Electronics
Engineers'
Handbook

OTHER McGRAW-HILL HANDBOOKS OF INTEREST

American Institute of Physics · American Institute of Physics Handbook
Baumeister · Marks' Standard Handbook for Mechanical Engineers
Beeman · Industrial Power Systems Handbook
Brady and Clauser · Materials Handbook
Burington and May · Handbook of Probability and Statistics with Tables
Condon and Odishaw · Handbook of Physics
Considine · Energy Technology Handbook
Coombs · Basic Electronic Instrument Handbook
Coombs · Printed Circuits Handbook
Croft, Carr, and Watt · American Electricians' Handbook
Dean · Lange's Handbook of Chemistry
Fink and Beaty · Standard Handbook for Electrical Engineers
Fink and Christiansen · Electronics Engineers' Handbook
Giacoletto · Electronics Designers' Handbook
Harper · Handbook of Components for Electronics
Harper · Handbook of Electronic Packaging
Harper · Handbook of Electronic System Design
Harper · Handbook of Materials and Processes for Electronics
Harper · Handbook of Thick Film Hybrid Microelectronics
Harper · Handbook of Wiring, Cabling, and Interconnecting for Electronics
Hicks · Standard Handbook of Engineering Calculations
Hunter · Handbook of Semiconductor Electronics
Ireson · Reliability Handbook
Jasik · Antenna Engineering Handbook
Juran · Quality Control Handbook
Kaufman and Seidman · Handbook of Electronics Calculations
Kaufman and Seidman · Handbook for Electronics Engineering Technicians
Korn and Korn · Mathematical Handbook for Scientists and Engineers
Kurtz and Shoemaker · The Lineman's and Cableman's Handbook
Machol · System Engineering Handbook
Maissel and Glang · Handbook of Thin Film Technology
Markus · Electronics Dictionary
McPartland · McGraw-Hill's National Electrical Code Handbook
Perry · Engineering Manual
Skolnik · Radar Handbook
Smeaton · Motor Application and Maintenance Handbook
Stout · Handbook of Microcircuit Design and Application
Stout and Kaufman · Handbook of Operational Amplifier Circuit Design
Truxal · Control Engineers' Handbook
Tuma · Engineering Mathematics Handbook
Tuma · Handbook of Physical Calculations
Tuma · Technology Mathematics Handbook
Williams · Electronic Filter Design Handbook

Electronics Engineers' Handbook

DONALD G. FINK   *Editor-in-chief*

Director Emeritus, Institute of Electrical and Electronics Engineers;
Fellow, IEEE; Member of the National Academy of Engineering;
Eminent Member, Eta Kappa Nu; Registered Professional Engineer;
Formerly Executive Director and General Manager, IEEE; Editor-in-
chief, Electronics; Vice President—Research, Philco Corporation;
President, Institute of Radio Engineers; Editor, Proceedings of the
IRE; Fellow of the Institution of Electrical Engineers (London)

DONALD CHRISTIANSEN   *Associate Editor*

Staff Director, Institute of Electrical and Electronics Engineers;
Editor, IEEE Spectrum; Fellow, IEEE; Member, IEEE Publications
Board; Member, Eta Kappa Nu; Member, New York Academy of
Sciences; Registered Professional Engineer; formerly Editor-in-chief
Electronics

Second Edition

McGRAW-HILL BOOK COMPANY
New York   St. Louis   San Francisco   Auckland   Bogotá
Hamburg   Johannesburg   London   Madrid   Mexico
Montreal   New Delhi   Panama   Paris   São Paulo
Singapore   Sydney   Tokyo   Toronto

Library of Congress Cataloging in Publication Data
Main entry under title:

Electronics engineers' handbook.

   "Companion volume to the Standard handbook for
electrical engineers . . . 11th edition." — Pref.
   Includes bibliographical references and index.
   1. Electronics — Handbooks, manuals, etc.
I. Fink, Donald G.   II. Christiansen, Donald.
TK7825.E34   1982   621.381'0202   81-3756
ISBN 0-07-020981-2   AACR2

 234567890   KPKP   898765432

ISBN 0-07-020981-2

*The editors for this book were Harold B. Crawford and
Geraldine Fahey and the production supervisor
was Thomas G. Kowalczyk. It was set in Trump
by University Graphics.
Printed and bound by The Kingsport Press.*

Contents

ELECTRONIC CIRCUITS AND FUNCTIONS

Contributors

Ronald T. Anderson   *ITT Research Institute* (SEC. 28)
Clarence M. Bailey, Jr.   *Bell Telephone Laboratories* (SEC. 22)
M. C. Bailey   *NASA Langley Research Center* (SEC. 18)
David K. Barton   *Raytheon Corporation* (SEC. 25)
James F. Bartram   *Raytheon Company* (SEC. 25)
J. C. Baumhauer, Jr.   *Bell Telephone Laboratories* (SEC. 22)
Gilmer L. Blankenship   *University of Maryland* (SEC. 5)
Ilan A. Blech   *Technion-Israel Institute of Technology, Haifa* (SEC. 6)
William Blood, Jr.   *Semiconductor Division, Motorola, Inc.* (SEC. 8)
Alan R. Bormann   *Motorola Semiconductor, Austin, Tex.* (SEC. 8)
D. A. Bosserman   *U.S. Army Research and Development Command* (SEC. 11)
Jenny Rosenthal Bramley   *U.S. Army Research and Development Command* (SEC. 11)
A. B. Brown, Jr.   *Bell Telephone Laboratories* (SEC. 22)
W. Calder   *The Foxboro Company* (SEC. 24)
L. V. Caldwell   *U.S. Army Research and Development Command* (SEC. 11)
David Cave   *Semiconductor Division, Motorola, Inc.* (SEC. 8)
R. L. Cerbone   *Bell Telephone Laboratories* (SEC. 22)
Peter W. Cheung   *Case Western Reserve University* (SEC. 26)
Dudley Childress   *Northwestern University* (SEC. 26)
Joseph L. Chovan   *General Electric Company* (SEC. 14)
Donald Christiansen   *Editor, IEEE Spectrum* (SEC. 28)
Wils L. Cooley   *West Virginia University* (SEC. 3)
Munsey E. Crost   *U.S. Army Electronics Research and Development Command* (SECS. 7 and 11)
William F. Croswell   *NASA Langley Research Center* (SEC. 18)
Jose B. Cruz, Jr.   *University of Illinois* (SEC. 5)
Dwight E. Davis   *Martin Marietta Aerospace, Orlando Division* (SEC. 28)
W. F. Davis   *Semiconductor Division, Motorola, Inc.* (SEC. 8)
N. A. Diakides   *U.S. Army Research and Development Command* (SEC. 11)
S. W. Director   *Carnegie Mellon University* (SEC. 27)
Sam Di Vita   *U.S. Army Communications Research and Development Command* (SEC. 7)
Sven H. Dodington   *International Telephone and Telegraph Company* (SEC. 25)
H. Bennett Drexler   *Martin Marietta Aerospace, Orlando Division* (SEC. 28)
B. Dudley   *Consultant* (SEC. 1)
H. W. Earle   *Bell Telephone Laboratories* (SEC. 22)
Myron D. Egtvedt   *Electronics Laboratory, General Electric Company* (SEC. 14)
Stanley L. Ehrlich   *Raytheon Company* (SEC. 25)
J. C. Engle   *Westinghouse Research Laboratories* (SEC. 15)
George K. Farney   *Varian Associates* (SEC. 9)
Joseph Feinstein   *U.S. Department of Defense* (SEC. 9)
Donald G. Fink   *Institute of Electrical and Electronics Engineers* (SEC. 20)
Lester H. Fink   *Systems Engineering for Power, Incorporated* (SEC. 5)

C. S. Fox   *U.S. Army Research and Development Command* (SEC. 11)

R. E. Franseen   *U.S. Army Research and Development Command* (SEC. 11)

Donald A. Fredenburg   *Raytheon Company* (SEC. 25)

Richard W. French   *ELEMEK, Incorporated* (SEC. 13)

G. C. Fritz   *Teletype Corporation* (SEC. 22)

Glenn B. Gawler   *General Electric Company* (SEC. 14)

E. A. Gerber   *U.S. Army Electronics Research and Development Command (retired)* (SEC. 7)

Robert A. Gerhold   *U.S. Army Electronics Research and Development Command (retired)* (SEC. 7)

Joseph M. Giannotto   *U.S. Army Communications Research and Development Command* (SEC. 7)

S. B. Gibson   *U.S. Army Research and Development Command* (SEC. 11)

Emanuel Gikow   *U.S. Army Electronics Research and Development Command* (SEC. 7)

John Goodrin   *Intel Corporation* (SEC. 8)

Kurt E. Gonzenbach   *Martin Marietta Aerospace, Orlando Division* (SEC. 28)

Thomas S. Gore, Jr.   *U.S. Army Electronics Research and Development Command (retired)* (SEC. 7)

J. M. Gotway   *Bell Telephone Laboratories* (SEC. 22)

R. D. Graft   *U.S. Army Research and Development Command* (SEC. 11)

W. T. Grant   *U.S. Army Research and Development Command* (SEC. 11)

Alan R. Grebene   *Exar Integrated Systems* (SEC. 3)

W. A. Guttierrez   *U.S. Army Research and Development Command* (SEC. 11)

L. Gyugyi   *Westinghouse Research Laboratories* (SEC. 15)

Edward B. Hakim   *U.S. Army Electronics Research and Development Command* (SEC. 7)

Harry W. Hale   *Iowa State University* (SEC. 12)

P. D. Hansen   *The Foxboro Company* (SEC. 24)

G. Burton Harrold   *General Electric Company* (SEC. 13)

Jack H. Heimann   *Raytheon Company* (SEC. 25)

T. M. Heinrich   *Westinghouse Research Laboratories* (SEC. 15)

Joseph P. Hesler   *AKF Design, Limited* (SECS. 13 and 14)

D. J. Horowitz   *U.S. Army Research and Development Command* (SEC. 11)

W. E. Hostetler   *Bell Telephone Laboratories* (SEC. 22)

R. M. Hunt   *Bell Telephone Laboratories* (SEC. 22)

Robert C. Huntington   *Semiconductor Division, Motorola, Inc.* (SEC. 8)

G. M. Janney   *Hughes Aircraft Company* (SEC. 11)

Paul G. A. Jespers   *Catholic University of Louvain, Belgium* (SEC. 16)

J. J. Jetzt   *Bell Telephone Laboratories* (SEC. 22)

A. E. Joel, Jr.   *Bell Telephone Laboratories* (SEC. 22)

V. I. Johannes   *Bell Telephone Laboratories* (SEC. 22)

C. A. Johnson   *U.S. Army Research and Development Command* (SEC. 11)

Edwin C. Jones, Jr.   *Iowa State University* (SEC. 12)

Peter G. Katona   *Case Western Reserve University* (SEC. 26)

A. Kennedy   *U.S. Army Research and Development Command* (SEC. 11)

Chang S. Kim   *Taihan Electric Wire Company, Limited* (SEC. 13)

Edwin W. Kimball   *Martin Marietta Aerospace, Orlando Division* (SEC. 28)

Raymond S. Kiraly   *Cleveland Clinic Foundation* (SEC. 26)

Richard C. Kirby   *International Radio Consultative Committee (CCIR), Geneva* (SEC. 18)

Wen H. Ko   *Case Western Reserve University* (SEC. 16)

Granino A. Korn   *University of Arizona* (SEC. 2)

Theresa M. Korn   *Tucson, Arizona* (SEC. 2)

Samuel M. Korzekwa   *General Electric Company* (SEC. 13)

Stanislaw Kus   *ITT Research Institute* (SEC. 28)

Joseph A. Kuzneski   *Raytheon Company* (SEC. 25)

Owen P. Layden   *U.S. Army Electronics Research and Development Command* (SEC. 7)

M. R. Lightner   *University of Illinois* (SEC. 27)

David Linden   *U.S. Army Electronics Research and Development Command (retired)* (SEC. 7)

R. E. Longshore   *U.S. Army Research and Development Command* (SEC. 11)

Harold W. Lord   *Consulting Engineer* (SEC. 13)

John W. Lunden   *General Electric Company* (SEC. 13)

Robert J. McFadyen   *General Electric Company* (SEC. 13)

Renville H. McMann, Jr.   *Thompson-CSF Laboratories* (SEC. 20)

Paul S. Malchesky   *Cleveland Clinic Foundation* (SEC. 26)

Gregory J. Malinowski   *U.S. Army Electronics Research and Development Command* (SEC. 7)

P. R. Manzo   *Science Applications, Inc.* (SEC. 11)

Daniel W. Martin   *Baldwin Piano & Organ Company* (SEC. 19)

Richard E. Matick   *International Business Machines Corporation* (SEC. 23)

James D. Meindl   *Stanford University* (SEC. 26)

Charles S. Meyer   *Semiconductor Division, Motorola, Inc.* (SEC. 8)

J. E. Miller   *U.S. Army Research and Development Command* (SEC. 11)

Floro Miraldi   *Case Western Reserve University and Cleveland Metropolitan General Hospital* (SEC. 26)

K. W. Mitchell   *Solar Energy Research Institute* (SEC. 11)

Eugene Mittelmann (deceased)   *Consulting Engineer* (SEC. 10)

Berton D. Moldow   *IBM Systems Research Institute* (SEC. 23)

J. Thomas Mortimer   *Case Western Reserve University* (SEC. 26)

J. W. Motto   *formerly Westinghouse Research Laboratories* (SEC. 15)

V. E. Munson   *Bell Telephone Laboratories* (SEC. 22)

Conrad E. Nelson   *General Electric Company* (SEC. 13)

Richard B. Nelson   *Varian Associates* (SEC. 9)

Gerald D. Netzband   *Martin Marietta Aerospace, Orlando Division* (SEC. 28)

Michael R. Neuman   *Case Western Reserve University* (SEC. 26)

W. E. Newell (deceased)   *Westinghouse Research Laboratories* (SEC. 15)

W. H. Ninke   *Bell Telephone Laboratories* (SEC. 22)

Harry N. Norton   *formerly Jet Propulsion Laboratory* (SEC. 10)

R. M. Oates   *Westinghouse Research Laboratories* (SEC. 15)

Neil V. Owen   *Martin Marietta Aerospace, Orlando Division* (SEC. 28)

Norman W. Parker   *Motorola Inc.* (SECS. 20 and 21)

C. M. Patel   *The Foxboro Company* (SEC. 24)

B. R. Pelly   *formerly Westinghouse Research Laboratories* (SEC. 15)

George F. Pfeifer   *General Electric Company* (SEC. 14)

P. F. Pittman   *Westinghouse Research Laboratories* (SEC. 15)

Robert Plonsey   *Case Western Reserve University* (SEC. 26)

J. H. Pollard   *U.S. Army Research and Development Command* (SEC. 11)

Philip T. Porter   *Bell Telephone Laboratories* (SEC. 22)

Noble Powell   *General Electric Company* (SEC. 14)

M. Prasad   *The Foxboro Company* (SEC. 24)

Isaac H. Pratt   *U.S. Army Electronics Research and Development Command* (SEC. 7)

Donald H. Preist   *Varian Associates* (SEC. 9)

I. Reingold   *U.S. Army Electronics Research and Development Command* (SECS. 7 and 11)

Charles W. Rhodes *Tektronix Inc.* (SEC. 20)

D. A. Richardson *The Foxboro Company* (SEC. 24)

Henry C. Rickers *Rome Air Development Center* (SEC. 28)

Raymond M. Roop *Semiconductor Division, Motorola, Inc.* (SEC. 8)

J. J. Rosinski *Bell Telephone Laboratories* (SEC. 22)

James M. Rugg *Integrated Circuits Division, Motorola, Inc.* (SEC. 8)

J. Salz *Bell Telephone Laboratories* (SEC. 22)

Allan Scott *Varian Associates* (SEC. 9)

E. J. Sharp *U.S. Army Research and Development Command* (SEC. 11)

R. R. Shurtz, II *U.S. Army Research and Development Command* (SEC. 11)

Paul Skitzki *Raytheon Company* (SEC. 25)

Bernard Smith *U.S. Army Electronics Research and Development Command* (SEC. 7)

Jack Spergel (deceased) *formerly U.S. Army Electronics Research and Development Command* (SEC. 7)

George M. Stamps *GMS Consulting* (SEC. 20)

Hans R. Stellrecht *Signetics Corporation* (SEC. 8)

Joseph L. Stern *Stern Telecommunication Corporation* (SEC. 21)

C. Stockbridge *Bell Telephone Laboratories* (SEC. 22)

S. Sugihara *The Aerospace Corporation* (SEC. 23)

George W. Taylor *U.S. Army Electronics Research and Development Command* (SEC. 7)

Stephen W. Tehon *General Electric Company* (SEC. 13)

John B. Thomas *Princeton University* (SEC. 4)

Francis T. Thompson *Westinghouse Research Laboratories* (SEC. 17)

R. K. Thompson *Bell Telephone Laboratories* (SEC. 22)

G. P. Torok *Bell Telephone Laboratories* (SEC. 22)

Richard W. Ulmer *Semiconductor Division, Motorola, Inc.* (SEC. 8)

E. W. Underhill *Bell Telephone Laboratories* (SEC. 22)

W. E. Vannah *The Foxboro Company* (SEC. 24)

John R. Vig *U.S. Army Electronics Research and Development Command* (SEC. 7)

R. E. Waddell *Bell Telephone Laboratories* (SEC. 22)

Pamela L. Walchli *Varian Associates* (SEC. 9)

Kurt F. Wallace *Ampex Corporation* (SEC. 20)

Claude E. Walston *International Business Machines Corporation* (SEC. 23)

Harold R. Ward *Raytheon Company* (SEC. 25)

Gunter K. Wessel *Syracuse University* (SEC. 13)

D. L. Whitson *Bell Telephone Laboratories* (SEC. 22)

James W. Wilbur *Rome Air Development Center* (SEC. 28)

Peter Wood *Westinghouse Research Laboratories* (SEC. 15)

Preface to the Second Edition

Since the first edition of this Handbook was published, the digital revolution has taken full command of electronics engineering. In 1975 digital methods were at the root of the computer and radar industries and were making large inroads into the field of telecommunications. Hardly to be imagined was that the vastly increased density of devices in integrated circuits and their sharply lower costs, then in store, would in less than a decade so extend the range of digital electronics that no aspect of commercial, industrial, or domestic life would remain untouched.

This sweeping change in the methods and outlook of electronics engineering has made a corresponding impression on the content of this new edition. Except for parts dealing with unchanging fundamentals, few sections of the earlier edition escaped the need for revision. The experts who have served as contributors have taken on the task of reviewing, deleting, adding to, and reorganizing their earlier work, particularly to incorporate new digital theory and technology. This occurred even in the section on mathematical formulas, which now includes standard notation and formulas for the Nyquist limits and other aspects of digitized signal transmission.

The earlier analogic systems of electronics have not been quiescent. For example, many startling improvements in video and audio engineering, which still deal primarily with analog signals, are fully covered in this new edition, as are the new arts of transduction and detection and production of radiant energy. Perhaps the majority of applications, once limited to analog signals, are now committed to digital methods, e.g., sound reproduction and television production. The digital filter now permits low-cost operations once prohibitively expensive or practically impossible in lumped-constant or distributed analog designs. This list could be extended indefinitely. Suffice it to say that digital methods have been embraced in this edition in those sections to which they apply as we go to press.

In addition to these revisions of previously published material, three entirely new subjects, dealing primarily with digital methods, have been added to the Handbook. The first is, of course, the microprocessor, on which the pervasive influence of electronics now so largely depends. This subject required the addition of 56 printed pages of material to Section 8, on integrated circuits. That section has also been expanded to include new material on VMOS technology, charge-coupled devices, voltage reference circuits, switching regulators, linear MOS circuits, data-conversion circuits, integrated-circuit filters, integrated injection logic, static and dynamic memory circuits, and logic arrays.

The second new section is on computer-aided design of electronic circuits, Section 27. Third is reliability of electronic components and systems, Section 28. This last is a vital new subject, as the ever increasing number of active elements in a given system puts new demands on the reliability of each one and requires great care in assessing the overall reliability of a system of hundreds of thousands of nominally highly reliable parts, the failure of any one of which might have disastrous results.

Central to the successful extension of digital electronics to many new areas of application are the new arts of telecommunication. Section 22 is a wholly new treatment by 21 members of the technical staff of the Bell Telephone Laboratories and one from the Teletype Corporation. This is the best handbook treatment of the subject yet composed, in my judgment. It is comprehensive, up to date, clearly written, well organized, and, as might be expected, quietly authoritative.

The editor and associate editor are deeply impressed with the competence, balance, and insight the 173 contributors have brought to this new edition, in taking into account one of the most dramatic and rapid periods of change that has occurred in the busy history of electronics. The result is a Handbook of 2,270 printed pages, more than a million words of text, 2,125 illustrations, and 3,285 bibliographic entries. To all those who have faithfully labored to make it possible, we give thanks.

The present Handbook can be considered a companion volume to the "Standard Handbook for Electrical Engineers," the 11th edition of which, under the editorship of the undersigned and H. W. Beaty, is now in print. The Electrical Handbook is devoted primarily to the techniques of electrical power engineering, i.e., "heavy current" generation, distribution, and application. Together the two Handbooks cover the whole field embraced, for example, by the Institute of Electrical and Electronics Engineers. Aside from the different focus of subject matter, the aim of the Electronics Handbook is the same as that of the Electrical Handbook: to contain in a single volume all pertinent data within its scope, to be accurate and comprehensive in technical treatment, to be used in engineering practice (as well as in study in preparation for such practice), and to be oriented toward application, with sufficient theoretical background to assure basic understanding of application requirements. The present Handbook is divided into four major parts: principles employed in electronics engineering (Sections 1 to 5), materials, devices, components and assemblies (Sections 6 to 11), electronic circuits and functions (Sections 12 to 18), and systems and applications (Sections 19 to 28).

The substantial effort made by all the contributors not only to cover their special fields comprehensively but to present their work in the most compact fashion consistent with informed and ready use is gratefully acknowledged. I wish particularly to welcome Associate Editor Donald Christiansen, well known to readers of this Handbook as editor of *IEEE Spectrum*. We look forward to continued collaboration as the art progresses to the point where still another edition is required.

Donald G. Fink
Editor-in-Chief

Basic Phenomena of Electronics

B. DUDLEY *Consultant, formerly staff member, Institute for Defense Analyses, Editor Technology Review, Massachusetts Institute of Technology; Senior Member, IEEE*

CONTENTS

Numbers refer to paragraphs

ELECTRONICS ENGINEERING

1. Electronics. Electronics is the field of science and engineering dealing with the release, transport, control, collection, and energy conversion of subatomic particles having mass and charge (such as electrons) acting in materials with known electromagnetic properties, e.g., vacuum, gases, or semiconductors. The charged particles are called *charge carriers*.

The phenomena of electronics depend upon the number of participating charge carriers, their dynamic activity, and the properties of the environment in which the charges act. The charge carriers are usually electrons, but they may be holes or positive or negative ions. The dynamic activity of charge carriers results from the force and recoverable energy needed to release them from atoms to produce their displacement, velocity, or acceleration in accordance with the principles of relativistic quantum mechanics. The properties of the environment depend upon the composition, structure, and changes in energy levels of atoms composing the substance through which charge carriers (or their fields) pass.

The basic principles of electronics are the same as those of electricity and magnetism. Electricity is any manifestation of energy conversion of charge carriers that initiates or yields forces producing displacement, velocity, or acceleration in the direction of their movement. Magnetism is any manifestation of the kinetic energy of charge carriers arising from or producing forces in a direction perpendicular to their motion. The principles of electronics and electromagnetism are built upon the physical entities of mass, length, time, electric charge (or current), temperature, amount of substance, and luminous intensity. All electromagnetic quantities are now expressed in the SI units (see Par. **1-201**).

The primary differences between electronics and electromagnetism lie in their applications. Compared with the traditional field of electromagnetism, electronics makes possible devices having much greater degree of control over the instantaneous, rather than the average, movement of charges during transport, and the control of charges can be exceedingly rapid. Active electron devices require an external source of power to maintain their electrodes at suitable operating voltages and currents. Electron devices are also, for the most part, nonlinear elements whose output voltages and current are disproportionately related to their input counterparts. At the expense of power from an external supply, many electron devices can provide at their output terminals an amplified version of the voltage, current, or power supplied to their input terminals.

Originally electronics dealt with the conduction of electricity in vacuum or gaseous tubes. Since the invention of the transistor in 1948 conduction through crystalline semiconductors (*solid-state* conduction) has virtually dominated the field, and thermionic electron tubes have played a role of diminishing importance except for applications requiring high power.[1]

2. Electronics Engineering. Electronics engineering is the branch of applied science concerned with active devices and systems, i.e., those requiring external power supply to function properly, in which a dominant role is played by the release, transport, control, collection, and energy conversion of elementary charges. It frequently deals with lower energy levels than electrical engineering does.

Electronics engineering is more closely allied with, and dependent upon, the composition, atomic structure, and mode of electric conduction in composite materials than electrical engineering.

ELECTRONIC PROPERTIES AND STRUCTURE OF MATTER

3. Elementary Particles. The charged elementary particles of principal interest in electronics are the electron and the proton, designated e^- and p^+, respectively. The mass, charge, and charge-to-mass ratios of these particles are as follows:

	Electron	Proton
Mass at rest, kg	9.1096×10^{-31}	1.6726×10^{-27}
Charge, C	-1.6022×10^{-19}	$+1.6022 \times 10^{-19}$
Charge-to-mass ratio, C/kg	1.7588×10^{11}	9.5791×10^7

The elementary particles whose existence has been experimentally verified or postulated on theoretical grounds are listed in Table 1-1.

4. Atomic Structure. The atoms of each element consist of a dense nucleus around which electrons travel in well-defined orbits, or shells. The total mass of the nucleons (protons and neutrons) is taken to be equal to the mass of the atom. The number of nuclear protons is equal to the *atomic number* Z of the element. The number of nucleons is equal to the *mass number* A of the atom, and $A - Z$ is the number of neutrons in the nucleus. Heavy atoms have more neutrons than protons; excess of neutrons over protons is important in determining the stability of atoms, i.e., their radioactive properties. Atoms having the same atomic number but different mass numbers have the same chemical properties but different atomic weights. They are called *isotopes* of the chemical element.

The diameter of the atomic nucleus is between 10^{-15} and 10^{-16} m, whereas the diameter of the outer orbiting electrons (the diameter of the atom) is of the order of 10^{-10} m.

The nucleus carries a positive charge equal to the atomic number Z of the element times 1.6 $\times 10^{-19}$ C, the charge of a proton. In the normal (un-ionized) atom there are Z orbiting electrons,

---

[1] Earlier developments in electron tubes and circuits are covered in the following works, which include extensive references: H. J. Van der Bijl, "Thermionic Vacuum Tube," McGraw-Hill, New York, 1920 (the first comprehensive and authoritative treatment of electron tubes and long a standard reference); E. L. Chaffee, "Theory of Thermionic Vacuum Tubes," McGraw-Hill, New York, 1933 (fundamentals of low-power, negative-grid tubes as amplifiers and detectors; uses nonstandard mathematical notation); H. J. Reich, "Theory and Applications of Electron Tubes," McGraw-Hill, New York, 1939 (physical principles of tubes and associated networks; includes treatment of gaseous devices); M.I.T. Dept. of Electrical Engineering, "Applied Electronics," Wiley, New York, 1943 (undergraduate course in electronics, electron tubes, associated circuits, and practical applications); K. R. Spangenberg, "Vacuum Tubes," McGraw-Hill, New York, 1948 (concerned largely with physical behavior and basic design of tubes; discusses ultra-high-frequency effects, electron bunching, electron optics, and cathode-ray and special tubes); F. A. Maxfield and R. R. Benedict, "Theory of Gaseous Conduction and Electronics," McGraw-Hill, New York, 1941 (fundamentals and applications of gaseous conduction, corona, sparking, glows, and arcs); J. D. Cobine, "Gaseous Conductors," McGraw-Hill, New York, 1941 (theory and applications of electrical discharges in gases); A. Guthrie and R. A. Wakerlin (eds.), "Characteristics of Electrical Discharges in Magnetic Fields," McGraw-Hill, New York, 1949 (Manhattan Project studies on the characteristics of electric discharges in gases and vapors, especially of uranium compounds, in magnetic fields).

Table 1-1. Elementary Particles

Family name	Name of particle	Symbol	Mass ($e^- = 1.0$)	Mass, MeV	Lifetime, s	Spin	Charge ($e^- = -1.0$)	Anti-particle
	Photon	γ	0	0	∞	1	0	γ
Electron	Electron	e^-	1	0.51098	∞	½	-1	e^+
	Electron neutrino	ν_e	0	0	∞	½	0	$\bar{\nu}_e$
Muon	Muon	μ^-	206.768	105.654	2.212×10^{-6}	½	-1	μ^+
	Muon neutrino	ν_μ	...	0	∞	½	0	$\bar{\nu}_\mu$
Meson	Pion, positive	π^+	273.18	139.59	2.55×10^{-8}	0	1	π^-
	Neutral	π^0	264.20	135.0	1.9×10^{-16}	0	0	π^0
	Kaon, positive	K^+	966.6	493.9	1.22×10^{-8}	0	1	K^-
	Neutral	K^0	974.2	497.8	1.0×10^{-10} 6.1×10^{-8}	0	0	$\bar{K}^0$
Baryons	Nucleon, proton	p^+	1,836.12	938.213	∞	½	1	$\bar{p}^+$
	Nucleon, neutron	n^0	1,838.65	939.507	1.013×10^3	½	0	$\bar{n}^0$
	Lambda	Λ^0	2,182.8	1,115.36	2.51×10^{-10}	½	0	$\bar{\Lambda}^0$
	Sigma, positive	Σ^+	2,327.7	1,189.40	8.1×10^{-11}	½	1	$\bar{\Sigma}^+$
	Neutral	Σ^0	2,332	1,191.5	$\sim 10^{-20}$	½	0	$\bar{\Sigma}^0$
	Negative	Σ^-	2,340.5	1,195.6	1.6×10^{-10}	½	-1	$\bar{\Sigma}^-$
	Xi, neutral	Ξ^0	2,566	1,311	1.5×10^{-10}	½	0	$\bar{\Xi}^0$
	Negative	Ξ^-	2,580	1,318	1.28×10^{-10}	½	-1	$\bar{\Xi}^-$

each with negative charge $e^- = -1.6 \times 10^{-19}$ C. At distances large compared with the atomic radius, the atom shows no net electric charge.

The extranuclear (electronic) structure of the atom is characteristic of the element. The orbiting electrons are arranged in successive *shells*. In order of increasing distance from the nucleus these shells are designated K, L, M, N, O, P, and Q. The number of electrons each shell can contain is limited. The electrons of the inner shells of complex atoms are tightly bound to the nucleus, and their paths can be altered only by high-energy particles, such as gamma rays. In the more complex atoms, electrons of the outer shells are relatively loosely bound to the nucleus. The outer shells account for the chemical and electrical properties of the elements.

5. Electron Orbits, Shells, and Energy States. Each orbiting electron in an atom has energy which is uniquely characterized by four *quantum numbers*. According to Pauli's exclusion principle, the wave functions describing the electrons must differ by at least one quantum number in the complete set required for their description.

An electron within an atom can be specified in terms of (1) a *principal quantum number n*, (2) an *azimuthal quantum number l*, (3) a *spatial quantum number m_l*, and (4) a *spin quantum number m_s or s*. The principal quantum number n specifies the shell in which an electron is located and hence principally specifies the energy state of the electron. Electrons lodged in the K, L, M, N, O, P, and Q shells have principal quantum numbers $n = 1, 2, 3, 4, 5, 6,$ or 7, respectively.

The azimuthal quantum number l specifies the angular orbital momentum of an electron in each orbital state in various subshells. Together with n, the value of l designates the eccentricity of an electron orbit; the smaller the value of l the greater the eccentricity of the orbits for any given shell. The magnitudes of l may be any integer from 0 to $n - 1$. Electrons whose values of l are 0, 1, 2, 3, 4, and 5, respectively, are referred to as the s, p, d, g, and f electrons. The number of electrons in a subshell is determined by restrictions on m_l and m_s imposed by Pauli's exclusion principle.

The spatial quantum number m_l specifies differently oriented orbits having the same general shape; it specifies the orientation of the magnetic field of the electron orbit. This quantity is the projection of l on the magnetic axis; it may have $\pm(2l - 1)$ integral values from $-l$ to $+l$ including 0.

The spin quantum number, m_s or s, specifies the direction of spin of an electron on its own axis. Corresponding to spin in opposite directions, the two spin quantum numbers are $+h/2$ and $-h/2$, where $h/2\pi$ is Planck's constant ($= 6.626 \times 10^{-34}$ J·s).

In a normal atom, orbiting electrons are arranged in the set of allowed states having the lowest total energy. As the complexity of atoms increases from hydrogen to uranium (the latter having 92 protons, and 146 neutrons), the electrons fill the shells and subshells by taking those states having the lowest total energy. Sometimes the energy state of an inner shell is less than that of a state in the outermost shell, and this accounts for the fact that some shells may begin to be filled before inner shells are totally filled.

6. Chemical Valence. The chemical properties of the elements are determined by the electrons in the outermost shell (valence electrons). Atoms with completely filled outer shells (the rare gases: helium, argon, krypton, zenon, and radon) are chemically inert. They contain eight electrons in their outer shells.

Atoms with a single electron in the outer shell (lithium, sodium, potassium, rubidium, cesium, franconium, and hydrogen) can easily lose their outer electron. They then become positive ions with completely filled shells.

Atoms with seven outer-shell electrons (the halogens: fluorine, chlorine, bromine, iodine, and astatine) readily pick up an electron from other atoms and become negative ions; they form molecules by sharing electrons and are said to have ionic bonding. Atoms with other numbers of outer-shell electrons tend to unite with other atoms in such ways that each atom has eight outer-shell electrons. Partially filled inner shells have an important bearing on the magnetic properties of the elements.

7. Conduction Electrons. When electrons are in close proximity in crystalline solids, the presence of nearby atoms affects their behavior and their energies are no longer uniquely determined. The single energy level of an electron in a free or isolated atom is thereby spread into a band, or range, of energy levels. Whether or not the band of allowed energies is completely filled with electrons determines its properties as an electric conductor or insulator.

The *conduction band* is a range of states in the free-energy spectrum of a solid in which electrons can move freely; i.e., the electrons must be capable of effecting transitions between energy states. The valence electrons in metals, for example, are not firmly attached to individual atoms but are free to travel within the crystal lattice. Such electrons are called *conduction electrons*. There is one such conduction electron per atom in silver, copper, gold, and the alkali metals, all of which are good conductors.

An insulator or dielectric is a material in which every energy level is filled and the electrons are unable to effect the transitions between states required for electric conduction.

8. Chemical Bonds and Compound Formation. Chemical bonds occur when the total energy of an aggregate is less with atoms near each other than separated. The charges of the atom play an important role in bonding, especially electrons in the outer shells.

Electrostatic or *ionic bonds* result from attractive forces between positive and negative ions or between pairs of oppositely charge ions. *Covalent bonds* occur when atoms share two or more electrons; i.e., shared electrons are attracted simultaneously to two atoms, and the resulting energy stability produces the bond. *Metallic bonds* are those in which the attractive forces result from the exchange interaction of the electron gas with the ionic lattice. *Van der Waals bonds* occur when molecules are formed, giving each atom an outer shell of eight atoms, as in an inert gas.

9. Energy Conversion. Energy in a system can be neither created nor destroyed (except in nuclear processes when energy is converted into its equivalent form, mass). However, one form of energy can be converted into other forms. Thus, the potential energy of a system can be converted into kinetic energy and vice versa, and the energy of particles of one kind can be converted into energy of particles of quite another kind. Heat, produced by the random motion of elementary particles and their aggregates, is the form of energy into which other forms are ultimately converted.

10. Energy Conservation and Mass Equivalence. According to classical physics, energy can be neither created nor destroyed; similarly, mass can be neither created nor destroyed. Relativistic physics identifies mass and energy according to the equation for total energy

$$mc^2 = m_0 c^2 + U_k$$

where $m_0 c^2$ is the rest energy of a body or particle and U_k is its kinetic energy. Hence the two conservation laws stated above become one and the same physical law.

11. Electromagnetic Effects. The dynamic behavior of elementary particles possessing both mass and charge produces electromagnetic phenomena. In free space the effects depend only on the nature and distribution of the charges and their motions, but such effects are greatly modified by atomic and molecular structure when charges move in material substances. Thus, the properties of materials modify the effects observed in free space.

The three major classes of electromagnetic materials are *conductors* or *semiconductors*, through which charges can flow more or less readily; *dielectrics* or *insulators*, through which charges are prevented from flowing; and *magnetic materials*, in which the motion of charges produces enhanced transverse forces.

12. Conduction Effects (see Par. **1-62**) Electric conduction is the effect produced in a substance or system having mobile charges by the application of an electric force such that the charged particles flow through the conductor in the direction of the applied force. The phenomenon is attributed to electrons in the outer shells of atoms that are so loosely bound that they can be released by small electric forces. In good conductors free electrons can also be released by chemical, thermal, or other kinds of forces.

13. Dielectric Phenomena. Displacement of electric charges occurs in a substance or system having bound charges. The application of an electric force produces a directed motion of charged particles in the direction of the applied force but of such limited extent that the charges do not separate from their parent atoms (see Par. **1-64**).

Dielectric phenomena are attributed to electrons in the outer shells of atoms which are so tightly bound that they cannot become mobile except through application of electric forces strong enough to destroy the dielectric properties of the material.

Under ordinary conditions, the electric flux density D in a dielectric is proportional to the electric field intensity E acting across the dielectric. But at very high frequencies, hysteresis effects occur in dielectric materials. The phenomenon is similar to that produced in magnetic materials at much lower frequencies (see Par. **1-118**). Hysteresis produces heat losses in dielectrics, just as heat losses occur in magnetic materials displaying hysteresis phenomena.

14. Magnetic Phenomena (see Pars. **1-81** to **1-110**). Magnetic phenomena include a number of effects of substances or systems in which the motions of charged particles set up forces transverse to the motion of charges.

The phenomena are attributed to the electric fields carried by moving charges. Any directed motion of charges produces magnetic effects. In magnetic materials the phenomena of magnetism arise from the orientation of orbiting electrons and, to a smaller extent, from electron spins in atoms, molecules, and crystal domains.

The interaction of matter with a magnetic field produced by moving charges arises from several mechanisms: (1) ferromagnetism, produced by exchange forces between atomic moments, (2) diamagnetism, produced by electron spins in antiparallel pairs in closed electronic shells, (3) paramagnetism, produced by the orbital or spin moments of electrons, or both, as well as by moments of free electrons, (4) antiferrimagnetism, produced by exchange forces between atomic moments, and (5) ferrimagnetism, produced by the moment resulting from two antiferromagnetic lattices.

15. Thermoelectric Phenomena. Three important relations between electricity and heat are the Seebeck effect, the Thomson effect, and the Peltier effect.

16. Work Function. Work function is a term applied to the amount of energy required to transfer electrons, ions, molecules, etc., from the interior of one substance across an interface boundary into an adjacent substance or space. It is commonly expressed by the transfer of an electron across a boundary in units of electronvolts (see Pars. **1-124** and **1-125**).

17. Bibliography on Electronic Properties and Structure of Matter

1. Allen, J. S.   "The Neutrino," Princeton University Press, Princeton, N.J., 1958.
2. Aller, L. H.   "Atmosphere of the Sun and Stars," 2d ed., Ronald, New York, 1963.
3. Allis, W. P., and M. A. Herlin   "Thermodynamics and Statistical Mechanics," McGraw-Hill, New York, 1952.
4. Arye, A. P.   "Fundamentals of Atomic Physics," Allyn and Bacon, Boston, 1975.
5. Beiser, A.   "Concepts of Modern Physics," McGraw-Hill, New York, 1967.
6. Bescancon, R. M.   "Encyclopedia of Physics," Reinhold, New York, 1974.
7. Bergman, P. G.   "Introduction to the Theory of Relativity," Prentice Hall, Englewood Cliffs, N.J., 1942.
8. Bjorken, J. D.   "Relativistic Quantum Mechanics," McGraw-Hill, New York, 1964.
9. Bjorken, J. D., and S. D. Drell   "Relativistic Quantum Fields," McGraw-Hill, New York, 1965.
10. Borowitz, S. and L. A. Bornstein   "Contemporary View of Elementary Physics," McGraw-Hill, New York, 1968.
11. Borth, V. H., and M. L. Bloom   "Physical Science: A Study of Matter and Energy," Macmillan, New York, 1972.
12. Bozworth, R. M.   "Ferromagnetism," Van Nostrand, New York, 1951.
13. Brown, G. E.   "Unified Theory of Nuclear Models," North-Holland, Amsterdam, 1974.
14. Cognac, B.   "Modern Atomic Physics," Halsted, Chicago, 1975.
15. Clark, F. M.,   "Insulating Materials for Design and Engineering Practice," Wiley, New York, 1962.
16. Clark, G. L.   "Applied X-Rays," McGraw-Hill, New York, 1940.
17. Clemmow, P. C., and J. P. Dougherty   "Electrodynamics of Particles and Plasmas," Addison-Wesley, Reading, Mass., 1969.
18. Condon, E. U.   "New Directions in Physics," Yale University Press, New Haven, Conn., 1972.
19. Cullin, R. E.   "Foundations of Microwave Engineering," McGraw-Hill, New York, 1966.
20. Cullity, B. D.   "Introduction to Magnetic Materials," Addison-Wesley, Reading, Mass., 1972.
21. Darmois, G.   "Matter, Electricity, and Energy," Walker, New York, 1957.
22. Davydor, A.   "Quantum Mechanics," Pergamon, New York, 1976.
23. Dicke, R. H., and J. P. Wittke   "Introduction to Quantum Mechanics," Addison-Wesley, Reading, Mass., 1960.
24. Eisberg, R., and R. Resnick   "Quantum Physics of Atoms, Molecules, Solids, Nuclei, and Particles," Wiley, New York, 1974.
25. Feynman, R. P.   "Quantum Mechanics and Path Integrals," McGraw-Hill, New York, 1965.
26. Finck, J. L.   "Thermodynamics from Classical and Generalized Standpoints," Princeton University Press, Princeton, N.J., 1955.
27. Garrison, S.   "Introduction to Quantum Physics," Wiley, New York, 1970.
28. Gasierwicz, S.   "Quantum Physics," Wiley, New York, 1974.
29. Harper, C. A.   "Handbook of Materials and Processes for Electronics," McGraw-Hill, New York, 1970.
30. Harper, C. A.   "Handbook of Plastics and Elastomers," McGraw-Hill, New York, 1975.
31. Heck, C.   "Magnetic Materials and Their Applications," Crane, Russack, New York, 1974.
32. Hemmenway, C. H., R. W. Henry, and M. Caulton   "Physical Electronics," 2d ed., Wiley, New York, 1967.
33. Hency, E. M., and W. Thirring   "Elementary Quantum Field Theory," McGraw-Hill, New York, 1962.
34. Hlawicskz, P.   "Introduction to Quantum Electronics," Academic, New York, 1971.
35. Heisinger, D. B.   "Nucleonics Fundamentals," McGraw-Hill, New York, 1959.
36. Houston, W. V.   "Principles of Quantum Mechanics," Dover, New York, 1959.
37. Hoyle, F., and J. V. Narliker   "Action at a Distance in Physics and Cosmology," Freeman, San Francisco, 1974.
38. "International Encyclopedia of Chemical Science," Van Nostrand, N.Y., 1964.

39. Lorentz, H. A., A. Einstein, and H. Minkowski "The Principles of Relativity," Dover. New York, 1952.

40. Marcuse, D. "Engineering Quantum Electrodynamics," Harcourt, New York 1970.

41. Merzberger, E. "Quantum Mechanics," Wiley, New York, 1970.

42. Nussbaum, A. "Electromagnetic and Quantum Properties of Materials," Prentice-Hall Englewood Cliffs, N.J., 1966.

43. Nussbaum, A. "Electronic and Magnetic Behavior of Materials," Prentice-Hall, Englewood Cliffs, N.J., 1967.

44. Page, L. "Introduction to Theoretical Physics," Van Nostrand New York, 1928.

45. Pierce, J. R. "Quantum Electronics," Doubleday, New York, 1956; Anchor Books, New York, 1966.

46. Rapp, D. "Quantum Mechanics," Holt, New York, 1971.

47. Richards, P. I. "Manual of Mathematical Physics," Pergamon, New York, 1959.

48. Richtmyer, F. K., and E. H. Kennard "Introduction to Modern Physics," 3d ed., McGraw-Hill, New York, 1942.

49. Romer, R. "Energy: An Introduction to Physics," Freeman, San Francisco, 1976.

50. Shortley, G., and D. Williams "Elements of Physics," Prentice-Hall, Englewood Cliffs, N.J., 1971.

51. Singer, J. R. "Advances in Quantum Electronics," Columbia University Press, New York, 1971.

52. Slater, J. C. "Modern Physics," McGraw-Hill, New York, 1955.

53. Smyth, C. "Dielectric Behavior and Structure," McGraw-Hill, New York, 1955.

54. Solomon, J. "The Structure of Matter," Wiley, New York, 1974.

55. Spraull, R. L. "Modern Physics," Halsted, Chicago, 1975.

56. Weyl, H. "Space, Time, Matter," Methuen, London, 1922; Dover, New York, 1951.

57. Weast, R. C. "Handbook of Physics and Chemistry," Chemical Rubber Co., Cleveland (annual editions).

58. Whitehead, S. "Dielectric Breakdown of Solids," Oxford University Press New York, 1951.

ELECTROSTATICS

18. Electric Charges. All electronic phenomena are expressible in terms of the masses and charges of elementary particles (see Par. **1-3**). Since electrically neutral substance has equal numbers of positive and negative charge carriers in its atoms, it exhibits no net electric charge at distances large compared with atomic dimensions. Thus, $Q^+ + Q^- = 0$ for bodies in the unelectrified state.

A body becomes electrically charged when charge carriers are transferred from one body to another. The negative charge Q^- acquired on one body is equal to the positive charge Q^+ lost from the other, so that $Q^+ = -Q^-$. The magnitude of any charge Q is equal to an integral number n of transported elementary charges e^- or p^+, and so $Q^- = ne^-$ or $Q^+ = np^+$.

Electronic phenomena are determined from the gravitational and electric forces acting between charge carriers. A force of attraction exists between their masses. Like charges show a mutual force of repulsion and unlike charges a force of attraction. For elementary particles the gravitation forces (due to their masses) are so very small compared with the electric forces that the masses of such particles need be considered only when charges are accelerated.

19. Coulomb's Law. The law of force between charges at rest can be stated as follows: The force of repulsion between like charges concentrated at points in an isotropic medium is proportional to the product of their charges, inversely proportional to the square of the distance between them, and inversely proportional to the dielectric coefficient of the medium in which the charges reside. The force acts in a straight line between the centers of the charges. If the charges are of like sign or polarity, the electric force is one of repulsion; if the charges are of opposite sign, the electric force is one of attraction.

In the rationalized system of units the Coulomb force is

$$F = \frac{Q_1 Q_2}{4\pi r_{12}^2 \epsilon} = \frac{Q_1 Q_2}{s\epsilon}$$

where Q_1, Q_2 = magnitudes of two concentrated charges, r_{12} = distance between centers of two charges, ϵ = dielectric coefficient (permittivity) of medium in which charges are situated, and s = surface area of sphere centered on one charge and just touching the other.

In SI units force is measured in newtons, distance in meters, charge in coulombs (1 C being the charge equal to 6.24×10^{18} elementary charges), and permittivity in coulombs^2 per newton-meter^2 (see Par. **1-24**).

20. Principle of Superposition. In any linear system, i.e., one in which the effect produced is proportional to the cause, each cause acting separately produces its separate effect.

When several causes act simultaneously, the resultant effect is the sum of the effects produced by each cause acting separately. Coulomb's law implies that the principle of superposition applies for electric charges. If the principle of superposition holds for any system, that system is linear. The total resultant effect at any point is then the vector sum of the individual component effects at that point.

21. Electric Field. The region in which an electric charge exerts a measurable force on another (above and beyond the gravitational force between the two) is called an *electric field*. Since the force has magnitude and direction, it is a vector field. The direction of the field is represented by lines designating the direction of the force which the field exerts on a small, isolated charge acting as a test body (see Par. **1-22**).

22. Test Body for Electric Field. An electric field can be detected and its direction and magnitude can be determined by measuring the force acting on an isolated charge used as a test body. A positive charge has been arbitrarily selected for this purpose. Since a charged test body has a field of its own and modifies the field to be measured, the electric field can be uniquely ascertained only if the charge on the test body is as small as possible and the polarity of the test charge is stated.

23. Lines and Tubes of Force. In an electric field, a *line of electric force* is a curve drawn so that at every point it has the direction of the force acting on a charge used as a test body; i.e., it has the same direction as that of the field.

A *tube of force* is obtained by drawing a number of lines of force through the boundary of any small closed curve. The lines then form a tubular surface which is not cut by any other lines of force.

24. Permittivity, Dielectric Coefficient, Electric Susceptibility. For a given distribution of charges, the forces between them depend on the environment in which they are located. The greatest force occurs when charges are in free space (vacuum); the presence of material bodies reduces the Coulomb force.

The *permittivity* of any dielectric can be expressed as the product of two terms. One accounts for the dielectric properties of material bodies. The other may be regarded as accounting for the dielectric properties of free space but is actually a factor that provides a consistent set of units for Coulomb's law when units of force, charge, and distance are set up arbitrarily.

If ϵ is the permittivity of any substance and ϵ_0 is the permittivity of free space (vacuum),

$$\epsilon = \epsilon_0 \epsilon_r$$

where ϵ_r is the *relative permittivity* of a material substance referred to that of vacuum. In general, ϵ_r is a complex number. Its quadrature (imaginary) term can usually be neglected, except at very high frequencies. The real component of ϵ_r is often called the *dielectric constant*. Since its value depends upon frequency and other factors, the term *dielectric coefficient* is preferable.

The permittivity can also be written as

$$\epsilon = \epsilon_0 (1 + \chi_e)$$

where χ_e is the *electric susceptibility* of a dielectric. The electric susceptibility is a numeric measure of the polarization or displacement of electrons in atoms or molecules of a dielectric. The dielectric properties of a material substance can be expressed in terms of relative permittivity ϵ_r or electric susceptibility χ_e in simple numerics. Such properties can also be expressed in terms of the absolute permittivity of the substance ϵ in coulombs^2 per newton-meter^2 or farads per meter.

In SI units the permittivity of free space is $\epsilon_0 = 8.854 \times 10^{-12}$ C^2/N·m^2 (or 8.854×10^{-12} F/m).

25. Surface and Volume Charge Density. A region of space containing a large number of closely spaced charges is equivalent to a continuous distribution of charges when viewed at a distance large compared with the separation between charges. The effects of such charges can be described in terms of charge density.

Surface charge density σ is defined as the ratio of the charge Q to the surface area s on which it is deposited, or $\sigma = Q/s$. The surface density at a point on the surface is

$$\sigma = dQ/ds$$

where dQ is the amount of charge on the surface element ds at the point where the surface density is determined. It is measured in coulombs per square meter.

Volume charge density ρ is the amount of charge Q per unit volume v, or $\rho = Q/v$. The volume charge density at a point is

$$\rho = dQ/dv$$

where dQ is the amount of charge in the element of volume dv at the point. It is measured in coulombs per cubic meter.

26. Electric Field Strength. The electric field strength (electric field intensity) at a given point is a vector quantity defined as the quotient of the force (that a small stationary test-body charge at that point will experience) to the charge as the charge approaches zero.

If an increment of force ΔF is produced on an increment of charge ΔQ at a point in an electric field, the electric field strength at the point is

$$E = \lim_{\Delta Q \to 0} \frac{\Delta F}{\Delta Q} = \frac{dF}{dQ}$$

Since F is a vector quantity whereas Q is a scalar, the field strength E is a vector quantity having the same direction as that of the force on the test body. In SI units, the electric field strength is measured in newtons per coulomb. The common unit is volts per meter.

27. Dielectric Strength. The dielectric strength is the maximum field strength that a dielectric can sustain without breakdown. It is determined experimentally by tests on materials and in SI units is expressed in volts per meter of thickness of the sample tested. The results obtained depend upon the method and conditions of the test as well as upon the thickness of sample being tested, being greater for thin rather than thick samples. Dielectric strength is usually expressed in volts per centimeter, volts per millimeter, kilovolts per centimeter, or volts per mil (1 mil = 0.001 in).

28. Field Strength Produced by Point Charges. Coulomb forces obey the principle of linear superposition in free space, as well as in many dielectrics. Therefore the electric field intensity at a point p produced by charges Q_1, Q_2, ... , Q_n whose distances from p are r_1, r_2, ... , r_n is the vector sum of the field strengths produced at the point by each charge individually. In a medium whose permittivity is ϵ, if the component field strengths are E_1, E_2, ... , E_n, the resultant field intensity of point charges is

$$E = E_1 + E_2 + \cdots + E_r = \frac{Q_1}{4\pi\epsilon r_1^2} + \frac{Q_2}{4\pi\epsilon r_2^2} + \cdots + \frac{Q_a}{4\pi\epsilon r_n^2}$$

In SI units, the field strength is in volts per meter when charge is specified in coulombs, distance in meters, and permittivity in the units given in Par. **1-24**.

29. Field Strength Produced by Distributed Charges. If charges are distributed continuously through a volume (instead of being located individually at discrete points), the total charge in volume v produced by charge density ρ is $Q = \int \rho \, dv$ and the field strength at a distance R (large compared with the dimensions of the volume in which charges are distributed) is given by the volume integral

$$E = \int \int \int_{\text{vol}} \frac{\rho \, dv}{4\pi\epsilon R^2}$$

In general ρ and R may be functions of the triple integral over the volume.

30. Electrostatic Potential and Potential Difference. In general, energy must be expended to move a charge through an electric field. The amount of energy required per unit of charge is called *electric potential*. While it is not possible to specify uniquely the *absolute potential* of any point in an electric field, the *difference in potential* between two points in the field can be determined when it is known how the electric field strength varies between the points.

The electrostatic potential difference between two points in an electric field resulting from a static distribution of electric charge is the scalar-product line integral of the electric field strength. It can be integrated along any path from one point to the other.

The potential difference between two points a and b in an electric field of average strength E is

$$V_{ab} = -\int_b^a E \cdot dl = -\int_b^a E \cos \theta \, dl$$

where θ is the angle between the direction of the electric field and the direction of the line element dl. The potential difference V_{ab} is taken to be positive if work is done in carrying a positive charge from b to a. In SI units potential difference is measured in volts (equivalent to joules per coulomb).

31. Field Strength and Potential Difference. Potential difference is the change in energy per unit charge as a charge is moved between two points in an electric field in the direction of the field. The energy is negative if work is done in carrying a positive charge between points a and b.

The differential potential difference can be expressed as

$$dV = -(F \cdot dl)/Q = -E \cdot dl$$

from which

$$E = -dV/dl$$

Hence, electric field strength is the negative rate of change of potential difference with respect to the distance a charge moves in an electric field; i.e., it is the gradient of the potential field. In SI units field strength is measured in volts per meter.

32. Voltage Rises and Voltage Drops. The amount of energy per charge between two points, whether due to separation of charges (as in electrostatics) or to changes of energy in dynamic processes, is commonly called voltage and is measured in volts.

A voltage drop exists when the energy of charges is diminished as charges do work. A voltage rise exists when the energy of charges is increased by some source of energy that does work on them. The total voltage rise is equal to the total voltage drop when voltages are measured in the same direction around a closed loop.

33. Potential Field of Point Charges. At any point the potential due to a number of charges establishing an electric field is the scalar sum of the potentials produced at that point by each charge acting alone.

If $Q_1, Q_2, \ldots, Q_n$ are the charges whose distances from a point at which the potential of the field is to be determined are $r_1, r_2, \ldots, r_n$, the potential at that point is

$$V = V_1 + V_2 + \cdots + V_n = \frac{1}{4\pi\epsilon}\left(\frac{Q_1}{r_1} + \frac{Q_2}{r_2} + \cdots + \frac{Q_n}{r_n}\right)$$

where ϵ is the permittivity of the medium in which charges reside.

34. Potential Field of Distributed Charges. If charges are distributed continuously through an element of volume dv in which the charge density is ρ, the potential at a point whose distance from dv is r is

$$V = \frac{1}{4\pi\epsilon}\int\int\int_{vol}\frac{\rho\,dv}{r}$$

where ϵ is the permittivity of the medium.

35. Electric Dipole and Dipole Moment. A combination of two point charges of equal magnitude and opposite polarity separated by a distance small compared with that at which the field of the dipole is to be determined is called an *electric dipole* (or a *doublet*). The dipole moment is the product of the magnitude of either charge of a dipole and the distance separating the two charges. In SI units dipole moment is measured in coulomb-meters; its direction is the direction of the line joining the two charges.

36. Field of a Dipole. Let a dipole consist of charges $+Q$ and $-Q$ separated by a distance d in a medium of permittivity ϵ. The center of the dipole is at the origin, and the field is to be determined at a point P (so remote from the origin that we can take $R_1 = r = R_2$ with negligible error) and at a direction measured by the angle θ. It is desired to determine the dipole field at point P, relative to that at the origin O (see Fig. 1-1).

The potential produced at point P is given by

$$V_{OP} = Qd/4\pi\epsilon r^2$$

whereas the electric field strength of the dipole at point P is

$$E_{OP} = (Qd/4\pi\epsilon r^3)(2\cos\theta + \sin\theta)$$

In SI units, V is in volts, E in volts per meter, Q in coulombs, r in meters, and permittivity ϵ in coulombs^2 per newton-meter^2.

37. Electric Polarization. At any point in a dielectric the electric polarization is the electric dipole moment per unit volume. It can also be expressed as the amount of bound charge per unit area perpendicular to the direction in which the charges of the dipole are displaced. If Q is the amount of bound charge displaced by distance d, the dipole moment is $p = Qd$. If $v = sd$ is the volume of a dielectric, d the distance between dipole charges, and s the area perpendicular to the direction in which bound charges move, the polarization is given by

$$P = p/v = Qd/v = Q/s$$

Electric polarization, in SI units, is measured in coulomb-meters per cubic meter (coulombs per square meter).

38. Electric Flux (Electric Displacement). The electric flux or displacement is a quantity associated with the amount of bound charge Q displaced in a dielectric which is subject to an electric field. The magnitude of the flux is proportional to the amount of bound charge displaced. In SI units the magnitude of the electric flux is equal to the amount of bound charge and is measured in coulombs. It is a scalar quantity.

39. Electric Flux Density. The electric flux density (also called the displacement density) at a point in an isotropic dielectric is a vector having the same direction as the electric field strength and a magnitude equal to the product of the electric field strength E and the permittivity of the medium $\epsilon = \epsilon_0\epsilon_r$, or

$$D = \epsilon E = \epsilon_0\epsilon_r E$$

Since $E = F/Q$ and, from Coulomb's law, $F = Q_1Q/\epsilon s$, where s is the closed surface surrounding charge Q, it follows that $D = Q/s$. D is measured in coulombs per square meter.

In nonisotropic media, ϵ becomes a tensor represented by a matrix, and D and E are not necessarily in the same direction.

40. Gauss' Theorem for Electrostatics. The electrostatic theorem of Gauss is a necessary consequence of Coulomb's law of electrostatics. It states that the amount of electric flux passing through any closed surface is proportional to (and in the rationalized systems of units is equal to) the amount of charge contained within the surface.

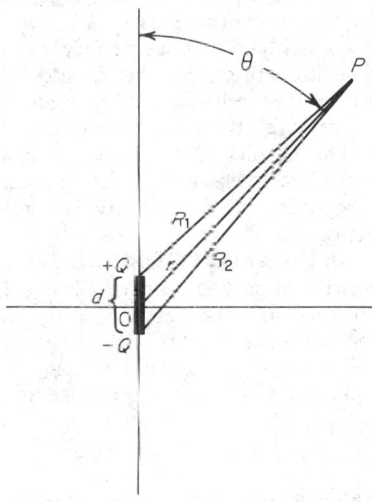

Fig. 1-1. Geometry for determining the field at a point P of an electric dipole consisting of two equal charges of opposite signs, $+Q$ and $-Q$, separated by a small distance d.

If D is the amount of electric flux density crossing any element of surface ds, θ is the angle between the direction of D and the normal to the surface element, and p is the electric charge density in volume element dv, the electric flux in rationalized units is

$$\Psi = \int_{\substack{closed \\ surface}} D \cos \theta \, ds = \int_{vol} \rho \, dv = Q$$

The surface integral is taken over a surface enclosing the volume.

41. Divergence of Electric Flux Density. The divergence of the vector electric flux density is a scalar quantity equal to the ratio of the flux passing through a closed surface to the volume contained therein, as the volume becomes vanishingly small, or

$$\text{div } D = \lim_{\Delta v \to 0} \frac{\int_{\substack{closed \\ surface}} D \cos \theta \, ds}{\Delta v}$$

If D is expressed in terms of rectangular components D_x, D_y, and D_z, the divergence is expressed as the sum of the partial derivatives of each component or

$$\text{div } D = \frac{\partial D_x}{\partial x} + \frac{\partial D_y}{\partial y} + \frac{\partial D_z}{\partial z}$$

42. Electric Field Vectors. The properties of electric fields in dielectrics can be expressed in terms of three field vectors: (1) electric field strength E, related to all charges, (2) electric flux density D, related only to free charges on the conducting surfaces at a boundary between dielectrics and conductors, and (3) polarization, or electric dipole moment per unit volume of dielectric P, related to the bound charges within a material dielectric. All three field vectors can be expressed in terms of surface charge density Q/s.

The three electric field vectors are connected by the general relation

$$D = \epsilon_0 E + P$$

For free space $P = 0$, and $D = \epsilon_0 E$. For dielectric materials

$$D = \epsilon E = \epsilon_0 \epsilon_r E \quad \text{and} \quad P = (\epsilon_r - 1)\epsilon_0 E$$

The last equations are generally true for homogeneous, isotropic dielectric media, and the relative permittivity is usually independent of electric field strength E. For certain dielectrics, called *ferroelectrics*, ϵ is a function of field strength E, and the ferroelectrics show properties of electric hysteresis (see Par. **1-13**).

43. Field Vectors at Dielectric Boundaries. The tangential components of electric field strength are continuous across the boundary, i.e., the same on both sides of the boundary. At the interface of a conductor and a dielectric the tangential components of the electric field are zero, i.e., they are the same.

The normal or perpendicular component of electric flux density changes at a boundary between two charged dielectrics by an amount equal to the surface charge density. The normal component of flux density is continuous across a charge-free boundary between the two dielectrics.

44. Potential Gradient. In an electric field, the maximum value of change of potential V with respect to distance l is equal to the magnitude of the electric field strength E and is obtained when the direction of the electric field strength is opposite that in which the potential increases most rapidly with distance.

Thus, the potential gradient at a point in an electric field is the limiting value of the ratio of a change in potential to a change in distance as the latter becomes vanishingly small:

$$\nabla V = \lim_{\Delta l \to 0} \frac{\Delta V}{\Delta l} = \frac{dV}{dl} = -E$$

The potential gradient is measured in volts per meter. A negative sign is used because a rise in potential requires the positive test charge to be moved in a direction opposite to that taken to be the direction of the electric field.

45. Force on a Charge in an Electric Field. The electric field strength is defined as the force per charge acting on a charge in an electric field, or $E = F/Q$. Therefore, the force acting on a charge Q in an electric field of intensity E is

$$F = QE$$

In SI units force is measured in newtons, charge in coulombs, and field strength in newtons per coulomb (volts per meter).

46. Work Done on Moving Charges. When a charge Q in an electric field of intensity E is moved a distance dl at an angle θ with respect to the direction of the field, the amount of work done is

$$dW = QE \cos \theta \, dl$$

and the total work done in moving the charge between two points a and b in the field is

$$W = -\int_b^a QE \cos \theta \, dl$$

In SI units work is in joules, charge in coulombs, and field strength in volts per meter.

47. Conservative Properties of Electrostatic Fields. No work is done in carrying a charge around a closed path in an electrostatic field, for

$$W = -\oint QE \cdot dl = -\left(\int_b^a QE \cdot dl - \int_a^b QE \cdot dl \right) = 0$$

This result is true for electrostatic fields but must be modified for time-varying fields. In a conservative field the work done in moving a particle between two points is independent of the path taken. A field satisfying the equation given above is said to be a *conservative field*.

48. Storage of Electric Charges. If charges Q^+ and Q^- are collected on conductors separated by free space or by a dielectric material, the mutual attraction between the two charges tends to hold the charges in place indefinitely. The charges are then said to be stored. They produce a stress in the intervening dielectric, measured by the electric field strength and a voltage difference exists between the two charge-bearing conductors

49. Capacitance and Elastance. Capacitance is the property of a system of conductors and dielectrics that permits the storage of electrically separated charges when a potential difference exists between the conductors. Elastance is the reciprocal of capacitance.

Elastance may be defined as the amount of energy U per charge Q required to transfer unit charge between two conductors separated by a dielectric. Quantitatively elastance is given by

$$S = \frac{U/Q}{Q} = \frac{V}{Q}$$

where V is the voltage difference required to transfer the charge Q between the two conductors. Capacitance is given by the inverse relation

$$C = 1/S = Q/V$$

50. Capacitors. A capacitor is a device consisting of conductors separated by a dielectric (which may be air or vacuum) for introducing capacitance into an electric circuit or system or for providing for the storage of electric charge.

51. Capacitance between Two Conductors. For systems of simple geometry, the capacitance between two conductors separated by a single, homogeneous, isotropic dielectric may

TABLE 1-2 Capacitance Relationships

Geometry	Capacitance C
Parallel plates, closely spaced	$\dfrac{\epsilon s}{l}$
Coaxial cylinders of length l	$\dfrac{2\pi\epsilon l}{\ln\,(r_2/r_1)}$
Concentric spheres	$\dfrac{4\pi\epsilon}{1/r_1 - 1/r_2}$
Isolated sphere	$4\pi\epsilon r$

be calculated in terms of the physical dimensions of the conducting electrodes and the permittivity of the dielectric substance (Table 1-2).

In Table 1-2 $\epsilon = \epsilon_0\epsilon_r$ is the permittivity of the dielectric in coulombs^2 per newton-meter^2, and capacitance is in farads when the dimensions are expressed in meters. The distance between parallel plates is l, and the plates have an area of s, whereas r_1 and r_2 are the radii of inner and outer conductors. The effects of fringing are neglected.

52. Energy of a Charged Capacitor. The total energy of the electric field in a capacitor whose capacitance is C when charged to potential difference V is

$$U = VQ$$

If the plates were initially at the same potential ($V = 0$), the average energy needed to charge the capacitor to voltage V is

$$U_{av} = \tfrac{1}{2}VQ = \tfrac{1}{2}V^2C = \frac{1}{2}\frac{Q^2}{C}$$

The energy is in joules if the charge is in coulombs and the potential difference in volts.

53. Energy Density of an Electric Field. In a dielectric of permittivity ϵ, in the form of a rectangular solid with surface area s and length l, a voltage difference V is needed to transfer a charge Q between the surfaces. The average energy stored in the dielectric in this charge transfer is

$$U_{av} = \tfrac{1}{2}VQ = \tfrac{1}{2}(El)(Ds) = \tfrac{1}{2}EDls = \tfrac{1}{2}EDv = \tfrac{1}{2}\epsilon E^2 v = \frac{1}{2}\frac{D^2}{\epsilon}v$$

where D is the flux density on the charge surfaces and $v = sl$ is the volume of the dielectric. Since $E = D/\epsilon$, the average energy density at a point in the field is

$$u_{av} = \frac{U_{av}}{v} = \tfrac{1}{2}\epsilon E^2 = \frac{1}{2}\frac{D^2}{\epsilon}$$

54. Bibliography on Electrostatics. See Par. **1-121.**

ELECTROKINETICS

55. Moving Charges. Mobile carriers of charge Q^+ and Q^- and masses m^+ and m^- situated in a field whose electric intensity is E are acted upon by a force

$$F = QE = (Q^+ + Q^-)E = Q^+E + Q^-E$$

Mobile carriers of opposite polarity move in opposite directions under the influence of the same electric field and have accelerations inversely proportional to their masses.

When a particle of mass m moves with velocity v, the force required to give it acceleration a is

$$F = \frac{d}{dt} mv = m\frac{dv}{dt} + v\frac{dm}{dt} = ma + v\frac{dm}{dv}\frac{dv}{dt} = \left(m + v\frac{dm}{dt}\right)a$$

The component of force ma lies in the direction of the acceleration, but the component $v\ dm/dt$ lies in the direction of the velocity v.

According to the theory of relativity, the mass of a moving particle is not constant but increases with its velocity v relative to the velocity of light c. If m_0 is the mass of a particle at rest relative to an observer, its effective transverse mass when moving with velocity v is

$$m_t = m_0/(1 - v^2/c^2)^{1/2}$$

and its effective longitudinal mass is

$$m_l = \frac{dm}{dt} = \frac{mv/c^2}{1 - v^2/c^2} = \frac{m_0}{(1 - v^2/c^2)^{3/2}}$$

In terms of the rest mass, the longitudinal component of force acting on the particle is

$$F_{\parallel} = \frac{ma}{1 - v^2/c^2} = \frac{m_0}{(1 - v^2/c^2)^{3/2}} a = m_l a$$

In accordance with the principle of relativity, this is true whether F and a are in the same or in opposite directions.

The transverse force acting on moving particles sets them moving in a circular path with radius R and is

$$F_{\perp} = ma = m\frac{v^2}{R} = \frac{m_0}{1 - v^2/c^2}\frac{v^2}{R}$$

The kinetic energy of the particle is

$$U_k = \int F \cos\theta\ dl = \int F\ v \cos\theta\ dt = c^2\ dm/dt$$

where F = force acting, $\int dl$ = distance moved, and θ = angle between F and dl. Thus kinetic energy is

$$U_k = \int c^2 \frac{dm}{dt}\ dt = c^2 \int_{m_0}^{m} dt = c^2 m - c^2 m_0 = U_t - U_p$$

where U_t = total energy and $U_p = m_0 c^2$ = potential energy of particle at rest. The total energy is

$$U_t = mc^2 = m_0 c^2 + U_k = U_p + U_k$$

In SI units, force is in newtons, distance in meters, velocity in meters per second, and energy in joules.

56. Speed and Mobility of Charge Carriers. For free charges moving in a vacuum with velocity small compared with that of light, the force acting is $F = ma = QE$, where m = mass of accelerated particle, a = its acceleration, Q = its charge, and E = intensity of electric field in which charge travels.

The drift speed acquired by the charge is

$$v = \int a\ dt = \int \frac{F}{m}\ dt = \int \frac{Q}{m} E\ dt = \frac{Q}{m} t'E = \mu E$$

where t' = mean free time and $\mu = Qt'/m$ = mobility of moving charge. The mobility of charge carriers in semiconductors is treated in Sec. **6**.

57. Electric Current. In a medium having mobile charges of volume density ρ in which the particles move across a surface of area s with average speed v_{av} traversing a mean free path length l in mean free time t, the electric current is the rate of charge transport given by

$$I = \rho s v_{av} = \frac{ne}{v} s v_{av} = \frac{Q}{v} s v_{av} = \frac{Q}{sl} s \frac{l}{t} = \frac{Q}{t}$$

where $v = sl$ = volume and $Q = ne$ = charge transported. The current is thus seen to be the directed, i.e., nonrandom, motion of n elementary particles of charge ϵ passing through a charge-bearing medium per unit of time.

If the charges do not move at constant speed but charge increment dQ passes through a surface s in time dt, the instantaneous value of the current is

$$i = dQ/dt$$

In SI units, current is measured in amperes, charge in coulombs, volume in cubic meters, area in square meters, speed in meters per second, volume charge density in coulombs per cubic meter, and time in seconds.

58. Current and Charge. Transfer of charge Q in time t at a steady rate produces a current $I = Q/t$, and $Q = It$. When an infinitesimal amount of charge dQ is transported in time interval dt, the current is $i = dQ/dt$ and the amount of charge transferred in the interval from t_1 to t_2 is

$$Q = \int_{t_1}^{t_2} i\, dt$$

Current i is measured in amperes when charge Q is in coulombs and time t is in seconds.

59. Current Density. Current density J is defined as the quotient of the amount of current ΔI transferred normal to a surface area Δs_n as the latter becomes vanishingly small. In the limit

$$J = dI/ds_n$$

If θ is the angle between the motion of charges and the normal to the surface across which charges pass, the effective surface element is $ds_n = \cos\theta\, ds$ and the current density is

$$J = \frac{dI}{\cos\theta\, ds}$$

J is a vector quantity measured in amperes per square meter in SI units.

60. Current Element. A quantity of charge remains constant whether stationary or in motion. Charges in motion pass through a conductor of length dl in time interval dt and have velocity $v = dl/dt$. Hence a quantity of moving charge can be expressed in terms of current as

$$Q = \int I\, dt = \int \frac{I}{v}\, dl$$

On the basis of this relation it is convenient to define a current element as

$$Qv = \int I\, dl$$

In SI units the current element is measured in ampere-meters, equivalent to coulomb-meters per second.

61. Sense of Electric Current. Mobile charges move in the direction of the electric force acting on them. Thus it is possible to have a current I^- due to the motion of mobile negative charges or a current I^+ due to motion of mobile positive charges. If charges of both polarity are mobile, the total current is the sum $I = I^- + I^+$.

When it is important to specify the direction of an electric current, the polarity of charge carriers constituting the current must be stated (see Par. **1-22**).

62. Conduction Current. Conduction current is measured by the amount of mobile charge transferred per unit time through a conductor. The transferred charges may be electrons (in metals and semiconductors), holes (in semiconductors), or ions (in gases and electrolytes).

The conduction current can be expressed as

$$I = \sigma\,(s/l)V = GV$$

where σ is a property of a unit volume of the conducting material called its *conductivity*, s is the area of the surface through which charges pass, and l is the distance along the conductor across which the difference of potential is V. The conductivity σ depends on properties of the conducting material (see Par. **1-69**). The quantity G is called the *conductance* of the conductor.

63. Convection Current. The convection current is usually regarded as the motion of charge density, as in the motion of electrons in formed beams in electron tubes. It is expressed as

$$I_{conv} = \rho v s$$

where ρ is the volume charge density of charge carriers moving with velocity v across a surface area s. In SI units current is in amperes, volume charge density is in coulombs per cubic meter, velocity is in meters per second, and area is in square meters.

64. Displacement Current. Electric fields of elementary charges extend considerably beyond the dimensions of charge carriers and can be expressed in terms of electric flux. The expression for current, $I = psv$, can be applied to show that a time-changing electric flux in a dielectric produces a *displacement current*.

The transfer of free charge dQ to conducting surfaces on opposite sides of a dielectric sets up flux $d\psi$ in the dielectric, and the charge density is $\rho = dQ/dv = d\psi/dv$. Since the volume v is the product of the surface area s and length l, and since $dv = s\, dl$, the displacement current is

$$I_{dis} = \rho sv = \frac{d\psi}{dv} sv = \frac{d\psi}{s\, dl} s \frac{dl}{dt} = \frac{d\psi}{dt}$$

Since $\psi = Ds$ and $D = \epsilon E$, the displacement current in the dielectric can be written in terms of electric field strength as

$$I_{dis} = \frac{d\psi}{dt} = \frac{d}{dt} Ds = s\frac{dD}{dt} = s\frac{d}{dt}\epsilon E = s\epsilon \frac{dE}{dt}$$

The displacement current is zero if the electric flux remains constant with respect to time.

65. Total Current. Total current is the sum of the conduction, convection, and displacement components in a circuit or system. The conduction and convection components are called *true currents* since they represent transfer of charges. Displacement current is a *virtual* current, arising from the time rate of change of electric flux of charges that do not pass through a dielectric.

66. Continuity of Current. Charges can neither be created nor destroyed. Thus, for steady flow of charges, the outward motion of positive charge through any closed surface must be equal to the decrease of positive charges within the surface. If Q denotes the charge inside a closed surface and J_n is the current density normal to the surface s across which charges move, the equation of continuity of current is

$$I = \int J_n \, ds = -\, dQ/dt$$

If ρ is the volume charge density, the divergence of steady current through a closed surface is the time rate of decrease of charge density, which is zero over a closed surface. For time-varying currents i, the displacement current $\partial\psi/\partial t$ must also be taken into account, and the divergence becomes

$$\text{div } (i + \partial\psi/\partial t) = 0$$

67. Conduction in Crystals. Because of their wave nature, electrons are believed to be able to pass through a perfect crystal without encountering any resistance. Opposition to motion of electrons arises from deviations in periodicity of the potential in which electrons move through a crystal. Such deviations may be due to (1) lattice vibrations, (2) vacancies, holes, or other defects in the lattice, (3) foreign impurity atoms, or (4) grain boundaries.

In a pure metal at finite temperature, the ions are not at rest but vibrate thermally about their positions of equilibrium. Such vibrations cause scattering of electrons, which accounts for resistance to motion of charges. The ion vibrations and scattering increase with temperature, and resistance also increases with rise in conductor temperature.

68. Current and Voltage Relations in Conductors. The electric current through a conductor may be regarded as the effect produced by the energy expended to keep charges moving through the conductor. For a given applied voltage V the current I depends upon the nature of the material of which the conductor is made and its dimensions. In most metals and other good conductors, the current can be expressed as

$$I = (\sigma s/l)V = GV = V/R$$

where $G = \sigma s/l$ = conductance, $R = V/I$ = resistance of conductor or system, σ = electric conductivity of conducting material, l = length of conductor across which voltage difference V exists when current I passes through it, and s = cross-sectional area of conductor, assumed constant.

In SI units current is measured in amperes, potential difference in volts, and conductance in amperes per volt, to which the name siemens is given. The resistance is measured in volts per ampere, or ohms.

Many cases occur in which the current through a conductor is not directly proportional to voltage across it. Such devices are said to be nonlinear, and the voltage-current relations are best shown by graphs relating V and I.

Frequently current depends upon some parameter other than voltage, such as temperature, pressure, incident radiation, etc.

69. Resistivity and Conductivity. *Conductivity* is a factor expressing the ease with which an electric current is able to pass through a conducting material. It is given by

$$\sigma = Il/Vs = J/E$$

where J = current density in conductor, E = electric field strength along conductor, s = area of conductor across which current passes, and V = voltage along length of conductor l. In SI units it is measured in (amperes per square meter) per (volt per meter), which is equivalent to siemens per meter.

Resistivity is a factor expressing the opposition to the motion of charges through a conductor of unit length and unit cross-sectional area in the system of units used. It is the reciprocal of conductivity and is given by

$$\rho = 1/\sigma = Vs/Il = E/J$$

where the symbols have the meaning given above. Resistivity is measured in (volts per meter) per (amperes per square meter) in SI units, or in ohm-meters.

70. Temperature Variation of Resistivity. The resistivity of highly conductive solid metals increases with temperature. Over a fairly wide temperature range, the increase in resistivity is proportional to the fractional change in absolute temperature.

If ρ_{T_0} is the resistivity of a conducting material at a reference temperature T_0, and if ρ_{T_1} is the resistivity of the same material at some other temperature T_1, then the resistivity at the non-reference temperature is

$$\rho_{T_1} = \rho_{T_0}[1 + (\kappa/\rho_{T_0}T_0)(T_1 - T_0)] = \rho_{T_0}(1 + \alpha \, \Delta T)$$

where $\alpha = \kappa/\rho_{T_0}T_0$ = temperature coefficient of resistivity at reference temperature T_0 and $\Delta T = T_1 - T_0$ is temperature change (K). The values for α are determined experimentally and are given in Sec. **6**. At room temperatures, the temperature coefficient of resistivity of most metallic conductors lies between 0.002 and 0.005.

71. Resistance. The resistance of a linear conductor, in which current is proportional to applied voltage, is regarded as the opposition offered to the transport of charges through it. It is expressed as the ratio of the constant voltage difference between the ends of a conductor, through which current passes, to the resulting current. Since the voltage V across a conductor is the amount of energy U expended per unit charge Q and the current through the conductor is the amount of charge Q transferred per unit time t, the resistance is given by

$$R = \frac{U/Q}{Q/t} = \frac{V}{I}$$

Resistance is measured in units of ohms, or volts per ampere.

The resistance of a nonlinear conductor, in which current is not directly proportional to applied voltage, is called the *variational* or *differential resistance*. It is measured as the limiting value of the quotient of a small change ΔV in voltage to the resulting small change ΔI in current as the latter becomes vanishingly small.

From another point of view, resistance is the factor by which the mean-square conduction current must be multiplied to give the corresponding power lost by dissipation as heat or put to some useful purpose.

72. Conductance. The conductance of a linear conductor, in which the voltage drop is proportional to the current, is regarded as the ease with which charges are transferred through it. It is expressed as the ratio of the current through the conductor to the voltage difference across it

$$G = \frac{Q/t}{U/Q} = \frac{I}{V}$$

Conductance is measured amperes per volt, or siemens.

73. Resistance of a Conductor. The resistance of a conductor of length l, uniform cross-sectional area s, and material resistivity ρ_0 at temperature T_0 is

$$R_0 = V/I = El/Js = \rho_0\, l/s$$

The resistance of the conductor is in ohms when l is in meters, cross-sectional area s is in square meters, resistivity ρ is in ohm-meters, and temperature T_0 is in kelvins.

74. Resistor. A resistor is a device the primary purpose of which is to introduce resistance into an electric circuit or system. The ideal resistor dissipates electric energy without storing electric or magnetic energy.

75. Work Done by Electric Current. Work is done on charge carriers when they are accelerated by an electric field. Conversely, the charges do work or dissipate energy when they are decelerated. Their energy is dissipated as heat (or other lost energy) or in work done by the system in which they are decelerated.

If a charge Q is decelerated and produces a steady voltage drop V when a steady current I passes through a conductor, the work done in time t is

$$W = U = VQ = VIt$$

If charges do not move at a uniform rate, the voltage and current vary with respect to time and can be represented by v and i, respectively. In this case, the amount of work done in the time interval from $t = t_1$ to $t = t_2$ is

$$W = U = \int_{t_1}^{t_2} vi\, dt$$

The work done is in joules when the potential difference is in volts, current in amperes, and time in seconds.

76. Energy Dissipated in a Resistor. The energy dissipated in a resistor is

$$U = VIt = (V^2/R)\, t = I^2Rt$$

for charges transported at steady rate, or

$$U = \int_{t_1}^{t_2} vi\, dt = \int_{t_1}^{t_2} \frac{v^2}{R} dt = \int_{t_1}^{t_2} i^2R\, dt$$

for charges transferred at a nonuniform rate.

77. Electric Power. If a steady current I and a steady voltage V produce work W in time interval t, the time rate of energy conversion (electric power) is

$$P = W/t = VIt/t = VI = V^2/R = I^2R$$

If the current and voltage vary with time and are expressed as i and v, the instantaneous power p in a circuit of resistance R is

$$p = vi = v^2/R = i^2R$$

In systems in which voltage and current vary with time according to simple harmonic motion, the average value of power is

$$P_{av} = \tfrac{1}{2}V_mI_m \cos\theta = VI \cos\theta$$

where the subscripts m = maximum or amplitude values of voltage and current whose effective values are V and I, $\cos\theta$ = power factor, and θ = phase difference between current and voltage.

78. Joule's Law of Heating Effect. Joule's law states that the rate at which heat is produced in an electric circuit of resistance R is equal to the product of the resistance and the square of the current. If I is the steady current through a resistor, the steady power dissipated in it is

$$P = I^2R$$

If i is the instantaneous value of time-varying current, the instantaneous power developed in resistor R is

$$p = i^2 R$$

Power is in watts when current is in amperes and resistance in ohms.

79. Faraday's Law of Electrolysis. In the process of electrolytic change, the mass of a substance deposited at one electrode or liberated at the other varies (1) directly as the quantity of electricity, (2) directly as the atomic weight of the substance deposited, and (3) inversely as its chemical valence. The Faraday electrolytic constant is the amount of charge required to liberate one gram atom of any univalent element. This quantity of electricity is 96,485 C/mol.

Faraday's law of electrolysis can be stated quantitatively as

$$m = \frac{a}{e} \frac{WQ}{v} = \frac{1}{F} \frac{WQ}{v}$$

where m = mass of substance deposited (g), a = mass of atom having unit atomic weight (g), Q = quantity of electricity passed through cell (C), W = atomic weight, v = valence of atoms deposited or liberated, e = charge of electron (C), and $a/e = 1/F$, where F = Faraday constant = 96,485 C/mol.

80. Bibliography on Electrokinetics. See Par. **1-121**.

MAGNETOSTATICS

81. Transverse Forces between Moving Charges. Mobile charges may be set into motion when acted upon by an electric field, and in motion, the charges carry their own electric fields with them. The three-dimensional field of charges can be resolved into a longitudinal component in the direction of motion and transverse components perpendicular to the direction of motion. The perpendicular components produce transverse, magnetic forces between moving charges. The forces also act between the conductors in which the charges travel. A *magnetostatic field* is produced when the charges move with uniform motion, a *magnetokinetic field* when the charges move nonuniformly.

The law of magnetic force between current elements is attributed to Biot and Savart and also to Ampère and Laplace. This law is as basic to magnetostatics and magnetokinetics as Coulomb's law is to electrostatics and electrokinetics.

82. Biot-Savart/Ampère-Laplace Law. Experiment and theoretical reasoning show that the increment of force between two parallel current elements[1] is

$$d(dF) = \frac{\mu I_1 \, dl_1 \, I_2 \, dl_2}{4\pi r_{12}^2} \sin \theta$$

where I_1 and I_2 are values of steady currents in conductors whose elements of length are dl_1 and dl_2, respectively, r_{12} is the straight-line distance between dl_1 and dl_2, μ is a property of the medium in which the current-carrying conductors reside, and θ is the angle between the direction of r_{12} and the perpendicular from a current element and the direction of the parallel conductors.

Experimentally it is not feasible to check the validity of the Biot-Savart law since differential current elements cannot be isolated. Currents flow in closed paths of finite length. The force between two closed current-carrying conductors can be measured experimentally. The force produced by currents in closed paths can be expressed mathematically by integrating the expression for $d(dF)$ over the total path lengths l_1 and l_2 by the double line integral

$$F = \oint \left(\oint \frac{\mu I_1}{4\pi r_{12}^2} \, dl_1 \right) I_2 \sin \theta \, dl_2$$

While this equation has theoretical importance, the integrations required are too difficult (except for conductors of simple geometry) for practical application.

---

[1] For discussion of the transverse force of charges moving in nonparallel paths, see M. Mason and W. Weaver, "The Electromagnetic Field," Dover, New York, 1929, or A. O'Rahilly, "Electromagnetic Theory: A Critical Examination of Fundamentals," 2 vols., Dover, New York, 1965.

If the first and second current elements contain N_1 and N_2 turns with currents I_1 and I_2, respectively, the force must be multiplied by N_1N_2. The force equation then takes the form

$$F = \oint \left(\oint \frac{\mu N_1 I_1}{4\pi r_{12}^2} \, dl_1 \right) N_2 I_2 \sin \theta \, dl_2 = \oint B_1 N_2 I_2 \sin \theta \, dl_2$$

The last expression on the right defines a field vector B, called the magnetic flux density (see Par. **1-87**).

83. Principle of Superposition. Magnetic forces are produced by the interaction of a pair of current-carrying conductors or their equivalent. One of these establishes a magnetic field, and the other exhibits a force reaction to the field. The magnitude of the magnetic field is proportional to the total current setting up the field, provided the permeability is constant. For a given field, the force on a current-carrying conductor is proportional to the current through it.

For magnetic fields, the principle of linear superposition applies when the permeability of the medium is constant, as it is for free space and most nonmagnetic substances. The permeability of iron and other ferromagnetic materials is not constant. In general, therefore, the principle of linear superposition does not apply for current-carrying conductors in the vicinity of ferromagnetic materials.

84. Magnetic Field. A magnetic field may be defined as a vector property of space or material bodies capable of exerting a force on a moving charge to accelerate the charge in proportion to its magnitude and its speed relative to that of the field. A magnetic field may be produced by a permanent magnet or by charges moving in a conductor with directed motion. In general the magnetic field may vary with time and is a vector function of the position at which a force acts in a transverse direction on the moving charge; it also depends upon the properties of the medium in which the charges move.

85. Test Body for Magnetic Field. The magnitude and direction of a steady magnetic field of flux density B can be determined by measuring the transverse force F acting on an isolated charge Q moving with velocity v at a point in the magnetic field. The magnetic flux density is in the direction of the force and has magnitude

$$B = F/Qv$$

The flux density is in units of teslas (webers per square meter) when force is in newtons, charge is in coulombs, and velocity is in meters per second.

The magnetic field can also be measured by the torque exerted on a small plane test coil of area s carrying a small current I if the coil is free to turn in any direction when situated at the point at which the magnetic field is to be assessed. The magnitude of the magnetic flux density is

$$B = \lim_{s \to 0} (T/Is)$$

where B = flux density (T), T = torque on the coil (N·m), I = current (A), and s = area of the test coil (m^2). At equilibrium, the direction of B is perpendicular to the plane of the coil in the sense that a right-hand screw would move if turned in the direction of current in the coil.

86. Permeability and Magnetic Susceptibility. The magnetic effect of material substances can be specified in terms of the *magnetic permeability* μ, which can be written as the product of two terms. One of the terms may be taken to express the magnetic properties of free space but is better regarded as a constant relating units in mechanical and electromagnetic systems. The other term accounts for the magnetic properties of material substances.

If μ is the permeability of any substance and μ_0 is the permeability of free space (or vacuum),

$$\mu = \mu_0\mu_r$$

where μ_r is the relative permeability of a material substance referred to that of vacuum. In SI units, $\mu_0 = 4\pi \times 10^{-7}$ H/m.

The permeability can also be written as the sum

$$\mu = \mu_0(1 + \chi_m)$$

where χ_m, a dimensionless quantity called the *magnetic susceptibility* of a substance, is a measure of the alignment of atoms or molecules in a magnetic material.

87. Magnetic Flux Density. Magnetic flux density can be looked upon from two different points of view. In terms of the force which the first current element $N_1 I_1 \, dl_1$ exerts on a second current element $N_2 I_2\ dl_2$, the physical significance of magnetic flux density can be expressed as

$$B_1 = \left(\frac{dF}{N_2 I_2 \sin\theta\ dl_2} \right)_{max} = \frac{dF_{max}}{N_2 I_2\ dl_2}$$

where F_{max} is the maximum force exerted on the second current element for which, in the Biot-Savart law, $\sin\theta = \sin 90° = 1$.

In terms of the magnetic field produced by a current element $N_1 I_1\ dl_1$, the magnetic flux density acting on a current-conducting element of length dl_2 located at a distance r_{12} from dl_1 is

$$B_1 = \oint \frac{\mu N_1 I_1\ dl_1}{4\pi r_{12}^2} = \frac{\mu N_1 I_1 l_1}{s}$$

where $s = 4\pi r_{12}^2$ is the surface area of a sphere whose center is at dl and whose surface just touches dl_2. In SI units, magnetic flux density is measured in teslas (webers per square meter).

88. Magnetic Flux. Magnetic flux ϕ through an area s is defined as the surface integral of the normal component of magnetic flux density over the area. Thus

$$\phi = \int_{surface} B \cos\theta\ ds = \int_{surface} B \cdot ds$$

where B is the magnetic flux density at the surface element ds and θ is the angle between the surface element ds and the direction of B at the element of area.

It is sometimes helpful to visualize the magnetic flux as the total quantity of lines of magnetic force set up around a current-carrying conductor in a medium whose permeability is μ. The magnetic flux produced by a current I in each of N closely spaced conductors of length l in a medium whose permeability is μ is

$$\phi = \mu N I l = Bs$$

In SI units magnetic flux is measured in webers, current in amperes, length in meters, and the permeability in henrys per meter.

89. Gauss' Theorem for Magnetic Flux. The theorem of Gauss for magnetic flux states that the amount of magnetic flux passing through any closed surface is equal to zero. This is interpreted to mean that as many lines of magnetic flux enter any closed surface as leave it.

If B is the amount of flux density crossing any element of surface ds and if θ is the angle between the direction B and the normal to the surface element. the magnetic flux is

$$\phi = \int_{\substack{closed \\ surface}} B \cos\theta\ ds = \int_{\substack{closed \\ surface}} B \cdot ds = 0$$

The theorem of Gauss and the statement of the divergence for magnetic flux are valid for time-varying magnetic fields.

90. Divergence of Magnetic Flux. The lines of magnetic flux around a current-carrying conductor are closed loops concentric about the conductor. For any closed surface surrounding a segment of such a conductor, as many lines of magnetic flux will cross the surface in an outward direction as recross it elsewhere in an inward direction.

The divergence of magnetic flux density is equal to the ratio of the flux passing through a closed surface to the vanishingly small volume contained therein, or

$$\text{div } B = \lim_{\Delta v \to 0} \frac{\int_{\substack{closed \\ surface}} B \cos\theta\ ds}{\Delta v} = 0$$

If B is expressed in terms of rectangular components, B_x, B_y, and B_z, each of which is a function of coordinates x, y, and z, the divergence is expressed as the sum of the partial derivatives of each component in the direction of its axis, or

$$\text{div } B = \frac{\partial B_x}{\partial x} + \frac{\partial B_y}{\partial y} + \frac{\partial B_z}{\partial z} = 0$$

This result applies for time-varying as well as for time-invariant magnetic fields.

91. Magnetic Field Strength. The magnetic effects of a current can be defined in terms of a quantity called *magnetic field strength H*. The magnetic field strength is a field vector specifying the amount of force F produced per unit of magnetic flux ϕ set up by a conductor of N closely spaced turns each carrying the same current I through a conductor of length l in a medium whose permeability is μ. From this point of view, $H = F/\phi$, but in terms of fluxes produced by current-carrying conductors, the force is

$$F = \phi_1\phi_2/4\pi r_{12}^2\mu = \phi_1\phi_2/s\mu$$

where $s = 4\pi r_{12}^2$ is the surface area of a sphere, as defined in Par. **1-87**. Hence

$$H = \phi_1\phi_2/\mu s\phi_1 = \phi_2/\mu s = B/\mu = NI/l$$

and the magnetic field strength (also called magnetizing force) is measured in ampere-turns per meter. The magnetizing force H is taken around a closed path of length l. This length is the circumference of a circular loop of magnetic lines centered on an element of the current-carrying conductor.

The magnetic field strength is a vector point function whose curl is the current density. It is proportional to magnetic flux density in regions free of magnetized substances.

92. Magnetization. Magnetization (the intensity of magnetization at a point in a medium) is defined as the intrinsic magnetic flux density at the point divided by the permeability of free space in the system of units employed. It is expressed as

$$M = (B - \mu_0 H)/\mu_0$$

The magnetization can be interpreted as the volume density of magnetic moment. It is a scalar quantity.

93. Ampère's Law for Magnetic Intensity. Ampère's circuital law (also referred to as Ampère's work law) can be derived from the Biot-Savart law of force. The circuital law states that the line integral of magnetic field intensity H about any closed path is equal to the current crossing the surface enclosed by the path. If H is the magnetic field intensity and I is the current crossing a surface bounded by a closed loop, then

$$\oint H \cdot dl = \int_{\text{surface}} J \cdot ds = I$$

In this result, positive current is defined as that in which charges flow in the direction of advance of a right-hand screw when it is turned in the direction in which the closed loop is traced. The application of Ampère's circuital law requires determination of the total current enclosed by a closed loop. The line integral is defined as the *magnetomotance* of the closed loop (see Par. **1-96**).

94. Relations between Magnetic Vectors. The magnetic field can be defined in terms of (1) the magnetic induction of flux density B, (2) the magnetic field strength H, or (3) the magnetization M of a magnetic substance.

The magnetic flux density, $B = F/NIl$, is the force per magnetic test body of a magnetic field acting on currents, whether the currents are *true currents*, i.e., due to charges moving with directed motion in a circuit, or *orbital currents*, due to the motions of orbiting electrons in magnetic substances.

The magnetic field strength, $H = NI/l$, is produced by the directed motion of free or mobile charges and applies only to "true" currents.

The magnetization M is the magnetic dipole moment per unit of volume. It applies to material substances under magnetic stress and is zero for free space.

For the three magnetic vectors, the following general relation holds:

$$B = \mu_0 H + M$$

For homogeneous, isotropic magnetic media of relative permeability μ_r, the magnetic field vectors are related by the equations

$$B = \mu_0\mu_r H = \mu H \qquad \text{and} \qquad M = (\mu_r - 1)H$$

95. Vectors at Magnetic Boundaries. The tangential components of magnetic field strength are continuous across the boundary of magnetic substance and are the same on either side of the boundary. The normal, or perpendicular, component of magnetic flux density is continuous between two media.

96. Magnetomotance. Magnetomotance V_m can be defined as the energy W per unit magnetic flux ϕ or the work done in moving a charge around a closed loop surrounding a current-carrying conductor.

In terms of the magnetizing force H produced by a current I flowing in N closely bunched conductors (each having the same current) and the length l_c of a closed loop surrounding the current-carrying conductor, the magnetomotance is

$$V_m = W/\phi = F \cdot l_c/\phi = F/\phi l_c = Hl = NIl_c/l_c = NI$$

The magnetomotance is measured in ampere-turns, the magnetic field strength H in ampere-turns per meter, and distance l_c in meters. The notion of magnetomotance is especially useful when dealing with magnetic circuits. In this case N is the number of turns of a winding, I is the current through the winding, and l_c is the mean length of path of magnetic flux in the magnetic circuit.

97. Magnetic Potential. The difference in magnetic potential between two points in a magnetic field is the integral of the dot product of the magnetic field strength H and the length of path l between the two points. Thus, if U_1 is energy at point 1 and U_2 is energy at point 2 in a magnetic field,

$$\int_1^2 H \cdot dl = U_2 - U_1$$

For a closed path which does not enclose current,

$$\oint H \cdot dl = 0$$

For a closed path which encloses N turns carrying current I

$$\oint H \cdot dl = NI = V_m$$

In SI units H is in amperes per meter, length l is in meters, current I is in amperes; N is a numeric representing the number of turns. The magnetic potential is specified in terms of amperes or ampere-turns.

98. Flux Linkages. When a closed line of magnetic flux links a closed turn of a conductor (or vice versa), the result is called a *flux link*. For an amount of magnetic flux ϕ linking N turns (closely spaced) of a conductor, the total number of flux linkages is

$$\Lambda = N\phi$$

In SI units flux linkages are measured in weber-turns and flux in webers.

99. Inductance. Inductance can be defined as the property of an electric circuit by virtue of which a varying current induces an electromotance in that circuit or a neighboring circuit. It can also be defined as the property of a conductor or circuit that establishes magnetic flux linkages.

The definition of inductance is restricted to circuits that are small compared with a wavelength of an oscillation, so that the charges flowing are essentially functions of time but not of spatial distribution.

100. Self-Inductance. Self-inductance is the property of an electric circuit by which a voltage is induced in the circuit by a change of current in the circuit. It is the property of a conductor or circuit for establishing magnetic flux linkages with its own current-carrying winding.

The coefficient of self-inductance L is

$$L = \frac{\partial \Lambda}{\partial I} = \frac{\partial}{\partial I} N\phi = \frac{N\phi}{I}$$

where $\Lambda = N\phi$ = number of flux linkages and I = current in conductor or winding. The SI unit of inductance is the henry (equal to weber-turns per ampere). The magnetic flux is in webers and current in amperes; the number of turns is a numeric.

101. Mutual Inductance. Mutual inductance is the property of two electric circuits whereby an electromotive force is induced in one circuit by a change of current in the other. It can also be defined in terms of the number of flux linkages set up in one circuit per unit of current in the other. Mutual inductance may be either positive or negative.

The coefficient of mutual inductance M between two windings a and b is given by

$$M = \partial \Lambda_b/\partial I_a$$

where Λ_b = total flux linkages in one winding and I_a = current in other winding.

The mutual inductance between two linear circuits is independent of the direction in which the flux linkages are taken. Hence, for linear systems,

$$M = M_{ab} = M_{ba} = \partial\Lambda_b/\partial I_a = \partial\Lambda_a/\partial I_b$$

102. Force on a Moving Charge in a Magnetic Field. A charged particle moving in a magnetic field is subject to a force in a direction perpendicular to its direction of travel and also perpendicular to the direction of the field. If Q is the magnitude of the charge, v its velocity, B the flux density of the magnetic field, and ϕ the angle between the direction of B and the direction of travel, the force on the particle is

$$F = QvB \sin \phi = Qv \times B$$

The magnetic field tends to accelerate the charge in a direction at right angles to its motion. The acceleration changes only the direction of the particle velocity, not its magnitude. The force F is in newtons when Q is in coulombs, v in meters per second, and B in teslas (webers per square meter).

103. Lorenz Force on a Moving Charge in a Combined Electric and Magnetic Field. If a charge Q is accelerated by an electric field of intensity E and moves with velocity v in a magnetic field of flux density B, the total force acting on it is the vector sum of the electric and magnetic forces, F_e and F_m, respectively, and is given by

$$F = F_e + F_m = QE + QvB \sin \phi = Q(E + v \times B)$$

where ϕ is the angle between the direction of motion imparted by the electric field and the direction of B. The force is in newtons when Q is in coulombs, E is in volts per meter, v is meters per second, and B is in teslas (webers per square meter).

104. Magnetic Force on Current-Carrying Conductor. The force F acting on an N-turn conductor of length $l = l_2 - l_1$ carrying current I when located in a magnetic field whose average flux density over the length of conductor is B is

$$F = \int_{l_1}^{l_2} NI(dl \times B) = BlIN \sin \theta$$

where θ is the angle between the direction of B and that of the conductor. The force is measured in newtons if current I is in amperes, length l is in meters, and the flux density B is in teslas (webers per square meter).

105. Torque on Current-Carrying Loop. If a small current-carrying loop is immersed in a magnetic field, equal and opposite forces act on opposite sides of the loop, tending to cause the loop to rotate as a result of the torque thus acting on it. If the loop is a rectangle of length l and width w and is located in a field of flux density B, and if ϕ is the angle between the plane of the loop and the direction of the magnetic field the torque is

$$T = 2F \, w/2 = IBlw \cos \phi = IAB \cos \phi$$

since $A = lw$ is the area of the loop. The torque is in newton-meters when the current is in amperes, flux density in webers per square meter, the loop dimensions l and w in meters, and A in square meters.

106. Magnetic Moment. The product of the current I and the area A of a current-carrying loop is called the *magnetic moment* of the loop and is designated by $m = IA$.

The torque acting on the loop is expressed as

$$T = mB \cos \phi = mB \sin \theta$$

In the last term, θ is the angle between the direction of the flux density B and the normal to the plane of the coil.

The torque is in newton-meters when B is in webers per square meter and m is in ampere-meters^2.

107. Magnetic Energy. The energy U in any system in which a force F acts for a distance l_c is $U = F \cdot l_c$. The magnetic force is given by the relation $F = \phi NII/s$, and the magnetic energy is then

$$U = F \cdot l_c = (\phi/s)(NII)l_c = \phi(NI)ll_c/s = \phi NI$$

since $s = ll_c$. The average energy is

$$U_{av} = \tfrac{1}{2}U = \tfrac{1}{2}\phi NI$$

The energy is in joules when magnetic flux is in webers, current in amperes, force in newtons, and distance in meters. N is a numeric, and area s is in square meters.

108. Magnetic Energy Density. Since the flux is $\phi = Bs$, $NI = Hl$ and l is the length of the volume $v = sl$, the average energy of a magnetic field is

$$U_{av} = \tfrac{1}{2}BHv = \tfrac{1}{2}\mu H^2 v = \tfrac{1}{2}(B^2/\mu)\, v$$

The energy density is the energy per unit volume, or

$$u_{av} = U_{av}/v = \tfrac{1}{2}\mu H^2 = \tfrac{1}{2}(B^2/\mu)$$

Energy density is in joules per cubic meter when H is in amperes per meter and B in webers per square meter.

109. Magnetic Energy Stored in Inductors. The energy stored in a winding of self-inductance L is

$$U = N\phi I = LI^2$$

The energy stored in a mutual inductance M whose coupled coils carry currents I_a and I_b is

$$U = N\phi_a I_a = (N\phi_a/I_b)I_a I_b = MI_a I_b$$

In each case the average energy is $U_{av} = \tfrac{1}{2}U$. Energy is in joules when currents are in amperes, flux in webers, and inductance in henrys.

110. Bibliography on Magnetostatics. See Par. **1-121**.

MAGNETOKINETICS

111. Electromagnetic Energy Conversions. In a loss-free, energy-storing system in which charges move with directed nonuniform motion, the sum of the instantaneous values of electric and magnetic energy must remain constant as required by the law of energy conservation. Accordingly, any change in the electric energy of the system, $du_e = d(vq)$, must be accompanied by an equal and opposite change in the magnetic energy of the system, $-du_m = -d(\Lambda i)$. Since charges carry their fields with them when they move, the transport of an elemental amount of charge dq in the electric portion of the system requires a change in flux linkages by an amount $-d\Lambda$ in the magnetic portion of the system.

For the system as a whole, the net change in the two forms of energy is given by

$$du_e = v\, dq = -du_m = -i\, d\Lambda = -\frac{dq\, d\Lambda}{dt}$$

112. Induced Electromotance. By applying the principle of energy conservation to charges in nonuniform motion, it can be shown that the conversion of magnetic to electric energy establishes a voltage

$$v = -d\Lambda/dt$$

where $d\Lambda/dt$ is the time rate of change of flux linkages. Flux linkages can be expressed as

$$\Lambda = N\phi = NBs = NBlw$$

where N = number of turns, ϕ = magnetic flux, B = flux density, and s = area across which flux linkages occur; s may be given in terms of a length l and width w. It is possible, therefore, to establish an *induced electromotance* by varying any one of these factors with respect to time.

Only two methods of producing time-varying flux linkages are in common use. In the flux-changing method, the magnetic flux in a fixed mechanical system changes with time, and the induced voltage is

$$v = -N\, d\phi/dt$$

In the flux-cutting method, usually the magnetic flux remains constant and a conductor cuts flux lines of the magnetic field perpendicular to lines of induction, with velocity v, to change flux linkages and generate a voltage

$$v = -NBlv$$

The induced voltage is in volts if the flux is in webers, flux density in webers per square meter, the length of the flux-cutting loop in meters, time in seconds, and the speed with which the conductor cuts lines of flux in meters per second.

113. Faraday's Law for Induced Electromotance. Faraday's law for electromagnetic induction in a nonmoving circuit states that the induced electromotance (voltage) is proportional to the time rate of change of magnetic flux linked with the circuit.

For stationary circuits this law can be stated as

$$\oint E \cdot dl = -\frac{d}{dt}\int_{surface} B \cdot ds = -\int_{surface}\frac{\partial B}{\partial t} \cdot ds$$

where E = electric field intensity along path length dl and B = magnetic flux density normal to surface ds.

If the voltage induced in a closed circuit is due to motion of the circuit with respect to the magnetic field, then

$$\oint E \cdot dl = \oint (v \times B) \cdot dl$$

If the electromotance is due to motion of the circuit with respect to the magnetic field as well as to time variations of flux density, then

$$\oint E \cdot dl = \oint (v \times B) \cdot dl - \int_{surface}\frac{\partial B}{\partial t} \cdot ds$$

The latter is the general case.

114. Voltage Induced in Inductors. The general equation for induced voltage can be used to derive the voltage induced in a self- or mutual inductance. The coefficient of self-inductance is defined as $L = N\phi/I$, so that the voltage induced in it is

$$v = -\frac{d}{dt}N\phi = -\frac{d}{dt}LI = -\left(L\frac{dI}{dt} + I\frac{dL}{dt}\right)$$

If the coefficient of inductance is time-invariant, as is often the case, the second term becomes zero and for a self-inductor

$$v = -L \, di/dt$$

The voltage induced is measured in volts when L is in henrys and the current change is in amperes per second.

The coefficient of mutual inductance M is defined to be $M = \Lambda_1/I_2$. Hence, the voltage v_2 induced in one winding by a current i_1 in another winding of a transformer or mutual inductor whose flux Λ links both circuits is

$$v_2 = -d\Lambda/dt = -(M \, di_1/dt + i_1 \, dM/dt)$$

If the mutual inductance is constant with respect to time,

$$v_2 = -M \, di_1/dt$$

The induced voltage is in volts when M is in henrys and the current change is in amperes per second.

115. Current Induced in Conductors and Semiconductors. Time-varying flux linkages occurring in conducting materials induce a voltage in the conductor, and current will flow in the conductor as a result. Such a current is called an *induced current*. If a well-defined path is provided, the induced current flows in a definite path and can be made to perform useful functions. But if the conducting path extends over substantial volume, many paths are provided for the transfer of charges. Many current paths then exist, and the currents may generate a considerable amount of heat.

116. Eddy (Foucault) Currents. Eddy, or Foucault, currents exist as a result of voltages induced in the body of a conducting mass by variations of the magnetic flux. The variation of magnetic flux is the result of a varying magnetic field or of the relative motion of the mass with respect to a magnetic field.

Eddy currents result in Joule heating of the conducting material. Losses in ferromagnetic materials (in which voltages are often induced in electric machinery) can be minimized by dividing the magnetically conducting substance into small segments (such as wires or laminations) to reduce the length of path in which a current exists. At high frequencies, finely divided particles of magnetic materials can be embedded in a nonconducting binder. The particles account for the magnetic properties of the material, but the nonconducting binder prevents large eddy currents from being generated.

117. Lenz's Law. The law of Lenz can be stated as follows: The current in a conductor as a result of an induced voltage is in such a direction that the change in magnetic flux due to it is opposite to the change in flux that caused the induced voltage. This is in accordance with the principle of energy conservation.

118. Magnetic Hysteresis. If the magnetic field applied to a ferromagnetic material is gradually increased from zero and the flux density B produced is plotted for various values of the magnetizing force H, a nonlinear relation between B and H is usually observed. The curve of initial BH values when a material is first magnetized is shown between O and A in Fig. 1-2,

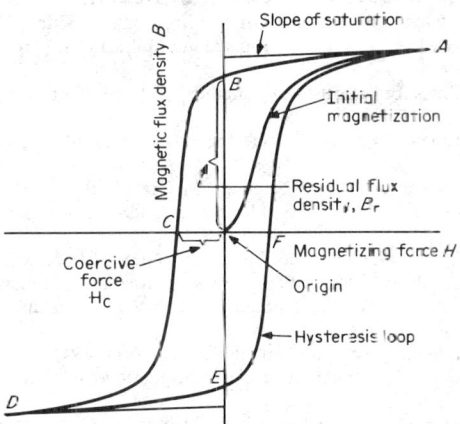

Fig. 1-2. Initial magnetization curve OA and cyclic curve $ABCDEFA$ of a ferromagnetic substance. The cyclic curve is known as a hysteresis loop.

the maximum value of B being called the *saturation flux density*. The curve OA is called the *initial* or *normal magnetization curve* of the material.

If the magnetic field is then slowly decreased, the flux density decreases, but not in proportion to the decrease in magnetizing field strength. When H is reduced to zero, the specimen retains a certain magnetic flux B_r called the *residual flux density*. To reduce the magnetic flux to zero, the field strength H must be increased in the reverse direction from the initial magnetization. The magnetic field strength H_c at which B becomes zero is called the *coercive force*.

As H is further increased in a negative direction, the flux density also increases with opposite polarity and again reaches a saturation value at high values of H. When the field strength decreases from its maximum negative value, the flux density again decreases but not proportionally. It has residual flux density when $H = 0$, and further increase in H in a positive direction is needed to reduce B to zero. For further increases in H the value of B increases to its saturation value, completing the loop.

The phenomenon causing values of B to lag behind values of H so that the increasing and decreasing fields differ in magnitude is called *hysteresis*. The loop traced out by a complete cycle of BH values is called the *hysteresis loop*.

119. Magnetic Effect of Displacement Current. The displacement current, treated in Par. **1-64**, occurs only in dielectrics and insulators, in which case conduction electrons are tightly bound to their atomic nuclei and are not able to move freely through the material. The displacement current I_d is found to produce a magnetic field intensity H which is proportional to the time rate of change of the electric field intensity E producing the displacement current. This situation is different from the case of conduction current, in which H and E are in time phase. Because H is proportional to dE/dt when a magnetic field is produced by displacement current and E is proportional to dH/dt, it is possible to generate electromagnetic waves in nonconducting materials.

120. Electromagnetic Oscillations and Waves. A time-varying electric field produces a time-varying current with its corresponding magnetic field. If the time variations of the electromagnetic field are restricted to a negligibly small region of space, e.g., a circuit, there are no spatial variations of the field to contend with and electrical oscillations are produced. If the time variations of the electromagnetic field extend over distances that are appreciable compared with a wavelength of oscillations, electromagnetic waves are produced along the system by the oscillations set up from the source of time-varying electromagnetic energy.

Electromagnetic waves are propagated with velocity $v = 1/\sqrt{\mu\epsilon}$, where μ is the permeability and ϵ is the permittivity of the medium in which the field exists. When the field is produced by conduction current, H is proportional to E, a field of induction is set up, and the energy of the field remains within the system. When the field is produced by displacement currents, H is proportional to dE/dt, a radiation field is produced, and electromagnetic energy may leave the system to be radiated into space or other dielectric.

121. Bibliography on Electricity and Magnetism

1. Alley, C. L., and K. W. Atwood   "Electronic Engineering," 3d ed., Wiley, New York, 1973.

2. Armstrong, R. L., and J. D. King   "The Electromagnetic Interactions," Prentice-Hall, Englewood Cliffs, N.J., 1973.

3. Benedict, R. R.   "Electronics for Scientists and Engineers," 2d ed., Prentice-Hall, Englewood Cliffs, N.J., 1976.

4. Bleaney, B. I., and B. Bleaney   "Electricity and Magnetism," Oxford University Press, Fair Lawn, N.J., 1957.

5. Booker, H. G.   "An Approach to Electrical Science," McGraw-Hill, New York, 1959.

6. Boylsted, R., and L. Nashelsky   "Electricity, Electronics, and Electromagnetism," Prentice-Hall, Englewood Cliffs, N.J., 1977.

7. Brophy, J. J.   "Basic Electronics for Scientists," 3d ed., McGraw-Hill, New York, 1977.

8. Carlin, H. J., and A. B. Giordano   "Network Theory," Prentice-Hall, Englewood Cliffs, N.J., 1964.

9. Clemmow, P. C., and J. P. Dougherty   "Electrodynamics of Particles and Plasmas," Addison-Wesley, Reading, Mass., 1969.

10. Cobine, J. D.   "Gaseous Conductors," McGraw-Hill, New York, 1941.

11. Corson, D., and P. Lorrain   "Introduction to Electromagnetic Waves and Fields," Freeman, San Francisco, 1962.

12. Cullwick, E. G.   "Electromagnetism and Relativity," Longmans Green, New York, 1957.

13. Dummer, P. A.   "Electronic Inventions and Discoveries," 2d ed., Pergamon, New York, 1959.

14. Durney, C. H., and C. C. Johnson   "Introduction to Modern Electromagnetism," McGraw-Hill, New York, 1969.

15. Elliott, R. S.   "Electromagnetics," McGraw-Hill, New York, 1966.

16. Foecke, H. A.   "Introduction to Electrical Engineering Science," Prentice-Hall, Englewood Cliffs, N.J., 1961.

17. Frank, N. H.   "Introduction to Electricity and Optics," 2d ed., McGraw-Hill, New York, 1950.

18. Guillemin, E. A.   "Network Theory," Wiley, New York, 1963.

19. Hallen, E.   "Electromagnetic Theory," Chapman Hall, London, 1962.

20. Harrington, R. F.   "Introduction to Electromagnetic Engineering," McGraw-Hill, New York, 1958.

21. Heaviside, O.   "Electromagnetic Theory," 3 vols., Chelsea, New York, 1971.

22. Javid, M., and P. M. Brown   "Field Analysis and Electromagnetics," McGraw-Hill, New York, 1963.

23. Johnson, C. C.   "Field and Wave Electrodynamics," McGraw-Hill, New York, 1965.

24. Kip, A. F.   "Fundamentals of Electricity and Magnetism," McGraw-Hill, New York, 1969.

25. Kline, M., and W. K. Irvin   "Electromagnetic Theory and Geometrical Optics," Interscience, New York, 1965.

26. Kraus, J. D.   "Electromagnetics," McGraw-Hill, New York, 1953.

27. Landau, L. D., and E. M. Lifschitz   "Classical Theory of Fields," Pergamon, New York, 1975.

28. Lorrain, P., and D. R. Corson   "Electromagnetic Fields and Waves," Prentice-Hall, Englewood Cliffs, N.J., 1970.

29. Magid, L. M.   "Electromagnetic Fields, Energy, and Waves," Wiley, New York, 1972.

30. Michaels, W. C.   "International Dictionary of Physics and Electronics," Van Nostrand, New York, 1961.

31. Nussbaum, A.   "Field Theory," Merrill, Columbus, Ohio, 1966.

32. Page, L., and N. I. Adams   "Principles of Electricity," Van Nostrand, New York, 1969.

33. Panofsky, W. K. H., and M. Phillips   "Classical Electricity and Magnetism," Addison-Wesley, Reading, Mass., 1962.

34. Portis, A. M.   "Electromagnetic Fields: Sources and Media," Wiley, New York, 1978.

35. Rao, N. N.   "Elements of Engineering Electromagnetics," Prentice-Hall, Englewood Cliffs, N.J., 1977.

36. Rojansky, V. B.   "Electromagnetic Fields and Waves," Prentice-Hall, Englewood Cliffs, N.J., 1971.

37. Robinson, F. N. H.   "Electromagnetism," Oxford University Press, Oxford, 1973.

38. Roters, H. C.   "Electromagnetic Devices," Wiley, New York, 1941.

39. Rocard, Y.   "Principles of Electricity and Magnetism," Pitman, New York, 1960.

40. Smythe, W. R.   "Static and Dynamic Electricity," 2d ed., McGraw-Hill, New York, 1950.

41. Schwarz, W. W.   "Intermediate Electromagnetic Theory," Krieger, Huntington, N.Y., 1973.

42. Stuart, R. D.   "Electromagnetic Field Theory," Addison-Wesley, Reading, Mass., 1965.

43. Sanders, K. F., and G. A. L. Reed   "Transmission and Propagation of Electromagnetic Waves," Cambridge University Press, Cambridge, 1978.

44. Seely, S.   "Introduction to Electromagnetic Fields," McGraw-Hill, New York, 1958.

45. Skilling, H. H.   "Electromechanics," Wiley, New York, 1962.

46. Shortley, G., and D. Williams   "Elements of Physics," Prentice-Hall, Englewood Cliffs, N.J., 1971.

RELEASE, TRANSPORT, CONTROL, COLLECTION, AND ENERGY CONVERSION OF CHARGE CARRIERS

122. Electronic Phenomena. Electronic applications depend upon the release, transport, control, collection, and energy conversion of charge carriers derived from material substances. To avoid duplication in treating these topics, only brief mention of the release of charge carriers is given here. More detailed treatment of emission processes is given in Sec. 6.

123. Charges Released from Materials. The mobile charges required to establish an electric current are released by applying force to, and expending energy on, material substances. Mobile charges may be released by heat, electromagnetic radiation, impact of particles, intense electric fields, and the tunneling effect, by which an electron crosses a potential barrier.

124. Thermionic Emission. Thermionic emission occurs when the electrons in a substance have enough thermal energy to overcome the restraining forces at the surface of the substance and so escape into the surrounding space. Emitted electrons tend to cluster at the emitting surface, which carries a positive charge equal to the negative charge of the emitted electrons. The resulting *space charge* limits release of additional electrons until those already released are removed from the vicinity of the emitter.

125. Photoelectric Emission. Photoelectric emission occurs when electrons are liberated from the surface of metals and compounds as a result of absorption of electromagnetic radiation in the ultraviolet, visible, or infrared portions of the spectrum. The emission current is proportional to the amount of radiant energy incident on the surface provided that photoelectric emission occurs for the spectral distribution of the incident radiation.

126. Secondary Emission. Secondary emission occurs when primary electrons impinge on a surface with velocities greater than a critical value for each material and, by their impact, release other electrons, called *secondary electrons*. The energy of primary electrons is transferred to electrons in the bombarded surface, some of which acquire sufficient energy to overcome the potential barrier restraining them to the surface. Electrons emitted in this process include primary electrons elastically reflected from the surface, primary electrons inelastically reflected with some loss of energy, and true secondary electrons, whose mean energy is of the order of 10 eV and is substantially independent of the energy of primary electrons impinging on the surface.

127. High-Field Emission. Field emission occurs when a high positive electric field exists at the surface of a cold metal-cathode surface. The electric field reduces the potential barrier at the surface of the metal, thereby lowering the energy needed to release electrons from the surface.

128. Ionization and Deionization. Ionization is the process by which a neutral atom or molecule acquires a positive or negative charge and thereby becomes an ion. Deionization is the process by which an ion loses its charge and becomes a neutral atom or molecule.

The minimum amount of energy required to free an electron occupying the highest energy state in an atom is known as the *ionization energy*. It is usually expressed in electronvolts. More than one electron can be removed from an atom, but the energy required to produce multiple ionization increases rapidly above that needed to remove the first electron. An important effect of ionization in gases is *space-charge neutralization*.

Ionizing agents include cosmic rays; natural radioactivity; x-rays; intense electric fields, giving rise to ionization by collision; high temperature, giving rise to thermal ionization; diffusion of charged particles; flames; chemical reactions and effects; photoelectric emission; thermionic emission; high-field emission; and secondary emission.

Deionizing agents include electric field in a gas, electric field at electrodes, diffusion by neutral molecules, recombination of ions in gases, and recombination of ions at surfaces. In a gas, volume recombination usually occurs when electrons become attached to neutral atoms to form negative ions, which then recombine with positive ions to form electrically neutral particles.

129. Unbound Electrons, Holes, and Electron-Hole Pairs. The highly conducting properties of common metals arise because electrons are not attached to any particular atom but form a mobile electron gas. In contrast, electrons in semiconductors are held together by covalent or valence bonds, along which electrons can move. Such bonds between atoms dominate the electrical behavior of semiconductor crystals.

At very low temperatures, valence electrons in a perfect crystal remain closely bound and the substance acts as a dielectric. As the temperature is increased, some of the valence bonds break, and bond disruption influences the motions of other bound electrons.

When a covalent bond in a semiconductor is broken, the vacancy left by the released electron behaves as though it were a newly formed mobile particle having positive charge whose magnitude is that of an electron and whose mass is slightly greater than that of an electron. The charge cavity thus formed is called a *hole*. Holes appear to move through the crystal structure much as a positive charge might be expected to move.

Actually, the hole itself does not move but becomes filled by an electron which leaves a hole elsewhere in the crystal to be filled subsequently by another electron leaving a hole behind it. The apparent motion of a positive hole in one direction is really produced by the release of electrons at other bonds and the motions of the newly released electrons in a direction opposite that in which the holes appear to travel. An electron filling a hole as an *electron-hole pair* neutralizes the local charge of both these particles in the crystal.

130. Charge Transport. Once mobile charges are released from their host material, they may be set into motion and given kinetic energy by means of an applied electric field. A magnetic field can be used to accelerate charge carriers in a direction perpendicular to that of their velocity.

131. Motion of Charges in Electromagnetic Fields. A particle of charge Q (C) located in a vacuum with electric field intensity E (V/m) is acted upon by a longitudinal accelerating force

$$F_\parallel = QE$$

If the particle moves in a magnetic field of flux density B (T) with velocity v (m/s), it is acted upon by a transverse force of

$$F_\perp = (v \times B)Q$$

If the particle is in an electromagnetic field, the total force acting upon it is the sum

$$F = F_\parallel + F_\perp = Q(E + v \times B)$$

In a vacuum, a particle of charge $-Q$, in coulombs, moving in an electric field E, whose intensity is in volts per meter, from point x_1 to x_2, in meters, undergoes a change in energy, in joules, of

$$U = -Q \int_{x_1}^{x_2} E \cdot dx = -Q(V_1 - V_2)$$

The change of kinetic energy, in joules, of a particle of rest mass m_0 moving with velocity v (m/s) in vacuum is

$$U = m_0 c^2 \left[\frac{1}{1 - (v/c)^2} - 1 \right]$$

if m_0 is in kilograms and $c = 3 \times 10^8$ km/s is the speed of light propagation.

A charged particle moving with velocity v (m/s) in a magnetic field whose flux density is B (T) travels in a circular orbit whose radius, in meters, is

$$r = mv/QB$$

where m = mass of particle (kg) and Q = charge (C).

132. Electric Conduction. Electric conduction is the transmission of charge carriers through, or by means of, electric conductors. The conductors may be in the solid, liquid, or gaseous state, each of which exhibits different mechanisms of charge transfer. The atomic structure of materials greatly influences the facility with which charges can be transmitted. The environment in which the conductor is located also modifies the conductivity, sometimes to a remarkable extent.

The conduction current passing through a surface s with drift velocity v is $I_{cond} = (Q/V)sv$ $= \rho sv$, where $\rho = Q/V$ is the volume density of mobile charge carriers. The current density is $J = \rho I/s = \rho v$, the drift velocity is $v = \mu E$, and $\mu = v/E$ is the mobility of charge carriers moving with drift speed v in an accelerating electric field of intensity E. Hence the current density can be expressed as

$$J = \rho v = \rho \mu E$$

In general, charge carriers may be of two kinds, one of charge Q^+ and mass m^+ and the other of charge Q^- and mass m^-. Both contribute to current, but positive and negative charges move in opposite directions, so that the conductivity is

$$\sigma = J/E = \rho^+ \mu^+ - \rho^- \mu^-$$

In SI units conductivity is measured in ampere-meters per volt or coulomb-meters per second. This unit is the same as the siemens per meter.

133. Superconductivity. At ordinary temperatures the electric conductivity of substances extends over a tremendous range, from about 1.6×10^{-8} $\Omega \cdot$cm for silver to at least 10^{16} $\Omega \cdot$cm for such dielectrics as quartz. The range is even greater at very low temperatures.

Superconductivity is a property of materials characterized by zero electric resistivity (infinite conductivity) and, ideally, zero permeability. More than a score of elements, including semiconductors and many alloys, become superconducting at temperatures of 20 K or less.

Superconductivity is found to occur in elements having two to five valence electrons outside the closed shell. It occurs at a *transition temperature* below which the material is superconducting and above which it is not. Superconductivity also depends upon the temperature and the magnetic field surrounding the conductor. It is explained on the hypothesis that the vibrating ions in a crystal lattice do not impede the flow of electrons at very low temperatures. At sufficiently low temperatures ions cease to vibrate, and a perfect, stationary crystal lattice freely allows the electrons to pass.

Currents once established in closed superconducting circuits have been found to continue undiminished for weeks but rapidly drop to zero when the temperature rises above the superconducting value.

134. Conduction in Metals. Metals, while free from net volume density of charge (of either polarity), contain approximately 10^{23} atoms per cubic centimeter. In ordinary metals (but not pure semiconductors) each atom is capable of releasing a free electron and leaving an immobile positive ion. Thus, highly conducting metals have a large volume density of unbound charges (electrons).

A voltage V applied between two points a and b of a metallic conductor of length l_{ab} produces a current I in the conductor through which charges pass and along which an electric field E is established. In terms of the electric field E, the voltage between a and b is

$$V_{ab} = - \int_a^b E \cdot dl = - El_{ba}$$

Since the current density is $J = \rho v$ and the drift velocity is $v = \mu E$ (Par. **1-132**), the current through the conductor is

$$I = \int_{surface} J \cdot ds = J \cdot s = \rho vs = \rho \mu Es = (\rho \mu) s E = (\sigma s/l) V$$

The conductivity σ is a constant for each metal and is independent of the current through it or the electric field intensity to which it is subjected.

135. Conduction in Semiconductors. Semiconductors are electronic conducting materials in which the conductivity depends, in large measure, upon impurities, lattice defects, and the temperature of the material. Two kinds of mobile charges are present simultaneously in a semiconductor, electrons and holes. They are oppositely charged and move in opposite directions in a given electric field. Holes and electrons have different masses and mobilities and are scattered in different ways.

The current density of charge carriers moving in a semiconductor is

$$J = e(\mu_e n + \mu_h p)E = (\sigma_e + \sigma_a)E$$

where e = electron charge, μ = mobility, subscripts e, h = electrons and holes, respectively, n = number of electrons per unit volume, p = number of holes per unit volume, σ_e = electron conductivity, and σ_h = hole conductivity.

The conductivity of semiconductors is highly temperature-dependent. The acceleration imparted to a mobile particle of mass m and charge Q is $a = F/m = QE/m$, where $F = EQ$ is the force acting upon the charge Q by the electric field intensity E. The velocity acquired by the charge is $v = QEt/m = \mu E$, where t is the square root of mean-square time during which the particles are accelerated. Thus the mobility is

$$\mu = \frac{Q}{m} t = \frac{Q}{m} \frac{l}{v}$$

where v is the thermal velocity and l is the mean length of the free path. Both v and l vary with temperature in a complicated way such that the conductivity increases with temperature.

136. Conduction in Liquids. According to their electric conductivity, liquids can be divided into three classes: those having conductivities of about 10^{-11} Ω/m (such as paraffin oils, xylol, etc.), which do not conduct and so serve as dielectrics or insulators; those having conductivities of about 10^{-4} Ω/m (alcohol and water), which are neither good conductors nor insulators; and those having conductivity of about 10 Ω/m (aqueous solutions of acids, bases, and salts), which are often regarded as good conductors.

The electric conductivity depends upon the number of ions per unit volume and the rate at which they move through the liquid. If there are n univalent molecules per unit volume of solution each with charge of an electron e, and if k is the fraction of molecules dissociated, then the number of charged particles is kn. The mobilities μ^+ and μ^- are the velocities per unit field strength, and the current density is

$$J = ken(\mu^+ + \mu^-)E$$

The conductivity is

$$\sigma = J/E = ken(\mu^- + \mu^-)$$

Electrolytic conduction follows Ohm's law approximately. The current is due to the movement of positive and negative ions (as well as electrons) that are abundant in ionized liquids. As ions move through the electrolyte, their transfer is evidenced by the deposit of neutralized ions as atoms at one electrode and the liberation of ions at the other electrode, whose atoms are used up as the electrode corrodes.

137. Conduction in Gases. Ionization is a dominant feature of electric conduction in gases; deionization is another. The different types of gas discharge depend upon the relative importance of ionization and deionization processes, which can be divided into self-maintaining and non-self-maintaining discharges. The first do not require an external source of ionization; the latter do.

The condition to be fulfilled for self-maintaining discharge is

$$\varepsilon^{ad} = (1 + r)/r$$

where a is the number of ion pairs formed by an electron, per meter drift toward the anode, d is the distance between emitter and collector electrodes, and r is the average number of secondary electrons emitted from the cathode for each new positive ion formed in the gas. The establishment of a self-maintaining discharge requires that, on the average, each electron leaving the cathode must initiate sufficient ionization by collision for the positive ions so formed to produce at least one electron.

138. Conduction in Plasmas. A plasma is a highly ionized gas that does not easily fit the category of a solid, liquid, or gas. The ionization of a plasma is maintained primarily by electron collisions and by photoionization. The concentrations of positive and negative ions are relatively high and roughly equal. Since a plasma is highly conducting, a relatively low voltage gradient exists across it. In general, the negative charge carriers are electrons. The neutral gas atoms, ions, and electrons are not necessarily in thermal equilibrium, and their velocities vary over a wide range.

Electrons in a plasma may vibrate longitudinally about their mean positions while the positive ions remain relatively fixed. This gives rise to electron plasma oscillations of frequency

$$f = (ne^2/\pi m_e)^{1/2}$$

where n = number of electrons producing charge density, e = charge of electron, and m_e = mass of electron. If λ is the wavelength of oscillation, the velocity of propagation is

$$v = \lambda(ne^2/\pi m_e)^{1/2}$$

The frequency of electron plasma oscillations may be in the vicinity of 10^9 Hz.

Positive ions may also oscillate longitudinally, but at a lower frequency than electrons, and the frequency is temperature-dependent.

139. Current in Dielectric and Insulating Materials. In dielectric and insulating materials, electrons are so closely bound to their parent atoms that they cannot become free without physically damaging the material. Consequently, so long as the dielectric properties are maintained, no conduction current can occur in dielectrics. But a varying voltage applied across the faces of a dielectric can produce a varying displacement of charges in the dielectric, and the directed displacement behaves much as a time-varying conduction current does (see Par. **1-64**).

If a time-varying electric field is applied across a dielectric, the displacement current in the dielectric is given as

$$I_{dis} = \frac{d\psi}{dt} = \frac{d}{dt}Ds = \varepsilon\frac{dD}{dt} = s\frac{d}{dt}\epsilon E = s\epsilon\frac{dE}{dt}$$

where ψ = electric flux displaced in dielectric, D = electric flux density, s = surface area of dielectric subject to displacement, ϵ = permittivity of dielectric, E = intensity of applied electric field, and t = time.

140. Mechanisms of Current Control. Current can be controlled by changing the forces acting on, and the energies expended upon, the moving charges constituting the current. The motions of charge carriers can be controlled by an electric field, a magnetic field, or some combination of these components in an electromagnetic field.

In practice, the environment in which charges move, including the properties of surrounding materials, has great influence on mechanisms for current control. The properties of the materials used in electron devices are themselves important in effecting certain controls, including unilateral conductivity, nonlinear voltage-current relations between electrodes, amplification, energy-conversion processes at input or output of the device, hysteresis of magnetic materials, and hysteresis of dielectric materials.

141. Control in Vacuum Tubes. The current in vacuum tubes can be controlled by electric fields established by electrode potentials, magnetic fields through which charges pass, and by both electric and magnetic fields.

Electric Control. When the current through a tube is limited by space charge, the current can be controlled by voltage applied to its electrodes, particularly one or more grids between emitter and collector. For such conditions, the current to the collector (plate) is of the form

$$I_b = K\left(\frac{C_{gk}}{C_{pk}} V_c + V_p\right)^{3/2} = K(\mu V_c + V_p)^{3/2}$$

where K = const depending upon tube geometry, C_{gk} = capacitance between control electrode (grid) and emitter (cathode), C_{pk} = capacitance between collector (plate) and emitter (cathode), $\mu = C_{gk}/C_{pk}$ = voltage amplification factor of tube (depends on its geometry), V_c = voltage applied to control electrode relative to cathode potential taken as reference value, and V_p = voltage applied to collector electrode measured with respect to emitter (cathode) voltage. The equation given above is for a tube having a single control electrode or grid. Similar equations can be developed for tubes having auxiliary grid electrodes. The current is usually controlled by a time-varying voltage applied to one control electrode; sometimes control is effected by varying the potentials of more than one control electrode.

Magnetic Control. The simplest electron tube having conduction characteristic controlled by a magnetic field is the early form of magnetron, invented by A. W. Hull. It consists of a filamentary electron-emitting cathode located on the axis of a cylindrical collector electrode. When the filament is heated to incandescence and a positive potential V is applied to the collector relative to the emitter, electrons flow from filament to plate. If the tube is surrounded by a magnetic field of intensity H, the electrons do not travel from filament to plate in straight lines but move in a circle of radius r. The electrons fail to strike the collector unless the magnetic field causes the electrons to move in a radius r equal to or greater than the distance between filament and cylindrical plate. The magnetic field that just causes electrons to graze the cylindrical collector is

$$H = (8mV/er)^{1/2}$$

where m = mass of electron, e = electron charge, r = radius of cylindrical collector, and V = positive voltage of collector relative to that of emitter. For values of H greater than that indicated, for a given accelerating voltage V, electrons do not strike the collector, and current in the tube abruptly drops to zero. Many other more sophisticated forms of magnetic control have been developed (see Sec. **9**).

142. Control in Gaseous Devices. Two methods of controlling current through gas tubes have been devised: those which control starting of a discharge by delaying the formation of the discharge until a desired time and those which control the starting of conduction by enhancing formation of the discharge at the desired time.

Devices of the first class can be controlled by a magnetic field, an external grid wound around the tube, or an electrode partially surrounding the collector electrode in a tube with cold cathode, as in the grid-glow tube.

Devices of the second class can be controlled by an igniter electrode in a tube having a mercury-pool cathode, application of voltage to an electrode insulated from the mercury-pool cathode, or one or more auxiliary electrodes to initiate discharge between cold electrodes in a gas.

In all cases, gas-discharge tubes operate by control of the *average current* through them rather than control of the instantaneous current, as in vacuum tubes. Since control periods are usually of the order of milliseconds, gas tubes are not generally useful for high-frequency applications.

Ionization of the gas neutralizes space charge within the device, so that the voltage drop across the gas tube is relatively small. Current density in vacuum tubes is proportional to the ½ power of the emitter-collector voltage; in high-pressure gas tubes, current density is proportional to the square of the cathode-anode voltage.

143. Control in Semiconductors. In semiconductor devices, such as transistors, control of current in the output electrode circuit is possible because the terminal currents depend upon voltages applied across *semiconductor junctions.* Emitter and collector currents are of nearly equal magnitude and increase exponentially with voltage between emitter and base. The current in the base is smaller than the emitter and collector currents, and it also depends to a marked extent upon the emitter-base voltage.

When a small, slowly varying voltage is applied across emitter and base, the concentration of holes in the base, at the edge of the emitter-junction and space-charge layer, increases. The distribution of holes in the base changes in such a way that the collector current increases and the base current decreases. The ratio of change in collector current to change in base current is called the *current gain* of the device. Complete discussions are presented in Secs. **7** and **8.**

144. Auxiliary Methods of Current Control. While current control is achieved basically by changing the electric or magnetic field, or both, auxiliary methods of control exist in which the field itself is controlled by some energy-converting device responding to pressure, temperature, radiation, or other nonelectromagnetic quantity. For example, the current through a phototube can be controlled by mechanically varying the amount of radiation falling upon its electron-emitting surface.

An important aspect of control in all types of electron devices is modifying the wave shape of an input signal as it reaches the output of the device. Such wave-shaping techniques include clipping, expanding or compressing signal amplitude, frequency multiplication, division, addition or subtraction, differentiation, integration, and intentional distortions of the signal (see Secs. **16, 17, 19,** and **20**). Further flexibility in control is provided by energy conversions at the input and output terminals of electron devices (see Pars. **1-147** and **1-148**).

145. Collection of Charges: Impact Phenomena. After being released, transported, and controlled, the mobile charges collected at an electrode convert their kinetic energy into other forms of energy upon impact. The kinetic energy of collected charges can be converted into useful current in an external circuit having the electrode as a terminal, heat energy that raises the temperature of the collector electrode, or radiant energy as the impacting electrons become absorbed by atoms of the collector electrode material, e.g., produce x-rays.

146. Electrode Heating. The collector electrode receiving current from a thermionically heated cathode must radiate power of amount

$$P = V_b I_b + A V_c I_c$$

where V_b = voltage at which collector is operated measured relative to emitter potential, I_b = current flowing to collector, A = coefficient whose value (less than 1.0) depends upon extent to which the collector encloses the emitter, V_c = voltage drop across emitter, and I_c = current taken by emitter electrode.

If the collector is cooled only by radiation, the rise in temperature at the surface of the anode is found, from the Stefan-Boltzmann law, to be of the form

$$\Delta T = \left(\frac{P}{sek} \right)^{1/4} = \left(\frac{V_b I_b + A V_c I_c}{sek} \right)^{1/4}$$

where P = amount of power to be dissipated, s = total area of radiating surface, k = the Stefan-Boltzmann constant, e = total radiation emissivity, and ΔT = rise in temperature (K). The other symbols have the meanings given in the equation for P, above. For metals, e has typical values of 0.1 at 1000 K and 0.2 at 2000 K. For radiation cooling alone, the permissible power dissipation is approximately 5 W/cm².

147. Energy Conversions at Input. In addition to the advantages of the control methods outlined above, the versatility and flexibility of electronic systems are greatly increased by their ability to respond to nonelectrical stimuli through the use of suitable transducers. A transducer is a device for converting energy from one form into another.

Nonelectrical forms of energy are often applied to the input of electronic systems by transducers able to convert energy into electrical form (see Secs. **9, 10, 11, 19,** and **20**). For example, sound is converted into electrical form by microphones or hydrophones (Sec. **19**), light is converted into electrical form by photoelectric materials and devices (Secs. **6, 7, 10,** and **11**), mechanical motion is converted into electric energy by selsyns (Sec. **17**), mechanical strain can be converted into changes of electrical resistance by strain gages (Sec. **10**), and so on. The electrical representation of the converted energy is then processed by the electronic system, as required, to yield output energy of the desired kind.

The design and use of transducing devices is treated in Sec. **10**. A summary on transducers, with bibliography also appears in pp. 3-76 to 3-83 of Ref. 12 of Sec. **11**.

148. Energy Conversions at Output. The output of an electronic system may be electrical in form but through the use of suitable transducers it can also be presented in nonelectrical form. For example, acoustical energy can be delivered from a loud speaker (see Sec. **19**), or mechanical energy can be obtained from an electronic system through the use of selsyns or servomechanisms connected to its output (see Secs. **10, 13, 17,** and **20**).

By making use of the luminescence produced by phosphors (see Sec. **11**) excited by electrons, ions, neutrons, photons, or electric fields (see Sec. **6**) light can be delivered at the output of electronic systems. Light-emitting diodes (see Sec. **7**) can provide light of different wavelengths. Light-emitting diodes (LEDs) are *pn* junctions of semiconductor materials biased in the avalanche breakdown region; they produce light from electricity. Lasers also emit light (see Secs. **6** and **11**). By means of cathode-ray and similar tubes (see Secs. **11, 20, 21,** and **25**) a wide range of visual images can be produced at the output of electronic systems.

X-rays are produced by allowing a stream of high-energy electrons strike a hard metal target (see Sec. **26**).

149. Bibliography on Release, Transport, Control, Collection, and Energy Conversion of Electron Devices

1. Adler, R. B., A. C. Smith, and R. K. Longini "Introduction to Semiconductor Physics," Wiley, New York, 1964.

2. Allison, J. "Electronic Engineering Materials and Devices," McGraw-Hill, New York, 1971.

3. Angelo, E. J. "Electronic Circuits: Unified Treatment of Vacuum Tubes and Transistors," McGraw-Hill, New York, 1964.

4. Anner, G. E. "Elementary Nonlinear Circuits," Prentice-Hall, Englewood Cliffs, N.J., 1974.

5. Beck, A. H. W. "Space Charge Waves and Slow Electromagnetic Waves," Pergamon New York, 1958.

6. Boylstad, R. L. "Electronic Devices and Circuit Theory," Prentice-Hall, Englewood Cliffs, N.J., 1978.

7. Bruining, H. "Physics and Applications of Secondary Electron Emission," McGraw-Hill, New York, 1954.

8. Chaffee, E. L. "Theory of Thermionic Vacuum Tubes," McGraw-Hill, New York, 1933

9. Chapman, S., and T. G. Cowling "Mathematical Theory of Non-Uniform Gases," Cambridge University Press, London, 1961.

10. Cobine, J. D. "Gaseous Conductors," McGraw-Hill, New York, 1941.

11. Compton, K. T., and I. Langmuir Electrical Discharges in Gases, I: Survey of Fundamental Processes, *Rev. Mod. Phys.*, 1930, Vol. 2, pp. 123–343; II: Fundamental Phenomena in Electrical Discharges, *Rev. Mod. Phys.*, 1931, Vol. 3, pp. 191–257.

12. Dekker, A. J. "Secondary Electron Emission," of Vol. 6, "Solid State Physics," Academic, New York, 1958.

13. DeWitt, D., and A. L. Rossoff "Transistor Electronics," McGraw-Hill, New York, 1957.

14. Dushman, S. Thermionic Emission, *Rev. Mod. Phys.*, 1930, Vol. 3, pp. 381–476.

15. Drummond, J. E. "Plasma Physics," McGraw-Hill, New York, 1941.

16. Gray, P. E., D. DeWitt, A. R. Boothroyd, and J. F. Gibbons "Physical Electronics and Circuit Models of Transistors," Wiley, New York, 1964.

17. Guthrie, A., and R. A. Wakerling "Characteristics of Electrical Discharges in Magnetic Fields," McGraw-Hill, New York, 1949.

18. Hemenway, C. L., R. W. Henry, and M. Caulton "Physical Electronics," 2d ed., Wiley, New York, 1967.

19. Hess, E. M. "Essentials of Semiconductor Circuits," Prentice-Hall, Englewood Cliffs, N.J., 1976.

20. Hutter, R. G. E. "Beam and Wave Electronics in Microwave Tubes," Van Nostrand, New York, 1960.

21. Kennard, E. H. "Kinetic Theory of Gases," McGraw-Hill, New York, 1938.

22. Llewellyn, F. B. "Electron Inertia Effects," Cambridge University Press, London, 1941.

23. Maxfield, F. A., and R. R. Benedict "Theory of Gaseous Conduction and Electronics," McGraw-Hill, New York, 1941.

24. Okress, E. "Crossed Field and Microwave Devices," Academic, New York, 1961.

25. Parr, G. "Cathode Ray Tube and Its Applications," Chapman & Hall, London, 1941.

26. Pierce, J. R. "Traveling Wave Tubes," Van Nostrand, New York, 1950.

27. Reich, H. J. "Theory and Application of Electron Tubes," McGraw-Hill, New York, 1939.

28. Spangenberg, K. "Vacuum Tubes," McGraw-Hill, New York, 1948.

29. Spielman, H. S. "Electronics Sourcebook for Teachers" (3 vols., many references), Hayden, New York, 1965.

30. Stanley, J. K. "Electrical and Magnetic Properties of Metals," American Society for Metals, Metals Park, Ohio, 1963.

31. Tannenbaum, B. S. "Plasma Physics," McGraw-Hill, New York, 1967.

32. Uman, M. F.   "Introduction to Physics of Electronics," Prentice-Hall, Englewood Cliffs, N.J., 1974.
33. Zworykin, V. K., and G. A. Morton   "Television: The Electronics of Image Transmission," Wiley, New York, 1940.

ELECTROMAGNETIC FIELDS

150. Equations for Electromagnetic Field Vectors. The basis of electromagnetic phenomena in the macroscopic (nonquantum) state is given by the four equations of Maxwell. The electric and magnetic intensites E and H are related to the electric and magnetic flux densities D and B and to the current density J. Maxwell's equations specify how D, B, and J depend upon the space and time variations of E and H, for fields in free space as well as in material substances.

The electromagnetic properties of a medium are specified in terms of auxiliary equations for the field vectors previously defined in this section, as follows:

$$D = \epsilon E = \epsilon_0 E + P = \epsilon_0(1 + \chi_e)E$$
$$B = \mu H = \mu_0 H + \mu_0 M = \mu_0(1 + \chi_m)H$$
$$J = \sigma E = \rho v$$

where ϵ = permittivity of medium, ϵ_0 = permittivity of free space, P = dielectric polarization, χ_e = electric susceptibility, μ = permeability of medium, μ_0 = permeability of free space, M = magnetization, χ_m = magnetic susceptibility, σ = electric conductivity of medium, ρ = volume charge density, and v = volume in which charge density ρ occurs.

These equations for the electromagnetic properties of a medium, together with the Lorentz force equation (see Par. **1-103**)

$$F = F_e + F_m = QE + Q(v \times B)$$

are used in applying the Maxwell field equations relating the field vectors, E, D, B, H, and J.

151. Maxwell's Equations. The four fundamental Maxwell relations between the electromagnetic field vectors are stated here in words as well as in integral and differential form, in the latter case in terms of the notation of vector analysis.

1. The total electric displacement passing through a closed surface is equal to the total charge within the volume enclosed (in free space, volume charge density is $\rho = 0$)

$$\int_{\text{surface}} D \cdot ds = \int_{\text{vol}} \rho \, dv \qquad \nabla \cdot D = \rho$$

2. The net magnetic flux passing through any closed surface is zero

$$\int_{\text{surface}} B \cdot ds = 0 \qquad \nabla \cdot B = 0$$

3. The magnetomotance around any closed path is equal to the sum of the conduction and convection current and to the time rate of change of electric displacement passing through the surface bounded by the closed path.

$$\oint H \cdot dl = \int_{\text{surface}} \left(J + \frac{\partial D}{\partial t} \right) \cdot ds \qquad \nabla \times H = J + \frac{\partial D}{\partial t}$$

4. The electromotance taken around any closed path is equal to the time rate of change of the magnetic displacement passing through the surface bounded by the closed path.

$$\oint E \cdot dl = -\int_{\text{surface}} \frac{\partial B}{\partial t} \cdot ds \qquad \nabla \times E = \frac{\partial B}{\partial t}$$

The equations at the left are the integral forms of Maxwell's relations over finite lengths and surface areas. The equations at the right are the differential forms of Maxwell's equations, expressing the field relationships at a point.

152. Restricted Applications of Maxwell's Equations. In the general forms given above, Maxwell's equations are difficult to apply except to extremely simple physical and geometrical conditions. The Maxwell equations can be reduced to less formidable equivalents under certain conditions for which some of the terms become zero or take simple forms. Modifications of the general form of the Maxwell equations are given for various special conditions in Table 1-3.

153. Electromagnetic Waves. The wave equation is a quantitative expression for the dynamic behavior of some physical quantity that varies with time at a particular point in space and varies with space coordinates at any instant of time. The Maxwell equations can be used to derive the wave equation for time-varying electromagnetic quantities for an infinite, homogeneous, linear, isotropic medium. For such a situation, the properties of the medium, as specified by ϵ, μ, σ, and ρ, are constant and have the same values in all directions over the region considered. For this case, $D = \epsilon E$, $B = \mu H$, $J_{cond} = \sigma E$, and $\rho = dQ/dv$.

The divergence equations of Maxwell play no role in specifying wave properties since they are time-invariant. But the curl equations are used in their general form. The differential forms

TABLE 1-3 Special Conditions for Maxwell's Equations

Condition	Modifications to Maxwell's equations
1. Electrostatics	$J = 0$ $\quad \partial D/\partial t = 0$ $\quad \partial B/\partial t = 0$ Only the electric-divergence equation is applicable
2. Stationary electromagnetic field; magnetostatics	$J = 0$ $\quad \partial D/\partial t = 0$ $\quad \partial B\partial t = 0$ Only electric and magnetic-divergence equations applicable
3. Quasi-stationary state for closed circuits	All Maxwell equations are applicable, but displacement current can usually be neglected, except in capacitors
4. Quasi-stationary state for open circuits	All Maxwell equations are applicable, but displacement current can sometimes be neglected
5. Electromagnetic field in vacuum or perfect dielectric	$J_{cond} = 0$ $\quad \rho = Q/v = 0$ $\quad \partial D/\partial t = J_{dis}$
6. Harmonic time variations of electromagnetic field of angular frequency ω	$\nabla \cdot D = \rho$ $\quad \nabla \times H = (\sigma + j\omega\epsilon)E$ $\nabla \cdot B = 0$ $\quad \nabla \times E = -j\omega\mu H$

of the Maxwell equations are of interest since it is helpful to know field variations at a point in space or in a material medium. By a process of elimination, the variables of the time-dependent curl equations can be separated, to yield the relations

$$\nabla^2 E = \mu\epsilon \frac{\partial^2 E}{\partial t^2} + \mu\sigma \frac{\partial E}{\partial t} + \frac{1}{\epsilon}\nabla\rho \qquad \nabla^2 H = \mu\epsilon \frac{\partial^2 H}{\partial t^2} + \mu\sigma \frac{\partial H}{\partial t}$$

Subject to the boundary conditions in any particular situation, these are the equations that must be satisfied by E and H in order for electromagnetic waves to be produced.

Important special cases, having considerable practical application, occur when $\rho = 0$ (as in free space) and when the time variations of the field vectors are expressible by the exponential $\epsilon^{j\omega t}$, where ω is a constant angular frequency in radians per second and t is the time. For this special case the wave equations are

$$\nabla^2 E = (\mu\epsilon\omega^2 - j\sigma\mu\omega)E = \gamma^2 E \qquad \nabla^2 H = (\mu\epsilon\omega^2 - j\sigma\mu\omega)H = \gamma^2 H$$

Solutions of these equations take the form

$$E = [f_1(1 - v_0 t)\epsilon^{-\gamma l} + f_2(1 + v_0 t)\epsilon^{+\gamma l}]\epsilon^{j\omega t}$$
$$H = [f_1(1 - v_0 t)\epsilon^{-\gamma l} + f_2(1 + v_0 t)\epsilon^{+\gamma l}]\epsilon^{j\omega t}$$

where γ is a propagation constant expressing the change in amplitude and phase as the disturbance travels in a medium with speed v_0. The expression f_1 represents a wave of E and H flowing in the $+l$ direction with speed v_0, whereas f_2 represents a wave of E or H flowing in the opposite $-l$ direction and having the same variation as f_1. As might be expected, E and H are related by the properties of the medium. This relation is usually expressed in terms of the intrinsic impedance of the medium (Par. **1-156**) for effective rather than instantaneous values of E and H.

154. Propagation Coefficient. When a homogeneous, isotropic medium whose properties are specified by $D = \epsilon E$, $B = \mu H$, $J_c = \sigma E$, and $\rho = 0$ is subject to harmonic time variations, the E and H field components are found to be proportional to $\gamma^2 = -(\mu\epsilon\omega^2 - j\sigma\mu\omega) = j\omega\mu(\sigma - j\omega\epsilon)$. The quantity γ^2 is given in terms of ϵ, μ, σ, and ρ specifying the physical properties of the medium and the factor ω expressing the time rate of change of the field vectors. The quantity γ relates the distance an electromagnetic wave travels in unit time, and is called the *propagation factor*.

The propagation factor can be expressed as a complex quantity

$$\gamma = \pm[j\omega\mu(\sigma - j\omega\epsilon)]^{1/2} = \pm j\omega\sqrt{\mu\epsilon}(1 - j\sigma/\omega\epsilon)^{1/2} = \alpha + j\beta$$

This result can be expanded by the binomial theorem. By using only the first few terms, we obtain an attenuation factor $\alpha \approx \frac{1}{2}\sigma(\mu/\epsilon)^{1/2}$ and a phase factor $j\beta \approx j\omega \sqrt{\mu\epsilon}[1 + \frac{1}{8}(\sigma/\omega\epsilon)^{1/2}]$.

The attenuation factor α is a measure of the change in amplitude of E or H per unit length of the propagation path, in units of nepers per meter (or in most engineering work, in decibels per meter).

The phase factor β is a measure of the change in phase per unit length as the field vectors E and H are propagated through the medium. The phase factor is measured in radians per meter.

155. Velocity of Wave Propagation. If the E and H vectors of an electromagnetic wave vary with time according to the function $\epsilon^{j\omega t}$, where $\omega = 2\pi f$ is the angular velocity in radians per second, the disturbance from equilibrium conditions at the source will be 0 when $t_0 = 0$. At some later time t, the disturbance at the source will have produced a relative phase change of $\theta = \omega t$. If the electromagnetic wave travels with speed $v = l/t$, the phase shift at the source can be expressed in terms of the distance l the wave has traveled. Since the phase shift relates time and distance, it follows that $\theta = \omega t = \omega l/v = (\omega/v)l = \beta l$. Here $\beta = \theta/l$ is the phase change per unit length of wave propagation whereas $\omega = \theta/t$ is the phase shift per unit time. The velocity of propagation is

$$v = l/t = \omega/\beta$$

The speed of propagation is related to the electromagnetic constants of a substance by the relation $v = 1/(\mu\epsilon)^{1/2}$. Since $\mu = \mu_0\mu_r$ and $\epsilon = \epsilon_0\epsilon_r$, the velocity can be expressed as

$$v = \sqrt{1/\mu\epsilon} = \sqrt{1/(\mu_0\epsilon_0)(\mu_r\epsilon_r)} = c/\sqrt{\mu_r\epsilon_r}$$

where $c = \sqrt{1/\mu_0\epsilon_0}$ is the speed of propagation in free space $\approx 300{,}000$ km/s or 186,300 mi/s.

The velocity with which waves travel in material substances is always less than their speed in free space or a vacuum since μ_r and ϵ_r always have values greater than 1 (except for paramagnetic materials). Propagation is rapidly attenuated in paramagnetic and other conducting materials.

156. Intrinsic Impedance of a Medium. The field vectors of an electromagnetic wave are interrelated by the *intrinsic impedance* of the medium, $Z_0 = E/H$, where E and H are the effective, or root-mean-square (rms), values of the electric and magnetic field intensities. By making use of the relation $\nabla \times E = j\omega\mu H$ for harmonic time variations of the field, it can be shown that

$$H = (E/Z_0)\,\epsilon^{-\gamma l}$$

where

$$Z_0 = \sqrt{\frac{j\omega\mu}{\sigma + j\omega\epsilon}} = \sqrt{\frac{\mu}{\epsilon}}\frac{1}{1 - j\sigma/\omega\epsilon}$$

The reference, or in-phase, component is the resistance

$$R_0 \approx \sqrt{\frac{\mu}{\epsilon}}\left[1 - \frac{3}{8}\left(\frac{\sigma}{\omega\epsilon}\right)^2 + \cdots\right]$$

and the quadrature component is the reactance

$$X_0 \approx \sqrt{\frac{\mu}{\epsilon}}\frac{\sigma}{2\omega\epsilon}$$

both of which are measured in ohms. For free space the properties of the medium are $\sigma = 0$, $\mu = \mu_0$, and $\epsilon = \epsilon_0$, and so the intrinsic impedance is

$$Z_0 = R_0 = \sqrt{\mu_0/\epsilon_0} = 379.7\ \Omega$$

157. Radiation from a Simple Dipole. The basic features of electromagnetic radiation can be illustrated by considering radiation from a current element $I\,dl$ located at the center of a system of spherical coordinates, in which the instantaneous current is $i = I\cos\omega t$.

Any disturbance created by the dipole at time t_0 will be observed at a point whose distance from the origin is r at time $t = t_0 + r/v$. That is, the effect observed at r at time t must have originated at the source at an earlier time $t_0 - r/v$, where v is the velocity of wave propagation.

At a remote point r at a distance very large compared to the dipole length l and at a frequency ω whose wavelength λ is also very large compared to the dipole length, the components of the radiated electromagnetic field are given, in spherical coordinates, as

$$E_\phi = 0 \qquad E_\theta = \frac{I\,dl\sin\theta}{4\pi\epsilon}\left(\underbrace{\frac{\sin\omega t}{\omega r^3}}_{\text{static}} + \underbrace{\frac{\cos\omega t}{r^2 v}}_{\substack{\text{induction}\\ \text{(quasi-steady)}}} - \underbrace{\frac{\omega\sin\omega t}{rv^2}}_{\text{radiation}}\right) \qquad H_r = 0 \qquad H_\theta = 0$$

$$E_r = \frac{I\,dl\cos\theta}{2\pi\epsilon}\left(\underbrace{\frac{\sin\omega t}{\omega r^3}}_{\substack{\text{induction}\\\text{(quasi-steady)}}} + \underbrace{\frac{\cos\omega t}{r^2 v}}_{\text{radiation}}\right) \qquad H_\phi = \frac{I\,dl\sin\theta}{4\pi}\left(\underbrace{\frac{\cos\omega t}{r^2}}_{\substack{\text{induction}\\\text{(quasi-steady)}}} - \underbrace{\frac{\omega\sin\omega t}{rv}}_{\text{radiation}}\right)$$

In these equations, $v = (1/\mu\epsilon)^{1/2}$ is the velocity of wave propagation.

158. Poynting's Vector. For electromagnetic energy flowing into or out of any closed region, at any instant the rate of flow is proportional to the surface integral of the vector product of the electric field strength and the magnetizing force. This product, known as *Poynting's vector*, is

$$S = E \times H$$

The energy density S is in watts per square meter when the electric field intensity E is in volts per meter and the magnetic field intensity H in amperes per meter.

159. Transmission of Electromagnetic Waves. Absorption can be regarded as the transformation of energy from its original form to some other form as waves pass through a material substance. When a beam of radiant energy passes through a transparent medium whose size is large compared with the wavelength of the radiation, some of the radiant flux is continuously absorbed in the medium.

The decrease in intensity of the radiant energy in passing through the medium, called *absorption*, depends on the thickness of the medium and is given quantitatively by Lambert's law.

If I_0 is the intensity of monochromatic energy entering an absorbing medium, the intensity of the emerging energy in passing through a transmission path of length x is

$$I = I_0 \epsilon^{-kx}$$

where k is a characteristic of the medium called the *absorbing coefficient*. The value of k is determined experimentally and, in general, is dependent upon the wavelength of the radiation. The intensity of the beam of radiant energy emerging from the medium of thickness x is a measure of the transmission properties of the medium and is given by

$$I_{em} = I_0 - I = I_0(1 - \epsilon^{-kx})$$

160. Reflection. Reflection is the change in direction of an electromagnetic wavefront, with a sudden change in its phase. The reflection coefficient is the ratio of the radiant energy reflected from a surface to the total amount of energy incident on the surface.

The laws of reflection depend upon the shape of the reflecting surface. If the reflecting surface is a smooth, flat plane, the laws of reflection of a plane wave, i.e., one having infinite radius of curvature, are (1) the reflected beam lies in the plane of the incident ray and the normal of the reflecting surface at the point of incidence, and (2) the angle of reflection, i.e , the angle between the reflected ray and the surface normal, is equal to the angle of incidence.

161. Refraction. The index of refraction of a substance is the ratio of the velocity of radiant energy in a vacuum to its velocity in the substance. The index of refraction varies with the wavelength of the radiation.

Snell's law states that if i is the angle of incidence, r the angle of refraction, v_1 the velocity of propagation in the first medium, and v_2 the velocity of propagation in the second medium, the index of refraction is

$$n = v_1/v_2 = (\sin i)/(\sin r)$$

Refraction is used to designate the change in the direction of a ray as it passes from one medium in which its velocity is v_1 to another in which its velocity is v_2.

162. Diffraction. Wave diffraction refers to the bending of rays around an obstacle such as a sharp edge or point over which radiant energy passes. It results from the spherical spreading of electromagnetic energy from a point source and indicates that the straight-line or rectilinear propagation of electromagnetic waves is an approximate concept.

If a source of radiant energy is small enough to be regarded as a point, the shadow of an object has its maximum sharpness. A certain amount of radiant energy from the source is always found, however, within the geometrical shadow as a result of diffraction at the edge of the shadow-casting object.

163. Wave Interference. The principle of linear superposition applies to radiation in a linear propagation medium. Hence, the intensity of two rays of monochromatic radiation in a linear medium is the sum or difference of their individual effects, taking account of their amplitude, frequency, and phase. If two beams have the same frequency, the additive and subtractive effects are said to produce interference when the beams lie in the same plane and travel substantially in the same path, large compared with the wavelength of the radiation. The phenomenon of cancellation and reinforcement of waves of the same frequency is called *wave interference.*

If the intensity of one beam of frequency $\omega = 2\pi f$ is

$$i_1 = I_1 \sin \omega t$$

and the properties of the second beam are given by

$$i_2 = I_2 \sin (\omega t + \phi)$$

where ϕ is the relative phase displacement between the two, then the net intensity of the two beams has the same frequency as that of the component rays and its amplitude is

$$I = (I_1^2 + I_2^2 + 2I_1 I_2 \cos \phi)^{1/2}$$

164. Dispersion and Scattering. Dispersion is the process by which radiation is separated in accordance with some characteristic (such as frequency, wavelength, or energy) into components which have different directions. The difference between the indices of refraction of any substance for two different wavelengths is a measure of dispersion for these two wavelengths and is called the *coefficient of dispersion* or the *dispersive power.*

If n_1 and n_2 are the refractive indices of radiation whose wavelengths are λ_1 and λ_2, respectively, within the spectral range considered, and if n_m is the refractive index for some specified wavelength λ_m between λ_1 and λ_2, the dispersive power for wavelength λ_m is the ratio

$$d = (n_1 - n_2)/(n_m - 1)$$

The wavelength λ_m is commonly taken to be the mean of λ_1 and λ_2.

Scattering is the process of diffusing, in various directions, electromagnetic radiation by reflection from molecules, atoms, electrons, or other particles.

165. Bibliography on Electromagnetic Fields and Waves

1. Adler, R. B., L. J. Chu, and R. M. Fano   "Electromagnetic Energy Transmission and Radiation," Wiley, New York, 1960.
2. Andrews, C. L.   "Optics of the Electromagnetic Spectrum," Prentice-Hall, Englewood Cliffs, N.J., 1960.
3. Becker, R., and F. Sauter   "Electromagnetic Fields and Interactions," 2 vols., Blaisdell, New York, 1964.
4. Budden, K. G.   "Wave Guide Mode Theory of Wave Propagation," Prentice-Hall, Englewood Cliffs, N.J., 1962.
5. Bronwell, A. B., and R. E. Beam   "Theory and Application of Microwaves," McGraw-Hill, New York, 1947.
6. Fano, R. M., L. J. Chu, and R. B. Adler   "Electromagnetic Fields, Energy, and Forces," Wiley, New York, 1960.
7. Frank, N. H.   "Introduction to Electricity and Optics," McGraw-Hill, New York, 1950.
8. Glasstone, S.   "Sourcebook of Space Science," Van Nostrand, New York, 1965.
9. Hayt, W. H.   "Engineering Electromagnetics," 3d ed., McGraw-Hill, New York, 1974.
10. Henney, K.   "Radio Engineering Handbook," 5th ed., McGraw-Hill, New York, 1959.
11. Hund, A.   "Phenomena in High Frequency Systems," McGraw-Hill, New York, 1936.
12. Hund, A.   "Short Wave Radiation Phenomena," 2 vols., McGraw-Hill, New York, 1952.
13. Javid, M., and P. M. Brown   "Field Analysis and Electromagnetics," McGraw-Hill, New York, 1963.
14. Jeans, J.   "Mathematical Theory of Electricity and Magnetism," Cambridge University Press, London, 1948.
15. Jordan, E. C.   "Electromagnetic Waves and Radiating Systems," Prentice-Hall, Englewood Cliffs, N.J., 1950.
16. King, R. W. P.   "Electromagnetic Engineering," McGraw-Hill, New York, 1945.
17. Kraus, J. D.   "Electromagnetics," McGraw-Hill, New York, 1953.
18. Laport, E. A.   "Radio Antenna Engineering," McGraw-Hill, New York, 1952.
19. Lorrain, P., and D. Corson   "Electromagnetic Fields and Waves," 2d ed., Freeman, San Francisco, 1970.
20. Montgomery, G. G., R. H. Dicke, and E. M. Purcell   "Principles of Microwave Circuits," Dover, New York, 1965.
21. Reintges, J. F., and G. T. Coate   "Principles of Radar," McGraw-Hill, New York, 1952.
22. Ramo, S., and J. R. Whinnery   "Fields and Waves in Modern Radio," Wiley, New York, 1944.
23. Sarbacher, R. I., and W. A. Edson   "Hyper and Ultrahigh Frequency Engineering," Wiley, New York, 1943.
24. Schelkunoff, S. A.   "Electromagnetic Waves," Van Nostrand, New York, 1943.
25. Seely, S.   "Introduction to Electromagnetic Waves," McGraw-Hill, New York, 1958.
26. Stratton, J. S.   "Electromagnetic Theory," McGraw-Hill, New York, 1941.
27. Smythe, W. R.   "Static and Dynamic Electricity," McGraw-Hill, New York, 1950.
28. Wait, J. R.   "Electromagnetic Waves in Stratified Media," Pergamon, Long Island City, N.Y., 1962.

THE ELECTROMAGNETIC SPECTRUM

166. Regions of the Spectrum. It is useful to divide the electromagnetic spectrum into regions exhibiting common properties useful to science and technology. Near the lower end of the frequency spectrum radiations are designated in terms of frequency bands which for the

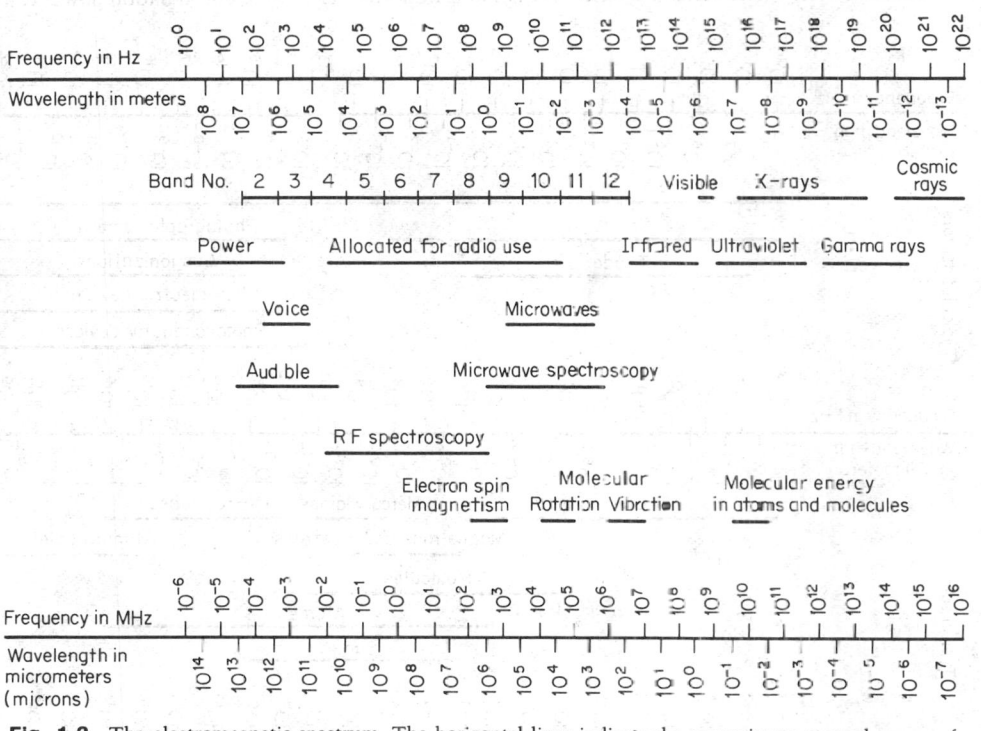

Fig. 1-3. The electromagnetic spectrum. The horizontal lines indicate the approximate spectral ranges of various physical phenomena and practical applications.

most part are useful in radio communication. The limits of frequency of bands used for radio communication are specified by governmental authorities.

Beyond the radio spectrum lie other frequency bands for which frequency assignments are not necessary. These bands include the infrared, visible, ultraviolet, x-ray, gamma-ray, and cosmic-ray portions of the spectrum. The relations between the various portions of the spectrum are shown in Fig. 1-3.

167. Sources and Detectors of Electromagnetic Energy (see Sec. 11). The greater portion of the spectrum is known to us only through the effects produced by energy sources and observed by detectors of radiant energy.

Sources and detectors are of two major classes: frequency-selective devices, responding to a band of frequencies that is small compared with the mean frequency of response, and broadband devices, capable of producing and responding to a range of frequencies that is large compared with the frequency of the band in which energy is generated or detected.

A wide variety of natural and artificial sources and detectors of radiant energy are available for use in different portions of the spectrum. Typical sources and detectors, together with the spectral region for which they are normally useful, are shown in Fig. 1-4.

168. Spectrum Utilization. The use to which the electromagnetic spectrum is put depends primarily upon the frequency (or wavelength) of the radiation and the propagation properties of the medium in which the waves travel. Because the electromagnetic spectrum covers a range of more than 22 decades, it is desirable to divide the spectrum into regions having similar properties.

Lowest-Frequency Bands (1 to 10^3 Hz). In this lowest-frequency portion of the spectrum, restricted bandwidths prevent use for communications, although proposals have been made to use this band for worldwide communications with submarines. Electric power is generated at frequencies of 50 or 60 Hz (occasionally at 25 Hz or other low frequency). For aviation use, electric power is used at 400 Hz, to reduce weight of iron-core apparatus.

This band also includes a portion of the audible (voice) frequencies, but electromagnetic energy is not radiated in this application.

Low Radio Frequencies (10^3 to 2×10^5 Hz). This band is particularly useful for long-distance communication where reliability of transmission is important and sufficient radiation power can

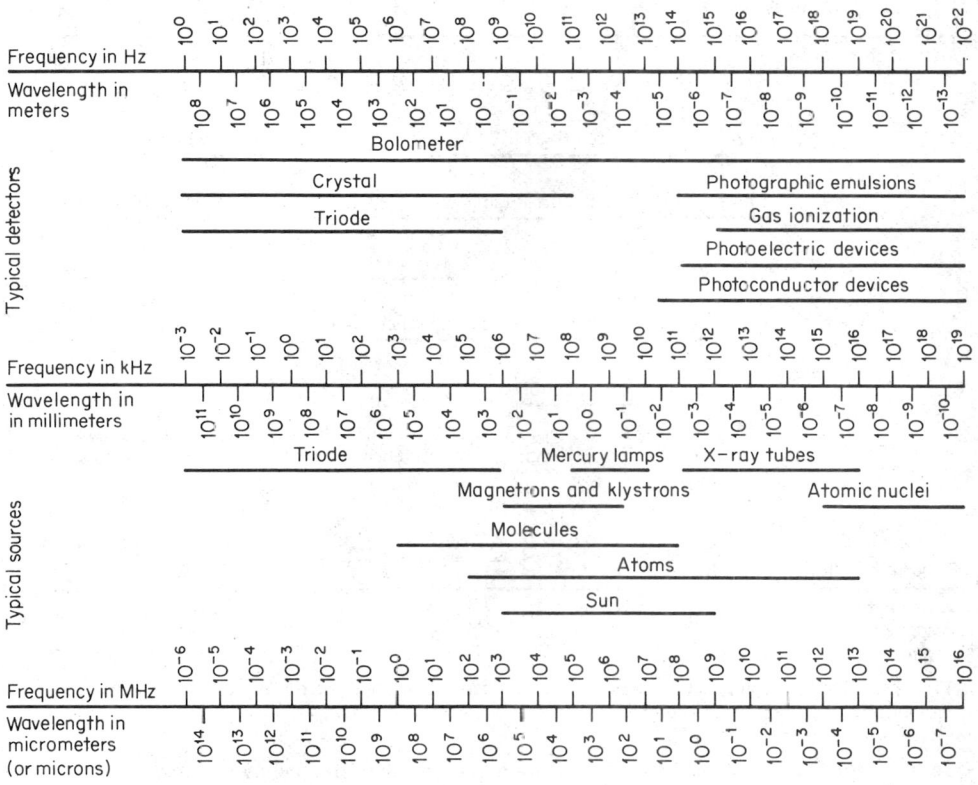

Fig. 1-4. Typical sources and detectors of electromagnetic energy.

be made available. This band is generally used for radio telegraphy. As frequency is decreased in this band, reliability and signal strength improve; there are fewer interruptions from diurnal, seasonal, and solar causes, but static and other radio noises tend to increase.

Medium Frequencies (2×10^5 to 2×10^6 Hz). The lower-frequency portion of this band is useful for services requiring reasonably stable transmission, day and night, over moderately long distances. Since appreciable atmospheric noise occurs in this band, substantial field strength is required for reliable communication. The upper-frequency portion of the band is characterized by relatively weak ground wave, but the sky wave prevailing at night is relatively large. This band is commonly used for radio broadcasting. The band is also used for fixed services, maritime mobile service, maritime navigation including loran, aeronautical radio navigation, and amateur communication.

High Frequencies (2×10^6 to 3×10^7 Hz). Useful but somewhat erratic long-range propagation is possible with low power in this frequency range. When the transmission path is entirely in darkness and the ionosphere is undisturbed, frequencies below a maximum usable frequency are propagated over long distances. Transmission is dependent upon vagaries of the ionosphere. Fading and multiple-path effects often limit speed of communication. The large interference range limits the number of emissions that can be simultaneously radiated at a given frequency. The band is used for fixed services, mobile services, amateur transmissions, broadcasting, maritime mobile service, and telemetering.

Very High and Ultra High Frequencies (3×10^7 to 3×10^9 Hz). Lumped circuits, useful at lower frequencies, give way to transmission lines and other distributed circuits in this band. The band is suitable for relatively short-distance communication for services transmitting a large amount of detail, including radar and television. Directive antenna systems of small size are economical and effective. If powerful transmitters and high-gain antennas are used, reliable long-distance propagation is possible by using waves scattered by turbulence in the troposphere. The band is used for fixed services, mobile services, space research, radio astronomy, telemetering and tracking, amateur transmissions, satellite communication, meteorological aids, radio location, radar, and television. It is also used for radio-frequency spectroscopy and in studies of magnetic resonance and nuclear quadruple resonance.

Microwaves (3×10^9 to 3×10^{11} Hz). Transition from circuit to optical techniques characterizes microwave bands. Electromagnetic waves in this band are short enough to be propagated by highly directive antennas and waveguides. Long-distance communication can be carried out by a series of automatic relay stations mounted at high elevations within line of sight of each other. This portion of the spectrum is used for high-definition radar, television, and similar services requiring extensive bandwidths to convey a considerable amount of information. Microwave spectroscopy, dealing with electron-spin resonance and molecular rotation, is carried out in this band.

Infrared Region (3×10^{11} to 4×10^{14} Hz). The infrared, visible, and ultraviolet portions of the spectrum have long been used for identifying molecules by their spectral emissions and as a subsequent means of qualitative analysis, for determining the geometry of simple molecules, and for quantitative analysis where spectral-line intensity is related to the concentration of substance in question. The low-frequency limit is not well defined and often extends into the microwave region of the spectrum.

Visible Spectrum (3.95×10^{14} to 7.90×10^{14} Hz). This is a relatively narrow part of the spectrum, covering only one octave, but it is extremely important. This band contains all the radiant energy visible to the human eye; it is commonly divided into six or seven regions identified with different spectral hues. Spectroscopy conducted in the visible spectrum is usually identified with molecular vibrations or with electronic transitions of outer electrons of atoms.

Ultraviolet Spectrum (7×10^{14} to 3×10^{16} Hz). Ultraviolet radiations have found bactericidal applications and, along with infrared radiations, are used in some specialized communications. In this portion of the spectrum electronic transitions of outer-shell electrons occur, and the spectrum measured is often referred to as the electronic spectrum.

X-Ray Spectrum (3×10^{16} to 10^{19} Hz). Extensive medical, biological, and industrial applications have been made of x-rays. X-rays have made important contributions to our understanding of the structure of matter. X-ray spectroscopy is a valuable tool in studying electronic transitions of inner-shell electrons occurring when high energy is absorbed.

Gamma-Ray Spectrum (10^{19} to 10^{22} Hz). This portion of the spectrum is useful in studying nuclear transitions. The upper part of this spectrum, from about 10^{20} Hz and beyond, is sometimes called the cosmic-ray spectrum.

169. Frequency Tolerances. Frequency tolerance is defined to be the maximum permissible departure of the center frequency of the band from an assigned frequency. Frequency tolerance is usually expressed in parts per million so that, for a specified frequency tolerance, the absolute deviation, in hertz, tends to be greater for stations operating on a higher frequency than those operating on lower frequencies. Sometimes, however, the maximum permissible tolerance is specified in absolute values of frequency, in hertz.

For stations of different classes of service, the permissible frequency tolerances also depend upon the peak power radiated from the antenna. In general, stations with low values of radiated power may have larger frequency tolerances than high-power stations for the same class of service.

170. Spurious Emissions. Spurious emissions are those occurring outside the frequency band assigned for a particular type of emission and quality of service. They do not contribute to the amount of information conveyed but often produce interference with other services. For these reasons, the amplitude of spurious emissions should be kept as low as possible. The absolute level of spurious-emission intensity depends upon the amount of power radiated. Accordingly, it may be specified in terms of absolute power levels or in terms of power levels relative to that of the radiated power. Spurious emissions may include harmonics, intermodulation products, cross modulation, and parasitic emission.

171. Standard-Frequency Transmissions. A number of stations throughout the world radiate waves whose carrier and modulation frequencies are very precisely established and so can be used for frequency standardization. The standard-frequency transmissions are usually radiated by government stations whose precision of frequency is established by national laboratories. In the United States authority for national standards resides in the National Bureau of Standards, and standard-frequency transmissions are made from stations in Fort Collins (near Boulder), Colo., and in Puunene, Maui, Hawaii. Transmissions suitable as frequency standards for use in North America and the Pacific are listed in Table 1-4.

Transmissions also include time signals and telegraphic transmissions regarding the ionospheric conditions to be expected in the near future. Details can be determined from the Frequency-Time Broadcast Services, National Oceanic and Atmospheric Administration, Boulder, Colo., 80302.

Table 1-4   Standard-Frequency Transmissions

Frequency	Error, parts in 10^{10}	Comment
440 Hz*	0.5	WWV, Fort Collins, Colo., WWVH, Hawaii
600 Hz*	0.5	WWV, Fort Collins, Colo., WWVH, Hawaii
1,000 Hz*	0.5	WWV, Fort Collins, Colo.
17.8 kHz	0.5	NAA, Cutler, Maine
18.6 kHz	0.5	NPG/NLK, Jim Creek, Wash.
20.0 kHz	0.5	WWVL, Fort Collins. Colo.
21.4 kHz	0.5	NSS, Annapolis, Md.
24.0 kHz	0.5	NBA, Balboa, Panama, C.Z.
26.1 kHz	0.5	NPM, Hawaii
60.0 kHz	0.5	WWVB, Fort Collins, Colo.
100.0 kHz	0.5	Loran C, Carolina Beach, N.C.
2.5 MHz	0.5	WWV, Fort Collins. Colo., WWVH, Hawaii
3.33 MHz	50	CHU, Ottawa, Canada
5.0 MHz	0.5	WWV, Fort Collins, Colo., WWVH, Hawaii
7.335 MHz	50	CHU, Ottawa, Canada
10.0 MHz	0.5	WWV, Fort Collins, Colo., WWVH, Hawaii
14.67 MHz	50	CHU, Ottawa, Canada
15.0 MHz	0.5	WWV, Fort Collins, Colo., WWVH, Hawaii
20.0 MHz	0.5	WWV, Fort Collins, Colo.
25.0 MHz	0.5	WWV, Fort Collins, Colo.

* Modulation frequencies; all others are carrier frequencies.

172. Bibliography on Spectrum Utilization

1. Andrews, C. L.   "Optics of the Electromagnetic Spectrum," Prentice-Hall, Englewood Cliffs, N.J., 1960.
2. Billmeyer, F. W., and M. Saltzman   "Principles of Color Technology," Wiley, New York, 1966.
3. Bransen, M. A.   "Infrared Radiation: A Handbook for Application," Plenum, New York, 1968.
4. Clark, G. L.   "Encyclopedia of X-Ray and Gamma Ray Spectroscopy," Van Nostrand, New York, 1963.
5. Chang, R.   "Basic Principles of Spectroscopy," McGraw-Hill, New York, 1973.
6. Illuminating Engineering Society   "I.E.S. Lighting Handbook," 5th ed., Illuminating Engineering Society, New York, 1972.
7. International Telephone & Telegraph Co.   "Reference Data for Radio Engineers," 6th ed., Howard Sams, Indianapolis, Ind., 1975.
8. Jamieson, J. A., R. H. McFee, G. N. Plass, R. H. Grube, and R. G. Richards   "Infrared Physics and Engineering," McGraw-Hill, New York, 1963.
9. Johler, J. R.   Propagation of the Low Frequency Radio Signal, *Proc. IEEE*, 1962, Vol. 50, No. 4, pp. 404–427.
10. Joint Technical Advisory Committee, IEEE   "Spectrum Engineering—Key to Progress," Institute of Electrical and Electronics Engineers, New York, 1968.
11. Kaelbe, E. F.   "Handbook of X-Rays," McGraw-Hill New York, 1967.
12. Koller, L. R.   "Ultraviolet Radiation," Wiley, New York, 1952.
13. Szymanski, H. A.   "IR: Theory and Practice of Infrared Spectroscopy," Plenum, New York, 1967.
14. Taylor, R. E.   "Radio Frequency Interference Handbook," National Aeronautics and Space Administration, Washington, 1971.
15. Wait, J. R.   "Electromagnetic Waves in Stratified Media," Pergamon, New York, 1962.
16. Watt, A. D.   "Very Low Frequency Radio Engineering," Pergamon, New York, 1967.
17. Walker, S., and B. P. Straughan   "Spectroscopy," 2 vols., 2d ed., Macmillan, New York, 1978.
18. Special Issue on Efficient Use of the Spectrum,   *Proc. I.E.E.E.*, Dec. 1980.

SPEECH, HEARING, AND VISION

173. Sensory Perceptions in Electronics Engineering. In designing electronic devices it is often necessary to employ data on the relations between physical stimuli and the corresponding sensations evoked in people. To do so in terms of concepts and parameters capable of objective and mathematical formulation is difficult, since human sensations depend upon the situation as a whole and a given situation cannot be split into quantified fragments without seriously modifying the human response. Nor is it possible to measure sensations or impressions objectively, for these depend upon individual experiences, which are unique.

Fortunately it is possible to establish procedures for carrying out operations in which the magnitudes of the measurements of physical quantities (stimuli) are related to the sensitivity, or acuity, of human senses. When properly carried out and interpreted such psychophysical measurements can be very useful in the design of electronic systems and devices.

174. Cognition. The objective relations existing between entities of the physical world and a person's subjective evaluation (cognition) of such relations are quite different and distinct.

The relationships between events described by the physical sciences and those described by the behavioral sciences must be examined with caution. There are no direct, immediate correlations between these fields in the sense that the relations between events of physical objects are of the same class as those represented by our subjective evaluations of them or even of our sensory perceptions of them. Hence, psychophysical data can be easily misinterpreted and misused. It is essential, therefore, that the engineer understand in some detail the relations between stimuli and the psychophysical data used in the design of systems and devices.

175. Evaluation of Physical Stimuli. In assessing human response to, and subjective evaluation of, various stimuli, there are three classes of evaluative concepts: *physical concepts,* which are quantitative and independent of the particular observer making measurements or observations; *psychological concepts,* which are subjective and uniquely related to individual experience (they have no quantitative significance because they cannot be expressed in operational terms); and *psychophysical concepts,* which can be made quantitative to the extent that they depend upon measurable aspects of the sense perceptions of the observer.

While different observers usually obtain different results when making psychophysical measurements of a given kind, it is often found that observations made with a large number of persons under identical measurement conditions tend to cluster around a set of values. Persons whose measured characteristics fall within small deviations of the mean of measurements on a large number of observers are regarded as having *normal sense responses.*

Particular sets of data may be adopted for general use by competent groups who examine, evaluate, and agree to accept the best data available as a standard. In this way, visibility curves, audibility curves, and similar data have been adopted and used in the design of systems or equipment.

176. Stimulus-Response Relations. The change in a given stimulus that produces a just detectable change in sensory response is called the *difference threshold.* Experiments show that the difference threshold is neither a fixed difference nor a fixed ratio of a change in stimulus but depends in a complicated way on the magnitude of the stimulus and technique of measurement.

The psychophysical response to stimuli activating different human senses can be expressed, roughly, by the law of psychology established by Weber and Fechner, which states that the least noticeable change of a stimulus is proportionally related to the magnitude of the stimulus. Thus, if Δp is the minimum recognizable change in the perception of a stimulus as its magnitude is changed from s_1 to s_2 over a range $\Delta s = s_2 - s_1$, then, by the Weber-Fechner law,

$$\Delta p = k \frac{s_2 - s_1}{s} = k \frac{\Delta s}{s}$$

where k is a constant of proportionality. By letting the increments become vanishingly small and integrating, the response is

$$p = k \log s + C$$

where C is a constant of integration. This equation states that the response in perception to a stimulus is proportional to the logarithm of the magnitude of the stimulus. Note that p represents a measure of the *perception* of a stimulus and is not a measure of subjective *sensation.*

The value of k varies considerably from sense to sense; it even varies for different magnitudes of a stimulus for the same sense. Some experiments tend to show that the Weber-Fechner law might more accurately describe situations if written

$$\Delta p = k \frac{\Delta s}{s + A}$$

where A is a small value of the stimulus related to the threshold value but not identical to it.

At small values of s, the constant A is a significant factor, but it becomes increasingly less important as values of s increase.

While the Weber-Fechner law is not beyond criticism, it is a pragmatically useful tool in dealing with perception and recognition processes. It provides a rationale for a variety of logarithmic response measurements commonly used in communications and electronics engineering.

177. Logarithmic Response Units. A number of logarithmic units have found extensive use in science and engineering. Most of them have been used to designate power ratios. J. W. Horton has suggested a definition of logarithmic units to relate ratios of any two physical quantities, in units of *logits*.

A relative change in a quantity can be written

$$R = r^m$$

where R is a ratio of two quantities that expresses the change in their relative magnitude, r is a standard ratio in relative magnitude, and m is the exponent indicating the power to which r must be raised to yield the given ratio R. From this relation

$$m = \log_r R = (\log R)/(\log r)$$

In many physical situations, a barely perceptible change occurs when the quantities have a ratio of about 1.25, and so it is convenient to take $r = 10^{0.1} = 1.2589 \cdots$. If this value is adopted

$$m = 10 \log R \qquad \text{(logits)}$$

and m is a numeric, expressed in units of logits. This definition can be used and applied to any physical unit.

178. Bel and Decibel. The *bel*, named in honor of Alexander Graham Bell, is defined as the common logarithm of the ratio of two powers, P_1 and P_2. Thus the number of bels N_B is

$$N_B = \log (P_2/P_1)$$

If P_2 is greater than P_1, N_B is positive, representing a gain in power; if $P_2 = P_1$, N_B is zero and the power level of the system remains unchanged; if P_2 is less than P_1, N_B is negative and the power level is diminished.

A smaller and more convenient unit for engineering purposes is the decibel, whose magnitude is $\frac{1}{10}$ B. Thus

$$N_{dB} = 10N_B = 10 \log (P_2/P_1)$$

In terms of electrical quantities, power may be expressed as $P = I^2R$ or $P = V^2/R$. Accordingly, the change in power level, in terms of changes in current and voltage, is

$$N_{dB} = 20 \log (I_2/I_1) + 10 \log (R_2/R_1) = 20 \log (V_2^2/V_1) + 10 \log (R_2/R_1)$$

The power-level change, expressed in decibels, is correctly given in terms of voltage and current ratios alone only for the special case for which $R_1 = R_2$.

179. Neper. The *neper* is defined to be the natural logarithm of the ratio of two currents, I_1 and I_2, or two voltages, V_1 and V_2. Thus, the number of nepers expressing a given change in voltage or current level is

$$N_{Np} = \ln (I_2/I_1) = \ln (V_2/V_1)$$

The neper can also be defined as one-half the natural logarithm of two power ratios measured under such circumstances that the resistance of the circuit remains the same as the level is changed.

180. Volume Unit. The *volume unit* is defined to be 10 times the common logarithm of the power ratio, P_2/P_1, where $P_1 = 1$ mW. If P_2 is expressed in watts,

$$N_{vu} = 10 \log (P_2/0.001) = 10 \log P_2 + 30$$

Note that the expression for volume unit specifies a reference level (1 mW) so that the number of volume units represents the actual power level in a circuit.

Logarithmic expressions for power ratios cannot be used to indicate the amount of power in a circuit or system unless and until some reference level or zero power level is stated.

181. Phon. The *phon* is a unit of loudness level. The level of a sound, in phons, is numerically equal to the intensity level (in decibels) of a pure 1-kHz tone which is judged by the listener to be of equivalent loudness.

182. Adaptive Processes. Adaptive processes include a wide range of adjustments to changes in environmental conditions. Specifically, the term is used to refer to the adjustment of sense organs, such as the eye or receptors of the skin, to the intensity or quality of stimulation such as light, temperature. or pressure prevailing at the moment, by changes in the sensitivity of the sense organs.

In visual processes, *dark adaptation* is the gradual adjustment and increased visual sensitivity to light of reduced intensity.

Light adaptation requires adjustment to highly intense visual excitation from previous excitation at low light intensities. *Visual adaptation* is a function of a number of variables including the wavelength and intensity of the stimulus before and after adaptation and the length of time the eye has been stimulated. Light adaptation usually occurs within a few minutes, but complete adaptation to the dark may require half an hour or more.

Similar effects occur in hearing. The ear appears to adapt rather quickly to new conditions of stimuli.

183. Components and Frequencies of Speech. On the average, the syllables of speech have a duration of about ⅛ s, and the interval between syllables is about ⅒ s. The frequency spectrum of speech sounds depends upon the resonant cavities formed by the throat,

TABLE 1-5   Representative Values of Human Peak Sound Pressure

Sound pressure, dB above 0.002 dyn/cm²	35	60	78
Fraction of speech lying above stated sound pressure, %	90	50	1

mouth, lips, and teeth as well as upon the syllables and words spoken. In general, the frequency spectrum for speech of women tends to be about an octave higher than that of men.

Frequencies below 200 Hz and above 7,000 Hz make little contribution to speech intelligibility. Consonants are more essential to speech intelligibility than vowels, but vowels account for the greater portion of the power generated in speech.

The dynamic power range of normally spoken speech is about 30 dB. Peaks of speech power average about 12 dB above the average level of speech, whereas the weakest speech sounds lie about 18 dB below average speech levels. For men, maximum power in speech is produced at frequencies near 400 Hz; the power decreases gradually with frequency up to about 4,000 Hz and then rapidly drops for high frequencies.

In large measure, the frequency spectrum to which an audio system should respond for the reproduction of speech depends upon the degree of naturalness required. Since it is also influenced by personal preferences and the noise and other interfering effects at the receiving point, it is likely to be a highly subjective choice. See also Sec. **19**.

184. Power Levels of Speech. Over several minutes the average power of connected speech is about 10 μW. For as loud shouting as possible, average speech sounds may reach levels of 1 mW (30 dB above normal speech levels). In faint whispers, speech power may drop to levels of 0.001 μW, 40 dB below the normal speech level. Hence, speech may vary over a dynamic range of 70 dB.

Measurements of the speech power levels of conversational speech for men s voices show that the average long-time speech power occurs in the octave between 500 and 1,000 Hz. Relative to the reference-level value of 10^{-13} W, the normal male voice produces sound levels of about 66 dB in the octave between 30 and 75 Hz. The level rises to about 80 dB in the octave between 500 and 1,000 Hz and falls to about 52 dB in the octave between 5 and 10 kHz.

185. Peak Factor Statistics of Speech. The natural reproduction of speech requires that peak values of speech power throughout the spectrum be taken into account as well as average levels. Table 1-5 gives a rough indication of peak sound pressures for normal speech.

186. Speech Intelligibility. The most common tests for speech intelligibility are articulation tests, in which a participant is asked to report, in writing, sounds heard during a test. The articulation index is the ratio of the number of correctly reported sounds or syllables to the total number presented to the observers. Such an index takes account of the speech clarity of the speaker as well as the speech perception of the listener.

Articulation tests are influenced by many factors, including the intensity of speech, background noise, the listener's familiarity with words or syllables spoken, the nature of the spoken sounds, the frequency spectrum of the transmission system, removal or absence of certain frequency bands in the transmitted spectrum, and removal of peak amplitudes of speech power by clipping or amplitude limiting.

If lower frequencies of speech are removed, the articulation index does not change markedly until frequencies above 500 Hz are removed. If low frequencies between 500 and 2,500 Hz are removed, the articulation index drops sharply. On the other hand, if high frequencies beyond 2,500 Hz are removed, 80% articulation remains, but removal of frequencies above 1,000 Hz leads to impractical communication systems since only 40% of the words spoken are correctly identified.

187. Audibility. To a significant extent the threshold of audibility depends upon how audibility is determined as a psychophysical response. The threshold of audibility for a specified sound is the minimum effective sound pressure capable of providing an auditory sensation, in the absence of noise, in a specified fraction of all observers, the specified fraction usually being taken to be the 1% of subjects having the most acute hearing. The audibility threshold is usually taken to be 0.0002 microbar, or 0.0002 dyn/cm^2.

188. Loudness. The human ear is responsive to frequencies from about 20 to 20,000 Hz, covering a range of 10 octaves, but the sensitivity varies greatly with frequency. The human ear is most sensitive to frequencies from about 2 to 5 kHz and least sensitive to sounds at the extreme frequencies of the audible range. The subjective response to sounds of increasing sound-pressure intensity is described by the loudness of sounds. The loudness of sounds depends upon their intensity and their frequency, and is specified quantitatively in terms of loudness levels, expressed in phons and measured at 1 kHz, which is arbitrarily chosen as the reference frequency (see Par. **1-181**). The loudness of a pure tone of a frequency other than 1 kHz is defined in terms of the sound-pressure level of a sound that appears to be as loud as the 1-kHz pure tone.

As shown in Fig. 1-5, the loudness levels of pure tones are a group of equal-loudness contours. In general the intensity level and the loudness level are described by the same number only at a frequency of 1 kHz. The loudness is a subjective evaluation; the intensity level is a physical measure.

189. Minimum Perceptible Changes in Intensity and Pitch. A change in sound-pressure level of about 1 dB can be detected for pure tones in the frequency range between 50 and 10,000 Hz by persons with normal hearing provided the sound level is 50 dB or more above the threshold value. If the sound-pressure level is less than 40 dB above the threshold value, changes of up to 3 dB are required to be perceptible. Under favorable conditions, changes in sound level as small as 0.3 dB can be detected by persons with normal hearing in the mid-audio-frequency range.

When the sound level is 40 dB or more above the threshold value, persons with normal hearing are able to detect changes in frequency of about 0.3 of 1% at frequencies of 1,000 Hz or higher. A frequency change of about 3 Hz is detectable at frequencies less than 1,000 Hz.

190. Space Perception in Hearing (Stereophony). The ability to localize the direction of a sound source depends, in part, upon facility in recognizing differences in loudness or sound intensity, phase in pure tones, and quality of sounds in the case of complex sounds for each of the two ears. In addition to these purely physical factors, localization of sound by binaural listening depends upon the ability to associate these differences with the direction of a sound source relative to the observer. Thus, experience is a factor in sound localization, quite independent of the hearing process itself.

Localization of sounds can be accomplished in different frequency ranges by different hearing mechanisms. At frequencies of about 300 to 400 Hz or less, the ability to evaluate the direction of a sound source depends upon detecting phase differences and usually is not well developed. Between 400 and 1,000 Hz, sound sources can be localized by detecting differences in phase of sound reaching the two ears. At frequencies above 1,000 Hz differences in the quality of sound reaching the two ears are used, and for frequencies greater than about 3,000 Hz, sound can be localized by differences in loudness at the two ears of an observer.

191. Luminous Energy. Radiant energy evaluated by the human eye is called *luminous energy*. It is measured in physical quantities similar to those used for radiant energy but is weighted, according to frequency or wavelength, by the properties of the normal human eye.

Quantity of light Q (or Q_v) is the quantity of radiant energy as evaluated by the normal human eye. In SI units it is specified in lumen-seconds.

Luminous flux (or Φ_v) is the time rate of flow of light. In SI units it is measured in lumens or candela-steradians. The lumen is the unit of luminous flux and is equal to the flux through a unit steradian from a uniform point source of one candela (candle). Alternatively, it is the flux on a unit surface all points of which are at unit distance from a uniform point source of one candela luminous intensity.

Luminous emittance or *exitance* M (or M_v) is the luminous flux density at a surface. In SI units it is measured in lumens per square meter.

Illuminance E (or E_v) is the density of luminous flux incident on a surface and is the ratio of the luminous flux to the area of the surface when the latter is uniformly illuminated. The SI unit is the lux.

Luminous intensity I (or I_v) of a source of light in a given direction is the luminous flux per solid angle in the direction in question. It is the luminous flux on a small surface normal to that

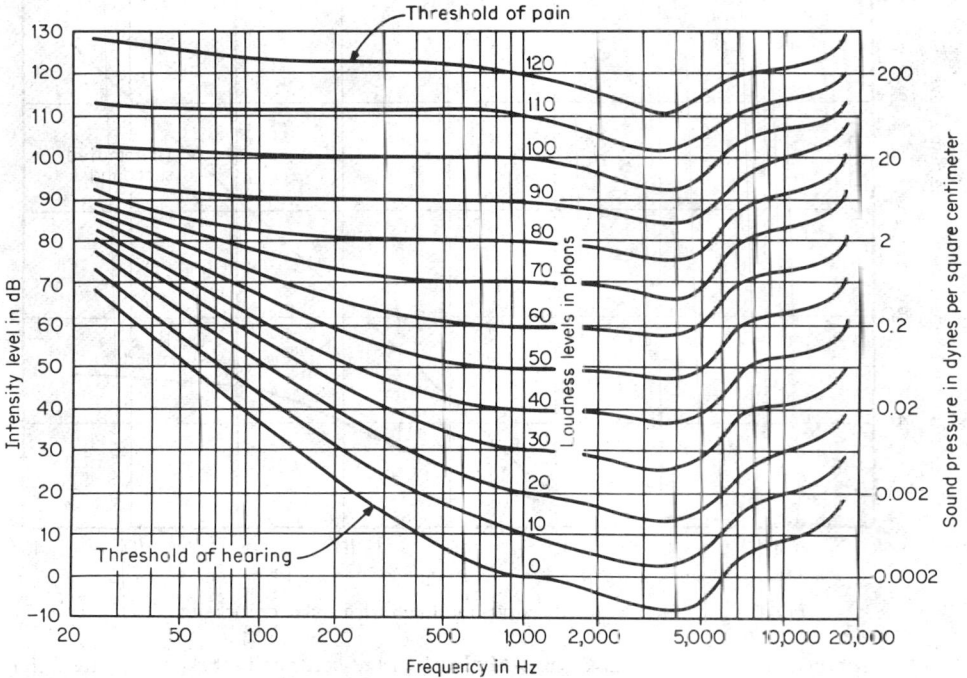

Fig. 1-5. Contours of equal loudness for the normal human ear. The curves are normalized at 1,000 Hz, and the corresponding loudness levels in phons are indicated at this frequency.

direction divided by the solid angle in steradians that the surface subtends at the source. In SI units (Par. **1-201**) it is measured in candelas, approximately equal to the older unit the *candle*.

The *candela* is the luminous intensity, in the perpendicular direction, of a surface of $1/600,000$ m^2 of a blackbody at a temperature of freezing platinum under a pressure of 101,325 N/m^2.

192. Spectral Sensitivity of the Normal Observer (see Sec. **20**). The relative sensitivity of human eyes varies with frequency of monochromatic light and also with intensity of the radiation. It also varies from person to person. For the so-called normal observer, the relative sensitivity of light is the function of wavelength shown in Fig. **20-1**. The curve shows the relative spectral response for daytime (photopic) vision in which the cones play the dominant visual role.

For photopic vision, maximum response occurs at 554 nm. For nighttime vision, the maximum response occurs at 507 nm. The shift of 47 nm is known as the *Purkinje shift* and occurs at light levels between 0.001 and 0.01 lm/ft^2, when the transition takes place between rod and cone vision.

193. Brightness and Brightness Sensitivity. Brightness is the visual sensation produced by the emission or reflection of light. It is a measure of the light emitted from a surface in a given direction per unit area per unit solid angle and can be measured in units such as lumens per square meter.

The sensitivity of the human eye to brightness is measured by the minimum change in illumination needed to produce a just perceptible difference. Brightness sensitivity of the human eye for changes in illumination is nearly constant for moderate light levels. The least perceptible change of brightness, or the brightness threshold, is about 1.6 to 1.8% for all colors at normal light levels.

At low light levels, the brightness sensitivity decreases, first in the red end of the spectrum and finally in the blue, which is the last hue to be perceived. At the lowest levels at which objects can be seen, all sensation of hue or color disappears.

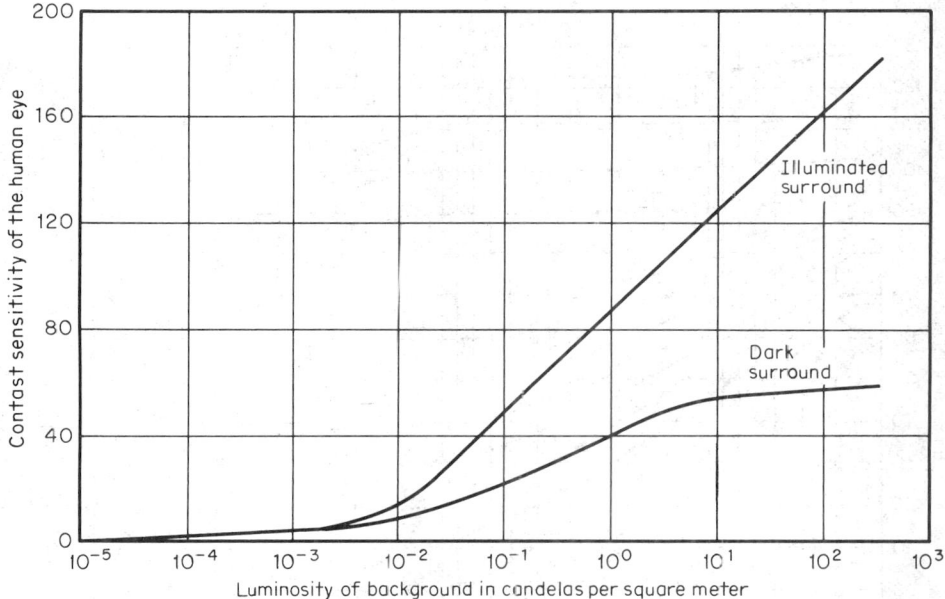

Fig. 1-6. Contrast-sensitivity curves for illuminated and dark background.

194. Contrast and Contrast Sensitivity. *Visual contrast* is the ratio of the difference in brightness between an object and its background to the brightness of the background. Such a definition is not entirely adequate since our normal notions of contrast vary depending upon the type and size of the object to be detected. Thus, it is common to divide contrast sensitivity into two classes: those in which small objects are to be contrasted and detected against their backgrounds and those in which the contrast between large contiguous surfaces is to be determined. The first of these involves variations of contrast with size of the object as well as contrast with illumination. Size is not a factor in determining contrast between large contiguous areas.

The *contrast sensitivity* is a measure of the ability to discriminate slight differences in contrast. It is the reciprocal of *visual contrast*, as defined above.

The contrast sensitivity for an object against its background when both are of the same color and are illuminated by white light is illustrated in Fig. 1-6. The reduced contrast sensitivity for objects having a dark surround may result from glare, which makes the object more difficult to detect.

195. Tonal Discrimination. As used in photography, television, and the visual arts, tone discrimination is the ability to discern a change in the brightness of an image reproduced from an original scene. If the original scene and its reproduction have the same values of brightness in corresponding elements, they have the same brightness range. In this case the tone discrimination is the same as the brightness discrimination.

The full range of brightness values cannot usually be reconstructed in a reproduction; i.e., the brightness range of the reproduced scene is less than that of the original. If the brightness of all elements in the reproduction is proportional to the brightness of the corresponding elements of the original, the brightness range of the reproduction is reduced. If no other property than brightness is modified, there is no loss of detail in the bright or dark portions of the reproduced image, although the detail may be more difficult to detect because of the reduced brightness range or contrast.

If the brightness of the image is not proportional to that of the original, then in addition to brightness compression, details may be lost in the reproduction, either in the shadows, the highlights, or both. Tone discrimination can be improved by increasing the contrast of the reproduction, but this is accompanied by brightness or amplitude distortion in the dark or light areas of the reproduced image.

196. Resolving Power. Resolving power is the ability to detect small objects and distinguish fine detail in the presence of large contrast between an object and its background. It varies

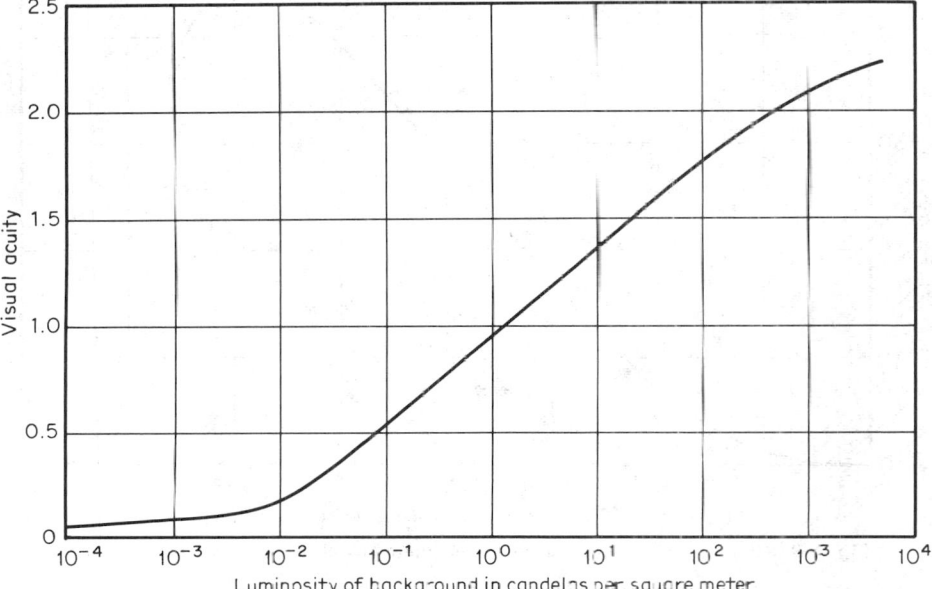

Fig. 1-7. Visual acuity as a function of background illumination.

widely with the type of test object, the spectral distribution of luminous energy, background luminosity, contrast between the object and its background, and (most important) the criteria used in making determinations of the quantities or operations being performed or measured.

The resolving power of the eye is sometimes taken to be a measure of the distinctness with which the images of two point sources of light can be detected separately. It is the minimum angular separation of two objects which appear distinct and separate.

197. Visual Acuity. Visual acuity is the ability to distinguish fine detail. It is usually expressed either as the ratio of the distance at which a line of letters on a Snellen test chart can be read by the observer being tested to the distance at which an observer with normal vision could read the line or as a visual efficiency rating whose percentage figures are related to the size of characters that can be read in each line of a chart such as that of the American Medical Association. Visual acuity may also be expressed as the reciprocal of the angle (measured in minutes of arc) subtended at the eye by a specified test object, such as two black dots on a white background.

Visual acuity increases with the luminosity to which the eye is adapted (Fig. 1-7). For values of luminosity less than about 10^{-2} cd/m², the increase in acuity is small.

The measurement of visual acuity depends upon the luminosity of the entire field, background as well as foreground. Visual acuity is greatest with white or nearly white light.

198. Flicker and Persistence of Vision. The eye does not respond instantly to a visual stimulus, nor does the sensation of vision disappear instantly when the stimulus is removed. Experiments show that the time an image is retained after the stimulus is removed is somewhat greater than the time needed to produce a visual sensation when the stimulus is suddenly applied.

This situation makes possible the viewing, in rapid succession, of related but slightly different scenes in such a way that a sensation of apparently continuous motion is conveyed. If the rate at which different images is presented is too small, the effect is accompanied by *flicker*. At sufficiently high projection rates, flicker disappears. The frequency at which flicker disappears depends upon the brightness of the field being viewed, as indicated in Fig. 1-8.

199. Depth Perception and Stereoscopy. The images of an object, especially one nearby, are seen in a slightly different manner because the pupils of the two eyes are not coincident. This dissimilarity of two simultaneous images enables the brain to fuse the two so that

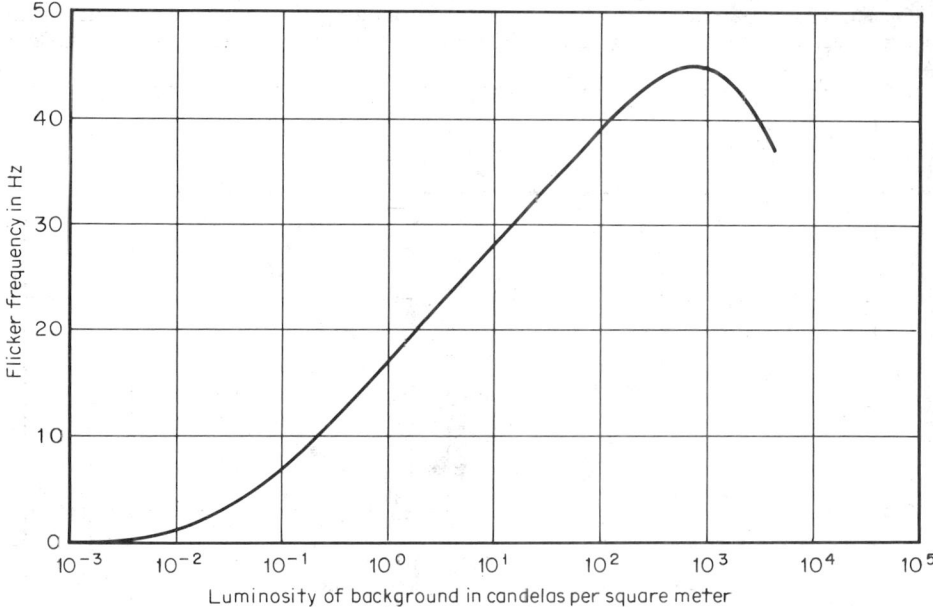

Fig. 1-8. Flicker frequency as a function of the luminosity of successively projected images.

a mental picture of a scene in three-dimensional space results. The perception of depth gained in this way is called *stereoscopic vision* and has its audio counterpart in stereophony.

200. Bibliography on Speech, Hearing, and Vision

PSYCHOLOGY AND COGNITION; SENSORY RESPONSES AND BEHAVIOR

1. Breger, L.   "Clinical Cognitive Psychology," Prentice-Hall, Englewood Cliffs, N.J., 1969.
2. Burns, W.   "Noise and Man," Lippincott, Philadelphia, 1969.
3. Candland, D. K.   "Psychology: The Experimental Approach," McGraw-Hill, New York, 1968.
4. Corsi, J. F.   "Experimental Psychology of Sensory Behaviour," Holt, New York, 1967.
5. Carterette, E. C., and M. P. Friedman   "Handbook of Perception," 7 vols., Academic, New York, 1978.
6. Fogel, L. J.   "Biotechnology," Prentice-Hall, Englewood Cliffs, N.J., 1963.
7. Goldenson, R. M.   "Encyclopedia of Human Behavior," 2 vols., Doubleday, New York, 1970.
8. Gulick, W. L.   "Hearing: Physiology and Psychophysics," Oxford University Press, London, 1971.
9. Harris, C. M.   "Handbook of Noise Control," McGraw-Hill, New York, 1957.
10. King, J. W., and L. A. Riggs   "Woodworth and Schlossberg's Experimental Psychology," Holt, New York, 1971.
11. Luce, R. D., R. R. Bush, and E. Galenter   "Handbook of Mathematical Psychology," Wiley, New York, 1965.
12. McCormick, E. J.   "Human Factors in Engineering," McGraw-Hill, New York, 1970.
13. Osgood, C. E.   "Method and Theory in Experimental Psychology," Oxford University Press, London, 1953.
14. Pieron, H.   "The Sensations," Muller, London, 1952.
15. Reeves, R. G.   "Manual of Remote Sensing," 2 vols., American Society for Photogrametry, Falls Church, Va., 1975.
16. Rubin, M. R., and G. I. Wallis   "Fundamentals of Visual Science," Thomas, Springfield, Ill., 1969.
17. Savage, C. W., and S. S. Stevens   Introspection and Behavioristic Interpretations, *Am. Psychol. Ass. Monogr.*, 1967, Vol. 80, No. 19, p. 8.
18. Stevens, S. S.   "Handbook of Experimental Psychology," Wiley, New York, 1951.
19. Stevens, S. S.   Psychophysics of Sensory Functions, *Am. Sci.*, 1960, Vol. 43, pp. 226–252.

20. Swets, J. A. "Signal Detection and Recognition by Human Observers," Wiley, New York, 1964.
21. Welman, B. B., and E. Nagel "Handbook of Clinical Psychology," Basic Books, New York, 1965.

LIGHT, OPTICS AND VISION

22. Bartlett, S. H. "Vision: A Study of Its Basis," Hafner, New York, 1963.
23. Born, M., and E. Wolff "Principles of Optics," 2d ed., Macmillan, New York, 1968.
24. Brindley, G. S. "Physiology of the Retina and the Visual Pathology," Williams & Wilkins, Baltimore, 1960.
25. Cornsweet, T. N. "Visual Perception," Academic, New York, 1970.
26. Davson, H. "The Eye," 4 vols., 2d ed., Academic, New York, 1969.
27. Ditchburn, R. W. "Light," 2d ed., Wiley, New York, 1963.
28. Hering, E. "Outline of a Theory of Light Sense," Harvard University Press, Cambridge, Mass., 1964.
29. Hurrich, L. M., and D. Jameson "Perception of Brightness and Darkness," Allyn & Bacon, Boston, 1966.
30. Illuminating Engineering Society "I.E.S. Lighting Handbook," 5th ed., New York, 1972.
31. Julesz, B. "Foundations of Cyclopean Perception," University of Chicago Press, Chicago, 1971.
32. LeGrand, Y. "Light, Color, and Vision," Wiley, New York, 1957.
33. Levi, L. "Applied Optics: A Guide to Optical System Design," Wiley, New York, 1969.
34. Linksz, A. "An Essay on Color Vision," Grune and Stratton, New York, 1964.
35. Ogle, K. N. "Researches in Binocular Vision," Hafner, New York, 1964.
36. Pirenne, M. H. "Vision and the Eye," Chapman & Hall, London, 1948.
37. Polyak, M. "The Vertebrate Visual System," University of Chicago Press, Chicago, 1957.
38. Rubin, M. L., and G. L. White "Fundamentals of Visual Science," Thomas, Springfield, Ill., 1969.
38. Shepard, J. J. "Human Color Perception," Elsevier, New York, 1960.
39. Smelzer, G. K. "The Structure of the Eye," Academic, New York, 1961.
40. Southall, J. P. C. "Mirrors, Prisms, and Lenses," 3d ed., Dover, 1964.

ACOUSTICS, SOUND, SPEECH, AND HEARING

41. Albers, V. M. "Underwater Acoustics Handbook," Pennsylvania State University Press, University Park, 1960.
42. Beranek, L. L. "Acoustics," McGraw-Hill, New York, 1954.
43. Beranek, L. L. "Acoustic Measurements," Wiley, New York, 1949.
44. Blake, M. P., and W. S. Mitchell "Vibration and Acoustics Measurement Handbook," Spartan, New York, 1972.
45. Carterette, E. C., and M. P. Friedman "Handbook of Perception," 3 vols., Academic, New York, 1974.
46. Deutsch, L. J. "Elements of Hearing Science," University Park Press, Baltimore, 1975.
47. Dittrich, F. L., and R. C. Exterman "Biophysics of the Ear," Thomas, Springfield, Ill., 1963.
48. Durrant, J. D. "Bases of Hearing Science," Williams & Wilkins, Baltimore, 1977.
49. Ensminger, D. "Ultrasonics," Dekker, New York, 1973.
50. Evans, E. F., and J. P. Wilson "Psychophysics and Physiology of Hearing," Academic, New York, 1977.
51. Goldman, R. "Ultrasonic Techniques," Reinhold, New York, 1966.
52. Gray, G. W., and C. W. Wise "The Bases of Speech," Harper, New York, 1959.
53. Green, D. M. "An Introduction to Hearing," Lawrence Erlbaum, Hillsdale, New York, 1976.
54. Grossman, S. D. "Textbook of Physiological Psychology," Wiley, New York, 1967.
55. Harmouth, H. F. "Acoustic Imaging with Electric Circuits," Academic, New York, 1979.
56. Hawley, M. E. "Speech Intelligibility and Speaker Recognition," Academic, New York, 1977.
57. Hirsh, I. J. "Measurement of Hearing," McGraw-Hill, New York, 1952.
58. Officer, C. B. "Introduction to the Theory of Sound Transmission, with Applications to the Ocean," McGraw-Hill, New York, 1958.
59. Rayleigh, J. W. S. "Theory of Sound," 2 vols., Dover, New York, 1965.
60. Richards, A. M. "Basic Experiments in Psychoacoustics," University Park Press, Baltimore, 1976.
61. Richardson, E. G. "Technical Aspects of Sound," 2 vols., Elsevier, New York, 1953.
62. Sanders, D. A. "Auditory Perception of Speech," Prentice-Hall, Englewood Cliffs, N.J., 1977.
63. Schubert, E. D. "Psychological Acoustics," Academic, New York, 1979.
64. Small, A. E. "Elements of Hearing Science," Wiley, New York, 1978.
65. Smith, C. A. "Handbook of Auditory and Vesticular Research Methods," Thomas, Springfield, Ill., 1976.
66. Stevens, S. S., and H. Davis "Hearing: Its Psychology and Physiology," Wiley, New York, 1938.
67. Von Bekesy, G. "Experiments in Hearing," McGraw-Hill, New York, 1960.
68. Yost, W. A. "Fundamentals of Hearing," Holt, New York, 1977.
69. Zemlin, W. R. "Speech and Hearing Science," Prentice-Hall, Englewood Cliffs, N.J., 1978 (many references).

ELECTRONIC QUANTITIES

201. Systems of Units. In 1960 the Eleventh General Congress on Weights and Measures promulgated the International System of Units (SI). This comprises a universal coherent system of units in which the following six quantities are considered to be basic: (1) meter of length, (2) kilogram of mass, (3) second of time, (4) ampere of current, (5) Kelvin degree of temperature, and (6) candela of luminous intensity. In 1967 the General Conference on Weights and Measures gave the name kelvin to the International Standard of temperature (which had previously been called the degree Kelvin) and assigned to this temperature unit the symbol K without the associated symbol ° for the degree.

In 1971 the amount of substance expressed in units of moles was adopted as the seventh basic quantity not derived from any other. The SI units are of convenient size for most branches of science and engineering, especially when used with suitable metric multipliers. It should be noted that the candela is a unit of radiant energy evaluated in accordance with the visual energy curve for a normal human observer and to this extent depends upon subjective factors.

The SI proposals are described in Refs. 3 and 4 of Par. **1-204**.

Definitions of the most important SI units given in the following paragraphs have been extracted from the records of the International Committee of Weights and Measures and the General Conferences of Weights and Measures.

Meter (m).   The *meter* is the length equal to 1,650,763.73 wavelengths in vacuum of the radiation corresponding to the transition between the levels $^2p_{10}$ and $^5d_5$ of the krypton 86 atom.

Kilogram (kg).   The *kilogram* is the unit of mass; it is equal to the mass of the international prototype of the kilogram. (The international prototype of the kilogram is a particular cylinder of platinum-iridium alloy which is preserved in a vault at Sèvres, France, by the International Bureau of Weights and Measures.)

Second (s).   The *second* is the duration of 9,192,631,770 periods of the radiation corresponding to the transition between the two hyperfine levels of the ground state of the cesium 133 atom.

Ampere (A).   The *ampere* is that constant current which, if maintained in two straight parallel conductors of infinite length, of negligible circular cross section, and placed 1 m apart in vacuum, would produce between these conductors a force equal to 2×10^{-7} N per meter of length.

Kelvin (K).   The *kelvin*, unit of thermodynamic temperature, is the fraction 1/273.16 of the thermodynamic temperature of the triple point of water.

Candela (cd).   The *candela* is the luminous intensity, in the perpendicular direction, of a surface of 1/600,000 m^2 of a blackbody at the temperature of freezing platinum under a pressure of 101,325 N/m^2.

Mole (mol).   The *mole* is that amount of substance of a system which contains as many elementary entities (molecules, atoms, or ions) as there are atoms in 0.012 kilogram of carbon 12. The nature of the elementary entities is to be specified.

Newton (N).   The *newton* is that force which gives to a mass of 1 kg an acceleration of 1 m/s^2.

Joule (J).   The *joule* is the work done when the point of application of 1 N is displaced a distance of 1 m in the direction of the force.

Watt (W).   The *watt* is the power which gives rise to the production of energy at the rate of 1 J/s.

Volt (V).   The *volt* is the difference of electric potential between two points of a conducting wire carrying a constant current of 1 A when the power dissipated between these points is equal to 1 W.

Ohm (Ω).   The *ohm* is the electrical resistance between two points of a conductor when a constant difference of potential of 1 V, applied between these two points, produces in this conductor a current of 1 A, this conductor not being the source of any electromotive force.

Coulomb (C).   The *coulomb* is the quantity of electric charge transported in 1 s by a current of 1 A.

Farad (F).   The *farad* is the capacitance of a capacitor between the plates of which there appears a difference of potential of 1 V when it is charged by a quantity of electricity equal to 1 C.

Henry (H).   The *henry* is the inductance of a closed circuit in which an electromotive force of 1 V is produced when the electric current in the circuit varies uniformly at a rate of 1 A/s.

Weber (Wb).   The *weber* is the magnetic flux which, linking a circuit of one turn, produces in it an electromotive force of 1 V as it is reduced to zero at a uniform rate in 1 s.

Lumen (lm).   The *lumen* is the luminous flux emitted in a solid angle of 1 sr by a uniform point source having an intensity of 1 cd.

202. Values of Physical Constants. Table 1-6 lists physical constants.

203. Symbols. Table 1-7 lists symbols, dimensions, and units for physical quantities.

TABLE 1-6 Values of Fundamental Physical Constants

Quantity	Symbol	Value[1]	Uncertainty, ppm
Permeability of vacuum	μ_0	$4\pi \times 10^{-7}$ H/m	0.004
Speed of light in vacuum	c	299 792 458(1.2) m/s	0.008
Permittivity of vacuum	$\epsilon_0 = (\mu_0 c^2)^{-1}$	8.854 187 82(7) $\times 10^{-12}$ F/m	0.82
Fine-structure constant	α	0.007 297 350 6(60)	2.9
Elementary charge	e	1.602 189 2(46) $\times 10^{-19}$ C	5.4
Planck constant	h	6.626 176(36) $\times 10^{-34}$ J/Hz	5.4
	$\hbar = h/2\pi$	1.054 588 7(57) $\times 10^{-34}$ J s	5.1
Avogadro constant	N_A	6.022 045(31) $\times 10^{23}$ mol^{-1}	5.1
Atomic mass unit	1 u = $(10^{-3}$ kg/mol)$/N_A$	1.660 565 5(86) $\times 10^{-27}$ kg	5.1
Electron rest mass	m_e	0.910 953 4(47) $\times 10^{-30}$ kg	5.1
Muon rest mass	m_μ	1.883 566(11) $\times 10^{-28}$ kg	5.6
Proton rest mass	m_p	1.672 648 5(86) $\times 10^{-27}$ kg	5.1
Neutron rest mass	m_n	1.674 954 3(86) $\times 10^{-27}$ kg	5.1
Ratio, proton mass to electron mass	m_p/m_e	1836.151 52(70)	0.38
Ratio, muon mass to electron mass	m_μ/m_e	206.768 65(47)	2.3
Specific electron charge	e/m_e	1.758 804 7(49) $\times 10^{11}$ C/kg	2.8
Faraday constant	$F = N_A e$	9.648 456(27) $\times 10^4$ C/mol	2.8
Magnetic flux quantum	$\phi_0 = h/2e$	2.067 850 6(54) $\times 10^{-15}$ Wb	2.6
Quantum of circulation	$h/2m_e$	3.636 945 5(60) $\times 10^{-4}$ J/Hz·kg	1.6
Rydberg constant	R_∞	1.097 373 177(83) $\times 10^7$ m^{-1}	0.075
Bohr radius	$a_0 = \alpha/4\pi R_\infty$	0.529 177 06(44) $\times 10^{-10}$ m	0.82
Electron Compton wavelength	$\lambda_C = \alpha^2/2R_\infty$	2.426 308 9(40) $\times 10^{-12}$ m	1.6
	$X_C = \lambda_C/2\pi = \alpha a_0$	3.861 590 5(64) $\times 10^{-13}$ m	1.6
Classical electron radius	$r_e = \mu_0 e^2/4\pi m_e = \alpha X_C$	2.817 938 0(70) $\times 10^{-15}$ m	2.5
Bohr magneton	$\mu_B = e\hbar/2m_e$	9.274 078(36) $\times 10^{-24}$ J/T	3.9
Nuclear magneton	$\mu_N = e\hbar/2m_p$	5.050 824(20) $\times 10^{-27}$ J/T	3.9
Electron magnetic moment	μ_e	9.284 832(36) $\times 10^{-24}$ J/T	3.9
Proton magnetic moment	μ_p	1.410 617 1(55) $\times 10^{-26}$ J/T	3.9
Proton Compton wavelength	$\lambda_{C,p} = h/m_p c$	1.321 409 9(22) $\times 10^{-15}$ m	1.7
Neutron Compton wavelength	$\lambda_{C,n} = h/m_n c$	1.319 590 9(22) $\times 10^{-15}$ m	1.7
Molar gas constant	R	8.314 41(26) J/mol·K	31
Molar volume, ideal gas (T_0 = 273.15 K, p_0 = 1 atm)	$V_m = RT_0/p_0$	0.022 413 83(70) m³/mol	31
Boltzmann constant	$k = R/N_A$	1.380 662(44) $\times 10^{-23}$ J/K	32
Stefan-Boltzmann constant	$\sigma = (\pi^2/60)k^4/\hbar^3 c^2$	5.670 32(71) $\times 10^{-8}$ W·m^{-2}·K^4	125
First radiation constant	$c_1 = 2\pi h c^2$	3.741 832(20) $\times 10^{-16}$ W·m²	5.4
Second radiation constant	$c_2 = hc/k$	0.014 387 86(45) m·K	31
Gravitational constant	G	6.672 0(41) $\times 10^{-11}$ N·m²/kg^{-2}	615
Gas constant	R	82.056 8(26) cm³·atm/mol·K	31

[1]The digits in parentheses following a numerical value represent the standard deviation of that value, in terms of the final listed digits. Abstracted by permission from Recommended Consistent Values of the Fundamental Constants, *CODATA BULLETIN* 11, December 1973.

TABLE 1-7 Standard Symbols for Quantities

Quantity	Symbol	Unit*
Space and time		
Angle, plane	$\alpha, \beta, \gamma, \theta, \phi, \psi$	rad
Solid	Ω, ω	sr
Length	l	m
Area	A, S	m²
Volume	V, v	m³
Time	t	s
Period	T	s
Time constant	τ, T	s
Frequency	f, ν	Hz
Speed of rotation	n	r/s
Angular frequency	ω	rad/s
Angular acceleration	α	rad/s²
Velocity	v	m/s
Speed of propagation of electromagnetic waves	c	m/s
Acceleration (linear)	a	m/s²
Attenuation coefficient	α	Np/m
Phase coefficient	β	rad/m
Propagation coefficient	γ	m⁻¹
Fields and circuits		
Electric charge	Q	C
Volume density of charge	ρ	C/m³
Electric field strength	E, K	V/m
Electrostatic potential, potential difference	V, ϕ	V
Voltage, emf	V, E, U	V
Electric flux	ψ	C
Electric flux density, (electric) displacement	D	C/m²
Permittivity	ϵ	F/m
Electric susceptibility	X_e, ϵ_i	Numeric
Electric polarization	P	C/m²
Electric dipole moment	p	C·m
(Electric) current	I	A
Current density	J, S	A/m²
Magnetic field strength	H	A/m
Magnetic (scalar) potential, magnetic potential difference	U, U_m	A
Magnetomotive force	$F, F_m, \mathcal{F}$	A
Magnetic flux	Φ	Wb
Magnetic flux density, magnetic induction	B	T
Magnetic flux linkage	Λ	Wb
(Magnetic) vector potential	A	Wb/m
Permeability	μ	H/m
Capacitance	C	F
Mechanics		
Mass	m	kg
Momentum	p	kg·m/s
Moment of inertia	I, J	kg·m²
Force	F	N
Weight	W	N

Quantity	Symbol	Unit
Torque	T, M	N·m
Work	W	J
Energy	E, W	J
Energy (volume) density	w	J/m³
Power	P	W
Efficiency	η	Numeric

Heat

Quantity	Symbol	Unit
Thermodynamic temperature	T, Θ	K
Temperature	t, θ	°C
Heat	Q	J
Temperature coefficient	α	K⁻¹
Thermal conductivity	λ, k	W/m·K

Radiation and light

Quantity	Symbol	Unit
Radiant intensity	I, I_e	W/sr
Radiant power	P, Φ, Φ_e	W
Radiant flux	Φ	W
Radiant energy	W, Q, Q_e	J
Radiance	L, L_e	W/sr·m²
Luminous intensity	I, I_v	cd
Luminous flux	Φ, Φ_v	lm
Quantity of light	Q, Q_v	lm·s
Luminance	L, L_w	cd/m²
Refractive index	n	Numeric

Quantity	Symbol	Unit
Elastance	S	F⁻¹
(Self-) inductance	L	H
Reciprocal inductance	Γ	H⁻¹
Mutual inductance	L_{ij}, M_{ij}	H
Coupling coefficient	k, κ	Numeric
Resistance	R	Ω
Resistivity, volume resistivity	ρ	Ω·m
Conductance	G	S
Conductivity	γ, σ	S/m
Reluctance	$R, R_m, \mathscr{R}$	H⁻¹
Permeance	$P, P_m, \mathscr{P}$	H
Impedance	Z	Ω
Reactance	X	Ω
Capacitive	X_C	Ω
Inductive	X_L	Ω
Quality factor	Q	Numeric
Admittance	Y	S
Susceptance	B	S
Active power	P	W
Reactive power	Q, P_q	var
Apparent power	S, P_s	VA
Power factor	$\cos \phi, F_p$	Numeric
Poynting vector	S	W/m²
Characteristic impedance	Z_0	Ω
Intrinsic impedance of a medium	η	Ω
Voltage standing-wave ratio	S	Numeric
Hysteresis coefficient	k_h	Numeric
Eddy current coefficient	k_e	Numeric
Phase angle	ϕ, θ	rad

*With a very few exceptions these are SI units. The International System (SI) is a coherent system based on (1) meter of length, (2) kilogram of mass, (3) second of time, (4) ampere of current, (5) kelvin of temperature, (6) candela of luminous intensity, and (7) mole of amount of substance; see Par **1-201**.

204. Bibliography on Units and Measurements

1. Institute of Electrical and Electronics Engineers   Letter Symbols for Quantities Used in Electrical Science and Electrical Engineering, ANSI Standard Y10.5-1968; IEEE Standard 280-1968, New York, 1968.

2. Institute of Electrical and Electronics Engineers   Letter Symbols for Units in Science and Technology, ANSI Standard Y-10.19; IEEE Standard 260-1969, New York, 1969.

3. Institute of Electrical and Electronics Engineers   Graphic Symbols for Electrical and Electronics Diagrams, IEEE Standard 315, 1975, New York, 1975.

4. American National Standards Institute   SI Units and Recommendations for the Use of Their Multiples and Certain Other Units, International Standards, LSO-1000-1973, New York, 1973.

5. ICSU-CODATA Central Office   Recommended Consistent Values of the Fundamental Physical Constants, *CODATA Bull.* 11, December 1973.

6. Institute of Electrical and Electronics Engineers   IEEE Standard and American National Graphic Symbols for Electrical and Electronic Diagrams, IEEE Standard 315, 1975. New York, 1975.

7. Institute of Electrical and Electronics Engineers   "IEEE Standard Dictionary of Electrical and Electronics Terms," IEEE Standard 100-1972, Wiley, New York, 1975.

8. Mechtly, E. A.   The International System of Units, NASA SP 7012, NASA, 1969.

9. Institute of Electrical and Electronics Engineers   Recommended Practice for Units in Published Scientific and Technical Work, IEEE Standard 268, 1973, New York, 1973.

10. Silsbee, F. L.   Establishment and Maintenance of Electrical Units, *Natl. Bur. Std. Cir.* 475, 1962.

11. Institute of Electrical and Electronics Engineers   Standard Metric Practice, IEEE Standard 268, 1976, New York.

12. Fink, D. G., and W. Beaty   "Standard Handbook for Electrical Engineers," 11th ed., McGraw-Hill, New York, 1979.

Mathematics: Formulas, Definitions, and Theorems Used in Electronics Engineering

GRANINO A. KORN *Professor of Electrical Engineering, The University of Arizona*

THERESA M. KORN

CONTENTS

Numbers refer to paragraphs

1. Introduction. This section contains a selection of reference material believed to be most useful for practicing electronics engineers. Topics generally understood by such workers, e.g., college algebra and plane and solid geometry, have been omitted; these are treated in the references listed in Par. **2-45**, notably in the "Mathematical Handbook for Scientists and Engineers," by G. A. Korn and T. M. Korn (McGraw-Hill, New York, 1968). The major portion of this section has been taken, by permission of the publisher, from that handbook and from the "Manual of Mathematics," by G. A. Korn and T. M. Korn (McGraw-Hill, New York, 1967).

TABLE 2-1 Derivatives of Frequently Used Functions

$f(x)$	$f'(x)$	$f^{(r)}(x) = \dfrac{d^r}{dx^r}[f(x)]$
x^a	ax^{a-1}	$a(a-1)(a-2)\cdots(a-r+1)x^{a-r}$
e^x	e^x	e^x
a^x	$a^x \log_e a$	$a^x(\log_e a)^r$
$\log_e x$	$\dfrac{1}{x}$	$(-1)^{r-1}(r-1)!\,\dfrac{1}{x^r}$
$\log_a x$	$\dfrac{1}{x}\log_a e$	$(-1)^{r-1}(r-1)!\,\dfrac{1}{x^r}\log_a e$
$\sin x$	$\cos x$	$\sin\left(x+\dfrac{\pi r}{2}\right)$
$\cos x$	$-\sin x$	$\cos\left(x+\dfrac{\pi r}{2}\right)$

$f(x)$	$f'(x)$	$f(x)$	$f'(x)$
$\tan x$	$\dfrac{1}{\cos^2 x}$	$\arcsin x$	$\dfrac{1}{\sqrt{1-x^2}}$
$\cot x$	$-\dfrac{1}{\sin^2 x}$	$\arccos x$	$-\dfrac{1}{\sqrt{1-x^2}}$
$\sec x$	$\dfrac{\sin x}{\cos^2 x}$	$\arctan x$	$\dfrac{1}{1+x^2}$
$\csc x$	$-\dfrac{\cos x}{\sin^2 x}$	$\text{arccot } x$	$-\dfrac{1}{1+x^2}$
$\sinh x$	$\cosh x$	$\sinh^{-1} x$	$\dfrac{1}{\sqrt{x^2+1}}$
$\cosh x$	$\sinh x$	$\cosh^{-1} x$	$\dfrac{1}{\sqrt{x^2-1}}$
$\tanh x$	$\dfrac{1}{\cosh^2 x}$	$\tanh^{-1} x$	$\dfrac{1}{1-x^2}$
$\coth x$	$-\dfrac{1}{\sinh^2 x}$	$\coth^{-1} x$	$\dfrac{1}{1-x^2}$
$\text{vers } x$	$\sin x$	x^x	$x^x(1+\log_e x)$

DIFFERENTIAL CALCULUS

2. Derivatives and Differentiation. Let $y = f(x)$ be a real, single-valued function of the real variable x throughout a neighborhood of the point x. The *(first, first-order) derivative* or *(first-order) differential coefficient of $f(x)$ with respect to x at the point x* is the limit

$$\lim_{\Delta x \to 0} \frac{f(x + \Delta x) - f(x)}{\Delta x} \equiv \lim_{\Delta x \to 0} \frac{\Delta y}{\Delta x} \equiv \frac{dy}{dx} \equiv \frac{d}{dx} f(x) \equiv f'(x) \equiv v' \tag{2-1}$$

The function $dy/dx \equiv f'(x)$ is a measure of the *rate of change of y with respect to x* at each point x where the limit (2-1) exists. On a graph of $y = f(x)$, $f'(x)$ corresponds to the *slope of the tangent*. Table 2-1 lists derivatives of frequently used functions.

3. Partial Derivatives. Let $y = f(x_1, x_2, \ldots, x_n)$ be a real single-valued function of the real variables $x_1, x_2, \ldots, x_n$ in a neighborhood of the point $(x_1, x_2, \ldots, x_n)$. The *(first-order) partial derivative of $f(x_1, x_2, \ldots, x_n)$ with respect to x_1 at the point $(x_1, x_2, \ldots, x_n)$* is the limit

$$\lim_{\Delta x_1 \to 0} \frac{f(x_1 + \Delta x_1, x_2, x_3, \ldots, x_n) - f(x_1, x_2, \ldots, x_n)}{\Delta x_1} \equiv \frac{\partial}{\partial x_1} f \equiv \frac{\partial y}{\partial x_1} \equiv f_{x_1}(x_1, x_2, \ldots, x_n) \tag{2-2}$$

The function $\partial y/\partial x_1 \equiv (\partial y/\partial x_1)_{x_1, x_2, \ldots, x_n} \equiv f_{x_1}(x_1, x_2, \ldots, x_n)$ is a measure of the *rate of change of y with respect to x_1* for fixed values of the remaining independent variables at each point $(x_1, x_2, \ldots, x_n)$ where the limit (2-2) exists. The partial derivatives $\partial y/\partial x_2, \partial y/\partial x_3, \ldots, \partial y/\partial x_n$ are defined in an analogous manner. Each partial derivative $\partial y/\partial x_k$ can be found by differentiation of $f(x_1, x_2, \ldots, x_n)$ with respect to x_k while the remaining $n - 1$ independent variables are regarded as constant parameters [*partial differentiation of $f(x_1, x_2, \ldots, x_n)$ with respect to x_k*].

4. Differentiation Rules. Table 2-2 summarizes the most important differentiation rules. The formulas of Table 2-2a and b apply to *partial differentiation* if $\partial/\partial x_k$ is substituted for d/dx in each case. Thus, if $u_i = u_i(x_1, x_2, \ldots, x_n)$ $(i = 1, 2, \ldots, m)$,

$$\frac{\partial}{\partial x_k} f(u_1, u_2, \ldots, u_m) = \sum_{i=1}^{m} \frac{\partial f}{\partial u_i} \frac{\partial u_i}{\partial x_k} \qquad k = 1, 2, \ldots, n \tag{2-3}$$

INTEGRALS AND INTEGRATION

5. Definite Integrals (Riemann Integrals). A real function $f(x)$ bounded on the bounded closed interval $[a, b]$ is *integrable over (a, b) in the sense of Riemann* if and only if the sum $\sum_{i=1}^{m} f(\xi_i)(x_i - x_{i-1})$ tends to a unique finite limit I for every sequence of partitions $a = x_0 < \xi_1 < x_1 < \xi_2 < x_2 \cdots < \xi_m < x_m = b$ as max $|x_i - x_{i-1}| \to 0$. In this case

$$I = \lim_{\max |x_i - x_{i-1}| \to 0} \sum_{i=1}^{m} f(\xi_i)(x_i - x_{i-1}) = \int_a^b f(x) \, dx \tag{2-4}$$

is the *definite integral of $f(x)$ over (a, b) in the sense of Riemann (Riemann integral)*. $f(x)$ is called the *integrand*; a and b are the *limits of integration*. Table 2-3 summarizes important properties of definite integrals.

$\int_a^b f(x) \, dx$ represents the *area* bounded by the curve $y = f(x)$ and the x axis between the lines $x = a$ and $x = b$; areas below the x axis are represented by negative numbers.

6. Indefinite Integrals. A given single-valued function $f(x)$ has an *indefinite integral* $F(x)$ in $[a, b]$ if and only if there exists a function $F(x)$ such that $F'(x) = f(x)$ in $[a, b]$. In this case $F(x)$ is uniquely defined in $[a, b]$ except for an arbitrary additive constant C *(constant of integration)*; one writes

$$F(x) = \int f(x) \, dx + C \qquad a \leq x \leq b \tag{2-5}$$

Note that $F(x) - F(a) \equiv F(x)]_a^x$ is uniquely defined for $a \leq x \leq b$.

7. Fundamental Theorem of the Integral Calculus. If $f(x)$ is *single-valued, bounded, and integrable on $[a, b]$ and there exists a function $F(x)$ such that $F'(x) = f(x)$ for $a \leq x \leq b$, then*

$$\int_a^x f(\xi) \, d\xi = F(x)]_a^x = F(x) - F(a) \qquad c \leq x \leq b \tag{2-6}$$

In particular, *if f(x) is continuous in* [a, b],

$$\frac{d}{dx} \int_a^x f(\xi) \, d\xi = f(x) \qquad a \le x \le b$$

and Eq. (2-6) applies.

NOTE: The fundamental theorem of the integral calculus enables one (1) to evaluate integrals by reversing the process of differentiation and (2) to solve differential equations by numerical evaluation of definite integrals.

TABLE 2-2   Differentiation Rules*

a. Basic rules

$$\frac{d}{dx} f[u_1(x), u_2(x), \ldots, u_m(x)] = \frac{\partial f}{\partial u_1} \frac{du_1}{dx} + \frac{\partial f}{\partial u_2} \frac{du_2}{dx} + \cdots + \frac{\partial f}{\partial u_m} \frac{du_m}{dx}$$

$$\frac{d}{dx} f[u(x)] = \frac{df}{du} \frac{du}{dx} \qquad \frac{d^2}{dx^2} f[u(x)] = \frac{d^2 f}{du^2} \left(\frac{du}{dx} \right)^2 + \frac{df}{du} \frac{d^2 u}{dx^2}$$

b. Sums, products, and quotients; logarithmic differentiation

$$\frac{d}{dx} [u(x) + v(x)] = \frac{du}{dx} + \frac{dv}{dx} \qquad \frac{d}{dx} [\alpha u(x)] = \alpha \frac{du}{dx}$$

$$\frac{d}{dx} [u(x)v(x)] = v \frac{du}{dx} + u \frac{dv}{dx} \qquad \frac{d}{dx} \left[\frac{u(x)}{v(x)} \right] = \frac{1}{v^2} \left(v \frac{du}{dx} - u \frac{dv}{dx} \right) \qquad v(x) \ne 0$$

$$\frac{d}{dx} \log_e y(x) = \frac{y'(x)}{y(x)} \qquad \text{logarithmic derivative of } y(x)$$

NOTE: To differentiate functions of the form $y = \dfrac{u_1(x)u_2(x) \, \cdots}{v_1(x)v_2(x) \, \cdots}$, it may be convenient to find the logarithmic derivative first.

$$\frac{d^r}{dx^r} (\alpha u + \beta v) = \alpha \frac{d^r u}{dx^r} + \beta \frac{d^r v}{dx^r} \qquad \frac{d^r}{dx^r} (uv) = \sum_{k=0}^{r} \binom{r}{k} \frac{d^{r-k} u}{dx^{r-k}} \frac{d^k v}{dx^k}$$

c. Inverse function given

If $y = y(x)$ has the unique inverse function $x = x(y)$, and $dx/dy \ne 0$,

$$\frac{dy}{dx} = \left(\frac{dx}{dy} \right)^{-1} \qquad \frac{d^2 y}{dx^2} = -\frac{d^2 x/dy^2}{(dx/dy)^3}$$

d. Implicit functions

If $y = y(x)$ is given implicitly in terms of a suitably differentiable relation $F(x, y) = 0$, where $F_y \ne 0$,

$$\frac{dy}{dx} = -\frac{F_x}{F_y} \qquad \frac{d^2 y}{dx^2} = -\frac{1}{F_y^3} (F_{xx} F_y^2 - 2 F_{xy} F_x F_y + F_{yy} F_x^2)$$

e. Function given in terms of a parameter

Given $x = x(t)$,

$$y = y(t) \qquad \dot{x}(t) \equiv \frac{dx}{dt} \ne 0 \qquad \dot{y}(t) \equiv \frac{dy}{dt} \qquad \ddot{x}(t) \equiv \frac{d^2 x}{dt^2} \qquad \ddot{y}(t) \equiv \frac{d^2 y}{dt^2}$$

$$\frac{dy}{dx} = \frac{\dot{y}(t)}{\dot{x}(t)} \qquad \frac{d^2 y}{dx^2} = \frac{\dot{x}(t)\ddot{y}(t) - \ddot{x}(t)\dot{y}(t)}{[\dot{x}(t)]^3}$$

*Existence of continuous derivatives is assumed in each case.

8. Integration of Polynomials

$$\int(a_n + a_{n-1}x + a_{n-2}x^2 + \cdots + a_0x^n)\, dx \equiv a_nx + \tfrac{1}{2}a_{n-1}x^2$$

$$+ \tfrac{1}{3}a_{n-2}x^3 + \cdots + \frac{1}{n+1}a_0x^{n+1} + C$$

TABLE 2-3   Properties of Integrals

a. Elementary properties

If the integrals exist,

$$\int_a^b f(x)\, dx = -\int_b^a f(x)\, dx \qquad \int_a^b f(x)\, dx = \int_a^c f(x)\, dx + \int_c^b f(x)\, dx$$

$$\int_a^b [u(x) + v(x)]\, dx = \int_a^b u(x)\, dx + \int_a^b v(x)\, dx \qquad \int_{-b}^{-a} \alpha u(x)\, dx = \alpha \int_a^b u(x)\, dx$$

b. Integration by parts

If $u(x)$ and $v(x)$ are differentiable for $a \leq x \leq b$, and if the integrals exist,

$$\int_a^b u(x)v'(x)\, dx = u(x)v(x) \Big]_a^b - \int_a^b v(x)u'(x)\, dx \qquad \text{or} \qquad \int_a^b u\, dv = uv \Big]_a^b - \int_a^b v\, du$$

c. Change of variable (integration by substitution)

If $u = u(x)$ and its inverse function $x = x(u)$ are single-valued and continuously differentiable for $a \leq x \leq b$, and if the integral exists,

$$\int_a^b f(x)\, dx = \int_{u(a)}^{u(b)} f[(x(u)]\frac{dx}{du}\, du = \int_{u(a)}^{u(b)} f[x(u)] \left(\frac{du}{dx}\right)^{-1} du$$

d. Differentiation with respect to a parameter

If $f(x, \lambda)$, $u(\lambda)$, and $v(\lambda)$ are continuously differentiable with respect to λ,

$$\frac{\partial}{\partial\lambda} \int_a^b f(x, \lambda)\, dx = \int_a^b \frac{\partial}{\partial\lambda} f(x, \lambda)\, dx$$

$$\frac{\partial}{\partial\lambda} \int_{u(\lambda)}^{v(\lambda)} f(x, \lambda)\, dx = \int_{u(\lambda)}^{v(\lambda)} \frac{\partial}{\partial\lambda} f(x, \lambda)\, dx + f(v, \lambda)\frac{\partial v}{\partial\lambda} - f(u, \lambda)\frac{\partial u}{\partial\lambda} \qquad \text{Leibnitz' rule}$$

provided that the integrals exist and in the case of improper integrals, converge uniformly in a neighborhood of the point λ.

The second case can often be reduced to the first by a suitable change of variables. Note also

$$\frac{\partial}{\partial\lambda} \int_a^\lambda f(x, \lambda)\, dx = \frac{1}{\lambda - a} \int_a^\lambda \left[f(x, \lambda) + (\lambda - a)\frac{\partial f}{\partial\lambda} + (x - a)\frac{\partial f}{\partial x} \right] dx$$

e. Inequalities

If the integrals exist,

$$f(x) \leq g(x) \text{ in } (a, b) \text{ implies } \int_a^b f(x)\, dx \leq \int_a^b g(x)\, dx$$

If $|f(x)| \leq M$ on the bounded interval (a, b), the existence of $\int_a^b f(x)\, dx$ implies the existence of $\int_a^b |f(x)|\, dx$, and

$$\left| \int_a^b f(x)\, dx \right| \leq \int_a^b |f(x)|\, dx \leq M(b - a)$$

9. Integration of Rational Fractions. Every rational integrand reduces to a sum of a polynomial and a set of partial fractions. Partial-fraction terms are integrated with the aid of the following formulas:

$$\int \frac{dx}{(x - x_1)^m} \equiv \begin{cases} -\dfrac{1}{(m - 1)(x - x_1)^{m-1}} + C & m \neq 1 \\ \log_e (x - x_1) + C & m = 1 \end{cases}$$

$$\int \frac{dx}{(x - a)^2 + \omega^2} \equiv \frac{1}{\omega} \arctan \frac{x - a}{\omega} + C$$

$$\int \frac{dx}{[(x - a)^2 + \omega^2]^{m+1}} \equiv \frac{x - a}{2m\omega^2[(x - a)^2 + \omega^2]^m} + \frac{2m - 1}{2m\omega^2} \int \frac{dx}{[(x - a)^2 + \omega^2]^m}$$

$$\int \frac{x \, dx}{[(x - a)^2 + \omega^2]^{m+1}} \equiv \frac{a(x - a) - \omega^2}{2m\omega^2[(x - a)^2 + \omega^2]^m} + \frac{(2m - 1)a}{2m\omega^2} \int \frac{dx}{[(x - a)^2 + \omega^2]^m}$$

10. Integrands Reducible to Rational Functions by a Change of Variables (Table 2-3c).

1. *If the integrand $f(x)$ is a rational function of* sin x *and* cos x, *introduce* $u = \tan (x/2)$, *so that*

$$\sin x = \frac{2u}{1 + u^2} \qquad \cos x = \frac{1 - u^2}{1 + u^2} \qquad dx = \frac{2 \, du}{1 + u^2}$$

2. *If the integrand $f(x)$ is a rational function of* sinh x *and* cosh x, *introduce* $u = \tanh (x/2)$, *so that*

$$\sinh x = \frac{2u}{1 - u^2} \qquad \cosh x = \frac{1 + u^2}{1 - u^2} \qquad dx = \frac{2 \, du}{1 - u^2}$$

NOTE: If $f(x)$ is a rational function of $\sin^2 x$, $\cos^2 x$, sin x cos x, and tan x (or of the corresponding hyperbolic functions), one simplifies the calculation by first introducing $v = x/2$, so that $u = \tan v$ (or $u = \tanh v$).

3. *If the integrand $f(x)$ is a rational function of* x *and either* $\sqrt{1 - x^2}$ *or* $\sqrt{x^2 - 1}$, reduce the problem to case 1 or 2 by the respective substitutions $x = \cos v$ or $x = \cosh v$.

4. *If the integrand $f(x)$ is a rational function of* x *and* $\sqrt{x^2 + 1}$, *introduce* $u = x + \sqrt{x^2 + 1}$, *so that*

$$x = \tfrac{1}{2}(u - 1/u) \qquad \sqrt{x^2 + 1} = \tfrac{1}{2}(u + 1/u) \qquad dx = \tfrac{1}{2}(1 + 1/u^2) \, du$$

5. *If the integrand $f(x)$ is a rational function of* x *and* $\sqrt{ax^2 + bx + c}$, reduce the problem to case 3 ($b^2 - 4ac < 0$) or to case 4 ($b^2 - 4ac > 0$) through the substitution

$$v = \frac{2ax + b}{\sqrt{|4ac - b^2|}} \qquad x = \frac{v\sqrt{|4ac - b^2|} - b}{2a}$$

6. *If the integrand $f(x)$ is a rational function of* x *and* $u = \sqrt{(ax + b)/(cs + d)}$, introduce u as a new variable.

7. *If the integrand $f(x)$ is a rational function of* x, $\sqrt{ax + b}$ *and* $\sqrt{cx + d}$, introduce $u = \sqrt{ax + b}$ as a new variable.

Many other substitution methods apply in special cases. Note that the integrals may not be real for all values of x.

Integrands of the form $x^n e^{ax}$, $x^n \log_e x$, $x^n \sin x$, $x^n \cos x$ ($n \neq -1$); $\sin^m x \cos^n x$ ($n + m \neq 0$); $e^{ax} \sin^n x$, $e^{ax} \cos^n x$ yield to repeated *integration by parts* (Table 2-3b).

11. Some Frequently Used Limits (Values of Indeterminate Forms)

$$\lim_{n \to \infty} (1 + 1/n)^n = e \approx 2.71828 \qquad n = 1, 2, \ldots \qquad \lim_{x \to 0} (1 + x)^{1/x} = e$$

$$\lim_{x \to 0} \frac{c^x - 1}{x} = \log_e c \qquad \lim_{x \to 0} x^x = 1$$

$$\lim_{x \to 0} \frac{\sin x}{x} = \lim_{x \to 0} \frac{\tan x}{x} = \lim_{x \to 0} \frac{\sinh x}{x} = \lim_{x \to 0} \frac{\tanh x}{x} = 1$$

$$\lim_{x \to 0} \frac{\sin \omega x}{x} = \omega \qquad -\infty < \omega < \infty$$

$$\lim_{x \to 0} x^a \log_e x = \lim_{x \to 0} x^{-a} \log_e x = \lim_{x \to \infty} x^a e^{-x} = 0 \qquad a > 0$$

FOURIER SERIES AND FOURIER INTEGRALS

12. Fourier Series

$$\tfrac{1}{2}a_0 + \sum_{k=1}^{\infty} (a_k \cos k\omega_0 t + b_k \sin k\omega_0 t) \equiv \sum_{k=-\infty}^{\infty} c_k e^{ik\omega_0 t} \qquad \omega_0 = \frac{2\pi}{T}$$

with

$$a_k = \frac{2}{T} \int_{-T/2}^{T/2} f(\tau) \cos k\omega_0 \tau \, d\tau \qquad b_k = \frac{2}{T} \int_{-T/2}^{T/2} f(\tau) \sin k\omega_0 \tau \, d\tau$$

$$c_k = c_{-k}^* = \tfrac{1}{2}(a_k - ib_k) = \frac{1}{T} \int_{-T/2}^{T/1} f(\tau) e^{-ik\omega_0 t} \, d\tau \qquad \begin{aligned} \omega_0 &= \frac{2\tau}{T} \\ k &= 0, 1, 2, \dots \end{aligned}$$

13. Properties of Fourier Transforms. Let

$$\mathcal{F}[f(t)] \equiv \int_{-\infty}^{\infty} f(t) e^{-2\pi i \nu t} \, dt \equiv c(\nu) \equiv F_F(i\omega) \equiv \sqrt{2\pi} C(\omega) \qquad \omega = 2\pi\nu$$

$$f(t) \equiv \int_{-\infty}^{\infty} c(\nu) e^{2\pi i \nu t} \, d\nu \equiv \int_{-\infty}^{\infty} F_F(i\omega) e^{i\omega t} \frac{d\omega}{2\pi} \equiv \frac{1}{\sqrt{2\pi}} \int_{-\infty}^{\infty} C\omega e^{i\omega t} \, d\omega$$

and assume that the Fourier transforms in question exist.

(a) $\mathcal{F}[\alpha f_1(t) + \beta f_2(t)] \equiv \alpha \mathcal{F}[f_1(t)] + \beta \mathcal{F}[f_2(t)]$ linearity

$$\mathcal{F}[f^*(t)] \equiv c^*(-\nu) \equiv F_F^*(-i\omega)$$

$$\mathcal{F}[f(\alpha t)] \equiv \frac{1}{\alpha} c\left(\frac{\nu}{\alpha}\right) \equiv \frac{1}{\alpha} F_F\left(\frac{i\omega}{\alpha}\right) \qquad \text{change of scale, similarity theorem}$$

$$\mathcal{F}[f(t + \tau)] \equiv e^{2\pi i \nu \tau} c(\nu) \equiv e^{i\omega \tau} F_F(i\omega) \qquad \text{shift theorem}$$

(b) *Continuity Theorem.* $\mathcal{F}[f(t, \alpha)] \to \mathcal{F}[f(t]$ as $\alpha \to a$ implies $f(t, \alpha) \to f(t)$ wherever $f(t)$ is continuous. Analogous theorems apply to Fourier cosine and sine transforms.

(c) *Borel's Convolution Theorem.* $\mathcal{F}[f_1(t)] \mathcal{F}[f_2(t)] \equiv \mathcal{F}[f_1(t) * f_2(t)]$, where

$$f_1(t) * f_2(t) \equiv \int_{-\infty}^{\infty} f_1(\tau) f_2(t - \tau) \, d\tau \equiv \int_{-\infty}^{\infty} f_1(t - \tau) f_2(\tau) \, d\tau$$

$$\mathcal{F}[f_1(t) f_2(t)] \equiv \int_{-\infty}^{\infty} c_1(\lambda) c_2(\nu - \lambda) \, d\lambda \equiv \int_{-\infty}^{\infty} c_1(\nu - \lambda) c_2(\lambda) \, d\lambda$$

$$\equiv \int_{-\infty}^{\infty} F_{F1}(i\lambda) F_{F2}[i(\omega - \lambda)] \frac{d\lambda}{2\pi} \equiv \int_{-\infty}^{\infty} F_{F1}[i(\omega - \lambda)] F_{F2}(i\lambda) \frac{d\lambda}{2\pi}$$

(d) *Parseval's Theorem.* if $\displaystyle\int_{-\infty}^{\infty} |f_1(t)|^2 \, dt$ and $\displaystyle\int_{-\infty}^{\infty} |f_2(t)|^2 \, dt$ exist, then

$$\int_{-\infty}^{\infty} \mathcal{F}^*[f_1(t)] \mathcal{F}[f_2(t)] \, d\nu = \int_{-\infty}^{\infty} f_1^*(t) f_2(t) \, dt$$

(e) *Modulation Theorem*

$$\mathcal{F}[f(t) e^{i\omega_0 t}] \equiv F_F[i(\omega - \omega_0)] = c(\nu - \nu_0)$$

$$\mathcal{F}[f(t) \cos \omega_0 t] \equiv \tfrac{1}{2}\{F_F[i(\omega - \omega_0)] + F_F[i(\omega + \omega_0)]\} \equiv \tfrac{1}{2}[c(\nu - \nu_0) + c(\nu + \nu_0)]$$

$$\mathcal{F}[f(t) \sin \omega_0 t] \equiv \frac{1}{2i} \{F_F[i(\omega - \omega_0)] - F_F[i(\omega + \omega_0)]\} \equiv \frac{1}{2i}[c(\nu - \nu_0) - c(\nu + \nu_0)]$$

(f) *Differentiation Theorem*

$$\mathcal{F}[f^{(r)}(t)] = (2\pi i \nu)^r \mathcal{F}[f(t)] \qquad r = 0, 1, 2, \dots$$

provided that $f^{(r)}(t)$ exists for all t, and that all derivatives of lesser order vanish as $|t| \to \infty$.

Table 2-4 Fourier Coefficients and Mean-Square Values of Periodic Functions $\left(\text{sinc } x \equiv \dfrac{\sin \pi x}{\pi x}\right)$

Periodic function, $f(t) = f(t + T)$	Fourier coefficients (for phasing as shown in diagram)	Average value $\langle f \rangle = \dfrac{a_0}{2}$	Mean-square value $\langle f^2 \rangle$
1. Rectangular pulses	$a_n = 2A\dfrac{T_0}{T}\,\text{sinc}\dfrac{nT_0}{T}$ $b_n = 0$	$A\dfrac{T_0}{T}$	$A^2\dfrac{T_0}{T}$
2. Symmetrical triangular pulses	$a_n = A\dfrac{T_0}{T}\,\text{sinc}^2\dfrac{nT_0}{2T}$ $b_n = 0$	$A\dfrac{T_0}{2T}$	$A^2\dfrac{T_0}{3T}$
3. Symmetrical trapezoidal pulses	$a_n = 2A\dfrac{T_0+T_1}{T}\,\text{sinc}\dfrac{nT_1}{T}\,\text{sinc}\dfrac{n(T_0+T_1)}{T}$ $b_n = 0$	$A\dfrac{T_0+T_1}{T}$	$A^2\dfrac{3T_0+2T_1}{3T}$
4. Half-sine pulses*†	$a_n = A\dfrac{T_0}{T}\left\{\text{sinc}\left[\dfrac{1}{2}\left(\dfrac{2nT_0}{T}-1\right)\right] + \text{sinc}\left[\dfrac{1}{2}\left(\dfrac{2nT_0}{T}+1\right)\right]\right\}$ $b_n = 0$	$\dfrac{2}{\pi}A\dfrac{T_0}{T}$	$A^2\dfrac{T_0}{2T}$
5. Clipped sinusoid $A = A_0\left(1-\cos\dfrac{\pi T_0}{T}\right)$	$a_n = \dfrac{A_0 T_0}{T}\left\{\text{sinc}\left[(n-1)\dfrac{T_0}{T}\right]\right.$ $\left. + \text{sinc}\left[(n+1)\dfrac{T_0}{T}\right] - 2\cos\dfrac{\pi T_0}{T}\,\text{sinc}\dfrac{nT_0}{T}\right\}$	$\dfrac{1}{\pi}A_0\left(\sin\dfrac{\pi T_0}{T}\right.$ $\left. -\dfrac{\pi T_0}{T}\cos\dfrac{\pi T_0}{T}\right)$	$\dfrac{1}{2\pi}A_0^2\left(\dfrac{\pi T_0}{T}-\dfrac{3}{2}\sin\dfrac{2\pi T_0}{T}\right.$ $\left. +\dfrac{2\pi T_0}{T}\cos^2\dfrac{\pi T_0}{T}\right)$
6. Triangular waveform	$a_n = 0$ $b_n = -\dfrac{A}{n\pi}\qquad n = 1, 2, \ldots$	$\dfrac{A}{2}$	$\dfrac{A^2}{3}$

*For $T_0 = \dfrac{T}{2} = \dfrac{\pi}{\omega}$, $f(t) = \dfrac{2}{\pi}A\left(\dfrac{1}{2}+\dfrac{\pi}{4}\cos\omega t+\dfrac{1}{3}\cos 2\omega t-\dfrac{1}{15}\cos 4\omega t+\dfrac{1}{35}\cos 6\omega t\pm\cdots\right)$ half-wave-rectified sinusoid.

†For $T_0 = T = \dfrac{2\pi}{\omega}$, $f(t) = -\dfrac{4}{\pi}A\left(\dfrac{1}{2}+\dfrac{1}{3}\cos 2\omega t-\dfrac{1}{15}\cos 4\omega t+\dfrac{1}{35}\cos 6\omega t\pm\cdots\right)$ full-wave-rectified sinusoid.

Table 2-5   Fourier Transform Pairs*

	$f(t) = \int_{-\infty}^{\infty} F(j\omega)e^{j\omega t}\frac{d\omega}{2\pi}$	$F(j\omega) = \int_{-\infty}^{\infty} f(t)e^{-j\omega t}\,dt$	
	$\text{rect}\,\dfrac{t}{T} = \begin{cases} 1 & (\lvert t \rvert < T/2) \\ 0 & (\lvert t \rvert > T/2) \end{cases}$	$T\,\text{sinc}\,\dfrac{\omega T}{2\pi} \equiv T\,\dfrac{\sin\dfrac{\omega T}{2}}{\dfrac{\omega T}{2}}$	
	$\text{sinc}\,\dfrac{t}{T} \equiv \dfrac{\sin\dfrac{\pi t}{T}}{\dfrac{\pi t}{T}}$	$T\,\text{rect}\,\dfrac{\omega T}{2\pi} = \begin{cases} 0 & \left(\lvert\omega\rvert < \dfrac{\pi}{T}\right) \\ T & \left(\lvert\omega\rvert > \dfrac{\pi}{T}\right) \end{cases}$	
	$\begin{cases} 1 - \dfrac{\lvert t \rvert}{T} & (\lvert t \rvert < T) \\ 0 & (\lvert t \rvert \geq T) \end{cases}$	$T\,\text{sinc}^2\,\dfrac{\omega T}{2\pi} \equiv T\left(\dfrac{\sin\dfrac{\omega T}{2}}{\dfrac{\omega T}{2}}\right)^2$	
	$e^{-\frac{\lvert t \rvert}{T}}$	$\dfrac{2T}{(\omega T)^2 + 1}$	
	$e^{-\frac{1}{2}\left(\frac{t}{T}\right)^2}$	$\sqrt{2\pi}\,T e^{-\frac{1}{2}(\omega T)^2}$	
	$\delta(t - T)$	$e^{-j\omega T}$	(Complex)
	$\cos\omega_0 t$	$\pi[\delta(\omega - \omega_0) + \delta(\omega + \omega_0)]$	
	$\sin\omega_0 t$	$\dfrac{\pi}{j}[\delta(\omega - \omega_0) - \delta(\omega + \omega_0)]$	(Imaginary)
	$\displaystyle\sum_{k=-\infty}^{\infty} \delta(t - kT)$ $\equiv \dfrac{1}{T}\displaystyle\sum_{i=-\infty}^{\infty} e^{2\pi j i\frac{t}{T}}$	$\dfrac{2\pi}{T}\displaystyle\sum_{i=-\infty}^{\infty}\delta\left(\omega - \dfrac{2\pi i}{T}\right)$ $\equiv \displaystyle\sum_{k=-\infty}^{\infty} e^{jk\omega T}$	

* Reprinted from G. A. Korn, "Basic Tables in Electrical Engineering," McGraw-Hill, New York, 1965, by permission.

VECTOR ALGEBRA

14. Vector Addition and Multiplication of Vectors by (Real) Scalars. Vectors, for example, **a**, **b**, are shown in boldface type.

$$\mathbf{a} + \mathbf{b} = \mathbf{b} + \mathbf{a}$$
$$\mathbf{a} + (\mathbf{b} + \mathbf{c}) = (\mathbf{a} + \mathbf{b}) + \mathbf{c} = \mathbf{a} + \mathbf{b} + \mathbf{c}$$
$$\alpha(\beta\mathbf{a}) = (\alpha\beta)\mathbf{a} \qquad (\alpha + \beta)\mathbf{a} = \alpha\mathbf{a} + \beta\mathbf{a}$$
$$\alpha(\mathbf{a} + \mathbf{b}) = \alpha\mathbf{a} + \alpha\mathbf{b}$$
$$(1)\mathbf{a} = \mathbf{a} \qquad (-1)\mathbf{a} = -\mathbf{a} \qquad (0)\mathbf{a} = 0$$
$$\mathbf{a} - \mathbf{a} = 0 \qquad \mathbf{a} + 0 = \mathbf{a}$$

15. Scalar Product (Dot Product, Inner Product). The *scalar product (dot product, inner product)* $\mathbf{a} \cdot \mathbf{b}$ [alternative notation $(\mathbf{ab})$] of two euclidean vectors $\mathbf{a}$ and $\mathbf{b}$ is the scalar $\mathbf{a} \cdot \mathbf{b} = |\mathbf{a}||\mathbf{b}| \cos \gamma$ where γ is the angle $\sphericalangle \, \mathbf{a}, \mathbf{b}$.

$$\mathbf{a} \cdot \mathbf{b} = \mathbf{b} \cdot \mathbf{a} \qquad \mathbf{a} \cdot (\mathbf{b} + \mathbf{c}) = \mathbf{a} \cdot \mathbf{b} + \mathbf{a} \cdot \mathbf{c} \qquad (\alpha \mathbf{a}) \cdot \mathbf{b} = \alpha(\mathbf{a} \cdot \mathbf{b})$$

$$\mathbf{a} \cdot \mathbf{a} = a^2 = |a^2| \geq 0 \qquad |\mathbf{a} \cdot \mathbf{b}| \leq |\mathbf{a}||\mathbf{b}| \qquad \cos \gamma = \frac{\mathbf{a} \cdot \mathbf{b}}{\sqrt{a^2 b^2}}$$

$$\mathbf{i} \cdot \mathbf{i} = \mathbf{j} \cdot \mathbf{j} = \mathbf{k} \cdot \mathbf{k} = 1 \qquad \mathbf{i} \cdot \mathbf{j} = \mathbf{j} \cdot \mathbf{k} = \mathbf{k} \cdot \mathbf{i} = 0$$

$$\mathbf{a} \cdot \mathbf{b} = (a_x\mathbf{i} + a_y\mathbf{j} + a_z\mathbf{k}) \cdot (b_x\mathbf{i} + b_y\mathbf{j} + b_z\mathbf{k}) = a_xb_x + a_yb_y + a_zb_z$$

$$a_x = \mathbf{a} \cdot \mathbf{i} \qquad a_y = \mathbf{a} \cdot \mathbf{j} \qquad a_z = \mathbf{a} \cdot \mathbf{k}$$

16. Vector (Cross) Product. The *vector (cross) product* $\mathbf{a} \times \mathbf{b}$ (alternative notation $[\mathbf{ab}]$) of two vectors $\mathbf{a}$ and $\mathbf{b}$ is the vector of magnitude

$$|\mathbf{a} \times \mathbf{b}| = |\mathbf{a}||\mathbf{b}| \sin \gamma$$

where γ is the angle $\sphericalangle \, \mathbf{a}, \mathbf{b}$. The direction of the vector is perpendicular to both $\mathbf{a}$ and $\mathbf{b}$ and such that the axial motion of a right-handed screw turning $\mathbf{a}$ into $\mathbf{b}$ is in the direction of $\mathbf{a} \times \mathbf{b}$.

17. Scalar Triple Product (Box Product)

$$\mathbf{a} \cdot (\mathbf{b} \times \mathbf{c}) \equiv [\mathbf{abc}] = [\mathbf{bca}] = [\mathbf{cab}] = -[\mathbf{bac}] = -[\mathbf{cba}] = -[\mathbf{acb}]$$

$$[\mathbf{abc}]^2 = [(\mathbf{a} \times \mathbf{b})(\mathbf{b} \times \mathbf{c})(\mathbf{c} \times \mathbf{a})] = a^2b^2c^2 - a^2(\mathbf{b} \cdot \mathbf{c})^2 - b^2(\mathbf{a} \cdot \mathbf{c})^2$$
$$- c^2(\mathbf{a} \cdot \mathbf{b})^2 + 2(\mathbf{a} \cdot \mathbf{b})(\mathbf{b} \cdot \mathbf{c})(\mathbf{a} \cdot \mathbf{c})$$

$$= \begin{vmatrix} \mathbf{a} \cdot \mathbf{a} & \mathbf{a} \cdot \mathbf{b} & \mathbf{a} \cdot \mathbf{c} \\ \mathbf{b} \cdot \mathbf{a} & \mathbf{b} \cdot \mathbf{b} & \mathbf{b} \cdot \mathbf{c} \\ \mathbf{c} \cdot \mathbf{a} & \mathbf{c} \cdot \mathbf{b} & \mathbf{c} \cdot \mathbf{c} \end{vmatrix} \qquad \text{Gram's determinant}$$

$$[\mathbf{abc}][\mathbf{def}] = \begin{vmatrix} \mathbf{a} \cdot \mathbf{d} & \mathbf{a} \cdot \mathbf{e} & \mathbf{a} \cdot \mathbf{f} \\ \mathbf{b} \cdot \mathbf{d} & \mathbf{b} \cdot \mathbf{e} & \mathbf{b} \cdot \mathbf{f} \\ \mathbf{c} \cdot \mathbf{d} & \mathbf{c} \cdot \mathbf{e} & \mathbf{c} \cdot \mathbf{f} \end{vmatrix}$$

$$[\mathbf{abc}] = \begin{vmatrix} a_x & b_x & c_x \\ a_y & b_y & c_y \\ a_z & b_z & c_z \end{vmatrix} \qquad > 0 \text{ if } \mathbf{a}, \mathbf{b}, \mathbf{c} \text{ are directed like right-handed cartesian axes}$$

TABLE 2-6 Relations Involving Vector (Cross) Products

a. Basic relations

$$\mathbf{a} \times \mathbf{b} = -(\mathbf{b} \times \mathbf{a})$$
$$\mathbf{a} \times \mathbf{a} = 0 \qquad \mathbf{a} \cdot (\mathbf{a} \times \mathbf{b}) = \mathbf{b} \cdot (\mathbf{a} \times \mathbf{b}) = 0$$
$$(\alpha \mathbf{a}) \times \mathbf{b} = \alpha(\mathbf{a} \times \mathbf{b}) \qquad \mathbf{a} \times (\mathbf{b} + \mathbf{c}) = \mathbf{a} \times \mathbf{b} + \mathbf{a} \times \mathbf{c}$$
$$[(\alpha + \beta)\mathbf{a}] \times \mathbf{b} = (\alpha + \beta)(\mathbf{a} \times \mathbf{b}) = \alpha(\mathbf{a} \times \mathbf{b}) + \beta(\mathbf{a} \times \mathbf{b})$$

b. In terms of any basis e_1, e_2, e_3

$$\mathbf{a} = \alpha_1\mathbf{e}_1 + \alpha_2\mathbf{e}_2 + \alpha_3\mathbf{e}_3 \qquad \mathbf{b} = \beta_1\mathbf{e}_1 + \beta_2\mathbf{e}_2 + \beta_3\mathbf{e}_3$$

$$\mathbf{a} \times \mathbf{b} = \begin{vmatrix} \mathbf{e}_2 \times \mathbf{e}_3 & \alpha_1 & \beta_1 \\ \mathbf{e}_3 \times \mathbf{e}_1 & \alpha_2 & \beta_2 \\ \mathbf{e}_1 \times \mathbf{e}_2 & \alpha_3 & \beta_3 \end{vmatrix}$$

c. In terms of right-handed rectangular cartesian components

$$\mathbf{i} \times \mathbf{i} = \mathbf{j} \times \mathbf{j} = \mathbf{k} \times \mathbf{k} = 0 \qquad \mathbf{i} \times \mathbf{j} = \mathbf{k} \qquad \mathbf{j} \times \mathbf{k} = \mathbf{i} \qquad \mathbf{k} \times \mathbf{i} = \mathbf{j}$$

$$\mathbf{a} \times \mathbf{b} = \begin{vmatrix} \mathbf{i} & a_x & b_x \\ \mathbf{j} & a_y & b_y \\ \mathbf{k} & a_z & b_z \end{vmatrix} = \mathbf{i}\begin{vmatrix} a_y & a_z \\ b_y & b_z \end{vmatrix} + \mathbf{j}\begin{vmatrix} a_z & a_x \\ b_z & b_x \end{vmatrix} + \mathbf{k}\begin{vmatrix} a_x & a_y \\ b_x & b_y \end{vmatrix}$$
$$= \mathbf{i}(a_yb_z - a_zb_y) + \mathbf{j}(a_zb_x - a_xb_z) + \mathbf{k}(a_xb_y - a_yb_x)$$

18. Other Products of More than Two Vectors

$$\mathbf{a} \times (\mathbf{b} \times \mathbf{c}) = (\mathbf{a} \cdot \mathbf{c})\mathbf{b} - (\mathbf{a} \cdot \mathbf{b})\mathbf{c} = \begin{vmatrix} \mathbf{b} & \mathbf{c} \\ \mathbf{a} \cdot \mathbf{b} & \mathbf{a} \cdot \mathbf{c} \end{vmatrix} \qquad \text{vector triple product}$$

$$(\mathbf{a} \times \mathbf{b}) \cdot (\mathbf{c} \times \mathbf{d}) = (\mathbf{a} \cdot \mathbf{c})(\mathbf{b} \cdot \mathbf{d}) - (\mathbf{a} \cdot \mathbf{d})(\mathbf{b} \cdot \mathbf{c}) = \begin{vmatrix} \mathbf{a} \cdot \mathbf{c} & \mathbf{b} \cdot \mathbf{c} \\ \mathbf{a} \cdot \mathbf{d} & \mathbf{b} \cdot \mathbf{d} \end{vmatrix}$$

$$(\mathbf{a} \times \mathbf{b})^2 = a^2 b^2 - (\mathbf{a} \cdot \mathbf{b})^2$$

$$(\mathbf{a} \times \mathbf{b}) \times (\mathbf{c} \times \mathbf{d}) = [acd]\mathbf{b} - [bcd]\mathbf{a} = [abd]\mathbf{c} - [abc]\mathbf{d}$$

VECTOR ANALYSIS: DIFFERENTIAL OPERATORS

19. The Operator ∇.

In terms of rectangular cartesian coordinates, the linear operator ∇ (*del* or *nabla*) is defined by

$$\nabla \equiv \mathbf{i}\frac{\partial}{\partial x} + \mathbf{j}\frac{\partial}{\partial y} + \mathbf{k}\frac{\partial}{\partial z}$$

Its application to a scalar point function $\Phi(\mathbf{r})$ or a vector point function $\mathbf{F}(\mathbf{r})$ corresponds formally to a noncommutative multiplication operation with a vector having the rectangular cartesian components $\partial/\partial x, \partial/\partial y, \partial/\partial z$; thus in terms of right-handed rectangular cartesian coordinates $x, y, z,$

$$\nabla\Phi(x, y, z) \equiv \text{grad } \Phi(x, y, z) \equiv \mathbf{i}\frac{\partial\Phi}{\partial x} + \mathbf{j}\frac{\partial\Phi}{\partial y} + \mathbf{k}\frac{\partial\Phi}{\partial z}$$

$$\nabla \cdot \mathbf{F}(x, y, z) \equiv \text{div } \mathbf{F}(x, y, z) \equiv \frac{\partial F_x}{\partial x} + \frac{\partial F_y}{\partial y} + \frac{\partial F_z}{\partial z}$$

$$\nabla \times \mathbf{F}(x, y, z) \equiv \text{curl } \mathbf{F}(x, y, z)$$

$$\equiv \mathbf{i}\left(\frac{\partial F_z}{\partial y} - \frac{\partial F_y}{\partial z}\right) + \mathbf{j}\left(\frac{\partial F_x}{\partial z} - \frac{\partial F_z}{\partial x}\right) + \mathbf{k}\left(\frac{\partial F_y}{\partial x} - \frac{\partial F_x}{\partial y}\right) \equiv \begin{vmatrix} \mathbf{i} & \dfrac{\partial}{\partial x} & F_x \\ \mathbf{j} & \dfrac{\partial}{\partial y} & F_y \\ \mathbf{k} & \dfrac{\partial}{\partial z} & F_z \end{vmatrix}$$

$$(\mathbf{G} \cdot \nabla)\mathbf{F} \equiv G_x\frac{\partial \mathbf{F}}{\partial x} + G_y\frac{\partial \mathbf{F}}{\partial y} + G_z\frac{\partial \mathbf{F}}{\partial z} \equiv \mathbf{i}(\mathbf{G} \cdot \nabla F_x) + \mathbf{j}(\mathbf{G} \cdot \nabla F_y) + \mathbf{k}(\mathbf{G} \cdot \nabla F_z)$$

TABLE 2-7   Rules* for Operations Involving the Operator ∇

a. Linearity

$$\nabla(\Phi + \Psi) = \nabla\Phi + \nabla\Psi \qquad\qquad \nabla(\alpha\Phi) = \alpha\nabla\Phi$$
$$\nabla \cdot (\mathbf{F} + \mathbf{G}) = \nabla \cdot \mathbf{F} + \nabla \cdot \mathbf{G} \qquad\qquad \nabla \cdot (\alpha\mathbf{F}) = \alpha\nabla \cdot \mathbf{F}$$
$$\nabla \times (\mathbf{F} + \mathbf{G}) = \nabla \times \mathbf{F} + \nabla \times \mathbf{G} \qquad\qquad \nabla \times (\alpha\mathbf{F}) = \alpha\nabla \times \mathbf{F}$$

b. Operations on products

$$\nabla(\Phi\Psi) = \Psi \nabla\Phi + \Phi \nabla\Psi$$
$$\nabla(\mathbf{F} \cdot \mathbf{G}) = (\mathbf{F} \cdot \nabla)\mathbf{G} + (\mathbf{G} \cdot \nabla)\mathbf{F} + \mathbf{F} \times (\nabla \times \mathbf{G}) + \mathbf{G} \times (\nabla \times \mathbf{F})$$
$$\nabla \cdot (\Phi\mathbf{F}) = \Phi\nabla \cdot \mathbf{F} + (\nabla\Phi) \cdot \mathbf{F}$$
$$\nabla \cdot (\mathbf{F} \times \mathbf{G}) = \mathbf{G} \cdot \nabla \times \mathbf{F} - \mathbf{F} \cdot \nabla \times \mathbf{G}$$
$$(\mathbf{G} \cdot \nabla)\Phi\mathbf{F} = \mathbf{F}(\mathbf{G} \cdot \nabla\Phi) + \Phi(\mathbf{G} \cdot \nabla)\mathbf{F}$$
$$\nabla \times (\Phi\mathbf{F}) = \Phi\nabla \times \mathbf{F} + (\nabla\Phi) \times \mathbf{F}$$
$$\nabla \times (\mathbf{F} \times \mathbf{G}) = (\mathbf{G} \cdot \nabla)\mathbf{F} - (\mathbf{F} \cdot \nabla)\mathbf{G} + \mathbf{F}(\nabla \cdot \mathbf{G}) - \mathbf{G}(\nabla \cdot \mathbf{F})$$
$$(\mathbf{G} \cdot \nabla)\mathbf{F} = \tfrac{1}{2}[\nabla \times (\mathbf{F} \times \mathbf{G}) + \nabla(\mathbf{F} \cdot \mathbf{G}) - \mathbf{F}(\nabla \cdot \mathbf{G}) + \mathbf{G}(\nabla \cdot \mathbf{F}) - \mathbf{F} \times (\nabla \times \mathbf{G})$$
$$- \mathbf{G} \times (\nabla \times \mathbf{F})]$$

*Note that vector equations involving $\nabla\Phi$, $\nabla \cdot \mathbf{F}$ and/or $\nabla \times \mathbf{F}$ have a meaning independent of the coordinate system used.

20. The Laplacian Operator. The *Laplacian operator* $\nabla^2 \equiv (\nabla \cdot \nabla)$ (sometimes denoted by Δ), expressed in terms of rectangular cartesian coordinates by

$$\nabla^2 \equiv (\nabla \cdot \nabla) \equiv \frac{\partial^2}{\partial x^2} + \frac{\partial^2}{\partial y^2} + \frac{\partial^2}{\partial z^2}$$

may be applied to both scalar and vector point functions by noncommutative scalar "multiplication," so that

$$\nabla^2\Phi \equiv \left(\frac{\partial^2}{\partial x^2} + \frac{\partial^2}{\partial y^2} + \frac{\partial^2}{\partial z^2} \right) \Phi \qquad \nabla^2\mathbf{F} \equiv \mathbf{i}\,\nabla^2 F_x + \mathbf{j}\,\nabla^2 F_y + \mathbf{k}\,\nabla^2 F_z$$

Note

$$\nabla^2(\alpha\Phi + \beta\Psi) = \alpha\,\nabla^2\Phi + \beta\,\nabla^2\Psi \qquad \text{linearity}$$

and

$$\nabla^2(\Phi\Psi) = \Psi\,\nabla^2\Phi + 2(\nabla\Phi) \cdot (\nabla\Psi) + \Phi\,\nabla^2\Psi$$

21. Repeated Operations. Note the following rules for repeated operations with the operator ∇:

$$\begin{aligned}
\text{div grad } \Phi &= \nabla \cdot (\nabla\Phi) = \nabla^2\Phi \\
\text{grad div } \mathbf{F} &= \nabla(\nabla \cdot \mathbf{F}) = \nabla^2\mathbf{F} + \nabla \times (\nabla \times \mathbf{F}) \\
\text{curl curl } \mathbf{F} &= \nabla \times (\nabla \times \mathbf{F}) = \nabla(\nabla \cdot \mathbf{F}) - \nabla^2\mathbf{F} \\
\text{curl grad } \Phi &= \nabla \times (\nabla\Phi) = 0 \\
\text{div curl } \mathbf{F} &= \nabla \cdot (\nabla \times \mathbf{F}) = 0
\end{aligned}$$

THE LAPLACE TRANSFORMATION

22. Introduction. The Laplace transformation associates a unique function $F(s)$ of a complex variable s with each suitable function $f(t)$ of a real variable t. This correspondence is essentially reciprocal one to one for most practical purposes (Par. **2-26**); corresponding pairs of functions $f(t)$ and $F(s)$ can often be found by reference to tables. The Laplace transformation is defined so that many relations between, and operations on, the functions $f(t)$ correspond to simpler relations between, and operations on, the functions $F(s)$ (Table 2-9). This applies particularly to the solution of differential and integral equations. It is thus often useful to transform a given problem involving functions $f(t)$ into an equivalent problem expressed in terms of the associated Laplace transforms $F(s)$ (*operational calculus* based on Laplace transformation or *transformation calculus*, Par. **2-29**).

23. Definition. The *one-sided Laplace transformation*

$$F(s) \equiv \mathfrak{L}[f(t)] \equiv \int_0^\infty f(t)e^{-st}\,dt \equiv \lim_{\substack{a \to 0 \\ b \to \infty}} \int_a^b f(t)e^{-st}\,dt \qquad 0 < a < b$$

associates a unique *result or image function* $F(s)$ of the complex variable $s = \sigma + i\omega$ with every single-valued *object or original function* $f(t)$ (t real) such that the improper integral exists. $F(s)$ is called the *(one-sided) Laplace transform* of $f(t)$. The more explicit notation $\mathfrak{L}[f(t);s]$ is also used.

The Laplace transform exists for $\sigma \geq \sigma_0$, and the improper integral converges absolutely and uniformly to a function $F(s)$ analytic for $\sigma > \sigma_0$ if

$$\int_0^\infty |f(t)|\,e^{-\sigma t}\,dt = \lim_{\substack{a \to 0 \\ b \to \infty}} \int_a^b |f(t)|\,e^{-\sigma t}\,dt \qquad 0 < a < b$$

exists for $\sigma = \sigma_0$. The greatest lower bound σ_a of the real number σ_0 for which this is true is called the *abscissa of absolute convergence* of the Laplace transform $\mathfrak{L}[f(t)]$.

Although certain theorems relating to Laplace transforms require only the existence (simple convergence) of the transforms, *the existence of an abscissa of absolute convergence will be implicitly assumed throughout the following sections.* Wherever necessary, it is customary to specify the region of absolute convergence associated with a relation involving Laplace transforms by writing $\sigma > \sigma_a$ to the right of the relation in question.

The region of definition of the analytic function

$$F(s) = \mathfrak{L}[f(t)] \qquad \sigma > \sigma_a$$

can usually be extended so as to include the entire s plane with the exception of singular points situated to the left of the abscissa of absolute convergence. Such an extension of the region of definition is implied wherever necessary.

24. Inverse Laplace Transformation. The inverse Laplace transform $\mathcal{L}^{-1}[F(s)]$ of a (suitable) function $F(s)$ of the complex variable $s = \sigma + i\omega$ is a function $f(t)$ whose Laplace transform is $F(s)$. Not every function $F(s)$ has an inverse Laplace transform.

25. The Inversion Theorem. Given $F(s) = \mathcal{L}[f(t)]$, $\sigma > \sigma_a$, then throughout every open interval where $f(t)$ is bounded and has a finite number of maxima, minima, and discontinuities,

$$f(t) = \frac{1}{2\pi i} \lim_{R \to \infty} \int_{\sigma_1 - iR}^{\sigma_1 + iR} F(s)e^{st}\, ds$$

$$= \begin{cases} \frac{1}{2}[f(t-0) + f(t+0)] & \text{for } t > 0 \\ \frac{1}{2}f(0+0) & \text{for } t = 0 \\ 0 & \text{for } t < 0 \end{cases} \qquad \sigma_1 > \sigma_a$$

In particular, for every $t > 0$ where $f(t)$ is continuous

$$f(t) = \frac{1}{2\pi i} \lim_{R \to \infty} \int_{\sigma_1 - iR}^{\sigma_1 + iR} F(s)e^{st}\, ds = f(t) \qquad \sigma_1 > \sigma_a \qquad (2\text{-}7)$$

The path of integration in Eq. (2-7) lies to the right of all singularities of $F(s)$. The inversion integral $f(t)$ reduces to $(1/2\pi i) \int_{\sigma_1 - i\infty}^{\sigma_1 + i\infty} F(s)e^{st}\, ds$ if the integral exists.

26. Uniqueness of the Laplace Transform and Its Inverse. The Laplace transform is unique for each function $f(t)$ having such a transform. Conversely, two functions $f_1(t)$ and $f_2(t)$ possessing identical Laplace transforms are identical for all $t > 0$ where both functions are continuous (Lerch's theorem). A given function $F(s)$ cannot have more than one inverse Laplace transform continuous for all $t > 0$.

Different discontinuous functions may have the same Laplace transform. In particular, the generalized unit step function defined by $f(t) = 0$ for $t < 0$, $f(t) = 1$ for $t > 0$ has the Laplace transform $1/s$ regardless of the value assigned to $f(t)$ for $t = 0$.

LINEAR DIFFERENTIAL EQUATIONS

27. Homogeneous Linear Equations with Constant Coefficients. The first-order differential equation

$$a_0 \frac{dy}{dt} + a_1 y = 0 \qquad a_0 \neq 0 \qquad (2\text{-}8)$$

has the solution

$$y = Ce^{-(a_1/a_0)t} \qquad C = y(0)$$

For $a_0/a_1 > 0$

$$y(a_0/a_1) = \frac{1}{e} y(0) \approx 0.37 y(0) \qquad y(4a_0/a_1) \approx 0.02 y(0)$$

a_0/a_1 is often referred to as the *time constant*.

The *second-order equation*

$$a_0 \frac{d^2y}{dt^2} + a_1 \frac{dy}{dt} + a_2 y = 0 \qquad a_0 \neq 0 \qquad (2\text{-}9)$$

has the solution

$$y = C_1 e^{s_1 t} + C_2 e^{s_2 t} \qquad (2\text{-}10a)$$

$$s_{1,2} = \frac{-a_1 \pm \sqrt{a_1^2 - 4a_0 a_2}}{2a_0} \qquad a_1^2 - 4a_0 a_2 \neq 0 \qquad (2\text{-}10b)$$

$$y = (C_1 + C_2 t)e^{-(a_1/2a_0)t} \qquad a_1^2 - 4a_0 a_2 = 0 \qquad (2\text{-}10c)$$

TABLE 2-8   Theorems Relating Volume Integrals and Surface Integrals

	Theorem	Vector formula	Sufficient conditions a. Throughout V	Sufficient conditions b. On S
1	Divergence theorem (Gauss' integral theorem)	$\int_V \nabla \cdot \mathbf{F}(\mathbf{r})\, dV = \int_S d\mathbf{A} \cdot \mathbf{F}(\mathbf{r})$	$\mathbf{F}(\mathbf{r}), \Phi(\mathbf{r})$ differentiable with continuous partial derivatives	Existence of integrals is sufficient
2	Theorem of the rotational	$\int_V \nabla \times \mathbf{F}(\mathbf{r})\, dV = \int_S d\mathbf{A} \times \mathbf{F}(\mathbf{r})$		
3	Theorem of the gradient	$\int_V \nabla \Phi(\mathbf{r})\, dV = \int_S d\mathbf{A}\Psi(\mathbf{r})$		
4	Green's theorems	$\int_V \nabla\Phi \cdot \nabla\Psi\, dV + \int_V \Psi \nabla^2\Phi\, dV = \int_S d\mathbf{A} \cdot (\Psi \nabla\Phi)$ $= \int_S \Psi\frac{\partial\Phi}{\partial n}\, dA$	$\Phi(\mathbf{r}), \Psi(\mathbf{r})$ differentiable with continuous partial derivatives; $\Phi(\mathbf{r})$ twice differentiable with continuous second partial derivatives	$\Psi(\mathbf{r})$ continuous, $\Phi(\mathbf{r})$ differentiable with continuous partial derivatives
5		$\int_V (\Psi\nabla^2\Phi - \Phi\nabla^2\Psi)\, dV = \int_S d\mathbf{A} \cdot (\Psi \nabla\Phi - \Phi \nabla\Psi)$ $= \int_S \left(\Psi\frac{\partial\Phi}{\partial n} - \Phi\frac{\partial\Phi}{\partial n}\right) dA$		
6	Special cases	$\int_V \nabla^2\Phi\, dV = \int_S d\mathbf{A} \cdot \nabla\Phi = \int_S \frac{\partial\Phi}{\partial n}\, dA$ (Gauss' theorem)	$\Phi(\mathbf{r}), \Psi(\mathbf{r})$ twice differentiable with continuous second partial derivatives	$\Phi(\mathbf{r}), \Psi(\mathbf{r})$ differentiable with continuous partial derivatives
7		$\int_V |\nabla\Phi|^2\, dV + \int_V \Phi\nabla^2\Phi\, dv = \int_S d\mathbf{A} \cdot (\Phi \nabla\Phi) = \int_S \Phi\frac{\partial\Phi}{\partial n}\, dA$		

If a_0, a_1, and a_2 are real, s_1 and s_2 become complex for $a_1^2 - 4a_0a_2 < 0$; in this case, Eq. (2-10a) can be written as

$$y = e^{\sigma_1 t}(A \cos \omega_N t + B \sin \omega_N t) = Re^{\sigma_1 t} \sin (\omega_N t + \alpha)$$

where
$$\sigma_1 = -a_1/2a_0 \qquad \omega_N = \sqrt{4a_0a_2 - a_1^2}/2a_0$$

are respectively known as the *damping constant* and the *natural (characteristic) circular frequency*. The constants C_1, C_2, A, B, R, and α are chosen so as to match given initial or boundary conditions (see also Fig. 2-1).

If $a_0a_2 > 0$, the quantity $\zeta = a_1/2\sqrt{a_0a_2}$ is called the *damping ratio*; for $\zeta > 1$, $\zeta = 1$, $0 < \zeta < 1$ one obtains, respectively, an *overdamped* solution (2-10a), a *critically damped* solution (2-10b), or an *underdamped (oscillatory)* solution (2-10c). In the latter case, the *logarithmic decrement* $2\pi\sigma_1/\omega_N$ is the natural logarithm of the ratio of successive maxima of $y(t)$.

Equation (2-9) is often written in the *nondimensional form*

$$\frac{1}{\omega_1^2}\frac{d^2y}{dt^2} + 2\frac{\zeta}{\omega_1}\frac{dy}{dt} + y = 0 \qquad \text{with } s_{1,2} = -\omega_1\zeta \pm \omega_1\sqrt{\zeta^2 - 1}$$

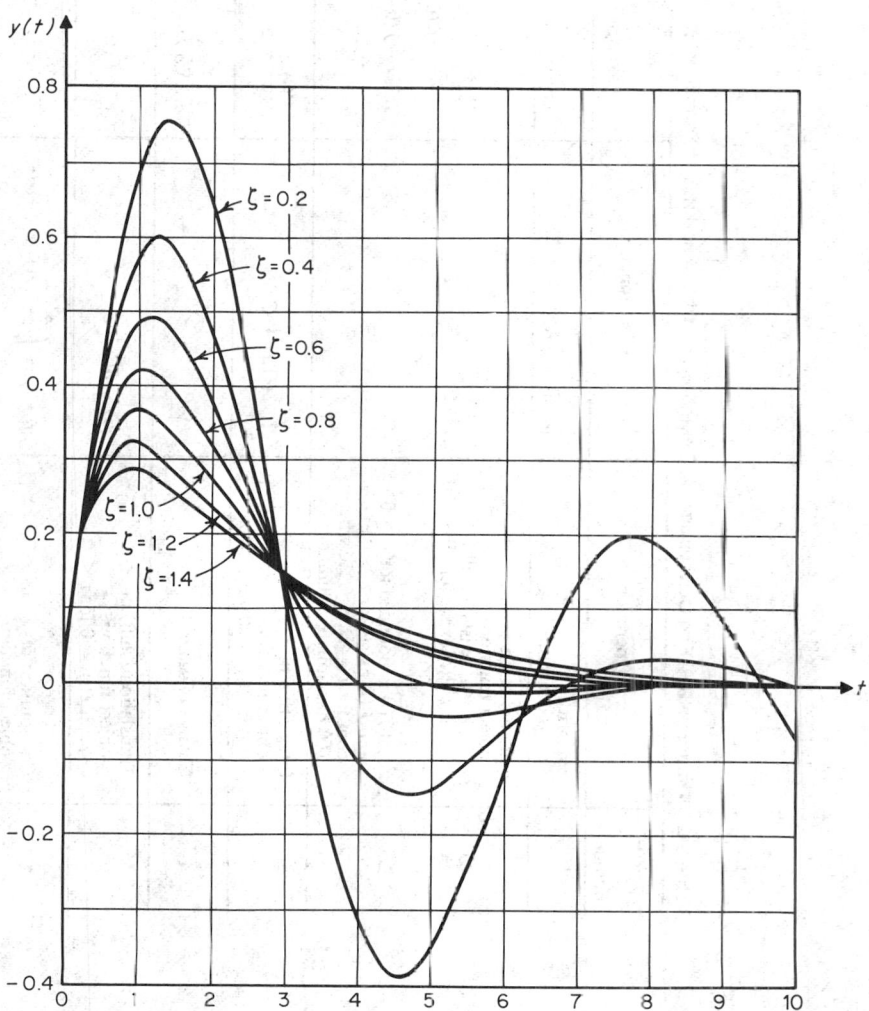

Fig. 2-1. Solution for the second-order differential equation

$$d^2y/dt^2 + 2\zeta\, dy/dt + v = 0$$

For $y(0) = 0$, $dy/dt_0 = 1$. Response is overdamped for $\zeta > 1$, critically damped for $\zeta = 1$, and underdamped for $\zeta < 1$.

Table 2-9 Theorems Relating Corresponding Operations on Object and Result Functions*†

Theorem number	Operation	Object function	Result function
1	Linearity (α, β constant)	$\alpha f_1(t) + \beta f_2(t)$	$\alpha F_1(s) + \beta F_2(s)$
2a 2b	Differentiation of object function‡ $\ldots$ if $f'(t)$ exists for all $t > 0$ $\ldots$ if $f^{(r)}(t)$ exists for all $t > 0$	$f'(t)$ $f^{(r)}(t)$ $\quad r = 1, 2, \ldots$	$sF(s) - f(0+0)$ $s^r F(s) - s^{r-1}f(0+0)$ $-s^{r-2}f'(0+0)$ $-\cdots - f^{(r-1)}(0+0)$
2c	$\ldots$ if $f(t)$ is bounded for $t > 0$, and $f'(t)$ exists for $t > 0$ except for $t = t_1, t_2, \ldots$, where $f(t)$ has unilateral limits	$f'(t)$	$sF(s) - f(0+0)$ $-\sum_k e^{-t_k s}[f(t_k+0) - f(t_k-0)]$
3	Integration of object function $\ldots$ if $f'(t)$ exists for $t > 0$	$\int_0^t f(\tau)\,d\tau + C$	$\dfrac{F(s)}{s} + \dfrac{C}{s}$
4	Change of scale	$f(at) \quad a > 0$	$\dfrac{1}{a}F\left(\dfrac{s}{a}\right)$
5	Translation (shift) of object function $\ldots$ if $f(t) = 0$ for $t \leq 0$	$f(t-b) \quad b \geq 0$	$e^{-bs}F(s)$
6	Convolution of object function¶	$f_1 * f_2 = \displaystyle\int_0^\infty f_1(\tau)f_2(t-\tau)\,d\tau$ $= f_2 * f_1$	$F_1(s)F_2(s)$

		$f(t)$	$F(s)$
7	Corresponding limits of object and result function (continuity theorem; α is independent of t and s)	$\lim_{\alpha\to a} f(t,a)$	$\lim_{\alpha\to a} F(s,a)$
8a	Differentiation and integration with respect to a parameter α independent of t and s	$\dfrac{\partial}{\partial\alpha} f(t,\alpha)$	$\dfrac{\partial}{\partial\alpha} F(s,\alpha)$
8b		$\displaystyle\int_{a_1}^{a_2} f(t,\alpha)\,d\alpha$	$\displaystyle\int_{a_1}^{a_2} F(s,\alpha)\,d\alpha$
9a	Differentiation of result function	$-t f(t)$	$F'(s)$
9b		$(-1)^r t^r f(t)$	$F^{(r)}(s)$
10	Integration of result function (path of integration situated to the right of the abscissa of absolute convergence)	$\dfrac{1}{t} f(t)$	$\displaystyle\int_s^\infty F(s)\,ds$
11	Translation of result function	$e^{at} f(t)$	$F(s-a)$

*From G. A. Korn and T. M. Korn, "Mathematical Handbook for Scientists and Engineers," McGraw-Hill, New York, 1961.

†The following theorems are valid whenever the Laplace transforms $F(s) = \mathcal{L}[f(t)]$ in question exist in the sense of absolute convergence.)

‡The abscissa of absolute convergence for $\mathcal{L}[f^{(r)}(t)]$ is 0 or a_a, whichever is greater.

¶The existence of $\int_{f_1} * f_2$ is assumed; absolute convergence of $\mathcal{L}[f_1(t)]$ and $\mathcal{L}[f_2(t)]$ is a sufficient condition for the absolute convergence of $\mathcal{L}[f_1 * f_2]$.

$\omega_1 = \sqrt{a_2/a_0}$ is called the *undamped natural circular frequency*; for weak damping ($\zeta^2 \ll 1$). $\omega_1 \approx \omega_N$.

To solve the *rth-order differential equation*

$$\mathbf{L}y \equiv a_0 \frac{d^r}{dt^r} y + a_1 \frac{d^{r-1}}{dt^{r-1}} y + \cdots + a_r y = 0 \qquad a_0 \neq 0 \qquad (2\text{-}11)$$

find the roots of the *rth-degree algebraic equation*

$$a_0 s^r + a_1 s^{r-1} + \cdots + a_0 = 0 \qquad \text{characteristic equation} \qquad (2\text{-}12)$$

obtained, for example, on substitution of a trial solution $y = e^{st}$. If the r roots $s_1, s_2, \ldots$ of the characteristic equation (2-12) are distinct, the given differential equation (2-11) has the general solution

$$y = C_1 e^{s_1 t} + C_2 e^{s_2 t} + \cdots + C_r e^{s_r t} \qquad (2\text{-}13a)$$

If a root s_k is of multiplicity m_k, replace the corresponding term in Eq. (2-13a) by

$$(C_k + C_{k1} t + C_{k2} t^2 + \cdots + C_{km_k-1} t^{m_k-1}) e^{s_k t} \qquad (2\text{-}13b)$$

The various terms of the solution (2-13b) are known as *normal modes* of the given differential equation. The r constants C_k and C_{kj} must be chosen so as to match given initial or boundary conditions.

If the given differential equation (2-11) is real, complex roots of the characteristic equation appear as pairs of complex conjugates $\sigma + i\omega$. The corresponding pairs of solution terms will also be complex conjugates and can be combined to form real terms:

$$t^m C e^{(\sigma + i\omega)t} + t^m C^* e^{(\sigma - i\omega)t} = t^m e^{\sigma t}(A \cos \omega t + B \sin \omega t)$$
$$= R t^m e^{\sigma t} \sin (\omega t + \alpha) \qquad (2\text{-}13c)$$

where A and B, or R and α, are new real constants of integration.

Given a *system of n homogeneous linear differential equations with constant coefficients*

$$\varphi_{j1}\left(\frac{d}{dt}\right) y_1 + \varphi_{j2}\left(\frac{d}{dt}\right) y_2 + \cdots + \varphi_{jn}\left(\frac{d}{dt}\right) y_n = 0 \qquad j = 1, 2, \ldots, n \qquad (2\text{-}14)$$

where the $\varphi_{jk}(d/dt)$ are polynomials in d/dt, each of the n solution functions $y_k = y_k(t)(k = 1, 2, \ldots, n)$ has the form (2-13a); the s_k are now the roots of the algebraic equation

$$D(s) \equiv \det [\varphi_{jk}(s)] = 0 \qquad \text{characteristic equation of system (2-14)} \qquad (2\text{-}15)$$

The constants of integration must again be matched to the given initial or boundary conditions.

28. Nonhomogeneous Equations. The general solution of the nonhomogeneous differential equation

$$\mathbf{L}y \equiv a_0 \frac{d^r y}{dt^r} + a_1 \frac{d^{r-1} y}{dt^{r-1}} + \cdots + a_r y = f(t) \qquad (2\text{-}16)$$

can be expressed as the sum of the general solution (2-13a) of the reduced equation (2-11) and any particular integral of Eq. (2-16).

29. Laplace-Transform Method of Solution. (See also Pars. **2-22** to **2-26**.) To solve a linear differential equation (2-16) with given initial values $y(0 + 0)$, $y'(0 + 0)$, $y''(0 + 0)$, $\ldots, y^{(r-1)}(0 + 0)$, apply the Laplace transformation to both sides, and let $\mathscr{L}[y(t)] \equiv Y(s)$, $\mathscr{L}[f(t)] \equiv F(s)$. The resulting linear *algebraic equation (subsidiary equation)*

$$(a_0 s^r + a_1 s^{r-1} + \cdots + a_r)Y(s) = F(s) + G(s)$$
$$G(s) \equiv y(0 + 0)(a_0 s^{r-1} + a_1 s^{r-2} + \cdots + a_{r-1})$$
$$+ y'(0 + 0)(a_0 s^{r-2} + a_1 s^{r-3} + \cdots + a_{r-2})$$
$$+ \cdots \cdots \cdots \cdots \cdots$$
$$+ y^{(r-2)}(0 + 0)(a_0 s + a_1) + a_0 y^{(r-1)}(0 + 0) \qquad (2\text{-}17)$$

is easily solved to yield the Laplace transform of the desired solution $y(t)$ in the form

$$Y(s) = \frac{F(s)}{a_0 s^r + a_1 s^{r-1} + \cdots + a_r} + \frac{G(s)}{a_0 s^r + a_1 s^{r-1} + \cdots + a_r} \qquad (2\text{-}18)$$

The second term represents the effects of nonzero initial values of $y(t)$ and its derivatives.

In the same manner, one applies the Laplace transformation to a system of linear differential equations

$$\varphi_{j1}\left(\frac{d}{dt}\right)y_1 + \varphi_{j2}\left(\frac{d}{dt}\right)y_2 + \cdots + \varphi_{jn}\left(\frac{d}{dt}\right)y_n = f_j(t) \qquad j = 1, 2, \ldots, n \quad (2\text{-}19)$$

to obtain

$$\varphi_{j1}(s)Y_1(s) + \varphi_{j2}(s)Y_2(s) + \cdots + \varphi_{jn}(s)Y_n(s) = F_j(s) - G_j(s) \qquad j = 1, 2, \ldots, n \quad (2\text{-}20)$$

where the functions $G_j(s)$ depend on the given initial conditions. The linear algebraic equations (2-20) are solved by Cramer's rule to yield the unknown solution transforms

$$Y_k(s) = \sum_{j=1}^{n} \frac{A_{jk}(s)}{D(s)} F_j(s) + \sum_{j=1}^{n} \frac{A_{jk}(s)}{D(s)} G_j(s) \qquad k = 1, 2, \ldots, n \quad (2\text{-}21)$$

where $A_{jk}(s)$ is the cofactor of $\varphi_{jk}(s)$ in the *system determinant* $D(s) \equiv \det [\varphi_{jk}(s)]$.

30. Sinusoidal Forcing Functions and Solutions: The Phasor Method. *Every system of linear differential equations (2-19) with sinusoidal forcing functions of equal frequency,*

$$f_j(t) \equiv B_j \sin (\omega t + \beta_j) \qquad j = 1, 2, \ldots, n \quad (2\text{-}22a)$$

admits a unique particular solution of the form

$$y_k(t) \equiv A_k \sin (\omega t + \alpha_k) \qquad k = 1, 2, \ldots, n \quad (2\text{-}22b)$$

In particular, *if all roots of the characteristic equation* $[D(s) = 0]$ *have negative real parts (stable systems), the sinusoidal solution (2-22b) is the unique steady-state solution obtained after all transients have died out.*

Given a system of linear differential equations (2-19) relating sinusoidal forcing functions and solutions (2-22), one introduces a reciprocal one-to-one representation of these sinusoids by corresponding complex numbers (vectors, phasors)

$$\begin{aligned} \vec{F}_j &= \frac{B}{\sqrt{2}} e^{i\beta_j} = \frac{B_j}{\sqrt{2}} \,\underline{/\beta_j} \qquad j = 1, 2, \ldots, n \\ \vec{Y}_k &= \frac{A_k}{\sqrt{2}} e^{i\alpha_k} = \frac{A_k}{\sqrt{2}} \,\underline{/\alpha_k} \qquad k = 1, 2, \ldots, n \end{aligned} \qquad (2\text{-}23)$$

The absolute value of each phasor equals the root-mean-square value of the corresponding sinusoid, while the phasor argument defines the phase of the sinusoid. *The phasors (2-23) are related by the (complex) linear algebraic equations (phase equations)*

$$\varphi_{j1}(i\omega)\vec{Y}_1 + \varphi_{j2}(i\omega)\vec{Y}_2 + \cdots + \varphi_{jn}(i\omega)\vec{Y}_n = \vec{F}_j \qquad j = 1, 2, \ldots, n \quad (2\text{-}24)$$

which correspond to Eq. (2-19) and can be solved for the unknown phasors

$$\vec{Y}_k = \sum_{j=1}^{n} \frac{A_{jk}(i\omega)}{D(i\omega)} \vec{F}_j \qquad k = 1, 2, \ldots, n \quad (2\text{-}25)$$

MATRICES AND LOGIC

31. Rectangular Matrices. An array

$$A \equiv \begin{bmatrix} a_{11} & a_{12} & \cdots & a_{1n} \\ a_{21} & a_{22} & \cdots & a_{2n} \\ \cdots & \cdots & \cdots & \cdots \\ a_{m1} & a_{m2} & \cdots & a_{mn} \end{bmatrix} \equiv [a_{ik}]$$

of real or complex numbers (scalars) a_{ik} will be called a real or complex *rectangular m × n matrix* whenever one of the matrix operations defined in Par. **2-32** is to be used. The elements a_{ik} are called *matrix elements;* the matrix element a_{ik} is situated in the ith row and in the kth column of the matrix; m is the number of rows, and n is the number of columns.

32. Basic Operations. Operations on matrices are defined in terms of operations on the matrix elements.

1. Two $m \times n$ matrices $A \equiv [a_{ik}]$ and $B \equiv [b_{ik}]$ are *equal* $(A = B)$ if and only if $a_{ik} = b_{ik}$ for all i, k.

2. The *sum of two $m \times n$ matrices* $A \equiv [a_{ik}]$ and $B \equiv [b_{ik}]$ is the $m \times n$ matrix

$$A + B \equiv [a_{ik}] + [b_{ik}] \equiv [a_{ik} + b_{ik}]$$

TABLE 2-10 Laplace Transform Pairs Involving Rational Algebraic Functions $F(s) = D_1(s)/D(s)$

Each formula holds for complex as well as for real polynomials $D_1(s)$ and $D(s)$, but the latter case is of greater practical interest. In this case the roots of $D(s) = 0$ are either real or they occur as pairs of complex conjugates, and the functions $f(t)$ are real. Note

$$(s - a)^2 + \omega_1^2 = [s - (a + i\omega_1)][s - (a - i\omega_1)] \quad \text{and} \quad K_1 \sin \omega t + K_2 \cos \omega t = \sqrt{K_1^2 + K_2^2} \, \sin(\omega t + \alpha) \qquad \alpha = \arctan K_2/K_1$$

No.	$F(s)$	$f(t)(t > 0)$	
1.1	$\dfrac{1}{s}$	1	
1.2	$\dfrac{1}{s - a}$	e^{at}	
1.3	$\dfrac{1}{s(s - a)}$	$Ae^{at} + K$	$A = \dfrac{1}{a} \quad K = -\dfrac{1}{a}$
1.4	$\dfrac{s + d}{s(s - a)}$		$A = \left(1 + \dfrac{d}{a}\right) \quad K = -\dfrac{d}{a}$
1.5	$\dfrac{1}{(s - a)(s - b)}$	$Ae^{at} + Be^{bt}$	$A = \dfrac{1}{a - b} \quad B = \dfrac{1}{b - a}$
1.6	$\dfrac{s + d}{(s - a)(s - b)}$		$A = \dfrac{a + d}{a - b} \quad B = \dfrac{b + d}{b - a}$
1.7	$\dfrac{1}{s(s - a)(s - b)}$	$Ae^{at} + Be^{bt} + K$	$A = \dfrac{1}{a(a - b)} \quad B = \dfrac{1}{b(b - a)} \quad K = \dfrac{1}{ab}$
1.8	$\dfrac{s + d}{s(s - a)(s - b)}$		$A = \dfrac{a + d}{a(a - b)} \quad B = \dfrac{b + d}{b(b - a)} \quad K = \dfrac{d}{ab}$
1.9	$\dfrac{s^2 + gs + d}{s(s - a)(s - b)}$		$A = \dfrac{a^2 + ga + d}{a(a - b)} \quad B = \dfrac{b^2 + gb + d}{b(b - a)} \quad K = \dfrac{d}{ab}$

1.10	$\dfrac{1}{(s-a)(s-b)(s-c)}$	$Ae^{at} + Be^{bt} + Ce^{ct}$
		$A = \dfrac{1}{(a-b)(a-c)}$ $\qquad B = \dfrac{1}{(b-a)(b-c)}$ $\qquad C = \dfrac{1}{(c-a)(c-b)}$
1.11	$\dfrac{s+d}{(s-a)(s-b)(s-c)}$	
		$A = \dfrac{a+d}{(a-b)(a-c)}$ $\qquad B = \dfrac{b+d}{(b-a)(b-c)}$ $\qquad C = \dfrac{c+d}{(c-a)(c-b)}$
1.12	$\dfrac{s^2+gs+d}{(s-a)(s-b)(s-c)}$	
		$A = \dfrac{a^2+ag+d}{(a-b)(a-c)}$ $\qquad B = \dfrac{b^2+bg+d}{(b-a)(b-c)}$ $\qquad C = \dfrac{c^2+cg+d}{(c-a)(c-b)}$
2.1	$\dfrac{1}{(s-a)^2+\omega_1^2}$	$Ae^{at}\sin(\omega_1 t + \alpha)$
		$A = \dfrac{1}{\omega_1}$ $\qquad \alpha = 0$
2.2	$\dfrac{s+d}{(s-a)^2+\omega_1^2}$	
		$A = \dfrac{1}{\omega_1}[(a+d)^2+\omega_1^2]^{1/2}$ $\qquad \alpha = \arctan\dfrac{\omega_1}{a+d}$
2.3	$\dfrac{1}{s[(s-a)^2+\omega_1^2]}$	
		$A = \dfrac{1}{\omega_1}\dfrac{1}{(a^2+\omega_1^2)^{1/2}}$ $\qquad \alpha = -\arctan\dfrac{\omega_1}{a}$ $\qquad K = \dfrac{1}{a^2+\omega_1^2}$
2.4	$\dfrac{s+d}{s[(s-a)^2+\omega_1^2]}$	$Ae^{at}\sin(\omega_1 t + \alpha) + K$
		$A = \dfrac{1}{\omega_1}\left[\dfrac{(a+d)^2+\omega_1^2}{a^2+\omega_1^2}\right]^{1/2}$ $\qquad \alpha = \arctan\dfrac{\omega_1}{a+d} - \arctan\dfrac{\omega_1}{a}$ $\qquad K = \dfrac{d}{a^2+\omega_1^2}$
2.5	$\dfrac{s^2+gs+d}{s[(s-a)^2+\omega_1^2]}$	
		$A = \dfrac{1}{\omega_1}\left[\dfrac{(a^2-\omega_1^2+ag+d)^2+\omega_1^2(2a+g)^2}{a^2+\omega_1^2}\right]^{1/2}$ $\qquad \alpha = \arctan\dfrac{\omega_1(2a+g)}{a^2-\omega_1^2+ag+d} - \arctan\dfrac{\omega_1}{a}$ $\qquad K = \dfrac{d}{a^2+\omega_1^2}$

Table 2-10 Laplace Transform Pairs involving Rational Algebraic Functions $F(s) = D_1(s)/D(s)$ (Continued)

No.	$F(s)$	$f(t)(t > 0)$	
2.6	$\dfrac{1}{(s - b)[(s - a)^2 + \omega_1^2]}$		$A = \dfrac{1}{\omega_1}\,\dfrac{1}{[(a - b)^2 + \omega_1^2]^{1/2}}$ $B = \dfrac{1}{(a - b)^2 + \omega_1^2}$ $\alpha = -\arctan\dfrac{\omega_1}{a - b}$
2.7	$\dfrac{s + d}{(s - b)[(s - a)^2 + \omega_1^2]}$	$Ae^{at}\sin(\omega_1 t + \alpha) + Be^{bt}$	$A = \dfrac{1}{\omega_1}\left[\dfrac{(a + d)^2 + \omega_1^2}{(a - b)^2 + \omega_1^2}\right]^{1/2}$ $B = \dfrac{b + d}{(a - b)^2 + \omega_1^2}$ $\alpha = \arctan\dfrac{\omega_1}{a + d} - \arctan\dfrac{\omega_1}{a - b}$
2.8	$\dfrac{s^2 + gs + d}{(s - b)[(s - a)^2 + \omega_1^2]}$		$A = \dfrac{1}{\omega_1}\left[\dfrac{(a^2 - \omega_1^2 + ag + d)^2 + \omega_1^2(2a + g)^2}{(a - b)^2 + \omega_1^2}\right]^{1/2}$ $B = \dfrac{b^2 + bg + d}{(a - b)^2 + \omega_1^2}$ $\alpha = \arctan\dfrac{\omega_1(2a + g)}{a^2 - \omega_1^2 + ag + d} - \arctan\dfrac{\omega_1}{a - b}$

$$2.9 \qquad \frac{1}{s(s-b)[(s-a)^2+\omega_1^2]}$$

$$A=\frac{1}{\omega_1}\frac{1}{(a^2+\omega_1^2)^{1/2}[(a-b)^2+\omega_1^2]^{1/2}}$$

$$B=\frac{1}{b[(b-a)^2+\omega_1^2]}\qquad K=-\frac{1}{b(a^2+\omega_1^2)}$$

$$\alpha=-\arctan\frac{\omega_1}{a-b}-\arctan\frac{\omega_1}{a}$$

$$2.10 \qquad \frac{s+d}{s(s-b)[(s-a)^2+\omega_1^2]}$$

$$Ae^{at}\sin(\omega_1 t+\alpha)+Be^{bt}+K$$

$$A=\frac{1}{\omega_1(a^2+\omega_1^2)^{1/2}}\left[\frac{(d+a)^2+\omega_1^2}{(a-b)^2+\omega_1^2}\right]^{1/2}$$

$$B=\frac{b+d}{b[(b-a)^2+\omega_1^2]}\qquad K=-\frac{d}{b(a^2+\omega_1^2)}$$

$$\alpha=\arctan\frac{\omega_1}{a+d}-\arctan\frac{\omega_1}{a-b}-\arctan\frac{\omega_1}{a}$$

$$2.11 \qquad \frac{s^2+gs+d}{s(s-b)[(s-a)^2+\omega_1^2]}$$

$$A=\frac{1}{\omega_1}\left\{\frac{(a^2-\omega_1^2+ag+d)^2+\omega_1^2(2a+g)^2}{(a^2+\omega_1^2)[(a-b)^2+\omega_1^2]}\right\}^{1/2}$$

$$B=\frac{b^2+hg+d}{b[(b-a)^2+\omega_1^2]}\qquad K=-\frac{d}{b(a^2+\omega_1^2)}$$

$$\alpha=\arctan\frac{\omega_1(2a+g)}{a^2-\omega_1^2+ag+d}-\arctan\frac{\omega_1}{a-b}-\arctan\frac{\omega_1}{a}$$

TABLE 2-10 Laplace Transform Pairs Involving Rational Algebraic Functions $F(s) = D_1(s)/D(s)$ (Continued)

No.	$F(s)$	$f(t) (t > 0)$	
2.12	$\dfrac{1}{[(s-a)^2 + \omega_1^2](s^2 + \omega_2^2)}$		$A = \dfrac{1}{\omega_1} \dfrac{1}{[(a^2 + \omega_1^2 - \omega_2^2)^2 + 4a^2\omega_2^2]^{1/2}}$ $\qquad \alpha = -\arctan \dfrac{2a\omega_1}{a^2 - \omega_1^2 + \omega_2^2}$ $\\[2mm] B = \dfrac{1}{\omega_2} \dfrac{1}{[(a^2 + \omega_1^2 - \omega_2^2)^2 + 4a^2\omega_2^2]^{1/2}}$ $\qquad \beta = \arctan \dfrac{2a\omega_2}{a^2 + \omega_1^2 - \omega_2^2}$
2.13	$\dfrac{s + d}{[(s-a)^2 + \omega_1^2](s^2 + \omega_2^2)}$	$Ae^{at}\sin(\omega_1 t + \alpha) + B\sin(\omega_2 t + \beta)$	$A = \dfrac{1}{\omega_1}\left[\dfrac{(a+d)^2 + \omega_1^2}{(a^2 + \omega_1^2 - \omega_2^2)^2 + 4a^2\omega_2^2} \right]^{1/2}$ $\\[2mm] \alpha = \arctan \dfrac{\omega_1}{a + d} - \arctan \dfrac{2a\omega_1}{a^2 - \omega_1^2 + \omega_2^2}$ $\\[2mm] B = \dfrac{1}{\omega_2}\left[\dfrac{d^2 + \omega_2^2}{(a^2 + \omega_1^2 - \omega_2^2)^2 + 4a^2\omega_2^2} \right]^{1/2}$ $\\[2mm] \beta = \arctan \dfrac{\omega_2}{d} + \arctan \dfrac{2a\omega_2}{a^2 + \omega_1^2 - \omega_2^2}$
2.14	$\dfrac{s^2 + gs + d}{[(s-a)^2 + \omega_1^2](s^2 + \omega_2^2)}$		$A = \dfrac{1}{\omega_1}\left[\dfrac{(a^2 - \omega_1^2 + ag + d)^2 + \omega_1^2(2a + g)^2}{(a^2 + \omega_1^2 - \omega_2^2)^2 + 4a^2\omega_2^2} \right]^{1/2}$ $\\[2mm] \alpha = \arctan \dfrac{\omega_1(2a + g)}{a^2 - \omega_1^2 + ag + d} - \arctan \dfrac{2a\omega_1}{a^2 - \omega_1^2 + \omega_2^2}$ $\\[2mm] B = \dfrac{1}{\omega_2}\left[\dfrac{(d - \omega_2^2)^2 + g^2\omega_2^2}{(a^2 + \omega_1^2 - \omega_2^2)^2 + 4a^2\omega_2^2} \right]^{1/2}$ $\\[2mm] \beta = \arctan \dfrac{g\omega_2}{d - \omega_2^2} + \arctan \dfrac{2a\omega_2}{a^2 + \omega_1^2 - \omega_2^2}$
3.1	$\dfrac{1}{s^2}$	t	

3.2	$\dfrac{1}{(s-a)^2}$	$(A+A_1 t)e^{at}$	$A = 0 \quad A_1 = 1$
3.3	$\dfrac{s+d}{(s-a)^2}$		$A = 1 \quad A_1 = a+d$
3.4	$\dfrac{1}{s^2(s-a)}$	$Ae^{at} + K + K_1 t$	$A = -\dfrac{1}{a^2} \quad K = -A \quad K_1 = -\dfrac{1}{a}$
3.5	$\dfrac{s+d}{s^2(s-a)}$		$A = \dfrac{a+d}{a^2} \quad K = -A \quad K_1 = -\dfrac{d}{a}$
3.6	$\dfrac{s^2+gs+d}{s^2(s-a)}$		$A = \dfrac{a^2+ag+d}{a^2} \quad K = 1-A \quad K_1 = -\dfrac{d}{a}$
3.7	$\dfrac{1}{s(s-a)^2}$	$(A+A_1 t)e^{at} + K$	$A = -\dfrac{1}{a^2} \quad A_1 = \dfrac{1}{a} \quad K = -A$
3.8	$\dfrac{s+d}{s(s-a)^2}$		$A = -\dfrac{d}{a^2} \quad A_1 = \dfrac{a+d}{a} \quad K = -A$
3.9	$\dfrac{s^2+gs+d}{s(s-a)^2}$		$A = \dfrac{a^2-d}{a^2} \quad A_1 = \dfrac{a^2+ag+d}{a} \quad K = 1-A$
3.10	$\dfrac{1}{(s-a)^2(s-b)}$	$(A+A_1 t)e^{at} + Be^{bt}$	$A = -\dfrac{1}{(a-b)^2} \quad A_1 = \dfrac{1}{a-b} \quad B = -A$
3.11	$\dfrac{s+d}{(s-a)^2(s-b)}$		$A = -\dfrac{b+d}{(a-b)^2} \quad A_1 = \dfrac{a+d}{a-b} \quad B = -A$

3. The *product of the m × n matrix A ≡ [a_{ik}] by the scalar α* is the m × n matrix

$$\alpha A \equiv \alpha[a_{ik}] \equiv [\alpha a_{ik}]$$

4. The *product of the m × n matrix A ≡ [a_{ij}] and the n × r matrix B ≡ [b_{jk}]* is the m × r matrix

$$AB \equiv [a_{ij}][b_{jk}] \equiv \left[\sum_{j=1}^{3} a_{ij}b_{jk} \right]$$

In every matrix product *AB* the number *n* of columns of *A* must match the number of rows of *B* (*A* and *B* must be *comformable*). The existence of *AB* implies that of *BA* if and only if *A* and *B* are square matrices; in general *BA ≠ AB*. Note

$$A + B = B + A \qquad\qquad A + (B + C) = (A + B) + C$$
$$\alpha(\beta A) = (\alpha\beta)A \qquad\qquad \alpha(AB) = (\alpha A)B = A(\alpha B)$$
$$A(BC) = (AB)C$$
$$\alpha(A + B) = \alpha A + \alpha B \qquad\qquad (\alpha - \beta)A = \alpha A + \beta A$$
$$A(B + C) = AB + AC \qquad\qquad (B + C)A = BA + CA$$

33. Identities and Inverses. Note the following definitions:

1. The *m × n null matrix (additive identity)* [0] is the m × n matrix all of whose elements are equal to zero. Then

$$A + [0] = A \qquad [0]A = [0]$$
$$[0]B = C[0] = [0]$$

where *A* is any m × n matrix, *B* is any matrix having n rows, and *C* is any matrix having m columns.

2. The *additive inverse (negative)* −A of the m × n matrix A ≡ [a_{ik}] is the m × n matrix

$$-A \equiv (-1)A \equiv [-a_{ik}]$$

with *A + (− A) = A − A = [0]*.

3. The *identity matrix (unit, matrix, multiplicative identity) I* of order *n* is the n × n diagonal matrix with unit diagonal elements:

$$I \equiv [\delta_k^i] \qquad \text{where}^1 \qquad \delta_k^i = \begin{cases} 0 & \text{if } i \neq k \\ 1 & \text{if } i = k \end{cases}$$

Then

$$IB = B \qquad CI = C$$

where *B* is any matrix having n rows, and *C* is any matrix having n columns. For any n × n matrix *A*

$$IA = AI = A$$

4. A (necessarily square) matrix *A* is *nonsingular (regular)* if and only if it has a (necessarily unique) *multiplicative inverse* or *reciprocal A⁻¹* defined by

$$AA^{-1} = A^{-1}A = I$$

Otherwise *A* is a *singular* matrix.

A finite n × n matrix A ≡ [a_{ik}] is nonsingular if and only if det (A) ≡ det [a_{ik}] ≠ 0; in this case *A⁻¹* is the n × n matrix

$$A^{-1} \equiv [a_{ik}]^{-1} \equiv [A_{ki}'(\det [a]_{ik})]$$

where *A_{ik}* is the cofactor of the element *a_{ik}* in the determinant det [a_{ik}].

34. Matrix Notation for Simultaneous Linear Equations. A set of simultaneous linear equations

$$\sum_{k=1}^{n} a_{ik}x_k = b_i \qquad i = 1, 2, \ldots, m$$

is equivalent to the matrix equation

---

[1]The symbol δ_k^i (or δ_i^k) is known as the *Kronecker delta*.

$$Ax = b \qquad \text{or} \qquad \begin{bmatrix} a_{11} & a_{12} & \cdots & a_{1n} \\ a_{21} & a_{22} & \cdots & a_{2n} \\ \cdots & \cdots & \cdots & \cdots \\ a_{m1} & a_{m2} & \cdots & a_{mn} \end{bmatrix} \begin{bmatrix} x_1 \\ x_2 \\ \cdot \\ x_n \end{bmatrix} = \begin{bmatrix} b_1 \\ b_2 \\ \cdot \\ b_m \end{bmatrix} \qquad (2\text{-}26)$$

The unknowns x_k may be regarded as components of an unknown vector such that the transformation (2-26) yields the vector represented by the b_i. If, in particular, the matrix $[a_{ik}]$ is nonsingular (Par. **2-33**), then the matrix equation (2-26) can be solved to yield the unique result

$$x = A^{-1}b$$

which is equivalent to Cramer's rule.

35. Boolean Algebras. A *boolean algebra* is a class $\mathscr{S}$ of objects $A, B, C, \ldots$ admitting two binary operations, denoted as *(logical) addition and multiplication*, with the following properties.

For all A, B, C in $\mathscr{S}$
1. $\mathscr{S}$ contains $A + B$ and AB closure
2. $A + B = B + A$
   $\quad AB = BA$ commutative laws
3. $A + (B + C) = (A + B) + C$
   $\quad A(BC) = (AB)C$ associative laws
4. $A(B + C) = AB + AC$
   $\quad A + BC = (A + B)(A + C)$ distributive laws
5. $A + A = AA = A$ idempotency
6. $A + B = B$ if and only if $AB = A$ consistency

In addition,
7. $\mathscr{S}$ contains elements I and 0 such that, for every A in $\mathscr{S}$,

$$A + 0 = A \qquad A0 = 0 \qquad AI = A \qquad A + I = I$$

8. For every element A, $\mathscr{S}$ contains an element $\tilde{A}$ (*complement* of A, also written $\overline{A}$ or $I - A$) such that

$$A + \tilde{A} = I \qquad A\tilde{A} = 0$$

In every boolean algebra

$$A(A + B) \equiv A + AB \equiv A \qquad \text{laws of absorption}$$
$$(\overline{A + B}) \equiv \tilde{A}\tilde{B} \qquad (\widetilde{AB}) \equiv \tilde{A} + \tilde{B} \qquad \text{dualization, or De Morgan's laws}$$
$$\tilde{\tilde{A}} \equiv A \qquad \tilde{I} = 0 \qquad \tilde{0} = I$$
$$A + \tilde{A}B \equiv A + B \qquad AB + AC + B\tilde{C} \equiv AC + B\tilde{C}$$

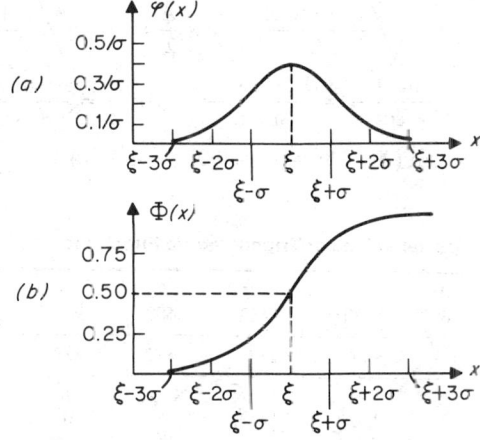

Fig. 2-2. (*a*) The normal frequency function; (*b*) the normal distribution function.

If $A + B = B$, one can write AB as $B - A$ (*complement of A with respect to B*). Two or more objects $A, B, C, \ldots$ of a boolean algebra are *disjoint* if and only if every product involving distinct elements of the set equals 0.

The symbols $\cup$ (cup) and $\cap$ (cap) are frequently employed to denote logical addition and multiplication in any boolean algebra, so that $A \cup B$ stands for $A + B$, and $A \cap B$ stands for AB.

FORMULAS FOR TRIGONOMETRIC AND HYPERBOLIC FUNCTIONS

36. Relations between the Trigonometric Functions. The basic relations

$$\sin^2 z + \cos^2 z = 1 \quad \text{and} \quad \frac{\sin z}{\cos z} = \tan z = \frac{1}{\cot z}$$

yield

$$\sin z = \pm\sqrt{1 - \cos^2 z} = \frac{\tan z}{\pm\sqrt{1 + \tan^2 z}} = \frac{1}{\pm\sqrt{1 + \cot^2 z}}$$

$$\cos z = \pm\sqrt{1 - \sin^2 z} = \frac{1}{\pm\sqrt{1 + \tan^2 z}} = \frac{\cot z}{\pm\sqrt{1 + \cot^2 z}}$$

$$\tan z = \frac{\sin z}{\pm\sqrt{1 - \sin^2 z}} = \frac{\pm\sqrt{1 - \cos^2 z}}{\cos z} = \frac{1}{\cot z}$$

$$\cot z = \frac{\pm\sqrt{1 - \sin^2 z}}{\sin z} = \frac{\cos z}{\pm\sqrt{1 - \cos^2 z}} = \frac{1}{\tan z}$$

37. Addition and Multiple-Angle Formulas. The basic relation

$$\sin (A + B) = \sin A \cos B + \sin B \cos A$$

yields

$$\sin (A \pm B) = \sin A \cos B \pm \cos A \sin B$$

$$\cos (A \pm B) = \cos A \cos B \mp \sin A \sin B$$

$$\tan (A \pm B) = \frac{\tan A \pm \tan B}{1 \mp \tan A \tan B} \qquad \cot (A \pm B) = \frac{\cot A \cot B \mp 1}{\cot A \pm \cot B}$$

$$\sin 2A = 2 \sin A \cos A$$

$$\cos 2A = \cos^2 A - \sin^2 A = 2 \cos^2 A - 1 = 1 - 2 \sin^2 A$$

$$\tan 2A = \frac{2 \tan A}{1 - \tan^2 A} \qquad \cot 2A = \frac{\cot^2 A - 1}{2 \cot A} = \tfrac{1}{2}(\cot A - \tan A)$$

$$\sin \frac{A}{2} = \pm\sqrt{\frac{1 - \cos A}{2}} \qquad \cos \frac{A}{2} = \pm\sqrt{\frac{1 + \cos A}{2}}$$

$$\tan \frac{A}{2} = \frac{\sin A}{1 + \cos A} = \frac{1 - \cos A}{\sin A} \qquad \cot \frac{A}{2} = \frac{\sin A}{1 - \cos A} = \frac{1 + \cos A}{\sin A}$$

$$a \sin A + b \cos A = r \sin (A + B) = r \cos (90° - A - B)$$

TABLE 2-11   Special Values of Trigonometric Functions

A (degrees)	0° 360°	30°	45°	60°	90°	180°	270°
A (radians)	0	$\pi/6$	$\pi/4$	$\pi/3$	$\pi/2$	π	$3\pi/2$
$\sin A$	0	$\frac{1}{2}$	$1/\sqrt{2}$	$\frac{1}{2}\sqrt{3}$	1	0	-1
$\cos A$	1	$\frac{1}{2}\sqrt{3}$	$1/\sqrt{2}$	$\frac{1}{2}$	0	-1	0
$\tan A$	0	$1/\sqrt{3}$	1	$\sqrt{3}$	$\pm\infty$	0	$\pm\infty$
$\cot A$	$\pm\infty$	$\sqrt{3}$	1	$1/\sqrt{3}$	0	$\pm\infty$	0

Table 2-12 Relations between Trigonometric Functions of Different Arguments

	$-A$	$90° \pm A$	$180° \pm A$	$270° \pm A$	$n360° \pm A$
sin	$-\sin A$	$\cos A$	$\mp \sin A$	$-\cos A$	$\pm \sin A$
cos	$\cos A$	$\mp \sin A$	$-\cos A$	$\pm \sin A$	$\cos A$
tan	$-\tan A$	$\mp \cot A$	$\pm \tan A$	$\mp \cot A$	$\pm \tan A$
cot	$-\cot A$	$\mp \tan A$	$\pm \cot A$	$\mp \tan A$	$\pm \cot A$

$$r = +\sqrt{a^2 + b^2} \qquad \tan B = \frac{b}{a}$$

$$\sin A \pm \sin B = 2 \sin \frac{A \pm B}{2} \cos \frac{A \mp B}{2}$$

$$\cos A + \cos B = 2 \cos \frac{A + B}{2} \cos \frac{A - B}{2}$$

$$\cos A - \cos B = -2 \sin \frac{A + B}{2} \sin \frac{A - B}{2}$$

$$\tan A \pm \tan B = \frac{\sin (A \pm B)}{\cos A \cos B} \qquad \cot A \pm \cot B = \frac{\sin (B \pm A)}{\sin A \sin B}$$

$$2 \cos A \cos B = \cos (A - B) - \cos (A + B)$$
$$2 \sin A \sin B = \cos (A - B) - \cos (A + B)$$
$$2 \sin A \cos B = \sin (A - B) + \sin (A + B)$$
$$2 \cos^2 A = 1 + \cos 2A \qquad 2 \sin^2 A = 1 - \cos 2A$$

38. Hyperbolic Functions

$$\sinh z = \frac{e^z - e^{-z}}{2} \qquad \cosh z = \frac{e^z + e^{-z}}{2} \qquad \tanh z = \frac{\sinh z}{\cosh z} \qquad \coth z = \frac{\cosh z}{\sinh z}$$

$$\operatorname{sech} z = \frac{1}{\cosh z} \qquad \operatorname{csch} z = \frac{1}{\sinh z} \qquad \cosh^2 z - \sinh^2 z = 1 \qquad \frac{\sinh z}{\cosh z} = \tanh z = \frac{1}{\coth z}$$

yield

$$\sinh z = \pm \sqrt{\cosh^2 z - 1} = \frac{\tanh z}{\pm \sqrt{1 - \tanh^2 z}} = \frac{1}{\pm \sqrt{\coth^2 z - 1}}$$

$$\cosh z = \pm \sqrt{1 + \sinh^2 z} = \frac{1}{\pm \sqrt{1 - \tanh^2 z}} = \frac{\coth z}{\pm \sqrt{\coth^2 z - 1}}$$

$$\tanh z = \frac{\sinh z}{\pm \sqrt{1 + \sinh^2 z}} = \frac{\pm \sqrt{\cosh^2 z - 1}}{\cosh z} = \frac{1}{\coth z}$$

$$\coth z = \frac{\pm \sqrt{1 + \sinh^2 z}}{\sinh z} = \frac{\cosh z}{\pm \sqrt{\cosh^2 z - 1}} = \frac{1}{\tanh z}$$

39. Formulas Relating Hyperbolic Functions of Compound Arguments

$$\sinh (A \pm B) = \sinh A \cosh B \pm \cosh A \sinh B$$
$$\cosh (A \pm B) = \cosh A \cosh B \pm \sinh A \sinh B$$

$$\tanh (A \pm B) = \frac{\tanh A \pm \tanh B}{1 \pm \tanh A \tanh B} \qquad \coth (A \pm B) = \frac{\coth A \coth B \pm 1}{\coth B \pm \coth A}$$

$$\sin 2A = 2 \cosh A \sinh A \qquad \cosh 2A = \cosh^2 A + \sinh^2 A$$

$$\tanh 2A = \frac{2 \tanh A}{1 + \tanh^2 A} \qquad \coth 2A = \frac{\coth^2 A + 1}{2 \coth A}$$

$$\sinh \frac{A}{2} = \pm \sqrt{\frac{\cosh A - 1}{2}} \qquad \cosh \frac{A}{2} = \pm \sqrt{\frac{\cosh A + 1}{2}}$$

$$\tanh \frac{A}{2} = \frac{\sinh A}{\cosh A + 1} = \frac{\cosh A - 1}{\sinh A} \qquad \coth \frac{A}{2} = \frac{\sinh A}{\cosh A - 1} = \frac{\cosh A + 1}{\sinh A}$$

$$\sinh A \pm \sinh B = 2 \sinh \frac{A \pm B}{2} \cosh \frac{A \mp B}{2}$$

$$\cosh A + \cosh B = 2 \cosh \frac{A + B}{2} \cosh \frac{A - B}{2}$$

$$\cosh A - \cosh B = 2 \sinh \frac{A + B}{2} \sinh \frac{A - B}{2}$$

$$\tanh A \pm \tanh B = \frac{\sinh (A \pm B)}{\cosh A \cosh B} \qquad \coth A \pm \coth B = \frac{\sinh (B \pm A)}{\sinh A \sinh B}$$

$$2 \cosh A \cosh B = \cosh (A + B) + \cosh (A - B)$$
$$2 \sinh A \sinh B = \cosh (A + B) - \cosh (A - B)$$
$$2 \sin A \cosh B = \sinh (A + B) + \sinh (A - B)$$
$$2 \cosh^2 A = 1 + \cosh 2A \qquad 2 \sinh^2 A = \cosh 2A - 1$$

40. Relations between Exponential, Trigonometric, and Hyperbolic Functions

$$e^{iz} = \cos z + i \sin z \qquad \cos z = \frac{e^{iz} + e^{-iz}}{2} \qquad \sin z = \frac{e^{iz} - e^{-iz}}{2i}$$

$$e^{-iz} = \cos z - i \sin z \qquad e^z = \cosh z + \sinh z \qquad e^{-z} = \cosh z - \sinh z$$

$$\cosh z = \frac{e^z + e^{-z}}{2} \qquad \sinh z = \frac{e^z - e^{-z}}{2}$$

$$\cos z = \cosh iz \qquad \cosh z = \cos iz \qquad \sin z = -i \sinh iz \qquad \sinh z = -i \sin iz$$
$$\tan z = -i \tanh iz \qquad \tanh z = -i \tan iz \qquad \cot z = i \coth iz \qquad \coth z = i \cot iz$$
$$\log_e z = \log_e |z| + i \arg (z) \qquad \log_e ix = \log_e x + (2n + \tfrac{1}{2})\pi i \qquad n = 0, \pm 1, \pm 2, \ldots$$
$$\log_e (-x) = \log_e x + (2n + 1)\pi i \qquad n = 0, \pm 1, \pm 2, \ldots$$

$$\arccos z = i \cosh^{-1} z \qquad \cosh^{-1} z = i \arccos z$$
$$\arcsin z = -i \sinh^{-1} iz \qquad \sinh^{-1} z = -i \arcsin iz$$
$$\arctan z = -i \tanh^{-1} iz \qquad \tanh^{-1} z = -i \arctan iz$$
$$\text{arccot } z = i \coth^{-1} iz \qquad \coth^{-1} z = i \text{ arccot } iz$$

$$\arccos z = i \log_e (z + i\sqrt{1 - z^2}) \qquad \cosh^{-1} z = \log_e (z + \sqrt{z^2 - 1})$$
$$\arcsin z = -i \log_e (iz + \sqrt{1 - z^2}) \qquad \sinh^{-1} z = \log_e (z + \sqrt{z^2 + 1})$$

$$\arctan z = -\frac{i}{2} \log_e \frac{1 + iz}{1 - iz} \qquad \tanh^{-1} z = \tfrac{1}{2} \log_e \frac{1 + z}{1 - z}$$

$$\text{arccot } z = -\frac{i}{2} \text{ lob}_e \frac{iz - 1}{iz + 1} \qquad \coth^{-1} z = \tfrac{1}{2} \log_e \frac{z + 1}{z - 1}$$

41. Power-Series Expansions

$$\frac{1}{1 - z} = 1 + z + z^2 + \cdots \qquad |z| < 1 \qquad \text{geometric series}$$

$$(1 + z)^p = 1 + \binom{p}{1} z + \binom{p}{2} z^2 + \cdots \qquad |z| < 1 \qquad \text{binomial series}$$

$$e^z = 1 + z + \frac{z^2}{2!} + \frac{z^3}{3!} + \cdots \qquad z \neq \infty$$

$$\sin z = z - \frac{z^3}{3!} + \frac{z^5}{5!} \mp \cdots \qquad \cos z = 1 - \frac{z^2}{2!} + \frac{z^4}{4!} \mp \cdots \qquad z \neq \infty$$

$$\sinh z = z + \frac{z^3}{3!} + \frac{z^5}{5!} + \cdots \qquad \cosh z = 1 + \frac{z^2}{2!} + \frac{z^4}{4!} + \cdots \qquad z \neq \infty$$

$$\log_e (1 + z) = z - \frac{z^2}{2} + \frac{z^3}{3} - \frac{z^4}{4} \pm \cdots \qquad |z| < 1$$

$$\arcsin z = z + \frac{1}{2}\frac{z^3}{3} + \frac{1\,3}{2\,4}\frac{z^5}{5} + \frac{1\,3\,5}{2\,4\,6}\frac{z^7}{7} + \cdots$$

$$\sinh^{-1} z = z - \frac{1}{2}\frac{z^3}{3} + \frac{1\,3}{2\,4}\frac{z^5}{5} - \frac{1\,3\,5}{2\,4\,6}\frac{z^7}{7} \pm \cdots \qquad |z| < 1$$

$$\arctan z = z - \frac{z^3}{3} + \frac{z^5}{5} \mp \cdots$$

$$\tanh^{-1} z = \tfrac{1}{2} \log_e \frac{1 + z}{1 - z} = z + \frac{z^3}{3} + \frac{z^5}{5} + \cdots \qquad |z| < 1$$

SIGNAL-ANALYSIS FORMULAS

42. Delta "Functions" and Dirac Combs. Limits (especially infinite series and indefinite integrals) appearing under an integral sign can often be represented in terms of the *Dirac delta "function"* $\delta(t)$ with the properties

$$\int_a^b f(\lambda)\delta(t - \lambda)\, d\lambda = \begin{cases} 0 & t < a \text{ or } t > b \\ \frac{1}{2}f(t + 0) & t = a \\ \frac{1}{2}f(t - 0) & t = b \\ \frac{1}{2}[f(t - 0) + f(t + 0)] & a < t < b \end{cases} \tag{2-27}$$

where $f(t)$ is of bounded variation in every finite interval. In view of the sampling property [Eq. (2-27)], sequences $f(k\,\Delta t)$ $(k = 0, \pm 1, \pm 2, \dots)$ of *sampled data* are often modeled with the aid of the *Dirac comb*

$$\mathrm{Comb}_{\Delta t}(t) = \Delta t \sum_{k=-\infty}^{\infty} \delta(t - k\,\Delta t) = \sum_{k=-\infty}^{\infty} e^{jk(2\pi/\Delta t)t} \tag{2-28}$$

Formulas using such *symbolic functions* are often a useful shorthand notation, but they must always be checked by other means, since $\delta(t)$ (= 0 everywhere, except ∞ at $t = 0$) and $\mathrm{Comb}_{\Delta t}(t)$ are not really properly defined functions.

43. Repetition Operator and Fourier Transforms. *Periodic repeated waveforms* can be represented with the aid of the *repetition operator* (rep_T)

$$\mathrm{rep}_T\, f(t) = \sum_{k=-\infty}^{\infty} f(t - k\,\Delta t) \tag{2-29}$$

and we have the useful symbolic (heuristic) Fourier-transform pairs

$$\left. \begin{array}{l} \mathscr{F}[\mathrm{rep}_T\, f(t)] = F(j\omega)\, \mathrm{Comb}_{2\pi/T}(\omega) \\[6pt] \mathscr{F}[f(t)\, \mathrm{Comb}_{\Delta t}(t)] = \mathrm{rep}_{2\pi/\Delta t}\, F(j\omega) \end{array} \right\} \quad F(j\omega) = \int_{-\infty}^{\infty} f(t)e^{-j\omega t}dt \tag{2-30}$$

The Fourier transform of the *impulse-sampler output* $f(t)\, \mathrm{Comb}_{\Delta t}(t)$ superimposes repeated versions of $F(j\omega)$; if $F(j\omega) = 0$ for $\omega \geq 2\pi B = \pi/\Delta t$ *(Nyquist circular frequency)* then the samples $f(k\,\Delta t)$ determine $F(j\omega)$ and thus $f(t)$ uniquely and

$$f(t) = \sum_{k=-\infty}^{\infty} f(k\,\Delta t)\, \frac{\sin[2\pi B(t - k\,\Delta t)]}{2\pi B(t - k\,\Delta t)} \tag{2-31}$$

(This is the *Nyquist-Kotelnikov-Shannon theorem for band-limited functions.*)
The Fourier transform of a *finite pulse train* is

$$\mathscr{F}\left[\sum_{k=-(n-1)/2}^{(n-1)/2} f(t - kT) \right] = F(j\omega)\, \frac{\sin(n\omega T/2)}{\sin(\omega T/2)} \tag{2-32}$$

The fraction on the right is a periodic function which approximates a Dirac comb as $n \to \infty$.

44. Digital Fourier Analysis: Effects of Sampling and Finite Record Length. Digital computers approximate the Fourier transform $F(j\omega)$ by the sampled-data approximation:

$$\hat{F}(j\omega) = \Delta t \sum_{k=-(n-1)/2}^{(n-1)/2} f(k\,\Delta t)e^{-j\omega k\,\Delta t} = \int_{-\infty}^{\infty} \hat{f}(t)e^{-j\omega t}\, dt \tag{2-33}$$

$$= \int_{-\infty}^{\infty} P[j(\omega - \lambda)]F(j\lambda)\, \frac{d\lambda}{2\pi}$$

$P(j\omega)$ is a finite-bandwidth *comb-filter frequency-response function*

$$P(j\omega) = T\, \mathrm{rep}_{2\pi/\Delta t}\, \frac{\sin(\omega t/2)}{\omega t/2} \tag{2-34}$$

Thus, the correct transform $F(j\omega)$ is replaced by the weighted average [Eq. (2-33)]. Errors due to *sampling (aliasing)* correspond to the repetitive nature of the weighting function [Eq. (2-34)]; the "sharpness" of each filter peak improves with increasing observation time T.

45. Bibliography

MATHEMATICAL REFERENCE BOOKS

Abramowitz, M., and I. A. Stegun   "Handbook of Mathematical Functions," National Bureau of Standards, Washington, 1964.

Bierens De Hann, D.   "Nouvelles Tables d'intégrales définies," Stechert, New York, 1939.

Byrd, P. F., and M. D. Friedman   "Handbook of Elliptic Integrals for Engineers and Physicists," 2d ed., Springer, New York, 1971.

Gröbner and Hofreiter   "Integral Tafel," 2d ed., Springer, Vienna, 1958.

Korn, G. A., and T. M. Korn   "Mathematical Handbook for Scientists and Engineers," 2d ed., McGraw-Hill, New York, 1968.

Luke, Y. L.   "Mathematical Functions and Their Approximations," Academic, New York, 1975.

Lindman, C. F.   "Examen des novelles tables d'intégrales définies de M. Bierens de Haan," 1944.

Luke, Y. I.   "Integrals of Bessel Functions," McGraw-Hill, New York, 1962.

Petit Bois, G.   "Tables of Indefinite Integrals," Dover, New York, 1961.

Ryshik, I. M., and I. S. Gradstein   "Tables of Series, Products, and Integrals," Academic, New York, 1964.

SHORT TABLES OF TRANSCENDENTAL FUNCTIONS

Abramowitz, M., and I. A. Stegun   "Handbook of Mathematical Functions," National Bureau of Standards, Washington, D.C., 1964.

Dwight, H. B.   "Tables of Integrals and Other Mathematical Data," 4th ed., Macmillan, New York, 1961.

Flügge, W.   "Four-Place Tables of Transcendental Functions," McGraw-Hill, New York, 1954.

Jahnke, E., and F. Emde   "Tables of Functions with Formulae and Curves," 6th ed., McGraw-Hill, New York, 1960.

STATISTICAL TABLES

Beyer, W. H.   "CRC Handbook of Tables for Probability and Statistics," 2d ed., Chemical Rubber Co., Cleveland, Ohio, 1968.

Burington, R. S., and D. C. May   "Handbook of Probability and Statistics," 2d ed., McGraw-Hill, New York, 1967.

Hald, A.   "Statistical Tables and Formulas," Wiley, New York, 1952.

Meredith, W.   "Mathematical and Statistical Tables," McGraw-Hill, New York, 1967.

Owen, D. B.   "Handbook of Statistical Tables," Addison-Wesley, Reading, Mass., 1962.

Pearson, E. S., and H. O. Hartley   "Biometrika Tables for Statisticians," Cambridge University Press, New York, 1956.

INDEXES TO NUMERICAL TABLES

Etherington, Harold (ed.)   "Nuclear Engineering Handbook," McGraw-Hill, New York, 1958.

Fletcher, A.   Guide to Tables of Elliptic Functions,   "Mathematical Tables and Other Aids to Computation," Vol. 3, No. 24, 1948.

——, J. C. P. Miller, and L. Rosenhead   "Index of Mathematical Tables," Addison-Wesley, Reading, Mass., 1962.

Greenwood, J. A., and H. O. Hartley   "Guide to Tables in Mathematical Statistics," Princeton University Press, Princeton, N.J. 1962.

Section 3

Circuit Principles

WILS L. COOLEY *Professor of Electrical Engineering, West Virginia University; Senior Member, IEEE*[1]

CONTENTS

Numbers refer to paragraphs

[1]The author acknowledges with gratitude the contributions of the late Professor Everard M. Williams to this section of the "Handbook" in the first edition, on which this revision is in part based.

ELECTRIC-CIRCUIT CONCEPTS

1. Electric-circuit theory deals with the behavior of electric apparatus (components, devices, equipment, systems, etc.) as measured by the values and fluctuations of the *circuit variables*, the voltages between various terminals and between various conductors and the currents at and in various terminals and conductors.

Electric-circuit theory has two principal divisions: *equivalent circuits* for such circuit components and devices as can be treated by circuit theory and the *mathematical analysis* of the equivalent circuits.

For digital systems, the final equivalent circuit constitutes a logic diagram. In either case, the equivalent circuit represents a concise summary of the elements of a mathematical model for the behavior of the apparatus.

2. Equivalent-circuit theory is also widely known as *network theory*; the term *network*, in fact, is often applied synonymously with *circuit*. However, simple configurations of apparatus are more commonly known as circuits, while more complicated configurations (many connections and many devices or elements) are more likely to be referred to as networks. In network *topology*, the field concerned with the geometric configurations of networks, the term *circuit* often designates a *loop*.

Equivalent circuits are classed as *linear* or *nonlinear*, *reciprocal* or *nonreciprocal*, and *unilateral* or *bilateral*. There are various (essentially synonymous) descriptions for the property of linearity, one of which is as follows: if the response of a circuit, circuit element, etc., to an input $f_1(t)$ is $g_1(t)$ and its response to an input $f_2(t)$ is $g_2(t)$, where $f_1(t)$ and $f_2(t)$ are arbitrary functions (including arbitrary amplitude), the circuit is *linear* if its response to an input $f_1(t) + f_2(t)$ is equal to $g_1(t) + g_2(t)$. This description is an illustration of the *superposition principle*.

A *bilateral* circuit is one in which the circuit behavior is not changed if the terminal connections of the elements are interchanged. A rectifier conducts current easily in one direction and with difficulty in the other and is therefore *unilateral*. Other typical unilateral devices are vacuum tubes and transistors. Linear circuits are necessarily bilateral.

Unless specifically stated otherwise, the treatment of analog equivalent circuits at every point in this section is limited to circuits which are linear. The question of linearity does not arise in the logic diagrams, which are mathematical models of digital circuits.

Reciprocal circuits are circuits which satisfy the reciprocity theorem (see Par. **3-5**). All linear bilateral equivalent circuits are reciprocal unless they contain *gyrators*.

3. Circuit Elements and Configurations. Equivalent circuits are composed of interconnected basic building blocks, e.g., resistances, capacitances, inductances, and voltage and current sources. The following, with examples given in Figs. 3-1 and 3-2, are common terms for describing circuits. *Elements*, or *circuit elements*, are the smallest units into which circuits can be subdivided; in Fig. 3-1, examples of *two-terminal* circuit elements are resistances R_1, R_2, . . ., R_6 and capacitance C_3; magnetically coupled inductors L_a and L_b are an example of a *four-terminal* circuit element.

Any point on an element, component, device, etc., to which a conductor is attached is called a *terminal*. Any portion of a circuit between two terminals, like that between terminals a and b, b and e, or f and b, is a *branch*. The portion between a and f is also a branch because the induced voltage in L_b caused by current in L_a acts in series with the circuit.

A point at which conductors from two or more elements join is sometimes called a *node*, but this term is more often restricted to points at which conductors from three or more branches join.

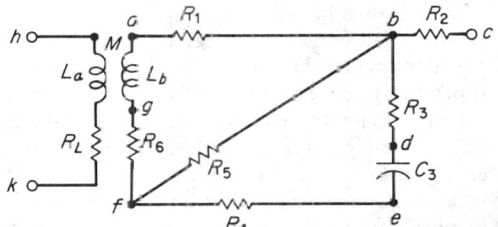

Any specified sequence of branches, e.g., in Fig. 3-1 from a to b to c or from a to b to d to e, is a *path*. Any path, such as a to b to f to a or a to b to d to e to f to a, which ends at the point at which it starts is a *closed path*, also known in circuit theory (in contrast with network toplogy) as a *loop*.

Fig. 3-1. Example of an equivalent circuit. Inductances L_a and L_b are magnetically coupled. Each is associated with one of the two separate parts in this circuit, i.e., with parts not physically connected with each other.

An equivalent circuit which can be laid out on a plane without any crossing of conductors is a *planar circuit*. Any loop in a circuit which does not enclose another loop, such as loops *abfga* and *bdefb* in Fig. 3-1, is also a *mesh*, also known as a *window* (the term *mesh* is sometimes limited to planar circuits).

Any element which receives energy from some source other than a signal generator is an *active element*. The term is also sometimes applied to any element which supplies energy to the rest of the circuit. Any element to which energy is supplied by other parts of the circuit is a *passive element*, also, in principle, a *load*, although the latter term is often applied only to a passive element to which energy is intentionally supplied.

Many circuits transmit, shape, amplify, or modify the energy (or signals) supplied by a source before delivering the energy (or signals) onward to a load. Each pair of terminals of such a circuit to which a source or load is connected is termed a *port*, an appropriate terminology because such terminal pairs acts as points of entry or exit from the circuit. In a circuit with but one port, the circuit itself is the load; circuits with two ports commonly have an input and output port. A circuit may have n ports; when only one port of a n-port circuit is an input, this port is also known as the *driving point* (see Fig. 3-2).

4. Linear-Circuit Analysis. Once an equivalent circuit is formulated, circuit theory provides a method for calculating the behavior of the currents and voltages. For a circuit with n elements, there are, to start with, $2n$ unknowns, since there is a voltage drop and current associated with each element. When the known voltage-drop-vs.-current characteristics of each element are taken into account, there remain n unknowns.

The n equations required to determine these unknowns are obtained by applying Kirchhoff's laws. *Kirchhoff's current law* (KCL) states that the algebraic sum of the instantaneous currents entering a node (in this case a node should be the common terminal of three or more branches) is zero. A necessary and sufficient condition for independence of the resulting current equations

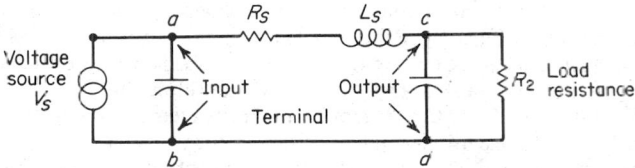

Fig. 3-2. Example of a complete equivalent circuit with source and load. Terminal pairs a, b and c, d are input and output terminals, or ports, respectively.

is that each equation contain at least one current which does not appear in another current-law equation. *Kirchhoff's voltage law* (KVL) states that the algebraic sum of the instantaneous voltages around any loop is zero. The number of independent voltage-law equations is equal to the number of meshes in the circuit.

In applying Kirchhoff's laws, two strategies are used in writing final circuit equations to minimize the number of simultaneous equations to be solved. In *loop (or mesh) analysis*, the currents in the various branches are treated as unknowns, and the voltage drops are written in terms of these currents, with only the minimum number of unknown currents necessary (on the basis of KCL) specified at each node. In *nodal analysis*, the branch voltages are treated as the unknowns, with only the minimum number of unknown voltages necessary (on the basis of KVL) specified in each loop.

The result of either strategy for writing the individual equations is a system of simultaneous equations which can be solved to obtain a single equation in a single current or voltage as a function of the parameters of the circuit elements and the source or sources. If the original equations have been written with the instantaneous voltages $v_i(t)$ or currents $i_i(t)$ as variables, this single equation is a differential equation *of order equal to the number of independent energy-storing circuit elements in the circuit*. If the original equations have been written with voltages or currents in *sinor* form (for which the properties of the elements necessarily appear in *phasor* form), the single equation is an algebraic equation.

Solution of the differential equation referred to above, with appropriate initial conditions (which depend upon the initial energy states of the energy-storing circuit elements), yields an equation in the dependent variable or variables (voltages or currents) as a function of time; it contains a *transient* component, which generally decays with time (unless the circuit is unstable), and a *steady-state* component, which continues indefinitely. Solution of the algebraic equation with sinor and phasor terms yields an equation for only the steady-state behavior of the dependent variable or variables. Except for first- or second-order equations, the differential-equation solutions for most circuits, however, are formidably difficult and rarely feasible without numerical values of the circuit elements, which makes it difficult to test various values of design parameters in order to obtain a desired performance. For high-order differential equations the most practicable solution is with an analog computer or an analog simulation on a digital computer. The sinor and phasor steady-state solution, on the other hand, is straightforward, albeit possibly algebraically complicated, regardless of the circuit complexity.

5. Network-Analysis Theorems. The analysis of linear circuits can be carried out using only Kirchhoff's current law and Kirchhoff's voltage law in conjunction with the defining relationships for each branch. The efforts can be significantly reduced, however, by the application of shortcuts in the form of *network theorems*.

(a) *The Thevenin theorem* states that insofar as the behavior of the circuit at its terminals is concerned, such a circuit can be replaced by a single sinor voltage source E in series with an impedance Z. The Thevenin equivalent source is illustrated in Fig. 3-3a; the value of the E is the open-circuit voltage developed between the terminals a and b, and Z can be found from the impedance observed between terminals a and b when all voltage sources are short-circuited and all current sources are open-circuited. In a dc case, E is a constant dc voltage and Z becomes a resistance R.

(b) *The Norton theorem* states that insofar as the circuit behavior at its terminals is concerned, such a circuit can be replaced by a sinor current source I in parallel with an impedance Z. The Norton equivalent source is shown in Fig. 3-3b; the value of I is equal to the short-circuit current between terminals a and b, and the impedance Z is equal to the Thevenin impedance Z and is found in the same way. In the dc case, the Norton source becomes a constant dc current, and Z becomes a resistance R.

In certain simple cases, in which Z reduces to a non-frequency-dependent parameter, the Thevenin and Norton theorems are applicable in transient as well as steady-state calculations.

(c) *Superposition Theorem.* In a linear network containing more than one independent source, the voltage across or the current through any element can be computed as follows.

1. Replace all the sources except one by their internal impedances (ideal current sources are replaced by open circuits, and ideal voltage sources by short circuits).

2. Compute the voltage across or the current through the element due to the single source.

3. Repeat steps 1 and 2 for each independent source in turn.

4. The sum of all the individual calculated voltages or currents provides the actual value of voltage or current for the complete circuit.

(d) *The compensation theorem*, also known as the *substitution theorem*, states that any branch in a circuit can be replaced by a substitute branch, so long as the branch voltage and current remain the same, without affecting the voltages and currents in other branches of the circuit.

(e) *Reciprocity Theorem*

$$I_j/V_K = I_K/V_j$$

In any linear, time-independent network with independent sources, the ratio of the current in a short circuit at one part of the network to the output of a voltage source at another part of the network does not change if the positions of the source and the short circuit are interchanged. Reciprocity also applies to a current source and the voltage developed across an open circuit.

(f) *The maximum-power-transfer condition*, sometimes termed the maximum-power-transfer theorem, describes the condition under which maximum possible power will be delivered to a load by a linear source. There are certain relations between Z_L and Z_i (Fig. 3-4) for which maximum power will be transferred from the source E to the load Z_L.

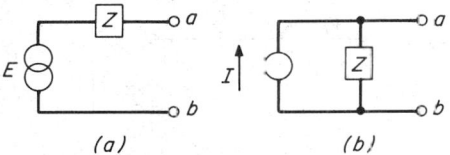

Fig. 3-3. (a) The Thevenin equivalent source and (b) the Norton equivalent source.

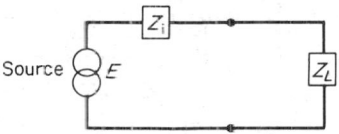

Fig. 3-4. Equivalent circuit of a source (represented by its Thevenin equivalent sinor voltage E and phasor impedance Z_i) connected to a load (represented by its impedance phasor Z_L).

If the source is dc and Z_i and Z_L are resistances R_i and R_L, respectively, maximum power transfer will occur for $R_i = R_L$. If the source is a steady-state sinusoidal voltage source and the load impedance Z_L can be varied in any way whatever, maximum power transfer occurs for a phasor value of Z_L given by $Z_L = Z_i^*$, where the phasor Z_i^* is the *complex-conjugate impedance* of the phasor Z_i; that is, if $Z_i = R_i + jX_i$, then $Z_L = R_i - jX_i$. *This yields the maximum possible power transfer under ac conditions.*

(g) *Miller's Theorem.* If an admittance Y is connected between the input and output terminals of a two-port network as shown in Fig. 3-5a, it can be replaced by $Y_1 = Y/(1 - K)$ and $Y_2 = Y(K - 1)/K$, referred to ground, without altering the properties of the circuit. K is defined as V_2/V_1. If an impedance Z is connected to ground as in Fig. 3-5b, it can be replaced by $Z_1 = Z(1 - A)$ and $Z_2 = Z[(A - 1)/A]$ without altering the properties of the circuit. $A = -I_2/I_1$.

(h) *Millman's Theorem* (see Fig. 3-6). Two or more nonideal voltage sources in parallel or nonideal current sources in series can be combined into a single equivalent source, where

$$E = \frac{E_1/Z_1 + E_2/Z_2}{1/Z_1 + 1/Z_2} \qquad Z = \frac{1}{1/Z_1 + 1/Z_2} \qquad I = \frac{I_1/Y_1 + I_2/Y_2}{1/Y_1 + 1/Y_2} \qquad Y = \frac{1}{1/Y_1 + 1/Y_2}$$

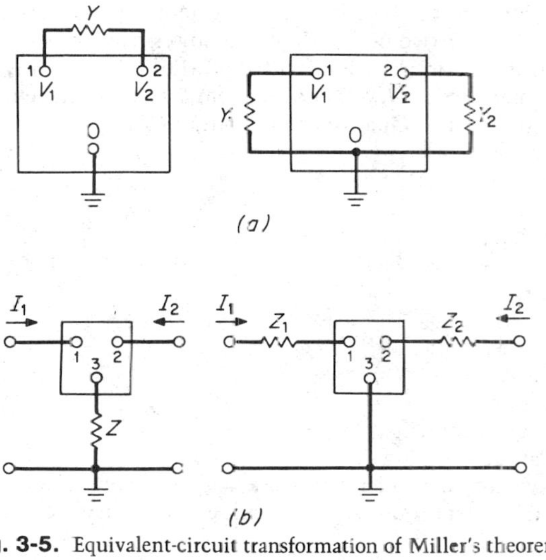

Fig. 3-5. Equivalent-circuit transformation of Miller's theorem.

(i) *Ohm's Law*

$$V = IR$$

When the current in a conductor is steady and there are no sources of emf within the conductor, the voltage V between terminals of the conductor is proportional to the current I; the proportionality constant R is called resistance.

(j) *Joule's Law*

$$P = i^2R = v^2/R$$

If a circuit contains only an ohmic resistance, that is, $R = V/I$ for all values of V, the instantaneous power dissipated in the resistance is equal to the square of the instantaneous current

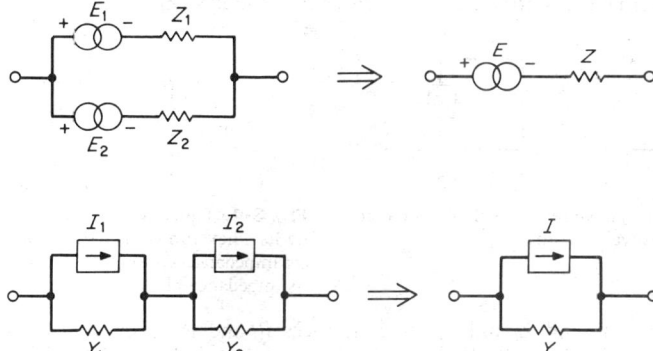

Fig. 3-6. Millman's reduction.

flow through the resistor times its resistance or, alternately, equal to the square of the instantaneous voltage applied divided by its resistance.

(k) *Tellegen's Theorem*

$$\sum_{k=1}^{n} i_k v_k = 0$$

In any lumped network which contains any type of elements, the sum of the products of all the instantaneous branch voltages times their respective instantaneous branch currents is instantaneous power, and therefore the result is that the sum of the instantaneous power in all parts of a circuit is zero, which is consistent with the principle of conservation of energy.

6. Dual Networks. The property of *duality* is extremely important in network analysis. One network is said to be the dual of another if they are described by the same equations except that all voltages have been replaced by currents, all currents by voltages, all resistances by conductances, etc. Figure 3-7 shows two networks which possess the property of duality.

For an arbitrary realizable network N and its dual network $\overline{N}$, if S is any true statement about N, then $\overline{S}$ is a true statement about $\overline{N}$. $\overline{S}$ is derived from S by replacing every electrical quantity or network description by its dual. Dual words and phrases are

Voltage	Current
Resistance	Conductance
Capacitance	Inductance
Impedance	Admittance
Reactance	Susceptance
Node	Loop
KVL	KCL
Short circuit	Open circuit
Charge	Flux
Tree branch	Chord
Node voltage	Loop current
Series addition	Parallel addition

Dual-network relationships are difficult to work with for circuits containing coupled inductors, transformers, and dependent sources. The circuit elements may be nonlinear and time-varying, however.

CIRCUIT ELEMENTS

7. Passive Lumped Parameters. Except for circuits in which the physical distance from one element to another is comparable to the wavelength λ of the signal being processed ($\lambda \approx c/f$, where c is the speed of light and f is the signal frequency), networks can be analyzed as an interconnected set of *lumped elements*. These elements form the branches of the circuit and have relatively simple defining equations relating the current through them to the voltage placed across them. If these relations do not change with time and do not supply energy to the

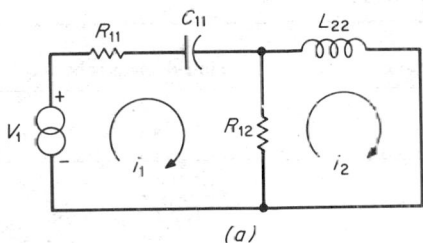

(a)

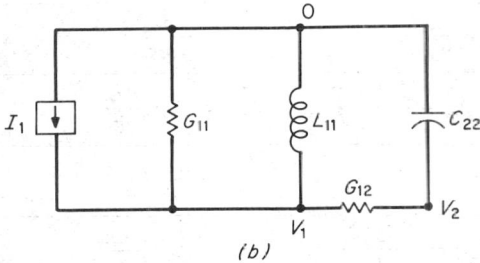

(b)

Fig. 3-7. Dual networks: (a) Loop analysis yields equations $V_1 = (R_{11} + R_{12} + 1/sC_{11})i_1 - R_{12}i_2$ and $0 = -R_{12}i_1 + (sL_{22} + R_{12})i_2$. (b) Node analysis yields equations $I_1 = (G_{11} + G_{12} + 1/sL_{11})V_1 - G_{12}V_2$ and $0 = -G_{12}V_1 + (sC_{22} + G_{12})V_2$.

network, they are constant and passive. Such passive lumped-constant elements form the major portion of most networks. The defining equations for them are given in Table 3-1, which also shows relationships for analyzing time-domain behavior and frequency-domain behavior, detailed in Pars. **3-17** and **3-21**.

Negative Parameters. Circuit elements, circuits, and devices which can be represented in the equivalent circuit by *negative resistance, negative inductance,* and/or *negative capacitance* are quite useful. Negative *incremental* resistance occurs in passive devices such as tunnel diodes and thermistors. Negative inductances and capacitances, however, can be realized only with circuits containing active devices.[1]

Quality Factor. Except for the resistance parameter in steady-state dc circuits, no circuit parameter is present as the sole parameter of a circuit element. The equivalent circuit of an inductor, for example, must include with its inductance some resistance to correspond to the power losses in the windings (and also in the core if a core is used). To describe the *quality* of coils and capacitors, a *quality factor* is used, defined as

$$Q_L = 2\pi f L/R(f) \qquad Q_c = 1/2\pi f C R(f)$$

in which L, C, and $R(f)$ are the equivalent series inductance, series capacitance, and series resistance measured at the frequency f for which the corresponding quality factor Q is specified.

8. Active Lumped Parameters. The equivalent circuits of active networks are constructed from the passive-circuit building blocks, defined in Table 3-1, the so-called voltage and current sources, various incremental parameters peculiar to each type of device, and/or from their graphical characteristics.

The active devices, such as transistors, electron tubes, converter diodes, and parametric diodes,

---

[1] J. L. Merrill, Theory of the Negative-Impedance Converter, *Bell Syst. Tech. J.*, January 1951, Vol. 30, No. 1, pp. 88–109.

are inherently nonlinear; their terminal characteristics are described by graphical data. For small-signal operation, they can be modeled by linear equivalent circuits, using incremental parameters. For large signals, their operation is represented by *load lines* on the graphical characteristics, with supplementary linear parameters, e.g., capacitances. The load-line treatment of active devices appears in Fig. **7-44**. The most significant element of the active-device equivalent circuit is the *dependent* or *controlled source*.

TABLE 3-1 Lumped-Constant Circuit Parameters

Parameter	Defining equation	Time-domain behavior	Frequency-domain behavior
Resistance, Ω	$R = \dfrac{V_{ab}}{I_{ab}}$	$v_{ab} = Ri_{ab}$	$V_{ab} = RI_{ab}$ that is, $Z = R + j(0)$
Conductance, S (reciprocal of resistance)	$G = \dfrac{I_{ab}}{V_{ab}}$	$i_{ab} = Gv_{ab}$	$I_{ab} = GV_{ab}$ that is, $Y = G + j(0)$
Incremental resistance, Ω	$R_{inc} = \left(\dfrac{dv_{ab}}{di_{ab}}\right)V_oI_o$	$i_{ab} \approx I_o + \dfrac{V_{ab} - V_o}{(R_{inc})v_o,I_o}$	$I_{ab} = \dfrac{(V_{ab})_{ac\ comp}}{(R_{inc})_{i=I_o}}$
Capacitance, F	$C = \dfrac{Q_a}{V_{ab}}$	$i_{ab} = \dfrac{C\,dv_{ab}}{dt}$	$I_{ab} = 2\pi fCV_{ab}$ that is, $jY = 2\pi fC$
Incremental capacitance	$(C_{inc})_{v_0} = \left(\dfrac{dq_{ab}}{dv_{ab}}\right)_{v=V_o}$	$i_{ab} \approx (C_{inc})_{v_0}\dfrac{dv_{ab}}{dt}$	$I_{ab} \approx j2\pi fC_{inc}(V_{ab})_{ac\ comp}$
Self-inductance, H	$L = \dfrac{\psi_{ab}}{I_{ab}}$	$v_{ab} = \dfrac{L\,di_{ab}}{dt}$	$V_{ab} = j2\pi fLI_{ab}$ that is, $Z = j2\pi fL$
Incremental inductance, H	$(L_{inc})_{I_0} = \left(\dfrac{d\psi_{ab}}{di_{ab}}\right)_{i=I_o}$	$v_{ab} = (L_{inc})_{I_0}\dfrac{di_{ab}}{dt}$	$V_{ab} = j2\pi f(L_{inc})_{I_0}(I_{ab})_{ac\ comp}$
Mutual inductance, H	$M_{12} = M_{21} = \dfrac{\psi_{ab}}{I_{cd}}$	$v_{ab} = L_1\dfrac{di_{ab}}{dt} + M_{12}\dfrac{di_{cd}}{dt}$	$V_{ab} = j2\pi f(L_1I_{ab} + MI_{cd})$
		$v_{cd} = L_2\dfrac{di_{cd}}{dt} + M_{21}\dfrac{di_{ab}}{dt}$	$V_{cd} = j2\pi f(L_2I_{cd} + MI_{ab})$

9. Dependent Sources. Any circuit which is capable of delivering more power to the load at the signal frequency than is incident upon it from external sources must incorporate an active device. In order to analyze circuits containing active devices, devices are replaced by an *equivalent circuit* of passive linear elements and one or more dependent sources. Dependent sources are legitimately sources and may be treated as such in any analysis, except for applying superposition. They are dependent, or controlled, however, in that the magnitude of their output voltage (or current) is determined by some other variable within the network (voltage or current). It is this dependency which allows one to analyze circuits containing active devices.

10. Small-Signal Representation of Active Devices. Every transistor or electron tube operated under small-signal conditions has two equivalent circuits. The first is a dc equivalent circuit, which deals with the various resistors, voltage supplies, and/or current supplies required to provide bias and to power the device so that it can function at the desired operating point. One step in the design of a small-signal amplifier (after the selection of the device and circuit configurations to be used) is a dc circuit calculation for the selection of the various resistors and power source or sources, using this dc equivalent circuit.

The second is an ac equivalent circuit, which includes the incremental ac parameters of the device, capacitances and inductances (where appropriate), and the ac circuit equivalents of the passive elements used for coupling, loads, isolation, etc.

A wide variety of small-signal ac equivalent circuits is found in practice, particularly for transistors. The following discussion summarizes the most commonly used, together with definitions of the incremental parameters.

11. Small-Signal Transistor Equivalent Circuits.

The symbols commonly used are e emitter; b base; c collector; v_{ce} (V_{ce}) variable (fixed) voltage, collector to emitter; v_{be} (V_{be}) variable (fixed) voltage, base to emitter; v_{cb} (V_{cb}) variable (fixed) voltage, collector to base; i_b (I_b) variable (fixed) current, base; i_e (I_e) variable (fixed) current, emitter; and i_c (I_c) variable (fixed) current, collector (see Fig. 3-8).

The h or *hybrid parameters* can be defined for the common-emitter, common-base, or common-collector circuits. For example, for the common-emitter transistor circuit, they are

$h_{11} = h_{ie}$ (input resistance) $= (\partial v_{be}/\partial i_b)_{V_{ce}=\text{const}}$

$h_{22} = h_{oe}$ (output admittance) $= (\partial i_c/\partial v_{ce})_{I_b=\text{const}}$

$h_{12} = h_{re}$ (feedback factor) $= (\partial v_{be}/\partial v_{ce})_{i_b=\text{const}}$

$h_{21} = h_{fe}$ (current amplification factor)

$\qquad = (\partial i_c/\partial i_b)_{V_{ce}=\text{const}}$

The corresponding h parameters for the common-base and common-collector amplifier circuits are designated as $h_{11} = h_{ib}$ and $h_{11} = h_{ic}$, respectively, and defined consistently with the circuits:

$\qquad h_{11} = h_{ib} =$ input resistance

(common-base circuit) $= (\partial v_{eb}/\partial i_e)_{V_{cb}=\text{const}}$

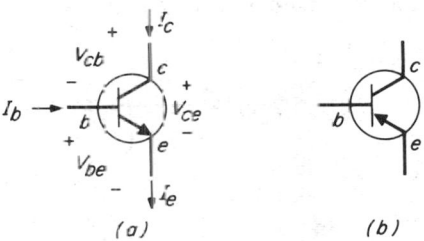

Fig. 3-8. Symbols for (a) npn and (b) pnp bipolar transistors.

Use of the h parameters leads to the equivalent circuit of Fig 3-9, useful for low-frequency calculations.

Other Transistor Parameters. α_0 is the short-circuit common-base current amplification at low frequencies (the symbol α, however, is also sometimes used for the transport *factor*, the fraction of minority carriers in the base which manage to arrive, through diffusion, in the collector region):

$$\alpha_0 = (-\partial i_c/\partial i_e)_{V_{cb}=\text{const}} \qquad (3\text{-}1)$$

α_f is the short-circuit common-base amplification at a frequency f

$$\alpha_f = \alpha_0/(1 + jf/f_\alpha) \qquad (3\text{-}2)$$

in which f_α is the frequency at which the magnitude $|\alpha_f|$ of α_f has fallen to 0.707 of its low-frequency value. f_α is known as the alpha cutoff frequency or common-base alpha cutoff frequency. The symbol f_{ab} is also used for this frequency.

f_T (also sometimes f_1) is the current-gain bandwidth product, also known as the common-emitter frequency f_{ae}. This is the frequency at which the common-emitter current gain of the device has dropped to unity.

β (beta) is a symbol used sometimes for h_{fe}, the current-amplification factor (common emitter).

These are the transistor parameters commonly given in the data sheets for commercial transistors. The h parameters are not convenient for the formulation of equivalent circuits for calculating high-frequency performance, however. For this purpose the T and pi parameters are useful. These follow, defined for the common-base amplifier, using the circuits of Fig. 3-10a and b.

Corresponding to a, as defined in the caption for Fig. 3-10c, an approximate current-amplification factor b for common-emitter operation is defined as $b = a/(1 - a)$. This in turn is approximately equal to h_{fe}, the hybrid-parameter current-amplification factor.

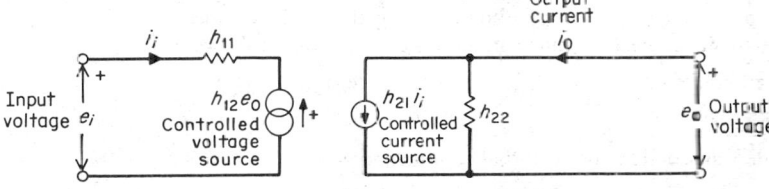

Fig. 3-9. Two-source equivalent circuit, using the hybrid (h) incremental parameters, a circuit valid for low-frequency small-signal transistor amplifiers. The feedback-voltage source $h_{12}e_0$ is often omitted. For a common-emitter amplifier, for example, $i_1 = i_b$, $h_{11} = h_{ie}$, $h_{22} = h_{oe}$ and $h_{21} = h_{fe}$.

Examples of some further equivalent circuits for high-frequency operation are shown in Fig. 3-11. The capacitance C_c is the collector-to-base capacitance, which incidentally varies with the collector voltage.

Since the device parameters normally given are the h parameters, the following equivalence relations are needed to determine the parameters in these equivalent circuits:

$$r_c = 1/h_{oe} \qquad r_e = h_{re}/h_{oe} \qquad r_b = h_{ie} = (1 + h_{fe})\, h_{re}/h_{oe} \tag{3-3}$$

At the highest frequencies, it is necessary to take into account the parameters of the leads in the transistors, resulting in even more complicated equivalent circuits.

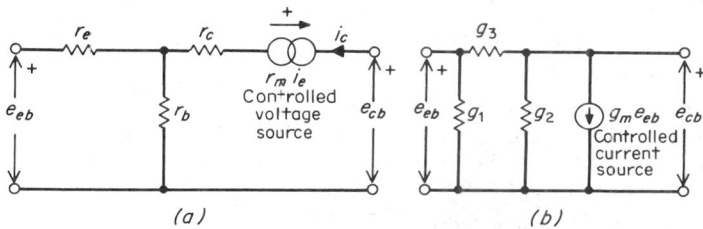

(a)   *(b)*

Fig. 3-10. *(a)* T equivalent circuit for the common-base transistor small-signal amplifier. The resistors r_e and r_c are the forward and backward resistance of the emitter-base and collector-base diodes, respectively. Resistor r_b represents a collector-emitter interaction effect, and $r_m i_e$ provides a voltage source to correspond to the amplification effect. This voltage source in series with r_c is often replaced with a current source in parallel with r_c, of magnitude $a i_e$, where $a = r_m/r_c$. The constant c is approximately equal to the α, defined in the text. *(b)* Pi equivalent circuit for the common-base amplifier. The conductances g_1 and g_2 correspond to the input and output diode conductances, and g_3 represents the interaction. The current source $g_m e_{cb}$ provides for the amplification effect.

The y Parameters. For high-frequency transistors, the y equivalent-circuit parameters are sometimes specified. The complete y equivalent for a bipolar-junction transistor is shown in Fig. 3-12.

12. Small-Signal Electron-Tube Circuits. As with transistors, there are two equivalent circuits to be considered in each electron-tube application. The first circuit is a dc steady-state circuit, in which only resistances and the electron tube are considered. The electron tube is represented by its actual terminal dc characteristics, usually expressed graphically; capacitors are open circuits, and inductors and/or transformer windings are replaced by their dc resistances. This dc equivalent circuit is designed to provide proper-operating quiescent dc electrode potentials (grid bias, screen and anode voltages, etc.).

The second, or ac, equivalent circuit determines the transfer function of the electron tube and its associated components under signal conditions. Large-signal calculations are developed by graphical plots on the charts of electron-tube characteristics and are not treated here. The following is a brief treatment of *small-signal linear equivalent circuits*.

The following symbols are used: e_{ak}, E_{ak} instantaneous anode-to-cathode voltage, dc anode-to-cathode supply voltage; e_{gk}, E_{gk} instantaneous grid-to-cathode voltage, dc grid-bias (grid-to-cathode) voltage; i_a, I_a instantaneous anode current, average anode current; C_{gk} capacitance grid to cathode; C_{ga} capacitance grid to anode; and C_{ak} capacitance anode to cathode. If the electron tube has four or five elements *(cathode, control grid, screen grid, suppressor grid, anode)*, the capacitances are defined as follows: C_{gk} ac measured capacitance grid to cathode with screen and suppressor (if present) at cathode potential; C_{ak} ac measured capacitance anode to cathode with screen and suppressor (if present) at ac cathode potential; and C_{ga} capacitance grid to anode with screen, suppressor (if used), and cathode tied together and grounded.

The small-signal electron-tube parameters are defined as follows:

Amplification factor: $\mu = (\partial e_{ak}/\partial e_{gk})_{i_a = \text{const}}$

Anode (plate) resistance: $r_a = (\partial e_{ak}/\partial i_a)_{e_{gk} = \text{const}}$

Grid-anode transconductance (mutual conductance):

$$g_m \text{ or } g_{ag} = (\partial i_a/\partial e_{gk})_{e_{ak} = \text{const}}$$

In general, each of the quantities μ, r_a, and g_m is a function of the actual values of i_a, e_{gk}, and e_{ak}, respectively. Figure 3-13 gives the commonly used small-signal equivalent circuits for common-cathode three-element (triode) electron tubes.

The performance of the triode electron-tube amplifier in the common-cathode mode is dominated at all but the lowest frequencies by the effect of the large difference between input and output voltages which is developed across the gride-anode capacitance C_{ga}. This results in an effectively greatly reduced input impedance (known as the *Miller effect*), so that small-signal operation of triodes in the common-cathode mode is restricted to very low frequencies.

13. Small-Signal Field-Effect-Transistor Equivalent Circuits. Field-effect transistors (FETs) are becoming useful for a wide range of circuit applications, including processing small analog signals. Figure 3-14 shows the FET equivalent circuit, which is topographically identical to the vacuum-tube circuit in Fig. 3-13.

14. Charge-Transfer Devices. Charge-transfer devices or *charge-coupled devices* (CCDs) are used in a wide variety of circuits. They work by passing along a charge which has been temporarily stored on the parasitic capacitance between the gate and substrate of MOSFETs (metal-oxide semiconductor field-effect transistor). These devices are interconnected in strings such that a pulse applied to a particular input (clock, strobe) causes the charge to be transferred from one stage to the next, as in Fig. 3-15. When this charge represents an analog signal, the device is called a *bucket brigade*. When this charge represents a digital signal, the device is called a *dynamic shift register*. These devices provide an effective way of producing a time delay within a signal path, since what is put in appears at the output n clock pulses later. The delay can be adjusted readily by adjusting the clock rate, but because of leakage of charge from the capacitance the devices will not function properly if clocked below some minimum rate.

15. Four-Layer Diodes. A large class of useful devices (especially for controlling large amounts of power) is based on the *pnpn* junction device, or four-layer diode, as shown in Fig. 3-16. This device behaves like a reverse-biased junction in one direction, but in the other direction it exhibits a breakover in its characteristic to a low-resistance region, as shown in Fig. 3-17. It is also called a *Shockley diode*.

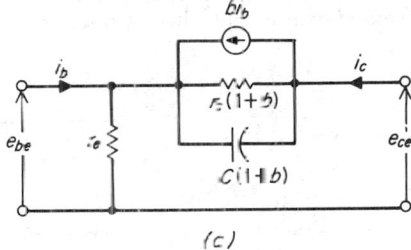

(c)

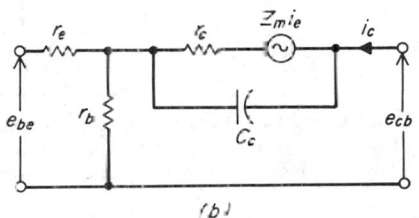

(b)

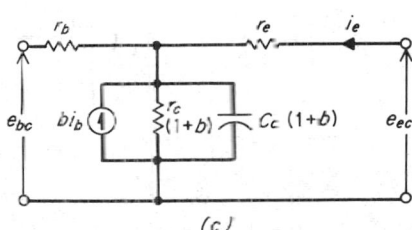

(c)

Fig. 3-11. Examples of high-frequency equivalent circuits for (a) common-emitter, (b) common-base, and (c) common-collector small-signal transistor amplifiers. The impedance Z_m in (b) is equal to αr_e.

The device will therefore withstand a large voltage until it *breaks over* and will then conduct current readily until its current is reduced below the *holding current*, at which point it regains its previous character. It can be said to exist in two distinct states, OFF and ON. The device can

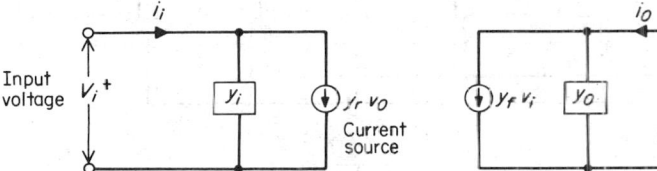

Fig. 3-12. The y-parameter equivalent circuit. For a common-base or common-emitter circuit, for example, the notation y_{ib} or y_{ie} respectively indicates the corresponding forward transfer admittance. The complex components of these admittances are often given in terms of capacitances because the capacitance values are frequency-independent.

be switched from OFF to ON not only by exceeding v_{BO} but also by injecting current into one of the interior regions (a *silicon-controlled rectifier*, SCR, or thyristor), shining light on the junctions (*light-activated* SCR, or photo-SCR), by raising its temperature, or by pulsing the voltage applied from anode to cathode. Back-to-back parallel devices can be packaged together to form DIACs (bilateral Shockley diodes) or TRIACs (bilateral SCRs). Other devices using the *pnpn* structure are the *silicon-controlled switch* (SCS) and *gate turn-off switch* (GTO).

SCRs have been developed which can handle considerably more power than transistors, but they are often rather hard to turn OFF once they are ON. The GTO and other improvements should soon remove this disadvantage.

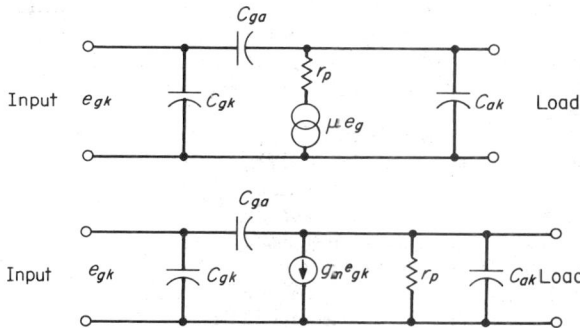

Fig. 3-13. Two alternative ac equivalent circuits for the common-cathode triode electron-tube amplifier. The load may, for example, be a resistor, transformer primary, tuned circuit, or the input to a subsequent stage. The incidental (stray) capacitances associated with the input and output must be added to C_{gk} and C_{ak}.

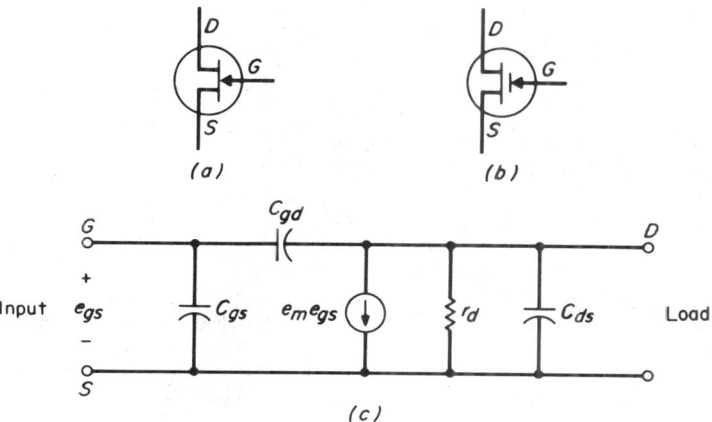

Fig. 3-14. (*a*) Symbols for junction-type field-effect transistors, (*b*) insulated gate field-effect transistors, and (*c*) the small-signal equivalent circuit.

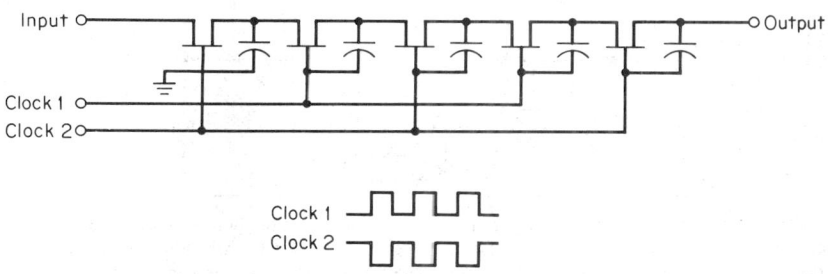

Fig. 3-15. A charge-transfer device.

16. Unijunction Transistor. This device (Fig. 3-18) has two ohmic contacts to a bar of semiconducting material and a third contact to its single junction. Its v-i characteristic displays a negative-resistance region, making it particularly useful for various oscillator circuits.

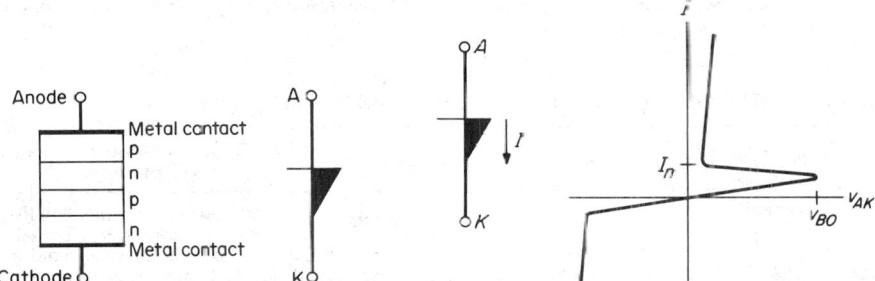

Fig. 3-16. The construction and symbol for a four-layer device.

Fig. 3-17. Voltage-current (v-i) characteristics of a four-layer diode.

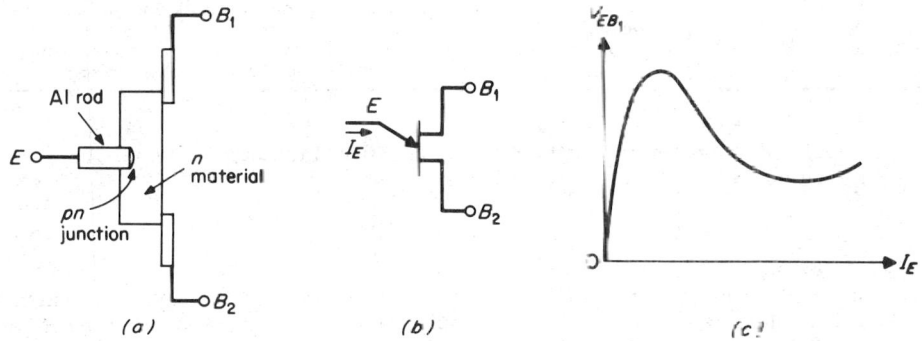

Fig. 3-18. Unijunction transistor: (a) construction; (b) symbol; (c) v-i characteristic.

NETWORK ANALYSIS

17. Time-Domain Analysis. Networks are constructed generally to perform some operation upon inputs that vary with time to produce various outputs which vary with time. Since time is a major independent variable, to understand many circuits requires one to ascertain what operations they perform on time-varying signals. Time-domain analysis is the straightforward technique of computing output functions of time from input functions of time and circuit characteristics. The conventional analysis of those equivalent circuits of interest to electronic engineers deals with three classes of systems: *power-supply systems*, in which the circuits deliver voltage and current at approximate values required to operate electronic equipment; *signal systems*, apparatus and networks associated with the origination, transmission, distribution, and reproduction of electric signals, e.g., telephone apparatus and the circuit portions of radio, television, and radar apparatus; and *control systems*, apparatus in which information is sensed (or fed in the form of commands) and used to control the application of energy, e.g., automatic controls for machine tools, automatic manufacturing processes, and automatic positioning systems.

In treating the equivalent circuits of *signal* and *control systems*, the engineer is basically concerned in determining the *output performance* of the circuit, i.e., the *output* which results from some *input*. Some examples of output characteristics of interest are shown in Table 3-2.

In time-domain analysis, then, a signal voltage is specified by its actual variation. The direct solutions of the circuit differential equations for the (transient) response of a circuit to the nonperiodic time-varying function which corresponds to a realistic information carrying signal, however, are seldom undertaken for any but the simplest circuits and signals. While the solutions are not impossible (since the advent of the internally programmed digital computer, any calculation of circuit behavior for numerically specified excitation and circuit elements is inherently feasible), considerable information concerning the response of a circuit to realistic signals can be inferred more simply from its response to one or more of the three basic *time-domain excitations*, the *step* (Fig. 3-19), the *impulse* (Fig. 3-20), and the *ramp* (Fig. 3-21) functions. Any other time-varying excitation can be approximated by a superposition of step functions of approximate amplitudes and starting instants. In accordance with the superposition principle, if the circuit is linear, its response to the series of steps which approximate the signal $h(t)$ can be determined by summing its responses to each step.

TABLE 3-2   Output Characteristics of Some Typical Circuit Types

Type of system	Typical input	Typical outputs or responses of interest
Analog signal	Analog voltage or current	Signal-to-noise ratios, fidelity, information rates, channel capacities, channel requirements
Digital signal	Digital voltage or current pulses	Logic behavior, error rates, information rates, channel capacities, spectrum requirements
Regulator	Step change in regulated output as a function of an input or load fluctuation	Response time, shape of response, limits of regulation, noise, stability
Control	Control instruction	Shape of response, response time, accuracy, stability, noise

The *impulse function* can be thought of as the derivative of the step function. When the impulse is defined as a $\delta(t)$ such that

$$\int_{t=0^-}^{t=0^+} \delta(t)\, dt = 1$$

it is a *unit impulse*, also known as a *Dirac function* or *delta function*. An impulse excitation of an electric circuit excites the circuit into its so-called *natural response*. Also, since the Fourier transform of the impulse yields a continuous frequency spectrum of constant amplitude from zero frequency to infinite frequency, the response of a circuit to an impulse can also be regarded as representation of its response to simultaneous excitations of a continuum of frequencies.

The *ramp function* can be regarded as the integral of a step function.

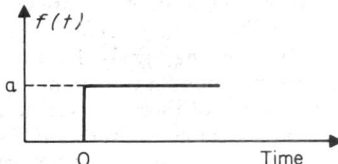

Fig. 3-19. The step function. For t smaller than 0, the amplitude is zero. For t greater than 0, the amplitude is a constant, a. If $a = 1$, it is a unit step function, frequently designated as $u(t)$.

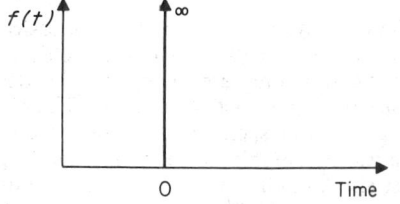

Fig. 3-20. The impulse function, of zero duration and infinite amplitude.

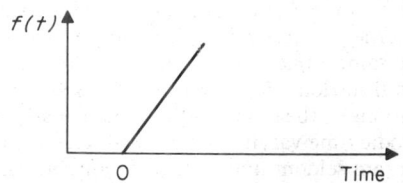

Fig. 3-21. The ramp function. For t greater than 0, $f(t) = kt$. When $k = 1$, that is, when the slope is unity, this is known as a unit ramp.

18. Laplace Transforms. Referring to Table 3-1, one can see that the time-domain behavior of many circuit elements is such that the application of Kirchhoff's laws to a network yields a series of integrodifferential equations. As already stated, such a system of equations is so difficult to solve that it is done only for simple circuits.

Fortunately, the time-domain analysis of circuits can be made strikingly simpler by a transformation of the variables involved, using the *Laplace transformation*. The Laplace transform is an *integral transformation*, since it transforms a function of one variable into a function of another variable through the process of integration. It is exceedingly important to linear-circuit analysis because it transforms the solution of simultaneous differential circuit equations with constant coefficients into an *algebra* problem. Although the transformation process is not always possible, all functions ordinarily encountered in practical networks can be handled using Laplace transforms. The Laplace transform converts a function of *time* into a function of a variable *s*, which is a *complex number* traditionally expressed as $s = \sigma + j\omega$ ($j = \sqrt{-1}$) and called the *complex frequency*

$$F(s) = \mathcal{L}[f(t)] = \int_0^\infty f(t)e^{-st}\, dt \tag{3-4}$$

$F(s)$ is in general a complex function. Note that the lower limit of integration is zero. It is assumed that $f(t) = 0$ for $t < 0$. If this is not the case, it will be necessary to redefine the time-axis origin or to use the *two-sided Laplace transform*.

It is also necessary to define the *inverse Laplace transform* to form the *Laplace-transform pair*. This transformation function is a complex integration which results in a time function

$$f(t) = \mathcal{L}^{-1}[F(s)] = \frac{1}{2\pi j}\int_{s=C-j\infty}^{C+j\infty} F(s)e^{st}\, ds \tag{3-5}$$

Since this integration is usually exceedingly difficult to carry out for any worthwhile function, it is not done, but tables are generated which express various common functions of time in terms of their Laplace transforms. These are called *Laplace-transform tables*. Table 3-3 shows some of the more common transform pairs. The Laplace transformation is *unique*; i.e., for any $F(s)$, there exists one and only one corresponding $f(t)$, and vice versa.

TABLE 3-3 Laplace-Transform Pairs

$f(t) \quad t \geq 0$	$F(s)$	$f(t) \quad t \geq 0$	$F(s)$
$af(t)$	$aF(s)$	$\cos \beta t$	$\dfrac{s}{s^2 + \beta^2}$
$f_1(t) \pm f_2(t)$	$F_1(s) \pm F_2(s)$	$e^{-\alpha t}\cos \beta t$	$\dfrac{s + \alpha}{(s + \alpha)^2 + \beta^2}$
$\dfrac{d}{dt}f(t) \equiv f'(t)$	$sF(s) - f(0+)$	$\dfrac{1}{a}(1 - e^{-\alpha t})$	$\dfrac{1}{s(s + a)}$
$\int f(t)\, dt \equiv f^{(-1)}(t)$	$\dfrac{F(s)}{s} + \dfrac{f^{(-1)}(0+)}{s}$	$\dfrac{1}{s_1 - s_2}(e^{s_1 t} - e^{s_2 t})$	$\dfrac{1}{(s - s_1)(s - s_2)}$
1 or unit step at $t = 0$	$\dfrac{1}{s}$	te^{-at}	$\dfrac{1}{(s + a)^2}$
t	$\dfrac{1}{s^2}$	t^n	$\dfrac{n!}{s^{n-1}}$
$\dfrac{1}{(n-1)!}t^{n-1}$ n a positive integer	$\dfrac{1}{s^n}$	$e^{-at}\sin \omega_1 t$	$\dfrac{\omega_1}{(s + a)^2 + \omega_1^2}$
e^{-at}	$\dfrac{1}{s + \alpha}$	$e^{-at}\cos \omega_1 t$	$\dfrac{s + c}{(s + a)^2 - \omega_1^2}$
$\sin \beta t$	$\dfrac{\beta}{s^2 + \beta^2}$	δ (delta function)	

The Laplace transform exhibits a number of useful properties:
1. It is linear. If α is a constant and $f_1(t)$ is some time function,

$$\mathcal{L}[\alpha f_1(t)] = \alpha \mathcal{L}[f_1(t)] = \alpha F_1(s)$$

It is also true that if there exists another time function $f_2(t)$, then

$$\mathcal{L}[f_1(t) + f_2(t)] = \mathcal{L}[f_1(t)] + \mathcal{L}[f_2(t)] = F_1(s) + F_2(s)$$

2. It can be integrated:

$$\int_0^1 f_1(t) \, dt = \frac{1}{s}[f_1(t)] = \frac{1}{s} F_1(s)$$

3. It can be differentiated:

$$\mathcal{L}\left[\frac{d}{dt} f_1(t)\right] = s\mathcal{L}[f_1(t)] - f_1(0) = sF_1(s) - f_1(0_+)$$

Note the extra term $f_1(0_+)$. This is an *initial condition* and is the numerical value of $f_1(t)$ at $t = 0_+$. Laplace transforms automatically handle initial conditions which occur in circuit analysis.

4. It can be differentiated two or more times by a repetitive process.

$$\frac{d^2}{dt^2} f_1(t) = s\left\{\frac{d}{dt}[f_1(t) - f_1(0_+)]\right\}$$

$$d^2 f_1(t) = s^2 F_1(s) - s\frac{d}{dt} f_1(0_+) - f_1(0_+)$$

5. It can be *shifted in time*:

$$\mathcal{L}[f(t)u(t-y)] = e^{-sy}F(s)$$

6. *Multiplication* of Laplace transforms is equivalent to *convolution* of their time functions.

$$\mathcal{L}[f_1(t) * f_2(t)] = F_1(s)F_2(s)$$

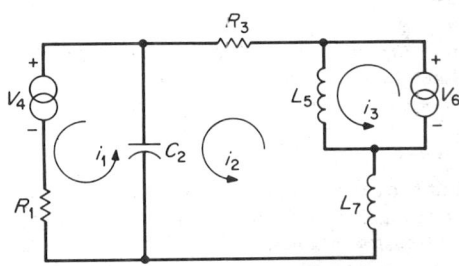

Fig. 3-22. A circuit containing three loops.

19. Example of Laplace Transformation. In the circuit of Fig. 3-22, the application of Kirchhoff's voltage law to all the loops yields a system of three equations. (Here it is assumed that at $t = 0$ there was no charge on the capacitor or current in any inductor.)

$$-V_4 = i_1R_1 + (1/C_2)\!\int i_1 \, dt - (1/C_2)\!\int i_2 \, dt$$
$$0 = i_2R_3 + (1/C_2)\!\int i_2 \, dt + L_7 \, di_2/dt + L_5 \, di_2/dt - (1/C_2)\!\int i_1 \, dt - L_5 \, di_3dt \qquad \text{(3-6)}$$
$$V_6 = L_5 \, di_3/dt - L_5 \, di_2/dt$$

Solving this circuit for $i_1(t)$, for instance, becomes a formidable task, but if the Laplace transformation is applied, the equations become

$$-V_4(s) = I_1(s)R_1 + I_1(s)/sC_2 - I_2(s)/sC_2$$
$$0 = I_2(s)R_3 + I_2(s)/sC_2 + I_2(s)sL_7 + I_2sL_5 - I_1(s)/sC_2 - I_3(s)sL_5 \qquad \text{(3-7)}$$
$$V_6(s) = I_3(s)sL_5 - I_2(s)sL_5$$

[Note that $I_1(s)$, $V_4(s)$, etc., are the *Laplace transforms* of time functions, i_1, V_4, etc.]

Although the complex number s is actualy a mathematical operator, it can be manipulated algebraically in linear circuits. Therefore, if one wishes to solve the equations for $I_1(s)$, some algebra yields

$$I_1 = \frac{-s^3C_2^2L_7V_4 - s^2C_2^2R_3 - s(C_2V_4 - C_2V_6)}{s^3C_2^2L_7 + s^2(C_2^2R_1R_3 + C_2L_7) + s(R_1C_2 + R_3C_2) - 1} \qquad \text{(3-8)}$$

$V_4(s)$ and $V_6(s)$ can (in almost every instance) be expressed as the ratio of two polynomials in s. The final expression for $I_1(s)$ is then, in general, the ratio of two polynomials in s

$$I_1(s) = \frac{\alpha_n s^n + \alpha_{n-1}s^{n-1} + \cdots + \alpha_1 s + \alpha_0}{\beta_m s^m + \beta_{m-1}s^{m-1} + \cdots + \beta_1 s + \beta_0} \qquad \text{(3-9)}$$

To determine $i_1(t)$ it is necessary to find the inverse transform of $I_1(s)$.

Partial-Fraction Expansion. Laplace-transform tables do not cover such complicated expressions, and so $I_1(s)$ must be broken down by *partial-fraction expansion.* Partial-fraction expansion requires that $I_1(s)$ be a *proper fraction,* that is, $m > n$. If it is not, first use long division to reduce the power of the numerator.

$$I_1(s) = \gamma_{m-n}s^{m-n} + \gamma_{m-n-1}s^{m-n-1} + \cdots + \gamma_0 + \frac{\xi_{n-1}s^{n-1} + \xi_{n-2}s^{n-2} + \cdots + \xi_0}{\beta_n s^n + \beta_{n-1}s^{n-1} + \cdots + \beta_n} \tag{3-10}$$

Then the fraction denominator must be factored:

$$I_1(s) = \gamma_{m-n}s^{m-n} + \cdots + \gamma_0 \frac{\xi_{n-1}s^{n-1} + \cdots + \xi_0}{\beta_n(s - s_1)(s - s_2) \cdots (s - s_n)} \tag{3-11}$$

where $s_1, s_2, \ldots, s_n$ are the roots of the denominator. $I_1(s)$ can be converted from this form into the *partial-fraction expansion*

$$I_1(s) = \gamma_{m-n}s^{m-n} + \cdots + \gamma_0 + \frac{K_1}{s - s_1} + \frac{K_2}{s - s_2} + \cdots + \frac{K_n}{s - s_n} \tag{3-12}$$

if all the s_i are different. If some roots are repeated ($s_2 = s_3$, for instance), the form is modified. The new form is best explained by example:

$$F(s) = \frac{\xi_5 s^5 + \xi_4 s^4 + \xi_3 s^3 + \xi_2 s^2 + \xi_1 s + \xi_0}{(s - s_1)(s - s_2)^3(s - s_3)^2} \tag{3-13}$$

$$F(s) = \frac{K_1}{s - s_1} + \frac{K_{21}}{(s - s_2)^3} + \frac{K_{22}}{(s - s_2^2)^2} + \frac{K_{23}}{(s - s_2)} + \frac{K_{31}}{(s - s_3)^2} + \frac{K_{32}}{(s - s_3)} \tag{3-14}$$

The unknown constants K_i can be evaluated by a simple process. To find K_1, for instance, simply multiply both sides of $F(s)$ by $s - s_1$ and evaluate the function for $s = s_1$

$$K_1 = \left.\frac{\xi_{n-1}s^{n-1} + \cdots + \xi_0}{\beta_n(s - s_2) \cdots (s - s_n)}\right|_{s=s_1} \tag{3-15}$$

To find the K's for a multiple root, one multiplies both sides by the highest power and then differentiates to find the K's associated with lower powers. For instance, in Eq. (3-14)

$$K_{21} = (s - s_2)^3 F(s)|_{s=s_1}$$

$$K_{22} = \frac{d}{ds}[(s - s_2)^3 F(s)]|_{s=s_1} \tag{3-16}$$

$$K_{23} = \frac{1}{2}\frac{d^2}{ds^2}[(s - s_2)^3 F(s)]|_{s=s_1}$$

The final result is that $I_1(s)$ becomes a series of simple terms such as $r_i s^i$, γ_0, and $K_{rp}/(s - s_r)^p$.

All these terms are documented in a complete table of Laplace transforms, so that $I_1(s)$ becomes the sum of a number of smaller, easier to find functions: $I_1(s) = I_{1a}(s) + I_{1b}(s) + \cdots + I_{1l}(s)$. Since each $I_{1k}(s)$ has a corresponding time function $i_{1k}(t)$, $i_1(t) = i_{1a}(t) + i_{1b}(t) + \cdots + i_{1l}(t)$.

A further discussion of Laplace transforms and more extensive tabulation of transform pairs can be found in Sec. **2**.

20. Poles and Zeros. Functions of s are often defined in terms of their *poles* and *zeros.* $I_1(s)$, for instance [Eq. (3-9)], has n roots in its numerator polynomial and m roots in its denominator. For the n values which correspond to the roots of the numerator, $I_1(s) \rightarrow 0$. These are called *zeros* of the function $I_1(s)$. For the m values which correspond to the roots of the denominator, $I_1(s) \rightarrow \infty$. These are called *poles* of $I_1(s)$. They can be located on a graph of the s plane ($s = \sigma + j\omega$).

This graph (Fig. 3-23) of the poles (x's) and zeros (0s) is equivalent to the polynomial representation. It is particularly useful in systems engineering (Sec. **5**) and filter design (Sec. **12**).

21. Frequency-Domain Analysis. Even though linear circuits can be solved in the time domain using Laplace-transform techniques, the resultant expressions are difficult to work with. The final result often consists of a sum of exponential time functions, with a separate term for each energy-storage element (capacitor or inductor) in the circuit, and it becomes difficult to comprehend the physical meaning of the expression.

It was discovered some time ago that equivalent information could be obtained in a form easier to understand by carrying out circuit calculations in the frequency domain. Often the signals which circuits process are periodic, occurring over and over again at regular intervals. In fact, one of the most common signals considered is of the form $g_1(t) = A_1 \sin 2\pi f_n t$, a sinusoid of amplitude A_1 (voltage or current) and frequency f_n (in hertz). For any linear circuit, the resultant output for this input is of the form $g_2(t) = A_2 \sin (2\pi f_n t - \theta_n)$, revealing an extremely simple relationship between input signal and output signal.

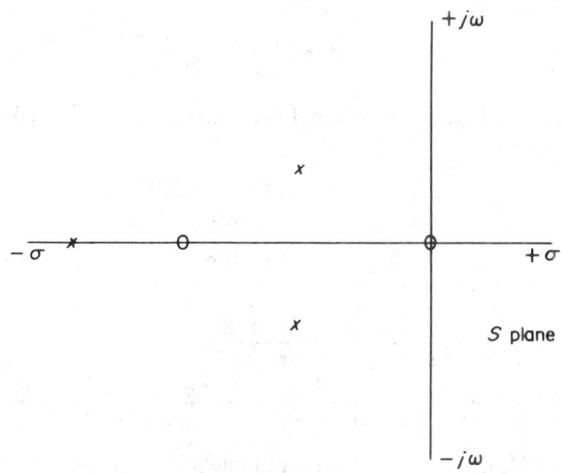

Fig. 3-23. Pole-zero representation of the ratio of two polynomials in s.

22. Fourier Series. The French mathematician Fourier showed that *any* realistic periodic signal can be represented as the sum of sine waves of various amplitudes (A's) and phases (θ's). Although his sum is an infinite series *(Fourier series)*, it converges quickly. By applying the techniques of Fourier series it is possible to decompose any periodic input into a sum of sine waves, simply calculate the circuit response to each component wave, and then synthesize the resulting output time function from the output series. The ease with which the circuit response to sine waves can be determined greatly outweighs the extra computation necessary to convert to the Fourier-series representation and back again.

The specific statement of Fourier's theorem is

$$f(x) = \tfrac{1}{2}a_0 + \sum_{m=1}^{\infty} (a_m \cos mx + b_m \sin mx)$$

Any function $f(x)$ which is well behaved in the interval $a < x < b$, that is, it has only a finite number of finite discontinuities, can be expressed in the interval as an infinite sum of sine and cosine functions.

The Fourier series for a periodic function $h(t)$, of period T, in seconds, is most conveniently determined by scaling (in time) the function to a function $f(x)$ with a period equal to 2π units of time. For the scaled function, the series in trigonometric form is

$$f(x) = a_0/2 + a_1 \cos x + a_2 \cos 2x + \cdots + a_n \cos nx$$
$$+ \, b_1 \sin x + b_2 \sin 2x + \cdots + b_n \sin nx \qquad (3\text{-}17)$$

where
$$a_n = \frac{1}{\pi} \int_0^{2\pi} f(x) \cos nx \, dx \qquad b_n = \frac{1}{\pi} \int_0^{2\pi} f(x) \sin nx \, dx \qquad (3\text{-}18)$$

In exponential form, the series is

$$f(x) = A_{-n}e^{-jnx} + \cdots + A_{-2}e^{-j2x} + A_{-1}e^{-jx}$$
$$+ \, A_0 + A_1 e^{jx} + A_2 e^{j2x} + \cdots + A_n e^{jnx} \qquad (3\text{-}19)$$

where
$$A_n = \frac{1}{2\pi} \int_0^{2\pi} f(x)e^{-jnx} \, dx \qquad (3\text{-}20)$$

The nth harmonic component of the function $f(x)$ is made up of the two terms $a_n \cos nx$ and $b_n \sin nx$, which can be combined into a single component by the identity

$$a_n \cos nx + b_n \sin nx = \sqrt{a_n^2 + b_n^2}(\sin nx + \theta_n) \tag{3-21}$$

where $\theta_n = \tan^{-1}(a_n/b_n)$. However, the series for many simple waveforms can be reduced to a series in sine terms only (or cosine terms only) by a judicious choice of the zero of the independent variable (x in the scaled series, t in the actual function). For example if the zero can be chosen so that the function $f(t)$, as in Fig. 3-24a and b, is an odd or even function the series contains only the sine or only the cosine terms, respectively.

23. Parseval's Theorem. One extremely useful shortcut resulting from Fourier analysis is Parseval's theorem, which enables one to calculate the *root-mean-square* (rms) or *effective* value of a periodic function directly from the Fourier series, without determining the time function. If $h(t)$ is a nonsinusoidal periodic function,

$$[h(t)]_{rms} =$$

$$\sqrt{\left(\frac{a_0}{2}\right)^2 + \left(\frac{a_1}{2}\right)^2 + \left(\frac{a_2}{2}\right)^2 + \cdots + \left(\frac{a_n}{2}\right)^2 + \left(\frac{b_1}{2}\right)^2 + \left(\frac{b_2}{2}\right)^2 + \cdots + \left(\frac{b_n}{2}\right)^2} \tag{3-22}$$

24. Fourier Transforms. The method used for Fourier-series analysis can be extended to include *nonperiodic* functions, or *transients*. In this case the infinite-series sum becomes a continuous function expressed as an integral, the *Fourier transform*. As mentioned previously, the time-function expression for any time-varying signal $h(t)$ is termed its *time-domain* representation. The mathematical equivalent of the same signal in the *frequency domain* is the sum of the infinite number of infinitesimal sinusoidal signals with the relative amplitude and phase at each frequency f_n given by the *Fourier transform* $F(\omega_n)$ of the signal $h(t)$, or

$$F(\omega_n) = \frac{1}{2\pi} \int_{-\infty}^{\infty} h(t)e^{-j\omega_n t} \, dt \tag{3-23}$$

where ω_n, the angular frequency, is equal to $2\pi f_n$.

In the above form of the Fourier integral, the relative amplitude of each component $F(\omega_n) \, d\omega$ is actually the amplitude of an exponential excitation, $F(\omega_n)e^{j\omega_n t} \, d\omega$, rather than the amplitude of a sinusoidal excitation, and the steady-state components are understood to exist in the spectrum space from $\omega_n = -\infty$ to $\omega_n = +\infty$. With sinusoidal excitation, no significance is attached to a negative frequency. In terms of the above $F(\omega)$, however, the relative component, in the conventional sinusoidal sense, at each angular frequency ω_n comprises the sum $F(+\omega_n)e^{+j\omega_n t} + F(-\omega_n)e^{-j\omega_n t}$. The exponential form of excitation, with frequencies from $\omega_n = -\infty$ to $\omega_n = +\infty$, is quite convenient analytically. It has the further advantage that in the form $e^{(+\sigma+j\omega)t}$ it can represent damped (for σ negative) trains of sine waves. In this representation, the coefficient $+\sigma + j\omega$ is known as the *complex frequency*. The plot of $F(\omega_n)$ as a function of frequency is known as the *frequency spectrum* of the signal $h(t)$.

As an example of an idealized signal spectrum, consider the pulse-excitation function $h(t)$ shown in Fig. 3-25. The Fourier transform $F(\omega)$ of this time function is $F(\omega) = 2/\omega \sin (\omega\tau_1/2)$, with the frequency spectrum plotted in Fig. 3-26. The plot shows the relative amplitude of the infinitesimal amplitude component at all angular frequencies ω_n. Any practical circuit to the input of which the pulse function of Fig. 3-25 is applied will necessarily cut off at some frequency less than infinity, and the resulting output pulse will have finite rise and decay times, also, probably an overshoot.

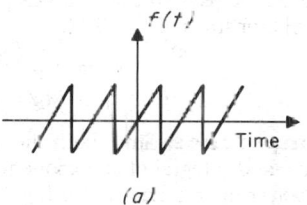

(a)

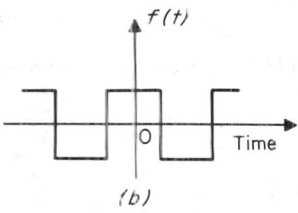

(b)

Fig. 3-24. (a) A sawtooth wave positioned so that it is an odd function; that is, $f(t) = -f(-t)$; (b) a square wave positioned so that it is an even function; that is, $f(t) = f(-t)$.

If the frequency and phase characteristics of a signal circuit are expressed as a *transfer function*, such as

$$T(\omega_n) = (A_2/A_1)e^{j\theta_n}$$

where A_2 is the amplitude of the output function

$$A_2 e^{j(\omega_n t + \theta_n)}$$

corresponding to an input $A_1 e^{j\omega_n t}$, the actual time-varying output $H(t)$ corresponding to an input $h(t)$ [for which the Fourier transform is $F(\omega_n)$] is given by the *inverse Fourier transform*

$$H(t) = \int_{-\infty}^{\infty} F(\omega_n)e^{j\omega_n t} T(\omega_n)\, d\omega \tag{3-24}$$

The procedure for determining the response of a linear circuit to a time-varying input function by the resolution of the input function into so-called *frequency components*, each of which is a steady-state sinusoid, and the calculation of the response by superposing the circuit response to each frequency taken separately is valid because the circuit is linear, in accordance with the definition of linearity given in Par. **3-2**. It is thus an application of the so-called principle of *superposition*, which applies to all linear systems.

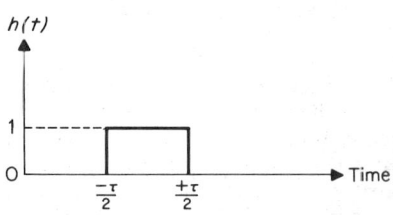

Fig. 3-25. Ideal pulse signal of duration τ_1 and unit amplitude.

25. Phasor Representation. Since many circuits are analyzed on the basis of their effects on a sine wave or sum of sine waves, methods have been developed to make this as easy as possible. Sinusoidal signals are usefully presented as *phasors*, in which the signal is represented as a line segment rotating counterclockwise about the origin in the complex plane, as in Fig. 3-27. The amplitude of the signal is represented by the length of the line. It completes f r/s, where f is the signal frequency. It is customary to show the several phasors representing various voltages and currents in a circuit on the same phasor diagram. The diagram is drawn to depict the stop-action relationship between the phasors at one particular time, usually when the input phasor is pointing directly to the right ($0°$ phase). At this point all sinusoidal signals can be represented as having a certain magnitude and phase ($M\angle\theta$) with respect to the reference input. Phasor notation ($M\angle\theta$) is an extremely compact way of representing information about circuit variables. Since Euler's relationship can be used in the complex plane, there are three equivalent ways of representing the phasor relationships of Fig. 3-27:

$$V_1 = M_1\angle 0° = M_1 e^{j0°} = M_1 \cos 0°$$
$$I_1 = M_2\angle\theta = M_2 e^{j\theta} = M_2 \cos\theta + jM_2 \sin\theta$$

26. Impedance. Since both the derivative and integral of a sinusoid are a sinusoid and the derivative and integral of an exponential function are an exponential function, powerful circuit calculations can be carried out using phasors. If V_1 is the voltage across some circuit element and I_1 is the current through it, its *impedance* can be expressed by $Z_1 = V_1/I_1$, in a manner analogous to Ohm's law. The ratio is actually taken by dividing the magnitude of V_1 by the magnitude of I_1 and subtracting the phase of I_1 from that of V_1:

$$Z_1 = V_1/I_1 = (M_1/M_2)\angle(0° - \theta) \tag{3-25}$$

It should be noted that whereas V_1 and I_1 are functions of time that have paused in their

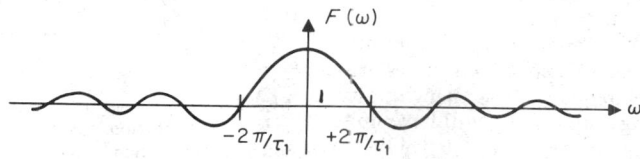

Fig. 3-26. Plot of $F(\omega) = 2/\omega \sin(\omega\tau_1/2)$, the Fourier transform of Fig. 3-25.

rotation, Z_1 is *not* a function of time but simply a complex number relating the two phasors. The impedance of a resistor is $Z_R = R\angle 0°$. The impedance of a capacitor is $Z_C = (1/C)\angle -90°$ $= 1/j2\pi fC$. The impedance of an inductor is $Z_L = L\angle + 90° = j2\pi fL$.

NETWORK CONCEPTS

27. Circuit-Analysis Techniques. Because the Laplace-transform representation of circuit variables used to make time-domain calculations is nearly identical to the impedance expressions derived from phasor analysis, the distinction between time-domain and frequency-domain analysis is often blurred. Many of the calculations are the same, and the information obtained can be transferred from one domain to the other. One can therefore describe general circuit-analysis techniques without specific reference to either domain.

28. Network Topology. Network topology is concerned with the interconnection of the elements of a network. We study how the branches are connected without really considering the elements themselves. Here we define a *branch* as a directed line segment, i.e., a line with an assigned direction. Each end of the line is called a node. Branches may be connected to each other only at their nodes. A group of branches is called a *network* or a *graph*. It is properly called a *directed graph* because all the lines have assigned directions. Although shown as a directed line segment only, a branch can represent a *circuit element*, such as a resistor, generator, capacitor, etc. It can also simply represent a wire. Branches are usually identified by numbering them (see Fig. 3-28).

A *path* is defined as the record of a journey through a network in which one follows branches from node to node, never passing through any particular node more than once. If a path exists between every pair of nodes in a network, the network is *connected*.

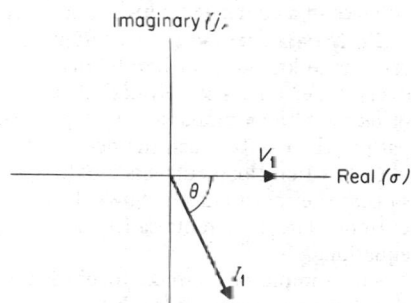

Fig. 3-27. Phasor representation of sinusoids.

29. Analysis of Connected Networks. A *subnetwork* can be identified as a part of a larger graph. Branches (with their nodes) which are not in the subnetwork are in the *complement* of the subnetwork. In a connected graph, it follows that a subnetwork will have at least one node in common with its complement (it will have no branches in common, however).

A *mesh* or *loop* is a path which ends upon the node from which it began. The branches of a mesh necessarily form a connected network (see Fig. 3-29). A *tree* is a subnetwork which contains all the nodes of the original connected network. It must be connected, and it must contain no meshes. A tree is not necessarily a path and in any event is not a closed path. Every connected network has at least one tree, which can be formed from the original network by removing one branch from each closed path subject to the constraint that the network remain connected. Branches which have been removed to form the tree are called the *chords* or *links* of that tree.

It can be shown by a construction technique[1] that for a connected network

$$T = N - 1 \qquad \text{and} \qquad C = B - T$$

where T is the number of tree branches, N the number of nodes, B the number of branches, and C the number of chords. It can also be demonstrated that every tree has at least two nodes to which *only one* tree branch is connected.

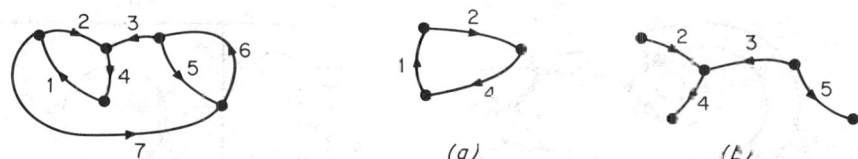

Fig. 3-28. Directed-graph representation of a simple network.

Fig. 3-29. A network (a) loop and (b) tree.

[1] Paul M. Chirlian, "Basic Network Theory," McGraw-Hill, New York, 1969.

We can use Kirchhoff's laws (see Par. **3-4**) to state some useful conditions in determining how much information is necessary to specify a circuit uniquely. We begin by realizing that Kirchhoff's current law (KCL) is applicable at every node and that Kirchhoff's voltage law (KVL) can be written around every mesh in the connected circuit. From this and the above definitions it can be stated that all branch voltages can be uniquely determined from the tree-branch voltages. Since the tree contains all the nodes of a network, it is possible to determine the voltage between any two nodes by applying KVL along a path connecting the nodes, thereby establishing any unknown branch voltage in terms of tree-branch voltages. Similarly, all the network currents can be uniquely determined from the chord currents. Remembering that there are at least two nodes of the tree which are connected to only one branch, it is clear that at least one node of the network contains only one current in addition to chord currents. This one current can then be determined by KCL. Let us now consider the tree subnetwork corresponding to the tree, with the branch we have just solved for removed from it. This new tree (and its network) has at least one node with only one unknown current (by previous arguments). Using KCL, we solve for this and remove its branch. It can be shown that all currents can be systematically found in this manner. These two developments tell us that there are no more than T independent branch voltages in a connected network nor more than C independent currents.

30. Mesh analysis is a systematic network-analysis procedure which chooses chord currents as unknowns, further defining them in such a way that KCL is automatically satisfied at every node. This is sometimes called *tie-set* analysis. Analysis rests on the selection of a number of *basic-mesh* or *fundamental-loop* units. A basic mesh is a loop comprising one chord and the path in the tree between the nodes of the chord (see Fig. 3-30). The number of basic mesh units is C. Fictitious loop currents are drawn through the C basic-mesh units choosing the direction so that the mesh current flows through the chord in the same direction as the original chord current. These currents satisfy KCL at every node, and it is unnecessary to write any KCL equations.

An example of a circuit topologically equivalent to Fig. 3-30 is shown in Fig. 3-31. Three simultaneous equations for this circuit can be written using KVL around each loop, assuming all initial conditions = 0:

$$-V_4 = i_1 R_1 + \frac{1}{C_2} \int i_1\, dt - \frac{1}{C_2} \int i_2\, dt$$

$$0 = i_2 R_3 + \frac{1}{C_2} \int i_2\, dt + L_7 \frac{di_2}{dt} + L_5 \frac{di_2}{dt} - \frac{1}{C_2} \int i_1\, dt - L_5 \frac{di_3}{dt} \qquad (3\text{-}26)$$

$$V_6 = L_5 \frac{di_3}{dt} - L_5 \frac{di_2}{dt}$$

These are the necessary and sufficient equations to solve for the three unknown currents (and hence all circuit unknowns).

For a given loop, all the voltage drops caused by the flow of its circulating current are called *self-voltage drops*, whereas drops caused by currents from other loops are called *mutual* or *coupled voltage drops*. Thus in Fig. 3-31 for loop 2, $i_2 R_3$, $(1/C_2)\int i_2\, dt$, $L_7\, di_2/dt$, and $L_5\, di_2/dt$ are all self-voltage drops, and $-(1/C_2)\int i_1\, dt$ and $L_5\, di_3/dt$ are coupled voltage drops.

31. Nodal analysis is a systematic network-analysis procedure which chooses nodal voltages as unknowns. One begins by arbitrarily picking a node which is labeled 0 and called the *reference node* or *datum node* or *ground*. All other circuit nodes are numbered systematically. Branch voltages are then determined in terms of the unknown node voltages by subtracting the

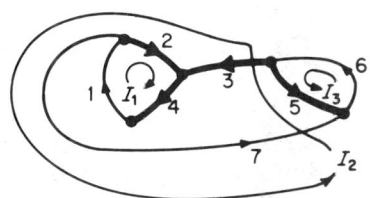

Fig. 3-30. Defining mesh currents. The heavy lines (2, 4, 3, and 5) indicate a tree.

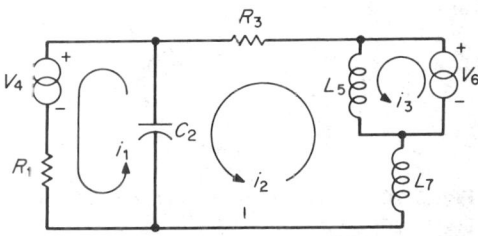

Fig. 3-31. An electric network topologically equivalent to Fig. 3-30.

voltage of the node at the tail of the branch arrow from the voltage of the node at the head of the branch arrow. The voltage of the reference node is defined as being zero.

Figure 3-32 shows such a network and circuit. Each of the four following equations is the result of applying KCL to each of the four nodes:

$$I_2 + I_7 = C_1 \frac{d}{dt} (V_{N_2} - V_{N_1}) \tag{3-27}$$

$$0 = -\frac{V_{N_2} - V_{N_3}}{R_4} - C_1 \frac{d}{dt} (V_{N_2} - V_{N_1}) \tag{3-28}$$

$$I_2 = \frac{V_{N_2} - V_{N_3}}{R_4} + \frac{V_{N_3} - V_{N_4}}{R_3} \tag{3-29}$$

$$0 = -\frac{V_{N_4} - V_{N_3}}{R_3} - \frac{1}{L_5} \int V_{N_4} \, dt - \frac{V_{N_4}}{R_6} \tag{3-30}$$

If a branch is connected between the node for which KCL is being written and the reference node, the current through this branch is due only to the voltage at the node of interest and is therefore called a *self-current*. The two currents represented by the second and third terms of Eq. (3-30) are self-currents. If the branch connects the node of interest with a node other than a reference node, the current in that branch is called a *coupled* or *mutual current*. All other terms in the above equations are of this type.

These four equations can be solved for the four unknown node voltages (and hence all circuit unknowns).

32. Choice of Analysis Type. The chief virtue of the mesh and nodal analysis procedures is that they are systematic and hence guarantee a solution (although not necessarily with the application of the least possible work). Because they are systematic approaches, they are mutually exclusive. The decision to choose one approach above the other is based on:

1. *Number of unknowns.* Mesh analysis will produce C simultaneous equations, whereas nodal analysis will produce $N - 1$. That method which produces the fewer equations is preferred.

2. *Type of sources.* Mesh analysis cannot be done with current sources in the network. Nodal analysis cannot be done with voltage sources. Thevenin or Norton equivalents can be found for sources of the undesired type, but this represents more calculations.

3. *Mutual inductance.* Nodal analysis can be used with coupled coils only by finding noncoupled equivalent circuits for the coils.

33. Cut Sets and Cut-Set Analysis. Nodal analysis is a special case of *cut-set analysis*. A *fundamental cut set* or *basic cut set* consists of one (and only one) branch of the network tree together with any links which must be cut to divide the network into two parts. An isolated node, if it is formed, is a *circuit part*. Removal of the cut set should divide the circuit into only two parts. A set of fundamental cut sets includes those cut sets formed by applying the cut-set division for each of the branches of the network tree. The number of cut sets is therefore equal to the number of tree branches of the network. Once the cut sets have been determined, KCL is applied to the interior of each. By convention, a positive current is one which crosses the cut-set boundary in the same direction as the tree branch. The circuit unknowns are the cut-set voltages, the voltage assigned to the node which is within the cut set.

Figure 3-33 shows the cut-set analysis of a circuit together with the resulting equations. A branch voltage is computed as the algebraic sum of all the *cut-set voltages* of the cut sets through which it passes. If the

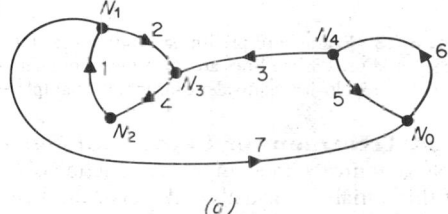

(a)

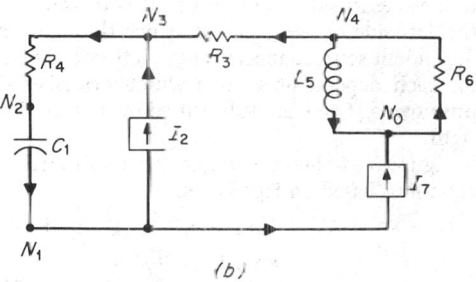

(b)

Fig. 3-32. Preliminary procedures for nodal analysis of (a) a directed graph and (b) its topological equivalent.

positive branch direction passes out of the cut set, the cut-set voltage is positive. If the branch passes into the cut set (in the same direction as the arrows on the boundary corners), it is taken negative. Application of KVL to Fig. 3-33 yields the following results: $V_{B_1} = V_2 - V_1$, $V_{B_2} = V_1$, $V_{B_3} = V_3$, $V_{B_4} = -V_2$, $V_{B_5} = V_3 - V_4$, $V_{B_6} = V_4 - V_3$, and $V_{B_7} = V_1 - V_3 - V_4$. Application of KCL yields $i_2 - i_1 + i_7 = 0$, $i_4 - i_1 = 0$, $i_3 - i_7 = 0$, and $i_5 + i_7 - i_6 = 0$. The circuit unknowns are V_1, V_2, V_3, and V_4. Also needed are branch equations for the unknown currents: $i_1 = C_1 \, dV_{B_1}/dt$, $i_2 = I_2$, $i_3 = 1/R_3 V_{B_3}$, etc. If these branch equations are substituted into the cut-set KCL, then by substituting V_1, V_2, V_3, V_4 for all branch voltages, one has four equations in four unknowns.

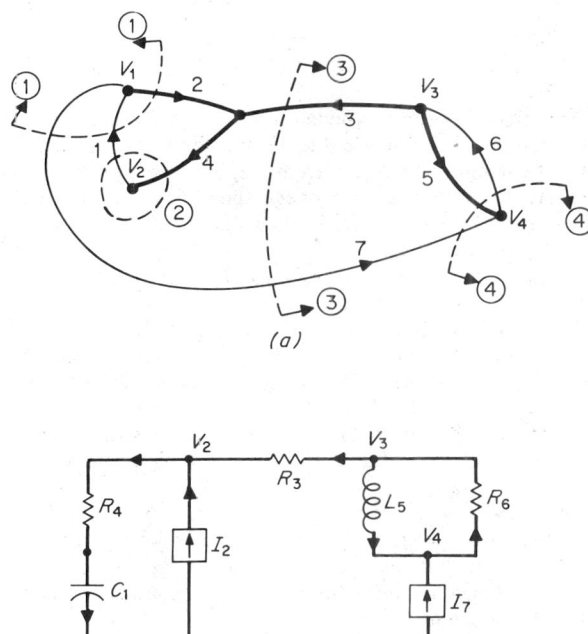

(a)

(b)

Fig. 3-33. Example of preliminary procedures for cut-set analysis. In (a) the heavy solid lines represent a tree, while the dashed lines are the cut-set boundaries. Each cut-set boundary is given a number and a direction. The topological equivalent for which example equations are given is shown in (b).

34. Treatment of Dependent Generators. A *dependent generator* is a current or voltage source whose output is a function of (depends on) some other circuit variable. The value of this variable is usually unknown, and hence the value of the generator output is unknown. These generators therefore require special treatment. Dependent generators are a necessary part of all transistor and vacuum-tube equivalent circuits. Analysis begins with the standard techniques described earlier, where it is the custom to write all the source voltages or currents on the left side of the equation, since they are known values. In spite of their being unknown, dependent sources appear on the left side. Then an additional series of equations is written (one for each dependent source) which describes the source dependency in terms of the circuit unknowns. These are substituted in, and the equations are regrouped with all unknowns on the right.

Figure 3-34 shows a simple two-mesh circuit with dependent voltage generators. The mesh equations based on Fig. 3-34 are

$$V_1 - r_m i_{ab} = i_1 R_2 + L_2 \, di_1/dt + (1/C)\smallint i_1 \, dt - (1/C)\smallint i_2 \, dt$$
$$\mu V_{ad} = L_1 \, di_2/dt + R_1 i_2 + (1/C)\smallint i_2 \, dt - (1/C)\smallint i_1 \, dt$$
$$i_{ab} = -i_2 \qquad V_{ad} = -V_1 + i_1 R_2 \tag{3-31}$$
$$V_1 = i_1 R_2 + L_2 \, di_1/dt + (1/C)\smallint i_1 \, dt - (1/C)\smallint i_2 \, dt - r_m i_2$$
$$-\mu V_1 = L_1 \, di_2/dt + R_1 i_2 + (1/C)\smallint i_2 \, dt - (1/C)\smallint i_1 \, dt - i_1 \mu R_2$$

35. Treatment of Nonlinear Elements. Analysis of networks containing nonlinear elements may be handled in a number of ways. Probably the most common approach is *linearization with small-signal analysis*. This technique approximates complex relationships as linear functions in the neighborhood of some operating point. Once the linearization has been accomplished, analysis can proceed according to the standard methods. If the circuit cannot be linearized, the nonlinear elements must be dealt with directly by developing an equation for voltage and current. Analysis can then be carried out almost as before but subject to a number of constraints:

1. It is not valid to use superposition, i.e., to say that if current i_1 produces voltage drop V_1 and i_2 produces drop V_2, then $i_1 + i_2$ will produce drop $V_1 + V_2$.

2. The circuit may contain a nonlinear element which is multivalued in its relationship. If the branch voltage cannot be determined uniquely from the branch current, the circuit cannot be analyzed using nodal or cut-set analysis.

3. If the branch current cannot be determined uniquely from the branch voltage, the circuit cannot be analyzed using loop or mesh analysis.

Figure 3-35 shows an example of circuit analysis with a nonlinear capacitance. The node equations are $I = V_A/R_1 + (V_A - V_B)/R_2$ and $0 = (V_A - V_B)/R_2 + f_c(V_b)|_{V_B} dV_B/dt$. The capacitor equations are $i = C dV_c/dt, C = f_c(V_c)|_{V_c}, i = f_c(V_c)|_{V_c} dV_c/dt$, and $V_c = V_B$.

If f_c is a fairly simple relationship, the equations

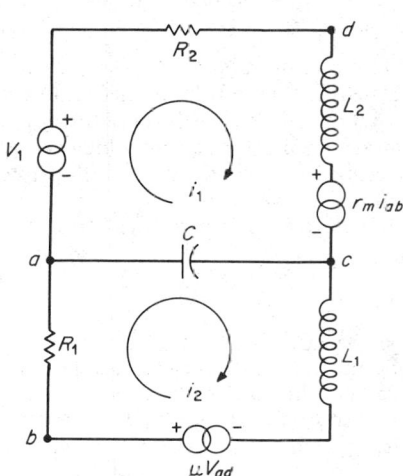

Fig. 3-34. A circuit with dependent sources.

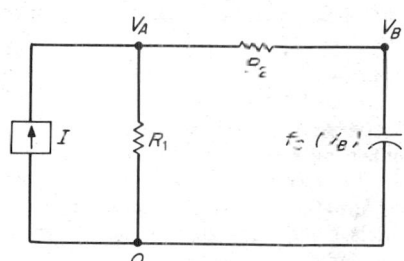

Fig. 3-35. Simple circuit containing a nonlinear capacitance.

can be solved analytically by a number of techniques. If the relationship is complex, solution must be accomplished by graphical techniques, state-variable techniques, or numerical analysis.

36. Matrix Formulation. Circuit-analysis techniques are systematic. Indeed, if the equations are set up according to the system definitions, algorithms exist to carry out the solution completely. The systematic form of the equations is called a *matrix formulation*. A *matrix* is an array of numbers, symbolically denoted by square brackets, with m rows and n columns:

$$\begin{bmatrix} z_{11} & z_{12} & \cdots & z_{1n} \\ z_{21} & z_{22} & \cdots & z_{2n} \\ \cdots & \cdots & \cdots & \cdots \\ z_{m1} & z_{m2} & \cdots & z_{mn} \end{bmatrix}$$

Formulation of the *loop-analysis matrix* representation of Fig. 3-32 is carried out by writing the source voltages in a single column matrix on the left. The rest of the equations are arranged to the left. All terms in Eq. (3-27) which contain i_1 are grouped as the z_{11} term. All terms of Eq. (3-28) which contain i_1 are grouped as the z_{21} term. All terms of Eq. (3-27) which contain current i_2 are grouped as the z_{12} term, and so on. The currents themselves are placed in a column matrix to the right of this matrix.

$$\begin{bmatrix} -V_4 \\ 0 \\ V_6 \end{bmatrix} = \begin{bmatrix} R_1 + (1/C_2)\!\int & -(1/C_2)\!\int & C \\ -(1/C_2)\!\int & R_3 + (1/C_2)\!\int + L_5\,d/dt + L_7\,d/dt & -L_5\,d/dt \\ 0 & -L_5\,d/dt & L_5\,d/dt \end{bmatrix} \begin{bmatrix} i_1 \\ i_2 \\ i_3 \end{bmatrix}$$

$$(3\text{-}32)$$

This representation is equivalent to the system of equations and can be manipulated in representative form $[V] = [Z][I]$.

Notice that the matrix is symmetric about its main diagonal. The main-diagonal elements are the self-voltage drops; all other elements are coupled-voltage drops. Notice also the unusual forms of terms like $R_1 + (1/C_2)\int$. These forms are called *operators (impedance operators)*. When this term multiplies i_1, for instance, it implies an operation

$$[R_1 + (1/C_2)\int]i_1 \Rightarrow R_1 i_1 + (1/C_2)\int i_1 \, dt$$

This notation is usually made more compact by the definition of the *differential operator p*, such that $pf(t) \equiv df(t)/dt$; in addition, the *integral operator* p^{-1} is defined as $p^{-1}f(t) \equiv f(t)/p = \int f(t) \, dt$. Thus $[R_1 + (1/C_2)\int]i_1 = (R_1 + 1/pC_2)i_1$.

By analogy, nodal (cut-set) analysis yields a matrix formulation for the circuit of Fig. 3-32.

$$
\begin{bmatrix} I_2 + I_7 \\ 0 \\ I_2 \\ 0 \end{bmatrix}
=
\begin{bmatrix}
-pC_1 & pC_1 & 0 & 0 \\
pC_1 - 1/R_4 & -pC_1 & pC_1 - 1/R_4 & 0 \\
0 & 1/R_4 & (1/R_3) - 1/R_4 & 1/R_3 \\
0 & 0 & 1/R_3 & -(1/R_3) - (1/pL_5) - 1/R_6
\end{bmatrix}
\begin{bmatrix} V_{N_1} \\ V_{N_2} \\ V_{N_3} \\ V_{N_4} \end{bmatrix}
$$

$$(3\text{-}33)$$

$[I] = [Y][V]$, where the terms of the $[Y]$ matrix are *admittance* operators.

37. n-Port Networks. All networks have sites at which they can be connected to the outside world, called *terminals*. A simple resistor has two terminals, a transistor has three, and a stereo amplifier often has more than ten. The most powerful general statement which can be made about an *m-terminal network* results from considering all the currents which flow into the network. Potentials can also be defined for each terminal (see Fig. 3-36).

A simple application of the principle of conservation of charge to the network yields

$$\sum_{k=1}^{m} i_k = 0$$

The network can therefore be treated as a giant node obeying KCL.

If it is possible to exhaustively separate the *m* terminals into *pairs*, such that for each pair (say i_3 and i_7) one can state $i_3 = -i_7$, these pairs are called *ports*. This can nearly always be done, thereby producing an *n-port network*. A transistor is a two-port network, although it has three terminals. Connecting an extra wire to one of the terminals provides the extra terminal without violating any network laws.

All useful networks can therefore be expressed as *n*-port networks with paired currents (see Fig. 3-37); *n*-port voltages can also be derived from the *m*-terminal potentials. If port 1 was formed by pairing terminal 3 with terminal 7, then $V_1 = E_3 - E_7$. Convention assigns the positive direction of a port voltage such that port current flows into the port at the positive terminal and out of the negative terminal.

It can be shown that every *n*-port network can be uniquely characterized by *driving-point functions*. These driving-point functions relate voltage and currents at each of the ports and may be any of a number of types, e.g., *impedances, admittances,* or *hybrid* parameters.

A typical set of impedance parameters for an *n*-port network is

$$
[Z] =
\begin{bmatrix}
z_{11} & z_{12} & \cdots & z_{1n} \\
z_{21} & & \cdots & \\
z_{31} & & \cdots & \\
\cdots & & & \\
z_{n1} & & \cdots & z_{nn}
\end{bmatrix}
$$

where z_{ii} is the voltage of port i divided by the current flowing in port i with all other ports open-circuited. This is the *self-impedance* of port i. The other impedances z_{ij}, $i \neq j$, are called *transfer impedances*. Z_{ij} is obtained by shorting port j, causing its port voltage to be zero. The ratio is then taken relating the voltage at port i to the current in port j with all other ports open.

By far the most frequently considered *n*-port network is the two-port. Further discussion will concern two-port networks, although the results apply to more complex *n*-ports.

38. Two-Port Networks. Two-port networks are commonly described in a number of ways. Table 3-4 shows these representations. They are also called *coupling networks, electric transducers, four-pole networks,* or *two-terminal-pair networks*.

The scattering-matrix representation seems unduly complicated but is of great use in distributed-parameter systems such as transmission lines (see Sec. **9**).

A two-port network may have a very large number of internal loops and nodes, but it can be completely defined by knowing four parameters, one of the sets described in Table 3-4. Thus, problems involving very complex networks are handled with ease if they can be represented as a combination of a small number of two-ports. In general, there are three types of manipulations which commonly arise in two-port networks.

Transfer Calculations. It is necessary to find one of the voltages (input or output) in terms of both currents, or it is necessary to find one of the currents in terms of the two voltages.

TABLE 3-4   Matrix Representations of Two-Port Networks

Representation		Deriving equations			
Impedance	$\begin{bmatrix} z_{11} & z_{12} \\ z_{21} & z_{22} \end{bmatrix}$	$z_{11} = V_1/I_1$ $I_2 = 0$	$z_{12} = V_1/I_2$ $V_2 = 0$	$z_{21} = V_2/I_1$ $V_1 = 0$	$z_{22} = V_2/I_2$ $I_1 = 0$
Admittance	$\begin{bmatrix} y_{11} & y_{12} \\ y_{21} & y_{22} \end{bmatrix}$	$y_{11} = I_1/V_1$ $V_2 = 0$	$y_{12} = I_1/V_2$ $I_2 = 0$	$y_{21} = I_2/V_1$ $I_1 = 0$	$y_{22} = I_2/V_2$ $V_1 = 0$
Transmission or chain	$\begin{bmatrix} A & B \\ C & D \end{bmatrix}$	$A = V_1/V_2$ $I_2 = 0$	$B = -V_1/I_2$ $V_2 = 0$	$C = I_1/V_2$ $I_2 = 0$	$D = -I_1/I_2$ $V_2 = 0$
Hybrid	$\begin{bmatrix} h_{11} & h_{12} \\ h_{21} & h_{22} \end{bmatrix}$	$h_{11} = V_1/I_1$ $V_2 = 0$	$h_{12} = V_1/V_2$ $I_1 = 0$	$h_{21} = I_2/I_1$ $V_2 = 0$	$h_{22} = I_2/V_2$ $I_1 = 0$
Inverse hybrid	$\begin{bmatrix} g_{11} & g_{12} \\ g_{21} & g_{22} \end{bmatrix}$	$g_{11} = I_1/V_1$ $I_2 = 0$	$g_{12} = I_1/I_2$ $V_1 = 0$	$g_{21} = V_2/V_1$ $I_2 = 0$	$g_{22} = V_2/I_2$ $V_1 = 0$
Scattering	$\begin{bmatrix} s_{11} & s_{12} \\ s_{21} & s_{22} \end{bmatrix}$				

$$s_{11} = \frac{(V_1/\sqrt{R_{01}}) - I_1\sqrt{R_{01}}}{(V_1/\sqrt{R_{01}}) + I_1\sqrt{R_{01}}} \qquad \frac{V_2}{\sqrt{R_{02}}} + I_2\sqrt{R_{02}} = 0$$

$$s_{12} = \frac{(V_1/\sqrt{R_{01}}) - I_1\sqrt{R_{01}}}{(V_2/\sqrt{R_{02}}) + I_2\sqrt{R_{02}}} \qquad \frac{V_1}{\sqrt{R_{01}}} + I_1\sqrt{R_{01}} = 0$$

$$s_{21} = \frac{(V_2/\sqrt{R_{02}}) - I_2\sqrt{R_{02}}}{(V_1/\sqrt{R_{01}}) + I_1\sqrt{R_{01}}} \qquad \frac{V_1}{\sqrt{R_{02}}} + I_2\sqrt{R_{02}} = 0$$

$$s_{22} = \frac{(V_2/\sqrt{R_{02}}) - I_2\sqrt{R_{02}}}{(V_2/\sqrt{R_{02}}) + I_2\sqrt{R_{02}}} \qquad \frac{V_1}{\sqrt{R_{01}}} + \frac{V_2}{\sqrt{R_{02}}} = 0$$

Transmission Calculations. It is necessary to find the voltage or current at one port in terms of the voltage and current at the other port.

Insertion Calculation. It is desired to ascertain the effect of inserting a two-port network into a system.

39. Two-Port Parameters and Transfer Functions. The term *transfer function*, used in the frequency-domain discussion above, represents the modification in the output imposed by a circuit on its input. If the input is a voltage or current, the output is a voltage or current, respectively, altered in amplitude and shifted in phase. In this context, the voltage-transfer function and current-transfer function are measures of the complete *two-port system*

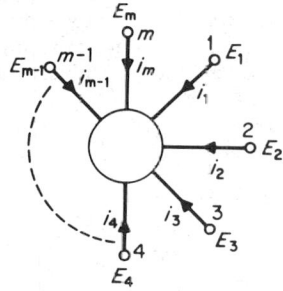

Fig. 3-36. An m-terminal network.

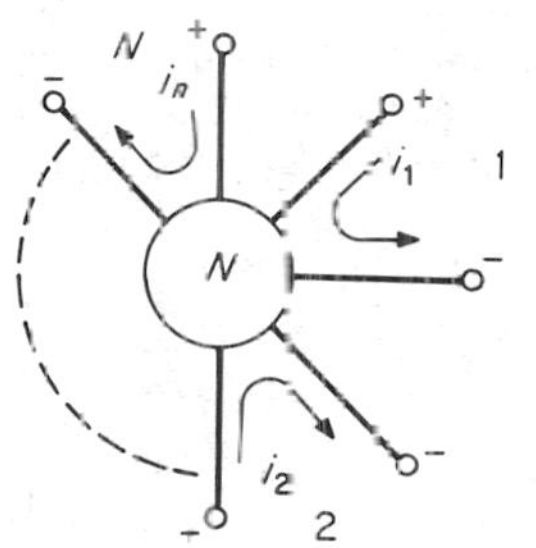

Fig. 3-37. An n-port network.

performance, starting with the *input signal source*, with its internal impedance, and ending with the output signal delivered to particular *load*, the effect of the impedance of which is included.

In another approach to circuit behavior, the properties of two-ports are described in such a way as to be independent of the impedance of the source and load connected to the input and output, respectively; in this latter context, which follows, transfer functions have a different meaning.

The relations between the various quantities in the circuit of Fig. 3-38, regardless of what may be connected at the input and output ports, can be shown to be given by any one pair of the following pairs of equations:

$$I_1 = y_{11}E_1 + y_{12}E_2 \qquad I_2 = y_{21}E_1 + y_{22}E_2 \qquad (3\text{-}34)$$
$$E_1 = z_{11}I_1 + z_{12}I_2 \qquad E_2 = z_{21}I_1 + z_{22}I_2 \qquad (3\text{-}35)$$
$$E_1 = h_{11}I_1 + h_{12}E_2 \qquad I_2 = h_{21}I_1 + h_{22}E_2 \qquad (3\text{-}36)$$
$$I_1 = g_{11}E_1 + g_{12}I_2 \qquad E_2 = g_{21}E_1 + g_{22}I_2 \qquad (3\text{-}37)$$
$$E_1 = AE_2 - BI_2 \qquad I_1 = CE_2 - DI_2 \qquad (3\text{-}38)$$
$$E_2 = DE_1 - BI_1 \qquad I_2 = CE_1 - AI_1 \qquad (3\text{-}39)$$

The coefficients in Eqs. (3-34) and (3-35) are known as the *short-circuit admittance* and *open-circuit impedance* parameters, respectively. The coefficients y_{21} (z_{21}) and y_{12} (z_{12}) are often termed the *forward-transfer admittance (impedance)* and *reverse-transfer admittance (impedance)*, respectively. The coefficients y_{11}, y_{22} and z_{11}, z_{22} are often referred to as *driving-point admittances* and *driving-point impedances*, respectively.

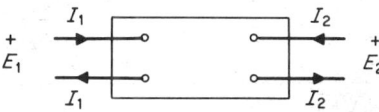

Fig. 3-38. Conventional representation of the two-port circuit. Since the input is understood to be connected to a source independent of the output, the instantaneous current I_1 entering the upper input terminal is equal and opposite to the instantaneous current I_1 leaving the lower input terminal. A similar situation occurs at the output port.

The coefficients A, B, C, and D are known simply as the *AECD* parameters or the *transfer* or *chain* parameters. The expressions $1/A$ and $-1/D$ are called the *open-circuit forward-voltage gain* and *short-circuit forward-current gain*, respectively. The expressions $-1/B$ and $1/C$ are called the *short-circuit forward-transfer admittance* and *open-circuit forward-transfer impedance*, respectively. It can be shown by the reciprocity theorem that $AD - BC = 1$.

40. Transfer Function. In addition to the various relationships which have been defined for two-port networks, others can also be formed. Two very important relationships are *voltage transfer function* and *current transfer function*. Consider a two-port connected as in Fig. 3-39 with one port defined as input and the other as output. Two dimensionless quantities can be defined: voltage transfer function V_2/V_1 and current transfer function I_2/I_1. Other possible ratios between one input and one output variable are also correctly called transfer functions. These have already been dealt with. They are z_{12} and z_{21}, transfer impedances, and y_{12} and y_{21}, transfer admittances.

41. Image Impedance. *Image impedance* or *iterative impedance* is the impedance which when connected to the output of a two-port causes the *driving-point impedance* (the input impedance) to be *identical* to the *terminating impedance* (that which has been connected to the output). The use of image impedances is particularly important in line-amplifier and filter design, in which it is desirable to insert a two-port in the middle of a network string. If the two-port is designed so that its image impedance is identical to the input of that which follows it, it can be inserted without changing *any* of the characteristics of the former network. This means that each two-port can be evaluated independently, and a network of many two-ports in series will have a transfer function which is the simple product of each individual transfer function.

Image-impedance terminations produce both maximum power transfer (subject to the constraints stated in Par. **3-5**) and minimum circuit calculations.

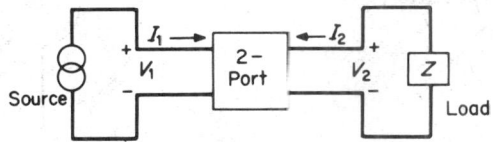

Fig. 3-39. Standard connections for a two-port.

42. Bode Plots. As mentioned earlier, when the ratio is taken of a two-port output parameter to an input parameter, the result is a transfer function. If one divides the output-voltage phasor by the input-voltage phasor, for instance, the result is the voltage gain A_V, which has both a magnitude $|A|$ and a phase θ. It is not a function of time, but it usually is a function of the frequency of the phasors used. This function may be expressed analytically or plotted graphically. Such a plot of the phase and magnitude of the transfer function is extremely useful, especially if the phase and logarithm of the magnitude are plotted vs. the logarithm of the frequency. This representation, known as a *Bode plot*, is particularly useful because it clearly shows the locations of the roots of the transfer function (see Par. **3-20**). With a little experience one can do a considerable amount of analysis and design of signal and control circuits directly from the Bode plot. A typical Bode plot is shown in Fig. 3-40.

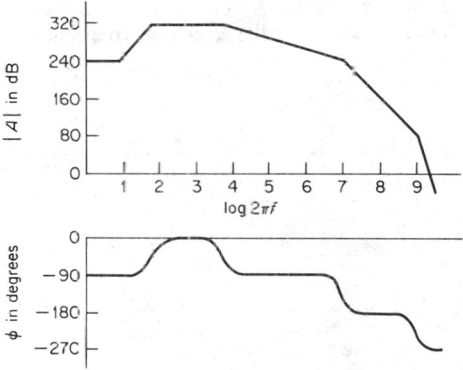

Fig. 3-40. A typical Bode plot.

43. Feedback Circuits. Many of the complex circuits used in modern electronics incorporate the concept of *feedback*, in which a portion of a circuit's output is returned to its input either to augment or to reduce the original input signal. If the feedback tends to increase the input amplitude, it is called *positive feedback*. If it tends to decrease the input amplitude, it is called *negative feedback*. Positive feedback increases the circuit gain at the expense of the range of frequencies it can amplify. Negative feedback enhances all the desirable properties of a circuit at the expense of reduced gain. Negative feedback is extremely useful for nearly all control circuits and many signal circuits. Figure 3-41 shows the general feedback configuration, and Table 3-5 shows the results of negative feedback on circuit performance.

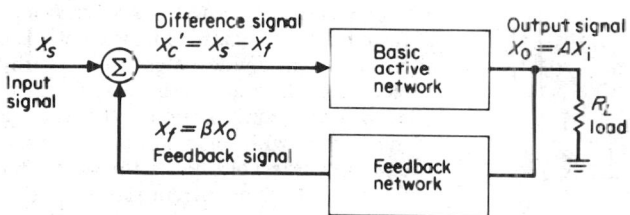

Fig. 3-41. Basic feedback-circuit configuration.

44. Control Theory. The basic feedback configuration shown in Fig. 3-41 can be applied for most situations where equipment is designed to hold a close tolerance between some desired quantity and the output of some large production system, often referred to as a *plant*. An example might be a home heating system, in which the actual room temperature is held as close as possible to the desired setting on the thermostat. In Fig. 3-41 the plant corresponds to the basic

TABLE 3-5   Effect of Negative Feedback on Circuit Performance

	Type of amplifier			
	Voltage	Transconductance	Current	Transresistance
Output resistance	Decreases	Increases	Increases	Decreases
Input resistance	Increases	Increases	Decreases	Decreases
Reduces	Voltage gain	Transconductance	Current gain	Transresistance
Bandwidth	Increases	Increases	Increases	Increases
Distortion	Decreases	Decreases	Decreases	Decreases

active network, the feedback network is called the *controller*, X_s is the desired quantity, and X_f is the actual quantity.

The object of any feedback control system is to maintain a close relationship between desired and actual quantities, including situations in which the desired value is changed suddenly or the system is subjected to strong outside influences. The analysis of feedback control systems is covered in detail in Sec. **17**.

45. Stability. A major concern of control-circuit analysis is that a configuration which produces negative feedback at one frequency will produce positive feedback at some other fre-

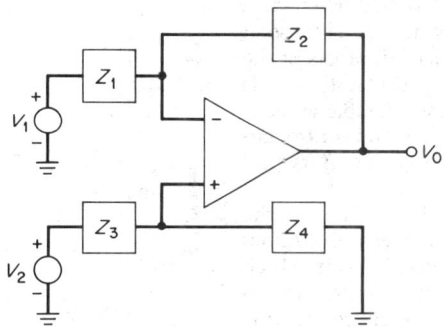

Fig. 3-42. A typical operational-amplifier circuit.

quency, causing the circuit to be unstable. An unstable circuit may oscillate wildly or perform poorly in some other manner and is to be avoided. Various stability tests have been developed to locate potentially unstable circuit conditions easily.

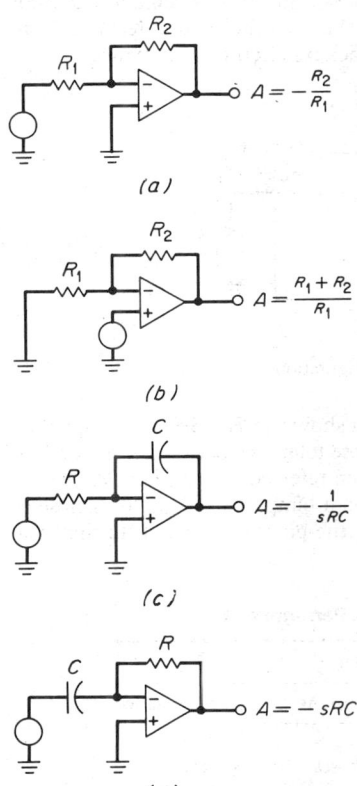

Fig. 3-43. Common operational-amplifier circuits: (a) inverting amplifier; (b) noninverting amplifier; (c) integrator; (d) differentiator.

46. Noise. Noise can be a problem in many circuits, especially those processing very small signals or handling heavy current in circuits adjacent to circuits using much smaller signals. Noise is classified as *extrinsic* or *intrinsic*, depending on whether it arises from outside or inside the circuit of concern. Extrinsic noise is *coupled* into the circuit by magnetic fields (from nearby current flow), electric fields (from nearby charged objects), by radiation, or by having some portion of the noise-generating circuit in common with some portion of the noise-receiving circuit. Intrinsic noise takes the form of random fluctuations in voltage or current caused by the particular characteristics of circuit devices at the molecular level. Thermal agitation of material (Johnson noise), processes at semiconductor junctions (shot noise), crystal defects (popcorn noise), and surface phenomena (contact noise) all contribute to unwanted signals in low-level circuits.

Noise can be reduced by a proper design which eliminates or reduces the *source* of the noise, reduces its coupling to the susceptible circuit, and reduces the susceptibility to the effects of noise. Each of these methods must be considered for its overall effect. Techniques which are usually useful include shielding, grounding, guarding, reduction of circuit impedance, reduction of power-supply impedance, filtering, decoupling, balancing, optical coupling, use of differential amplifiers, isolation and neutralizing transformers, modulation, coding, limitation of system bandwidths, and the use of transient suppressors.

47. Operational Amplifiers. Many modern electronic circuits use operational amplifiers, which are high-gain voltage amplifiers. If the amplifier has two nongrounded input terminals, it is a *differential amplifier*, since it amplifies the voltage difference between the two inputs. Operational amplifiers are nearly always used in feedback circuits, in such a way that the properties of the circuit are controlled by the circuit parameters and not by the characteristics of the operational amplifier.

In analyzing an operational-amplifier feedback circuit, one uses the properties of high gain and high input impedance to establish a *virtual* short between the amplifier input terminals. This means that the voltage between the input terminals is essentially zero, but no current flows in them. This allows the circuit to be solved by writing one or two node equations. Referring to Fig. 3-42, one can use the virtual short to work equations:

$$V_- = V_+ \qquad \frac{V_1 - V_-}{Z_1} = \frac{V_- - V_0}{Z_2} \qquad \frac{V_2 - V_+}{Z_3} = \frac{V_+}{Z_4}$$

which reduce to

$$V_0 = V_1 \frac{-Z_2}{Z_1} + V_2 \frac{Z_4}{Z_1} \frac{Z_1 + Z_2}{Z_3 - Z_4} \tag{3-40}$$

Equation (3-40) can be used to construct a multitude of transfer functions with the appropriate substitutions for Z_1 through Z_4. Figure 3-43 shows some common operational-amplifier circuits and their transfer functions.

CHARACTERISTICS OF SPECIFIC NETWORK CONFIGURATIONS

A number of networks are both very common and illustrative of many fundamental concepts.

48. Frequency- and Time-Domain Properties of Single-Port Circuits. The frequency- and time-domain properties of examples of simple equivalent circuits of resistors, capacitors, and/or inductors in one-port (two-terminal) configurations are shown in Table 3-6. The time-domain response given is the response current to a step function of voltage E_0, for cases a, b, and c, and the response voltage to a step function of current I_0 for case d. The frequency-domain description comprises the terminal (phasor) impedance Z_{ab} and its phase angle θ_{ab}.

Time Constant. The characteristic exponential time-domain response of the RL and RC circuits has led to the extremely important concept of the *time constant*. The time constants L/R (for the RL circuit) and RC (for the RC circuit) represent the amount of time, in seconds, before the current reaches within a certain fraction of its steady-state value; for the RL circuit, this is the time required for a rise to about 63% of the final value and for the RC circuit a drop to about 37% of the initial value. The time-constant concept, i.e., the time which must elapse before a physical phenomenon comes close to its final value, is widely employed not only in circuits but also in dealing with mechanical and thermal effects.

The RC circuit is often used to produce a controlled time delay, e.g., in analog timers.

Resonance. The RLC circuit displays *resonances* in the frequency domain. In the series RLC circuit, the minimum impedance and in-phase resonances occur when $2\pi fL = 1/2\pi fC$ or for a frequency $f = 1/2\pi \sqrt{LC}$; the natural frequency of oscillation f_n in the time domain $f_n = \sqrt{(1/LC) - R^2/4L^2}/2\pi$ differs from the other resonant frequencies only slightly for small values of R. The series RLC circuit can, in principle, be used as a tuned circuit to select a signal of a particular frequency (or a signal with a narrow frequency spectrum); such use is limited because few practical electronic circuits are equivalent to true voltage sources. The selective properties of the series resonant circuit can be represented by the normalized universal[1] resonance curve shown in Fig. 3-44, based on the approximation that

$$\frac{\text{Current at frequency } f}{\text{Current at resonance } f_r} = \frac{1}{1 + (a/Q) + ja(2 + a/Q)/(1 + a/Q)} \approx \frac{1}{1 + j2a} \tag{3-41}$$

where

$$a = Q \frac{f - f_r}{f_r} = Q \frac{\text{cycles off resonance}}{\text{resonant frequency}}$$

$a/Q \ll 1$ for quite practical values of Q, and Q is a constant, independent of frequency.

The parallel RLC circuit of Table 3-6 is of considerably more importance than the series RLC circuit; it is widely used as a *frequency-discriminating circuit*, and its characteristics for the case in which it is excited with a current source often correspond to realistic situations. Typical resonance characteristics are shown in Fig. 3-45. The maximum-impedance resonant frequency and

[1]Originally derived by F. E. Terman.

"in-phase" resonant frequency differ from each other, and neither coincides with the *nominal frequency* $f_r = 1/2\pi\sqrt{LC}$ or the natural frequency $f_n = \sqrt{(1/LC) - R^2/4L^2}/2\pi$. For physically realizable high values of Q, the differences between the various resonant frequencies become insignificant. The normalized universal resonance curve of Fig. 3-44 can be applied with reasonable accuracy to the parallel resonant circuit excited with a current source by connecting the ordinate of the curve to the ratio of voltage developed at frequency f to that developed at f_r.

TABLE 3-6   Frequency- and Time-Domain Properties

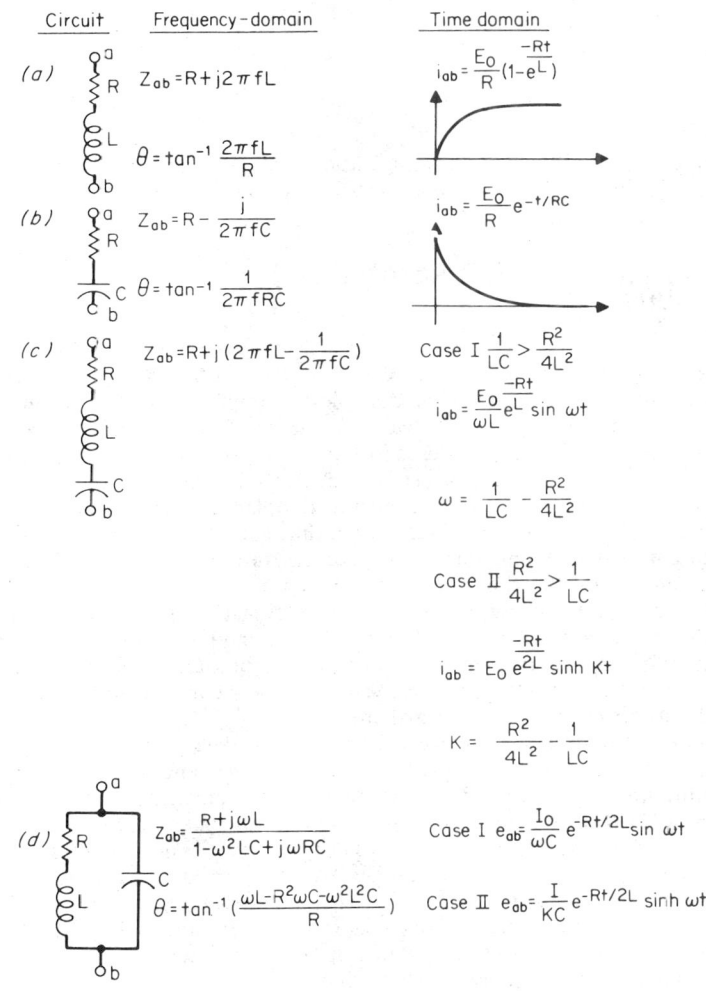

49. Frequency- and Time-Domain Properties of Two-Port Circuits.
The *RL* circuit and, particularly, the *RC* circuit can appear in equivalent circuits as two-port networks; the simple two-port configurations are shown in Fig. 3-46.

The amplitudes and phase angles of the transfer functions of the *RC* networks appear in Fig. 3-47 (the independent variable is normalized as $K = 2\pi fRC$ so that the curves are of universal application). The time-domain responses for a typical pulse signal are given in Fig. 3-48.

Some limiting conditions are of special interest. For $K \gg 1$, the frequency-domain transfer function of the *RC* network of Fig. 3-46c becomes very nearly $E_{out}/E_{in} = 1/K$; this corresponds to an output inversely proportional to the frequency of the input. This is a time derivative of the input; the *RC* network must be followed by a distortionless amplifier of appropriate gain to increase the proportionality factor to unity.

The equivalent circuits of substantially all *RC* interstage coupling networks in electronic amplifiers reduce to the $K \ll 1$ configuration and condition of Fig. 3-46a at their low-frequency cutoff and the $K \gg 1$ configuration and condition at their high-frequency cutoff; hence, the time-domain signal distortions of such amplifiers can be interpreted in the light of such differentiating and integrating effects.

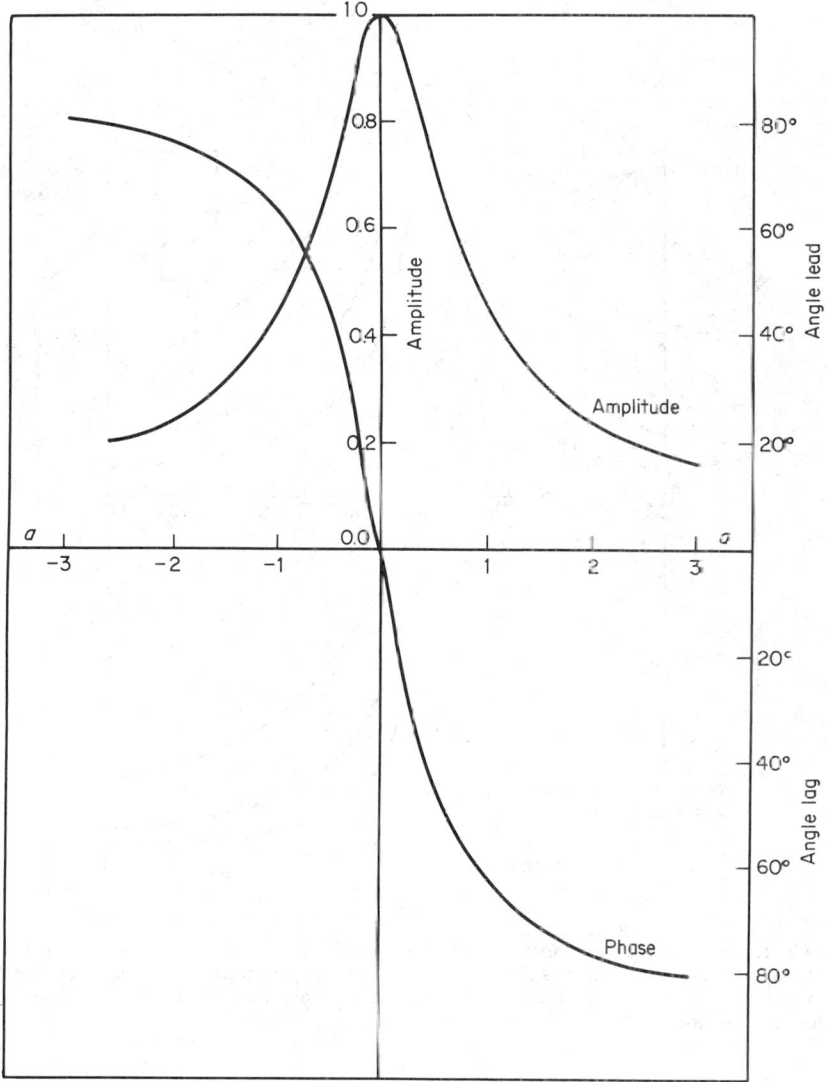

Fig. 3-44. Universal resonance curve for constant Q. Ratio of current and phase angle at frequency f to current at f_r as a function of normalized variable $a = Q(f - f_r)/f_r$, $f_r = 1/2\pi\sqrt{LC}$.

50. Coupling and Coupling Networks. Coupling networks are used to interconnect devices, to interconnect apparatus and devices, etc. Their functions include (1) isolation of dc components, a necessary feature for cascading electron-tube or transistor amplifiers (except for transistors in a complementary configuration), (2) shaping the (frequency-domain) amplitude or phase-angle transfer function, (3) impedance matching, and (4) waveform preservation or correction, i.e., direct control (as an alternative to the indirect frequency-domain approach) of time-domain characteristics. In each case, the circuit function depends not only upon the equivalent-circuit parameters of the coupling network itself but also upon the equivalent-circuit parameters of the devices or apparatus being coupled (since it is rare in electronics engineering to encounter devices or apparatus which are perfect current or voltage sources).

Coupling Networks for Audio and Video Amplifiers.   Audio and video amplifiers are essentially low-pass amplifiers, the response of which excludes a narrow range of frequencies from zero frequency (dc) up to some low-frequency cutoff. Equivalent circuits for some examples of such amplifiers appear in Fig. 3-49.

Figure 3-49c is a simple example of a compensated video-amplifier coupling circuit in which

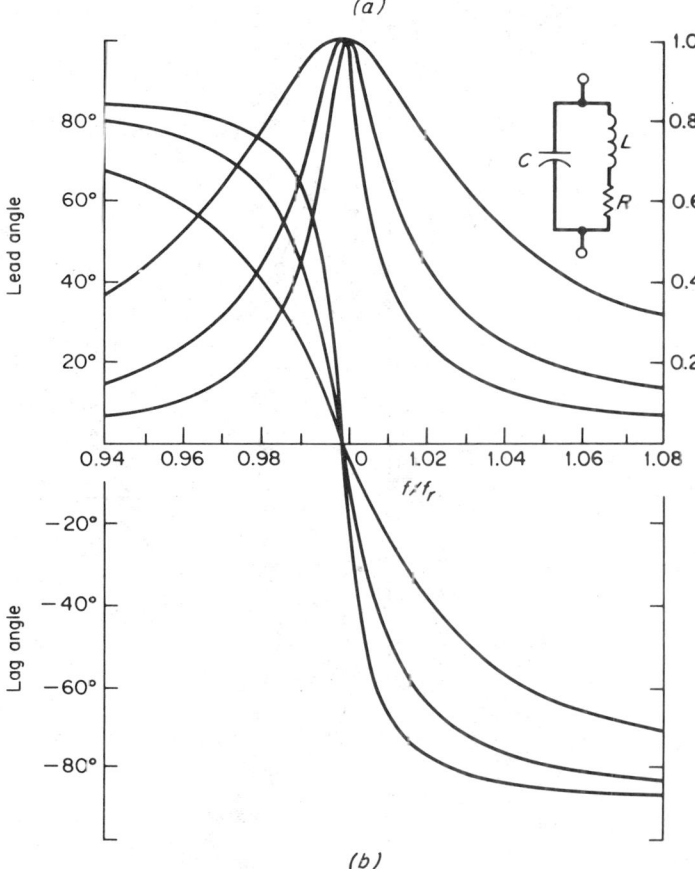

Fig. 3-45. (*a*) Parallel-circuit resonance curve (for constant Q). The magnitude of the impedance relative to the maximum impedance is plotted vs. the normalized frequency parameter f/f_r for various values of Q. (*b*) The phase angle of the impedance relative to the phase at resonance is plotted vs. the normalized frequency parameter f/f_r for various values of Q.

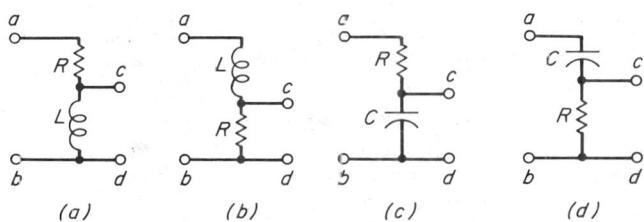

Fig. 3-46. Arrangements of RL and RC circuits as two-port networks. Cases (*a*) and (*b*) occur infrequently in equivalent circuits because an inductor can seldom be realistically treated as pure, i.e., without an equivalent resistance. On the other hand, (*c*) and (*d*) represent realistic equivalent circuits for a number of important cases.

an inductor L_L is inserted to compensate for the effect of the shunt capacitances C_{pk} and C_{gk} and increase the cutoff frequency.

Coupling Networks for Bandpass Amplifiers (IF and RF Amplifiers). Since most electronic amplifiers are used to process signals rather than a single-frequency sinusoidal ac voltage or current, their transfer functions must be essentially bandpass in order to provide for the reproduc-

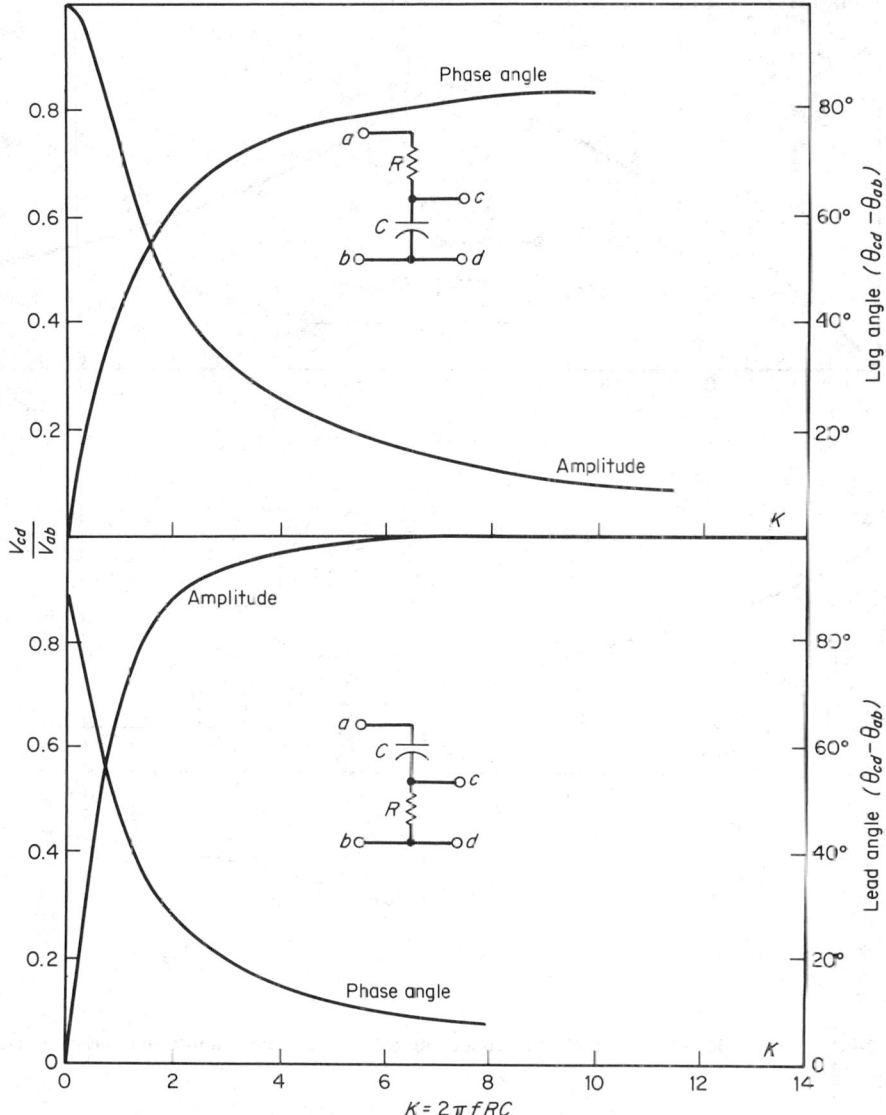

Fig. 3-47. Amplitude and phase characteristics of series RC networks. The plots show the magnitude of the ratio of output voltage V_{cd} to input voltage V_{ab} as well as the phase shift between output and input $\theta_{cd} - \theta_{ab}$.

tion of both the signal carrier and its side frequencies. At the same time the passband must be restricted in its frequency width to reject the components of *adjacent* (in frequency) unwanted signals.

The coupling networks in such amplifiers are universally of the tuned type, comprising capacitors and inductors in various configurations. Figure 3-50 shows some examples of simple bandpass coupling networks. The coupling arrangement of Fig. 3-50a has an optimum response to a single frequency and can be expected to attenuate side frequencies. The overall amplifier response can be converted to a bandpass response by staggering the resonant frequencies of successive stages. The coupling network of Fig. 3-50b lends itself better to a bandpass response. The characteristic is highly dependent upon the coefficient of coupling between the primary (L_1) and secondary (L_2) inductors. Examples of the effect of coupling coefficient upon the frequency-domain amplitude and phase components of the transfer functions of the double-tuned circuit are shown in Fig. 3-51. By using different coefficients of coupling in successive stages, a relatively flat amplitude transfer function is quite readily obtainable.[1]

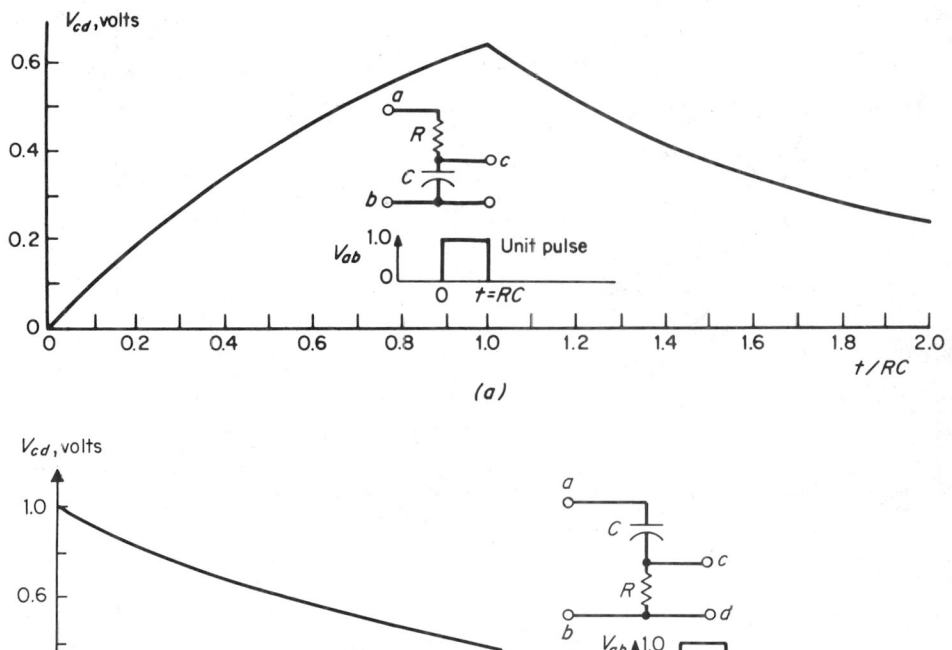

(a)

(b)

Fig. 3-48. Time-domain response of RC series-circuit output for input pulse of unit amplitude and duration RC seconds.

51. Filters. A filter is a two-port network (in general) which has been designed to transmit freely sinusoidal signals within one or more frequency bands and to attenuate sinusoids of other frequencies substantially. Many coupling networks perform a filter function by shaping the overall transfer function of the amplifier within which they are used.

Filters are usually characterized by their *voltage transfer function*. They are classed according to their voltage transfer function as low-pass, high-pass, bandpass, and band-stop, as sketched in Fig. 3-52.

[1] For a detailed treatment of tuned amplifiers see R. F. Shea (ed.), "Amplifier Handbook," McGraw-Hill, New York, 1966, Chap. 24, pp. 24-1 to 24-21.

A *low-pass filter* passes steady-state sinusoids of all frequencies from zero (continuous, or dc excitation) up to some cutoff frequency f_1. Ideally, such a filter would pass no sinusoids of frequency higher than f_1; in practice, the attenuation characteristic for unwanted signals falls off above the cutoff frequency with a finite slope. It is physically impossible (indeed undesirable) to obtain an infinitely steep cutoff above f_1 because such a cutoff entails an infinite phase shift. A *bandpass filter* passes all frequencies from some f_2 to f_3, and a *high-pass filter* passes all frequencies above some f_4. A *band-stop filter* passes all frequencies except those between frequency f_1

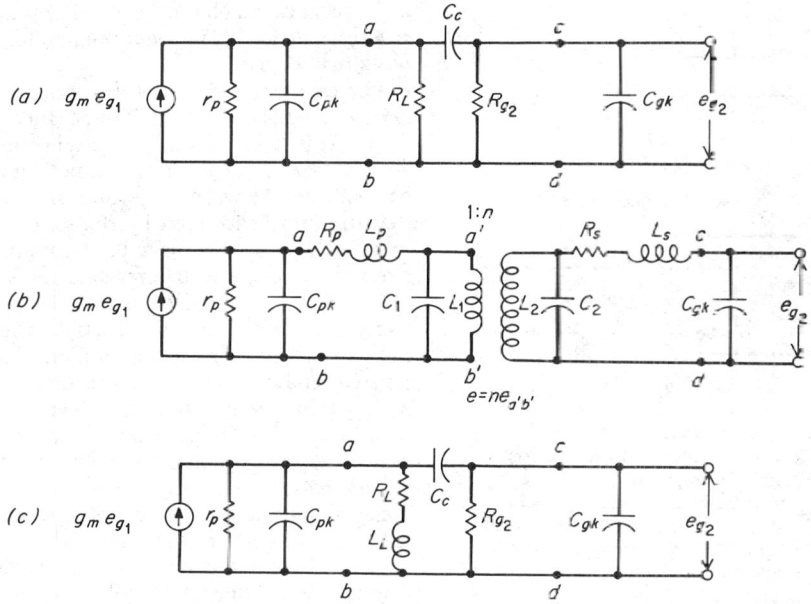

Fig. 3-49. Equivalent circuits for the interstage coupling in three examples of electron-tube low-pass (excluding dc) amplifier. In each case the portion of the circuit to the left of the terminals a and b is the equivalent circuit of the input electron tube, where e_{g1} is the signal at its grid. The two-port network appearing between terminals a and b (input) and c and d (output) is the coupling network, and the circuit to the right of terminals c and d is the equivalent circuit of the input to the next stage.

and f_2. (Figure 3-52 shows a second stop band between f_3 and f_4.) All the attenuation characteristics outside the desired band have a finite slope for the same reason as that of the low-pass filter. The synthesis of filters to obtain desired characteristics is discussed in Sec. 12.

52. Phase-Locked Loops. Figure 3-53 shows a typical phase-locked-loop (PLL) system available on a single integrated-circuit chip. It consists of a *voltage-controlled oscillator* (VCO) (varying the dc voltage varies the frequency of the sine-wave oscillator), a *phase-sensitive detector* (the output is a dc voltage proportional to the phase *difference* between its two inputs), a high-gain amplifier, and a low-pass filter. It is connected in a loop arrangement so that any phase (frequency) difference between the VCO and the input is minimized. Since the VCO output is a pure sine wave, output 1 is a sine wave *locked* to the predominant frequency of the input, thus causing the PLL to function as a high-Q resonant filter. If output 2 is used, a dc voltage is generated proportional to the VCO *frequency* (and hence the input frequency), causing the circuit to act as an FM demodulator or detector.

53. Microelectronic Configurations. The circuit parameters that can be constructed in microelectronic configurations are limited to resistances and capacitances, excluding, for practical purposes, inductors. Tuned circuits, however, can be achieved by the use of piezoelectric (electromechanical) components. A single-tuned circuit, for example, can be realized with a piezoelectric resonator with an electric-to-mechanical transducer at the input and a mechanical-to-electric transducer at the output. A double-tuned electric coupling network can be realized with two loosely coupled mechanical elements.

In microelectronic devices, neither very low nor very high values of resistance can be fabricated. The equivalent circuits of resistors include significant and unavoidable shunt capacitances, which at high frequencies must be treated as distributed elements.

54. Impedance Matching. As pointed out in Par. **3-5**, *maximum transfer of power from a source to a load takes place when the source impedance and load impedance are complex conjugates*. In certain cases in which a source and its load do not have complex-conjugate impedances (because their design properties are dictated by other factors, for example) and an impedance match is desired, an *impedance-matching device* or *impedance-matching network* may be used.

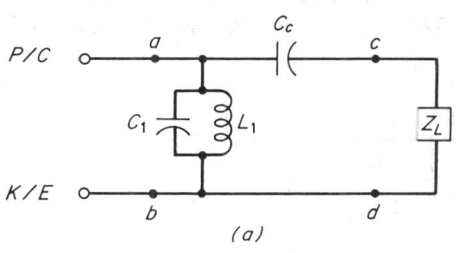

(a)

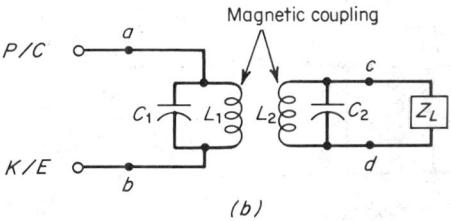

Magnetic coupling

(b)

Fig. 3-50. Simplified schematic diagrams of (a) single-tuned and (b) double-tuned bandpass coupling networks. The impedance Z_L represents the impedance presented by the load or by the input to a following stage. The corresponding equivalent circuits would include the parameters of the input and output electron-tube or transistor input and the resistances of the inductors L_1 and L_2. P/C refers to the plate of an electron tube or collector of a transistor. K/E refers to the cathode of an electron tube or emitter of a transistor.

The essential problem is illustrated in Fig. 3-54. A generator of internal impedance $Z_g = R_g + jX_g$ is to be coupled to a load of impedance $Z_L = R_L + jX_L$. A device or network is to be connected between the source and load so that the impedance seen by the source is $Z'_L = R_g - jX_g$; that is, R_L is to be transformed to a resistance R_g and X_L to a reactance $-X_g$.

For an audio or video source with a relatively constant, largely resistive internal impedance when the load is similarly largely resistive and constant, a transformer is used with a transformation ratio n given by $n = \sqrt{R_L/R_G}$. Such transformers are of the untuned conventional ferromagnetic-core closely coupled winding types and can be designed to operate reasonably well from a signal-handling point of view over a frequency range such that the ratio of high-frequency cutoff to low-frequency cutoff can approach 1,000:1.

For a high-frequency bandpass application in which the bandwidth is a small fraction of the center frequency, a *tuned impedance-matching network* is feasible. Examples of such networks are shown in Fig. 3-55.

The components of the selected network (one of those shown in Fig. 3-55 or one of a wide variety of alternatives) are selected so that the impedance Z_{ab}, with the actual load ($Z_L = R_L + jX_L$) connected between terminals c and d, has the desired resistive and reactive components. Analytically, the design values are determined by solving two simultaneous equations, such as (in the case of Fig. 3-55c, for example)

$$R_{ab} = \frac{X_1^2 R_L + X_1 X_L R_L + X_1 X_2 R_L}{(X_1 - X_L - X_2)^2 + R_L^2} \qquad X_{ab} = \frac{X_1^2(X_1 + X_2) - X_1 R_L^2}{(X_1 - X_L - X_2)^2 + R_L^2} \qquad (3\text{-}42)$$

with $R_{ab} = R_g$ and $X_{ab} = -X_g$, the specified internal impedance components of the source. (In this particular network, physically realizable values of X_1 and X_2 restrict the application to cases in which $R_L < R_g$; for $R_L > R_g$ the network of Fig. 3-55d can be used.) The two-element impedance-matching networks have uniquely determined values of components since there are two variables and two conditions to be met. With networks having three or more components, the components can be chosen from a range for convenience, on account of some particular component availability or to obtain some desired phase shift. When the load impedance is a pure resistance, charts of values of the components[1] are available.

55. Analog-Digital Interfaces. Since nearly all information in the physical world resides in the relative magnitudes of infinitely variable continuous quantities while calculations are usually carried out using pure numbers to represent worldly quantities, it often becomes necessary to design interfaces to convert from analog (infinitely variable) to digital (represented

---

[1]See for example F. E. Terman, "Radio Engineers' Handbook," McGraw-Hill, New York, 1943, pp. 206–215.

by a finite number string) quantities. These interfaces are called *converters*. *Analog-to-digital* converters (A/D) produce a number representation of the analog input, while *digital-to-analog* converters (D/A) produce a continuous output representing a number string at the input. They are an important part of the ability of a digital computer to interact with the physical world.

D/A converters accept a digital input as logical pulses on parallel input lines and output a

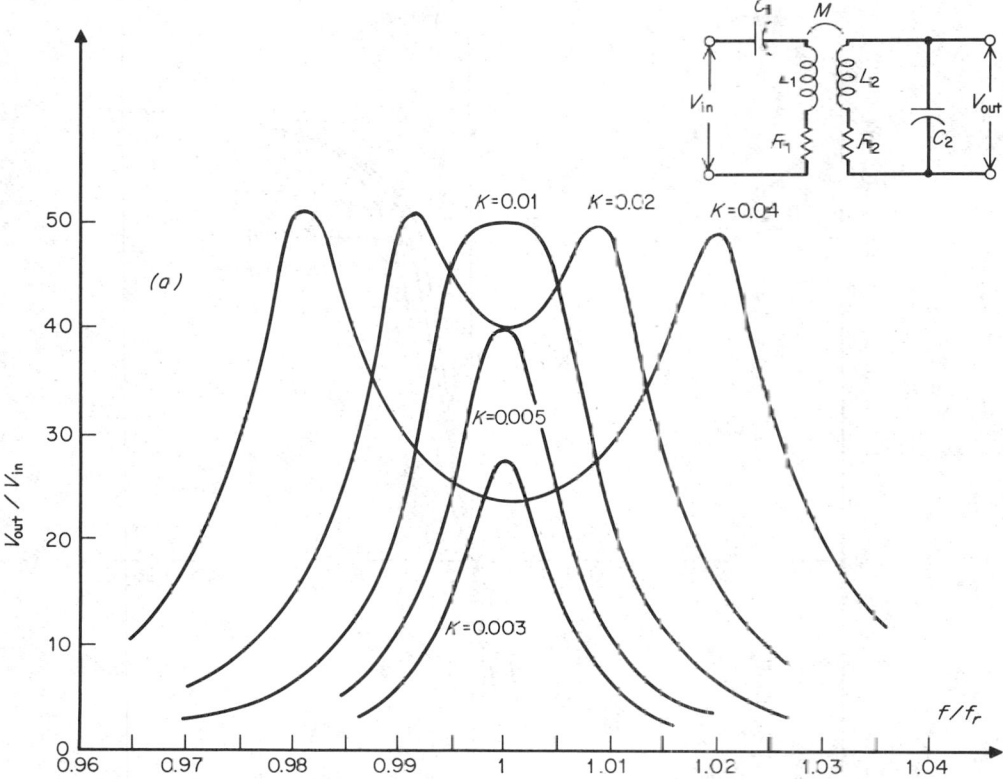

Fig. 3-51. Frequency-response characteristics of a double-tuned circuit. The magnitude ratio (*a*) and the phase difference (*b*) between the output voltage and the input voltage are plotted vs. a normalized frequency parameter f/f_r for several values of coupling coefficient $K = M/\sqrt{L_1 L_2}$. See following page for Fig. 3-51 (*b*).

continuous waveform which represents the input numbers. This conversion is usually accomplished by either a *binary weighted-resistor* network or a *matched-resistor-ladder* network. In either case several currents are weighted according to the *significance* they represent (their place in the number) and summed at the input of an operational amplifier. The matched-resistor-ladder network is usually more accurate, especially if the digital number contains eight or more digits.

A/D converters produce digital output (usually in parallel) from a sampled analog waveform. There are three basic techniques for performing the conversion. The most basic (and slowest) technique is to begin a digital counter at zero and count up. A D/A converter generates the analog representation of the accumulated count, which is compared with the analog signal to be converted. When the two signals are equal, the counter is stopped and its contents appear at the output of the A/D.

A second type of A/D uses essentially the same hardware except that the counter performs a *successive approximation* instead of a uniformly increasing count. It counts up its most significant digit first and compares the two signals. If the generated signal exceeds the signal to be converted, the counter resets this digit to zero and proceeds to increment the adjacent digit (of lesser significance). If the generated signal does not exceed the signal to be converted, the counter holds the digit and proceeds to the adjacent digit. Although the logic of the successive-approximation method is more complicated, it is significantly faster than the first method.

A third type of A/D is quite fast and relatively unaffected by noise in the signal to be digitized. It is known as *dual-slope integration*, in which the time required to integrate the unknown signal up to a fixed limit is compared with the time required to integrate a reference signal to the same limit. This method is extremely useful for digital voltmeters or other general purpose A/D converters, since a scale change can be made easily and without affecting the percentage accuracy of the converter.

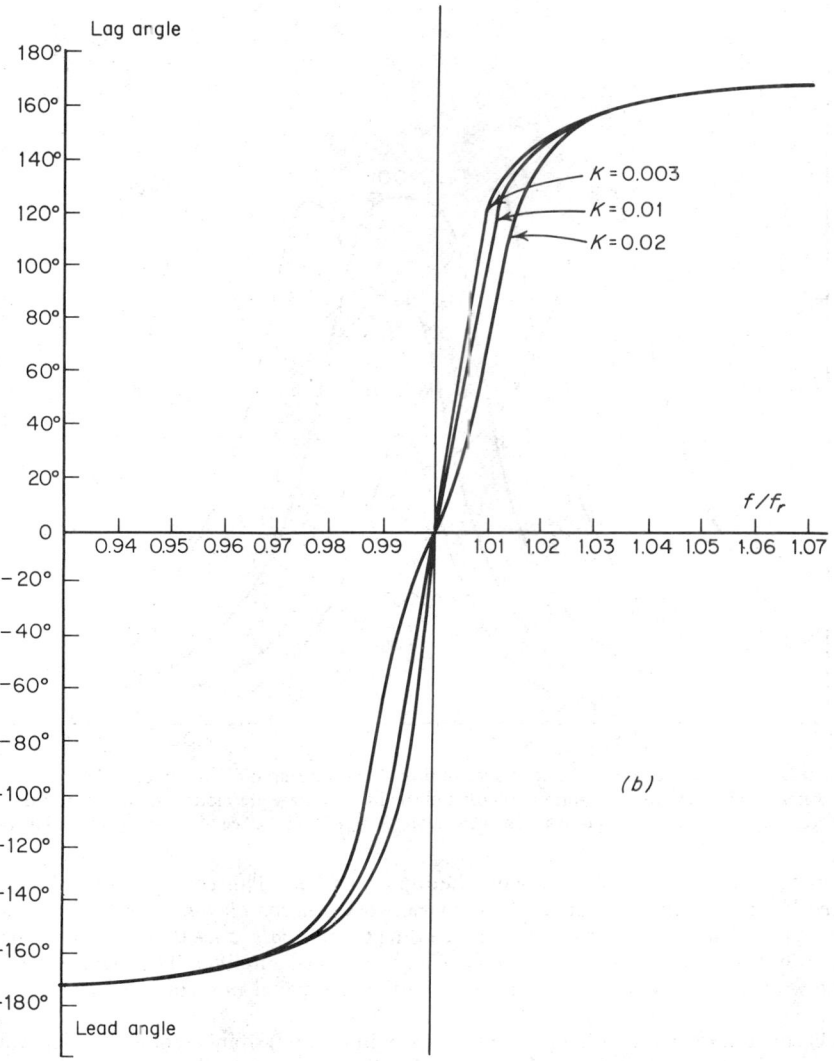

Fig. 3-51. (b) Phase angles for various values of k.

CIRCUITS WITH DISTRIBUTED PARAMETERS[1]

56. Distributed-Parameter Concepts. At low frequencies, where the dimensions of the circuit components are small compared with the wavelengths of the signals considered, the *lumped-parameter* concept provides an adequate description for analysis. The *displacement-current* term in Maxwell's equations is negligible compared with the *conduction-current* term, and the variations of the magnetic and electric fields in space can be neglected. At very high frequencies, however, this is not the case, since the dimensions of the device are comparable to the

[1]The author received valuable assistance in preparing this material from Andrew T. Perlik.

wavelengths of the propagating signals. We associate the energy stored in the magnetic field with the *distributed inductance* of the structure, the energy stored in the electric field with the *distributed capacitance*, and the power loss with the *distributed resistance*. The concept of distributed parameters is useful together with the concepts of equivalent voltage and current because it enables us to apply many of the techniques and properties of low-frequency analysis to high-frequency structures.

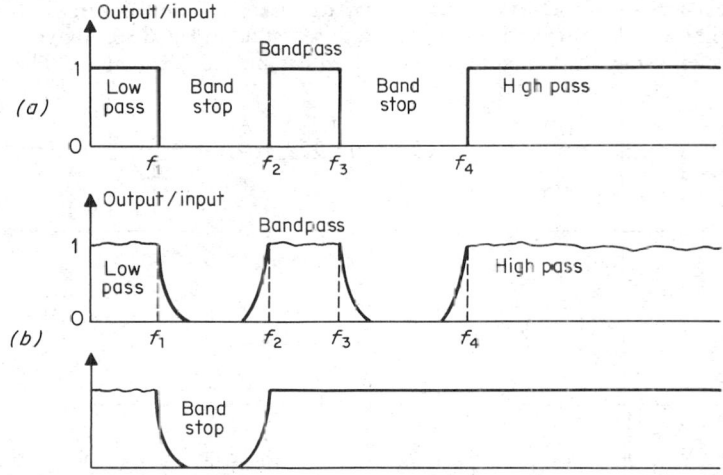

Fig. 3-52. Filter responses: (*a*) ideal responses and (*b*) examples of actual responses.

A *transmission line* is a structure consisting of two or more parallel conductors which guides *transverse electromagnetic* (TEM) *waves*. TEM waves have their electric and magnetic fields perpendicular to the direction of propagation. A *waveguide* is a hollow structure with conducting walls which supports transverse electric (TE) and transverse magnetic (TM) waves. A waveguide is a more desirable structure to use than a transmission line for applications where the wavelength is less than 10 cm. Transmission lines and waveguides are treated in detail in Sec. **9**.

57. Modes and Boundary Conditions. The term *mode* is used to describe the electric and magnetic *field pattern* in a waveguide. Specific modes are identified by specifying the type

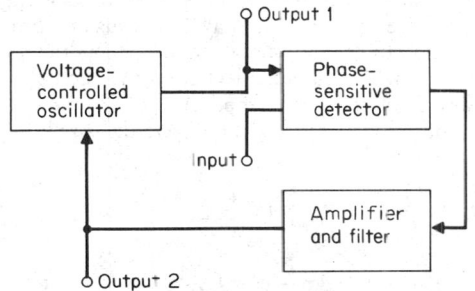

Fig. 3-53. A typical phase-locked-loop system.

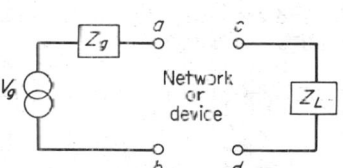

Fig. 3-54. General form of the impedance-matching configuration. The network or device ideally transforms Z_g so that the driving-point impedance Z_{ab} = $R_g - jX_g$. $Z_g = R_g + jX_g$, $Z_L = R_L + jX_L$.

of wave, for example, TM, and by specifying the number of relative maxima occurring in the field configuration of the cross section with subscripts. For a *rectangular waveguide* (rectangular cross section) TE$_{mn}$ denotes that the electric field is transverse to the direction of propagation and that the electric field has m relative maxima occurring along the width of the cross section and n relative maxima occurring along the height of the cross section.

For *circular waveguides* the subscript m denotes the number of relative maxima occurring in the radial field component in the angular direction, and the subscript n denotes the total number of relative maxima and minima of the angular field component in the radial direction.

The *boundary conditions* on the electric and magnetic fields at the boundary between two different media are derived by applying Maxwell's equations to elemental volumes containing the boundary or closed contours cutting the boundary. For perfect conductors, i.e., resistanceless conductors, the *normal component of the magnetic field intensity* $\mathbf{H}$ is zero and the *tangential component of the electric field* $\mathbf{E}$ is zero at the surface of the conductor.

58. Calculation of Distributed Parameters. A two-wire transmission line is shown together with its distributed-parameter representation in Fig. 3-56. A given transmission line is characterized by the values of its *distributed parameters* r, l, g, and c. These quantities

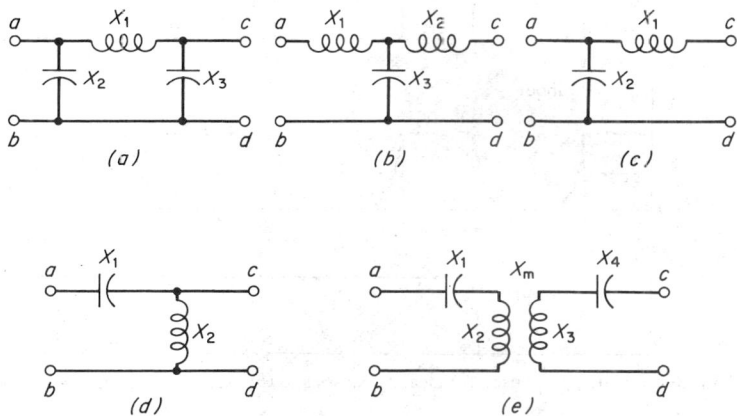

Fig. 3-55. Examples of impedance-matching networks.

are given per unit length of the line. Formulas are available for substantially all simple two-wire transmission-line configurations;[1] however, for any particular configuration these parameters are, in general, frequency-dependent because of the skin effect.

The *skin effect* is the phenomenon of increasing current-density concentration in the surface layers of the conductors as frequency increases; it arises because the reactance associated with the possible current paths is smallest near the conductor surface. The *skin depth* is given by $\sqrt{\sigma\pi\mu f}$, where f is the frequency of the current, μ the permeability of the medium, and σ its conductivity; it is the depth at which the current density has decreased to $1/\varepsilon$ (37%) of its value at the surface of the conductor, where ε is used for the exponential to avoid confusion. This effect is not restricted to the conductors of transmission lines only but occurs in all conductors, including the conductors of circuit elements and those which interconnect circuit elements.

In Fig. 3-56, the voltage $e(x)$ is attenuated because of the series impedance $Z_s = r\,dx + j\omega l\,dx$, and the current $i(x)$ is attenuated by the shunt admittance $g\,dx + j\omega c\,dx$. Applying Kirchhoff's laws yields

$$e(x + dx) - e(x) = -\left[r\,dx\,i(x) + l\,dx\,\frac{\partial}{\partial t}\,i(x) \right]$$

$$i(x + dx) - i(x) = -\left[g\,dx\,e(x) + c\,dx\,\frac{\partial}{\partial t}\,e(x) \right]$$

(3-43)

Dividing by dx and taking the limit as $dx \to 0$ gives

$$\frac{\partial}{\partial x}\,e(x) = -ri(x) - l\frac{\partial}{\partial t}\,i(x) \qquad \frac{\partial}{\partial x}\,i(x) = -ge(x) - c\frac{\partial}{\partial t}\,e(x)$$

(3-44)

which are the equations in terms of the voltage and current as a function of the distance from some reference position.

This equivalent circuit gives excellent results in comparison to experimental data for low-frequency power transmission, communication, and some high-frequency applications for lines

[1] R. W. P. King, "Transmission Line Theory," Dover, New York, 1965, Chap. 1.

of uniform spacing and conductor size in which end effects are neglected. The *distributed series resistance r* in ohms per unit length is the resistance of a unit length of the two conductors. The *distributed shunt conductance g* in siemens per unit length is the equivalent conductance of a unit length of the lines caused by dielectric losses in the medium between the conductors. The *distributed series inductance l* in henrys per unit length is calculated by assuming that a unit current travels along an infinite extent of transmission line, going in one conductor and returning in the other, and calculating the resulting flux linkages in the section of unit length. The *distributed shunt capacitance c* in farads per unit length is calculated by assuming a constant voltage *V* between the two conductors, extended to infinity in either direction from the unit section, and determining the corresponding charge *q* in coulombs on the unit section

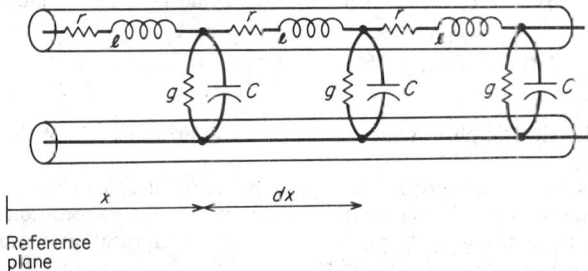

Reference
plane

Fig. 3-56. Two-wire transmission line and parameters.

Voltage and current definitions are introduced on an equivalent basis and are of value because many of the circuit-analysis techniques valid at low frequencies are also applicable at microwave frequencies.

Propagating *waveguide modes* have the following properties:[1] (1) power transmitted is given by an integral involving the transverse electric and transverse magnetic fields only; (2) in a loss-free guide supporting several *modes* of propagation, the power transmitted is the sum of that contributed by each mode individually; (3) the transverse fields vary with distance along the guide according to a propagation factor $\varepsilon^{\pm j\beta z}$ only; (4) the transverse magnetic field is related to the transverse electric field by a simple constant, the *wave impedance* of the mode. The equivalent voltage and current for a waveguide supporting N modes can be written using superposition. The impedance-matrix description for a two-port waveguide for each propagating mode is

$$\begin{bmatrix} V_1 \\ V_2 \end{bmatrix} = \begin{bmatrix} Z_{11} & Z_{12} \\ Z_{21} & Z_{22} \end{bmatrix} \begin{bmatrix} I_1 \\ I_2 \end{bmatrix} \tag{3-45}$$

59. Delay Lines. Transmission lines used to obtain pulse time delays are one class of structure known as *delay lines*. The drawback is that the line must be rather long even for small time delays since the electromagnetic waves propagate at a speed close to the speed of light. Special compact low-velocity lines have been developed to avoid this inconvenience. The most common type is a coaxial line, in which the inner conductor is a helix. The vast majority of the so-called "electric" delay lines are artificial transmission lines consisting of lumped capacitors and inductors. The limitations of physically realizable amplitude- and phase-transfer functions are such that the practical delays obtained do not exceed the order of a few pulse periods. Longer time delays are achieved with *acoustic delay lines*, employing acoustic wave propagation and electromechanical transducers at the input and output.

60. Pulse-Forming Lines. A particularly useful application of the transient phenomena in transmission lines is the *pulse-forming line*, an example of which appears in Fig. 3-57. In this example, the transmission line is slowly charged, by means of a voltage source in series with a relatively high resistance $R_s \gg R_0$, to an initial steady-state voltage V_0. When the load is connected by closing the switch S, the line develops a load voltage equal to $V_0/2$ and a wave of voltage with an amplitude $-V_0/2$ travels back toward the source. Since the source appears as an open circuit to the traveling wave, it is reflected as a voltage wave $-V_0/2$, which brings the resultant voltage to zero. The net effect is that the line delivers a rectangular voltage pulse to the load of amplitude $V_0/2$ and of duration equal to the round-trip time delay of the line. The

[1] R. E. Collin, "Fundamentals for Microwave Engineering," McGraw-Hill, New York, 1966, Chap 4, p. 145.

charging source can alternatively be connected at the load end; in this case, the pulse-forming line is a one-port device. One-port "artificial" pulse-forming lines synthesized with lumped capacitors and inductors are called *Guillemin lines*.

61. Scattering-Matrix Description. One difficulty that arises in the impedance-matrix formulation is that of measuring the impedance parameters. The quantities that are directly measurable are the amplitudes and phase angles of the reflected waves. Since the field equations are linear, the reflected-wave amplitudes are linearly related to the incident-wave amplitudes. In matrix form we write $V^- = AV^+$ The equivalent voltages are chosen so that the *characteristic impedance* is 1. Thus the *scattering matrix* S is *symmetric* for *reciprocal structures*, and the power transmitted to each port is $\frac{1}{2}V_n^+ V_n^{+*}$. For a given structure the impedance matrix has definite values which are, of course, a function of frequency. If the terminal plane of the nth port is changed by l_n, the resulting scattering matrix is easily calculated:

$$[S'] = \begin{bmatrix} e^{-j\theta_1} & \cdots & \cdots \\ \cdots & e^{-j\theta_n} & \cdots \\ & & e^{-j\theta_N} \end{bmatrix} [S] \begin{bmatrix} -e^{-j\theta_1} & \cdots \\ & \cdots \\ & e^{-j\theta_N} \end{bmatrix} \tag{3-46}$$

where $\theta_i = \beta_i l_i$ is the electric phase shift and S is the scattering matrix of the original configuration (see Par. **3-36**).

Several properties of the scattering matrix can be easily derived. For a lossless structure $\sum_{n=1}^N S_{ns} S_{nr}^* = \delta_{sr}$, where δ_{sr} is the *Kronecker delta* function. If the most common microwave structure, the two-port, in addition to being lossless is also reciprocal, it is easily shown that $|S_{11}| = |S_{22}|$ and $|S_{12}| = \sqrt{1 - |S_{11}|^2}$ With $S_{11} = |S_{11}|e^{j\theta_1}$, $S_{22} = |S_{22}|e^{j\theta_2}$, and $S_{12} = (1 - |S_{11}|^2)^{1/2} e^{j\phi}$ we get

$$\phi = (\theta_1 + \theta_2 + \pi)/2 \pm n\pi$$

If the waveguide is not lossless, the power dissipation is given by $P_{\text{DISS}} = \frac{1}{2}V^+(I - SS^*)V^{+*}$.

62. Transmission-Matrix Description. When two-port microwave circuits are *cascaded*, it is convenient to represent each circuit by a *transmission matrix*, the transmission matrix of the entire structure being the product of the individual transmission matrices (see Par. **3-38**). When the independent variables are chosen to be the *port voltage* V_1 and the *port current* I_1, the transmission matrix is called the *voltage-current transmission matrix*. This approach is invaluable in the study of periodic structures.

$$\begin{bmatrix} V_2 \\ I_2 \end{bmatrix} = [T] \begin{bmatrix} V_1 \\ I_1 \end{bmatrix} \tag{3-47}$$

63. Resonators. At high frequencies short-circuit or open-circuit transmission lines can be used as *resonant circuits*. It is usually assumed that such lines are air-filled since dielectric-filled lines introduce additional losses resulting in a lower Q (see Par. **3-7**). For a short-circuited line of length L, with series resistance r Ω per unit length, l H per unit length, and c F per unit length, the input inpedance is given by $Z_{\text{in}}(\omega) = rL + j\omega lL$, which is analogous to the result for the lumped-parameter series RLC circuit. Typical values for Q are from 100 to 10,000. *Microwave cavities can also be used as resonators and are of practical importance at frequencies above 1,000 MHz.*

The *resonant cavity* is a completely enclosed structure with a resonant frequency given by $f_{mnl} = c\sqrt{(l/2d)^2 + (n/2b)^2 + (m/2a)^2}$, where c is the velocity of light, a, b, and d the dimensions of the structure, and l, m, n are integers. Each set of integers l, m, n corresponds to a different resonant frequency.

64. Tuning. When a load is connected to a source through a transmission line or waveguide, the impedance of both the source and the load is matched to the impedance of the line. This reduces the sensitivity to frequency variations. Matching the source to the loaded waveguide or transmission line for maximum power transfer is not practical since a small change in the frequency of the source changes the electric length of the propagating structure, which greatly modifies the effective load impedance. In addition, if the load is not matched to the propagating structure, *standing waves* are present and additional losses are incurred. When the propagating structure is a transmis-

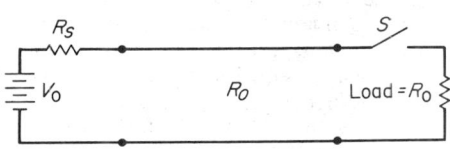

Fig. 3-57. Schematic diagram of an application of a transmission line as a pulse-forming line. In practical applications, a charging inductor is generally used in place of the charging resistor R_s.

sion line, impedance matching is accomplished with tuning stubs. A *tuning stub* is a short-circuited transmission line of variable length. Tuning is accomplished by adjusting the length of the stub. When the losses are small, the effect of the tuning stubs can be studied on the Smith chart (see Par. **9-2**).

65. Antennas and Radiators. An *antenna* is an impedance-matching device used to absorb or radiate electromagnetic waves. Its purpose is to match the impedance of the *propagating medium* (usually air or free space) to the source. The field configuration produced by a radiating antenna is separated into two parts, the *induction field*, which attenuates rapidly in the vicinity of the antenna, and the *radiating field*, which attenuates as $1/r^2$.

To transmit energy to the propagating medium most effectively the antenna must appear as a resistive load to the source at the frequency to be propagated. This is the *radiation resistance* R_r; it and the ohmic resistance R_0 are the only resistances that are usually considered, the loss due to eddy currents, leakage, etc., being neglected.

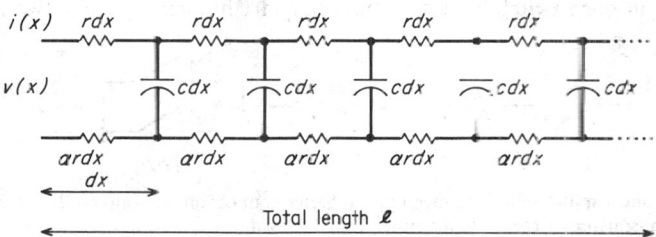

Fig. 3-58. Equivalent circuit for a general *RC* ladder

Antennas capable of functioning over a wide range of frequencies are called *aperiodic*, and those designed for a particular frequency, *tuned*. The geometry of the antenna determines the radiation pattern of the propagating signal in space.

Three-dimensional plots of the radiation pattern in space are difficult to construct. The normal procedure is to plot the radiation pattern for vertical planes, the *elevation patterns*, and horizontal planes, the *azimuth patterns*.

The two most important properties in antenna design are the *gain* and the *directivity*. Antenna *gain* is defined as $G = 4\pi P_r/P_T$, where P_T is the total power delivered to the antenna and P_r is the power radiated per unit solid angle in a given direction. The *directivity* is defined as $D = P_r/P_A$, where P_A is the average power radiated per unit solid angle.

When antennas are separated at distances comparable to the wavelength of the signals propagated, the mutual coupling becomes significant. The composite radiation pattern has properties that are not achievable with single-element radiators.

At microwave frequencies a variety of antenna types and arrays can be designed. The *horn radiator*, for example, is easier to couple to a waveguide than a *dipole antenna*, has a large power capacity, and provides more control over the radiation pattern. The design criteria involve a trade-off between the physical size of the radiator and the phase change it introduces. For details of antenna design, see Sec. **18**.

66. Distributed Parameters in Integrated Circuits. The technology of integrated circuits arose from the need to construct small, reliable, low-cost components for modern electronic systems (see Sec. **8**). The components are mounted on a common insulating or semiconducting substrate. The latter is more desirable because active elements can be synthesized more easily, but the use of a semiconducting substrate reduces the isolation between the circuit components. As a result, even at low frequencies, the distributed resistance and capacitance are important and affect the circuit operation. These distributed effects are not always detrimental; in particular, the *RC ladder* is a useful structure. A greater phase shift can be achieved over conventional lumped-ladder networks.

The equivalent circuit for a general type of *RC* ladder is shown in Fig. 3-58, where r and c are the resistance and capacitance per unit length. The factor α accounts for the fact that the substrates may have different resistivities. If the voltages and currents are assumed to be sinusoidal, we derive, using Kirchhoff's laws,

$$\frac{d^2v}{dx^2} = j\omega(1 + \alpha)rcv \qquad \frac{di}{dx} = j\omega cv \qquad (3\text{-}48)$$

One application for the distributed *RC* ladder network is the *notch filter*.

DIGITAL CIRCUITS

67. Switching Networks. A switching network has, in general, n pairs of terminals, some of which are designated as inputs and some as outputs. In contrast to the networks considered above, however, it is assumed that each of the terminals of a switching circuit can exist only in a *finite number of states*; i.e., each terminal can take on only a finite number of values, which are discrete and identifiable. Since there are a finite number of terminals, it is theoretically possible to list all combinations of terminal values in a table, called a *truth table* or *state table*. Since the relationships between input states and output states can be identified in the truth table, it serves the same function as the transfer functions identified for analog n-port networks.

A switching network is composed only of *switching elements*. A switching element is a *resistive branch element* which exists only in states $R = 0$ and $R = \infty$.[1] Each individual switch element may be considered either by its *transmission properties* or *hindrance properties*. If the switch is *closed*, it will transmit signals freely and hence has a transmission of 1 and hindrance of 0. Conversely an *open* switch has 0 transmission and 1 hindrance. These two properties are

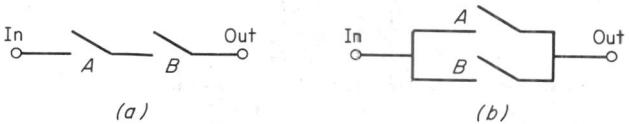

Fig. 3-59. Series and parallel switch connection. (*a*) Series connection transmits only when A and B are both closed. (*b*) Transmission is complete for either A or B closed.

never mixed in identifying switches in the same network. Switching networks are usually designed using transmission properties. Thus, a switch is defined as closed by a 1 and open by a 0.

Switches may be connected either in series or parallel. If two switches are in series, they will not transmit unless both are closed. If they are in parallel, transmission will be complete if *either* is closed. Figure 3-59 shows the results of series and parallel connections.

68. Uses of Switching Networks. Switching networks were first used to connect things together. An example of a network of this type is the telephone switching system, which by closing the proper switches in a large network can connect a caller to any other telephone in the system. Here one is concerned with finding a pathway through a maze, and a good deal of theory has been built up on this subject (see Sec. **22**).

Switching networks are also used to do *logical manipulations*, using *boolean algebra* (see Sec. **23**). This is one of the primary functions of digital computers, especially those used to control processes of manufacture.

Switching networks are also used to do arithmetic. Switching networks add, subtract, multiply, and divide in the binary system (see Sec. **23**). This is also a primary function of many digital computers. Because the arithmetic is done with 0 and 1, like the logical manipulations, sometimes there is little distinction made between the two, and arithmetic becomes a type of logical manipulation.

Finally, switching networks are used to store information. Many switches are designed to remain in either the 0 or 1 state for extended periods of time. One can therefore set a particular pattern of 1s and 0s in a switch network, and it will be remembered for future reference. This is the primary function of computers which keep government and bank records, etc.

The *state* of a mechanical switch is determined by the position of its contacts, and the state exists whether any voltage or current is applied to the switch or not. In order to use the switch state as a computational variable, it is necessary to ascertain its position. To determine its resistance or otherwise assess position, a voltage or current must be applied. If a current source or voltage source is permanently associated with a switch element, the current through or the voltage across the switch element will reflect its state and may be considered as its *state value*. Since most electronic elements require permanent sources of voltage and current for their proper operations, it becomes convenient to consider a switch in terms of its output voltage or current.

69. Gate Elements. A gate element, or decision element, is a network with one or more input ports and a single output port. The state value of every port of a gate is either the voltage or the current associated with the port. The present output state is uniquely determined by the

[1]Paul E. Wood, "Switching Theory," McGraw-Hill, New York, 1968.

present input state. The use of voltage (current) as a state value at every port of a gate is called *voltage* (current) *logic*.

Gate elements are the basic building units in the construction of switching networks. They usually perform a simple and generally useful functional transformation.

A very small number of different gate operations are necessary to carry out any logic functional relationship. The set of operations AND, OR, and NOT are a *sufficient set*, for instance. Other sets are also possible. Table 3-7 shows three common sets of logic gate types together with

TABLE 3-7   Gate Sets and Their Symbols

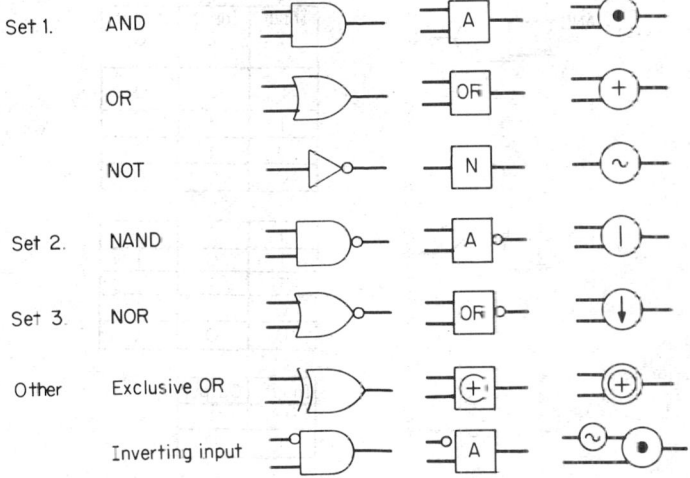

their usual electrical symbols. It also shows the exclusive-or (XOR) function and an inverting input.

Set 1, 2, or 3 is necessary and sufficient to completely implement operations. They are therefore *minimal* and *complete*. More efficiency may result from mixing sets or using the XOR in addition to the usual set gates. In general, however, sets are not mixed, and design is carried out in one set.

It can be shown that there are 18 different minimal complete gate (or operation) sets which have two members. There are 6 minimal complete sets with three members, 4 with four members, and none with more than four.[1] As Table 3-7 implies, the NOR and NAND operations are each minimal sets in themselves. They are called *universal operations* because all logic functions can be performed by one type of gate. This makes NOR and NAND logic easy to manufacture and maintain since only a single type of gate circuit is used.

A small circle associated with a port is the usual symbol for *inversion*. If it is associated with the output, it provides the *logical complement* of the normal output. If it is associated with an input, this indicates that the complement of the input is taken before it enters the main-gate circuitry.

If the state variable x associated with a port is defined so that $x = 1$ for a more positive voltage or current value and $x = 0$ for a less positive voltage or current value, the gate is called a *positive-logic gate*. Otherwise, it is a *negative-logic gate*.

As mentioned earlier, a switching network is described by its truth table. These tables actually serve to define the gate elements. Table 3-8 shows the truth tables for various gate elements.

Switching networks are further classified as being either *combinational* or *sequential*. A combinational network is a switching network whose present output port values (states) depend only on the present input port values.

Gate networks which contain no loops (a gate is never encountered twice along any network path) is called a *feedforward* network. A feedforward gate network is always a combinational network.

A switching network whose present output port values (states) depend on past input port states is called a sequential network.

[1]See ibid., pp. 65–66.

Logic and arithmetic operations are generally considered combinational. The process of switching lines together or remembering information requires sequential networks. Sequential networks are built up from combinational networks by the introduction of *time-delay elements* or *memory elements* and by network interconnections which involve loops.

70. Digital Hardware. To be useful a logical design must be implemented by physical devices. Gates can be built using springs and levers, hydraulic valves, electromechanical relays,

TABLE 3-8   Truth Tables for Two-Input Gates

Gate   Truth table

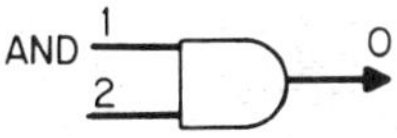

AND gate symbol

Input 1	Input 2	Output O
1	0	0
1	1	1
0	1	0
0	0	0

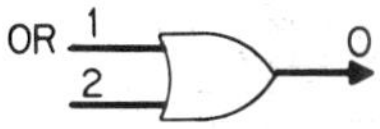

OR gate symbol

1	2	O
1	0	1
1	1	1
0	1	1
0	0	0

NOT gate symbol

Input	Output
0	1
1	0

NAND gate symbol

1	2	O
1	0	1
1	1	0
0	1	1
0	0	1

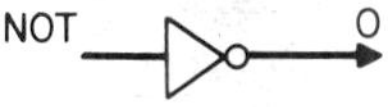

NOR gate symbol

1	2	O
1	0	0
1	1	0
0	1	0
0	0	1

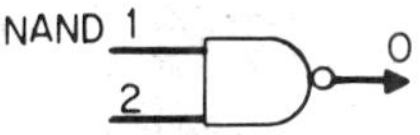

XOR gate symbol

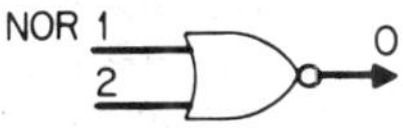

1	2	O
1	0	1
1	1	0
0	1	1
0	0	0

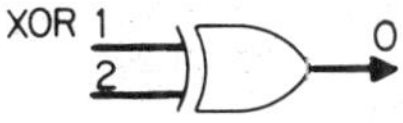

AND gate symbol

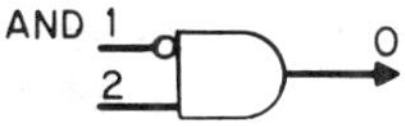

1	2	O
1	0	0
1	1	0
0	1	1
0	0	0

or other devices. The most common gates are electronic, however, in which circuit components are interconnected to carry out the logic or memory functions covered earlier. Gates can be made from resistors and diodes, resistors and transistors (RTL), diodes and transistors (DTL), multiple-emitter transistors (TTL), MOSFETs (MOS), complementary MOSFETs (CMOS), multiple-collector transistors (IIL or I^2L), and various other devices. Each type of construction, called a *logic family*, has certain characteristics in addition to its basic capacity to perform logical operations. Table 3-9 shows several logic families and outlines their basic characteristics. CMOS is very popular because of its low power dissipation, while TTL is reasonably fast and very flexible.

TABLE 3-9 Comparison of Logic Families

Family	Basic gate	Fan-out	Power dissipated per gate, mW	Noise immunity	Propagation delay per gate, ns	Maximum clock rate, MHz	Number of functions	Packing density
RTL	NOR	5	12	Nominal	12	8	High	Very poor
DTL	NAND	8	8	Good	30	12	Fairly high	Very poor
TTL	NAND	10	10	Good	10	35	Very high	Nominal
LTTL	NAND	20	1	Good	33	3	High	Nominal
HTTL	NAND	10	22	Good	6	50	Nominal	Nominal
STTL	NAND	10	19	Good	3	125	Very high	Nominal
LSTTL	NAND	20	2	Good	9.5	45	Very high	Nominal
ECL	OR-NOR	25	40	Good	2	60	High	Poor
MOS	NAND	20	1	Nominal	100	2	Low	Good
CMOS	NOR-NAND	>50	0.1–1	Excellent	30	10	High	Nominal
PMOS	NAND	...	0.1	Fair	700	1	Custom	Good
I^2L	NOR	...	0.001–0.01	Poor	25–250	...	Custom	Very high

71. Combinational Circuit Interconnections. The major characteristic of combinational networks is that they contain *no loops*. Gates are interconnected for the purpose of carrying out logical or arithmetical manipulations of a more complex nature than a single gate can provide.

The central problem of a switching network is the *design problem*, in which some desired function is to be implemented in terms of a number of interconnected gates. Superficially, this process is quite easy, because it can be shown that any boolean expression can be implemented directly in terms of circuit hardware; i.e., if the boolean expression is written in a *standard* or *canonical form*, simple algorithms exist for translating the expression into gate hardware (see Sec. **23**).

There are two canonical forms, the *canonic sum*, or *disjunctive canonical form*, and the *canonic product*, or *conjunctive canonical form*.

The *canonic-sum form* is such that the function is expanded into a number of product terms (AND) in which each term is expressed as a product of all variables or their complements (where no product term has the same variable appearing twice) and these product terms are connected by the OR operation.[1] This is also known as a *sum of minterms*, where a minterm is a term which is the product of all variables or their complements.

The *canonic-product form* is such that the function is expanded into a number of sum terms (OR) in which each term is expressed as a sum of all variables or their complements (where no sum term has the same variable appearing twice) and these sum terms are connected by the AND operation. This is also known as a *product of maxterms*, where a maxterm is a term which is the sum of all variables or their components. Designating a circuit from the canonic forms is a straightforward procedure, as Fig. 3-60 shows.

Problems of design of switching circuits are not concerned with whether or not a particular function can be realized but with how it can be realized with minimal cost or other additional constraints.

72. Minimum-Complexity Combinational Networks. The important design aim of reducing network complexity usually leads to lower cost and greater ease of construction. Minimum complexity may have several meanings, some of which are in opposition. A *minimally complex network* may be defined as:

[1]See R. E. Miller, "Switching Theory," Vol. 1, Wiley, New York, 1965.

1. That which contains the minimum number of gate elements
2. That which utilizes a certain set of gate elements (see Table 3-8) but which contains the minimum number of elements from that set
3. That which has the fewest number of interconnections
4. That which can be wired with the fewest crossovers on a circuit board
5. That which minimizes the total cost of components used in its construction
6. That which is easiest to maintain and repair
7. That which operates at the highest speed
8. That which has the highest reliability

These and many other criteria may be of major importance in designing a switching network. It is known how to optimize the design in terms of some of these criteria but not others. Several minimization techniques will be discussed.

73. Map Techniques. The prime goal of *map minimization methods* is to produce a circuit realization which uses a minimum number of gate elements. The *map*, or *Karnaugh map* (Fig. 3-61), is an alternate way of presenting the data contained in the truth table. A map of function $f(x_1, x_2, \ldots, x_n)$ of n variables contains 2^n squares, one for every possible value of the n-tuple $(x_1, x_2, \ldots, x_n)$. If it is desired that some particular value of the n-tuple produce an output from the circuit, a 1 is placed in the corresponding square. If no output is desired, a 0 goes in the square, and if one does not care what the output is, a d is placed in the square. A d can be counted as either a 1 or 0, depending on which condition produces greater minimization.

Minimization by the map method uses two basic laws of boolean algebra:

$$x + \bar{x}y = x + y \quad \text{and} \quad xy + \bar{x}y = y \tag{3-49}$$

It is therefore necessary to arrange the map to make it easy to apply these laws by inspection. This is done by arranging the squares so that any two adjacent squares differ by only one of the variables in the n-tuple. Figure 3-61 shows an example, where special ordering is

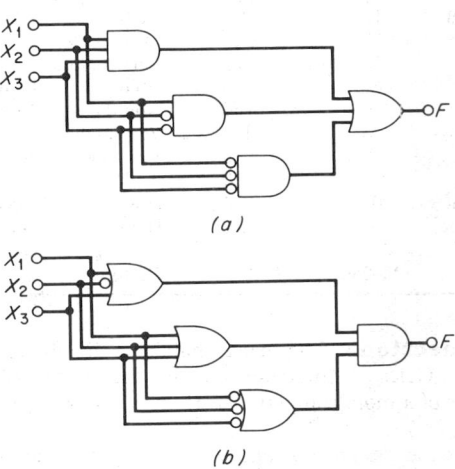

Fig. 3-60. Realization of canonical forms: (a) realization of $F = (x_1 \cdot x_2 \cdot x_3) + (x_1 \cdot \bar{x}_2 \cdot \bar{x}_3) + (\bar{x}_1 \cdot \bar{x}_2 \cdot \bar{x}_3)$; (b) realization of $F = (x_1 + \bar{x}_2 + x_3) \cdot (x_1 + x_2 + x_3) \cdot (\bar{x}_1 + \bar{x}_2 + \bar{x}_3)$.

required in the 4 × 4 map to achieve properly adjacent squares.

Once the map is drawn, the output function is entered in all the squares. If one proceeds to minimize using the 1 and d entries, the result will be a minimal sum-of-products form, similar to the canonical sum. This is also called the *minterm canonical form* or *standard sum*. If one proceeds using the 0 and d entries, the result will be a minimal product-of-sums form, similar to the canonical product. This is also called the *maxterm canonical form* or *standard product*.

Minimization is accomplished by combining the map entries into groups. A *group* is a large square or rectangle (twice as long as wide) which contains squares of all the same value. For

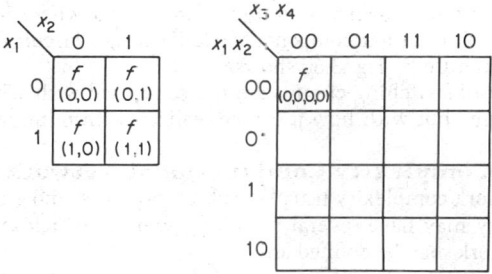

Fig. 3-61. 2 × 2 and 4 × 4 Karnaugh maps.

instance, to use the sum-of-products form, one groups squares containing 1s and d's. The highest order of simplification is achieved when the 1s are combined into *as few groups as possible*, each of which contains as many adjacent cells as possible. Because of the form of the map, squares along its two sides are adjacent, as well as those on the top and bottom.

Figure 3-62 shows such a grouping carried out in a 4-tuple. In Fig. 3-62b it can be seen by

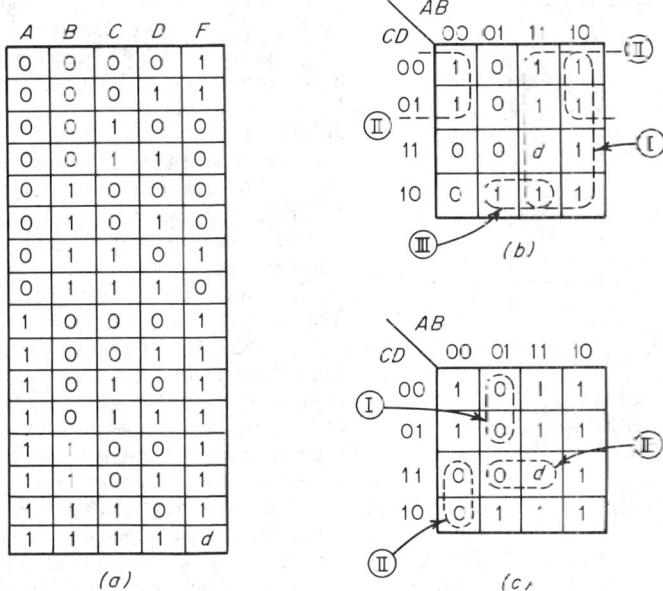

Fig. 3-62. Map minimization: (a) truth-table representation of a function F of four variables; (b) sum-of-minterms minimization; (c) product-of-maxterms minimization.

looking at group I that $F = 1$ (or d) for all 4-tuples where $A = 1$. Therefore, group I $= A$. Group II is such that $F = 1$ for all occasions where B and C both $= 0$, and so on:

$$\text{Group I} = A \qquad \text{Group II} = \overline{BC} \qquad \text{Group III} = BC\overline{L}$$

It is also possible to group the zeros, as in Fig. 3-62c. (In this case the d is considered a zero because it aids grouping: it need not be included if greater minimization is achieved by leaving it out.) Grouping of zeros makes use of two other boolean algebra rules:

$$(x + y)(x + \overline{y}) = x \qquad \text{and} \qquad (x + y)(x + z + y) = (x + y)(x + z) \qquad (3\text{-}50)$$

The zero groups represent maxterms. Group I in Fig. 3-62c is such that $F = 0$ only if $A, \overline{B}$, and C are each zero. The value of D does not matter. Similar conditions hold for groups II and III.

$$\text{Group I} = A + \overline{B} + C \qquad \text{Group II} = A + C + \overline{C} \qquad \text{Group III} = \overline{B} + \overline{C} + \overline{D}$$

Based on these two approaches, there are two minimum networks, shown in Fig. 3-63a and b. Clearly, if AND and OR gates are about the same price, Fig. 3-63a is preferred.

74. Prime Implicants. When a Karnaugh map is used to find a minimal representation, one tries to combine adjacent 1-squares into larger groups. Each group that can be made which is not properly contained in a larger group is a graphical example of a *prime implicant*. Thus in Fig. 3-62b, for instance, there are three prime implicants: A, $\overline{BC}$, and $BC\overline{D}$.

For large networks the map would become quite unwieldy, and so prime implicants are searched for directly from the truth table. The search process consists of the following steps:

1. Form the canonical sum-of-product representation of the function to be minimized.

2. Examine all product terms and apply the reduction $xy + x\overline{y} = x$ as many times as possible. All the new product terms so formed will have one less variable than the original terms.

3. Take the new set of product terms and repeat step 2 on this new set of terms. When no further reductions are possible, all the product terms that were generated by steps 1 and 2 and which cannot be further reduced are the prime implicants associated with the function to be minimized.

4. The set of prime implicants is then inspected to choose a minimal set that can be ORed together to represent the function.[1]

A similar process can be carried out with the canonical product of sums.

75. Cubical Representation. This approach is another geometric representation similar to the Karnaugh map, in which the boolean function of *n* variables is represented by an *n*-dimensional *unit cube*. Minimization is accomplished by combining cubes into higher-order

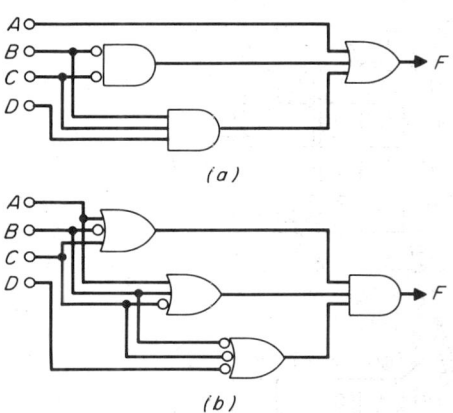

(a)

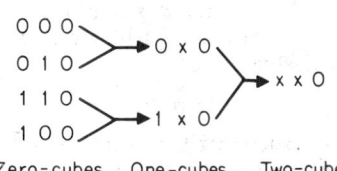

(b)

Fig. 3-63. Realization of the reduction of (*a*) Fig. 3-62*b* and (*b*) Fig. 3-62*c*.

cubes. A 0-*cube* is a single expression, equivalent to a 1-square on the Karnaugh map. A 1-*cube* is a pair of expressions (0-cubes) differing by only one variable. Together the two 0-cubes form a 1-cube. Two 1-cubes can form a 2-cube if they have one component which differs.

In Fig. 3-64, the *X* represents the variable which differs, called the *free component*. The 1s and 0s which remain are called *bound components*. The two 0-cubes which combine are called the *faces* of the resultant 1-cube, and so on. See Miller[2] for an explanation of minimization using cubical complexes.

76. Design Using Other Gate Sets. The techniques mentioned above all produce realizations utilizing AND, OR, and NOT operations. These are not necessarily the most desirable functions (indeed, NAND or NOR logic is usually much preferred because of the inherent stability and ease of construction of the gate circuitry). It therefore becomes necessary to translate the results of the minimization technique into other sets, most commonly *inverting* gates, NAND, NOR, NOT. The NAND operation can be represented in two ways, as shown in Fig. 3-65*a*. The NOR operation can also be represented in two ways (see Fig. 3-65*b*). Each gate produces a *complementary*, or inverting, signal. If a signal passes through two inverting gates in series, it is twice complemented and regains its original character. An algorithm for generating NAND or NOR logic can be stated.

A NAND (NOR) gate performs an OR (AND) operation on its complemented input variables if it is at an *odd* number of inversion levels; whereas an AND (OR) operation will be performed on its uncomplemented input variables if it is at an *even* number of inversion levels.

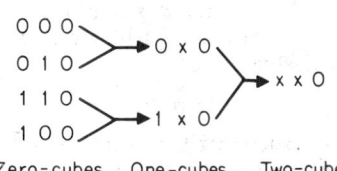

Zero-cubes   One-cubes   Two-cube

Fig. 3-64. Cubical minimization.

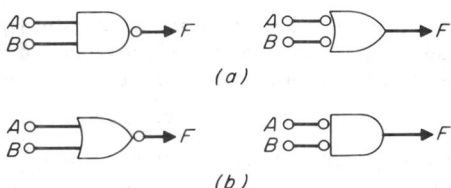

(a)

(b)

Fig. 3-65. Symbols for gates performing (*a*) the NAND operation and (*b*) the NOR operation.

The foregoing statements lead to the following set of rules for obtaining the logic expression of the output signal of an interconnected array of inverting gates:

1. Consider the gate from which the output signal will be obtained as the first (odd) level of inversion, the preceding gates as the second (even) level, etc.
2. Consider all NAND gates in odd levels to perform the OR operation.
3. Consider all NAND gates in even levels to perform the AND operation.
4. Consider all NOR gates in odd levels to perform the AND operation.
5. Consider all NOR gates in even levels to perform the OR operation.

---

[1]From Taylor L. Booth, "Digital Networks and Computer Systems." Copyright © 1971 by John Wiley and Sons, Inc. Reprinted by permission.
[2]"Switching Theory," pp. 145–156.

6. All input variables entering gates in odd levels should appear complemented in the logic expression of the output signal.

7. All input variables entering gates in even levels should appear uncomplemented in the logic expression of the output signal.[1]

77. Arithmetic Circuits. Binary arithmetic can be carried out using logic gates. Figure 3-66 shows an example of a logic circuit used to add binary numbers. Each digit of number A is added to the corresponding digit of number B along with a carry digit (if present) from the right

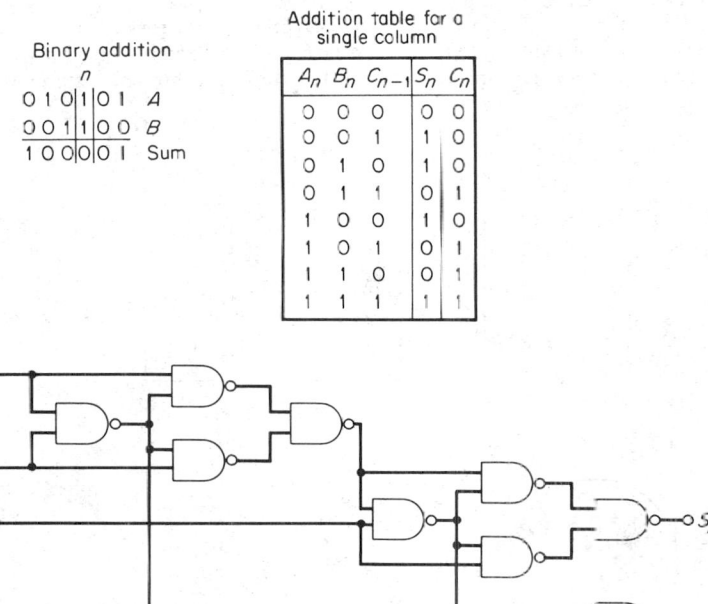

Binary addition

$$n$$
$$0\ 1\ 0\ 1\ 0\ 1\quad A$$
$$0\ 0\ 1\ 1\ 0\ 0\quad B$$
$$1\ 0\ 0\ 0\ 0\ 1\quad \text{Sum}$$

Addition table for a single column

A_n	B_n	C_{n-1}	S_n	C_n
0	0	0	0	0
0	0	1	1	0
0	1	0	1	0
0	1	1	0	1
1	0	0	1	0
1	0	1	0	1
1	1	0	0	1
1	1	1	1	1

Fig. 3-66. NAND realization of addition network.

adjacent column. The result is the sum of the digits in that column plus a carry to be forwarded to the left adjacent column. The table which summarizes the results is indistinguishable from a logical truth table, and hence the adder can be implemented with logic gates.

Logic circuits can be designed which carry out binary subtraction, multiplication, and division also.

78. Sequential-Circuit Interconnections. Sequential circuits possess the property that their output states depend not only upon present input states but upon past values of inputs. This characteristic is achieved by including *delay elements* or *memory elements* into the circuit. Boolean algebra and the design techniques mentioned previously cannot express delay or memory, and so a different representation must be used. It is necessary to specify *input states*, *output states*, and *internal states* to describe a *sequential machine*, where the internal state describes the condition of all the memory or delay elements in the circuit. The input and internal states together are known as the *total state*. The present total state determines both the present output state and the next internal state. If the number of possible internal states is finite, the machine (the behavior representation of the network) is called a *finite-state sequential machine*.

A finite-state sequential machine can be uniquely and completely specified by its *state table* or a *state diagram*. The state table indicates the next state and present output for every possible combination of present inputs and internal states, as shown in Fig. 3-67a. The state diagram for the same machine is shown in Fig. 3-67b. It shows transitions from the present to next state as *directed line segments*. The segments are labeled to show the inputs which caused the transition and the present output.

[1] From R. L. Morris and J. R. Miller (eds.), "Designing with TTL Integrated Circuits," McGraw-Hill, New York, 1971, p. 115.

79. Synchronous Sequential Machines. The most elementary form of sequential network is composed of AND, OR, NOT, and delay elements. A delay element is a one-input, one-output network such that its output state at time τ is equal to its input state at time $\tau - d$, where d is called the *delay time*. With these elements, one can form a synchronous sequential machine such that:

1. Input changes occur only at instants in time iD, where i is a positive integer and D is a standard time interval.
2. All delay elements have delays equal precisely to integer multiples of D.
3. Gate elements and interconnections are delay-free.
4. All input-, state-, and output-variable changes occur instantaneously.[1]

Any *finite-memory-span sequential machine* can be realized by a feedforward network with delay. A finite-memory-span machine is one in which the present state (and output) can be

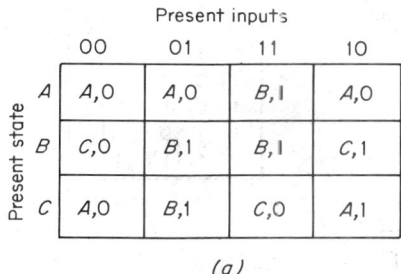

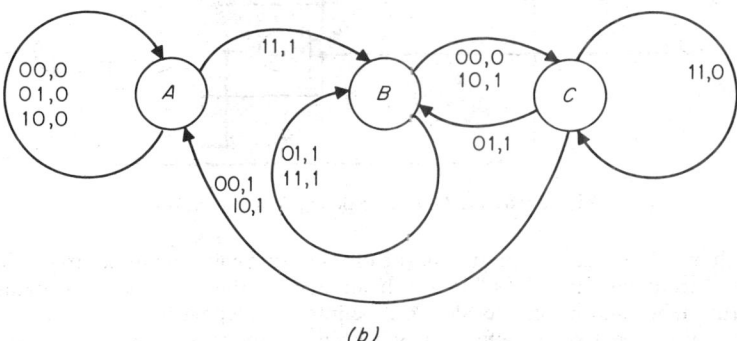

(b)

Fig. 3-67. (a) State table and (b) state diagram for sequential machine M.

determined from knowledge of present and a finite number of past inputs alone, without knowledge of the original internal state of the machine. Such a machine can be realized with a network which can be reduced in complexity by techniques analogous to those used for combinational-circuit reduction. The following stages are necessary in this process:

Stage 1: Description of Desired Network Operation. A complete set of specifications must be prepared describing the operation of the network. All inputs and outputs must be identified, and the relationship between the quantities must be defined in a consistent manner.

Stage 2: Determination of State Table. Using the specification established at stage 1, an initial state table or state diagram is defined for the network. The state table is checked to make sure that it satisfies all design criteria.

Stage 3: State-Table Minimization. In the process of developing a state table to satisfy a given set of operational requirements an unnecessarily large number of states may be introduced. Since the number of information storage elements in a circuit increases as the number of states increases, it is often desirable to remove redundant states from the state table.

Stage 4: State Assignments. The information contained in the state table must be encoded into binary form. This is not a unique process, and the encoding used can considerably influence the complexity of the resulting circuit. The result of this stage is to transform the state table into a *transition table*.

[1]From P. Wood, "Switching Theory," McGraw-Hill, New York, 1968, p. 279.

Stage 5: Network Realization. Once a transition table has been constructed and a decision made concerning the type of storage elements to be used, the logic expressions relating the input and present state to the output can be obtained.[1]

Stages 1, 2, and 4 cannot usually be carried out by a straightforward algorithm, and therefore no more details of the procedure will be mentioned here.

A *memory element* may be distinguished from a *delay element*. A memory element produces delays of any desired length. Whereas the delay element produces an output equal to its input value d s earlier, the memory element is designed to hold as its output the most recently received input, regardless of the length of time involved.

80. Asynchronous Sequential Machines. Any sequential machine that violates one of the four synchronous criteria is considered *asynchronous*. Since the criteria are rather ideal, most real machines are asynchronous. Such machines are difficult to represent and construct.

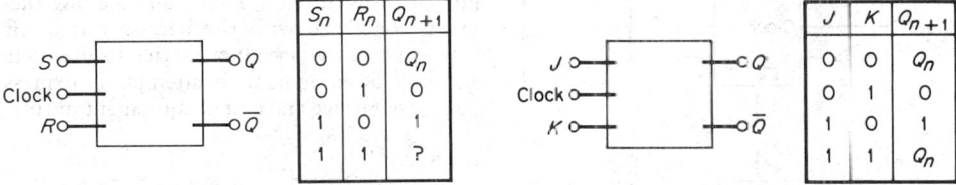

S_n	R_n	Q_{n+1}
0	0	Q_n
0	1	0
1	0	1
1	1	?

Fig. 3-68. Description of S-R flip-flop.

J	K	Q_{n+1}
0	0	Q_n
0	1	0
1	0	1
1	1	$\overline{Q_n}$

Fig. 3-69. Description of J-K flip-flop.

Usually, therefore, pseudo-synchronous behavior is achieved by incorporating a *clock*. *Clocked sequential systems* form the bulk of useful large-scale switching networks. The clock is a sequential machine that produces a series of pulses, usually occurring at regular intervals. The other network inputs and outputs can change at any time except during a clock pulse, at which time a new total state is defined. Thus, network activity exhibits the characteristics of a synchronous sequential network, and the clocked system can be designed in the same manner. In order to operate the clocked sequential system, it is necessary to distribute the clock pulse to virtually every element to assure synchronism.

81. Flip-Flops. Simple sequential circuit elements are called *flip-flops*. There are several types in widespread use. These are set-reset (S-R), J-K, and J-K master-slave. They can be used as memory elements, delay elements, or as a *toggle*. The *state table* of a *clocked* S-R flip-flop is shown in Fig. 3-68.

The output Q changes according to the levels present at S and R *during* the time the clock pulse is present, or in synchronism with the clock, and can be treated as a synchronous circuit.

If S and R are both 1, the output of the flip-flop is indeterminate. To avoid this problem, the J-K flip-flop is used. Figure 3-69 shows its state table.

The J-K flip-flop may oscillate when J and K are both 1 unless the clock pulse is extremely short. To prevent this the J-K master-slave flip-flop can be used. Whereas other flip-flops change states as soon as possible after the clock goes "high" (1), the master-slave initiates its change as the clock goes high but cannot complete it until the clock returns low, thereby preventing oscillation. J-K flip-flops are often provided with asynchronous inputs also. These can be used to *preset* or *clear* the flip-flop at any time *between* clock pulses, while the J and K inputs only act *during* the clock pulse.

Delay and toggle functions are common. They can be implemented from the J-K flip-flop, as shown in Fig. 3-70.

Flip-flops are implemented from standard gates using feedback loops. Figure 3-71 shows an S-R flip-flop implemented with four NAND gates.

82. Shift registers are rows of interconnected flip-flops in which stored data can be removed from all locations to the right (or left) adjacent locations by the application of clock pulses. Such networks can be used to store or delay data transmission temporarily, perform some arithmetic operations, count events, or derive timing signals from the clock. Figure 3-72 shows a simple shift register.

83. Traffic and Loading. The previous concepts dealt with techniques used to simplify the physical design of a switching network. The network is designed to carry out a certain

[1]From Taylor L. Booth, "Digital Networks and Computer Systems." Copyright © 1971 by John Wiley & Sons, Inc. Reprinted by permission.

operation or to combine its inputs in a certain way to produce an output. Since the operation takes a certain time to perform, there is a limit to the number of operations the circuit can perform in a given time. It certainly cannot meaningfully operate on two different sets of inputs at exactly the same time, for instance. Such situations regularly occur in telephone switching systems and computer systems. The system is called upon to perform an operation while it is busy working on some other operation and hence cannot respond. It then becomes necessary for those demanding services to (1) wait until facilities are available, (2) try again later, (3) share the system in some manner, or (4) use a different identical system.

The first three alternatives generally involve deterioration of service; i.e., the operation is not carried out as quickly as if the individual were the only user of the system. Providing multiple systems can reduce the deterioration of service in the multiple-demand situation but at the cost of maintaining the extra equipment necessary to provide faster service. *Traffic and loading theory* is concerned with the necessary trade-off between *quality* (speed) of service to the users and *cost* of equipment. It attempts to express quality of service and cost of equipment in com-

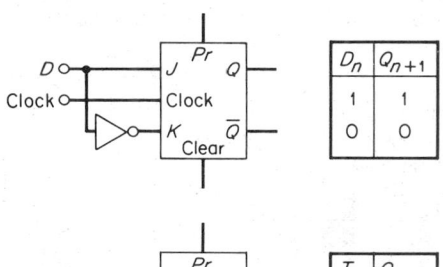

Fig. 3-70. *D-* and *T*-type flip-flops.

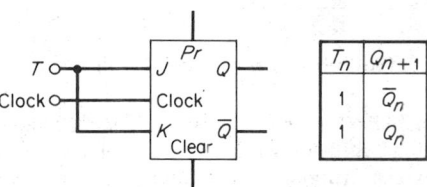

Fig. 3-71. Implementation of *S-R* flip-flop.

parable units (dollars, for instance) and *optimize* the system to provide the best service at the lowest cost. The optimization depends largely upon the traffic patterns the switching system is required to handle. Variations in traffic patterns are of several kinds.

(a) *Variation in Demand with Time.*   The system may be required to provide service under constant demand, or demands may come in rush-hour patterns. Nonconstant demand may vary predictably or randomly.

(b) *Variation in Holding Time.*   Holding time refers to the length of time necessary to perform the service demanded. This can be fixed, vary in well-defined ways, or be randomly distributed.

(c) *Priority Considerations.*   Some systems allow the users to have priority assignments, in which a high-priority user can usurp the place of a lower-priority user.

Given the above variations in traffic, a theory has been developed which estimates (for a given situation) the expected quality of service to the user. The mathematics involved is that of probability. The calculations involved are not difficult; it is the determination of criteria for service which poses the greatest problem.

84. Queuing theory is a subset of traffic theory which has to do with *waiting for service.* If a user demands service and the facilities are busy, two things can happen: the user is lost, i.e., the user must try again later, or is put on a waiting list and remains within the system awaiting service. Queuing theory deals with the characteristics of the waiting list (queue). It may be formed in a number of ways.

(a) *First In, First Out.*   When the facility becomes available, the user who has been waiting longest is the next to receive service.

(b) *First In, Last Out.*   Here the last person to request service gets it. This technique may seem "unfair," but it has wide use in the storage of data, which do not mind the wait. Sometimes it is called a *push-down stack.*

(c) *Random Order of Service.*   As the facility becomes available, the next user is picked at random from a pool of waiting customers.

(d) *Nearest-Neighbor Service.*   This is the type of service usually on elevators, in which service is rendered to the waiting customer nearest to the location of the system at the time it becomes free.

(e) *Priority Queue.* Here service is given on the basis of importance of the user or the loudness of the user's demands.

Unless there is a priority consideration, it would appear that a type *a* queue would provide the best service to its customers. Unfortunately, formation of this queue requires the system to *remember* the order in which the customers have demanded service, which in itself can require a sophisticated logic network. It is much cheaper for the system to form a type *c* or *d* queue. If this is done, calculation of the quality of service (waiting time) yields a probability function. Often the parameter of interest is average waiting time for service. This is a common criterion of service, but it can result in an unusually long wait for a particular customer. Therefore, an estimation of the maximum waiting time for service is also made. It is these waiting times which

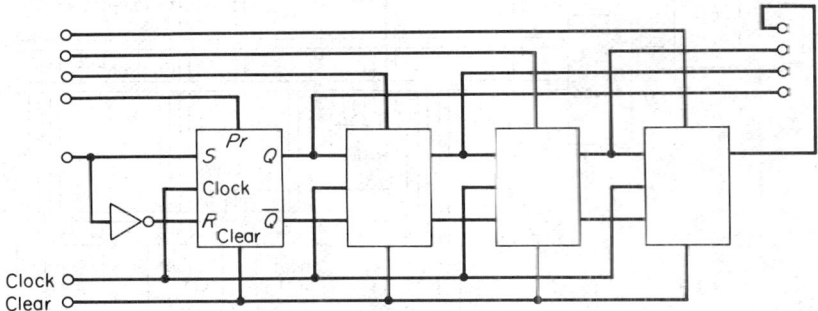

Fig. 3-72. A 4-bit shift register.

are compared to the dollar cost of additional service equipment to determine the required system size.

Scheduling Theory. Another important aspect of designing optimum switching systems concerns *routing*, or *scheduling*, of service. In completing a long-distance telephone call or a complex computer calculation, it becomes necessary to connect subsystems together temporarily to achieve the final result. Routing and scheduling theory concerns itself with the determination of the optimum patterns of interaction of these subsystems such that they are utilized as fully as possible without causing bottlenecks in the overall system. One tries to avoid as much as possible the situation where one part of the overall system is consistently limiting the quality of service.

85. Stored-Program Methods. In all the previous discussion, it has been implicit that the networks mentioned are designed to do a specific job by connecting various circuit elements together with wires, and indeed this is very often the case. Classical telephone switching circuits and many special-purpose computers are examples of large systems which are *hard-wired.* General-purpose digital computers and the newest telephone systems function in a very different manner, however. In these, the switching function is a *stored program*, and the switching network can be changed without moving a single physical wire of the system, i.e., by changing the program.

86. Programmable Networks. Figure 3-73 shows a basic programmable network. Input *A* is a *data line*, and input *L* is called the *control line*. As can be seen from the truth table, when *L* is low (0), the output γ is simply the input *A*, whereas when *L* is high (1), the output γ is the complement of *A*. It is therefore possible to construct networks which perform various functions depending on the signals applied to various control lines.

A *stored-program system* is divided into two main units, a *control unit* and a *processing unit*. The processing unit carries out the switching, logic, or arithmetic computations. It is constructed so that it can carry out a number of basic operations, such as AND, OR, NOT. The control unit tells the processing unit which operations to perform, in what order, what variables to consider as inputs, and

L	A	γ	
0	0	0	} A
0	1	1	
1	0	1	} $\bar{A}$
1	1	0	

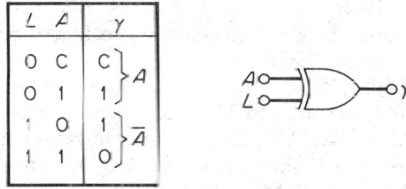

Fig. 3-73. A simple programmable circuit.

what to do with the result. In order to do this, of course, the processing unit must have access to all inputs and all output lines, as well as the ability to perform any operation with any combination of variables. It must also be totally controllable by the control unit.

The control unit is basically a memory unit. The programmer enters a set of instructions into the memory in a binary coded form. During the operation, the control unit proceeds from one memory location to the next in an organized manner. At every location, the instruction contained within that location is sent to the processing unit, controlling its activity for a short time. In this way an extensive, easily altered switching system can be built up (see Sec. **23**).

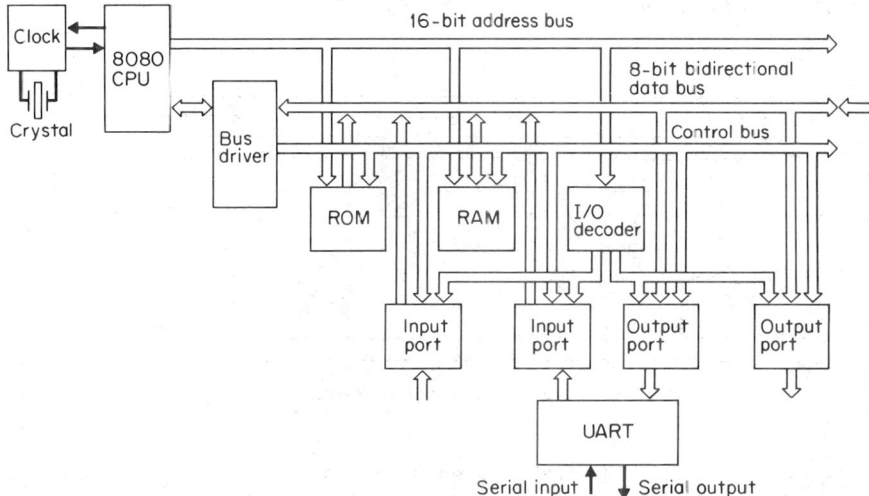

Fig. 3-74. Basic block diagram for the Intel 8080 Microprocessor.

87. Microprocessor Circuits. (See also Sec. **8**.) Large-scale integration has made it possible to construct complex programmable circuits on a single silicon chip, allowing the design of complete general-purpose digital computers on a single printed-circuit board. Figure 3-74 shows one such device based on the Intel 8080 integrated circuit.

The 8080 chip itself is the central processing unit. It can be *programmed* by sorting instructions in its memories, read-only memory (ROM) and random-access memory (RAM). It stores and retrieves data from locations which it selects by a signal on the *address bus*, sends these data along the *data bus*, and performs various operations controlled by signals on the *control bus*. Each of these three lines is multiplexed so that the various modules can communicate without interference. Also included is a *decoder*, which controls the multiplexing function. Decoders can also be used with the output ports to convert to decimal representation for easier interpretation by the user. Finally, a universal asynchronous receiver-transmitter (UART) is used to communicate data to other digital circuits. This device is a buffer which provides enough temporary storage to allow conversion of data from serial to parallel or parallel to serial.

88. Bibliography

Beckmann, P.   "Introduction to Elementary Queuing Theory and Telephone Traffic," Golem Press, Boulder, Colo., 1968.

Belevitch, V.   "Classical Network Theory," Holden-Day, San Francisco, 1968.

Bode, H. W.   "Network Analysis and Feedback Amplifier Design," Van Nostrand, New York, 1945.

Bohn, E. V.   "Introduction to Electromagnetic Fields and Waves," Addison-Wesley, Reading, Mass., 1968.

Booth, T. L.   "Digital Networks and Computer Systems," 2d ed., Wiley, New York, 1978.

Brenner, E., and M. Javid   "Analysis of Electric Circuits," McGraw-Hill, New York, 1959.

Casasent, D.   "Digital Electronics," Quantum, New York, 1974.

———   "Electronic Circuits," Quantum, New York, 1973.

Chirlian, P. M.   "Basic Network Theory," McGraw-Hill, New York, 1968.

———   "Electronic Circuits: Physical Principles, Analysis, and Design," McGraw-Hill, New York, 1971.

Collin, R. E.   "Foundations for Microwave Engineering," McGraw-Hill, New York, 1966.

Cowan, J. D., and H. S. Kirschbaum   "Introduction to Circuit Analysis," Merrill, Columbus, Ohio, 1961.

Cruz, J. B., and M. E. Van Valkenburg   "Introductory Signals and Circuits," Blaisdell, Waltham, Mass., 1967.

Desoer, C. A., and E. S. Kuh   "Basic Circuit Theory," McGraw-Hill, New York, 1969.

Hunt, W. T., Jr., and R. Stein   "Static Electromagnetic Devices," Allyn and Bacon, Boston, Mass., 1963.

Karni, S.   "Network Theory: Analysis and Synthesis," Allyn and Bacon, Boston, Mass., 1966.

King, R. W. P.   "Transmission Line Theory," McGraw-Hill, New York, 1955.

Kraus, J. D.   "Antennas," McGraw-Hill, New York, 1950.

Lin, H. C.   "Integrated Electronics," Holden-Day, San Francisco, 1967.

Manning, L. A.   "Electrical Circuits," McGraw-Hill, New York, 1966.

Matick, R. E.   "Transmission Lines for Digital and Communication Networks," McGraw-Hill, New York, 1969.

McCluskey, E. J.   "Introduction to the Theory of Switching Circuits," McGraw-Hill, New York, 1965.

Middendorf, W. H.   "Analysis of Electric Circuits," Wiley, New York, 1956.

Miller, R. E.   "Switching Theory," Vol. 1, "Combinational Circuits," Wiley, New York, 1965.

Millman, J.   "Microelectronics," McGraw-Hill, New York, 1979.

Mittleman, J.   "Circuit Theory Analysis," Hayden, New York, 1964.

Morris, R. L., and J. L. Miller (eds.)   "Designing with TTL Integrated Circuits," McGraw-Hill, New York, 1971.

Motchenbacher, C. D., and F. C. Fitchen   "Low-Noise Electronic Design," Wiley, New York, 1973.

Newcomb, R. W.   "Active Integrated Circuit Synthesis," Prentice-Hall, Englewood Cliffs, N.J., 1968.

Ott, H. W.   "Noise Reduction Techniques in Electronic Systems," Wiley, New York, 1976.

Popović, B. D.   "Introductory Engineering Electromagnetics," Addison-Wesley, Reading, Mass., 1971.

Ramo, S., J. R. Whinnery, and T. VanDuzer   "Fields and Waves in Communication Electronics," Wiley, New York, 1967.

Rubin, M., and C. E. Haller   "Communication Switching Systems," Reinhold, New York, 1966.

Ruston, H., and J. Bordogna   "Electric Networks: Functions, Filters, Analysis," McGraw-Hill, New York, 1966.

Seely, C.   "Electronic Circuits," Holt, Rinehart and Winston, New York, 1958.

Shea, R. F. (ed.)   "Amplifier Handbook," McGraw-Hill, New York, 1966.

Skilling, H. H.   "Electrical Engineering Circuits," 2d ed., Wiley, New York, 1965.

Stewart, J. L.   "Circuit Theory and Design," Wiley, New York, 1956.

Terman, F. E.   "Radio Engineer's Handbook," McGraw-Hill, New York, 1943.

Thourel, L.   "The Antenna," Wiley, New York, 1960.

Van Valkenburg, M. E.   "Introduction to Modern Network Synthesis," Wiley, New York, 1960.

Weinberg, L.   "Network Analysis and Synthesis," McGraw-Hill, New York, 1962.

Williams, E. M., and J. B. Woodford   "Transmission Circuits," Macmillan, New York, 1957.

Wood, P. E., Jr.   "Switching Theory," McGraw-Hill, New York, 1968.

Section 4

Information, Communication, Noise, and Interference

JOHN B. THOMAS *Professor of Electrical Engineering, Princeton University; Fellow, IEEE*

CONTENTS

Numbers refer to paragraphs

1. Introduction. Attempts began in the 1920s to develop a quantitative theory of information measure and to apply this measure to communication systems. One of the pioneering efforts was that of Hartley in 1928, who defined the information rate of a communication system as the logarithm of the number of possible messages that could be sent through the system, assuming that all messages were equally likely.

During World War II Norbert Wiener was largely responsible for the development of a general philosophy of communication and control called *cybernetics*, formalizing the concept that both desirable signals and undesirable signals (noises) could be defined in probabilistic terms as random processes. His work was well known to initiates by the end of World War II but did not become readily available until 1948.

Drawing on Wiener's concepts and taking into account the effect of noise and message probabilities, C. E. Shannon produced two classic papers in 1948. He introduced the concepts of entropy and channel capacity in communication systems and related them through the coding theorems. Wiener and Shannon might be considered the creators of modern communication and information theory.

The principal problem in most communication systems is the transmission of information in

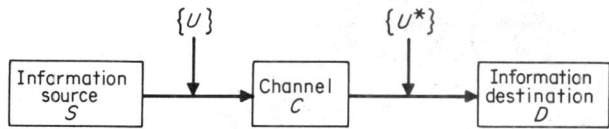

Fig. 4-1. Basic communication system.

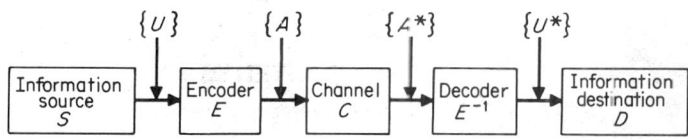

Fig. 4-2. Communication system with encoding and decoding.

the form of messages or data from some originating *information source S* to some *destination* or *receiver D*. The method of transmission is frequently by means of electric signals more or less under the control of the sender. These signals are transmitted via a channel C, as shown in Fig. 4-1. The set of messages sent by the source will be denoted by $\{U\}$. If the channel were such that each member of U were received exactly, there would be no communication problem. However, due to channel limitations and noise, a corrupted version $\{U^*\}$ of $\{U\}$ is received at the information destination. It is generally desired that the distorting effects of channel imperfections and noise be minimized and that the number of messages sent over the channel in a given time be maximized.

These two requirements are interacting, since, in general, increasing the rate of message transmission increases the distortion or error. However, some forms of message are better suited for transmission over a given channel than others, in that they can be transmitted faster or with less error. Thus it may be desirable to modify the message set $\{U\}$ by a suitable *encoder E* to produce a new message set $\{A\}$ more suitable for a given channel. Then a decoder E^{-1} will be required at the destination to recover $\{U^*\}$ from the distorted set $\{A^*\}$. A typical block diagram of the resulting system is shown in Fig. 4-2.

2. Self-Information and Entropy. Information theory is concerned with the relative frequency of occurrence of messages and not with their meaning. In the model communication system given in Fig. 4-1, we presume that each member of the message set $\{U\}$ is expressible by means of some combination of a finite set of symbols called an *alphabet*. Let this source alphabet be denoted by the set $\{X\}$ with elements $x_1, x_2, \ldots, x_M$, where M is the size of the alphabet. The notation $p(x_i)$, $i = 1, 2, \ldots, M$, will be used for the probability of occurrence of the ith symbol x_i. In general the set of numbers $\{p(x_i)\}$ can be assigned arbitrarily so long, of course, as

$$p(x_i) \geq 0 \qquad i = 1, 2, \ldots, M \tag{4-1}$$

and
$$\sum_{i=1}^{M} p(x_i) = 1 \tag{4-2}$$

A measure of the amount of information contained in the ith symbol x_i will be defined based solely on the probability $p(x_i)$. Thus, for reasons that will become clearer later, the *self-information* $I(x_i)$ of the ith symbol x_i is defined as

$$I(x_i) = \log 1/p(x_i) = -\log p(x_i) \tag{4-3}$$

This quantity is a decreasing function of $p(x_i)$ with the endpoint values of infinity for the impossible event and zero for the certain event.

It follows directly from Eq. (4-3) that $I(x_i)$ is a discrete random variable, i.e. a real-valued function defined on the elements x_i of a probability space. Of the various statistical properties of this random variable $I(x_i)$, the most important is the expected value, or mean, given by

$$E\{I(x_i)\} = H(X) = \sum_{i=1}^{M} p(x_i)I(x_i) = -\sum_{i=1}^{M} p(x_i) \log p(x_i) \tag{4-4}$$

This quantity $H(X)$ will be called the *entropy* of the distribution $p(x_i)$. If $p(x_i)$ is interpreted as the probability of the ith state of a system in phase space, then this expression is identical to the entropy of statistical mechanics and thermodynamics. Furthermore, the relationship is more than a mathematical similarity. In statistical mechanics, entropy is a measure of the disorder of a system; in information theory, it is a measure of the uncertainty associated with a message source.

In the definitions of self-information and entropy, the choice of the base for the logarithm is arbitrary, but of course each choice results in a different system of units for the information measures. The most common bases used are base 2, base e (the natural logarithm), and base 10. When base 2 is used, the unit of $I(\cdot)$ is called the *binary digit* or *bit*; this base is most convenient when dealing with the binary case, i.e., the case where $M = 2$ and the alphabet consists of only two symbols, for example 0 and 1. When base e is used, the unit is the *nat*; this base is often used because of its convenient analytical properties in integration, differentiation, etc. The base 10 is encountered only rarely; the unit is the *Hartley*.

3. Entropy of Discrete Random Variables. The more elementary properties of the entropy of a discrete random variable can be inferred from a simple example. Consider the binary case, where $M = 2$, so that the alphabet consists of the symbols 0 and 1 with probabilities p and $1 - p$ respectively. It follows from Eq. (4-4) that

$$H_1(X) = -[p \log_2 p + (1 - p) \log_2 (1 - p)] \quad \text{(bits)} \tag{4-5}$$

This equation can be plotted as a function of p, as shown in Fig. 4-3, and has the following interesting properties:

(1) $H_1(X) \geq 0$.
(2) $H_1(X)$ is zero only for $p = 0$ and $p = 1$.
(3) $H_1(X)$ is a maximum at $p = 1 - p = \frac{1}{2}$.

From the graph of Fig. 4-3 and from Eq. (4-4) it can be inferred or proved that the entropy $H(X)$ has the following properties for the general case of an alphabet of size M:

(1) $H(X) \geq 0$. $\hspace{3.5cm}$ (4-6)
(2) $H(X) = 0$ if and only if all of the probabilities are zero except for one, which must be unity. $\hspace{2cm}$ (4-7)
(3) $H(X) \leq \log_b M$. $\hspace{2.8cm}$ (4-8)
(4) $H(X) = \log_b M$ if and only if all the probabilities are equal so that $p(x_i) = 1/M$ for all i. $\hspace{2.3cm}$ (4-9)

4. Mutual Information and Joint Entropy. The usual communication problem concerns the transfer of information from a source S through a channel C to a destination D, as shown in Fig. 4-1. The source has available for forming messages an alphabet X of size M. A particular symbol x_i is selected from the M possible symbols and is sent over the channel C. It is the limitations of the channel that produce the need for a study of information theory.

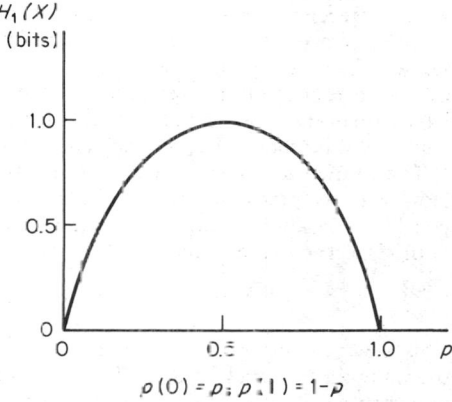

Fig. 4-3. Entropy in the binary case.

$H_1(X)$
(bits)

1.0

0.5

0

0   0.5   1.0 p

$p(0) = p; p(1) = 1 - p$

The information destination has available an alphabet Y of size N. For each symbol x_i sent from the source, a symbol y_j is selected at the destination. Two probabilities serve to describe the state of knowledge at the destination. Prior to the reception of a communication, the state of knowledge of the destination about the symbol x_j is the *a priori* probability $p(x_i)$ that x_i would be selected for transmission. After reception and selection of the symbol y_j, the state of knowledge concerning x_i is the conditional probability $p(x_i|y_j)$, which will be called the *a posteriori* probability of x_i. It is the probability that x_i was sent given that y_j was received. Ideally this a posteriori probability for each given y_j should be unity for one x_i and zero for all other x_i. In this case an observer at the destination is able to determine exactly which symbol x_i has been sent after the reception of each symbol y_j. Thus the uncertainty which existed previously and which was expressed by the a priori probability distribution of x_i has been removed completely by reception. In the general case it is not possible to remove all the uncertainty, and the best that can be hoped for is that it has been decreased. Thus the a posteriori probability $p(x_i|y_j)$ is distributed over a number of x_i but should be different. If the two probabilities are the same, then no uncertainty has been removed by transmission or no information has been transferred.

Based on this discussion and on other considerations that will become clearer later, the quantity $I(x_i; y_j)$ is defined as the information gained about x_i by the reception of y_j, where

$$I(x_i; y_j) = \log_b \left[p(x_i|y_j)/p(x_i) \right] \tag{4-10}$$

This measure has a number of reasonable and desirable properties.

Property 1.   The information measure $I(x_i; y_j)$ is symmetric in x_i and y_j; that is,

$$I(x_i; y_j) = I(y_j; x_i) \tag{4-11}$$

Property 2.   The mutual information $I(x_i; y_j)$ is a maximum when $p(x_i|y_j) = 1$, that is, when the reception of y_j completely removes the uncertainty concerning x_i:

$$I(x_i; y_j) \le - \log p(x_i) = I(x_i) \tag{4-12}$$

Property 3.   If two communications y_j and z_k concerning the same message x_i are received successively, and if the observer at the destination takes the a posteriori probability of the first as the a priori probability of the second, then the total information gained about x_i is the sum of the gains from both communications:

$$I(x_i; y_j, z_k) = I(x_i; y_j) + I(x_i; z_k|y_j) \tag{4-13}$$

Property 4.   If two communications y_j and y_k concerning two *independent* messages x_i and x_m are received, the total information gain is the sum of the two information gains considered separately:

$$I(x_i, x_m; y_j, y_k) = I(x_i; y_j) + I(x_m; y_k) \tag{4-14}$$

These four properties of mutual information are intuitively satisfying and desirable. Moreover, if one begins by requiring these properties, it is easily shown that the logarithmic definition of Eq. (4-10) is the simplest form that can be obtained.

The definition of mutual information given by Eq. (4-10) suffers from one major disadvantage. When errors are present, an observer will not be able to calculate the information gain even after the reception of all the symbols relating to a given source symbol, since the same series of received symbols may represent several different source symbols. Thus the observer is unable to say which source symbol has been sent, and the best he can do is to compute the information gain with respect to each possible source symbol. In many cases it would be more desirable to have a quantity which is independent of the particular symbols. A number of quantities of this nature will be obtained in the remainder of this section.

The mutual information $I(x_i; y_j)$ is a random variable just as the self-information $I(x_i)$ was. However, two probability spaces X and Y are involved now, and several ensemble averages are possible. The *average mutual information* $I(X; Y)$ is defined as a statistical average of $I(x_i; y_j)$ with respect to the joint probability $p(x_i; y_j)$; that is,

$$I(X; Y) = E_{XY}\{I(x_i; y_j)\} = \sum_i \sum_j p(x_i, y_j) \log \left[p(x_i|y_j)/p(x_i) \right] \tag{4-15}$$

This new function $I(X; Y)$ is the first information measure defined which does not depend on the individual symbols x_i or y_j. Thus it is a property of the whole communication system and will turn out to be only the first in a series of similar quantities used as a basis for the characterization of communication systems. This quantity $I(X; Y)$ has a number of expected properties.

It is nonnegative; it is zero if and only if the ensembles X and Y are *statistically independent*; and it is symmetric in X and Y so that $I(X; Y) = I(Y; X)$.

A source entropy $H(X)$ was given by Eq. (4-4). It is obvious that a similar quantity, the destination entropy $H(Y)$, can be defined analogously by

$$H(Y) = -\sum_{j=1}^{N} p(y_j) \log p(y_j) \tag{4-16}$$

This quantity will, of course, have all the properties developed for $H(X)$. In the same way the *joint* or *system entropy* $H(X, Y)$ can be defined by

$$H(X, Y) = -\sum_{i=1}^{M} \sum_{j=1}^{N} p(x_i, y_j) \log p(x_i, y_j) \tag{4-17}$$

If X and Y are *statistically independent* so that $p(x_i, y_j) = p(x_i)p(y_j)$ for all i and j, then Eq. (4-17) can be written

$$H(X, Y) = H(X) + H(Y) \tag{4-18}$$

On the other hand, if X and Y are not independent, Eq. (4-17) becomes

$$H(X, Y) = H(X) + H(Y|X) = H(Y) + H(X|Y) \tag{4-19}$$

where $H(Y|X)$ and $H(X|Y)$ are *conditional entropies* given by

$$H(Y|X) = -\sum_{i=1}^{M} \sum_{j=1}^{N} p(x_i, y_j) \log p(y_j|x_i) \tag{4-20}$$

and

$$H(X|Y) = -\sum_{i=1}^{M} \sum_{j=1}^{N} p(x_i, y_j) \log p(x_i|y_j) \tag{4-21}$$

These conditional entropies each satisfy an important inequality

$$0 \le H(Y|X) \le H(Y) \tag{4-22}$$

and

$$0 \le H(X|Y) \le H(X) \tag{4-23}$$

It follows from these last two expressions that Eq. (4-15) can be expanded to yield

$$I(X; Y) = -H(X, Y) + H(X) + H(Y) \ge 0 \tag{4-24}$$

This equation can be rewritten in the two equivalent forms

$$I(X; Y) = H(Y) - H(Y|X) \ge 0 \tag{4-25}$$

or

$$I(X|Y) = H(X) - H(X|Y) \ge 0 \tag{4-26}$$

It is also clear, say from Eq. (4-24), that $H(X, Y)$ satisfies the inequality

$$H(X, Y) \le H(X) + H(Y) \tag{4-27}$$

Thus the joint entropy of two ensembles X and Y is a maximum when the ensembles are independent.

At this point is may be appropriate to comment on the meaning of the two conditional entropies $H(Y|X)$ and $H(X|Y)$. Let us refer first to Eq. (4-26). This equation expresses the fact that the average information gained about a message, when a communication is completed, is equal to the average source information less the average uncertainty that still remains about the message. From another point of view, the quantity $H(X|Y)$ is the average additional information needed at the destination after reception to completely specify the message sent. Thus $H(X|Y)$ represents the information lost in the channel. It is frequently called the *equivocation*. Let us now consider Eq. (4-25). This equation indicates that the information transmitted consists of the difference between the destination entropy and that part of the destination entropy that is not information about the source; thus the term $H(Y|X)$ can be considered a *noise entropy* added in the channel.

INFORMATION SOURCES

5. Message Sources. As shown in Fig. 4-1, an information source can be considered as emitting a given message u_i from the set $\{U\}$ of possible messages. In general, each message u_i will be represented by a sequence of symbols x_j from the source alphabet $\{X\}$, since the number of possible messages will usually exceed the size M of the source alphabet. Thus sequences of symbols replace the original messages u_i, which need not be considered further. When the source alphabet $\{X\}$ is of finite size M, the source will be called a *finite discrete source*. The problems of concern now are the interrelationships existing between symbols in the generated sequences and the classification of sources according to these interrelationships.

A random or stochastic process x_t, $t \in T$, can be defined as an indexed set of random variables where T is the *parameter set* of the process. If the set T is a sequence, then x_t is a stochastic process with discrete parameter (also called a *random sequence* or series). One way to look at the output of a finite discrete source is that it is a discrete-parameter stochastic process with each possible given sequence one of the ensemble members or realizations of the process. Thus the study of information sources can be reduced to a study of random processes.

The simplest case to consider is the *zero-memory source*, where the successive symbols obey the same fixed probability law so that the one distribution $p(x_i)$ determines the appearance of each indexed symbol. Such a source is called *stationary*. Let us consider sequences of length n, each member of the sequence being a realization of the random variable x_i with fixed probability distribution $p(x_i)$. Since there are M possible realizations of the random variable and n terms in the sequence, there must be M^n distinct sequences possible of length n. Let the random variable x_i in the jth position be denoted by x_{i_j} so that the sequence set (the message set) can be represented by

$$\{U\} = X^n = \{x_{i_1}, x_{i_2}, \cdots, x_{i_n}\} \qquad i = 1, 2, \ldots, M \tag{4-28}$$

The symbol X^n is sometimes used to represent this sequence set and is called the *nth extension of the zero-memory source X*. The probability of occurrence of a given message u_i is just the product of the probabilities of occurrence of the individual terms in the sequence so that

$$p\{u_i\} = p(x_{i_1})p(x_{i_2}) \cdots p\{x_{i_n}\} \tag{4-29}$$

Now the entropy for the extended source X^n is

$$H(X^n) = - \sum_{X^n} p\{u_i\} \log p\{u_i\} = nH(X) \tag{4-30}$$

as expected. Note that, if base 2 logarithms are used, then $H(X)$ has units of bits per symbol, n is symbols per sequence, and $H(X^n)$ is in units of bits per sequence. For a zero-memory source, all sequence averages of information measure are obtained by multiplying the corresponding symbol by the number of symbols in the sequence.

6. Markov Information Source. The zero-memory source is not a general enough model in most cases. A constructive way to generalize this model is to assume that the occurrence of a given symbol depends on some number m of immediately preceding symbols. Thus the information source can be considered to produce an mth-order Markov chain and is called an *mth-order Markov source*.

For an mth-order Markov source, the m symbols preceding a given symbol position are called the *state* s_j of the source at that symbol position. If there are M possible symbols x_i, then the mth-order Markov source will have $M^m = q$ possible states s_j making up the *state set*

$$S = \{s_1, s_2, \cdots, s_q\} \qquad q = M^m \tag{4-31}$$

At a given time corresponding to one symbol position the source will be in a given state s_j. There will exist a probability $p(s_k|s_j) = p_{jk}$ that the source will move into another state s_k with the emission of the next symbol. The set of all such conditional probabilities is expressed by the *transition matrix* T, where

$$T = [p_{jk}] = \begin{bmatrix} p_{11} & p_{12} & \cdots & p_{1q} \\ p_{21} & p_{22} & \cdots & p_{2q} \\ \cdots & \cdots & \cdots & \cdots \\ p_{q1} & p_{q2} & \cdots & p_{qq} \end{bmatrix} \tag{4-32}$$

A *Markov matrix* or *stochastic matrix* is any square matrix with nonnegative elements such that the row sums are unity. It is clear that T is such a matrix since

$$\sum_{j=1}^{q} p_{ij} = \sum_{j=1}^{q} p(s_j|s_i) = 1 \qquad i = 1, 2, \ldots, q \tag{4-33}$$

Conversely, any stochastic matrix is a possible transition matrix for a Markov source of order m, where $q = M^m$ is equal to the number of rows or columns of the matrix.

A Markov chain is completely specified by its transition matrix T and by an initial distribution vector π giving the probability distribution for the first state occurring. For the zero-memory source, the transition matrix reduces to a stochastic matrix where all the rows are identical and are each equal to the initial distribution vector π, which is in turn equal to the vector giving the source alphabet a priori probabilities. Thus, in this case we have

$$p_{jk} = p(s_k|s_j) = p(s_k) = p(x_k) \qquad k = 1, 2, \ldots, M \tag{4-34}$$

For each state s_i of the source an entropy $H(s_i)$ can be defined by

$$H(s_i) = -\sum_{j=1}^{q} p(s_j|s_i) \log p(s_j|s_i) = -\sum_{k=1}^{M} p(x_k|s_i) \log p(x_k|s_i) \tag{4-35}$$

The source entropy $H(S)$ in information units per symbol is the expected value of $H(s_i)$; that is,

$$H(S) = -\sum_{i=1}^{q}\sum_{j=1}^{q} p(s_i)p(s_j|s_i) \log p(s_j|s_i) = -\sum_{i=1}^{q}\sum_{k=1}^{M} p(s_i)p(x_k|s_i) \log p(x_k|s_i) \tag{4-36}$$

where $p(s_i) = p_i$ is the *stationary state probability* and is the ith element of the vector $\mathbf{P}$ defined by

$$\mathbf{P} = [p_1 \quad p_2 \quad \cdots \quad p_q] \tag{4-37}$$

It is easy to show, as in Eq. (4-8), that the source entropy cannot exceed $\log M$, where M is the size of the source alphabet $\{X\}$. For a given source, the ratio of the actual entropy $H(S)$ to the maximum value it can have with the same alphabet is called the *relative entropy* of the source. The *redundancy* η of the source is defined as the positive difference between unity and this relative entropy:

$$\eta = 1 - \frac{H(S)}{\log M} \tag{4-38}$$

The quantity $\log M$ is sometimes called the *capacity* of the alphabet.

CODES AND CODING

7. Noiseless Coding. The preceding discussion has emphasized the information source and its properties. We now consider the properties of the communication channel of Fig. 4-1. In general, an arbitrary channel will not accept and transmit the sequence of x_i's emitted from an arbitrary source. Instead the channel will accept a sequence of some other elements a_i chosen from a *code alphabet* A of *size* D, where

$$A = \{a_1, a_2, \ldots, a_D\} \tag{4-39}$$

with D generally smaller than M. The elements a_i of the code alphabet are frequently called *code elements* or *code characters*, while a given sequence of a_is may be called a *code word*.

The situation is now describable in terms of Fig. 4-2, where an encoder E has been added between the source and channel. The process of *coding*, or *encoding*, the source consists of associating with each source symbol x_i a given code word, which is just a given sequence of a_i's. Thus the source emits a sequence of x_i's chosen from the source alphabet X, and the encoder emits a sequence of a_i's chosen from the code alphabet A. It will be assumed in all subsequent discussions that the code words are distinct, i.e., that each code word corresponds to only one source symbol.

Even though each code word is required to be distinct, sequences of code words may not have this property. An example is code A of Table 4-1, where a source of size 4 has been encoded in binary code with characters 0 and 1. In code A the code words are distinct, but sequences of code words are not. It is clear that such a code is not *uniquely decipherable*. On the other hand, a given sequence of code words taken from code B will correspond to a distinct sequence of source symbols. An examination of code B shows that in no case is a code word formed by adding characters to another word. In other words, no code word is a *prefix* of another. It is clear that this is a *sufficient* (but not necessary) condition for a code to be uniquely decipherable. That it is not necessary can be seen from an examination of codes C and D of Table 4-1. These codes are uniquely decipherable even though many of the code words are prefixes of other words. In these cases any sequence of code words can be decoded by subdividing the sequence of 0s and 1s to the left of every 0 for code C and to the right of every 0 for code D. The character 0 is the first (or last) character of every code word and acts as a comma; therefore this type of code is called a *comma code*.

In general the channel will require a finite amount of time to transmit each code character. The code words should be as short as possible in order to maximize information transfer per unit time. The average length L of a code is given by

$$L = \sum_{i=1}^{M} n_i p(x_i) \tag{4-40}$$

where n_i is the length (number of code characters) of the code word for the source symbol x_i and $p(x_i)$ is the probability of occurrence of x_i. Although the average code length cannot be computed unless the set $\{p(x_i)\}$ is given, it is obvious that codes C and D of Table 4-1 will have

Table 4-1. Four Binary Coding Schemes

Source symbol	Code A	Code B	Code C	Code D
x_1	0	0	0	0
x_2	1	10	01	10
x_3	00	110	011	110
x_4	11	111	0111	1110

NOTE: Code A is not uniquely decipherable; codes B, C, and D are uniquely decipherable; codes B and D are instantaneous codes; and codes C and D are comma codes.

a greater average length than code B unless $p(x_4) = 0$. Comma codes are not optimal with respect to minimum average length.

Let us encode the sequence $x_3 x_1 x_3 x_2$ into codes B, C, and D of Table 4-1 as shown below:

Code B:	110011010
Code C:	011001101
Code D:	110011010

Codes B and D are fundamentally different from code C in that codes B and D can be decoded word by word *without examining subsequent code characters* while code C cannot be so treated. Codes B and D are called *instantaneous codes* while code C is noninstantaneous. The instantaneous codes have the property (previously mentioned) that no code word is a prefix of another code word.

The aim of noiseless coding is to produce codes with the two properties of *unique decipherability* and *minimum average length* L for a given source S with alphabet X and probability set $\{p(x_i)\}$. Codes which have both these properties will be called *optimal*. It can be shown that if, for a given source S, a code is optimal among instantaneous codes, then it is optimal among all uniquely decipherable codes. Thus it is sufficient to consider instantaneous codes. A *necessary* property of optimal codes is that source symbols with higher probabilities have shorter code words; i.e.,

$$p(x_i) > p(x_j) \Rightarrow n_i \leq n_j \tag{4-41}$$

The encoding procedure consists of the assignment of a code word to each of the M source symbols. The code word for the source symbol x_i will be of length n_i; that is, it will consist of n_i code elements chosen from the code alphabet of size D. It can be shown that a necessary and sufficient condition for the construction of a uniquely decipherable code is the *Kraft inequality*

$$\sum_{i=1}^{M} D^{-n_i} \leq 1 \tag{4-42}$$

8. Noiseless-Coding Theorem. It follows from Eq. (4-42) that the average code length L, given by Eq. (4-40), satisfies the inequality

$$L \geq H(X)/\log D \tag{4-43}$$

Equality (and minimum code length) occurs if and only if the source-symbol probabilities obey

$$p(x_i) = D^{-n_i} \qquad i = 1, 2, \ldots, M \tag{4-44}$$

A code where this equality applies is called *absolutely optimal*. Since an integer number of code elements must be used for each code word, the equality in Eq. (4-43) does not usually hold.

However, by using one more code element, the average code length L can be bounded from above to give

$$H(X)/\log D \le L \le H(X)/\log D + 1 \qquad (4\text{-}45)$$

This last relationship is frequently called the *noiseless-coding theorem*.

9. Construction of Noiseless Codes. The easiest case to consider occurs when an absolutely optimal code exists; i.e., when the source-symbol probabilities satisfy Eq. (4-44). Note that code B of Table 4-1 is absolutely optimal if $p(x_1) = \frac{1}{2}$, $p(x_2) = \frac{1}{4}$, and $p(x_3) = p(x_4) = \frac{1}{8}$. In such cases, a procedure for realizing the code for arbitrary code-alphabet size $(D \ge 2)$ is easily constructed as follows:

1. Arrange the M source symbols in the set x_i in order of decreasing probability.

2. Arrange the D code elements in an arbitrary but fixed order, that is $a_1, a_2, \ldots, a_D$.

3. Divide the set of symbols x_i into D groups with equal probabilities of $1/D$ each. This division is always possible if Eq. (4-44) is satisfied.

4. Assign the element a_1 as the first digit for symbols in the first group a_2 for the second, and a_i for the ith group.

5. After the first division each of the resulting groups contains a number of symbols equal to D raised to some integral power if Eq. (4-44) is satisfied. Thus a typical group, say group i, contains D^{k_i} symbols, where k_i is an integer (which may be zero). This group of symbols can be further subdivided k_i times into D parts of equal probabilities. Each division decides one additional code digit in the sequence. A typical symbol x_i is isolated after q divisions. If it belongs to the i_1 group after the first division, the i_2 group after the second division, etc., then the code word for x_i will be $a_{i_1} a_{i_2} \cdots a_{i_q}$.

An illustration of the construction of an absolutely optimal code for the case where $D = 3$ is given in Table 4-2. This procedure ensures that source symbols with high probabilities will have short code words and vice versa, since a symbol with probability D^{-n_i} will be isolated after n_i divisions and thus will have n_i elements in its code word, as required by Eq. (4-44).

The code resulting from the process just discussed is sometimes called the *Shannon-Fano* code. It is apparent that the same encoding procedure can be followed whether or not the source probabilities satisfy Eq. (4-44). The set of symbols x_i is simply divided into D groups with probabilities as nearly equal as possible. The procedure is sometimes ambiguous, however, and more

Table 4-2. Construction of an Optimal Code; $D=3$

Source symbols x_i	A priori probabilities $p(x_i)$	Step			Final code		
		1	2	3			
x_1	$\frac{1}{3}$	1			1		
x_2	$\frac{1}{9}$	0	1		0	1	
x_3	$\frac{1}{9}$	0	0		0	0	
x_4	$\frac{1}{9}$	0	-1		0	-1	
x_5	$\frac{1}{27}$	-1	1	1	-1	1	1
x_6	$\frac{1}{27}$	-1	1	0	-1	1	0
x_7	$\frac{1}{27}$	-1	1	-1	-1	1	-1
x_8	$\frac{1}{27}$	-1	0	1	-1	0	1
x_9	$\frac{1}{27}$	-1	0	0	-1	0	0
x_{10}	$\frac{1}{27}$	-1	0	-1	-1	0	-1
x_{11}	$\frac{1}{27}$	-1	-1	1	-1	-1	1
x_{12}	$\frac{1}{27}$	-1	-1	0	-1	-1	0
x_{13}	$\frac{1}{27}$	-1	-1	-1	-1	-1	-1

NOTE: Average code length $L=2$ code elements per symbol; source entropy $H(X) = 2 \log_2 3$ bits per symbol

$$L = \frac{H(X)}{\log_2 3}$$

Table 4-3. Construction of a Huffman Code; $D = 2$

Source symbols x_i	A priori probabilities $p(x_i)$	Final code	Reduction 1 (Step 5)	Reduction 2 (Step 4)	Reduction 3 (Step 3)	Reduction 4 (Step 2)	Reduction 5 (Step 1)
x_1	0.40	0	0 — 0.40	0 — 0.40	0 — 0.40	0 — 0.40	1 — 0.60
x_2	0.20	111	111 — 0.20	111 — 0.20	10 — 0.24	11 — 0.36	0 — 0.40
x_3	0.12	101	101 — 0.12	110 — 0.16	111 — 0.20	10 — 0.24	
x_4	0.08	1101	100 — 0.12	101 — 0.12	110 — 0.16		
x_5	0.08	1100	1101 — 0.08	100 — 0.12			
x_6	0.08	1001	1100 — 0.08				
x_7	0.04	1000					

Average code length $L = 1(0.40) + 3(0.20) + 3(0.12) + 4(0.08) + 4(0.08) + 4(0.08) + 4(0.04)$
$= 2.48$ code elements/symbol

than one Shannon-Fano code may be possible. The ambiguity arises, of course, in the choice of approximately equiprobable subgroups.

For the general case where Eq. (4-44) is not satisfied, a procedure due to Huffman guarantees an optimal code, i.e., one with minimum average length. This procedure for code alphabet of arbitrary size D is as follows:

1. As before, arrange the M source symbols in the set x_i in order of decreasing probability.
2. As before, arrange the code elements in an arbitrary but fixed order, that is, $a_1, a_2, \ldots, a_D$.
3. Combine (sum) the probabilities of the D least likely symbols and reorder the resulting $M - (D - 1)$ probabilities; this step will be called *reduction* 1. Repeat as often as necessary until there are D ordered probabilities remaining. *Note:* For the binary case ($D = 2$), it will always be possible to accomplish this reduction in $M - 2$ steps. When the size of the code alphabet is arbitrary, the last reduction will result in exactly D ordered probabilities if and only if

$$M = D + n(D - 1)$$

where n is an integer. If this relationship is not satisfied, add *dummy* source symbols with zero probability. The entire encoding procedure is followed as before, and at the end the dummy symbols are thrown away.

4. Start the encoding with the last reduction which consists of exactly D ordered probabilities; assign the element a_1 as the first digit in the code words for all the source symbols associated with the first probability; assign a_2 to the second probability; and a_i to the ith probability.
5. Proceed to the next to the last reduction; this reduction consists of $D + (D - 1)$ ordered probabilities for a net gain of $D - 1$ probabilities. For the D new probabilities, the first code digit has already been assigned and is the same for all of these D probabilities; assign a_1 as the second digit for all source symbols associated with the first of these D new probabilities; assign a_2 as the second digit for the second of these D new probabilities, etc.
6. The encoding procedure terminates after $1 + n(D - 1)$ steps, which is 1 more than the number of reductions.

As an illustration of the Huffman coding procedure, a binary code is constructed in Table 4-3.

COMMUNICATION CHANNELS

10. Channel Capacity. The average mutual information $I(X; Y)$ between an information source and a destination was given by Eqs. (4-25) and (4-26) as

$$I(X; Y) = H(Y) - H(Y|X) = H(X) - H(X|Y) \geq 0 \tag{4-46}$$

The average mutual information depends not only on the statistical characteristics of the channel but also on the distribution $p(x_i)$ of the input alphabet X. If the input distribution is varied until Eq. (4-46) is a maximum for a given channel, the resulting value of $I(X; Y)$ is called the *channel capacity* C of that channel; i.e.,

$$C = \max_{p(x_i)} I(X; Y) \tag{4-47}$$

In general, $H(X)$, $H(Y)$, $H(X|Y)$, and $H(Y|X)$ all depend on the input distribution $p(x_i)$.

Hence, *in the general case*, it is not a simple matter to maximize Eq. (4-46) with respect to $p(x_i)$.

All the measures of information that have been considered in this treatment have involved only probability distributions on X and Y. Thus, for the model of Fig. 4-1, the joint distribution $p(x_i, y_j)$ is sufficient. Suppose the source [and hence the input distribution $p(x_i)$] is known; then it follows from the usual conditional-probability relationship

$$p(x_i, y_j) = p(x_i)p(y_j|x_i) \tag{4-48}$$

that only the distribution $p(y_j|x_i)$ is needed for $p(x_i|y_j)$ to be determined. This conditional probability $p(y_j|x_i)$ can then be taken as a description of the information channel connecting the source X and the destination Y. Thus, a *discrete memoryless channel* can be defined as the probability distribution

$$p(y_j|x_i) \qquad x_i \in X \text{ and } y_j \in Y \tag{4-49}$$

or, equivalently, by the *channel matrix D*, where

$$D = [p(y_j | x_i)] = \begin{bmatrix} p(y_1|x_1) & p(y_2|x_1) & \cdots & p(y_N|x_1) \\ p(y_1|x_2) & p(y_2|x_2) & \cdots & p(y_N|x_2) \\ \cdots\cdots\cdots\cdots\cdots\cdots\cdots \\ p(y_1|x_M) & \cdots & \cdots & p(y_N|x_M) \end{bmatrix}$$ (4-50)

A number of special types of channels are readily distinguished. Some of the simplest and/or most interesting are listed.

(a) *Lossless Channel.* Here $H(X|Y) = 0$ for all input distribution $p(x_i)$, and Eq. (4-47) becomes

$$C = \max_{p(x_i)} H(X) = \log M$$ (4-51)

This maximum is obtained when the x_i are equally likely, so that $p(x_i) = 1/M$ for all i. The channel capacity is equal to the source entropy, and no source information is lost in transmission.

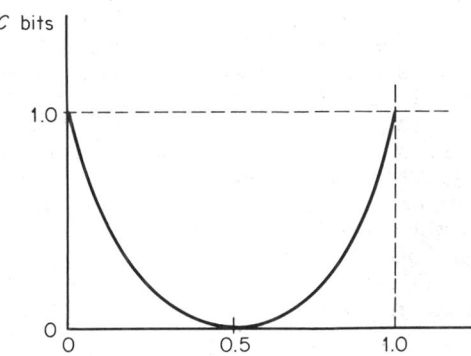

Fig. 4-4. Capacity of the binary symmetric channel.

(b) *Deterministic Channel.* Here $H(Y|X) = 0$ for all input distributions $p(x_i)$, and Eq. (4-47) becomes

$$C = \max_{p(x_i)} H(Y) = \log N$$ (4-52)

This maximum is obtained when the y_j are equally likely, so that $p(y_j) = 1/N$ for all j. Each member of the X set is uniquely associated with one, and only one, member of the destination alphabet Y.

(c) *Symmetric Channel.* Here the rows of the channel matrix D are identical except for permutations, *and* the columns are identical except for permutations. If D is square, rows and columns are identical except for permutations. In the symmetric channel, the conditional entropy $H(Y|X)$ is independent of the input distribution $p(x_i)$ and depends only on the channel matrix D. As a consequence, the determination of channel capacity is greatly simplified and can be written

$$C = \log N + \sum_{j=1}^{N} p(y_j | x_i) \log p(y_j | x_i)$$ (4-53)

This capacity is obtained when the y_j are equally likely, so that $p(y_j) = 1/N$ for all j.

(d) *Binary Symmetric Channel (BSC).* This is the special case of a symmetric channel where $M = N = 2$. Here the channel matrix can be written

$$D = \begin{bmatrix} p & 1-p \\ 1-p & p \end{bmatrix}$$ (4-54)

and the channel capacity is

$$C = \log 2 - G(p)$$ (4-55)

where the function $G(p)$ is defined as

$$G(p) = -[p \log p + (1-p) \log (1-p)]$$ (4-56)

This expression is mathematically identical to the entropy of a binary source as given in Eq. (4-5) and plotted in Fig. 4-3 using base 2 logarithms. For the same base, Eq. (4-55) is shown as a function of p in Fig. 4-4. As expected, the channel capacity is large if p, the probability of correct transmission, is either close to unity or to zero. If $p = \frac{1}{2}$, there is no statistical evidence which symbol was sent and the channel capacity is zero.

DECODING

11. Decision Schemes. A decision scheme or decoding scheme $\mathcal{B}$ is a partitioning of the Y set into M disjoint and exhaustive sets $B_1, B_2, \ldots, B_M$ such that when a destination symbol y_k falls into set B_i, it is decided that symbol x_i was sent. Implicit in this definition is a *decision rule*

$d(y_j)$, which is a function specifying uniquely a source symbol for each destination symbol. Let $p(e|y_j)$ be the probability of error when it is decided that y_j has been received. Then the *total error probability* $p(e)$ is

$$p(e) = \sum_{j=1}^{N} p(y_j)p(e|y_j) \tag{4-57}$$

For a given decision scheme $\mathcal{B}$, the conditional error probability $p(e|y_j)$ can be written

$$p(e|y_j) = 1 - p[d(y_j)|y_j] \tag{4-58}$$

where $p[d(y_j)|y_j]$ is the conditional probability $p(x_i|y_j)$ with x_i assigned by the decision rule; i.e., for a given decision scheme $d(y_j) = x_i$. The probability $p(y_j)$ is determined only by the source a priori probability $p(x_i)$ and by the channel matrix $D = [p(y_j|x_i)]$. Hence, only the term $p(e|y_j)$ in Eq. (4-57) is a function of the decision scheme. Since Eq. (4-57) is a sum of nonnegative terms, the error probability is a minimum when each summand is a minimum. Thus, the term $p(e|y_j)$ should be a minimum for each y_j. It follows from Eq. (4-58) that the minimum-error scheme is that scheme which assigns a decision rule

$$d(y_j) = x^* \qquad j = 1, 2, \dots, N \tag{4-59}$$

where x^* is defined by

$$p(x^*|y_j) \geq p(x_i|y_j) \qquad i = 1, 2, \dots, M \tag{4-60}$$

In other words, each y_j is decoded as the *a posteriori most likely* x_i. This scheme, which minimizes the probability of error $p(e)$, is usually called the *ideal observer*.

The ideal observer is not always a completely satisfactory decision scheme. It suffers from two major disadvantages: (1) For a given channel D, the scheme is defined only for a given input distribution $p(x_i)$. It would be preferable to have a scheme which was insensitive to input distributions. (2) The scheme minimizes average error but does not bound certain errors. For example, some symbols may always be received incorrectly. Despite these disadvantages, the ideal observer is a straightforward scheme which does minimize average error. It is also widely used as a standard with which other decision schemes may be compared.

Consider the special case where the input distribution is $p(x_i) = 1/M$ for all i, so that all x_i are equally likely. Now the conditional likelihood $p(x_i|y_j)$ is

$$p(x_i|y_j) = \frac{p(x_i)p(y_j|x_i)}{p(y_j)} = \frac{p(y_j|x_i)}{Mp(y_j)} \tag{4-61}$$

For a given y_j, that input x_i is chosen which makes $p(y_j|x_i)$ a maximum, and the decision rule is

$$d(y_j) = x\dagger \qquad j = 1, 2, \dots, N \tag{4-62}$$

where $x\dagger$ is defined by

$$p(y_j|x\dagger) \geq p(y_j|x_i) \qquad i = 1, 2, \dots, M \tag{4-63}$$

The probability of error becomes

$$p(e) = \sum_{j=1}^{N} p(y_j)\left[1 - \frac{p(y_j|x\dagger)}{Mp(y_j)}\right] \tag{4-64}$$

This decoder is sometimes called the *maximum-likelihood* decoder or decision scheme.

It would appear that a relationship should exist between the error probability $p(e)$ and the channel capacity C. One such relationship is the *Fano bound*, given by

$$H(X|Y) \leq G[p(e)] + p(e) \log(M - 1) \tag{4-65}$$

and relating error probability to channel capacity through Eq. (4-47). Here $G(\cdot)$ is the function already defined by Eq. (4-56). The three terms in Eq. (4-65) can be interpreted as follows:

$H(X|Y)$ is the equivocation. It is the average additional information needed at the destination after reception to completely determine the symbol that was sent.

$G[p(e)]$ is the entropy of the binary system with probabilities $p(e)$ and $1 - p(e)$. In other words, it is the average amount of information needed to determine whether the decision rule resulted in an error.

$\log(M - 1)$ is the maximum amount of information needed to determine which among the remaining $M - 1$ symbols was sent if the decision rule was incorrect; this information is needed with probability $p(e)$.

EFFECTS OF NOISE

12. The Noisy-Coding Theorem. The concept of channel capacity was discussed in Par. **4-10**. Capacity is a fundamental property of an information channel in the sense that it is possible to transmit information through the channel at any rate less than the channel capacity with arbitrarily small probability of error. This result, which will be stated more precisely shortly, is called the *noisy-coding theorem* or *Shannon's fundamental theorem for a noisy channel*.

A proof of this theorem for an arbitrary channel is difficult and quite beyond the level of this treatment. However, a heuristic, i.e., nonrigorous, discussion is relatively straightforward and affords considerable insight into the general problem. We begin by stating the assumptions made in order that confusion over what has been proved will be at a minimum.

1. The *source* will consist of a set $U = \{u_1, u_2, \ldots, u_m\}$ of m messages u_k. It will be assumed that these are *independent* and *equally likely*. This last assumption is a strong one, and it may be necessary to form this message set by encoding an original source alphabet $X = \{x_1, x_2, \ldots, x_M\}$ into long sequences; i.e., each sent message u_k may consist of a sequence of x_i's of length p, where p may be large. Now the source entropy $H(U)$ is given by

$$H(U) = \log m \tag{4-66}$$

2. Each message u_k will be *encoded* into a code word consisting of a sequence of n binary digits for transmission over the binary channel. The number of possible code words of length n is 2^n; the number of messages u_k is m; thus the inequality

$$2^n \geq m \quad \text{or} \quad n \geq \log_2 m \tag{4-67}$$

must be satisfied. The code words will be assigned at random; i.e., the probability is $m/2^n$ that a given n-place code word (of the 2^n possible code words) will be chosen to represent one of the set of m messages.

3. The *channel* will be taken to be the zero-memory binary symmetric channel (BSC) with probability p of correct transmission of a binary digit and probability of error $q = 1 - p$.

4. The *decision scheme* or *decoder* will operate as follows. At the destination, the received messages (there are 2^n possible) will be compared with all m of the messages that could have been sent. The message u_k will be considered sent which differs from the received message in the least number of binary digits. This decoder can be shown to be the ideal observer (actually the maximum-likelihood decoder since, in this particular case, the source probabilities are equally likely).

Consider a given code word of length n. The probability that exactly r digits will be altered in transmission (while $n - r$ digits are not altered) is just the probability of $n - r$ successes and r failures in n Bernoulli trials. This is the binomial distribution and is given by

$$p\{r \text{ errors}\} = \binom{n}{n - r} p^{n-r} q^r \quad r = 0, 1, 2, \ldots, n \tag{4-68}$$

The mean number of errors is

$$E\{r\} = nq = n(1 - p) \tag{4-69}$$

Thus, the average number of binary digits altered by noise in the BSC will be nq.

There are $\binom{n}{r}$ code words (out of the total of 2^n) that differ *in exactly r digits* from a given code word. Therefore, the number K of code words that differ *in nq digits or less* from a given code word is given by the sum

$$K = \sum_{r=0}^{nq} \binom{n}{r} = \binom{n}{0} + \binom{n}{1} + \cdots + \binom{n}{nq - 1} + \binom{n}{nq} \tag{4-70}$$

For $q < \frac{1}{2}$, each term in this series for K is larger than the preceding term, so that K is bounded by

$$K \leq (nq + 1) \binom{n}{nq} = (nq + 1) \frac{n!}{(nq)!(n - nq)!} \tag{4-71}$$

For large n, the factorial is given approximately by Stirling's formula

$$n! \approx \sqrt{2\pi}\, n^{n+1/2} e^{-n}$$

With this substitution, Eq. (4-71) can be written

$$K \le \sqrt{(nq + 1)^2/2\pi npq}\; q^{-nq} p^{-np} \tag{4-72}$$

Thus, on the average, a given sent code word can be received as any one of K code words differing in nq digits or less from the sent word. An error will result if any message u_k, other than the one originally transmitted, is represented by one of this group of K code words.

Recall that the code words were assigned at random from the set of 2^n possible code words. Therefore, the expected number of messages that could be changed into the original message by errors in the BSC and hence be confused with this original message is

$$L = Km/2^n \tag{4-73}$$

It follows from Eq. (4-72) that L is bounded by

$$L \le m\sqrt{(nq + 1)^2/2\pi npq}\; 2^{-nC_{s2}} \tag{4-74}$$

where C_{s2} is the capacity of the BSC and is given by

$$C_{s2} = 1 + p \log_2 p + q \log_2 q \tag{4-75}$$

For finite m and C_{s2}, the limit of L for large n is zero. Thus, for sufficiently long code words, the probability that a message will be decoded incorrectly approaches zero. Now let the number of messages m be fixed by the relation

$$m = 2^{nC_{s2}}/n \tag{4-76}$$

Then Eq. (4-74) becomes
$$L \le \sqrt{(nq + 1)^2/2\pi n^3 pq} \tag{4-77}$$
and, as before,
$$\lim_{n \to \infty} L = 0 \tag{4-78}$$

The channel input entropy per message is given by Eq. (4-66). This entropy per code digit is

$$H'(U) = (\log_2 m)/n \quad \text{(bits/digit)} \tag{4-79}$$
or, from Eq. (4-76),
$$H'(U) = C_{s2} - (\log n)/n \tag{4-80}$$
and
$$\lim_{n \to \infty} H'(U) = C_{s2} \quad \text{(bits/digit)} \tag{4-81}$$

Thus, for sufficiently long message sequences, information can be conveyed in the BSC at a rate approaching channel capacity and with arbitrarily small probability of error.

The noisy-coding theorem can be stated more precisely as follows: "Consider a discrete memoryless channel with nonzero capacity C; fix two numbers H and ϵ such that

$$0 < H < C \tag{4-82}$$
and
$$\epsilon > 0 \tag{4-83}$$

Let us transmit m messages $u_1, u_2, \ldots, u_m$ by code words each of length n binary digits. The positive integer n can be chosen so that

$$m \ge 2^{nH} \tag{4-84}$$

In addition, at the destination the m sent messages can be associated with a set $V = \{v_1, v_2, \ldots, v_m\}$ of received messages and with a decision rule $d(v_j) = u_j$ such that

$$p[d(v_i)|v_i] \ge 1 - \epsilon \tag{4-85}$$

i.e., decoding can be accomplished with a probability of error that does not exceed ϵ." The foregoing discussion of this theorem for the BSC is not a formal proof, of course, but merely a plausibility argument. However, a rigorous proof is quite involved and somewhat difficult. There is a converse to the noisy-coding theorem which states that it is not possible to produce an encoding procedure which allows transmission at a rate greater than channel capacity with arbitrarily small error.

ERROR CORRECTION

13. Error-Correcting Codes. The codes considered in Par. **4-7** were designed for minimum length in the noiseless-transmission case. For noisy channels, the coding theorem of Par.

4-12 guarantees the existence of a code which will allow transmission at any rate less than channel capacity and with arbitrarily small probability of error. However, the theorem does not provide a constructive procedure to devise such codes. Indeed, it implies that very long sequences of source symbols may have to be considered if reliable transmission at rates near channel capacity are to be obtained. In this section, we consider some of the elementary properties of simple *error-correcting codes;* i.e., codes which can be used to increase reliability in the transmission of information through noisy channels by correcting at least some of the errors that occur so that overall probability of error is reduced.

The discussion will be restricted to the BSC, and the notation of Par. **4-12** will be used. Thus, a source alphabet $X = \{x_1, x_2, \ldots, x_M\}$ of M symbols will be used to form a message set U of m

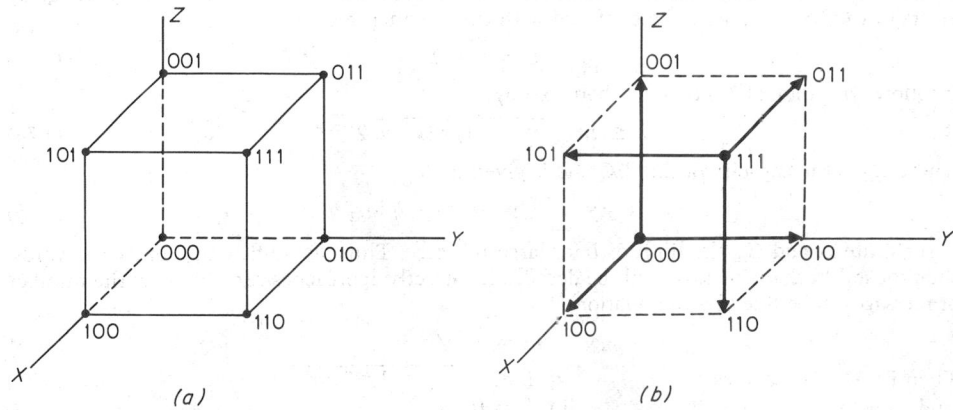

Fig. 4-5. Representation of binary sequences as the corners of an n-cube, $n = 3$: (a) the eight binary sequences of length 3; (b) shift in sequences 000 and 111 from a single error.

messages u_k, where $U = \{u_1, u_2, \ldots, u_m\}$. Each u_k will consist of a sequence of the x_i's. Each message u_k will be encoded into a sequence of n binary digits for transmission over the BSC. At the destination, there exists a set $V = \{v_1, v_2, \ldots, v_{2n}\}$ of all possible binary sequences of length n. The inequality $m \leq 2^n$ must hold. The problem is to associate with each sent message u_k a received message v_j so that $p(e)$, the overall probability of error, is reduced.

In the discussion of the noisy-coding theorem, a decoding scheme was used that examined the received message v_j and identified it with the sent message u_k, which differed from it in the least number of binary digits. In all the discussions of Par. **4-13** it will be assumed that this decoder is used. Let us define the *Hamming distance* $d(v_j, v_k)$ between two binary sequences v_j and v_k of length n as that integer which is the number of digits in which v_j and v_k disagree. Thus, if the distance between two sequences is zero, the two sequences are identical. It is easily seen that this distance measure has the following elementary properties:

$$d(v_j, v_k) \geq 0 \text{ with equality if and only if } v_j = v_k \tag{4-86}$$
$$d(v_j, v_k) = d(v_k, v_j) \tag{4-87}$$
$$d(v_j, v_l) \leq d(v_j, v_k) + d(v_k, v_l) \tag{4-88}$$
$$d(v_j, v_k) \leq n \tag{4-89}$$

The decoder we use is a *minimum-distance* decoder. As mentioned in Par. **4-12** the ideal-observer decoding scheme is a minimum-distance scheme for the BSC.

It is intuitively apparent that the sent messages should be represented by code words that all have the greatest possible distances between them. Let us investigate this matter in more detail by considering all binary sequences of length $n = 3$; there are $2^n = 2^3 = 8$ such sequences, viz.,

$$\begin{array}{cccc} 000 & 001 & 011 & 111 \\ & 010 & 110 & \\ & 100 & 101 & \end{array}$$

It is convenient to represent these as the eight corners of a unit cube, as shown in Fig. 4-5a, where the x axis corresponds to the first digit, the y axis to the second, and the z axis to the third. Although direct pictorial representation is not possible, it is clear that binary sequences of length n greater than 3 can be considered as the corners of the corresponding n-cube.

Suppose that all eight binary sequences are used as code words to encode a source. If any binary digit is changed in transmission, an error will result at the destination since the sent message will be interpreted incorrectly as one of the three possible messages that differ in one code digit from the sent message. This situation is illustrated in Fig. 4-5b for the code words 000 and 111. A change of one digit in each of these code words produces one of three possible other code words.

Figure 4-5b suggests that only two code words, say 000 and 111, should be used. The distance between these two words, or any other two words on opposite corners of the cube, is 3. If only one digit is changed in the transmission of each of these two code words, they can be correctly distinguished at the destination by a minimum-distance decoder. If two digits are changed in each word in transmission, it will not be possible to make this distinction.

It is clear that this reasoning can be extended to sequences containing more than three binary digits. For any $n \geq 3$, single errors in each code word can be corrected. If double errors are to be corrected without fail, there must be at least two code words with a minimum distance between them of 5; thus, for this case, binary code words of length 5 or greater must be used.

In the light of the previous discussion, it can be seen that the error-correcting properties of a code depend on the distance $d(v_j, v_k)$ between code words. Specifically, single errors can be corrected if all code words employed are at least a distance of 3 apart, double errors if the words are at a distance of 5 or more from each other, and, in general, q-fold errors can be corrected if

$$d(v_j, v_k) \geq 2q + 1 \qquad j \neq k \tag{4-90}$$

Of course errors involving less than q digits per code word can also be corrected if Eq. (4-90) is satisfied. If the distance between two code words is $2q$, there will always be a group of binary sequences which are in the middle, i.e., a distance q from each of the two words. Thus, by the proper choice of code words, q-fold errors can be *detected* but not *corrected* f

$$d(v_j, v_k) = 2q \qquad j \neq k \tag{4-91}$$

Let us now consider the maximum number of code words r that can be selected from the set of 2^n possible binary sequences of length n to form a code that will correct all single, double, ..., q-fold errors. In the example of Fig. 4-5, the number of code words selected was 2. In fact, it can be shown that there is no single-error-correcting code for $n = 3, 4$ containing more than two words. Suppose we consider a given code consisting of the words $\ldots, u_k, u_j, \ldots$. All binary sequences of distance q or less from u_k must "belong" to u_k, and to u_k only, if q-fold errors are to be corrected. Thus, associated with u_k are all binary sequences of distance $0, 1, 2, \ldots, q$ from u_k. The number of such sequences has been given previously by Eq. (4-70) and is

$$\binom{n}{0} + \binom{n}{1} + \binom{n}{2} + \cdots + \binom{n}{q} = \sum_{i=0}^{q} \binom{n}{i} \tag{4-92}$$

Since there are r of the code words, the total number of sequences associated with all the code words is

$$r \sum_{i=0}^{q} \binom{n}{i}$$

This number can be no larger than 2^n, the total number of distinct binary sequences of length n. Therefore the following inequality must hold:

$$r \sum_{i=0}^{q} \binom{n}{i} \leq 2^n \qquad \text{or} \qquad r \leq \frac{2^n}{\displaystyle\sum_{i=0}^{q} \binom{n}{i}} \tag{4-93}$$

This is a *necessary* upper bound on the number of code words that can be used to correct all errors up to and including q-fold errors. It can be shown that it is not sufficient.

Let us consider the eight possible distinct binary sequences of length 3. Suppose we add one binary digit to each sequence in such a way that the total number of 1s in the sequence is *even* (or odd, if you wish). The result is shown in Table 4-4. Note that all the word sequences of length 4 differ from each other by a distance of at least 2. In accordance with Eq. (4-91), it should be possible now to *detect single errors* in all eight sequences. The detection method is straightforward. At the receiver, count the number of 1s in the sequence; if the number is odd, a single error (or, more precisely, an odd number of errors) has occurred; if the number is even, no error (or an even number of errors) has occurred. This particular scheme is a good one if only single errors are likely to occur and if only detection (rather than correction) is desired. The added digit is called a *parity-check digit*, and the scheme is a very simple example of a *parity-check code*.

14. Parity-Check Codes. More generally, in parity-check codes, the encoded sequence consists of n binary digits of which only $k < n$ are *information digits* while the remaining $l = n - k$ digits are used for error detection and correction and are called *check digits* or *parity checks*. The example of Table 4-4 is a single-error-detecting code, but, in general, q-fold errors can be detected and/or corrected. As the number of errors to be detected and/or corrected increases, the number l of check digits must increase. Thus, for fixed word length n, the number of information digits $k = n - l$ will decrease as more and more errors are to be detected and/or corrected. Also the total number of words in the code cannot exceed the right side of Eq. (4-93) or the number 2^k.

Parity-check codes are relatively easy to implement. The simple example given of a single-error-detecting code requires only that the number of 1s in each code word be counted. In this light, it is of considerable importance to note that these codes satisfy the noisy-coding theorem.

TABLE 4-4   Parity-Check Code for Single-Error Detection

Message digits	Check digit	Word	Message digits	Check digit	Word
000	0	0000	110	0	1100
100	1	1001	101	0	1010
010	1	0101	011	0	0110
001	1	0011	111	1	1111

In other words, it is possible to encode a source by parity-check coding for transmission over a BSC at a rate approaching channel capacity and with arbitrarily small probability of error. Then, from Eq. (4-84), we have

$$2^{nH} = 2^k \tag{4-94}$$

or H, the rate of transmission, is given by

$$H = k/n \tag{4-95}$$

As $n \to \infty$, the probability of error $p(e)$ approaches zero. Thus, in a certain sense, it is sufficient to limit a study of error-correcting codes to the parity-check codes.

As an example of a parity-check code, consider the simplest nondegenerate case where l, the number of check digits, is 2 and k, the number of information digits, is 1. This system is capable of single-error detection and correction, as we have already decided from geometric considerations. Since $l + k = 3$, each encoded word will be three digits long. Let us denote this word by $a_1 a_2 a_3$, where each a_i is either 0 or 1. Let a_1 represent the information digit and a_2 and a_3 represent the check digits.

Checking for errors is done by forming two independent equations from the three a_i, each equation being of the form of a modulo-2 sum, i.e., of the form

$$a_i \oplus a_j = \begin{cases} 0 & a_i = a_j \\ 1 & a_i \neq a_j \end{cases}$$

Take the two independent equations to be

$$a_2 \oplus a_3 = 0 \quad \text{and} \quad a_1 \oplus a_3 = 0$$

for an *even-parity* check. For an odd-parity check, let the right sides of both of these equations be unity. If these two equations are to be satisfied, the only possible code words that can be sent are 000 and 111. The other six words of length 3 violate one or both of the equations.

Now suppose that 000 is sent and 100 is received. A solution of the two independent equations gives, for the received word,

$$\left. \begin{array}{l} a_2 \oplus a_3 = 0 \oplus 0 = 0 \\ a_1 \oplus a_3 = 1 \oplus 0 = 1 \end{array} \right\} \to 01$$

The check yields the binary check number 1, indicating that the error is in the first digit a_1, as indeed it is. If 111 is sent and 101 received, then

$$\left.\begin{array}{l} a_2 \oplus a_3 = 0 \oplus 1 = 1 \\ a_1 \oplus a_3 = 1 \oplus 1 = 0 \end{array}\right\} \Rightarrow 10$$

and the binary check number is 10, or 2, indicating that the error is in a_2.

In the general case, a set of l independent linear equations is set up to derive a binary checking number whose value indicates the position of the error in the binary word. If more than one error is to be detected and corrected, the number l of check digits must increase, as discussed previously.

In the example just treated, the l check digits were used only to check the k information digits immediately preceding them. Such a code is called a *block code*, since all the information digits and all the check digits are contained in the block (code word) of length $n = k + l$. In some encoding procedures, the l check digits may also be used to check information digits appearing in preceding words. Such codes are called *convolutional* or *recurrent codes.* A parity-check code (either block or convolutional) where the word length is n and the number of information digits is k is usually called an (n, k) *code.*

15. Other Error-Detecting and Error-Correcting Codes. Unfortunately, a general treatment of error-detecting and error-correcting codes requires that the code structure be cast in a relatively sophisticated mathematical form. The commonest procedure is to identify the code letters with the elements of a finite (algebraic) field. The code words are then taken to form a vector subspace of n-tuples over the field. Such codes are called *linear codes* or, sometimes, *group codes.* Both the block codes and the convolutional codes mentioned in the previous paragraph fall in this category.

An additional constraint often imposed on linear codes is that they be *cyclic.* Let a code word $\bar{a}$ be represented by

$$\bar{a} = (a_0, a_1, a_2, \ldots, a_{n-1})$$

Then the ith *cyclic permutation* a^{-i} is given by $\bar{a}^i = (a_i, a_{i+1}, \ldots, a_{n-1}, a_0, a_1, \ldots, a_{i-1})$. A linear code is cyclic if, and only if, for every word $\bar{a}$ in the code, there is also a word $\bar{a}^i$ in the code. The permutations need not be distinct and, in fact, generally will not be. The eight code words

$$\begin{array}{llll} 0000 & 0110 & 1001 & 1010 \\ 0011 & 1100 & 0101 & 1111 \end{array}$$

constitute a cyclic set. Included in the cyclic codes are some of those most commonly encountered such as the Bose and Ray-Chaudhuri (BCH) codes and shortened Reed-Muller codes. A more detailed discussion is beyond the scope of this treatment, but will be found in the references, Par. **4-42**.

CONTINUOUS CHANNELS

16. Continuous-Amplitude Channels. The preceding discussion has concerned discrete message distributions and channels. Further, it has been assumed, either implicitly or explicitly, that the time parameter is discrete, i.e., that a certain number of messages, symbols, code digits, etc., are transmitted per unit time. Thus, we have been concerned with *continuous-amplitude, discrete-time* channels and with messages which can be modeled as *discrete random processes* with *discrete parameter.* There are three other possibilities, depending on whether the process amplitude and the time parameter have discrete or continuous distributions.

In Pars. **4-16** to **4-18** we consider the *continuous-amplitude, discrete-time* channel, where the input messages can be modeled as *continuous random processes* with *discrete parameter.* It will be shown later that continuous-time cases of engineering interest can be treated by techniques which amount to the replacement of the continuous parameter by a discrete parameter. The most straightforward method involves the application of the sampling theorem to band-limited processes. In this case the process is sampled at equispaced intervals of length $1/2W$, where W is the highest frequency of the process. Thus the continuous parameter t is replaced by the discrete parameter $t_k = k/2W$, $k = \ldots, -1, 0, 1, \ldots$.

Let us restrict our attention for the moment to continuous-amplitude, discrete-time situations. The discrete density $p(x_i)$, $i = 1, 2, \ldots, M$, of the source-message set is replaced by the continuous density $W_x(x)$, where, in general, $-\infty < x < \infty$, although the range of x may be restricted in particular cases. In the same way, other discrete densities are replaced by continuous densities. For example, the destination distribution $p(y_j)$, $j = 1, 2, \ldots, N$, becomes $W_y(y)$, and the joint distribution $p(x_i, y_j)$ will be $W_2(x, y)$.

A reasonable place to begin the study of continuous distribution is with the definition of source entropy as given by Eq. (4-4) for the discrete case:

$$H(X) = - \sum_i p(x_i) \log p(x_i) \tag{4-96}$$

The continuous density $W_x(x)$ can be approximated by

$$p(x_i) \approx W_x(x_i) \, \Delta x_i \tag{4-97}$$

where $p(x_i)$ is approximately the probability that the continuous random variable x with density $W_x(x)$ lies in an interval Δx_i which includes x_i. In the limiting case as $\Delta x_i \to 0$, this relationship will become exact, and Eq. (4-97) becomes

$$H(X) = \lim_{\Delta x_i \to 0} \left\{ - \sum_i W_x(x_i) \, \Delta x_i \log \left[W_x(x_i) \, \Delta x_i \right] \right\} \tag{4-98}$$

The logarithm can be expanded to yield

$$H(X) = - \lim_{\Delta x_i \to 0} \sum_i W_x(x_i) \log \left[W_x(x_i) \right] \Delta x_i - \lim_{\Delta x_i \to 0} \sum_i W_x(x_i) \, \Delta x_i \log \Delta x_i \tag{4-99}$$

The first term of this expression can be considered as a limiting form of the integral $\int_{-\infty}^{\infty} W_x(x)$ $\log W_x(x) \, dx$, while the second term tends to infinity, e.g., if the intervals Δx_i are equal,

$$- \lim_{\Delta x \to 0} \sum_i W_x(x_i) \, \Delta x \log \Delta x = - \lim_{\Delta x \to 0} \log \Delta x \int_{-\infty}^{\infty} W_x(x) \, dx$$

$$= - \lim_{\Delta x \to 0} \log \Delta x = \infty \tag{4-100}$$

This approach suggests that the entropy of a continuous distribution $W_x(x)$ might be defined in analogy with the discrete case by the first term of Eq. (4-99)

$$H(X) = - \int_{-\infty}^{\infty} W_x(x) \log W_x(x) \, dx \tag{4-101}$$

This definition is not completely satisfactory for a number of reasons having to do with the properties of this new $H(X)$.

(a) $H(X)$ May Be Negative, Positive, or Zero. In the discrete case, it was shown that $H(X)$ was nonnegative. This is no longer necessarily true. For example, let $W_x(x)$ be uniformly distributed in the interval $(0, 1/a)$. Then we have

$$H(X) = - \int_0^{1/a} a \log a \, dx = - \log a \begin{cases} > 0 & a < 1 \\ = 0 & a = 1 \\ < 0 & a > 1 \end{cases} \tag{4-102}$$

(b) $H(X)$ Depends on the Coordinate System. In the interest of generality, consider the set of random variables $x_1, x_2, \ldots, x_n$ with joint distribution

$$W_{\bar{x}}(x_1, x_2, \ldots, x_n) = W_x(\bar{x}) \tag{4-103}$$

Consider the entropy

$$H(X) = - \int_{-\infty}^{\infty} \cdots \int_{-\infty}^{\infty} W_{\bar{x}}(\bar{x}) \log W_x(\bar{x}) \, d\bar{x} \tag{4-104}$$

where $d\bar{x} = dx_1 \, dx_2 \cdots dx_n$. Let us transform to a new coordinate system (set of random variables) $y_1, y_2, \ldots, y_n$ with joint distribution

$$W_{\bar{y}}(y_1, y_2, \ldots, y_n) = W_y(\bar{y})$$

and entropy $H(\bar{Y})$, where

$$H(\bar{Y}) = - \int_{-\infty}^{\infty} \cdots \int_{-\infty}^{\infty} W_{\bar{y}}(\bar{y}) \log W_{\bar{y}}(\bar{y}) \, d\bar{y} \tag{4-105}$$

The densities $W_{\bar{x}}(\bar{x})$ and $W_{\bar{y}}(\bar{y})$ are related by

$$W_{\bar{y}}(\bar{y}) = W_{\bar{x}}(\bar{x})J(\bar{x}|\bar{y})$$

where $J(\bar{x}|\bar{y})$ is the Jacobian of the transformation from $\bar{x}$ to $\bar{y}$. This last expression together with

$$d\bar{y} = J(\bar{y}|\bar{x})\, d\bar{x}$$

can be substituted into Eq. (4-105) to yield

$$H(\bar{Y}) = -\int_{-\infty}^{\infty} \cdots \int_{-\infty}^{\infty} W_{\bar{x}}(\bar{x})J(\bar{x}|\bar{y}) \log\,[W_{\bar{x}}(\bar{x})J(\bar{x}|\bar{y})]J(\bar{y}|\bar{x})\, d\bar{x}$$

On using the relationship

$$J(\bar{x}|\bar{y})J(\bar{y}|\bar{x}) = 1$$

and expanding the logarithm, we obtain

$$H(\bar{Y}) = -\int_{-\infty}^{\infty} \cdots \int_{-\infty}^{\infty} W_{\bar{x}}(\bar{x}) \log W_{\bar{x}}(\bar{x})\, d\bar{x} - \int_{-\infty}^{\infty} \cdots \int_{-\infty}^{\infty} W_{\bar{x}}(\bar{x}) \log J(\bar{x}|\bar{y})\, d\bar{x}$$

or

$$H(\bar{Y}) = H(\bar{X}) - E_{\bar{x}}\{\log J(\bar{x}|\bar{y})\} \tag{4-106}$$

where $E_{\bar{x}}\{\cdot\}$ indicates the expectation operation with respect to the $\bar{x}$ distribution. Thus the entropy of a continuous distribution changes with the coordinate system. As an example, consider the linear transformation

$$y_i = \sum_{j=1}^{n} a_{ij}x_j \qquad i = 1, 2, \ldots, n$$

In this case the Jacobian is

$$J(\bar{y}|\bar{x}) = \begin{vmatrix} a_{11} & a_{12} & \cdots & a_{1n} \\ a_{21} & a_{22} & \cdots & a_{2n} \\ \cdots & \cdots & \cdots & \cdots \\ a_{n1} & a_{n2} & \cdots & a_{nn} \end{vmatrix} = \|a_{ij}\|$$

where $\|a_{ij}\|$ is the magnitude of the determinant $|a_{ij}|$. Now Eq. (4-106) becomes

$$H(\bar{Y}) = H(\bar{X}) + \log \|a_{ij}\|$$

In one dimension, where $y = ax$, this becomes

$$H(Y) = H(X) + \log |a|$$

(c) $H(X)$ *Is Invariant to Translation.* Consider only the simple translation

$$y = x + b$$

The entropy $H(Y)$ is

$$H(Y) = -\int_{-\infty}^{\infty} W_y(y) \log W_y(y)\, dy = -\int_{-\infty}^{\infty} W_x(y - b) \log W_x(y - b)\, dy$$

After the change of variable $x = y - b$, we have

$$H(Y) = -\int_{-\infty}^{\infty} W_x(x) \log W_x(x)\, dx = H(X)$$

as expected. Actually this result follows immediately from Eq. (4-106) since the Jacobian of a translation is unity.

It is clear that joint and conditional entropies can be defined in exact analogy to the discrete case discussed in Par. **4-4**. If the joint density $W_2(x, y)$ exists, and if

$$W_x(x) = \int_{-\infty}^{\infty} W_2(x, y)\, dy \qquad \text{and} \qquad W_y(y) = \int_{-\infty}^{\infty} W_2(x, y)\, dx$$

then the joint entropy $H(X, Y)$ is given by

$$H(X, Y) = -\int\int_{-\infty}^{\infty} W_2(x, y) \log W_2(x, y) \, dx \, dy \tag{4-107}$$

and the conditional entropies $H(X|Y)$ and $H(Y|X)$ are

$$H(X|Y) = -\int\int_{-\infty}^{\infty} W_2(x, y) \log \frac{W_2(x, y)}{W_y(y)} \, dx \, dy \tag{4-108}$$

and

$$H(Y|X) = -\int\int_{-\infty}^{\infty} W_2(x, y) \log \frac{W_2(x, y)}{W_x(x)} \, dx \, dy \tag{4-109}$$

The average mutual information follows from Eq. (4-15) and is

$$I(X; Y) = -\int\int_{-\infty}^{\infty} W_2(x, y) \log \frac{W_x(x) W_y(y)}{W_2(x, y)} \, dx \, dy \tag{4-110}$$

Although the entropy of a continuous distribution can be negative, positive, or zero, the average mutual information $I(X; Y)$ is nonnegative as in the discrete case. Consider two continuous densities $p(x) \geq 0$ and $q(x) \geq 0$, where

$$\int_{-\infty}^{\infty} p(x) \, dx = \int_{-\infty}^{\infty} q(x) \, dx = 1$$

A well-known inequality is

$$\log_a z \geq \frac{1}{\ln a}\left(1 - \frac{1}{z}\right) \qquad z \geq 0$$

It follows that

$$\int_{-\infty}^{\infty} p(x) \log_a \frac{p(x)}{q(x)} \, dx \geq \frac{1}{\ln a} \int_{-\infty}^{\infty} p(x)\left[1 - \frac{q(x)}{p(x)}\right] dx = 0$$

with equality if and only if $p(x) = q(x)$. It can be seen immediately from Eq. (4-110) that $I(X; Y) \geq 0$ with equality when x and y are statistically independent, i.e., when $W_2(x, y) = W_x(x) W_y(y)$.

17. Maximization of Entropy of Continuous Distributions. The entropy of a discrete distribution is a maximum when the distribution is uniform, i.e., when all outcomes are equally likely. In the continuous case, the entropy depends on the coordinate system, and it is possible to maximize this entropy subject to various constraints on the associated density function. The maximization itself is the so-called isoperimetric problem of the calculus of variations. For our purposes, the procedure is as follows. It is desired to find $y = y(x)$ so that the integral

$$I = \int_a^b F(x, y) \, dx \tag{4-111}$$

is an extremum subject to the constraints that

$$\int_a^b F_1(x, y) \, dx = c_1$$
$$\cdots \cdots \cdots \tag{4-112}$$
$$\int_a^b F_n(x, y) \, dx = c_n$$

where the c_i are preassigned constants and the F_i are functions determined by the problem. The function y is found by solving

$$\frac{\partial F}{\partial y} + \lambda_1 \frac{\partial F_1}{\partial y} + \cdots + \lambda_n \frac{\partial F_n}{\partial y} = 0 \tag{4-113}$$

The λ_i are constants (called *undetermined multipliers*) whose values are found by substituting the y found from Eq. (4-113) into the set of equations given by Eq. (4-112). We consider now several of the commonest and most useful entropy maximizations.

(a) *The Maximization of H(X) for a Fixed Variance of x.* We wish to maximize

$$H(X) = -\int_{-\infty}^{\infty} W_x(x) \log W_x(x) \, dx$$

subject to the constraints that

$$\int_{-\infty}^{\infty} W_x(x) \, dx = 1 \tag{4-114}$$

and

$$\int_{-\infty}^{\infty} x^2 W_x(x) \, dx = \sigma^2 \tag{4-115}$$

Let us form the expression

$$F_0 = F + \lambda_1 F_1 + \lambda_2 F_2 = W_x \log W_x + \lambda_1 W_x + \lambda_2 x^2 W_x$$

where W_x has been used as a simplified notation for $W_x(x)$. Equation (4-113) now becomes

$$\frac{\partial F_0}{\partial y} = 0 = 1 + \log W_x + \lambda_1 + \lambda_2 x^2$$

or

$$W_x(x) = e^{-(\lambda_1+1)} e^{-\lambda_2 x^2} \tag{4-116}$$

The constants λ_1 and λ_2 can be found by substituting Eq. (4-116) into Eqs. (4-114) and (4-115). The result is

$$W_x(x) = (1/\sqrt{2\pi}\sigma)e^{-x^2/2\sigma^2} \qquad -\infty < x < \infty \tag{4-117}$$

Thus, for fixed variance, the normal distribution has the largest entropy. It is clear that

$$\ln W_x(x) = -\tfrac{1}{2}\ln 2\pi\sigma^2 - x^2/2\sigma^2$$

and that the entropy is this case is

$$H(X) = \tfrac{1}{2}\ln 2\pi\sigma^2 + \tfrac{1}{2}\ln e = \tfrac{1}{2}\ln 2\pi e\sigma^2 \tag{4-118}$$

This last result will be of considerable use later. Note that the normal distribution need not have a zero mean since entropy is invariant to a translation. For convenience, the natural logarithm has been used, and the units of H are nats.

(b) *The Maximization of H(X) for a Limited Peak Value of x.* In this case, the single constraint is

$$\int_{-M}^{M} W_x(x) \, dx = 1 \tag{4-119}$$

and $W_x(x)$ is found from

$$\frac{\partial}{\partial W_x}(-W_x \ln W_x + \lambda_1 W_x) = 0 \qquad \text{or} \qquad W_x(x) = e^{\lambda_1 - 1} = \text{const}$$

This result can be used in Eq. (4-119) to obtain the uniform distribution

$$W_x(x) = \begin{cases} 1/2M & |x| \le M \\ 0 & |x| > M \end{cases}$$

The associated entropy is

$$H(X) = -\int_{-M}^{M} \frac{1}{2M} \log \frac{1}{2M} \, dx = \log 2M \tag{4-120}$$

(c) *The Maximization of H(X) for x Limited to Nonnegative Values and a Given Average Value.* The constraints are

$$\int_{0}^{\infty} W_x(x) \, dx = 1 \tag{4-121}$$

and

$$\int_0^\infty x W_x(x) \, dx = \mu \tag{4-122}$$

and $W_x(x)$ is found from

$$\frac{\partial}{\partial W_x}(-W_x \ln W_x + \lambda_1 W_x + \lambda_2 x W_x) = 0 \quad \text{or} \quad W_x(x) = e^{\lambda_1 - 1} e^{\lambda_2 x}$$

As before, λ_1 and λ_2 can be eliminated through Eqs. (4-121) and (4-122) to yield

$$W_x(x) = \begin{cases} 0 & x < 0 \\ (1/\mu) e^{-(x/\mu)} & x \geq 0 \end{cases}$$

The entropy associated with this distribution is

$$H(X) = \ln \mu + 1 = \ln \mu e \tag{4-123}$$

18. Gaussian Signals and Channels. Let us assume that the source symbol x and the destination symbol y are jointly gaussian, so that the joint density $W_2(x, y)$ is

$$W_2(x, y) = \frac{1}{2\pi\sigma_x\sigma_y\sqrt{1-\rho^2}} \exp\left\{-\frac{1}{2(1-\rho^2)}\left[\left(\frac{x}{\sigma_x}\right)^2 - 2\rho\frac{xy}{\sigma_x\sigma_y} + \left(\frac{y}{\sigma_y}\right)^2\right]\right\} \tag{4-124}$$

where σ_x^2 and σ_y^2 are the variances of x and y, respectively, and ρ is the correlation coefficient given by

$$\rho = \frac{E\{xy\}}{\sigma_x\sigma_y} \tag{4-125}$$

The univariate densities of x and y are given, of course, by

$$W_x(x) = \frac{1}{\sqrt{2\pi}\sigma_x} \exp\left[-\frac{1}{2}\left(\frac{x}{\sigma_x}\right)^2\right] \qquad -\infty < x < \infty \tag{4-126}$$

and

$$W_y(y) = \frac{1}{\sqrt{2\pi}\sigma_y} \exp\left[-\frac{1}{2}\left(\frac{y}{\sigma_y}\right)^2\right] \qquad -\infty < y < \infty \tag{4-127}$$

Let us make use of Eq. (4-110) to calculate the average mutual information $I(X; Y)$. We have

$$\frac{W_2(x, y)}{W_x(x)W_y(y)} = \frac{1}{\sqrt{1-\rho^2}} \exp\left\{-\frac{\rho^2}{2(1-\rho^2)}\left[\left(\frac{x}{\sigma_x}\right)^2 - \frac{2xy}{\rho\sigma_x\sigma_y} + \left(\frac{y}{\sigma_y}\right)^2\right]\right\}$$

and

$$I(X; Y) = -\tfrac{1}{2}\ln(1-\rho^2) \int\int_{-\infty}^{\infty} W_2(x, y) \, dx \, dy$$

$$- \frac{\rho^2}{2(1-\rho^2)} \int\int_{-\infty}^{\infty} W_2(x, y)\left[\left(\frac{x}{\sigma_x}\right)^2 - \frac{2xy}{\rho\sigma_x\sigma_y} + \left(\frac{y}{\sigma_y}\right)^2\right] dx \, dy \tag{4-128}$$

This expression can be rewritten

$$I(X; Y) = -\tfrac{1}{2}\ln(1-\rho^2) - \frac{\rho^2}{2(1-\rho^2)}(1 - 2 + 1)$$

or

$$I(X; Y) = -\tfrac{1}{2}\ln(1-\rho^2) \tag{4-129}$$

Thus the average mutual information in two jointly gaussian random variables is a function only of the correlation coefficient ρ and varies from zero to infinity.

The noise entropy $H(Y|X)$ can be written

$$H(Y|X) = H(Y) - I(X; Y) = \tfrac{1}{2}\ln 2\pi e\sigma_y^2(1-\rho^2) \tag{4-130}$$

Suppose that x and y are jointly gaussian as a result of independent zero-mean gaussian noise n being added in the channel to the gaussian input x, so that

$$y = x + n \tag{4-131}$$

In this case the correlation coefficient ρ becomes

$$\rho = \frac{E\{x^2 + nx\}}{\sigma_x \sigma_y} = \frac{\sigma_x^2}{\sigma_x \sigma_y} = \frac{\sigma_x}{C_y} \qquad (4\text{-}132)$$

and the noise entropy is

$$H(Y|X) = \tfrac{1}{2} \ln 2\pi e \sigma_n^2 \qquad (4\text{-}133)$$

where σ_n^2 is the noise variance given by

$$\sigma_n^2 = E\{n^2\} = \sigma_y^2 - \sigma_x^2 \qquad (4\text{-}134)$$

In this situation, Eq. (4-129) can be rewritten as

$$I(X; Y) = \tfrac{1}{2} \ln (1 + \sigma_x^2/\sigma_n^2) \qquad (4\text{-}135)$$

It is conventional to define the signal power as $S = \sigma_x^2$ and the noise power as $N = \sigma_n^2$ and to rewrite this last expression as

$$I(X; Y) = \tfrac{1}{2} \ln (1 + S/N) \qquad (4\text{-}136)$$

where S/N is the signal-to-noise power ratio.

In analogy with Par. **4-10**, channel capacity C for the continuous-amplitude, discrete-time channel is

$$C = \max_{W_x(x)} I(X; Y) = \max_{W_x(x)} [H(Y) - H(Y|X)] \qquad (4\text{-}137)$$

Suppose the channel consists of an additive noise which is a sequence of independent gaussian random variables n each with zero mean and variance σ_n^2. In this case the conditional probability $W(y|x)$ at each time instant is normal with variance σ_n^2 and mean equal to the particular realization of x. The noise entropy $H(Y|X)$ is given by Eq. (4-133) and Eq. (4-137) becomes

$$C = \max_{W_x(x)} [H(Y)] - \tfrac{1}{2} \ln 2\pi e \sigma_n^2 \qquad (4\text{-}138)$$

If the input power is fixed at σ_x^2, then the output power is fixed at $\sigma_y^2 = \sigma_x^2 + \sigma_n^2$ and $H(Y)$ is a maximum if $y = x + n$ is a sequence of independent gaussian random variables. The value of $H(Y)$ is

$$H(Y) = \tfrac{1}{2} \ln 2\pi e(\sigma_x^2 + \sigma_n^2)$$

and the channel capacity becomes

$$C = \tfrac{1}{2} \ln (1 + \sigma_x^2/\sigma_n^2) = \tfrac{1}{2} \ln (1 + S/N) \qquad (4\text{-}139)$$

where S/N is the signal-to-noise power ratio. Note that the input x is a sequence of independent gaussian random variables and this last equation is identical to Eq. (4-136). Thus for additive independent gaussian noise and an input power limitation, the discrete-time continuous-amplitude channel has a capacity given by Eq. (4-139). This capacity is realized when the input is an independent sequence of independent, identically distributed gaussian random variables.

BAND-LIMITED CHANNELS

19. Band-Limited Transmission. In this section, messages will be considered which can be modeled as continuous random processes $x(t)$ with continuous parameter t. The channels which transmit these messages will be called *amplitude-continuous, time-continuous* channels. Specifically attention will be restricted to signals (random processes) $x(t)$ which are *strictly band-limited*.

Suppose a given arbitrary (deterministic) signal $f(t)$ is available for all time. Is it necessary to know the amplitude of the signal for every value of time in order to characterize it uniquely? In other words, can $f(t)$ be represented (and reconstructed) from some set of *sample values* or samples ... , $f(t_1)$, $f(t_0)$, $f(t_1)$, ... ? Surprisingly enough, it turns out that, under certain fairly reasonable conditions, a signal can be represented exactly by samples spaced relatively far apart. The reasonable conditions are that the signal be strictly band-limited.

A (real) signal $f(t)$ will be called *strictly band-limited* $(-2\pi W, 2\pi W)$ if its Fourier transform $F(\omega)$ has the property

$$F(\omega) = 0 \qquad |\omega| > 2\pi W \tag{4-140}$$

It is clear that this spectrum could be extended into a periodic frequency function with period $4\pi W$. In other words, a new function $F_e(\omega)$ can be defined by

$$F_e(\omega) = \sum_{n=-\infty}^{\infty} F(\omega + n4\pi W) \tag{4-141}$$

This function is periodic with period $4\pi W$.

For reasonably well-behaved $F(\omega)$, the periodic function $F_e(\omega)$ can be expanded in a Fourier series with period $4\pi W$, and, in the interval $-2\pi W \le \omega \le 2\pi W$, this Fourier series will converge to $F(\omega)$; that is,

$$F(\omega) = \sum_{k=-\infty}^{\infty} F_k e^{-jk\omega/2W} \qquad |\omega| \le 2\pi W \tag{4-142}$$

where $j = \sqrt{-1}$ and F_k is the Fourier coefficient given by

$$F_k = \frac{1}{4\pi W} \int_{-2\pi W}^{2\pi W} F(\omega) e^{jk\omega/2W} \, d\omega \tag{4-153}$$

Since $F(\omega)$ is band-limited as described by Eq. (4-140), its inverse Fourier transform is

$$f(t) = \frac{1}{2\pi} \int_{-2\pi W}^{2\pi W} F(\omega) e^{j\omega t} \, d\omega \tag{4-144}$$

If the *Nyquist instants* are defined as the set of times

$$\{t_n\} = \{t_n | t_n = n/2W \qquad n = \ldots, -1, 0, 1, \ldots\} \tag{4-145}$$

then it is clear that $f(t_n)$ is given from Eq. (4-144) as

$$f(t_n) = f\left(\frac{n}{2W}\right) = \frac{1}{2\pi} \int_{-2\pi W}^{2\pi W} F(\omega) e^{jn\omega/2W} \, d\omega \tag{4-146}$$

A comparison of this last equation with Eq. (4-143) shows that the Fourier coefficient F_k is related to the sample value $f(k/2W)$ by

$$F_k = (1/2W)f(k/2W) \tag{4-147}$$

If the sample values $f(n/2W)$ are given for all time, then the Fourier series

$$F(\omega) = \frac{1}{2W} \sum_{k=-\infty}^{\infty} f\left(\frac{k}{2W}\right) e^{-jk\omega/2W} \qquad |\omega| \le 2\pi W \tag{4-148}$$

determines $F(\omega)$ exactly and hence $f(t)$ through the inverse Fourier transform given by Eq. (4-144). This completes the proof of the existence of a sampling theorem: a function $f(t)$, strictly band-limited $(-2\pi W, 2\pi W)$ rad/s, is uniquely and exactly determined by its sample values spaced $1/2W$ apart throughout the time domain. Of course, there are an infinite number of such samples.

The reconstruction of $f(t)$ from its sample values is obtained on substituting Eq. (4-148) into Eq. (4-144):

$$f(t) = \frac{1}{2\pi} \int_{-2\pi W}^{2\pi W} \frac{1}{2W} \sum_{k=-\infty}^{\infty} f\left(\frac{k}{2W}\right) e^{-jk\omega/2W} e^{j\omega t} \, d\omega \tag{4-149}$$

20. Sampling Theorem. We can interchange the order of summation and integration and evaluate the integral to obtain the *sampling representation*

$$f(t) = \sum_{k=-\infty}^{\infty} f\left(\frac{k}{2W}\right) \frac{\sin(2\pi Wt - k\pi)}{2\pi Wt - k\pi} \tag{4-150}$$

This expression is sometimes called the *Cardinal series* or *Shannon's sampling theorem*. Together with Eq. (4-146) it relates the discrete time domain $\{k/2W\}$ with sample values $f(k/2W)$ to the continuous time domain $\{t\}$ of the function $f(t)$.

The interpolation function

$$k(t) = (\sin 2\pi Wt)/2\pi Wt \tag{4-151}$$

has a Fourier transform $K(\omega)$ given by

$$K(\omega) = \begin{cases} 1/4\pi W & |\omega| < 2\pi W \\ 0 & |\omega| > 2\pi W \end{cases} \tag{4-152}$$

Also the shifted function $k(t - k/2W)$ has the transform

$$\mathcal{F}\{k(t - k/2W)\} = K(\omega)e^{j\omega k/2W} \tag{4-153}$$

Therefore, each term on the right side of Eq. (4-150) is a time function which is strictly band-limited $(-2\pi W, 2\pi W)$. Note also that

$$k\left(t - \frac{k}{2W}\right) = \frac{\sin(2Wt - k\pi)}{2\pi Wt - k\pi} = \begin{cases} 1 & t = t_k = k\pi/2W \\ 0 & t = t_n, \quad n \neq k \end{cases} \tag{4-154}$$

Thus this sampling function $k(t - k/2W)$ is zero at all Nyquist instants except t_k, where it equals unity.

Suppose that a function $h(t)$ is not strictly band-limited to at least $(-2\pi W, 2\pi W)$ rad/s and an attempt is made to reconstruct the function using Eq. (4-150) with sample values spaced $1/2W$ s apart. It is apparent that the reconstructed signal [which is strictly band-limited $(-2\pi W, 2\pi W)$, as already mentioned] will differ from the original. Moreover a given set of sample values $\{f(k/2W)\}$ could have been obtained from a whole class of different signals. Thus it should be emphasized that the reconstruction of Eq. (4-150) is unambiguous only for signals strictly band-limited to at least $(-2\pi W, 2\pi W)$ rad/s. The set of different possible signals with the same set of sample values $\{f(k/2W)\}$ is called the *aliases* of the band-limited signal $f(t)$.

Let us now consider a signal (random process) $x(t)$ with *autocorrelation function* given by

$$R_x(\tau) = E\{x(t)x(t + \tau)\} \tag{4-155}$$

and *power spectral density*

$$\varphi_x(\omega) = \int_{-\infty}^{\infty} R_x(\tau)e^{-j\omega\tau}\,d\tau \tag{4-156}$$

which is just the Fourier transform of $R_x(\tau)$. The process will be assumed to have zero mean and to be strictly *band-limited* $(-2\pi W, 2\pi W)$ in the sense that the power spectral density $\varphi_x(\omega)$ vanishes outside this interval; i.e.,

$$\varphi_x(\omega) = 0 \qquad |\omega| > 2\pi W \tag{4-157}$$

It has just been shown that a deterministic signal $f(t)$ band-limited $(-2\pi W, 2\pi W)$ admits the *sampling representation* of Eq. (4-150). It can also be shown that the random process $x(t)$ admits the same expansion; i.e.,

$$x(t) = \sum_{k=-\infty}^{\infty} x\left(\frac{k}{2W}\right)\frac{\sin(2\pi Wt - k\pi)}{2\pi Wt - k\pi} \tag{4-158}$$

The right side of this expression is a random variable for each value of t and converges mean square to the process $x(t)$ for each t. The proof is straightforward. Since $\varphi_x(\omega)$ vanishes outside $(-2\pi W, 2\pi W)$, its Fourier transform $R_x(\tau)$ has a sampling representation

$$R_x(\tau) = \sum_{k=-\infty}^{\infty} R_x\left(\frac{k}{2W}\right)\frac{\sin(2\pi Wt - k\pi)}{2\pi Wt - k\pi} \tag{4-159}$$

We define a partial sum $x_N(t)$ by

$$x_N(t) = \sum_{k=-N}^{N} x\left(\frac{k}{2W}\right)\frac{\sin(2\pi Wt - k\pi)}{2\pi Wt - k\pi}$$

Then, after some manipulation, we find that

$$\lim_{N \to \infty} E\{|x(t) - x_N(t)|^2\} = R_x(0) - 2R_x(0) + R_x(0) = 0$$

and $x_N(t)$ converges mean square to $x(t)$. Thus the process $x(t)$ with continuous time parameter t can be represented by the process $x(k/2W)$, $k = \ldots, -2, -1, 0, 1, 2, \ldots$, with discrete time parameter $t_k = k/2W$. For band-limited signals or channels it is sufficient, therefore, to consider the discrete-time case and to relate the results to continuous time through Eq (4-158).

Suppose the continuous-time process $x(t)$ has a spectrum $\varphi_x(\omega)$ which is *flat and band-limited* so that

$$\varphi_x(\omega) = \begin{cases} N_0 & |\omega| \leq \omega_b \\ 0 & |\omega| > \omega_b \end{cases} \tag{4-160}$$

Then the autocorrelation function can be found as the inverse Fourier transform of $\varphi_x(\omega)$, namely,

$$R_x(\tau) = \frac{1}{2\pi} \int_{-\infty}^{\infty} \varphi_x(\omega) e^{j\omega\tau} \, d\omega \tag{4-161}$$

In this special case, Eq. (4-161) becomes

$$R_x(\tau) = 2N_0 W(\sin 2\pi W\tau)/2\pi W\tau \tag{4-162}$$

This function passes through zero at intervals of $1/2W$ so that

$$R_x(k/2W) = 0 \quad k = \ldots, -2, -1, 1, 2, \ldots \tag{4-163}$$

Thus, samples spaced $k/2W$ apart are uncorrelated *if the power spectral density is flat and band-limited* $(-2\pi W, 2\pi W)$. *If the process is gaussian, the samples are independent.* This implies that continuous-time band-limited $(-2\pi W, 2\pi W)$ gaussian channels, where the noise has a flat spectrum, have a capacity C given by Eq. (4-139) as

$$C = \tfrac{1}{2} \ln (1 + S/N) \quad \text{(nats/sample)} \tag{4-164}$$

Here N is the variance of the additive, flat, band-limited gaussian noise and S is $R_x(0)$, the fixed variance of the input signal. The units of Eq. (4-164) are on a per sample basis. Since there are $2W$ samples per unit time, the capacity C' per unit time can be written as

$$C' = W \ln (1 + S/N) \quad \text{(nats/s)} \tag{4-165}$$

The ideas developed thus far in this section have been somewhat abstract notions involving information sources and channels, channel capacity, and the various coding theorems. We now look more closely at conventional channels. Many aspects of these topics fall into the area often called *modulation theory.*

MODULATION

21. Modulation Theory. As discussed in Pars. **4-1** to **4-3** and shown in Fig. 4-1, the central problem in most communication systems is the transfer of information originating in some source to a destination by means of a channel. It will be convenient in this section to call the sent message or intelligence $a(t)$ and to denote the received message by $a^*(t)$, a distorted or corrupted version of $a(t)$.

The message signals used in communication and control systems are usually limited in frequency range to some maximum frequency $f_m = \omega_m/2\pi$ Hz. This frequency is typically in the range of a few hertz for control systems and moves upward to a few megahertz for television video signals. In addition the bandwidth of the signal is often of the order of this maximum frequency so that the signal spectrum is approximately low-pass in character. Such signals are often called *video signals* or *baseband signals.* It frequently happens that the transmission of such a spectrum through a given communication channel is inefficient or impossible. In this light, the problem may be looked upon as the one shown in Fig. 4-2, where an encoder E has been added between the source and the channel. However, in this case, the encoder acts to *modulate* the signal $a(t)$, producing at its output the *modulated wave* or *signal* $m(t)$.

Modulation can be defined as the modification of one signal, called the *carrier,* by another, called the *modulating signal.* The result of the modulation process is a modulated wave or signal. In most cases a frequency shift is one of the results. There are a number of reasons for producing modulated waves, of which the major ones are given.

(a) *Frequency Translation for Efficient Antenna Design.* It may be necessary to transmit the modulating signal through space as electromagnetic radiation. If the antenna used is to radiate an appreciable amount of power, it must be large compared with the signal wavelength. Thus translation to higher frequencies (and hence to smaller wavelengths) will permit antenna structures of reasonable size and cost at both the transmitter and receiver.

(b) *Frequency Translation for Ease of Signal Processing.* It may be easier to amplify and/or shape a signal in one frequency range than in another. For example, a dc signal may be converted to ac, amplified, and converted back again.

(c) *Frequency Translation to Assigned Location.* A signal may be translated to an assigned frequency band for transmission or radiation, e.g., in commercial radio broadcasting.

(d) *Changing Bandwidth.* The bandwidth of the original message signal may be increased or decreased by the modulation process. In general, decreased bandwidth will result in channel economies at the cost of fidelity. On the other hand, increased bandwidth will be accompanied

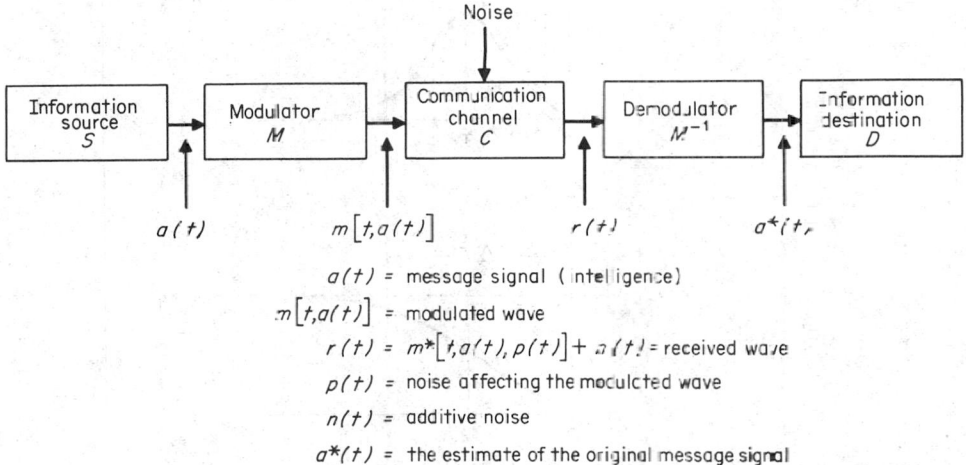

$$a(t) = \text{message signal (intelligence)}$$
$$m[t,a(t)] = \text{modulated wave}$$
$$r(t) = m^*[t,a(t),p(t)] + n(t) = \text{received wave}$$
$$p(t) = \text{noise affecting the modulated wave}$$
$$n(t) = \text{additive noise}$$
$$a^*(t) = \text{the estimate of the original message signal}$$

Fig. 4-6. Communication system involving modulation and demodulation

by increased immunity to channel disturbances, as in wide-band frequency modulation, for example.

(e) *Multiplexing.* It may be necessary or desirable to transmit several signals occupying the same frequency range or the same time range over a single channel. Various modulation techniques allow the signals to share the same channel and yet be recovered separately. Such techniques are given the generic name of *multiplexing.* As discussed later, multiplexing is possible in either the frequency domain (frequency-domain multiplexing FDM) or in the time domain (time-domain multiplexing TDM). As a simple example, the signals may be translated in frequency so that they occupy separate and distinct frequency ranges as mentioned in item (b) above.

Thus it can be seen that, in a general way, the process of modulation can be considered as a form of encoding used to match the message signal arising from the information source to the communication channel. At the same time it is generally true that the channel itself has certain undesirable characteristics resulting in distortion of the signal during transmission. A part of such distortion can frequently be accounted for by postulating noise disturbances in the channel. There noises may be additive and may also affect the modulated wave in a more complicated fashion, although it is usually sufficient (and much simpler) to assume additive noise only. Also, the received signal must be decoded (demodulated) to recover the original signal.

In view of this discussion, it is convenient to change the block diagram of Fig. 4-2 to that shown in Fig. 4-6. The waveform received at the demodulator (receiver) will be denoted by $r(t)$, where

$$r(t) = m^*[t, a(t), p(t)] + n(t) \tag{4-166}$$

where $a(t)$ is the original message signal, $m[t, a(t)]$ is the modulated wave, $m^*[t, a(t), p(t)]$ is a corrupted version of $m[t, a(t)]$, and $p(t)$ and $n(t)$ are noises whose characteristics depend on the channel. Unless it is absolutely necessary for an accurate characterization of the channel, we will assume that $p(t) \equiv 0$ to avoid the otherwise complicated analysis that results.

The aim is to find modulators M and demodulators M^{-1} which make $a^*(t)$ a "good" estimate of the message signal $a(t)$. It should be emphasized that M^{-1} is not uniquely specified by M; for example, it is not intended to imply that $MM^{-1} = 1$. The form of the demodulator, for a given modulator, will depend on the characteristics of the message $a(t)$ and the channel as well as on the criterion of "goodness of estimation" used.

We now take up a study of the various forms of modulation and demodulation, their principal characteristics, their behavior in conjunction with noisy channels, and their advantages and disadvantages. We begin with some preliminary material on signals and their properties.

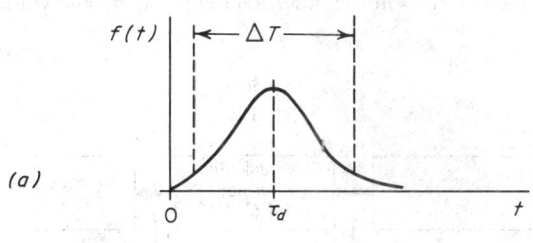

(a)

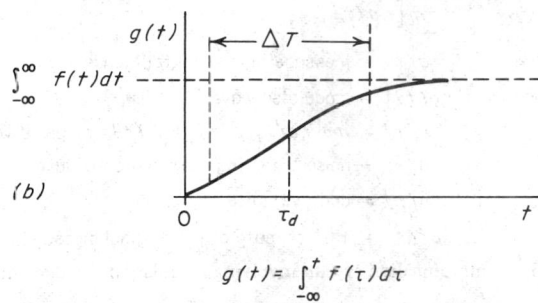

(b)

$$g(t) = \int_{-\infty}^{t} f(\tau)\, d\tau$$

Fig. 4-7. Duration and delay: (a) typical pulse; (b) integral of pulse.

22. Elements of Signal Theory. The real time function $f(t)$ and its Fourier transform form a Fourier transform pair given by

$$F(\omega) = \int_{-\infty}^{\infty} f(t) e^{-j\omega t}\, dt \qquad (4\text{-}167)$$

and

$$f(t) = \frac{1}{2\pi} \int_{-\infty}^{\infty} F(\omega) e^{j\omega t}\, d\omega \qquad (4\text{-}168)$$

It follows directly from Eq. (4-167) that the transform $F(\omega)$ of a real time function has a real part which is even and an imaginary part which is odd.

Consider the function $f(t)$ shown in Fig. 4-7a. This might be a pulsed signal or the impulse response of a linear system, for example. The time ΔT over which $f(t)$ is appreciably different from zero is called the *duration* of $f(t)$, and some measure, such as τ_d, of the center of the pulse is called the *delay* of $f(t)$. In system terms, the quantity ΔT is the system *response time* or *rise time*, and τ_d is the system delay. The integral of $f(t)$, shown in Fig. 4-7b, corresponds to the step-function response of a system with impulse response $f(t)$.

If the function $f(t)$ of Fig. 4-7 is nonnegative, the new function

$$\frac{f(t)}{\int_{-\infty}^{\infty} f(t)\, dt}$$

is nonnegative with unit area. We now seek measures of duration and delay which are both meaningful in terms of communication problems and also mathematically tractable. It will be clear that some of the results we obtain will not be universally applicable and, in particular, must be used with care when the function $f(t)$ can be negative for some values of t. However, the results will be useful for wide classes of problems.

Consider now a frequency function $F(\omega)$, which will be assumed to be real. If $F(\omega)$ is not real, either $|F(\omega)|^2 = F(\omega)F(-\omega)$ or $|F(\omega)|$ can be used. Such a function might be similar to that shown in Fig. 4-8a. The radian frequency range $\Delta\Omega$ (or the frequency range ΔF) over which $F(\omega)$ is appreciably different from zero is called the *bandwidth* of the function. Of course, if the function is a *bandpass* function, such as that shown in Fig. 4-8b, the bandwidth will usually be taken to be some measure of the width of the positive-frequency (or negative-frequency) part of the function only. As in the case of the time function previously discussed, we may normalize to unit area and consider

$$\frac{F(\omega)}{\int_{-\infty}^{\infty} F(\omega)\,d\omega}$$

Again this new function is nonnegative with unit area.

Let us consider the Fourier pair $f(t)$ and $F(\omega)$ and change the time scale by the factor a, replacing $f(t)$ by $af(at)$ so that both the old and the new signal have the same area, i.e.

$$\int_{-\infty}^{\infty} f(t)\,dt = \int_{-\infty}^{\infty} af(at)\,dt \tag{4-169}$$

For $a < 1$, the new signal $af(at)$ is stretched in time and reduced in height; its "duration" has been increased. For $a > 1$, $af(at)$ has been compressed in time and increased in height; its "duration" has been decreased. The transform of this new function is

$$\int_{-\infty}^{\infty} af(at)e^{-j\omega t}\,dt = \int_{-\infty}^{\infty} f(x)e^{-j(\omega/a)x}\,dx = F\left(\frac{\omega}{a}\right) \tag{4-170}$$

The effect on the bandwidth of $F(\omega)$ has been the opposite of the effect on the duration of $f(t)$. When the signal duration is increased (decreased), the bandwidth is decreased (increased) in the same proportion. From the discussion, we might suspect that more fundamental relationships hold between properly defined durations and bandwidth of signals.

23. Duration and Bandwidth—Uncertainty Relationships. It is apparent from the discussion above that treatments of duration and bandwidth are mathematically similar although one is defined in the time domain and the other in the frequency domain. Several specific measures of these two quantities will now be found, and it will be shown that they are

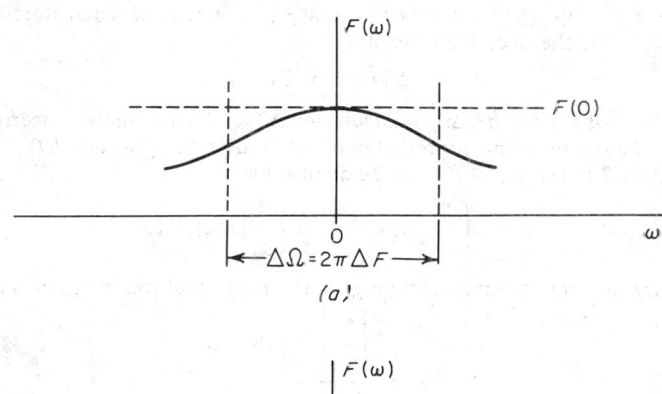

(a)

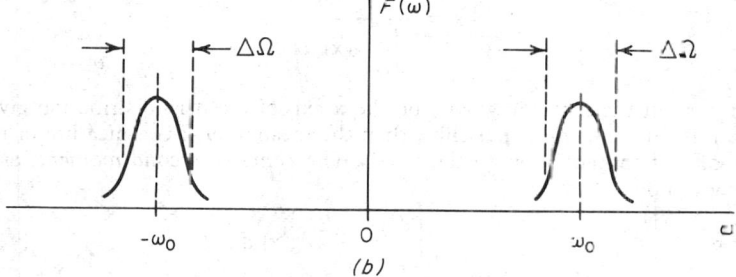

(b)

Fig. 4-8. Illustrations of bandwidth: (a) typical low-pass frequency function; (b) typical bandpass frequency function.

intimately related to each other through various so-called *uncertainty relationships*. The term "uncertainty" arises from the *Heisenberg uncertainty principle* of quantum mechanics, which states that it is not possible to determine simultaneously and exactly the position and momentum coordinates of a particle. More specifically, if Δx and Δp are the uncertainties in position and momentum, then

$$\Delta x \, \Delta p \geq h \tag{4-171}$$

where h is a constant. A number of inequalities of the form of Eq. (4-171) can be developed relating the duration ΔT of a signal to its (radian) bandwidth $\Delta \Omega$. The value of the constant h will depend on the definitions of duration and bandwidth.

(a) *Equivalent Rectangular Bandwidth* $\Delta \Omega_1$ *and Duration* ΔT_1.   The *equivalent rectangular bandwidth* $\Delta \Omega_1$ of a frequency function $F(\omega)$ is defined as

$$\Delta \Omega_1 = \frac{\displaystyle\int_{-\infty}^{\infty} F(\omega) \, d\omega}{F(\omega_0)} \tag{4-172}$$

where ω_0 is some characteristic center frequency of the function $F(\omega)$. It is clear from this definition that the original function $F(\omega)$ has been replaced by a rectangular function of equal area, width $\Delta \Omega_1$, and height $F(\omega_0)$. For the low-pass case ($\omega_0 \equiv 0$), it follows from Eqs. (4-167) and (4-168) that Eq. (4-172) can be rewritten

$$\Delta \Omega_1 = \frac{2\pi f(0)}{\displaystyle\int_{-\infty}^{\infty} f(t) \, dt} \tag{4-173}$$

where $f(t)$ is the time function which is the inverse Fourier transform of $F(\omega)$.

The same procedure can be followed in the time domain, and the *equivalent rectangular duration* ΔT_1 of the signal $f(t)$ can be defined by

$$\Delta T_1 = \frac{\displaystyle\int_{-\infty}^{\infty} f(t) \, dt}{f(t_0)} \tag{4-174}$$

where t_0 is some characteristic time denoting the center of the pulse. For the case where $t_0 \equiv 0$, it is clear, then, from Eqs. (4-173) and (4-174) that equivalent rectangular duration and bandwidth are connected by the uncertainty relationship

$$\Delta T_1 \, \Delta \Omega_1 = 2\pi \tag{4-175}$$

(b) *Second-Moment Bandwidth* $\Delta \Omega_2$ *and Duration* ΔT_2.   An alternative uncertainty relationship is based on the second-moment properties of the Fourier pair $F(\omega)$ and $f(t)$.

The total energy $\mathcal{E}$ of the signal $f(t)$ can be defined by

$$\mathcal{E} = \int_{-\infty}^{\infty} f^2(t) \, dt = \frac{1}{2\pi} \int_{-\infty}^{\infty} |F(\omega)|^2 \, d\omega \tag{4-176}$$

For an arbitrary real nonnegative function $g(x)$, the *mean* $\bar{x}$ of the function is given by

$$\bar{x} = \frac{\displaystyle\int_{-\infty}^{\infty} x g(x) \, dx}{\displaystyle\int_{-\infty}^{\infty} g(x) \, dx} \tag{4-177}$$

This quantity is just the center of gravity on the x axis of the mass distribution given by the function $g(x)$. If this function is pulselike, then the mean $\bar{x}$ of x is a measure of the pulse's location or point of concentration on the x axis. The *centered second moment*, or *variance*, $(\Delta x)^2$ of x is given by

$$(\Delta x)^2 = \frac{\displaystyle\int_{-\infty}^{\infty} (x - \bar{x})^2 g(x) \, dx}{\displaystyle\int_{-\infty}^{\infty} g(x) \, dx} \tag{4-178}$$

This quantity is a measure of the dispersion of the function $g(x)$ about the mean $\bar{x}$.

A small value for $(\Delta x)^2$ indicates that most of the area under the $g(x)$ curve is concentrated near the point $x = \bar{x}$. It is apparent that, at least for functions $g(x)$ which are nonnegative, the quantity $(\Delta x)^2$ can serve as a measure of the bandwidth or duration of $g(x)$.

This second moment may be used as a measure of bandwidth and duration by replacing the function $g(\cdot)$ in the time domain by

$$g(\cdot) \approx |f(t)|^2$$

and in the frequency domain by

$$g(\cdot) \approx (1/2\pi)|F(\omega)|^2$$

Thus a *second-moment bandwidth* $\Delta\Omega_2$ can be defined by

$$\mathcal{E}(\Delta\Omega_2)^2 = \frac{1}{2\pi} \int_{-\infty}^{\infty} (\omega - \bar{\omega})^2 |F(\omega)|^2 \, d\omega \qquad (4\text{-}179)$$

and a *second-moment duration* ΔT_2 by

$$\mathcal{E}(\Delta T_2)^2 = \int_{-\infty}^{\infty} (t - \bar{t})^2 |f(t)|^2 \, dt \qquad (4\text{-}180)$$

It is obvious that, without loss of generality, the total energy $\mathcal{E}$ can be set equal to unity; i.e.,

$$\mathcal{E} = \frac{1}{2\pi} \int_{-\infty}^{\infty} |F(\omega)|^2 \, d\omega = \int_{-\infty}^{\infty} |f(t)|^2 \, dt = 1$$

provided all results are divided by the actual value of $\mathcal{E}$. In other words, replace $g(\cdot)$ by $g(\cdot)/\mathcal{E}$.

In order to simplify Eqs. (4-179) and (4-180) let us note that any t translation of $f(t)$, a linear change of variable, will not effect the value of $|F(\omega)|^2$ in Eq. (4-179). Similarly any ω translation in Eq. (4-179) will not affect the value of $|f(t)|^2$ in Eq. (4-180). Thus it can be assumed without loss of generality that $\bar{t}$ and $\bar{\omega}$ are zero. The derivative of $f(t)$ can be written from Eq. (4-168) as

$$f'(t) = \frac{1}{2\pi} \int_{-\infty}^{\infty} j\omega F(\omega) e^{j\omega t} \, d\omega \qquad (4\text{-}181)$$

and we can form

$$\int_{-\infty}^{\infty} f'(t) f^{*\prime}(t) \, dt = \frac{1}{2\pi} \int_{-\infty}^{\infty} \omega^2 |F(\omega)|^2 \, d\omega \qquad (4\text{-}182)$$

From these results and the assumption that $\mathcal{E} = 1$, the bandwidth product $\Delta\Omega_2 \, \Delta T_2$ can be written as

$$(\Delta\Omega_2)^2 (\Delta T_2)^2 = \int_{-\infty}^{\infty} tf(t) tf^*(t) \, dt \int_{-\infty}^{\infty} f'(t) f^{*\prime}(t) \, dt \qquad (4\text{-}183)$$

We now apply a form of the Schwarz inequality and, after some manipulation, obtain

$$\Delta\Omega_2 \, \Delta T_2 \geq \tfrac{1}{2} \qquad (4\text{-}184)$$

This expression is a second uncertainty relationship connecting the bandwidth and duration of a signal. Many other such inequalities can be obtained.

24. Continuous Modulation. As discussed in Par. **4-21**, modulation can be defined as the modification of one signal, called the *carrier*, by another, called the *modulation*, *modulating signal*, or *message signal*. In this section we will be concerned with situations where the carrier and the modulation are both continuous functions of time. Later we will treat the cases where the carrier and/or the modulation have the form of pulse trains.

For our analysis, Fig. 4-6 can be modified to the system shown in Fig. 4-9, where the message is sent through a modulator (or transmitter) to produce the modulated continuous signal $m[t, a(t)]$. This waveform is corrupted by additive noise $n(t)$ in transmission so that the received (continuous) waveform $r(t)$ can be written

$$r(t) = m[t, a(t)] + n(t) \qquad (4\text{-}185)$$

The purpose of the demodulator (or receiver) is to produce some best estimate $a^*(t)$ of the original message signal $a(t)$. As pointed out in Par. **4-21**, a more general model of the transmission medium would allow corruption of the modulated waveform itself so that the received signal was of the form of Eq. (4-165). For example, in ionospheric radio-wave propagation, multiplicative disturbances might result due to multipath transmission or fading so that the received signal was of the form

$$r_1(t) = p(t)m[t, a(t)] + n(t) \tag{4-186}$$

where both $p(t)$ and $n(t)$ are noises. However, we shall not treat such systems, confining ourselves to the simpler additive-noise model of Fig. 4-9.

25. Linear, or Amplitude, Modulation. In a general way, *linear (or amplitude) modulation* (AM) can be defined as a system where a *carrier wave* $c(t)$ has its amplitude varied linearly by some *message signal* $a(t)$. More precisely, a waveform is linearly modulated (or

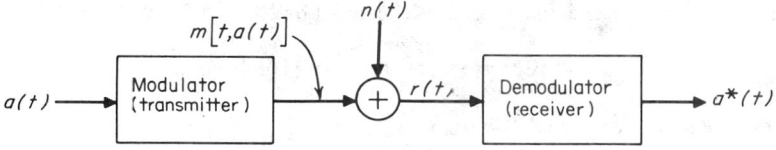

a(t) = message signal (intelligence)

$m[t,a(t)]$ = modulated signal

r(t) = $m[t,a(t)]$ + n(t) = received signal

n(t) = additive noise

a*(t) = the estimate of the original message signal

Fig. 4-9. Communication-system model for continuous modulation and demodulation.

amplitude-modulated) by a given message $a(t)$ if the partial derivative of that waveform with respect to $a(t)$ is independent of $a(t)$. In other words, the modulated signal $m[t, a(t)]$ can be written in the form

$$m[t, a(t)] = a(t)c(t) + d(t) \tag{4-187}$$

where $c(t)$ and $d(t)$ are independent of $a(t)$. Now we have

$$\frac{\partial m[t, a(t)]}{\partial a(t)} = c(t) \tag{4-188}$$

and $c(t)$ will be called the *carrier*. In most of the cases we will treat, the waveform $d(t)$ will either be zero or will be linearly related to $c(t)$. It will be more convenient, therefore, to write Eq. (4-187) as

$$m[t, a(t)] = b(t)c(t) \tag{4-189}$$

where $b(t)$ will be either

$$b_1(t) \equiv 1 + a(t) \tag{4-190}$$

or

$$b_2(t) \equiv a(t) \tag{4-191}$$

Also, at present it will be sufficient to allow the carrier $c(t)$ to be of the form

$$c(t) = C \cos(\omega_0 t + \theta) \tag{4-192}$$

where C and ω_0 are constants and θ is either a constant or a random variable uniformly distributed on $(0, 2\pi)$.

Whenever Eq. (4-190) applies, it will be convenient to assume that $b_1(t)$ is nearly always nonnegative. This implies that if $a(t)$ is a deterministic signal, then

$$a(t) \geq -1 \tag{4-193}$$

It also implies that if $a(t)$ is a random process, the probability that $a(t)$ is less than -1 in any finite interval $(-T, t)$ is arbitrarily small; i.e.,

$$p[a(t) < -1] \leq \epsilon \ll 1 \qquad -T \leq t \leq T \tag{4-194}$$

The purpose of these restrictions on $b_1(t)$ is to ensure that the carrier is not overmodulated and that the message signal $a(t)$ is easily recovered by simple receivers.

We can take the Fourier transform of both sides of Eq. (4-189) to obtain an expression for the frequency spectrum $M(\omega)$ of the general form of the linear-modulated waveform $m[t, a(t)]$:

$$M(\omega) = \frac{1}{2\pi} \int_{-\infty}^{\infty} B(\omega - v)C(v) \, dv \qquad (4-195)$$

where $B(\omega)$ and $C(\omega)$ are the Fourier transforms of $b(t)$ and $c(t)$ respectively. If the carrier $c(t)$ is the sinusoid of Eq. (4-192), then

$$C(\omega) = \pi C e^{j\theta}\delta(\omega - \omega_0) + \pi C e^{-j\theta}\delta(\omega + \omega_0) \qquad (4-196)$$

and Eq. (4-195) becomes

$$M(\omega) = (C/2)e^{j\theta}B(\omega - \omega_0) + (C/2)e^{-j\theta}B(\omega + \omega_0) \qquad (4-197)$$

Thus, for a sinusoidal carrier, linear modulation is essentially a symmetrical frequency translation of the message signal through an amount ω_0 rad/s, and no new signal components are generated. On the other hand, if $c(t)$ is not a sinusoid, so that the spectrum $C(\omega)$ has nonzero width, then $M(\omega)$ will represent a spreading and shaping as well as a translation of $B(\omega)$.

Suppose that the message signal $a(t)$ [and hence $b(t)$] is low-pass and strictly band-limited (0, ω_s) rad/s. Then the frequency spectrum of $b(t)$ obeys $B(\omega) = 0$, $|\omega| > \omega_s$, and $B(\omega - \omega_0)$ and $B(\omega + \omega_0)$ do not overlap. The spectrum $B(\omega - \omega_0)$ occupies only a range of positive frequencies while $B(\omega - \omega_0)$ occupies only negative frequencies.

The *envelope* of the modulated waveform $m[t, a(t)]$ is the magnitude $|b(t)|$. If, as mentioned earlier, the function $b(t)$ is restricted to be nonnegative almost always, then this envelope becomes just $b(t)$. In such cases, an *envelope detector* will uniquely recover $b(t)$ and hence $a(t)$. We now consider the common forms of simple amplitude, or linear, modulation.

26. Double-Sideband Amplitude Modulation (DSBAM). In this case the function $b(t)$ is given by Eq. (4-190) so that

$$m[t, a(t)] = C[1 + a(t)] \cos (\omega_0 t + \theta) \qquad (4-198)$$

The transform of $b_1(t)$ is just

$$B_1(\omega) = \mathcal{F}[1 + a(t)] = 2\pi\delta(\omega) - A(\omega) \qquad (4-199)$$

where $A(\omega)$ is the Fourier transform of $a(t)$ and $\delta(\omega)$ is the Dirac delta function. It follows, therefore, from Eq. (4-197) that the frequency spectrum of $m[t, a(t)]$ is given by

$$M(\omega) = \frac{C}{2} e^{j\theta}[2\pi\delta(\omega - \omega_0) + A(\omega - \omega_0)] + \frac{C}{2} e^{-j\theta}[2\pi\delta(\omega + \omega_0) + A(\omega + \omega_0)] \qquad (4-200)$$

Depending on the form of $a(t)$, a number of special cases can be distinguished, but in any case the spectrum is given by Eq. (4-197).

The simplest case is where $a(t)$ is the *periodic function* given by

$$a(t) = \eta \cos \omega_s t \qquad \omega_s < \omega_0 \text{ and } 0 \le \eta \le 1 \qquad (4-201)$$

Then the frequency spectrum $A(\omega)$ is

$$A(\omega) = \pi\eta\delta(\omega - \omega_s) + \pi\eta\delta(\omega + \omega_s) \qquad (4-202)$$

For convenience, let us take the phase angle θ to be zero so that

$$M(\omega) = \pi C[\delta(\omega - \omega_0) + \delta(\omega + \omega_0) + \frac{\eta}{2} \delta(\omega - \omega_0 - \omega_s) + \frac{\eta}{2} \delta(\omega - \omega_0 + \omega_s)$$

$$+ \frac{\eta}{2} \delta(\omega + \omega_0 - \omega_s) - \frac{\eta}{2} \delta(\omega + \omega_0 + \omega_s)] \qquad (4-203)$$

This spectrum is illustrated in Fig. 4-10 together with the corresponding modulated waveform

$$m[t, a(t)] = C(1 + \eta \cos \omega_s t) \cos \omega_0 t \qquad (4-204)$$

Note that this last equation can also be written as

$$m[t, a(t)] = C \cos \omega_0 t + (C/2)\eta \cos (\omega_0 - \omega_s)t + (C/2)\eta \cos (\omega_0 + \omega_s)t \qquad (4-205)$$

$$\underbrace{\hspace{2cm}}_{\text{carrier}} \qquad \underbrace{\hspace{2.5cm}}_{\text{lower sideband}} \qquad \underbrace{\hspace{2.5cm}}_{\text{upper sideband}}$$

In this form it is easy to distinguish the *carrier*, the *lower sideband*, and the *upper sideband*.

Suppose that $a(t)$ is not a single sinusoid but is periodic with period P. Then it can be expanded in a Fourier series and each term of the series treated as in Eq. (4-203).

27. Suppressed-Carrier DSBAM. If $b(t)$ is given by Eq. (4-191) so that

$$m[t, a(t)] = Ca(t) \cos (\omega_0 t + \theta) \tag{4-206}$$

then the carrier is suppressed and Eq. (4-200) reduces to

$$M(\omega) = (C/2)e^{j\theta} A(\omega - \omega_0) + (C/2)e^{-j\theta} A(\omega + \omega_0) \tag{4-207}$$

The principal advantage of DSBAM-SC over DSBAM is that the carrier is not transmitted in the former case, with a consequent saving in transmitted power. The principal disadvantages relate to problems in generating and demodulating the suppressed-carrier waveform.

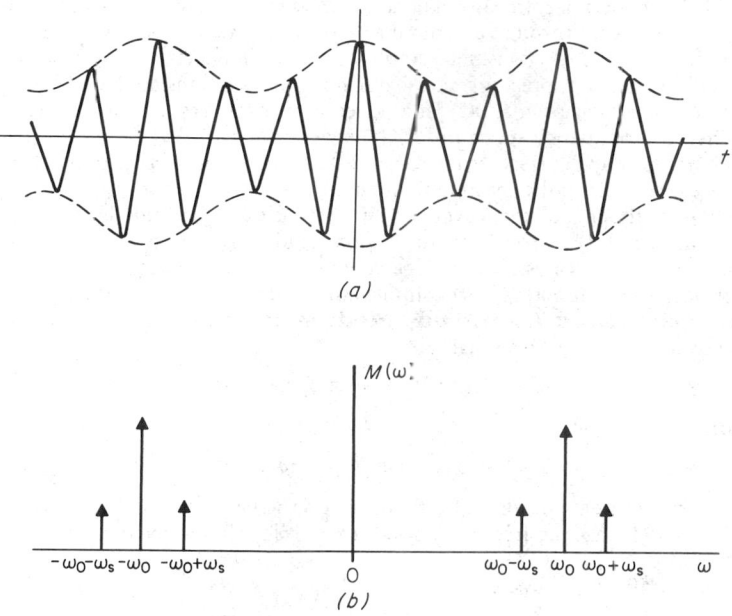

(a)

(b)

Fig. 4-10. (*a*) Double-sideband AM signal and (*b*) frequency spectrum.

28. Vestigial-Sideband AM (VSBAM). It is apparent from Eq. (4-200) that the total information regarding the message signal $a(t)$ is contained in either the upper or lower sideband in conventional DSBAM. In principle it is only necessary to transmit one of these sidebands and the other could be eliminated, say by filtering, with a consequent reduction in bandwidth and transmitted power. Such a procedure is actually followed in single-sideband amplitude modulation (SSBAM), discussed next. However, completely filtering out one sideband requires an ideal bandpass filter with infinitely sharp cutoff or an equivalent technique. A policy intermediate between the production of DSBAM and SSBAM is vestigial sideband amplitude modulation (VSBAM), where one sideband is attenuated much more than the other with some saving in bandwidth and transmitted power. The carrier may be present or suppressed.

The principal use of VSBAM has been in commercial television. The video (picture) signal is transmitted by VSBAM with a consequent reduction in the total transmitted signal bandwidth and in the frequency difference that must be allowed between adjacent channels.

29. Single-Sideband AM (SSBAM). This important type of AM can be considered as a limiting form of VSBAM when the filter for the modulated waveform is an ideal filter with infinitely sharp cutoff so that one sideband, e.g., the lower, is completely eliminated. The modulation and demodulation of SSBAM are relatively complicated and are treated in more detail in the references, Par. **4-42**.

30. Bandwidth and Power Relationships for AM. It is clear that the bandwidth of a given AM signal is related in a simple fashion to the bandwidth of the modulating signal since AM is essentially a frequency translation. If the modulating signal $a(t)$ is assumed to be low-pass and to have a bandwidth of W Hz, then the bandwidth ΔF of the modulated signal $m[t, a(t)]$ must satisfy

$$W \leq \Delta F \leq 2W \qquad (4\text{-}208)$$

From the previous discussions, it is obvious that the upper limit of $2W$ Hz holds for double-sideband modulation and the lower limit of W Hz for single-sideband. For the concept of bandwidth to be meaningful for a single-frequency sinusoidal modulating signal, it is assumed, of course, that a low-frequency sinusoid of frequency W Hz has a bandwidth W. Actually, what is really being assumed is that this sinusoid is the highest-frequency component in the low-pass modulating signal.

The only case where intermediate values of the inequality of Eq. (4-208) are encountered is in VSBAM. In principle any bandwidth between W and $2W$ Hz is possible. In this respect SSBAM can be considered as a limiting case of VSBAM. From a practical point of view, however, it is difficult (or expensive) to design a filter whose gain magnitude drops off sharply.

Power relationships are also simple and straightforward for AM signals. Consider Eq. (4-198), which is a general expression for the modulated AM waveform $m(t)$. Let the average power in this signal be denoted by P_{av}. The phase angle θ will be fixed if $a(t)$ is deterministic and will be taken to be a random variable uniformly distributed on $(0, 2\pi)$ if $a(t)$ is a random process. Also, it will be assumed that $a(t)_{av}$ is zero. It is clear that P_{av} is given by

$$P_{av} = (C^2/2)[1 + a^2(t)_{av}] \qquad (4\text{-}209)$$

where $a^2(t)_{av}$ is the average power in $a(t)$. The first term, $C^2/2$, is the carrier power, and the second term, $(C^2/2)a^2(t)_{av}$, is the signal power in the upper and lower sidebands. If $|a(t)| \leq 1$ to prevent overmodulation, then at least half the transmitted power is carrier power and the remainder is divided equally between the upper and lower sideband.

In double-sideband, suppressed-carrier operation, the carrier power is zero and half the total power exists in each sideband. The fraction of information-bearing power is ½. For single-sideband, suppressed-carrier systems, all the transmitted power is information-bearing, and, in this sense, SSB-SC has maximum transmission efficiency.

The disadvantages of both suppressed-carrier and single-sideband operation lie in the difficulties of generating the signals for transmission and in the more complicated receivers required. Demodulation of suppressed-carrier AM signals involves the reinsertion of the carrier or an equivalent operation. The local generation of a sinusoid at the exact frequency and phase of the missing carrier is either difficult or impossible unless a pilot tone of reduced magnitude is transmitted with the modulated signal for synchronization purposes or unless some nonlinear operation is performed on the suppressed-carrier signal to regenerate the carrier term at the receiver. In SSBAM, not only is the receiver more complicated, but transmission is considerably more difficult. It is usually necessary to generate the SSB signal at a low power level and then to amplify with a linear power amplifier to the proper level for transmission. On the other hand, DSB signals are easily generated at high power levels so that inefficient linear power amplifiers need not be used.

31. Angle (Frequency and Phase) Modulation. In angle modulation, the carrier $c(t)$ has either its phase angle or its frequency varied in accordance with the intelligence $a(t)$.

The result is not a simple frequency translation, as with AM, but involves both translation and the production of entirely new frequency components. In general, the new spectrum is much wider than that of the intelligence $a(t)$. The greater bandwidth may be used to improve the signal-to-noise performance of the receiver. This ability to *exchange* bandwidth for signal-to-noise enhancement is one of the outstanding characteristics and advantages of angle modulation.

In the form of angle modulation which will be called *phase modulation* (PM) the phase of the carrier is varied linearly with the intelligence $a(t)$. Thus the modulated signal is given by

$$m(t) = C \cos [\omega_0 t + \theta + k_p c(t)] \qquad (4\text{-}210)$$

where k_p is a constant and the *modulation index* $\oslash_m$ is defined by

$$\oslash_m = \max |k_p a(t)| \quad \text{(rad)} \qquad (4\text{-}211)$$

In *frequency modulation*, the instantaneous frequency is made proportional to the intelligence $a(t)$. The modulated signal is given by

$$m(t) = C \cos \left[\omega_0 t + k_f \int_{-\infty}^{t} a(\tau) \, d\tau \right] \qquad (4\text{-}212)$$

where k_f is a constant. The *maximum deviation* $\Delta\omega$ is given by

$$\Delta\omega = \max |k_f a(t)| \qquad (\text{rad/s}) \tag{4-213}$$

and, as before, a *modulation index* $\oslash_m$ by

$$\oslash_m = \max \left| k_f \int_{-\infty}^{t} a(\tau)\, d\tau \right| \tag{4-214}$$

In general, the analysis of angle-modulated signals is difficult even for simple modulating intelligence. We will consider only the case where the modulating intelligence $a(t)$ is a sinusoid. Let $a(t)$ be given by

$$a(t) = \eta \cos \omega_s t \qquad \omega_s < \omega_0 \tag{4-215}$$

Then the corresponding PM signal is

$$m(t) = C \cos(\omega_0 t + k_p \eta \cos \omega_s t) \tag{4-216}$$

where the phase angle θ has been set equal to zero. In the same way, the FM signal is

$$m(t) = C \cos [\omega_0 t + (k_f \eta/\omega_s) \sin \omega_s t] \tag{4-217}$$

Let us now consider the FM signal of this last equation. Essentially the same results will be obtained for PM. The equation can be expanded to yield

$$m(t) = C \cos(\oslash_m \sin \omega_s t) \cos \omega_0 t - C \sin(\oslash_m \sin \omega_s t) \sin \omega_0 t \tag{4-218}$$

where the modulation index $\oslash_m$ is given by $\oslash_m = k_f \eta/\omega_s$. Sinusoids with sinusoidal arguments give rise to Bessel functions, and Eq. (4-218) can be expanded and rearranged to obtain

$$m(t) = \underbrace{C J_0(\oslash_m) \cos \omega_0 t}_{\text{carrier}} + C \sum_{n=1}^{\infty} J_{2n}(\oslash_m)[\underbrace{\cos(\omega_0 + 2n\omega_s)t}_{\text{USB}} + \underbrace{\cos(\omega_0 - 2n\omega_s)t}_{\text{LSB}}]$$

$$+ C \sum_{n=1}^{\infty} J_{2n-1}(\oslash_m)\{\underbrace{\cos[\omega_0 + (2n-1)\omega_s]t}_{\text{USB}} - \underbrace{\cos[\omega_0 - (2n-1)\omega_s]t}_{\text{LSB}}\} \tag{4-219}$$

where $J_m(x)$ is the Bessel function of the first kind and order m. This expression for $m(t)$ is relatively complicated even though the modulating intelligence is a simple sinusoid. In addition to the carrier, there are an infinite number of upper and lower sidebands separated from the carrier (and from each other) by integral multiples of the modulating frequency ω_s. Each sideband, and the carrier, has an amplitude determined by the appropriate Bessel function. When there is more than one modulating sinusoid, the complexity increases rapidly.

In the general case, an infinite number of sidebands exist dispersed throughout the whole frequency domain. In this sense, the bandwidth of an FM (or PM) signal is infinite. However, outside of some interval centered on ω_0, the magnitude of the sidebands will be negligible; this interval may be taken as a practical measure of the bandwidth. An approximation can be obtained by noting that $J_n(\oslash_m)$, considered as a function of n, decreases rapidly when $n < \oslash_m$. Therefore only the first $\oslash_m$ sidebands are significant. If the highest-frequency sideband of significance is $\oslash_m \omega_s$, then the bandwidth is given approximately by

$$BW \approx 2\omega_s \oslash_m \qquad (\text{rad/s}) \tag{4-220}$$

A more accurate rule of thumb is the slightly revised expression

$$BW \approx 2\omega_s(\oslash_m + 1) \qquad (\text{rad/s}) \tag{4-221}$$

which may be considered a good approximation when $\oslash_m \geq 5$.

DIGITAL DATA TRANSMISSION AND PULSE MODULATION

32. Digital Transmission. The first half of the twentieth century saw the development of a theory of communication based on analog signals. Starting in the early 1950s, however, the widespread availability of digital computers and the concurrent development of sampled-data theory lead to a *digital* revolution which is now well under way. The technology ranges from the digital wristwatches available in every drug and department store to the sophisticated data-handling techniques of the synchronous satellites used in worldwide communications.

Actually, most of the theoretical foundations for digital data-communication-system design have already been covered in this chapter. The information-theoretic material of Pars. **4-1** to **4-15** is set in the natural context of digital signals; i.e., it relates directly to discrete-time and discrete-amplitude signals. The sampling theorem of Pars. **4-19** and **4-20** relates continuous-time band-limited signals to discrete-time samples of the signals and forms the natural connection between sampled data and their continuous-time counterpart. There remains the problem of *quantization*, by which a continuum of amplitude values is converted into a finite number of preassigned possible levels. However, we first discuss briefly a form of modulation which stands somewhere between the continuous modulation of Pars. **4-24** to **4-31** and the pulse modulation systems used in digital data transmission.

33. Pulse-Amplitude Modulation (PAM). Previously modulation schemes were considered which operated with sinusoidal carriers. Any other continuous waveform could have been used, although analysis of the resulting modulated signal might have become very complicated. Here systems are considered where the carrier is no longer continuous but consists of a pulse train, some parameter of which is suitably modified by the modulating intelligence. It will be seen that there are natural applications for such modulation schemes and that they may provide striking improvements in noise immunity. One of the simplest of such systems is PAM, where the amplitude of a pulsed carrier $p(t)$ is varied by the modulating signal $a(t)$. Here the modulated signal is given by

$$m(t) = m[t, a(t)] = a(t)p(t) \tag{4-222}$$

Let us assume that the carrier $p(t)$ is a periodic pulse train with basic pulse shape $f(t)$ and period T. Since the pulse train is periodic, it can be expanded in a Fourier series as

$$p(t) = \sum_{k=-\infty}^{\infty} P_k e^{jk\omega_0 t} \tag{4-223}$$

where $\omega_0 = 2\pi/T$ and the Fourier coefficient P_k is given by

$$P_k = \frac{1}{T} \int_{-T/2}^{T/2} p(t)e^{-jk\omega_0 t} \, dt \tag{4-224}$$

The PAM signal of Eq. (4-222) can be rewritten with the help of Eq. (4-223) as

$$m(t) = a(t)p(t) = \sum_{k=-\infty}^{\infty} P_k a(t)e^{jk\omega_0 t} \tag{4-225}$$

with Fourier transform $M(\omega)$ given by

$$M(\omega) = \sum_{k=-\infty}^{\infty} P_k \mathscr{F}\{a(t)e^{jk\omega_0 t}\} = \sum_{k=-\infty}^{\infty} P_k A(\omega - k\omega_0) \tag{4-226}$$

where $A(\omega)$ is the Fourier transform of $a(t)$. Suppose now that $a(t)$ is band-limited $(-B/2, B/2)$ rad/s, where $B < \omega_0$. Then it is clear from Eq. (4-226) that the PAM signal $m(t)$ has a spectrum where the basic spectral shape is repeated periodically throughout the frequency domain; but each spectral pulse $A(\omega - k\omega_0)$ is weighted by the appropriate Fourier coefficient P_k. The value of this coefficient depends on the amplitude, shape, and spacing of the carrier pulse train $p(t)$.

In practice, the width D of the basic pulse $p(t)$ is small compared with the pulse spacing T. The general effect of the modulating process is to sample the intelligence $a(t)$ at a fixed interval of T s throughout the time domain. If $a(t)$ is strictly band-limited $(-B/2, B/2)$ rad/s, the sampling theorem of Par. **4-20** ensures that $a(t)$ can be reconstructed exactly from its samples spaced $2\pi/B$ s apart throughout the time domain. Thus a *necessary* condition for exact recovery of $a(t)$ from $a(t)p(t)$ is that

$$T \leq 2\pi/B \quad \text{or} \quad BT \leq 2\pi \tag{4-227}$$

It is clear from Eq. (4-226) that this condition is the necessary restriction to prevent overlapping of the repeated spectra of the PAM signal $m(t)$. (Note that ω_0 has been defined as $2\pi/T$.)

In the demodulation of PAM signals, two cases must be distinguished. If the Fourier coefficient P_0 is not zero, the modulating intelligence $a(t)$ can be recovered by low-pass filtering. The resulting output is just the original intelligence $a(t)$ weighted by the constant P_0. When P_0 is zero but P_1 is not, the intelligence $a(t)$ can be recovered by some form of synchronous demodulation.

As a practical matter, the exact waveform given by Eq. (4-225) is not generally used in PAM. More specifically, the tops of the pulse train are not usually shaped by the modulating signal. Instead they are kept flat, and the pulse height is determined by the value of $a(t)$ at some point in the pulse-length interval D. Thus the PAM signal is not exactly $a(t)p(t)$, and the resulting spectrum $M(\omega)$ does not yield $A(\omega)$ with a constant scale factor. Instead there is some distortion of the spectrum of $A(\omega)$, depending on the exact shape of the pulse train $p(t)$. In many cases this distortion can be kept negligibly small, and in other cases it can be substantially removed by a low-pass equalizing filter in the PAM receiver.

34. Quantizing and Quantizing Error. The PAM system just discussed involves converting a time-continuous modulating intelligence $a(t)$ into a time-discrete form by *sampling*. The sampling theorem guarantees that $a(t)$ can be reconstructed exactly from its samples $a(k/2W)$, $k = \dots, -1, 0, 1, \dots$, spaced throughout the time domain, provided $a(t)$ has a spectrum $A(\omega)$ which is strictly band-limited $(-2\pi W, 2\pi W)$ rad/s. In a similar fashion, an amplitude-continuous signal is converted into an *amplitude-discrete* or *digital* signal by quantizing in the amplitude domain into a finite number of fixed distinguishable amplitude levels, each a distance Q apart.

The process of quantizing is irreversible since, regardless of how small the quantization level Q is taken to be, an unresolvable uncertainty of $\pm Q/2$ is associated after quantizing with each amplitude value. Thus a *quantization noise* N_q is inevitably associated with all quantized signals. This noise can be made as small as desired by choosing enough quantization levels or, equivalently, by making each quantization level small enough, but it cannot be eliminated. In the pulsed carrier systems discussed subsequently, it turns out that the noise added in transmission can be almost completely eliminated at the receiver. In other words, the type of noise interference that is added externally (say in the channel) is negligible, and the principal source of contamination is the quantization noise.

The quantization levels are often taken to be equal; i.e., the spacing between amplitude levels is uniform. Nonuniform quantizing can be used to favor small amplitudes where noise has a greater effect at the expense of large amplitudes. However, for this discussion, it will be assumed that the differences in amplitude levels are all equal to Q. Let us assume that the actual amplitude of the signal being sampled is equally likely to lie anywhere within the particular quantization level. In other words, the quantizing error in an amplitude interval is taken to be a uniformly distributed random variable q with density $p_q(x)$ given by

$$p_q(x) = \begin{cases} 0 & |x| > Q/2 \\ 1/Q & |x| \le Q/2 \end{cases} \tag{4-228}$$

The quantization noise N_q can be written as

$$N_q = \int_{-Q/2}^{Q/2} x^2 p_q(x)\, dx = \frac{Q^2}{12} \tag{4-229}$$

The average value of the quantizing error is zero since the distribution of Eq. (4-228) has zero mean. In terms of 1-V quantization levels the rms error voltage is $1/\sqrt{12}$ V. Suppose that the peak-to-peak value of the signal to be quantized is limited to $\pm A$. Then the quantization level Q can be written as

$$Q = 2A/d \tag{4-230}$$

where d is the number of quantization levels. The signal-to-quantization-noise power ratio for a sampled signal $a*(t)$ can then be written as

$$\frac{S}{N_q} = \frac{12[a*(t)]^2_{\text{av}}}{Q^2} = 3d^2 \frac{[a*(t)]^2_{\text{av}}}{A^2} \tag{4-231}$$

where the quantity $[a*(t)]^2_{\text{av}}/A^2$ is the normalized sampled signal power.

35. Signal Encoding. The result of amplitude quantizing and time sampling a message signal is a sequence of numbers, i.e., a *digital signal*. We now consider advantageous ways to represent this sequence. Each number could be represented by a pulse of height proportional to that number. The result would be a quantized PAM pulse train. However, there can be great advantages in representing *each* number by a *sequence* of numbers which are allowed to assume fewer values than the original set. Such a system of representation (Pars. **4-7** to **4-9**) is called a code. The principal advantages of encoding a message set are (1) that it is easier to discriminate in the presence of noise between a few possible numbers (pulse heights) than between many and (2) that an error in the value of a given code number will affect only part of the information contained in the original sample value. In general, an m-ary code will consist of m possible levels. Then a sequence of n symbols of this code can represent m^n possible message levels since there are $M = m^n$ distinguishable code groups of length n. As an example, suppose that the original message is quantized at eight levels. The message sequence is to be encoded into a binary code consisting of two levels or symbols 0 and 1. In this case $m = 2$ and, since M must equal 8, we have

$$M = m^r \quad \text{or} \quad 8 = 2^3$$

Thus each binary code word, representing one of the possible quantization levels, must consist of three binary digits. A logical encoding is

Quantization level	Code word		Quantization level	Code word
0	000		4	100
1	001		5	101
2	010		6	110
3	011		7	111

Of the possible m-ary codes that could be used, the binary code ($m = 2$) is the most common, for at least two reasons: (1) the binary code offers the least number of choices, namely, two, in decoding, and (2) binary systems are most easily implemented electronically since the two possible states can be made to correspond to on-off, open-closed, or conducting-nonconducting conditions in circuits and systems.

It is clear from the previous discussion that the process of amplitude quantization, time sampling, and encoding can be used to convert a continuous band-limited message signal into a pulse train consisting of equally spaced pulses of two heights, e.g., zero and unity. The steps by which this is accomplished are as follows: (1) a low-frequency signal $a(t)$ band-limited $(-2\pi W, 2\pi W)$ is quantized into M amplitude levels and sampled at an interval of $1/2W$ s. The result is a sequence of numbers, a digital signal; (2) each of these numbers is then encoded in a binary code; (3) the corresponding binary sequence becomes a pulse train consisting of pulses of height unity and zero to correspond to 1 and 0, respectively. This binary sequence can now be transmitted directly, in which case it will be called a *baseband* signal, or it can be used as a modulating signal in a *pulse-code-modulation* (PCM) system. The steps followed up to this point are as follows:

1. A continuous signal $a(t)$ is sampled and quantized.
2. The quantized sample values are encoded into a pulse train $p(t)$.
3. The pulse train $p(t)$ may be transmitted directly or used as the modulating signal in any appropriate modulation scheme.

The great advantage of this system is its inherent resistance to external noise corruption. If the external noise is not too large, it is clear that the original pulse $p(t)$ can be recovered *exactly* from the received pulse $p^*(r)$ since only a knowledge of the *presence* or *absence* of a pulse and not the pulse *shape* itself is needed at each pulse position to reproduce the original pulse train exactly. Thus, if the signal is large enough compared with corrupting noise, there will be *no error.* Furthermore, if it is desired to transmit the pulse train for a long distance, it can be regenerated exactly as often as desired at repeater stations along the way. In other words, no noise is added in transmission in the large-signal case. This situation is in striking contrast to the analog schemes already studied. Thus, in a properly designed large-signal (encoded) digital-data system, all error that results will arise from the original quantization noise, discussed previously.

The digital-data system can be considered a coded wide-band system which trades complexity and bandwidth for noise immunity. In general, if there are d quantization levels, and if binary pulses are used, each code word must contain n pulses, where

$$d = 2^n \quad \text{or} \quad n = \log_2 d \tag{4-232}$$

Thus there must be $2nW$ pulses/s in the pulse train $p(t)$ and any two adjacent pulses must be clearly distinguishable. If the bandwidth of the system transmitting the pulse train is made too small, the pulses will be broadened and will interfere with each other, so that errors in identification will result. In fact, the uncertainty relationships previously developed bear directly on this problem. Thus Eq. (4-184) relates the bandwidth $\Delta\Omega$ and the duration ΔT of a pulse by

$$(\Delta\Omega)(\Delta T) \geq \tfrac{1}{2} \qquad (4\text{-}233)$$

Since there must be $2nW$ pulses/s, each pulse cannot exceed a width $\Delta T \leq 1/2nW$ s.

It follows directly from Eq. (4-233) that the system bandwidth $\Delta\Omega$ is bounded from below by

$$\Delta\Omega \geq nW \qquad (\text{Hz}) \qquad (4\text{-}234)$$

It is clear that the digital-data signal requires a transmission bandwidth which is at least n times that of the original intelligence, where n is determined by Eq. (4-232) and depends on the num-

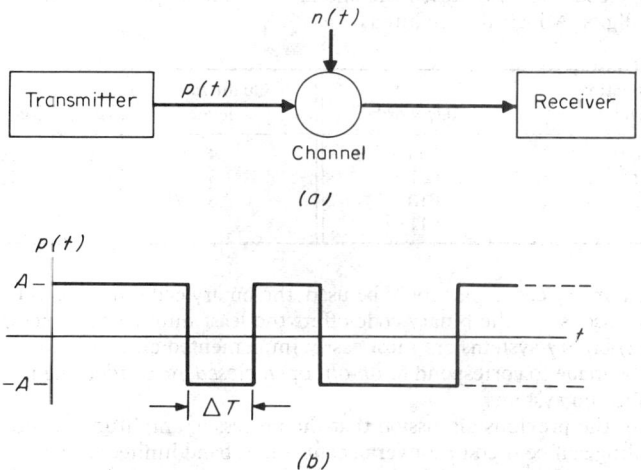

Fig. 4-11. Baseband digital-data transmission: (a) system structure and (b) baseband-encoded sequence.

ber of quantization levels desired. If the number of quantization levels is taken to be small (so that n is small), the quantization noise will increase as shown by Eqs. (4-229) and (4-230). The output signal-to-noise power ratio due to quantization noise is given by Eq. (4-231), where $[a^*(t)]_{av}^2$ is the power in the quantized signal and must be calculated. For reasonable signal distributions and uniform quantizers it can be shown that the signal-to-noise ratio increases approximately as 2^{2n}.

In many practical situations, equipment inadequacies and/or low-signal conditions may create situations where transmission noise as well as quantization noise is added. In other words, some pulses in the coded pulse train will be incorrectly identified. This problem is most conveniently formulated in terms of a *probability of error*, which gives the average rate at which incorrect pulse identification occurs. This criterion was discussed in Par. **4-10** and is directly applicable here.

36. Baseband Digital-Data Transmission. Consider the system of Fig. 4-11, where the baseband signal (encoded sequence) is denoted by $p(t)$. To avoid bias (dc level) problems, it may be convenient to send a sequence of $+1$s and -1s instead of $+1$s and 0s. A noise $n(t)$ is added in the transmission channel and the problem at the receiver is to determine whether a $+1$ or a -1 was transmitted in each time slot of length ΔT.

To obtain a general idea of receiver performance, it will be assumed that the noise $n(t)$ is a zero-mean gaussian process, as discussed in Par. **4-20** with power spectral density [see Eq. (4-156)] taken to be the constant $N_0/2$. Such a process is called a *white gaussian noise*. It will also be assumed that the *synchronization* problem has been solved; i.e., the receiver knows when each pulse in $p(t)$ starts and ends.

A general theory of detection and estimation can be used to specify an optimal receiver structure; however, a reasonable (and quasi-optimal) way to proceed is to use an *integrate-and-dump* system, where, at the end of each pulse occurrence time, the random variable

$$V = \int_{\Delta T} [p(t) + n(t)]\, dt \qquad (4\text{-}235)$$

is compared with zero. If $V > 0$, it is decided that $+A$ was sent, and if $V < 0$, that $-A$ was sent. It is clear that

$$V = \begin{cases} ADT + N & \text{if } +A \text{ was sent} \\ -ADT + N & \text{if } -A \text{ was sent} \end{cases}$$

where N is a normal random variable (result of a linear operation on a gaussian process) with zero mean and variance σ^2 given by

$$\text{Var } N = \sigma^2 = E\{N^2\} = N_0 \, \Delta T/2 \tag{4-236}$$

It is obvious that an error can occur in either one of two ways: (1) $+A$ is sent and $ADT + N < 0$ or (2) $-A$ is sent and $-ADT + N > 0$. From symmetry, these two errors are equally likely, and the total error probability P_e is

$$P_e = \int_{A \, \Delta T}^{\infty} f_N(x) \, dx \tag{4-237}$$

where $f_N(x)$ is the normal density function

$$f_N(x) = (1/\sqrt{2\pi}\sigma)e^{-x^2/2\sigma^2} \qquad -\infty < x < \infty \tag{4-238}$$

and $\sigma^2 = N_0 \, \Delta T/2$ has already been defined. The expression for P_e can be rewritten, after a change in variable, as

$$P_e = \frac{1}{\sqrt{\pi}} \int_k^{\infty} e^{-y^2} \, dy \tag{4-239}$$

where $k^2 = A^2 \, \Delta T/N_0$. Using tables or numerical integration, one can use this last equation to plot P_e vs. k, as shown in Fig. 4-12. Note that k^2 is a measure of signal-to-noise power ratio. Naturally, as k increases, the probability of error P_e decreases.

The preceding calculations are intended to give a general idea of error-probability calculations and receiver design. For voluminous details, see the references listed at the end of this section under Digital-Data Transmission.

37. Pulse-Code Modulation (PCM). Thus far, no consideration has been given to the actual transmission of the baseband signal. It is clear that the pulse train cannot be radiated from an antenna of any reasonable size. As in the analog case, some frequency-translation technique must be employed to make radiation practical and/or to allow frequency stacking of more than one baseband signal. A number of systems are used for such *digital signaling.* Some of the commonest follow.

(a) *Binary On-Off Keying (OOK).* This type of binary signaling is almost self-explanatory. A high-frequency sinusoidal signal is switched on and off so that on periods correspond to 1 and off periods to 0 in the PCM wave. In practice. the pulsed sinusoids do not start or stop abruptly but exhibit a transient buildup and decay.

(b) *Binary Frequency-Shift Keying (FSK).* In this form of digital signaling, a continuous wave is sent which shifts between two frequencies, one representing the symbol 1 in the PCM wave and the other representing 0.

(c) *Binary Phase-Shift Keying (PSK).* Here the frequency is kept constant, but the phase is shifted 180° whenever the basic signal changes from 0 to 1.

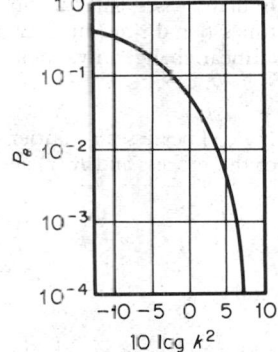

Fig. 4-12. Probability of error plotted against signal-to-noise ratio for an integrate-and-dump receiver.

NOISE AND INTERFERENCE

38. General. In a general sense, *noise* and *interference* are used to describe any unwanted or undesirable signals in communication channels and systems. Since in many cases these signals are random or unpredictable, some study of random processes is a useful prelude to any consideration of noise and interference.

39. Random Processes. A random process $x(t)$ is often defined as an indexed family of random variables where the *index* or *parameter* t belongs to some set T; that is, $t \in T$.

The set T is called the *parameter set* or *index set* of the process. It may be finite or infinite, denumerable or nondenumerable; it may be an interval or set of intervals on the real line, or it may be the whole real line $-\infty < t < \infty$. In most applied problems, the index t will be time, and the underlying intuitive notion will be that of a random variable developing in time. However, other parameters such as position, temperature, etc., may also enter in a natural manner.

It follows from the definition that there are at least two ways to view an arbitrary random process: (1) as a set of random variables: this viewpoint follows from the definition; for each value of t, the random process reduces to a random variable; and (2) as a set of functions of time. From this viewpoint, there is an underlying random variable each realization of which is a time function with domain T. Each such time function is called a *sample function* or *realization* of the process.

From a physical point of view, it is the sample function which is important, since this is the quantity that will almost always be observed in dealing experimentally with the random process. One of the important practical aspects of the study of random processes is the determination of properties of the random variable $x(t)$ for fixed t on the basis of measurements performed on a single sample function of the process $x(t)$.

A random process is said to be *stationary* if its statistical properties are invariant to time translation. This invariance implies that the underlying physical mechanism producing the process is not changing with time. Stationary processes are of great importance for two reasons: (1) they are common in practice or approximated to a high degree of accuracy (actually, from the practical point of view, it is not necessary that a process be stationary for all time but only for some observation interval which is long enough to be suitable for a given problem); (2) many of the important properties of common stationary processes are described by first and second moments. Consequently, it is relatively easy to develop a simple but useful theory (*spectral theory*) to describe these processes. Processes which are not stationary are called *nonstationary*, although they are also sometimes referred to as *evolutionary* processes.

The *mean* $m(t)$ of a random process $x(t)$ is defined by

$$m(t) = E\{x(t)\} \tag{4-240}$$

where $E\{\cdot\}$ is the mathematical expectation operator defined in Par. **4-4**. In many practical problems, this mean is independent of time. In any case, if it is known, it can be subtracted from $x(t)$ to form a new "centered" process $y(t) = x(t) - m(t)$ with zero mean.

The *autocorrelation function* $R_x(t_1, t_2)$ of a random process $x(t)$ is defined by

$$R_x(t_1, t_2) = E\{x(t_1)x(t_2)\} \tag{4-241}$$

In many cases, this function depends only on the time difference $t_2 - t_1$ and not on the absolute times t_1 and t_2. In such cases, the process $x(t)$ is said to be *at least wide-sense stationary*, and by a linear change in variable Eq. (4-241) can be written

$$R_x(\tau) = E\{x(t)x(t + \tau)\} \tag{4-242}$$

If $R_x(\tau)$ possesses a Fourier transform $\varphi_x(\omega)$, this transform is called the *power spectral density* of the process and $R_x(\tau)$ and $\varphi_x(\omega)$ form the Fourier-transform pair

$$\varphi_x(\omega) = \int_{-\infty}^{\infty} R_x(\tau)e^{-j\omega\tau} \, d\tau \tag{4-243}$$

and

$$R_x(\tau) = \frac{1}{2\pi} \int_{-\infty}^{\infty} \varphi_x(\omega)e^{j\omega\tau} \, d\omega \tag{4-244}$$

For processes which are at least wide-sense stationary, these last two equations afford a direct approach to the analysis of random signals and noises on a power ratio or mean-squared-error basis. When $\tau = 0$, Eq. (4-244) becomes

$$R_x(0) = E\{x^2(t)\} = \frac{1}{2\pi} \int_{-\infty}^{\infty} \varphi_x(\omega) \, d\omega \tag{4-245}$$

an expression for the normalized power in the process $x(t)$.

As previously mentioned, in practical problems involving random processes, what will generally be available to the observer is not the random process but one of its sample functions or realizations. In such cases, the quantities that are easily measured are various time averages, and an important question to answer is: Under what circumstances can these time averages be related to the statistical properties of the process?

We define the *time average* of the random process x(t) by

$$A\{x(t)\} = \lim_{T \to \infty} \frac{1}{2T} \int_{-T}^{T} x(t)\, dt \tag{4-246}$$

The *time autocorrelation function* $\mathcal{R}_x(\tau)$ is defined by

$$\mathcal{R}_x(\tau) = A\{x(t)x(t + \tau)\} = \lim_{T \to \infty} \frac{1}{2T} \int_{-}^{-} x(t)x(t + \tau)\, dt \tag{4-247}$$

It is intuitively reasonable to suppose that, for stationary processes, time averages should be equal to expectations; e.g.,

$$E\{x(t)x(t + \tau)\} = A\{x(t)x(t + \tau)\} \tag{4-248}$$

A heuristic argument to support this claim would go as follows. Divide the parameter t of the random process x(t) into long intervals of T length. If these intervals are long enough (compared with the time scale of the underlying physical mechanism, the statistical properties of the process in one interval T should be very similar to those in any other interval. Furthermore, a new random process could be formed in the interval (0, T) by using as sample functions the segments of length T from a single sample function of the original process. This new process should be statistically indistinguishable from the original process, and its ensemble averages would correspond to time averages of the sample function from the original process.

The foregoing is intended as a very crude justification of the condition of *ergodicity*. A random process is said to be *ergodic* if time averages of sample functions of the process can be used as approximations to the corresponding ensemble averages or expectations. A further discussion of ergodicity is beyond the scope of this treatment, but this condition can usually be assumed to exist for stationary processes. In this case, time averages and expectations can be interchanged at will, and, in particular,

$$E\{x(t)\} = A\{x(t)\} = u = \text{const} \tag{4-249}$$

and

$$R_x(\tau) = \mathcal{R}_x(\tau) \tag{4-250}$$

40. Classification of Random Processes. A central problem in the study of random processes is their classification. From a mathematical point of view, a random process x(t) is defined when all n-dimensional distribution functions of the random variables $x(t_1)$, $x(t_2)$, ..., $x(t_n)$ are defined for arbitrary n and arbitrary times t_1, t_2, ..., t_n. Thus classes of random processes can be defined by imposing suitable restrictions on their n-dimensional distribution functions. In this way, we can define the following (and many others):

(a) *Stationary processes,* whose joint distribution functions are invariant to time translation.

(b) *Gaussian (or normal) process,* whose joint distribution functions are multivariate normal.

(c) *Markov processes,* where given the value of $x(t_1)$, the value of $x(t_2)$, $t_2 > t$, does not depend on the value of $x(t_3)$, $t_0 < t_1$; in other words, the future behavior of the process, given its present state, is not changed by additional knowledge about its past.

(d) *White noise,* where the power spectral density given by Eq. (4-243) is assumed to be a constant N_0. Such a process is not realizable since its mean-squared value (normalized power) is not finite; i.e.,

$$R_x(0) = E\{x^2(t)\} = \frac{N_0}{2\pi} \int_{-\infty}^{\infty} d\omega = \infty \tag{4-251}$$

On the other hand, this concept is of considerable usefulness in many types of analysis and can often be postulated where the actual process has an approximately constant power spectral density over a frequency range much greater than the system bandwidth.

Another way to classify random processes is on the basis of a model of the particular process. This method has the advantage of providing insights into the physical mechanisms producing the process. The principal disadvantage is the complexity that frequently results. On this basis, we may identify the following (natural) random processes.

(e) *Thermal noise* is caused by the random motion of the electrons within a conductor of nonzero resistance. The mean-squared value of the thermal-noise voltage across a resistor of resistance $R\ \Omega$ is given by

$$\overline{v^2} = 4kTR\ \Delta f \quad \text{(volts}^2) \tag{4-252}$$

where k is Boltzmann's constant, T is the absolute temperature in kelvins, and Δf is the bandwidth of the measuring equipment.

(f) *Shot noise* is present in any electronic device (vacuum tube, transistor, etc.) where electrons move across a potential barrier in a random way. Shot noise is usually modeled as

$$x(t) = \sum_{i=-\infty}^{\infty} f(t - t_i) \tag{4-253}$$

where the t_i are random emission times and $f(t)$ is a basic pulse shape determined by the device geometry and potential distribution.

(g) *Partition noise* results where the electrons in an electron beam can travel to two or more electrodes. Random fluctuations in the number of electrons reaching each electrode are the basis for this noise. It is frequently the predominant noise in multielectrode vacuum tubes such as the pentode.

(h) *Defect noise* is a term used to describe a wide variety of related phenomena which manifest themselves as noise voltages across the terminals of various devices when dc currents are passed through them. Such noise is also called current noise, excess noise, flicker noise, contact noise, or $1/f$ noise. The power spectral density of this noise is given by

$$\varphi_x(\omega) = kI^\alpha/\omega^\beta \tag{4-254}$$

where I is the direct current through the device, ω is radian frequency, and k, α, and β are constants. The constant α is usually close to 2, and β is usually close to 1. At a low enough frequency this noise may predominate due to the $1/\omega$ dependence.

41. Man-Made Noise. The noises just discussed are more or less fundamental and are caused basically by the noncontinuous nature of the electronic charge. In contradistinction to these are a large class of noises and interferences which are more or less man-made and hence, in principle, under our control. The number and kinds here are too many to list and the physical mechanisms usually too complicated to describe. However, to some degree, they can be organized into three main classes.

(a) *Interchannel Interference.* This includes the interference of one radio or television channel with another, which may be the result of inferior antenna or receiver design, variations in carrier frequency at the transmitter, or unexpectedly long-distance transmission via scatter of ionospheric reflection. It also includes crosstalk between channels in communication links and interference caused by multipath propagation or reflection. These types of noises can be removed, at least in principle, by better equipment design, e.g., by using a receiving antenna with a sufficiently narrow radiation pattern to eliminate reception from more than one transmitter.

(b) *Hum.* This is a periodic and undesirable signal arising from the power lines. Usually it is predictable and can be eliminated by proper filtering and shielding.

(c) *Impulse Noise.* Like defect noise, this term describes a wide variety of phenomena. Not all of them are man-made, but the majority probably are. This noise can often be modeled as a low-density shot process or, equivalently, as the superposition of a small number of large impulses. These impulses may occur more or less periodically, as in ignition noise from automobiles or corona noise from high-voltage transmission lines. On the other hand, they may occur randomly, as in switching noise in telephone systems or the atmospheric noise from thunderstorms. The latter type of noise is not necessarily man-made, of course. This impulse noise tends to have an amplitude distribution which is decidedly nongaussian, and it is frequently highly nonstationary. It is difficult to deal with in a systematic way because it is ill-defined. Signal processors which must handle this type of noise are often preceded by limiters of various kinds or by noise blankers which give zero transmission if a certain amplitude level is surpassed. The design philosophy behind the use of limiters and blankers is fairly clear. If the noise consists of large impulses of relatively low density, the best system performance is obtained if the system is limited or blanked during the noisy periods and behaves normally when the (impulse) noise is not present.

42. References. The large number of textbooks listed on the following pages contain extensive references to the periodical literature.

GENERAL

1. Lawson, J. W., and G. E. Uhlenbeck   "Threshold Signals," McGraw-Hill, New York, 1950.
2. Davenport, W. B., and W. L. Root   "An Introduction to Random Signals and Noise," McGraw-Hill, New York, 1958.

3. Lee, Y. W.   "Statistical Theory of Communication," Wiley, New York, 1960.

4. Middleton, D.   "An Introduction to Statistical Communication Theory," McGraw-Hill, New York, 1960.

5. Javid, M., and E. Brenner   "Analysis, Transmission, and Filtering of Signals," McGraw-Hill, New York, 1963.

6. Golomb, S. W.   "Digital Communication with Space Applications," Prentice-Hall, Englewood Cliffs, N.J., 1964.

7. Bennett, W. R., and J. R. Davey   "Data Transmission," McGraw-Hill, New York, 1965.

8. Lathi, B. P.   "Signals, Systems, and Communication," Wiley, New York, 1965.

9. Papoulis, A.   "Probability, Random Variables, and Stochastic Processes," McGraw-Hill, New York, 1965.

10. Wozencraft, J. M., and I. M. Jacobs   "Principles of Communication Engineering," Wiley, New York, 1965.

11. Schwartz, M., W. R. Bennett, and S. Stein   "Communication Systems and Techniques," McGraw-Hill, New York, 1966.

12. Cooper, G. R., and C. D. McGillem   "Methods of Signal and System Analysis," Holt, Rinehart and Winston, New York, 1967.

13. Lucky, R. W., J. Salz, and E. J. Weldon   "Principles of Data Communication," McGraw-Hill, New York, 1968.

14. Sakrison, D. J.   "Communication Theory: Transmission of Waveforms and Digital Information," Wiley, New York, 1968.

15. Thomas, J. B.   "An Introduction to Statistical Communication Theory," Wiley, New York, 1969.

16. Schwartz, M.   "Information Transmission, Modulation, and Noise," 2d ed., McGraw-Hill, New York, 1970.

INFORMATION THEORY

1. Shannon, C. E., and W. Weaver   "The Mathematical Theory of Communication," University of Illinois Press, Urbana, 1949.

2. Bell, D. A.   "Information Theory and Its Engineering Applications," Pitman, London, 1952.

3. Goldman, S.   "Information Theory," Prentice-Hall, Englewood Cliffs, N.J., 1953.

4. Brillouin, L.   "Science and Information Theory," Academic, New York, 1956.

5. Khinchin, A. I.   "Mathematical Foundations of Information Theory," Dover, New York, 1957.

6. Feinstein, A.   "Foundations of Information Theory," McGraw-Hill, New York 1958.

7. Kullback, S.   "Information Theory and Statistics," Wiley, New York, 1959.

8. Fano, R. M.   "Transmission of Information," M.I.T. Press, Cambridge, Mass. 1961.

9. Reza, F. M.   "An Introduction to Information Theory," McGraw-Hill, New York, 1961.

10. Wolfowitz, J.   "Coding Theorems of Information Theory," Prentice-Hall, Englewood Cliffs, N.J., 1961.

11. Abramson, N.   "Information Theory and Coding," McGraw-Hill, New York, 1963.

12. Raisbeck, G.   "Information Theory," M.I.T. Press, Cambridge, Mass., 1964.

13. Ash, R.   "Information Theory," Wiley-Interscience, New York, 1965.

14. Billingsley, P. B.   "Ergodic Theory and Information," Wiley, New York, 1965.

15. Jelinek, F.   "Probablistic Information Theory," McGraw-Hill, New York, 1968.

16. Gallager, R.   "Information Theory and Reliable Communication," Wiley, New York, 1968.

CODING THEORY

1. Peterson, W. W.   "Error-Correcting Codes," Wiley, New York, 1961.

2. Berlekamp, E. R.   "Algebraic Coding Theory," McGraw-Hill, New York, 1968.

3. Lin, S.   "An Introduction to Error-Correcting Codes," Prentice-Hall, Englewood Cliffs, N.J., 1970.

4. Blake, I., and R. C. Mullin   "An Introduction to Algebraic and Combinatorial Coding Theory," Academic, New York, 1976.

DIGITAL-DATA TRANSMISSION

1. Viterbi, A. J., and J. K. Omura   "Principles of Digital Communication and Coding," McGraw-Hill, New York, 1979.

2. Berger, T. "Rate Distortion Theory," Prentice-Hall, Englewood Cliffs , N.J., 1971.

3. Oppenheim, A. V., and R. W. Shafer "Digital Signal Processing," Prentice-Hall, Englewood Cliffs, N.J., 1975.

4. Spilker, J. J. "Digital Communications by Satellite," Prentice-Hall, Englewood Cliffs, N.J., 1977.

NOISE

1. Van der Ziel, A. "Noise," Prentice-Hall, Englewood Cliffs, N.J., 1954.

2. Smullin, L. D., and H. A. Haus "Noise in Electron Devices," Wiley, New York, 1959.

3. Bell, D. A. "Electrical Noise: Fundamentals and Physical Mechanism", Van Nostrand, London, 1960.

4. Bennett, W. R. "Electrical Noise," McGraw-Hill, New York, 1960.

5. MacDonald, D. K. C. "Noise and Fluctuations," Wiley, New York, 1962.

Section 5

Systems Engineering

LESTER H. FINK *President, Systems Engineering for Power, Incorporated; Fellow, IEEE*

GILMER L. BLANKENSHIP *Associate Professor, University of Maryland; Member, IEEE (linear systems; control theory, in part)*

JOSÉ B. CRUZ, JR. *Professor, University of Illinois; Fellow, IEEE (game theory)*

Since everything, then, is cause and effect, dependent and supporting, mediate and immediate, and all is held together by a natural though imperceptible chain which binds together things most distant and most different, I hold it equally impossible to know the parts without knowing the whole and to know the whole without knowing the parts in detail.
— PASCAL "Pensées," No. 72

CONTENTS
Numbers refer to paragraphs

INTRODUCTION

1. The Discipline of Systems Engineering. The history of science and technology is one of the successive emergence of new disciplines from older ones. In consequence, it is difficult to delineate a set of separate peer disciplines which jointly span scientific knowledge and engineering practice. In its early stages, engineering as a discipline distinguished civil from military; later, mechanical and chemical were recognized. Early in this century, electrical engineering was distinguished from mechanical, and since World War II, nuclear engineering in turn has emerged. More recently, computer engineering has gained recognition.

Against this background systems engineering is an emergent allied discipline, having its own domain of knowledge and range of application. Only when it is viewed in this light and accepted on its engineering merits can it make its proper contribution to theory and practice. The point can also be made negatively: systems engineering is *not* just all good engineering; more emphatically it is *not* just elaborated common sense. It is a powerful tool only in the hands of those who recognize it as a hard, mathematical discipline, distinguished from its peers by its focus on the nature and effects of interconnections between objects — objects which are otherwise studied in their own frames of reference.

One of the difficulties in recognizing and accepting systems engineering as an independent discipline is just the fact that all the concrete objects with which it deals are individually already the objects of study, in their own frame of reference, of other, older disciplines. An electrical engineer feels that he knows what there is to know about transformers or transistors; a mechanical engineer is expert in the characteristics of turbines and engines; a chemical engineer understands chemical process reactors, and a civil engineer understands building frame stresses.

In contrast, systems engineers must discern and understand the constraints and requirements to which each of these objects, individually and collectively, is subjected when they are jointly incorporated in a functioning system. Moreover, systems engineers must discern and understand the characteristics of these elements, individually and in tradeoffs, so that the overall system can fulfill its purpose.

These problems are real; they transcend the provinces of the experts on the individual objects composing the systems. They cannot be answered, or even properly posed, in the framework of any of the component objects themselves. Above all they cannot be answered on the basis of "common sense" or even "good engineering in general," any more than an integrated circuit can be designed in terms of such generalities.

How do we distinguish a system from a "component" which is itself a functioning assemblage of elements? Perhaps because of the nested, hierarchical nature of the physical world, almost every object can be considered as a system. This makes it necessary to distinguish "between the study of *a system (complex) object* and *the systems study of such an object*" (Ref. 5, p. 118, Par. **5-34**).

The study of an object contained within a system may proceed (1) by decomposing the object into its elements, weighing, measuring and analyzing them, and reassembling the object as the sum of its parts or (2) it may content itself with an input-output "black-box" analysis, characterizing the system by its response to stimuli or by its transformation of inputs.

In contrast, the systems study of an object is concerned with functioning structure: if it uses analysis by decomposition (reductionism), it carefully retains the functioning internal structure of the object system while paring away detailed characteristics of the interconnected elements of which the object system is composed.

Conversely, systems study of an object system cannot properly be satisfied with a purely input-output analysis, since this provides no insight into the internal functioning structure of the object system which is the proper objective of systems analysis.

2. Systems Theory. To place our treatment of systems engineering in perspective, it is necessary to comment briefly on the wider range of *systems theory*. There has been much discussion of general systems theory[56]† during the last decade, but it is by no means clear that this concept has any real meaning other than collectively. It is important to note that systems can be of profoundly different types, that the problems posed by these different types are themselves as profoundly different, and that how these problems are addressed and the resources brought to bear in their solution (or transformation, or amelioration, or whatever) must be carefully adapted to their nature and characteristics.

---

†Superior numbers correspond to references in Par. **5-34**.

In his presidential lecture at the 1980 meeting of the American Association for the Advancement of Science, Kenneth E. Boulding[7] made the point that

> Within the scientific community there is a great variety of methods, and one of the problems which science still has to face is the development of appropriate methods corresponding to different epistemological[†] fields. Methods that work in one field do not necessarily work in another. There has to be constant critique and evaluation of the methods themselves. Perhaps one of the greatest handicaps to the growth of knowledge in the scientific community has been the uncritical transfer of methods which have been successful in one epistemological field into another where they are not really appropriate.

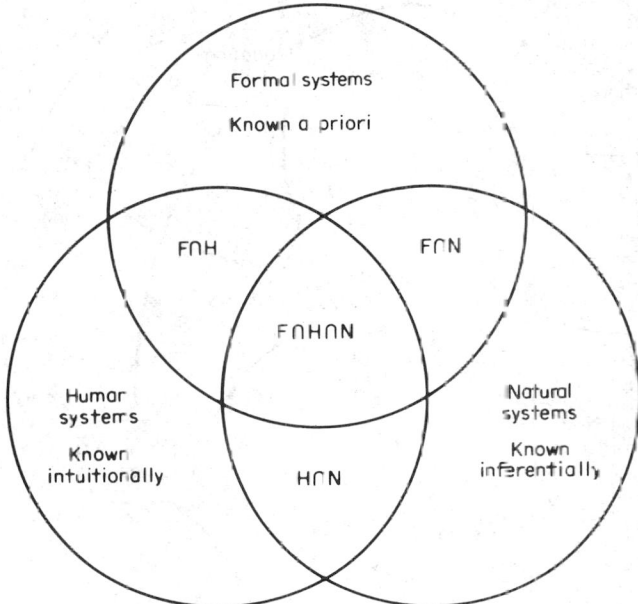

Fig. 5-1. Types of systems.

Boulding's caution is certainly appropriate, even within the relatively narrow field of systems theory.

3. Formal, Natural, and Human Systems. Many discussions of systems engineering refer indifferently to varieties of systems which are fundamentally different in their nature, which are, in fact and in Boulding's terminology, epistemologically different. We suggest there are at least three such types of systems, which we denote respectively as *formal*, *natural*, and *human* systems. A simple trefoil Venn diagram (Fig. 5-1) represents these three types and their pairwise and triple intersections.

By a *formal system* we mean a set of simultaneous mathematical equations. Any such set can be viewed as constituting a system all of whose characteristics are implicit in the equations, and of which our knowledge is complete, a priori. By a *natural system* we mean a natural unconstrained physical process: a growing tree, an ocean, the formation of a snowflake, a planet's atmosphere, and so on. Of such systems, whatever knowledge we have is at best inferential and tentative, attained indirectly by observation of the observable effects of that system's functioning.

Finally, by a *human system*, we mean a social system comprising mutually interacting people. Our knowledge here can best be described as intuitional. We can never fully understand the functioning of such systems without involving ourselves either overtly or at least sympathetically. For example, anthropologists have often found it necessary to participate in the life of the tribes they study, even at the risk of affecting them.

The significance of the intersections (Fig. 5-1) of these three types follows directly. *Formal-natural systems*, for instance, are natural processes constrained by formal design, e.g., chemical processing plants, television sets, internal combustion engines. *Formal-human systems*, for

[†]Relating to the nature and limits of human knowledge.

instance, are groups of people interacting within some formal structure, e.g., the political system of the United States, a college faculty, a Boy Scout troop. The important triple intersection *(formal-natural-human)* involves all three elements, e.g., a manufacturing company with a physical plant designed, operated, and managed by people.

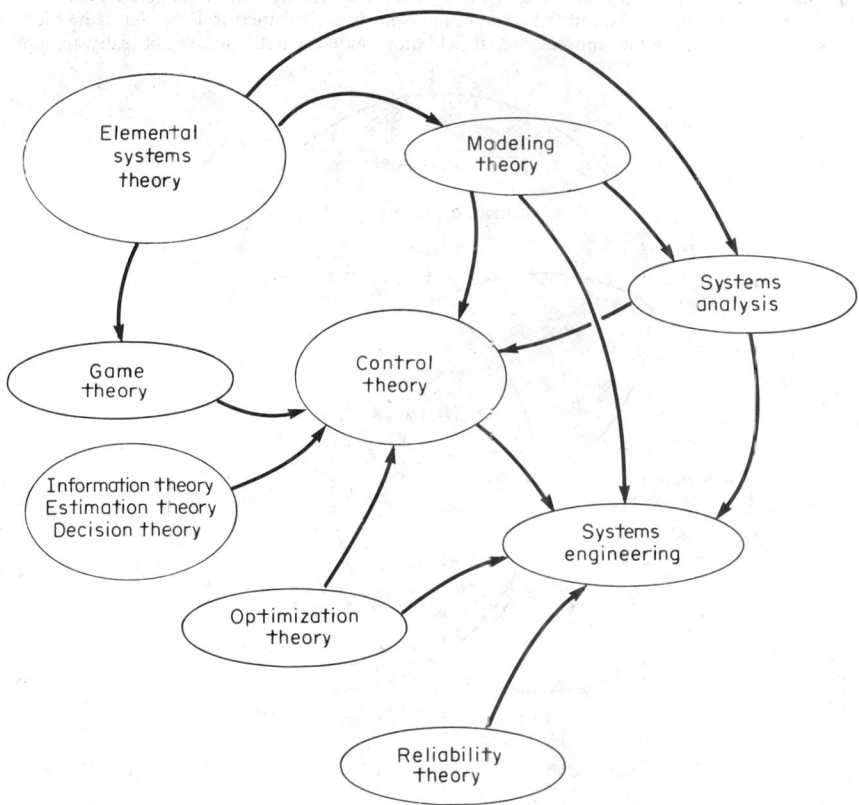

Fig. 5-2. Domain of systems engineering.

It is important to note that our understanding of these three types of systems is basically epistemologically different. Boulding's caution applies: we cannot uncritically transfer methods successful in the study of formal or natural systems to the study of human systems. Unfortunately such caution has not been widely observed, and much confusion and controversy has resulted. Not all of systems theory, even good systems theory, is relevant or useful to the solution of any systems problem, and the systems engineer must be careful not to be misled into bringing a solution technique to bear on a problem for which it is not fitted.[4]

In this section on systems engineering, to keep our discussion brief and to minimize confusion, we constrain our treatment to *formal* or *formal-natural systems,* i.e., physical engineering systems. The discussion given here cannot with impunity be considered equally valid for systems characterized by significant human participation, which would require much more extensive treatment.

4. Essential Elements of Systems Engineering. If we attempt a constructive definition of systems engineering, we can cite several narrower disciplines which are mutually related by their focus on systems-type problems and which collectively provide a coherent foundation for the discipline of systems engineering. Our own candidates for such a set of disciplines would include:

- Elemental systems theory
- Modeling theory
- Linear and nonlinear systems analysis
- Optimization theory

- Control theory
- Theory of games
- Information, estimation, and decision theories
- Reliability theory

Added to this list would be, at a more basic level, the useful branches of mathematics necessary to mastering almost any area of science or engineering, including probability and mathematical statistics, linear algebra, functional analysis, category theory, as well as more hardware-oriented supporting disciplines, including computer science and instrumentation.

Figure 5-2 indicates the mutual relationships we see to exist between these areas with regard to their respective contributions to systems engineering. In our view, control theory plays a central role (1) by virtue of having been the first engineering discipline to focus on "systems study," necessitated by the nature of feedback and (2) as being more mature. Unfortunately, space does not permit even a cursory treatment of all the areas cited above, but this should not be taken as an indication of lesser importance. In the pages which follow, we attempt to characterize five of the areas, from the standpoint of their relevance and contribution to systems engineering as we have defined it, and we conclude with an example.

ELEMENTAL SYSTEMS THEORY

5. Abstract Systems. There have been continuing efforts to provide a set-theoretic base for systems theory. The initial hurdle is formulation of a definition of the concept "system" that conforms to our intuitive understanding, is general enough to span the range of our interests, and yet is specific enough to provide a meaningful base for study. Blauberg et al.[5] survey a wide range of attempts at such a definition and suggest a formulation of their own which seems promising.

They suggest that three facets of our intuitive understanding of systems must be captured: *wholeness*, which reflects the primacy of the system as a whole over its elements; *hierarchy*, which reflects not only the potential for successive decompositions of systems into subsystems but also the role of a given system as a subsystem within a larger framework; and *relativity*, which reflects the inadequacy of any single delineation of a system and the consequent necessity of resorting to classes of descriptions, any one of which can capture only certain aspects of the wholeness and hierarchy of the system.

On the basis of this perception, they suggest the following definition of the system concept: a system S represents a class of sets, $S = \{M_s^i, L_s^j, K_s^k\}$, where $\{M_s^i\}$, $i = \alpha, \beta, \gamma, \ldots$, is a subclass of sets representing decompositions of the initial object, the system, into elements; $\{L_s^j\}$, $j = a, b, c, \ldots$, is a subclass of sets formed as a result of further divisions of the initial object; and $\{K_s^k\}$, $k = 1, 2, 3, \ldots$, is a subclass of sets each of which includes the system object under study as a definite element. It appears to us that this definition could provide a very interesting starting point for a formal theory of systems.

6. Multilevel Systems. One of the most extensive efforts to date toward the development of such a theory has been carried out by a group at Case Western Reserve University, particularly M. D. Mesarović. The latter's own work, developed at length in Mesarović et al.,[42] provides a basically descriptive, deterministic formalism based on a very simple notion: a system S is a relation on abstract sets X and Y, $S \subseteq X \times Y$; if S is a function, $S:X \rightarrow Y$, is called a *functional system*.

The major impact of Mesarović's work has been in the general acceptance of his ad hoc delineation of basic types of hierarchies and his classification of alternative principles for coordination of functional subsystems by a higher-level decision unit.

Mesarović considers systems containing multiple decision units that operate within some framework. Whenever the framework is other than horizontal, i.e., whenever there is some element of primacy-dependency in the interunit relationships, he suggests that in general (1) higher-level units are concerned with larger portions or broader aspects of overall system behavior, (2) the decision periods of higher-level units are longer than those of lower-level units, (3) higher-level units are concerned with slower aspects of overall system behavior, (4) higher-level problems are less structured, exhibit greater uncertainty, and therefore are more difficult to formalize or quantify.

He elaborates three basic notions of level: *strata*, representing levels of description or abstraction; *layers*, representing levels of decision complexity; and *echelons*, representing organizational levels.

This classification is *not* to be understood as a partition. Not only is the possibility of a system's being described by more than one class not excluded, but all three notions may be involved in the description of any one actual hierarchical system, the case where only one notion is applicable being an exception rather than the rule.

It is interesting to note that this classification introduces, on an ad hoc basis, the essential qualities which Blauberg et al. deemed it necessary to capture in their basic definition of the notion of system, wholeness, hierarchy, and particularly relativity, to the latter two of which

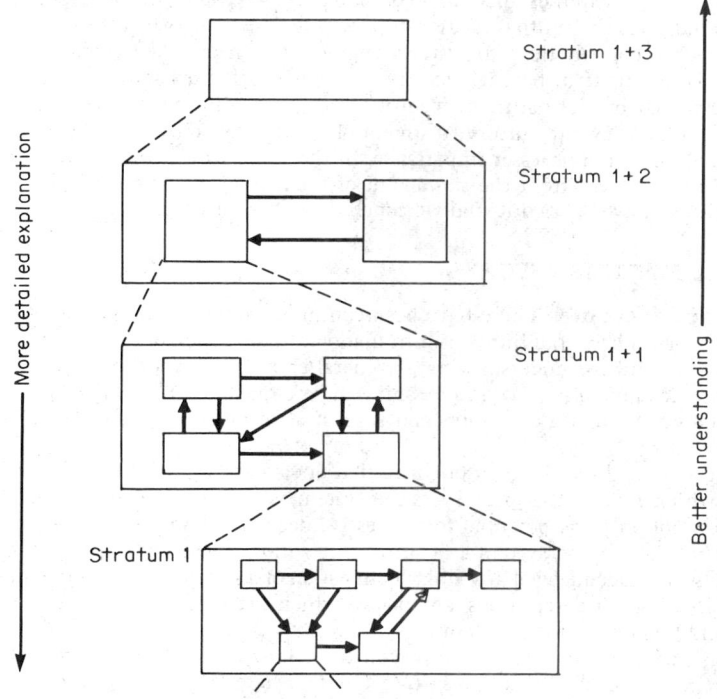

Fig. 5-3. Multistrata system. *(From M. D. Mesarcvić, Proc. IEEE, January 1970, p. 113.)*

Mesarović's basic definition is completely transparent. He states that the three types were conceived heuristically, strata with reference to modeling problems, layers with reference to decision-problem decomposition, and echelons with reference to the interrelationship of decision units within a system.

Figure 5-3, depicting the relationship between strata, suggests how modeling is naturally approached in practical situations; i.e., a subsystem within the system considered at one level becomes a component within a subsystem at a higher level. Moreover, the principles or laws used to characterize the system on any one stratum cannot generally be derived from the principles used on other strata. Instead each stratum has its own set of terms, concepts, and principles. For example, the principles of computation or programming considered by computer science at one level of abstraction are not derivable from the physical laws governing the behavior of a computer, which are considered on a lower stratum, and vice versa.

Figure 5-4 illustrates the concept of layers involving increasing degrees of decision complexity. In this example, the lowest layer involves *selection* of a preferred course of action *m*, that is, control. The selection is on the basis of process data and algorithms or criteria provided by the higher layers. As suggested in the figure, this selection layer might be subdivided into a lower, regulating, layer and an upper, optimizing, layer. The second layer in this example involves *learning* or *adaptation*, which is used to improve knowledge about the system and environment, through updating models, tracking disturbances, etc. The uppermost layer in Fig. 5-4 involves *self-organization*, where decisions are made with regard to the structure of the lower layers, i.e., the strategies to be used in solution of lower-layer problems. Mesarović points out that different methods and techniques are appropriate at the several layers. For example, feedback control and numerical optimization may be used on the selection layer, statistical and logical techniques at the adaptive layer, and heuristic techniques at the self-organizing layer.

Figure 5-5 illustrates the most familiar of the hierarchical structures, which Mesarović desig-

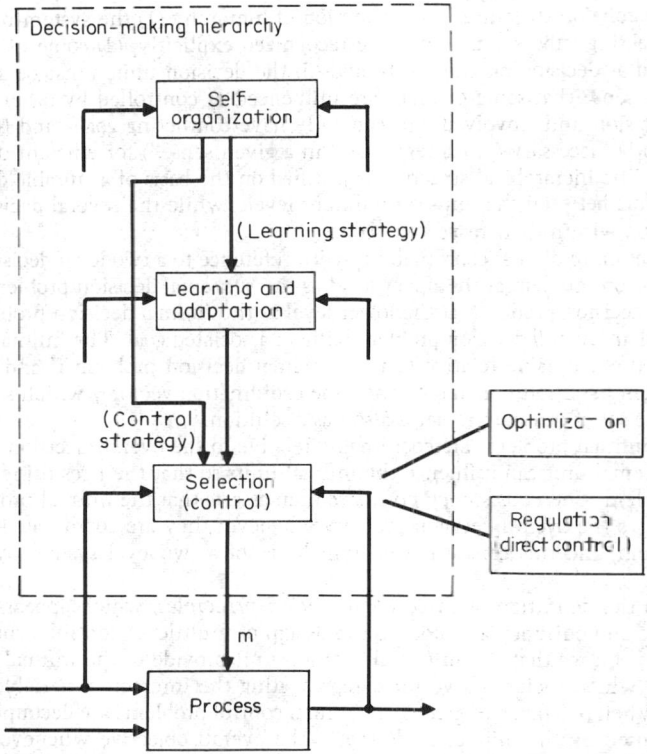

Fig. 5-4. Multilayer system. *(From M. D. Mesarović, Proc. IEEE, January 1970, p. 114.)*

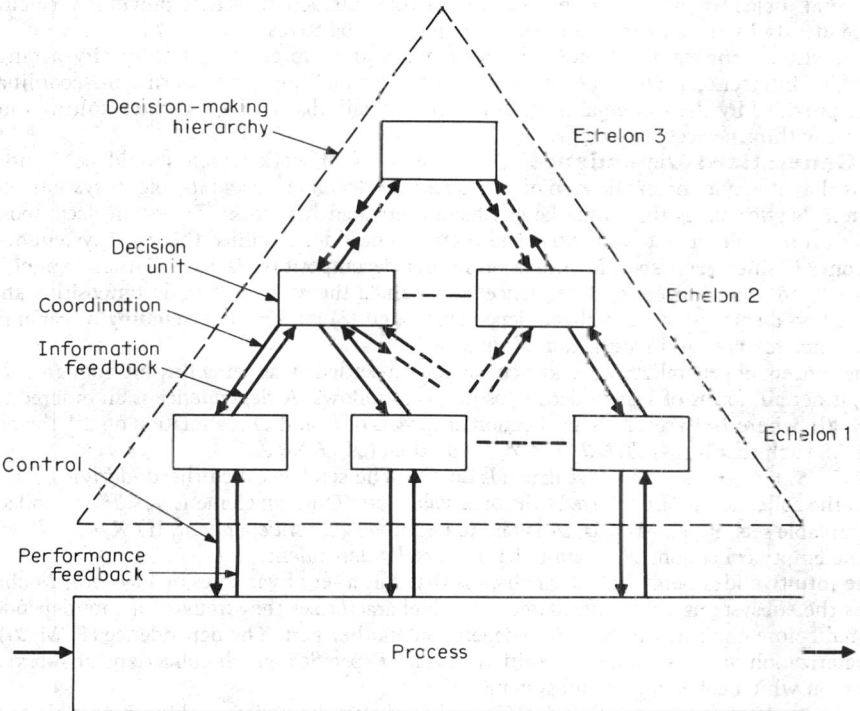

Fig. 5-5. Multiechelon system. *(From M. D. Mesarović, Proc IEEE, January 1970, p. 114.)*

nates as a *multiechelon* structure. For this notion of hierarchy (1) the system must consist of a family of interacting subsystems which are recognized explicitly, (2) some of the subsystems must be defined as decision-making units, and (3) the decision units must be arranged hierarchically, in the sense that some of them are influenced or controlled by other decision units. The several decision units involved will generally have conflicting goals, and Mesarović holds such conflict to be necessary ("to a degree and in a given sense") for efficient operation of the overall system. The hierarchical structure is justified on the basis of a suitable division of decision-making effort between the units on different levels, while the several decision units have freedom of action within their respective scopes.

Mesarović goes on to discuss coordinability with reference to a two-level decision framework. Denote the decision problem at the upper level as the *supremal* decision problem with an associated goal, the decision problems at the lower level as the *infimal* decision problems with associated goals, and an *overall* decision problem with an associated goal. The infimal decision problems are deemed coordinable relative to the supremal decision problem if and only if (1) the supremal problem has a solution and (2) for some coordination vector γ which solves the supremal problem the set of infimal problems also has a solution.

Further, the infimal problems are coordinable relative to the overall decision problem if and only if the supremal unit can influence the infimal units so that their resulting action satisfies the overall problem. The *consistency postulate* then asserts that the infimal problems are coordinated relative to the overall decision problem whenever they are coordinated relative to the supremal problem, and this constitutes coordinability of a two-level system by the supremal decision problem.

This leads to the definition of three *coordination principles*, which appear to have found some acceptance and currency as guides to the design of multilevel control structures. *Interaction prediction* requires that the supremal decision unit provide to the infimal units a coordination vector γ which includes a vector α representing the interaction variables between the infimal units. When α is taken as given, the infimal control problems are decoupled. The resulting controls chosen by the infimal units satisfy the overall objective whenever the resulting interactions are those predicted by the supremal unit.

Interaction decoupling requires each infimal unit to view the interactions as an additional control set and to select any interaction which is desirable for its own local objective. The infimal control decision, based also on the coordination vector provided by the supremal unit, satisfies the overall objective whenever the resulting actual interaction is that previously calculated independently by the supremal unit to satisfy its own objectives.

Interaction estimation, a looser concept, requires the supremal unit to specify a range of acceptable interactions. The controls chosen by the infimal units, considering the coordination vector provided by the supremal unit, satisfy the overall objective when the resulting interactions lie within the acceptable range.

7. Generalized Dependence. One other body of work which should be mentioned here is that of Naylor on explication of the concept of *dependence* and its role in *system decomposition*. Naylor states that three beliefs have motivated his work: (1) system decomposition plays a central role in systems theory, and there is a need for a unified theory of system decomposition; (2) since each specific means of system decomposition is based either explicitly or implicitly on some concept of dependence, any unified theory of system decomposition should be based on a concept of generalized dependence; and (3) a deep understanding of generalized dependence requires an investigation of *duality*.

The concept of generalized dependence Naylor has settled on as being flexible enough to cover most, if not all, forms of system decomposition is as follows. A *dependence* is an ordered triple $(E, M, \mathcal{D})$, where E is a set, M is a collection of subsets of E, and $\mathcal{D}$ is a relation on 2^M, the power set of M, such that if $(\mathcal{A}; \mathcal{B}) \in \mathcal{D}$ and $\mathcal{B} \subseteq \mathcal{B}'$, then $(\mathcal{A}; \mathcal{B}') \in \mathcal{D}$.

If $(\mathcal{A}; \mathcal{B}) \in \mathcal{D}$, we say that "$\mathcal{A}$ depends on $\mathcal{B}$." The set E is called the *underlying set*. The sets in the collection M are referred to as *observable sets*. Thus, an element $\mathcal{A} \in 2^M$ is a collection of observable sets. If $(\mathcal{A}; M) \notin \mathcal{D}$, $\mathcal{A}$ is said to be *universally independent*. If $(\mathcal{A}; \phi) \in \mathcal{D}$, where ϕ is the empty collection, $\mathcal{A}$ is said to be *universally dependent*.

The intuitive idea behind the formalism is that E is a set of variables of a system, M characterizes the subsystems being considered, and $\mathcal{D}$ characterizes the strength of interdependence required before one part can be said to depend on another part. The dependence $(E, M, \mathcal{D})$ is a characterization of a system decomposition in that it specifies which collections of subsystems depend on which collections of subsystems.

For a given system, any or all of E, M, and $\mathcal{D}$ can often be varied, making it possible to vary system decomposition. For example, one may start with a very simple $\mathcal{D}$ to obtain an initial system decomposition and then alter $\mathcal{D}$ to obtain a more detailed decomposition. Indeed, one can imagine various hierarchical taxonomies of system decomposition based on comparison of dependences. For instance, $(E, M, \mathcal{D}_1)$ and $(E, M, \mathcal{D}_2)$, where $\mathcal{D}_1 \subseteq \mathcal{D}_2$. Naylor's work appears to be of fundamental importance and could come to play a key role in the evolution of systems theory.

MODELING THEORY

8. Models. By and large, models constitute the realm of discourse within which systems engineering is carried on. The previous discussion of elemental systems theory was largely a search for a suitable covering model for systems theory in terms of which the subject could be studied and discussed. More pragmatically, a model is a prerequisite to the use of analytical methods in engineering design.

9. Definition and Role of Modeling. By *modeling* we mean any deliberate intelligible cognitive activity aimed at abstracting, and reproducing in some convenient realm of discourse, features of an object or system (the prototype) of interest to the modeler. The activity is deemed cognitive in that it is directed toward achieving an understanding of the inherent nature and characteristics of the prototype, of the results of the future evolution of the prototype, or its autonomous response to possible exogenous stimuli.

Of late, models are becoming generally recognized as indispensible tools for effective understanding of the behavior of complex systems. Yet mathematical modeling is still, at best, an art. There is no comprehensive, consistent body of theory which constitutes a theory of modeling. There is not even any assortment of theoretical notions which collectively could provide an adequate foundation for existing modeling methodologies. Perhaps worst of all, there is little available theoretical guidance for exploration and development of important areas for which as yet, there are not even adequate heuristic modeling procedures.

System and subsystem models are integral elements in the design and subsequent functioning of modern control systems. A controller sees a controlled plant in terms of a model and interprets plant data in relation to that model. The problems involved in clarifying this relationship of mathematical control to physical plant are epistemological, but they cannot be ignored if a sound body of theory is to be developed.

Engineering systems modeling — indeed *systems analysis* — is a blending of physical and mathematical theory. It is a sterile activity if either is left out. In the sciences, models are sought which illuminate natural phenomena. The objective is to strip away all that is not essential so that our observations of reality can be characterized and understood in terms of some ultimate simplicity. In this sense, a model explains those phenomenological patterns of interest in terms of a set of easily understood elements. In this context a model is a *theory* constituting a set of propositions or laws from which facts exhibited in nature can be deduced.

This last notion illustrates what has come to be called the *scientific method*. The scientific method of establishing an understanding of any physical phenomenon is generally identified as consisting of three phases: (1) initial observation, (2) formulation of a theory, and (3) prediction of new observations and experimentation. Moreover, the completion of the last stage frequently suggests refinements to the theory, and the process is repeated. The emphasis on observation has its roots in the empiricist philosophy which has been at the heart of modern science.

These concepts are reflected in the systems engineer's perception of a model. In engineering systems analysis, the model represents the hypotheses about how that part of the world of interest operates, at least in regard to those aspects of concern. The ultimate use of a model in systems analysis is for decision making, and its final evaluation is in terms of the decisions made. However, the model is much more than simply a means to an end. The modeling process is very much at the heart of systems analysis. It is during this process that the issues at hand are clarified and made explicit.

One useful classification of models distinguishes three types: (1) *naïve models* (the past trends of a single variable are used to project future behavior of that variable); (2) *simple correlative models* (past observations are used to correlate several interrelated variables in order to forecast future trends); and (3) *causal models* (the response of certain variables due to changes in others is predicted).

Although simple correlative models may be considerably more complex than naïve models in mathematical structure, both are obtained from routine manipulation of past data and are not based on any underlying theory of behavior. They are meaningful only under the assumption that no changes occur to the object of interest that would alter its behavioral patterns. Causal models, on the other hand, require some understanding of underlying cause-and-effect relationships and consequently are more difficult to obtain. They are extremely valuable in systems engineering, because they permit evaluation of changes to the system itself.

A fundamental distinction can be drawn between the first two model types and causal models. Naïve and simple correlative models are *descriptive*, whereas causal models are *explanatory*. A descriptive model answers such questions as "How?" or "How much?" An explanatory model treats the question "Why?"

There is, of course, a wide gradation of causal models. On one extreme, the only phenomenological knowledge about the system may be the causal classification of inputs and outputs. On the other extreme, a priori knowledge constitutes a complete theory concerning the operation of the system providing detailed causal relationships between all relevant variables. In the former case, the model is completed by postulating the form of the mathematical statements relating inputs and outputs and determining unspecified coefficients in order to achieve a best characterization of known input-output pairs. Causal models constructed in this way are known as *black-box*, *input-output*, or *inductive models*. Modeling at the other extreme has been referred to by various names, e.g., *mathematicophysical*, *physicochemical*, *first principle*, *theoretical*, and *deductive*.

Models are frequently developed which are a mixture of these two extremes. For example, physical theory is often used to establish the mathematical structure of a model, after which input-output techniques are used to establish numerical values for model parameters. In other situations theoretical analysis is employed to partition the system of interest into an interconnection of subsystems, some of which are subsequently modeled using input-output techniques.

10. Theoretic Base. There have been a few attempts at developing a rigorous base for modeling in the form of an axiomatic system. Ziegler[63] has developed a modeling framework within which deterministic systems can be discussed and has paid special attention to the question of relating system structure and behavior. Corynen[12] essays a more general framework than Ziegler's, encompassing the modeling of stochastic objects and statistical issues associated with simulation. His definition of modeling involves three essential ingredients: *imitation*, the pertinent features which a model has in common with its prototype; *approximation*, the degree to which those features are present in the model; *interpretation*, a mapping relation from the model performance set into the prototype performance set.

These performance sets include all possible values of the features of interest. Approximation is discussed in terms of topologies on these sets. Model is defined formally in terms of a requirement that, given corresponding objects in the prototype and in a candidate model, a neighborhood of the image (of the prototype object) in the performance set of the prototype contains the mapping, under the interpretation relation, of the image (of the model object) in the performance set of the model.

Corynen's modeling paradigm is illustrated in Fig. 5-6. The sets S and S' are the *prototype objects* and *model candidates*, respectively. The maps ϕ and ϕ' are *feature extraction maps* which associate, with each $s \in S$ and $s' \in S'$, a feature or performance $v = \phi(s) \in V$ and $v' = \phi'(s') \in V'$, respectively. These performance or feature-extraction maps make it possible to express the aspects or features of the prototypes and models which are important to the modeler.

Associated with these objects, maps, and feature spaces are various types of errors, uncertainties, ambiguities, tolerances, and approximations. To capture these aspects Corynen introduces *approximation structures*. Finally, he defines an *interpretation relation* $H \subset V' \times V$. If we are dealing with an activity where models are used as explanatory or predictive devices, H is used to make inferences in V about the prototypes from model features or outputs in V'. This is called the *model-use activity*.

If we are dealing with the process of modeling itself, where simplified or otherwise transformed objects and features are obtained from the prototypes and their features, H is used to associate features in V' with the original features in V. This is called the *model-development activity*. In both cases, the satisfaction of a modeling relation is determined by what modelers want from their models compared with what they actually get from them. The formal definition of model is slightly different for each case. Formally, in a model-development activity, the *prototype approximation structure* (PAS) is a mapping

$$\mathcal{A}:S \times V \times I \to 2^V$$

where S is the set of prototype objects, V is the set of prototype performances or features, I is a partially ordered set $\langle |I|, \alpha_I \rangle$ with elements $|I|$ and partial order α_I on $|I|$, and 2^V is the power set of V', with V' the model candidate performances or features.

This mapping can be interpreted as follows. If $v \in V$ is the feature of $s \in S$, $\mathcal{A}(s, v, i)$ is the set of model features which are *approximations* of v at level $i \in I$. The basic idea is that i reflects the modeler's standard of "closeness" between the model and the original object, and if the model candidate meets or exceeds this standard, the candidate qualifies as a model at tolerance level i.

A similar structure is associated with the model candidates for a model-development activity. The *model approximation structure* (MAS) is a mapping

$$\mathcal{A}' : S' \times V' \times I' \to 2^V$$

where S' is the set of model candidates, V' is the set of model features, and I' is the partially ordered set $\langle |I'|, \alpha_{r} \rangle$. The usual interpretation of $\mathcal{A}$ and $\mathcal{A}'$ will be that $\mathcal{A}(s, v, i)$ are the approximates tolerated at level i and $\mathcal{A}'(s', v', i')$ are the approximates obtained at level i'.

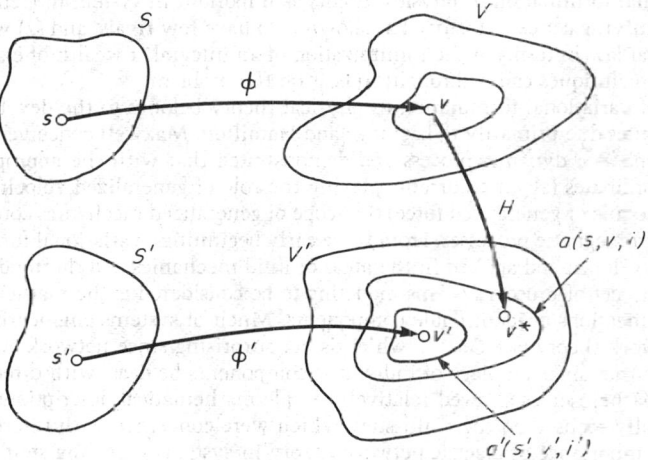

Fig. 5-6. Important elements associated with the concept "model."

In many cases the approximation of the performances is not dependent upon the systems with which they are associated, and the arguments s and s' in $\mathcal{A}$ and $\mathcal{A}'$ can be dropped. Corynen has shown that this approximation framework captures topological, statistical, fuzzy, and combined notions of closeness and confidence in a natural way.

Whether some $s' \in S'$ is a model of $s \in S$ is not only a matter of degree but also depends on the modeler's point of view. This aspect is captured by introducing a *modeling criterion* $\mathcal{C} = \langle \phi, \phi', \mathcal{A}, \mathcal{A}' \rangle$, where all symbols have been defined above. The modeler is then represented by his *point of view* $\mathcal{M} = \langle \mathcal{C}, H \rangle$, where H is his interpretation. We call $\mathcal{M}$ itself the *modeler*. This leads to Corynen's definition: Given a modeler $\mathcal{M} = \langle \mathcal{C}, H \rangle$, tolerance and obtainment levels i and i', respectively (Fig. 5-6), then $s' \in S'$ is a *model* of $s \in S$ at tolerance level i with strength i' relative to $\mathcal{M}$ if and only if

$$\mathcal{A}'(s', \Phi'(s'), i') \subset \mathcal{A}(s, \Phi(s), i)$$

Important issues discussed and clarified within this framework include conditions under which transitivity of the modeling relationship can be asserted and conditions under which partial (submodel) validation can provide validation of the full model.

11. Modeling of Dynamic Physical Processes. Notwithstanding the hopes and claims of the advocates of a general systems theory,[56] it seems clear that in large part specific concepts and techniques used in the practice of modeling depend for their validity on the nature of the system being modeled. Dangers which follow upon losing sight of this truism are illustrated by Berlinski's eloquent jeremiad.[4] Accordingly, to discuss the modeling process at a less abstract level, let us focus on dynamic physical processes. Two aspects are important. The first concerns the development of equations describing the elementary components, and the second has to do with the interconnection of the component models so that they form a meaningful whole, representative in all relevant respects of the prototype process.

The foundations of process systems modeling are the laws of conservation of mass, energy, and momentum. With appropriate interpretation, all the conservation equations take the form:

Accumulation of conserved quantity X in system
= net influx of X across system boundaries + net generation of X within system

Add to these an infinite variety of arrangements, constitutive relations and boundary conditions, and there arises the diversity of processes that systems engineers are concerned with.

In beginning an analysis, model builders must decide at what level of internal physical detail to direct their efforts. In this regard, conservation laws are classified as *microscopic*, i.e., the system of interest is considered to be a continuum, and differential balance equations are formulated using differential constitutive relations, or *macroscopic*, i.e., spatial variations of system parameters are ignored by (usually) integrating the balance equations over a spatial volume and employing algebraic (no spatial gradients) constitutive relations. A microscopic formulation results in a description in terms of partial differential equations with time and position the independent variables, whereas a macroscopic formulation results in ordinary differential equations with time the independent variable and internal detail sacrificed for simplicity.

The variational formulation of physical theory is important in system modeling for two reasons: (1) as a unifying concept, history has shown it to have few rivals, and (2) with the restatement of physical law in terms of the minimization of an integral, a wealth of exact and approximate solution techniques can be brought to bear on the problem.

The origin of variational formulation in physical theory belongs to the development of generalized mechanics due primarily to Lagrange and Hamilton. Maxwell conceived of electromagnetic phenomena as a dynamic process and demonstrated that with the appropriate choice of generalized coordinates (electric currents playing the role of generalized velocities and electromotive force the role of generalized force) the scope of generalized mechanics could be expanded to include electromagnetic processes. From these early beginnings, variational formulations have been further developed and applied in the areas of fluid mechanics and thermodynamics.

The second aspect of process systems modeling to be considered is the mathematical description of interconnections of identifiable components. Much of systems engineering has its roots in electric-network theory (see Sec. **3**), which is not surprising since network analysis required from the outset that an assemblage of individual components be dealt with directly. Moreover, the components themselves allowed relatively simple mathematical descriptions, so that attention was naturally focused on the real issues, which were concerned with interconnection and topology. The importance of electric-network theory in systems modeling stems directly from the fact that conservation laws, fundamental to all process modeling, are easily related to the structure of the system in the electric-network context. The prime example of this relationship is Kirchhoff's formulation of his topological theorems for electric networks.

Kirchhoff's laws were discovered some 100 years after d'Alembert had restated Newton's second law to show that the sum of all forces acting on a body is zero. It was Maxwell, about 30 years later, who first stated the analogy of force to voltage. The force-current analogy was not formulated until the 1930s.

Firestone[21] distinguished between two types of physical variables, *across variables*, e.g., velocity, voltage, pressure, and temperature, which are measured across two spatially different points, and *through variables*, e.g., force, current, fluid flow, or heat flow, which are measurable at a single point. The product of each pair has the units of power.

This point of view led to identifying ideal elements of systems each of which interacts with other elements via a limited number of *energy ports*, a port being a point at which power can be transmitted. Each energy port transmits a single type of power, and the power transmission is instantaneously described by a pair of signal variables:

$$e_f \downarrow i_f$$

Multiport: $\xleftarrow{\tau}{\omega}$ dc motor $\xleftarrow{e_a}{i_a}$

Elements are coupled by interconnections of their energy ports, the interconnection being called a *bond*:

$$e_f \downarrow i_f$$

Bond graph: $\xleftarrow{P}{Q}$ pump $\xleftarrow{\tau}{\omega}$ dc motor $\xleftarrow{e_a}{i_a}$ battery

Thus a *bond graph* is a collection of multiport elements bonded together. In the general sense it is a linear graph whose nodes are multiport elements and whose branches are bonds. Associated with a given port are three direct and three integral quantities. *Effort* and *flow* are assumed to be scalar functions of time. *Power* is the scalar product of effort and flow, the direction of positive power being indicated by the arrow. *Momentum* and *displacement* are integrals of effort and flow, respectively, and *energy* is the integral of power. From a sufficiently detailed bond graph, state equations can be written using standard techniques. Lucid expositions are given, for example, in Karnopp and Rosenberg[31] and in Takahashi et al.[55]

From the outset bond-graph theory embraced the domain of applicability of linear graphs, and it has rapidly expanded into nontraditional areas. Some of the important contributions are nonlinear mechanics, fluid mechanics, thermodynamics, chemical networks, and electric fields.

12. Parsimony and Validation. Certain qualitative issues in modeling are as important as they are difficult. Two in particular must be faced by any responsible systems engineer, *parsimony* and *validation*. We mentioned at the outset that in the sciences the object is to characterize and understand our observations of reality in terms of an ultimate simplicity. This is important also to effective modeling. It is necessary that only essential phenomena be characterized and that such characterization be accomplished with utmost efficiency. A more detailed model is not necessarily a better model, even in the absence of error; it is often a worse one.

Every model, by definition, is a reduced-order, aggregated, or abstracted representation of its prototype. Otherwise we would be dealing, not with a model, but with a duplicate of the system. Accordingly a successful model is one which, among other things, abstracts from the prototype only those factors which are fully relevant (necessary and sufficient) to the inquiry in view. Every unnecessary feature introduces a likely source of error in the structure of the model, the data with which it is represented, and the interpretation of any simulation or analysis. Recognition of the overriding need for parsimony, however, provides no guidance to how it can be achieved or even recognized. Nevertheless, the development of efficient, useful, understandable, credible models of complex, large-scale systems is virtually impossible in the absence of an effective grasp of how to achieve parsimony in modeling. Parsimony, then, is a matter of basic engineering wisdom in analyzing the system to be modeled.

A related, quantitative concept, which cannot substitute for engineering wisdom but which can be a valuable adjunct, is that of model reduction, on which there is a considerable literature. Perhaps one of the oldest ideas to be exploited for the purpose of constructing simplified dynamic models, and still one of the most valuable, is the notion that if attention need be paid only to a limited speed range or dynamic behavior, it is often possible to achieve a significant reduction in the order of the system of differential equations required to characterize the process. This is so because it is frequently possible to eliminate consideration of the dynamics of those elements which respond much faster or slower than the phenomena of interest.

This idea is quantified by the concept of singular perturbation.[33] Consider a simple situation in which a dynamic system has two groups of time constants, one relatively faster than the other. In addition, suppose that there is a small parameter μ such that as μ shrinks to zero, the fast time constants become infinitely fast; i.e., the time constants shrink to zero.

This can be illustrated by a simple linear system

$$\dot{x} = Ax + Bz \qquad \mu\dot{z} = Cx + Dz \tag{5-1}$$

In the limit, as $\mu \to 0$, this reduces to

$$\dot{x} = Ax + Bz \qquad 0 = C\bar{x} + D\bar{z}$$

yielding

$$\bar{z} = -D^{-1}C\bar{x} \tag{5-2}$$

If this is substituted into Eq. (5-1), we have

$$\dot{\bar{x}} = (A - BD^{-1}C)\bar{x}$$

which is a smaller model representing only the "slow" modes of the original model. Actually, however, the substitution of Eq. (5-2) into (5-1) involves some sleight of hand, since x and z differ from $\bar{x}$ and $\bar{z}$ by their fast parts. Kokotović[34] and his colleagues have developed convenient techniques for dealing with this discrepancy. Letting η represent the fast part of z, that is, $\eta = z - \bar{z}$, we have

$$\eta = z + D^{-1}Cx$$

Repeated substitutions of this sort into Eq. (5-1) will, in the limit, decouple the fast and slow systems

$$\dot{\xi} = \bar{A}\xi + 0 \qquad \eta = 0 + \bar{D}\eta$$

which can be studied separately and from which the original variables can be recovered.

The techniques of singular perturbations and the similar techniques of weak coupling (together referred to as *asymptotic techniques*) are proving of considerable value in the analysis and design of controls for large-scale and complex systems.

Compared with the ontological issue of parsimony, an even thornier issue is the epistemological problem of model *validation*. Without a clear understanding of the relationship between a model and its prototype, it is not clear how necessary and sufficient conditions for validation can be established (or even how "validation" can be defined unambiguously). The problem of how a model, e.g., a mathematical system, relates to its prototype, e.g., a physical or a social system, is rarely addressed. In consequence, discussions of validation are diverse and inconclusive.

A model is one of many possible hypotheses about the behavior of a system, and its validity rests jointly on its correlation with observations and the strength of its theoretical base. Moreover, as with all physical theory, a model can at most be discredited; it cannot be demonstrated to be uniquely and absolutely true. Model validity, then, can only mean that sufficient evidence has been put forth to judge the model adequate for the intended purposes.

Three considerations have been identified for the validation of models. The first is input-output comparison, a topic for which a considerable literature is available. Special consideration must be given to those cases where system observations are nonexperimental; i.e., the system can be observed only during normal operation. The second is the soundness of the overall theoretical structure, and the third is sensitivity analysis. If system behavior is unreasonably sensitive to small changes in specific parameters, this could be an indication that model structure or assumptions are not properly formulated. At any rate, unvalidatable models are at best dangerous constructs, and if any good is to come of our attempts to model complex systems, we need a procedure based on a clear understanding of how we come by our knowledge of such systems and how that knowledge can be verified.

In summary, we emphasize that the theory of modeling is still embryonic and that while the practice of modeling physical processes has achieved a considerable level of maturity, there are still unresolved issues of basic importance which limit wider usefulness. Nevertheless, modeling is an indispensible basis for systems engineering, without which that discipline cannot exist.

LINEAR SYSTEMS

13. Linear systems are a large and important area of control and systems engineering which we can only sketch here. Our survey touches on *classical frequency-domain methods* and their relationship to "modern" state-space methods for the analysis of linear dynamic systems. We treat the *state-space realization problem*, some of the *structure theory* of linear systems, and *multivariable frequency-response methods*. The references contain more detailed treatments of these topics and related ones. We do not discuss linear optimization techniques or linear programming. Linear stochastic systems are discussed briefly in Par. **5-24**.

14. Classical Frequency-Domain Methods. The appearance of Nyquist's classic paper[46] in 1932 began a period of about 30 years during which the analysis and synthesis of linear systems was dominated by frequency-response methods. Nyquist's result showed the great power of complex-variable methods for the treatment of feedback systems. The elegantly simple representation of linear time-invariant systems introduced by Bode was so compelling that it, together with the Nyquist and Evans representations, formed a conceptual framework for circuit and control system engineers educated during that period. To a certain extent this phenomenon has persisted in undergraduate engineering education and in industrial practice.

Linear systems in continuous time t can be represented by the convolution integral

$$y(t) = \int_{-\infty}^{\infty} g(t, r)u(r) \, dr \tag{5-3}$$

where u is the input, y is the output, and g is the impulse response (or weighting pattern or Green's function).

In the classical theory of Nyquist, Bode, and others a single-input, single-output linear time-invariant system is represented by its transfer function $G(s)$, the Fourier-Laplace transform of $g(t)$. Only the case when $G(s)$ is a rational function $G(s) = q(s)/p(s)$, p and q polynomials, is treated. One can assume that $p(s)$ is a monic polynomial, so that $p(s) = s^n + a_{n-1}s^{n-1} + \cdots + a_0$; and we shall consider only $q(s) = b_m s^m + b_{m-1}s^{m-1} + \cdots + b_0$, with $m \le n$ (the "proper" case), and assume that p and q have no common factors. The roots of $q(s) = 0$ are the *zeros* of $G(s)$, and those of $p(s) = 0$ are the *poles*.

The time response of Eq. (5-3) is essentially determined by the poles of $G(s)$; that is, if p_i, i

$= 1, 2, \ldots, d$ are the distinct roots of $p(s) = 0$, and if m_i, $i = 1, 2, \ldots d$, are their multiplicities, then the inverse Laplace transform of $G(s)$ is

$$g(s) = \sum_{i=1}^{d} \sum_{j=0}^{m_i} c_{ij} t^j e^{p_i t}$$

The c_{ij} are determined by the method of residues. If Re (real part) $p_i < 0$ for each i, then $G(s)$ is stable. If, in addition, Re $q_i < 0$, $G(s)$ is minimum phase.

Now consider a negative-feedback system

$$y = G\varepsilon \qquad x = Fy \qquad \varepsilon = u - x = u - FG\varepsilon \tag{5-4}$$

with u the input, y the output, x the feedback signal, ε the error, G the plant, and F the feedback compensator. The classic Nyquist criterion deals with the case when $G = G(s)$ is a rational transfer function and $Fy = ky$ is a constant feedback gain. Suppose that $G(s)$ has n, poles in the right half of the complex plane; then the closed-loop system of Eq. (5-4) is stable if and only if the Nyquist locus of $G(j\omega)$, $-\infty < \omega < \infty$, encircles the critical point $(1/k + j0)$ n, times in the counterclockwise direction.

This result can be proved by the principle of the argument or Cauchy's integral theorem from the theory of complex variables. This theorem shows that the (normalized) integral of a function of a complex variable about a closed contour in the complex plane is equal to the net number of singularities (poles minus zeros) of the function enclosed by the contour. For rational functions one can understand the result best by expanding the function in powers of $L'(s - p_i)$, where p_i is a pole, and integrating the terms along circular contours.

The integer which results following a division by 2π is the number of times the phase angle of the term passes through 2π as the contour is traced in a specific (counterclockwise) direction. Nyquist's theorem is an application of this procedure to contour integrals of a transfer function when the contour encloses the right-half complex plane, Re $s = \geq 0$. The rational case is easy. The irrational case is a deep result, established first by Desoer[17] in 1965.

For systems which are open-loop-stable, one can define the gain margin as the smallest number g_m for which $g_m G(j\omega)$ passes through the critical point; the phase margin is defined similarly. One can use the Nyquist plot to design systems $FG(s)$, $F(s)$ the compensator, which are stable with (approximately) given gain and phase margins. This procedure is typical of classical frequency-response design methods, i.e., trial-and-error procedures based on geometric representations of the system which partially modify the system function to realize some performance standard. The latter may be context-dependent and heuristic or based on simple parameters (effective damping and natural frequency) defined in terms of low- (second) order systems. A very complete treatment of these methodologies can be found in D'Azzo and Houpis[16] and Newton et al.[45]

The process of modifying the system compensation to meet performance standards is a kind of scenario analysis. In contrast, "modern" control systems design based on state-space models relies on specification of a somewhat artificial performance index and an almost mechanical optimization algorithm to select the system structure which maximizes this index.

15. State-Space Realization Problem. "Modern" linear-system theory is based on the model

$$\begin{aligned}
\dot{x}(t) &= A(t)x(t) + B(t)u(t) \\
y(t) &= C(t)x(t) + D(t)u(t) \\
x(t_0) &= x_0 \qquad t_0 \leq t \leq T
\end{aligned} \tag{5-5}$$

in which $u(t)$ is the vector of inputs or controls, $y(t)$ is the vector of outputs, A, B, C, D are time-varying matrices which parameterize the system, and $x(t)$ is the vector of state variables for the system. If we define the transition matrix $X(t, t_0)$ as the fundamental solution of

$$\dot{X}(t, t_0) = A(t)X(t, t_0) \qquad X(t_0, t_0) = I$$

then Eq. (5-5) is solved by

$$y(t) = D(t)u(t) + C(t)X(t, t_0)x_0 + \int_{t_0}^{t} C(t)X(t, \tau)B(\tau)u(\tau) \, d\tau \tag{5-6}$$

We call $D(t)u(t)$ the feedthrough component and $y_0(t) = C(t)X(t, t_0)x_0$ the free response, and the integral is the input-output response. Since the latter is of the form (5-3), we call $C(t)X(t, \tau)B(\tau)$ the *weighting pattern*.

A key observation regarding this model is that the nth-order differential equation

$$y^{(n)}(t) + a_{n-1}y^{(n-1)}(t) + \cdots + a_0y(t) = bu(t) \tag{5-7}$$

can be written in the form of Eq. (5-5) by defining the matrices

$$A = \begin{bmatrix} 0 & 1 & 0 & \cdots & 0 \\ 0 & 0 & 1 & \cdots & 0 \\ \cdots & \cdots & \cdots & \cdots & 1 \\ -a_0 & -a_1 & & \cdots & -a_{n-1} \end{bmatrix} \qquad E = \begin{bmatrix} 0 \\ 0 \\ \vdots \\ b \end{bmatrix} \qquad C = [1, \quad 0, \quad \cdots \quad 0]$$

which are $n \times n$, $n \times 1$, and $1 \times n$, respectively. The power of representing general dynamical systems by sets of first-order differential equations was realized by H. Poincaré, who introduced the idea of considering the evolution of dynamical system variables in terms of the trajectory of a point in an n-dimensional space. His famous treatise on celestial mechanics (1892) together with A. M. Lyapunov's study of stability (1907) are built on this idea, of which MacFarlane[39] provides an excellent historical survey. Evidently the state-space approach which dominates "modern" control theory is much older than classical linear-system theory!

The extra set of variables $x(t)$ in Eq. (5-5), which are only implicit in the input-output formula, have the effect of accounting for the past history of the system. In Eq. (5-5) or (5-7) the vector $x(t)$ has a finite number (n) of components; hence, the system has a finite-dimensional "memory." This property of the state model of linear systems was decisive in the use of digital computers with limited storage as controllers. On the other hand, the emergence of large, reliable digital computers as aids to engineering design in the 1950s made possible the consideration of complex multivariable nonstationary models of physical processes. The aerospace problems which dominated the engineering work of the 1960s were especially amenable to these kinds of models. The profound change of methods for designing linear systems which occurred between 1955 and 1965 was at once fortunate and inevitable.

Returning to the systems of Eqs. (5-3) to (5-5), one can ask a more basic question: Given a linear system in input-output form, when can one construct a state-space realization given by Eq. (5-5) of the system? This problem was solved, in the finite-dimensional case, by E. G. Gilbert and R. E. Kalman in the early 1960s. Referring to Eq. (5-6), one sees that the fundamental matrix X satisfies

$$X(t, r) = X(t, q)X(q, r)$$

for all t, q, r; indeed this semigroup property is synonymous with the state concept.

Using this, one sees immediately that a matrix-valued weighting pattern $G(t, r)$ has a finite-dimensional state-space realization (A, B, C) if and only if $G(t, r) = H(t)F(r)$ for some matrices H, F. Then $A = 0$, $B = F$, $C = H$ does the job. Notice that one can add additional (zero) columns to H and rows to F and produce a realization of arbitrarily large dimension. Clearly, the issue of minimality is the interesting one.

Consider the stationary case, A, B, C constant, in Eq. (5-5) (and $D = 0$). Then one can solve for y as a function of u in terms of Laplace transforms. Let $X(s)$ be the Laplace transform of $x(t)$, $U(s)$ that of $u(t)$, etc.; then

$$X(s) = (Is - A)^{-1}BU(s) + (Is - A)^{-1}x(0)$$
$$Y(s) = C(Is - A)^{-1}BU(s) + Y_0(s)$$

provided that s is not an eigenvalue of A. Evidently, $G(s) = C(Is - A)^{-1}B$ is the transfer function of the system in state-space form. From this one can see that a matrix transfer function $G(s)$ has a realization, (A, B, C) constant, only if all its elements are proper rational functions. With some effort one can show that this is also a sufficient condition.[8]

16. Structure Theory of Linear Systems. It is clear that one can produce such realizations of arbitrarily high degree, so that the realization is by no means unique. However, given a realization (A, B, C), if one makes the change of coordinates $x \rightarrow Px$ (P nonsingular) in the state space, one finds that $(A, B, C) \rightarrow (PAP^{-1}, PB, CP^{-1})$ and that

$$G(s) = C(Is - A)^{-1}B = CP^{-1}(Is - PAP^{-1})^{-1}PB$$

Therefore, the input-output behavior of a linear system is invariant under a change of coordinates in the state space. This is another form of nonuniqueness of realizations.

A natural question is: Does there exist a realization of minimal dimension (of the state)? For the time-invariant rational $G(s)$ this question is readily answered in the affirmative, modulo the selection of coordinates in the state space. The degree of a minimal realization is called the

McMillan degree, and it can be computed from $G(s)$ by a simple test.[8] In summary, a matrix of proper rational functions as the model of a time-invariant multivariable linear system has an infinite number of state-space realizations; all realizatons of minimal degree are related by a change of coordinates in the state space.

Realizations have a number of other structural properties, including *controllability* and *observability*. In essence, controllability is a measure of the coupling between the input variables and the state, and observability is a measure of the coupling between states and outputs. More precisely, a realization (A, B, C) is controllable if given any two state values x_0 x_1 there exists a time $t_1 \geq t_0$ and a control $u(t)$, $t_0 \leq t \leq t_1$, such that $x(t_0) = x_0$ is steered to $x(t_1) = x_1$.

The *constant* pair (A, B) is controllable if and only if the matrix $[B \quad AB \quad \cdots \quad A^{n-1}B]$ ($n \times nm$) has full rank n. Similarly, a realization (A, B, C) is observable if there is a $t_1 \geq t_0$ such that the initial state $x(t_0)$ can be uniquely determined from the output $y(t)$ [and the input $u(t)$] over $t_0 \leq t \leq t_1$. The *constant* pair (A, C) is observable if and only if the pair (A^T, C^T) is controllable.

These concepts, while of little interest in themselves, are decisive as structural features of realizations. Unnecessary states in nonminimal realizations must be either uncontrollable, unobservable, or both. Alternatively, minimal realizations must be simultaneously controllable and observable. This is a basic classification theorem for linear dynamic systems

Further information and more advanced topics in linear-system theory can be found in Kwakernaak and Sivan[35] and Wonham.[62]

17. Multivariable Frequency-Response Methods. Over the past decade there has been a great resurgence of interest in frequency-response methods, especially for multi-input, multioutput systems. According to MacFarlane,[39] this renaissance was brought on in part by difficulties in applying state-space models to industrial processes. Intellectual curiosity possibly played as large a role. In either case, key initial contributions were made by Kalman in 1964 and by Rosenbrock in 1966. Since that time the literature has grown explosively. The book by Rosenbrock[51] is a major contribution.

The basic idea in designing a feedback compensator for a multivariable system is to reduce the problem as far as possible to designing a series of compensators for single-loop systems. If one could introduce a precompensator in cascade with the given plant so that the resulting system were diagonal, i.e., consisted of noninteracting blocks, the ideal would be at hand.

Rosenbrock's basic contribution was to relax this requirement to permit weak interaction by allowing the transfer matrix to be diagonally dominant. A square matrix is diagonally dominant if the magnitude of the diagonal elements exceeds the sum of the magnitudes of the elements in the same row. (One can also use the column sums.) Using this and Gershgorin's theorem for locating the eigenvalues of a matrix, Rosenbrock was able to develop a method, called the *inverse Nyquist array*, which approximates the decoupled design.

Briefly, this method permits one to select compensators for a series of single-loop control systems subject to certain constraints expressed as geometric bounds on polar plots or scalar transmittances in a modified version of the original diagonally dominant system. The compensators are selected to preserve the diagonal-dominance property of the resulting cascaded system. The single-loop systems are to be designed by classical trial-and-error methods.

Rival methods due to MacFarlane, Mayne, and others are described by MacFarlane.[38] The situation does not seem to be entirely stable; e.g., there are a number of competing definitions for the zeros of a multivariable transfer function.[38] In addition, there is a very complete and intuitively appealing geometric theory for the design of linear systems due to Wonham and coworkers.[62]

CONTROL THEORY

18. Introduction. In Par. **5-1** we asserted that systems engineering deals with the understanding of systems as such, for which an understanding of the components is necessary but not sufficient. From this point of view, perhaps the epitome of systems engineering, in the sense of exemplar or archetype, is *control theory*, and the essence of control theory is found in the concept of *feedback*.

The basic concern of control is to provide, wholly or in part, automatic regulation or guidance of dynamic systems (or devices) under steady-state and transient conditions. Almost invariably satisfactory achievement of such results involves the use of feedback, and control did not begin to emerge as a discipline until the implications of the use of feedback began to be understood.

Automatic control can be, and sometimes is, effected without the use of feedback, e.g., clock-controlled timing of traffic lights to control the flow of traffic, and very often control regimes incorporate elements of feedforward as well as feedback, but such uses are little more than glosses on the vast body of control theory and practice. In general, the use of feedback makes possible (1) the establishment and maintenance of stability in system operation, (2) the decrease of operating sensitivity to design limitations, to plant parameter variations and plant nonlinearities, and to noise, and (3) the adaptation of system performance to unknown or time-varying plant characteristics.[13]

Classical control theory, which still dominates most control engineering, developed from analysis of the control problem reduced to its simplest dimensions: the analysis of linear, deterministic, stationary, single control loops operating in steady state. Continuing subsequent advances in control theory have resulted in successive removal of most of these constraints.[47] Classical techniques were extended very successfully to satisfy transient as well as steady-state requirements. Modern control from the early 1950s, via reversion to time-domain analysis, effected multiloop and time-varying system design; stochastic control has enabled us to face up to pervasive uncertainty; and current research deals (with increasing success) with problems involving decentralized and multilevel systems.

The most intractable problem is that of nonlinearity.[26] Despite continuing advances in nonlinear systems analysis, the most powerful and most useful tools of control theory still are generally those of linear theory. Undoubtedly this in large part results from, and is a testimonial to, the contributions of feedback techniques, not only in providing tight regulation of controlled system variables around operating points (thus confining operation within regions where local linearization closely approximates actual conditions) but also in often providing some degree of linearization of plant characteristics.

Control design involves some formal statement of performance criteria, a formal model of the plant (or system) to be controlled, provision for measurement or estimation of relevant variables, appropriate analysis techniques, and finally synthesis procedures. Of these, we will focus here on the last, as being central to control theory proper.

19. Root Locus. Since World War II, when control theory first emerged as an engineering discipline and entered its classical era, control engineering at the applied level has been dominated by artful use of frequency-domain analysis techniques, notably those formulated by Bode and Nyquist. Undoubtedly, the great bulk of all control loops now in operation were designed by use of such methods. More recently, since the 1970s, a few sophisticated control systems have been designed via the "modern" or "optimal" state-space techniques, which captured the preponderance of research attention during the 1960s. For expository purposes, however, and very often for design purposes as well, the most useful tool for discussion of control-loop synthesis is the *root-locus* technique, introduced by W. R. Evans in 1948 and sometimes referred to abroad as the *Evans locus.*

This technique, once the basic concept has been grasped, is captivating in its simplicity and awe-inspiring in its sufficiency. As a classical technique — perhaps the capstone of classical techniques — it deals with linear time-invariant single-loop systems and provides a direct bridge from the characteristics of system elements in terms of their transfer functions to the characteristics of the closed-loop feedback system they constitute. Significantly, also, the root locus can readily be interpreted in terms of frequency-domain characteristics on the one hand or time-domain characteristics on the other.

One of the beautiful aspects of the technique is that, depending on the quality of available information and the extent of one's interest, it can be carried out at any level of precision, from back-of-the-envelope approximation to multidigit computer accuracy, always yielding profound insight into system behavior and any potential for its modification. An exposition can be found in almost any undergraduate control textbook,[11,56] and while the pages of "rules" stated by some authors may seem stultifying at first glance, they quickly become transparent and self-evident.

If the transfer function of a linear system is denoted by $G(s)$, which represents a ratio of polynomials in s, $kN(s)/D(s)$, the transfer function of that same system within a unity-feedback loop (Fig. 5-7) will be $G(s)/[1 + G(s)]$. Thus, the roots of the denominator, i.e., the system poles which determine the basic dynamic characteristics of the system, are constrained by the relationship

$$G(s) = -1$$

Root locus exploits this relationship for all it is worth.

Appreciation of the root-locus technique depends on understanding the significance to system transient characteristics of system pole positions on the complex plane. Since system transfer functions are functions of the complex variable s, the system poles and zeros must be plotted on the complex plane. Figure 5-8 shows graphically the significance of a pole's position on this plane to a system's transient response to an impulse. Since we are dealing with linear systems, the response of a system having a multiplicity of poles is given by superposition of the individual responses; the relative magnitudes are determined by the relative distances between the system poles and the system zeros.

In terms of the numerator and denominator of $G(s)$, the closed-loop transfer function can be written $kN(s)/[kN(s) + D(s)]$. Thus the roots of the numerator, the system zeros, are not affected by the feedback loop. As the scalar gain parameter k is increased from zero toward infinity, the closed-loop system poles, always constrained by the relationship $G(s) = -1$, move along clearly defined loci from the positions of the open-loop poles (with which they coincide for $k = 0$) toward the position of the system zeros (some of which may lie at infinity in clearly defined directions). Thus, given the appropriate easily constructed root locus, fixing the loop gain k will fix the dynamic characteristics of the closed-loop system, which may be very different from those of the corresponding open-loop system.

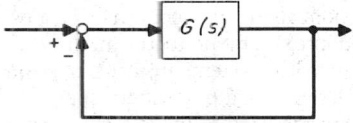

Fig. 5-7. Single unity-feedback loop.

While consideration only of unity feedback with no dynamics in the feedback loop and of variation only of the loop-gain parameter may seem of limited usefulness, it can easily be shown that any linear feedback loop can be written in an equivalent unity-feedback form, and any one parameter can play the role of the loop gain. The real restrictions are those common to classical techniques: linear time-invariant deterministic single-loop systems, although the last can be relaxed with ingenuity.

A few examples will show the power and usefulness of the technique and simultaneously clarify effects of feedback on stability which were frustrating to control pioneers. The lowest imaginable level of even negative feedback around a system as simple as third order, with no left-half-plane zeros and at least two integrators (poles at the origin) (Fig. 5-9a), will be absolutely destabilizing! Furthermore, even without the poles at the origin, some minimum value of gain will be sufficient to guarantee instability (Fig. 5-9b). Open-loop unstable systems, stabilized via negative feedback, may be only conditionally stable (Fig. 5-10a and b).

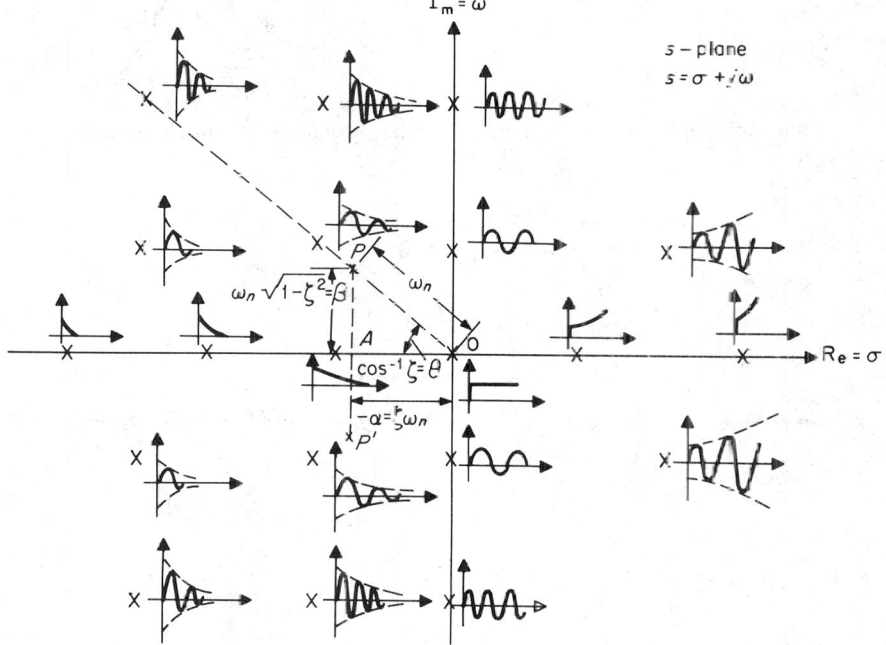

Fig. 5-8. Significance of root location in the complex plane. *(From Y. Takahashi, M. J. Rabins, and D. M. Auslander, "Control and Dynamic Systems," Addison-Wesley, 1970, p. 330.)*

Critical to the successful application of such techniques, however, is an understanding of modeling. Systems (or devices) which are poorly modeled will be analyzed incorrectly. All models are lower-order simplifications; the only unsimplified model is the prototype itself. However, models which omit relevant states or modes will inevitably lead to error. A system requiring three states for adequate representation but modeled by two (Fig. 5-11a) will appear absolutely stable under negative feedback, but the system will not have got the message (Fig. 5-11b).

20. Linearization. The success and much of the continuing power of linear control theory is due in part to the technique of *linearization* and the concept of regulation under feedback. Linearization, usually a straightforward application of common techniques such as series expansion about an operating point with higher-order nonlinear terms discarded, provides linear approximations to the nonlinear models which invariably result from attempts to represent physical systems in terms of basic laws of conservation of mass, momentum, etc. Analysis and design based on such an approach has been found to work well for the control of plants which are required to operate near a given steady-state operating point for long periods of time. Such

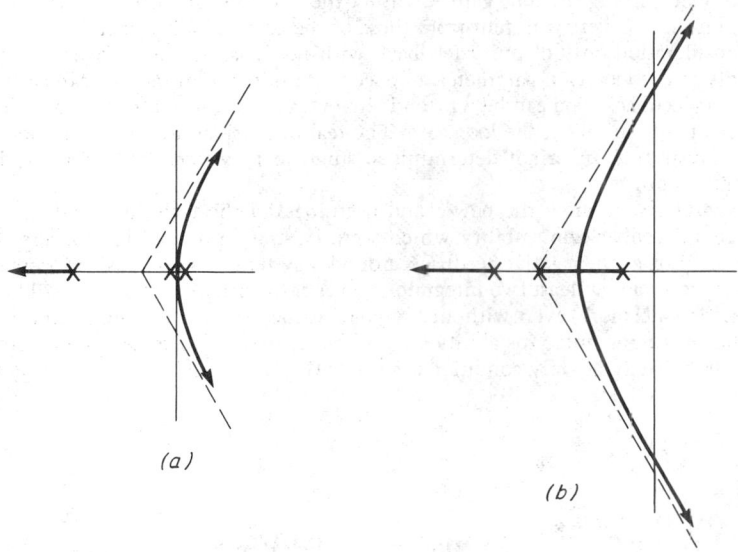

(a)

(b)

Fig. 5-9. Third-order system stability: (a) two poles at origin; (b) no poles at origin.

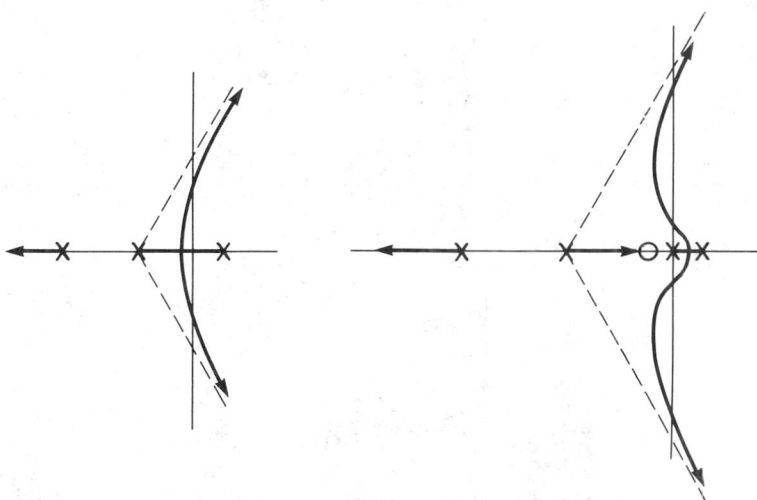

Fig. 5-10. Conditional stabilization.

plants are *regulated* to maintain control parameters and plant conditions within a small range of their scheduled values in the face of persistent but small disturbances.

Large disturbances driving the plant outside the region where its behavior approximates that of the linearized model will seriously degrade performance of the normal control and may require the intervention of a human operator. However, the range of usefulness of regulating controls based on linearized models can be dramatically extended by linearizing the model, not around a single operating point but around an operating trajectory, and using linear feedback

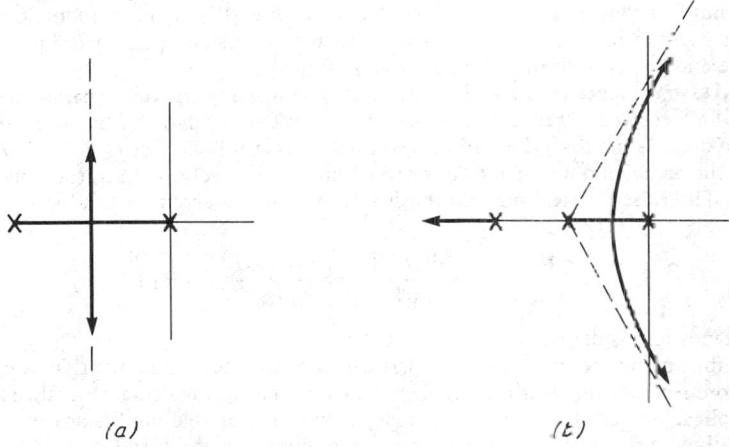

(a)   (b)

Fig. 5-11. Model adequacy: (a) second-order model; (b) third-order model.

control to keep the plant within the vicinity of the trajectory as the operating point varies over a wide range; such *tracking control* usually involves a combination of feedforward with feedback control.

21. Compensation. The basic idea of feedback is intuitive and simple. From the perspective of a human operator attempting any control action, whether that of positioning a lamp on a table, steering an automobile, or any of the innumerable actions we take continually and instinctively, our action is almost invariably tempered by our continuing observation of any discrepancy between intent and status thus far. This is negative feedback: the control action is a function of the difference between the desired output and the actual output.

For simple position control in the absence of disturbing forces, a control signal proportional to error may often be adequate, but the situation is rarely that simple. Obviously, even the simple act of holding an object at a given elevation or holding a spring in a certain degree of compression requires a continuous force at zero error in the position, so that "proportional" control would require a continuous error input in order to provide a constant control effort.

Engineers early learned to circumvent this difficulty by adding a *reset* term to the control: integrating any output error until the resultant corrective force was sufficient to drive the error to zero and keep it there. Later it was found possible to improve the performance of sluggish systems by the cautious addition of a *rate* term: a derivative term proportional to the rate at which the output error was increasing.

The resulting three-mode control (proportional, integral or reset, and derivative or rate) has become standard in the process industries. Adjustment of the relative magnitudes of these three components of the control signal to attain the most desirable system performance has been more of an art than a science (with some heuristic rules of thumb[10]. From the standpoint of root locus, adjustment can be analyzed in terms of the addition of one or a few poles and zeros to the system, providing the control engineer with additional degrees of freedom, enough indeed to make considerably improved performance possible.

The full extent of the power of feedback did not become apparent, however, until the development of state-space formalism. In 1967 it was shown[61] that controllability of an open-loop system is equivalent to the possibility of assigning an arbitrary set of poles to the transfer matrix of the closed-loop linear state-feedback system: The pair (A, B) is controllable if and only if for every choice of the set Λ there is a matrix C such that $A + BC$ has Λ for its set of eigenvalues.

This is a most remarkable fact; so remarkable that we must be careful to note its limitations. It speaks only to linear systems which are controllable and all of whose states are available for feedback. This property of arbitrary pole assignment cannot be guaranteed for output feedback with the same generality (Fig. 5-12), but it has been shown[32] that such is possible for almost all systems under certain conditions: $n < r + m + \nu - 1$, $r > \mu$, $m \geq \nu$; n, r and m being respectively the number of states, of inputs, and of outputs, and ν and μ being respectively the controllability and observability indices. Such pole assignment, when possible, means that the system involved can be given, via feedback, not only any desired degree of stability but also any desired causal response consonant with its order.

Full treatment of basic control topics is contained in standard undergraduate texts, including Eveleigh[19] and Takahashi et al.[55] Clark[11] contains an unusually lucid introduction to classical control, and Elgerd[18] has a comparably lucid introduction to state-space techniques. Something of the range and power of feedback is indicated in Cruz.[13]

22. Sensitivity Reduction. Sensitivity of system characteristics to parameter variations, noise, etc., has received continuing and widespread attention, dating from the early work of Bode.[6] As so often is the case, this earliest formulation is still basic. Let $G = G(S, \alpha)$ and $G_0 = G(S, \alpha_0)$ be the actual and nominal transfer functions, respectively, with α_0 the nominal parameter vector. Then the classical or Bode sensitivity function is given by the logarithmic partial derivative

$$S_\alpha(s) \equiv \left. \frac{\partial \ln G}{\partial \ln \alpha} \right|_{\alpha_0} = \left. \frac{\partial G/G}{\partial \alpha/\alpha} \right|_{\alpha_0} = \left. \frac{\partial G}{\partial \alpha} \right|_{\alpha_0} \frac{\alpha_0}{G_0} \qquad \text{for } \alpha = \alpha_j$$

where $\equiv$ stands for "is defined as."

Bode's definition applies only to small perturbations in linear time-invariant single-variable systems. More recently, Perkins (in Ref. 13) and Cruz[13] have provided a generalized sensitivity operator applicable to arbitrary parameter changes in multivariable nonlinear time-varying systems. Since they deal directly with the comparison of open- and closed-loop systems, we begin with a definition of *nominal equivalence*. Two system configurations are said to be nominally equivalent if, for nominal parameter values, the outputs due to the same inputs are equal.

If we take

$$\Delta Y_o \equiv Y_o(s, \alpha) - Y_o(s, \alpha_n) \qquad \Delta Y_c \equiv Y_c(s, \alpha) - Y_c(s, \alpha_n)$$

as the output difference of equivalent open- and closed-loop systems due to a plant-parameter change, their comparison sensitivity function is defined as

$$S_p(s, \alpha) \equiv \Delta Y_c(s, \alpha)/\Delta Y_o(s, \alpha)$$

It can be shown that the Bode function is equal to the comparison sensitivity function for linear time-invariant systems with differentially small plant-parameter variations.

While it is not true that feedback reduces parameter sensitivity in any event (Ref. 23, p. 246), feedback remains a powerful tool for such reduction. For single-input single-output control loops, Bode's absolute-value integral theorem states that if the open-loop transfer function $F_o(s)$ is stable and $1 + F_o(s)$ has no right-half-plane zeros, and if the number of poles of $F_o(s)$ exceeds the number of zeros by at least 2, then the average value of the logarithm of the (Bode) sensitivity function is zero:

$$\int_0^\infty \ln |S_p^G(j\omega)| \, d\omega = 0$$

This says that if the controller is chosen such that $|S_p^G| < 1$ holds for a frequency range of interest, what is gained in the range below ω is lost in the range above.

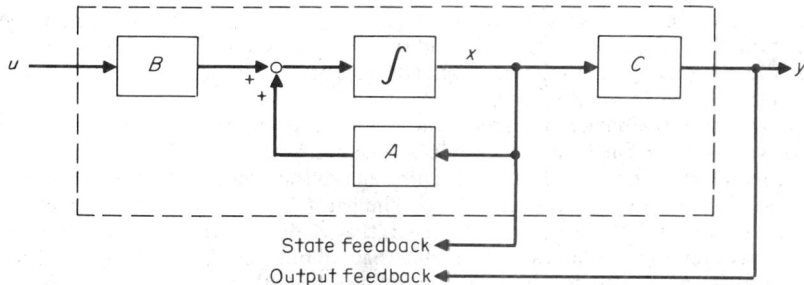

Fig. 5-12. State vs. output feedback.

It also implies that choice of controller characteristics to satisfy sensitivity requirements fore-closes on choice of control quality. This can be circumvented by introducing additional degrees of freedom into the structure. For example, the Bode sensitivity of the system in Fig. 5-13 depends only on $L(s)$, so that control quality can be determined by proper choice of $R(s)$ and $H(s)$. Furthermore, proportional state feedback of all states results in a transfer function not sat-isfying the absolute-value integral theorem requirement of pole excess greater than 1, so that in each case the closed-loop control can be less sensitive than open-loop control over the entire frequency range.

For full treatment of sensitivity, see Cruz[13] and Frank.[23]

23. Deterministic Optimal Control. Awareness of the power of feedback design and compensation leads naturally to experimentation with techniques for exploiting that power and

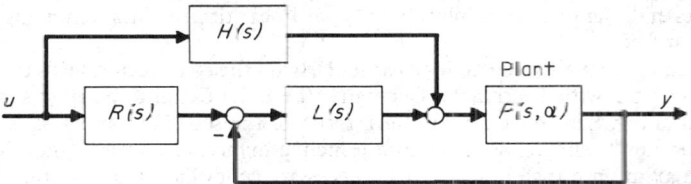

Fig. 5-13. Feedforward compensation. *(From P. M Frank, "Introduction to System Sensitivity Theory," Academic, 1978, p. 53.)*

doing the job "right." The first attempts, in the 1950s, were in terms of analytic design, wherein system parameters, within a fixed structure, were set to minimize a function of the error between desired and actual system response to a specified input signal.[45] The chosen input could be a deterministic or a random function of time or frequency, and a variety of figures of merit were used, among the more popular being the integral squared error

$$\int_0^\infty \varepsilon^2(t)\, dt$$

and the mean squared error

$$\lim_{T\to\infty} \frac{1}{2T} \int_{-T}^{T} \varepsilon^2 t\, dt$$

Analytic design techniques, which were based on frequency-domain transfer-function analy-sis, foundered probably as much on mathematical complexity as on anything and were overtaken by the development of deterministic optimal control, based on burgeoning state-space time-domain analysis.

Deterministic optimal control assumes known plant dynamics

$$\dot{x} = f(x, u)$$

with the state vector x and/or the control vector u subject to equality or inequality constraints. The performance index is a functional in both state and control vectors

$$J = \int_{t_0}^{t_f} F(x, u)\, dt$$

which is to be optimized subject to the imposed constraints.

Static optimization applies to plants which are normally operated under steady-state condi-tions. In such, the plant dynamic equations

$$\dot{x} = f(x, u)$$

reduce to

$$f(x^0, u^0) = 0$$

and the performance index reduces to a constant function

$$J' = F(x^0, u^0)$$

to be optimized. Depending on the nature of the constraints, this can be done analytically, by the method of Lagrange multipliers (when the governing constraints are equalities), by linear programming, or more generally by gradient techniques.

Consideration of non-steady-state performance introduces considerably more difficulty. Since the system is not in the steady state, the derivative of the vector state equation is no longer zero and we are forced back to consideration of the performance-index functional in integral form. Treatment of this case is possible in principle because we are considering deterministic situations: the demands to be made on the plant are known, as are the plant dynamics, and the control input can be specified as a function of time to yield the desired plant performance.

However, it is well to recognize at the outset that this is really a very unusual situation, i.e., perfect knowledge of the plant and its environment over the time horizon of interest, from t_0, the lower limit of integration, to t_f, the upper.

As in the case of static optimization, the appropriateness of various solution techniques depends on the nature of the governing constraints. If they are equalities, the classical calculus of variations provides a solution satisfying necessary conditions via the Euler-Lagrange equations, which are generally amenable to solution only for linear time-invariant systems and quadratic performance indices.

Introduction of inequality constraints, particularly on the state vector, leads to the *maximum principle*, associated with Pontryagin. Generally, the Euler-Lagrange equations lead to an optimum control only when this exists in the *interior* of the set U of admissible controls, where constraints are not binding. The maximum principle involves controls at the *boundary* of the set U. The maximum principle states that a necessary condition for a minimum of the performance index is that a scalar function, akin to a Hamiltonian,

$$H(x, p, u, \lambda_0) = \lambda_0 F(x, u) - \langle p, f(x, u) \rangle$$

be maximized with respect to the control vector u at all times during the duration of the problem. [Here x and p represent the state and costate vectors, and $F(\cdot)$ represents the time cost function.] The significance of this principle is that the difficult problem of minimization of a functional

$$J = \int F \, dt$$

is replaced by the more tractable problem of minimizing a scalar function H.

The answer is obtained by solving a matrix Riccati-type equation

$$P = -PA - A^T P + PBR^{-1}B^T P - Q$$

where all the matrices may be time-varying, A and B are the usual system and control matrices, and R and Q are the weighting matrices from the cost functional to be minimized. Since the terminal conditions to be satisfied are given, $P(t_f)$ is known and the equation must be integrated backward in time to yield $P(t)$. The optimal-control law is then given by

$$u(t) = K(t)x(t) = -R^{-1}(t)B^T(t)P(t)x(t)$$

It is important to note, however, that the resultant optimal control is a function of initial *and final* boundary conditions and is *basically* open-loop *in nature* [since the feedback gain matrix $K(t)$ must be prespecified as a function of time] even though it appears in feedback form.

An alternative but related solution to the optimal-control problem is provided by Bellman's dynamic programming via the principle of optimality.[36] This principle states that any portion of the optimal trajectory is also an optimal trajectory. Dynamic programming deals with the discretized version of the optimal-control problem. In this version, optimal control becomes an N-state decision process, at each of which stages the decision to be made must be optimal over the remaining (future) stages, regardless of the path by which the current stage was reached. For an n-dimensional N-stage problem where each state may take on discrete values, an exhaustive search over all possible trajectories for the optimum would involve approximately r^{nN} decisions. Application of the principle of optimality reduces this to approximately Nr^2, since it deals with N single-stage decision problems. Nevertheless, the dimensionality of the problem remains unrealistically forbidding for many actual problems.

Deterministic optimal control seems to us to represent a period during which control theory forsook for a time its attachment to physical considerations to retreat into a mathematical realm where it reexamined its nature and gathered material for a new assault on reality. The retreat was not intentional (although one is left somewhat uneasy by the repeated emphasis in Kalman et al. that "control theory does not deal with the real world"[30]), but apparently the atmosphere generated by the advances in the late 1950s and early 1960s was irresistibly heady.

Excellent treatments of deterministic optimal control are contained in Athans[2] and Sage and White.[52]

24. Stochastic Estimation and Control Algorithms. It is natural and traditional to model complex phenomena by random processes. This organizes empirical data so as to incorporate implicit information in the data explicitly into the model. Therefore stochastic models contain more, not less, information than deterministic models of the same complexity. Unfortunately, design methods for stochastic models have not been perfected, and the existing theory has been applied only to a small but important class of problems.

The design of stochastic, dynamic control systems has two important aspects: (1) the problem of information and (2) the superiority of feedback controls. In the first case, the designer must consider not only performance objectives and admissible control laws but also various information-processing structures. If noise is present in the system, measurements of variables will be distorted and must be filtered in conjunction with the control process.

Second, noise in the model describes unpredictable disturbances in the physical system. It follows that a responsive controller must base its actions on current information about the physical process; i.e., it must be in feedback form. Open-loop controllers are inferior. In deterministic control problems, open-loop and feedback controllers produce the same performance, and information processing is irrelevant.

Design of a stochastic control system therefore requires the simultaneous design of coupled information processing and control structures. Disjoint designs almost surely result in total structures which are suboptimal. At present, there are no general techniques (like the Pontryagin maximum principle) for treating these combined problems. However, in one special case, the design of the filter and the controller can be separated into classical problems.

This result is rooted in the work of Wiener, Kolmogorov, Wold, and others in the 1940s on the estimation of stationary random processes. Briefly, given two jointly stationary processes $\{x(t), y(t) -\infty < t < \infty\}$, the Wiener filtering problem is to determine a linear time-invariant filter (transfer function) which estimates current (future, past) values $x(t)$ of the message process from past and present values of the observed process $\{y(s), s \leq t\}$. That is, one must choose a function $g(t)$ so that $\hat{x}(t) = \int_{-\infty}^{t} g(t - \tau)y(\tau)\,d\tau$ estimates $x(t)$ for all t in an optimal fashion.

Given the correlation functions $Ex(t + r)x(r) = R_x(t)$, $Ex(t + r)y(r) = R_{xy}(t)$ and $R_y(t)$, or their Fourier transforms,

$$S_x(j\omega) = \int_{-\infty}^{\infty} e^{j\omega t} R_x(t)\,dt$$

etc. (power spectral density functions), the $g(t)$ which minimizes

$$\bar{e}^2 = E|x(t) - \hat{x}(t)|^2$$

satisfies the Wiener-Hopf equation

$$R_{xy}(t) + \int_{0}^{\infty} g(u)R_y(t - u)\,du = 0$$

Since g must be realizable $[g(u) = 0, u < 0]$, this equation cannot be solved by Fourier transforms and one must use the method of spectral factorization developed by Wiener and Hopf.

In 1960, Kalman and Bucy reformulated and solved Wiener's problem in a way which naturally incorporated nonstationary vector-valued random processes, thereby circumventing two major obstacles in the Wiener theory. Kalman's filter was also recursive in form and so required minimum on-line memory in the implementation of the estimator. These features have led to the widespread use of the Kalman-Bucy algorithm. Kalman and Bucy postulated the model

$$\begin{align}
\dot{z}(t) &= A(t)z(t) + B(t)w(t) & x(t) &= C(t)z(t) & (5\text{-}8)\\
y(t) &= x(t) + v(t) & t_0 \leq t \leq T & & z(t_0) = z_0
\end{align}$$

where w and v are vector-valued uncorrelated white noise with $E\,v(t)v^T(u)\} = S_v\delta(t - u)$, etc. Assuming that the matrices A, B, C are known is equivalent to assuming that the correlation functions of x and y are known. Thus, the model is compatible with the one postulated by Wiener; it is, however, more flexible in digital computations.

The linear, least-mean-square estimator of $x(t)$ given $y(s)$, $s \leq t$, satisfies

$$\frac{d}{dt}\hat{z}(t) = A(t)\hat{z}(t) + K(t)[y(t) - C(t)z(t)]$$

$$\hat{x}(t) = C(t)\hat{z}(t) \qquad \hat{z}(t_0) = E\{z_0\}$$

where $K(t) = Q(t)C^T(t)S_v^{-1}$ and the error covariance matrix $Q(t)$ satisfies

$$\dot{Q}(t) = A(t)Q + QA^T(t) + B(t)S_w B^T(t) - QC^T(t)S_v^{-1}C(t)Q$$
$$Q(t_0) = \text{covar } z_0$$

The estimator is recursive in form; i.e., to compute $\hat{z}(t + \Delta t)$ one need only process the current estimate $\hat{z}(t)$ and observation $y(t)$. The equation for Q depends only on the data (A, B, C, S_w, S_v) and can be solved off line by numerical integration. If A, B, C are constant and A is stable, then over an infinite time interval (x, z, y) in Eq. (5-8) are stationary processes. In this case, the steady-state filter is

$$\frac{d\hat{z}}{dt} = (A - \bar{K}C)\hat{z} + \bar{K}y(t) \qquad \hat{x}(t) = C\hat{z}(t)$$
$$K = \bar{Q}CS_v^{-1} \qquad 0 = A\bar{Q} + \bar{B}QA^T + BS_w B^T - \bar{Q}C^T S_v^{-1}C\bar{Q}$$

The algebraic equation for $\bar{Q}$ has a unique positive definite symmetric solution under natural conditions on the data which can be found as the limit of the solution of the Riccati differential equation (among various methods). The matrix $A - \bar{K}C$ is stable in this case. In integrated form the Kalman-Bucy filter is

$$\hat{x}(t) = \int_{-\infty}^{t} Ce^{(A-\bar{K}C)(t-u)}\bar{K}y(u)\, du$$

The combined problem of filtering and control is encountered in

$$\min_u E\left[\int_0^{t_1} y^T(t)y(t) + u^T(t)u(t)\, dt + y^T(t_1)F_y(t_1)\right]$$
$$\dot{z}(t) = Az(t) + Bw(t) + Gu(t) \qquad x(t) = Cz(t)$$
$$z(0) = z_0 \qquad y(t) = x(t) + v(t)$$

where A, B, C, F, G are (time-varying) matrices, $F = F^T \geq 0$, and w, v are white noise. The information-controller structure determines the problem. The controls may be chosen as functionals of z (perfect observations); they may be memoryless functions $\{u(t) = f[t, z(t)]\}$ or functionals of past observations $\{u(t) = h[t; y(s), s \leq t]\}$, depending on the application.

Suppose that one has the latter situation and that the disturbances (w, v, z_0) are gaussian. Then (x, y, z) are gaussian, and the control problem has a very special structure. The optimal-control law is

$$u^*(t) = -G^T P(t)\hat{z}(t)$$

where
$$\dot{P} = -(A^T P + PA + CC^T) + PG^T GP$$
$$P(T) = F \qquad 0 \leq t \leq T$$
$$\frac{d\hat{z}}{dt} = [A - K(t)C]\hat{z}(t) + Gu^*(t) + K(t)y(t)$$

With $K(t)$ as above, this solution has the following features:

1. The control law is the same as the optimal law in the deterministic (perfect-observations) problem. This is the *certainty-equivalence principle.*

2. The control law operates on the state estimate $\hat{z}$, which is computed in a filter whose parameters are independent of the controller. This is the *separation principle.*

3. The optimal cost decomposes into the effects of initial disturbances ($z_0 \neq 0$), state driving noise w, and estimation errors.

In nonlinear or nongaussian cases one can expect certainty equivalence to fail, e.g., when the cost is nonquadratic, followed soon after by the separation property as assumptions are relaxed. General algorithms for partially observed stochastic control problems are not yet available.

Davis[15] and Åström[1] describe the linear quadratic gaussian problem in detail. The December 1971 issue of the *IEEE Transactions on Automatic Control* contains a number of useful papers on related subjects. Advanced dynamic-programming results for perfectly observed stochastic control problems are discussed in Fleming and Rishel.[22] Filtering for nonlinear processes is an active area of current research; however, few practical algorithms are available. A heuristic algorithm, the *extended Kalman filter,* which uses sequential linearization of the dynamics, is described in Jazwinski.[29]

25. Analysis of Nonlinear Systems. In the last 20 years limited analytical methods for nonlinear systems like the describing function and phase portraits for second-order systems have been supplanted by a number of results of greater power and generality. These include (1) input-output stability theory for feedback systems; (2) advanced Lyapunov methods; and (3) accurate approximation methods for nonlinear systems.

Consider a feedback system with input u, output y, return-difference error ε, plant G, and feedback function F:

$$y = G\varepsilon \qquad \varepsilon = U - Fy = u - FG\varepsilon \qquad (5\text{-}9)$$

Two basic principles provide a stability theory for such systems. First, if the system FG attenuates signals in a suitable sense, then

$$\|\varepsilon\| \leq \|u\| + \|FG\| \cdot \|\varepsilon\|$$

or

$$\|\varepsilon\| \leq (1 - \|FG\|)^{-1}\|u\| \qquad \text{and} \qquad \|y\| \leq \|G\|(1 - \|FG\|)^{-1}\|u\|$$

Hence, if G is bounded on the set of admissible signals and $\|FG\| < 1$ on the same signal space, the system converts bounded inputs into bounded outputs with an upper bound on the transmission and is, in this sense, stable.

This simple result provides a direct design or analysis method for such systems. If G is a given plant, any feedback function F which results in a causal, unique relationship between (ε, y) and u and which satisfies $\|FG\| < 1$ results in a stable system.

For example, suppose that the set of signals is

$$L_2 = \left\{ f(t) \text{ with } \int_0^\infty f^2(t)\, dt < \infty \right\}$$

Then

$$\|f\| = \left(\int_0^\infty f^2(t)\, dt \right)^{1/2}$$

and such signals have finite energy. Suppose

$$(G\varepsilon)(t) = \int_0^t g(t - \tau)\varepsilon(\tau)\, d\tau \qquad \text{and} \qquad (Fy)(t) = f(t, y(t))$$

a memoryless, time-varying nonlinearity. Then on L_2, $\|G\| = \max |\hat{g}(j\omega)|$, where $\hat{g}(j\omega)$ is the Fourier transform of $g(t)$. Suppose that there are positive constants a, b such that $a \leq f(t, y)/y \leq b$ for all y and t. Then on L_2 we have $\|Fy - \frac{1}{2}(a + b)y\| \leq \frac{1}{2}(b - a)\|y\|$. The condition $\|FG\| < 1$ yields (after a simple transformation) the circle criterion for stability of the feedback system: the system is stable on L_2 if the Nyquist locus of $\hat{g}(j\omega)$ in the complex plane does not encircle or intersect the disk centered on the real axis at $-\frac{1}{2}(1/a + 1/b)$ with radius $\frac{1}{2}(1/a - 1/b)$. This simple graphical condition in the frequency domain is completely in the spirit of the Nyquist criterion. It reduces to the latter if $a = b$, the case of a constant feedback gain.

The second basic principle makes use of passivity. Intuitively, a system G is passive if it is lossless or dissipates energy. On the space L_2 we define

$$\langle y, x \rangle = \int_0^\infty y(t)x(t)\, dt \qquad \text{and} \qquad \langle y, Gy \rangle = \int_0^\infty y(t)(Gy)(t)\, dt$$

A system is passive in this sense if $\langle y, Gy \rangle \geq 0$ for every y in L_2. It is strictly passive if the inequality is strict for every $y \neq 0$. For example, a convolution, $y(t) = \int_0^t g(t - r)x(r)\, dr$ is passive if $\mathrm{Re}\, \hat{g}(j\omega) \geq 0$ for every ω, that is, if it is positive real, a term from network theory.

The general stability principle is that the feedback, or lossless, interconnection of passive systems is passive and is stable if it is strictly passive. For example, let G be the convolution above, and let $(Fy)(t) = f(y(t))$, a time-invariant nonlinearity with $0 < f(y)/y < k$ for all y. Then the feedback system of Eq. (5-9) on L_2 is stable (strictly passive) if for some $a \geq 0$

$$\mathrm{Re}\, [(1 + j\omega a)\hat{g}(j\omega) + 1/k] \geq 0 \qquad \text{for all } \omega$$

This result was discovered by Popov in 1960. Like the circle criterion, it is a frequency-domain condition which can be tested graphically. It is a sufficient condition for stability which is more accurate than the circle criterion when both apply. It also reduces to the Nyquist criterion when f is linear.

Both the circle criterion and the Popov criterion are sufficient conditions for *absolute stability*; any function f which satisfies the sector conditions produces a stable feedback system. They may or may not produce sharp estimates for specific systems. Other, more complex results produce sharper estimates for more tightly constrained systems, e.g., linear periodic systems. Instability conditions can also be obtained by similar arguments; these provide necessary conditions for stability. More information on this theory, including its multivariable versions, can be found in Desoer and Vidyasagar[17] and Willems[59].

For nonlinear (feedback) systems described by differential (or difference) equations one can use the classical theorems of Lyapunov to check for stability. Suppose that $\dot{x}(t) = F(x(t))$ has an equilibrium $[F(x) = 0]$ at $x = 0$. If one can find a real-valued function $V(x)$ which is positive definite, $V(0) = 0$, and which satisfies

$$\dot{V}(x(t)) = \text{grad } V(x(t)) \cdot f(x(t)) < 0 \tag{5-10}$$

then the solutions $x(t) \to 0$ as $t \to \infty$. A region D containing $x = 0$ in which grad $V(x) \cdot f(x) < 0$ is a *domain of attraction* of the equilibrium $x = 0$; any solution with $x(0)$ in D approaches $x = 0$ as $t \to \infty$.

The basic problem of Lyapunov stability theory is to find V which produces the largest possible D. Evidently, one must solve a variational problem to find V. Aside from a few exceptional cases (including the circle and Popov criteria) there are no general procedures for producing V functions. It should be noted, however, that if a given system is asymptotically stable with domain D and the solutions approach the equilibrium fast enough, there is a Lyapunov function V which is positive definite and satisfies Eq. (5-10) on D.

This result, which establishes necessary and sufficient stability conditions, was published in 1949, but it is little known in the engineering community. More unfortunate is the fact that it is nonconstructive; one must solve the system to construct the V function. Recent years have seen the application of Lyapunov methods to stability problems in several areas, including electric power systems, and the development of Lyapunov methods for large-scale systems. This work has not been exhaustive, however, and one cannot judge the potential for future applications. Willems[60] provides a useful introduction to the subject, which also brings out its connections with the input-output stability theory.

In the past decade the extensive structure theory for linear systems has been extended to the class of so-called bilinear systems

$$\dot{x}(t) = Ax(t) + \sum_{i=1}^{m} B_i x(t) u_i(t) + Cv(t) \qquad y(t) = Dx(t) \tag{5-11}$$

where A, B_i, C, D are matrix parameters, v and u_i are input controls or disturbances, y is the output, and x is the state. This system is linear in x and u_i separately but not jointly. The structure theory available includes algebraic tests for controllability and observability reminiscent of those for linear systems ($B_i = 0$). Realization of certain kinds of input-output maps (u_i, v) $\to y$ in the form of Eq. (5-11) is possible using these structural determinants. An important result is that in certain cases the nonlinear system

$$\dot{X}(t) = f(X(t)) + g[X(t)u(t)], \; Y(t) = h[(X(t)]$$

can be approximated to an arbitrary degree of accuracy over a finite interval by a system in the form of Eq. (5-11). This substantially extends the validity of linearization and completes in a limited way research started by Wiener and Volterra. More information on Eq. (5-11) can be found in Mayne and Brockett[40] and Mohler and Ruberti.[43]

GAME THEORY: OBJECTIVE FUNCTIONS IN LARGE-SCALE SYSTEMS

26. Introduction. Thus far we have had little to say about the choice of the objective, or cost functions, which are fundamental to determining control actions. A distinguishing characteristic of large-scale systems is the presence of two or more decision centers where controls or policies are determined. Associated with each decision center is at least one goal or objective function which is to be optimized. Typically, the values of the objective functions depend on the state of the system and on the individual controls or policies of the decision makers.

In certain situations, the values may depend explicitly on the controls of the other decision makers. Because of the interdependences of the cost functions on the state and on the control inputs of several decision makers, it is generally impossible unilaterally to optimize, i.e., mini-mize or maximize, a cost function of a decision maker with respect to its own control or decision inputs alone. The strategy for optimization in the multiple-decision-maker case is intrinsically different from that in the single-decision-maker case as extensively developed in the optimal-control literature. The foundation of this new field of multiple decision making in large-scale systems is rooted in several relevant areas: game theory, differential games, team theory, and decentralized control of large-scale systems.

Game theory was originated by von Neumann[57] in the context of static (nondynamic) situa-tions whereby a set of choices by a set of players results in a payoff function for each of the players. The simplest situation occurs when there are only two players and the payoff to one player is the negative of the payoff to the other player. Such games are called *zero-sum games*. The early development of this theory focused on the existence of equilibrium solutions to such games. Von Neumann showed that if the players randomize their strategies and the payoff func-tions are expectations of functions convex in one control variable and concave in the other, a solution exists. This theory has been extended to n-person non-zero-sum games and also to dif-ferential games.

Differential games are dynamic-system extensions of static games. The bulk of the early lit-erature on differential games is on zero-sum games. Isaacs[27] created this field. Pursuit-evasion games are the principal examples of zero-sum differential games, but the connotation of a dif-ferential game as a zero-sum game is so well entrenched in the literature that it is best to use a different label for non-zero-sum games. We will simply refer to them as the control of systems with multiple decision makers.

Typically, a dynamic system with multiple decision makers is described by a vector state equation

$$\dot{x} = f(x, u_1, \ldots u_N)$$

where x is the state, u_1 is the control vector of decision maker 1, and f is a vector function of x, $u_1, u_2, \ldots, u_N$. Each decision maker has an output measurement

$$y_i = h_i(x, u_1, \ldots, u_N)$$

The controls u_i are to be chosen as functions of y_i. Associated with each decision maker is a cost function

$$J_i = \phi_i(x(T)) + \int_0^T L_i(x, u_1, \ldots, u_N) \, dt$$

where ϕ_i is a scalar function of $x(T)$ and L_i is a scalar function of $x, u_1, \ldots u_N$. In contrast to ordinary optimal-control problems, it is generally not possible to minimize J_i with respect to all the u_j's. Accordingly, optimality has to be defined differently.

27. Nash Strategy. Several solution concepts have been advanced in the literature. One is the Nash or Cournot strategy. Let γ_i denote the strategy for decision maker i. The set $(\gamma_1^*, \gamma_2^*, \ldots, \gamma_N^*)$ is said to be a Nash strategy if

$$J_i(\gamma_1^*, \gamma_2^*, \ldots, \gamma_i^*, \ldots, \gamma_N^*) \leq J_i(\gamma^*, \ldots, \gamma_i^*, \ldots, \gamma_i^*) \text{ for all } i = 1 \ldots, N \qquad (5\text{-}12)$$

That is, if $\gamma_1^*, \ldots, \gamma_{i-1}^*, \gamma_{i+1}^*, \ldots, \gamma_N^*$ have been chosen by the $N - 1$ players and decision maker i examines the cost J_i, no strategy γ_i will result in a cost function lower than that corresponding to γ_i^*. No single decision maker can gain anything by deviating from the Nash strategy, provided that all the other $N - 1$ decision makers are using their Nash strategies. If two or more decision makers cooperate, it is conceivable that the alliance could be better off than sticking to the Nash strategy, but the situation is generally not stable.

Conceptually, one can imagine that the N decision makers have a nonoptimal initial strategy choice and that each is allowed to revise its strategy one at a time. For example, γ_1 could be chosen to minimize J_1 assuming that the other γ_i's are equal to the initial strategies. With the initial choice of γ_1 replaced by the minimizing γ_1, decision maker 2 is then given an opportunity to change its strategy. The new γ_2 is chosen to minimize J_2 given that $\gamma_3, \ldots, \gamma_N$ are equal to the initial guesses and γ_1 is equal to the previously optimized γ_1. The process is continued until γ_N has been optimized. We then have a new set of controls $(\gamma_1, \ldots, \gamma_N)$.

The process is repeated by reoptimizing γ_1 and then the rest of the γ_i's, one at a time. If the sequential optimization converges, we arrive at a Nash equilibrium point $(\gamma^*, \ldots, \gamma_N^*)$, which may not be unique. When such a set is reached, it will satisfy Eq. (5-12). Necessary conditions for the Nash strategy can be obtained by introducing Hamiltonians or via the Hamilton-Jacobi-Bellman equation of dynamic programming. When the dynamic system is linear, and when the cost functions are quadratic, linear Nash strategies are obtained from coupled matrix Riccati equations.[44]

28. Pareto-Optimal Strategy. The Nash solution concept is natural to adopt when cooperation between the decision makers cannot be agreed upon. However, if collaboration can be achieved, it is possible that two or more of the decision makers can move away from the Nash strategy so that at least one cost function is reduced and no cost function is increased. If such a possibility occurs, the Nash strategy is said to be dominated by another strategy.

One can continue in this manner to seek strategies that dominate previously obtained strategies until we arrive at a noninferior strategy. A noninferior strategy is one whereby any deviation which results in a reduction in one cost function is accompanied by an increase in another cost function. The set of noninferior strategies is generally infinite. A noninferior solution is also called a *Pareto-optimal solution*.[50]

A Pareto-optimal solution can be obtained as follows. Consider a set of numbers $\alpha_1, \ldots, \alpha_N$ such that

$$\alpha_i \geq 0 \qquad \text{for } i = 1, \ldots, N \text{ and } \sum_{i=1}^{N} \alpha_i = 1 \qquad (5\text{-}13)$$

and form

$$J = \sum_{i=1}^{N} \alpha_i J_i$$

Choose $u_1, \ldots, u_N$ so as to minimize J. Such a set of u_i's is a Pareto solution. For each choice of $\{\alpha_i\}$ we obtain a different Pareto solution, and thus we generate an infinite set of Pareto-optimal solutions. Under certain conditions this procedure will generate all the Pareto-optimal solutions for all choices of α_i satisfying Eq. (5-13). There may be other solutions.[50] For recent developments, see Lin.[37]

29. Min-Max Strategy. In the Pareto-optimal concept, it is assumed that the decision makers collaborate with each other completely to ensure that whenever it is possible to move to another choice of strategies such that at least one will benefit without hurting anyone, such a move will be carried out. The Nash solution concept assumes that cooperation cannot be enforced, but it is assumed that each decision maker is attempting to minimize its own cost function.

A more severe degree of noncooperation is complete antagonism, whereby other decision makers are assumed to choose controls which maximize your cost. For example, for decision maker 1, for a choice of u_1, it is conceivable that $u_2, \ldots, u_m$ are chosen so as to maximize J_1

$$J_1(u_1, \bar{u}_2, \bar{u}_3, \ldots, \bar{u}_N) \geq J_1(u_1, \ldots, u_N) \qquad \text{for all } u_1, u_2, \ldots, u_N$$

In such a situation, the best choice for u_1 is $\check{u}_1$, where

$$J_1(\check{u}_1, \bar{u}_x, \ldots, \bar{u}_N) \leq J_1(u_1, \bar{u}_2, \ldots, \bar{u}_N) \qquad \text{for all } u_1$$

This choice of $\check{u}_1$ is pessimistic in the sense that it is computed under the worst contingencies of choices for $u_2, \ldots, u_N$ which are not under the control of decision maker 1. Similarly, $\check{u}_2$ is chosen to minimize J_2 under the assumption that $u_1, u_3, \ldots, n_N$ are chosen to create the worst condition for J_2. Thus each $\check{u}_i$ is obtained from

$$\min_{u_i} \max_{\substack{u_j \\ j \neq i}} J(u, \ldots, u_N) = J_1(\bar{u}_1, \ldots, \check{u}_i, \ldots, \bar{u}_N)$$

If each decision maker i chooses $\check{u}_i$ in the above manner, the resulting set $(\check{u}_1, \ldots \check{u}_N)$ will result in costs which are not as large as the min-max costs. For example,

$$J_1(\check{u}_1, \check{u}_2, \ldots, \check{u}_N) \leq J_1(\check{u}_1, \bar{u}_2, \ldots, u_N)$$

Similarly,

$$J_i(\check{u}_1, \check{u}_2, \ldots, \check{u}_N) \leq \min_{u_i} \max_{\substack{u_j \\ j \neq 1}} J_i(u_1, \ldots, u_N)$$

Min-max strategies are conservative strategies, but they guarantee the lowest cost possible under the worst possibilities for the choices of the other decision makers. For more details see Nash.[44]

30. Leader-Follower Strategies. In certain situations it is possible for a decision maker to declare its strategy in advance. When there are only two decision makers, the one who declares in advance is called the *leader* and the other one is called the *follower*. Such games were first considered by von Stackelberg in the context of a nondynamic two-player economic market situation.[58] Suppose decision maker 1 is the leader and decision maker 2 the follower. For each choice u_1 the follower chooses $u_2 = T_2(u_1)$ such that

$$J_2(u_1, T_2(u_1)) \leq J_2(u_1, u_2) \tag{5-14}$$

for all u_1 and u_2. The problem for the leader is to choose u_1^* such that

$$J_1(u_1^*, T_2(u_1^*)) \leq J_1(u_1, T_2(u_1)) \tag{5-15}$$

for all u_1. The pair $(u_1^*, u_2^*) = (u_1^*, T_2(u_1^*))$ is called a *Stackelberg strategy pair*.

It is interesting to compare the Stackelberg strategy to the Nash strategy for the leader. Compare Eqs. (5-12) and (5-14); when u_1 is chosen as u_1^N, it is clear that $u_2^N = T_2(u^N)$. Thus, from Eq. (5-15),

$$J_1(u_1^*, T_2(u_1^*)) \leq J_1(u_1^N, T_2(u_1^N)) \quad \text{or} \quad J_1(u_1^*, u_2^*) \leq J_1(u_1^N, u_2^N)$$

Whenever a decision maker has the option of declaring its strategy first, it is advantageous for it to do so since it will be at least as good as playing Nash. This will not be true for the follower, since being follower in a Stackelberg game, while sometimes being better than playing Nash, may sometimes be worse.

The concept of Stackelberg has been generalized to the dynamic-system case.[4,14,48,54] Various information structures have been studied. In multiple-decision-maker problems the information structure influences the solution, unlike the single-decision-maker situation. In general the open-loop solution is different from the state-feedback solution. The control may involve the past states also, leading to a variety of different solutions.[3,48,49]

Necessary conditions for open-loop Stackelberg solutions were known for a long time, but the closed-loop Stackelberg remained elusive until recently.[48] Furthermore, in general, the Stackelberg strategy does not obey the principle of optimality.[54] The desirable properties of the principle of optimality make it useful to introduce a modified Stackelberg strategy which obeys that principle. The feedback Stackelberg strategy and the Stackelberg equilibrium strategy were introduced for this purpose. It has been generalized also to the M-level leader-follower strategies with M players and to coordination strategies with one leader and several followers.[14,54]

In the coordination of multiple decision makers with one leader, various solution concepts may be imposed on the followers. Suppose cooperation cannot be guaranteed. One might examine the conditions for Nash equilibrium between the followers but a leader-follower relationship between the coordinator and the rest. Let us consider two followers and a leader. Let u_0 be the control of the leader with cost function J_0 and u_1, u_2 the control of the followers with cost functions J_1, J_2. For each u_0 decision makers 1 and 2 choose $T_1(u_0)$ and $T_2(u_0)$, which satisfy

$$J_1(u_0, T_1(u_0), T_2(u_0)) \leq J_1(u_0, u_1, T_2(u_0))$$

and

$$J_2(u_0, T_1(u_0), T_2(u_0)) \leq J_2(u_0, T_1(u_0), u_2)$$

Then u_0 is chosen as u_0^s so as to satisfy

$$J_0(u_0^s, T_1(u_0^s), T_2(u_0^s)) \leq J_0(u_0, T_1(u_0) T_2(u_s))$$

The coordinator's cost function may be

$$J_0(u_0, u_1, u_2) = \alpha J_1(u_0, u_1, u_2) + (1 - \alpha) J_2(u_0, u_1, u_2)$$

In this case, the coordinator's control is chosen to achieve a limited or constrained Pareto optimality for the two remaining decision makers. The constraint is that the followers have to choose $T_2(u_0)$ and $T_3(u_0)$, respectively. The coordinator then examines how to vary all controls by varying only u_0, thus varying $T_2(u_0)$ and $T_3(u_0)$.

A recent intriguing development in leader-follower strategies is the consideration of information structures for which the leader attains its team solution.[3,49,50] That is to say, the resulting cost function for the leader is equal to the minimum of the leader's cost function with respect to the controls of the leader and the follower. Suppose that

$$J_1(u_1^0, u_2^0) \leq J_1(u_1, u_2)$$

for all u_1 and u_2. Thus (u_1^0, u_2^0) achieves a global minimum for J_1; that is, it is the team solution for the leader. Suppose that u_1^0 is such that

$$J_2(u_1^0, T_2(u_1^0)) \leq J_2(u_1^0, u_2) \qquad \text{and} \qquad T_2(u_1^0) = u_2^0$$

Then (u_1^0, u_2^0) is a Stackelberg solution which is also the team solution.

Let us demonstrate that this is possible by a simple example.[25] Consider a simplified electricity-pricing problem for a utility company whose profit function is

$$J_1 = R(q) - \tfrac{1}{2}Cq^2$$

where C is a positive constant, $\tfrac{1}{2}Cq^2$ is the cost for generating electricity q, and $R(q)$ is the revenue. The function $R(q)$ is to be determined so as to maximize J_1. To make a profit, J_1 should be positive. The aggregate of the customers is represented by a follower who wishes to maximize

$$J_2 = \tfrac{1}{2}S[\bar{q}^2 - (q - \bar{q})^2] - R(q)$$

where S and $\bar{q}$ are positive constants. The function J_2 is consumer's surplus, which is the consumer's satisfaction less the cost of electricity. There is a capacity constraint

$$q \leq \hat{q}$$

and a regulation constraint

$$R(q) - \tfrac{1}{2}Cq^2 \leq kq$$

where $\hat{q}$ and k are positive constants. Suppose that the leader's information is q. Thus we assume the information structure to be such that although $R(q)$ is declared in advance, q is actually available at the time $R(q)$ is implemented. Clearly the team solution is

$$R(q) = kq + \tfrac{1}{2}Cq^2 \tag{5-16}$$
$$q = \hat{q} \qquad \text{or} \qquad R(\hat{q}) = k\hat{q} + \tfrac{1}{2}C\hat{q}^2$$

For a given $R(q)$, the follower maximizes J_2 to obtain

$$-S(q - \bar{q}) - dR/dq = 0 \qquad \text{or} \qquad Sq = S\bar{q} - dR/dq \tag{5-17}$$

Now we want to choose $R(q)$ to satisfy Eq. (5-16). Denoting dR/dq by p, we have from Eq. (5-17)

$$q = (S\bar{q} - p)/S = \hat{q} \qquad \text{or} \qquad dR/dq = p = S(\bar{q} - \hat{q}) \tag{5-18}$$

Since

$$dR/dq \geq 0 \qquad \bar{q} \geq \hat{q}$$

any $R(q)$ which satisfies Eqs. (5-18) and (5-16) would be a candidate leader strategy provided that Eq. (5-17) provides a global maximum for the consumer. For example,

$$R(q) = S(\bar{q} - \hat{q})q + k\hat{q} + \tfrac{1}{2}C\hat{q}^2 - S\hat{q}(\bar{b} - \hat{q})$$

is a Stackelberg strategy which achieves a team solution for the leader.

The above example indicates that it is possible for the leader virtually to force the follower to cooperate in optimizing the leader's objective function. The key to this possibility is the choice of the leader strategy as a function of the follower's strategy, although the leader declares first. Thus, the follower actually plays first.

Recent results on leader-follower strategies indicate that modeling and control should be developed jointly rather than sequentially. Certain ill-posedness can result if the traditional modeling and control sequence is used when neglecting fast phenomena in the model.[24]

SYSTEMS ENGINEERING: AN EXAMPLE

31. Introduction. The domain of systems engineering is even broader than that of other engineering disciplines, and we have been able to touch briefly on only a few representative aspects of the field. Because of this breadth, at least in part, systems engineering as practiced is rarely, if ever, fully documented, and we know of no published sources which illustrate the discipline in action with any degree of completeness or fidelity. Individual practitioners, of course, if queried would each cite their own favorite (undocumented) examples of cases which come closest to their idea of good practice.

32. Dynamic Performance of a Power Plant. Our example will be that of the modeling and simulation of the dynamic performance of a complete power plant in parallel with the design of the plant before its construction.[9,41] To our knowledge this was the first time this had ever been attempted. The plant was to have been the first full-scale high-temperature gas-cooled reactor.

Any power plant is a large undertaking which involves diverse groups to design the many systems involved. Here, five major groups were involved: the utility, the reactor manufacturer, the turbine manufacturer, the controls vendor, and the architect-engineering firm. The utility had separate contracts with each of the other firms, the architect-engineer being responsible for the so-called "balance of plant," i.e., equipment other than the reactor and turbine.

It is normal practice in such a case for each of the contractors to work independently on the basis of specifications, scope of supply, and other project management documents, final design

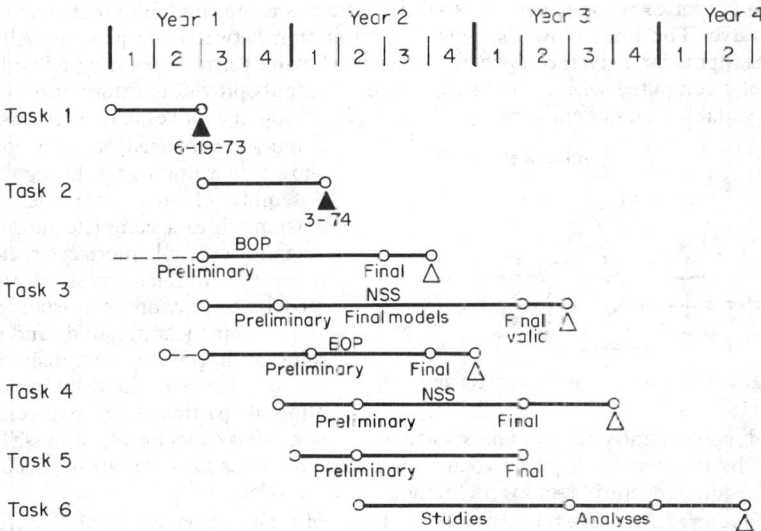

Fig. 5-14. Project task schedule: task 1, scope of work; task 2, model definition; task 3, component models and validation; task 4, subsystem models and validation; task 5, overall plant model; task 6, model utilization.

review and acceptance being the utility's responsibility. This time, however, the utility involved had pioneered in the development of plant-modeling techniques and simulation and had demonstrated that it was possible to construct reliable, nonlinear dynamic plant models on the basis of steady-state design data alone.[20] They recognized that the proposed plant had many unique design features and that the potential benefits of applying their newly developed techniques of plant modeling and simulation during the design phase would more than justify the estimated costs of such an effort. Accordingly a unique team was formed, drawing from all the firms involved to bring together engineers with expertise in the realities of power-system and power-plant operation, the design and operation of nuclear, mechanical, and electrical equipment, the design and analysis of control systems, mathematical modeling, computer simulation, and data analysis.

33. Task Organization. The project was organized in six tasks, shown in Fig. 5-14 along with the associated schedule. Task 1 was completed by the utility as a basis for a cost estimate and as a means for convincing the several contractors, to whom such an effort was unheard of and suspect, of the mutual benefits to be realized. The results were gratifying, and generous cooperation was achieved, the team working as a unit and not at arm's length via intercompany communication channels.

Task 2 involved definition of the basic structure of the proposed plant model. The utility had developed six modeling norms which in their experience characterized successful process models. These were that the model be *deductive*, of *minimum order*, i.e., parsimonious, *nonlinear*, *explicit*, i.e., not contain any implicit algebraic relationships between variables, thoroughly *validated*, and, once validated, *inviolable*, i.e., not be altered without explicitly resatisfying each of the other norms.

Strict guidelines were established, including a standard nomenclature for assigning variable names. The complete plant diagram (too large and complex to reproduce here) was reviewed to eliminate all equipment not essential to the modeling purpose. This simplified plant schematic diagram, still too detailed for feasible simulation, was subjected to a second level of simplification to yield a "model schematic diagram" with redundant components eliminated, retaining the minimum number necessary to predict the behavior of important plant variables.

The next step involved preparation of a model solution diagram. In deductive modeling, each component is analyzed using fundamental physical laws: defining equations are developed to describe the pertinent static and dynamic characteristics, and these equations constitute the model of that component. In developing the overall plant model by interconnecting the component models according to the physical plant arrangement of equipment (as simplified), it is necessary to maintain a consistent causality so that the resultant set of equations is solvable.

Accordingly for each component a causality diagram was prepared, like that shown in Fig. 5-15 for a valve. The lines show the flow of *information* between components. All variables required as inputs for a given component model are shown by arrows pointing toward its block, and variables computed within the block appear as outputs providing information to adjacent blocks. A variable (such as enthalpy of a fluid passing through a valve) may often pass through

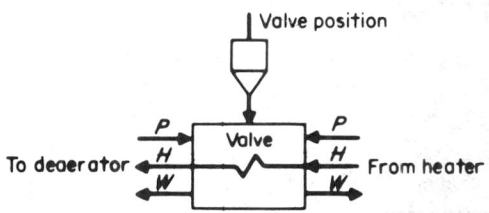

Fig. 5-15. Causality diagram for valve.

a block unchanged; such a condition is shown by a line going through the block. Assembly of such component causality diagrams into a complete model solution diagram with all interfaces matched confirms the well-posedness of the overall model; i.e., it assures consistency.

The nomenclature guide and the model solution diagram became major coordinating tools between the team members, enabling all participating engineers to interface models conveniently and to understand and review models developed by others. The model as defined by this process required about 180 differential equations and about 1,000 algebraic equations. Fourteen control loops were included in the model.

Task 3 required full development and validation of each component model, including controls. This was possible since most components were individually standard, so that the models could be validated against existing equipment in other plants. Figure 5-14 shows task 3 as providing parallel efforts for modeling of the nuclear steam supply (NSS) and the balance of plant equipment (BOP).

Task 4 required assembling the component models into the two major subsystems, NSS and BOP. At this stage, validation against an existing plant was not possible, so that reliance was placed on component validity and matching of designed steady-state characteristics. Particular attention was paid to the interface between the two subsystems; the interface variables, their numeric value, and their order of causality had to be determined to ensure the compatibility of the two models.

Finally, task 5 was assembly of these two models into the overall plant model and final reduction of the equations into a convenient form for numeric solution during simulation runs.

In the third quarter of the third year, work on the plant was suspended and subsequently terminated. One month later work on the model was completed, after steady-state verification at rated conditions and a reactor scram transient. After 2½ years of effort, the simulation project was on schedule and the project team was completing the overall model development and beginning to use the model. The overall model was verified at rated reactor conditions, and a reactor scram transient was completed.

These accomplishments were quite significant in that for the first time an intercompany working group had developed a mathematical model of a complete power plant which was suitable for dynamic analysis. One of the original objectives of this project was to identify and correct during the design phases normal operation and control problems which were related to the dynamic interaction of the two plant subsystems, NSS and BOP. These interactions would not have been identifiable by any of the traditional analyses. The reactor scram transient identified two such conditions. both of which would have contributed to start-up delays if left undetected and therefore uncorrected.

It is rare but not unheard of for design flaws to slip through the interface between companies, engineering disciplines, or even between system design documents. Occasionally, these flaws escape detection even after repeated design reviews by expert engineers because reviewers tend to use the same system documentation. An error subtle enough to be missed once is likely to be missed repeatedly. The procedure described in this example was found to be extremely valuable in:

- Identifying subtle flaws in normal control logic and operating procedures early in the design effort
  - Spawning suggestions for improving the design of the process and controls
  - Coordination of overall design particularly at data interfaces

This example is only partially illustrative of systems engineering as we have defined it. The example emphasized modeling, and drew heavily also on systems analysis and on control synthesis (not related here). Other examples could be cited which would emphasize game theory, optimization, or reliability. The point, however, is not the particular tools used in a given case but that in the design and analysis of complex systems they must be studied as systems, with attention to, and understanding of, the interactions of their components. Systems engineering brings together the tools for such efforts and makes them available.

34. References

1. Aström, K. J. "Introduction to Stochastic Control Theory," Academic, New York, 1970.

2. Athans, M., and P. L Falb "Optimal Control," McGraw-Hill, New York, 1965.

3. Basar, T., and H. Selbuz Closed Loop Stackelberg Strategies with Applications in Optimal Control of Multilevel Systems, *IEEE Trans. Autom. Control*, April 1979, Vol. AC-24, pp. 166–179.

4. Berlinski, D. "On Systems Analysis: An Essay Concerning the Limitations of Some Mathematical Methods in the Social, Political, and Biological Sciences," M.I.T Press, Cambridge, Mass., 1976.

5. Blauberg, I. V., V. N. Sadovsky, and E. G. Yudin "Systems Theory: Philosophical and Methodological Problems," Progress, Moscow, 1977.

6. Bode, H. W. "Network Analysis and Feedback Amplifier Design," Van Nostrand, Princeton, N.J., 1945.

7. Boulding, K. E. Science: Our Common Heritage, *Science*, Feb. 22, 1980, Vol. 207, No. 4433, pp. 831–836.

8. Brockett, R. W. "Finite Dimensional Linear Systems," Wiley, New York, 1970.

9. Broer, W. T. F., K. O. Jaegtnes, J. P. McDonald, and R. W. McNamara Fulton Station Plant Dynamic Simulation, *Joint IEEE/ASME Power Generator Conf.*, 1974

10. Caldwell, W. I., G. A. Coon, and L. M. Zoss "Frequency Response for Process Control," McGraw-Hill, New York, 1959.

11. Clark, R. N. "Introduction to Automatic Control Systems," Wiley, New York, 1962.

12. Corynen, G. C. "A Mathematical Theory of Modeling and Simulation," Ph.D. dissertation, University of Michigan, 1975.

13. Cruz, J. B. Jr. (ed.) "Feedback Systems," McGraw-Hill, New York, 1972.

14. Cruz, J. B., Jr. Leader-Follower Strategies for Multilevel Systems, *IEEE Trans. Autom. Control*, April 1978, Vol. AC-23, No. 2, pp. 244–255.

15. Davis, M. H. A. "Linear Estimation and Stochastic Control," Wiley-Halstead, New York, 1977.

16. D'Azzo, J. J., and C. H. Houpis "Feedback Control Systems Analysis and Synthesis," 2d ed., McGraw-Hill, New York, 1980.

17. Desoer, C. A., M. Vidyasagar "Feedback Systems: Input-Output Properties," Academic, New York, 1975.

18. Elgerd, O. I. "Control Systems Theory," McGraw-Hill, New York, 1967.

19. Eveleigh, V. W. "Introduction to Control System Design," McGraw-Hill, New York, 1975.

20. Fink, L. H., and H. G. Kwatny Model Simulates Reheat Unit Dynamics, *Elec. World*, Sept. 1, 1971, Vol. 176, No. 5, pp. 30–33.

21. Firestone, F. A. A New Analogy between Mechanical and Electrical Systems, *J. Acoust. Soc. Am.*, 1933, Vol. 4, pp. 249–267.

22. Fleming, W. A., and R. W. Rishel "Deterministic and Stochastic Optimal Control," Springer-Verlag, New York, 1975.

23. Frank, P. M. "Introduction to System Sensitivity Theory," Academic, New York, 1978.

24. Gardner, B. F., and J. B. Cruz, Jr. Well-Posedness of Singularly Perturbed Nash Games, *J. Franklin Inst.*, November 1978, Vol. 306, pp. 355–374.

25. Ho, Y.-C., P. B. Liu, and R. Muralidharan Information Structure, Stackelberg Games and Incentive Controllability, *IEEE Trans. Autom. Control*, April 1981, Vol. AC-25.

26. Holtzman, J. M.   "Nonlinear System Theory: A Functional Analysis Approach," Prentice-Hall, Englewood Cliffs, N.J., 1970.

27. Isaacs, R.   "Differential Games," Wiley, New York, 1965.

28. James, H. M., N. B. Nichols, and R. S. Phillips (eds.)   "Theory of Servomechanisms," Dover, New York, 1965.

29. Jazwinski, A. H.   "Stochastic Processes and Filtering Theory," Academic, New York, 1970.

30. Kalman, R. E., P. L. Falb, and M. A. Arbib   "Topics in Mathematical Systems Theory," McGraw-Hill, New York, 1969.

31. Karnopp, D. C., and R. C. Rosenberg   "System Dynamics: A Unified Approach," Wiley, New York, 1975.

32. Kimura, H.   A Further Result on the Problem of Pole Assignment by Output Feedback, *IEEE Trans. Autom. Control*, June 1977, Vol. AC-22, No. 3, pp. 458–463.

33. Kokotović, P. V.   Feedback Design of Large Linear Systems, in J. B. Cruz, Jr. (ed.), "Feedback Systems," McGraw-Hill, New York, 1972.

34. Kokotović, P. V., J. J. Allemong, J. R. Winkelman, and J. H. Chow   Singular Perturbation and Iterative Separation of Time Scales, *Automatica*, January 1980, Vol. 15, No. 1, pp. 23–33.

35. Kwakernaak, H., and R. Sivan   "Linear Optimal Control Systems," Wiley-Interscience, New York, 1972.

36. Larson, R. E., and J. L. Casti   "Principles of Dynamic Programming," Dekker, New York, 1978.

37. Lin, J. G.   Multiple-Objective Optimization by a Multiplier Method Proper of Equality Constraints, I: Theory, *IEEE Trans. Autom. Control*, August 1979, Vol. AC-24, pp. 567–573.

38. MacFarlane, A. G. J.   "Frequency-Response Methods in Control Systems," IEEE Press, New York, 1979.

39. MacFarlane, A. G. J.   The Development of Frequency-Response Methods in Automatic Control, *IEEE Trans. Autom. Control*, April 1979, Vol. AC-24, No. 2, pp. 250–265.

40. Mayne, D. Q., and R. W. Brockett (eds.)   "Geometric Methods in System Theory," Reidel, Dordrecht, 1973.

41. McDonald, J. P.   Nuclear Plant Dynamic Simulation: A Design Analysis Tool, *Proc. 1977 Joint Autom. Control Conf.*, pp. 381–387.

42. Mesarović, M. D., D. Macko, and Y. Takahara   "Theory of Hierarchical, Multilevel Systems," Academic, New York, 1970.

43. Mohler, R., and A. Ruberti (eds.)   "Theory and Applications of Variable Structure Systems," Academic, New York, 1972.

44. Nash, J. F.   Noncooperative Games, *Ann. Math.*, 1951, Vol. 54, No. 2, pp. 286–295.

45. Newton, G., L. Gould, and J. Kaiser   "Analytical Design of Linear Feedback Controls," Wiley, New York, 1964.

46. Nyquist, H.   Regeneration Theory, *Bell Syst. Tech. J.*, 1932, Vol, 11, pp. 126–147.

47. Padulo, L., and M. A. Arbib   "System Theory: A Unified State Space Approach to Continuous and Discrete Systems," Saunders, Philadelphia, 1974.

48. Papavassilopoulos, G. P., and J. B. Cruz, Jr.   Nonclassical Control Problems and Stackelberg Games, *IEEE Trans. Autom. Control*, April 1979, Vol. AC-24, No. 2, pp. 155–166.

49. Papavassilopoulos, G. P., and J. B. Cruz, Jr.   Sufficient Conditions for Stackelberg and Nash Strategies with Memory, *J. Optimization Theory Appl.*, June 1980, Vol. 31, pp. 233–260.

50. Pareto, V.   "Cours d'économie politique," Rouge, Lausanne, 1896.

51. Rosenbrock, H. H.   "Computer-Aided Control System Design," Academic, New York, 1974.

52. Sage, A. P., and C. C. White III   "Optimum Systems Control," 2d ed., Prentice-Hall, Englewood Cliffs, N.J., 1977.

53. Schultz, D. G., and J. L. Melsa   "State Functions and Linear Control Systems," McGraw-Hill, New York, 1967.

54. Simaan, M., and J. B. Cruz, Jr.   Additional Aspects of the Stackelberg Strategy in Nonzero-Sum Games, *J. Optimization Theory Appl.*, 1973, Vol. 11, No. 6, pp. 613–626.

55. Takahashi, Y., M. J. Rabins, and D. M. Auslander   "Control and Dynamic Systems," Addison-Wesley, Reading, Mass., 1972.

56. von Bertalanffy, L.   "General System Theory," Braziller, New York, 1968.

57. von Neumann, J., and O. Morgenstern   "Theory of Games and Economic Behavior," 2d ed., Princeton University Press, Princeton, N.J., 1947.

58. von Stackelberg, H.   "The Theory of the Market Economy," Oxford University Press, Oxford, 1952.

59. Willems, J. C.   "The Analysis of Feedback Systems," M.I.T. Press, Cambridge, Mass., 1971.

60. Willems, J. L.   "Stability Theory of Dynamical Systems," Wiley-Interscience, New York, 1970.

61. Wonham, W. M.   On Pole Assignment in Multi-Input Controllable Linear Systems, *IEEE Trans. Autom. Control*, December 1967, Vol. AC-12, No. 6, pp. 660–665.

62. Wonham, W. M.   "Linear Multivariable Control: A Geometric Approach," Springer-Verlag, New York, 1974.

63. Ziegler, B. P.   "Theory of Modeling and Simulation," Wiley, New York, 1976.

Section 6

Properties of Materials

ILAN A. BLECH *Professor, Department of Materials Engineering, Technion-Israel Institute of Technology, Haifa*

CONTENTS

Numbers refer to paragraphs

CONDUCTIVE AND RESISTIVE MATERIALS

1. Volume Electrical Conductivity. The volume conductivity is the ratio between the electric current density and the electric field strength,

$$\sigma = J/E \tag{6-1}$$

where σ = volume electrical conductivity (S/m), J = electric current density (A/m^2), and E = electric field strength (V/m). The conductivity of metals is independent of current density. Noncubic materials have an anisotropic electrical conductivity.

The volume conductivity is related to the conductance through

$$\sigma = Gl/A \tag{6-2}$$

where G = conductance (S), l = conductor length (m), and A = conductor cross section (m^2).

2. Mass Conductivity. Mass conductivity is defined as

$$\sigma_m = Gl^2/m \tag{6-3}$$

where σ_m = mass conductivity (S·m^2/kg), G = conductance (S), l = conductor length (m), and m = conductor mass (kg).

3. Volume Resistivity. Volume resistivity is the electric field required to produce a unit current density

$$\rho = E/J \tag{6-4}$$

where ρ = resistivity ($\Omega \cdot$m), J = current density (A/m^2), and E = electric field strength (V/m). The volume resistivity is the reciprocal of volume conductivity. The relation between the volume resistivity and electrical resistance is

$$\rho = RA/l \tag{6-5}$$

where R = resistance (S), A = conductor cross section (m^2), and l = conductor length (m). Table 6-1 lists electrical resistivities of elements. Table 6-2 and Fig. 6-1 show the electrical resistivity of alloys and compounds.

4. Mass Resistivity. The mass resistivity is defined as

$$\delta = Rm/l^2 \tag{6-6}$$

where δ = mass resistivity ($\Omega \cdot$kg/m^2), m = conductor mass (kg), and l = conductor length (m).

5. International Annealed Copper Standard (IACS). The conductivity of annealed copper at 20°C has been chosen as a standard, and a value of 100% is assigned to it. The mass resistivity of annealed copper is 0.00015328 $\Omega \cdot$kg/m^2, and the volume resistivity is 1.7241 $\times$ 10^{-8} $\Omega \cdot$m.

TABLE 6-1 Physical Properties of Pure Metals[60]

Metal	mp, °C	bp, °C	Density at 20°C, g/cm^3	Thermal conductivity 0–100°C, W/m·K	Resistivity at 20°C, $\mu\Omega \cdot$cm	Temp. coeff. of resistivity, 0–100°C, K$^{-1} \times 10^{3a}$	Coefficient of expansion, 0–100°C, K$^{-1} \times 10^{6a}$
Aluminum	660.1	2520	2.70	238	2.67	4.5	23.5
Antimony	630.5	1590	6.68	23.8	40.1	5.1	8–11
Barium	729	2130	3.5	. . .	60[b]	. . .	18
Beryllium	1287	2470	1.848	194	3.3	9.0	12
Bismuth	271	1564	9.80	9	117	4.6	13.4
Cadmium	320.9	767	8.64	103	7.3	4.3	31
Calcium	839	1484	1.54	125	3.7	4.57	22
Cerium	798	3430	6.75	11.9	85.4	8.7	8
Cesium	28.5	670	1.87	36.1[c]	20	4.8	97
Chromium	1860	2680	7.1	91.3	13.2	2.14	6.5
Cobalt	1492	2930	8.9	96	6.34	6.6	12.5
Copper	1083.4	2560	8.96	397	1.694	4.3	17.0
Gallium	29.7	2205	5.91	41.0[c]	[d]	. . .	18.3
Germanium	937	2830	5.32	56.4	. . . .	. . .	5.75
Gold	1063	2860	19.3	315.5	2.20	4.0	14.1
Hafnium	2227	4600	13.1	22.9	32.2	4.4	6.0
Indium	156.4	2070	7.3	80.0	8.8	5.2	24.8
Iridium	2454	4390	22.4	146.5	5.1	4.5	6.8
Iron	1536	2860	7.87	78.2	10.1	6.5	12.1
Lead	327.4	1750	11.68	34.9	20.6	4.2	29.0
Lithium	181	1342	0.534	76.1	9.29	4.35	56
Magnesium	649	1090	1.74	155.5	4.2	4.25	26.0
Manganese	1244	2060	7.4	7.8	160(α)	. . .	23
Mercury	−38.87	357	13.546	8.65	95.9	1.0	61
Molybdenum	2615	4610	10.2	137	5.7	4.35	5.1
Nickel	1455	2915	8.9	88.5	6.9	6.8	13.3
Niobium	2467	4740	8.6	54.1	16.0	2.6	7.2
Osmium	3030	5000	22.5	87.5	8.8	4.1	4.57
Palladium	1552	2960	12.0	75.5	10.8	4.2	11.0
Platinum	1769	3830	21.45	71.5	10.58	3.92	9.0
Potassium	63.2	759	0.86	104[c]	6.8	5.7	83
Radium	700	1500	5				
Rhenium	3180	5690	21.0	47.6	18.7	4.5	6.6
Rhodium	1966	3700	12.4	149	4.7	4.4	8.5
Rubidium	38.8	688	1.53	58.3[c]	12.1	4.8	9.0
Ruthenium	2310	4120	12.2	116.3	7.7	4.1	9.6
Silicon	1412	3270	2.34	138.5	. . .	. . .	7.6
Silver	960.8	2163	10.5	425	1.63	4.1	19.1
Sodium	97.8	883	0.97	128	4.7	5.5	71
Strontium	770	1375	2.6	. . .	23[b]	. . .	100
Tantalum	2980	5370	16.6	57.55	13.5	3.5	6.5
Tellurium	450	988	6.24	3.8	1.6 × 10^{5b}	. . .	[e]
Thallium	304	1473	11.85	45.5	16.6	5.2	30
Thorium	1755	4290	11.5	49.2	14	4.0	11.2
Tin	231.9	2625	7.3	73.2	12.6	4.6	23.5
Titanium	1667	3285	4.5	21.6	54	3.8	8.9
Tungsten	3400	5555	19.3	174	5.4	4.8	4.5
Uranium	1132	4400	19.05(α), 18.89(β)	28	27	3.4	[f]
Vanadium	1902	3410	6.1	31.6	19.6	3.9	8.3
Zinc	419.5	911	7.14	119.5	5.96	4.2	31
Zirconium	1852	4400	6.49	22.6	44	4.4	5.9

[a]By convention, the signs on orders of magnitude are reversed when they are moved from column to stub (leftmost column) or from column to column heading in a table; i.e., the first value here is read 4.5 × 10^{-3} K^{-1}. The convention is not always observed by different authors, but it has been followed throughout Sec. 6.
[b]0°C.   [c]Solid.
[d]17.4 || a axis, 8.1 || b axis, 54.3 || c axis.   [e]1.7 || c axis, 27.5 $\perp$ c axis.
[f]α-Uranium at 25–300°C 23 || a axis, −3.5 || b axis, 17 || c axis; β-uranium at 20–720°C, 4.6 || c axis, 23 $\perp$ c axis.

TABLE 6-2   Electrical Resistivity of Some Alloys and Compounds[51,52]

Material	Conductivity, % IACS	Resistivity, $\mu\Omega \cdot cm$	Material	Conductivity % IACS	Resistivity, $\mu\Omega \cdot cm$
92.5 Ag–7.5 Cu	85	2			
60 Ag–40 Pd	8	23	TaN	1.2	135
97 Ag–3 Pt	50	3.5	ZrN	12.6	13.6
90 Pt–10 Ir	7	25	TiN	7.9	21.7
96 Pt–4 W	5	36	VN	2	85.9
70 Pd–30 Ag	4.3	40	TaC	5.7	30
90 Au–10 Cu	16	10.8	ZrC	2.7	63.4
75 Au–25 Ag	16	10.8	SiC	1–17	100–200
78.5 Ni–20 Cr–1.5 Si	1.6	108.05	WC	14	12
71 Ni–29 Fe	9	19.95	MoSi$_2$	4.5	37
80 Ni–20 Cr	1.5	112.2	CbSi$_2$	26	6.5
Carbon steel 0.65% C	9.5	18	ZrB$_2$	19	9
Electrical Si sheet steels	3–9.5	18–52	LaB$_6$	11.5	15
Stainless steel, type 302	3	72	TiB$_2$	8.6	20
Type 316	2.5	74			

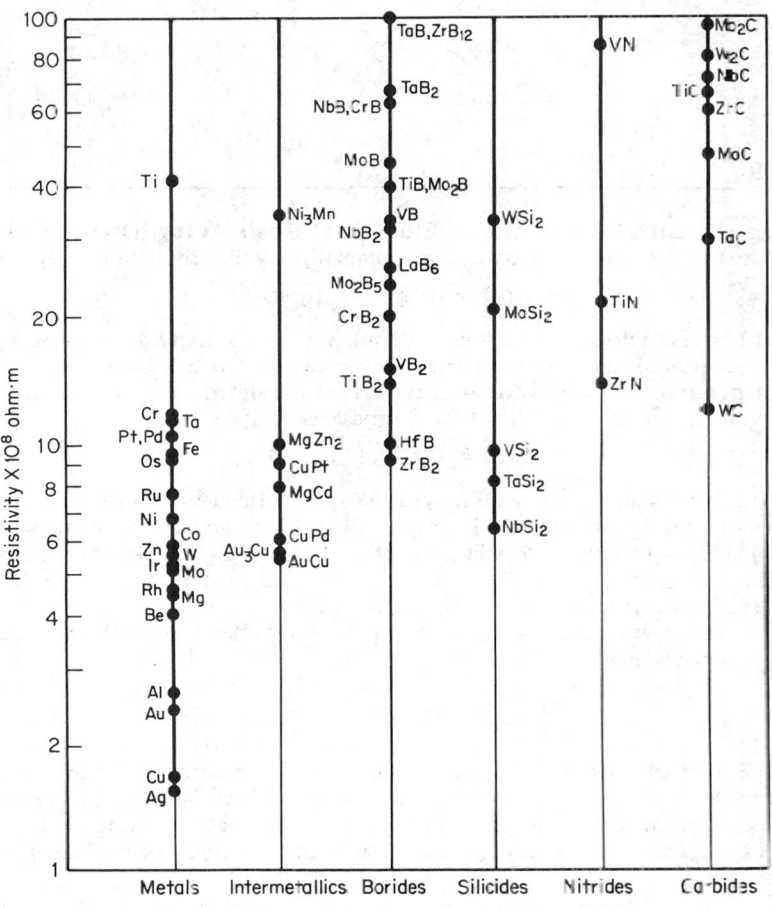

Fig. 6-1. Electrical resistivity of various materials.[54]

6. Sheet Resistivity. The resistance of a conductor or resistor in sheet form is, according to Eq. (6-5),

$$R = \frac{\rho l}{db} = \frac{\rho}{d}\frac{l}{b}$$ (6-7)

where d, l, and b are the conductor thickness, length, and width, respectively; ρ/d is called *sheet resistivity* and is measured in ohms per square. l/b is the number of squares. The resistance of a sheet is its sheet resistivity multiplied by the number of squares.

7. Surface Resistivity. At high frequencies, electric current is mainly conducted near the surface of the conductor. The depth at which the current density falls to $1/e$ of its value at the surface is called the *skin depth* δ. The surface resistivity R_s is the dc sheet resistivity of a conductor having a thickness of one skin depth

$$R_s = \rho/\delta = 1/\sigma\delta$$ (6-8)

where ρ = electrical resistivity ($\Omega \cdot m$), δ = thickness (m), and σ = electrical conductivity (Ω/m).

The skin depth and therefore R_s are functions of the ac frequency. Values of R_s are given in Table 6-3.

TABLE 6-3   High-Frequency Resistivity Characteristics of Several Metals[63]

Material	dc resistivity ρ (relative to copper)	Skin depth δ at 2 GHz, μm	Surface resistivity $(R_s \times 10^7)/f^{1/2}$, $(\Omega/\square)s^{1/2}$
Ag	0.95	1.4	2.5
Cu	1.0	1.5	2.6
Au	1.36	1.7	3.0
Al	1.6	1.9	3.3
W	3.2	2.6	4.7
M_o	3.3	2.1	4.7
Ni	5.1	0.31	5.5
Cr	7.6	4.0	7.2

8. Temperature Coefficient of Electrical Resistivity. Over a narrow range of temperature the electrical resistivity changes approximately linearly with temperature

$$\rho(t_2) = \rho(t_1)[1 + \alpha_{t_1}(t_2 - t_1)]$$ (6-9)

where $\rho(t_1)$ = resistivity at temperature t_1 ($\Omega \cdot m$), $\rho(t_2)$ = resistivity at temperature t_2 ($\Omega \cdot m$), and α_{t_1} = temperature coefficient of electrical resistivity (K^{-1}) (see Table 6-1).

9. Temperature Coefficient of Electrical Resistance. The electrical resistance changes with temperature similarly to the electrical resistivity

$$R(t_2) = R(t_1)[1 + \alpha_{t_1}(t_2 - t_1)]$$ (6-10)

where $R(t_2)$ = resistance at temperature t_2 (Ω), $R(t_1)$ = resistance at temperature t_1 (Ω), and α_{t_1} = temperature coefficient of electrical resistance (K^{-1}). It is customary to take $t_1 = 20°C$. Equation (6-10) holds in a range of temperatures of about 0 to 100°C. The resistance of most conductors increases with temperature. The resistance of carbon, semiconductors, several thin-film resistors, and electrolytes can decrease with temperature.

When the temperature of reference t_1 is changed to some other value t, the coefficient α_{t_1} will change to a new value α_t

$$\alpha_t = \frac{\alpha_{t_1}}{1 + \alpha_{t_1}(t - t_1)}$$ (6-11)

Equation (6-10) does not take into account changes in the dimensions of the conductor with temperature. Such changes depend on the temperature coefficient of linear expansion, which is generally much smaller than the temperature coefficient of electrical resistance. If dimensional changes are neglected, the coefficient of electrical resistance equals the coefficient of electrical resistivity.

10. Resistivity-Temperature Constant. The change of resistivity of a material per kelvin is called the resistivity-temperature constant. For example, the resistivity of copper changes at the rate of 68 $p\Omega \cdot m/K$ irrespective of the temperature.

11. Matthiessen's Rule. Electrical resistivity originates from two main sources: thermal scattering of the electrons (*thermal resistivity*) and scattering of electrons from imperfection in the lattice (*residual resistivity*). According to Matthiessen's rule,[1]* the total electrical resistivity ρ is the sum of thermal resistivity ρ_T and the residual resistivity ρ_R

$$\rho = \rho_R + \rho_T \tag{6-12}$$

At very low temperatures, $\rho_T \ll \rho_R$, and the electrical resistivity is mainly the residual resistivity. The residual resistivity arises from point defects (vacancies, solute atoms or interstitials), line defects (dislocations), surface defects (stacking faults or grain boundaries), and volume defects (voids or second phase). The increase in resistivity due to vacancies is approximately 10 $n\Omega \cdot m/$ (at %) of concentration.[2] The residual resistivity due to dislocations is, roughly,[3-*] $\rho_R = 10^{-25}C_D$ $\Omega \cdot m$, where C_D is the dislocation density in meters per cubic meter. The effect of impurities on residual resistivity is discussed under Nordheim's rule (Par. **6-13**).

12. Resistivity Ratio. The resistivity ratio is the ratio of electrical resistivity at 298 K to that at 4.2 K. This ratio is a measure of the perfection of the conductor. Higher ratios are obtained for more perfect materials since their residual resistivity at 4.2 K is smaller. Ratios up to 100,000 have been obtained.

13. Nordheim's Rule. According to Nordheim's rule, the addition of solute of concentration X to a metal results in an increase in the residual resistivity of

$$\rho_R(X) = CX(1 - X) \tag{6-13}$$

where $\rho_R(X)$ = increase in residual resistivity ($\Omega \cdot m$), C = const [$\Omega \cdot m/$(at %)], and X = solute concentration (at %).

For small values of X, Eq. (6-13) can be written

$$\rho_R(X) = CX \tag{6-14}$$

The increase in residual resistivity is due to electron scattering by the disordered solute atoms.

*Superior numbers correspond to numbered references in Par. **6-257**.

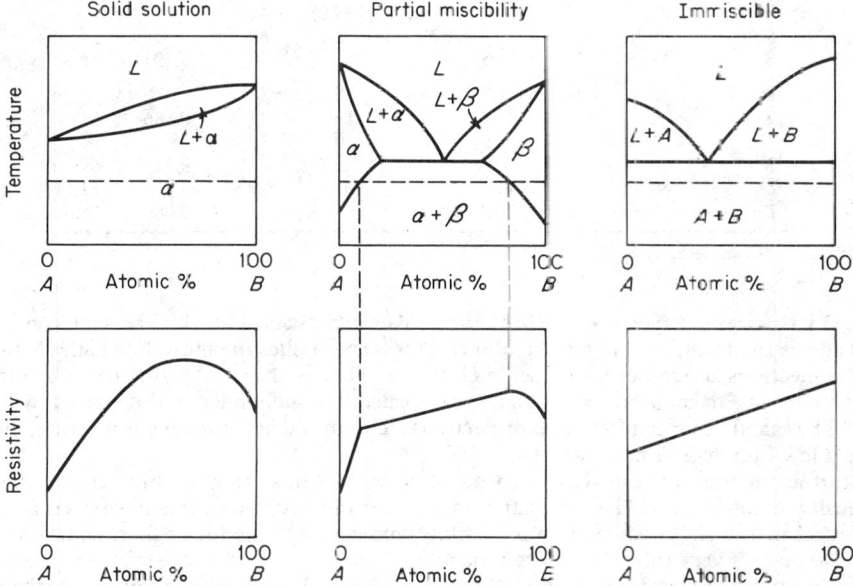

Fig. 6-2. Typical resistivity of binary alloys as a function of composition. α and β are solid solutions rich in A and B, respectively. L denotes liquid.[65]

The appearance of intermediate phases or ordered solutions generally lowers the resistivity of the alloy. Schematic diagrams of the electrical resistivity vs. composition are given in Fig. 6-2. The highest increase in resistivity is obtained in binary systems with complete miscibility.

For further information see Refs. 7 and 8. Values of C are given in Table 6-4.

TABLE 6-4   Maximum Solid Solubility c_{max} and Atomic Resistivity Increase C [N·m/(at %)] for Cu, Ag, Au, and Al[8,65,66]

| | Base metal | | | | | | | |
| | Cu | | Ag | | Au | | Al | |
Solute	c_{max}	C	c_{max}	C	c_{max}	C	c_{max}	C
Ag	4.9	1.4	....	....	*	3.6	23.8	115
Al	19.6	12.5	20.34	5	6	18.7		
As	....	68	8.8	85	....	80		
Au	*	55	*	36				
Be	16.4	62						
Bi	...	...	3	73	...	(65)		
Cd	...	30	42.2	3.8	32.5	6.3	0.11	0.57
Co	12.0	63.5	...	...	23.5	61		
Cr	0.8	(36)	...	...	(23)	42.5	0.40	77
Cu	...	...	14.1	0.77	*	4.5	24.8	7.85
Fe	4.5	93	....	....	75	79	0.025	53
Ga	19.9	14.2	18.7	23.6	13	22	....	2.5
Ge	11.8	37.9	9.6	55	3.2	52		
In	11	10.6	20	17.8	12.6	13.9		
Ir	...	57						
Mg	7	6.5	29.3	19.5	(25)	13	16.26	4.9
Mn	*	29	47	16	...	24.1	0.90	59.7
Ni	*	12.5	...	...	*	7.9	0.023	17.7
P	3.5	6.7						
Pb	....	...	2.8	46.5	...	(39)		
Pd	*	8.9	*	4.4	*	4.1		
Pt	*	21	40.5	16	*	10.1		
Rh	(30)	44	...	...	(9.0)	41.5		
Sb	6	54	7.2	72.5	1.1	68		
Si	11.25		39.5	...	...	...	1.59	6.5
Sn	9.1	28.8	11.5	43.6	6.8	33.6		
Ti	...	...	...	...	(11)	129	0.566	51
Tl	...	...	7.5	22.7	0.9	19		
V	...	...	...	...	...	...	0.32	69
Zn	38.3	3.2	40.2	6.4	31	9.5	66.4	2.11
Zr	...	...	...	...	...	...	0.085	58.5

*Complete.

14. Thickness Effect on Electrical Resistivity. The thickness of conducting films affects the resistivity. When the film thickness approaches the mean free path of the conduction electrons, a significant number of electrons collide with the film surfaces. The increase in resistivity due to collision with surfaces was treated by Sondheimer[9] and reviewed by Campbell.[10] The calculated mean free paths of electrons are 10 to 100 nm at room temperature, increasing to a few hundreds of nanometers at −200°C.

A further increase in resistivity is encountered when films are very thin, so that they are physically discontinuous. The conduction mechanisms in very thin films have been reviewed by Neugebauer.[11] The conduction in these films proceeds by tunneling or thermionic emission. The resistance of very thin films is often nonohmic.

The resistivity of thin films is normally higher than bulk even when films are thicker than the mean free path of electrons. This increase is due to imperfections in the films such as grain boundaries, included oxides, and voids. Films deposited at higher substrate temperatures normally show values closer to bulk.

15. Thickness Effect on Temperature Coefficient of Electrical Resistivity. The increase of resistivity in very thin films results in great part from geometrical separation of the conducting islands. The resistivity is governed by processes such as tunneling or thermionic emission. An increase in temperature will lower the resistivity since conduction mechanisms such as tunneling and thermionic emission are assisted by the increased temperature. The temperature coefficient of electrical resistivity of very thin films will therefore be negative for most materials.

16. Thermal Conductivity. The heat flow rate along a rod is proportional to its cross section and to temperature gradient dT/dX along the rod (see Fig. 6-3)

$$q = -\lambda A \, dT/dX \tag{6-15}$$

where dT/dX = temperature gradient (K/m), A = cross section (m²), and λ = thermal conductivity (W/m·K). The thermal conductivity is a function of temperature (examples shown in Fig. 6-4). Values of λ are listed in Table 6-1.

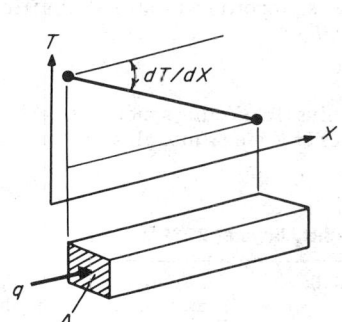

Fig. 6-3. Heat flow due to a temperature gradient.

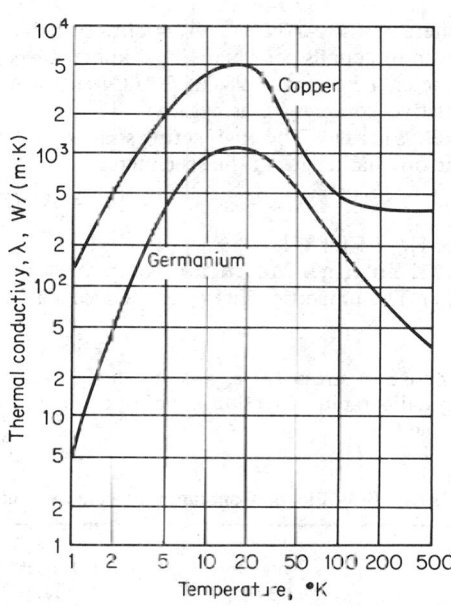

Fig. 6-4. Thermal conductivity of copper and germanium as a function of temperature.[57]

17. Wiedemann-Frantz Ratio. The Wiedemann-Frantz ratio L is defined as

$$L = \lambda/\sigma T \tag{6-16}$$

where λ = thermal conductivity (W/m·K), σ = electrical conductivity (S/m), and T = temperature (K). This ratio, approximately constant for most pure metals, is 20 000 to 30,000 (μV/K)².

18. Temperature Coefficient of Linear Expansion. Materials change their dimensions with temperature. The temperature coefficient of linear expansion α_L is defined by

$$\alpha_L = \frac{1}{l}\frac{dl}{dT} \tag{6-17}$$

where l = any linear dimension (e.g., length of rod) of material (m) and T = temperature (K). If the length l_0 of a rod at temperature T_0 is known, the length of the rod at temperature T can be calculated from Eq. (6-17). When α_L is constant,

$$l = l_0[1 + \alpha_L(T - T_0)] \tag{6-18}$$

Values of α_L are given in Table 6-1.

A more exact relation between l and l_0 involves a power series of the temperature difference.

$$l = l_0[1 + \alpha_L(T - T_0) + \beta_L(T - T_0)^2 + \cdots] \tag{6-19}$$

19. Volume Coefficient of Thermal Expansion. The volume coefficient of thermal expansion α_V is defined as

$$\alpha_V = \frac{1}{V}\frac{dV}{dT} \tag{6-20}$$

where V = volume (m³) and T = temperature (K). Since α_L and α_V are small, $\alpha_V \approx 3\alpha_L$.

20. Specific Heat. Specific heat at constant pressure c_p (J/K·kg), is defined as the amount of heat needed to raise the temperature 1 kg of material by 1 K at constant pressure. Specific heat at constant volume c_v (J/K·kg), is defined as the amount of heat needed to raise the temperature 1 kg of material by 1 K at constant volume. The difference between c_p and c_v is small in solids.

21. Tensile Stress. The engineering tensile stress is defined as the applied force per unit area

$$\sigma = F/A_0 \tag{6-21}$$

where σ = stress (N/m²), A_0 = cross section (m²) (see Fig. 6-5), and F = force (N). Stress is often given in pounds per square inch, kilograms per square millimeter, or kilograms per square centimeter; 1 kg/cm² = 98,066.5 Pa (N/m²); 1 lb/in² = 6,894.8 Pa. Tensile stress is defined to be positive; compression is negative.

22. Strain. The engineering strain ϵ_x is defined as the elongation of the material divided by the original length of the specimen.

$$\epsilon_x = \Delta l/l_0 = (l - l_0)/l_0 \tag{6-22}$$

See Fig. 6-5 for l, l_0.

23. Young's Modulus. For small strains, the stress is proportional to the strain (Hooke's law). The proportionality constant is called *Young's modulus,*

$$\sigma = \epsilon_x E \tag{6-23}$$

where σ = stress (Pa), ϵ_x = strain, and E = Young's modulus (Pa). Young's modulus depends on crystallographic direction, even for cubic materials. Values of Young's modulus are compiled in Table 6-5.

TABLE 6-5   Elastic Constants for Polycrystalline Metals and Alloys at 20°C[80]

Metal	Poisson's ratio	Bulk modulus K, GPa	Young's modulus E, GPa	Rigidity modulus G, GPa
Aluminum	0.345	75.2	70.6	26.2
Chromium	0.210	160.2	279	115.3
Copper	0.343	137.8	129.8	48.3
Iron, soft	0.293	169.8	211.4	81.6
Cast*	0.27	109.5	152.3	60
Lead*	0.44	45.8	16.1	5.59
Magnesium	0.291	35.6	44.7	17.3
Molybdenum	0.293	261.2	324.8	125.6
Nickel, unmagnetized, soft*	0.312	177.3	199.5	76
Hard*	0.306	187.6	219.2	83.9
Niobium	0.397	170.3	104.9	37.5
Silver	0.367	103.6	82.7	30.3
Steel, mild	0.291	169.2	211.9	82.2
Stainless†	0.283	166	215.3	83.9
Tantalum	0.342	196.3	185.7	60.2
Tin	0.357	58.2	49.9	18.4
Titanium	0.361	108.4	120.2	45.6
Tungsten	0.280	311	411	160.6
Tungsten carbide	0.22	319	534.4	219
Vanadium	0.365	158	127.6	46.7
Zinc	0.249	69.4	104.5	41.9

*Approximate values for materials of variable composition.
†Composition 0.2% C, 0.5% Si, 0.7% Mn, 2.0% Ni, 18.0% Cr by weight.

24. Poisson's Ratio. When a material is stressed in the x direction (Fig. 6-5), it deforms also in the transverse directions. The width b_0 will contract to a value b upon elongation from l to l_0. The transverse strain is $\epsilon_y = (b - b_0)/b_0$. Poisson's ratio ν is defined as

$$\nu = -\ \epsilon_y/\epsilon_x \tag{6-24}$$

Values of Poisson's ratio are compiled in Table 6-5.

25. Shear. The shear stress is defined as the shear force F per unit area A_0 (see Fig. 6-6)

$$\tau = F/A_0 \tag{6-25}$$

where τ = shear stress (Pa), F = shear force (N), and A_0 = area (m²). The shear strain γ is defined from Fig. 6-6 as

$$\gamma = \Delta/l_0 \tag{6-26}$$

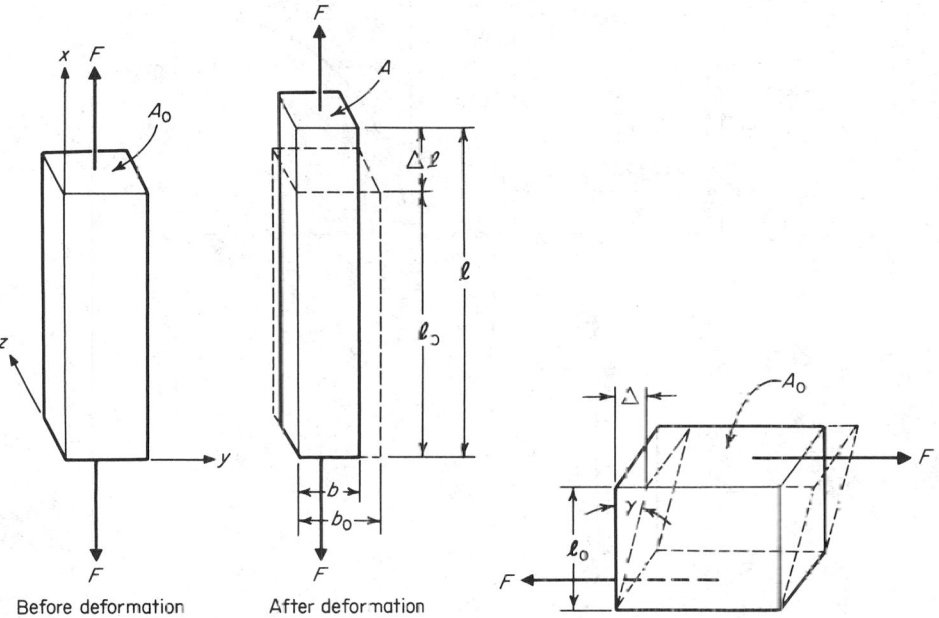

Before deformation   After deformation

Fig. 6-5. Uniaxial tensile force and deformation Δl.   **Fig. 6-6.** Shear force F and shear strain γ.

26. Shear Modulus. For small strains the shear stress is proportional to the shear strain

$$\tau = \gamma G \tag{6-27}$$

where τ = shear stress (Pa), γ = shear strain, and G = shear (or rigidity) modulus (Pa). (See Table 6-5.)

27. Compressibility. A material of volume V will contract by ΔV upon application of pressure Δp. The compressibility is the fractional volume change per unit of applied pressure,

$$-\Delta V/V = \beta \Delta p \tag{6-28}$$

where V = volume (m³), ΔV = volume change (m³), Δp = applied pressure (Pa), and β = compressibility (m²/N). Values of $K = 1/\beta$ (*bulk modulus*) are compiled in Table 6-5.

28. Stress-Strain Diagrams. Typical stress-strain diagrams are shown in Fig. 6-7. Figure 6-7a, typical of low-carbon steels, displays a distinct *elastic region* up to a sharp *yield point* σ_y. The elastic deformation is reversible; i.e., it can be removed by removal of the stress. Beyond the yield stress (or flow stress) the *plastic region* starts. The plastic deformation is not reversible; it remains after removal of the stress. The stress increases with deformation in the plastic region (*strain hardening*) until it reaches the *ultimate tensile strength* (UTS).

Figure 6-7b shows a typical diagram for ductile metals such as aluminum or copper. No distinct yield point is seen. It is customary to define an *offset yield strength* by plotting a line parallel to the slope at the origin at an offset of $\epsilon = 0.2\%$ (see Fig. 6-7*b*) and noting its intersection with the stress-strain curve. Figure 6-7*c* shows a typical stress-strain diagram for a *brittle material*. The material breaks without any appreciable plastic deformation.

The ductility of materials is characterized by the final *elongation* of the sample after fracture

$$\text{Elongation} = [(l - l_0)/l_0](100) \tag{6-29}$$

where l_0 = original length (m), l = length after fraction (m), and by the final *reduction* in the cross section at fracture

$$\text{Reduction} = [(A_0 - A)/A_0](100) \tag{6-30}$$

where A_0 = original cross section (m²) and A = cross section after fracture (m²). See Table 6-6 for compilation of mechanical properties.

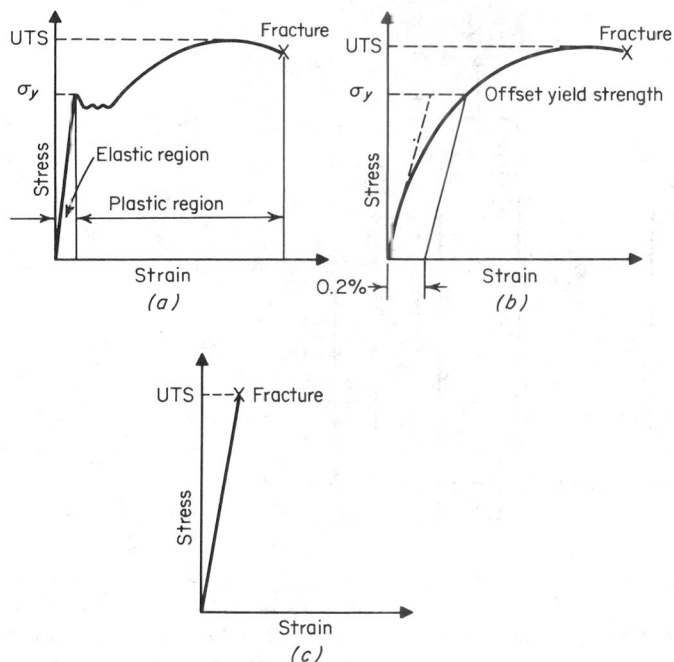

Fig. 6-7. Schematic stress-strain diagrams: (*a*) sharp yield point; (*b*) ductile material without sharp yield point; (*c*) brittle material; σ_y = yield point. UTS = ultimate tensile strength.

29. Hardness. Hardness measures the resistance of the material to indentation. Hardness tests measure the plastic deformation (the size or depth) of an indentation. Brinell hardness tests use spheres as indenters; the Vickers test uses pyramids. Rockwell tests use cones or spheres. Microhardness tests for specimens are also available, using the Knoop method with miniature pyramid indenters. Hardness values are given in Table 6-6.

30. Notch Toughness. Notch toughness is measured by the energy necessary to fracture a standard notched specimen. The notch toughness is a function of temperature. Many materials display a transition from high toughness (ductility) to low toughness (brittleness). The transition temperatures for low-carbon steels are in the region of -20 to $-70°C$.

31. Fatigue. Under cycling loading, materials tend to break after a large number of cycles even when stressed below their ultimate strength. Figure 6-8 shows the fracture stress vs. number-of-cycles curve (*S-N* curve) for steel. Ferrous metals tend to exhibit a lower limit of stress under which they do not fail due to fatigue. This limit is the *endurance limit*. Nonferrous metals do not show a definite endurance limit. *S-N* diagrams are usually reported for fully reversing stress.

Fatigue fracture starts with fine cracks that propagate from the surface through the specimen. The fatigue strength is therefore very sensitive to surface finish and tensile-stress risers such as keyholes and notches.

If a steady tensile stress is imposed on the alternating stress, the fatigue life will decrease. A compressive stress will increase fatigue life. Prestressing or hardening of the surface also results in an increased life. Fatigue strength is lower in corrosive environments (called *corrosion fatigue*).

32. Creep. Materials will continue to deform under a given stress at high temperatures. As a function of time, three distinct stages occur. The first stage, primary creep, is characterized by a decelerating creep rate; the second stage is the steady creep region; during the third stage the creep rate accelerates and fracture finally results. The stress-rupture test records the rupture times at a given stress and temperature.

TABLE 6-6   Mechanical Properties of Metals and Alloys at Room Temperature[61,62]

Material	Composition	Condition	Yield point or 0.2% proof stress, MPa	Ultimate tensile stress, MPa	Elongation on 2 in, %	Hardness*
Ag	99.9	Annealed 600–650°C	76†	137	50	26 VHN
		Hard	...	380	4	90 VHN
Al	99.95	Rolled rod 0	...	55	61	17 B
Au	99.99	Soft, cast	0	121	30	33 B
		Hard, 60% red.	212	228	4	58 B
Co	99.9	Soft	190	240	4–8	124 B
		Hard	...	675	2–8	165 B
Cu	99.997	Annealed	340	351	60	
		Rod, cold-drawn	34	213	14	37 RB
Ni	>99.0	Annealed	138	482	40	100 B
		Cold-drawn	482	654	25	170 B
Pt	99.99	Annealed	...	120–130	25–40	38–40 VHN
		50% cold-rolled	180	200	3	92 VHN
Pd	99.9	Annealed	35	190	40	37 VHN
		50% cold-drawn	...	320	1.5	106 VHN
Ta	99.98	Annealed	180	200	36	90 VHN
	99.95	Cold-rolled	330	410	5	160 VHN
W		Swaged, recrys.	195	405	16	(200 VHN)
	99.9	Swaged	...	1,750	1–4	450–490 VHN
Aluminum alloys:						
1100	1% Si	O	27.5	69	45	19 B
1100		H18	124	131	15	35 B
3003	1–1.5% Mn	O	41	110	40	28 B
3003		H18	186	200	10	55 B
5056	4.5–5.6% Mg	O	152	290	35	65 B
5056		H38	345	415	15	100 B
7075	1.2–2% Cu, 2.1–2.9% Mg, 5.1–5.6% Z	O	104	227	...	60 B
7075		T6	505	570	...	150 B
Copper alloys:						
OHFC, copper		Wire, soft	...	241	35	
		Wire, hard	...	380	1.5	
Gilding metal	5% Zn	Annealed, strip	69	234	45	46 RF
		Extra hard	390	435	4	73 RB
Red brass	15% Zn	Annealed	70–130	280–320	48	
		Half hard	350	405	12	65 RB
		Extra hard	435	555	4	83 RB
Yellow brass	35% Zn	Annealed	100–157	326–376	54–65	58–78 RF
		Half hard	426	526	8	80 RB
		Extra hard	44	605	5	87 RB
Phosphor bronze	5% Sn	Annealed	150–220	340–390	57–48	33–46 RB
		Hard	560	575	8	89 RB
		Extra spring	710	725	3	98 RB
Beryllium copper	1.9% Be	Annealed	...	430–550	35	45–78 RB
		HT (cold worked and precipitation-hardened)	1,070	1,420	2	42 RC
Steels:						
C1010	0.08–0.13 C	Hot-rolled	188	340	28	95 B
		Cold-drawn	318	383	20	105 B
C1080	0.77–0.88%	Hot-rolled	440	810	10	229 B
12% Mn steel	12% Mn	Tempered 600°F	1,480	1,520	10	
Stainless steel	9% Ni, 19% Cr	Annealed	340	590	60	160 B
Type 304		Cold-rolled	1,110	1,280	...	400 B

*VHN = Vickers hardness number; B = Brinell; RB = Rockwell B; RC = Rockwell C; RF = Rockwell F.
†0.01% proof stress.

33. Cold Working. Plastic deformation at room temperature is called cold working. Hardening and strengthening of a material by plastic deformation are called *work hardening*. Wire drawing or other cold-working processes raise the yield strength of metals. Plastic deformation may introduce a preferred crystallographic orientation, sometimes referred to as *texture*. Since some physical properties such as elastic constants and magnetic permeability are anisotropic, the texture strongly affects them.

34. Solute Effects on Mechanical Properties. A solute generally raises the yield and ultimate tensile strength of a material. Controlled precipitation of the solute from supersaturated solutions brings an even more drastic improvement in mechanical properties. The precipitation is achieved by first heat treating above the solubility limit to dissolve all the solute (solution treatment), then quenching to room temperature, and finally heat treating at 100 to 200°C to form very fine precipitates. The strengthening due to precipitation is often referred to as *age hardening* or *precipitation hardening*. Figure 6-9 shows the change in mechanical properties with aging time for an Al alloy.

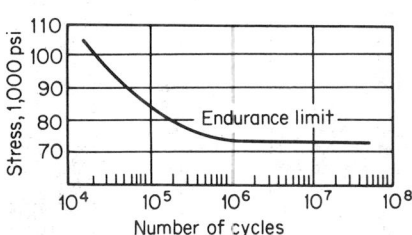

Fig. 6-8. Stress vs. number of cycles to failure for a hot-worked 4340 steel bar.[68]

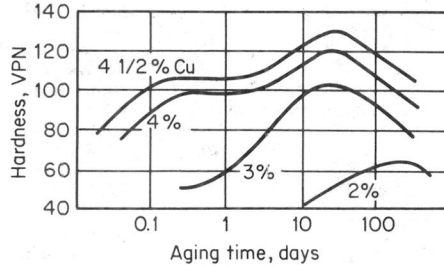

Fig. 6-9. Hardness of various Al–Cu alloys aged at 130°C, as a function of time.[69]

35. Annealing. Softening of materials by exposure to high temperature for long times is referred to as annealing.

36. Stress Corrosion Cracking. Rapid corrosion of highly cold-worked materials is known as stress corrosion cracking. Examples are brass in mercury or steel in sodium hydroxide.

37. Melting Point. The temperature at which an element or compound transforms from solid to liquid under equilibrium conditions is the melting point (mp). The melting point is a function of the external pressure. See Table 6-1 for values of melting points.

38. Boiling Point. The temperature at which an element or compound transforms from liquid to vapor under equilibrium conditions is the boiling point (bp). The boiling point is a function of the external pressure. See Table 6-1.

39. Allotropic Forms. Different crystallographic forms of the same element or compound are called allotropic forms. For example, iron can exist in a body-centered cubic (bcc) form or a face-centered cubic (fcc) form.

40. Equilibrium Phase Diagrams. Phase diagrams show which phase or phases are most stable in a given system of components at a given temperature and composition. Most diagrams are binary (for two elements) or ternary. Sections through diagrams with more than three components are sometimes available.

Figure 6-10 is an example of a binary diagram with complete *miscibility*. Unlike the pure materials, the alloy does not show a sharp melting point. The melting occurs over a wide temperature range. Figure 6-11 shows a system with partial miscibility. This is called a *eutectic* diagram. Except for the eutectic composition (38.1 wt % Pb) the alloys do not have a sharp melting point. Figure 6-12 shows a binary diagram of gold and tin. This diagram shows a series of eutectics, peritectics, and *intermetallic compounds*.

41. Diffusion. Diffusion is mass transport in a solid, liquid, or gas. The driving force for diffusion is generally chemical, originating from concentration gradients. The diffusion flux J in a binary alloy is proportional to the chemical gradient of the solute (Fick's law)

$$J = -D \, dC/dX \tag{6-31}$$

where C = concentration (atoms/m³), X = direction of the material flux (m), J = flux (atoms/m²·s), and D = diffusion coefficient (m²/s).

The diffusion coefficient is temperature-dependent and can be approximated by

$$D = D_0 e^{-Q/kT} \tag{6-32}$$

where D_0 = preexponential constant (m²/s), Q = activation energy (J, normally given in cal/mol or eV), k = Boltzmann's constant (J/K), and T = temperature (K). The diffusion coefficient is generally concentration-dependent. The diffusion coefficients are generally larger for structurally imperfect solvents. Thus heat treatment, mechanical working, and radiation damage appreciably increase the diffusion coefficient. Diffusion in thin films, which normally are structurally imperfect, is much faster than in bulk materials.

Diffusion distance, or solute penetration, can be roughly estimated by the formula $X = \sqrt{Dt}$, where X = diffusion distance (m) and t = diffusion time (s).

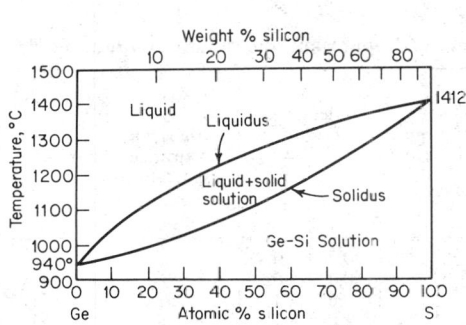

Fig. 6-10. Germanium-silicon phase diagram.[66]

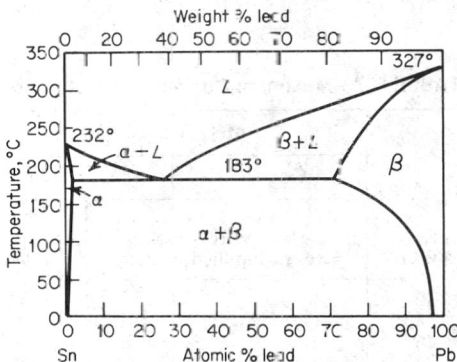

Fig. 6-11. Lead-tin phase diagram. α and β are the tin- and lead-rich solid solutions, respectively; L denotes liquid.[66]

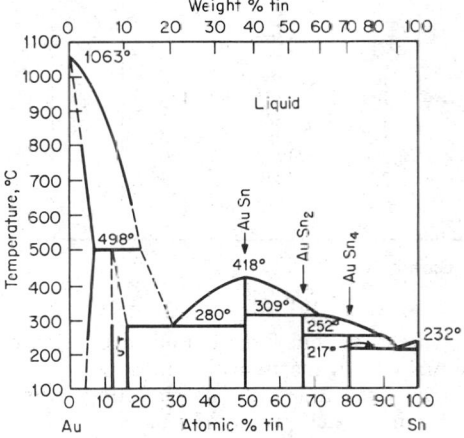

Fig. 6-12. Gold-tin phase diagram.[66]

42. Electromigration. An electric current density J induces in bulk conductors an atomic flux J_a (atoms/m²·s). The atomic flux is created by either simple electrostatic forces or by a momentum transfer from the electrons. Huntington[1] proposed the relation

$$J_a = (ND/kT)Z^* e\rho J \tag{6-33}$$

where N = density of metallic ions (m⁻³), D = diffusion coefficient of vacancies (m²/s), which can be expressed in terms of a constant D_0 and an activation energy Q,

$$D = D_0 e^{-Q/kT}$$

J = current density (A/m^2), k = Boltzmann's constant (J/K), ρ = resistivity of the conductor ($\Omega \cdot$m), T = temperature (K), and Z^*e = *effective charge* of the metallic ions (C).

The effective charge on the metallic ions can be positive or negative, depending on whether the electrostatic field which creates an atomic flux toward the cathode is larger than the momentum transfer which creates a flux toward the anode.

Electromigration in thin films proceeds much faster than in bulk conductors since the diffusion processes in thin films are generally much faster than bulk values.[13,14]

43. Current-Carrying Capacity. The current-carrying capacities of wires, cables, and bus bars are limited by heating effects produced by the current. The permissible temperature rise of bus bars is around 30°C above an ambient of 40°C. Current densities for copper buses are about 10^6 A/m^2 ($= 10^2$ A/cm^2). Aluminum current densities are 75% of those permitted in copper. Current-carrying capacities of various wires and cables are given in Table 6-7. Figure 6-13 shows the current-carrying capacity of copper conductors on printed-circuit boards.

TABLE 6-7   Maximum Current Capacity (Amperes) of Copper and Aluminum Conductors[*70]

	MIL-W-5088				National Electrical Code	Underwriters Laboratory		American Insurance Association	500 c mils/A
	Copper		Aluminum						
Size, AWG	Single-wire	Wire-bundled	Single-wire	Wire-bundled		60°C	80°C		
30	...	...	...	...	...	0.2	0.4	...	0.20
28	...	...	...	...	...	0.4	0.6	...	0.32
26	...	...	...	...	...	0.6	1.0	...	0.51
24	...	...	...	...	...	1.0	1.6	...	0.81
22	9	5	...	...	...	1.6	2.5	...	1.28
20	11	7.5	...	...	...	2.5	4.0	3	2.04
18	16	10	...	...	6	4.0	6.0	5	3.24
16	22	13	...	...	10	6.0	10.0	7	5.16
14	32	17	...	...	20	10.0	16.0	15	8.22
12	41	23	...	...	30	16.0	26.0	20	13.05
10	55	33	...	...	35	...	...	25	20.8
8	73	46	58	36	50	...	...	35	33.0
6	101	60	86	51	70	...	...	50	52.6
4	135	80	108	64	90	...	...	70	83.4
2	181	100	149	82	125	...	...	90	132.8
1	211	125	177	105	150	...	...	100	167.5
0	245	150	204	125	200	...	...	125	212.0
00	283	175	237	146	225	...	...	150	266.0
000	328	200	...	...	275	...	...	175	336.0
0000	380	225	...	...	325	...	...	225	424.0

*For further information consult Ref. 73.

The current-carrying capacity of thin films is limited by electromigration (see Par. **6-42**). Thanks to the good thermal conduction from the films, it is possible to conduct currents in excess of 10^9 A/m^2. At higher current densities irregularities in the metal transport will create holes in the conductors, leading to their eventual failure.[14]

The lifetime of thin-film conductors depends strongly on their composition, structural perfection, width, thickness, length, temperature, and temperature gradients. There is no good set of tabulated lifetime data that encompasses all the above variables.

FORMS OF CONDUCTORS

44. Wire. Wires are the most common form of conductors for power transmission, smaller electric signals, or as resistors. Wires are either solid or stranded. Most wires are round; occasionally square or rectangular conductors are used, such as integrated-circuit external leads.

Insulated wires are most often used in electronic circuits. The insulation provides electrical insulation and mechanical and chemical protection. There are several wire gages:

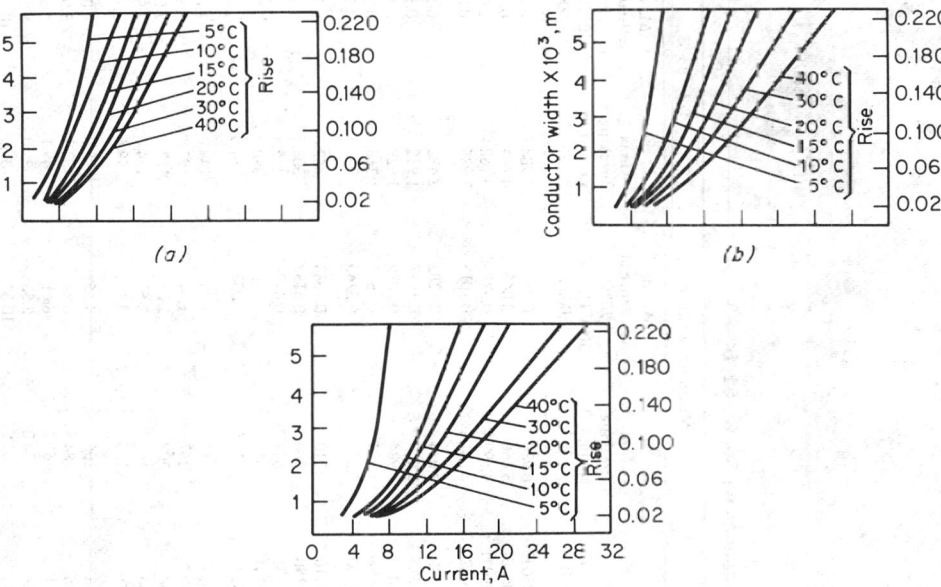

Fig. 6-13. Temperature rise vs. current for (a) 1-oz copper; (b) 2-oz copper; (c) 3-oz copper.[u]

AWG (American B&S): This wire gage divides the range from 0.0050 in (AWG No. 36) to 0.4600 in (AWG No. 0000) into 39 intervals. Sizes progress in geometrical fashion with a ratio of 1.1229322 between adjacent gages.

BWG (Birmingham wire gage): An English designation based on the number of drawings necessary to produce the wire.

Metric wire gage: 0.1 mm is No. 1, 0.2 mm is No. 2, etc.

Steel wire gage: Used exclusively for steel wire.

Wire size: A direct way to specify wire size is to give the diameter in mils (1 mil = 0.001 in).

Circular mil: A unit of area that equals the cross-sectional area of a 1-mil-diam. wire. 1 cmil = 0.7854 mil².

Table 6-8 shows the properties of copper wire.

45. Stranded Wire. A group of wires used as a single wire is called stranded wire. Due to their high flexibility, stranded wires are the most commonly used conductors.

Stranded conductors are based on a 1-wire or a 3-wire core. The total number of wires is $3n(n + 1) + 1$ for a 1-wire core (that is, 1, 7, 19, . . .) and $3n(n + 2) + 3$ for a 3-wire core (that is, 3, 12, . . .), $n = 0, 1, 2, \ldots$. There is no sharp distinction between stranded wire and cable. Tables of stranded wires are given in Table 6-9.

46. Cables. Cables are stranded conductors or a combination of stranded conductors. Cables can be made from one material or a combination of materials. Cables can be bare or insulated. There are various geometrical configurations for cables:

Concentric-lay cables: Successive layers of helically laid wires.

Bunch-stranded conductors: A group of conductors bunched together in no particular geometrical form. If conductors are parallel, they are referred to as parallel strand, and the strands are usually twisted together.

Rope lay: Made of helically laid successive layers of stranded conductors. The stranded conductors can be either bunch-stranded or concentric-lay-stranded.

Annular conductors: Helically stranded wires over a central rope, copper helix or twisted I beam.

Expanded ACSR: Steel core covered with a filler material and helically laid hard-drawn aluminum wires.

Composite conductors: Made up from two different conductors to obtain desired ratios between mechanical and electrical properties.

47. Coaxial Cable. A wide variety of cables having a central conductor covered with a

TABLE 6-8 Weight, Breaking Strength, and DC Resistance of Copper Wire Based on ASTM Specifications B1-56, B2-52, B3-63

Size, AWG	Diam., in	Area		Weight		Hard (B1-56)		Medium (B2-52)		Soft (B3-63)	
		cmils	in²	lb/1,000 ft	lb/mil	Min breaking strength,* lb	Max dc resistance at 20°C (68°F),† Ω/1,000 ft	Min breaking strength,* lb	Max dc resistance at 20°C (68°F),† Ω/1,000 ft	Max breaking strength,‡ lb	Max dc resistance at 20°C (68°F),† Ω/1,000 ft
4/0	0.4600	211,600	0.1662	640.5	3,382	8,143	0.05045	6,980	0.05019	5,983	0.04901
3/0	0.4096	167,800	0.1318	507.8	2,681	6,720	0.06362	5,666	0.06330	4,744	0.06182
2/0	0.3648	133,100	0.1045	402.8	2,127	5,519	0.08021	4,599	0.07980	3,763	0.07793
1/0	0.3249	105,600	0.08291	319.5	1,687	4,518	0.1022	3,731	0.1016	2,985	0.09825
1	0.2893	83,690	0.06573	253.3	1,338	3,688	0.1289	3,024	0.1282	2,432	0.1239
2	0.2576	66,360	0.05212	200.9	1,061	3,002	0.1625	2,450	0.1617	1,928	0.1563
3	0.2294	52,620	0.04133	159.3	841.1	2,439	0.2050	1,984	0.2039	1,529	0.1971
4	0.2043	41,740	0.03278	126.3	667.1	1,970	0.2584	1,584	0.2571	1,213	0.2485
5	0.1819	33,090	0.02599	100.2	528.8	1,590	0.3260	1,265	0.3243	961.5	0.3135
6	0.1620	26,240	0.02061	79.44	419.4	1,280	0.4110	1010	0.4088	762.6	0.3952
7	0.1443	20,820	0.01635	63.03	332.8	1,030	0.5180	806.7	0.5153	605.1	0.4981
8	0.1285	16,510	0.01297	49.98	263.9	826.1	0.6532	644.0	0.6498	479.8	0.6281
9	0.1144	13,090	0.01028	39.61	209.2	660.9	0.8241	513.9	0.8199	380.3	0.7925
10	0.1019	10,380	0.008155	31.43	166.0	529.3	1.039	410.5	1.033	314.0	
11	0.0907	8,230	0.00646	24.9	131	423	1.31	327	1.30	249	1.26
12	0.0808	6,530	0.00513	19.8	104	337	1.65	262	1.64	197	1.59
13	0.0720	5,180	0.00407	15.7	82.9	268	2.08	209	2.07	157	2.00
14	0.0641	4,110	0.00323	12.4	65.7	214	2.63	167	2.61	124	2.52
15	0.0571	3,260	0.00256	9.87	52.1	170	3.31	133	3.29	98.6	3.18
16	0.0508	2,580	0.00203	7.81	41.2	135	4.18	106	4.16	78.0	4.02
17	0.0453	2,050	0.00161	6.21	32.8	108	5.26	84.9	5.23	62.1	5.05
18	0.0403	1,620	0.00128	4.92	26.0	85.5	6.64	67.6	6.61	49.1	6.39
19	0.0359	1,290	0.00101	3.90	20.6	68.0	8.37	54.0	8.33	39.0	8.05
20	0.0320	1,020	0.000804	3.10	16.4	54.2	10.5	43.2	10.5	31.0	10.1

21	0.0285	812	0.000638	2.46	13.0	43.2	13.3	34.4	13.2	24.6	12.8
22	0.0253	640	0.000503	1.94	10.2	34.1	16.9	27.3	16.8	19.4	16.2
23	0.0226	511	0.000401	1.55	8.16	27.3	21.1	21.9	21.0	15.4	20.3
24	0.0201	404	0.000317	1.22	6.46	21.7	26.7	17.5	26.6	12.7	25.7
25	0.0179	320	0.000252	0.970	5.12	17.3	33.7	13.9	33.5	10.1	32.4
26	0.0159	253	0.000199	0.765	4.04	13.7	42.7	11.1	42.4	7.94	41.0
27	0.0142	202	0.000158	0.610	3.22	10.9	53.5	8.87	53.2	6.33	51.4
28	0.0126	159	0.000125	0.481	2.54	8.64	67.9	7.02	67.6	4.99	65.3
29	0.0113	128	0.000100	0.387	2.04	6.97	84.5	5.68	84.0	4.01	81.2
30	0.0100	100	0.0000785	0.303	1.60	5.47	108	4.48	107	3.14	104
31	0.0089	79.2	0.0000622	0.240	1.27	4.35	136	3.6	135	2.49	131
32	0.0080	64.0	0.0000503	0.194	1.02	3.63	169	2.90	168	7.01	162
33	0.0071	50.4	0.0000396	0.153	0.806	2.79	214	2.30	213	1.58	206
34	0.0063	39.7	0.0000312	0.120	0.634	2.20	272	1.82	270	1.25	261
35	0.0056	31.4	0.0000246	0.0949	0.501	1.75	344	1.44	342	0.985	331
36	0.0050	25.0	0.0000196	0.0757	0.400	1.40	431	1.16	429	0.785	415
37	0.0045	20.2	0.0000159	0.0613	0.324	1.13	533	0.944	530	0.636	512
38	0.0040	16.0	0.0000126	0.0484	0.256	0.898	674	0.750	671	0.503	648
39	0.0035	12.2	0.00000962	0.0371	0.196	0.691	880	0.577	876	0.385	847
40	0.0031	9.61	0.00000755	0.0291	0.154	0.543	1,120	0.455	1,120	0.302	1,080
41	0.0028	7.84	0.00000616	0.0237	0.125	...	1,380	...	1,370	0.246	1,320
42	0.0025	6.25	0.00000491	0.0189	0.0999	...	1,730	...	1,720	0.196	1,660
43	0.0022	4.84	0.00000380	0.0147	0.0774	...	2,230	...	2,220	0.152	2,140
44	0.0020	4.00	0.00000314	0.0121	0.0639	...	2,700	...	2,680	0.126	2,590

*No. 19 AWG and smaller, based on Anaconda data.

†based on nominal diameter and ASTM resistivities.

‡No requirements for tensile strength are specified in ASTM B3-63. Values given here based on Anaconda data.

dielectric material, a metallic shield, and a jacket conductor are available. The shield is generally made of braided copper, metal tape, or solid shield. Properties of coaxial cables are given in Par. **9-3**.

48. Bar. Bus bars of rectangular cross section are used in general for carrying high electric currents. Occasionally, tubular or angled bars are used. Tubular conductors are used for high-voltage applications. These conductors have a small skin-effect ratio, larger current-carrying capacity, and smaller corona losses.

49. Strip and Sheet. Conductors in strip (or foil) form can sometimes be found in interconnections of hybrid circuits. Flat flexible conductors are used for interconnections on printed-circuit boards. These wires are flat conductors laminated between layers of plastic insulation. The flexible wiring can be single- or double-layered. Shielded flat wiring has also been used.

Printed-circuit boards also contain flat conductors. The boards are produced by etching, electroplating, stamping, or molding the conductor (almost invariably copper) on an insulating rigid plastic board. Copper-cladding thicknesses are given in Table 6-10.

TABLE 6-9   Properties of Stranded Conductors[71] (MIL-W-81044)

Size desig- nation, AWG	Nominal conductor area, c mils	No. of strands	Allow- able no. of missing strands	Nominal diam. of individual strands, in.	Max. diam. of stranded conductor, in.	Max. resistance of finished wire at 20°C, Ω/1,000 ft			
						Tin- coated copper	Silver- plated copper	Nickel- plated copper	Silver- plated high- strength copper alloy
30	112	7	0	0.0040	0.013	107.0	101.0	109.0	116.0
28	175	7	0	0.0050	0.016	67.6	62.9	68.3	72.2
26	304	19	0	0.0040	0.021	39.3	36.2	40.1	41.5
24	475	19	0	0.0050	0.026	24.9	23.2	25.1	26.6
22	754	19	0	0.0063	0.033	15.5	14.6	15.5	16.8
20	1,216	19	0	0.0080	0.041	9.70	9.05	9.79	10.4
18	1,900	19	0	0.0100	0.052	6.08	5.80	6.08	6.65
16	2,426	19	0	0.0113	0.060	4.76	4.54	4.76	5.23
14	3,831	19	0	0.0142	0.074	2.99	2.87	3.00	3.30
12	7,474	37	0	0.0142	0.102	1.58	1.48	1.59	1.70
10	9,361	37	0	0.0159	0.118	1.27	1.20	1.27	1.38
8	16,983	133	0	0.0113	0.176	0.700	0.661	0.680	0.760
6	26,818	133	0	0.0142	0.218	0.436	0.419	0.428	0.483
4	42,615	133	0	0.0179	0.272	0.274	0.263	0.269	0.302
2	66,500	665	2	0.0100	0.345	0.179	0.169	0.174	0.194
0	104,500	1,045	3	0.0100	0.432	0.114	0.105	0.109	0.123

TABLE 6-10   Copper-Cladding Thickness Tolerances[70]

Nominal thickness		Nominal weight		Tolerance			Sheet resistivity, mΩ/□ (based on 1.724 μΩ·cm)
in	mm	oz/ft²	g/mm²	By weight, %	By gage		
					in	mm	
0.0007	0.0178	½	1.5	±10	±0.0002	±0.0051	0.9685
0.0014	0.0355	1	3.06	±10	+0.0004 −0.0002	+0.0102 −0.0051	0.4843
0.0028	0.0715	2	6.12	±10	+0.0007 −0.0003	+0.0178 −0.0076	0.2421
0.0042	0.1065	3	9.18	±10	±0.0006	±0.0152	0.1614
0.0056	0.1432	4	12.24	±10	±0.0006	±0.0152	0.1211
0.0070	0.1780	5	15.30	±10	±0.0007	±0.0178	0.0968
0.0084	0.2130	6	18.36	±10	±0.0008	±0.0204	0.0807
0.0098	0.2460	7	21.42	±10	±0.001	±0.0254	0.0692
0.014	0.3530	10	30.6	±10	±0.0014	±0.0355	0.0484
0.0196	0.4920	14	43.2	±10	±0.002	±0.0508	0.0346

50. Thin-Film Conductors. Some of the interconnections in microelectronics are made of thin films. The films are generally vacuum-deposited and are about 1 μm thick.

In integrated circuits the most widely used conductor is aluminum. Other conductors such as Ti–Pt–Au or Ti–Pd–Au are also occasionally used.[15,16] In hybrid circuits, thicker films are used, up to 25 μm. These films are either vacuum-deposited, electroplated, chemically vapor-deposited, or screen-printed and fired. Thin-film resistors, vacuum-deposited or directly diffused in the semiconductor, are also used.

51. Joining of Conductors. *Solders.* Solders are elements or alloys with low melting temperatures used for joining two or more conductors. Solders are available in various forms: wires, wires with a flux core, sheets (preforms), and balls.

Conductive Adhesives. Epoxies filled with conductive metals such as silver and gold are available for joining conductors. These glues are cured at relatively low temperatures.

Welding. Metals are joined by a process of melting the parent metals.

Brazing. Metals are joined by melting nonferrous brazing alloys (brazing alloys melt above 80°F but lower than the parent metals).

Spot Welding. Metals are joined by application of pressure and electric current.

Thermocompression. Metals are joined by application of pressure and heat.

SPECIFIC CONDUCTOR MATERIALS

52. Copper. Copper is used very extensively for electrical conductors. It has a very high electrical and thermal conductivity and can easily be formed into wires, tubes, and sheet. Copper has a high resistance to corrosion, forming an oxide on its surface.

Copper is normally supplied as electrolytic tough pitch (ETP), containing 99.95% Cu, with about 0.04% oxygen and approximately 0.01% other impurities. This material is used for electrical wires, bus bars, switches, and terminals.

Other forms of commercially pure copper are deoxidized low- or high-residual phosphorus (DLP and DHP, respectively). The OFHC, oxygen-free high-conductivity copper (99.95% Cu) is an electrolytic copper free from cuprous oxide. This material is used for electrical conductors of various forms, bus bars, and waveguides.

For electrical connector and switch components, the free-machining copper is used. Both tellurium and lead additions are found. Additions of silver (zinc or chromium) to copper increases the resistance to creep and also raises the softening temperature after cold working.

When copper is used for contacts, the oxidation of copper is occasionally not desirable. The copper can be coated with a precious metal such as silver to prevent oxidation. Nickel-plated copper is also used.

When high resistance to sticking, arcing, or welding is desired, copper-tungsten or copper-graphite mixture is used.

Copper alloys are also commonly used for plugs, connectors, and other contacts.

Yellow brass (65% Cu–35% Zn) has improved mechanical properties but reduced corrosion resistance and electrical conductivity (28% IACS).

Phosphor bronze is used where good resistance to wear is desired in electric contacts (98.75% Cu–1.25% Sn). Bronzes with low additions of tin are desired since they have only a small loss of electrical conductivity.

Nickel silver (55% Cu–27% Zn–18% Ni) is used for telephone equipment, resistance wire, and contacts. It has 5.5% (IACS) conductivity.

Beryllium copper (97.9% Cu–1.9% Be–0.2% Ni or Co) alloy is used where a relatively high electrical conductivity (18% IACS cold-worked, 30% IACS annealed) and high mechanical strength are desired, as in spring contacts. This alloy also has good corrosion resistance.

Tin bronze (88% Cu–8% Sn–4% Zn) is used as collectors for electric generators.

When copper containing cuprous oxides is annealed in a hydrogen-bearing atmosphere (above about 500°C), the hydrogen diffused through the metal reduces the oxide and forms steam, which produces cracks in the metal. This cracking is termed *hydrogen embrittlement*. The embrittlement does not occur in OFHC copper due to the absence of oxides in this material.

Mechanical and electrical properties of copper wires are given in Tables 6-9 and 6-10. Electrical properties of copper alloys are given in Table 6-11.

53. Aluminum. Aluminum has a high electrical and heat conductivity. It is malleable, very strong compared with its weight, and has good reflectivity and good corrosion resistance.

Aluminum is used for electrical conductors. EC alloy (99.45% Al) is used for bus bars, wires, and stranded conductors. For cables it is used with a central steel-core reinforcement. Aluminum-alloy conductors are also used for cables. Aluminum-alloy bus bars (6061: 1% Ag, 0.6% Si, 0.25% Cu, 0.25% Cr) are also in use. Alloying increases the mechanical strength and normally reduces electrical conductivity. Some alloys can be age-hardened to obtain optimum mechanical properties.

TABLE 6-11   Electrical Properties of Copper Alloys[61]

Alloy	Electrical conductivity, % IACS	Electrical resistivity, $\mu\Omega \cdot cm$	Temperature coefficient of electrical resistivity, K^{-1}
Pure copper	103.06	1.67	0.00404
Electrolytic copper (ETP)	101	1.71	0.00397
Oxygen-free copper (OF)	101	1.71	
Free-machining copper (1.0% Pb)	98	1.76	
Gilding metal, 5% Zn	56	3.10	0.00231
Red brass, 15% Zn	37	4.70	0.0016
Cartridge brass, 30% Zn	28	6.20	0.00148
Yellow brass, 35% Zn	27	6.40	
Phosphor bronze, grade A, 5% Sn	15	11	
Cupro Nickel, 30% Ni	4.6	37.00	0.00048
Beryllium copper	15–18	9.6–11.5* 5.7–7.8†	

*Cold-worked.
†Precipitation-hardened.

In thin-film or fine-wire form it is used in the microelectronic industry. In thin-film form it is prone to electrical failures due to electromigration (see Par. **6-42**).

Aluminum is used for heat sinks, radiators, and for reflective coatings. Aluminum can be anodized to give it very good corrosion protection.

Properties of aluminum wires are listed in Table 6-12. Resistivities of aluminum alloys are given in Table 6-13.

54. Silver. Silver has the highest electrical conductivity at room temperature. Silver also has excellent heat conduction. Mechanically, it is malleable. Silver has good corrosion resistance but poor resistance to tarnishing.

Silver is used for electric contacts, usually with lower current (up to 20 A) and voltages. Silver is used mainly in electrodeposited form for plugs, sockets, rotary switches, and occasionally slip rings. For many applications silver alloys are used. The alloys are harder and less prone to wear. Silver-copper alloys from sterling silver (92.5% Ag–7.5% Cu) to the eutectic alloy (72% Ag–28% Cu) are used. Ag–Cd are also used.

Silver can be used in screened conductors for hybrid circuits and as fillers in conducting low-curing-temperature adhesives. Silver is also used as a component in brazes: silver-copper or silver-copper-zinc alloys are extensively used as high-temperature brazes.

In the presence of humidity and electric fields, silver will migrate in the form of fine threads between silver conductors.[17] This ionic migration occurs between conductors on organic or ceramic surfaces and eventually causes electrical shorts.

55. Gold. Gold has excellent conductivity, similar to that of aluminum, as well as excellent corrosion and oxidation resistance. Gold is a very soft material: it can easily be fabricated to very small dimensions by cold working.

Gold is used in high-frequency conductors employed in corrosive environments and as a plated layer over plugs. Gold is used for fine-wire interconnections and integrated and hybrid circuits. Gold plating is extensively used on semiconductor packages, leads, and on circuit boards. Gold and glass frit are used as conductors on thick-film hybrid circuitry. Gold in thin-film form is used on thin-film hybrids and on beam-lead integrated circuits as the conductor material. Conductive adhesives occasionally contain gold.

Gold alloys are used for rotary switches and telephone relays. In particular, 60% Au–25% Ag–6% Pt is a useful alloy. Au–Ag alloys are also useful for low-current electric contacts.

Au–Sn, Au–Si, and Au–Ge are some of the gold-base solders used whenever an intermediate

TABLE 6-12   Dimensions, Weight, and DC Resistance at 20°C (68°F) of Aluminum Wire

Conduc-tor size, AWG	Diam., mils	Area cmils	Area in²	Dc resistance,* Ω/1,000 ft	Weight,† lb Per 1,000 ft	Weight,† lb Per Ω	Length, ft/Ω
2	257.6	66,360	0.05212	0.2562	61.07	238.4	3,903
3	229.4	52,620	0.04133	0.3231	48.43	149.9	3,095
4	204.3	41,740	0.03278	0.4074	38.41	94.30	2,455
5	181.9	33,090	0.02599	0.5139	30.45	59.26	1,946
6	162.0	26,240	0.02061	0.6479	24.15	37.28	1,544
7	144.3	20,820	0.01635	0.8165	19.15	23.47	1,225
8	128.5	16,510	0.01297	1.030	15.20	14.76	971.2
9	114.4	13,090	0.01028	1.299	12.04	9.272	769.7
10	101.9	10,380	0.008155	1.637	9.556	5.836	610.7
11	90.7	8.230	0.00646	2.07	7.57	3.66	484
12	80.8	6,530	0.00513	2.60	6.01	2.31	384
13	72.0	5,180	0.00407	3.28	4.77	1.45	305
14	64.1	4,110	0.00323	4.14	3.73	0.914	242
15	57.1	3,260	0.00256	5.21	3.00	0.575	192
16	50.8	2,580	0.00203	6.59	2.33	0.361	152
17	45.3	2,050	0.00161	8.29	1.89	0.228	121
18	40.3	1,620	0.00128	10.5	1.49	0.143	95.5
19	35.9	1,290	0.00101	13.2	1.19	0.0899	75.8
20	32.0	1,020	0.000804	16.6	0.942	0.0568	60.2
21	28.5	812	0.000638	20.9	0.748	0.0357	47.8
22	25.3	640	0.000503	26.6	0.589	0.0222	37.6
23	22.6	511	0.000401	33.3	0.470	0.0141	30.0
24	20.1	404	0.000317	42.1	0.372	0.00884	23.8
25	17.9	320	0.000252	53.1	0.295	0.00556	18.8
26	15.9	253	0.000199	67.3	0.233	0.00346	14.9
27	14.2	202	0.000158	84.3	0.186	0.00220	11.9
28	12.6	159	0.000125	107	0.146	0.00136	9.34
29	11.3	128	0.000100	133	0.118	0.000883	7.51
30	10.0	100	0.0000785	170	0.0920	0.000541	5.88

*Conductivity = 61.0% IACS.   †Density = 2.703 g/cm³ (0.09765 lb/in³).
SOURCE: Based on ASTM Specifications B230-60, B262-61, and B323-61.

TABLE 6-13   Electrical Properties of Aluminum Alloys[62]

Alloy and heat treatment*	Composition	Electrical conductivity, % IACS	Electrical resistivity, μΩ·cm
Aluminum	99.996% Al	64.9≈	2.655
EC (O and H 19)	99.45% Al	62	2.8
1100 (O)	1% Si, 0.2% Cu, 0.05% Mn, 0.1% Zn	59	2.9
2011 (T3)	5–6% Cu, 0.4% Si, 0.7% Fe, 0.3% Zn	36	4.8
3003 (O)	1.0–1.5% Mn, 0.6% Si, 0.7 Fe, 0.2% Cu, 0.1% Zn	50	3.4
5056 (H38)	4.5–5.6% Mg, 0.3% Si, 0.4% Fe, 0.1% Zn, 0.05–0.2% Mn, 0.05–0.2% Cr	27	6.4
6061 (T4 and T6)	0.8–1.2% Mg, 0.4–0.8% Si, 0.7% Fe, 0.15% Mn, 0.15–0.4% Cu, 0.15–0.35% Cr, 0.25% Zn, 0.15% Ti	40	4.31
7075 T6	5.1–6.1% Zn, 2.1–2.9% Mg, 1.2–2% Cu, 0.5% Si, 0.7% Fe, 0.3 Mn, 0.18–0.4% Cr, 0.2% Ti	30	5.74

*For heat treatment designations see, for example, K. R. Van Horn (ed.), "Aluminum," Vol. 1, ASM The American Society for Metals, Metals Park, Ohio, 1967, p. 112.

melting temperature is desired. Au–Si is frequently used to solder integrated circuits to the gold-plated packages.

56. Platinum. Platinum is a precious metal with very good corrosion resistance and a high melting point; it is ductile and can be easily formed. Platinum is used for resistance thermometers and thermocouples.

Platinum is also used for electric contacts, brushes, and precision potentiometer wires. Another use is in contacts to silicon and as part of the thin-film conductors on beam-lead integrated circuits.[15]

Platinum-palladium, rhodium, or ruthenium alloys are used as electric contacts. Platinum and platinum–10% rhodium wires are frequently used for high-temperature thermocouples.

57. Palladium. Palladium is a precious metal with very good corrosion resistance and high melting point. It can be formed to various shapes and can be electroplated. It resembles platinum in appearance.

Palladium has wide use as telephone relay contacts because it is less prone to erosion, is relatively noise-free when used as contacts and is more economical than platinum.

A number of palladium alloys are used for relay applications: palladium-ruthenium, palladium-silver-nickel, palladium-silver-copper. For sliding contacts, palladium-silver-gold-platinum and palladium-silver-copper-platinum-gold-zinc alloys are sometimes used.

Palladium has been suggested as a replacement for platinum in beam-lead circuits.

58. Nickel. Nickel-coated copper conductors can be used at temperatures up to 300°C. There is no enhanced corrosion at defective areas of the coating. Nickel can be electroplated or cladded. Nickel plating is frequently used as a coating under the final gold plating on conductors. Nickel is also an important constituent in resistance alloys.

59. Tungsten. Tungsten has a very high melting point and a relatively high electrical conductivity. Tungsten is a hard and brittle material. It tends to develop oxide films in air.

Since it has good resistance to erosion, arcing, and welding, it is used for vibrators, voltage regulators, and other low-current repetitive contacts. Tungsten has been suggested as a possible contact metal on integrated circuits. Tungsten is also used as heating wires and filaments for incandescent lamps.

60. Molybdenum. Molybdenum has a lower melting point than tungsten and a higher tendency for oxide formation. The electrical conductivity of molybdenum is similar to that of tungsten. It is not attacked but is wetted by mercury, and it is therefore used extensively in mercury switches.

Molybdenum is also used as heating wires for high-temperature furnaces.

61. Sintered Materials. In electric contacts sintered powders are sometimes used. The advantage of a composite powder is the added control over the physical properties of the conductors. Among the more common sintered powders are Ag–CdO, Ag–graphite, Ag–W, Ag–Mo, Ag–WC, and Cu–W.

62. Rhodium. Rhodium is a hard, high-melting metal with relatively low electrical resistance and good corrosion resistance. Rhodium is used for sliding contacts and can be electroplated on silver or other base material. Rhodium is also used in thermocouple wires as an addition to platinum.

63. Contact Materials. See Ref. 61, Vol. 1, pp. 801–816, Par. **6-257**.

64. Materials for Wires and Cables. Aside from pure copper and aluminum, other wire and cable construction are available, as follows.

Copper-Clad Steel. Steel wires covered with copper provide high strength with some conductivity loss (30 to 40% IACS). At high frequencies these wires have the same conductivity as solid-copper wires since most of the current is conducted near the surface.

Aluminum-Clad Steel. This is used for communication or signal wires and cables. Again, the steel imparts the strength while the aluminum serves as the conductor.

Galvanized Steel. Galvanized-steel conductors are coated-steel wires with relatively low electrical conductivity and high mechanical strength. Galvanized-steel wires are not commonly used.

Copper Alloys. For high-strength wires, cadmium-copper, zirconium-copper, chromium-copper, and cadmium-chromium-copper alloys have been used.

Metal-Coated Copper. Tin is used as a protective coating. The thickness of the coat can vary from 0.5 to 5 μm. In stranded cables the tin is prefused, overcoated, or top-coated on bare strands. Silver coating is used for high-temperature (up to 200°C) application. Silver-coated-copper wire is susceptible to enhanced corrosion at discontinuities, cracks, and pinholes in the coating. Nickel-coated copper can be used up to 300°C. Nickel has poor solderability.

65. Fusible Alloys. Alloys of low-melting-temperature materials such as lead, cadmium, bismuth, and tin have even lower melting temperatures than their pure components. Examples of low-melting-temperature alloys are given in Ref. 61, Vol. 1, p. 864.

66. Resistance Metals. Resistor materials have several requirements. The resistivity should be 50 to 150 $\mu\Omega\cdot$cm. The thermoelectric potential against copper should be small. The wires should be solderable, and the temperature coefficient of electrical resistance has to be very small. The wires have to be stable metallurgically and chemically. Stress relieving helps in achieving stability.

Most precision resistors are wire-wound nickel alloys: manganin or constantan. If high frequency and high stability are required, film-type resistors are used. These resistors are composed of a resistive film on a substrate. The least accurate resistors are the composition type, such as carbon resistors.

Resistor inks are used on thick-film hybrid circuits. These inks or pastes are screened on the surface of a ceramic substrate and subsequently fired. Thin-film hybrids or integrated circuits use different materials as resistors. Table 6-14 shows the composition and properties of thick- and thin-film resistors.

TABLE 6-14   Thick- and Thin-Film Resistors[64]

		Selected thick-film resistors	
Supplier*	Series	Temperature coefficient of resistivity, ppm/K, −55 to 125°C	Sheet resistivity per square
Ceramalloy	100	. . .	100 Ω–50 kΩ
	500	0 ±50	100 Ω–1 MΩ
	800	0 ±200	10 Ω–10 MΩ
	2000	0 ±200	3 Ω–100 MΩ
EMCA	5500	0 ±50	100 Ω–100 kΩ
	5100	0 ±100	10 Ω–10 MΩ
ESL	2700	200–350	10 kΩ–10 MΩ
	3800	50	1 Ω–10 MΩ
Thermistor	PTC2600	4,000	5–100 Ω
		1,200	1 kΩ
Plessey	SRD	200	10 Ω–10 MΩ
	SRA	100	10 Ω–100 MΩ
Du Pont	1500	±50	100 Ω–1 MΩ
	4500	+250	1.5 Ω–1 MΩ

	Thin-film resistors	
Material	Sheet resistivity, Ω/□	Tolerance, %
Diffused silicon	50–250	±10
NiCr	40–400	±10
Ta, Ta–N	200	
Cr–SiO	100–10,000	±20
Si–Cr	2,000–20,000	±15 to ±30
MoSi$_2$	200	±10

*Ceramalloy, West Coshocken, PA 19428; EMCA, Mamaroneck, NY 10543; ESL, Pennsauken, NJ 08110; Plessey, South Melville, NY 11746; Du Pont EMD, Wilmington, DE 19898.

67. Carbon. Carbon occurs in two crystalline forms, graphite and diamond. Graphite is the stable form at room temperature. Carbon is also found in an amorphous (noncrystalline) form. Carbon or graphite is used with binders to form parts for electric sliding contacts such as motor brushes. Graphite has a low shearing stress parallel to its basal plane which makes it a solid lubricant. Diamond is sometimes used as a substrate material for some unique applications where extremely good heat conduction is required.

68. Soldering Alloys. Soldering alloys are low-melting-temperature conducting alloys which join two conductors by wetting their surfaces and then solidifying to a mechanically strong solid.

Solders lose their mechanical strength above solidus temperature and become sufficiently fluid at about 50°C above the liquidus (see Fig. 6-11). The solder should be strong and creep- and corrosion-resistant and should be a reasonable electrical conductor.

TABLE 6-15   Selected Thick-Film Solders and Brazes[64]

Mfgr.	No.	Composition	mp, °C
EMCA	6099	58 Bi–42 Sn	138
		62 Sn–36 Pb–2	
Du Pont	8956	Ag	179
Ceramalloy	6000	10 Sn–90 Pb	179
	6003	60 Sn–40 Pb	188
Du Pont	8522	60 Sn–40 Pb	188
Ceramalloy	60807	50 In–50 Pb	216
Du Pont	9577	80 Pb–20 In	275
	8511	80 Au–20 Sn	280
EMCA	512	80 Au–20 Sn	280
Ceramalloy	6079	80 Au–20 Sn	280
	6000	10 Sn–90 Pb	300
Du Pont	8513	88 Au–12 Ge	356
Ceramalloy	60193	72 Ag–28 Cu	780

Lead-tin solders are most commonly used for electronic applications. The eutectic composition (63% Sn–37% Pb) has good mechanical properties combined with high wettability. Electrical conductivity decreases from 13% IACS for tin to 7% IACS for lead. Antimony additions increase the strength of lead-tin alloys. To minimize scavenging of silver (when soldering onto thin silver films), silver is added to saturate the solders. Gold is unintentionally dissolved into the solder when soldering gold surfaces. Gold-tin intermetallics seem to impart brittleness to the joints. Aluminum, cadmium, and zinc are detrimental to the solder. *Tin* melts at 232°C and has good wetting. Tin transforms into a gray brittle allotropic form at 13°C, although alloying additions will retard this occurrence. *Gold-silicon* or *gold-germanium* solders are used for soldering semiconductor dice to gold-plated surfaces. *Tin-antimony* is an intermediate-temperature solder which is not, however, suitable for brass. *Lead-silver-(tin)* is used for higher temperatures, but has poor corrosion resistance. *Lead-5% indium* has good wetting at high temperatures. *Aluminum-zinc* is used for aluminum soldering.

69. Conductive Coatings and Films. Conductive films on printed-circuit boards are primarily of copper. The copper may be gold-plated or tinned.

On hybrid circuits conductors are formed by screen printing or electroplating (thick films) or by vapor deposition (thin films). A variety of conductive pastes are available (see Table 6-16). The conductive pastes containing the conductor and a binder are screen-printed on dielectric substrates (commonly alumina) and subsequently fired to remove the binder. The resulting conductor thickness is about 10 to 25 μm. Screened conductors are normally wider than 100 μm.

Another way of achieving a thick conductor is by electroplating. In this method a thick conducting layer is electroformed through a photoresist pattern (pattern plating) on a previously deposited thin film. An alternative method is provided by electroplating on a previously deposited metallic film without use of a pattern and subsequently etching the thick conductors. Gold is most commonly used as the conductor layer when electroforming is performed. Copper is occasionally used.

Thin conducting films are used on semiconductor devices. Al is the most commonly used film. The film thickness is about 1 to 2 μm (sheet resistivity of 0.015 to 0.03 Ω/□). Al thin films are susceptible to electromigration and corrosion. The reliability of thin films has been the subject of a number of studies.[16,18]

Alternative conductors are Mo-Au films. Mo serves as an adherence layer between the gold and the silicon or silicon oxide, and the gold is the main conductor. The Mo–Au films are not commonly used. Still another combination of films is Ti–Pt–Au. This system, developed at Bell Laboratories, consists of a titanium adherence layer, followed by a platinum diffusion barrier (to prevent gold diffusion into silicon), topped by the final gold conducting layer.[15] This system is relatively corrosion-free (electromigration in these films has not yet been extensively studied).

TABLE 6-16 Selected Thick-Film Conductors[64]

Material	Engelhard* no.	Sheet† resistivity, Ω/□	Adhesion strength, lb/in²	EMCA No.	Sheet resistivity, mΩ/□	Adhesion‡ strength, lb/in²	Line definition, mi	Bondability§			
								Die	Ultrasonic	Solder	Thermocompression
Au¶	A-1456	0.004	4,000	3264	<2	Superior	3	E	E	F	E
Au	A-3319	0.002	>5,000	212B	<2	Good	5	E	E	P	E
Pt–Au	A-2496	0.05	>5,000	180	20	>3,000	3	G	F	E	F
Pd–Ag	A-2475	0.03	4,500	6290	40	. . .	10	G	G	E	F
Pt–Ag	A-3058	0.006	5,000	4112	30	>3,000	10	G	G	E	F
Ag	A-3061	0.005	7,000	92	10	>750	10	F	G	F	F
Pd–Au	A-1927	0.060	3,000	7225							
Ni	A-2964	0.100	3,000	6500	. . .	>1,500	5				

*East Newark NJ 07029.
†0.012 in thickness.
‡0.1- by 0.1-in square.
§E = excellent, G = good, F = fair, P = poor.
¶Alloyed Au for good high-temperature Au–Al bond aging available from several companies: EMCA 7287-3, ESL 8882, Ceramalloy 4393, and Du Pont.

Other conductive films have been reported as suitable for semiconductor devices: tungsten, tungsten-gold, tungsten-titanium-gold.[18]

70. Shielding. It is often desirable to shield part of the circuit from electromagnetic fields. The shields can *absorb, reflect,* or *degrade* (by multiple internal reflections) the electromagnetic energy. The most commonly used shields are braided copper. Tapes of copper or aluminum and copper or aluminum-coated *lossy* foils are also used. Solid shields can be found where complete shielding is necessary. Conductive plastics, yarns, and metal sprays also serve as shielding materials.

Magnetic shielding is obtained by high-permeability materials such as Mumetal.

71. Radar-Absorbing Materials. Frequently, as in radar or antenna applications, it is necessary to absorb rf energy without reflection. This can be performed by *resonant* or *broadband* absorbing materials.[19] Examples of resonant absorbers are seen in Fig. 6-14a and b. Resonant absorbers have a certain effective bandwidth in which the reflections are strongly attenuated by interference, similar to light absorption in antireflection optical coatings.

Broadband absorbing materials are divided into several kinds. In $\mu = \epsilon$ *absorbers*, the normal incidence reflection coefficient is zero when the relative magnetic permeability μ equals the relative electrical permittivity ϵ_r. Ferrites with $\mu_r \approx \epsilon_r$ are used for this purpose (Fig. 6-14c). *Inhomogeneous layers* (Fig. 6-14d) are materials with an ϵ varying from a low value at the surface ($\epsilon \approx 1$) to a high value of ϵ at the back of the layer. Geometric transition absorbers are pyramids or wedges of synthetic sponge rubber or plastic foam loaded with a material such as carbon (Fig. 6-14e). Sometimes these absorbers are constructed so that the absorption is increased toward the back of the layer. *Low-density absorbers* with $\epsilon \approx \epsilon_0$ have a small reflection (Fig. 6-14f). Such materials are low-density plastic foams with some carbon particles as absorbers or lossy fibers.

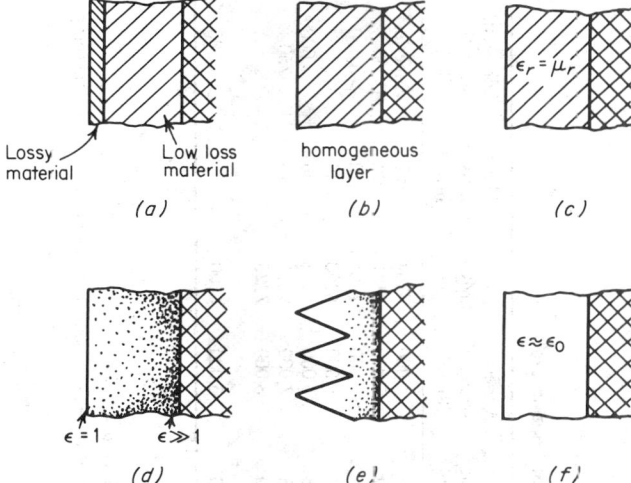

Fig. 6-14. Various configurations of radar-absorbing materials: (*a*) resonant absorber, Salisbury screen; (*b*) resonant absorber, homogeneous layer; (*c*) $\mu = \epsilon$ absorber; (*d*) inhomogeneous layer; (*e*) geometric transition absorber; (*f*) low-density absorber.

72. Bibliography on Conductive and Resistive Materials

Harper, C. A. (ed.) "Handbook of Materials and Processes for Electronics," McGraw-Hill, New York, 1970.

Gerritsen, A. N. Metallic Conductivity, Experimental Part, in S. Flügge (ed.), "Handbuch der Physik," Vol. 19, Springer, Berlin, 1922, pp. 137–226.

Meadea, G. T. "Electrical Resistance of Metals," Plenum, New York, 1965.

Parker, E. S. "Materials Data Book," McGraw-Hill, New York, 1967.

Lyman, T. (ed.) "Metals Handbook," The American Society for Metals, Cleveland, 1961.

DIELECTRIC AND INSULATING MATERIALS

73. Permittivity of Empty Space. The permittivity of empty space is defined through the force equation between two charges in empty space.

$$F = Q_1Q_2/4\pi\epsilon_0 r^2 \tag{6-34}$$

where Q_1, Q_2 = charges (C), r = distance between the charges (m), F = force (N), ϵ_0 = permittivity of empty space and ϵ_0 = 8.85 pF/m.

74. Permittivity. The force [Eq. (6-34)] is modified if the charges are separated by a dielectric medium.

$$F = Q_1Q_2/4\pi\epsilon r^2 \tag{6-35}$$

where ϵ = permittivity of the dielectric (F/m).

75. Relative Permittivity (Dielectric Constant). The permittivity of a dielectric relative to that of empty space is called the relative permittivity

$$\epsilon_r = \epsilon/\epsilon_0 \tag{6-36}$$

where ϵ_r = relative permittivity (dielectric constant), ϵ = permittivity (F/m), and ϵ_0 = permittivity of empty space (F/m).

76. Electric Flux Density. The electric flux density in empty space is defined as

$$D_0 = \epsilon_0 E \tag{6-37}$$

where ϵ_0 = permittivity of empty space (F/m), E = electric field (V/m), and D_0 = electric flux density (C/m²).

The electric flux density in a dielectric is

$$D = \epsilon E \tag{6-38}$$

where ϵ = permittivity of the dielectric.

77. Electric Polarization. The electric polarization P is the addition in electric flux density in a dielectric material to the density in free space.

$$P = D - \epsilon_0 E \tag{6-39}$$

where P = polarization (C/m²), D = electric flux density (C/m²), ϵ_0 = permittivity of free space (F/m), and E = electric field strength (V/m). The polarization is the total dipole moment induced in a unit volume of the dielectric.

78. Electric Susceptibility. The electric susceptibility x_e is the ratio between the polarization P and the electric flux density in empty space $\epsilon_0 E$

$$x_e = P/\epsilon_0 E = \epsilon/\epsilon_0 - 1 \tag{6-40}$$

79. Polarization Processes. The polarization within dielectric materials is determined by the displacement of charges. Four sources for polarization exist in the material: *electronic polarization*, due to displacement of electronic charges; *ionic polarization*, due to displacement of ions; *dipole polarization*, due to reorientation of permanent dipoles; and polarization by *space charges*, due to macroscopic displacement. Schematic diagrams of these four mechanisms are given in Fig. 6-15.

Since the polarization process takes place in a finite time at any given temperature, it is expected that ϵ_r will be frequency- and temperature-dependent. A typical frequency dependence of the dielectric constant is seen in Fig. 6-16. The series of inflections in the curve occurs at the relaxation times for the

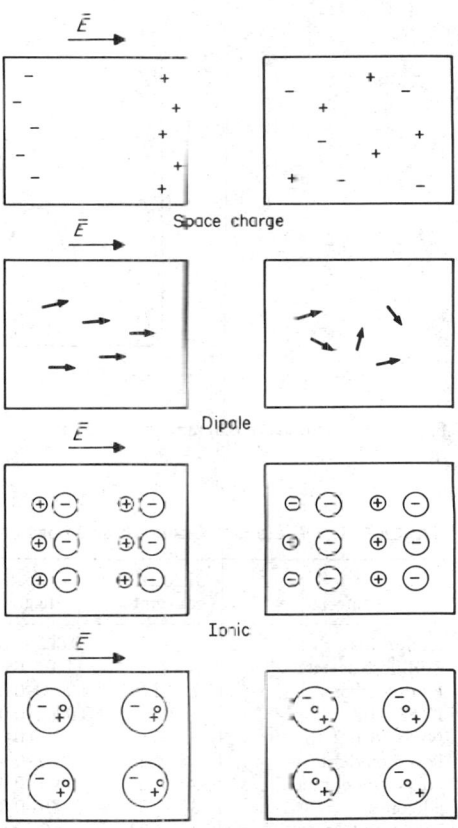

Fig. 6-15. Four types of polarization. On the right are schematic representations of the material without field; on the left, the material is under electric field.

various polarization processes. For example, the dipoles of the material represented in Fig. 6-16 can follow audio frequencies but are incapable of following infrared frequencies. The dielectric constant at infrared frequencies is decreased by the dipole component and contains only the ionic and electronic components. Table 6-17 lists the dielectric constant at 1 MHz for several materials.

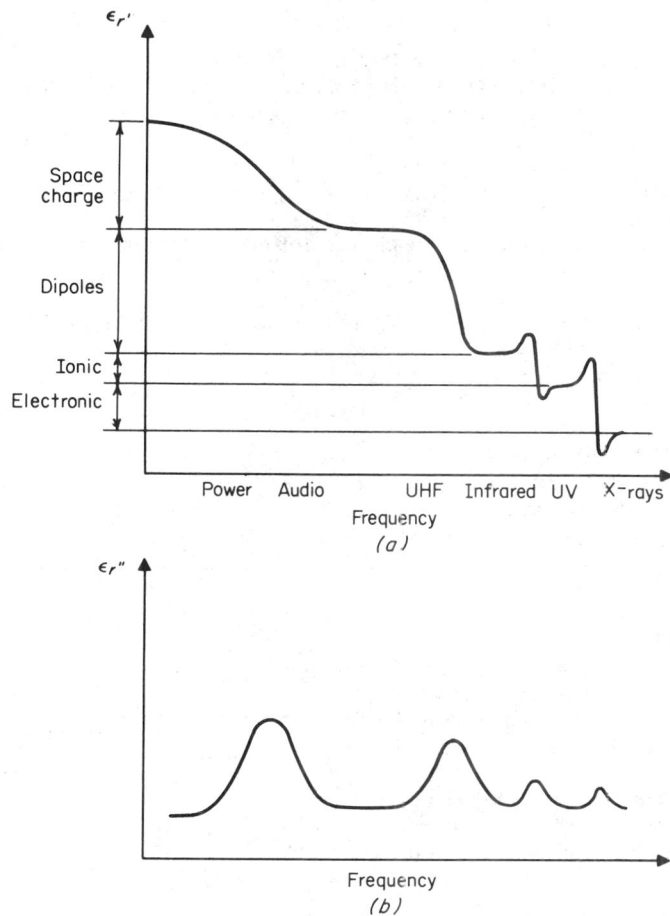

Fig. 6-16. Schematic diagram of (a) real and (b) imaginary parts of the dielectric constant as a function of frequency.[72]

TABLE 6-17   Dielectric Constant and Loss Factor for Several Dielectrics at 1 MHz

Material	Dielectric constant	Dissipation factor	Material	Dielectric constant	Dissipation factor
ABS resins	2.4–3.2	0.005–0.016	Cordierite	4.02–6.23	0.001–0.009
Cellulose acetate	3.5–7.5	0.01–0.06	Forsterite	6.2–6.5	0.0002–0.0004
Fluorocarbons	2.1–3.6	0.0003–0.0015	Alumina	8.0–10.0	0.0001–0.0009
PTFE, FEP	2.0	0.0002–0.0003	Beryllia		
Nylon 6 and 10	3.5–3.6	0.04	Soda-lime glass	7.2	0.009
Polypropylene	2.20–2.28	0.0002–0.002	Borosilicate	4.1–4.9	0.0006–0.005
Polystyrene, model	2.45–4.0	0.0001–0.002	Fused quartz	3..8	0.00001
Silicones	3.4–4.3	0.001–0.004	Mica, muscovite	5.4–8.7	0.0001–0.0004
Polystyrene, foam	1.02–1.24	<0.0005	Phlogopite	6.5	0.0001–0.0003
Hard rubber	2.95–4.80	0.007–0.028	Glass-bonded mica	6.3–9.3	0.0011–0.0025
Pyroceram 9606	5.58	0.0015	Boron nitride	4.15	0.0002

80. Dielectric Constant. The dielectric constant is generally described as a complex quantity

$$\epsilon_r = \epsilon_r' + j\epsilon_r'' \tag{6-41}$$

The imaginary part ϵ_r'' is related to dielectric losses.

A typical frequency dependence of ϵ_r'' is shown in Fig. 6-16, and a typical temperature dependence of δ_r'' in Fig. 6-17.

In an ideal lossless dielectric, alternating current leads the voltage by 90°. In the presence of dielectric losses the current leads the voltage by 90° − δ, and the power losses are proportional to tan δ, which is called the *loss tangent* or *dissipation factor*. The loss tangent is related to the dielectric constant

$$\tan \delta = \epsilon_r''/\epsilon_r' \tag{6-42}$$

When, in addition to the polarization processes, Joule heating occurs by leakage currents, the total energy dissipated per unit volume is

$$W = \sigma E^2 + \epsilon_0 \epsilon_r \omega E^2 \tan \delta \tag{6-43}$$

where W = losses per unit volume (W/m²), σ = conductivity of the dielectric (S/m), ϵ_0 = permittivity of empty space (F/m), ϵ_r = dielectric constant, E = electric field (V/m), tan δ = loss tangent due to polarization processes, and ω = angular frequency (s⁻¹).

The Joule heating is often negligible, and the power losses are proportional to ϵ_r, tan δ (often referred to as the *loss factor*). Table 6-17 shows the loss factor for various materials.

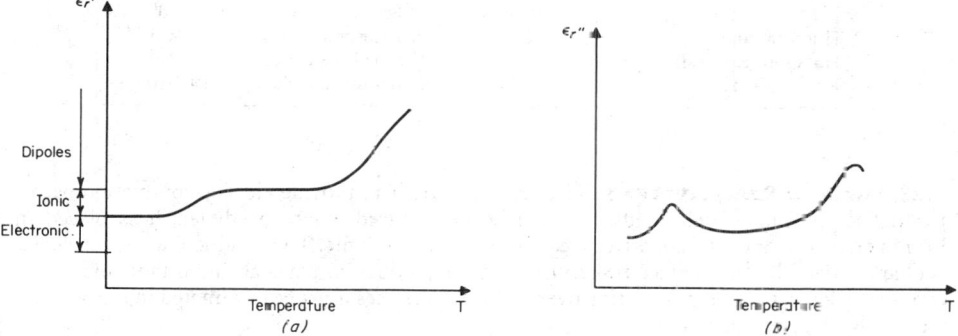

Fig. 6-17. Schematic diagram of (a) real and (b) imaginary parts of the dielectric constant as a function of temperature.

81. Conduction in Dielectrics. The volume resistivity of dielectrics is defined as for conductors, i.e., as the electric field strength required to produce a unit current density. The resistivity is time-, temperature-, and field-dependent. The dc component of the conduction follows Ohm's law roughly up to fields of 10⁷ V/m and then rises rapidly with increasing fields. The conduction in dielectrics is ascribed to electronic and ionic processes.

Electronic conduction can be electrode- or bulk-controlled. At high fields electrode-controlled processes (important in thin dielectric films) can be thermionic with Schottky lowering of the barrier, field emission, and temperature-aided field emission. The electronic conduction in the bulk is usually considered to proceed by hopping processes. The hopping can occur by tunneling between localized states or by thermal activation (Poole-Frenkel, or Poole, effect). As a consequence of such processes, the conductivity is found to increase as some exponential function of field and/or temperature. In addition, space charges produced by the carriers can determine the conductivity at higher current densities and increasing thicknesses. The ionic conductivity occurs by the motion of the charged ions in the electric field. The ionic conduction is a linear function of the field at low fields but increases rapidly at high fields. The ionic conductivity increases exponentially with the inverse absolute temperature.

Low-field resistivities of various dielectrics are given in Table 6-18. Tabulated values of resistivities should be treated with caution since the resistivity depends on fields, trapped space charges, defects, impurities, surface leakage currents, and time effects.

82. Carrier Mobility. The mobility of charge carriers in an electric field is defined as their average drift velocity in a unit electric field

$$\mu = V_{drift}/E \tag{6-44}$$

where μ = mobility ($m^2/V \cdot s$), V_{drift} = drift velocity (m/s), and E = electric field strength (V/m).

If conduction is occurring by charges of one sign only, the conductivity σ is related to the mobility as

$$\sigma = \mu N q \tag{6-45}$$

where σ = conductivity (S/m), μ = mobility ($m^2/V \cdot s$), N = density of charge carriers (m^{-3}), and q = charge on a carrier (C).

TABLE 6-18   Resistivities of Dielectrics

Material	Volume resistivity, $\Omega \cdot m$	Material	Volume resistivity, $\Omega \cdot m$
Cellulose acetate, molding	$10^8 - 10^{11}$	Silicones	$10^9 - 10^{11}$
Epoxy, cast resin	$10^{14} - 10^{15}$	Alumina	$10^9 - 10^{12}$
Methyl methacrylate,		Beryllia	$> 10^{13}$
cast	$> 10^{13}$	Forsterite	$> 10^{12}$
Mica	$10^{12} - 10^{15}$	Cordierite	$> 10^{12}$
Nylon 6 and 12	$10^{12} - 10^{13}$	Boron nitride	10^{11}
Fluorocarbons	$10^{13} - 2 \times 10^{16} +$	Pyroceram 9606	2×10^{14}
Polyester, cast resin	10^{12}	Soda-lime glass	10^3
Rubber, band	2×10^{13}	Borosilicate glass	$10^{14} - 10^{15} +$

83. Surface Resistivity. Surfaces of dielectrics can provide electric-conduction paths in parallel to the bulk. Such conduction can be pronounced when the dielectric is placed in a humid environment. The conduction can be electronic or ionic. The conductance of the surfaces is characterized by the *surface resistivity.* Surface conduction plays an important role in semiconductor devices.[20] Charges drifting over dielectric surfaces must be eliminated to prevent interference with device operation.

84. Piezoelectric and Pyroelectric Coefficients. Piezoelectricity is the appearance of electric polarization with applied stress. The *piezoelectric constants* d_{ij} are defined as the partial derivative of the electric displacement in the i direction with respect to the stress applied in the j direction at constant field

$$d_{ij} = (\partial D_i / \partial \sigma_j)_E \tag{6-46}$$

The piezoelectric effectivity K_{ef} is defined as

$$K_{ef} = d_{ij}/\sqrt{\epsilon_{ij}} \tag{6-47}$$

where ϵ_{ij} is the ij tensor element of the dielectric constant. K_{ef} is insensitive to the crystal directions. The appearance of polarization can come about by thermal expansion. Such an effect is called *pyroelectricity.*

The *pyroelectric coefficient* P is defined as the partial derivative of the displacement with respect to the temperature at constant electric field

$$P = (\partial D / \partial T)_E \tag{6-48}$$

85. Breakdown Processes and Fields.* Electrical breakdown is a sequence of often rapid processes leading to a change from an insulating to a conducting state. Breakdown develops in a number of successive stages: (1) the initiating stage, increasing the conductivity, leading to (2) instability with current runaway; (3) voltage collapse with discharge of the electrostatic

*The author thanks Prof. N. Klein for his valuable suggestions and for preparation of the paragraphs on electrical breakdown.

energy stored in the specimen through the breakdown channel; and (4) settling down to the low-voltage, high-current state.

Insulating media cannot sustain indefinite voltages, and breakdowns occur at electric fields varying from 10^5 to more than 10^9 V/m. It has been customary to assign to insulating media a dielectric strength which denotes the breakdown field. Dielectric strength is not a particularly useful concept. In most cases the breakdown field is found to be strongly dependent on such parameters as the time of voltage application to breakdown, thickness, temperature, electrode material, frequency, geometry, micro- and macroscopic defects, chemical changes, and the presence of interfaces.

Information on breakdown properties can be given in breakdown characteristics. For thin insulating films such characteristics show that the time to breakdown decreases as some exponential function of increasing field and that breakdowns occur over a wide range of fields. Increase in thickness generally decreases the breakdown field, while increase in temperature not only decreases but also increases the breakdown field. For further details see Refs. 21 to 28.

86. Breakdown in Gases. Breakdown in gases is produced by impact ionization of the molecules. Instability and current runaway arise due to the positive feedback of secondary ionization processes. Two kinds of conducting states after breakdown are observed: the glow discharge at voltages roughly above 70 V and arc discharge at much lower voltages. As sparking in air is possible above 330 V, electronic equipment can be easily exposed to the harmful effects of spark discharges. The breakdown field of gases can be enhanced by increase in pressure and by the use of gases such as sulfur hexafluoride or Freon. Breakdown in gases is observed from fields less than 10^5 to 10^8 V/m. High-intensity laser radiation can also cause gas breakdown

87. Breakdown in Liquids. Several mechanisms were proposed for liquid breakdown, such as thermal breakdown or impact ionization in the liquid and field emission from the cathode. It has been suggested in recent years that breakdown can be initiated also by the formation of tiny gas bubbles in the high fields in the liquid, leading to completion of the event by gas discharges. A further mechanism was observed in impure liquids with conducting particles in suspension. Such particles coalesce into conducting bridges between the electrodes, causing breakdown. The breakdown field of liquids extends from less than 10^7 to 5×10^8 V/m. Lower values are found in impure liquids and in those containing dissolved gases.

88. Breakdown in Solids. Breakdown in solids is caused mainly by thermal, electronic, and internal and external discharge mechanisms. In a solid specimen that mechanism prevails which causes breakdown quicker or at a lower voltage. The discharge of the electrostatic energy stored in the specimen through the conducting channel determines mainly whether destruction arises. It is usually found that destruction occurs at applied fields larger than 10^7 V/m. At lower fields with no destruction, the processes lead to switching events.

Thermal breakdown is caused at low frequencies by the Joule heat of the leakage current and at very high frequencies, mainly by dielectric losses. At low frequencies the Joule heat leads to thermal instability and current runaway due to rapid, often exponential, rise of the conductivity with temperature. The breakdown fields decrease rapidly with increasing temperature and time of voltage application. Thermal instability has been observed from 10^5 to 10^9 V/m.

Electronic breakdown can be caused by a variety of processes such as impact ionization, field emission, double injection, and, possibly, by insulator-to-metal transition. Breakdown in insulators is usually ascribed to the interaction of impact ionization, charge carrier injection into the insulator, and the effect of space charges. In thin films breakdowns are found to occur randomly in space and in time, involving only a very small part of the insulator. The typical range of electronic breakdown in insulators is from 10^7 to more than 10^9 V/m. In semiconductors electronic breakdown is initiated either by impact ionization or by tunneling processes at fields of 10^6 to 10^8 V/m. It should be remarked that impact ionization is observed in semiconductors at liquid-helium temperatures at fields less than 10^3 V/m; these processes lead to switching but not to breakdown.

Breakdown by internal discharges is possible once the voltage is sufficient to cause sparking in the cavities in the solid. Thus, this kind of sparking does not occur in thin films. As sparking at atmospheric pressure can start at fields less than 10^6 V/m, the cavities are the weak links of thick insulators. Discharges in cavities can cause erosion by sputtering, chemical reactions, local melting, and evaporation and can promote breakdown in the adjacent solid. These local destructive processes can be fast at high voltages and high frequencies but very slow at lower voltages and on direct current. The breakdown field is a decreasing function of increasing time, and breakdowns occurring after years of operation are known to be caused by internal discharges.

The effect of internal discharges is relatively small on dc operation, but the effect becomes marked in ac systems, increasing with frequency. When internal discharges are present, breakdown of thick insulators may arise readily above fields of 10^6 V/m, and the breakdown progresses through the insulator in branched channels in the form of *treeing*.

TABLE 6-19   Thermal Conductivity of Dielectrics at Room Temperature

Material	Thermal conductivity, W/m·K	Material	Thermal conductivity, W/m·K
Diamond	658	Soda-lime glass	0.8
BeO	167	Cast epoxies	0.17–1.3
Graphite	117–200	Wood	0.5–2.5
Bonded SiC	42	Melamines	0.3–0.7
MgO	42	Mica	0.3–0.7
Al_2O_3	33	Phenolics (molded)	0.17–0.67
TiC	33	Polyethylenes	0.33
BN	28	Nylons	0.17–0.25
Mullite	6.3	TFE fluorocarbons	0.25
Titania	3.3–4.2	Elastomers	0.08–0.17
Porcelain	1.7	Foams	0.003–0.04
Fused silica glass	1.5		

Breakdown fields are lowest when breakdown is produced by external discharges. Such discharges occur along the interface of an insulator with a gaseous or a liquid medium. These discharges, and also those occurring in cavities, are often denoted as *corona* discharges. The development of electric discharges along surfaces is enhanced by the interaction of processes in the solid and in the gas or liquid. Favorable conditions for impact ionization and charge-carrier emission produce sparking and lead to *arc tracking* on the surface of the solid. Tracking produces conducting channels in the form of treeing. Conditions for tracking are especially favorable on impure surfaces and in humid atmospheres, and the lowest design fields, less than 10^6 V/m, in electronic equipment are assigned to surfaces.

All the breakdown processes can be enhanced and the lifetime of the insulator shortened when the insulator is subject to *thermal aging* by chemical processes.

89. Thermal Properties. The thermal conductivity of dielectrics is defined as in conductors (see Par. **6-16**). Values of thermal conductivity are given in Table 6-19 and in Fig. 6-18. Dielectrics are in general poor heat conductors, although a few notable exceptions are diamond and BeO. Crystalline dielectrics are generally better heat conductors than amorphous dielectrics. Porous dielectrics have a smaller heat conduction than the dense materials.

90. Potential Distribution in Dielectrics. *Uniform Dielectrics.* The electrostatic field in dielectrics is the voltage difference divided by the spacing. Nonuniform electrostatic fields can be calculated, a common case being the coaxial geometry

$$E = V/[r \ln (R_2/R_1)] \qquad (6\text{-}49)$$

where E = electrostatic field (V/m), V = applied voltage (V), r = radius at which value of E is sought (m), R_2 = outer radius of dielectric (m), and R_1 = inner radius of dielectric (m). The maximum field occurs at $r = R_1$.

Composite Dielectrics. If the dielectric is composed of two substances parallel to the electrodes of thickness t_1, t_2, then on ac operation the electric fields are related as

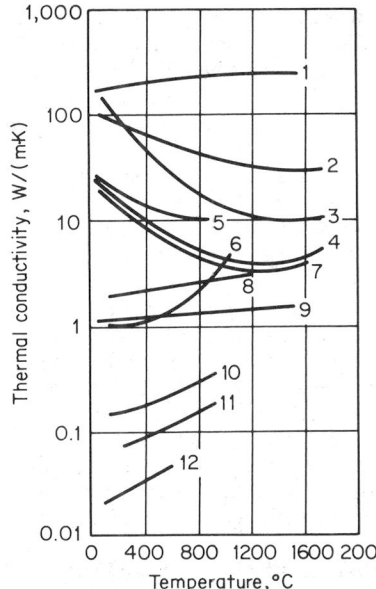

Fig. 6-18. Thermal conductivity of some ceramic materials as a function of temperature:[72] (1) platinum; (2) graphite; (3) pure dense BeO; (4) pure dense MgO; (5) bonded SiC; (6) clear fused silica; (7) pure dense Al_2O_3; (8) fine-clay refractory; (9) dense stabilized ZrO_2; (10) 2800°F insulating fire brick; (11) 2000°F insulating fire brick; (12) powdered MgO.

$$E_1/E_2 = \epsilon_{r2}/\epsilon_{r1} \qquad (6\text{-}50)$$

where E_1, E_2 = electric fields (V/m), and ϵ_{r1}, ϵ_{r2} = dielectric constants. The material with the lower dielectric constant will have the larger electric field.

In a static (dc) electric field,

$$E_1/E_2 = \rho_1/\rho_2 \qquad 6\text{-}51)$$

where ρ_1, ρ_2 = resistivities of the two dielectrics ($\Omega \cdot$m). In this case the larger field will occur in the dielectric with the higher resistivity.

If two dielectrics with different resistivities are placed in a field, space charges will accumulate in the interface between the layers. This space-charge accumulation has been used at high fields for memory application in MNOS (metal nitride oxide semiconductor) devices.[*] Here, accumulated charge was first produced in desired locations (writing) and then sensed at a later time (reading).

Nonuniform Dielectrics. Real dielectrics may have nonuniformities such as voids, changes in chemical composition, and also conducting inclusions. These nonuniformities act in a manner similar to that described above. The accumulation of space charge at nonuniformities can cause additional dielectric losses, which can reduce the breakdown field. Reduction of nonuniformities, such as the impregnation of porous dielectrics by oil, increases the dielectric strength. Intentional nonuniformities have been used in programmable memories where a semiconductor was buried in the insulator. A charge is injected through the dielectric into the buried layer. The stored charge can be subsequently sensed if desired.

SPECIFIC DIELECTRIC AND INSULATING MATERIALS

91. Gases. At low temperatures and low electric fields, gases have practically no electrical conduction. Ionization by electromagnetic radiation or particles can ionize gases and largely increase their conductivity. The dielectric constant of gases is slightly larger than unity. Breakdown occurs at high fields due to electron multiplication. The electrical breakdown in uniform fields is a function of pressure times spacing (Fig. 6-19). The dielectric strength tends to increase with increasing molecular weight of the gases. Figure 6-19 shows that electrical breakdown can occur at voltages as low as a few hundred volts. Electrical breakdown can occur even in vacuum. This phenomenon, termed *vacuum breakdown*, is probably related to field emission from the cathode.

92. Liquids. *Oils.* Oils are used to provide electrical insulation by replacing the air in certain systems or to impregnate porous insulators. The properties of interest in the oils are breakdown voltage, electrical conductivity, heat conductivity, viscosity, flammability, and vapor pressure.

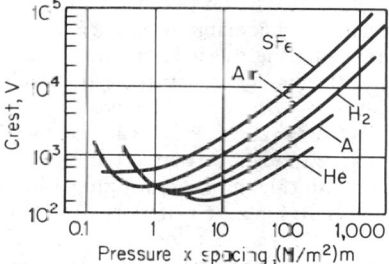

Fig. 6-19. Pressure-spacing dependence of the dielectric strength of gases (Pashen's curves).

The oils are usually mineral hydrocarbons derived from petroleum deposits. The oils contain aliphatic and aromatic compounds. The ratio of these compounds affects the properties of the oil. Properties of mineral oils are given in Table 6-20. Mineral oils can be chlorinated. The chlorination raises the viscosity and boiling point and reduces the dielectric constant.

Oils dissolve gases to 10 to 100% by volume, and water to 100 ppm at 100% relative humidity. Breakdown voltage of oils depends on the oil purity; i.e., water and particulate matter will reduce their dielectric strength. Oils can deteriorate with time due to oxidation leading to sludge and acidity. Electrode materials such as copper can enhance the oil oxidation. Oils are more stable in N_2 atmosphere. Oxidation inhibitors can also be used. The properties of oils are usually maintained by filtering (see *IEEE Stand. Publ. 64*).

Chlorinated Aromatics (Askarel Liquids). These liquids are used at low frequencies and are lossy at high frequency. Two main types are used: (1) chlorinated diphenyl liquids for capacitors and (2) mixtures of highly chlorinated diphenyl with trichlorobenzene or tetrachlorobenzene for transformers. Additives are generally used in these liquids to avoid the effects of hydrogen chloride production in the liquid on discharge or arcing.

Fluorocarbon Liquids. Fluorocarbon liquids are nonflammable aliphatic compounds with low permittivities and conductivities and are chemically inert. They have been used for large transformers or filling electronic apparatus.

Silicon Fluids. These have high-stability, low-dielectric losses and high dielectric strength. They can be obtained at various levels of viscosity, from 10^{-7} to several m^2/s. Their working range is between 40 and 200°C.

Ester Fluids. These are used occasionally for high-frequency capacitors. They are prone to hydrolysis and have poor thermal stability.

TABLE 6-20   Characteristic Properties of Insulating Liquids[73]

| | Mineral oil | | | |
	Transformer	Cable and capacitor	Solid cable	Transformer Askarel
Specific gravity	0.88	0.885	0.93	1.56–1.57
Viscosity, Saybolt seconds at 37.8°C	57–59	0.100	100	52–56*
				41–45†
Flash point, °C	135	165	235	
Fire point, °C	148	185	280	None
Pour point, °C	−45	−45	−5	<−32°C*
				<−44°C†
Specific heat	0.425	0.412	. . .	0.251
Coefficient of expansion	0.00070	. . .	0.00075	0.0007
Thermal conductivity, cal/cm·s·°C	0.39	. . .	. . .	0.30
Approx. dielectric strength,‡ kV	30	. . .	. . .	>35
Permittivity at 25°C	2.2	. . .	. . .	4.5–4.7
Resistivity, TΩ·cm	1–10	50–100	1–10	0.1

*A mixture of 60% hexachlorobiphenyl and 40% trichlorobenzene.
†A mixture of 45% hexachlorobiphenyl and 55% trichloro- and tetrachlorobenzenes.
‡ASTM D877.

93. Liquid Crystals. Liquid crystals are used for optical displays. Their optical properties change considerably by application of weak electric fields. Common liquid-crystal displays operate using the twisted nematic mode (TNM). In this mode the device rotates any optical beam by 90°; application of an electric field changes the orientation pattern of the nematic liquid and reversibly destroys this optical rotation.[29]

94. Plastics. Plastics are the most widely used dielectric materials in the electrical and electronics industry. There are numerous plastic materials available with a spectrum of electrical, mechanical, and chemical properties. Plastics can be divided into elastomers and thermosetting and thermoplastic materials.

Elastomers. Elastomers are natural or synthetic rubberlike materials which have outstanding elastic characteristics. Their elastic-deformation range is normally a few hundred percent. Elastomers are principally used where their outstanding mechanical properties are sought, i.e., for damping, sealing, or gasketing. Designation and applications of elastomers are given in Table 6-21, and properties are given in Table 6-22.

Thermosetting Plastics. These are materials which are cured and hardened to a desired form at room temperature or at a higher temperature. The chemical change on curing is permanent, and the material cannot be softened by reheating. Classification and properties of thermosetting plastics are given in Table 6-23, and properties of plastic materials are given in Table 6-24.

Thermoplastic Plastics. Thermoplastic materials do not cure or set upon heating. They soften and can be shaped by molding into any desired form. Thermoplastics can be repeatedly resoftened by heating. Table 6-25 lists the applications of several thermoplastics. Properties of thermoplastics are given in Table 6-26.

95. Fillers. Fillers in form of particles, fibers, or platelets are added mainly to thermosetting material to improve their properties. Fillers can improve the dimensional stability, stiffness, hardness, and tensile and impact strength, as well as improve chemical and heat resistance. Graphite and molybdenum disulfide are used to improve lubrication properties. Fillers can be organic, such as cotton, paper, or cellulose, or inorganic, such as glass, asbestos, mica, clay, carbonates, or oxides.

96. Laminates. Laminates are made by impregnating glass cloth, paper, cotton or other fibers with various thermosetting compounds such as phenolics, melamines, silicones, or epoxies. The laminate is cured by heat and pressure. Laminates with thermoplastics are not common. Laminates find a very important use as the boards for printed circuits. Application characteristics of several copper-clad laminates are listed in Tables 6-27 and 6-28.

97. Plastic Films, Sheets, and Tapes. Most thermoplastics are available in sheet, film, or tape form. Thin films are similar in most properties to bulk materials. The dielectric strength

TABLE 6-21 Designations and Application Information for Elastomers[20]

Elastomer designation			
ASTM D1418	Trade name or common name	Chemical type	Major application considerations
NR	Natural rubber	Natural polyisoprene	Excellent physical properties; good resistance to cutting, gouging, and abrasion; low heat, ozone, and oil resistance. The best electrical grades are excellent in most electrical properties at room temperature.
IR	Synthetic natural	Synthetic polyisoprene	Same general properties as natural rubber; requires less mastication in processing than natural rubber.
CR	Neoprene	Chloroprene	Excellent ozone, heat, and weathering resistance; good oil resistance, excellent flame resistance. Not so good electrically as NR or IR. However, the combination of generally good electricals for jacketing application, coupled with all the other good properties, gives this elastomer broad use for electrical wire and cable jackets.
SBR	GRS. Buna S	Styrene-butadiene	Good physical properties; excellent abrasion resistance; not oil-, ozone-, or weather-resistant. Electrical properties generally good but not specifically outstanding in any area.
NBR	Buna N, nitrile	Acrylonitrile-butadiene	Excellent resistance to vegetable, animal, and petroleum oils; poor low-temperature resistance. Electrical properties not outstanding; probably degraded by molecular polarity of acrylonitrile constituent.
IIR	Butyl	Isobutylene-isoprene	Excellent weathering resistance; low permeability to gases; good resistance to ozone and aging; low tensile strength and resilience. Electrical properties generally good but not outstanding in any area.
IIR	Chlorobutyl	Chloroisobutylene-isoprene	Same general properties as butyl.
BR	Cis-4	Polybutadiene	Excellent abrasion resistance and high resilience; used principally as a blend in other rubbers.
	Thiokol (PS) (Thiokol chemical)	Polysulfide	Outstanding solvent resistance; widely used for potting of electrical connectors.
R	EPR	Ethylene propylene	Good aging, abrasion, and heat resistance; not oil-resistant. Good general-purpose electrical properties.
R	EPT	Ethyl propylene terpolymer	Good aging, abrasion, and heat resistance; not oil-resistant. Good general-purpose electrical properties.
CSM	Hypalon (HYP) (Du Pont)	Chlorosulfonated polyethylene	Excellent ozone, weathering, and acid resistance; fair oil resistance; poor low-temperature resistance. Not outstanding electrically, but has some special-application uses based on other properties.
SIL	Silicone	Polysiloxane	Excellent high- and low-temperature resistance; low strength; high compression set. Among the best electrical properties in the elastomer grouping. Especially good stability of dielectric constant and dissipation factor at elevated temperatures.
	Urethane (PU)	Polyurethane diisocyanate	Exceptional abrasion, cut, and tear resistance; high modulus and hardness; poor moist-heat resistance. Generally good general-purpose electrical properties. Some special high-quality electrical grades available from formulators.
	Viton (FLU) (Du Pont)	Fluorinated hydrocarbon	Excellent high-temperature resistance, particularly in air and oil. Not outstanding electrically.
ABR	Acrylics	Polyacrylate	Excellent heat, oil, and ozone resistance; poor water resistance. Not outstanding for or widely used in electrical applications.

of thin films is higher, however, than that of bulk. Several tapes are available with adhesive backing. Properties of films are given in Table 6-29.

TABLE 6-22   Comparison of Properties of Rubbers and Elastomers[70]*

Material	Dielectric constant†	Power factor $\times 10^2$†	Volume resistivity, Ω-cm	Surface resistivity, Ω	Dielectric strength, V/mil
Natural rubber	2.7–5	0.05–0.2	10^{15}–10^{17}	10^{14}–10^{15}	450–600
Styrene-butadiene rubber	2.8–4.2	0.5–3.5	10^{14}–10^{16}	10^{13}–10^{14}	450–600
Acrylonitrile-butadiene rubber	3.9–10.0	3–5	10^{12}–10^{15}	10^{12}–10^{15}	400–600
Butyl rubber	2.1–4.0	0.3–8.0	10^{14}–10^{16}	10^{13}–10^{14}	400–800
Polychloroprene	7.5–14.0	1.0–6.0	10^{11}–10^{12}	10^{11}–10^{12}	100–500
Polysulfide polymer (Thiokol)	7.0–9.5	0.1–0.5	10^{11}–10^{12}	. . .	250–325
Silicone	2.8–7.0	0.10–1.0	10^{13}–10^{17}	10^{13}	300–700
Chlorosulfonated polyethylene (Hypalon)	5.0–11.0	2.0–9.0	10^{13}–10^{17}	10^{14}	400–600
Polyvinylidene fluoride copolymer-hexafluoropronylene (Viton A)	10.0–18.0	3.0–4.0	10^{13}	. . .	250–700
Polyurethane (Adiprene)	5.0–8.0	3.0–6.0	10^{10}–10^{11}	. . .	450–500
Ethylene-propylene terpolymer (Nordel)	3.2–3.4	0.6–0.8	10^{15}–10^{17}	. . .	700–900

*Thiokol produced by Thiokol Corp.; other trade names those of DuPont.
†At 1 MHz.

98. Plastic Foams. Plastic foams are cellular forms of urethanes, polystyrenes, vinyls, polyethylenes, polypropylenes, phenolics, epoxies, and a variety of other plastics. The foams are used for electrical, thermal, or acoustical insulators (see also radar-absorbing materials, Par. **6-71**). Properties of plastic foams are given in Table 6-30.

99. Plastic Coatings. Plastic coatings can be thermosetting, thermoplastic, or elastomeric. The coating properties are similar to the bulk properties. In addition, the coatings have characteristic properties such as coatability, adhesion, and wetting. In general, coatings can be applied by dipping, brushing, spraying, roller, screening, impregnation, fluidized beds, or vacuum deposition.

Varnishes are used for impregnation (such as transformer coils) or for coating electronic equipment. Varnishes may contain solvents for reducing the viscosity and then be air-dried or baked. Solventless coatings consist generally of a resin and a curing agent. The resins are most commonly epoxies, polyesters, or polyurethanes. Such coatings have, generally, a shorter working life than the solvent-based coatings.

Printed-circuit boards are coated with conformal coating by dipping the boards in thixotropic coatings. Such coatings are up to 1 mm thick. The coatings improve the protection against corrosion, electrical resistance degradation, and moisture. Coatings provide some mechanical protection. Coatings are sometimes filled to improve their properties. Flexible silicone coatings are used in packaging semiconductor devices. The flexible coating is applied directly on the semiconductor die surface after it is assembled on the package, and subsequently the entire assembly is covered with an outer shell by transfer molding or potting with silicone or epoxy.

The inner flexible coating, often referred to as junction coating, has to provide flexibility, electrical insulation, and some moisture resistance. The purity of this coating is essential to minimize the corrosion of the metallization and interconnecting fine wires which are used in this technology. Coatings are also used for wire and coil insulation. The coatings are generally high-molecular-weight thermoplastic or thermosetting. Some applications of coatings are shown in Table 6-31.

100. Paper. Paper is widely used as electrical insulation. Unimpregnated cellulose paper (kraft paper) is used for cable, small transformers, and capacitors. The dielectric constant, up to 6, and dissipation factor (as low as 0.001 for dry paper) are functions of humidity, temperature, and density. The dielectric strength of kraft paper is 6 to 12 MV/m. Press boards of cellulose can also be found up to 0.5-in thickness with dielectric strength of 2 to 12 MV/m.

TABLE 6-23 Application Information for Thermosetting Plastics[70]

Material	Major application considerations	Common available forms
Alkyds	Excellent dielectric strength, arc resistance, and dry insulation resistance; low dielectric constant and dissipation factor; good dimensional stability; easily molded	Compression moldings, transfer moldings
Aminos (melamine formaldehyde and urea formaldehyde)	Available in an unlimited range of light-stable colors; exhibit hard glossy molded surface and good general electrical properties, especially arc resistance; excellent chemical resistance to organic solvents and cleaners and household-type cleaners	Compression moldings, extrusions, transfer moldings, laminates, film
Diallyl phthalates (DAP) (allylics)	Unsurpassed among thermosets in retention of properties in high-humidity environments; have among the highest volume and surface resistivities in thermosets; low dissipation factor and heat resistance to 400°F or higher; excellent dimensional stability; easily molded	Compression moldings, extrusions, injection moldings, transfer moldings, laminates
Epoxies	Good electrical properties, low shrinkage, excellent dimensional stability, and good to excellent adhesion, extremely easy to compound, using nonpressure processes, for providing a wide variety of end properties; useful over a wide range of environments	Castings, compression moldings, extrusions, injection moldings, transfer moldings, laminates, matched-die moldings, filament windings, foam
Phenolics	Among the cheapest, most widely used thermoset materials; excellent thermal stability to over 300°F generally, and over 400°F in special formulations; can be compounded to a broad choice of resins, fillers, and other additives	Castings, compression moldings, extrusions, injection moldings, transfer moldings, laminates, matched-die moldings, stock shapes, foam
Polyesters	Excellent electrical properties and low cost; extremely easy to compound using nonpressure processes; like epoxies, can be formulated for either room-temperature or elevated-temperature use; not equivalent to epoxies in environmental resistance	Compression moldings, extrusions, injection moldings, transfer moldings, laminates, matched-die moldings, filament windings, stock shapes
Silicones (rigid)	Excellent electrical properties, especially low dielectric constant and dissipation factor, which change little up to 400°F and over; nonrigid silicones are covered in elastomers and embedding-material sections	Castings, compression moldings, transfer moldings, laminates
Urethanes (rigid foams)	Low-weight plastics; excellent electrical properties, which are basically variable as a function of density; easy to use for foam-in-place and embedding applications; flexible urethane foams and nonrigid high-density urethanes are covered in sections on foams, elastomers and embedding materials	Castings, coatings

Synthetic-fiber papers are also used, having improved range of working temperatures and higher dielectric strengths. Paper is also used with metallization for capacitors.

101. Impregnated Papers. Liquid impregnation of papers considerably improves their dielectric strength and thermal stability. The impregnating liquids are mineral oils or chlori-

TABLE 6-24 Thermosetting Molding Materials[73]

		Diallyl phthalate, SPI prefix DAP			Epoxy		Melamine SPI prefix MF		
	ASTM no.	Glass-fiber filler	Mineral filler	Synthetic-fiber filler	Glass-fiber filler	Mineral filler	α-Cellulose filler	Asbestos filler	Glass-fiber filler
Electrical properties									
Volume resistivity, Ω·cm	D257	10^{16}	10^{13}	10^{16}	10^{14}	10^{14}	10^{14}	10^{12}	10^{11}
Dielectric strength, V/mil, short-time	D149	450	420	400	400	400	400	430	300
Step-by-step	D149	400	400	410	400	400	300	320	240
Dielectric constant, 60 Hz	D150	4.3	5.2	5.0	5.0	5.0	9.5	10.2	11.1
1 kHz	D150	4.4	5.3	3.9	5.0	5.0	9.2	9.0	
1 MHz	D150	4.5	4.0	3.6	5.0	5.0	8.4	6.7	7.5
Dissipation factor, 60 Hz	D150	0.01	0.03	0.026	0.01	0.01	0.030	0.07	0.14
1 kHz	D150	0.004	0.03	0.004	0.01	0.01	0.015	0.07	0.07
1 MHz	D150	0.009	0.02	0.012	0.01	0.01	0.027	0.041	0.013
Arc resistance, s	D495	180	190	1.30	180	190	180	180	180
Mechanical properties									
Specific gravity	D792	1.78	1.68	1.39	2.0	2.0	1.52	2.0	2.0
Specific volume, in³/lb	D792	17.2	16.8	20.7	15.4	14.2	18.2	13.8	13.8
Tensile strength, lb/in²	D638, D651	11,000	8,700	6,800	30,000	15,000	13,000	7,000	10,000
Elongation, %	D638	...	...	...	4	...	0.9	0.45	
Tensile modulus, lb/in² $\times 10^{-5}$*	D638	22	22	6.0	30.4	...	14	19.5	24
Compressive strength, lb/in²	D695	35,000	32,000	30,000	40,000	40,000	45,000	30,000	35,000
Flexural strength, lb/in²	D790	19,000	11,000	19,000	60,000	15,000	16,000	11,000	23,000
Impact strength, ft-lb/in of notch	D256	10.0	0.45	8.0	10.0	0.4	0.35	0.4	6.0
Hardness, Rockwell	D785	M108-110	M100-103	M108-115	M100-110	M100-110	M110-125	M110	
Thermal properties									
Thermal conductivity, (cal/s·cm·K) $\times 10^{4}$*	C177	5-10	7-25	5-6	4-10	4-30	7-10	13-17	11-0
Thermal expansion, K^{-1} $\times 10^{5}$*	D696	3.6	4.2	6.0	3.5	5.0	4.0	2.0-4.5	1.5-1.7
Maximum-use temp, °F	...	350	350	300	300	300	210	250	300
Heat-distortion temp, °F at 264 lb/in²	D648	350	320	300	250	250	360	265	400

Chemical properties

		0.35	0.5	0.2	0.2	0.04	0.6	0.14	0.21
Water absorption, % in 24 h	D570	0.35	0.5	0.2	0.2	0.04	0.6	0.14	0.21
Effect of weak acids	D543	None	None	None	None	None	None	None to slight	None
Of strong acids	D543	Slight	Slight	Slight	Negligible	None	Decomposes	Decomposes	Decomposes
Of weak alkalies	D543	None to slight	None to slight	None	None	None	None	Very slight attack	None
Of strong alkalies	D543	Slight	Slight	Slight	None	Slight	Attacked	Slight attack	None to slight
Of organic solvents	D543	None	None	None	None	None	None	None	None

Trade names

Acme, Cosmic, Dapon, Diall, Poly-Dap, RX	Bakelite, Dri-Coat, Eccomold, EMC, Epiall, Eposet, Fibercore, Formitt, High Strength, Hysol, Plenco, Polyset, Scotchply, Smooth-on, Trevarno, Unipoxy	Amres, Cymel, Diaron, Melmac, Permelite, Plenco, Resimene, Resloom, Syr-U-Tex

*See convention footnote a in Table 6-1.

TABLE 6-24 Thermosetting Molding Materials (Continued)

		Phenolic, SPI prefix PF			Polyester, SPI prefix EA		Silicone, SPI prefix S		Urea formaldehyde SPI prefix UF
	ASTM no.	Woodflour and cotton flock filler	Asbestos filler	Glass-fiber filler	Glass-fiber filler	Mineral filler	Glass-fiber filler	Mineral filler	α-Cellulose filler
Electrical properties									
Volume resistivity, Ω·cm	D257	10^{13}	10^{13}	10^{12}	10^{15}	10^{14}	10^{14}	10^{14}	10^{13}
Dielectric strength, V/mil, short-time	D149	400	350	400	420	450	400	400	400
Step by step	D149	375	300	270	390	350	300	380	300
Dielectric constant, 60 Hz	D150	13	50	7.1	7.3	7.5	5.2	3.6	9.5
1 kHz	D150	9.0	30	6.9	4.68	6.2	5.0	...	7.5
1 MHz	D150	6.0	10	6.6	6.4	5.5	4.7	6.3	6.8
Dissipation factor, 60 Hz	D150	0.05	0.1	0.05	0.011	0.009	0.004	0.004	0.035
1 kHz	D150	0.04	0.1	0.02	...	0.02	0.0035	...	0.025
1 MHz	D150	0.03	0.4	0.012–0.026	0.008	0.015	0.002	0.002	0.25
Arc resistance, s	D495	Tracks	120	120	180	150	250	420	150
Mechanical properties									
Specific gravity	D792	1.45	1.9	1.95	2.3	2.30	2.0	2.82	1.52
Specific volume, in³/lb	D792	17.8	11.9	14.1	...	5.4	13.8	...	18.2
Tensile strength, lb/in²	D638,D651	10,000	7,500	18,000	10,000	8,000	5,000	3,500	13,000
Elongation, %	D638	0.8	0.50	0.2	...	...	...	...	1.0
Tensile modulus, lb/in² $\times 10^{-5}$*	D638	17	30	33	25	26	...	...	15
Compressive strength, lb/in²	D695	36,000	35,000	70,000	30,000	25,000	15,000	18,000	45,000
Flexural strength, lb/in²	D790	12,000	14,000	60,000	20,000	10,000	14,000	8,000	18,000
Impact strength, ft·lb/in of notch	D256	0.60	3.5	18	16.0	0.50	15	0.35	0.40
Hardness, Rockwell	D785	M96-120	M95-115	M95-100	...	...	M84	M71-95	M110-120
Thermal properties									
Thermal conductivity, (cal/s·cm·K) $\times 10^4$**	C177	4.7	8–22	9–14.5	10–16	15–25	7.51–7.54	11–13	7–10
Thermal expansion, $K^{-1} \times 10^6$*	D696	3.0–4.5	0.8–4	0.8–1.6	2.5–3.3	3.5–5	0.8	2–4	2.2–3.6
Maximum-use temp, °F	...	350	350	350	300	300	>600	>600	170
Heat-distortion temp, °F at 264 lb/in²	D648	260	300	300	>400	350	>900	>900	260

Chemical properties:

	Test method							
Water absorption, % in 24 h	D570	0.7	0.5	1.2	0.28	0.5	0.2	0.13	0.8
Effect of weak acids	D543	None			Slight	None	None to slight	None	None to slight
Of strong acids	D543	Oxidizing acids			Attacked	Attacked	Slight	Slight	Decomposed if surface attacked
Of weak alkalies	D543	Slight			Slight to attacked	Attacked	None to slight	None to slight	Slight to marked
Of strong alkalies	D543	Attacked			Slight to attacked	Decomposes	Slight to marked	Slight to marked	Decomposes
Of organic solvents	D543	None			None	None	Attacked by some	Attacked by some	None to slight

Trade names

	Amres, Aroclreen, Bakelite, Catalin, Celcron, Durez, Fiberite, Haveg, Mouldrite, Nestorite, Permelite, Plenco, Plyophen, Resinox, Snap-Cure, Synvaren, Synvorite, Tetra-Flex, Tybon, Varcum
	Alpon, Co-Rezyn, Durez, Fibercore, Formadall, Glasdramatic, Glaskyd, Glasrin, Glastic, Haysite, Insulstruct, Mobaloy, Parr, Plaskon, Politen, Polyglas, Premix, Resistrac, Rosite, Trevarno
	Dow Corning
	Amres, Arodure, Beetle, Mouldrite, Nestorite, Plaskon, Resimene, Resloom, Sylplast, Syn-U-Flex, Synvarol, Tetra-Ria, Tybon

*See convention footnote *a* in Table 6-1.

TABLE 6-25   Application Information for Thermoplastics[70]

Material	Major application considerations	Common available forms
ABS (acrylonitrile-butadiene-styrene)	Extremely tough, with high impact resistance; can be formulated over a wide range of hardness and toughness properties; special grades available for plated surfaces with excellent pull-strength values; good general electrical properties but not outstanding for any specific electrical applications	Blow moldings, extrusions, injection moldings, thermoformed parts, laminates, stock shapes, foam
Acetals	Outstanding mechanical strength, stiffness, and toughness properties, combined with excellent dimensional stability; good electrical properties at most frequencies, which are little changed in humid environments up to 125°C	Blow moldings, extrusions, injection moldings, stock shapes
Acrylics (polymethyl methacrylate)	Outstanding properties are crystal clarity and resistance to outdoor weathering; excellent resistance to arcing and electrical tracking	Blow moldings, castings, extrusions, injection moldings, thermoformed parts, stock shapes, film, fiber
Cellulosics	There are several materials in the cellulosic family, such as cellulose acetate (CA), cellulose propionate (CAP), cellulose acetate butyrate (CAB), ethyl cellulose (EC), and cellulose nitrate (CN); widely used plastics in general but not outstanding for electronic applications	Blow moldings, extrusions, injection moldings, thermoformed parts, film, fiber, stock shapes
Chlorinated polyethers	Good electrically, but most outstanding properties are corrosion resistance and good physical and thermal stability by thermoplastic standards	Extrusions, injection moldings, stock shapes, film
Ethylene-vinyl acetates (EVA)	Excellent flexibility, toughness, clarity, and stress-crack resistance; somewhat like a tough synthetic rubber or elastomer; not widely used in electronics; comparatively low resistance to heat and solvents	
Fluorocarbons: Chlorotrifluoroethylene (CTFE)	Excellent electrical properties and relatively good mechanical properties; somewhat more stiff than TFE and FET fluorocarbons but does have some cold flow; widely used in electronics but not quite so widely as TFE and FEP. Useful to about 400°F	Extrusions, isostatic moldings, injection moldings, film, stock shapes
Fluorinated ethylene propylene (FEP)	Very similar properties to those of TFE, except useful temperature limited to about 400°F; easier to mold than TFE	Extrusions, injection moldings, laminates, film
Polytetrafluoroethylene (TFE)	Electrically one of the most outstanding thermoplastic materials; exhibits very low electrical losses and very high electrical resistivity; useful to over 500°F and to below −300°F; excellent high-frequency dielectric; among the best combinations of mechanical and electrical properties but relatively weak in cold-flow properties; nearly inert chemically, as are most fluorocarbons; very low coefficient of friction, nonflammable	Compression moldings, stock shapes, film
Polyvinylidine fluoride (PVF$_2$)	One of the easiest of the fluorocarbons to process; stiffer and more resistant to cold flow than TFE; good electrically; useful to about 300°F; a major electronic application is wire jacketing	Extrusions, injection moldings, laminates, film
Nylons (polyamides)	Good general-purpose for electrical and nonelectrical applications; easily processed; good mechanical strength, abrasion resistance, and low coefficient of friction; some nylons have limited use due to moisture-absorption properties	Blow moldings, extrusions, injection moldings, laminates, rotational moldings, stock shapes, film, fiber

TABLE 6-25 Application Information for Thermoplastics[70] (*Continued*)

Material	Major application considerations	Common available forms
Parylenes (polyparaxylylene)	Excellent dielectric properties and good dimensional stability; low permeability to gases and moisture; produced as a film on a substrate, from a vapor phase; such vapor-phase polymerization is unique in polymer processing; used primarily as thin films in capacitors and dielectric coatings; numerous polymer modifications exist	Film coatings
Phenoxies	Tough, rigid, high-impact plastic; has low mold shrinkage, good dimensional stability, and very low coefficient of expansion for a thermoplastic; useful for electronic applications below about 175°F; useful in adhesive formulations	Blow moldings, extrusions, injection moldings, film
Polyallomers	Thermoplastic polymers produced from two monomers; somewhat similar to polyethylene and polypropylene, but with better dimensional stability, stress-crack resistance, and surface hardness than high-density polyethylene; electronic application areas similar to polyethylene and polypropylene; one of the lightest commercially available plastics	Blow moldings, extrusions, injection moldings, film
Polyamide-imides and polyimides	Among the highest-temperature thermoplastics available, having useful operating temperatures between about 400°F and about 700°F or higher; excellent electrical properties, good rigidity, and excellent thermal stability; low coefficient of friction; polyamide-imides and polyimides are chemically similar but not identical in all properties; they are difficult to process, but are available in molded and block forms, and also as films and resin solutions	Films, coatings, molded and/or machined parts, resin solutions
Polycarbonates	Excellent dimensional stability, low water absorption, low creep, and outstanding impact-resistance thermoplastics; good electrical properties for general electronic packaging application; available in transparent grades	Blow moldings, extrusions, injection moldings, thermoformed parts, stock shapes, film
Polyethylenes and polypropylenes (polyolefins or polyalkenes)	Excellent electrical properties, especially low electrical losses; tough and chemically resistant, but weak to varying degrees in creep and thermal resistance; there are three density grades of polyethylene: low (0.910–0.925), medium (0.926–0.940), and high (0.941–0.965); thermal stability generally increases with density class; polypropylenes are generally similar to polyethylenes but offer about 50°F higher heat resistance	Blow moldings, extrusions, injection molding, thermoformed parts, stock shapes, film, fiber, foam
Polyethylene terephthalates	Among the toughest of plastic films with outstanding dielectric strength properties; excellent fatigue and tear strength and resistance to acids, greases, oils, solvents; good humidity resistance; stable to 135–150°C	Film, sheet, fiber
Polyphenylene oxides (PPO)	Excellent electrical properties, especially loss properties to above 350°F and over a wide frequency range; good mechanical strength and toughness; a lower-cost grade (Noryl) somewhat similar properties to PPO but with a 75–100°F reduction in heat resistance	Extrusions, injection moldings, thermoformed parts, stock shapes, film
Polystyrenes	Excellent electrical properties, especially loss properties; conventional polystyrene is	Blow moldings, extrusions, injection

TABLE 6-25   Application Information for Thermoplastics[70] (*Continued*)

Material	Major application considerations	Common available forms
	temperature-limited, but high-temperature modifications exist, such as Rexolite or Polypenco cross-linked polystyrene, which are widely used in electronics, especially for high-frequency applications; polystyrenes are also generally superior to fluorocarbons in resistance to most types of radiation	moldings, rotational moldings, thermoformed parts, foam
Polysulfones	Excellent electrical properties and mechanical properties to over 300°F; good dimensional stability and high creep resistance; flame-resistant and chemical-resistant; outstanding in retention of properties upon prolonged heat-aging, as compared with other tough thermoplastics	Blow moldings, extrusions, injection-mold thermoformed parts, stock shapes, film sheet
Vinyls	Good low-cost general-purpose thermoplastic materials but not specifically outstanding electrical properties; greatly influenced by plasticizers; many variations available, including flexible and rigid types; flexible vinyls, especially polyvinyl chloride (PVC), widely used for wire insulation and jacketing	Blow moldings, extrusions, injection moldings, rotational moldings, film sheet

nated hydrocarbons. Solid impregnants such as wax or resins are also used. Impregnated paper is widely used for capacitors and transformers.

102. Wood. Wood is not commonly used as electrical insulation. Some use remains as poles of electrical transmission lines. Wood has a dielectric constant of 2.5 to 7.7 and resistivity of 10^{10} to 10^3 Ω-cm. The dielectric strength is about 4×10^5 V/m.

103. Fabrics. Both natural and synthetic organic fibers are used for insulation. Inorganic fibers such as glass and asbestos are also used. Fibers can be used in nonwoven form such as laminates.

Unimpregnated fibers are used in low-voltage applications such as thermocouple wires. Impregnated fabrics are coated with elastomers and provide an improved dielectric strength and thermal endurance. These flexible fabrics are used for cable, coil, or bus bars.

104. Mica. Mica is available as muscovite or phlogopite. The muscovite, or ruby mica, is $KAl_3Si_3O_{10}(OH)_2$, while the phlogopite, or amber mica, is $KMg_3AlSi_3O_{10}(OH)_2$.

Mica can be cleaved into very thin sheets. Muscovite can be used up to about 500°C and heat-treated phlogopite up to 800°C. The resistivity of mica in the range 0 to 300°C is between 10^8 and 10^{14} Ω·m. Dissipation factors for muscovite are 0.0001 to 0.0004 and for phlogopite 0.004 to 0.07 between 60 and 10^6 Hz. The dielectric constant is about 6.5 to 8.7 for muscovite and 5 to 6 for phlogopite. The dielectric strength is about 10 to 20 MV/m for 0.0001- to 0.0003-in-thick material in air. Synthetic mica can be obtained as a compacted aggregate. It has electrical properties similar to bulk muscovite but can be heated to 1100 to 1200°C. Mica paper, i.e., mica splittings, often oriented platelets, embedded in resins are very stable insulators. Mica is used for capacitors interlayered with tin foil, silvered, or as mica paper. Glass-bonded mica, Mycalex, for example, is a very stable insulator, possessing good arc resistance and high dielectric strength (20 MV/m).

105. Ceramics. Ceramics are extensively used as electric insulators or capacitors in the electrical and electronics industries. The uses of ceramics are summarized in Table 6-32. Melting points of some ceramics are given in Fig. 6-20. A more detailed description of representative ceramics is given below.

Porcelain.   Most porcelain insulators consist of mullite and quartz embedded in a glassy matrix. High-tension porcelain is completely vitrified, normally covered with a glaze to improve mechanical properties and corona resistance.

Porcelain can be formed in the soft state, then dried, glazed, and fired. It is used for low-frequency application since it has a high dielectric loss factor due to its lossy glass content. Porcelains have relatively low electrical resistivities due to their high content of mobile alkali ions. Properties of porcelain are given in Table 6-33.

TABLE 6-26 Thermoplastic Molding Materials[73]

	ASTM no.	Acetal	ABS	Acrylic	Cellulose acetate	Cellulose acetate butyrate	Cellulose propionate	Chlorinated polyether	Chlortri-fluoro-ethylene	Nylon (polyamide)	Poly-carbonate
SPI prefix		...	ABS	MM	CA	CAB	CP	...	HH	PA	...
					Electrical properties						
Arc resistance	D495	129	90	No track	200	...	180	...	>360	140	120
Dielectric constant,	D150										
60 Hz		3.8	3.0	4.0	7.5	6.4	4.0	3.1	2.8	5.5	3.2
1 MHz	...	3.8	3.0	3.5	7.0	6.3	4.0	3.0	2.7	4.9	3.0
1 GHz	...	3.8	3.0	3.2	7.0	6.2	3.6	2.9	2.5	4.7	3.0
Dissipation factor,	D150										
60 Hz		0.004	0.003	0.04	0.01	0.02	0.01	0.01	0.001	0.01	0.0009
1 GHz	...	0.004	0.005	0.02	0.01	0.05	0.01	0.01	0.09	0.03	0.01
Dielectric strength, V/mil step by step	D149	400	350	350	200	250	300	400	450	320	364
Volume resistivity, $\Omega\cdot$cm	D257	10^{14}	10^{16}	10^{14}	10^{13}	10^{14}	10^{15}	10^{15}	10^{18}	10^{15}	10^{12}
					Mechanical properties						
Tensile strength, lb/in²	D651,D638	9,000	7,000	11,000	8,500	6,900	7,800	6,000	6,000	14,000	9,500
Tensile modulus, lb/in² $\times 10^{5*}$	...	4.1	3.5	4.5	4.0	2.0	2.2	1.6	3.0	3.8	3.5
Elongation, %	...	15	...	10	70	88	100	160	250	320	100
Compressive strength, lb/in²	D695	10,000	7,000	18,000	36,000	22,000	22,000	...	7,400	13,000	12,500
Flexural strength, lb/in²	D790	14,000	10,500	17,000	16,000	9,300	11,100	5,000	9,300	...	13,500
Impact strength, ft·lb/in of notch	D256	1.4	7.0	0.5	5.2	6.3	11.5	0.4	2.7	4.0	16
Hardness, Rockwell	D785	R120	M110	M105	R125	R115	R122	R100	R95	R118	R118
					Thermal properties						
Heat-distortion temp at 264 lb/in²	D648	255	200	210	190	202	228	285	258	167	280
Maximum-use temp, °F	...	185	210	190	220	220	220	290	390	250	250
Coefficient of thermal expansion, $K^{-1} \times 10^{5*}$	D696	8.1	11	9	16	17	17	8	7	8	7

*See explanation of convention in footnote a, Table 6-1.

TABLE 6-26 Thermoplastic Molding Materials[73] (Continued)

	ASTM no.	Acetal	ABS	Acrylic	Cellulose acetate	Cellulose acetate butyrate	Cellulose propionate	Chlorinated polyether	Chlortri-fluoro-ethylene	Nylon (polyamide)	Poly-carbonate
SPI prefix		...	ABS	MM	CA	CAB	CP	...	HH	PA	...
Thermal conductivity, cal/s·cm·K	C177	5.5	3.0	0.0	8.0	8.0	8.0	3.1	5.3	5.9	4.6
Flammability, in/min	D635	1.1	...	1.2	...	...	1.3	No burn	No burn	No burn	No burn
							Chemical properties				
Water absorption, % in 24 h	D570	0.25	0.45	0.4	6.5	2.2	2.8	0.01	0.00	1.88	0.15
Not resistant to	D543	Strong acids	Oxidizing acids, ketones, esters, chlorinated solvents	Ketones, esters, aromatic and chlorinated solvents	...	Strong bases, ketones, esters, aromatic and chlorinated solvents	...	Oxidizing acids	Chlorinated solvents	Strong acids, phenol	Alkalies, aromatic and chlorinated solvents
						Trade names					
		Betalux Celcon Delrin Dielax Formaldafil Thermocomp	Abson Cycolac Cyclon Kralastic Lustran Royalite Sulivac Triform Tybrene	Acrapon Acrydass Acrylite Acrylux Aerysol Areset Bovick Glopaque Implex Interpole Kydex Lacite Lactrelite Oraglos Plexiglas Stsvol XT Polymer Zerlon	Cal-Stix Celanese Joda Kodacel Plasticel Yenite	Acelon Cabulite Joda Tenite Uvex	Fortice l Tenite	Penton	Kel-F Halon	Capran Catalin Firestone Fosta-Nylon Glastil Moleculoy Monocast Nylafil Nylux Plaskon Spencer Thermocomp X-Tal Zytel	Dupilon Lexan Merlon Penntube IV Polycarbafil Thermocomp Zelux

	ASTM no	Polyethylene, low-density PE	Polyethylene, med-density PE	Polyethylene, high-density PE	Polypropylene	Polystyrene PS	Polysulfone	Polyphenylene oxide	Phenoxy	Polyvinyl chloride VC	SAN	Tetrafluoroethylene HH
SPI prefix		PE	PE	PE	…	PS	…	…	…	VC	…	HH
Electrical properties												
Arc resistance	D495	140	200	200	185	100	122	75	…	80	150	<200
Dielectric constant,	D150											
60 Hz	…	2.4	2.4	2.4	2.6	3.4	3.1	2.6	4.1	3.6	3.4	2.1
1 MHz	…	2.4	2.4	2.4	2.6	3.2	3.1	2.6	4.1	3.3	2.5	2.1
1 GHz	…	2.4	2.4	2.4	2.6	3.1	3.1	2.6	3.8	3.4	3.1	2.1
Dissipation factor,	D150											
60 Hz	…	<0.0005	<0.0005	<0.0005	<0.0005	0.0004	0.0008	0.0004	0.001	0.007	0.004	<0.0002
1 MHz	…	<0.0005	<0.0005	<0.0005	<0.0005	0.0004	0.001	…	0.002	0.009	0.007	<0.0002
1 GHz	…	<0.0005	<0.0005	<0.0005	<0.0005	0.0004	0.005	0.0009	0.03	0.006	0.001	<0.0002
Dielectric strength, V/mil, step by step	D149	420	500	550	450	300	400	400	400	375	300	430
Volume resistivity, Ω·cm	D257	10^{16}	10^{16}	10^{16}	10^{16}	10^{16}	10^{17}	10^{13}	10^{13}	10^{16}	10^{16}	10^{18}
Mechanical properties												
Tensile strength, lb/in²	D651, D638	2,300	3,500	5,500	5,500	6,800	10,200	11,000	9,500	9,000	12,000	4,500
Tensile modulus, lb/in¹ × 10^{-5}*	…	0.35	0.55	1.5	2.3	4.5	3.6	3.8	3.8	6.0	5.6	0.58
Elongation, %	…	800	600	100	700	80	100	80	100	40	3.5	400
Compressive strength, lb/in²	D695	…	…	3,200	8,000	9,000	14,000	13,000	12,000	13,000	17,000	1,700
Flexural strength, lb/in²	D790	…	7,000	1,000	8,000	10,000	15,400	15,000	14,000	16,000	19,000	…
Impact strength, ft·lb/m of notch	D256	No break	<16	20	1.5	8	1.3	1.9	12	20	0.5	3.0
Hardness, Rockwell	D785	…	…	…	R110	R100	R120	R123	R123	…	M90	…
Thermal properties												
Heat-distortion temp at 264 lb/in²	D648	105	120	120	145	205	…	375	…	164	215	<250
Maximum-use temp, °F	…	212	250	250	320	175	345	…	300	175	205	550
Coefficient of thermal expansion, K^{-1} × 10^6*	D696	18	16	13	10	21	6	3	6	18	8	10
Thermal conductivity, cal/·cm,K	C177	8.0	10.0	12.4	2.8	3.0	6.2	4.5	8.4	7.0	2.9	6

*See explanation of convention in footnote *a*, Table 6-1.

TABLE 6-26 Thermoplastic Molding Materials (*Continued*)

	Polyethylene, low-density	Polyethylene, med-density	Polyethylene, high-density	Polypropylene	Polystyrene	Polysulfone	Polyphenylene oxide	Phenoxy	Polyvinyl chloride	SAN	Tetrafluoroethylene
SPI prefix	PE	PE	PE	...	PS	...	...	...	VC	...	HH
Flammability, in/min	1.04	1.04	1.04	1.04	1.0	No burn	No burn	No burn	No burn	2.0	No burn
Chemical properties											
Water absorption, % in 24 h	<0.01	<0.01	<0.01	0.03	0.6	0.2	0.06	0.13	0.4	0.3	0.00
Not resistant to	Oxidizing acids, aromatic and chlorinated solvents	Oxidizing acids, aromatic and chlorinated solvents	Oxidizing acids, aromatic and chlorinated solvents		Oxidizing acids, aromatic chlorinated solvents	Aromatic solvents	Aromatic and chlorinated solvents		Ketones, esters, aromatic solvents	Acids, ketones, esters, chlorinated solvents	
Trade names	Agilene, Althon, Bakelite, Cao X-L, Dylan, El-Rex, Epolene, Ethylux, Excelite, Fortiflex, Hi-Fax, Marlex, Molulene, Petrothene, Poly-Eth			Acrotut, Avison, Bakelite, Chevron, Escon, Marlex, Petrothene, Pro-Fax	Bakelite, Biax, Dylene, El-Rex, Exenglo, E-Z Flow, Fostarene, Gilco, Grace, Hypac, Kardel, Lustrex, Shell, Sofar, Styrafil, Styroflex, Styroflux, Styron	Bakelite	PPO	Bakelite	Bakelite, Biacar, Dacovin, Durelene, Esamgia, Ethyl, Excelon, Exon, Geon, Insular, Kenron, Kohinor, Marvinol, Nalgon, Opalon, Pliovic, Resinite, Rucoblend, Secron, Trulon, Tygon, Vicoa, Vygen, Gyran, Gyran	Acrylafil, Bakelite, Catalin, Kralac, Lustran, Plaxacrin, Tyril	Halon, Teflon

ASTM no: D635 (Flammability), D570 (Water absorption), D543 (Not resistant to)

TABLE 6-27 Properties of NEMA Copper-Clad[74]
See Table 6-28

Property	Conditioning procedure by ASTM methods D618*	NEMA grade							
		XXXP	XXXPC	FR-2	FR-3	FR-4	FR-5	G-10	G-11
Peel strength (min), lb/in width, 1-oz copper, after solder dip	A	6	6	6	7	7	7	7	7
After elevated temp	E-1/140†	6	6	6	7	7	7	7	7
2-oz copper, after solder dip	A	7	7	7	9	9	9	9	9
After elevated temp	E-1/140†	7	7	7	9	9	9	9	9
Volume resistivity (min), mΩ·cm	C-96/35/90	10,000	10,000	10,000	100,000	100,000	100,000	100,000	100,000
Surface resistance (min), mΩ	C-96/35/90	1,000	1,000	1,000	1,000	1,000	1,000	5,000	5,000
Dielectric breakdown parallel to laminations (min), step by step, kV	D-48/50	15	15	15	30	30	30	30	30
Dielectric constant (avg max) at 1 MHz‡	D-48/50	5.3	5.3	5.3	5.0	5.8	5.8	5.8	5.8
Dissipation factor (avg max) at 1 MHz‡	D-48/50	0.05	0.05	0.05	0.045	0.045	0.045	0.045	
Flexural strength (avg min), lb/in², lengthwise, $\frac{1}{32}$ to $\frac{3}{32}$ in thick	A	12,000	12,000	12,000	20,000	55,000	55,000	55,000	55,000
$\frac{1}{8}$ and $\frac{1}{4}$ in thick						50,000	50,000	50,000	50,000
Crosswise, $\frac{1}{32}$ to $\frac{3}{32}$ in thick	A	10,500	10,500	10,500	16,000	45,000	45,000	45,000	45,000
$\frac{1}{8}$ and $\frac{1}{4}$ in thick						40,000	40,000	40,000	40,000
Condition A value retained at elevated temp. lengthwise, %, $\frac{1}{32}$ in thick	E-1/150, T-150					50	50	50	50
$\frac{1}{8}$ and $\frac{1}{4}$ in thick							40		40
$\frac{1}{4}$ and $\frac{1}{2}$ in thick							50		50
Flammability (avg max), time to extinguish, s	A			15	15	15	15		
Water absorption (avg max), %, $\frac{1}{32}$ in thick		1.30	1.30	1.30	0.65	0.80	0.80	0.80	0.80
$\frac{1}{16}$ in thick		1.00	0.75	0.75		0.35	0.35	0.35	0.35
$\frac{3}{32}$ in thick		0.85	0.65	0.65	0.50	0.25	0.25	0.25	0.25
$\frac{1}{8}$ in thick		0.75	0.55	0.55	0.50	0.20	0.20	0.20	0.20

*Methods of Conditioning Plastics and Electrical Insulating Materials for Testing (ASTM Designation D618).
†For grades XXXP, XXXPC, FR-2, and FR-3, use condition E-1/120.
‡Applies only to $\frac{3}{32}$ and $\frac{1}{8}$-in thicknesses.

Steatite.   Steatite ceramic bodies are low-dielectric-loss porcelains. Steatite is used for variable capacitors, switches, coil forms, resistor shafts, spacers, and bushings. Steatite is mainly used for low-tension, high-frequency applications. Natural steatite talc is often referred to as *lava*. It is

TABLE 6-28   Description of NEMA Grades Listed in Table 6-27

NEMA grade	Resin	Reinforcement	Description
XXXP	Phenolic	Paper	Hot-punching grade for general use
XXXPC	Phenolic	Paper	Room-temperature punching for boards with small close holes
G-10	Epoxy	Glass cloth	General-purpose glass base; excellent electrical and physical properties; good moisture resistance
G-11	Epoxy	Glass cloth	Similar to G-10 but more thermally stable
FR-2	Phenolic	Paper	Similar to XXXPC but flame-retardant
FR-3	Epoxy	Paper	Room-temperature punching; flame-retardant; better physical properties than XXXPC
FR-4	Epoxy	Glass cloth	Similar to G-10 but flame-retardant
FR-5	Epoxy	Glass cloth	Similar to G-11 but flame-retardant

readily machinable in the soft state and can subsequently be fired to form a hard product. Properties of steatite are given in Table 6-33.

Cordierite.   Cordierite has a low thermal expansion coefficient and a high resistance to thermal shock. Cordierite is used as electrical insulation where thermal stresses are important, as in heating apparatus, thermocouple insulators, etc. See Table 6-33 for physical properties.

Forsterite.   The main advantages of forsterite are its ease of firing, low dielectric loss at high frequencies, and high resistivity. Forsterite is used in small microwave tubes. It has a high thermal expansion coefficient and poor thermal shock resistance. See Table 6-33 for physical properties.

Alumina (Al_2O_3).   Alumina ceramics are composed of fine crystalline Al_2O_3 particles and a glass matrix. High-purity alumina can be >99.9% Al_2O_3. Al_2O_3 can also be found as single crystals, termed sapphire.

Alumina ceramics have a high mechanical and dielectric strength, as well as a high resistivity and low dielectric loss. Alumina ceramics are very stable chemically, retaining their properties over a wide temperature and frequency range. The properties of alumina are affected by its purity. For example, ionic impurities cause higher dielectric losses in impure alumina. Porosity causes a pronounced degradation in thermal conductivity. High-purity, high-density alumina can be obtained by sintering fine-grained powders at high temperatures. Alumina can be metallized by vacuum deposition or by screen printing metal oxides, and firing. Alumina is used for circuit breakers, spark plugs, resistor cores, substrates for hybrid microelectronic circuits, microwave integrated circuits, integrated circuits, and power transistors. See Table 6-33 and Fig. 6-21 for properties of alumina. Sapphires are used as substrates where high heat dissipation and a very flat surface are needed.

Beryllia (BeO).   Beryllia is a unique material having a very high heat conductivity with very low electrical conductivity. At room temperature BeO thermal conductivity is about that of aluminum (Fig. 6-18). BeO has good thermal shock resistance and good mechanical characteristics.

BeO is used where high heat dissipation is essential, as in high-power transistor bases, microwave windows, high-power klystrons, and lasers. Inhalation of BeO powder is hazardous. Properties are given in Fig. 6-21.

Magnesia (MgO).   MgO has poor shock resistance but shows little electrical conduction at high temperatures. MgO has found extensive use as a basic refractory in steelmaking, while in electronics it is used for heaters and thermocouple insulators.

Zirconia (ZrO_2).   Zirconia has poor thermal conductivity and shock resistance. It is usually available with yttrium or calcium, which stabilize the zirconia. Changes in electrical conduction of zirconia have been used to measure oxygen partial pressure. ZrO_2–MgO combinations have been used for high-temperature heating elements.

Carbides.   Silicon carbide is the most widely known carbide. SiC forms a protective-surface oxide layer which permits its use up to 1700°C in oxidizing and 2200°C in inert atmosphere.

TABLE 6-29 Insulating Films[73]

Film base	ASTM No.	Cellulose acetate	Cellulose triacetate	Polytrifluoro chloroethylene copolymers	Polyurethane elastomer	FEP fluorocarbon	Polyethylene	Polyvinyl chloride	Ionomer	Polyimide
Method of processing	...	Cast, extruded	Cast	Extruded	Cast, calendered extruded	Extruded	Extruded	Cast, calendered	Extruded	Rolls
Forms available	...	Rolls, sheets, tapes	Rolls, sheets	Rolls	Rolls, sheets	Rolls, sheets, tapes	Rolls, sheets, tapes	Rolls, sheets, tapes	Rolls, sheets, tapes	Rolls
Thickness range, in	...	0.0005-0.250	0.0008-0.020	0.0005-0.030	0.005-0.030	0.0005-0.020	0.00075-0.010	0.00075-0.08	0.001-0.010	0.001-0.005
Maximum width, in	...	60	45	54	52	48	240	84	60	18
Area factor, in^2/lb·mil	...	22,000	21,400	13,000	22,000	12,900	29,500	23,000	29,400	19,400
Specific gravity, g/cm^3	D792-60T	1.31	1.31	2.15	1.26	2.15	0.940	1.50	0.94	1.42
Tensile strength, lb/in^2	D882-61T	16,400	16,000	8,000	9,000	3,000	3,500	5,600	5,000	25,000
Elongation, %	D882-61T	70	40	150	650	300	650	500	450	70
Bursting strength, 1 mil thickness, Mullen points	D774-63T	40	70	42	...	11	...	20	...	75
Tearing strength, g	D689-62	10	10	26	...	125	...	1,400	...	8
Tearing strength, lb/in	D689-62	415	395	900	600	600	300	490	70	
Folding endurance, c	D643-63T	2,000	4,000	...	...	4,000	>100,000	...	...	10,000
Water absorption, % in 24 h	D570-63	9	2.4-5	0.00	0.55-0.77	<0.01	<0.01	...	0.4	
Dielectric constant, 1 kHz	D150-62	3.6	3.2	2.7	7.5	2.0	2.2	6.0	2.1	2.9
1 MHz	D150-62	3.2	3.3	2.4	7.1	2.0	2.2	4.0	2.4	3.6
1 GHz	D150-62	3.2	3.2	2.3			2.2		2.4	3.4
Dissipation factor, 1 kHz	D150-62	0.10	0.10	0.027	0.060	0.0002	0.0003	0.16	0.007	0.003
1 MHz	D150-62	0.10	0.10	0.017		0.0007	0.0003	0.14	0.007	0.010
1 CHz	D150-62	0.094	0.094	0.004			0.0003		0.007	
Dielectric strength, V/mil	...	5,000	3,700		500	5,000	500	600	1,000	7,000
Volume resistivity, Ω·cm	...	10^{13}	10^{15}	10^{18}	10^{11}	10^{19}	10^{16}	10^{14}	10^{16}	10^{18}
Orientation possible	...	No	No	No	Yes	Yes	Yes	Yes	Yes	No

TABLE 6-29 Insulating Films[73] (Continued)

Film base	ASTM No.	Cellulose acetate	Cellulose triacetate	Polytrifluoro chloroethylene copolymers	Polyurethane elastomer	FFP-fluorocarbon	Polyethylene	Polyvinyl chloride	Ionomer	Polyimide
Trade names		Auburn Acelon Bexfilm Bexoid Cadeo Campco Inceloid Kodacel Lumarith Midlon Saplasco Tenite Vucpak	Rexfilm Celanese Kodacel	Kel-F Polyfluoron	Adiprene Estane Perflex Texin	Teflon	Arnar Amerifilm Ampacet Auburn Boltaron Clysar Conolean Crown-Seal Dynafilm Ger-Pae Hypac Katharon Midlon Plicose Polython Prohi Seilon Vis-Queen Zee Zendel	Agilide Amerifilm Arnar Boltaflex Cadeo Clopane Colovin Dorn Fabray Fabtex Ger-Pac Katharon Korrseal Krene Monosol Pantex Randfilm Resinite Reynolon Ruccam Saplasco Touchstone Ultron Velon Vylene Watascal	Surlyn	Kapton

Film base	ASTM No.	Polypropylene	Polyvinyl fluoride	Polyester	TFE tetrafluoroethylene	Vinylidene chloride	Fluorohalocarbon	Vinylidene fluoride	Polyamide
Method of processing	...	Extruded	Extruded	Cast	Skived, extruded	Extruded, cast	Extruded	Extruded	Extruded, cast
Forms available		Rolls, sheets, tapes	Rolls	Rolls, sheets, tapes	Sheets, tapes	Rolls, tapes	Rolls	Rolls	Rolls
Thickness range, in	0.0005-0.020	0.0005-0.02	0.00015-0.014	0.0005-0.01	0.0004-0.02	0.005-0.02	0.001-0.010	0.0005-0.02	0.0005-0.02
Maximum width, in	...	60	138	120	38	40	54	16	54
Area factor, in²/lb·mil	...	31,300	17,200	22,600	12,800	16,800	13,000	15,000	24,5000
Specific gravity, g/cm³	D792 60T	0.9	1.38	1.4	2.2	1.61	2.2	1.76	1 13
Tensile strength, lb/in²	D882-61T	10,000	18,000	30,000	4,000	20,000	11,000	6,500	11,500
Elongation, %	D882-61T	1,000	250	120	350	80	150	300	400
Bursting strength, 1 mil thickness, Mullen points	D774-63T	...	70	66	...	...	31	20	90
Tearing strength, g	D689-62	45	40	15	100	...	150	60	1,200
Tearing strength, lb/in	D689-62	...	1,400	1,300	...	...	29	...	>250,000
Folding endurance, c	D643-63T	>100,000	47,000	14,000	...	50,000	>100,000	75,000	...
Water absorption, % in 24 h	D570-63	0.005	0.5	0.8	0.00	0.00	0.00	0.04	9.5
Dielectric constant, 1 kHz	D1531-62	2.1	8.5	3.1	2.1	6.0	2.6	7.72	3.8
1 MHz	D1531-62	2.1	7.4	3.0	2.1	5.0	2.3	6.43	3.7
1 GHz	D1531-62	...	...	2.8	2.1	4.0	...	2.98	3.4
Dissipation factor, 1 kHz	D1531-62	0.0003	0.009	0.0047	0.0002	0.045	0.039	0.019	0.010
1 MHz	D1531-62	0.0003	...	0.016	0.0002	0.075	0.037	0.159	0.016
1 GHz	D1531-62	0.0003	...	0.003	0.0002	0.08	...	0.11	0.025
Dielectric strength, V/mil	...	3,000	...	7,000	430	5,000	3,000	1,280	1,700
Volume resistivity, Ω·cm	...	10^{16}	10^{13}	10^{18}	10^{18}	10^{16}	10^{17}	10^{14}	...
Orientation possible	...	Yes	No	Yes	No	Yes	No	No	No
Trade names	...	Bexphone, Clysar, Dynafilm, Electro-film, Hypac, Midlon, Olefane, Profax, Propylux, Udel, Vypro	Tedlar	Celanar, Kodar, Mylar, Scotchpar, Scotchpak, Videne	Halon, Teflon	Cryovac, Saran	Aclar	Kynar	Cadco, Califilm, Capran, Plastex, Sapalaso

SiC has electrical resistivity of about 1 mΩ·m at room temperature. SiC semiconductor devices capable of performing up to 500°C have been made. The main uses of SiC are in cutting tools and heating elements.

Nitrides. *Boron nitride* is used as a high-temperature insulator. It is readily machinable and needs no firing. BN absorbs water and must therefore be kept dry. BN powder can be used up to 3000°C in nitrogen.

Silicon nitride is used in the integrated circuits primarily as a sodium diffusion barrier on the SiO_2. It has been used for double dielectric memories (MNOS). Si_3N_4 has an electrical resistivity of 10 GΩ·m at 200°C. Films can be formed by chemical vapor deposition, sputtering, and vapor deposition. Such films have been used as oxidation masks in integrated circuits.

106. Ferroelectric Ceramics. Ferroelectric materials have a **D-E** curve similar to the **B-H** hysteresis curve in ferromagnetic materials. This nonlinear nature of the induction originates from domain structure in the ferroelectrics. Each domain consists of parallel dipoles. When the field is applied, favorable domains grow at the expense of their neighbors, and very high polarization can be obtained with low fields. Ferroelectrics have a Curie temperature (1200°C for $BaTiO_3$) above which the ferroelectric behavior vanishes. The main materials used as ferroe-

TABLE 6-30 Properties of Rigid Plastic Forms

Property	ASTM no.	ABS	Cellulose acetate	Epoxy	Syntactic epoxy*		Phenolic	Poly-ethylene	Poly-propylene
Density per cubic foot	...	31	6-8	5-8	36	42	2-4	4	5
Thermal conductivity, Btu·in/h·ft²·°F	177	0.58	0.31-0.32	0.24-0.28	4.56	...	0.20-0.22	0.92	0.27
Coefficient of thermal expansion, °F⁻¹ × 10⁵†	D696	9.7	2.5	2.3	4.5	...	0.5	4.18	
Water absorption, % vol	C272	0.6	13-17	...	...	1.5	<3	0.22	
Flammability, in/min	D1692	1.04	4.9	...	...	...	SE	...	1.6
Dielectric const at 1 MHz	....	1.59	1.10-1.12	2.0	1.55	...	...	1.48	
Dissipation factor at 1 MHz	....	0.007	0.003	0.005	0.01	...	...	0.0003	
Max recorded service temp, °F	....	200	350	400-500	300	300	270	195	230
Strength, lb/in², tensile	D1623	1,400	170	50-200	3,300	4,600	20-55	1,000	170
Compressive (10%)	D1621	....	125-150	60-90	9,600	13,400	20-90	800	55
Flexural	D790	2.4	150	200-800	3,800	6,000	25-65	1,900	230
Shear	C273	....	140	....	3,800	4,400	15-30		
Modulus of elasticity, 1,000 lb/in², in tension	D1623	2.4	....	....	....	610			
In compression	D1621	....	5.5-13	2.1-6.5	373	480	....	....	1.2
In flexion	D790	9	5.5	2.5-6	...	480	....	....	9.6
In shear	C273	....	....	....	...	...	0.4-0.75		
Hardness (Shore D)	....	60	....	....	80-85				

Property	ASTM no.	Polyvinyl chloride	Polystyrene (expanded)		Urea		Urethane		
Density per cubic foot	....	3	2	6	0.8-1.2	2-3	4-7	18-25	
Thermal conductivity, Btu·in·ft²·°F	177	0.15-0.20	0.20-0.28	0.20-0.25	0.18-0.21	0.11-0.23	0.15-0.28	0.29-0.52	
Coefficient of thermal expansion, °F⁻¹ × 10⁵†	D696	2	2.7-4	2.7-4	...	3-4	4	4	
Water absorption, % vol	C272	0.1	<0.1	<0.1	...	3-4	1.5	0.2	
Flammability, in/min	D1692	...	2-8	2-8	SE	NB	NB	NB	
Dielectric const at 1 MHz	...	...	1.02-1.24						
Dissipation factor at 1 MHz	...	...	<0.0005						
Max recorded service temp, °F	...	180	175	175	120	200-250	250-300	300-400	
Strength, ln/in², tensile	D1623	100-200	50-55	120	...	20-70	90-250	700-1,300	
Compressive (10%)	D1621	70-100	25-30	100-150	5	20-50	65-275	1,200-2,000	
Flexural	D790	120-160	55-75	-00-300	17	60-100	200-350	700-2,000	
Shear	C273	60-80	35	150	...	20-30	60-130	7,600	
Ultimate tension elong., %	D1623	5-20	5	2					
Modulus of elasticity, 1,000 lb/in², in tension	D1623	3-4	740	6,100					
In compression	D1621	3-4	0.55-2	3-6	...	0.3-0.6	1.5-4.5	10-40	
In flexion	D790	3-4	1.3-3.8	5-15	0.7	0.8-0.9	0.8-15	12-100	
In shear	C273	2-2.5	1.15-1.6	3	...	0.17-0.21	0.5-1.5	3-9	

*Glass microballoons filled with epoxy. †See foothote *a*, Table 6-1.

lectric ceramics are titanates, zirconates, niobates, tantalates, and stannates Properties of some ferroelectrics are listed in Table 6-34.

Ceramic capacitors can be divided into two classes:

Class I: Capacitors which have a definite temperature coefficient of capacitance. The dielectric constants of such capacitors (usually containing TiO_2) are up to 500. The capacitors have loss factors up to 0.004. Class I capacitors are suited for resonator or oscillator circuits where good capacitance stability is required.

Class II: Capacitors which are highly nonlinear materials ($BaTiO_3$, for example) having very high dielectric constants (typically 500 to 10,000). These capacitors are not stable and have high losses. Class II capacitors are used in coupling, filtering, or bypass applications where capacitance stability is not important but miniaturization is essential.

All ferroelectric materials are piezoelectric and pyroelectric.

107. Piezoelectric and Pyroelectric Materials. Piezoelectric materials are non-centrosymmetric. Quartz, $BaTiO_3$, and rochelle salt are common piezoelectric materials. Piezo-electrics are used for ultrasonic cleaning, delay lines, strain gages, thickness monitors, acceler-ometers, etc. Properties of some piezoelectric and pyroelectric materials are given in Table 6-34. A description of acoustic devices such as surface acoustic wave devices is given in Sec. **13** and Ref. 30.

108. Electrolytic Films. Electrolytic capacitors are formed by anodizing aluminum or tantalum foils to form Al_2O_3 or Ta_2O_5 dielectric films. Electrolytically formed oxide is between 0.01 and 1.0 μm thick. Tantalum oxide capacitors are used also in thin-film hybrid circuits where

TABLE 6-31   Applications of Organic Coatings[77][*]

Coating	Distinguishing characteristics	Max. continuous use temp., °F
Thermosetting coatings:		
Acrylics	Excellent resistance to ultraviolet and weathering	250
Alkyds	Most widely used general-purpose coating	200–250
Epoxies	Excellent chemical resistance and good insulation coating; widely used as circuit-board coatings	400
Phenolics	Good chemical resistance against alkalies and good insulation coating	350-400
Phenolic-oil varnishes	Electrical impregnating varnishes	250
Polybutadienes	Good electrical insulation properties	450
Polyesters	Good electrical insulation properties	200
Silicones	Excellent high-temperature electric insulation properties	500
Thermoplastic coatings:		
Acrylics	Excellent resistance to ultraviolet and weathering	180
Cellulosics	Fast-dry commercial lacquers	180
Fluorocarbons	Excellent chemical resistance, excellent electrical insulation properties, even at high temperatures	400–500
Penton chlorinated polyethers	Primarily used as chemically resistant coatings for equipment; coatings have good electrical properties, however	250
Phenoxies	Low coefficient of expansion	180
Polyimides	High-temperature coating having excellent insulation properties; widely used as high-temperature magnet-wire enamel	600–700
Polyurethanes	Good electrical insulation coatings, widely used as circuit-board coatings	250
Vinyls	There are many varieties; some vinyls are outstanding in resistance to inorganic and plating chemicals; vinyl plastisols are convenient for dip coating of electrical parts	150

[*]In addition to the above basic types of coatings, there are many specialty coatings, such as ablative coatings, thermal control coatings, flame-retardant coatings, fungus-resistant coatings, electrically conductive coatings, and magnetic coatings. In addition, many elastomers can be applied as coatings (e.g., neoprene). Most of the specialty coatings are specifically filled or otherwise modified variations of basic thermosetting, thermoplastic, or elastomeric coating polymers. Hence, the possible coatings available are as unlimited as the broad range of plastics and polymers.

TABLE 6-32 Uses of Technical Ceramics[64]

Applications	Types of ceramics
Electrical	
Low- and high-tension insulation:	
Domestic fittings, power generation, and transmission	Electrical porcelain, zircon porcelain
Insulation at elevated temperatures:	
Electric fires, ovens, and low-temperature kilns	Aluminous porcelain, cordierite
Fireproof cables	Magnesia
Insulation at high temperatures:	
Thermocouple sheaths, furnace muffles, various furnace parts	Mullite, fused silica, alumina
Sparking plug insulators	Alumina
Electronic	
High-frequency insulation:	
Rods, tubes, plates, coil formers, valve parts	Steatite, zircon porcelain, alumina
Capacitor dielectrics:	
Trimmer capacitors	Steatite
High-permittivity dielectrics	Rutile, titanates
Transmitter capacitor dielectrics	Rutile, magnesium titanate
Nonlinear dielectrics:	
Dielectric amplifiers, memory units, accelerometers, electromechanical transducers, e.g., gramophone pickup crystals, ultrasonic generators	Barium titanate, lead zirconate titanate
Insulation and good heat conduction:	
Transistors and integrated-circuit packages and hybrid-circuit substrates	Alumina, beryllia

TABLE 6-33 Some Typical Properties of Ceramic Dielectrics[72]

Material	Vitrified products			
	High-voltage porcelain	Alumina porcelain	Steatite	Forsterite
Typical applications	Power-line insulation	Spark plug cores, thermocouple insulation, protection tubes	High-frequency insulation, electrical appliance insulation	High-frequency insulation, ceramic-to-metal seals
Specific gravity, g/cm^3	2.3–2.5	3.1–3.9	2.5–2.7	2.7–2.9
Water absorption, %	0.0	0.0	0.0	0.0
Coefficient of linear thermal expansion, °K^{-1} (20–700°C)	5.0–6.8 × 10^{-6}	5.5–8.1 × 10^{-6}	8.6–10.5 × 10^{-6}	11 × 10^{-6}
Safe operating temperature, °C	1000	1350–1500	1000–1100	1000–1100
Thermal conductivity, cal/cm·s·K	0.002–0.005	0.007–0.05	0.005–0.006	0.005–0.010
Strength, lb/in^2, tensile	3000–8000	8000–30,000	8000–10,000	8000–10,000
Compressive	25,000–50,000	80,000–250,000	65,000–130,000	60,000–100,000
Flexural	9000–15,000	20,000–45,000	16,000–24,000	18,000–20,000
Impact strength, ft·lb; ½-in rod	0.2–0.3	0.5–0.7	0.3–0.4	0.03–0.04
Modulus of elasticity, lb/in^2	7–14 × 10^6	15–52 × 10^6	13–15 × 10^6	13–15 × 10^6
Thermal shock resistance	Moderately good	Excellent	Moderate	Poor
Dielectric strength, V/mil; ¼-in-thick specimen	250–400	250–400	200–350	200–300
Resistivity Ω·cm, at room temp	10^{12}–10^{14}	10^{14}–10^{15}	10^{13}–10^{15}	10^{13}–10^{15}
Te value, °C	200–500	500–800	450–1000	Above 1000
Power factor at 1 MHz	0.006–0.010	0.001–0.002	0.0008–0.0035	0.0003
Dielectric constant	6.0–7.0	8–9	5.5–7.5	6.2
L grade (JAN Spec. T-10)	L-2	L-2–L-5	L-3–L-5	L-6

vapor-deposited tantalum is anodized to form the oxide on its surface.[51] The oxide forms at 1.6 nm/V. Formation voltages range between 30 and 250 V. The capacitance density is 0.004 to 0.0005 F/m², and the dielectric constant is 21.2. The temperature coefficient of capacitance can be up to 250 ppm/K. Dissipation factor ranges from 0.002 to 0.01 at frequencies less than 10 kHz. Leakage currents are between 0.01 and 0.1 A/F. Manganese oxide (MnO_2) is used on Ta_2O_5

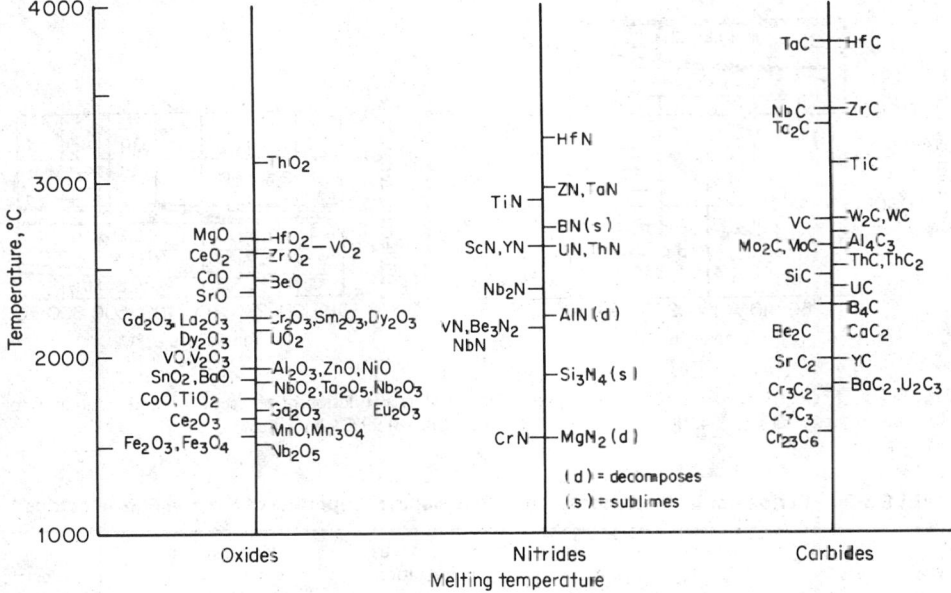

Fig. 6-20. Melting points of oxides, nitrides, and carbides.[75]

	Semivitreous and refractory products						
Zircon porcelain	Lithia porcelain	Titania, titanate ceramics	Low-voltage porcelain	Cordierite refractories	Alumina, aluminum silicate refractories	Massive fired talc, pyrophyllite
Spark plug cores, high-voltage high-temperature insulation	Temperature, stable, inductances, heat-resistant insulation	Ceramic capacitors, piezoelectric ceramics	Switch bases, low-voltage wire holders, light receptacles	Resistor supports, burner tips, heat insulation, arc chambers	Vacuum spacers, high-temperature insulation	High-frequency insulation, vacuum tube spacers, ceramic models
3.5–3.8 0.0	2.34 0.0	3.5–5.5 0.0	2.2–2.4 0.5–2.0	1.6–2.1 5.0–15.0	2.2–2.4 10.0–20.0	2.3–2.8 1.0–3.0
3.5–5.5 × 10⁻⁶ 1000–1200	1 × 10⁻⁶ 1000	7.0–10.0 × 10⁻⁶ ...	5.0–6.5 × 10⁻⁶ 900	2.5–3.0 × 10⁻⁶ 1250	5.0–7.0 × 10⁻⁶ 1300–1700	11.5 × 10⁻⁶ 1200
0.010–0.015 10,000–15,000 80,000–150,000 20,000–35,000 0.4–0.5 20–30 × 10⁶ Good	... ... 60,000 8000 0.3 ... Excellent	0.008–0.01 4000–10,000 4,000–120,000 10,000–22,000 0.3–0.5 10–15 × 10⁶ Poor	0.004–0.005 1500–2500 25,000–50,000 3500–6000 0.2–0.3 7–10 × 10⁶ Moderate	0.003–0.004 1000–3500 20,000–45,000 1500–7000 0.2–0.25 2–5 × 10⁶ Excellent	0.004–0.005 700–3000 15,000–60,000 1500–6000 0.17–0.25 2–5 × 10⁶ Excellent	0.003–0.005 2500 20,000–30,000 7000–9000 0.2–0.3 4–5 × 10⁶ Good
250–350 10¹³–10¹⁵ 700–900 0.0006–0.0020 8.0–9.0 L-4	200–300 ... ... 0.05 5.6 L-3	50–300 10⁸–10¹⁵ 100–400 0.0002–0.050 15–10,000	40–100 10¹²–10¹⁴ 300–400 0.010–0.020 6.0–7.0	40–100 10¹²–10¹⁴ 400–700 0.004–0.010 4.5–5.5	40–100 10¹²–10¹⁴ 400–700 0.0002–0.010 4.5–6.5	80–100 10¹²–10¹⁵ 600–900 0.0008–0.010 5.0–6.0

capacitors as a protective layer under the counterelectrodes. This layer permits large-area capacitors to be used. Electrolytic SiO_2 can be formed on silicon semiconductor devices as a dielectric. However, the anodic silicon oxide has not found a wide application in the microelectronic industry.

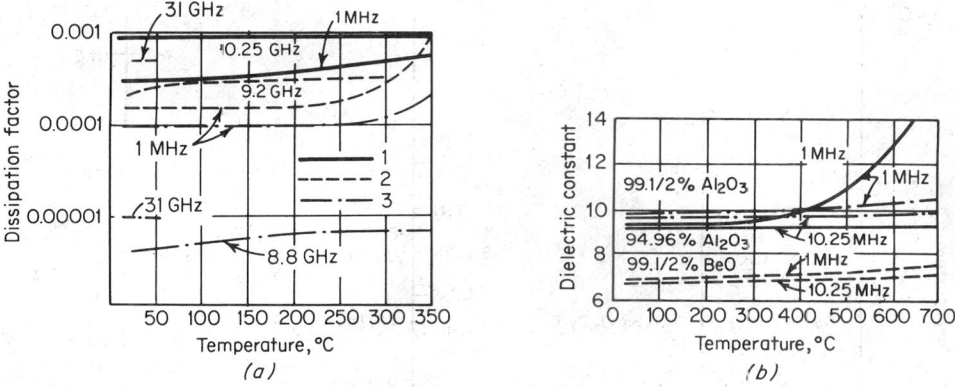

Fig. 6-21. Dielectric constant (b) and dissipation factor (a) as a function of temperature and frequency for alumina and beryllia.[76] (1) 94.9% alumina; (2) 99.5% alumina; (3) 99.5% beryllia.

TABLE 6-34   Ferroelectric, Piezoelectric, and Pyroelectric Properties of Several Ferroelectrics[64]

Material	Critical temperature T_c, K	Spontaneous polarization		Piezoelectric efficiency K_{eff}, pC/N receiving mode	Pyroelectric coefficient p, $\mu C/m^2 \cdot K$
		p_s, mC/m^2	T, K		
TGS and KDP types:					
$NaK_4H_4O_6 \cdot H_2O$	297	2.3	278	26.2	20
Rochelle salt,					
$(CH_2NH_2COOH)_3H_2SO_4$	322	28	293		
KH_2PO_4	123	53.3	96		
$Pb_2KNb_5O_{15}$	658	280	293		
Perovskites:					
$BaTiO_3$	393	260	293	. . .	330
$Bi_4Ti_3O_{12}$	948	. . .	. . .	1.47	
$LiNbO_3$	1470	3,000	. . .	6.33	40
Smectic liquid crustal:					
DOBAMBC	368	1	333		
Polyvinylidene fluoride					
(PVDF)	Unknown	25	293	0.33	24

109. Glass. Glass is widely used in the electronic industry as a structural member, such as tube envelopes, hermetic seals to metal or ceramic, protective coating on hybrid and integrated circuits, insulating layers and crossovers in microelectronics, and capacitors. Semiconducting chalcogenide glasses have been reported as suitable for switching and memory application.

Glass is usually a noncrystalline solid having no long-range order, a state referred to as *vitreous*. Glasses have a very pronounced short-range order, for example, in vitreous SiO_2, where most silicon atoms have four oxygen neighboring atoms at distances typical to silicon-oxygen bond length.[32] Glasses can be devitrified (recrystallized) in order to modify their properties.

Most glasses are composed of a variety of oxides with hundreds of glass compositions available. Table 6-35 lists the composition of some typical glasses. Common additions to glass are alkali oxides to lower the melting point. These oxides add an undesirable ionic conductivity. In this respect, the potassium, with the larger ionic radius, is less detrimental than sodium. Alkali-earth oxides improve the weathering resistance of low-melting glasses. Aluminum is added to increase strength; PbO lowers the melting point; B_2O_3 lowers the melting point and results in chemical resistance without affecting the electrical resistivity. Volume resistivity of glass decreases with temperature (Fig. 6-22). Surface conductivity of glass becomes pronounced above about 50% relative humidity (Fig. 6-22). Chemical reduction of the lead in lead glasses can result in an electrically conductive layer on the surface of the glass. This occurrence has to be avoided when lead solder glasses are used as insulators or sealing glasses.

The dielectric strength of glasses is a function of thickness, temperature, time, frequency, and testing medium. The dielectric constants of glasses vary from 3.8 for fused silica to 2,000 for glasses containing ferroelectric materials. Typical dielectric properties of glasses are given in Table 6-36.

TABLE 6-35 Typical Composition of Some Glasses

Constituent oxides, wt %:	Fused silica	Silica glass (7900)*	Soda lime (0080)	Lead silicate (0010)	Boro-silicate (7740)
SiO_2	99.5	96.3	73.6	63.0	80.5
B_2O_3	. . .	2.9	. . .	0.2	12.9
Na_2O	. . .	0.2	16.0	7.5	3.8
K_2O	. . .	0.2	0.6	6.0	0.4
CaO	. . .	. . .	5.2	0.3	
MgO	. . .	. . .	3.6	0.2	
Al_2O_3	. . .	0.4	1.0	0.5	2.2
PbO	. . .	. . .	—	21.0	
SO_3	. . .	. . .	0.2		
Thermal expansion coefficient $\times 10^{-7}$,† K^{-1}	5.5	8.0	92	93	33

*Corning's number in parentheses.
†See footnote a, Table 6-1.

110. Fused Quartz and Fused Silica. Fused quartz and fused silica are vitreous forms of SiO_2. Fused quartz is prepared from rock crystal or white quartzite sand, and fused silica (translucent fused quartz) is made from white silica sand. Fused quartz and silica have very low expansion coefficients and poor heat conduction (see Fig. 6-18). They are mechanically strong and have a high softening point.

Vitreous SiO_2 is used extensively in thin-film form in the microelectronics industry. Thermally oxidized silicon provides a thin (1-μm) layer of SiO_2 which is used as insulation and as diffusion barriers. Sputtered SiO_2 has been used as dielectric layer for two-level metallizations. Chemically vapor deposited SiO_2 is also used as protective layers. Properties of fused silica are given in Table 6-36.

111. Glass Ceramics. Glass ceramics are glasses which are devitrified about 100°C below their softening point to form a very fine network of crystalline phase. The glass ceramics can be shaped at the soft state and then hardened by crystallization. They are mechanically strong and have a low thermal expansion coefficient and therefore have a high thermal shock resistance. Glass ceramics retain their properties at higher temperatures than glasses. Properties of a glass ceramic, Pyroceram,* are given in Table 6-36. Another machinable glass ceramic, Macor,* has a maximum use temperature of 1000°C (no load) and a coefficient of thermal expansion of 9.4 ppm/K.

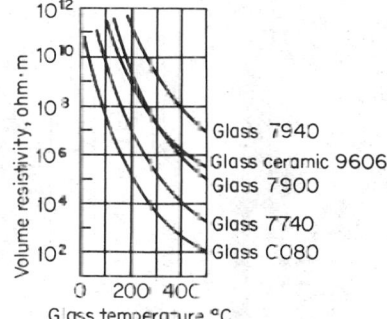

Fig. 6-22. Resistivity of several glasses as a function of temperature.[73]

112. Thick-Film Dielectrics. The dielectric films can be divided into crossover insulators, capacitors, overglazes, and solder-glass pastes. These films are formed as a screenable paste which are screened on substrates and subsequently fired to leave the desired dielectric layer. Table 6-37 lists some of the available pastes and their properties.

*Trademark of Corning Glass Works, Corning, N.Y.

TABLE 6-36 Comparison of Properties of Pyroceram, Glass, and Ceramic[70]

Property	Pyroceram 9605	Pyroceram 9606	Pyroceram 9607	Pyroceram 9608	Fused silica 7940	Vycor 7900	Pyrex 7740	Lime glass 0080	High-purity alumina (93%+)	Steatites MgO-SiO$_2$	Forsterite 2MgO-SiO$_2$
	Pyroceram				Glass				Ceramic		
Specific gravity, 25°C	2.62	2.61	2.40	2.50	2.20	2.18	2.23	2.47	3.6	2.65-2.92	2.8
Water absorption, %	0.00	0.00	0.00	0.00	0.00	0.00	0.00	0.00	0.00	0-0.03	0-0.01
Gas permeability	0	0	0	0	0	0	0	0	0	......	0
					Thermal						
Softening temp,[a] °C	1350	1260	...	1250	1584	1500	820	696	1700	1349	1349
Specific heat, 25°C, cal/g	0.185	0.185	...	0.190	0.176	0.178	0.186	0.200	0.181		
Mean, 25-400°C	0.230	0.230	...	0.235	0.223	0.224	0.233	0.235	0.241		
Thermal conductivity, mean, 25°C, W/m·k	4.19	3.64	...	1.97	1.17	...	1.09	...	21.8-24.3	2.6-2.72	4.19
Linear coefficient of thermal expansion[b] 25-300°C, × 10^{-6}	14	57	-7	4-20	5.5	8	32	92	73[d]	81.5-99[d]	99[d]
					Mechanical						
Modulus of elasticity, lb/in^2 × 10^{-6}	19.8	17.3	...	12.5	10.5	9.6	9.5	10.2	40	15	
Poisson's ratio	...	0.245	...	0.25	0.17	0.17	0.20	0.24	0.32		
Modulus of rupture (abraded), 0.001 lb/in^2	...	20	...	16-23	...	5-9	6-10	...	40-50	20[e]	19[e]
Hardness, Knoop, 100 g	...	698	...	703	...	532	481	...	1,880		
500 g	720	619	...	588	...	477	442	...	1,530		

Electrical

	Electrical										
Dielectric constant, 1 MHz, 25°C	(6.1)	5.58	...	6.78	3.78	3.8	4.6	7.2	8.81	5.9	6.3
300°C	(6.3)	5.60	...	...	...	3.9	5.9	...	...	...	
500°C	...	8.80	...	...	...	...	...	...	...	...	
10 GHz, 25°C	(6.1)	5.45	...	6.45	3.78	3.8	4.5	6.71	9.03	5.8	5.8
300°C	(6.1)	5.51	...	6.65	3.78	...	...	...	8.79		
500°C	(6.1)	5.53	...	6.78	3.78	...	...	...	9.03		
Dissipation factor, 1 MHz, 25°C	(0.0017)	0.0015	...	0.0030	...	0.0005	0.0046	0.009	0.00035	0.0013	0.0003
300°C	(0.014)	0.0154	...	...	...	0.0042	0.0130	...	0.012		
500°C									0.0015		
10 GHz, 25°C	(0.0002)	0.00033	...	0.0068	...	0.0009	0.0085	0.017	0.0015	0.0014	0.0010
300°C	(0.0008)	0.00075	...	0.0115	...						
500°C											
Loss factor, 1 MHz, 25°C	(0.010)	0.008	...	0.02	...	0.0019	0.0212	0.065	0.0031	0.0077	0.0019
300°C	(0.078)	0.086	...	...	...	0.0164	0.0566	...	0.108		
500°C											
10 GHz, 25°C	(0.001)	0.002	...	0.045	...	0.0036	0.0282	0.114	0.0132	0.0082	0.0058
300°C	(0.005)	0.004	...	0.077	...				0.108		
500°C	(0.015)	0.008	...	0.27	...				0.019		
Volume resistivity $\log_{10}$, $\Omega\cdot$cm, 250°C	10.1	10	...	8.1	12.0	9.7	8.1	6.4	14.0[f]	14[g]	14[g]
350°C	8.7	816	...	6.8	9.7	8.1	6.6	5.1	12.95[h]	14[g]	

[a]Softening temperature. method of evaluation for Pyroceram comparable with ASTM C 24-56, for glass, ASTM C 338-54T, for ceramics, ASTM C 24-56.
[b]Expansion coefficients depend on heat treatment. See note a, Table 6-1. [d]20–300°C.
[c]Unabraded values. [e]100°C. [f]20°C. [g]300°C.

TABLE 6-37   Selected Thick-Film Dielectrics[64]

Type and trade name	Dielectric strength, kV/cm	Thermal coefficient, K^{-1}, $\times 10^{6*}$	Dielectric constant	Remarks
Multilayer dielectric:				
EMCA 3186	>200	6.6	16	
Ceramalloy 9114	...	...	6–8	
Du Pont 9429	...	...	9–12	
ESL 4608	...	...	8–10	Alkali-free, low-porosity general-purpose
Overglazes:				
EMCA 99	>200	8	15	
ESL 4612	...	...		Permanent green, for use in hermetic packaging, overglazing
Solder glasses:				
EMCA 92	>200	11.5	13	
ESL 4017C	...	Matches alumina, beryllia	...	Firing temp 400–500°C, vitreous
Dielectric capacitors:				
Du Pont 8299	...	...	10–20	
8229	...	...	500	
8456	...	...	2,000	
ESL 4903	...	...	6–10	
4515	...	...	500 ± 150	
4520	...	...	$1,750 \pm 250$	
Ceramalloy 9160	...	...	50	
9170	...	...	250	

*See footnote *a*, Table 6-1.

113. Bibliography on Dielectric Materials

Harper, C. A. (ed.)   "Handbook of Electronic Packaging," McGraw-Hill, New York, 1969.

Kohl, W. H.   "Handbook of Materials and Techniques for Vacuum Devices," Reinhold, New York, 1967.

Kingery, W. D.   "Introduction to Ceramics," Wiley, New York, 1960.

Megaw, H. D.   "Ferroelectricity in Crystals," Methuen, London, 1957.

Von Hippel, A. R.   "Molecular Science and Molecular Engineering," M.I.T., Wiley, New York, 1959.

Shand, E. B.   "Glass Engineering Handbook," McGraw-Hill, New York, 1958.

Lang, S. B.   "Sourcebook of Pyroelectricity," New York, 1974.

Jaffe, B., W. R. Cook, and H. Jaffe   "Piezoelectric Ceramics," Academic, New York, 1971.

Broadhurst, H. G., and G. T. Davies (eds.)   In "Piezo and Pyroelectric Properties of Electrets," "Topics in Modern Physics," Springer, Berlin, 1980.

MAGNETIC MATERIALS

114. Magnetic Field. Forces beyond electrostatic forces are exerted between moving charges. It is convenient to speak of a magnetic field set up by the moving charge and forces which appear on a second moving charge when it traverses through the field.

The magnetic field in a long thin coil is

$$H = nI \tag{6-52}$$

where H = magnetic field strength (A/m), n = number of turns per meter (m^{-1}), and I = electric current through the coil (A).

115. Magnetic Pole Strength. The magnetic pole strength of a material is analogous to the charge in electrostatics. For example, the force on a magnetic pole of strength m in a magnetic field of strength H is

$$F = mH \tag{6-53}$$

where F = force (N), m = magnetic pole strength (Wb), and H = magnetic field strength (A/m).

116. Magnetization. The magnetization I is defined as the magnetic moment per unit volume.

117. Magnetic Flux Density. In empty space the magnetic flux density (or induction) B is given by

$$B = \mu_0 H \tag{6-54}$$

where B = magnetic flux density (Wb/m², T) and μ_0 = magnetic permeability of empty space = 0.4π μH/m.

In a material the flux density is

$$B = \mu H \tag{6-55}$$

where μ = magnetic permeability of the material (H/m).

The magnetic flux density can also be written as a sum of the flux in empty space $\mu_0 H$ and the flux density due to magnetization I

$$B = \mu_0 H + I \tag{6-56}$$

118. Relative Magnetic Permeability. The ratio between the permeability of a material and that of empty space is the relative permeability,

$$\mu_r = \mu/\mu_0 \tag{6-57}$$

where μ_r = relative permeability. It follows that

$$B = \mu_0 \mu_r H \tag{6-58}$$

In analogy to the electrostatics, μ_r is written in a complex form for lossy materials

$$\mu_r = \mu'_r - j\mu''_r \tag{6-59}$$

119. Magnetic Susceptibility. The magnetic susceptibility is the magnetization per unit magnetic field

$$\chi = I/H \tag{6-60}$$

where χ = magnetic susceptibility (H/m). From Eqs. (6-60), (6-55), and (6-56) it follows that

$$\mu = \mu_0 + \chi \tag{6-61}$$

See also Sec. **1-86.** The relative magnetic susceptibility $\bar{\chi}$ is defined as

$$\bar{\chi} = \chi/\mu_0 \tag{6-62}$$

120. Magnetic Energy. The magnetic energy stored in a unit volume by increasing the magnetic flux density from B_1 to B_2 is

$$\text{Magnetic energy density} = \int_{B_1}^{B_2} H \, dB \quad \text{J/m}^3 \tag{6-63}$$

121. Classification of Magnetic Materials. The atomic origin of magnetism stems from the orbital motion and spin of electrons. The main classes of magnetic materials (see Fig. 6-23) are as follows.

Diamagnetic Materials. These materials have a small negative susceptibility $\bar{\chi} \approx -10^{-5}$. The magnetism originates from induced currents opposing the external field.

Paramagnetic Materials. These materials have a $\bar{\chi}$ of $+10^{-5}$ to 10^{-3}. The magnetism stems from partial alignment of existing spins, randomly oriented by thermal agitation in the absence of external fields.

Ferromagnetic Materials. Ferromagnetic materials have spins aligned parallel to each other. The material is divided into *magnetic domains*, each domain having a net magnetization even without an external field. This magnetization is called *spontaneous magnetization*. A bulk sample will generally not have a net magnetization since the spontaneous magnetization in the various domains will cancel each other.

Application of a small magnetic field will cause growth of favorable domains resulting in high magnetization and high $\bar{\chi}$ (up to 10^6). Above a critical temperature, Curie temperature, these materials become paramagnetic.

Ferrimagnetic Materials. These materials have two kinds of magnetic ions with unequal spins, oriented in an antiparallel fashion. The spontaneous magnetization can be regarded as the two opposing and unequal magnetizations of the ions on the two sublattices. Ferrimagnetic materials become paramagnetic above a Curie temperature.

Antiferromagnetic Materials. The materials have an antiparallel arrangement of equal spins resulting in very low $\overline{\chi}$ similar to paramagnetic materials. The spin arrangement of antiferromagnetic materials is not stable above a critical temperature (Neél temperature).

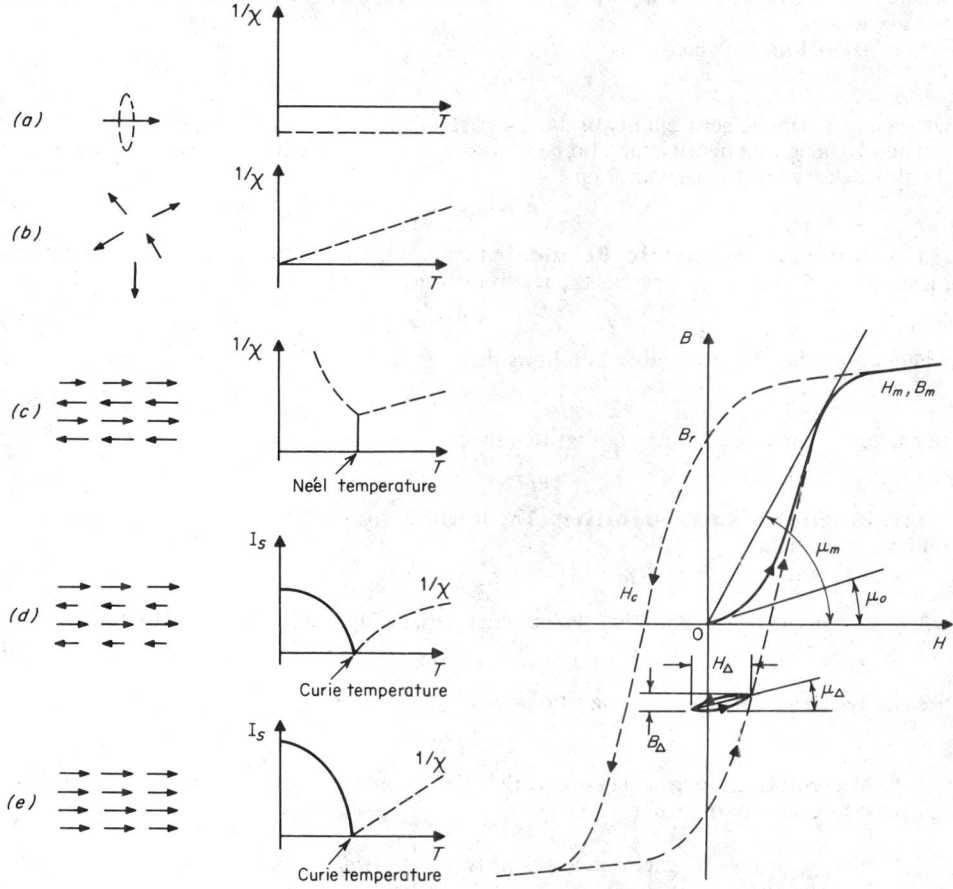

Fig. 6-23. Magnetic susceptibility and saturation magnetization for several magnetic materials: (a) diamagnetic; (b) paramagnetic; (c) antiferromagnetic; (d) ferrimagnetic; (e) ferromagnetic. On the left are schematic representations of spin arrangements within the materials.

Fig. 6-24. Magnetization curve and hysteresis loop of a ferromagnetic material.

122. Magnetization of Ferromagnetic Materials. The magnetic induction of a ferromagnetic material is seen in Fig. 6-24 as a function of the applied magnetic field. The induction starts from 0 at zero field and reaches *maximum induction* B_m and *maximum field* H_m at saturation. The magnetization reaches an upper limit called the *saturation induction* B_s. When the field is decreased, the induction will follow a curve with higher values than the original curve. At $H = 0$ there remains an induction B_r *residual induction*, or *remanence*. The maximum residual induction (when materials are fully magnetized) is the *retentivity*. To remove the retentivity, a negative magnetic field is applied and the induction is completely removed at H_c, the *coercive force*, or its maximum, the *coercivity*. The process of removal of residual induction is often referred to as *demagnetization*, and the portion of curve between B_r and H_c is called the *demagnetization curve*. Application of higher negative fields will eventually saturate the material. Reversing the field again will complete the *B-H* curve and bring it again to the maximum field H_m and maximum induction B_m. The entire curve (Fig 6-24) is called *hysteresis loop*.

123. Ferromagnetic Domains. Ferromagnetic materials are composed of domains which are spontaneously magnetized in a specific direction. The external magnetic field merely moves domain walls and at higher fields changes the direction of the magnetization in the domains. Domain theory explains the initial low slope of the hysteresis curve as a reversible wall movement, the second steep part of the curve as irreversible wall movement, and the final convergence toward saturation as magnetization rotation (see Fig. 6-25).

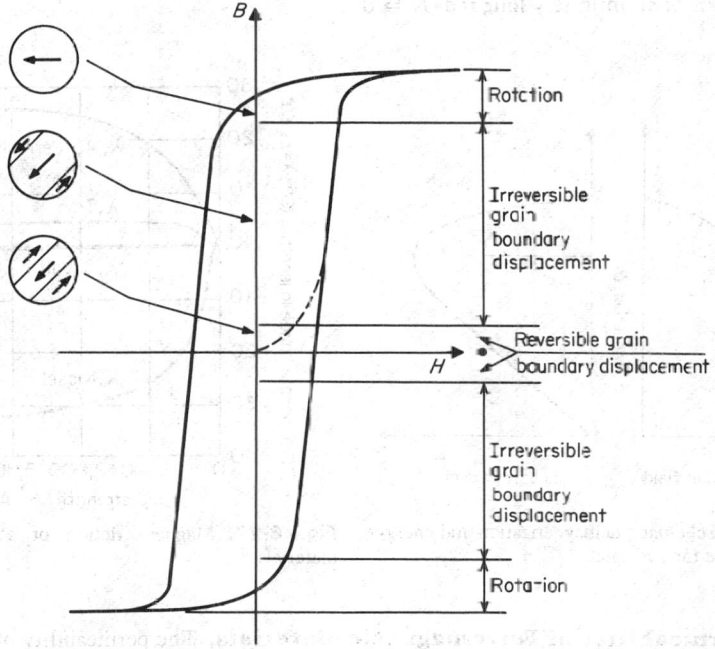

Fig. 6-25. Ferromagnetic-domain configuration at various stages of magnetization.

124. Energy-Product Curve. The *B-H* curve at any point on the demagnetization curve is termed the energy-product curve. Figure 6-26 shows a typical demagnetization curve. The *BH* product reaches a maximum value BH_{max}. This value is a good criterion for permanent magnetic materials; i.e., higher BH_{max} values are obtained with higher-quality magnets.

125. Hysteresis Losses. The work per unit volume required to magnetize a ferromagnetic material to saturation is $\int_0^{B_m} H\,dB$. In one cycle of the hysteresis loop there is a loss of energy per unit volume which equals the area within the hysteresis loop.

Materials are called *hard magnetic materials* when they have a high coercivity and a large hysteresis loop. *Soft magnetic materials* are low-loss, low-coercive-force, high-permeability materials.

126. Eddy Current Losses. The varying magnetic flux in a material induces an emf, which in turn creates eddy currents. The Joule heating due to the currents is an energy loss, termed *eddy current loss*. At low frequencies, when the flux penetration is complete, the eddy current losses are proportional to the square of the frequency f^2 and inversely proportional to the electrical resistivity. At high frequency and constant flux amplitude, the losses become proportional to $f^{3/2}$. Eddy current losses are greatly reduced if cores are made from laminated sheets so as to confine the currents within the relatively high resistance sheets. Substantial reduction of eddy current losses can be achieved by using ferromagnetic material having very high electrical resistivities.

127. Total Core Losses. The sum of all losses in a core is termed total core loss. It should be noted that the losses are normally given for a sinusoidally varying magnetic field under symmetrically cycling conditions. Corrections have to be made for deviations from sinusoidal condition.

128. Demagnetizing Factor. If a material is magnetized, free poles are produced in it which create a field opposing the external applied field. This field is termed the demagnetizing field H_d

$$H_d = NI/\mu_c \tag{6-64}$$

where I = magnetization (Wb/m²), μ_0 = permeability of empty space (H/m), and N = demagnetizing factor. For fields perpendicular to sheets or thin plates, $N = 1$, and for applied fields along the axis of an infinitely long rod, $N = 0$.

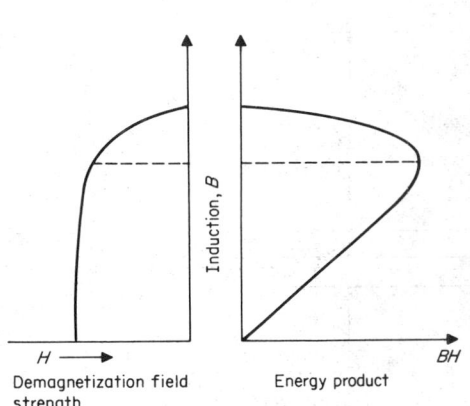

Fig. 6-26. Schematic demagnetization and energy-product curve for a magnet.

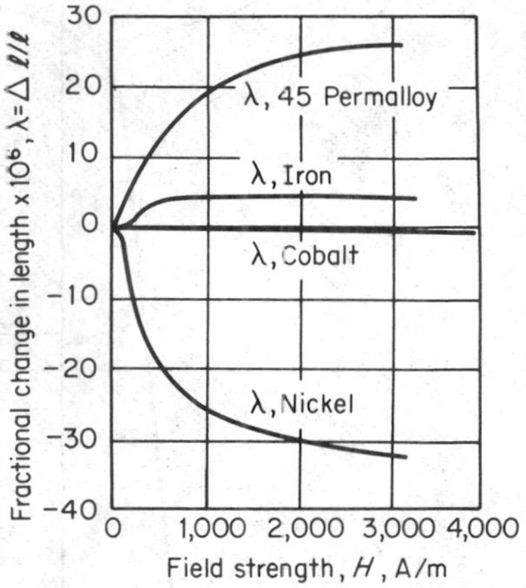

Fig. 6-27. Magnetostriction of some common materials.[80]

129. Permeability of Ferromagnetic Materials. The permeability of a ferromagnetic material B/H is not constant (see Fig. 6-24). When the material is first magnetized, the initial slope of the B-H curve is called *initial permeability* μ_0. The maximum slope from the origin to the B-H curve is the *maximum permeability* μ_m. If a biasing field H_b is held constant and a small alternating field H_Δ is applied, then $B_\Delta/H_\Delta = \mu_\Delta$, the *incremental permeability*. *Differential permeability* is dB/dH, the local slope of the hysteresis curve. Permeability measured under ac excitation is termed *ac permeability*.

130. Magnetostriction. The change in a material's dimension due to magnetization is called magnetostriction. Changes in linear dimensions are called *linear* magnetostriction, while volume changes are referred to as *volume* magnetostriction. The change in length in the direction of the magnetization is called *Joule* magnetostriction. The symbol λ is generally used for the value of the fractional change in length of saturation. Figure 6-27 shows the magnetostriction field strength for iron, nickel, and cobalt.

131. Magnetic Anisotropy. The internal energy of a ferromagnetic or ferrimagnetic crystal depends on the direction of the spontaneous magnetization with respect to the crystal axes; i.e., it is anisotropic. This anisotropy is called *magnetocrystalline anisotropy*. It follows that the ease of magnetization of crystals depends on the direction of magnetization. For example, cobalt shows a uniaxial anisotropy, the easy and stable direction of spontaneous magnetization being its c axis. Iron has an easy axis of magnetization in the <100> directions, while nickel has easy <111> directions (Fig. 6-28). Polycrystalline materials with strong preferred orientation also show magnetic anisotropy.

In thin rods or thin films there is a tendency for the spontaneous magnetization to align itself in the direction of the rod axis or in the film plane, respectively. This form of anisotropy is called *shape anisotropy*.

Anisotropy can also be induced by strain due to the magnetostriction energy. In some alloy systems, that is, Ni–Fe (see Par. **6-141**), heat treatment in the presence of a magnetic field results in a uniaxial magnetic anisotropy.

132. Aging Coefficient. The percentage change in magnetic properties of materials resulting from temperature aging is called the aging coefficient. Typical aging treatments are 100 h at 150°C and 600 h at 100°C.

133. Magnetoresistance. The change in electrical resistance due to the application of a magnetic field is called magnetoresistance.

134. Magnetocaloric Effect. A change in temperature of the material due to a change in the magnetic field is called the magnetocaloric effect.

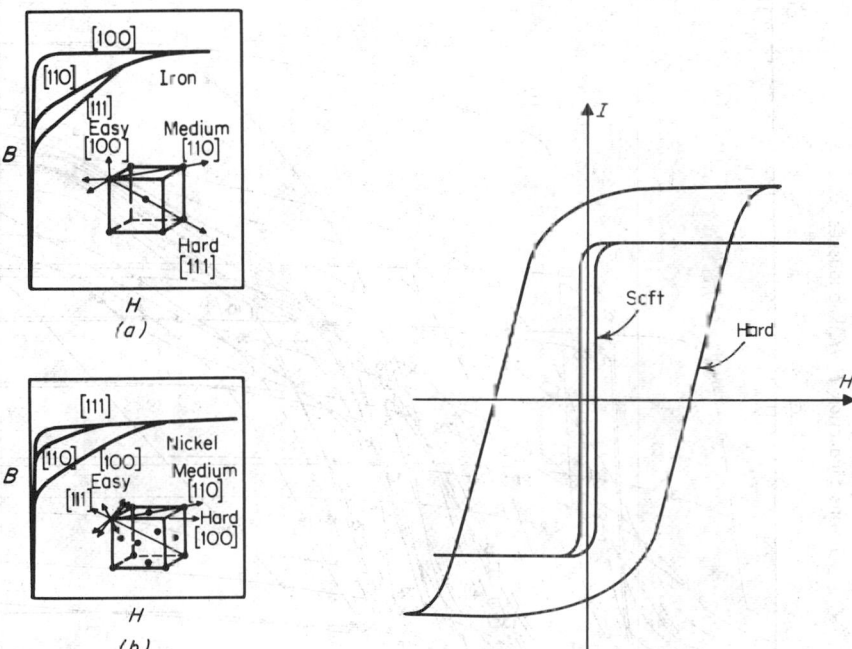

Fig. 6-28. Magnetization vs. applied field for field-directions along the [100], [110], and [111] directions for (a) iron and (b) nickel single crystals.[81]

Fig. 6-29 Hysteresis loops for soft and hard magnetic materials.

135. Magnetooptical Effects. Spontaneously magnetized material can affect the polarization of light transmitted through the materials (Faraday effect) or reflected from their surface (Kerr effect). These are called magnetooptical effects.

SPECIFIC MAGNETIC MATERIALS

136. Retentive and Nonretentive Materials. Magnetic materials can be classified into *nonretentive soft* materials and *retentive hard* materials (see Fig. 6-29). The nonretentive materials are low-loss materials used in transformers, motor or generator cores, electromagnetic apparatus, and memories. These materials can be further classified according to the frequency range for which they are used, from nonalloyed iron for low-frequency uses, silicon-iron, Permalloy, ferrites, and finally garnets for the highest-frequency use. Hard materials are high-loss, high-retentivity materials with a high-energy product used for permanent magnets. Hard magnetic alloys can be further classified into alloys undergoing martensitic transformation, precipitation-hardened alloys, ordered-hardened alloys, and powder magnets.

Magnetic materials with special properties are also often desired, for example, square-loop materials for memory application and constant-permeability materials or alloys capable of operating at elevated temperatures.

DC magnetization curves of various magnetic materials are given in Fig. 6-30.

NONRETENTIVE MAGNETIC MATERIALS

137. Iron. Iron is used for dc applications such as electromagnet cores and relays. Iron contains nonmetallic impurities which reduce its permeability and increase the hysteresis losses. *Hydrogen annealing* at high temperature can remove the impurities and greatly increases the permeability. Another common annealing procedure which increases permeability is an 800°C anneal followed by a slow (5°C/min) cool. Impurities such as carbon or nitrogen in iron cause magnetic aging, i.e., deterioration of magnetic properties with time.

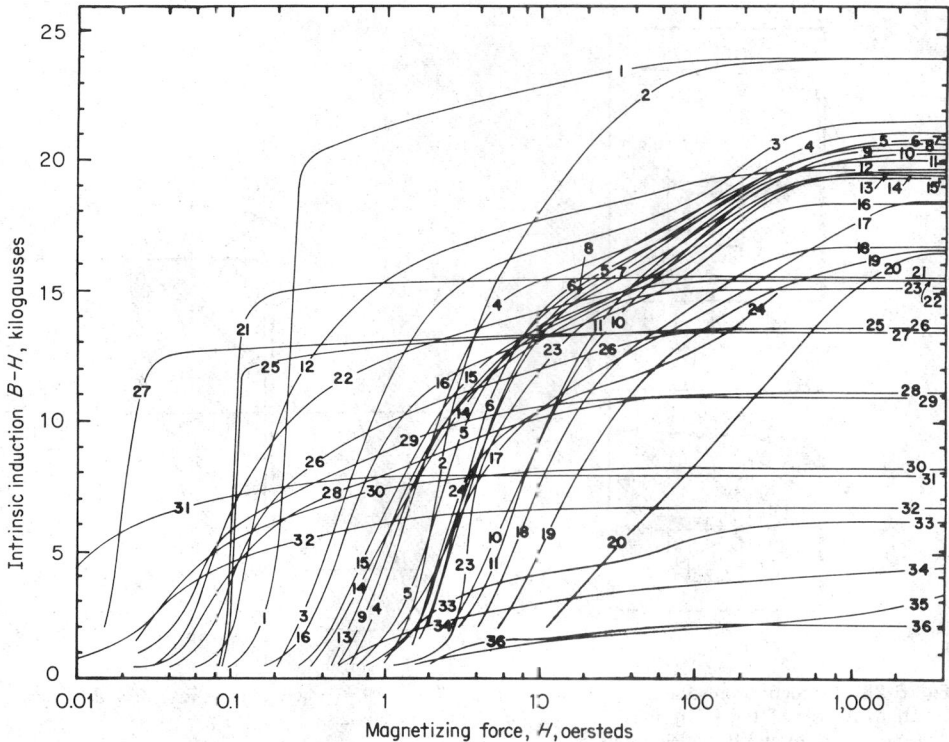

Fig. 6-30. DC magnetization curves for various magnetic materials: (1) Supermendur; (2) vanadium Permendur (50% Co-2% V); (3) pure iron, annealed; (4) ingot iron, annealed; (5) hot-rolled low-carbon sheet steel; (6) M-50; (7) gold-rolled low-carbon strip steel; (8) cold-drawn carbon steel, annealed; (9) M-43 (0.5% Si) and M-36 (1.5% Si); (10) steel castings, as cast; (11) carbon-steel forgings, annealed; (12) 3% Si strip, oriented; (13) M-27 (2.75% Si); (14) M-22 (3.25% Si); (15) M-19 (3.75% Si); (16) M-14 4.50% Si); (17) malleable iron castings, as cast; (18) type 416 stainless steel, annealed; (19) nodular cast iron; (20) gray iron, as cast; (21) Deltamax, oriented; (22) 4750 alloy; (23) Perminvar (45% Ni-25% Co); (24) powdered iron, sintered and annealed; (25) Monimax, oriented; (26) Monimax, nonoriented; (27) 65 Permalloy, oriented; (28) Sinimax; (29) 78 Permalloy; (30) M.1-Permalloy; (31) Superalloy; (32) Mumetal; (33) pure nickel, annealed; (34) soft magnetic ferrite (Fe–Ni–Zn–V); (35) 30% Ni–Fe temperature compensator alloy; (36) Monel, annealed.[61]

Iron is available as electrolytic iron, which is a very pure form containing typically 0.006% C, 0.005% Si, 0.005% P, and 0.005% S. Very high permeabilities can be obtained in electrolytic iron by hydrogen annealing. The use of electrolytic iron is limited mainly by economic considerations. *Ingot iron* is highly refined iron of typical composition 0.08% C, 0.015% Mn, 0.16% Si, 0.06% P, 0.01% S, and 1.2 to 2.8% slag which is finely dispersed in the iron. Annealed ingot-iron bars are used for relay cores. *Low-carbon steels* have lower permeability and higher hysteresis losses and higher aging coefficient due to the carbon content. *Cast steels* are used where strength is required. *Gray cast iron* containing about 3% C and 1 to 3% Si can be cast to form complex shapes. *Ductile cast iron* is superior to gray cast iron in magnetic properties. Ductile cast iron contains the carbon in the form of nodular graphite which imparts the ductility to the material. *Malleable cast iron* is magnetically superior to ductile cast iron.

138. Iron-Silicon Steels. The useful magnetic properties of iron-silicon alloys were reported at the end of the nineteenth century by Hadfield Silicon steels, known as electrical steels, are very widely used for low- and intermediate-frequency applications. Their main advantages are low core losses, high maximum permeability, increase in electrical resistivity (lower eddy current loss), and reduction of magnetic aging. The effect of silicon additions on the coercive force, hysteresis and core losses, and maximum relative permeability is seen in Fig. 6-31.

The disadvantages of silicon additions are the decrease in saturation magnetization and the increase in brittleness. The increase in brittleness limits the amount of Si to 1 to 3% for rotating machinery and 3 to 4.5% for transformers. Si increases the yield and ultimate tensile strength of steel and improves the high-temperature mechanical properties. Table 6-35 lists the typical applications and properties of silicon steels. Conventional material is supplied in four classes: hot-rolled, fully processed; cold-rolled, fully processed; hot-rolled, semiprocessed; and cold-rolled, semiprocessed. Silicon steels can also be obtained in grain-oriented form. This product is rolled to obtain the easy [100] magnetization direction parallel to the rolling direction. Grain-oriented silicon steels have very low core losses. Very thin oriented silicon steels for higher-frequency applications are also available in tape form. Tapes with thickness of 0.001, 0.002, and 0.004 in are available for pulse transformers, reactors, magnetic amplifiers, and small power transformers. Core laminations are frequently coated to provide electrical insulation between the laminations. These coatings can be organic or inorganic.

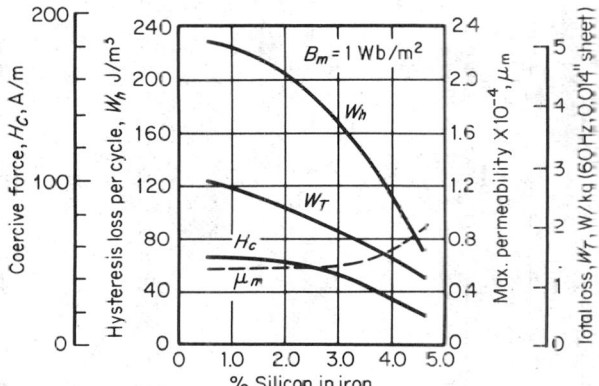

Fig. 6-31. Magnetic properties of hot-rolled commercial silicon-iron sheet as a function of silicon content. Some of the changes are due to higher annealing temperatures of the materials having higher silicon content.[80]

139. Iron-Cobalt Alloys. Iron-cobalt alloys are used when a high magnetic saturation is required. Vanadium is added to the iron-cobalt alloys to permit easy processing. When annealed under a magnetic field, vanadium Permendur has a high maximum permeability and the highest residual induction of any other alloy with very low coercive force. Iron-cobalt alloy is one of the highest in positive magnetostriction ($\lambda = 60$ to 70×10^{-6}). Other alloys are Hyperco 35% Co, 0.5% Cr, and the balance Fe; and alloys with the addition of several percent Cr: stainless invar.

140. Iron-Aluminum and Iron-Aluminum-Silicon Alloys. Iron-aluminum alloys have high electrical resistivities, high maximum permeabilities, and low coercivity. Commercial alloys are available, such as the 16% Al, Alperm, and the 13% Al, Alfer (used as a magnetostriction material). Iron-aluminum alloys can be cold-rolled. Grain-oriented material can also be produced. Iron-aluminum-silicon alloys have good ac properties but are brittle. A commercial alloy of about 10% Si and 6% Al (Sendust) is available in sheet and powder form.

141. Iron-Nickel Alloys. Iron-silicon alloys have high permeabilities at high fields. However, for low-field applications which are common in higher-frequency operations, these alloys do not have a high enough permeability. The iron-nickel alloys have a high initial permeability which makes them suitable for low-field, high-frequency operations. Figure 6-32 shows the initial permeability of iron-nickel alloys as a function of nickel content. The maximum of the permeability occurs around 78% Ni, between the composition which corresponds to zero magnetocrystalline anisotropy ($\approx 75\%$ Ni) and the one corresponding to $\lambda_{111} = 0$ (just below 80% Ni). Another permeability maximum occurs at about 45% Ni. At this composition the alloys

TABLE 6-38 Properties of Laminated or Wound Cores of Silicon Steel[81]

Grade	General applications	Nominal Si content, cold-rolled, %	Max core loss at 60 Hz, W/lb						Saturation induction B_s		sp gr, g/cm³	ρ, μΩ·cm
			10,000B			15,000B						
			0.014 in	0.0185 in	0.025 in	0.014 in	0.0185 in	0.025 in	kG	Wb/m²		
M-6	Power and distribution transformers, large turbogenerators, small power and audio transformers	3.25	...	...	...	0.66	...	...	19.7	1.97	7.65	50
M-7		3.25	...	...	...	0.73	...	...	19.7	1.97	7.65	50
M-19	Large rotating machines, communication transformers	3.00	0.72	0.83	0.97	1.75	2.00	2.35	19	1.90	7.65	47
M-22	Stators of induction motors and high-efficiency rotating equipment	3.00	0.82	0.94	1.10	2.00	2.20	2.60	19	1.90	7.65	47
M-27	High-efficiency motors, small transformers	3.00	1.01	1.14	1.30	2.46	2.66	3.05	19	1.90	7.65	47
M-36	Routing machines, including ac-dc motors	1.75	1.17	1.35	1.70	3.00	3.20	3.85	20	2.00	7.75	37
M-43	Intermittent-duty rotating machines, pole pieces, relays	0.50	1.30	1.55	1.98	3.50	3.70	4.60	20.5	2.05	7.76	28

have higher saturation magnetization, which permits the use of smaller volumes of material in core construction. Composition and properties of iron-nickel alloys are given in Table 6-39. 78 *Permalloy* is a 78% Ni–22% Fe alloy. Permalloy has high initial and maximum permeabilities and low losses. High permeabilities are obtained by a high-temperature heat-treatment (preferably in hydrogen), cooling to 600°C, followed by a rapid quench. The quench prevents the formation of the low-permeability ordered phases.

The maximum and differential permeabilities and remanence (i.e., loop squareness) can be increased appreciably by inducing a uniaxial isotropy in iron-nickel alloys. This anisotropy is technically accomplished by rolling or by annealing in a magnetic field. Permalloys have high squareness and very low coercivities.

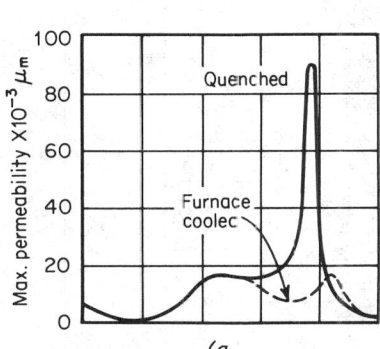 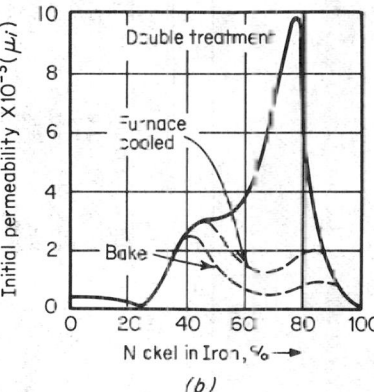

Fig. 6-32. (a) Maximum permeability and (b) initial permeability vs. composition of nickel-iron alloys.[82]

Permalloy has a higher saturation density than 78 Permalloy. This alloy is available under several trade names: 4750 Allegheny and Carpenter 5549 alloy. *Supermalloy* has very high initial and maximum permeabilities. Its composition is 79% Ni 5% Mo, and 16% Fe. Mo Permalloy results from the addition of small amounts of Mo, Cr, or Cu to Permalloy (for example, 4% Mo Permalloy) to improve the permeability, increase the resistivity, and eliminate the necessity of quenching from 600°C. 2% Mo Permalloy is also pulverized and used as magnetic powder cores. Uniaxial anisotropy has been introduced in Mo Permalloy also by a radiomagnetic treatment, i.e., bombarding with high-energy electrons at 100°C under a magnetic field.[33] Iron-nickel alloys possessing uniaxial anisotropy are also used in thin-film form for high-speed memory application (in the 1- to 10-ns range). These films are prepared by electroplating on wires or by vacuum deposition in the presence of a magnetic field. *Deltamax* is a recrystallized sheet with a (001)[100] texture of 50% iron-nickel alloy. Deltamax reaches saturation at very low fields. Deltamax is also available in tape form. *Isoperm* is a specially treated 50% Fe-Ni alloy with a constant permeability. Its linear magnetization curve is useful for circuit transformers. *Thermoprem* is a temperature-sensitive alloy containing 30% nickel with a Curie point just above room temperature. The alloy decreases in magnetization with temperature and is suitable for temperature-compensating elements.

Tape-wound cores of nickel-iron alloys are also available. Tapes range from 0.004 in (0.1 mm) to 0.000125 in (3.125 μm) in thickness. Typical properties of tape-wound and laminated cores are given in Tables 6-40 and 6-41.

Other nickel-iron alloys are as follows. *Mumetal*, containing 77% Ni, 4.8% Cu 1.5% Cr, 14.9% Fe, has a high initial permeability, especially after hydrogen annealing. It is used as a magnetic shielding for electronic equipment and for magnetic amplifier cores. *Perminvar*, containing 25% Co, 45% Ni, 30% Fe, has a constant permeability over a wide range of inductions. This alloy must be annealed for long periods at 400 to 450°C. Perminvar can be made, by magnetic annealing, to have a square hysteresis loop. *Superinvar*, containing 31% Ni, 4 to 6% Co, is a magnetic alloy with a very small thermal expansion coefficient. *Elinvar*, containing 36% Ni, 12% Co, has constant elastic moduli, which are temperature-independent.

TABLE 6-39 Magnetic Properties of Soft Magnetic Alloys[83]

Material	Composition	Heat treatment,* °C	$\bar{\mu}_a$	$\bar{\mu}_{max}$	H_c A/m	H_c Oe	I_s Wb/m²	I_s G	θ, °C	ρ, $\mu\Omega\cdot$cm	σ, g/cm³
Iron	0.2 (imp.)	950	150	5,000	80	1.0	2.15	1,710	770	10	7.88
Purified iron	0.05 (imp.)	1480(H₂); 880	10,000	200,000	4	0.05	2.15	1,710	770	10	7.88
Silicon-iron	4 Si	800	500	7,000	40	0.5	1.97	1,570	690	60	7.65
Silicon-iron (oriented)	3 Si	800	1,500	40,000	8	0.1	2.00	1,590	740	47	7.67
Silicon-iron (cubic)	3 Si	…	…	116,000	5.6	0.07	2.00	1,590	740	47	7.67
Silicon-iron (cubic)	3 Si	…	…	65,000	6.4	0.08	2.00	1,590	740	47	7.67
Aluminum-iron	3.5 Al	1,100	500	19,000	24	0.3	1.90	1,510	750	47	7.5
Alfer	13 Al	…	700	3,700	53	0.66	1.20	955	…	90	6.7
Alperm	16 Al	600Q	3,000	55,000	3.2	0.04	0.80	637	400	140	6.5
16 Alfenol	16 Al	600Q	600–4,100	4,000–90,000			0.80	637	…	153	6.5
78 Permalloy	78.5 Ni	1,050; 600Q	8,000	100,000	4	0.05	1.08	860	600	16	8.6
Supermalloy	5 Mo, 79 Ni	1300C	100,000	1,000,000	0.16	0.002	0.79	629	400	60	8.77
Cr Permalloy	3.8 Cr, 78 Ni	1000	12,000	62,000	4	0.05	0.80	637	420	65	8.5
Mumetal	5 Cu, 2 Cr, 77 Ni	1175(H₂)	20,000	100,000	4	0.05	0.65	517	…	62	8.58
Hipernik	50 Ni	1200(H₂)	4,000	70,000	4	0.05	1.60	1,270	500	45	8.25
50 Isoperm	50 Ni	1100CR	90	100	480	6	1.60	1,270	500	40	8.25
Deltamax	50 Ni	1075	500	200,000			1.55	1,230	…	45	8.25
Thermoperm	30 Ni	1000					0.20	159			
Permendur	50 Co	800	800	5,000	160	2.0	2.45	1,950	980	7	8.3
45–25 Perminvar	25Co, 45 Ni	1000, 400	400	2,000	95	1.2	1.55	1,230	715	19	
7–70 Perminvar	7 Co, 70 Ni	1000, 425	850	4,000	48	0.6	1.25	995	650	16	8.6

*Q = quenched; C = controlled cooling rate; CR = severely cold-rolled; H₂ = annealed in pure hydrogen.

TABLE 6-40 High-Permeability Laminated or Wound Cores of Silectron† and Nickel Alloys[81]

	Silectron cores, 0.012 in	Silectron laminations, 0.014 in	Mumetal, 0.014 in	Molybdenum Permalloy, 0.014 in	4750, 0.014 in	Deltamax, 0.002 in	Supermalloy, 0.004 in
General applications	Power distribution transformers	Power and audio transformers	Low induction filters and audio transformers		Servo and synchro motors, audio transformers	Magnetic amplifiers	Magnetic amplifiers, specialty transformers
Nominal composition, %:							
Molybdenum	···	···	···	4.00	···	···	5.00
Silicon	3.25	3.25					
Nickel	···	···	77.0	79.00	48.00	50.00	79.00
DC permeability:							
μ_{max}	50,000	···	100,000	200,000	80,000	100,000	700,000
B at μ_{max}	8,000	···	2,500	3,000	5,000	12,000	3,000
μ at 40B	4,000	···	25,000	30,000	8,000	500	75,000
μ at 100B	6,500	···	30,000	45,000	12,000	2,000	80,000
AC permeability, 60 Hz, at 40B	3,500		20,000	26,000*	8,000*	500	70,000
At 200B	6,500		30,000	32,000*	13,500*	1,000	90,000
At 2,000B	15,000		40,000	54,000*	30,000*	20,000	160,000
Saturation induction B_s, G	19,700	19,700	7,500	8,000	15,500	16,000	7,800
Specific gravity, g/cm³	7.65	7.65	8.5	8.74	8.20	8.25	8.77
Electrical resistivity, $\mu\Omega\cdot$cm	50	50	60	55	45	45	65

*Minimum values.
†Allegheny Ludlum Steel Co.

TABLE 6-41 High-Frequency Silicon and Nickel Cores[81]

Grade	Monimax	Powder cores	Rotosil, 7 mils	Silectron			Ultrathin nickel irons		
				4 mils	2 mils	1 mil	Square Permalloy	Delta max	Super malloy
General applications	Pulse transformers, high-frequency transformers	Loading coils, filters	High-frequency rotating machinery	Pulse transformers, high-frequency transformers, magnetic amplifiers			Magnetic amplifiers, computer cores		
Nominal composition, %:									
Nickel	47.00	81.00	...	...	...	...	79.00	50.00	79.00
Silicon	...	...	3.25	3.25	3.25	3.25	...	...	...
Molybdenum	3.00	2.00	...	...	...	...	4.00	...	5.00
Nominal frequency range	Audio range	Audio to low rf	400–800 Hz	400 Hz audio range	Pulse 0.5–10	Pulse under 0.50	Pulse at repetition rates up to 1 MHz		
Specific gravity	8.25	...	7.65	7.65	7.65	7.65	8.74	8.25	8.77
Saturation induction B_s, G	14,500	7,800	19,700	19,700	19,700	19,700	8,000	16,000	8,000
Electrical resistivity, $\mu\Omega \cdot cm$	65 min	High	50	50	50	50	55	45	65

142. Powder Cores. For higher-frequency applications powdered ferromagnetic materials are used. These powdered materials are compacted with an insulating binder, thereby reducing considerably the eddy currents. The reduction in eddy current losses comes at the expense of the permeability, which is normally around 100. Powder grain size varies from 10 to 100 μm, approximately. Powder cores are made of pure carbonyl iron, Sendust, or Permalloy. At low frequencies and for high stability, Permalloy cores are preferable. At audio frequencies all three powders are suitable, while at high frequencies iron and Sendust are useful. At very high frequency iron cores are used. Properties of powder cores are given in Table 6-42.

TABLE 6-42 Magnetic Properties of Pressed Powder and Ferrite Cores[83]

Material	Composition	Treat-ment*	$\bar{\mu}_a$	H_c A/m	H_c Oe	I_s Wb/m²	I_s G	Curie temp., °C	ρ, $\Omega \cdot m$
Carbonyl iron powder	100 Fe	P	20	1,200	15	1.56	1,240	770	10^8
Mo Permalloy powder	2 Mo, 81 Ni	P 650	125	...	...	0.70	560	480	10^4
Sendust powder	5 Al, 10 Si	P, 800	80	100	1.25	0.45	360	500	
Mn–Zn ferrite	50 Mn, 50 Zn	1,150	2,000	8	0.1	0.25	200	110	1
Cu–Zn ferrite	40 Cu, 60 Zn	1,000	1,100	40	0.5	...	...	90	10^3
Cu–Mn ferrite	40 Cu, 60 Mn	1,250	...	80	1.0	0.29	230		
Ni–Zn ferrite	30 N , 70 Zn	1,050	80	...	...	0.40	320	130	10^{10}
Mg–Zn ferrite	50 Mg, 50 Zn, 50 Mg, 50	...	500	80	1.0	0.26	207	120	
Mg–Mn ferrite	Mn	1,400	...	40	0.5	0.27	215	130	

*P = pressed; numerical values are Celsius sintering temperatures.

143. Ferrites. At higher frequencies nonconducting ferrimagnetic materials are used as core materials. The ferrimagnetic materials have smaller saturation inductions (up to 0.5 Wb/m²) and very small core losses due to their high electrical resistivities. The ferrites are generally of the spinel, magnetoplumbite, or garnet types. The permeability of some common ferrites is shown in Fig. 6-33 as a function of frequency. It can be seen that the real part of the permeability decreases with increasing frequency, while the imaginary part, which is related to losses, reaches a maximum and then decreases. In garnets the losses at specific dc bias magnetic fields become important. Large losses are encountered at specific fields. Ferrites can be classified also according to their use: high permeability, low frequency; low loss, high frequency; microwave materials; square-loop materials. Table 6-43 lists some properties of ferrites.

Impurities tend to decrease the initial permeability, and insoluble inclusions increase the coercive force. Pores at grain boundaries have little effect on the initial permeability. Pores within grains decrease the initial permeability. Pores both at grain boundaries and in grains can affect the maximum permeability, coercivity, and hysteresis losses.

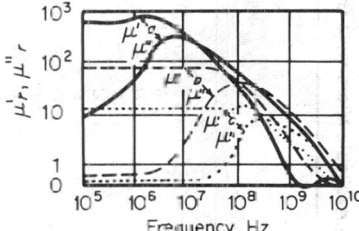

Fig. 6-33. Permeability spectra at room temperature of three commercial ferrites: (a) $Ni_{0.36}Zn_{0.54}Fe_2O_4$; (b) $Ni_{0.64}Zn_{0.36}Fe_2O_4$; (c) $NiFe_2O_4$. The μ scale is logarithmic for μ greater than 1 and linear for μ less than 1.[84]

144. Spinels. The ferrimagnetic spinel structure is obtained by replacing the Mg in $MgAl_2O_4$ with a divalent metal Me^{II} and Al with Fe^{III}. The divalent metal can be Mg, Ni, Co, Cu, Fe, Zn, Mn, or Cd. The spinel has 8 occupied tetrahedral sites and 16 occupied octahedral sites. The spins of the ions are oppositely directed on the two kinds of sites. The two sites can be occupied by Me^{II} and Fe^{III} in the following fashion:

$$Me_\delta^{II}Fe_{1-\delta}^{III}[Me_{1-\delta}^{II}Fe_{1-\delta}^{III}]O_4$$

$Me_\delta^{II}Fe_{1-\delta}^{III}$ ions are on the tetrahedral sites, while $Me_{1-\delta}^{II}Fe_{1-\delta}^{III}$ ions are on the octahedral site. Normal spinel has $\delta = 1$, and Me^{II} is found only on the tetrahedral sites. Inverse spinel has $\delta = 0$, and the Fe^{III} ions are equally divided between tetrahedral and octahedral sites, thus canceling their net magnetic divalent metal ions in the octahedral sites.

TABLE 6-43 Properties of Commercially Available Microwave Garnets and Ferrites[54]

Supplier*	Material code	Composition	$4\pi M_s$, G	Saturation magnetization M_s, mT	Curie temp. T_c, K	Resonance line width† ΔH, A/m	Landé g factor	Coercive force H_c, A/m	Dielectric constant ϵ	Dielectric loss factor tan $\delta_c \times 10^3$‡	Remannent induction B_r, mT
						Garnets					
T	G-113	YIG	1,780	14.2	280	3,581	1.97	35.8	15.0	<2	127.7
C	CG-1800	YIGk	1,800	14.3	280	<6,366	1.98	...	15.8	15	
C	CG-1400A	YAl	1,400	11.1	250	5,570	1.98	...	15.5	15	78.4
T	G-1210	YAl	1,200	9.55	220	3,183	1.98	31.8	14.8	<2	
C	CG-1200A	YAl	1,200	9.55	250	<4,775	1.99	...	15.7	15	
T	G-250	YAl	250	1.99	105	<3,581	2.02	35.8	13.8	<2	12.3
C	CG-250A	YAl	250	1.99	115	2,387	2.03	...	14.1	15	
C	CG-1600G	YGd	1,600	12.7	280	<7,958	1.98	...	15.8	15	
T	G-1001	YGd	1,200	9.55	280	5,968	1.99	59.7	15.2	<2	71.7
C	CG-720G	YGd	720	5.73	280	17,507	2.01	...	16.2	15	
C	G-1400	YGdAl	1,400	11.1	265	3,979	1.98	39.8	15.1	<2	91.8
C	CG-1400GA	YGdAl	1,400	11.1	270	<7,162	1.98	...	15.6	15	
T	G-500	YGdAl	550	4.38	180	5,173	2.00	51.7	14.4	<2	28
C	CG-550GA	YGdAl	550	4.38	185	4,775	2.02	...	14.6	15	
T	G-4256	YHO	1,600	12.7	280	5,570	1.98	55.7	15.1	2	98.6
						Ferrites					
T	TT1-2000	Mg	2,000	15.9	290	19,894	1.98	198	12.4	<2.5	138.5
C	C-145	MgZn	2,750	21.9	270	10,743	2.02	...	12.6	3	128.8
T	TT1-390	MgMn	2,150	17.1	320	42,972	2.04	430	12.7	<2.5	
C	C-115	MgMnAl	1,300	10.4	180	6,764	2.02	...	11.8	2	
T	TT71-4800	LiZn	4,800	38.2	400	19,099	2.01	191	14.5	<25	284.3
C	C-10	Ni	3,000	23.9	580	33,423	2.40	...	11.8	6	
C	C-21	NiAl	2,150	17.1	510	21,486	2.38	...	13.2	5	
C	C-50	NiAl	500	3.98	190	23,873	1.47	...	9.9	2	
T	TT2-113	NiAl	500	3.98	120	11,937	1.54	119	9.0	<8	14
T	TT2-111	Ni	5,000	39.8	375	12,732	2.11	127	12.5	<10	195.6
C	C-48	NiZn	5,250	41.8	410	14,324	2.06	...	13.0	10	
C	C-9	NiCo	3,150	25.1	585	10,345	2.22	...	13.0	7	
C	C-30	NiAlC	1,330	10.6	495	22,680	2.47	...	12.7	5	
T	TT2-116	NiAlO	1,400	11.1	425	20,690	2.40	207	12.3	<10	51.2

*C = Countis Industries, San Luis Obispo, CA 93401; T = Trans-Tech, Gaithersburg, MD 20760.
†For C entries g is effective and tan δ is measured at 9.0 GHz. Line widths are measured on thin slabs at one-half the peak loss in decibels. For T entries line width at −3 dB; B_r and H_c at 2 kHz with $H_{app} = 5H_c$. Saturation magnetization at 23°C.
‡See footnote a, Table 6-1.

Ni–Zn ferrites are most commonly used in the frequency range 0.1 to 200 MHz. Figure 6-33 shows μ'_r, μ''_r for such ferrites as a function of frequency. Ni–Zn ferrites have high initial permeability and low coercive force and are used for recording tapes.

Mn–Zn ferrites have a very small coercive force and very high initial permeability as well as a high-saturation magnetization. These ferrites are used in the 1-kHz to 1.5-MHz range and are lossy at higher frequencies.

Ni–Co ferrites have been used as magnetostrictive materials.

Square-loop ferrites (Mn–Cu, Li–Ni, and Mn–Mg ferrites) and doped YIG exhibit high permeability but low coercive forces. Such materials are used for high-speed memory cores.

145. Hexagonal Ferrites. The magnetoplumbites are the most common materials in this group. The mineral is of the formula $PbFe_{7.5}Mn_{3.5}Al_{0.5}Ti_{0.5}O_{19}$. The hexagonal ferrites show a considerable magnetocrystalline anisotropy, which makes them suitable for hard magnets.

The ferrites are denoted M, U, Z, Y, X, and W, where

$$M = BaFe_{12}O_{19} \qquad U = Ba_4Me_2Fe_{36}O_{60} \qquad Z = Ba_6Me_4Fe_{48}O_{82}$$
$$Y = Ba_2Me_2Fe_{12}O_{22} \qquad X = Ba_2Me_2Fe_{28}O_{46} \qquad W = BaMe_2Fe_{16}O_{17}$$

Me can be Ni, Mg, Co, Fe, Zn, Mn, or Cu.

Hexagonal ferrites can be used at very high frequencies.

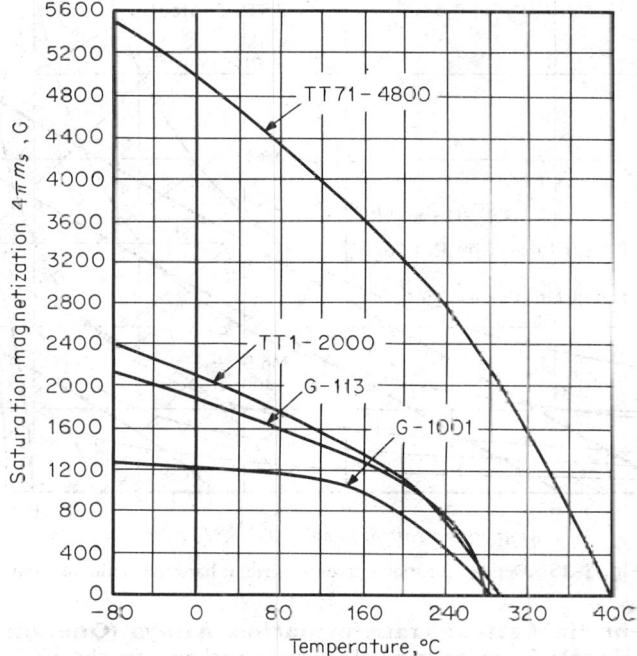

Fig. 6-34. Saturation magnetization for several garnets and ferrites.[86] Material designation corresponds to Table 6-45.

146. Garnets. The highest-frequency ferrites available are garnets. Garnets are of the $Me_3^{II}Me_2^{III}Si_3^{IV}O_{12}$ type, where Me^{II} can be replaced by Ca, Fe, Mg, or Mn and the Me^{III} by Al, Cr, Fe, or Mn. Magnetic iron magnets are derived by substituting $Me^{III}Fe^{III}$ for $Me^{II}Si^{IV}$. Me^{III} can be Y, Sm, or Lu. The unit cell contains eight units of $Me_3^{III}Fe_5^{III}O_{12}$.

The garnet has 24 tetrahedral Fe^{III} sites (d sites), 16 octahedral Fe^{III} sites (a sites), and 24 dodecahedral Me^{III} sites (c sites).

Partial substitution of other trivalent ions such as aluminum is also possible. YIG, for example, is $Y_3Fe_5O_{12}$ and has a very low hysteresis loss at microwave frequencies. Saturation magnetization of garnets as a function of temperature is given in Fig. 6-34. The peculiar shape of the saturation magnetization occurs because it is a sum of the magnetization due to the c sublattice and the a-d sublattice. The magnetization at high temperature is due to the Fe^{III} ions only. Properties of some microwave materials are given in Table 6-43.

RETENTIVE MAGNETIC MATERIALS

147. Retentive materials are characterized by a high energy product, making them suitable for permanent magnets. These materials have high remanences and coercive forces. Retentive materials are composed of a fine structure, which is responsible for the large coercive force.

There are four classes of retentive materials: (1) martensitic lattice-transformation alloys, (2) precipitation-hardened alloys, (3) ordered alloys, and (4) fine-particle magnets.

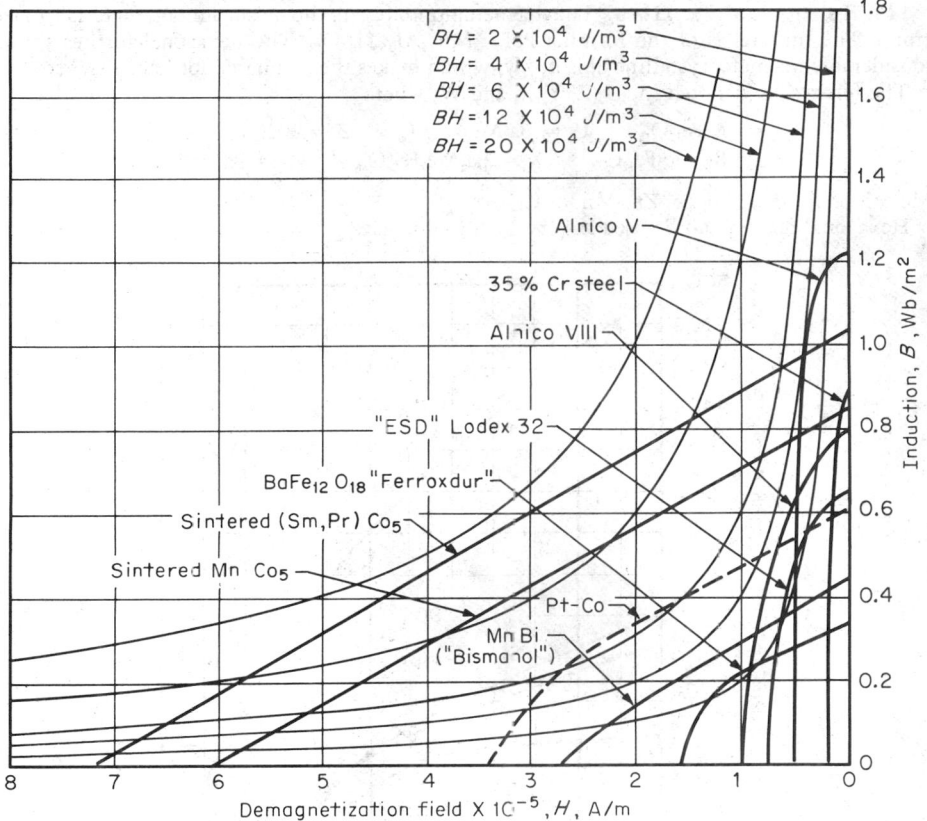

Fig. 6-35. Demagnetization curves for several hard magnetic materials.[54]

148. Martensitic Lattice-Transformation Alloys (Quench- and Work-Hardened Alloys). These are iron-rich alloys quenched from the fcc γ phase to form a martensitic structure. It is a very fine platelike structure which is created from the γ phase by a rapid shear transformation. This fine structure is both mechanically and magnetically hard. High-carbon steels containing W, Cr, Co, Al, or V are most commonly used. Properties and composition of some of these steels are given in Table 6-44, and the demagnetizing curves in Fig. 6-35. In some alloys the martensitic transformation can be induced by work hardening.

149. Precipitation-Hardened Alloys. These alloys develop a very fine structure due to precipitation upon heat treatment. This fine structure results in very high remanence, coercive force and very high energy product.

Alnico is the best-known alloy in this group. Its maximum energy product can be increased by directional solidification in the <100> directions. Solution treatment at 1300°C followed by a 800°C heat treatment will product further improvements. The 800°C treatment will produce a network of fine α'-Fe, Co-rich precipitates with an α-Ni–Al–rich matrix. The α' has a higher magnetization. Annealing in the presence of a magnetic field results in elongated α' particles which improve the magnetic hardness due to shape anisotropy of the α'. Long heat-treatments at 580°C further increase the "hardness" of Alnico.

Alnico magnets can be cast to complex shapes. There are various kinds of Alnico alloys, and their magnetic properties are listed in Table 6-44. Demagnetization curves are given in Fig. 6-35. The remanence of a magnet is affected by temperature. Alnico 5 suffers a structural change at 550°C but will lose part of the remanence at lower temperatures.

TABLE 6-44   Properties of Permanent-Magnet Materials[84,87]

	Required magnetizing force H_s, kA/m	Residual induction B_r, T	Coercive force H_c, kA/m	Peak energy product $(B_dH_d)_{max}$, kJ/m^3
Cast Alnico:				
1	159	0.71	32	10
2	199	0.72	43	13
3	199	0.67	36	11
4	279	0.52	56	10
5	239	1.24	51	43.8
5DG	279	1.26	53	49.7
5-7	279	1.30	58	57.7
6	318	1.02	61	29.8
8A	637	0.85	127	40
8B	796	0.75	147	41.8
Sintered Alnico:				
2	199	0.68	41	12
5	239	1.05	47	28
6	318	0.86	63	24
8	637	0.76	123	36
Platinum-cobalt	1,671	0.60	334	60
P-6 alloy†	24	1.40	5	4
Lodex:				
30	477	0.40	99	13
31	477	0.62	91	27
32	398	0.73	75	27
33	398	0.80	58	25
1095 C steel	24	0.86	4	1.4
5% W steel	24	1.03	6	2.5
3½% Cr steel	24	0.90	5	2.3
17% Co steel	60	0.90	14	5.2
36% Co steel	80	0.96	18	7.5
Cunife	191	0.54	44	12
Cunico	239	0.34	53	6
Barium ferrite (isotropic)	796	0.22	127	7
Oriented barium ferrite:				
1	796	0.305	191	17.5
2	796	0.37		26
Remalloy (Comol)	80	1.00	18	9
Rare-earth–cobalt Vacomax:‡				
140	1,500	0.85–0.94	550–720	125–170
145 (Co₅Sm)	2,500	0.85–0.94	600–720	140–170
200	1,500	1.0 –1.05	710–800	190–220

*Properties are based on relatively simple geometry. In the case of antisotropic materials, these properties refer to the preferred direction of magnetization.
†When heat-treated at 600°C for 2 h.
‡Trade name of Vacuumschmelze GmbH, West Germany.

Alnico magnets are mechanically hard and brittle. To compromise these properties with the magnetic hardness, the ductile alloys were developed. *Cunife* is a ductile alloy, containing 20% Ni, 60% Cu, and 20% Fe, which has still a reasonable magnetic hardness. *Cunico* can be machined before heat treatment but not after. *Vicalloy* is a Co–V–Fe alloy which is ductile before heat treatment. *Silmanal* is a ductile Ag–Mn–Al alloy having a very high coercive force. *Remalloy* or *Comol* are Co or Mo ductile age-hardened iron alloys. Properties of age-hardened permanent magnets are listed in Table 6-44.

150. Ordered Alloys. These include the Fe–Pt and Co–Pt alloys which exhibit ordered superlattices on aging at 700°C. These alloys have very high coercive forces and the highest energy product. For properties see Table 6-44.

151. Fine-Particle Magnets. Fine-particle magnets are produced from compacted metallic powders or ceramic magnetic materials. Metallic-power magnets with high energy products have been produced from iron, 70% Fe–30% Co alloys, Alnico, Remalloy, and others. The properties of powder magnets have been further improved by the use of elongated-single-domain (ESD) powders which consist of elongated fine particles (formed electrolytically on a liquid-metal electrode) which impart very high retentivities to the magnets. Rare-earth–Co sintered magnets have the highest BH product and coercive force. More information on those magnets can be found in the last two items in Par. **6-153** and in Fig. 6-35.

An outstanding ceramic permanent-magnet material is the $BaFe_{12}O_{19}$ (Ferroxdure), belonging to the group of hexagonal ferrites. The ceramic magnets have a much higher ohmic resistance than metallic-powder magnets. They have high coercive forces, are chemically inert, and usually have a lower cast. Ceramic magnets can operate at high frequency and are sometimes used as resonance oscillators at the 50-GHz range. Ceramic magnets have been used with plastic or elastomeric binders. Such construction allows machining of the magnets to the desired shape. Properties of ceramic magnets are listed in Table 6-44.

152. Magnetic-Bubble Memories. These special retentive magnetic devices use the stability of miniature circular magnetic domains (bubbles) in garnets. The magnetic domains are generated in a garnet layer grown over a gadolinium gallium garnet (GGG) metallized and patterned with Permalloy and aluminum conductors. The bubbles travel along the Permalloy pattern by action of a rotating magnetic field. Several technologies for forming bubble memories have been reported.[34–36]

153. Bibliography on Magnetic Materials

Chikazumi, S.   "Physics of Magnetism," Wiley, New York, 1966.

Bozorth, R. M.   "Ferromagnetism," Van Nostrand, Princeton, N.J., 1951.

Tebble, R. S., and D. J. Craik,   "Magnetic Materials," Wiley-Interscience, New York, 1969.

Berkowitz, A. E., and E. Kneller (eds.)   "Magnetism and Metallurgy," Academic, New York, 1969.

Von Aulock, W. H. (ed.)   "Handbook of Microwave Ferrite Materials," Academic, New York, 1965.

Smit, J. (ed.)   "Magnetic Properties of Materials," McGraw-Hill, New York, 1971.

Strnat, K. J.   Rare Earth–Cobalt Magnets, *2nd Int. Workshop Rare Earth Cobalt Magnets Their Appl., University of Dayton*, 1976.

Eschenfelder, A. H.   "Magnetic Bubble Technology," Springer, New York, 1980.

SEMICONDUCTOR MATERIALS

154. Elemental and Compound Semiconductors. Semiconductors are usually materials which have energy-band gaps smaller than 2 eV. An important property of semiconductors is the ability to change their resistivity over several orders of magnitude by doping. Semiconductors have electrical resistivities between 10^{-5} and 10^7 $\Omega \cdot m$. Semiconductors can be crystalline or amorphous.

Elemental semiconductors are single-element semiconductor materials (except for intentional minute additions) such as silicon or germanium. *Compound semiconductors* are materials containing more than one element, for example, III-V compounds such as GaAs, AlAs, or GaP.

155. Energy Bands and Gaps. According to band theory, discontinuities occur in energy levels at certain values of electronic wave vectors. The discontinuities are called energy gaps, or band gaps. Figure 6-36 shows the typical energy-band occupancy for metals, semiconductors, and insulators. In metals there is an incompletely filled conduction band (or overlapping bands) which allows free movement of electrons. Semiconductors have a filled *valence band* with a relatively small band gap separating it from the *conduction band*. At any finite temperature a small number of electrons are found at the bottom of the conduction band, and a number of unfilled states, or *holes*, are found in the valence band. Electrical conduction is possible by the motion of electrons or holes. Insulators have large energy gaps with filled valence bands and empty conduction bands resulting in extremely small conductivity.

156. Intrinsic and Extrinsic Semiconductors. Semiconductors with a low impurity concentration having electrical properties characteristic of the pure semiconductor material are called *intrinsic* semiconductors. Semiconductors with electrical properties modified by impurities are called *extrinsic* semiconductors.

157. Charge Transport. A perfect semiconductor is essentially an insulator at absolute zero temperature, since its conduction band is vacant. As the temperature increases, electrons are thermally excited from the valence into the conduction band, and the conductivity increases. The free electrons in the conduction band and the holes in the valence band participate in electrical conduction. The electrons are negative charge carriers, while the holes are positive carriers. Actual crystals exhibit purely *intrinsic conduction* (i.e., independent of impurity concentration) only at high temperatures. At lower temperatures the impurity effect can be quite large.

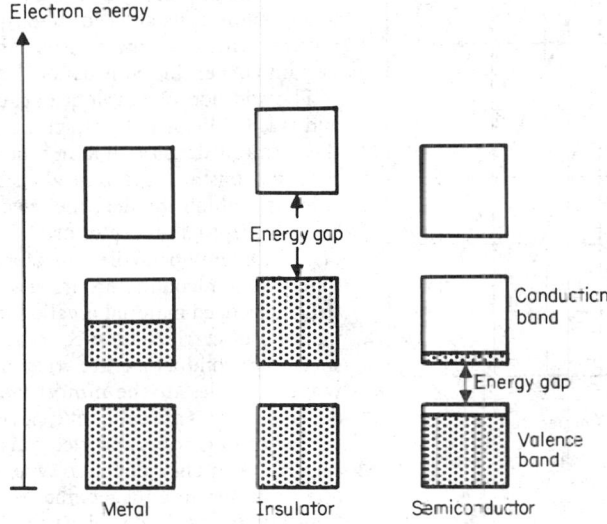

Fig. 6-36. Energy-band diagrams for metals, insulators, and semiconductors.

158. Carrier Concentration in Intrinsic Semiconductors. The concentration of electrons n and of holes p in intrinsic semiconductors can be calculated based on a simple model of the band structure (assuming the distance between the Fermi level and the band edges to be large in comparison with kT)

$$n = N_c \exp[-(E_c - E_f)]/kT \qquad p = N_v \exp[-(E_f - E_v)]/kT \qquad (6\text{-}65)$$

where n and p = electron and hole densities (m^{-3}).

$$N_c = 2\frac{(2\pi m_n^* kT)^{3/2}}{h^2}$$

can be thought of as an effective density of states at the bottom edge of the conduction band.

$$N_v = 2\left(\frac{2\pi m_p^* kT}{h^2}\right)^{3/2}$$

is the effective density of states at the top edge of the valence band. E_f, E_c, E_v are the energies corresponding to the Fermi level, conduction-band edge, and valence-band edge, respectively. m_n^*, m_p^* are the effective masses of electrons and holes, respectively (kg), k = Boltzmann's constant (J/K), h = Planck's constant (J·s), and T = temperature (K).

The product of the electron and hole densities is

$$np = N_c N_v \exp[-(E_c - E_v)]/kT = N_c N_v \exp(-E_g/kT) \qquad (6\text{-}66)$$

where the energy gap $E_g = E_c - E_v$. The np product depends on the energy gap and not on the Fermi level. For intrinsic semiconductors $n = p = n_i$ and

$$np = n_i^2 \qquad (6\text{-}67)$$

$$n_i = (N_c N_v)^{1/2} \exp(-E_g/kT) \qquad (6\text{-}68)$$

Figure 6-37 shows n_i as a function of T for Si, Ge, and GaAs. It should be noted that N_c and N_v are also slightly temperature-dependent.

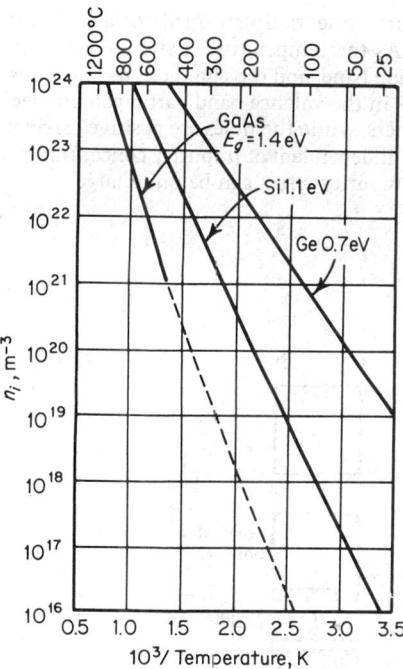

Fig. 6-37. Intrinsic carrier concentration in Si, Ge, and GaAs as a function of temperature.[88-90]

159. Donor and Acceptor Impurities.

Impurities can be added intentionally to semiconductors to modify their electrical properties. This addition is termed *doping*.

Silicon doped with pentavalent substitutional impurities such as P, As, or Sb will have a higher electron density due to the easily ionized fifth electrons of the impurities. The hole concentration in such a silicon crystal is reduced since the np product remains constant [Eq. (6-66)]. Dopants which increase the electron density are called *donors*. Dopant concentration is denoted N_p.

The addition of trivalent dopants such as B, Al, and Ga to silicon will attract electrons and reduce the electron density. Again, since the np product remains constant, the hole density must increase. Dopants which reduce the electron density are called *acceptors*. Acceptor concentration is denoted N_A. The energy-band diagrams for donor- and acceptor-doped semiconductors are shown in Fig. 6-38.

Donor-doped material is called *n*-type because the majority of carriers are electrons. The electrons in such semiconductors are the *majority carriers*, while the holes are the *minority carriers*. Acceptor-doped materials are called *p*-type. Here the holes are the majority carriers and electrons are the minority carriers. A highly doped *n*-type semiconductor is often referred to as degenerate and denoted by n^+. Similarly, a highly doped *p*-type is denoted p^+.

160. Carrier Concentration in Extrinsic Semiconductors.

To calculate the carrier concentration in extrinsic semiconductors we use the electric charge neutrality equation. For fully ionized impurities,

$$p_n + N_D = n_n + N_A \qquad p_p + N_D = n_p + N_A \qquad (6\text{-}69)$$

where p_n and n_n = hole and electron concentration in *n*-type semiconductors (m^{-3}), p_p and n_p = hole and electron concentrations in *p*-type semiconductors (m^{-3}), and N_p and N_A = donor and acceptor concentrations (m^{-3}).

Using Eqs. (6-69) and (6-67) and assuming $n_i \ll |N_D - N_A|$, we get the carrier concentration in semiconductors of

n-type:

p-type:

$$\text{*n*-type: } n_n = N_D - N_A \qquad p_n = n_i^2/(N_D - N_A)$$

$$\text{*p*-type: } p_p = N_A - N_D \qquad p_n = n_i^2/(N_D - N_A) \qquad (6\text{-}70)$$

It can be seen from Eqs. (6-70) that a reduction in charge carriers (and conductivity) can be obtained in materials with low $N_D - N_A$. This lowering can be done, for example, by intentionally adding donors into a *p*-type material. Such a material is called *compensated intrinsic*.

The carrier concentration determines the position of the Fermi level. Figure 6-39 shows the Fermi level as a function of donor and acceptor concentrations.

161. Impurity Energy Levels. The doping impurities have characteristic energy states within the energy gap. Impurities with energy states close to the band edges are called *shallow states*, and those closer to the center of the gap are called *deep states*. Figure 6-40 shows the energy states of various impurities in Si, Ge, and GaAs. Some impurities have several energy states within the band gap.

162. Surface States. Oxidized semiconductors are of interest in most semiconductor devices. Such semiconductors have localized electronic states both in the oxide and at the semiconductor interface. The density of states on freshly cleaved surfaces is about 10^{19} m^{-2}. Oxidized surfaces show densities of 10^{15} to 10^{16} m^{-2}. The density of the states can be reduced by proper manufacturing techniques.

Surface states have been divided into slow and fast states, depending upon the speed of their interaction with the semiconductor space-charge region.

163. Conduction in Semiconductors. An electric field applied to a semiconductor will induce a drift on the charge carriers. At low fields, the drift velocity v_d is proportional to the field E:

$$v_d = \mu/E \tag{6-71}$$

where the proportionality constant μ is the mobility.

The conductivity in semiconductors is given by

$$\sigma = e(n\mu_n + p\mu_p) \tag{6-72}$$

where σ = conductivity, n, p = electron and hole density, respectively (m^{-3}), μ_p, μ_n = mobilities of electrons and holes, respectively ($m^2/V \cdot s$), and e = electronic charge (C).

The mobility of electrons and holes is affected by two main scattering mechanisms, *impurity* and *lattice scattering*. The mobility due to impurity scattering μ_i is proportional to $T^{3/2}$, while the mobility due to lattice scattering μ_L is proportional to $T^{-3/2}$. The total mobility μ is related to the individual components as

$$1/\mu = 1/\mu_i + 1/\mu_L \tag{6-73}$$

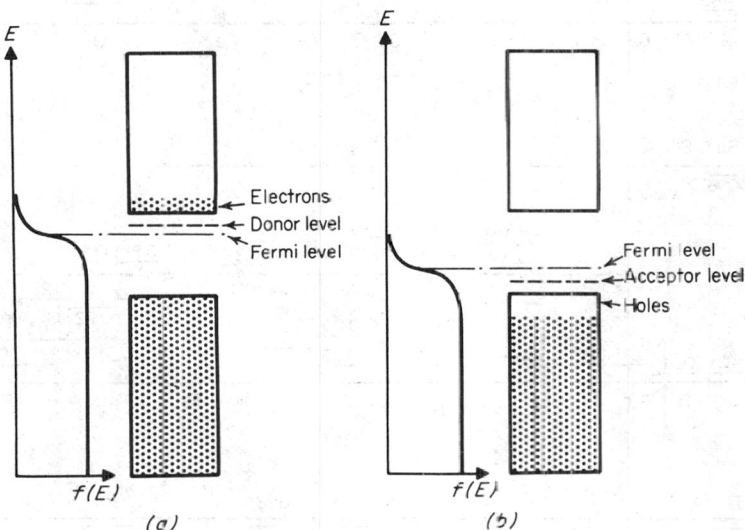

Fig. 6-38. Band diagrams for (a) n-type and (b) p-type semiconductors. The Fermi-Dirac distribution functions $f(E)$ are also shown.

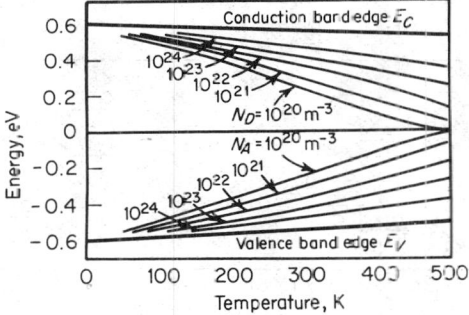

Fig. 6-39. The Fermi level in Si as a function of impurity concentration.[20]

The mobility of electrons and holes in Si, Ge, and GaAs is shown as a function of impurity concentration in Fig. 6-41. The resistivity of these semiconductors as a function of impurity concentration is shown in Fig. 6-42.

Crystal imperfections also scatter electrons. This scattering is small and has only a slight temperature dependence. The mobilities near semiconductor surfaces are decreased to a fraction of the bulk mobilities. This decrease is in part due to extra scattering of the carriers at the surfaces.

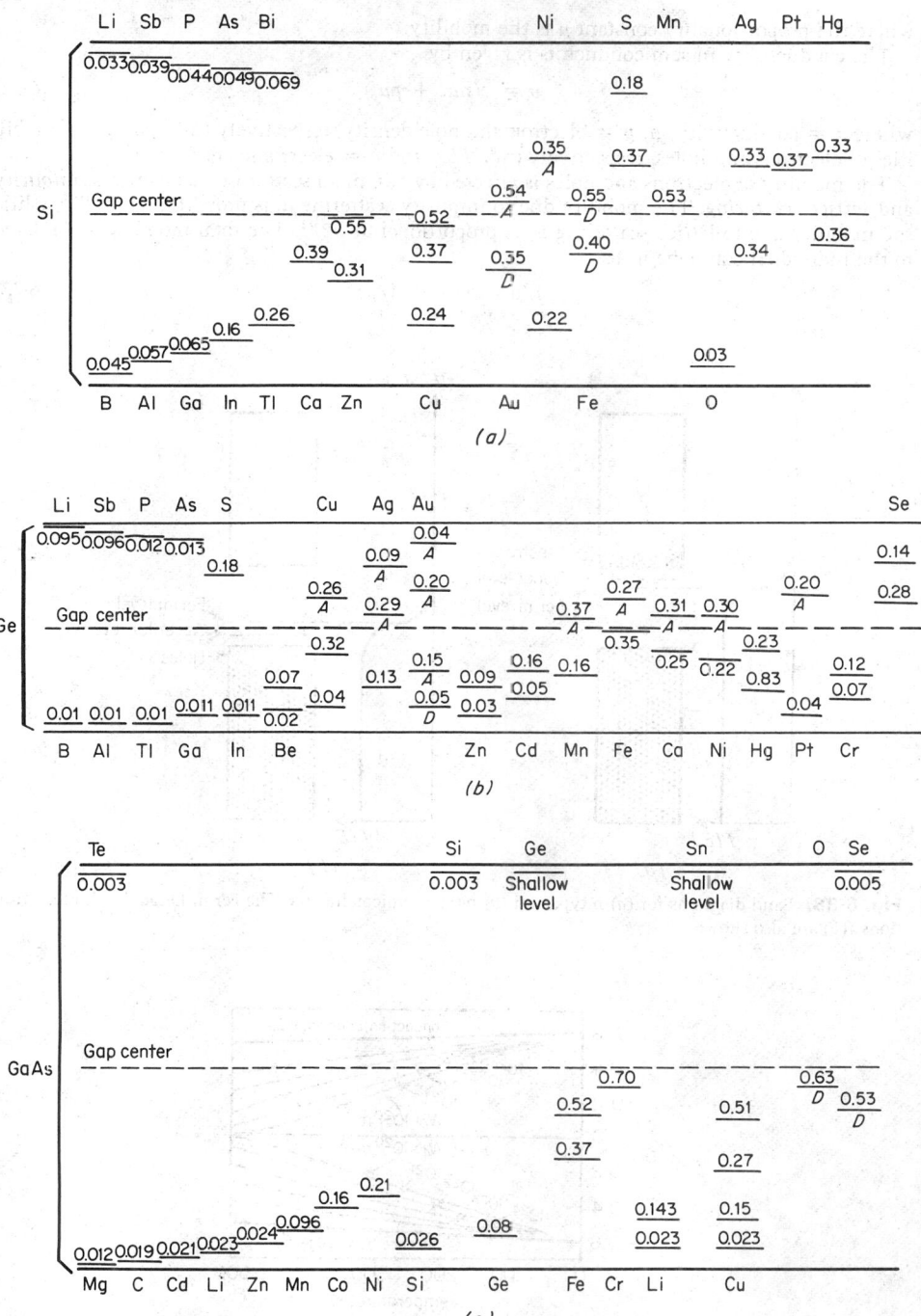

Fig. 6-40. Impurity energy levels in (a) Si, (b) Ge, and (c) GaAs.[74]

164. Carrier Diffusion. Carrier concentration gradients cause a current flow in addition to the normal flow by electric fields. The current density for electrons or holes in a concentration gradient along the x axis is

$$J_n = eD_n\frac{dn}{dx} \qquad J_p = -eD_p\frac{dp}{dx} \qquad (6\text{-}74)$$

where J_n, J_p = electron or hole current densities (A/m²), e = electronic charge (C), D_n, D_p = diffusion coefficients for electrons or holes (m²/s), and n, p = electron or hole concentrations (m⁻³).

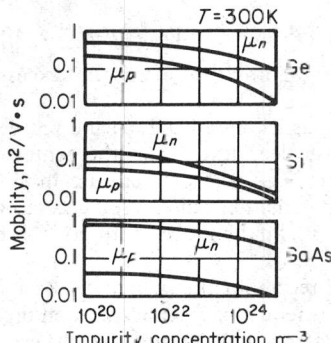

Fig. 6-41. Drift mobility of Ge and Si and Hall mobility of GaAs vs. impurity concentration at 300 K.[74]

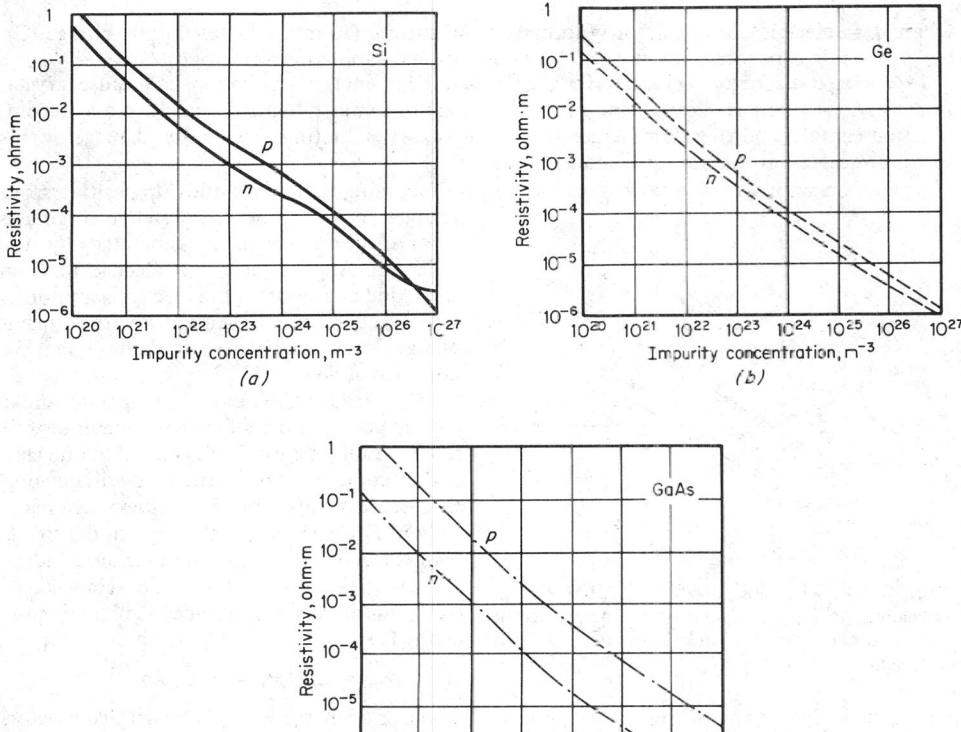

Fig. 6-42. Resistivity vs. carrier concentration at room temperature for (a) Si, (b) Ge, and (c) GaAs.[91-93]

The diffusion coefficients are related to mobilities according to Einstein's relation,

$$D_n = (kT/e)\mu_n \qquad D_p = (kT/e)\mu_p \tag{6-75}$$

where k = Boltzmann's constant (J/K), μ_n, μ_p = electron or hole mobilities (m²/V·s), and T = temperature (K).

165. Carrier Generation and Recombination. *Generation* of electron-hole pairs can be accomplished by means other than thermal, e.g., by illuminating the semiconductor. A semiconductor with excess electron-hole pairs will return to equilibrium by *recombination* of holes and electrons. An n-type semiconductor with p_n minority carriers at equilibrium and p_{n0} carriers at time $t = 0$ (just after illumination removal) will have a concentration of p carriers at time t

$$p = p_m + (p_{n0} - p_n)e^{-t/\tau_p} \tag{6-76}$$

where τ_p = hole lifetime (s). (Similarly, τ_n is the excess electron lifetime in p-type semiconductors.)

Recombination can be accomplished by a band-to-band process or with the aid of impurities or imperfections. Impurities with deep levels near the center of the energy gap are efficient recombination-generation centers, e.g., gold or platinum in silicon or copper in germanium. Defects produced by high-energy radiation can also serve as recombination centers and decrease the minority-carrier lifetime. τ_n, τ_p should not be confused with *relaxation time* for conduction, also denoted by τ.

Surfaces are sites of enhanced recombination which causes a decrease in the minority-carrier concentration at the surface. This reduction in concentration produces strong concentration gradients which create hole and electron flow toward the surface. The diffusion flux toward the surface is balanced by the recombination at the surface. For example, the current density for holes is given by

$$J = e(p_{n0} - p_a)S \tag{6-77}$$

where J = electric current density normal to the surface (A/m²), e = electronic charge (C), p_{n0}, p_n = hole concentration (m⁻³), and S = surface recombination velocity (m/s).

166. High-Energy Irradiation Effects. High-energy irradiation can cause crystal imperfections (such as vacancies and interstitialcies) in semiconductors which serve as recombination centers, and therefore reduce the minority-carrier lifetime. Irradiation damage of this sort is annealed out even at room temperature.

Irradiation with x-rays, γ-rays, electrons, or other ionizing radiation builds up positive space charges in SiO₂. The space charge originates from electron-hole pairs generated by the radiation. Application of an electric field on the oxide can separate the electrons and holes upon generation, accentuating the space-charge buildup. Irradiation damage can be annealed at about 300°C.

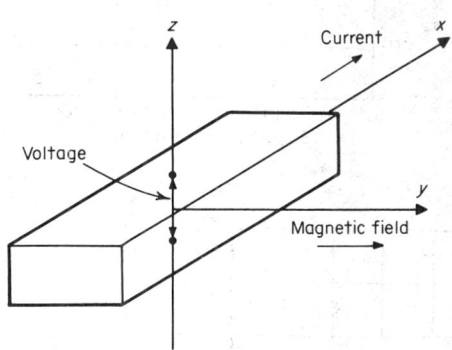

167. Hall Effect. A magnetic field applied across a current-carrying material will force the moving carriers to crowd to one side of the conductor. An electric field will develop as a result of this crowding, shown schematically in Fig. 6-43, where the current density J_x is in the x direction and the magnetic induction B is in the y direction. The Hall coefficient R_H is for an n-type semiconductor, defined as

Fig. 6-43. Schematic diagram of Hall voltage developing in the z direction relative to an electric current in the x direction and a magnetic field in the y direction.

$$R_H = E_H/J_xB_y = 1/(-en) \tag{6-78}$$

where n = density of electrons (m⁻³) and e = electronic charge (C). p-type semiconductors show a positive Hall coefficient.

168. Photoconductive Effects. Photons with high enough energies can generate electron-hole pairs. The excess carriers will cause an increase in conductivity, called *photoconduc-*

tivity. Illuminated nonbiased junctions can develop a voltage across the junction, called *photo-voltaic effect*. Light emission from semiconductors is discussed in Par. **6-229**.

169. Thermoelectric Effects. Several semiconductors have large thermoelectric power. The thermal and electrical properties of some semiconductors allow their use in thermoelectric heating and refrigeration. See also Pars. **27-14** to **27-18**.

170. Acoustoelectric Effect. The acoustoelectric effect is an electric current generated by longitudinal acoustic waves. The waves transmit some of their momentum to electrons and force a small current to flow.

JUNCTION PHENOMENA

171. pn Junctions. When a *p*-type semiconductor is brought into contact with an *n*-type semiconductor, holes will diffuse from the *p*-type to the *n*-type material, while electrons will diffuse in the opposite direction. The displacement of charges creates an electric field which opposes the diffusion. The Fermi level stays constant across a junction in equilibrium.

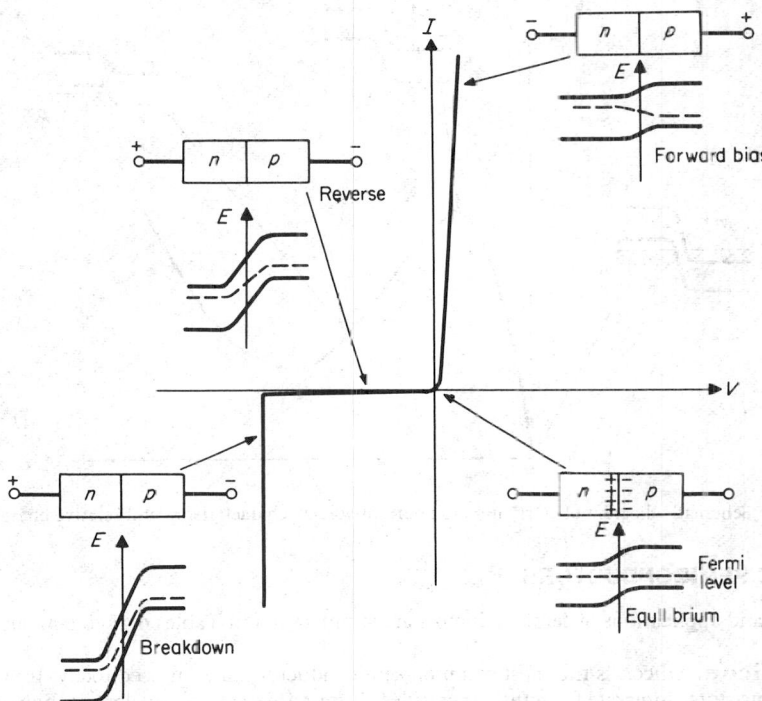

Fig. 6-44. Schematic diagram of *pn* junction current-voltage characteristics and related energy levels.

Reverse biasing a *pn* junction (that is, *p* is negatively biased with respect to *n*) will tend to create an electron flow in the *n*-type material and a hole flow in the *p*-type material away from the junction. Such currents resulting from electron-hole-pair generation are small. Forward biasing a junction will tend to move the majority carriers toward the junction, where their recombination can create a large forward current.

The *I-V characteristic* of a *pn* junction is shown in Fig. 6-44. Note that at high reverse bias the junction suffers an electrical *breakdown*. The breakdown can be due to avalanche processes (multiplication of electron-hole pairs by collision of fast-moving charge carriers) or by *zener breakdown* (due to tunneling between the two sides of the junction). *Soft breakdown* is a very undesirable occurrence in semiconductor junctions. This kind of breakdown is characterized by high-reverse-bias currents at voltages smaller than breakdown voltage. Copper and iron are notorious for producing "soft" junctions in silicon. Junctions can be "hardened" by a gettering process with phosphorus or nickel.

Junctions cannot be operated above a temperature which causes the semiconductor to be intrinsic because the rectification properties vanish.

172. $p^+ n^+$ Junctions. Highly doped p^+n^+ junctions are used in tunnel diodes. The energy-band diagram and I-V characteristic are shown in Fig. 6-45. The forward biasing of the diode will move the conduction states of the n^+ material close to the states of the p^+ valence band, and a large tunnel current will result. Higher forward bias will decrease the tunneling current since it will separate the n^+ and p^+ states.

173. Metal-Semiconductor Junctions. Metal semiconductor junctions can be either rectifying (nonohmic) or nonrectifying (ohmic) junctions. Low-resistance ohmic junctions are needed as contacts to the various p or n regions in semiconductors. It is customary to dope the semiconductor contacts heavily to achieve low-resistance ohmic junction. Rectifying metal semiconductor junctions are often used as high-frequency diodes (Schottky barrier diodes).

174. Heterojunctions. Occasionally, junctions are produced between two kinds of semiconductors, for example, GaAs and Ge or GaAs–GaAlAs (for lasers). Such junctions are called heterojunctions.

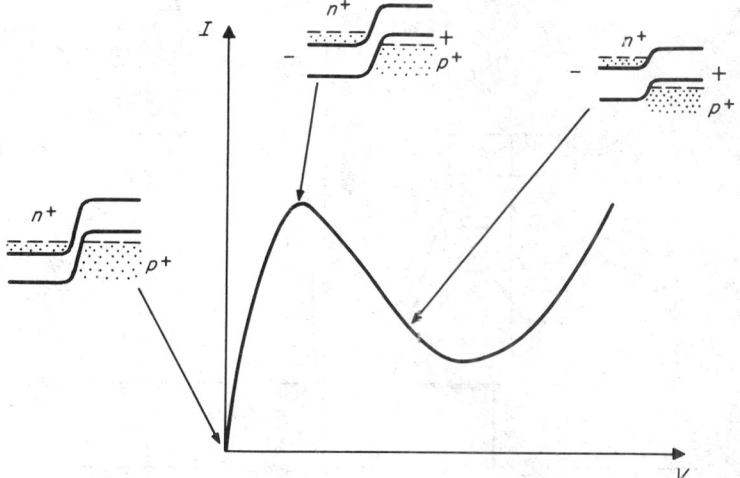

Fig. 6-45. Schematic diagram of p^+n^+ junction current-voltage characteristics and relative energy levels.

SPECIFIC SEMICONDUCTORS

Properties and applications of semiconductors are summarized in Tables 6-45 to 6-47 and Figs. 6-37 to 6-51.

175. Silicon. Silicon is the most common semiconductor material used today. It is used for diodes, transistors, integrated circuits, memories, infrared detection and lenses, light-emitting diodes (LED), photosensors, strain gages, solar cells, charge transfer devices, radiation detectors, and a variety of other devices.

Silicon belongs to group IV in the periodic table. It is a gray brittle material with a diamond cubic structure (lattice parameter = 0.543 nm). The energy gap of silicon is 1.1 eV, this value permitting the operation of Si semiconductor devices at higher temperatures than germanium. Silicon is conventionally doped with P, As, and Sb donors and B, Al, and Ga acceptors. Doping is achieved by thermal diffusion, ion implantation, or transmutation.

Silicon is mainly processed by the planar technology. Silicon (n^+) single crystals are sawed into slices, and then polished mechanically and chemically. An epitaxial n layer is deposited onto the n^+ substrate; then a predeposition of p-type dopant is performed on the base region. Diffusion of the predeposited dopant drives the doping impurities into the silicon. The n-type emitter is next predeposited and diffused, and the desired npn structure is achieved. Reduction of minority-carrier lifetime is obtained by diffusing gold or platinum throughout the silicon slice. Phosphorus or nickel is also generally diffused into the silicon to obtain stable hard junctions. Silicon is also used in polycrystalline form as a thin-film conductor.

The metal oxide semiconductor (MOS) technology[20] is another common technology used for producing field effect transistors (FET) and large-scale memories and circuits (LSI).

TABLE 6-45 Summary of Properties of Lightly Doped Semiconductors[*]

Group	Semiconductor	EG, eV	Mobility, m²/V·s		m_n/m_0	m_p/m_0	a, nm	mp, °C	ϵ_s	Density, g/cm³
			Electron	Hole						
IV	C	5.3	0.1800	0.1600	...	...	0.356	3800	5.8	3.51
	Si	1.1	0.1350	0.0475	0.23	0.12	0.543	1417	11.7	2.33
	Ge	0.7	0.3900	0.1900	0.03	0.08	0.566	937	16.0	5.33
	SiC	2.8	0.0400	0.0050	0.60	1.20	0.436	2830	10.0	3.22
III-V	AlAs	2.2	0.0180	...	...	...	0.566	1600	8.5	3.79
	AlP	3.0	0.0080	...	...	...	0.546	1500	11.6	2.38
	AlSb	1.6	0.0200	0.0420	0.30	0.40	0.614	1050	10.1	4.26
	BN	4.6	...	...	...	...	0.362	3000	7.1	2.20
	BP	6.0	...	0.0300	...	...	0.454	1250	11.6	2.97
	GaAs	1.4	0.8500	0.0400	0.07	0.09	0.565	1237	10.4	5.32
	GaP	2.3	0.0110	0.0075	0.12	0.50	0.545	1465	8.5	4.13
	GaSb	0.7	0.4000	0.1400	0.20	0.39	0.610	712	14.0	5.60
	InAs	0.4	3.3000	0.0460	0.03	0.02	0.606	942	11.7	5.66
	InP	1.3	0.4600	0.0150	0.07	0.69	0.507	1070	10.3	4.78
	InSb	0.2	8.0000	0.0750	0.01	0.18	0.648	525	15.6	5.77
II-VI	CdS	2.6	0.0340	0.0018	0.21	0.80	0.583	1750	5.4	4.84
	CdSe	1.7	0.0600	...	0.13	0.45	0.605	1350	10.0	5.74
	CdTe	1.5	0.0300	0.0065	0.14	0.37	0.648	1098	11.0	5.86
	ZnS	3.6	0.0120	0.0005	0.40	...	0.541	1850	5.2	4.09
	ZnSe	2.7	0.0530	0.0016	0.10	0.60	0.567	1515	8.4	5.26
	ZnTe	2.3	0.0530	0.0900	0.10	0.60	0.609	1238	9.0	5.70
IV-I	PbS	0.4	0.0600	0.0200	0.25	0.25	0.594	1077	17.0	7.50
	PbSe	0.3	0.1400	0.1400	0.33	0.31	0.615	1062	23.6	8.10
	PbTe	0.3	0.6000	0.4000	0.22	0.29	0.646	904	30.0	8.16
II-IV	Mg_2Ge	0.7	0.0530	0.0110	...	...	0.634	1115	...	1.94
	Mg_2Si	0.8	0.0370	0.0065	...	0.46	0.675	1102	...	3.66
	Mg_2Sn	0.4	0.0210	0.0150	...	...	...	778	...	6.21
II-V	Cd_3As_2	0.1	...	1.5000	0.05	...	0.647	721	...	6.92
	CdSb	0.5	0.0300	0.1000	0.16	0.10	...	456	...	5.53
	Zn_3As_2	0.9	...	0.0350	0.15	...	...	1015	...	6.33
	ZnSb	0.5	0.0010	0.0045	...	...	...	546	...	4.75
V-VI	As_2Se_3	1.6	0.0015	0.0080	0.36	...	1.440	608	...	6.00
	As_2Te_3	1.0	0.0170	0.0400	0.32	0.21	1.045	360	...	7.70
	Bi_2Te_3	0.2	1.0000	0.0400	...	...	1.168	580	...	5.81
	Sb_2Se_3	1.2	0.0015	0.0045	...	0.34	...	612	...	6.50
	Sb_2Te_3	0.3	...	0.0270	...	...	...	620	...	...
V-VIII	$PtSb_2$	0.1	0.0200	0.1400	...	1.23	0.643	1240	...	5.67
III-VI	In_2Se_3	1.3	0.0030	...	...	...	0.615	890	...	5.78
	In_2Te_3	1.0	0.0340	...	0.70	...	...	667	...	...

[*]EG = energy gap; m_N/m_0 = effective energy mass; m_p/m_0 = effective hole mass; a = lattice constant; ϵ_s dielectric constant.

TABLE 6-46. Applications of Semiconductors[94]

Effect	Cause	Application	Semiconductors
Transistor effect.....	Current multiplication	Amplifier, etc.	Si, Ge
Tunnel effect	pn junction in degenerate semiconductor	High-frequency switch and storage element, oscillator, amplifier	Si, Ge, GaAs
Avalanche effect	Carrier generation, hot electrons	Cryogenic switches, high-frequency generation, high-frequency amplification	Si, Ge, GaAs
Gunn effect.........	Hot electrons in semiconductors with two different band minima	High-frequency generation, high-frequency amplification	GaAs, InP, CdTe
Piezo effect	Polar cohesion in semiconductor	Electroacoustic amplifier	GaAs, CdS, CdSe
Piezoresistance......	Disformation of band structure by pressure	Pressure indicator	Si, Ge
Varactor effect	Voltage-dependent space charge and capacity at pn junction	Parametric amplification, frequency multiplication, tuning diode	Si, Ge, GaAs
Pair generation......	Carrier generation by light or irradiation at a pn junction	Photocell, solar cell, particle counter	Si, Ge, GaAs, Se
Electroluminescence.	Radiative carrier recombination at pn junction	Light displays, generation of incoherent light by injection	GaAs, GaP, InAs, InSb, SiC
Laser effect.........	Radiative carrier recombination by injection to degeneracy	Laser diode, generation of coherent light by injection	GaAs, InAs, InSb
Galvanomagnetic effect.............	Influence of magnetic field or carrier motion	Hall generator, field plate	Si, InAs, InSb
Plasma waves.......	Interaction between charge carriers and electromagnetic waves in a magnetic field	Gyrator	InSb

TABLE 6-47. Properties of Ge, Si, and GaAs (at 300 K, in Alphabetical Order)[95]

Property	Ge	Si	GaAs
Atoms/cm^3	4.42×10^{22}	5.0×10^{22}	2.21×10^{22}
Atomic weight	72.6	28.08	144.63
Breakdown field, V/cm	$\sim 10^5$	$\sim 3 \times 10^5$	$\sim 4 \times 10^5$
Crystal structure	Diamond	Diamond	Zincblende
Density, g/m^3	5.3267	2.328	5.32
Dielectric constant	16	11.8	10.9
Effective density of states, in conduction band N_c, cm^{-3}	1.04×10^{19}	2.8×10^{19}	4.7×10^{17}
in valence band N_v, cm^{-3}	6.1×10^{18}	1.02×10^{19}	7.0×10^{18}
Effective mass $m*/m_0$, electrons	$m_l^* = 1.6$	$m_l^* = 0.97$	0.068
	$m_t^* = 0.082$	$m_t^* = 0.19$	
Holes	$m_{lh}^* = 0.04$	$m_{lh}^* = 0.16$	0.12,
	$m_{hh}^* = 0.3$	$m_{hh}^* = 0.5$	0.5
Electron affinity x, V	4.0	4.05	4.07
Energy gap, eV, at 300 K	0.303	1.12	1.43
Intrinsic carrier concentration, cm^{-3}	2.5×10^{12}	1.6×10^{10}	1.1×10^7
Lattice constant, nm	0.565748	0.543086	0.56534
Linear coefficient of thermal expansion, $(\Delta l/L)/\Delta T$, K^{-1}	5.8×10^{-6}	2.6×10^{-6}	5.9×10^{-6}
Melting point, °C	937	1420	1238
Minority-carrier lifetime, s	10^{-3}	2.5×10^{-3}	$\sim 10^{-8}$
Mobility (drift), cm^2/V·s, μ_n (electrons)	3.900	1.500	8.500
μ_p (holes)	1.900	600	400
Raman phonon energy, eV	0.037	0.063	0.035
Specific heat, J/g·°C	0.31	0.7	0.35
Thermal conductivity at 300 K, W/cm·K	0.64	1.45	0.46
Thermal diffusivity, cm^2/s	0.36	0.9	0.44
Vapor pressure, torr	10^{-3} at 1270°C 10^{-8} at 800 °C	10^{-3} at 1600°C 10^{-8} at 930°C	1 at 1050°C 100 at 1220°C
Work function, V	4.4	4.8	4.7

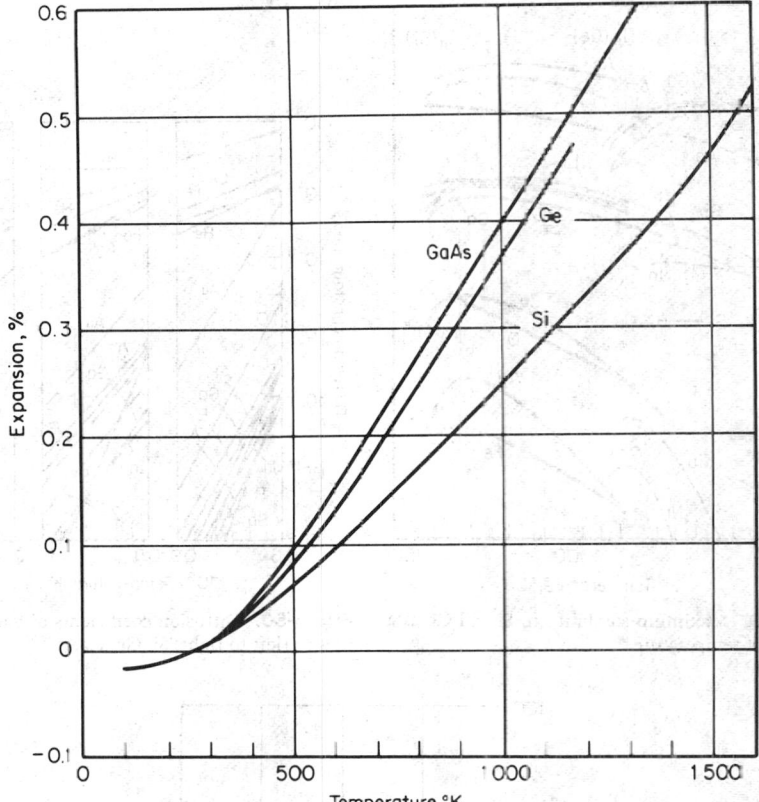

Fig. 6-46. Thermal expansion of Si, Ge, and GaAs.[94]

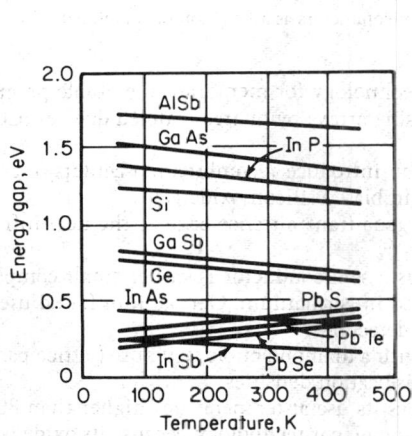

Fig. 6-47. Energy gap of several semiconductors as a function of temperature.[94]

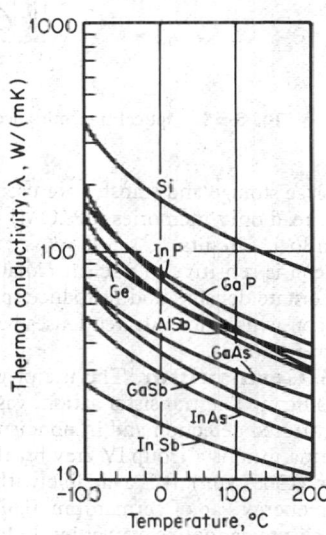

Fig. 6-48. Thermal conductivity of various near-intrinsic semiconductors as a function of temperature.[94]

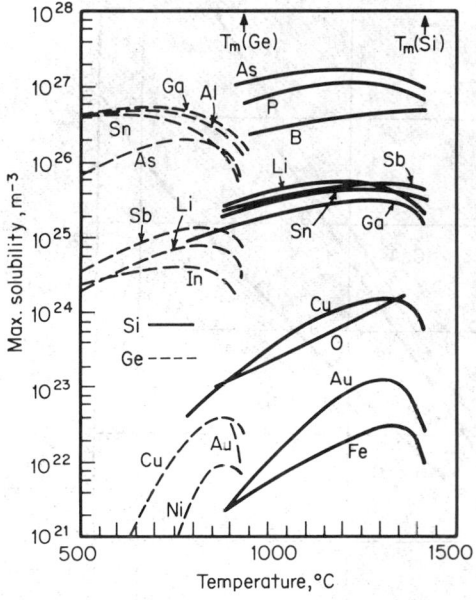

Fig. 6-49. Maximum solubility in Si and Ge as a function of temperature.[94]

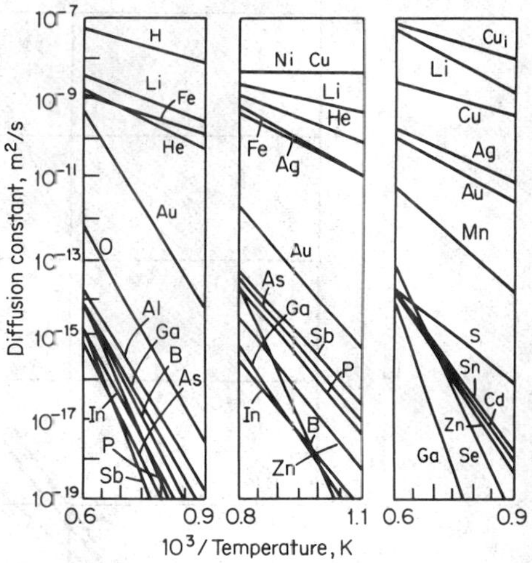

Fig. 6-50. Diffusion coefficients of various impurities in (left to right) Si, Ge, GaAs.[96,97]

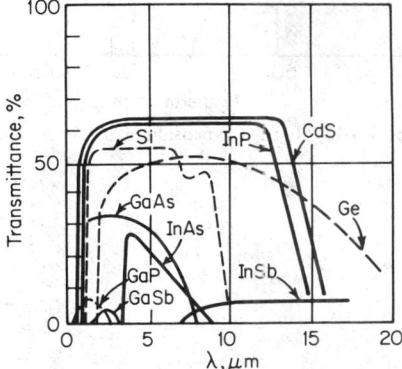

Fig. 6-51. Optical transmittance of several semiconductors as a function of wavelength.[94]

Charge storage and transfer are used in silicon technology for memories in erasable programmable read only memories (EPROM), in photosensing arrays by charge coupled devices (CCD), and in logic circuits.

Silicon is sensitive to nuclear radiation which can introduce recombination centers, increase surface state density, and introduce space changes in biased silicon oxide gates.

Silicon is used for infrared lenses because it has good transmittance edge in the near-infrared region (Fig. 6-51).

176. Germanium. The use of germanium as a semiconductor is rather small compared with silicon. The transistor action was first observed in germanium. Germanium found uses in near-infrared detection and in nondispersive x-ray detectors.

Germanium is a group IV gray brittle material with a diamond cubic structure (lattice parameter = 0.5657 nm). It is available with very low dislocation densities.

The energy gap of germanium, 0.67 eV, precludes its use at temperatures higher than 80°C. Germanium cannot be conveniently fabricated by the planar technology because its oxide is not stable.

Doped germanium crystals, cooled to liquid-helium temperature, serve as infrared detectors. Li-doped crystals are used at this temperature for nondispersive x-ray detection.

177. Diamond and Graphite. Diamond is a transparent extremely hard cubic material (lattice parameter = 0.356 nm). Diamond has a very high band gap ($\sim$ 5.3 eV) at room temperature, making it essentially an insulator. Several p-type diamonds have been found, however, with resistivities of 0.1 to 1 $\Omega \cdot$m. Diamonds have not been used commercially as semiconductors, although they have found some use as heat sinks since they have the highest observed heat conductivity.

Graphite has a hexagonal structure (a = 0.2461 nm, c = 0.6703 nm). It has a high electrical resistivity perpendicular to the basal plane but is metallic along some directions. It can be made an n- or p-type semiconductor by doping.

178. Selenium. Selenium is an element in the VI period group. It can be found in the monoclinic (α), monoclinic (β), and the hexagonal (γ) allotropic forms. Se can also be found in a noncrystalline form. All forms are semiconducting.

Selenium is used as a rectifier material, for photovoltaic cells, and for xerographic printing. As a rectifier it is prepared as a thin crystalline layer of doped Se on steel or nickel electrodes with a Sn or Cd counterelectrode. Se films are used also as photoelectric light meters. For xerography a vitreous Se layer is used on a plate. This layer can be electrostatically charged and then preferentially discharged by selective illumination. The electrostatic image can be transmitted to a paper by dusting the plate with fine powder and transferring the image to the paper.

179. Silicon Carbide. Silicon carbide has a large energy gap ($\sim$ 3.0 eV) and was consequently tried for use as a high-temperature semiconductor. SiC can be prepared by chemical vapor decomposition, from the melt, or by sublimation. SiC is a very hard material which decomposes at 2380°C. It is found in cubic form (a = 0.436 nm) and in a variety of hexagonal polytypes (a = 0.30806 nm, C = multiples of 0.252 nm). The band gap changes slightly with the polytypic form. SiC can be prepared both n-type (normally by nitrogen doping) or p-type (by aluminum doping). Mobilities are low ($\sim$ 0.01 m^2/V·s). The acceptor level is 0.25 eV above the valence band, causing incomplete acceptor ionization at room temperature. Large variations in electrical properties occur in p-type SiC when the temperature is increased due to changes in the degree of acceptor ionization.

180. Gallium Arsenide. Gallium arsenide is an important device material. Gallium arsenide has a large band gap (1.47 eV) at room temperature and high mobility (0.85 m^2/V·s). It is used for Schottky barrier diodes, light-emitting diodes, Gunn diodes, and injection lasers. GaAs is a gray brittle material with the zincblende structure (a = 0.565 nm).

GaAs can be obtained with low dislocation density, although it often contains a large number of other defects. p-type material is normally obtained by Zn doping, while n-type is Te-doped. Deviation from stoichiometry has also the effect of adding donor or acceptor levels.

181. Indium Antimonide. Indium antimonide has a small band gap and a very high electron mobility. It is useful for infrared detectors, infrared filter material, and as magnetoresistance and Hall effect devices.

InSb has the zincblende structure with a lattice parameter of 0.648 nm. InSb becomes body-centered tetragonal above 23 kbar, and its properties become metallic. InSb crystals normally contain precipitates and dislocations.

The band gap is about 0.18 eV at 298 K and 0.23 eV at 80 K. Electron mobility is 7.7 m^2/V·s at 290 K and larger than 50 m^2/V·s at 77 K. Sn(Na), S, Se, and Te have been used as n-type dopants, while Zn, Cd, Mg, Hg, Ag, Au, Al, Ge, and Mn have been used as p-type dopants.

The magnetoresistance effect is quite large: 1 Wb/m^2 can raise the resistance of InSb (5 $\times$ 10^{12} acceptors/m^3) by 1.7%. The optical absorption edge for InSb is about 7 pm at 290 K. Donor doping shifts the absorption edge to lower wavelength.

Tunnel diodes, transistors, and laser diodes have also been made with InSb. Photosensing arrays based on charge injection (C10) have been prepared from InSb.

182. Gallium Phosphide. Gallium phosphide has a zincblende structure with a lattice parameter of 0.545 nm and 2.3-eV band gap. It is used for electroluminescent diodes which can emit either green or red light. Red light is obtained with zinc or cadmium oxide dopants acting as isoelectronic pairs. Green light is emitted when the dopant atoms are separated.

183. Isomorphous Systems. Isomorphous systems are solid solutions between III-V compounds. Examples are G(P, As), used for light-emitting diodes; (In, Ga)Sb, (In, Al)Sb, (Ga, In)As, and (Ga, Al)As, used for continuous-wave injection lasers.

184. Cadmium Compounds. *Cadmium sulfide* is the best-known II-VI compound. Its main use is in photodetectors. Cadmium sulfide is a brittle material which sublimes under atmospheric pressure. It has two allotropic forms: hexagonal (Wurtzite structure) with a = 0.41368 nm and C = 0.67163 nm and cubic (ZnS structure) with a = 0.582 nm. The hexagonal form is the stable one.

The color of the pure material is pale yellow. CdS has an energy gap of 2.4 eV and can be prepared only as an n-type semiconductor. CdSe and CdTe have smaller band gaps. n-type and p-type CdTe can be obtained by Ga and Ag (or Sb) doping.

CdS is a most sensitive photoconductor in the 0.5- to 0.6-μm range, while CdSe is sensitive at 0.7 to 0.75 μm, and CdTe at about 0.85 μm. Ternary HgCdTe is used for infrared detection. A common composition is $(Hg_{0.8}Cd_{0.2})$ Te with a band gap of 0.1 eV at 77 K, sensitive up to about 12 μm.

185. Lead Compounds. Lead sulfide, selenide, and telluride have three possible applications: diodes and transistors at low temperatures, infrared detectors, or thermoelectric applications.

PbTe diodes have been operated at 4 K, where the mobility may be greater than 10^2 $m^2/V \cdot s$. PbS infrared detectors have a long-wavelength threshold of 3 μm with a maximum specific detectivity at 2 μm. PbTe has a threshold at 5 μm; PbSe at about 6 μm. These II-VI compounds are very sensitive to deviations from stoichiometry. For example, sulfur-deficient PbS is p-type, while excess sulfur makes PbS n-type. PbSnTe ternary is another infrared detector material.

187. Metal Oxides. Metal oxides normally have large energy gaps but can be conductive when they deviate from stoichiometry or when suitably doped. Mobilities are very small. Semiconductivity has been found in a variety of oxides: ZnO, MgO, transition metal oxides, and ferrites. Sintered mixtures of transition metal oxides have been used as thermistors (temperature-sensitive resistors).

188. Organic Semiconductors. One of the most studied organic semiconductors is the anthracene C_6H_4: (CH_2): C_6H_2. Semiconducting polymers have also been studied. Organic semiconductors have very low mobilities.

189. Amorphous Semiconductors.* The past few years have seen a great deal of research and development activity in the area of amorphous semiconductors. This has been motivated, in part, by the scientific interest in understanding such structurally disordered solids, which lack the crystalline periodicity underlying much of solid-state theory. However, the decisive factor in the present level of interest in these materials stems, in the traditional way, from technological considerations.

Uses of amorphous solids can be roughly categorized in two classes: applications in which they can function as well as crystalline solids but are much easier to use, and applications in which their potentially unique properties are exploited. The most obvious example of the first class is ordinary window glass, composed of amorphous oxide insulators (largely SiO_2). The existing technologies based on amorphous semiconductors also belong to this category. The largest-scale application is as the light-sensitive element in electrophotography,[37] as currently used in document-copying machines. The semiconducting chalcogenide glasses, Se and As_2Se_3, are employed here as large-area photoconductive films. Chalcogenide glasses (amorphous solids which contain one or more of the group VI elements S, Se, or Te) are also used as infrared-transmitting windows and lenses.

Applications of the second type, while not yet technologically significant, have provided much impetus to recent work. Glasses, while they can normally persist indefinitely in their amorphous state, are thermodynamically metastable. Thus, for example, the methods of preparing such solids typically employ rapid quenching to bypass crystallization: either from the melt, for bulk glasses relatively easy to prepare, such as SiO_2 or As_2S_3; or from the vapor, for amorphous films of more difficult glass formers such as Ge and Si. The possibility of reversible amorphous-crystalline transformation is a unique feature of semiconducting glasses which underlies some promising applications.[38] Such transformations could be induced electrically, optically, or thermally. Switching between a high- and a low-resistance condition in tellurium-based glasses such as $Te_{0.8}Ge_{0.2}$ probably involves such a mechanism, perhaps the phase separation of Te-rich conducting filaments.[39] Optically induced phase transitions may also find applications in image handling or optical mass memories.[40] This area, the investigation of structural transformations (catalyzed by various means) in amorphous semiconductors is a rapidly developing one; the scope of the eventual technological usefulness of such phenomena is very much an open question at this time.

190. Bibliography on Semiconductor Materials
Sze, S. M. "Physics of Semiconductor Devices," Wiley Interscience, New York, 1969.
Azaroff, L. V., and J. J. Brophy "Electronic Processes in Materials," McGraw-Hill, New York, 1963.

*Dr. R. Zallen prepared the material on amorphous semiconductors; his contribution is gratefully acknowledged.

Hannay, N. B. "Semiconductors," Reinhold, New York, 1960.

Wolf, H. "Semiconductors," Wiley Interscience, New York, 1971.

Hogarth, C. A. (ed.) "Materials Used in Semiconductor Devices," Interscience, New York, 1965.

Grove, A. S. "Physics and Technology of Semiconductor Devices," Wiley, New York, 1967.

Madelung, O. "Grundlagen der Halbleiterphysik," Springer, Berlin, 1970.

ELECTRON-EMITTING MATERIALS

191. Work Function and Surface-Potential Barrier. The potential energy of electrons increases when electrons are removed from the material. This increase is due to the work needed to detach the electron from the positively charged surface. The potential energy of the electrons far from the surface is called surface-potential-energy barrier. The difference between the Fermi level (in the material) and the surface potential energy at absolute zero of the temperature is called the work function (see Fig. 6-52a).

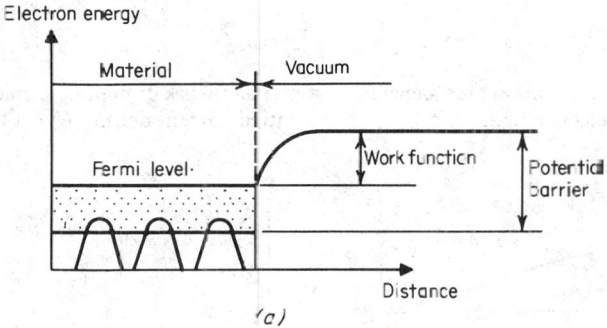

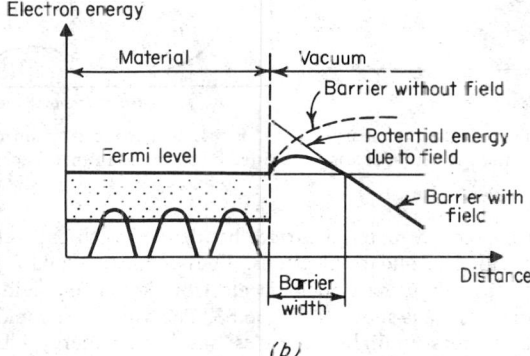

Fig. 6-52. Electron energy near a surface of a material (a) without electric field and (b) with an electric field normal to the surface.

192. Thermal Emission from Metals. Thermal emission of electrons, *thermionic emission*, is an electron current leaving the surface of a material due to the thermal activation. Electrons with sufficient thermal energy to overcome the surface-potential barrier escape from the material's surface. The thermally emitted electron current increases with temperature, since more electrons have the energy necessary to leave the material.

The current density of the thermal emission follows the Richardson-Dushman equation.

$$J_s = AT^2 e^{-e\phi/kT} \tag{6-79}$$

where J_s = emitted current density (A/m²), T = absolute temperature (K), e = electronic charge (C), e_ϕ = work function (eV), A = Richardson's constant (theoretial value = 1.2 MA/m²·K).

The value of ϕ can be written

$$\phi = \phi_0 + \alpha T \qquad \phi_0, \alpha = \text{const} \tag{6-80}$$

Substituting Eq. (6-80) in (6-79) gives

$$J_s = A_0 T^2 e^{-e\phi_0/kT} \tag{6-81}$$

A plot of $\ln(J_s/T^2)$ vs. $1/T$ yields the values of ϕ_0 and $A_0 e^{-e\phi_0/kT}$ by measuring the slope of the curve and its intercept at $1/T = 0$.

193. Schottky Effect. The thermal electron emission can be increased by applying an electric field to the cathode. The electric field lowers the surface-potential barrier, as seen in Fig. 6-52b enabling more electrons to escape the material. This field-assisted emission is called Schottky effect.

194. Field Emission. When large electric fields ($\sim$ 1 GV/m) are applied to the material, the surface-barrier width decreases. When the barrier becomes very thin ($\sim$ 10 nm), electrons can tunnel through it quantum-mechanically. This phenomenon is called field emission. The current density of emitted electrons is almost unaffected by the temperature and is primarily field-dependent. The emitted current density is given by the Fowler-Nordheim equation, which has the form

$$J = C_1 E^2 e^{-C_2/E} \tag{6-82}$$

where C_1, C_2 = constants which depend on material's work function, surface potential, and Fermi level; E = electric field (V/m); and J = emitted current density (A/m^2).

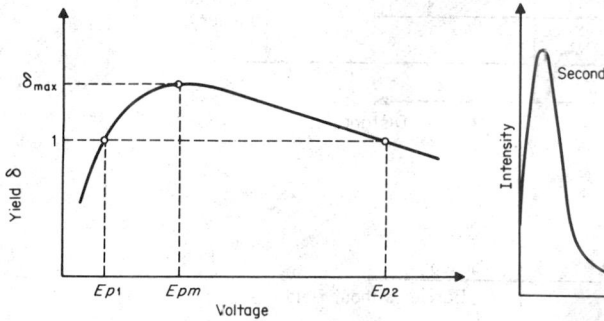

Fig. 6-53. Schematic diagram of the secondary-electron yield as a function of the incident electron voltage.

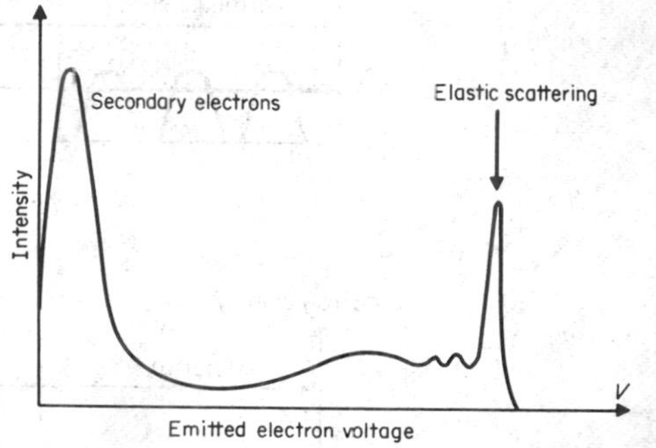

Fig. 6-54. Schematic diagram of the intensity of emitted electrons as a function of voltage.

195. Secondary Emission. A material surface bombarded with energetic electrons will emit low-energy electrons called secondary electrons. The secondary yield δ is defined as the number of secondary electrons per primary incident electron. Secondary field as a function of primay-electron accelerating voltage is shown in Fig. 6-53. The yield rises rapidly, goes through a broad maximum δ_{max}, and drops at higher energies, since high-energy electrons penetrate deeper in the material and the secondaries generated there are unable to reach the material surface with enough energy to be emitted.

196. Energy Distribution of Secondary Electrons. Figure 6-54 shows schematically the energy distribution of emitted electrons. There are two distinct energy regions which are populated: primary electrons with energies close to the primary energy and secondary electrons with energies about 10 eV.

197. Effect of Surface Contamination. Since most secondary electrons have energies larger than the work function, their yield is relatively insensitive to monolayer-type absorbed films. However, films of few to tens of nanometers play an important role, since most electrons are emitted from this depth. Specifically, some oxide films may increase appreciably the electron yields, while absorbed hydrocarbons may reduce yields. This sensitivity to surface cleanliness accounts for some of the discrepancies of reported secondary-emission yields.

198. Effect of Angle of Incidence. The yield of secondary electrons increases with the increase of angle of incidence of the primary electrons. The angular distribution of the secondary electrons is, roughly, a cosine distribution and is hardly affected by the primary incidence angle.

199. Photoemission. Electrons excited by photons can acquire enough energy to surmount the surface-potential barrier. Electron emission due to the illumination of surfaces is called photoemission. The maximum kinetic energy of electrons emitted by photons with frequency v is, according to Einstein's equation,

$$KE_{max} = hv - e\phi \tag{6-83}$$

where h = Planck's constant ($J \cdot s$), v = frequency of impinging radiation (s^{-1}), e = electronic charge (C), $e\phi$ = work function (eV), and KE_{max} = maximum kinetic energy of emitted electrons (J). The work function can be found experimentally by the use of Eq. (6-83).

Photoemission spectra from insulators or semiconductors are more complicated than those of metals. Electrons can be excited from different energy states and emitted. The photocurrent may have a series of maxima as a function of photon energy corresponding to excitation of electrons from various impurity levels. To emit electrons, the photons must first produce energetic electrons; then the electrons have to travel through the material to the surface; and finally the electron has to overcome the surface-potential barrier.

In metals, the efficiency of energetic electron production is low due to the high reflectivity of light, and moreover, the motion of electrons through the solid is very inefficient, due to electron-electron scattering. In semiconductors and insulators the production and motion of energetic electrons are much easier.

Photoemission due to very energetic photons may produce secondary electrons (i.e., production of primary electrons, which in turn excite secondary electrons). Yields exceeding unity can be achieved by such a process. Electrons can be emitted by low-energy photons when two or three photons transmit their energy to a single electron.

200. Photoelectric Efficiency (Yield). The photoelectric yield, or quantum yield, is the number of emitted electrons per incident photon. The photoelectric yield is termed also photoelectric (or quantum) efficiency. It is also customary to use the photocurrent per watt incident radiation in milliamperes per watt or microamperes per lumen, the latter having reference to a tungsten lamp of a given temperature. The microampere per lumen values are useful when comparing photocathodes with similar spectral response and for applications where incandescent lamps are useful as light sources.

SPECIFIC ELECTRON-EMITTING MATERIALS

201. Thermionic Emitters. Thermionic emitters can be classified as pure-metal emitters, monolayer-type emitters, oxide emitters, compound emitters, and alloy emitters.

202. Pure-Metal Emitters. Thermionic emitters are metals with high melting temperatures, low vapor pressure, and preferably low work function. Tungsten and tantalum are the metals most suited for thermal emission. Thermal-emission characteristics of several metals are given in Table 6-48.

TABLE 6-48.   Thermal Emission of Different Cathodes[98]

Cathode form and material	Work function, eV	Constant A, $\mu A/m^2 \cdot K^2$	Cathode form and material	Work function, eV	Constant A, $\mu A/m^2 \cdot K^2$
Metal:			Oxide:		
W	4.5	0.60	BaO,SrO	0.95	10^{-3}–10^{-4}
Zr	4.1	3.3	Thoria	2.5	0.03
Re	4.8	1.0	CsO	0.75	10^{-3}–10^{-4}
Ta	4.3	1.2	Compound:		
Mo	4.2	0.55	TiC	3.35	0.25
Th	3.4	0.6	SiC	3.5	0.64
Ba	2.5	0.6	UC	2.9	0.33
Cs	1.9	1.6	ThC$_2$	3.5	5.5
Film:			LaB6	2.7	0.29
W–Cs	1.5	0.03	ThO$_2$	2.6	0.05
W–Ba	1.6	0.015			
W–Th	2.7	0.04			
Ta–Th	2.5	0.015			
W–O–Ba	1.3	0.015			

203. Monolayer Emitters. Tungsten filaments containing small amounts of thoria have much better emission characteristics than pure tungsten. The work function of thoriated tungsten is reduced from 4.5 eV for pure W to 2.6 eV for the thoriated materials. The improved electron emission was found on addition of other electropositive elements such as Cs, Ba, and La. This lowering is probably due to formation of dipole monolayers of these elements on the tungsten, facilitating the escape of electrons. A monolayer of electronegative oxygen increases the work function considerably.

Additions of 0.5 to 2% thoria to tungsten (with other additions of alumina, SiO_2, Na_2O) improves the emission by several orders of magnitude.[41] Thoria prevents tungsten grain growth,

TABLE 6-49.   Secondary Electron Yields from Insulators, Compounds, and Semiconductors[99]

Group	Substance	δ_{max}	E_{pm}
Semiconductive elements .	Ge (single crystal)	1.2–1.4	400
	Si (single crystal)	1.1	250
	Se (amorphous)	1.3	400
	Se (crystal)	1.35–1.40	400
	C (diamond)	2.8	750
	C (graphite)	1	250
	B	1.2	150
Semiconductive compounds	Cu_2O	1.19–1.25	400
	PbS	1.2	500
	MoS_2	1.10	
	MoO_2	1.09–1.33	
	WS_2	0.96–1.04	
	Ag_2O	0.98–1.18	
	ZnS	1.8	350
Intermetallic compounds .	$SbCs_3$	5–6.4	700
	SbCs		
	Initial	5.7	600
	After 50 h bombardment	4.0	400
	SbCs	1.9	550
	$BiCs_3$	6–7	1,000
	Bi_2Cs	1.9	1,000
	GeCs	7	700
	Rb_3Sb	7.1	450
Insulators..............	LiF (evaporated layer)	5.6	
	NaF (layer)	5.7	
	NaCl (layer)	6–6.8	600
	NaCl (single crystal)	14	1,200
	NaBr (layer)	6.2–6.5	
	NaBr (single crystal)	24	1,800
	NaI (layer)	5.5	
	KCl (layer)	7.5	1,200
	KCl (single crystal)	12	
	KI (layer)	5.5	
	KI (single crystal)	10.5	1,600
	RbCl (layer)	5.8	
	KBr (single crystal)	12–14.7	1,800
	BeO	3.4	2,000
	MgO (layer)	4	400
	MgO (single crystal)	23	1,200
	MgO film on Mg-Ag alloy	12	600
	BaO (layer)	4.8	400
	BaO	3–9	500–700
	BaO-SrO (layer)	5–12	1,400
	Common phosphors:		
	P-1	2.7	750
	P-2	3.4	750
	P-3	3.9	1,000
	P-4	3.7	700
Insulators..............	Al_2O_3 (layer)	1.5–9	350–1,300
	SiO_2	2.4	400
	Mica	2.4	300–384
Glasses................	Technical glasses	2–3	300–420
	Pyrex	2.3	340–400
	Quartz glass	2.9	420

NOTE: For E_{pm}, E_{p1}, E_{p2} see Fig. 6-60.

and therefore improves its high-temperature mechanical properties. Barium monolayers reduce the work function to 1.6 eV. See Table 6-49 for properties of monolayer emitters. Carburization of thoriated tungsten filaments reduces considerably the rate of thorium evaporation and prolongs the filament lifetime. Carburized filaments have peak emission yields of about 30 kA/m².

204. Oxide Emitters. Several types of oxide emitters are found. The dispenser-type cathodes are essentially porous tungsten or molybdenum-tungsten cathodes in which the emitting films are dispensed to the surface of the cathode from the interior. The L *cathode* is a porous tungsten disk behind which barium-strontium carbonate is located.[42] A filament heats the carbonates, which decompose, and the barium and strontium migrate to the outer surface to form the emitting film. The impregnated cathode[43] uses a single pressed disk of barium calcium aluminates and tungsten rather than a tungsten bloc and a separate source of alkaline-earth carbonates. Such cathodes are capable of emitting up to 10^5 A/m² pulsed emission and can be operated up to 1150°C.

Another dispenser cathode is made by pressing Mo-W alkaline-earth compounds.[44]

Dispenser cathodes can be poisoned by metal vapors; therefore care must be taken not to expose the cathodes to metal vapors.

Nickel-base pressed and sintered cathodes are widely used. The emitter is a disk of sintered Ni, ZrH, and Ba-Sr-Ca carbonate. The ZrH is used as a reducing agent. The top surface is again a monolayer of Ba continuously replenished from the sintered body.[45]

Another type of emitter is the oxide-coated cathode[46] made of Ba-Sr carbonate layers on nickel. This layer is heated in a vacuum to form a BaO + SrO oxide layer. Apparently, the emitting surface of such cathodes is also a monolayer of Ba. These cathodes are also formed by triple carbonates containing Ba, Sr, and Ca carbonates. The carbonates are applied in paste form or by transfer tape to the cathode sleeve. Nickel or nickel-based alloys are almost exclusively used as sleeve materials. Properties of oxide emitters are given in Table 6-48.

205. Compound Emitters. Emitters such as uranium carbide and zirconium carbide have been studied. Lanthanum hexaboride has been recently introduced as emitters in several configurations.[47] Emission densities of up to 70 A/m² have been obtained with LaB₆. See Table 6-48 for properties of compound emitters.

206. Alloy Emitters. Emitters have also been produced by using alloys of W, Ta, and Mo with mobilizers such as Ti, Zr, or Hf. The alloys are impregnated with activators such as La, Ce, Gd, Th, or Ca to form a monolayer-type cathode.[48]

207. Secondary Electron Emitters. Secondary emitters can be classified according to the maximum yield δ and the accelerating voltage needed for maximum yield. Table 6-49 lists the maximum yield of various cathode materials. The listing includes metals, alloys, oxides, and glass. Large discrepancies in values of δ_{max} can be found in the literature, mainly due to the variations in the emitting-surface condition.

Materials with high maximum yield are used for electron multiplier surfaces, called *dynodes*. Silver-magnesium alloys are used after oxidation. The surfaces are oxidized to form MgO. High secondary yields can be obtained with these emitters. Pure magnesium oxide has a very high yield (over 20). Nickel-beryllium and nickel-magnesium alloys can also be oxidized to form secondary emitters. Copper-beryllium is used extensively as a dynode material.

Occasionally, the secondary emission has to be suppressed. Materials with low secondary emission such as graphite can be used for this purpose.

208. Photoelectric Emitters. Photoelectric emitters are divided into metallic emitters and various types of semiconductor emitters.

209. Metallic Emitters. Metallic emitters have lost their importance since the advent of high-efficiency insulating and semiconducting emitters.

Photoelectric emission is confined to a layer of about 10 atomic layers at the surface of metals.

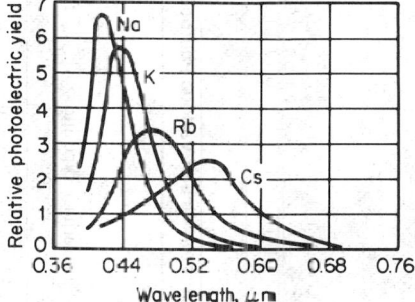

Fig. 6-55. Spectral response of alkali metals.[100]

Metallic surfaces are very sensitive to absorbed surface layers. These layers can form in a few seconds, even at pressures of 10^{-7} torr. The presence of surface layers casts doubt on many of the earlier experimental results. The alkali metals are the only metals with appreciable efficiency in the visible-light region. The spectral response of alkali metals is given in Fig. 6-55.

TABLE 6-50 Characteristics of Various Photocathodes[103]

Number[a]	Photocathode type and envelope	Conversion factor[b] k, lm/W	Typical luminous sensitivity[c] s_typ, μA/lm	Maximum luminous sensitivity[d] s_max, μA/lm	λ_{max}, nm	Typical radiant sensitivity[e] σ_{typ}, mA/W	Typical quantum efficiency[f], %	Typical photocathode dark emission[g] at 25°C, pA/cm²	μA/m²
S-1	Ag-O-Cs, time-glass bulb	93.9	25	60	800	2.35	0.36	900	9,000
S-3	Ag-O-Rb, lime-glass bulb	286	6.5	20	420	1.86	0.55		
S-4	Cs-Sb, time-glass bulb	977	40	110	400	39.1	12		
S-5[h]	Cs-Sb, 9741 glass bulb	1252	40	80	340	501[h]	18[h]	0.2	2
S-8	Cs-Bi, lime-glass bulb	755	3	20	365	2.26	0.77	0.3	3
S-9	Cs-Sb, semitransparent lime-glass bulb	683	30	110	480	20.5	5.3	0.13	13
S-10	Ag-Bi-O-Cs, semitransparent lime-glass bulb	508	40	100	450	20.3	5.6	70	700
S-11	Cs-Sb, semitransparent, lime-glass bulb	804	60	110	440	48.2	14	3	30
S-13	Cs-Sb, semitransparent fused-silica bulb	795	60	80	440	47.7	13	4	40
S-17	Cs-Sb lime-glass bulb, reflecting substrate	664	125	160	490	83	21	1.2	12
S-19[i]	Cs-Sb, fused-silica bulb		40	70		22[i]	11[i]	0.3	3
S-20	Sb-K-Na/Cs, (multialkali), semitransparent, lime-glass bulb	428	150	250	420	64.2	18	0.3	3
S-21	Cs-Sb, semitransparent 9741 glass bulb	779	30	60	440	23.4	6.6		

[a] The S number is the designation of the spectral-response characteristic of the device and includes the transmission of the device envelope.

[b] k is the conversion factor from amperes per lumen to amperes per watt at the wavelength of peak sensitivity.

[c] s is the luminous sensitivity for the photocathode for 2870 color-temperature test lamp. For a multiplier phototube, output sensitivity is μs, where μ is the amplification of the multiplier phototube.

[d] Care must be used in converting s_{max} to a σ_{max} figure. Photocathodes having maximum lumen sensitivity frequently have more red sensitivity than normal, and the formula cannot be applied without reevaluation of the spectral response for the particular maximum-sensitivity device.

[e] σ is the radiant sensitivity at the wavelength of maximum response.

[f] 100% quantum efficiency implies one photoelectron per incident quantum, or $e/h\nu = \lambda/1{,}239.5$, where λ is expressed in nanometers. Quantum efficiency at λ_{max} is computed by comparing the radiant sensitivity at λ_{max} with the 100% quantum-efficiency expression above.

[g] Most of these data are obtained from multiplier phototube characteristics. For tubes capable of operating at very high gain factors, the dark emission at the photocathode is taken as the output dark current divided by the gain (for the equivalent minimum anode-dark-current input multiplied by cathode sensitivity). On tubes where other dc dark-current sources are predominant, the dark noise figure may be used. In this case, if all the noise originates from the photocathode emission, it can be shown that the photocathode dark emission in amperes is approximately $0.4 \times 10^{18} \times$ (equivalent noise input in lumens times cathode sensitivity in amperes per lumen). The data shown are all given per unit area of the photocathode.

[h] The S-5 spectral response is suspected to be in error. The data tabulated conform to the published curve, which is maximum at 340 nm. Present indications are that the peak value should agree with that of the S-4 curve (400 nm). Typical radiant sensitivity and quantum efficiency would then agree with those for the S-4 response.

[i] No value for k or λ_{max} is given because the spectral-response data are in question. The values quoted for σ and typical quantum efficiency are typical only of measurements made at the specific wavelength 283.7 nm and not at the wavelength of peak sensitivity as for the other data.

210. Semiconductor Emitters. Cs₃Sb.Cs₃Sb cathodes have a much higher efficiency than the alkali metals in the visible-light region. These cathodes are prepared by evaporating Sb on glass and exposing it to Cs vapor. The resulting structure is approximately Cs₃Sb. Additional sensitivity can be obtained by partially oxidizing the cathode. Both transparent and opaque cathodes can be fabricated. Transparent cathodes can be further improved by using a manganese oxide substrate. It is not possible to assign an exact spectral response curve for the Cs₃Sb cathodes. Instead, an S number is assigned to various cathodes, and typical response curves have been agreed on to classify them. Table 6-50 describes some of these cathodes. The spectral response of Cs₃Sb can be seen in Fig. 6-56.

Changes in the performance of photoemitters with time are referred to as *fatigue*. Fatigue can be due to excessive exposure to the sun, temperatures above 100°C, chemical changes, nonhermetic sealing, or electron and ion bombardment.

Cathodes of K₃Sb, Rb₃Sb, Na₃Sb. K₃Sb photocathodes are prepared by evaporating Sb on glass and exposing it to glow discharge in oxygen followed by K vapor. Spectral responses of the K₃Sb, Rb₃Sb, and Na₃Sb cathodes can be seen in Fig. 6-56.

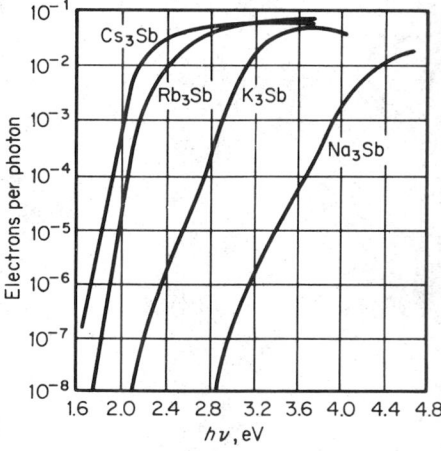

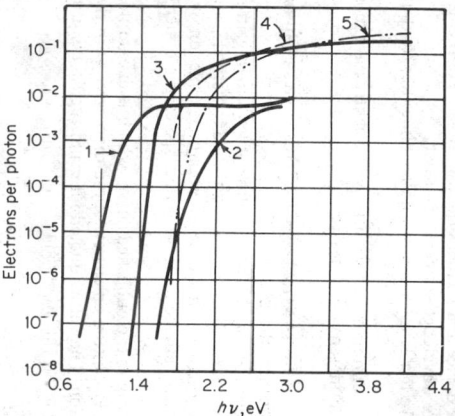

Fig. 6-56. Spectral response of Cs₃Sb, Rb₃Sb, K₃Sb, and Na₃Sb.[101]

Fig. 6-57. Spectral response of (1) AgOCs, normal; (2) AgOCs, silver-deficient; (3) Cs(Na₂KSb); (4) K₂Cs₃b; (5) Na₂KSb.[102]

Multialkali Cathodes. Compounds of several alkali metals and antimonium have very high quantum yields. Na-K-Sb and Cs-Na-K-Sb cathodes (designated S-20) are examples of such cathodes. Na-K-Sb cathodes are formed by exposing K₃Sb films to Na. Sb and K are added to the film to obtain maximum sensitivity. The resulting compound is approximately Na₂KSb. Cs-Na-K-Sb cathodes are even more complicated. Superficial oxidation and manganese oxide substrates have no beneficial effect on these cathodes. Spectral response of Na₂KSb and (Cs)Na₂KSb cathodes is seen in Fig. 6-57.

211. Silver-Oxygen-Cesium Photocathodes. The Ag-O-Cs cathodes are fabricated by first evaporating, or using other deposition methods for formation of a silver base. The silver is then oxidized, normally, by a glow discharge in oxygen. The Cs vapor is then deposited on the oxidized silver. The entire process requires care and skill. Occasionally, additional silvering and superficial oxidation are also performed. The resulting structure of Ag-O-Cs is probably Ag-Cs₂O-Cs. The spectral response of Ag-O-Cs is seen in Fig. 6-57. Another interesting cathode is the Bi-Ag-O-Cs combination, which has a panchromatic response.

212. Ultraviolet-Sensitive Photocathodes. Cs₂Te and Rb₂Te are sensitive at the 0.2- to 0.35-μm-wavelength range. Cs₂Te cathodes are prepared by evaporating Te on a metallic or quartz substrate followed by exposure to Cs vapor at elevated temperature. In the 0.1- to 0.2-μm-wavelength range the alkali halides (with the exception of some fluorides) reach a quantum yield of 0.1 electron per photon. In this range the silver halides and the oxides of the alkaline-earth metals (MgO, CaO, SrO, and BaO) have also high quantum yields. The alkali halides have high quantum yields also below 0.1 μm.

213. Special Cathodes. GaAs crystals with a Cs surface layer have infrared applications.[49] A variety of elemental and compound semiconductors have also been investigated, but none are used as photoemissive cathodes.

214. Bibliography on Electron-Emitting Materials

Kohl, V. H. "Handbook of Materials and Techniques for Vacuum Devices," Reinhold, New York, 1967.

Bruining, H. "Physics and Applications of Secondary Electron Emission," McGraw-Hill, New York, 1954.

Sommer, A. H. "Photoemissive Materials," Wiley, New York, 1968.

Feuerbacher, B., et al. (eds.) "Photoemission and Electronic Properties of Surfaces," Wiley, New York, 1978.

RADIATION-EMITTING MATERIALS

215. Blackbody Radiation. Solids or liquids at finite temperatures radiate a continuous spectrum of electromagnetic energy. This radiation is called *thermal radiation.* The power radiated at different wavelengths is a function of temperature.

A body which completely absorbs all radiation falling on it (and, according to Kirchhoff's law, which also emits all the radiation) is called a *blackbody.* The power distribution of the energy radiated from blackbodies is given by Planck's formula

$$W_\lambda \, d\lambda = \frac{C_1 \lambda^{-5} \, d\lambda}{\exp(C_2/kT) - 1} \qquad (6\text{-}84)$$

where λ = wavelength (m), T = temperature (K), $W_\lambda \, d\lambda$ = energy emitted per unit area between λ and $\lambda + d\lambda$ (W/m²), $C_1 = 3.740 \; 10^{-16} \; \text{W}\cdot\text{m}^2$, and $C_2 = 1.4385 \times 10^{-2} \; \text{m}\cdot\text{K}$.

Figure **11-1** shows the emitted energy from a blackbody at various temperatures. Higher temperatures produce radiation with shorter wavelengths.

216. Relative Luminosity and Luminous Efficiency. The human eye has a brightness sensation which is wavelength-dependent. A standard luminosity curve which gives the relative brightness sensation for a standard observer of monochromatic light of different wavelengths is given in Fig. **20-1**. It should be noted that this curve is not intended to predict retinal response under every lighting condition. At 555 nm, 1 W of radiant flux has a luminous flux of 685 lm.

The luminous efficiency is defined as luminous flux divided by radiant flux. For a blackbody radiator the luminous efficiency can be calculated by dividing the total luminous flux by the total radiant flux. Figure 6-58 shows the luminous efficiency of a blackbody as a function of its temperature. The maximum efficiency is about 94 lm/W. Nonideally, blackbodies have smaller radiant emittance. They are characterized by their emissivity, which is the ratio of their emittance to that of a blackbody.

217. Coherent and Incoherent Radiation. A wavetrain which preserves its wavelength and phase is called *time-coherent.* Two time-coherent sources with identical wavelengths can cause interference since there is a definite relation between their phases. Wavetrains which have no phase relationship are called *incoherent waves.* Incoherent waves do not produce interference phenomena. Sources of incoherent radiation can produce interference phenomena by spatially

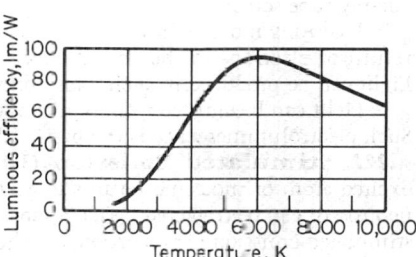

Fig. 6-58. Luminous efficiency of a blackbody as a function of temperature.[104]

separating the radiation to produce two (or more) wavetrains which are space-coherent. Interference occurs because there is a definite phase relationship between the separated beams originating from the same source. Most sources of light are time-incoherent. Coherent radiation persisting over small periods of time can be obtained from lasers.

218. Radiation from Heated Solids. Metal filaments heated below their melting point are used as radiation emitters in incandescent lamps. To obtain maximum luminous efficiency from such lamps, it is best to operate the filaments at temperatures of about 6000 K. Carbon has the highest melting point (3600°C), but it evaporates excessively above 1850°C. Incandescent lamps have been made from tungsten (mp 3380°C), tantalum (mp 2900°C), and osmium (mp 2700°C). Another type of heated-solid-light source is the arc lamp. In open arc lamps the carbon electrode emits radiation due to its high temperature.

219. Radiation from Combustion. Radiation from flames produced by combustion arises mainly by heated fine carbon particles in the flame. Another source of radiation is the combustion of Al or Zr wires in photographic flashbulbs. The light originates from heated oxide particles obtained by the combustion of the metal.

220. Radiation from Excited Gases, Vapors, and Plasmas. An electric field applied to a gas can ionize or excite the gas atoms. The ionization (removal of electrons) or excitation (momentary raising of the energy of an electron) is produced by the impact of energetic electrons. Electromagnetic radiation is produced when the excited electrons return to lower energy levels.

Every gas has a series of possible energy levels which its electrons can occupy. Since there are specific energy changes possible for the electrons in a given gas, there will also be only a number of characteristic wavelengths emitted from the gas.

Discharge lamps use the light emitted from the characteristic lines of noble gases, mercury vapor, or sodium vapor at low pressures. Radiation can also be obtained from discharge in high pressure. The spectrum from high-pressure arcs does not show sharp distinct spectral lines but resembles the continuous blackbody radiation. The luminous part of the arc is confined to a central column of the gas, which may reach 6000 K.

221. Luminescence. Luminescence is the emission of light from a substance, above the thermal emission, due to external excitations. Luminescent materials are often classified according to the rise or decay time of the emitted light. Materials which emit light almost at the instant of excitation and have a rapid decay of emission with the removal of excitation are called *fluorescent* materials. Materials which have an afterglow, i.e., a long decay time for the emitted light, are called *phosphorescent* materials, or *phosphors*.

The emission of light from a substance can be excited by a variety of methods:

Chemical reactions. The slow oxidation of organic or inorganic materials can produce some light *(chemiluminescence)*. Such reactions are responsible for the light emitted by fireflies and glowworms *(bioluminescence)*.

Electromagnetic radiation. Light can be produced by illuminating a substance with visible or ultraviolet light *(photoluminescence)*. Shorter wavelength, such as x-rays or γ radiation, can also produce light emission.

High-energy particles. Ions (protons, alpha particles) can produce visible radiation *(ionoluminescence)*. Radioactive materials can excite light emission from solids by providing alpha particles, which in turn cause the light emission *(radioluminescence)*.

Electrons or electric fields. Electrons moving in a gas can excite its atoms, which emit light upon return to their excited state (glow, or arc, discharge).

Electrons impinging on a substance can produce light *(cathodoluminescence)*, e.g., the cathode-ray-tube screen.

Electrons moving in a solid can excite its electrons, and light emission can occur upon return of electrons to the vacant energy states. Such a process is called *electroluminescence*. Light can be produced in such a fashion when an ac field is applied to phosphorus.

Light can be emitted from semiconductors due to recombination processes in a *pn* junction. Such electroluminescence is the basis of light-emitting diodes.

222. Stimulated Emission (Laser Radiation). Light can be emitted when an excited atom or molecule returns to a low-energy state. This light emission can occur spontaneously or can be triggered by radiation. A triggered emission is called stimulated emission. The stimulated emission is proportional to the incident radiation and is coherent with respect to it.

Normally, materials contain a small number of excited atoms, and consequently, incident radiation causes negligible stimulated emission. In lasers (light amplification by stimulated emission of radiation) the population of excited states is inverted by pumping, i.e., by external excitation, for example by a flash lamp, and sizable stimulated emission can be obtained. Figure 11-18 shows a schematic diagram of a solid-state laser. A flash lamp excites the atoms in the material, and the emitted light from some excited atoms stimulates radiation. The two ends of the laser are silvered so that only stimulated radiation parallel to the rod axis is reflected back and forth within the laser cavity. This increases the radiation field within the cavity, thus further increasing the stimulation emission. When the losses in the cavity are overcome, a coherent parallel, and highly monochromatic beam emerges from the laser.

Laser action can also be obtained in semiconductors. The injection of minority carriers at a *pn* junction can invert the carrier concentration there. Stimulated emission from this inverted population can occur.

Laser operation can be divided into three modes: *continuous wave; pulses* of microseconds to milliseconds, a duration similar to that of the pumping pulses; and *phase-locked pulses* in which the radiation emerges from the laser in short intense pulses down to picoseconds.

SPECIFIC RADIATION-EMITTING MATERIALS

223. Metal Filaments. Incandescent lamps use a resistance-heated filament of tungsten, tantalum, osmium, or platinum, tungsten being the most common filament material.

Tungsten has the highest melting point (3380°C) of all metals. Since incandescent lamps need to be operated at the highest possible temperature, tungsten is the best choice as the filament. A typical spectral emittance from a tungsten incandescent lamp is shown in Fig. 6-59.

Tungsten filaments are produced by first forming bars by powder metallurgy, swaging them to wire form, and finally drawing them to produce fine wires. Coiled filaments are produced by winding the fine wires and then etching the mandrel.

Tungsten filaments evaporate slowly, causing local hot spots at the thinner areas. These hot spots enhance further evaporation, followed by additional thinning, and finally failure. When the evaporated tungsten settles on the glass, it reduces light transmission. Filament evaporation can be reduced by filling the bulbs with argon gas. Normally, some nitrogen is added to the argon to reduce its tendency to ionization. Krypton has been used to increase the lamp efficiency since its heat conduction is smaller than that of argon.

At the high operating temperatures (2500°C) there is an excessive tungsten grain growth and a loss of strength. The grain growth can be reduced by addition of thoria

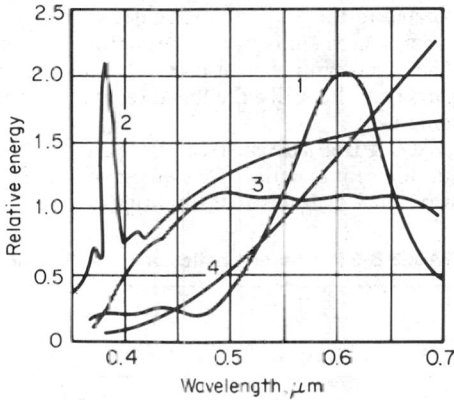

Fig. 6-59. Spectral energy distribution of emitted lights from (1) white fluorescent lamp, (2) carbon arc, (3) sunlight, (4) 50-W incandescent lamp.[105]

(2%), which prevents grain boundary motion. Another approach is provided by additions of silica or alumina to enhance grain growth; a stable grain configuration is produced shortly after the filament is lit the first time.[50]

Failures due to tungsten evaporation have been drastically reduced by halogen-filled lamps. In the quartz iodine lamp the tungsten vapors reach the hot quartz envelope and combine with iodine to form volatile tungsten compounds which are reduced to metallic tungsten on contact with the hot filament. The evaporated tungsten is continuously carried back from the envelope to the filament, reducing considerably the evaporation loss and envelope blackening.

224. Noble Gases. Noble gases are used in discharge lamps. The light is derived from either the typical resonance radiation or the continuum. The noble gases are He, Ne, Ar, Kr, Xe, and Rn.

Xenon. Xenon lamps are used for optical equipment, film projectors, high-power large-area lights, and electronic flashtubes. Xenon spectral distribution is close to sunlight distribution. Xenon emits a sizable power in continuous ultraviolet radiation. In addition, xenon lamps emit both characteristic and continuous radiation in the infrared. Xenon lamps normally have quartz envelopes because of their high operating temperatures. Xenon flashlight spectral distribution is more continuous.

Neon. Neon lamps are used as indicator lamps and advertising signs. The indicator lamps use the negative glow, while the advertising signs use a long positive glow. Neon lamps often contain some argon to permit lower striking voltage.

225. Sodium Vapor. Low-pressure sodium-vapor lamps are used as low-cost sources for street lighting, for instance. The low-pressure lamps are filled with neon (at a few millimeters of mercury) and a small addition of argon and/or xenon. The neon serves both to start the discharge and also to enhance electron collision with sodium atoms. The small addition of argon serves to lower the starting voltage. Low-pressure sodium vapor has discrete resonance lines at 0.5890 and 0.5896 μm. At 270°C the sodium pressure is 3 to 4 mmHg. High-pressure lamps operate at several hundred degrees Celsius. The spectrum of the high-pressure lamps tends to be continuous.

226. Mercury Vapor. Mercury-vapor lamps can be low, medium, high, and extra-high pressure. The low-pressure lamps are low-power, small sources for instrument-panel illumination or ultraviolet excitation of fluorescent materials. Another kind of low-pressure mercury lamp is the familiar fluorescent lamp. It has hot cathodes which are filled with low-pressure argon gas and a drop of mercury. The excitation of the mercury produces the 0.2537-μm resonant radiation in the ultraviolet. This radiation reaches the tube walls, which are phosphor-coated, and excites the phosphor to emit visible light. Medium-pressure mercury lamps ($\sim$ 1 atm) are not used to a large extent. High-pressure (2 to 10 atm) lamps are used for such applications as street illumination. The high-pressure lamps have a very high temperature at the central portion of the discharge. Lamps are double-walled, the inner envelope being quartz and the outer one clear or pearl-finished glass. The outer envelope can be coated with magnesium fluorogermanate phosphor for color correction. Extra-high-pressure (20 atm) mercury lamps are also available.

Mercury-iodide discharge lamps are high-pressure mercury lamps with a metal iodide incorporated in the gas. The iodide decomposes in the high-temperature region, releasing the metal vapor, which emits light. The additional light emission improves the color balance of the lamp. The metal can never reach the lamp walls and damage them, since on approaching the cooler parts near the walls, the metal recombines and forms the iodides, which are not harmful to the lamp walls.

227. Phosphors. (See also Pars. **11-77** and **11-86**.) Phosphors are inorganic or organic luminescent materials. Most phosphors are crystalline inorganic crystals (host crystals) with substitutional or interstitial impurities (activators) which increase the light emission of the host

TABLE 6-51   Characteristics of Some Scintillators in Current Use[105]

Phosphor	Density, g/cm	Max. emission wavelength, nm	Absolute efficiency, %	Decay time, μs
ZnS–Ag	4.1	470	23	10
NaI–Tl	3.7	410	13	0.23
CsI–Tl	4.5	410–580	6–7	1.5
LiI–Eu	4.06	475	4	0.94
High-silica glass and Ce	2.5	390	1	0.05
Lithium silicate glass and Ce	2.5	390	1	0.05
BaPt(CN)$_4 \cdot 4H_2O$	2.1	550	9	1.8
CaWO$_4$	6.1	430	2	3
Anthracene	1.25	445	4	30
Naphthalene	1.15	385	1	0.08
trans-Stilbene	1.16	410	2	0.0064
p-Terphenyl, in toluene	1	320–400	1.6	0.002–0.003
In xylene	1	320–400	1.1	0.003–0.003
POPOP + p-terphenyl, in toluene	1	444	2.2	0.01
In PVT	1	440	2	0.01

crystal (or of another impurity) or introduce new emission lines. Some impurities, such as Ni in ZnS or Ag, are "poisons" and tend to suppress light emission. Phosphors are compounds which have oxygen or fluorine anions (O-dominated), sulfur or selenium (S-dominated) mixed compounds, alkali halide crystals, or organic compounds.

The decay times of light emission of crystals may follow an exponential decay e^{-at}, a power law t^{-n}, or a mixed behavior.

Phosphors Excited by Photons. Long-wave ultraviolet. Little use is made of these phosphors, mainly due to the lack of economical sources in this wavelength region.

Shortwave ultraviolet. These phosphors are used in fluorescent lamps. Fluorescent lamps contain low-pressure mercury vapor which emits 0.2537-μm radiation. This radiation is then exciting visible radiation from a phosphorescent coating inside the lamp envelope. Coatings for fluorescent lamps include $3Ca_3(PO_4)_2 \cdot Ca(F, Cl)_2$:Sbo, Mn with Sb^{3+} at type II Ca^{2+} sites and locally compensated by O^{2-} at the nearest-neighbor halide site and the Mn^{2+} at type I and type II Ca^{2+} sites. Mixtures of three phosphors are used today, e.g., strontium chlorophosphate:Eu^{2+} for blue, willemite for green, and yttrium oxide:Eu^{3+} for red.[107]

X-Rays. For x-ray fluoroscopy: hexagonal, (Zn:Cd)S:Ag, is used at voltages smaller than 100 kV and tetragonal, CaWO$_4$:[W], rhombohedral, BaSO$_4$:Pb, or cubic, BaFCl for higher voltages. Short persistent luminescent crystals are used for scintillation counters, for example, cubic, NaI:Tl, hexagonal, ZnO:[Zn], or rhombohedral, BaSO$_4$:Pb. The microsecond delay time of such crystals allows high counting rates (see Table 6-51).

Phosphors Excited by Electrons. These phosphors are used for cathode-ray tubes, television screens, and electron microscopes. Table **11-13** lists the common phosphors. Short-persistent phosphors are P15, P5, P11, P3, and P1. Long-persistent phosphors are P12, P2, P14, and P7. (See Fig. 6-60.)

Phosphors Excited by Ions and Neutrons. The main use of these phosphors is in luminescent coating for dials. $RaBr_2$ or $RaSO_4$ is mixed with hexagonal ZnS:Cu. The decaying radium excites the ZnS to emit visible radiation. Neutrons can excite phosphors indirectly by mixing the phosphors with materials that emit charged particles when irradiated with neutrons. The neutrons cause charged-particle emission, which in turn excite the phosphors to emit light.

Phosphors Excited by Electric Fields. ZnS phosphors are mainly used for electroluminescent devices which operate by application of an ac field on the powder. Electroluminescent lamps are prepared by sandwiching the phosphor between two conducting layers which serve as the electrodes. The phosphor can be embedded in an organic material or in an enamel ("ceramic" lamps). The top electrodes can be conducting glass or plastic. The construction employs a bottom metal plate, a reflective conducting enamel followed by the phosphor, top electrode, and a transparent cover. A list of electroluminescent materials is given in Table 6-51.

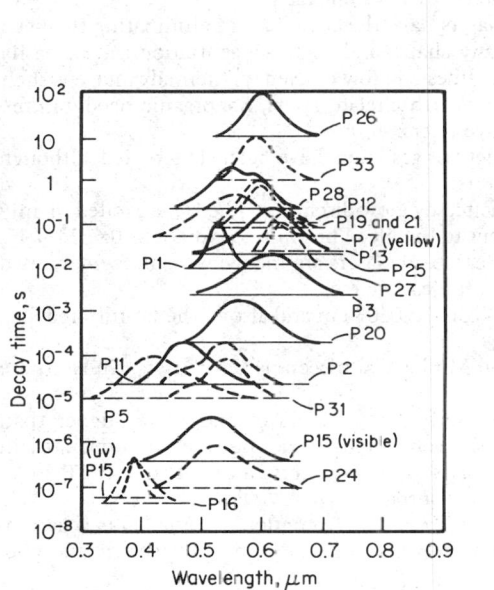

Fig. 6-60. Spectral emission curve of some commercial cathodoluminescent phosphors.[106]

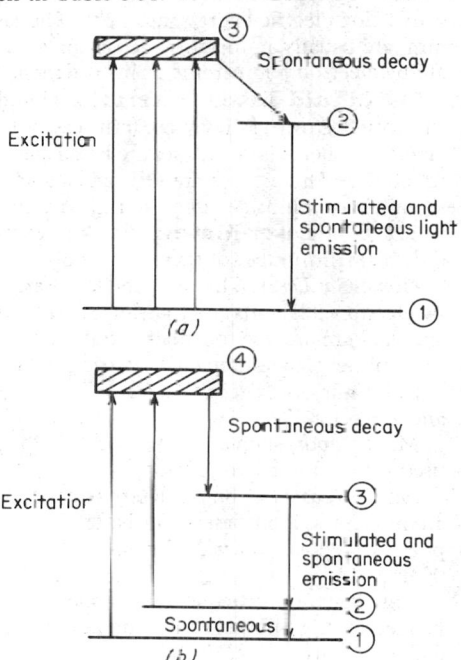

Fig. 6-61. Schematic energy-level diagram for (a) three-level and (b) four-level lasers.

Another type of electroluminescent lamp is the forward biased *pn semiconductor junction.* The GaAs, GaP light-emitting diodes operate on this principle. Radiation is emitted due to recombination of charge carriers in the semiconductor junction.

Avalanche in *reverse-biased pn junctions* causes hole-pair generation followed by light emission. Again GaAs is an example for this type of electroluminescence.

Light emission can also occur due to electrons supplied by *tunneling* into the phosphor.

228. Laser Materials. Lasers can be classified according to the type of lasing material: solid lasers (actually, insulating solids incorporating ions), semiconductor lasers, liquid lasers, and gas lasers.

Solid lasers use solid materials which contain impurity ions that can be excited to high levels. Figure 6-61a shows a simplified energy diagram for ruby (Cr-doped Al_2O_3 crystals). This diagram is typical of three-level solid lasers. The Cr ions are excited to high energy level 3, and then return to level 2, which must be overpopulated with respect to level 1. The spontaneous radiation emitted by the return from level 2 to 1 stimulates more transfers from level 2 to 1 and further radiation. This situation is termed population inversion and is mandatory for lasing action.

Population inversion is even easier to achieve in a four-level material (Fig. 6-61b). Most solid lasers have this type of energy-level diagram. These lasers contain transition metals, rare-earth metals, or actinide ions in glass, garnets, tungstate, or fluorides. Crystalline imperfections of laser crystals have an effect on the minimum power (threshold) needed to operate the laser because the scattering from imperfections must be overcome to obtain amplification. Lasers are also difficult to operate at high temperatures since it is harder to obtain population inversion. Table **11-2** lists some of the solid-laser materials.

Ruby. The laser crystal is Al_2O_3 with addition of 0.05 wt % Cr_2O_3. The Cr addition is responsible for the pink color of the crystals. Ruby is a common solid-laser material. These lasers give powerful pulses which can be used for material processing, triggering chemical reactions, holography, and other applications.

Neodymium Crystal. 0.5 to 2% Nd ions in crystals such as $CaWO_4$, $SrWO_4$, $SrMoO_4$, $Ca(NbO_3)$, and $Y_3Al_5O_{12}(YAG)$ have been used for emission in the near-infrared region. Out of the above-mentioned materials, the YAG is a very useful material because it has better optical and mechanical properties. YAG lasers can be operated either in continuous or pulsed modes. They are useful for machining, communication, and ranging.

229. Semiconductor-Laser Materials. Semiconductor lasers are efficient devices for converting electric energy into light. The lasers, generally of linear dimensions smaller than 1 mm, are usually pumped by carrier injection at a *pn* junction. Other possible pumping modes are by electron bombardment, by avalanche breakdown, or optically.

230. Liquid-Laser Materials. Liquid lasers have the advantage of eliminating the need for crystal growing. They conform easily to any shape, and their concentration can be easily changed. Their disadvantages are broad spectral lines and low efficiency; their advantage is their tunability. Three types are distinguishable: rare-earth chelate lasers, nonorganic neodymium–selenium oxychloride lasers, and the organic-dye lasers.

231. Gas-Laser Materials. A large variety of gas lasers have been constructed, although only a small number of these can be discussed here.

Noble-Gas Lasers. Helium-neon lasers are excited by glow discharge. The He is excited mainly by electron collisions and transfers its excitations to the Ne. The emitted radiations 0.6328, 1.15, and 3.39 μm are due to radiative transfers in Ne. He–Ne lasers are inexpensive and widely used in alignment, measurements, holography, communication, etc.

Helium lasers emit in the near infrared 1.9543 and 2.0603 μm and also in the far infrared 95.8 and 216 μm.

Metal vapors such as Cd, Hg, Cs, Cu, Pb, and Mn have also been used for lasers. Cu lasers are used for pumping dye lasers.

Ion-Gas Lasers. Ion-gas lasers operate at powers several orders of magnitude higher than atomic lasers. Ionic lasers can be operated in the continuous mode, but are often operated in pulses to reduce power dissipation in the laser parts. Ionic-gas lasers can be used at the 0.26- to 0.7-μm-wavelength range with considerable power output.

Ionic lasers most frequently use ions of Ne, Ar, Kr, or Xe. Excitation of ionic lasers is accomplished by high-power gas discharges. Ion lasers are very convenient in photoelectric and photochemical work.

Noble-gas ions are chemically very reactive, similar to the halogens; therefore special precaution is necessary to protect the inner parts of such lasers. The noble-gas ion lasers are the best-known ion lasers. However, a large number of other ion lasers have been reported, among them Hg, halogens, O, N, and vaporized elements such as B, C, Si, Mn, Cu, Zn, Ge, As, Cd, In, Sn, and Pb.

Molecular Lasers. The best-known laser of this group is the CO_2 laser. This laser contains CO_2, N_2, and a large proportion of He, and sometimes also H_2O. The N_2 molecules are excited by electrons and transfer their excitation to the CO_2. The He improves the discharge characteristics of the gas mixture and helps relaxation of the CO_2 molecules to the ground state. H_2O gas helps to remove the energy of the CO_2 molecules in the terminal laser levels. Lasers operating at tens of kilowatts are available. These lasers can have very high output power and can be operated continuously or in a pulsed mode. These lasers have applications where a high-power output is required.

Another version of the pulsed CO_2 laser, the so-called TEA (transversely excited at atmospheric pressures), is capable of providing very high output pulses of the order of hundreds of joules.

The most powerful CO_2 laser is the gas-dynamic laser. The population inversion is obtained in this laser by letting a mixture of gases, including CO_2, expand from a region of high temperature and high pressure into a low-temperature and low-pressure region.

232. Chemical Lasers. In chemical lasers the population inversion is brought about by chemical means. An example is the population inversion created by mixing hydrogen and chlorine to produce excited HCl. Other types are photodissociated CS_2–O_2 mixtures.

233. Bibliography on Radiation-Emitting Materials

Leverenz, H. W.   "Luminescence of Solids," Dover, New York, 1968.
Ivey, H. F.   Electroluminescence and Related Effects, *Adv Electron. Electron Opt*, supplements, 1963.
Goldberg, P. (ed)   "Luminescence of Inorganic Solids," Academic, New York, 1966.
Hewitt, H., and A. S. Vause   "Lamps and Lighting," Arnold, London, 1966.
Lengyel, B. A.   "Lasers," Wiley-Interscience, New York, 1971.

OPTICAL AND PHOTOSENSITIVE MATERIALS

234. Index of Refraction. The index of refraction of a material is the ratio of the light velocity in empty space to the velocity in the material

$$n = c/v \tag{6-85}$$

where c = velocity of light in empty space (m/s), v = velocity of light in the material (m/s), and n = index of refraction. The index of refraction for nonmagnetic materials is

$$n = \epsilon_r^{1/2} \tag{6-86}$$

where ϵ_r = dielectric constant. The index of refraction is a function of the wavelength. This effect is called *dispersion*. Figure 6-62 shows the dispersion in several glasses.

235. Dispersive Power. The dispersive power ω is defined by the indexes of refraction in the blue (n_F), yellow (n_D), and red (n_C) portions of the spectrum,

$$\omega = (n_F - n_C)/(n_D - 1) \tag{6-87}$$

The dispersive power is a measure of the changes of the index of refraction of the material over the whole visible range, relative to the mean deviation of this index from unity.

236. Birefringence. Optically anisotropic crystals having two indexes of refraction are called birefringent crystals or double refracting crystals.

237. Dichroic Crystals. Doubly refracting materials with high absorption for one polarized component are called dichroic. Dichroic crystals can serve as light polarizers since they can be made to transmit only one polarization direction.

238. Refraction and Reflection. When light waves arrive at the boundary between two media, part enter the media, changing direction, while part are reflected. The angles of incidence and refracted rays are related.

$$n_1 \sin \phi_1 = n \sin \phi_2 \qquad \phi_1 = \phi_3 \tag{6-88}$$

where n_1, n_2 = indexes of refraction, and ϕ_1, ϕ_2, ϕ_3 = angles of incidence, refraction, and reflection, respectively, measured from the normal to the surface

239. Transmittance and Reflectance. For a monochromatic light of wavelength λ the transmittance T_λ of a material is defined as

$$T_\lambda = P_T/P_0 \tag{6-89}$$

where P_T = radiant flux transmitted through the material and P_0 = radiant flux of the incident beam.

Reflectance of a monochromatic light of wavelength λ, R_λ, from a material is defined as

$$R_\lambda = P_R/P_0 \tag{6-90}$$

where P_R = reflected radiant flux and P_0 = radiant flux reflected from 100% reflecting (completely diffusing) sample.

Transmittance and reflectance can be defined also with respect to sensors for nonmonochromatic light. The transmittance of light with respect to a sensor with a given sensitivity is

$$T = \frac{\int P_T s \, d\lambda}{\int P_0 s \, d\lambda} \tag{6-91}$$

where T = transmittance and s = sensitivity of the sensor. In effect, the transmittance is the total transmitted luminous flux divided by the total incident luminous flux.

The reflectance, similarly, is

$$R = \frac{\int P_R s \, d\lambda}{\int P_0 s \, d\lambda} \tag{6-92}$$

where R = reflectance.

240. Transmission and Reflection Density. The transmission density D_T is defined as

$$D_T = \log(1/T) = -\log T \qquad (6\text{-}93)$$

The reflection density D_R is defined as

$$D_R = \log(1/R) = -\log R \qquad (6\text{-}94)$$

where T and R are the transmittance and reflectance, respectively.

241. Exposure. Exposure is defined as the amount of energy incident on a unit area of photographic emulsion.

242. Characteristic Curve of Emulsions. The density of the emulsion as a function of exposure is shown in Fig. 6-63. This curve is called the *H-D curve* (Hurter and Driffield). The slope of this curve is called *gamma*. Emulsion speed can be defined by this curve. Exposures beyond the maximum of the curve begin to decrease the film density. This phenomenon is called *solarization* or *reversal*.

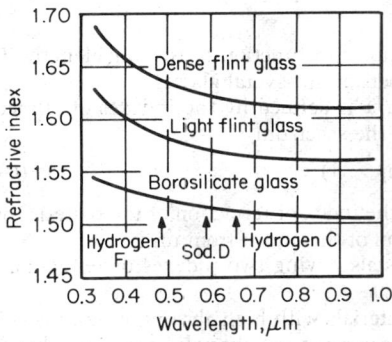

Fig. 6-62. Refractive index as a function of wavelength for various glasses.[72]

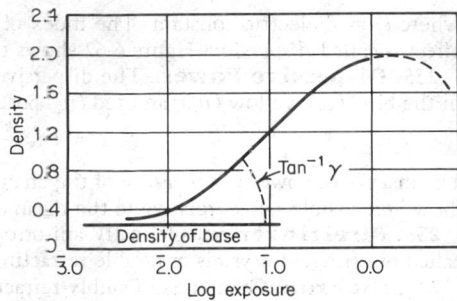

Fig. 6-63. Schematic *H-D* curve for a photographic emulsion.

243. Spectral Sensitivity of Photographic Emulsions. The sensitivity s of a photographic emulsion at a given wavelength λ is

$$s = (1/U_\lambda) \times \text{cons} \qquad (6\text{-}95)$$

where U_λ = incident energy per unit emulsion area necessary to produce a given density D.

244. Image Formation on Photographic Emulsions. Photographic emulsions are in most cases minute silver halide crystals dispersed in a gelatine carrier. The gelatine carrier separates the silver halide crystals and may contain additional materials such as sensitizers, desensitizers, and fogging or antifogging materials.

Although the theory of image formation is not yet completely founded, it is generally accepted that the incident light releases electrons from the halide ion and these electrons are captured by *sensitivity specks*. Interstitial silver ions can move toward the negatively charged specks and become neutralized. Sufficient neutral silver atoms render a grain developable. These minute chemical changes of the emulsion on exposure are called *latent image*. The latent image has not yet been detected directly but can be made visible by the development process. The development process causes chemical decomposition almost exclusively in the grains in which silver atoms collected at sensitivity specks. The development process multiplies the chemical decomposition by a factor as high as 10^8.

245. Development and Fixing. During development, the developing agent differentially reduces the light-exposed silver halide grains to metallic silver. This development process is called *chemical developing*. Long development time will affect all the crystals in the emulsion. Most developers contain reducing agents such as hydroquinone or p-methylaminophenol sulfate (for black and white) and pyrogallol or derivatives of p-phenylenediamine (for color photography).

In addition, the developer contains sulfite, which serves as a preservative against oxidation and also a solvent for silver halide. Bromine is frequently added to the developer to promote adsorption onto silver halide grain surface. This retards the development of very weakly or unexposed grains, i.e., prevents fog. An alkali is added to the developer to control the proper pH of the solution. Other additions such as wetting agents are also common.

The fixing process dissolves the undeveloped silver halide grains. The photographic picture is then composed of the remaining dark silver particles.

246. Reversal Process. In this process the emulsion is first developed and then the silver is dissolved by a strong oxidizing agent. The silver remains only in the unexposed areas of the film. The film is now exposed and developed, and the result is a direct positive. Reversal process is used for 16-mm amateur motion picture films.

SPECIFIC OPTICAL AND PHOTOSENSITIVE MATERIALS

247. Optical Glasses. Glasses for optical instruments have to fulfill stringent requirements, those being that the glass have a specified refractive index and dispersion; it must have minimal variations in its optical properties; and it must be strain-free to minimize birefringence and have high dimensional stability and a very smooth surface finish.

Optical glasses are discussed in great detail by Morley.[51]

*Fiber Optics.** Transmittal of light through long optical fibers is rapidly gaining importance in communications. The optical fibers must have very low losses and ideally should transmit the light coherently. There are two types of optical fibers.

Graded Index. The fiber is made of many layers of different refractive indexes with the highest in the center. The fibers are prepared by first chemically vapor depositing silicon and germanium containing glasses of increasing refractive index inside a quartz tube. Then the tube is made to collapse and then pulled to form the fibers. The resulting fiber has a parabolic refractive index.

Step Index. The fiber is made of ultrapure quartz cone cladded with a polymer of a lower refractive index or with a halide-doped low-refractive-index glass.

248. Filters. Glasses can be obtained in different colors to serve as optical filters. The various colors result by compound additions:

    Green: chromium oxide, iron oxide, vanadium oxide
    Blue: cobalt oxide, copper oxide
    Purple: manganese oxide
    Amber: carbon or sulfur
    Red: selenium and cadmium sulfide, cuprous oxide, elemental gold
    Opal glass: fluorine compounds (cryolide)

Filter glass is normally not of optical grade. A large selection of filters can be obtained with various transmittance spectra.

Filtering can also be obtained by multiple dielectric coating on glass. The multiple coating removes certain light wavelengths by interference.

249. Mirrors. Reflecting of glass surfaces can be significantly improved by dense metallic coatings. Selective reflectance can be achieved by incorporation of gold or nickel in glass. For example, visible light can be transmitted; infrared radiation is reflected by certain mirrors. Interference coatings can also serve as selective reflectors.

250. Light-Sensitive Glasses (UV, Visible, IR).[52-54] Light-sensitive glasses can be divided into the following two types.

Photosensitive Glasses. The color of these glasses can be permanently affected by ultraviolet (UV) radiation. Such glasses are commonly lithium silicates. The glasses contain metallic ions such as silver and gold and small additions of cerium oxide and common reducing agents such as tin oxide.

The UV light is applied at room temperature, and the color changes are observable only after a heat treatment above the glass-annealing range. The color is produced by elemental silver or gold precipitation and growth.

Photosensitive glasses can be made into photomachinable glasses by slight variations in composition. In these glasses lithium metasilicate is nucleated by the silver or gold particles, and crystals can be grown by heat treatment. The lithium metasilicate phase is etched considerably faster in diluted hydrofluoric acid than the matrix. These chemically machined glasses are used, for example, for fluidic controlled systems and engravings.

*The author wishes to thank Dr. T. M. Weinberg of Fibronics for his contribution to this section.

The photomachined glasses may be further heated at slightly higher temperatures to produce lithium disilicate crystals. These crystals form an interlocking network of considerable mechanical strength similar to that of glass-ceramic bodies.

Photochromic Glasses[53,54] These normally contain silver halides and change their transmittance reversibly with incident-light intensity. Their transmittance decreases upon strong illumination and is regained when the illumination is decreased. Such glass is used for windows, sunglasses, and optical memories.

251. Polarizing Materials. Tourmaline is an example of a dichroic crystal which has a pale yellow color for one polarization direction and becomes almost entirely opaque to the other direction.

Polarizing filters, Polaroid, have been developed by E. H. Land.[55] Originally, these filters contained aligned crystallites of herapathite. Polaroid filters have later been made with long-chain polymers such as polyvinylene or polyvinyl alcohol treated with iodine. The polymer chains are aligned to produce the necessary optical anisotropy.

252. Monochrome Film Emulsions. A variety of monochrome photographic emulsions are available. Emulsion speeds vary over three orders of magnitudes. The resolution varies from less than 50 to 2,000 lines/mm. The contrast obtained in films also varies, and the appropriate emulsion must be chosen for each application.

The granularity changes from coarse to extremely fine. Emulsions can also be made with different spectral sensitivity in the visible ultraviolet and infrared. Emulsions for nuclear, x-ray, and electron microscope work are also available.

The spectral sensitivities of unsensitized emulsions show some differences in response. For example, silver chloride is UV-sensitive, while silver bromide is blue-sensitive. The differences in these curves are due to differences in spectral absorption of those materials; i.e., yellow silver bromide absorbs blue light and therefore is sensitive to blue, while silver chloride absorbs only in the UV range, etc.

The emulsions can be either chemical-sensitized with sulfur-containing organic compounds or optical-sensitized with dyes such as cyanines. The sensitizers increase the range of emulsion sensitivity. Supersensitizing of emulsion is obtained by addition of several dyes. Desensitizing of an emulsion can be used to decrease the sensitivity of a film to some desired wavelengths (such as yellow, to allow film developing in bright illumination). For spectral sensitivity of film types see Ref. 56.

Orthochromatic emulsions are green-sensitive, panchromatic emulsions are sensitive to the entire visible range, and nonsensitized emulsions are blue-sensitive.

253. Photoresists. Photoresists are organic compounds whose structure can be changed upon exposure in the UV region. Photoresists can be preferentially dissolved after the exposure. Photoresists remaining in the areas which received the illumination are called *negative resists*. Resists remaining in the unilluminated areas are *positive resists*.

Negative resists contain polymers, sensitizers, and solvents. Such resists are kept from polymerizing. The light promotes the polymerization with the aid of the sensitizer. The polymerized photoresist remains insoluble during the development. Positive resists, when illuminated, render the illuminated area soluble.

Table 6-52 lists some of the common photoresists.

Photoresists have found use in chemical-machining metal foils, printed-circuit boards, and very extensively in microelectronics.

254. Other Recording Media. In addition to photoresists and photographic films, there are a variety of image-recording techniques. Electrophotographic recording includes techniques such as xerography, or elastrostatically produced deformations in plastic films. *Dielectric recording* (electret) is another recording technique in which the image is transferred to paper as in xerography. Other techniques use metallic oxide reduction, electrolytic recording, diazo-recording photochromism, and photochemical recording. A review of recording techniques is found in Refs. 57 to 59, Par. **6-257**.

255. Color-Film Emulsions. Color-film emulsions are usually of the multilayer substractive type. The color films are prepared from several layers of emulsion. For example, Kodachrome has a top layer of blue-sensitive emulsion followed by a yellow filter. Underneath is an orthochromatic (mainly green-sensitive), and the bottom is a panchromatic, mainly red-sensitive, emulsion. The processing of such an emulsion is quite involved. It consists in developing, exposing to red light from the back side, then dye-coupler developing (i.e., developing which includes dyeing during the process). More development and dye coupling are necessary before the final image is obtained.

TABLE 6-52   Photoresist Materials[64]

Supplier and name	Type*	Comment
Du Pont:		
Riston	N	Supplied as a film for plating or etching
Kodak:		
KPR	N	For copper and copper-base alloys
KPR3	N	For dip coating, electroplating
KPR4	N	For roller coating
KOR	N	Same as KPR with greater sensitivity
KMER	N	For most metallic surfaces
KFTR	N	For microelectronic applications
Microneg. 747, 752	N	High purity, electronic applications
Shipley:		
AZ1300 series 1350, 1350J, 1370, 1375, 1350B, 1360J	P	Various thicknesses (1375 high step coverage, 1350B for photomask fabrication)
AZ2400 series 2400, 2415, 2430	P	For microelectronic applications (2430 high step coverage, 2415 photomask fabrication)
AZ1400 series	P	Striation-free
AZ111 series	P	General semiconductor fabrication
Philip A. Hunt:		
Waycoat MPR	P	For photomask fabrication
JC, SC	N	For semiconductor industry
450	N	Chemical milling
204/206	P	Striation-free
HNR	N	High-resolution semiconductor work
Mead:		
PBS, PMMA	P	Electron resists
COP	N	Electron resists

*N = negative; P = positive.

Ansco, Agfacolor, and Ektachrome are emulsions that can be developed in a simpler manner because the dye couplers are contained in the emulsion. Other emulsion types are Kodacolor, Ektacolor, and a Du Pont process.

256. Bibliography on Optical and Photosensitive Materials

Mees, C. E. "The Theory of Photographic Process," Macmillan, New York, 1942.

Evans, R. M., W. T. Hanson, and W. L. Brewer "Principles of Color Photography," Wiley, New York, 1953.

Deforest, W. S. "Photoresist," McGraw-Hill, 1975.

257. References

1. Matthiessen, A., and C. Vogt Proc. R. Soc., 1863, Vol. 12, p. 652; Ann Phys. Chem., 1864, Vol. 122, p. 19.

2. Thompson, M. W. "Defects and Radiation Damage in Metals," Cambridge University Press, New York, 1969, p. 27.

3. Clarebrough, L. M., M. E. Hargreaves, and M. H. Loretto Phil. Mag., 1962, Vol 1, p. 115.

4. Panseri, C., and T. Federighi Phil. Mag., 1958, Vol. 3, p. 1223.

5. Yoshida, S., T. Kino, M. Kiritani, S. Kabemoto, H. Maeta, and Y. Shimomura J. Phys. Soc., 1963, Vol. 18, Suppl. 2, p. 98.

6. Cotterill, R. M. J. Phil. Mag., 1963, Vol. 8, p. 1937.

7. Washburn, E. W. (ed.) "International Critical Tables of Numerical Data," Vol. 6, McGraw-Hill, New York, 1929, p.135.

8. "Landolt-Bornstein," Zahlenwerte und Funktionen aus Naturwissenschaften und Technik, Vol. 6, Springer, Berlin, 1959.

9. Sondheimer, E. H. The Mean Free Path of Electrons in Metals, Adv. Phys., 1952, Vol. 1, p. 1.

10. Campbell, D. S. "The Use of Thin Films in Physical Investigations," Academic, New York, 1966, p. 299.

11. Neugebauer, C. A. in B. Schwartz and N. Schwartz (eds.), "Measurement Techniques for Thin Films," Electrochemical Society, New York, 1967, pp. 191–220.

12. Huntington, H. B., and A. R. Grone J. Phys. Chem. Solids. 1961, Vol. 20, p. 76.

13. Rosenberg, R., and L. Berenbaum   *Appl. Phys. Lett*, 1968, Vol. 12, p. 201.
14. Blech, I. A., and E. S. Meieran   *J. Appl. Phys.*, 1968, Vol. 40, p. 485.
15. Lepselter, M. P.   *Bell Syst. Tech. J.*. 1966, Vol. 45, p. 233.
16. Wilson, R. W., and L. E. Terry   *Proc. IEEE*, 1969, Vol. 57, p. 1580.
17. Kohman, G. T., et al.   *Bell Syst. Tech. J.*, 1955, Vol. 34, p. 115.
18. Cunningham, J. A., C. R. Fuller, and T. Haywood   *IEEE Trans.* rel.-R-19, 1970, p. 182.
19. Ruck, G. T., D. E. Barrick, W. D. Stuart, and C. K. Krichbaum   "Radar Cross-section Handbook," Plenum, New York, 1970.
20. Grove, A. S.   "Physics and Technology of Semiconductor Devices," Wiley, New York, 1967, pp. 346–350.
21. Whitehead, S.   "Dielectric Breakdown of Solids," Oxford University Press, London, 1953.
22. O'Dwyer, J. J.   "The Theory of Dielectric Breakdown in Solids," Oxford University Press, London, 1964.
23. Klein, N.   Electrical Breakdown in Solids, *Adv Electron. Phys.*, 1969, Vol. 26, pp. 309–424.
24. Mason, J. H.   Dielectric Breakdown in Solid Insulation, *Prog. Dielect.* (London), 1959, Vol. 1, pp. 1–58.
25. Adamczewski, I.   "Ionization, Conductivity and Breakdown in Dielectric Liquids," Taylor & Francis, London, 1969.
26. Llewellyn-Jones, F.   "Ionization, Conductivity and Breakdown in Gases" and "Ionization Avalanches and Breakdown," Methuen, London, 1957 and 1967.
27. Von Engel, A.   "Ionized Gases," 2d ed., Oxford University Press, London, 1965.
28. Frohman-Bentchkowski, D., and M. Lenzliner   *J. Appl. Phys.*, 1969, Vol. 40, p. 3307.
29. Schadt, M., and W. Helfrich   *Appl. Phys. Lett.* 1971, Vol. 18, p. 127.
30. Special issue on Acoustics, *Proc. IEEE*, May 1976, Vol. 64, No. 5.
31. Berry, R. W., P. M. Hall, and M. T. Harris   "Thin Film Technology," Van Nostrand, Princeton, N.J., 1968.
32. Warren, B. E.   "X-Ray Diffraction," Addison-Wesley, Reading, Mass., 1969.
33. Hadfield, R. A.   "Metallurgy and Its Influence on Modern Progress," Van Nostrand, Princeton, N.J., 1926.
34. Kestigian, M., A. B. Smith, and W. R. Bekebred,   *J. Appl. Phys.*, 1979, Vol. 50, p. 2161.
35. Bobeck, A. H., and E. Della Torre   "Magnetic Bubbles," North-Holland, Amsterdam, 1975.
36. Voegeli, I. O., B. A. Calhoun, L. I. Rosier, and J. C. Slonczewski   *AIP Conf. Proc.*, 1975, Vol. 24, 617.
37. Dessauer, J. H., and E. H. Clark (eds.)   "Xerography and Related Processes," Focal Press, New York, 1965.
38. Ovshinsky, S. R., and H. Fritzche   *Met. Trans.*, 1971, Vol. 2, p. 641.
39. Sie, C., P. Dugan, and S. C. Moss   *4th Int. Conf. Amorphous Liquid Semiconductors*, Ann Arbor, Mich., August, 1971.
40. Feinleib, J., J. de Neuville, S. C. Moss, and S. R. Ovshinsky   *Appl. Phys. Lett.*, 1971, Vol. 18, p. 254.
41. Langmuir, I.   *J. Franklin Inst.*, 1934, Vol. 217, p. 543.
42. Lemmens, H. J., M. J. Jansen, and R. Loosjes   *Philips Tech. Rev.*, 1959, Vol. 11, p. 341.
43. Levi, R.   *J. Appl. Phys.*, 1952, Vol. 24, p. 233.
44. Copola, P. P., and R. C. Hughes   *Proc. IRE*, 1956, Vol. 44, p. 351.
45. Hadley, C. P., W. G. Rudy, and A. J. Stoeckert   *J. Electrochem. Soc.*, 1958, Vol. 105, p. 395.
46. Hermann, G., and S. Wagener   "The Oxide Coated Cathode," Champan & Hall, London, 1951.
47. Lafferty, J. M.   *J. Appl. Phys.*, 1951, Vol. 22, p. 299.
48. Albert, M. J., and M. A. Atta   *Metals Mater.*, February 1967, p. 43.
49. Gobell, G. W., and F. G. Allen   *Phys. Rev.*, 1965, Vol. 137, p. 245A.
50. Hewitt, H., and A. S. Vause   "Lamps and Lighting," Arnold, London, 1966, p. 168.
51. Morley, G. W.   "Properties of Glass," 2d ed., *ACS Monog.* 77, 1954.
52. Lillie, H. R.   *Glass Technol.*, 1960, Vol. 1, p. 115.
53. Araujo, R. J., and S. D. Stookey   *Glass Ind.*, December 1967, p. 687.
54. Smith, G. P.   *J. Photo. Sci.*, 1970, Vol. 18, No. 2.
55. Land, E. H.   *J. Opt. Soc. Am.*, 1951, Vol. 41, p. 957.

56. "Kodak Data Book of Applied Photography," Vol. A. Kodak Ltd., London.

57. Robillard, J. J.   *Photogr. Sci. Eng.*, 1964, Vol. 8, p. 18.

58. Kosar, J., and W. Clark (eds.)   "Wiley Series in Photographic Sciences and Technology," Wiley, New York, 1965.

59. Soule, H. V.   "Electro-Optical Photography at Low Illumination Levels," Wiley, New York, 1968.

60. Smithels, C. J.   "Metals Reference Book," 5th ed. Butterworths, London, 1976.

61. Lyman, T. (ed.)   "Metals Handbook," 8th ed., The American Society for Metals, Cleveland, 1961.

62. Parker, E. (ed.)   "Material Data Book," McGraw-Hill, New York, 1967.

63. Caulton, M.   NEREM   *Rec.*, November 1968.

64. Compiled from several sources.

65. Gerritsen, A. N.   Metallic Conductivity, Experimental Part, Vol. 19, p. 137, in S. Flugge (ed.), "Handbuch der Physik," Springer, Berlin, 1956.

66. Hansen, M.   "Constitution of Binary Diagrams," McGraw-Hill, New York, 1958.

67. Goldsmith, A., et al. (eds.)   "Handbook of Thermophysical Properties of Solid Materials," Pergamon, Oxford, 1961.

68. Garwood, M. F., H. H. Zurburg, and M. A. Erickson   "Correlation of Laboratory Tests and Service Performance," American Society for Metals, Cleveland, 1951.

69. Hardy, H. K., and T. J. Heal   *Prog. Metal Phys.*, 1954, Vol. 5, p. 195.

70. Harper, C. A. (ed.)   "Handbook of Electronic Packaging," McGraw-Hill, New York, 1969, pp. 1–52.

71. After Murphy, E. J., and S. D. Morgan   *Bell Syst. Tech. J.*, 1935, Vol. 16 p. 493.

72. Kingery, W. D.   "Introduction to Ceramics," Wiley, New York, 1960.

73. Fink, D. G. (ed.)   "Standard Handbook for Electrical Engineers," McGraw-Hill, New York, 1978.

74. Harper, C. A. (ed.)   "Handbook of Materials and Processes for Electronics," McGraw-Hill, New York, 1970.

75. Hague, J. R., et al.   "Refractory Ceramics for Aerospace," American Ceramic Society, Columbus, Ohio, 1964.

76. American Lava Corporation, Chattanooga, Tenn.

77. Licari, J. J., and E. R. Brands   *Mach. Des. Mag.*, 1967.

78. "Properties of Selected Commercial Glasses" Corning Glass Works, Corning, N.Y., 1971.

79. Corning Glass Works, Corning, N.Y.

80. Bozorth, R. M.   "Ferromagnetism," Van Nostrand, Princeton, N. J., 1951.

81. "Electrical Materials Handbook," Allegheny Ludlum Steel Co., Pittsburgh, 1961.

82. Bozorth, R. M.   *Rev. Mod. Phys.* 1953, Vol. 25, p. 42.

83. Chikazumi, S.   "Physics of Magnetism," Wiley, New York, 1964.

84. Smit, J. (ed.)   "Magnetic Properties of Materials," McGraw-Hill, New York, 1971, p. 77.

85. Tebble, R. S., and D. J. Craik   "Magnetic Materials," Wiley-Interscience, New York, 1969.

86. Trans Tech Co., Gaithersburg, Md.

87. General Electric "Permanent Magnet Manual."

88. Hall, R. N., and J. H. Racette   Diffusion and Solubility of Copper in Extrinsic and Intrinsic Germanium, Silicon, and Gallium Arsenide, *J. Appl. Phys.*, 1964, Vol. 35, p. 379.

89. Morin, F. J., and J. P. Maita   Electrical Properties of Silicon Containing Arsenic and Boron, *Phys. Rev.*, 1954, Vol. 96, p. 28.

90. Morin F. J., and J. P. Maita   Conductivity and Hall Effect in the Intrinsic Range of Germanium, *Phys. Rev.*, 1954, Vol. 94, p. 1525.

91. Sze, S. M., and J. C. Irvin   *Solid State Electron.*, 1968, Vol. 11, p. 599.

92. Curtiss, D. B.   *Bell Syst. Tech. J.*, 1961, Vol. 40, p. 509.

93. Irvin, J. C.   *Bell Syst. Tech. J.*, 1962, Vol. 41, p. 387.

94. Wolf, H.   "Semiconductors," John Wiley-Interscience. New York, 1971.

95. Sze, S. M.   "Physics of Semiconductor Devices," Wiley-Interscience, New York, 1969.

96. After Burger, R. M., and R. P. Donovan (eds.)   "Fundamentals of Silicon Integrated Device Technology," Vol. 1, Prentice-Hall, Englewood Cliffs, N.J., 1967.

97. Kendall, D. L.   *Stanford Univ. Dept. Mater. Sci. Rep. 65-29*, August 1965.

98. Müller, H. O.   *Z. Phys.*, 1937, Vol. 104, p. 475.

99. Gray, D. E. (ed.)   "American Institute of Physics Handbook." McGraw-Hill, New York, 1957.

100. Rose, R. M., L. A. Shepard, and J. Wulff "The Structure and Properties of Materials," Vol. 4, Wiley, New York, 1966.

101. Spicer, W. E. *Phys. Rev.*, 1958, Vol. 112, p. 114, and *RCA Rev.*, 1958, Vol. 19, p. 555.

102. Sommer, A. H. "Photoemissive Materials," Wiley, New York, 1968.

103. Kazan, B., and M. Knoll "Electronic Image Storage," Academic, New York, 1968.

104. Sears, F. W. "Optics," Addison-Wesley, Reading, Mass., 1949.

105. Garlick, G. F. J. in P. Goldberg (ed.), "Luminescence of Inorganic Solids," Academic, New York, 1966, p. 708.

106. Lengyel, B. A. "Lasers," Wiley-Interscience, New York, 1971.

107. Williams, F. in B. Di Bartolo (ed.), "Luminescence of Inorganic Solids," Plenum, New York, 1977.

Discrete Circuit Components

E. A. GERBER   Director, Electronic Components Laboratory (ret.), U.S. Army Electronics Command Fort Monmouth; Consultant. U.S. Army Electronics Technology and Devices Laboratory, Electronics Command, Fort Monmouth, New Jersey; Fellow, IEEE

THOMAS S. GORE, JR.   Deputy Chief, Frequency Control and Signal Processing Devices Technical Area (ret.);* Member, IEEE

EMANUEL GIKOW   Leader, Filter Devices Team (ret.); Senior Member, IEEE; Licensed Professional Engineer, New Jersey (deceased)

JOSEPH M. GIANNOTTO   Electronics Engineer, Test Measurements and Diagnostics System Division, Center for Tactical Computer Systems, U.S. Army Communications Research and Development Command, Fort Monmouth, N.J.; Senior Member, IEEE

JOHN R. VIG   Chief, Frequency Control Branch; Member, IEEE

SAM DI VITA   Ceramics and Glass Engineer, Fiber Optics Team, Multichannel Division, U.S. Army Communications Research and Development Command, Fort Monmouth, N.J.; Fellow, American Ceramic Society

GEORGE W. TAYLOR   Leader, Special Devices and Technology Team; Beam, Plasma and Display Division; Member, IEEE

BERNARD SMITH   Electronics Engineer, Special Devices and Technology Team; Beam, Plasma and Display Division

I. REINGOLD   Director, Beam, Plasma and Display Division; Fellow, IEEE; Fellow, Society for Information Display; Licensed Professional Engineer, New Jersey

MUNSEY CROST   Project Engineer, Displays and Peripherals Team; Beam, Plasma and Display Division; Member, American Physical Society; Member, Society for Information Display

*Unless stated otherwise, all authors are with the U.S. Army Electronics Technology and Devices Laboratory, Electronics Research and Development Command, Fort Monmouth, N.J.

GREGORY J. MALINOWSKI *Electronics Engineer, Microwave and Signal Processing Devices Division*

EDWARD B. HAKIM *Leader, Reliability and Failure Analysis Team, Microelectronics Division; Member, IEEE*

DAVID LINDEN *Director, Power Sources Division (ret.); Fellow, American Institute of Chemists; Member. American Chemical Society; Member, Electrochemical Society*

JACK SPERGEL *Chief Applications Engineer, General Cable Corporation, Colonia, N.J.; Senior Member, IEEE (deceased)*

ROBERT A. GERHOLD *Deputy Director, Microelectronics Division (ret.); Consultant; Senior Member, IEEE*

OWEN P. LAYDEN *Leader, Hybrid Microcircuits and Packaging Team, Microelectronics Division; Member, IEEE*

ISAAC H. PRATT *Research Physical Scientist, Hybrid Microcircuits, Microelectronics Division; Member, IEEE; Member, American Physical Society*

CONTENTS

Numbers refer to paragraphs

Resistors

BY THOMAS S. GORE, JR.

1. Fundamentals. A resistor is a device that introduces resistance into an electronic circuit and as such is used for setting biases, controlling gain, fixing time constants, matching and loading circuits, and for voltage division, heat generation, and other related functions. Resistance is a fundamental property of a conductor, as shown:

$$R = \rho l / A$$

where R = resistance (Ω), ρ = specific resistance or resistivity of the conductor material ($\Omega \cdot cm$), l = length of conductor (cm), and A = cross-sectional area (cm^2). Resistance controls either the voltage or the flow of current in an electronic circuit and in so doing produces dissipation of power. Thus

$$\text{Resistance} = R = V/I \quad \text{Ohm's law}$$
$$\text{Power dissipated in resistor} = P = V^2/R = I^2R$$

where V = voltage (V), I = current (A), and P = power (W).

Resistance-Temperature Characteristic. The magnitude of change in resistance due to temperature is usually expressed as a percentage per degree Celsius or parts per million per degree Celsius. If the changes are linear over the operating temperature range, the parameter is known as *temperature coefficient;* if nonlinear, the parameter is known as *resistance-temperature characteristic.*

Hot-spot temperature is the maximum temperature measured on the resistor body due to both internal heating and the ambient operating temperature and is usually the maximum temperature at which the resistor is derated to zero power.

Noise is an unwanted voltage fluctuation generated within the fixed resistor. Total noise of a resistor always includes (1) Johnson noise, which is dependent only on resistance value and temperature of the resistance element, due to thermal agitation, and (2) noise caused by current flow, cracked bodies, and loose end caps or leads, depending on the type of resistor element and construction. For variable resistors, noise may also be caused by jumping of the moving contact over turns and by an imperfect electrical path between the contact and resistance element.

Maximum Working Voltage $(V = \sqrt{PR})$. This quantity represents the maximum voltage stress (dc or rms) that can be applied to the resistor. Its value is a function of the materials used, the physical dimensions, and the quality of performance desired.

High-Frequency Effects. For most resistors, the lower the resistance value, the less total impedance the resistor exhibits at high frequency. Resistors are not generally tested for total impedance at frequencies above 120 Hz; therefore this characteristic is not controlled. For the best high-frequency performance, the ratio of resistor length to the cross-sectional area should be a maximum. Dielectric losses are kept low by proper choice of the resistor base material. When dielectric binders are used, their total mass is kept to a minimum.

Film-type resistors have the best high-frequency performance (see Fig. 7-1). The effective dc resistance for most resistance values remains fairly constant up to 100 MHz and decreases at higher frequencies. In general, the higher the resistance value, the greater the effect of frequency.

2. Fixed Resistors. Typical characteristics of fixed resistors are listed in Table 7-1.

Precision Resistors. The precision resistor, available in metal-film or wire constructions, is designed for use in circuits requiring close tolerance, long-term resistance stability, low noise,

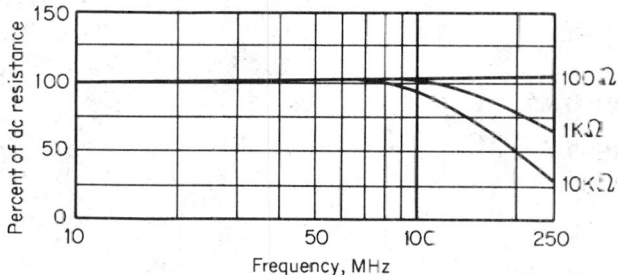

Fig. 7-1. Resistance vs. frequency, film resistances, ½ W.

and low temperature coefficient. The wire-wound variety is comparatively large and available in a limited resistance range only; however, it is the most stable of all resistors. The inductive *L* and capacitive *C* effects of the wire-wound units make them unsuitable for use above 50 kHz even when specially wound to reduce the associated inductance and capacitance. Wire-wound resistors usually exhibit an increase in resistance with higher frequencies because of skin effects. The units are usually low-power devices. The metal-film resistor element is normally nichrome, tin oxide, or tantalum nitride and is not so stable as the wire-wound unit but is less inductive. Metal-film resistors are available either in hermetically sealed or molded-phenolic cases.

Semiprecision Resistors. The semiprecision resistor is designed for circuits requiring long-term temperature stability. This resistor is normally smaller than the precision unit and is less expensive. The units are used primarily for current-limiting or voltage-dropping functions in circuit applications.

General-Purpose Resistors. These types are small, of inexpensive composition (carbon with binder), most frequently used in electronic circuit applications. They are for use in circuits that are not critical of initial tolerances (e.g., 5% or more) or of long-term stability that may approach 20% under full rated power. These resistors should not be used where a low temperature coefficient of resistance and low noise levels are desired. Composition resistors exhibit little change in effective dc resistance up to frequencies of about 100 kHz. Resistance values above 0.3 MΩ begin to show decreasing resistance at approximately 100 kHz; above frequencies of 1 MHz, all resistance values exhibit decreased resistance.

Power Resistors. These resistors are available in both wire-wound and film constructions. The wire-wound units are designed for precision and general-purpose applications. The use of tapped resistors should generally be avoided because the insertion of taps weakens the resistors mechanically and lowers their effective power ratings. The power resistors are normally rated at a 25°C ambient. When necessary to operate these resistors above this temperature, a power-rating correction factor with reference to 25°C must be applied, as specified in applicable specification. The film types have the advantage of good performance at high frequencies, especially as dummy loads, and have higher resistance values than wire-wound types for a given size. The power resistors are used in power supplies, control circuits, and voltage dividers where an operational stability of 5% is satisfactory.

3. Special Resistors. *High-Megohm Resistors.* The high-megohm resistors (10^8 to 10^{15} Ω) are used in test instrumentation, photocell circuits, and other special applications. The resistance element is hermetically sealed in a glass or glazed-ceramic case. Extreme care must be taken in handling and mounting the resistor to prevent creation of leakage paths on the resistor surface.

Chip Resistor.   The chip resistor, a relatively new device, is used in hybrid microelectronic circuits. These units are available in either thick- or thin-film construction. The thick-film element is screened onto a ceramic or glass substrate; the thin-film element is vacuum-deposited. The desired resistance is obtained by scribing, sandblasting, or otherwise adjusting the resistor

TABLE 7-1   Characteristics of Typical Fixed Resistors

Resistor types	Resistance range	Watt range (full rating at indicated temp)*	T, °C	Operating temp range, °C, −55 to:	Resistance temp characteristics, ppm/°C	Terminations
Precision:						
Wire-wound	0.1 Ω–1.2 MΩ	⅛–¾	125	+145	±10	Axial leads, printed circuit
Metal film	10 Ω–5 MΩ	½₀–½	125	125	±25	Axial leads
Semiprecision:						
Metal oxide	10 Ω–1.5 MΩ	¼–2	70	150	±200	Axial leads
Cermet	10 Ω–1.5 MΩ	½₀–½	125	175	±200	Axial leads
Deposited carbon	10 Ω–5 MΩ	⅛–1	70	165	+200, −500	Axial leads
General-purpose:						
Composition	2.7 Ω–100 MΩ	⅛–2	70	130	±1500	Axial leads
Power:						
Wire-wound:						
Tubular	0.1–180 kΩ	1–210	25	275	±260	Radial tab, axial leads
Chassis mount	1.0–38 kΩ	5–30	25	275	±50	Axial leads, radial tab
Precision	0.1–40 kΩ	1–10	25	275	±20	Axial leads
Film	20 Ω–2 MΩ	7–1000	25	225	±500	Radial tab, ferrule
Chip resistor	1–22 MΩ	...		125	±25 to ±200	Beam lead, tab, solder

*When resistors are operated above the temperature at which full wattage is listed, the wattage must be derated in accordance with applicable specifications.

element to tolerance. A general comparison of the characteristics of most common thick- and thin-film planar resistors is shown in Table 7-2.

Chassis-Mounted Power Resistor.   This is a wire-wound unit housed in a metal case for mounting directly on the chassis, thereby permitting use of the chassis as a heat sink and allowing the size of the resistor to be considerably smaller than conventional resistors of equivalent wattage rating.

High-Voltage Resistors.   These are designed to fulfill the special requirement for high-voltage, high-resistance units capable of dissipating moderate power. Such resistors are rated from 5 to 20 kV, have a resistance range of 2,000 Ω to 1,000 MΩ, and are rated from 5 to 20 W. The resistor is noninductive and is used primarily in high-voltage bleeder circuits, high-voltage voltage dividers, and high-voltage networks.

4. Variable Resistors. *Precision Resistors (Potentiometers).*   The variable resistor is designed with either film, wire-wound, or conductive plastic elements and is available in single or multiturn and single multisection units. The electrical output in terms of applied voltage is linear or follows a specified mathematical law (*taper*) with respect to the angular position of the contact arm. The resolution (the tracking of the actual resistance to the theoretical resistance) is a very important operating parameter when the resistor is used in servo applications requiring precise electrical and mechanical output and quality performance. Such resistors are most frequently used in computers, flight-control instrumentation, and radar-system circuitry.

General-Purpose Variable Resistors (Potentiometers).   These units (Fig. 7-2) have composition, cermet, or wire-wound elements and are used principally as gain or volume controls, voltage dividers, or current controls in circuits. The resistors are available with linear taper *A*, clockwise taper *B*, or counterclockwise taper *C* (see Fig. 7-3). The resistors can be ganged, and a rear-mounted switch can be attached to the end resistor.

Rheostat (Power).   These are wire-wound resistors used as speed controls for motors, ovens, and heater controls and in applications where adjustments in voltage or current levels are required, such as voltage dividers and bleeder circuits.

TABLE 7-2   General Comparison of Typical Thick- and Thin-Film Planar Resistor Types

Characteristics	Printed carbon	Cermet	Tin oxide	Nickel chromium	Tantalum nitride
Resistance range	4 Ω–20 MΩ	10 Ω–5 MΩ	10 Ω–100 kΩ	15 Ω–150 kΩ	15 Ω–150 kΩ
Resistance tolerance, %	±5, ±10	±1, ±5	±1, ±5	±2, ±5	±0.5, ±2
Temp coeff of resistance, ppm/°C	−1500	±300	±100	±25, ±50	±25
Typical noise coeff, μV/ V per frequency decade	2	<1.0*	<0.1	<0.1	<0.1
Typical voltage coeff, ppm/V	500	<100*	<25	5	5
Typical power rating, mW	200	125	125	50	50
Change in resistance after load life, %†	< −7	<0.5	<0.5	<0.25	<0.25
Resistivity, Ω/square	50–5 × 10⁶	10–10⁵	20–400	25–300	25–300
Typical power dissipation, W/cm² of resistor surface	8	8	8	8	2

*Varies with resistivity.
†Based on 1,000 h at rated wattage and 85°C ambient, no heat sink.

Trimmer Resistors (Trim Pots).   Screw-lead actuated trim pots, in single or multiturn design, are available in composition, cermet, or wire-wound construction. They are used principally to control the low-current or bias voltages in transistor circuits and as matching, balancing, and adjusting circuit variables in critical circuit applications.

5. Reference Specifications.   Established Reliability (ER) Specifications are a new series of Military (MIL) Specifications that define the failure-rate levels for selected styles of resistors.

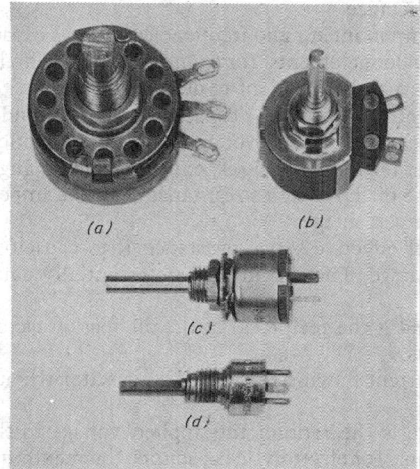

(a)   (b)

(c)

(d)

Fig. 7-2. Potentiometers: (*a*) composition ½ W; (*b*) wire-wound 1½ W; (*c*) composition ½ W; (*d*) cermet ¼ W.

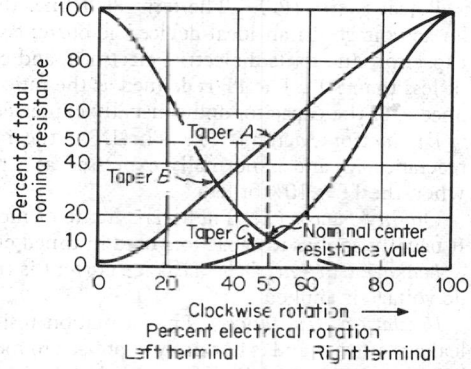

Fig. 7-3. Taper (percentage change in resistance vs. rotation).

The predicted failure rate is based on qualification-approval life testing and continuous-production life testing by the resistor manufacturer. Table 7-2 lists characteristics of film-type planar resistors.

Capacitors

BY THOMAS S. GORE, JR.

6. Fundamentals. A capacitor consists basically of two conductors separated by a dielectric or vacuum so as to store a large electric charge in a small volume. The capacitance is expressed as a ratio of electric charge to the voltage applied $C = Q/V$, where Q = charge (C), V = voltage (V), C = capacitance (F). A capacitor has a capacitance of one farad when it receives a charge of one coulomb at a potential of one volt. The electrostatic energy in watt-seconds or joules stored in the capacitor is given by

$$J = \tfrac{1}{2}CV^2$$

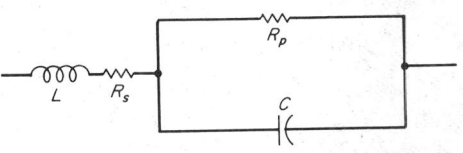

Fig. 7-4. Equivalent circuit of a capacitor; R_s = series resistance due to wire leads, contact terminations, and electrodes (Ω); R_p = shunt resistance due to resistivity of dielectric and case material (Ω), and to dielectric losses; and L = stray inductance due to leads and electrodes (H).

Dielectric. Depending on the application, the capacitor dielectric may be air, gas, paper (impregnated), organic film, mica, glass, or ceramic, each having a different dielectric constant, temperature range, and thickness.

Equivalent Circuit. In addition to capacitance, a practical capacitor has an inductance and resistance. An equivalent circuit useful in determining the performance characteristics of the capacitor is shown in Fig. 7-4.

Equivalent Series Resistance (ESR). The ESR is the ac resistance of a capacitor reflecting both the series resistance R_s and the parallel resistance R_p at a given frequency so that the loss of these elements can be expressed as a loss in a single resistor R in the equivalent circuit.

Capacitive Reactance. The reactance of a capacitor is given by

$$X_c = 1/2\pi fC = 1/\omega C \quad (\Omega)$$

where $f = \omega/2\pi$ is the frequency in hertz.

Impedance (Z). In practical capacitors operating at high frequency, the inductance of the leads must be considered in calculating the impedance. Specifically,

$$Z = \sqrt{R^2 + (X_L - X_C)^2}$$

where R is the ESR, and X_L reflects the inductive reactance.

The effects illustrated in Fig. 7-4 are particularly important at radio frequencies where a capacitor may exhibit spurious behavior due to these equivalent elements. For example, in many high-frequency tuned circuits the inductance of the leads may be sufficient to detune the circuit.

Power Factor (PF). The term PF defines the electrical losses in a capacitor operating under an ac voltage. In an ideal device the current will lead the applied voltage by 90°. A practical capacitor, due to its dielectric, electrode, and contact termination losses, exhibits a phase angle of less than 90°. The PF is defined as the ratio of the effective series resistance R to the impedance Z of the capacitor and is usually expressed as a percentage.

Dissipation Factor (DF). The DF is the ratio of effective series resistance R to capacitive reactance X_C and is normally expressed as a percentage. The DF and PF are essentially equal when the PF is 10% or less.

Quality Factor Q. The Q is a figure of merit and is the reciprocal of the dissipation factor. It usually applies to capacitors used in tuned circuits.

Leakage Current, DC. Leakage current is the current flowing through the capacitor when a dc voltage is applied.

Insulation Resistance. The insulation resistance is the ratio of the applied voltage to the leakage current and is normally expressed in megohms. For electrolytic capacitors, the maximum leakage current is normally specified.

Ripple Current or Voltage. The ripple current or voltage is the rms value of the maximum allowable alternating current or voltage (superimposed on any dc level) at a specified frequency at which the capacitor may be operated continuously at a specified temperature.

Surge Voltage.   The surge voltage applicable to electrolytic capacitors is a voltage in excess of the rated voltage which the capacitor will withstand for a specified limited period at any temperature.

7. Fixed Capacitors. Table 7-3 shows the types of capacitors generally used over the frequency range from direct current to 10 GHz.

Precision Capacitors.   Capacitors falling into the precision category (see Table 7-4) are generally those having exceptional capacitance stability with respect to temperature, voltage, frequency, and life. They are available in close capacitance tolerances and have low-loss (high-Q) dielectric properties. These capacitors are generally used at radio frequencies in tuner, rf filter,

TABLE 7-3   Useful Frequency Range of Capacitors
Dashed lines indicate fringe areas of application

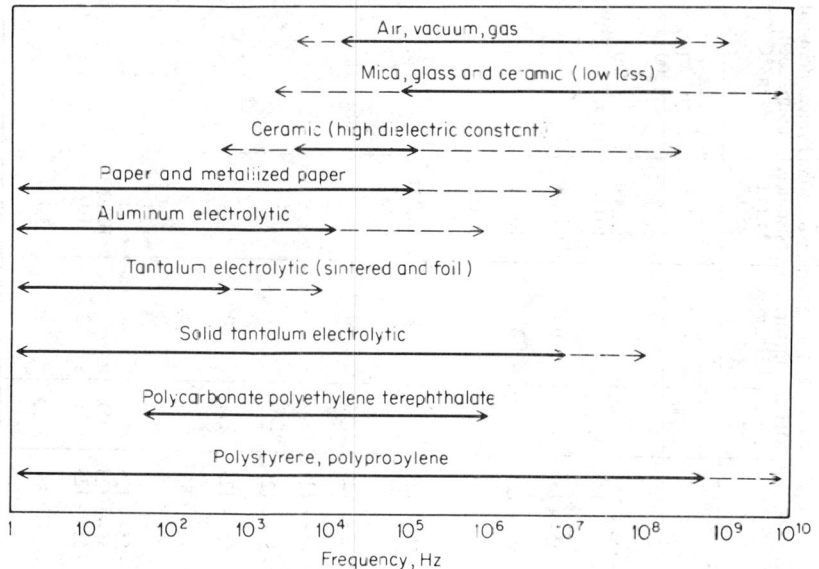

SOURCE: Adapted from G. W. Dummer and Harold M. Nordenberg, "Fixed and Variable Capacitors," McGraw-Hill Book Company, New York, 1960.

coupling, bypass, and temperature-compensation applications. Typical capacitor types in this category are mica, ceramic, glass, and polystyrene. The polystyrene capacitor has exceptionally high insulation resistance, low losses, low dielectric absorption, and a controlled temperature coefficient for film capacitors.

Semiprecision Units.   Paper- and plastic-film capacitors listed in Table 7-4 with foil or metallized dielectric constitute a large portion of the present applications. These capacitors are nonpolar and generally fall between the low-capacitance precision types, such as mica and ceramic, and the high-capacitance electrolytics.

General-Purpose.   Electrolytic aluminum and tantalum capacitors and the large-usage general-purpose (high-K) ceramic capacitors are so grouped in Table 7-4 because both have broad capacitance tolerances, are temperature-sensitive, and have high volumetric efficiencies (capacitance-volume ratio). They are primarily used as bypass and filter capacitors where high capacitance is needed in small volumes, and with guaranteed minimum values. These applications do not require low dissipation factors, stability, or high insulation resistance found in precision and semiprecision capacitors. On a performance-vs.-cost basis, the general-purpose capacitors are the least expensive of the groups. High-capacitance aluminum electrolytic capacitors have been designed for computer applications featuring low equivalent series resistance and long life.

Suppression Capacitors.   The feed-through capacitors listed in Table 7-4 are three-terminal devices designed to minimize effective inductance and to suppress rf interference over a wide frequency range. For heavy feed-through currents, applications in 60- and 400-Hz power supplies, paper or film dielectrics are normally used. For small low-capacitance, low-current units, the ceramic and button-mica feed-through high-frequency styles are used.

Capacitors for Microelectronic Circuits.   Table 7-5 lists representative styles of discrete miniature capacitors electrically and physically suitable for microelectronic circuit use (filtering,

TABLE 7-4 Characteristics of Typical Fixed Capacitors

Capacitor type	Typical capacitance range	Typical voltage-rating range, V. dc	Operating temp. range, °C	Q, 1 MHz, min.	% dissipation factor, 1 kHz, max.	Min. insulation resistance, MΩ at 25°C	Typical nominal temp. coeff., ppm/°C	Max. capacitance change over temp. range, %	Terminations
Precision Mica	1–91,000 pF	100–2,500	−55–+125	1,200	...	7,500	−20, +100, ±200	...	Axial leads, Radial leads, tab
Glass	1–10,000 pF	300–500	−55–+125	1,500	...	10,000	+140	...	Axial leads
Ceramic	1–1,100 pF	150–500	−55–+85	1,000	...	10,000	+100 to −750*	±3	Axial, radial leads
Film polystyrene	1,000–222,000 pF	200–600	−55–+85	...	0.1	100,000	...	...	Axial leads
Semiprecision Plastic Film	1,000 pF–10 µF	30–1,000	−55–+125	...	0.5	10,000	...	+15 −10	Axial leads
Paper-plastic film (metallized)	4,700 pF–10 µF	50–400	−55–+125	...	1	10,000	...	+20 −10	Axial leads
General-purpose Ceramic (hi-K)	10–100,000 pF	50–200	−55–+125	...	2	10,000	...	±15	Axial, radial leads
Tantalum oxide, sintered, solid electrolyte, polar	4,700 pF–330 µF	6–100	−55–+125	...	6†	...	...	+8 −10	Axial leads
Foil polar	1–580 µF	10–300	−55–+125	...	...	...	...	+50 −40	Axial leads
Sintered (wet), polar	3.6–560 µF	4.0–85	−55–+125	...	...	...	...	+20 −64	Axial leads
Aluminum oxide, dry, polar	3.3–1,000 µF	7–250	−55–+125	...	20†	...	...	+25 −30	Axial
Computer-grade polar	150–120,000 µF	5–450	−40–+85	...	...	...	...	...	Solder lugs
Suppression Feed through Paper	10,000 pF–3.0 µF	100–600	−55–+125	...	...	...	...	±15	Axial
Mica, button	5–2,400 pF	500	−55–+125	1,000	...	100,000	...	±100	Special
Ceramic	100–1,500 pF	500–1,500	−55–+125	...	2	10,000	...	±15	Axial leads, solder lugs
Bypass paper	10,000–25,000 pF	100–500‡	−55–+85	...	...	3,000	...	±25	Solder lugs screw type

* +100, 0, −330, −750, preferred values.
† pF at 120 Hz.
‡ ac/dc.

TABLE 7-5 Characteristics of Typical Capacitors for Microelectronic Circuits

Type of capacitor	Typical capacitance range	Typical working voltage range, V, dc	Temperature range, °C	Dissipation factor, %	Minimum insulation resistance, MΩ at 25°C	Temp. coeff., ppm/°C	Max.-capacitance change over temp. range, %	Termination
Chip:								
Ceramic, temperature-compensating	1 pF-0.027 µF	50-200	-55-+125	0.1 at 1 MHz	50,000	0 to -750	...	Metallized
Ceramic, general-purpose	390 pF-0.47 µF	25-200	-55-+125	3.0 at 1 kHz	10,000	+200	±15	Metallized
Tantalum oxide (beam-lead), dry	100-3,000 pF	12-50	-55-+85	0.6 at 1 kHz	10,000			Beam lead
Tantalum oxide Polar, solid electrolyte	0.1-47 µF	3-35	-55-+85	4.0 at 120 Hz	...	...	±10	Metallized
Metallized, film (metal case)	0.1-10 µF	50-100	-55-+85	2.5 at 1 kHz	5,000	...	±10	Axial
Variable ceramic trimmer	Min 1-3 nF, max. 5-25 pF	25	-55 +125	0.2 at 1 MHz	10,000	...	±5-±15	Printed circuit

coupling, tuning, bypass, etc.). The chip capacitor, widely used in hybrid circuits, is available in single-wafer or multilayer (monolithic) ceramic, or in tantalum constructions, both offering reliable performance, very high volumetric efficiency, and a wide variety of capacitance ranges at moderate cost. Temperature-compensating ceramic chips are used where maximum capacitance stability or predictable changes in capacitance with temperature are required.

General-purpose ceramic and solid electrolytic tantalum chips are used for coupling and bypass applications where very high capacitance and small size are necessary. The ceramic chips are unencapsulated and leadless, with suitable electrodes for soldering in microcircuits. The beam-leaded tantalum chips are attached by pressure bonding. The tantalum chip is also available in a multiple-unit assembly in the dual-in-line package for use on printed-circuit boards.

Transmitter Capacitors.   The principal requirements for transmitter capacitors are high rf power-handling capability, high rf current and voltage rating, high Q, low internal inductance, and very low effective series resistance. Mica, glass, ceramic, and vacuum- or gas-filled capacitors

TABLE 7-6   Transmitter Capacitors (Fixed)

Type of capacitor	Typical capacitance range, pF	Peak voltage range, kV, ac	Current rating, A
Mica	47–100,000	0.25–35	0.51–75 at 1 MHz
Glass	22–150,000	0.5–6	0.33–25 at 1 MHz
Gas-filled	500–2,000	5–14	120–150 at 16 MHz
Vacuum-filled	500–1,000	2–35	30–200 at 16 MHz

are used primarily for transmitter applications. Glass dielectric transmitter capacitors have a higher self-resonating frequency and current rating than comparable mica styles. The ceramic capacitors offer moderately high rf current ratings, operating temperatures to 105°C, and high self-resonant frequencies. The gas or vacuum capacitor is available in a wide range of capacitance and power ratings and is used where very-high-power, high-voltage, and high-frequency circuit conditions exist. These units are also smaller than other transmitting types for comparable ratings. The circuit performance of the transmitter capacitor is highly dependent on the method of mounting, lead connections, operating temperatures, air circulation, and cooling, which must be considered for specific applications. Typical ratings are listed in Table 7-6.

8. Variable Capacitors. Variable capacitors are used for tuning and for trimming or adjusting circuit capacitance to a desired value.

Tuning Capacitor (Air-, Vacuum-, or Gas-Filled).   The parallel plate (single or multisection style) is used for tuning receivers and transmitters. In receiver applications, one section of a multisection capacitor is normally used as the oscillator section, which must be temperature-stable and follow prescribed capacitance-rotation characteristics. The remaining capacitor sections are mechanically coupled and must track to the oscillator section. The three most commonly used plate shapes are semicircular (straight-line capacitance with rotation), midline (between straight-line capacity and straight-line frequency), and straight-line frequency (which are logarithmically shaped). For transmitter applications, variable vacuum- or gas-filled capacitors of cylindrical construction are used. These capacitors are available in assorted sizes and mounting methods. Typical ratings are:

			rf current, A	
	Capacitance, pF	Peak rf voltage, kV	At 50 MHz	At 16 MHz
Gas-filled:				
Minimum	7–50	3.5	35	
Maximum	500–3,000	14.0		150
Vacuum:				
Minimum	3–30	7.5		36
Maximum	100–5,000	15		125

9. Special Capacitors. *High-Energy Storage Capacitors.*   Oil-impregnated paper and/or film dielectric capacitors have been designed for voltages of 1,000 V or higher for pulse-forming networks. For lower voltages, special electrolytic capacitors can be used which have a low inductance and equivalent-series resistance.

Commutation Capacitors. The widespread use of SCR devices (see Par. **7-79**) has led to the development of a family of oil-impregnated paper and film dielectric capacitors for use in triggering circuits. The capacitor is subjected to a very fast rise time (0.1 ms) and high current transients and peak voltages associated with the switching.

Reference Specifications. Established Reliability (ER) Specifications are a new series of Military (MIL) Specifications that define the failure-rate levels for selected styles of capacitors. The predicted failure rate is based on qualification-approval life testing and continuous-production life testing by the capacitor manufacturer.

Inductors and Transformers

BY EMANUEL GIKOW

10. Introduction. Among the many families of electronic parts, inductive components are generally unique in that they must be designed for a specific application. Whereas resistors, capacitors, meters, switches, etc., are available as standard or stock items, inductors and transformers have not usually been available as off-the-shelf items with established characteristics. With recent emphasis on miniaturization, however, wide varieties of chokes have become available as stock items. Low inductance values are wound on a nonmagnetic form, powdered iron cores are used for the intermediate values of inductance, and ferrites are used for the higher-inductance chokes. High-value inductors use both a ferrite core and sleeve of the same magnetic material. Wide varieties of inductors are made in tubular, disk, and rectangular shapes. Direct-current ratings are limited by the wire size or magnetic saturation of the core material.

11. Solenoids. Inductance can be calculated readily from the geometry of the inductors. The nomograph in Fig. 7-5 can be used as a design guide for calculating the inductance of single-

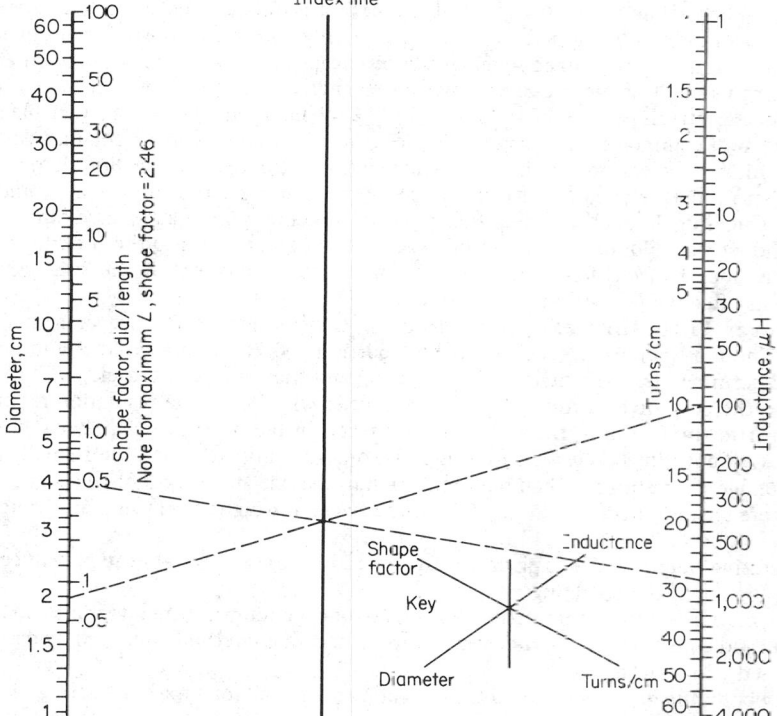

Fig. 7-5. Single-layer solenoid design chart. (*Adapted from K. Henney, "Radio Engineering Handbook,"* 5th ed., McGraw-Hill Book Company, New York, 1959.)

layer air-core solenoids. As shown by the key, if three known values are connected, the fourth unknown can be read off at the intersection. For example: If $L = 100 \ \mu H$, $d = 2$ cm, shape factor $= 0.5$, turns/cm $= 28$, total turns $= 112$.

12. Distributed Capacitance. The distributed capacitance between turns and windings has an important effect upon the characteristics of the coil at high frequencies. At a frequency determined by the inductance and distributed capacitance, the coil becomes a parallel resonant circuit. At frequencies above self-resonance, the coil is predominantly capacitive. For large-valued inductors the distributed capacitance can be reduced by winding the coil in several sections so that the distributed capacitances of the several sections are in series.

13. Toroids. The toroidal inductor using magnetic-core material has a number of advantages. Using ungapped core material, the maximum inductance per unit volume can be obtained with the toroid. It has the additional advantages that leakage flux tends to be minimized and shielding requirements are reduced. The inductance of the toroid and its temperature stability are directly related to the average incremental permeability of the core material. For a single-layer toroid of rectangular cross section without air gap,

$$L = 0.0020 N^2 b \mu_d \ln (r_2/r_1)$$

where $L =$ inductance (μH), $b =$ core width (cm), $\mu_d =$ average incremental permeability, $r_1 =$ inside radius (cm), $r_2 =$ outside radius (cm), and $N =$ total number of turns.

Incremental permeability is the apparent ac permeability of a magnetic core in the presence of a dc magnetizing field. As the dc magnetization is increased to the point of saturating the magnetic core, the effective permeability decreases. This effect is commonly used to control inductance in electronically variable inductors.

Inductors with magnetic cores are often gapped to control certain characteristics. In magnetic-cored coils the gap can reduce nonlinearity, prevent saturation, lower the temperature coefficient, and increase the Q.

14. Adjustable Inductors. The variable inductor is used in a number of circuits: timing, tuning, calibration, etc. The most common form of adjustment involves the variation of the effective permeability of the magnetic path. In a circuit comprised of an inductor and capacitor, the slug-tuned coil may be preferred over capacitive tuning in an environment with high humidity. Whereas the conventional variable capacitor is difficult and expensive to seal against the entry of humidity, the inductor lends itself to sealing without seriously inhibiting its adjustability. For example, a slug-tuned solenoid can be encapsulated, with a cavity left in the center to permit adjustment of the inductance with a magnetic core. Moisture will have little or no effect on the electrical performance of the core. The solenoid inductor using a variable magnetic core is the most common form of adjustable inductor. The closer the magnetic material is to the coil (the thinner the coil form), the greater the adjustment range will be. Simultaneous adjustment of both a magnetic core and magnetic sleeve will increase the tuning range and provide magnetic shielding. In another form, the air gap at the center post of a cup core can be varied to control inductance. Not only is the result a variable inductor, but also the introduction of the air gap provides a means for improving the temperature coefficient of the magnetic core by reducing its effective permeability.

15. Power Transformers. Electronic power transformers normally operate at a fixed frequency. The most popular frequencies are 50, 60, and 400 Hz. Characteristics and designs are usually determined by the voltampere (VA) rating and the load. For example, when used with a rectifier, the peak inverse voltage across the rectifier will dictate the insulation requirements. With capacitor input filters, the secondary must be capable of carrying higher currents than with choke input filters. With the increased use of semiconductors, and their small size, there is considerable interest in a concomitant reduction in the size and weight of all other component parts. There are a number of ways by which the size and weight of power transformers can be reduced, as follows:

Operating frequency. For a given voltampere rating, size and weight can be reduced as some inverse function of the operating frequency.

Maximum operating temperature. By the use of high-temperature materials and at a given VA rating and ambient temperature, considerable size reduction can be realized by designing for an increased temperature rise.

Ambient temperature. With a given temperature rise and for a fixed VA rating, transformer size can be reduced if the ambient temperature is lowered.

Regulation. If the regulation requirements are made less stringent, the wire size of the windings can be reduced, with a consequent reduction in the size of the transformer.

16. Audio Transformers. These transformers are used for voltage, current, and impedance transformation over a nominal frequency range of 20 to 20,000 Hz. The equivalent circuit
is the starting point for the basic analysis of the transformer frequency response. Figure 7-6 is
the equivalent circuit of a transformer.

A prime consideration in the design, size, and cost of audio transformers is the span of the
frequency response. In wide-band audio transformers the frequency coverage can be separated

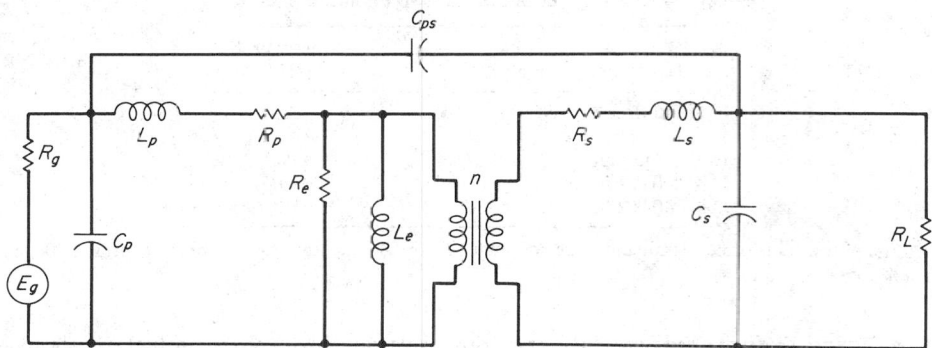

Fig. 7-6. Equivalent circuit of a broadband transformer; E_g = generator voltage, R_g = generator impedance,
C_p = primary shunt and distributed capacitance, L_p = primary leakage inductance, L_s = secondary leakage
inductance, C_s = secondary shunt and distributed capacitance, R_p = primary winding resistance, R_e = equivalent resistance corresponding to core losses, L_e = equivalent magnetizing (open-circuit) inductance of primary, n = ideal transformer primary-to-secondary turns ratio, C_{ps} = primary-to-secondary capacitance (interwinding capacitance), R_s = secondary winding resistance, R_L = load impedance.

into three nearly independent ranges for the purpose of analysis. Thus, in the high-frequency
range (Fig. 7-7) leakage inductance and distributed capacitance are most significant. In the low-
frequency region (Fig. 7-8) the open-circuit inductance is important. At the medium-frequency
range, approximately 1,000 Hz, the effect of the transformer on the frequency response can be
neglected. In the above discussion, the transformation is assumed to be that of an ideal
transformer.

17. Miniaturized Audio Transformers. *Frequency Response.* The high-frequency response (Fig. 7-7) is dependent on the magnitude of the leakage inductance and distributed capacitance of the windings. These parameters decrease as the transformer size decreases.
Consequently, miniaturized audio transformers generally have an excellent high-frequency
response. On the other hand, the small size of these transformers results in increased loss and in
a degradation of the low-frequency response, which is dependent on the primary open-circuit
inductance.

Air Gap. The primary open-circuit inductance is proportional to the product of the square
of the turns and the core area, and inversely proportional to the width of the gap. As the transformer size is reduced, both the core area and number of turns must be reduced. Consequently,
if the open-circuit inductance is to be maintained, the air gap must be reduced. Table 7-7 shows
the reduction in open-circuit inductance (OCL) as the air gap increases. A butt joint has of necessity a finite air gap; as a result the full ungapped inductance is not realized. An interlaced structure of the core, where the butt joints for each lamination are staggered, most closely approximates an ungapped core. The substantially higher inductance, 120 H, is shown in Table 7-7.
The problem with taking advantage of this effect is that as the air gap is reduced, the allowable

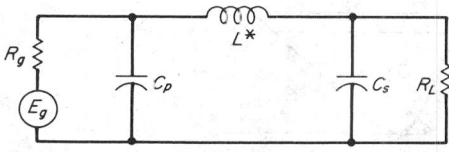

L* is the leakage inductance of both the primary,
and of the secondary referred to the primary

Fig. 7-7. Equivalent circuit of audio transformer at
high frequencies (~15,000 Hz).

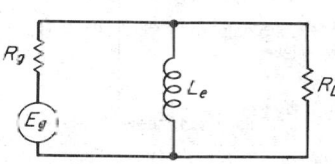

Fig. 7-8. Equivalent circuit of audio transformer at
low frequencies (~300 Hz).

amount of unbalanced direct current flowing in the transformer winding must be lowered to prevent core saturation.

Operating Voltage Level. To avoid core saturation, which will result in distortion, the voltage level of operation must be lowered as core size is reduced. Existence of a dc magnetizing field will further reduce the operating voltage.

TABLE 7-7   Effect of Air Gap on Inductance*

Lamination assembly	Primary OCL, H
Interlace 3 × 3	120
Butt joint:	
No gap	32
0.001-in gap	20
0.002-in gap	12

*Data of test transformer: lamination size and material, EE 24-25, alloy 4750; number of turns, 1,000; core area $\frac{1}{16}$ in^2

18. Pulse Transformers. Pulse transformers used in high-voltage on high-power applications, above 300 W peak, are generally used in modulators for radar sets. Their function is to provide impedance matching between the pulse-forming network and the magnetron. Prime concern is transformation of the pulse with a minimum of distortion. Lower-power pulse transformers fall into two categories: those used for coupling or impedance matching similar to the high-power pulse transformers, and blocking oscillator transformers used in pulse-generating circuits. Pulse widths for such transformers most commonly range from about 0.1 to 20 μs.

Assuming the pulse transformer is properly matched and the source is delivering an ideal rectangular pulse, a well-designed transformer should have small values of leakage inductance and distributed capacitance. Within limits dictated by pulse decay time the open-circuit inductance should be high. Figure 7-9 shows the pulse waveform with the various types of distortions that may be introduced by the transformer.

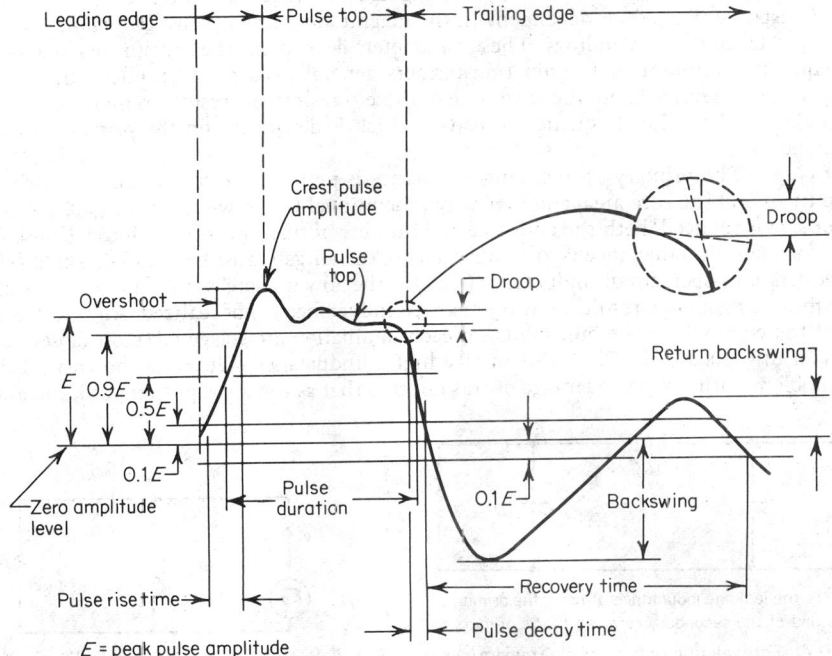

Fig. 7-9. Pulse waveform.

19. Broadband RF Transformers. At the higher frequencies, transformers provide a simple, low-cost, and compact means for impedance transformation. Bifilar windings and powdered-iron or ferrite cores provide optimum coupling. The use of cores with high permeabilities at the lower frequencies reduces the number of turns and distributed capacitance. At the upper frequencies the reactance increases, even though the permeability of the core may fall off.

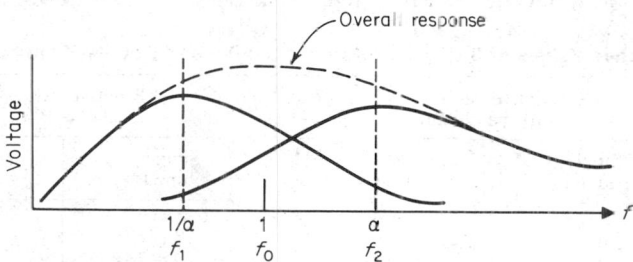

Fig. 7-10. Response of a staggered pair, geometric symmetry.

20. Transmission-Line Transformers. Where dc isolation is not a factor, transmission-line transformers can be made to provide polarity reversal, balanced to unbalanced, 4:1 impedance transformation. By hooking these transformers in tandem, higher impedance transformations are possible. Very broad bandwidths have been attained with a single transformer, as high as 0.1 to 1,000 MHz within ± 1 dB. Successful transformers were made using ferrite toroidal cores.

21. Inductive Coupling. There are a variety of ways to use inductive elements for impedance-matching or coupling one circuit to another. Autotransformers and multiwinding transformers which have no common metallic connections are a common method of inductive coupling. In a unity-coupled transformer, N_1 = number of primary turns, N_2 = number of secondary turns, k = coefficient of coupling, M = mutual inductance, n = turns ratio, L_1 = primary open-circuit inductance, L_2 = secondary open-circuit inductance, I_1 = primary current, I_2 = secondary current, E_1 = primary voltage, E_2 = secondary voltage, Z_1 = primary impedance with matched secondary, Z_2 = secondary impedance with matched primary. Transformer relationships for unity-coupled transformer, $k = 1$, assuming losses are negligible.

$$n = N_2/N_1 = E_2/E_1 = I_1/I_2 = Z_2/Z_1 \qquad M = \sqrt{L_1 L_2}$$

22. Single-Tuned Circuits. Single-tuned circuits are most commonly used in both wideband and narrow-band amplifiers. Multiple stages which are cascaded and tuned to the same frequency are synchronously tuned. The result is that the overall bandwidth of the cascaded amplifiers is always narrower than the single-stage bandwidth. The shrinkage of bandwidth can be avoided by stagger tuning. A stagger-tuned system is a grouping of single-tuned circuits where each circuit is tuned to a different frequency. For a flat-topped response the individual stages are geometrically balanced from the center frequency. In Fig. 7-10, which illustrates the response of a staggered pair, f_0 = center frequency of overall response, and f_1, f_2 = resonant frequency of each stage. The frequencies are related as follows:

$$f_0/f_1 = f_2/f_0 = \alpha$$

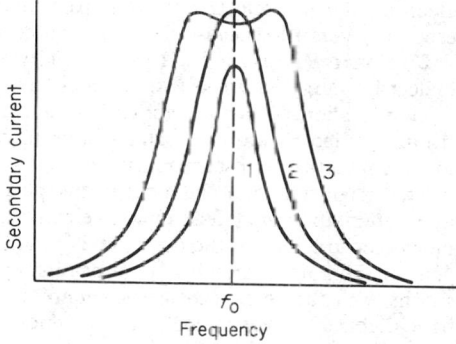

Fig. 7-11. Variation of secondary current and gain with frequency and with degree of coupling: 1 = undercoupled; 2 = critically coupled; 3 = overcoupled.

23. Double-Tuned Transformers. One of the most widely used circuit configurations for i.f. systems in the frequency range of 250 kHz to 50 MHz is the double-tuned transformer. It consists of a primary and secondary tuned to the same frequency and coupled inductively to a degree dependent on the desired shape of the selectivity curve. Figure 7-11 shows the variation of secondary current vs. frequency.

24. Bandwidth. A comparison of the relative 3-dB bandwidth of multistage single- and double-tuned circuits is shown in Table 7-8. Most significant is the lower skirt ratio of the double-tuned circuit, i.e., relative value of the ratio of the bandwidth at 60 dB (BW_{60}) to the bandwidth at 6 dB (BW_6).

25. Fabrication Techniques. *Control of Inductance.* In a single-layer winding, if the end turns are spaced, a finer adjustment of inductance can be made by repositioning the final

TABLE 7-8   Relative Values of 3-dB Bandwidth for Single- and Double-Tuned Circuits

No. of stages	Relative values of 3-dB bandwidth		Relative values of BW_{60}/BW_6	
	Single-tuned	Double-tuned*	Single-tuned	Double-tuned
1	1.00	1.00	577	23.9
2	0.64	0.80	33	5.65
3	0.51	0.71	13	3.59
4	0.44	0.66	8.6	2.94
6	0.35	0.59	5.9	2.43
8	0.30	0.55	5.0	
10	0.27	0.52	4.5	

* Based upon identical primary and secondary circuits critically coupled.

turns. In a universal winding, the slight sponginess of the winding permits a final adjustment of inductance by squeezing it parallel to the coil axis or perpendicular to the axis.

Q Adjustment. It is sometimes difficult and uneconomical to design a coil to specific Q values. Accordingly, if the coil is designed to a somewhat higher Q than is needed, a shunt resistor can be used to reduce the effective Q of the coil.

Poled Ferroelectric Ceramic Devices

BY J. M. GIANNOTTO

26. Introduction. The usefulness of ferroelectrics rests on two important characteristics, asymmetry and high dielectric constant. Poled ferroelectric devices are capable of doing electric work when driven mechanically or mechanical work when driven electrically. In poled ferroelectrics, the piezoelectric effect is particularly strong. From the design standpoint, they are especially versatile because they can be used in a variety of ceramic shapes.

27. Basic Properties. Piezoelectricity is the phenomenon of coupling between elastic and dielectric energy. Piezoelectric ceramics have gained wide use in the low-frequency range up to a few megahertz over the strongly piezoelectric nonferroelectric single crystals such as quartz, lithium sulfate, lithium niobate, lithium tantalate, and zinc oxide. High dielectric strength and low manufacturing cost are prime factors for their usefulness.

The magnitude and character of the piezoelectric effect in a ferroelectric material depend upon orientation of applied force or electric field with respect to the axis of the material. With piezoelectric ceramics, the polar axis is parallel to the original dc polarizing field. In all cases the deformations are small when amplification by mechanical resonance is not involved. Maximum strains with the best piezoelectric ceramics are in the range of 10^{-3}. Figure 7-12 illustrates the basic deformations of piezoelectric ceramics and typical applications.

28. Transducers. The use of barium titanate as a piezoelectric transducer material has been increasingly replaced by lead titanate zirconate solid-solution ceramics since the latter offer higher piezoelectric coupling, wider operating temperature range, and a choice of useful variations in engineering parameters. Table 7-9 gives the characteristics of different compositions.

The high piezoelectric coupling and permittivity of PZT-5H have led to its use in acoustic devices such as phonograph pickups, where its high electric and dielectric losses can be tolerated. For hydrophones or instrument applications PZT-5A is a better choice, since its higher Curie point leads to better temperature stability. The low elastic and dielectric losses of the PZT-8 composition at high drive level point to its use in high-power sonic or ultrasonic transducers.

Basic deformations   Typical applications

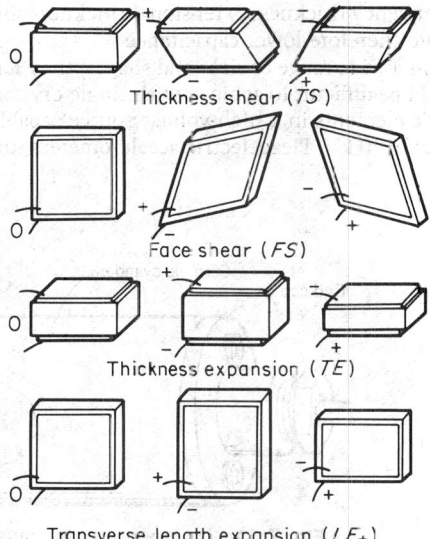

Thickness shear (TS)

Delay-line transducers; ultrasonic transducers; accelerometers, high-frequency resonators.

Face shear (FS)

Headphones; microphones; twister "Bimorph" phonograph cartridges.

Thickness expansion (TE)

Delay-line transducers; ultrasonic testing and cleaning transducers; high-frequency resonators.

Transverse length expansion (LE_t)

Sonar transducers and hydrophones; "Bimorph" phonograph cartridges; headphones; microphones; tweeters; clock drivers; heart pacers.

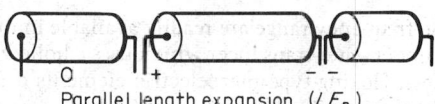

Parallel length expansion (LE_p)

Sonar radiating transducers; ultrasonic bonders and welders; ultrasonic solder cleaners

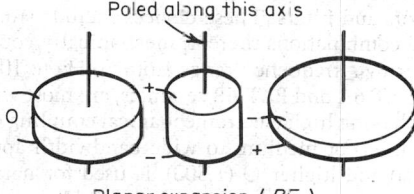

Poled along this axis

"Bimorphs" and "Unimorphs" for labs; tweeters for cameras, radios and simple alarms; transmitter and receiver for intrusion alarms.

Planar expansion (PE_t)

Fig. 7-12. Basic piezoelectric action depends upon the type of material used and the geometry. Generally, two or more of these actions are present simultaneously. TE and TS are high-frequency (greater than 1 MHz) modes and FS, LE_t, LE_p, and PE_t are low-frequency (less than 1 MHz) modes. The thickness controls the resonant frequency for the TE and TS modes, the diameter for PE_t, and the length for the LE_t and LE_p modes. Typical applications illustrate selections based on performance characteristics and cost. *(Electro-Technology.)*

TABLE 7-9 Ceramic Compositions

	k_{33}	k_p	$\varepsilon_{33}^T / \varepsilon_c$	Change in N_1 −60 to +85°C %	Change in N_1 per time decade, %
PZT-4	0.70	0.58	1,300	4.8	+1.5
PZT-5A	0.705	0.60	1,700	2.6	+0.2
PZT-5H	0.75	0.65	3,400	9.0	+0.25
PZT-6A	0.54	0.42	1,050	<0.2	<0.1
PZT-8	0.62	0.50	1,000	2.0	+1.0
Na$_{0.5}$K$_{0.5}$NbO$_3$	0.605	0.46	500	?	?
PbNb$_2$O$_6$	0.38	0.07	225	3.3	?

k_{33} = coupling constant for longitudinal mode.
k_p = coupling constant for radial mode.
ε_{33}^T = permittivity parallel to poling field, stress-free condition.
N_1 = frequency constant (resonance frequency × length).

The very low mechanical Q of lead metaniobate has encouraged its use in ultrasonic flaw detection, where the low Q helps the suppression of ringing. The high acoustic velocity of sodium potassium niobate is of advantage in high-frequency thickness-extensional thickness-shear transducers, since this allows greater thickness and therefore lower capacitance.

Since ceramic materials can be fabricated in a wide range of sizes and shapes, they lend themselves to designs for applications which would be difficult to achieve with single crystals. Figure 7-13 illustrates the use of simple piezoelectric elements in a high-voltage source capable of generating an open-circuit voltage of approximately 40 kV. Piezoelectric accelerometers suitable for

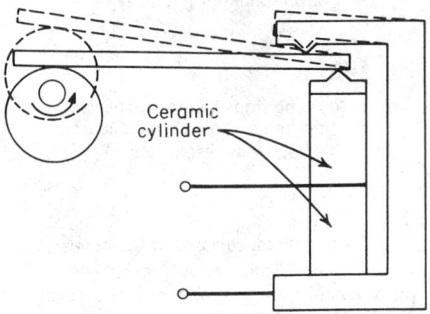

Fig. 7-13. High-voltage generator.

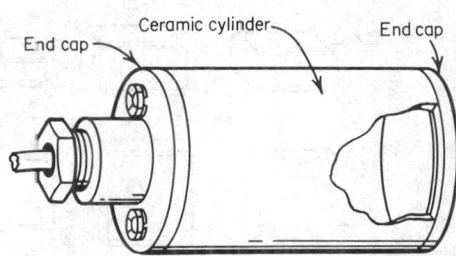

Fig. 7-14. Underwater sound transducer.

measuring vibrating accelerations over a wide frequency range are readily available in numerous shapes and sizes. Figure 7-14 shows a typical underwater transducer which uses a hollow ceramic cylinder polarized through the wall thickness. Flexing-type piezoelectric elements can handle larger motions and smaller forces than single plates.

29. Resonators and Filters. The development of temperature-stable filter ceramics has spurred the development of ceramic resonators and filters. These devices include simple resonators, multielectrode resonators and cascaded combinations thereof, mechanically coupled pairs of resonators, and ceramic ladder filters, covering a frequency range from 50 Hz to 10 MHz.

Two lead titanate-zirconate compositions, PZT-6A and PZT-6B ceramics, are most widely used for resonator and filter applications. PZT-6A, having high electromechanical coupling coefficient (45%) and moderate mechanical Q (400), is used for medium to wide bandwidth applications, while PZT-6B, with moderate coupling (21%) and higher Q (1,500), is used for narrow bandwidths. The compositions exhibit a frequency constant stable to within $\pm 0.1\%$ over a temperature range from -40 to $+85°C$. The frequency characteristics increase slowly with time at less than 0.1% per decade of time.

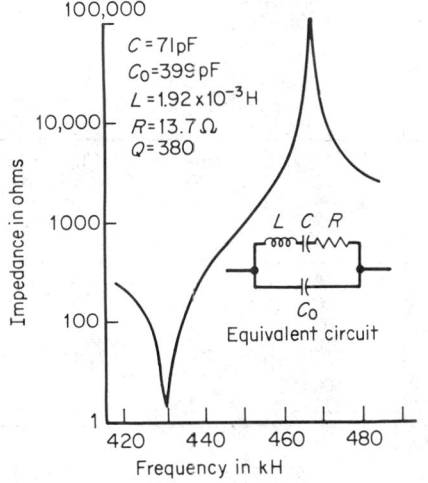

Fig. 7-15. Impedance response of a fundamental radial resonator.

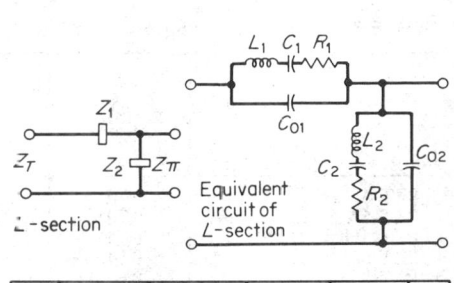

Fig. 7-16. Ceramic L section.

Disk	C (pF)	C_0 (pF)	L (henrys)	R (ohms)	Q
1	6	153	2.04×10^{-2}	137	425
2	12	324	1.05×10^{-2}	62	478

30. Ceramic Resonators. A thin ceramic disk with fully electroded faces, polarized in its thickness direction, has its lowest excitable resonance in the fundamental radial mode. The impedance response of such a disk and its equivalent circuit are shown in Fig. 7-15. Ceramic resonators can be used in various configurations for single-frequency applications or combined in basic L-sections to form complete bandpass filter networks (see Fig. 7-16). Table 7-10 presents typical performance characteristics for ferroelectric ceramic filter elements.

TABLE 7-10 Ferroelectric Ceramic Filter Elements

Center frequency	50 Hz to 10 MHz
Bandwidth	½ to 10% of center frequency
Impedance	A few to 50,000 Ω
Temperature stability	±0.1% from −55 to +85°C
Operating temperature	To 125°C
Aging stability	+0.06% center frequency per decade of time

Quartz Crystal Devices

BY JOHN R. VIG

31. Introduction. Piezoelectric crystal devices are used primarily for precise frequency control and timing. The piezoelectric material used for most applications is quartz. A quartz crystal acts as a stable mechanical resonator, which, by its piezoelectric behavior and high Q, determines the frequency generated in an oscillator circuit. Bulk-wave resonators are available in the frequency range from about 1 kHz to 200 MHz. Surface-acoustic-wave (SAW) and shallow-bulk-acoustic-wave devices can be made to operate at well above 1 GHz. SAW devices are discussed in Sec. 9; this section deals with bulk-wave resonators.

In the manufacture of the different types of quartz resonators, wafers are cut from the mother crystal along precisely controlled directions with respect to the crystallographic axes. The properties of the device depend strongly on the angles of cut. After shaping to required dimensions, metal electrodes are applied to the quartz wafer, which is mounted in a holder structure. The assembly, called a *crystal unit* (or *crystal* or *resonator*) is sealed hermetically.

To cover the wide range of frequencies, different cuts, vibrating in a variety of modes, are used. Above 1 MHz, the AT or BT cuts, which vibrate in the thickness shear mode, are commonly used. For high-precision applications, the SC cut (also thickness shear mode) has important advantages over the AT and BT cuts. AT-, BT-, and SC-cut crystals can be manufactured for fundamental-mode operation at frequencies up to about 40 MHz. Above 40 MHz, *overtone* crystals are generally used. Such crystals operate at a selected harmonic mode of vibration. Below 1 MHz, tuning forks, X-Y and NT bars (flexure mode), +5° X cuts (extensional mode), or CT and DT cuts (face shear mode) are commonly used.

32. Equivalent Circuit. The circuit designer treats the behavior of a quartz crystal unit by considering its equivalent circuit (Fig. 7-17). The mechanical resonance in the crystal is rep-

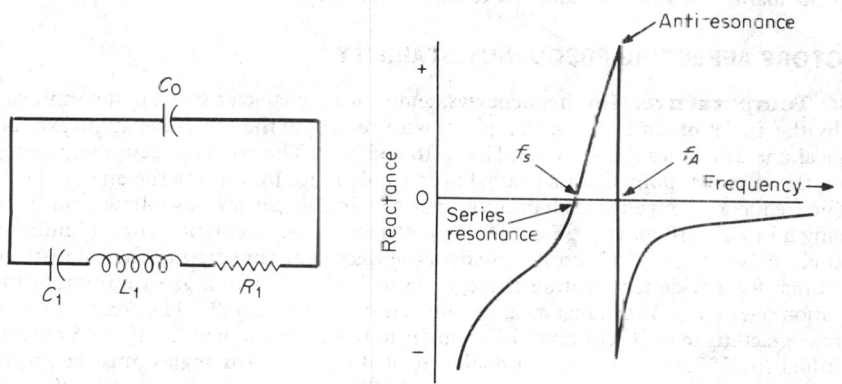

Fig. 7-17. Equivalent circuit and frequency-reactance relationship

resented by L_1, C_1, and R_1. Because it is a dielectric with electrodes, the device also displays an electrical capacitance C_0. The parallel combination of C_0 and the motional arm, C_1-L_1-R_1, represents the equivalent circuit. As shown in Fig. 7-17b, the reactance of this circuit varies with frequency.

The Q values ($Q^{-1} = 2\pi f_s R_1 C_1$) of quartz-crystal units are much higher than those attainable with other circuit elements. In general-purpose units, the Q is usually in the range of 10^4 to 10^6. The intrinsic Q of quartz is limited by internal losses. For AT-cut crystals, the intrinsic Q has been experimentally determined to be 16×10^6 at 1 MHz; it is inversely proportional to frequency. At 5 MHz, for example, the intrinsic Q is 3.2×10^6.

33. Oscillators. The commonly used crystal oscillator circuits fall into two broad categories. In series-resonance oscillators, the crystal operates at series resonance, i.e., at f_s. In parallel-resonance or antiresonance oscillators, the crystal is used as a positive reactance; i.e., the frequency is between f_s and f_A. In this latter mode of operation the oscillator circuit provides a load capacity to the crystal unit. The oscillator then operates at the frequency where the crystal unit's reactance cancels the reactance of the load capacitor. If the load capacitance is changed, the oscillator frequency will change. An important parameter in this connection is the capacitance ratio $r = C_0/C_1$. Typically, the value of r is a few hundred for fundamental-mode crystals. It is larger by a factor of n^2 for nth-overtone crystals.

When a load capacitor C_L is connected in series with a crystal, the series-resonance frequency of the combination is shifted from f_s by a Δf which is related to the other parameters by

$$\frac{\Delta f}{f_s} \approx \frac{C_0}{2r(C_0 + C_L)}$$

For a typical fundamental-mode AT-cut crystal unit with $r = 250$, $C_0 = 5$ pF, and $C_L = 30$ pF, the shift is 286 ppm. If such a crystal unit is to remain stable to, for example, 1×10^{-9}, the load reactance due to C_L and the other circuit components must remain stable to within 1.2×10^{-4} pF. The frequency can also be "tuned" by intentionally changing C_L. For the above example, a change of 1 pF in C_L shifts the frequency by nearly 10 ppm.

34. Filters. Quartz crystals are used as selective components in crystal filters. With the constraint imposed by the equivalent circuit of Fig. 7-17, filter design techniques can provide bandpass or band-stop filters with prescribed characteristics. Crystal filters exhibit low insertion loss, high selectivity, and excellent temperature stability.

The main difference between a filter and oscillator crystal is the requirement that the filter crystal have only one strong resonance in its region of operation, with all other responses (unwanted modes) attenuated as much as possible. The application of energy-trapping theory can provide such a response. If electrode size and thickness are selected in accordance with that theory, the energy of the main response is trapped between the electrodes, whereas the unwanted modes are untrapped and propagate toward the edge of the crystal resonator, where their energy is dissipated. It is possible to manufacture AT-cut filter crystals with greater than 40-dB attenuation of the unwanted modes relative to the main response.

35. Crystal-Unit Standardization. Standardization exists concerning dimensions and performance characteristics of crystal units. The principal documents are the U.S. Military Standards, the standards issued by the Electronics Industry Association (EIA), and by the International Electrotechnical Commission (IEC). The generally used designations are those of the Military Standards, such as HC-6 and HC-18 for the commonly used crystal enclosures.

FACTORS AFFECTING FREQUENCY STABILITY

36. Temperature. The frequency-vs.-temperature characteristics are determined primarily by the angles of cut of the crystal plates with respect to the crystallographic axes of quartz. Typical characteristics are shown in Figs. 7-18 and 7-19. The points of zero temperature coefficient, the *turnover points*, can be varied over a wide range by varying the angles of cut.

The frequency-temperature characteristic of AT- and SC-cut crystals follow a third-order law, as shown in Fig. 7-18 for the AT cut. A slight change in the orientation angle (7 minutes in the example shown in Fig. 7-18) greatly changes the frequency-temperature characteristic. Curve 1 is optimal for a wide temperature range (-55 to $105°C$). Curve 2 gives minimum frequency deviation over a narrower range near the inflection temperature T_i. The frequency-vs.-temperature characteristic of SC-cut crystals is similar to the curves shown in Fig. 7-18 except that T_i is shifted to $95°C$. The SC cut is a doubly rotated cut; i.e., two angles must be specified and controlled during the manufacturing process. The SC cut is therefore more difficult to manufacture with predictable frequency-vs.-temperature characteristics than the singly rotated cuts, such as the AT and BT cuts. Many of the crystal cuts have a parabolic frequency-vs.-temperature characteristic, as shown in Fig. 7-19 for the BT, CT, and DT cuts. The turnover temperatures of these cuts can also be shifted up or down by changing the angles of cut.

To achieve the highest stability, oven-controlled oscillators are used. In such oscillators, the crystal unit and the temperature-sensitive components of the oscillator are placed in a stable

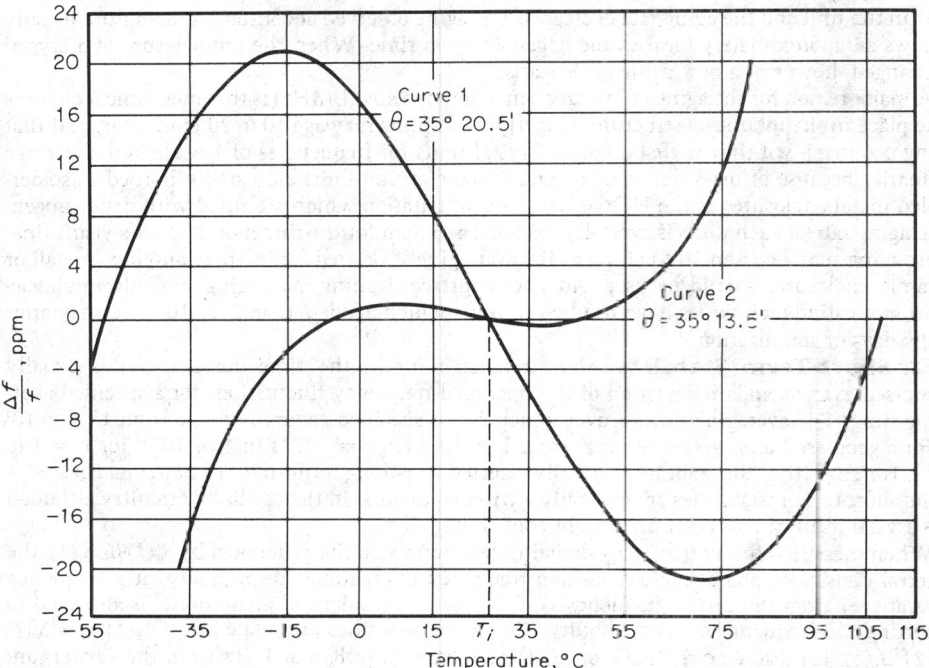

Fig. 7-18. Frequency-temperature characteristics of AT-cut crystals.

oven whose temperature is set to the crystal's turnover temperature. Oven-controlled oscillators are bulkier and consume more power than other types. In addition, when an oven-controlled oscillator is turned on, one must wait about 10 min for the oscillator to stabilize. During warm-up, the thermal stresses in the crystal can produce significant frequency shifts. This thermal-transient effect means that the typical warm-up time of an oscillator is several minutes longer than the time it takes for the oven to stabilize. The thermal-transient effect is absent in SC-cut crystals. In oscillators which use SC-cut crystals, the warm-up time can therefore be much shorter.

In temperature-compensated crystal oscillators (TCXOs) the crystal's frequency-vs.-temperature behavior is compensated by varying a load capacitor. The output signal from a temperature sensor, e.g., a thermistor network, is used to generate the correction voltage applied to a varactor. Digital techniques are capable of providing better than 1×10^{-7} frequency stability from -40 to $+75°C$. TCXOs are smaller and consume less power than oven-controlled oscillators and require no lengthy warm-up times. A major limitation on the stabilities achievable with TCXOs is the thermal hysteresis exhibited by crystal units (see Par. **7-39**).

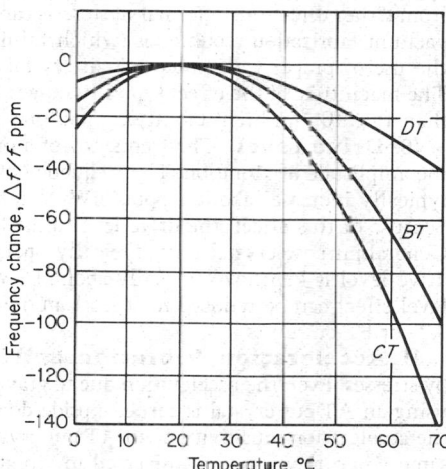

Fig. 7-19. Parabolic frequency-temperature characteristic of some crystal cuts.

37. Aging. Aging, the gradual change in a crystal's frequency with time, can be due to several causes. The main causes are mass transfer to or from the resonator surfaces (due to adsorption and desorption of contamination) and stress relief within the mounting structure or at the interface between the quartz and the electrodes. The observed aging is the sum of the aging produced by the various mechanisms and may be positive or negative. Aging is also sometimes referred to as *drift* or *long-term stability*.

The aging rate of a crystal unit is highest when it is new. As time elapses, stabilization occurs within the unit and the aging rate decreases. The aging observed at constant temperature usually follows an approximately logarithmic dependence on time. When the temperature of a crystal is changed, however, a new aging cycle starts.

A major reason for the aging of low-frequency units (below 1 MHz) is that mechanical changes take place in the mounting structure. Properly made units may age 10 to 20 ppm/year, half that aging occurring within the first 30 days. Crystal units for frequencies of 1 MHz and above age primarily because of mass transfer. General-purpose crystal units are usually housed in solder-sealed metal enclosures of the HC-6 or HC-18 configuration which are filled with dry nitrogen. The aging rate of such units is typically specified as 5 ppm for the first month; over a year's time their aging may be from 10 to 60 ppm. If lower aging is desired, units in clean glass, metal, or ceramic enclosures should be used. Advanced surface-cleaning, packaging, and ultrahigh-vacuum fabrication techniques have resulted in units which age less than 1×10^{-10} per day after a few days of stabilization.

38. Short-Term Stability. Short-term stability in the time domain $\sigma(\tau)$ is usually expressed as the standard deviation of the fractional frequency fluctuations for a specified averaging time. The averaging times τ over which $\sigma(\tau)$ is specified generally range from 10^{-3} to 10^3 s. For a good oscillator, $\sigma(\tau)$ may range from 1×10^{-9} for $\tau = 10^{-3}$ s to 1×10^{-12} for $\tau = 1$ to 10^3 s. For $\tau > 10^3$ s, the stability is usually referred to as *long-term stability* or aging. For $\tau < 1$ s, the short-term instabilities are generally attributed to noise in the oscillator circuitry although the crystal itself can also be a significant contributor.

When measured in the frequency domain, short-term stability is denoted by $S_\phi(f)$ or $S_y(f)$, the spectral density of phase fluctuations and frequency fluctuations, respectively, at a frequency separation f from the carrier frequency ν. $\mathcal{L}(f)$, the single-sideband phase noise, is also used in reporting the frequency-domain stability. The three quantities are related by $f^2 S_\phi(f) = \nu^2 S_y(f) = 2 f^2 \mathcal{L}(f)$. For a low-noise oscillator, $\mathcal{L}(f)$ may be -115 dB(c) at 1 Hz from the carrier, and -150 dB(c) at 1 kHz from the carrier.

39. Thermal Hysteresis. When the temperature of a crystal unit is changed and then returned to its original value, the frequency will generally not return to its original value. This phenomenon, called *thermal hysteresis* or lack of *retrace*, can be caused by the same mechanisms as aging, i.e., mass transfer due to contamination and/or stress relief.

For a given crystal unit, the magnitude of the effect depends on the magnitude and direction of the temperature excursion and on the thermal history of the unit. The effect tends to be smaller in oven-controlled oscillators, where the operating temperature of the crystal is always approached from below, than in TCXOs, where the operating temperature can be approached from either direction. Thermal hysteresis can be minimized through the use of clean, ultrahigh-vacuum fabrication techniques (which minimize the mass-transfer contribution) and through the use of properly mounted SC-cut crystals (which minimize the stress-relief contributions). The magnitude of the effect typically ranges from several ppm in general-purpose crystals to less than 1×10^{-9} in high-stability crystals operated in oven-controlled oscillators.

40. Drive Level. The frequency of a crystal unit also depends on the drive level, i.e., on the amplitude of vibration. The frequency of an AT-cut crystal, at sufficiently high drive level, typically increases about 1 ppm/mW; that of a BT cut decreases by about the same amount. Because of this effect, the drive level must be specified when specifying a crystal's frequency. General-purpose crystal units typically operate at 2 mW. For high-precision crystal units, the drive level is kept low, typically about 1 μW. In properly designed SC-cut crystals, the drive-level effect can be reduced to at least an order of magnitude below the magnitude observed in AT- or BT-cut units.

41. Acceleration, Vibration, and Shock. The frequency of a crystal unit is affected by stresses. Even the acceleration due to gravity produces measurable effects. When an oscillator using an AT-cut crystal is turned upside down, the frequency typically shifts about 4×10^{-9}; the acceleration sensitivity of an AT-cut crystal is typically 2×10^{-9} g^{-1}. The sensitivity is the same when the crystal is subjected to vibration; i.e., the time-varying acceleration due to the vibration modulates the frequency at the vibration frequency with an amplitude of 2×10^{-9}

g^{-1}. In the frequency domain, the vibration sensitivity manifests itself as vibration-induced sidebands that appear at plus and minus the vibration frequency away from the carrier frequency. The acceleration sensitivity of SC-cut crystals can be made to be substantially less than that of comparably fabricated AT- or BT-cut crystals.

Shock places a sudden stress on the crystal. During shock, the crystal's frequency changes due to the crystal's acceleration sensitivity. If during shock the elastic limits in the crystal's support structure or in its electrodes are exceeded, the shock can produce a permanent frequency change. Crystal units made with chemically polished crystal plates can withstand shocks in excess of 20,000g. Such crystals have been successfully fired from howitzers.

42. Radiation. The degree to which ionizing radiation affects the frequency of a crystal unit depends primarily on the quality of quartz used to fabricate the device. When resonators made of natural quartz are subjected to steady-state ionizing radiation, the frequency is changed permanently by approximately a few parts in 10^{11} per rad. To minimize the effect of such radiation, high-purity cultured quartz should be used. Pulse irradiation produces a transient frequency shift due to the thermal-transient effect. This effect can be minimized by using SC-cut crystals.

Insulators

BY SAM DI VITA

43. General. Ceramics and plastics are the principal materials for electronics insulation and mounting parts. Ceramic materials are outstanding in their resistance to high temperature, mechanical deformation, abrasion, chemical attack, electrical arc, and fungus attack. Ceramics also possess excellent electrical insulating properties and good thermal conductivity and are impervious to moisture and gases. These properties of ceramics are retained throughout a wide temperature range and are of particular importance in high-power applications such as vacuum-tube envelopes and spacers, rotor and end-plate supports for variable air capacitors, rf coil forms, cores for wire-wound resistors, ceramic-to-metal seals, and feed-through bushings for transformers.

The properties of plastics differ rather markedly from ceramics over a broad range. In a number of properties, plastics are more desirable than ceramics. These include lighter weight; better resistance to impact, shock, and vibration; higher transparency; and easier fabrication with molded-metal inserts (however, glass-bonded-mica ceramic material may be comparable with plastic in this latter respect).

CERAMIC INSULATORS

44. Linear Dielectric Radio Insulators. Ceramic insulators are linear dielectrics having low loss characteristics, which are used primarily for coil forms, tube envelopes and bases, and bushings, which all require loss factors less than 0.035 when measured at standard laboratory conditions. Dielectric loss factor is the product of power factor and dielectric constant of a given ceramic. Military Specification MIL-I-10B, Insulating Compound Electrical Ceramic Class L, covers low-dielectric-constant (12 or under) ceramic electrical insulating materials, for use over the spectrum of radio frequencies used in electronic communications and in allied electronic equipments. In this specification the "grade designators" are identified by three numbers, the first representing dielectric loss factor at 1 MHz, the second dielectric strength, and the third flexural strength (modulus of rupture).

Table 7-11 lists the various types of ceramics and their grade designators approved for use in the fabrication of Military Standard ceramic radio insulators specified in MIL-I-23264A, Insulators—Ceramic, Electrical and Electronic, General Specification For. This specification covers only those insulators characterized by combining the specific designators required for the appropriate Military Standard insulators used as standoff, feed-through, bushing, bowl, strain, pin, spreader, and other types of insulators. As a typical example of one military standard for insulators, a feed-through is shown in Fig. 7-20.

Currently, Grade L-242 is typical of porcelain, L-422 of steatite, L-523 of glass, L-746 of alumina, L-442 of glass-bonded mica, and L-834 of beryllia.

TABLE 7-11 Property Chart of Insulating Ceramic Materials Qualified under MIL-I-10

Class L ceramics	Grade designators	Power factor, 1 MHz	Dielectric constant, 1 MHz	Dielectric loss factor, 1 MHz	Dielectric strength, V/mil	Flexural strength, lb/in.²
Steatite:						
Unglazed	L-523	0.00069-0.0010	6.42-6.03	0.0041-0.0063	230	21,200
Glazed	L-543	0.0008-0.0014	5.73-6.14	0.005-0.008	330	28,600
Porcelain:						
Glazed	L-232	0.0076-0.0099	5.42-6.01	0.041-0.059	249	13,600
Zircon:						
Unglazed	L-433	0.0012-0.0014	8.14-8.22	0.010-0.012	259	22,900
Glazed	L-413	0.00119	8.92	0.011	191	29,800
Alumina:						
Unglazed	L-746	0.0001-0.0008	8.14-	0.0009	500	68,000
Glass:						
Borosilicate (Pyrex)	L-622	0.00074	4.19	0.0031	226	16,000
High silica (Vycor)	L-541	0.0017	3.78	0.0065	363	8,520
Glass-bonded mica:						
Unglazed	L-442	0.0017-0.0018	7.08-7.44	0.012-0.013	382	18,600
Forsterite:						
Glazed	L-723	0.0003	6.37	0.002	200	20,000
Cordierite:						
Unglazed	L-321	0.0049	4.57	0.022	245	13,000
Wallastonite:						
Unglazed	L-621	0.0004	6.49	0.003	293	13,700
Beryllia	L-834	0.00015	6	0.0009	295	29,000

Applications, electronic and electrical:

Electronic and electrical
Bushings
Coil forms
Capacitor leads
Electronic packages
Envelopes, tubes
Insulators, antenna
Insulators, cyclotron
Insulators, spark plug
Insulators, thermocouple
Insulators, tube element
Housing lamp
Magnetron parts
Printed-circuit boards
Radomes
Resistor bases
Supports, tube element
Shafts, condenser
Substrates
Terminals
Tube windows
Transformer bushings
Tuner-coupling arms

45. Good Practice Design. The data compiled in Table 7-12 represent good design practice for ceramic radio insulators but do not necessarily imply that closer tolerances or any important dimension or special designs cannot be produced by special handling.

46. Surface Finishes. Surface finish is the deviation of the heights and depths of surface irregularities from a central reference line.

An *applied* finish results from a modification of the as-fired condition by grinding, lapping, polishing, tumbling, or glazing. Glazed (vitreous coating) surfaces provide the smoothest surfaces, finishes of 1 μin (25 nm) or better being fairly common. Composition specifically designed to have smooth as-fired finishes can be made having surface finishes of 5 μin (125 nm) or smoother, whereas composition designed for other applications may run up to 50 μin (1,500 nm) surface finish.

47. High-Thermal-Shock-Resistant Ceramics. Lithia porcelain is the best thermal-shock-resistant ceramic because of its low (close to zero) coefficient of thermal expansion. It is followed in order by fused quartz, cordierite, high-silica glass, porcelain, steatite beryllium

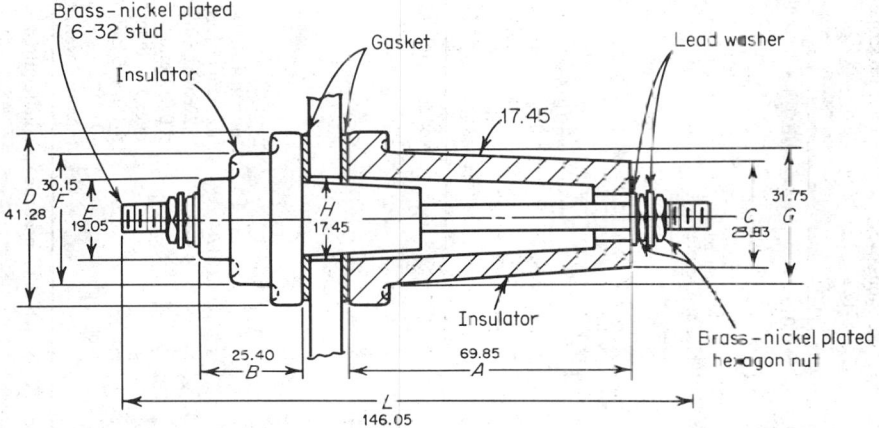

Fig. 7-20. Feed-through insulator as per MIL-23264/13A, Type-NL 422B34-046 (dimensions in millimeters).

oxide, alumina, and glass-bonded mica. Those materials find wide use for rf coil forms, cores for wire-wound resistors, stator supports for air dielectric capacitors, coaxial cable insulators, standoff insulators, capacitor trimmer bases, tube sockets, relay spacers, and base plates for printed radio circuits.

48. High Thermal Conductivity. High-purity beryllium oxide is unique among homogeneous materials in possessing high thermal conductivity comparable with metal, together with excellent electrical insulating properties.

Care must be exercised in the use of beryllium oxide because its dust is highly toxic. Although it is completely safe in dense ceramic form, any operation that generates dust, fumes, or vapors is potentially very dangerous.

Some typical uses of beryllium oxide are:
1. Heat sinks for high-power rf amplifying tubes, transistors, and other semiconductors
2. Printed-circuit bases
3. Antenna windows and tube envelopes
4. Substrates for vapor deposition of metals
5. Heat sinks for electronic chassis or subassemblies

49. Mounting of Ceramic Parts. Parts to be mounted on a flat surface should be designed with one, two, or three mounting bosses that can be ground flat after firing to prevent breakage during the mounting operation. If two bosses are used, they should be spaced 180°, and if three bosses are used they should be spaced 120°. When screws are used, they should be properly secured with corrosion-resistant lock washers or, if this is not practical, secured with weatherproof cement. Ceramic insulators should be cushioned with fungus-proof resilient gasket, fish paper, cork, lead, or other shock-absorbing material.

TABLE 7-12 Good Design Practice for Ceramic Radio Insulator: Dimension Tolerances
1 in = 2.54 cm

Material	Type of dimensions	Dimensions up to and including 12½ in	Dimensions over 12½ in	Cylindrical shapes (OD and ID), in	Hole center,* in	Hole diameter,† in
Glass or glass-bonded mica‡	Noncritical and critical	±0.010	±0.015	±0.015	12½ diam and less, ±0.005; over 12½ diam, ±0.007	0.500 diam and less, ±0.005; over 0.500 diam ±1%
	Wall	Not applicable	Not applicable	±0.005	Not applicable	Not applicable
Porcelain	Noncritical	±3% with ±⅜ max, ±3/32 min	±1.5%	Not applicable	Not applicable	Not applicable
	Critical: Unglazed	±1.5% with ±0.015 min	±1.5%	±1.5% with ±0.015 min	±1.5% with ±0.015 min	±1.5% with ±0.010 min
	Thickness	±0.015	Not applicable	±0.015	Not applicable	Not applicable
	Glazed	±2% with ±0.1875 max	±1.5%	±1.5% + 0.015 with ±0.020 min	±1.5% with ±0.015 min	±1.5% with 0.020 min
Steatite and other fired ceramics (except porcelain)	Noncritical	±2% with ±⅛ max, ±3/64 min	±1%	Not applicable	Not applicable	Not applicable
	Critical: Unglazed	±1% with ±0.005 min	±1%	±1%	Less than 0.500 diam ± 0.005; 0.500 diam and greater ±1%	Less than 0.500 diam ± 0.005; 0.500 diam and greater ±1%
	Thickness	±0.010	Not applicable	±0.010 (nominal wall)	Not applicable	Not applicable
	Glazed	±2% + 0.012	±2% + 0.012	1 in diam or less, ±2% with ±0.012 min; over 1 in diam, ±1% + 0.010	Less than 0.500 diam ± 0.005; 0.500 diam and greater ±1%	Less than 0.500 diam. +0.015; 0.500 diam and greater ±3%
	Thickness	±1% + 0.012	Not applicable	±0.010 (nominal wall)	Not applicable	Not applicable

*If pin gages are used for inspecting hole-center spacings, the design of the gage should be such as to meet the specific requirements. If pin gages are used, consideration should be given to the tolerance between hole centers as well as hole-diameter tolerance.

†Holes leading from the glazed surfaces should have the tolerances for glazed surfaces applied except when specified that the glaze should be removed from the hole, in which case the tolerances for unglazed surfaces should apply. Holes which are not perfect circles should have the same tolerances applied to the minor axis only.

‡The flatness tolerance should be ±0.0015 in/in. When angles or V cuts are required in the edges of flat pieces, the tolerance on such angular dimensions should be ±1°.

PLASTIC INSULATORS

50. General. The term *plastics* usually refers to a class of synthetic organic materials (resins) which are solid in finished form but at some stage in their processing are fluid enough to be shaped by application of heat and pressure. The two basic types of plastic are *thermoplastic resins*, which, like wax or tar, can be softened and resoftened repeatedly without undergoing a change in chemical composition, and *thermosetting resins* which undergo a chemical change with application of heat and pressure and cannot be resoftened.

51. Choice of Plastics. Some of the differences between plastics that should be considered when defining specific needs are degree of stiffness or flexibility; useful temperature range; tensile, flexural, and impact strength; intensity, frequency, and duration of loads; electrical strength and dielectric losses; color retention under environment; stress-crack resistance over time; wear and scratch resistance; moisture and chemical resistance at high temperature; gas permeability; weather and sunlight resistance over time; odor, taste, and toxicity; and long-term creep properties under critical loads.

52. Reinforced Plastics. These comprise a distinct family of plastic materials which consist of superimposed layers of a synthetic resin-impregnated or resin-coated filler. Fillers such as paper, cotton fabric, glass fabric or fiber, glass mats, felted asbestos, nylon fabric — either in the form of sheets or macerated — are impregnated with a thermosetting resin (phenolic, melamine, polyester, epoxy, silicone). Heat and pressure fuse these materials into a dense insoluble solid and nearly homogeneous mass, which may be fabricated in the form of sheets or rods or in molded form.

53. Mounting Plastic Parts. Choice of assembly method depends on the strength needed, contour to the parts, appearance demands, and mold design. Parts may be attached to each other or to other materials by bolts or screws, but care must be exercised to assure a tight fit without crushing or damaging the plastic. Bolts with large heads and washers should be used to distribute the damping force. Parts must fit snugly; loose fit allows movement, causing rapid wear and deterioration of the part by abrasion. Load-bearing parts are often made with metal inserts to hold the bolts and distribute the stress.

Power and Receiving Tubes

BY G. W. TAYLOR AND B. SMITH

54. Introduction. Power and receiving tubes are active devices using either the flow of free electrons in a vacuum or electrons and ions combined in a gas medium. The vast majority of uses for low-power electron tubes are found in the area of high vacuum with controlled free electrons. The source of free electrons is a heated material that is a thermionic emitter (cathode). The control element regulating the flow of electrons is called a *grid*. The collector element for the electron flow is the *anode*, or *plate*. Power tubes, in contrast to receiving-type tubes, handle relatively large amounts of power, and for reasons of economy, a major emphasis is placed on efficiency. The traditional division between the two tube categories is at the 25-W plate-dissipation rating level.

Receiving tubes (and transistors, see Pars. **7-80** to **7-90**) provide the essential active-device function in electronic applications. The general uses of receiving tubes cover nearly all functions of radio and television receivers, low-power transmitters, telephone systems, industrial control, and measurement devices. Most receiving-type tubes produced at the present time are for replacement use in existing equipments. Most of the electronic functions of the new equipment designs are being handled by solid-state devices.

Power tubes are widely used as high-power-level generators and converters in radio and television transmitters, radar, sonar, manufacturing-process operations, and medical and industrial x-ray equipment.

55. Classification of Types. Power and receiving-type tubes can be separated into groups according to their circuit function or by the number of electrodes they contain. Table 7-13 illustrates these factors and compares some of the related features. The physical shape and location of the grid relative to the plate and cathode are the main factors that determine the amplification factor μ of the triode. The μ values of triodes generally range from about 5 to over 200. The mathematical relationships between the three important dynamic tube factors are

Amplification factor $\mu = \Delta e_b / \Delta e_{c1}$

Dynamic plate resistance $r_p = \Delta e_b / \Delta i_b$ e_b, e_{c1}, i_b = const

Transconductance Sm or $Gm = \Delta i_b / \Delta e_{c1}$

where e_b = total instantaneous plate voltage, e_{c1} = total instantaneous control grid voltage, and i_b = total instantaneous plate current. Note that $\mu = Gmr_p$. Figure 7-21 shows the curves of

TABLE 7-13   Tube Classification by Construction and Use

Tube type	No. of active electrodes	Typical use	Relative features and advantages
Diode	2	Rectifier	High back resistance
Triode	3	Low- and high-power amplifier, oscillator, and pulse modulator	Low cost, circuit simplicity
Tetrode	4	High-power amplifier and pulse modulator	Low drive power, low feedback
Pentode	5	Low-power amplifier	High gain, low drive, low feedback, low anode voltage
Hexode, etc. ...	6 or more	Special applications	Multiple-input mixers, converters

plate and grid current as a function of plate voltage at various grid voltages for a typical triode with a μ value of 30.

The tetrode, a four-element electron tube, is formed when a second grid (screen grid) is mounted between grid 1 (control grid) and the anode (plate). The plate current is almost independent of plate voltage. Figure 7-22 shows the curves of plate current as a function of plate voltage at a fixed screen voltage and various grid voltages for a typical power tetrode.

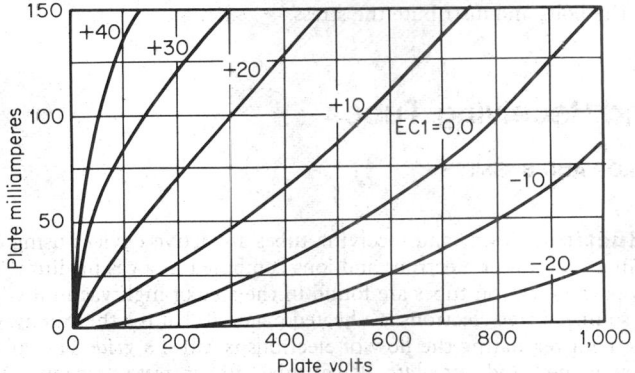

Fig. 7-21. Typical triode plate characteristics.

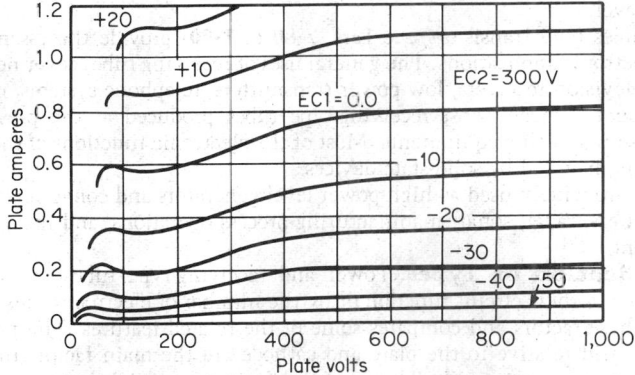

Fig. 7-22. Typical tetrode plate characteristics.

56. Basic Construction. *Cathodes* used in most power and receiving tubes obtain the energy required for electron emission from heat. The two types of cathodes based upon the method of heating are directly heated (filament types) and indirectly heated. The three types of emitting surfaces most commonly used are thoriated tungsten, alkaline-earth oxides, and tungsten barium aluminate–impregnated emitters. While most receiving tubes and many power tubes use an oxide cathode because of its high efficiency, i.e., ratio of emission current per watt

TABLE 7-14 Typical Characteristics of Emitters

Emitter	Heating method	Operating temp, °C	Emission density, A/cm²	
			Average	Peak
Oxide	Direct and indirect	700–820	0.100–0.5	0.100–20
Thoriated tungsten	Direct	1600–1800	0.04–0.43	0.04–10
Impregnated tungsten	Direct and indirect	900–1175	0.5–8.0	0.5–12

of heating power, thoriated tungsten and tungsten-impregnated cathodes are more tolerant to ion bombardment and are used in many high-power, high-voltage tubes. In some applications which previously used either oxide-coated cathodes or thoriated tungsten cathodes, the impregnated cathode is being used because of its ruggedness and high average emission capability. The characteristics of these three emitting surfaces are given in Table 7-14.

Grids. The most widely used grid construction in receiving-type tubes involves helically wound fine wire of nickel alloy supported on two side rods. Much of the mechanical strength of this construction is derived from the strength of the lateral grid wires themselves. Grids of high electrical performance require finer wires of low mechanical strength and are not good for use as a support. A structure that is used in modern, close-spaced, high-performance receiving tubes is the frame grid. Extremely fine lateral wire is wound under tension on a rigid open frame with close spacing between grid wires and between tube electrodes, thereby improving tube performance. Grid wire of 12 μm diameter or less is practical using this construction technique.

The grids of power tubes are made to very close tolerances and must maintain their shape and spacing at elevated temperatures. They must also withstand shock and vibration. For this reason, most grids in power tubes are made with tungsten or molybdenum, which have good hot strength. The grids are also required not to emit either primary or secondary electrons. Gold plating of the wires is the most widely used technique to inhibit primary emission in tubes with oxide cathodes. The maximum safe operating temperature for gold plating is on the order of 550°C. Many special coatings have been developed for specific applications that are effective in reducing grid emission.

Anodes. The anode (plate) usually encloses the other electrodes, and in most cases it is basically cylindrical in cross section. The major exception is the planar construction used in high-frequency types. The anode of most receiving tubes and many older medium-power tubes is completely enclosed by the tube envelope. Stamped or formed sheet nickel is generally used in receiving tubes, although some tubes have been made with aluminum-clad steel. The most commonly used material for power tubes with enclosed internal anodes is zirconium-clad molybdenum. In most power tubes, the anode is a part of the envelope, and since the outer surface is external to the vacuum, it can be cooled directly (see Fig. 7-23). The material normally used for external anodes is copper.

Envelopes of most receiving tubes and power tubes of older design are glass. Glass is easy to form, readily available, and low in cost. Ceramic material is used as the envelope material in power tubes of recent design and in some receiving tubes for special appli-

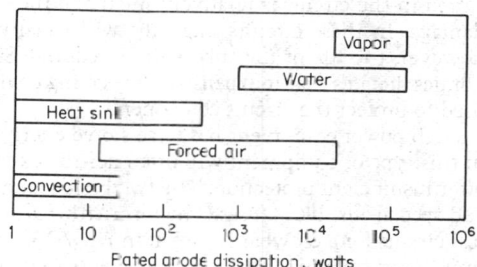

Fig. 7-23. Anode cooling methods.

cations. Ceramic envelopes are more expensive than glass; however, higher-temperature processing improves the vacuum, resulting in longer tube life. Metal envelopes have been used in the manufacture of receiving tubes, such as types 6L6, 12SQ7, 12SK7. Receiving tubes made with iron (or steel) envelopes possessed advantages with respect to electrostatic shielding and were less subject to mechanical damage. However, this is an expensive construction and not widely used today. An exception is the *nuvistor* construction, which uses a ceramic-metal envelope to house a small cantilever-supported cylindrical structure. Examples of this construction are the types 6DS4 and 6CW4.

57. Life Expectancy. Life expectancy is one of the most important factors to be considered in the use of tubes. Whereas it is impossible to predict the life for an individual tube, accurate average life expectancy can be determined for a group of tubes run under closely controlled conditions. Two types of failures are encountered.

Catastrophic failure is the premature demise of the electron tube long before its statistical life expectancy. Most of these failures are in the mechanical category, i.e., broken filament, poor weld, cracked insulators, etc. These failures generally occur early in life and can be reduced by careful design, selection of materials, and rigorous control of manufacturing techniques. These types of failures have been dramatically reduced in modern tubes.

Wear-out failure results from gradual deterioration of electrical characteristics. Modern tube technology has virtually overcome manufacturing problems that result in fast wear-out. The one remaining consideration for achieving ultimate reliability, i.e., the proper application of the tube, is the responsibility of the user. A modern tube, properly used, can be reasonably expected to perform for many thousands of hours.

The cathode is the heart of any tube, and cathode temperature, together with current loading, is of major importance in determining tube life. Oxide cathodes are used in almost all receiving tubes and in the majority of modern power tubes. Figure 7-24 is a plot of the tube wear-out life capability of oxide cathodes as a function of cathode current density. This curve assumes that the tubes are properly cooled, the cathode temperature is adjusted to the optimum value required by the current demand, the heater voltage is regulated to 1%, and the tube is protected from arc damage. Thoriated tungsten emitters, which are used in some high-voltage power tubes, have very long life. This type emitter, operating at 1650°C with a filament-voltage variation of less than 1%, has a life capability in excess of 25,000 h. An increase of 5% in filament voltage will reduce life approximately 50%. Impregnated-cathode life is proportional to the quantity of active material in the matrix and the operating temperature. Life tests conducted on these cathodes have shown that they are capable of operating in excess of 40,000 h at 1000°C and at least 5,000 h at 1075°C.

58. Cooling Methods. Cooling of the tube envelope, seals, and anode, if external, is a major factor affecting tube life. The data sheets provided by tube manufacturers include with the cooling requirements a maximum safe temperature for the various external surfaces. The temperature of these surfaces should be measured in the operating equipment. The temperature can be measured with thermocouples, optical pyrometers, a temperature-sensitive paint such as Tempilaq, or temperature-sensitive tapes.

The envelopes and seals of most tubes are cooled by convection of air around the tube or by using forced air. The four principal methods used for cooling external anodes of tubes are by air, water, vapor, and heat sinks. Other cooling methods occasionally used are oil, heat pipes, refrigerants, such as Freon, and gases, such as sulfahexafluoride. Figure 7-23 shows the range of anode dissipation generally used for the four principal methods used for anode cooling.

59. Protective Circuits. Arcs can damage or destroy electron tubes, which may significantly increase the equipment operating cost. In low- or medium-power operation, the energy stored in the circuit is relatively small and the series impedance is high, which limits the tube damage. In these circuits, the tube will usually continue to work after the fault is removed; however, the life of the tube will be reduced. Since these tubes are normally low in cost, economics dictates that inexpensive slow-acting devices, e.g., circuit breakers and common fuses, be used to protect the circuit components.

High-power equipment has large stored energy, and the series impedance is usually low. Arcs in this type of equipment will often destroy expensive tubes, and slow-acting protective devices offer insufficient protection. The two basic techniques used to protect tubes are *energy diverters* and special *fast-blow fuses*. The term *crowbar* is commonly used for energy diverters. The typical circuit for a crowbar is shown in Fig. 7-25. In the event of a tube arc, the trigger-fault sensor unit "fires" to a crowbar, which is a very-low-impedance gas-discharge device. The firing time can be less than 2 μs. The low impedance in the crowbar arm is in shunt with the Z_2 and tube arm and diverts current from the tube during arcs. The impedance Z_2, which is in series with the tube, is used to ensure that most of the current is diverted through the crowbar. The value of Z_2 is primarily limited by efficiency considerations during normal operation. The impedance Z_1 is required to limit the fault current in the storage condenser to the maximum current rating of the condenser. The impedance Z_g is the internal power-supply impedance. Devices used as crowbars are thyratrons, ignitrons, triggered spark gaps, and plasmoid-triggered vacuum gaps.

60. Fast Fuses. Two types of fast-blow fuses are used to protect power tubes. They are *exploding-wire* and *exothermic* fuses. Exploding-wire fuses require milliseconds to operate and are limited in their ability to protect the tube. Exothermic fuses, although faster-acting than exploding wires, are significantly slower than crowbars in clearing the fault current in the tube. A second disadvantage of fuses is that after each fault the power supply must be turned off and the fuse replaced. For this reason, fuses are limited to applications where the tubes seldom arc. The major advantage of fuses is low cost.

61. Special Application Considerations. *Noise.* The electronic noise generated with an electron tube is a basic limitation to the magnitude of amplification achievable. Resistor and tube noise can never be avoided; however, much can be done by proper circuit design and choice of tube types to approach the minimum attainable noise. Random division of current between electrodes contributes to the noise in multielectrode

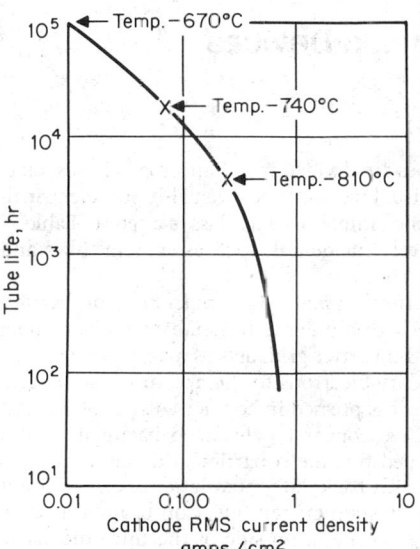

Fig. 7-24. Tube life vs. current density.

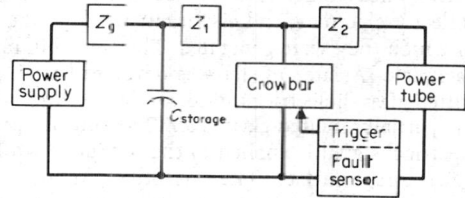

Fig. 7-25. Energy-diverter circuit.

tubes and makes pentodes 3 to 5 times as noisy as the same tube connected as a triode. Mixer tubes are also noisy, the triode converter being the lowest in noise. Mechanical noise results from some aspect of tube construction or use. Vibration of elements may cause microphonic noise. The design of the control grid is the most important factor in microphonic effects. Hum can come from poor heater design or construction.

Frequency. Transit time is a large factor in considering the upper frequency limitation of electron tubes. A finite time is taken by the electron to traverse the space from the cathode, through the grid, and on to the plate. As the frequency of operation is raised, a frequency is finally reached at which electron transit-time effects become significant. The point of onset is dependent on accelerating voltages to the grid and anode planes and the respective spacings. Closer-spaced tubes, in particular, with close spacing in the grid-to-cathode region, have reduced transit-time effects.

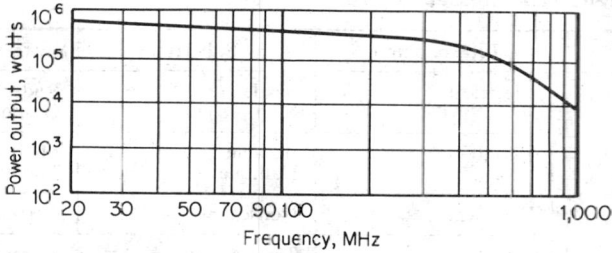

Fig. 7-26. Continuous-wave output-power capability.

Lead inductance and *capacitance* act to reduce the upper frequency limit of electron tubes. Modern high-frequency electron tubes have short structures, short lead-electrode connections, and a minimum of unnecessary or parasitic capacitance.

Power limitations are interrelated with cathode emission capability, electrode sizes, and design-form factors, materials, and methods of cooling. There is also a power-frequency relationship in that, as the frequency of required operation is increased, closer spacing and smaller-sizes electrodes are used. This reduces the power-handling capability for tube designs having the same form factor. Figure 7-26 illustrates the maximum continuous-wave power-output capability as a function of frequency for high-power gridded tubes.

Cathode-Ray Storage and Conversion Devices

BY I. REINGOLD AND M. CROST

62. Introduction. Electronic charge-storage tubes are divided into four broad classes: electrical-input–electrical-output types, electrical-input–visual-output types, visual-input–electrical-output types, and visual-input–visual output types. An example of each class is cited in Table 7-15. Tubes under these classes in which storage is merely incidental, such as camera tubes and image converters, are classed under conversion devices. See also Sec. **11**.

63. Electrical-Input Devices. *Electrical-Output Types.* The *radechon, or barrier-grid storage tube,* is a single-electron-gun storage tube with a fine-mesh metal screen in contact with a mica storage surface. The metal screen, called the barrier grid, acts as a very-close-spaced collector electrode, and essentially confines the secondary electrons to the apertures of the grid in which they were generated. The very thin mica sheet is pressed in contact with a solid metal backplate. A later model was developed with a concave copper bowl-shaped backplate and a fritted-glass dielectric storage surface. A similarly shaped fine-mesh barrier grid was welded to the partially melted glass layer. The tube is operated with backplate modulation; i.e., the input electrical signal is applied to the backplate while the constant-current electron beam scans the mica storage surface. The capacitively induced voltage on the beam side of the mica dielectric is neutralized to the barrier-grid potential at each scanned elemental area by collection of electrons from the beam and/or by secondary electron emission (equilibrium writing). The current involved in recharging the storage surface generates the output signal.

In a single-electron-gun *image-recording storage tube* (Fig. 7-27) information can be recorded and read out later. The intended application is the storage of complete halftone images, such as a television picture, for later readout. In this application, write-in of information is accomplished by electron-beam modulation, in time-sharing sequence with the readout mode. Reading is accomplished nondestructively, i.e., without intentionally changing the stored pattern, by not permitting the beam to impinge upon the storage surface. The electron-gun potentials are readjusted for readout so that none of the electrons in the constant-current beam can reach the storage surface, but divide their current between the storage mesh and the collector in proportion to the stored charge. This process is called *signal division.*

TABLE 7-15   Storage-Tube Classes

Type of input	Type of output	Subclass	Representative example
Electrical	Electrical	Single-gun Multiple-gun	Radechon Graphechon
	Visual	Bistable Halftone	Memotron Tonotron
Visual	Electrical	Nonscanned	Correlatron
	Visual	Time-sharing	Storage image tube

One type of double-ended, multiple-electron-gun *recording storage tube* with nondestructive readout operates by recording halftone information on an insulating coating on the bars of a metal mesh grid, which can be penetrated in both directions by the appropriate electron beams. Very high resolution and long storage time with multiple readouts, including simultaneous writing and reading, are available with this type of tube. Radio-frequency separation or signal cancellation must be used to suppress the writing signal in the readout during simultaneous operation.

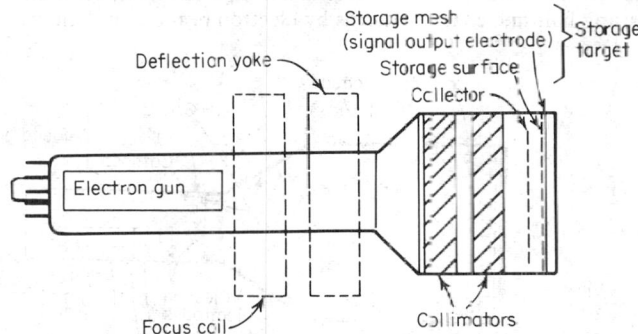

Fig. 7-27. Electrical-signal storage tube, basic structure. (*Hughes Aircraft Corp.*)

Figure 7-28 is a representative schematic drawing of a double-ended, multiple-electron-gun *membrane-target storage tube* with nondestructive readout, in which the writing beam and the reading beam are separated by the thin insulating film of the storage target. In this case there is a minimal interference of the writing signal with the readout signal in simultaneous operation, except for capacitive feed-through, which can readily be canceled. Writing is accomplished by beam modulation, while reading is accomplished by signal division. Very high resolution and long storage time with multiple readouts are available with this group of tubes.

The *graphechon* differs from the two groups of tubes just described in that its storage target operates by means of electron-bombardment-induced conduction (EBIC). Halftones are not available from tubes of this type in normal operation. The readout is of the destructive type, but since a large quantity of charge is transferred through the storage insulator during writing, a multitude of readout copies can be made before the signal displays noticeable degradation. In simultaneous writing and reading, signal cancellation of the writing signal in the readout is generally accomplished at video frequencies.

Visual-Output Types. This class comprises the *display storage tubes (DSTs)* or *direct-view storage tubes (DVSTs)*. Figure 7-29 shows a schematic diagram of a typical DVST with one electrostatic focus-and-deflection writing gun (other types may have electromagnetic focus and/or deflection) and one flood gun, which is used to provide a continuously bright display and may also establish bistable equilibrium levels. The storage surface is an insulating layer deposited on the bars of the metal-mesh backing electrode. The view screen is an aluminum-film-backed phosphor layer.

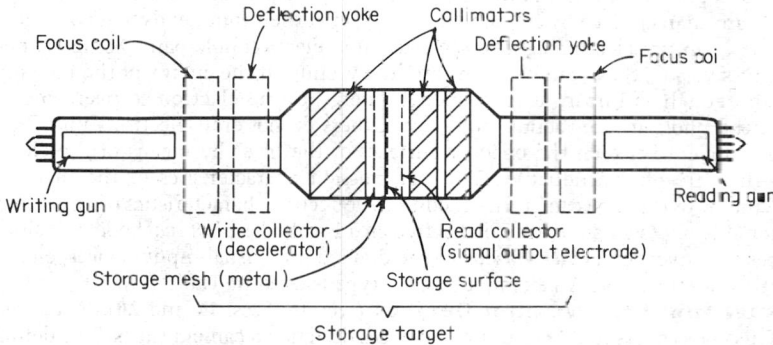

Fig. 7-28. Typical double-ended scan converter, basic structure. (*Hughes Aircraft Corp.*)

The *memotron* is a DST that operates in the bistable mode; i.e., areas of the storage surface may be in either the cutoff condition (flood-gun-cathode potential) or in the transmitting condition (collector potential), either of which is stable. The focusing and deflection are electrostatic. Normally, the phosphor remains in the unexcited condition until a trace to be displayed is written into storage; then this trace is displayed continuously until erased, so long as the flood beam is maintained in operation.

The *tonotron* is typical of a large number of halftone DSTs. These operate in the nondestructive readout mode, with the storage surface at or below flood-cathode potential. Writing is accomplished by depositing halftone charge patterns by electron-beam modulation.

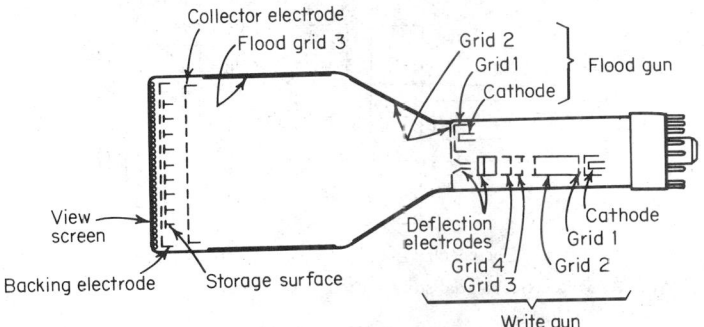

Fig. 2-29. Cross-sectioned view of direct-view storage tube. (*Westinghouse Electric Corp.*)

64. Visual-Input Devices. *Electrical-Output Types.* The *correlatron* is a storage tube that receives a visual input to a photoemissive film, focuses the photoelectron image upon a transmission-grid type of storage target, where it is stored, and later compares the original image with a similar image. A single total output current is read out for the entire image, with no positional reference. The purpose of this comparison is to ascertain whether the first and second images correlate.

Visual-Output Types. In the *storage image tube*, a positive electron image can be stored upon the insulated bars of the storage mesh. Then, if the photocathode is uniformly flooded with light to produce a flood electron cloud, a continuously bright image of the stored charge pattern can be obtained upon the phosphor screen in the following section of the tube. A high degree of brightness gain can be achieved with this type of tube, or a single snapshot picture of a continuous action can be "frozen" for protracted study.

CONVERSION DEVICES

65. Principles of Conversion Devices. The conversion devices discussed receive images in visible or infrared radiation and convert them by internal electronic processes into a sequence of electrical signals or into a visible output image. Some of these devices may employ an internal storage mechanism, but this is generally not the primary function in their operation. These tubes are characterized by a photosensitive layer at their input ends that converts a certain region of the quantum electromagnetic spectrum into electron-hole pairs. Some of these layers are photoemissive; i.e., they emit electrons into the vacuum if the energy of the incoming quantum is high enough to impart at least enough energy to the electron to overcome the work function at the photosurface-vacuum interface. Others do not emit electrons into the vacuum but conduct current between the opposite surfaces of the layer by means of the electron-hole pairs; i.e., they are photoconductive. The transmission characteristics of the material of the entrance window of the tube can greatly modify the effective characteristics of any photosurface. If the material is photoemissive, the total active area is called a *photocathode*. See also Sec. **11**.

The types of conversion devices discussed are divided into visual-input devices, electrical-output and visual-output types. An example of each type is cited in Table 7-16.

66. Visual-Input Conversion Devices (see also Secs. **11** and **20**). *Electrical-Output Types.* This class covers the large group of devices designated camera tubes. The defining attributes of the *vidicon* are a photoconductive rather than photoemissive image surface or target, a direct readout from the photosensitive target rather than by means of a return beam, and a much smaller size than the above camera tubes. The original vidicons used coincident electromagnetic deflection and focusing. Many later versions employ either or both electrostatic focusing and deflection.

A very important group of tubes is the *semiconductor diode-array vidicons*. In place of the usual photoconductive target, these tubes include a very thin monolithic wafer of the semiconductor, usually single-crystal silicon. On the beam side of this wafer a dense array of junction photodiodes has been generated by semiconductor diffusion technology. These targets are very sensitive compared with the photoconductors, and they have very low leakage, low image lag, and low blooming from saturation.

Visual-Output Types.   The *image tube,* or *image amplifier*, with input in the visible spectrum is used principally to increase the light level and dynamic range of a very low light-level image to a level and contrast acceptable to a human observer, a photographic plate, or a camera

TABLE 7-16   Conversion Devices

Type of input	Type of output	Subclass	Representative example
Visual	Electrical	Photoemissive Photoconductive	Image orthicon Vidicon
	Visual	Photoemissive	Image amplifier

tube. The image tube consists basically of a photoemissive cathode, a focusing and accelerating electron-optical system, and a phosphor screen.

Since the photocathode would be illuminated by light returning from the phosphor screen, the internal surface of the phosphor is covered with a very thin film of aluminum that can be penetrated by the high-energy image electrons. The aluminum film also serves as the tube anode and as a reflector for the light that would otherwise be emitted back into the tube.

Semiconductor Diodes and Controlled Rectifiers

BY G. J. MALINOWSKI

67. Semiconductor Materials and Junctions. Transistors and diodes are fabricated from semiconductor materials, a form of matter situated between metals and insulators in their ability to conduct electricity. Typical values of electrical resistivity of conductors, semiconductors, and insulators are 10^{-6} to 10^{-5}, 10 to 10^4, and 10^{12} to 10^{18} $\Omega \cdot$cm, respectively. By far the most widely used semiconducting materials are germanium and silicon. Other semiconductor materials such as gallium arsenide, selenium, cadmium sulfide, and copper oxide have electrical properties that make them useful in special applications.

Unlike the vacuum tube, where current flow arises from the motion of charge carriers within a vacuum, semiconductor devices develop current flow from the motion of *charge carriers* within a crystalline solid. The conduction process in semiconductors is most easily visualized in terms of silicon and germanium. The atoms of each of these elements have four electrons in the outer shell (valence shell). These electrons are normally bound in the crystalline lattice structure. Some of these valence electrons are free at room temperature, and hence can move through the crystal; the higher the temperature, the more electrons are free to move. Each vacancy, or hole, left in the lattice can be filled by an adjacent valence electron. Since a hole moves in a direction opposite to that of an electron, a hole may be considered as a positive-charge carrier. Electrical conduction is due to the motion of holes and electrons under the influence of an applied field.

Intrinsic (pure) semiconductors exhibit a negative coefficient of resistivity, since the number of carriers increases with temperature. Conduction due to thermally generated carriers, however, is usually an undesirable effect, because it limits the operating temperature of the semiconductor device.

At a given temperature, the concentration of thermally generated carriers is related to the energy gap of the material. This is the minimum energy (stated in electronvolts) required to free a valence electron (1.1 eV for silicon and 0.7 eV for germanium). Silicon devices perform at higher temperatures because of the wider energy gap.

The conductivity of the semiconductor material can be altered radically by doping with minute quantities of *donor* or *acceptor impurities*. Donor (*n*-type) impurity atoms have five valence electrons, whereas only four are accommodated in the lattice structure of the semiconductor. The extra electron is free to conduct at normal operating temperatures. Common donor impurities include phosphorus, arsenic, and antimony. Conversely, acceptor (*p*-type) atoms have three valence electrons; a surplus of holes is created when a semiconductor is doped with them. Typical acceptor dopants include boron, gallium, and indium.

In an *extrinsic (doped) semiconductor*, the current-carrier type introduced by doping predominates. These carriers, electrons in *n*-type material and holes in *p*-type, are called *majority carriers*. Thermally generated carriers of the opposite type are also present in small quantities and are referred to as *minority carriers*. Resistivity is determined by the concentration of majority carriers.

Lifetime is the average time required for excess minority carriers to recombine with majority carriers. Recombination occurs at "traps" caused by impurities and imperfections in the semiconductor crystal. Semiconductor junctions are formed in material grown as a single continuous crystal to obtain the lattice perfection required, and extreme precautions are taken to ensure exclusion of unwanted impurities during processing. However, in some applications short lifetime is desired, and in such cases gold doping is used to achieve this.

Carrier mobility is the property of a charge carrier which determines its velocity in an electric field. Mobility also determines the velocity of a minority carrier in the diffusion process. High mobility yields a short transit time and good frequency response.

68. pn Junctions. If *p*- and *n*-type materials are formed together, a unique interaction takes place between the two materials, and a *pn* junction is formed. In the immediate vicinity of this junction (in the *depletion region*), some of the excess electrons of the *n*-type material diffuse into the *p* region, and likewise holes diffuse into the *n*-type region. During this process of recombination of holes and electrons, the *n* material in the depletion region acquires a slightly positive charge and the *p* material becomes slightly negative. The space-charged region thus formed repels further flow of electrons and holes, and the system comes into equilibrium. Figure 7-30 shows a typical *pn* junction.

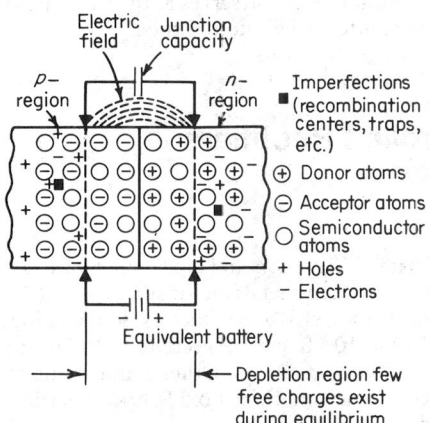

Electric field   Junction capacity   *p*-region   *n*-region   Imperfections (recombination centers, traps, etc.)   ⊕ Donor atoms   ⊖ Acceptor atoms   ○ Semiconductor atoms   + Holes   − Electrons   Equivalent battery   Depletion region few free charges exist during equilibrium

Fig. 7-30. *pn* junction. (*General Electric Co.*)

To keep the system in equilibrium, two related phenomena constantly occur. Due to thermal energy, electrons and holes diffuse from one side of the *pn* junction to the other side. This flow of carriers is called *diffusion current*. When a current flows between two points, a potential gradient is produced. This potential gradient across the depletion region causes a flow of charge carriers, drift current, in the opposite direction to the diffusion current. As a result, the two currents cancel at equilibrium, and the net current flow is zero through the region. An energy barrier is erected such that further diffusion of charge carriers becomes impossible without the addition of some external energy source. This energy barrier formed at the interface of the *p*- and *n*-type materials provides the basic characteristics of all junction semiconductor devices.

69. Fabrication. Silicon diodes can be made by using any of the various technologies employed in the fabrication of transistors, including alloying, growing, diffusion, or ion implantation (see Par. **7-82**). The planar epitaxial process is widely employed. See also Sec. **8**.

70. pn Junction Characteristics. When a dc power source is connected across a *pn* junction, the quantity of current flowing in the circuit is determined by the polarity of the applied voltage and its effect upon the depletion layer of the diode. Figure 7-31 shows the classical condition for a reverse-biased *pn* junction. The negative terminal of the power supply is connected to the *p*-type material, and the positive terminal to the *n*-type material. When a *pn* junction is reverse-biased, the free electrons in the *n*-type material are attracted toward the positive terminal of the power supply and away from the junction. At the same time, holes from the *p*-type material are attracted toward the negative terminal of the supply and away from the junction. As a result, the depletion layer becomes effectively wider, and the potential gradient increases to the value of the supply. Under these conditions the current flow is very small because no electric field exists across either the p or n region.

Figure 7-32 shows the positive terminal of the supply connected to the p region and the negative terminal to the n region. In this arrangement, electrons in the p region near the positive terminal of the supply break their electron-pair bonds and enter the supply, thereby creating new holes. Concurrently, electrons from the negative terminal of the supply enter the n region and diffuse toward the junction. This condition effectively decreases the depletion layer, and the energy barrier decreases to a small value. Free electrons from the n region can then penetrate

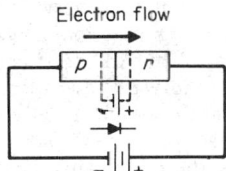

Fig. 7-31. Reverse-biased diode.

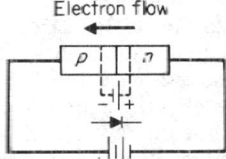

Fig. 7-32. Forward-biased diode.

the depletion layer, flow across the junction, and move by way of the holes in the p region toward the positive terminal of the supply. Under these conditions, the pn junction is said to be *forward-biased.*

A general plot of voltage and current for a pn junction is shown in Fig. 7-33. Here both the forward- and reverse-biased conditions are shown. In the forward-biased region, current rises rapidly as the voltage is increased and is quite high. Current in the reverse-biased region is usually much smaller and remains low until the breakdown voltage of the diode is reached. Thereupon the current increases rapidly. If the current is not limited, it will increase until the device is destroyed.

Junction Capacitance. Since each side of the depletion layer is at an opposite charge with respect to each other, each side can be viewed as the plate of a capacitor. Therefore a pn junction has capacitance. As shown in Fig. 7-34, junction capacitance changes with applied voltage.

DC Parameters of Diodes. The most important of these parameters are as follows:

Forward voltage V_F is the voltage drop at a particular current level across a diode when it is forward-biased.

Breakdown voltage BV is the voltage drop across the diode at a particular current level when the device is reverse-biased to such an extent that heavy current flows. This is known as *avalanche.*

Reverse current I_R is the current specified at a voltage less than BV when the diode is reverse-biased.

AC Parameters of Diodes. The most important of these parameters are as follows:

Capacitance C_0 is the total capacitance of the device, which includes junction and package capacitance. It is measured at a particular frequency and bias level.

Rectification efficiency R_E is defined as the ratio of dc output (load) voltage to the peak of the input voltage, in a detector circuit. This provides an indication of the capabilities of the device as a high-frequency detector.

Forward recovery time t_{fr} is the time required for the diode voltage to drop to a specified value after the application of a given forward current.

Reverse recovery time t_{rr} is specified as the time between the application of reverse voltage and the point where the reverse current has dropped to a spec-

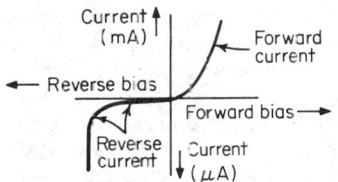

Fig. 7-33. Voltage-current characteristics for a pn junction.

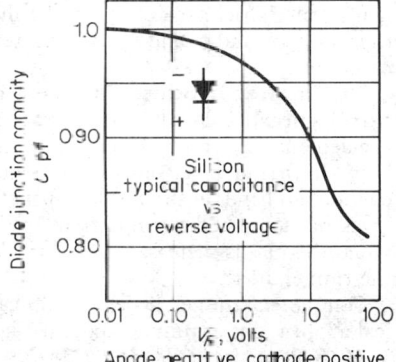

Fig. 7-34. Diode junction capacitance vs. reverse voltage. (*General Electric Co.*)

ified value. It is further necessary to have the diode forward-biased before application of the reverse-voltage pulse.

Stored charge C_s can be used to characterize reverse recovery time of diodes.

Transient thermal resistance provides data on the instantaneous junction temperature as a function of time with constant applied power. This parameter is valuable in ensuring reliable operation of diodes in pulse circuits.

71. Small-Signal Diodes. Small-signal diodes are the most widely used discrete semiconductor devices. The capabilities of the general-purpose diode as a switch, demodulator, rectifier, limiter, capacitor, and nonlinear resistor suit it to many applications.

The most important characteristics of all small-signal diodes are forward voltage, reverse breakdown voltage, reverse leakage current, junction capacitance, and recovery time.

72. Silicon Rectifier Diodes. Silicon rectifier diodes are *pn* junction devices that have up to several hundred amperes forward-current-carrying capability. An ideal rectifier has an infinite reverse resistance, infinite breakdown voltage, and zero forward resistance. The silicon rectifier approaches these ideal specifications in that the forward resistance is only a few tenths of an ohm, while the reverse resistance is in the megohm range.

Since silicon rectifiers are primarily used in power supplies, thermal dissipation must be adequate. To avoid excessive heating of the junction, the heat generated must be efficiently transferred to a heat sink. The relative efficiency of this heat transfer is expressed in terms of the thermal resistance of the device. For stud-mounted rectifiers, the thermal-resistance range is typically 0.1 to 1°C/W.

73. Zener Diodes. Externally the zener diode is similar in appearance to other silicon rectifiers, electrically it is capable of rectifying alternating current. While it serves in a variety of applications, its primary use is as a voltage reference or regulator element. Its ability to maintain a desired operating voltage is limited by the temperature coefficient and the impedance of the zener device.

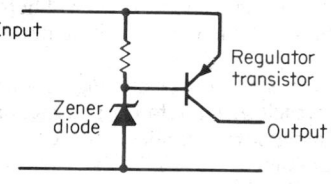

The voltage-reference and regulation performance of the zener diode is based on the avalanche characteristics of the *pn* junction. When a source of voltage is applied to the diode in the reverse direction (anode negative), a reverse current I_r is observed. As the reverse potential is increased beyond the knee of the current-voltage curve, avalanche-breakdown current becomes well developed. This occurs at the zener voltage V_z. Since the resistance of the device drastically drops at this point, it is necessary to limit the current flow by means of an external resistor. Avalanche breakdown of the operating zener diode is not destructive as long as the rated power dissipation of the junction is not exceeded.

Fig. 7-35. Transistor voltage regulator using zener diode for reference.

The zener diode can be used to control the reference voltage of a transistor power supply, as shown in Fig. 7-35. The zener diode also finds use in audio or rf applications where a source of stable reference voltage is required. The design of zener diodes permits them to absorb overload surges and thereby serves the function of protecting circuitry from transients.

74. Varactor Diodes. The varactor diode is a *pn* junction device that has useful nonlinear voltage-dependent variable-capacitance characteristics. Varactor diodes are useful in microwave oscillators when employed with the proper filter and impedance-matching circuitry. Varactor diodes are also used for sensitive microwave amplification in parametric amplifiers. The voltage-dependent capacitance effect in the diode permits its use as an electrically controlled tuning capacitor in radio and television receivers, to replace conventional mechanical variable capacitors.

75. Tunnel Diodes. The tunnel diode is a semiconductor device whose primary use arises from its negative conductance characteristic. In a *pn* junction, a *tunnel* effect is obtained when the depletion layer is made extremely thin. Such a depletion layer is obtained by heavily doping both the *p* and *n* regions of the device. In this situation it is possible for an electron in the conduction band on the *n* side to penetrate, or tunnel, into the valence band of the *p* side. This gives rise to an additional current in the diode at a very small forward bias, which disappears when the bias is increased. It is this additional current that produces the negative resistance of the tunnel diode.

Commercial tunnel diodes are fabricated from germanium, silicon, and gallium arsenide. Typical applications of tunnel diodes include oscillators, amplifiers, converters, and detectors.

76. Schottky Barrier Diodes. This diode (also known as the surface-barrier diode, metal-semiconductor diode, and hot-carrier diode) consists of a rectifying metal-semiconductor junction in which majority carriers carry the current flow. When the diode is forward-biased, the carriers are injected into the metal side of the junction, where they remain majority carriers at some energy greater than the Fermi energy in the metal this gives rise to the name *hot carriers*. The diode can be switched to the OFF state in an extremely short time (in the order of picoseconds). No stored minority-carrier charge exists.

The reverse dc current-voltage characteristics of the device are very similar to those of conventional *pn* junction diodes. The reverse leakage current increases with reverse voltage gradually, until avalanche breakdown is reached.

Schottky barrier diodes usually consist of silicon or gallium arsenide semiconductor material onto which gold, platinum, palladium, or silver is deposited by evaporation techniques or is ion-implanted through the surface.

Schottky barrier diodes used in detector applications have several advantages over conventional *pn*-junction diodes. They have a lower noise and better conversion efficiency, and hence have greater overall detection sensitivity.

77. Diode Light Sensors (Photodiodes). When a semiconductor junction is exposed to light, photons generate hole-electron pairs. When these charges diffuse across the junction, they constitute a photocurrent. Junction light sensors are normally operated with a load resistance and a battery which reverse-biases the junction. The device acts as a source of current which increases with light intensity.

Silicon sensors are used for sensing light in the visible and near-infrared spectra. They can be fabricated as phototransistors in which the collector-base junction is light-sensitive. Phototransistors are more sensitive than photodiodes because the photon-generated current is amplified by the current gain of the transistor.

78. Light-Emitting Diodes (LEDs). These devices have found wide use in visual displays, isolators, and as digital storage elements.

LEDs are principally manufactured from gallium arsenide. When biased in the avalanche-breakdown region, *pn* junctions emit visible light at relatively low power levels. In many applications LED displays are powered by a multiplexing network to reduce their power requirements. LEDs are capable of providing light of different wavelengths by varying their construction. They can be used individually or in matrix arrays. Large quantities of LEDs are made for the calculator and electronic-game markets.

79. Silicon Controlled Rectifiers (SCRs). A silicon controlled rectifier is basically a four-layer *pnpn* device that has three electrodes (a cathode, an anode, and a control electrode called the *gate*). Figure 7-36 shows the junction diagram and voltage-current characteristics for an SCR. See also Sec. **15.**

When an SCR is reverse-biased (anode negative with respect to the cathode), it is similar in characteristics to that of a reverse-biased silicon rectifier or other semiconductor diode. In this bias mode, the SCR exhibits a very high internal impedance, and only a very low reverse current flows through the *pnpn* device. This current remains small until the reverse voltage exceeds the reverse breakdown voltage; beyond this point, the reverse current increases rapidly.

During forward-bias operation (anode positive with respect to the cathode), the *pnpn* structure of the SCR is electrically bistable and may exhibit either a very high impedance (OFF state) or a very low impedance (ON state). In the forward-blocking state (OFF), a small forward current, called the *forward* OFF-*state current*, flows through the SCR. The magnitude of this current is approximately the same as that of the reverse-blocking current that flows under reverse-bias conditions. As the forward bias is increased, a voltage point is reached at which the forward current increases rapidly, and the SCR switches to the ON state. This voltage is called the *forward breakover voltage*.

When the forward voltage exceeds the breakover value, the voltage drop across the SCR abruptly decreases to a very low value, the forward ON-state voltage. When the SCR is in the ON state, the forward current is limited primarily by the load impedance. The SCR will remain in this state until the current through the SCR decreases below the holding current and then reverts back to the OFF state.

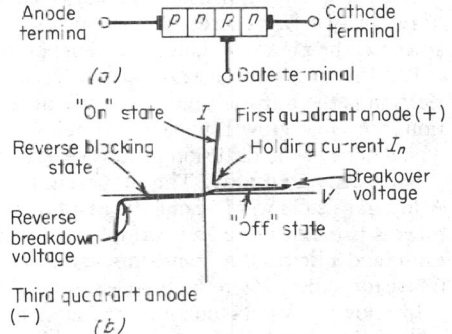

Fig. 7-36. (*a*) SCR junction diagram. (*b*) typical SCR characteristics.

The breakover voltage of an SCR can be varied or controlled by injection of a signal at the gate. When the gate current is zero, the principal voltage must reach the breakover value of the device before breakover occurs. As the gate current is increased, however, the value of breakover voltage becomes less until the device goes to the ON state. This enables an SCR to control a high-power load with a very-low-power signal.

Silicon controlled switches (SCS) are basically SCRs with a second gate. This second gate can be used to turn the device on or off, depending upon the circuitry employed and the polarity of the applied signal. SCSs are basically low-current devices and have found application in low-power digital circuits.

Transistors

By E. B. HAKIM

BIPOLAR TRANSISTORS

80. Semiconductor *pn* Junctions. See Pars. **7-68** to **7-70**.

81. Transistor Action. A bipolar transistor consists of two junctions in close proximity within a single crystal. An *npn* transistor is shown in Fig. 7-37. In normal bias conditions, the emitter-base junction is forward-biased and the collector-base junction is reverse-biased. For FET and unijunction transistors, see Par. **7-89**.

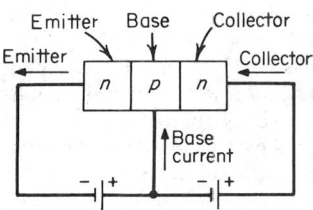

Fig. 7-37. An *npn* junction transistor.

Forward bias of the emitter-base junction causes electrons to be injected into the base region, producing an excess concentration of minority carriers there. These carriers move by diffusion to the collector junction, where they are accelerated into the collector region by the field in the depletion region of the reverse-biased collector junction. Some of the electrons recombine before reaching the collector. Current flows from the base terminal to supply the holes for this recombination process. Another component of current flows in the emitter-base circuit because of the injection of holes from the base into the emitter.

Practical transistors have narrow bases and high lifetimes in the base to minimize recombination. Injection of holes from the base into the emitter is made negligible by doping the emitter much more heavily than the base. Thus the collector current is less than, but almost equal to, the emitter current.

In terms of the emitter current I_E, the collector current I_C is

$$I_C = \alpha I_E + I_{CBO}$$

where α is the fraction of the emitter current that is collected and I_{CBO} is due to the reverse-current characteristic of the collector-base junction. Increase of I_{CBO} with temperature sets the maximum temperature of operation.

High-frequency transistors are fabricated with very narrow bases to minimize the transit time of minority carriers across the base region. Although germanium has a higher carrier mobility, silicon is the preferred material because of its availability and superior processing characteristics.

82. Transistor Fabrication. Today the most widely used technique for transistor fabrication is the diffusion process. Without the development of this technique of junction formation, the rapid growth made in all fields of transistor electronics would have been impossible. The details of the diffusion process appear in Par. **7-84**.

83. Alloy Process. The alloy technique preceded the development of the diffusion process. Alloy transistors were among the first types manufactured and are still widely used, because the process is inexpensive and performance of the devices is good at low frequencies. Both germanium and silicon-alloy transistors, for example, are used in applications such as audio stages of transistor radios, where high power at low frequency is required.

In alloy transistors, donor or acceptor impurities are applied directly to the top and bottom surfaces of a carefully prepared wafer of semiconductor material. The impurities on what will eventually become the collector are made to cover more area than on the emitter side to improve the current-gain characteristics. The assembly is brought up to such a temperature that the impurities alloy into the semiconductor wafer, whereupon a saturated liquid solution of both materials is formed on both sides. This assembly is then permitted to cool. The resulting alloy fronts form abrupt (step) junctions with the base. The final electrical characteristics of the alloy transistor are determined in part by the area the impurities cover and their depth of penetration into the wafer.

84. Diffusion Process. The diffusion process has many advantages over the alloy process, i.e., precise control of junction areas and layer width, nonuniform-resistivity regions to provide for a variety of electrical characteristics, graded junctions instead of abrupt junctions, and a variety of geometries to optimize current handling and frequency response. There are many variations of this process, but it basically involves exposing a semiconductor wafer of predetermined resistivity to a gaseous flow of impurities in a furnace. The gaseous-impurity atoms thus diffuse into the semiconductor surface, forming a pn junction. The time, temperature, gas flow, and concentration (all of which must be accurately controlled) determine the junction depth, width, resistivity, impurity concentration, and general electrical characteristics of the pn junction.

A modification in the diffusion process which led to the development of integrated circuits, large-scale integration, and other advanced technologies in semiconductor devices is the *planar process*. The term planar refers to a device in which each of the junctions, emitter-base and collector-base, is brought to a common plane surface. This structure is distinguished from the mesa structure, in which the junctions are terminated at the edge of the layers constituting the device. The planar structure is illustrated in Fig. 7-38. The significance of the planar process is that the pn junctions are terminated and protected beneath a silicon oxide layer. Thus many of the surface problems associated with other types of transistor fabrication techniques, i.e., high leakage currents and poor low-current dc gain, are eliminated.

To improve switching speed, operating frequency, dc characteristics, collector voltage ratings, power dissipation, and reliability, the *epitaxial* collector was introduced to the planar transistor devices, as shown in Fig. 7-38. The epitaxial process provides a means of growing a very thin high-purity single crystal layer of semiconductor material on a very heavily doped crystal wafer of the same type. The epitaxial process can be used for epitaxial-base devices, as well as collector fabrication. The topography and geometry of transistors take many shapes. In power transistors, the geometry is chosen to favor current-handling capability, whereas in small-signal transistors, high-speed operation is the design goal.

BIPOLAR TRANSISTOR OPERATION

85. Circuit Models of the Transistor. Performance of the transistor as an active circuit element is analyzed in terms of various small-signal equivalent circuits. The low-frequency T-equivalent circuit (Fig. 7-39) is closely related to the physical structure. This circuit model is used here to illustrate the principle of transistor action. Carriers are injected into the base region by forward current through the emitter-base junction. A fraction α (near unity) of this current is collected. The incremental change in collector current is determined essentially by the current generator αi_e, where i_e is the incremental change of emitter current. The collector resistance r_c in parallel with the current generator accounts for the finite resistance of the reverse-biased collector-base junction. The input impedance is due to the dynamic resistance r_e of the forward-biased emitter-base junction and the ohmic resistance r_b of the base region.

The room temperature value of r_e is about $26/I_E$ Ω, where I_E is the dc value of emitter current in milliamperes. Typical ranges of the other parameters are as follows: r_b varies from tens of ohms to several hundred ohms; α varies from 0.9 to 0.999; and r_c ranges from a few hundred

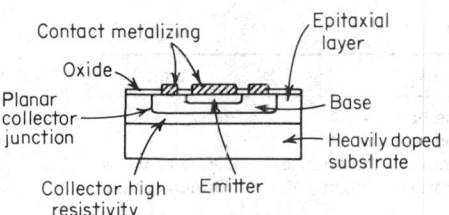

Fig. 7-38. Double-diffused epitaxial planar transistor structure.

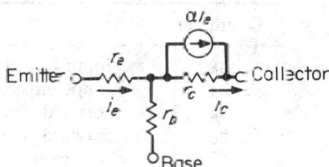

Fig. 7-39. Common-base T-equivalent circuit.

ohms to several megohms. The symbolic representations of an *npn* and *pnp* transistor are shown in Fig. 7-40. The direction of conventional current flow and terminal voltage for normal operation as an active device are indicated for each. The voltage polarities and current for the *pnp* are reversed from those of the *npn*, since the conductivity types are interchanged.

Transistors may be operated with any one of the three terminals as the common, or grounded, element, i.e., common base, common emitter, or common collector. These configurations are shown in Fig. 7-41 for an *npn* transistor.

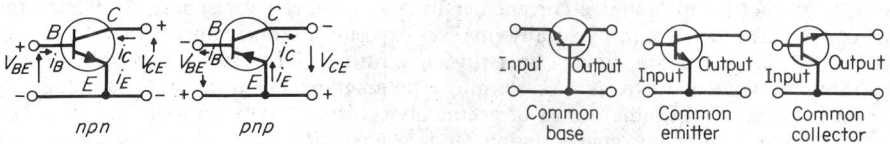

Fig. 7-40. Transistor symbols.   **Fig. 7-41.** Circuit connections for *npn* transistor.

Common Base. The transistor action shown at the left in Fig. 7-41 is that of the common-base connection whose current gain (approximately equal to α) is slightly less than 1. Even with less than unity current gain, voltage and power amplification can be achieved, since the output impedance is much higher than the input impedance.

Common Emitter. For the common-emitter connection, only base current is supplied by the source. Base current is the difference between emitter and collector currents and is much smaller than either; hence current gain I_c/I_b is high. Input impedance of the common-emitter state is correspondingly higher than it is in the common-base connection.

Common Collector. In the common-collector connection, the source voltage and the output voltage are in series and have opposing polarities. This is a negative-feedback arrangement, which gives a high input impedance and approximately unity voltage gain. Current gain is about the same as that of the common-emitter connection. The common-base, common-emitter, and common-collector connections are roughly analogous to the grounded-grid, grounded-cathode, and grounded-plate (cathode-follower) connections, respectively, of the vacuum tube.

h Parameters. Low-frequency performance of transistors is commonly specified in terms of the small-signal *h* parameters listed in Table 7-17. In the notation system used, the second subscript designates the circuit connection (*b* for common-base and *e* for common-emitter). The forward-transfer parameters (h_{fb} and h_{fe}) are current gains measured with the output short-circuited. The current gains for practical load conditions are not greatly different. The input parameters h_{ib} and h_{ie}, although measured for short-circuit load, approximate the input impedances of practical circuits. The output parameters h_{ob} and h_{oe} are the output admittances.

The current gain of the common-base stage is slightly less than unity; common-emitter current gains may vary from ten to several hundred. Input impedance and output admittance of the common-emitter stage are higher than those of the common-base circuit by approximately h_{fe}. Nomenclature and units for *h* parameters are given in Table 7-18.

TABLE 7-17   Transistor Small-Signal *h* Parameters

	Input parameter	Transfer parameter	Output parameter
Common-base	$h_{ib} \approx r_e + (1 - \alpha)r_b$	$h_{fb} \approx \alpha$	$h_{ob} \approx 1/r_c$
Common-emitter	$h_{ie} \approx r_e/(1-\alpha)+r_b$	$h_{fe} \approx \alpha/(1-\alpha)$	$h_{oe} \approx 1/r_c(1-\alpha)$

TABLE 7-18   *h*-Parameter Nomenclature

Parameter	Nomenclature	Unit
h_{ib}	Input impedance (common-base)	Ω
h_{ie}	Input impedance (common-emitter)	Ω
h_{fb}	Forward-current transfer ratio (common-base)	Dimensionless
h_{fe}	Forward-current transfer ratio (common-emitter)	Dimensionless
h_{ob}	Output admittance (common-base)	S
h_{oe}	Output admittance (common-emitter)	S

Although matched power gains of the common-base and common-emitter connections are about the same, the higher input impedance and lower output impedance of the common-emitter stage are desirable for most applications. For these reasons the common-emitter stage is more commonly used. For example, the voltage gain of cascaded common-base stages cannot exceed unity unless transformer coupling is used.

The common-collector circuit has a higher input impedance and lower output impedance than either of the other connections. It is used primarily for impedance transformation.

High-Frequency Limit. The current gain of a transistor decreases with frequency, principally because of the transit time of minority carriers across the base region. The frequency f_T at which h_{fe} decreases to unity is a measure of high-frequency performance. Parasitic capacitances of junctions and leads also limit high-frequency capabilities. These high-frequency effects are shown in the modified equivalent circuit of Fig. 7-42. The maximum frequency f_{max} at which the device can amplify power is limited by f_T and the time constant $r_b' C_c$, where r_b' is the ohmic base resis-

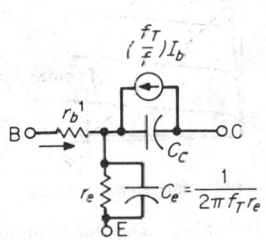

Fig. 7-42. High-frequency common-emitter equivalent circuit.

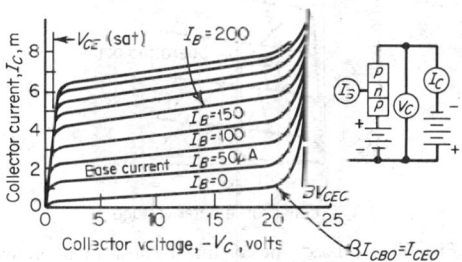

Fig. 7-43. V_C-I_C characteristic for grounded-emitter junction transistor.

tance and C_c is that portion of the collector-base junction capacitance which is under the emitter stripe. Values of f_T greater than 2 GHz and f_{max} exceeding 10 GHz are obtained by maintaining very thin bases (<30 μm) and narrow emitters (<3 μm).

86. Transistor Voltampere Characteristics. The performance of a transistor over wide ranges of current and voltage is determined from static characteristic curves, e.g., the common-emitter output characteristics of Fig. 7-43. Collector current I_C is plotted as a function of collector-to-emitter voltage V_C for constant values of base current I_B. Maximum collector voltage for grounded emitter is limited by either punch-through or avalanche breakdown, whichever is lower, depending on the base resistivity and thickness. When a critical electric field is reached, avalanche occurs due to intensive current multiplication. At this point current increases rapidly with little increase in voltage. The common-emitter breakdown voltage BV_{CEO} is always less than the collector-junction breakdown voltage BV_{CBO}. Another characteristic evident from Fig. 7-43 is the grounded-emitter saturation voltage $V_{CE,sat}$. This parameter is especially important in grounded-emitter switching applications.

Two additional parameters, both related to the emitter junction, are BV_{EBO} and V_{BE}. The breakdown voltage emitter to base with the collector open-circuited BV_{EBO} is the avalanche-breakdown voltage of the emitter junction. The base-to-emitter forward voltage of the emitter junction V_{BE} is simply the junction voltage necessary to maintain the forward-bias emitter current.

The leakage current I_{CBC} in the common-base connection is the reverse current of the collector-base junction; common-emitter leakage is higher by the factor $1/(1 - \alpha)$ because of transistor amplification. In either case, the leakage current increases exponentially with temperature. Maximum junction temperatures are limited to about 100°C in germanium and 250°C in silicon. The locus of maximum power dissipation is a hyperbola on the voltampere characteristic curve. Power dissipation must be decreased when higher ambient temperatures exist. Large-area devices and physical heat sinks of high thermal dissipation are used to extend power ratings (see also Par. **7-88**).

Dynamic variations of voltage and current are analyzed by a load line on the characteristic curves, as in vacuum tubes. For a linear transistor amplifier with load resistance R_L, the output varies along a load line of slope $-1/R_L$ about the dc operating point (Fig. 7-44). Since the minimum voltage $V_{CE,sat}$ is quite low, good efficiencies can be obtained with low values of supply voltage. The operating point on the V_{CE}-I_C coordinates is established by a dc bias current in the input circuit. Transistor circuits should be biased for a fixed emitter current rather than a fixed base current to maintain a stable operating point, since the lines of constant base current are variable between devices of a given type and with temperature.

The common-emitter circuit can also be used as an effective switch, as shown by the load line of Fig. 7-45. When the base current is zero, the collector circuit is effectively open-circuited and only leakage current flows in the collector circuit. The device is turned on by applying base current I_{BI}, which decreases the collector voltage to the saturation value.

87. Transistor Applications. Representative circuit functions are discussed in Secs. **13** to **17**. RC-coupled amplifiers are used for audio and video applications. Bandwidths in excess of 250 MHz can be achieved with high-frequency transistors. Transistors are used in class B push-

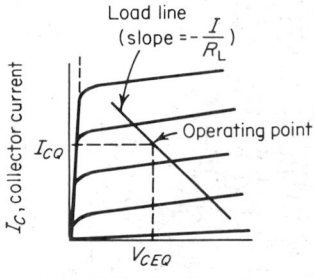

Fig. 7-44. Load line for linear-transistor-amplifier circuit.

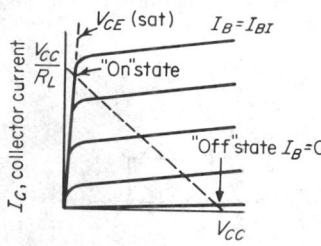

Fig. 7-45. Switching states for common-emitter circuit.

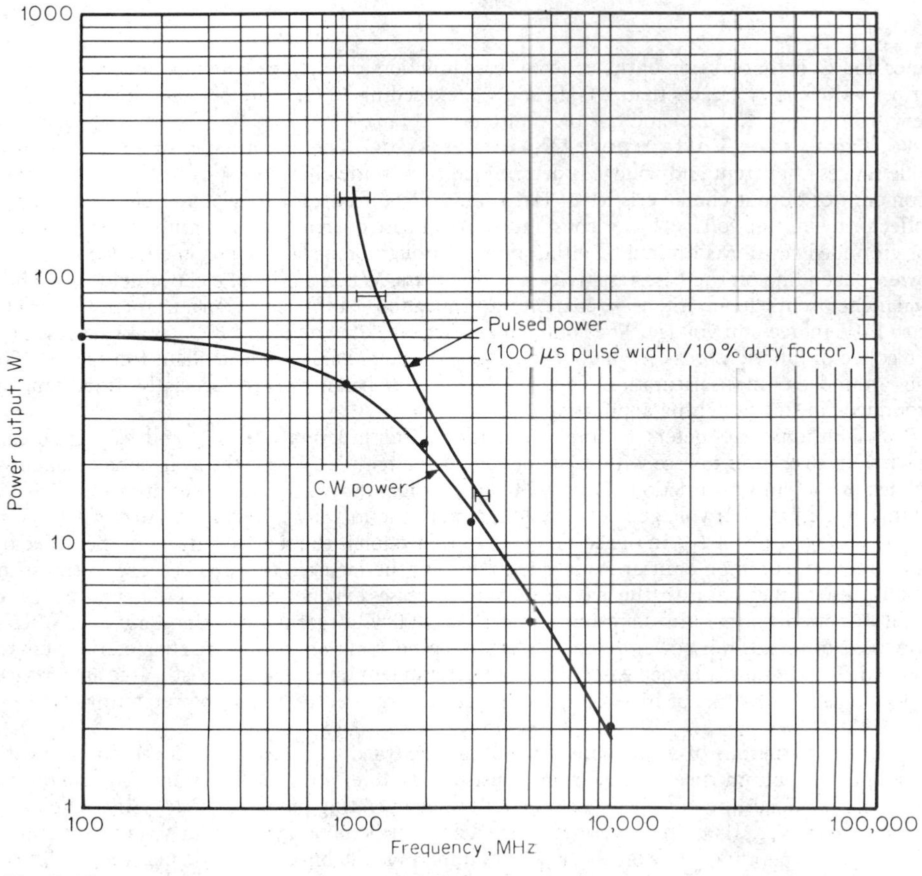

Fig. 7-46. Power output vs. frequency of available bipolar silicon rf transistors. (*Courtesy R. A. Gilson.*)

pull amplifiers for high-power linear applications. Since the transistor is a high-current, low-voltage device, low-impedance loads such as speakers and servomotors can be driven without a matching transformer. Use of complementary stages provides versatility of source and load connections not possible with vacuum tubes. The npn and pnp transistors conduct during alternate half cycles, since they are forward-biased for opposite polarities of the input signal.

High-frequency applications of silicon transistors include amplifiers, oscillators, and mixers in communications systems. They provide useful power gains at frequencies as high as 10 GHz, with noise performance superior to that of vacuum tubes. High-frequency transistors used in mobile transmitters can supply as much as 50 W at 500 MHz.

Shown in Fig. 7-46 is a plot of power-output-vs.-frequency capability of silicon rf power transistors available in the late 1970s. Of significance is the rapid falloff of power above 1 GHz.

88. Transistor Reliability. The life of a transistor in any system should be inherently greater than other electronic components used in the system, but manufacturing defects, poor circuit design, hostile environments, or improper troubleshooting mean that the life of a transistor may be limited to hours. In the early days of semiconductors, reliability was determined by putting a number of devices on various life tests. Today, this is impractical because of the hundreds of thousands of hours required to induce failure in high-quality parts. Various accelerated test programs are used in conjunction with long-term life tests to determine failure rates.

Most test programs involve temperature as a stress, to predict the behavior of transistors over a period of time. Temperature is used since the failure rate is related to an exponential function involving the inverse of temperature. Figure 7-47 is an example of data obtained from an accel-

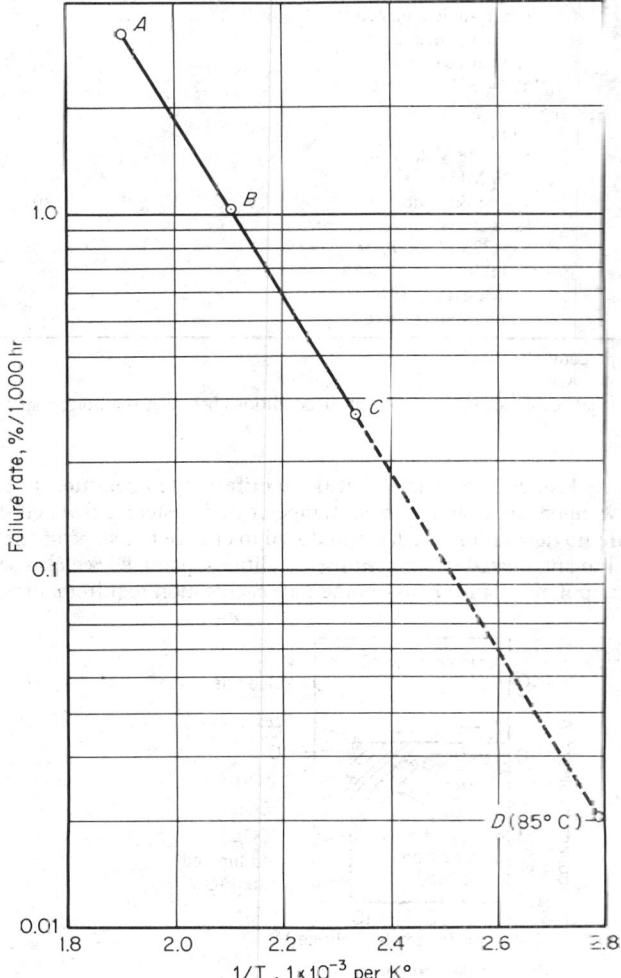

Fig. 7-47. Typical failure-rate plot as a function of temperature.

TABLE 7-19   Typical Military Group B Mechanical-Environmental Specification Requirement for RF Power Transistors

Subgroup	Examination or test	Conditions	LTPD*	Symbol	Min	Max	Unit
1	Physical dimensions	MIL-STD-750	20				
2	Soldering temperature cycling	−65 to +200°C; 5 c†	10				
3	Shock	Nonoperating 5 blows in each orientation X_1, Y_1, Y_2, and Z_1 (total of 20 blows)	10				
4	RF hot-spot thermal resistance‡	$V_{CC} = 35$ V $f = 225$ MHz $T_C = 100°C$ Broadband circuit; 50-Ω termination	13	θ_{rf}‡	...	5.0	°C/W
5	Storage life	$T_{stg} = 200°C$ 1,000 h	10				
6	RF operation life	$T_C = 85°C$ $V_{CE} = 25$ V $P_0 = 25$ W, cw $f = 225$ MHz, $t = 1,000$ h $R_L = 50$ Ω	10				
7	VSWR test	VSWR = 10:1 Phase angle adjusted to achieve max device power dissipation $T_C = 55°C$, $t = 1$ h $P_0 = 25$ W, cw, into 50 Ω $V_c = 28$ V dc, $f = 225$ MHz	20				
	Peak rf voltage capability	$V_{CC} = 30$ V dc $P_0 = 25$ W $f = 225$ MHz, $B_L = 50$ Ω Increase V_{peak}, rf by increasing rf drive	...	$V_{(peak)}$rf	120	...	V

*LTPD = lot tolerance percent defective.
†MIL-STD-202, Method 102A.
‡θ_{rf} is defined as the thermal resistance under rf operating conditions between the hottest spot on the chip and the case.

erated test. The points *A*, *B*, and *C* are obtained at three different temperatures (junction temperature if the devices were operated, and ambient temperature if tested at elevated storage conditions only). If there are no new failure modes introduced to change the slope of the curve, the line can be extrapolated to anticipated equipment-use conditions, point *D*. See also Sec. **8**.

A typical military group B, mechanical-environmental specification requirement, is shown in

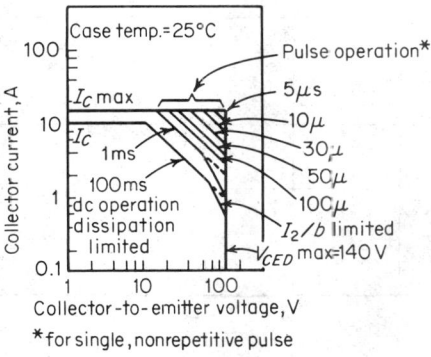

Fig. 7-48. Safe-operating-area chart.

Table 7-19. LTPD, lot tolerance percent defective, assures that a lot will not be accepted that has a proportion defective equal to or greater than the specified LTPD.

Second breakdown, a phenomenon common in power transistors (both audio and rf), is a potentially destructive mode of operation. This condition is caused by localized hot-spotting and thermal runaway. To prevent devices from being operated at conditions which can cause second breakdown, manufacturers specify on their data sheets "safe operating areas." One type of safe operating area is illustrated in Fig. 7-48. In conjunction with safe area, the device *thermal resis-*

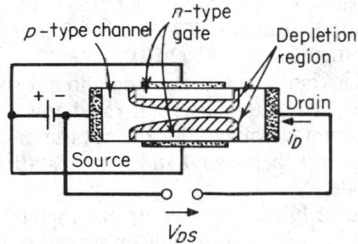

Fig. 7-49. A *p*-channel junction field-effect transistor.

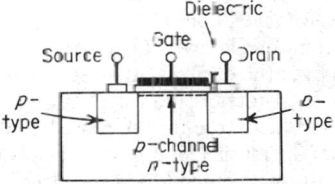

Fig. 7-50. A *p*-channel MOS transistor.

tance θ_{JC}, junction to case, is used to avoid hot-spotting. The value of thermal resistance is usually determined by the use of a temperature-sensitive parameter V_{BE} whose value as a function of temperature is predictable. However, this technique gives an average indication of the junction temperature. For greater reliability assurance *hot-spot thermal resistance* θ_{BS} is becoming popular. This technique requires measurement of the surface temperature under various operating conditions. The most common tools used for these measurements are the infrared radiometer, thermal graphic phosphors, and liquid crystals.

FIELD-EFFECT AND UNIJUNCTION TRANSISTORS

89. Field-Effect (Unipolar) Transistors. There are two general types of field-effect transistors (FET): junction (JFET) and insulated-gate (IGFET). The IGFET has a variety of structures, known as the metal-insulator-semiconductor (MISFET) and the metal-oxide-semiconductor (MOSFET). See Sec. **8**.

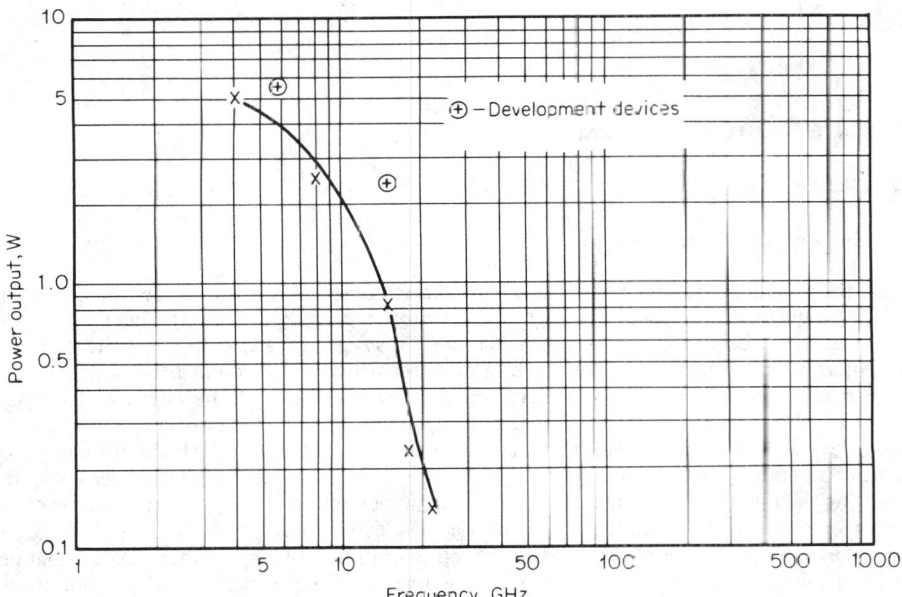

Fig. 7-51. Power output vs. frequency of available GaAs FETs. (*Courtesy of R. A. Gilson.*)

The cross section of a p-channel JFET is shown in Fig. 7-49. Channel current is controlled by reverse-biasing the gate-to-channel junction so that the depletion region reduces the effective channel width. The input impedance of these devices is high because of the reverse-biased diode in the input circuit. In fact, the voltampere characteristics are quite similar to those of a vacuum tube. Another important feature of the junction FET is the excellent low-frequency noise characteristics, which surpass those of either the vacuum tube or conventional (bipolar) transistor.

The cross section of the p-channel MOSFET (or MOS transistor) is shown in Fig. 7-50. This device operates in the depletion mode. For zero gate voltage, there is no channel, and the drain current is small. A negative voltage on the gate repels the electrons from the surface and produces a p-type conduction region under the gate. Compared with the JFET, the MOS transistor has a wider gain-bandwidth product and a higher input impedance (>100 GΩ).

Power MOSFETs offer the major advantage over bipolar transistors in that they do not suffer from second breakdown. Their safe dc operating areas are determined by their rated power dissipation over the entire drain-to-source voltage range up to rated voltage. This is not the case for bipolar transistors, as indicated in Fig. 7-48. The superiority in power handling capability of MOSFETs is also true for the pulsed-power operating mode.

Further operational and reliability advantages of FETs over bipolar devices are obtained by the use of gallium arsenide (GaAs) in place of silicon (Si). The advantages include enhanced switching speeds resulting from electron velocity twice that of Si; lower operating voltages and lower ON resistance resulting from a fivefold greater electron mobility; and 350°C maximum operating temperature versus 175°C. Figure 7-51 depicts the power output vs. frequency of currently available GaAs power MISFETs.

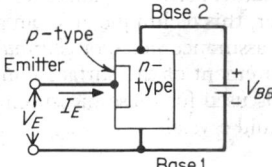

Fig. 7-52. Unijunction transistor.

90. Unijunction Transistors. A unijunction transistor is shown in Fig. 7-52. The input diode is reverse-biased at low voltages owing to IR drop in the bulk resistance of the n-type region. When V_E exceeds this drop, carriers are injected and the resistance is lowered. As a result, the IR drop and V_E decrease abruptly. The negative-resistance characteristic is useful in such applications as oscillators and as trigger devices for silicon controlled rectifiers.

Batteries and Fuel Cells

BY D. LINDEN

PRINCIPLES OF OPERATION

91. Electrochemical Principles and Reactions. A battery is a device which converts the chemical energy contained in its active materials directly into electric energy by means of an oxidation-reduction electrochemical reaction. This type of reaction involves the transfer of electrons from one material to another. In a nonelectrochemical reaction this transfer of electrons occurs directly, and only heat is involved. In a battery (Fig. 7-53) the negative electrode or anode is the component capable of giving up electrons, being oxidized during the reaction. It is separated from the oxidizing material, which is the positive electrode or cathode, the component capable of accepting electrons. The transfer of electrons takes place in the external electric circuit connecting the two materials and in the electrolyte, which serves to complete the electric circuit in the battery by providing an ionic medium for the electron transfer.

92. Fuel Cells. The operation of the fuel cell is similar to that of a battery except that one or both of the reactants are not permanently contained in the electrochemical cell but are fed into it from an external supply when power is desired. The fuels are usually gaseous or liquid (compared with the metal anodes generally used in batteries), and oxygen or air is the oxidant.

93. Components of Batteries. The basic unit of the battery is the cell. A battery consists of one or more cells, connected in series or parallel, depending on the desired output voltage or capacity. The cell consists of three major components: the anode (the reducing material or fuel), the cathode or oxidizing agent, and the electrolyte which provides the necessary internal ionic conductivity. Electrolytes are usually liquid, but some batteries employ solid electrolytes which are ionic conductors at their operating temperatures. In addition, practical cell design requires a separator material, which serves to separate the anode and cathode electrodes mechanically. Electrically conducting grid structures or materials are often added to each electrode to reduce internal resistance. The containers are of many types, depending on the application and its environment.

94. Theoretical Cell Voltage and Capacity. The theoretical capacity (amperehours) of a battery system is determined by its active materials; the maximum electric energy (watt-hours) corresponds to the free-energy change of the reaction. The theoretical voltage and ampere-hour capacities of a number of electrochemical systems are given in Table 7-20. The voltage is determined by the active materials selected; the ampere-hour capacity is determined

TABLE 7-20　Characteristics of Batteries

System	Anode	Cathode	Typical voltage, V	Capacity[a] Wh/kg	Capacity[a] Wh/liter
Primary:					
Leclanche	Zn	MnO_2	1.2	65	175
Magnesium	Mg	MnO_2	1.5	90	180
Alkaline MnO_2	Zn	MnO_2	1.15	65	200
Mercury	Zn	HgO	1.2	80	370
Mercad	Cd	HgO	0.85	45	175
Silver oxide	Zn	Ag_2O	1.5	120	465
Zinc-air	Zn	Air (O_2)	1.1	350	920[c]
Li-organic					
electrolyte	Li	SO_2	2.8	300	475
	Li	MnO_2	2.8	200	500
Solid	Li	PbI_2/PbS	1.9	100	400
		I_2 (poly-2-vinylpyri-			
Electrolyte	Li	dine)	2.8	125	400
Secondary:					
Lead-acid	Pb	PbO_2	2.0	37	70
Edison	Fe	Ni oxides	1.2	29	60
Nickel-cadmium	Cd	Ni oxides	1.2	33	60
Silver-zinc	Zn	AgO	1.5	90	180
Silver-cadmium	Cd	AgO	1.05	65	125
Zinc-nickel oxide	Zn	Ni oxides	1.6	55	110
Nickel-hydrogen	H_2	Ni oxides	1.2	60	65
Zinc-chlorine	Zn	Cl_2	2.0	10[h]	
High-	Na	S	1.8	50[h]	
temp	Li	FeS_x	1.4	20[h]	
Reserve					
Cuprous chloride[d]	Mg	CuCl	1.4	60	80
Silver-chloride[d]	Mg	AgCl	1.5	100	180
Zinc–silver oxide	Zn	AgO	1.5	30	75[e]
Thermal	Ca	[b]	2.6	10	20[f]
NH_3-activated	Mg	m-DNB	1.7	22	60[g]

[a] Delivered capacity when discharged at normal temperatures (20°C) at normal discharge rates.
[b] Several different cathodes used.
[c] Weight of air not considered in computation of watt-hours.
[d] Water-activated.
[e] Automatically activated; high rate discharge, 2- to 10-min rate
[f] Fused salt; heat-activated; high rate discharge; 2- to 10-min rate.
[g] Four-minute discharge rate.
[h] Projected.
NOTE: An expanded table appears in "Handbook of Batteries and Fuel Cells," McGraw-Hill, New York (in press).

by the amount (weight) of available reactants. One gram-equivalent weight of material will supply 96,500 C, or 26.8 Ah, of electric energy.

95. Factors Influencing Battery Voltage and Capacity. In actual battery practice, only a small fraction of the theoretical capacity is realized. This is due not only to the presence of nonreactive components (containers, separators, electrolyte) that add to the weight of the battery but also to many other factors that prevent the battery from performing at its theoretical level. Factors influencing the voltage and capacity of a battery are as follows.

Voltage Level. When a battery is discharged in use, its voltage is lower than the theoretical voltage. The difference is caused by *IR* losses due to cell resistance and polarization of the active

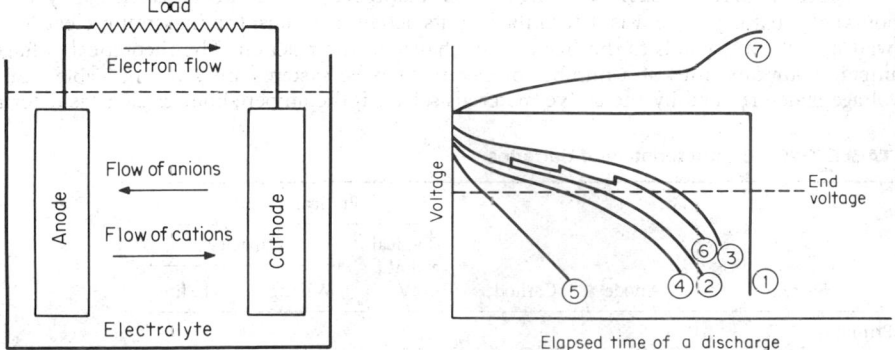

Fig. 7-53. Electrochemical operation of a battery.   **Fig. 7-54.** Battery discharge characteristics.

materials during discharge. This is illustrated in Fig. 7-54. The theoretical discharge curve of a battery is shown as curve 1. In this case, the discharge of the battery proceeds at the theoretical voltage until the active materials are consumed and the capacity fully utilized. The voltage then drops to zero. Under load conditions, the discharge curve is similar to curve 2. The initial voltage is lower than theoretical, and it drops off as the discharge progresses.

The Current Drain of the Discharge. As the current drain of the battery is increased, the *IR* loss increases, the discharge is at a lower voltage, and the service life of the battery is reduced (curve 5). At extremely low current drains it is possible to approach the theoretical capacities (in the direction of curve 3). In a very long discharge period, chemical self-deterioration during the discharge becomes a factor and causes a reduction of capacity.

Voltage Regulation. The voltage regulation required by the equipment is most important. As is apparent by the curves in Fig. 7-54, design of equipment to operate to the lowest possible end voltage results in the highest capacity and longest service life. Similarly, the upper voltage limit of the equipment should be established to take full advantage of the battery characteristics. In some applications, where only a narrow voltage range can be tolerated, voltage regulators may have to be employed to take full advantage of the battery's capacity. If a storage battery is used in conjunction with another energy source, which is permanently connected in the operating circuit, allowances must be made for the voltage required to charge the battery, as illustrated in curve 7, Fig. 7-54. The charging voltage must just exceed the maximum voltage on charge.

The Type of Discharge (Continuous, Intermittent, etc.). When a battery stands idle after a discharge, certain chemical and physical changes take place which can result in voltage recovery. Thus the voltage of a battery which has dropped during a heavy discharge will rise after a rest period, giving a sawtooth discharge. Curve 6 of Fig. 7-54 shows the characteristic of a battery discharged intermittently at the same drain in curve 2. The improvement resulting from the intermittent discharge depends on the current drain, length of the recovery period, discharge temperature, end voltage, and the particular battery system and design employed. Some battery systems, during inactive stand, develop a protective film on the active-material surface. These batteries, instead of showing a recovery voltage, may momentarily demonstrate a lower voltage after stand until this film is broken by the discharge. This is known as a *voltage delay.*

Temperature of the Battery during Discharge. The temperature at which the battery is discharged has a pronounced effect on its service and voltage characteristics. This is due to the reduction in chemical activity and the increase in battery internal resistance at lower temperatures. Curves 3, 2, 4, and 5 of Fig. 7-54, also represent discharges at the same current drawn but at progressively reduced temperatures. The specific characteristics vary for each battery system and discharge rate, but generally best performance is obtained between 20 and 40°C. At higher temperatures, chemical deterioration may be rapid enough during discharge to cause a loss in capacity.

Effect of Size on Capacity. Battery size influences the voltage characteristics by its effect on current density. A given current drain may be a severe load on a small battery, giving a discharge similar to curve 4 or 5 (Fig. 7-54), but be a mild load to a larger battery and give a discharge similar to curve 3. It is often possible to obtain more than a proportional increase in the service life by increasing the size of the battery. The absolute value of current, therefore, is not the key influence, although its relation to the size of the battery, i.e., the current density is important.

The Age and Storage Condition of the Battery. Batteries are a perishable product and deteriorate as a result of chemical action that proceeds during storage. The type of battery, design, temperature, and length of storage period are factors which affect the shelf life of the battery. Since self-discharge proceeds at a lower rate at reduced temperatures, refrigerated storage extends the shelf life. Refrigerated batteries should be warmed before discharge to obtain maximum capacity.

CLASSIFICATION OF CELLS AND BATTERIES

96. General Characteristics. The many and varied requirements for battery power and the multitude of environmental and electrical conditions under which they must operate necessitate the use of a number of different battery types and designs, each having superior performance under certain discharge conditions. The key theoretical and performance characteristics of the major primary and secondary batteries and fuel cells are listed in Table 7-20.

97. Types of Batteries. Batteries are generally identified as primary or secondary.

Primary batteries are not capable of being easily recharged electrically and hence are used or discharged a single time and discarded. Many of the primary batteries, in which the electrolyte is contained by absorbent or separator materials (i.e. there is no free or liquid electrolyte), are termed *dry cells.*

Secondary batteries are those which are capable of being recharged electrically, after discharge, to their original condition by passing current through them in the opposite direction to that of the discharge current. They are electric-energy storage devices and are known also as storage batteries or accumulators.

Reserve batteries are primary types in which a key component is separated from the rest of the battery before activation. In the inert condition, the battery is capable of long-term storage. Usually, the electrolyte is the component that is isolated; in other systems, such as the thermal battery, the battery is inactive until it is heated, melting a solid electrolyte, which then becomes conductive.

PRIMARY BATTERIES

98. General. A number of different types of primary batteries are used widely in civilian, industrial, and military applications. They are a convenient, usually relatively inexpensive, lightweight source of power for the portable electric devices. The general advantages of primary batteries are reasonably good shelf life; high energy densities at low to moderate rates; little, if any, maintenance; and ease of use. Typical characteristics, applications, or uses of these batteries are shown in Tables 7-20 and 7-21.

99. Leclanche Battery (Zn–MnO$_2$). The Leclanche dry cell, known for over 100 years, is still the most widely known and used of all the dry-cell batteries.

The most common construction (Fig. 7-55) uses a cylindrical zinc container as the negative electrode and manganese dioxide as the positive element, with electrical connection through a carbon electrode. A paste or paper separator separates the two electrodes. Another common design is the flat cell, used only in multicell batteries, which offers better volume utilization. The Lechanche cell is fabricated in a number of sizes of varying diameter (or cross section) and height.

Service Life. Typical discharge curves for the Leclanche dry cell are shown in Fig. 7-56. Note that the capacity varies with the different discharge conditions (higher capacities are available at lower current or power densities) to the point where shelf deterioration during the long discharge causes a loss of capacity. Performance is usually better under intermittent discharge, as the cell is given an opportunity to recover. This effect is particularly noticeable at the heavier discharge loads.

Effect of Temperature. Leclanche cells operate best at normal temperatures (20°C). The energy output of the cell increases with higher operating temperatures, but prolonged exposure to high temperatures (50°C) will cause rapid deterioration. The capacity of the Leclanche cell

TABLE 7-21   Major Characteristics and Applications of Primary Batteries

System	Characteristics	Applications
Zinc-carbon (Leclanche) (zinc-MnO₂)	Popular common low-cost primary battery, available in variety of sizes	Flashlight, portable radios and electronics, toys, novelties, instruments, etc.
Magnesium (Mg-MnO₂)	High-capacity primary battery, long shelf life	Military receiver-transmitters, aircraft emergency transmitters
Mercury (Zn-HgO)	Highest capacity (by volume) of conventional types, flat discharge, good shelf life	Hearing aids, medical (heart pacers), photography, detectors, receiver-transmitters, military sensor and detection equipment
Alkaline (Zn-alkaline electrolyte-MnO₂)	Good low-temperature and high-rate performance, moderate cost	Cassettes and tape recorders, calculators, radio and TV—popular for high-drain primary-battery application
Silver-zinc (Zn-AgO)	Highest capacity (by weight) of conventional types, flat discharge, good shelf life	Hearing aids, photography, electric watches, missiles and space application (larger sizes)
Lithium (lithium-SO₂)	New battery system—recent development; highest-performance primary battery, excellent low-temperature performance, long shelf life	Will have wide, general-purpose application when available. First uses will be military and special civilian applications needing high-capacity and low-temperature performance
Solid electrolyte	Extremely long shelf life, low-power battery	Medical electronics, memory circuits, fusing

falls off rapidly with decreasing temperature and is essentially inoperative below −20°C. The effects are more pronounced at heavier current drains. For best operation at low ambient temperatures, the Leclanche cell should be kept warm by some appropriate means. A vest battery has proved effective, for example, in military applications. This battery is worn under the user's clothing and uses body heat to maintain the battery at a satisfactory operating temperature.

Shelf Life. Figure 7-57 shows the capacity retention (shelf life) of the Leclanche battery at different temperatures. These data point out the advantage of storing batteries at low temperatures for preserving their capacity.

100. Zinc Chloride Cell. A recent modification of the Leclanche cell is the zinc chloride electrolyte cell. The construction is similar to the conventional carbon-zinc cell but the electrolyte contains only zinc chloride, without the saturated solution of ammonium chloride. The zinc chloride cell is a high-performance cell with improved high-rate and low-temperature performance and a reduced incidence of leakage. A comparison of the performance of the zinc chloride cell with the conventional cell is presented in Fig. 7-58.

101. Magnesium Dry-Cell Batteries (Mg-MnO₂). The magnesium battery was developed for military use and has two principal advantages over the zinc dry cell: (1) it has twice the capacity or service life of an equivalently sized zinc cell, and (2) it can retain this capacity during storage, even at elevated temperatures (Fig. 7-59). The construction of the magnesium dry cell is similar to that of the cylindrical zinc cell,

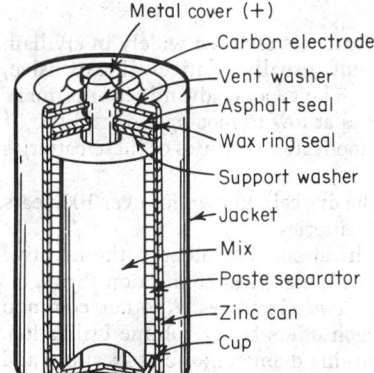

Metal cover (+)
— Carbon electrode
— Vent washer
— Asphalt seal
— Wax ring seal
— Support washer
— Jacket
— Mix
— Paste separator
— Zinc can
— Cup
— Star bottom
Metal bottom (−)

Fig. 7-55. Cross section of a Leclanche cylindrical cell.

except that a magnesium can is used instead of the zinc container. The magnesium cell has a mechanical vent for the escape of hydrogen gas, which forms as a result of a parasitic reaction during the discharge of the battery. Magnesium batteries have not been fabricated successfully in flat-cell designs.

The good shelf life of the magnesium battery results from a film which forms on the inside of the magnesium can, preventing corrosion. This film, however, is responsible for a delay in the battery's ability to deliver full output voltage after it is placed under load. The delay is usually less than 0.3 s but can be longer at low temperatures and high current drains.

102. Zinc–Mercuric Oxide Battery (Zn–HgO). The zinc–alkaline–mercuric oxide battery is noted for its high capacity per unit volume, a relatively constant output voltage during its discharge, and good storageability.

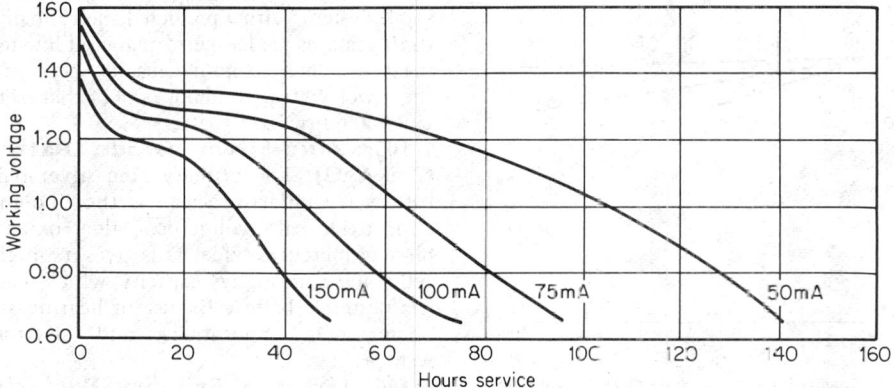

Fig. 7-56. Typical discharge curves for a Leclanche dry cell (size D).

The zinc–mercuric oxide cell is constructed in a sealed but vented structure, with the active materials balanced to prevent formation of hydrogen in a discharged battery. Three basic structures are used: the wound-anode, the flat-pressed powder-anode, and the cylindrical-pressed powder-electrode types. The cell is available in a number of sizes, from the miniature 16-mAh to the largest 14-Ah size.

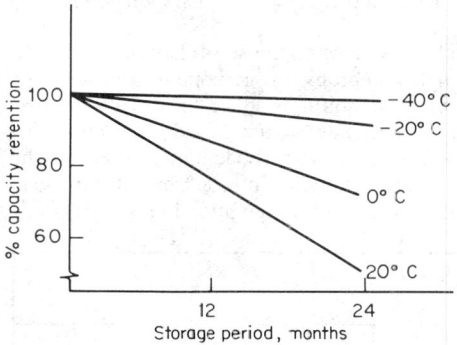

Fig. 7-57. Capacity retention of Leclanche cells.

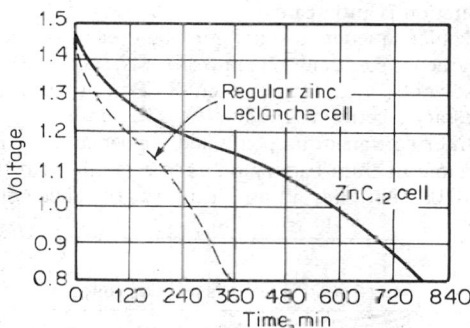

Fig. 7-58. Comparative performance of zinc chloride and Leclanche cells.

The general discharge characteristics of the mercury battery are shown in Fig. 7-60. The mercury battery is suited to use at normal and elevated temperatures. The low-temperature performance is poor, particularly under heavy loads, and its use at temperatures below 0°C is not recommended, except in special low-temperature designs.

The mercury cell has good storage characteristics. While this varies with cell size, capacity retention is about 90% after 2 years storage at 20°C and over 80% after 1 year at 45°C.

103. Alkaline–MnO$_2$ Cell (Zn–MnO$_2$). The zinc–alkaline–MnO$_2$ cell uses the same electrochemically active materials, zinc and manganese dioxide, as the Leclanche cell but differs in construction (Fig. 7-61) and in the use of highly conductive potassium hydroxide electrolyte which result in a lower internal resistance. The advantage on low-rate or intermittent discharge is marginal, but on high and continuous drain conditions, the alkaline cell can deliver from 2 to 10 times the ampere-hour capacity of the Leclanche cell. Its performance at low temperatures is superior to other commercially available dry batteries, operating at temperatures as low as $-25°C$. The electrolyte undergoes no change during the discharge, maintaining its high conductivity throughout the cell's life. It thus differs from the Leclanche cell, whose resistance increases during the discharge. Typical discharge curves are given in Fig. 7-62. The shelf life of the alkaline–MnO_2 cell is moderately superior to that of the Leclanche cell. Capacity retention is about 90% after 1 year of storage.

104. Cadmium–Mercuric Oxide (Cd–HgO). The cadmium–alkaline-mercuric oxide battery is similar in design to the zinc–mercuric oxide battery. The substitution of cadmium for zinc lowers the cell voltage but gives a very stable system, with a predicted shelf life of up to 10 years, as well as performance at low temperatures. Its watt-hour capacity, because of the lower voltage, is about 60% of that of the zinc–mercuric oxide battery.

105. Zinc–Silver Oxide Battery (Zn–AgO). The primary zinc–silver oxide battery is similar in design to the zinc–mercuric oxide battery but uses silver oxide in place of mercuric oxide. This gives it a higher cell voltage and energy capacity, which makes it a desirable battery for use in hearing aids, photographic applications, and electronic watches.

106. Lithium Primary Batteries. The lithium battery is a recent development with a number of advantages over other primary-battery systems. Lithium is an attractive anode because of its reactivity, light weight, and high voltage; cell voltages range between 2 and 3.6 V, depending on the cathode material.

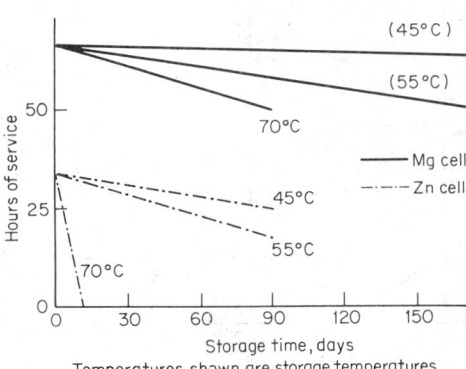

Fig. 7-59. Comparison of magnesium and Leclanche (zinc) dry cells.

The advantages of the lithium battery include high energy density, of the order of 250 Wh/kg and 400 Wh/l; high power density; flat discharge characteristics; excellent service over a wide temperature range, down to $-40°C$ or below; good shelf life; up to 5 years without refrigeration is anticipated.

Nonaqueous solvents must be used as the electrolyte because of the solubility of lithium in aqueous solutions. Organic solvents, such as acetonitrile and propylene carbonate, and inorganic solvents, such as thionyl chloride, are typical. A compatible solute is added to provide the necessary electrolyte conductivity. A number of different materials (sulfur dioxide, carbon monofluoride, vanadium pentoxide, copper sulfide) are used as the active cathode material. Hence the name lithium battery refers to a family of battery systems, ranging in size from 100 mAh to 10,000 Ah; they all use lithium as the anode but differ widely in design and chemical action.

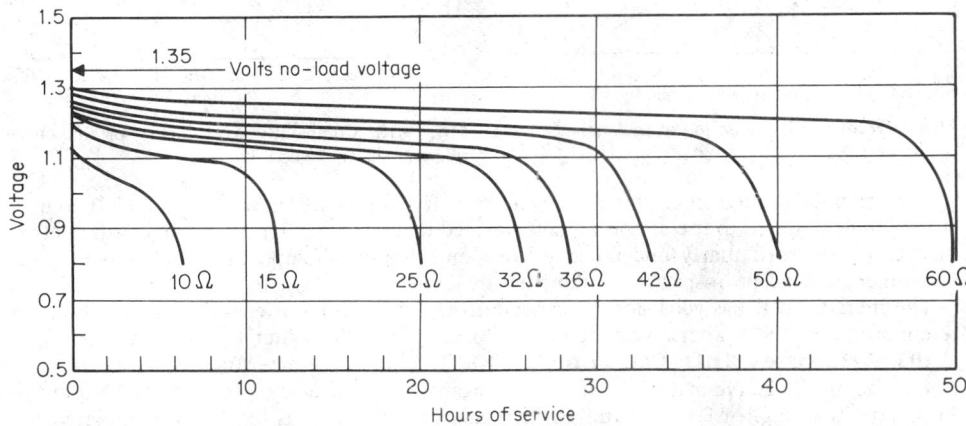

Fig. 7-60. Discharge curves for the Zn–HgO cell.

107. Lithium-Sulfur Dioxide Cell. This lithium primary cell uses sulfur dioxide for the cathode material and an electrolyte consisting of acetonitrile and lithium bromide. The cell is typically fabricated in a cylindrical hermetically sealed structure, as shown in Fig. 7-63. A jelly-roll construction is used, made by spirally winding strips of lithium ribbon, a polypropylene separator, and the cathode electrode (a Teflon-carbon mix pressed on an aluminum screen). This design provides the high surface area and low cell resistance which are necessary to obtain high-current and low-temperature performance.

The good shelf life of the lithium–SO_2 cell is attributed to the protective film formed by the initial reaction of lithium and SO_2, which prevents further reaction or loss of capacity on stand. During discharge, the SO_2 is depleted and the cell pressure reduced.

108. Lithium-Thionyl Chloride Cell. A more recent advance is the lithium–thionyl chloride cell, which has an energy density 25% greater than the lithium–SO_2 cell. Figure 7-64 compares the lithium primary cells with other primary systems.

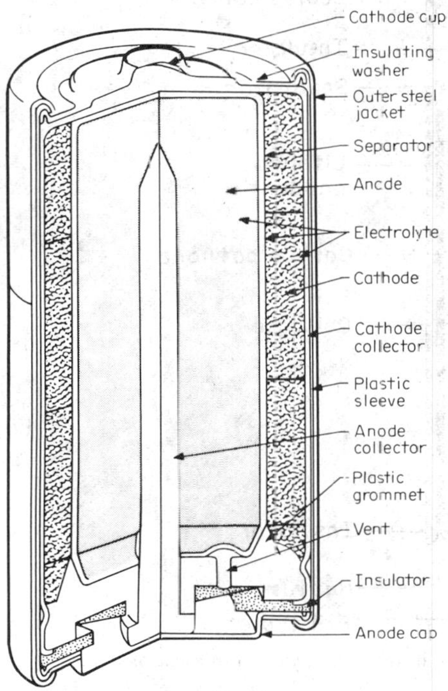

Labels on Fig. 7-61 (top to bottom): Cathode cup, Insulating washer, Outer steel jacket, Separator, Anode, Electrolyte, Cathode, Cathode collector, Plastic sleeve, Anode collector, Plastic grommet, Vent, Insulator, Anode cap

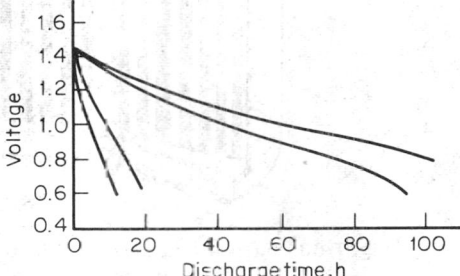

Fig. 7-61. Internal construction of an alkaline–MnO_2 cell. (*Duracell International Inc.*)

Fig. 7-62. Discharge curves of an alkaline–MnO_2 cell at 20°C.

109. Solid Cathode Cells. Another type of lithium anode primary cell uses solid (rather than gaseous or liquid) materials for the cathode. These cells have the advantage of not being pressurized, but they do not have the high-rate capability. They have been designed generally for low-rate applications in crimped-sealed flat or button cells. Larger cylindrical cells, in both high-rate and low-rate jelly-roll configurations, have been constructed. A number of different solid cathode materials are being used in these batteries. Figure 7-65 presents typical discharge curves for several of these cells. The discharge of the solid cathode cell is not as flat as that of the Li–SO_2 cell, but the performance approaches that of the Li–SO_2 cell at the lower discharge rates.

110. Air Batteries. Primary batteries using air as the depolarizer can provide high energy densities since they do not contain an active cathode material. Zinc-air batteries, using bulky zinc anodes, a carbon-air cathode, and potassium hydroxide electrolyte in heavy glass or hard-rubber containers, have been used for many years in railway signal systems and marine applications. They have high energy density but low power output. Miniature designs, employing thin Teflon-coated electrodes in button configurations, are now available. These cells deliver up to twice the service of a comparable zinc-mercuric cell when used in hearing aids, hand-held calculators, and similar applications. Moisture loss must be prevented; this requires careful packaging during storage and limits the life after the battery has been put in service.

111. Solid Electrolyte Batteries. Most batteries depend on the ionic conductivity of liquid electrolytes for their operation. The solid-electrolyte batteries depend on the ionic conductivity of an electronically nonconductive salt in the solid state, for example, Li^+ ion mobility in lithium iodide. Cells using these solid electrolytes are low-power (microwatt) devices but have an extremely long shelf life and can operate over a wide temperature range. The absence of liquid eliminates corrosion and gassing and permits the use of a hermetically sealed cell. The solid electrolyte batteries are used in medical electronics (in devices such as heart pacemakers), for memory circuits, and other such applications requiring a long-life, low-power battery.

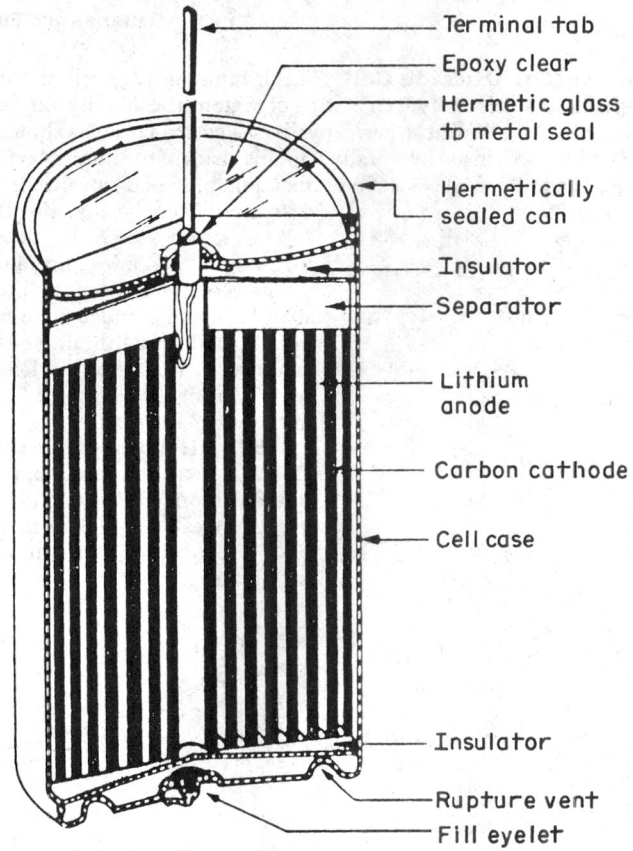

Terminal tab
Epoxy clear
Hermetic glass to metal seal
Hermetically sealed can
Insulator
Separator
Lithium anode
Carbon cathode
Cell case
Insulator
Rupture vent
Fill eyelet

Fig. 7-63. Internal structure of a lithium–sulfur dioxide cell. (*Duracell International, Inc.*)

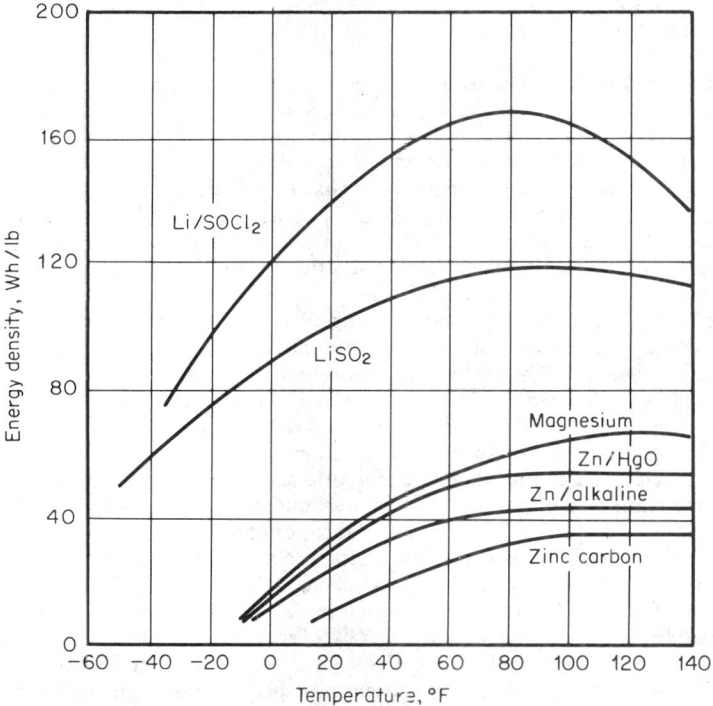

Fig. 7-64. Comparative performance of primary battery systems.

Several types of solid electrolyte batteries are being marketed using different solid electrolytes and active materials. Of special significance are the high energy densities (5 to 10 Wh/in³) achieved with the Li-anode solid electrolyte battery.

112. Recharging Primary Batteries. Most primary batteries can be recharged for a small number of cycles under controlled conditions. With Leclanche cells the charging must be done at a slow rate and on freshly discharged cells which have not been discharged below 1.0 V

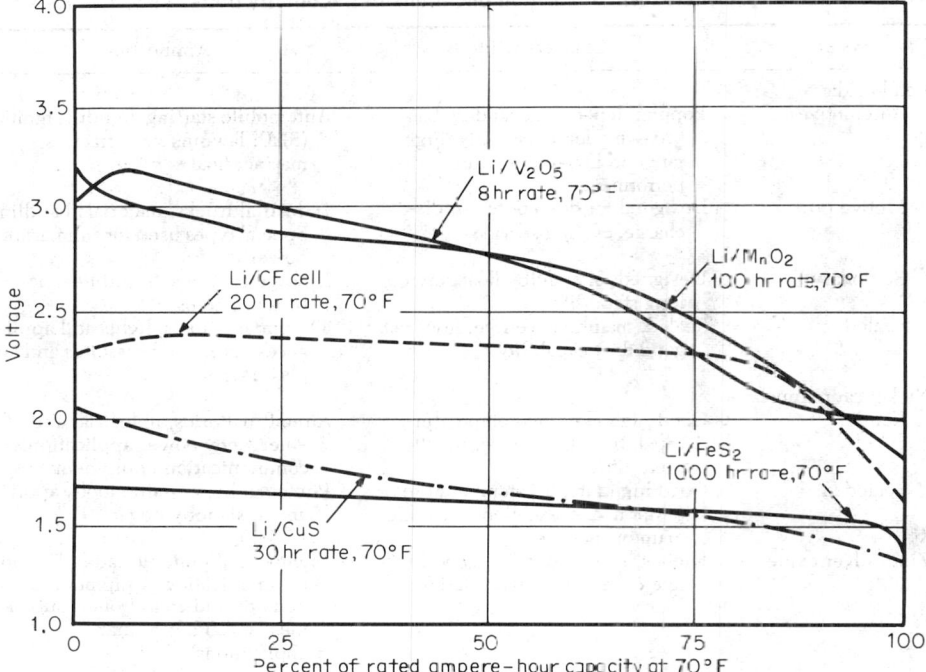

Fig. 7-65. Typical discharge characteristics of lithium–solid cathode cells.

per cell. The cells must be returned to service soon after recharging since recharged cells have poor shelf life.

Recharging dry cells is usually impractical and can be hazardous, particularly with cells that are not properly vented to eliminate the gases that form during the charge.

SECONDARY (STORAGE) BATTERIES

113. General. Secondary batteries are characterized, in addition to their ability to be recharged, by high power density, high discharge rate, flat discharge curves and good low-temperature performance. Their energy densities are usually lower than those of primary batteries. Tables 7-20 and 7-22 list the characteristics and applications of secondary batteries.

The applications of secondary batteries fall into two major categories: (1) applications where the secondary battery is used essentially as a primary battery but recharged after use secondary batteries are used in this manner for convenience (as in hand-held calculators or electronic flash units), for cost savings (as they can be recharged rather than replaced), or for power drains beyond the level of primary batteries; (2) applications where the secondary battery is used as an energy-storage device, being charged by a prime energy source and delivering its energy to the load on demand. Examples are automotive and aircraft systems, emergency no-fail and standby power sources, and hybrid applications.

114. Lead-Acid Battery. This is the most widely used and economical secondary battery. It uses sponge lead for the negative electrode, lead oxide for the positive, and a sulfuric acid solution for the electrolyte. As the cell discharges, the active material is converted into lead sulfate and the sulfuric acid solution is diluted; i.e., its specific gravity decreases. On charge the reverse actions take place. The state of charge of the battery can be determined by measuring the specific gravity.

The lead-acid battery is manufactured in many sizes, ranging from small plastic-encased batteries of less than 1 Ah capacity to large automotive and stationary units with hundreds of ampere-hours of capacity. The most common construction is the *pasted-plate design*, where the lead oxides are pasted onto a flat antimonial lead grid. Maintenance-free designs, in which the electrolyte is immobilized and loss of water is minimized, have been introduced and are particularly advantageous in portable applications.

TABLE 7-22   Major Characteristics and Applications of Secondary Batteries

System	Characteristics	Applications
Lead-acid:		
Automotive	Popular, low-cost secondary battery—moderate capacity, high-rate and low-temperature performance	Automobile starting, lighting, ignition (SLI); lawnmowers, tractors, marine, float service
Motive power	Designed for deep 6- to 9-h discharge, cycling service	Industrial trucks, materials handling. Special types used for submarine power
Stationary	Designed for standby float service, long stand life	Emergency power—utilities, no-break systems
Sealed	Sealed, maintenance-free, low cost, good float capability	TV, portable tools, lights and appliances, radios and cassettes and tape players
Nickel-cadmium:		
Vented	Good high-rate, low-temperature capability; flat voltage, excellent cycle life	Aircraft batteries, industrial and emergency-power applications, communication equipment
Sealed	Good high-rate, low-temperature performance, excellent cycle life, maintenance-free	Photography, portable tools, appliances, standby power
Zinc–silver oxide	Highest energy density, good high-rate capability, low cycle life	Lightweight portable radio, TV, and communication equipment; torpedo propulsion, drones, submarines, and other military applications

The general performance characteristics of the lead-acid battery are given in Fig. 7-66. Several limitations of the lead-acid battery are its poor low-temperature characteristics (particularly its inability to accept a charge at low temperatures), its loss of capacity on stand (self-discharge), and its relatively weak mechanical structure. A lead-acid battery left in a discharged condition for more than 6 months will become "sulfated" and difficult to recharge.

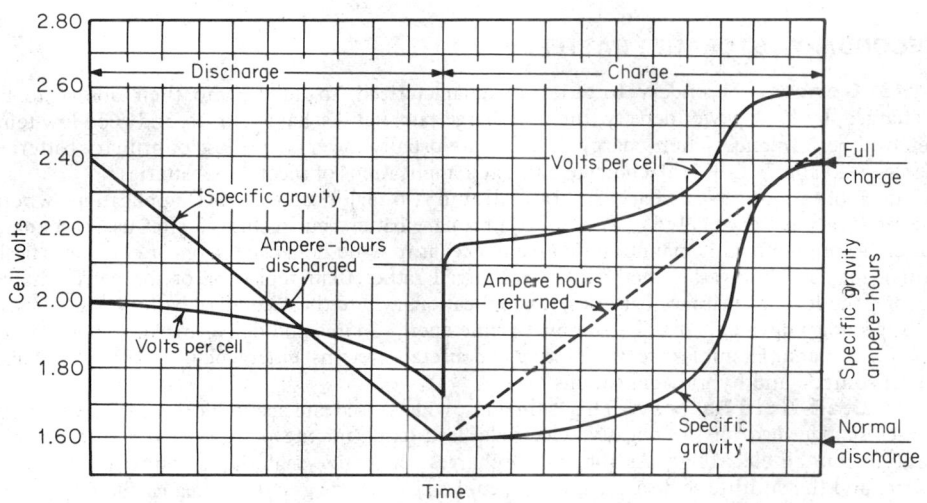

Fig. 7-66. Typical voltage and gravity characteristics of a lead-acid battery (constant-rate discharge and charge).

Since they cannot be maintained indefinitely in a charged condition, lead-acid batteries are usually stored *dry-charged*. In this condition they can retain their charge for as long as 2 years and can be put into service by adding the acid electrolyte. Recently, a water-activated battery, which contains the sulfuric acid in dry form and only requires the addition of water, has been introduced.

A lead-acid battery can be charged at any rate that does not produce excessive gassing or high temperatures. The most common practice for recharging a fully discharged battery is to start the charge at the $C/5$ rate (amperes, corresponding to one-fifth the rated capacity of the battery) for 5 h, which will return about 80 to 85% of the battery's capacity. The charging current is then tapered to about one-half to complete the charge. In an emergency, fast or boost charging is used. In this type of charge, the current should not exceed the C rate or the battery can suffer damage.

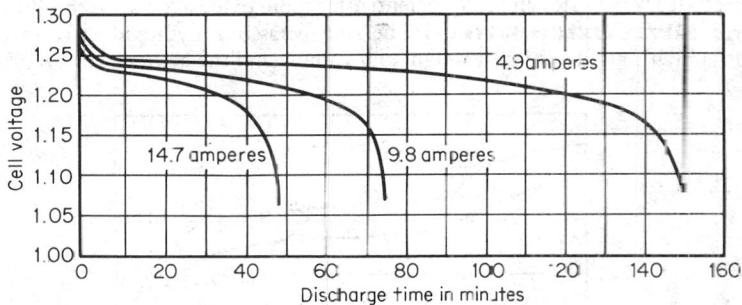

Fig. 7-67. Discharge curves for vented sintered-plate nickel-cadmium battery (12.5 Ah).

The battery can also be *float-* or *trickle*-charged when it is continuously connected to an electrical system. The current should be regulated at a low level to maintain the battery in a charged condition (sufficient just to replace capacity loss due to stand) and to prevent overcharging. The battery manufacturers can supply detailed charging instructions.

Maintenance-free batteries use a small percentage of alloys of calcium or other metals to give the lead grid the necessary physical rigidity and a corresponding lesser amount of the antimony normally used. The low-antimony grid construction considerably reduces the self-discharge due to local chemical action and reduces overcharge and gassing, thus significantly increasing the life of the battery. Such calcium-alloy cells are best suited for standby or float service rather than deep-discharge-cycling use. Small sealed cells, using the lead-acid system in cylindrical or rectangular shape, are available for portable applications. These cells use a gelled or immobilized electrolyte. They can be discharged in any position and are maintenance-free; i.e., there is no need to replace electrolyte or add water.

115. Nickel-Cadmium Batteries. The major alkaline secondary battery is the nickel-cadmium battery, noted for high power capability, long cycle life, good low-temperature performance, ruggedness, and reliability. It uses nickel oxide for the positive electrode, cadmium for the negative, and a solution of potassium hydroxide for the electrolyte.

The *pocket-type vented batteries* are used primarily for standby service or emergency lighting, with the battery maintained in a fully charged condition by trickle charging. Their long life and low cost make them ideally suited for these applications.

The *sintered-plate vented battery* is a more recent development. It is more expensive than the pocket type but gives better performance at high discharge rates and at low temperatures due to its lower resistance. The characteristics of the vented sintered-plate nickel-cadmium battery are given in Fig. 7-67.

The nickel-cadmium battery requires little maintenance and can take considerable abuse. Unlike the lead-acid battery, it can be stored in either the charged or discharged condition without damage. The self-discharge rate is very low, tapering off to about 3% per month, with a capacity loss of less than 50% during a year's storage at normal temperatures. Cycle life is excellent (over 1,000 c on deep discharges; many more on partial discharges). The battery can be recharged rapidly. Reasonable overcharging has no detrimental effect, and float charging for extended periods of time is permissible, although overheating the battery should be avoided.

The nickel-cadmium battery is also manufactured in a sealed construction as button or larger cylindrical or rectangular cells up to 25 Ah capacity. These cells are completely sealed but incorporate a design feature to prevent pressure buildup caused by gassing during charge. A safety vent is used in some cells to prevent rupture by excessive gas pressure due to malfunction, overcharge, or abuse.

Sealed cells have many advantages for portable applications, e.g., freedom from maintenance (other than charging), operation in any position, long storage and cycle life, good performance over a wide temperature range, and rugged construction.

116. Nickel-Zinc Batteries. The zinc–nickel oxide system has had limited use because of its poor cycle life, which is characteristic of the zinc electrode. Recently, however, this type of battery has received attention because of its high energy density, which is twice that of the lead-acid and nickel-cadmium systems. This fact, plus its moderate cost, makes the zinc-nickel battery a prime candidate for use in electrically propelled vehicles. Recent improvement in limiting the degradation of the zinc electrode has extended the life to 200 to 250 c, corresponding to 30,000 mi in typical electric vehicle use. The nickel-zinc system should be cost-competitive with other secondary batteries and have potential for more extensive use in the future.

117. Zinc–Silver Oxide Battery. The zinc–potassium hydroxide–silver oxide battery is noted for its high capacity per unit weight and volume and the ability to deliver this capacity

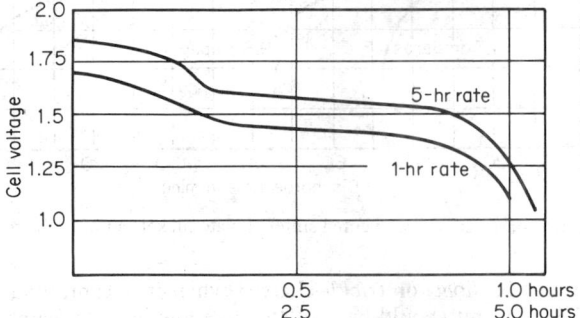

Fig. 7-68. Discharge curves for the zinc–AgO battery.

at high current drains (Fig. 7-68). It is useful in applications where high energy density is a prime requisite. The battery is not recommended for general storage battery use because its cost is high, its performance at low temperatures falls off more markedly than other storage batteries, and its cycle life is limited. Charging can be accomplished efficiently by normal methods, but it is necessary to limit overcharging due to the growth of zinc dendrites, which can short-circuit the cell internally.

118. Other Alkaline Secondary Batteries. A large number of electrochemical systems other than those described above can perform as secondary batteries. Of these, the *iron-nickel oxide* Edison cell was an important battery for traction and standby applications. In new designs it is now being investigated for electric vehicles, which can take advantage of its excellent cycle life. The *cadmium–silver oxide* battery, while expensive, has better cycle-life and low-temperature performance than the zinc–silver oxide battery but has a lower energy density.

The *nickel–hydrogen secondary battery* combines the nickel electrode of the nickel-cadmium cell with the fuel-cell hydrogen-gas-diffusion electrode. The resultant battery has a long life, capable of several thousand deep discharges and an energy density of 65 Wh/kg, twice that of the nickel-cadmium cell. The *silver-hydrogen battery*, using the silver electrode of the zinc–silver oxide battery, has an energy density of about 100 Wh/kg and a life of about 500 c.

119. Advanced Secondary Batteries. A number of electrochemical systems are being investigated as candidates for a low-cost efficient rechargeable battery with an energy density exceeding 200 Wh/kg at high power densities. Such batteries are needed for new applications, such as electric vehicles and utility load leveling.

Four different types of batteries are being considered: (1) high-temperature systems using molten-salt electrolytes; (2) high-temperature systems using ceramic or glass electrolytes; (3) organic-electrolyte ambient-temperature systems; and (4) aqueous zinc-halogen and metal-air systems. Figure 7-69 plots the anticipated performance characteristics of these new batteries and compares them with several existing battery systems. Most of these batteries use the alkali metals, lithium or sodium, as the active negative material because of their light weight, high voltage, and high specific energy with nonaqueous electrolytes to avoid the solubility of the alkali metals in aqueous solutions.

RESERVE AND SPECIAL BATTERIES

120. Reserve Batteries. Many batteries use highly active materials to obtain the high-energy content and power levels that are required, and are designed in a reserve construction to withstand long or severe environmental storage conditions. These batteries are used primarily to deliver high power for very short periods of time.

An important reserve battery is the *automatically activated zinc–silver oxide battery* used for missiles. The electrolyte of this battery is contained in a copper tube, sealed by a metal-foil diaphragm and activated by firing a gas squib which forces the electrolyte into the battery. The battery is fully discharged within 10 to 20 min.

The most prominent water or electrolyte manually activated batteries are the magnesium–

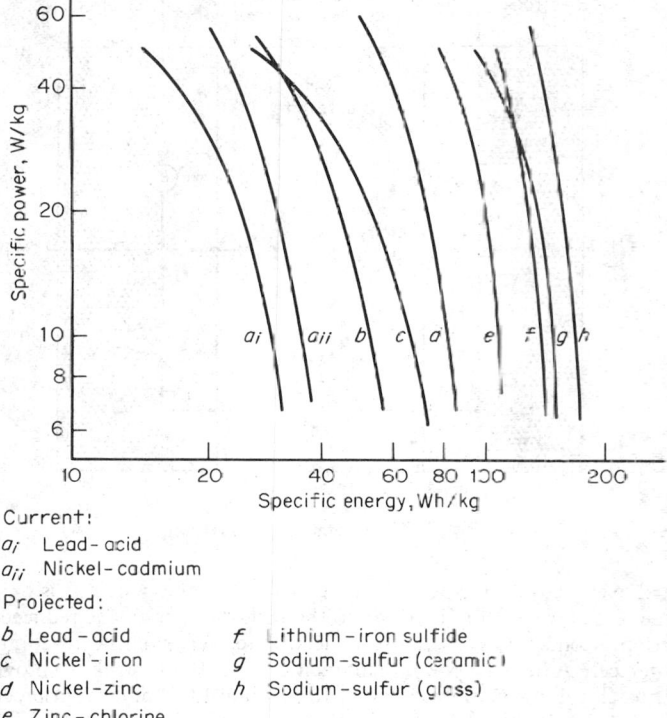

Current:
  a_i Lead–acid
  a_{ii} Nickel–cadmium

Projected:
  b Lead–acid f Lithium–iron sulfide
  c Nickel–iron g Sodium–sulfur (ceramic)
  d Nickel–zinc h Sodium–sulfur (glass)
  e Zinc–chlorine

Fig. 7-69. Projected performance of advanced secondary batteries. (*Data from Yao and Landgrebe, "New Battery Techologies," Battery Council, 1977.*)

silver chloride and magnesium–cuprous chloride systems. These are used for weather balloons or marine applications (sonobuoys, radios) to take advantage of their good low-temperature capability and high energy content. The battery is used within a few hours after activation.

The *thermal, or heat-activated, battery* employs a solid nonconducting electrolyte when the battery is inactive. The cell is activated by heating it sufficiently high to melt the electrolyte, making it conductive and permitting the flow of current. The battery is used in applications requiring its full capacity in minutes or less.

The *liquid-ammonia battery,* using liquid ammonia as the electrolyte, is an interesting system, being operable in the cold as well as normal temperatures, with little change in cell voltage and energy output. It is useful at the 1- to 20-min rate. The major problem is the volatilization of gas at the higher temperatures.

FUEL CELLS

121. General. Potentially the most efficient of the modern approaches to electric-power generation is the fuel cell. It converts liquid or gaseous fuels electrochemically into electric energy without the thermal losses typical of the internal-combustion engine and the efficiency limits of the Carnot cycle. The main advantage of the fuel cell is its high conversion efficiency (electrical output/heat content of the fuel). This can range as high as 50 to 55%, compared with 35% for conventional power plants.

The fuel cell can be adapted to cover a wide range of output power from portable units, 5 W or smaller in size, where ease of operation, low maintenance, and silence are important characteristics, to large stationary plants delivering megawatts of power where the high efficiency over the range to full load and reduced pollution problems are significant.

The operation of a fuel cell is shown graphically in Fig. 7-70. The basic fuel cell consists of an anode (−) and cathode (+) separated by a conducting electrolyte, such as a potassium hydroxide

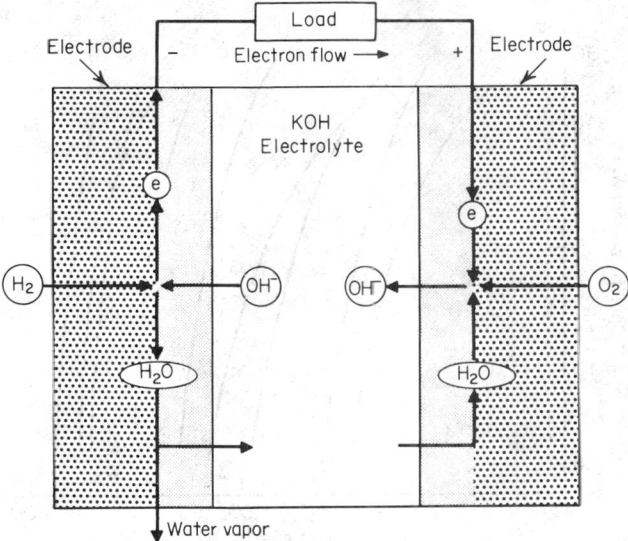

Fig. 7-70. Operation of a fuel cell.

solution. A fuel, e.g., hydrogen, is fed to the negative electrode, where it is oxidized, releasing electrons to the load. Oxygen (or air) is fed to the cathode, where it is reduced. The circuit is completed by ionic conduction through the electrolyte. Water is the reaction product of the hydrogen-oxygen cell. A fuel-cell power plant includes a fuel cell "stack" to provide the required output voltage (each cell operates in the range from 0.5 to 1.0 V dc), a fuel processor to convert the fuel into hydrogen-rich gas, and a power converter.

The ultimate objective is a direct fuel cell using conventional hydrocarbon fuels. A number of technological advances are required before such a device can be achieved. Current approaches, therefore, are concentrated on (1) fuels, such as hydrogen, hydrazine, and methanol, which are readily oxidized electrochemically and (2) indirect systems in which hydrocarbon fuels are first converted into hydrogen. Typical approaches to fuel-cell design are illustrated in Fig. 7-71.

Fuel cells can also be classified by the electrolyte, temperature of operation, oxidant or fuel. Low-power fuel cells operate from the ambient temperature to 70°C; the molten carbonate electrolyte systems in the range 600 to 700°C, and the solid-oxide electrolyte systems at about 1000°C. Higher operating efficiencies (as well as the reduction or elimination of the need for expensive catalysts) are achieved at the higher operating temperatures.

The fuel-cell systems built in 1965–1970 for the Gemini and Apollo space missions were among the first successful fuel cells demonstrated. They used hydrogen as the fuel and oxygen as the

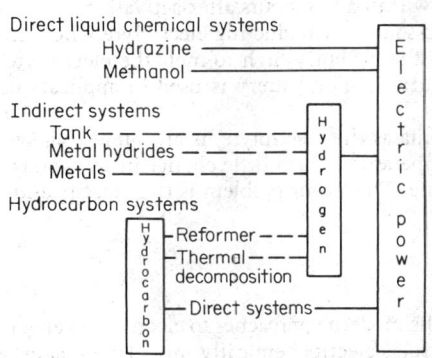

Fig. 7-71. Approaches to fuel cells.

oxidizer. The Gemini system, designed as a 1-kW system, employed an ion-exchange membrane-separator–electrolyte system. The larger 1.5-kW Apollo system was based on a potassium hydroxide electrolyte.

For ground applications, air-breathing, rather than pure oxygen systems, have been considered. In the smaller units (up to 250 W) hydrogen is supplied in a pressure vessel or generated from a metal hydride. In the larger systems (500 W and above), hydrocarbon fuels are used and hydrogen is generated by steam reforming or thermocatalytic cracking of the fuel.

Prototypes of low-power fuel-cell systems in the 1- to 100-W range using hydrogen as fuel have been built but have not shown sufficient advantage over battery systems to warrant serious consideration, except possibly for long-term unattended operation.

122. Future Directions. The major emphasis of the fuel-cell program is directed toward application in the electric utility systems, primarily for handling peak loads, in place of the conventional electric generators. For this application the fuel cell is more efficient, should be less

TABLE 7-23   Fuel-Cell Applications*

Type of plant	Size	Fuel	Cost per kW	Efficiency %	Electrolyte	Availability
On-site (TARGET)	25–200 kW	Natural gas	$500	30–40	Phosphoric acid	Early 1980s
Dispersed (load-following)	10–25 MW	Naphtha	350	37–39	Phosphoric acid	Early 1980s
	10–25 MW	Petroleum	350	45–47	Molten carbonate	Late 1980s
Central (base load)	150–600 MW	Coal	800	45–47	Molten carbonate	1990s

*Source: Arnold P. Pickett, Fuel Cell Power Plants, *Scientific American*, December 1978, p. 76.

costly, and can be located close to the point of use (since it is environmentally acceptable), minimizing transmission losses. The current fuel-cell program for utility applications is summarized in Table 7-23.

The plants listed in this table will operate on fossil fuels, such as natural gas or low-sulfur-content light distillates. They will contain three sections: the reformer section, for converting the fuel into hydrogen-rich gas; the power section, using the phosphoric acid fuel cell initially; and an inverter, to convert the dc output of the cells into alternating current. The overall conversion efficiency is estimated at 40%. Later plants will use the molten-carbonate high-temperature electrolyte system at 600°C. These cells will use porous nickel-electrode structures and will require no platinum catalyst. The fuel used can thus be broadened to include coal and oil.

Relays and Switches

BY J. SPERGEL

123. Introduction. The primary function of electromechanical components is the transmission and control of electric current accomplished by mechanical contacting and actuating devices. In recent years, solid-state (nonmechanical) switching devices have come into wide use and their applications are extending rapidly.

124. Relay Types. The simplified diagram of a relay shown in Fig. 7-72 illustrates the basic elements that constitute an electromagnetic relay (EMR). The most common EMR types are as follows:

General-purpose. Design, construction, operational characteristics, and ratings are adaptable to a wide variety of uses.

Latch-in. Contacts lock in either the energized or deenergized position until reset either manually or electrically.

Polarized (or polar). Operation is dependent upon the polarity of the energizing current. A permanent magnet provides the magnetic bias.

Differential. Functions when the voltage, current, or power difference between its multiple windings reaches a predetermined value.

Telephone. An armature relay with an end-mounted coil and spring-pickup contacts mounted parallel to the long axis of the relay coil. Ferreeds are also widely used for telephone cross-point switches.

Stepping. Contacts are stepped to successive positions as the coil is energized in pulses; they may be stepped in either direction.

Interlock. Coils, armature, and contact assemblies are arranged so that the movement of one armature is dependent upon the position of the other.

Sequence. Operates two or more sets of contacts in a predetermined sequence. (Motor-driven cams are used to open and close the contacts.)

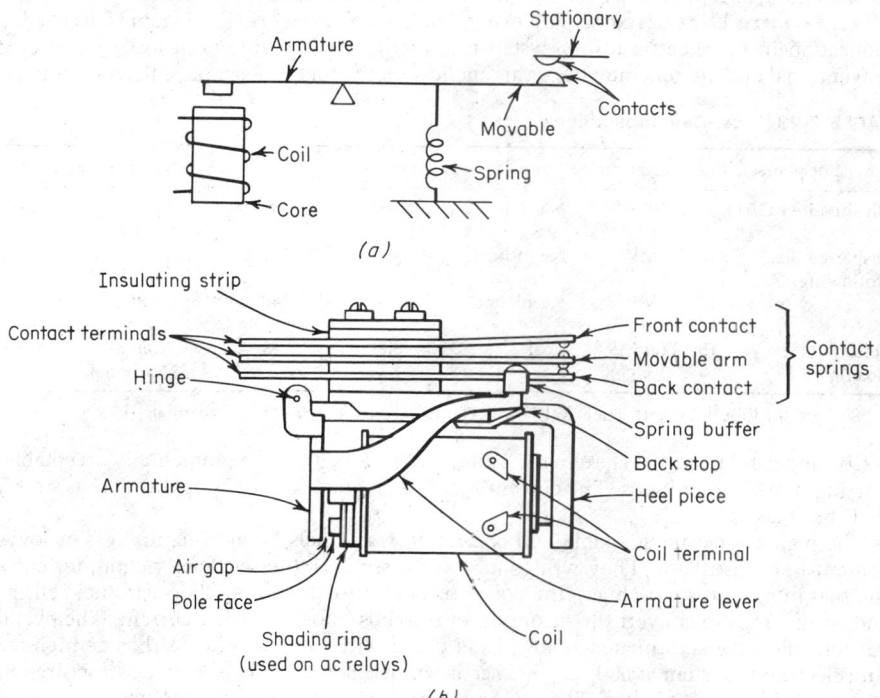

Fig. 7-72. (*a*) Simplified diagram of a single-pole, double-throw, normally open relay; (*b*) structure of a conventional relay. (*From "Electronics Components Handbook," McGraw-Hill Book Company, New York, 1958.*)

Time-delay. A synchronous motor is used for accurate long time delay in opening and closing contacts. Armature-type relay uses a conducting slug or sleeve on the core to obtain delay.

Marginal. Operation is based on a predetermined value of coil current or voltage.

125. Performance Criteria. The design or selection of a relay should be based on the following circuit-performance criteria:

Operating frequency. Electrical operating frequency of relay coil

Rated coil voltage. Nominal operating voltage of relay coil

Rated coil current. Nominal operating current for relay

Nonoperate current (or voltage). Maximum value of coil current (or voltage) at which relay will not operate

Operate voltage (or current). Minimum value of coil voltage (or current) at which switching function is completed

Release voltage (or current). Value of coil voltage (or current) at which contacts return to the deenergized position

Operate time. Time interval between application of power to coil and completion of relay-switching function

Release time. Time interval between removal of power from coil and return of contacts to deenergized position

Contact bounce. Uncontrolled opening and closing of contacts due to forces within the relay

Contact chatter. Uncontrolled opening and closing of contacts due to external forces such as shock or vibration

Contact rating. Electrical load on the contacts in terms of closing surge current, steady-state voltage and current, and induced breaking voltage

Figure 7-73 illustrates some of the contacting characteristics during energization and deenergization. Figure 7-74 illustrates the effect on relay operation of changing coil current. Detailed discussion and analysis of relay design parameters, as well as data on magnetic-core materials, winding coils, and general formulas for temperature of electromagnets, are provided by Peek and Wagar (see reference, Par. **7-189**, Relays and Switches).

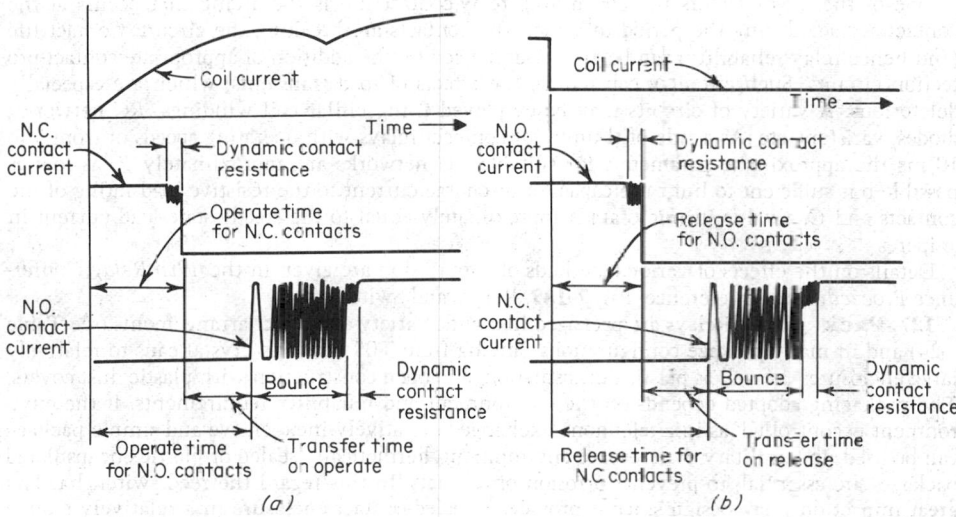

Fig. 7-73. Typical oscillograph pictures of contacting characteristics during (a) energization and (b) deenergization of a relay. (*Automatic Electric, Northlake, Ill.*)

126. General Design and Application Considerations. The dynamic characteristics of the moving system, i.e., armature and contact assembly, are primarily determined by the mass of the armature and depend upon the magnet design and flux linkage. Typical armature configurations are clapper or balanced armature, hinged or pivoted lever about a fixed fulcrum; rotary armature; solenoid armature; and reed armature. Contact and restoring-force springs are attached or linked to the armature to achieve the desired make and/or break characteristics. The types of springs used for the contact assembly and restoring force are generally of the cantilever, coil, or helically wound spring type. Primary characteristics for spring materials are modulus of elasticity, fatigue strength, conductivity, and corrosion resistance. They should also lend themselves to ease of manufacture and low cost. Typical materials for springs are spring brass, phosphor bronze, beryllium copper, nickel silver, and spring steel.

Contacts. These include stationary and moving conducting surfaces that make and/or break the electric circuit. The materials used depend on the application; the most common are palladium, silver, gold, mercury, and various alloys. Plated and overlaid surfaces of other metals such as nickel or rhodium are used to impart special characteristics such as long wear and arc resistance or to limit corrosion.

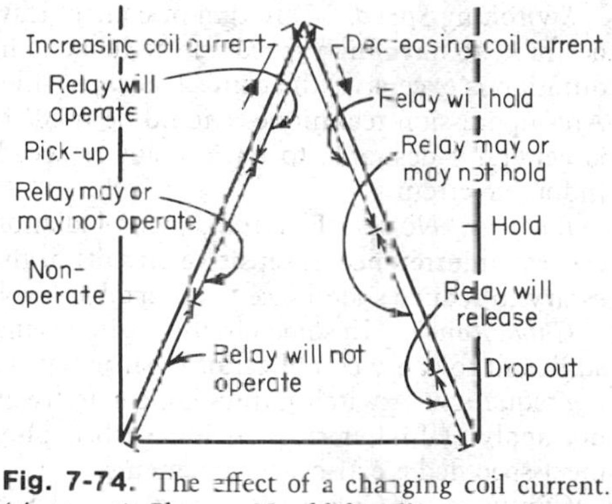

The heart of the relay is the contact system that is typically required to make and/or break millions of times and provide a low, stable electrical resistance. The mechanical design of the relay is aimed principally at achieving good con-

Fig. 7-74. The effect of a changing coil current. (*Automatic Electric, Northlake, Ill.*)

tact performance. Because of the numerous operations and arcing often occurring during operation, the contacts are subject to a wide variety of hazards that may cause failure, such as:

Film formation. Effect of inorganic and organic corrosion, causing excessive resistance, particularly at dry-circuit conditions.

Wear erosion. Particles in contact area which can cause bridging between small contact gaps.

Gap erosion. Metal transfer and welding of contacts.

Surface contamination. Dirt and dust particles on contact surfaces can prevent achievement of low resistance between contacts and may actually cause an open circuit.

Cold welding. Clean contacts in a dry environment will self-adhere or cold-weld.

One of the major factors in determining relay-contact life is the arcing that occurs at the contact surface during the period in which the contacts are breaking the circuit. Contact life (and hence relay reliability) can be greatly enhanced by the addition of appropriate contact-protection circuits. Such circuitry can reduce the effects of load transients, which are especially deleterious. A variety of circuits may be employed using bifilar coil windings, RC networks, diodes, varistors, etc. As a rule of thumb, for compact relays with operating speeds of from 5 to 10 ms the approximate parameters for suitable RC networks are approximately R, as low as possible but sufficient to limit the capacitor discharge current to the resistive load rating of the contacts and C, a value in microfarads approximately equal to the steady-state load current in amperes.

Details on the effects of various methods of suppression are given in the *19th Relay Conference Proceedings* (see reference, Par. **7-189**, Relays and Switches).

127. Packaging. Relays are packaged in a wide variety of contact arrangements (see Table 7-24) and in many package configurations ranging from T05 cans and crystal cans to relatively large enclosures, as well as plastic encapsulation, and open construction with plastic dust covers. The packaging adopted depends on the environment and reliability requirements. If the environment is controlled, as in a telephone exchange, a relatively inexpensive and simple package can be used. In a military or aerospace environment, hermetically sealed or plastic-encapsulated packages are essential, to prevent corrosion of contacts. In this regard the reed switch has had great impact on relay design since it provides a sealed-contact enclosure in a relatively simple and inexpensive package. It has become widely used in the telephone and electronics industry because it has been able to extend relay life to millions of cycles. A comparison between the reed and typical conventional relays is given in Table 7-25.

With the growth of the microelectronic technology, the development of solid-state relays (SSR) has been achieved and is making an impact in the relay industry. Table 7-26 highlights the important differences in performance between EMRs and SSRs.

SWITCHES

128. Introduction. Switches are electromechanical devices used to make or break an electric circuit by manual or mechanical operation. Switches are available in many types for many functions.

129. Selection Criteria. Some of the more important criteria follow.

Switching Speed. Duration of contact travel during *make* or *break* function. It is generally desirable to have high speed during make to minimize the duration of an arc or flashover. If the duration is excessive, the contact surface will deteriorate and welding of the contacts may occur. Arc-suppression techniques should be used if higher currents are anticipated. During break, it is generally desirable to have a slower speed to minimize transients, particularly for dc and inductive circuits.

Electrical Noise. Electromagnetic radiations may occur during make and break of switches, causing interference in sensitive circuits or high-gain amplifiers. Suppression of arc may be necessary to reduce such noise to acceptable levels.

Capacitance. In some circuits switches may look like a capacitor and its capacitance may be sufficient to cause complications, particularly in frequencies of 60 to 100 MHz.

Frequency. Switch ratings are generally given for direct current, 60 and 400 Hz, and would not apply at higher frequencies. Such applications may necessitate a special switch design to meet specified electrical requirements.

Contact Snap-Over and Bounce Time. Snap-over time is the time a contact separates from a normally closed position and travels to a normally open position and makes contact with the circuit. Bounce time is the interval between initial contact and steady contact during which the

Table 7-24 Nomenclature for Basic Contact Forms

Form	Description	Symbol	Form	Description	Symbol
A	Make or SPSTNO		J	Make, make, break or SPST (M-M-B)	
B	Break or SPSTNC		K	Single pole, double throw, center off or SPDTNO	
C	Break, make or SPDT (B-M)		L	Break, make, make or SPST (B-M-M)	
D	Make, break or SPDT (M-B)		U	Double make, contact on arm or SPSTNODM	
E	Break, make, break or SPDT (B-M-B)		V	Double break, contact on arm or SPSTNCDB	
F	Make, make or SPST (M-M)		W	Double break, double make, contact on arm or SPDTNC-NO(DB-DM)	
G	Break, break or SPST (B-B)		X	Double make or SPSTNODM	
H	Break, break, make or SPST (B-B-M)		Y	Double break or SPSTNCDB	
I	Make, break, make or SPST (M-B-M)		Z	Double break double make or SPDTNC-NO(DB-DM)	

Poles, single (SP), double (DP); throws, single (ST), double (DT); normal position, open (NO), closed (NC); double make (DM), double break (DB).

Table 7-25 Estimated Load-Life Capability of Two-Pole Miniature Relays

	Dry reed, 0.125 A	Conventional crystal can, 5 A	Miniature power, 10 A
λ at 1×10^6 (%/10^4)	0.002	1.2	0.80
$R_{(.999)}$ (10^3 operation)	700	2.8	6.0
$R_{(.90)}$ (10^3 operation)	10,000	60	120.0

λ = failure rate in %/10,000 h operation; $R_{(.999)}$ = operating life with 99.9% probability; $R_{(.90)}$ = operating life with 90% probability.

Table 7-26 Relative Comparison of Electromagnetic Relays (EMR) vs. Solid-State Relays (SSR)

Characteristic	EMR	SSR	Advantage
Life	From 100,000 to millions of cycles. Reed contacts are outstanding.	No moving parts. When properly designed should last life of equipment.	SSR
Isolation	Infinite dielectric isolation.	Not dielectrically isolated; however, several techniques are available to achieve up to 10 k$M\Omega$.	EMR
EMI (RFI)	Can generate EMI by switching of its coil, thereby requiring special isolation (i.e., shielding).	Noise generated is negligible compared with EMR.	SSR
Speed	Order of milliseconds.	Up to nanoseconds.	SSR
Operate power	Uses more power than SSR.	Lower power requirements but requires continuous standby power.	SSR
Contact voltage drop	Relatively low voltage drop because of low contact resistance.	High voltage drop which is dissipated into heat.	EMR
Thermal power dissipation	Primarily concerned with dissipating coil power.	Higher voltage drop develops appreciable heat to be dissipated.	EMR

contact bounces as a result of impact of a moving contact on a stationary contact. These times should be kept to a minimum because the time intervals and/or the contacting instability may influence critical or sensitive circuits.

130. Switch Configurations. Switches are available in numerous configurations and packages to meet special equipment designs. These are often *ganged* or assembled in matrices. A variety of unique switching capabilities can be achieved by using a permanent magnet in combination with reed capsules. In addition, many pushbutton switches are available with illuminated faces.

Discrete Component Modules

BY R. A. GERHOLD

131. General. Physical implementation of a circuit schematic into a functional component assembly requires selection of the discrete component parts, mechanical support, electrical interconnection, provision for dissipation of heat generated, and adequate protection against anticipated environmental factors. Design of such a functional assembly must also consider maintenance, cost, and reliability.

132. Definitions. *Circuit packaging.* The branch of electronic design concerned with the physical assembly and electrical interconnection of elements into circuits, and with location, mounting, assembly, protection, and interconnection of circuits into subassemblies, assemblies, and equipments to meet performance, operational, and maintenance requirements.

Module, equipment. An assembly or subassembly of an equipment which displays dimensional regularity and separable repetition within a given equipment. It is generally designed to be handled as a single unit to facilitate supply and installation, operations and/or maintenance. It can be either repairable or nonrepairable.

Printed wiring. A pattern of interconnecting wiring formed on a common insulating base.

133. Planar Assembly. Printed wiring on rigid laminated plastic boards provides a planar interconnection wiring array. Discrete components are mounted on the printed-wiring boards; their leads are bent at right angles and inserted into predrilled or prepunched holes. Packages for microelectronics are mounted similarly (see Pars. **7-176** to **7-185**). Soldering of the leads to the printed wiring provides both electrical connection and mechanical support of the components in the completed printed-wiring assembly.

134. Embedment as a modular-assembly concept can be applied to assemblies of components in free form which are electrically interconnected but require encapsulating resin to provide mechanical support and environmental protection. Encapsulation or embedment material should be applied to individual components only as part of a controlled production process for that component. Module design should be completely verified for the particular encapsulation or embedment materials and processes to be employed. In particular, extreme temperature, aging, and temperature-cycling tests are required to verify adequacy of the design.

135. Printed Wiring. The printed-wiring conductor is usually a metal such as copper, and the *insulator* is usually a laminated plastic sheet. In the *subtractive process*, the starting point is a copper-clad laminate; the desired configuration is printed with an acid resist on the foil surface. Etching then removes the unwanted surface areas, leaving a foil pattern.

136. In additive processes the printed-wiring pattern is built up by plating on the insulator surface. Plating techniques are also employed to produce *through connections* between circuits on opposite sides of a plastic board.

Thickness of the conductor foil and width of individual printed-wiring lines are determined by the *current-carrying capacity* required, as indicated in Fig. 7-75.

137. Printed-wiring-board material and *thickness* are determined mainly by mechanical considerations of strength and stiffness vs. cost and fabricability. Some thin materials are used where flexible printed wiring is desired.

138. Insulation resistance and breakdown voltage between adjacent conductors are severely degraded by surface contamination in the presence of high humidity. *Spacing*

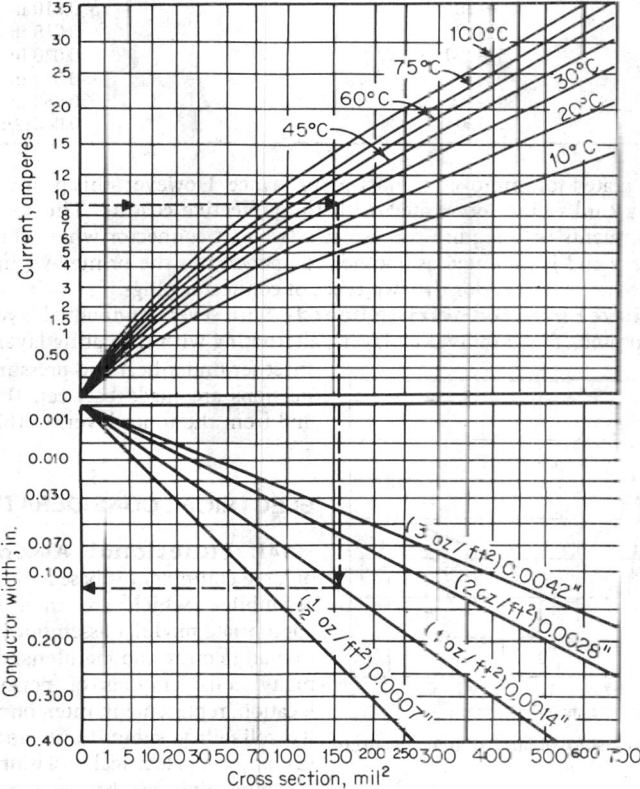

For four standard thicknesses of copper.

(For use in determining current carrying capacity and sizes of etched copper conductors for various temperature rises above ambient.)

Fig. 7-75. Printed-wiring conductor thickness and width.

requirements between conductors for various classes cf application and voltages are indicated in Table 7-27. Applications for military field equipment require that all printed-wiring boards be *conformally coated.*

139. Connection. The above requirements also relate to the distinction between the two broad classes of connectors for printed-wiring assemblies. In the card-edge, or one-part, connector, the plug is printed on the end of the printed-wiring board as part of the printed wiring. The

TABLE 7-27   Spacing Requirements

Voltage between conductors, dc or ac peak, V	Minimum spacing
Uncoated printed-wiring boards, sea level through 10,000 ft:	
0–150	0.025 in
151–300	0.050 in
301–500	0.100 in
>500	0.0002 in/V
Over 10,000 ft:	
0–50	0.025 in
51–100	0.060 in
101–170	0.125 in
171–250	0.250 in
251–500	0.500 in
>500	0.001 in/V
Conformal coated printed-wiring boards, all altitudes:	
0–30	0.010 in
31–50	0.015 in
51–100	0.020 in
101–300	0.030 in
301–500	0.060 in
>500	0.00012 in/V

surface may be plated for improved contact performance. However, this printed-plug portion of a printed-wiring card cannot be treated with the protective coating. The solution for humid military environments is to require the use of a two-part connector where the plug portion is formed separately and is mounted as another component on the printed-wiring board. Appropriate receptacles are provided for the two types of connector plugs.

140. Multilayer printed-wiring boards with several individual layers of conductive patterns are also used. These individual layers, alternating with insulating layers, are laminated together under heat and pressure. Through connections are made between the various layers and from the inner layers to the outer surfaces.

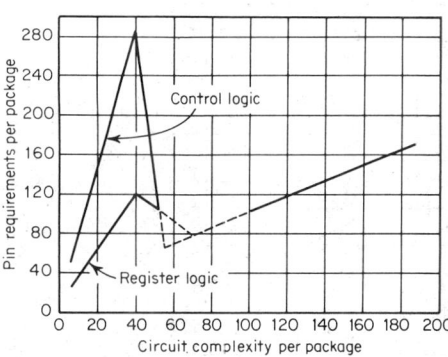

Fig. 7-76. Relationships of pin requirements to circuit complexity.

ELECTRICAL CONSIDERATIONS

141. Functional Assemblies. Design of large equipment is based on major equipment assemblies, which are then subdivided into appropriate modular assemblies for convenience in manufacture and maintenance. This involves many considerations of performance, failure location, replacement, interconnection complexity, reliability, subunit test capability, standardization and electrical isolation and shielding. Obvious divisions between modules correlate with functional circuit groupings but frequently lead to unequal complexities, whereas uniformity is desirable for manufacturing.

142. Connector Pins. Another prime consideration is the number of connector pins required, per module and overall. As shown in Fig. 7-76, the number of pins required per package may vary widely as circuit complexity is increased. Beyond some maximum point, the number of pins per package frequently falls to a minimum, which serves to indicate a desirable division point between packages. Provision must also be made for ease of testing, installation, repair, replacement, and troubleshooting.

143. Standard Modules. Standard modules, of specific design, are available for use throughout an entire system. Thus a variety of systems can be built and changes incorporated simply by changing interconnecting wiring. However, modules built specifically for the indi-

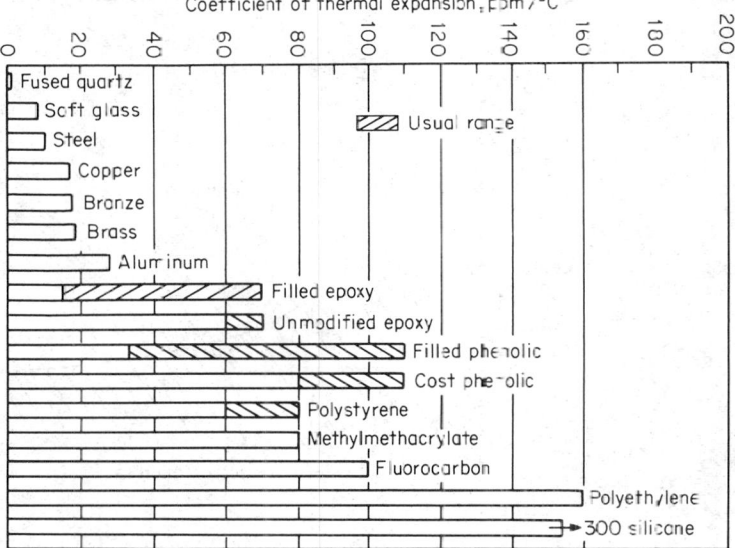

Fig. 7-77. Coefficients of thermal expansion. [*After C. A. Harper (ed.), "Handbook of Electronic Packaging," McGraw-Hill Book Company, New York, 1969.*]

vidual purposes usually provide simplified wiring, smaller size, and improved electrical performance. These advantages must be weighed against the cost, reliability, and availability which may accrue from the use of the standard modules.

Consideration must also be given to the need to minimize interference between stages, excess propagation time, pickup on signal lines, surface-leakage currents under humidity, changes in components by heat of soldering, resistance of joints in series, and effect of component hot spots on operation. Voltage distribution and grounding systems can be series, tree, loop, and point types. Multilayered printed-wiring boards facilitate use of multiplaned grounding systems and of several layers of voltage distribution. Additional electrostatic and electromagnetic shielding may be used.

ENVIRONMENTAL CONSIDERATIONS

144. Humidity. Consideration must be given to the effect of volumetric absorption of moisture, to surface absorption or condensation of moisture, and to possibilities of moisture entrapment within the case or container or between closely spaced surfaces. Aside from performance changes, moisture greatly accelerates corrosion of metals, particularly where voltage differences are present. Even in sealed containers, temperature drops (such as occur in high-altitude flight) may increase the relative humidity to the dew point and result in condensed moisture on the circuitry. Localized heating from power-dissipating components may aid in crying off surfaces.

145. Temperature. Temperature considerations involve expected environmental temperatures, plus additional heat rise within the equipment due to power dissipation. Hot-spot temperatures of critical components must be maintained within acceptable limits. Low temperature, particularly with thermal cycling, can cause severe mechanical degradation due to cracking. Figure 7-77 indicates the wide range of coefficients of thermal expansion of commonly used materials.

LIFE-CYCLE CONSIDERATIONS

146. Initial design and production cost is but one aspect of the overall life cycle of a piece of equipment. For complex equipment, the cost of locating faults during the operating life of the equipment, of replacing failed modular assemblies, and of repairing such replaced assemblies can add significantly to total life-cycle cost. For military applications, this additional cost may run several times the initial cost of the equipment.

147. Reliability. The failure rate of a particular equipment is directly related to the sum of the failure rates of its individual component parts. This failure-rate total is also affected by

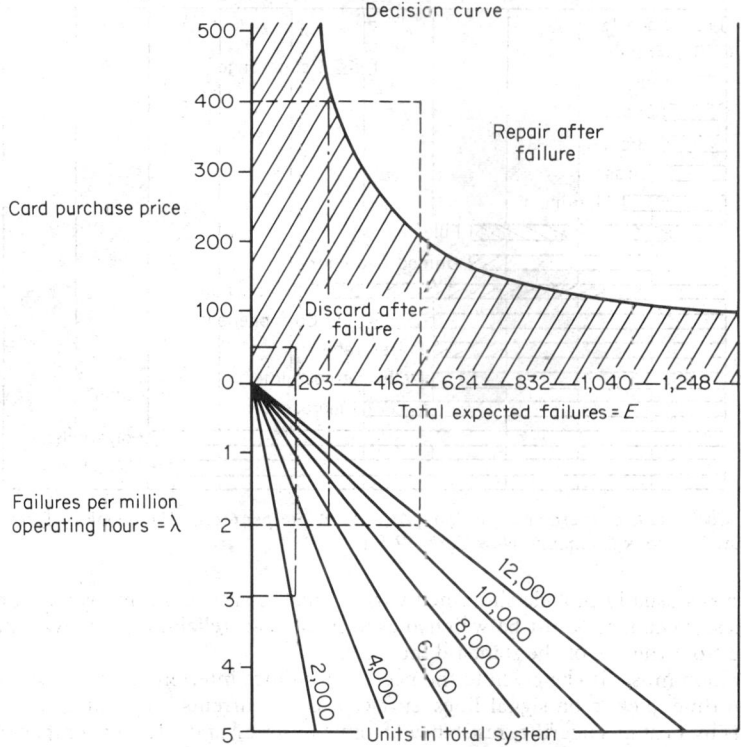

Fig. 7-78. Repair-vs.-discard decision curve.

any redundancy of circuitry and by the degree of criticality of application (i.e., the severity of the failure criteria) of the parts in the individual circuits.

148. Repair vs. Discard. In each application, there is a cost value for the modular-component assembly below which it is more economic to discard a failed assembly and to replace it with a new one, rather than to locate and repair the failure. Figure 7-78 indicates the type of decision curve that applies to such an analysis. The curve itself is a function of the cost of repair time, repair parts, test equipment, transportation, failure analysis, and quantities involved. Such a curve varies widely with the specific type of equipment and its manner of application. When Fig. 7-78 is entered at the lower left, the known or anticipated failure rate applied to the total number of units in the system leads to the total number of expected failures per year. The decision to repair or discard after failure is determined by the intersection point between this total number of expected failures and the card or assembly purchase price. If this intersection falls above the decision curve (in the white area), it indicates that enough failues are anticipated to warrant setting up a cost-effective procedure for repairing such failures when they occur. If the intersection point falls below the decision curve (in the shaded area), it indicates that not enough failures are expected to support a repair procedure that would be cost-effective compared with the cost of discarding failed assemblies.

Hybrid Microcircuits

BY I. PRATT

FABRICATION AND ASSEMBLY

149. Introduction. Hybrid technology provides the means for assembling all major categories of electronic components in relatively small enclosures to fill system size requirements which cannot be met by more conventional packaging techniques. Analog and digital circuits as well as microwave modules are made in hybrid form. The enclosure contains an insulating substrate with deposited networks, generally conductors and resistors, to which semiconductor devices, integrated circuits, and passive elements are attached in chip form.

The development of the transistor and associated minicomponents accelerated miniaturization and the growth of electronic applications. New complex products requiring large numbers of components resulted in interconnection problems which were resolved mainly by the development of monolithic integrated-circuit technology and the thick- and thin-film technologies defined below.

Interconnections formed as part of the silicon in monolithic integrated circuits are supplemented by thin- and thick-film patterns on the insulating substrate. Since not all system electrical requirements can be met solely by only one or the other of these technologies, integration of the several methods into one package has been achieved by the hybrid-microcircuit technology.

150. Definitions. *Monolithic technology.* Science of formation of electronic elements (by diffusion, ion implantation, or evaporation) on or within a single semiconductor chip, e.g., silicon, to perform an electronic-circuit function. Chips are normally 9 to 14 mils (225 to 350 μm) thick and up to 0.06 in^2 (0.40 cm^2) in area (see Sec. **8**).

Thin-film technology. Science of the formation of electronic elements or networks (by vacuum evaporation, sputtering, or anodization) on a supporting substrate. Film thicknesses are less than 5 μm (5000 nm) and usually on the order of 0.03 to 1.0 μm.

Thick-film technology. Science of the formation of electronic elements or networks (by printing liquid or paste through a screen or mask and then firing) on a supporting substrate. Film thicknesses are usually 10 μm or greater.

151. Design and Fabrication. From analysis of the circuit schematic diagram, thick- and/or thin-film materials are chosen, along with the discrete components, which include uncased active and passive chips. The geometry which defines the dominant parameter of each film-type circuit and the position of each component are determined next. After establishing the deposition sequence for the individual layers of film materials, master patterns for each layer are prepared. From them, photoreduced transparencies are made for photolithographic fabrication of the masks used to deposit or define the film geometry on the substrate, after which the chip parts are attached and interconnected to the film circuit. Maximum circuit yield with related cost effectiveness necessitates an optimum number of add-on components per substrate, a limitation on the range of resistance and capacitance values and tolerances, a minimum use of inductors and transformers, and a system circuit design divided into testable subfunctions.

152. Testing. Film depositions and assembly procedures with adequate process and quality controls ensure desired performance and reliability. Hybrid microcircuits, normally packaged in hermetically sealed enclosures to satisfy environmental and life requirements, can be subjected to a variety of screening procedures, as defined under the International Society of Hybrid Microelectronics (ISHM) Hybrid Microelectronics Standard Specification Guidelines. These screening procedures, extracted from military specifications (MIL-M-38510) and standards (MIL-STD-883) for microelectronics, can be carried out to ensure that the circuits achieve levels of quality and reliability needed for the intended application. Specification and quality control at the material and process level and final product assurance are defined in control documents by each manufacturer.

153. Applications. A typical hybrid application is a pulse-restorer circuit used as an unattended repeater in a two-way pulse-code-modulation transmission system. The physical layout of this linear circuit on an alumina substrate includes two-layer thick-film gold conductors, thick-film resistors, and chip resistors, diodes, capacitors, and inductors. A typical digital processor circuit for use in electronic countermeasure (radar-jamming) systems includes six thick-film conductor layers interconnecting over 100 chips, including LSI microprocessors, RAMs, ROMs, and other integrated-circuit types. Similarly, a microwave phase-shifter circuit used repetitively in a phased-array radar antenna, consists of a gold thick-film microstrip distributed network and includes 30 thick-film capacitors and 15 *pin* diodes connected through the substrate to a thick-film ground plate on the back side of a 2½-in-diam substrate.

FILM COMPONENTS

154. Networks. Thick- and thin-film networks involve the deposition of passive circuit elements in a predetermined geometric pattern on the surface of an insulating substrate (Fig. 7-79). Deposited thin films can be made with precision and stability for the diverse requirements of linear circuits and can be fabricated with finer lines than thick films.

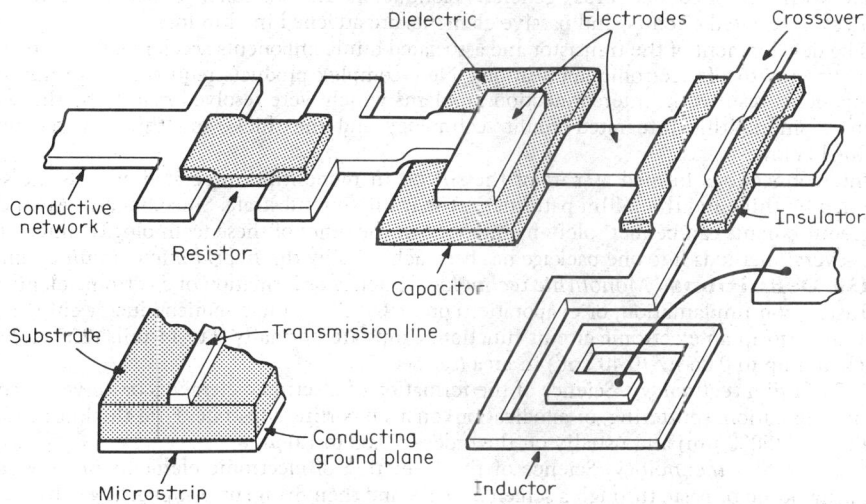

Fig. 7-79. Hybrid-microcircuit film components.

155. Resistive Elements. Resistor films are described in terms of sheet resistance (ohms per square), which is actually the resistance of a square of the film material, and varies inversely with film thickness. The value of a resistor of a given film is determined by its length-to-width ratio (Fig. 7-80).

156. Capacitive Elements. Deposited capacitors are composed of two layers of conductive material (electrodes) separated by a dielectric layer. The capacitance C is a function of area, dielectric constant, and dielectric thickness

$$C = \epsilon_0 K A / t$$

where ϵ_0 = permittivity of free space, A = area of electrode, K = dielectric constant, and t = dielectric thickness. Anodic tantalum oxide capacitors, an integral part of the well-established thin-film hybrid-microcircuit technology, are the most thoroughly studied and developed. The use of thick-film capacitors is relatively limited, compared with the extensive use of thick-film resistors and conductors. Thick-film capacitors are difficult to fabricate to close tolerances, require large substrate areas to achieve high capacitance values, and are plagued with dielectric limitations.

157. Inductive Elements. Flat spiral inductors are made by printing or defining conductor patterns of an appropriate spiral design directly onto the substrate. Such configurations exhibit relatively low Q factors and also require extensive substrate area to achieve high values since the inductance is directly proportional to the overall size and the number of turns.

158. Microstrip. Microstrip is a form of transmission line. It is the basic building block for hybrid microwave microcircuits. Microstrip is a planar structure consisting of the dielectric substrate, a conducting strip for the conductor pattern on one side of the substrate, and a conducting ground plane on the other. The transmission line is required to provide matched interconnections between various passive components, including resonators and filters, and integral parts of phase shifters, isolators, and circulators.

SUBSTRATES

159. The substrate of the hybrid microcircuit supports the deposited film and the mounted chip components. Substrate properties are important since film properties depend upon the characteristics of the substrate surface, particularly for thin-film components Thermal conductivity, surface finish, dielectric constant, thermal coefficient of expansion, and ability to withstand processing temperature are significant for different circuit types and applications.

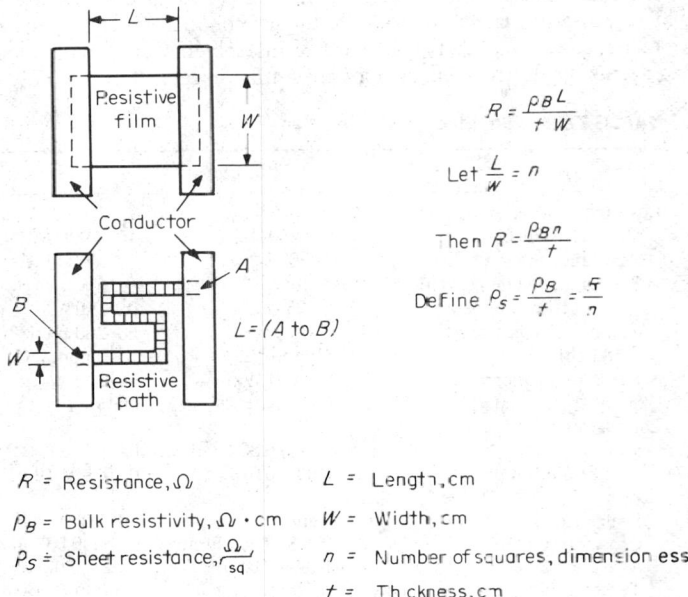

$$R = \frac{\rho_B L}{t\,W}$$

$$\text{Let } \frac{L}{W} = n$$

$$\text{Then } R = \frac{\rho_B n}{t}$$

$$\text{Define } \rho_S = \frac{\rho_B}{t} = \frac{R}{n}$$

$L = (A \text{ to } B)$

R = Resistance, Ω L = Length, cm

ρ_B = Bulk resistivity, $\Omega \cdot$ cm W = Width, cm

ρ_S = Sheet resistance, $\frac{\Omega}{sq}$ n = Number of squares, dimensionless

t = Thickness, cm

Fig. 7-80. Design formulas for resistive microcircuit elements.

The most popular substrate is alumina, which can withstand the high temperature required for thick-film processing. Alumina is available with surface quality for thin-film processing and can be obtained in a wide range of sizes and forms. The ceramic substrates (Table 7-28) are available as 96% Al_2O_3 for thick-film applications, 99.5% Al_2O_3 for thin-film applications, and as single crystal Al_2O_3 (sapphire) for special applications. Other substrates include beryllia (BeO), which is used in high-power applications, based on its combination of high thermal conductivity and high electrical resistivity, and glasses for use where maximum surface smoothness is

TABLE 7-28   Properties of Substrate Materials

| Material | sp gr | Surface finish, μin | | Dielectric constant | Thermal conductivity, cal·cm/ °C·cm²·s | Dissipation factor | Tensile strength, lb/in² |
		As fired	Polished				
Al_2O_3, 96%	3.76	<25	10	9.5	0.063	0.001	20,000
99.5%	3.89	<10	3	10.1	0.078	0.0002	30,000
Sapphire	3.98	. . .	1	9.53	0.10	<0.0001	58,000
BeO	2.88	<20	5	6.4	0.8	0.0001	20,000
Glass 7900 96% silica	2.18	0.25	. .	3.9	0.003	0.0005	
Epsilam-10	2.98	. . .	. .	10.4	0.00089	0.002	1,400
Porcelainized steel	7.86(Fe)*	9–10	. . .	6.8	†	0.0016	35,000

*Porcelainized steel 0.050 in thich is approximately 3 times as heavy as 0.025-in alumina.
†Steel = 0.13; insulator = 0.002.

required. A relatively new substrate type is a ceramic-filled Teflon compound (Epsilam-10), which combines the physical properties of plastic with the dielectric properties of alumina. When clad with copper, this substrate is particularly suitable for microwave microstrip applications. Another new substrate for hybrid-microcircuit applications, porcelain-enameled steel, is of particular interest where larger and more rugged packaging areas are preferred.

FILM MATERIALS

160. Thick Films. Specially formulated inks or pastes (Table 7-29) are applied to ceramic substrates by screen-printing techniques. Resistor and conductor pastes consist of a finely divided conductive phase, a finely divided glass phase, resins, solvents, and small amounts of surfactants and wetting agents. Dielectric compositions are similar except that the finely divided phase is a

TABLE 7-29   Film Electrical Properties

	Thin film	Thick film
Resistors:		
Sheet resistance	10–250 $\Omega/\square$	10–1,000 M$\Omega/\square$
Resistance range	5–300 kΩ	10–1,000 MΩ
Resistance tolerance, initial	$\pm 10\%$	$\pm 15\%$
With trimming	0.005–1.0%	0.1–2.0%
Temp coefficient of resistance	5–50 ppm/°C	50–200 ppm/°C
Matching	1 ppm/°C	± 10 ppm/°C
Power dissipation	25–100 W/in^2	50–200 W/in^2
Stability per year	0.1%	0.5%
Conductors:		
Sheet resistance	0.015–0.30 $\Omega/\square$	0.0017–0.060 $\Omega/\square$
Line resolution	0.001–0.010 in	0.007–0.010 in
Capacitors		
Dielectric constant K	6 and 25	10–2,000
Capacitance per area	0.008–0.05 μF/cm^2	0.003–0.037 μF/cm^2
Range of values	1 pF – 0.02 μF	2 pF – 0.02 μF
Temp coefficient of capacitance	$+100$ ppm/°C	± 50 to $-180,000$ ppm/°C
Allowable stress	2 MV/cm	0.2 MV/cm (500 V/mil)
Operating voltage	20 V	100 V

dielectric material. The films are dried after printing and then fired in air to a precise time-temperature profile for 15 min to 1 h at 750 to 1000°C.

The main classes of standard conductor compositions are noble metals, including Au, PtAu, PdAu, Ag, and PdAg. More recently some copper- and nickel-based pastes are being used, which require firing in inert or reducing atmospheres.

The resistive components of resistor pastes may comprise any one of the following: ruthenium oxide (the most popular), palladium–palladium oxide–silver compositions, mixtures of precious metals, thallium oxide, indium oxide, tungsten–tungsten carbide, or tantalum–tantalum oxide.

Dielectrics constitute materials with either low or high permittivity, the latter largely based on ferroelectric ceramic barium titanate with various additives and the former based on glasses and glass-ceramic mixtures.

161. Thin Films. Thin films are normally deposited in vacuum by electron-beam evaporation or sputtering. They serve as conductors, resistors, and dielectrics similar to the thick-film materials. In contrast to the thick films, however, which are applied as bulk materials, thin-film growth on a substrate involves the formation of independently nucleated particles, or islands, which grow together to form a continuous film as the deposition continues. The physical and electrical properties of deposited thin films are affected by the nature of the substrate, rate of film deposition, and the thickness, structure, and composition of the film. Thin-film conductors (Table 7-29) are normally gold films; resistors are either nickel-chromium, tantalum nitride, or silicon carbide films; and dielectrics or insulators include silicon monoxide, silicon dioxide, and tantalum oxide.

162. Resistor Trimming. Most film resistors, although designed carefully and processed with close controls, must be trimmed to a specified tolerance during production since the inaccuracy ($\pm 15\%$) inherent in the fabricating process cannot be tolerated in most circuit designs. The most popular methods of trimming are laser and abrasive trimming for thick-film resistors and laser and anodic trimming for thin films.

Laser trimming systems, specifically designed for trimming resistor elements, use yttrium aluminum garnet (YAG) or carbon dioxide (CO_2) lasers with the associated electronic computer control. Laser trimming permits high-circuit-density design because of the fine-spot diameter of the laser beam (0.001 to 0.01 mil). Systems have been developed to achieve maximum trimming speeds for high circuit throughput and to provide for sophisticated functional trimming.

Abrasive trimming (trimming by blasts of fine alumina particles) is an older method but still popular in the hybrid-microcircuit industry. It is reliable, relatively cheap, achieves adequate accuracies (0.5%), and avoids formation of microcracks in the substrate, which may occur with a laser.

Anodization, another trimming technique, is an electrochemical oxidation process used to grow an oxide in certain thin-film metals. When a tantalum-type resistor is anodized, the metal becomes thinner since a portion is converted into an oxide, thereby increasing its resistance. In principle, the technique has infinite resolution; however, standard trimming is $\pm 1\%$, with precision adjustments capable of $\pm 0.02\%$ to as low as $\pm 0.001\%$. The technique allows resistor adjustment independent of resistor geometry and also provides a passivating layer over the resistor surface.

DISCRETE CHIP DEVICES

163. Discrete Devices. A variety of discrete active and passive components, generally in chip form, are available for hybrid-microcircuit application and are used to provide functions difficult to achieve in film form or to improve circuit layout efficiency.

164. Active Devices. Most active devices are available from semiconductor manufacturers in wafer or chip form. Of continuing concern to the hybrid manufacturer is reduction in damage during handling of the unused chips and achievement of realistic circuit yields as the number, types, and complexity of the devices increase.

165. Resistors. Discrete-chip-resistor use is limited to special conditions where very low or high values are needed, to eliminate extra printing sequence, or to conserve circuit space. The number of circuits to be produced will influence whether the chips are used or not. A wide variety of ceramic and silicon chips of the thick, thin, and diffused types are available. The last type, which uses the bulk resistivity of silicon, is available in very small sizes but finds limited application because of the relatively poor resistor characteristics.

166. Capacitors. The limited availability of capacitance per unit area of the thick- and thin-film techniques and the difficulty in printing thick-film capacitors to close tolerances make chip capacitors attractive for hybrid-microcircuit applications. Capacitor-chip types include silicon MOS, porcelain, ceramic, tantalum, and glass.

ASSEMBLY

167. Assembly. *Assembly Techniques (Device and Wire Bonding).* The mechanical and electrical bonding of dice or chips and other discrete devices is a basic process in the assembly of the hybrid microcircuit, following fabrication of the thick- and thin-film components. Bonding may include eutectic alloy, solder, epoxy, wire, tape, flip-chip, or beam-lead technologies. Epoxy bonding is the most common, and eutectic bonding is a widely used mechanical attachment process. Wire bonding is the most extensively used electrical interconnect process.

168. Eutectic Bonding. Essentially any type of semiconductor chip can be back-bonded to the thick- or thin-film substrate by eutectic bonding. The technique uses the silicon from the chip and the gold on the substrate to form a silicon-gold eutectic alloy at about 370°C. Bonding is also done with preforms of gold-silicon or with gold-tin eutectic alloys when a lower (280°C) bonding temperature for heat-sensitive devices is desired.

169. Solder Bonding. Solder bonding is similar to eutectic bonding and is used for mounting active and passive devices to the hybrid substrate. Various soft-solder preforms of tin- and lead-based systems are used with melting temperatures of the order of 180 to 300°C. Solders are available as pastes and are specifically designed for thick- and thin-film circuits; they can be applied by screen printing, brushing, dipping, or with a syringe to provide bonding alloys for device attachment.

170. Epoxy Bonding. As the complexity of the hybrid microcircuit increases, it is desirable to keep the substrate at a lower temperature during device mounting to avoid degradation of the active devices' characteristics during assembly. The use of polymer adhesives (epoxy compounds) for chip bonding is desirable from such a manufacturing standpoint because of the relatively low temperature (<150°C) required for curing the epoxy, the relative ease of application, and the strength and reliability of the bonds. Epoxies exist in conductive and nonconductive forms, with gold or silver fillers in the conductive types. The use of organic or polymeric material inside a hybrid-microcircuit package has been of concern from the point of view of reliability, but the development of improved materials specifically for electronic applications has led to the general consensus that if properly selected, controlled, and applied, microcircuit epoxy adhesives are satisfactory for most hybrid-microcircuit bonding applications.

171. Wire Bonding. After the chip has been bonded to the substrate, fine aluminum or gold wires are used to make electrical contact between the metallized chip pads and the conductive network on the substrate (Fig. 7-81a). The most commonly used wire diameter is 1.0 mil;

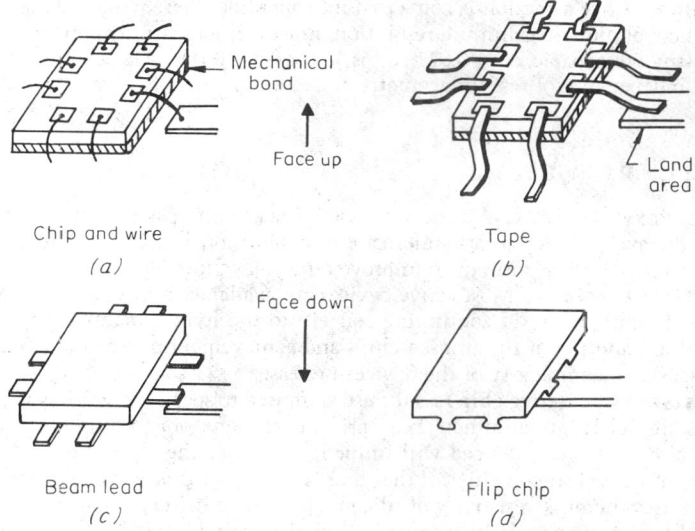

Fig. 7-81. Methods of bonding chips to substrates.

available diameters range from 0.7 to 20 mils. Wires are bonded one at a time, each wire requiring two bonds to be made. Wire-bonding techniques include thermocompression bonding (pressure and heat), ultrasonic bonding (friction of wire results in melting and alloying of wire), and thermosonic bonding (a combination of the two processes). Poor physical connections and chemical reactions between the wire and the thin-film pads on the die and between the wire and the thin- or thick-film land areas on the substrate can occur with improper bonding controls. This makes wire bonding one of the weakest parts of the hybrid-microcircuit assembly process.

172. Flip-Chip Technology. A flip chip (Fig. 7-81d) performs both electrical and mechanical connection after the chip is inverted and bonded face down to the substrate interconnection pattern. A raised metallic bump of solder is made on each of the chip mounting pads, the bump corresponding to the conductive land areas on the substrate. Bonding is normally carried out by reflow solder techniques, the bonds being made simultaneously, eliminating one wire and its two bonds per connection. Disadvantages include inability to make visual inspection of the bonds and limitations in removing heat since the chip is not in intimate contact with the substrate surface. More importantly, the availability of flip chips is limited for most hybrid manufacturers because of the reluctance of the semiconductor manufacturers to convert to this technology; the bump-processing steps add to the total processing cost of the wafer.

173. Beam-Lead Technology. Gold beams (3 mils wide, 5 mils long, spaced on 10-mil centers) are formed as an inherent part of the beam-lead integrated-circuit chip during processing of the silicon wafer (Fig. 7-81c). The metallurgy and sealed junction used for the wafer processing make the beam-leaded device one of the most reliable chip configurations. The beams are exposed and extend over the edge of the chip after the chip's separation from the wafer. The chip is inverted (face down), and several beam leads are attached simultaneously to the substrate. The chip remains slightly elevated above the substrate and therefore, like the flip chip, is limited in power dissipation.

174. Tape Bonding. A laminated tape of gold-plated copper foil acts as a carrier vehicle for semiconductor chips (Fig. 7-81b). The foil, etched in the form of leads, is bonded to metallurgical bumps on the chip pads, after which the chip and the attached leads are excised from the tape and bonded to the substrate. The technology lends itself to mass production, is an established method for individually packaging integrated circuits, and has been under active development for hybrid-microcircuit application since 1976. In addition to elimination of the wire bonding, the technology permits pretesting and burn-in of chips on the tape before substrate mounting. These pretest and burn-in techniques are difficult and usually avoided in the previously described electrical bonding techniques.

Microelectronics Packaging

BY OWEN P. LAYDEN

175. Introduction. Protective packages for microelectronics (both hybrid and monolithic) include (1) hermetic packages, using metal, glass, or ceramic and means of sealing, such as solder or welding, and (2) nonhermetic structures using plastic encapsulation, molding, potting, or polymer sealing. A trade-off must be made between using hermetic packages which afford a maximum of protection at a higher cost, and using nonhermetic packaging which offers less protection at lower cost.

HERMETIC PACKAGES

176. Hermetic Packages. A hermetic package is the most effective means for protecting hybrid microcircuits (or integrated circuits, as discussed in Sec. 8) from moisture and the effects of a harsh environment. The most common hermetic package for hybrid microcircuits is made of metal and uses glass-to-metal feed-throughs for electrical connection to other elements. Figure 7-82 shows various types of packages in use.

177. Cover Seals. Sealing of these packages is typically accomplished by soldering or welding. Most metal packages can be soldered, but special designs are required for welding. The major considerations in selecting the sealing method are availability of equipment and the cost of the

Fig. 7-82. Hermetic packages: 1, 2, 3 = uniwall type, weld or solder sealing; 4 = platform type, solder sealing; 5, 11 = TO type, weld sealing; 6, 7, 8, 10, 13, 14 = butterfly type, weld or solder sealing; 9 = special pack, solder sealing; 12 = flatpack, solder sealing. (*Tekform Products Co., Anaheim, Calif.*)

hybrid circuit. For low-cost circuits, where it is not economically feasible to open the package for circuit repair, welding should be considered for sealing. Welding, either seam welding or cold welding, is fast and repeatable and normally results in a high yield. It is usually impractical and uneconomical to remove the welded cover; i.e., delidding and rework is not feasible.

For high-cost circuits it is advantageous to be able to remove the cover and repair the circuit. The solder seal is used primarily in such applications. Solder sealing can be accomplished by machine or by hand. Problems of solder sealing include variations in solder material, lack of proper lid to package configuration, and temperature restrictions.

178. Pin Seals. The metal-to-metal seal of a hermetic package is a minor problem, compared with the glass-to-metal seal used for the package pins or leads to make electrical interconnections to outside elements. The sealing problem arises from the different thermal coefficient of expansion (TCE) between metals and glasses. To solve this problem a metal-and-glass matched combination of Kovar (TCE $= 47 \times 10^{-7}$) and glass (TCE $= 46 \times 10^{-7}$) is used. An oxide layer is grown on the metal surface to act as a barrier and to fuse with the glass to provide the seal. A properly fabricated glass-to-metal seal will provide a reliable hermetic seal over a wide temperature range. The packages shown in Fig. 7-82 are of matched seal construction, made with borosilicate glass sealing the pins and body of ASTM F-15 alloy (Kovar).

179. Hermetic Chip Carriers. Hermetic chip carriers (HCC) are square leadless packages for packaging one or more integrated circuits. They are made with the same materials and processes used for years in making ceramic dual-in-line packages (DIPs). Normal chip-mounting methods are used with either gold or aluminum wire for chip-to-carrier interconnection. The package center-to-center lead spacing is usually 40 mils for interfacing with a ceramic substrate or 50 mils for use on printed-circuit boards. When used for hybrid-microcircuit packaging (as contrasted with single-chip packaging), a thin substrate can be used for chip-to-chip interconnection. The chip carrier is mounted, usually with solder, on printed-circuit boards or ceramic substrates. When used on printed-circuit boards, a thermal mismatch exists between the ceramic of the HCC and the copper or board material. It is good practice to eliminate this problem by interfacing to the board through connectors especially designed for this purpose. This is unnecessary when the HCC is mounted on a ceramic substrate since there is no thermal mismatch between the two ceramic surfaces.

Figure 7-83 shows the typical construction of a three-layer bottom-contact chip carrier compared with a DIP with the same lead count. The chip carrier is essentially the center portion of the DIP with the two rows of metal leads replaced by contact pads on all four sides. As in the DIP, screen-printed metallized traces extend from the internal bonding pads to the external contacts. In a typical mounting arrangement the chip carrier will consume less area than DIPs by a factor of 3. Chip carriers used as replacement for chip-and-wire assembly (see Par. **7-171**) typically use more area by a factor of 2.

180. Hermeticity Testing. The hermeticity of the sealed package is determined by gross- and fine-leak tests in accordance with Method 10.4.2 of MIL-STD-883B. The fine-leak test, with limits from 1×10^{-7} to 5×10^{-8} atm·cm³/s of helium, depending on package volume, is a procedure whereby a tracer gas, helium, is forced under pressure into the package. Its leak rate from the package is then measured by a mass-spectrometer leak detector. The gross-leak test consists of immersion of the package, after pressurizing, in a fluorocarbon fluid heated to 125°C. A gross leak exists if air escapes from the package, resulting in a stream of air bubbles in the fluid. Testing of nonhermetic packages (Pars. **7-181** to **7-185**) is accomplished in part by using the 85°C/85% RH test. The packaged microcircuits are subjected to an environment of 85°C temperature and 85% relative humidity and electrically tested after a period of time.

NONHERMETIC PACKAGING

181. Nonhermetic Packaging. Various methods and materials are used in the nonhermetic packaging of hybrid microcircuits. The materials used allow moisture to permeate the enclosure and thus provide protection inferior to hermetic packaging. The nonhermetic packaging is advantageous when low cost is of utmost importance or where environments are less rigorous, as in the commercial market compared with the high-reliability military electronics market.

182. Ceramic Lid. One method of nonhermetic packaging makes use of a ceramic lid, or cover, used with an assembled ceramic substrate. The hybrid microcircuit is assembled on a ceramic substrate using chip-and-wire techniques. The semiconductor devices are covered with a silicone junction coating, and leads for external connection are attached to the substrate metallization with solder. The packaging is completed by sealing the ceramic lid to the substrate with an epoxy preform.

183. Epoxy Shell. Another method involves conformal casting of the entire chip-and-wire substrate with high-purity, semiconductor-grade silicone material. The assembled substrate is then inserted in an epoxy shell and sealed with a liquid epoxy. This shell supports the substrate but does not touch the conformal coating. The back of the substrate is epoxy-filled and cured as the final packaging step. Both methods make use of preformed covers or shells for microcircuit protection.

184. Conformal Coating. Conformal coating is used extensively for thick-film resistor and resistor-capacitor networks and has been adopted for inexpensive chip-and-wire microcircuits in which restricted size is of prime importance. The active chips are coated with a silicone

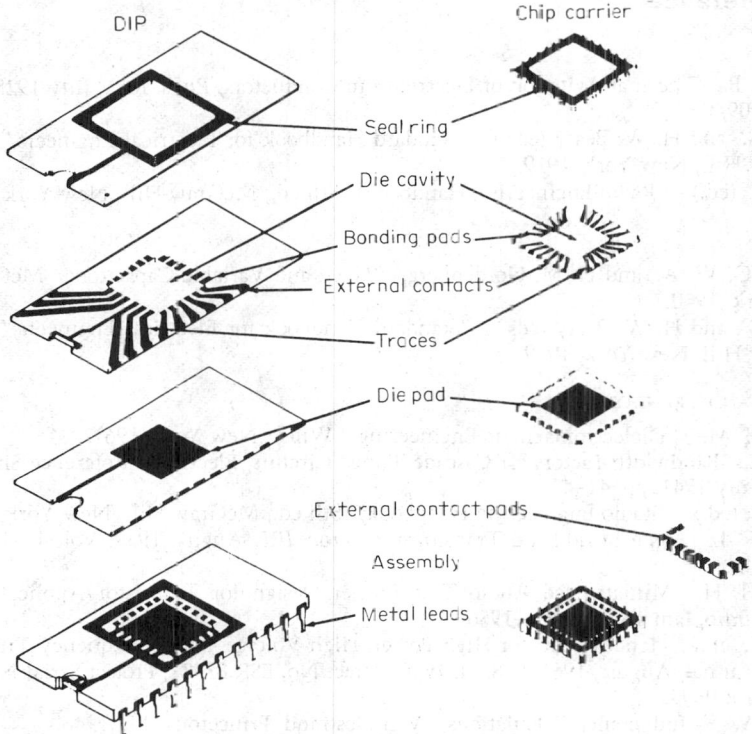

Fig. 7-83. Exploded view of chip carrier and dual-in-line (DIP) package. (*Proc. Nat. Electronics Packaging and Production Conference, March 1977.*)

junction coating normally dispensed with a syringe. The circuits are then dipped in a phenolic compound, which, after curing, acts as a mechanical protection for the microcircuit and has a thermal coefficient of expansion closely matched to that of the ceramic substrate. The microcircuit is again dipped in epoxy as the final processing step.

185. Plastic Chip Carriers. A new method of nonhermetic packaging for integrated circuits is the use of a plastic chip carrier. This premolded carrier consists of a metal lead frame and a molded insulating body. The microcircuit assembly technique (normally chip and wire) is the same as in the hermetic chip carrier (Par. **7-179**).

The unit is then encapsulated. One method uses a silicone-gel encapsulation compound. After encapsulation the leads of the premolded carrier are formed down the side and under the carrier, allowing for insertion into a printed-circuit board.

ELECTRICAL CONSIDERATIONS

186. Electrical Connections. The package must provide current paths between the hybrid microcircuit and external elements to interconnect dc power and analog and digital signals. Parasitic capacitance, inductance, and resistance can degrade the electrical performance of the circuit.

187. Parasitic capacitance is of two types: lead-to-lead and lead-to-ground. Capacitance is a direct function of wire diameter and the relative permittivity of the separating dielectric and is inversely related to the separation distance. The parasitic capacitance is greatest where the leads pass through the package walls.

188. Parasitic inductance generally appears as a series inductance in the leads and includes a resistive component which degrades electrical performance. Lead inductance and resistance should be kept low by using as short a lead length as possible. The lead inductance can resonate with the parasitic capacitance, producing spurious resonances, while the lead resistance can decrease analog circuit gain and give rise to improper *RC* time constants for digital signals.

189. References

RESISTORS

Johnson, J. B.   Thermal Agitation of Electricity in Conductors, *Phys. Rev.*, July 1928, Vol. 32, pp. 97–109.
Fink, D. G., and H. W. Beaty (eds.)   "Standard Handbook for Electrical Engineers," 11th ed., McGraw-Hill, New York, 1979.
Henney, K. (ed.)   "Radio Engineering Handbook," 5th ed., McGraw-Hill, New York, 1959.

CAPACITORS

Dummer, G. W. A., and H. M. Nordenberg   "Fixed and Variable Capacitors," McGraw-Hill, New York, 1960.
Fink, D. G., and H. W. Beaty (eds.)   "Standard Handbook for Electrical Engineers," 11th ed., McGraw-Hill, New York, 1979.

INDUCTORS AND TRANSFORMERS

Schlicke, H. M.   "Dielectromagnetic Engineering," Wiley, New York, 1961.
Dean, C. E.   Bandwidth Factors for Cascade Tuned Circuits, Electronic Reference Sheet, *Electronics*, July 1941, pp. 41–42.
Henney, K. (ed.)   "Radio Engineering Handbook," 5th ed., McGraw-Hill, New York, 1959.
Ruthroff, C. L.   Some Broad-Band Transformers, *Proc. IRE*, August 1959, Vol. 47, No. 8, pp. 1337–1342.
Kajihara, H. H.   Miniaturized Audio Transformer Design for Transistor Applications, *IRE Trans. Audio*, January-February 1956.
Howe, J. G., et al.   Final Report for High Power, High Voltage, Audio Frequency Transformer Design Manual, August, 1964, U.S. Navy, Contract No. BSR 87721, Project Serial No. SR008-03-02, Task 9599.
Grover, F. W.   "Inductance Calculations," Van Nostrand, Princeton, N.J., 1946.
U.S. Department of Commerce   *Circ. Natl. Bur. Stand.* C74, 1952.

FERROELECTRIC DEVICES

Jaffe, Hans   Piezoelectric Applications of Ferroelectrics, *IEEE Trans. Electron Devices*, June 1969, Vol. ED-1G, No. 6, pp. 557–561.
Berlincourt, Don A.   Piezoelectric Transducers, *Electro-Technol.*, January 1970, pp. 33–38.

QUARTZ CRYSTAL DEVICES

Proc. Annu. Symp. Frequency Control 1956–present.   Electronics Industries Assoc., National Technical Information Service, Springfield, Va.
Hafner, E.   Crystal Resonators, *IEEE Trans. Sonics Ultrason.*, October 1974, Vol. SU-21, pp. 220–237.
Buchanan, J. P.   Handbook of Piezoelectric Crystals for Radio Equipment Designers, *WADC Tech. Rep.* 54-243, National Technical Information Service, AD 110448, October 1956.
Bennet, R. E.   "Quartz Resonator Handbook," National Technical Information Service, AD 251289, 1960.

STANDARDS

Military Standards MIL-C-3098, MIL-H-10056, MIL-O-55310, IEEE Standards 176, 177, and 178, IEC Publ. 122, 283, 302, 314, 368, and EIA Publ. RS-192-A, RS-417, and *Component Bull.* 6.

INSULATORS

Inorganic Dielectrics Research (a history of 23 years of ceramic dielectric research sponsored by the U.S. Army Electronics Command, Fort Monmouth, N.J., at Rutgers University), *Eng. Res. Bull.* 50, December 1969

Military Specification MIL-I-10, Insulating Compound, Electrical Ceramic, Class L; U.S. Army Electronics Command, ATTN:AMSEL-PPEM-2, Fort Monmouth, N.J., December 1966.

Military Specification MIL-L-23264A. Insulators, Ceramic, Electrical and Electronic General Specification for Use by the Dept. of the Army, Navy, and Air Force, July 1968.

Military Specification MIL-S-55620, Substrates Ceramic for Deposition Thin Film Microcircuits; Commanding General, U.S. Army Electronic Command, ATTN:AMSEL-PPEM-2, Fort Monmouth, N.J., January 1969.

American Standards Association B46.1-1962 Standard for Surface Finishes.

Standards of the Alumina and Steatite Ceramic Manufacturers Association, New York, 1964.

"Modern Plastics Encyclopedia," McGraw-Hill, New York, 1970-1971, Vol. 47, No. 10a.

POWER AND RECEIVING TUBES

Doolittle, H. D.   Vacuum Power Tubes for Pulse Modulation, *Machlett Cathode Press*, 1964, Vol. 21, No. 1.

Kohl, W. H.   "Handbook of Materials and Techniques for Vacuum Devices," Reinhold, New York, 1967.

Millman, J., and S. Seely   "Electronics," McGraw-Hill, New York, 1951.

Schneider, S., and G. W. Taylor   Transients in High-Power Modulators, *IEEE Trans. Electron Dev.*, 1966, Vol. ED-13, No. 12.

Spangenberg, K. R.   "Vacuum Tubes," McGraw-Hill, New York, 1948.

Terman, F. E.   "Electronic and Radio Engineering," 4th ed., McGraw-Hill, New York, 1955.

Eimac Division of Varian   "Care and Feeding of Power Grid Tubes," San Carlos, Calif., 1967.

"Receiving Tube Manual," RCA, Harrison, N.J., 1973.

CATHODE-RAY STORAGE AND CONVERSION DEVICES

Kazan, B., and M. Knoll   "Electronic Image Storage," Academic, New York, 1968.

IEEE Trans. Electron Dev. September 1971, Vol. ED-18, No. 9.

IRE Standards on Electron Tubes   "Methods of Testing," Pt. 8, Camera Tubes; Pt. 10, Cathode-Ray Charge Storage Tubes, IEEE, New York, 1962.

Crowell, M. H., and T. M. Labuda   Silicon Diode Array Camera Tube, *Bell Syst. Tech. J.*, May-June 1969, Vol. 48, No. 5, pp. 1481–1528.

SEMICONDUCTOR DIODES AND SCRs

Cleary, J. F.   "G.E. Transistor Manual," General Electric Company, 1964.

Gutzwiller, F. W.   "G.E. Silicon Controlled Rectifier Manual," General Electric Company, 1967.

Everitt, W. L.   "Semiconductor Controlled Rectifiers," Prentice-Hall, Englewood Cliffs, N.J., 1964.

"RCA Tunnel Diode Manual," Radio Corporation of America, 1963.

"RCA Silicon Power Circuits Manual," Radio Corporation of America, 1969.

Lindmayer, J., and C. Y. Wrigley   "Fundamentals of Semiconductor Devices," Van Nostrand, Princeton, N.J., 1965.

TRANSISTORS

Phillips, A. B.   "Transistor Engineering," McGraw-Hill, New York, 1962.

Gartner, W. W.   "Transistors: Principles, Design and Applications," Van Nostrand, Princeton, N.J., 1960.

Joyce, M., and K. K. Clarke   "Transistor Circuit Analysis," Addison-Wesley, Reading, Mass., 1961.

Cleary, J. F.   "G.E. Transistor Manual," General Electric Company, 1964.

Tenk, J. D.   "Handbook for Transistors," Prentice-Hall, Englewood Cliffs, N.J., 1975.

Cooper, W. D.   "Solid-State Devices: Analysis and Application," Reston, Reston, Va. 1974.

BATTERIES AND FUEL CELLS

Falk, S. Uno, and Alvin J. Salkind   "Alkaline Storage Batteries," Wiley, New York, 1969.

Mantell, C. L.   "Batteries and Energy Systems," McGraw-Hill, New York, 1970

Heise, G. W., and N. C. Cahoon "The Primary Battery," Vol. 1, Wiley, New York, 1971.

Fleischer, A. and J. L. Lander "Zinc-Silver Oxide Batteries," Wiley, New York, 1971.

Proc. Power Sources Conf. 1956–1970, PSC Publications Committee, Red Bank, N.J.

"Application Engineering Handbook," General Electric Company, Gainesville, Fla., 1971.

Eveready Battery Applications Engineering Data, Union Carbide Corp., 1971.

Fink, D. G., and H. W. Beaty (eds.) "Standard Handbook for Electrical Engineers," 11th ed., McGraw-Hill, New York, 1979.

RELAYS AND SWITCHES

Peek, R. L., and H. N. Wagar "Switching Relay Design," Van Nostrand, Princeton, N.J., 1955.

Henney, K., C. Walsh, and H. Mileaf "Electronics Components Handbook," Vol. 1, Chap. 5, Relays, and Vol. 3, Chap. 2, Connectors, McGraw-Hill, New York, 1958.

Harper, C. A. (ed.) "Handbook of Electronic Packaging," Chap. 6, Connectors, and Chap. 8, Packaging with Conventional Components, McGraw-Hill, New York, 1969.

Proc. Relay Conf. 1953–1971, National Association Relay Manufacturers, and Oklahoma University, Scottsdale, Ariz.

Proc. Holm Sem. Elec. Contact Phenomena, 1955–1971, Illinois Institute of Technology and ITT Research Institute, Chicago.

Harper, C. A. (ed.) "Handbook of Wiring, Cabling and Interconnecting for Electronics," Chap. 3, Hook-Up Wiring and Connector Systems, and Chap. 4, Coaxial Cable and Connector Systems, McGraw-Hill, New York, 1972.

DISCRETE COMPONENT MODULES

Harper, C. A. (ed.) "Handbook of Electronic Packaging," McGraw-Hill, New York, 1969.

Keonjian, E. "Microelectronics," McGraw-Hill, New York, 1963.

MIL-STD-275C, Printed Wiring for Electronic Equipment, Jan. 9, 1970.

ON HYBRID MICROCIRCUIT TECHNOLOGY

Maissel, L. I., and R. Glang "Handbook of Thin Film Technology," McGraw-Hill, New York, 1970.

Harper, C. A. "Handbook of Thick Film Hybrid Microelectronics," McGraw-Hill, New York, 1974.

Hamer, D. W., and J. V. Biggers "Thick Film Hybrid Microcircuit Technology," Wiley-Interscience, New York, 1972.

Agnew, J. "Thick Film Technology: Fundamentals and Applications in Microelectronics," Hayden, New York, 1973.

Berry, R. W., P. M. Hall, and M. T. Harris "Thin Film Technology," Van Nostrand, Princeton, N.J., 1968.

International Society for Hybrid Microelectronics "Hybrid Microelectronics Standard Specification Guidelines," Montgomery, Ala., 1975.

Coulton, M. Substrate Materials and Processes for Microwave Applications, *Proc. 29th Electron. Components Conf. May 1979*, pp. 126–131.

Comeforo, J. E. Selecting Ceramic Substrates For Semiconductor Devices, *Insulation/Circuits*, December 1976, pp. 2–6.

MICROELECTRONIC PACKAGING

Riel, M., and R. Chalman Specifying Special IC Packages, *Electron. Packaging Prod.*, May 1975.

Fogiel, M. "Modern Microelectronics," Research and Education Association, New York, 1972.

Markstein, H. Chip Carrier Update, *Electron. Packaging Prod.*, April 1979.

Burch, M. L., and B. M. Hargis Ceramic Chip Carrier: The New Standard in Packing, *NEPCON Proc. 1977*.

Older, B., and R. A. Bly Non-Hermetic Packaging Techniques for Hybrids, *Electron. Packaging Prod.*, June 1979.

Harding, M. L. Non-Hermetic Packaging for Commercial Hybrid Microcircuits, *Proc. Symp. Plastic Encapsulated/Polymer Sealed Semiconductor Dev. Army Equip.*, May 1978.

Integrated Circuits and Microprocessors

ALAN B. GREBENE *Vice President, Engineering, Exar Integrated Systems, Inc., Sunnyvale, Calif.; Fellow, IEEE*

HANS H. STELLRECHT *Signetics Corporation, Sunnyvale, Calif.; Member, IEEE*

WILLIAM F. DAVIS *Senior Member, Technical Staff, Motorola Semiconductor, Mesa, Ariz.*

CHARLES S. MEYER *VHSIC Technical Program Manager, Motorola Semiconductor, Phoenix, Ariz.*

ROBERT C. HUNTINGTON *Linear CCD Project Manager, APRDL, Motorola Semiconductor, Phoenix, Ariz.*

DAVID CAVE *Integrated Circuits Design Manager, Linear Integrated Circuits, Motorola Semiconductor, Mesa, Ariz.*

RICHARD W. ULMER *Section Manager, Design Engineering, Logic and Special Functions, MOS Integrated Circuits Division, Motorola Semiconductor, Austin, Tex.*

ALAN R. BORMANN *MOS Memory Design Manager, MOS Integrated Circuits Division, Motorola Semiconductor, Austin, Tex.*

JAMES M. RUGG *Device Technology Manager, Advanced MOS Product Development, Integrated Circuits Division, Motorola Semiconductor, Phoenix, Ariz.*

WILLIAM BLOOD, JR. *Manager, Bipolar LSI Applications and Product Planning, Motorola Semiconductor, Phoenix, Ariz.*

RAYMOND M. ROOP *Leader, New Process/Product Development Group, Linear Integrated Circuits, Motorola Semiconductor, Phoenix, Ariz.*

JOHN GOODRIN *Product Marketing Engineer, Intel Corp., Santa Clara, Calif.*

CONTENTS

Numbers refer to paragraphs

INTRODUCTION*

1. Overview. The concept of an integrated circuit, having all required circuit elements built into a single monolithic block or chip, together with the development of the planar process,† led to today's complex digital and analog integrated circuits (ICs). Medium-scale integration, large-scale integration (LSI), and now very large scale integration (VLSI) have been successive objectives of IC developers. Single chips containing 100,000 to 200,000 transistors or more and memory chips of 1 million bits seem to be within the grasp of present technology.

Integrated circuits have cost, size, weight, and reliability advantages over conventional circuitry. From a structural standpoint, ICs are generally classified in one of three categories: monolithic, thin- or thick-film, or hybrid circuits.

Monolithic circuits are the fundamental form of IC, in which the entire circuit function is formed within a monolithic body of semiconductor material. Both active and passive circuit elements are an integral part of the substrate. *Thin- or thick-film* circuits are fabricated by depositing resistive and/or conductive films on an insulating substrate and by imposing patterns on them to form an electric network. *Hybrid* circuits are a natural extension of the first two classes. They contain passive and active devices, as well as monolithic ICs assembled and interconnected on a common insulating substrate.

Functional classifications of ICs include digital, linear, and microwave. *Digital* ICs include various types and classes of logic circuits, as well as semiconductor memories. *Linear* ICs include amplifiers, regulators, multipliers, modulators, and other types of signal-processing circuitry. *Microwave* ICs cover the frequency range from 0.5 to 15 GHz and rely to a great extent on hybrid-circuit technology.

IC and microprocessor technology involves a wide variety of disciplines, from physical chemistry and metallurgy to solid-state device and circuit theory.

IC FABRICATION TECHNOLOGY

2. Introduction. The planar technology is the principal method used in the fabrication of semiconductor devices for hybrid and monolithic ICs. As applied to semiconductor devices,

---

*Paragraphs **8-1** to **8-100** were written by the authors except John Goodrin, who was responsible for Pars. **8-101** to **8-162**. The editors express their additional thanks to David Hodges, University of California, Berkeley, Frank A. Brunot, Signetics Corporation, Michael Peak and Robert Patterson, Intel Corp. — D.G.F, D.C.

†J. A. Hoerni, U.S. Patent 3,025,589, assigned to Fairchild Camera and Instrument Corp., New York, 1960.

the term planar means that the devices and components fabricated by this process extend below the surface of one plane of a silicon substrate; the surface structure of the semiconductors remains essentially unaltered during the fabrication process. Silicon is especially well suited for this process because of the relative ease with which an insulating oxide layer can be formed at the surface. This oxide layer can act as a barrier to the diffusion of dopants used for the device

fabrication of ICs. Thus, by selectively etching openings into the oxide layer, components can be formed by localized diffusions of impurities.

Fabrication of an IC requires a sequence of several independent processing steps: (1) material preparation, (2) epitaxial growth, (3) surface passivation, (4) photolithography, (5) junction formation, and (6) film deposition. The sequence of these basic processing steps is shown in Fig. 8-1.

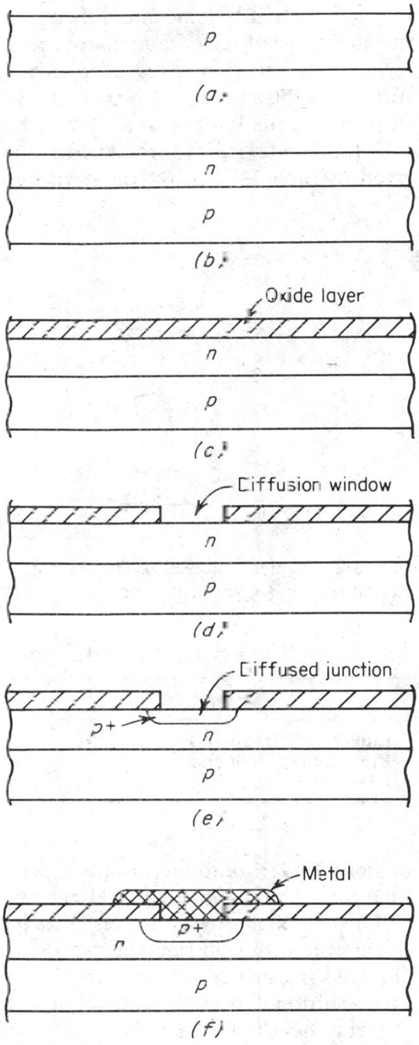

3. Material Preparation. The basic material preparation starts with growing the single-crystal semiconductor material. The grown crystal is sliced into thin *wafers*, which are lapped, polished, and chemically etched to provide a smooth, defect-free semiconductor surface.

Crystal Growing. The most common method used for the growth of single crystals for IC fabrication is the Czochralsky *pulling* technique. In this method an induction-heated melt of high-purity silicon with the desired doping impurity is prepared in an inert atmosphere. A small seed crystal of silicon is then dipped into the melt, rotated, and very slowly withdrawn from the melt. If temperature conditions of the melt are properly maintained as the seed is withdrawn, it grows as a single oriented crystal bar.

Wafer Preparation. After the single crystal has been grown, it is sliced into wafers. The crystal is first oriented according to crystallographic planes by x-ray techniques. A diamond saw is used to slice the oriented crystal into wafers. The wafers are then lapped to the required thickness and polished with fine-graded diamond polish. Often a chemical polish is used as the final preparation step to remove any mechanical damage at the surface. The final dimensions of the prepared silicon wafer are a diameter of 1 to 3 in and thickness of approximately 250 to 400 μm.

4. Epitaxial Growth. Epitaxy, derived from the Greek word meaning "arranged upon," is a growth technique in which the crystal structure of a silicon substrate is extended by arranging new layers of atoms in precisely the same lattice structure as the substrate. The additional layers are formed by vapor-phase deposition in an epitaxial reactor. By

Fig. 8-1. Basic processing steps in planar technology: (a) p-type starting material after material preparation; (b) epitaxial growth; (c) surface passivation; (d) photolithography; (e) junction formation; (f) metallization.

controlling the deposition rates and introducing controlled amounts of impurities into the carrier-gas stream, the thickness and the resistivity of the deposited layer can be controlled.

Different systems have been developed for the growth of epitaxial layers. The most common systems today are open-tube systems, where silicon is deposited on the surface of heated substrates by the reaction[1]* of hydrogen with silicon tetrachloride ($SiCl_4$) or by the decomposition[2] of silane (SiH_4). Figure 8-2 shows a simplified diagram of a typical epitaxial system using both

*Superior numbers correspond to the numbered references in Par. 8-163.

horizontal and vertical reactors. The resistivity of the deposited layers can be controlled by introducing controlled amounts of p- or n-type impurities into the carrier gas.

$SiCl_4$ is an inexpensive source of silicon; it is easy to purify and nontoxic but requires relatively high temperatures, which causes considerable redistribution of previously diffused impurities.

An alternative epitaxial growth process which has gained considerable importance is based on the *pyrolytic decomposition of silane* (SiH_4) into Si and H_2. The disadvantage of this process is that SiH_4 is highly toxic and difficult to handle. However, this disadvantage is compensated by the advantage of lower-temperature processing.

Since the purity and crystal perfection of the epitaxial film are strongly dependent on the surface condition of the substrate, final cleaning of the substrate wafers is very important. This cleaning is usually done in the reaction chambers with vapor-phase hydrochloric acid.

Typical epitaxial-layer thicknesses and resistivity ranges for integrated circuit applications are given in Table 8-1. Unlike the diffusion process, epitaxial growth proceeds by uniform addition

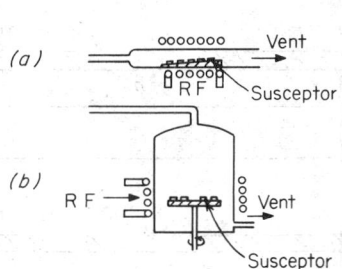

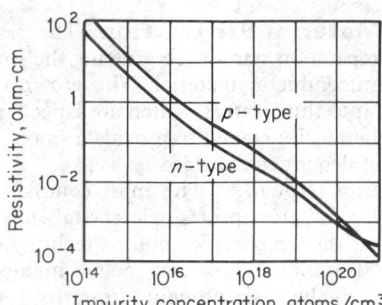

Fig. 8-2. Typical epitaxial reactor systems: (a) horizontal reactor; (b) vertical reactor.

Fig. 8-3. Resistivity of silicon as a function of uniform impurity concentration at 300 K.[4]

Table 8-1. Typical Epitaxial-Layer Thickness and Sheet Resistivity

Circuit type	Thickness, μm	Resistivity range, $\Omega \cdot cm$
Linear, low frequency	8–15	1–6
Linear, high frequency	6–10	0.2–1
Digital	5–8	0.22

of atomic layers onto the substrate. Thus the dopant impurities are relatively uniformly distributed through the epitaxial layer and show no significant concentration gradient.

The epitaxial growth process takes place at temperatures comparable with those encountered during the diffusion steps. Consequently, when the epitaxial-layer resistivity is different from the background resistivity, some impurity redistribution can take place across the epitaxial-layer–substrate interface, due to the exchange of impurities by diffusion.[3]

Figure 8-3 shows the resistivity of uniformly doped p- or n-type silicon substrate or epitaxial layers as a function of impurity concentration.[4]

5. Surface Passivation. The passivation of the silicon surface by an insulating dielectric layer is one of the principal features of the planar process. Passivation is normally achieved by thermal oxidation of silicon (Par. **8-6**). The thin layer of silicon dioxide obtained in this manner serves four functions:

1. It serves as a diffusion mask and allows selective diffusion into the silicon through the openings etched into the silicon dioxide.

2. It protects the diffused junctions from impurity contamination.

3. It serves as a surface insulator which separates devices and metal interconnections.

4. It can serve as a dielectric film for monolithic capacitors.

6. Thermal Oxidation. Thermal oxidation[5,6] of the silicon wafers is usually carried out in a diffusion furnace at temperatures between 900 and 1200°C. If a pure oxygen ambient is used, the oxide layer is formed on the silicon surface through the basic chemical reaction

$$Si + O_2 \rightarrow SiO_2$$

This reaction is significantly accelerated in the presence of water vapor and proceeds in accordance with the reaction

$$Si + 2H_2O \rightarrow SiO_2 + 2H_2$$

Figure 8-4 shows typical growth rates of SiO_2 as a function of temperature for the cases of dry and wet oxygen (steam).[7]

7. Masking Properties of SiO_2. The diffusion coefficients of most dopants in SiO_2 are between two and four orders of magnitude smaller than in silicon. Therefore an SiO_2 layer of proper thickness can serve as a diffusion barrier against these dopants. Figure 8-5 shows the typical oxide thickness required to mask against boron or phosphorus diffusion, as a function of diffusion times and temperatures.[8]

Ionic Contamination. Some positively charged ions such as sodium (Na^+) or hydrogen (H^+) have relatively large diffusion coefficients in SiO_2 at low temperatures. This can be especially detrimental to integrated circuit devices containing lightly doped p-type regions.

8. Silicon Nitride Passivation. Silicon nitride (Si_3N_4) is more resistant to ionic contamination than SiO_2. It can be used as an additional passivating layer to prevent surface contaminants from reaching the SiO_2 layer. This is especially true of bipolar circuits designed to operate at low current levels and for devices such as MOS transistors which rely heavily on surface phenomena. The silicon nitride layer can be formed by pyrolytic deposition at temperatures of 800 to 1000°C, or by rf or reactive-sputtering techniques. Often an additional silicon dioxide layer is deposited over the silicon nitride to facilitate photomasking.

Deposited Oxides. The high temperatures needed for the growth of thermal oxides cause redistribution of impurities within the previously diffused layers. To avoid this, passivating

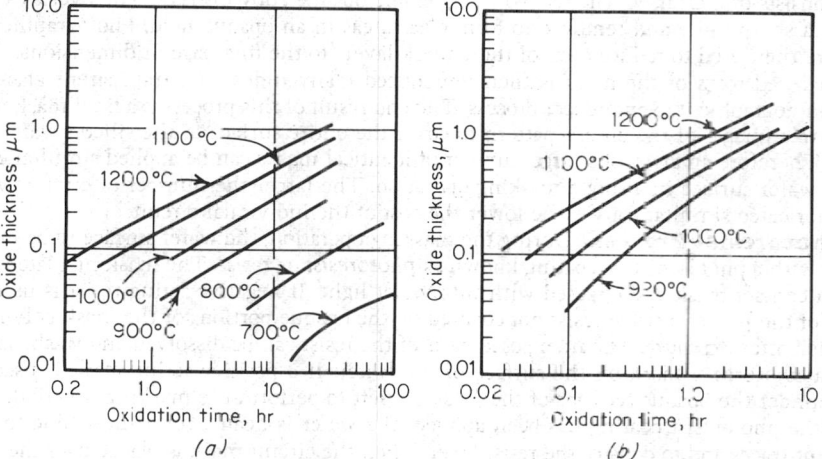

Fig. 8-4. Typical growth rates of SiO_2 as a function of temperature: (a) dry O_2; (b) oxygen plus water vapor. *(Adapted from Ref. 7.)*

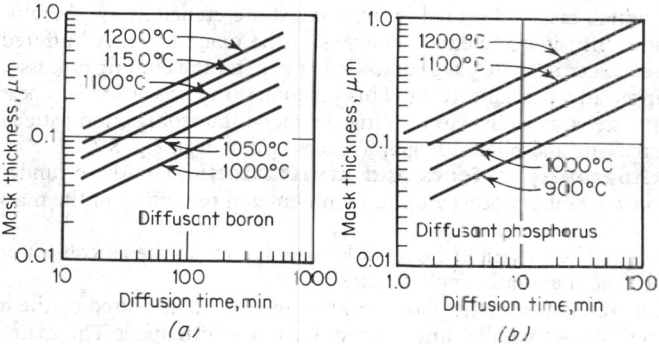

Fig. 8-5. Masking oxide thickness for: (a) boron and (b) phosphorus.[8]

oxide layers can be formed by low-temperature deposition processes such as pyrolytic decomposition of oxysilane, vapor-phase reactions, anodizing, sputtering, or plasma oxidation.[9]

9. Postmetal Passivation. Sometimes it is desirable to form a protective dielectric coating over the IC surface after the metal interconnections have been formed. This can be done by pyrolytic deposition of SiO_2 or silicon glass, commonly referred to as *s-glass*. This postmetal passivation involves additional processing steps, but in return it provides surface protection for the interconnections and for the thin-film devices on the surface. In addition it allows the use of multiple-layer metallization, which in turn greatly simplifies the circuit layout. A structural diagram of a monolithic circuit chip using postmetal passivation is shown in Fig. 8-6.

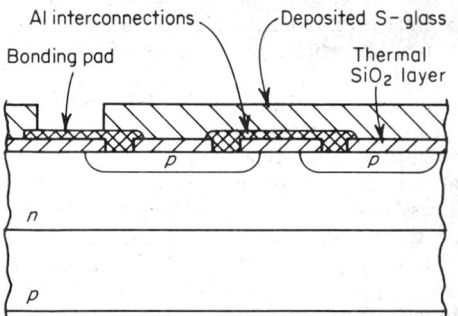

Fig. 8-6. Postmetal passivation of wafer surface by pyrolytic deposition of silicon glass.

10. Photolithography. Initial design and layout of an IC are carried out on a scale several hundred times larger than the final desired dimensions of the circuit pattern. The initial layout of an integrated circuit is normally done on a magnification scale in the range of 250:1 to 500:1. This layout is a composite of different mask patterns corresponding to the different masking steps associated with the fabrication process. The dimensional accuracy of the final circuit depends on the dimensional stability of this master layout.

Each mask pattern is cut into a dimensionally stable plastic laminate layer, called a *rubylith*, which consists of a clear Mylar base with a peelable opaque ruby overlay. The overlay can be cut with a sharp knife and removed to form clear areas in an opaque field. Photographic techniques are then used to reduce each of these mask layers to the final circuit dimensions.

The reduced form of the mask is then reproduced many times on a transparent glass slide using a photographic step-and-repeat process. The end result of this process is a final mask which has multiple images of the circuit pattern to cover the entire surface of the silicon wafer to be masked. Therefore an array of a large number of identical masks can be applied simultaneously over the wafer surface in a single masking operation. The larger the number of circuits which can be fabricated simultaneously, the lower the cost of the individual circuit.

11. Photoresist Process. During the masking operation, the wafer surface to be masked is coated with a photosensitive coating known as *photoresist* or *resist*. The masking plate is then used as a contact mask and exposed with ultraviolet light. If negative-acting resist is used, the portions of the photosensitive resist not covered by the opaque portions of the mask polymerize and harden after exposure. The unexposed parts of the resist can be dissolved and washed away, leaving a *photoresist mask* on the surface of the wafer. If a positive-acting resist is used, the portions under the opaque sections of the mask are left to perform the masking function.

After the photoresist coating has been applied, the wafer is heated for a short time to drive out solvent traces and to densify the resist layer. Then the circuit mask is placed over the wafer and aligned to its desired position by two-directional micrometer adjustments. This alignment process requires the use of a microscope.

12. Etching. The photomasking step is followed by an etching step. In this step, the parts of the SiO_2 layer which are not covered by photoresist are etched away, thereby exposing the bare silicon to form diffusion or contact windows in the oxide layer. A buffered hydrofluoric acid solution is used to etch the SiO_2 at a controlled rate. After the etching process, the remaining photoresist is stripped in a special solvent. This photomasking procedure is repeated for all the remaining diffusion steps, as well as in forming the metal-interconnection pattern. A flowchart of standard mask making and photolithography steps is shown in Fig. 8-7.

13. Photolithography Defects and Limitations. The three fundamental limitations of the photolithography process are the alignment and resolution of the mask patterns and etching defects.[10]

Mask Alignment. The pattern of each mask layer applied to the silicon-wafer surface must align properly with the previously applied patterns.

Resolution. The resolution of the photomasking step can be measured by the minimum line width needed to resolve two parallel lines spaced one line width apart. The main limitations of mask resolution are due to the finite grain size in the photographic emuls.ons and to the diffraction of light at the mask edges.

Etching Defects. Etching defects result from irregular thickness or adherence properties of the photoresist or the nonuniform etching properties of the oxide. These defects can lead to poorly defined oxide edges and may increase the size of resulting diffusion windows.

14. Junction Formation. To form a *pn* junction in an IC, a controlled number of impurities must be introduced into the silicon substrate. This can be done by either *diffusion* or by *ion-implantation techniques*, as described below.

15. Diffusion. The diffusion process is presently the most widely used method of introducing controlled amounts of impurities into the silicon substrate. It is a relatively well understood and highly reproducible process which readily lends itself to the batch-processing advantages of the planar technology, where many silicon wafers can be processed in a single operation.

In the crystalline silicon lattice, the impurities can move by two different diffusion mechanisms: *substitutional* and *interstitial diffusion* [11] In substitutional diffusion, the impurity atoms

Table 8-2.   *p*- and *n*-Type Dopants for Silicon-Device Fabrication

p-Type	*n*-Type
Boron (B) Aluminum (Al) Gallium (Ga) Indium (In)	Phosphorus (P) Arsenic (As) Antimony (Sb)

propagate through the crystal lattice by jumping from one lattice site to the next, thus substituting for the original host atom. The number of available lattice sites to which the impurity atom can jump is given by the number of vacancies in the lattice. In interstitial diffusion the impurity atoms move through the crystal lattice by jumping from one interstitial site to the next. Since there are five interstitial voids in a unit cell of the silicon lattice and only a few are occupied by point defects, an interstitial impurity can diffuse through the lattice at a much faster rate.

The impurities which are diffused into silicon to determine the type and the resistivity of various regions of the semiconductor material are elements from group III or V of the periodic table, for *p*- or *n*-type doped regions, respectively. Some of these dopants are listed in Table 8-2.

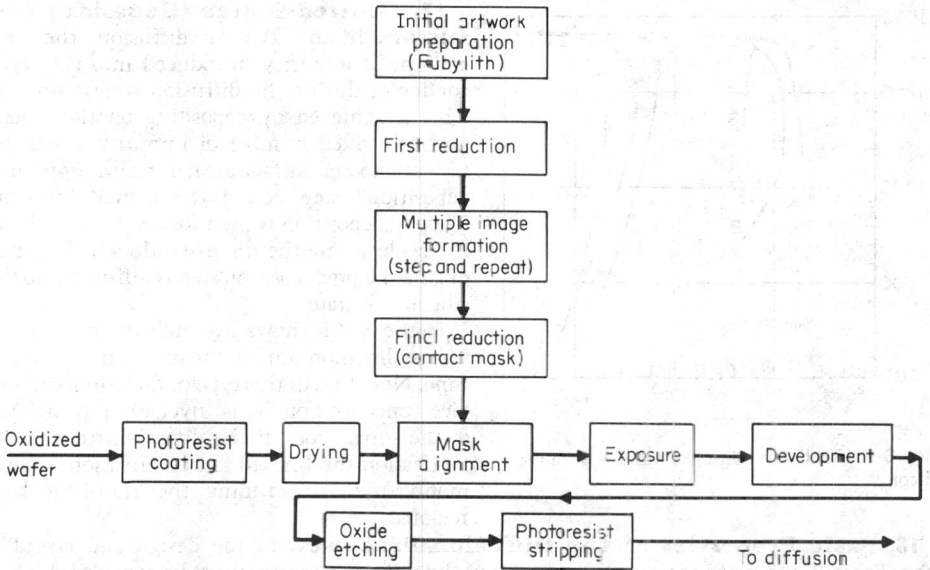

Fig. 8-7. Flowchart of mask-making and photolithography steps.

For IC fabrication, B, P, As, and Sb are the most common dopants. All these dopants are substitutional impurities. Figure 8-8 shows the diffusion coefficients of silicon dopants as a function of temperature.[12]

16. Constant-Source Diffusion. In this type of diffusion the impurity concentration at the semiconductor surface is maintained at a constant level throughout the diffusion cycle. The constant impurity level on the surface is determined by the temperature and the carrier-gas flow rate of the diffusion furnace. In most constant-source diffusion systems, it is convenient to let the surface concentration N_0 be determined by the solid-solubility concentration limit of the particular dopant in silicon. As shown in Fig. 8-9, the solid-solubility limit is also a strong function of the temperature.[13]

A typical set of constant-source impurity distribution profiles formed by complementary-error-function diffusions is shown in Fig. 8-10 as a function of the diffusion time. Note that, in this type of diffusion, the total amount of impurities diffused into silicon increases indefinitely with time. Similarly, the impurity concentration at any point within the material (except at the surface) is a monotonically increasing function of time; therefore, as t goes to infinity, the entire wafer would have a uniform concentration of N_0. If the diffused impurity type is different from the resistivity type of the substrate material, a junction is formed at the points where the diffused impurity concentration is equal to the background concentration already present in the substrate. These junction depths are shown as points x_1, x_2, and x_3, respectively, for the diffusion profiles of the figure. In the fabrication of monolithic circuits, constant-source diffusion is commonly used for the isolation and the emitter diffusion steps.

17. Limited-Source (Gaussian) Diffusion. In this type of diffusion, the total amount of impurity introduced into the semiconductor during the diffusion step is limited. This is achieved by depositing on the silicon surface a fixed number of impurity atoms per unit of exposed surface area, during a short "predeposition" step before the actual diffusion. This predeposition is then followed by a "drive-in" cycle where the impurity already deposited during the predeposition step is diffused into the silicon substrate.

Figure 8-10b shows a sketch of the limited-source diffusion profile, for increasing values of time. Note that in this type of diffusion, the surface concentration N_0 is inversely proportional to the square root of the diffusion time. In IC fabrication the limited-source diffusion is commonly used in forming the transistor base regions.

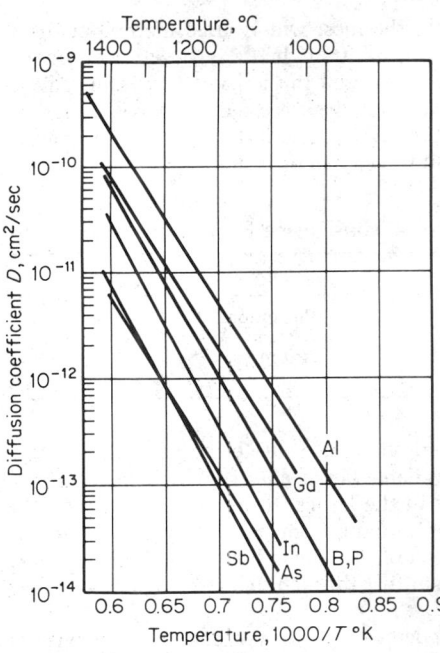

Fig. 8-8. Diffusion coefficients of various dopants in silicon.

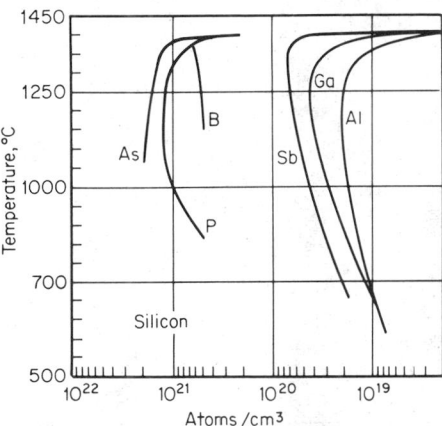

Fig. 8-9. Solid solubilities of various dopants in silicon.[16]

18. Basic Properties of the Diffusion Processes. In the design and layout of monolithic ICs three fundamental properties of the diffusion process must be considered:

(a) *All diffusions proceed simultaneously.* The impurities introduced in an earlier diffusion step continue to diffuse during the subsequent diffusion cycles. Therefore, when calculating the total effective diffusion time for a given impurity profile, one must often consider the effects of the subsequent diffusion cycles. The effects of the subsequent diffusions on a given impurity profile can be estimated by defining an effective product equal to the sum of the products of the diffusion coefficients and the respective times of diffusion.

$$Dt \approx D_1 t_1 + D_2 t_2 + D_2 t_3 + \cdots$$

Thus, for example, in the planar-device fabrication, the emitter region of a bipolar transistor is formed by a diffusion process which succeeds the base diffusion step. Therefore the effective Dt product of the base region contains a finite contribution from the emitter diffusion step.

(b) For a given surface and background concentration, the *junction depths* x associated with the two separate diffusions having different diffusion times and temperature are related as the square root of the ratio of the respective products.

$$x_1/x_2 = \sqrt{D_1 t_1/D_2 t_2}$$

(c) The diffusion proceeds *sideways from a diffusion window* as well as downward. In considering the lateral dimensions of the planar devices, these lateral diffusion effects must be taken into account. Figure 8-11 shows the lateral diffusion effects for various concentration ratios.[14]

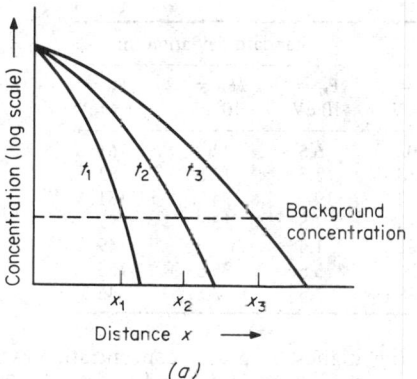 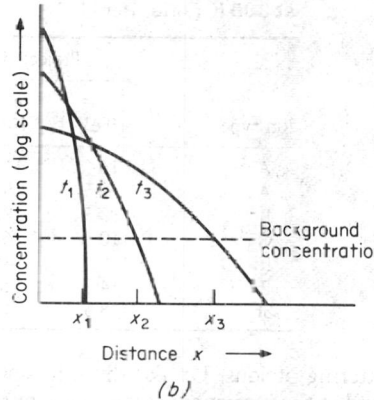

(a)   (b)

Fig. 8-10. (a) Constant-source and (b) limited-source diffusion profiles as a function of time.

Figure 8-11a gives the constant concentration contours for the case of a constant-source diffusion. The diffusion contours of Fig. 8-11b correspond to the limited-source case. Typical side diffusion effects are about 75 to 85% of the vertical penetration for most impurity concentration levels encountered in IC fabrication.

19. Ion Implantation. Ion implantation provides an alternative method for introducing impurities into a semiconductor.[15,16] In contrast to diffusion processes, the number of implanted ions is controlled by the external system, rather than by the physical properties of the substrate. This allows a precise monitoring of the dose of the implanted layers, even at temperatures at which normal diffusion is insignificant. The implanted dopant concentration is not limited by solubility considerations; and a much wider variety of dopant elements can be used. An important side benefit of this is that implanted layers can be formed without affecting previously diffused device structures.

In the ion-implantation process, the desired impurities are introduced into the semiconductor lattice by bombarding the surface with high-energy impurity ions. The implantation operation takes place in vacuum. Impurity ions are accelerated from an ion source, and a mass spectrometer is used to separate the undesired impurities from the beam. The ion beam is then focused to a small area (typically smaller than ¼ in²) and is scanned across the semiconductor wafer which serves as the target.

The impurity ions penetrate the target lattice at typical energy levels used for implantation applications, in the 50- to 150-keV range. The depth of penetration is a function of the kinetic energy of the impurity ions and the mass of the target atoms, the lattice spacing and the crystal plane facing the incident ion beam.

Because of better reproducibility, ions are usually implanted with the beam at a small angle with respect to the <111> crystal plane of the semiconductor target. This prevents channeling because the semiconductor appears amorphous to the incident beam, thereby resulting in better reproducibility.[16]

The semiconductor surface can be masked against the implanted ions by using a metal layer (such as aluminum) or a thick oxide layer as a mask. Thus ion-implanted regions can readily be patterned at the silicon surface in the same manner as diffused regions, using photomasking techniques.

20. Range Distribution. The range is the total distance the ion travels within the target before coming to rest. The projection of this distance on the direction of incidence is called the *projected range* X_p. A typical range distribution in an amorphous substrate is approximately gaussian in shape and can be characterized by a mean range and a finite range distribution about this mean value, as shown in Fig. 8-12. Typical value of the mean range for 100-keV ions is approximately 0.1 μm. Figure 8-13 shows X_p as a function of the implantation energy E_{p0} for different impurity ions. The standard deviation σ_R associated with the range distribution is also shown in the figure.[17]

Table 8-3 shows the projected range and standard deviation of some important dopants in silicon. The impurity profile in a semiconductor after ion implantation is determined by random

TABLE 8-3   Projected Range and Standard Deviation of Selected Ions as a Function of Initial Kinetic Energy E_{p0} in Silicon for Nonchanneling Conditions at 300 K (After Ref. 17)

Ion type	Projected range, nm			Standard deviation, nm		
	$E_{p0} =$ 10 eV	$E_{p0} =$ 10^2 eV	$E_{p0} =$ 1 MeV	$E_{p0} =$ 10 eV	$E_{p0} =$ 10^2 eV	$E_{p0} =$ 1 MeV
Al	16	145	1,302	6.5	42	161
As	9.5	58.5	573	2.5	12.5	82
B	38.5	398	2,323	19	94	181
Ga	10	61	613	2.5	13.5	89
In	9	46.5	382	1.5	7.5	49
P	14.5	123	1,176	5.5	35.5	153
Sb	8.5	45.5	368	1.5	7.5	46.5

scattering of ions. Unlike the diffusion processes, the highest impurity concentration is not found at the semiconductor surface, but at a distance X_p from the surface, as shown in Fig. 8-12. Figure 8-14 shows a typical gaussian impurity profile obtained by implantation with different ion doses.

21. Crystal Damage. Ion bombardment of a crystalline semiconductor usually results in structural damage to the crystal lattice of the target material. After an incident ion has dissipated its kinetic energy upon entering the target, it typically lodges in an interstitial position of the target lattice. During its penetration, the incident ion displaces many target atoms and creates a large number of interstitials and vacancies. The extent of this damage depends on the properties of the impurity ions and the target, the ion energy, the impurity concentration, and the ambient conditions. At sufficiently high doses, a noncrystalline or amorphous layer is formed. The crystal damage can be almost completely removed by a high-temperature annealing heat treatment. If the damage is confined to isolated disordered regions, anneal temperatures of 200 to 300°C are sufficient to remove most of the damage.

22. Applications. The spatial distribution of implanted impurities can be determined by channeling-effect measurements, and their electrical behavior can be measured using Hall effect techniques. After the implantation step, only a small fraction of the implanted atoms are electrically active.

Ion implantation offers a wide variety of appli-

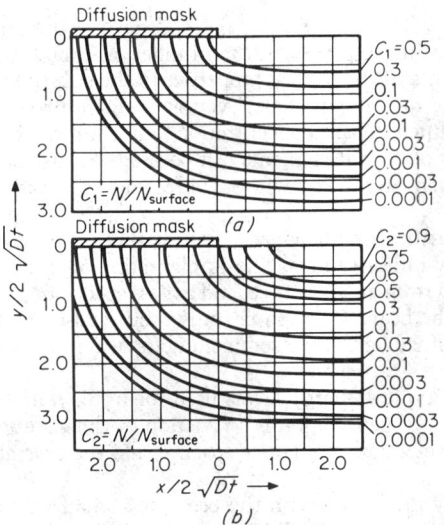

Fig. 8-11. Lateral diffusion profiles for (a) constant-source and (b) limited-source diffusions.[14]

cations in the fabrication[19] of monolithic devices. Some of its most important applications are in fabricating complementary bipolar or MOS transistors, implanted resistors, and MOS transistors with self-aligned gate structures and in adjusting the threshold voltage of MOS transistors.

23. Isolation Techniques. In monolithic ICs electrical isolation between the devices on the same substrate is achieved by fabricating them in electrically isolated regions of the substrate known as isolation *pockets* or *tubs.* The electrical separation of these pockets can be achieved by one of three methods: reverse-biased junctions (junction isolation), dielectric barrier layers (dielectric isolation), or beam-lead-connected components (air isolation).

24. Junction Isolation. Junction-isolation technique is the most common and economical method used in monolithic circuits. Electrical isolation is achieved by separating the individual components by reverse-biased pn junctions. This method requires few additional process steps and wastes relatively little surface area, but parasitic effects are created which can affect circuit performance. Figure 8-15 shows the cross section of a junction-isolated pocket containing an npn bipolar transistor. In fabricating this structure the basic planar process is used. Before the epitaxy step, a selective n^+ diffusion is made into the p-type substrate. This buried layer provides a low-resistivity current path from the collector contact to the active collector area directly below the base region.

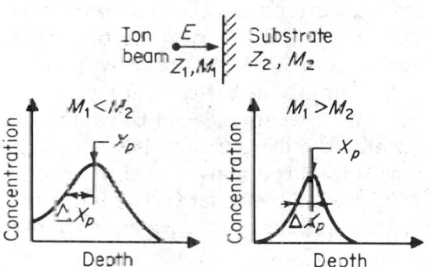

Fig. 8-12. Depth distribution of implanted atoms in an amorphous target where the ion mass is less or greater than the mass of the substrate atoms. To a first approximation the mean depth X_p depends on ion mass M_1 and incident energy E, whereas the relative width $\Delta X_p/C_p$ of the distribution depends primarily on the ratio of the mass of the ions to the mass of the substrate atoms M_2.[16]

Following the epitaxy and surface oxidation steps, an *isolation mask* is applied to the wafer surface, which opens the diffusion windows outlining the isolation pockets. The p-type isolation walls are then diffused from the wafer surface, through the n-type epitaxial layer into the p-type substrate. This isolation diffusion forms a continuous p-type wall around a selected region of the n-type epitaxial layer. When the substrate and the isolation walls are at negative dc poten-

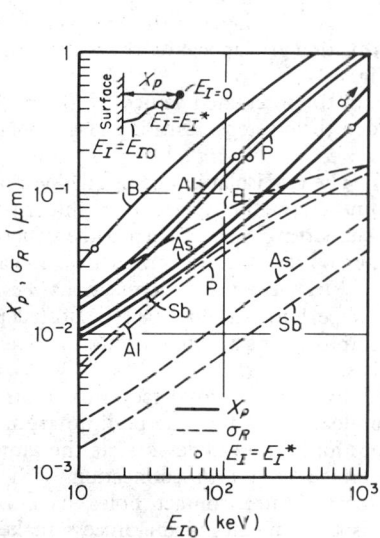

Fig. 8-13. Projected range X_p and standard deviation σ_R vs. ion energy at semiconductor surface E_{I0} for various impurities in silicon at 300 K. (*After Ref. 17.*)

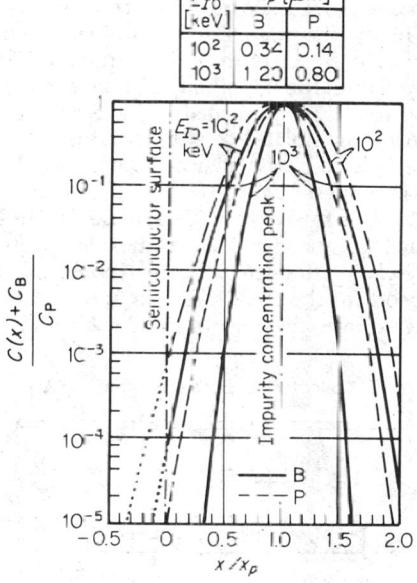

Fig. 8-14. C (x) at a distance from surface x and location of impurity concentration peak X_p for boron and phosphorus in silicon at 300 K. (*After Ref. 17.*)

tial with respect to the n-type pocket, the reverse-biased pn junction surrounding the pocket electrically isolates it from the rest of the wafer.

Isolation diffusion is a relatively noncritical diffusion step, with the basic requirement that the final depth of the isolation wall be greater than the epitaxial layer thickness. After completion of the isolation diffusion, the transistor base and the emitter regions are formed by respective diffusions into the n-type pocket, which serves as the collector of the npn transistor.

25. Dielectric Isolation. In certain applications, the parasitic junction capacitances or leakage currents associated with the junction isolation methods may not be acceptable. In such cases, a superior electrical isolation is obtained by insulating each pocket with a dielectric layer, as shown in Fig. 8-16. Normally, thermally grown SiC_2 is used as the dielectric material.

In forming the dielectrically isolated pockets on the wafer surface, a number of alternative fabrication techniques can be used. Figure 8-17 shows a typical sequence of fabrication steps in forming the dielectrically isolated single-crystal silicon pockets or islands. Starting with an n-type substrate, a nonselective n^+ layer is diffused into the wafer surface. Following the initial n^+ diffusion, the wafer surface is oxidized, and a mirror-image mask of the desired isolation grid pattern is applied to the wafer, to remove the oxide along the isolation grid. The exposed silicon

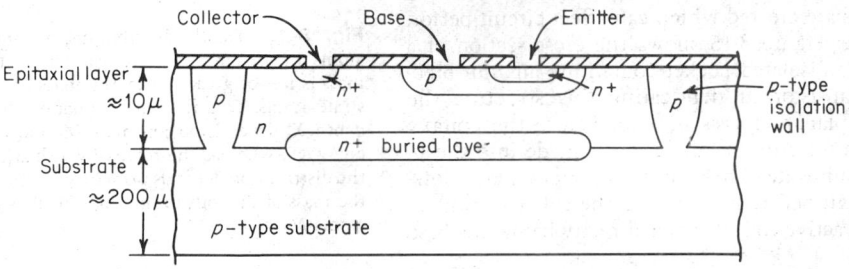

Fig. 8-15. Structural diagram of a pn-junction-isolated pocket containing an npn transistor.

surface is then etched by a potassium hydroxide (KOH)-based etch. If a $<111>$-oriented silicon crystal is used as the substrate, this preferential etching results in the formation of a V-shaped isolation groove, or moat, on the wafer surface, as shown in Fig. 8-17b. After the preferential etching step, the exposed silicon is reoxidized, and a thick layer of polycrystalline silicon is deposited over the oxide layer, as shown in Fig. 8-17c.

Finally, the original wafer is flipped around and the bottom surface of Fig. 8-17c now corresponds to the top of the device structure. Then the single-crystal n layer of thickness d_2 is backlapped until the tips of the isolation moats forming the isolation grid appear on the wafer surface. This results in an isolated n-type pocket as shown in Fig. 8-17d.

After the isolated pockets are formed, the fabrication of the integrated devices in the pockets is completed by a sequence of conventional masking and diffusion steps, resulting in the isolated-device structure of Fig. 8-16.

26. Air Isolation (Beam-Lead Technology). The two isolation techniques described above use silicon and silicon dioxide as a support medium to give the IC structural integrity. In air isolation thick metallic interconnections are used as supporting "beams" to provide structural support for the circuit, as well as serve as electrical interconnections.[21] These beams are cantilevered over the edge of the devices and also serve as external bonding connections. Besides providing superior isolation, this type of structure greatly simplifies the interconnection and packaging of ICs.

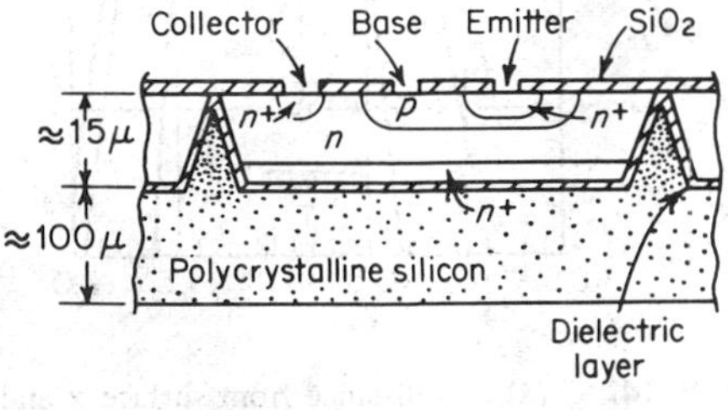

Figure 8-18 shows a cross section of an air-isolated beam-lead structure. The preliminary fabrication steps for beam-lead devices are the same as those described earlier in connection with the planar process. After contact holes have been etched, a special metallization process makes it possible to realize good ohmic contacts and provide good adherence to the SiO_2 layer. Normally, multiple metal layers are required for this operation. Initially, platinum silicide ohmic contacts

Fig. 8-16. Structural diagram of a dielectrically isolated pocket containing an npn transistor.

are formed; then titanium and platinum layers are sputtered on the SiO_2 surface; and finally the heavy gold beam leads are electroformed on the platinum base. Electroforming is a special form of electroplating where material is built up in selective areas only.

After the metallization pattern has been applied, the wafer is turned over and the isolation-etch-masking pattern is applied in registry with the metalization pattern on the other side of the wafer. The unmasked areas are then etched away, leaving discrete silicon mesas containing active and passive devices interconnected by the beam leads.

The beam-lead technology has several advantages over conventional processes. Higher device yields can be realized because of chemical separation rather than the scribe-and-break technique normally used; mounting and bonding of the circuit are greatly simplified; and higher reliability is obtained. The major advantage is the improvement in the electrical performance due to the reduction of the parasitic capacitances between the active device areas.

27. Thin- and Thick-Film Deposition. The electrical interconnection of integrated components is achieved by evaporation of conductive thin films on the wafer surface. Resistor and capacitor structures can also be formed on the passivated silicon surface by deposition of resistive or dielectric thin-film layers. A wide variety of thick and thin films are available in ICs.

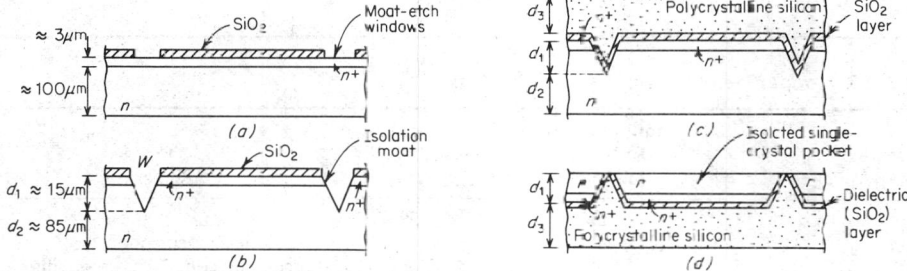

Fig. 8-17. Sequence of processing steps in dielectric isolation.

The term *thin film* is used to describe approximate film thickness of 1 μm or less, compared with the larger geometry and thicker films associated with hybrid ICs. Tables 8-4 and 8-5 list some common thin-film materials and their electrical properties.[22]

High-conductivity metallic films are used for circuit interconnections. Table 8-6 lists some of the films used in forming the circuit interconnections. Among these, aluminum (Al) is the most

TABLE 8-4   Thin-Film Materials in Microelectronics

Interconnections and/or terminals	Insulation and/or passivation	Encapsulation	Resistive
Al	$SiO_2 \cdot P_2O_5$	SiO_2	Ta
Al alloys (Cu, Si)	SiO_2	$SiO_2 \cdot P_2O_5$	TaN_2
Cu	Si_3N_4	Al_2O_3	$Cr \cdot SiO$
Mo–Au	Al_2O_3	Parylene	NiCr
		$PbO \cdot B_2O_3 \cdot SiO$	
Ti–Ag	BN		SnO_2
Cr–Ag	SiO		Kanthal
Pt–Au			
Cr–CuAu			
Pb–Sn			

Capacitive	Semiconducting		Processing	Development
SiO_2	Si	InAs	SiO_2	Si_3N_4–SiO_2
Ta_2O_5	Se	InSb	Si	P_2O
HfO_2	Te	PbS	Photopolymer	Nb_2O_5
ZrO_2	SiC	PbTe	Mo	Ge–Si–Te
$PbTiO_3$	GaAs	CdS	Cr	ZnO
	GaP	CdSe		Polymer
	AlN	ZnSe		

often used thin-film material because of its high electrical conductivity and good adherence to the SiO_2 surface. In forming the resistor patterns, resistive thin films such as tantalum (Ta), nickel-chromium (Ni–Cr) alloys, and tin oxide (SnO_2) are the most commonly used materials.

Figure 8-19 shows the cross sections of some thin-film components used in various types of ICs.

Table 8-5. Classification of Typical Microelectronic Thin-Film Material by Resistivity

Type	Example	Resistivity, $\Omega \cdot cm$
Dielectric	SiO_2 $SiTiO_2$	10^{18} 10^{14}
Resistor	NiCr $CrSiO_2$	10
Semiconductor	CdS PbTe	10^{-3}
Conductor	Al Mo Pt	10^{-6}

TABLE 8-6   Properties of Metals Used as Thin Films

Metal	Limiting current density I_1, MA/cm^2	Resistivity, $\mu\Omega \cdot cm$	Remarks
Silver	40	1.59	Poor adhesion
Copper	40	1.67	Poor adhesion, corrosion
Gold	70	2.35	Silicon eutectic 370°C, poor adhesion
Aluminum	5	2.65	Silicon eutectic 577°C, electromigration
Aluminum + Cu	20		
Magnesium		4.45	Extremely reactive
Rhodium		4.51	Poor adhesion
Iridium		5.3	Poor adhesion
Tungsten	200	5.6	Difficult etching
Molybdenum	100	5.7	Corrosion susceptibility
Platinum		9.8	Poor adhesion
Titanium		55	

28. Deposition Techniques. A variety of deposition techniques are available for forming thick- or thin-film layers on a dielectric substrate. Some of these processes are briefly outlined below.

Vacuum Evaporation. The passivated substrate together with the source of the material to be evaporated, is placed in a bell jar under high vacuum (10^{-5} to 10^{-6} torr). The material to be evaporated is heated by an electrical element until it vaporizes. Under the high vacuum the mean free path of the vaporized molecules is comparable with the dimensions of the bell jar; therefore the vaporized material radiates in all directions in the bell jar. Some of the vaporized material then deposits on the substrate, which is placed some distance from the source to ensure uniformity of deposition. The substrate is also maintained at an elevated temperature to provide a good adhesion of the deposited film.

Both conductive and resistive films can be deposited by vacuum evaporation. Aluminum, gold, and silver are among the conductive films formed in this manner. Nickel-chromium resistors can also be deposited by vacuum-evaporation techniques, except that high power densities required to vaporize the source require electron-beam bombardment rather than thermal heating of the source material to be used.

Cathode Sputtering. Sputtering process takes place in a low-pressure gas atmosphere. A glow discharge is formed by applying a high voltage (typically 5,000 V) between the cathode and the anode sections of the sputtering apparatus. The cathode is coated with the material to be evaporated, and the substrate is attached to the anode or placed in the glow-discharge region. Normally, an inert gas, such as argon (Ar), is used as the sputtering medium. The Ar^+ ions generated by the glow discharge accelerate toward the cathode, due to the negative cathode potential.

When these high-energy ions impinge on the cathode, they cause the atoms or the molecules of the cathode to break away, or sputter, from the cathode surface. Then a part of these cathode particles which float away are intercepted by the substrate and are deposited in the form of a thin layer. Under the low-vacuum conditions used in sputtering, the mean free path of the source atoms is much shorter than the source-to-substrate spacing. Therefore the deposition rates in the sputtering process are much slower than in vacuum evaporation.

Vapor-Phase Deposition. In vapor- or gas-phase deposition, halide compounds of the material to be deposited are chemically reduced, and the resulting metal atoms are deposited on the substrate. This basic deposition process closely resembles the epitaxial-growth step of the planar process. Vapor deposition is particularly useful for obtaining thick layers of deposited films (up to 20 μm). It is

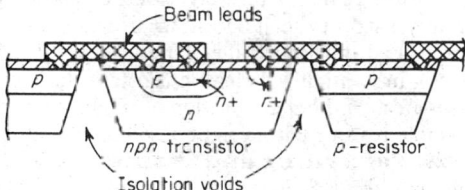

Fig. 8-18. Cross section of an air-isolated IC with beam-lead interconnections.

commonly used for forming aluminum oxide ($Al_2O_3 \cdot SiO_2$) dielectric layers or tin oxide (SnO_2) resistive films. The sheet resistance of SnO_2 films can be controlled by introducing group III or group V ions (such as In or Sb) to increase or reduce the sheet resistance. In this manner, a sheet resistivity range of 100 to 5,000 $\Omega/\square$ can be obtained.

Plating Techniques. Two basic kinds of plating techniques are used in forming metallic films on semiconductor substrate: electroplating and electroless plating. In electroplating, the substrate to be plated is placed at the cathode terminal of the plating apparatus and is immersed in an electrolytic solution. An electrode made up of the metal to be plated serves as the anode. When a direct current is passed through the solution, the positive-charged metal ions which dissolve into the solution from the anode migrate and plate at the cathode. This method of plating is often used for forming conductive films of gold or copper.

In electroless plating, simultaneous reduction and oxidation of a chemical agent are used in

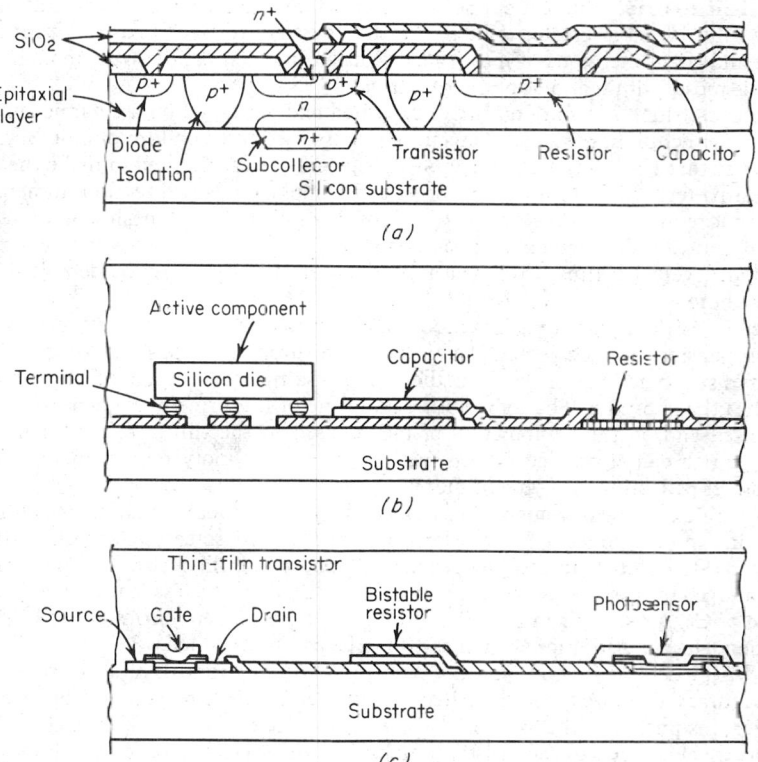

Fig. 8-19. Cross-sectional diagram of the three types of microelectronic circuits using thin films: (*a*) monolithic silicon IC; (*b*) hybrid silicon thin-film circuit; (*c*) all-thin-film IC.

forming a free metal atom or molecule. Since this method does not require electrical conduction during the plating process, it can be used with insulating substrates. Nickel, copper, and gold are among the most common metals that can be deposited in this manner.

Anodization.   Anodization is an electrochemical process for converting a metal into its oxide. In the fabrication of thin-film devices, anodization can be used to replace deposited dielectrics by forming an anodized layer of metal oxide to serve as the dielectric. Two basic metals commonly used in the anodization process are tantalum and titanium. Anodization techniques are often used in forming dielectric layers of Ta_2O_5 on tantalum films to fabricate capacitors. Since Ta_2O_5 has a higher dielectric constant than SiO_2, a larger capacitance value can be obtained per unit area. Another application of anodization methods is in trimming tantalum resistors to obtain a stable resistor with a tight absolute-value tolerance.

29. Patterning and Etching of Thin Films. With minor modifications, the basic photomasking and etching techniques can be used in patterning the thin-film components. One significant exception is the case of multiple thin-film layers (such as aluminum interconnections over thin-film resistors), where additional care must be taken in the choice of the etchant to ensure that the bottom film is not damaged by the patterning of the top layer.

In Al, SnO_2, Ta, or $Al_2O_3 \cdot SiO_2$ layers patterning and etching can be achieved by direct photoresist techniques. In very thin (30 to 50 nm) nickel-chromium films, an *inverse metal-masking* technique can be used. In this process, a thin layer of metal film (typically copper) is deposited and etched by the inverse of the desired final metal pattern. Then the desired thin-film layer is deposited on this inverse metal pattern. In the final etching step the inverse metal pattern of the initial metal film is etched away, taking with it the layer of desired metal deposited on it, and only the portion of the desired metal layer which directly adheres to the substrate is left behind.

In a simpler version of the inverse metal-masking technique, a photoresist layer may be used in place of the first metal layer, on which the desired metal can be deposited. Then the photoresist is etched and cleared away to leave behind the desired metal pattern, which adheres directly to the substrate. Since photoresist is an organic polymer, it cannot withstand exposure to high temperatures. Therefore inverse photoresist techniques are limited to thin-film processes where the substrate temperature is maintained below 250°C.

30. Thick Films. No clear-cut definition exists for separating the thick and the thin films; generally, film layers of greater than 10 μm thickness are referred to as thick films. Since the structure and the dimensions of most thick films are not compatible with monolithic circuits, their use is mostly limited to hybrid structures.

The resistive thick films are normally deposited and patterned on a ceramic substrate using silk-screening techniques. The silk-screening process used in thick-film circuit fabrication is an adaptation of that used in the graphic arts. This technique can be used to make the metal-interconnection pattern for hybrid circuits or in other applications where close dimensional tolerances are not required. A screen made of silk or fine stainless-steel mesh is mounted on a rigid frame and coated with a photographic emulsion or a resist. The screen is then exposed through a mask and developed, thus leaving clear areas where the film is to be deposited and opaque areas elsewhere.

The screen is then placed on the circuit substrate, and a thin layer of the film material in viscous suspension with powdered glass is squeezed through the open areas of the screen onto the ceramic substrate. The substrate is then fired in a furnace to drive off liquids and to form the bond to the substrate. The temperature and time of the firing cycle depend on the type of ceramic used and on the combination of metals used in the thin-film material. Some of the materials which can be applied by this method are carbon, molybdenum-manganese, gold-palladium, silver-palladium, and tin-antimony.

Resistive thick films are composed of a suspension of conductive metal particles in a ceramic matrix, with an organic resin as the filler. Following the silk-screening operation, the substrate is fired at 1400 to 1600°C to ensure a permanent adhesion of the resistive film. Because of their ceramic and metal base, these resistors are known as *cermets*. Typical cermet structures are composed of Cr, Ag, or PbO in an SiO or SiO_2 matrix. By proper control of the original resistor composition, sheet resistivities as high as 10 kΩ/$\square$ can be obtained.

Since this process in general does not lend itself to tight control of sheet resistance, where resistor accuracy is desired, the deposited resistor values must be adjusted by trimming techniques. For this purpose either abrasive or laser trimming methods can be used.

The electrical contacts to the resistive thick films are normally made by gold or nickel interconnections which are deposited by electroless plating techniques and patterned by photographic etching.

Figure 8-20 is a flowchart of process steps associated with the fabrication of a thick-film hybrid IC.[23]

31. Interconnections. *Ohmic Contacts.* The basic requirement for the conductive films used for interconnections is that they should make good ohmic contact with the diffused components or other metallic films deposited on the device surface. A good ohmic contact is defined as one that exhibits a linear voltage-current relationship and introduces a negligible amount of series resistance.

The exposure of contact areas to the ambient atmosphere often results in the formation of parasitic oxide layers over the chip areas to be interconnected. Therefore, to provide good ohmic contact, the interconnecting metal must be chemically active so that it can be alloyed or sintered through these parasitic oxide layers. The most commonly used interconnection metal is aluminum, which can be readily alloyed into the silicon substrate, to form ohmic contacts.

In conventional monolithic circuit fabrication, the alloying of the Al interconnections into silicon is the last step of the planar process. It is normally accomplished by a short heat treatment in an inert atmosphere, typically about 10 min at 500°C.

A troublesome metal-interconnection problem can occur in devices which employ two active dissimilar metals in their interconnection scheme. At the interface of two dissimilar metals, parasitic intermetallic compounds and oxides can form. A typical example of this is the intermetallic gold-aluminum compounds forming between the aluminum bonding pads and the gold wires which may be used to connect the chip to the package terminals.

Interconnections in Monolithic Circuits. Aluminum is the most commonly used interconnection metal in monolithic ICs. In other types of ICs, such as hybrid or beam-lead devices, another conductive metal layer, such as gold or molybdenum, may be used for interconnections. The deposition and patterning techniques for these metal film layers are discussed in Pars. **8-28** and **8-29**.

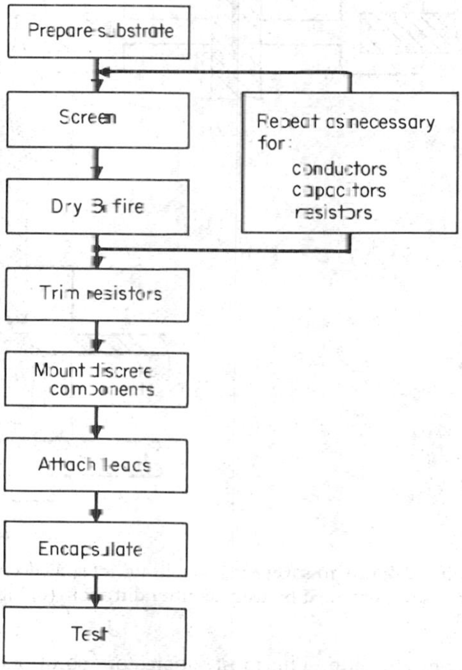

Crossovers. If the interconnection cannot be accomplished in one plane, different methods of producing crossovers are available without additional processing steps. For example, the area over a diffused resistor can be used for a metal crossover, as shown in Fig. 8-21a. The emitter diffusion can be used if a low-value resistance is permissible for the interconnection. This is shown in Fig. 8-21b. Figure 8-21c shows how the geometry of an npn transistor can be modified to facilitate a crossover near the device. By use of these crossover methods it is possible to interconnect most ICs in one plane. However, if very high packing densities of devices are required, or if stray capacitances must be minimized multilayer metallization can be used.

Fig. 8-20. Typical thick-film process sequence.

Multilayer Metallization. Multilayer interconnections are formed by depositing alternate layers of patterned interconnections separated by dielectric layers (Fig. 8-22a). The dielectrics used in this application must have sufficient dielectric strength, volume resistivity, and thermal and mechanical stability to withstand the processing operations used to form the interconnect metallization. The first dielectric layer is usually the thermal oxide used to passivate the silicon surface. Subsequent dielectric layers must be formed by deposition techniques. Most multilayer metal processes use aluminum as the interconnection metal. The multiple metal-interconnection layers are then interconnected by means of the *via holes* cut in the separating dielectric layer, as shown in Fig. 8-22b.

Reliability of Interconnections. The reliability of interconnections is a serious problem in microcircuits, largely because of the sharp oxide steps which must be covered by the deposited metal layers. In many cases, the step in the oxide thickness may be significantly greater than the thickness of the metal-interconnection layer. This is particularly true in multilayer metal systems and can lead to the formation of *microcracks* in the metal layer at the edges of contact windows or at sharp oxide steps.

Figure 8-22c shows a structural diagram indicating the possible location of microcracks in the vicinity of a contact window. It can be shown that if the step height is an appreciable part of the film thickness, cracking due to geometrical shadowing effects can be expected from any source

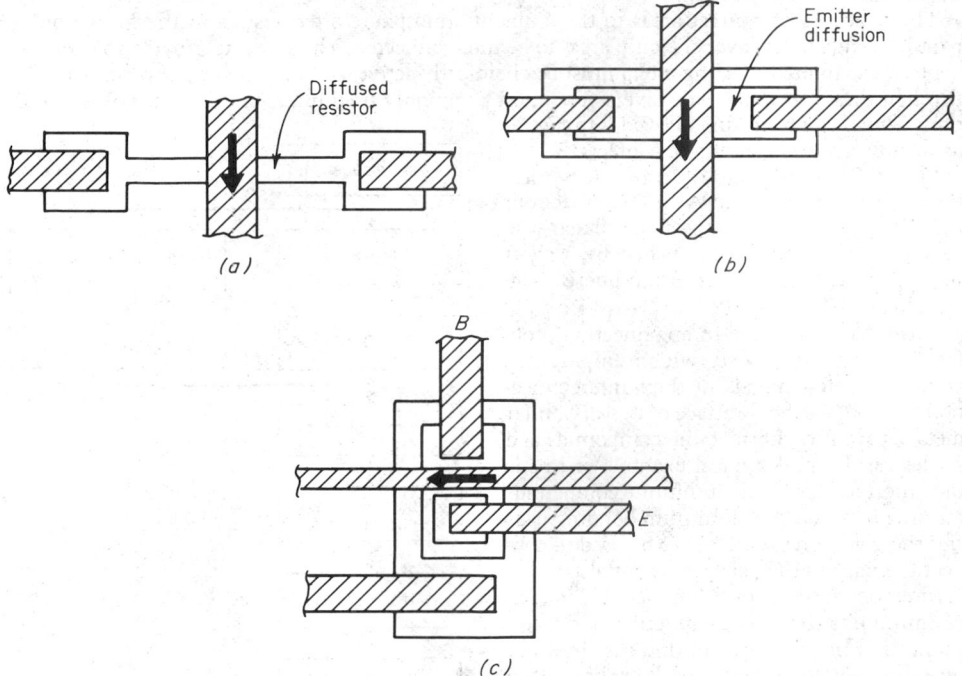

Fig. 8-21. Crossovers in monolithic integrated circuits. (a) interconnection crossing diffused resistor; (b) crossover obtained by using emitter diffusion; (c) interconnection passing over base region of *npn* transistor.

configuration. The most satisfactory solution to this problem appears to be a reduction of the height of the oxide step or tapering of the oxide, as shown in Fig. 8-22d.

Various other failure mechanisms can occur in interconnection systems. Some of these are *electromigration* effects at high current densities, or formation of parasitic intermetallic compounds. These failure mechanisms will be discussed further, in Sec. 28.

32. Testing and Assembly. The basic fabrication processes covered in the preceding sections deal with the fabrication of the *intrinsic* circuit only. Before the circuit is functional, it has to be assembled in a circuit package and electrically tested to guarantee certain performance requirements. Figure 8-23 shows a flowchart of the complete IC processes, including the testing and the assembly steps.

The basic fabrication steps outlined in the figure can be briefly described as follows. After the wafer fabrication is completed, each finished wafer undergoes a *wafer-sort* process where each and every circuit on the wafer is contacted and electrically tested using miniaturized probes. Units that do not meet the desired electrical specifications are marked on the wafer, to be discarded. The tested wafer is then *scribed* by using a diamond-tipped cutting tool, and the individual circuit chips, or dice, are separated. Chips which have passed the wafer-sort step are assembled in packages, where the circuit terminals are brought out to the package leads. After the assembly operation, the packaged circuits undergo a final-test step, which completes the fabrication cycle.

Testing and assembly steps lack the batch-processing advantages of the basic fabrication steps because every circuit needs to be tested and assembled individually. For this purpose a larger number of automated test and assembly techniques have been developed.

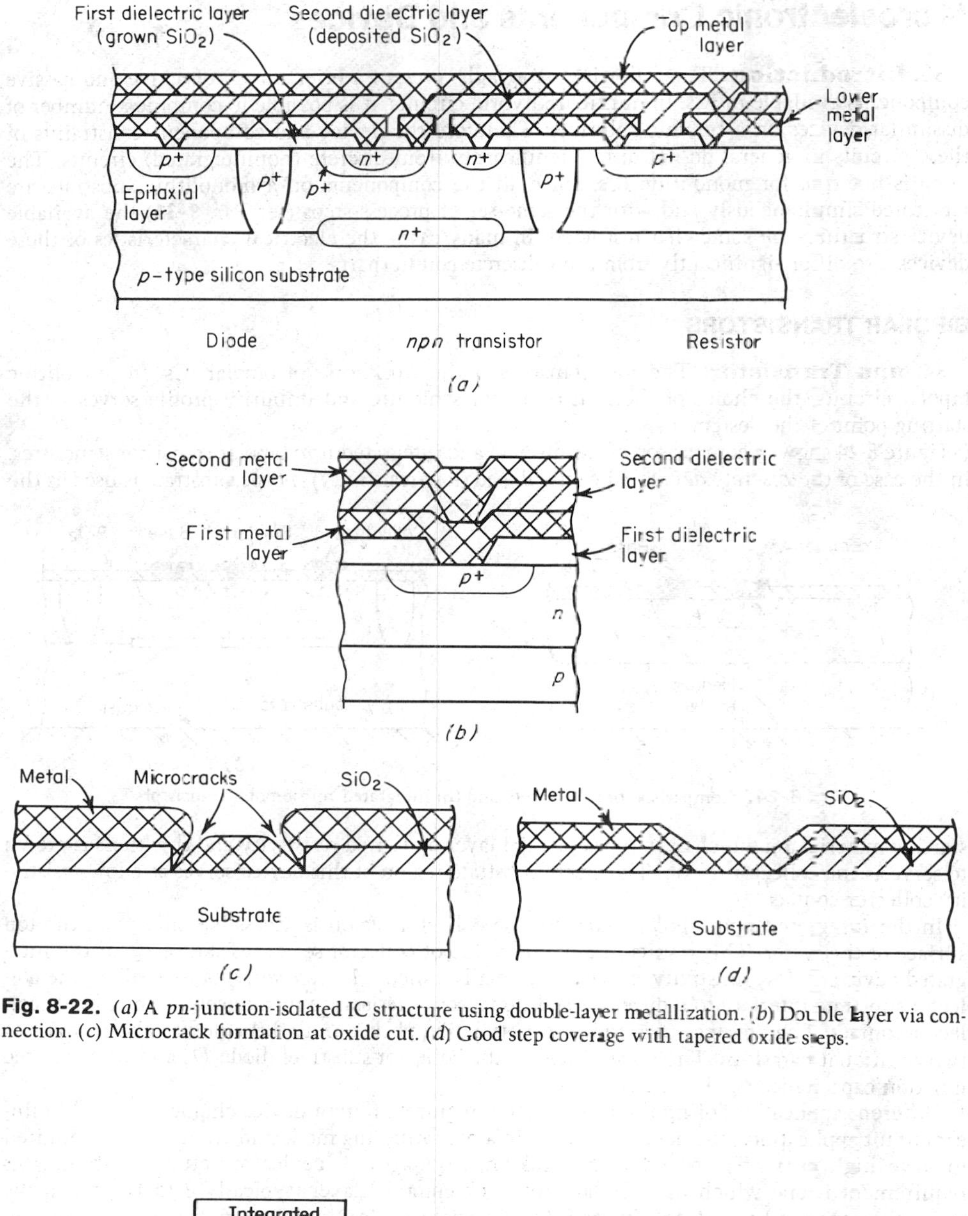

First dielectric layer Second dielectric layer Top metal
(grown SiO$_2$) (deposited SiO$_2$) layer

Lower metal layer

$p+$ $p+$ $n+$ $n+$ $p+$ $p+$

$p+$ $p+$

Epitaxial layer

$n+$

p-type silicon substrate

Diode npn transistor Resistor

(a)

Second metal layer Second dielectric layer

First metal layer First dielectric layer

$p+$

n

p

(b)

Metal Microcracks SiO$_2$ Metal SiO$_2$

Substrate Substrate

(c) (d)

Fig. 8-22. (a) A pn-junction-isolated IC structure using double-layer metallization. (b) Double layer via connection. (c) Microcrack formation at oxide cut. (d) Good step coverage with tapered oxide steps.

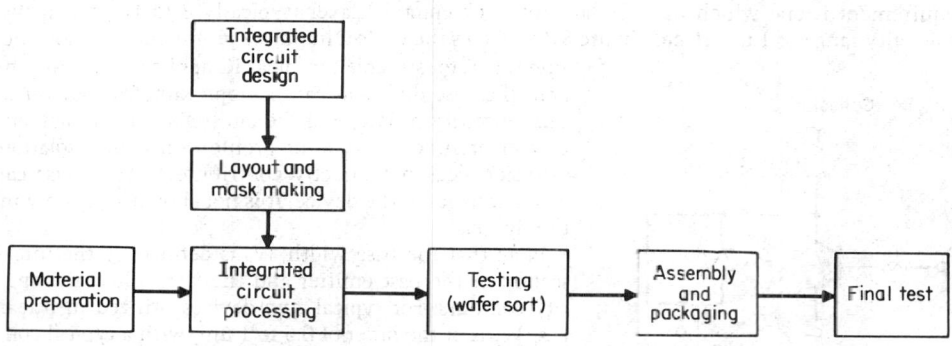

Integrated circuit design

Layout and mask making

Material preparation → Integrated circuit processing → Testing (wafer sort) → Assembly and packaging → Final test

Fig. 8-23. Flowchart showing testing and assembly steps.

Microelectronic Components and Devices

33. Introduction. The following paragraphs present a brief survey of active and passive components available in ICs. In multichip hybrid circuits, it is possible to combine a number of dissimilar devices on a ceramic substrate. Therefore the device and component constraints of these circuits in general do not differ significantly from discrete (nonintegrated) circuits. The same is not true for monolithic ICs. Since all the components on a monolithic substrate are fabricated simultaneously and with the same set of process steps (see Par. **3-31**) the available device structures are somewhat restricted. In many cases, the electrical characteristics of these devices also differ significantly from their discrete counterparts.

BIPOLAR TRANSISTORS

34. npn Transistor. The *npn* transistor is the workhorse of bipolar ICs. In monolithic bipolar circuits, the choice of the *npn* transistor structure and impurity profile serves as the starting point of the design.

Figure 8-24 shows a comparison of the discrete and integrated *npn* bipolar transistor structures. In the case of the *discrete device*, a heavily doped (low-resistivity) n-type substrate is used as the

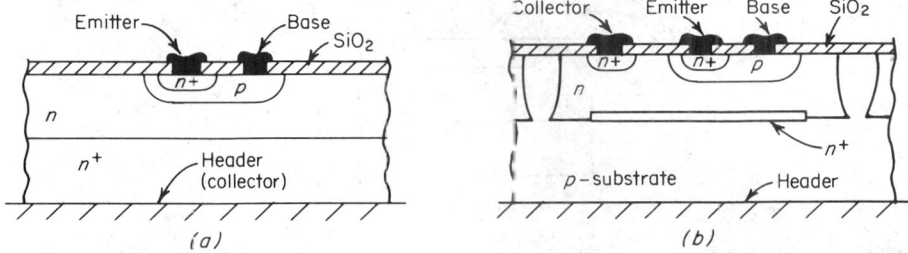

Fig. 8-24. Comparison of (*a*) discrete and (*b*) integrated *npn* bipolar transistors.

starting material, on which an n-type epitaxial layer of desired resistivity and thickness is grown to serve as the collector region. The entire substrate region of this device serves as a low-resistivity collector contact.

In the *integrated npn bipolar transistor* the collector region is accessible only from the top surface of the wafer. This results in a higher value of collector series resistance r_{cs} for the integrated device. A low-resistivity n^+ layer (normally called a *buried layer*) is used along the collector-substrate interface to reduce this additional series resistance introduced by the topside collector contact. The presence of a pn junction isolation pocket around the device also introduces two additional parasitics. These parasitics are the collector-substrate diode D_{ss} and its associated junction capacitance C_{ss} (Fig. 8-25).

Different applications of *npn* transistors often require different device characteristics.[24] In linear-circuit applications, the devices are used in a nonsaturating mode and are in general required to have high current gain and high breakdown voltages. A device structure satisfying this requirement is one which uses a relatively thick epitaxial layer (typically 8 to 12 μm) in the resistivity range of 1 to 5 $\Omega \cdot$cm. Figure 8-26a shows the typical impurity profile for a monolithic *npn* transistor suitable for linear IC applications.[25] In general, the base diffusion can be approximated by a *gaussian* impurity profile, and the emitter by a complementary error function type of profile. Since the isolation diffusion does not directly contribute to the electrical characteristics of the device, it is not shown separately in the figure.

Note that the base width W_B is defined by the intersection of the base-emitter and the base-collector impurity contours. For typical *npn* devices utilized in linear ICs, W_B is of the order of 0.5 to 1 μm, with a typical control tolerance of ± 0.1 μm.

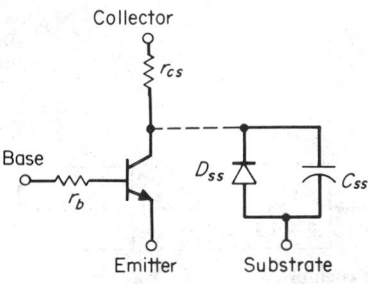

Fig. 8-25. Inherent parasitics associated with a junction-isolated transistor.

In digital circuit applications where high switching speeds and low saturation voltages are needed, and where low breakdown voltages (≤ 20 V) can be tolerated, a somewhat different impurity profile is used.[26] A typical example of this is shown in Fig. 8-26b. A high-speed digital circuit transistor normally uses a thinner epitaxial collector region and shallow base and emitter diffusions. This choice is made to reduce the parasitic *sidewall* capacitances associated with the side portions of diffused regions.

In saturating logic circuits, a small quantity of gold is also diffused into the device structure. Gold atoms diffuse very rapidly through the entire silicon lattice and serve as recombination centers for free carriers in silicon. The net effect of gold doping is the reduction of minority-

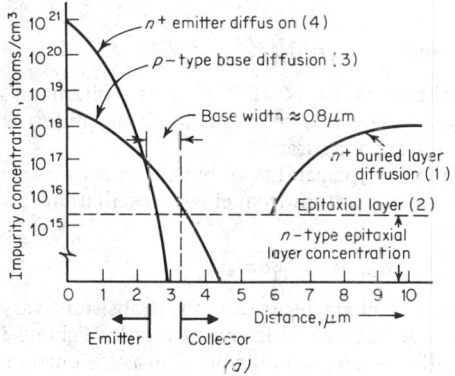

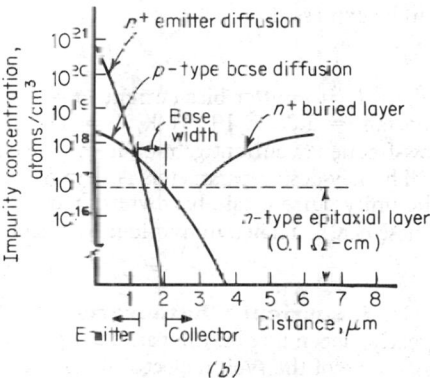

Fig. 8-26. Typical impurity profile for (a) linear-circuit transistor and (b) digital IC transistor.

carrier lifetime, which in turn reduces the *storage time* or the turnoff delay associated with switching an *npn* transistor from saturated operation to OFF condition.

35. npn Device Layout. Figure 8-27 shows the lateral geometry and the structural cross section of a typical small-signal *npn* transistor, having the impurity profile shown in Fig. 8-26a. Note that the vertical dimensions of the structural cross section are not drawn to scale.

The common-emitter current gain β is a function of collector I_c. Figure 8-28 shows the β-vs.-I_c characteristics for a typical small-signal *npn* transistor, having the geometry of Fig. 8-27 and the impurity profile of Fig. 8-26a. Current-gain characteristics of various other types of monolithic bipolar transistors of comparable size and geometry are also shown in the figure for comparison.

The degradation of β at low current levels is due primarily to parasitic surface-recombination effects. Low-current β can be improved by additional surface-passivation steps (such as nitride passivation) and by minimizing the emitter periphery-to-area ratio.[27]

At high current levels, transistor current gain degrades, due to two separate effects: (1) decrease of emitter efficiency and (2) current crowding at the emitter periphery. The emitter efficiency can be improved by increasing the emitter area, and the current-crowding effects can be minimized by increasing emitter periphery-to-area ratio. This can be done using multiple emitter *stripes* in the device layout. Figure 8-29 shows the practical layout for an integrated power transistor. Note that the device has three emitter stripes, surrounded by interdigitated base contacts and a low-resistivity collector contact that surrounds the active-device region.

36. Frequency-Response Parameters. For small-signal applications, the frequency response of bipolar transistors can be

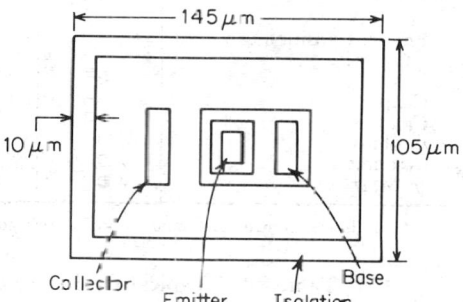

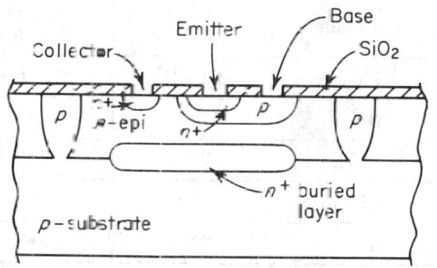

Fig. 8-27. Lateral geometry and the structural cross section of a small-signal *npn* transistor (vertical dimension not to scale).

closely approximated by the hybrid-π model shown in Fig. 8-30. In the circuit model, the parameters r_b, C_c, and g_0 are the parasitics inherent in any bipolar transistor structure, either discrete or integrated. The base spreading resistance r_b is the physical resistance of the path traversed by the base current from the base contact to the active base region, directly below the emitter. C_c is the collector-base junction capacitance, which is directly proportional to collector-base junction area. The finite output conductance g_0 comes about as a result of base-width modulation.

Collector substrate capacitance C_{ss} and the collector series resistance r_{cs} are the two additional parasitics particularly associated with integrated transistor structures (see Figs. 8-24 and 8-25).

The device transconductance g_m and the dynamic resistance r_{π} of the base emitter junction can be expressed as

$$g_m = qI_E/kT = I_E/V_T \quad \text{and} \quad r_{\pi} = \beta_0/g_m$$

where I_E = emitter bias current, q = electronic charge = 1.6×10^{-19} C, k = Boltzmann's constant = 1.38×10^{-23} J/K, T = temperature (K), $V_T = kT/q$ = thermal voltage, and β_0 = low-frequency current gain of device in common-emitter configuration.

The key design parameters describing the high-frequency capability of bipolar transistors are the unity current-gain-bandwidth frequency f_t and the maximum frequency of oscillation f_{max}. These can be related to hybrid-model parameters as

$$2\pi f_t = g_m/(C_{\pi} + C_c) \quad \text{and} \quad f_{max} = \sqrt{f_t/8\pi r_b C_c}$$

37. Electrical Characteristics. The electrical characteristics of IC transistors vary greatly, depending on the particular choice of device geometry and impurity profile. Table 8-7 lists some of the typical electrical parameters of small-geometry npn bipolar transistors encoun-

Table 8-7.   Typical Electrical Parameters of Integrated npn Bipolar Transistors*

	2.5 Ω·cm collector (no gold)	0.5 Ω·cm collector (no gold)	0.1 Ω·cm collector (gold-doped)	Measurement condition
Low-frequency current gain β	150	50	25	$I_C = 0.1$ mA
	150	60	50	$I_C = 1$ mA
	80	60	50	$I_C = 10$ mA
LV_{CEO}, V	60	35	22	
BV_{CBO}, V	85	45	25	$I_0 = 10$ μA
BV_{EBO}, V	7.0	6.7	6.7	$I_0 = 10$ μA
V_{BE}, V	0.65	0.69	0.73	$I_E = 1$ mA
$V_{CE.sat}$, V	1.2	0.5	0.26	$I_C = 5$ mA
Emitter transition cap. C_{TE}, pF	4	6	9	Forward-bias condition 5 V reverse bias
C_{ss}, pF	0.5	0.8	1.5	
	0.6	1	1.5	5 V reverse bias
C_{ES}, pF	1.8	2.9	4.6	5 V reverse bias
r_b, ohms	100	80	60	
r_{cs}, ohms	200	75	15	
f_T, MHz	400	440	550	$I_C = 5$ mA, $V_{CE} = 5$ V

* For device geometry and impurity profiles refer to Figs. 8-26 and 8-27.

Table 8-8.   Typical Design Tolerances for Integrated npn Transistors

Device parameter	Typical value*	Typical design tolerance	Temperature coefficient	Matching tolerance	Temperature tracking tolerance
V_{BE}	0.65 V	± 20 mV	-2 mV/°C	± 1 mV	± 10 μV/°C
β	150	$\pm 25\%$	$+3000$ ppm/°C	$\pm 8\%$	± 100 ppm/°C
BV_{EBO}	7.0 V	± 0.3 V	$+3$ mV/°C	± 25 mV	± 100 μV/°C
BV_{CBO}	85 V	± 15 V	$+60$ mV/°C	± 2 V	± 3 mV/°C
r_b	100 Ω	$\pm 20\%$	$+1500$ ppm/°C	$\pm 3\%$	± 100 ppm/°C
r_{cs}	200 Ω	$\pm 20\%$	$+2500$ ppm/°C	$\pm 3\%$	± 200 ppm/°C

* An impurity profile similar to that of Fig. 8-26a is assumed.

tered in monolithic circuits. For comparison a device geometry similar to that shown in Fig. 8-27 is assumed. The devices listed in the first and last columns of the table correspond, respectively, to linear and digital device structures having impurity profiles similar to those shown in Fig. 8-26.

Some of the process tolerances and temperature coefficients associated with the key *npn* device parameters are listed separately in Table 8-8. Note that even though the absolute-value tolerances of these device parameters are relatively poor by discrete-device standards, their matching and thermal-tracking properties are far superior to those of discrete devices.

38. *pnp* Transistors. In monolithic IC structures, high-performance *pnp* and *npn* devices are not readily compatible. To fabricate a high-performance *pnp* transistor on the same substrate

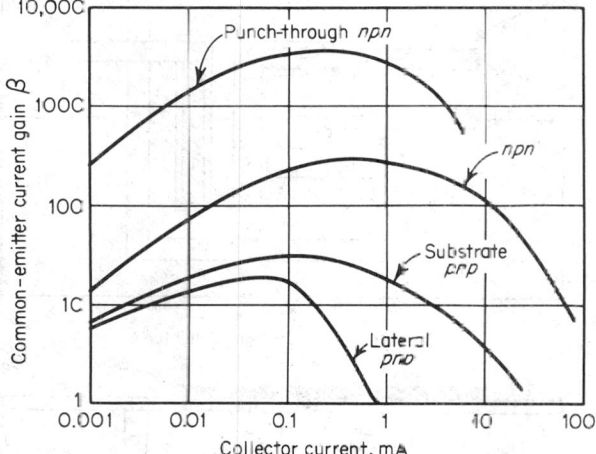

Fig. 8-28. Typical β-vs.-I_c characteristics of various IC transistors of equal emitter area. (For comparison an emitter area equal to that shown in Fig. 8-27 is assumed.)

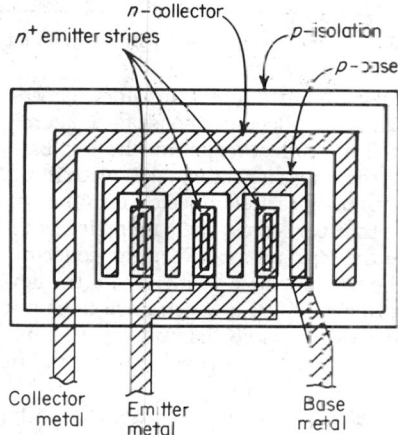

Fig. 8-29. Lateral geometry of a high-current *npn* transistor.

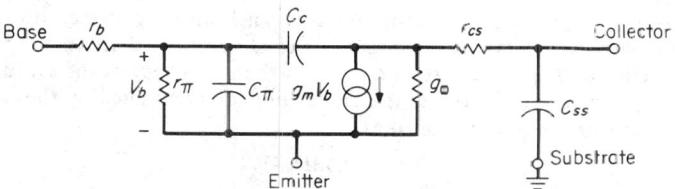

Fig. 8-30. Small-signal equivalent circuit for an integrated bipolar transistor.

with an *npn* bipolar transistor usually requires additional critical diffusion steps. In certain applications where the *pnp* performance is not critical, it is possible to fabricate functional *pnp* transistors using the same process steps as for *npn* devices. The two types of *pnp* transistors which can be fabricated in this manner are the lateral-*pnp* and the substrate-*pnp* transistors.

39. Lateral-*pnp* Transistors. The simplest *pnp* transistor structure which can be fabricated simultaneously with the *npn* bipolars is the lateral-*pnp* transistor, which requires no additional masking or diffusion steps.[28] Figure 8-31 shows the plane view and the structural diagram of a lateral-*pnp* transistor. The base region of the device is formed by the *n*-type epitaxial

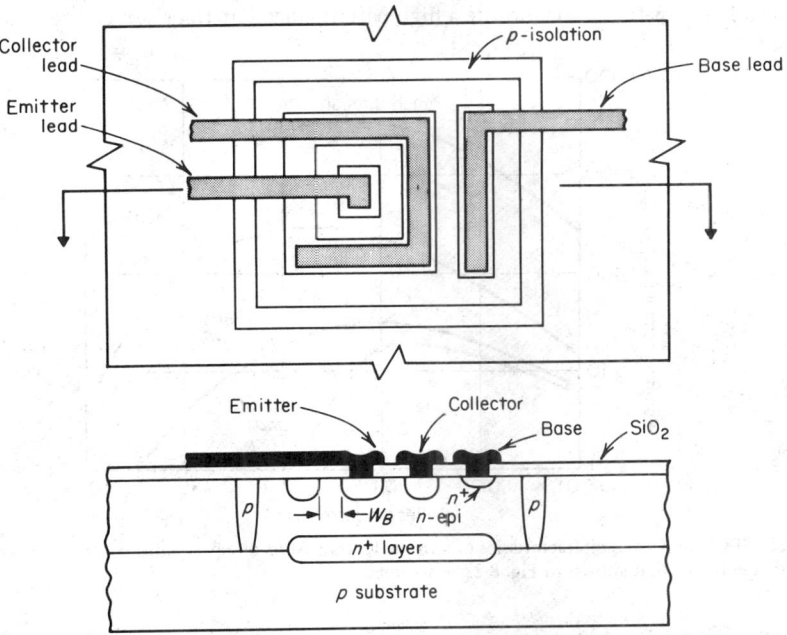

Fig. 8-31. Lateral-*ppn* transistor with wraparound collector.

layer, which serves as the collector of the *npn* transistors. The *p*-type base diffusion of the *npn* is used to form the emitter and the collector regions of the lateral *pnp*; the *n*⁺ emitter diffusion of the *npn* is used to form the *n*⁺ contact region for the *pnp* base.

In such a device structure, the transistor action takes place in the lateral direction, i.e., parallel to the device surface. The minority carriers injected into the base diffuse laterally toward the collector region. The carrier transport across the base region is the most efficient at or near the surface of the device, where the separation between the collector and the emitter is minimum. This minimum spacing is the effective base width W_B for the device. Due to masking-tolerance and voltage-breakdown requirements, W_B is constrained to be of the order of 6 to 12 μm. This value for the effective base width is more than an order of magnitude larger than that of a vertical-*npn* transistor. Consequently, the current gain and the frequency characteristics of lateral *pnps* are inferior to those of *npn* devices.

As shown in Fig. 8-31, the lateral-*pnp* structure uses the *n*⁺ buried layer at the episubstrate interface. In the case of a lateral *pnp*, this *n*⁺ layer serves a dual function; it avoids parasitic *pnp* action between the lateral-*pnp* emitter and the substrate, and it provides a low-resistivity path for *pnp* base current.

The low-frequency current gain β of the lateral *pnp* is limited by poor emitter efficiency, wide base width, and surface recombination effects. Typical values of β are in the range of 5 to 20. Typical dependence of β on the collector current of the lateral *pnp* is shown in Fig. 8-28. Frequency-response characteristics of the lateral *pnp* are dominated by the base transit time. Its cutoff frequency can be approximated as[29]

$$f_t \approx 2.43 D_p / W_B^2$$

where D_p is the diffusion constant of holes in the base region. Typical values of f_t for lateral-*pnp* structures are of the order of 2 to 5 MHz.

The low current gain of a lateral *pnp* can be improved by combining it with a monolithic *npn* transistor, to form the composite transistor structure shown in Fig. 8-32. The polarity and the current gain of such a composite device are equivalent to those of a single *pnp* device having an effective current gain $\beta_t = \beta_p \beta_n$, where β_p and β_n are the current gains of the *pnp* and the *npn* devices. However, the frequency response of the composite device is determined by the lateral *pnp*.

40. Substrate-*pnp* Transistors. A functional *pnp* transistor can also be formed by using the base region of the *npn* transistor as the emitter, the n-type epitaxial layer as the base, and the p-type substrate as the collector. Such a device is called a *substrate pnp* and has the typical layout and the cross section shown in Fig. 8-33. The n^- buried layer is not present in this device structure. Since the epitaxial-layer thickness is now directly related to the effective base width W_B of the *pnp*, a tighter control of the epitaxial-layer thickness is necessary (typically to $\pm 1 \ \mu m$).

The collector of the substrate *pnp* is formed by the p-type substrate, which is common to the rest of the circuit and is at all times ac-grounded. Therefore the substrate *pnp* is available only in the grounded-collector configuration.

The current gain and the frequency response of the substrate *pnp* are also limited by its relatively large base width ($W_B \approx 6$ to $10 \ \mu m$) and poor emitter efficiency. The typical values of β_0

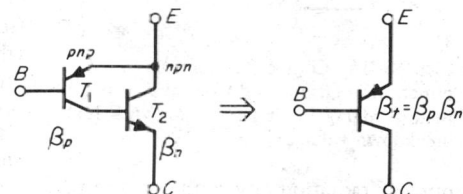

Fig. 8-32 Composite connection of complementary transistors.

for the device are in the range of 5 to 30. Since the entire bottom surface of the emitter is electrically active, the substrate *pnp* can handle a higher amount of current than the lateral *pnp* of comparable geometry. Typical values of f_t are in the 10- to 30-MHz range.

41. Punch-Through Transistors. In certain analog circuits, such as the input stages of operational amplifiers, it is necessary to have very high input impedance and low input bias currents. For such an application, β_0 of a typical integrated *npn* transistor is not high enough, since the device design requires a compromise between the current-gain and the voltage-break-down requirements.

The current gain can be greatly increased at the expense of the collector-emitter breakdown voltage by using a very narrow base structure ($W_B \approx 200$ nm). Such a transistor structure can provide current gains of 1,000 to 3,000 at collector current levels of 10 to 20 μA, with a collector-emitter voltage of 0.5 V (see Fig. 8-28). However, the collector-emitter voltage breakdown of such a device structure is limited to a 2- to 3-V range because of the collector-base depletion layer punching through the active base region into the emitter. The term *punch-through transistor* is derived from this phenomenon.

Punch-through transistors can be fabricated simultaneously with the conventional *npn* bipo-

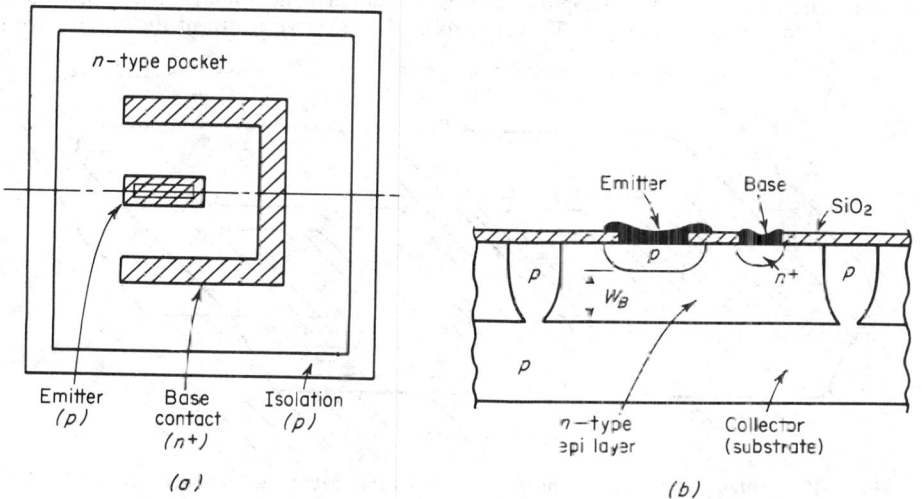

Fig. 8-33. Structural diagram of a substrate-*pnp* transistor: (*a*) plane view; (*b*) cross section.

lar transistors, with the addition of one extra masking and diffusion step. The emitter of the punch-through transistor is partially diffused before the emitter diffusion cycle for conventional *npn* transistors. Due to this initial diffusion step, the emitter of the punch-through transistor is diffused to a slightly greater depth than the conventional *npn* device. Thus, if both devices have the same base diffusion depth, the resulting base width W_B of the punch-through device is significantly smaller (250 vs. 700 nm) than that of a conventional *npn* transistor on the same silicon substrate.

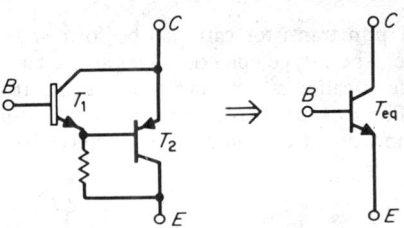

Fig. 8-34. Composite connection of a punch-through *npn* and a lateral *pnp* to simulate a high-gain *npn* transistor with high breakdown voltage.

In circuit applications, punch-through transistors are often used together with a conventional IC transistor which can provide the necessary voltage protection for it. Figure 8-34 shows a composite connection of a punch-through transistor T_1 with a lateral-*pnp* transistor T_2 to form an equivalent *npn* transistor T_{eq} which has the effective β of the punch-through *npn* but the breakdown voltage of the lateral *pnp*.

42. High-Performance Complementary Transistors. In the design of high-performance analog circuits, particularly for operation within a nuclear radiation environment, the performance characteristics of the lateral- or the substrate-*pnp* transistors are not acceptable. For these specialized applications, a number of high-performance complementary bipolar structures have been developed. Each of these structures requires additional processing steps beyond those required for the basic *npn* transistor. Their use is therefore limited to design applications where the added fabrication cost or complexity can be justified.

Figure 8-35 shows a practical *pnp-npn* device structure using dielectric isolation techniques.[30] In fabricating complementary bipolar devices, the use of dielectric isolation is preferred over conventional isolation because it provides access to the back of the device structure during a portion of the fabrication steps. In fabricating the device structure of Fig. 8-35, the process steps follow the basic sequence of steps associated with dielectrically isolated devices, as discussed in Par. **8-25** (see Fig. 8-17).

FIELD-EFFECT TRANSISTORS

43. Terminology. The field-effect transistor (FET) is a voltage-controlled device in which the current conduction between the *source* and the *drain* regions is controlled or modulated by means of a control voltage applied to the *gate* terminal. Depending on the physical structure of the gate region, the FETs are classified as junction-gate (JFET) or insulated-gate (IGFET) devices.

44. Junction-Gate Field-Effect Transistor (JFET). *Principle of Operation.* Figure 8-36 shows the cross section of an integrated JFET with an *n*-type channel region. In such a device structure, reverse bias is applied to the gate-channel depletion layer and causes the layer to extend into the channel region. This modulates the effective width of the conductive path

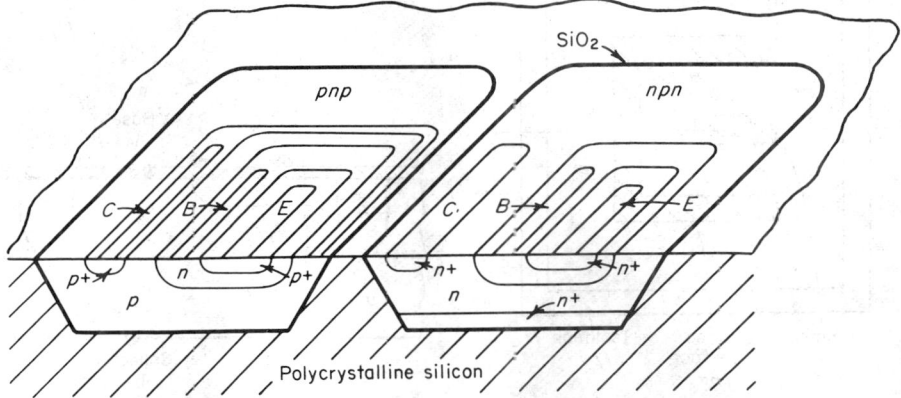

Fig. 8-35. A high-performance complementary bipolar transistor structure using dielectric isolation.

between the source and the drain. Assuming that the impurity concentration in the gate region is much higher than that of the channel and that the channel has a uniform n-type impurity concentration N_D, the thickness d of the gate-channel depletion layer can be related to the reverse bias V_{GC} across the gate-channel junction as

$$d \approx (2\epsilon V_{GC}/qN_D)^{1/2}$$

where ϵ is the dielectric constant of silicon and q is the electronic charge.

The reverse bias across the gate-channel junction, which would cause d to extend into the channel to deplete, or *pinch*, the entire channel, is known as the *pinch-off voltage* V_p. For a uniformly doped channel having a half-width a (see Fig. 8-36), V_p can be written

$$V_p = qN_Da^2/2\epsilon$$

At low values of drain voltage, the JFET operates as a voltage-controlled resistor, with an effective source-drain resistance R given as

$$R = R_0(1 - V_G/V_p)^{-1/2}$$

where R_0 is the bulk resistance of the channel with zero depletion layer.

At any given gate bias $V_G > V_p$, as the drain voltage is increased, the drain current also increases, thus increasing the ohmic drop along the channel. At any point along the channel, this voltage drop adds to the net bias across the gate-channel interface, thus causing the depletion region to extend farther into the channel, in the vicinity of the drain.

Consequently, for drain voltages in excess of the pinch-off voltage, a space-charge region is formed near the drain end of the channel. This space-charge layer then causes I_D to reach a saturation level and be relatively insensitive to the further increase of the drain potential. This is known as the pinched operation of the FET, where the device functions as a voltage-controlled current source. For operation below the drain-current saturation (that is, $V_D > V_p - V_G$), the drain-current-voltage characteristics can be approximated as

$$I_D = \frac{2azV_p}{3\rho L}\left(\frac{3V_D}{V_p} - \frac{2V_D + V_G}{V_p} + \frac{2V_G}{V_p}\right)^{3/2}$$

where z is the dimension of the channel measured normal to a and L.

Figure 8-37 shows the typical I_D-vs.-V_D characteristics of an n-channel JFET, with the gate bias V_G as a parameter. The region of validity of the equation for I_D corresponds to the area to the left of the dashed line, where the net voltage across the gate-drain junction is less than the pinch-off voltage.[31] For higher values of the drain voltage, the first-order theory predicts a total saturation of I_D. However, in practical devices, I_D still exhibits a slight increase with increasing V_D, thus leading to a nonzero value of dynamic drain conductance g_d. This effect comes about due to the modulation of the effective channel length L by the space-charge region near the drain, in a manner analogous to the base-width modulation effect in bipolar transistors.[32]

The saturation value of the drain current I_{DS} can be expressed as

$$I_{DS} \approx \frac{V_p}{3R_0}\left[1 - \frac{V_G}{V_p} + 2\left(\frac{V_G}{V_p}\right)^{3/2}\right]$$

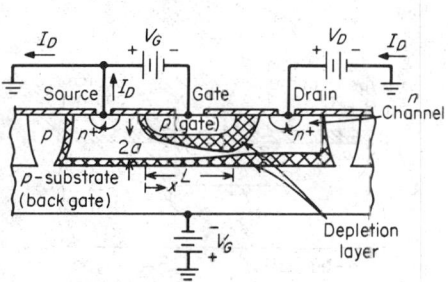

Fig. 8-36. Structural diagram of an n-channel JFET. (Crosshatched areas denote the gate-channel depletion layer.)

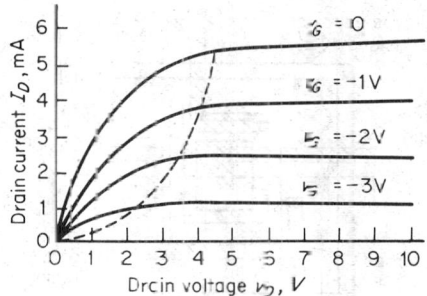

Fig. 8-37. Current-voltage characteristics for a typical JFET.

Similarly, the device transconductance g_m for pinched operation can be written

$$g_m = \frac{\partial I_D}{\partial V_G} = \frac{1}{R} = \frac{1 - \sqrt{V_G/V_p}}{R_0}$$

45. JFET High-Frequency Characteristics. The frequency performance of the JFET can be closely approximated by the ac equivalent circuit of Fig. 8-38 for device operation in its pinched region. In the figure, R_s represents the parasitic bulk resistance in series with the source contact. For typical integrated JFET structures (see Fig. 8-39) R_s is of the order of 50 to 80 Ω. C_{gs} and C_{gd} are the gate-source and gate-drain capacitances; g_d is the dynamic output conductance due to channel-length modulation effects. In the device layout, the drain area is made as small as possible to minimize the C_{gd}, because this capacitance provides parasitic coupling between the drain and the gate terminals and reduces the frequency capability of the FET in a manner similar to C_c of bipolar transistors.

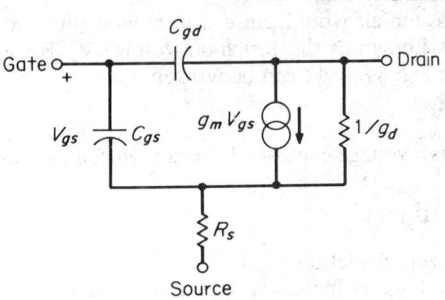

Fig. 8-38. AC small-signal equivalent circuit for a JFET.

A useful figure of merit for high-frequency capability of an FET is its transconductance cutoff frequency f_c, which is inversely proportional to the total capacitance C_g, seen looking into the gate terminal. For a uniform-channel device with a channel resistivity ρ_c, f_c can be related to the device dimensions as

$$f_c = \frac{g_m}{2\pi C_g} = \frac{\rho_c}{2\pi}\left(\frac{a}{L}\right)^2$$

In a typical integrated JFET structure with a 5 $\Omega \cdot$cm channel resistivity, having a channel half-width of 2.5 μm and a length of 10 μm, f_c is of the order of 150 to 200 MHz.

46. JFET Device Structures. For monolithic integrated circuits, the most useful JFET structures are those which are compatible with the npn bipolar technology and can be fabricated simultaneously with npn transistors. Figure 8-39 shows the layout and the structural diagram of an n-channel JFET which uses the n-type epitaxial collector region of the npn transistor as the channel of the FET. Similarly, the p-type base diffusion of the npn transistor is used to form the control gate with the source and the drain contacts formed by the n^+ emitter diffusions for the npn transistor.

To obtain a narrow channel width without degrading the npn bipolar breakdown characteristics, a p^+ subepitaxial layer is diffused under the FET-gate region, as well as under the isolation walls. Then, during the isolation diffusion, this subepitaxial p^+ layer outdiffuses into the epitaxial layer and reduces the effective channel width of the n-channel FET. In such a structure, the p-type substrate is also a part of the FET gate. However, since the substrate is common to the rest of the circuit, it cannot be used as a control terminal. Therefore only the top gate func-

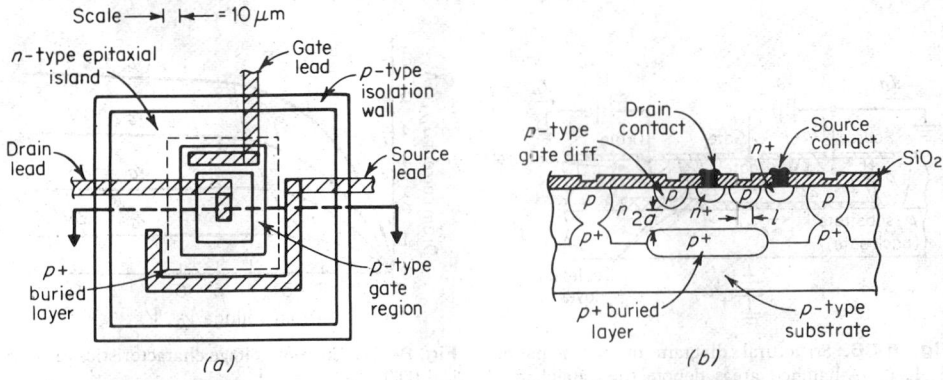

Fig. 8-39. Layout and structural diagram of a typical integrated n-channel JFET.

tions as the control electrode, and the bottom gate is at all times connected to a fixed negative potential, with respect to the rest of the device.

Figure 8-40 shows two additional JFET structures which are also compatible with the npn bipolar technology. The structure of Fig. 8-40a is a p-channel device with a diffused channel region. The p-type channel region can be formed either by npn base diffusion or by an additional p-diffusion step, resulting in a deeper junction structure with higher gate-channel breakdown voltage.

The device structure of Fig. 8-40b is particularly suitable for high-frequency applications and uses dielectric isolation. The gate and the source and drain contacts are again formed by the npn base and emitter diffusions. The channel is formed by a wedge-shaped groove etched under the gate region.

One of the major drawbacks of JFET devices is the strong dependence of device parameters on channel geometry, and particularly on the channel half-width a. Table 8-9 lists the dependence of some of the significant JFET parameters on channel dimensions. Due to tolerances associated with the epitaxial-growth and diffusion steps, the absolute tolerance associated with the channel width is of the order of $\pm 15\%$. This can result in absolute tolerances of ± 25 and $\pm 40\%$ for V_p and I_{DS}.

Table 8-10 lists some of the typical parameter values and tolerances associated with the JFET structure of Fig. 8-39, assuming a channel resistivity of 5 $\Omega \cdot$ cm and a channel half-width of 2.5 μm and a substrate bias of -6 V.

BIFET Technology.   BIFET technology[33] is the merging of JFET device structures using bipolar technology. A typical p-channel JFET compatible with bipolar technology is shown in Fig. 8-41. The p-type source and drains are produced by diffusing the bipolar base into an n-type epitaxial island, which serves as the bottom gate. The n epitaxial material is also isolated elsewhere

Table 8-9.   Dependence of Junction-Gate FET Parameters on Channel Dimensions

Device parameter	Dependence on channel dimensions	
	Half-width a	Length L
Saturation current I_{DS}	a^3	$1/L$
Transconductance g_m	a	$1/L$
Cutoff frequency	a^2	$1/L^2$
Drain conductance beyond pinch-off, g_d	a^3	$1/L^2$
Pinch-off voltage V_p	a^2	
"On" resistance R_{on}	$1/a$	$\dfrac{L}{}$
Input capacitance C_{in}		L

TABLE 8-10   Typical Device Parameters and Tolerances Associated with the JFET Structure of Fig. 8-39, with $\rho_c = 5 \ \Omega \cdot$ cm and $a = 2.5 \ \mu$m and Substrate Bias of -6V

Device parameter	Typical value	Absolute tolerance	Matching tolerance	Temperature coefficient, ppm/K
Pinch-off voltage V_p	5 V	± 1.2 V	± 0.15 V	-600
Drain saturation current I_{DS}	4 mA	± 2 mA	± 0.5 mA	$-4,000$
Transconductance g_m	900 μS	$\pm 20\%$	$\pm 8\%$	$-4,000$
"On" resistance R_{on}	500 Ω	$\pm 15\%$	$\pm 5\%$	$+4,000$

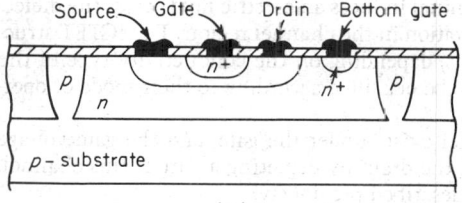
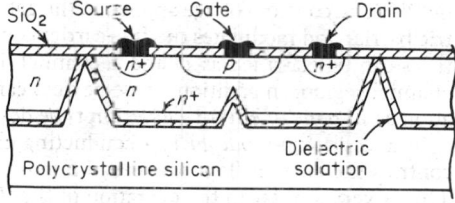

(a)   (b)

Fig. 8-40. Other possible JFET structures compatible with monolithic ICs.

for other bipolar functions such as collector regions of *npn* transistors, base regions of lateral *pnp*-transistors, etc. The actual *p* channel is formed by ion implanting *p*-type material over the entire source-drain area at an appropriate subsurface level. This technique connects the channel to the source and drain and permits the use of a lightly doped *n*-type ion-implanted top gate over the channel surface. This top gate enhances transconductance of the device by also depleting

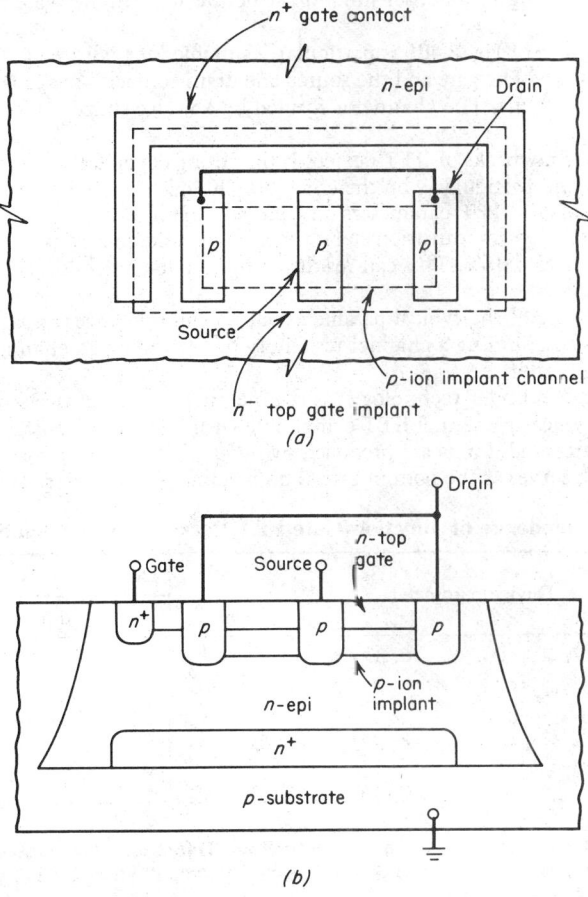

Fig. 8-41. Ion-implanted JFET: (*a*) lateral geometry; (*b*) cross section.

the channel from the upper junction region and improves frequency response by reducing the series gate resistance. Placement of the lightly doped top gate over the source-drain area is not critical since the richly doped *p*-type material will not invert. The n^+ buried layer, located under the device in the n epitaxial material, is used to reduce the series resistance of the bottom gate. Various channel width-to-length ratios Z/L can be provided to optimize either transconductance g_m or drain-source saturation current I_{DSS}.

47. Insulated-Gate Field-Effect Transistor (IGFET). *Principle of Operation.* In the insulated-gate FET structure a thin dielectric barrier is used to isolate the gate and the channel.[34,35] The control voltage applied to the gate terminal induces an electric field across the dielectric barrier and modulates the free-carrier concentration in the channel region. The IGFET structures can be classified as *p*- and *n*-channel devices, depending on the conductivity type of the channel region. In addition, these devices can also be classified according to their mode of operation as *enhancement-* or *depletion-type* devices.

In a *depletion-mode FET*, a conducting channel exists under the gate, and the gate voltage controls the current flow between the source and the drain by depleting a part of this channel. This is very similar to the operation of the JFET described previously.

In the *enhancement-mode IGFET*, no conductive channel exists between the source and the drain applied gate voltage. As a gate bias of proper polarity is applied and increased beyond a threshold value V_T, a localized inversion layer is formed directly below the gate, which serves as a conducting channel between the source and the drain electrodes. If the gate bias is increased further, the resistivity of induced channel is reduced and the current conduction from the source to the drain is enhanced.

Figure 8-42 shows a cross-sectional diagram of a p-channel enhancement-mode IGFET. The particular polarities of the gate and the drain bias for the proper operation of the device are also identified on the figure. Silicon dioxide (SiO_2) is the most common gate dielectric, but other types of gate-dielectric materials such as silicon nitride (Si_3N_4) or aluminum oxide (Al_2O_3) are also used in a number of device structures. The thickness of the gate dielectric is usually much less than the oxide layers normally used for masking or surface passivation.

The enhancement-mode IGFET is preferred over its depletion-mode counterpart because it is a *self-isolating* device, does not require tight control of diffusion cycles, and can be fabricated by a single diffusion step forming the source and the drain pockets. Since all the active regions of the IGFET are reverse-biased with respect to the substrate, adjacent devices fabricated on the

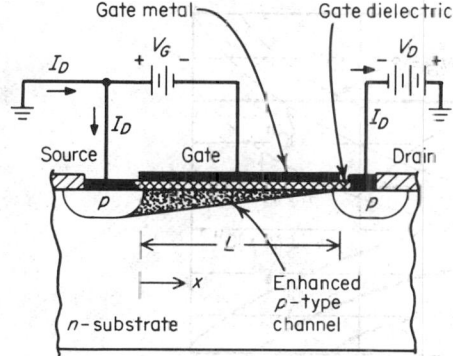

Fig. 8-42. Cross section of a p-channel enhancement-mode IGFET.

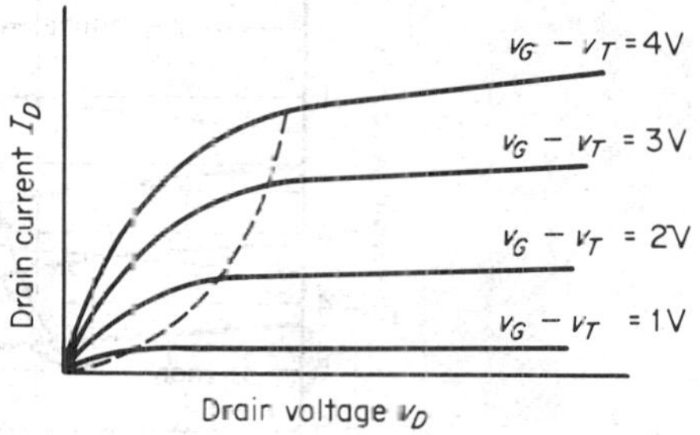

Fig. 8-43. Typical current voltage characteristic of a p-channel enhancement-mode IGFET.

same substrate are electrically isolated, without requiring a separate isolation diffusion. Because of this advantage, IGFET devices offer a much higher packing density per unit area of silicon surface than the bipolar transistors.

If the drain voltage is increased in the polarity shown in Fig. 8-42, a finite drain current I_D flows through the induced channel. This current also causes an ohmic drop along the channel which subtracts from the net gate voltage ($V_G - V_T$). Since the apparent sheet conductance at any point along the channel is proportional to this net gate voltage, the voltage gradient due to I_D causes the channel to deplete along its length. Thus an increase of V_D causes the drain current to increase, which in turn causes the channel to deplete, or pinch off, near the drain For values of the drain voltage in excess of the net gate bias, a space-charge layer is formed at the drain end of the channel, since the net gate bias, i.e., the applied gate voltage minus the voltage drop along the channel, at this point is no longer sufficient to maintain an induced channel. This leads to a saturation of drain current similar to the case of JFET, for values of $V_D \geq V_G - V_T$, and results in a set of drain-current–vs.–drain-voltage characteristics shown in Fig. 8-43.

The threshold voltage V_T is one of the key design parameters in IGFET structures, since it appears in most of the device equations. V_T is a complex function of the gate capacitance, the Fermi level in the silicon substrate, the work-function difference between the gate conductor and silicon, and excess charge built up at the silicon-dielectric interface. The excess charge is created by the presence of surface states at the semiconductor-dielectric interface.[35] Figure 8-44 shows a plot of the threshold voltage as a function of impurity concentration for various values of surface state concentration N_{SS}. For illustration purposes, a gate oxide thickness of 100 nm with aluminum gate electrode is assumed.[36] The threshold voltage decreases linearly with decrease of work-function difference between the gate electrode and silicon. For example, V_T would be reduced by approximately 1.5 V if polycrystalline silicon, rather than aluminum, were used as the gate electrode. This has led to the development of so-called *silicon-gate* IGFET structures, which offer significantly lower threshold (~ 1.2 V) compared with conventional devices ($V_T \approx 3$ V).

48. IGFET High-Frequency Characteristics. For small-signal operation, the frequency performance of an IGFET can be closely approximated by the equivalent circuit of Fig. 8-45 for $V_D > V_G - V_T$. It should be noted that except for the source-substrate and the drain-substrate capacitances C_{ss} and C_{ds} this equivalent circuit is identical with that of Fig. 8-38. Normally, the IGFET is used in the grounded-source configuration, with the substrate either ac-grounded or short-circuited to the source. In this latter case, C_{ss} is short-circuited out, and C_{gs} appears between the drain and the ground terminal. As in the JFET case, C_{gd} and C_{gs} refer to the components of the gate capacitance associated with the source and the drain electrodes. The drain output conductance g_d is due to the modulation of the effective channel length L by the drain space-charge layer width. Although not shown explicitly in Fig. 8-45, the parasitic bulk resistances R_S and R_D are still present in the IGFET devices and have the same degradation effects on the device performance as discussed earlier in connection with junction-gate FET transistors.

To optimize the high-frequency performance of an IGFET, it is necessary to reduce the channel length L and to minimize the parasitic *gate overlap capacitance* due to physical overlap of

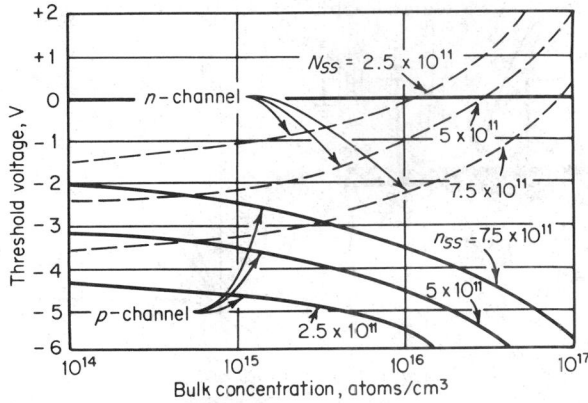

Fig. 8-44. Threshold voltage vs. bulk doping level and surface density N_{ss}.

gate dielectric over the source and drain regions. In practical MOS structures, $L \geq 5 \ \mu m$, due to gate alignment tolerances and punch-through breakdown limitations between the source and the drain.

49. IGFET Integrated Device Structures. The most commonly used IGFET structure is the p-channel enhancement-mode device. Figure 8-46 shows a typical layout and structural cross section for such a device. The resistivity of the n-type substrate is in the range of 5 to 8 $\Omega \cdot cm$. In the fabrication of the device, p-type beds, forming the source and the drain, are diffused first. The separation between these beds defines the effective channel length L. In considering the final dimensions of the channel, the side diffusion of p-type source and drain pockets must be taken into consideration. After the pocket diffusion, a relatively thick oxide layer (~ 1 to $1.4 \ \mu m$) is grown over the device surface. This oxide is later etched to form the contact areas to the device.

The thin gate dielectric can be formed by either etching back the oxide layer on the gate to the desired thickness (~ 100 to 150 nm) or by etching and regrowing a new layer of gate dielectric. In certain applications other gate dielectrics such as silicon nitride (Si_3N_4) or aluminum oxide (Al_2O_3) can also be used in conjunction with SiO_2 to form a multilayer gate structure. The

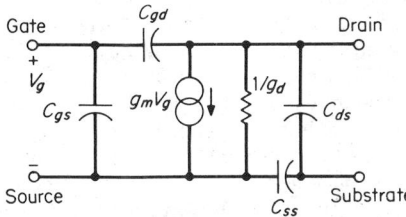

Fig. 8-45. AC equivalent circuit for an IGFET.

device structure of Fig. 8-46 is then completed by depositing the metal interconnections and the gate electrode. In an enhancement-mode IGFET, the gate electrode must overlap the source and the drain pockets to form a continuous channel.

The crystal orientation of n-type starting material has a significant effect on the threshold voltage of the device. A <100>-oriented silicon crystal exposes a smaller number of incomplete interatomic bonds at the crystal surface than the <111>-oriented crystal. This reduces the charge accumulation at the silicon-dielectric interface and reduces the threshold voltage V_T. Typical ranges of values of V_T for different device structures are listed in Table 8-11.

Compared with bipolar transistors, enhancement-mode IGFET structures have two distinct advantages for monolithic fabrications: (1) They are self-isolating; i.e., the adjacent devices on the same substrate are electrically isolated without requiring separate isolation pockets. (2) They can be fabricated without requiring any critical diffusion step.

Complementary IGFET devices can be fabricated on the same monolithic substrate, with only a moderate increase in process complexity.[36,37] Figure 8-47 shows the structural cross section of a pair of n- and p-channel IGFET devices on the same silicon substrate. In fabricating complementary IGFET structures, a relatively deep (~ 5 to 7 μm) p-type island, or pot, is diffused into the n-substrate. This p-type pot forms the isolated island or which n-channel IGFET is later formed. Following the pot diffusion, the p- and n-type beds are diffused to form the source and drain regions of respective devices; and finally, the gate dielectric layer is grown or deposited, and the aluminum interconnections are deposited to complete the structure.

An alternative device structure which offers a low-threshold voltage and higher packing density than conventional IGFETs is the *silicon-gate* IGFET. In such a device structure, p-type doped polycrystalline silicon layer is used as the gate electrode, instead of aluminum. This reduces the work-function difference between the gate electrode and the silicon surface and

Table 8-11. Typical Threshold-Voltage Ranges for Various IGFET Structures

Device type	Crystal orientation	Threshold voltage range, V
Conventional p-channel	$\langle 111 \rangle$ $\langle 100 \rangle$	3.5-5 2.0-3.0
Silicon-gate p-channel	$\langle 111 \rangle$ $\langle 100 \rangle$	1.5-2.0 0.4-1.2

decreases V_T.[38] The sequence of process steps in fabricating the silicon-gate FET is shown in Fig. 8-48. The polycrystalline silicon layer forming the gate electrode has very low sheet resistivity. Therefore it also works as an additional layer of interconnection between the devices. This allows the active devices to be placed close to each other on the chip, without metal-crossover problems.

Ion implantation also provides an added degree of flexibility to IGFET fabrication.[39,40] Figure 8-49 shows the cross section of an ion-implanted-device structure. In this type of structure, the source and drain beds are located relatively far apart, and the actual channel length L is defined at a later step, by ion implantation. After the gate dielectric is grown, aluminum gate electrode is deposited and etched to define the gate region. Then a p-type layer is implanted through the thin gate dielectric. During this implantation step, the aluminum gate electrode works as a mask; therefore the region directly below the gate is not implanted. In this manner, the channel is self-defined by the outline of the gate metal. Since no mask alignment is necessary in defining the channel, this is known as a *self-aligning* structure.

VMOS Devices.   Short-channel silicon MOS field-effect devices can be fabricated by using an

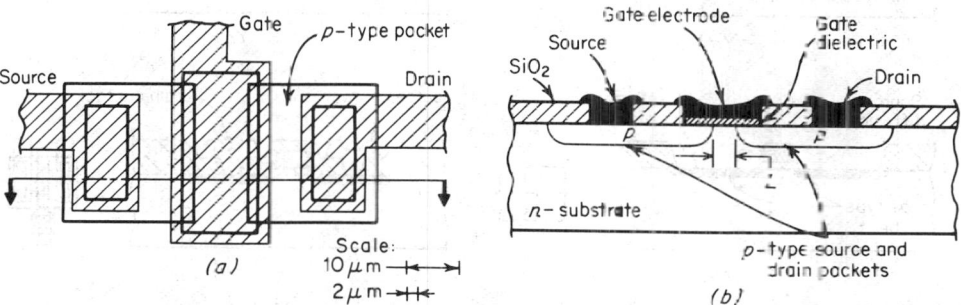

Fig. 8-46. Lateral geometry and cross section of a p-channel enhancement-mode IGFET (vertical dimensions not to scale).

anistropic etch such as potassium hydroxide to etch a V groove through an *npn* or a *pnp* structure in <100> silicon. The resulting groove is defined by <111> planes and is inclined 54.7° to the horizontal. An *n*-channel device is shown in Fig. 8-50.

The device characteristics of VMOS devices are similar to those of other short-channel MOS devices. The velocity-saturation effect of mobility due to high electric fields in the short channel causes the drain current in the saturation region to be proportional to the gate voltage rather than the gate voltage squared.[41] In VMOS transistors if the drain region is lightly doped, as it would be for high breakdown, the carrier concentration can be enhanced by the portion of the gate extending into the drain, thus affecting ON resistance. The lower ON resistance per unit area of VMOS devices compared with long-channel devices allows them to be used as high-current switches although the voltage drop is not as low as for bipolar devices. VMOS transistors have the following advantages over bipolar devices used as power switches:

1. No stored charge and therefore a rapid turn-off characteristic

2. A negative temperature dependence of output current which eliminates thermal runaway

3. No second breakdown

4. High input impedance and therefore high current gain

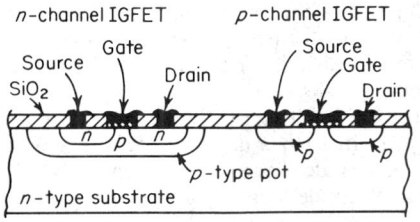

Fig. 8-47. Complementary IGFET structures on a monolithic substrate.

Charge-Coupled Devices. The charge-coupled device (CCD) is a unique and versatile semiconductor structure,[42,43] invented in 1969 by W. S. Boyle and G. E. Smith, of Bell Telephone Laboratories. Circuits using CCDs are truly integrated since these devices cannot be breadboarded using discrete components. The basic concept of a CCD involves shifting mobile charge (minority carriers) between temporary storage sites (potential-energy wells) at (or somewhat beneath) the semiconductor surface. Figure 8-51 is a repeating cross-sectional slice of one type of a CCD, the *surface-channel CCD.* In this structure charge moves along the silicon surface (at the silicon–silicon dioxide interface) under the control of the ϕ_1 and ϕ_2 clocks in the following manner. With ϕ_1 high and ϕ_2 low, any existing signal charge (electrons for a *p*-type substrate) resides under those ϕ_1 electrodes with the thinner associated dielectric (the *storage plates*). This is because the *surface potential* (the electrostatic potential of the silicon surface with respect to the silicon bulk) is more positive at these locations due to the higher applied voltage and the larger value of gate oxide capacitance. No change occurs when ϕ_2 switches to the high level, but when ϕ_1 subsequently drops, the signal charge moves to the right and relocates beneath the neighboring ϕ_2 storage plates due to the effects of self-induced drift, diffusion, and fringing fields. Movement to the right is ensured by using thicker oxide beneath the *transfer plates* so that a potential barrier prevents reverse flow.

The cross section of Fig. 8-51 shows the repeating portion of a CCD shift register but does not

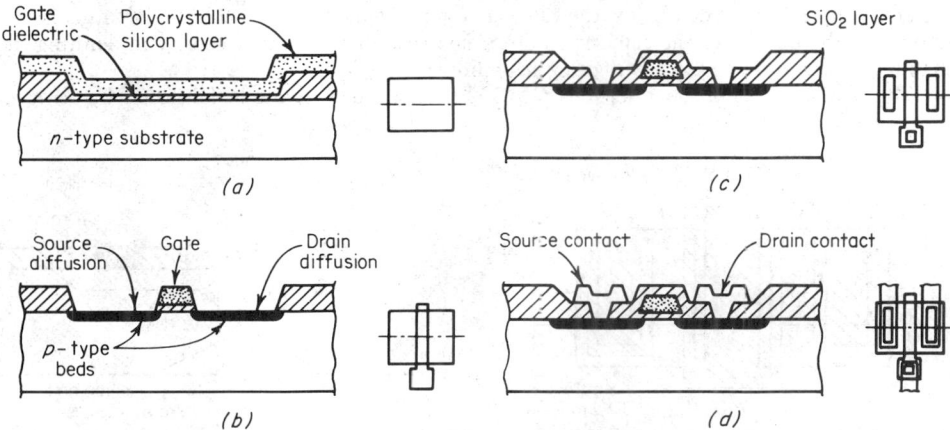

Fig. 8-48. Sequence of process steps in the fabrication of a silicon-gate IGFET: (*a*) gate oxide and polycrystalline silicon are grown; (*b*) source and drain beds are diffused; (*c*) source and drain contacts are defined; (*d*) aluminum interconnections are deposited and etched.

show the two ends. Signal charge is injected at the input end by using an n diffusion as the source and several gates as the launching control. An n diffusion is also used as the sensing means at the output of the shift register. There are a number of variations of the surface channel CCD of Fig. 8-51, including different clocking approaches and the use of ion implantation to create or augment the potential barrier. One of the key parameters of a CCD is its *transfer efficiency*, which measures the completeness of the charge transfer from one site to the next. Since the interaction of the signal charge with interface states tends to reduce transfer efficiency, the

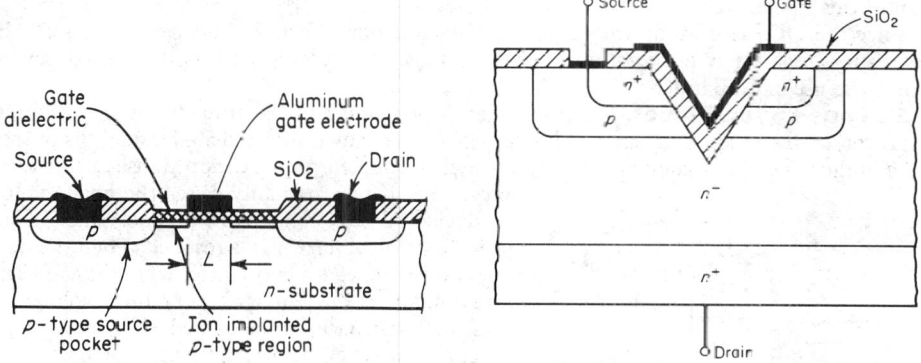

Fig. 8-49. Ion-implanted IGFET.

Fig. 8-50. An n-channel VMOS device.

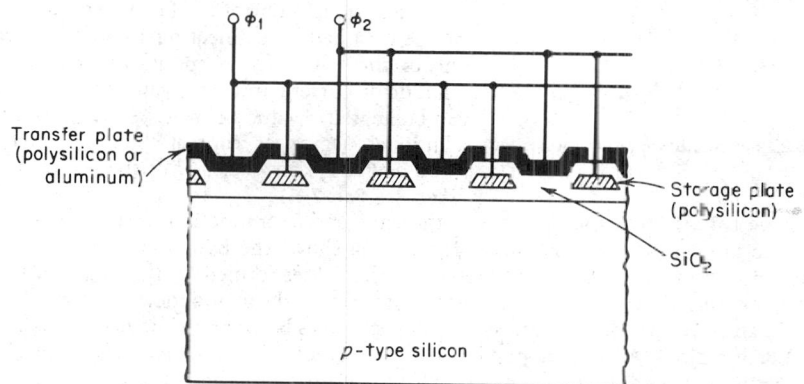

Fig. 8-51. Cross-sectional slice of a surface-channel CCD.

buried-channel CCD was created to move the temporary storage sites away from the surface and into the semiconductor bulk. This is accomplished by using a thin ion-implanted or epitaxial n-type layer on the p-type substrate. The resulting improvement in transfer efficiency occurs at the expense of a somewhat reduced charge-handling capability.

One of the interesting aspects of CCDs is the wide variety of applications for which they are used. Although it now appears that their use in mass memories will not be as extensive as originally thought, they still are quite useful for certain classes of storage where a fast serial data rate is required. They can also implement logic operations although again this use is limited. Their most promising areas of application are as imaging and analog devices. As imagers they can be area or linear sensors for video camera or facsimile equipment; their analog applications range from electronically variable delay lines to transversal filters.

OTHER ACTIVE DEVICES

50. Unijunction Transistors (UJT). The unijunction transistor (UJT) is a single junction device with two ohmic base contacts. Figure 8-52 shows a simplified diagram of a UJT, along with its IC realization. The device exhibits a voltage-controlled negative-resistance characteristic as a result of the conductivity modulation within the high-resistivity base region. For a given base bias V_B, negligible anode current I_i flows until the anode terminal is raised to a sufficiently positive voltage level V_p to cause the junction to be forward-biased. The minority carriers injected into the base region cause the effective value of R_A to decrease, due to conductivity modulation, thus further forward-biasing the junction. This results in a negative-resistance characteristic, where I_j increases rapidly, without requiring V_j to rise, and the device switches to its ON state. Similarly, it can be returned to its stable OFF state by reducing the anode voltage to below V_B. The voltage-controlled negative-resistance characteristics of the UJT make it useful for relaxation-oscillator application. However, due to excessive charge-storage effects in the high-resistivity base region, the switching speed of the UJT is several orders of magnitude lower than that of the bipolar transistor.

Since the UJT depends on the conductivity-modulation effects for its operation, its *IV* characteristics are strongly temperature-dependent. This effect, along with its slow speed, severely limits the use of UJT in ICs.

51. Four-Layer Diodes. The four-layer diode *(pnpn)* and its three-terminal version, the thyristor, each contain four semiconductor regions. As shown in Fig. 8-53, for analysis purposes a *pnpn* diode can be decomposed into a set of cross-coupled *pnp* and *npn* transistors. It can be shown that if one gradually raises the anode voltage, the device initially remains in a nonconductive state, until a turn-on voltage V_{BO} is reached. Then the device suddenly switches to its ON state, where it can carry a large amount of current with very little voltage drop across it. The device can be turned off by reducing the current to a level below the critical current level, known as the *holding current*. This is shown as the current level I_H in Fig. 8-53.

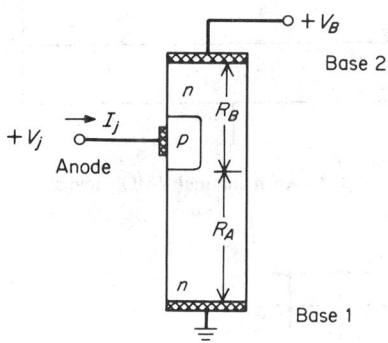

Fig. 8-52. Schematic representation of a UJT.

An additional control of the turn-on characteristics can be obtained by connecting a *gate* electrode to any one of the two center regions. By injecting a small amount of current into this gate, the turn-on voltage can be kept at a value below V_{BO}, giving the device an additional degree of control. Such a three-terminal *pnpn* device is known as a *thyristor*, or *controlled rectifier*. See Par **7-51**.

52. Integrated Diodes. Any one of the semiconductor junctions forming the monolithic circuit structure can be used as a diode. Figure 8-54 shows the basic diodes associated with an integrated *npn* transistor. D_{BE} and D_{BC} represent the diodes formed by the base-emitter and base-collector junctions; D_{CS} is the collector-substrate diode in junction-isolated circuits. The resistors R_b and R_{ct} represent the parasitic bulk resistances between the device terminals and the actual diode junctions. A parasitic *pnp* transistor is formed when the diode D_{BC} is forward-biased and D_{CS} is reverse-biased. The current gain of this parasitic *pnp* can be greatly reduced by using an n^+ buried layer between the diode junction and the p-type substrate.

The diode current I_D is exponentially related to the voltage V_D applied across the diode as

$$I_D = I_0 (e^{qV_D/kT} - 1)$$

where I_0 is the reverse saturation current and is proportional to the junction area. For a typical integrated diode the forward (I_D, V_D) characteristics predicted by this equation are valid over six

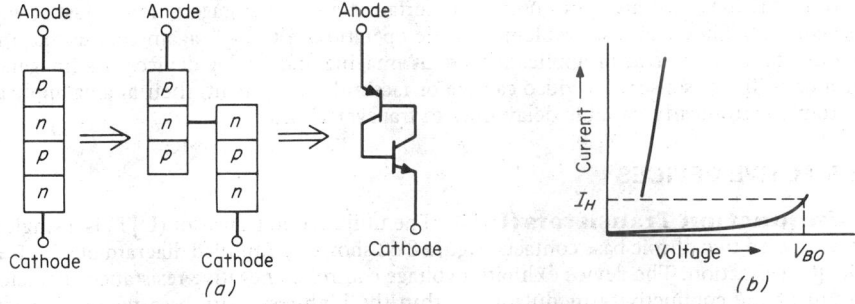

Fig. 8-53. The four-layer diode: (a) transistor equivalent circuit; (b) current-voltage characteristics.

orders of magnitude in current. For a 1-mil^2 base-emitter junction area, this range covers current values of 10 nA to 10 mA.

Under forward bias, the junction diode exhibits a differential conductance g_d given as

$$g_d = \partial I_D/\partial V_D = q I_D/kT$$

The diode forward voltage V_D shows a strong negative temperature coefficient. This thermal drift is a highly predictable and repeatable effect. For the case of the base-emitter diode, which is the most commonly utilized diode connection, the temperature coefficient of V_D falls within the narrow range of -1.9 to -2.1 mV/°C.

Only two of the diodes, D_{BE} and D_{BC}, associated with the npn structure of Fig. 8-54 are readily suitable for circuit applications. The collector-substrate diode D_{CS} is not as useful, since its cath-

Table 8-12. Typical Characteristics of Integrated Circuit Diodes

Diode connection	Series resistance	Reverse breakdown	Storage time ($I=2$ mA)	Parasitic pnp
	Low ($\approx r_b$)	Low ($\approx 6 - 9$ V)	High (≈ 60 ns)	No
	Low ($\approx r_{cs} + r_b$)	Low ($\approx 6-9$ V)	Low (≈ 15 ns)	No
	High ($\approx r_b + r_{cs}$)	High > 30 V	High (≈ 30 ns)	Yes
	High ($\approx r_b + r_{cs}$)	High > 30 V	High (≈ 50 ns)	Yes
	High ($\approx r_b + r_{cs}$)	Low ($\approx 6 - 9$ V)	High (≈ 100 ns)	Yes

ode, the substrate, is common to the rest of the circuit. The semiconductor junctions which make up the npn bipolar transistor can be interconnected as a diode in any one of the five possible configurations shown in Table 8-12. The series bulk resistances associated with each diode are in general the most significant parasitics. In this table these parasitic resistances are expressed in terms of the parameters of the npn bipolar equivalent circuit of Fig. 8-30. For most circuit applications where the low reverse breakdown is not a problem, the diode connection is preferred since this configuration has the least series resistance and the lowest storage time.

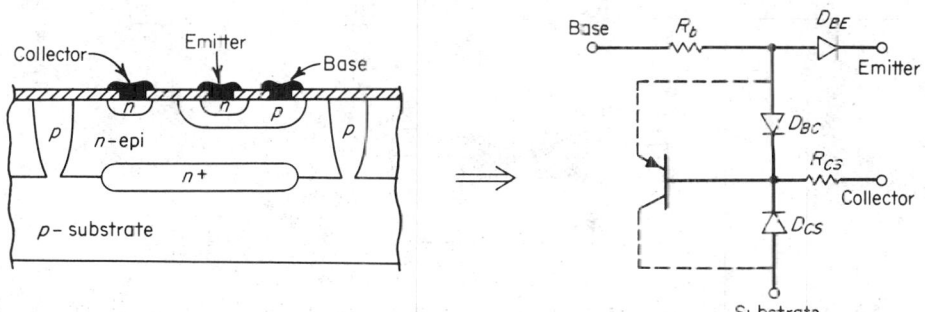

Fig. 8-54. Possible diodes available in a monolithic npn structure.

53. Avalanche Diodes. The avalanche-breakdown characteristics of junctions can be used for voltage reference or dc level-shift purposes. Typical breakdown voltages associated with each of the five diode connections are listed in Table 8-12. The base-emitter breakdown, which falls within the 6- to 9-V range, is the most commonly used avalanche diode because its breakdown voltage is compatible with the voltage levels available in most ICs. Each of these diodes has a reverse-breakdown resistance approximately equal to r_b of Fig. 8-30.

The avalanche-breakdown voltage BV_{EB} associated with the base-emitter junction shows a typical positive temperature coefficient in the range of 2 to 5 mV/°C. Since the thermal drifts of the diode forward voltage V_D and BV_{EB} are in opposite directions, it is possible to partially compensate the thermal drift of an avalanche-breakdown diode by connecting a forward-biased diode in series with it.

54. Schottky Barrier Diodes. When a metal and a semiconductor material are brought into contact, an electrostatic barrier is formed at the interface which causes the metal-semiconductor interface to have rectifying properties.[44] Fig-

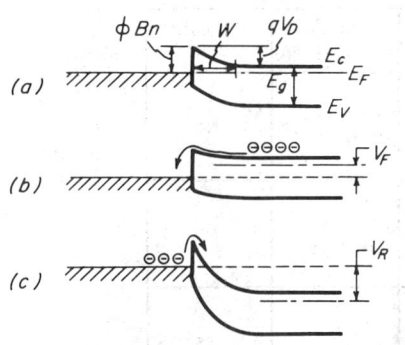

ure 8-55 shows the thermal-equilibrium diagrams of Schottky barriers on n- and p-type semiconductors, where E_c and E_v are the conduction and the valence band edges, E_g is the band gap, and E_F is the Fermi level. The bending of the energy bands at the interface occurs in such a manner as to retard the flow of majority carriers into the metal. Thus, in the case of the n-type semiconductor, an energy barrier of height ϕ_{Bn} is present at the contact. It is found that the barrier height is essentially independent of semiconductor doping level.

A metal-semiconductor junction having an energy diagram similar to that of Fig. 8-55 acts as a rectifying junction, with the metal as the anode. When a forward bias V_F is applied to the interface, the barrier height is lowered and current flows freely. If a reverse voltage V_R is applied, the barrier height increases and the current flow is blocked.

Fig. 8-55. Energy-band diagrams for a Schottky barrier junction under various bias conditions: (*a*) thermal equilibrium; (*b*) forward bias; (*c*) reverse bias.

Compared with conventional pn junction diodes, Schottky barrier-diode characteristics have the following significant differences:

1. In Schottky barrier diodes current flow is by majority carriers, rather than by minority-carrier diffusion. Thus switching speeds of Schottky diodes are not limited by storage-time delays.

2. For a given diode area and forward-current level, Schottky diodes have a smaller forward-voltage drop V_F. (For a typical forward current of 10 μA, V_F is about 0.25 V for a Schottky diode and about 0.55 V for a comparable pn junction diode.)

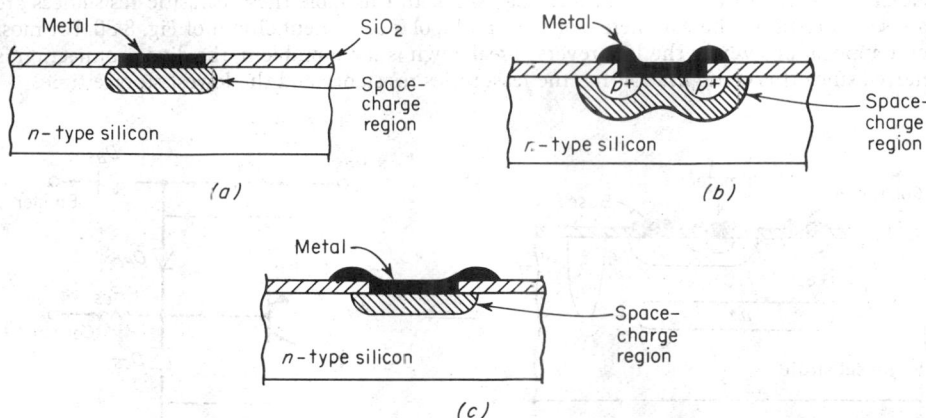

Fig. 8-56. Three basic Schottky barrier structures: (*a*) without metal overlap; (*b*) with p-guard ring; (*c*) with metal overlap.

These two properties of Schottky barrier diodes make them well suited for various switching or clamping applications.

Figure 8-56 shows typical Schottky diode structures suitable for ICs.[45] The structure of Fig. 8-56a is the simplest diode structure, formed by placing a metal such as aluminum in contact with lightly doped n-type silicon ($N_D < 10^{17}$ atoms/cm^3). The positive fixed surface charge at the silicon surface causes the space-charge region to narrow along the periphery of the diode. This results in leaky, or "soft," reverse characteristics for the diode. In the diode structure (Fig. 8-56b) this problem is avoided by putting a p^+ guard ring around the diode, but the ring adds additional shunt capacitance to the diode.

Schottky barrier diodes can be fabricated simultaneously with ohmic contacts by using conventional aluminum evaporation or sputtering techniques. The nature of the contact (ohmic or rectifying) is determined by the resistivity of the semiconductor region directly below the metal.

The most important application of Schottky diodes in ICs is to perform clamping functions for npn bipolar transistors. In this application, the Schottky barrier diode is connected in parallel with the collector-base junction of an npn transistor to prevent the transistor from going into

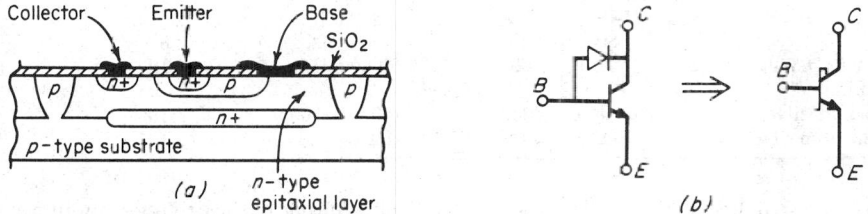

Fig. 8-57. Schottky-clamped npn transistor: (a) structural diagram; (b) equivalent circuit.

heavy saturation and thereby reducing the switching time. Figure 8-57 shows the structural diagram and the electrical equivalent of a Schottky clamped npn transistor. The Schottky diode is incorporated in the structure with little increase in device size or processing complexity.

RESISTORS IN INTEGRATED CIRCUITS

55. Integrated Circuit Resistors. IC resistors can be classified into two general categories: (1) semiconductor resistors and (2) deposited (film) resistors. The semiconductor resistors make use of the bulk resistivity of doped semiconductor regions to obtain a desired resistance value. Depending on their structure, semiconductor resistors can also be subdivided into four groups: diffused, bulk, pinched, and ion-implanted resistors. Deposited, or film, resistors are formed by depositing resistive films on an insulating substrate. These films are then etched and patterned to form the desired resistive network. Depending on the thickness and the dimensions of the deposited films, film resistors are classified into two groups: thin- and thick-film resistors.

To obtain a given resistor L/W ratio with efficient use of the chip area often requires the resistor to have a folded or zigzag structure. This can be achieved by making square or round bends in the lateral layout of the resistor. The effective length of a resistor around such a bend can be readily calculated by numerical-integration techniques. Figure 8-58 shows the effective resistor length around round and square bends for some common resistor layouts.

56. Diffused Semiconductor Resistors. The diffused resistor[46] is formed by the bulk resistivity of a diffused region in a semiconducting substrate. In monolithic circuits based on the npn bipolar technology, both the p-type base and the n-type emitter diffusions can be used to form a diffused resistor. Because of its higher breakdown and R_s values, p-type base diffusion is almost always preferred over the emitter diffusion.

Figure 8-59 shows the typical geometry and cross section of a p-type base-diffused resistor. The sheet resistance R_s is normally dictated by the npn transistor base-diffusion specifications and is typically 80 to 250 $\Omega/\square$.

The lower limit of the resistor width W is normally set by the photomasking tolerances and is approximately 5 to 7 μm for production circuits. Best matching is obtained for the values of W in the range of 20 to 50 μm. Table 8-13 shows the typical properties of base-diffused resistors, along with other types of integrated resistor structures. The matching tolerance shown in the table corresponds to a 10-μm-wide resistor structure.

The diffused resistor structure has a distributed capacitance associated with it, due to the reverse-biased junction which outlines the resistor. For a diffused resistor of total value R_1, the high-frequency equivalent circuit looks as shown in Fig. 8-60, where C_1 is the total distributed capacitance associated with R_1. The effect of C_1 is to shunt the ac signal to ground and cause excessive phase lag. Figure 8-60b shows the typical frequency response of a diffused resistor in terms of the magnitude of its driving-point impedance.[47] Assuming that the capacitance C_1 is

TABLE 8-13 Typical Characteristics of Integrated Circuit Resistors

Resistor type	Sheet resistivity, per □	Temp. coef., ppm/°C	Absolute tolerance, %	Matching tolerance, %
Semiconductor:				
Diffused	0.80 to 250 Ω	1,200 to 2,000	±12	±1.5
Pinched	3 Ω to 10 kΩ	3,000 to 5,000	+80 to −50	±10
Bulk	1 Ω to 10 kΩ	3,500 to 6,000	±30	±5
Ion-implanted	500 Ω to 20 kΩ	100 to 1,200	±6	±1
Thin-film:				
Tantalum	10 Ω to 1 kΩ	±100	±5	±1
Ni-Cr	40 to 400 Ω	±100	±5	±1
SnO₂	80 to 4,000 Ω	0 to −1,500	±8	±2
Cermet (Cr-SiO)	30 to 2,500 Ω	±50 to ±150	±10	±2
Thick-film:				
Palladium-silver	100 Ω to 100 kΩ	±150 to −500	±10	±2
Ruthenium-silver	10 Ω to 10 MΩ	±200	±10	±2

uniformly distributed along R_1, one can show that the magnitude of the driving-point impedance z_{in} is down by approximately 3 dB at the frequency f_1 given as

$$f_1 \approx 1/3R_1C_1 \approx 1/3R_sC_0L^2$$

where C_0 is the capacitance per unit area of the resistor junction.

Bulk Resistors. The bulk resistance of the n-type epitaxial layer can be used in some applications to form a noncritical high-value resistor. This can be done by using a device structure as shown in Fig. 8-61. The resulting structure, known as a *bulk resistor*, has a sheet resistivity $R_s = \rho_e/d$.

Pinched Resistors. The sheet resistivity semiconductor region can be increased by reducing its effective cross-sectional area. In a pinched-resistor structure, this technique is used to obtain a high-value sheet resistance from an ordinary diffused or bulk resistor. Figure 8-62 shows the structural diagram of a base-pinch resistor formed by placing an n^+ type emitter diffusion over the p-type diffused resistor.

The current-voltage characteristics of a pinched resistor are similar to those of a JFET since the resistor body is analogous to the FET channel, with the surrounding junction area functioning as the gate. Consequently, for increasing voltages, the pinched-resistor current I_p saturates at a value I_0, as shown in Fig. 8-62. In base-pinch resistors, the resistor breakdown voltage is quite

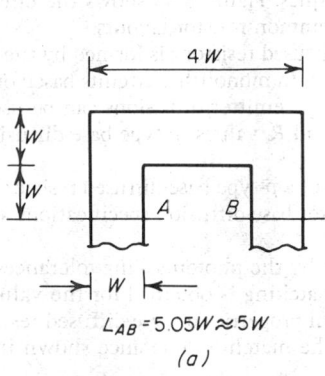

$$L_{AB} = 5.05W \approx 5W$$

(a)

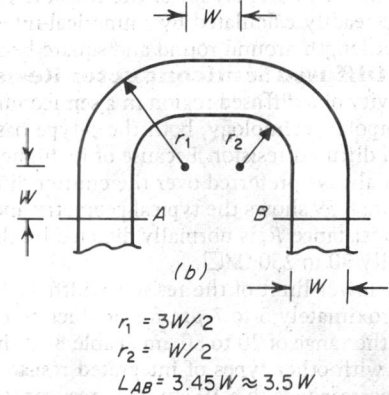

(b)

$$r_1 = 3W/2$$
$$r_2 = W/2$$
$$L_{AB} = 3.45W \approx 3.5W$$

Fig. 8-58. Calculation of effective resistor length around corners: (a) square corners; (b) round corners.

low, i.e., equal to transistor base-emitter breakdown BV_{EBO} (6 to 8 V). Typical matching and tracking tolerances of base-pinch resistors are listed in Table 8-13

Ion-Implanted Resistors. Ion-implantation techniques can be utilized to form resistor structures on the semiconductor surface. With this technique the impurities are introduced into the silicon lattice by bombarding the wafer surface with high-energy ions. The implanted ions lie within a very shallow layer (typically of the order of 0.1 to 0.8 μm) along the silicon surface. Thus, for the similar doping levels, the implanted layers yield a sheet resistivity which is roughly 20 times higher than a correspondingly doped diffused layer of 2- to 4-μm thickness.[40,48]

57. Thin-Film Resistors. Resistive films can be deposited and patterned on a dielectric substrate by using the deposition and photomasking techniques outlined in Par. **8-31**. In some

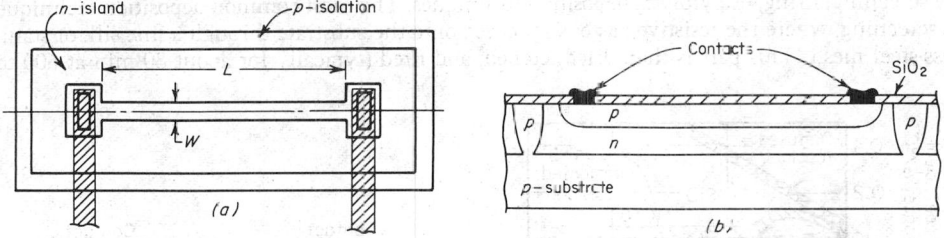

Fig. 8-59. Basic diffused resistor structure: (*a*) lateral geometry; (*b*) cross section.

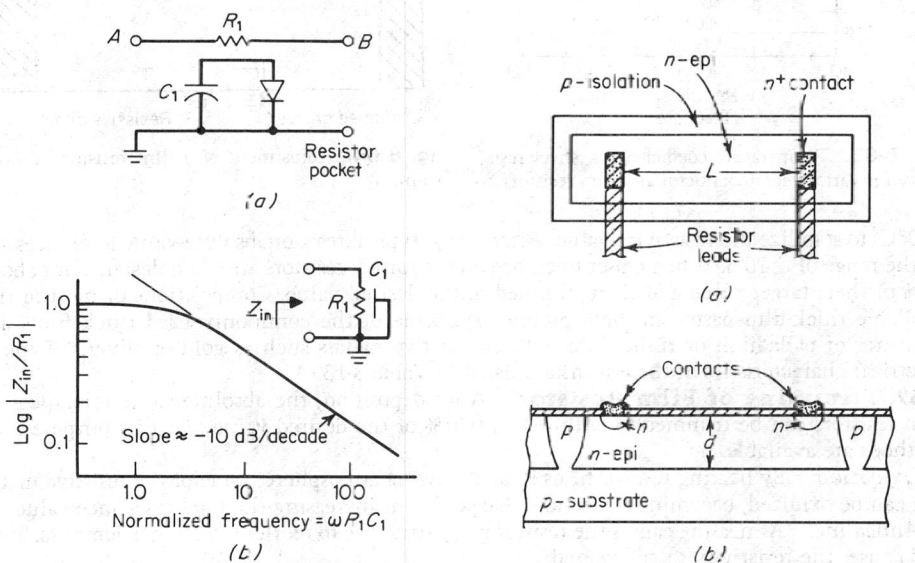

Fig. 8-60. High-frequency characteristics of diffused resistors: (*a*) equivalent circuit; (*b*) typical frequency response.[51]

Fig. 8-61. Bulk resistor: (*a*) lateral geometry; (*b*) cross section.

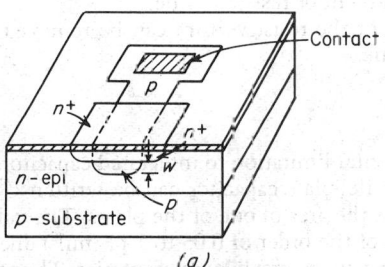

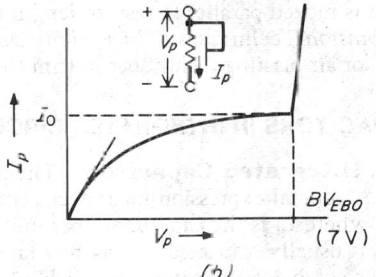

Fig. 8-62. Structural diagram and electrical characteristics of a base-pinch resistor.

cases, besides the basic deposition and patterning steps, an additional dielectric deposition is necessary to stabilize the resistor structure by sealing it off from the ambient atmosphere.[49]

Compared with the diffused resistors, thin films offer the following advantages:

1. Low temperature coefficient
2. Tighter absolute-value control
3. Lesser parasitics
4. Higher sheet resistivity

Table 8-13 and Fig. 8-63 summarize the basic properties of some of the thin-film resistors used in ICs. Among those listed, tantalum (Ta), nickel-chromium (Ni–Cr), and tin oxide (SnO_2) are by far the most common.

58. Thick-Film Resistors. Thick-film resistors are deposited on an insulating substrate by screening, firing, or pyrolytic deposition techniques. The most common deposition technique is screening, where the resistive paste is squeezed onto the substrate through a fine silk or stainless-steel mesh. This part is then dried, etched, and fired (typically for about 30 min at 600 to

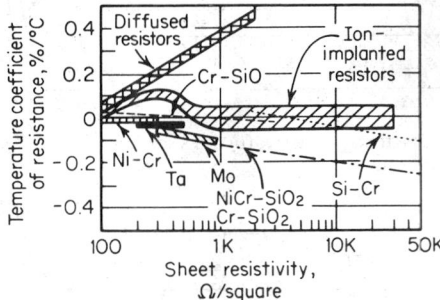

Fig. 8-63. Temperature coefficient vs. sheet resistivity for various semiconductor and film resistors.

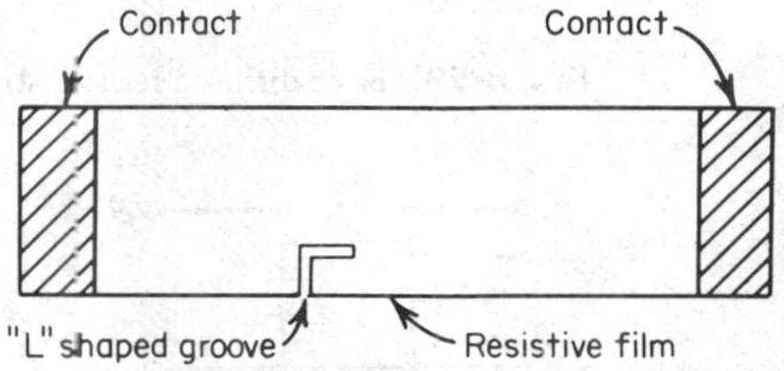

Fig. 8-64. Adjustment of a film resistor by laser trimming.

900°C) to stabilize the resistivity value. After firing, typical resistor absolute-value tolerances are in the range of ±10%. When closer tolerances are required, resistors are often designed for about 80% of their target value and then trimmed to the desired value. Compositions of most of the available thick-film pastes are held proprietary. Some of the commonly used thick films are mixtures of palladium or ruthenium with conductive metals such as gold or silver.[49] Typical electrical characteristics of these films are listed in Table 8-13.

59. Trimming of Film Resistors. After deposition, the absolute-value tolerances of film resistors can be trimmed to within 1 to 0.01% of the desired value. For this purpose four methods are available.

Oxidation. By heating resistor films in an oxidizing atmosphere, some of the resistive material can be oxidized, becoming nonconductive, and thus increasing the total resistance value.

Annealing. Annealing causes the resistor grain structure to reorient itself in a denser fashion and causes the resistivity to be lowered.

Laser Trimming. By selectively evaporating a small portion of a film resistor, its effective resistance can be increased. A common technique for this purpose is the L-shaped groove cut into the resistor, as shown in Fig. 8-64. A focused laser beam (spot size about 1 μm) is first moved perpendicular to the resistor for coarse trimming; as the resistance approaches its final value, the beam is moved parallel to resistor length for fine adjustment of resistor value.

Abrasion Techniques (Thick Films Only). A part of the resistive film can be removed by sand- or air-blasting techniques to trim the resistor value.

CAPACITORS IN INTEGRATED CIRCUITS

60. Integrated Capacitors. The most fundamental limitation to integrated capacitors is size. A general expression for the capacitance of a parallel-plate capacitor can be written $C = C_0 A$, where C_0 is the capacitance per unit area and A is the area of one of the plates. The value of C_0 is usually restricted to a narrow range (typically, of the order of 0.05 to 1 pF/mil^2) due to the type of dielectric materials available in ICs and their voltage-breakdown properties. Thus the area requirement increases quite rapidly with the required capacitor value.

There are two basic classes of capacitor structures available in integrated circuits: *junction* and *thin-film capacitors.*

61. Junction Capacitors. Application of a reverse bias across a semiconductor junction forces the mobile carriers to move away from the immediate vicinity of the junction. This creates a depletion layer of thickness x_d across the junction, and causes the junction to act as a parallel-plate capacitor with a plate separation x_d.[50]

In general, the voltage dependence of most junction capacitances that can be fabricated with the IC processes can be described by an expression of the form

$$C_0 = K_1(1/V_R)^n$$

where $n = 0.33$ to 0.5, K_1 = constant of proportionality depending on impurity concentration levels in immediate vicinity of the junction, and V_R = total reverse bias. The $n = 0.5$ case corresponds to the step-junction structure, and $n = 0.33$ corresponds to a linearly graded junction. A detailed family of capacitance-vs.-voltage curves is available in the literature.[51]

Figure 8-65 shows the family of C_0-vs.-V_R curves for a step junction, for various values of N. Figure 8-66 shows the three separate sets of junction capacitors available in bipolar integrated

Table 8-14. Typical Values of Capacitance per Unit Area Associated with Integrated Transistor Junctions

Applied voltage, V	Typical junction capacitance, pF/mil²			
	C_{EB}	C_{BC}	C_{cs} without n^+ layer	C_{cs} with n^+ layer
0	0.9	0.26	0.12	0.17
5	0.65	0.11	0.04	0.06
10		0.08	0.025	0.035

circuits. Also shown in the figure is the resultant interconnection of these three capacitors, along with the ideal diodes in shunt with them to indicate the bias polarity requirements for each capacitor. Note that each capacitance also has a finite bulk resistance in series with it. The collector-substrate capacitance C_{ss} has only a very limited application since one of its terminals, the substrate, is common to the rest of the circuit and represents an ac ground point in any junction-isolated-device structure. The remaining capacitances C_{EB} and C_{BC} can be eliminated when not needed by omitting the emitter or the base diffusion steps. Typical values of C_0 associated with each of these junctions are listed in Table 8-14 for a device impurity profile similar to that shown in Fig. 8-26a.

62. Thin-Film Capacitors. The thin-film capacitor is a direct miniaturization of the conventional parallel-plate capacitor. It is composed of two conductive layers separated by a dielectric. In ICs it can be fabricated in either one of the two forms shown in Fig. 8-67: (a) using a metal-oxide semiconductor (MOS) structure; (b) using a thin dielectric film between two conducting metal layers. The MOS structure is the most common thin-film capacitor in monolithic circuits because it is readily compatible with the conventional processing technology and does not require multiple metallization layers.

The capacitance per unit area of a thin-film capacitor is equal to the ratio of the permittivity ϵ_x to the thickness T_x of the dielectric layer. In MOS capacitors either SiO_2 or silicon nitride (Si_3N_4) can be used as the dielectric layer. Although SiO_2 is more readily available, Si_3N_4 is preferred whenever the extra processing steps can be justified, since it offers a higher dielectric constant (see Table 8-15). The minimum thickness of the dielectric layer is set by the yield and process control and breakdown requirements, and typically about 50 to 250 nm is preferred.

Unlike their junction counterparts, thin-film capacitors are not a function of the mag-

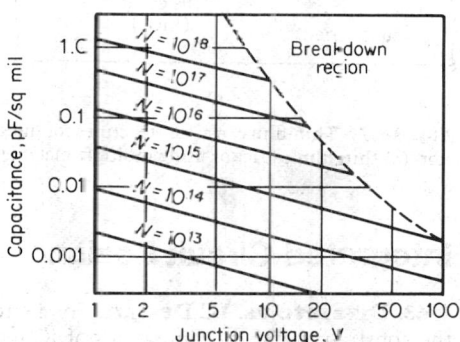

Fig. 8-65. Capacitance per unit area vs. junction voltage for a step junction (N = concentration of the lighter-doped side).

nitude or the polarity of the voltage applied across them and can offer higher capacitance per unit area with lesser parasitics. However, thin-film capacitors fail by the breakdown of dielectric layer when their voltage rating is exceeded. This is a destructive irreversible failure mechanism; therefore additional care should be taken in their current application to provide overvoltage protection.

Table 8-15 gives a summary of the electrical characteristics of thin-film capacitors available in ICs. Note that the first two on the list are the MOS type of structures, and the last two columns correspond to the multilayer metal structure of Fig. 8-67b.

Table 8-15.   Thin-Film Capacitor Characteristics

	Dielectric material			
	SiO_2	Si_3N_4	Al_2O_3	Ta_2O_5
Capacitance, pF/mil²	0.25–0.4	0.5–1.0	0.3–0.5	2–3.5
Relative dielectric constant	2.7–4.2	3.5–9	4–8.5	24–28
Breakdown voltage, V	50	50	20–40	20
Absolute tolerance, %	± 20	± 20	± 20	± 20
Matching tolerance, %	± 3	± 3	± 5	± 5
Temperature coef., ppm/°C	15	4–10	300	200–500
Q, at 10 MHz	25–80	20–100	10–100	10–100

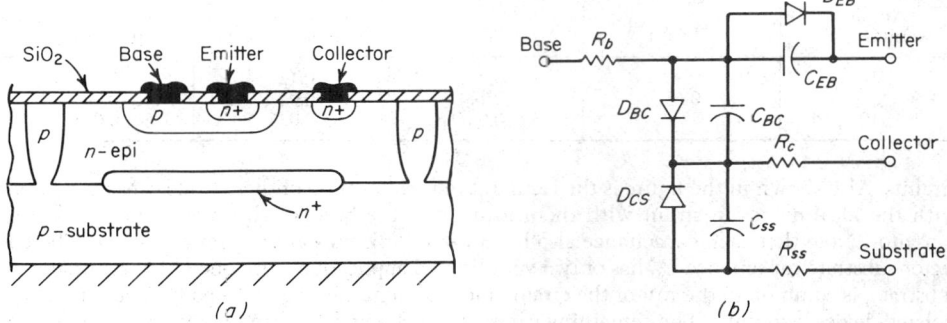

Fig. 8-66. Junction capacitances in bipolar ICs: (a) physical structure; (b) equivalent circuit of junction capacitances.

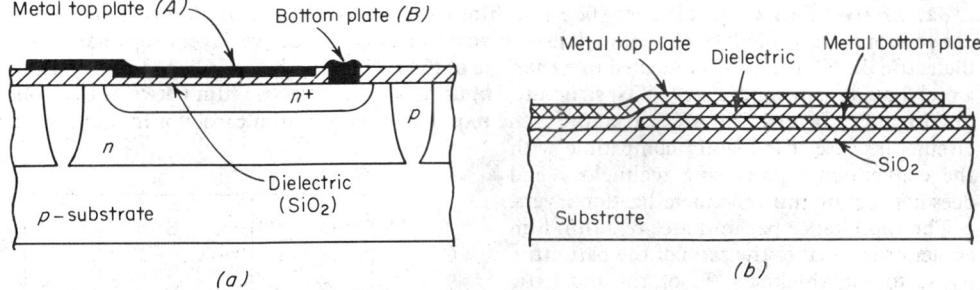

Fig. 8-67. Thin-film capacitor structures for integrated circuits: (a) metal-oxide semiconductor (MOS) capacitor; (b) thin-film capacitor using multiple metal layers.

Integrated-Circuit Design

63. Discrete vs. IC Design. For a circuit designer trained in the use of discrete circuits, the constraints and the limitations of IC technology may pose a difficult challenge. The constraints are the most severe in the case of monolithic circuits. Some of the basic limitations of monolithic components are:

Poor absolute-value tolerances
Poor temperature coefficients

Limitations on component values
Lack of integrated inductors
Limited choice of compatible active devices
Limited power-handling capability

On the other hand, IC fabrication methods offer a number of advantages to the circuit designer:

Availability of a large number of active devices
Good matching and tracking of component values
Close thermal coupling
Control of device layout and geometry

Through efficient use of these advantages it is often possible to design ICs that exceed the performance of similar discrete-component circuits.

LINEAR INTEGRATED CIRCUITS

64. Basic Building Blocks. In the design of linear ICs repetitive use is made of basic circuit configurations which utilize the matching and tracking properties of monolithic devices. These basic circuit configurations form a set of useful building blocks, which in turn serve as

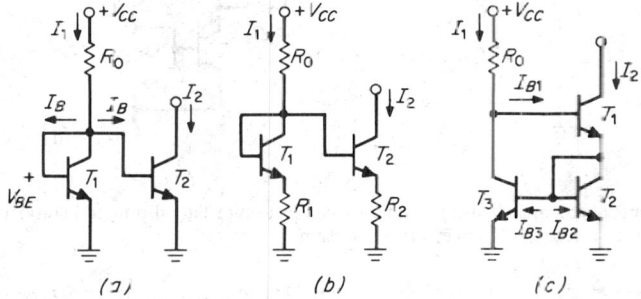

Fig. 8-68. npn constant-current stages: (a) diode-biased; (b) diode-resistor-biased; (c) internal-feedback constant-current stages.

the starting points for the design of larger and more complex functional circuits. Several of these are described in the following paragraphs.

65. Constant-Current Stages. In the design of monolithic circuits, constant-current stages or constant-current sources are often utilized for biasing purposes. They can also be used as *active loads*, to simulate large-value resistors in high-gain amplifier stages.[52] Figure 8-68 shows some common constant-current stages, using npn bipolar transistors. In each case, a reference current I_1 applied to one branch of the circuit generates an equal current I_2 at the output of the stage, over a wide range of temperature or bias conditions.

The circuit of Fig. 8-68a relies solely on the close matching of base-emitter voltage V_{BE} between two monolithic transistors. The emitter current I_E of a transistor is related to V_{BE} as

$$V_{BE} = V_T \ln(I_E/I_0)$$

where $V_T = kT/q$ is the thermal voltage* and I_0 is the base-emitter reverse leakage current which is proportional to emitter area. For transistor current gain much greater than unity, I_E can be replaced by collector current I_C. Then, for a pair of transistors T_1 and T_2 biased as shown in Fig. 8-68a, the collector currents I_{C1} and I_{C2} of the respective transistors are related as $I_{C1}/A_1 = I_{C2}/A_2$, where A_1 and A_2 are the effective emitter areas of T_1 and T_2. If $A_1 = A_2$, the input and output currents can be related as

$$I_2 = I_1 - 2I_B \approx I_1$$

Figure 8-68b shows an alternative constant-current stage, where the current level is determined by a resistor ratio, given as

$$I_2 = I_1 R_1/R_2 + V_T/R_2 \ln (I_1/I_2)$$

For $I_1 R_1 \gg V_T$, only the first term of the equation is dominant, and $I_2 \approx I (R_1/R_2)$. By setting $R_1 = 0$ in this circuit one can also obtain a constant-current stage for low current levels.[53]

*At 300 K, $V_T = 0.0259$ V.

An improved version of the diode-biased constant-current stage is shown in Fig. 8-68c. This circuit uses internal feedback through the connection between T_2 and T_3 to reduce the effects of transistor base currents.[54] The input and output currents are related as

$$I_2 = I_1 + (I_{B3} + I_{B2} - 2I_{B1})$$

If transistor current gains are well matched, the last term is negligible and independent of the absolute value of β.

Lateral-*pnp* transistors can also be used in designing constant-current stages. Figure 8-69 shows a composite connection of *npn* and lateral-*pnp* transistors for such an application. Note that, in each case, the high-β characteristics of the *npn* devices are used to minimize the effects of low

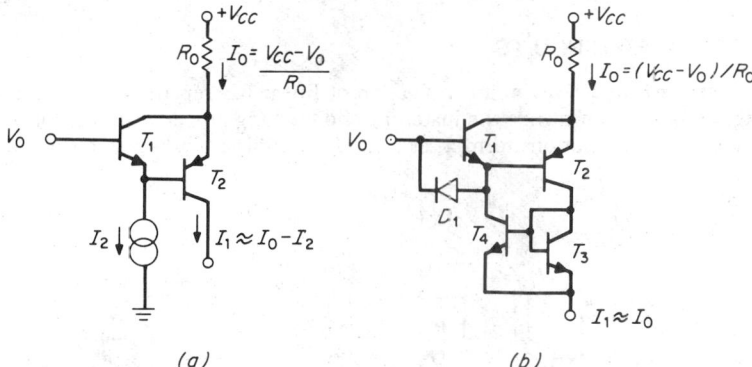

(a)   *(b)*

Fig. 8-69. Current-source circuits using composite connection of lateral-*pnp* and conventional-*npn* transistors: (*a*) basic current-source circuit; (*b*) improved version.

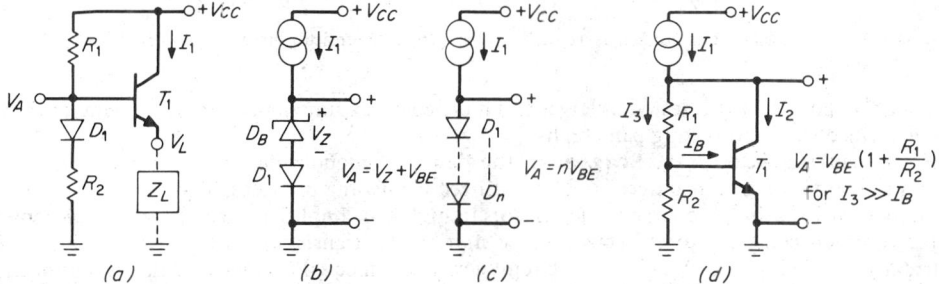

(a)   *(b)*   *(c)*   *(d)*

Fig. 8-70. Voltage-source configurations for linear ICs: (*a*) common-collector stage; (*b*) temperature-compensated avalanche diode; (*c*) diode string; (*d*) shunt-feedback diode amplifier.

β values associated with the lateral *pnp*. Assuming the *npn* β is sufficiently high, the circuit performance becomes independent of *pnp* gain characteristics. The simplified circuit in Fig. 8-69a shows a method of generating a constant current I_1 independent of *pnp* characteristics. The circuit shown in Fig. 8-69b is an improved version of this circuit in which the current source I_2 of the previous circuit is replaced by a diode-biased *npn* constant-current stage formed by T_3 and T_4.[55]

66. Voltage Sources. In many circuit applications, it is necessary to establish a low-imped-ance point within the circuit which can serve as an internal voltage supply. Such a voltage reference point should have both a very low ac impedance and a very stable dc voltage level, insensitive to power-supply and temperature variations. Usually, however, either the low imped-ance or the dc voltage stability is of prime importance. The circuits which primarily fulfill the low-impedance requirements are known as *voltage sources*, whereas those specifically designed to provide a constant voltage independent of the supply or the temperature changes are called *voltage references*.

Figure 8-70 shows typical voltage-source configurations often used in IC design. In the circuit of Fig. 8-70a the low output impedance of an emitter follower is used to simulate a low-imped-ance source, the diode D_1 providing the temperature compensation for the V_{BE} drift of T_1. The circuits of Fig. 8-70b and c use a temperature-compensated avalanche diode and a diode string, respectively, to provide a low-impedance voltage output. The circuit of Fig. 8-70d uses a shunt-feedback transistor amplifier stage to amplify the transistor V_{BE} drop. This circuit provides an efficient substitute for the diode string of the circuit of Fig. 8-70c if a large number of diodes are involved.[56]

67. Voltage References. In the design of analog circuits, such as the low-drift amplifiers or voltage regulators, it is often necessary to establish an internal voltage reference in the circuit. Unlike the case of voltage sources, the main emphasis in a voltage reference stage is on the thermal stability of the reference voltage. Temperature-stability requirements for a reference voltage are typically ≤ 100 ppm/°C. Temperature coefficients of most monolithic circuit components or devices are much greater than this, but by making use of the matching and tracking properties of integrated components and the close thermal coupling on the chip it is possible to compensate the thermal drifts to a few parts per million. Figure 8-71 shows one circuit configuration which can be used for such an application.

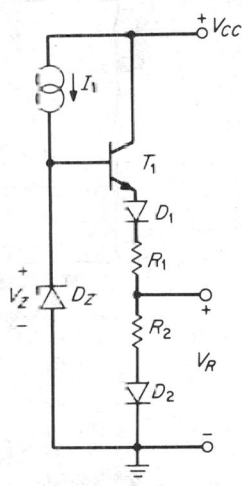

Buried Zener. There are basically two devices that can be used as voltage references in the bipolar IC technology, the forward-biased diode and the zener diode. In the past, the emitter-base zener breakdown voltage of 6 to 7 V was commonly used. A lower zener voltage of 5 to 6 V was achieved by diffusing the rich n^+ emitter material into the richly doped p-type isolation material. However, since the zener breakdown would occur at the richly doped surface near mobile oxide contaminants, the long-term stability was noticeably poor. For this reason, steps were taken to push the zener breakdown into the n epitaxial material. One such device is shown in Fig. 8-72. This type of *buried zener* makes use of the p^+-type isolation diffusion and n^--type buried layer. The n^+ buried layer prevents the isolation diffusion from inverting the p-type material, thus forming the zener junction far below the surface.

Fig. 8-71. Circuit configuration for generating a temperature-independent voltage reference V_R.

The long-term stability of this device is excellent when compared with that of a surface zener diode. Unfortunately, the actual zener tolerance is poor because of the poor control usually associated with isolation and buried-layer diffusion. The zener diode shown in Fig. 8-73 is produced at subsurface levels using ion implantation and can achieve much better control over initial tolerance while maintaining excellent long-term stability. This results from ion implanting a p-type region in contact with the n^+ emitter diffusion and then ion implanting an n^- region on top of the p-type implant at the junction. This forces the zener breakdown to occur below the surface at the pn^+ interface, achieving excellent stability and better initial tolerance with the well-controlled ion implants. Contact is made to the p-type ion implant with standard base diffusions. These implanted zener voltages typically range from 5 to 6 V and find extensive use as IC voltage-regulator reference elements, provided the input voltages are always in excess of the zener voltage. Unfortunately, there are many IC applications where the input voltage will decrease below the lowest achievable zener voltage, yet voltage regulation is still required.

Band-Gap Reference. A low-voltage reference has been developed for low-voltage applica-

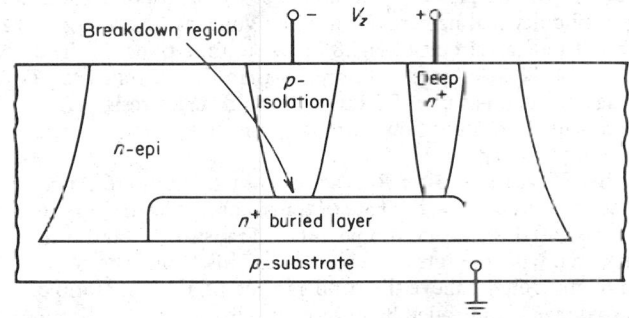

Fig. 8-72. Buried zener diode using p isolation and n^- buried-layer junction.

tions using the forward-biased base-emitter junction characteristics in a circuit configuration called the *band-gap reference*. This circuit in its basic form is shown in Fig. 8-74. It is well known that the temperature coefficient (TC) of a base-emitter junction is inversely related to the current density in the device. Because of resistor R_1, the emitter current in transistor T_1 is less than that flowing in the emitter of diode D_1. Thus the temperature coefficient of the base-emitter junction of transistor T_1 is greater than that of diode D_1. This fact produces a positive TC across

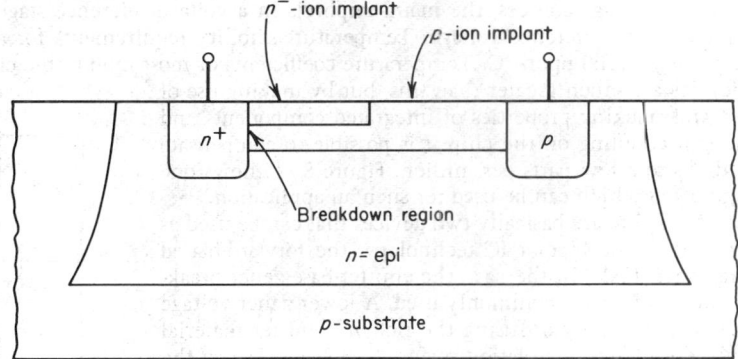

Fig. 8-73. Ion-implanted buried zener diode.

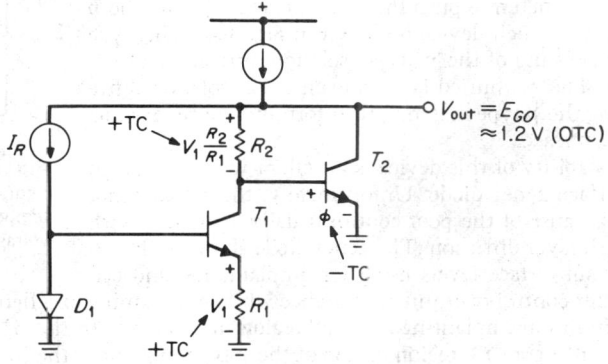

Fig. 8-74. Basic low-voltage band-gap voltage reference.

resistor R_1, which is scaled across resistor R_2 by the R_2/R_1 ratio. By carefully adjusting this resistor ratio so that the positive TC across R_2 will exactly balance the negative TC at the base of transistor T_2 the output voltage will exhibit a zero temperature coefficient. Notice that by increasing the area of the base-emitter junction of transistor T_1 a greater current can be made to flow in the transistor for the same TC at its emitter, thus effectively reducing the R_2/R_1 ratio. The resulting output voltage at which this TC cancellation will occur is the band-gap potential of silicon, or about 1.2 V. The primary limitations of this particular band-gap circuit are poor output impedance, difficulty scaling zero TC to other voltages, and a large R_2/R_1 resistor ratio.

By using the basic band-gap circuit of Fig. 8-75 these limitations can be overcome. The base-emitter area of transistor T_1 is designed to be greater than that of transistor T_2. Thus the voltage across resistor R_1 again has a positive TC but will scale across resistor R_2 as the ratio $2R_2/R_1$ because twice the resistor R_1 current flows through resistor R_2 due to transistor mirror M_1. This reduces the required R_2/R_1 ratio by 2.

Again, the positive TC across resistor R_2 is adjusted by this ratio to cancel the negative TC of transistor T_2 to produce a zero TC at the base of transistor T_2. Again, this zero TC voltage at the base of transistor T_2 is the silicon band-gap potential. Transistor T_3 and resistors R_A and R_B form a negative-feedback loop with the reference elements. This allows the zero TC band-gap voltage to be level-shifted to any voltage above the band-gap potential by the ratio of R_A/R_B. In addition, because the output voltage is controlled by negative feedback, the output impedance is very low. For a band-gap output voltage, the input voltage to this reference can be as low as 2 V, making this reference ideally suited for low-voltage applications. Zero TC reference voltages as low as

100 to 200 mV are presently being developed with ground potential apparently being the ultimate aim.

MOS (IGFET) Voltage References.   An on-chip NMOS voltage-reference circuit[57-59] (Fig. 8-76) has been developed using the difference in the threshold of two transistors T_2 and T_3, which differ only in the implants they receive. The transistors show very similiar sensitivities to substrate bias and temperature, so that the difference in thresholds is determined exclusively by the implants. The implanted ions are immobile in the crystal structure at temperatures below 900°C and are sealed in by high-integrity thermal gate oxide. By implanting T_2 to make it an enhancement type and T_3 to make it a depletion type the magnitudes of the thresholds add when the threshold difference is taken. In this way, a reference voltage of several volts can be produced.

The manufacturing spread of the reference value showed a larger variation than tolerable for the PCM codec application. To compensate, a buffer amplifier[60] with gain-trimming polysilicon* fuses was used to adjust the final value to 3.15 ± 0.002 V. This value was chosen to fall within a range compatible with the chosen power-supply voltages yet be related to telephone system voltages. The overall variation in the reference voltage with operating conditions and time is equivalent to a gain variation of less than 0.07 dB. Accelerated life tests indicate a 40-year drift equivalent of 0.01 dB.

*Doped silicon having small grain size.

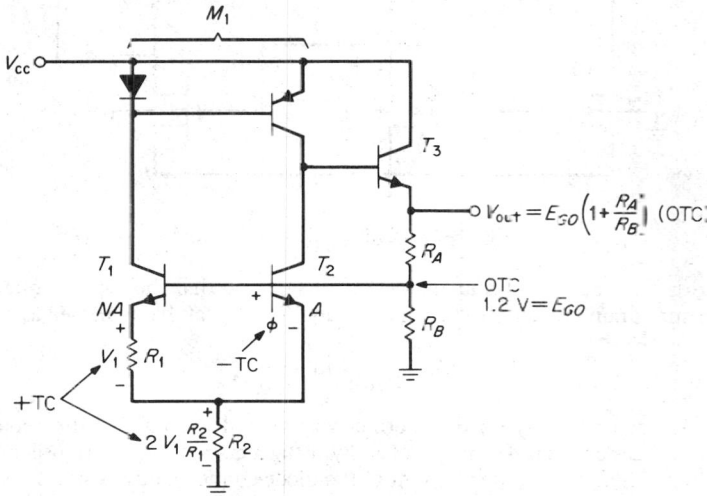

Fig. 8-75. Improved low-voltage band-gap voltage reference.

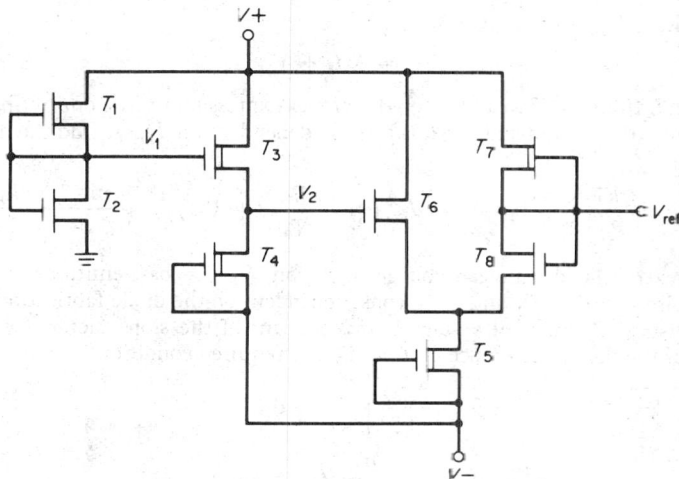

Fig. 8-76. MOS voltage reference.[5]

Another voltage reference[61] which can be constructed in standard CMOS technology (Fig. 8-77) consists of five MOS transistors operated in weak inversion and is used to generate a voltage V_{R2} across resistor R_2. This voltage has a positive temperature coefficient and is compensated by the negative-temperature coefficient of the base-emitter voltage V_{BE} of a diode-connected transistor.[62]

The current mirrors[63] T_1 to T_4 form a closed loop with an initial loop gain greater than 1 such that the current in both branches increases until equilibrium is achieved when the gain is reduced to 1 by the voltage V_{R1} across resistance R_1.

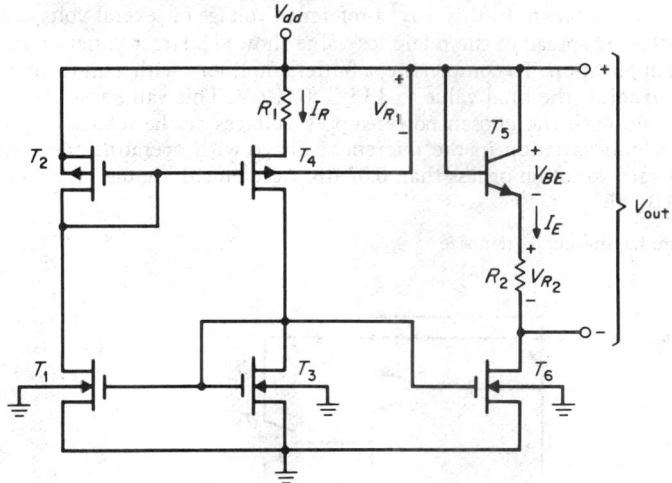

Fig. 8-77. CMOS voltage reference.[61]

Assuming that the devices operate in weak inversion and that the supply voltage V_{DD} is high enough to ensure drain current saturation of T_1 and T_4, V_{R1} can be expressed as

$$V_{R1} = \frac{kT}{q} \ln \left(\frac{S_4 \, S_1}{S_2 \, S_3} \right) \tag{8-1}$$

The voltage V_{R1} depends only on the thermal voltage and the ratio of the geometrical shape factors of the devices and is independent of n. By using a current mirror transistor T_6 to provide current multiplication, the emitter current of the diode-connected transistor T_5 is

$$I_E = \frac{S_6}{S_3} \frac{kT}{q} \frac{1}{R_1} \ln \left(\frac{S_4 \, S_1}{S_2 \, S_3} \right) \tag{8-2}$$

The output voltage V_{out} is then given by

$$V_{out} = R_2 I_E + V_{BE} \tag{8-3}$$

Substituting for I_E from Eq. (8-2) and using the proper expression for the emitter-base voltage of a diode-connected transistor (assuming I_E is linearly dependent on T), we find the output voltage to be

$$V_{out} = \frac{kT}{q} \frac{R_2}{R_1} \frac{S_6}{S_3} \ln \frac{S_4 S_1}{S_2 S_3} + V_{GO} \left(1 - \frac{T}{T_0} \right) + V_{BEO} \frac{T}{T_0} + \frac{mkT}{q} \ln \frac{T_0}{T} \tag{8-4}$$

where V_{GO} = extrapolated band-gap voltage of silicon, V_{BEO} = base-emitter voltage of diode-connected transistor at $T = T_0$, and m = const dependent on the diode fabrication and temperature characteristics. The output voltage is independent of the slope factor. For temperature independence of the voltage reference at $T = T_0$, the required condition is

$$\left. \frac{\partial V_{out}}{\partial T} \right|_{T=T_0} = 0 \tag{8-5}$$

or

$$\frac{R_2}{R_1} \frac{S_6}{S_3} \ln \frac{S_4 S_1}{S_2 S_3} = \frac{q}{kT_0} (V_{GO} - V_{BEO}) + m \tag{8-6}$$

This equation in conjunction with Eq. (8-2) and the diode exponential relationship can be used to select the circuit element values. Substituting Eq. (8-6) into Eq. (8-4) results in

$$V_{out} = V_{GO} + \frac{mkT}{q}\left(1 + \ln\frac{T_0}{T}\right)$$

(8-7)

and the output voltage at $T = T_0$ is

$$V_{out}(T_0) = V_{GO} + \frac{mkT_c}{q}$$

(8-8)

Trimming can easily be accomplished for the zero-temperature coefficient at $T = T_0$. In principle, once the circuit is trimmed so that its output voltage is equal to that given by Eq. (8-8) at $T = T_0$, a zero temperature coefficient is achieved at that temperature.

68. DC-Level-Shift Stages. Since large-value coupling capacitors are not available in monolithic circuits, all broadband gain stages must be dc-coupled. In a bipolar gain stage using an npn transistor, the output dc level is always higher than the dc level of the input. Therefore, if a number of such gain stages are cascaded, the output dc level rapidly builds up toward the positive supply voltage. This in turn limits amplitude and the linearity of the available output swing. Ideally, such a dc-level buildup can be avoided by using complementary pnp-npn gain stages. However, the pnp transistors available in monolithic form have relatively poor frequency-response and current-gain characteristics.

Figure 8-78 shows some practical dc-level-shift stages commonly used in monolithic design. A level-shift stage serves as a unilateral buffer between the successive gain stages; therefore it is required to have a high input impedance and a relatively low output impedance to prevent interstage loading. A possible exception to this is the resistor-current-source level-shift circuit of Fig. 8-78d, which has an output impedance approximately equal to R_1.

In certain low-frequency applications, the composite npn-lateral-pnp connection of Fig. 8-79 can also be used as a level-shift stage. As described earlier (see Fig. 8-32), this composite connection of the npn-pnp transistors is equivalent to a single transistor which has the polarity of the pnp and the β of the npn. Hence, a level-shift stage such as shown in Fig. 8-79 can also provide a voltage gain $A_V \approx -(R_2/R_1)$.

69. Differential Gain Stages. The differential gain stage is a balanced amplifier circuit, designed to amplify only the *difference* between the two input signals.[64,65] Figure 8-80 shows the basic differential gain stage in its simplest form. The types of inputs which are applied simultaneously to both inputs of the circuit are called *common-mode* (CM) inputs. Thanks to the symmetry of the circuit, CM inputs cause the current and voltage levels within both branches of the circuit to vary in an identical level. Thus, for common-mode inputs, the differential gain stage has no voltage gain and can be described by the equivalent circuit of Fig. 8-80b. On the other hand, if a set of *differential-mode* (DM) signals is applied to the inputs, the current and voltage levels in the circuit vary in a differential, e.g., antisymmetric, manner and a net output voltage is produced. For differential input signals, the response of the amplifier can be described by the equivalent circuit of Fig. 8-80c.

Figure 8-81 shows some of the differential gain stage configurations commonly used in inte-

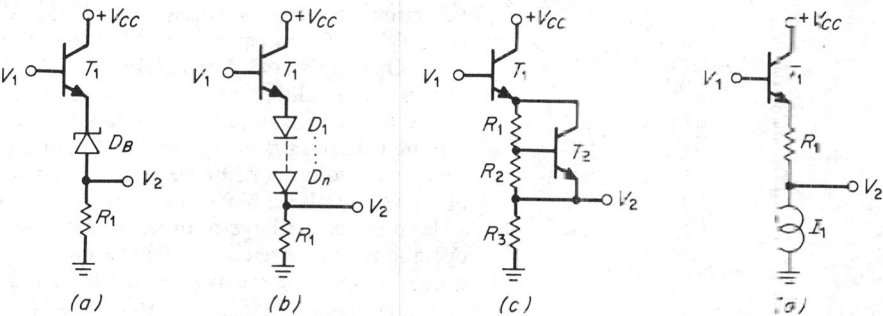

Fig. 8-78. DC level-shift stages: (a) avalanche diode; (b) diode string; (c) diode amplifier; (d) resistor-current-source combination.

grated circuits. The expression for differential voltage gain, $A_V (= V_{out}/V_{in})$, is also shown in the figure for each circuit configuration.

In spite of the inherent matching advantages of monolithic devices, small differences exist between two identical transistors on the same chip. These mismatches result in a finite *offset* between the currents in the two branches of the circuit, under zero input condition. In a generalized differential amplifier stage, as shown in Fig. 8-82, the V_{BE} and β mismatches of active devices can be related to the collector-current offset ΔI_c as[56]

$$\frac{\Delta I_c}{I_1} = \frac{\Delta V_{BE} + (R_B I_1/\beta_1)(1 - \beta_1/\beta_2)}{V_T + R_E I_1 + R_B I_1/\beta_1} \tag{8-9}$$

where $\Delta I_c = I_1 - I_2$ and $\Delta V_{BE} = V_{BE1} - V_{BE2}$.

If significant resistor mismatches were present in the circuit, the collector-current mismatch expression of Eq. (8-9) could be made to include them by replacing the ΔV_{BE} term in the equation by an equivalent offset voltage ΔV_{io}, given as

$$\Delta V_{io} = V_{BE} + \Delta R_E I_1 + \Delta R_B I_1/\beta \tag{8-10}$$

where ΔR_E and ΔR_B are the emitter and the base resistor mismatches. Even though it is fairly complicated, Eq. (8-10) gives a good insight into the circuit parameters which contribute to the current unbalance in a differential stage.

70. Active Loads. In a number of circuit applications, it is necessary to obtain very high voltage gain from an amplifier stage. This can often be achieved by using active devices to simulate high-value load resistors. Figure 8-83 shows a simplified circuit schematic of an *npn* common-emitter amplifier stage, using a lateral-*pnp* transistor as an active load. In this case, the lateral *pnp* is biased as a constant-current stage. The voltage gain of the stage can be expressed as

$$A_v = -g_m r_0$$

where g_m is the transconductance of T_1 and r_0 is the combined collector impedance of T_1 and T_2.

71. Class B Output Stages. In most high-gain amplifier circuits it is necessary to have a sizable output-current capability without high quiescent-power dissipation. This requires the use of class AB or class B type of output stages. Figure 8-84 shows some output circuit configurations commonly used in IC design.

The circuit of Fig. 8-84a is an all-*npn* class B amplifier stage. This circuit configuration, often referred to as the *totem-pole* topology, was initially developed for digital ICs but is readily adaptable to analog ICs. The operation of the circuit can be briefly described as follows. For positive swing of the input voltage, T_2 is in its active region, and the load current flows toward the negative supply through D_2 and T_2. The diode drop across D_2 assures that T_1 is off during this half cycle. For negative-going input signals, T_2 is cut off and the load current is supplied from T_1. For standby condition, with no ac signal at the input, T_1 stays off, and T_2 is biased near cutoff with a small standby current I_1 equal to V_{CC}/R_1.

Figure 8-84b shows a more conventional class B output stage using a substrate-*pnp* transistor, along with a conventional *npn*. The substrate *pnp* has a relatively low β but a good high-current capability. The diodes D_1 and D_2 are diode-connected transistors whose V_{BE} drops match those of T_1 and T_2, respectively. Therefore, in the quiescent state, a bias current I_2 flows through both T_1 and T_2 such that $I_2 = mI_1$, where m is the emitter area ratio of T_1 to D_1 or T_2 to D_2. Therefore this circuit operates as a class AB amplifier since both T_1 and T_2 can be on in standby condition.

72. Operational Amplifiers. An "ideal" operational amplifier can be defined as a voltage-controlled voltage-amplifier circuit which offers infinite voltage gain with an infinite input impedance and zero output impedance. The advantage of such an idealized block of gain is that it is possible to perform a large number of mathematical operations or generate a number of circuit functions, by applying passive feedback around the amplifier. Figure 8-85 shows some of the basic functional applications of an operational amplifier. If the input and the output impedance levels of the amplifier are respectively high and low

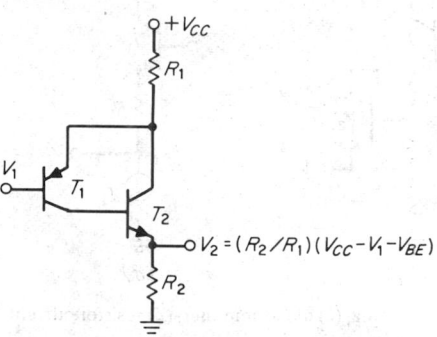

Fig. 8-79. *pnp-npn* level-shift stage.

compared with the source and feedback impedances, and if the voltage gain is sufficiently high, the resulting amplifier performance becomes determined solely by the external feedback components.

Because of its versatility as a general-purpose building block, the operational amplifier is the most widely accepted class of linear ICs. Other widely used classes of linear ICs closely related to the operational amplifiers are the *voltage comparators* and the *sense amplifiers*. Voltage comparators compare the amplitude and the polarity of an input signal and produce a large dc level shift at the output if the input exceeds the reference level.

Operational amplifiers possess finite voltage gain (typically of the order of 80 to 100 dB) and large but finite input impedance (typically above 400 kΩ).

The circuit model of an actual integrated operational amplifier is shown in Fig. 8-86, where I_B indicates the finite input bias currents, dc sources V_{io} and I_{io} represent the voltage and current

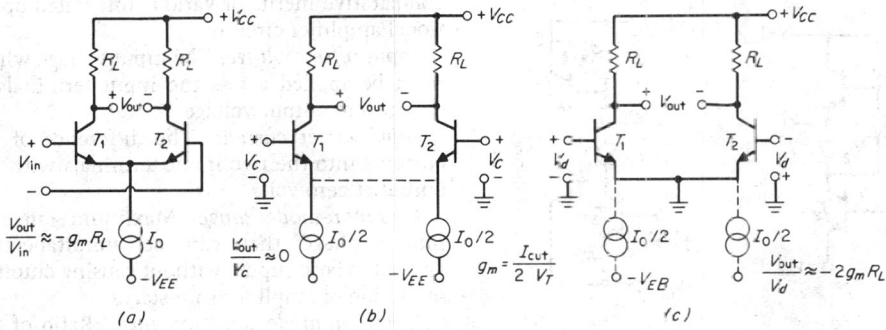

Fig. 8-80. Differential gain stages: (a) basic circuit; (b) common-mode equivalent; (c) differential-mode equivalent.

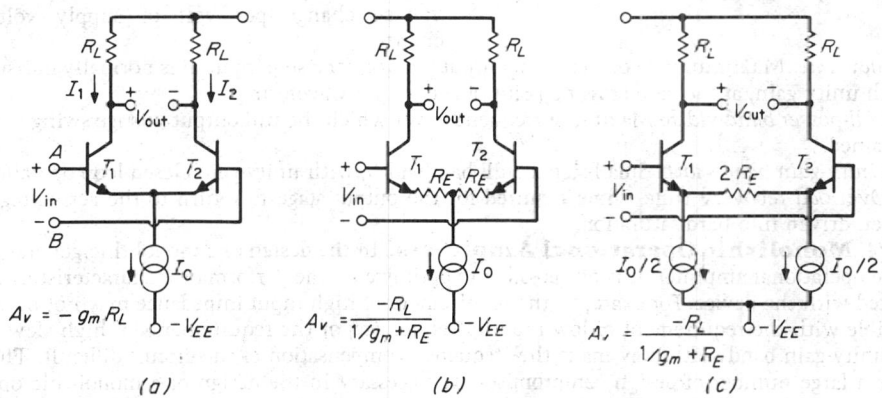

Fig. 8-81. Common differential gain stages.

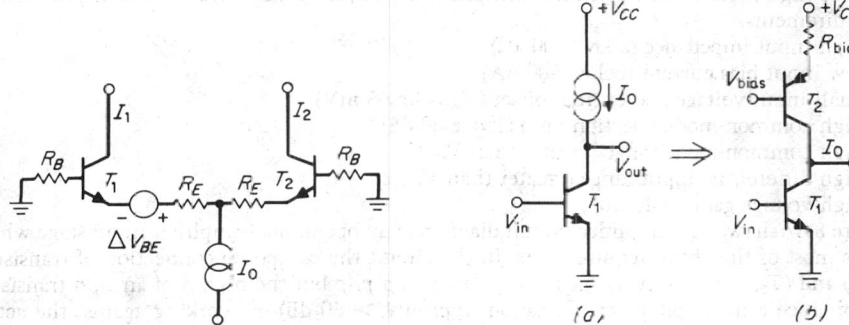

Fig. 8-82. Offset sources in a current-biased differential amplifier.

Fig. 8-83. Active loads: (a) equivalent circuit; (b) implementation.

offsets associated with the circuit, and A is the voltage amplification factor. Due to nonzero values of V_{io} and I_{io}, the output voltage $V_{out} \neq 0$ for input voltage $V_{in} = 0$.

Neglecting the nonzero output impedance, the overall voltage gain $A_V (= V_{out}/V_{in})$ can be expressed as

$$A_V = \frac{V_{out}}{V_{in}} = \frac{R_F}{R_S} \frac{1}{1 + (1/A)[1 + (R_F/R_S) + R_F/R_{in}]}$$

where R_{in} is the input impedance of the amplifier. For large values of R_{in}, as $A \to \infty$, A_V becomes determined solely by the external components, as $A_V = -R_F/R_S$.

73. Operational-Amplifier Terminology. Since the operational amplifier has become a universal building block for circuit and system design, a number of widely accepted design terms have evolved which describe the comparative merits of various integrated operational amplifier circuits.

Input offset voltage: The input voltage which must be applied across the input terminals to obtain zero output voltage.

Input offset current: The difference of the currents into the two input terminals with the output at zero volts.

Common-mode range: Maximum range of input voltage that can be simultaneously applied to both inputs without causing cutoff or saturation of amplifier gain stages.

Common-mode rejection ratio: Ratio of the differential open-loop gain to the common-mode open-loop gain.

Supply-voltage rejection ratio: Input offset voltage change per volt of supply voltage change.

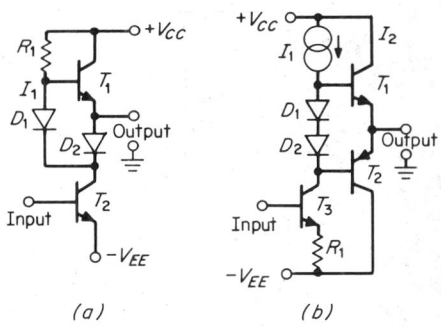

(a)   *(b)*

Fig. 8-84. Class B output stages: (a) all-npn totem-pole output stage; (b) complementary output stage using substrate pnp.

Slew rate: Maximum rate of change of output voltage, for a step input. It is normally measured with unity gain, at the zero crossing point of the output waveform.

Full-power bandwidth: Maximum frequency over which the full output voltage swing can be obtained.

Unity-gain bandwidth: Small-signal 3-dB bandwidth, with unity-gain closed-loop operation.

Overload recovery time: Time required for the output stage to return to the active region, when driven into hard saturation.

74. Monolithic Operational Amplifiers. In the design of a monolithic general-purpose operational amplifier, it is not possible to optimize all the performance characteristics associated with the device. For example, the requirement of high input impedance may not be compatible with the requirement of low input offset voltage; or the requirements of high slew rate or unity-gain bandwidth may make the frequency compensation of the circuit difficult. Therefore a large number of design compromises are necessary in the design of a monolithic operational amplifier.

In the design of integrated operational amplifiers, the input stage is often the most critical stage. In a high-performance operational-amplifier circuit, the input stage must fulfill the following requirements:

High input impedance (above 500 kΩ)
Low input bias current (below 500 nA)
Small input voltage and current offset (V_{io} below 5 mV)
High common-mode rejection ratio (above 60 dB)
High common-mode range (greater than $V_{CC}/2$)
High differential-input range (greater than $V_{CC}/2$)
High voltage gain (≥ 40 dB)

Figure 8-87 shows the simplified circuit diagram of an operational-amplifier input stage which satisfies most of the above requirements. In the circuit the composite connection of transistors (T_1, T_3) and (T_2, T_4) effectively have the polarity of a pnp but the high β of an npn transistor. They produce a high-voltage amplification (typically > 60 dB) by working against the active loads formed by current sources T_8 and T_9. A common-mode feedback loop formed by T_5, T_6 and current source I_B is used to set the operating point of the pnp transistors.[66]

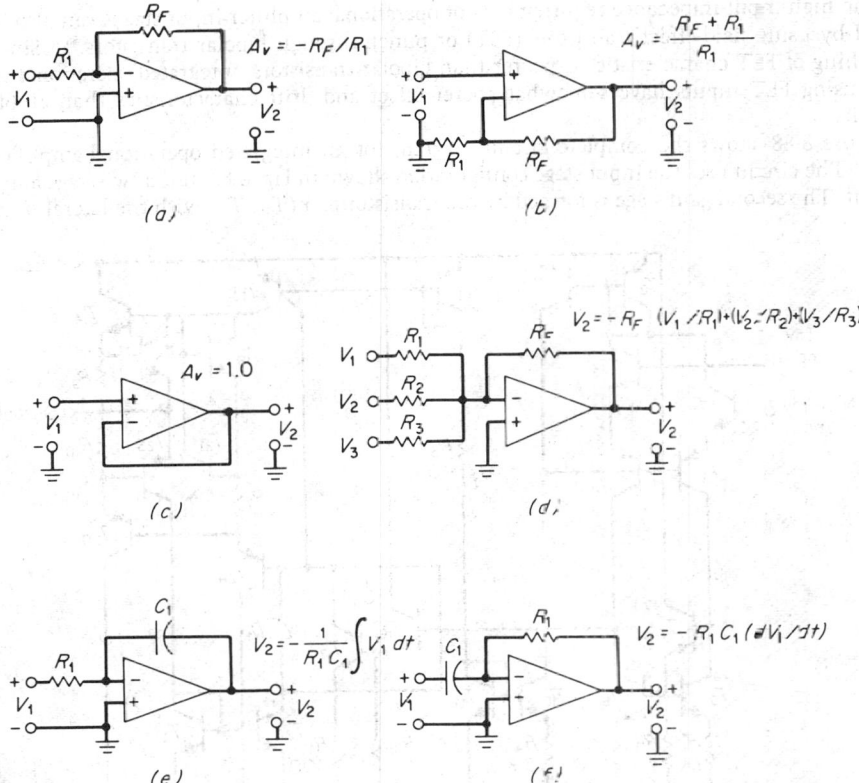

Fig. 8-85. Functional applications of an operational amplifier: (a) inverting amplifier; (b) noninverting amplifier; (c) voltage follower; (d) summing amplifier; (e) integrator; (f) differentiator.

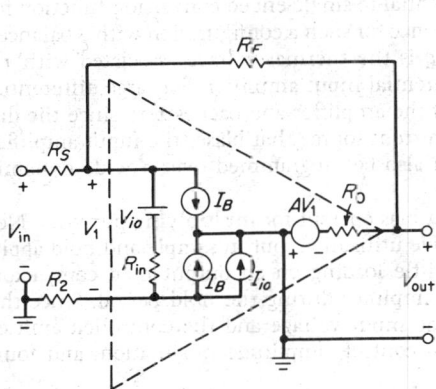

Fig. 8-86. Equivalent circuit of an operational amplifier showing the effect of finite input impedance, bias currents, and current and voltage offsets.

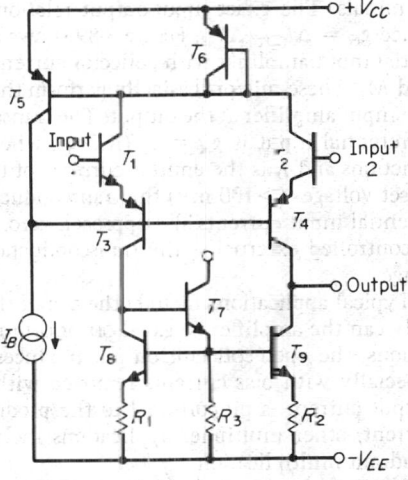

Fig. 8-87. Operational-amplifier input-stage configuration using active loads.

The high-input-impedance requirements of operational-amplifier input stages can also be satisfied by using field-effect transistors (FET) or punch-through bipolar transistors.[56,57] Since the matching of FET characteristics is poorer than bipolar transistors, integrated operational amplifiers using FET inputs have somewhat poorer offset and drift characteristics than all-bipolar circuits.

Figure 8-88 shows the complete circuit diagram for an integrated operational-amplifier circuit.[66] The circuit uses the input-stage configuration shown in Fig. 8-87. It is a two-stage amplifier circuit. The second gain stage is formed by the transistor pair (T_{16}, T_{17}) with the lateral pnp, T_{13},

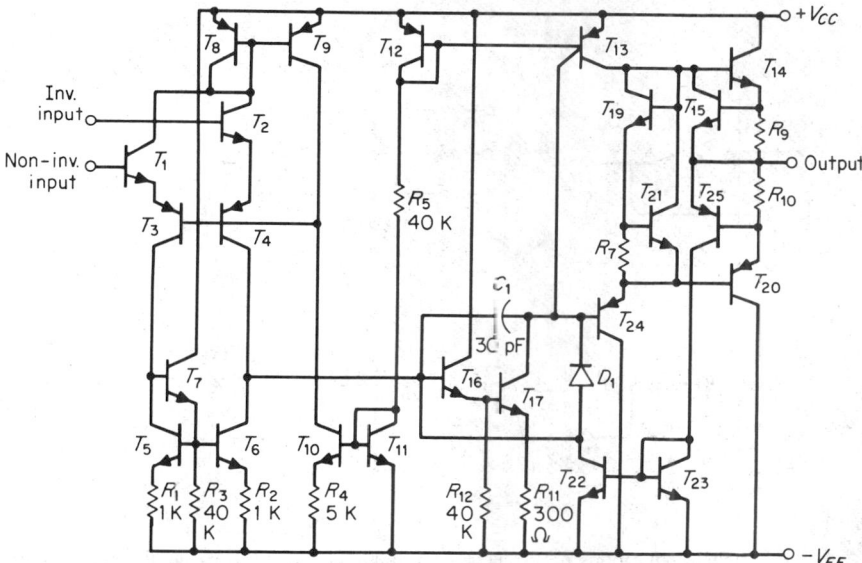

Fig. 8-88. Circuit of a monolithic operational amplifier using internal compensation and overload protection.

serving as an active load. The circuit can be internally frequency-compensated by connecting a 30-pF capacitor across this gain stage. The output circuit uses a class B stage (see Fig. 8-84b), with T_{20} designed as a substrate-pnp transistor.

Transconductance Operational Amplifier.[67]   A transconductance operational amplifier produces an output current that is proportional to the voltage difference at the differential input terminals. The exact input-output relationship is determined by the amplifier's transconductance $g_m = \Delta I_{OUT}/\Delta V_{in}$. Figure 8-89 shows a basic transconductance amplifier that uses a differential input amplifier with collector currents mirrored to the output by current mirrors M_1, M_2, and M_3. These mirrors basically perform the differential-to-single-ended conversion function for the input amplifier at the output. The transconductance for such a configuration with a balanced differential input is $g_m = I_E/(KT/q)$, where KT/q is the thermal voltage associated with pn junctions and I_E is the emitter currents of the differential input amplifier. For large differential offset voltages (>100 mV) the transconductance of the amplifier approaches zero since the differential input currents also approach zero. If the current source that biases the input amplifier is controlled externally, the transconductance can also be programmed over a wide dynamic range.

Typical applications include the use of this input bias current for multiplying purposes. Not only can the amplifier be gated OFF or ON for selective utilization, but in sample and hold applications, the open-collector output produces very little loading on an output hold compactor, especially with bias currents removed within the amplifier during the hold period. Since the output current is proportional to the product of the input voltage and the controlled emitter current, other multiplier applications include gain control, amplitude modulation, and four-quadrant multiplication.

BIFET Operational Amplifiers.   BIFET operational amplifiers are a class of circuits that use the BIFET technology. Historically, the performance of operational amplifiers has been limited by the standard bipolar technology. The large dynamic range usually required of the input common-mode voltage and of the output-voltage swing means that large dc level shifting in the amplifier must be accomplished; npn transistors were generally used for level shifting ac signals to higher dc voltages, and pnp transistors were generally used to level shift ac signals to lower dc voltages. Both types of level shifting are common in the classical two-stage operational amplifier. Unfortunately, the frequency performance of bipolar lateral-pnp transistors is poor relative to that of npn transistors. As a result, most bipolar operational amplifiers with a large dynamic range had limited frequency-response capability, usually with 1- to 3-MHz unity-gain band-

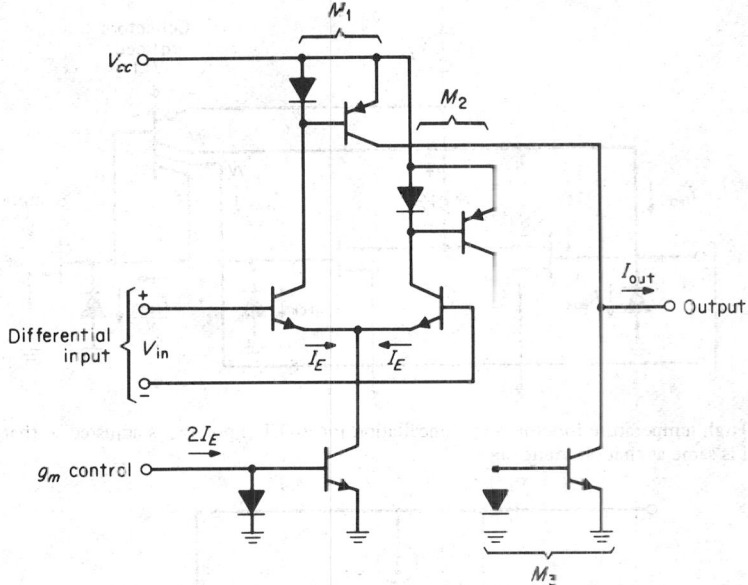

Fig. 8-89. Basic transconductance amplifier

widths. The use of substrate-pnp transistor followers in the output stage further aggravated the poor frequency response.

Classically, the transconductance g_m of the input amplifier equals the unity-gain bandwidth BW times the frequency-compensation capacitance C_c. Thus, for given bipolar bandwidths, the input transconductance was limited by the affordable size of the compensation capacitance. This transconductance limitation also limited the slew rate of the operational amplifier since both are directly related to the limited input tail current I_0. Several bipolar techniques have been developed to separate the bandwidth–slew-rate interdependence but not without inducing input offset-voltage trade-offs.

Another characteristic of bipolar operational amplifiers is their relatively high input bias current, typically in the 100- to 500-nA region. Some superbeta operational amplifiers, however, could achieve 1- to 2-nA input values.

The introduction of the JFET to the bipolar IC operational-amplifier technology produced improvements in bandwidth, slew rate, and input bias current. The ability of JFETs to level shift ac signals to lower dc voltage levels at higher frequencies with less excess phase shift was a significant advantage over the conventional lateral-pnp transistor. This fact, in and of itself, allowed BIFET operational amplifiers to achieve higher unity-gain stable bandwidths. In some cases, the conventional substrate-pnp follower has been replaced by JFETs in the output stage to increase bandwidths further. Thus for a given input transconductance, the higher bandwidths allowed reduction of the compensation capacitance.

It is characteristic of JFET amplifiers to be biased at a substantially higher quiescent current for a given transconductance than bipolar amplifiers. Thus for a given input transconductance, the quiescent tail current of a BIFET differential input amplifier is greater than that of a bipolar input stage.

As a result, higher slew rates can be achieved with BIFET operational amplifiers, since not only is there more input tail current available to slew the compensation capacitance but the capacitance itself is generally reduced.

The input bias current of BIFET operational amplifiers is typically in the 40- to 100-pA range at 25°C. This low input current dominantly results from the large n-epitaxial-gate-to-p-substrate reversed-biased junction of the JFET. Less dominant gate-to-channel reverse-biased junction leakage also contributes to the input current. At higher temperatures these junction leakages increase, causing a corresponding increase in input bias current. Techniques have been devel-

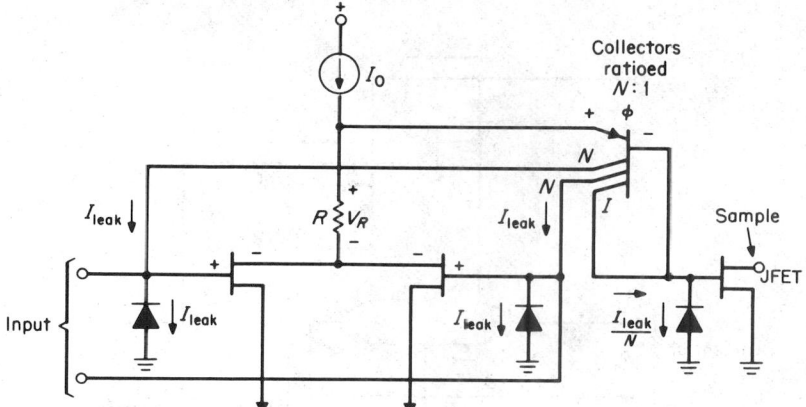

Fig. 8-90. High-temperature input-leakage cancellation for BIFET input. V_R is adjusted so that the sample gate potential is same as that on input gates.

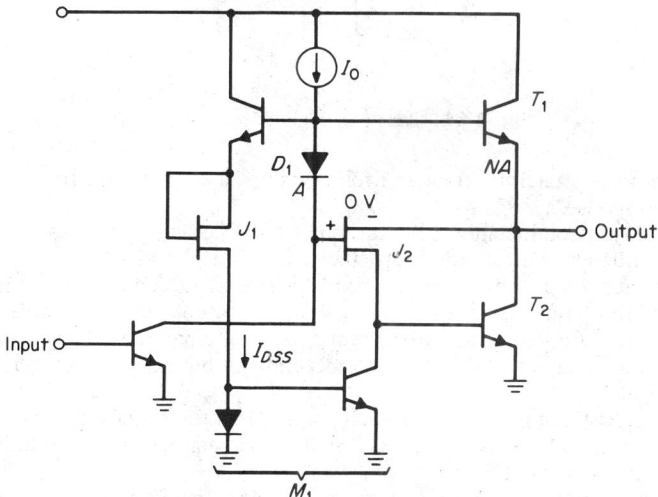

Fig. 8-91. JFET output stage for BIFET operational amplifiers.

oped which sample an n-epitaxial-to-p-substrate junction leakage current and add a scaled amount to the input, reducing input bias current by canceling input leakages. Usually the sampled junction is bootstrapped to the input-gate potential to ensure that the canceling leakage will track over all ranges of input-gate voltages. An example of this technique is shown in Fig. 8-90.

Figure 8-91 demonstrates the use of JFETs in the output stage. JFET J_1 is used as a current source of value I_{DSS} since the gate is shorted to the source. Current mirror M_1 reflects this current through an identical JFET J_2, forcing the gate-to-source voltage to this device to be zero as well. Since the voltage across the base emitter of the diode D_1 and transistor T_1 is the same, the quiescent value of output current through transistor T_1 will be controlled by the drive current I_0 and the ratio of emitter-base junction area of transistor T_1 to diode D_1. JFETS J_1 and J_2 are both tracking the output voltage to ensure that the gate-to-source voltage of JFET J_2 will remain zero, independent of the output voltage. For negative output swings, JFET J_2 provides the drive to the pull-down transistor T_2, and current source I_0 provides the drive to transistor T for positive output swings. This combination provides higher-frequency response with less excess phase than when using the conventional substrate-pnp output follower.

A simplified schematic of a typical BIFET operational amplifier is shown in Fig. 8-92. The differential input current mirror is often provided with emitter degeneration resistors to facilitate trimming the input offset voltage of the amplifier near zero by adjusting the resistor ratio. This is usually accomplished by allowing an external user adjust or an internal trim using laser technology.

75. Voltage Regulators.[68] The function of a voltage regulator is to maintain a constant output-voltage level, irrespective of the changes of the input voltage or the output current. Volt-

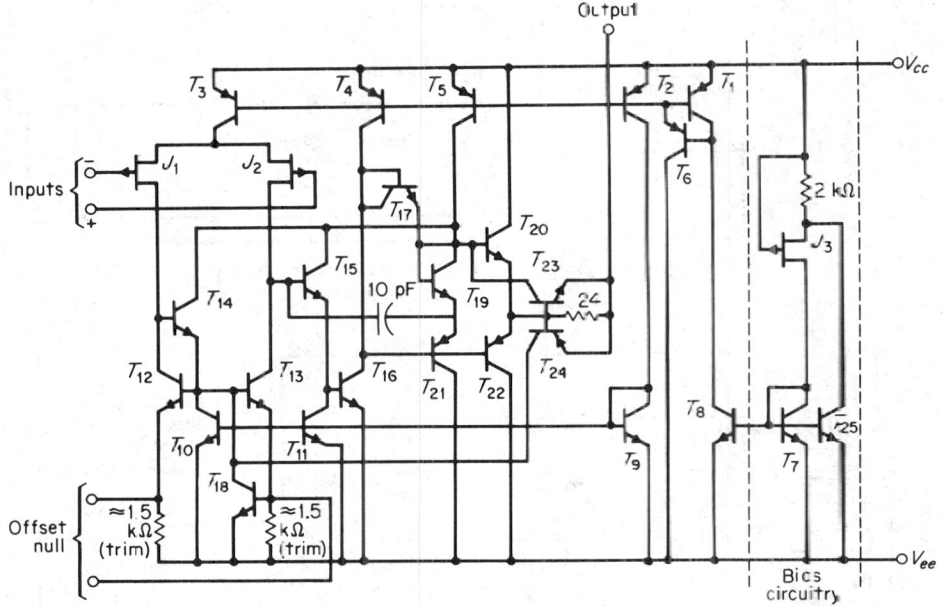

Fig. 8-92. Typical BIFET operational amplifier.

age-regulator circuits suitable for monolithic integration are *series*-type regulators which are connected in series between the load and the unregulated supply line. The key design parameters for a voltage regulator are its *line-* and *load*-regulation characteristics. In addition, the output voltage level is required to have very low temperature coefficient and to be protected from burning out under accidental output short-circuit conditions.

An integrated voltage-regulator circuit in general consists of three main parts: the reference voltage element, the error amplifier, and the series pass element. Figure 8-93 shows a simplified circuit diagram for an integrated regulator circuit where each of these basic sections is identified. The voltage reference V_R must be insensitive to supply voltage or temperature changes. It can be either external or internal to the monolithic design. For most applications the open-loop voltage gain A for the error amplifier is of the order of 60 to 70 dB, which can be obtained from a single high-gain differential stage using active (current-source) loads. The pass element is normally a power transistor connected in common-collector configuration. In most cases a Darlington connection of two transistors can be used as the pass device to reduce the output impedance and to provide adequate buffering of the error amplifier from the load.

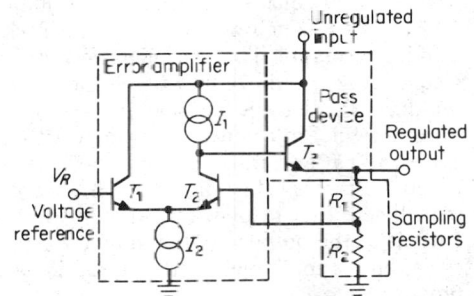

Fig. 8-93. Simplified diagram for a series-type voltage regulator.

Switching Regulators.[69,70]   As power requirements increase, the low efficiency (as low as 30%) of series pass regulators can no longer be ignored from either a cost or a thermal standpoint. Switching regulators whose efficiency is 70 to 80% do have some limitations: poorer load and line regulation, slower transient response, and much higher complexity. Figure 8-94 shows a typical half-bridge output configuration; the control circuit alternately saturates discrete power

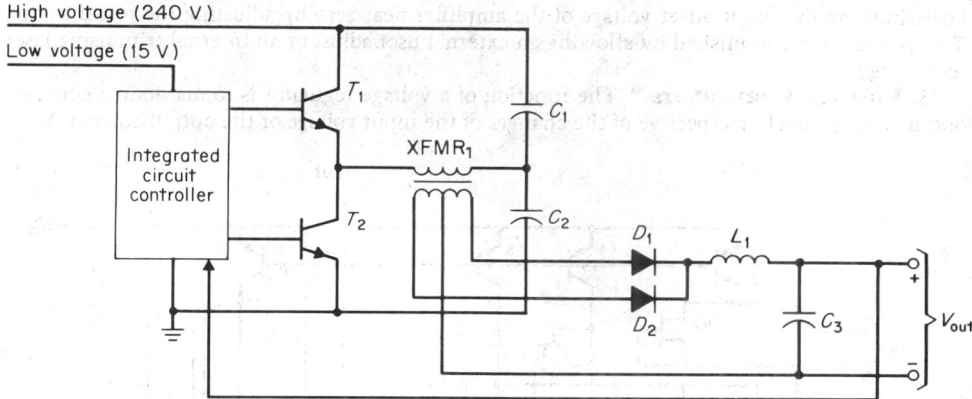

Fig. 8-94. Output portion of half-bridge switching regulator.

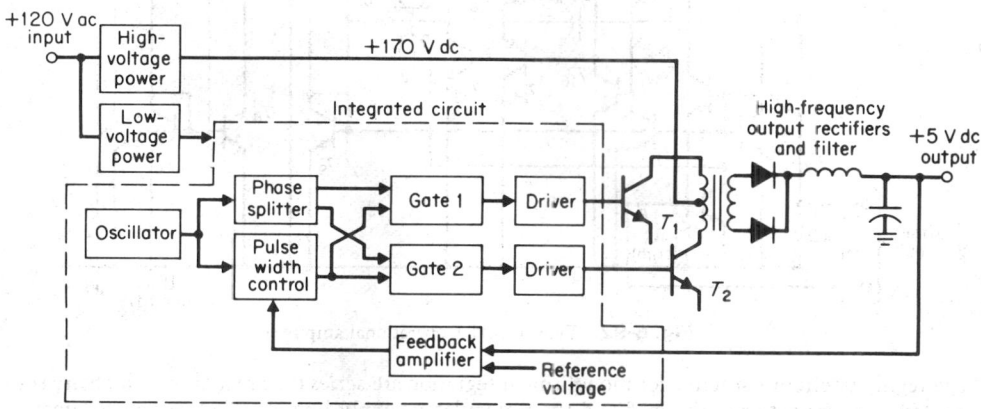

Fig. 8-95. Half-bridge switching regulator.

transistors T_1 and T_2. When T_1 is saturated, current begins to pass through the primary side of $XFMR_1$, with C_1 and C_2 providing the path to ground. The secondary current passes through diode D_1 and is filtered by L_1 and C_3. During the second half cycle T_1 turns off. After a delay, determined by load conditions, T_2 saturates, reversing the primary current, while the secondary current passes through diode D_2 to be filtered and stored. During regulation the average secondary current is equal to the load current and is controlled by the duty cycle of T_1 and T_2. Figure 8-95 shows a switching regulator with block diagram of the IC controller and a lower-voltage half-bridge output. The controller consists of a fixed-frequency oscillator for setting the operating frequency (typically 20 to 200 kHz), a phase splitter to drive the outputs out of phase, a pulse-width control in conjunction with output gates for controlling the output duty cycle, a feedback-error amplifier to compare the output with a reference voltage, and a pair of output drivers to drive the power transistors. Figure 8-96 shows typical waveforms of the regulator. When the output voltage is below the reference voltage, the output devices are on nearly 100%, excluding a small dead time to allow for switching times of the external devices, thus delivering maximum current to the output. As the output voltage rises and overshoots, the outputs go to 0% duty cycle, but as the output settles to the proper voltage, the duty cycle adjusts to maintain this proper output.

76. Wide-Band Amplifiers. Lack of large-value coupling and bypass capacitors and matching transformers makes the design of wide-band integrated amplifiers a difficult problem.

In the integrated design, the ac design and performance of the circuit cannot be considered as a separate problem from the dc bias considerations. The availability of a large number of well-matched active devices again provides a distinct advantage for the designer. For example, one can now use compound connections of two or more devices to replace a single transistor for improved high-frequency performance, or since the devices are well matched, their parasitics can be neutralized by proper circuit layout or interconnections.[71] Among the monolithic devices, the npn bipolar transistor has the best high-frequency capability. For this reason almost all wide-band amplifiers use npn devices in the signal path; the lateral- or substrate-pnp devices, when used, are confined to the biasing applications.

Compound Devices.   The gain-bandwidth product of an amplifier stage can be significantly improved by using a multiple number of active devices as a direct replacement for a single transistor.

Figure 8-97a shows the simplified circuit diagram of a common-collector–common-emitter stage. In this compound device configuration, T_1 effectively buffers the gain stage T_2 from the

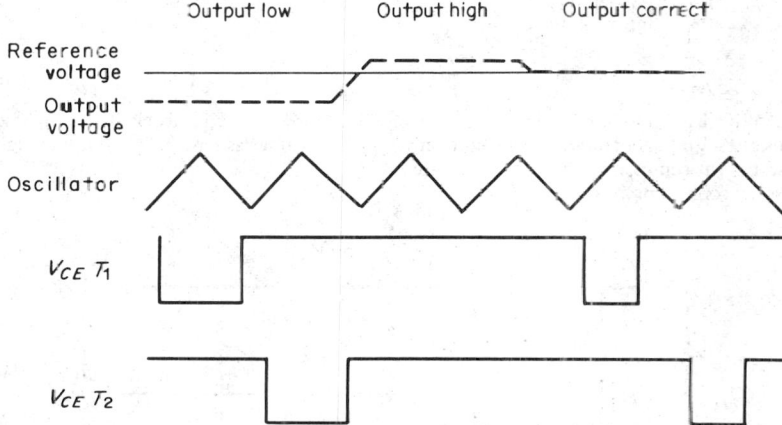

Fig. 8-96. Waveforms for a typical switching regulator.

input. Therefore capacitive loading at the input due to the Miller capacitance of T_2 is greatly reduced. This results in an improved gain-bandwidth product for the compound device over a single common-emitter stage, particularly for large values of source resistance R_s.

The cascode, or common-emitter–common-base stage of Fig. 8-97b, also forms a useful compound-device topology suitable for wide-band amplifier design. The improvement in the high-frequency performance is obtained by the impedance mismatch between the transistors T_1 and T_2. The collector load of T_1 is formed by the input impedance of T_2. Since T_2 is operated in common-base configuration, its input impedance is very low. This in turn keeps the voltage gain across T_1 low, and minimizes the capacitive loading at the input due to the base-collector capacitance of T_1. The advantage of the cascode configuration over a simple common-emitter stage is particularly significant at large values of R_L.

Feedback Stages.   Feedback is one of the most commonly used broadbanding techniques. It allows one to exchange gain for bandwidth in amplifier performance. Feedback can be applied locally (around each stage) or it can be applied around the overall amplifier. Figure 8-98 shows the two basic single-stage amplifier configurations using local feedback. Series feedback stage offers high input and low output impedance; therefore it operates best as a voltage amplifier. Shunt feedback stage is best suited to current amplification because of its low input and high output impedance. In cascading feedback stages, one normally alternates between series and shunt feedback. In this manner the shunt feedback stage can provide the necessary low-impedance drive for the series feedback stage; and conversely, the series feedback output can provide the high-impedance drive for the shunt feedback stage succeeding it.

Figure 8-99 shows the circuit diagram of a practical wide-band amplifier circuit which uses a cascade of differential series and shunt feedback stages to obtain broadband performance. In this circuit, the output emitter followers Q_5 and Q_6 also function as buffer stages and provide low output impedance in spite of shunt feedback applied around the second gain stage. The circuit gain can be adjusted by interconnecting the taps A and B and A', B' at the input stage. The monolithic circuit offers a bandwidth of 70 MHz with a voltage gain of 35 dB.

Figure 8-100 shows some of the basic wide-band amplifier configurations using overall rather than local feedback. Approximate voltage or current-gain expression for each of these stages is also shown in the figure. Stability considerations limit the practical number of gain stages which can be included in a feedback loop to three.

77. Communication Circuits. *Balanced Modulators.* A circuit configuration which is often used in modulator or mixer applications is the cross-coupled differential stage shown in Fig. 8-101. The differential configuration of the circuit makes it particularly suited to integra-

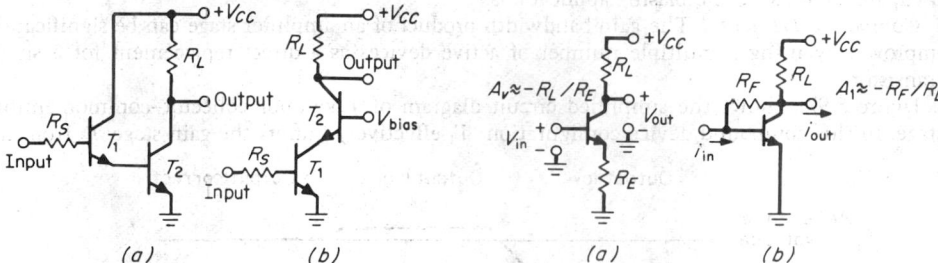

(a)   (b)   (a)   (b)

Fig. 8-97. Wide-band amplifier stages using compound transistors: (a) common-collector–common-emitter cascade; (b) common-emitter–common-base cascade (cascode connection).

Fig. 8-98. Basic feedback stages suitable for integration: (a) series feedback; (b) shunt feedback.

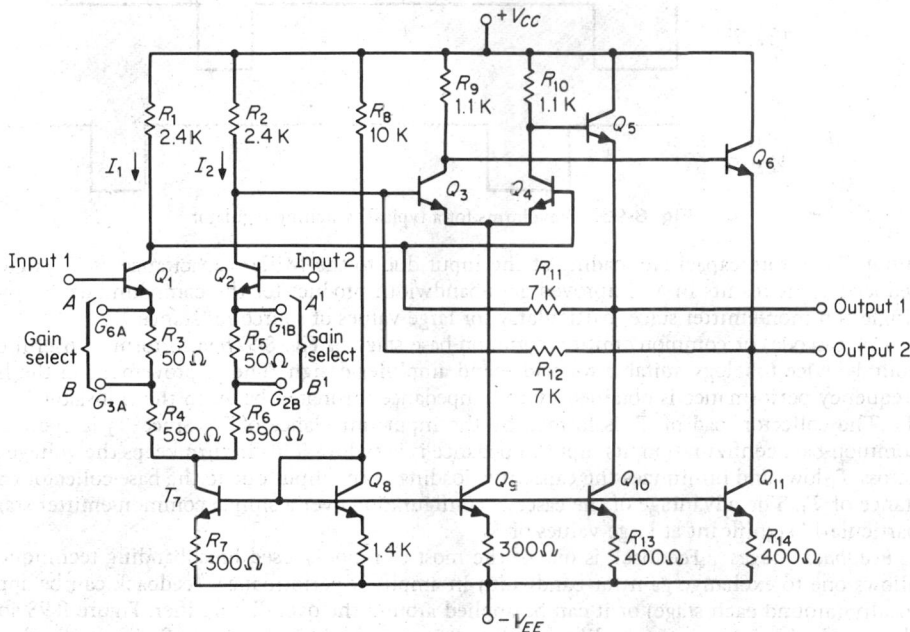

Fig. 8-99. Wide-band differential amplifier using a cascade of series and shunt feedback stages.

tion.[72] Because of the differential symmetry of the circuit, it provides an output signal only if both of the inputs are present simultaneously. Because of this property, it is called a balanced modulator. In the normal operation of the circuit, the high-level constant-amplitude carrier signal $V_1(t)$ is applied to the bases of the cross-coupled transistors T_1 through T_4. For high-level inputs $[V_1(t) > 25 \text{ mV, rms}]$ these transistors function as a set of synchronous single-pole, double-throw switches, as shown in the equivalent circuit model. The low-level modulation input is applied to the bases of T_5 and T_6 and is effectively *chopped* by the carrier signal. The output voltage $V_{out}(t)$ can be expressed as

$$V_{out}(t) = (R_L/R_E) \, v_2(t) \, S(t)$$

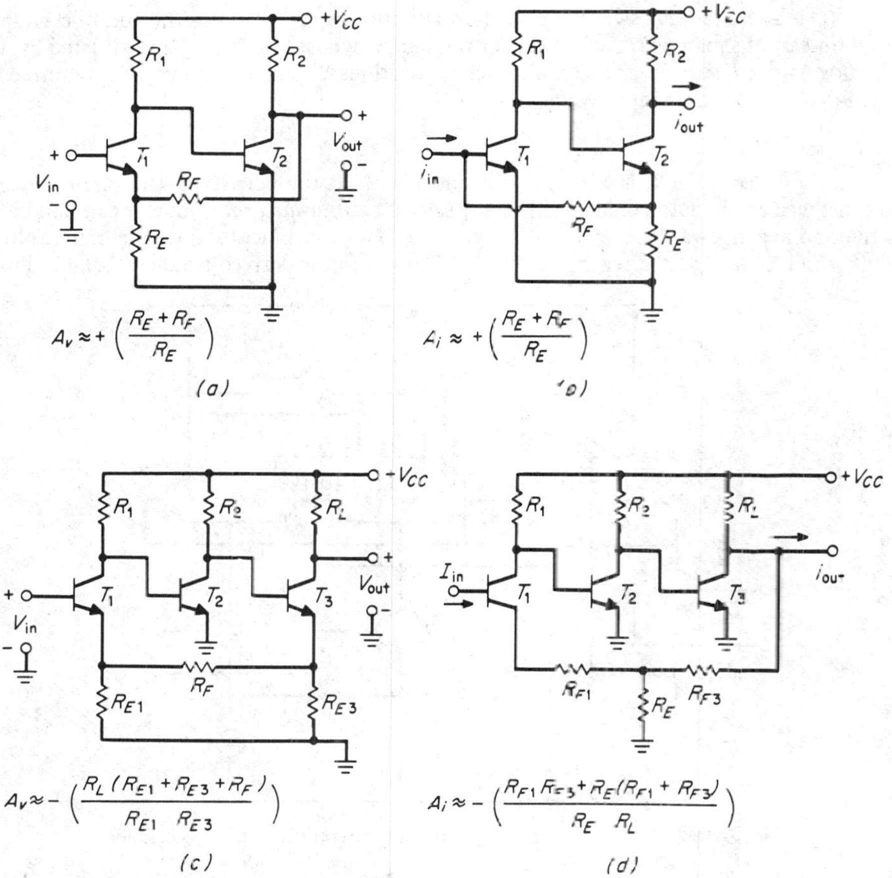

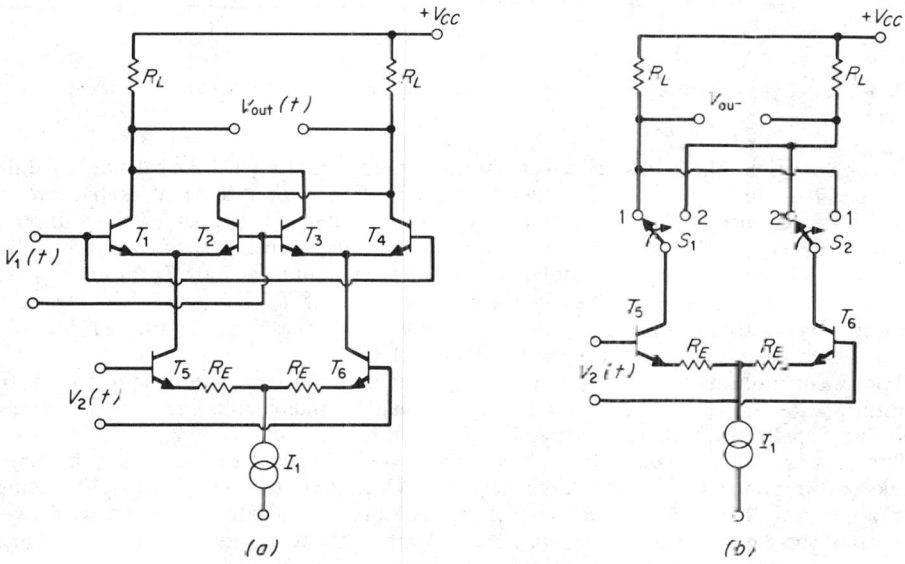

+VCC
R1
R2
T1
T2
RF
RE
Vin
Vout
iin
iout
RL
T3
RE1
RE3
Iin
RF1
RF3
Av ≈ + ((RE + RF) / RE)
Ai ≈ + ((RE + RF) / RE)
Av ≈ − ((RL (RE1 + RE3 + RF)) / (RE1 RE3))
Ai ≈ − ((RF1 RF3 + RE (RF1 + RF3)) / (RE RL))
(a)
(b)
(c)
(d)

Fig. 8-100. Basic feedback pairs and triples for wide-band amplifier design: (*a*) series-shunt pair; (*b*) shunt-series pair; (*c*) series-series triple; (*d*) shunt-shunt triple.

+VCC
RL
Vout (t)
T1
T2
T3
T4
V1 (t)
T5
T6
RE
V2(t)
I1
S1
S2
(a)
(b)

Fig. 8-101. (*a*) Balanced modulator circuit and (*b*) its equivalent circuit

where $S(t)$ is a square-wave switching waveform of unit amplitude. Thus the output corresponds to a symmetrical square wave at the carrier frequency, whose amplitude is modulated by $V_2(t)$.

Analog Multipliers.   In analog-multiplier applications, an analog output V_o is required to be related to the two sets of input voltages V_x, V_y as

$$V_o = KV_x V_y$$

where the gain constant K is normally chosen to be ⅒. If the polarity of the input voltages is also conserved at the output, the multiplier is said to be *four-quadrant*, since it can handle both positive and negative values of V_x and V_y. Figure 8-102 shows a practical four-quadrant multiplier circuit which is well suited for integration.[73,74] This circuit is derived from the balanced-modu-

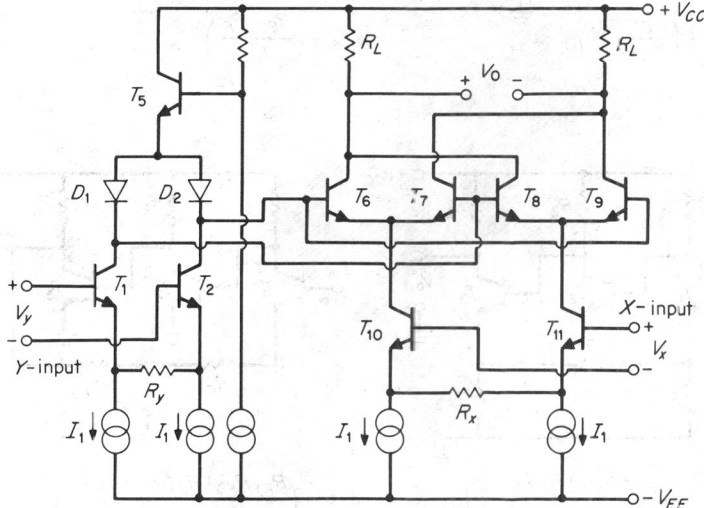

Fig. 8-102. Four-quadrant analog multiplier circuit suitable for integration.

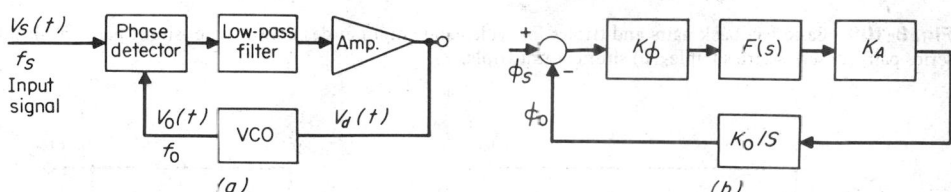

Fig. 8-103. The phase-locked loop: (*a*) functional block diagram; (*b*) linearized model as a negative feedback system.

lator configuration. The X-input signal effectively corresponds to $V_2(t)$ of the balanced modulator (see Fig. 8-101), and is processed linearly by the circuit. The Y-input signal V_y is first converted to a differential current at the collectors of T_1 and T_2, and causes a *logarithmic* voltage difference across diodes D_1 and D_2. This voltage difference is then exponentially related to the collector currents of the doubly balanced modulator section of the circuit formed by T_6 through T_9. If the diodes D_1 and D_2 are well matched to base-emitter diodes of T_6 through T_9, the logarithmic nonlinearity exactly cancels the exponential nonlinearity in the Y-signal path and thus allows the output to be a linear product of the two input voltages.

The analog multiplier can be used in a wide variety of analog computation applications, such as multiplication, division, square-root extraction, squaring, and obtaining root-mean-square values. These applications are well covered in the literature.[73,74]

Phase-Locked Loop Circuits.   The phase-locked loop (PLL) is one of the fundamental circuit blocks of communication systems. It is a frequency-selective circuit comprising a phase comparator, a low-pass filter, and an amplifier interconnected to form a feedback system as shown in Fig. 8-103. Some of the basic applications of a PLL are frequency-selective AM or FM demodulation, signal conditioning, frequency synchronization, and frequency synthesis.[75]

The basic principle of operation of a PLL can be briefly explained as follows. With no signal input to the system, the error voltage $V_d(t)$ of Fig. 8-103a is equal to zero. Then the voltage-controlled oscillator (VCO) operates at a set frequency, $\omega_0 = 2\pi f_0$, known as its *free-running* frequency. When an input signal $V_s(t)$ is applied to the system, the phase comparator compares the phase and frequency of the input with the VCO frequency and generates an error voltage, related to the frequency and phase difference between the two signals.

This error voltage is filtered, amplified, and applied to the control terminals of the VCO. In this manner, the control voltage $V_d(t)$ forces the VCO frequency to vary in a direction which reduces the frequency difference between f_0 and the input signal. If the input frequency f_s is sufficiently close to f_0, the feedback nature of the PLL causes the VCO to synchronize, or *lock*, with the incoming signal. Once in lock, the VCO frequency is identical with the input signal, except for a finite phase difference. This net phase difference ϕ_0 is necessary to generate the corrective error voltage V_d to shift the VCO frequency from its free-running value to the input signal frequency f_s and thus keep the PLL in lock. This self-correcting ability of the system also allows the PLL to *track* the frequency changes of the input signal once it is locked.

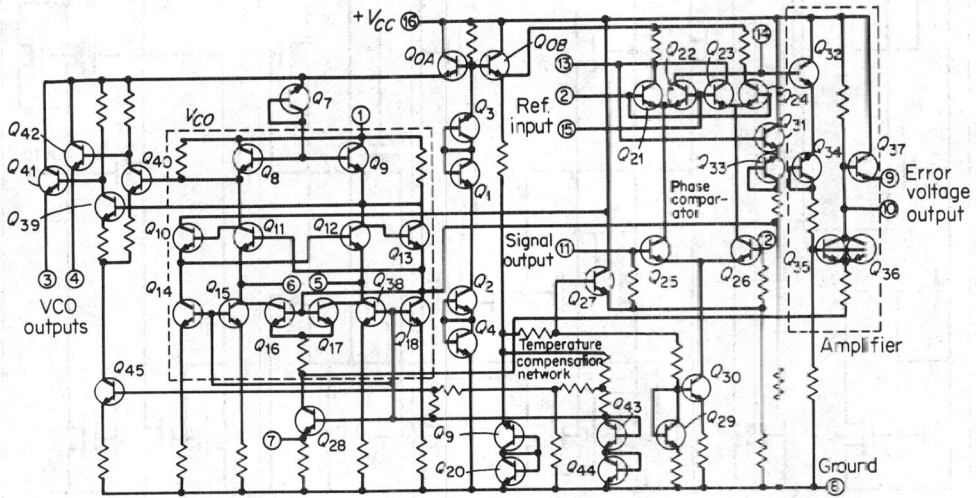

Fig. 8-104. Circuit schematic for a monolithic PLL circuit (Signetics NE562).

The range of frequencies over which the PLL can maintain lock with an input signal is defined as the *lock range* of the system. This is always larger than the band of frequencies over which the PLL can acquire lock with an incoming signal This latter range of frequencies is known as the *capture range* of the system. Thus the PLL has a high degree of frequency selectivity since it responds only to a narrow band of frequencies falling within the lock and capture ranges of the system. When the PLL is in lock, it can be approximated and analyzed as a linear feedback system, as shown in Fig. 8-103b, where $F(s)$ is the low-pass-filter transfer function.

The basic building blocks forming the PLL are well suited for integration. Figure 8-104 shows the circuit diagram of a monolithic phase-locked loop circuit, designed for FM demodulation and frequency synthesis applications.[76] The circuit uses a balanced-modulator type of phase comparator (see Fig. 8-101).

The VCO section is designed as an emitter-coupled multivibrator. The feedback path of the PLL is closed externally by connecting the VCO outputs to the *reference input* terminals of the phase comparator section. The idling frequency of the VCO is determined by a single external capacitor connected across oscillator terminals 5 and 6 in the figure The low-pass-filter section can be implemented by connecting a capacitor across the phase comparator outputs (terminals 13 and 14).

78. Linear Circuit Functions Using MOS Technology. *MOS Amplifiers and Comparators.*[77]   High-gain amplifiers are easily achieved with bipolar-transistor IC techniques because of the large voltage gain with a single amplifying stage. However, only recently has it been demonstrated that stable operational amplifiers with sufficient gain can be made with conventional CMOS and NMOS techniques. Figure 8-105 illustrates the complexity of an NMOS operational amplifier developed at the University of California at Berkeley; similar designs are used extensively in the Model 2912 pulse-code modulation filter. The 2912 has a wide common-mode voltage range and a large output-voltage swing. Typical of MOS devices that occupy small die areas compared with the space taken up by bipolar devices, this circuit requires only one-third the area of an equivalent bipolar-transistor operational amplifier (300 mil²). Although its open-loop gain is about 1,000 (far below that possible with bipolar-transistor operational amplifiers), it is ample for many applications. The unity-gain bandwidth is about 2 MHz, and power consumption is 10 mW. Figure 8-106 illustrates a CMOS operational amplifier like that used on the MC14413 PCM filter for similar filter interstage purposes. The complementary metal gate

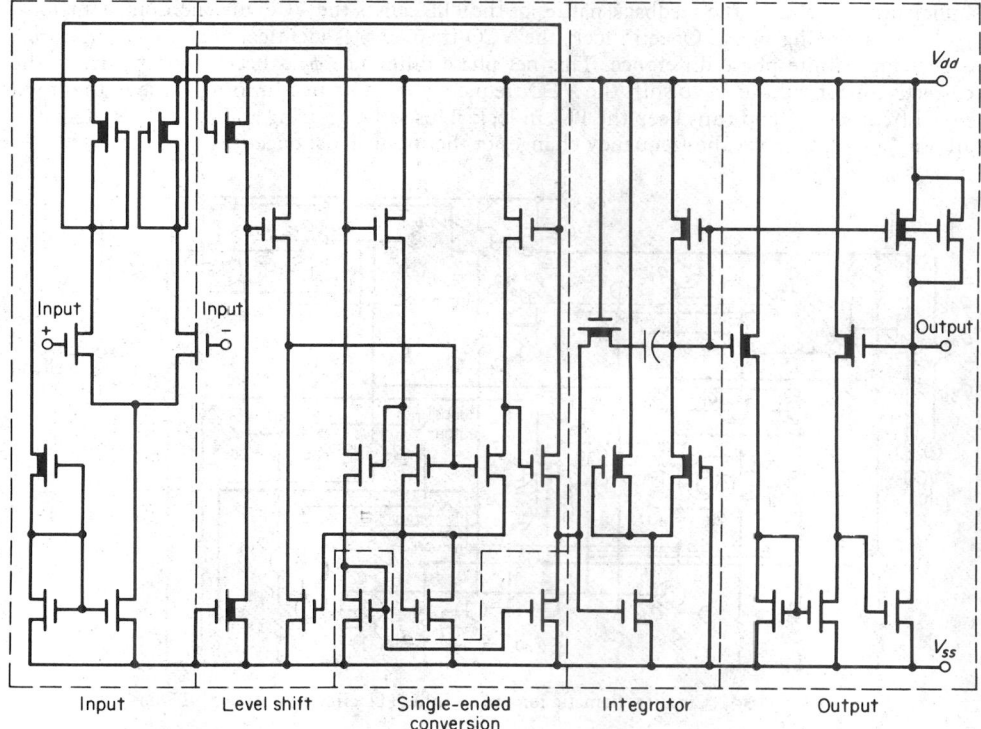

Fig. 8-105. NMOS operational amplifier.

p- and n-channel transistors allow simple circuit configurations to deliver performance (90 dB loop gain, 300 kHz unity-gain bandwidth) along with small die area (200 mils²) and low power consumption (0.5 mW).

The equivalent input-offset voltage for a nominally symmetrical MOS differential amplifier is typically 10 to 20 mV, about 10 times greater than that for bipolar-transistor amplifiers. This can be a severe limitation on precision comparators and other circuits. Fortunately circuit techniques exist to deal with this problem (Fig. 8-107). Using the technique in the figure, the equivalent input-offset voltage can be reduced by a factor of 10 or more if one uses electronically controlled analog switches. A small MOS transistor, with typical ON resistance of a few thousand ohms and OFF leakage current of less than 1 pA, performs well in this role. The relatively large ON resistance is entirely acceptable in conjunction with on-chip capacitance values of 100 pF and less, since the resulting RC time constant is under 1 μs. In contrast, current switching rather than voltage switching is used in bipolar-transistor analog circuits because of the inherent suitability of emitter-coupled transistor pairs as current switches.

Signal-Processing Functions   Many signal-processing functions can be realized with resistors and operational amplifiers. Addition, subtraction, multiplication, and division by fixed coefficients, as well as comparison of two voltages are examples. These same functions can sometimes be realized with capacitors, analog switches, and operational amplifiers (Fig. 8-108). One limitation in using capacitive summers and dividers is that the operations must be completed faster than the 10 to 100 ms it takes to accumulate significant error voltage from transistor leakage currents. Nevertheless, capacitors are more useful than resistors as precision passive elements in analog MOS LSI circuits. They dissipate no power under a steady-state voltage. In addition their precision capabilities for sample-and-hold functions can often be combined with precision-ratio voltage or charge divisions for high-level integration of functions in sampled-data analog systems. And they offer signal inversion at negligible cost by the simple inversion of a clock phase.

One problem with monolithic circuits having precision capacitor ratios is parasitic capacitance. Often it can equal the desired capacitance value of the circuit. For example, the parasitic capacitance from the top plate of capacitor C_1 in Fig. 8-109 can seriously degrade circuit performance.

To minimize parasitic capacitance, circuit designs with voltage-driven nodes (or with nodes at

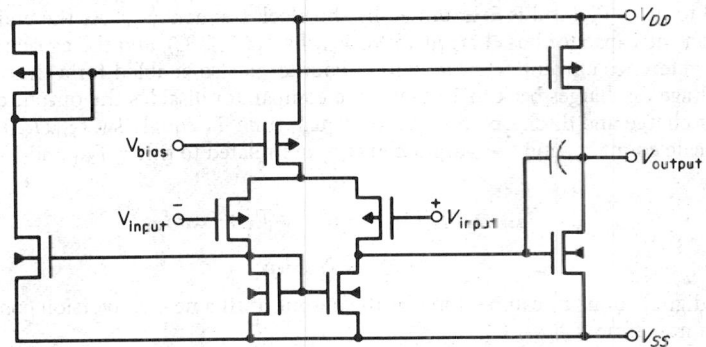

Fig. 8-106. CMOS operational amplifier used in PCM filter.

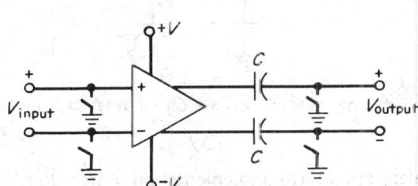

Fig. 8-107. Offset-nulled MOS differential amplifier. When switches are closed during the initialization cycle, the amplifier offset voltage is sampled and held on capacitors. Effective offset voltage is zero when switches are opened. *(After Ref. 77.)*

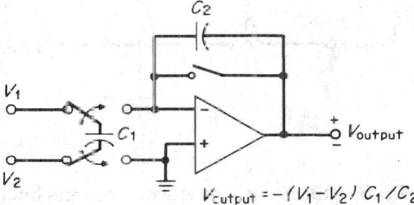

Fig. 8-108. Summing and multiplication configuration. Feedback switch is initially closed. When opened, the input switches are thrown to the right. Output voltage is then a function of the capacitance ratio and input voltage difference. *(After Ref. 77.)*

virtual ground potential) are generally used. Examples of voltage-driven nodes include those driven by supply voltages, ground potential, input voltages, and low-impedance amplifier outputs. Virtual-ground nodes can be obtained at inputs to operational-amplifier integrators.

79. Data-Conversion Circuits. *Analog-to-Digital (A/D) Converters.*[78] The A/D converter function provides a well-defined digital-output representation of the value of an analog input. Although other digital formats are used, the *natural binary representation* is the most common. Using binary, an n-bit A/D converter implies a resolution of 2^n possible output codes over the useful analog input range. An 8-bit A/D converter, for example, would have 256 digital output states, representing the full analog input range. Figure 8-110 is an ideal transfer curve of an A/D converter where each analog increment is the least significant bit (LSB) of analog value equal to the full-scale value divided by $2^n - 1$. Each digital increment represents one of the 2^n digital words. The accuracy of an A/D converter is measured against this ideal graph. Notice that the analog value must stay within ½ LSB of its ideal value or an output-bit change would occur. Thus an 8-bit A/D converter will be ½ LSB accurate out of 256 bits or 0.2% accurate. Linearity refers to an accuracy band in which all transitions will occur.

Flash A/D Converter. Figure 8-111 demonstrates the basic principle behind the flash A/D converter, so named because of its high conversion speed.

A precision voltage reference is divided by a precision resistor divider to provide individual comparator reference voltages. As the input voltage increases, the comparators switch in turn, and the encode logic provides an appropriate digital output as a function of the state of the comparator input voltages. On an n-bit A/D converter, $2^n - 1$ comparators and $2^n - 1$ corresponding resistor taps generally must be used. An 8-bit converter would require, for example,

255 comparators and 255 resistor taps. If a 10-mV LSB was allowed, the required input common-mode voltage would have to be at least 2.56 V. This limitation, however, is usually tolerated in view of the fast conversion times inherent in the parallel operation, limited only by the dynamics of the comparator and encode logic. Any output digital encoding can be used, including nonlinear varieties which weight the comparator reference voltages.

Dual-Ramp A/D Converter. Figure 8-112 demonstrates the popular dual-slope A/D converter. When a conversion command is given, an input current I_{IN}, which is proportional to the analog input, charges the ramp capacitor C for a reference time T_{REF}. The full-scale count is proportional to time T_{REF} and is determined by the clock frequency f_C. At the end of time T_{REF}, the voltage on the capacitor has charged above V_{REF} by $I_{IN}(T_{REF}/C)$, and the capacitor is switched to discharge reference current I_{REF}. The output counter is also enabled for a time T_X when the capacitor voltage discharges back to V_{REF}, and the comparator disables the output counter. Since the capacitor charge and discharge voltages are equal, time T_X equals $I_{IN}(T_{REF}/I_{REF})$.

The full-scale count n_{FS} and the output count n_X are related to times T_{REF} and T_X by the clock frequency f_C

$$T_{REF} = n_{FS}/f_C \quad \text{and} \quad T_X = n_X/f_C$$

Thus
$$n_X = I_{IN} N_{FS}/I_{REF} \tag{8-11}$$

The output digital count n_X can be stored at the output until a new conversion command begins to generate a new time T_X.

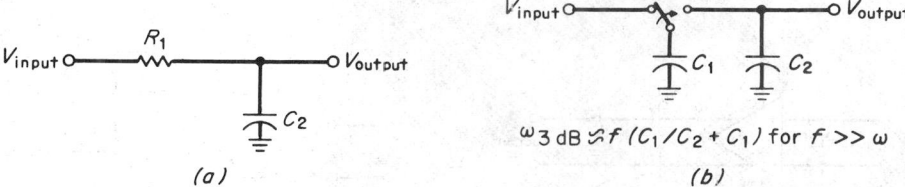

$$\omega_{3\,dB} \cong f\,(C_1/C_2 + C_1)\ \text{for}\ f \gg \omega$$

(a)   *(b)*

Fig. 8-109. Switched-capacitor low-pass filter: (a) equivalent circuit; (b) implementation. *(After Ref. 77.)*

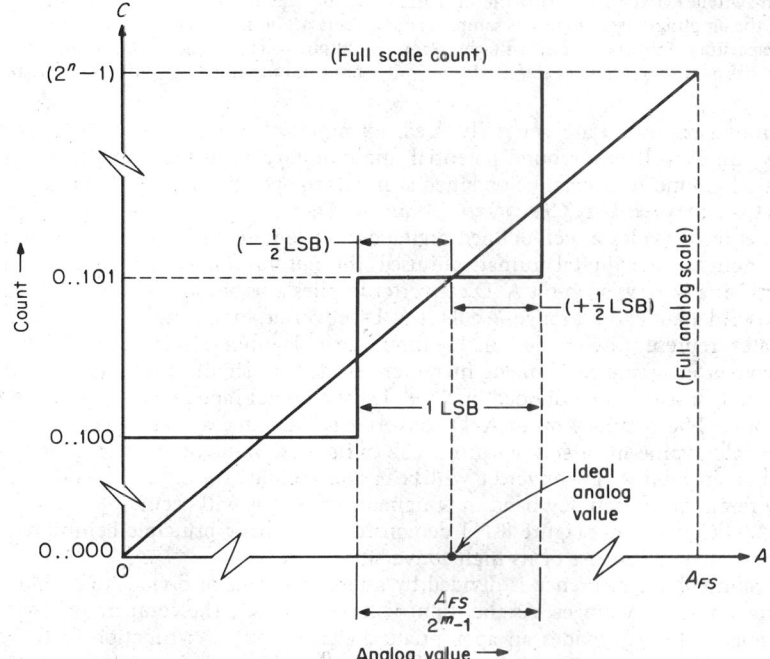

Fig. 8-110. Ideal transfer curve for an A/D or D/A converter.

This type of conversion is used where high accuracy is desired, since the output count is not a function of clock frequency or ramp capacitance and accurate current references can be provided. However, the conversion speed is generally slow, approximately equal to $T_{REF} + T_X$.

Staircase A/D Converter. The n-bit staircase A/D converter shown in Fig. 8-113 increments a digital counter from zero, converts the count into an analog output with a D/A converter, and compares this analog output with the analog input using a comparator. When the two analog voltages are equal, the increment sequence stops and the digital word in the counter represents the value of the analog input; $2^n - 1$ increments would occur at full scale. This technique is used where medium speed and accuracy are required. Accuracy is limited by the offset voltage and common-mode rejection of the comparator and the inherent accuracy of the D/A converter. Speed is limited to the speed of the comparator, the time for the increments to occur, and the settling time of the D/A converter.

Successive-Approximation (S/A) A/D Converter. The n-bit successive-approximation A/D converter (Fig. 8-114) first generates an n-bit half-scale word of 2^{n-1} with a special S/A register, converts this word into the half-scale analog output using a D/A converter, and compares this analog output with an analog input using a comparator. If the analog input is greater than half scale, a second word is generated by saving the original MSB and setting the next MSB, which increases the D/A converter output to three-fourths scale. However, if the analog input is less

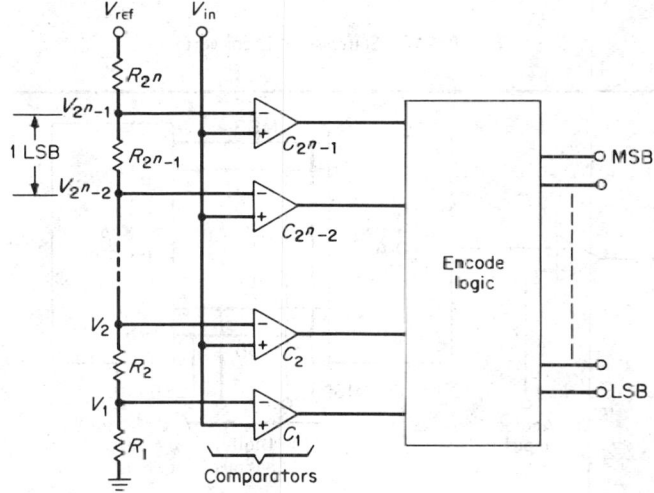

Fig. 8-111. Basic flash A/D converter.

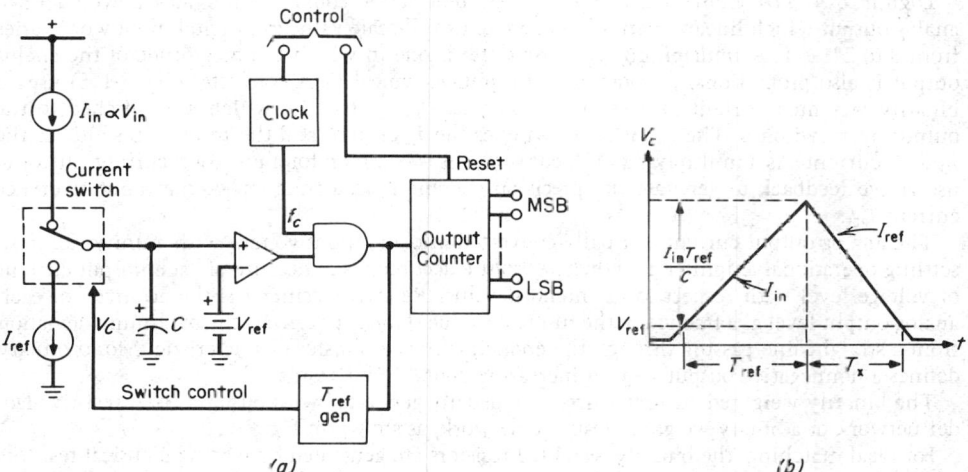

(a)   (b)

Fig. 8-112. Dual-ramp A/D converter: (a) basic block diagram; (b) capacitor waveform.

than half scale, the MSB is unset and only the next MSB is set, which decreases the D/A converter output to one-fourth scale. This process continues n times, producing a binary search that results in an n-bit word representing the analog input. The digital output can be in a serial or parallel form. Speed of conversion is faster than the stairstep approach, since only n steps are required, but bit decisions based on preceding bit valves are sensitive to noise-induced bit errors.

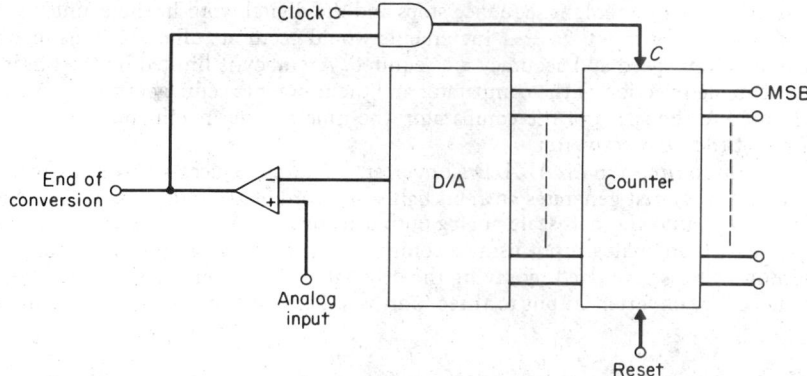

Fig. 8-113. Staircase A/D converter.

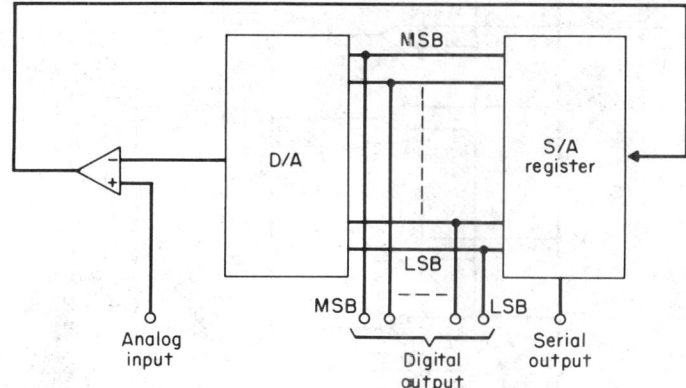

Fig. 8-114. Successive-approximation A/D converter.

Digital-to-Analog Converters (D/A).[79] An n-bit D/A converter provides a well-defined analog output which linearly varies from zero to its full-scale value as a digital input word varies from 0 to $2^n - 1$. A *multiplying D/A converter* is one in which the magnitude of the analog output is also proportional to some analog input. A typical D/A converter (Fig. 8-115) uses n binarily weighted current sources of I_0, $I_0/2^1$, $I_0/2^2$, ..., $I_0/2^{n-1}$, which are switched to the output by n switches. The input MSB switches the I_0 current and the input LSB switches the $I_0/2^{n-1}$ current. As a multiplying D/A converter, it is common for a precision-current mirror to use active feedback to generate the precision current I_0 as a function of the input reference current I_{REF}.

The analog output current is usually converted into an output voltage with a unity-gain fast-settling operational amplifier as shown. *Absolute accuracy* is a measure of each output current or voltage level with respect to its intended value. *Relative accuracy* is the accuracy of each analog output level as a fraction of the full-scale value. *Linearity error* is the maximum deviation from a straight line passing through the endpoints of the transfer characteristic. *Monotonicity* defines a nonnegative output step for increasing digital input words.

The binarily weighted current sources are usually generated with either a R-$2R$ resistive-ladder network or a binary weighted resistive network, as shown in Fig. 8-116.

For ideal matching, the binarily weighted resistors are generated by placing identical resistors in parallel. This technique, although accurate, requires $2^n - 1$ identical resistors. Current division in the R-$2R$ ladder requires only $5 + 3(n - 2)$ resistors and can be best understood by noticing that the current at node n is split into equal parts through the $2R$ resistors. Since at node $n - 1$ the resistance also appears as a $2R$ resistor in parallel with $2R$ of resistance, the current again splits evenly at node $n - 1$. This process continues down the ladder to generate the n binary currents, as shown. Although the R-$2R$ ladder has a decided cost advantage with fewer resistors, in the binarily weighted technique any one of the weighted currents can be trimmed without affecting the ratio of the remaining currents.

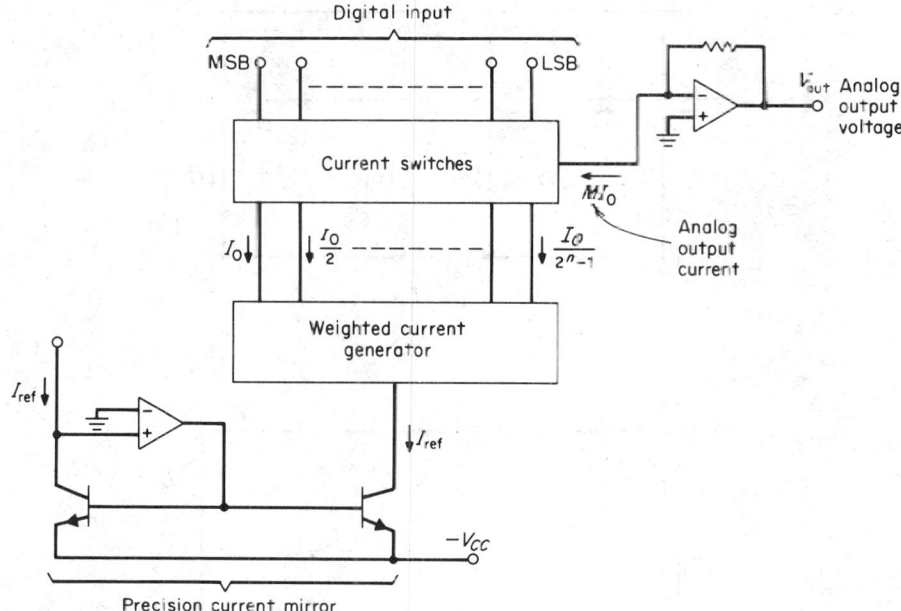

Fig. 8-115. Basic multiplexing D/A converter.

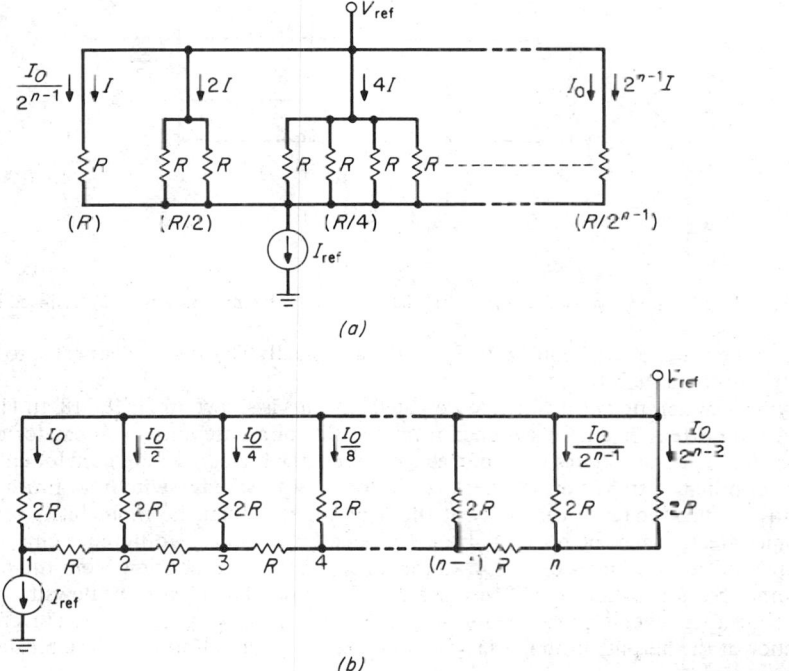

(a)

(b)

Fig. 8-116. Resistive-ladder networks for weighted current generation: (*a*) binary weighted; (*b*) R-2R.

When extracting accurate current from the resistive networks it is important for the reference potential to be identical at each terminating resistor end. This can be accomplished either by binarily scaling the emitter areas of the current source, as shown in Fig. 8-117a, or by using feedback termination techniques, as shown in Fig. 8-117b. In Fig. 8-117b the termination voltages will be identical if the currents I_X are all identical and the transistors T_1 to T_N have identical

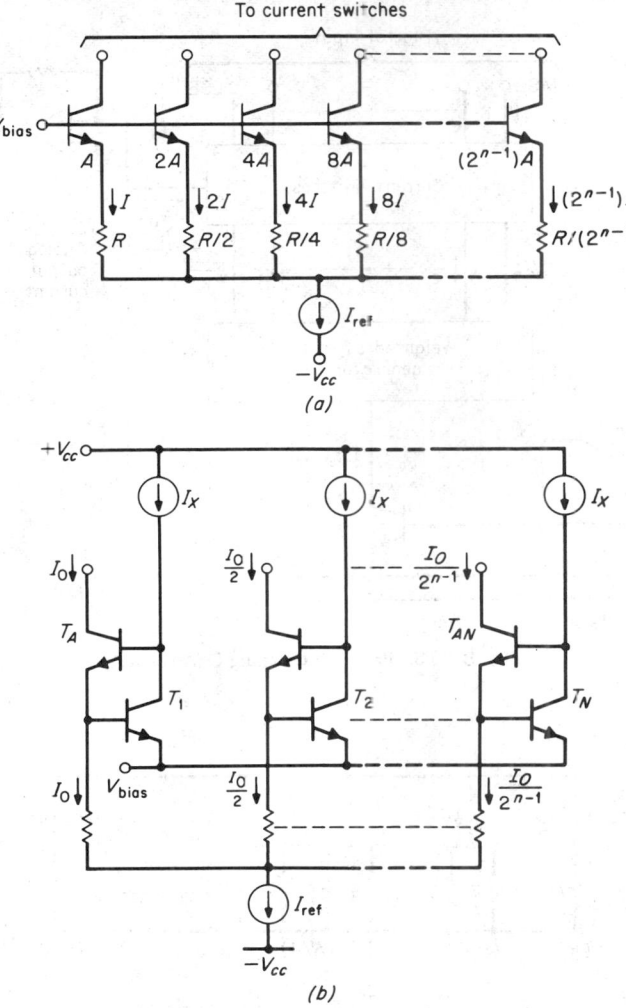

Fig. 8-117. Resistive-ladder current extraction: (a) binary emitter scaling; (b) feedback technique.

betas and emitter areas. Transistors T_A to T_{AN} are usually Darlington-connected to optimize current transfer through them.

Typical switching of the binarily weighted currents is shown in Fig. 8-118. In Fig. 8-118a, the input tail current is simply switched from one side of a differential comparator to the output. Figure 8-118b demonstrates an improved technique in which diodes are used for current steering. This eliminates transistor emitter-to-collector losses, speeds switching times, and allows improved dynamic-range capability at the logic input. When the input bit is zero, current I_L becomes greater than the binary current I_B. Diode D_{1n} clamps the difference current $(I_L - I_B)$ to ground and reverse biases diode D_{2n} so the binary current cannot flow from the output. When the input bit is 1, the current I_L becomes 0, and the total binary current flows from the output. By connecting several stages as shown, the diodes D_2 to D_{2n} perform the current-summation function at the output, although the output dynamic range is limited by the breakdown voltage of these diodes.

Typical applications of multiplying D/A conversion include programmable power supplies, gains, current sources, and pulse generators. Other applications include polarity switching, digital attenuators, panel meter readouts, digital subtraction and summation, sample and hold, analog division by a digital word, analog quotient of two digital words, analog product of two digital words, and A/D conversion.

MOS A/D and D/A Converters. The need for converters has been heightened with the advent of single-chip LSI microcomputers. In some cases it is desirable to include the data converter on the same chip with the microcomputer. A variety of circuit techniques for doing so exists.

One very simple technique works well for relatively slow A/D converters and is useful for making serial or slope A/D converters (Fig. 8-119). A minimum number of analog circuit ele-

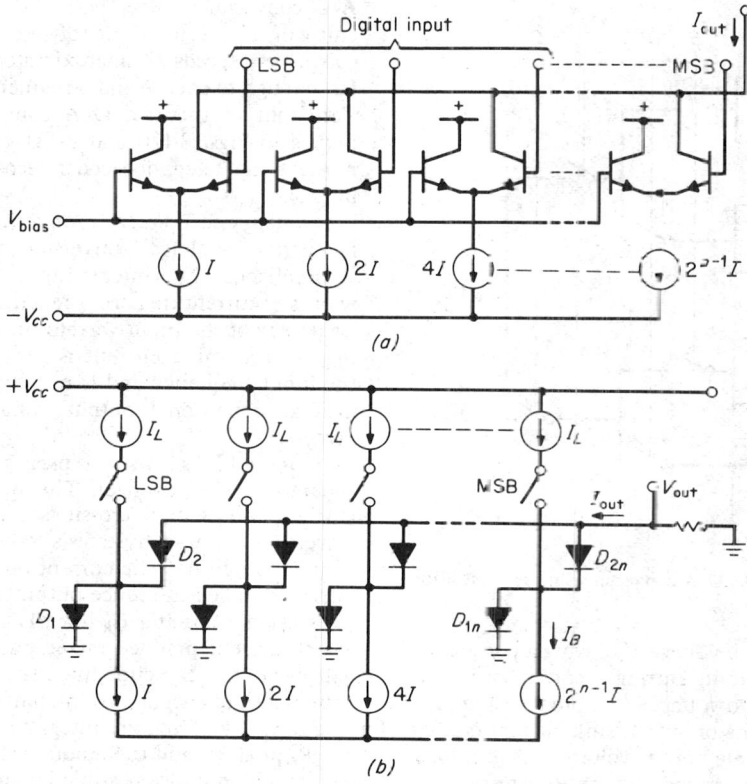

Fig. 8-118. D/A current-switching techniques: (a) differential switching; (b) diode switching.

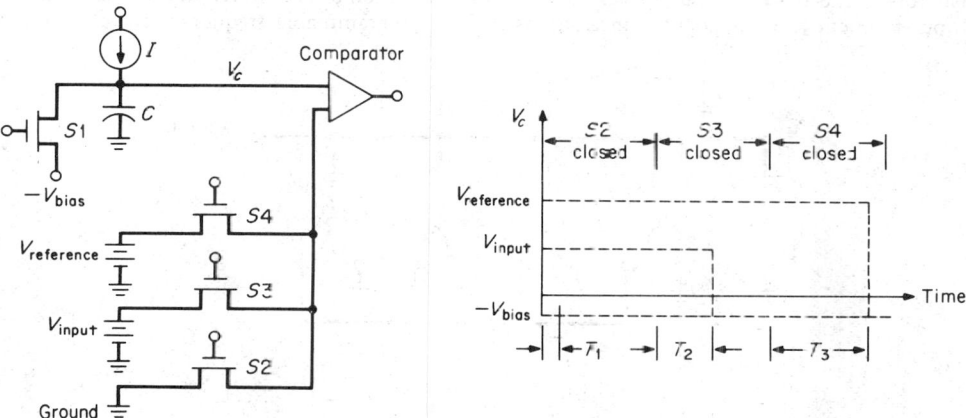

Fig. 8-119. Serial or slope A/D converter.

ments are employed, leading to a very small chip area. The subtractions and division to obtain the digital output are performed by microcomputer or special logic. The overall error of such a circuit is on the order of 0.1%, and conversion rates of 10 to 100 s^{-1} are achievable. Such specifications are adequate for many data-acquisition and instrumentation applications. Slope A/D converters are commercially available from several suppliers in LSI IC form.

For faster A/D and D/A conversion, several other techniques exist in MOS LSI technology. Series strings of resistors (Fig. 8-120) and binary-ratioed capacitor arrays (Fig. 8-121) are two techniques for D/A converter circuits. With either approach, a D/A converter circuit with a conversion speed of 1 μs or less can be made. When any of these approaches is combined with a comparator and successive-approximation logic, a high-speed successive-approximation A/D converter results (Fig. 8-122). Such a circuit can have 8 to 10 bits of resolution with conversion speeds of approximately 20 μs — fast enough to encode audio-frequency signals. Variations of the two D/A converter techniques in Figs. 8-120 and 8-121 are used in most of the telephone coder-decoder (codec) units.

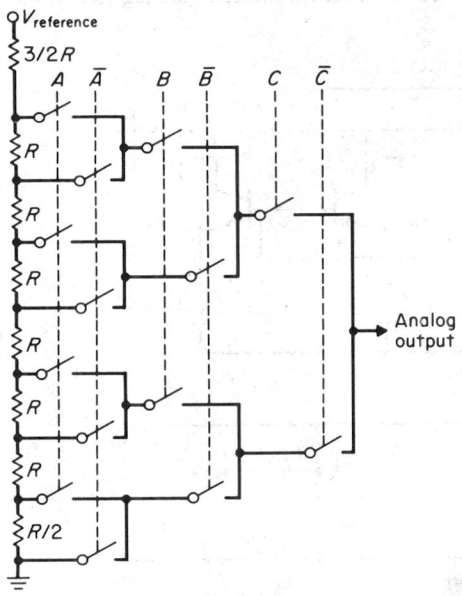

Fig. 8-120. D/A converter using series resistors for speed-up.

Frequency-to-Voltage Converters (F/V). Frequency-to-voltage conversion is usually accomplished by integrating fixed pulse widths of current that occur repetitively at the frequency of the input waveform. The resulting average value current is proportional to the input frequency and is usually converted into a proportional output voltage in the process.

Figure 8-123 shows a typical method for generating F/V conversion. The input comparator C_A senses zero crossings of the input waveform and uses hysteresis to ensure well-defined switching. The current pulses are frequency-doubled to reduce output ripple after integration. Capacitor C_R linearly ramps from voltage V_1 to voltage V_2 during a portion of the first half cycle to produce a fixed pulse width of output current. During a portion of the second half cycle, the capacitor linearly ramps from voltage V_2 to voltage V_1 to produce another fixed pulse width of output current. Thus two fixed pulse widths of output sink current occurred for one input cycle and are integrated as shown. The full-scale output voltage is determined by the $I_O R_F$ product, and the amount of frequency ripple is determined by the $R_F C_F$ product. Two-pole integration is also possible by changing the RC network configuration about the integration amplifier for improved ripple rejection. Comparators connected to the output voltage switch as a function of their reference voltage and the input frequency. Thus popular applications include programmable frequency switches. Over

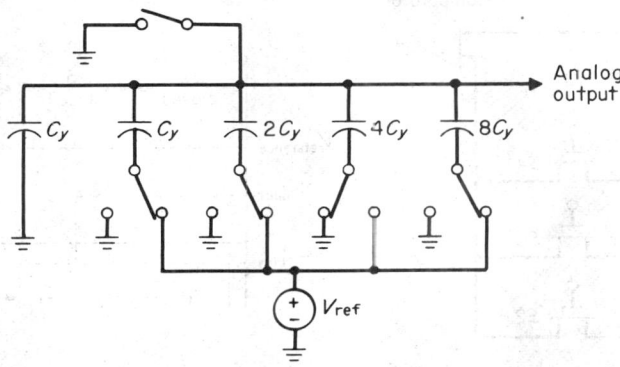

Fig. 8-121. D/A converter using binary-ratioed capacitor arrays.

and under speed sensing, tachometers, breaker-point dwell meters, and speed governors are just a few of the many applications for this type of circuit.

Voltage-to-Frequency Converters (V/F).[80] In general, most IC V/F converters use voltage-to-current converters to provide current for modulating a current-controlled oscillator using a capacitor as the timing element. If both charge and discharge currents are proportional to the

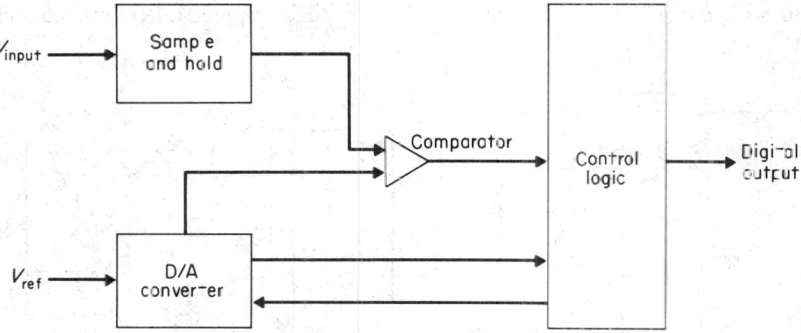

Fig. 8-122. High-speed successive-approximation A/D converter.

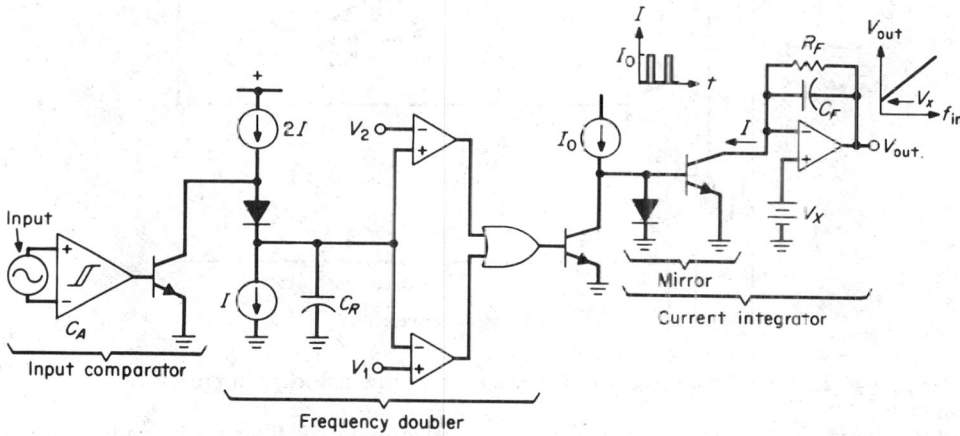

Fig. 8-123. Typical F/V converter.

input voltage, buffered output waveforms may be triangular, square, or both. An example of this type of IC V/F converter is shown in Fig. 8-124. Voltage-to-current conversion is accomplished with the use of a unity-gain feedback voltage follower. The input voltage V_{IN} will be regulated across resistor R_X, producing a current of $I_O = V_{IN}/R_X$ which is supplied to the current mirror M_1 through transistor T_X. The I_O charge current is mirrored directly to the capacitor through diode D_1. The capacitor discharge current I_O is pulled from the capacitor through diode D_2 with the use of mirrors M_1 and M_2. Current steering at the capacitor can be accomplished using the four-diode matrix and the switching transistor T_A.

During the charge mode, transistor Q_A is OFF, allowing current I_O to ramp the capacitor linearly to voltage V_H. At this point, the comparator C_F sets the RS flip-flop, which turns ON transistor Q_A, thus allowing current I_O to discharge the capacitor linearly to voltage V_L. At this point, the RS flip-flop is reset by comparator C_L, turning OFF transistor Q_A to begin a new cycle. Since the frequency of this oscillator is a direct function of current I_O and current I_O is a direct function of the input voltage V_{IN}, V/F conversion is accomplished. Notice that either a buffered square wave occurs at the flip-flop $\overline{Q}$ output or a triangular waveform is available at the output of the unity-gain buffer, as shown. Typical applications include FM modulation, signal generation, function generation, frequency-shift keying, and tone generation.

80. Integrated-Circuit Filters. A variety of newly emerging techniques is providing means of integrating filters capable of meeting high-quality filter specifications. Performance of these filters typically equals and in some particulars exceeds performance available with passive *RLC* filters.

At the present state of IC technology, all truly integrated filters are digital filters in which a continuous-time signal is sampled with a fixed periodicity and the analog amplitudes, the samples, are either linearly manipulated and transformed or digitally coded and processed before transformation. Digital filters which process sampled data linearly are classified as either *infinite-impulse-response* (IIR) *recursive filters* or *finite-impulse-response* (FIR) *nonrecursive filters*,

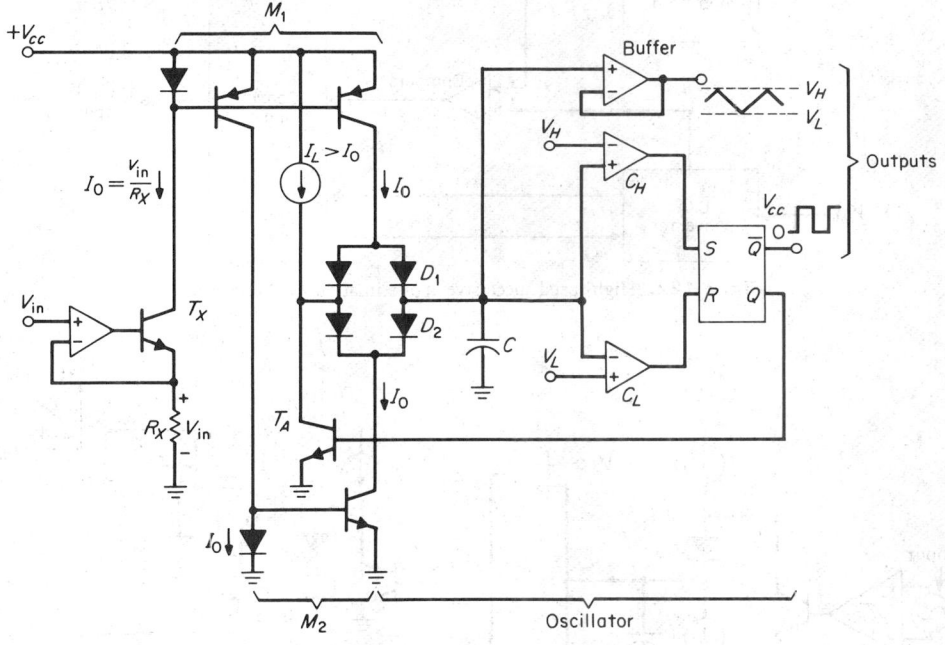

Fig. 8-124. V/F converter.

according as the filter does or does not use recursion in internal loops in processing the analog samples.

Infinite-Impulse-Response Integrated Filters. Design of an IIR filter traditionally involves derivation first of the desired analog filter function in continuous time and Laplace transform–related frequency domain and then converting them into the corresponding difference equations in discrete time and the Z-transform-related Z plane. The resulting Z-function defines the configuration of the IIR filter.

At present the only practical approach to IIR filter integration is provided by MOS LSI technology using switched capacitors. Switched-capacitor implementation replaces each resistor critical to the transfer function with a capacitor and two switches.[81,82] In Fig. 8-125a, when the switch is to the left, the capacitor charges to voltage V_1. When the switch is thrown to the right, the capacitor discharges to V_2. The net charge moving from input to output during one cycle (clock period) is $Q = C_1(V_2 - V_1)$. If the clock frequency is f, the average current flow I is

$$I = C_1(V_2 - V_1)f \quad \text{or} \quad (V_2 - V_1)/I = 1/C_1 f \qquad (8-12)$$

When the clock frequency is high with respect to the input-signal frequency, the effects of sampling can be ignored and Fig. 8-125b can be considered as a resistor replacement of value $R_1 = 1/fC_1$.

Figure 8-126a shows a simple continuous-time integrator whose transfer function is $T = -1/SR_1C_2$; Fig. 8-126b is its discrete-time replacement when $C_1 = 1/fR_1$.

This principle is used in implementing an active ladder ("leapfrog") *RLC* filter section[83] or can be extended to provide direct-form implementation of a second-order filter section.[84] Higher-order filters are realized by cascading or paralleling first- or second-order sections.

Finite-Impulse-Response Filters. The direct form for a FIR nonrecursive filter is shown in Fig. 8-127, in which $f_s = 1/T$ is the frequency with which the input continuous-time function $V_{IN}(t)$ is sampled. (This form is often called a *tapped-delay-line filter* or a *transversal filter*.) The sampling section is followed by M delay stages, each of which delays the associated sample by the period T. At each node K the signal is multiplied by the weighting coefficient $h(K)$, and the products are summed to provide the filter output in discrete time:

$$V_{OUT}(nT) = \sum_{k=1}^{M} h(K)V_K(nT) \qquad (8\text{-}13)$$

At $t = nT$ the delay associated with node K is $(K - 1)T$. Therefore,

$$V_K(nT) = V_{IN}[nT - (K - 1)T]$$

and

$$V_{OUT}(nT) = \sum_{k=1}^{M} h(K)V_{IN}[nT - (K - 1)T] \qquad (8\text{-}14)$$

This represents the linear convolution of the sequences $h(K)$ and the sampled input signal. Where $h(K)$ is the sequence of coefficients defining the impulse response of the filter, which is determined as the inverse Fourier transform of the desired filter response in the frequency domain, $V_{OUT}(nT)$ is the desired filter output in discrete time. Design of an FIR filter may consist of the derived impulse response modified by one of several available window functions. Computerized design programs using a different approach also are available.[85]

Implementation of this filter form is readily accomplished using a charge-transfer device, either a charge-coupled device (CCD) or a bucket-brigade device (BBD). Figure 8-128 is a schematic of one form of a CCD filter in which the weighting coefficients, i.e., the sequence $h(K)$ which defines the filter impulse response, are defined by the locations of splits in successive sensing gates. In this

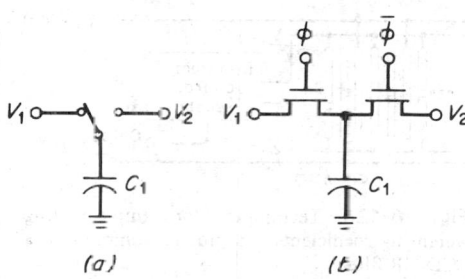

Fig. 8-125. Switched-capacitor replacement of resistors: (a) circuit technique; (b) MOS implementation.

structure a succession of charge samples which are linearly related to the sampled input voltage are propagated down the channel in the same manner as in a CCD shift register. As a charge packet moves into a potential well beneath a sense gate, a charge is induced on each gate segment which is proportional to the integral of the displacement current times the gate-segment length. With gate segments connected as shown, the filter output for a given discrete time increment is

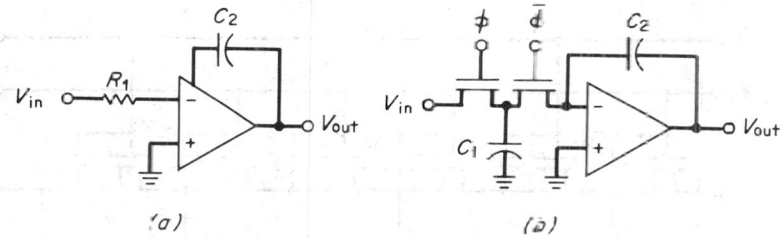

Fig. 8-126. Active integrator: (a) continuous-time and (b) discrete-time switched capacitor with MOS switches and integrated MOS operational amplifier.

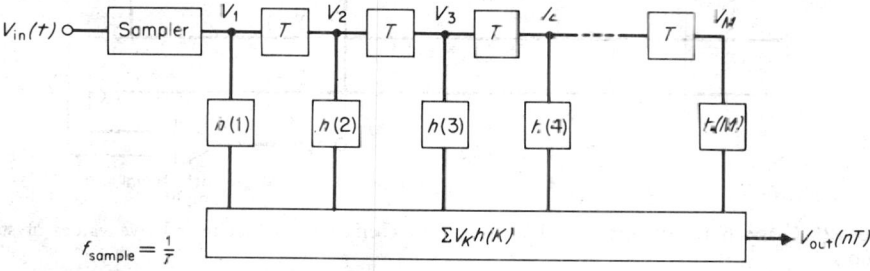

Fig. 8-127. Finite-impulse-response nonrecursive filter.

obtained by detecting the net charge difference between the two buses using a charge-sensitive differential amplifier,[86] usually an operational amplifier having a MOSFET input differential stage.

A BBD consists essentially of an array of capacitors and clocked switches. One of several methods of defining weighting coefficients in a basic BBD is shown in the simplified schematic of Fig. 8-129. A succession of charge packets representing the sampled input signal are injected serially into the line and conducted sequentially from one storage capacitor C_s to the next by appropriately clocking the ϕ_1 and ϕ_2 lines. Coefficient weights are set by the relative values of the storage capacitors. The resulting node voltages drive the associated source followers, which can be used as current sources to drive the output buses. These buses drive a differential amplifier to derive the output signal. The polarity of a coefficient is established by connecting its output to either the inverting or noninverting input to the amplifier.

Integration of both CCD and BBD structures uses MOS technology. In the CCD, overlapping gates are required for the charge-transfer process. In present technology this usually consists of a polycrystalline silicon first gate and either a polysilicon or metal second gate. Historically, integration of charge-transfer-device filters consisted of integrating only the CCD or BBD structures used with off-chip amplifier sensing circuits. However, with the recent maturing of MOS linear-circuit technology, fully integrated transversal filters in both CCD and BBD forms have appeared.

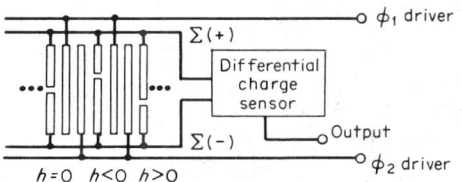

Fig. 8-128. Technique for implementing weighting coefficients and product summing in a CCD FIR filter.

FIR filters in symmetrical form provide truly linear phase response at frequencies up to the Nyquist frequency (one-half sampling frequency), an advantage not available in other filter forms. In addition they provide very narrow transition bands, typically of the order of 150 dB/octave. However, in common with all sampled signal systems, they have a limitation imposed by aliasing; i.e., their frequency response is periodic, with the sampling frequency as period.

Switched-Capacitor Filters. One solution to making monolithic frequency filters is to use switched capacitors. They were not implemented in MOS form until 1976 because of the lack of suitable MOS operational amplifiers. Figure 8-130a shows one switched-capacitor configuration for a high-Q bandpass filter. The center frequency and Q are determined by capacitance ratios and clock frequency, parameters that are easily controllable to within a few tenths of 1% without trimming or adjustment. Such a filter is as precise as those of active RC designs that are trimmed. Higher-order, switched-capacitor filters are possible when second-order sections are cas-

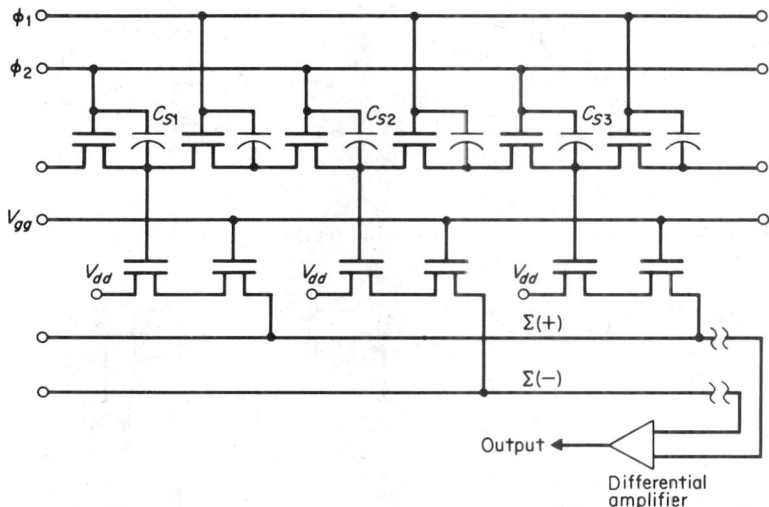

Fig. 8-129. Basic BBD FIR filter in which weighting coefficients are set by relative values of storage capacitors.

caded if an active-ladder configuration is used (Fig. 8-130b). With this technique Chebyshev filters and elliptic filters have been made.

Because switched-capacitor filters are sampled-data circuits, care must be taken to band-limit incoming signals to one-half the clock frequency to avoid foldover or antialiasing distortion. For audio applications, clock frequencies are in the range of 32 to 512 kHz; at the high end of this range, continuous-time band-limiting filters of simple construction can be used.

Another MOS LSI filter limitation is the electrical noise present in MOS amplifiers. At present such noise is at least an order of magnitude higher than the noise for bipolar-transistor amplifiers

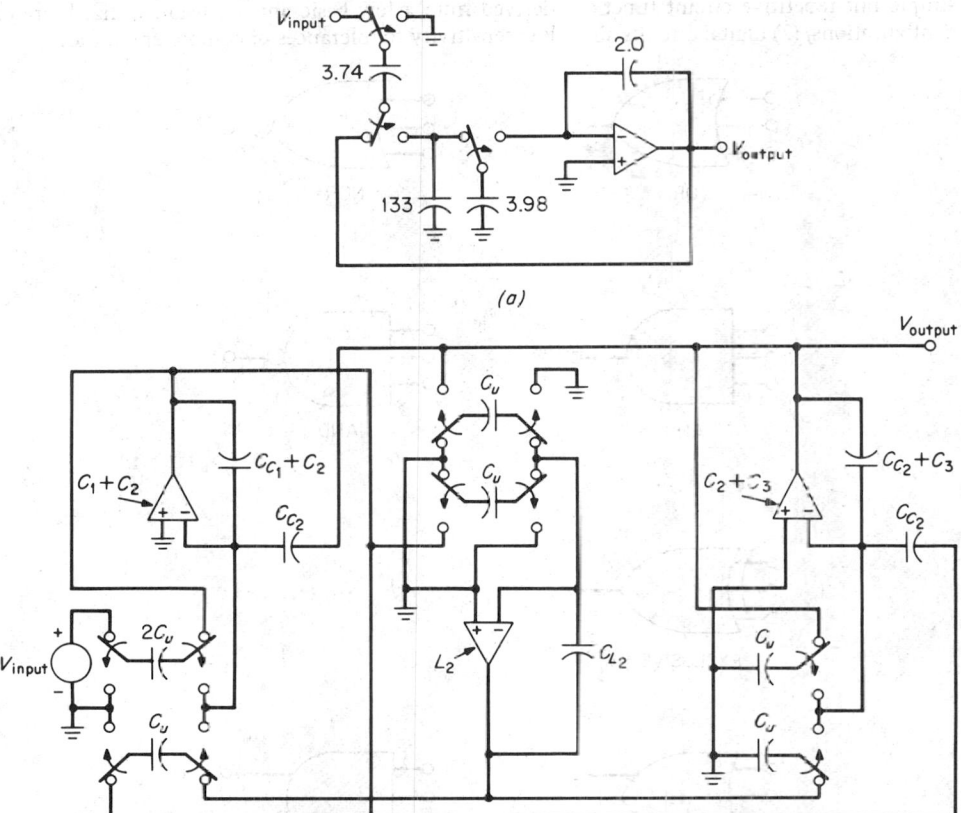

Fig. 8-130. Switched-capacitor configurations: (a) high-Q bandpass filter; (b) active-ladder configuration.

with junction field-effect input transistors (the equivalent rms noise voltage, referred to the amplifier input and integrated over the audio-frequency band, is in the range of 10 to 50 μV for MOS amplifiers). When several MOS amplifiers are used in a filter, such as a switched-C elliptic model, the overall signal-to-noise ratio is about 80 to 90 dB. This is adequate for telephony applications but insufficient for high-quality stereo-music systems.

Active Filters. Linear-active-filter design and synthesis techniques are conceptually well suited to integration, but from a practical point of view, the component tolerance and gain-sensitivity requirements often impose significant limitations in their application to ICs. The requirement of tight component absolute-value tolerance for linear active filters stems from one fundamental fact: the performance characteristics are a very strong function of the system natural frequencies of "poles" which are determined by the RC products and the overall loop gain in the feedback circuit. The absolute value of the loop gain can be desensitized at low frequencies by using local feedback around gain stages. However, the absolute-value control of a product of two dissimilar circuit elements such as a resistor and a capacitor requires a tight control of the absolute value of each element, and does not benefit from the matching and tracking between "similar" monolithic components. In many cases, this drawback restricts linear active filters to hybrid rather than monolithic ICs where additional trimming of component values is possible after circuit fabrication. Unfortunately, this also sacrifices some of the inherent batch-processing advantages of ICs.

A large number of design and synthesis techniques have been developed for designing active filters and are well covered in the literature.[87-91]

DIGITAL CIRCUITS

81. Digital Integrated Circuits. The greatest impact of ICs has been in the area of digital circuits, for the following reasons: (1) digital systems use large quantities of relatively simple but repetitive circuit functions derived from a few basic and well-standardized circuit configurations; (2) digital circuits show low sensitivity to tolerances of component values.

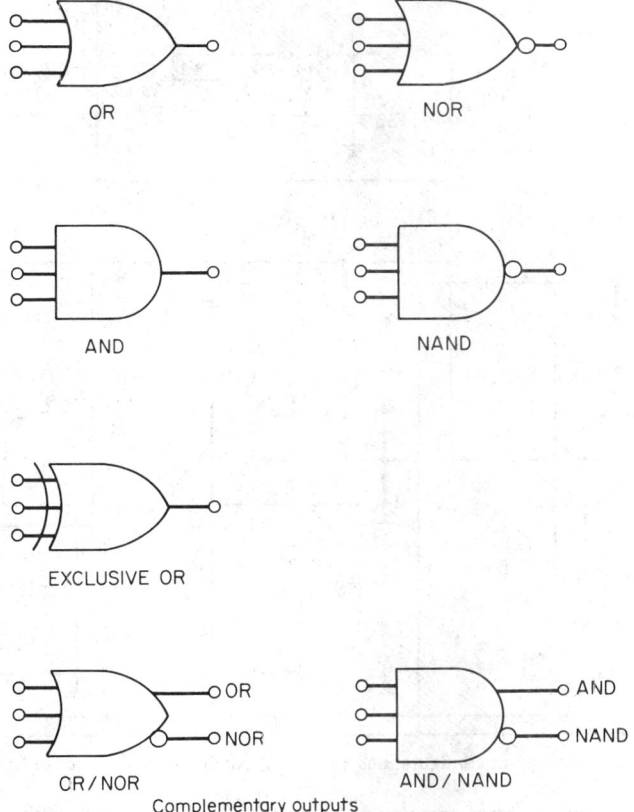

OR

NOR

AND

NAND

EXCLUSIVE OR

OR

NOR

OR / NOR

AND

NAND

AND / NAND

Complementary outputs

Fig. 8-131. Standard symbols for logic gates.

The logic gate is the fundamental circuit block in digital ICs, and it is customary to use the basic gate circuit for comparison between various logic families. Standard symbols for various gate types are shown in Fig. 8-131.

Eight major circuit families are used in digital ICs.[92-95] The basic characteristics and applications of each of these are briefly described below.

82. Resistor Transistor Logic (RTL). As the name implies, this logic family contains only resistors and transistors. The schematic diagram of a three-input RTL gate is shown in Fig. 8-132 together with its logic symbols.

The operation of the circuit in Fig. 8-132a is as follows. When input A goes to a high level, current will flow through R_1 and turns T_1 ON. The input voltage must be high enough to supply sufficient base current to drive T_1 into saturation. The base resistors must be large enough to prevent current hogging, which could be caused by V_{BE} mismatches between the transistors T_1, T_2, and T_3.

If any one or combination of the inputs goes to a high level, the respective transistors conduct, and current flow through R_L causes the output to go to a low state. For *positive logic*, the circuit performs the NOR function; i.e., if one or more inputs are high, the output is low. For *negative logic*, the circuit performs the NAND function; i.e., when all inputs are at a low level, each of the transistors is turned off, and the output voltage is at high level.

The ouput dc level at the high state of the RTL gate depends on the effective load resistance seen at the output port. This load resistance is in turn determined by the fan-out requirements. *Fan-out* is defined as the number of gate inputs which are driven by the output. Figure 8-132c shows the output voltage as a function of fan-out for a typical RTL gate with a load resistance of 640 Ω and R_B = 450 Ω, as indicated by the equivalent circuit. Temperature dependence of device parameters and the absolute-value tolerances associated with resistor values limit the "worst-case" fan-out capability of practical RTL circuits to five.

82. Diode-Transistor Logic (DTL). The diode-transistor logic gate circuit of Fig. 8-133 avoids some of the inherent shortcomings of RTL, such as current hogging or poor noise immunity, by using diodes in series with each of the input terminals. In the circuit of Fig. 8-133a, if one or more of the inputs is pulled low, the base drive to the output transistor is shunted to ground and T_1 stays in the OFF state. T_1 is turned ON only when all the logic inputs are high.

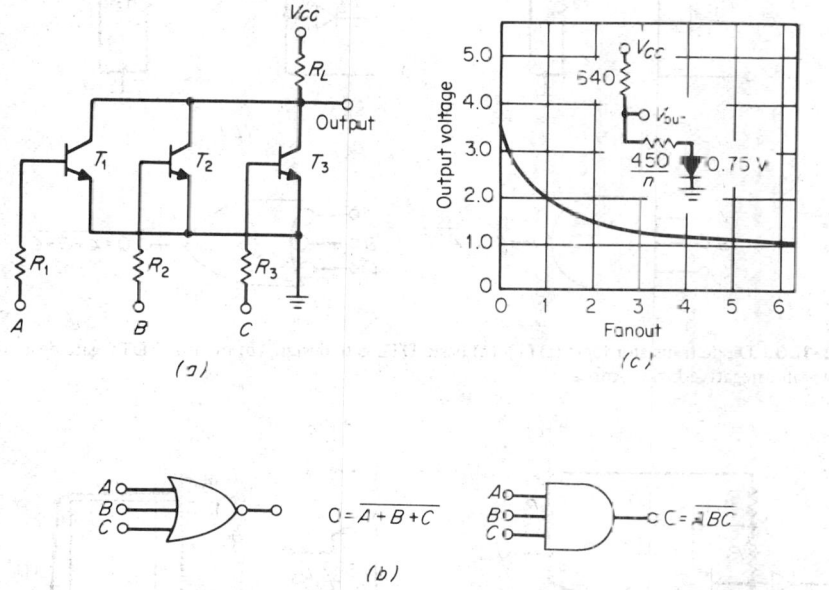

(a)

(c)

(b)

Fig. 8-132. Resistor-transistor logic (RTL): (a) three-input RTL gate design; (b) positive and negative logic symbols; (c) output voltage vs. fan-out and equivalent circuit for a typical RTL device (n fan-out).

The transistor acts simply as an inverting amplifier, while the logic function is performed by the diode network. The output voltage for the low level is $V_{CE_{sat}}$ (normally 0.2 V). Additional diodes may be added at the X input for increasing fan-in capability. The input threshold voltage and noise margin is determined by the saturation voltage of the transistors driving the input diodes and the forward voltage across diodes D_4 and D_5. Thus noise margin can be selected by the number of diodes in series with D_4. The pulldown resistor R_2 provides a discharge path for stored charge in the output transistor in saturation, thus reducing turnoff delay. This pulldown resistor also improves the noise immunity by keeping the transistor turned off during short positive-input transients.

A modified version of the basic DTL gate is obtained if one of the two series diodes is replaced by a transistor, as shown in Fig. 8-133b. In this design the gain of the transistor T_2 is used to reduce the required input power and to enhance the performance characteristics of the basic circuit. In the modified circuit the input current drawn through an input diode is reduced by approximately one-third. Since the transistor T_2 acts as an emitter follower, the available base drive for the output transistor T_1 is increased by almost a factor of 2. This reduces the minimum current-gain requirement for the output transistor and doubles the output fan-out capability.

DTL circuits lend themselves well to IC fabrication. For example, the input diodes can be placed into a single isolation pocket, because they share a common anode. This conserves area, especially in circuits with a large number of inputs. Another desirable feature of DTL logic is that the input impedance is high, for a high-level input, thus effectively unloading the driving circuit for positive signals. Figure 8-133c gives the logic symbols for DTL logic, for both positive and negative conventions.

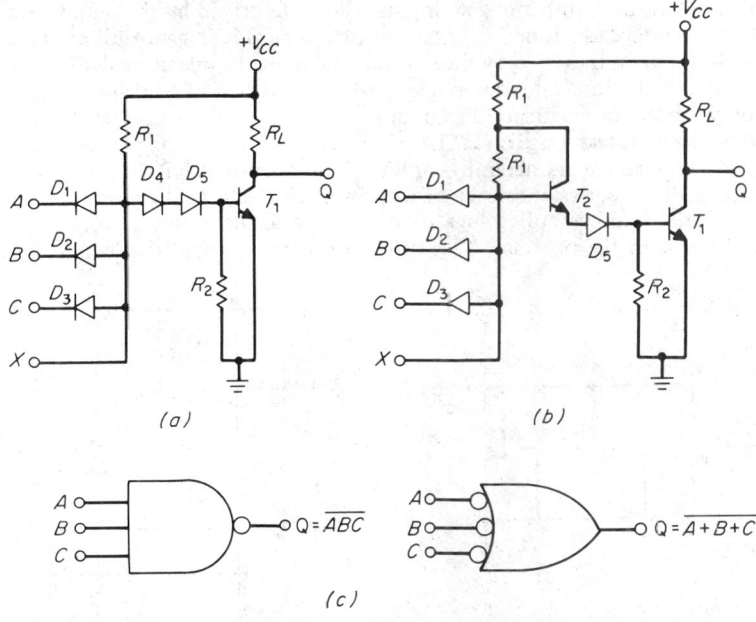

(a)   (b)

(c)

Fig. 8-133. Diode-transistor logic (DTL): (a) basic DTL gate design; (b) modified DTL gate design; (c) DTL positive- and negative-logic symbols.

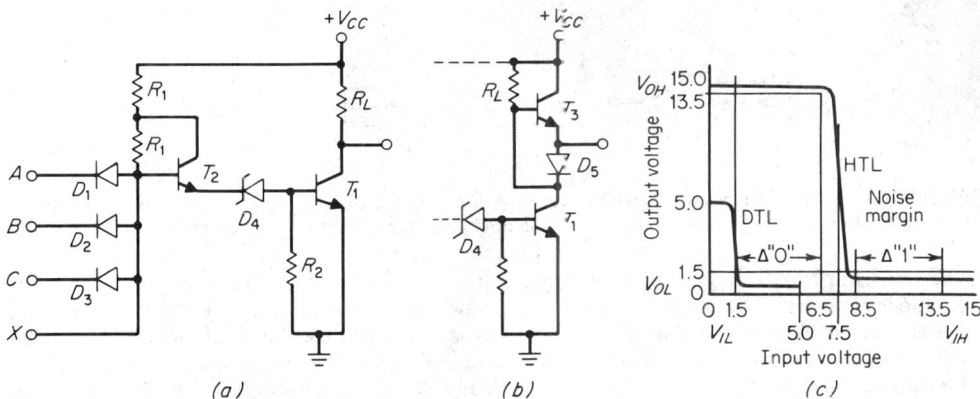

(a)   (b)   (c)

Fig. 8-134. High-threshold logic (HTL): (a) HTL gate design with passive pull-up; (b) active pull-up for HTL gate of part (a); (c) worst-case noise margins for HTL and DTL, V_{cc} = 15 V.

83. High-Threshold Logic (HTL). Many digital logic applications in noisy environments require circuitry with considerably greater noise immunity than that available from the standard logic families. High-threshold logic (HTL) has been designed for such applications. HTL designs are characterized by higher supply voltages (approximately 15 V), higher noise immunity, and higher thresholds than other logic families. These characteristics are usually obtained by adding a zener-diode voltage drop to the normal diode voltage drops in DTL designs. The circuit form is basically the same as DTL except for the zener diode, which replaces a conventional diode, as shown in Fig. 8-134a. To prevent excessive power dissipation at the higher supply voltages, the resistor values in the circuit are increased several fold over DTL circuits.

To reduce the output impedance of the HTL gate, an active pull-up configuration as shown in Fig. 8-134b can be used. With transistor T_1 turned off, current flows through R_1 into the base of the pull-up transistor T_3. This makes a large amount of current available to move the output voltage close to V_{CC}. When transistor T_1 is turned on, the load current flows through the diode D_5, thereby turning off the emitter-base junction of T_3. A disadvantage of the active pull-up circuit is that the implied AND function can no longer be used. Except for the higher voltage levels, the basic HTL gate operation is the same as the DTL.

The threshold voltage is dependent on the zener voltage of D_4 and on the base-emitter voltage of the output transistor T_1. The zener diode is a reverse-biased base-emitter junction and has a process-dependent voltage drop of approximately 6.9 V. This, together with the typical V_{BE} turn-on voltage of the output transistors, gives a 7.5-V gate threshold. Figure 8-134c shows the typical worst-case noise margins for HTL as compared with DTL.

The design shown in Fig. 8-134a and b allows a worst-case voltage noise immunity of 5 V. The thresholds are fairly insensitive to temperature changes since the positive temperature coef-

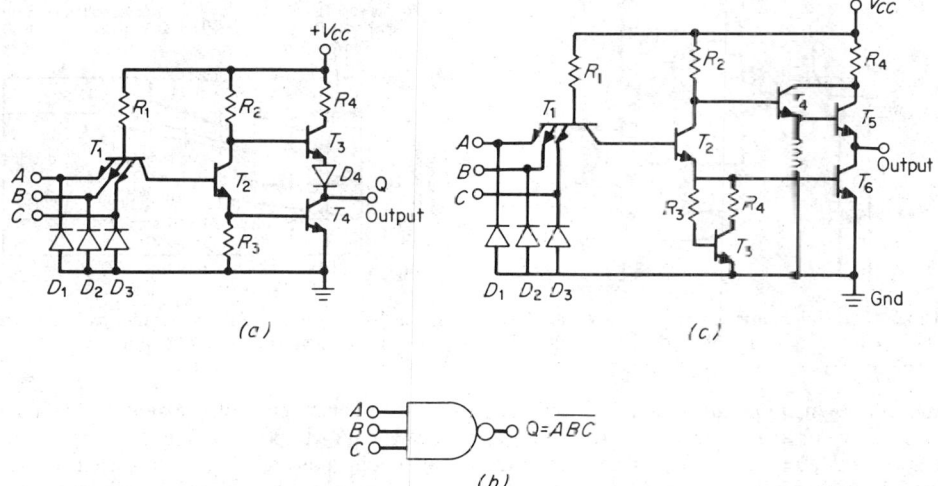

Fig. 8-135. Transistor-transistor logic (TTL): (a) typical medium-speed TTL gate; (b) TTL NAND gate for positive convention; (c) high-speed TTL circuit.

ficient of the zener diode is approximately compensated by the negative temperature coefficient of the base-emitter diode drop.

84. Transistor-Transistor Logic (TTL). TTL is one of the most commonly used logic families, because of its versatility and high speed. TTL is basically an extension of the DTL logic family, but it has higher noise immunity and output current capability. TTL logic circuits are classified as medium- and high-speed types.

Figure 8-135 shows the operation of the basic TTL gate. The input transistor T_1 performs the same function as the input diodes in a DTL circuit. For normal operation the clamp diodes D_1, D_2, and D_3 are reverse-biased and can be neglected. If any one of the inputs is at a low level, current flows through R_1, causes T_1 to conduct and keep collector of T_1 at a low voltage level, and thus prevents T_2 from conducting. Thus T_2 stays OFF and the output reads high.

When all the inputs are high, the base-emitter junction of T_1 is reverse-biased. Under this condition T_1 is OFF; yet its base-collector junction becomes conducting and provides the base current for T_2. Thus T_2 becomes conducting, and the output level reads low.

The output circuitry of Fig. 8-135, known as the *totem-pole* configuration, provides a higher output-current drive capability than the RTL or DTL type of circuit. The output-drive transistor T_2 is known as a *phase-splitter* stage because it provides a simultaneous in-and-out-of-phase drive to bases of T_4 and T_3. When T_2 is OFF, T_3 functions as an emitter follower and the output level is two V_{BE} below V_{CC}, corresponding to a high reading. If T_2 is ON or saturated T_4 conducts and the output level is equal to the saturation voltage $V_{CE,sat}$ of T_4 (~0.2 V). The basic three-input TTL gate shown in Fig. 8-135 performs the NAND function; i.e., the output will be low only when all the inputs are high.

Medium-Speed TTL. The medium-speed TTL gate uses the basic circuit configuration shown in Fig. 8-135a. One major advantage of the circuit is the high current-drive capability of the totem-pole output stage, which can provide fast rise times under capacitive loading conditions. The turnoff delay of TTL is shorter than that of a comparable DTL gate, since T_1 turns on during the "turnoff" period and rapidly drains the excess charge from the base of the phase-splitter transistor T_2. While T_2 is turning off, there is a shot time during which both T_3 and T_4 are simultaneously ON. This results in a brief low-impedance state across the power supply lines, which can cause a transient current spike. Total propagation delays associated with a medium-speed TTL gate are of the order of 10 to 15 ns.

High-Speed TTL. The basic circuit configuration for a high-speed TTL gate is shown in Fig. 8-135c. This circuit uses a lower resistance value for the base pulldown resistors to minimize parasitic RC time constants. The lower totem-pole transistor T_6 also has a lower base drive impedance provided by T_3, which increases its current-handling capability. The addition of the active

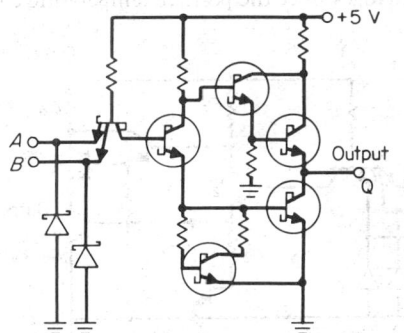

High-speed TTL designs conventional design (6 ns, 0.7 mW/MHz)

Active pull-down with Schottky (3.5 ns, 0.6 mW/MHz)

Active pull-down (6 ns, 0.4 mW/MHz)

Average power dissipation, mW

50.0
40.0
30.0
20.0
10.0
0

Medium-speed TTL designs original TTL design (10 ns, 0.35 mW/MHz)

5400/7400 design (13 ns, 0.3 mW/MHz)

5.0   15.0   25.0   35.0
Frequency, MHz

Fig. 8-136. Schematic diagram of active-pulldown TTL circuit employing Schottky diode clamping technique.

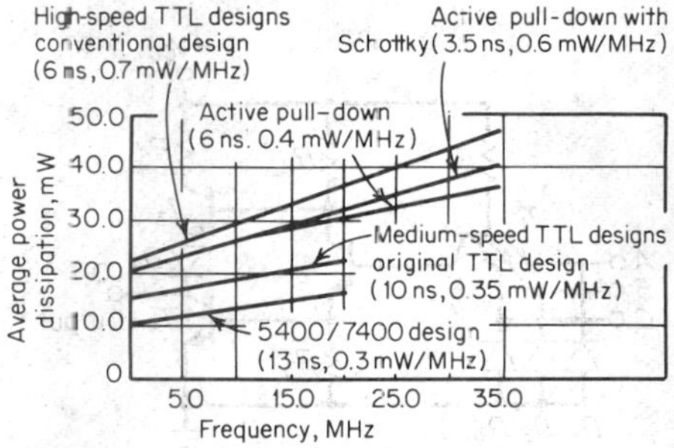

Fig. 8-137. Average TTL power dissipation vs. frequency for typical gate, with 50% duty cycle.

base-pulldown transistor T_3 also reduces the temperature dependence of the switching characteristics by compensating for the temperature drift of the V_{BE} drop of T_6. The Darlington connection of T_4 and T_5 on the upper part of the output totem pole reduces the transient current spike by reducing the turnoff delay. Typical values of propagation delay for high-speed TTL circuits are in the range of 6 to 10 ns.

Schottky Clamped TTL. In saturated logic circuits, storage-time delay associated with saturated transistors is the most significant limitation on switching speed. Storage-time delay can be eliminated by preventing the transistors from saturating. This can be accomplished by placing a clamping diode across the base-collector junction. If the clamping diode has a lower turn-on voltage than that of the collector-base junction diode, it will keep the collector-base junction from being forward-biased. Such a low turn-on voltage can be obtained by using a Schottky barrier diode as described in Par. **8-54** (see Figs. 8-55 to 8-57). The clamped npn transistor is formed by placing an integrated Schottky barrier diode in parallel with the base-collector junction, as shown in Fig. 8-57. A metal electrode is then connected across the p-type base and to the lightly doped n-type collector region, where it forms a rectifying contact. The Schottky diode turns on at a forward bias of about 0.4 V and keeps the transistor in its active region.

Figure 8-136 shows the schematic diagram of a high-speed TTL circuit using Schottky clamped transistors. The Schottky clamped transistors are identified with the electrical symbol defined in Fig. 8-57. Typical propagation delay for Schottky clamped TTL circuits is 3.5 to 6 ns. Figure 8-137 gives a comparison of speed and power-dissipation properties of various types of TTL circuits.[95]

85. Emitter-Coupled Logic (ECL). In emitter-coupled logic (also called current-mode logic) the transistors are switched between OFF and ACTIVE states without going into saturation. The transistors are kept out of saturation by limiting the total current flow through them. Since the transistors never go into saturation, storage-time delays are eliminated, and the overall propagation delay can be significantly reduced. Figure 8-138 shows the basic circuit configuration and the logic symbol for a three-input ECL gate. If all three inputs are low (approximately -1.6 V), the emitter voltage V_3 will be one diode drop below V_{BB} (approximately -1.9 V). Under this condition, none of the input transistors conduct; thus the base of T_6 is at ground potential, and the NOR output at the emitter of T_6 is one V_{BE} below ground.

If one or more of the inputs is high (-0.75 V), current will flow through R_1, causing the output of emitter follower T_6 to drop to about -1.65 V. This corresponds to a NOR function, since the output is *not* high if one or more of the inputs is high. The output at the emitter of T_7 is 180° out of phase with the NOR output; thus it corresponds to a logical OR function. The internal reference voltage V_{EB} determines the switching level of the gate. The emitter-follower outputs provide a low output impedance of about 15 Ω. The input circuit of the gate is similar to that used in most operational amplifiers. Its input impedance is of the order of 100 kΩ.

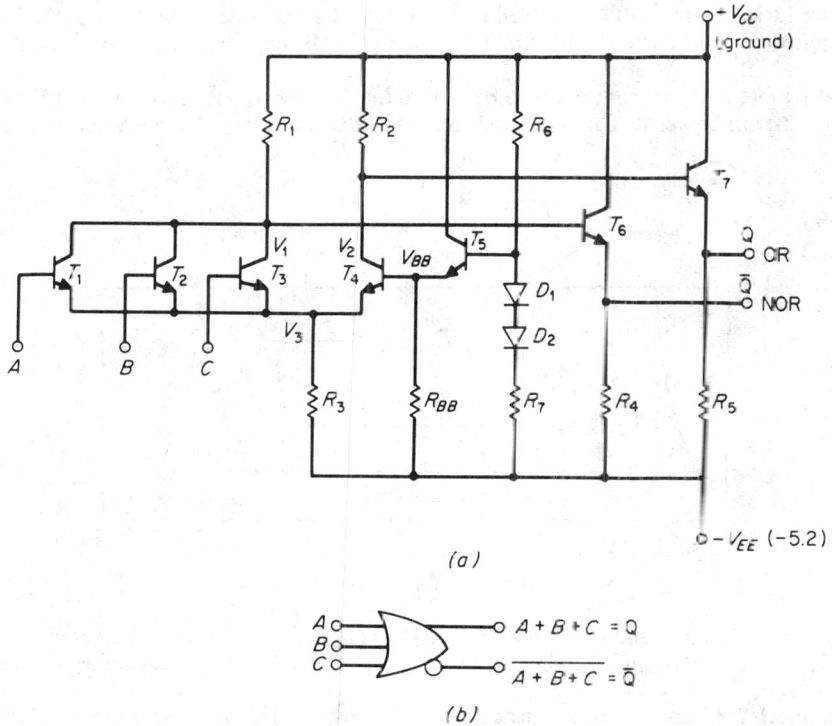

(a)

(b)

Fig. 8-138. Emitter-coupled logic (ECL): (a) emitter-coupled logic gate with complementary outputs; (b) ECL gate symbol.

Recent versions of ECL circuits exhibit the best speed-power products and have the shortest propagation delay of any logic form. In contrast to other logic families, the current drain is essentially constant regardless of frequency. The ECL circuit is capable of very-high-frequency operation, due to the use of nonsaturating transistors and small voltage swings. Propagation delays of 1 to 2 ns can be achieved using ECL circuits. To use the maximum-speed advantages of ECL, it is necessary to use special circuit-board layout and termination techniques. The main disadvantages of ECL are high power dissipation and low noise immunity. Its low swing levels make interfacing with other logic families difficult.

86. Integrated Injection Logic (I^2L).[96] Integrated injection logic is a dense common-emitter npn logic where the load and drive gate currents are provided by the injection of minority carriers. A basic gate structure is shown in Fig. 8-139. With the n epitaxial and p substrate usually grounded, current supplied to the p-type injector emits holes which diffuse through the n epitaxial material until they are collected by the p-type base of the I^2L gate. This base current forward biases the p-base–n-epitaxial junction, causing the gate's n-epitaxial emitter to inject electrons which are collected by the gate's n^+-type collector.

Since the p- and n^+-type diffusions are the standard bipolar base and emitter diffusions, respectively, this logic is compatible with other types of bipolar circuitry. However, the npn gate uses these diffusions in an inverted manner; hence the name inverse npn transistor. The equivalent circuit of Fig. 8-140 shows a lateral-pnp transistor providing base drive to the npn gate. In practice, the collector of the pnp transistor is the same p region as the base of the npn transistor.

Figure 8-141 shows a typical I^2L gate array with a geometrical plan view and its equivalent circuit. Since each npn gate receives identical injector current I_0, the inverse beta of each gate must be greater than 1 if the base drive of any given gate is expected to sink the base drive of the following gate. The inverse betas of these gates are typically optimized by minimizing the ratio of hole current received by the base-to-hole current reinjected back into the n epitaxial. Reinjected hole current is minimized by the placement of a rich n^+ diffusion around the base as shown. This also helps to minimize hole-current interaction between adjacent gates, since each gate can collect or inject holes in the common n epitaxial material. Optimizing the distance W to the p-type injector also improves inverse beta since the injector acts as a collector for reinjected holes. For a given injector current, a large width W will produce poor low-current inverse beta since little current will be received by the gates. Too narrow a W will force a large undesirable reinjected hole current collected by the injector itself, again forcing a reduction in inverse beta.

Usually more than one collector is diffused into the p-type base as shown, giving I²L a fan-out capability that improves its flexibility and functional density. Typical applications include flip-

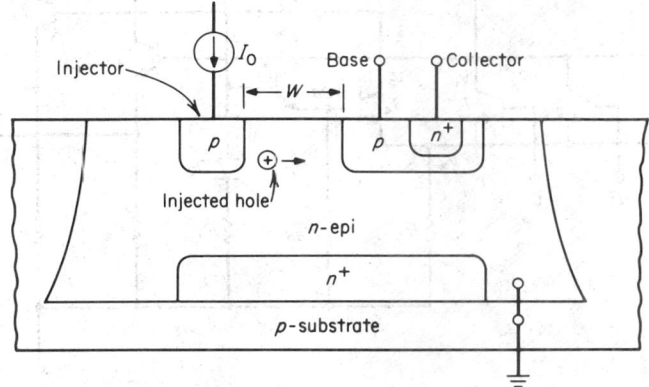

Fig. 8-139. Basic I²L gate structure.

flops, counters, dividers, encoders, and decoders. More important is building I²L compatibly with analog circuitry to give the IC designer analog and high-density digital capability on the same die.

87. Metal-Oxide Semiconductor (MOS) Logic. The small size and relatively simple device structure of MOS transistors are very attractive for digital circuit design. These devices are also known as insulated-gate field-effect transistors (IGFET) as discussed in detail in Par. **8-47** (see Figs. 8-43 to 8-49).

MOS logic circuits offer three significant advantages over the bipolar logic families: high component density, low power dissipation, and high fan-out capability. The last advantage comes about because of the high input impedance associated with the gate terminal of MOS devices. MOS logic circuits also have the disadvantages of low operating speeds and low current-drive capability. In addition, MOS logic circuits often require two power supplies for proper operation.

Because of its simplicity and ease of fabrication, the p-channel enhancement-mode MOS transistor is one of the most widely used devices in MOS logic circuits. In IC terminology, this device is often referred to as the PMOS transistor. For PMOS devices the threshold voltage V_T (the gate-source bias at which a conductive channel is formed between the source and the drain) is in the range of 2 to 5 V. This threshold voltage is controlled by the thickness of the gate dielectric, the crystal orientation of the semiconductor and the impurity concentration in the channel region (see Fig. 8-44 and Table 8-11).

The ON resistance of MOS devices depends on the applied gate voltage and on the channel dimensions. For conventional small-geometry PMOS devices, the ON resistance is of the order of several kilohms. Device performance can be significantly improved by using advanced fabrication techniques such as silicon-gate or ion-implantation technologies. Sili-

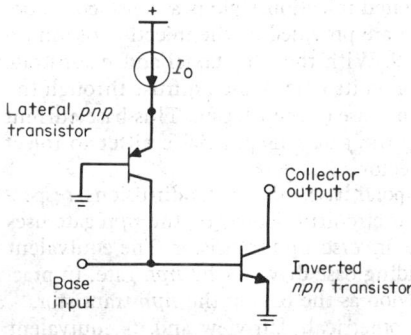

Fig. 8-140. Equivalent circuit of I²L gate structure.

con-gate-device structures shown in Fig. 8-48 can reduce the threshold voltage to below 1 V. Ion-implanted MOS transistors such as shown in Fig. 8-49 also provide low threshold voltages and minimize the gate-drain overlap capacitance to increase switching speeds.

Figure 8-142 shows the basic three-input NAND gate using MOS transistors and the corresponding symbols for positive and negative logic. Transistors T_1, T_2, and T_3 are used as input gates, and the output is low only when all the inputs are high. The lower device, T_4, acts as a constant-

TABLE 8-16   **Typical Performance Characteristics and Supply-Voltage Requirements for Various MOS Logic Circuit Types**

	p-channel MOS (PMOS)			Complementary MOS (CMOS)		n-channel MOS (NMOS)	
	Metal gate						
Process	High threshold	Low threshold	Silicon gate, low threshold	Metal gate	Silicon gate	Metal gate	Silicon gate
Threshold voltage, V	−3 to −5	−1.5 to −2.5	−1.7 to −2.5	1.5 to 2.5	0.5 to 1.0	0.7 to 1.5	0.7 to 1.0
Depletion threshold	Not used	+3 to +5	+3 to +5	NA	NA	−3 to −4	−1 to −2, −3 to −5
Typical power-supply voltages, V	$V_{DD} = -13$ $V_{GG} = -27$ $V_{SS} = 0$	$V_{DD} = -5$ $V_{GG} = -12$ $V_{SS} = +5$	$V_{DD} = -5$ $V_{GG} = -12$ $V_{SS} = +5$	3 to 18	1.5 to 16	5 to 15	5
Maximum frequency, MHz	2.0	3.0	8	1 to 15	3 to 25	5.0	10
Power dissipation, mW/gate	1.5	0.7	0.7	0.01	0.001	100	40

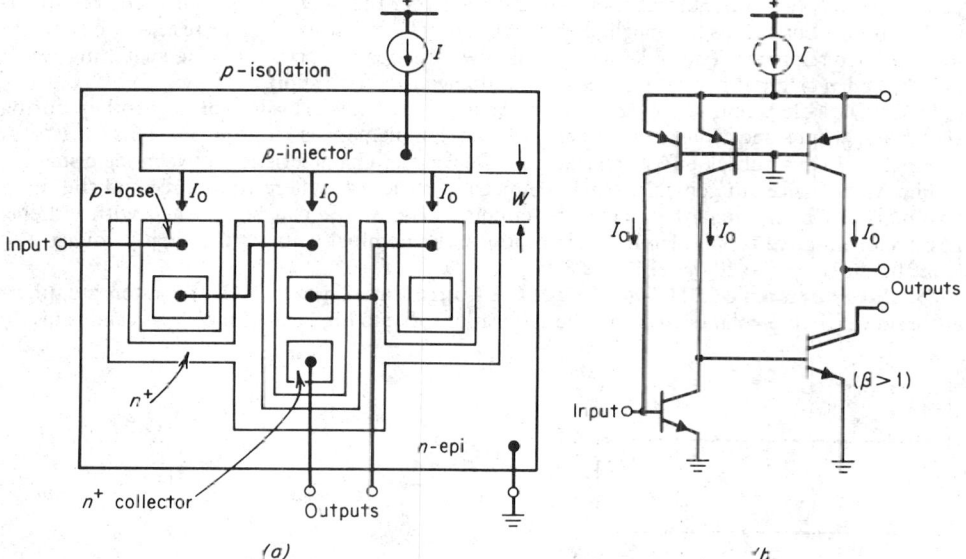

(a)   (b)

Fig. 8-141. I^2L gate array: (a) top view; (b) equivalent circuit.

current source. The input impedance is capacitive and can be considered a dc open circuit. The output characteristic is resistive. The resistance to ground is approximately 2 kΩ for each device that is turned on for a device geometry similar to that shown in Fig. 8-46. With the devices off, the output sees an impedance of about 25 kΩ to V_{DD}. The load MOS device is kept turned on with a negative potential applied to V_{GG}.

Typical power-supply requirements and other pertinent electrical characteristics of PMOS logic circuits are listed in Table 8-16. PMOS logic permits high packing density of circuit functions and is therefore well suited for large complex repetitive functions such as shift registers and memories. Computer-aided design and layout techniques are often used to minimize the design time and expense for complex MOS logic circuits.

88. Complementary Metal-Oxide Semiconductor (CMOS) Logic. Significant improvements in switching speed and power dissipation can be made if both n- and p-channel devices are used in a logic circuit. CMOS transistors can be fabricated on a monolithic substrate using the device structure of Fig. 8-47.

Figure 8-143a shows a typical circuit connection for a CMOS three-input NOR gate. The circuit symbols for p- and n-channel MOS devices are defined in Fig. 8-143b. Note that the direction of the arrow from channel to substrate corresponds to the polarity of the diode formed by the channel-substrate interface. The logic symbol for the circuit is given in Fig. 8-143c.

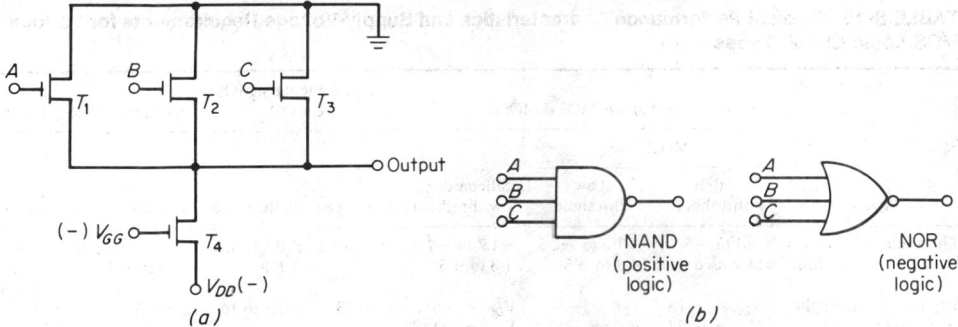

Fig. 8-142. p-channel MOS logic: (a) basic three-input PMOS NAND gate; (b) logic symbols for basic PMOS gate.

In CMOS logic circuits, the complementary devices are connected in series between the power supplies V_{DD} and V_{SS}, with PMOS devices located adjacent to the positive supply voltage V_{DD}, as shown in Fig. 8-143. At any given time, only one device is turned ON, and its complementary counterpart is OFF. Thus, at steady-state conditions, either the p- or the n-channel devices are OFF under all logic conditions, and negligible current flows from V_{DD} to V_{SS}. This results in extremely low power dissipation. The only substantial power dissipation occurs during switching when both p- and n-channel devices may be ON simultaneously, for a short duration.

In CMOS logic circuits, the devices also function as "active loads" for each other during switching, where one tends to turn ON while its complement is turning OFF. This creates an internal positive-feedback effect and sharpens the transfer characteristics between logic states.

The MOS devices in general have a symmetrical structure, where the source and the drain terminals can be interchanged. Thus the circuit of Fig. 8-143a can be operated with either a positive or negative supply. For most applications, the nominal value of the supply voltage V_{DD} is in the range of 3 to 20 V (see Table 8-16).

89. Comparison of MOS and Bipolar Logic Families.[97] MOS transistors are inherently slower than bipolar devices in logic applications, due to high impedance levels and parasitic

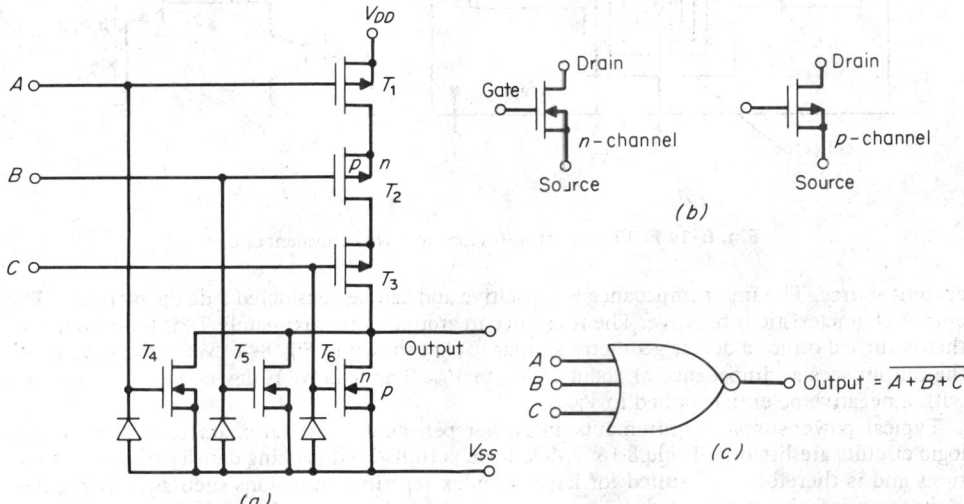

Fig. 8-143. Complementary MOS logic circuit: (a) three-input NOR gate using CMOS transistors; (b) circuit symbols for n- and p-channel MOS transistors; (c) logic symbol for three-input NOR gate.

capacitances associated with them. The speed-power product of bipolar circuits is normally 10 to 100 times greater than MOS circuits.

One significant advantage of MOS devices is their small size.[75] Figures 8-144 and 8-145 give some indication of the size comparison between an MOS and a bipolar transistor. Table 8-17 compares the fundamental IC logic families.

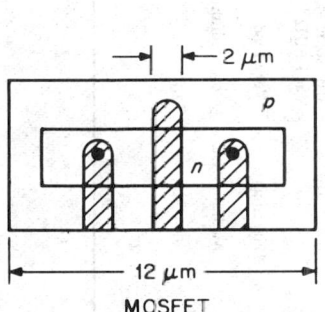

MOSFET

Fig. 8-144. Planar layout of MOS device.

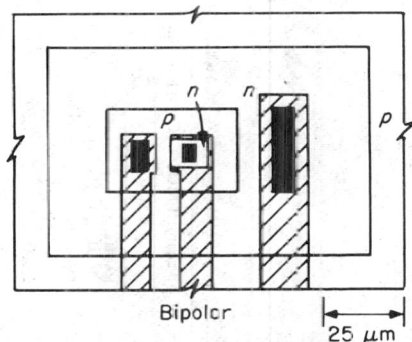

Bipolar

Fig. 8-145. Planar layout of bipolar device.

INTEGRATED-CIRCUIT MEMORIES

90. Semiconductor Memories.[98] Semiconductor memories are rapidly becoming highly competitive with magnetic data-storage devices. Semiconductor memories are also easier to interface with the external drive and sense circuitry than their magnetic counterparts. This advantage often leads to significant reduction in overall memory-system complexity.

Semiconductor memory devices fall into two groups: bipolar and MOS. Bipolar memories offer higher speeds and signal levels, which are directly compatible with the remaining bipolar logic circuits most often used in computer design. MOS memories, on the other hand, are less expensive to fabricate and offer higher bit densities. In addition, the power consumption of MOS circuits is usually lower, particularly in the case of CMOS circuitry. The high impedance of MOS devices also makes dynamic circuits possible, in which the information is temporarily stored as a charge on a capacitor and replenished periodically. The dynamic approach often allows a higher functional density than the static approach.

Depending on their functional use, semiconductor memories can also be classified as random-access read-write memories (RAM) or the read-only memories (ROM). The basic difference between these two memory types is that the bit pattern of the stored information is fixed in the read-only memory, while it can be changed during normal operation in the random-access memory.

91. Bipolar Memory Cells. The bistable flip-flop, made up of two cross-coupled-inverter stages, is the most widely used basic memory cell for both bipolar and static MOS memory circuits.[99,100] This cell is inherently fast, simple to design, and insensitive to process variations. Figure 8-146a shows a basic bipolar memory cell made up of npn transistors and collector load resistors. In this circuit one transistor is normally ON, which keeps the other transistor turned OFF. When an external signal is used to force the OFF transistor into the ON state, the ON transistor turns off.

Thus the flip-flop can have two stable states and will remain in either of them until an external signal is used to change its state. These two stable states can be interpreted as stored logic 1 and 0. This type of memory cell is generally used for random-access read-write memories. To get information in and out of the circuit — writing and reading — a gating arrangement is used. This gating can be achieved with dual-emitter transistors, as shown in Fig. 8-146b. One of the emitters of each transistor is tied to a common word line, while the other two emitters are each tied to one of the bit lines.

A large memory array can be formed by interconnecting many such flip-flops. Any particular memory cell can be selected or addressed, for writing or reading. To read the contents of a cell, the word-line voltage is raised. This causes the flip-flop current, which normally flows through the word line, to transfer to one of the bit lines. The signal current is then detected by a current-sensing amplifier.

TABLE 8-17   Comparison of IC Logic Families

Parameter	DTL	HTL	TTL 12-ns	TTL 6-ns	TTL 4-ns	ECL 1-ns	PMOS	CMOS	I²L	ISL
Circuit form	Diode-transistor	Diode-zener transistor	Transistor-transistor	Transistor-transistor	Transistor-transistor	Current mode emitter-coupled	p-channel MOS	Complementary CMOS	Integrated-injection logic	Integrated-Schottky logic
Positive-logic function of basic gate	NAND	NAND	NAND	NAND	NAND	OR/NOR	NAND	NOR or NAND	NAND	NAND
Wired positive-logic function	Implied AND	Implied AND (A·0·1)	AND-OR INVERT	AND-OR INVERT	AND-OR INVERT	Implied OR (all functions)	None	None	Open collector	Open collector
Typical Z_O high-level, Ω	6 or 2,000	15 or 1,500	70	10	50	6	2,000	1,500	N.A.	N.A.
Low-level	R_{sat}	R_{sat}	R_{sat}	R_{sat}	R_{sat}	6 Ω or 21 mA	25 kΩ	1.5 kΩ	R_{sat}	R_{sat}
Fan-out	8	10	10	10	20 or 10	N.A.	20	50+	4	4
Specified temp. range, °C	−55 to 125 0 to 75	−30 to 75	−55 to 125 0 to 70	−55 to 125 0 to 75	−55 to 175 0 to 75	0 to 75	−55 to 125 0 to 75	−55 to 125	−55 to 125	−55 to 125
Supply voltage, V	5.0 ± 10%	15 ± 1	5.0 ± 10% 5.0 ± 5%	5.0 ± 10% 5.0 ± 5%	5.0 ± 10% 5.0 ± 5%	−5.2 ± 5%	−27 ± 2 −13 ± 1	4.5 to 16	N.A.	1.5 min
Typical power dissipation per gate, mW	8 or 12	55	12	22	1	25 + load	0.2 to 10	0.01 static ≈ 1 at 1 MHz	0.05	0.2
Immunity to external noise	Good	Excellent	Very good	Very good	Very good	Good	Nominal	Very good	Nominal to good	Nominal
Noise generation	Medium	Medium	Medium high	High	Medium	Low	Medium	Low medium	Low	Low
Propagation delay per gate, ns	30	90	12	6	4	1	300	70	20	4
Typical clock rate for flip-flops, MHz	12 to 30	4	15 to 30	30 to 60	60 to 100	400	2	5	8	35 to 40

SOURCE: After Ref. 95.

For writing, a cell is similarly selected by the word line; then, unbalancing the voltage at the two bit lines forces the memory cell into the desired state. For unselected cells, the word-line voltage is low, forcing the cell current through the word line. For this condition, no signal current flows through the bit line, and the cell content is not sensed. A change in voltage on the bit line will have no effect on the state of the cell. Because of its simplicity and small area, this cell structure is used in many bipolar memory circuits for read-write applications.

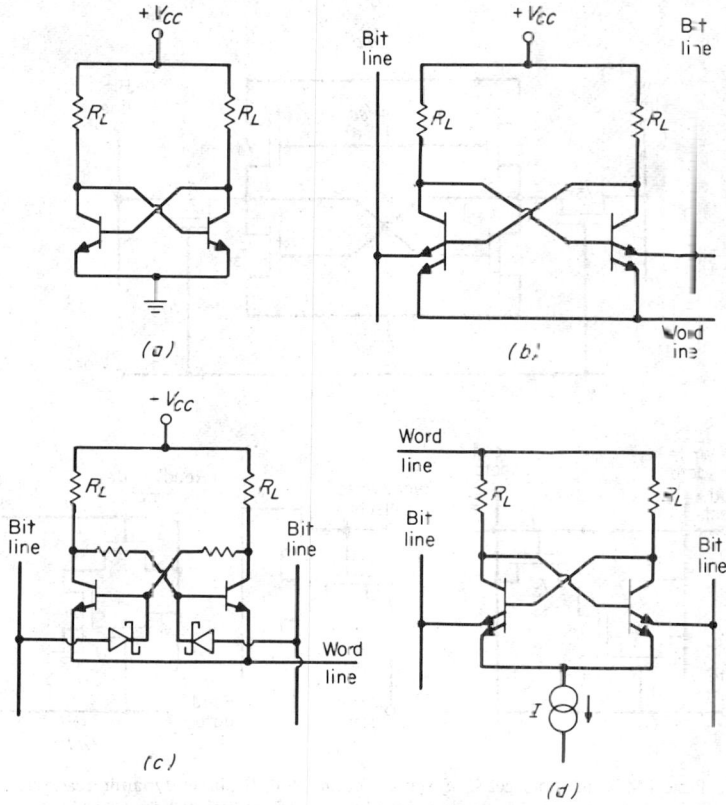

Fig. 8-146. Basic bipolar-memory-cell structures: (a) basic bipolar flip-flop memory cell; (b) bipolar memory cell with multiemitter transistors; (c) bipolar memory cell with Schottky diode gating (d) emitter-coupled memory cell.

A Schottky diode (see Figs. 8-56 and 8-57) can also be used to address the basic flip-flop, as shown in the cell configuration of Fig. 8-146c. This cell can be selected for reading by lowering the word-line voltage. Signals can then be detected on the bit lines through the Schottky diode. For writing, a large current through the Schottky diode simultaneously turns on the OFF transistor and forces the previously ON transistor to turn off by increasing the load current and decreasing the base drive. This cell has the advantage of small size, low power operation, and high speed, because the collector resistor R_C can be made large without degrading the access speed of the memory cell.

Figure 8-146d shows a memory-cell structure based on the emitter-coupled logic (ECL) configuration. In this type of memory cell, bit selection is achieved by raising the word-line voltage. The write-read operation of the ECL memory cell is much as in the dual emitter cell shown in Fig. 8-146d. In the ECL memory cell, the voltage across the selected cell is higher than across the unselected cell, making a large sense current available. Since the voltage across the supply terminals of unselected cells can be made quite low, standby power dissipation can also be kept low.

92. MOS Memory Cells. Because of the self-isolating feature of MOS devices, MOS memory cells can be packed much closer together than comparable bipolar memory cells. The relative sizes of MOS and bipolar devices are shown in Fig. 8-144.

Static Memory. In static-memory cells, information can be stored indefinitely, provided the bias supplies are not turned off. The basic bipolar memory cells of Fig. 8-146 all fall into this category. In the case of MOS circuits, the basic cross-coupled flip-flop configuration can also be used for static-memory applications.

A circuit configuration suitable for this application is shown in Fig. 8-147a. In the figure, transistors T_1, T_2 and T_3, T_4 make up the basic bistable circuit. Transistors T_5 and T_6 provide the gating function for the cell. To read the cell, T_5 and T_6 are turned on by the word line, and the data on T_1 and T_3 are transferred to the bit lines.

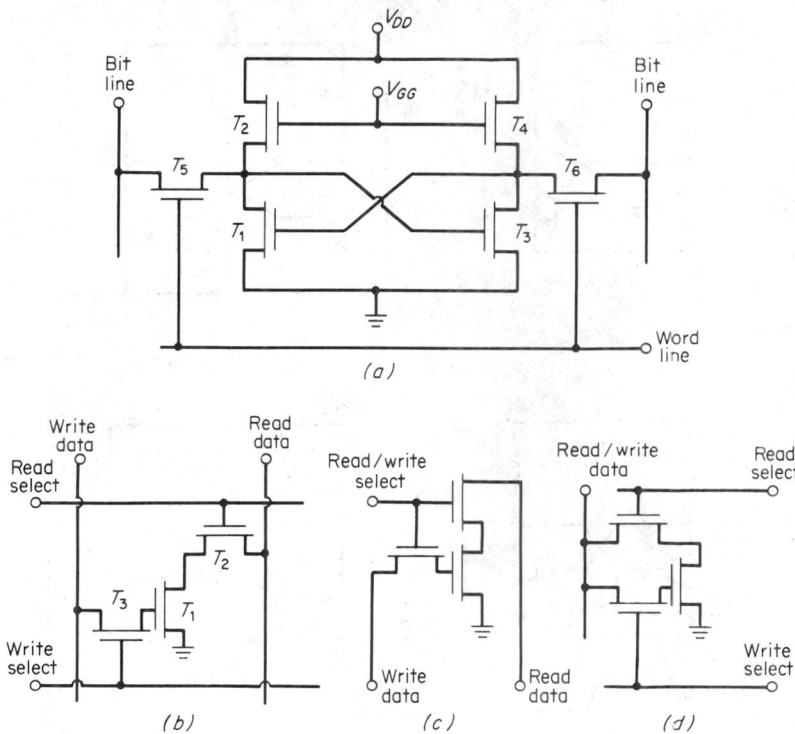

Fig. 8-147. Basic PMOS memory cells: (a) static memory cell; (b) basic dynamic read-write memory cell; (c), (d) simplified cell structures using combined read-write and data select lines.

Writing can be achieved by again turning on T_5 and T_6 and then forcing the cell into the desired logic state by applying the proper voltage to the bit lines.[101] If p-channel MOS transistors are used, V_{DD} and V_{GG} are negative supply voltages. The basic MOS flip-flop storage cell using six transistors per bit is used in many read-write MOS memories.

Dynamic (Charge-Storage) Memory Cells. The dynamic MOS memory cells overcome the two major disadvantages of bistable flip-flop cells, namely, high-power-dissipation and large-chip-area requirements. In these types of memory cells, one makes use of the high-impedance and low-leakage-current properties of MOS devices, which permit using the parasitic device capacitances for temporary charge storage. For example, the gate of a MOS transistor can be used as storage node. For a p-channel device, the presence of a sufficient amount of negative charge on its gate will turn the device on. If this charge is removed, the device is off. This can be interpreted as a logic 1 or 0 condition. If provisions are made to supply and remove charge from the gate, and for sensing the presence or absence of charge, a memory cell can be obtained.[102,103]

Figure 8-147b shows a circuit configuration of such a charge-storage memory cell. Transistor T_1 is used as the charge-storage device. Transistor T_2 provides a connection between T_1 and the read-data terminal, when the read-select line is activated. This permits interrogation of the cell for its content. Transistor T_3 permits access to the charge-storage node for write purposes. This is accomplished by activating the write-select line. Since charge can be stored for a limited time only, a means for periodic refreshing of the memory data must be provided. This is also done through T_3 by reinforcing the charge condition of the storage node from a refresh amplifier.

Read and write select functions or data lines can be combined as shown in Fig. 8-147c and d, respectively. These differences can save interconnections at the expense of more complex drive circuitry. In general, dynamic read-write memory cells achieve higher performance (less than 100 μW/bit power dissipation in the active mode) and higher packing densities than static memory cells. As a result of these advantages, dynamic memory structures can be competitive with either core or plated-wire memories.

93. Functional Classification of Semiconductor Memories. Depending on their functional application, semiconductor memories can be divided into three basic categories: read-only memories, random-access memories, and shift registers.

94. Read-Only Memories (ROM). This type of memory contains a "permanent" set of information written into the memory. Under ordinary operation, the content of the memory remains unchanged. Semiconductor ROMs can be divided into three groups:

1. Mask programmed ROMs

2. Fusible-link ROMs, in which a permanent and irreversible change in the memory interconnection pattern is caused by either an electrical pulse or by mechanical means

3. Alterable ROMs, in which a reversible change in active-device characteristics is induced electrically

In both the mask-programmed memories and the fusible-link memories, the data in the individual memory cells are "written" at the time the memory is fabricated. Since the ROM cells are very small, such a memory takes up much less chip area than a RAM of comparable bit count.

Mask-programmed ROMs are frequently used to store microprograms for code conversion, process control, or for character generation. In these cases a custom bit pattern is specified by the customer for a special application. Typically, the customer specifies the input-output requirements of the custom ROM on a coding sheet. This information is then used in the fabrication process to generate custom interconnection masks for the IC.

Small size and high functional density of MOS devices make them particularly useful for read-only memories. In most cases, the additional circuitry required for address decoding and input buffering is included on the same chip with the memory matrix.

Figure 8-148 shows a block diagram of an MOS read-only memory which is organized as a 16-word matrix, with 2 bits per word.[104] The basic storage elements are the MOS transistor cells. Since the memory has 32 transistor locations, it can be coded with 32 bits of information by inserting or deleting these transistors at the desired locations.

Fusible-link ROM circuits are programmed after the monolithic circuit fabrication and packaging process is completed. This programming is achieved by fusing, or burning out certain critical interconnections at the desired storage locations. It can be done by the customer with the help of special programming equipment. These memories are also called *field-programmable ROMs* (FROM). One disadvantage of this type of memory is that they cannot be reprogrammed, which prevents complete functional testing before delivery and also makes it impossible to modify the content of the memory in case of an error.

Alterable ROM Circuits. In these circuits the memory contents can be reprogrammed by inducing a change in the active-device characteristics. An example of such an "alterable" active device is the floating-gate avalanche injection metal-oxide semiconductor (FAMOS) structure, shown in Fig. 8-149a.[105] This device is essentially a *p*-channel silicon-gate MOS transistor (see Fig. 8-48) with no electrical contact made to the gate. The operation of this memory cell depends on the charge transport of electrons to the floating gate by avalanche injection from either the source or the drain *pn* junctions. An applied high-voltage pulse causes an accumulation of charge on the gate, which induces a conductive inversion layer from source to drain. When the applied voltage is removed, no discharge path is available to the accumulated electrons, since the gate is surrounded by thermal oxide. Thus this injected charge becomes trapped at the gate electrode, and maintains an induced channel under the gate until a high-voltage pulse of opposite polarity is applied to alter the memory setting. The circuit diagram of a FAMOS memory cell is shown in Fig. 8-149. The programming of a memory bit is accomplished by simultaneously selecting the *X*- and *Y*-select lines.

Electrically Programmable Read-Only Memory (EPROM). The ultraviolet erasable EPROM[106] provides nonvolatile read-only memory which can be electrically programmed after it has been fabricated and assembled in its package. This programming can be done by the customer.

Device Structure. The design of an EPROM centers on a unique memory cell using the floating-gate structure shown in Fig. 8-150. The device is very similar to a conventional *N*-channel MOS transistor except that it has an extra floating-gate electrode between the control gate and the active channel. It is this floating gate (floating in the sense that it is isolated electrically from all other nodes) that provides the memory capability of the device.

Programming.   Information can be stored in the EPROM cell by accumulating an excess negative charge on the floating gate. This is done through hot-electron channel injection (Fig. 8-151). A programming voltage V_p of about $+25$ V is applied to both the control gate and drain terminals of the EPROM device. Under these conditions the memory cell is biased in a heavily conducting state with an electron-inversion layer carrying current from source to drain. When this electron flow enters and crosses the drain depletion layer, it sees a very high electric field. A small fraction of the carriers gain enough energy from this field to overcome the oxide potential barrier at the silicon surface and drift to the floating gate. This injection builds up an excess

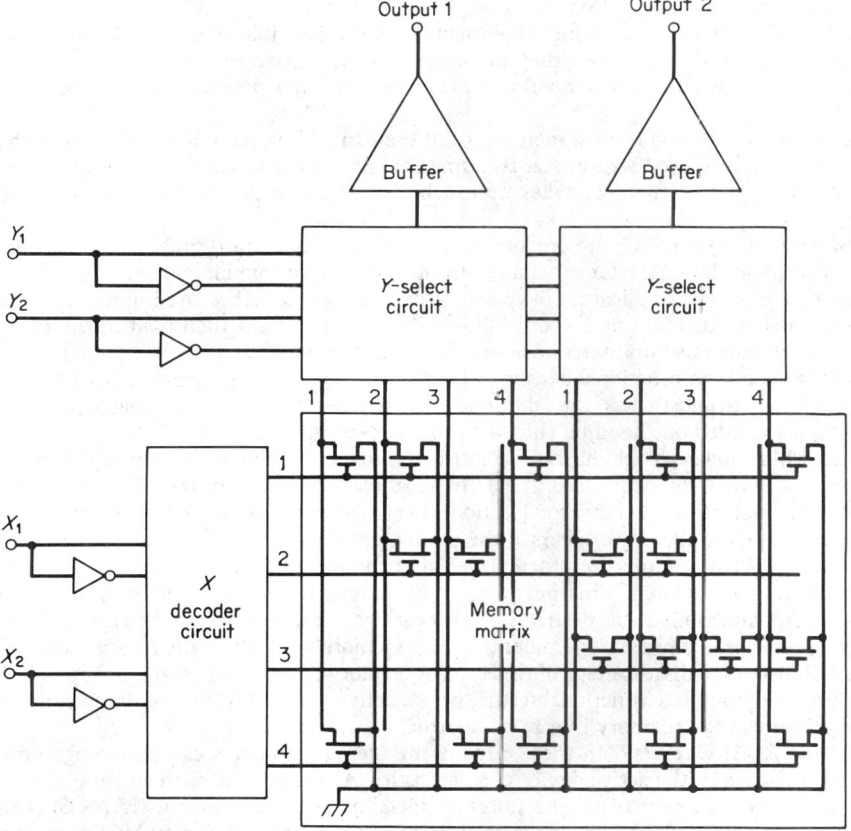

Fig. 8-148. Block diagram of a 16-word, 2 bits/word MOS ROM. Logic 1 in a cell is represented by a transistor at the intersection of the decoded X and Y address lines.[134]

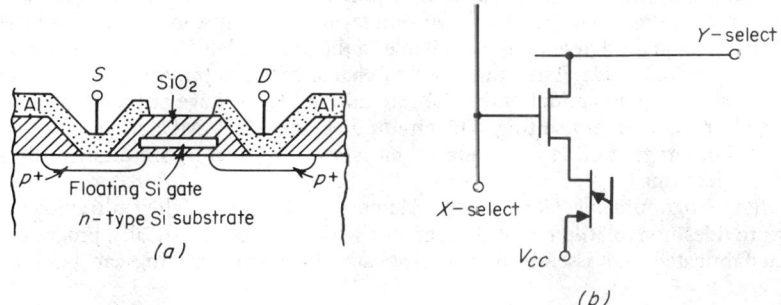

Fig. 8-149. Floating-gate avalanche-injection MOS transistor (FAMOS): (a) device cross section; (b) typical memory cell.

negative charge on the gate, and the overall device threshold becomes more positive. Based on the capacitive coupling shown in Fig. 8-152, typical initial thresholds are about +2 V; the programmed thresholds may typically be greater than +10 V.

Reading. This EPROM cell can be sensed (Fig. 8-153) by placing the control-gate voltage at an intermediate voltage of +5 V. If the device is unprogrammed, this +5 V is greater than the

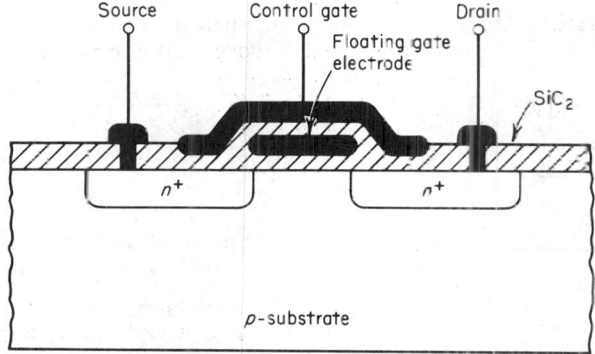

Fig. 8-150. Cross section of EPROM cell.

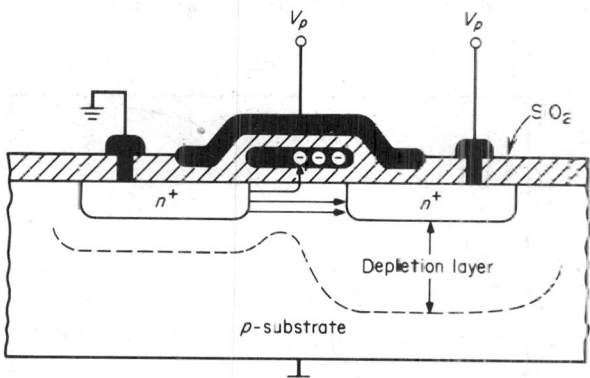

Fig. 8-151. Hot-electron channel injection.

threshold voltage and the device will conduct current. This state is generally defined as a stored binary 1. On the other hand if the device is programmed, the threshold voltage is greater than the read voltage ($V_{T,final} > V_{CG}$) and the device does not conduct. This state is defined as a stored binary 0.

Erasing. Once an EPROM has been programmed, it retains the data indefinitely; estimates are that the charge leakage from a floating gate would take 10 to 100 years to discharge a cell in normal operation. The primary method for erasing or discharging a cell is to expose it to ultraviolet radiation for about 20 min. This radiation reaches the top surface of the die through a transparent lid on the semiconductor package. Some of this ultraviolet radiation reaches the electrons on the floating gate, giving them enough energy to overcome the potential barrier of the oxide surface and to drift back to the silicon substrate. At the end of such an ultraviolet erasure, the excess electron density on the floating gate returns to zero, and the memory-cell thresholds return to their original low-threshold state $V_{T,initial}$. This cycle of programming followed by ultraviolet erasure can be repeated up to several hundred times.

Memory Array. When connected in a memory circuit (Fig. 8-154), the individual EPROM cells are configured as an array with gates tied to form word lines and drains tied to form data lines. Individual memory elements are accessed by raising one word line to a positive voltage and then selecting the desired data line through a column decoder. In this manner any bit in the memory array can be uniquely selected for either programming or reading.

The simplicity of design and layout and the inherent density advantage of the single-device-per memory bit have made the ultraviolet EPROM a very popular and economical alternative to ROMs.

95. Random-Access Read-Write Memories. Random-access read-write memories, often simply called random-access memories (RAM), are used where it is required to change the bit location during routine operation of the memory. Various typical static and dynamic RAM cells are shown in Figs. 8-146 and 8-147. Fast- and medium-speed RAMs are generally used for scratch-pad memory applications, while low-speed RAMs are used for main storage or in peripheral buffers.

Static Random-Access Memory (SRAM). The static random-access read-write memory (SRAM)[107] is called static because it will retain data indefinitely as long as the power supplies are

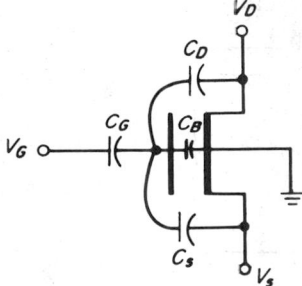

$$V_T = (C_T/C_G) V_{T1} - (C_D/C_G) V_D - (C_S/C_G) V_S - (Q_S/C_G)$$
$$\approx (C_T/C_G) V_{T1} - (Q_S/C_G)$$

where V_{T1} = threshold of first polytransistor; Q_S = stored charge on floating gate; and C_T = total capacitance of floating gate = $C_D + C_B + C_S$

Fig. 8-152. EPROM threshold calculation.

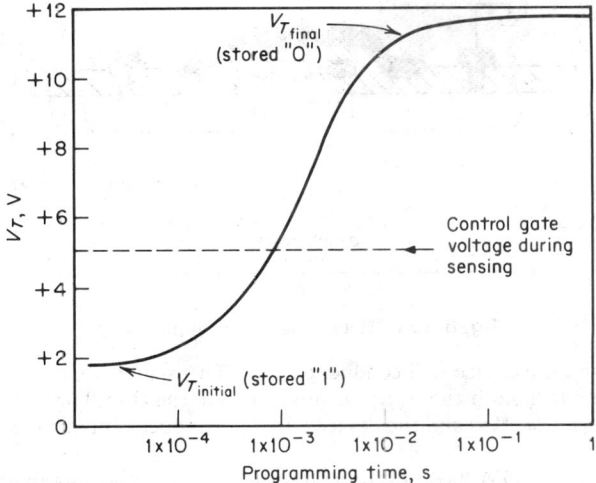

Fig. 8-153. EPROM programming characteristics.

not disconnected. In its IC form the SRAM is composed of address input buffers, word-line decoders, static memory cells, bit-line decoders, and sense amplifiers and output buffers. The static RAM can be manufactured in either bipolar or MOS technologies, and its cell is basically a cross-coupled latch with provisions made for an *XY* selection scheme. The most popular and widely used static RAMs today are those manufactured in the *n*-channel MOS technology. Figure 8-155 shows the schematic of a typical MOS static RAM cell.

Operation of Cell. In the SRAM cell L_1 and L_2 are load devices that provide the conduction path to keep the higher-voltage internal nodes charged to a high level. L_1 and L_2 can be either MOS transistors or extremely high resistance polysilicon resistors. Although they add extra steps to the manufacturing, polysilicon resistors provide two important advantages: they provide a physically smaller cell (thereby allowing more cells to be placed on one die), and they reduce the power consumed by the memory cell to very low levels. Transistors T_1 and T_3 are the cross-coupled pair and form the regeneration in the cell. T_1 and T_3 are very high gain compared with

L_1 and L_2. Transistors T_2 and T_4 are the access transistors. The gain of these transistors must be less than half the gain of transistors T_1 and T_3 for the cell to be stable during the selection or deselection of the cell. To write data into the memory cell the word-line voltage is raised to a high level. Then complementary voltages are impressed on the bit-line and $\overline{\text{bit-line}}$ nodes by the data input circuitry. Transistors T_2 and T_4 are now turned on and couple the voltage on the bit-line and $\overline{\text{bit-line}}$ bar onto the memory-cell nodes A and B. This completes a write operation into the cell; the word-line voltage can be lowered, and the data input circuitry can cease to control the voltage on the bit-line and $\overline{\text{bit-line}}$ nodes. To read data from the cell the word-line voltage is

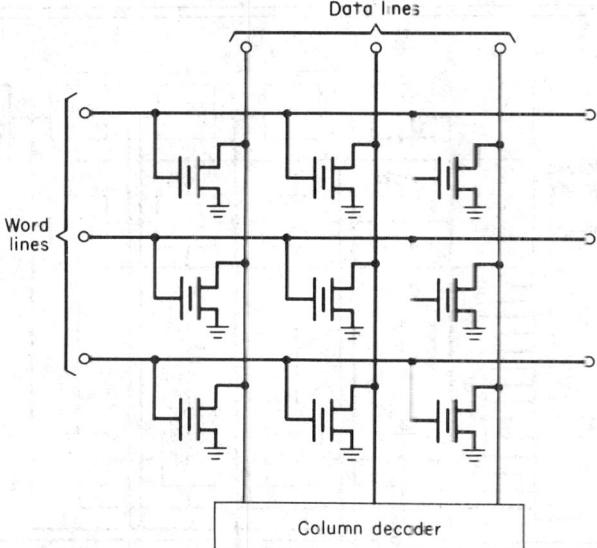

Fig. 8-154. EPROM memory array.

again raised to a high level. Transistors T_2 and T_4 are now turned on and, depending on the state of the cell, either T_1 or T_3 is turned on. A conductive path is now formed by the series combination of transistors T_1 and T_2 or T_3 and T_4. This in turn lowers the voltage on either the bit line or $\overline{\text{bit line}}$, respectively, while the opposite node rises due to the bit-line and $\overline{\text{bit-line}}$ load devices. This complementary voltage is then passed through the bit-line decoder and amplified.

Figure 8-156 shows the basic organization for a 16-word by 4-bit static bipolar RAM system.[108] The basic cell structures can be designed using any one of the bipolar memory-cell circuits shown in Fig. 8-106. Reading the contents of a particular word is accomplished by applying a signal to the appropriate word-select input and detecting the outputs of the sense amplifiers (S/ A) at terminals D_0 to D_3. Writing is accomplished by enabling word and write lines and applying a signal to the proper write amplifier.

Dynamic Random-Access Memory (DRAM). The dynamic random-access read-write memory (DRAM)[137] is called dynamic because data are stored only temporarily and must be continually rewritten or refreshed. The data are stored in the form of charged capacitors and are necessarily temporary because of the parasitic leakage currents in ICs. The attribute of the DRAM cell that makes its transitory nature tolerable is its small size. Memory-cell sizes in current DRAMs range from 450 μm² in the 16,384 × 1 bit DRAM to 160 μm² in the more advanced 65,536 × 1 bit DRAM.

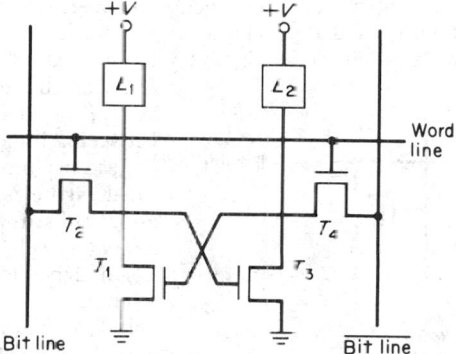

Fig. 8-155. Typical MOS static RAM cell.

The basic DRAM cell consists of one transistor and one capacitor connected as shown in Fig. 8-157. T_1 is the select transistor and allows data to be read into or out of the memory storage node A. C_S is the storage capacitor and is connected between the memory-storage node A and the supply voltage V, which can be any steady-state voltage.

Writing information into the cell is straightforward. To accomplish a write operation the word line is raised to a high voltage and T_1 is turned on fully. Then the bit line is driven either to a

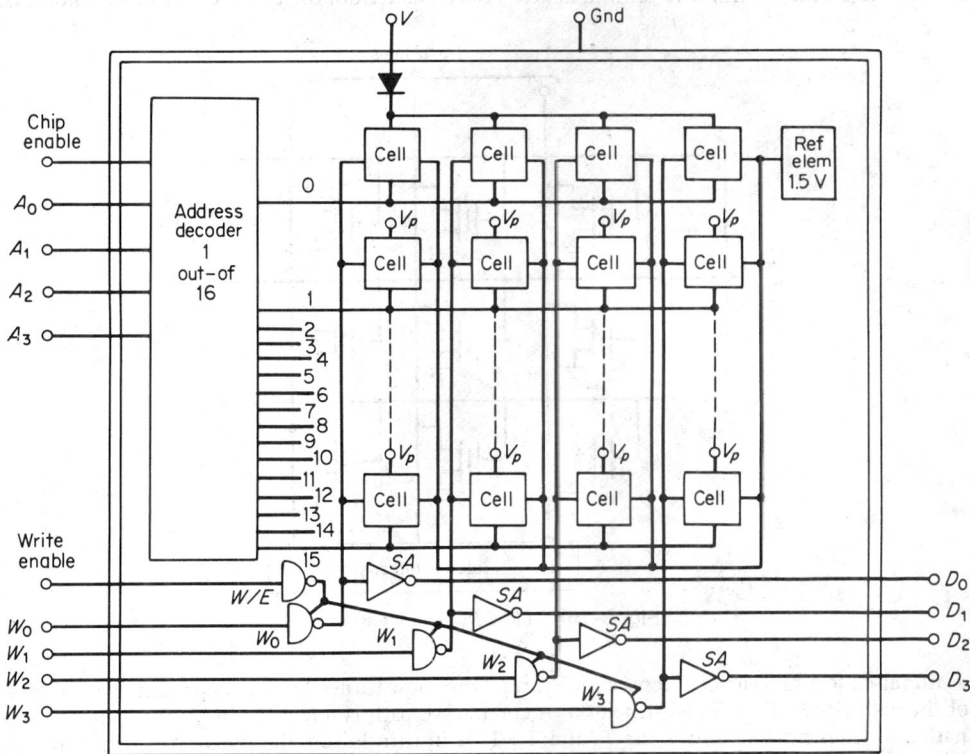

Fig. 8-156. Typical organization diagram of a 16-word by 4-bit (64-bit total) bipolar RAM.

high or low voltage. Because T_1 is on, memory-storage node A will be charged to approximately the same voltage as the bit line. Then the word line is returned to a low voltage, turning off T_1 and isolating node A from the bit line. The memory cell now has the information stored on node A in the form of the charge (or lack of charge) stored on C_S. This charge will slowly leak away because of the junction leakage of node A. Current cell-storage capacitor (C_S) sizes are typically 0.050 pF, and leakage currents are typically 2.5pA at 80°C. Therefore, the voltage on node A will decay about 2 V in 40 ms at 80°C. To ensure that data are not lost by this voltage decay on the storage node, the data must be refreshed before the voltage on node A decays past the point for discrimination between a 1 and a 0.

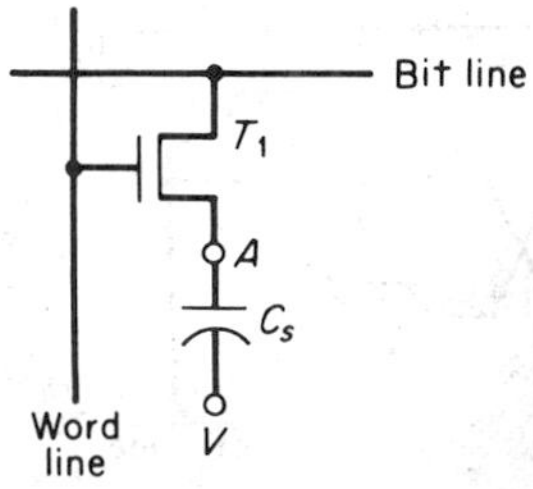

Reading information from the memory cell is more complicated. Figure 8-158 shows a simplified schematic of a cell and the associated sensing circuitry. In the schematic T_1 and C_S make up a memory cell, T_R and C_R make up a reference cell, and C_B and C_{RB}, designed to be equal, are the capacitances of the bit line and reference bit line, respectively. C_R is designed to be equal to one-half C_S. Before a read operation begins, the word line and reference word line are at a low voltage and the bit line and reference bit line have been charged equally to the supply voltage ($+V$) and then disconnected from the supply voltage. Node B in the reference cell has also been charged to ground potential. To begin the

Fig. 8-157. Dynamic random-access read-write memory cell.

read operation the word line and reference word line are charged to the supply voltage. This turns T_R and T_1 on. Since node B was initially charged to ground potential, charge from the reference bit line will redistribute and charge C_R. The amount of voltage change on the reference bit line will then be

$$\Delta V_{RBL} = \frac{VC_S}{2C_B - C_S} \qquad C_R = C_S/2 \qquad C_{RB} = C_B$$

If node A stored a 1 (high voltage), the bit-line voltage would not change. In this instance the sense amplifier will have a differential voltage on its inputs equal to the change in voltage on the reference bit line. The sense amplifier will then be activated and amplify that voltage differential. If node A stored a 0 (ground potential), node A would be charged to a higher voltage due to the redistribution of the charge from C_B. The amount of voltage change on the bit line will then be

$$\Delta V_{BL} = VC_S/(C_B + C_S)$$

The sense amplifier is then presented with a voltage differential equal to

$$V_{diff} = \Delta V_{RBL} - \Delta V_{BL} = \frac{VC_S}{2C_B - C_S} - \frac{VC_S}{C_B + C_S}$$

In general $C_S \ll C_B$; then

$$V_{diff} = -\frac{VC_S}{2C_B}$$

The sense amplifier will then be activated and amplify that voltage differential.

When the sense amplifier has completed its function, the data will automatically be rewritten into the cell and a bit-line decoder will route the information to the output-buffer circuitry.

A potential charge-loss mechanism that affects DRAMs is α-particle radiation. An α particle impinging on the surface of a DRAM penetrates about 25 um deep into the silicon. During its trajectory through the silicon the α particle generates roughly 1.2 million electron-hole pairs. Some percentage of the electrons generated will find their way to the surface and be collected by memory-cell nodes. If the number collected by any one node is large, the voltage on the node may be reduced sufficiently to cause a misread or soft failure. One source of the α particles is the IC packaging material, and efforts are being made to reduce the α radiation to a minimum. Another possible solution is to coat the surface of the IC with a material that will stop the α particle from getting to the silicon surface.

Figure 8-159 shows the organization diagram for a dynamic MOS RAM. Reading and writing occur for all cells in one row simultaneously. Since only 1 bit at a time is available for writing, an internal read operation is used to transfer the data to the refresh amplifier before writing. In this manner, the refresh amplifier contains data corresponding to the contents of the row into which writing takes place.[101]

96. Shift Registers. A shift register is an arrangement of an arbitrary number of storage cells in a row and is used primarily for temporary storage of digital information. Some common applications of shift registers are in serial-data entry and serial-data output, as well as in serial-to-parallel converters. The serial-in–serial-out shift register can perform similarly to a high-speed drum memory; however, unlike the mechanical drum memory, it can be stopped instanta-

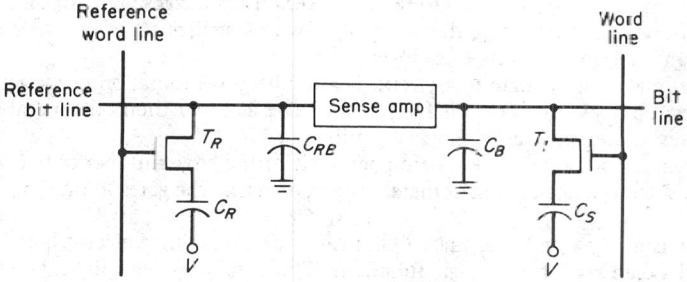

Fig. 8-158. Simplified schematic of a dynamic RAM with associated sensing circuitry.

neously. Serial-to-parallel converters are often used in accumulators, where the data are entered in serial fashion, e.g., from a keyboard, and then acted upon in parallel, as by an adder.

The shift registers can be designed for either static or dynamic operation. Static shift registers make use of the basic flip-flop circuits of Figs. 8-146 and 8-147 for data-storage purposes. Dynamic shift registers operate in the same manner as the dynamic RAM circuits, where each bit of information is constantly refreshed and recycled. Figure 8-160 shows the circuit diagram of a section of a dynamic MOS shift-register circuit. The circuit operates with two-phase clock pulses, $V_{\phi 1}$ and $V_{\phi 2}$. After each clock pulse, one bit of information is inverted and transferred half a cell to the right; thus, after two clock pulses, the contents of each cell are shifted over to the next one.

97. Array Circuit Types. LSI logic arrays fall into two basic categories, *field-programmable* and *mask-programmable*,[109,110] each with important advantages and disadvantages. All field-programmable types have a fixed number of OR gates, AND gates, and sometimes flip-flops per circuit. Programming is handled through *fusible links* similar to those of a PROM. Advantages are that several custom logic functions can be developed from one IC part number by blowing the fuse links. Changes are easily incorporated by programming a new fuse pattern. Programming commonly requires special equipment, although some of the smaller array circuits use standard PROM programming hardware. Disadvantages of the fusible-link approach relate to the fixed pattern of inputs, outputs, and logic functions on chips. For example, an LSI circuit incorporating a 10-bit shift register cannot be built if the field-programmable circuit is limited to eight flip-flops. Similarly an LSI part requiring multiphase clocking cannot be built on an array which connects all flip-flops to a common clock line.

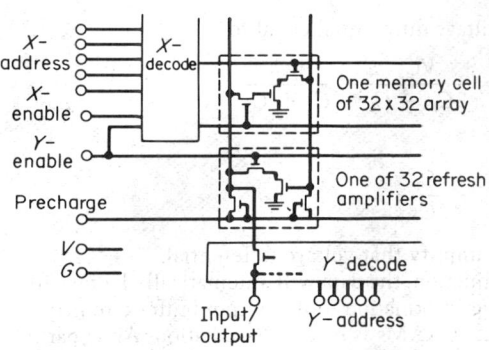

Fig. 8-159. Organization diagram of a dynamic MOS random-access memory system.

Mask-programmable arrays, commonly known as *gate arrays*, and *macrocell arrays* and sometimes called *master slices*, offer greater logic flexibility. Gate arrays have a large number of gates on chips which can be connected to build any logic function, subject only to package pins and the number of gates. By interconnecting gates as flip-flops, adders, multiplexers, etc., these circuit elements can be placed in any combination anywhere on a chip.

The *macrocell array* carries the concept one step further by subdividing the array into MSI-complexity blocks, called macrocells, rather than individual gates. The designer works with a library of macrocell functions and need not implement everything from simple gates.

Disadvantages of mask-programmable arrays center around mask programming. Every circuit option requires a custom metal pattern on top of a standard semiconductor diffusion set. While having only custom metal is much simpler than a custom circuit designed from scratch with a full semiconductor mask set, it is more complex than programming with fuses at the user's location.

Arrays approaching 1,200 gates have more logic power per package, speed-power products are generally better than with the mask programmable circuits, and circuit performance, especially with ECL arrays, is superior to other forms of digital logic.

The main purpose of logic arrays is to replace several IC packages on a circuit board with one LSI circuit. One way of comparing the array types is to visualize them as a miniature printed-circuit board and see how each replaces logic.

The *fusible-link arrays* equate to a circuit board with a combination of OR gates, AND gates, and sometimes flip-flops. The board is fully wired. The designer then cuts metal (fusible links) so that the gates implement various logic equations.

The *gate array* is the same circuit board populated with a large number of two- or three-input gate packages. The designer than adds metal, interconnecting the gates to implement the desired function.

The analogy can be extended to macrocell arrays by visualizing a circuit board full of empty IC sockets and a data book full of logic functions. The designer selects logic functions from the data book (macrocell library), puts them in the IC sockets, and interconnects the ICs through routing channels.

A typical macrocell array has 85 logic functions in a macrocell library and a total of 106 cell positions on the chips.

Array Terminology. The following definitions will be helpful:

Gate array: An IC containing a number of uncommitted gates which are interconnected with dedicated metal patterns to form a custom-circuit logic function. Gate arrays vary between 100 and 1,200 gates and may use CMOS, I^2L, TTL, or ECL circuit technologies.

Macrocell: An array subsection performing a higher level logic function than a basic gate. Macrocells normally relate to MSI complexity circuits such as flip-flops, decoders, multiplexers, adders, etc., and make design easier than using gates. Macrocells can be several gates in a gate array or the basic building block in a more advanced macrocell array.

Macrocell array: An array circuit built around macrocell functions rather than individual gates. Macrocell arrays offer more logic power, lower power dissipation, and easier design interface than an equivalent gate array.

Macrocell library: A set of predefined macrocell functions available to the logic designer. Many different MSI and SSI equivalent macrocell functions can be characterized in the macrocell library.

Array option: A specific logic-function metal pattern over a standard array diffusion set. The array diffusion set and macrocell library may be available to anyone, but specific array options may be proprietary to a given customer.

Logic-diagram interface: The metal patterns from a user-supplied logic diagram, produced by a semiconductor supplier. While feasible for a small array, this interface becomes very labor-intensive for a large array and is impractical if the array is intended as a wide-customer-base industry standard.

Calma interface: A good interface for customers with semiconductor design experience and Calma graphics equipment. The designer begins with Calma tapes containing fixed metal such as power buses, bonding pads, gate or macrocell metal, etc., and then adds customizing metal. Resulting tapes are then used directly to produce the finished array metal.

CAD interface: A computer-aided design (CAD) interface is the most practical for most array customers. The option designer interfaces by means of a terminal to provide all option development information. The designer can select macrocell functions from the library and place them in cell locations on the chip. Macrocells are then interconnected through dedicated horizontal and vertical routing channels. The CAD system can handle certain design decisions such as first- or second-layer metal selection, metal widths, channel spacing, via interconnects, etc. Simulation programs available through the CAD interface can be used to help minimize design errors so that the array options function properly on the first pass.

98. Microwave Integrated Circuits. Microwave integrated circuits (MICs) are designed to operate at frequencies beyond the capabilities of conventional ICs. They cover the frequency range from 0.5 to 15 GHz. MICs can be fabricated with either monolithic or hybrid technology. The small size of monolithic circuits is a significant advantage for microwave appli-

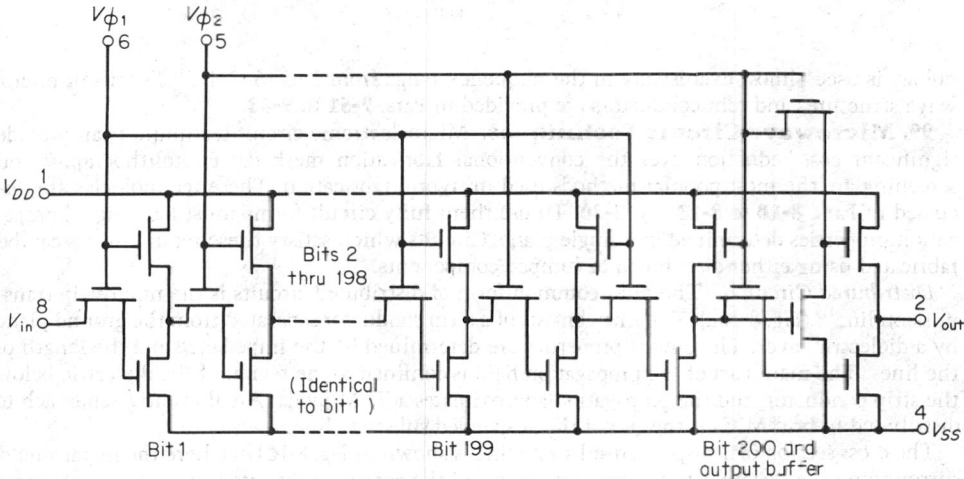

Fig. 8-160. Circuit diagram of one section of a 200-bit dynamic shift register.

cations because it minimizes the lead-inductance problem associated with discrete circuits. However, the use of monolithic IC technology for microwave applications has been somewhat limited because many of the microwave circuit components and devices are not readily compatible with the monolithic technology.

The hybrid IC technology, on the other hand, permits the use of a wider variety of devices and thus overcomes many of the difficulties of the monolithic approach. To date, hybrid tech-

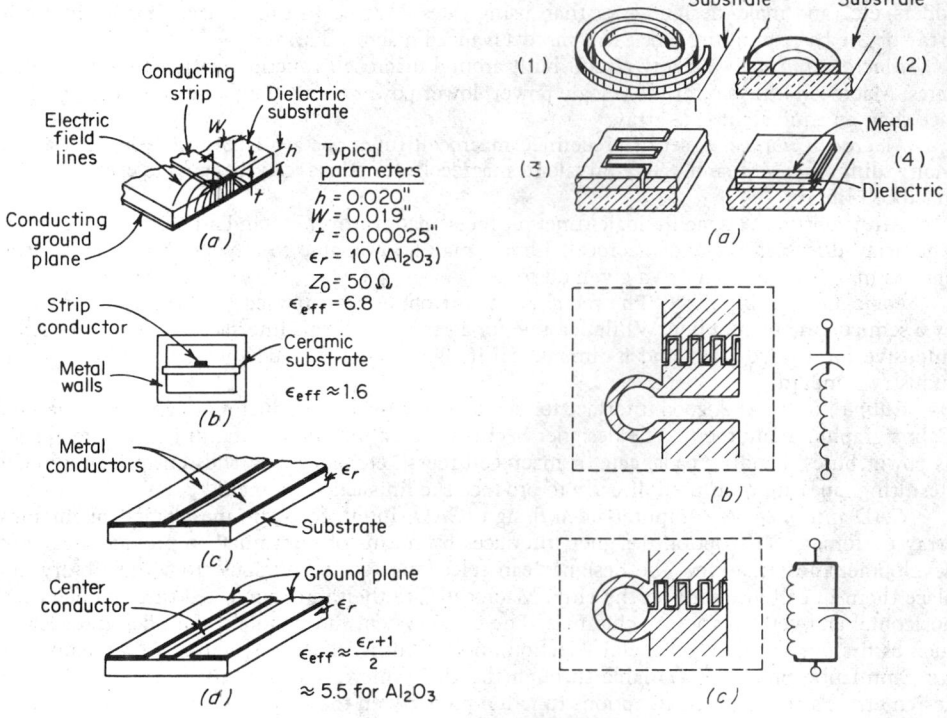

Fig. 8-161. Commonly used MIC transmission lines: (a) microstrip transmission line; (b) suspended substrate line; (c) slot line; (d) coplanar waveguide.

Fig. 8-162. Lumped components for microwave ICs: (a) typical structures; (1) spiral inductor, (2) strip inductor, (3) interdigitated capacitor, (4) metal-oxide-metal capacitor; (b) series LC circuit using strip inductor and interdigitated capacitor; (c) parallel LC circuit using strip inductor and interdigitated capacitor.

nology is used almost exclusively in the frequency range from 1 to 15 GHz.[111] Details of microwave structures and semiconductors are provided in Pars. **9-51** to **9-63**.

99. Microwave Circuit Techniques. Microelectronic circuit techniques can provide significant cost reduction over the conventional fabrication methods. Photolithography and screening are the most popular methods used in hybrid fabrication. These technologies are discussed in Pars. **8-10** to **8-13** and **8-30**. To use them fully circuit forms must have signal-propagation properties determined in a single plane. Circuits which satisfy these requirements can be fabricated using either distributed or lumped components.

Distributed Circuits. The most common form of distributed circuits is the microstrip transmission line[112] (Fig. 8-16a). The line consists of a strip conductor separated from the ground plane by a dielectric layer. The circuit properties are determined by the impedance and the length of the lines. The main part of the propagation field is confined to the region of the dielectric below the strip conductor, and the propagation approximates a TEM mode. An alternative approach to distributed hybrid MICs is the use of the suspended substrate line.

The cross section of a suspended substrate line is shown in Fig. 8-161b, where the metal shield surrounding the system acts as a ground plane and the ceramic substrate serves as a mechanical support for the suspended line. Two other types of distributed lines used in hybrid MICs are also shown in the figure. The finite slot line of Fig. 8-161c is a geometrical and electrical dual of the coplanar waveguide line of Fig. 8-161 d. In both these lines, the wave propagation mode is not TEM, and there are longitudinal as well as transverse magnetic rf fields.

Lumped Circuits. This class of MICs uses lumped circuit elements such as resistors, capacitors, or inductors whose values are independent of frequency within the frequency band of interest. For electrical components to act as lumped elements, it is necessary that their physical dimensions be much smaller than the wavelength of the electric signal they are supposed to handle. The small size of IC components is a distinct advantage in this application since they can maintain their lumped characteristics up to much higher frequencies than their discrete counterparts.

Some of the lumped inductor, conductor, and capacitor structures used in fabricating hybrid microwave ICs[112] are shown in Fig. 8-162a. Typical lumped LC circuits[113] which can resonate in the frequency range of 4 to 12 GHz are shown in Fig. 8-162b and c.

Most hybrid MIC structures require a three-layer sandwich of metal-dielectric-metal on a dielectric substrate. Figure 8-163 shows the cross section and the equivalent circuit of such a multilayer circuit structure.[114] First a layer of chromium-copper-chromium is deposited on the substrate. The thin layers of chromium (or titanium) on either side of the copper (or gold) conductive layer are necessary to ensure proper adhesion to the dielectric surfaces. After the bottom metal layer is deposited and patterned, an SiO_2 film is deposited as the dielectric layer; finally, the top metal layer is deposited and etched.

100. Materials for MICs. *Substrates.* Substrates for MICs should have low dielectric loss. The surface finish is important because it determines the definition of the circuit pattern, the yield in thin-film MOS capacitors, and rf conductor loss. The relative dielectric constant ϵ_r should be in the range of 8 to 16. Heat conductivity is important where high-power devices are used. Some properties and applications[112] of substrates that are commonly used for MIC applications are listed in Table 8-18.

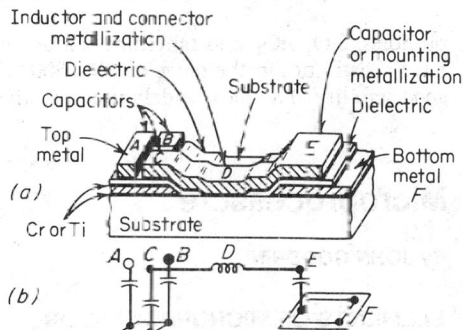

Fig. 8-163. Multilayer metal dielectric films for MICs (a) Cross section of integrated structure; (b) equivalent circuit contacts.[114]

Conductors. Important considerations for MIC conductors are the rf resistance and skin depth, deposition technique, substrate adherence, and thermal expansion during processing. Conductors can be divided into four categories, as shown in Table 8-19. The categories range from good conductors with poor substrate adherence to relatively poor conductors with good substrate adherence. The metals of the first two categories are usually deposited by vacuum evaporation, or electron-beam heating. Molybdenum and tungsten, in the third category, are

TABLE 8-18 Properties of Substrate Materials Used in Microwave Integrated Circuits

Material	Surface roughness Δ, μm	ϵ_r	K, $W/cm^2 \cdot K$	MIC applications
Alumina: 99.5%	2–8	10	0.3	Microstrip, suspended
96%	20	9	0.28	substrate
85%	50	8	0.20	
Sapphire	1	9.3–11.7	0.4	Microstrip, lumped element
Glass	1	5	0.01	Lumped element quasi-monolithic MICs
Quartz (fused)	1	3.3	0.01	Microstrip, lumped element
Beryllia	2–50	6.6	2.5	Compound substrates
Rutile	10–100	100	0.02	Microstrip, slot-line coplanar
Ferrite/garnet	10	13–16	0.03	Microstrip, coplanar compound substrates, nonreciprocal components
GaAs (high resistivity)	1	13	0.3	High-frequency microstrip, monolithic MICs
Si (high resistivity)	1	12	0.9	Monolithic MICs

refractive materials, and vacuum evaporation using electron-beam heating is required for deposition. Sputtering works with all these conductors and is especially useful for category III materials. The resistivity of the metals listed in the table is normalized to that of copper (~ 1.7 $\mu\Omega \cdot cm$).

Dielectrics and Resistors. Isolation and capacitor dielectrics for MIC applications must be reproducible, withstand high voltages, and be able to undergo processing without developing

TABLE 8-19 Characteristics of Conductor Materials for Microwave Integrated Circuits

Material	DC resistivity ρ (relative to Cu)	Skin depth δ at 2 GHz, μm	α, thermal expansion, μK^{-1}	Adherence to dielectric
I (Ag, Cu, Au, Al)	0.95–1.6	1.4–1.9	15–26	Poor
II (Cr, Ta, Ti)	7.6–48	4.0–10.5	8.5–9.0	Good
III (Mo, W)	3.3	2.6	6.0, 4.6	Fair
IV (Pt, Pd)	6.2	3.6	9–11	

pinholes. SiO, SiO$_2$, and tantalum pentoxide are the most widely used. Resistive films should have resistivities in the range of 10 to 500 $\Omega/\square$, a low-temperature coefficient of resistance, and good stability. The most widely used resistive films are nichrome and tantalum.

Microprocessors

By JOHN GOODRIN

ELEMENTS OF MICROPROCESSOR ARCHITECTURE, PROGRAMMING, DEVELOPMENT, AND IMPLEMENTATION

101. Introduction. These paragraphs describe the basic architecture of a microprocessor and the interrelationship of the microprocessing unit (MPU) with the forms of data and instruction storage (memory devices), input and output interfaces, and peripherals. They also describe data flow, program execution, register manipulation, addressing, microprocessor software, and microprocessor hardware and software development and integration.[115–125]

102. General Overview. In discussing microprocessors or microprocessor-based systems, there are three common elements (Fig. 8-164): one or more devices that function as a *processing unit*, *storage devices* for instructions to the processor and data to be processed, and a group of *sending and receiving interfaces* to connect the processor with the outside world. The outside world consists of a myriad of I/O devices that transmit data into the MPU system from the human interface, e.g., teletype, keyboard, paper-tape reader, and display or store data such as a CRT, LEDs, magnetic tape, discs, and line printers.

103. Definitions. Following are brief definitions of basic microprocessor and microprocessor system terms; through conventional usage they are often interchanged as they possess the same common generic elements.

Microprocessor. A microprocessor is one or more LSI devices that perform the function of a *central processing unit* (CPU) or *microprocessor unit* (MPU). A microprocessor is a logic device used in digital electronics. The MPU is a complex logic element which performs arithmetic, logic, and control operations and which is generally packaged as a single IC. Its operation is synchronized by internal or external clocking, or both, from pulses generated by a free-running oscillator or crystal. The device accesses external memory, its I/O interface, I/O devices, and peripherals through a group of parallel conductors which carry informational signals. The MPU can control and service peripheral devices which share an external bus, but peripheral control and I/O operations have a great effect on the computational and performance (speed) efficiency of the MPU. The primary function of the MPU is to process large quantities of data quickly. This has led to the development of specialized processors which handle I/O and peripheral control and service, serving as "slave" processors to the main or host processor in a multiprocessing environment. Discussions of microprocessor concepts in this section are based on the above assumptions.

Microcomputer. A microcomputer is a single LSI device acting as a complete computer system. The device integrates a microprocessor, memory, and I/O interface on a single chip. Microcomputer architectural definition varies as to the solution the device is designed to implement, e.g., a controller or a minimum chip system.

Single-Board Computer. A single-board computer (SBC) is a combination of a microprocessor, memory, and I/O on a single printed-circuit board that acts as a complete computer system. SBCs

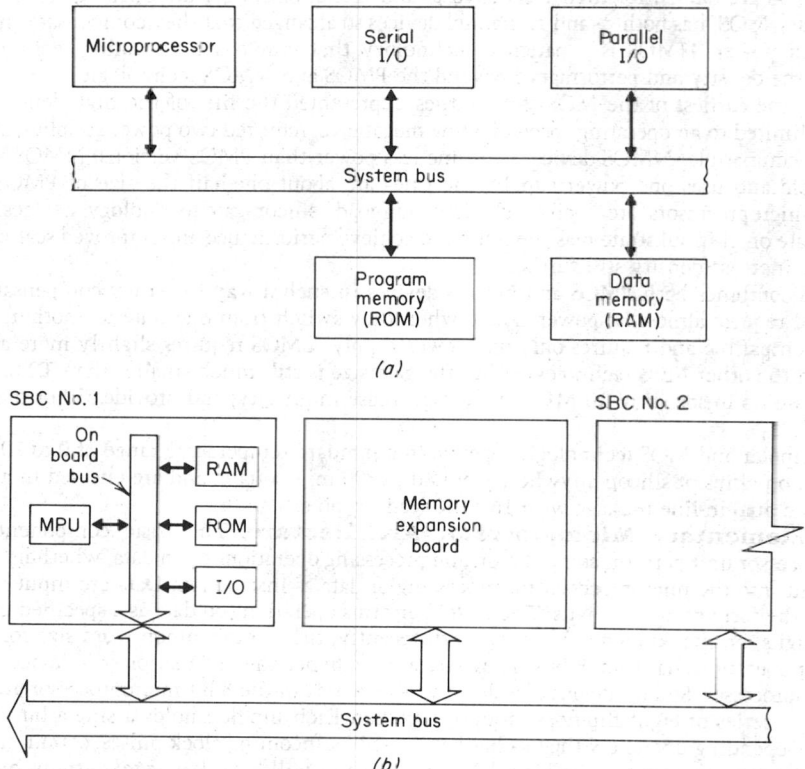

Fig. 8-164. Elements are common in (*a*) microprocessor serving as an MPU or (*b*) single board computer in a multiprocessing environment.

provide economical self-contained computer-based solutions, including specialized modules such as a high-speed mathematics unit.

Microcomputer System. A microcomputer system is the combination of a microcomputer, mechanical enclosure, and power distribution functioning as a complete computer system.

There are two basic technologies associated with microprocessors, bipolar and MOS. Bipolar is based on conventional transistors, while MOS technologies are based on a newer type of transistor constructed from metal-oxide semiconductors.

Before the advent of MOS technology, the most widely used logic had been *transistor-transistor logic* (TTL). TTL is a modification of the older diode-transistor logic (DTL), and it remains compatible with DTL. The main advantage of TTL is its high operating speed, ranging from 5 MHz to more than 50 MHz. It has a good drive capability with a typical fan-out of 10, so that it can be used with flexibility in complex circuits. It has good internal and external noise immunity.

TTL logic is normally classified in three categories, reflecting a power-speed trade-off: low-power, high-speed, and high-speed Schottky-clamped. There are also three-stated circuits, with stable states at a logical 0 and 1 and a disabled state that prevents data transfer. *Three-state logic* is used to interface several devices to a bus. When a device is in a disabled state, it is effectively disconnected.

Emitter-coupled logic (ECL) is another popular bipolar logic that has the advantages of high speed and flexibility, providing speeds in excess of 75 MHz. A fan-out of 25 allows it to drive many gates. Both the output and its complement are usually available, so that inverters can be eliminated or avoided. ECL circuits generate very little noise in switching. The disadvantages of

ECL are high power dissipation, poor external noise immunity, and need for a negative voltage supply. Consequently, both high and low voltages are negative and must be interfaced to make them compatible with the positive voltages of TTL and CMOS.

MOS technologies offer some advantages over bipolar technologies. MOS consumes less power, and MOS gates are smaller, providing greater density.

MOS technologies can be classified into PMOS, NMOS, HMOS, and CMOS. PMOS and standard NMOS are named for their respective *p*- and *n*-channels; they are duals. CMOS, or complementary MOS, has both *p*- and *n*-channel devices so arranged that they compensate and draw minimum power. HMOS is a patented technology that employs high-scaling techniques to increase the density and performance beyond the PMOS and NMOS technologies.

PMOS, the earliest of the MOS technologies, represented the first of the high-density chips but was limited to an operating speed of a few megahertz, required two power supplies, and was not TTL-compatible. NMOS devices consume less power than PMOS versions. NMOS is TTL-compatible and uses one power supply; the gates are about one-half the size of PMOS gates. HMOS microprocessors are *n*-channel, depletion-load, silicon-gate-technology devices. They incorporate on-chip substrate-bias generation to achieve performance and improved scaling techniques to increase density still further.

CMOS combines both PMOS and NMOS devices in such a way that they compensate each other and require almost no power except when they switch from one state to another. CMOS is TTL-compatible and requires only one power supply. CMOS requires slightly more area per gate than the other MOS technologies, but the gate size is still much smaller than TTL. CMOS can operate up to speeds of 20 MHz, has a high noise immunity, and provides 50 or more fan-outs.

Both bipolar and MOS technologies operate in a standard temperature range of 0 to 70°C, are produced on chips of silicon anywhere from 40 to 300 mils square, and are encased in a plastic or ceramic dual-in-line package with 16 to 40 leads or pins.

104. Elementary Microprocessor Architecture. The basic components of a microprocessor unit perform computation and processing operations upon data, whether they are data input into the microprocessor for processing or data as instruction. Data are input into the MPU in the form of binary digits. The MPU performs operations on data as a specified group of binary digits, or bits, known as a *word*. Until recently, the most common word size for microprocessor operations has been 8 bits, known as a *byte,* hence the designation of a device as an 8-bit microprocessor. Storage devices for data bytes or words in the 8-bit microprocessor are called registers, a series of eight flip-flops grouped together. Each flip-flop holds a single bit of information. Depending upon how the bus has been gated by incoming clock pulses, certain flip-flops will have a *high-level* output, or logical 1, while the rest will have *low-level* outputs, or logical 0. This results in a binary number.

The basic architecture of an elementary MPU is shown in Fig. 8-165. The elements are described in the following paragraphs.

105. Arithmetic Logic Unit. The arithmetic logic unit (ALU) performs arithmetic and logic operations on data determined by signals from various control lines. The ALU performs binary addition, subtraction, multiplication, and division for mathematical calculations. It also performs a variety of logical comparisons through boolean operations, complements words, i.e., inverts, and shifts a data word either right or left 1 bit at a time.

106. Controller-Sequencer. The control logic of the MPU manages (controls) and supervises (sequences) the activities of the processor by producing a variety of control signals to various input pins to carry out an instruction. The control logic performs *clock generation, read and write data control* to and from memory and I/O, *address latching* (holding the location of data contents), *bus control,* and *reset* functions.

107. Accumulator. The accumulator is the basic functional register of the MPU. During arithmetic and logic operations it performs a dual function (Fig. 8-166). Before an operation it holds an operand. After, it holds the resulting sum, difference, or logical answer. It is a special-purpose register, either architecturally defined or user-specified, that acts as the transfer focal point. It serves as the transfer point from almost all register-to-memory and memory-to-register operations. It serves as the binary adder. It can be complemented, tested, and shifted to perform most of the MPU's basic functions, whether arithmetic or logic.

108. Data Register. The data register is the temporary storage location for data received from the data bus. Also called the *instruction register* (IR), the storage location holds a byte of data while its function is being decoded. The register also holds a byte while the word is being stored in a memory location.

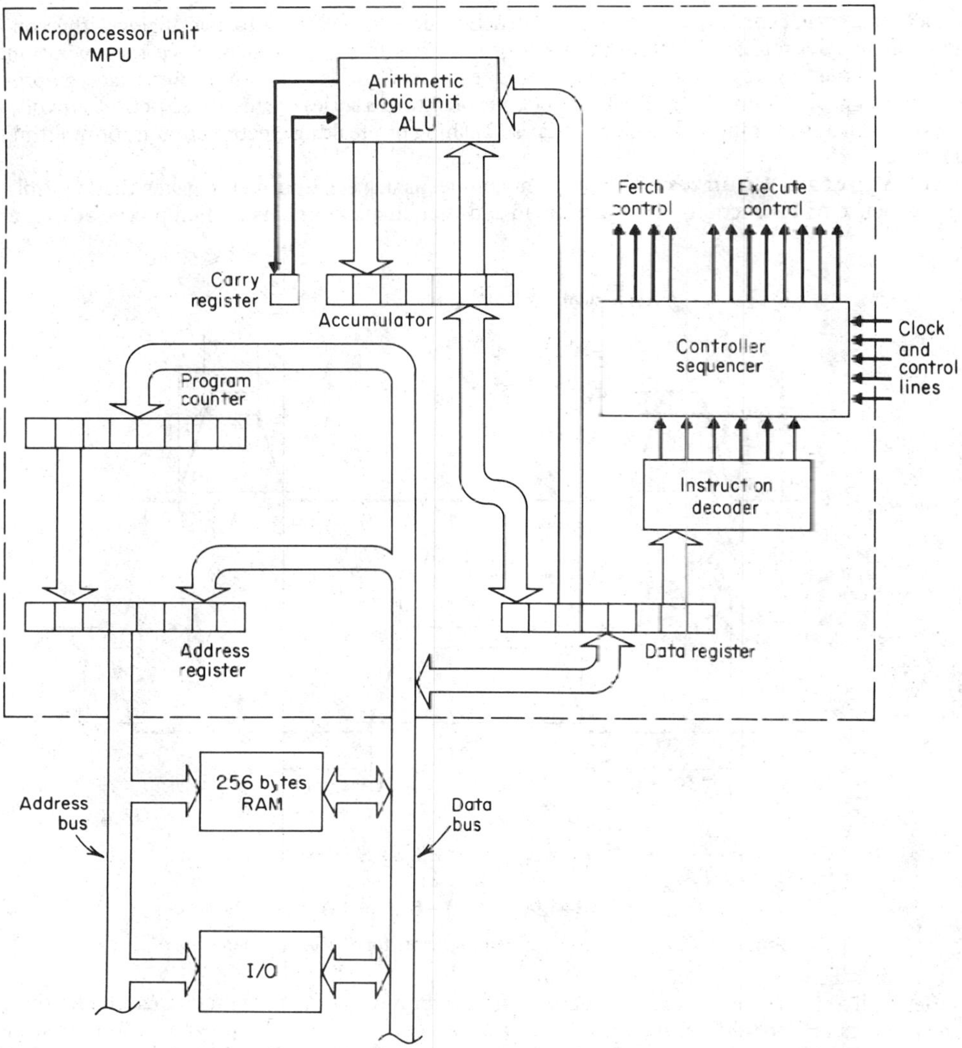

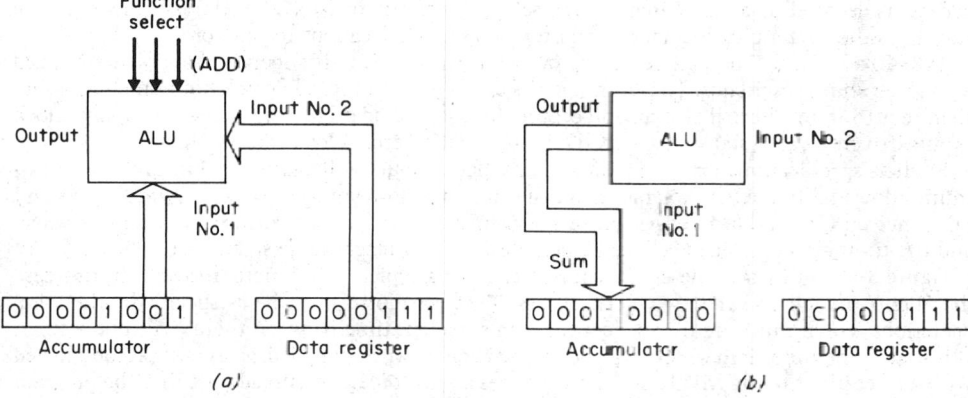

Fig. 8-165. Architecture of an elementary MPU.

Fig. 8-166. Accumulator operation: (a) accumulator holds initial operand and (b) result.

109. Instruction Decoder. The instruction decoder is a circuit that analyzes the contents of the data register to determine what operation is to be performed. Once the *operation code (op code)* is recognized, the controller-sequencer can supply the appropriate preprogrammed signals for operation. Each instruction consists of a series of *microinstructions*, involving such tasks as opening and closing gates, establishing or breaking electric circuits, or resetting status flags, etc.

110. Program Counter. The program counter is a special-purpose register that controls the sequence of instructions. For example, to add two binary numbers a microprocessor uses a

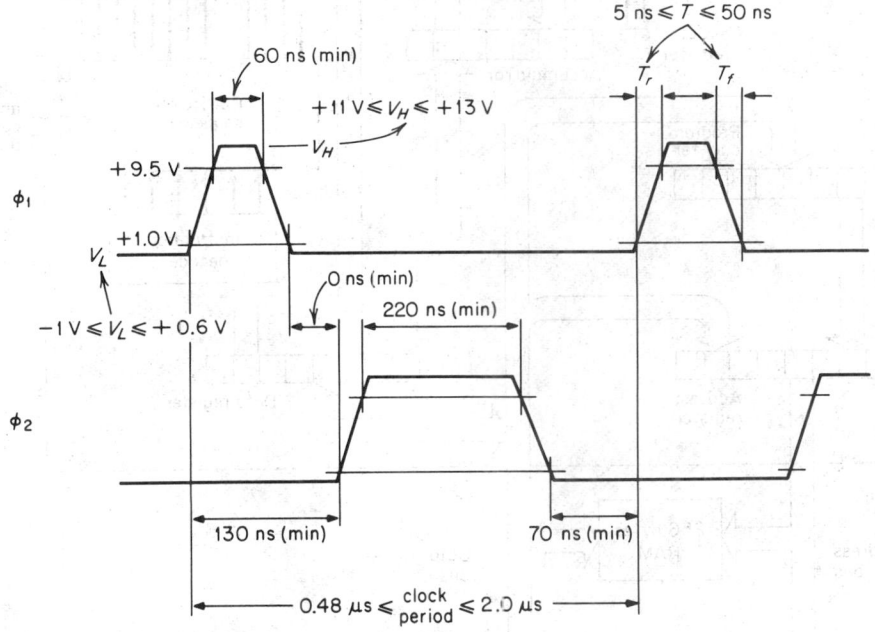

Power supply voltages = +12 V, +5 V, −5 V (and 0 V)

Fig. 8-167. Clock-waveform requirements for 8080A microprocessor.

series of five instructions. The instructions are a subroutine of a computer program. As these instructions are stored in sequentially accessed memory locations, the program counter is initialized to the first location in memory. As each op code or operand is processed, the register is incremented. It always acts as a pointer to the location in memory from which the next instruction or data word will come.

111. Address Register. The address register is a temporary storage device that holds the location or address of a data word in memory or I/O being used in a current operation. The address is decoded by a logic circuit at the selected memory or I/O. The address decoding logic translates the contents of the address register into specific locations in memory.

112. Clocking. The precise timing of events in the MPU is supervised and synchronized by a free-running oscillator. Today's microprocessors have an internal clock built on-chip, rather than requiring an external clocking procedure. Industry standard microprocessors can have clock inputs driven by a crystal, an *LC*-tuned circuit, or an external clock source.

Explicit specifications for a microprocessor's clock inputs will include minimum and maximum allowable frequency; tolerances on the high and low voltage levels, maximum rise and fall times on the waveform edges, pulse-width tolerance if the waveform is not a square wave, and the timing relationships between clock phases if two-clock phase signals are needed.

Figure 8-167 illustrates the clock requirements for a typical 8-bit microprocessor, in this case the 8080A, which requires two clock phases. This waveform must be as specified; otherwise, operations which must occur in sequence, with sufficient time between them, will not occur.

Initiation of the microprocessor's operation when power is applied is usually accomplished with a reset input to the MPU. Activating the reset will force a certain address into the program counter. Common reset mechanisms look for the rising edge on the reset input to initiate operation. An *RC* circuit keeps the input, as shown in Fig. 8-168, from rising until some time after the voltage of the device itself has reached 5 V. The actual values of resistance and capacitance required depend upon how fast the power-supply output reaches 5 V. The Schmitt trigger is either contained on-chip or is an external clock-driven chip.

Some microprocessors include automatic clearing of scratch-pad registers upon reset or require a user-designed initialization routine. If the MPU has a stack pointer pointing to a stack in RAM, loading the initial address into the stack pointer is part of the initialization routine.

113. Control Lines. The control lines in the MPU carry the signals that control the actual processing. For example, these lines may activate the parallel conductors, called a bus, so that data placed on them will flow to the proper destination. The control lines supervise the order in which microinstructions are carried out after the op code has been decoded by the logic circuit.

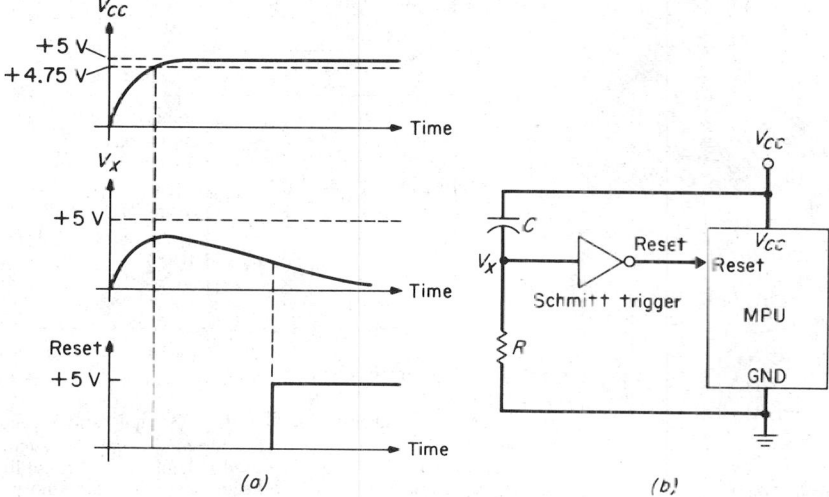

Fig. 8-168. Start-up circuit: (*a*) waveforms; (*b*) circuit.

Using the clock inputs, the control circuitry maintains the proper sequence of the processing task. The signals on control lines enable the various device functions which can be active high or active low. Table 8-20 describes a typical microprocessor's functional pin definitions, showing how they are activated by various signals.

114. Data and Address Buses. To complete the function of an MPU, a microprocessor must access memory and peripheral devices. This is accomplished by placing data on a bus, either an *address bus* or *data bus*, depending upon the function of the operation. The standard 8-bit microprocessor requires an 8-line parallel bus for each function. The trend (although it complicates the system design and timing) is to multiplex the address or data bus to reduce the number of pin connections. To do so, the microprocessor first selects the address which must be latched in a register or buffer. This process is described below.

The common approach to address-bus or data-system-bus organization is illustrated in Fig. 8-169. The MPU communicates with ROM, RAM, and I/O ports over a common bus consisting of (1) eight bidirectional data-bus lines to transfer one data word between the MPU and the addressed device. The direction of data flow will be defined by the operation. During a write, or load-to-memory, instruction the data will flow from the MPU to the addressed device. During a read, or load-from-memory, instruction data will flow from the selected device to the MPU. (2) Two unidirectional control lines, READ and WRITE, functionally inverted since when they are high (logic 1), they are not being used. READ is driven low by the MPU to tell the addressed device to put data onto the data bus. WRITE is driven low by the MPU to alert the selected device that it will receive a data word from the data bus for storage.

A bidirectional data bus implies that the data lines can carry data in either direction, to or from the MPU. Any one line can be driven in either direction. This is implemented by connecting only devices with three-state outputs to the data bus. An input port is eight three-state buffers plus gating to select the input port based on a port address supplied by the MPU and to read data from. Figure 8-170 shows schematic and logic symbol of the input port.

TABLE 8-20   Typical MPU Functional Pin Definitions (Continued)

Symbol	Function
$A_8–A_{15}$ (output, 3-state)	Address bus: the most significant 8 bits of memory address or the 8 bits of the I/O address, 3-stated during HOLD and HALT modes and during RESET
AD_{0-7} (I/O, 3-state)	Multiplexed address or data bus: lower 8 bits of memory address (or I/O address) appear on bus during first clock cycle (T state) of machine cycle; it becomes data bus during second and third clock cycles
ALE (output)	Address latch enable: occurs during first clock state of machine cycle and enables address to get latched into on-chip latch of peripherals; falling edge of ALE can also be used to strobe status information; ALE never 3-stated
S_0, S_1, and IO/$\overline{\text{M}}$ (output)	Machine cycle status:

IO/$\overline{\text{M}}$	S_1	S_0	Status
0	0	1	Memory write
0	1	0	Memory read
1	0	1	I/O write
1	1	0	I/O read
0	1	1	Op code fetch
1	1	1	Interrupt acknowledge
*	0	0	Halt
*	X	X	Hold
*	X	X	Reset

* = 3-state (high impedance)
X = unspecified
S_1 can be used as an advanced R/$\overline{\text{W}}$ status. IO/$\overline{\text{M}}$, S_0, and S_1 become valid at beginning of machine cycle and remain stable throughout cycle; falling edge of ALE may be used to latch state of these lines

$\overline{\text{RD}}$ (output, 3-state)	READ control: a low level on $\overline{\text{RD}}$ indicates that selected memory of I/O device is to be read and that data bus is available for data transfer, 3-stated during HOLD and HALT modes and during RESET
$\overline{\text{WR}}$ (output, 3-state)	WRITE control: a low level on $\overline{\text{WR}}$ indicates that data on data bus are to be written into selected memory or I/O location; data are set up at trailing edge of $\overline{\text{WR}}$; 3-stated during HOLD and HALT modes and during RESET
READY (input)	If READY is high during read or write cycle, it indicates that memory or peripheral is ready to send or receive data; if READY is low, CPU will wait an integral number of clock cycles for READY to go high before completing read or write cycle; READY must conform to specified data and hold times
HOLD (input)	HOLD indicates that another master is requesting use of address and data buses; upon receiving HOLD request CPU will relinquish use of bus as soon as current bus transfer has been completed; internal processing can continue; processor can regain bus only after HOLD is removed; when the HOLD is acknowledged, address, data, $\overline{\text{RD}}$, $\overline{\text{WR}}$, and IO/$\overline{\text{M}}$ lines are 3-stated
HLDA (output)	HOLD ACKNOWLEDGE: indicates that CPU has received HOLD request and will relinquish bus in next clock cycle; HLDA goes low after HOLD request removed; CPU takes bus one-half clock cycle after HLDA goes low
INTR (input)	INTERRUPT REQUEST: used as general-purpose interrupt; sampled only during next-to-last clock cycle of instruction and during HOLD and HALT states; if it is active, program counter (PC) will be inhibited from incrementing and $\overline{\text{INTA}}$ will be issued; during this cycle RESTART or CALL instruction can be inserted to jump to interrupt-service routine; INTR is enabled and disabled by software; it is disabled by RESET and immediately after interrupt is accepted
$\overline{\text{INTA}}$ (output)	INTERRUPT ACKNOWLEDGE: used instead of (and has same timing as) $\overline{\text{RD}}$ during instruction cycle after INTR is accepted; can be used to activate 8259 interrupt chip or some other interrupt port

TABLE 8-20   Typical MPU Functional Pin Definitions (*Continued*)

Symbol	Function
RST 5.5, RST 6.5, RST 7.5 (inputs)	RESTART INTERRUPTS: These three inputs have same timing as INTR but cause an internal RESTART to be inserted automatically; have higher priority than INTR; may be individually masked out using SIM instruction
TRAP (input)	Trap interrupt is a nonmaskable RESTART interrupt; it is recognized at the same time as INTR or RST 5.5–7.5; unaffected by any mask or interrupt enable; has highest priority of any interrupt
RESET IN (input)	Sets program counter to zero and resets interrupt enable and HLDA flip-flops; data and address buses and control lines are 3-stated during RESET; because of asynchronous nature of RESET, processor's internal registers and flags may be altered by RESET with unpredictable results; RESET IN is Schmitt-triggered input, allowing connection to an RC network for power-on RESET delay; CPU is held in reset condition as long as RESET IN is applied
RESET OUT (output)	Indicates CPU is being reset; can be used as system reset; signal is synchronized to processor clock and lasts integral number of clock periods
X_1, X_2 (input)	X_1 and X_2 are connected to a crystal, LC, or RC network to drive internal clock generator; X_1 can also be external clock input from logic gate; input frequency is divided by 2 to give processor's internal operating frequency
CLK (output)	Clock output for use as a system clock; period of CLK is twice the X_1, X_2 input period
SID (input)	*Serial input data line:* data on this line are loaded into accumulator bit 7 whenever RIM instruction is executed
SOD (output)	*Serial output data line:* output SOD is set or reset as specified by SIM instruction
V_{cc}	+5-V supply
V_{ss}	Ground reference

115. Instruction and Data Movement in the MPU. An add subroutine is a good example of how data and instructions move within the basic elements of the MPU. Table 8-21 shows the hypothetical instruction and operand, mnemonics for programming, and the op code or operand binary representations. An add instruction to a microprocessor consists of LOAD THE ACCUMULATOR, ADD, and HALT instructions, plus two operands as data bytes.

TABLE 8-21   Sample Program

Name	Mnemonic	Op code	Description
Load accumulator	LDA	1000 0110	Load contents of next memory location into accumulator
Add	ADD	1000 1011	Add contents of next memory location to present contents of accumulator; place sum in accumulator
Halt	HLT	0011 1110	Stop all operations

116. Accessing Memory. To read the contents of a memory location, the microprocessor places the address on the address bus. The address decode logic translates the address. The MPU then issues a READ signal to the memory. The memory responds by placing the contents on the data bus. A memory is always unconcerned about what operation the MPU is performing. Whenever it receives a READ signal, it will send a selected data word to the MPU.

117. Executing a Subroutine: Fetch Phase. Figure 8-171 shows the basic registers and elements of the MPU with the corresponding segment of memory containing the add instruction. The binary contents represents the op code or operand to be accessed.

The program counter is first initialized to the first location in memory to be accessed. Corresponding to the program, the MPU fetches (retrieves) the first address in response to control signals from the control logic. The address from the program counter is latched into the address register, and the program counter is initialized to the next byte to be accessed in memory, as shown in Fig. 8-172*a* and *b*. The address is then placed on the address bus for decoding (Fig. 8-172*b*). The normal latching–data-out procedure is for the control signal to be active on two inputs of the address latch, STR and OE. Strobing the STR input latches the data.

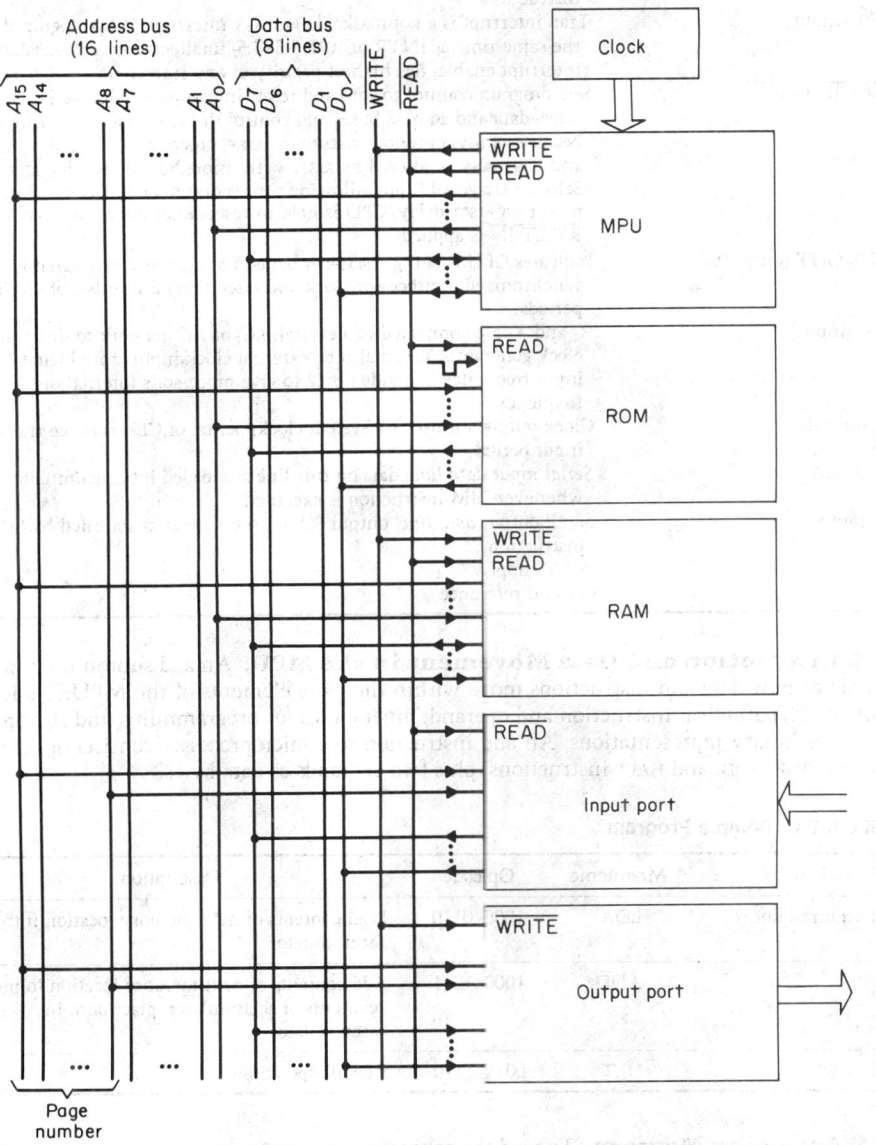

Fig. 8-169. Address and data bus organization.

The memory device responds by placing the contents of the decoded address on the data bus. The contents, an op code, are latched in the instruction register, where they are decoded. The instruction decoder translates the op code as a *load accumulator* instruction (LDA), which is then converted into the proper control signals by the controller-sequencer (Fig. 8-172*b*), during the execution phase of the instruction when the LDA instruction operand is read into the accumulator. This procedure encompasses the fetch phase of operation. Each operation contains both a fetch and an execution phase of operation.

118. The Execution Phase. The process for retrieving the operand is the same, but when the operand is decoded and placed in the data register it is transferred to the accumulator. This completes the execution of the LDA instruction.

The computer then performs a fetch of the next (ADD) instruction. When the add op code is translated, the controller-sequencer activates the ALU to perform the binary addition by adding the contents of the data or temporary register to the accumulator. The accumulator will retain

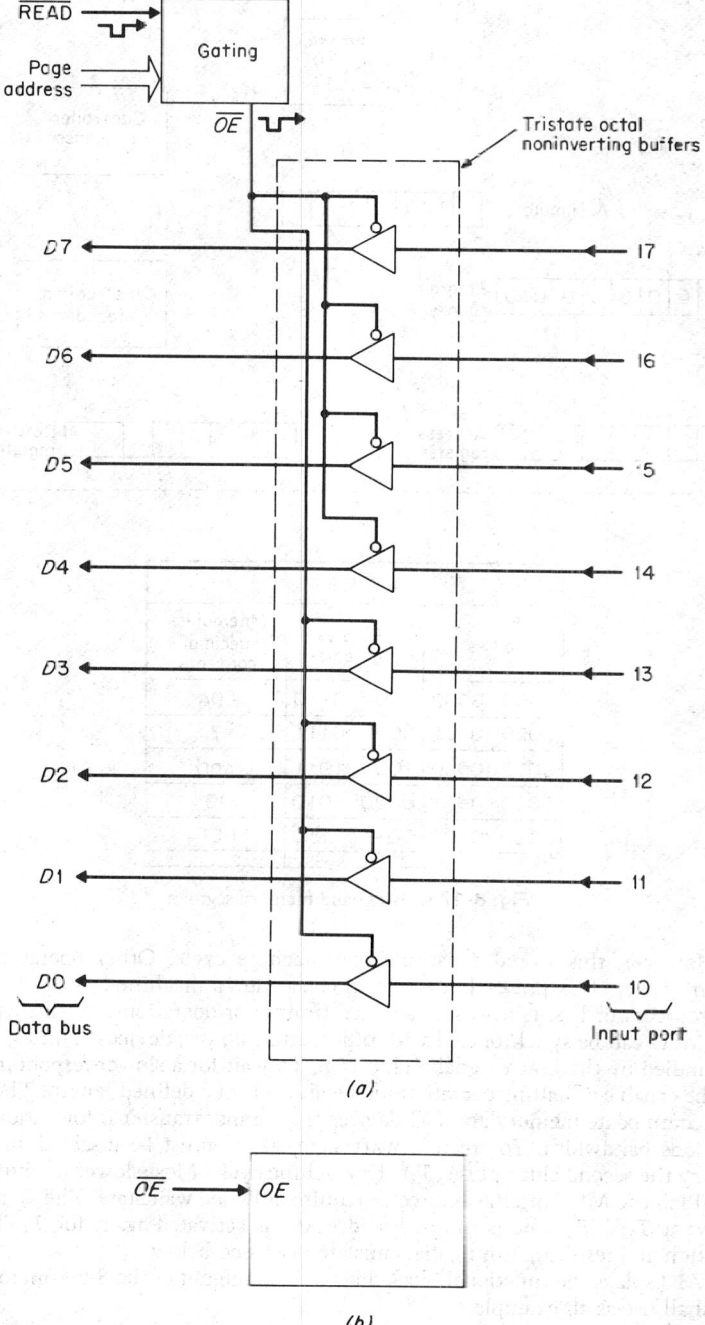

(a)

(b)

Fig. 8-170. Input port: (a) circuit; (b) symbol.

the result of the addition. The halt instruction signals the end of this particular program sequence.

119. Timing. Each microprocessor has a specified basic timing, called a *machine cycle*. The machine cycle is a consistent sequence of clock pulses needed to perform an MPU function. In a standard microprocessor, an op-code fetch may take three to six clock pulses to complete the

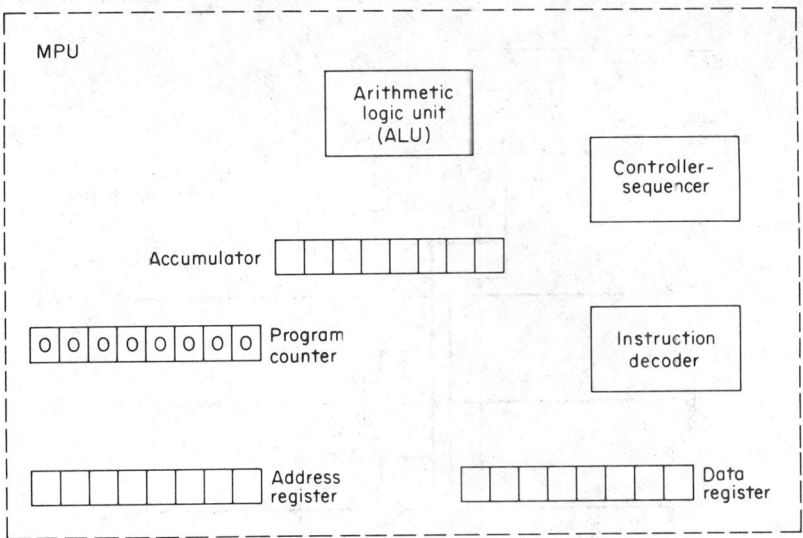

MEMORY		
Address	Binary contents	Mnemonics decimal contents
0000 0000	1000 0110	LDA
0000 0001	0000 0111	7
0000 0010	1000 1011	ADD
0000 0011	0000 1010	10
0000 0100	0011 1110	HLT

Fig. 8-171. MPU and memory segment.

operation. However, this would constitute one machine cycle. Other operations, reads and writes, require fewer clock pulses, but each still constitutes a machine cycle.

Each microprocessor has its own standardized timing for operations. With this standardized timing, the MPU can be synchronized with other, often slower, devices. Timing with external devices is handled by the READY signal (Fig. 8-173). To wait for a slower-responding device, the MPU must be capable of halting operation for a wait cycle of a defined length. The READY signal is used to accommodate memory and I/O devices that cannot transfer information at the maximum MPU bus bandwidth. To create a WAIT state, READY must be disabled and driven low, usually during the second clock pulse (T_2) of a machine cycle. Most slower memory and peripherals with which the MPU interfaces directly require only one wait state. The system ready line is driven low at T_2, a T_w state is inserted, and READY is activated again for T_3. The hardware implementation and resulting timing diagrams are discussed below.

Figure 8-173 displays the functional block diagram and pin-out of the 8-bit microprocessor, the 8085A, we shall use as an example.

Table 8-20 provides sample functional pin definitions, many of which will be expanded in subsequent subsections.

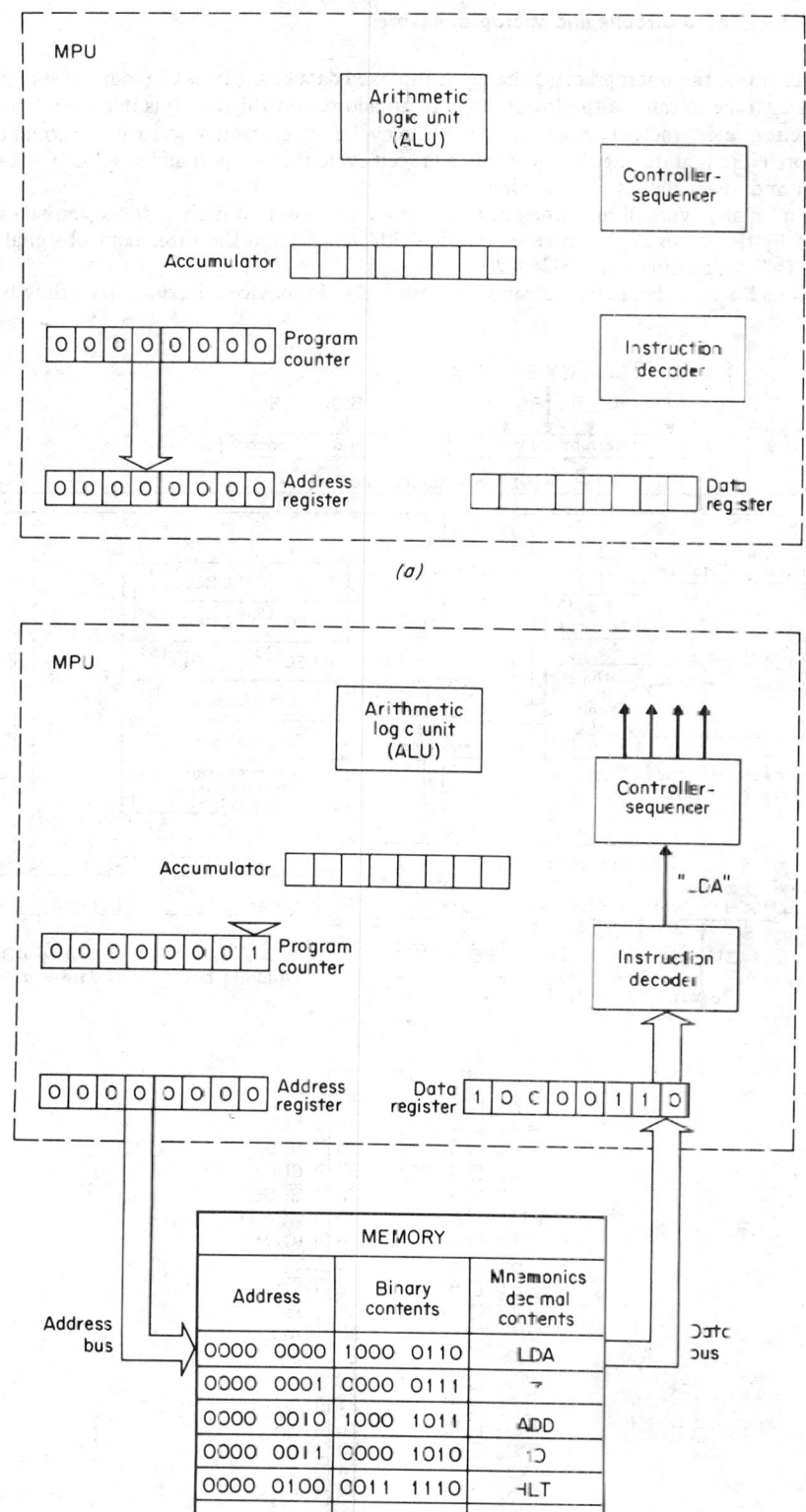

Fig. 8-172. Fetch phase: (a) address is latched; (b) program counter initialized, address placed on address bus, decoded address placed on data bus, op code decoded and converted.

In our example the microprocessor has a multiplexed data bus. The ALE (address latch enable) is used as a strobe to sample the lower 8-bits of an address on the data bus. Figure 8-174 shows an instruction fetch, memory read, and I/O write cycle, which would occur in a normal output instruction. Note that during the I/O write and read cycle the I/O port address is copied on both the upper and lower halves of the address.

There are many types of machine cycles in a microprocessor. Which of the seven takes place is defined by the status of the three status lines ($IO/\overline{M}, S_1, S_2$) and the three control signals (RD, WR, and INTA), as shown in Table 8-20.

The status lines can be used as advanced controls, e.g., for device selection, since they become

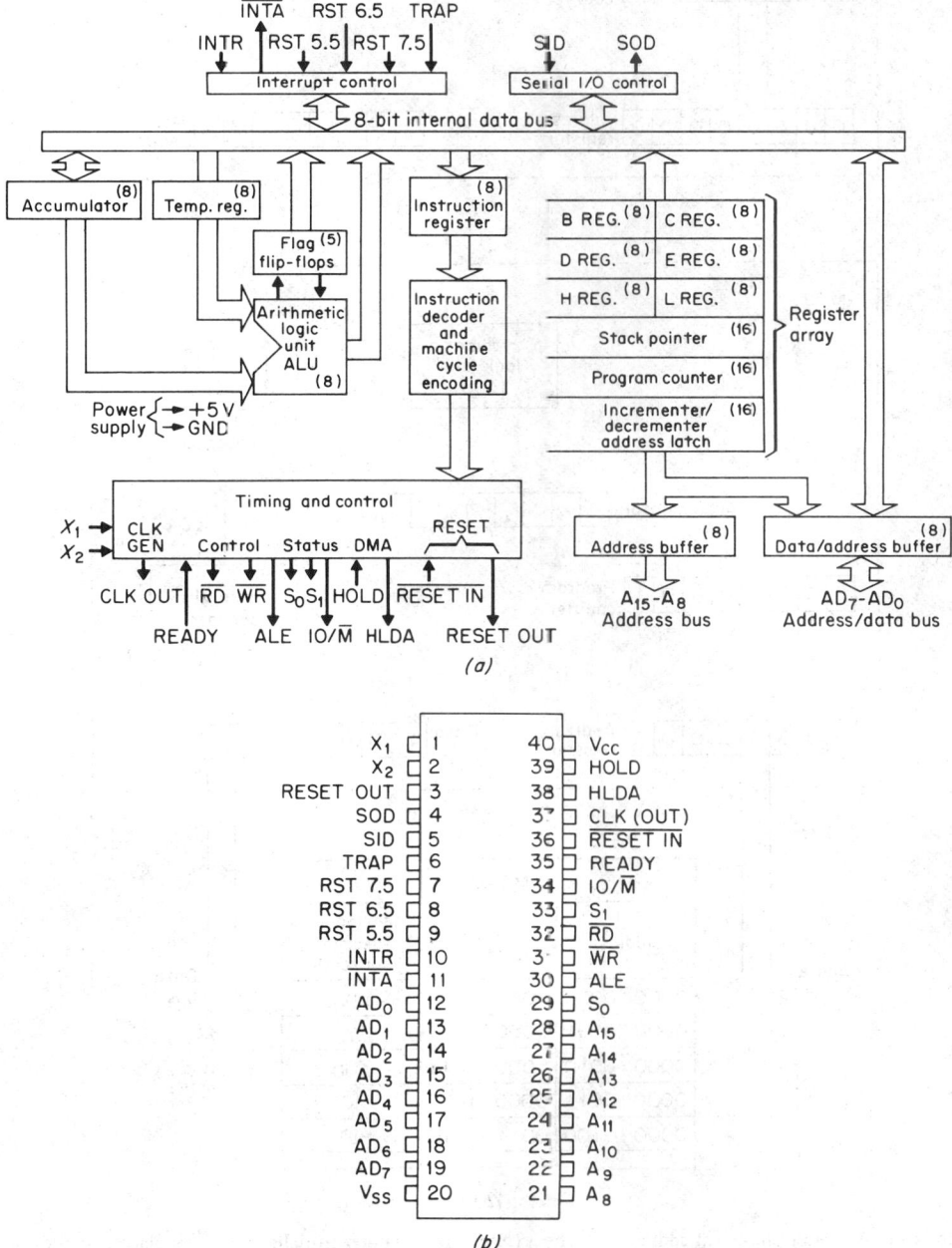

(a)

(b)

Fig. 8-173. (*a*) Functional block diagram and (*b*) pin-out.

active at the start of each machine cycle, at the T_1 state. Since control lines RD and WR become active later, when data transfer is to take place, they are used as command lines.

For a typical microprocessor a machine cycle normally consists of three T states with the exception of the op-code fetch.

Because all current microprocessors are geared for the highest performance possible for their application environments, timing with external devices such as slower memories or peripheral

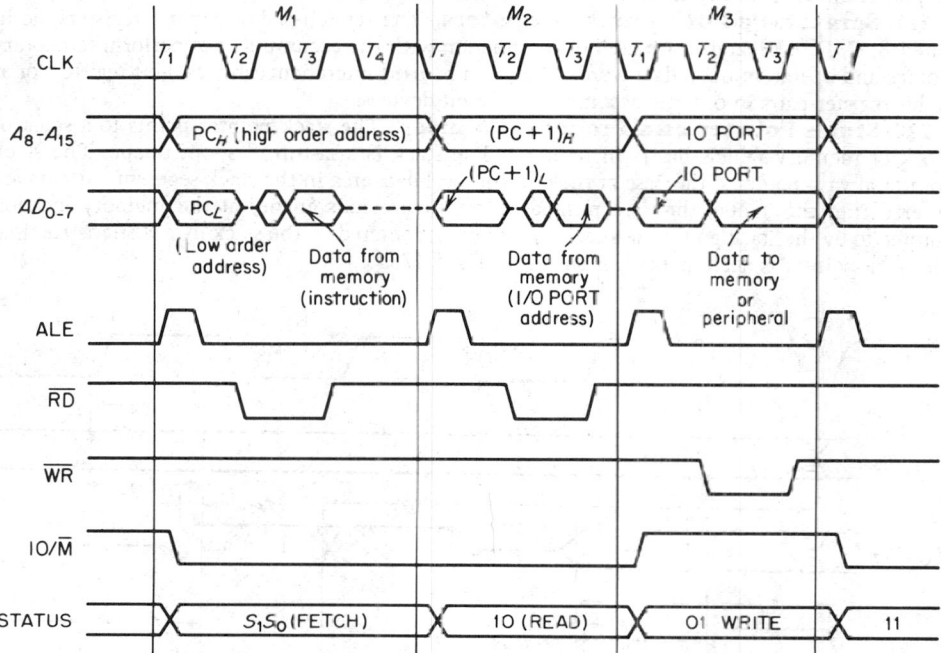

Fig. 8-174. Basic system timing for the 8085A.

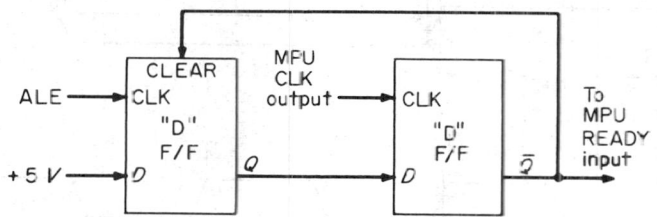

Fig. 8-175. Flip-flop circuit for generating a wait state.

devices creates processing delays. To solve the problem wait states are inserted in the machine cycle through the use of simple circuits such as D flip-flops. Figure 8-175 shows such a circuit that generates one wait state.

In this case, the D flip-flops should be chosen so that CLR is triggered on the rising edge of the signal and CLEAR is low-level-active. The READY line is used to extend the read and write pulse lengths so that the microprocessor can be used with devices slower than itself. Figure 8-176 shows bus timing with and without wait states.

If the microprocessor-based system includes an electromechanical device, such as a printer or stepper motor which must be driven, the MPU will have to generate pulses with a specified width. The interval is likely to be measured in milliseconds and may be repeated over and over. One method for generating the long pulses has been to implement a one-shot, as shown in Fig. 8-177 a.

One-shot circuits have now been replaced by *programmable timers* (Fig. 8-177 b), which can be either on-chip (as with microcomputers) or external, interrupt the MPU for the specified length of time, and generate output pulses and time intervals.

If there are numerous requests for data transfers in the system, the MPU may be supported by a *direct-memory-access* (DMA) controller, which permits the accessing of memory without involving MPU. The DMA controller exercises the processor's HOLD input, causing the MPU to relinquish control of its buses to the DMA controller which moves the data. The DMA controller can move data faster than an MPU can with I/O routines.

120. Architecture of an Industry Standard Microprocessor. Using the functional block diagram of an industry standard microprocessor in Fig. 8-173, the following paragraphs detail the standard microprocessor architectural elements not yet described.

121. Scratch-Pad or General Registers. The scratch-pad or general registers, designated B, C, D, E, H, and L, are used in conjunction with the accumulator to perform temporary storage and operation upon data words. They can be used interchangeably as 8-bit registers or as 16-bit register pairs in order to accommodate 16-bit devices.

122. Stack Pointer and Program Counter. The stack pointer points to a stack or block of memory which has been reserved. The stack is a last-in, first-out queue. The stack pointer always points to the base of the current local data area in the stack segment. As data are entered into the queue, they are pushed on top of previous entries at the memory location pointed to by the stack pointer. As each byte of data is entered on the stack by a PUSH instruction, the stack pointer is decremented, as shown in Fig. 8-178.

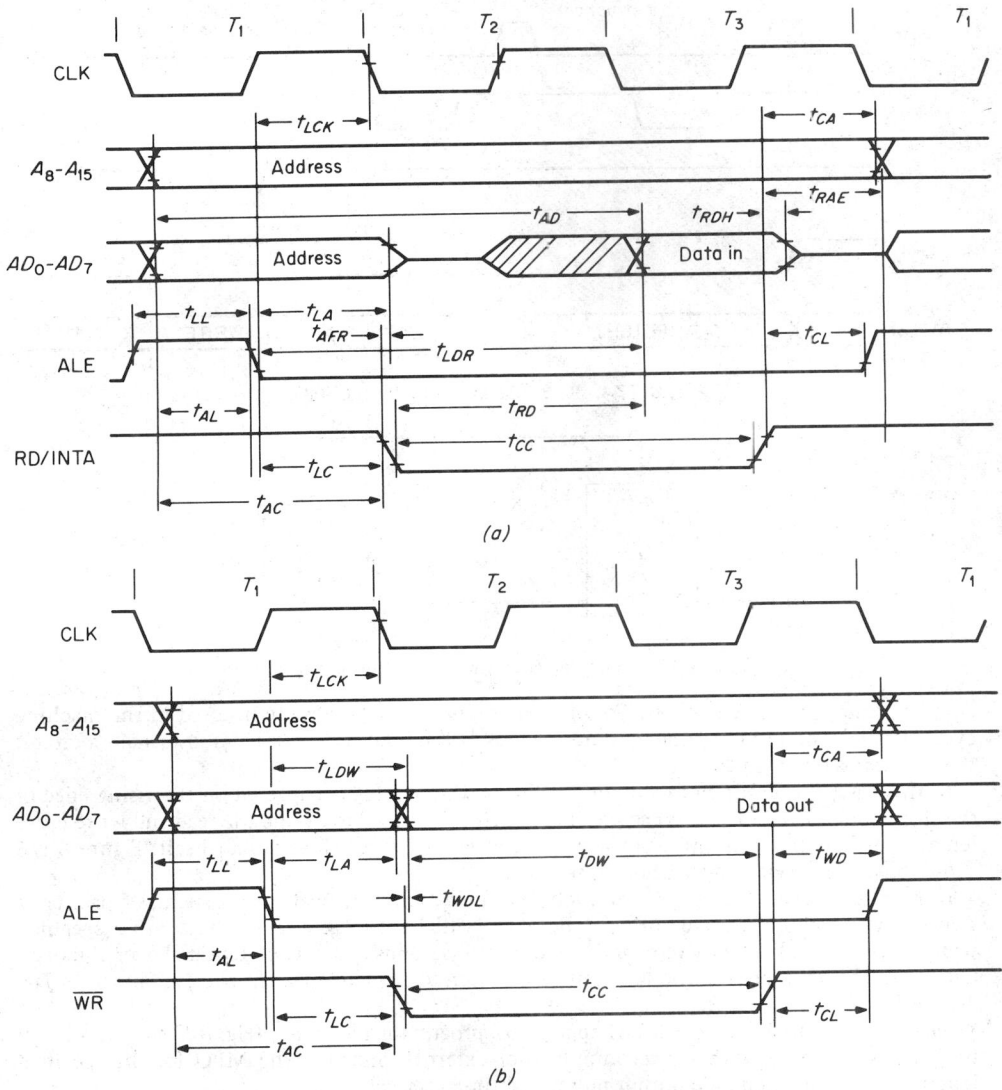

Fig. 8-176. Bus timing with and without wait: (*a*) read operation and (*b*) write operation.

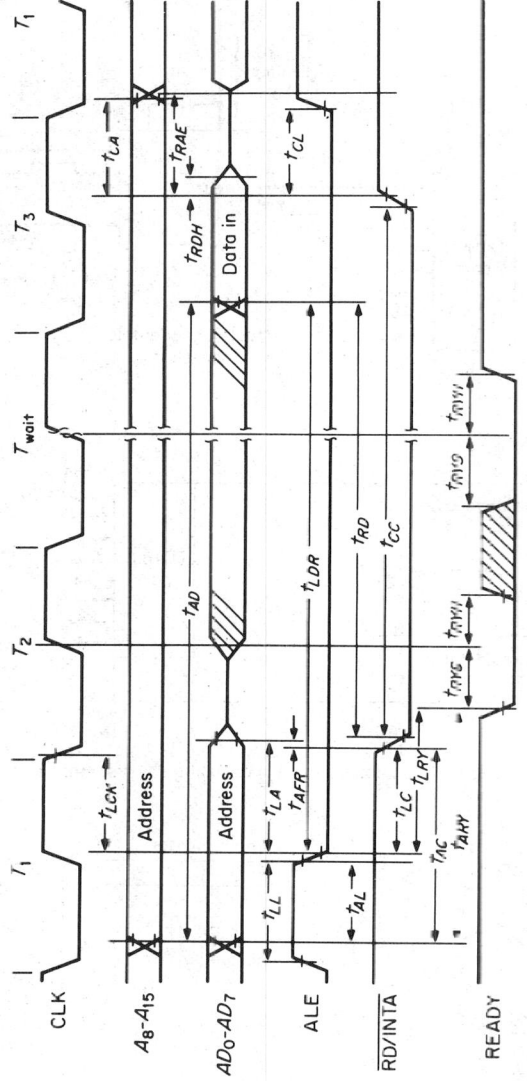

Note 1: Ready must remain stable during setup and hold times.

(c)

Fig. 8-170. (Continued). (c) Read operation with wait cycle.

Conversely, the stack pointer is incremented as data are brought off the stack to the MPU with a POP instruction. In this manner, the data byte available to the MPU is the last byte placed on the stack.

Typically, the microprocessor uses a stack located in data memory to handle subroutine return addresses automatically during subroutine call and return instructions. Alternatively, during an interrupt-service routine, any of the MPU register pairs desired can be explicitly set aside onto

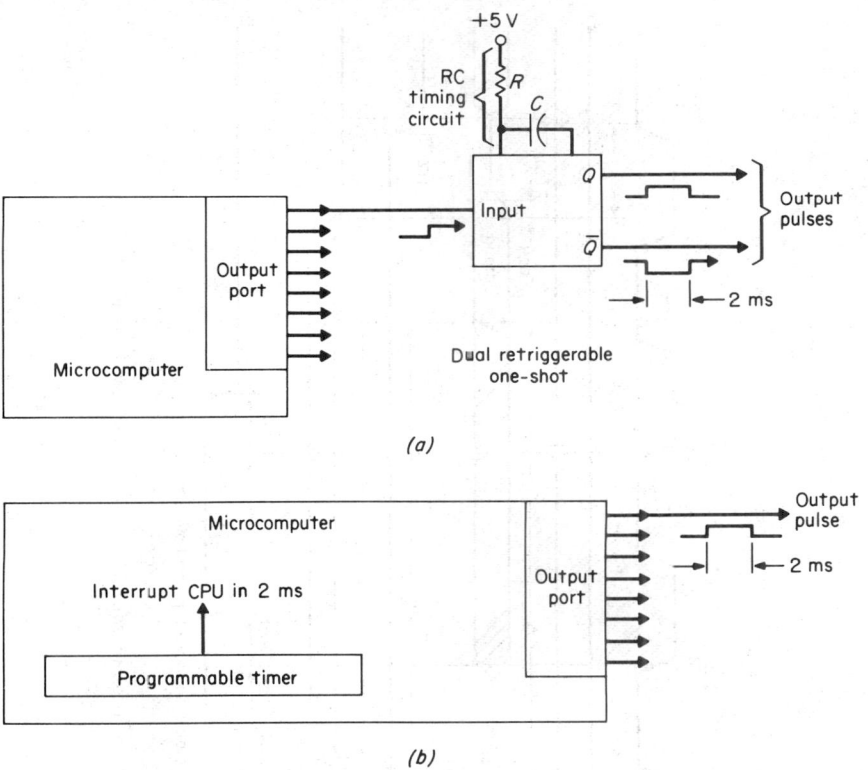

(a)

(b)

Fig. 8-177. Generating a long pulse: (*a*) one-shot; (*b*) programmable timer.

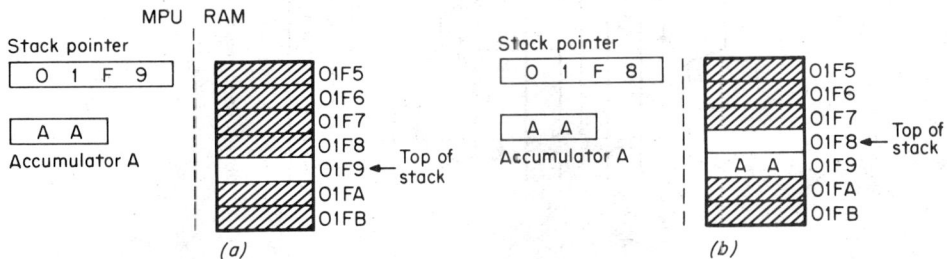

(a)   *(b)*

Fig. 8-178. Stack-pointer operation: (*a*) before data entry and (*b*) after data entry.

the stack with a PUSH instruction and subsequently restored with a POP instruction. In fact, any time data must be set aside to free a processor register temporarily, a 1-byte PUSH instruction accomplishes it.

When 2 bytes of data are pushed onto the stack from two registers, the high-order byte goes first. The stack pointer is decremented before each data transfer. Consequently, the stack pointer always points to the last byte of data entered onto the stack.

123. Flags. Flags can be used to alter a normal instruction. The five-flag set (1 bit each) contains the CARRY, ZERO, SIGN, PARITY, and AUXILIARY CARRY status bits. The CARRY indicates whether a carry or borrow has been generated in a binary operation. This is necessary if the microprocessor is to perform multiple-precision binary arithmetic. The ZERO flag indicates whether or not the result was zero, allowing the comparison of two values for equality. The SIGN flag indicates the setting of the leftmost bit of the result, which is later interpreted as the sign (− or +) of the binary number. A PARITY flag indicates whether the result is even (0) or odd (1) parity, to permit testing for transmission errors. Single-bit errors can occur through low-level electrostatic discharge or line fluctuations. One bit for each data word is used as a check bit. An even-parity construct results when, by adding the checking bit, the result of the 1 digits is even. With three active bits, a checking bit of a logical 1 creates an even result. The test system is the same for creating odd parity. The parity flag indicates whether the result is even or odd.

An AUXILIARY CARRY flag sometimes augments the flag set and indicates whether a carry was generated in the four lower-order bits. In conjunction with a decimal-adjustment instruction, the AUXILIARY CARRY makes it possible to perform packed BCD addition.

124. Interrupts. With devices external to the MPU requesting and receiving data, interrupt signals are used to halt current processing temporarily. Interrupt handling as it generally applies to common 8-bit microprocessors is handled as follows; some additional features for interrupt handling will be discussed in later paragraphs.

There are two ways to handle an interrupt situation: (1) a program can be written which accepts the first character from a serial input device, waits until the next character is ready by checking the ready status of the input device then accepts the next character, proceeding in this fashion until the entire string has been received; this method is referred to as *polled I/O*; (2) the device controller can interrupt the MPU when a character is ready for input, forcing a branch from a background application program to a special interrupt-service routine. The advantage of using interrupts is that the MPU can process data while it is waiting for inputs instead of wasting time polling.

The MPU will receive an interrupt-request-signal input from either a requesting device or an interrupt controller. The INTR input is used as a general-purpose interrupt. It is sampled only during the next to the last clock cycle of an instruction and during HOLD and HALT states. If INTR is active, the PC is inhibited from incrementing and an interrupt acknowledge (INTA) signal is issued. During this cycle a RESTART or call instruction can be inserted that results in branching from the current executing program to an interrupt-service routine located in a designated memory segment.

RST is a special-purpose 1-byte CALL instruction designed primarily for use with interrupts. RST pushes the contents of the program counter onto the stack to provide a return address and then jumps to one of eight predetermined addresses. A 3-bit code carried in the op code of the RST instruction specifies the jump address.

When a device requests interrupt service and interrupts are enabled, the processor acknowledges the request and prepares the data lines to accept an instruction from the interrupt control device.

An external device, known as an interrupt controller, may handle priority interrupts to the microprocessor. The controller functions as an overall manager of an interrupt-driven system environment. It accepts requests from the peripheral equipment, determines which request has priority, ascertains whether the incoming request has higher priority than one being currently serviced, and issues an interrupt to the CPU. The address of the interrupt-service routine is typically contained in the controller's vectoring table, which is transferred to the PC after the controller places a CALL instruction on the microprocessor's data bus in response to INTA pulses from the MPU.

125. Serial Data Control. Microprocessors interface with a variety of devices (modems, teletypes, and magnetic-tape cassettes) which transmit and receive data in serial streams. The interface logic converts serial data to parallel data and vice versa. A programmable communication interface is used to handle this conversion and samples for data transmission, both synchronous and asynchronous.

Serial protocol involves three possible serial communication techniques. Two of the most widely used are *synchronous* and *asynchronous* serial communication. The third, *isosynchronous*, is a hybrid.

Asynchronous communication is best suited for data transmission between two devices where data are sent at low speed in intermittent, small groupings. A typical application is data entry through a teletype or CRT keyboard. Since the receiving device has no way of knowing when data will be sent, the asynchronous format requires *framing information* for each character transmitted. This enables the receiver, such as a universal synchronous-asynchronous receiver-transmitter (USART), to detect a valid data signal properly.

The first bit, the *start bit*, is always a logical 0. The start bit is followed by the *data bits* forming the character. The data bits are then followed by one or more *stop bits*, which are logical ls. Stop bits ensure that the start of the next character will cause a transition on the communication line.

This transition is important. While the receiver waits on a dead line for possible transmission, it constantly samples for the leading edge of the start bit to occur. Sampling is performed by the USART much faster than the transmission baud rate. For example, a receiver may sample the signal edge at 8 times the baud rate. If the baud rate is 100 (100 bits/s), the USART would sample for the leading edge every ⅛₀₀ s. Once the edge is detected, however, the receiver must ensure that it is receiving a valid signal and not a transmission caused by line noise. Upon detecting a possible start bit, the receiver steps off a half-bit time to see that the bit is a logical 0. Once the start bit is detected, the receiver times a 1-bit time sampling for the remaining data and stop bits. Through this routine, the receiver synchronizes one on each character transmitted.

In a synchronous protocol there is clocking between the two systems. The transmitter generates the clock and transmits data on the leading edge of the clock pulse. The receiver uses this clock to read in the serial data stream.

With synchronous protocol, bit synchronization is not necessary, as data are expected continuously. Still, the receiver must establish a reference indicating where data begin and end. With a synchronous protocol, characters are grouped into records, and *framing characters* are added to the record. These framing characters are SYN (or synch) characters. The USART uses the SYN characters to determine the boundaries of the message. SYN characters consist of a synchronization pattern, a pattern not likely to occur during a normal message. They are generated by the transmitter. The receiver samples these SYN characters bit by bit to establish its reference.

The isosynchronous format of serial communication retains the clock interconnect of the synchronous protocol, but it does not generate SYN characters. As with the asynchronous format, a start bit is generated. As the protocol uses clock synchronization, the repetitive samplings of the asynchronous format are eliminated. The hybrid protocol reduces the amount of MPU time devoted to message recognition, eliminates software overhead required by framing characters, and implements a greatly simplified protocol.

In receiving and transmitting the serial protocol in any form, the USART strips (or inserts) the framing characters and bits from the serial data stream and converts the data into a parallel format to be placed on the data bus. Figure 8-179 shows the elementary function of a serial I/O interface.

As far as the MPU is concerned, the USART consists of the data bus buffer, a control, and status registers. The receive data and transmit data buffers lie passively in the path of received and transmitted data and do not need direct access. The USART receives the serial-data-in signal and transmits the serial-data-out control signal. The MPU can service the transmissions on an interrupt-driven basis.

126. Memory. A microprocessor accesses the instructions of the MPU structure and the data to be manipulated from devices referred to as memory. Program instructions may be stored in memory devices separate from memory used for data storage. The following paragraphs outline the structure and implementation of each.

127. Volatile and Nonvolatile Memory. For different applications, memory comes in three forms: *volatile* or *dynamic storage*, in which data must be constantly reapplied or refreshed and in which data are erased when power is no longer applied; *nonvolatile memory*, in which data remain permanently, even when power is not applied, and *static memory*, which does not need to be refreshed but which loses its data when power is no longer applied.

By the nature of their functions and cost of manufacture, nonvolatile memories are used basically for program data storage, while data input for processing is stored in dynamic or volatile memories. Static memories are used widely for data storage, with a higher cost for increased throughput. With the advent of *magnetic-bubble storage devices*, nonvolatile memory is finding widespread application as mass storage medium. Bubble memories are expected to increase in density comparable to the higher densities of earlier semiconductor memories.

128. Program Data Memories. Program memories are available in several forms (also see Par. **8-94**).

ROMs.   ROMs are *read-only memories*, used to retain the final microprocessor instructions for operation. The contents are mask-programmed by the manufacturer where the final chip is made.

PROMs.   PROMs are *programmable (field) read-only memories*. The contents are programmed by the user with a special PROM programmer in conjunction with a microprocessor development system.

EPROMs. EPROMs are *electrically programmable read-only memories* that can be erased under ultraviolet light. This permits the device to be reprogrammed over and over. In contrast, PROMS permit only previously unprogrammed bits to be programmed later.

The choice of which to use depends on the application. EPROMs are today overwhelmingly used during the prototype development of an instrument. ROMs, with their significantly lower cost in large quantities, are used in products shipped to the end user in quantity. PROMs provide a middle ground.

PROMs provide a more reliable instrument than EPROMs as the program cannot be erased accidentally. They are off-the-shelf devices that can be purchased in any quantity, allowing the

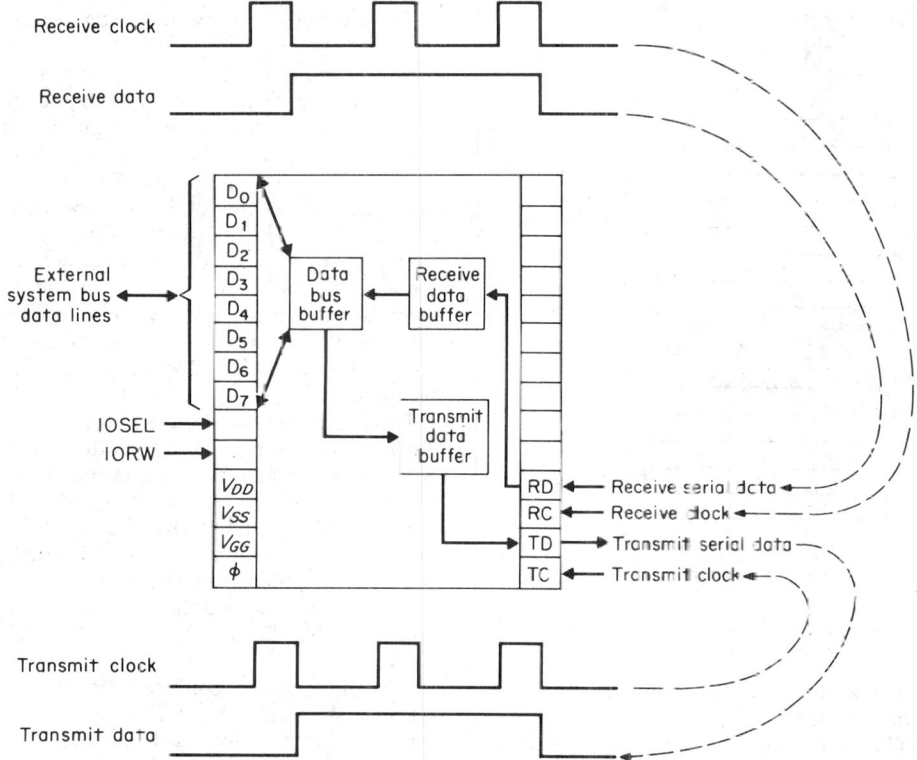

Fig. 8-179. Serial I/O interface.

user to maintain a low inventory. ROMs are usually ordered in quantities of 100 minimum and carry a masking charge. The use of a PROM also encourages minor modifications to an instrument to meet the needs of a relatively small group of end users.

The use of ROM chips with the bus structure is identical to the use of an input port. ROM chips are normally designed with three-state outputs. As shown in Fig. 8-18C, the page address must be decoded together with the RD signal to enable the contents of the addressed word to be placed on the data bus upon request by the MPU.

Since the output port does not drive the data bus, it is not three-stated. The incoming data must be latched until a subsequent WR signal to the port occurs. An output port can be implemented with *D* flip-flops. Each flip-flop remembers its state at the moment the clock input goes from low to high until the clock input goes from low to high again.

PROMs and EPROMs are programmed with a PROM programmer, a device that permits data entry through a keyboard, paper-tape reader, development system, or another PROM. The programmer puts successive pulses onto each bit at the recommended rate and recommended voltage.

Programming an EPROM is a multistep procedure. The device is powered normally as shown in Fig. 8-181. When the WE pin is raised to +12 V, the outputs become data inputs. Both data and addresses are applied using normal logic levels, as shown in Fig. 8-182. Then, successive addresses are written into, using the abnormal PROGRAM pulse in Fig. 8-181 as a write pulse.

129. Data Memories. The common reference to data memories is RAM (a misnomer), standing for *random-access memory*, a name which has stuck. Data memories are normally read-write memories which may be either MOS or bipolar. In either case, the basic storage structure is one or more transistors. Bipolar RAMs have fast access times, but each memory cell requires a larger area and dissipates more power than a MOS memory cell. MOS RAMs may be either static or dynamic.

The normal RAM circuits include address decoders to select the desired memory cell or cells, a chip-select (CS) input to enable the RAM addressing, and read-write circuits. A bit driver amplifies the data that are to be written into memory. Sense amplifiers detect and amplify the data that are read from memory. Sometimes buffers are needed to compensate for differences between memory and external circuit levels.

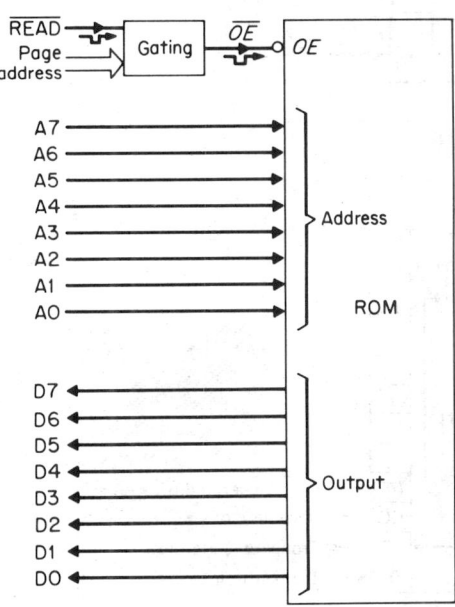

Fig. 8-180. ROM chip implementation.

RAMs are organized as modules that are arrays of LSI circuits. Two addressing schemes are commonly used, word and bit organization. Word organization, as shown in Fig. 8-183, requires just one decoder. The memory is arranged as W words of B bits. The decoder selects one of the W words. To read from memory, all bits of the selected word are sensed and read by the sense amplifiers. To write a word into memory, the location is selected by the decoder, and then data are written into the selected word through the B-bit drivers.

Large memories often use bit organization. Two decoders are needed to address $W = 2^n$ cells. Each decoder has $n/2$ inputs and can select one of $2^{n/2}$ lines. A 1,024-word by 1-bit memory is addressed by two five-input decoders, each of which drives 32 lines. One decoder selects the X address of the desired memory cell; the other selects the Y address. This scheme is illustrated in Fig. 8-184 for a 16-bit RAM.

Bipolar or static RAM cells are constructed with bipolar transistor memory cells, the earlier emitter-coupled cell and the Schottky-diode-coupled cell, as shown in Fig. 8-185.

The two transistors of the multiemitter cell form a flip-flop that stores a binary digit of 0 or 1, according to which device is on or off. Normally the word line is low and current is conducted over one of the inner emitters. When the word is selected, the W line goes high. Since the bit lines (D and $\overline{D}$) are normally at about $+0.5$ V, the flip-flop current is transferred to one of the outer emitters and the inner emitter junctions are reverse-biased. To write to the cell, the W line is raised in potential, and current is applied to either D or $\overline{D}$ to force the flip-flop to the desired state.

Static MOS memory uses a flip-flop per memory cell, as shown in Fig. 8-186. Transistors T_1 and T_2 are flip-flop transistors, and transistors T_3 and T_4 provide the load resistors. The flip-flop is coupled to the bit lines D and $\overline{D}$ through two transistors, T_5 and T_6, controlled by the W line, which is normally off. Activating the W line turns on the coupling transistors, enabling the cell to be impressed on the D lines for reading or enabling a change in the flip-flop state in response to voltage applied for writing.

Figure 8-187 shows a possible three-transistor dynamic RAM cell. The bit is stored as a charge on the gate-to-substrate capacitance of T_1, while T_2 and T_3 serve as gating switches for reading and writing. Since the capacitance-stored charge decays with time, the memory needs refreshing periodically, by reading the cell content and rewriting into it.

More recent dynamic RAMs use single transistor cells (Fig. 8-188). Dynamic RAMs such as 2117 or 2118 multiplex the incoming address bits onto seven address pins, have their own on-chip decoding logic and gating, and have address latching provided by two TTL clocks, row address strobe (RAS), and column address strobe (CAS). The noncritical timing requirements for the two TTL clocks allow multiplexing to be used without sacrificing speed.

Basically, there are two elements in a single transistor cell: the storage capacitor, which holds a charge quantity appropriate to a logical 1 or 0, and the transistor itself, which acts as a switch, dumping the capacitor's charge onto the bit sense line for sensing. In the 16K-bit* dynamic RAM, the capacitor element is placed beneath the transistor, using two polysilicon interconnections. The first layer serves as one capacitor element, and the second serves as a gate and interconnection.

*K (not to be confused with $k = 1,000$) stands for $2^{10} = 1,024$, the closest in binary to 1,000.

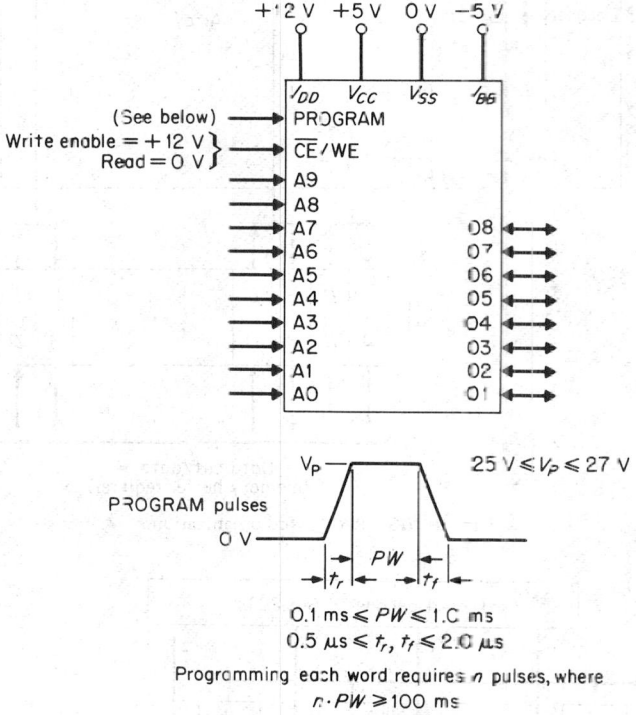

$25 \text{ V} \leqslant V_p \leqslant 27 \text{ V}$

$0.1 \text{ ms} \leqslant PW \leqslant 1.0 \text{ ms}$
$0.5 \text{ μs} \leqslant t_r, t_f \leqslant 2.0 \text{ μs}$

Programming each word requires n pulses, where
$$n \cdot PW \geqslant 100 \text{ ms}$$
PROGRAM input current $\approx 10 \text{ mA}$

Fig. 8-181. EPROM programming characteristics.

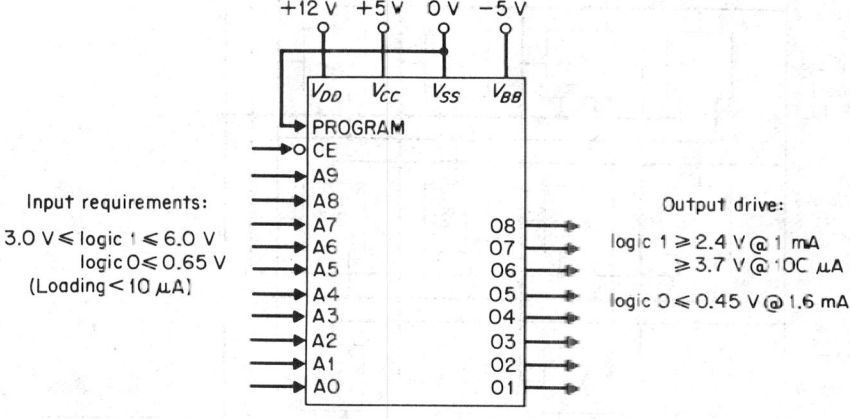

Chip enable to output delay $\leqslant 120 \text{ ns}$
Address to output delay $\leqslant 450 \text{ ns}$

Fig. 8-182. Read-mode characteristics.

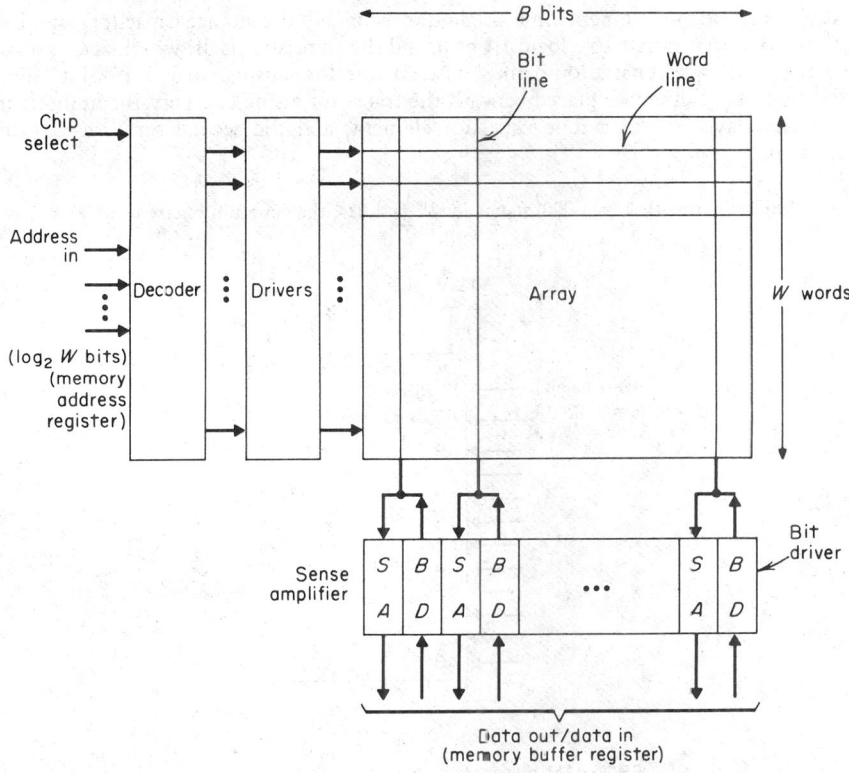

Fig. 8-183. RAM word organization.

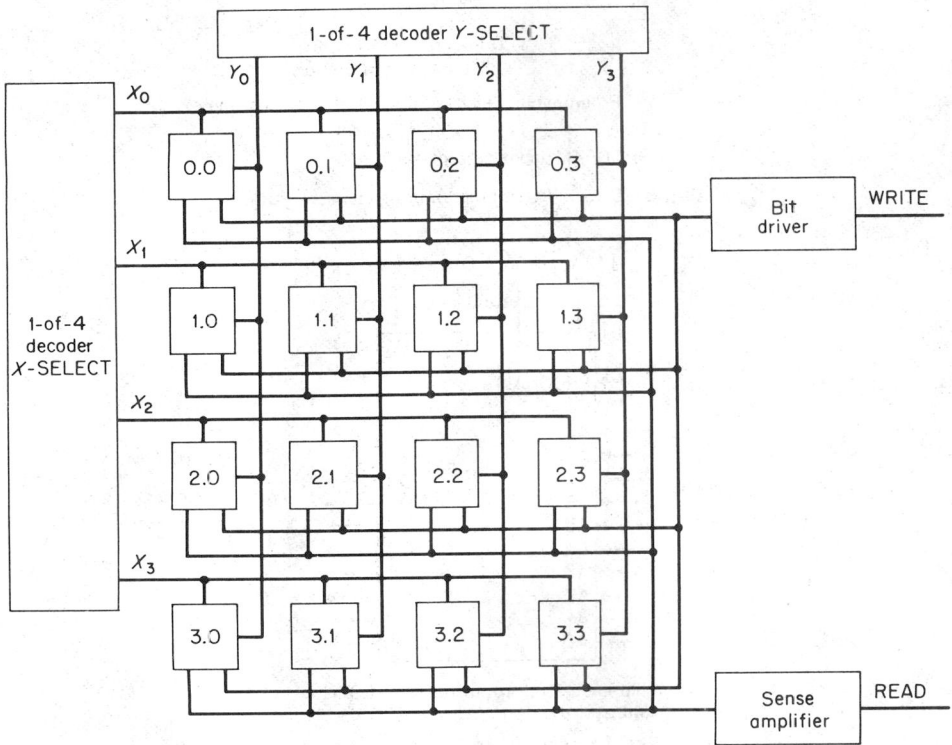

Fig. 8-184. 16-bit RAM organization.

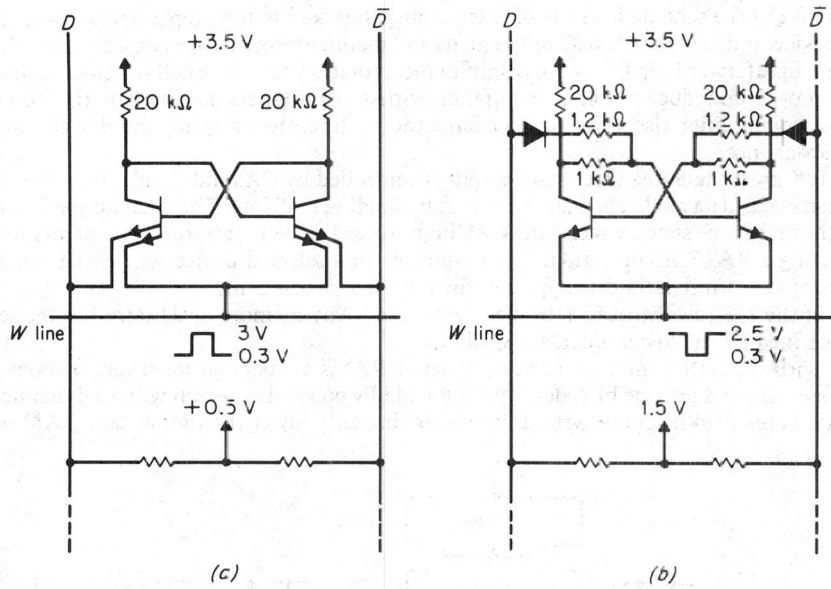

(c)   (b)

Fig. 8-185. Bipolar memory cell: (a) multiemitter; (b) Schottky diode.

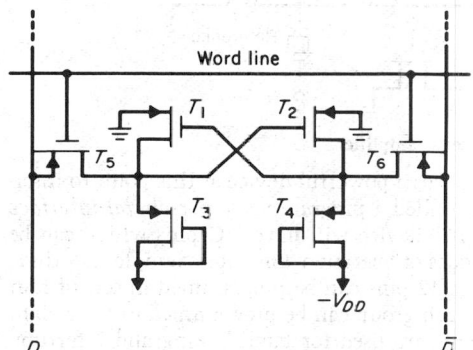

Fig. 8-186. Static MOS RAM memory cell.

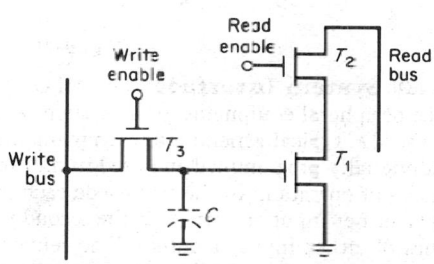

Fig. 8-187. Three-transistor dynamic MOS RAM cell.

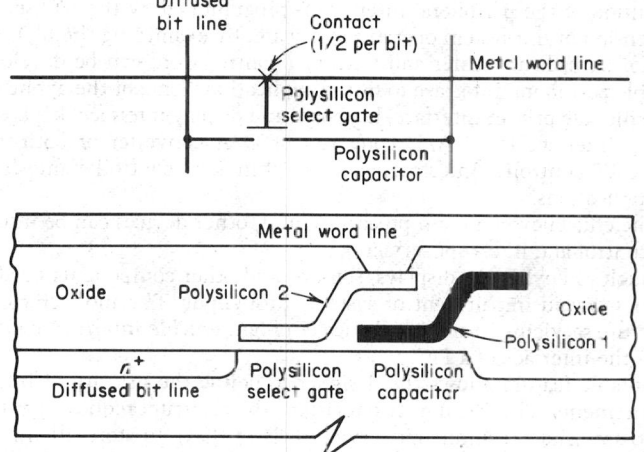

Fig. 8-188. Single-transistor RAM.

Figure 8-189 is a schematic of the data sense amplifier, key to the proper operation of the 16K RAM device. It detects small (200-mV) signals and maintains low power consumption. The otherwise straightforward flip-flop sense amplifier incorporates a reference cell to detect such a small signal. In order to reduce power consumption, with 256 amplifiers on the chip, the load device of the sense amplifier also serves to precharge the bit line, thus minimizing device count and input capacitance.

The 16K dynamic RAM three-state output is controlled by CAS independent of RAS. After a valid read cycle, data are latched on the output by holding CAS low. The data-out pin is returned to a high-impedance state by returning CAS high. A read cycle is performed by maintaining WE high during a RAS/CAS operation. The output pin of a selected device will remain in a high-impedance state until valid data appear at the output at access time.

Each of the 128 rows must be refreshed every 2 ms. Any memory cycle refreshes the selected row as defined by the lower-order (RAS) address.

Even with the refreshing requirement, dynamic RAMs are popular for several reasons. They are inexpensive and provide high density. The stand-by power dissipation when a dynamic RAM is neither being read from nor written into is significantly lower than for a static RAM device.

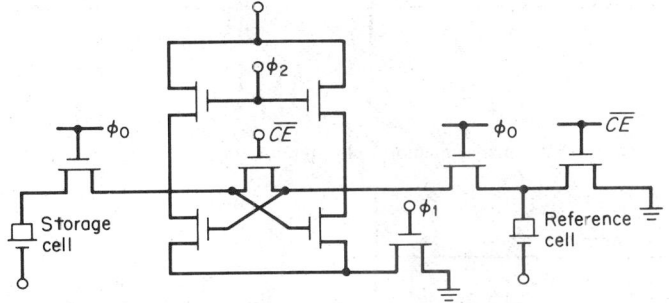

Fig. 8-189. Data sense amplifier.

130. System Interface. The MPU system needs a powerful device at this point to interface peripheral equipment to the system. A device called a *programmable peripheral interface* is used. A typical general-purpose programmable I/O device will have I/O pins which can be individually programmed and used in multiple modes of operation. One, for example, has three modes of operation. In the first mode each group of 12 pins can be programmed in sets of four to be either input or output. In the second mode, each group can be programmed to have eight lines of either input or output. The remaining pins are used for handshaking and interrupt-control signals. The third mode of operation is a bidirectional bus mode, which uses eight lines for a bidirectional bus and five lines, borrowing one from the other group, for handshaking. Through this process and flexibility, almost any I/C device can be interfaced without external logic. As mentioned before, every peripheral device has a service routine associated with it. The functional definition of the peripheral interface is programmed by the I/O service routine and becomes an extension of the system operating software. By examining the I/O device's interface characteristics for both data transfer and timing, a control word can be developed to initialize the programmable peripheral interface to fit the application. Some of the applications of such an interface device include printer interface, keyboard and display interface, keyboard and terminal address interface, interface to an A/D converter or D/A converter or both, basic floppy-disc interface, basic CRT controller interface, and a machine-tool controller interface. Figure 8-190 shows typical applications.

To increase the efficiency of a main processing unit, other devices can be interfaced to offload the peripheral control and interrupt servicing.

I/O devices such as keyboards, displays, sensors, and other components need servicing in an efficient manner to avoid impairment of system throughput. The most efficient manner is to offload the interrupt servicing to another device, a *programmable interrupt controller* (PIC). Figure 8-191 shows the interface of a PIC.

The programmable feature allows the designer to define the priority of interrupts based on the system requirements. The PIC uses two registers, the interrupt request register (IRR), to store requesting interrupts and the interrupt service register (ISR), to store all interrupts currently being serviced. A *priority resolver* is a logic circuit that determines the highest requesting priority and strobes it into the ISR during the INTA pulse. In this manner, the processor does not have to stop to evaluate and decide priority for each service interrupt.

Some devices work with the processor in a *slave* capacity, directly offloading normal tasks of the *host*, or there can be an additional *master* to share and enhance the system performance. A problem may arise here if the device performing the slave or multimaster function operates at a slower speed. All devices would request bus control at some time, with the slower devices affecting the performance of the main processor. The impact on timing and performance can outweigh the offloading benefits. A system bus addresses this problem by providing a common element for communication between a wide variety of system modules, including SBCs, memory, digital and analog I/O expansion boards, and peripheral controllers.

System bus signal lines can be grouped in the following categories: address lines, bidirectional data lines, multilevel interrupt lines, bus control, timing, and power-supply lines. The address and data lines are driven by three-state devices, while the interrupt and some other control lines are open-collector-driven.

Modules that use a system bus have a master-slave relationship. The bus master module can drive the command and address lines and control the bus. A bus slave cannot control the bus. Memory and I/O expansion boards are examples of bus slaves.

Bus arbitration occurs when more than one master requests control of the bus at the same time. A bus clock is usually provided by one of the bus masters and may be derived independently from the processor clock. The bus clock provides a timing reference for resolving bus contention between multiple requests. For example, a processor and a DMA module may both request control of the bus. This feature allows masters with different speeds to share resources on the same bus. Actual transfers via the bus, however, proceed asynchronously with respect to the bus clock. The transfer speed depends on the transmitting and receiving devices only. The bus design prevents slow master modules from being handicapped in their attempts to gain control of the bus. Once a bus request is granted, single or multiple read-write transfers can proceed.

A primary application of the system bus is for high-speed data transfers by a DMA controller used with a main processor.

The DMA controller can offload the processing function by allowing external devices to transfer information to or from the system memory directly. The DMA controller uses a hold-request (HRQ) signal input to the processor to request control of the system bus. Figure 8-152 shows a convenient method for configuring the DMA controller with a processor.

The multimode DMA controller issues a HRQ to the processor whenever there is at least one valid DMA request from a peripheral device. When the processor replies with a HLDA signal, the DMA takes control of the bus. The address for the first transfer operation comes out in 2 bytes, the least significant 8 bits on the address outputs and the most significant bits on the data bus. The contents of the data bus are latched into the 8-bit latch to complete the full 16 bits of the address bus.

The power of a DMA device is considerable. After being initialized by the program, a standard DMA can transfer a block of data up to 64K bytes without further intervention by the CPU. A DMA controller also resolves the priority, much like the PIC, for peripheral requests.

In three standard modes of operation the DMA controller can transfer data from a peripheral to memory, transfer data from memory to peripheral, and perform data-transfer verification procedures.

An MPU or DMA channel interfaces with other masters on a common system bus through the use of a bus arbiter. A bus arbiter provides bus-arbitration logic, timing logic, and output-drive logic.

For arbitration, the master sends a bus-request signal to the bus arbiter. A request-strobe (RSTB) input latches the request signal into the arbiter. During the next two falling edges of the bus clock (BCLK), used to synchronize the resolution circuitry asynchronously to the MPU clock, the controller sends out a bus-request signal based on the priority-resolving technique implemented. When priority is granted to the master, the master outputs a BUSY signal that locks the master onto the bus. When BUSY is active, the address and data enable (ADEN) goes active to enable the bus drivers, which amplify the transmission control signals.

One of the control inputs from the microprocessor is sent to the controller decode-control logic, which in turn activates memory and I/O read and write request signals.

131. I/O and Peripheral Devices. The following paragraphs briefly describe I/O and peripheral devices which can be used outside the microprocessor to enhance the MPU system's abilities.

132. Dynamic RAM Controller. A dynamic RAM controller can be interfaced to a microprocessor system as shown in Fig. 8-193. It provides all the signals necessary to use dynamic RAM, including multiplexed addresses and address strobes, as well as refresh, access, and arbitration.

The controller distinguishes three operational cycles: read, write, and refresh. The *refresher timer* is a sample timer that indicates to the controller when the refresh cycle is to be initiated. The address counter contains the address of the row to be refreshed. The counter is incremented after every cycle. The multiplexer is designed to provide the dynamic RAM array with row address, column address, and refresh address at the proper times.

The *timing and control block* executes one of the three execution cycles at the request of the arbiter. It provides the RAM array with WE, CAS, and RAS signals. It provides the MPU with the transfer and system acknowledge signals.

133. Arithmetic Processing Unit. An APU is a monolithic LSI device that enhances the mathematical capability of the microprocessor by providing high performance in floating-point arithmetic and floating-point trigonometric functions. All transfers, including operand, result, status, and command information, take place over an 8-bit bidirectional data bus. Operands are pushed onto an internal stack, and commands are issued to perform operations on the data in the stack. Results are then retrieved from the stack.

Transfers to and from the APU may be handled by an associated processor using conventionally programmed I/O or by the DMA controller. At the end of the operation, the APU issues an end of execution signal to the MPU to be used by the processor as an interrupt. Figure 8-194 shows the block diagram of the APU.

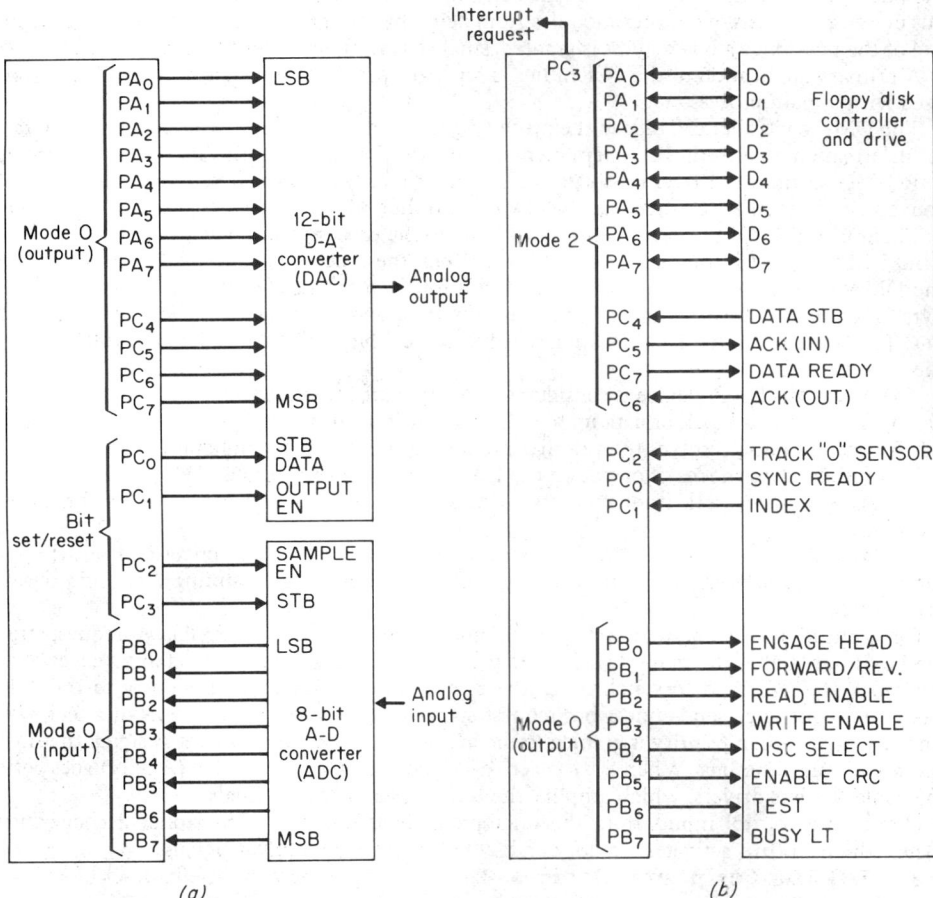

(a) *(b)*

Fig. 8-190. Programmable peripheral interface applications: *(a)* D/A; A/D; *(b)* basic floppy-disc interface.

134. Floating-Point Processor (FPU). A floating-point processor performs like an APU, providing single-precision (32-bit) and double-precision (64-bit) add, subtract, multiply, and divide functions.

135. Programmable HDLC/SDLC Protocol Controller. An HDLC/SDLC controller is a microcomputer peripheral-interface device which supports the International Standards (ISO) *high-level-data-link-control* (HDLC) and IBM *synchronous-data-link-control* (SDLC) protocols. It minimizes MPU software by supporting a frame-level instruction set and by hardware implementation of the low-level tasks associated with frame assembly-disassembly and data integrity.

Both protocols are bit-oriented, code-dependent, and ideal for full duplex communication. Some common applications include terminal-to-terminal, terminal-to-MPU, MPU-to-MPU, satellite communication, packet switching, and other high-speed data links.

The controller relieves the MPU of many of the tasks associated with constructing and receiving frames. A frame is a single communication element which can be used for both link-control and data-transfer purposes. The element consists of six fields which must be constructed: opening and closing 8-bit flags, 8-bit address and control fields, a variable-length information field, and a 16-bit error-checking field. The controller allows either DMA or non-DMA data transfers. It accepts a command from the MPU, executes the command, and provides an interrupt and result back to the MPU.

136. Floppy-Disc Controller. One of the most common peripherals a microprocessor will interface with is the floppy disc. A high performance LSI controller chip is available to interface both single- and double-density disc drives. Floppy-disc controllers interface up to four

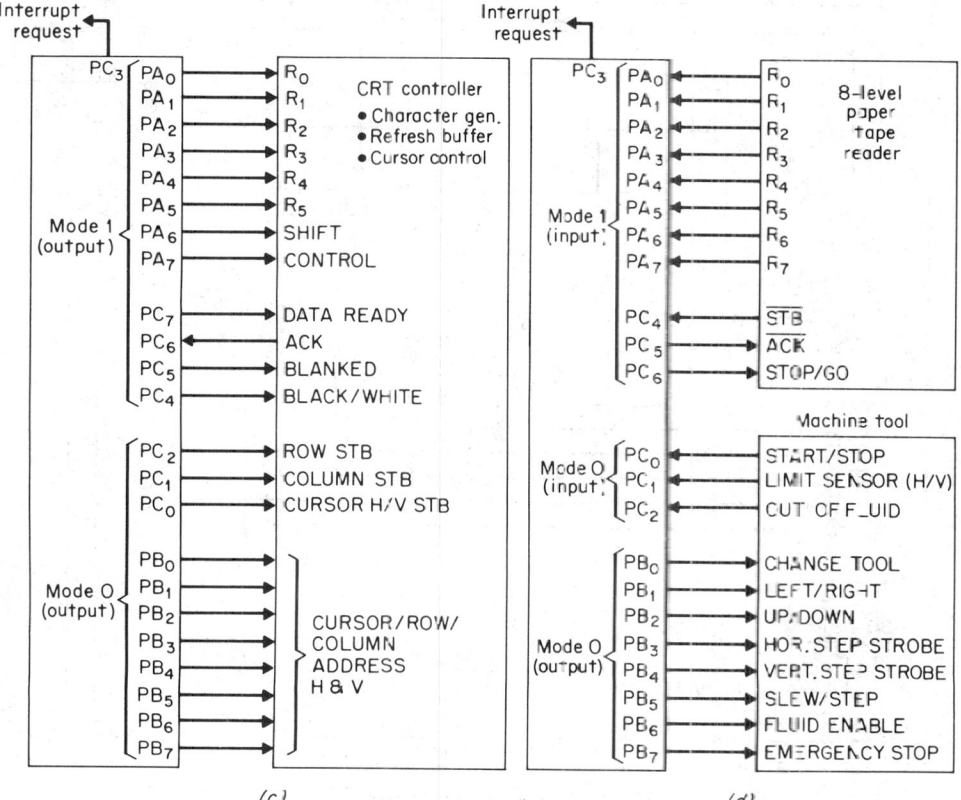

(c)   *(d)*

Fig. 8-190. *(Continued)* (c) CRT-controlled interface; (d) machine-tool interface.

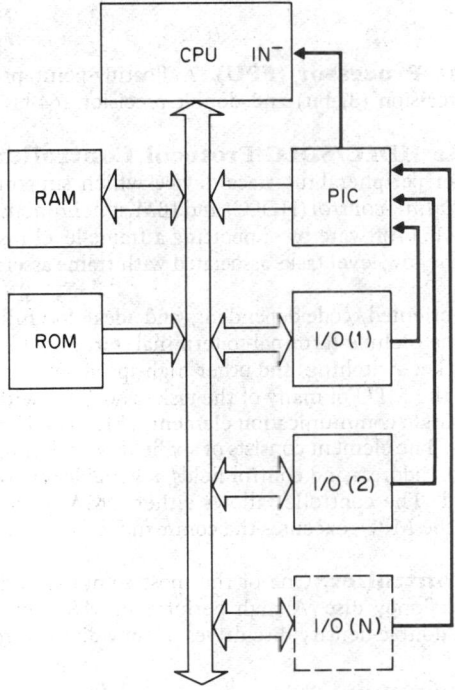

Fig. 8-191. A programmable interface system configuration.

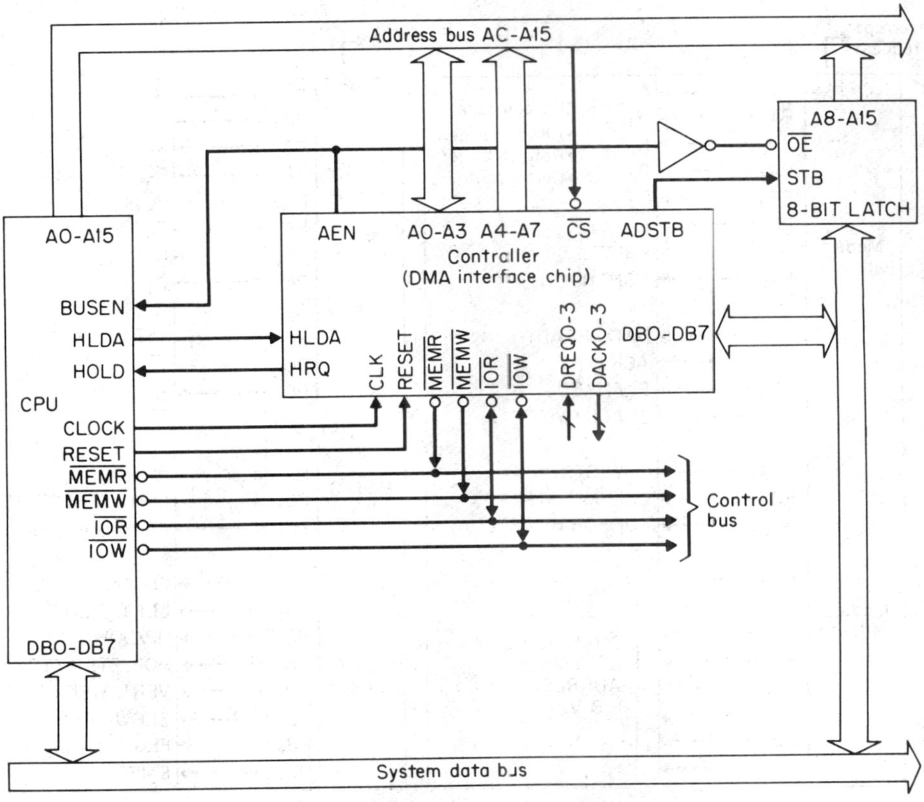

Address bus AC-A15

A8-A15
$\overline{OE}$
STB
8-BIT LATCH

AO-A15

AEN   AO-A3 A4-A7 $\overline{CS}$ ADSTB
Controller
(DMA interface chip)

BUSEN
HLDA   HLDA
HOLD   HRQ

CPU
CLOCK
RESET
$\overline{MEMR}$
$\overline{MEMW}$
$\overline{IOR}$
$\overline{IOW}$

CLK
RESET
$\overline{MEMR}$
$\overline{MEMW}$
$\overline{IOR}$
$\overline{IOW}$
DREQO-3
DACKO-3
DBO-DB7

Control
bus

DBO-DB7

System data bus

Fig. 8-192. DMA interface configuration.

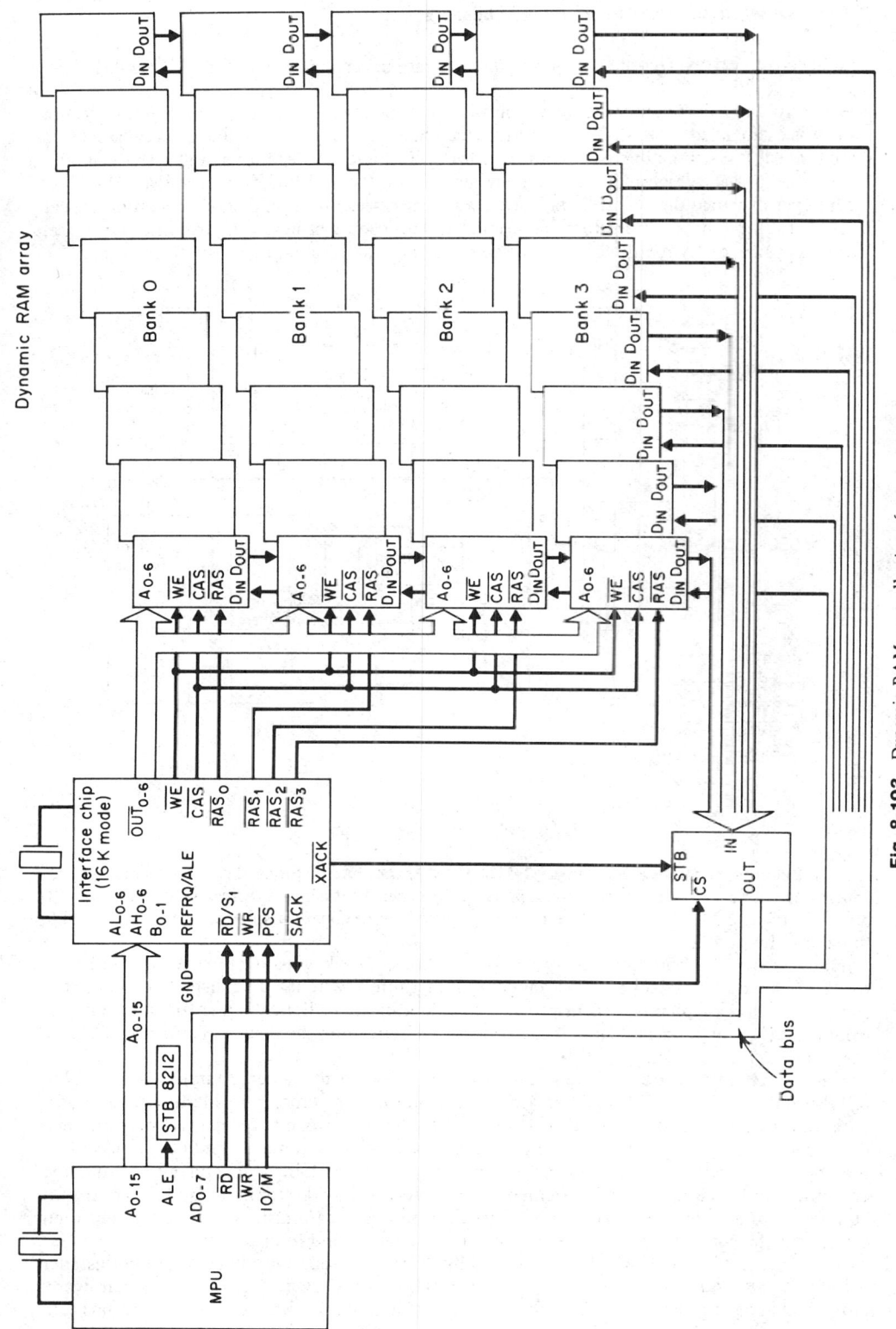

Fig. 8-193. Dynamic RAM controller interface.

double-density drives. Figure 8-195 shows the interface of a floppy disc to the MPU and the disc drives.

A controller chip supports the quadrupling of mass storage (double-density) to industry disc standards established by IBM. The four drives can be either single- or double-density, single- or double-sided. It is compatible with both the IBM 3740 single-density format and system 34 double-density format, which require frequency modulation (FM) and modified FM (MFM).

High-performance disc controllers feature automatic seek, read, and write operation, select lines for multiple heads, programmable loading and unloading of heads, control lines for phase-locked loops generally required for disc-drive circuitry and operate on a +5-V power supply.

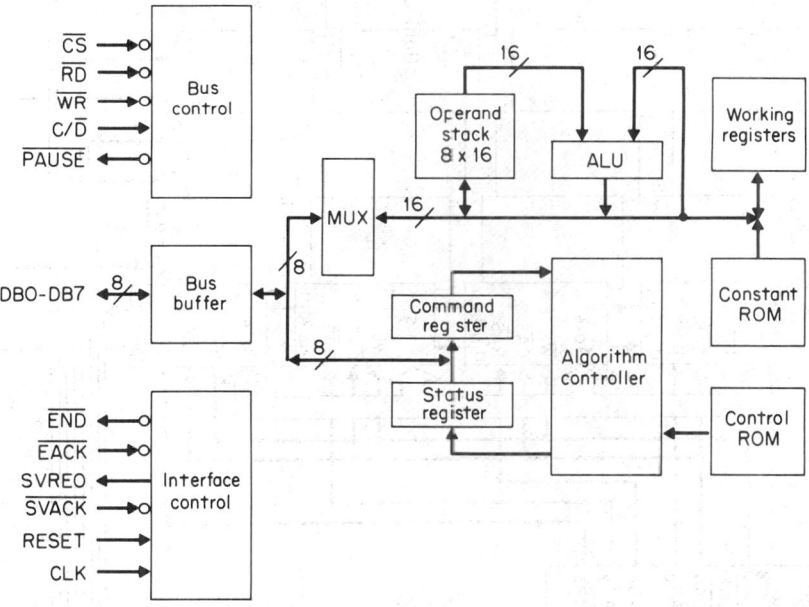

Fig. 8-194. APU block diagram.

137. Programmable Keyboard-Display Interface. Since data and display are an integral part of many microprocessor designs, an interface that can control these functions without affecting the MPU is often needed. A programmable keyboard-display interface is available to handle this function.

The keyboard (or console) usually has two sections. The keyboard proper can interface to regular typewriter-style keyboards or random-toggle thumb switches. The display section of the console drives the alphanumeric displays or a bank of indicator lights. These functions off-load from the MPU the tasks of scanning the keyboard or refreshing the display. Figure 8-196 shows the interface arrangement.

138. GPIB Interface Chips. The accepted standard for general-purpose interface-bus (GPIB) configuration is IEEE Standard 488, Digital Interface for Programmable Instrumentation. This interface is primarily intended as a method of allowing computer control of peripheral devices such as chemical analysis equipment, spectrographic equipment, and other specialized scientific or laboratory-type equipment. The interface allows both input to, and output from, the computer, so that equipment measurements can be processed by the computer and the controlled equipment instructed how to correct for departures from standard results. The equipment then carries out the computer's instructions, and another analysis and input results.

An interface to the IEEE 488 bus is handled by three ICs, used to implement a *talker-listener* function, a *bus-controller* function, and a *bidirectional transceiver*. The last can be hardware-programmed to support either the talker-listener or the combined talker-listener and bus-controller functions. Figure 8-197 shows the system diagram with the three chips mentioned, together with an optional DMA controller.

139. Data Encryption. Electronic-funds-transfer (EFT) applications as well as other electronic banking and data-handling operations require data to be encrypted. A *data-encryption unit (DEU)* is a microprocessor peripheral device designed to encrypt and decrypt blocks of data.

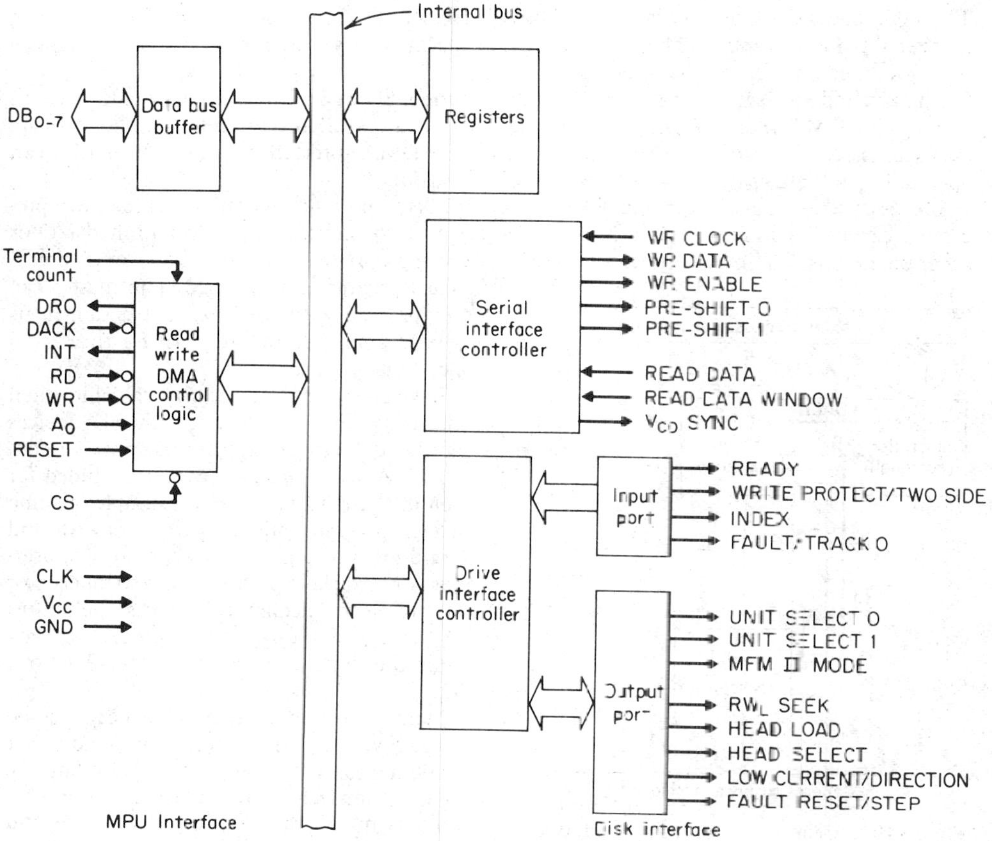

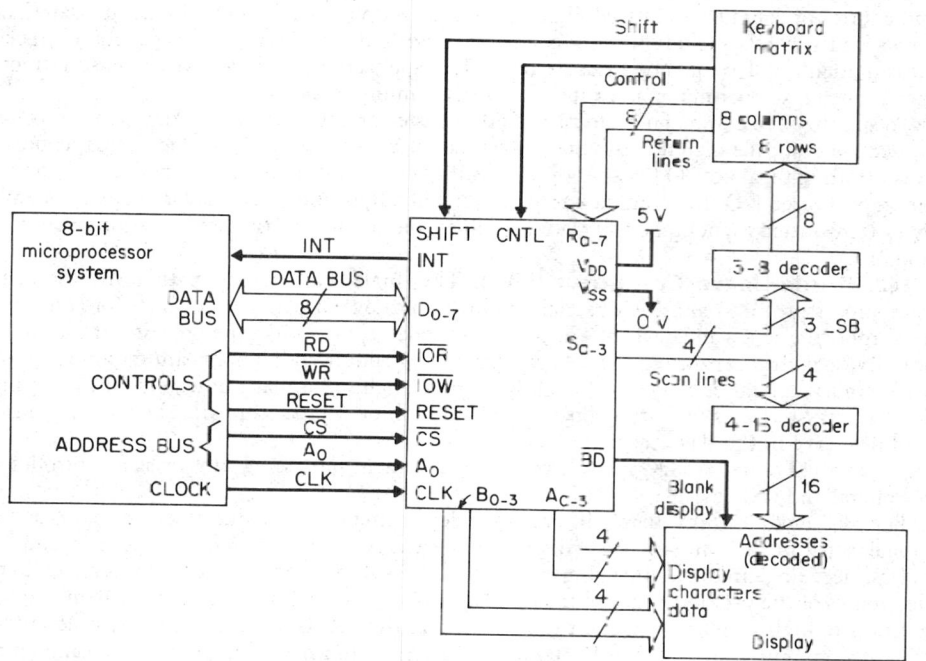

Fig. 8-195. Floppy-disc controller interface.

Fig. 8-196. Keyboard-display interface.

The DEU operates on text words using a specified key to produce cipher words. The operation is reversible for decrypting. The algorithm used is permanently stored in the DEU, but the key is user-defined and can be changed at any time.

The key and message are transferred to and from the DEU in 8-bit bytes by way of the system data bus. A DMA interface and interrupt output ports are available to minimize the software overhead associated with data transfer. By using the DMA interface two or more DEUs can operate in parallel.

140. Slave Processing. Where execution speed is important a technique called slave processing can be used. This means that one master processor is used to conduct limited I/O or other operations. The master processor can signal the appropriate slave unit whenever its predesigned program can be used. The master can now proceed with the next instruction in its own program without concern for the slaved operation.

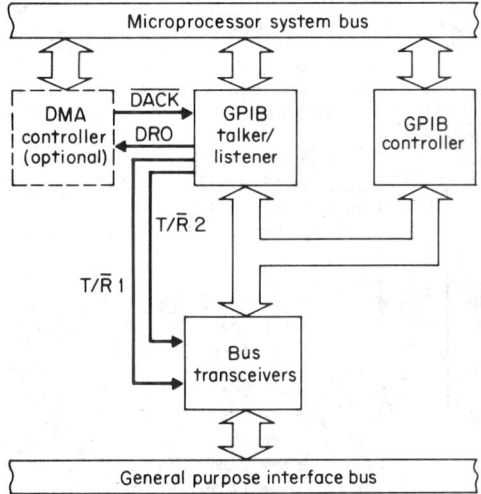

Fig. 8-197. General-purpose interface-bus system diagram.

Slave processing is especially valuable when relatively slow electromechanical devices such as disc drives or printers must be controlled. A slave processor can be included for each of these devices. As an example, assume that the program calls for both a disc write and a hard print by a printer. By providing a user-programmed slave processor to control each of the two electromechanical devices, the master can be allowed to continue processing data while the two slaves perform the I/O control duties.

Many I/O interfacing problems can also be solved by using slave processors, and what would appear at first glance to be a relatively high-cost answer to high-speed requirements really is not since many interface chips and systems can be handled by slave processing without holding up the master processor.

141. Industrial Digital Processor. In the industrial environment, the measurement and control of common industrial digital input and output signals is required. These measurements and controls include sensing change of state, pulse counting, pulse generation, period measurement, and frequency measurement. The applications include switch sensing, motor-speed control, stepper-motor actuation, and serial communications.

To simplify the command protocol with the master processor, an industrial digital processor is used. One of nine digital I/O functions can be selected at any time for measuring, counting, or controlling separate I/O lines. Additional utility commands allow reading or setting the condition of unused I/O lines. Simplex serial input and output modes can assemble or disassemble bytes transmitted asynchronously over TTL lines, including insertion and deletion of start and stop bits.

142. Single-Board Computer (SBC). The single-board computer typically includes a microprocessor, ROM and RAM, serial I/O lines, and several parallel I/O ports for connection to peripheral circuits and devices. The SBC is expandable, much like the processor itself. Instead of individual chip expansion, the SBC is expanded by adding additional board modules containing memory, digital or analog I/O facilities, special high-speed math units, etc. SBCs are interfaced to a system bus with components such as additional processors and the like. Expansion modules serve in the slave capacity.

As with LSI chips, an SBC module can serve as a main processing unit or as a controller for peripheral and I/O devices.

The SBC module is invaluable in prototype development, as it reduces component costs and special software techniques. For example, unnecessary additional memory storage and I/O address decoding circuits can be eliminated. The SBC can be programmed and interfaced to the electronics of the prototype unit so that field experience can be gained before the manufacturer commits to high-volume production. The domain of the SBC is in the medium-volume production area, ranging from 1,000 to 10,000 units. The SBC eliminates the problem of maintaining future inventory, maintenance and installation documentation, and software support. A typical

SBC module contains 50 or more ICs, meeting and exceeding the common requirements of the medium-range manufacturing production.

The SBC module solves the problems involved with the modularity needs of both hardware and software. A modular system is composed of identifiable, reproduceable components that can be mated together to achieve an overall configuration that meets a specific need. Although somewhat complicated by the competing requirements of different modules, computer design for special applications is often no more than selection and interconnection of the appropriate hard-

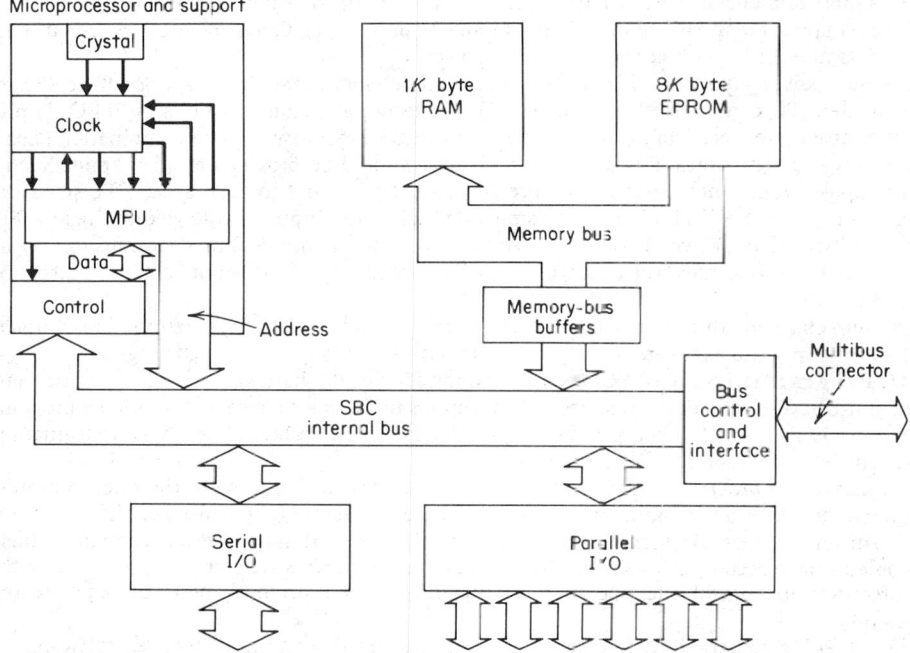

Fig. 8-198. Block diagram of typical single-board computer.

ware and software. In a specific system, it is possible to design a special printed-circuit card to hold all the required components, but such a system is inflexible. Another approach to modularity would be to partition the system into a number of arbitrary modules, each containing a half-dozen ICs. Each card would then have to be connected to its neighbor cards. These interconnections take the form of edgecard connectors mounted on a mother board; each module is inserted into a specific connector, and intermodule signal transfers take place along wires or traces on the mother board between edge connectors. Unfortunately, the connectors are expensive and the least reliable component in electronic systems. Furthermore, if a module is to communicate with several other modules, e.g., a microprocessor board that communicates with 10 or more memory storage modules, each module will consume valuable space with additional logic to coordinate and buffer those signals.

Large-scale integration (LSI) encourages the installation of many devices in close proximity with many on-board connections. Paradoxically, LSI devices also require wide signal pathways to external equipment.

Many of the signals that appear at the edge connector(s) of a 40-pin microprocessor, for example, are connected only to immediately adjacent ICs (system clocks, control-signal-derivation logic, signal buffers, etc.). It is practical to cluster these components together, but the entire system requires storage and I/O facilities to be functional. It is good engineering practice and makes economic sense to provide some of that capacity on the same card with the microprocessor. In the interests of flexibility, some means must also be provided to access external modules for additional resources required for a particular application. Enough memory and I/O are included with a microprocessor in an SBC module to satisfy a large percentage of all potential applications, yet the SBC unit also includes off-card signals that permit the addition of up to 20 other modules to create powerful computer configurations for special needs.

The typical single-board computer shown in Fig. 8-193 includes seven major elements: (1)

microprocessor and support, including the processor itself, clock, and control signal interfacing; (2) 1K bytes RAM; (3) 8K bytes EPROM; (4) serial I/O for connecting the module to serial peripherals or to a data communications channel; (5) parallel I/O, with up to 48 bits of I/O capacity, much of which may be used bidirectionally; (6) memory and I/O decoding and control logic; and (7) bus control and interface to permit communications with other modules used to expand the system capacity.

This particular configuration includes 60 ICs and several discrete components.

143. Power Supplies and Chassis. To provide a complete MPU system, some form of power supply and chassis is required. Typically, these are off-the-shelf items.

Power supplies for an SBC system should be able to provide regulated dc output power at +12, +5, −5, and −12 V, with access to ac input power.

A typical power supply provides sufficient current-delivery capabilities to support the SBC (or an equivalent PC card) fully loaded with I/O line terminators and drivers and EPROMs plus residual capacity for most combinations of up to three SBC memory, I/O, or combination expansion boards. Current limiting and voltage protection should be provided at all outputs. Such a power supply would include logic to sense ac power failure and to generate a TTL signal for powerdown control. A TTL high signal is provided when the input ac voltage drops below 90% of its nominal value. DC voltages remain within 5% of their nominal values for 3 ms minimum after the ac low signal goes true (changes state). The frequency of the input ac may be either 50 or 60 Hz.

A system chassis with from one to eight card-cage slots and a backplane is required to complete the SBC system. Card cages are usually available with power supplies and cooling fans.

144. Programming and Software. To implement the hardware functions, a microprocessor must have instructions. A series of instructions to the microprocessor is called a program. A *program* is an algorithm composed of instructions placed in a logical sequence that unambiguously solves a problem in a finite number of steps.

A *collection of programs*, and in some cases a single program, residing in the microcomputer or entered into the microcomputer in some manner are referred to generically as *software*. There are two different types. *Applications programs* are software written to solve a particular problem or problems or to accomplish specific tasks; and *systems programs* are software produced by the manufacturer to control the operation of the devices and to implement the applications program(s).

The following paragraphs define common terms associated with microprocessor software.

145. Object Program or Code. A microprocessor operates on data in some binary form. An object program or code is a program in the language of the device and directly executable by the computer. This kind of code is also often called *machine code*.

146. Source Program or Code. A source program is a program written in a software language that must first be translated into executable object code. Such a language might be in assembly language (mnemonics or symbolic representation of machine code), or it might be in a high-level language that is closer in structure and syntax to an understandable conversational or written language.

147. Instruction Set. A microprocessor instruction set breaks down conveniently, from a programming point of view, into data movement, data manipulation, decision and control, and I/O instructions. A typical medium-scale MPU has a set of 72 basic instructions and can range up to about 200. These instructions may include binary and decimal arithmetic functions as well as logical, shift, rotate, load, store, branch, interrupt, and stack-manipulation functions. Most of the instructions have several variations, and most can be used with several memory-addressing modes.

These instructions are encoded in a binary form, usually in hexadecimal format. In this form, instructions are said to be in *machine language*. Instruction codes, data, and the names of data written in mnemonic form using combinations of letters and numbers are said to be written in *assembly language*, which must be translated by programs called assemblers into machine language before they can be executed.

148. Assembly Language. An assembly language is the *mnemonic form of language* used to develop a source program for a microprocessor. An assembly language is easier for a programmer to use, read, and understand than the hexadecimal machine code. It reduces the possible errors associated with writing a program as a series of binary 1s and 0s or hexadecimal numbers. Assembly language and high-level languages as well permit the programmer to annotate a program with comments to explain and document the purpose and operation of the instructions. The mnemonic codes used in assembly language refer directly to operations of the MPU. The use of an assembly language gives the programmer strong control over the operation of the device, and since the mnemonics directly represent binary object code, the language is translated more directly, producing a benefit in efficiency and speed of execution compared with high-level languages, which often use *pseudo code* or *tokens* needing translation by the computer into executable code before execution.

149. High-Level Languages. High-level languages are programming languages closer to forms of communication used by people. These languages give the programmer more control of a problem or procedure and are known as *problem-oriented* and *procedure-oriented* languages for that reason. They are not as efficient as assembly language for the control of devices, but they were not primarily intended to turn individual bits on and off or effect register-to-register data movements. What they do accomplish is to increase programmer productivity and further reduce potential programming errors inherent in designing a program in machine code or assembly language.

Just as an assembly language must be translated by an assembler into executable object code, a high-level language must also be compiled or interpreted into object code. This may be done at run time by solid-state circuitry called an *interpreter*, but most higher-level languages use a program called a *compiler* to compile a separate file of executable machine code.

When microprocessors first appeared in 1972, there was no choice: assemblers, cross-assemblers, and later PL/M (a high-level compiled language) were the only tools available. BASIC was the next high-level language used for the Intel 8080 (in 1975). A subset of FORTRAN IV appeared in 1976. Since microcomputers became available in kit form, system software has been written in assembly language and applications programs in PL/M or BASIC. More recently it has become possible to buy applications programs in a variety of languages: FORTRAN, COBOL, PASCAL, FORTH, BASIC in several versions, and several other languages are now used with a wide range of both 8- and 16-bit processors.

150. Basic. BASIC (*Beginner's All-Purpose Symbolic Instruction Code*) was developed primarily because FORTRAN concepts were difficult to learn. Today, BASIC is available in a wide variety of enhanced versions and is used for many applications programs. It can be used for programming a whole range of microprocessors. As computer languages go, BASIC is not a sophisticated language and has no inherent structure for specific applications, but it has the advantage of being easy to use. Because it usually is interpreted at run time instead of being compiled before the first execution of a program, it is very interactive with the programmer or user. It is a good language to use for programmer productivity (statements are in the form of PRINT, RUN, GO TO, LIST, etc., and are easily read by the programmer), unless the program needs primitive controls over I/O devices. Most BASIC implementations allow the programmer to CALL segments of machine code where these controls are required or execution speed is an important consideration.

151. Fortran. The name is short for *formula translator*, and FORTRAN is an algebraic language with static arrays. The first widely successful high-level language, it is designed for both numerical analysis and I/O control. Symbols that start with I,J,K,L,M, or N are automatically assumed by FORTRAN to be integers; all others are floating-point numbers called *reals*. For highly precise numbers, the programmer may declare that twice as many bits are to be used to store a particular value. (A few BASIC implementations also have this feature.)

FORTRAN has a powerful subroutine-invocation scheme that permits the creation of separate subprograms merged later during a system-integration phase. Each subroutine is generally compiled separately from the main program, and then all the subroutines are linked together by part of the systems software called a *linker loader* (or sometimes a linking loader). Each subroutine may have arguments declared. When a subroutine is invoked, these values are effectively replaced by the values supplied in the calling program, sometimes called the *host program*.

152. Cobol. COBOL is a higher-level language designed for use in business-applications programming. It supports files of multilevel use. COBOL is especially good at handling many small pieces of similar data, the kind of file found in commercial and industrial data-handling programs. It is oriented toward reading and writing complicated data records and arranging them in storage for easy processing.

153. Pascal. PASCAL was designed as a tool for teaching programming methodology and structure and is currently BASIC's only challenger as the major teaching language. Pascal has a variety of programming and data structures. It is unlikely to replace BASIC as the hobbyist's or home computer language because its use involves a time-consuming compilation step not needed with BASIC (which is interpreted rather than compiled).

154. Other Languages. The following is a brief description of some of the other available high-level programming languages used with microprocessors and their applications within microcomputers.

ALGOL is a block-structured algebraic language with recursion and structured-programming primitives. It is widely used in Europe and for publication of algorithms.

APL is designed primarily for efficient handling of *n*-dimensional arrays and is both very powerful and concise. It is widely used in engineering, mathematics instruction, business, and digital hardware design.

C is a structured language designed to give efficient control of machine resources while making device-independent programming possible.

FORTH is an extendable, stack-oriented language, developed for data acquisition and radio-telescope orientation at an astronomical observatory. Its program listings are hard to follow. Users may define their own instructions in FORTH.

LISP is used widely in artificial-intelligence research, symbolic mathematics, and computing theory. Data and programs are represented as list structures using Church's lambda calculus extended with a variety of program and data structures to be a general-purpose language.

MUMPS is a text-oriented language with built-in data-base facilities and string-and-pattern-matching features. It is widely used in hospitals and large organizations for unified accounting systems.

PILOT was created for computer-aided instruction programming. It has facilities for interactive input and output, pattern matching, and conditional branching. It is extended with various data types and structures. PILOT was intended to allow noncomputer people (schoolteachers) to write usable interactive programs intended to teach and test students using small, inexpensive microcomputers designed for home use.

PL/1 is a large general-purpose programming language often used for systems programming.

RPG-II (report generator) is widely used in formatting of reports. It has some arithmetic and file-handling capabilities. Today, however, much of the software inherent in word-processing systems has even greater capabilities.

155. Binary Loaders and Relocatable Loaders. When a user generates an object program, it still has to be read into memory for execution. This is accomplished by a binary loader as part of the system software. The loader reads the object code lines and stores the successive words into the appropriate memory locations. In the simplest sense, for an absolute object program, where every instruction in the object file is assumed to have a fixed and final memory location, a loader reads the words into successive memory locations. Where the system permits simultaneous programs to be stored, called, and executed, the actual locations in memory may change. Therefore, without a relocatable loader, the absolute object program might have to be retranslated for each use.

Relocatable object programs are always assembled or compiled as if the first location assigned were zero. Every instruction that refers to an address in memory is marked so that it can be adjusted when the program is loaded. The relocatable binary loader starts with the knowledge of the address at which the program is to be loaded; this address is then added to an offset at every one of the marked addresses.

A relocatable loader also provides the ability to take two or more separate pieces of a program or binary object file and load them into contiguous memory. A composite program is pieced together to avoid the need for recompiling if a small error is found. The final program then can be "bound" together by a *binder*, another part of system software, to form the final program object-code unit.

156. Linking Editor. To handle all the different memory locations and references and offsets, a linkage editor is required. In using a linkage editor, each program module has all its important symbols defined in tables by symbolic names. All the points in the module that may be referenced to or from another module are named and given relative (as opposed to absolute) addresses inside the module.

As an object program is read by the linkage editor, all relocation is performed and references are linked into the appropriate memory locations. In addition to software techniques required for linking, there are two associated tables needed to store the symbolic references. As this requires substantial memory space, linkage editors are usually found only on large microprocessor development systems.

157. Operating Systems. An operating system (OS) is an easy-to-use sophisticated software system that operates on the main processor. The OS extends the MPU architecture to provide a structured, efficient environment for applications programs. The OS includes facilities for concurrent program execution, resource and information sharing, servicing asynchronous events, and interactive control over system resources and utilities. In addition, it provides major real-time facilities and includes priority-based system-resource allocation; means for concurrently monitoring and controlling multiple external events, real-time clock control, and interrupt handling and task dispatching.

158. Utility Program. Both assembler support and library support usually come in the form of a software package called a utility program that can be loaded into a microcomputer development system via disc, magnetic tape, or punched paper tape. An assembler support package for the large processor usually contains the assembler, linking and loading facilities, a PROM programming facility, and additional enhancements such as a utility to translate object code into hexadecimal format.

The utility programs of a library range from pseudo operations that complement and augment a microprocessor's instruction set to routines that support digital processing control, elementary statistics functions, and math functions with single, double, and extended single- and double-point precision.

The hierarchy of a support library may be similar to the one shown in Fig. 8-199, displaying the level of complexity and the interrelationship of the programs.

As these routines and subroutines can be used over and over for different programs or called in different languages, they are set up as a library to be handled by the operating system as needed. As each is called, it is linked into the object program as if it were another object file.

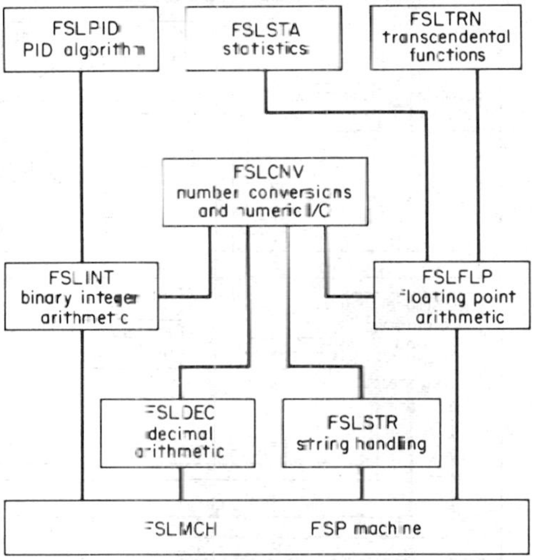

159. Assemblers, Compilers, and Cross-Compilers. Language translators come in two basic forms, an assembler to translate assembly mnemonics into an object program for the loader or a compiler for high-level languages (except BASIC, which is interpreted, not compiled).

A useful characteristic of an assembler is the evaluation of expressions at assembly time, which permits the programmer to specify certain parameters algebraically instead of numerically or symbolically. Then, when the program is assembled, the assembler evaluates the algebraic expressions and inserts the correct values into the machine language (object) code. The process requires the variables to be specified ahead of time, but it permits the programmer to alter them by changing their specification only once, rather than each time a value is used in the program. This saves time and helps in debugging the program later.

Inherent in the assembler are pseudo operations, or assembler instructions. They do not assemble into object code directly but control the assembly of other instructions. Compilers are more remote from machine language than assemblers. Often a single compiler statement constitutes several machine-language instructions. The compiler operates on a high-level language such as FORTRAN to produce the executable machine code or in some cases an intermediate *p-code* or *pseudo code*. The compiler chooses register assignments, assigning memory areas to programs and subroutines, linking labels in one program with addresses in another etc. The compiler produces a diagnostic run of the program, showing errors, called *compile-time errors*, as they pertain to the target or host processor. With a compiler, programmers know very little about the device for which they are programming.

A *cross-compiler* or *cross-assembler* program is an assembler or compiler that compiles or assembles programs run on one device for the purpose of transporting the program to another device for execution.

160. Basic Addressing Methods. Until now, we have discussed software tools that implement instructions from the programmer of the device or the operator of the system. Software directs the hardware functions of the MPU or external devices. In the instruction set of the processor, the basic instruction to perform an operation is called the *operation code*, often shortened to op code and expressed in mnemonic assembly-language form. The *operand* is the data that will be manipulated or a number acting as a pointer to the address in which the data are stored.

Often the data to be operated on are stored in memory. When multibyte numeric data are

Fig. 8-199. Hierarchy of a library.

used, the data, like instructions, are stored in successive memory locations, with the least significant byte first, followed by those of increasing significance. Several different methods, called *modes*, may be used to address memory locations that contain the desired data.

Direct Mode.   Bytes 2 and 3 of the instruction contain the exact memory address of the data, with the low-order bits stored in the address represented by byte 2 and the high-order bits at the address contained in byte 3.

Register Mode.   The instruction specifies the register on register pair where the data are located.

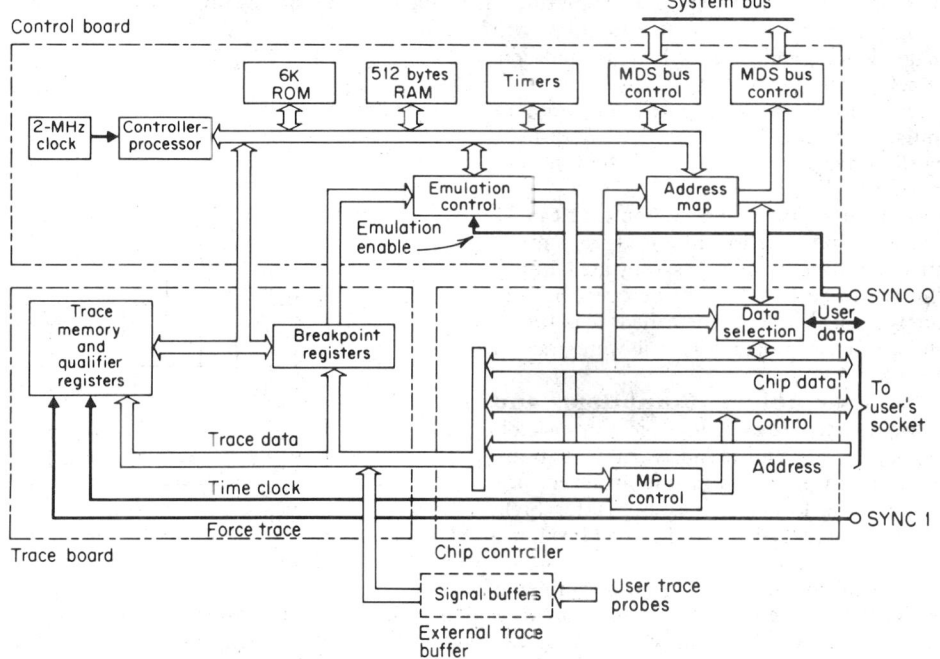

Fig. 8-200. Emulator block diagram.

Register Indirect Mode.   The instruction specifies a register pair that contains the memory address where the data are located (high-order bits in the address specified by the first register in the pair, low-order bits in that specified by the second register).

Immediate Mode.   The instruction contains the data themselves. This is either an 8-bit quantity (up to 255 decimal) or a 16-bit quantity, with the most significant byte stored first and the least significant byte last.

Unless directed by an interrupt or branch instruction, the execution of instructions proceeds through consecutively increasing memory locations as controlled by the program counter. A branch instruction can specify the address of the next instruction to be executed by altering the contents of the program counter in two ways:

1. *Direct mode.* The branch instruction contains the address of the next instruction to be executed. With the single exception of the RST (return from subroutine) instruction, byte 2 of the branch instruction contains the low-order address byte, and byte 3 contains the high-order byte of the address of the branch target.

2. *Register indirect mode.* The branch instruction indicates a register pair containing the address of the next instruction to be executed. The high-order byte of the address is in the first register of the pair, and the lower-order byte is in the second.

The RST instruction is a special 1-byte call instruction generally used during interrupt sequences. RST includes a 3-bit field; program control is transferred to the instruction whose address is 8 times the contents of this 3-bit field.

161. Microprocessor Development. To implement a microprocessor or an MPU system with a myriad of external logic and interfaces and to program its operation a microcomputer development system is required. In the early days of digital processing systems, breadboards were used to test the various functions and interconnections of subcomponents. Then, after the hardware was designed, the software was developed and integrated. The process was both tedious and time-consuming.

Today, MPUs and MPU-based systems are designed in a total interactive environment that permits simultaneous hardware and software development debugging, and integration. This environment is the microprocessor development system.

The development system is used to design, debug, and integrate both hardware and software products from the initial prototype state to the final production unit. The system allows interaction between the system and the development personnel at every stage to ensure that problems can be corrected immediately. A development system provides the firm base for systematic and orderly project development. A development system typically provides a variety of high-level and assembly programming languages. The operating system provides for the easy combination of software modules even if coded by different programmers. Also available are tests and debugging tools, including in-circuit emulators, for simultaneous hardware and software debugging and integration.

162. In-Circuit Emulation. To provide simultaneous hardware and software debugging and integration, in-circuit emulators are used. The key device providing this function is the in-circuit emulator module. It is connected to a development system and the emulated MPU socket via ribbon cables and a buffer box. The emulator uses both its own and the system's memory, as well as additional on-line storage, memory, and available user-supplied memory to emulate the system MPU, peripherals, and other associated hardware under the prototype software operation.

Figure 8-200 shows the block diagram for an in-circuit emulator that emulates the processor to be used in the final design.

The in-circuit emulator for a target processor provides the symbolic definition of all of the MPU's registers, interrupt bits, and flags. Additional references include the lower-order bits containing the number of clock pulses that elapse during emulation, the high-order bits of the timer-counter, the address of the last instruction emulated, and the number of frames of valid trace data.

163. References

FABRICATION TECHNOLOGY

1. H. C. Theurer   Epitaxial Silicon Films by Hydrogen Reduction of $SiCl_4$ *J. Electrochem. Soc.*, Vol. 108, pp. 649–653, 1961.

2. Research Triangle Institute   "Integrated Silicon Device Technology: Epitaxy," ASD-TDR 63–316, Vol. 9, pp. 42–47, 1965.

3. A. S. Grove, A. Roder, and C. T. Sah   Impurity Distribution in Epitaxial Growth, *J. Appl. Phys.*, Vol. 36, p. 802, 1965.

4. J. C. Irvin   Resistivity of Bulk Silicon and of Diffused Layers in Silicon, *Bell System Tech. J.*, Vol. 41, pp. 387–410, 1962.

5. B. E. Deal   The Oxidation of Silicon in Dry Oxygen, Wet Oxygen and Steam *J. Electrochem. Soc.*, Vol. 110, p. 527, 1963.

6. A. S. Grove   "Physics and Technology of Semiconductor Devices," Chap. 2, Wiley, New York, 1967.

7. B. E. Deal and A. S. Grove   General Relationships for the Thermal Oxidation of Silicon, *J. Appl. Phys.*, Vol. 36, pp. 3370–3378, 1965.

8. S. K. Ghandhi   "The Theory and Practice of Microelectronics," Chap. 6, Wiley, New York, 1968.

9. Research Triangle Institute   "Integrated Silicon Device Technology: Oxidation," ASD-TDR 63–316, Vol. 7, p. 76, 1965.

10. Research Triangle Institute   "Integrated Silicon Device Technology: Photoengraving," ASD-TDR 63–316, Vol. 3, 1964.

11. C. S. Fuller and J. A. Ditzenberger   Diffusion of Donor and Acceptor Elements in Silicon, *J. Appl. Phys.*, Vol. 27, pp. 544–553, 1956.

12. Research Triangle Institute   "Integrated Silicon Device Technology: Diffusion," ASD-TDR 63–316, Vol. 4, 1964.

13. F. A. Trumbore   Solid Solubilities of Impurity Elements in Germanium and Silicon, *Bell System Tech. J.*, Vol. 39, pp. 205–234, 1960.

14. D. P. Kennedy and R. R. O'Brian   Analysis of the Impurity Atom Distribution near the Diffusion Mask for a Planar P-N Junction, *IBM J. Res. Dev.*, Vol. 9, No. 3, pp. 179–186, 1965.

15. J. T. Burrill, W. J. Kind, S. Harrison, and P. McNally   Ion Implantation as a Production Technique, *IEEE Trans. Electron Devices*, Vol, ED-14, No. 1, pp. 10–17, January 1967.

16. J. W. Mayer, L. Ericksson, and J. A. Davies   "Ion Implantation in Semiconductors," Academic, New York, 1970.

17. H. T. Wolf   "Semiconductors," Chaps. 2–5, Wiley-Interscience, New York, 1971.

18. H. H. Stellrecht, D. S. Perloff, and J. T. Kerr   Precision Ladder Networks Using Ion Implanted Resistors, *WESCON Tech. Program Rec.*, 1971.

19. J. D. MacDougall and K. E. Manchester   Implanted Components in Microcircuits, *Proc. Electron. Conf.*, Vol. 25, pp. 140–145, December 1969.

20. H. A. Waggener, R. C. Krogness, and A. L. Tyler   Anisotropic Etching for Forming Isolation Slots in Silicon Beam-Leaded Integrated Circuits, *Proc. IEEE Int. Electron Devices Conf.*, Washington, October 1967.

21. M. P. Lepselter   Beam Lead Technology, *Bell System Tech. J.*, Vol. 45, No. 2, February 1966.

22. L. V. Gregor   Thin Film Processes for Microelectronic Application, *Proc. IEEE*, Vol. 59, pp. 1390–1403, October 1971.

23. T. C. Reissing   An Overview of Today's Thick Film Technology, *Proc. IEEE*, Vol. 59, pp. 1448–1454, October 1971.

COMPONENTS AND DEVICES

24. A. B. Phillips   "Transistor Engineering," Chaps 8–17, McGraw-Hill, New York, 1962.

25. A. B. Grebene   "Analog Integrated Circuit Design," Chap. 2, Van Nostrand, New York, 1972.

26. Motorola Semiconductor Products Div., Engineering Staff   "Integrated Circuits: Design Principles and Fabrication," Chap. 7, McGraw-Hill, New York, 1965.

27. C. A. Bittmann, G. H. Wilson, R. J. Whittier, and R. K. Waits   Technology for the Design of Low-Power Circuits, *IEEE J. Solid State Circ.*, Vol. SC-5, No. 1, pp. 29–37, February 1970.

28. H. C. Lin, T. B. Tan, G. Y. Chang, and B. Van Der Leest   Lateral Complementary Transistor Structure for Simultaneous Fabrication of Functional Blocks, *Proc. IEEE*, Vol. 49, pp. 1491–1495, 1964.

29. H. C. Lin   "Integrated Electronics," Chap. 8, Holden-Day, New York, 1967.

30. B. Polata   Compatible High Performance Complementary Bipolar Transistors for Integrated Circuits, *IEEE Int. Electron Devices Meet.*, Washington, 1969.

31. R. Cobbold   "Theory and Applications of Field-Effect Transistors," Wiley, New York, 1970.

32. A. B. Grebene and S. K. Ghandhi   General Theory for Pinched Operation of Junction-Gate FET, *Solid State Electron.*, Vol. 12, pp. 573–589, 1969.

33. R. W. Russell and D. D. Culmer   Ion-Implanted JFET-Bipolar Monolithic Analog Circuits, *Int. Solid-State Circ. Conf. Dig. of Tech. Pap.*, February 1974, pp. 140–141.

34. H. K. J. Ihantola and J. L. Moll   Design Theory of a Surface Field-Effect Transistor, *Solid State Electron.*, Vol. 7, pp. 423–430, 1964.

35. A. S. Grove   "Physics and Technology of Semiconductor Devices," Chaps. 8 and 9, Wiley, New York, 1967.

36. H. R. Camenzind   "Designing Integrated Systems," Chap. 11, Van Nostrand, New York, 1972.

37. M. H. White and T. R. Cricchi   Complementary MOS Transistors, *Solid State Electron.*, Vol. 9, pp. 991, 1966.

38. L. Vadasz, A. S. Grove, T. A. Rowe, and G. E. Moore   Silicon-Gate Technology, *IEEE Spectrum*, October 1969, pp. 28–35.

39. R. W. Bower, H. G. Dill, K. G. Aubuchon, and S. A. Thompson   MOS Field-Effect Transistors Formed by Gate-Masked Ion Implantation, *IEEE Trans. Electron Devices*, Vol. ED-15, pp. 757–761, 1968.

40. J. D. MacDougall and K. E. Manchester   Implanted Components in Microcircuits, *Proc. Natl. Electron. Conf.*, Vol. 25, pp. 140–145, December 1969.

41. T. J. Rodgers and James D. Meindl   VMOS: High Speed TTL Compatible MOS Logic, *IEEE J. Solid State Circ.*, Vol. SC-9, pp. 239–249, October 1974.

42. C. H. Sequin and M. F. Tomprett   "Charge Transfer Devices," Academic, New York, 1975.

43. W. S. Boyle and G. E. Smith   Charge-Coupled Devices: A New Approach to MIS Device Structures, *IEEE Spectrum*, July 1971, pp. 18–27.

44. D. Khang and M. P. Lepselter   Planar Epitaxial Schottky Barrier Diodes, *Bell Syst. Tech. J.*, Vol. 44, p. 1525, 1965.

45. A. Y. C. Yu   The Metal-Semiconductor Contact: An Old Device with a New Future, *IEEE Spectrum*, March 1970, pp. 83–89.

46. J. C. Irvin   Resistivity of Bulk Silicon and Diffused Layers in Silicon, *Bell Syst. Tech. J.*, Vol. 41, pp. 287–410, 1962.

47. A. B. Grebene   A Practical Method for Reducing the Effects of Parastic Capacitances in Integrated Circuits, *Proc. IEEE*, Vol. 55, No. 2, pp. 235–236 1967.

48. H. H. Stellrecht, D. S. Perloff, and J. T. Kerr   Precision Ladder Networks Using Ion Implanted Resistors, *WESCON Tech. Program Rec.*, 1971.

49. M. Fogiel   "Microelectronics," Chaps. 3 and 4, Research and Education Association, New York, 1968.

50. A. B. Phillips   "Transistor Engineering," Chap. 5, McGraw-Hill, New York, 1962.

51. H. Lawrence and R. M. Warner   Diffused Junction Depletion Layer Calculation, *Bell Syst. Tech. J.*, Vol. 39, pp. 389–404, 1960.

INTEGRATED-CIRCUIT DESIGN

52. H. R. Camenzind and A. B. Grebene   An Outline of Design Techniques for Linear Integrated Circuits, *IEEE J. Solid State Circ.*, Vol. SC-4, pp. 110–122, June 1969.

53. R. J. Widlar   Some Circuit Design Techniques for Linear Integrated Circuits, *IEEE Trans. Circ. Theory*, Vol. CT-12, pp. 586–590, December 1965.

54. G. R. Wilson   A Monolithic Junction FET-NPN Operational Amplifier, *Int. Solid State Circ. Conf. Dig. Tech. Pap.*, Vol. 11, pp. 20–21, February 1968.

55. J. A. Mattis and H. R. Camenzind   A New Phase-Locked Loop with High Stability and Accuracy, *Signetics Corp. Appl. Note*, September 1970.

56. A. B. Grebene   "Analog Integrated Circuit Design," Chap. 4, Van Nostrand Reinhold, New York, 1972.

57. M. E. Hoff, Jr., J. Huggins, and B. M. Warren   An NMOS Telephone Codec for Transmission and Switching Applications, *IEEE J. Solid State Circ.*, Vol. SC-14, February 1979.

58. M. Hoff   MOS Reference Voltage Circuit, U.S. Pat. 4,100, 437, July 11, 1978.

59. M. Tobey, D. Giuliani, and B. Ashkin   Flat Band Voltage Reference, U.S. Pat. 3,975,648, Aug. 17, 1976.

60. D. Senderowicz, D. A. Hodges, and P. R. Gray   High-Performance NMOS Operational Amplifier, *IEEE J. Solid-State Circ.*, Vol. SC-13, pp. 760–766, December 1978.

61. G. Tzanateas, C. A. T. Salama, and Y. P. Tsividis   A CMOS Bandgap Voltage Reference, *IEEE J. Solid State Circ.*, Vol. SC-14, pp. 655–657, June 1979.

62. Y. P. Tsividis and R. W. Ulmer   CMOS Reference Voltage Source. *Int. Solid-State Circ. Conf.*, pp. 49–50, 1978.

63. E. Vittoz and J. Fellrah   CMOS Analog Integrated Circuits Based on Weak Inversion Operation, *IEEE J. Solid State Circ.*, Vol. SC-12, pp. 224–231 June 1977.

64. R. D. Middlebrook   "Differential Amplifiers," Wiley, New York, 1953.

65. L. J. Giacoletto   "Differential Amplifiers," Wiley, New York, 1970

66. D. Fullagar   A New High Performance Monolithic Operational Amplifier, *Fairchild Semiconductor Appl. Brief*, May 1968.

67. H. A. Witt Linger   Applications of the CA3080 and CA3080A High-Performance Operational Transconductance Amplifiers, *RCA Appl. Note*, ICAN-6668.

68. R. J. Widlar   New Developments in IC Voltage Regulators, *IEEE J. Solid State Circ.*, Vol. SC-6, pp. 2–7, February 1971.

69. Intersil, Inc.   "Switchmode Converter Topologies: Make Them Work for You!", Cupertino, Calif., 1980.

70. Signetics Corp.   Signetics Switched Mode Power Supply; Push/Pull Regulator, *Appl. Note* AN130, Sunnyvale, Calif.

71. G. W. Haines C_c Compensated Transistors, *IEEE Int. Electron Devices Meet.*, Washington, October 1966.

72. A. Bilotti   Applications of a Monolithic Analog Multiplier, *IEEE J. Solid State Circ.*, Vol. SC-3, pp. 373–380, December 1968.

73. B. Gilbert   A Precise Four-Quadrant Multiplier with Subnanosecond Response, *IEEE J. Solid State Circ.*, Vol. SC-3, No. 4, pp. 365–373, December 1968.

74. E. Renschler   Theory and Application of a Linear Four-Quadrant Monolithic Multiplier, *IEEE Mag.*, Vol. 17, No. 5, May 1969.

75. A. B. Grebene and H. R. Camenzind Frequency Selective Integrated Circuits Using Phase-Lock Techniques, *IEEE J. Solid State Circ.*, Vol. SC-4, pp. 216–225, August 1969.

76. A. B. Grebene The Monolithic Phase-Locked Loop: A Versatile Building Block, *IEEE Spectrum*, March 1971, pp. 38–49.

77. D. A. Hodges, P. R. Gray, and R. W. Broderson MOS Amplifiers and Comparators, *IEEE Spectrum*, February 1979.

78. D. H. Sheingold "Analog-Digital Conversion Handbook," Analog Devices, Inc., Norwood, Mass., pp. 1–50, 1972.

79. D. F. Hoeschele "Analog to Digital to Analog Conversion Techniques," pp. 108–120, Wiley, New York, 1968.

80. P. Pinter and D. Timms Voltage to Frequency Converter: IC Versions Perform Accurate Data Conversion (and Much More) at Low Cost, *EDN*, Vol. 22, pp. 153–157, Sept. 5, 1977.

81. G. M. Jacobs, D. J. Allstot, R. W. Broderson, and P. R. Gray Design Techniques for MOS Switched Capacitor Ladder Filters, *IEEE Trans. Circ. Syst.*, Vol. CAS-25, pp. 1014–1020, December 1978.

82. J. T. Caves, M. A. Copeland, C. F. Rahim, and D. S. Rosenbaum Sampled Analog Filtering Using Switched Capacitors as Resistor Equivalents, *IEEE J. Solid State Circ.*, Vol. SC-12, pp. 592–599, December 1977.

83. P. J. Allstot, R. W. Broderson, and P. R. Gray MOS Switched Capacitor Ladder Filters, *IEEE J. Solid State Circ.*, Vol. SC-13, pp. 806–814, December 1978.

84. I. A. Young and D. A. Hodges MOS Switched-Capacitor Analog Sampled-Data Direct-Form Recursive Filters, *IEEE J. Solid State Circ.*, Vol. SC-14, pp. 1020–1033, December 1979.

85. L. R. Rabiner and B. Gold "Theory and Application of Digital Signal Processing," Prentice-Hall, Englewood Cliffs, N.J., 1975.

86. R. D. Baertsch, W. E. Engeler, H. S. Goldberg, C. M. Puckette, and J. S. Tiemann The Design and Operation of Practical Charge-Transfer Transversal Filters, *IEEE J. Solid State Circ.*, Vol. SC-11, pp. 65–73, February 1976.

87. S. K. Mitra (ed.) "Active Inductorless Filters," IEEE Press, New York, 1971.

88. R. W. Newcomb "Active Integrated Circuit Synthesis," Prentice-Hall, Englewood Cliffs, N.J., 1968.

89. L. P. Huelsman (ed.) "Active Filters: Lumped, Distributed, Integrated, Digital, and Parametric," McGraw-Hill, New York, 1970.

90. A. V. Oppenheim and R. W. Schafer "Digital Signal Processing," Prentice-Hall, Englewood Cliffs, N.J., 1975.

91. D. D. Buss, D. R. Collins, W. H. Bailey, and C. R. Reeves Transversal Filtering Using Charge-Transfer Devices, *IEEE J. Solid State Circ.*, Vol. SC-8, pp. 138–146, April 1973.

92. D. Christiansen Integrated Circuits in Action: In Search of the Ideal Logic Scheme, *Electronics*, Vol. 40, pp. 149, Mar. 6, 1967.

93. L. S. Garrett Integrated Circuit Digital Logic Families, Part I, *IEEE Spectrum*, October 1970, pp. 46–58.

94. L. S. Garrett Integrated Circuit Digital Logic Families, Part II, *IEEE Spectrum*, November 1970, pp. 63–72.

95. L. S. Garrett Integrated Circuit Digital Logic Families, Part III, *IEEE Spectrum*, December 1970, pp. 30–42.

96. H. H. Borger and S. H. Wiedmann Terminal-Oriented Model for Merged Transistor Logic (MTL), *IEEE J. Solid State Circ.*, Vol. SC-9, pp. 211–217, October 1974.

97. R. M. Warner, Jr. Comparing MOS and Bipolar Integrated Circuits, *IEEE Spectrum*, June 1967, pp. 50–58.

98. D. A. Hodges "Semiconductor Memories," IEEE Press, New York, 1971.

99. M. E. Hoff, Jr. Application Considerations for Semiconductor Memories, *IEEE Mag.*, June 1970, pp. 62–69.

100. L. L. Vadasz, H. T. Chua, and A. S. Grove Semiconductor Random Access Memories, *IEEE Spectrum*, May 1971, pp. 40–48.

101. J. S. Schmidt Integrated MOS Random-Access Memory, *Solid State Design*, January 1965, pp. 21–25.

102. L. Boysel, W. Chan, and J. Faith Random-Access MOS Memory Packs More Bits to the Chip, *Electronics*, Vol. 43, pp. 109–115, February 1970.

103. W. M. Regitz and J. A. Garp Three-Transistor-Cell 1020-Bit 500-ns MOS RAM, *IEEE J. Solid State Circ.*, Vol. SC-5, pp. 182–186, October 1970.

104. M. R. McCoy Semiconductor Memories, pp. 47–58, Chap. 6 in J. Eimbinder (ed.), "Semiconductor Memories," Wiley-Interscience, New York, 1971.

105. D. Frohman-Bentchkowsky A Fully-Decoded 2048-Bit Electrically Programmable FAMOS Read-Only Memory, *IEEE J. Solid State Circ.*, Vol. SC-6, No. 5, October 1971.

106. Y. Tarui, Y. Hayashi, and K. Nagai Electrically Programmable Non-Volatile Semiconductor Memory, *IEEE J. Solid State Circ.*, Vol. SC-7, No. 5, pp. 369, October 1972.

107. G. Luecke, J. P. Mize, and W. N. Carr "Semiconductor Memory Design and Application," pp. 115–123, McGraw-Hill, New York, 1973.

108. M. G. Snyder Semiconductor Memories, pp. 159–168, Chap. 6 in J. Eimbinder (ed.), "Semiconductor Memories," Wiley-Interscience, New York, 1971.

109. J. Prioste MECL 10,000 Macrocell Array Preliminary Design Manual, Motorola Semiconductor Products, Inc., Phoenix, Ariz., 1979.

110. J. Prioste Macrocell Approach Customizes Fast VLSI, *Electronic Design*, June 7, 1980, pp. 159–166.

111. H. Sobol Applications of Integrated Circuit Technology to Microwave Frequencies, *Proc. IEEE*, Vol. 59, No. 8, pp. 1200–1211, August 1971.

112. M. Caulton Film Technology in Microwave Integrated Circuits, *Proc. IEEE*, Vol. 59, No. 10, pp. 1481–1489, October 1971.

113. C. S. Aitchison et al. Lumped Circuit Elements at Microwave Frequencies, *IEEE Trans. Microwave Theory Tech.*, Vol. MTT-19, p. 928, December 1971.

114. M. Caulton et al. Status of Lumped Elements in Microwave Integrated Circuits: Present and Future, *IEEE Trans. Microwave Theory Tech.*, Vol. MTT-19, No. 7, pp. 588–599, July 1971.

MICROPROCESSORS

115. S. P. Morse "The 8086 Primer," Hayden, Rochelle Park, N.J., 1978.

116. S. P. Morse "Intel Microprocessors: 8008 to 8086," Intel Corporation, Santa Clara, Calif., 1978.

117. C. A. Ogdin "Programming and Interface Single Board," Software Technique, Inc., Alexandria, Va., 1978.

118. C. A. Ogdin "Software Design for Microcomputers," Prentice-Hall, Englewood Cliffs, N.J., 1978.

119. A. Osborne "An Introduction to Microcomputers," Adam Osborne and Associates, Berkeley, Calif., 1978.

120. J. B. Peatman "Microcomputer-Based Design," McGraw-Hill, New York, 1977.

121. A. Ralston and C. L. Meek "Encyclopedia of Computer Science," Petrocelli/Charter, New York, 1976.

122. C. Sippl "Microcomputer Dictionary and Guide," Matrix Publishers, Newport Beach, R.I., 1975.

123. M. E. Sloan "Computer Hardware and Organization," Science Research Associates, Inc., Chicago, 1976.

124. B. Soucek "Microprocessors and Microcomputers," Wiley, New York, 1976.

125. T. W. Pratt "Programming Language: Design and Implementation," Prentice-Hall, Englewood Cliffs, N.J., 1975.

Section 9

UHF and Microwave Devices

JOSEPH FEINSTEIN *Varian Associates, Palo Alto, Calif., Fellow, IEEE; Currently Director of Electronics and Physical Sciences, U. S. Department of Defense*

RICHARD B. NELSON *Varian Associates, Palo Alto, Calif.; Fellow, IEEE*

GEORGE K. FARNEY *Varian Associates, Beverly, Mass.; Fellow, IEEE*

DONALD H. PREIST *Varian Associates, Carlos, Calif.; Fellow, IEEE*

ALLAN SCOTT *Varian Associates, Palo Alto, Calif.; Member IEEE*

PAMELA L. WALCHLI *Varian Associates, Palo Alto, Calif.; Member IEEE*

CONTENTS
Numbers refer to paragraphs

Passive Microwave Components

BY JOSEPH FEINSTEIN

1. Introduction. While the physical concepts and mathematical theory underlying electromagnetic-wave propagation in confined structures were developed at the end of the nineteenth century, the practical utilization of wavelengths shorter than 1 m began during World War II. A wide variety of devices is now available for the transmission, sampling, filtering, and impedance matching of UHF and microwave power. Because of the short wavelengths at these frequencies (10 cm at 3 GHz, 1 cm at 30 GHz), most of these components use distributed elements and obtain specific reactances by judicious use of short-circuited lengths of transmission line. However, the trend toward microcircuitry has led recently to the introduction of lumped elements in some low-power, low-Q applications. In addition, the use of strip and microstrip transmission lines marks a return to open-wire media, with the attendant radiation loss kept low by close spacing and the presence of the dielectric filler.

Except for special applications (such as rotating joints for antenna feeds where a cylindrical member is essential), the use of rectangular waveguide, dimensioned to transmit in the dominant (lowest-order) mode, is standard for high-power transmission. Coaxial cable is used for short-distance runs where the advantage of its flexibility outweighs its higher attenuation. Ridged waveguide is useful for designing matching sections and for providing very-wide-bandwidth single-mode transmission. Oversized cylindrical guide operated in the circular electric mode is finding use in millimeter-wave-carrier telephony, where its extremely low attenuation justifies the special precautions which must be taken to avoid mode conversions.

Reciprocal and Nonreciprocal Components. Of the components which have become standard in this field, perhaps the most unusual are ferrite devices. When biased with the proper magnetic field, ferrites act as nonreciprocal elements with respect to microwave transmission in an appropriate frequency band. This behavior allows isolation of a signal source from reflections and the separation of incident from reflected power along the same transmission line (by means of a ferrite device called a *circulator*). It is also possible to vary the phase of a transmitted wave by adjusting the magnetizing field on the ferrite. Such phase shifters are capable of handling high powers with low loss.

All other types of microwave components, such as hybrid junctions and directional couplers, are reciprocal in their action. The latter are employed for power division and for signal sampling. A wide variety of transmission components such as variable attenuators, matched-load terminations, and slotted lines are used in microwave measurements and design.

High-Q resonators are formed from completely enclosed short-circuited sections of waveguide, with slit or loop coupling. Resonators using lengths of strip line or coaxial line provide smaller size but at lower Q. Finally, lumped elements such as varactor diodes or YIG spheres provide electrically tunable resonators for microwave oscillators and receivers.

2. Transmission-Line Relationships. The basic transmission-line equations are derived for distributed parameters R (series resistance), G (shunt conductance), L (series inductance), and C (shunt capacitance), all defined per unit length of line. Some useful relations are shown in Table 9-1. For zero losses ($R, G = 0$) one obtains the ideal line expression shown on the right. Table 9-2 gives some equations that are useful for relating the measured voltage standing wave ratio (VSWR) to wave transmission and reflection.

TABLE 9-1 Summary of Transmission-Line Equations

l = length of transmission line

Quantity	General line expression	Ideal line expression
Propagation constant	$\gamma = \alpha + j\beta = \sqrt{(R + j\omega L)(G + j\omega C)}$	$\gamma = j\omega\sqrt{LC}$
Phase constant β	Im γ	$\beta = \omega\sqrt{LC} = 2\pi/\lambda$
Attenuation constant α	Re γ	0
Impedance, characteristic	$Z_0 = \sqrt{\dfrac{R + j\omega L}{G + j\omega C}}$	$Z_0 = \sqrt{\dfrac{L}{C}}$
Input	$Z_{-l} = Z_0 \dfrac{Z_r + Z_0 \tanh \gamma l}{Z_0 + Z_r \tanh \gamma l}$	$Z_{-l} = Z_0 \dfrac{Z_r + jZ_0 \tan \beta l}{Z_0 + jZ_r \tan \beta l}$
Of short-circuited line, $Z_r = 0$	$Z_{sc} = Z_0 \tanh \gamma l$	$Z_{sc} = jZ_0 \tan \beta l$
Of open-circuited line, $Z_r = \infty$	$Z_{oc} = Z_0 \coth \gamma l$	$Z_{oc} = -jZ_0 \cot \beta l$
Of line an odd number of quarter wavelengths long	$Z = Z_0 \dfrac{Z_r + Z_0 \coth \alpha l}{Z_0 + Z_r \coth \alpha l}$	$Z = \dfrac{Z_0^2}{Z_r}$
Of line an integral number of half wavelengths long	$Z = Z_0 \dfrac{Z_r + Z_0 \tanh \alpha l}{Z_0 + Z_r \tanh \alpha l}$	$Z = Z_r$
Voltage along line	$V_{-l} = V_i (1 + \Gamma_0 e^{-2\gamma l})$	$V_{-l} = V_i (1 + \Gamma_0 e^{-2j\beta l})$
Current along line	$I_{-l} = I_i (1 - \Gamma_0 e^{-2\gamma l})$	$I_{-l} = I_i (1 - \Gamma_0 e^{-2j\beta l})$
Voltage reflection coefficient	$\Gamma = \dfrac{Z_r - Z_0}{Z_r + Z_0}$	$\Gamma = \dfrac{Z_r - Z_0}{Z_r + Z_0}$

The nomographs in Figs. 9-1 and 9-2 provide a convenient means for determining these quantities. The transmission-line impedance transformations of Table 9-1 can be seen graphically on the Smith chart shown in Fig. 9-3. Circles of constant reflection coefficient are concentric with the center of the chart ($R/Z_0 = 1.0$). The VSWR magnitudes are given by their horizontal intercepts. Radial lines from the center are loci of constant phase angle, read against the degree or fractional wavelength scales at the circumference of the chart. Normalized values of impedance seen as one moves along the line can be read at each radial intercept from the resistance and reactance grids on which the circle of constant VSWR has been drawn. This chart is also of value in matching reactances by providing a graphical determination of the range of impedance transformations available from a stub tuner by changes in line length.

Although L and C for a waveguide cannot be defined uniquely, because its electric and magnetic field patterns are not purely transverse, it is possible to construct a useful equivalent circuit for the dominant mode. This is based on a capacitance defined from the total electric field energy storage across the guide cross section and on an impedance which is related to power flow down the guide. The relations given in Tables 9-1 and 9-2 can then be used for the dominant-mode

TABLE 9-2   Some Miscellaneous Relations in Low-Loss Transmission Lines

Equation	Explanation
$r = \dfrac{1 + \mid \Gamma \mid}{1 - \mid \Gamma \mid}$	r = VSWR
$\mid \Gamma \mid = \dfrac{r - 1}{r + 1}$	$\mid \Gamma \mid$ = magnitude of reflection coefficient
$\Gamma = \dfrac{R - Z_0}{R + Z_0}$	Γ = reflection coefficient (real) at a point in a line where impedance is real (R)
$r = \dfrac{R}{Z_0}$	$R > Z_0$ (at voltage maximum)
$r = \dfrac{Z_0}{R}$	$R < Z_0$ (at voltage minimum)
$\dfrac{P_r}{P_i} = \mid \Gamma \mid^2 = \left(\dfrac{r - 1}{r + 1} \right)^2$	P_r = reflected power P_i = incident power
$\dfrac{P_t}{P_i} = 1 - \mid \Gamma \mid^2 = \dfrac{4r}{(r + 1)^2}$	P_t = transmitted power
$\dfrac{P_b}{P_m} = \dfrac{1}{r}$	P_b = net power transmitted to load at onset of breakdown in a line where VSWR = r exists P_m = same when line is matched, $r = 1$
$\dfrac{\alpha_r}{\alpha_m} = \dfrac{1 + \Gamma^2}{1 - \Gamma^2} = \dfrac{r^2 + 1}{2r}$	α_m = attenuation constant when $r = 1$, matched line α_r = attenuation constant allowing for increased ohmic loss caused by standing waves
$r_{max} = r_1 r_2$	r_{max} = maximum VSWR when r_1 and r_2 combine in worst phase
$r_{min} = \dfrac{r_2}{r_1} \quad r_2 > r_1$	r_{min} = minimum VSWR when r_1 and r_2 are in best phase
$\mid \Gamma \mid = \dfrac{\mid X \mid}{\sqrt{X^2 + 4}}$	Relations for a normalized reactance X in series with resistance Z_0
$\mid X \mid = \dfrac{r - 1}{\sqrt{r}}$	
$\mid \Gamma \mid = \dfrac{\mid B \mid}{\sqrt{B^2 + 4}}$	Relations for a normalized susceptance B in shunt with admittance Y_0
$\mid B \mid = \dfrac{r - 1}{\sqrt{r}}$	

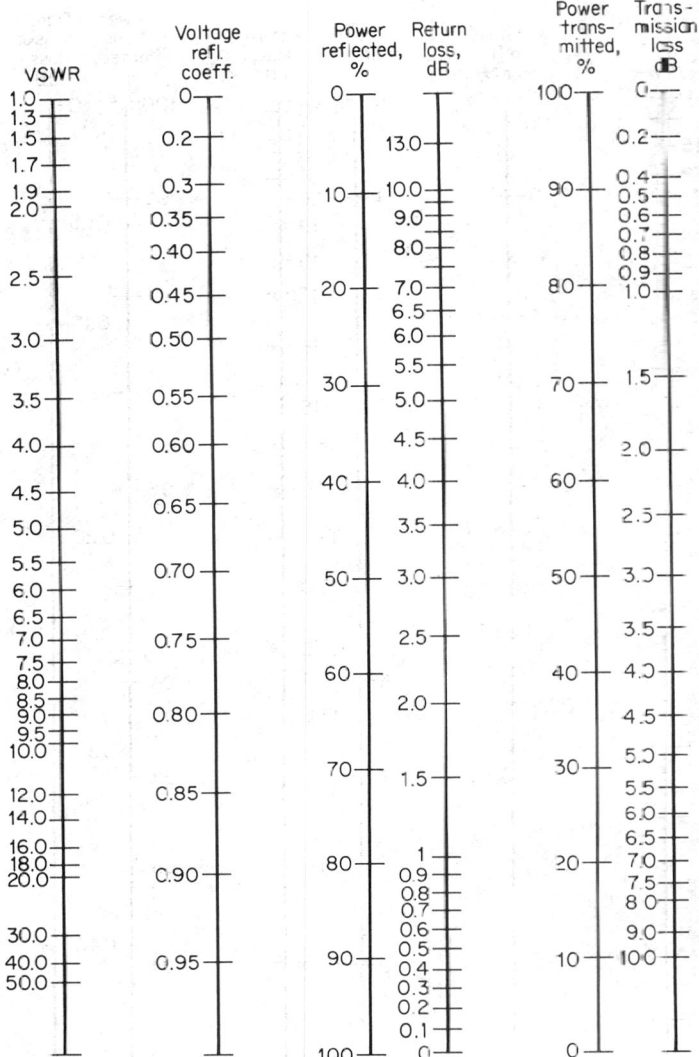

Fig. 9-1. Nomograph for transmission and reflection of power at high voltage standing wave ratios (VSWR).

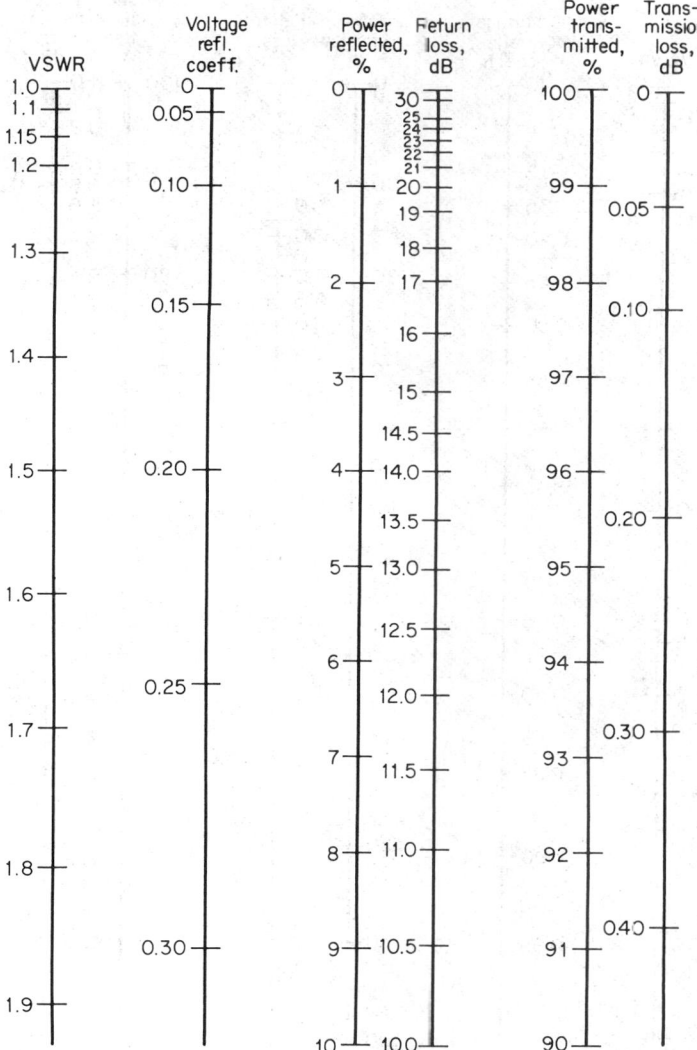

Fig. 9-2. Nomograph for low VSWR.

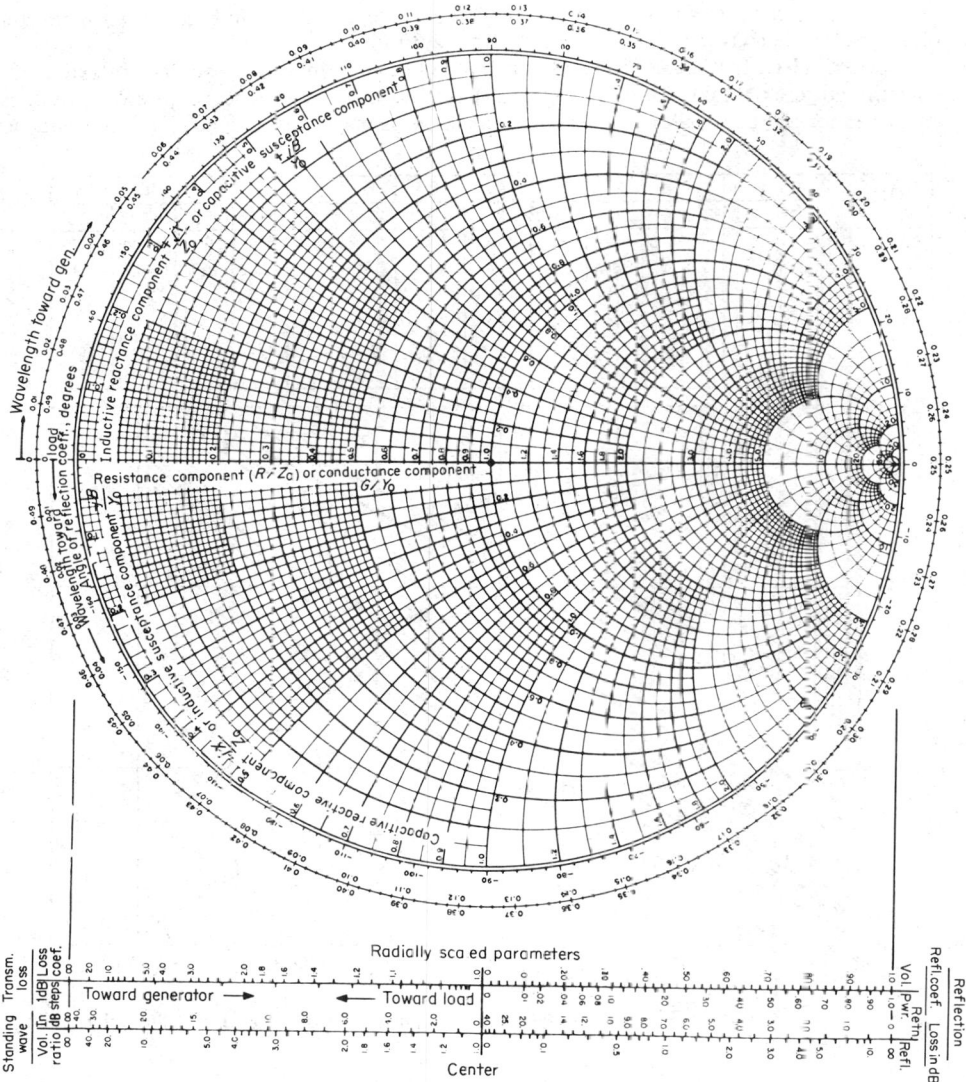

Radially scaled parameters

Fig. 9-3. The Smith chart.

guide. When nondominant-mode transmission is possible, the situation becomes far more complex, since mode conversion can occur at any discontinuity.

3. Coaxial-Cable Data. The attenuation and power-handling capability of flexible 50-Ω polyethylene coaxial cable are shown in Fig. 9-4 as a function of frequency. Figure 9-5 gives these quantities for rigid coaxial of various outer diameters as marked on the curves. The frequency

USASI C83.2 cable group	1	2	3	4	5	6	7	8	9	10	11
Approx size O.D. inches	0.080	0.110	0.160	0.200	—	0.330	0.415	0.415	0.550	0.725	0.870
RG-( )/U	178,A,B	174	122	55,A,B	75Ω	5,A,B	115,A	8,A	14,A	Teflon	17,A
Cable designation	196,A	188,A		58,A,C	cables	21A,B		9,A,B	217	cables	218
	(Teflon)			142A,B		143,A		213			
				223		212		214			
						222		225			
						304					

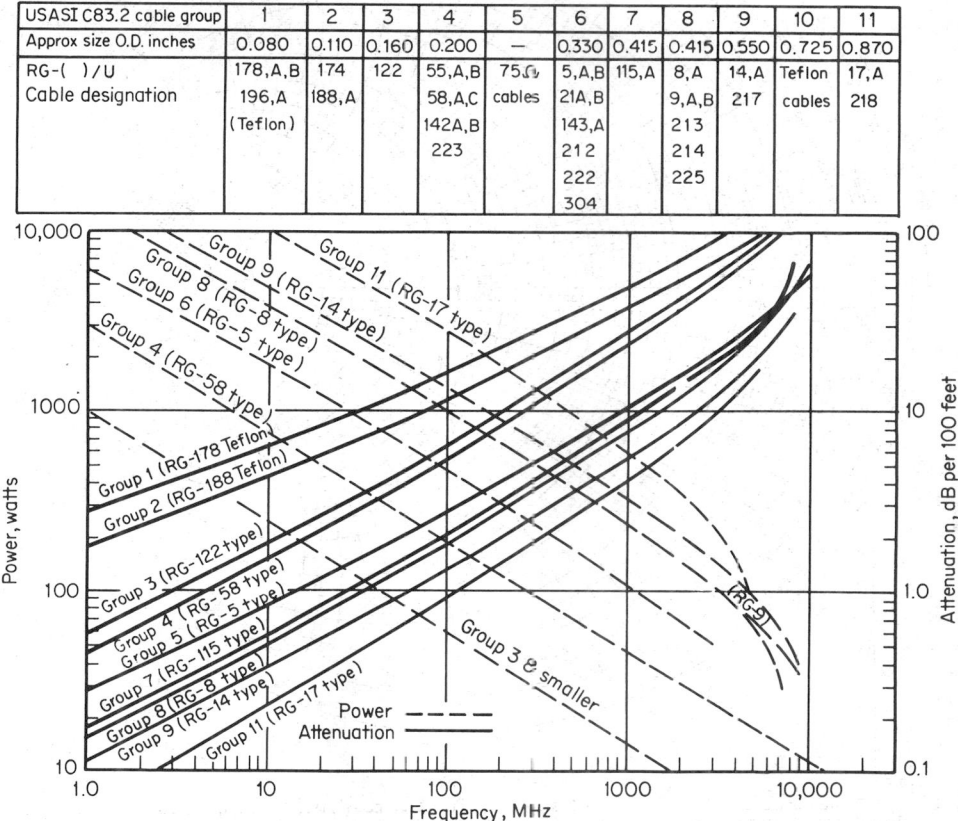

Fig. 9-4. Power rating and attenuation of flexible polyethylene coaxial cables. For Teflon cables multiply power ratings by 5.

scale on this figure extends only to 3 GHz because such high-power transmitting coaxial is generally not used at higher frequencies. Figure 9-6 gives the variation of impedance with dimensions for a standard coaxial line (curve 1) as well as for several other forms of inner-conductor TEM-mode configurations.

Care must be taken to avoid operation of a coaxial line at wavelengths where it becomes possible for additional modes to propagate. This occurs when the mean circumference between the inner and outer cylinders forming the transmission path equals a full wavelength; a stable standing wave is then possible in a circumferential direction. The higher attenuation and reduced power-carrying capability which accompany the small dimensions necessary to avoid higher-order modes generally lead to the choice of waveguide as a transmission medium at frequencies above 3 GHz.

4. Waveguide Data. Table 9-3 gives the standard dimensions, flange-coupling codes, attenuation, and power-handling capability for dominant-mode (TE_{10}) rectangular waveguide. The electric field pattern of this mode is a half sinusoid across the transverse guide dimension with its maximum at the center of the broad wall. The frequency range shown for each guide size follows conventional practice in avoiding operation too close to the cutoff wavelength λ_c (= $2a$, where a is the broad dimension of the guide), on the low-frequency side, or too close to the next higher mode (usually, $\lambda'_c = a$), on the high-frequency side.

When higher power must be transmitted than is possible because of breakdown in air with

Table 9-3. Rectangular Waveguide Data and Fittings

EIA WG Desig. WR	Recommended operating range for TE₁₀ mode — Frequency, GHz	Wavelength, cm	Cutoff for TE₁₀ mode — Frequency, GHz	Wavelength, cm	Range in $\frac{2\lambda}{\lambda_c}$	Range in $\frac{\lambda_g}{\lambda}$	Theoretical cw power rating, lowest to highest frequency, MW	Theoretical attenuation, lowest to highest frequency, dB/100 ft	Material alloy	JAN WG RG ( )/U	JAN flange Choke UG ( )/U	JAN flange Cover UG ( )/U	EIA WG Desig. WR	Inside	Tolerance	Outside	Tolerance	Wall thickness, nominal
2300	0.32–0.49	93.68–61.18	0.256	116.84	1.60–1.05	1.68–1.17	153.0–212.0	0.051–0.031	Alum.	—	—	—	2300	23.000–11.500	±.020	23.250–11.750	±.020	0.125
2100	0.35–0.53	85.65–56.56	0.281	106.68	1.62–1.06	1.68–1.18	120.0–173.0	0.054–0.034	Alum.	—	—	—	2100	21.000–10.500	±.020	21.250–10.750	±.020	0.125
1800	0.41–0.625	73.11–47.96	0.328	91.44	1.60–1.05	1.67–1.18	93.4–131.9	0.056–0.038	Alum.	201	—	—	1800	18.000–9.000	±.020	18.250–9.250	±.020	0.125
1500	0.49–0.75	61.18–39.97	0.393	76.20	1.61–1.05	1.62–1.17	67.6–93.3	0.069–0.050	Alum.	202	—	—	1500	15.000–7.500	±.015	15.250–7.750	±.015	0.125
1150	0.64–0.96	46.84–31.23	0.513	58.42	1.60–1.07	1.82–1.18	35.0–53.8	0.128–0.075	Alum.	203	—	—	1150	11.500–5.750	±.015	11.750–6.000	±.015	0.125
975	0.75–1.12	39.95–26.76	0.605	49.53	1.61–1.08	1.70–1.19	27.0–38.5	0.137–0.095	Alum.	204	—	—	975	9.750–4.875	±.010	10.000–5.125	±.010	0.125
770	0.96–1.45	31.23–20.67	0.766	39.17	1.60–1.06	1.66–1.18	17.2–24.1	0.201–0.136	Alum.	205	—	—	770	7.700–3.850	±.005	7.950–4.100	±.005	0.125
650	1.12–1.70	26.76–17.63	0.908	33.02	1.62–1.07	1.70–1.18	11.9–17.2	0.317–0.212 / 0.269–0.178	Brass / Alum.	69 / 103	—	417A / 418A	650	6.500–3.250	±.005	6.660–3.410	±.005	0.080
510	1.45–2.20	20.67–13.62	1.157	25.91	1.60–1.05	1.67–1.18	7.5–10.7	0.388–0.385	Brass / Alum.	104	—	435A / 437A	510	5.100–2.550	±.005	5.260–2.710	±.005	0.080
430	1.70–2.60	17.63–11.53	1.372	21.84	1.61–1.06	1.70–1.18	5.2–7.5	0.501–0.330	Brass / Alum.	105	—	—	430	4.300–2.150	±.005	4.460–2.310	±.005	0.080
340	2.20–3.30	13.63–9.08	1.736	17.27	1.58–1.05	1.78–1.22	3.1–4.5	0.877–0.572 / 0.751–0.492	Brass / Alum.	112 / 113	—	553 / 554	340	3.400–1.700	±.005	3.560–1.860	±.005	0.080
284	2.60–3.95	11.53–7.59	2.078	14.43	1.60–1.05	1.67–1.17	2.2–3.2	1.102–0.752 / 0.940–0.641	Brass / Alum.	48 / 75	54A / 585	53 / 584	284	2.840–1.340	±.005	3.000–1.500	±.005	0.080
229	3.30–4.90	9.08–6.12	2.377	11.63	1.56–1.05	1.62–1.17	1.6–2.2	2.08–1.44	Brass / Alum.	49 / 95	148B / 406A	149A / 407	229	2.290–1.145	±.005	2.418–1.273	±.005	0.064
187	3.95–5.85	7.59–5.12	3.152	9.510	1.60–1.08	1.67–1.19	1.4–2.0	1.77–1.12	Brass / Alum.	187	—	—	187	1.872–0.872	±.005	2.000–1.000	±.005	0.064
159	4.90–7.05	6.12–4.25	3.711	8.078	1.51–1.05	1.52–1.19	0.79–1.0	2.87–2.30	Brass / Alum.	50 / 106	343A / 440A	344 / 441	159	1.590–0.795	±.004	1.718–0.923	±.004	0.064
137	5.85–8.20	5.12–3.66	4.301	6.970	1.47–1.05	1.48–1.17	0.56–0.71	2.45–1.94	Brass / Alum.	137	—	—	137	1.372–0.622	±.004	1.500–0.750	±.004	0.064
112	7.05–10.00	4.25–2.99	5.259	5.700	1.49–1.05	1.51–1.17	0.35–0.46	4.12–3.21 / 3.50–2.74	Brass / Alum.	51 / 68	52 / 137A	51 / 138	112	1.122–0.497	±.004	1.250–0.625	±.004	0.064
90	8.20–12.40	3.66–2.42	6.557	4.572	1.60–1.06	1.68–1.18	0.20–0.29	6.45–4.48 / 5.49–3.83	Brass / Alum.	52 / 67	40A / 136A	39 / 135	90	0.900–0.400	±.003	1.000–0.500	±.003	0.050
75	10.00–15.00	2.99–2.00	7.868	3.810	1.57–1.05	1.64–1.17	0.17–0.23	9.51–8.31	Brass / Alum. / Silver	91	541	419	75	0.750–0.375	±.003	0.850–0.475	±.003	0.050
62	12.4–18.00	2.42–1.66	9.486	3.160	1.53–1.05	1.55–1.18	0.12–0.16	6.14–5.36	Brass / Alum. / Silver	107	—	—	62	0.622–0.311	±.0025	0.702–0.391	±.0025	0.040
51	15.00–22.00	2.00–1.36	11.574	2.590	1.54–1.05	1.58–1.18	0.080–0.107	20.7–14.8	Brass / Alum. / Silver	53 / 121	596 / 598	595 / 597	51	0.510–0.255	±.0025	0.590–0.335	±.003	0.040
42	18.00–26.50	1.66–1.13	14.047	2.134	1.56–1.06	1.60–1.18	0.043–0.058	27.6–12.6	Brass / Alum. / Silver	66	—	—	42	0.420–0.170	±.0020	0.500–0.250	±.002	0.040
34	22.00–33.00	1.36–0.91	17.328	1.730	1.57–1.05	1.62–1.18	0.034–0.048	13.3–9.5	Brass / Alum. / Silver	96	600	599	34	0.340–0.170	±.0020	0.420–0.250	±.003	0.040
28	26.50–40.00	1.13–0.75	21.081	1.422	1.59–1.05	1.65–1.17	0.022–0.031	—	Brass / Alum. / Silver	97	—	—	28	0.280–0.140	±.0015	0.360–0.220	±.002	0.040
22	33.00–50.00	0.91–0.60	26.342	1.138	1.60–1.05	1.61–1.17	0.014–0.020	21.9–15.0	Brass / Silver	96	600	103	22	0.224–0.112	±.0010	0.304–0.192	±.002	0.040
19	40.00–60.00	0.75–0.50	31.357	0.956	1.57–1.05	1.63–1.16	0.011–0.015	31.0–20.9	Brass / Silver	97	—	—	19	0.188–0.094	±.0010	0.268–0.174	±.002	0.040
15	50.00–75.00	0.60–0.40	39.863	0.752	1.60–1.06	1.67–1.17	0.0063–0.0090	52.9–39.1	Brass / Silver	98	—	385	15	0.148–0.074	±.0010	0.228–0.154	±.002	0.040
12	60.00–90.00	0.50–0.33	48.350	0.620	1.61–1.06	1.68–1.18	0.0042–0.0060	93.3–52.2	Brass / Silver	99	—	387	12	0.122–0.061	±.0005	0.202–0.141	±.002	0.040
10	75.00–110.00	0.40–0.27	59.010	0.508	1.57–1.06	1.61–1.18	0.0030–0.0041	152–99	Silver	138	—	99	10	0.100–0.050	±.0005	0.180–0.130	±.002	0.040
8	90.00–140.00	0.333–0.214	73.840	0.406	1.64–1.05	1.75–1.17	0.0018–0.0026	163–137	Silver	136	—	—	8	0.080–0.040	±.0003	0.156 dia	±.001	—
7	110.00–170.00	0.272–0.176	90.840	0.330	1.64–1.06	1.77–1.18	0.0012–0.0017	308–193	Silver	135	—	—	7	0.065–0.0325	±.00025	0.156 dia	±.001	—
5	140.00–220.00	0.214–0.136	115.750	0.259	1.65–1.05	1.78–1.17	0.00071–0.00107	384–254	Silver	137	—	—	5	0.051–0.0255	±.00025	0.156 dia	±.001	—
4	170.00–260.00	0.176–0.115	137.520	0.218	1.61–1.05	1.69–1.17	0.00052–0.00075	512–348	Silver	139	—	—	4	0.043–0.0215	±.00020	0.156 dia	±.001	—
3	220.00–325.00	0.136–0.092	173.280	0.173	1.57–1.06	1.62–1.18	0.00035–0.00047	—	Silver	—	—	—	3	0.034–0.0170	±.00020	0.156 dia	±.001	—

these restrictions on dimensions, pressurization of the air or a sulfur hexafluoride gas fill is generally used. The increase in power rating obtained is shown in Fig. 9-7.

Unlike the case for the TEM mode of a coaxial or parallel-wire line, the guide wavelength λ_g departs from the free-space wavelength λ_0 and varies with frequency, producing phase distortion in a wide-band signal. The relationship is

$$\lambda_g = \lambda_0 / \sqrt{1 - (\lambda_0/\lambda_c)^2}$$

5. Ridge Waveguide. To obtain broader bandwidth in a single mode, as well as increased flexibility in the choice of impedance, ridge waveguide is generally employed. The variation of cutoff wavelength with dimensions for single and double ridges (always inserted on the broad

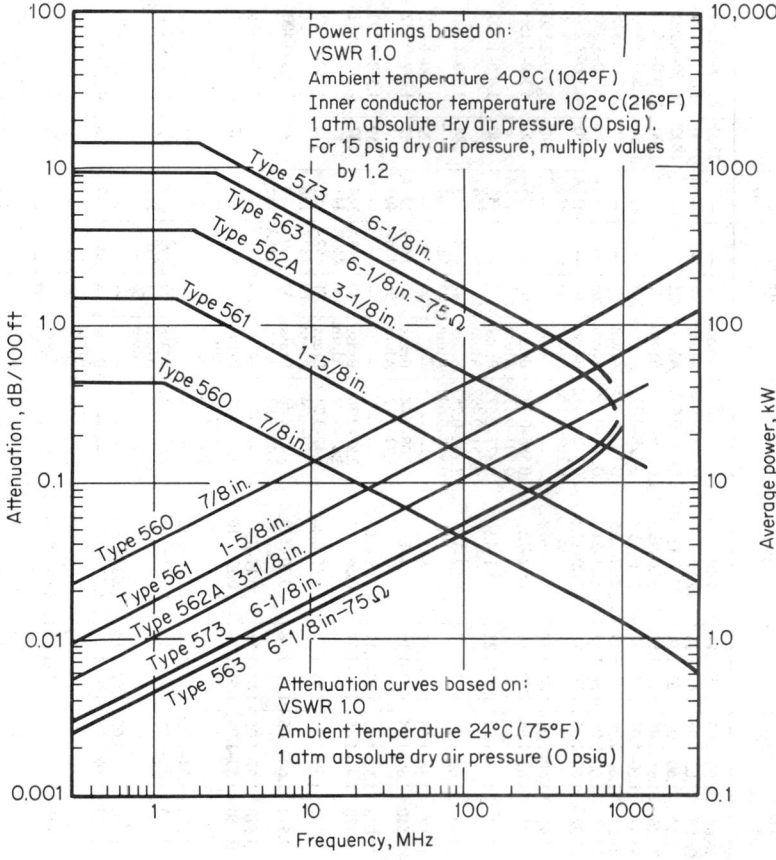

Fig. 9-5. Power rating and attenuation for rigid coaxial cables.

wall of a rectangular guide to give capacitive loading) is given in Fig. 9-8. The correction factor to these curves for nonstandard b/a ratio is given in Fig. 9-9. A large increase in bandwidth is made possible by this type of guide.

6. Circular-Mode Transmission. The circular electric (TE$_{01}$) mode is currently employed in long-distance broadband communications at millimeter wavelengths because of its low-loss transmission property. The electric field pattern and the wall currents form concentric rings in this unusual mode. Conversion to lower-order modes occurs if the cross section is even slightly elliptical or if the guide axis is curved. Various forms of mode suppression are employed; a common technique consists in fabricating the cylindrical wall from a tightly wound helix with loss between the turns to damp out modes with axial components of current. Attenuation as low as a few decibels per mile has been achieved.

7. Strip and Microstrip Transmission Lines. Strip transmission lines constitute an increasingly useful medium for short-distance, low-power applications. Strip line consists of a printed conductor between two ground planes, typically formed from copper-clad polyethylene sheets. Electrically, such lines have properties similar to coaxial transmission lines. The higher-order-mode-free operating frequency range depends upon the width of the strip and the placement of mode-suppression screws. Physical characteristics of standard 50-Ω line are given in Table 9-4. The attenuation of these lines over a range of impedance and frequency is shown on Fig. 9-10, and their power-handling capability in Fig. 9-11.

Table 9-4.   Physical Characteristics of 50-Ω Strip Line

Standard Designation	Material thickness,* in.	Copper weight,† oz.	Strip width, in.	Screw spacing, in. Long.‡	Lat.§	Upper frequency limit.¶ GHz
MPC-062-2	0.062	2	0.083 ± 0.0015	0.375	0.200	7.5
MPC-125-2	0.125	2	0.182 ± 0.003	0.375	0.300	5.0
MPC-187-2	0.187	2	0.280 ± 0.004	0.375	0.400	3.6
MPC-250-2	0.250	2	0.380 ± 0.005	0.375	0.500	2.8

* Material thickness measured over copper cladding.
† Copper thickness: 1 oz, 0.0014 in.; 2 oz, 0.0028 in.
‡ Type 4-40 screws recommended.
§ Lateral screw spacing measured from edge of strip to center of screw.
¶ Recommended for this lateral spacing from circuits. For higher-frequency limits, the specified spacing should be reduced.

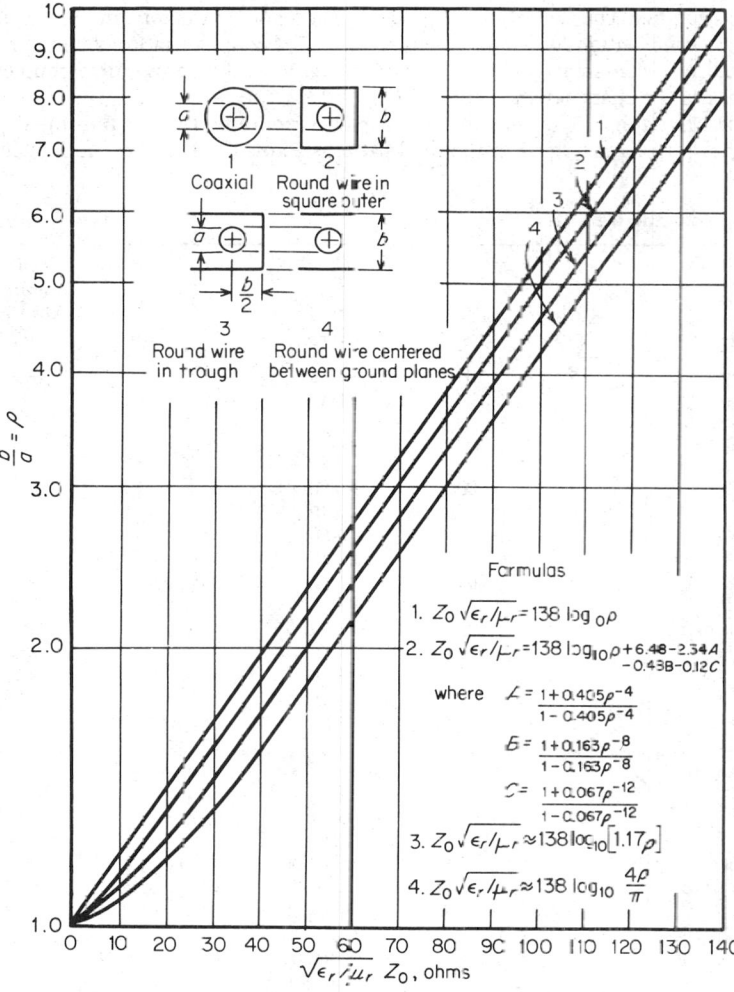

Fig. 9-6. Characteristic impedance of coaxial lines.

Microstrip transmission line is a miniaturized version of strip line best suited to circuit integration of semiconductor devices. Polished substrates are employed, with a precious metal deposited or etched away to form the line conductor. A wide variety of dielectric substrates is available, with typical properties shown in Table 9-5. The most common material for high-quality circuits is alumina; quartz is chosen when extreme dimensional stability is required. The variation of characteristic impedance with microstrip parameters is shown in Fig. 9-12. The attenuation of this type of line is greater than that for strip line, while its power-handling capacity is less.

8. Reciprocal Circuit Components. The basic circuit theory for microwaves is identical with that for lower frequencies, but the physical form of the components is usually quite different from conventional lumped elements because of the distributed nature of microwave electric and magnetic field patterns.

Resonant circuits are generally formed from short-circuited lengths of transmission line. The dimensions for such a resonator are selected from a mode chart of the type shown in Fig. 9-13 for the right circular cylinder. The Q which is obtained for some typical cavity resonators in hollow-cylinder, coaxial, and rectangular configurations is shown in Fig. 9-14.

A metal post, screw, or dielectric discontinuity inserted in a transmission line acts as a lumped circuit element. Typical examples of capacitive and inductive equivalent circuits in waveguide are shown in Fig. 9-15.

Directional Couplers. Directional couplers are employed to sample power for measurement purposes. By spacing a series of holes at quarter-wavelength intervals one obtains phase addition of the propagation field coupled from one guide into another for the forward direction of power flow and phase cancellation for the reverse direction. The coupling ratios obtained for holes in the broad wall of rectangular guide are shown in Fig. 9-16. The cross-guide coupler, another popular type, has the characteristics shown in Fig. 9-17.

Impedance Matching. The smooth flow of power from one type of transmission medium into another requires a matching of the field patterns across the boundary to launch the wave

TABLE 9-5   Dielectric Materials

Material	Dielectric constant	Loss Factor	Useful Temp range, °C	Flexibility	Coeff of thermal expansion $\times 10^6$, °C	Surface finish, μ in*
Woven TFG	2.55	.0015	−60 to +200	Good	18.5	N/A
Microfiber TFG, Duroid 5870	2.33	.0005	−60 to +200	Good	5	N/A
Duroid 5880	2.2	.0006	−60 to +200	Good	32	N/A
Polystyrene	2.53	.0003	−60 to +100	Very poor	7	N/A
Reinforced	2.62	.002	−60 to +100	Poor	5.7	N/A
Polyphenelene oxide (PPO)	2.55	.0016	−60 to +200	Fair	29	N/A
Polyolefin	2.32	.00015	−60 to +100	Excellent	4.4	N/A
Quartz Teflon	2.47	.0006	−60 to +200	Good	18.5	N/A
Polymide, Micaply 5032	4.8	.01	−60 to +250	Good	9	N/A
Epsilam-10	10.0	.002	−60 to +150	Good	11–23	N/A
99.5% alumina	9.9	.00008	Up to 500	Very poor	7.5	< 3
Quartz (fused silica)	3.78	.0001	Up to 500	Very poor	0.55	< 1
Sapphire	9.4 11.6	.00008	Up to 500	Very poor	7.7 8.3	< 1
99.5% BeO	6.6	.0004	Up to 500	Very poor	7.8	3–7
Boron nitride	4.4	.0003	Up to 500	Poor	1–2	N/A

*N/A = Not available.

into the second medium with a minimum of reflection. Coaxial line is generally matched into rectangular waveguide by extending the center conductor of the coaxial line through the broad wall of the guide, parallel to the electric field lines across the guide. Alternatively, the center conductor can be formed into a loop and oriented to couple the magnetic field of the guide mode.

The higher-order-mode field patterns, which are necessary for analytical reasons to match the transition completely, represent lumped reactances provided no propagation of these modes is possible, i.e., that they are cut off. Compensating susceptances can then be introduced in the form of tunable stubs designed by means of the Smith chart (Fig. 9-3) but with an empirical fine adjustment to cancel reflections. If an impedance transformation is required in addition to the pattern match described, quarter-wave transformers are generally used. They consist of $\lambda_g/4$ lengths of line designed for an intermediate impedance equal to the square root of the product of the two impedances to be matched.

For broader bandwidth, multiple quarter-wave sections are called for, leading in the limit to a smooth taper of the dimension(s) so that the fractional change in impedance with distance is constant. The resultant reflection coefficients for various forms of tapers are compared as a function of taper length in Fig. 9-18. Figure 9-19 shows the bandwidth obtainable from a two-section (quarter-wave) transformer for given VSWR (denoted by S on the curves).

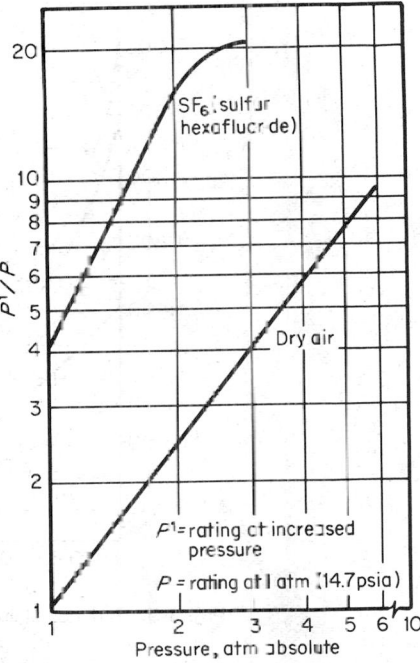

Fig. 9-7. Power rating of pressurized transmission lines

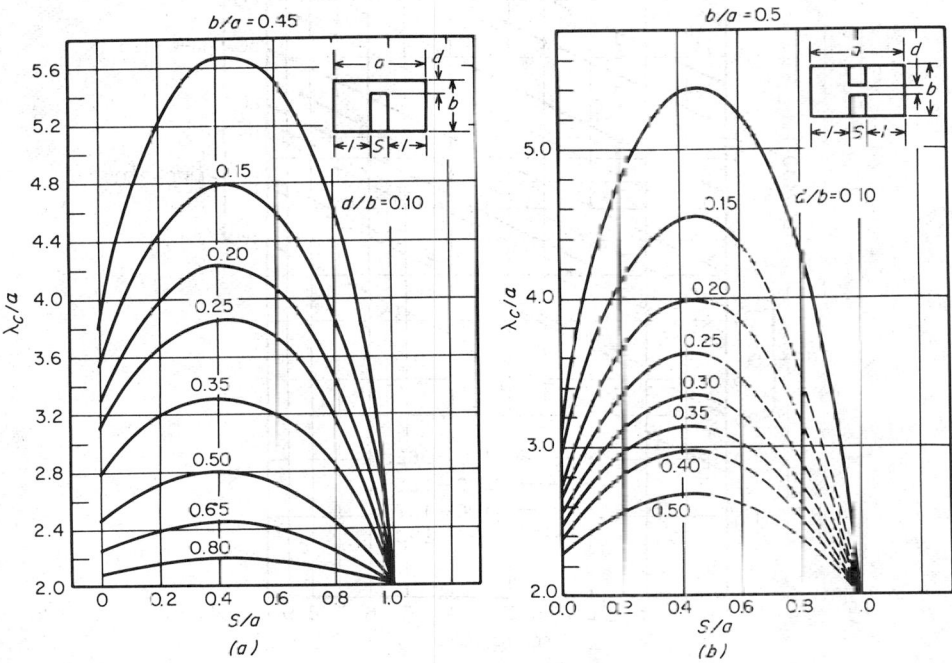

Fig. 9-8. Cutoff wavelength of ridged waveguide: (a) single ridge (b) double ridge

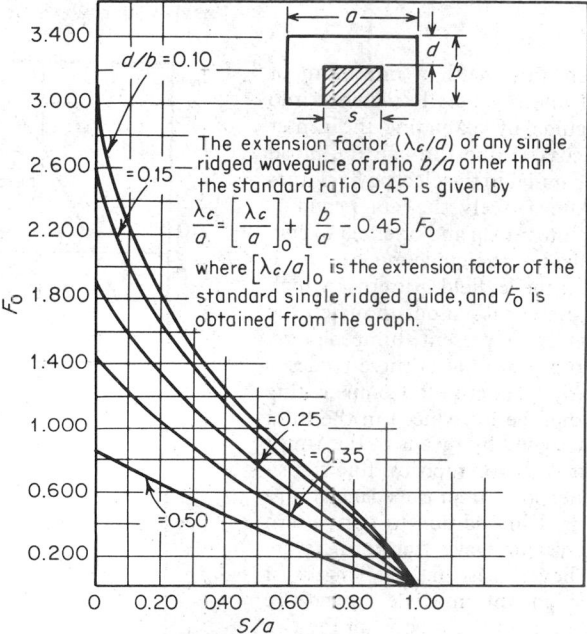

Fig. 9-9. Variation of cutoff wavelength for nonstandard b/a ratios.

The extension factor (λ_c/a) of any single ridged waveguide of ratio b/a other than the standard ratio 0.45 is given by

$$\frac{\lambda_c}{a} = \left[\frac{\lambda_c}{a}\right]_0 + \left[\frac{b}{a} - 0.45\right] F_0$$

where $\left[\lambda_c/a\right]_0$ is the extension factor of the standard single ridged guide, and F_0 is obtained from the graph.

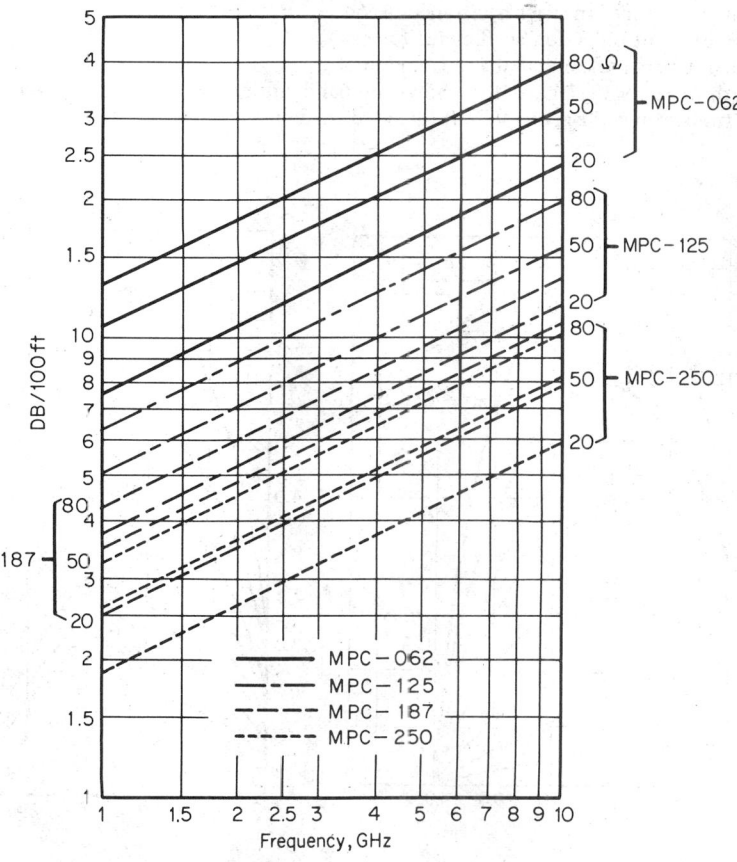

Fig. 9-10. Attenuation characteristics of strip line.

Filters. Frequency filters are used to separate the components of a composite waveform for signal-processing purposes or to suppress rf interference (RFI) which results from the spurious output of transmitters. The latter problem has only recently become serious at microwave frequencies as this area of the spectrum has become congested. High-power capability is required for such filters, leading generally to the use of waveguide structures. A section of waveguide

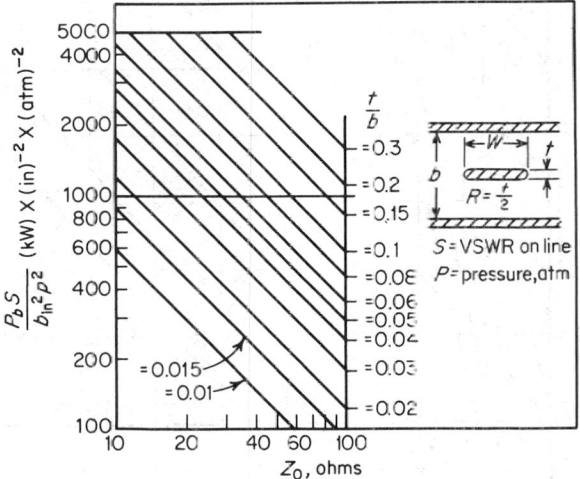

Fig. 9-11. Power-carrying capacity of strip line.

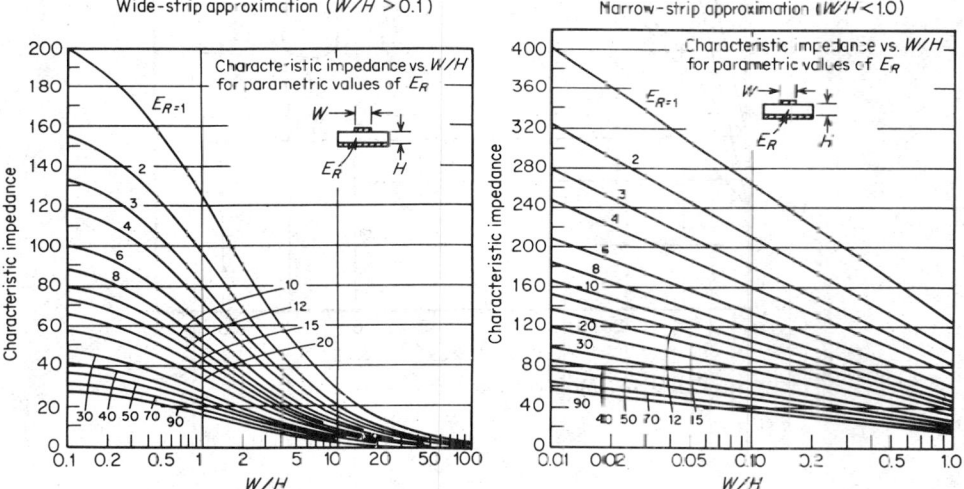

Fig. 9-12. Characteristic impedance of microstrip line: wide-strip approximation (*left*); narrow-strip approximation (*right*).

beyond cutoff constitutes a simple high-pass reflective filter. Loading elements in the form of posts, irises, or stubs are employed to supply the reactances required for conventional lumped-constant-filter design.

The desired skirt steepness and stopband attenuation determine the number of sections, as at lower frequencies. A disk-loaded coaxial line is generally used as a low-pass high-power filter. Insertion loss of reflective filters is typically 0.1 to 0.2 dB, with stopband attenuation of the order of 50 dB. Absorption filters avoid the reflection of unwanted energy by incorporating lossy material in secondary guides which are coupled through leaky walls (typically small sections of guide beyond cutoff in the passband). These filters are effective primarily against harmonics.

For low-power applications, strip-line filters are widely used because of their compact size and low cost. Typical dimensions for a low-pass filter of this type are shown in Fig. 9-20.

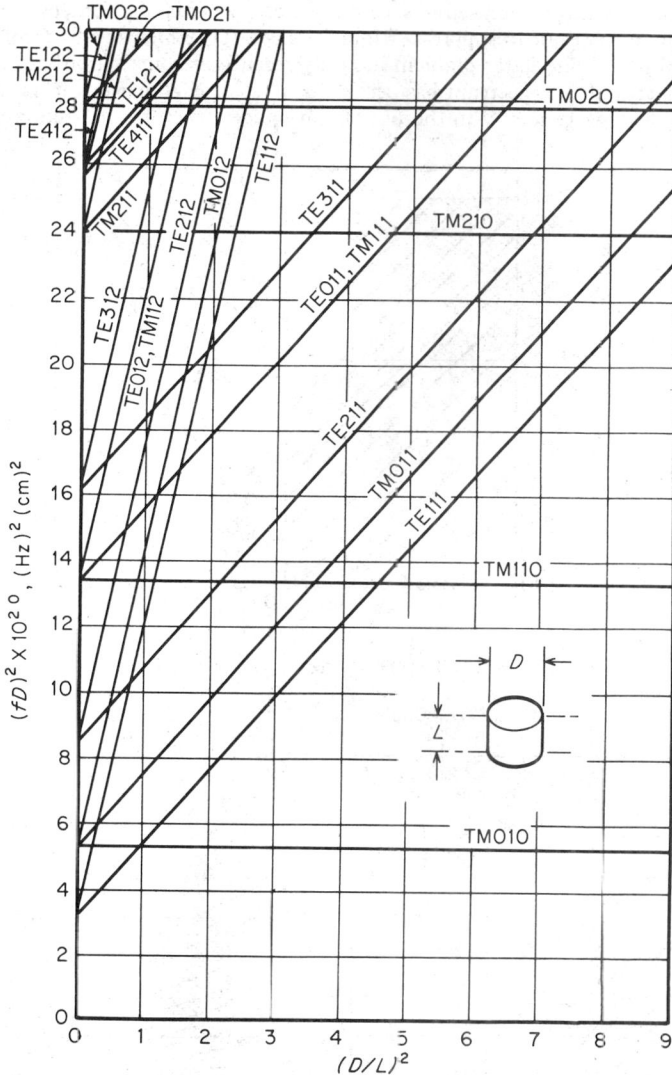

Fig. 9-13. Mode chart for right circular cylinder.

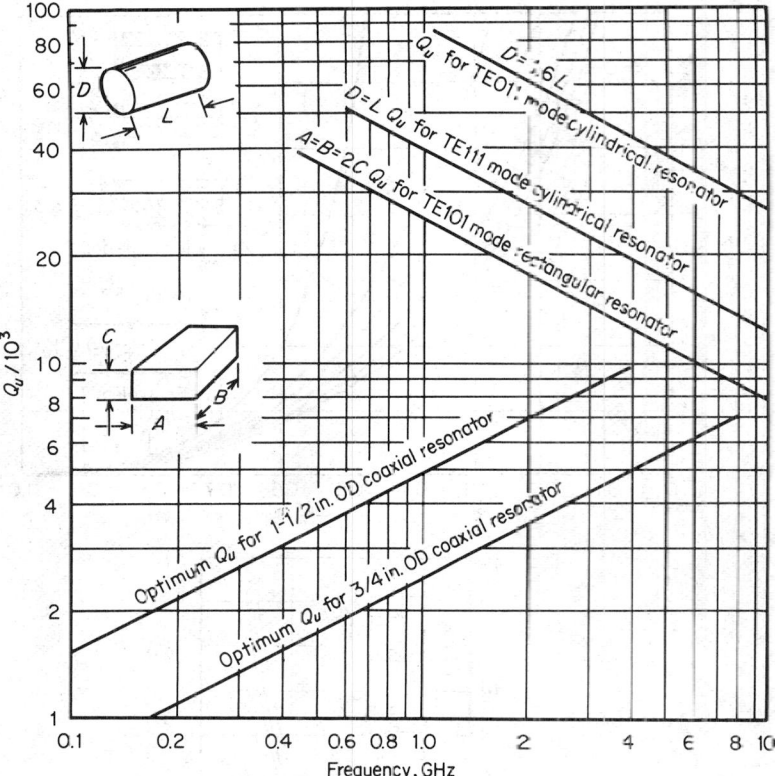

Fig. 9-14. Q for typical cavity resonators.

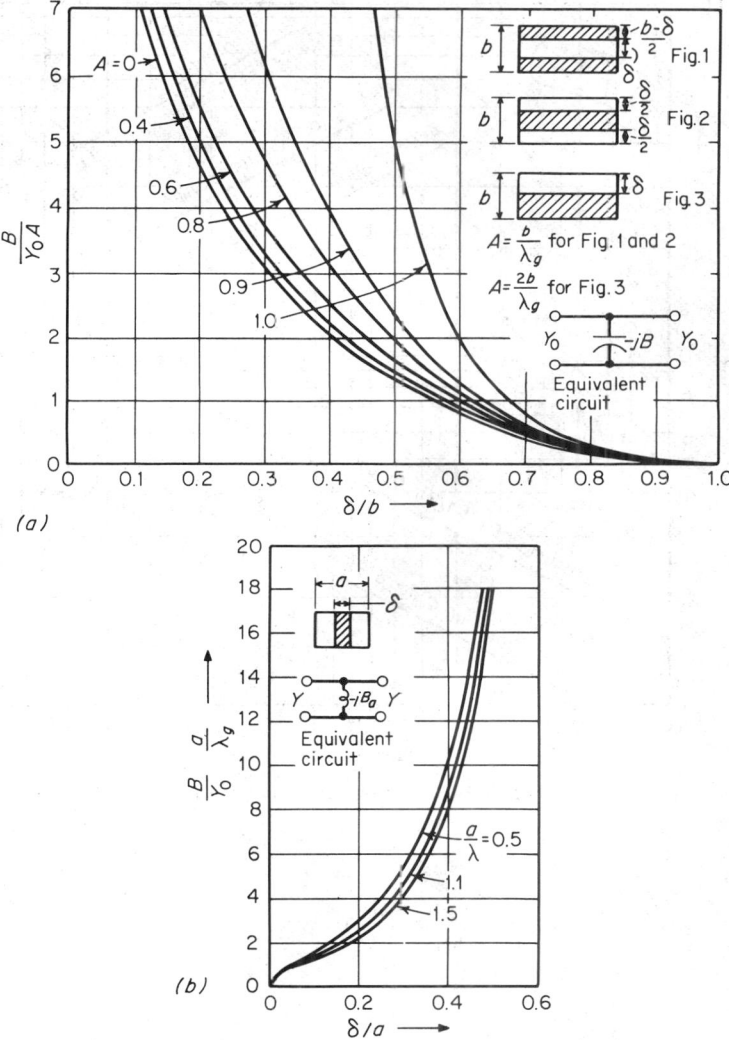

(a)

(b)

Fig. 9-15. Susceptance of waveguide elements: (a) capacitive irises; (b) inductive, centered thin vane.

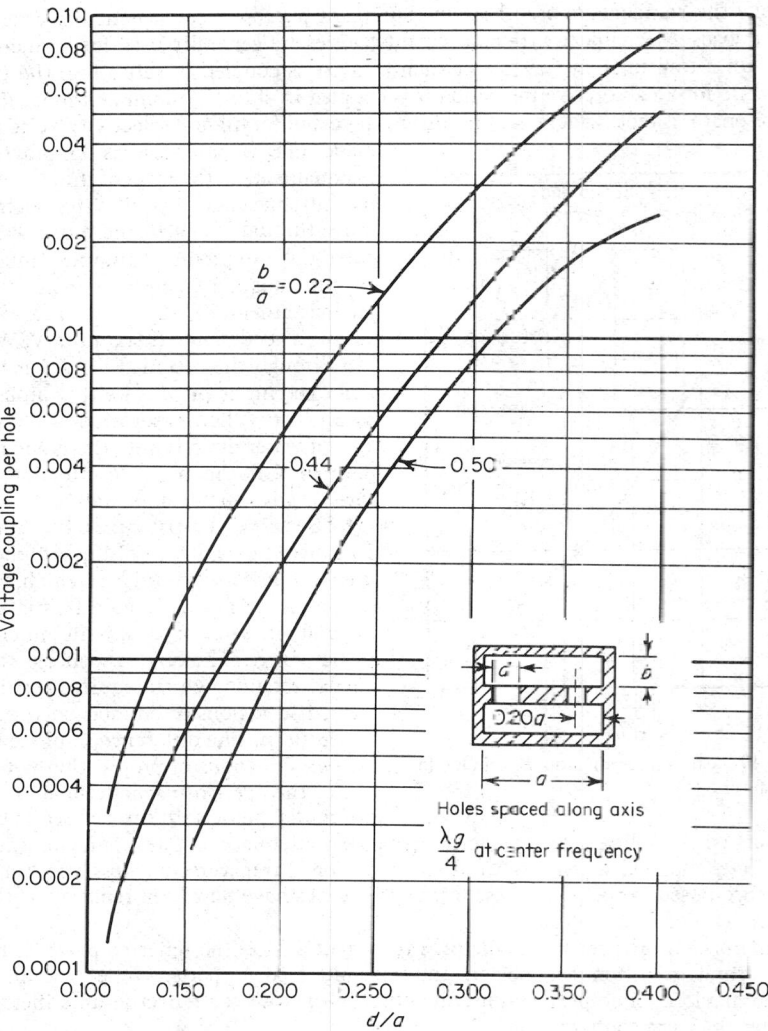

Fig. 9-16. Broad-wall directional-coupler ratios.

Among the components useful for measurement purposes are wavemeters, attenuators, and matched loads. Mechanically tuned resonant cavities are generally used for frequency determination. For a transmission wavemeter such a cavity is coupled in series into the transmission path, while for an absorptive indication it is coupled in shunt. A dominant-mode (TE_{111}) cylindrical resonator is most widely used for this purpose, but for highest selectivity a circular electric-mode (TE_{011}) resonator is employed. Variable attenuators take the form of thin absorptive material introduced tangential to the electric field typically through a slot in the broad wall of rectangular guide to produce minimum reflection. Load terminations are tapered attenuators designed to produce at least 40 dB of return loss while maintaining a good match (maximum VSWR less than 1.2) through the specified band. For high power, water cooling is provided either around a loaded ceramic or Teflon absorber or by introducing a tube of water directly into the guide to act as the absorber. Calorimetric determination of power is possible with such water loads.

9. Ferrite Components. *Isolators.* Ferrite isolators owe their usefulness to the large ratio of reverse loss to forward loss which occurs when the proper dc magnetic field bias is applied to a material chosen for a specific microwave frequency range. The original concept was that of a Faraday rotation, which required positioning a ferrite rod in regions of circular polarization of the field pattern. The difference in phase velocity for the two directions of propagation is then a function of the applied magnetic field. The resultant differential phase shift can be used in conjunction with 3-dB power splitters to provide nonreciprocal attenuation. The nonreciprocal change in field pattern obtained when a ferrite slab is positioned transversely across a rectangular guide has also been used in the past. These ferrite geometries have now been replaced by the junction configuration described below.

Circulators. In the ferrite circulator, a three-port device, the reflected power is brought out from a separate port from the incident and transmitted power ports, where it can be monitored, absorbed in a load, or employed as useful output in the case of a reflective amplifier. Figure 9-21 illustrates the basic configuration.

Circulators have also found use as duplexers, providing greater than 20 dB isolation between

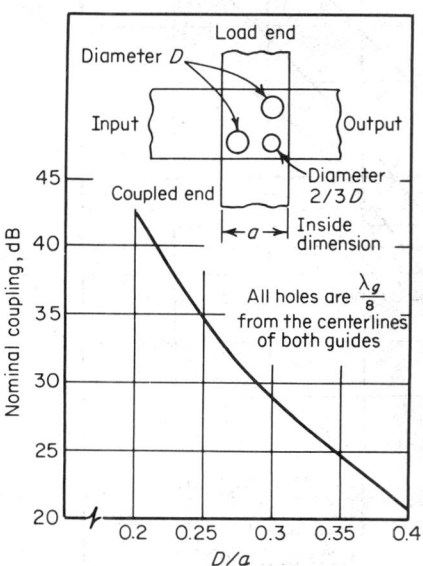

Fig. 9-17. Cross-guide directional coupler for signal sampling.

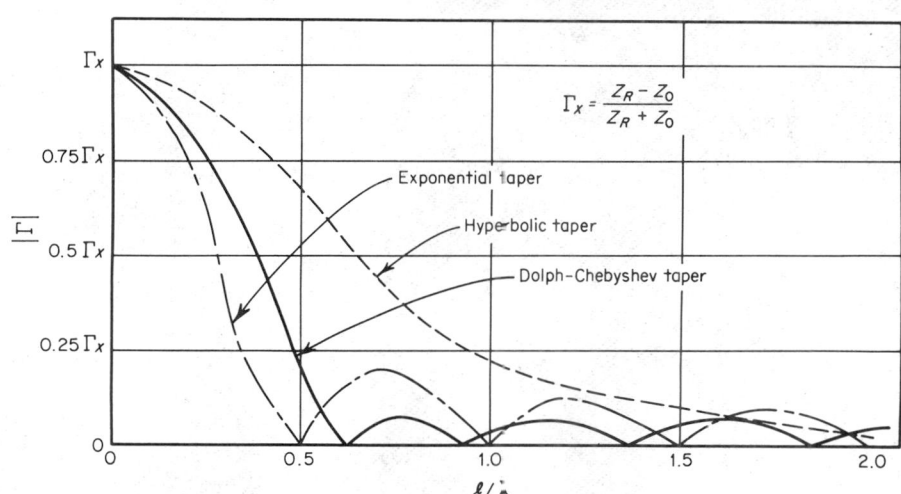

Fig. 9-18. Use of tapers for impedance matching.

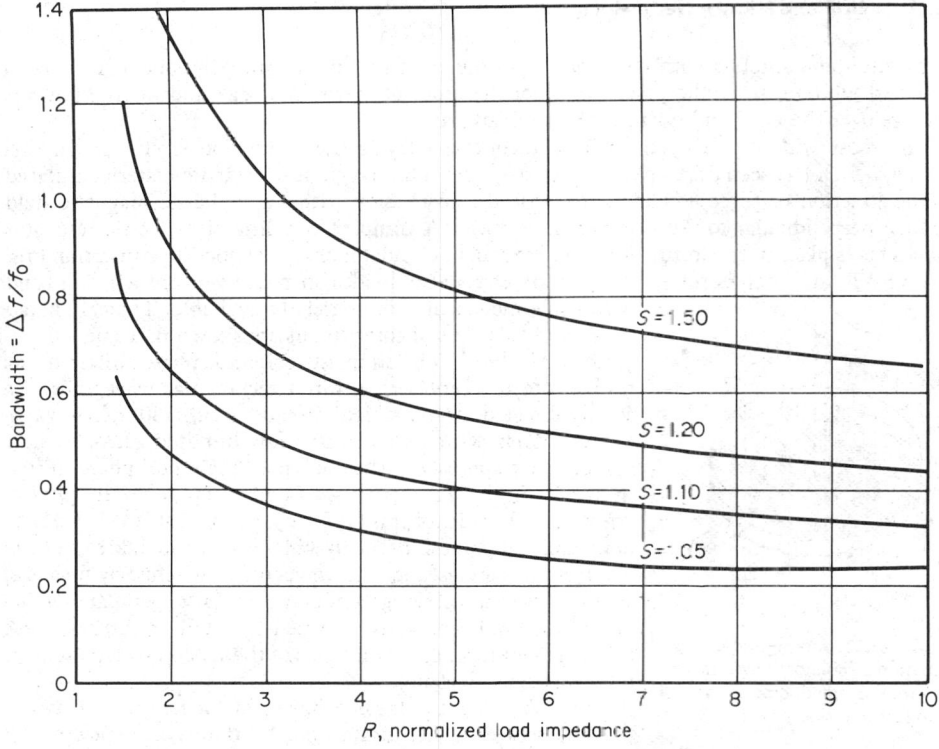

Fig. 9-19. Bandwidth and VSWR of quarter-wave transformers.

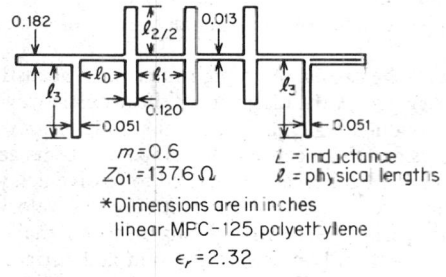

$m = 0.6$
$Z_{01} = 137.6\ \Omega$

L = inductance
ℓ = physical lengths

*Dimensions are in inches
linear MPC-125 polyethylene
$\epsilon_r = 2.32$

Dimensions

Frequency f_c	ℓ_0 $0.084\lambda_{\epsilon_r}$	ℓ_1 $0.100\lambda_{\epsilon_r}$	$\ell_{2/2}$ $0.144\lambda_{\epsilon_r}$	ℓ_3 $0.200\lambda_{\epsilon_r}$
1 GHz	0.6653	0.7920	1.093	1.548
2 GHz	0.3326	0.3960	0.533	0.756
3 GHz	0.2220	0.2640	0.333	0.493

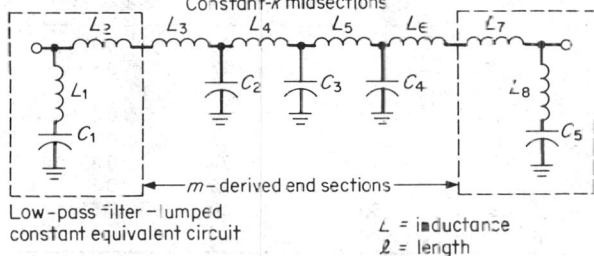

Constant-k midsections

m-derived end sections

Low-pass filter – lumped
constant equivalent circuit

L = inductance
ℓ = length

Fig. 9-20. Strip-line filter design.

transmitter and antenna with insertion loss of the order of ¼ dB. Ferrite components have been designed which can handle power levels of the order of megawatts peak and many kilowatts average over the UHF and microwave frequency bands.

The most widely employed ferrite component today is the Y-junction circulator. In this device, which has been developed for both strip line and waveguide, the ferrite material is placed at the junction of three equiangularly spaced transmission paths, and the dc magnetic field applied perpendicular to the plane of these paths. A complex coupling of dielectric resonator modes takes place in the ferrite, leading to excellent circulator characteristics (low insertion loss, low VSWR, and high port-to-port isolation). Octave-bandwidth strip-line versions and full-band waveguide models are commercially available. Typical characteristics of this class of components are shown in Table 9-6.

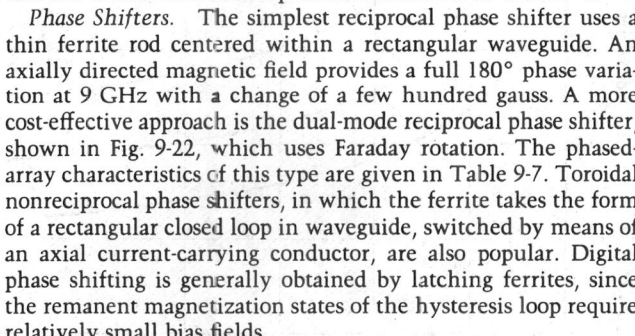

Phase Shifters. The simplest reciprocal phase shifter uses a thin ferrite rod centered within a rectangular waveguide. An axially directed magnetic field provides a full 180° phase variation at 9 GHz with a change of a few hundred gauss. A more cost-effective approach is the dual-mode reciprocal phase shifter, shown in Fig. 9-22, which uses Faraday rotation. The phased-array characteristics of this type are given in Table 9-7. Toroidal nonreciprocal phase shifters, in which the ferrite takes the form of a rectangular closed loop in waveguide, switched by means of an axial current-carrying conductor, are also popular. Digital phase shifting is generally obtained by latching ferrites, since the remanent magnetization states of the hysteresis loop require relatively small bias fields.

Fig. 9-21. Configuration of junction circulator, consisting of a circular conductor between two ferrite disks.

Limiters. The nonlinear behavior of ferrites at high power levels is used in ferrite limiters. Such limiters are replacing TR gas-discharge tubes in radar today. Peak powers of tens of kilowatts at X band can be handled, but with considerable spike-energy leakage, which requires a follow-on solid-state *(pin)* diode stage. Typical insertion loss is about 0.5 dB, with 30 dB of flat limiting and about 6 dB of spike limiting for the ferrite alone. The diode stage increases the insertion loss to about 1 dB but reduces all leakage an additional 30 dB. The recovery time is very short, typically 100 ns, and is determined primarily by the diode section. Characteristics of such diodes are discussed further in Pars. **9-51** and **9-59**.

10. Acoustic-Wave Devices. Although interest in acoustic waves at UHF and microwave frequencies was triggered by the discovery of the acoustic-wave amplifier, a device employing a piezoelectric semiconductor such as cadmium sulfide to provide interactive gain between its drifting charge carriers and the electric field component of its acoustic wave, practical applications of these waves have thus far been restricted to passive delay lines and signal processors. Acoustic waves owe their usefulness to their relatively low velocity, typically 10^{-5} times the electromagnetic velocity, permitting relatively long electric-signal delay times to be obtained in a physically small space. Both bulk-mode propagation and surface waves have been used, the latter gaining in popularity because of the relative ease of access to intermediate points along the propagation path for structure shaping and tapping.

Transducers. The transducers designed for the two types of acoustic modes are physically

TABLE 9-6   Characteristics of Typical Junction Circulators

Circulator type	Center frequency, GHz	Bandwidth, %	Isolation, dB	Insertion loss, dB	Power capacity Avg, W	Peak kW
Waveguide	12.4–18	. . . .	20	0.3	20	1
	18–26.5	. . . .	17	0.5	15	0.5
Strip transmission line	4–8	. . . .	20	0.4	35	
	12.4–18	. . . .	18	0.5	25	0.25
Switching	2.9	8.9	26	0.35		15
	35	5	15	0.5		15
Lumped constant	1.2	30	20	0.6		
	0.4–0.5	30	20	0.4		

SOURCE: *Microwave J.*, November 1978.

TABLE 9-7 Characteristics of Reciprocal Dual-Mode Phasers

Phaser	Power capacity, kW		Center frequency, GHz	Band-width SWR <1.20, %	Insertion loss, dB	Phase shift, deg		Switching speed, µs	Cooling	Material	Cost
	Peak	Average				Analog	Latching				
X band	1	0.1	9.1	10	0.6 ± 0.2*	0-650	0-500	20-40†	Air-cold plate	Ferrite	Low
S band	150	1.5	3.2	3	0.9 ± 0.1*	0-200		CV‡	Liquid ± 2°F	DTCG§	High
	8	1.5	3.2	3	1.1 ± 0.2*	0-1050		CV‡	Liquid ± 2°F	DTCG§	Medium
K_a band	>0.5	0.5	35	6	2.0 ± 0.2*	0-800	0-360	5†	Air	NiZn ferrite	Low

*Modulation.
|Latching.
‡Continuously varying.
§Doped temperature-compensated garnet.
SOURCE: *Microwave J.*, November 1978.

quite different, but both contain electrodes spaced either a quarter or half an acoustic wavelength in a piezoelectric material. Microwave bulk-wave transducers consist of multiple films of ZnO or CdS, each approximately $\lambda/4$ thick, separated by similar films of gold or other nonpiezoelectric material deposited on the bulk medium to provide electric-to-acoustic wave coupling and impedance transformation.

Surface acoustic waves (SAW) are generally excited by interdigital transducers. An interdigital transducer, illustrated in Fig. 9-23, consists of two sets of interleaved metal electrodes, called *fingers*, deposited on the piezoelectric substrate. To generate a wave an rf potential is applied between the adjacent sets of fingers, which are spaced by a distance equal to one-half wavelength at the transducer design frequency. A typical 100-MHz transducer on LiNbO$_3$ has aluminum fingers 0.2 μm thick by 9 μm wide with 9-μm gaps.

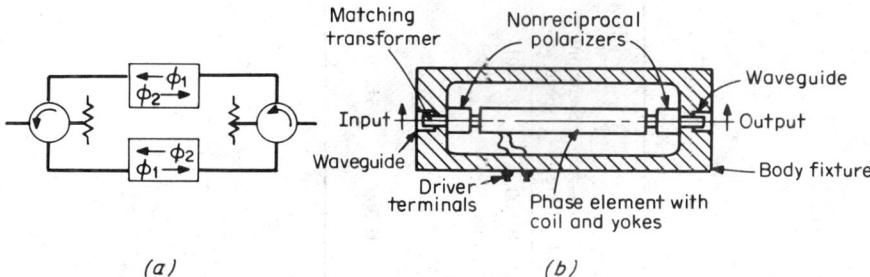

Fig. 9-22. Dual-mode reciprocal phase shifter: (a) equivalent circuit schematic; (b) typical configuration.

The wave excited by the rf potential between a pair of fingers travels at the surface-wave velocity. By the time the wave arrives midway between the next pair of fingers, the rf excitation potential has reversed sign, and the wave excited by the second pair of fingers will be in phase with the wave from the first pair. Thus the excitation due to the second pair is added to the excitation from the first, and so on. The mechanism is reciprocal, and hence the transducer that excites a wave will also detect it.

The transducer has a fractional bandwidth of $1/N$, where N is the number of finger pairs. Electrically, the transducer is represented by a capacitance shunted by a radiation resistance which depends on the choice of finger length.

A surface-wave transducer is a three-port device, i.e., with one electric and two elastic ports. Figure 9-24 shows the conversion loss as a function of frequency from a 50-Ω source to one of the two acoustic outputs for three different transducers on a lithium niobate surface. These calculated curves show that for this particular case the use of five finger pairs provides the widest bandwidth and smallest conversion loss. The attenuation of the wave,

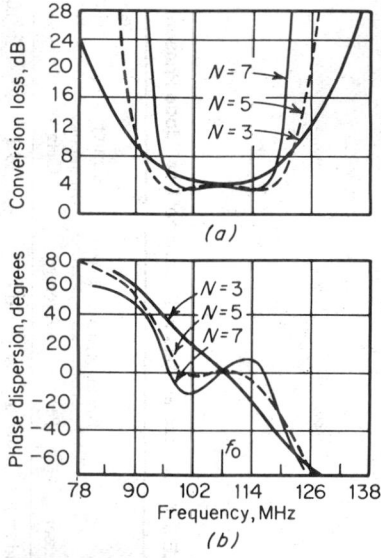

(a)

(b)

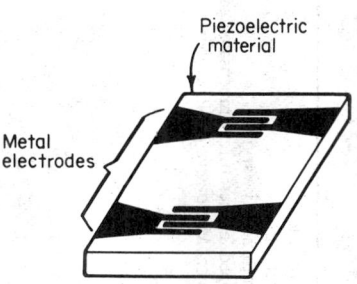

Fig. 9-23. Piezoelectric surface-wave microwave device with interdigital electrodes. *(IEEE Spectrum, August 1971.)*

Fig. 9-24. Coupling between electric port and one acoustic port for transducers with three, five, and seven interdigital periods: (a) Theoretical conversion loss; (b) phase dispersion.

once it has been launched is given in Fig. 9-25 for lithium niobate. This quantity generally varies as the square of the frequency. Surface-wave amplification, obtained by coupling to the carriers in a thin film of semiconducting material mounted adjacent to the surface, has been used to compensate this attenuation but is still in an experimental stage.

The lowest operating frequency of acoustic surface-wave devices is limited by the allowable size to typically 10 MHz. At present the upper operating frequency is limited by fabrication difficulties to about 1 GHz. A typical 500-MHz transducer on quartz has interleaved metal fingers 1.5 μm wide by 3 mm long, separated by 1.5-μm gaps. Usual photolithography techniques can be employed by make transducers up to 600 MHz routinely. Transducers with an operating

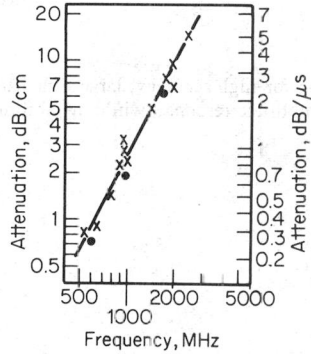

Fig. 9-25. Surface-wave attenuation of Y-cut lithium niobate. (*After Armstrong. WESCON, 1971.*)

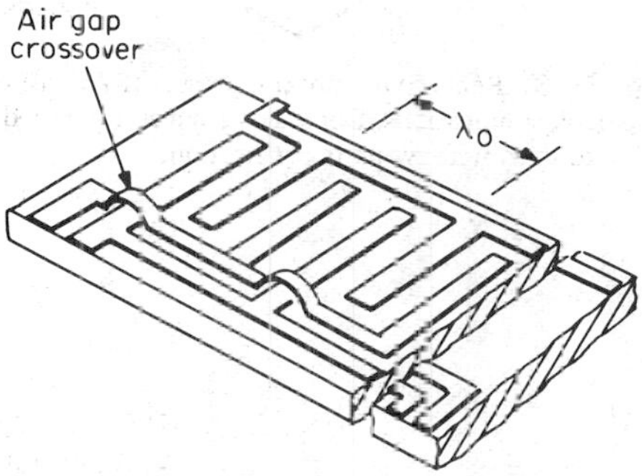

Fig. 9-26. Three-phase unidirectional transducer. Each electrode is 120° out of phase, so that the acoustic waves add in one direction and cancel in the other. (*Microwave J., December 1977.*)

frequency up to 3.5 GHz have been made by using a scanning electron microscope to expose the photoresist in the photolithographic process.

Bandpass Filters. Bandpass filtering is the principal commercial application of SAW technology. Such devices are replacing *LC* filters in television receiver i.f. circuits. Minimum stopband rejection of the order of 60 dB with in-band response flat to ±0.1 dB and phase deviation from linearity of only a few degrees are typical of these filters. The major drawback of SAW devices, their high insertion loss (of the order of 15 dB), can be reduced to less than 3 dB by using a three-phase unidirectional transducer structure, as shown in Fig. 9-26.

Dispersive Filters. Dispersive filters are used primarily for pulse compression. A schematic of the metallization pattern in Fig. 9-27 illustrates the linear increase in delay that results from

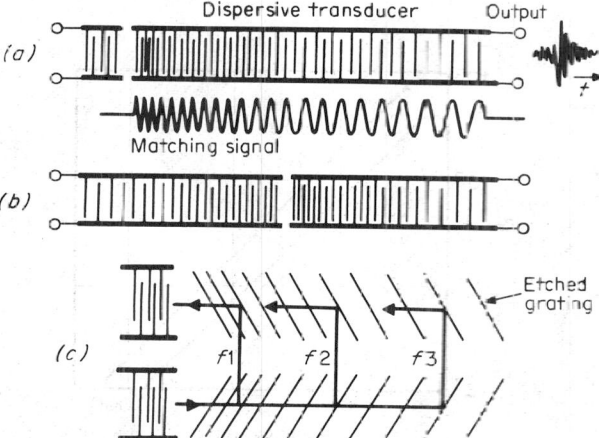

Fig. 9-27. Basic forms of matched SAW filter for chirp signals: (a) delay line with dispersion designed in one transducer; (b) dispersive delay line with dispersion in both transducers; (c) reflective array compressor. (*Proc. IEEE, May 1976, Vol. 64, No. 5.*)

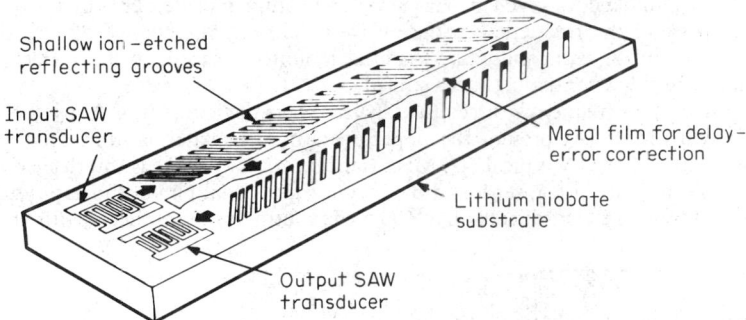

Fig. 9-28. Reflective-array-compressor (RAC) filter configuration for high-accuracy, large time-bandwidth dispersive applications. Depths of grooves are varied to equalize amplitude response, while metal film adjusts delay to within a few parts per million.

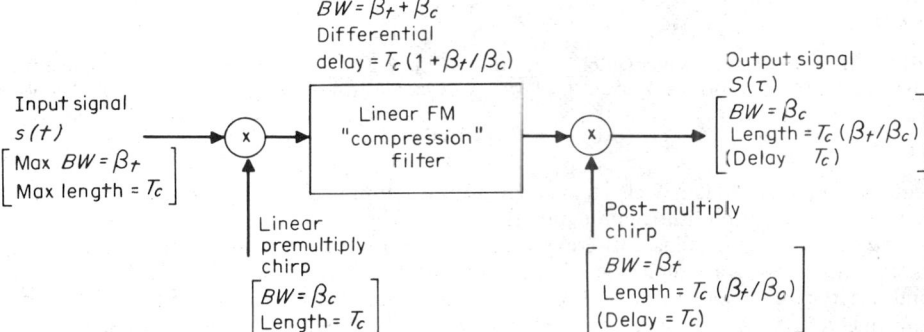

Fig. 9-29. Algorithm for chirp-transform cell, in which the output is the Fourier transform of the input.

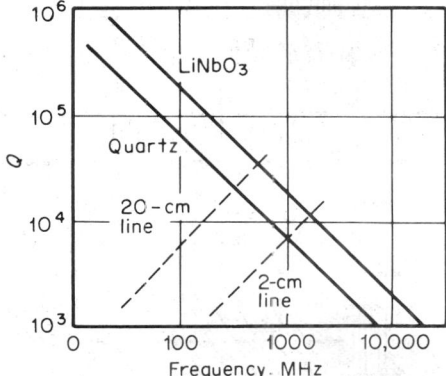

Fig. 9-30. Q factors of SAW materials. Solid lines show intrinsic Q without air loading; dashed lines show the limit to Q imposed by a single transit in a finite length of delay line (*Proc. IEEE, May 1976, Vol. 64, No. 5.*)

the monotonically decreasing finger spacing. A linear frequency-modulated signal of bandwidth B and pulse duration T is thereby converted into a single pulse compressed by the ratio BT. Side-lobe suppression of better than 40 dB is obtained for time-bandwidth (BT) products up to 500. For larger products, up to 10,000, reflective array compressors (RAC), shown in Fig. 9-28, are used.

Chirp Transforms. The linear FM, or chirp response, described above, can be used to obtain a Fourier transform of an input signal with the algorithm shown in Fig. 9-29. The input signal is multiplied by a linear frequency-modulated waveform (chirp) followed by a convolution of the product with a filter whose impulse response is also a chirp but with a frequency-time relation of opposite slope. When only the power spectrum of the input signal is required, the last step is unnecessary.

SAW Resonators. SAW resonators have the advantages of small size and low cost over conventional electromagnetic structures. The Q's obtainable are shown in Fig. 9-30. Reflective configurations are used to obtain the advantage of long transit length. Such resonators are used for narrow-band filters and for oscillators at frequencies below 1 GHz.

11. Bibliography on Passive Microwave Components

Ramo, S., and J. R. Whinnery   "Fields and Waves in Modern Radio," Wiley, New York, 1953.
Reich, H. J., J. G. Skalnik, P. F. Ordung, and H. L. Krauss   "Microwave Principles," Van Nostrand, New York, 1957.
Ginzton, E. L.   "Microwave Measurements," McGraw-Hill, New York, 1957.
Saad, T. S. (ed.)   "Microwave Engineers' Handbook," Artech House, 1971.
Young, L. (ed.)   "Microwave Filters, Impedance-Matching Networks and Coupling Structures," McGraw-Hill, New York, 1972.
Special Issue on Surface Acoustic Waves   *Proc. IEEE*, May, 1976, Vol. 64, No. 5.

Planar Microwave Tubes and Circuits

BY DONALD H. PREIST

12. Development of Planar Tubes. *Triodes* for rf power generation at frequencies above about 1 GHz were developed just before World War II in Europe and the United States for both continuous-wave (cw) and pulsed operation. Cylindrical and planar electrode configurations were used with oxide-coated cathodes in Britain. In the United States the "lighthouse tube" used planar geometry and an internal anode. All used metal and glass envelopes. Continuous-wave rf powers of the order of 1 to 10 W at 1 GHz and peak pulse powers of the order of kilowatts were obtained in coaxial-line circuits in several industrial laboratories.

The need for centimeter-wave equipment for military purposes stimulated development of microwave tubes and associated circuitry during World War II, particularly in the United States, and by 1946 the 2C39 type triode with external anode emerged as the most popular design. The version with an aluminum oxide ceramic envelope (3CX100A5) was produced in very large quantities and with minor internal modifications is still in use today. Before the emergence of the power klystron in the 1950s, these tubes and combinations of them (e.g., in an *annular circuit*) produced the highest rf power outputs obtainable above 1 GHz.

The power gain of rf amplifiers with planar triodes is sufficient to make practical the use of multistage amplifiers driven by crystal-controlled oscillators with high-frequency stability. For example, the TD2 microwave radio relay system operating at 4 GHz in service since 1951, uses a planar triode of advanced design (type 416A/1553) in broadband circuitry for television and multichannel telephone transmission.[1]* Planar triodes are preferred today in some applications (airborne radar and navigation equipment and space communications) because of their small size and/or low cost relative to other technical approaches. They have been improved considerably in performance over the years, and are manufactured in Europe, the United States, and Japan. Triodes of small size using cylindrical electrodes ("pencil triodes") are available for similar applications.

Tetrodes (4X150A, 4X150G, 1949, and later models; 6816, 7213, and others) also give useful performance at 1 GHz and above, especially in pulse service, but are less widely used than the triodes because of greater difficulty in designing and fabricating the associated cavity circuitry

*Superior numbers correspond to numbered references, Par. **9-23**.

and the tube socket. These difficulties increase rapidly with frequency, and the improvement in power gain over the planar triode decreases rapidly. In general, therefore, tetrodes are preferred below 1 GHz and triodes above 1 GHz today.

13. Performance vs. Frequency. Power output and power gain of amplifiers with a single tube fall off with increasing frequency, for several basic reasons. Physical size imposes an upper limit related directly to wavelength and inversely to frequency. Ideally, the rf voltages between electrodes should be uniform, but this cannot be realized unless the major electrode dimensions are much less than one quarter wavelength (radius of planar triodes, axial length of cylindrical electrode triodes or tetrodes). The radius cannot be increased indefinitely because of the tendency to introduce circumferential variations of electric field which can exist when circumference exceeds about one wavelength. These factors cause maximum tube electrode area, and therefore power output, to fall off inversely with the square of the frequency.

In addition, *electron-transit-time effects* cause a further falloff unless interelectrode spacings, mainly grid to cathode, and grid-wire diameters are scaled inversely with frequency. At 1 GHz, the grid cathode spacing must not exceed a few mils, or transit time will result in excessive loading of the drive source, significantly reducing power gain of amplifiers, causing back-heating of the cathode by electron bombardment and reduction of conversion efficiency due to increase of the *conduction angle* (fraction of the rf cycle during which electrons arrive at the anode).

Another cause of reduced performance with increasing frequency is the increase in rf displacement currents drawn by interelectrode capacitances, resulting in increased i^2R heating of the grid wires, cathodes, and connecting leads, including the seal areas between metal leads and the insulating envelope.

For these reasons, the performance of a given tube falls off very rapidly above a specified frequency, and in practice an upper limit of frequency (sometimes called f_1) is designated by manufacturers above which the dc power input to the anode circuit must be reduced to prevent overheating. Below this frequency, performance stays roughly constant as the frequency is reduced, since the basic limiting factors are then cathode emission density and voltage breakdown between electrodes, which are not functions of frequency.

14. Pulsed Operation. In pulsed operation higher frequencies can be reached than with cw operation because the higher voltages, used intermittently, result in increased electron velocities and decreased transit times with higher conversion efficiencies. At a given frequency the rf power output during a short pulse can be as much as 100 times the cw output. This is due partly to the higher voltages but mainly to the greatly increased emission density available from the cathodes during short pulses. The usable cathode current depends on the pulse length and the duty factor (pulse duration divided by pulse-to-pulse time interval, see Fig. 9-31). The cathode temperature, and therefore the heater voltage, may also be different for pulsed operation. Lifetime of tubes depends on other operating conditions besides cathode temperature, and tube manufacturers should advise about these conditions if maximum life is to be obtained, whether in cw or pulsed operation.

Examples of performance vs. frequency from available tubes are given in Fig. 9-32 for cw and pulsed operation. The curves show maximum rf power output, pulsed and cw, considered to be obtainable from tubes, using present technology if designed for narrow-band operation at the frequency chosen, and assuming zero circuit losses and a tube life of 1,000 h or more. Pulse lengths of a few microseconds are assumed.

Higher peak power can be obtained if the plate voltage is pulsed (rather than if dc plate voltage is used with grid pulsing) because the higher probability of internal arcing requires a lower dc voltage than the allowable pulse voltage. For longer pulses, cathode current must be reduced, and therefore less rf power output can be obtained. Typical operating conditions for a few planar triodes are given in Table 9-8.

15. Multitube Arrangements. To obtain more power output, several tubes can be used together if care is taken to suppress *moding*, which tends to occur when some dimension of the combining circuitry used becomes large in terms of the wavelength. For example, using an annular circuit of 14 planar triodes can produce 500 W cw output at 1 GHz.[2] This approach to higher power has lost popularity at microwave frequencies with the advent of the klystron, traveling-wave tube, and crossed-field amplifier, all capable of very much higher power and competitive in price with multiple triodes because of the high cost of the combining circuitry required.

16. Efficiency and Gain. Typical plate-conversion efficiencies, assuming zero circuit losses, narrow-band operation, and optimum conditions for planar triodes, fall from about 60%

at 1 GHz to about 30% at 5 GHz when pulsed, or to about 5% at 5 GHz cw. Circuit losses reduce these numbers, depending on the circuit design.

The power gain obtainable from amplifiers depends on circuit design (see below) but seldom exceeds 12 dB at 1 to 2 GHz, falling off with increasing frequency. For this reason self-excited oscillators are used at the higher frequencies. The bandwidth of such amplifiers also depends on circuit design but seldom exceeds a few percent of the center frequency.

17. Design and Construction of Planar Tubes. The envelopes of planar tubes are almost always of ceramic insulation with penetrating metal members supporting the electrodes.

TABLE 9-8 Planar Tubes

Type	Frequency, GHz	Service	Plate voltage, kV	Plate current, A	Rf output W	Rf drive, W	Plate efficiency %	Gain, dB	Heater power, W
3CX100A5, 7289	3.0	Plate-pulsed oscillator	3.5	3.0	2,000	...	19	...	6.0
8892, 18651	5.0	Grid-pulsed oscillator	2.0	2.0	1,000	...	25	...	4.5
8756	2.35	Cw amplifier	1.25	0.15	60	3.0	33	13	5.1
416B*	4.0	Cw amplifier	0.3	0.03	0.5	...	...	6–9	

*See Ref. 4, Par. **9-23**.

High purity alumina (above 99% pure) is most commonly used, but beryllia is used when the added performance made possible by its increased thermal conductivity (10 times) and lower dielectric constant (%) justifies the increased cost (5 to 10 times). The metal members are usually of a material (such as Kovar) with thermal expansion coefficient close to that of the insulator, to minimize thermal stresses in fabrication and operation. They are coated with a highly conductive metal, e.g., copper or silver, to reduce i^2R losses. The metal members are shaped either as disks or disks with cylindrical projections.

The cathodes are usually oxide-coated (Ba, Sr, Ca on Ni), indirectly heated. There has been considerable developmental effort to increase emission density, to increase power output (since size is limited as noted above). Increased electric field strength, which also contributes to higher power, is related to cathode performance because the condition of the cathode surface often determines the voltage at which internal arcing occurs. Because rf displacement currents and emission currents can cause significant cathode heating, attempts are

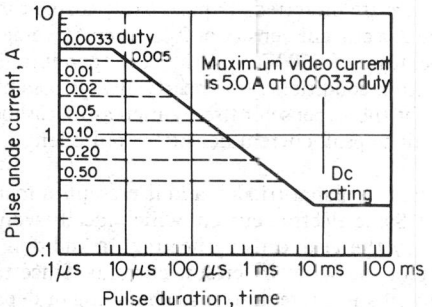

Fig. 9-31. Derating of peak pulse current with increasing pulse length. *(EIMAC Division of Varian Associates.)*

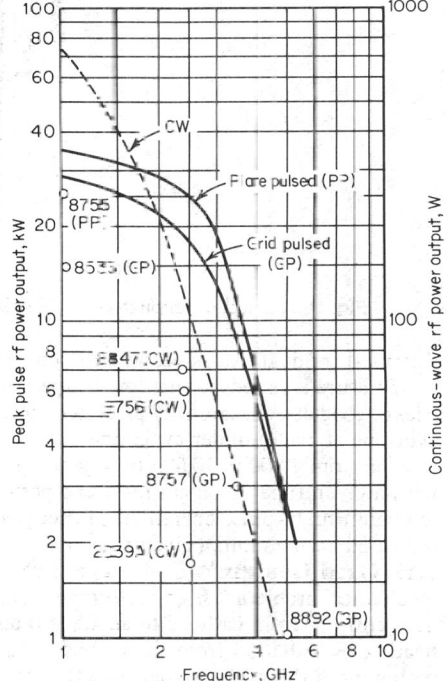

Fig. 9-32. Power output vs. frequency of typical planar triode and manufacturer's ratings of typical tubes.

made to reduce cathode-coating resistance. Mixing Ni powder with carbonates of Ba, etc., sometimes followed by application of pressure, is beneficial. Other successful approaches include the *dispenser cathode*, a metal sponge impregnated with Ba compounds. A recent development uses osmium doping.

18. Cathode Life. An objective of cathode development is to improve emission density and electric conductivity without decreasing tube life or increasing rate of evaporation of emitter material, which deposits on other electrodes and may actually block the perforations in very

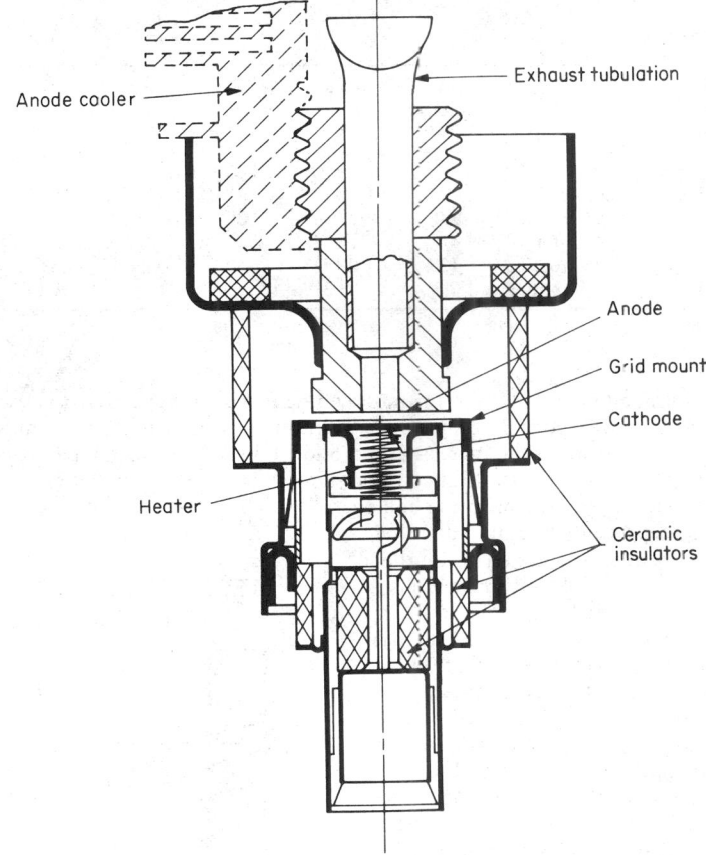

Fig. 9-33. Section through type 7289 planar triode *(EIMAC Division of Varian Associates.)*

fine mesh grids if excessive. The life tends to decrease (and evaporation to increase) as cathode temperature is raised, so that low-temperature emitters are preferred. Typical oxide-coated cathodes of good design, carefully processed, may operate at a current density of 0.25 A/cm² (averaged over one rf cycle) under cw conditions, at a temperature of 750°C, and the corresponding life may be from 2,000 to 10,000 h, depending on operating conditions. It is nearly always possible, with a given tube, to obtain increased performance at the expense of life by increasing cathode temperature. In pulse operation it is necessary to reduce peak current density as pulse length is increased, to maintain a given tube life.

19. Grid Design.[3,4] Grid design is very important in planar triodes, and it presents a major mechanical problem if high performance is required. Some electron current which ideally would flow entirely from cathode to anode is intercepted by the grid, causing heating, in addition to heating by radiation from the cathode and heating of the wires by circuit currents. Since the periphery of the grid is colder than the center, thermal stresses tend to cause buckling or departure from planeness if not counteracted. In some designs the counteracting means is by prestressing, in others by a rigid cellular supporting structure

Increasing the grid temperature not only tends toward deformation but also leads to emission of primary electrons, which affects performance adversely. Grids are usually made of wires of a refractory metal (W, Mo) for high tensile strength and thermal conductivity, coated with gold or other material to inhibit grid emission.

To obtain high performance, the grid must have the thinnest possible wires, spaced as close as possible to the cathode. This not only minimizes electron-transit-time losses but also maximizes amplifier power gain per stage, which is limited to approximately the tube-amplification factor because of the *grounded-grid* circuitry invariably used. Wires are typically 1 to 0.4 mil in diameter, and closest grid-to-cathode spacing is 2 to 3 mils.

20. Anode Design. Anodes are usually of Cu, the main function being to conduct the heat of electron bombardment to an external heat sink, usually air-cooled. A design problem exists because thermal expansion may change anode-to-grid spacing and capacitance, detuning the resonant circuit needed for efficient energy transfer to the load.

The internal structural details of the type 7289 planar triode are shown in Fig. 9-33.

21. Circuits for Planar Triodes. Planar triodes for 1 GHz and above are used in resonant-cavity circuits, of waveguide, coaxial line, or strip line. They confine the rf fields to the interior, minimizing radiation loss, and provide convenient and efficient means of presenting the correct impedance between tube terminals. Since the tube itself usually provides the major part of the shunt-capacitance reactance of the resonant circuit and also has substantial series inductive reactance, the tube and associated circuits must be considered together during the design process.

The basic amplifier circuit used is the *grounded-grid (common-grid, grid-separation, or cathode-driven)* type. At low frequency the equivalent circuit is shown in Fig. 9-34a. It is characterized by relatively low power gain, large input bandwidth, and high stability, resulting from the inherent large negative feedback due to the loading of the rf drive source by the in-phase rf component of the pulsating anode current.

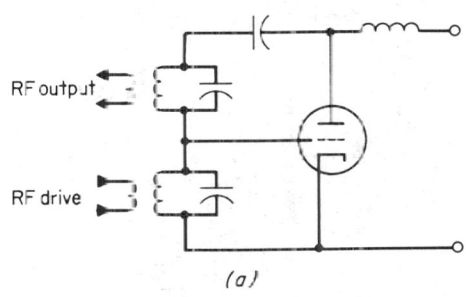

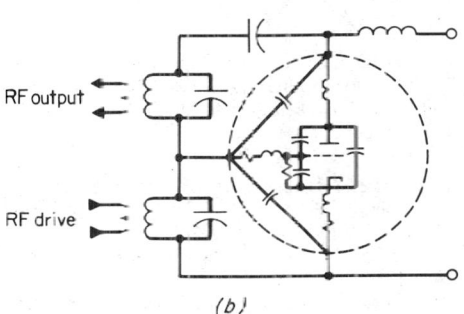

Fig. 9-34. Grounded-grid amplifier equivalent circuits: (a) at low frequencies; (b) at microwave frequencies. Cathode-heating and grid-bias circuits not shown. (*EIMAC Division of Varian Associates.*)

At microwave frequencies the equivalent circuit is made more complicated (Fig. 9-34b) by (1) the dominating influence of circuit elements inside the tube envelope and the capacitance of the envelope itself; (2) the effect of contact fingers between tube and circuit, which may have significant reactance; (3) the distributed reactance of cavity resonators and the tube itself; and (4) electron transit time, which causes resistive loading and phase shifts of considerable magnitude.

An example of the physical arrangement of tube, cavities, and coupling elements is shown in Fig. 9-35. This is an untuned amplifier. Tuned amplifiers require variation of cavity inductance with sliding contacts or noncontacting capacitive plungers, both of which introduce added circuit losses. The design of planar triode amplifiers has always been difficult and highly specialized and is more so today because of demands for large bandwidth and untuned circuits, which stem from similar characteristics obtained from solid-state amplifiers used to drive the tube amplifiers.

In general, a single stage will provide power gain of 5 to 10 dB, depending on bandwidth, frequency, and drive level. Stages can be cascaded to obtain more gain. Interstage coupling may use either waveguide or coaxial-line elements. A good example of waveguide technique is used in the TD2 transmitter operating at 4 GHz. Three stages have 18 dB gain over a 20-MHz bandwidth between points 0.1 dB down.[1]

The use of double- and triple-tuned circuits to improve bandwidth is discussed in Refs. 3 and 5, Par. **9-23**. In some applications where reduced frequency stability can be tolerated and small size is paramount, oscillators are used, e.g., in airborne altimeters. Useful rf power can be obtained from tubes as oscillators at frequencies higher than the maximum frequency for amplification with useful power gain. Circuits used are derived from the basic amplifier form in coaxial or strip-line construction, with the addition of positive feedback. An example is shown in Fig. 9-36.

22. Economics of Planar Triodes. Although planar triodes are outclassed technically in power capability by velocity-modulated tubes and in life potential by transistors and other solid-state devices, they continue to find favor with new-equipment designers for special applications where their performance is adequate, such as airborne IFF and navigational aids, both civil and military, and for space missions. This is not only because of their small size and weight and high tolerance to aircraft environment but because of their low first cost and replacement cost, resulting from high production rates and high yields in modern manufacturing plants and price competition in the industry. It is to be expected that triodes will be preferred for the power-frequency range shown in Fig. 9-32 in the foreseeable future.

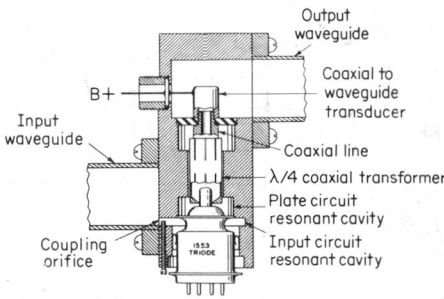

Fig. 9-35. Planar triode in waveguide cavity for TD-2 system amplifier. *(A. A. Roetken, K. D. Smith, and R. W. Friis, Bell Syst. Tech. J., October 1951, p. 1041.)*

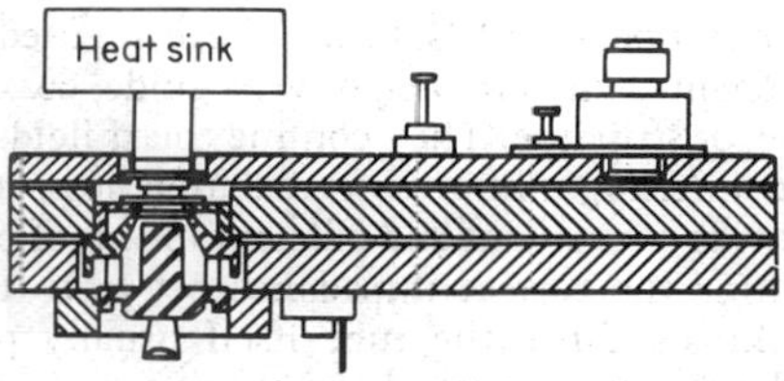

Fig. 9-36. Oscillator using strip-line construction with a planar triode. *(From U.S. Patent 3,596,130 to Melvin D. Clark.)*

23. References on Planar Tubes

1. Roetken, A. A., K. D. Smith, and R. W. Friis   The TD-2 Microwave Radio Relay System, *Bell System Tech. J.*, October 1951, p. 1041.

2. Preist, D.   Annular Circuits for High Power Multiple Tube RF Generators at VHF and UHF, *Proc. IRE*, May 1950.

3. Morton, J. A., and R. M. Ryder   Design Factors of the B.T.L. 1553 Triode, *Bell System Tech. J.*, October 1950, p. 496.

4. Bowen, A. E., and W. W. Mumford   A New Microwave Triode: Its Performance as a Modulator and as an Amplifier, *Bell System Tech., J.*, October 1950, p. 531.

5. Beggs, J. E., and N. T. Lavoo   High Performance Experimental Power Triodes, *IEEE Trans.*, May 1966, ED-13, No. 5, p. 502.

6. Gurewitsch, A. M., and J. R. Whinnery   Microwave Oscillators Using Disk Seal Tubes, *Proc. IRE*, May 1947.

7. "Klystrons and Microwave Triodes," Radiation Laboratory Series, Vol. 7, McGraw-Hill, New York, 1948.

Klystrons

BY RICHARD B. NELSON

24. Introduction. For high frequencies, linear-beam tubes overcome the transit-time limitations of grid-controlled tubes by accelerating the electron stream to high velocity before it is modulated. Modulation is accomplished by varying the velocity, with consequent drifting of electrons into bunches to produce rf space current. The rf circuits for coupling signals to and from the electron beam are generally integral parts of the tube. Two basic types are important today, klystrons and traveling-wave tubes (see Pars. **9-34** to **9-39**). Different versions are used as oscillators and amplifiers.

In a klystron, the rf circuits are resonant cavities which act as transformers to couple the high-impedance beam to low-impedance transmission lines. The frequency response is limited by the impedance-bandwidth product of the cavities but can be increased by stagger tuning and by multiple-resonance filter-type cavities.

Table 9-9 lists the principal types of linear-beam tubes, with typical power and bandwidth values.

25. Reflex Klystrons. In the reflex klystron a single resonator is used to modulate the beam and extract rf energy from it, making the tube simple and easy to tune. The beam passes through the cavity and is reflected by a negatively charged electrode to pass through again in the reverse direction. With proper phasing determined by applied voltages, oscillating modes occur for n + three-quarters-cycle transit time between passes through the cavity. The frequency can be modulated by varying voltage on the reflector (which draws no current). Figure 9-37 shows a schematic of a reflex klystron.

Most reflex tubes have two grids to concentrate the electric field of the cavity in the hole through which the beam passes. Some are tuned by deforming the cavity envelope to vary the spacing between grids, and hence the capacity loading. A tuning range up to 1.4:1 is thus obtainable.

Reflex tubes requiring stable output, even under shock and vibration, are tuned by a second resonant cavity outside the vacuum envelope, which is tightly coupled to the evacuated cavity through an iris. A capacitive post in the outer cavity tunes the circuit. The tuning range is typically 1.1:1.

Figure 9-38 shows the power-supply circuit, and Fig. 9-39 the operating parameters of a typical reflex klystron, the VA-244E.

Reflex klystrons are used as test signal sources, receiver local oscillators, pump sources for parametric amplifiers, and low-power transmitters for FM line-of-sight relays. Among microwave devices they are cheap and highly reliable. Reflex-tube frequencies cover the entire microwave range from 1 to 100 GHz. Current use is mainly from 4 to 40 GHz.

26. Two-Cavity Klystron Oscillators. In all klystrons except the reflex, the beam goes through each cavity in succession, and so there is no feedback. The tube is a buffered amplifier, with each stage isolated from those upstream. Electromagnetic feedback may be provided to make an oscillator.

The specialized two-cavity oscillator has a coupling iris in the wall between the cavities. This tube is more efficient and more powerful than the reflex klystron. It can be frequency-modulated by varying the cathode voltage around the center of the oscillating mode but requires more modulator power than a reflex klystron.

Two-cavity oscillators are used where moderate power, stable frequency, and low sideband noise are needed. Examples are the transmitter source in Doppler navigators, pumps and parametic amplifiers, and master oscillators for cw Doppler radar illuminators. To improve stability, the tubes are usually fixed-frequency or have at most a few percent tuning range.

Figure 9-40 shows the power and frequency of cataloged reflex klystron tubes. Figure 9-41 shows operating characteristics of a typical high-power, low-noise tube.

TABLE 9-9   Linear-Beam Tubes

Klystrons		Traveling-wave tubes		
Type	Watts	Type	Watts	Frequency range
Oscillators				
Reflex	0.01–1	Helix BWO	0.01–0.1	2:1
Two-cavity	1–100	Power BWO	0.1–1.0	1.4:1
Extended interaction (mm wave)	0.1–1			
Amplifiers				
Two-cavity	1–10	Low-noise helix	0.001–0.01	2:1
Multicavity, cw	10^3–10^5	Medium-power helix	1–1,000	2:1
Pulse	10^3–10^7	Ring-bar, pulse	10^3–10^5	1.3:1
		Coupled cavity, cw	10^3–10^4	1.1:1
		Pulse	10^3–10^6	1.1:1
Hybrid Twystron, pulse			10^6–10^7	1.1:1

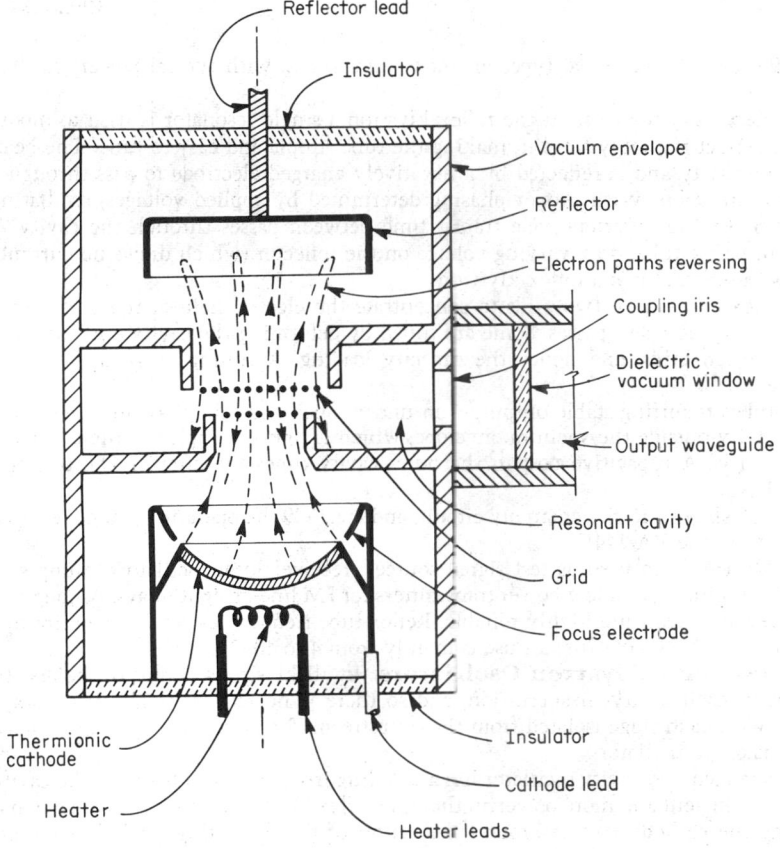

Fig. 9-37. Schematic cross section of a reflex oscillator. *(Varian Associates.)*

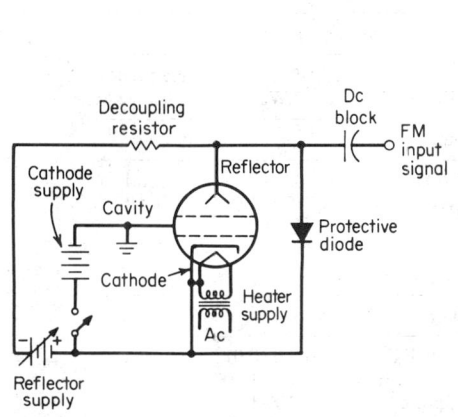

Fig. 9-38. Supply circuits for a reflex klystron. The reflector bias should be turned on before the cathode voltage. The protective diode prevents transient positive excursions of the reflector.

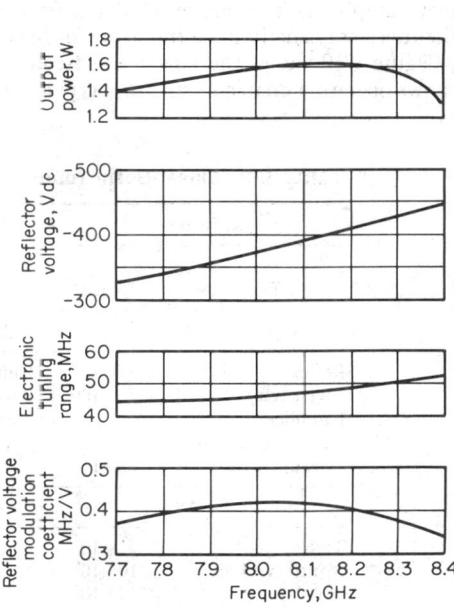

Fig. 9-39. Operating parameters of reflex klystron VA-244E: beam voltage, 750 V dc; heater voltage, 6.3 V; beam current, 78 mA dc; heater current, 0.75 A.

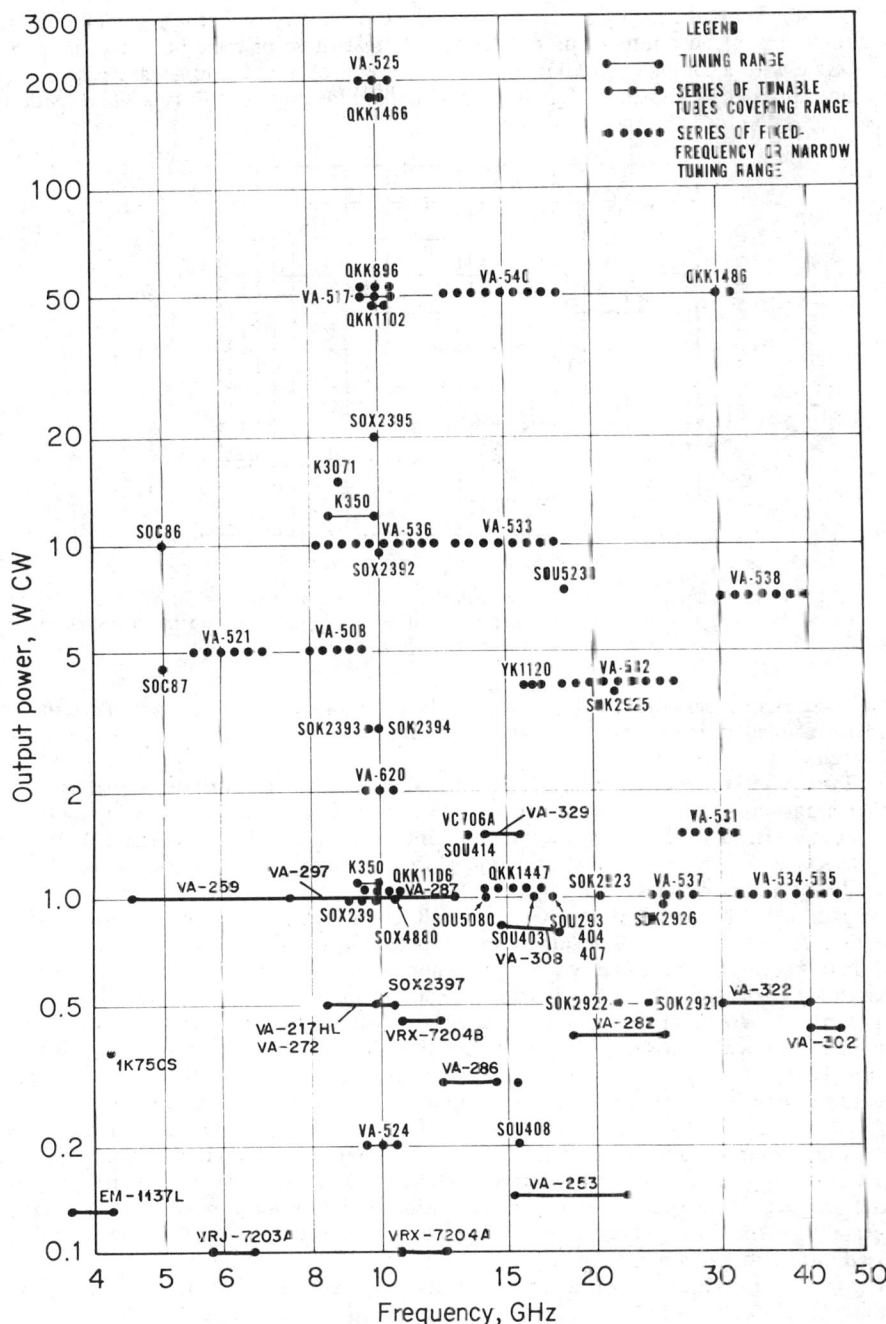

Fig. 9-40. Power and frequency of cataloged klystron tubes.

27. Extended-Interaction Oscillators. At millimeter wave frequencies the losses in klystron cavities make it hard to build up the impedance necessary to oscillate with the very small low-current beams required.

If a series of cavities are coupled together and interact sequentially with the beam in the proper phase, the total interaction impedance increases directly with the number of cavities. The circuits of extended-interaction oscillators resemble those of traveling-wave tubes. Since they operate with a complete standing wave (at the cutoff of the traveling-wave passband), the tubes can be classed as klystrons. Various names are used for tubes of this type. The Laddertron

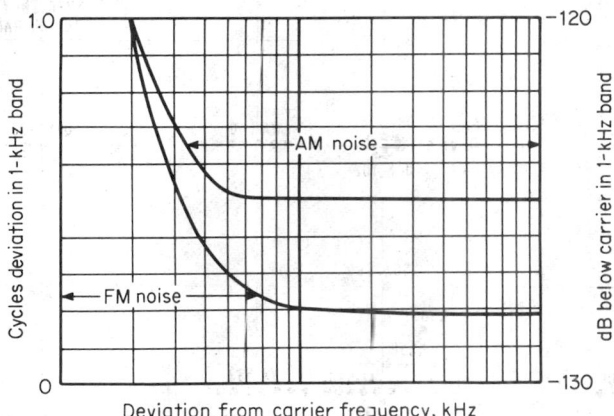

Fig. 9-41. Two-cavity oscillator noise characteristics, type VA-517: frequency = 10,000 ± 0.125 GHz, output power = 75 W, beam voltage = 10 kV dc, beam current = 90 mA dc, temperature coefficient = 200 kHz/°C, beam voltage coefficient = 1.8 kHz/V, coolant flow, water = 0.5 gal/min.

uses a ladder-shaped periodic circuit and a flat-ribbon electron beam. *Multicavity klystron oscillators* use coupled cavities and cylindrical beams. Most of the extended-interaction tubes are still experimental.

28. Two-Cavity Amplifiers. In the simplest klystron the driving signal is coupled through a transmission line to the input cavity. The cavity voltage produces velocity modulation of the beam. After a single drift space, the resultant density modulation induces current in the output resonator, from which power is extracted through another transmission line. The beam is usually focused electrostatically.

The gain of a two-cavity klystron is about 10 dB. Use is limited because more gain is desired in high-power tubes and solid-state amplifiers are available at low powers.

29. Multicavity Amplifiers. Downstream from the input cavity, cascaded intermediate cavities are inserted. They have no external coupling and are driven by the rf beam current and in turn remodulate the beam velocity. Figure 9-42 shows the structure schematically.

Each cavity tuned to the signal frequency adds about 20 dB of gain to the 10 dB of a two-cavity klystron. Net gain up to 60 dB is practical. The penultimate cavity is usually detuned to a higher frequency to improve efficiency by about 5%. Other intermediate cavities are often detuned, or *staggered*, to increase bandwidth, at the expense of gain. Up to eight cavities have been used.

Focusing. In multicavity amplifiers the electron beam is long and requires focusing forces to keep it uniformly small. A series of electrostatic lenses is used in a few tubes where light weight or stray magnetic fields are important. Most klystrons use a uniform magnetic field parallel to the beam. Permanent magnets are used for rf power below 5 kW. At higher power the tubes are inserted interchangeably in electromagnets.

Tuning. The cavities are tuned mechanically in several ways, depending on the range required. Some examples follow:

Fixed-frequency. Set on frequency by permanent deformation of a cavity wall.

3% tuning. One wall of the cavity is a thin diaphragm which is forced in and out by a mechanism outside the vacuum.

10% tuning. A movable interior cavity wall is not part of the vacuum wall but is moved through a flexible bellows.

(The above three methods vary the inductance of the cavity.)

   25% *tuning.* A paddle inside the cavity moves perpendicular to the beam and adds capacity across the interaction gap.

   35% *tuning.* (1) A combination of the above inductive and capacitive tuners inside the cavities. (2) The cavities extend outside the vacuum envelope, through dielectric windows and have movable walls with sliding-contact fingers.

   Cooling. Klystrons are cooled by air or liquid for powers up to 5 kW. Higher-power cooling is by boiling water (used in the United States only in television transmitters) or recirculating liquid.

   Performance. The klystron is a true linear amplifier from zero signal level up to 2 or 3 dB below saturated output. Figure 9-43 shows the gain characteristic. The efficiency at maximum output is 35 to 55%. Figure 9-44 shows the variation of saturated power and efficiency with applied beam voltage. The data are from the VA-884D, a 14-kW cw broadband amplifier designed for ground-to-satellite communication. The characteristics are typical of any well-designed klystron.

   Rf modulation is applied to the input drive signal. Amplitude modulation is usually limited to the linear portion of the gain characteristic, with consequent loss of efficiency, because the beam power is always full on. For frequency modulation the drive power is set for saturated output.

   Pulse modulation is obtained by applying a negative rectangular voltage pulse to the cathode instead of dc voltage. The rf drive (saturation value) is usually pulsed on for a slightly shorter time than the beam pulse.

   Bandwidth of a klystron is roughly proportional to the fifth root of the beam power and the ⅖ power of the perveance. Perveance is typically 1.0 to 2.0 $\mu A/V^{3/2}$.

   Bandwidth is increased by stagger tuning and sometimes by multiple-resonance filter-type out-

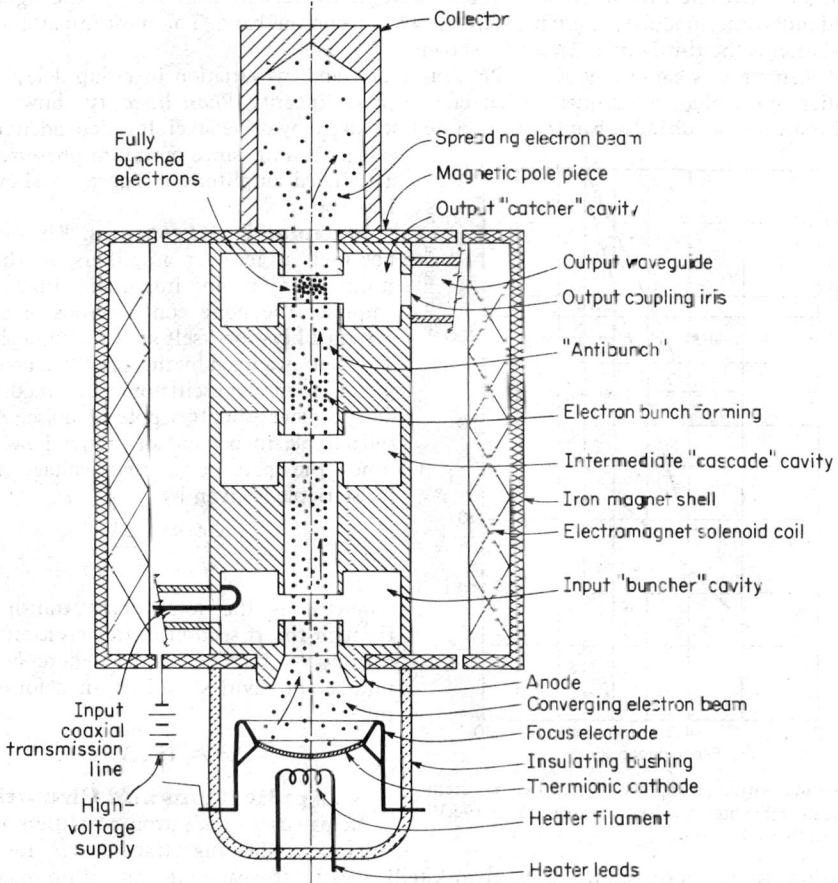

Fig. 9-42. Cross section of cascade klystron amplifier. *(Varian Associates,*

put cavity circuits. Figure 9-45 shows a stagger-tuned band characteristic. Figure 9-46 shows the interchange between gain and bandwidth for various degrees of stagger tuning.

Distortion.   Amplitude distortion is reasonably predictable because the nonlinearity of the gain characteristic is quite similar for all klystrons. Figure 9-47 shows the increase in output of the second harmonic with drive power. Other harmonics are similar but with rapidly decreasing

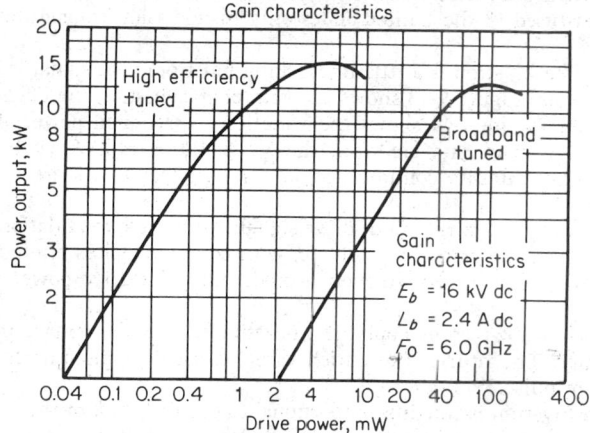

Fig. 9-43. Typical gain, output power and drive power characteristics.

amplitude above the third harmonic. When multiple carriers are combined in the signal, the second-order intermodulation products are always outside the band. The most important intermodulation is the third order $(2f_2 - f_1)$, shown in Fig. 9-48.

FM distortion is caused by AM-to-PM conversion and by variation in group delay due to deviation of the phase-vs.-frequency characteristic from linearity. *Phase linearity*, shown in Fig. 9-49 for a very broadband X-band tube, is quite unchanged by drive level. It is dependent on the stagger tuning, since optimum phase requires a rounded amplitude response, as shown on Fig. 9-50.

30. Noise in Klystrons. Klystrons are not used as receiver amplifiers, so that the *noise figure* is not important. In a power amplifier the noise contributions of a well-designed klystron itself are usually negligible. They are swamped by the amplified noise output of the master oscillator and by modulation due to power-supply ripple or noise. Amplitude ripple in percent of carrier power is ⅝ times the percentage beam-voltage ripple. Phase ripple is given by

$$\frac{\Delta\phi}{\phi} = \frac{1}{2}\frac{\Delta V}{V}$$

where ϕ is the total phase transit angle through the rf section of the klystron. Typically, $\phi \approx 10(N - 1)$ rad, where N is the number of cavities, and so, in a four-cavity tube,

$$\Delta\phi \approx 15\,\Delta V/V$$

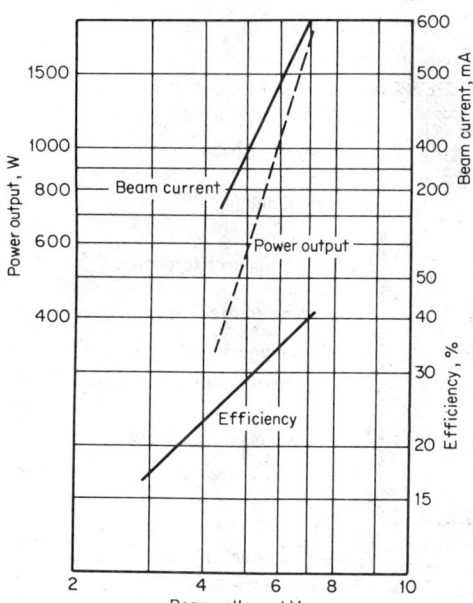

Fig. 9-44. Saturated rf power output, beam current and beam efficiency vs. beam voltage for a 15-kW klystron (VA-884 series).

31. Applications and Circuits. The principal uses of klystron amplifiers include microwave heating, transmitters for radar, television, tropospheric scatter, ground to satellite, earth to spacecraft, and illuminators for guided missiles.

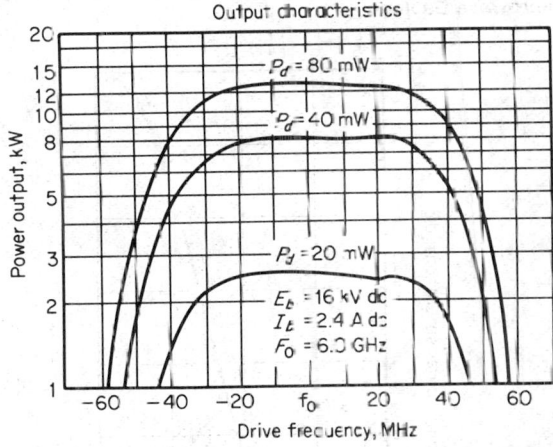

Fig. 9-45. Gain and output vs. frequency characteristics under saturated and unsaturated rf drive conditions.

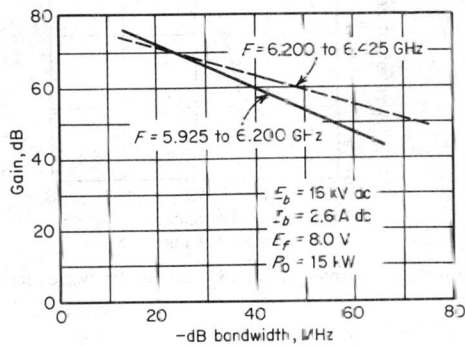

Fig. 9-46. Type VA-884 series klystron gain-bandwidth characteristics.

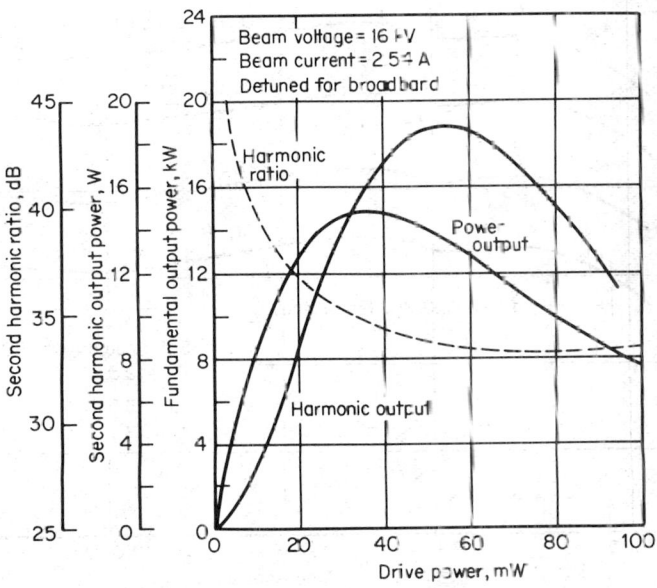

Fig. 9-47. Harmonic output of a typical klystron.

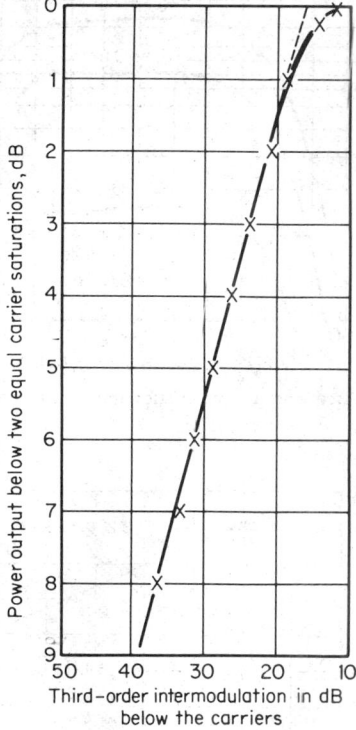

Fig. 9-48. Third-order intermodulation distortion under two equal carrier conditions.

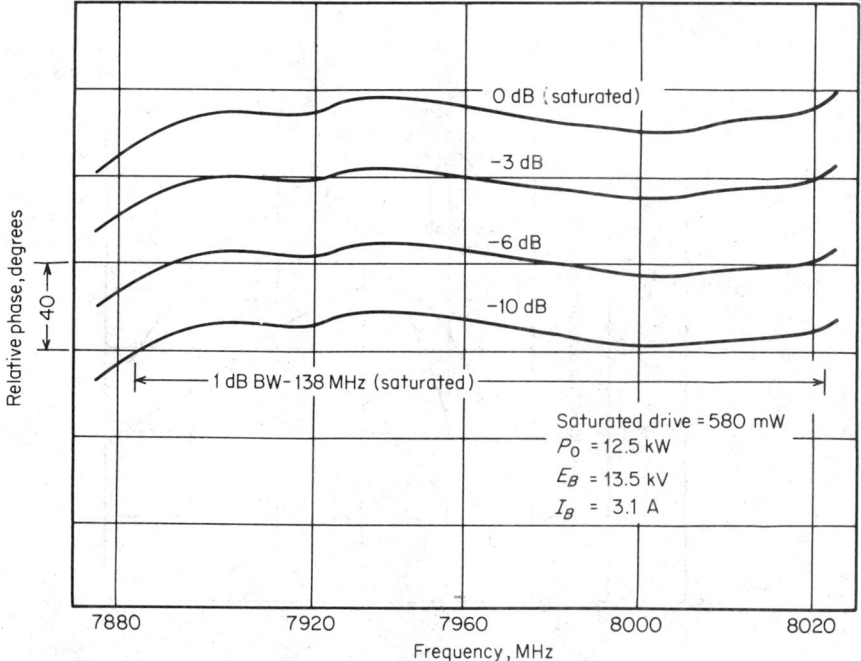

Fig. 9-49. Phase linearity of type VKX-7753 klystron. The deviation from linearity is unaffected by the drive level.

A typical power-supply schematic for a cw amplifier is shown in Fig. 9-51. Protective devices often required for klystron amplifiers include sensors to monitor cooling-air velocity, cooling-water flow, collector overtemperature, cathode-heating time delay, cathode overcurrent, body overcurrent, and output waveguide arcing.

32. Klystron Types. In Figs. 9-52 and 9-53 cataloged cw and pulsed klystrons are arranged by power and frequency coverage.

33. References on Klystrons

1. Harrison, A. E. "Klystron Tubes," McGraw-Hill, New York, 1947.
2. Pierce, J. R., and W. C. Shepherd "Reflex Oscillators," *Bell System Tech. J.*, July 1947, Vol. 26, No. 3, p. 460.
3. Beck, A. H. W. "Velocity Modulated Thermionic Tubes," Macmillan, New York, 1948.
4. Warnecke, R. R., and P. R. Guenard "Le Tube éctronique à commande par modulation de vitesse," Gauthier-Villars, Paris, 1951.
5. Hutter, R. G. E. "Beam and Wave Electronics in Microwave Tubes," Van Nostrand, New York, 1960.
6. Moreno, T. "High Power Axial Beam Tubes," *Adv. Electron. Electron Phys.*, 1961, Vol. 14, p. 299.

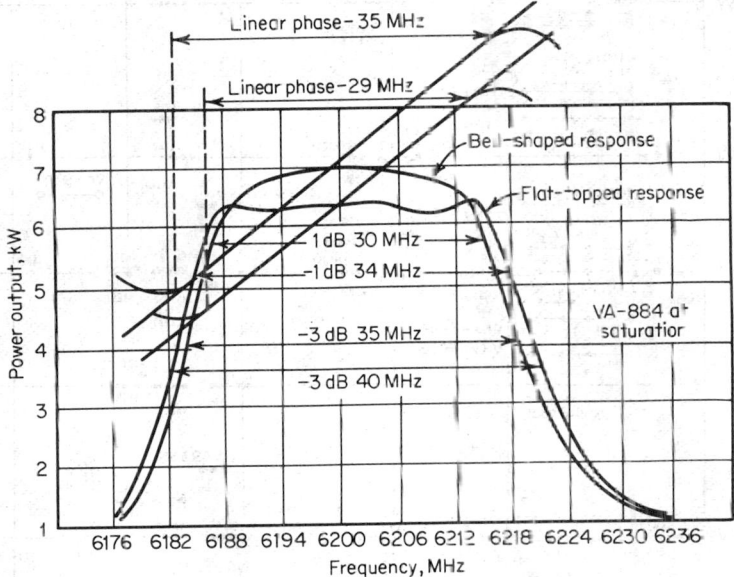

Fig. 9-50. Response curves for bell-shaped and flat-topped tuned amplifiers.

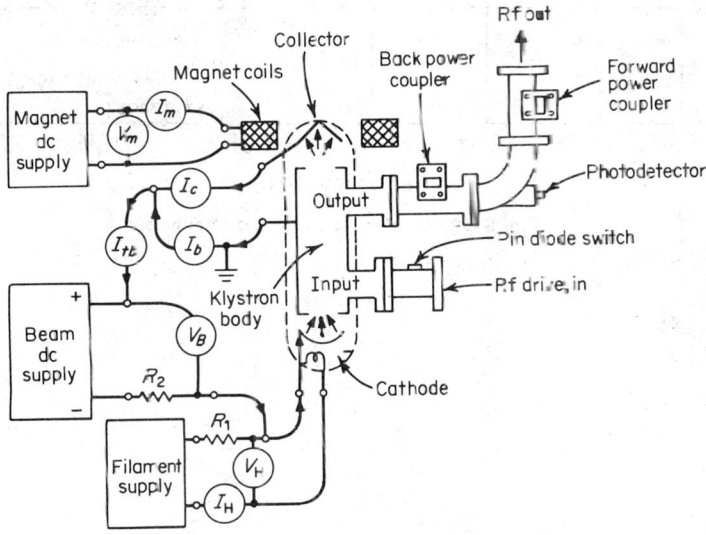

Fig. 9-51. Circuits for a cw klystron amplifier.

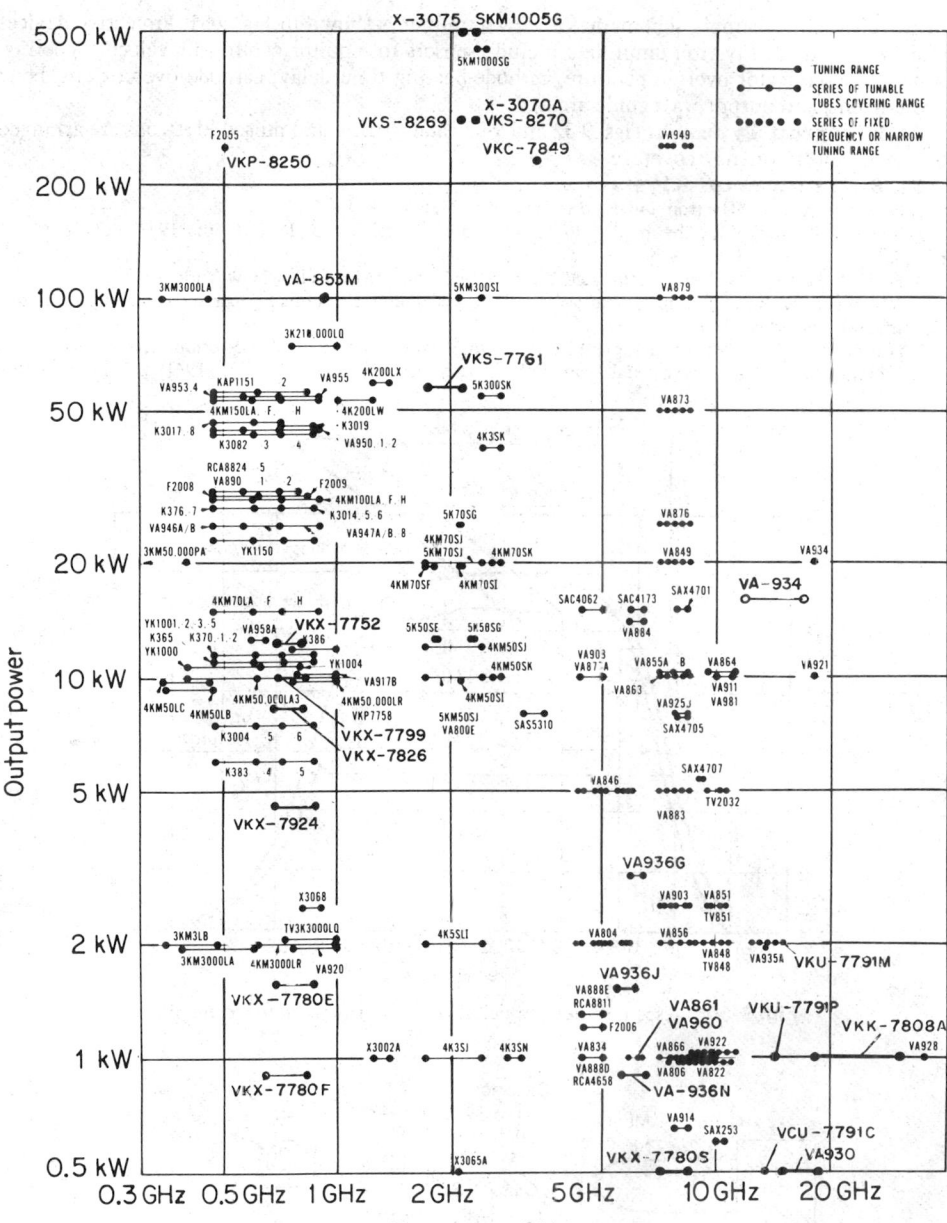

Fig. 9-52. Cataloged cw klystron amplifiers.

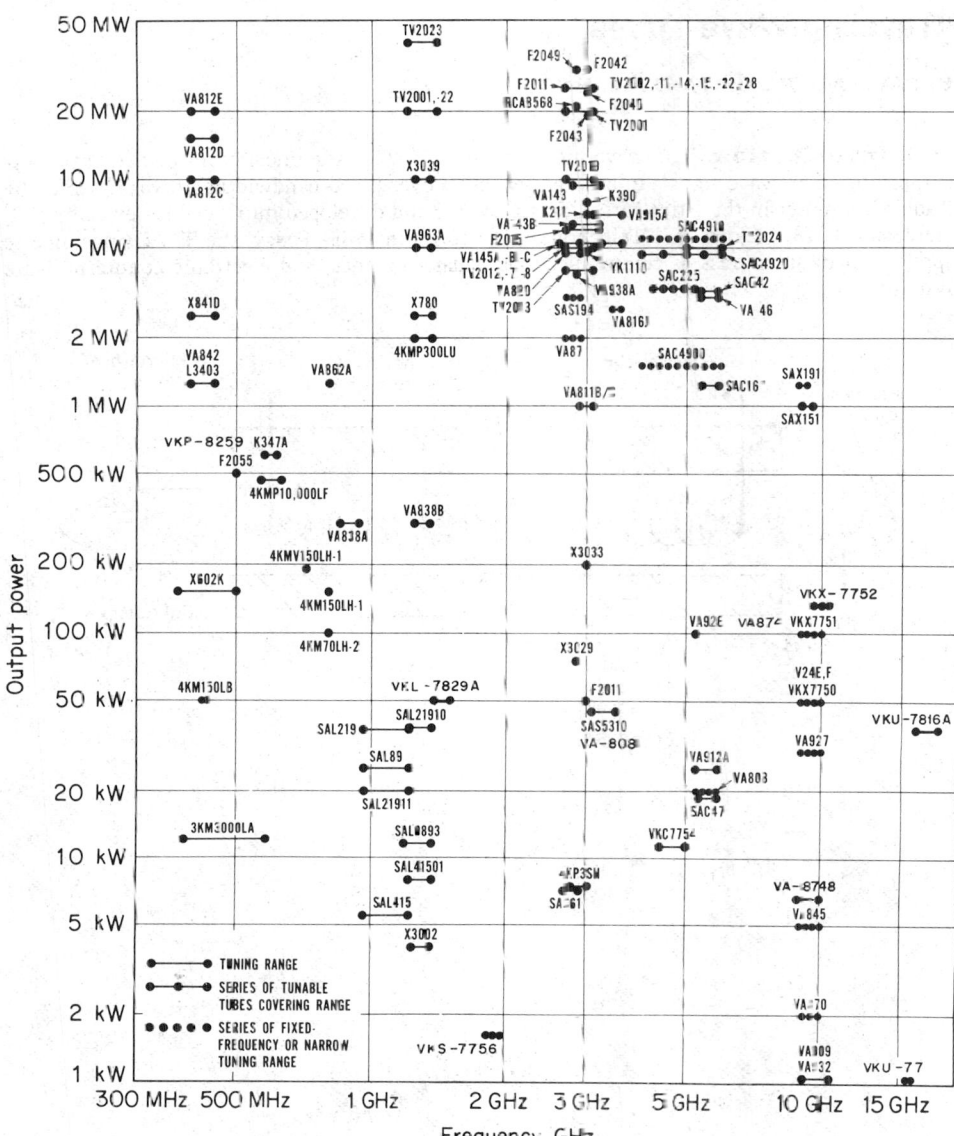

Fig. 9-53. Cataloged pulsed klystron amplifiers.

Traveling-Wave Tubes

BY PAMELA L. WALCHLI AND ALLAN SCOTT

34. Introduction. The traveling-wave tube (TWT) is a linear-beam device capable of amplifying microwave signals to high power levels over broad bandwidths. It was invented by Rudolf Kompfner in the latter part of World War II and developed into a viable device by J. R. Pierce and L. M. Field at Bell Telephone Laboratories in 1945. Today, the TWT finds diverse application in such areas as communications, radar, guidance, and electronic countermeasure systems.

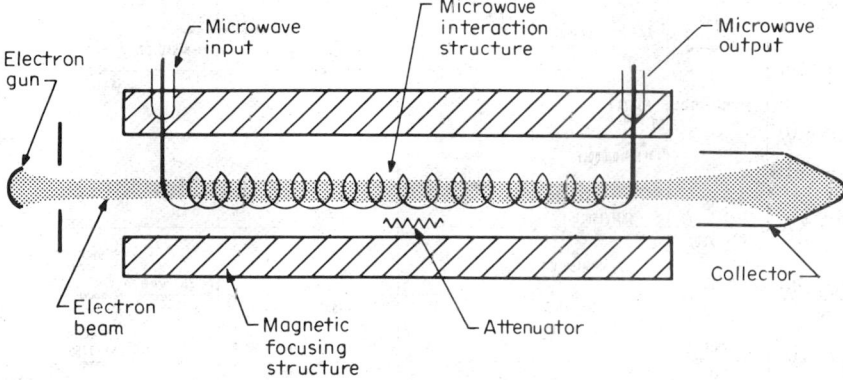

Fig. 9-54. Basic elements of a typical TWT.

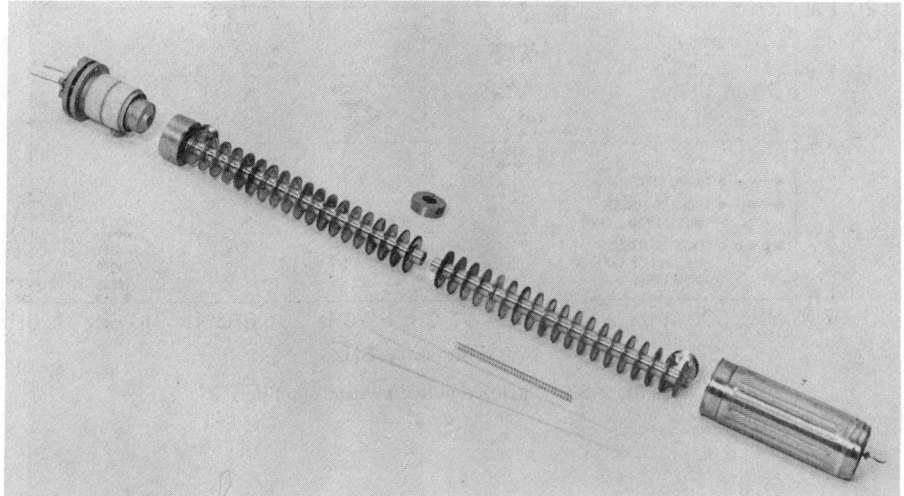

Fig. 9-55. Exploded view of internal structure of a TWT. *(Varian Associates.)*

Basic Structure.   All traveling TWTs use four basic elements: (1) an electron gun, (2) an rf interaction circuit, (3) a magnetic electron-beam focusing system, and (4) a collector to dissipate the remaining beam power.

The most pronounced difference between the various types of TWTs lies in the rf interaction structure used. A schematic representation of a typical TWT is shown in Fig. 9-54. At the left is the electron gun, which forms the beam; at the center are the rf interaction structure (in this case, a helix) and the magnetic beam-focusing system; on the right is the collector which absorbs the spent beam power. A partially exploded view of a TWT is shown in Fig. 9-55, with the gun at the left, collector at the right, and focusing structure in the center. In the foreground are a helix and three support rods, which form part of the interaction structure. One of the focusing magnets shows in the background.

Theory of Operation. The purpose of the interaction structure is to slow the rf signal so that it travels at the same speed as the electron beam. Electrons enter the structure during both positive and negative phases of an rf cycle. Those entering during a positive phase are accelerated; those entering during a negative phase are decelerated. The electrons which experience a velocity increase catch up with the electrons which have been slowed down, forming electron bunches. These bunches produce an alternating current superimposed on the dc beam current. The alternating current induces growth of the rf circuit wave, which, in turn, forms tighter electron bunches and thus a larger component of alternating current.

Growth of the wave on the circuit occurs because the velocity at which the beam is traveling forces the electron bunches to enter a decelerating phase of the rf field. In the decelerating field, the electrons are slowed, transferring their energy to the rf wave. This cycle is limited by one or more severs and ultimately by the extraction of the rf power through the output connector. The sever absorbs the rf power that has been built up on the circuit but does not affect the ac component of current in the beam.

The modulated beam drifts through the sever region and induces a new rf wave in the next circuit section, where the interaction process begins again. The purpose of the sever is to absorb reflected power, which travels in a backward direction on the circuit. The reflected power arises from an imperfect match between the rf circuit and the output connector. Without the sever, regenerative oscillations would be induced.

At any given frequency, a certain level of drive power will cause the maximum degree of bunching and thus the greatest amount of output power. This condition is known as *saturation.* For small drive signals, typical TWTs have 40 to 70 dB of gain.

35. The Electron Gun. The electron gun forms a high-current-density pencil beam of electrons, part of whose energy is converted to rf power through interaction with a wave traveling along the rf circuit. In the typical electron gun, electrons are emitted from a spherical cathode and converged to the required beam size by focusing electrodes. The final converged beam size is maintained through the interaction structure by either a permanent magnet or electromagnet focusing field.

On-off switching modulation of the electron beam is accomplished by applying a pulse to one of the following four electrodes: (1) cathode, (2) anode, (3) focus electrode, or (4) grid.

Cathode and Anode Pulsing. In the first method, the cathode is pulsed negatively with respect to the grounded anode, requiring both the full beam voltage and current to be switched. Alternatively, anode modulation involves switching the full beam voltage between cathode potential and ground, but the current switched is just that intercepted on the anode, usually only a few percent of the full beam voltage.

Focus-Electrode Pulsing. The focus electrode, which normally operates at or near cathode potential, can be biased negatively with respect to the cathode in order to turn the beam off. The voltage swing required is usually one-third or less of the full cathode voltage. Since the focus electrode draws no current, reduction in power requirements is significant.

Grid Pulsing. Switching power requirements are minimized with grid modulation. A grid structure, to which the modulating voltage is applied, is placed directly in front of the cathode surface. The amount of voltage swing needed is typically only one-twentieth or less of the full beam voltage.

Some common grid structures in use are shown in Fig. 9-56. The grid in Fig. 9-56a is a simple, single intercepting grid. To turn the beam on, a voltage positive with respect to the cathode is applied to the grid, drawing current from the full cathode surface. The grid webs intercept the current drawn from the cathode area directly behind the grid. This interception limits the duty cycle at which the tube can be run.

Current drawn by the grid can be minimized by schemes such as those shown in Fig. 9-56b and c. The structure in Fig. 9-56b is composed of two grids, the one nearest the cathode surface being operated at cathode potential and the outer one being biased positively. The inner grid, identical in pattern to the outer grid, prevents emission from the cathode surface directly behind it, effectively eliminating intercepted current on the control grid. This inner grid is referred to as a *shadow grid.* Beam optics are improved when the shadow grid is attached directly to the cathode surface, as in Fig. 9-56c. This kind of structure has a trade name Unigrid.

A simple grid has application in smaller guns, where high amplification factors are not required. The grid is operated at cathode potential and therefore intercepts no current. To turn off the beam, a voltage negative with respect to the cathode is applied to the grid. Amplification factors around 10 are typical for this kind of electron gun.

36. Magnetic Beam Focusing. Without a focusing system the electron beam would spread due to the mutually repulsive forces on like-charged particles, causing the electrons eventually to be intercepted on the rf circuit. A magnetic beam-focusing system is the most widely used and is usually implemented in one of three ways: (1) electromagnet, (2) permanent magnet, or (3) periodic permanent magnet.

Electromagnet Focusing. Electromagnet focusing is used primarily on very high power coupled-cavity tubes. Tight beam focusing is required in these tubes because significant interception on the rf circuit is intolerable at the power levels in question. Disadvantages of this kind of

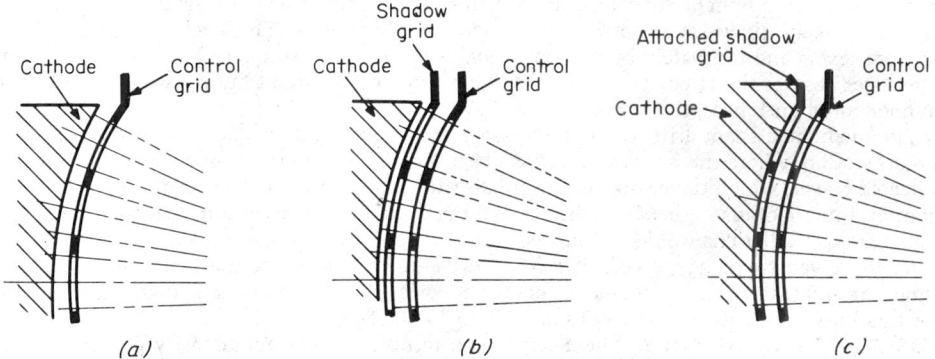

Fig. 9-56. Grid structures used in TWTs: (a) simple intercepting grid; (b) double grid, with shadow grid operated at cathode potential; (c) unigrid type, with shadow grid attached directly to cathode.

focusing are size, weight, and consumption of power, but they all can be reduced somewhat by wrapping the windings of the solenoid directly on the tube body. Solenoidal focusing is illustrated in Fig. 9-57a.

Permanent-Magnet Focusing. Permanent-magnet focusing is possible where the interaction structure is short, e.g., in low-gain or millimeter-wave tubes. It can be used in place of solenoidal focusing in these kinds of tubes. This focusing system is shown in Fig. 9-57b.

Periodic-Permanent-Magnet Focusing. Periodic-permanent-magnet focusing is used on almost all helix TWTs and most coupled-cavity tubes. A periodic-permanent-magnet (PPM) structure is shown in Fig. 9-57c. The magnets are arranged with alternate polarity in successive cells. In helix TWTs the pole pieces (with nonmagnetic spacers) may form the tube's vacuum envelope, or the pole pieces and spacers may be slipped over a stainless-steel tube which serves the same purpose. In coupled-cavity TWTs the cavity walls themselves are the pole pieces.

This kind of focusing provides a major reduction of tube size and weight, along with the elimination of the requirement for a magnet power supply. The drawback of this scheme is that the electron beam ripples with a periodicity of the length of one magnet cell. This increases beam interception on the rf circuit and thus generally limits the use of PPM focusing to lower average-power TWTs.

37. The Interaction Circuit. The fundamental principle of operation of a TWT is that an electron beam moving at approximately the same velocity as an rf wave traveling along a circuit gives up energy to the rf wave. Since the rf wave travels at the speed of light, a method must be found to slow the forward progress of the wave down to roughly the same velocity as that of an electron beam. The beam speed in a TWT typically falls between 10 and 50% of the velocity of light, corresponding to cathode voltages of 4 to 120 kV. The two structures which accomplish the slowing of the rf wave are the helical and coupled-cavity circuits.

Helix Circuits. A helix (Fig. 9-58) is supported inside the vacuum envelope by three or more ceramic support rods, which also conduct heat away from the helix. A helix interaction structure is used where bandwidths of an octave or more are required, since over this range the velocity of the signal carried by the helix is almost constant with frequency. For greater than octave-bandwidth operation, the variation of velocity with frequency can be modified by the introduction of metal loading segments near the helix, causing the phase velocities of a wider range of frequencies to be more nearly in synchronism with the beam velocity.

The helix provides satisfactory performance over the range of frequencies from 500 MHz up to 40 GHz. However, the typical helix circuit is limited in average power-handling capability to a few hundred watts. Peak power levels above several kilowatts cannot, in general, be achieved due to circuit instabilities. Higher peak power levels can be obtained by elimination of these oscillations with the use of a special type of helix circuit consisting of the superposition of a helix wound in a right-hand sense on a helix wound in a left-hand sense. Two practical implementations of this configuration are the ring-loop and ring-bar circuits (Fig. 9-59). Peak powers of hundreds of kilowatts are attainable, but average power capability is no better than that of the conventional helix circuits, since the structures are supported in a like manner. Because the ring-bar and ring-loop circuits are dispersive, the maximum bandwidth of a tube using them is only one-third octave.

Coupled-Cavity Circuits. Because of its superior ability to dissipate heat, the coupled-cavity structure is capable of both high peak and average power over moderate bandwidths. Coupled-cavity tubes find application from 2 GHz up to nearly 100 GHz. Bandwidths of 10% are typical, although tubes with 40% bandwidth have been developed.

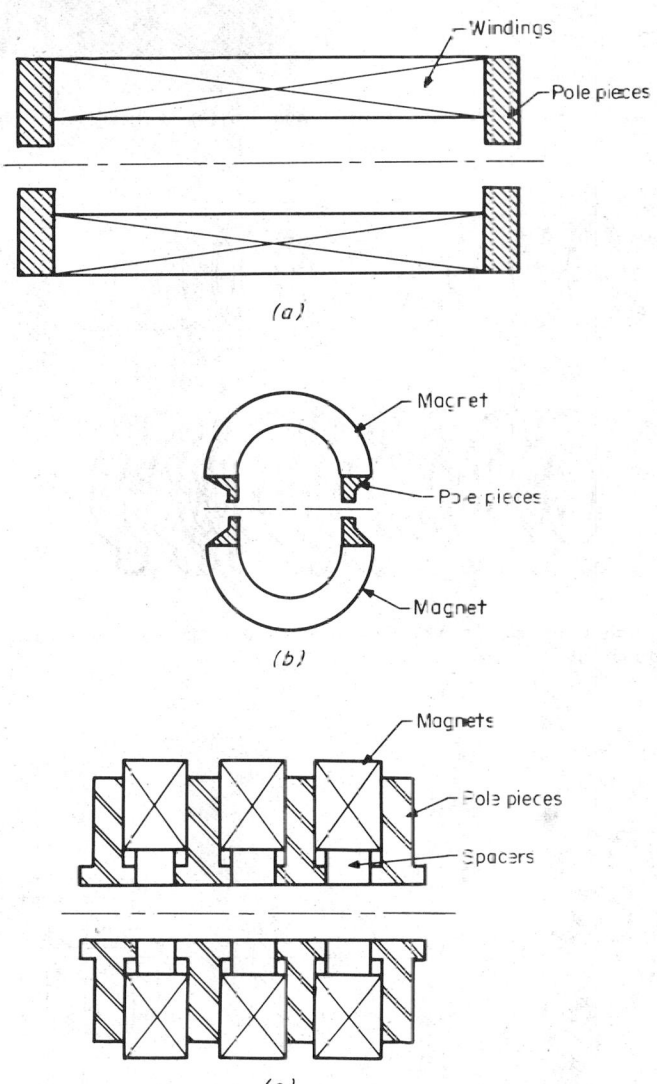

Fig. 9-57. Magnetic focusing arrangements: (*a*) solenoidal type; (*b*) permanent-magnet type; (*c*) periodic permanent-magnet structure.

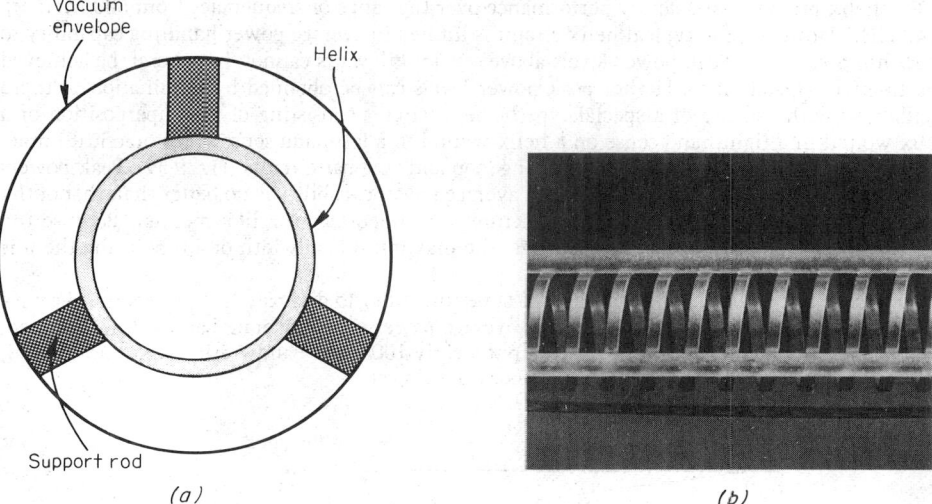

Fig. 9-58. Helix circuit: (a) end view; (b) side view.

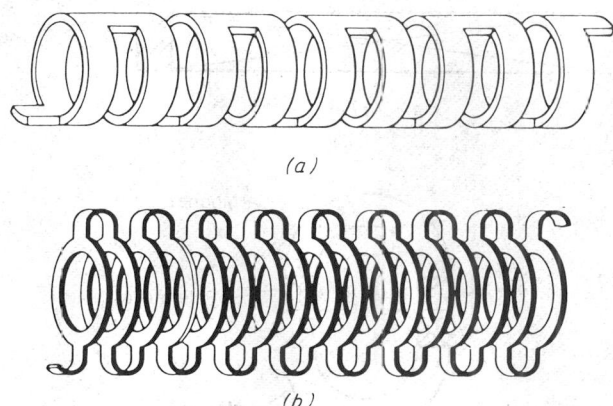

Fig. 9-59. Structures composed of two helixes superimposed in opposite senses of rotation: (a) ring-loop circuit; (b) ring-bar circuit.

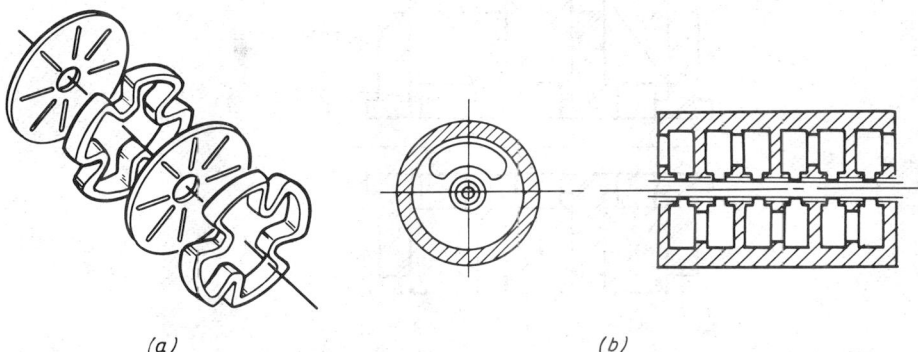

Fig. 9-60. Coupled-cavity circuits: (a) forward fundamental circuit ("cloverleaf"); (b) single-slot space harmonic circuit.

The coupled-cavity circuit consists of resonant cavities coupled through slots cut in the cavity end walls, resembling a folded waveguide. This arrangement results in a bandpass filter network that is highly dispersive, limiting the tube bandwidth. The two most common kinds of coupling schemes are illustrated in Fig. 9-60. The structure in Fig. 9-60a is a forward fundamental circuit, also called a cloverleaf circuit from the shape of its cavities. It is used primarily on extremely high peak power coupled-cavity tubes or in the output section of a hybrid kylstron TWT, known as a Twystron. Typical performance of a tube with a forward fundamental circuit is 3 MW peak and 5 kW average at S band. Figure 9-60b illustrates the more commonly used coupled-cavity structure, the single-slot space harmonic circuit. A peak power of 50 kW and an average power of 5 kW at X band are typical for space harmonic TWTs.

38. The Collector. The function of the collector is to collect the electron beam after it has passed through the interaction structure, dissipating the remaining beam energy. During interaction electrons give up various amounts of energy, and some actually gain energy. Typically, the slowest electrons lose no more than 50% of their original energy; the fastest gain at most 20%, the remainder being distributed between these extremes. If this TWT had an interaction efficiency of 20%, the average electron would possess 80% of its original energy.

Single-Stage Collectors. The overall efficiency of the TWT can be increased by operating the collector at a voltage lower than the full beam voltage, a practice known as *collector depression*. This introduces a potential difference between the interaction structure and the collector through which the electrons pass. The amount by which a single-stage collector can be depressed is limited by the remaining energy of the slowest electrons; i.e., the potential drop can be no greater than the amount of energy of the slowest electrons or they will be turned around and reenter the interaction structure, causing oscillations.

Multistage Collectors. Efficiency can be increased still more by introducing multiple depressed-collector stages. This method provides for the collection of the slowest electrons on one stage, while allowing those with more energy to be collected on other stages depressed still further. Figure 9-61a and b represents the configuration of power supplies to operate a gridded TWT with a single-stage and a multistage depressed collector. respectively. Calculations of the overall efficiency of such TWTs are shown in the following table, assuming a beam power of 5 kW (10 kV, 0.5 A) and an interaction efficiency of 15%:

	Voltage, kV	Current, A	Power, W
Single-stage collector			
Helix supply	10	0.025	250
Collector supply	5	0.475	2,375

$$\text{Overall efficiency} = \frac{750 \text{ W}}{250 \text{ W} + 2375 \text{ W}} = 29\%$$

	Voltage, kV	Current, A	Power, W
Multistage collector			
Helix supply	10	0.025	250
Collector stage, 1	5	0.23	1,150
2	2.5	0.15	375
3	1	0.085	85
4	0	0.01	0

$$\text{Overall efficiency} = \frac{750 \text{ W}}{250 \text{ W} + 1150 \text{ W} + 375 \text{ W} + 85 \text{ W}} = 40\%$$

Cooling of helix TWTs is accomplished by clamping the tube to a metal baseplate, mounted in turn, on an air- or liquid-cooled heat sink. Coupled-cavity tubes below 1 kW average power are cooled convectively by drawing air over the entire tube length. Higher-power coupled-cavity tubes are cooled by circulating liquid over the tube body and collector.

A finished, packaged helix TWT is shown in Fig. 9-62a. An air-cooled coupled-cavity tube in its final package is shown in Fig. 9-62b.

39. References on Traveling-Wave Tubes.

1. Pierce, J. R. "Traveling-Wave Tubes," Van Nostrand, New York, 1950.

2. Gittins, J. "Power Traveling Wave Tubes," Elsevier, New York, 1965.

3. Gewartowski, J. W., and H. A. Watson "Principles of Electron Tubes," Van Nostrand, Princeton, N.J., 1965.

4. Staprans, A., E. E. McCune, and J. A. Ruetz High Power Linear Beam Tubes, *Proc. IEEE*, March 1973, Vol. 61, No. 3.

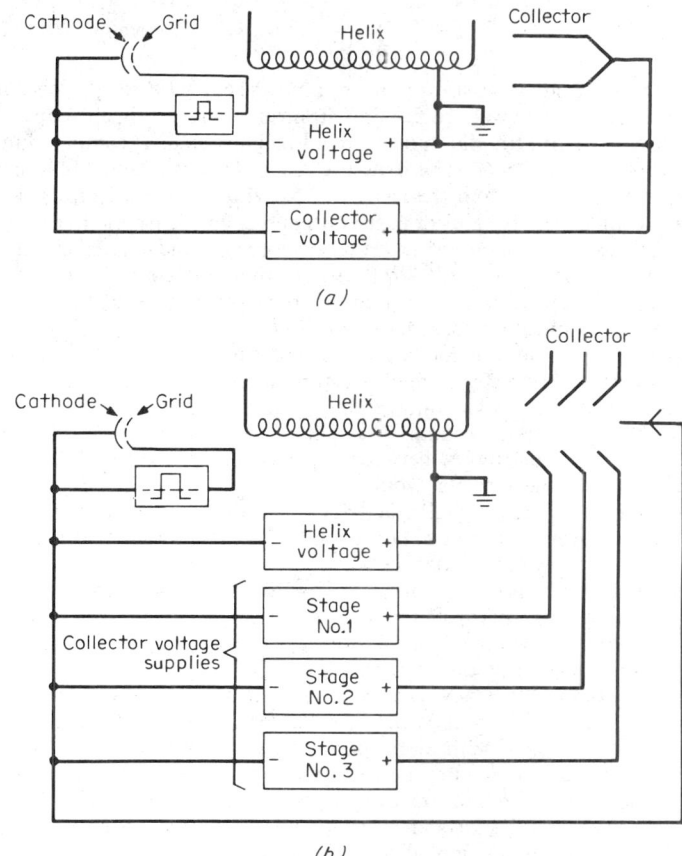

Fig. 9-61. Power supplies for TWTs: (a) single collector TWT; (b) multistage depressed collector TWT.

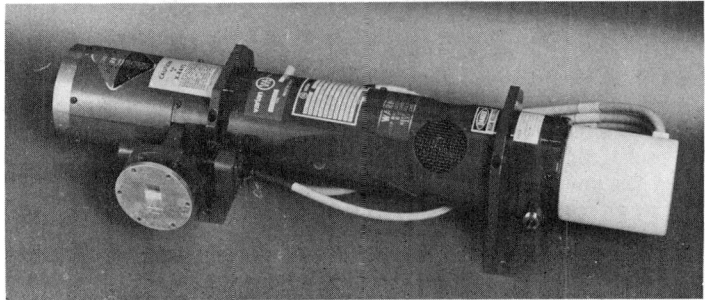

Fig. 9-62. Packaged traveling-wave tubes: (a) liquid-cooled type; (b) air-cooled type. *(Varian Associates.)*

Crossed-Field Tubes

BY GEORGE K. FARNEY

40. Crossed-Field Interaction Mechanism. A crossed-field microwave tube is a device that converts dc electric power into microwave power using an electronic energy-conversion process similar to that used in a magnetron oscillator. These devices differ from beam tubes in that they are potential-energy converters rather than kinetic-energy converters. The term *crossed field* is derived from the orthogonality of the dc electric field supplied by the source of

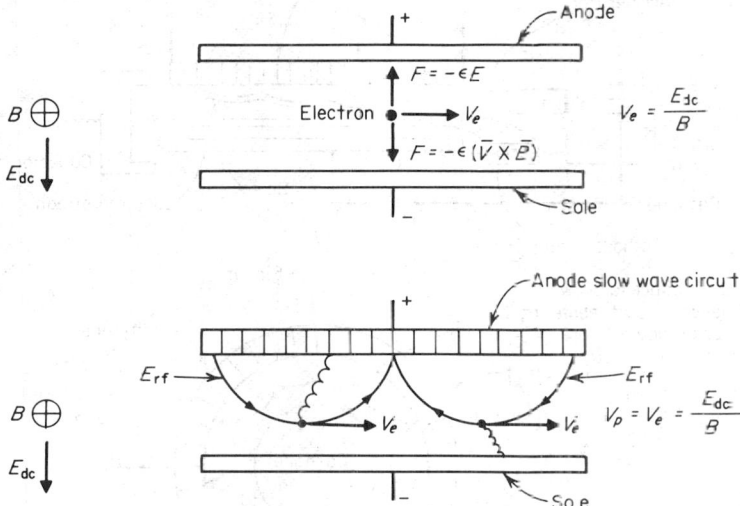

Fig. 9-63. Forces exerted on a moving electron in a crossed-field environment.

dc electric power and the magnetic field required for beam focusing in the interaction region. Typically, the magnetic field is supplied by a permanent-magnet structure. These tubes are sometimes called *M tubes.*

The electronic interaction is illustrated schematically in Fig. 9-63. Electrons moving to the right in the figure experience electric field deflection forces ($f_e = -\epsilon E$) toward the electrically positive anode, while the magnetic deflection forces ($f_m = -\epsilon V \times B$) resulting from the motion of the negatively charged electron in the orthogonal magnetic field cause deflection toward the negative electrode. This electrode is also called the *sole.*

The forces are balanced when an electron is traveling in a parallel direction between the electrodes with a velocity numerically equal to the ratio of the dc electric field to the magnetic field ($V_e = E/B$). Any alteration of the electron velocity leads to an unbalanced condition. Reduction of the electron forward motion causes the magnetic deflection force to become less, and the electron trajectory is deviated toward the positive electrode. Conversely, an increase of velocity causes a greater magnetic deflection force, which causes trajectory deviation toward the negative electrode.

Electronic interaction with a traveling wave occurs when the positive electrode is an rf-guiding slow-wave circuit whose phase velocity for the traveling wave is numerically equal to the ratio of the dc electric field to the magnetic field ($V_p = E/B$). Under these conditions synchronous interaction occurs between the rf fields on the slow-wave circuit and the stream of electrons traveling in the interaction region.

Two general kinds of motion result, as illustrated schematically in Fig 9-63, where a moving frame of reference is shown traveling from left to right at a velocity equal to the phase velocity of the circuit wave, so that the instantaneous rf fields are seen as stationary. The electronic motion resulting from interaction with the tangential components of the additional rf electric fields depends on the location of the electron relative to the phase of the rf fields of the slow-wave circuit. Those located so that their forward motion is retarded by the rf electric field are slowed, and the energy they lose is transferred to the rf wave on the circuit. These slower-moving electrons are subsequently accelerated toward the anode by the dc electric field, and their velocity is increased to the synchronous condition. The energy-exchange cycle can then be repeated. Electrons moving in this phase of the rf field pattern transfer energy to the rf wave on the circuit while maintaining nearly constant kinetic energy. The energy transfer results from the loss of potential energy of the electrons as they move to the anode.

Electrons located in the alternate phase of the rf field pattern are accelerated by the rf field and move away from the anode. The intensity of the slow wave decreases exponentially with distance away from the slow-wave circuit so that the magnitude of this interaction decreases.

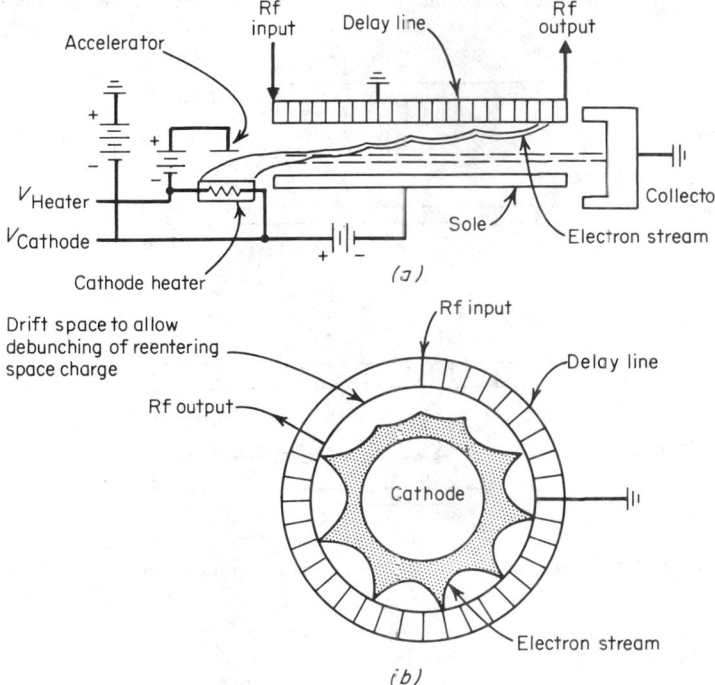

Fig. 9-64. (*a*) Linear injected beam and (*b*) reentrant emitting-sole crossed-field amplifier.

The result is the transfer of dc electric power to microwave power on the slow-wave circuit, with the phase-sorted electrons in the crossed-field interaction region providing the necessary coupling mechanism. Electron current thus flows to the anode only in the region of suitably phased rf electric fields.

The components of the rf field which are perpendicular to the forward motion of the electrons exert forces which phase-lock the electron near the center of the pattern. These regions are called *spokes* because of the similarity, in a magnetron oscillator, to the spokes in a rotating wheel. The phase locking of the sorted space charge relative to the traveling rf wave on the slow-wave circuit reduces the effect of power-supply variations on the electron trajectories. The details of the electron trajectories are extremely complex and have been calculated only approximately, using sophisticated computer techniques.

It is an important fundamental of crossed-field interaction that very high electronic conversion efficiency can be obtained because the kinetic energy of the electrons lost as heat upon ultimate impact with the slow-wave circuit can be designed to be a small fraction of the total potential energy transferred from the power supply. The ideal electronic conversion efficiency is given by $\eta = 1 - V_0/V$, where η is efficiency, V_0 is the synchronous voltage, and V is the cathode-to-anode voltage. Large ratios of V/V_0 lead to high efficiencies.

Many microwave devices have been investigated which are variations of this basic concept, but a majority have not proceeded beyond laboratory evaluation. Devices in practical use fall into two broad categories, which differ primarily in the method by which the electron current is delivered to the interaction region. These are illustrated schematically in Fig. 9-64. In one class, called the *injected-beam crossed-field tubes*, the electron stream is produced by an electron gun located external to the interaction region, rather like a traveling-wave tube. In the second class, the *emitting-sole tubes*, the electron current for interaction is produced directly within the interaction region by secondary electron emission, which results when some electrons are driven to the negative electrode and allowed to strike it. In the latter devices, this electrode is formed from a material capable of delivering copious secondary-emission electrons.

41. Slow-Wave Circuits for Crossed-Field Tubes. Electron current in crossed-field interaction moves toward the slow-wave circuit rather than through the circuit as in beam tubes. This leads to the use of open circuits that present an rf waveguiding surface to the electron stream. Maximum energy conversion efficiency is usually obtained when the current is intercepted on the slow-wave circuit; so the structures must withstand the thermal stress associated with electron bombardment. Electronic interaction can occur using either forward-wave or backward-wave traveling-wave circuits, as well as with circuits supporting a standing wave. Example of circuits suitable for use in forward-wave interaction are various *meander lines, helix-derived structures, bar* and *vane structures,* which are capacitively loaded by ground planes, and *capacitively strapped bar circuits.*

A helix-coupled vane circuit and a ceramic-mounted meander line are shown in Fig. 9-65a. The most common backward-wave circuits are derivatives of the interdigital line and strapped-bar and vane circuits. Examples of a choke-supported interdigital line and a strapped-bar circuit are shown in Fig. 9-65b. Traveling-wave circuits are used mostly in amplifiers. Standing-wave circuits are resonant and used typically in magnetron oscillators. The most commonly used standing-wave circuits are composed of arrays of quarter-wave resonators that may or may not be strapped for improved oscillating-mode stability.

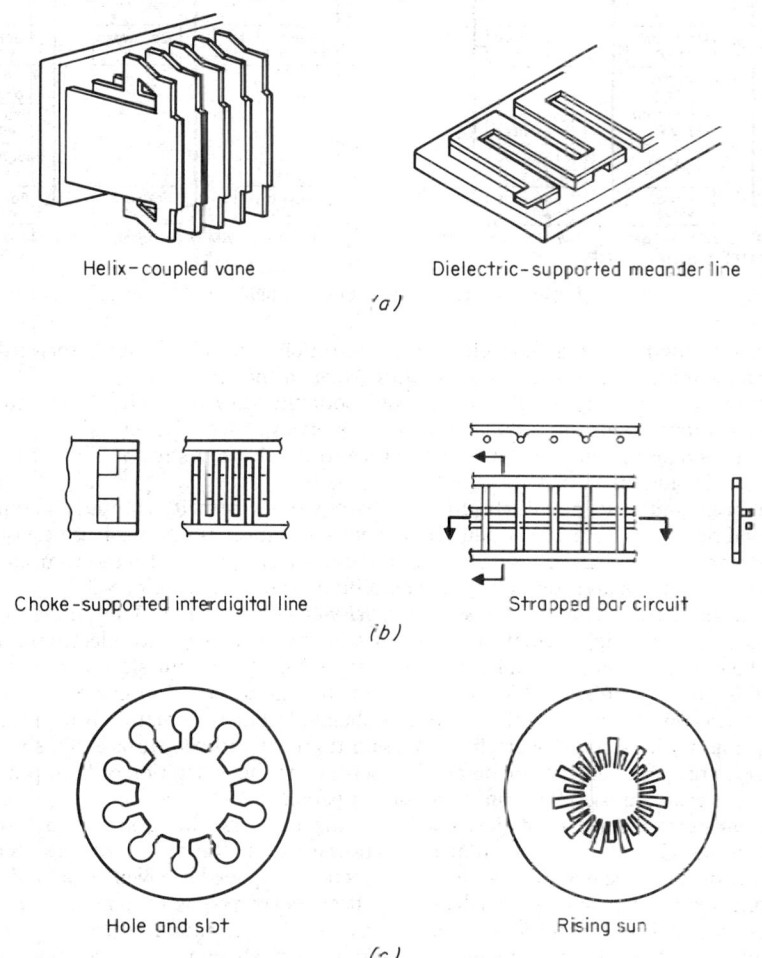

Helix–coupled vane

Dielectric–supported meander line

(a)

Choke–supported interdigital line

Strapped bar circuit

(b)

Hole and slot

Rising sun

(c)

Fig. 9-65. Slow-wave circuits for crossed-field tubes.

Variations of these circuits include *hole-and-slot resonators* and *rising-sun anodes*. Examples are shown in Fig. 9-65c. Cooling of vane structures for high average power is obtained by heat conduction along the vanes to the back wall of the anode to a heat sink, which may be liquid- or forced-air-cooled. Bar structures are cooled by passage of liquid coolant through the tubular bars of the slow-wave circuit.

42. Crossed-Field Family Tree. The current crossed-field tube types can be classified in a family tree (Fig. 9-66), divided into two major divisions as *injected-beam* and *emitting-sole* tubes. Subdivisions relate to *linear* and *circular format, reentrant* and *nonreentrant* electron

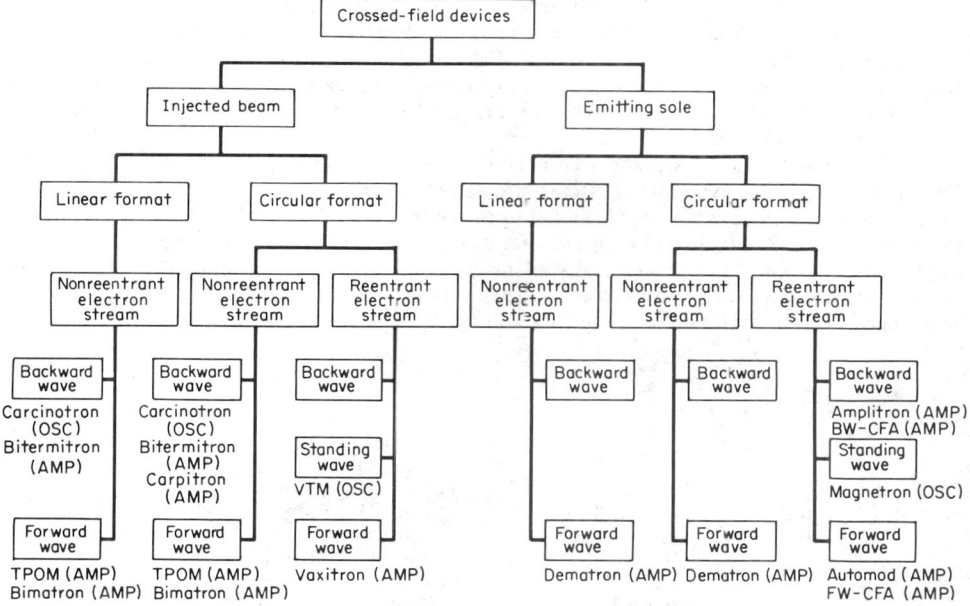

Fig. 9-66. Family tree of crossed-field tubes.

stream, and to whether the rf-wave electronic interaction takes place with a *forward-traveling wave*, a *backward-traveling wave*, or a *standing wave* on the circuit.

Magnetron oscillators are single-port devices. Both the slow-wave circuit and the electron stream are reentrant; i.e., the circular geometry is always used. Traveling-wave crossed-field oscillators are single-port devices but use a nonreentrant electron stream. They use either the linear or circular format.

Injected-beam and emitting-sole amplifiers are two-port devices with rf input and output ports. They are fabricated in both linear and circular format. Linear tubes must use a nonreentrant electron stream. Some circular-format amplifiers use a reentrant electron stream and some do not. Both forward-wave and backward-wave amplifiers have been developed.

43. Crossed-Field Oscillators. *Conventional Magnetrons.* The conventional magnetron is an emitting-sole, circular-format, reentrant-stream device with electronic interaction between the circulating current and a π-mode, rf standing wave on the slow-wave circuit. Oscillation builds up from noise contained initially in thermionic-emission current from a heated cathode. During operation interaction current is obtained from a circulating hub of space charge supplied primarily by secondary electron emission from the cathode surface. This is illustrated in Fig. 9-67. Large peak currents are obtainable, permitting the generation of high peak power at lower voltages than are used for beam tubes of comparable peak power.

Pulsed magnetrons have been developed covering frequency ranges from a few hundred megahertz to 100 GHz. Peak power from a few kilowatts to several megawatts has been obtained with typical overall efficiencies of 30 to 40%, depending upon the power level and frequency range. Continuous-wave magnetrons have also been developed with power levels of a few hundred watts, in tunable tubes, at an efficiency of 30%. As much as 25 kW cw has been obtained for a 915-MHz fixed-frequency magnetron at efficiency greater than 70%. Figure 9-68 illustrates typical performance.

Pulsed magnetrons are used primarily in radar applications as sources of high peak power. Low-power pulsed magnetrons find applications as beacons. Magnetrons operate electrically as a simple diode, and pulsed modulation is obtained by applying a negative rectangular voltage pulse to the cathode with the anode at ground potential. Voltage values are less critical than for beam tubes, and line-type modulators are often used to supply pulsed electric power. Tunable cw magnetrons are used in electronic countermeasure applications. Fixed-frequency magnetrons are used as microwave heating sources.

Mechanical tuning of conventional magnetrons is accomplished by moving capacitive tuners, near the anode straps or capacitive regions of the quarter-wave resonators, or by inserting symmetrical arrays of plungers into the inductive portions. Tuner motion is produced by a mechanical connection through flexible bellows in the vacuum wall. Tuning ranges of 10 to 12% bandwidth are obtained for pulsed tubes and as much as 20% for cw tubes.

44. Coaxial Magnetrons. The frequency stability of conventional magnetrons is affected by variations in the microwave load impedance (frequency pulling) and by cathode current fluctuations (frequency pushing). When the mode control becomes marginal, the tube may occasionally fail to produce a pulse. The coaxial magnetron minimizes these effects by using the anode geometry shown in Fig. 9-69. Alternate cavities are slotted to provide coupling to a surrounding coaxial cavity. π-mode operation of the vane structure provides in-phase currents at the coupling slots which excite the TE_{011} circular electric coaxial mode. The unique rf field pattern of the circular electric mode permits effective damping of all other cavity modes with little effect on the TE_{011} mode, and oscillation in other cavity modes is thereby prevented. Additional resistive damping is used adjacent to the slots but removed from the vanes to prevent oscillations in unwanted modes associated with rf energy stored in the vanes and slots that does not couple to the coaxial cavity.

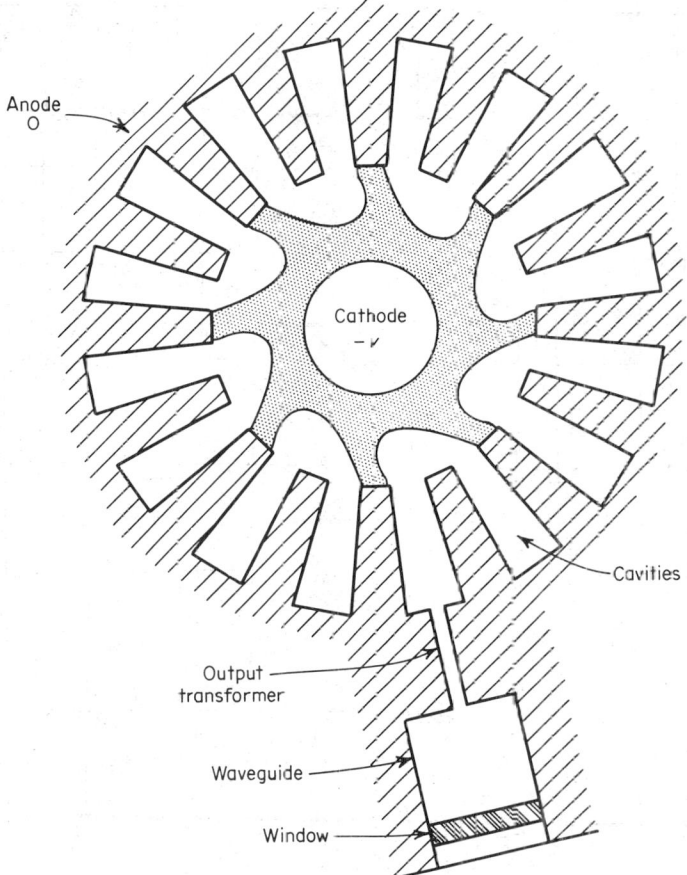

Fig. 9-67. Conventional magnetron structure.

The oscillation frequency is controlled by the combined vane system and resonant cavity. Sufficient energy is stored in the TE_{011} cavity to provide a marked frequency-stabilizing effect on the oscillation frequency. Hence the coaxial magnetron is much less subject to frequency pushing and pulling than conventional magnetrons, and it exhibits fewer missed pulses. Tunable versions of this tube type are tuned by a movable end plate in the coaxial cavity similar to a tunable coaxial wavemeter. This is illustrated in Fig. 9-70. The larger resonant volume for energy storage leads to a slower buildup time for oscillation than in conventional magnetrons. This causes greater statistical variation in the starting time for oscillation (leading-edge jitter).

These factors are compared in Table 9-10 for the SFD-349 coaxial magnetron, which was designed as an improved retrofit for the 7008 conventional magnetron. The operating efficiency of the SFD-349 was deliberately degraded to meet retrofit requirements. Typically, coaxial magnetrons operate with an efficiency of 40 to 50% or higher.

Spurious Noise.   The circulating space charge in the hub of both conventional and coaxial magnetrons contains wide-band noise-frequency components that can couple to the output. In conventional magnetrons this spurious noise can couple directly to the output waveguide. Spurious noise power measured in a 1-MHz bandwidth is typically greater than 40 to 50 dB below the carrier. The coaxial cavity in the coaxial magnetron provides some isolation between the

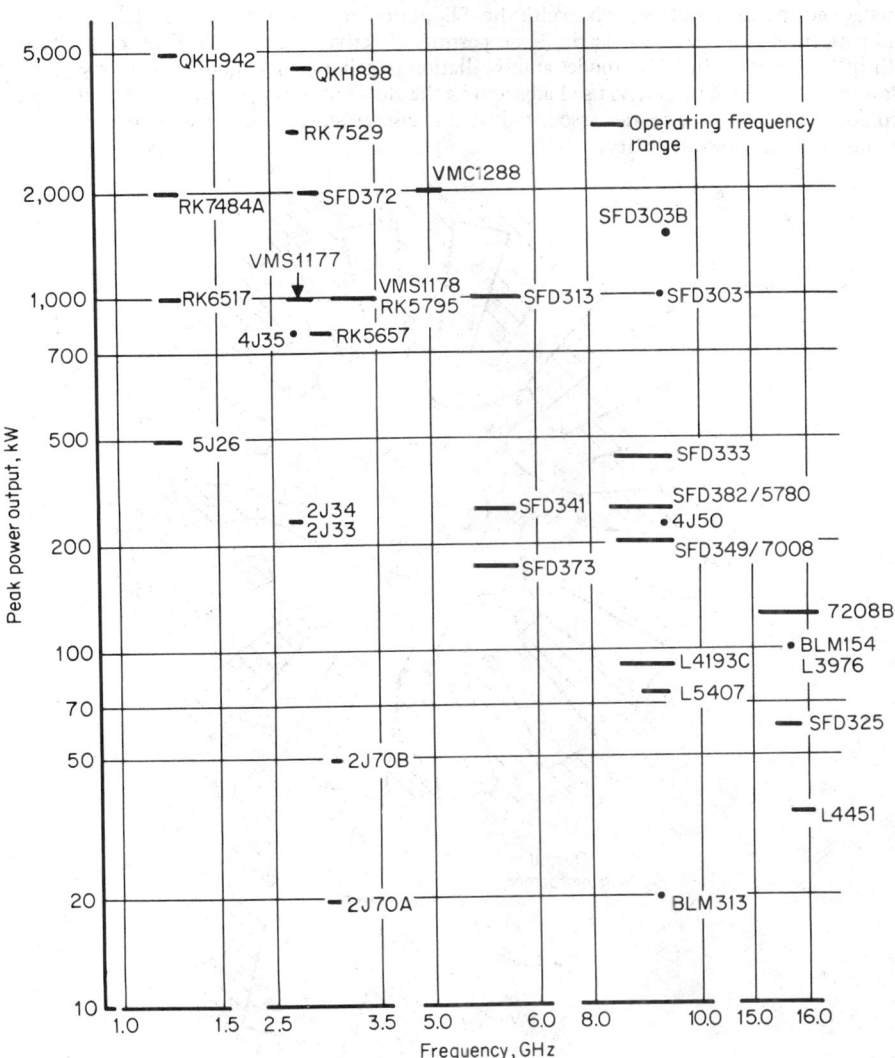

Fig. 9-68. Performance of typical conventional magnetrons.

spurious noise coupled to the vanes and the output waveguide. The spurious-noise power from coaxial magnetrons is typically 10 to 20 dB lower than conventional magnetrons or comparable peak power level.

45. Frequency-Agile Magnetrons. To improve radar-signal detection and electronic countermeasures, rapid frequency-changing signal sources have been developed. Frequency-agile conventional magnetrons are available with rapidly rotating capacitive tuners (*spin-tuned magnetrons*) or hydraulic-driven, *mechanically tuned* tubes. The operational advantages of the coaxial magnetron are preserved in frequency-agile dither-tuned and gyro-tuned coaxial magnetrons.

Dither-tuned magnetrons use a mechanically tuned coaxial magnetron with an integral motor and *resolver* to provide high-speed, narrow-band frequency-agile operation. Mechanical linkage between the rotating motor and the tuning plunger provides approximately sinusoidal tuning of the magnetron frequency. Mechanical limitations imposed by acceleration forces determine the attainable tuning range and tuning rates. A voltage output from the resolver is made proportional to the magnetron frequency and is used to adjust the receiver local oscillator to track the rapidly tuned frequency of the magnetron. X-band tubes, 200 kW, with narrow-band dither-tuned frequency ranges of 30 to 50 MHz, are tuned at rates of 200 Hz. Wider-band frequency excursions of 250 to 500 MHz are dithered at rates of 25 to 40 Hz. Some of these tubes are equipped with servo motors for tuning. Frequency can be set electronically to provide rapid changes of fixed-frequency operation or can be dither-tuned with various shapes of frequency-tuning curves. These tubes are called *Accutune* magnetrons.

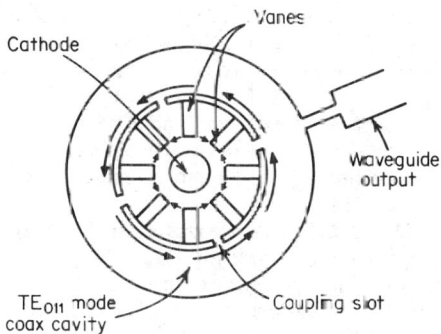

Fig. 9-69. Coaxial magnetron coupling.

Gyro-tuned coaxial magnetrons use several rotating dielectric ceramic paddles in the stabilizing coaxial cavity, which cause frequency variation as they are rotated in a plane normal to the rf electric field of the TE_{011} mode. The anode vane system of the tube is surrounded by a ceramic cylinder bonded to the ends of the coaxial cavity to form the vacuum wall for the electronic interaction region. The stabilizing cavity, containing the tuning, is outside of this vacuum wall and is pressurized with sulfur hexafluoride, to inhibit arcing or corona caused by high rf fields. The tuner drive motor and frequency readout generator are also located within the pressurized section of the magnetron. The symmetry and inherently low rotational mass of the dielectric paddles result in a mechanism in which tuning speed and rf frequency excursion are essentially independent. It is therefore possible to attain higher tuning rates and relatively broader frequency excursion simultaneously than is achieved currently with dither tuning. K band, 60 kW peak power, gyro-tuned magnetrons obtain frequency excursions of 300 MHz at 200-Hz tuning rates.

Electronic tuning of magnetrons has been demonstrated in laboratory models by several methods, including auxiliary electron beams, multipactor discharges, *pin* diodes and ferrite phase shifters; i.e., no mechanical moving parts were employed. Tuning rates as large as several megahertz per microsecond have been achieved. These offer the possibility of magnetron-type radars

TABLE 9-10 Comparison of the 7008 Magnetron and the SFD-349 Coaxial Magnetron

	7008	SFD-349
Efficiency, %	38	38
Leading-edge jitters, rms, ns	1.2	1.5
Pushing factor, kHz/A, specified	500	100
Typical	200	50
Pulling factor (VSWR 1 5), MHz	15	5
Spectra side lobes, dB	8–9	12–13
Missing pulses, %	1	0.01
Pulse-frequency jitter, rms, kHz	60	5
Life, h, specified	500	1,250
Typical	700–800	3,000–3,500

with intrapulse-compression signal-processing capabilities. In 1980, however, these tubes had not yet been employed in actual operating systems.

46. Voltage-Tunable Magnetrons (VTMs). The voltage-tunable magnetron uses a circular-format, reentrant-stream injected beam which interacts with a standing wave on a low-Q resonant structure. A hollow electron beam is launched axially from an electron gun into the interaction region. The hollow beam begins to rotate in the annular region between the anode and sole as it enters the interaction region. This is shown schematically in Fig. 9-71. The anode consists of an interdigital line which capacitively loads the resonant cavity in a manner similar to the grids in a klystron cavity. Oscillating energy in the resonant cavity provides electric fields of opposite polarity between the fingers of the interdigital line, which creates a π-mode standing-wave pattern. Varying the voltage between the anode and sole changes the circumferential velocity of the circulating electron stream by changing the ratio E/B. The oscillation frequency is determined by the circumferential velocity of the electrons interacting with the π-mode field pattern, and voltage tuning is obtained. Voltage-tuning ranges of more than 10% bandwidth have been obtained. Some achieved power levels are indicated in Fig. 9-71. Pulse power levels of 2 kW at 5% duty cycle have been achieved at X band.

Low-power VTMs (milliwatts) have low-noise power output that is at least 80 dB/MHz below the carrier power level. These tubes find application as receiver local oscillators and as signal sources in swept-frequency generators. High-power VTMs (watts) are used in electronic countermeasure applications to generate FM noise by applying a noise-modulated voltage between the sole and anode.

47. M-Carcinotrons and M-Backward-Wave Oscillators (MBWOs). M-carcinotrons and M-backward-wave oscillators use a circular- or linear-format, nonreentrant injected beam which interacts with a traveling wave on a dispersive backward-wave circuit. Voltage-tunable oscillations are generated as in linear-beam backward-wave oscillators. Voltage tuning is obtained by varying the anode-to-sole voltage. Tubes have been developed in frequency ranges between P and K_u band. Continuous-wave power greater than 100 W has been generated at X band with 30% voltage-tunable frequency range. Several hundred watts of cw power has been generated at lower frequencies with comparable tuning range. The primary use for these tubes is for noise-jamming signal sources in electronic countermeasure equipment.

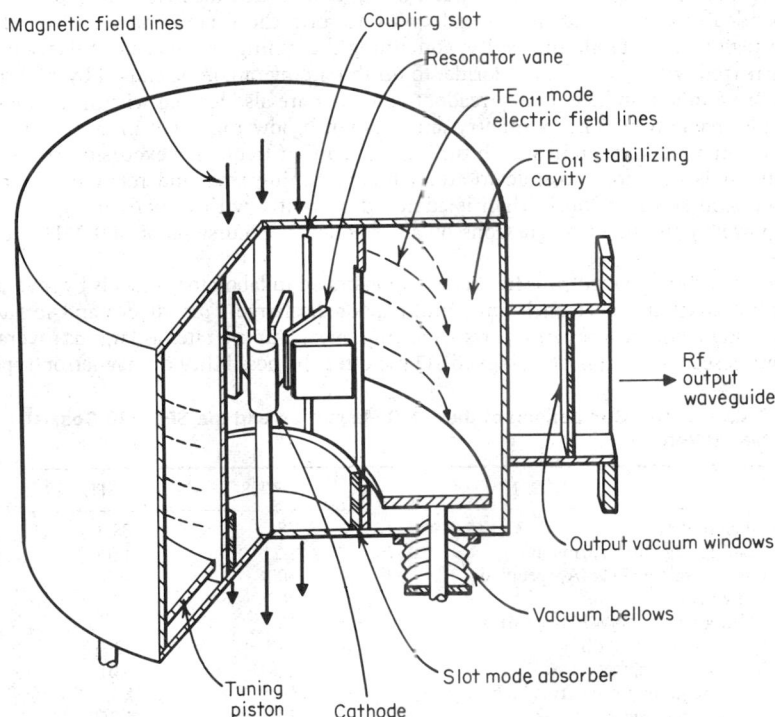

Fig. 9-70. Schematic of coaxial magnetron.

48. Cross-Field Amplifiers (CFAs). *Injected-Beam Types.* Nonreentrant injected-beam crossed-field amplifiers have been developed using linear- and circular-format geometry. Current for interaction is obtained from a heated emissive cathode located external to the interaction region as in a TWT (see Pars. **9-34** to **9-38**). The power-supply requirements for these tubes are greater than for emitting-sole oscillators and amplifiers, both in the number of voltages and stability required. An accelerator electrode, positive with respect to the cathode, draws current from the cathode surface. Magnetic forces deflect the current toward the interaction region over the surface of the cathode.

The presence of nonuniform space charge between the accelerator electrode and the cathode surface introduces nonuniformities in the intervening electric field. This can cause nonuniform

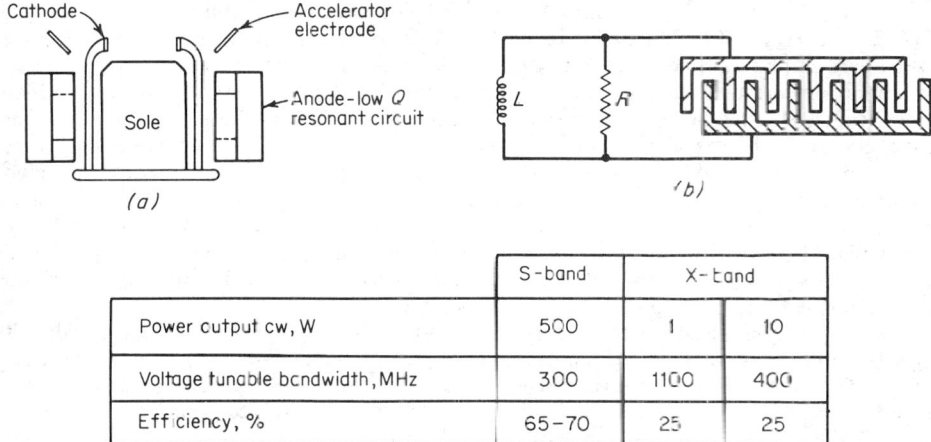

Cathode — Accelerator electrode

Sole — Anode—low Q resonant circuit

L R

(a)

(b)

	S-band	X-band	
Power output cw, W	500	1	10
Voltage tunable bandwidth, MHz	300	1100	400
Efficiency, %	65–70	25	25

Fig. 9-71. Schematic of voltage-tunable magnetron: (a) structure; (b) equivalent circuit.

cathode emission densities and perturbation of the electron launch trajectories. Noise generation and beam instabilities can result from an inadequate electron gun. Variation of the voltage between anode and sole causes variation in beam velocity. Low-current beams in low-power, high-gain amplifiers require power-supply stability comparable with that of TWTs to minimize output signal power and phase variation. Low-gain, high-power, high-current amplifiers are less sensitive to beam voltage variation by a factor of 3 to 5.

Backward-wave amplifiers (*Bitermitron, Carpitron*) are voltage-tunable, narrow-bandwidth devices that are essentially injected frequency-locked M-carcinotrons and MBWOs. The achievable power levels and frequency ranges are comparable. Amplification of 15 to 20 dB has been obtained at 30 to 35% efficiency. Noise power measured far from the carrier in Carpitron tubes is more than 65 dB/MHz below the carrier.

Extensive development efforts have been devoted to nonreentrant forward-wave amplifiers (*TPOM, Bimatron*) for both cw and pulsed application. Continuous-wave amplifiers with a few hundred watts output have been developed with gain of 20 to 30 dB and with efficiencies of 20 to 35%. Proper control of the electron stream at large values of gain requires the beam to be physically close to the anode at synchronism. Operation at low values of V/V_0 (3 to 6), together with increased insertion loss for tubes with greater circuit length for greater gain, leads to lower efficiency values. Like TWTs, the attainable bandwidth is dependent upon the electron-beam optics and the dispersiveness of the slow-wave circuit. Half-octave bandwidth has been obtained at constant-voltage settings, and full-octave bandwidth with adjustment of the anode-to-sole voltage. Stable operation at gain in excess of 20 dB requires circuit severs or distributed attenuation as in TWTs. Useful gain in excess of 30 dB is difficult because of excessive noise buildup in the electron stream.

A major source of noise comes from modulation of the potential minimum in front of the cathode emissive surface. Closely spaced grids located in front of the cathode have been employed to minimize this effect by electrostatic shielding of the potential minimum region. The grids are sets of parallel wires perpendicular to the magnetic field. The magnetic field deflection of the electrons minimizes beam interception. Grids are also used to modulate the emitted cathode current, permitting a cw amplifier to be pulsed to higher-peak-power output while maintaining the same average power level. Dual-mode operation is complicated by the necessary change in electron-beam optics. Peak-to-cw power ratios as great as 10:1 may be feasible by using multiple-element depressed collectors to preserve efficiency in the two operational modes.

Narrow-band (10%), pulsed, high-peak-power (5 MW) injected-beam amplifiers, which operate with efficiencies greater than 50%, have been developed for use in radar application. These tubes have gain of 11 to 15 dB. The lower gain values (greater input signal level) permit the use of less critical beam optics, shorter slow-wave circuits, and greater ratios of V/V_0.

Reentrant-stream forward-wave injected-beam amplifiers (*Vaxitron*) use an interaction geometry similar to that of the voltage-tunable magnetron, with the low-Q resonant structure replaced by a broadband, two-port, nondispersive forward-wave circuit. These tubes have not yet been fully developed, but experimental versions have demonstrated 30-dB gain with efficiencies in excess of 50% when constructed with multiple-element depressed collectors.

49. Emitting-Sole Crossed-Field Amplifiers. Electron current for emitting-sole crossed-field amplifiers is obtained from the sole electrode in the interaction space by electron-beam back bombardment, as in the magnetron oscillator. Unlike the magnetron, these amplifiers do not require thermionic emission to initiate current flow. Current flow can be started in an emitting-sole CFA by the admission of an rf signal to the input of the slow-wave circuit when the proper magnetic field and anode-cathode voltage are present in the interaction region. Amplifiers with rf-induced current flow are called *cold-cathode* amplifiers, regardless of cathode temperature, provided there is no thermionic emission from the cathode. In the absence of an rf input signal, these amplifiers remain quiescent even with full operational voltage applied. The details of the starting mechanism of rf-induced current flow are not fully understood, but the phenomenon is reliable in a properly designed amplifier.

Radio-frequency-induced current flow permits several modulation techniques for pulsed emitting-sole CFAs. These include cathode-pulsed CFAs, dc-operated CFAs with combination of dc voltage and a pulsed turn-off voltage, and dc-operated CFAs with only dc voltages applied. Cathode-pulsed amplifiers, like magnetron oscillators, use pulse modulators to supply the required dc electric power during amplification. Direct-current-operated amplifiers obtain electric input power for amplification from a dc power supply. Electron current flow is initiated by an rf input signal and is terminated at the end of the rf input signal either by a voltage pulse or a dc bias voltage applied to a quench electrode.

Cold-cathode starting for cathode-pulsed amplifiers is assured by correct temporal alignment so that the rf input pulse bridges the cathode voltage pulse. The rf signal is present on the slow-wave circuit as the applied voltage pulse increases to synchronous value. Current flow is initiated, and amplification occurs during the voltage pulse and ceases upon removal.

Both forward- and backward-wave cathode-pulsed CFAs are available. They use the circular-format reentrant-stream geometry illustrated in Fig. 9-72a. A reentrant electron stream is advantageous because electrons which have delivered only part of their available potential energy to

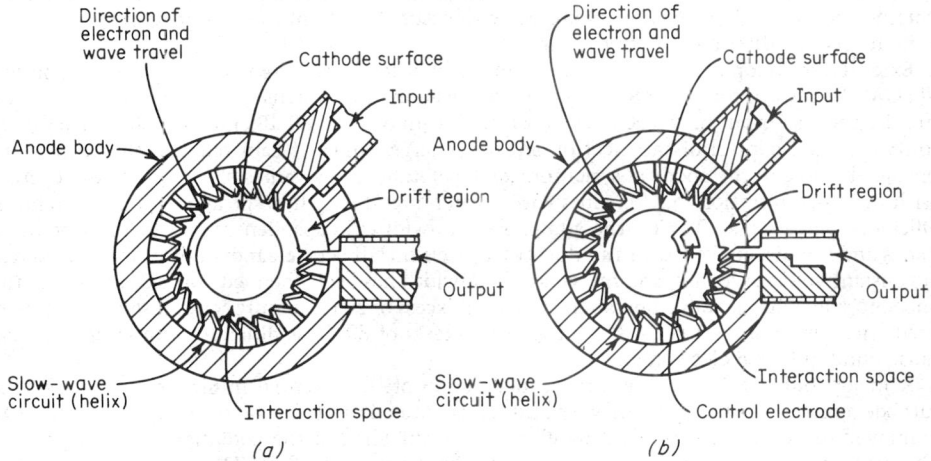

Fig. 9-72. Diagrams of reentrant-stream crossed-field amplifiers: (*a*) cathode-pulsed; (*b*) with control electrode turn-off.

the circuit wave as they leave the interaction region can reenter for further participation, thereby leading to higher electronic conversion efficiency. (Overall amplifier efficiencies of 45 to 50% or more are common.) The leaving phase-sorted electrons contain rf modulation. Some reentrant-stream amplifiers use a circuit geometry with the rf input and output ports spatially separated by a sufficiently long circuit-free region (called the *drift space*) for internal space-charge forces to cause dispersal of the electron spokes. This removes the rf modulation from the reentrant stream while preserving the reentrant-stream efficiency advantage. Other reentrant backward-wave amplifiers (*Amplitrons*) use the modulated electron stream to create a regenerative amplifier. A minimal drift space is used, so that the electron spokes reenter the interaction region before the modulation is dispersed. By suitable design of the shorter drift-space dimensions, the modulated stream reenters with positive phase to enhance the interaction.

Greater electronic conversion efficiency (amplifier efficiency in excess of 70% has been obtained) can be obtained at the expense of lower gain-bandwidth product than can be obtained with amplifiers which remove the reentrant modulation. The use of regenerative amplification is not feasible with a forward-wave amplifier because, at a fixed operating voltage, the simultaneous regenerative gain at frequencies other than the drive signal could lead to unwanted auxiliary oscillations or selective peaks in spurious-noise output power. This is avoided with a backward-wave amplifier because the dispersive slow-wave circuits require different voltages to obtain adequate amplification of separated frequencies.

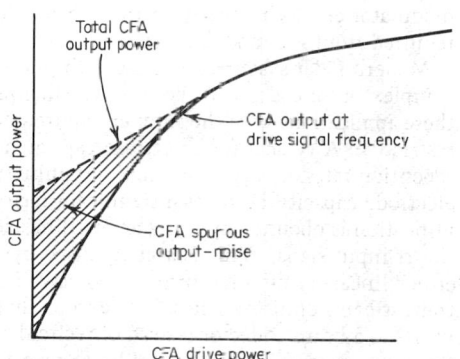

Fig. 9-73. Emitting-sole crossed-field power output.

Spurious Noise.   Dispersal by space-charge forces of the non-phase-locked electrons in a drifting spoke can lead to a rapid buildup of broadband spurious-noise components in the reentering electron stream. This is prevented from becoming severe by use of a sufficiently large rf input signal to lock out noise-signal growth. CFA power output as a function of rf drive signal is illustrated in Fig. 9-73. Reduction of the rf input signal leads to a reduced amplifier output signal at higher values of gain but is accompanied by an increased relative amount of broadband noise-power output. (Reentrant emitting-sole CFAs with terminated input have been used for high-efficiency broadband noise generators.) Adequate lockout of the noise power is obtained at reduced signal gain when the amplifier is driven well into saturation. The rapid growth of noise at small drive signals precludes the use of distributed attenuation and of circuit severs for emitting-sole amplifiers.

Relatively short slow-wave circuits are used with the minimum attainable insertion loss between the rf input and output connections. These amplifiers are called *transparent tubes*. Reflected signals from a load mismatch travel backward through the amplifier to the rf input, where they can be reflected, possibly leading to oscillation. Judicious use of ferrite isolators and circulators at the rf input and/or output minimizes this effect. The overall stable gain of emitting-sole CFAs is limited to 20 dB or less (typically 13 to 15 dB) because of the requirement for a sufficiently large rf input signal for adequate lockout of spurious-noise power and the need to limit gain to avoid oscillations caused by multiply reflected signals. Transparency is often used to advantage in radar systems employing the final amplifier in the transmitter chain in a non-operating feedthrough mode to provide coarse programming of the output power level.

Bandwidth Characteristics.   Cathode-pulsed forward-wave amplifiers offer 10 to 15% instantaneous bandwidth at a fixed value of pulsed voltage. Backward-wave amplifiers provide only 1 to 2% instantaneous bandwidth under comparable conditions but can accommodate 10% bandwidth by adjustment of cathode voltage. The static impedance of both tube types varies as a function of frequency. The constant voltage-vs.-frequency characteristics of forward-wave amplifiers is readily accommodated by a hard-tube cathode modulator, providing nearly constant power output across the frequency band of the amplifier. Constant power output vs. frequency for a variable-voltage backward-wave amplifier is nearly achieved with a constant-current modulator. For restricted bandwidth (4 to 6%) this condition is approximated by using a line-type modulator.

Modulation Requirements.   A simplification in modulator requirements is obtained with a broadband, dc-operated, rf-triggered CFA. The termination of the rf input signal after rf turn-on leaves uncontrolled circulating space charge that can generate cw spurious output signals of magnitude as large as half the amplified signal output. To avoid this, a *control electrode* isolated from the cathode is mounted as part of the cathode structure and is located in the drift space. This location minimizes interference with amplifier performance (Fig. 9-72b). This electrode is pulsed positive with respect to the cathode, coincident in time with removal of the rf input signal. The circulating space charge in the interaction space is collected upon the control electrode, and the cathode current flow is terminated. Modulator requirements are simplified because the pulse voltage required for the turn-off electrode is typically one-quarter to two-thirds of the anode-cathode voltage and the peak collected current is less than one-half of the peak cathode-current flow during amplification. The duration of the collection time for the circulating current is approximately one transit time for electron flow around the interaction region. Typically, this is a small fraction of the time duration of the amplified signal. Consequently, the modulator energy required for the control electrode per amplifier pulse is much less than that required from the modulator for full-cathode-pulsed amplifiers.

Modern radar systems often use high pulse-repetition frequencies, very short pulse durations, complex pulse codes with possible variable pulse widths, and/or burst modes of operation. For these applications even the simpler requirements for the control electrode modulator can become restrictive. A typical interelectrode capacitance for a turn-off electrode is 50 to 70 pF. At high repetition rates or very short amplifier pulses the energy consumed per pulse charging the interelectrode capacity becomes a significant factor affecting the transmitter design. A further simplification is obtained by using an amplifier geometry that leads to complete self-modulation by the rf input signal. This is accomplished by a nonreentrant, dc-operated, emitting-sole CFA of either linear- or circular-format geometry (*Dematron*). These CFAs function similarly to reentrant-stream, emitting-sole CFAs, except that electron current leaving the interaction region passes to a beam collector instead of recirculating for further participation. Obtainable gross-performance characteristics are similar to reentrant-stream amplifiers, but the overall efficiency is limited to about 35% because partially interacted electron current exiting from the interaction region is usually collected at full beam voltage. Complete rf-controlled modulation of the beam current is provided because removal of the rf input signal permits the electron current to pass unidirectionally through the interaction region to leave for removal from further participation by the beam collector.

Self-Modulation.   Reentrant-stream amplifiers that provide complete self-modulation by the rf input signal have been demonstrated in the laboratory. These dc-operated amplifiers (*AutoMod*) use an isolated turn-off electrode in the cathode structure located within the interaction region rather than in the drift space. A dc bias voltage, positive with respect to the cathode, is applied to the turn-off electrode. This electrode acts as a beam collector similarly to a normal control electrode located in the drift space when a positive turn-off voltage pulse is applied. During rf turn-on and normal amplification the fringing rf electric fields from the slow-wave circuit perturb the circulating electron trajectories in the vicinity of the dc-biased electrode sufficiently for enough electrons to pass through the collector region to provide adequate reentrancy for normal amplification. Upon removal of the rf input signal the bias electrode becomes totally effective for collecting circulating current, and the amplifier turns off automatically. Complete rf modulation is obtained at the expense of some power loss due to partial beam collection by the bias electrode during normal amplification. Overall efficiency of 40 to 45% has been demonstrated, which is midway between that achievable by Dematrons and control-electrode-modulated amplifiers.

Self-Emitting Cathodes.   A variety of materials are used for secondary-electron-emission cathodes in rf-triggered amplifiers. The selection is based on the amplifier drive signal level, peak power output, and the intended operating voltage for the tube. Materials used include pure metals, such as aluminum, beryllium, and platinum, as well as a variety of composite materials, such as dispenser cathodes and cermets. Dispenser cathodes and pure-platinum cathodes are suitable for drive signal levels in excess of 10 kW. Amplifiers with drive signals from a few hundred watts to 10 kW are better accommodated with metals supporting oxide surface layers such as aluminum and beryllium. Oxide layers are susceptible to erosion under electron-beam bombardment, and some CFAs employ a low-level background pressure of pure oxygen supplied from a suitable reservoir to rejuvenate and extend the active cathode life.

Frequency Range.   Emitting-sole amplifiers have been developed at frequency ranges extending from VHF to K_u band, with experimental models at lower and higher frequencies. Examples of peak power levels available include 100 kW at K_u band, 1 MW at X band, and 3 MW at S band. Average power levels vary from 200 W at K_u band to several kilowatts at lower frequencies. Laboratory models have demonstrated as much as 400 kW of cw power at S band. The performance of typical cataloged amplifiers is shown in Fig. 9-74.

Noise Power.   Noise-power output from a reentrant emitting-sole CFA with a space-charge-dispersing drift space measured in a 1-MHz bandwidth far from the carrier is typically greater

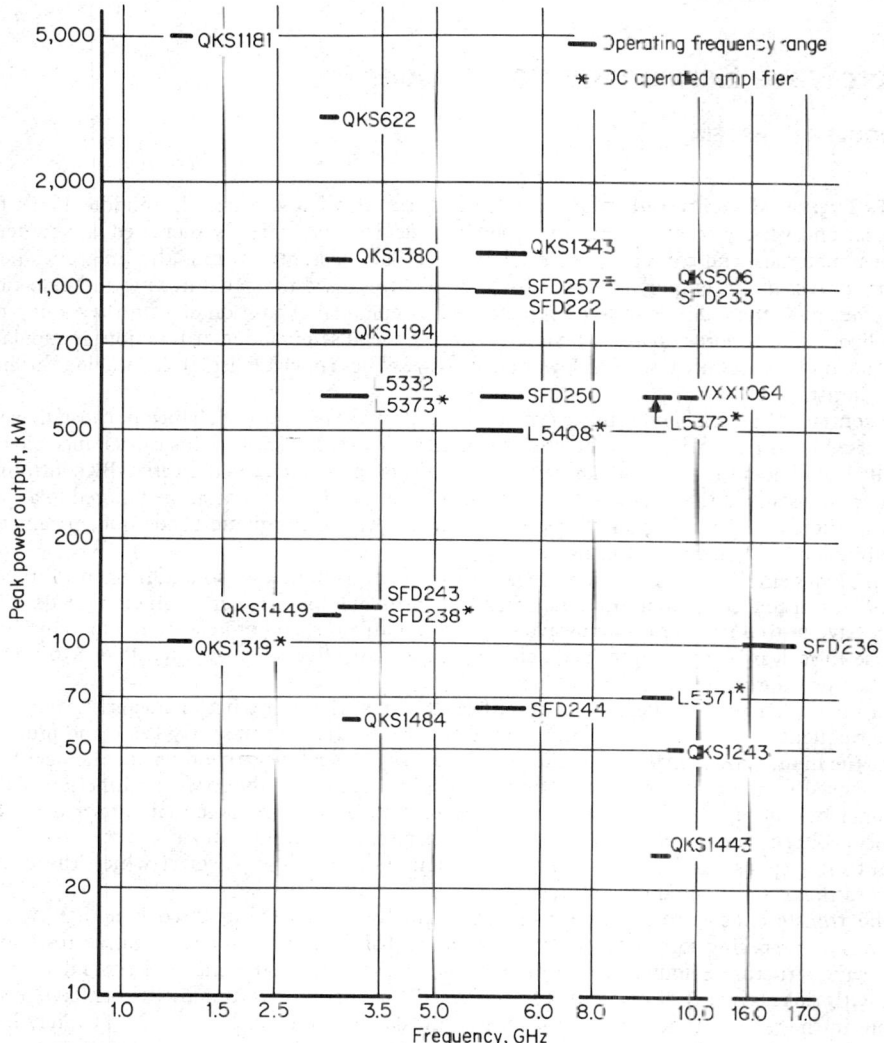

Fig. 9-74. Performance of cataloged emitting-sole, crossed-field amplifiers.

than 30 dB below the carrier power level. Noise levels greater than 40 dB below the carrier are not uncommon. Noise power close to the carrier frequency tends to be less, due to the phase-lockout phenomenon associated with high drive-signal levels. The maximum power density of the main lobe of the spectrum of a pulsed CFA is typically greater than 40 dB above the radiated interspectral line noise when both are normalized to a 1-MHz bandwidth. Ratios of more than 50 dB are not uncommon. Phase locking of the space charge by the drive signal also minimizes phase variation due to the voltage change and drive-signal variation. Saturated amplifiers with 13 to 15 dB gain have output phase variations of 3 to 8° for a 1% change in anode-cathode voltage. A comparable phase change occurs with a 1-dB variation in drive-signal level.

Applications.   The primary use for emitting-sole CFAs is for transmitter tubes in coherent radars. CFAs have been used in pulse compression radars and pulse-coded and phased-coded radars, as well as phased-array radars. Lightweight low-voltage cathode-pulsed amplifiers are attractive for airborne applications. High-power dc-operated CFAs are used in ground-based radars.

50. Bibliography on Crossed-Field Tubes

Collins, G. B. "Microwave Magnetrons," McGraw-Hill, New York, 1948.

Okress, E. "Crossed-Field Microwave Devices," Vols. 1 and 2, Academic, New York, 1961; "Microwave Power Engineering," Vol. 1, Academic, New York, 1968.

Microwave Semiconductor Devices

BY JOSEPH FEINSTEIN

51. Types of Microwave Semiconductor Devices. Since the middle 1950s the number and variety of microwave semiconductor devices have greatly increased as new techniques, materials, and concepts have been applied. The oldest structure is the tungsten-silicon *point-contact diode*, employing a metal-whisker contact, used for signal mixing and detection. More recently, these diodes have been fabricated by epitaxial deposition of a thin layer of p-on-p^+ silicon. The *Schottky barrier diode*, a rectifying metal-semiconductor junction, is supplanting the point contact because of its lower noise figure. These devices display a variable-resistance characteristic.

In contrast, the *variable-reactance (varactor) diode* makes use of the change in capacitance of a reverse-biased pn junction as a function of applied voltage. Physically, this capacitance change results from widening the depletion layer as the reverse-bias voltage is increased. By controlling the doping profile at the junction, the functional forms of this relation can be tailored to a specified application. Typical applications of varactor diodes are harmonic generation, parametric amplification, and electronic tuning.

pin diodes employ a wide intrinsic region which permits high-power-handling capability and offers an impedance at microwave frequencies controllable by a lower frequency (or dc) bias. They have proved useful for microwave switches, modulators, and protectors. In changing from reverse to forward bias the *pin* diode changes electrically from a small capacitance to a large conductance, approximating a short cicuit.

For microwave power generation or amplification, a negative-resistance characteristic at microwave frequencies is necessary. Beginning with the *tunnel diode* in the early 1960s and progressing to the higher-power *IMPATT diodes* and *Gunn diodes*, such negative-resistance devices have experienced rapid development. The tunnel diode utilizes a heavily doped pn junction which is sufficiently abrupt for electrons to be able to tunnel through the potential barrier near zero applied voltage. Because this is a majority-carrier effect, the tunnel diode is very fast acting, permitting response in the millimeter-wave region. The very low power at which the tunnel diode saturates has limited its usefulness.

The *transferred-electron oscillator* (TEO), originally named for its discoverer, J. B. Gunn, depends on a specific form of quantum-mechanical band structure for its negative resistance. This band structure is found in gallium arsenide, the semiconductor material generally associated with this class of device, and in a few other III-V compounds. *Gunn* oscillators have power output in the tens to hundreds of milliwatts at low dc operating voltage (9 to 28 V). They have a wide range of tunability and reasonably low AM and FM noise.

The IMPATT (*impact avalanche transit-time) diode owes its negative resistance to the classical effects of phase shift introduced by the time lag between maximum field and maximum avalanche current multiplication and by the transit time through the device of this current. These effects can occur in all semiconductor pn junctions under sufficient reverse bias to initiate breakdown. While IMPATT diodes were originally developed as silicon devices (still the predominant type), gallium arsenide IMPATTs have recently come into prominence because of their higher power (several watts) and superior noise characteristics. IMPATT diodes find applications in point-to-point microwave communications transmitters and as pumps for parametric amplification.

Combination modes are also possible. Since reverse breakdown in a diode may occur due to a tunneling mechanism as well as to avalanching, TUNNETT mode oscillators are possible. In such a mode the injected current is in phase with the applied rf voltage so that the negative resistance is supplied purely by transit-time effects. This leads to a lower conversion efficiency for a TUNNETT compared with an IMPATT but also to much lower noise because of the absence of avalanche multiplication. Devices can also be operated in a mixed tunneling and avalanche transit-time mode (MITATT), which partakes of the properties of both. Intermediate noise and power properties are also obtained from BARITT (barrier-injection transit-time) diodes.

The trapped-plasma-avalanche triggered-transit (TRAPATT) mode, on the other hand, uses class C operation to obtain high peak power capability at good efficiency (about 25%), but this use requires complex circuit matching.

Bipolar transistors have penetrated the microwave region through the refinement of fabrication techniques based on the planar photolithography technology. By reducing the emitter width and base thickness to micrometer dimensions (and by paralleling stripes of structure to maintain power capability), transit times and charging times (resistance-capacitance product) can be kept low enough to provide useful devices to frequencies of about 10 GHz where 1 W of power is available. Such transistors are also used in low-noise preamplifiers up to frequencies of about 4 GHz.

Field-effect transistors (FETs) are microwave solid-state devices which have progressed rapidly during the past several years. Using the higher mobility and saturated velocity of GaAs, amplifier gain has been obtained at frequencies as high as 40 GHz. The GaAs FET now dominates low-noise applications in the intermediate microwave region of the spectrum and is also taking over power-amplifier designs from the negative-resistance diodes discussed above.

The *ruby maser* (acronym for microwave amplification through stimulated emission of radiation), while historically very important in demonstrating the feasibility of the inverted-population principle, has been replaced by the *cooled parametric amplifier* in applications requiring ultralow noise reception. This obsolescence has been a consequence of the low power saturation and narrow bandwidth of the maser, together with the liquid-helium temperature necessary for its operation. It is used only in a few radio astronomy installations.

Acoustic-wave amplifiers came into prominence in electronics through the discovery that amplification could be obtained by coupling the charge carriers in a semiconductor to an acoustic wave propagating in a piezoelectric material. Although a great deal of experimental work continues on such amplifiers, the practical applications of acoustic waves have been in passive devices using the slow propagation velocity (relative to electromagnetic waves) and compressed wavelengths of acoustic waves. *Microwave delay lines* and *signal-processing devices* making use of these properties are discussed in Par. **9-10**.

52. Microwave Mixer and Detector Diodes. Crystal rectifiers used as mixers have good noise performance in comparison with thermionic tubes at microwave frequencies. Noise figures lower than 6 dB are now possible at K_u band using Schottky barrier diodes. These improvements have resulted from better control of semiconductor purity, epitaxial material to achieve low series resistance, and photolithographic techniques to achieve small-area Schottky diodes.

Mixing is defined as the conversion of a low-power-level signal from one frequency to another by combining it with a higher-power (local-oscillator) signal in a nonlinear device. In general, mixing produces a large number of sum and difference frequencies. Usually, the difference frequency between signal and local oscillator (the intermediate frequency) is of interest and is at a low power level.

In *microwave detection*, solid-state devices are used as nonlinear elements to accomplish direct rectification of an applied rf signal. The sensitivity of a detector is usually much less than that of a superheterodyne receiver, but the detector circuit is simple and easy to adjust. Recently this sensitivity has been improved considerably by the use of a wide-band low-noise microwave preamplifier preceding the detector.

Tunnel diodes, back diodes, point-contact diodes, and Schottky barrier diodes are majority-carrier devices that have nonlinear resistive characteristics and are useful for mixing and detecting. Because of greater susceptibility to rf burnout, circuit complications, and fabrication difficulties, tunnel and back diodes have not found as wide acceptance as mixers and detectors at microwave frequencies. The point-contact (pressure contact between metal and semiconductor) and Schottky barrier diodes (formed by deposition of metal on semiconductor surface) are the primary mixer and detector microwave devices. However, back diodes are used in selected applications such as broadband low-level detectors and Doppler mixers.

Point-contact, Schottky barrier, and back diodes are used as mixers and detectors from UHF to millimeter frequencies. The construction and current-voltage characteristics of these diodes are shown in Fig. 9-75. The point contact is fabricated by a metal whisker forming a rectifying junction in contact with the semiconductor. The Schottky is formed generally by an evaporated metal contact, and the back diode by an alloyed junction.

53. Point-Contact Diodes. The point-contact diode is the oldest structure. Until 1965 point-contact diodes were fabricated using moderately low resistivity material with the rectifying contact established by touching the semiconductor surface with a metal whisker (normally, tungsten for silicon and phosphorus bronze for germanium and GaAs).

Since that time the semiconductor epitaxial deposition has been applied to point-contact diodes (as well as to Schottky diodes) to maximize the frequency cutoff (minimize the R_sC_j product).

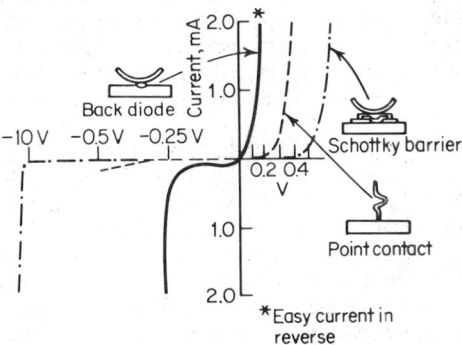

Fig. 9-75. Current-voltage characteristics of back, point-contact, and Schottky barrier diodes. (*Proc. IEEE, August 1971.*)   **Fig. 9-76.** Equivalent circuit of varistor diode.

This is a significant consideration since conversion loss is directly proportional to the product of diode series resistance R_s and junction capacitance C_j, as shown in Fig. 9-76. Further, the conversion loss L can be shown to be related to semiconductor properties as

$$L \approx R_s C_j \approx W\epsilon^{1/2}/N^{1/2}\mu$$

where ϵ is the semiconductor dielectric constant and W, N, and μ are the epitaxial-layer thickness, carrier concentration, and mobility, respectively. Present-day silicon point-contact diodes are mainly fabricated from p-on-p^+ expitaxial silicon (p layer 0.25 to 0.5 μm thick).

Such devices are generally used only in the millimeter-wavelength range because of inherent reproducibility problems and because they have poorer noise figures than Schottky diodes. Nevertheless they are more resistant to rf burnout than the Schottky devices.

54. Schottky Barrier Diodes. The Schottky barrier diode is a rectifying metal-semiconductor junction formed by plating, evaporating, or sputtering a variety of metals on n- or p-type semiconductor materials. Schottky diodes are fabricated from n-on-n^+ epitaxial material (n layer 0.5 to 1.0 μm thick), where the p and n layers are optimized in thickness and carrier concentration for minimum conversion loss and maximum rf burnout or power-handling capability. Generally, n-type silicon and n-type GaAs are used. Due to their higher cutoff frequency, GaAs devices are preferred in applications above X-band frequencies. This results from the higher mobility of electrons in GaAs than in silicon. Although in practice this advantage is not as significant as predicted, a conversion-loss improvement of 0.5 dB at K_u band is readily obtainable with GaAs, compared with silicon.

Schottky diodes are fabricated by a planar technique. A SiO_2 layer (1 μm thick) is thermally grown or deposited on the semiconductor wafer, and windows are etched in the SiO_2 by photolithography techniques. Schottky junctions are formed by evaporation, sputtering, or plating techniques. Metal on the oxide is removed by a second photo step. Junction diameters as small as 5 μm are made by this technique. Figure 9-77*a* shows a silicon diode intended for operation at X band, and Fig. 9-77*b* shows a GaAs diode for millimeter waves.

55. Mixer-Diode Parameters. Figure 9-78 gives the noise figure for a variety of mixer diodes as a function of local oscillator power at 16 GHz. Below this frequency, silicon Schottky diodes exhibiting a minimum noise figure of 5.5 dB are generally used, while GaAs is preferred at higher frequencies.

The *tangential signal sensitivity* (TSS) is the most widely used criterion for detector performance. It indicates the ability of the detector to detect a signal against a noise background and also includes the noise properties of the diode and video amplifier. Figure 9-79 gives the TSS as a function of microwave frequency for good-quality detector diodes. This quantity varies with the square root of the amplifier bandwidth, since square-law response is obtained at low signal levels. As a rough basis of comparison, the TSS rating corresponds to a signal-to-noise ratio of about 2.5.

Typical mixer- and detector-diode mounts are shown in Fig. 9-80. They are designed to be compatible with a particular type of microwave circuitry, waveguide, coaxial, or strip line. The

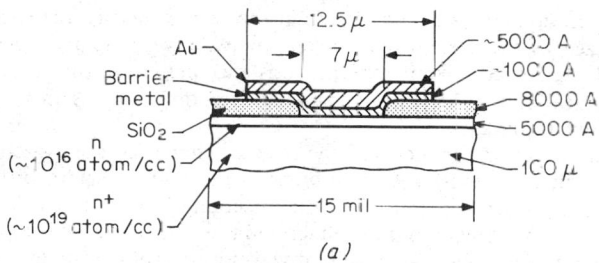

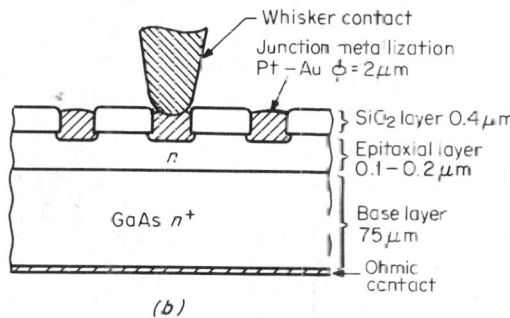

Fig. 9-77. Microwave diodes: (a) silicon diode for operation at X band; (b) GaAs diode for the millimeter-wave region of the spectrum.

encapsulations shown in Fig. 9-80a and b are generally used up to about 12 GHz. The coaxial package in Fig. 9-80c permits operation up to about 30 GHz, while the waveguide wafer in Fig. 9-80d is employed at millimeter wavelengths. Diodes for strip-line circuitry are generally unencapsulated, employing oxide protection and gold beam leads for attachment.

56. Varactor Diodes. The active element of a varactor diode consists of a semiconductor wafer containing a pn junction of a well-defined geometry, usually formed by diffusion. Varactor

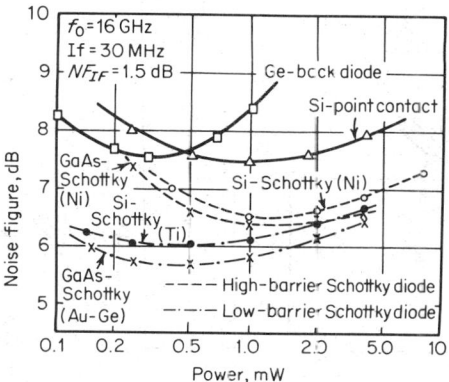

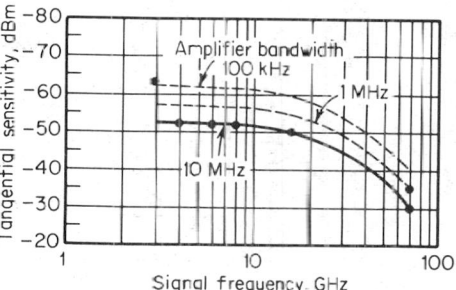

Fig. 9-78. Noise figure vs. local oscillator power for point-contact diodes and n-type Schottky barrier diodes.

Fig. 9-79. Tangential sensitivity vs. signal frequency of detector diodes in 1971. (*After H. A. Watson, "Microwave Semiconductor Devices and Their Circuit Applications," McGraw-Hill, New York, 1969.*)

diodes are normally operated under reverse bias, where the junction resistance (ordinarily 10 MΩ or more) is negligible in comparison with the microwave capacitive reactance of the junction. Therefore the equivalent circuit of the reverse-biased varactor wafer at microwave frequencies is simply a capacitance and resistance in series. The equivalent circuit of a forward-biased varactor at microwave frequencies is generally more complicated, since it must include the diffusion capacitance of the injected carriers, as well as the effect of these carriers on the conductance of the semiconductor material.

For forward bias the diode current increases exponentially with the applied voltage, and for reverse bias a small saturation current I_s flows. When the reverse bias is increased to the avalanche breakdown voltage V_B, the diode reverse current increases very rapidly, since it is limited only by the small diode resistance and any external resistance present in the circuit.

Figure 9-81 shows typical dc current-voltage characteristics, microwave series resistance, and 1-MHz current-voltage characteristics.

Charge-Storage Effects.   In some applications the injected charge-storage capacitance is more important than the capacitance variation association with the varying width of the depletion region. Charge-storage capacitance is produced by the injection of minority carriers during the forward-biased excursion of the varactor pump voltage and the withdrawal of this charge during the reverse-biased portion of the cycle. The resultant waveform is shown in Fig. 9-82. Efforts to

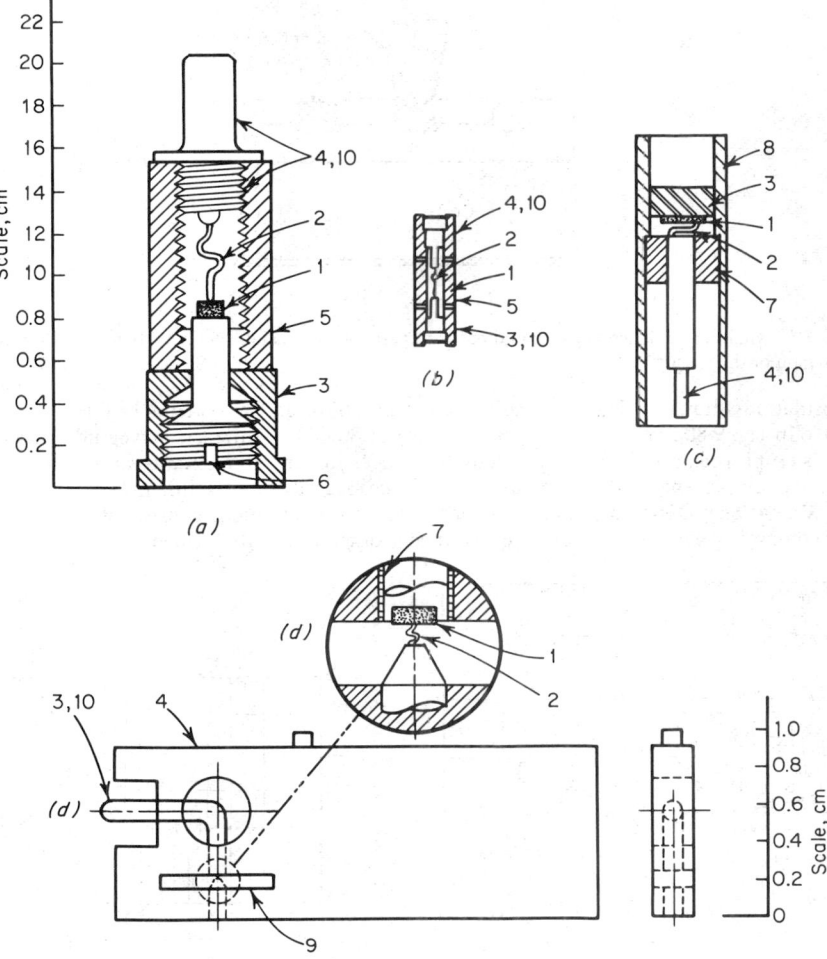

Fig. 9-80. Mixer and detector diode encapsulations: (1) semiconductor; (2) whisker; (3) wafer contact; (4) external whisker contact; (5) ceramic case; (6) adjustment screw; (7) insulating spacer; (8) outer conductor; (9) waveguide rf input port; (10) connection to i.f. amplifier or detector. *(After H. A. Watson, "Microwave Semiconductor Devices and Their Circuit Applications," McGraw-Hill, New York, 1969.)*

maximize charge-storage effects in microwave diodes have led to a class of devices known as *snapback*, or *step-recovery*, *diodes*. They feature steep doping profiles and narrow junctions, to give fast recovery of injected charge, typically in a transition period of a few tenths of a nanosecond, yielding a high harmonic content. However, this design results in lower breakdown voltage, reducing the power capability of the diode. Frequency multipliers use these diodes where high-order multiplication (above eight) and circuit simplicity are desired, at the expense of power output.

57. Varactor Frequency Multipliers. Figure 9-83 gives the characteristics of diffused epitaxial GaAs and Si varactor diodes employed in a conventional tripler (4 to 12 GHz) frequency multiplier. Because of its higher cutoff frequency, GaAs gives higher efficiency but lower maximum output power than Si.

58. Varactor Parametric Amplifiers (Paramps). The most commonly used parametric amplifier is the one-port nondegenerate configuration in which a circulator is employed to separate output from input signal. Typical bandwidths of practical amplifiers with single-tuned signal and idler circuits at 20 dB gain are of the order of 2 or 3%. To achieve wider bandwidth, gain is lowered, and an additional tuning resonator is added to broaden and flatten the passband. The low gain necessitates cas-

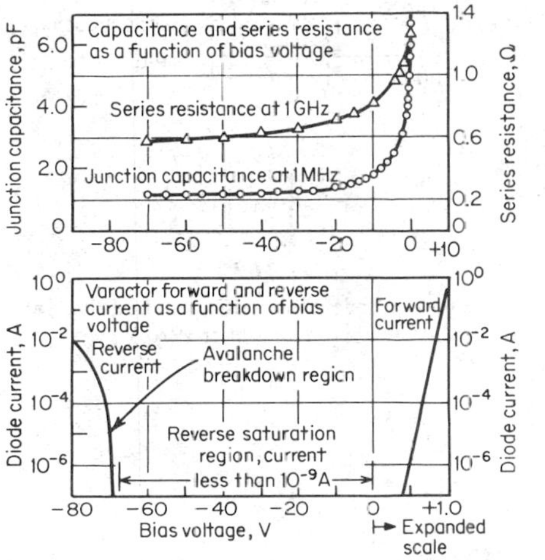

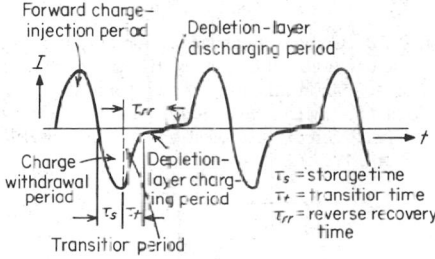

Fig. 9-81. Varactor junction capacitance, series resistance, forward current, and reverse current as functions of bias voltage. (*After H. A. Watson, "Microwave Semiconductor Devices and Their Circuit Applications," McGraw-Hill, New York, 1969.*)

Fig. 9-82. Current waveform of sinusoidally switched step-recovery diode. (*After H. A. Watson, "Microwave Semiconductor Devices and Their Circuit Applications," McGraw-Hill, New York, 1969.*)

cading paramp stages to make the noise contribution of subsequent devices negligible. For example, at 4 GHz, three double-tuned 10-dB stages are cascaded to give a 500-MHz 30-dB-gain low-noise receiver.

Most varactors currently used in paramps are made of gallium arsenide, in the form of epitaxial diffused junction diodes, with a dynamic cutoff frequency of about 100 GHz. These give, typically, noise temperatures of 120 K uncooled in the 7-GHz satellite communications band and 80 K at 3 to 4 GHz. When cooled by a 17-K cryogenic refrigerator, these noise temperatures drop to 30 and 20 K, respectively. GaAs Schottky barrier-type varactor diodes have recently appeared with a cutoff frequency above 500 GHz. When pumped in the 60-GHz range, amplifiers employing such diodes have exhibited uncooled noise temperatures of the order of 60 K at 7 GHz.

Mounting. Typical varactor diodes are mounted in hollow dielectric (usually alumina) cylinders with Kovar or copper end caps, as shown in Fig. 9-84. A flexible connection is made to the diode mesa from one end cap by means of a thin gold strap. The equivalent circuit shown represents the coupling between the diode junction region and the package surface.

59. pin Diodes. *pin* diodes consist of heavily doped *p* and n regions separated by a layer of high-resistivity intrinsic material. Typical construction of such a diode is shown in Fig. 9-85, and Fig. 9-86 illustrates a metal-ceramic package employed at microwave frequencies. Under zero and reverse bias this type of diode has a very high impedance, whereas at moderate forward current it has a very low impedance. This permits its use as a switch in microwave transmission lines. Generally, the diode is placed in shunt across a strip line allowing unimpeded transmission when reverse-biased but short-circuiting the line to produce almost total reflection when forward-biased by as little as 1 V. The wide intrinsic layer permits high microwave peak power to be controlled since the breakdown voltage is of the order of 1 kV. Very little power is dissipated by the diode itself because reflection switching is employed.

pin diodes can be used as limiters, replacing TR tubes for peak powers smaller than 100 kW. At higher peak power, these diodes are useful, following the TR box to eliminate any spike leakage, although if fast response is required (less than 1 μs) a varactor diode is used. In attenuator applications the diode behaves like a current-controlled resistance in parallel with the capacitance of the intrinsic region.

Electrically controllable, rapid-acting microwave phase shifters are finding increasing use in phased-array systems. *pin* diodes are employed to switch lengths of transmission line, providing

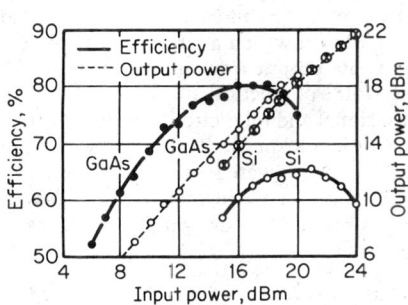

Fig. 9-83. Conversion efficiency and output power of varactors used as frequency triplers, with bias adjusted to maximum output at each point. *(After H. A. Watson, "Microwave Semiconductor Devices and Their Circuit Applications," McGraw-Hill, New York, 1969.)*

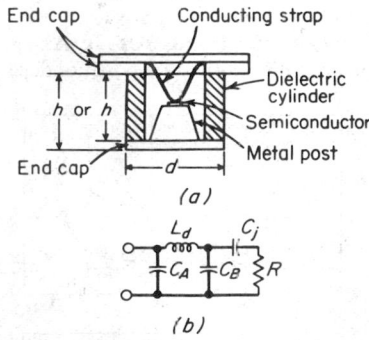

Fig. 9-84. Diode construction and equivalent circuit. *(Microwave J., November 1970.)*

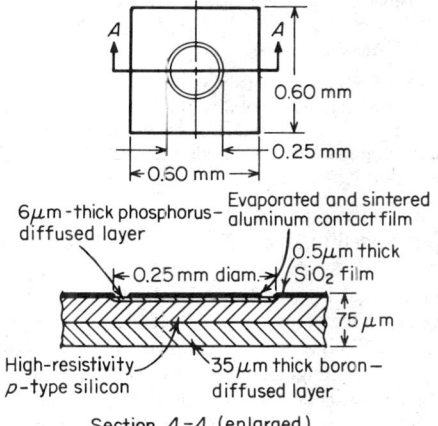

Fig. 9-85. Typical planar *pin* microwave wafer.

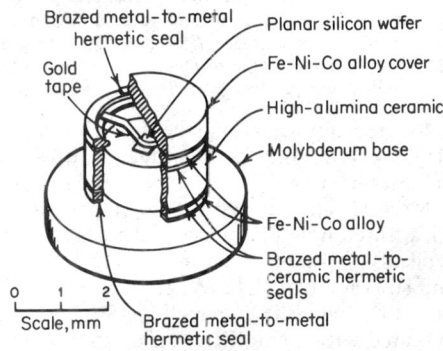

Fig. 9-86. Typical *pin* diode in microwave package.

digital increments of phase in individual transmission paths, each capable of carrying many kilowatts of peak power.

60. Microwave Bipolar Transistors. To achieve operation at microwave frequencies, individual transistor dimensions must be reduced to the micrometer range. To maintain current and power capability, various forms of internal paralleling on the chip are employed. These geometries fall into three general types, as shown in Fig. 9-87, interdigitated fingers forming emitter and base, overlay groupings of emitter and base stripes, and a mesh or matrix of emitter and base spots. All microwave transistors are now planar in form, and almost all are of the silicon *npn* type.

Construction and Fabrication. High-quality microwave transistors are built up on an *n*-type epitaxial layer of the order of 1 Ω·cm resistivity (deposited on lower-resistivity silicon) to keep the collector depletion layer narrow. Following a first thermal oxidation of the silicon, the base area is opened by masking and photoetching operations, and diffusion of p-type dopant (typically boron) proceeds. More recently ion implantation has replaced diffusion as the means of achieving base widths of less than 0.1 nm.

Following base formation, a second oxide is deposited on the slice, generally by thermal decomposition of a gas or liquid in a low-temperature process designed to minimize the possibility of further base diffusion. The emitter is then defined through photomasks, and the appropriate dopant (phosphorus or preferably arsenic) diffused in. The emitter diffusion is quite critical, since too shallow or too deep a penetration can result in no transistor action at all. Metal contacts are then evaporated or sputtered to interconnect the elements of the device

Figure 9-88 illustrates the steps in the fabrication process. Because of its superior properties, silicon nitride is replacing silicon oxide as a passivating agent when extremely shallow junctions are to be protected. Reduction of strip width to increase frequency response has been made possible by improved lithography and etching, leading to state-of-the-art line definition of 1 nm.

Power Capability.   Most power microwave transistors include some form of integral emitter resistors to aid in equalizing the current over the distributed emitter structure. The *overlay transistor* utilizes an integral diffused resistor as part of each emitter stripe, while *thin-film resistors* are deposited as part of the contacts on interdigitated devices. Guard rings are employed to raise the voltage breakdown limit of the collector to about half that of bulk silicon.

A figure of merit can be defined for the transistor in terms of the base resistance r_b', the collector capacitance C, and the emitter-to-collector signal delay time τ_{ec}:

$$(\text{Power gain})^{1/2}(\text{bandwidth}) \approx 1/4\pi(r_b' C \tau_{ec})^{1/2}$$

τ_{ec} is composed of the transit time of carriers across the base and collector depletion layer plus the charging time of the emitter-base junction capacitance and the collector capacitance. The maximum frequency of oscillation is obtained by setting the gain equal to unity in this expression.

Noise.   The principal sources of noise in a microwave transistor are shot noise associated with the emitter and collector current and thermal noise in the base resistance. It is increase in the latter and reduction in current gain which are responsible for the deterioration of noise figure with frequency. The current state of the art in the bipolar-transistor noise figure is about 2 dB at 1 GHz, increasing to 5 dB at 6 GHz. The power performance of silicon bipolar transistors is compared with that of GaAs FETs (Par. **9-61**) in Fig. 9-89.

The input-output isolation of the transistor, as well as its unconditional stability in properly designed circuits, makes it the component

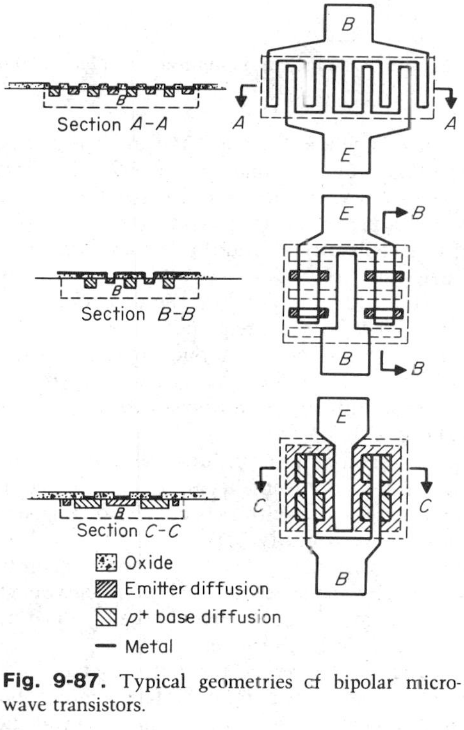

Section A-A

Section B-B

Section C-C

Oxide

Emitter diffusion

p+ base diffusion

Metal

Fig. 9-87. Typical geometries of bipolar microwave transistors.

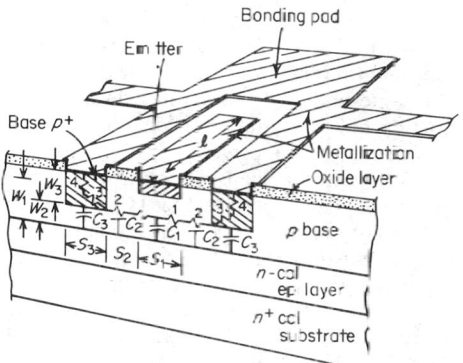

Bonding pad

Emitter

Base p+

Metallization

Oxide layer

p base

n-col

ep layer

n+ col

substrate

Fig. 9-88. Cross section of planar transistor, showing (1)–(3) r_b' and (4) contact resistance R_c (*Proc. IEEE, August 1971*.)

of choice in those applications where its power output or noise is competitive. Most circuit configurations require that the collector be isolated from the ground plane. A ceramic (commonly beryllium oxide because of its superior thermal characteristics) is employed in the package to provide the necessary insulation. At the higher microwave frequencies, conventional packaging results in a loss of gain, primarily because of lead inductance. Direct mounting of a chip in a microwave integrated circuit is called for in this application.

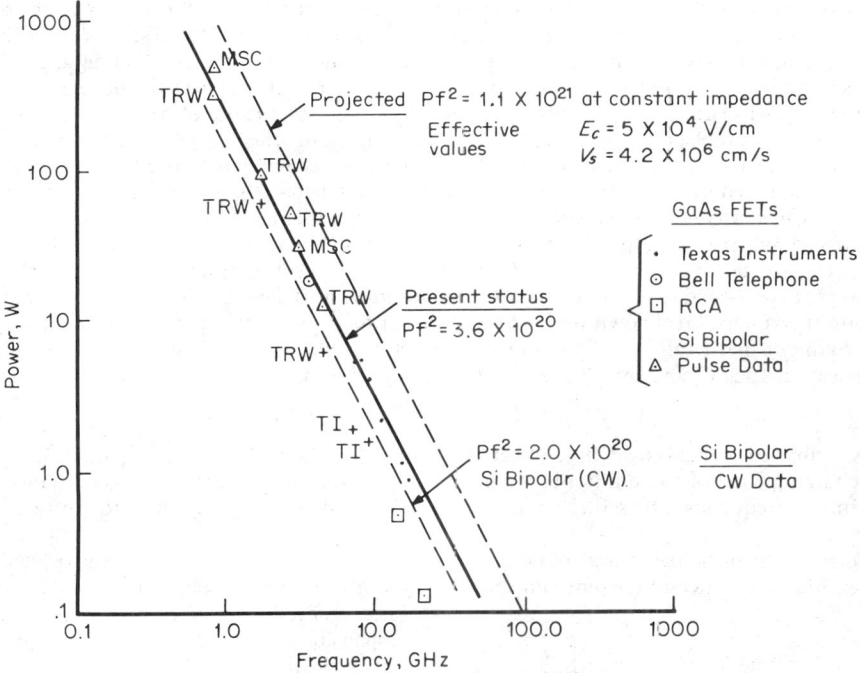

Fig. 9-89. Performance of microwave power transistors as of 1979 and projected improvement. *(Trans. IEEE, May 1979, Vol. MTT 27, No. 5.)*

61. Field-Effect Transistors. The FET device is generally considered to be the most important solid-state microwave development in the 1970s. The schematic in Fig. 9-90 illustrates its principle of operation. The flow of charge carriers from source to drain is controlled by the potential applied to the gate electrode. This flow takes place in a thin layer of n-type GaAs, usually fabricated by vapor-phase epitaxy. The gate is generally a Schottky barrier formed by deposition of an appropriate metal, leading to the acronym MESFET.

The electron current from source to drain is determined by the depth of the depletion layer formed under the gate by the negative potential applied to it. The frequency response of this type of amplifier goes up as the gate length is reduced, while its power output increases linearly with gate width. Typical gate widths range from ¼ to 1 μm, while gate lengths vary from a fraction of a millimeter for low-noise preamps to several millimeters for power devices. About 1 W of microwave power per millimeter is a widely used figure of merit.

Noise performance is shown in Fig. 9-91 for narrow-band amplifiers. Broadband (octave frequency coverage) performance is 1 to 2 dB poorer. Further improvements are to be expected in this rapidly moving field. The variation of gain and noise figure with bias current of a typical FET is shown in Fig. 9-92. Efficiencies for power devices are typically 30%.

Dual-Gate FETs.   By introducing a second gate, as shown in Fig. 9-93, functional capability can be improved. Mixing can be provided, for example, by applying the local oscillator voltage to the second gate. Gain control and improved isolation from feedback can also be obtained by the use of the second gate. Figure 9-94 illustrates the variable-gain feature.

62. Transferred-Electron (Gunn) Devices. The negative resistance that leads to oscillation for this class of device is a consequence of the band structure of certain semiconductors. An upper conduction band must exist in which carriers have lower mobility than in the initially occupied lower band. The transfer of carriers, generally electrons, from the lower to the

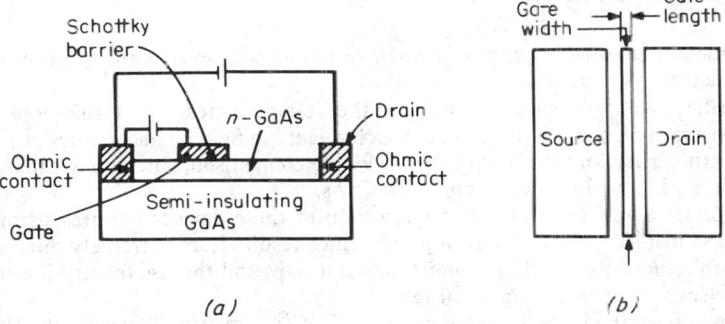

Fig. 9-90. (a) Schematic and (b) dimensions of a single-gate Schottky barrier GaAs field-effect transistor. (Microwave J., November 1978.)

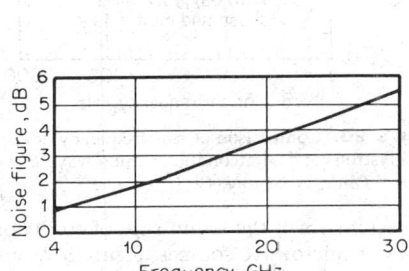

Fig. 9-91. Best reported noise figure as a function of frequency for Schottky GaAs FETs. (Microwave J., November 1978.)

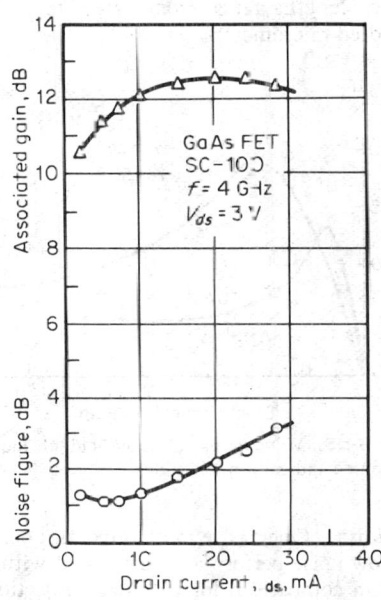

Fig. 9-92. Variation of noise and gain with drain current of a packaged GaAs FET. (IEEE Trans., June 1978, Vol. ED-25, No 6.)

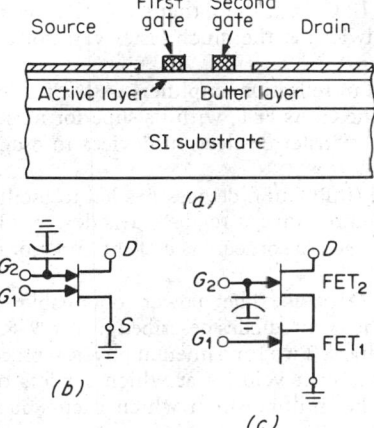

Fig. 9-93. Dual-gate FET: (a) cross section; (b) symbolic representation; (c) equivalent configuration of two single-gate FETs. (Trans. IEEE, June 1978, Vol. ED-25, No. 6.)

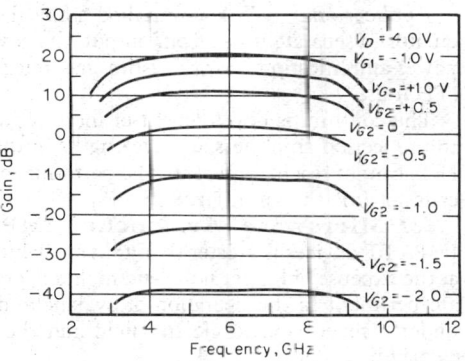

Fig. 9-94. Variation in gain vs. frequency as a function of the bias voltage V_{G2} applied to the second gate of a dual-gate FET amplifier. (Microwave J., November 1978.)

upper conduction band takes place as a result of lattice collisions as the electric field strength across the material is increased.

The transfer leads to a reduced current as the voltage increases and therefore represents a negative resistance on the v-i curve. The velocity–electric field characteristics of two materials that exhibit this behavior are shown in Fig. 9-95. By comparison, silicon has a monotonic characteristic which lies well below the curve for GaAs.

Fabrication of Gunn Devices.   The fabrication of these devices requires stringent control over the GaAs material. Good electrical performance results from extremely pure and uniform material with a minimum of deep donor levels and traps and the use of very-low-loss contacts. Modern devices use an n-type epitaxial layer of GaAs grown from the liquid or vapor phase on n^+ bulk material. Typical carrier concentrations range from 10^{14} to 10^{16} cm^{-3}, while device lengths range from a few to several hundred micrometers.

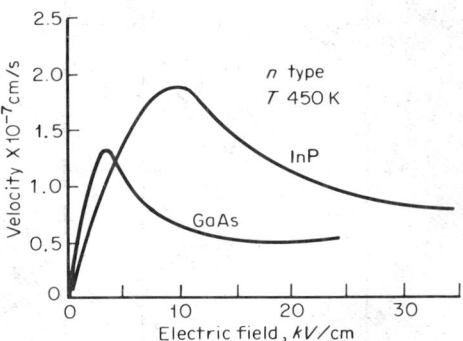

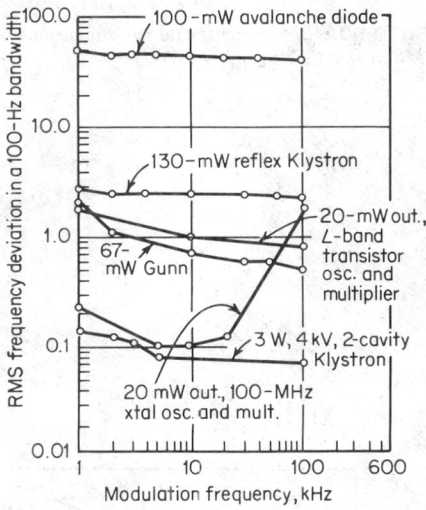

Fig. 9-95. Velocity vs. electric field of indium phosphide and gallium arsenide.

Fig. 9-96. Comparison of rms frequency deviation in klystrons and semiconductor microwave devices. (*After Fank, WESCON, 1971.*)

Noise.   One of the more important operating characteristics of the Gunn type of oscillator is its low-noise performance compared with other types of microwave sources. Figure 9-96 gives such a comparison for FM noise. Injection locking with a low-power source which has been cavity-stabilized (since FM noise is inversely proportional to loaded Q) can be used to reduce noise further.

Electronic Tuning.   Electronically tuned Gunn oscillators are available employing YIG spheres, varactor diodes, and *pin* diodes as the tuning element, in addition to mechanically tuned devices. Typical ranges covered are 8 to 12 and 12 to 18 GHz at the rather slow YIG magnetic-tuning speed of 100 Hz and of the order of 10% bandwidth at the much faster varactor rate of up to 10 MHz.

Applications.   While Gunn diodes have been used in reflection amplifiers with a ferrite circulator to separate input from output, the advent of the GaAs FET, with its superior noise and power-amplification characteristics, has relegated the transferred-electron devices to oscillator (TEO) use.

The continuing development of indium phosphide (InP) Gunn devices has led to oscillators and reflection amplifiers at wavelengths in the 3- to 5-mm range, a region GaAs devices (TE or FET) cannot reach at present. The performance advantage is a consequence of the superior properties of InP, shown in Fig. 9-95.

63. Microwave Avalanche (IMPATT) Diodes. The power obtainable from IMPATT devices is greater than that available from the Gunn diodes described in Par. **9-62** but at the expense of higher noise and higher operating voltage. Two fundamental physical processes are pertinent to the operation of avalanche diodes: the drift velocity at which carriers travel under a reverse-biased electric field and the avalanche multiplication which occurs at sufficiently high fields.

Read Diode.   An understanding of the dynamic operating characteristics of IMPATT diodes can be best obtained by considering the operation of the Read diode. The structure, field distribution, and current waveforms for a reverse-biased Read diode are shown in Fig. 9-97. The Read diode consists of two regions: a narrow avalanche region (p region), in which carrier multiplication by impact ionization occurs, and a drift region (i region), in which the carriers drift at saturated or field-independent velocities and where no impact ionization occurs.

The negative resistance or conductance of a Read diode is attributed to phase shift between the current through the diode and the voltage across it. This phase shift consists of two components. There is a phase delay of the current caused by the avalanche multiplication process and by the finite transit time of the holes drifting through the drift region. The phase delay caused by the avalanche process results from the condition that the rate of generation of electron-hole pairs in the avalanche region is proportional to both the electric field and the density of electrons and holes that are already present.

If a small rf voltage of sufficiently high frequency is assumed to be superimposed along with a dc voltage near breakdown across the diode, the rate of generation of electron-hole pairs will exceed the rate at which the pairs leave the avalanche region. Thus, both the density of carriers and the current will grow exponentially with time as long as the rf and dc voltages add to give a total field above breakdown. When the rf voltage changes sign and subtracts from the dc voltage, causing the field to fall below breakdown, the generation rate starts decreasing, causing the current to decrease.

Thus the current generated by the avalanche process has its maximum when the rf field goes through zero. That is, as shown in Fig. 9-97, the avalanche process contributes a 90° inductive phase lag to the current generated in the avalanche region. This current is then injected into the drift region of the diode. The current induced in the external circuit by this drifting charge is shown at the bottom of Fig. 9-97. It is obvious that the fundamental component of the external current is more than 90° out of phase with the rf voltage, causing the resistance or conductance of the diode to be negative. If the diode is to operate as a stable oscillator, the negative conductance of the diode must decrease with increasing rf voltage. The rf voltage across the diode will grow until the admittance of the diode is balanced by the admittance of the diode mount and microwave circuit.

The effect of the transit time on the frequency range of operation can be understood by assuming that the current generated in the avalanche region is generated in a sharp pulse. The precise time in the cycle when this pulse is generated is a function of the sum of the dc current and the rf voltage, as discussed previously. The phase delay caused by the transit time is proportional to $\omega\tau$, where τ is the transit time in seconds. For a given diode the transit time τ remains constant but the phase delay decreases with decreasing frequency. At a sufficiently low frequency the phase delay of the transit time plus the phase lag of the avalanche process are not sufficient to have the fundamental component of the external current lag the rf voltage by more than 90°. Therefore, below a certain cutoff frequency, the conductance of the diode becomes positive.

IMPATT Diode. At present the semiconductor materials used in commercial IMPATTs are silicon and gallium arsenide. The latter is operated in this case at much higher electric fields than corresponds to the region of negative differential mobility utilized in the Gunn effect. The basic structure of a typical *pn* junction silicon IMPATT is shown in Fig. 9-98.

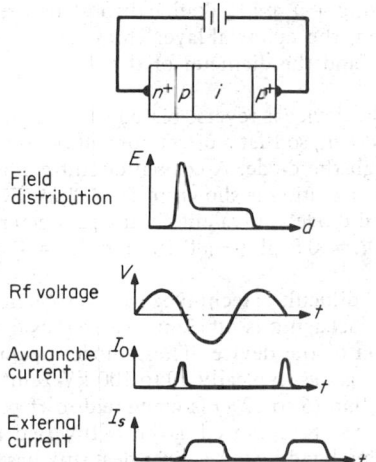

Fig. 9-97. Waveforms of the Read diode. *(IEEE Trans. MTT, November 1970.)*

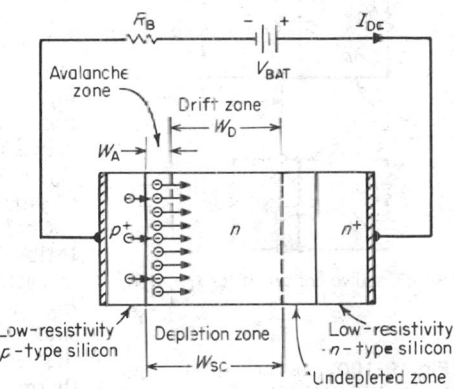

Fig. 9-98. pn junction of IMPATT diode under reverse bias. *(After Cowley, WESCON, 1971.)*

Significant improvement in performance is obtained, however, by modifying the doping profile so that the high-field avalanche region is narrowly confined and the electric field is optimized separately for the drift region. Such profiles, called high-low, low-high-low, and double-

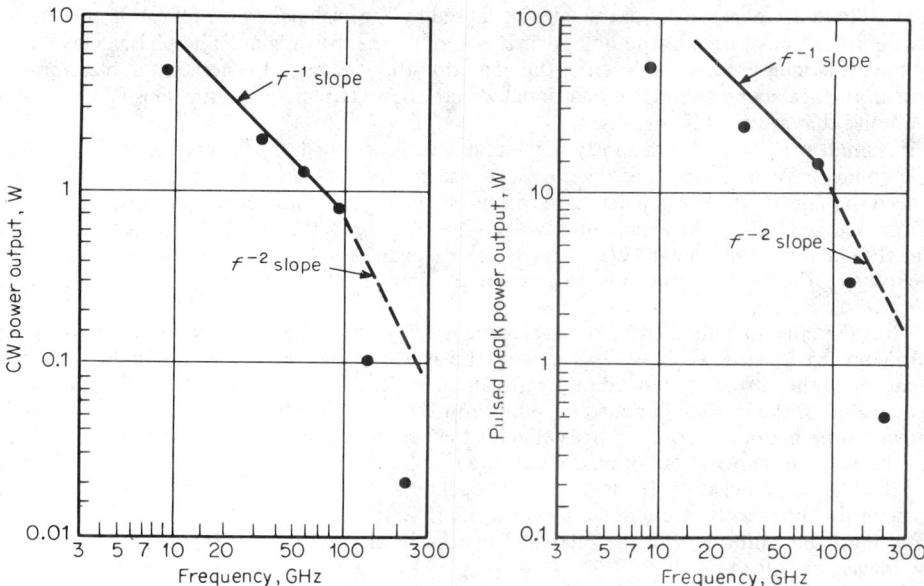

Fig. 9-99. Power vs. frequency of cw (*left*) and pulsed (*right*) IMPATT oscillators. (*Microwave J., June 1979.*)

drift, yield efficiencies of 20 to 35% compared to 6 to 10% and 10 to 15% obtained from flat-profile Si and GaAs respectively.

GaAs IMPATTs are preferred for frequencies up to 30 GHz because of their performance advantages. Si IMPATTs are used in the millimeter wave range and currently give the highest power available from a solid-state device in this frequency region. Power-vs.-frequency performance in the early 1980s is shown in Fig. 9-99.

Fabrication. The IMPATT structure is fabricated by first growing a thin epitaxial layer of n-type silicon on a heavily doped n-type (n^+) substrate and then adding a p layer by growing it epitaxially or by ion implantation. Finally a thin platinum contact layer is formed by a diffusion process, giving a p^+pnp^+ double-drift structure. For 94-GHz operation, the epitaxial-layer thickness is of the order of 0.5 μm, and the diameter of the diode is of the order of 50 μm.

In operation the device is reverse-biased past the point of avalanche breakdown, so that a direct current of 50 to 500 mA flows through the diode. A coaxial circuit configuration of IMPATT operation is shown in Fig. 9-100. Typical values for X-band diode equivalent-circuit parameters are $R_D = -2\ \Omega$, $X_D = 0.5$ pF ($-j30\ \Omega$), and $L_p = 0.6$ nH ($+j40\ \Omega$).

Cooling. A difficult technological problem in IMPATT diode packaging is efficient heat removal from the active portion of the device. These diodes operate at high dc power densities, typically 10 to 100 kW/cm^2, and since only a fraction (6 to 12%) is converted to rf power, the remainder must be removed as heat. Inverted mesa thermocompression bonding to a copper heat sink has been employed for this purpose. A better approach electroplates the heat sink at the wafer stage before individual diodes

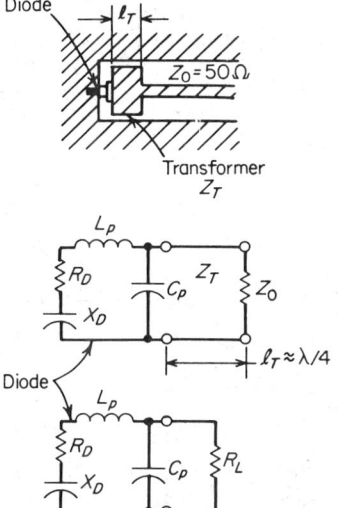

Series equivalent circuit for $\ell_T = \lambda/4$

$$R_L = \frac{Z_T^2}{Z_0}$$

Fig. 9-100. Low-Q coaxial circuit for use with IMPATT diodes as amplifiers or oscillators. (*After Cowley, WESCON, 1971.*)

have been fabricated. Diamond has also been employed as a heat sink in experimental devices using inverted chips.

Performance Limitations. A major limitation on the use of IMPATTs in system configurations arises from their high noise. Figure 9-101 compares the FM noise of several types. This characteristic has tended to restrict system applications to transmitters. Another effect that can

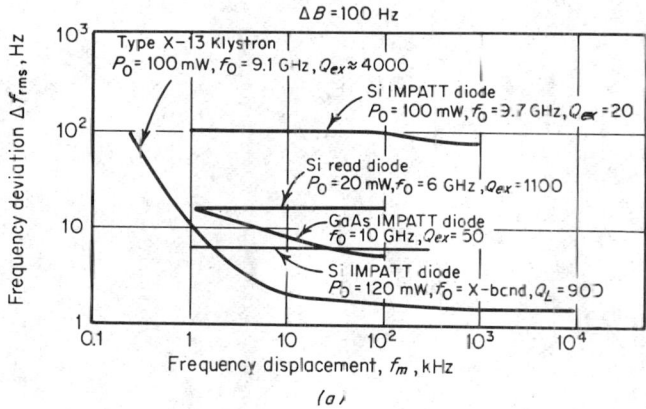

(a)

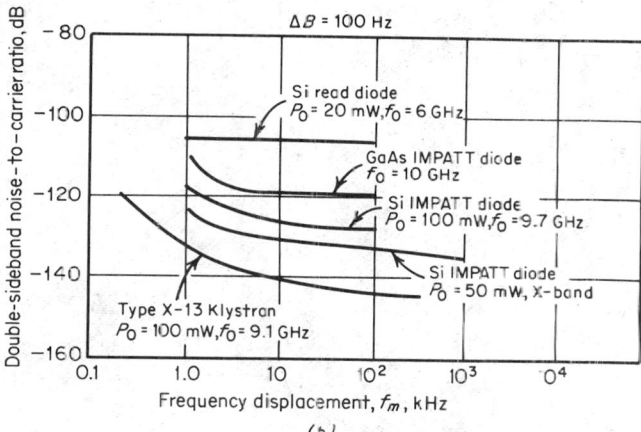

(b)

Fig. 9-101. Noise data for IMPATT diodes: (*a*) FM noise; (*b*) AM noise. (*IEEE Trans MTT, November 1970.*)

cause problems is the downward thermal drift of oscillation frequency during a pulse. Various types of stabilizing circuits, employing phase and injection locking, are necessary to provide coherence. Alternatively the frequency chirping can be used by tailoring the current pulse waveform.

64. References on Microwave Semiconductor Devices

Watson, H. A. "Microwave Semiconductor Devices and Their Circuit Applications," McGraw-Hill, New York, 1969.

Special Issue on Microwave Semiconductor Devices *IEE Trans. Electron Devices*, June 1978, Vol. ED-25, No. 6.

Special Issue on Solid State Microwave Techniques *IEEE Trans. Microwave Theory Tech*, May 1979, Vol. MTT-27, No. 5.

Microwave J., June 1979, Vol. 22, No. 6.

Transducers and Sensors*

HARRY N. NORTON *Member, Technical Staff, Jet Propulsion Laboratory, California Institute of Technology; Author, "Handbook of Transducers for Electronic Measurement Systems," "Sensor and Analyzer Handbook," and "Biomedical Sensors—Fundamentals and Applications"*

CONTENTS

Numbers refer to paragraphs

---

*The editors acknowledge the assistance given by the late Dr. Eugene Mittelmann in revising and expanding this section.

TRANSDUCER CHARACTERISTICS AND SYSTEMS

1. The Transducer Concept. Measurements by direct comparison with a reference standard having the same characteristics as those of the quantity measured are called *direct measurements*, but most measurements yield results in a more indirect way. They are based on knowledge of the relationship between the quantity to be measured and the response of the measuring instrument or system influenced by it. Moving-coil meters are an example: under the influence of the applied voltage or current (or both) a mechanical torque is generated. The pointer attached to the moving coil acquires its final position indicative of the electrical quantity measured, the mechanical torque being balanced by a spring. Thus an electrical quantity is first translated into a torque, which in turn is translated into a position on a scale calibrated in units of the quantity measured.

In this example, the electrical quantities are translated into mechanical ones. In contrast, *transducers* translate nonelectrical quantities into electric signals.

The term transducer has been applied to a variety of devices, including measuring instruments, acoustic-energy transmitters, signal converters, and phonograph cartridges. With the recent vast increase in the development and use of electronic measuring systems, however, instrumentation engineers found it necessary to devise a more limited definition of transducer as a device used for measurement purposes.

The Instrument Society of America (ISA) published its Standard S37.1 in 1969. This standard, Electrical Transducer Nomenclature and Terminology, defines a *transducer* as "a device which provides a usable output in response to a specified measurand." The *measurand* is "a physical quantity, property or condition which is measured." The *output* is "the electrical quantity, produced by a transducer, which is a function of the applied measurand." Only the last of these three definitions applies specifically to electrical transducers. It could apply equally well to transducers with pneumatic output if the word "electrical" in the definition were omitted. Only electrical transducers are covered in this Handbook.

ISA S37.1-1969 also applies to the construction of transducer nomenclature (see Table 10-1). When used in titles or for indexes, the sequence shown in the table should be used, e.g., "transducer, acceleration, potentiometric, ±5g." When the nomenclature is used in the text, the opposite of the sequence shown in the table should be used, e.g., "A 0- to 8-cm dc output reluctive displacement transducer was installed on the actuator."

2. Transducer Classes and Elements. Thermocouples are representatives of a transducer class in which transduction takes place in a single component. An electric signal is generated between the terminals of the two dissimilar wires forming the thermocouple when they are exposed to a temperature difference. This is the simplest class of transducer. It is also *self-generating*, the output signal being produced without an additional power source. Piezoelectric transducers are also of the self-generating type, electric charge or potential being generated when the crystal is exposed to stress.

Reference may be made to *sensing elements* (sensors) in distinction to the transducer elements proper, but sensing elements are actually transducer elements themselves; i.e., they perform the first step in a multistep translation process. In the Bourdon-tube type of pressure transducer, pressure is first translated into a deflection of the Bourdon tube, then into the movement of the wiper arm on a resistive potentiometer, resulting in an electric-signal output determined by the

TABLE 10-1 Construction of Typical Transducer Nomenclature and Examples of Modifiers

Main noun	First modifier, measurand, examples	Second modifier, restricts measurand, examples	Third modifier, electrical transduction principle, examples	Fourth modifier[a], sensing element, special features or provisions, examples	Range, examples	Unit, examples
Transducer	Acceleration	Absolute	Capacitive	AC output	0 to 1,000	A
	Air speed	Angular	Electromagnetic	Amplifying	±5	°C
	Attitude	Differential	Inductive	Bellows	−100 to +500	cm
	Attitude rate	Gage	Ionizing	Bondable	−430 to −415	cm/s
	Current	Infrared	Photoconductive	Bonded		deg
	Displacement	Intensity	Photovoltaic	Bourdon tube		°F
	Flow rate	Linear	Piezoelectric	Capsule[b]		ft/s
	Force	Mass	Potentiometric	DC output		g
	Heat flux	Radiant	Reluctive	Diaphragm		Hz
	Humidity	Relative	Resistive	Digital output		in/s
	Jerk	Surface	Strain gage	Discrete-increment		in
	Light	Total	Thermoelectric	Dual-output		K
	Liquid level	Volumetric		Exposed-element		kg
	Mach number			Frequency output		lb/min
	Nuclear radiation			Gyro		m
	Pressure			Integrating		mmHg
	Speed[c]			Self generating		N
	Sound pressure			Semiconductor		% RH
	Strain			Servo[d,e]		lb/in²
	Temperature			Switch		kPa
	Torque			Toothed-rotor		mbar
	Velocity[f]			Triaxial		rad/s
				Turbine		
				Ultrasonic		
				Unbonded		
				Vibrating-element[g]		
				Weldable		

[a]Nomenclature may include two of these terms.

[b]Preferred to "aneroid."

[c]Scalar quantity.

[d]Preferred to "force balance" or "null balance."

[e]When this modifier is used, the third modifier ("transduction principle") may be omitted.

[f]Vector quantity.

[g]When this modifier is used together with "frequency output," the third modifier may be omitted.

position of the potentiometer wiper arm. Many transducer types make use of such a multistep process before the output electric signal becomes available.

The term *translator chain* was coined by B. Ziebolz for such operations. In microprocessor operations for process control, e.g., in the chemical and automotive industries, it is customary to refer to the complete transducer system as a *sensor*. (For a list of sensors for automotive applications, see Table 10-7, Par. **10-98**.)

Excitation. Most transducers require an external power source to carry out the translation process. Whenever the transducing element forms part of a closed electric circuit, the use of an external power source becomes mandatory. Examples are resistive temperature measurements in a bridge circuit or displacement measurements by means of linear differential transformers.

3. Signal Conditioning. The widespread acceptance of industrial standards specific to a particular industry or application demands that the transducer output be in a specific form and within a specific range. Process industry standards, for instance, demand a transducer output signal either in the form of a direct current in the range from 4 to 20 mA or a direct voltage between 0 and 5 V. Transducers delivering control signals to microprocessors often require a digital output.

The process and the steps involved to provide an output signal in whatever specific form required are referred to as *signal conditioning.* The equipment may include ac or dc amplifiers, rectifiers, demodulators, circuits for square-root extraction, logarithmic amplifiers, etc., depending on the laws governing the relationship between the measurand and the desired output signal. For example, the radiation emanating from the surface of a body is an exponential function of its temperature. Thus to obtain a linear temperature scale at the output, logarithmic amplifiers are typically involved in the signal-conditioning process. In addition, voltage or current regulation circuits are used in many instances to ensure accuracy and repeatability.

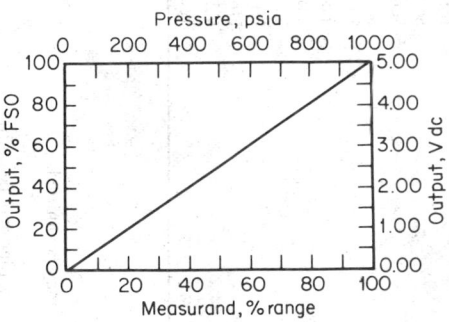

Fig. 10-1. Output-measurand relationship of ideal linear-output transducer as exemplified for a dc-output transducer. *(From Ref. 41 by permission of Prentice-Hall, Inc.)*

When a transducer system includes all the above-mentioned elements and circuits and is packaged in a single housing, it is referred to as an *integrally conditioned transducer.* Recent examples are integrated-circuit semiconductor transducers for temperature and pressure measurements contained in a single dual-in-line pin housing.

4. Transfer Function. Every transducer design can be characterized by an ideal or theoretical output-measurand relationship (*transfer function*). This relationship is capable of being described exactly by a prescribed or known *theoretical curve* (see Fig. 10-1), stated in terms of an equation, a table of values, or a graphical representation. This applies primarily to the *static characteristics* of a transducer, i.e., the output-measurand relationship for a steady-state or very slowly varying measurand. It can also apply to the transducer *dynamic characteristics*, i.e., the output-measurand characteristics for a relatively rapidly fluctuating measurand. However, this dynamic behavior is described by relationships other than the transducer's theoretical curve.

5. Transducer Errors. Because of a variety of factors, the behavior of a real transducer is nonideal. These factors include production variations, as well as the use of nonideal materials, production methods, ambient conditions during manufacture, and testing methods. It must also be recognized that many trade-offs enter into the design of a marketable transducer and that our knowledge (the *state of the art*) is limited with regard to producing an ideal transducer design and then compensating it perfectly for aging effects and a variety of environmental conditions the transducer may be subjected to during its operation.

Hence the measurand value indicated by the transducer may often differ from the true measurand value or the specified theoretical value. The algebraic difference between the indicated and the true value is the *error*.

6. Error Band. A convenient manner of determining or specifying transducer errors is to state them in terms of the band of maximum (or maximum allowable, for a specification) deviations from a specified reference line or curve. This band is defined as the *error band* The static *error band* (see Fig. 10-2) is that error band obtained (or obtainable) by means of a *static calibration*, which is performed under "room conditions" (controlled room temperature, humidity, and atmospheric pressure) and in the absence of any vibration, shock, or acceleration (unless one of these is the measurand) by applying known values of measurand to the transducer and recording corresponding output readings. Other types of error band are applicable under somewhat different (and rigorously specified) conditions.

7. Dynamic Characteristics. When a step change in a measurand is applied to a transducer, the transducer output does not instantaneously indicate the new measurand level. Examples of such step changes are mechanical shock, a sudden pressure rise when a solenoid valve opens, or a temperature transducer is rapidly immersed in a very cold liquid. The lag between the time the measurand reaches its new level and the corresponding steady (final) transducer output reading is defined in various ways. The time required for the output change to reach 63%

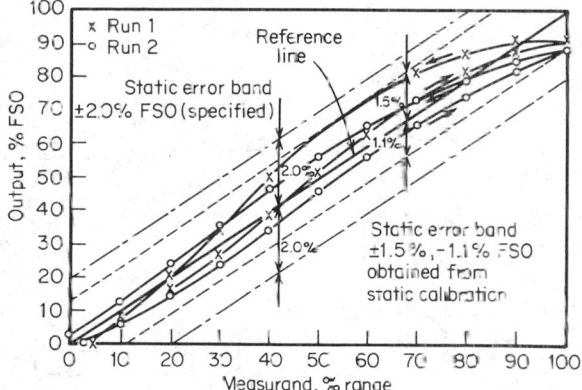

Fig. 10-2. Static error band referred to terminal line (error scale 10.1). (From Ref. 41 by permission of Prentice-Hall, Inc.)

of its final value is the *time constant* of the transducer. The time required to reach a different specified percentage of this final value (say 90 or 98%) is the *response time*. The time in which the output changes from a small to a large specified percentage of the final value (usually from 10 to 90%) is the *rise time*. The output may rise beyond the final value before it stabilizes at that value. This *overshoot* depends on the *damping* characteristics of the transducer.

When the measurand fluctuates (sinusoidally) over a stated frequency range, the transducer output may not be able to indicate the correct amplitude of the measurand over these excursions. An example is the mechanical vibration (vibratory acceleration) of an engine housing. The output may be somewhat higher at certain measurand frequencies but usually drops off as the frequency increases until the output is essentially zero. The change with measurand frequency of the output-measurand amplitude ratio is the *frequency response* of the transducer, always stated for a specified frequency range. The above characteristics, as well as other transducer dynamic characteristics, are defined in the terminology section of ISA Standard S37-1-1969.

8. Environmental Characteristics. In most applications transducers are used only under the controlled room conditions of the facility where they are calibrated and where various static performance characteristics are determined. The external conditions to which a transducer is exposed not only while operating but also during shipping, storage, and handling can contribute additional errors, such as temperature error, acceleration error, or attitude error. Such environmental conditions (which can also include corrosive atmosphere, salt-water immersion, or nuclear radiation) may even cause a permanent deterioration or malfunction in the transducer.

9. Transduction Principles. The most essential determinant of any one transducer type is its transduction principle. How the electrical output is originated affects most other characteristics of the transducer. The most frequently used transduction principles are described below and illustrated in Fig. 10-3. It should be noted that photovoltaic, piezoelectric, and electromagnetic transduction are used in *self-generating* transducers, whereas all other transduction methods illustrated require some sort of external excitation power.

Photovoltaic Transduction. The measurand is converted into a change in the voltage generated when a junction between certain dissimilar materials is illuminated. Used primarily in optical sensors, this principle has also been employed in transducers incorporating mechanical-displacement shutters to vary the intensity of a light beam between a built-in light source and the transduction element.

Piezoelectric Transduction.   The measurand is converted into a change in the voltage E or electrostatic charge Q generated by certain crystals when mechanically stressed by compression

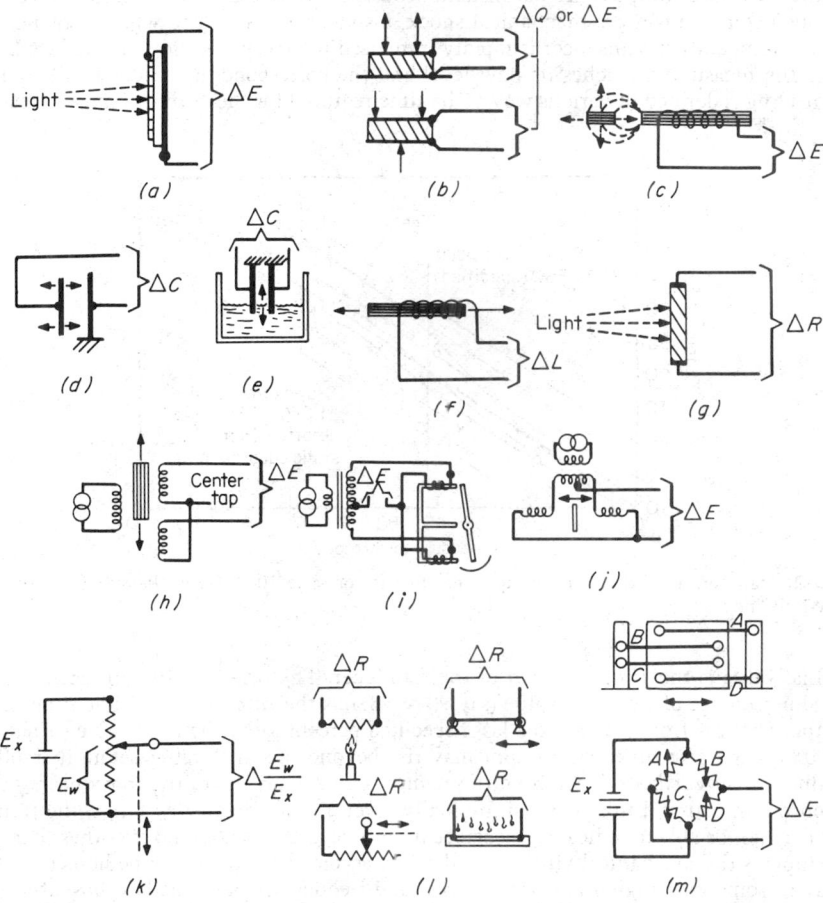

Fig. 10-3. Transduction principles: (*a*) photovoltaic; (*b*) piezoelectric; (*c*) electromagnetic; (*d*), (*e*) capacitive; (*f*) inductive; (*g*) photoconductive; (*h*), (*i*), (*j*) reluctive; (*k*) potentiometric; (*l*) resistive; (*m*) strain gage. (*From Ref. 41 by permission of Prentice-Hall, Inc.*)

or tension forces or by bending forces. Either natural or synthetic crystals (usually ceramic mixtures) are used in such transduction elements.

Electromagnetic Transduction.   The measurand is converted into a voltage (electromotive force) induced in a conductor by a change in magnetic flux, usually due to a relative motion between a magnetic material and a coil having a ferrous core (electromagnet).

Capacitive Transduction.   The measurand is converted into a change of capacitance. This change occurs typically either by having a moving electrode move to or from a stationary electrode or by a change in the dielectric between two fixed electrodes.

Inductive Transduction.   The measurand is converted into a change of the self-inductance of a single coil.

Photoconductive Transduction.   The measurand is converted into a change in conductance (resistance change) of a semiconductive material due to a change in the illumination incident on the material. This transduction is implemented in a manner similar to that explained for the case of photovoltaic transduction, above.

Reluctive Transduction. The measurand is converted into an ac voltage change by a change in the reluctance path between two or more coils while ac excitation is applied to the coil system. This transduction principle applies to a variety of circuits, including the differential transformer and the inductance bridge.

Potentiometric Transduction. The measurand is converted into a change in the position of a movable contact on a resistance element. The displacement of the contact (wiper arm) causes a change in the ratio between the resistance from one end of the element to the wiper arm and the end-to-end resistance of the element. In its most common applications the resistance ratio is used in the form of a voltage ratio when excitation is applied across the resistance element.

Resistive Transduction. The measurand is converted into a change of resistance. This change is typically effected in a conductor or semiconductor by heating or cooling, by the application of mechanical stresses, by sliding a wiper arm across a rheostat-connected resistive element, or by drying or wetting electrolytic salts.

Strain-Gage Transduction. The measurand is converted into a resistance change, due to strain, usually in two or four arms of a Wheatstone bridge. This principle is a special version of resistive transduction. However, the output is always given by the bridge-output voltage change. In the typical configuration illustrated in Fig. 10-3 the upward arrows indicate increasing resistance, and the downward arrows decreasing resistance, in the respective bridge arms for sensing link motion toward the left.

10. Transducers in Process Control. (See also Sec. 24.) The availability of an output signal responding to a measurand makes transducers key elements in the many areas of industrial process control. It permits direct comparison of a transducer output with an adjustable electric reference signal, calibrated in units of the measurand. Both the transducer output and the adjustable reference signal are fed into the input terminals of a differential amplifier. When any deviation of the transducer output from the adjustable reference signal (*set point*) occurs, an error signal is supplied to the controller, which in turn causes the control element to acquire such a position that the measurand again causes the transducer to deliver an output equal to the adjustable reference signal. This control process can be carried out in several alternative modes, e.g., on-off or proportional.

TRANSDUCERS FOR SOLID-MECHANICAL QUANTITIES

11. Terminology. *Acceleration,* a vector quantity, is the time rate of change of velocity with respect to a reference system. When the term acceleration is used alone, it usually refers to *linear acceleration a,* which is then related to linear (translational) velocity v, and time t by $a = dv/dt$. *Angular acceleration* α is related to angular (rotational) velocity ω and time t by $\alpha = d\omega/dt$. Mechanical *vibration* is an oscillation wherein the quantity, varying in magnitude with time so that this variation is characterized by a number of reversals of direction, is mechanical in nature. This quantity can be stress, force, displacement, or acceleration; however, in measurement technology the term vibration is usually applied to *vibratory acceleration* and sometimes to *vibratory velocity.* Mechanical *shock* is a sudden nonperiodic or transient excitation of a mechanical system.

12. Acceleration transducers (accelerometers) are used to measure acceleration as well as shock and vibration. Their sensing element is the *seismic mass,* restrained by a spring. The motion of the seismic mass in this acceleration-sensing arrangement is usually damped (see Fig. 10-4a). Acceleration applied to the transducer case causes motion of the mass relative to the case. When the acceleration stops, the mass is returned to its original position by the spring (see Fig. 10-4b). This displacement of the mass is then converted into an electrical output by various types of transduction elements in *steady-state acceleration transducers* whose frequency response extends down to essentially 0

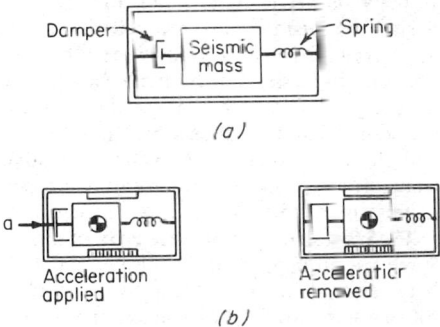

Fig. 10-4. Basic operating principle of an acceleration transducer: (a) spring-mass system; (b) displacement of seismic mass. (*From Ref. 41 by permission of Prentice-Hall, Inc.*)

Hz. In piezoelectric accelerometers the mass is restrained from motion by the crystal transduction element, which is thereby mechanically stressed when acceleration is applied to the transducer. Such *dynamic acceleration transducers* do not respond appreciably to acceleration fluctuating at a rate of less than 5 Hz. They are normally used for vibration and shock measurements.

Capacitive and photoelectric accelerometers have been produced at various times, and vibrating-element accelerometers (in which the mass, as it tends to move, applies tension to a wire or ribbon, thereby changing the frequency at which the wire can oscillate) have been used in some aerospace programs. However, the most commonly used steady-state acceleration transducers are the potentiometric, reluctive, strain-gage, and servo types. For vibration and shock measurement the piezoelectric accelerometers are most frequently used because of their inherently high fre-

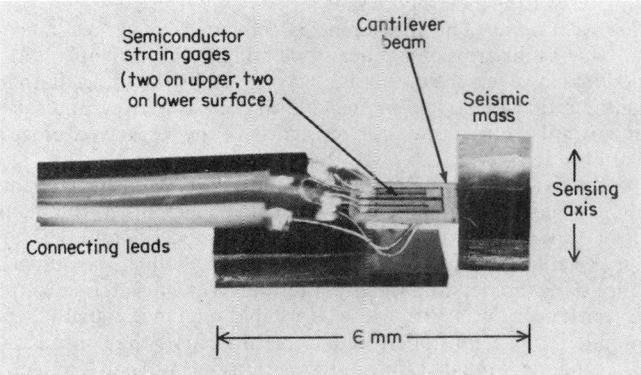

Fig. 10-5. Semiconductor-strain-gage acceleration transducer. *(Entran Devices, Inc.)*

quency-response capability; some miniature semiconductor-strain-gage accelerometers are also used for these measurements since they can respond to fairly high acceleration frequencies.

13. Potentiometric accelerometers usually employ a mechanical linkage to amplify the motion of the seismic mass so as to produce the necessary extent of wiper-arm travel over the resistance element. The mass is supported by flexural springs or a cantilever spring in some models. In others it slides on a central coaxial shaft, restrained by calibrated coil springs. Magnetic, viscous, or gas damping is normally used in potentiometric accelerometers, primarily to reduce output noise due to wiper-arm whipping and transient wiper-contact resistance changes. Overload stops keep the wiper arm from moving beyond the resistance-element ends in the presence of acceleration beyond the range of the accelerometer.

14. Reluctive accelerometers require ac excitation power having a frequency greater than the upper limit of the transducer's frequency response. When moderately high frequency response is needed, the inductance-bridge version has been found most suitable. In a typical design the seismic mass is attached to a spring-restrained ferromagnetic armature plate, pivoted at its middle and placed above two coils so that the small seesaw motion of the plate, due to acceleration action on the mass, causes a decrease of inductance in one coil and an increase of inductance in the other. Since the coils are in opposite bridge arms, these inductance changes are additive and produce a bridge output voltage double that obtainable from having only one coil change its inductance. When a relatively low frequency response is needed, a differential-transformer synchro or microsyn transduction circuit can be used to convert the seismic-mass displacement into the required electrical output.

15. Strain-gage accelerometers are very popular and exist in several design versions. Some use unbonded metal wire stretched between the seismic mass and a stationary frame or between posts on a cross-shaped spring to whose center the seismic mass is attached and whose four tips are attached to a stationary frame. Other designs use bonded-metal wire, metal foil, or semiconductor gages bonded to one or two elastic members deflected by the displacement of the seismic mass. The transducer shown in Fig. 10-5 has two semiconductor strain gages bonded to the upper surface and two to the lower surface of a flat cantilever beam to whose tip the seismic mass is attached. The gages are usually connected as a four-active-arm bridge.

16. Servo accelerometers are closed-loop force-balance, torque-balance, or null-balance transducers. The displacement of the seismic mass is detected by a position-sensing element, usually reluctive or capacitive, whose output is the error signal in the servo system. This signal is amplified and fed back to a torquer or restoring coil so that the restoring force is equal and opposite to the acceleration-induced force. The coil or torquer is attached to the seismic mass and returns the mass to its original position, when the feedback current is sufficient so that the position error signal is reduced to zero. The current, which is proportional to acceleration, passes through a resistor. The *IR* drop across the resistor is the accelerometer output voltage, proportional to the acceleration.

17. Piezoelectric accelerometers exist in several design versions, two of which are illustrated in Fig. 10-6. Both contain a seismic mass which applies a force, due to acceleration, to a piezoelectric crystal. With acceleration acting perpendicular to the base, an output is generated by the crystal due to compression force in one design and to shear force in the other. Crystal materials include quartz and several ceramic mixtures such as titanates, niobates, and zirconates.

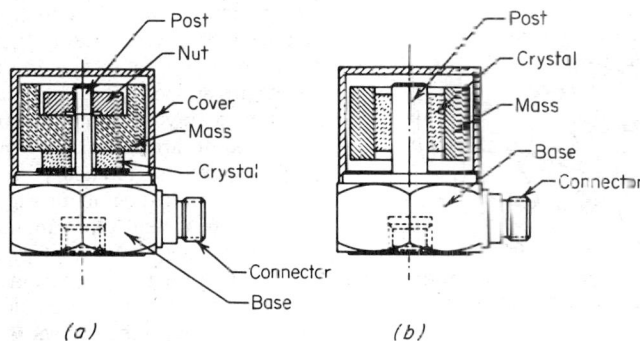

Fig. 10-6. Piezoelectric acceleration transducers: (*a*) single-ended compression type; (*b*) shear type. (*Endevco, Division of Becton, Dickinson & Co.*)

Ceramic crystals are used more frequently than natural crystals. They gain their piezoelectric characteristics by exposure to an orienting electric field during cooling after they are fired at a high temperature. If they are subsequently heated, as during transducer operation at elevated temperature, they can lose their piezoelectric qualities if that temperature is above the *Curie point*, which varies between about 100 and 600°C, depending on the materials used in the crystal. Piezoelectric accelerometers almost invariably require some signal-conditioning circuitry to provide a usable output since they have a relatively low output amplitude and a very high output impedance. In some designs, the necessary conditioning circuitry is included in the transducer. For most such accelerometers, a separate charge or voltage amplifier is needed, connected to the transducer by a thin shielded coaxial cable of special low-noise construction to avoid noise pickup from within the cable itself.

18. Criteria for selection of an acceleration transducer are primarily the required acceleration range and frequency response. They are mutually dependent; e.g., a typical $\pm 2g$ potentiometric accelerometer design will have an upper frequency limit for flat response of about 12 Hz, whereas a $\pm 20g$ accelerometer of the same design can have an upper frequency-response limit of about 40 Hz. As frequency-response requirements increase, the reluctive, servo, metal-strain-gage, semiconductor-strain-gage, and piezoelectric transducers successively become candidates for selection. The best accuracy characteristics are provided by servo accelerometers, which are also most suitable for low-range ($\pm 0.2g$ or lower) applications.

19. Attitude and Attitude-Rate Transducers. Attitude is the relative orientation of a vehicle or an object represented by its angles of inclination to three orthogonal reference axes. Attitude rate is the time rate of change of attitude.

The sensing methods employed by attitude transducers are best categorized by the kind of reference system to which the orientation to be measured is related. The *inertial* reference system is provided by a *gyroscope (gyro)* in which a rotating member will continue turning about a fixed axis as long as no forces are exerted on the member and the member is not accelerated. *Gravity* reference is used to establish a vertical reference axis. This principle is applied in *pendulum-type transducers*, in which a weight is attached to a wiper arm and a potentiometric ment is attached to the case, so that an output change is obtained when the object to which the case is mounted, deflects from a vertical position.

A *magnetic reference axis* can be established by the poles of a magnetic field which remains fixed in position. This reference system is employed by certain navigational transducers related to the compass. *Flow-stream reference* refers to the direction of fluid flow past an object moving within that fluid, a reference system employed in *angle-of-attack* transducers mounted well forward of the nose of high-speed aircraft and rockets, so that the flow stream sensed is not altered in direction by the vehicle itself.

Optical reference systems are used by electrooptical transducers mounted (in a known attitude) so as to sense a remote light source or a light-dark interface whose position is known. This establishes a reference axis between the object on which the transducer is mounted and the target sensed by the transducer. *Optically referenced* transducers include such aerospace (primarily spacecraft) devices as the sun sensor, Canopus sensor, star tracker, and horizon sensor, as well as military target-locating equipment.

20. Gyros are the most widely used attitude and attitude-rate transducers. The operating principle of the gyro is illustrated in Fig. 10-7. A fast-revolving rotor turns about the *spin axis* of the gyro. This axis, which remains fixed in space as long as the rotor revolves, establishes the inertial reference axis. The rotor shaft ends are supported by a *gimbal* frame which is free to pivot about the gimbal axis. The pivot points are part of the gyro housing structure, which is attached to the object whose changes in attitude about the gimbal axis are to be measured. An angular-displacement transduction element (pick-off) is then used to provide an output proportional to attitude. A simple example of such an element is a wiper arm, attached to the gimbal frame at the pivot point, wiping over a ring-shaped potentiometric resistance element attached to the inside of the case. Potentiometric transduction as well as reluctive (especially synchro) and, occasionally, capacitive and photoconductive transduction are used in most gyros.

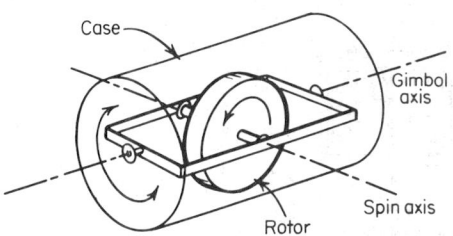

Fig. 10-7. Basic single-degree-of-freedom gyro. *(From Ref. 41 by permission of Prentice-Hall, Inc.)*

Gyro attitude transducers (free gyros) are often designed as two-degree-of-freedom gyros, i.e., those providing an output for each of two of a vehicle's three attitude planes (pitch, yaw, and roll, or x, y, and z axes). The design illustrated in Fig. 10-8 provides an inner gimbal for one axis and an outer gimbal for the other axis, with a separate pick-off for each axis. The caging mechanism (symbolized by the hand) is used to lock the inner gimbal to a reference position until the spin axis is to start serving as inertial reference axis. At this point the gyro is uncaged (after the rotor has come up to speed). AC or dc motors are commonly used to turn the rotor. Some gyros use a clock spring, wound before each use, or a pyrotechnic charge which, when activated, forces a stream of combustion gases into a small turbine.

Rate gyros are attitude-rate transducers. They provide an output proportional to angular velocity (time rate of change of attitude). The operating principle of the rate gyro (see Fig. 10-9) is similar to that of the single-degree-of-freedom free gyro, except that the gimbal is elastically restrained and its motion is damped. The output is representative of gimbal deflection about the output axis in response to attitude-rate changes about the input axis. The deflection of the gimbal (precession) is caused by the torque T applied to it. The applied torque is the product of the instantaneous attitude rate about the input axis and the angular momentum of the gyro.

When selecting a gyro, attention must be paid not only to the usual characteristics (weight, size, range, linearity, repeatability, threshold, etc., and dynamic characteristics for rate gyros) but also to *drift*, the amount of precession of the spin axis from its intended position due to internal unwanted torques, and the time period (after spin-motor runup or after uncaging) during which measurements must be obtained continuously.

21. Displacement and Position Transducers. *Position* is the spatial location of a body or point with respect to a reference point. *Displacement* is the vector representing a change in position of a body or point with respect to a reference point. Displacement transducers are used to measure linear and angular displacements, as well as to establish position from a displacement measurement.

The sensing element of most displacement transducers is the *sensing shaft* with its coupling device, which must be of a design suitable to make the motion of the sensing shaft truly representative of the motion of the measured point (driving point). A spring-loaded sensing shaft (without coupling device) is used for some applications. A number of *noncontacting* transducer designs are also in use. These require no coupling or sensing shaft. Various transduction principles are employed in displacement transducers.

22. Capacitive Displacement Transducers. In these devices a linear or angular motion of the sensing shaft causes a change in capacitance either by relative motion between one or more moving (*rotor*) electrodes and one or more stationary (*stator*) electrodes or by moving a sleeve of insulating material, having a dielectric constant different from that of air, between two stationary electrodes.

23. Inductive displacement transducers can be of the coupled or the noncontacting types. Coupled designs contain a coil whose self-inductance is varied as a nonmagnetic sensing shaft moves a magnetically permeable core gradually into or out of the central hollow portion

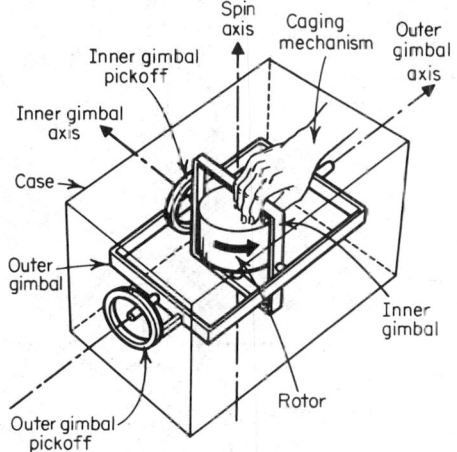

Fig. 10-8. Two-degree-of freedom gyro. (*Conrac Corp.; from Ref. 41 by permission of Prentice-Hall, Inc.*)

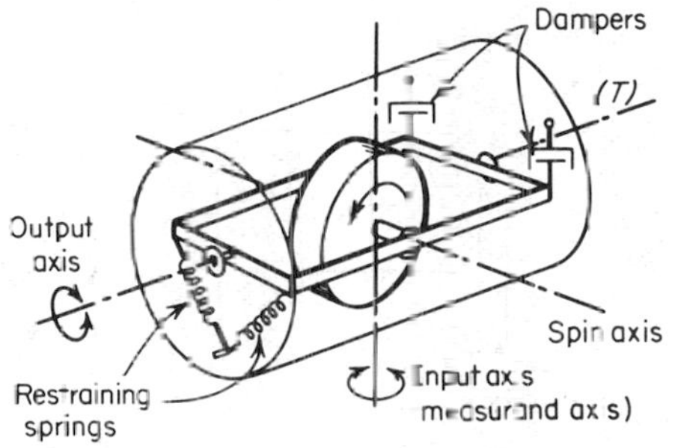

Fig. 10-9. Basic-rate gyro. (*From Ref. 41 by permission of Prentice-Hall, Inc.*)

of the coil. Some designs incorporate an additional coil (balancing coil) having a fixed inductance value equal to the inductance of the transduction coil at a predetermined "zero" position of the sensing shaft. The two coils are connected as two arms of an inductance bridge. This two-coil principle is used in some noncontacting displacement transducers in which the transduction coil has a stationary core but changes its inductance with the distance between itself and a moving ferromagnetic object.

24. Photoconductive displacement transduction is employed in at least one coupled displacement transducer design and in several noncontacting displacement measuring systems. In the coupled version the sensing shaft moves a shutter (a plate with a small slit) between a light source and either a potentiometric arrangement of photoconductive and conductive material or two photoconductive sensors connected in a bridge circuit and mounted so as to decrease the light incident on one sensor as the other sensor receives more light. Noncontacting photoconductive sensors usually require an optical reflector mounted to the measured object. Various optical configurations are used to obtain an output from the photoconductive element as the intensity, the phase, or the position of the reflected light beam changes. The *laser interferometer* is included in this group of displacement-sensing devices.

25. Potentiometric displacement transducers are widely used because of their relative simplicity of construction and their ability to provide a high-level output. All these designs use a sensing shaft. The wiper arm is either attached directly to the shaft (but insulated from it) or mechanically connected to it through an amplification linkage. Straight potentiometric resistance elements are used in linear displacement transducers, circular or arc-shaped elements in angular displacement transducers. The elements are usually wire-wound, but conductive plastic, carbon film, metal film, or ceramic-metal mixtures (cermets) are also used. Some transducers have two or more wiper-element combinations moved by the same sensing shaft. A good sliding seal is needed at the point where the sensing shaft enters the transducer case, to protect the internally exposed resistance elements from atmospheric contaminants and moisture.

Reluctive displacement transducers are as commonly used as the potentiometric types. The reluctive transduction circuits employed in linear- and angular-displacement transducers are illustrated in Fig. 10-10. Only the linear-variable differential transformer (LVDT) and the inductance-bridge circuits are used for linear-displacement measurements. Many winding configurations exist for the LVDT transducers; one manufacturer offers 12 different "off-the-shelf" con-

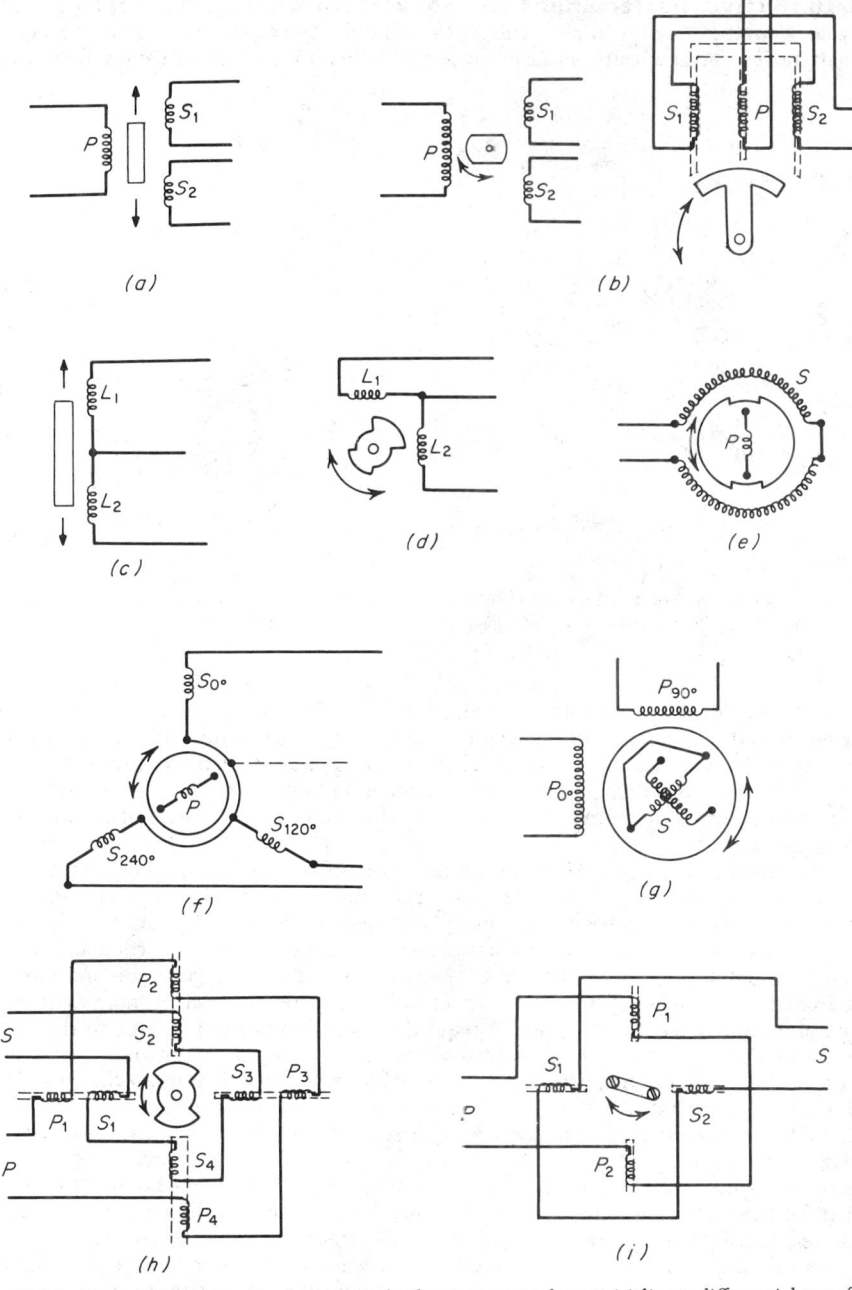

Fig. 10-10. Transduction circuits of reluctive displacement transducers: (a) linear differential transformer; (b) angular differential transformer; (c) linear inductance bridge; (d) angular inductance bridge; (e) induction potentiometer; (f) synchro; (g) resolver; (h) microsyn; (i) shorted-turn signal generator. *(From Ref. 41 by permission of Prentice-Hall, Inc.)*

figurations, including several with two separate secondary windings. Alternating-current excitation is required for all reluctive transducers. However, some designs are available with integral ac/dc output conversion and, in some cases, also integral dc/ac excitation conversion. Synchro-type transducers are often connected to a synchro-type receiver, which indicates the measured angle directly, e.g., on a dial.

A few strain-gage displacement transducers have been designed for the measurement of small linear and angular displacement. The gages are usually attached to the top and bottom surfaces of a cantilevered or end-supported beam which is deflected by the displacement.

26. Digital-output displacement transducers (encoders) are frequently referred to as *linear encoders* and *angular* or *shaft-angle encoders*, respectively. These consist essentially of a strip (for linear displacements) or a disk (for angular displacements), coded so as to provide a digital readout for discrete (sometimes very small) displacement increments and a reading head. Three types of encoders are in common use:

1. Brush-type encoders, in which the reading head is a sliding contact (brush) which wipes over a partly conductive, partly nonconductive coded pattern.

2. Photoelectric encoders, in which the reading head consists of a light-source assembly on one side of the disk or strip and a corresponding light-sensor assembly facing it on the other side of the disk or strip; the coded pattern is partly translucent, partly opaque.

3. Magnetic encoders, with a magnetic reading head and a partly magnetized, partly non-magnetized coded pattern.

Incremental encoders have a simple, alternately ON and OFF coded pattern. They provide an output in the form of number of *counts* between the start and end of the displacement. Hence the start position must be known if the end position is to be determined in absolute terms. *Absolute encoders* have a code pattern such that a unique digital word is formed for each discrete displacement increment. Various codes are used for this purpose, such as the binary, binary-coded-decimal (BCD), and the Gray code.

Among displacement-transducer *selection criteria,* the most critical are range, resolution, starting force, overtravel, and type and magnitude of full-scale output. Accuracy and dynamic characteristics, type of available excitation supply, and freedom from contamination by the ambient atmosphere or other fluids need to be considered as well, for all transducer applications.

27. Force, Torque, Mass, and Weight Transducers. *Force* is the vector quantity necessary to cause a change in momentum. *Mass* is the inertial property of a body, a measure of the quantity of matter in the body and of its resistance to change in its motion. *Weight* is the gravitational force of attraction; where gravity exists, it is equal to mass times acceleration due to gravity. *Torque* is the moment of force, the product of force and the perpendicular distance from the line of action of the force to the axis of rotation (lever arm).

28. Force transducers (load cells) are used for force measurements (compression, tension, or both) as well as for weight determinations in any locality where gravity exists and the gravitational acceleration g is known. The standard g (on earth) is 9.80665 m/s^2. Mass can be determined from weight, which is expressed in force units. A mass of 1 kg, for example, "weighs" 2.205 lb (pounds force) on earth. Torque is measured by *torque transducers.*

The sensing elements of force and torque transducers usually convert the measurand into a mechanical deformation of an elastic element. This deformation, in terms of either local strains or gross deflection, is then converted into a usable output by a suitable transduction element. Bending beams (cantilever, end-supported, or end-restrained), solid rings or frames (*proving rings*), and solid or hollow rectangular or cylindrical columns are the most commonly used force-sensing elements. Special solid or notched shafts are used as torque-sensing elements.

29. Piezoelectric force transducers are used for dynamic compression-force measurements. A typical design has the shape of a thick washer. The annular piezoelectric crystal segments are sandwiched between two hollow cylindrical columns. Bidirectional force measurements can be obtained by preloading this *force washer.* An amplifier is used to boost the low-level output signals.

30. Reluctive force transducers use proving-ring sensing elements in most design versions. The deflection of the proving ring is converted into an ac output by an inductance-bridge or differential-transformer transduction element. An entirely different design uses the permeability changes due to stresses in a laminated column to vary the voltage induced by a primary winding in a secondary winding.

31. Strain-gage force transducers are the most widely used type. Bonded-metal foil and metal wire gages predominate, but unbonded wire gages and bonded semiconductor gages are used in some designs. Columns and proving rings are the usual sensing elements. The shear-web sensing element of the force transducer shown in Fig. 10-11 is related to the column, but is reported to offer greater transduction efficiency.

32. Torque transducers are mostly of the reluctive, photoelectric, or strain-gage type. The last is more widely used than the first two. The metal-foil strain gages in the transducer shown in Fig. 10-12 are located on the sensing shaft, which is enclosed in a cylindrical *torque sensor* housing. The leads from the gages are carried through the shaft to slip rings. Brushes ride on the slip rings to provide stationary external connections. The brush assembly can be lifted off the slip rings to increase brush life during periods when torque is not monitored. The speed-sensing provisions are described in Par. **10-36**. In some strain-gage torque transducers the slip rings and brushes are replaced by a rotary transformer.

33. Reluctive torque transducers use changes in shaft permeability, due to torque-induced stresses in the shaft, to change the voltage coupled from a primary winding to two secondary windings. *Photoelectric torque transducers* use two incremental-encoder disks, one

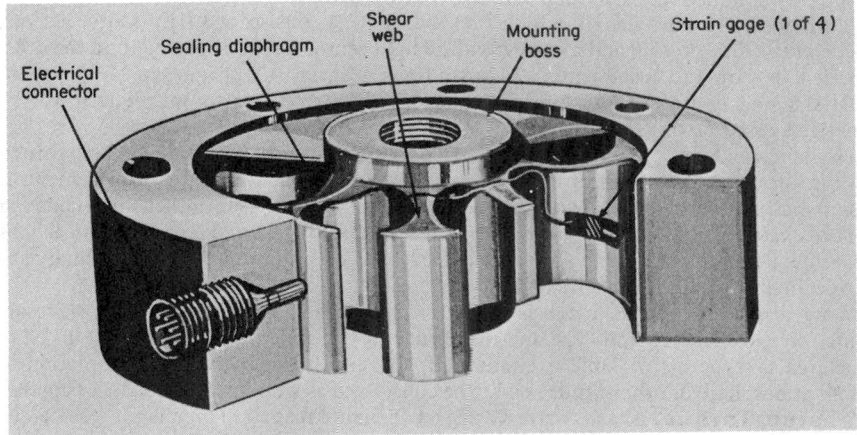

Fig. 10-11. Strain-gage force transducer. *(Interface. Inc.)*

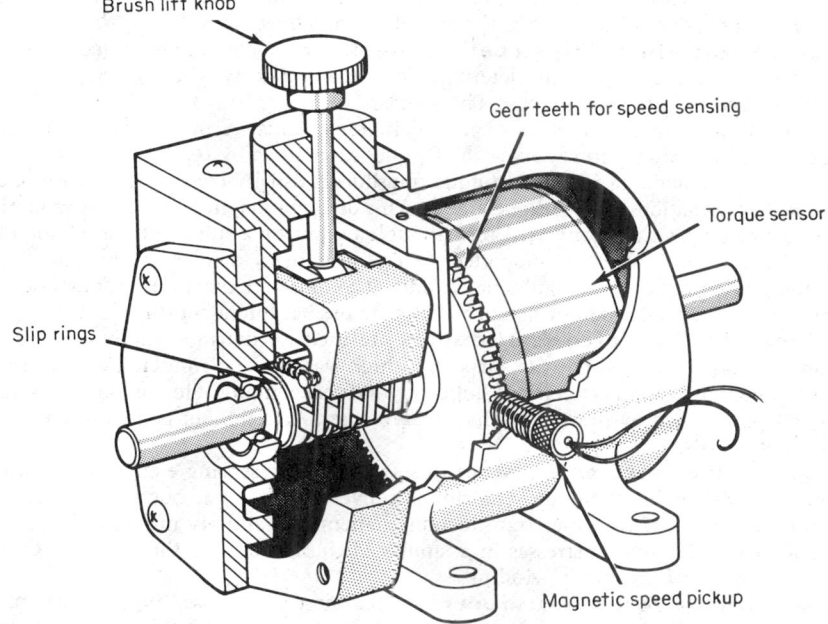

Fig. 10-12. Strain-gage torque transducer. *(Lebow Associates, Inc.)*

on each end of the shaft, to change the illumination on a light sensor when one disk undergoes a small angular deflection, due to torque, relative to the other disk.

Selection criteria include the usual range, accuracy, excitation, and output characteristics, case configuration and dimensional constraints, overload rating, the thermal environment, and, for torque transducers, maximum shaft speed and proximity of any magnetic fields that may cause reading errors. A frequent application of force transducers is in automatic weighing systems.

34. Speed and Velocity Transducers. *Speed* (a scalar quantity) is the magnitude of the time rate of change in displacement. *Velocity* (a vector quantity — magnitude and direction) is the time rate of change of displacement with respect to a reference system. *Velocity trans-ducers* are almost invariably linear-velocity transducers, whereas speed transducers are normally angular-speed transducers (*tachometers*).

35. Velocity transducers are usually of the electromagnetic type, exemplified by a coil in which a permanent-magnet core moves freely. The core has a sensing-shaft extension, and the shaft is attached to the object whose (usually oscillatory) velocity is to be measured. The rate at which lines of magnetic flux from the core are cut by the coil turns determines the amount of electromotive force generated in the coil; hence the output is proportional to the velocity of the measured point. In some designs the coil moves within a fixed magnetic field instead.

36. Tachometers are also predominantly of the electromagnetic type. Such angular-speed transducers as the *dc tachometer generator*, the *ac induction tachometer*, and the *ac permanent-magnet tachometers (ac magneto)* are electric generators. Their output amplitude increases with angular (rotational) speed. In the case of the ac magneto, the output frequency also increases with speed. The output of a *toothed-rotor tachometer* also varies in both amplitude and frequency, but the frequency variation is much greater than the amplitude variation and represents the angular speed much more accurately.

The speed-sensing gear teeth and sensing coil (pickup) incorporated in the torque transducer of Fig. 10-12 constitute a toothed-rotor tachometer. A pulse is generated in the electromagnetic sensing coil every time a ferromagnetic tooth passes by it. Since there are 60 teeth on the gear shown in the illustration, 60 pulses per revolution are provided by the sensing coil. By counting the pulses over a fixed time interval the angular speed can be determined with very close accuracy. *Photoelectric tachometers* provide the same degree of accuracy, typically by chopping a beam between a light source and a light sensor into equidistant pulses by an incremental-encoder disk attached to the sensing shaft. The pulse-frequency output can also be converted into a dc output voltage if the degraded accuracy, resulting from the conversion, can be tolerated.

Selection criteria include, besides range and accuracy characteristics, the mounting position and required frequency response for velocity transducers, and the type of available readout or signal-conditioning and telemetry equipment in the case of tachometers.

37. Strain Transducers. *Strain* is the deformation of a solid resulting from *stress*, the force acting on a unit area in a solid. Strain is measured as the ratio of dimensional change to the total value of the dimension in which the change occurs. Essentially, all strain transducers are resistive and are referred to as *strain gages*. Their essential characteristic is their sensitivity (*gage factor*), the ratio of the unit change in resistance to the unit change in dimension (length).

38. Strain gages employ either a conductor or semiconductor, of small cross-sectional area, suitable for mounting to the measured surface so that it elongates or contracts with that surface and changes its resistance accordingly. A typical metal-wire gage is shown in Fig. 10-13. Other types of metal gages are made of thin metal foil, die-cut or etched into the required pattern and deposited on an insulating substrate through a pattern mask by bombardment or evaporative methods. The metals used in strain gages are usually copper-nickel alloys; other alloys such as nickel-chromium, platinum-tungsten, and platinum-iridium are also used. Semiconductor gages are usually made from thin doped-silicon wafers or blocks.

Strain gages can be *bare* (*surface-transferable*, free-filament), bonded to an insulating carrier sheet on one side only, or completely *encapsulated* in a bondable (usually plastic) or weldable (metal) carrier, the latter insulated internally from the gage. Bare gages are normally supplied with a strippable insulating substrate (*carrier*). Since two or more gages are normally used to obtain a strain measurement, for temperature-compensation, linearity-com-

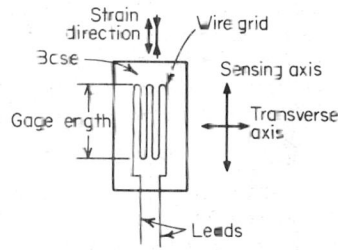

Fig.   10-13. Bondable   wire-grid strain gage. (*From Ref. 41 by permission of Prentice-Hall, Inc.*)

pensation, and output-multiplication puposes, strain-gage *rosettes* are sometimes used, combining two, three, or four gages, mutually aligned as to their strain-sensing axes, on one carrier.

Selection criteria involve the desired type and size (always including gage length and width), type, and material of connecting leads and spacing between them on the gage itself, type of carrier or encapsulation, gage resistance, gage factor, transverse sensitivity tolerances, allowable overload (*strain limit*), and maximum excitation current for a given application. Semiconductor gages must often be shielded from illumination, which can cause reading errors. Proper methods of attachment and of connection into a Wheatstone bridge circuit are very critical for strain gages.

TRANSDUCERS FOR FLUID-MECHANICAL QUANTITIES

39. Density Transducers. *Density* is the ratio of the mass of a homogeneous substance to a unit volume of that substance. Density transducers (*densitometers*) are used for the determination of the density of fluids (gases, liquids, and slurries). They are, however, not related to densitometers used to measure optical density, as of a photographic image, or to equipment used to determine spectral density (e.g., power spectral density).

Three methods are primarily used for density sensing.

Sonic density sensing is achieved by an arrangement of piezoelectric sound (usually ultrasound) transmitters and receivers producing outputs proportional to the speed of sound in the fluid and to the acoustic impedance of the fluid. Since acoustic impedance varies with the product of speed of sound and density, a signal proportional to density can be derived from the transducer and signal conditioning system.

Radiation density sensing relies upon the attenuation, due to density, of the radiation passing from a radioisotope source, on one side of the fluid-carrying pipe or vessel, to a radiation detector on the opposite side.

Vibrating-element density sensing employs a simple mechanical structure, such as a cylinder or a plate, electromagnetically set into vibration at its resonant frequency. This frequency changes with density, and an output is produced, proportional to density, which is related directly to the square of the period of vibration. The transducer illustrated in Fig. 10-14 uses a vibrating plate in its sensing head. The plate is installed as the end-supported beam in a circular support structure containing a crystal detector close to one of the attachment points of the beam. A magnetostrictive drive sets the beam into vibration at its resonant frequency. This frequency and its variations due to density change are converted into a usable output by the crystal detector and built-in preamplifier. Additional methods, used to infer density from other measurements, have also been employed in measurement systems.

40. Flow Transducers. *Flow* is the motion of a fluid. *Flow rate* is the time rate of motion expressed either as fluid volume per unit time (*volumetric flow rate*) or as fluid mass per unit time (*mass flow rate*). Transducers used for flow measurement (*flowmeters*) generally measure flow rate. Most flowmeters measure volumetric flow rate, which can be converted to mass flow rate by simultaneously measuring density and computing mass flow rate from the two measurements. Some flowmeters measure mass flow rate directly. Flow-sensing elements can be categorized as follows:

1. *Differential-pressure flow-sensing elements.* Sections of pipe provided with a restriction or curvature which produces a pressure differential ΔP proportional to flow rate across two points of the device (see Fig. 10-15). The output of a differential pressure transducer whose input ports are connected to these two points is representative of flow rate through the sensing element. Known relationships of ΔP vs. flow rate exist for each type of element.

2. *Mechanical flow-sensing elements.* Freely moving elements, e.g., turbine or propeller, or mechanically restrained elements e.g., a float in a vertical tapered tube, a spring-restrained plug, a hinged or cantilevered vane, whose displacement, deflection, or angular speed is proportional to flow rate.

3. *Flow sensing by fluid characteristics.* Certain transduction elements can be so designed and installed that they will interact with the moving fluid itself and produce an output relative to flow rate. The heated wire of a *hot-wire anemometer* transfers more of its heat to the fluid as the flow rate increases, thereby causing the resistance of the heated wire to decrease. When small amounts of radioisotope tracer material are added to the fluid, a radiation detector close to the moving fluid will respond with increasing output as the flow rate increases (*nucleonic flowmeter*).

In the *fluid-conductor magnetic flowmeter* an increasing electromotive force is induced in an electrically conductive fluid, flowing through a transverse-magnetic field, as the flow rate increases. In the *thermal flowmeter* two thermocouple junctions are immersed in the moving fluid, one upstream, the other downstream, from an electric heater immersed in the same fluid,

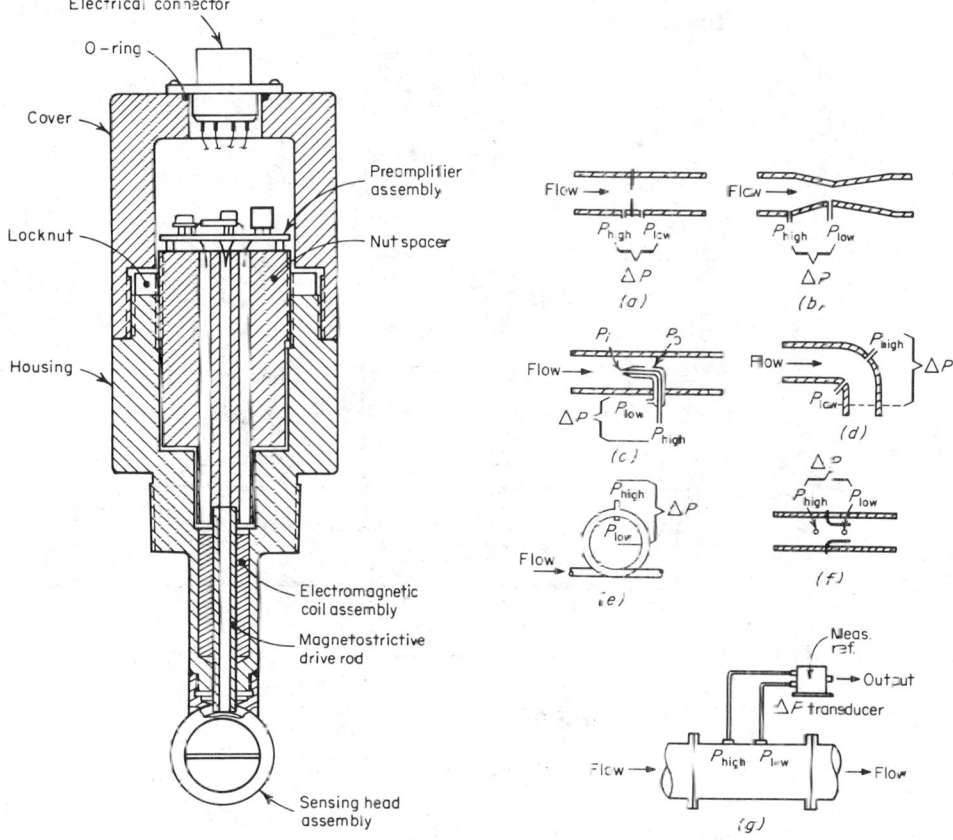

Fig. 10-14. Vibrating-beam density transducer. *(ITT Barton.)*

Fig. 10-15. Flow measurement using differential pressure-sensing elements (e) orifice plate; (b) venturi tube; (c) pitot tube; (d) centrifugal section (elbow); (e) centrifugal section (loop); (f) nozzle; (g) measurement of differential pressure due to flow rate. *(From Ref. 41 by permission of Prentice-Hall, Inc.)*

and the two junctions are connected as a differential thermocouple, the output of the latter increasing with mass flow rate. In a similar device, the *boundary-layer flowmeter*, only the portion of the fluid immediately adjacent to the inside wall of the pipe is heated and thermally sensed.

41. Turbine Flowmeter. The turbine flowmeter (see Fig. 10-16) is among the most widely used flow-rate transducers. Its operating principle is similar to that of the toothed-rotor tachometer described in Par. **10-36**. The bladed rotor (turbine) rotates at an angular speed proportional to volumetric flow rate. Rotational friction is reduced as much as possible by special bearing design. As each magnetic rotor blade cuts the magnetic flux of the pickup coil's pole piece, a pulse is induced in the pickup coil (sensing coil). A frequency meter is used to display the frequency output of the flowmeter, or a frequency-to-dc converter can be utilized to provide a dc voltage increase with flow rate. The rotor blades can be so machined that the variable-frequency ac voltage across the sensing coil terminals is virtually sinusoidal. This permits use of an FM demodulator as a frequency-to-dc converter. The number of turbine blades, the pitch of the blades, and the internal geometry of the flowmeter determine the range of output frequencies for a given flow-rate range.

42. Oscillating-Fluid Flowmeter. In this device the fluid is first forced into a swirling motion, then passes through a venturi-like cavity at a point of which the flow oscillates about the axis of the flowmeter. A fast-response resistive temperature transducer at that point provides an output in terms of frequency of resistance changes. This frequency, proportional to flow rate

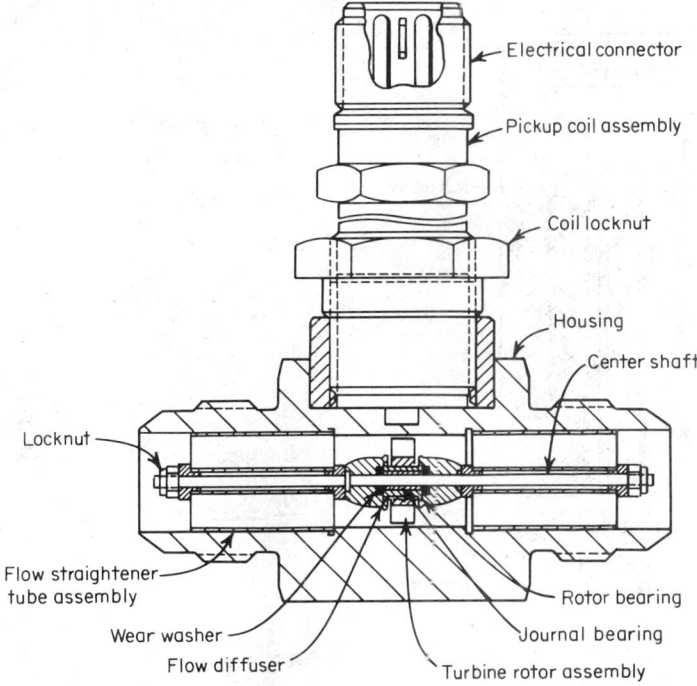

Fig. 10-16. Turbine flowmeter. *(ITT Barton.)*

and converted into voltage variations, can then be displayed on a counter after it has been amplified, filtered, and wave-shaped. Strain or pressure sensors are also used in such flowmeters.

43. Other flowmeter designs include the *ultrasonic flowmeter*, typically using pairs of piezoelectric transducers to establish sonic paths. Changes in flow rate produce corresponding changes in the propagation velocity of sound along the path. *Strain-gage flowmeters* use cantilevered vanes or beam-supported drag bodies which deflect or displace due to fluid flow. The strain in the deflecting beam is then transduced by strain gages. A few types of *angular-momentum mass flowmeters* have been developed in which the fluid either imparts angular momentum to a circular tube through which it flows or receives angular momentum by a rotating impeller. The angular momentum is then used to cause an angular displacement or a torque in a mechanical member, either of which can be transduced to provide an output proportional to mass flow.

Selection criteria involve, first, a choice of either a flowmeter alone or a complete flow-rate or flow-measuring system which can include signal conditioning and display equipment and, when required, a flow totalizer. Among the essential flowmeter characteristics are the (mass or volumetric) flow-rate range, the properties and type(s) of the measured fluid (gas, liquid, mixed-phase, slurry), the nominal and maximum pressure and temperature of the fluid, the configuration, mechanical support, weight and provisions for correction of the flowmeter, the required time constant, and the output, as well as accuracy, specifications. The sensitivity of a turbine flowmeter is usually expressed as the *K factor*, stated in hertz (or cycles) per gallon, per liter, per cubic foot, or per cubic meter. Attention must also be paid to the length of straight pipe upstream and downstream of the flowmeter and the necessity for flow straighteners other than those incorporated in the transducer itself.

44. Humidity and Moisture Transducers. *Humidity* is a measure of the water vapor present in a gas. It is usually measured as relative humidity or dew point temperature, sometimes as absolute humidity. *Relative humidity*, which is temperature-dependent, is the ratio of the water-vapor pressure actually present to water-vapor pressure required for saturation at a given temperature; it is expressed in percent (% RH). The *dew point* is the temperature at

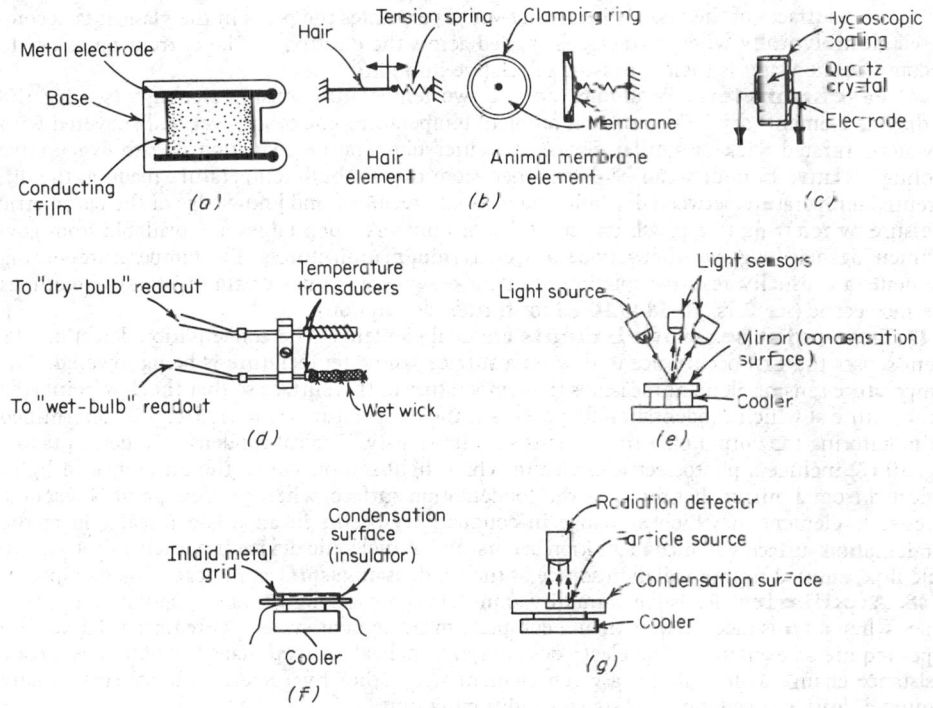

Fig. 10-17. Sensing elements of humidity transducers: (*a*) resistive, (*b*) mechanical, (*c*) oscillating crystal, (*d*) psychrometric, (*e*) photoelectric. (*f*) resistive, (*g*) nucleonic; (*e*) to (*g*) are dew-point sensors. (*From Ref. 41 by permission of Prentice-Hall, Inc.*)

which the saturation water-vapor pressure is equal to the partial pressure of the water vapor in the atmosphere. Hence any cooling of the atmosphere, even a slight amount below the dew point, produces water condensation. The relative humidity at the dew point is 100% RH. *Moisture* is the amount of liquid adsorbed or absorbed by a solid; it is also the amount of water adsorbed, absorbed, or chemically bound in a nonaqueous liquid. Humidity and moisture measurements are made by one of three methods: hygrometry, psychrometry, and dew-point determination.

45. Hygrometers. The hygrometer is a device which can measure humidity directly, with a single sensing element; it is usually calibrated in terms of relative humidity. Three types of hygrometric sensing elements are shown in Fig. 10-17. In the *resistive* humidity-transducer sensing element a change in ambient relative humidity produces a change in resistance of a conductive film between two electrodes. Carbon powder in a binder material has been used for such films, but hygroscopic salts, also in a binder material, are more common. Lithium chloride has been the most popular hygroscopic salt in such applications. The *mechanical* hygrometric element is the oldest type. It uses a material, such as human hair or animal membrane, which changes its dimension with humidity. The resulting displacement on an attaching point on the material is then transduced into an output proportional to humidity. The *oscillating-crystal* hygrometric element consists of a quartz crystal with a hygroscopic coating, so that the total crystal mass changes as water is adsorbed on, or desorbed from, the coating. When the crystal is connected into an oscillator circuit, the oscillator output frequency will change with changes in humidity.

Several other types of hygrometric sensing elements have also been developed. In the *aluminum oxide element* an impedance (resistance and capacitive reactance) change occurs with changes in humidity. The *Brady array* also provides an ac output when excited with alternating current (at about 1 kHz). However, it differs from other devices in that it consists of an array of semiconducting crystal matrices which look electrically neutral to the water molecule. Vapor pressure then allows the molecules to drift in and out of the interstices, creating an exchange of energy within the structure. The *porous-glass-disk* hygrometric element has electrodes plated on the two surfaces of the disk. When water vapor permeates the pores in the glass, it is decomposed electrolytically when a voltage is applied across the electrodes. The current necessary to decompose the water is then a measure of relative humidity.

46. Psychrometers. Psychrometers use two temperature-sensing elements (see Fig. 10-17d). One element, *dry bulb*, measures ambient temperature; the other, *wet bulb*, covered with a water-saturated wick or similar device, measures temperature reduction due to evaporative cooling. Relative humidity can be determined from the dry-bulb temperature reading, the differential temperature between dry-bulb and wet-bulb readings, and knowledge of the barometric pressure by referring to a *psychrometric table* of numbers. Such tables are available from government agencies, e.g., weather service, as well as from manufacturers. The temperature-sensing elements are usually resistive (platinum- or nickel-wire windings or thermistors), sometimes thermoelectric (see Pars. **10-58** to **10-65** for further description).

47. Dew-point sensing elements are dual elements. The condensation-detection element senses the first occurrence of dew on a surface whose temperature is being lowered. The temperature-sensing element measures the temperature of this surface so that the dew point (the temperature at which condensation first occurs as the temperature is lowered) can be determined by monitoring the output of both elements simultaneously. Typical condensation detectors (see Fig. 10-17) include a photoelectric device in which light sensors detect the difference in light, reflected from a mirror that serves as the condensation surface, when the dew point is reached; a resistive element in which a change in conductivity occurs in an inlaid metal grid at the condensation surface when condensation occurs; and a nucleonic device in which a drop in particle flux, emitted from a radiation source at the condensation surface, indicates the dew point.

48. Auxiliaries. Resistive humidity transducers are generally more popular than other types when a transducer, rather than a complete measurement system, is required. Almost all types require ac excitation. The electrodes are spiral, helical, or loop-shaped to obtain as large a resistance change as feasible for a given element size. Other hygrometric transducers usually require at least an excitation and signal conditioning unit.

Psychrometric transducers are typically complemented by a signal conditioning and readout system. A small blower (*aspirator*) is often included to blow the ambient air over the two sensing elements so that a faster response can be obtained.

Dew-point humidity transducers require, as a minimum, a cooler (thermoelectric coolers are often used), its associated control circuit, and a power conditioning circuit, as well as the two sensing elements. However, several designs are miniaturized and require sufficiently little signal conditioning.

Selection Criteria. Humidity transducer applications should first be examined to see whether relative humidity or dew point is to be measured. Relative humidity can, of course, also be inferred from psychrometric and dew-point readings but not without a look-up table or calculations. Among performance characteristics the measurement range is the most important; measurement accuracy can usually be improved when only a partial range needs to be measured. Other important characteristics include the temperature and the chemical properties of the ambient atmosphere or the measured material.

49. Liquid-Level Sensing. A large variety of sensing approaches and transducer types have been developed for the determination of the level of liquids and quasi liquids, e.g., slurries and powdered or granular solids, in open or enclosed vessels (tanks, ducts, etc.). Not only is the knowledge of the *level* itself important, but other measurements can be inferred from level. If the tank geometry and dimensions are additionally known, the *volume* of the liquid can be determined. If, additionally, the density of the liquid is known, its *mass* can be calculated.

Level is generally sensed by one of two methods: obtaining a discrete indication when a predetermined level has been reached (*point sensing*) or obtaining an analog representation of the level as it changes (*continuous sensing*). Point sensing is also used when it is only desired to establish whether a liquid or a gas exists at a certain point, e.g., in a pipe. The different level-sensing methods can be classified into those lending themselves primarily to point sensing, to continuous sensing, or both. It should be understood, of course, that point-sensing systems are usually simpler and cheaper than continuous sensing systems and should be used when only a discrete indication has to be obtained. Even when two or more discrete levels must be established in one vessel, the use of two or more point sensors may be preferable to a continuous sensing system. On the other hand, electronic circuitry can be used to provide one or more discrete level indications from a continuous sensing system.

50. Point level-sensing methods are usually aimed at indicating the interface between a liquid and a gas, sometimes the interface between two different liquids. Three methods are illustrated in Fig. 10-18. *Heat-transfer sensing* is used by two types of sensors: the resistive

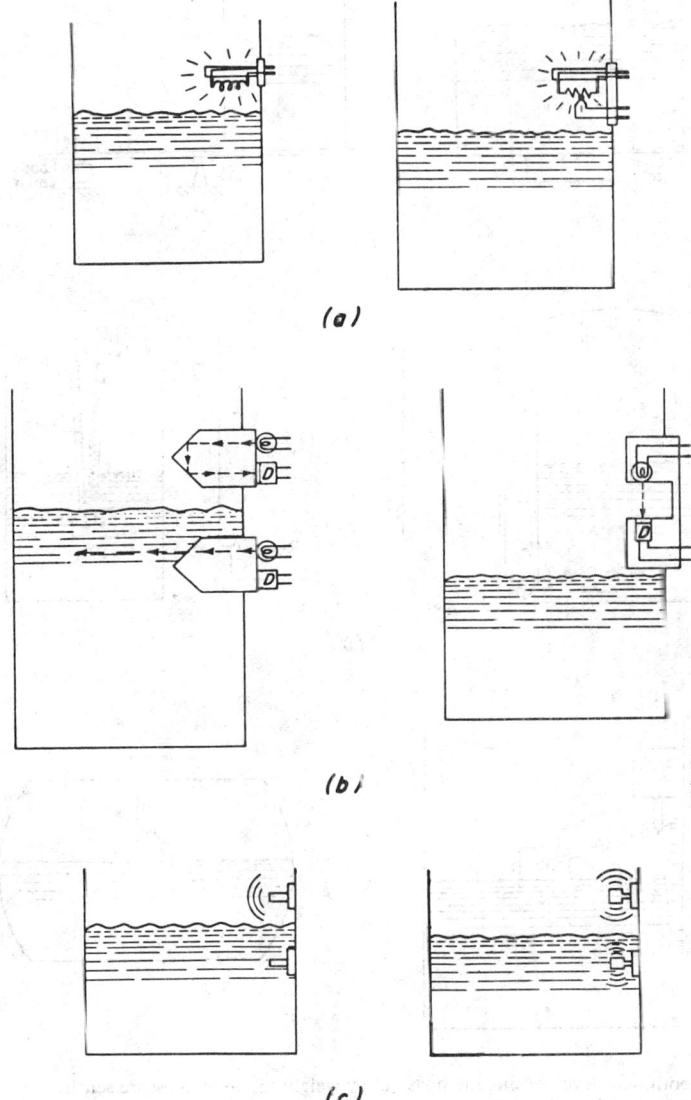

(a)

(b)

(c)

Fig. 10-18. Point level-sensing methods: (a) by heat-transfer rate, (b) by optical means, (c) by oscillation damping. *(From Ref. 41 by permission of Prentice-Hall, Inc.)*

sensor (wire-wound or thermistor) is heated to some degree by the current passing through it so that its resistance changes due to cooling when contacted by the liquid; the thermoelectric sensor detects the cooling, upon liquid contact, of a wire-wound heater it is in thermal contact with. *Optical sensing* relies either on the presence or absence of reflection of a light beam from the interface between a prism surface in contact with gas (reflection) or liquid (no reflection) or on the greater attenuation of a light beam when it passes through liquid on its way to a light sensor. In *damped-oscillation sensing* the mechanical vibration of an element, excited into such vibration electrically, is either stopped (in a magnetostrictive or piezoelectric element) or reduced in amplitude (e.g., in an oscillating-paddle element) due to acoustic damping or viscous damping, respectively, when the measured fluid changes to a liquid.

51. Continuous Level Sensors. Three classic continuous level-sensing methods are illustrated in Fig. 10-19. The level, volume, or mass of a liquid in a tank of known geometry can

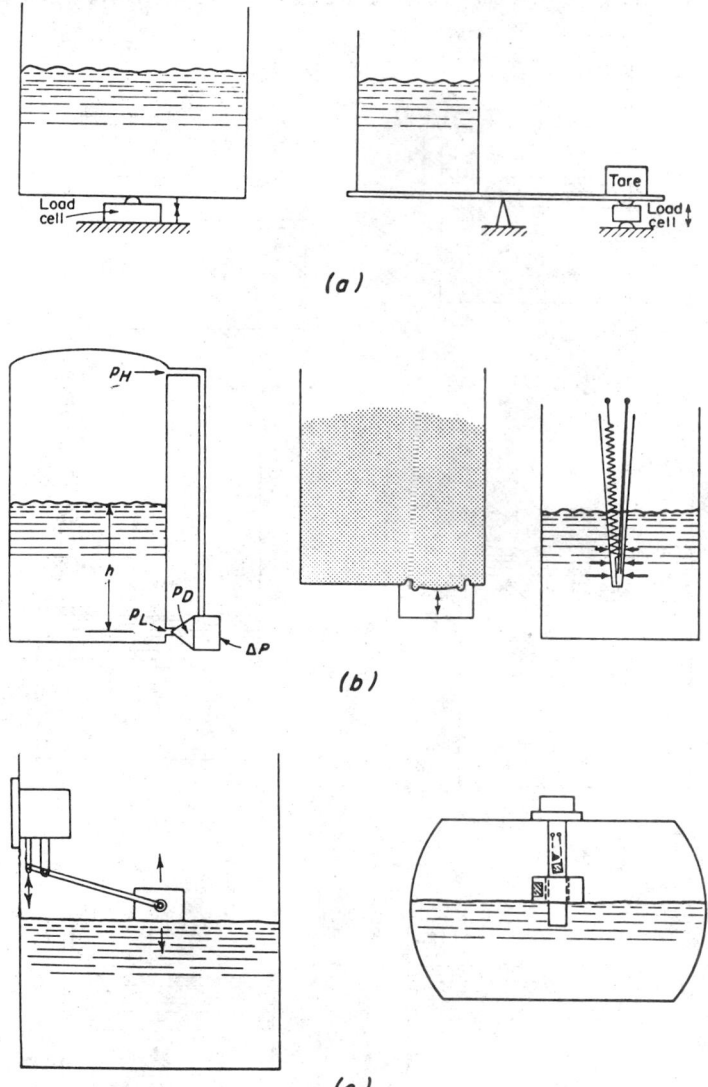

Fig. 10-19. Continuous level-sensing methods: (a) by weighing, (b) by pressure sensing, (c) by float. (*From Ref. 41 by permission of Prentice-Hall, Inc.*)

be determined by *weighing* the tank continuously, as by means of a load cell (force transducer), and subtracting the tare weight of the tank or compensating for the tare weight. *Pressure sensing* relies on the pressure (*head*) developed at the base of a liquid column. This pressure increases with the column height and hence with level above the point at which pressure is sensed. The differential pressure P_D, measured by the differential-pressure transducer, on the tank shown in the illustration, is equal to the difference in pressures between the bottom and top of the tank $(P_L - P_H)$. The level h of the liquid above the bottom sensing point is then given by $h = (P_L - P_H)/w$, where w is the specific weight of the liquid. Pressure acting at the bottom of the tank can also be sensed by a diaphragm built into the tank bottom, used as pressure-sensing element. A third method uses the increasing pressure near the bottom of a tank, as the level rises, to compress electrically conductive strips against a resistive element, gradually shorting out the resistance (and therefore decreasing this resistance) as the level increases. The *level-sensing float* mechanically actuates a transduction element, usually a potentiometer, sometimes a reluctive element or one or more magnetic reed switches. A radically different method (not illustrated) is *cavity-resonance sensing*, where electromagnetic oscillations are excited (from a coupling element at the tank top) within the gaseous cavity enclosed by the liquid surface and the upper tank walls, and the change in resonant frequency, as the liquid surface changes in location, becomes a measure of liquid level.

Several methods are equally useful for point and continuous level sensing (see Fig. 10-20). *Conductivity level sensing* is usable with even mildly conductive liquids. The resistance between two electrodes (the tank wall may serve as one of the two) changes continuously (or suddenly, in the case of the point-sensor version) as the liquid level rises or falls.

Dielectric-variation sensing is used primarily for nonconductive liquids, which then play the role of dielectric materials between two (sometimes four) concentric electrodes which are used (and electrically connected) as plates of a capacitor. The capacitance changes continuously (or suddenly, for the point sensor) as the vertical distance h of the level changes. If it is necessary to compensate for changes in the liquid's characteristics during measurement, a reference capacitor, always submerged, can be employed so that the ratio of the capacitance change equals the ratio of the measured level to the vertical dimension of the reference capacitor $\Delta C / \Delta C_R = h/h_R$).

Sonic level sensing uses ultrasound either emitted from a sound projector and detected by a sound receiver or emitted and detected by a single sound transceiver operating alternately in the transmit and receive mode. An echo-ranging technique is commonly used, the liquid-gas interface (the liquid level) acting as the target. The difference in attenuation or travel time of the beam of sound between liquid or gas in its path can also be used for sonic level sensing, especially for point sensing.

Radiation sensing is a nucleonic sensing method employing usually one or more radioisotope sources and radiation detectors to indicate level changes by virtue of the changes in attenuation of the radiation due to level changes. The attenuation in the liquid is caused mainly by absorption. Such nucleonic methods have also been used to study density profiles and the location and extent of vortices in tanks and of gas bubbles in pipes.

52. Liquid-level transducers, in their most common configuration, are probes, flange- or boss-mounted through the tank or duct wall. Some pipe-wall-mounted transducers are so designed that their sensing end is flush with the inside of the wall, to prevent obstructions to flow. Nucleonic transducer systems are attached to the outside of the wall.

The transduction principle of liquid-level transducers is given by the sensing technique employed. Dielectric-variation sensing demands capacitive transducers, using ac excitation having a frequency between 400 Hz and 200 kHz. Magnetostrictive and piezoelectric transducers, whose probe tip oscillates at a frequency in the vicinity of 40 kHz, find their application in the sonic, as well as the damped-oscillation, sensing techniques. Ionization-type, as well as solid-state, transducers are used in nucleonic systems. Photoelectric transducers are used in optical sensing systems. Potentiometric and reluctive transduction elements are found in float-actuated liquid-level transducers. Resistive transducers are used for heat-transfer sensing and, in a somewhat different form, for conductivity and pressure sensing. Thermoelectric elements are found in some heat-transfer sensors. Vibrating-element (notably vibrating-paddle) transducers find their use in damped-oscillation sensing systems.

Selection criteria involve, first of all, the choice of one or more point-level sensors or a continuous level sensor. After this choice has been made, together with an evaluation of end-to-end system requirements, the characteristics of the measured liquid are of primary importance. These include its conductivity, viscosity, temperature, chemical properties and, for installation in pipes or ducts, its flow rate and pressure. The transducer must also be designed and installed in such a manner as to prevent false level indications due to slosh, spray, and splash and to adherence of liquid to the transducer with falling level.

53. Pressure and Vacuum Transducers. *Pressure* is force acting on a surface; it is measured as force per unit area, exerted at a given point. *Absolute pressure* is measured relative to zero pressure, *gage pressure* relative to ambient pressure, and *differential pressure* relative to a *reference pressure* or a range of reference pressures. A perfect *vacuum* is zero absolute pressure. Vacuum measurement, however, is the measurement of very low pressures.

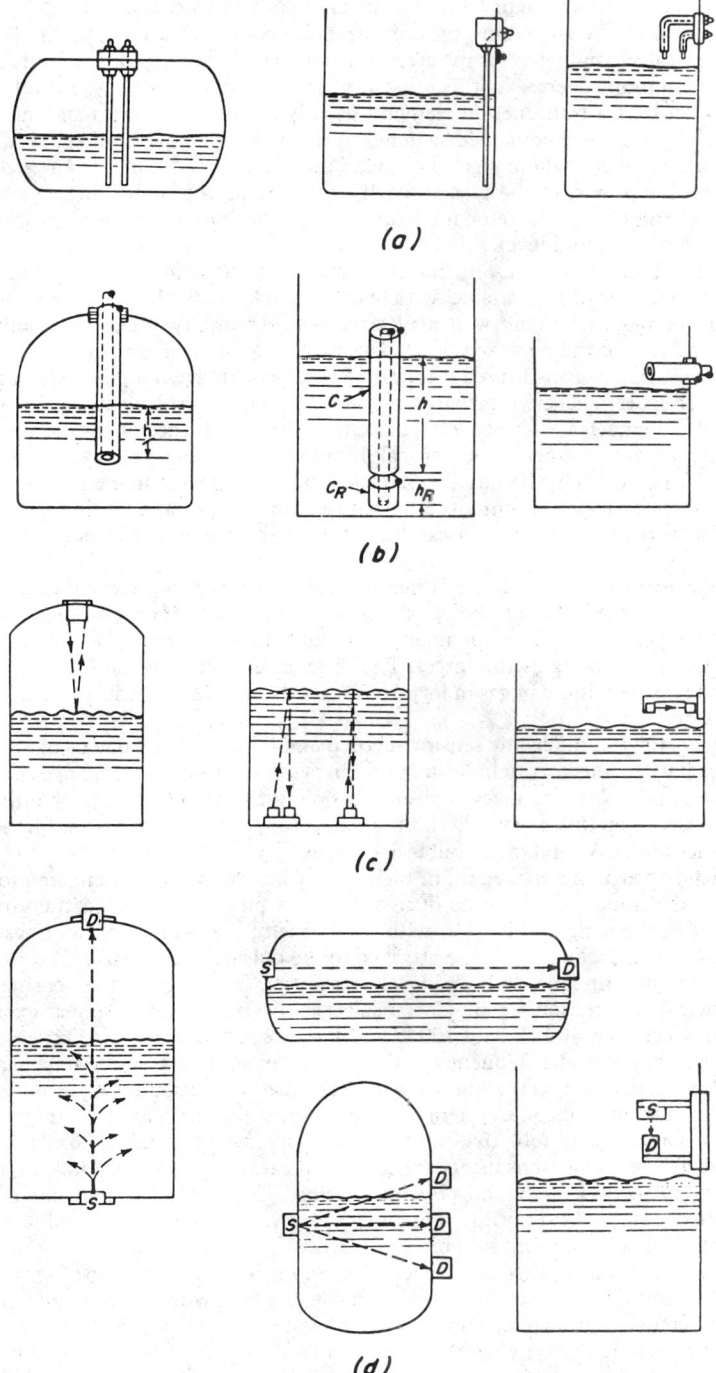

Fig. 10-20. Continuous- and point-level sensing: (a) conductivity; (b) dielectric variation; (c) sonic sensing; (d) radiation sensing. (From Ref. 41 by permission of Prentice-Hall, Inc.)

54. Pressure-sensing elements are almost invariably mechanical in nature (see Fig. 10-21). They can be described generally as thin-walled elastic members which deflect when the pressure on one side of their wall is not balanced by a pressure on the opposite side. The former pressure is the measured pressure; the latter is either a vacuum or near vacuum (for absolute-pressure transducers), the ambient atmosphere (for gage-pressure transducers), or some other pressure (for differential-pressure transducers).

The *diaphragm* is a circular plate fastened around its periphery so that its center will deflect when pressure is applied to it. It can be flat or, when a greater deflection is required, contain a

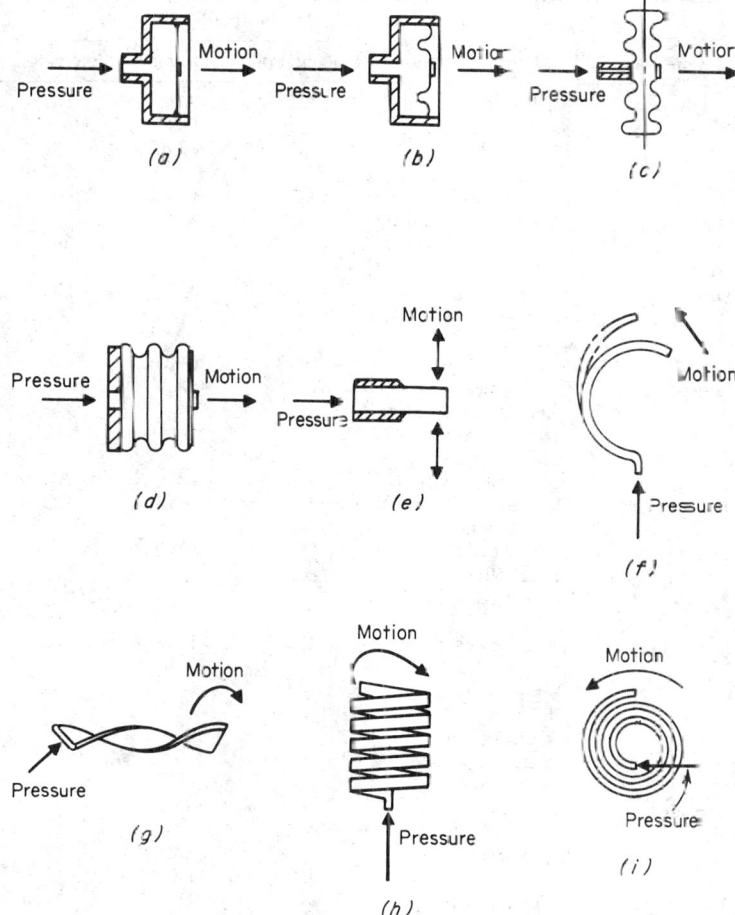

Fig. 10-21. Pressure-sensing elements: (a) flat diaphragm; (b) corrugated diaphragm; (c) capsule; (d) bellows; (e) straight tube; (f) C-shaped Bourdon tube; (g) twisted Bourdon tube; (h) helical Bourdon tube; (i) spiral Bourdon tube. (*From Ref. 41 by permission of Prentice-Hall, Inc.*)

number of concentric corrugations which increase the effective area upon which the force (pressure) can act. Two corrugated diaphragms, welded, brazed, or soldered together around their periphery, form a *capsule* sensing element (aneroid). Two or more capsules can be fastened together so that the pressure acts on all. The displacement obtainable at the end of such a multiple-capsule element nearly equals the displacement of one capsule multiplied by the number of capsules in the assembly. The *bellows* sensing element is typically made from a thin-walled tube formed into deep convolutions and sealed at one end, whose displacement can then be made to act on a transduction element. In the *straight-tube* sensing element, again sealed at one end, applied pressure causes an expansion of the tube diameter. This expansion, though slight, can be converted into a usable output by a transduction element.

The *Bourdon tube* is one of the most widely used sensing elements, particularly for pressure ranges higher than 2 MPa (about 300 lb/in^2). The Bourdon tube, elliptical in cross section and sealed at its tip, tends to straighten from its curved, twisted, helical, or spiral shape, thus causing the tip to deflect sufficiently to act upon a transduction element. The number of turns or twists in a Bourdon tube tends to multiply the tip travel.

55. Pressure transducers, using the sensing elements described above, provide their outputs by means of a large variety of transduction elements (see Table 10-2). Many designs are

TABLE 10-2　Pressure Transducers

Transduction	Sensing elements	Type variations	Normal		Optional	
			Excitation	Output	Excitation	Output
Capacitive	Diaphragm	AC bridge unbalance	AC	AC	DC	DC
		Variable ionization	AC	DC		
		RF-tank-circuit detuning	AC	Freq.		
Inductive	Bellows	LC tank circuit	AC	Freq.		
	Diaphragm Bourdon tube	AC bridge unbalance	AC	AC		
Piezoelectric	Diaphragm	Natural crystal Synthetic crystal Ceramic	None	AC	DC	Amplified ac
Potentiometric	Capsule Bourdon tube	Wire-wound element	(AC)	(AC)		
		Continuous-resolution element (metal film, cermet, plastic, carbon film)	DC	DC		
Reluctive	Diaphragm Bourdon tube	Inductance bridge	AC	AC	AC	DC
		Differential transformer			DC	DC
Strain gage	Diaphragm	Unbonded gages, metal wire Bonded gages, metal wire	(AC)	(AC)		
	Straight tube	Bonded gages, metal foil; bonded gages, semiconductor; diffused semiconductor gages; evaporated-metal gages	DC	DC	DC	Amplified dc
Servo type	Capsule	Null balance	AC	AC		Encoder (digital)
	Bellows	Force balance	AC DC	DC DC		Synchro Potentiometric
Vibrating element	Diaphragm	Vibrating wire Vibrating diaphragm	AC	Freq.		
	Straight tube	Vibrating cylinder	AC	Freq.		

available with integrally packaged output- and excitation-conditioning circuitry. Certain designs, notably potentiometric, reluctive, and strain-gage transducers, are more prevalent than other types. Piezoelectric transducers are usable only for dynamic pressure measurements. Inductive transducers are subject to severe temperature effects and are not used extensively.

A *potentiometric pressure transducer* is illustrated in Fig. 10-22. The dual-capsule sensing element transfers its displacement to a lever-type wiper arm by means of a pushrod. The wiper then slides over the curved resistance element. Capsule elements are commonly used in such transducers for pressure ranges up to 2.5 MPa (about 360 lb/in^2).

Reluctive pressure transducers use either the inductance bridge circuit or, primarily when only the normal ac output is required, the differential-transformer circuit. When inductance bridge transducers use a diaphragm sensing element, the magnetic diaphragm itself, positioned between two coils, acts as the armature which increases the inductance of one coil while decreasing the inductance of the other coil. When inductance bridge transducers use a Bourdon tube sensing element, a flat armature plate, positioned over two coils, tilts more toward one coil than toward the other as the Bourdon tube tip rotates slightly with applied pressure. In differential-transformer transducers the sensing-element displacement is used to move a magnetic core within the transformer.

Most *strain-gage pressure transducers* use a diaphragm sensing element, although at least one good design uses a straight tube. Most designs have a four-active-arm strain-gage bridge, with the gages either on the diaphragm or on a beam actuated by the diaphragm.

When the sensing-element displacement is not sufficient for a given transduction element, a mechanical amplification linkage can be inserted between the two elements. Special design considerations apply to differential-pressure transducers when the measured fluid (at one of the two

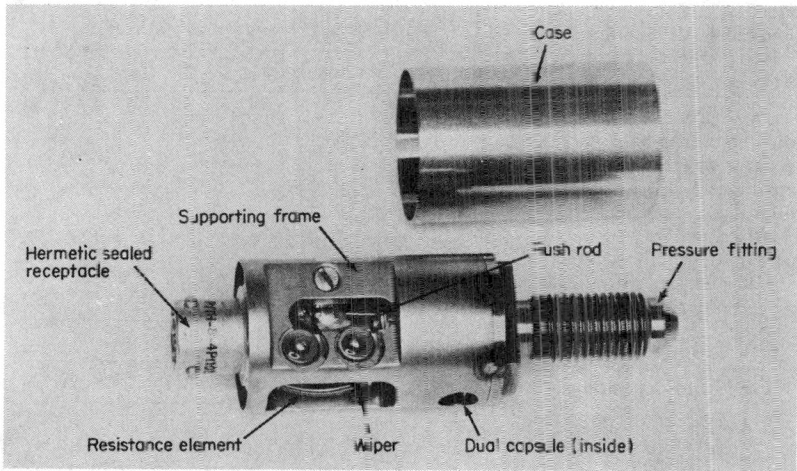

Fig. 10-22. Potentiometric pressure transducer. (Bourns, Inc.)

pressure ports) must not come in contact with the transduction element. One solution to this problem has been to fill the affected inside portion of the transducer with a *transfer fluid*, sealed off by a thin *membrane* to which the measured fluid can be applied safely. Gage-pressure transducers have the inside of their case (which usually acts as the *reference cavity*) vented to the outside through a small hole (*gage vent*), equipped with a fine-mesh screen, a porous plug, or another filter to prevent internal contamination.

Flush-diaphragm transducers are designed for high-frequency-response applications where use of tubing, or even the cavity formed by a mounting boss, may reduce response; these transducers are so designed that the diaphragm is flush (when installed) with the inside surface of the pipe wall (or other wall) through which they are mechanically fastened.

Specification characteristics of pressure transducers deserve particular attention since pressure is one of the two most common measurands (the other is temperature). Table 10-3 lists those characteristics which should be considered when preparing a specification for a pressure transducer. Not all these characteristics need always be specified; some can be omitted when sufficient knowledge of the application permits.

56. Vacuum transducers are an important subgroup of pressure transducers, though bearing little resemblance to them with regard to design and operation. The pressure constituting a practical dividing line between pressure and vacuum measurement is not well defined. Some pressure transducers are usable for very low pressure measurement. Generally, however, pressure measurements extending substantially below 133 Pa (= 1 torr) can be considered as vacuum measurements.

Vacuum transducers (see Table 10-4) exist in two major categories, given by their transduction principles.

Thermoconductive vacuum transducers measure pressure as a function of heat transfer by the measured gas. As the number of gas molecules within the transducer decreases, the quantity

TABLE 10-3   Specification Characteristics for Pressure Transducers*

Mechanical design characteristics	
Specified by user and manufacturer	Stated by manufacturer
Configuration and dimensions (shown on drawing)	Sensing element
Mountings (shown on drawing)	Transduction-element details
Type and location of pressure ports (shown on drawing)	Materials in contact with measured fluids
	Dead volume
Type and location of electrical connections (shown on drawing)	Type of damping (including type of damping oil if used)
Nature of pressure to be measured, including range	
Measured fluids	
Case sealing (explosion proof, burstproof, or waterproof enclosure)	
Isolation of transduction element	
Mounting and coupling torque	
Weight	
Identification	
Nameplate location (shown on drawing)	

Electrical design characteristics
Excitation (nominal and limits)
Power rating (optional)
Input impedance (or element resistance)
Output impedance
Insulation resistance (or breakdown-voltage rating)
Wiper noise (in potentiometric transducers)
Output noise (in dc output transducers)
Electrical connections and wiring diagram
Integral provisions for simulated calibration (optional)

Performance characteristics	
Individual specifications	Error-band specifications
Range	Range
Endpoints	Full-scale output (nominal)
Full-scale output	End points (defining reference line)
Creep (optional)	Resolution (where applicable)
Resolution (where applicable)	Static error band
Linearity	Reference-pressure range†
Hysteresis	Warm-up period (optional)
Repeatability	Frequency response
Friction error	
Zero balance	
Zero shift	
Sensitivity shift	
Warm-up period (optional)	
Reference-pressure range†	
Reference-pressure effects†	
Frequency response	
Operating temperature range	Operating temperature range
Temperature error or thermal zero shift and thermal sensitivity shift	Temperature error band
Temperature-gradient error	Temperature-gradient error
Ambient-pressure error	Ambient-pressure error band
Acceleration error	Acceleration error band
Vibration error	Vibration error band

Performance after exposure to:	Performance during and after exposure to:
Shock (triaxial)	High sound-pressure levels
Humidity	Sand and dust
Salt spray or salt atmosphere	Ozone
	Nuclear radiation
	High-intensity magnetic fields
	Etc.

*By permission of Prentice-Hall, Inc.
†Applies to differential-pressure transducers only.

heat transferred from a heated filament, through the gas, and to the case of the transducer, will decrease proportionally. The *Pirani gage*, as well as the *thermocouple gage* (which may use a thermopile instead of a single junction), both use this principle. A basic thermocouple gage is illustrated in Fig. 10-23.

Ionizing vacuum transducers measure pressure as a function of gas density by measuring ion current. Since different gases have different densities, the calibration of such a transducer will usually differ as well. The gas is usually ionized by electrons, except in one type using alpha particles for this purpose.

In thermionic vacuum transducers the electrons are emitted by a filamentary cathode, and positive ions are collected at the anode. Various modifications of the original triode type have helped to extend its lower range limit from 10^{-8} to 10^{-10} torr (*Bayard-Alpert gage*, by reducing internal x-ray effects) to 10^{-11} torr (*Nottingham gage*, by reducing electrostatic-charge effects) and to 10^{-12} torr (*Schuemann modification*, by virtually eliminating x-ray effects). The ion current, representative of pressure, is in the microampere region.

Several ionizing vacuum transducer types, whose electrons are emitted from either hot or ion-bombarded cold cathodes, use a magnetic field to increase the electron path length by forcing this path to be helical so that the probability of electron collisions with gas molecules is increased (*magnetron gages*). The hot-cathode versions include the *Lafferty gage*. The *Philips* (or *Penning*) *gage* and the *Redhead gage* are examples of the cold-cathode versions.

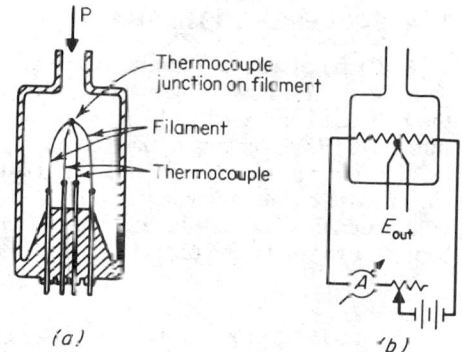

Fig. 10-23. Thermoelectric thermoconductive vacuum transducer: (*a*) transducer; (*b*) typical circuit. (*From Ref. 41 by permission of Prentice-Hall, Inc.*)

Selection criteria for vacuum transducers (and any necessary ancillary equipment for them) are primarily the required measuring range; secondarily, size, weight, ruggedness, and complexity. Considerations for the selection of a pressure transducer are primarily range, type of excitation and output, accuracy and frequency response; secondarily, the properties of the measured fluid and environmental conditions.

57. Ranges of Pressure Transducers. Some of the pressure transducers described in this section are available for ranges up to 100 MPa (about 15,000 lb/in²). Special sensing devices

TABLE 10-4. Vacuum Transducers

Sensing element	Common name	Output	Nominal range, torr
	Thermoconductive transduction		
Heated filament	Pirani gage	Resistance change	$10^{-3}-1$
	Thermocouple gage	Thermoelectric emf	$10^{-3}-1$
	Ionizing transduction		
Thermionic, triode	Bayard-Alpert gage	Direct current	$10^{-10}-10^{-3}$
Photomultiplier		Direct current	$10^{-15}-10^{-3}$
Hot cathode, magnetic field	Hot-cathode magnetron gage (Lafferty gage)	Direct current	$10^{-14}-10^{-5}$
Cold cathode, magnetic field	Cold-cathode magnetron gage (Philips gage, Penning gage)	Direct current	$10^{-7}-10^{-3}$
With flash filament		Direct current	$10^{-12}-10^{-3}$
With auxiliary cathode	Redhead gage	Direct current	$10^{-13}-10^{-4}$
Radioactive, alpha particles	Alphatron	DC voltage (from amplifier)	$10^{-5}-1,000$
Beta particles		DC voltage (from amplifier)	$10^{-3}-1,000$

have been designed for pressures up to 7 GPa (about 1 million lb/in²). Pressure transducers are also used to measure altitude (a known nonlinear relationship exists between atmospheric pressure and altitude above sea level), water depth [pressure increases at the rate of approximately 1 kPa/m (0.44 lb/in² ft) when descending below the water surface], and air speed (by measuring the difference between impact pressure, obtained from a pitot tube, and static pressure, while in flight).

TRANSDUCERS FOR THERMAL QUANTITIES

58. Temperature Transduction. The *temperature* of a body or substance is (*a*) its potential of heat flow, (*b*) a measure of the mean kinetic energy of its molecules, and (*c*) its thermal state considered with reference to its power of communicating heat to other bodies or substances. *Heat* is energy in transfer, due to a difference in temperature, between a system and its surroundings or between two systems, substances, or bodies. Heat energy is transferred by one or more of the following methods of *heat transfer:* (*a*) *conduction*, by diffusion through solid material or stagnant liquids or gases; (*b*) *convection*, by the movement of a liquid or gas between two points; and (*c*) *radiation*, by electromagnetic waves.

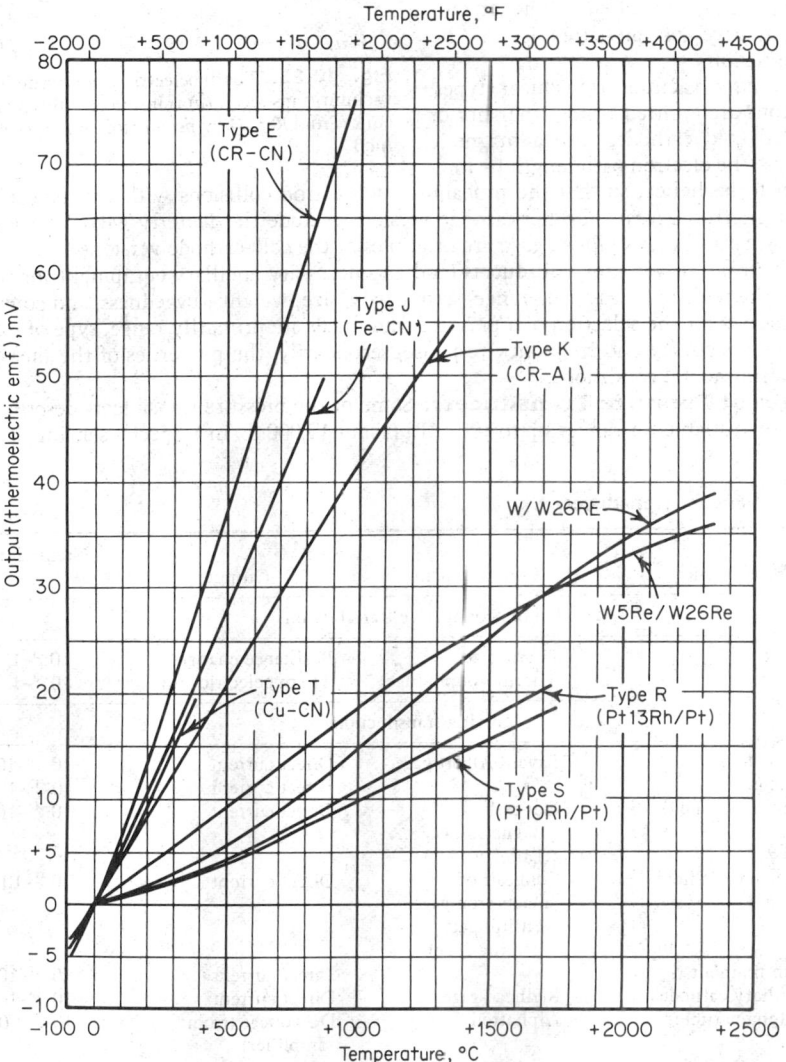

Fig. 10-24. Thermocouple output vs. temperature characteristics (reference junction at 0°C).

The sensing elements of temperature transducers typically act as transduction elements as well. The two most commonly used sensing-transduction elements are the *thermoelectric* element (*thermocouple*) and the *resistive* element (*resistance thermometer*). Among other sensing-transduction elements the only one which has found commercial acceptance is the *oscillating-crystal* element, essentially a quartz crystal (connected into an oscillator circuit) which has a substantial and highly linear temperature coefficient of frequency.

59. Thermocouples. A thermocouple is an electric circuit consisting of a pair of wires of different metals joined together at one end (*sensing junction*) and terminated at their other end in such a manner that the terminals (*reference junction*) are both at the same and known temperature (*reference temperature*). Connecting leads from the reference junction to some sort of load resistance (an indicating meter or the input impedance of other readout or signal-conditioning equipment) complete the thermocouple circuit. Both these connecting leads can be of copper or some other metals different from the metals joined at the sensing junction. Due to the *thermoelectric effect (Seebeck effect)*, a current is caused to flow through the circuit whenever the sensing junction and the reference junction are at different temperatures. In practice, the reference junction is either held at a known constant temperature (e.g., at 0°C) or is electrically compensated for variations from a preselected temperature.

The electromotive force (*thermoelectric emf*), which causes current flow through the circuit, is dependent in its magnitude on the sensing-junction wire materials, as well as on the temperature difference between the two junctions. Commonly used wire materials are Chromel (CR) and Alumel (AL) (both registered trade names of Hoskins Mfg. Co., Detroit, Mich.), Constantan (CN, an alloy of 53% copper and 45% nickel), copper (Cu), iron (Fe), platinum (Pt), an alloy of platinum and (either 10 or 13%) rhodium (Rh), tungsten (W), tungsten-rhenium (Rh) alloys (5 or 26% rhenium content is typical), nickel (Ni), and ferrous nickel alloys.

The characteristics of certain combinations of wire materials, such as their thermoelectric emf vs. temperature characteristics, their accuracy tolerances, and wire-insulation color codes, were standardized by ANSI Standard C96.1 (which is based on ISA Recommended Practice RP1) in such a manner that materials of different brand names can be used as long as the characteristics assigned to a specific type of thermocouple are maintained.

The names of the wire materials constituting, in their combination, a thermocouple sensing junction are now listed only as typical examples. Thus typical materials of a *type K* thermocouple are Chromel and Alumel. The ANSI Standard favors the use of type-letter designations in lieu of the names of the two metals used. Figure 10-24 shows the thermoelectric emf obtainable from various types of thermocouples when the reference temperature is held at 0°C.

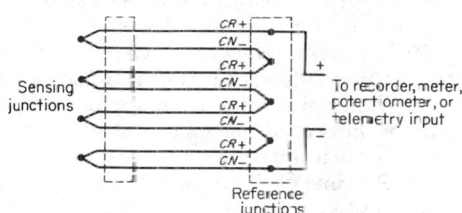

Fig. 10-25. Thermopile schematic diagram: CR-CN combination shown as example. *(From Ref. 41 by permission of Prentice-Hall, Inc.)*

Thermopiles (see Fig. 10-25) consist of several sensing junctions of the same material pairs, in close proximity to each other and connected in series so as to multiply the output obtainable from a single sensing junction. The isothermal reference junctions are usually also in close proximity to each other to assure an equal temperature for each reference junction.

60. Resistive temperature-sensing elements are either conductive or semiconductive. Conductive elements are usually wire-wound, sometimes made of metal foil or film. Elements wound in high-purity annealed platinum wire are best suited for most applications. Other metal-wire elements are wound of nickel or nickel alloy. Copper-wire elements are rarely used any more. Tungsten-wire elements have shown some promising characteristics but are generally considered too difficult to manufacture and too brittle to stay reliable.

A platinum-wire element has been used to define the International Practical Temperature Scale from −183 to +630°C, and it is expected that this upper limit will be extended to the melting point of gold (+1063°C). The resistance vs. temperature curve of such an element follows a well-defined theoretical relationship, making most points on the curve calculable within very close tolerances when only a few measured points have been established. Repeatabilities within about 0.01 K have been obtained at temperatures up to the gold point. Semiconductive resistive temperature-sensing elements include thermistors, germanium, and silicon crystals, carbon resistors, and gallium arsenide diodes. Thermistors have a nonlinear and negative temperature coefficient of resistance and an empirical resistance vs. temperature relationship.

61. Temperature transducers are classified into two general categories: surface-temperature transducers, which are cemented, welded, bolted, or clamped to a surface whose temperature is to be measured, and immersion probes, which are immersed into stagnant or moving fluids to measure their temperature. The fluid can be in a pipe, a duct, a tank, or other enclosed vessel, where the immersion probe is mounted through a pressure-sealed opening. It can also be freely moving, even at almost imperceptible rates of motion, e.g., an open body of water, an outdoor or indoor atmosphere.

Thermoelectric temperature transducers have the same sort of sensing junction, whether they are intended for surface temperature measurement or as immersion probes. The junctions between the two dissimilar-metal wire pairs are made by butt-welding the wire ends, by crossing them and welding them, by coiling one wire end around the other, or twisting the two ends about each other, then welding, brazing, or soldering the junction, or by welding both wire ends, in very close proximity to each other, to a metallic surface or to the metallic inside of an immersion-probe tip.

For surface measurements, the junctions are soldered, brazed, or welded to a surface (if it is metallic) or cemented to it (if it is not). If it is cemented, care must be taken to have the junction in solid thermal contact with the measured surface. Taping a junction to a surface is poor practice, since even a very small gap between junction and surface can introduce considerable errors. For immersion measurements, thermocouples are often produced with an integral sheath or inserted into a sealed immersion sheath (*thermowell*).

Junctions for thermoelectric immersion probes can be grounded (metallic contact from junction to sheath or thermowell) or isolated (ungrounded). In some cases, exposed junctions, at the tip of a probe, are immersed in the fluid without use of an integral sheath or thermowell. If terminals or connectors must be used between the sensing junction and the reference junctions, the terminals as well as the *extension wires* must be made of the same types of metals as used for the junction.

Thermocouples are usually made from two-conductor insulated cable, rarely from reels of individual bare-wire materials. The cables have a variety of insulation, over each conductor as well as over the conductor pair, and can be shielded or unshielded. Useful for many applications is thin (2 to 10 mm outside diameter) metal-sheathed, ceramic-insulated thermocouple cable.

Differential thermocouples can be used when the measurement objective is to measure the temperature difference between two points. In this case the sensing junction at the other measured point replaces the reference junction. The first wire of the first junction and the second wire of the second junction must still be brought to isothermal terminals; however, it is not necessary that the temperature of these terminals be known.

62. Resistive Temperature Transducers. Electrically conductive surface-temperature transducers are usually small and flat enough not to be influenced by convective heat transfer but only by conductive transfer from the measured surface. After installation they may be coated or covered to minimize any radiative heat transfer to them. The sensing element is usually a metal wire either wound around a thin insulating "card" or a coiled wire cemented to the base (see Fig. 10-26). Some metal-foil transducers (encapsulated or *free-grid*) are in the shape of a zigzag pattern. All designs are aimed at exposing the maximum sensing surface to the conductive heat transfer in an area of minimum size.

Resistive metal-wire *immersion probes*, most commonly with a platinum-wire element but sometimes with elements of nickel or nickel-alloy wire, are widely used for industrial and scientific fluid-temperature measurements. The probe-type transducer, illustrated in Fig. 10-27, has a ceramic encapsulated (coated) element in a perforated protective sheath so that it is usable for a variety of measured fluids over a wide temperature range. For applications in relatively stagnant fluids an unencapsulated (*exposed*) element is used to provide a shorter time constant. Some fluids require an element completely *enclosed* within a metallic well but with good thermal contact between well and element. The threaded mounting allows for compression sealing by means of a gasket or O ring between the housing and the mounting boss.

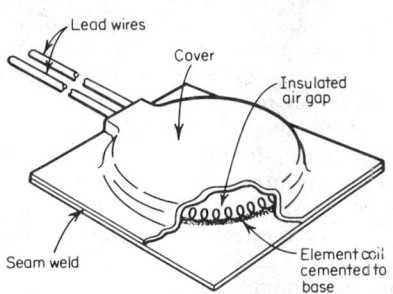

Fig. 10-26. Platinum-wire resistive surface-temperature transducer. (*Rosemount Engineering Co.*)

Lead wires
Cover
Insulated air gap
Seam weld
Element coil cemented to base

63. Thermistors are used for surface-temperature as well as fluid measurements. Because of their nonlinear (essentially negative exponential) resistance-vs.-temperature characteristics, they are particularly useful when a large resistance change is needed for a narrow range of temperature. Where a short time constant is required, a glass-coated thermistor bead, as small as 0.3 mm in diameter, can be suspended on its 0.03-mm-diameter precious-metal-alloy leads. Where somewhat more ruggedness is required, a glass-encapsulated bead about 1.5 mm around the tip and 4 mm long can be used. Excitation power must be kept low to avoid errors due to self-heating. Thermistor-type temperature transducers are available in a large variety of configurations, some of which are illustrated in Fig 10-28

64. Germanium thermometers are made of germanium crystals with highly selected and controlled impurities (dopants). They are intended primarily for cryogenic temperature measurements (below −195°C). Carbon resistors have also been used for such applications, as have

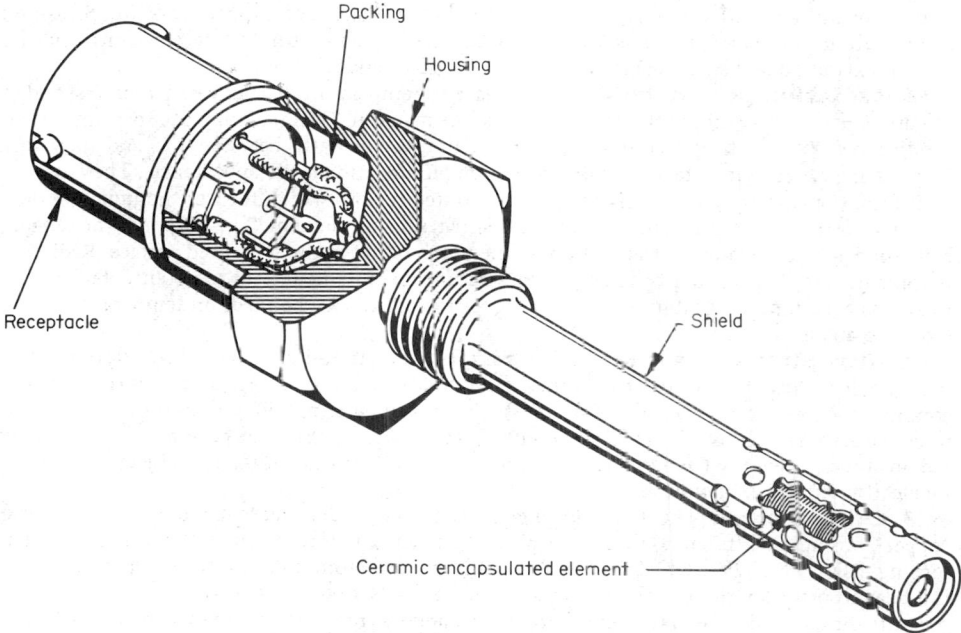

Packing

Housing

Receptacle

Shield

Ceramic encapsulated element

Fig. 10-27. Platinum-wire resistive immersible-probe temperature transducer. *(Rosemount Engineering Co.)*

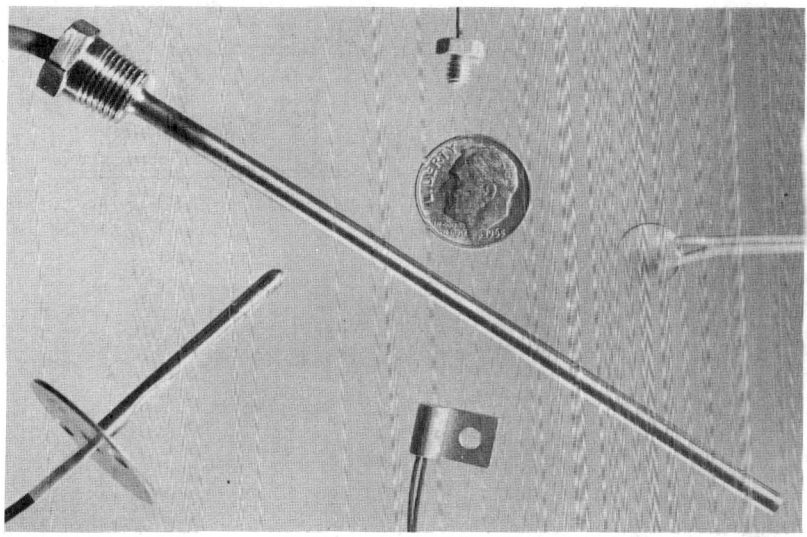

Fig. 10-28. Typical thermistor-transducer configurations. *(Fenwal Electronics, Inc.)*

gallium-arsenide junction diodes, which can be used to somewhat higher temperatures. Silicon-wafer transducers have been used for surface-temperature measurements in the range -50 to 275°C, where their resistance-vs.-temperature characteristics are similar to those of some metal wires.

65. Quartz-crystal temperature transducers use oscillating-crystal sensing elements in such a manner that the change of oscillator frequency with temperature is nearly linear over a range from about -50 to 250°C. They are usually furnished with associated electronics and readout equipment. This tends to limit their usability for general telemetry application without, however, detracting from their advantages in laboratory applications.

The selection of a temperature transducer is more complex than the selection of most other types of transducers. The objective is to select a design whose sensing element will attain the temperature of the measured material within the time available to make the measurement. Among primary selection criteria are, then, the characteristics and properties of the measured solid or fluid, the measuring-range limits, the required response time (time constant), and the type of excitation and signal conditioning available or intended to be used.

66. Radiation pyrometers are noncontacting temperature transducers which respond to radiative heat transfer from the measured surface or material. This radiation occurs primarily in the infrared portion of the electromagnetic spectrum (wavelengths between 0.75 and 1000 μm). Typical radiation pyrometers resemble a motion-picture camera in appearance. They use an optical lens or mirror system (sensitive in the infrared region) which focuses the radiation on a thermoelectric or resistive (usually photoconductive) sensing element. The output of the sensing element can be correlated, by calibration, to the temperature of the measured surface. Radiation pyrometers are used primarily for high-temperature measurements (up to about 3500°C), but have also been found useful for noncontacting measurements in the medium temperature range (down to about -50°C).

67. Heat-Flux Transducers. Two basic types of transducers have been developed to measure *heat flux*, heat transfer in terms of the total amount of thermal energy (heat flux is commonly expressed in W/cm² or Btu/ft²·s). The *calorimeter* provides an output proportional to convective as well as radiant thermal energy (*total heat flux*). The *radiometer* responds to radiant thermal energy (*radiant heat flux*) only. Virtually all heat-flux transducers have thermoelectric sensing elements.

68. Calorimeters. The *foil calorimeter* (*membrane calorimeter*, Gardon gage) acts as a copper-Constantan differential thermocouple. When heat flux is received by the thin Constantan sensing disk (Fig. 10-29a) which is metallurgically bonded around its rim to a copper heat sink, the heat absorbed by the membrane is transferred radially to the heat sink. This causes a temperature difference between the center of the disk and its rim. A thin copper wire is attached to the bottom surface of the disk, at its exact center, thus forming one copper-to-Constantan sensing junction. The copper-to-Constantan contact around the rim of the disk forms the other junction. The output of the calorimeter is then proportional to the energy absorbed. When heat flux must

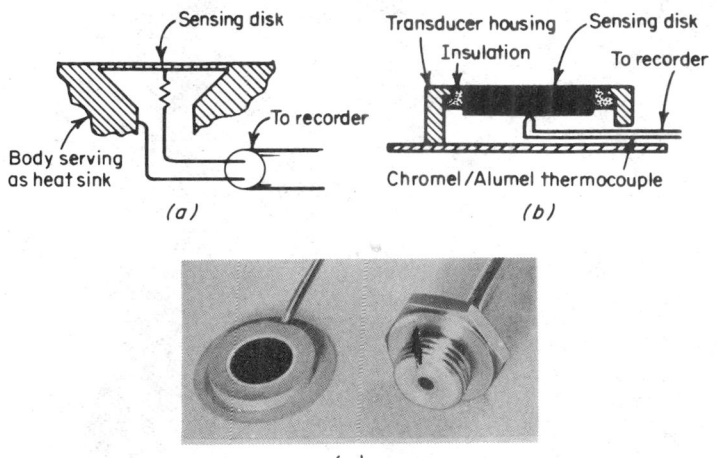

Fig. 10-29. Calorimeters: (a) foil; (b) slug; (c) typical appearance. (Hy-Cal Engineering)

be measured over long periods of time, the foil calorimeter can be provided with tubing and an internal flow path so that it can be water-cooled.

The *slug calorimeter (slope calorimeter)* uses a relatively thick thermal-mass sensing disk with an external high-emissivity (black) coating, which is thermally insulated from the transducer housing (Fig. 10-29b). A thin-wire thermocouple is attached to the bottom of the disk (slug), at its center. When heat flux is received by the slug, an output signal is produced by the thermocouple. The signal is proportional to the temperature rise of the slug. Two of the many available slug-calorimeter configurations are illustrated in Fig. 10-29c.

69. Radiometers. A typical *radiometer* is, essentially, a foil calorimeter with a *window* (usually of quartz or synthetic sapphire) mounted over the sensing disk so that a disk can receive radiant heat flux but no convective heat flux. The cavity formed by window, transducer housing, and sensing disk is usually sealed, but provisions for gas purging of this cavity can be made to prevent window clouding when the radiometer is to be used in a contaminating atmosphere. Radiometers can also be water-cooled. The sensitivity of a radiometer can be increased by using a differential (multijunction) thermopile instead of the two-junction differential thermocouple.

TRANSDUCERS FOR ACOUSTIC QUANTITIES

70. Terminology. *Sound* is an oscillation in pressure, stress, particle displacement, etc., in an elastic or viscous medium. *Sound sensation* is the auditory sensation evoked by the oscillations associated with sound. *Sound pressure* is the total instantaneous pressure at a given point, in the presence of a sound wave, minus the static pressure at that point. *Sound pressure level* (SPL or L_p) is 10 times the logarithmic ratio of the mean-square sound pressure p to a mean-square reference pressure p_{ref}. It is normally expressed as SPL $= 20 \log (p_{rms}/p_{ref,rms})$; see Fig. 10-30. The reference pressure is usually specified as 2×10^{-4} μbar, sometimes as 1 μbar (0.1 Pa). *Sound level* is a weighted sound-pressure-level reading obtained with a meter complying with a standard, e.g., ANSI Standard S1.4, Specification for General-Purpose Sound Level Meters.

71. Sound-Pressure Transducers. The sensing element of a sound-pressure transducer is almost invariably a diaphragm (see Pars. **10-53** to **10-55**). The reference cavity behind the diaphragm is vented to the ambient atmosphere by means of a small hole in the transducer case

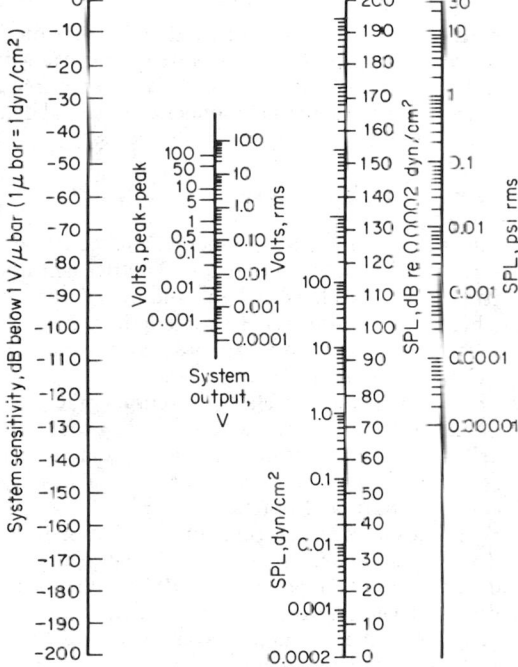

Fig. 10-30. Nomograph for sound-pressure-level calculations. (*From Ref. 41 by permission of Prentice-Hall, Inc.*)

so that static pressures on both sides of the diaphragm are equalized and only sound pressure is sensed.

A perforated cap over the diaphragm protects the diaphragm mechanically and, by its shape and geometry of perforations, provides some control over the transducer's directivity characteristics.

Sound-pressure transducers can be described, essentially, as special-purpose gage-pressure transducers. *Capacitive sound-pressure transducers* (sometimes called *condenser microphones*) use the sensing diaphragm as one electrode of a capacitor and a rigidly supported back plate, insulated from the rest of the structure but provided with a connecting lead or terminal, as the other electrode. A dc polarization voltage, applied across the two electrodes through a high-series resistance, maintains a constant charge on them. Capacitance changes due to diaphragm deflection cause changes in the voltage across the electrodes. The transducer output is first fed to an emitter follower so as to reduce the output impedance to a workable value. The output is then amplified. The emitter-follower (or cathode-follower) circuitry is sometimes built into the transducer case to keep the coupling path short. A shielded cable connects the transducer to the amplifier.

Piezoelectric and, to a limited extent, *inductive pressure transducers* have also been designed as sound-pressure transducers. Some piezoelectric designs have sealed cases, primarily to protect the internal components from atmospheric moisture and contaminants. The absence of a gage vent, however, necessitates correction of output readings when the transducer is used at low ambient pressures (e.g., high altitudes). Piezoelectric transducers do not require an excitation power supply but do require an amplifier.

The primary performance characteristics of sound-pressure transducers are range, output, frequency response, and directivity (directional response). Output is usually expressed as sensitivity or sensitivity level, sometimes as full-scale output for a stated range of sound pressures or sound-pressure levels. The nomograph shown in Fig. 10-30 can be used to correlate output characteristics stated in various ways.

72. Sound-level meters are complete, self-contained measuring systems, typically battery-operated and portable. A sound-level meter consists of a sound-pressure transducer (microphone), amplifier, standardized weighting networks, a calibrated attenuator, and an indicating meter. The sound-level range is always referred to a sound pressure of 10^{-4} μbar. The weightings denote different frequency-response characteristics of the measuring system. Referred to merely as A, B, or C, they are defined in a national standard as, for the United States, ANSI Standard S1.4.

73. Underwater sound detectors are used either for listening (*hydrophone*) or, in conjunction with an *underwater sound projector*, in sonar (*sound navigation and ranging*) systems. The transmitting and receiving function in a sonar system are frequently combined in a single device (sound transceiver). In the sonar field underwater sound detectors, as well as projectors and transceivers, are commonly referred to as *transducers*.

TRANSDUCERS FOR OPTICAL QUANTITIES

74. Terminology. *Light* is a form of radiant energy, an electromagnetic radiation whose wavelength is between approximately 100 and 0.01 μm. By strict definition, only visible light (0.4 to 0.76 μm wavelength) can be considered as light, and infrared or ultraviolet light is then termed *radiation*. The light spectrum, in terms of wavelength, frequency, photon energy, and blackbody temperature (all interrelated by physical laws), is illustrated in Fig. 10-31, with the visible-light spectrum (color spectrum) brought out in detail.

The transduction elements of light transducers (light sensors, photocells, photosensors, photodetectors, light detectors) also act as sensing elements since they convert electromagnetic radiation into a usable electrical output. Four transduction principles are commonly used: photovoltaic, photoconductive, photoconductive junction, and photoemissive (see Fig. 10-32). Section **11** provides detailed data on optical sensors and systems.

75. Photovoltaic light sensors are self-generating in that their output voltage is a function of the illumination of a junction between two dissimilar materials. These materials are semiconductive, either nonmetallic or metal compounds (Fig. 10-32a). Photons (particles of light) first pass through a thin conductive layer, and then impinge on the junction, causing an electron flow across the junction area so that the conductive layer becomes the negative terminal of the sensor. Various materials constitute the conductive and semiconductive portions of a photovoltaic light sensor.

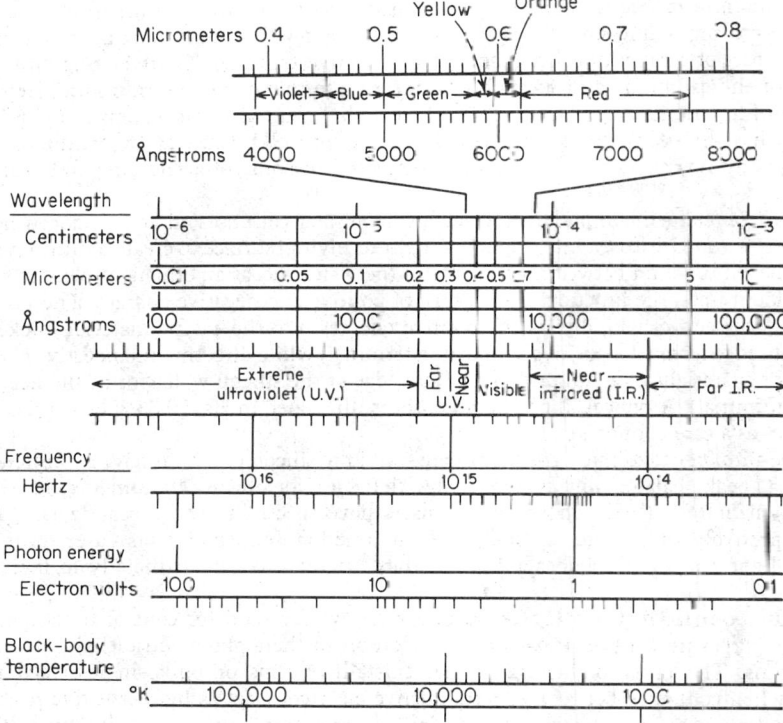

Fig. 10-31. The light spectrum. (*From Ref. 41 by permission of Prentice-Hall, Inc.,*

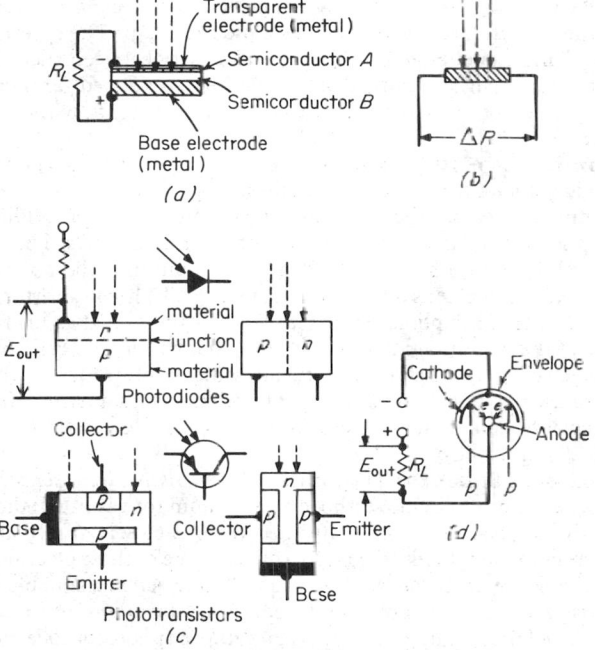

Fig. 10-32. Basic methods of light transduction: (*a*) photovoltaic, (*b*) photoconductive, (*c*) photoconductive junction semiconductor, (*d*) photoemissive. (*From Ref. 41 by permission of Prentice-Hall, Inc.*)

The *selenium cell* consists of an iron base (conductive, positive terminal) with a very thin selenium coating (semiconductive), which is in contact with cadmium oxide (semiconductive), which is part of a cadmium film (conductive, negative terminal). A silver ring on the outside surface of the cadmium film forms the connecting terminal for the cadmium electrode. The cadmium film and its oxide layer are so thin as to be essentially transparent to light. This type of sensor has also been referred to as *barrier-layer photocell* because the cadmium oxide acts as barrier to any reverse flow of electrons. The spectral response of a selenium cell peaks around 0.57 μm.

The *silicon cell* (silicon photovoltaic cell, silicon solar cell) uses an arsenic-doped n-type silicon wafer. Boron is diffused into the upper (light-receiving) surface to create a thin p-type silicon layer. The pn junction between the layer and the wafer acts as a permanent electric field. Photons incident upon the junction cause a flow of positive and negative charges. The pn junction, acting as an electric field, directs the positive charges into the p-type material (nickel plating around its edge forms the positive connecting terminal) while directing the negative charges into the n-type material (solder around the bottom edge of the silicon wafer forms the negative connecting terminal). A typical silicon photosensor is illustrated in Fig. 10-33. The third connecting pin serves as a case connection.

Germanium photovoltaic light sensors are similar to silicon types but have a different spectral response (a peak near 1.55 μm, as compared with 0.8 μm for silicon). *Indium-arsenide* (InAs) and *indium-antimonide* (InSb) photovoltaic sensors have spectral-response peaks near 3.2 and 6.8 μm, respectively. They are single-crystal pn junction semiconductors, used primarily for infrared-light sensing. In such applications they are often cooled artificially to increase their sensitivity.

76. Photoconductive light sensors are widely used for control functions, such as automatic exposure control in cameras, in addition to their photometric (light-measurement) applications. The photoconductors are polycrystalline films or bulk single-crystal materials which, when contained between two conductive electrodes, act as light-sensitive resistors (Fig. 10-32b) whose resistance decreases as incident illumination increases. *Cadmium sulfide* (CdS) and *cadmium selenide* (CdSe) are popular because of their spectral-response peaks in the visible-light region (approximately 0.6 μm for CdS, 0.72 μm for CdSe) and because of their relatively high output without artificial cooling. Two typical designs are shown in Fig. 10-34. Some photoconductive sensors use mixed CdS–CdSe crystals to obtain a response peak around 0.66 μm.

Lead sulfide (PbS) and *lead selenide* (PbSe) photoconductive cells are used for infrared-light sensing because their spectral-response peaks are close to 2.2 μm. The spectral-response curve of PbSe, however, is shallow enough to provide good sensitivity between about 1.8 and 3.6 μm, and its time constant is less than one-tenth that of PbS. *Mercury-doped-germanium* (HgGe) photoconductive sensors have been used for far-infrared measurements while being cooled to cryogenic temperatures.

77. Photoconductive Junction Sensors. In these devices the resistance across the junction in a semiconductor device changes as a function of incident light (Fig. 10-32c). Increasing incident illumination causes the junction photocurrent to increase. This category of photosensors includes *photodiodes* and *npn* as well as *pnp phototransistors*. They are made of silicon, with a spectral peak near 1.0 μm, except when germanium must be used to raise the spectral peak to about 1.6 μm. Time constants of photodiodes and phototransistors are less than 1 μs (compared with 10 μs for PbSe photoconductive sensors; 100 to 700 μs for PbS photoconductive sensors; 10 ms for CdSe photoconductive cells; and 100 ms for CdS cells but around 20 μs for silicon photovoltaic cells). Transistors also provide some amplification of the light-induced signal. Both types are usually sealed into a standard transistor can, sometimes furnished with a lens or window. A special silicon photodiode design has also been developed for ultraviolet measurements (0.06- to 0.25-μm region).

78. Photoemissive Sensors. The earliest *photoemissive light sensor* was the phototube, in which electrons are emitted by the cathode of a vacuum (or gas-filled) diode tube when photons impinge on the cathode surface (Fig. 10-32d). When a closely spaced anode is at a positive potential with respect to the cathode, the anode collects some of these electrons, and the resulting anode current can produce an output voltage as an *IR* drop across a suitable load resistor (R_L).

The photoemissive principle is now employed mostly by *photomultiplier tubes*, in which additional electrodes (*dynodes*) are placed between cathode (*photocathode*) and anode to amplify the electron current by means of secondary emission. An additional dynode is located behind the anode. A voltage divider network is used to apply a successively higher voltage to each of the dynodes as they approach the anode in proximity. Various spectral-response peaks can be obtained, depending on the photocathode material and material of that portion of the sealed envelope directly in front of the cathode (the window). Photomultiplier tubes are particularly useful in the visible-light and ultraviolet regions. They can provide very high sensitivities without artificial cooling.

Selection criteria for light sensors are, primarily, spectral-response characteristics and sensitivity and, secondarily, operating temperatures, relative complexity of associated circuitry, ruggedness, and cost. Where the spectral-response must be limited to only a portion of a sensor's basic response capabilities, a window can be placed in front of the sensing surface. Windows are optical filters which have spectral-response characteristics of their own, depending

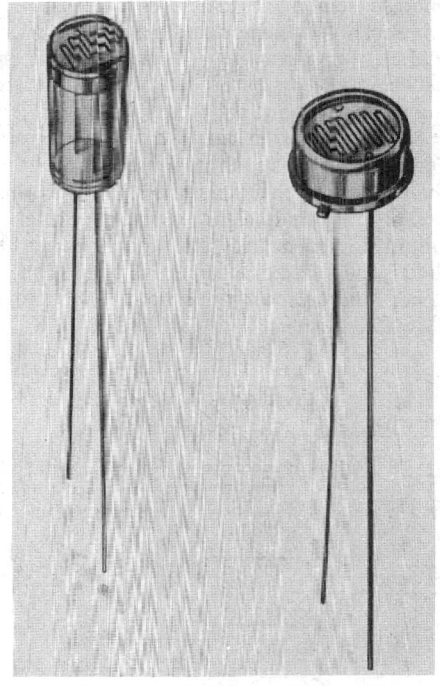

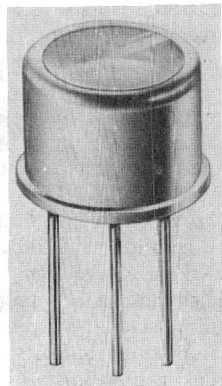

Fig. 10-33. Silicon photoelectric cell. *(Solar Systems Div., Tyco.)*

Fig. 10-34. Photoconductive (CdS or CdSe) cells. *(Clairex Corp.)*

on their material. Typical window spectral responses are 0.2 to 1.4 μm (quartz crystal or fused silica), 0.4 to 1.2 μm (borosilicate glass), 0.15 to 1.6 μm (cultured sapphire), 0.11 to 1.8 μm (lithium fluoride), 0.12 to 11 μm (calcium fluoride), and 0.25 to 70 μm (cesium iodide).

79. Spectrometers and Colorimeters. Light sensors in conjunction with light sources are used in a number of measuring devices other than for photometry. In optical *spectrometers* the incident light is passed through a *monochromator*, a grating or prism whose angular displacement relative to the incoming light beam can be closely correlated with the single wavelength of the light beam it sends on to the light sensor, whose spectral response is additionally known and selected for a specific range of wavelengths. Some monochromators consist simply of a set of windows, each having a different (and narrow) spectral response. The output of the light sensor then shows the light intensity of each of the sampled portions (groups of wavelengths) of the spectrum. The motion of the monochromator can be mechanized so that a given spectrum is scanned at a known rate over a known time interval. Wavelength can then be determined from the time counted from the start of the scan.

The *colorimeter* is similar to an optical spectrometer except that it responds to light of known characteristics (emanating from a built-in or associated light source) which is reflected by the surface whose color is to be determined.

80. Turbidity and Opacity Meters. In *turbidimeters, nephelometers,* and *opacity meters* a fixed amount of light from a light source is made to pass through the measured fluid and onto a light sensor whose output varies with changes in the transmittance of the measured liquid or gaseous medium.

TRANSDUCERS FOR NUCLEAR RADIATION

81. Terminology. *Nuclear radiation* is the emission of charged and uncharged particles and of electromagnetic radiation from atomic nuclei. *Charged particles* include (*a*) *alpha parti-

cles, helium-atom nuclei consisting of two protons and two neutrons and having a double positive charge; (b) *beta particles*, negative electrons or positive electrons (positrons) emitted when *beta decay* occurs in a nucleus, a radioactive transformation by which the atomic number is changed by +1 or −1 while the mass number remains unchanged; (c) *protons*, positively charged elementary particles of mass number 1. *Uncharged particles* are typified by the *neutron*, an uncharged elementary particle of mass number 1. Nuclear *electromagnetic radiation* includes (a) *gamma rays*, electromagnetic radiation quanta resulting from quantum transitions between two energy levels of a nucleus; (b) *x-rays*, quanta of electromagnetic radiation originating in the extranuclear part of the atom.

82. Ionization and Scintillation. Two basic transduction methods are used to convert nuclear radiation into a usable electrical output. When *ionizing* transduction is employed, the transduction element acts also as the sensing element. This method relies upon the production of ion pairs in a gas or solid by incidence of nuclear radiation. The two ions of an ion pair are subatomic particles, one with a positive charge, the other with a negative charge. An electric field is then used to separate the positive and negative charges to produce an electromotive force. When *photoelectric* transduction is employed, a *scintillator* material is used as a sensing element which converts the nuclear radiation into light. A light sensor is then utilized as the transduction element.

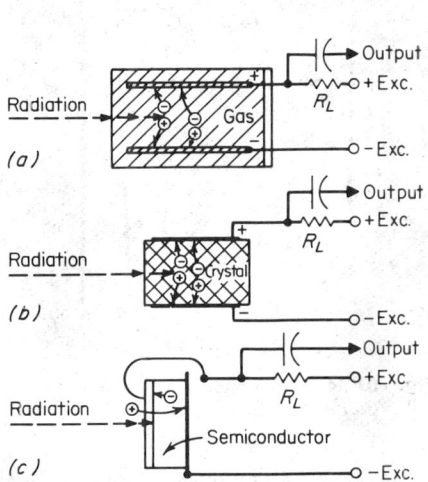

Ionization occurs in gases and solids when charged particles pass atoms at a high enough velocity to separate one of the outer electrons from them. The resulting ion pair consists of an ion and a secondary electron. The same incoming particle can cause such ion pairs to be produced a number of times before its energy is expended. Uncharged particles can produce ion pairs by collision with atomic nuclei. Nuclear electromagnetic radiation can produce ion pairs by removing secondary electrons from atoms with which the radiation interacts.

When ion pairs are produced in a gas (argon, neon, xenon, hydrogen), the electric field between a cathode and an anode in a gas-filled envelope, with a dc potential applied across the two electrodes, causes charge separation (see Fig. 10-35a). The *ionization current*, the current due to flow of charges to electrodes of opposite polarities, can then be monitored as the average IR drop across the load resistor R_L. This principle is used in three types of nuclear radiation sensors.

Fig. 10-35. Transduction of nuclear radiation. (a) Ionization in gas; (b) ionization in a solid crystal; (c) ionization in a solid semiconductor. (*From Ref. 41 by permission of Prentice-Hall, Inc.*)

83. Ionization Chambers. The ionization chamber consists typically of a metallic outer cylinder (cathode) with a metallic rod or wire (anode) mounted along the axis of the cylinder. One end of the cylinder is the sealed base of the sensor which provides the cathode terminal and a well-insulated anode terminal. The other end of the cylinder is also sealed, either by a thin mica, nylon, or metal window or by a metal disk when a window is provided in a portion of the cathode. In windowless ionization chambers the thin-walled cathode itself acts as a window. The gases or gas mixtures, contained within the ionization chamber at a pressure below, above, or equal to atmospheric pressure, are selected on the basis of ionization potential and energy per ion pair produced. The material and thickness of the window control the types of radiation to which the ionization chamber will respond with the necessary output. Tungsten is the most commonly used anode material.

84. Proportional counters are ionization chambers whose anode potential is above a certain limiting value. Above this voltage value, different for different fill gases, *gas amplification* occurs within the chamber, the production of additional ions by the *avalanche effect*. The incident radiation produces an ion pair, including a secondary electron. Each secondary electron produces more ion pairs, including more secondary electrons, and they, in turn, produce additional ion pairs, etc. The gas amplification can be up to about 10,000. For each incident particle, the total amount of ionization, which dictates the magnitude of the output current pulse, is proportional to the energy of radiation.

85. Geiger Counter (Geiger-Mueller Tube). The Geiger counter is a proportional counter to whose fill gas a *quenching agent* has been added and whose anode voltage is higher than it would be for a proportional counter of similar design. An incident radiation particle (or other ionizing event related to radiation) causes a discharge within the fill gas that spreads along the entire length of the anode. The quenching vapor (a halogen such as chlorine or bromine, or alcohol) quenches the discharge by preventing the production of secondary electrons by positive ions at the cathode.

Because of the discharge phenomenon, the magnitude of the output pulse is independent of the type of energy of the radiation measured. After the anode voltage is reached at which the ionization counter starts operating as a Geiger counter, a further increase causes no change in operation until, finally, a voltage is attained at which the quenching action begins to fail. The anode voltage is therefore chosen somewhere around the center of this plateau. Since no such plateau characteristic exists for ionization chambers or proportional counters, a well-regulated power supply must be used for these sensors to avoid inadvertent calibration changes. In its typical operation a Geiger counter produces an output pulse with a rise time of less than 1 μs and a duration of several microseconds. The pulse then decays because of quenching. After the decay the counter is inoperative over a *dead time* of 50 to 150 μs and recovers gradually over an additional but shorter *recovery time*.

86. Ionization in Solid Crystals. (See Fig. 10-35b.) Crystals have been used for the transduction of nuclear radiation in some cases. Such crystals must have a near-perfect crystal structure. Charge separation and the resulting ionization current flow are effected by the electric field established by two electrodes of opposite polarity on opposite crystal surfaces.

Ionization in solid semiconductors (see Fig. 10-35c), such as silicon or germanium, has been used in many useful nuclear radiation sensors. The positive-negative charge pair, resulting from an ionizing event due to incident radiation, is separated by the permanent electric field estab-lished by the junction between the *p* and *n* materials. Although *intrinsic* semiconductors (those having negligible impurity concentrations) have been used in some radiation-detection applica-tions, *extrinsic* semiconductors (those having significant controlled impurity concentrations) make up the bulk of semiconductor radia-tion sensor designs because of their more favorable transduction characteristics. Two major types of extrinsic-semiconductor radiation sensors have been developed in various designs: the surface-barrier type and the diffused-junction type.

Surface-barrier radiation sensors (see Fig. 10-36) usually consist of a *p*-type layer of sil-icon dioxide formed on one surface of a wafer of *n*-type single-crystal silicon. The very thin oxide layer is then covered with a vacuum-evaporated film to serve as electric contact.

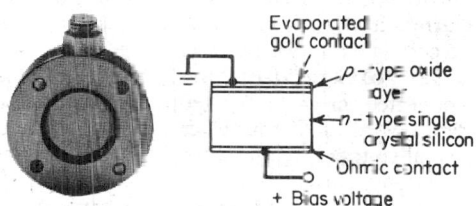

Fig. 10-36. Surface-barrier-type semiconductor radiation sensor. *(ORTEC, Inc.)*

Diffused-junction radiation sensors typically consist of a *p*-type single-crystal silicon wafer into one surface of which a shallow diffusion of *n*-type material (e.g., phosphorus) has been made. Some such sensors use a *p*-type diffusion into an *n*-type base instead.

In a *pin junction* radiation sensor the *n*- and *p*-type semiconductors are separated by an intrin-sic semiconductor. The foremost example of such a device is the *lithium-drifted* radiation detec-tor, in which lithium ions are made to form an intrinsic region slightly below the sensor surface.

87. Photoelectric transduction of nuclear radiation is used in *scintillation counters*. The scintillator material — inorganic or organic crystals, solid plastics, or plastics in liquid solution — is placed in front of a light sensor, usually a photomultiplier tube, in a single enclosure which is lightproof (except at the scintillator's sensing surface) and acts as a magnetic shield. The scintillator converts nuclear radiation into light (photons), which is then converted into an electrical output by the light sensor.

88. Neutron detection involves the interaction of neutrons with a *conversion material* which emits particles other than neutrons when neutrons impinge on them. In gas-filled sensors the conversion material (a stable isotope of one of several selected elements) can be applied either as a fill gas or as internal coating. In semiconductor sensors the material is usually in the form of a film or foil placed in front of the sensing surface. In scintillation counters the material can be a constituent of the scintillator material.

Selection criteria of nuclear radiation transducers involve primarily the type of radiation to be measured and its expected energy levels; secondarily, the presence of other nuclear or light radiation which should not influence the measurement. Special circuitry has been developed to improve the performance of radiation transducers. Since the transducer output is almost invariably in pulse form, logic circuitry has been used extensively to isolate the measured radiation from other prevailing but unwanted events also capable of causing ionization or scintillation.

TRANSDUCERS FOR ELECTRICAL QUANTITIES

89. Terminology. *Current measurements* can be made by various devices acting as transducers. A *series resistance*, inserted into the conductor where current is to be measured, provides a usable voltage across it due to the *IR* drop caused by the resistance. When the series resistance is relatively low, it is referred to as a *shunt*.

If it is necessary to keep the measurement circuit electrically isolated from the measured circuit (which is always desirable), a differential amplifier can be used, with its input terminals closely coupled across the series resistance so that its output terminals can be isolated from the measured circuit, and so that a signal sufficient for telemetry can be provided without inserting too high a series resistance into the measured circuit. This method is usable for both ac and dc currents. Other isolating current transducers are the *saturable reactor*, an adjustable inductor in which the input-current–vs.–output-voltage relationship is adjusted by controlled magnetomotive forces applied to the core, and the *Hall effect* current transducer, in which an output-voltage change is produced by measured-current-originated electromagnetic effects on a semiconductor placed in a magnetic field. The *current transformer* can be used to convert the measured current, with circuit isolation, into an output current or voltage.

90. Electrometers. Small currents (down to 10^{-15} A) can be measured by means of an *electrometer tube* or by special semiconductor devices such as an amplifier with varactor diodes, metal-oxide semiconductor field-effect transistors (MOSFET), or junction field-effect transistors. All these devices require output amplification to provide signals suitable for remote measurement. When current must be measured on the high-voltage secondary side of a transformer, the current in the low-voltage primary can often be measured instead, and a suitable calibration used to correlate the two currents.

91. Voltage Monitors. DC and ac voltages can be monitored by means of voltage divider connected across the two terminals to be measured. A voltage divider consists of two resistors in series, with the output taken across only one of the two resistors when a signal lower than the actual voltage is required by the measurement system. AC voltages can also be measured by use of a transformer.

92. Frequency Converters and Dividers. Frequency can be converted into a voltage signal by use of a tuned discriminator or of integrating circuitry. When a digital signal is required, the measured frequency can be passed through an electronic "gate" which is "opened" for a fixed period of time. The pulse count over the gated period is then indicative of the frequency. If the measuring system can accept a frequency, but one much lower than the measured frequency, a *frequency divider* circuit can be used to provide an output frequency which is a fixed fraction of the input frequency.

93. Other Electrical Transducers. *Power measurements* are usually derived from simultaneous but separate voltage and current measurements. Power (especially at microwave frequencies) is also measured by using a portion of the power to raise the temperature of a resistive temperature transducer (e.g., a thermistor), then measuring the temperature change, which can be correlated to measured power by a suitable calibration.

Conductivity, typically of a liquid material, is measured by a *conductivity cell*, a chamber containing two electrodes of selected materials, connected into a bridge circuit.

Inductance, capacitance, and resistance characteristics are converted into a usable electrical output as described for inductive, capacitive, and resistive transduction elements in Par. **10-9**.

SOLID-STATE SENSORS

94. Solid-State Transducers. Some versions of solid-state transducers are described in Pars. **10-75** to **10-77** and **10-82** to **10-88**. A large variety of new transducers using semiconductor properties have been developed. A number of them are used in biomedical engineering to determine the condition of life systems in research and in clinical applications. See Sec. **26**.

Another widespread application is in the control of automotive engines; solid-state sensors and integrated circuits were universally used in all gasoline-powered passenger cars manufactured in the United States beginning with the 1981 production year and are destined to become more sophisticated as the requirements for fuel economy and pollution control are tightened. Automotive sensors and the associated integrated circuits are described in Pars. **10-98** and **10-99**.

Arrays of small solid-state sensors are used to detect and act on the properties of two-dimensional measurands, a form of image sensing that is becoming more and more detailed. One example is a television camera whose sensitive surface is an array of charge-coupled photodiode solid-state elements capable of resolving more than 100,000 picture elements. For detailed description of these devices, see the references listed in Par. **10-100**.

95. Sensors Applied to Integrated Circuits. An important recent development is the production of silicon sensors (microsensors) on the same chip with a microprocessor. Many large-scale applications of microprocessors in the consumer field (automobiles, cameras, washing machines, microwave ovens, refrigerators, and furnaces) are currently using the transduction properties of silicon and compatible materials to determine light values, temperature, pressure, humidity, force, acceleration, and gas flow.

The analog output of the transducer, or sensor, must be converted into digital form by an analog-to-digital converter built into the microprocessor chip. Silicon microsensors of this type are more reliable than their discrete transistor counterparts. Moreover, they can be verified periodically by the microprocessor.

The principal disadvantage of silicon as a transducer is its limited temperature range, roughly -100 to $+200°C$. Mechanical problems with bonding, etc., may also pose problems in the severe environment of an automobile engine. However, as noted in Par. **10-97**, measurement of gas flow, oxygen content, and mechanical position and temperature are routinely carried out in solid-state automotive control systems (see Figs. 10-43 to 10-45).

96. Types of Semiconductor Transducers. Several types of semiconductors used recently in transducers (silicon and compatible materials) have been applied to the following seven areas of measurement.

Photoconductive Sensors. For wavelengths longer than about 1.1 μm, silicon is photoconductive. For other ranges, including the visible, lead sulfide, cadmium sulfide, or indium antimonide can be evaporated on a silicon wafer. The resistance of the wafer decreases and increased current from an external power supply flows when the incident radiant energy increases.

Photovoltaic Junction Transducers. Silicon photodiodes and phototransistors of both the npn and pnp types are used in photovoltaic applications. The photodiode is reversed-biased so that incident light modulates the diode leakage current. The lateral photoeffect, in which illumination of a photoconductive sensor causes a photovoltage parallel to the junction as well as across it, may also be used.

Piezoresistive Transducers. Piezoresistive wires have long been used in strain gages to measure force, vibration, acceleration, and pressure. For application in silicon systems, p-type monolithic silicon is suitable for strain gages. It displays a linear relationship between stress and resistance up to 2.5% resistivity variation with only 0.5% nonlinearity. It does not display hysteresis under repeated applications of stress, and it obeys Hooke's law over a broad range.

Piezoelectric Transducers. Whereas silicon is not piezoelectric, zinc oxide and indium antimonide are, and they can be deposited on silicon wafers for use in strain gages.

Temperature Transducers. Silicon temperature sensors operate by virtue of the increase with temperature of the base-emitter voltage of a junction transistor operating at constant collector current. Typically a planar npn transistor can be used for this purpose, directly fabricated on a silicon chip. Resistors with a positive temperature coefficient can also be fabricated on chips to serve as temperature transducers.

Dew-Point Transducers. (See also Par. **10-47**.) Dew-point sensors operate by cooling a moisture-carrying gas until condensate forms and determining the temperature at which it forms. In solid-state dew-point indicators, the condensate detector may be a capacitor formed by depositing aluminum on each side of a layer of SiO_2. This capacitor is placed (see Fig. 10-37) on top of a Peltier cooling device. On the upper surface of the capacitor a temperature-sensitive transistor (see preceding paragraph) measures the temperature at which the formation of moisture changes the value of the capacitance.

Hall-Effect Transducers. The Hall effect, by which the concentration of electrons and holes in n and p materials can be measured, is used for precise control in the manufacture of semiconductor devices. The Hall effect can be measured by applying a magnetic field perpendicular to the surface of a semiconducting plate. This produces a transverse electric field which can be measured as a voltage in the transverse direction.

97. Practical Forms of Semiconductor Transducers. As of the early 1980s, several forms of semiconductor transducers have been developed in practical form. The following examples are taken from Ref. 34.

Position Detector. A position detector developed at the Delft University of Technology (Fig. 10-38) operates by the photovoltaic junction effect. A layer of boron-doped p-type silicon is dif-

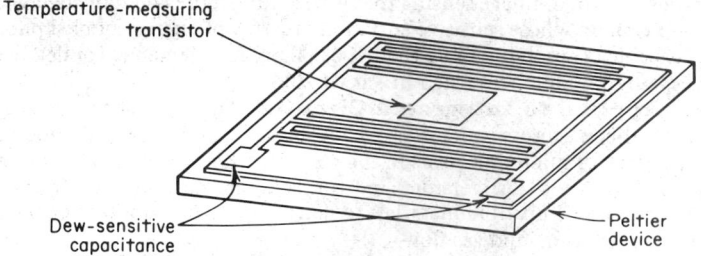

Fig. 10-37. A solid-state dew-point transducer, which measures the dew-point temperature at the instant condensate forms and changes the capacitance between the aluminum fingers.[34]

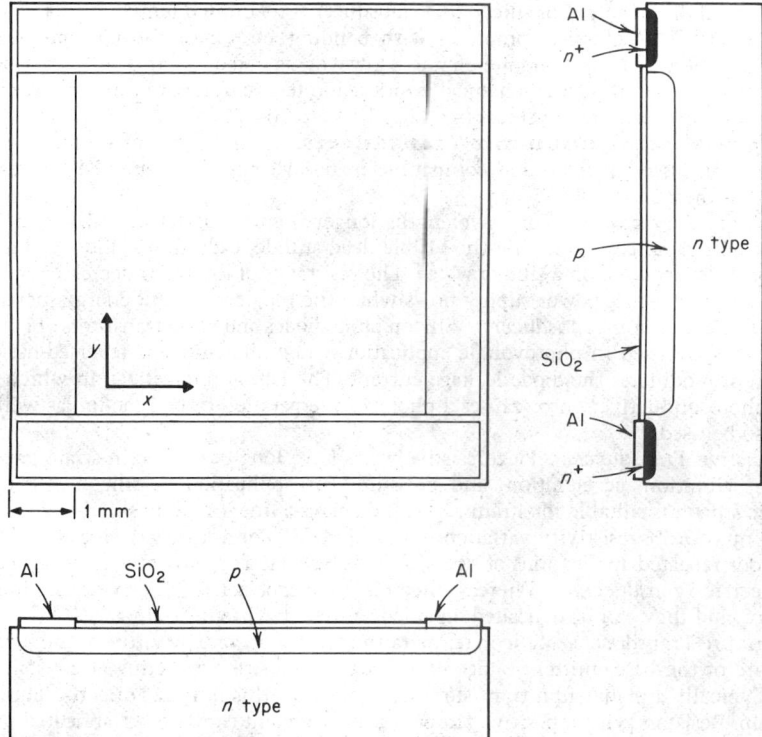

Fig. 10-38. Top and cross-sectional views of a position detector fabricated on silicon by integrated-circuit methods. The device measures 8 by 6 mm.[34]

fused on a substrate of phosphor-doped n-type silicon. This element is framed by contact strips deposited on n^+ islands. The shorter edge strips are evaporated on the p layer, while the longer ones are placed on the substrate, adjacent to the layer. The position of light impinging on the xy coordinate, as shown in the figure, is to be determined. The current between the x contacts depends only on the x coordinate; that between the y contacts only on the y coordinate. The voltages on the contacts are kept constant.

Calculations indicate that the "center of gravity" on the light spot is measured linearly by the contact currents, so that very precise position measurements of unfocused spots can be determined as long as the whole area of the spot is within the sensitive area. Such position indicators are useful in photolithography, pattern recognition, and target seeking for missiles; another application is in adjusting laser-beam systems in video-disc players.

Pressure Transducer. A pressure transducer (Fig. 10-39) was commercially available in 1980, applied to process controllers, and to measurement of oil pressure and exhaust pressure in automotive engine manifolds. It consists of a diaphragm set in a ring clamp and anchored to four silicon strain gages. Two of the gages are placed near the center of the diaphragm and two near the edges. The gages form the arms of a Wheatstone bridge. The bridge becomes unbalanced to a degree depending on the applied stress. The size and thickness of the diaphragm is adjusted to meet different sensitivity- and response-range requirements.

This transducer is fabricated by integrated-circuit methods; p silicon is diffused onto an n epitaxial layer, grown on an n^+ substrate. The n^+ substrate is selectively removed by spark machining and by an electrochemical process. The remaining epitaxial layer forms an integral unit with the clamped ring and the diaphragm.

Silicon Accelerometer. An accelerometer developed at Stanford University (Fig. 10-40) consists of two parts: a piezoelectric strain gage in the form of a 15-μm-thick silicon cantilever and

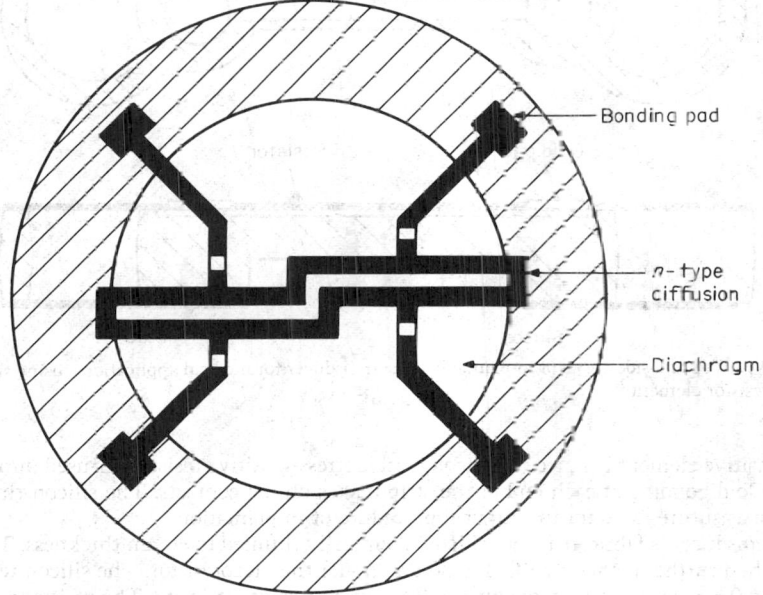

Fig. 10-39. Pressure transducer consisting of four silicon strain-gage elements in a Wheatstone bridge.[34]

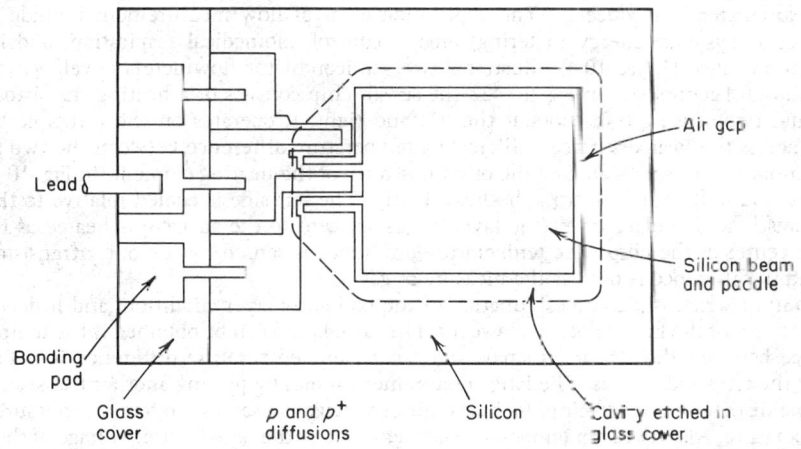

Fig. 10-40. Accelerometer, measuring 2 by 3 by 0.6 mm, constructed of silicon, using integrated-circuit technology.[34]

a 200-μm-thick silicon mass which serves as the inertial reference, attached to the movable end of the cantilever. The device is glass-covered, the glass being etched to permit motion of the cantilever. It measures accelerations perpendicular to the glass face, which bend the cantilever, producing stresses proportional to the acceleration.

Temperature compensation is provided by a second gage on a nearby, unstressed piece of thin silicon. This accelerometer is fabricated by etching and diffusion. p-type strain elements and p^+ contact pads are diffused on the top surface, which is then covered with a thick layer of SiO$_2$. The bottom surface is etched with KOH through a mask. This device, measuring 2 by 3 mm, will measure accelerations from 10^{-1} to 10^3 m/s^2.

Implantable Force Transducer.   Figure 10-41 illustrates a transducer developed at Stanford University and made of materials suitable for implantation in biomedical applications. The

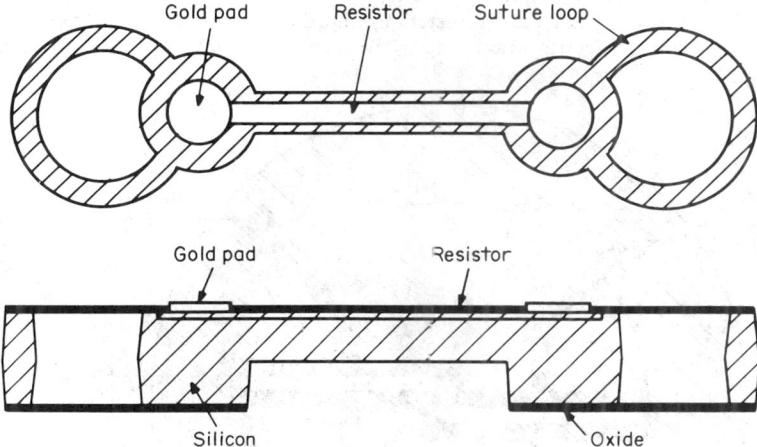

Fig. 10-41.   Top and side views of implantable force transducer for medical applications, using silicon stress-sensitive resistor element.[34]

stress-sensitive element is a piece of silicon, with a stress-sensitive resistor diffused into its middle section. Gold bonding at each end of the strip makes ohmic contacts. The silicon rings at each end serve as suture loops for the surgical procedure of implantation.

The transducer is fabricated from silicon chemically thinned to 60 μm thickness. The bottom layer is then further etched by KOH to 30 μm, under the piezoresistor. The silicon-resistor strip will stand 0.5 g·cm bending moment and 90 g tension before fracture. The resistance measured across the device is proportional to stress, without hysteresis, over the range of force up to 10% of the breaking stress.

Gas Flowmeter Transducers.   The application areas of flow measurement include pumps, fuel-injection systems, energy metering, process control, biomedical respiration, and infused-fluid motion rates. Figure 10-42 illustrates two semiconductor flowmeters developed at Delft University of Technology. In Fig. 10-42a the sensor chip consists of a heating transistor in the center and two sensing transistors at the left and right. It operates on the principle that gas flowing across the heated surface will create a temperature difference between the two temperature sensors, one downstream and the other upstream of the heating element. In Fig. 10-42a the gas flows from left to right, across the heated chip. The left side is cooled relative to the right side because the boundary layer (the layer of gas adjacent to the surface) is heated as it passes over the center of the chip. The temperature difference is sensed by the outer transistors, and their output difference is used as the measure of gas flow.

The output signal is a complex function of the boundary-layer condition, and it depends on the square root of the flow velocity. However, a linear relation can be obtained if the temperature difference between the sensors is small and a constant temperature difference is maintained between the chip and the gas. The latter requirement is met by placing another transistor in the flow some distance from the chip. Its base-emitter voltage is used as a reference measure of the gas temperature. Also, the chip temperature is measured by the base-emitter voltage of the downstream sensor transistor on the chip. An operational amplifier amplifies the difference between these two voltages, and the result is used to drive the heat transistor in the center of the chip, thus maintaining a constant temperature difference between gas and chip. The outputs of the temperature-sensing transistors on the chip are applied to an operational amplifier, which generates an output voltage proportional to the voltage difference between them. When this output voltage is squared, it provides a linear measure of the gas velocity.

In Fig. 10-42b, a Wheatstone bridge is formed by four p-type resistors diffused onto a substrate. Gas flowing over the chip cools the resistors perpendicular to the flow more than the resistors

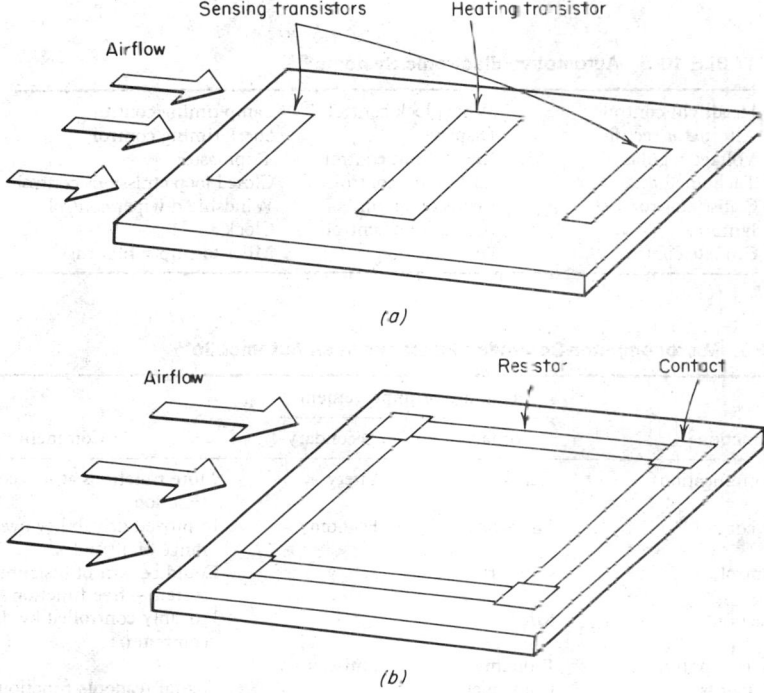

(a)

(b)

Fig. 10-42. Two forms of silicon gas flowmeters. (a) Two temperature-measuring transistors on each side of a heating transistor detect the change in gas temperature as it flows over the chip. (b) Four temperature-sensitive transistors form a Wheatstone bridge, the different effects of the lateral and longitudinal flow causing bridge unbalance proportional to gas velocity.[34]

parallel to it. The corresponding difference in the resistances causes the bridge to become unbalanced, providing a signal across its diagonal that is proportional to the flow velocity. A similar principle can be used to measure fluid velocity.

Dew-Point Transducer. The dew-point sensor previously mentioned (Fig. 10-37) was developed at Delft University of Technology. It consists of a planar npn transistor placed in the center of a silicon chip, its base-emitter voltage being a measure of the temperature of the chip. The transistor is surrounded by a fingerlike pattern of 15-μm-wide aluminum conductors deposited on a 1-μm-thick layer of SiO$_2$. The conductive strips form a capacitor of about 20 pF. The chip is mounted on a Peltier-type cooling device. When vapor is precipitated on the chip, the capacitance increases to 40 to 80 pF (since pure water has a permittivity of about 20). When the sudden change in capacitance is detected, the output (base-emitter voltage) of the heat-sensitive transistor measures the dew-point temperature. The accuracy is about $\pm0.2°C$, a very small error in such measurements, attributable to the chip's small size and the close proximity of the temperature sensor to the dew collector. Hysteresis effects can be minimized by operating the temperature sensor in a constant low-impedance mode. This can be achieved by regulating the current through the Peltier device to keep the impedance and hence the amount of water at a constant low level.

This detector is insensitive to contaminants and can be integrated into a single chip with other elements. Applications include automotive engines to enhance fuel economy and the regulation of humidity in climate-control systems.

98. Transducers in the Automotive Industry. During the 1970s a large-scale effort was mounted to develop sensors and associated computing systems suitable for use in automobiles. Strict Federal standards for fuel economy and pollution control were the motivation. In 1977 Cadillac introduced MISAR, the first microprocessor-controlled engine control system. By 1981 virtually all gasoline-powered passenger cars produced in the United States were equipped with more advanced versions of microprocessor engine control, and many other functions were carried out by computer methods. Table 10-5 lists the major electronic automotive functions and Table 10-6 lists those under microcomputer control.

TABLE 10-5   Automotive Electronic Systems[46]

Headlight control	Wheel-lock control	Lamp-timing control
Alternator rectifier	Displays	Spark-timing control
Voltage regulator	Knock-limit control	Tripmaster
Tachometer	Carburetor control	Closed-loop emissions control
Cruise control	Intrusion alarm	Windshield-wiper control
Ignition	Air-cushion control	Clock
Climate control	Fuel injection	Miles-to-empty fuel gage

TABLE 10-6   Microcomputer-Controlled Functions in an Automobile[46]

Function	Advantage or improvement		Comment
	Primary	Secondary	
Audio communications	Comfort	Safety	More functions at less cost per function
Carburetor control	Emissions	Economy	Improved drivability over a wider range of altitudes
Climate control	Comfort	Safety	Could be part of instrument system — free function
Crash protection	Safety		Probably controlled by dash computer
Cylinder select control	Economy	Emissions	
Dash instruments	Convenience		Digital readouts functional and attractive
Dynamic ride control	Comfort	Safety	
Engine control	Emissions	Economy	Maximizes drivability within emission and economy constraint
Miles-to-empty	Convenience	Safety	
On-board diagnostics	Convenience	Safety	Helps relieve anxiety about mechanic's analysis
Security control	Safety	Comfort	Permits unique security system and code for each driver
Speed control	Comfort	Economy	
Transmission control	Economy	Drivability	
Wheel status	Safety		Low-tire indicator will improve safety and tire wear

One of the biggest problems in reaching the universal use of computers in automobiles was the development of transducers, primarily sensors, rugged enough to meet the severe environmental conditions of the engine compartment and reliable enough to avoid potentially disastrous breakdowns. By the early 1980s most of the problems had been solved. Table 10-7, compiled in 1977, lists the sensors then being considered for automotive use, with the principles of operation, advantages, and disadvantages.

Figure 10-43 shows the major elements of a computer system envisaged for total engine control during the 1980s. The most important element (shown at the bottom), responsible in large part for achieving large-scale production of standard microprocessor chips, is the alterable instruction program, in which a programmable read-only memory (PROM) is used to adapt a standard microprocessor to specific combinations of engine and vehicle characteristics.

In its many interconnections Fig. 10-43 shows one of the major areas of design attention, centralized vs. distributed control of engine functions. The centralized system of Fig. 10-43 involves a complicated wiring harness, with attendant maintenance problems. Distributed control, on the other hand, uses a number of smaller dedicated microprocessor chips with their associated sensors and actuators. Much simpler wiring results, but the costs of the individual computer elements must be competitive, on a total performance-cost basis with the centralized system.

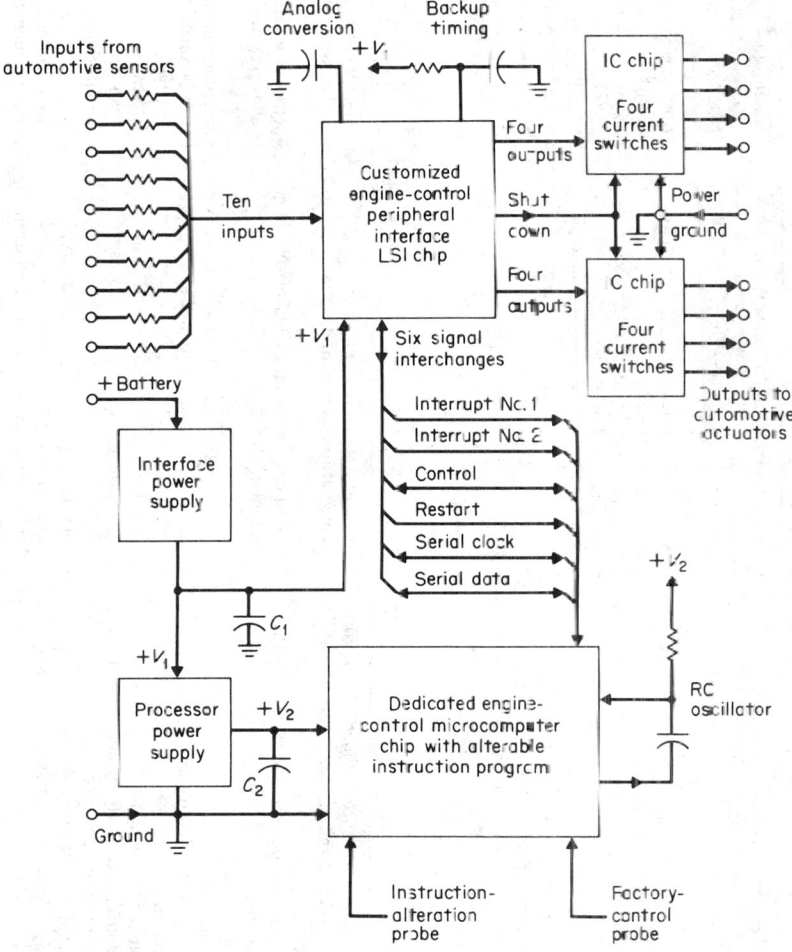

Fig. 10-43. Basic elements of a comprehensive automotive electronic engine control system, as envisaged for the 1980s. This centralized type of control requires a complicated wiring harness.[46]

99. Examples of Microprocessor-Controlled Engines. The principal requirements of computer-controlled engine systems relate to the opposed effects of fuel economy and pollution control, but many secondary benefits must also be kept in mind, particularly driveability, i.e., the ease and comfort with which the driver controls the vehicle and feels its response. These requirements may be met in large part by (1) optimized control of ignition timing, (2) optimization of air-to-fuel ratio in carburetion, (3) injection of air into the exhaust to reduce pollutants, (4) exhaust-gas recirculation (EGR) to use fuel more efficiently, (5) the prevention of untimely detonation (engine knock), and (6) diagnosis of faults not only in the engine but in the sensor-computer-actuator system itself.

A particularly important aspect of fuel economy is reaching the stochiometric ratio, 14.7:1, between air and fuel at which the three-way catalyst is most efficient in reducing exhaust emission pollutants to acceptable levels. Maintenance of this optimum ratio requires sensing temperature, pressure, oxygen content, throttle position, and gas flow.

TABLE 10-7 Transducers and Sensors Considered for Automotive Applications[46]

Device	Principle of operation	Advantages	Disadvantages
Temperature			
Thermocouple	EMF generated at junction of two dissimilar materials by Thompson effect	Largest temperature range, very high temperatures	Small signal, low sensitivity, lowest accuracy, unstable, expensive
Thermistor	Change in resistance with temperature in semiconducting metal oxides	Largest signal, high sensitivity, low cost, modest accuracy	Nonlinear, limited temperature range, part-to-part variation
Resistance temperature detector (RTD)	Change in resistance with temperature in metallic conductors	Wide temperature range, highest accuracy, very stable, linear	Modest signal, high cost, long thermal time constant
Semiconductor diode	Change in V_{be} with temperature in forward-biased diode	Good accuracy, highly stable, modest cost, manufactured using standard semiconductor processing methods	Small signal, part-to-part variation, nonlinear, requires external signal conditioning
Thyristor temperature switch	Thermal breakdown of thyristor gate at set temperature and voltage	Capable of handling large currents	Set point depends on doping profile of junction, limited temperature range
Spreading resistance sensor (SRS)	Change in resistance of extrinsic silicon with temperature	Manufactured using common planar processing methods, fast thermal response, comparable to RTDs in performance, potentially low cost	Requires highly uniform substrate material, high-resolution photolithography
Pressure			
Potentiometric	Change in resistance of slide-wire resistor coupled to diaphragm structure or Bourdon tube	Low cost, relatively insensitive to temperature, large output signal	Low accuracy, low sensitivity, hysteresis, noise, wear
Reluctive (LVDT)	Change in reluctance due to motion of core slug coupled to diaphragm structure or Bourdon tube	Modest cost, temperature insensitive, large output ac signal	Sensitive to shock and vibration, ac output, slow response time
Bonded strain gage	Change in resistance due to pressure-differential-induced strain in diaphragm structure	High sensitivity, low hysteresis, can have differential output, ratiometric to supply voltage	Small output signal, temperature sensitive, can be unstable due to bond degradation, costly
Capacitive	Change in capacitance due to deflection of diaphragm constituting one plate of parallel-plate capacitor	Highest sensitivity, can be used in harsh environments, temperature insensitive	Typically very low capacitance, difficult to manufacture, expensive in most applications
Piezoresistive strain gage	Change in resistance due to pressure-induced strain in integral diaphragm structure	Potentially low cost, high sensitivity, low hysteresis, avoids problems of bonded strain gage, can be manufactured using planar processing technology	Very temperature sensitive, low output signal, difficult to package, possible environmental problems
Position			
Potentiometric	Change in resistance of slide-wire resistor coupled to mechanical motion	Low cost, relatively temperature insensitive, large output signal	Low accuracy, low sensitivity, hysteresis, noise, wear

Type	Principle	Advantages	Disadvantages
Reluctive	Change in Q of coil due to motion of core slug (LVDT) or proximity of metallic material	Modest cost, temperature insensitive, large output signal, good accuracy	Sensitive to shock and vibration (LVDT), requires metallic material, ac output
Optical	Interruption or reflection of optical signal to optoelectronic detector	Noncontacting, can sense nonmetallic materials, fast response time	Subject to contamination by dirt, oil, and other opaque materials
Hall effect	Hall effect in semiconducting element generates output voltage proportional to field strength	Noncontacting, not subject to interference by dirt and oil, can be manufactured in integrated form using semiconductor methods	Requires external magnet, temperature sensitive, small output signal, difficult to package
Wiegand effect	Domain reversal in specially processed ferrous materials, generates large voltage pulse	Noncontacting, temperature insensitive, requires no external power supply, large output signal	Pulse output only, poorly understood phenomenon, requires external magnet
Magnetodiode	Change in V_{be} or I_c of pin or Schottky diode due to minority-carrier recombination at magnetic-barrier layer structure	Noncontacting, large output signal, can be incorporated in discrete devices such as IJFET's for switching	Poorly understood phenomenon, requires external magnet

Fluid flow

Type	Principle	Advantages	Disadvantages
Turbine or vane	Momentum transfer from fluid flow to rotating turbine or movable vane	Mass flow, linear output can provide digital signal (pulse train)	Requires calibration for specific fluid, can restrict flow
Vortex shedding	Bluff body induces turbulent vortices in flow that can be sensed as pressure or velocity changes	Simple, no moving parts, digital output (pulse train)	Provides velocity flow only
Thermal transfer	Heat transfer by fluid from heating element, can be sensed as change in resistance of heater element or by separate temperature sensor	Simple mass-flow sensor, no moving parts	Highly nonlinear, requires calibration for specific fluid
Differential pressure	Pressure differential across orifice in flow stream	Uses conventional pressure sensors in differential mode	Provides velocity flow only, highly nonlinear

Environmental

Type	Principle	Advantages	Disadvantages
Semiconducting metal-oxide gas	Change in conductivity due to surface adsorption of gaseous elements	Simple solid-state device, simple signal conditioning, low cost	Only partially selective, affected by several gaseous elements, part-to-part variation
Alumina humidity	Change in impedance of anodized alumina thin film due to adsorption of water vapor	Simple solid-state device, can be incorporated in conventional semiconductor devices	Unstable, part-to-part variation, high-temperature limitations
Brady array humidity	Change in conductivity of semiconducting oxide element due to adsorption of water vapor	High sensitivity, fast response	Expensive, part-to-part variation, complex processing of signal
MOSFET gas	Change in flat-band voltage of MOSFET due to surface adsorption of gaseous elements	Standard semiconductor device, highly sensitive, easily measured, can be incorporated in standard IC products	Sensitive to many different chemical elements, phenomenon poorly understood
Surface junction gas	Change in semiconductor junction properties due to adsorption at surface junction	Standard semiconductor device, highly sensitive, easily measured, can be incorporated in standard IC products	Sensitive to many different chemical elements, effect difficult to control

Initially, attempts to introduce electronic engine control, going back to the Volkswagen fuel injection system in 1967, were based on analog outputs of sensors. The size and complexity of the automobile electrical system, including the many functions potentially involved (Tables 10-5 and 10-6), plus the rapidly reduced costs of microprocessor chips made the digital approach much more attractive, and digital methods have been used since 1975 in all engine control applications. This development has culminated in the use of 16-bit microprocessors in the early 1980s. For example, the Ford PROCO system of stratified-charge fuel-injection, put into pilot production in 1981, used the Intel 16-bit model 8016 microcomputer.

Shown in Fig. 10-44 are the basic elements of the EEC-III Ford engine control system, introduced in 1980. The EEC-III incorporated four custom large-scale integrated-circuit chips in the

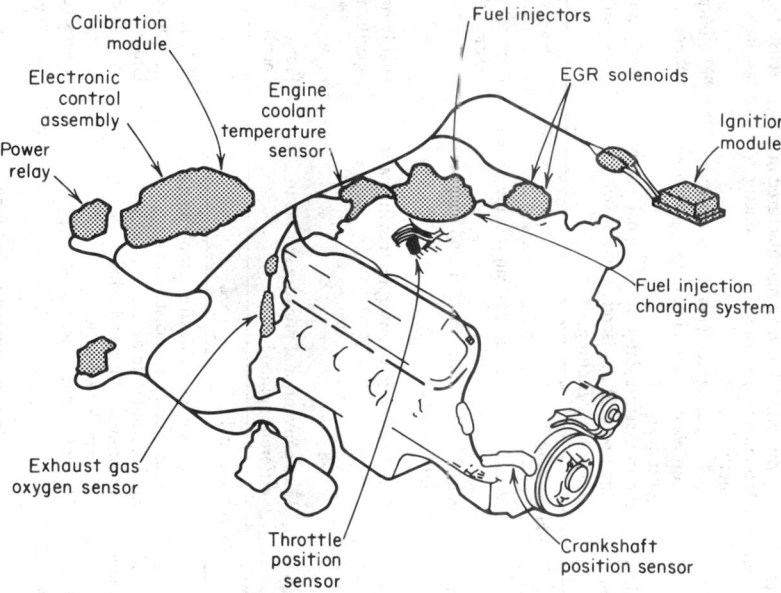

Fig. 10-44. Elements of the Ford model EEC-III engine control system. Read-only memory chips can be plugged into the base module to adapt the system to different engine-vehicle combinations.[47]

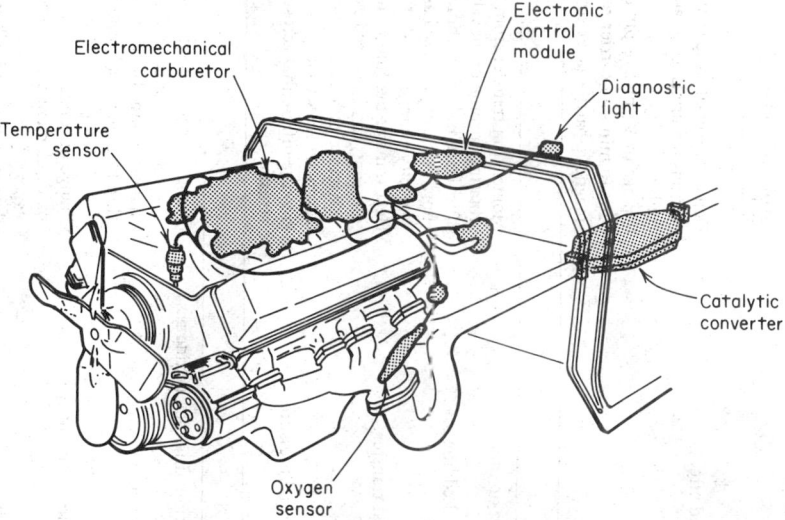

Fig. 10-45. Elements of the General Motors C-4 engine control system, used in nearly all GM cars produced for the 1981 model year. The feedback control of carburetion is shown, but many other functions can be performed.[47]

control assembly, with much faster processing time than was available in previous models. It provides two methods of controlling the air-to-fuel ratio (1) through feedback control of the carburetor and (2), for certain engines, electronic fuel injection. The fuel-injection control module has 11 inputs, 10 outputs and a memory of 8,000 words, built into replaceable chips in which the memory content is adapted to particular engine-vehicle combinations.

The C-4 engine control system, used on almost all 1981 General Motors gasoline-powered cars is shown in Fig. 10-45. This incorporates closed-loop control of carburetion. An oxygen sensor connected to the control module indicates the presence or absence of oxygen. This signal, with another from the coolant temperature sensor, is used to compute the proper air-to-fuel ratio. The desired mixture is then called out by the control module to the electromechanical carburetor. The C-4 system can be used on four-, six-, and eight-cylinder engines, with displacement from 1.6 to 6 L. The basic system can be expanded to incorporate a number of additional functions, e.g., throttle-body fuel injection, electronic spark timing, idle speed control, exhaust-gas recirculation, choke control, secondary air control, evaporative emissions purge control, knock limiting, and transmission control.

100. Bibliography

1. Ahmad, W., and M. Khan   A Digital Technique for Measurement of Differential Pressure and Flow, *IEEE Trans. Indus. Electron. Control Instrum.*, Vol. IECI-25, No. 1, February 1978.

2. Allan, R.   New Applications Open Up for Silicon Sensors, *Electronics*, Nov. 6 1980, pp. 113–122.

3. Aronson, M. H. (ed.)   "Temperature Measurement and Control Handbook," *Instruments Publishing Co.*, Pittsburgh, 1961.

4. Bollinger, L. E.   Transducers for Measurement, Part IV, Fluid Flow, *ISA J.*, November 1964, Vol. 11, No. 11, pp. 64–69.

5. Bradspies, R. W.   Bourdon Tubes, *Giannini Controls Tech. Note*, Conrac Corp., Duarte, Calif., 1961.

6. Brock, T. E., and C. J. Moon (eds.)   "A Bibliography on Hot-Wire Anemometry," British Hydromechanics Research Association, Cranfield, Bedford, October 1965.

7. Canfield, E. B.   "Electromechanical Control Systems and Devices," Wiley, New York, 1965.

8. Cerni, R. H., and L. E. Foster   "Instrumentation for Engineering Measurement," Wiley, New York, 1962.

9. Considine, D.   "Process Instrumentation and Control Handbook," McGraw-Hill, New York, 1957.

10. Considine, D.   "Handbook of Applied Instrumentation," McGraw-Hill, New York, 1964.

11. Corruccini, R. J.   Interpolation of Platinum Resistance Thermometers, 20 to 373.15°K, *Rev. Sci. Instrum.*, 1960, Vol. 31, pp. 637–640.

12. Dean, M., III (ed.)   "Semiconductor and Conventional Strain Gages," Academic, New York, 1962.

13. Dearnaley, G., and D. C. Northrop   "Semiconductor Counters for Nuclear Radiation," 2d ed., Barnes and Noble, New York, 1966.

14. Doebelin, E. A.   "Measurement Systems: Application and Design," McGraw-Hill, New York, 1966.

15. Dozer, B. E.   Liquid Level Measurement for Hostile Environments, *Instrum. Technol.*, February 1967, Vol. 14, No. 2, pp. 55–58.

16. Draper, C. S., W. McKay, and S. Lees   "Instrument Engineering," Vols. I and III, McGraw-Hill, New York, 1952.

17. Eisenmann, W. L.   Properties of Photodetectors, *Nav. Ordn. Lab. Photodetec. Ser. NOLC Rep.* 637, Corona, Calif., Feb. 15, 1966.

18. Fagenbaum, J.   Controlling the Synfuel Process, *IEEE Spectrum*, November 1980, pp. 31–36.

19. Frazine, D. F.   The Design and Construction of Thin Film Radiation Thermopiles, *Arnold Eng. Dev. Center Rep. AEDC-TR-66-38*, USAF-AFSC, Arnold Air Force Sta., Tenn., 1966.

20. Frederick, J. R.   "Ultrasonic Engineering," Wiley, New York, 1965.

21. Grabbe, M., S. Ramo, and D. E. Wooldridge (eds.)   "Handbook of Automation, Computation and Control," Vol. 3, Chap. 20, Wiley, New York, 1961.

22. Habibullah, B., H. Singh, K. L. Soo, and L. C. Ong   A New Digital Speed Transducer, *IEEE Trans. Indust. Electron. Control Instrum.*, April 1978, Vol. IECI-25, No. 4, pp. 339–342.

23. Harris, C. M., and C. E. Crede (eds.)   "Shock and Vibration Handbook," 3 vols., McGraw-Hill, New York, 1961.

24. Harrison, T. R.   "Radiation Pyrometry and Its Underlying Principles of Radiant Heat Transfer," Wiley, New York, 1960.

25. Holzbock, W. G. "Instruments for Measurement and Control," 2d ed., Reinhold, New York, 1962.

26. Hughes, W. G. Attitude Control: Gyros and Sensors, R. Aircraft Estab. Rep. ESRO-TM-32, Farnborough, England, 1966.

27. Jones, T. O. Some Recent and Future Automotive Electronics Developments, Science, March 1977.

28. Jones, T. O. et al. Automotive Microprocessors, Proc. Electron. (Turin), 1977, Vol. 4.

29. Keast, D. N. "Measurements in Mechanical Dynamics," McGraw-Hill, New York, 1967.

30. Lion, K. S. "Instrumentation in Scientific Research," McGraw-Hill, New York, 1959.

31. Lloyd, C., and A. A. Giardini Measurement of Very High Pressures, Acta IMEKO 1964 Pap. 21-USA-267, Budapest, 1964.

32. Merriam, J. D., W. L. Eisenman, and A. B. Naugle Properties of Photodetectors, Nav. Ord. Lab. NOLC Rep. 621, Corona, Calif., Apr. 1, 1965.

33. Merill, J. J. (ed.) "Light and Heat Sensing," Macmillan, New York, 1963.

34. Middelhoek, S., J. Angell, and D. J. W. Norlag Microprocessors Get Integrated Sensors, IEEE Spectrum, February 1980, Vol. 17, No. 2, pp. 42–46.

35. National Semiconductor Corporation "Pressure Transducer Handbook," Santa Clara, Calif., 1977.

36. National Semiconductor Corporation "Transducers for Pressure and Temperature," Santa Clara, Calif., 1977.

37. Neubert, H. K. P. "Instrument Transducers," Oxford, London, 1963.

38. Nokes, M. C. "Radioactivity Measuring Instruments: A Guide to Their Construction and Use," Philosophical Library, New York, 1959.

39. Norton, H. N. Error Band Concept Defines Transducer Performance, Ground Support Equip., January 1963.

  a. Norton, H. N. "Handbook of Electronic Sensing and Analyzing Devices," Prentice-Hall, Englewood Cliffs, N.J., 1981.

  b. Norton, H. N. "Introduction to Biomedical Sensors," Noyes, Park Ridge, N.J., 1981.

  c. Norton, H. N. (ed.) "Sensor Selection Guide," Elsevier-Sequoia, Lausanne, 1982.

40. Norton, H. N. Specification Characteristics of Pressure Transducers, Instrum. Control Syst. December 1963, Vol. 36.

41. Norton, H. N. "Handbook of Transducers for Electronic Measuring Systems," Prentice-Hall, Englewood Cliffs, N.J., 1969. (See Ref. 90.)

42. Oliver, B. M., and J. M. Cage "Electronic Measurement and Instrumentation," McGraw-Hill, New York, 1971.

43. Oliver, F. "Practical Instrumentation Transducers," Hayden, New York, 1971.

44. Ostrovskij, L. A. "Elektrische Messtechnik" ("Electrical Measurement Technology"), (Ger. trans. by D. Hoffman), VEB Verlag Technik, Berlin, 1969.

45. Pitman, G. R., Jr. (ed.) "Inertial Guidance," Wiley, New York, 1962.

46. Puckett, G., J. Marley, and J. Grapp Automotive Electronics, II, IEEE Spectrum, November 1977, Vol. 14, No. 11, pp. 37–45.

47. Rivard, J. G. Microprocessors Hit the Road, IEEE Spectrum, November 1980, Vol. 17, No. 11, pp. 44–47.

48. Roehrig, J. R. "High Vacuum Measuring Instrumentation and Methodology," U.S. Dept. Commerce, Office Tech. Serv. Rep. FDL-TDR-64-68 (AFSC-R&TD-AFFDL), 1964.

49. Ruskin, J. M., Jr. Thermistors as Temperature Transducers, Data Syst. Eng., February 1964, Vol 19, No. 2, pp. 24–27.

50. Savet, P. H. "Gyroscopes: Theory and Design," McGraw-Hill, New York, 1961.

51. Schweppe, J. L., et al. "Methods for the Dynamic Calibration of Pressure Transducers," Natl. Bur. Stand. Monogr. 67, Washington, 1963.

52. Schuster, G. M. On the Use of a Resonant Diaphragm as an FM Pressure Transducer, IEEE Trasn. Indus. Electron. Control Instrum., February 1978, Vol. IECI-25, No. 1.

53. Sharpe, J. "Nuclear Radiation Measurement," Temple, London, 1960.

54. Sheingold, D. H. "Transducer Interface Handbook," Analog Devices, Norwich, Mass., 1980.

55. Spaulding, Carl P. "How to Use Shaft Encoders," DATEX Division of Conrac Corp., Monrovia, Calif., 1965.

56. Special Issues of the IEEE Trans. Indus. Electron. Control Instrum., Vol. IECI-16, No. 1, July 1969; Vol. IECI-17, No. 2, August, 1970; Vol. IECI-19, No. 2, August 1971.

57. Spink, L. K. "Principles and Practice of Flow Meter Engineering," 9th ed., The Foxboro Co., Foxboro, Mass., 1967.

58. Stempel, F. C., and D. L. Rall Applications and Advancements in the Field of Direct Heat Transfer Measurements, *Instru. Soc. Am. Prepr.* 8.1.63, Pittsburgh, 1963.

59. Stiltz, H. L. (ed.) "Aerospace Telemetry," Prentice-Hall, Englewood Clifs, N.J., 1961.

60. Streeter, V. L. (ed.) 'Fluid Dynamics," McGraw-Hill, New York, 1948.

61. Taylor, J. M. "Semiconductor Particle Detectors," Butterworth, London, 1963.

62. Truxal, J. G. (ed.) "Control Engineers' Handbook," McGraw-Hill, New York, 1958.

   a. Tyson, F. C. "Industrial Instrumentation," Prentice-Hall, New York, 1943.

63. Van Der Pyl, L. M. Bibliography on Diaphragms and Aneroids, *ASME Pap.* 60-WA 122, 1960.

64. Vanzetti, R. "Practical Applications of Infrared Techniques, Wiley, New York, 1972.

   a. Volk, J. A. Gyroscopes, *Data Syst. Eng.*, February 1964, Vol. 19, No. 2 p. 28–31.

65. Werner, F. D. Time Constant and Self-Heating Effect for Temperature Probes in Moving Fluids, *Rosemount Eng. Co. Bull.* 106017, Minneapolis, Minn., 1960.

66. Wexler, A. (ed.) "Humidity and Moisture Measurement and Control in Science and Industry," Vols. 1–3, Reinhold, New York, 1965.

67. Wolber, W. G. Prime Sensor for Automotive Engine Control, *IEEE Trans. Vehic. Technol.*, 1977, Vol. VT-26, No. 2.

68. Wolber, W. G. New Sensors for Automotive Engine Control Instrumentation, *Instrum. Technol.*, August 1978, pp. 47–53.

69. Wolf, W. L. "Handbook of Military Infrared Technology," Office of Naval Research, Washington, 1966.

70. Zeisler, F. L. Transducers for Automotive Control Systems, *Automot. Eng. Congr. Soc. Automot. Eng.* 1973, Pap. 730130.

71. Ziebolz, H. "Analysis and Design of Translator Chains," 2d ed., Askania Regulator Co., Chicago, 1951.

72. Temperature: Its Measurement and Control in Science and Industry," Vol. 2, Reinhold, New York, 1955.

73. IRE Standards on Nuclear Techniques: Definitions for the Scintillation Counter Field, 1960. Institute of Electrical and Electronics Engineers, 60 IRE 13.S1, New York, 1960.

74. "Telemetry Transducer Handbook," *WADD Tech. Rep.* 61–67, Vols. I and II, ASD, Air Force Systems Command, USAF, Wright-Patterson AFB, Ohio, 1961.

75. "Strain Gage Handbook," *BLH Electronics, Bull.* 4311A, Waltham, Mass., 1962.

76. "Temperature: Its Measurement and Control in Science and Industry," Vol. 3, Parts 1–3, Reinhold, New York, 1962.

77. Synchro and Resolver Evaluation and Test Equipment (editorial survey), *Electro-Technol.*, May 1963, Vol. 71, No. 5, pp. 201–210.

78. Thermistor Definitions and Test Methods, Electronic Industries Association, EIA Standard RD-275, Washington, 1963.

79. Specifications and Tests for Piezoelectric Acceleration Transducers, ISA Stand. RP37.2, Pittsburgh, Pa., 1964.

80. "Standard Gyro Terminology" (rev.), Aerospace Industries Association, Washington, D.C., September 1964.

81. "Strain-Gages, Bonded Resistance," Aerospace Industries Association, National Aerospace Standard NAS 942, Washington, 1964.

82. Temperature Measurement Thermocouples, American National Standards Institute, ANSI C96.1, New York, 1964.

83. Load Cells, Bonded Strain Gage, General Specification for U.S. Naval Weapons Center, Specification NOTS-PD-101, China Lake, Calif., 1966.

84. "Photoconductive Cell Manual," Clairex Corp., New York, 1966.

85. Specifications and Tests of Potentiometric Pressure Transducers for Aerospace Testing, ISA Stand. S37.6, Pittsburgh, 1967.

86. Dynamic Response Testing of Process Control Instrumentation, ISA Stand. S25, Pittsburgh, 1968.

87. Electrical Transducer Nomenclature and Terminology, ISA Stand. S37.1, Pittsburgh, 1969.

88. "Specifications and Tests for Piezoelectric Pressure and Sound-Pressure Transducers," ISA Stand. S37.10, Pittsburgh, 1969

89. "Specifications and Tests for Strain-Gage Pressure Transducers," ISA Stand S37.3, Pittsburgh, 1970.

90. Norton, H. N. "Sensor and Analyzer Handbook," Prentice-Hall, Englewood Cliffs, N. J., 1982 (replaces Ref. 41).

Sources and Sensors of Infrared, Visible, and Ultraviolet Energy

JENNY ROSENTHAL BRAMLEY *Staff Scientist, Night Vision and Electro-Optics Laboratory, U.S. Army Electronics Research and Development Command; Fellow, IEEE, American Physical Society, Washington Academy of Sciences; Member, Optical Society of America, American Mathematical Society, Society for Information Display, Pattern Recognition Society, Sigma Xi*

C. S. FOX *Laser Division, Night Vision and Electro-Optics Laboratory, U.S. Army Electronics Research and Development Command; Member, IEEE, Laser Institute of America, Society of Photo-Optical Instrumentation Engineers, Sigma Xi, Illuminating Engineering Society*

J. E. MILLER *Laser Division, Night Vision and Electro-Optics Laboratory, U.S. Army Electronics Research and Development Command*

D. J. HOROWITZ *Advanced Concepts Division, Night Vision and Electro-Optics Laboratory, U.S. Army Electronics Research and Development Command*

R. R. SHURTZ II *Laser Division, Night Vision and Electro-Optics Laboratory, U.S. Army Electronics Research and Development Command; Member, Research Society of America, Sigma Xi, Sigma Pi Sigma*

E. J. SHARP *Night Vision Research Division, Night Vision and Electro-Optics Laboratory, U.S. Army Electronics Research and Development Command*

P. R. MANZO *Electro-Optics Technology Division, Science Applications, Inc.; Member, American Physical Society*

G. M. JANNEY *Space Sensors Division, Hughes Aircraft Company; Member, Optical Society of America, IEEE*

The Editor wishes to acknowledge with gratitude the care and depth of knowledge contributed by Dr. Jenny Bramley, who coordinated the revision and extension of Sec. 11. — D.G.F.

S. B. GIBSON *Systems Integration Division, Night Vision and Electro-Optics Laboratory, U.S. Army Electronics Research and Development Command; Member, Research Society of America, Illuminating Engineering Society, American Ordnance Society*

N. A. DIAKIDES *Visionics Division, Night Vision and Electro-Optics Laboratory, U.S. Army Electronics Research and Development Command; Member, American Chemical Society, Electrochemical Society, IEEE*

M. E. CROST *Beam, Plasma and Display Division, Electronics Technology and Devices Laboratory, U.S. Army Electronics Research and Development Command; Member, Society for Information Display, American Physical Society, American Association for the Advancement of Science, American Institute of Physics*

I. REINGOLD *Beam, Plasma, and Display Division, Electronics Technology and Devices Laboratory, U.S. Army Electronics Research and Development Command; Fellow, IEEE, Society for Information Display; Licensed Professional Engineer, State of New Jersey*

D. A. BOSSERMAN *Systems Integration Division, Night Vision and Electro-Optics Laboratory, U.S. Army Electronics Research and Development Command*

C. A. JOHNSON *Director's Staff, Night Vision and Electro-Optics Laboratory, U.S. Army Electronics Research and Development Command; Member, IEEE*

R. D. GRAFT *Night Vision Research Division, Night Vision and Electro-Optics Laboratory, U.S. Army Electronics Research and Development Command; Member, American Physical Society, IEEE*

R. E. FRANSEEN *Systems Engineering Division, Night Vision and Electro-Optics Laboratory, U.S. Army Electronics Research and Development Command; Member, IEEE, American Society of Mechanical Engineers; Associate Member, American Society of Heating, Refrigerating, and Air-Conditioning Engineers*

W. A. GUTIERREZ *Night Vision Research Division, Night Vision and Electro-Optics Laboratory, U.S. Army Electronics Research and Development Command*

J. H. POLLARD *Infrared Technology Division, Night Vision and Electro-Optics Laboratory, U.S. Army Electronics Research and Development Command; Member, Sigma Xi*

A. J. KENNEDY *Infrared Technology Division, Night Vision and Electro-Optics Laboratory, U.S. Army Electronics Research and Development Command; Member, IEEE, American Physical Society*

L. V. CALDWELL *Infrared Technology Division, Night Vision and Electro-Optics Laboratory, U.S. Army Electronics Research and Development Command*

W. T. GRANT *Infrared Technology Division, Night Vision and Electro-Optics Laboratory, U.S. Army Electronics Research and Development Command; Member, Society of Photo-Optical Instrumentation Engineers*

R. E. LONGSHORE *Infrared Technology Division, Night Vision and Electro-Optics Laboratory, U.S. Army Electronics Research and Development Command; Member, American Vacuum Society, American Society for Testing of Materials, American Chemical Society—Division of Colloidal Surface Chemistry, American Association of Physics Teachers, Sigma Pi Sigma, Phi Kappa Phi, Sigma Tau Epsilon*

K. W. MITCHELL *Photovoltaic Compound Semiconductor Program, Solar Energy Research Institute; Member, IEEE, American Physical Society, Sigma Xi, American Scientific Affiliation*

CONTENTS

Numbers refer to paragraphs

Lamps, Luminous Tubes, and Other Noncoherent Electric Radiation Sources

BY CLIFTON S. FOX

GLOSSARY AND BASIC CONCEPTS

1. Generation of Light. Light is produced by the transitions of electrons from states of higher energies to states of lower energies. The law of conservation of energy is satisfied in these transition processes[1,2]* by the emission of a photon, or quantum of light, whose energy corresponds to the difference in energy of the initial and final energy states of the electron.

2. Blackbody Radiation. A blackbody is defined as a body which, if it existed, would absorb all and reflect none of the radiation incident upon it. It is thus a perfect absorber and a

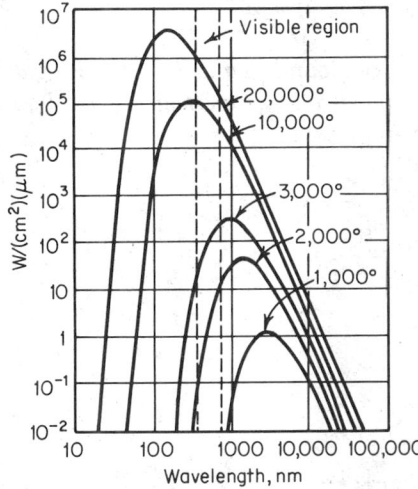

Fig. 11-1. Blackbody distribution curves for several values of temperature in kelvins.

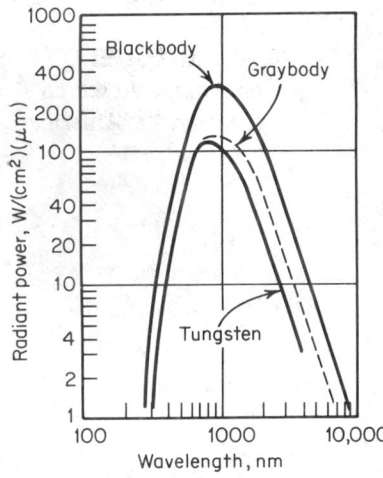

Fig. 11-2. Spectral distribution of tungsten at a color temperature compared with blackbody and graybody of the same temperature.

perfect emitter. The blackbody curves for several values of T are plotted on a logarithmic scale in Fig. 11-1.

3. The total emissivity ϵ of a thermal radiator at a given temperature is the ratio of the total radiation output of that radiator to that of a blackbody of the same temperature.

4. The spectral emissivity $\epsilon(\lambda)$ of a thermal radiator is defined as the ratio of the output of the source at the wavelength λ to that of a blackbody at the same wavelength and operating temperature.

5. Graybody Radiation. If the emissivity of a thermal radiator is a constant less than 1 for all wavelengths, the radiator is called a graybody.

6. Selective Radiation. A thermal radiator whose spectral emissivity is not constant but is a function of wavelength is called a selective radiator.

7. The color temperature of a thermal radiator is the temperature of a blackbody chosen such that its output is the closest possible approximation to a perfect color match with the thermal radiator. Figure 11-2 shows the spectral distribution of a tungsten filament operating at a color temperature of 3000 K compared with a blackbody of the same temperature and a graybody whose emissivity is the same as tungsten in the visible.

8. The candela (cd) is the unit of luminous intensity. Luminous intensity is the amount of luminous flux per unit solid angle in a given direction. This is measured as the luminous flux on a target normal to the direction of incidence divided by the solid angle (measured in steradians, abbreviated sr) subtended by the target as viewed from the source.

9. The lumen (lm) is the unit of luminous flux. It is equal to the flux in a unit solid angle from a uniform point source of 1 cd intensity.

*Superior numbers correspond to numbered references, Par. **11-27**.

10. The luminous efficacy of a light source is the measure of light-producing efficiency of the source. It is the ratio of the total luminous flux output to the total input power of the source. Luminous efficacy is measured in lumens per watt.

11. Radiative efficiency of a light source is the ratio (in percent) of total output power of the source measured in watts to the input power to the source.

12. Incandescence. Emission of radiation due to the temperature of the source.

13. Luminescence. Emission of radiation due to causes other than temperature of the source.

14. Fluorescence. Luminescence stimulated by radiation, not continuing more than about 10^{-8} s after the stimulating radiation is cut off.

15. The most commonly used light sources are the tungsten filament, electric discharge and electroluminescent lamps, and solid-state or light-emitting diodes. The first source is incandescent; the others are luminescent.

TUNGSTEN-FILAMENT LAMPS

16. Filament. The higher the operating temperature of a solid filament, the higher the percentage of its radiation which falls in the visible portion of the electromagnetic spectrum. Tungsten, with its high melting point (3653 K), low vapor pressure, and other favorable characteristics, is the most frequently used filament material. In higher-power incandescent lamps (generally above 40 W) an inert gas instead of vacuum surrounds the filament to reduce the evaporation rate of the tungsten.

17. Lamp Types. Tungsten filament lamps[4-6] are divided into the following categories: general-service lamps; high- and low-voltage lamps; series-burning lamps; projector and reflector lamps; showcase lamps; spotlight, floodlight, and projection lamps; halogen-cycle lamps; and infrared lamps.

Tungsten halogen-cycle lamps have a quartz envelope and use a halogen fill, usually iodine, to keep the bulb clean by chemical reaction with sublimated tungsten. This reaction provides a high-lumen maintenance throughout the life of the lamp by redepositing evaporated tungsten on the filament instead of on the bulb. *Infrared lamps* are tungsten-filament lamps which operate at low filament temperature.

18. Interrelationship of Lamp Parameters. The quantities voltage, current, resistance, temperature, watts, light output, efficacy, and life of a filament lamp are interrelated, and one cannot be changed without changing the others. Figure 11-3 shows how these quantities change typically as a function of the voltage for large gas-filled lamps.

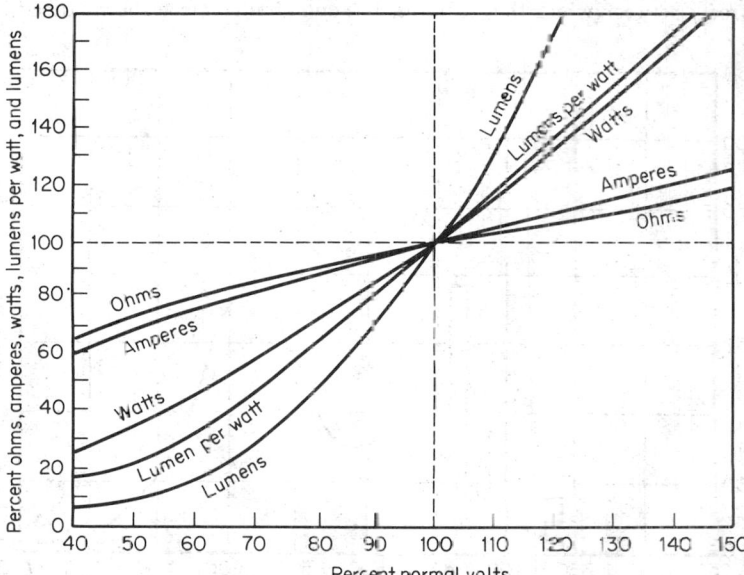

Fig. 11-3. Interrelation of lamp parameters for large tungsten-filament lamps. (*From GE Tech. Pamph. TP 110, GE Large Lamp Department, Nela Park, Cleveland, Ohio, 1969.*)

Some useful exponential relations frequently applied to incandescent filament lamps, where capital letters indicate normal rated values, are

$$\frac{\text{life}}{\text{LIFE}} = \left(\frac{\text{VOLTS}}{\text{volts}}\right)^d \qquad \frac{\text{lumens}}{\text{LUMENS}} = \left(\frac{\text{volts}}{\text{VOLTS}}\right)^k$$

$$\frac{\text{LM/W}}{\text{lm/w}} = \left(\frac{\text{VOLTS}}{\text{volts}}\right)^g \qquad \frac{\text{watts}}{\text{WATTS}} = \left(\frac{\text{volts}}{\text{VOLTS}}\right)^n$$

For approximate calculations the following average exponents may be used: $d = 13$, $g = 1.9$, $k = 3.4$, and $n = 1.6$.

ELECTRIC-DISCHARGE LAMPS

19. Fluorescent lamps[4,5,7] are electric-discharge lamps in which light is produced through the excitation of phosphors by the ultraviolet energy from a mercury arc. The lamp usually consists of a phosphor-coated tubular bulb with electrodes sealed into each end and containing mercury vapor at low pressure along with an inert starting gas such as argon or an argon-neon mixture. Various colors, grades of white light, and even black light[8] (near ultraviolet) are obtained by the choice of available phosphors. Cool white is the most widely used white-light fluorescent tube (Fig. 11-4). The deluxe fluorescent lamp has extra red output which enhances the natural appearance of red objects.

Lamp Types. Fluorescent lamps are classified as *hot-cathode* or *cold-cathode* type. There are three classes of hot-cathode fluorescent lamps: *preheat*, *instant-start*, and *rapid-start*. Preheat lamps allow preheating of the cathodes for a few seconds before striking the mercury arc. Instant-start lamps require no preheat because sufficient voltage is applied between the electrodes to strike the arc very quickly. Rapid-start lamps have continuously heated cathodes requiring a lower voltage than instant-start lamps. This feature also allows dimming and flashing of the lamp.

Life and Efficiency. The main advantages of fluorescent tubes are high efficiency and long life. A typical cool-white fluorescent tube renders 22% of its input energy in the form of visible light. Typical rated life for a 40-W fluorescent tube operated 3 h per start is 6,000 to 12,000 h.

20. Mercury Lamps.[4,5,9] Mercury vapor discharge lamps of the wall-stabilized variety are used primarily for general lighting. Mercury lamps with additives such as the metal halide lamp and sodium vapor lamp and mercury compact arcs are treated in separate succeeding paragraphs. Most mercury vapor lamps have two bulbs. The inner bulb, called the *arc tube*, contains the arc and the electrodes between which the arc burns. The outer bulb protects the arc tube from drafts

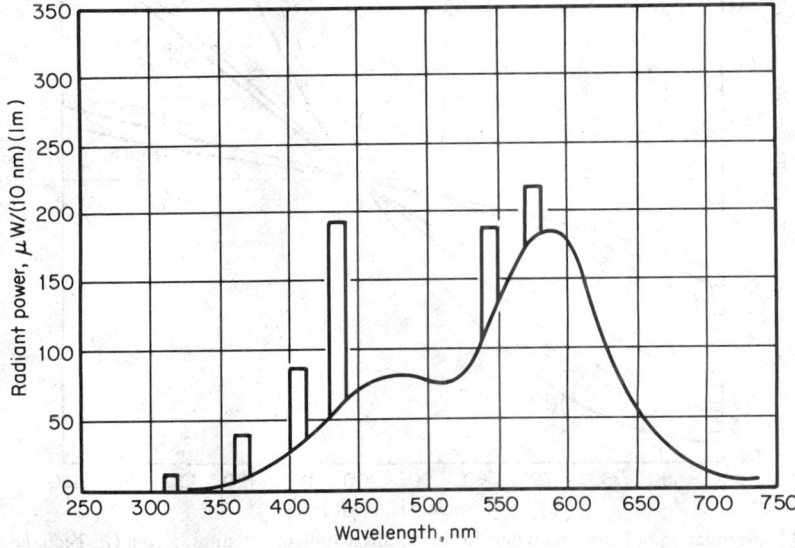

Fig. 11-4. Spectrum of a typical cool-white fluorescent lamp.[7]

and stabilizes the operating temperature. Mercury lamps for general lighting are available in input power sizes from 50 to 3,000 W.

In addition to mercury in the arc tube, an easily ionized inert gas such as argon is present to facilitate starting. The arc is generally ignited through use of a starting electrode and current-limiting starting resistor. An arc is first struck between the starting electrode and the adjacent main electrode. The heating and additional ionization resulting from this arc allow the large arc to form between the main electrodes.

Electrical and Radiation Characteristics.   Figure 11-5 shows the spectrum of a typical clear mercury lamp. The color-rendering properties of a clear mercury lamp are only fair, due to the

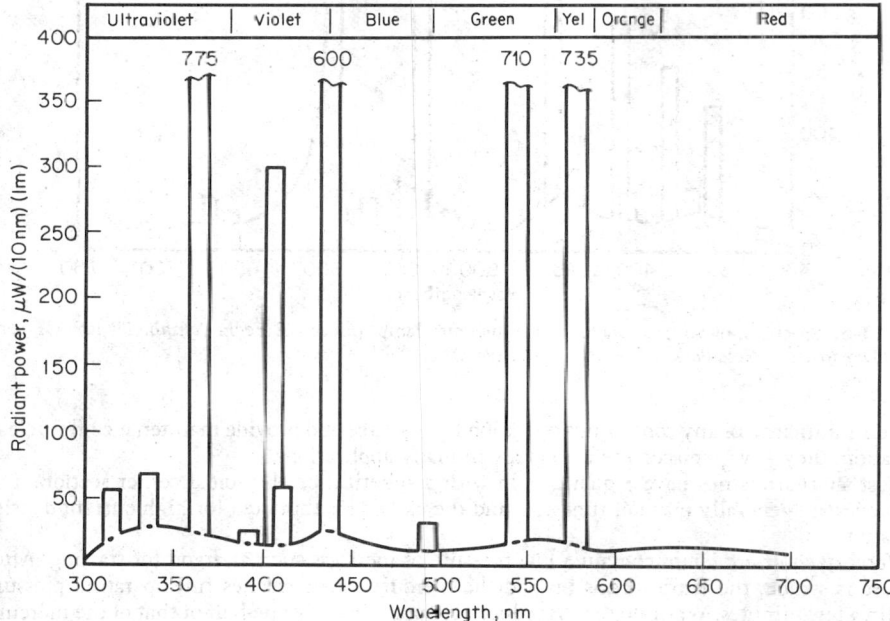

Fig. 11-5. Spectrum of a typical clear mercury lamp.[9]

line structure in the blue end of the spectrum. A clear mercury lamp of 1,000 W input power has a typical initial luminous efficacy of 56 lm/W.

Life.   General lighting lamps of 100 to 1,000 W input power have mean lifetimes in excess of 24,000 h, based on 5 h burning time per start and operation from the correct ballast.

21. Metal halide–mercury vapor lamps[4,5,9] are very nearly the same as regular mercury lamps except that additives such as the iodides of sodium, thallium, and indium are contained in the arc tube for the purpose of improving color rendition and efficiency (Fig. 11-6). The typical initial luminous efficacy of a metal halide–mercury lamp of 1,000 W input power is 90 lm/W. Typical life is 10,000 to 15,000 h.

22. High-pressure sodium vapor lamps[4,9] are presently the most efficient source of artificial light (Fig. 11-7). A typical 1,000-W high-pressure sodium lamp has an initial luminous efficacy of 140 lm/W. The theoretical efficiency of white light in the visible region, assuming 100% conversion of power into a continuum output, is 220 lm/W. Typical life of high-pressure sodium lamps is 12,000 to 20,000 h.

High-pressure sodium lamps require a high-transmission ceramic envelope such as alumina to contain the alkali metal at high temperature and an alkali-resistant high-temperature metal seal. The corrosive effects of sodium at high operating temperature prohibit the use of quartz and other glasses as an arc-tube material. Xenon is used as the readily ionized starting gas. When the xenon starting gas is ionized, the arc is struck, producing heat, and the vapor pressure starts to rise.

23. Arc-Light Sources. (a) *Short-arc or compact-arc lamps*[4,5] are of the enclosed-bulb type. Most are made for dc operation, which results in long life and good arc stability. They have high operating pressure, a comparatively short arc gap, and very high luminance (photometric brightness). The arc length may vary from about ⅓ to 17 mm. Since these lamps have the

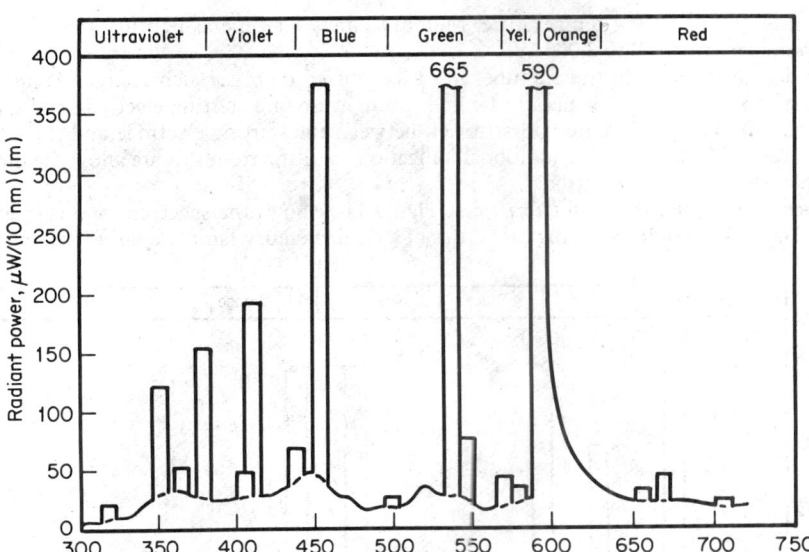

Fig. 11-6. Spectrum of a typical metal halide–mercury lamp. *(From GE Tech. Pamph. TP 109, GE Large Lamp Department, Nela Park, Cleveland, Ohio, 1969.)*

highest luminance of any continuous-operation light source and provide maintenance-free, clean operation, they have replaced the carbon arc in many applications.

Most short-arc lamps have a quartz bulb with a spherical or ellipsoidal center section. The electrodes are generally made of tungsten, and the cathode is thoriated for high-current-density operation.

Mercury short-arc lamps contain a low pressure of inert gas such as argon for starting. After the arc is struck, the lamp warms up and the mercury vapor reaches full operating pressure within a few minutes. Warmup time is reduced to approximately one-half of that of the mercury arc if xenon at 1 atm or more pressure is added. The spectra of mercury and mercury-xenon

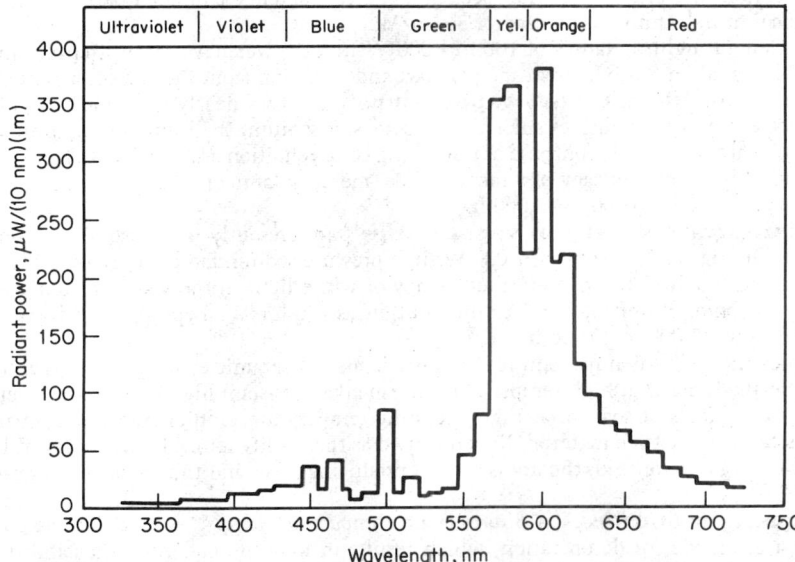

Fig. 11-7. Spectrum of a typical high-pressure sodium lamp. *(From GE Tech. Pamph. TP 109, GE Large Lamp Department, Nela Park, Cleveland, Ohio, 1969.)*

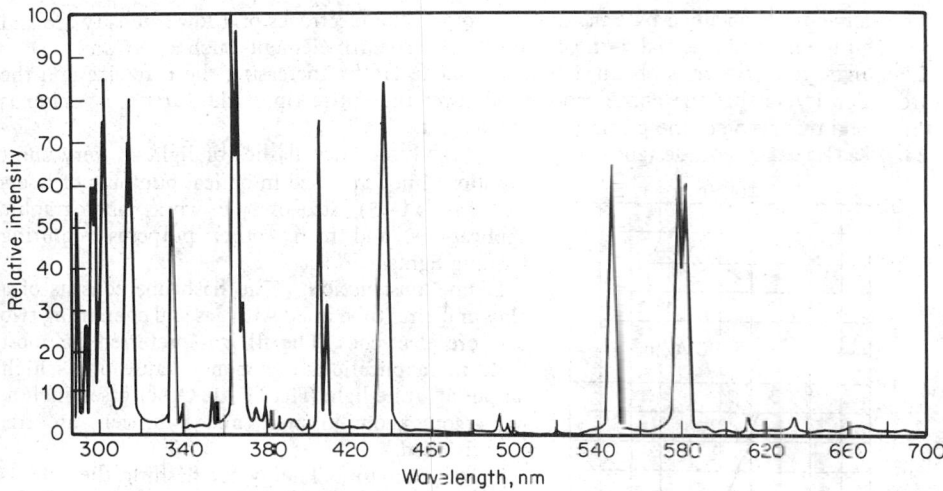

Fig. 11-8. Spectrum of a typical mercury-xenon short-arc lamp.

lamps (Fig. 11-8) are essentially the same. The luminous efficacy of these lamps is about 50 lm/W for a 1,000-W lamp. These lamps are available in a power range from 30 to 5,000 W.

Xenon short-arc lamps are filled to a cold pressure of several atmospheres with high-purity xenon gas. Operating pressure is roughly double the cold pressure. These lamps do not have as long a warmup time as the metal vapor types; 80% of the light output is obtained within 1 s of startup. Xenon has excellent color-rendering characteristics due to its continuous spectrum (Fig. 11-9) in the visible region. Luminous efficacy at 5,000 W is approximately 45 lm/W.

(b) *Carbon arcs* are of three basic types: the low-intensity arc, the flame arc, and the high-intensity arc.

The *low-intensity arc* has as its source of light the incandescent tip of the positive carbon which is maintained near the sublimation point of carbon (3700°C). The heat supplied to the positive carbon is from high-current-density electron bombardment originating from the negative carbon.

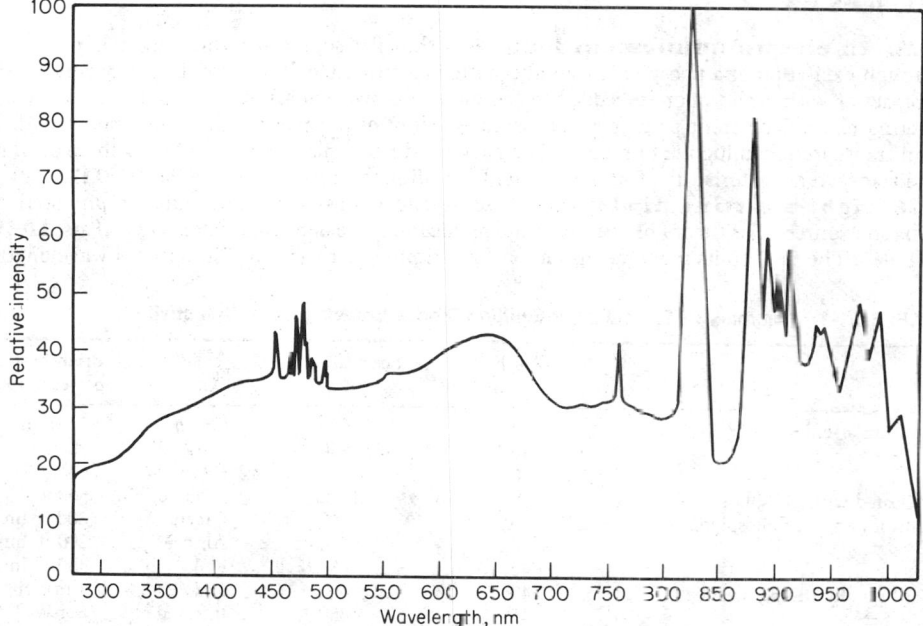

Fig. 11-9. Spectrum of a typical xenon short-arc lamp.

The *flame arc* is obtained by enlarging the core of the electrodes of a low-intensity arc and replacing the removed material with compounds of rare-earth elements such as cerium.

The *high-intensity arc* is obtained from the flame arc by increasing the core size and the current density so that the anode spot spreads over the entire tip of the carbon. A crater is formed, and this becomes the primary source of light.

24. Flashtubes[4] are designed to produce high-luminance flashes of light of very short duration. They are used in optical pumping of lasers (see Par. **11-39**), stroboscopic work, photographic applications, and many other purposes requiring flashing lights.

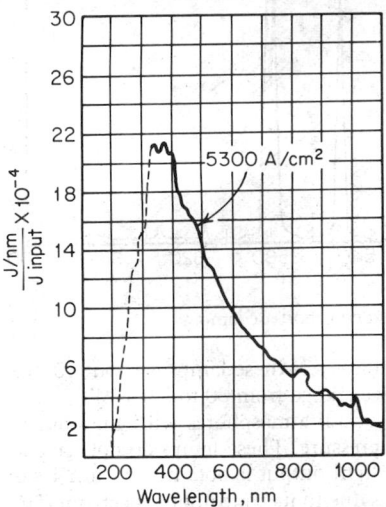

Lamp Construction. The flashtube consists of a glass or quartz tube filled with gas and containing two or more electrodes. The fill gas preferred for most flashtube applications is xenon because of its high output of white light (Fig. 11-10). Other gases, including argon, neon, krypton, and hydrogen, are frequently used.

Driving Circuit. Energy for flashing the tube is usually stored in a capacitor. This energy is determined by the equation $E = CV^2/2$, where $E =$ energy (J), $C =$ capacitance (μF), and $V =$ voltage on capacitor (kV). The duration of the flash usually depends on the resistance of the discharge and the capacitance of the storage capacitor, the duration being approximately $3RC$, with R in ohms and C in farads. For short pulses of 1 μs or less duration, frequently the inductance of the tube or circuit is the dominant factor over the resistance in determining pulse length.

Fig. 11-10. Typical spectrum of a xenon flashtube.

Many circuits have been used to flash lamps. One basic method is to hold off a voltage higher than the self-breakdown voltage of the lamp with an electronic switch, such as a thyratron, silicon-controlled rectifier, or spark gap, and trigger the switch when a flash is desired.

ELECTROLUMINESCENT LAMPS AND OTHER SOURCES

25. An electroluminescent lamp[4] is a thin flat source in which light is produced through excitation of a phosphor by an alternating electric field. The lamp basically consists of a capacitor with a phosphor embedded in the dielectric material sandwiched between two conducting plates. The front plate is a transparent sheet of either plastic, glass, or ceramic with a thin transparent conductive film on it. The back conductive plate may be a metal sheet or film or a transparent material like the front plate. For additional information see Par. **11-117**.

26. Light-emitting diodes (LEDs) (solid-state lamps) are semiconductor *pn*-junction radiation sources. For theory of operation and applications see Semiconductor Lasers, Pars. **11-59** to **11-63**. The LED emits spontaneous rather than stimulated radiation. The spectral wavelength

TABLE 11-1 Summary of Typical Light-Emitting Diodes Showing Color Selectivity

Type	Bandwidth at ½ power, nm	Spectral peak, nm	Color	Brightness or power output
Gallium arsenide	3	540	Green	80 ft·lm
	40	900	Infrared	10 mW
	60	980	Infrared	500 mW
Coated with phosphor	60	1000	Infrared	500 μW
Gallium arsenide phosphide	40	560	Green	300 ft·lm
	40	610	Amber	200 ft·lm
	40	680	Red	450 ft·lm
Gallium aluminum arsenide	40	800	Red	1 mW
	40	850	Infrared	5 mW
Silicon carbide	80	590	Yellow	150 ft·lm

bandwidth is on the order of 40 to 100 nm and the radiance is much lower than that of the laser diode. These noncoherent emitters selectively produce radiation throughout the visible and near-infrared spectral ranges, as shown in Table 11-1. The LED has gained wide acceptance in the electronics field for photoelectric systems, visual displays, and electro-optical components.[10]

27. References on Noncoherent Sources

1. McNally, J. R. Atomic Spectra Including Zeeman Effect and Stark Effect, in E. U. Condon and Hugh Odishaw (eds.), "Handbook of Physics," 2d ed., McGraw-Hill, New York, 1967.

2. Ditchburn, R. W. "Light," 2d ed., Interscience, New York, 1963.

3. Pivovonsky, M., and M. R. Nagel "Tables of Blackbody Radiation Functions," Macmillan, New York, 1961.

4. Kaufman, J. E. "IES Lighting Handbook," 5th ed., Illuminating Engineering Society, New York, 1972.

5. "Westinghouse Lighting Handbook," Westinghouse Electric Corporation, Bloomfield, N.J., 1969.

6. Incandescent Lamps, *General Electric Pamph.* TP-110R, Cleveland, Ohio, 1977.

7. Fluorescent Lamps, *General Electric Pamph.* TP-111R, Cleveland Ohio, 1978.

8. Black Light, *General Electric Pamph.* TP-215, Cleveland, Ohio, 1969.

9. High Intensity Discharge Lamps, *General Electric Pamph.* TP-109R, Cleveland, Ohio, 1975.

10. Directory of GaAs Lite Products, Monsanto Co., Electronics Special Products Pamphlet, Cupertino, Calif.

11. Opto-Electronics Products Directory, *Electro-Opt. Syst. Des.,* May 1971, Vol. 3, No. 5.

Lasers (Coherent Sources)

LASER FUNDAMENTALS

BY JAMES E. MILLER AND DANIEL J. HOROWITZ

28. Laser Light Compared with Nonlaser Light. The main characteristic of laser light[1-8]* is coherence, although laser light is usually more intense, more monochromatic, and more highly collimated than light from other sources.

Coherence is the property wherein corresponding points on the wavefront are in phase. A coherent beam can be visualized as an ideal wave whose spatial and time properties are clearly defined and predictable. Ordinary noncoherent light consists of random and discontinuous phases of varying amplitudes. The noncoherent beam has an average intensity and a predominant wavelength, but it is basically a superposition of different waves. The characteristic grainy appearance of laser light is due to interference effects which result from coherence.

Intensity of laser light can be very high. For example, power densities of over 1,000 MW/cm² can be obtained. A beam of such intensity can cut through and vaporize materials.

A *laser beam* is often highly *monochromatic* and highly *collimated* both varying with the type of laser.

29. Stimulated vs. Spontaneous Emission. The laser operates on the principle of stimulated emission, an effect which is rarely observed except in connection with lasers.

Spontaneous emission is the usual method whereby light is emitted from excited atoms or molecules. Assume that the laser material has energy levels which can be occupied by electrons, that the lowest level or ground state is occupied, and that the next upper level is unoccupied. An excitation process can then raise an electron from the ground state to this upper state. The electron, after a variable time interval, returns to the ground state and emits a photon whose direction and phase of the associated wave are random and whose energy corresponds to the energy difference between the states. The upper-level lifetime may be comparatively short (less than 10 ps) or it may be long (greater than 1 μs), in which case the level is referred to as *metastable* and the light emission is *fluorescence.*

Stimulated Emission. When the electron is in the upper level, if a light wave of precisely the wavelength corresponding to the energy difference strikes the electron in the excited state, the light stimulates the electron to transfer down to the lower level and emit a photon. This photon is emitted precisely in the same direction, and its associated wave is in the same phase as that of the incident photon. Thus a traveling wave of the proper frequency is produced, passing through the excited material and growing in amplitude as it stimulates emission.

*Superior numbers correspond to numbered references, Par. **11-34**.

30. Pumping and Population Inversion. The process of exciting the laser material (raising electrons to excited states) is referred to as *pumping*. Pumping can be done optically using a lamp of some kind, by an electric discharge, a chemical reaction, or in the case of the semiconductor laser, by injecting electrons into an upper energy level by means of an electric current.

A *population inversion* is necessary to initiate and sustain laser action. Normally, the ground state is almost entirely occupied and the upper level or levels, assuming they are more than a few tenths of an electronvolt above the ground state at room temperature, are essentially unoccupied. When the upper level has a greater electron population than the lower level, a population inversion is said to exist. This inverted population can support lasing, since a traveling wave of the proper frequency can stimulate downward transitions and the associated energy can be amplified.

31. Optical Resonators.[9] The addition of a positive-feedback mechanism to a lasing medium permits it to serve as an oscillator.

The Fabry-Perot resonator, which provides optical feedback, consists of two parallel mirrors, the rear mirror fully reflecting and the front mirror partly reflecting and partly transmitting at the laser wavelength. The light reflected back from the front and rear mirrors serves as positive feedback to sustain oscillation, and the light transmitted through the front mirror serves as the laser output. Laser action is started by spontaneously emitted light with the proper direction to travel down the axis of the laser rod and be reflected on itself from the end mirrors. The two mirrors form an optical cavity which can be tuned by varying the spacing of the mirrors. The laser can operate only at wavelengths for which a standing-wave pattern can be set up in the cavity, i.e., for which the length of the cavity is an integral number of half wavelengths. Mirrors may be separate from the laser rod or deposited on its end faces.

Spectral modes, i.e., a multiplicity of radiation patterns, are permitted within the cavity. Longitudinal modes, determined by the spacing of the mirrors, occur at each wavelength for which a standing-wave pattern can be set up in the cavity. Transverse modes vary not only in wavelength but also in field strength in the plane perpendicular to the cavity axis. The longitudinal, or axial, mode structure determines the spectral characteristics of the laser, i.e., coherence length and spectral bandwidth, while the transverse-mode structure determines the spatial characteristics, e.g., beam divergence and beam energy distribution.

If no attempt is made to control the mode of the laser, many longitudinal modes will be seen in the output. These modes lie within the natural spectral line width of the laser transition (Fig. 11-11). The transverse modes depend on the physical geometry of the cavity and are denoted similarly to waveguide modes at microwave frequencies. The lowest-order mode, characterized by a narrow, diffraction-limited beam spread, is the transverse electromagnetic (TEM_{00}) mode. Higher-order modes have multilobe intensity patterns and wider beams.

Mode control, or selection, is achieved by a variety of methods, e.g., varying the mirror curvatures, restricting the beam by apertures in the cavity, using resonant reflectors which reflect only a narrow band of wavelengths, and using Q-switching techniques.

32. Generation of Laser Pulses.[10] The typical output of an optical laser consists of a series of spikes occurring during the major portion of the time the laser is pumped (Fig. 11-12). These spikes result because the inverted population is being alternately built up and depleted. *Q switching* (Q spoiling) is a means of obtaining all the energy is a single spike of very high peak power. As an example, an ordinary laser might generate 100 mJ over a time interval of 100 μs for a peak power (averaged over this time interval) of 1,000 W. The same laser Q-switched might emit 80 mJ in a single 10-ns pulse for a peak power of 8 MW. The term Q switching is used by analogy to the Q of an electric circuit. Lowering the Q of the optical cavity means the laser cannot oscillate, and a large inverted population builds up. When the cavity Q is restored, a single "giant pulse" (see Fig. 11-12) is generated. This high-peak-power pulse is useful in optical ranging and communication and in producing nonlinear effects in materials. (See Par. **11-74.**)

Q switches use five techniques to inhibit laser action: the mechanical, or movable-optical-element, Q switch; the saturable organic-dye absorber; the electro-optic-effect crystal; the magneto-optic-effect crystal; and the acousto-optic-effect crystal (see Par. **11-74**).

All Q switches except the saturable absorber operate by deflecting radiation out of the optical cavity to prevent laser oscillation. The saturable absorber, called a *passive* Q switch because no external signal is required, consists of an organic compound usually in a liquid solution placed between the laser medium and one of the cavity mirrors. The dye is initially very highly absorbing at the laser wavelength, and the laser is isolated from one mirror. The absorbing transition in the dye is saturated as its ground state is depopulated, causing the dye to become transparent, or "bleach." The rate of bleaching is a strongly nonlinear function of incident radiation, and as the population inversion increases, the dye rapidly switches to a highly transparent state to allow lasing to occur.

Mode Locking (Phase Locking). This technique leads to even shorter pulses. A free-running laser output consists of a time average of many longitudinal and transverse modes with no fixed phase and amplitude relationship. If the laser is constrained to oscillate in only one transverse mode, there are still many longitudinal modes spaced in frequency at intervals of $c/2L$ (L = cavity length) which contribute randomly to the output. The time dependence of the output will be controlled by constraining these modes to maintain a fixed phase and amplitude relationship. A narrow pulse in the time domain requires a wide bandwidth in the frequency

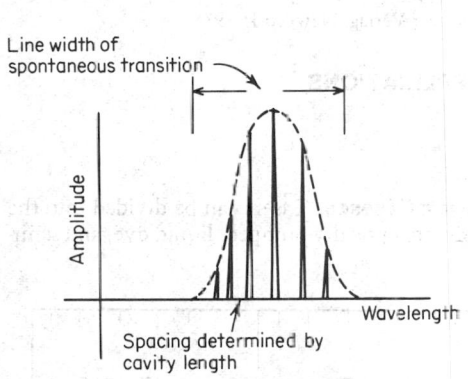

Fig. 11-11. Typical spectral modes of an optically pumped laser.

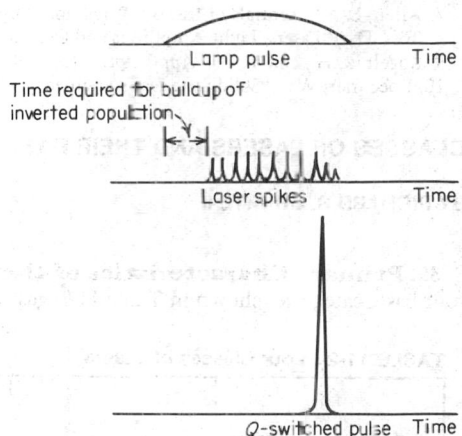

Fig. 11-12. Typical output of an optically pumped laser for Q-switched and non-Q-switched operation.

domain. The output of a mode-locked laser is a series of pulses, approximately $1/\Delta\nu$ wide, where $\Delta\nu$ is the laser gain bandwidth. These pulses are spaced at intervals of the cavity round-trip transit time.

In practice, Q-switch devices are also used to mode-lock lasers. For example, the saturable-dye Q switch can be used as follows. In the initial phase of amplification of spontaneous radiation after the laser material has been pumped, many modes are permitted within the bandwidth, and at this initial low-intensity level all are equally absorbed by the dye. As the amplification continues, natural-mode selection occurs as certain cavity modes are favored. These modes grow in intensity and become sufficient to begin to saturate, or bleach, the dye. These few pulses then are passed by the dye while the remaining weaker pulses are absorbed.

The allowed (passed) pulses are also narrowed in time (and thus broadened in frequency-mode content) because the tails of the peaks are more heavily absorbed due to the nonlinear absorption of the dye. Finally, the intensity of a pulse becomes sufficiently high to saturate the dye completely so that the amplification becomes nonlinear. At this point, all other pulses have been suppressed and the intensity of the remaining pulse rises quickly. The passage of this pulse through the optical resonator results in the output of a train of narrow pulses until the population inversion is depleted. An optical switch located outside the cavity is often used to select a 1-ps pulse out of this train of pulses.

33. Frequency Selection. Methods of frequency (or wavelength) selection vary with the type of laser. The following techniques are typical. For solid-state lasers, the output frequencies are limited to a few sharp closely spaced lines. The wavelength of these lines depends on the active ion in the laser, and the specific wavelength is chosen by a highly selective dielectric mirror or a dispersive prism used as part of the optical cavity. Liquid lasers have broad emission bands, several tens of nanometers wide. The output wavelength can be continuously tuned within this band by an intracavity prism or diffraction grating.

Semiconductor lasers have narrow emission bands determined by the constituents. For a given semiconductor, the output wavelength can be varied by changing the temperature. Gas lasers emit discrete lines at wavelengths depending on the gas. Specific lines are chosen by a diffraction grating or prism. Some gas lasers are pumped by a second gas laser. The wavelength of this type of laser is controlled by the wavelength of the pump laser as well as the cavity elements described previously.

34. References for Laser Fundamentals

1. Corcoran, V. J. "Introduction to Lasers," Optical Physics and Engineering Series, Plenum, New York, 1978.

2. O'Shea, D. C., et al. "Introduction to Lasers and Their Applications," Addison-Wesley, Reading, Mass., 1977.

3. Beesley, M. J. "Lasers and Their Applications," Halsted, New York, 1976.

4. Yariv, A. "Introduction to Optical Electronics," Holt, New York, 1976.

5. Lengyel, B. A. "Lasers," Wiley-Interscience, New York, 1971.

6. Arecchi, F. T., and E. O. Shultz-Dubois "Laser Handbook," Vols. I and II, North-Holland, Amsterdam, 1972.

7. Allen, L. "Essentials of Lasers," Pergamon, Elmsford, N.Y., 1969.

8. Röss, D. "Lasers, Light Amplifiers and Oscillators," Academic, New York, 1969.

9. Kogelnik, H., and T. Li Appl. Opt. **5**(10):1550–1567 (1966).

10. Koechner, W. "Solid-State Laser Engineering," Springer-Verlag, New York, 1976.

CLASSES OF LASERS AND THEIR BASIC APPLICATIONS

BY RICHARD R. SHURTZ II

35. Primary Characteristics of the Laser Classes. Lasers can be divided into the four basic categories shown in Table 11-2: gas, solid-state optically pumped, liquid dye, and semi-

TABLE 11-2   Four Classes of Lasers

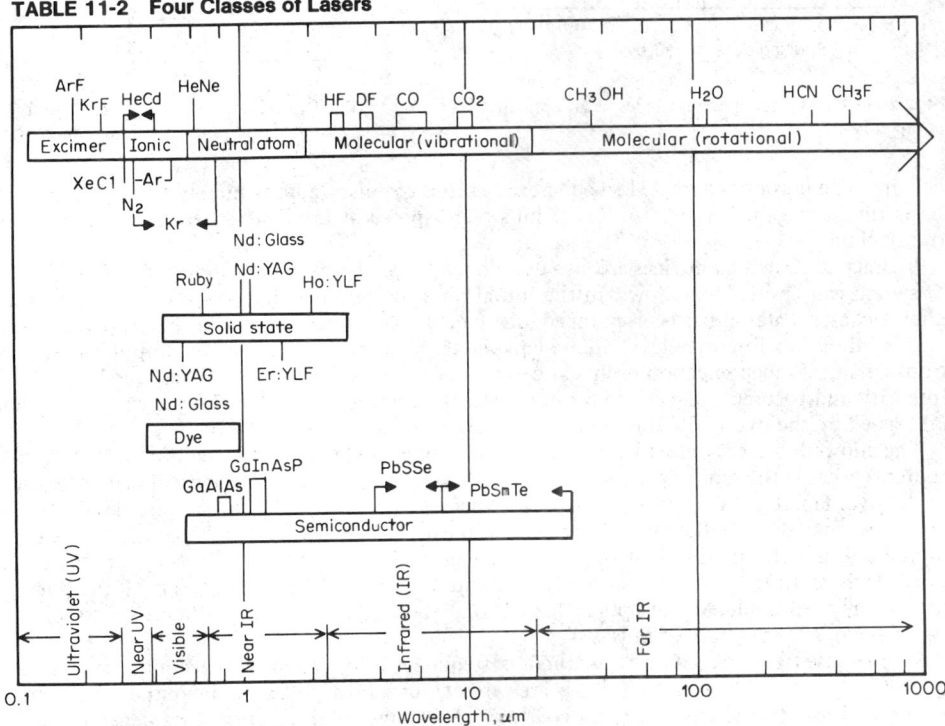

conductor. Combined, these lasers cover the spectral region from ultraviolet (0.1 μm) to far infrared (1000 μm), available wavelengths most densely populating the visible to infrared region.

The rapid development of new lasers over the past two decades has created a dynamic and continually changing ensemble of devices available to the applications engineer. This set has stabilized as the most useful lasing systems have become known commodities and commercially available. In this section we quantify this established set of commercial devices[1,2]* and outline

*Superior numbers correspond to references in Par. **11-38**.

the primary applications base of each laser type.[3-12] The reader should be able to identify the laser system best suited for a particular application and then refer to the appropriate paragraph for more specific information.

Gas lasers, shown in the top bar of Table 11-2, have the broadest spectral coverage.

The solid-state optically pumped laser category spans the visible to near infrared region. This is the only laser class that can be Q-switched or cavity-dumped. The electronic transitions are pumped either by flash lamps or by semiconductor diodes.

The family of *optically pumped organic-dye lasers* extends in wavelength from 0.4 to 1 μm. The major distinction of the category is its continuous tunability over the entire visible spectrum. Because of their broadband spectral output, these lasers can generate subpicosecond pulses when mode-locked. Their primary use is in spectroscopy and photochemistry.

Semiconductor lasers are pumped by the injection of excess electrons and holes into a thin semiconductor layer. Radiation is produced when the excess carriers recombine, producing photon energies equal to the band-gap energy. Lead salt laser diodes operate from 4 to 30 μm, continuously tunable either by varying the drive current or the temperature. They are used primarily for spectroscopy and must be cooled to nominally 77 K during operation. The shorter-wavelength semiconductor lasers are formed from III-V compounds: gallium arsenide (GaAs), gallium aluminum arsenide (GaAlAs), and gallium indium arsenide phosphide (GaInAsP). These devices, either pulsed or cw, operate at room temperature.

Single III-V compound laser diodes are used for fiber-optic communication, integrated optical processing, and rangefinding. Arrays of diodes are used for low-power infrared illuminators.

36. Characteristics of commercially available pulsed and cw lasers are summarized in Tables 11-3 and 11-4. The ranges shown are taken from tables published in the laser buyers' guides.[1,2] A few key lasers have been selected to illustrate the essential properties of each laser category, as discussed in the previous paragraphs.

37. Laser applications can be divided into several general categories: measurement of spatial parameters, material heating and/or removal, nondestructive probing of resonant phenomena, communications, optical processing, laser-induced chemical reactions, and weapons.[3-12] Selected laser applications are shown in Table 11-5. These areas are rapidly expanding as more engineers become familiar with the properties unique to the laser.

38. Selected References on Classes of Lasers and Their Basic Applications. Information on commercially available lasers can be found in the following buyers' guides:

1. "Laser Focus Buyers' Guide," Advanced Technology Publications, Inc., 1001 Watertown St., Newton MA 02165; published annually.

2. "The Optical Industry and Systems Directory," The Optical Publishing Co.. Inc., P.O. Box 1146, 59 Bartlett Ave., Pittsfield MA 01201; published annually.

Information on laser applications is found in the following references. Books before 1974 have not been included because this subject becomes rapidly dated.

3. Muncheryan, H. M. "Laser Technology," 2d ed., Sams, Indianapolis, 1979.

4. Ready, J. F. "Industrial Applications of Lasers," Academic, New York, 1978.

5. Hiefte, G. M. (ed.) "New Applications of Lasers to Chemistry," American Chemical Society Washington, 1978.

6. Röss, M. (ed.) "Laser Applications," Vols. 2–3, Academic, New York, 1974, 1977.

7. Mooradian, A., T. Jaeger, and P. Stokseth (eds.) "Tunable Lasers and Applications," Springer-Verlag, Berlin, 1976.

8. Kock, W. E. "Engineering Applications of Lasers and Holography," Plenum, New York 1975.

9. Harry, J. E. "Industrial Lasers and Their Application," McGraw-Hill, New York, 1974.

10. Jacobs, S., M. Sargent III, and M. O. Scully (eds.) "High Energy Lasers and Their Applications," Addison-Wesley, Reading, Mass., 1974.

11. Beesley, M. J. "Lasers and Their Applications," Taylor and Francis, London, 1976.

12. Ready, J. F. (ed.) "Lasers in Modern Industry," Society of Manufacturing Engineers, Dearborn, Mich., 1979.

SOLID OPTICALLY PUMPED LASERS

By EDWARD J. SHARP

39. Basic Principles. A solid optically pumped ionic laser consists of a solid material (Par. **11-42**) with optical gain, situated inside a resonator formed by two or more mirrors. Spontaneous emission in the material is amplified by stimulated emission. The resonator provides the feed-

TABLE 11-3 Properties of Commercially Available Pulsed Lasers

Laser	Type	Wavelength, μm	Output, J		Pulse per second	Pulse length, μs	Beam diam, mm	Beam divergence, mrad
			TEM_{00}	Multimode				
XeCl	Excimer	0.308	......	0.015-80	1-90	0.005-0.01	6 × 18-10 × 25	4-2 × 5
Nitrogen	Molecular gas	0.337-0.427	400-800 kw	0.002-0.01	1-100	0.005-0.01	6 × 20-10 × 25	1 × 7-5 × 10
Dye	Organic solution	0.39-1		10^2-10^6 W	1-10^3	10^{-5}-0.8		
Ruby	Solid state	0.694	0.02-5	0.3-50	0.1-1	10^{-5}-10^4	1.5-20	0.4-8
GaAlAs	Semiconductor	0.8-0.9		0.1-3,000 W	1-10^4	5×10^{-2}-1	......	100-300
Nd:Yag	Solid state	1.06	5×10^{-3}-5	0.1-400	1-100	10^{-5}-10^4	1.5-75	0.3-12
Nd:Glass	Solid state	1.06	10^{-3}-60	0.3-10^3	0.02-5	10^{-5}-100	2-120	0.1-10
Ho:YLF	Solid state	2.06		0.001-0.01	10-10^3	0.07-100	3	5
CO_2	Molecular gas	9-11	0.1-10^4	0.2-10^4	1-10^4	0.1-10^7	1-350	0.3-10
HCN	Molecular gas	337	0.001	......	1-100	30	10	40

TABLE 11-4 Properties of Commercially Available Continuous Wave (cw) Lasers

Laser	Type	Wavelength, μm	Power, W		Beam diam, mm	Beam divergence, mrad
			TEM_{00}	Multimode		
Argon (Ar)	Ionized gas	0.33-0.5145	4×10^{-3}-18	4×10^{-3}-22	0.8-2	0.5-1.5
Krypton (Kr)	Ionized gas	0.33-0.799	0.5-10	0.5-6	1-2	0.6-2
He-Cd	Ionized gas	0.325-0.442	10^{-3}-4×10^{-2}	1.5×10^{-3}-6×10^{-2}	0.82-1.5	0.5-1.6
Dye	Organic solution	0.39-1	.....	0.1-1	0.5-0.7	1.4-2
He-Ne	Neutral gas	0.6328	10^{-3}-2	10^{-2}-1.3	0.45-30	0.71-2.1
GaAlAs	Semiconductor	0.8-0.89	.....	10^{-3}-10^{-2}	.....	5-610
Nd:YAG	Solid state	1.06	0.2-18	0.4-100	0.8-7	1.5-15
Ho:YLF	Solid state	2.06	.....	5	3	25
CO_2	Molecular gas	9-11	1-10^4	4-10^4	1-70	1-10
HCN	Molecular gas	337	0.01-1	.....	10	40

TABLE 11-5 Selected Laser Applications

Application	Technique	Mode of operation*	Desired laser property	Possible laser
Measurement:				
Distance, short range	Interferometric	cw	Coherence	He–Ne
Long range	Time of flight	P	Short pulse length	Q-switch Nd:YAG, GaAs
Shape, thickness	Refraction measurement	cw	Collimation, transparency	He–Ne
Diameter	Interference	cw	Coherence	He–Ne
Overall dimension	Obscuration	cw	Collimation	He–Ne
Imperfections	Interference, obscuration	cw	Coherence, collimation for transparency	He–Ne or other cw
Alignment	Beam used as reference	cw	Collimation, visibility	He–Ne
Atmospheric monitoring:				
Wind, turbulence	Doppler-shifted particulate backscatter	P†	Short pulse, narrow bandwidth	CO_2, Nd:YAG
Particulate	Backscatter	P	Short pulse	CO_2, Nd:YAG
Gaseous pollution	Raman-shifted backscatter	P	Short pulse, narrow bandwidth	CO_2, Nd:YAG
	Absorption	cw	Tunable, narrow bandwidth	PbSnTe diode, dye
	Fluorescence	cw	Tunable, narrow bandwidth	PbSnTe diode, dye
Material removal:				
Drilling	Energy concentration	P	1–10 MW/cm²	Ruby, Nd:YAG, Nd:glass
Cutting	Energy concentration	P, cw	1–10 MW/cm²	CO_2, Nd:YAG
Film vaporization	Energy concentration	P, cw	100 kW/cm²	CO_2, Nd:YAG, argon, xenon
Scribing wafers	Energy concentration	P, cw	100 kW/cm²	CO_2 Nd:YAG
Illumination (infrared)	Pulse-gated for particulate penetration	P	Narrow pulse, efficient	GaAlAs diode arrays
Heating:				
Welding	Energy concentration	P, cw	1–10 MW/cm²	Nd:YAG, CO_2
Heat treating, annealing	Energy concentration	P, cw	Moderate to high power	Nd:YAG, CO_2, ruby
Fusion	Energy concentration	P	100 TW per 50-μm particle	Nd:glass, CO_2
Communications:				
Fiber optic	Digital, analog	P, cw	Bandwidth; linear output	cw GaAlAs; cw GaInAsP
Space	Analog	cw	Stability, collimation	CO_2
Optical information:				
Processing	Integrated optics; three-dimensional optics	cw	Coherence	cw GaAlAs; coherent cw gas
Storage	Holography	cw	Coherence	He–Ne, argon
Chemical:				
Laser-induced reactions	Selective breaking of bonds	P, cw	Tuning to molecular resonance	Variety of moderate-power lasers
Spectroscopy	Absorption; Raman emission	cw	Narrow bandwidth, tunable	Organic dye, PbSn Te diode
Isotope separation, e.g., uranium	Chemical isolation of excited species	cw	Laser tuned to excite one species	Organic dye, PbSnTe diode, others

TABLE 11-5 Selected Laser Applications (*Continued*)

Application	Technique	Mode of operation*	Desired laser property	Possible laser
Medical:				
Skin treatment	Laser-induced reactions	P, cw	Tuned to resonant absorption	Ruby, tunable dyes, ultraviolet lasers
Surgery	Cutting, cauterization	P, cw	Collimation, power	Nd:YAG, CO_2, argon
Eye repair	Retina repair	P, cw	Collimation, power	Ruby, argon
Destructive weapons	Energy concentration	P, cw	High average power	CO_2, HF

*cw = continuous-wave, P = pulsed.   †For range.

back for the optical amplifier, resulting in a laser oscillator. The optical arrangement for a typical optically pumped laser is shown in Fig. 11-13.

These devices generally consist of a pump reflector enclosing the laser rod and optical pump in one of a number of efficient ways which is usually determined by the application or laser property to be exploited.[1-3]* The laser rod is pumped by flash lamp (Par. **11-24**) for pulsed operation and usually by arc lamps (Par. **11-23**) for cw (continuous-wave) operation, or in some cases an appropriate laser can be selected for the optical pump.

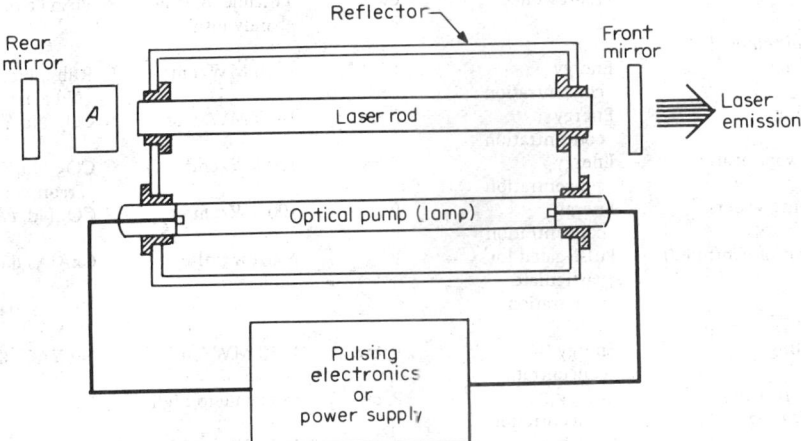

Fig. 11-13. Typical optically pumped laser configurations. Component *A* can represent a Q switch, polarizer, or other optical element.

Two important conditions must be met: the wavelength region of the pump emission must overlap a significant portion of the absorption spectrum of the active ion or ions which are incorporated in the laser rod, and the optical pump must not damage the laser rod. Figure 11-14 illustrates the overlap of pump emissions and the absorption bands of a typical Nd^{3+} four-level laser scheme where pumping is accomplished by xenon-lamp pumping and LED laser pumping.

40. Laser Components. *Optical Pumps.*   The general technique of optical pumping[4] has many uses. To achieve population inversion linear, low-pressure flash lamps are the most common pumping device. Typically they have tungsten electrodes and are filled to a few atmospheres pressure with xenon or krypton. For pulsed operation xenon is commonly used but the lines of krypton radiate at wavelengths more favorable than xenon for pumping a neodymium laser at low input levels. Thus it is a more efficient cw pump.[5] At higher input levels, the blackbody spectrum becomes stronger than the emission lines, and this advantage is lost. Lasers such as argon[6] and LEDs[7] have also been successfully used as optical pumps.

Pulsing Electronics.   For pulsed operation the energy to be discharged into a flash lamp is stored in a capacitor and discharged through the lamp via a choke whose value is selected to give the desired pulse width with minimum ringing.[8] The spectrum of the lamp discharge depends

*Superior numbers correspond to references in Par. **11-48**.

upon the electrical parameters, pressure, and fill gases, so that it can be tailored to match the absorption bands of the active ion to a certain extent.[9]

Coolants.   Lasers that operate cw or at a high repetition rate need to be cooled, since most of the energy expended in the flash lamp is converted into heat and would quickly cause overheating and or thermal cracking of the laser rod. Air and liquids have been successfully used as

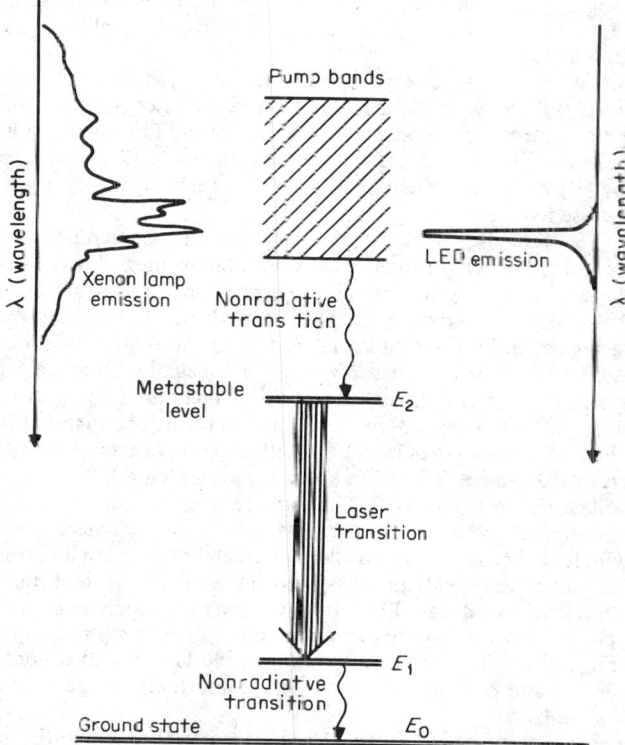

Fig. 11-14. A typical four-level Nd^{3+} laser, in which the Nd^{3+} ions are optically excited by pumping the absorption bands of the ions with (a) a xenon lamp and (b) an LED laser.

coolants, liquids being used when cooling requirements are more severe. Distilled or deionized water is often used, and mixtures of water and ethylene glycol are used for operation below 0°C. The coolant circuit includes a pump for circulation and a radiator, often with a fan, to dissipate the heat.

Pumping Reflectors.   To obtain maximum use of the pump light, it must be efficiently collected and coupled into the rod. The reflectors used to do this are usually elliptical in cross section, with a rod at one focus and the lamp at the other. Sometimes two lamps are used in a double elliptical cavity. Often a round cross section is used, resulting in an afocal system. In general, the smallest cross section is best, and the highest efficiency is obtained with a close-coupled arrangement, where the rod and lamp are almost in contact and the reflector encloses them closely. However, in this close-coupled configuration the thermal distortion of the rod due to the asymmetrical pumping could be a serious problem.

Mirrors.   Multilayer dielectric mirrors can be deposited on the laser-rod ends, which are finished flat and parallel. These mirrors are designed for specific reflectivities at the operating wavelengths and eliminate the problem of alignment. The rear-mirror reflectivity is normally 100%, but the front-mirror reflectivity is selected for the best performance, determined by laser-material properties, rod size, and operating power level. Where separate mirrors are used (see Fig 11-13) the rod ends should be antireflection-coated or cut at Brewster's angle. Separate or external mirrors must be mounted in holders that permit fine adjustment about two axes to achieve the required parallelism.

41. Q-Switched Lasers.   Q switching, a technique used for generating large output bursts of radiation from laser devices,[10] is accomplished by effectively blocking the optical path to one of the mirrors for most of the time during which the rod is being pumped, causing the rod to store energy. The Q switch then quickly restores the optical path to the mirror and a giant pulse (see Fig. 11-12) is generated. The four main type of Q switches are the rotating mirror, acousto-optical, electro-optical, and saturable absorber.

Rotating Sector, Prism, or Mirror. The rotating prism[11] or mirror[12,13] requires careful mechanical design to provide rotation at the required speeds (30,000 r/min is common) and still maintain alignment. They are generally less reliable than other methods of Q switching, but the extinction ratio is infinite and the insertion loss is negligible, enabling this type of Q switch to attain high efficiencies.

Electro-optic Q Switches. An electro-optic crystal, or liquid Kerr cell, can be placed in the optical resonator, usually between the rear mirror and the laser rod (position A in Fig. 11-13) along with an appropriate polarizer to effect the cavity losses. The polarization of the light passing through the electro-optic medium is changed by the application of an applied electric field and is rejected by the polarizer. When the voltage is removed, the cavity losses are reduced, permitting giant-pulse radiation.[3,14–17]

Acousto-optic Q Switch. An acousto-optic Bragg scattering switch deflects the beam, which is then used to provide feedback during the buildup time. In these devices a standing-wave pattern of ultrasonic waves is set up in a suitable cell positioned in the resonator. The refraction resulting from the passage of a plane-parallel light beam through this ultrasonic field is sufficient to provide shuttering action for infrared lasers, leading to giant-pulse radiation.[18,19]

Saturable Absorber Q Switch. These switches are usually bleachable dyes[15,20–22] which undergo decreased absorption at high light intensities. They are known as passive Q switches since they do not require any electrical or mechanical control. The material is opaque until the fluorescence of the laser rod bleaches it and the optical path is restored for a Q-switched pulse.

42. Laser Ions and Hosts. The current list of ions which can be incorporated into a solid host and made to lase numbers about 20. The total number of different wavelengths available from these ions is approximately 150. A complete listing is available.[23] The solid host is any crystal or glass which can accommodate trivalent or divalent rare-earth and iron-group ions. The iron-group ions in which laser oscillation has been achieved are divalent nickel and cobalt and trivalent chromium and vanadium. The rare-earth ions in which laser oscillation has been achieved are divalent samarium and dysprosium and trivalent europium, praseodymium, ytterbium, neodymium, erbium, thulium, and uranium. Table 11-6 lists these ions and the emission wavelength of the ion and host, as well as the mode and temperature of operation. A list by transition levels is available.[24]

43. Characteristics of the Hosts. The host material to which the active ion (dopant) is incorporated can be either crystalline or glass and should exhibit the following properties:

High thermal conductivity
Ease of fabrication
Hardness (to prevent degradation of optical finishes)
Resistance to solarization or radiation-induced color centers
Chemical inertness, i.e., not water-soluble
High optical quality, which implies uniformity of refractive index and absence of voids, inclusions, or other scattering centers

44. Crystalline hosts offer as advantages in most cases their hardness, high thermal conductivity, narrow fluorescence line width, and, for some applications, optical anisotropy. Crystalline hosts usually have as disadvantages their poor optical quality, inhomogeneity of doping, and generally narrower absorption lines.

45. Glass laser hosts are optically isotropic and easy to fabricate, possess excellent optical quality, and are hard enough to accept and retain optical finishes. In most cases glasses can be more heavily and more homogeneously doped than crystals, and, in general, glasses have broader absorption bands and exhibit longer fluorescence decay times. The primary disadvantages of glass are its broad fluorescence line widths (leading to higher thresholds), its significantly lower thermal conductivity (a factor of 10 leading to thermally induced birefringence and distortion when operated at high pulse-repetition rates or high average powers), and its susceptibility to solarization (darkening due to color centers which are formed in the glass as a result of the ultraviolet radiation from the flash lamps). These disadvantages limit the use of glass laser rods for cw and high-repetition-rate lasers. Table 11-7 gives a brief listing of glass lasers.

46. Sensitized Lasers. Laser performance and efficiency can be enhanced through the technique of energy transfer. A second ion (sensitizer) is incorporated into the host in addition to the laser ion (activator) to accomplish this effect. The sensitizer may be a color center. Pump energy is absorbed by the sensitizer and is transferred to the activator, which then emits this energy at the laser wavelength. A list of sensitized lasers is given in Table 11-8.

47. Miniature Lasers. Several host materials that incorporate Nd as a constituent rather than a dopant, such as $NdAl_3(BO_3)_4$, neodymium aluminum borate (NAB),[53,54] $LiNdP_4O_{12}$, lithium neodymium tetraphosphate (LNP),[95] and NdP_5O_{14}, neodymium pentaphosphate (NdPP),[35] have been studied recently, and since concentration quenching is not appreciable in these materials, they are excellent candidates for miniaturized lasers.[96] These materials typically have 30 times the Nd content of Nd^{3+}:YAG, yet the lifetime of the $^4F_{3/2}$ state is only quenched by a factor of 2.

TABLE 11-6   Laser Ions and Hosts

λ, nm	Ion	Host	Mode	Temp, K	Transition	Ref.
325.5	Ce^{3+}	$LiYF_4$	P	300	$5d \rightarrow {}^2F_{7/2}$	25
551.2	Ho^{3+}	CaF_2	P	77	${}^5S_2 \rightarrow {}^5I_8$	26
598.5	Pr^{3+}	LaF_3	P	77	${}^3P_0 \rightarrow {}^3H_6$	27
611.3	Eu^{3+}	Y_2O_3	P	220	${}^5D_0 \rightarrow {}^7F_2$	28
694.3	Cr^{3+}	Al_2O_3	P, cw	300	${}^2E(A) \rightarrow {}^4A_2$	29
750.0	Cr^{3+}	$BeAl_2O_4$	P, cw	300	Vibronic	30
854.8	Er^{3+}	CaF_2	P	77	${}^4S_{3/2} \rightarrow {}^4I_{13/2}$	31
941.0	Nd^{3+}	$CaY_2Mg_2Ge_3O_{12}$	cw	300	${}^4F_{3/2} \rightarrow {}^4I_{9/2}$	6
946.0	Nd^{3+}	$Y_3Al_5O_{12}$	P	230	${}^4F_{3/2} \rightarrow {}^4I_{9/2}$	32
1029.6	Yb^{3+}	$Y_3Al_5O_{12}$	P	77	${}^2F_{5/2} \rightarrow {}^2F_{7/2}$	33
1040.0	Pr^{3+}	$Ca(NbO_3)_2$	P	77	${}^1G_4 \rightarrow {}^3H_4$	34
1047.7	Nd^{3+}	$LiNdP_4O_{12}$	cw	300	${}^4F_{3/2} \rightarrow {}^4I_{11/2}$	7
1051.5	Nd^{3+}	NdP_4O_{14}	P	300	${}^4F_{3/2} \rightarrow {}^4I_{11/2}$	35
1053.0	Nd^{3+}	$LiYF_4$	P	300	${}^4F_{3/2} \rightarrow {}^4I_{11/2}$	36
1054.0	Nd^{3+}	$Ba_2MgGe_2O_7$	P	300	${}^4F_{3/2} \rightarrow {}^4I_{11/2}$	37
1058.0	Nd^{3+}	$CaWO_4$	cw	300	${}^4F_{3/2} \rightarrow {}^4I_{11/2}$	38
1058.9	Nd^{3+}	$CaY_2Mg_2Ge_3O_{12}$	P, cw	300	${}^4F_{3/2} \rightarrow {}^4I_{11/2}$	6, 39
1059.7	Nd^{3+}	$NaLa(MoO_4)_2$	P	300	${}^4F_{3/2} \rightarrow {}^4I_{11/2}$	40
1061.0	Nd^{3+}	$CaMoO_4$	cw	300	${}^4F_{3/2} \rightarrow {}^4I_{11/2}$	41
1062.9	Nd^{3+}	$Ca_5(PO_4)_3F$	P	300	${}^4F_{3/2} \rightarrow {}^4I_{11/2}$	42
1063.3	Nd^{3+}	$Y_3Ga_5O_{12}$	P	300	${}^4F_{3/2} \rightarrow {}^4I_{11/2}$	43
1064.1	Nd^{3+}	YVO_4	P	300	${}^4F_{3/2} \rightarrow {}^4I_{11/2}$	44
1064.5	Nd^{3+}	$YAlO_3$	P	300	${}^4F_{3/2} \rightarrow {}^4I_{11/2}$	45
1064.8	Nd^{3+}	$Y_3Al_5O_{12}$	P, cw	300	${}^4F_{3/2} \rightarrow {}^4I_{11/2}$	43
1070.0	Nd^{3+}	$La_2Be_2O_5$	cw	300	${}^4F_{3/2} \rightarrow {}^4I_{11/2}$	46
1076.0	Nd^{3+}	La_2O_2S	P	300	${}^4F_{3/2} \rightarrow {}^4I_{11/2}$	47
1079.5	Nd^{3+}	$YAlO_3$	P	300	${}^4F_{3/2} \rightarrow {}^4I_{11/2}$	43, 48
1111.0	Nd^{3+}	$Y_3Al_5O_{12}$	P	300	${}^4F_{3/2} \rightarrow {}^4I_{11/2}$	49
1116.0	Tm^{2+}	CaF_2	cw	27	${}^2F_{5/2} \rightarrow {}^2F_{7/2}$	50
1121.3	V^{3+}	MgF_2	P	77	${}^3T_2 \rightarrow {}^3T_1$	51
1314.4	Ni^{2+}	MgO	P	77	${}^3T_2 \rightarrow {}^3A_2$	52
1358.0	Nd^{3+}	$Y_3Al_5O_{12}$	P	300	${}^4F_{3/2} \rightarrow {}^4I_{13/2}$	53
1623.0	Ni^{2+}	MgF_2	F	77	${}^3T_2 \rightarrow {}^3A_2$	52
1660.2	Er^{3+}	$Y_3Al_5O_{12}$	P	77	${}^4I_{13/2} \rightarrow {}^4I_{15/2}$	33
1663.0	Er^{3+}	$YAlO_3$	P	300	${}^4S_{3/2} \rightarrow {}^4I_{9/2}$	54
1750.0	Co^{2+}	MgF_2	P	77	${}^4T_2 \rightarrow {}^4T_1$	51
1821.0	Co^{2+}	$KMgF_3$	P	77	${}^4T_2 \rightarrow {}^4T_1$	51
1833.0	Nd^{3+}	$Y_3Al_5O_{12}$	P	300	${}^4F_{3/2} \rightarrow {}^4I_{15/2}$	55
1911.0	Tm^{3+}	$CaWO_4$	P	77	${}^3H_4 \rightarrow {}^3H_6$	56
2013.2	Tm^{3+}	$Y_3Al_5O_{12}$	P	300	${}^3H_4 \rightarrow {}^3H_6$	33
2059.0	Ho^{3+}	$CaWO_4$	P	77	${}^5I_7 \rightarrow {}^5I_8$	57
2097.5	Ho^{3+}	$Y_3Al_5O_{12}$	cw	77	${}^5I_7 \rightarrow {}^5I_8$	33
			P	300		
2165.0	Co^{2+}	ZnF_2	P	77	${}^4T_2 \rightarrow {}^4T_1$	58
2171.0	Ho^{3+}	BaY_2F_8	P	300	Vibronic	59
2358.8	Dy^{2+}	CaF_2	cw	77	${}^5I_7 \rightarrow {}^5I_8$	60
			P	145		
2407.0	U^{3+}	SrF_2	P	90	${}^4I_{11/2} \rightarrow {}^4I_{9/2}$	61
2556.0	U^{3+}	BaF_2	P	20	${}^4I_{11/2} \rightarrow {}^4I_{9/2}$	62
2613.0	U^{3+}	CaF_2	P	300	${}^4I_{11/2} \rightarrow {}^4I_{9/2}$	63
2691.0	Er^{3+}	CaF_2	P	300	${}^4I_{11/2} \rightarrow {}^4I_{13/2}$	64

TABLE 11-7   Glass Lasers

Dopant	Glass	λ, nm	Mode	Temp, K	Transition	Ref.
Gd^{3+}	Li–Mg–Al–Si	312.5	P	77	$^6P_{7/2} \rightarrow {}^8S_{7/2}$	86
Ho^{3+}	Li–Mg–Al–Si	1950.0	P	77	$^5I_7 \rightarrow {}^8I_8$	87
Nd^{3+}	Nd-ultraphosphate	1051.0	P	300	$^4F_{3/2} \rightarrow {}^4I_{11/2}$	35
Nd^{3+}	K–Ba–Si	1060.0	P	300	$^4F_{3/2} \rightarrow {}^4I_{11/2}$	88
Nd^{3+}	Barium crown	1060.0	cw	300	$^4F_{3/2} \rightarrow {}^4I_{11/2}$	89
Nd^{3+}	Ba–Cs–Si	920.0	P	300	$^4F_{3/2} \rightarrow {}^4I_{9/2}$	90
Nd^{3+}	La–Ba–Th–B	1370.0	P	300	$^4F_{3/2} \rightarrow {}^4I_{13/2}$	91
Yb^{3+}	Li–Mg–Al–Si	1015.0	P	77	$^5F_{5/2} \rightarrow {}^5F_{7/2}$	92

TABLE 11-8   Sensitized Lasers

Active ion	Sensitizing ion	Host	λ, nm	Transition	Mode	Temp, K	Ref.
$Cr^{3+}(Pr)$	Cr^{3+}	Al_2O_3	704.0		P	77	65–67
Er^{3+}	Yb^{3+}	Silicate glass	1542.6		P	300	68
Er^{3+}	Color center	CaF_2	1530.8	$^4I_{13/2} \rightarrow {}^4I_{15/2}$	P	4	69
Ho^{3+}	Er^{3+}	Er_2O_3	2121.0	$^5I_7 \rightarrow {}^5I_8$	P, cw	77	70
Ho^{3+}	Er^{3+}	$CaMoO_4$	2070.0	$^5I_7 \rightarrow {}^5I_8$	P	77	71
Ho^{3+}	Cr^{3+}	$Y_3Al_5O_{12}$	2097.5	$^5I_7 \rightarrow {}^5I_8$	P	77	72
Ho^{3+}	Er^{3+}, Tm^{3+}	$Y_3Fe_5O_{12}$	2086.0	$^5I_7 \rightarrow {}^5I_8$	P	77	73
Ho^{3+}	$Er^{3+}, Tm^{3+}, Yb^{3+}$	$Y_3Al_5O_{12}$	2128.8	$^5I_7 \rightarrow {}^5I_8$	P	295	74
Ho^{3+}	$Er^{3+}, Tm^{3+}, Yb^{3+}$	$Y_3Al_5O_{12}$	2122.7	$^5I_7 \rightarrow {}^5I_8$	cw	85	74
Ho^{3+}	$Er^{3+}, Tm^{3+}, Yb^{3+}$	CaF_2	2060.0	$^5I_7 \rightarrow {}^5I_8$	P	298	75
Ho^{3+}	Er^{3+}, Tm^{3+}	$LiYF_4$	2065.4	$^5I_7 \rightarrow {}^5I_8$	P	300	76
Ho^{3+}	Yb^{3+}	Silicate glass	2080.0	$^5I_7 \rightarrow {}^5I_8$	P	80	77
Ho^{3+}	Er^{3+}	$LiYF_4$	2066.0	$^5I_7 \rightarrow {}^5I_8$	P	77	78
Nd^{3+}	Cr^{3+}	$Y_3Al_5O_{12}$	1061.2	$^4F_{3/2} \rightarrow {}^4I_{11/2}$	P, cw	300	79
Nd^{3+}	Cr^{3+}	$YAlO_3$	1064.0	$^4F_{3/2} \rightarrow {}^4I_{11/2}$	P	300	80
Nd^{3+}	Ce^{3+}	CeF_3	1060.0	$^4F_{3/2} \rightarrow {}^4I_{11/2}$	P	90	81
Nd^{3+}	Mn^{3+}	Phosphate glass	1060.0	$^4F_{3/2} \rightarrow {}^4I_{11/2}$	P		82
Nd^{3+}	UO_2	Barium crown glass	1060.0	$^4F_{3/2} \rightarrow {}^4I_{11/2}$			83
Tm^{3+}	Er^{3+}	Er_2O_3	1934.0	$^3H_4 \rightarrow {}^3H_6$	P	77	84
Tm^{3+}	Er^{3+}	$CaMoO_4$	1911.5	$^3H_4 \rightarrow {}^3H_6$	P	77	71
Tm^{3+}	Cr^{3+}	$Y_3Al_5O_{12}$	2019.0	$^3H_4 \rightarrow {}^3H_6$	P	295	72
Yb^{3+}	Nd^{3+}	Borate glass	1018.0	$^2F_{5/2} \rightarrow {}^2F_{7/2}$	P	77	85

48. References on Solid Optically Pumped Lasers

1. Patek, K.   "Lasers," Iliffe, London, 1967.
2. Smith, W. V., and P. P. Sorokin   "The Laser," McGraw-Hill, New York, 1966.
3. Yariv, A.   "Quantum Electronics," 2d ed., Wiley, New York, 1975.
4. Cohen-Tannoudji, C., and A. Kastler   Optical Pumping, in E. Wolf (ed.), "Progress in Optics," Vol. V, North-Holland, Amsterdam, 1966.
5. Read, T. B.   *Appl. Phys. Lett.*, **9**:342 (1966).
6. Birnbaum, M., A. W. Tucker, and C. L. Fincher   *IEEE J. Quantum Electron.*, **QE-13**:101D (1971).
7. Saruwatari, M., T. Kimura, and K. Otsuka   *Appl. Phys. Lett.*, **29**:291 (1976).
8. Markiewicz, J. P., and J. L. Emmett   *IEEE J. Quantum Electron.*, **QE-2**:707 (1966).
9. Liberman, I.   Incoherent Optical Sources, in R. J. Pressley (ed.), "Handbook of Lasers," Chemical Rubber Co., Cleveland, Ohio, 1961.
10. DeMaria, A. J., W. H. Glenn, Jr., M. J. Brienza, and M. E. Mack   *Proc. IEEE*, **57**:2 (1969).
11. Woodbury, E. J.   *IEEE J. Quantum Electron.*, **QE-3**:509 (1967).
12. Kovacs, M. A., G. W. Flynn, and A. Javan   *Appl. Phys. Lett.*, **8**:62 (1966).
13. Smith, R. G., and M. F. Galvin   *IEEE J. Quantum Electron.*, **QE-3**:406 (1967).
14. McClung, F. J., and R. W. Hellwarth   *J. Appl. Phys.*, **33**:828 (1962).
15. Mocker, H. W., and R. J. Collins   *J. Appl. Phys. Lett.*, **7**:270 (1965).
16. Bass, M., and K. Andringa   *IEEE J. Quantum Electron.*, **QE-3**:621 (1967).
17. Ammann, E. O., and J. M Yarborough   *Appl. Phys. Lett.*, **20**:117 (1972).
18. DeMaria, A. J., R. Gagosz, and G. Barnard   *J. Appl. Phys.*, **34**:453 (1963).
19. Scott, W. C., and M. deWit   *Appl. Phys. Lett.*, **20**:141 (1972).
20. DeMaria, A. J., D. A. Stetser, and H. Heynau   *Appl. Phys. Lett.*, **8**:174 (1966).
21. Kafalas, P., J. I. Masters, and E. M. E. Marray   *J. Appl. Phys.*, **35**:2349 (1964).

22. Frantz, L. M., and J. S. Nodvik *J. Appl. Phys.*, **34**:2346 (1963).

23. Weber, M. J. Insulating Crystal Laser, in R. J. Pressley (ed.) "Handbook of Lasers," Chemical Rubber Co., Cleveland, Ohio, 1971.

24. Chesler, R. B., and J. E. Geusic Solid State Ionic Laser, in "Laser Handbook," Vol. II, North-Holland, Amsterdam, 1972.

25. Ehrlich, D. J., P. F. Moulton, and R. M. Osgood, Jr. *Optic Lett.*, **4**:184 (1979).

26. Voronko, U. K., A. A. Kaminskii, V. V. Osiko, and A. N. Prokhorov *JETP Lett.*, **1**:3 (1965).

27. Solomon, R., and L. Mueller *Appl. Phys. Lett.*, **3**:135 (1963).

28. Chang, N. C. *J. Appl. Phys.*, **34**:3500 (1963).

29. Maiman, T. H. *Brit. Commun. Electron.*, **7**:674 (1960).

30. Walling, J. C., H. P. Jenssen, R. C. Morris, E. W. O'Dell, and O. G. Peterson *Optic Lett.*, **4**:182 (1979).

31. Voronko, Y. K., and V. A. Sychugov *Phys. Stat. Sol.*, **25**:K119 (1968).

32. Wallace, R. W., and S. E. Harris *Appl. Phys. Lett.*, **28A**:111 1969.

33. Johnson, L. F., J. E. Geusic, and L. G. Van Uitert *Appl. Phys. Lett.*. **7**:127 (1965).

34. Ballman, A. A., S.P. S. Porto, and A. Yariv *J. Appl. Phys.*, **34**:3155 1963.

35. Weber, H. P., T. C. Damen, G. G. Danielmeyer, and B. C. Tofield *Appl. Phys. Lett.*, **22**:534 (1973).

36. Harmer, A. L., A. Linz, and D. R. Gabbe *J. Phys. Chem. Solids*, **30**:1438 1969.

37. Alan, M., K. H. Gooen, B. DiBartolo, A. Linz, E. Sharp, L., Gillespie, and G. Janney *J. Appl. Phys.*, **39**:4738 (1968).

38. Johnson, L. F., G. D. Boyd, K. Nassau, and R. R. Soden *Proc. IRE*, **50**:213 (1962).

39. Sharp, E. J., J. E. Miller, D. J. Horowitz, A. Linz, and V. Belruss *J. Appl. Phys.*, **45**:4974 (1974).

40. Morozov, A. M., M. N. Tolstoi, P. P. Feotilov, and V. N. Shapovalov *Opt. Spectrom.*, **22**:224 (1967).

41. Duncan, R. C. *J. Appl. Phys.*, **36**:874 (1965).

42. Ohlmann, R. C., K. B. Steinbruegge, and R. Mazelsky *Appl. Opt.*, **7**:905 (1968).

43. Geusic, J. E., H. M. Marcos, and L. G. Van Uitert *Appl. Phys. Lett*, **4**:182 (1964).

44. Bagdasarov, K. W., G. A. Bogomolova, A. A. Kaminskii, and V. I. Popov *Sov. Phys. Dok.*, **13**:516 (1968).

45. Weber, M. J., M. Bass, K. Andringa, R. R. Monchamp, and E. Comperchio *Appl. Phys. Lett.*, **15**:342 (1969).

46. Morris, R. C., C. F. Cline, R. F. Begley, M. Dutoit, P. J. Harget, H. F. Jenssen, T. S. LaFrance, and R. Webb *Appl. Phys. Lett.*, **27**:444 (1975).

47. Alves, R. V., R. A. Buchann, K. A. Wickersheim, and E. A. C. Yates *J. Appl. Phys.*, **42**:3043 (1971).

48. Bagdasarov, K. S., and A. A. Kaminskii *JETP Lett.*, **9**:303 (1969)

49. Smith, R. G. *IEEE J. Quantum Electron.*, **QE-4**:505 (1968).

50. Kiss, Z. J., and R. C. Duncan *Appl. Phys. Lett.*, **3**:23 (1963).

51. Johnson, L. F., H. J. Guggenheim, and R. A. Thomas *Phys. Rev*, **149** 179 (1966).

52. Johnson, L. F., R. E. Dietz, and H. J. Guggenheim *Phys. Rev. Lett*, **11**:318 (1963).

53. Deserno, U., D. Ross, and G. Zeidler *Phys. Lett.*, **28A**:422 (1968).

54. Weber, M. J., M. Bass, G. A. DeMars, and K. Andringa *IEEE J. Quantum Electron.*, **QE-6** 654 (1970).

55. Wallace, R. W. *IEEE J. Quantum Electron.*, **QE-7**:203 (1971).

56. Johnson, L. F., G. D. Boyd, and K. Nassau *Proc. IRE*, **50**:86 (1962).

57. Johnson, L. F., G. D. Boyd, and K. Nassau *Proc. IRE*, **50**:87 (1962).

58. Johnson, L. F., R. E. Dietz, and H. J. Guggenheim *Appl. Phys. Lett.*, **5**:21 (1964).

59. Johnson, L. F., and H. J. Guggenheim *IEEE J. Quantum Electron*, **QE-10**:442 (1974).

60. Kiss, Z. J., and R. C. Duncan *Proc IRE*, **50**:1531 (1962).

61. Porto, S. P. S., and A. Yariv *Proc. IRE*, **50**:153 (1962).

62. Porto, S. P. S., and A. Yariv *Proc. IRE*, **50**:1542 (1962).

63. Boyd, G. D., R. J. Collins, S. P. S. Porto, A. Yariv, and G. W. Hargraves *Phys. Rev. Lett.*, **8** 269 (1962).

64. Robinson, M., and D. P. Devor *Appl. Phys. Lett.*, **10**:167 (1967).

65. Soffer, M. B., and R. H. Hoskins *Appl. Phys. Lett.*, **6**:200 (1965).

66. Schawlow, A. L., and G. E. Devlin *Phys. Rev. Lett.*. **6**:96 (1961).

67. Powell, R. C., B. DiBartolo, B. Birang, and C. S. Naiman *Phys. Rev.*. **155**:296 (1967)

68. Snitzer, E., and R. Woodcock *Appl. Phys. Lett*, **6**:45 (1965).

69. Forrester, P. A., and D. F. Sampson *Proc. Phys. Soc.*, **88**:199 (1966)

70. Soffer, B. H., and R. H. Hoskins *IEEE J. Quantum Electron.*, **QE-2**:253 (1966)

71. Johnson, L. F., L. G. Van Uitert, J. J. Rubin, and R. A. Thomas *Phys. Rev.*, **133**:A494 (1964).

72. Johnson, L. F., J. E. Geusic, and L. G. Van Uitert *Appl. Phys. Lett.*. **7**:127 (1965).

73. Johnson, L. F., J. P. Remeika, and J. F. Dillon, Jr. *Phys. Lett.*, **21**:37 (1966).

74. Johnson, L. F., J. E. Geusic, and L. G. Van Uitert *Appl. Phys. Lett.* **8**:200 (1966).

75. Robinson, M., and D. P. Devor *Appl. Phys. Lett.* **10**:167 (1967).

76. Chicklis, E. P., C. S. Naiman, R. C. Folweiler, D. R. Gabbe, H. P. Jenssen, and A. Linz *Appl. Phys. Lett.*, **19**:119 (1971).

77. Gandy, H. W., R. J. Ginther, and J. F. Weller *Appl. Phys. Lett.*, **6**:237 (1965).

78. Remski, R. L., L. T. James, K. H. Gooen, B. DiBartolo, and A. Linz *IEEE J. Quantum Electron.*, **5**:214 (1969).

79. Kiss, Z. J., and R. C. Duncan *Appl. Phys. Lett.*, **5**:200 (1964).

80. Bass, M., and M. J. Weber *Appl. Phys. Lett.*, **17**:395 (1970).

81. O'Connor, J. R., and W. A. Hargraves *Appl. Phys. Lett.*, **4**:208 (1964).

82. Melamed, N. T., C. Hirayama, and E. K. Davis   *Appl. Phys. Lett.*, **7**:170 (1965).
83. Melamed, N. T., and C. Hirayama   *Appl. Phys. Lett.*, **6**:431 (1965).
84. Soffer, B. H., and R. H. Hoskins   *Appl. Phys. Lett.*, **6**:200 (1965).
85. Pearson, A. D., and S. P. S. Porto   *Appl. Phys. Lett.*, **4**:202 (1964).
86. Gandy, H. W., and R. J. Ginther   *Appl. Phys. Lett.*, **1** 25 (1962).
87. Gandy, H. W., and R. J. Ginther   *Proc. IRE*, **50**:2113 (1962).
88. Snitzer, E.   *Phys. Rev. Lett.*, **7**:444 (1961).
89. Young, C. G.   *Appl. Phys. Lett.*, **2**:151 (1963).
90. Robinson, C. C., R. Shaw, and R. F. Woodcock   American Optical Corp., Interim Report, Contract DAAK02-70-C-0009, USAMERDC, Ft. Belvoir, Va., August 1970.
91. Maurer, P. B.   *Appl. Opt.*, **3**:153 (1964).
92. Etzel, H. W., H. W. Gandy, and R. J. Ginther   *Appl. Opt.*, **1**:534 (1962).
93. Winzer, G., P. G. Möckel, and W. W. Krühler   *IEEE J. Quantum ELectron.*, **QE-14**:840 (1978).
94. Chinn, S. R., and H. Y. P. Hong   *Opt. Comm.*, **15**:345 (1975).
95. Otsuka, K., T. Yamada, M. Saruwatari, and T. Kimura   *IEEE J. Quantum Electron.*, **QE-11**:330 (1975).
96. Budin, J. P., M. Neubauer, and M. Rondot   *IEEE J. Quantum Electron.*, **QE-14**:831 (1978).

LIQUID LASERS

BY PATRICK R. MANZO

49. Liquid lasers use a liquid as the laser medium in place of a large single crystal or a gas (Fig. 11-15). Their properties are intermediate between those of gaseous and solid lasers. They are easy to prepare in large samples with excellent optical quality, and for certain types their energy output can be as high as 10^8 W peak and several tens of watts average. The wavelength coverage available for certain types of liquid lasers is considerably greater than that of both the solid and the gaseous lasers. There are two types of liquid lasers in common use.

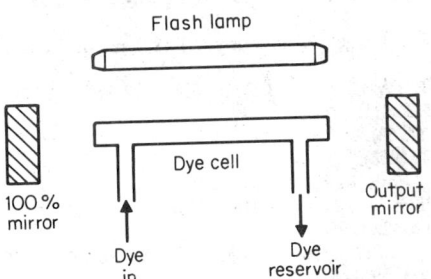

Flash lamp

Dye cell

100 % mirror

Output mirror

Dye in

Dye reservoir

Fig. 11-15. Typical dye-laser configuration.

The *aprotic liquid laser* consists of a rare-earth salt dissolved in an inorganic solvent which does not contain hydrogen. Energy from the excitation source is absorbed by the solvent and then transferred to the rare-earth ion, which lases. The absence of hydrogen in solution greatly increases the efficiency of this energy-transfer process because it lessens the possibility of this energy being transferred into vibrations of the molecule. The output wavelength is that of the rare-earth salt (see Table 11-2). In system gain, in output power levels, and in overall efficiencies, the aprotic liquid lasers are comparable with the solid-state lasers. They are capable of extremely high peak powers (10^8 W) and sustained high average powers (50 to 100 W). The principal aprotic liquid laser materials are Nd^{3+} in $SeOCl_3$ with $SnCl_4$ or in $PoCl_3$ with $SnCl_4$.

The *dye laser* uses highly fluorescent organic molecules as the laser medium; unlike the aprotic liquid, these molecules do not contain the rare-earth salts. The radiative transition in the dye laser does not originate at a metastable energy level, as in most other lasers, but takes place instead between two singlet electronic states of the molecule. Consequently it is strongly allowed and extremely short-lived — approximately 10 ns compared with several hundreds of microseconds for the rare-earth ions. This requires quite different excitation techniques to achieve laser action. The first organic dye lasers required Q-switched lasers or flash lamps with extremely fast rise times in order to invert the population fast enough to achieve laser action. However, techniques[4]* now available alleviate the necessity for such fast excitation sources. Rhodamine 6G and several other dyes have been made to operate in a continuous mode[3] with average power as high as 10 W.

The organic dyes exhibit excellent optical quality in solution, and they are extremely easy to prepare and handle. The absorption and fluorescence bands of the molecule are extremely broad (several tens of nanometers) due to the large number of vibrational and rotational energy levels associated with each electronic energy level. The laser output can be broadband (30 nm wide) or

*Superior numbers correspond to numbered references, Par. **11-50**.

very narrow (0.001 nm). Table 11-9 illustrates the broad wavelength coverage possible by select-ing the appropriate dye. In each case the wavelength is at approximately the center of the emis-sion bandwidth.

The very short lifetime of the excited state means that sources must be of very high intensity to cause the dyes to lase. Therefore, the sources used to pump the dyes include high-intensity flash lamps and other lasers. Distortion in the dye due to thermal inhomogeneities requires that the dye be mixed rapidly and exchanged in order to minimize beam spreading and losses during

TABLE 11-9   Most Common Organic-Dye Laser Materials

Dye	Solvent	Lasing wavelength, nm
p-Terphenyl	Cyclohexane	341
p-Quaterphenyl	Dimethyl sulfoxide	371
p,p'-Diphenylstilbene	Benzene	408
9,10-Diphenylanthracene	Cyclohexane	432.5
Acridone	Ethanol	437
9-Aminoacridine hydrochloride	Ethanol	458.5
4-Methyl-7-hydroxycoumarin	Water	450
7-Diethylamino-4-methyl coumarin	Ethanol	460
7-Hydroxycoumarin	Water	460
Trypaflavin	Ethanol	505
Acriflavin hydrochloride	Ethanol	510
Fluorescein	Aqueous alkaline	518
Na-fluorescein	Water	527
Eosin	Ethanol	540
Rhodamine 6G	Ethanol	590
Uranine 6	Ethanol	560
Rhodamine B	Ethanol	620
Acridine red	Ethanol	615
3,3'-Diethyloxadicarbocyanine iodide	Methanol	658
3,3'-Diethyl-2,2'-thiadicarbocyanine iodide	Acetone	711
3,3'-Diethyloxytricarbocyanine iodide	Ethanol	708.5
Cryptocyanine	Glycerin	745
Naphthalene green	Glycerin	756
Malachite green	Isoamyl alcohol	750
Chloro-aluminum phthalocyanine	Dimethyl sulfoxide	751.5
3,3'-Diethylthiatricarbocyanine iodide	Ethanol	807.5
	Methanol	835
Methylene green	Sulfuric acid	823
Toluidine blue	Sulfuric acid	848
Phthalocyanine (metal-free)	Sulfuric acid	863
1,1'-Diethyl-2,2'-quinotricarbocyanine iodide	Acetone	898
1,1'-Diethyl-4,4'-quinotricarbocyanine iodide	Acetone	1000

cw and repetition-rate operation. The line width of typical dye materials is homogeneously broadened, and therefore all the stored energy in the upper laser level can be channeled into a spectrally narrow emission line, which can be tuned over the gain profile of the active medium.

Flashlamp-pumped dye lasers have demonstrated average powers as high as 100 W at high repetition rates. Single pulse energies to 10 J per pulse and peak powers to 10 MW have also been demonstrated. Present flashlamp-pumped systems are limited to approximately 1% effi-ciency while the efficiency of laser-pumped dyes is somewhat lower. Average power from a few milliwatts to over 10 W has been achieved in cw operation of laser dyes using pump laser con-figurations such as the one shown in Fig. 11-16a. Frequency-stabilized cw dyes have achieved a line width of less than 1 MHz and long-term stabilities of a few hundred hertz.

With an appropriate selection of dyes, this type of laser can be directly tuned from 340 to 1200 nm. This wavelength range has been extended into the ultraviolet and the infrared by frequency mixing of dye-laser radiation in nonlinear optical materials. Tunable ultraviolet radiation down to 196 nm has been obtained by second-harmonic and sum-frequency generation in nonlinear crystals. In the long-wavelength range the generation of tunable near and middle infrared, as the difference frequency of two laser oscillators, has been successful out to 12.7 μm. It is also possible to produce multiple wavelengths from the same or combinations of dyes. Figure 11-16b illustrates one scheme for doing so.

It is possible to produce dye laser pulses with a duration of less than 1 ps. Durations as short as 0.3 ps have been reported. In addition, the broad emission bands of dyes make them the only sources capable of producing mode-locked picosecond pulses at high repetition rates which are wavelength tunable over several tens of nanometers.

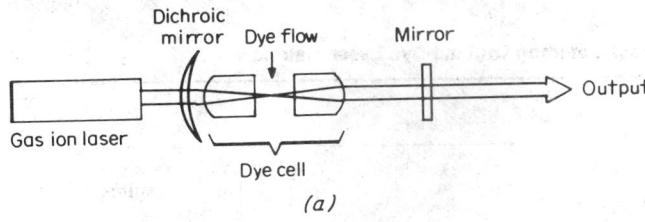

(a)

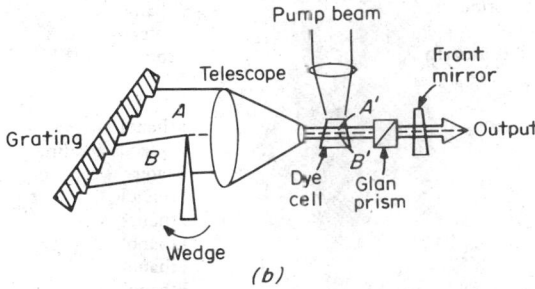

(b)

Fig. 11-16. (*a*) Schematic of a continuous-wave dye laser; (*b*) double-wavelength dye laser.[9] The lower portion of the beam is shifted toward the red by the wedge.

50. References on Liquid Lasers

1. Lempicki, A., H. Samelson, and C. Brecher   *Appl. Opt., Suppl.,* **2**:205 (1965).

2. Lempicki, A., and H. Samelson   Organic Laser Systems, in A. K. Levine (ed.). "Lasers," Vol. 1, pp. 181–252, Dekker, New York, 1966.

3. Peterson, O. G., S. A. Tuccio, and B. B. Snavely   *Appl. Phys. Lett.* **17**:245 (1970).

4. Snavely, B. B.   Flashlamp Excited Organic Dye Lasers, *Proc. IEEE,* **57**(8):1374–1390 (August 1969).

5. Sorokin, P. P.   Organic Lasers, *Sci. Amer.,* **220**:30–40 (February 1969).

6. Schafer, F. P.   Liquid Lasers, pp. 369–424 in F. T. Arecchi and E. O. Schulz-Dubois (eds.), "Laser Handbook," Vol. 1, North-Holland, Amsterdam, 1972.

7. Bass, M., T. F. Deutsch, and M. J. Weber   Dye Lasers, pp. 269–345 in A. K. Levine and A. J. DeMaria (eds.), "Lasers," Vol. 3, Dekker, New York, 1971.

8. Wallenstein, R.   Pulsed Dye Lasers, pp. 289–358 in M. L. Stitch (ed.), "Laser Handbook," Vol. 3, North-Holland, Amsterdam, 1979.

9. Lotem, H. and R. T. Lynch, Jr.   *Appl. Phys. Lett.* **27**:344 (1975).

GAS LASERS

BY GARETH M. JANNEY

51. General Characteristics. Gas lasers can best be characterized by their variety. The laser medium may be a very pure, single-component gas or mixture of gases. It may be a permanent gas or a vaporized solid or liquid. The active species in a gas laser may be a neutral atom, an ionized atom, or a molecule (including excimers, which are "stable" molecules only in an excited state). The operating pressures range from a fraction of a torr to atmospheric pressure, and the operating temperature from −196 to 1600°C. Excitation methods include electric discharges (glow, arc, pulsed, rf, dc), chemical reactions, supersonic expansion of heated gases (gas dynamic), and optical pumping. The average output power of useful gas lasers ranges from a few microwatts to tens of kilowatts (10 orders of magnitude), and the peak power ranges from a fraction of a watt to 100 MW. The range of output wavelengths extends from 0.16 to 774 μm at discrete wavelengths.

52. Multiple Wavelengths. Most gas-laser materials have a number of distinct laser transitions (i.e., different wavelengths). For example, the neon atom has more than 100, and the argon ion has more than 30. Lasers using these materials can operate with a multiwavelength output, or one transition at a time can be selected by a simple adjustment (rotating a prism or diffraction grating in the optical cavity). Table 11-10 (bottom) shows several commercially available lasers which provide more than one wavelength of operation.

53. Electrically Excited Electric-Discharge Gas Lasers. Most gas lasers are excited by electric discharges. Electrons which have been accelerated by an electric field transfer energy to the gas atoms and molecules by collisions. These collisions may excite the upper laser level directly. Indirect excitation is also possible by cascading from higher-energy levels of the same atom (or molecule) or resonant-energy transfer from one atom (or molecule) to another by collision.

Typical Configuration (Fig. 11-17).   The gas is contained in a glass tube having an electrode near either end. The ends are sealed by windows mounted at Brewster's angle to minimize reflections at the windows (for one plane of polarization). An optical cavity is formed by two mirrors (usually both are concave), at least one of which is partially transmitting When an electric discharge is produced in the tube between the electrodes, the gas atoms or molecules are excited and laser action starts. Listed below are some typical parameters for several common types of electric-discharge gas lasers.

Laser species	Gas mixture	Pressure, torr	Current density, A/cm^2
Ne (neutral atom)	He	1.0	0.05–0.5
	Ne	0.1	
Ar (ion)	Ar	0.3	100–2,000
CO_2	He	5–10	0.01–0.1
	N_2	1.5	
	CO_2	1.0	

The transverse-electric-discharge configuration (Fig. 11-18) is used for some gas lasers, especially for high-average-power, fast-flowing gas lasers. It provides high electric fields at practical voltages while retaining long paths of excited gas.

A pair of long electrodes (or linear arrays of electrodes) is located parallel to the optical axis, within the envelope which contains the gas. The discharge current flows transverse to the optical axis. Lasers employing this configuration include high peak power, pulsed N_2, Ne, and CO_2; high-average-power, fast-flowing CO_2 lasers; and excimer lasers (XeF, KrF, etc.).

E-Beam Lasers.   High-energy high-current electron beams are generated in an evacuated high-voltage electron gun. The electron beam passes through a thin metal foil into a chamber containing the gas mixture. As in electric-discharge lasers, some of the kinetic energy of the electrons is transferred directly or indirectly to the laser species (atom or molecule). Electron beams are sometimes used in conjunction with transverse electric discharges.

54. Chemical Lasers. Chemical lasers derive their energy from the free-energy change of a chemical reaction. The chemical reaction may be initiated by some other source of energy, such as light or electric discharges. Since the chemical reaction consumes the reactants, a flowing system is necessary for repetitive-pulsed or cw operation. Figure 11-19 shows an arrangement used to produce high-power, cw laser output from hydrogen fluoride (HF) and other molecules.

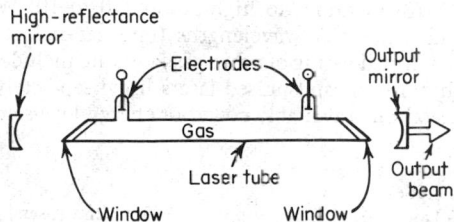

Fig. 11-17. Electric-discharge gas laser.

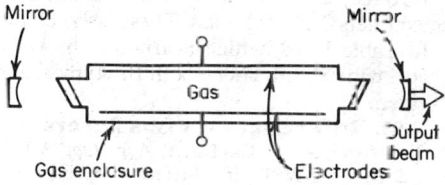

Fig. 11-18. Transverse-electric-discharge gas laser.

A series of chemical reactions is required to sustain the laser operation in practice, but in simplified form the reaction can be expressed as

$$F_2 + H_2 \rightarrow 2HF + \Delta E$$

where F_2 = reactant 1, H_2 = reactant 2, and ΔE = free-energy change, some of which is in the form of vibrationally excited HF molecules. Laser action occurs on vibrational transitions of the HF molecule in the wavelength region 2.6 to 2.9 μm. Chemical lasers of this general type are potential sources of high average power (multikilowatt).

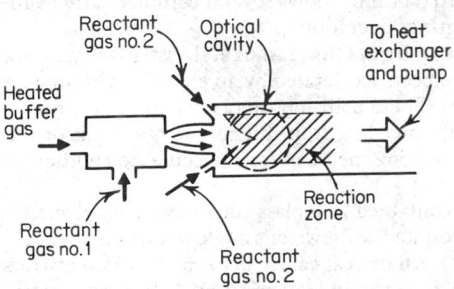

Fig. 11-19. Fast-flowing chemical laser.

55. Gas Dynamic Lasers. The expansion of a hot, high-pressure mixture of CO_2 and N_2 through a supersonic nozzle results in a lowering of the gas temperature in a time which is short compared with the vibrational relaxation time of the CO_2 molecule. A differential relaxation time between the upper and lower laser levels results in a population inversion for a short distance downstream from the supersonic nozzle, and laser operation in this region is possible. Average output power of 60 kW at 10.6 μm has been obtained from such a device.

56. Properties of the Gas-Laser Output Beam. *Wavelength and Frequency.* It is customary in laser terminology to refer to a laser transition by its wavelength (in micrometers or nanometers) and to discuss the fine structure of the transition in terms of frequency. Wavelength λ and frequency ν are related by $\lambda\nu = c$, where c is the velocity of light. The wavelength of a laser transition, as well as the wavelength interval or width of the transition over which optical gain exists, is a property of the laser atom or molecule. The transition width is also related to gas temperature and pressure. Typical widths of common gas lasers are:

Laser	λ, μm	Line width, MHz
HeNe	0.6328	1,700
	1.15	920
	3.39	310
Ion lasers	0.5	2,500–3,000
CO_2 low-pressure	10.6	60
High-pressure	10.6	800

Oscillation can occur at discrete frequencies (cavity modes) at frequency spacings of $\Delta\nu = c/2L$, where L is the separation of cavity mirrors. Depending on the nature of the laser medium and the geometry of the optical cavity, oscillation may occur at a number of different modes, within the line width of the transition. For example, for helium-neon at 0.6328 μm, with a cavity-mirror spacing of 100 cm, the mode spacing is 150 MHz, and 11 (axial) modes could oscillate. Single-axial-mode oscillation is generally achieved by reducing L so that only one axial mode lies within the line width of the laser transition. A stable optical cavity is required for both frequency and amplitude stability.

57. Available Gas Lasers. Of the many commercially available gas lasers (over 35 gas-laser species), four groups are most common: (1) low-power (few milliwatts) visible cw, including helium-neon (HeNe) and helium-cadmium (HeCd); (2) medium-power (few watts), visible, cw including argon-ion (Ar^+) and krypton-ion (Kr^+); (3) medium- to high-power (10 to 10^4 W) infrared, cw, primarily carbon dioxide (CO_2) and ultraviolet wavelengths (rare gas–halogen excimers, XeF, KrF, etc.). These lasers along with a selected listing of other gas lasers are included in Table 11-10, which is arranged by wavelength, with cw and pulsed lasers listed separately. For many of the lasers, both the typical and the maximum available power or energy levels are given.

58. References on Gas Lasers

1. Bloom, A. L.   Gas Lasers, *Appl. Opt.,* **5**:1500–1514 (1966).

2. Bennett, W. R., Jr.   Inversion Mechanisms in Gas Lasers, *Appl. Opt.,* Suppl. 2, 1965, Chem. Lasers, p. 34.

TABLE 11-10   Representative Gas Lasers*

λ, μm	Laser type	Continuous-wave (cw) Power, W Typical	Max	Comment
0.3252	HeCd	10^{-3}–10^{-2}		Metal vapor, electric discharge
0.33–0.53	Ar	0.5–5	20	High-current electric discharge
0.33–1.09	Ar/Kr	0.5	6	High-current electric discharge
0.4416	HeCd	1–5×10^{-3}		Metal vapor electric discharge
0.46–0.65	HeSe	10^{-3}		Metal vapor electric discharge
0.6328	HeNe	10^{-3}	10^{-1}	Most common gas laser
1.15	HeNe	10^{-3}		
1.315	I_2		10^2	Experimental chemical laser
2.6–3.0	HF	5–50	10^4	Chemical laser
3.39	HeNe	10^{-3}	10^{-2}	
3.5	HeXe	5×10^{-3}		
3.6–4.0	DF	2–100	10^4	Chemical laser
5–6.5	CO	2–10	50	
9–11	CO_2	1–100	10^4	
27–374	H_2O	1–10×10^{-3}		
34–388	NH_3	10^{-3}		Optically pumped by CO_2 or CO laser
311, 337	HCN	10^{-2}	1.0	
37–1217	Methanol	1–10×10^{-4}		

λ, μm	Laser type	Pulsed lasers Pulse energy, J Typical	Max	Comment
0.157	F_2	10^{-2}		
0.193	ArF	10^{-1}		
0.222	KrCl	10^{-2}		
0.248	KrF	10^{-1}	10^2	e beam for high energy
0.308	XeU	10^{-1}		
0.337	N_2	10^{-3}–10^{-2}		
0.351	XeF	10^{-1}		
0.458–0.53	Ar	10^{-8}–10^{-6}		
0.5105, 0.5782	Cu	10^{-4}–10^{-3}		
1.5	Ba	10^{-3}		
2.8–3	HF	1	2×10^2	Chemical, electrically initiated
3.6–4.1	DF	10^{-1}–1		Chemical, electrically initiated
9–11	CO_2	10^{-1}–10^2	10^4	e beam for high energy
118.8	Methanol	10^{-3}		
496.1	Methyl fluoride	10^{-4}		

Laser type	Multiple-wavelength gas lasers Wavelength, μm			Comment
Helium-neon	0.5939	0.6352	1.152	Rotate prism to select one wavelength at a time
	0.6046	0.6401	1.162	
	0.6118	0.7305	1.177	
	0.6294	1.080	1.199	
	0.6328	1.084	3.39	

TABLE 11-10 Representative Gas Lasers* (Continued)

Multiple-wavelength gas lasers

Laser type	Wavelength, μm			Comment
	Argon	Krypton	Xenon	
Noble-gas ions	0.3511	0.4619	0.4955	Rotate prism to select lines from one
	0.3638	0.4762	0.5007	gas; change gas in tube to change set
	0.3795	0.5208	0.5160	of wavelengths; mixtures of gases
	0.4579	0.5682	0.5260	can be used to extend range;
	0.4765	0.6471	0.5353	simultaneous oscillation on many
	0.4965	0.6764	0.5395	lines possible
	0.5017		0.5959	
	0.5145			
	0.5287			
Carbon dioxide	9.1– 11.3			Several groups of closely spaced lines in this wavelength region; line separations within groups ~0.02 μm; rotate diffraction grating to select lines one at a time

*Wavelength ranges are sets of discrete wavelengths and simultaneous output at more than one wavelength may or may not be possible.

3. Geusic, J. E., W. B. Bridges, and J. I. Pankove Coherent Optical Sources for Communications, Proc. IEEE, 58:1419–1439 (1970).
4. Emmett, J. L. Frontiers of Laser Development, Phys Today, March 1971, pp. 24–33.
5. "Laser Focus Buyers Guide," Advanced Technology Publications, Newton, Mass., 1980.
6. Bloom, A. L. "Gas Lasers," Wiley, New York, 1968
7. Sinclair, D. C., and W. E. Bell "Gas Laser Technology," Holt, New York, 1969.

SEMICONDUCTOR LASERS AND LEDS

BY RICHARD R. SHURTZ II

59. Semiconductor lasers and light-emitting diodes (LEDs) range in wavelength from 0.58 to 34 μm, the most efficient materials emitting near 1 μm.[1,2]* The fundamental light-producing mechanism in the semiconducor is the recombination of excess conduction-band electrons and valence-band holes. The wavelength of the emitted radiation is related to E_g, the bandgap of the material, by $\lambda = 1.24/E_g$, with λ in micrometers and E_g in electronvolts. The most effective method of pumping the semiconductor is the injection of excess carriers across a junction (see Table 11-11).

Injection lasers can be divided into two operational regimes — high-peak-power pulsed operation and low-power cw operation. The former is used primarily for ranging, infrared illumination, and intrusion alarms, the latter for fiber-optic communication, holographic readout, optical scanning, and storage-disk information retrieval.

LEDs can be fabricated either for high brightness, as needed in visual displays, photoelectric systems, and electro-optical components, or for high radiance, as needed in fiber-optics communication systems. Though LEDs are incoherent radiation sources they are included here since their structure is very similar to that of semiconductor lasers.

60. Semiconductor Lasers. Typical *broad-area pulsed* and *stripe-geometry cw*[3–12] semiconductors lasers are shown in Fig. 11-20. The distinguishing feature is the constricted current flow and the resulting reduced waveguide width W in the stripe-geometry case. This configuration provides improved heat sinking for cw operation, reduced emitting area for efficient fiber-optics coupling, and improved longitudinal-mode control for a narrow spectral bandwidth. The width W is in the 5- to 30-mil range for broad area and 0.4- to 1-mil range for stripe geometry. Since the front and rear Fabry-Perot surfaces are cleaved facets, they are always aligned and easily formed. Distributed-feedback (DFB) or grating Fabry-Perot reflectors are also used for single-longitudinal-mode (SLM) operation.[13–15]

*Superior numbers correspond to references in Par. **11-64**.

The rear surface is usually coated with a high-reflectivity metal, sandwiched between insulating layers, so that all light emerges from the partially reflecting front surface. The sides of the laser are sawed or highly absorbing to quench off-axis lasing in the broad-area case. The light propagates through a dielectric waveguide formed in the crystal using either doping or compositional variations. Typical ranges are 0.2 to 4 μm for the waveguide thickness d and 10 to 15 mils for the cavity length L, which can be selected to maximize the power efficiency.

TABLE 11-11   Emission Wavelengths of Selected III-V and II-VI Heterostructures

Composition			Bandgap eV	Emission wavelength, μm	Temp, K	Fabricated yet?
Active layer	Confinement layer	Substrate				
$(Al_xGa_{1-x})_{0.51}In_{0.49}P*$	$(Al_yGa_{1-y})_{0.51}In_{0.49}P*$	GaAs	1.8–2.15	0.58–0.69	300	No
$Ga_xAs_{1-x}P$	$In_{1-y}Ga_yP$	GaAs†	1.38–1.85	0.67–0.9	300	Yes
$Al_xGa_{1-x}As$	$Al_yGa_{1-y}As$	GaAs	1.38–1.85	0.7–0.9	300	Yes
$GaAs_ySb_{1-y}$	$Al_xGa_{1-x}As_ySb_{1-y}$	GaAs†	0.67–1.45	0.86–1.8	300	Yes
$In_xGa_{1-x}As$	$In_{1-y}Ga_yP$	GaAs†	0.9–1.38	0.9–1.37	300	Yes
$(Al_xGa_{1-x})_{0.47}In_{0.53}As$	$(Al_yGa_{1-y})_{0.47}In_{0.53}As$	InP	0.8–1.35	0.92–1.6	300	No
$Ga_xIn_{1-x}P_yAs_{1-y}$	InP	InP	0.74–1.35	0.9–1.7	300	Yes
$(Al_xGa_{1-x})_yIn_{1-y}Sb$	AlSb	GaSb†	0.6–1	1.24–2.07	300	No
InAs	$GaAs_{0.08}Sb_{0.92}$	InAs, GaSb†	0.360–0.420	3–3.44	0–300	No
$InAs_{0.91}Sb_{0.09}$	GaSb	GaSb	0.287–0.347	3.57–4.32	0–300	No
$InAs_{0.82}Sb_{0.18}$	AlSb	GaSb†	0.23–0.29	4.28–5.39	0–300	No
$PbS_{1-x}Se_x$	$PbS_{1-x}Se_x†$	PbS†	0.15–0.31	4–8.5	4–100	Yes
$Pb_{1-x}Sn_xTe$	$Pb_{1-x}Sn_xTe†$	PbTe†	0.04–0.19	6.5–34	4–100	Yes

*$0 < x < 1, 0 < y < 1$ used to indicate compositional degree of freedom for all quaternary and ternary entries.
†Not lattice-matched to active layer.

In application, diodes are mounted on a heat sink, traditionally a transistor header for discrete diodes or on a beryllium oxide block for arrays.

The planar waveguide can be formed, for example, by a GaAs ($n = 3.60$) layer sandwiched between two $Ga_{0.8}Al_{0.2}As$ ($n = 3.45$) layers of lower index of refraction, one doped n and the other p. In use, the junction laser is positively biased, p side positive. Excess electrons are injected from the n layer and excess holes from the p layer (each with nearly 100% injection efficiency) into the active waveguide region, where they recombine, via stimulated emission, producing the desired gain. The active or waveguiding region can be lightly doped, either n or p. The high injection efficiencies are caused by the band-edge discontinuities introduced by the GaAs–

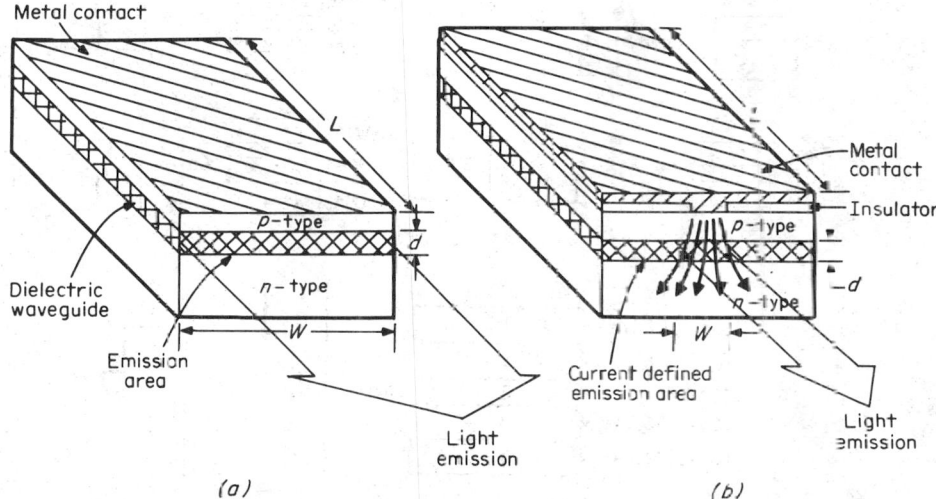

Fig. 11-20. Semiconductor laser structures: (a) broad-area geometry is used for high-power pulsed operation, where heat sinking is not critical and where maximum emission area is desired; (b) stripe geometry is used where heat sinking is required for cw operation and where reduced emission area is desired for efficient fiber coupling.[1]

GaAlAs heterojunctions. Thus in this example the heterojunctions provide both the optical and carrier confinement needed for efficient room-temperature lasing action.

At low current levels, the electrons recombine with holes, spontaneously emitting radiation in all directions. At higher current levels the excess carrier density is high enough to produce an inverted population, yielding a positive gain. The lasing threshold is reached when a light pulse can make a round trip in the Fabry-Perot cavity without attenuation, i.e.,

$$R_2 R_1 \exp\left[(g - a)2L\right] = 1$$

where R_1, R_2 = the reflectances at cavity ends, g = gain per unit length, a = the absorption per unit length, and L = cavity length. The prime causes of absorption are defect scattering and free-carrier absorption.

Laser Operational Characteristics.   Room-temperature threshold current densities of GaAlAs heterostructure lasers vary between 500 and 10,000 A/cm², with higher thresholds seen in high-peak-power pulsed diodes. Low-threshold cw devices have waveguide thicknesses less than 1 μm, while higher threshold pulsed devices have waveguide thicknesses varying between 2 and 4 μm. External differential quantum efficiencies vary between 30 and 50% at room temperature for GaAlAs heterojunction structures. CW devices emit in the 10-mW range, and pulsed devices typically emit 1 W per mil of junction width W at a 0.1% duty cycle. Operational lifetimes of defect-free devices exceed 10^4 h.[16,17]

61. Semiconductor Laser Configurations.   *Homojunction Laser.*[18–20]   This standard laser structure is formed by diffusing zinc into a thin n-type crystal layer grown on an oriented substrate (see Fig. 11-21a). These homojunction lasers have room-temperature threshold current densities which vary from 40 to 100 kA/cm² and a beam spread of about 12° × 15° full angle. Room-temperature efficiencies are less than 1%, and efficiencies at 77 K are as high as 60%. This structure is primarily used in high-power arrays when cooled.

Single-heterostructure lasers,[21–24] developed to reduce the room-temperature threshold, have a GaAlAs layer located about 2 μm from the GaAs homojunction on the p side[6,7] (see Fig. 11-21b). Because the GaAlAs has a larger bandgap and a lower index of refraction than the GaAs, the injected electrons are confined to a narrow recombination region and the optical modes to an asymmetrical waveguide formed by the pn junction and the GaAs-GaAlAs heterojunction. This structure has a typical threshold current density in the 8 to 12 kA/cm² range and has made room-temperature operation of the laser practical. The beam spread of this device is about 12° × 30° full angle.

The double-heterostructure laser[25–27] has a second GaAlAs layer added at the pn junction to provide a symmetrical waveguide (see Fig. 11-21c). The fraction of modes confined to the active region is much higher for this structure. Threshold currents can be reduced to arbitrarily low

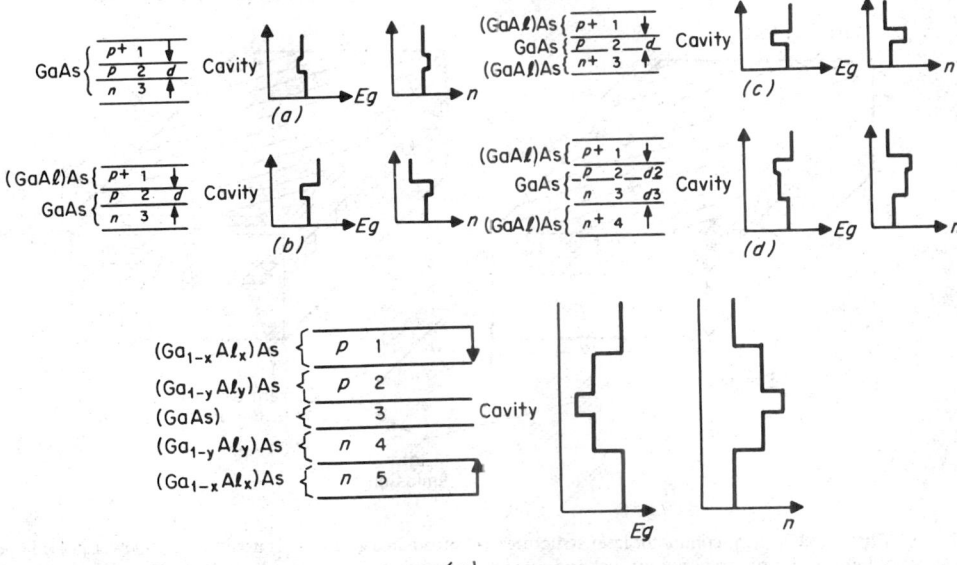

Fig. 11-21.   (a) Homojunction laser; (b) single-heterojunction laser; (c) double-heterojunction laser; (d) large-optical-cavity laser; (e) separate-confinement heterojunction laser.[1]

values by reducing the heterojunction separation and increasing the aluminum concentration in the outside layers for improved waveguiding. The first *cw room-temperature laser* was fabricated this way. Threshold current density is 5 kA/cm$^2 \cdot \mu$m of waveguide thickness.

Large-Optical-Cavity (LOC) Laser.[28-32] Both the single- and double-heterostructure lasers undergo *catastrophic degradation* at low current levels, where the reflecting surfaces of the Fabry-Perot cavity are destroyed. The destruction is caused by the high-optical-flux densities in the waveguide, which must be no wider than 2 μm for effective carrier confinement. The large-optical-cavity laser (LOC) was developed to solve this problem (Fig. 11-20d). The optical cavity here is allowed to be wider than the recombination region, thus lowering the flux densities and raising the threshold for catastrophic degradation. By optically coating the front surface, this threshold can be raised even higher.

The LOC lasers have power outputs as high as several watts per mil emitting-junction width, as opposed to about 1 W/mil for the single and double heterostructures with 2-μm waveguides. Typical threshold current densities can be obtained in the 4 to 12 kA/cm^2 region. This device emits two beams of light separated by 50 to 90° because it operates in a high-order transverse mode. On occasion a small amount of aluminum may be added to layer 3 to increase the hole-injection efficiency into the active region 2, and to force lasing in the fundamental-order mode for a reduced beam width.

Separate-Confinement Heterojunction Laser[33-36] *(SCH).* This device is designed to decouple the optical and carrier confinement functions by using four heterojunctions (Fig. 11-21e), the inner two for carrier confinement and the outer two for optical confinement. In practice, layer 3 must be thin (< 0.5 μm) to force the mode into layers 2 and 4. Low-threshold, moderate-beam-divergent (compared with DH) lasers can be made with this structure.

62. Materials.[1,2] Over 25 binary, ternary, or quaternary semiconductor materials have lased. Room-temperature operation has occurred only where heterojunctions have been incorporated into the structure for optical and carrier confinement.

Variable-bandgap ternary compounds are formed by combining two binary compounds which have one constituent in common (such as GaAs and AlAs). The bandgap of the mixture varies between the gaps of the two individual compounds, e.g., between 1.4 and 2.2 eV for GaAlAs as the aluminum arsenide concentration is increased. This compound produces devices with low defect densities and long operational lifetimes, because its lattice constant is nearly independent of composition. Hence heterojunction-caused strain and dislocations are negligible.

Other commonly used ternary compounds, such as GaInAs, GaInP, GaAsP, InAsP, and GaAsSb, lack this feature. To circumvent this problem, *quaternary compounds* have been developed with two degrees of compositional freedom so that a single lattice constant can be maintained over a range of bandgaps. Common quaternary compounds are InGaAsSb, InGaAsP, and AlGaAsSb.

Table 11-11 lists the selected active-region III-V compositions which have been demonstrated (or may be demonstrated). Except where indicated, the III-V compound lasers *are fully lattice-matched* (i.e., across all heterojunctions and with substrate) throughout the emission ranges shown. When the substrate lattice constant does not match the active region, buffer or transitional layers are commonly used to shift the lattice constant gradually.

The II-VI lead-salt diodes are current- and temperature-tunable over broad ranges using the same composition. The III-V compound device tunability is achieved principally by compositional variations.* All diodes emitting at wavelengths longer than 2 μm must be cooled to depopulate the conduction band thermally.

63. Light-emitting diodes (LEDs) are similar to injection lasers in all respects except for absence of feedback mechanism or Fabry-Perot resonator. Their spontaneous recombination bandwidth is 20 to 40 nm. In *high-brightness surface emitters,*[37-44] bulk reabsorption and surface reflectance are two factors which reduce diode efficiency. Reabsorption is minimized by capping the light-producing region with a material of slightly larger bandgap than the emitted photon energy. Reflectance is reduced by encapsulation, surface shaping (e.g., hemispherical), or anti-reflective coatings. Common materials are GaAs:Si, GaAs:Ge, GaAsP, GaAsP:N, GaAlAs, InGaP, and InAlP. Selected commercially available high-brightness diodes are shown in Table 11-1.

High-radiance LED[45-49] technology is similar to that used for heterojunction lasers. Both surface and edge emitters are commonly employed. A surface emitter specifically designed for fiber coupling is shown in Fig. 11-22a. It is a standard double-heterojunction configuration in which a hole has been etched through the absorbing substrate. The light-emitting region is defined by current injected by the circular contact. The emission is lambertian.

---

*III-V bandgaps are also temperature-tunable; e.g., GaAs emission varies 0.25 nm/K.

A restricted edge-emitting diode is shown in Fig. 11-22b. The device is a double heterostructure with the emitting region again defined by the etched contact. The emitting region is situated close to the edge in order to minimize reabsorption. The restricted edge emitter has very broad beam width (120° full width at half power, abbreviated by FWHP) in the plane of the junction and can have a relatively narrow one (for example, 30° FWHP) in the plane perpendicular. The latter beam width depends on the waveguiding nature of the double heterojunction waveguide. Devices with radiance as high as 1,000 $W/cm^2 \cdot sr$ provide improved fiber coupling efficiency over lambertian emitters.

Frequency response of LEDs is of prime importance for information-transfer systems. Short recombination lifetimes and hence fast response are achieved when the diodes are operated in

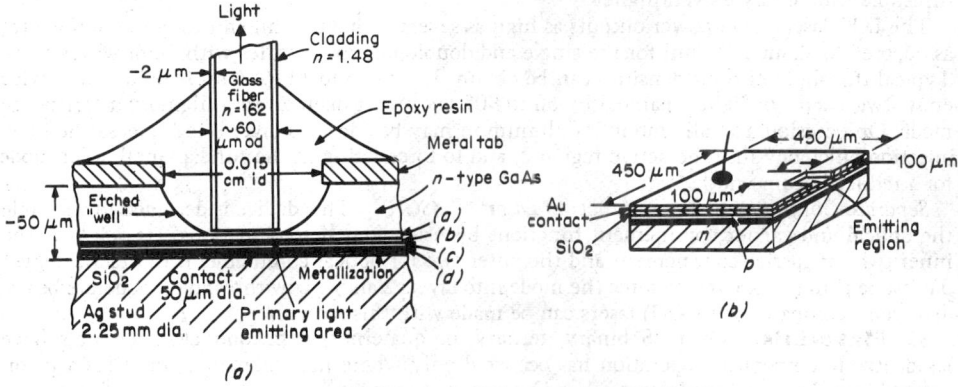

Fig. 11-22. Emitting diodes: (a) fiber-coupled DH electroluminescent diode;[46] (b) restricted edge-emitting diode.[47]

the high-injected-carrier-density regime. Hence narrow recombination layers and high drive currents are desired. Lifetimes as low as 1.2 ns and modulation rates as high as 250 MHz (MTF = 0.5) have been achieved using this technique in restricted edge emitters. For comparison DH lasers have been modulated at frequencies as high as 1 GHz.

64. References on Semiconductor Lasers and LEDs

SELECTED REVIEW BOOKS

1. Kressel, H., and J. K. Butler   "Semiconductor Lasers and Heterojunction LEDs," Academic, New York, 1977.
2. Casey, H. C., Jr., and M. B. Panish   "Heterostructure Lasers," Pts. A and B, Academic, New York, 1978.

SELECTED STRIPE GEOMETRY LASER REFERENCES

3. Yonezu, H., I. Sakuma, K. Kobayashi, T. Kamejima, M. Ueno, and Y. Nannichi   *Jpn. J. Appl. Phys.*, **12:** 1585 (1973).
4. D'Asaro, L. A.   *J. Lumin.*, **7:**310 (1973).
5. Yonezu, H., Y. Matsumoto, T. Shinohara, I. Sakuma, T. Suzuki, K. Kobayashi, R. Lang, Y. Nannichi, and I. Hayashi   *Jpn. J. Appl. Phys.*, **16:**209 (1977).
6. Aiki, K., M. Nakamura, T. Kuroda, J. Umeda, R. Ito, N. Chinone, and M. Maeda   *IEEE J. Quantum Electron.*, **QE-14:**89 (1978).
7. Namizaki, H., H. Kan, M. Ishii, and A. Ito   *J. Appl. Phys.*, **45:**2785 (1974).
8. Susaki, W., H. Namizaki, H. Kan, and A. Ito   *J. Appl. Phys.*, **44:**2893 (1973).
9. Tsukada, T., H. Nakashima, J. Umeda, S. Nakamura, N. Chinone, R. Ito, and O. Nakada   *Appl. Phys. Lett.*, **20:**344 (1972).
10. Chinone, N., R. Ito, and O. Nakada   *J. Appl. Phys.*, **47:**785 (1976).
11. Tsukada, T.   *J. Appl. Phys.*, **45:**4899 (1974).
12. Lee, T. P., and A. Y. Cho   *Appl. Phys. Lett.*, **29:**164 (1976).

SELECTED DISTRIBUTED FEEDBACK (DFB) LASER REFERENCES

13. Casey, H. C., Jr., S. Somekh, and M. Ilegems   *Appl. Phys. Lett.*, **27:**142 (1975).
14. Aiki, K., M. Nakamura, and J. Umeda   *IEEE J. Quantum Electron.*, **QE-12:**597 (1976).
15. Reinhart, F. K., R. A. Logan, and C. V. Shank   *Appl. Phys. Lett.*, **27:**45 (1975).

SELECTED OPERATIONAL-LIFETIME REFERENCES

16. Hartman, R. L., and R. W. Dixon   *Appl. Phys. Lett.*, **26:**239 (1975).
17. Ladany, I., M. Ettenberg, H. F. Lockwood, and H. Kressel   *Appl. Phys. Lett.*, **30:**87 (1977).

SELECTED HOMOJUNCTION-LASER REFERENCES

18. Nathan, M. I., W. P. Dumke, G. Burns, F. H. Dill, Jr., and G. Lasher Appl. Phys. Lett., 1:63 (1962).
19. Nathan, M. I. Appl. Opt., 5:1514 (1966).
20. Hall, R. N., G. E. Fenner, J. D. Kingley, T. J. Soltys, and R. O. Carlson Phys. Rev Lett., 9: 366 (1962).

SELECTED SINGLE HETEROSTRUCTURE LASER REFERENCES

21. Kressel, H., H. Nelson, and F. Z. Hawrylo J. Appl. Phys., 43:561 (1972).
22. Hayashi, I., M. B. Panish, and P. W. Foy IEEE J. Quantum Electron., QE-5:211 (1969).
23. Panish, M. B., I. Hayashi, and S. Sumski IEEE J. Quantum Electron., QE-5:210 (1959).
24. Kressel, H., and H. Nelson RCA Rev., 30:106 (1969).

SELECTED DOUBLE-HETEROJUNCTION-LASER REFERENCES

25. Panish, M. B., I. Hayashi, and S. Sumski Appl. Phys. Lett., 16:326 (1970).
26. Hayashi, I., M. B. Panish, P. W. Foy, and S. Sumski Appl. Phys. Lett., 17:109 (1970)
27. Alferov, Zh. I., V. M. Andreev, D. Z. Garbuzov, Yu. V. Zhilyaev, E. P. Morozov, E. L Portonoi, and V. G. Triofim Sov. Phys. Semicond., 4:1573 (1971).

SELECTED LOC-LASER REFERENCES

28. Lockwood, H. F., H. Kressel, H. S. Sommers, Jr., and F. Z. Hawrylo Appl. Phys. Lett., 17:499 (1970).
29. Kressel, H., H. F. Lockwood, and F. Z. Hawrylo Appl. Phys. Lett., 18:43 (1971).
30. Kressel, H., H. F. Lockwood, and F. Z. Hawrylo Appl. Phys. Lett., 43:561 (1972).
31. Paoli, T. L., B. W. Hakki, and B. I. Miller J. Appl. Phys., 44:1276 (1973).
32. Krupka, D. C. IEEE J. Quantum Electron., QE-11:390 (1975)

SELECTED SEPARATE-CONFINEMENT HETEROSTRUCTURE-LASER REFERENCES

33. Thompson, G. H. B., and P. A. Kirkby IEEE J Quantum Electron., QE-9:311 (1973)
34. Panish, M. B., H. C. Casey, Jr., S. Sumski, and P. W. Foy Appl. Phys. Lett., 22:590 (1973).
35. Thompson, G. H. B., and P. A. Kirkby Electron Lett., 9:295 (1973).
36. Casey, H. C., Jr., M. B. Panish, W. O. Schlosser. and T. L. Paoli J. Appl. Phys., 45:322 (1974).

SELECTED HIGH-BRIGHTNESS LED REFERENCES

37. Rupprecht, H., J. M. Woodall, K. Konnerth, and D. G. Pettit Appl. Phys. Lett., 9:221 (1966).
38. Ladany, I. J. Appl. Phys., 42:654 (1971).
39. Rupprecht, H., J. M. Woodall, and G. D. Pettit Appl. Phys. Lett., 11:81 (1967).
40. Woodall, J. M., H. Rupprecht, and W. Reuter J. Electrochem. Soc., 116:899 (1969).
41. Kressel, H., F. Z. Hawrylo, and N. Almeleh J. Appl. Phys., 40:2224 (1969).
42. Linden, K. J. J. Appl. Phys., 40:2325 (1969).
43. Kressel, H., and M. Ettenberg Appl. Phys. Lett., 23:511 (1973).
44. Rosztoczy, F. E., F. Ermahis, I. Hayashi, and B. Schwartz J. Appl Phys., 41:264 (1970).

SELECTED HIGH-RADIANCE LED REFERENCES

45. Wittke, J. P., M. Ettenberg, and H. Kressel RCA Rev., 37:159 (1976).
46. Burrus, C. A., and R. W. Dawson Appl. Phys. Lett., 17:97 (1970).
47. Kressel, H., and M. Ettenberg Proc. IEEE, 63:1360 (1975).
48. Ettenberg, M., H. Kressel and J. P. Wittke IEEE J. Quantum Electron., 12:360 (1976)
49. LeBailly, J. in M. Balkanski and P. Lallemand (eds.) "Photonics," Gauthier-Villars, Paris, 1975.

APPLICATION OF SEMICONDUCTOR LASERS

BY STEVE B. GIBSON

65. Applications of semiconductor radiation sources[1]* can be divided into two major categories, *signaling* and *illumination*. The choices include laser diodes, laser diode stacks, and laser diode arrays, as well as noncoherent light-emitting diodes (LEDs) (see Par. **11-63**). As shown in Table 11-12, the *laser source* becomes the choice when high radiance, high peak power, and/or high average radiant power are required in conjunction with narrow projected beam angles.

66. The wavelength selectivity of laser diodes ranges from 0.5 to 34 μm. Gallium aluminum arsenide and gallium arsenide phosphide laser diodes cover the spectral range from

*Superior numbers correspond to numbered references, Par. **11-73**.

TABLE 11-12 Diode-Source Application Selection Chart

Optics	Application	Type of operation	Room temperature devices				Cryogenic laser array
			Emitter	Laser diode	Laser diode stack	Laser array	
Signaling applications							
None	Paper tape reader	Cw	X				
	Card reader	Cw	X				
	Shaft encoder	Cw	X				
	Keyboard	Cw or coded	X				
	Circuit isolator coupler—"dc transformer"	Mod	X				
Fiber optics	Data transmission	Mod	X	X			
	Line finder/edge sensor	Cw or pulse	X	X	X		
	Intrusion alarm	Mod or pulse	X	X	X		
Single-lens	Remote control signaling	Mod		X			
	Voice communications	Pulse	...	X	X		
	Ranging	Pulse	...	X	X	X	
Illumination systems							
Single-lens	Illuminator for gated viewers	Pulse	...	...	...	X	X
	Target designator	Pulse or cw	...	X	X		X

0.8 to 0.89 μm. Gallium arsenide lasers cryogenically cooled to 77 K emit at 0.855 μm. At room temperature these laser diodes emit at 0.905 μm. The shift in spectral output with temperature is approximately 0.25 nm/°C. The best sensors available with spectral sensitivity in this wavelength region are the S-25 photoemissive detector and the silicon and gallium arsenide solid-state photodetectors (see Tables 11-17 and 11-26, Pars. **11-121** and **11-164**, respectively).

67. The electronic and optical-emission characteristics, as shown in Table 11-14, vary widely for operating temperatures of 77 K and room temperature. The average radiant power output and power dissipation must be calculated based on peak power output, duty factor, and power efficiency. In selecting a laser diode source it is of utmost importance to consider the effect of operating temperature on diode performance. Other important parameters shown in Table 11-14 are the width of the diode and the peak power output per unit of junction width. These two parameters are useful in determining the number of lasers in a diode stack or array and operating currents to obtain required radiant power outputs.

68. Designing a semiconductor laser system requires the characterization of four subelements: (1) laser diode source, (2) thermal dissipation element or heat sink, (3) collection and projection optics, and (4) electronic power supply.

69. Laser Diode Source. In signaling applications the peak radiant power output is the major parameter, while average power is more important in illumination systems. In room temperature pulsed applications, where the peak power output exceeds 15 W or the average power is above 15 mW, it is advantageous to use diode stacks or arrays. A diode stack is composed of from two to five lasers placed one upon another on the same heat sink. An array is composed of many laser diodes, ranging[3] from 5 to 1,000. The size of such a 200-laser diode array source using straightforward arraying techniques would be approximately 0.25 by 0.25 in. The average power of this source operating at room temperature with a duty factor of 0.01% is 200 mW. The power dissipation is 3.8 W, based on an external power efficiency of 5%. This heat can be dissipated by an air-cooled radiator or a thermoelectric cooling unit.

70. Thermal Dissipation. Temperature rise at the pn junction causes shift in spectral output, increase in threshold current, and reduction in average radiant power output. An electronic design which minimizes losses due to ohmic contacts and impedances is the first step in reducing the temperature-rise problem. Single diode-laser sources are usually placed on copper heat sinks designed to limit the temperature rise to less than 20°C. Closely packed multiple-diode arrays and stacks present severe thermal problems because of the high thermal flux density. To increase the area of the heat sink, a three-dimensional design in the shape of an inverted V is employed.[4] This design compromises between the effective optical size of the source and the heat sink.

The present approach to improved thermal heat sinking is the use of fiber optics in arraying multiple-diode lasers. The laser diodes are mounted on individual heat sinks, and the radiant output is piped through the fibers into a small emitting area. The large separation distances between stacks of diodes permit much larger heat sinks to reduce the thermal flux density. This technique is applicable to signaling applications where the laser diodes are operated individually and the radiation is piped into a fiber-optic cable. Other cooling systems include forced-air or liquid-cooled radiators, thermoelectric coolers, cryogenic refrigerators, and liquid-nitrogen dewars.

71. Electronic Power Supply for Diode Lasers. A variety of pulse power supplies (Table 11-13) has been developed for driving laser diodes and diode arrays. The major problem areas are impedance matching due to the nonlinear characteristics of the diodes, the high threshold voltages required by series-connected diodes and the protection of the diode from reverse current surges. Each approach requires a trigger circuit, electronic switch, protective circuits, and a primary power source.[4] The switching elements are usually SCRs and transistors, but gas tubes and electromechanical devices are also available.

The limitations on voltage and/or current of the SCR and transistor switching elements require that the laser diodes in a multiple-diode array be electrically connected in series-parallel circuits. As the number of diodes connected in series increases, the operating voltage and ohmic resistance rapidly increase to values causing impedance mismatching problems. As shown in Table 11-13, the transistor switching elements have the advantage of higher efficiency and continuous control over pulse width and pulse repetition rate. A high-efficiency delay-line SCR pulse power supply design has been developed which uses an exponentially varying impedance transformation.[8] This pulse power supply exhibits an overall charge-discharge power efficiency of 80%.

72. Fiber-optics communication[8-13] is potentially the broadest application for semiconductor lasers and LEDs. Research is currently directed toward emitters in the 1.1- to 1.4-μm

TABLE 11-13 Comparative Summary of the Types of Pulse Power Supplies for Laser Diode Sources

Type	Peak current range, A	Peak supply voltage range, V	Pulse-width control	Pulse repetition rate, kHz	Pulse rise time, ns	Reliable operational lifetime, h	Relative size	Relative efficiency, %
								Characteristics
Delay line SCR	100	1,000	Fixed-step adjustable	10	50	500	Moderate	75
Capacitor SCR	100	1,000	Fixed-step adjustable	10	10	500	Moderate	65
Pulse transformer SCR	300	1,000	Fixed	10	40	1,000	Moderate	70
Avalanche transistor	30	100	Continuously variable	10	2	1,000	Small	90
Transistor	60	100	Continuously variable	30	25	1,000	Small	90
Mechanical relay	100	1,000	Fixed	1	10	100	Large	80
Gas tube	1,000	1,000	Fixed	1	10	50	Moderate	75

range, where fibers exhibit the lowest loss and dispersion. Three material systems have been used to produce cw double heterojunction lasers in this region: AlGaAsSb/GaAsSb GaInAsP/InP, and GaInP/GaInP. GaAlAs heterostructure devices, emitting in the 0.8- to 0.85-μm range, are currently the industry's mainstay.

The selection of an emitter, LED, or laser depends upon the required information bandwidth and the fiber length. Key factors to consider are fiber attenuation, pulse broadening, and coupling

TABLE 11-14   Comparison of Laser Performance at Cryogenic and Room Temperatures

	Cryogenic temp, 77 K	Room temp, 27°C
Driving current, A	4	25
Threshold current A	0.7	7
Radiant power output (peak), W	2.5	6
Pulse duration, μs	2	0.2
Duty factor, %	2	0.1
Power efficiency, %	40	4
Driving voltage, V	1.6	9
Wavelength, nm	845	905
Diode width, mils	6	6
Power output per unit width, W/mil	0.4	1

efficiency. The first and last factors determine diode power and radiance. Pulse broadening dictates the broadest acceptable spectral bandwidth, assuming that fiber dispersion is the limiting factor. For long-distance applications the laser is the preferred source because of its narrow spectral bandwidth (0.016–2.0 nm) and high radiance. For short-distance lower data-rate applications the LED may be preferred because of its reduced temperature sensitivity and simpler construction.

73. References on Applications of Semiconductor Lasers

1. Optoelectronic Product Director, *Electro-Opt. Syst. Des.*, May 1971, Vol. 3, No. 5.

2. Vallese, L. M.   Temperature Dependence of Semiconductor Laser Characteristics, *Solid State Technol.*, January 1971, Vol. 14, No. 1.

3. Glicksman, R.   Technology and Design of GaAs Laser and Non-coherent IR Emitting Diodes, Pt. II, *Solid State Technol.*, October 1970, Vol. 13, No. 10.

4. Gibson, S. B., S. E. Smathers and A. J. Repasy   Optics and Pulse Power Supply for Laser Diode Array Sources, *1970 Proc. Electro-Opt. Design Conf.*, Industrial and Scientific Conference Management, Inc., Chicago, 1970.

5. Smathers, S. E.   Computer Designed Optical Integrating Devices and Projection Systems for Semiconductor Laser Arrays, *1971 Proc. Electro-Opt. Design Conf.*, Industrial and Scientific Conference Management, Inc., Chicago, 1971.

6. Millman, J., and H. Taub   "Pulse, Digital, and Switching Waveforms Devices and Circuits for Their Generation and Processing," McGraw-Hill, New York, 1965.

7. Repasy, A. J., and J. L. Peeler   A Microminiature Gallium Aluminum Arsenide Laser Diode Pulser, *1970 Gov. Microcir. Appl. Conf.*, October 1970.

8. Barnoski, M. K. (ed.)   "Fundamentals of Fiber Optic Communications," Academic, New York, 1976.

9. Arnaud, J. A.   "Beam and Fiber Optics," Academic, New York, 1976

10. Marcuse, D.   "Light Transmission Optics," Van Nostrand Reinhold, New York, 1972.

11. Miller, S. E., A. E. J. Marcatili, and T. Li   *Proc. IEEE*, 1973, Vol. 61, p. 1703.

12. Kressel, H. (ed.)   Semiconductor Devices for Optical Communication, *Topics Appl. Phys.*, 1980, Vol. 39.

13. Miller, S. E., and A. G. Chynoweth (eds.)   "Optical Fiber Telecommunications," Academic, New York, 1979.

Electro-Optics and Nonlinear Optics

BY EDWARD J. SHARP

74. Basic Principles. The optical properties of many materials can be altered by a strong electric field, and if the alterable properties can be made to interact with the light propagating through the material, the material serves as a transducer from electric to optical signals. When a material is subjected to electromagnetic radiation, the electric field E associated with the radiation induces a polarization P in the material. If the values of the optical electric field are large, as from a laser source, the induced optical polarization is represented by a series expansion as

$$P = \epsilon_0(\chi^{(1)}E + \chi^{(2)}E^2 + \chi^{(3)}E^3 + \cdots)$$

where the relationship between the polarization and the applied field is most generally of tensor form. Here $\chi^{(1)}$ is the linear optical susceptibility of the medium and ϵ_0 is the permittivity of free space.

The well-known optical phenomena of reflection and refraction arise from the first term of the optical polarization, which is linear in the electric field amplitude. The terms of the induced optical polarization having the nonlinear susceptibilities $\chi^{(2)}$ and $\chi^{(3)}$, which are quadratic or cubic with respect to the electric field amplitudes, give rise to a wide variety of interesting and important optical phenomena.[1-8]*

The linear electro-optic, or Pockels effect,[9-11] dc-optical rectification,[11-13] second harmonic generation,[14,15] and parametric generation[16-20] are important nonlinear optical phenomena that arise from the quadratic polarization term. Third harmonic generation,[21,22] the quadratic electro-optic, or Kerr, effect,[5,23] stimulated Raman,[24,25] Brillouin,[26] and Rayleigh[27] scattering, and two-photon absorption[28,29] arise from the cubic polarization term. Several of these effects are of great importance to the rapidly advancing fields of optical processing and laser applications. In general the effects arising from the cubic term are weaker and until now were somewhat less important from an applications point of view.

Linear Electro-Optic (Pockels) Effect. The linear electro-optic effect occurs only in acentric (lacking a center of inversion) crystals, where the form of the electro-optic tensor is determined by the point group symmetry of the crystal.[30-34] In 20 of the 21 acentric point groups a strong electric field will cause a change in the index refraction which is proportional to the field. This effect affords a convenient and widely used method of controlling the intensity and phase of propagating radiation. Modulation techniques[35] permit a wide variety of devices[36] including spatial,[37,38] temporal,[39] and spectral[40] modulators for use in the rapidly expanding area of optical processing.

Second Harmonic Generation (SHG). SHG was first demonstrated in a quartz crystal[15] when the frequency of a ruby laser, $\lambda = 0.694$ μm, was doubled, producing light at $\lambda = 0.3470$ μm. This nonlinear effect is of great importance since highly efficient lasers are being developed, and the frequency doubling process has been refined to the point where 100% conversion efficiencies have been achieved,[41] in the sense that the converted output at 0.532 μm is equal to the output of a Nd-doped YAG laser at 1.064 μm without the nonlinear crystal. A prerequisite for high-efficiency SHG is phase matching,[42,43] or the condition is imposed that the doubled frequency pass through the nonlinear medium with the same velocity as the fundamental frequency.

DC-Optical Rectification. SHG is intimately related to optical rectification, whereby radiation at a particular frequency beats with itself to produce a dc or low-frequency polarization in the medium.[12] Symmetry requires that the optical-rectification coefficient and the linear electro-optic coefficients be equal, and in fact experiments have been carried out demonstrating this in the material ADP.[13]

Parametric Generation. In genral, parametic interactions involve three or more waves and have given rise to such phenomena as amplification at optical frequencies,[44] both continuous and pulsed optical oscillation, frequency up-conversion and tunable coherent radiation sources. Typically a nonlinear material is used to cause interaction of a high-frequency electromagnetic wave, called the *pump*, with two lower-frequency electromagnetic fields to amplify them. After feedback is provided and the gain exceeds the losses, oscillation occurs and results in emitted radiation at the two lower frequencies, the sum of which equals the pump frequency. These frequencies can be tuned[45,46] by index matching in a number of ways, but normally this is accomplished through temperature control. The parametric up-converter,[47,48] or frequency converter, permits a signal frequency and a strong laser beam to combine in a nonlinear crystal to generate an output beam at the sum frequency. Up-converting an infrared signal[49] to the visible or near infrared permits the use of detectors with better sensitivities.

Third Harmonic Generation (THG). Optical third harmonic generation was first observed in calcite[21] using a pulsed ruby laser, and the output signal corresponded to approximately 10^4 third harmonic photons per laser pulse. Phase-matched THG has been observed in cholesteric liquid crystals[22] and alkali-metal vapors.[23]

The Quadratic Electro-optic (Kerr) Effect. All crystal systems exhibit a quadratic electro-optic effect, and in centric crystals it is the lowest-order effect. For the Kerr effect the induced

---

*Superior numbers correspond to references in Par. **11-75**.

optical birefringence is a quadratic function of the applied electric field, and much higher voltages are required than in the linear electro-optic effect. The Kerr effect is strongest in liquids with anisotropic molecules, e.g., nitrobenzene. Here the electro-optic effect is used for high-speed light modulators and is attributed to molecular reorientation in the presence of the applied field.[5,32,36]

Two-Photon Absorption. Real two-photon absorption[29] was observed in ruby-laser-illuminated CaF_2:Eu^{2+}. The observed intensity of the transmitted blue light was found to vary quadratically with the intensity of the incident ruby radiation. Two-photon excitation has been observed in a vapor of atomic cesium.[30]

Stimulated Raman, Brillouin, and Rayleigh Scattering. Stimulated Raman emission was first achieved by placing a cell of nitrobenzene in a cavity of a ruby giant-pulse laser.[25] The material gain provided by the nitrobenzene resulted in stimulated emission at the Stokes frequency in the near infrared. The phenomenon of stimulated Brillouin scattering is the scattering of an optical beam by an acoustic wave which is produced by the optical beam itself.[27] Stimulated Raman and Brillouin scattering has been observed in liquids,[25,50] gases,[51,52] and solids.[53,27] The stimulated Rayleigh scattering process associated with the alignment of molecules is responsible for line broadening of emission from nonlinear liquids at high laser power levels.[28,54] Stimulated Raman scattering occurs only when the optical field intensity exceeds a threshold value, yet the intensities required by laser beams are significantly lower due to beam self-focusing in the nonlinear medium.[55-59]

Nonlinear optical effects arising from the quadratic and cubic terms of the induced optical polarization can be used in a wide range of new optical devices, yet some of the nonlinearities such as self-focusing, two-photon absorption, and stimulated scattering impose an upper limit on the ability of the material to handle increased laser powers. In addition, very little work has been done to date on electro-optic and nonlinear materials for use in the infrared or millimeter regions of the spectrum.

Although the reference list in Par. **11-75** is long, it is only a representative, clear witness of the intensive efforts made in the last 15 years to exploit the laser in communication and optical processing technology. Many nonlinear and electro-optic materials have been characterized, and the data on the more important materials have been collected and prepared in various tables according to the nonlinear effect exploited. In addition to Refs. 16, 33, and 36, Refs. 50 to 69 contain tables of pertinent electro-optical and nonlinear optical data with appropriate references to the original work. In many cases it is advisable to consult the original source for material data since there is considerable variation in the lists of coefficients in these references, resulting primarily from differences in the quality of materials studied by different workers.

Discussions of the theory concerning electro-optical and nonlinear optical phenomena are found in Refs. 1 to 8, 11, 65, and 70 to 72.

75. References on Electro-optics and Nonlinear Optics

1. Franken, P. A., and J. F. Ward   Rev. Mod. Phys., **35**:23 (1963).
2. Baldwin, G. C.   "An Introduction to Nonlinear Optics," Plenum , New York, 1969.
3. Bloembergen, N.   "Nonlinear Optics," Benjamin, New York, 1965.
4. Ovander, L. N.   Soviet Phys. Usp., **8**:337 (1965).
5. Minck, R. W., R. W. Terhune, and C. C. Wang   Proc. IEEE, **54**:1357 (1966).
6. Zernike, F., and J. E. Midwinter   "Applied Nonlinear Optics," Wiley, New York, 1973.
7. Pershan, P. S.   Nonlinear Optics, Progr. Optics, Vol. 5 (1965).
8. Lines, M. E., and A. M. Glass,   "Principles and Applications of Ferroelectrics and Related Materials," Oxford University Press, London, 1977.
9. Billings, B. H.   J. Opt. Soc. Am., **39**:797, 802 (1949).
10. Carpenter, R. O. B.   J. Opt. Soc. Am., **49**:225 (1950).
11. Armstrong, J. N., N. Bloembergen, J. Ducuing, and P. S. Pershan   Phys. Rev., **127**:1918 (1962).
12. Bass, M., P. A. Franken, J. F. Ward, and G. Weinreich   Phys. Rev. Lett., **9**:446 (1962).
13. Ward, J. F.   Phys. Rev., **143**:569 (1966).
14. Kleinman, D. A.   Optical Harmonic Generation in Nonlinear Media in "Laser Handbook," Vol. 2, North-Holland, Amsterdam, 1972.
15. Franken, P. A., A. E. Hill, C. W. Peters, and G. Weinreich   Phys. Rev. Lett., **7**:118 (1961).
16. Smith, R. G.   Optical Parametric Oscillators, in "Laser Handbook," Vol. 1, North-Holland, Amsterdam, 1972.
17. Knoll, N. M.   Phys. Rev., **127**:1207 (1962).
18. Kinston, R. H.   Proc. IRE, **50**:472 (1962).
19. Akhmanov, S. A., and R. V. Khokhlov,   Sov. Phys. JETP., **16**:252 (1963).
20. Giordmaine, J. A., and R. C. Miller,   Phys. Rev. Lett., **14**:973 (1965).
21. Terhune, R. W., P. D. Maker, and C. M. Savage,   Phys. Rev. Lett., **8**:404 (1962).
22. Shelton, J., and Y. Shen,   Phys. Rev. Lett., **26**:538 (1971).
23. Miles, R. B., and S. E. Harris,   IEEE J. Quantum Electron., **QE-9**:470 (1973).

24. Mayer, G., and F. Gires   *C. R.*, **258**:2039 (1964).

25. Woodbury, E. J., and W. K. Ng   *Proc. IRE*, **50**:2367 (1962).

26. Eckhardt, G., R. W. Hellworth, F. J. McClung, S. E. Schwarz, D. Weiner, and E. J. Woodbury,   *Phys. Rev. Lett.*, **9**:455 (1962).

27. Chiao, R. Y., C. H. Townes, and F. P. Stoicheff   *Phys. Rev. Lett.*, **12**:592 (1964).

28. Bloembergen, N., and P. Lallemand   *Phys. Rev. Lett.*, **16**:81 (1966).

29. Kaiser, W., and C. G. B. Garrett   *Phys. Rev. Lett.*, **7**:229 (1961).

30. Abella, I. D.   *Phys. Rev. Lett.*, **9**:453 (1962).

31. Nye, J. F.   "Physical Properties of Crystals," Oxford University Press, London, 1969.

32. Fowles, G. R.   "Introduction to Modern Optics," 2d ed., Holt, New York, 1975.

33. Yariv, A.   "Optical Electronics," 2d ed., Holt, New York, 1976.

34. Born, M., and E. Wolf   "Principles of Optics," 2d ed., Pergamon, New York, 1964.

35. Denton, R. T.   Modulation Techniques, in "Laser Handbook," Vol. 1, North-Holland, Amsterdam, 1972.

36. Kaminow, I. P., and E. H. Turner   *Appl. Opt.*, **5**:1612 (1966); *Proc. IEEE*, **54**:1374 (1966).

37. Ninomiya, Y.   *IEEE J. Quantum Electron.*, **QE-9**:791 (1973).

38. Meyer, R. A.   *Appl. Opt.*, **11**:613 (1972).

39. Smith, D. L. and D. T. Davis   *IEEE J. Quantum Electron.*, **QE-10**:138 (1974).

40. Pinnow, D. A., R. L. Abrams, J. F. Lotspeich, D. M. Henderson, T. K. Plant, R. R. Stephens, and C. M. Walker,   *Appl. Phys. Lett.*, **34**:391 (1979).

41. Geusic, J., H. Levinstein, S. Singh, R. Smith, and L. Van Uitert   *Appl. Phys. Lett.*, **12**:306 (1968).

42. Maker, P. D., R. W. Terhune, M. Nisenhoff, and C. M. Savage,   *Phys. Rev. Lett.*, **8**:21 (1962).

43. Giordmaine, J. A.   *Phys. Rev. Lett.*, **8**:19 (1962).

44. Wang, C. C., and G. W. Racette,   *Appl. Phys. Lett.*, **6**:169 (1965).

45. Falk, J., and J. E. Murray   *Appl. Phys. Lett.*, **14**:245 (1969).

46. Kreuzer, L. B.   *Appl. Phys. Lett.*, **10**:336 (1967).

47. Midwinter, J. E., and J. Warner   *J. Appl. Phys.*, **38**:519 (1967).

48. Warner, J.   *Appl. Phys. Lett.*, **12**:222 (1968).

49. Evtuhov, V., B. H. Soffer, and D. M. Tseng   *Appl. Opt.*, **11**:2998 (1972).

50. Garmire, E., and C. H. Townes   *Appl. Phys. Lett.*, **5**:84 (1964).

51. Minck, R. W., R. W. Terhune, and W. G. Rado   *Appl. Phys. Lett.*, **3**:181 (1963).

52. Hagenlocker, E. E., and W. G. Rado,   *Appl. Phys. Lett.*, **7**:236 (1965).

53. Eckhardt, G., D. P. Bortfeld, and M. Geller   *Appl. Phys. Lett.*, **3**:137 (1963).

54. Mash, D. I., V. V. Morozov, V. S. Starunov, and I. L. Fabelinskii   *JETP Lett.*, **2**:25 (1965).

55. Chiao, R. Y., E. Garmire, and C. H. Townes   *Phys. Rev. Lett.*, **13**:479 (1964).

56. Hercher, M.   *J. Opt. Soc. Am.*, **54**:563A (1964).

57. Kelley, P. L.   *Phys. Rev. Lett.*, **15**:1005 (1965).

58. Garmire, E., R. Y. Chiao, and C. H. Townes   *Phys. Rev. Lett.*, **16**:347 (1966).

59. Bjorkholm, J. E., and A. Ashkin   *Phys. Rev. Lett.*, **32**:129 (1974).

60. Cook, W. R., Jr., and H. Jaffe,   Magneto-Electro-and Elasto-Optic Constants, in "American Institute of Physics Handbook," McGraw-Hill, New York, 1972.

61. Kurtz, S. K.   Nonlinear Optical Materials, in "Laser Handbook," Vol. 1, North-Holland, Amsterdam, 1972.

62. Wemple, S. N.   Electro-Optic Materials, in "Laser Handbook," Vol. 1, North-Holland, Amsterdam, 1972.

63. Kaminow, I. P., and E. H. Turner,   Linear Electro-Optic Materials, in R. J. Pressley (ed.), "Handbook of Lasers," Chemical Rubber Co., Cleveland, 1970.

64. Lynch, C. T. (ed.)   "Handbook of Materials Science," Vol. III, Chemical Rubber Co., Cleveland, 1975.

65. Yariv, A.   "Quantum Electronics," 2d ed., Wiley, New York, 1975.

66. Singh, S.   Non-linear Optical Materials, in R. J. Pressley (ed.), "Handbook of Lasers," Chemical Rubber Co., Cleveland, 1970.

67. Rez, I. S.   *Soviet Phys. Usp.*, **10**:759 (1968).

68. Byer, R. L.   Parametric Oscillators and Nonlinear Materials, in P. G. Harper and B. S. Wherrett (eds.), "Nonlinear Optics," Academic, New York, 1977.

69. Zernike, F.   Nonlinear Optical Coefficients, in "American Institute of Physics Handbook," McGraw-Hill, New York, 1972.

70. Kelley, P. L. J.   *Phys. Chem. Solids*, **24**:607 (1963).

71. Ward, J. F.   *Rev. Mod. Phys.*, **37**:1 (1965).

72. Cheng, H., and P. B. Miller   *Phys. Rev.*, **134**:A683 (1964).

Phosphor Screens

BY NICHOLAS A. DIAKIDES

76. Phosphor screens are used to convert electron energy into radiant energy in image tubes, cathode-ray tubes, and storage cathode-ray tubes. These screens are composed of a thin layer of luminescent crystals, phosphors, which emit light when bombarded by electrons (cathodoluminescence).

THEORY OF OPERATION AND DEPOSITION METHODS

77. Cathodoluminescence occurs when the energy of the electron beam is transferred to electrons in the phosphor crystal, raising their energy levels. When the electron returns to its initial state (ground state) following the excitation, it releases a quantum of light energy. The property of light emission during excitation is termed *fluorescence*, and that immediately after excitation is removed is *phosphorescence*. The latter emission process is also referred to as *persistence* or *decay characteristic*.

78. Phosphor materials are highly purified inorganic crystals containing traces of other elements which serve as activators and, in combination with the host crystals, promote the phenomenon of luminescence.[1-4]* The activator determines the luminous efficiency of the phosphor and may also affect other phosphor characteristics such as persistence and spectral emission. The most commonly used activators are metals such as copper, silver, manganese, and chromium. Typical characteristics of various standard phosphors which are used in image tubes and cathode-ray tubes are shown in Table 11-15.

79. Standardization of phosphor types has been accomplished by registration of the various phosphors with the Joint Electron Device Engineering Council (JEDEC) of the Electronic Industries Association.[5] Registered phosphors are designated by a number series known as P numbers, such as P1, P2, etc., to P45 (see Table 11-15).

80. Phosphor-Screen Deposition Methods. Phosphor screens are usually fabricated by depositing very small crystal grains on a glass faceplate to form a thin phosphor layer. The most common method used by industry for fabrication of phosphor screens is sedimentation.[6,7] known as *phosphor settling*. This method allows the phosphor crystals to settle from a liquid suspension under the influence of gravity.

81. Electrophoretic Screens. Another method for depositing phosphor screens makes use of the effect of particle motion under the influence of an electric field and is termed *electrophoretic* or *cataphoretic screening*.[8-10] Here the phosphor powder is first fractionated to select the desired particle-size range. The phosphor is then placed in suspension in an electrolytic solution. Under suitable conditions of electrolyte, phosphor suspension, electrode geometry, and electrical uniformity of the substrate, the resulting phosphor deposit is very uniform and compact and has higher resolution characteristics than settled screens.

82. Monolayer Screens.[11,12] The method consists of forming an ultrathin thermoplastic film on a glass faceplate. Phosphor particles are applied on the film surface using a high-velocity airbrush while the thermoplastic is being heated to near its melting point. The thickness of the thermoplastic determines the density of the phosphor layer. To form a monolayer screen, the film thickness must be approximately equal to the average grain size of the phosphor powder. The major advantage of this technique is that it provides means of controlling the screen thickness with good reproducibility.

83. Transparent Screens. For certain applications where optimum image resolution is required, transparent phosphor screens prepared by vacuum evaporation are sometimes employed.[13,14] Activators may be coevaporated or diffused into the host material subsequent to the deposition. This is accomplished usually by heat treatment following the deposition.

Thin-film phosphor screens are also fabricated by the vapor-reaction method.[15] They are converted to the vapor state during processing and react with the gaseous atmosphere to provide film growth of the desired phosphor material.

84. Aluminizing,[16,17] or coating the phosphor-screen surface with a thin (100- to 140-nm) metallic film (usually aluminum), prevents the charging of the phosphor and permits accurate control of the primary electron energy. The reflecting aluminum surface redirects the backward emission of light from the phosphor toward the observer and almost doubles (see Fig. 11-23) the effective light output of the phosphor screen. Another important function of the aluminum backing is protection of the phosphor screen from ion-bombardment damage.

85. Image-Forming Properties. The resolution and brightness output of phosphor screens are two important parameters influencing the image quality of the screen. Resolution is usually assessed by various methods, one of which is the modulation transfer function (MTF). The MTF is defined as the ratio of the modulation at each spatial frequency present in the final image to the modulation of the same frequency in the object scaled to the final image size. The

*Superior numbers correspond to numbered references, Par. **11-92**.

TABLE 11-15   Phosphor Characteristics

Type	Color*		Persistence†	Intended use
	Fluorescent	Phosphorescent		
P1	YG	YG	M	Oscillography; radar
P2	YG	YG	M	Oscillography
P3	YO	YO	M	
P4	W	W	MS	Direct-view TV
P5	B	B	MS	Photographic
P6	W	W	S	
P7	B	Y	MS(B), L(Y)	Radar
P8	Replaced by P7			
P9	Not registered			
P10	Dark-trace screen		VL	Radar
P11	B	B	MS	Photographic
P12	O	O	L	Radar
P13	RO	RO	M	Radar
P14	B	YO	MS(B), M(YO)	Radar
P15	UV	G	UV(VS), G(S)	Flying-spot scanners
P16	UV	UV	VS	Flying-spot scanners; photographic
P17	B	Y	S(B), L(Y)	Oscillography; radar
P18	W	W	M–MS	Projection TV
P19	O	O	L	Radar
P20	YG	YG	M–MS	Storage tubes
P21	RO	RO	M	Radar
P22	W(R, B, G)	W(R, B, G)	MS	Tricolor TV
P23	W	W	MS	Direct-view TV
P24	G	G	S	Flying-spot scanner
P25	O	O	M	Radar
P26	O	O	VL	Radar
P27	RO	RO	M	Color TV monitor
P28	YG	YG	L	Radar
P29	P2 and P25 stripes			Radar; indicators
P30	Canceled			
P31	G	G	MS	Oscillography; bright TV
P32	PB	YG	L	Radar
P33	O	O	VL	Radar
P34	BG	YG	VL	Radar; oscillography
P35	G	B	MS	Oscillography
P36	YG	YG	VS	Flying-spot scanner
P37	B	B	VS	Flying-spot scanner; photographic
P38	O	O	VL	Radar
P39	YG	YG	L	Radar
P40	B	YG	MS(B), L(YG)	Low repetition rate (P12 and P16)
P41	UV	O	VS(UV), L(O)	Radar with light trigger

*Colors: B = blue; P = purple; G = green; O = orange; Y = yellow; R = red; W = white; UV = ultraviolet.

†Persistence to 10% level; VS = <1 μs; S = 1 to 10 μs; MS = 10 μs to 1 ms; M = 1 to 100 ms; L = 100 ms to 1 s; VL > 1 s.

SOURCE: M. I. Skolnik (ed.), "RAdar Handbook," pp. 6–3, McGraw-Hill, New York, 1970.

MTF indicates the ability of the phosphor screen to produce image detail. Figure 11-24 shows a typical MTF performance of high-resolution screens prepared by the various methods discussed previously. The corresponding brightness output characteristics are shown in Fig. 11-25.

PHOSPHOR-SCREEN CHARACTERISTICS

86. Spectral Emission. Phosphors are commercially available with cathodoluminescent emission over the entire visible band, including ultraviolet and near-infrared. Typical absolute spectral characteristics of commercial phosphors are shown in Fig. 11-26.

The visible effectiveness of a phosphor is measured by comparing its spectral-emission curve with a standard visibility function (eye-response curve). This efficiency is usually stated in lumens per radiated watt and is referred to as *lumen equivalent*. It can be calculated[18] using the eye response per wavelength interval ν and the phosphor output per wavelength interval P_λ:

$$\text{Lumen equivalent} = 680 \int_0^\infty \nu P_\lambda \, d\lambda \qquad (11\text{-}1)$$

The constant 680 is the lumen content for 1 W of radiation at 555 nm (the wavelength of maximum eye response).

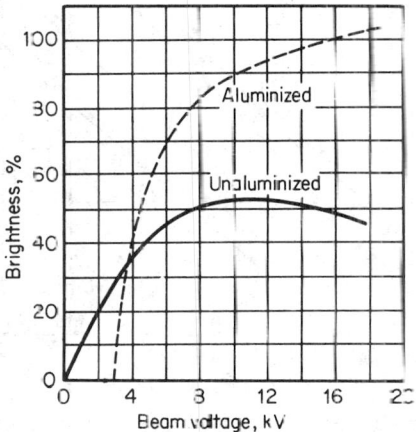

Fig. 11-23. Light output of aluminized and unaluminized phosphor screens. *(From D. G. Fink, "Television Engineering Handbook," McGraw-Hill, New York, 1957.)*

87. Brightness depends on various factors,[19] such as the type of phosphor used, accelerating voltage, electron-beam current, duration of excitation, and screen deposition method and is determined by[20]

$$B = K I (V - V_0) \qquad (11\text{-}2)$$

where B = output brightness (lm), K = constant defined by both the *luminous efficiency* (lm/W radiated energy) and *efficiency factor* of the specific phosphor, I, V = average beam current and voltage, respectively, and V_0 = voltage drop across aluminized screen, which is a function of the thickness of the aluminum film. Phosphor brightness as a function of voltage for a P20 phosphor screen is shown in Fig. 11-27.

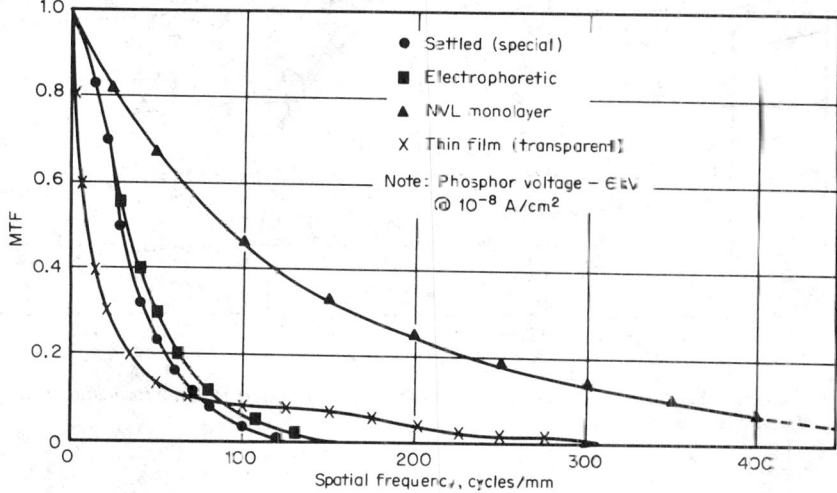

Fig. 11-24. Modulation-transfer-function (MTF) characteristics of high-resolution phosphor screens prepared by various methods.[11]

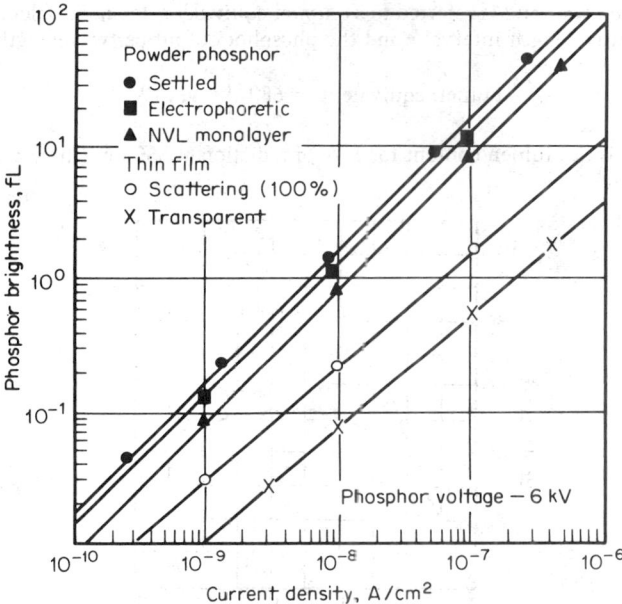

Fig. 11-25. Brightness output of high-resolution phosphor screens prepared by various methods. Brightness is expressed as a function of current density at constant phosphor voltage.[11]

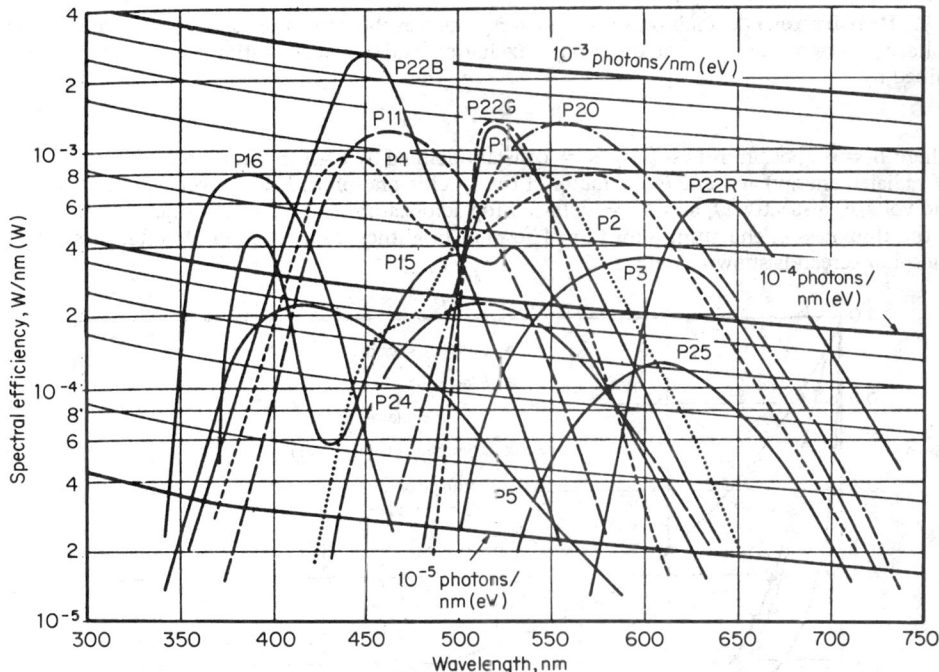

Fig. 11-26. Spectral characteristics of commercial phosphors. *(From "Reference Data for Radio Engineers," 5th ed., Sams, Indianapolis, Ind., 1968.)*

Luminous efficiency (LE) is usually measured with a calibrated eye-corrected photometer which measures the phosphor brightness of a known area on the screen. Both the electron-beam current and voltage are controlled. The luminous efficiency can be computed from

$$LE = (1.076 \times 10^{-3}B)/VI_d \qquad (11\text{-}3)$$

where B = phosphor brightness (fL), I_d = beam-current density (A/cm²), V = electron voltage (V); 1.076×10^{-3} is a conversion factor required to change the brightness measurement from foot-lamberts to lumens per square centimeter.

Conversion efficiency is a measure of the ability of the phosphor screen to convert input electric energy into emitted radiant energy. The conversion-efficiency percentage of the phosphor is simply luminous efficiency [Eq. (11-3)] divided by the lumen equivalent [Eq. (11-1)].

88. Persistence of phosphors is generally characterized by approximately exponential decay of the form $e^{-\alpha t}$ or of the power law t^{-n} or combinations of these forms. The decay characteristic of a phosphor is a function of numerous variables such as anode voltage, current density of the electron beam, duration of excitation, and pulse repetition rate. For this reason, empirical evaluation is required for the selection of phosphor persistence in visual display devices.

Persistence requirements for flicker-reduced displays are shown in Table 11-16. Phosphors are rated according to the persistence of phosphorescence as *short* (<1 s), medium (<2 s), and long (>1 min).

89. Long-Persistence Phosphors. *Cascade* (P7) and *dark-trace* (P10) phosphor screens are used where very long persistence is desired. Cascade phosphors are composed of two layers, the top layer emitting ultraviolet radiation. During operation, the electron beam excites this layer, which in turn excites the second layer (long-persistence photoluminescent phosphor). The dark-trace phosphor consists of a layer of potassium chloride crystals which exhibit the phenomenon of induced absorption bands. In operation, the screen is viewed by white light and shows darkening of the surface under electron bombardment.

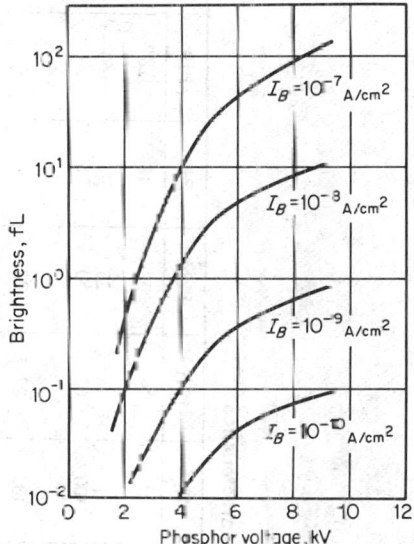

Fig. 11-27. Phosphor brightness as a function of voltage.

90. Rise Time. The rise of luminescence depends on the composition of the luminescent material, manufacturing process, crystal size, impurity content, and method of excitation. The period of luminescence from the beginning of the exciting pulse to the time of the phosphor light output reaches a value of 90% of maximum brightness is defined as rise time.

91. Fast Rise and Decay Phosphors. Phosphors exhibiting rapid rise and decay characteristics are usually approximately 50% less efficient in brightness output than the conventional P20 phosphor. Phosphors of this sort that are available commercially are yttrium aluminate, cerium-activated (P46); yttrium silicate, cerium-activated (P47), and an experimental RCA phosphor (G500). The decay of these phosphors is in the range of 70 to 150 ns at the 10% level, and the conversion efficiency is approximately 5 to 10 lm/W.

92. References on Phosphors

1. Curie, D., and G. F. J. Garlick "Luminescence in Crystals," Wiley, New York, 1963.
2. Leverenz, H. W. "An Introduction to Luminescence of Solids," Dover, New York, 1968.
3. Goldberg, P. "Luminescence of Inorganic Solids," Academic, New York, 1962.
4. Kallman, H. P., and G. M. Spruch "Luminescence of Organic and Inorganic Materials," Wiley, New York, 1962.
5. Optical Characteristics of Cathode-Ray Tube Screens, *JEDEC Electron Tube Council Publ.* 16B, 1971.
6. Sadowski, M. *RCA Rev.*, **95**:112 (1957).
7. Pakswer, S., and P. J. Intiso *J. Electro Chem. Soc.*, **99**:146 (1952).
8. Cerulli, N. F. Method of Electrophoretic Deposition of Luminescent Materials and Product Resulting Therefrom, U.S. Patent 2,851,408, Sept. 9, 1958.
9. Linden, B. R. *Adv. Electron. Electron Phys.*, **16**:311 (1962).

TABLE 11-16 Special Phosphor Characteristics

Phosphor	P1	P7A	P7N	P19	P19	P26	P31
Refresh rate for flicker-free display, Hz	32	29	27	18	28	17	55
Low-level persistence, s	1	15	25	500	160	700	0.6
Burn resistivity relative to P1 with an assigned weight of 100 (raster)	100	62	50	18	50	19	47
Burn resistivity relative to P1 with an assigned weight of 100 (spot)	100	27	22	9	42	12	30
Buildup	1.35	1.63	1.66	2.83	2.00	3.10	1.45
Light output relative to P1 with an assigned weight of 100 (5 kV, 10 mA)	100	58	32	1	24	34	100
Light output relative to P1 with an assigned weight of 100 (10 kV, 6.3 mA)	100	115	44	32	16	21	185
Spot size, mm	0.254	0.433	0.416	0.345	0.330	0.330	0.254
Apparent resolution relative to P1 with an assigned weight of 100	100	74	90	52	100	78	100
Short-time excitation magnitude of required pulse	28	25	30	65	50	60	20
Rise time, ms	20	20	20	120	60	100	0.2
Fall time, ms	28	60	55	230	80	350	1.2

SOURCE: H. R. Luxenberg and R. L. Kuehn, "Display Systems Engineering," p. 264, McGraw-Hill Book Company, New York, 1968.

10. Birks, J. B.   Electrophoretic Deposition of Insulating Materials, *Progr. Dielectrics.* 1:273 (1959).

11. Diakides, N. A.   Phosphors, *Proc. Soc. Photo-Opt. Instrum. Eng.*, 42:83–92 (August 1973)

12. Lehmann, W.   Method of Forming a Uniform Layer of Luminescent Material on a Surface U.S. Patent 2,798,821, July 9, 1957.

13. Feldman, C., and M. O'Hara   *J. Opt. Soc. Am.*, 47:300 (1957).

14. Koller, L. R.   Thin Film Phosphors, *Electrochem. Soc. Meet.*, Washington, D.C., May 13, 1957.

15. Studer, F. J., D. A., Cusano, and A. H. Yound   *J. Opt. Soc. Am.*, 41:559 (1951).

16. McGee, J. D., R. W. Airey, and M. Asian   *Adv. Electron. Electron Phys.*, 22A:571–581 (1966).

17. Sadowsky, M. J.   *J. Soc. Motion Picture Telev. Eng.*, 70:81 (February 1961).

18. Moon, P.   "The Scientific Basis of Illuminating Engineering," Dover, New York, 1961.

19. Poole, H. H.   "Fundamentals of Display Systems," pp. 335–338, Spartan, Washington, 1966.

20. Sherr, S.   "Fundamentals of Display System Design," p. 68, Wiley, New York, 1970.

Cathode-Ray Tubes

BY MUNSEY E. CROST AND IRVING REINGOLD

INTRODUCTION

93. Generation of Radiation. The cathode-ray tube (CRT) produces visible or ultra-violet radiation by bombardment of a thin layer of phosphor material by an energetic beam of electrons. The great preponderance of applications involves the use of a sharply focused electron beam directed time-sequentially toward relevant locations on the phosphor layer by means of externally controlled electrostatic or electromagnetic fields. In addition, the current in the electron beam can be controlled or modulated in response to an externally applied varying electric signal.

DESIGN FEATURES

94. General Principles. The generalized modern cathode-ray tube consists of an electron-beam-forming system, electron-beam deflecting system, phosphor screen and evacuated envelope (see Figs. 11-28 and 29).

The electron beam is formed in the electron gun, where it is modulated and focused, and then travels through the deflection region, where it is directed toward a specific spot or sequence of

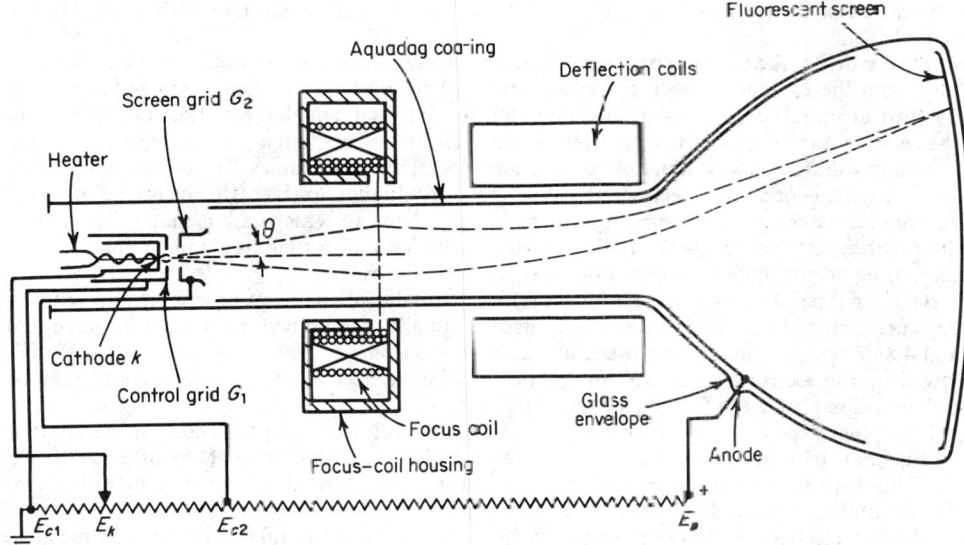

Fig. 11-28. Generalized schematic of a cathode-ray tube with electromagnetic focus and deflection. (*From "Cathode-Ray Tubes," Vol. 22, p. 47, Radiation Laboratory Series. McGraw-Hill, New York, 1948.*)

spots on the phosphor screen. At the phosphor screen the electron beam gives up some of the energy of the electrons in producing light or other radiation, some in generating secondary electrons, and the remainder in producing heat.

95. Electron-Gun Heater-Cathode. Almost all the CRT electron guns now available have indirectly heated cathodes in the form of a small capped nickel sleeve or cylinder with an insulated coiled tungsten heater inserted from the back end. Most heaters now operate at 6.3 V ac with 600 mA current. Since this is wasteful for the emission current required, some tubes with 300 mA current are available. Very-low-power heaters operate with 1.5 V and 140 mA.

96. Modulating Grid. The cathode assembly is mounted on the axis of the modulating or control grid cylinder, or simply grid, which is a metal cup of low-permeability or stainless steel about ½ in in diameter and ⅜ to ½ in long. A small aperture on the order of 10 mils diameter

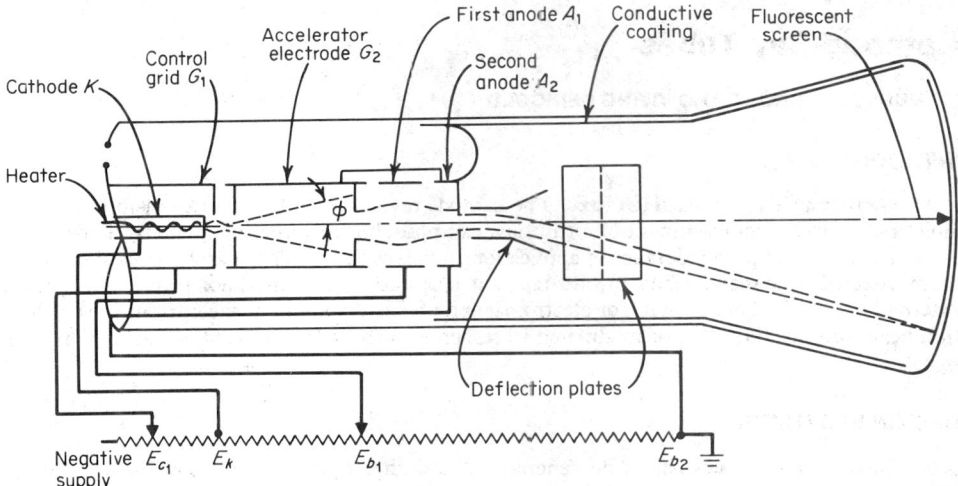

Fig. 11-29. Generalized schematic of a cathode-ray tube with electrostatic focus and deflection. An einzel focusing lens is depicted. *(From "Cathode-Ray Tubes," Radiation Laboratory Series, Vol. 22, p. 47, McGraw-Hill, New York, 1948.)*

is punched or drilled in the cap. The grid G_1 voltage is usually negative with respect to the cathode K.

97. Anodes, Accelerators, and Electrostatic Lenses. To obtain any electron current from the cathode through the grid aperture, there must be another electrode beyond the aperture at a positive potential great enough for its electrostatic field to penetrate the aperture to the cathode surface. Since there are a multitude of electrode-gun designs, this next electrode may have many designs, a wide range of voltages, and several different names. In a simple accelerating lens, in which successive electrodes have progressively higher voltages, this electrode may also be used for focusing the electron beam upon the phosphor, in which case it may be designated the *focusing* (or *first) anode* A_1. This is usually a cylinder, longer than its diameter and probably containing one or more disk apertures.

Another type of gun employs a screen grid G_2, usually in the form of a short cup with an aperture facing the grid aperture. The voltage is usually maintained unadjusted between 200 and 400 V positive. In an electrostatically focused electron gun, the screen grid is usually followed by the focusing anode. In a magnetically focused electron gun the screen grid may be followed directly by the final anode (Fig. 11-28).

In another type of electrostatically focused electron gun in widespread use (Fig. 11-29), the grid is followed immediately by a long apertured cylinder at the voltage of the principal anode A_2. This is called the *accelerator* or *preaccelerator*. It is followed in sequence by either two short cylindrical electrodes or apertured disks.

The last electrode and preaccelerator are connected within the tube. The set of three electrodes constitutes an *einzel lens*. By proper design of the einzel lens, the focal condition can be made to occur when the voltage on the central element is zero or a small positive voltage with reference to the cathode.

98. Electromagnetic Focusing Lenses. The focusing systems previously described use electrostatic electron lenses, but a large and important class of CRTs uses external magnetic components for focusing. The magnetic field imparts no kinetic energy to the electron, since it always acts in a direction perpendicular to the velocity.

The common method of magnetic focusing of CRTs employs a short magnetic lens which operates by means of the radial inhomogeneity of the magnetic field and can have both the object and image points distant from the lens. The typical short magnetic lens or focus coil for a CRT (see Fig. 11-28) consists of a large number of turns of fine wire with a total resistance of several hundred ohms, wound on a bobbin of insulating paper or plastic. The bobbin and coil are almost totally enclosed in a soft-iron shell, except for an annular gap of about ⅜ in at one end of the core tubing.

99. Deflection of the Electron Beam. There are two basic methods of deflection of the electron beam in a CRT, by a transverse electrostatic field and by a transverse electromagnetic field.

100. Electrostatic Deflection. In electrostatic deflection, metallic deflection plates are used in pairs within the neck of the CRT (see Fig. 11-29).

The simplest deflection plates are merely flat rectangular parallel plates facing each other, with the electron beam directed along the central plane between them. The deflection plates are located in the field-free space within the second-anode region, and the plates are essentially at second-anode voltage when no deflection signal is applied. Deflection of the electron beam is accomplished by establishing an electrostatic field between the plates.

The well-made modern electrostatic-deflection CRT does not exhibit excessive deflection defocusing until the beam deflection angle off axis exceeds the neighborhood of about 20°. Most electrostatic-deflection CRTs are used to display electric waveforms as a function of time.

To display the electric waveform it is necessary to generate a sweep representing passage of time and to superimpose on this an orthogonal deflection representing signal amplitude This is most readily accomplished by the use of two pairs of deflection plates. The second pair of deflection plates must have an entrance window large enough to accept the maximum deflection of the beam produced by the first pair. This requires that although the plates may be close enough together at the entrance to afford high deflection sensitivity, they must also have an appreciable width, which results in high capacitance. The plates must also diverge, to accommodate their own deflection of the beam.

To obtain an acceptable deflection sensitivity the plates must be made long, and consequently the capacitance is increased.

Electrostatic deflection CRTs are particularly suited for the display of arbitrary waveforms, as opposed to electromagnetic-deflection CRTs, because the deflection plates generally have capacitances with reference to each other and to all other electrodes of the order of 10 pF or less.

101. Electromagnetic-Deflection Systems. In contrast to electrostatic-deflection systems, the deflection components in electromagnetic-deflection systems are almost universally disposed outside the tube envelope, rather than inside the vacuum. Since the neck of the CRT beyond the electron gun is free of obstructions, a larger-diameter electron beam can be used in the magnetic-deflection CRTs than in the electrostatic-deflection CRTs, which permits a much greater beam current to the phosphor screen and consequently a much brighter picture than if electrostatic deflection were used. In fact, included deflection angles of 110° (55° off axis) are commonly used in television picture tubes without excessive spot defocusing. As is apparent, large deflection angles permit CRTs to be made with shorter bulb sections for any given screen size.

102. The electromagnetic deflection yoke is most suitable for repetitive types of scan, such as raster scans (parallel-line scans sweeping out a rectangular area), or plan-position-indicator (PPI) scans (radial line scan directed outward from the center to cover a circular area). In recent years, however, electromagnetic deflection has been used more and more frequently for random address. The principal problem with random deflection is the inductance of the deflection coils. For any specific field strength an ampere-turns product must be established. Therefore a low inductance implies a high current, which may be difficult to supply, especially with large bandwidth. Normally for each axis the yoke includes two coils, each bent into a saddle shape and extending halfway around the CRT neck.

103. PPI Deflection. For PPI deflection one common arrangement is to have the single-axis yoke rotated physically by an external motor or self-synchronous repeater driven by the radar antenna. In this arrangement, a constant-amplitude triggered linear-sawtooth wave of current is supplied to the yoke. A second common method employs a stationary yoke with two orthogonal deflection axes. One axis receives a current waveform of the linear sawtooth with its amplitude coefficient varying according to the algebraic sine of the antenna rotation angle, and the other axis receives a similar waveform, varying according to the algebraic cosine of the rotation angle.

104. Comparison of Electrostatic (ES) and Electromagnetic (EM) Deflection Systems. ES deflection can display information with extremely large bandwidth, while the bandwidth with EM deflection is limited by yoke inductance and amplifier capabilities. ES deflection is limited by unacceptable deflection defocusing to angles off axis in the neighborhood of 20°, while EM deflection can be used to angles as great as 55° off axis. Greater beam current, and consequently higher brightness of the trace, is usually possible with EM deflection than with ES deflection, because of the larger usable beam diameter.

With a given deflection voltage, in ES deflection the magnitude of deflection is inversely proportional to the anode voltage. In EM deflection, with a given deflection current, the angle of deflection is inversely proportional to the square root of the anode voltage.

105. Drift-Space and Postdeflection Acceleration (PDA). Beyond the deflection region the electron beam may travel in a field-free space until it impinges on the phosphor screen. This condition implies that the anode, or A_2, voltage is the most positive or ultimate voltage applied in the CRT. However, in a large percentage of ES deflection CRTs and in some specialized EM deflection CRTs, additional acceleration voltage is applied to the electron beam beyond the deflection region. This configuration is called postdeflection acceleration (PDA).

106. Postdeflection Accelerator. In ES deflection tubes the postdeflection accelerator electrode (also called the *postaccelerator* or *third anode* A_3) generally takes the form of a wide graphite band around the inside of the envelope funnel just behind the faceplate and connected to the aluminum film if the phosphor is aluminized.

Another type of postdeflection accelerator, the spiral accelerator, can be used with a considerably higher A_3/A_2 voltage ratio without detrimental effects caused by the localized electron lens. This accelerator consists of a high-resistance narrow circumferential spiral stripe of graphite of low screw pitch painted over a substantial length of the inside of the envelope funnel. The ends of the spiral are electrically connected to the A_2 and A_3 terminals. In operation the spiral accelerator requires a small continuous direct current to establish a nearly uniform potential variation from the A_2 to the A_3 voltage. In effect, this constitutes a thick electron lens, and since there are no abrupt changes in potential, the trace-distortion effects are much smaller than in the thin lens.

107. Cathode-Ray-Tube Envelopes. The cathode-ray-tube envelope consists of the faceplate, bulb, funnel, neck, base press, base, faceplate safety panels, shielding, and potting. Not all CRTs will incorporate each of these components, of course.

The *faceplate* is the most critical component of the envelope, since the display on the phosphor must be viewed through it. Most faceplates are now pressed in molds from molten glass and are trimmed and annealed before further processing. Some specialized CRTs for photographic recording or flying-spot scanning use optical-quality glass faceplates sealed to the bulb section in such a way as to produce minimum distortion.

To minimize the return scattering of ambient light from the white phosphor, many CRT types, especially for television applications, use a neutral-gray-tinted faceplate. While the display information will be attenuated as it makes a single pass through this glass, ambient light will be attenuated both going in and coming out, thus squaring the attenuation ratio and increasing contrast.

Certain specialized CRTs have faceplates made wholly or partially of fiber optics, which may have extraordinary characteristics, such as high ultraviolet transmission. A fiber-optic region in the faceplate permits direct-contact exposure of photographic or other sensitive film without the necessity for external lenses or space for optical projection.

108. Bulb Section. The bulb section of the CRT is the transition section necessary to enclose the full deflection volume of the electron beam between the deflection region and the phosphor screen on the faceplate. In most CRTs, this is a roughly cone-shaped molded-glass component.

Instead of an ordinary glass bulb, many of the larger TV-type CRTs have metal cone sections made of a glass-sealing iron alloy. The metal cones are generally lighter in weight than the corresponding glass sections.

109. Funnel Section. The junction region of the bulb of a CRT with the neck section is critical as to geometry, since tubes made with these separate sections are intended for electromagnetic deflection, and this region is just where the deflection yoke is intended to be located.

110. Tube Neck. The neck diameter of a CRT depends to a great extent upon the type of deflection used and the intended application of the CRT. In general, the ES-deflection CRTs have large neck diameters, while the EM types have small diameters.

111. Implosion Protection. As an additional safety factor, many of the larger TV-type CRTs are manufactured with a separate glass *implosion panel*, contoured to match the faceplate curvature, permanently bonded to the outside surface of the faceplate by means of a transparent rubbery silicone adhesive filler. This combination offers protection to the viewer in much the same manner as laminated safety glass in automobile windows. The implosion panel may be etched on its front surface to produce a ground-glass effect to avoid specular reflections from ambient light, and/or it may include neutral-gray attenuation.

CATEGORIES OF CATHODE-RAY TUBES

112. Oscilloscope Tubes. For oscilloscopic applications the general requirements on a CRT include a sharp, bright, rapidly deflectable, single-line trace with a minimum of deflection defocusing or astigmatism. The rapidity of deflection and the fact that arbitrary waveforms must be displayed dictate the use of ES deflection, at least for the vertical direction. For general use, both horizontal and vertical axes employ ES deflection. Since the included deflection angle must be small, usually less than 45°, to preserve good spot size and shape, these CRTs are relatively long compared with the face diameter.

The phosphors generally used for oscilloscope CRTs are P1 (green, medium persistence) or P2 (yellow-green, medium persistence, but with a much longer, low-level "tail" than P1, see Par. **11-87**).

113. Radar Display Tubes. Except for the *A-scope* radar display, which is essentially the same as an oscilloscope display, most radar displays consist of a two-dimensional coordinate display with beam-intensity modulation. Since the coordinate scans are mathematically regular and at preselected rates, EM deflection is generally used, inasmuch as this permits greater deflection angles and consequently shorter tubes to be used for a given face diameter.

Especially in filtered radar displays, it is often necessary to include alphanumeric characters, symbols, and vectors in the display along with the radar information. Shaped-beam tubes, such as the Charactron, or a multiple-beam tube, in which one beam is devoted to the tracing of the characters or symbols and the other to the plan-position-indicator (PPI) display, are used for this purpose.

Long-persistence phosphors are generally used in CRTs for radar displays, since it is desirable to be able to see the radar situation in the entire area covered at any given time.

114. Television Picture Tubes. *Monochrome Tubes.* Since the standards for television transmission in the United States call for 30 frames of two interlaced fields each per second, producing the effect of 60 pictures per second, which is above the flicker fusion frequency for all light levels, there is no stringent limitation on phosphor persistence for monochrome TV picture tubes so long as the persistence does not cause picture smearing. The white luminescence used for most applications is achieved by a mixture of phosphors rather than any single component. Several white-luminescing combinations, all designated P4 have been in common use, namely, the all-silicates, the silicate-sulfide mixture, and the all-sulfides.

Color Tubes. Many types of full-color CRTs have been developed for television use, but the shadow-mask tube is in most widespread use. This type of CRT uses a cluster of three electron guns in a wide neck, one gun for each of the colors red, green, and blue. All the guns are aimed at the same point at the center of the shadow mask, which is an iron-alloy grid with an array of perforations in triangular arrangement, generally spaced 0.025 in between centers for entertainment television. For high-resolution studio-monitor or computer-graphic readout monitor applications, color CRTs with shadow-mask aperture spacing as small as 0.012 in center-to-center are now readily available. This triangular arrangement of electron guns and shadow-mask apertures is known as the *delta-gun configuration.* Phosphor dots on the faceplate just beyond the shadow mask are arranged so that after passing through the perforations, the electron beam from each gun can strike only the dots emitting one color.

Because of the close proximity of the phosphor dots to each other and the strict dependence on angle of penetration of the electrons through the apertures to strike phosphor dots of the desired color, close attention must be paid to shielding the CRT from extraneous ambient magnetic fields and to degaussing of the shield and shadow mask, which is usually carried out automatically when the equipment is switched on or off. All three beams are deflected simultaneously by a single large-diameter deflection yoke, which is usually permanently bonded to the

CRT envelope by the tube manufacturer. The three phosphors together are designated P22, individual phosphors of each color being denoted by the numbers P22R, P22G, and P22B. Most of the present color CRTs are made with rare-earth-element-activated phosphors, because of their superior colors and brightness compared with previously used phosphors.

Two other classes of multicolor CRTs are of current interest, those with parallel-stripe phosphors and those with voltage-penetration phosphors. In the *parallel-stripe class* of CRTs, such as the Japanese *Trinitron*, sets of very fine stripes of red-, green-, and blue-emitting phosphors are deposited in continuous lines repetitively across the faceplate, generally in a vertical orientation. The Trinitron, unlike conventional color CRTs, has a single electron gun which emits three electron beams across a diameter perpendicular to the orientation of the phosphor stripes. This type of gun, also used by some United States CRT manufacturers, is called the *in-line gun*. Each beam is directed to the proper color stripe by means of the internal beam-aiming structure and a slitted *aperture grille*.

The *Lawrence tube*, or *Chromatron*, is another example of the parallel-stripe-phosphor class of color CRT. It employs a single electron beam, and color selection is accomplished solely by control voltages applied between the interdigitated combs of wires constituting the grille itself.

In the voltage-penetration type of phosphor screen, two or three unstructured layers of phosphors emitting different colors are deposited upon each other, sometimes with a nonluminescing, transparent barrier layer between them for better color differentiation. A second important structure consists of individual phosphor grains built up in layers, called *onionskin phosphors*. The core phosphor is generally green-emitting. This is surrounded by a nonemitting layer which in turn is surrounded by an outer red-emitting layer. With both types of phosphor screen, a single electron beam is employed, and the resultant color of the screen is determined by preselected beam-accelerating voltages, which are changed to control the depth of beam penetration into the phosphor layers.

Intermediate colors are produced by the visual combination of the first color, resulting from the penetration of the first layer by the electron beam, with varying intensities of the second color, as more electrons penetrate into the second color-emitting layer. The range of colors producible is thus limited by the hues of the two phosphor layers, which must lie on the same side of the CIE chromaticity diagram and as close to the spectral locus as possible. A major problem associated with voltage-penetration color CRTs is the change in deflection sensitivity and focus of the electron beam as the screen voltage is changed to change the color displayed. This usually dictates operation at only a few preset screen voltages (and therefore colors) where the deflection amplifications and focus voltages are also preset to correspond. However, a type of voltage-penetration CRT has recently been developed in which a constant-potential mesh grid very close to the phosphor screen separates the deflection and focusing space from the color-adjusting space. In this type of CRT a maximum residual deflection error of 1 to 2% can be automatically compensated for by means of a large weak electron lens formed between electrode bands deposited on the inside surface of the bulb.

115. Recording Tubes. Cathode-ray tubes for recording or transcribing information on photographic or otherwise sensitized film are usually of the very-high-resolution (vhr) or ultrahigh resolution (uhr) types. The great majority of these types have nominal faceplate diameters of 4 or 5 in. The spot diameters of the vhr and uhr tubes range from approximately 0.0015 in down to 0.00033 in.

The displayed information is transferred to the recording medium either by an external focusing optical system or by direct contact with a fiber-optic faceplate, requiring no focusing.

116. Computer-Terminal Display Tubes. CRTs for computer display are very similar to tubes used in high-resolution video monitors, but since the display is principally alphanumeric and vector-graphical, the linearity of the beam-modulation characteristics is less important. Well-focused round spots with minimum spot growth or deflection aberrations from the center to the useful edges of the display area are required. High legibility is of primary importance, implying high contrast. White-emitting phosphors are not necessary, so that highly efficient, high-visual-response phosphors emitting in the yellow or green spectral regions are applicable. Most of these CRTs are made with rectangular faceplates.

116a. References on Cathode-Ray Tubes

1. "Cathode-Ray Tube Displays," MIT Radiation Laboratory Series, Vol. 22, McGraw-Hill, New York, 1953.

2. Luxenberg, H. R., and R. L. Kuehn (eds.) "Display Systems Engineering," McGraw-Hill, New York, 1968.

3. Poole, H. "Fundamentals of Display Systems," Spartan, Washington, 1966.

4. IRE Standards on Electron Tubes, Methods of Testing (1962), Pt. 2, Cathode-Ray Tubes, Institute of Electrical and Electronics Engineers, New York, 1962.

5. Special Issue on Information Display, *IEEE Trans. Electron Dev.*, September 1971, Vol. ED 13, No. 9.

6. Crost, Munsey E.   Display Devices and the Human Observer, *Proc. Interlab. Sem. Component Technol.*, Pt. 1, *R&D Tech. Rep.* ECOM-2865, U.S. Army Electronics Command, Fort Monmouth, N.J., August 1967, pp. 365–429.

7. Sherr, S.   "Fundamentals of Display System Design," Wiley, New York, 1970.

Electroluminescent Displays

By DAVID A. BOSSERMAN

117. Early Developments. The possibility of using electroluminescent (EL) displays has been a moot point for a number of years (they were not considered of sufficient importance to warrant inclusion in the first edition of this Handbook).

Destriau[1]* discovered intrinsic electroluminescence (EL) in ZnS among other materials. Sylvania's powdered ZnS EL panel dominated EL work in the 1950s and early to mid-1960s, when most of the projects were cancelled or deferred because of life and/or crosstalk-contrast problems.[2] Attempts were made to correct crosstalk and contrast problems by placing addressable electronic components at each pixel site to serve as turn-on or threshold devices and to provide address-state memory, needed for bright displays where the addressing duty cycle is a factor. These attempts ran into manufacturing difficulties concerning the uniformity and reproducibility of the pixel circuit elements.[3]

118. Thin-Film Display. In 1972, Sigmatron, working on the thin-film EL (TFEL) [also called light-emitting film (LEF) and AC-TFEL], reported an *XY* matrix with inherent threshold and an optional light-absorbing layer in the otherwise transparent thin-film structure which enhanced the display contrast when viewed in bright ambient illumination.[4] Shortly thereafter, Sharp Corporation in Japan achieved a breakthrough with a panel life of over 10,000 h coupled with a continuous ac excitation brightness of over 3,500 cd/m² (over 1,000 fL).[5] Many laboratories now work on applying the TFEL structure to various types of matrix-addressed displays. Sharp has since reported on a memory phenomenon intrinsic to the TFEL structure which can be used to freeze a frame or to increase the brightness of line-at-a-time addressed displays.[6] Rockwell International has reported on a 500-pixel/in TFEL display.[7]

The early successes of TFEL displays are due to the thin-film planar sandwich structure, consisting of electrode-dielectric-phosphor-dielectric-electrode, where at least one electrode is transparent. Since the thickness of this structure is less than 1 nm, high field strengths can be achieved with modest voltages across the structure. This high field strength produces hot electrons through tunneling from relatively deep traps. The hot electrons are fired across the ZnS film, exciting the color centers associated with the dopant,[7] usually Mn. This process is relatively temperature-independent, which translates into a wide operational temperature range.[8] The details of the tunneling and subsequent retrapping of activator electrons are being studied in various laboratories to determine the best structure for enhancing the various desirable operational parameters such as threshold, brightness, memory, and uniformity.

Much of the work on TFEL has been funded by the U.S. Department of Defense and is not yet available commercially. The low-cost production processes inherent in the planar TFEL structure make them eminently suitable for the commercial and consumer markets. However, a prerequisite is the development of high-voltage, monolithic drive chips and inexpensive and reliable interconnect techniques.[8] In view of the very rapid advances in this field, it would be advisable to consult the most recent issues of *SID Digest* and *SID Proceedings* for state-of-the-art information.

119. References on Electroluminescent Displays

1. Destriau, G.   *J. Chem. Phys.*, **33** 620 (1936).
2. Mito, S.   1977 *SID Int. Symp. Dig.*, p. 86.
3. Tannas, L., Jr.   *Proc. SID*, **19**(4):193 (1978).
4. Ketchpel, R.   1972 *SID Int. Symp. Dig.*, p. 166.
5. Inguchi, T., et al.   1974 *SID Int. Symp. Dig.*, p. 84.
6. Suzuki, C. et al.   1976 *SID Int. Symp. Dig.*, p. 50.
7. Ketchpel, R.   *Proc. SID*, **19**(3):97 (1978).
8. Allan, R.   *Electronics*, Mar. 13, 1980, p. 127.

*Superior numbers correspond to references in Par. **11-119**.

GENERAL REFERENCES

9. *SID Digest of Technical Papers*, 1972, 1974, 1976–1978.
10. *Proceedings of the SID*, 1976–1978.

Photoemissive Electron Tubes, Image Converters, and Intensifiers

BY CLARENCE A. JOHNSON

PHOTOEMISSIVE ELECTRON TUBES

120. General Characteristics. In their various configurations, photomissive electron tubes are important devices for sensing, detecting, imaging, processing, amplifying, and displaying photon information. Devices are available with spectral sensitivity in the far-ultraviolet, visible, or near-infrared regions. They have linear characteristics over wide signal-input ranges. Their good signal-to-noise characteristics make them very useful in low-photon-level detection.

121. Photoelectron and Secondary Electron Emitters. The front element of the photoemissive electron tube is a photocathode which converts photons into electrons. Photons are absorbed in the photocathode material and excite mobile photoelectrons which diffuse toward the photocathode-vacuum interface and escapes into the vacuum if the work function of the surface is low enough. Once in the vacuum, the photoelectrons can be collected, amplified, and/or displayed.

Photocathodes for practical applications are formed by evaporation, oxidation of metals, and recently by epitaxial growth techniques. The lack of sufficient control over evaporation and oxidation processes, along with incomplete understanding of the physics of these photocathodes, has led to various proprietary recipes for their preparation. Solid-state theory, ultrahigh vacuum techniques, and good quality semiconductor materials have led to the development of highly sensitive negative-electron-affinity photoemitters (Fig. 11-30). Operationally stable GaAs photocathodes with average sensitivities of 1200 μA/lm and up to 2000 μA/lm are now routinely utilized in photomultiplier and image-intensifier tubes.

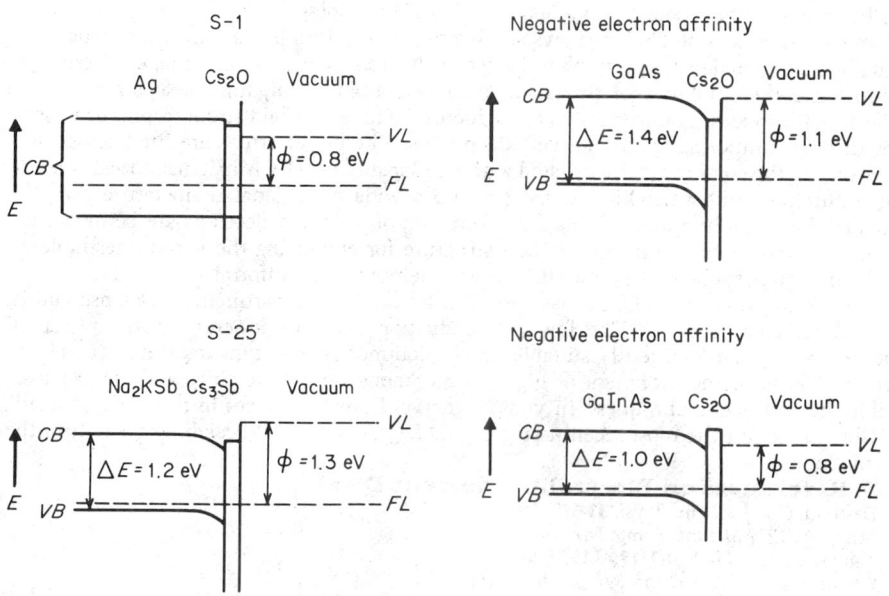

Fig. 11-30. Simplified energy-level diagrams for the photocathode: *(top left)* S-1; *(bottom left)* S-25; *(top right)* GaAs cathode (negative-electron-affinity case); *(bottom right)* GaInAs cathode with optimized 1.06-nm photoresponse (negative-electron-affinity case); E = energy; CB = conduction band; VB = valance band; VL = vacuum level; FL = Fermi level. $R(eV) = 1.2397\ \lambda^{-1}$ (nm), electron affinity $EA = CB$ in the bulk minus VL.

Photocathodes can operate either in a transmission or reflection mode. For the transmission mode a transparent substrate is required because free-standing films of the required thickness are not obtainable, except for the silicon photocathode. The substrate can be an insulator like glass, sapphire, LiF, or a wide-bandgap semiconductor like GaP, AlGaAs, or GaAs.

For reviews and comprehensive literature, see Refs. 1 to 7 Par. **11-129**. The limitation in the long-wavelength threshold is presently around 1.3 μm. In the future, longer-wavelength thresholds will be obtainable from tunnel emitters and heterojunction emitters.[15]* Typical photocathodes and their properties are listed in Table 11-17. Their cathode sensitivity is given in microamperes per lumen, measured with a standardized tungsten light source of 2854 K color temperature. As the evaluation with this light source tends to favor near-infrared-sensitive photocathodes, the quantum efficiencies at a specified wavelength are also given. Complete spectral-yield data for several selected photocathodes are shown in Fig. 11-31.

122. Secondary Emission.[8-14] Every good photoemitter is a potentially good secondary-electron emitter. Secondary-electron emitters are operated in reflection mode, except for porous potassium chloride layers and silicon films. The processes involved are the same except for the excitation. For a compilation of secondary emitters, see Table 11-18. The values are for normal incidence of the primary-electron beam. For a good signal-to-noise ratio the multiplication factor of the first dynode of the phototube must be as high as possible. This requires the use of GaP + Cs_2O or Si + Cs_2O as dynode material. Time lag is a problem in the zero- and negative-electron-affinity secondary emitter because of the long travel time of the excited electrons in the emitter. This effect can be offset by the use of fewer dynodes for a given amplification.

123. Phototubes and Photomultipliers. Phototubes and photomultipliers are glass, ceramic, or metal vacuum bottles with a transmission or reflection photocathode in end-on or

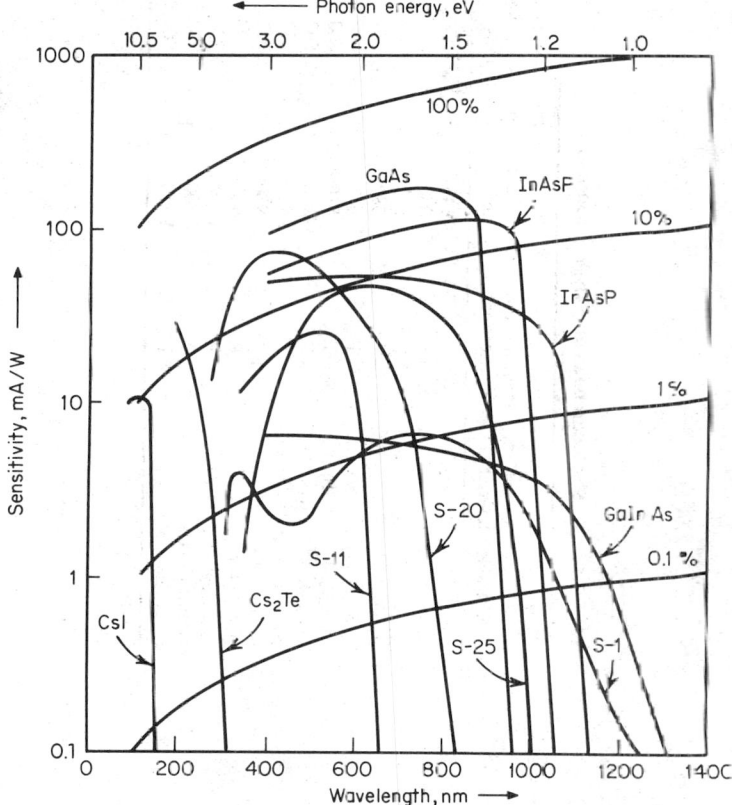

Fig. 11-31. Spectral response curves for various photocathodes. Sensitivity to incident light in milliamperes per watt is plotted on the ordinate. The curved lines with percentage numbers are quantum efficiencies.

*Superior numbers correspond to numbered references, Par. **11-129**.

TABLE 11-17 Photocathodes and Their Properties

Cathode Material	Operation (transmission or reflection)	Threshold, nm	Max. white-light sensitivity, μA/lm	Quantum efficiency %	at λ, nm	Dark current, A/cm²
CsI	T	145	.....	11	120	10^{-16}–10^{-17}
Cs₂Te	T	330	.....	8	250	10^{-16}–10^{-17}
K₃Sb	T	550	60	24	380	10^{-16}–10^{-17}
Rb₃Sb	T	580	25	8	380	10^{-14}–10^{-15}
Cs₃Sb (S-11)	T	670	60	15	440	10^{-15}
Na₂KSb + Cs (S-20)	T	870	300	20	400	10^{-15}
Na₂KSb + Cs₃Sb (S-25)	T	950	560	9	550	10^{-10}–10^{-13}
Ag + Cs₂O (S-1)	T	1,150	80	0.1	1,060	10^{-10}–10^{-13}
GaAs + Cs₂O	R, T	930	2,062	35	700	10^{-16}
InAsP + Cs₂O	R	960	1,200	12	940	10^{-15}
GaInAsP + Cs₂O	R	1,100	1,200	7	1,060	10^{-14}
GaInAs + Cs₂O	R	1,300	150	0.3	1,060	10^{-12}
Si + Cs₂O	R	1,150	500	0.5	1,060	10^{-9}

side-window configuration and an anode (in the case of the phototube) or discrete or continuous dynodes in front of an anode (in the case of photomultipliers). Discrete dynode structures, listed in order of improving time response, include venetian blind, box and grid, circular cage, and linear multiplier (Fig. 11-32). Collection-efficiency considerations determine location and shape of the photocathode-to-first-dynode region. The typical photocathode diameter for phototubes

TABLE 11-18 Secondary Emitters

Material	Secondary-emission yield δ_{max}	Primary energy E_{mp} at δ_{max}, V
Pt	1.8	700
KCl	7.5	1,200
BeO	12	700
MgO	24	1,200
GaP + Cs$_2$O	240	9,000
Si + Cs$_2$O	950	20,000

and photomultipliers is ¼ to 2 in. The number of dynodes in photomultipliers varies between 5 and 16. A typical current-amplification value is 1×10^6. The cc supply voltage lies between 1 and 3 kV, with maximum values of 300 V between cathode and first dynode and 200 V between consecutive dynodes. Ruggedized versions resistant to shock, vibration, acceleration, and high temperature are available.

124. Microchannel electron multipliers are continuous dynodes consisting of glass with a high lead oxide content, made slightly conductive through a hydrogen firing process. The glass tube has electrodes at the entrance and exit. The entrance can be conical or straight, while the main section is straight, bent, or spiraled. The output current has to be a tenth or less of the strip current; otherwise the multiplier operates in a saturated mode. The amplification depends on the length-to-diameter ratio of the multiplier, the axial field strength, and the secondary-electron-emitter material.

IMAGE CONVERTERS AND INTENSIFIERS

125. Principle of Operation. An image tube converts an image in one spectral region directly into an image in another spectral band, usually with an increase in intensity. If the primary purpose is to convert the spectral region of the image (near infrared or ultraviolet to

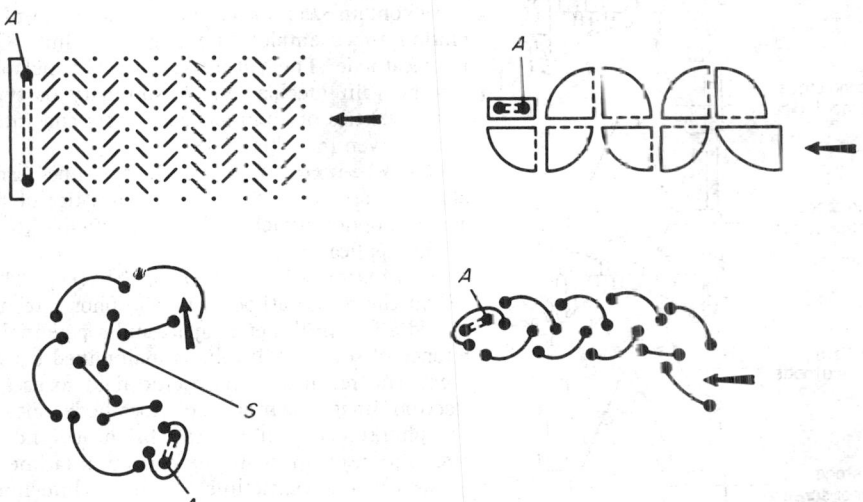

Fig. 11-32. Multiplier structures. Venetian blind (upper left), box and grid (upper right), circular cage (lower left), and linear (lower right). The arrows show the direction of the incident electrons. A = anode; S = shield.

visible), the tube is called an *image-converter tube* (Fig. 11-33). If the primary purpose is to intensify the image without regard to the spectral conversion, it is called an *image-intensifier tube* (Fig. 11-34). Normally, a tube designed to do both is categorized as an image-intensifier tube.

Direct-view image tubes like that in Fig. 11-33 use a photoemissive input surface (Par. **11-121**) to form an electron image from 4 to 20 kV to accelerate and focus the image, and a phosphor

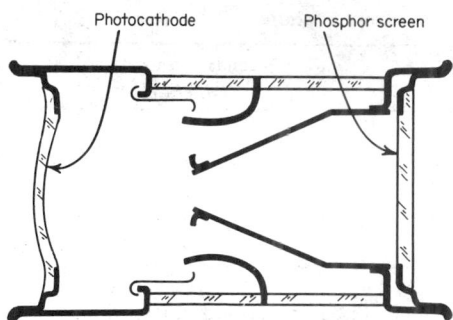

Photocathode   Phosphor screen

Fig. 11-33. Image-converter tube.

screen (Par. **11-76**) to display the image, all contained in a single vacuum envelope.[16] The exception to this arrangement is the solid-state image converter, which has no vacuum envelope (Fig. 11-35). The images formed by these devices are two-dimensional and are converted or amplified at all points simultaneously rather than being scanned as in a television system.

Image tubes are manufactured in various configurations, differing in size, photocathode, electron focusing, gain mechanism, phosphor screen, magnification, etc. By far, the largest number

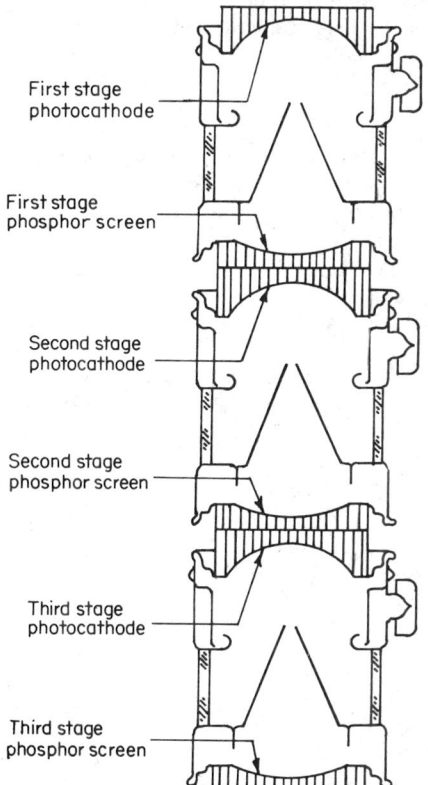

First stage photocathode

First stage phosphor screen

Second stage photocathode

Second stage phosphor screen

Third stage photocathode

Third stage phosphor screen

Fig. 11-34. Typical three-stage electrostatic-focus image intensifier.

of image tubes have been fabricated for night-vision applications. They are generally labeled zero, first, second, and third generation tubes. Zero generation tubes are simple one-stage image converters (Fig. 11-33). First generation tubes (Figs. 11-34 and 11-43) and second generation tubes (Figs. 11-37 and 11-42) achieve significantly higher gain by cascading two or more single converter stages (first generation) or through the use of a microchannel plate (second generation). For night-vision applications both tube types contain S25 photocathodes, whereas third generation tubes employ the negative-affinity GaAs photocathode. They also use microchannel plates for the gain mechanism. A summary of typical characteristics of common converters and intensifiers is given in Table 11-19.

126. Electron Optics. Image tubes can be classified according to the characteristics of their electron optics, which make them suitable for particular applications.

(a) *Proximity-Focused Tubes*[16,20] (Fig. 11-36). When the photocathode and phosphor screen are spaced a few millimeters apart and a potential difference of several kilovolts is maintained between them, the resulting electrostatic field focuses the electron image from the photocathode onto the phosphor screen with increased brightness and with a resolution of up to 40 line pairs/mm (a line pair consists of one black line and one white line of equal width).

A more useful version of this tube type is the microchannel wafer intensifier (Fig. 11-37), which uses a microchannel array as the electron multi-

TABLE 11-19   Typical Characteristics of the Most Common Image Converters and Intensifiers

Tube type*	Cathode Type	Diameter, mm	Center magnification	Focus†	Center resolution, line pairs per mm	Gain‡	Equivalent background input, lm/cm²
Zero generation	S1	25	0.76	El	50	15 CI§	2.5×10^{-7}
	S25	18	0.97	El	64	65	2×10^{-11}
	S25	25	0.92	El	63	63	2×10^{-11}
	S25	40	0.95	El	57	65	2×10^{-11}
	S25	40	1.00	Mag	70	75	5×10^{-12}
First generation	S25	25	0.82	El	32	35K	2×10^{-11}
	S25	40	0.82	El	32	35K	2×10^{-11}
	S25	40	1.00	Mag	45	35K	5×10^{-12}
Second generation	S25	18	1.00	Prox in Prox out	32	10K	2×10^{-11}
	S25	25	0.97	El in Prox out	32	25K	2×10^{-11}
Third generation	GaAs	18	1.00	Prox in	32	20K	2×10^{-11}
	GaAs	25	1.00	Prox out	32	20K	2×10^{-11}

*U.S. Tube Manufacturers: ITT, Roanoke, VA; Ft. Wayne, IN; Litton, Tempe, AZ; Varian, Palo Alto, CA; Ni-Tec, Niles, IL; RCA, Lancaster PA; Varo, Garland, TX.

†El = electrostatic; Mag = magnetic; Prox in = proximity focus from cathode to MCP; Prox out = proximity focus from MCP to phosphor screen.

‡Unless otherwise noted, gain listed is luminous gain. The gains listed for first, second, and third generation tubes are the gains generally used in night-vision devices. The maximum gain for these tubes is significantly higher.

§CI = conversion index — the ratio of the output flux from the phosphor screen in lumens to the input flux of 1 lumen from a 2854° C temperature source modified by a Corning 2540 infrared filter with a filter factor of 10.8%.

plier. These tubes are used in devices such as night-vision goggles, hand-held viewing devices, and rifle scopes.

Since these tubes do not invert the images formed, fiber-optic twists or inverters (fiber bundles twisted 180° to invert the images) are used where an inverted image is necessary. Typical performance characteristics of this tube type are listed in Table 11-19.

(b) *Electrostatic-Focus Tubes*   See Fig. 11-34.

1. *Infrared image converter.* The single-stage electrostatic-focus infrared converter tubes (Fig. 11-33) are usually diode fixed-focus types with curved-glass photocathode and flat-glass phosphor-screen faceplates.[30] However, there are also triode tubes which require a focusing voltage. They are made in photocathode sizes from 18 to 25 mm, and their magnification ranges from 0.50 to 0.75. These tubes have mainly been used in military-weapon sights, driving binoculars, and armored-vehicle night-vision systems, as well as for photographic-film inspection and in scientific apparatus such as ultraviolet and infrared microscopes. Figure 11-38 and Table 11-19 contain typical characteristics of this tube type.

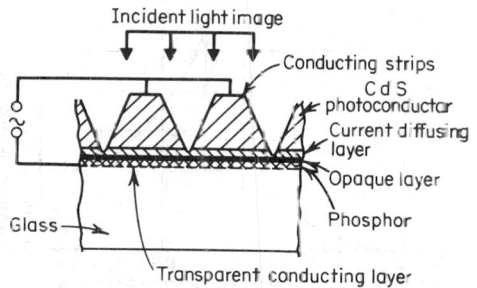

Fig. 11-35. Solid-state image converter.

2. *Image intensifiers.*[18,19] The preponderance of the imaging tubes in use today fall into this category. Most use modular construction and have fiber-optic faceplates on input and/or output ends (Fig. 11-34). The electron lenses in the electrostatic-focus types invert the image, and both the object plane and image plane of electrostatic-focus electron optics are curved.

Fiber optics often are used in electrostatic-focus tubes. They characteristically have magnifications on the order of 0.75 to 1.00, but there are versions available with variable magnification (zoom) (Fig. 11-39). In addition, there are versions made with fixed magnifications other than unity, i.e., demagnification down to 0.3X and magnification up to 10X. Included in this grouping are the multistage tubes[19] (Fig. 11-34), which are cascaded via fiber optics for increasing useful gain.

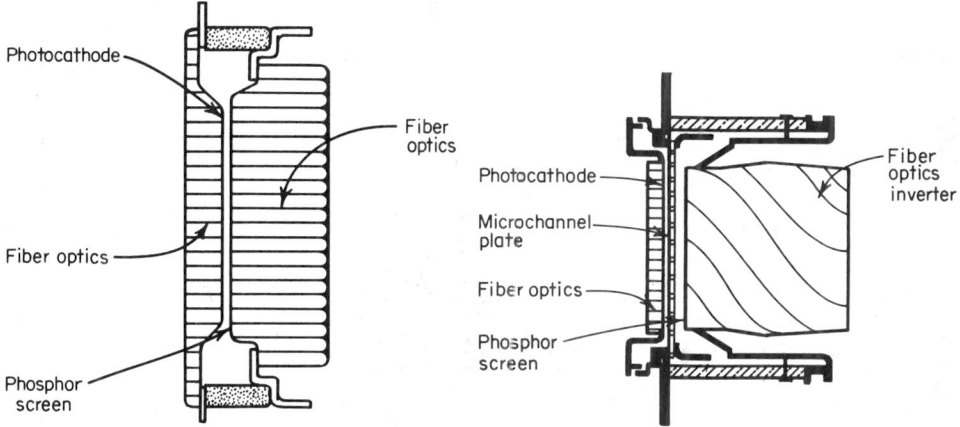

Fig. 11-36. Wafer-type proximity-focus tube with fiber-optics output.

Fig. 11-37. Wafer-type microchannel tube with 180° fiber-optics twist for image inversion.

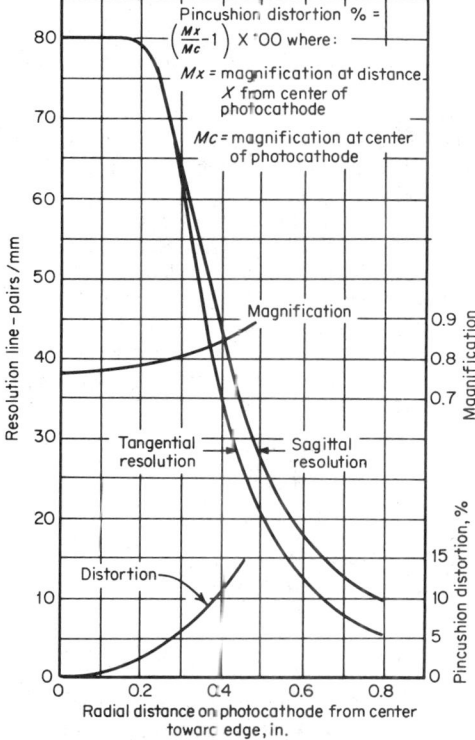

Fig. 11-38. Resolution, magnification, and distortion curves for a typical image-converter tube.

Some versions have automatic brightness control (ABC) (see Figs. 11-40 and 11-41 for typical performance characteristics of regular and ABC-type multistage intensifiers), which extends their useful range to light levels two orders of magnitude higher than conventional versions. Typical characteristics of this tube type are listed in Table 11-19.

3. *The electrostatic-focus microchannel intensifier* (Fig. 11-42) uses electrostatic input focusing and proximity output focusing. This tube, like the proximity-focus wafer tube, uses a microchannel array (to be described) as an amplifier. It is used in military night sights because of its compact size, high resolution, and high gain.

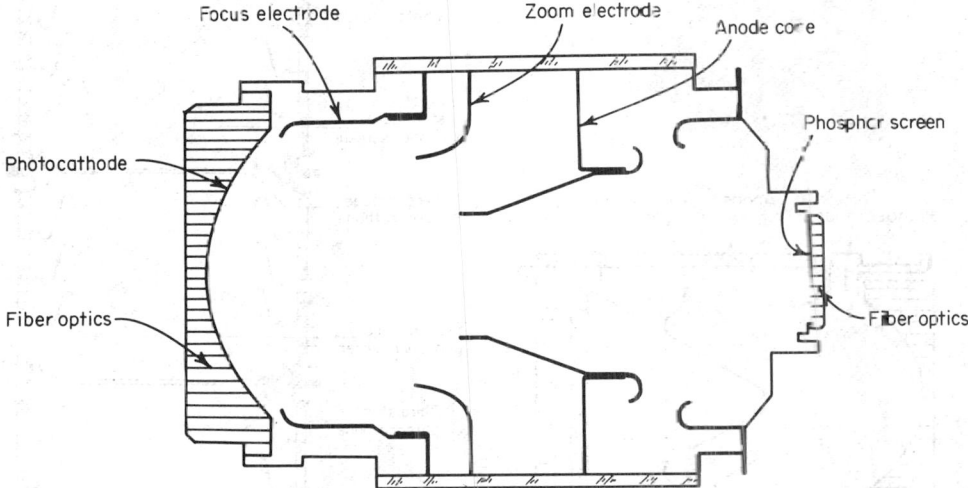

Fig. 11-39. Electrostatic-focus variable-magnification tube.

(c) *Magnetic-Focus Image Intensifier.*[16,20] Magnetically focused intensifiers (Fig. 11-43) are furnished as two-, three-, or four-stage tubes. These multistage devices are usually built into one vacuum envelope[21] and are focused with toroidal permanent magnets or electromagnets. These tubes are characterized by their high resolution, lack of distortion, low background, and high gain in the three- and four-stage types. The bulk and weight of their focusing magnets limit the portability of systems using this tube type, but it has found great favor for use on astronomical telescopes, where its low background (EBI, 5×10^{-12} cm^{-2}), high resolution (45 line pairs/mm for three stages), and absence of distortion make possible the long exposures necessary in astronomical photography.[17] Typical characteristics are listed in Table 11-19.

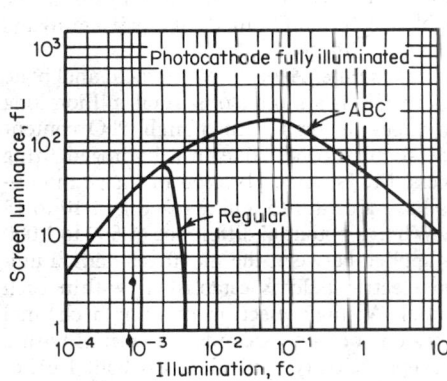

Fig. 11-40. Screen luminance vs. incident illumination for a three-stage assembly with and without automatic brightness control.

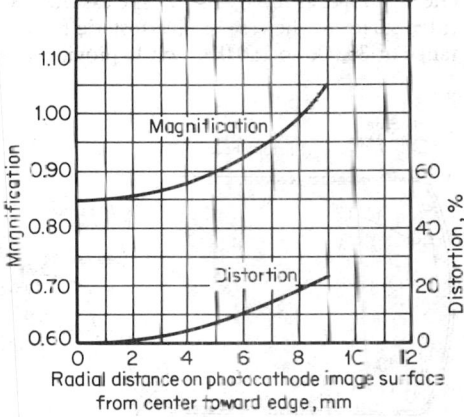

Fig. 11-41. Typical magnification and distortion characteristics of a three-stage assembly.

127. Gain Mechanism. *Electron Acceleration.* Electrons emitted from a photocathode with an energy of only a few electronvolts are given tremendous acceleration by an electrostatic field of several kilovolts. These electrons then strike the phosphor screen, which emits visible light; i.e., substantial gain is available.

Demagnification. The image brightness in an electrostatic-focus tube varies inversely as the square of the magnification; the lower the magnification, the higher the brightness. The electron

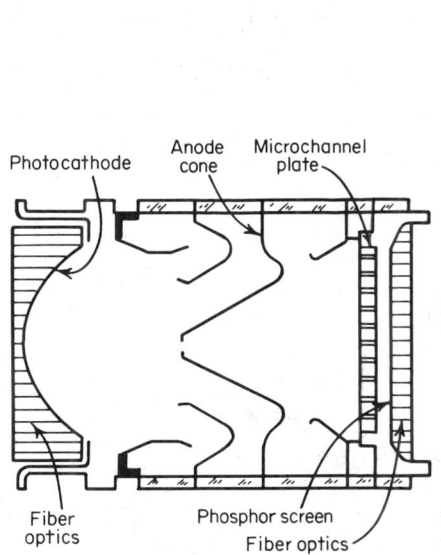

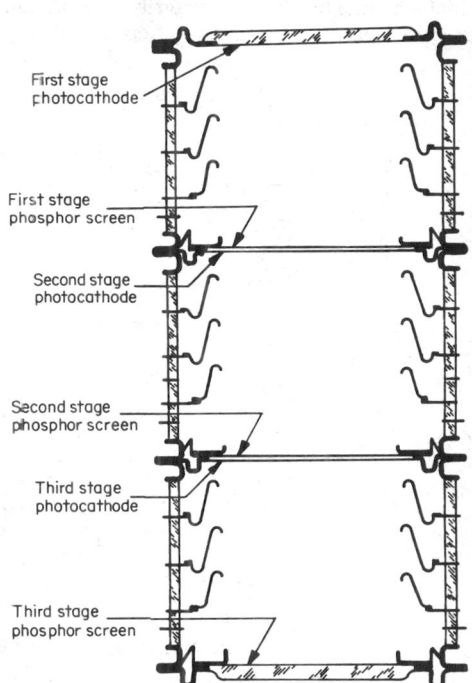

Fig. 11-42. Typical electrostatic-focus microchannel inverter tube.

Fig. 11-43. Three-stage magnetic-focus image intensifier.

acceleration in combination with demagnification concentrates the electron flux at the screen and thus increases the gain. The x-ray image-intensifier tube (Fig. 11-44) falls in this category.

Cascading. When several image-tube modules are placed in tandem or optically coupled to increase the gain, the method is called cascading. The cascading is accomplished by coupling several modules having fiber-optic faceplates, as in Fig. 11-34. Several stages can also be built into one envelope, as in Fig. 11-43. The cascaded assembly is called an *image intensifier* and contains three stages in the case of electrostatic focus, because three stages yield the desired gain (in the range of 30,000 to 150,000) while providing reasonable resolution (25 to 40 line pairs/mm) and an inverted image.

Microchannel Array. A microchannel plate array (MCP) is a mosaic of several million long thin tubes of glass with a high PbO content made slightly conductive in a hydrogen firing process. MCPs currently used in image intensifiers have a center-to-center spacing of 10 to 15 μm. When a potential difference of 500 to 1,000 V is applied between the metalized ends, a uniform electric field is established within each channel. A single electron entering a channel creates one or more secondary electrons with a transverse velocity equivalent to about 1 eV of energy. This transverse velocity carries it across the channel, while the electric field accelerates it along the channel. A sufficient amount of

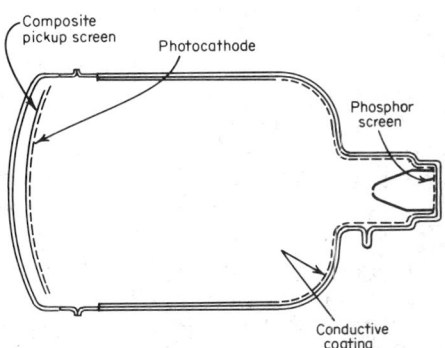

Fig. 11-44. X-ray image-intensifier tube.

energy is imparted to the typical electron so that, on the average it will generate more than one secondary upon collision with the opposite wall. Thus a cascading action is started which can produce electron gains in excess of 100,000. The microchannel array is used as the gain mechanism for the wafer tube (Fig. 11-37) and the inverter-type intensifier (Fig. 11-41).

Transmission Secondary-Electron Multiplication (TSEM). When electrons are accelerated into a thin foil at high energy and under proper conditions of film thickness and electron energy,

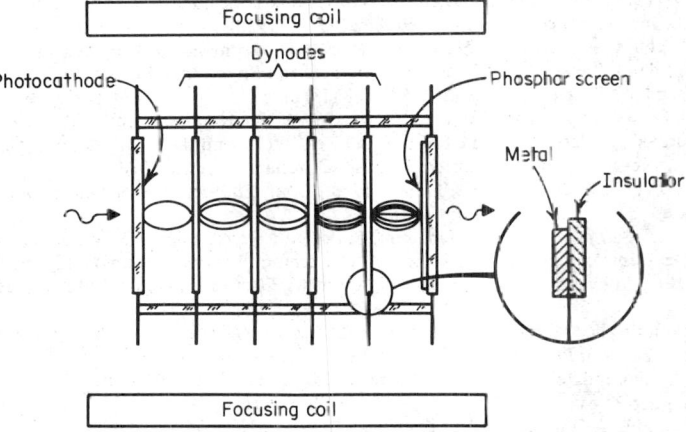

Fig. 11-45. Transmission secondary-electron-multiplication tube.

secondary electrons can be collected on the opposite, or transmission side. Figure 11-45 illustrates a tube using this principle, with several such films or dynodes.

128. Special Tube Types. *X-Ray Image Intensifiers* (Fig. 11-44). The composite target (composed of an x-ray phosphor deposited on an aluminum supporting substrate) of the x-ray image intensifier converts an x-ray image into a visible image which is intensified by demagnification and displayed on a phosphor screen for direct viewing. X-ray image intensifiers can also be coupled to television camera tubes by using fiber optics for remote viewing. X-ray image intensifiers are used mainly for medical diagnosis and structural inspection.

Image Magnifier (Fig. 11-46). The image magnifier is an electrostatic-focus image converter with a magnification usually of 5X and 10X. The gain in such tubes is low because, as previ-

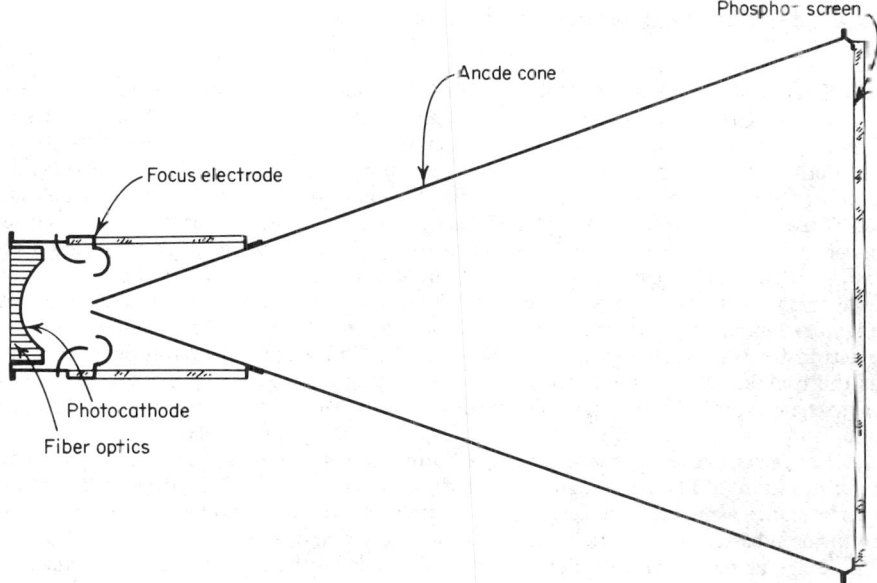

Fig. 11-46. Image-magnifier tube.

ously explained, it varies inversely as the square of the magnification; thus there is lower gain in the image magnifier. Magnifier tubes have typical gains of 6 to 8 for the 5× type and 2 to 3 for the 10× type. To increase the gain in magnifier tubes, additional unity-magnification stages are coupled to the input by means of fiber optics.

129. References on Photoemissive Tubes, Converters, and Intensifiers

1. Goerlich, P.   Recent Advances in Photoemission, *Adv. Electron. Electron Phys.*, **11**:1 (1959); and M. Berndt and P. Gorlich, *Phys. Stat. Sol.*, **3**:963 (1963).
2. Sommer, A. H.   "Photoemissive Materials," Wiley, New York, 1968.
3. Van Laar, J., and J. J. Scheer   *Philips Tech. Rev.*, **29**:54 (1970).
4. Schagen, P., and A. A. Turnbull   New Approaches to Photoemission at Long Wavelength, and C. H. A. Syms, GaAs Thin Film Photocathodes, *Adv. Electron. Electron Phys.*, **28A**:93, 399 (1969).
5. Bell, R. L., and W. E. Spicer   *Proc. IEEE*, **58**:1788 (1970).
6. Rome, M.   Photoemissive Cathodes, I, and C. H. A. Syms, Photoemissive Cathodes, II, in L. M. Biberman and S. Nudelman (eds.), "Photoelectronic Imaging Devices," Vol. 1, pp. 147, 161, Plenum, New York, 1971.
7. Bell, R. L.   "Negative Electron Affinity Devices," Clarendon, Oxford, 1973.
8. Martinelli, R. U.   *Appl. Phys. Let.*, **17**:313 (1970), *Conf. Photoelectric Secondary Electron Emission*, Minneapolis, Aug. 18–19, 1971, Pap. D1.
9. Dekker, A. J.   Secondary Electron Emission, *Solid State Phys.*, **6**:251 (1958); O. Hachenberg and W. Brauer, Secondary Electron Emission from Solids, *Adv. Electron. Electron Phys.*, **11**:413 (1959).
10. Bruining, H.   "Physics and Applications of Secondary Electron Emission," McGraw-Hill, New York, 1954.
11. RCA Photomultiplier Manual, RCA Electronic Components, *Tech. Ser.* PT-61, Harrison, N.Y., 1970.
12. Mark, D.   "Basics of Phototubes and Photocells," Rider, New York, 1956.
13. Kollath, R.   Sekundärelektronen-Emission fester Körper bei Bestrahlung mit Elektronen, in S. Flügge (ed.), "Handbuch der Physik," Vol. 21, p. 232, Springer, Berlin, 1956.
14. Simon, R., and B. F. Williams   *IEEE Trans. Nucl. Sci.*, **15**:167 (1968).
15. Milnes, A. G., and D. L. Feucht   "Heterojunctions and Metal Semi-Conductor Junctions," Academic, New York, 1972.
16. Grivet, P.   "Electron Optics," Pergamon, New York, 1965.
17. Soule, Harold V.   "Electro-Optical Photography at Low Illumination Levels," Wiley, New York, 1968.
18. Morton, G. A.   Image Intensifiers and the Scotoscope, *Appl. Optics*, June 1964.
19. *Proc. Image Intensifier Symp.*, Office of Scientific and Technical Information NASA, NASA SP-2, Oct. 1961.
20. Spangenberg, Karl R.   "Vacuum Tubes," McGraw-Hill, New York, 1948.
21. Kohl, Walter H.   "Materials and Techniques for Electron Tubes," Reinhold, New York, 1960.

Television Camera Tubes

BY RONALD D. GRAFT AND RICHARD E. FRANSEEN

130. Types of Television Camera Tubes. Modern television camera tubes are available in a wide variety of sizes and types to meet a multitude of applications. *Antimony trisulfide* and *lead oxide vidicons* are the workhorse television tubes, accounting for perhaps 70% of all applications. Other tube types such as the *SEC vidicon, image isocon, silicon vidicon,* and *silicon intensifier vidicon* provide special capabilities not generally required in routine applications.

131. Theory of Operation. Television camera tubes can be conveniently discussed in terms of the three major subassemblies shown in Fig. 11-47. The *image section* uses a photoemissive surface and electron optics to convert an optical image into an electron image which is focused upon the surface of the storage target to create a corresponding electric-charge image. The *storage target* integrates, or stores, the focused electric charge before readout and erasure by an electron beam generated in the *scan section*. The low-velocity electron beam repetitively scans the back surface of the target, thereby generating a time-varying electric signal proportional to the magnitude of the spatial charge distribution. Most vidicon-tube types forgo an image section, using a photoconductive target to perform the transducing function.

132. Performance. The major tube performance parameters are *sensitivity, modulation transfer function* (MTF), *limiting resolution, dark current,* and *lag*. Excluding dark current and lag, performance parameters are determined by limitations and conditions existing in each of the three major subassemblies. Dark current is exclusively determined by the electrical properties of the storage target; lag, or incomplete erasure, depends on the charge transport characteristics and capacitance of the target and the electron-velocity distribution in the scanning beam.

The sensitivity of a camera tube is the ratio of output-signal current to uniform tube-faceplate illumination. Some tubes, such as the antimony trisulfide vidicon, do not have a unique sensitivity for all illuminations; they are nonlinear or subunity gamma tubes. *Gamma* is the slope of a log-log plot of the light-transfer characteristic, as shown in Fig. 11-48. Sensitivity depends on the spectral-energy distribution of the illumination; therefore, the nature of the illumination must always be stated as part of the sensitivity parameter.

Electron gain is an important performance parameter for tubes with electron-image sections. It is defined as the ratio of output-signal current to uniformly illuminated photocathode current. It is a dimensionless parameter independent of the photocathode response characteristics. Electron gain is dependent upon target material and photoelectron energy and can usually be varied by changing the applied photocathode potential. Electron gain can be increased several orders of magnitude by fiber-optically coupling one or more image-intensifier tubes ahead of the camera tube (see Par. 11-127, under Cascading). The increased sensitivity is achieved at the cost of

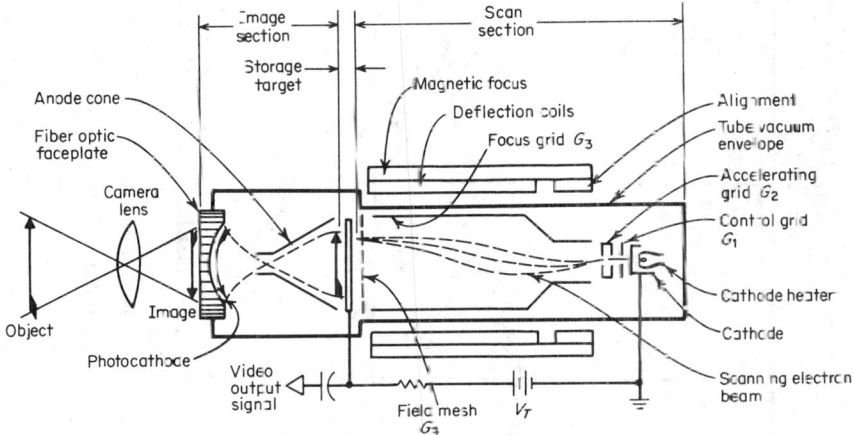

Fig. 11-47. Television-camera-tube schematic.

reduced MTF. In some tubes, additional electron gain is achieved by adding an electron-multiplier structure into the image section.

High-quality pictures require high *signal-to-noise ratios* (S/N). For large uniform areas in the picture, the S/N is the ratio of the peak-to-peak signal current to the rms noise current. The chief noise sources are the preamplifier and, in high-gain tubes, a shot-noise contribution from the photocathode current.

A full description of noise sources and a complete S/N analysis can be found in Ref. 2, Par. **11-140**.

The modulation transfer function (MTF) and limiting resolution[2]* are related parameters useful in specifying the image quality. The real part of the MTF is a measure of the modulation depth present in the output signal when a spatially varying sine-wave pattern is imaged on the tube. The limiting resolution is the highest spatial frequency just resolved by the camera tube-monitor-eye combination. In practice, *amplitude-response* data are commonly quoted for square-wave patterns (black-and-white bar pattern), as in Fig. 11-49. If desired, these data can be converted into sine-wave response via a well-known transformation.[1] It is common television practice to state the spatial frequency in TV lines per raster height. The MTF of several cascaded elements is the product of the MTF of each individual element. As a result, the MTF of a multielement tube can be discussed in terms of the MTF of each element. The limiting resolution is commonly used to evaluate the image capabilities at reduced light levels; see Fig. 11-50.

Dark current is an apparent signal current resulting from target leakage currents between successive scans by the electron beam. In the dark, the rear surface of an ideal vidicon-type target is clamped to near cathode potential; the front surface is maintained at some fixed positive potential by an applied voltage. If target current flows between successive scans, beam current is subsequently deposited on the rear surface, creating an apparent signal current (dark current).

*Superior numbers correspond to numbered references, Par. **11-140**.

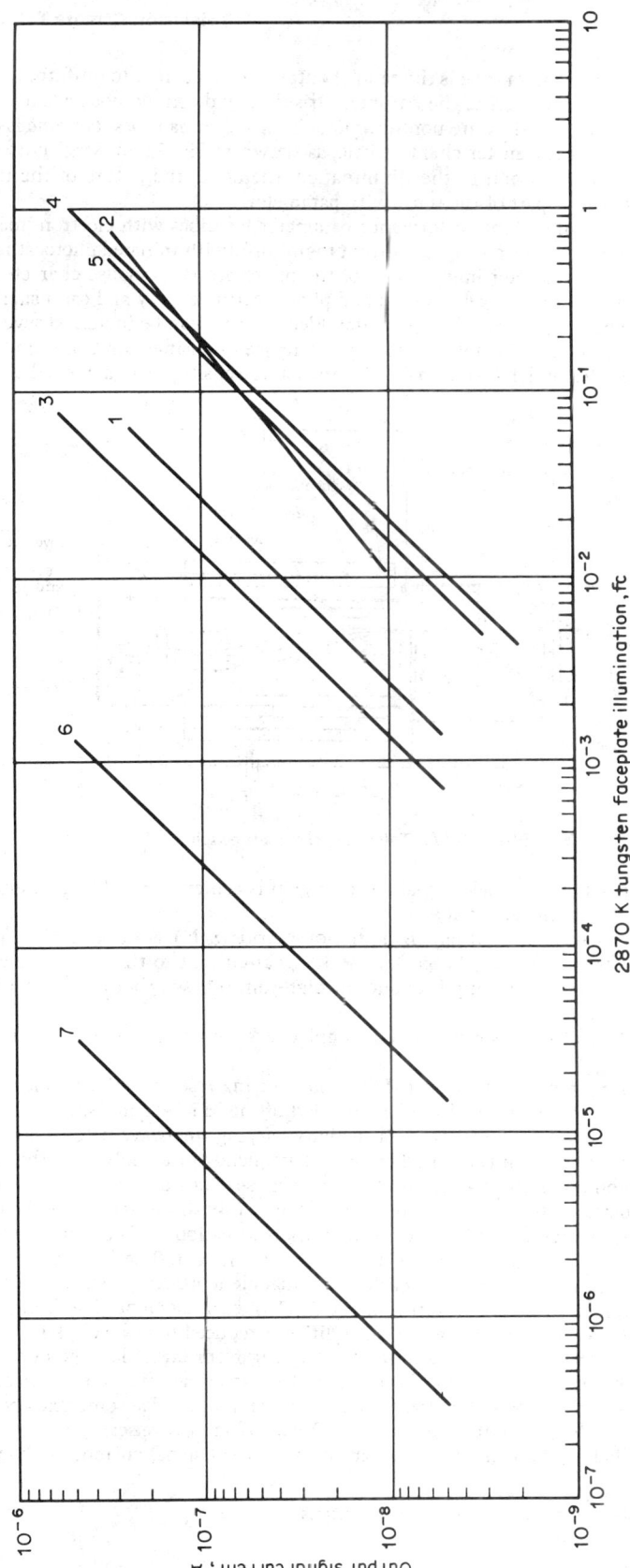

Fig. 11-48. Light-transfer characteristics of camera tubes (see Table 11-21 for key to tube identification).

Output signal current, A

2870 K tungsten faceplate illumination, fc

Lag arises from the failure of the beam to return the rear surface of a charged target to cathode potential after each scan. In viewing a dynamic scene, objectionable image smear will be noted.

133. Image Sections. Many camera tubes use an image section identical with the direct-view image intensifiers discussed in Par. **11-125**, except that the photoelectrons are imaged on the charge-storage target instead of a light-emitting phosphor. Image sections are either magnetically focused, as image orthicons generally are, or electrostatically focused, as silicon-intensifier and SEC vidicons generally are. Image orthicons and image isocons employ an additional mesh element in the image section to collect secondary electrons emitted by the target. The light-

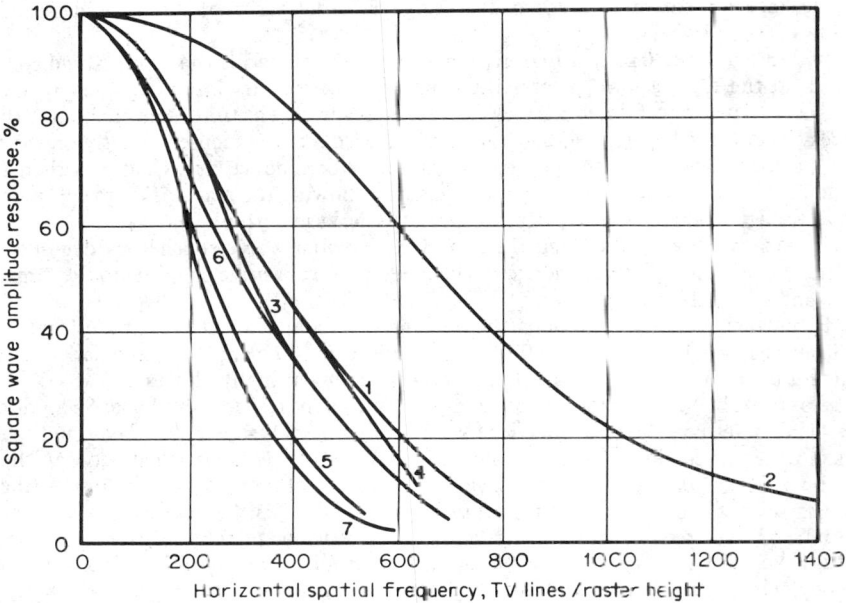

Fig. 11-49. Square-wave amplitude-response characteristics (see Table 11-21 for key to tube identification).

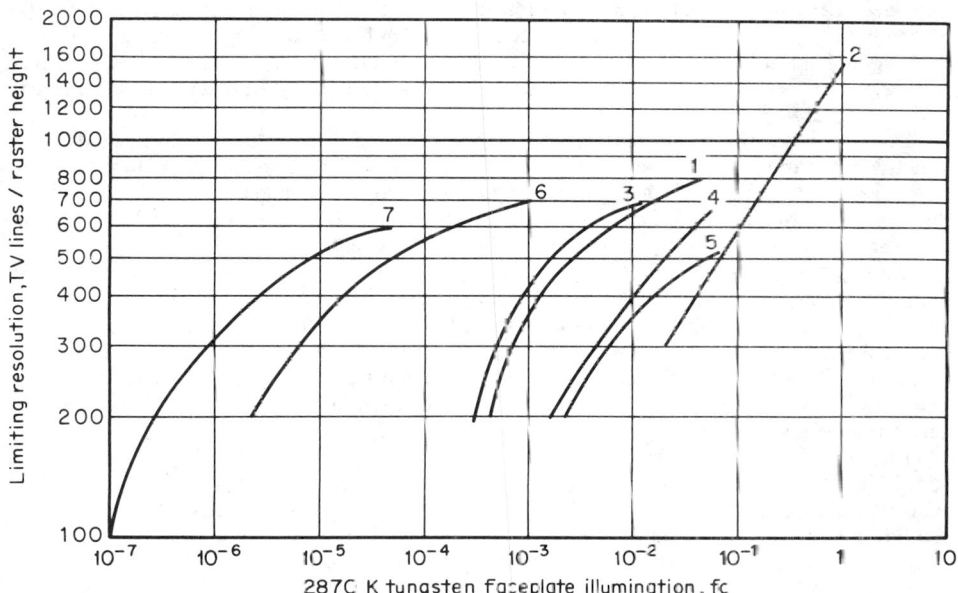

Fig. 11-50. Limiting resolution as a function of faceplate illumination (see Table 11-21 for key to tube identification).

input window (faceplate) of the image section or vidicon is used to restrict the spectral bandwidth to which the photoconductive layer is sensitive and may serve to support substrates for sensing layers. The materials being used are mainly glasses, high-silica glasses, fused quartz, and sapphire. Fiber-optic faceplates with concave back surfaces serve as faceplates for most electrostatic image sections and as the faceplate for vidicons to which image intensifiers are coupled. Optical systems with very short back focal length may also require fiber-optic faceplates.

134. Storage Target Materials. Target-material parameters must be carefully selected and controlled to obtain high image quality, low blemish count, and reproducible electrical characteristics. The target must store an electric charge image for at least $\frac{1}{30}$ s with small lateral charge leakage to ensure high image quality. Target materials most commonly used are described as follows.

Antimony trisulfide (Sb_2S_3), a high-resistivity photoconductor, is the standard vidicon target material. In the dark, a fixed potential is maintained across the thin-film target by a voltage applied to a transparent front-surface electrode; the scanning electron beam maintains the rear surface at near cathode potential. Light from the imaged scene is focused directly on the target, creating electron-hole pairs, thereby increasing the bulk conductivity. Charge is then free to migrate through the target between successive scans, allowing the rear surface to approach some positive voltage whose magnitude depends upon the leakage current.

At the next beam scan, the original rear-surface potential is reestablished by deposited beam current. The deposited charge generates a video signal via capacitive coupling to the target front surface and external circuitry.

Under normal operating conditions, the photon-conversion efficiency (light sensitivity) of the Sb_2S_3 layer is approximately 20% (550 nm) and nonlinear.[3,9] The efficiency and dark current are dependent upon the applied target voltage, both increasing at high voltages.

Lead oxide (PbO) is a photoconductive target,[6,10] in many respects similar to Sb_2S_3, discussed above. The target consists of a three-layered structure: a SnO_2 signal-plate layer and two PbO layers of different conductivity. A nearly intrinsic PbO layer is sandwiched between the SnO_2 layer and a p-type polycrystalline PbO layer. The surface of the intrinsic layer nearest the SnO_2 is n type, so that the structure forms a *pin* layer. The SnO_2 is maintained at a fixed positive potential, and the target rear surface driven to near cathode potential by the scanning beam. The structure then acts like a reverse-biased diode with small reverse current and high collection efficiency. The image is stored as positive charge (holes) in the p-type crystals, which are about 1 μm square and 0.1 μm thick, acting as discrete storage sites.

The PbO target circumvents three major shortcomings of the Sb_2S_3 vidicon: high dark current, low sensitivity, and high lag. The low dark current and improved sensitivity are a result of the *pin* layer structure, since reverse leakage currents are low and the intrinsic layer provides a strong field region for efficient carrier collection. Low lag is due to the absence of photoconductive lag and the low target capacitance. The current transfer characteristics are nearly linear, with a gamma between 0.8 and 0.9. The light sensitivity is target-voltage-dependent up to saturation near 20 V.

Recently developed photoconductive surfaces include a number of selenium-based heterojunction structures.[11,12] The operation of these surfaces is similar to that of PbO; they have been developed to overcome PbO deficiencies. Broader spectral response, improved resolution, increased sensitivity, and better manufacturing reproducibility are frequently claimed. Tubes using these materials are primarily intended for color applications.

The *silicon-diode-array* target is a product of modern microelectronics. Approximately 10^6 individual pn junctions are formed on a typical 120-mm² target area. The diodes are back-biased by an applied positive voltage and the scanning action of the electron beam. Incident photons or electrons generate electron-hole pairs which diffuse to the diode depletion region. The holes are swept into the p region, discharging the diode and reducing the back bias. On a subsequent scan, the electron beam returns the diode side to cathode potential, generating a video signal. The target is shown schematically in Fig. 11-51.

Potassium chloride (KCl) targets are used in secondary-electron conductivity (SEC) tubes. A highly porous KCl layer is deposited on a thin aluminum conducting plate. When an electron image from a photoemissive surface is focused on the target, the electrons penetrate the conducting coating, creating secondary electrons within the KCl layer. The secondary electrons are drawn through the porous structure by an applied electric field, creating a stored-charge analog of the optical image. Reading by the scanning electron beam is identical with that discussed previously.

Thin-film target materials include ionic and electronic conducting glasses and MgO. The targets are commonly used in the image orthicon and image isocon. Although differing in detail, operation of the three target types is similar. A scanning electron beam drives the thin-film target to near cathode potential. An electron image from the photocathode is focused on the target. The energy of the incident electrons is selected to provide a net secondary-electron-emission ratio greater than unity. A closely spaced mesh, located on the target write side, collects the secondary electrons emitted by the target.

The target becomes positively charged in proportion to the light intensity, the maximum excursion being determined by the mesh potential. The positive-potential pattern is then erased from the opposite reading side of the target by the scanning beam, and the video signal appears

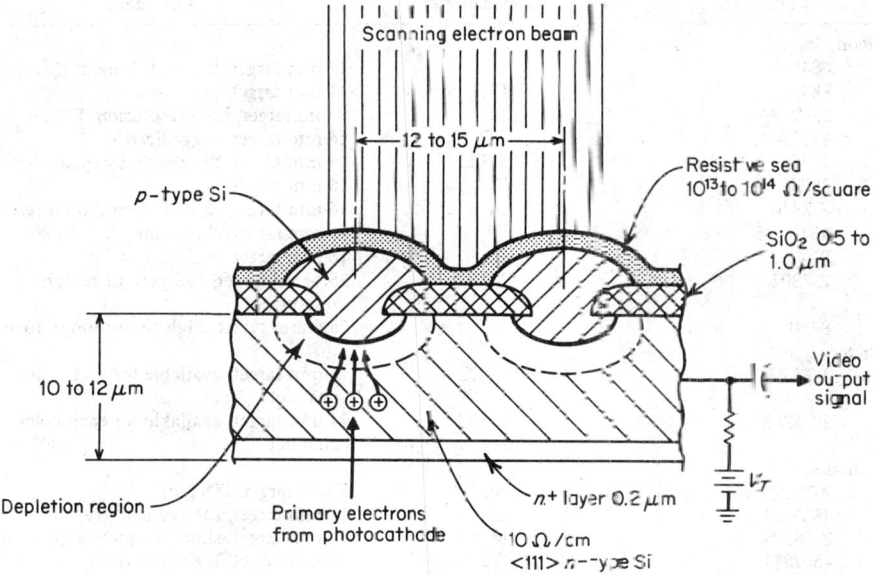

Fig. 11-51. Silicon-diode-array target.

as modulation in the scattered and reflected components of the electron beam. A multiplier structure provides an electron gain of about 500 to the return beam before coupling to an external circuit.

135. Scanning Sections. The function of the scanning section is to generate, focus, and deflect an electron beam. The electron gun is similar in design and operation to the CRT gun described in Par. **11-94**. Magnetic, electrostatic, or combined magnetic-electrostatic fields are used in focusing and deflection. Cameras designed for minimum size and power frequently employ all-electrostatic tubes. Cameras designed for maximum performance employ all-magnetic or magnetic-focus electrostatic-deflection tubes.

Return-beam camera tubes such as the image orthicon and image isocon generate an output signal via modulation of the return beam from the storage target. The return electron beam is composed of scattered and reflected components, both influenced by the electrostatic field and the magnetic-focusing field. The image orthicon collects and amplifies both components via an electron-multiplier structure. The image isocon uses special steering plates to separate the scattered and reflected components. Only the scattered component enters the multiplier and is used for signal generation. Since the scattered component is a minimum in the dark areas of the scene, the effects of beam shot noise are reduced. Return-beam tubes are generally large high-performance devices.

136. Broadcast Service Tubes. Television camera tubes suitable for broadcast service must have good sensitivity, high reliability, extremely low blemish count, low lag, and high image quality. For color applications using three-tube cameras, careful matching of the spectral response of the individual tubes is required. The introduction of lead oxide vidicons in the 1960s resulted in the gradual replacement of image orthicons, until now approximately 90% of live-color cameras employ the lead oxide vidicon. Conventional Sb_2S_3 vidicons are also useful in broadcast service, particularly for film-pickup applications. A representative listing of broadcast television camera tubes is given in Table 11-20. A limited amount of performance data is given in Figs. 11-48 to 11-50 and in Table 11-21.

137. Closed-Circuit Service Tubes (see Table 11-20). Single-tube color cameras and black-and-white cameras employing simple and less expensive Sb_2S_3 vidicons account for nearly all applications. Special requirements, such as high sensitivity or high resolution, may dictate alternative tube types.

A limited amount of performance data is given in Figs. 11-48 to 11-50 and in Table 11-21.

TABLE 11-20   Television Camera Tubes

Tube type	Principal applications*	Comment
Vidicon, Sb_2S_3:		
4848	CC	11-mm target, low cost, E focus
8844	CC	11-mm target, low cost
Z-7969A	M	9-mm target, high resolution, FPS gun
4503A	CC, M	16-mm target, ruggedized
4589	S, CC	16-mm target, fiber-optic faceplate
8507	CC	16-mm target
Z-7820	M, S	16-mm target, short FPS gun, high resolution
4542	SP	16-mm target, slow scan
BC4809	B	16-mm target
Z-7802	M, S	25.4-mm target, FPS gun, ultrahigh resolution
8480	B	25.4-mm target, high resolution, E focus
PbO:		
BC4892	B, CC	16-mm target, available for each color channel
BC4392	B, CC	21-mm target, available for each color channel
Si diode:		
Z-79695	M	9-mm target, FPS gun
4833/U	CC	11-mm target, high sensitivity
Z-7820S	M	16-mm target, short FPS gun, high resolution
4532/U	CC	16-mm target, high sensitivity
C23262A/P2	M	16-mm target, all-ceramic construction
WX32834	M	16-mm target
Other photoconductors:		
BC4390	B	11-mm Se–As–Te photoconductive target
BC4909	B	16-mm version of BC4390
S4076	CC, S	16-mm ZnSe–CdTe photoconductive target
E5001	CC	16-mm CdSe-based photoconductive target
WX-5157	CC, S	16-mm Se-based photoconductive target, x-ray-sensitive
Silicon-intensifier:		
C21199/P1	M	12-mm photocathode, miniature envelope
4804/H	CC, S	16-mm photocathode
WL31960	CC, S	16-mm photocathode
4826/H	M, S	25-mm photocathode
WL31792	M, S	25-mm photocathode
C21145/H	M, S	40-mm photocathode
WX32719	M, S	40-mm photocathode
C21146	M, S	80-mm photocathode
WX33697	M, S	80-mm photocathode
Two-stage:		
4849/H	M, S	16-mm photocathode, very low light level
SEC:		
WL-30893	S	25-mm photocathode
WL-31381	M	40-mm photocathode
WX-31958	S	50-mm photocathode, magnetic image section
WX-32193	S	85-mm photocathode, magnetic image section
Image orthicon:		
4536	B	41-mm photocathode
Image isocon:		
4807	M, S	33-mm photocathode

*CC = closed circuit, M = military, S = scientific, SP = special-purpose, B = broadcast.

138. Scientific and Military Service. Scientific and military applications frequently require tube operation under extreme conditions not normally encountered in studio or closed-circuit applications. Wide temperature and light-level extremes and high vibration and corrosive environments are frequently specified. In many instances, extraordinary performance and compact size are demanded. Typical applications include airborne target acquisition and surveillance, ground-based perimeter and battlefield surveillance, satellite mapping, and tactical-missile guidance. A listing of tube types useful in space and military service is provided in Table 11-20. Some tube performance data are shown in Figs. 11-48 to 11-50 and in Table 11-21.

TABLE 11-21   Representative Camera-Tube Performance

Key no.	Tube type*	Model	Lag,† %	Dark current ‡ nA
1	Vidicon, other	S4076	20	10
2	Vidicon, Sb₂S₃	Z7820	20	20
3	Vidicon Si diode	4532/U	8	8.5
4	Vidicon, other	BC4909	2	0.03
5	Vidicon, PbO	4892	2.5	3
6	Silicon-intensifier vidicon	4804/H	7	7
7	Silicon-intensifier vidicon (two-stage)	4849/H	8	7

*See Table 11-20 for additional description.

†Lag, or residual signal, is measured as a percentage of the typical original signal ⅕₀ s or three fields after exposure.

‡Measured under typical operating conditions, 30°C.

139. Special-Service Tubes. A number of camera tubes (Table 11-20) are designed for special applications. Included in this category are x-ray- and infrared-sensitive devices (see Par. **11-125**), slow-scan and charge-storage tubes, and image-dissector tubes. Slow-scan tubes are useful for remotely located cameras requiring a data link to the monitor wherein the bandwidth reduction significantly lowers the data-link cost

140. References on Television Camera Tubes

1. Coltman, J. W. *J. Opt. Soc. Am.*, **44**:468 (1954).

2. Biberman, L. M., and S. Nudelman (eds.) "Photoelectronic Imaging Devices," Vols. 1 and 2, Plenum, New York, 1971.

3. Weimer, P. K., S. V. Forgue, and R. R. Goodrich *RCA Rev.*, **3**:306 (1951).

4. Crowell, M. H., and E. F. Labuda, *Bell Syst. Tech. J.*, **48**:1481 (May–June 1969).

5. Gordon, E. I., and M. H. Crowell, *Bell Syst. Tech. J.*, **47**:1855 (November 1968).

6. DeHaan, E. F., A. van der Drift, and P. P. M. Schampers *Philips Tech. Rev.*, **25**:133 (1963/1964).

7. Kazan, B., and M. Knoll "Electronic Image Storage," Academic, New York, 1968.

8. Zworykin, V. K., and G. A. Morton "Television," 2d ed., Wiley, New York, 1954.

9. Weimer, P. K., J. V. Forgue, and R. R. Goodrich The Vidicon Photoconductive Camera Tube, *RCA Rev.*, **12**(1):306–313 (1951).

10. De Haan, E. F., A. van der Drift, and P. P. M. Schampers The Plumbicon: A New Television Camera Tube, *Philips Tech. Rev.*, **25**(6,7):133–155 (1963–1964).

11. Neuhauser, R. G. *J. SMPTE*, **87** (March 1978).

12. Shimizu, K., et al. *IEEE Trans.* **ED18**:1058 (1971).

Photoconductive and Semiconductor Junction Detectors

BY WILLIAM A. GUTIERREZ

INTRODUCTORY REMARKS

141. Photoconductors and junction devices constitute an important class of solid-state photodetectors which can operate somewhere within the 0.2- to 2-μm spectral region. These detectors convert electromagnetic energy directly into electric energy via the photoconductivity effect that occurs in semiconductors.

142. Figures of merit[1,2]* are given in Par. **11-162**.

*Superior numbers correspond to numbered references, Par. **11-151**.

PHOTOCONDUCTORS

143. Operation. The simplest photoconductor detector is a bar of relatively low conductivity n- or p-type semiconductor (in bulk or thin-film form) with ohmic contacts at its ends (Fig. 11-52). The photoconductor varies its electrical resistance in accordance with the light wavelength and intensity it receives. Its operation depends on the photoconductivity[3,4] which occurs in semiconducting materials. Electrons in bound states in the valence band (intrinsic) or in for-

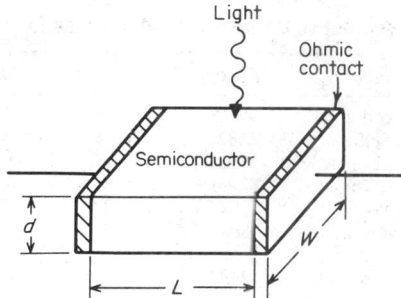

Fig. 11-52. Diagram of a photoconductor detector.

bidden-gap levels (extrinsic) absorb the energy of the incident photons and are excited into the free states in the conduction band, where they remain for a characteristic lifetime. Electric conduction may take place either by the electrons in the conduction band or by the positive holes vacated in the valence band. The electrical resistance of the material thus decreases on illumination, and this resistance change can be translated into a change in the current that flows through the output circuit.

144. The performance of photoconductor detectors is measured not only in terms of D^* (Par. **11-162**) but also in terms of photoconductivity gain, response time, dark current, spectral response, and temperature coefficient. For a photoconductor in which the conductivity is dominated by one carrier (either holes or electrons) the gain is given by the ratio of free-carrier lifetime to the transit time of this carrier. It can also be expressed as

$$\text{Gain} = \tau \mu V / L^2$$

where τ = free-carrier lifetime, μ = mobility, V = applied voltage, and L = spacing between ohmic contacts, as shown in Fig. 11-52. The maximum D^* and spectral dependence of commercially available CdS and CdSe photodetectors are shown in Fig. 11-53. Table 11-22 lists typical gains, response times, and dark current.

Properties of Specific Photoconductors.   The long-wavelength threshold for photoconductivity is usually determined by the bandgap of the material according to the relationship

$$\lambda_c = 1.24/Eg$$

where λ_c = threshold wavelength (μm) and Eg = bandgap (eV).

Bandgap values of materials commonly used as photodetectors in the 0.1- to 2-μm region are given in Table 11-23. The normalized spectral response of some of them is shown in Fig. 11-54.

TABLE 11-22   Parameters of Various Photoconductive Detectors

Photodetector	Gain	Response time, s	Dark current
Photoconductor	10^5	10^{-3}	1–10 mA
pn junction	1	10^{-11}	1–10 μA
Metal-semiconductor	1	10^{-11}	
Avalanche diode	10^4	10^{-10}	
Point contact	1	...	1–3 mA
Heterojunction photodiode	1	...	High
Phototransistor	10^2	10^{-8}	1 nA
Photofet	10^2	10^{-7}	1 μA

SEMICONDUCTOR JUNCTION DETECTORS

145. Semiconductor junction detectors (photodiodes and phototransistors) differ from photoconductive detectors in that their operation depends essentially on a reverse-biased diode whose leakage current is varied by electron-hole pairs generated near or at the depletion region by light absorption. Their response time is characteristically short.[5-7]

146. Basic Classes of Photodiode Detectors. Photodiodes fall into two general categories, the *depletion-layer type* and the *avalanche type*. The distinguishing feature between them is the existence of a gain mechanism.

(a) *The depletion-layer photodiode*[8] family includes the *pn*-junction diode, the *pin* diode, the Schottky barrier (metal-semiconductor) diodes, the point-contact diode, and the heterojunction diode.

1. *pn-junction diode.* Figure 11-55 is a diagram of a *pn*-junction diode. The junction is reverse-biased,[9] and the diode is illuminated either at the *n* or *p* region, away from the depletion region (Fig. 11-55a), or right at the depletion region (Fig. 11-55b). Their built-in field enables them to be operated in the photovoltaic mode (i.e., no externally applied bias); however, the photoconductive mode, with a fairly large reverse bias, is usually the more common mode of operation.

2. *pin photodiode.* Figure 11-56 shows a cross-sectional diagram of a typical *pin* photodiode.[10] The sensitivity range and frequency response of this type of diode depend principally on the thickness of the intrinsic layer (which defines the depletion layer). Light passes through the *p* region before it arrives at the depletion region, where it excites hole-electron pairs that are very quickly swept out by the large electric field present.

3. *Metal-semiconductor photodiode.* Figure 11-57 is a cross-sectional diagram of a metal-semiconductor (Schottky barrier) photodiode.[11] In this case, light passes through a thin (~10 nm) metal film with a suitable antireflection coating to minimize large absorption and reflection losses. As with the *pn* and *pin* diodes, the photogenerated electron-hole pairs in the semiconductor give rise to an output-signal current.

4. *Point-contact photodiode.* Figure 11-58 shows a diagram of a point-contact detector.[12] Light is incident onto the Schottky barrier through an etched cavity in the semiconductor. This detector is extremely fast because of small dimensions and low capacitance.

5. *Heterojunction photodiode.* A depletion-layer photodiode can be constructed by forming a junction between two semiconductors of different bandgaps. Figure 11-59 shows a photodiode[13] made up of n^- GaAs and p^- Ge. Light is absorbed almost completely in the low-bandgap material. Large dark currents could arise due to spontaneous electron-hole generation in the depletion region from a large density of interface states. This could have deleterious effects on the signal-to-noise ratio at low light levels.

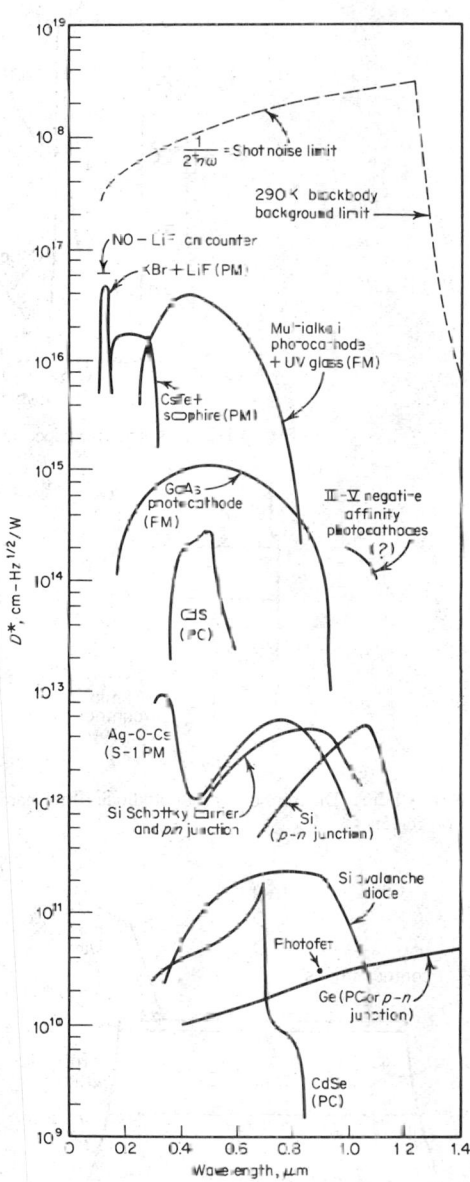

Fig. 11-53. Detectivity vs. wavelength for various photoconductors. PC = photoconductive mode. PM = photomultiplier mode.

TABLE 11-23 Bandgap Values for Photoconductor Materials

Material	Threshold λ_c, μm	Band gap, eV
CdS	0.52	2.4
CdSe	0.73	1.7
ZnS	0.33	3.7
ZnSe	0.48	2.6
GaAs	0.89	1.4
InP	1.03	1.2
Ge	1.77	0.7
Si	1.13	1.1

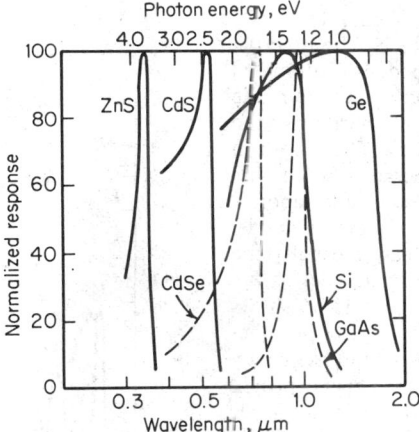

Fig. 11-54. Normalized spectral response of some photoconductors.

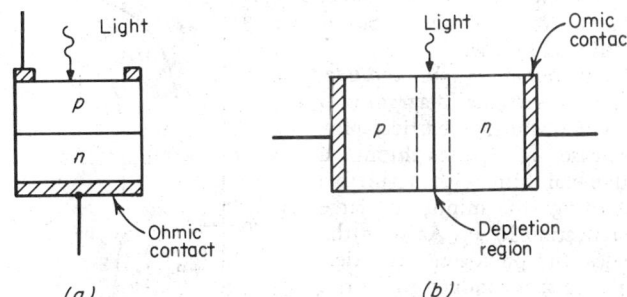

Fig. 11-55. Diagram of a *pn* photodiode illuminated (*a*) away from the depletion region or (*b*) at the depletion region.

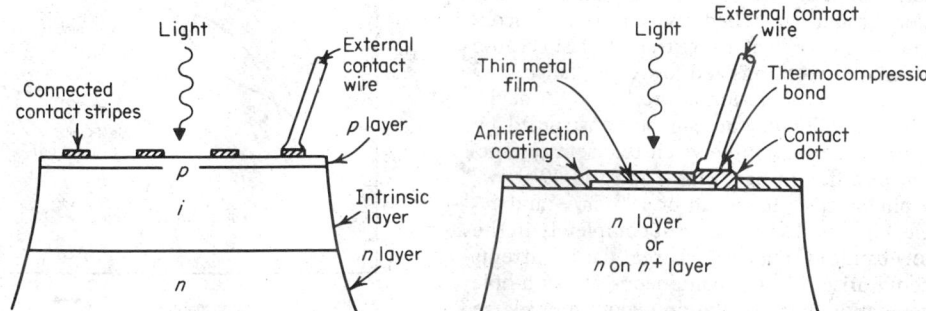

Fig. 11-56. Cross section of a *pin* photodiode.

Fig. 11-57. Cross section of a metal-semiconductor photodiode.

(b) *Avalanche Photodiodes.* Depletion-layer photodiodes operated at higher reverse-biased voltages give an increased output signal. This is due to internal carrier multiplication via the avalanche effect. If the field in the depletion region can impart an energy equal to or greater than the bandgap energy to an electron, this electron can create another hole-electron pair by collision, and this pair can be accelerated to create an additional pair, and so on. This gives rise

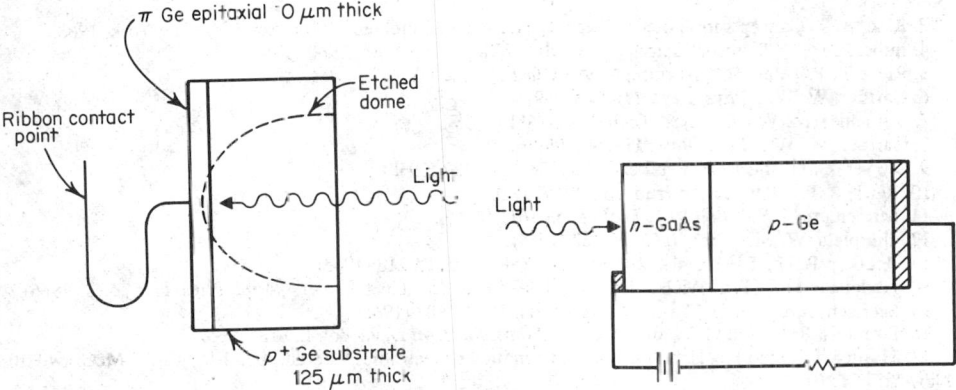

Fig. 11-58. Diagram of a point-contact photodiode.

Fig. 11-59. Diagram of a heterojunction photodiode with applied reverse bias.

to carrier multiplication and to internal gain in the photodiode. The avalance photodiode is therefore the counterpart of the photomultiplier tube, and its multiplication factor M is

$$M = K(1 - V/V_B)^{-1}$$

where K is a constant and V_B is the breakdown voltage.

Figure 11-60 shows two types of avalanche photodiodes[14-16] with guard rings. The guard ring prevents a high-field breakdown region from reaching the surface.

147. Performance of Photodiodes. The spectral dependence of D^*, gain, speed of response, and dark current of the various photodiodes are shown in Table 11-21 and Fig. 11-53.

148. Phototransistor. A *pnp* or *npn* junction transistor can act as a photodetector, with the possibility of large internal gain.[17,18] A *npn* structure, for example, is usually operated as a two-terminal device with the base floating and the collector positively biased.

Phototransistors are generally fabricated of Ge or Si in the same manner as conventional transistors, except that a lens or window is provided in the transistor to admit light at the base or base-collector junction. Response time of 10^{-8} and peak sensitivities of 30 A/lm are possible. Gains of several hundred have been attained with dark currents as low as nanoamperes.

149. Photofet, SMS Photodetector, and *pnpn* Device. The *photofet*, or *photosensitive field-effect transistor*, combines a photodiode and high-impedance amplifier in one device to achieve photodetection with large gain.[19]

A *semiconductor-metal-semiconductor* (SMS) device is essentially a Schottky barrier device with gain.[20]

A silicon controlled rectifier (SCR),[21] or *pnpn device* can be used as a photosensitive switch, with photogenerated current taking the place of the usual gate current.

150. D^* Comparison of Photoconductive Devices with Photoemissive Devices. Representative D^* values for various commercial types of radiation detectors in the 0.2- to 2-μm range are shown in Fig. 11-53. The D^* values given for the III-V negative-electron-

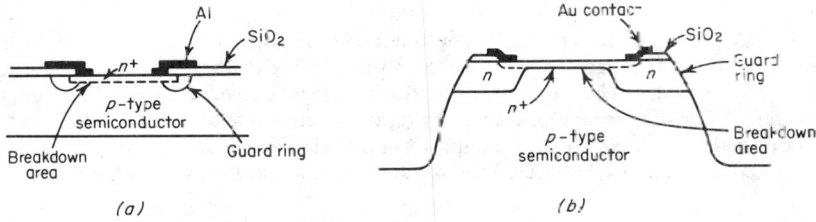

(a)   (b)

Fig. 11-60. Cross section of avalanche photodiode with guard ring: (a) planar type; (b) mesa type.

affinity photoemitter represent an estimate, since these detectors are not yet fully developed. Because of rapid improvements being made in III-V photoemitters, the D^* values can be expected to increase significantly.

151. References on Photoconductive and Semiconductor Devices

1. Ross, M. "Laser Receivers: Devices, Techniques, Systems," Wiley, New York, 1966.
2. Kruse, P. W., L. D. McGlauchlin, and R. B. McQuistan "Elements of Infrared Technology," Wiley, New York, 1962.
3. Rose, A. "Concepts in Photoconductivity and Applied Problems," Interscience, New York, 1963.
4. Bube, R. H. "Photoconductivity of Solids," Wiley, New York, 1960.
5. Riesz, R. P. *Rev. Sci. Instrum.*, **33**:994 (1962).
6. Gartner, W. W. *Phys. Rev.*, **116**:84 (1959).
7. Schneider, M. V. *Bell Syst. Tech. J.*, **45**:1611 (1966).
8. Gartner, W. W. *Phys. Rev.*, **116**:84 (1959).
9. Sawyer, D. E., and R. H. Rediker *Proc. IRE*, **46**:1122 (1958).
10. Reitz, R. P. *Rev. Sci. Instrum.*, **33**:994 (1962).
11. Schneider, M. V. *Bell Syst. Tech. J.*, **45**:1611 (1966).
12. Sharpless, W. M. *Proc. IEEE*, **52**:207 (1964).
13. Rediker, R. H., T. M. Quist, and B. Lax *Proc. IEEE*, **51**:218 (1963).
14. Anderson, L. K., P. G. McMullin, L. A. D'Asaro, and A. Goetzberger *Appl. Phys. Lett.*, **6**:62 (1965).
15. Baertsch, R. D. *IEEE Trans. Electron Dev.*, **Ed-13**:987 (1966).
16. Emmons, R. B., and G. Lucovsky *IEEE Trans. Electron Dev.*, Vol. ED-13, 1966.
17. Hunter, L. P. (ed.) "Handbook of Semiconductor Electronics," 2d ed., pp. 4-16, 16-17, McGraw-Hill, New York, 1970.
18. Schuldt, S. B., and P. W. Kruse *J. Appl. Phys.*, **39**:5573 (1968).
19. Shipley, M. *Solid State Design*, **5**:28 (1964).
20. Reynolds, J. H. *Trans. Metal. Soc. AIME*, **239**:326 (1967).
21. Gentry, F. E., F. W. Gutzwiller, N. H. Holonyak, and E. E. Van Zastrow "Semiconductor Controlled Rectifiers," Prentice-Hall, Englewood Cliffs, N.J., 1964.

Charge Transfer Device (CTD) Imagers

BY JOHN H. POLLARD, ANDREW J. KENNEDY, AND LARRY V. CALDWELL

152. Monolithic solid-state self-scanned image sensors of the charge-coupled-device (CCD) and charge-injection-device (CID) types are extremely versatile and can be applied to high- and low-light-level imaging, optical character recognition, and facsimile reproduction.[1-3*] Solid-state devices offer advantages over vidicons for imaging applications: they provide lag-free, burn-free imaging and operate at low power in a self-scanned mode; they are lightweight and have high sensitivity. Advanced metal-oxide-semiconductor (MOS) technology is employed to fabricate the closely spaced single- or multiple-capacitor imaging elements, called *pixels*. Linear or area configuration of the pixels, with the appropriate on-chip scanning circuit and low-noise preamplifier, constitute the focal-plane image sensor in a camera system.

153. CCD Imaging. The principle of operation is based on the transfer of photogenerated minority-carrier charge packets to adjacent pixels. Sequentially clocked voltage pulses applied to adjacent pixels create potential wells, which transfer charge packets to adjacent positions across the surface. Each charge packet corresponds to a pixel-sized optical image element. Thus, the moving charge packets are transferred to a low-capacitance output diode that has typical values on the order of 0.2 pF to obtain the video signal. This is one of the major advantages of the CCD imagers and produces high signal-to-noise output signals. Two implementations of CCD imagers are possible, one using charge transport in a surface channel and the other in a buried or bulk channel.

In the surface-channel device (SCCD) the conduction channel is at the semiconductor-oxide interface and allows the signal charge packets to interact with interface states. The resultant interface-state trapping can be reduced by introducing a bias charge 5 to 20% of full well, referred to as "fat zero." The performance is limited by temporal noise due to interface state trapping and a spatial or fixed pattern noise due to nonuniform fat-zero insertion.

For a buried-channel device (BCCD) the conduction channel is ion-implanted approximately

---

*Superior numbers correspond to references in Par. **11-158**.

0.5 μm into the bulk with respect to the semiconductor-oxide interface. The charge packets interact with low density (of the order of 10^{11} cm^{-3}) bulk states, but the bulk trapping noise is insignificant even for signals as low as approximately 10 electrons per pixel per frame at $-20°$C. The charge transfer efficiency is also improved over the surface-channel device because of the larger fringing fields between the pixels present in the channel.

CID Imaging. The CID is an *XY* addressed matrix of MOS capacitor pairs in which each capacitor pair constitutes a pixel. Charge transfer occurs only between the capacitor pairs within a pixel, and the photogenerated charge packets are not transferred to a common output. The charge packets are stored by biasing at least one of the capacitors. When the capacitor pair of a pixel is simultaneously pulsed to zero, the potential well collapses and the stored minority carrier charge is injected into the substrate, where most of the carriers recombine. This injected charge provides the video signal. A reverse-biased epitaxial junction in the substrate collects the unrecombined minority carriers to prevent them from diffusing back into the same pixel or into neighboring pixels, which would degrade the resolution.

An alternative low signal-to-noise readout technique uses an output circuit that senses the

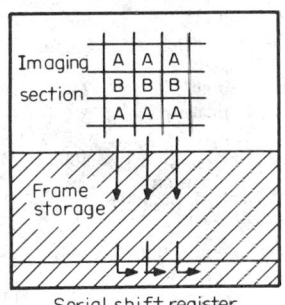

Fig.   11-61. Frame-storage charge transfer device.

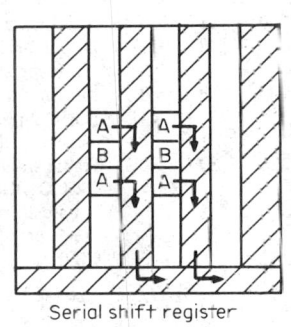

Fig.   11-62. Interline charge transfer device.

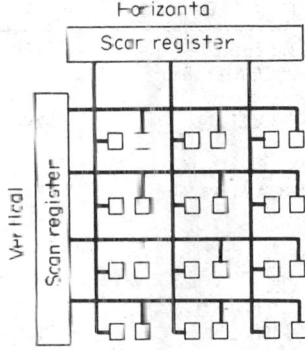

Fig.   11-63. Charge injection area imager.

relative magnitude of the stored charge in a given pixel by transferring the charge between the capacitor pairs. Even with this improvement, the effective output capacitance of a CID is still approximately an order of magnitude higher than that of a CCD. This is a definite disadvantage for low-light-level imaging, but the superior antiblooming control and the random-access features provided by the CID scheme are useful in many other applications.

154. Charge-Transfer-Device Architecture. Charge-coupled-device area imagers have been fabricated in two main configurations. One uses a frame-storage mode and the other an interline transfer mode. The former has nearly 100% optically sensitive imaging area with transparent electrodes or backside illumination but can suffer from optical transfer smearing effects. For certain applications not requiring TV readout rates, e.g., astronomical observations and space telescopes, full frame imagers can be used.

The frame-storage device is illustrated in Fig. 11-61. Charge is integrated in the imaging section and then rapidly moved into the frame-storage section for subsequent readout through the serial shift register. In this mode both frontside and backside illumination is possible. Interlaced readout is obtained by imaging under different electrode sets, giving alternate readout of frame A and frame B. The shaded regions are opaque.

The interline transfer device is shown in Fig. 11-62. The charge is integrated in the photosensitive areas and is shifted into opaque columns for subsequent transfer to the readout serial shift register. In this mode only frontside illumination is possible. Interlacing is obtained by integration under different electrodes in subsequent frames (*A* and *B* in the figure). Shaded regions are opaque.

In the charge-injection-device area imager (Fig. 11-63) a particular column is allowed to float, and the change in potential after a row transfer under that column provides the pixel signal readout. Area imaging is obtained by suitable scanning of the horizontal and vertical registers. The device has the advantage of random access and is also free of optical smearing effects. Transparent electrodes allow about 100% optically active areas with a sensitivity that depends on the readout scheme.

155. Low-Light-Level Imaging. Image-intensifier technology can be used with charge transfer devices for low-light-level applications, and two methods are currently being developed. In the EBS-CCD approach (electron-bombarded silion–CCD), the phosphor screen, in a proximity-focused and/or inverter-type image intensifier, is replaced by a backside thinned and accumulated silicon CCD. The impinging 8- to 15-kV photoelectrons produce high gains in the approximately 10-μm-thick CCD substrate. The maximum theoretical gain is $V/3.5$, where V is the accelerating voltage of the incident electrons.

In this mode of operation the effective preamplifier noise of a cooled direct-view CCD is reduced by the approximately 1,000 to 2,500 electron gain in the CCD, and low-light-level imaging has been demonstrated[5] under overcast night-sky illuminance levels of 8 μlx. The sec-

TABLE 11-24   Characteristics of Commercial CTD Area Imagers

Device	Pixel elements V	H	Pixel size, μm V	H	Architecture	Output	Resolution* V	H
Fairchild: CCAID 488	488	380	18 [18	30 12]	Interline transfer BCCD 2-phase shift register	Single floating gate	488	284
CCD 211	244	190	18 [18	30 14]	Interline transfer BCCD 2-phase shift register	Single floating gate	244	142
RCA SID 52501	512	320	30	30	Frame transfer SCCD/BCCD 3-phase shift register	Floating diffusion with reset	480	240
GE: TN 2500	244	248	36	46	CID	Preinjection readout	244	191
TN 2200	128	128	46	46	CID	Sequential readout	128	128
Hughes HCCI 100A	100	100	25	25	Full frame 4-phase	Surface gate output amplifier	100	100
Sony SiCCD color camera	492	245	14 [14 [photosites]	36 6]	Interline transfer BCCD 2-phase		350 (2 chips offset horizontally)	280

*Lines per picture height.

ond method uses image intensifiers, fiber optically coupled directly to either a frontside-illuminated CCD or CID area array. This hybrid approach allows the selection of conventional image intensifiers based on an optimum trade-off between gain, MTF, noise figure, and charge-transfer-device noise.

Both methods of low-light-level imaging are in a preliminary stage of development, and the technological progress will depend on the requirements for future applications.

156. Selection of Suitable CTD Imagers. To date a considerable research and development effort has produced steady progress in performance characteristics and array size. The first commercially available CTD imagers and TV-compatible cameras were produced by RCA, Fairchild, and General Electric. Most of the imagers were in a more or less 500 × 300 pixel format with the CCD pixels about 25 × 25 μm and the CIDs having pixels about twice that area but with fewer pixel elements. High-sensitivity large area arrays were custom fabricated. For example, Texas Instruments produced thinned, backside-illuminated, 100 × 160 and 400 × 400 pixel CCDs for night vision and space applications. Further advances at TI led to the development of 500 × 500 BCCDs with 15 × 15 μm pixels, followed by the development of 800 × 800 pixel CCDs for space application. Table 11-24 shows some of the charge transfer device area imagers commercially available at the start of the 1980s. Figure 11-64 shows the spectral responsivity of typical charge-coupled device imagers, and Fig. 11-65 shows the response of charge injection device imagers (the corresponding quantum efficiency is indicated).

157. Future Trends. A recent approach for future silicon CCDs is the implementation of a virtual-phase scheme that requires essentially a single gate for a 490 × 328 pixel imager.[5] The comparative simplicity of this approach promises significant improvement in device fabrication yield. The CTD's ability to perform sophisticated on-chip signal conditioning will be further exploited in both intrinsic silicon and narrower bandgap intrinsic semiconductors for infrared CCD (IRCTD) staring focal-plane-array imagers.

Extrinsic silicon (Si:X) IRCCDs with indium doping cover the 3- to 5-μm and, with gallium doping, the 8- to 14-μm spectral regions. Advanced focal-plane sensors fabricated of this material system will integrate the detector and signal-conditioning sections monolithically.

Monolithic IRCTDs fabricated of intrinsic narrow-bandgap semiconductors represent a longer-range possibility. The potentially useful materials for this application are the binary ternary,

---- Fairchild 488 X 380 interline transfer
---- RCA 512 X 320 frame transfer
········ Hughes 100 X 100 full frame

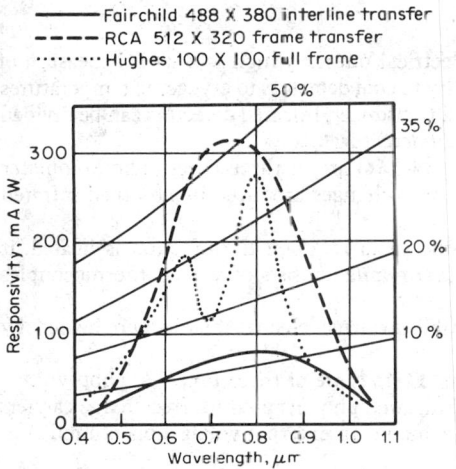

------ GE-TN2500 244×248
------ GE-TN2200 128 × 128

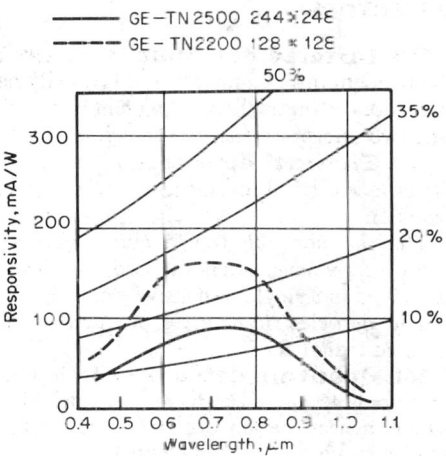

Fig. 11-64. Spectral responsivity of typical charge-coupled imagers. Corresponding quantum efficiencies are shown.

Fig. 11-65. Spectral response of charge injection imagers. Corresponding quantum efficiency in percent is shown.

and quaternary III-V, II-VI, and IV-VI, such as GaSb (E_g = 0.7 eV, λ_c = 1.8 μm), InAs (E_g = 0.4 eV at 77 K, λ_c = 3.1 μm), InSb (E_g = 0.23 eV at 77 K, λ_c = 5.4 μm), and $Hg_{0.8}Cd_{0.2}Te$ (E_g = 0.09 eV at 77 K, λ_c = 13.8 μm).

A shorter-term IRCCD development problem is the application of silicon CCDs in a hybrid configuration as multiplexers for photovoltaic (Hg, Cd)Te detector arrays. A variety of other near-term applications are also being explored, e.g. the CCD's ability to perform the time-delay and integration (TDI) function in serial-parallel scanned systems.

Schottky barrier gate CCDs with the GaAs/GaAlAs[6] and GaSb/GaAlAsSb systems represent another direct-view visible CCD or IRCCD monolithic focal-plane-sensor approach. For this scheme the p-type binary layer forms the absorber, and the generated electrons are injected through the heterojunction into the n-type ternary or quaternary wide-bandgap Schottky barrier charge transfer layer. The two main advantages of this approach are that high quantum efficiency CCDs can be fabricated on suitable bandgap materials without native oxides and that non-MOS CCDs are less susceptible to the total dose effects of ionizing radiation.[7–8]

158. References for CTD Imagers

1. Sequin, C. H., and M. F. Tompsett Charge Transfer Devices, "Advances in Electronics and Electron Physics," Suppl. 8, Academic, New York, 1975.

2. Buss, D. D., and M. F. Tompsett (eds.) Special Issue on Charge-Transfer Devices, IEEE Trans. Electron Dev., February 1976, Vol. ED-23.

3. Chamberlain, S. G., and M. Kuhn (eds.) Joint Special Issue on Optoelectronic Devices and Circuits IEEE Trans. Electron Dev., February 1978, Vol. ED-25.

4. Caldwell, L. V., J. J. Boyle, A. J. Kennedy, W. Tardiff, and S. Tomarchio Low Light Level CCD-TV, Proc. 5th Int. Conf. CCDs, Edinburgh, September 1979.

5. Hynecek, J. Virtual Phase CCD Technology, Proc. IEDM, Washington, December 1979, pp. 611–614.

6. Deyhimy, I., J. S. Harris, Jr., R. C. Eden, R. J. Anderson, and D. D. Edwall   An Ultra High Speed GaAs CCD, *Proc. IEDM, Washington, December 1979*, pp. 619–621.

7. Killiany, J.   Radiation Effects on Silicon Charge Coupled Devices, *IEEE Trans. Components, Hybrids, Manuf. Tech., December 1978*, Vol. CHMT-1, pp. 353–365.

8. Kennedy, A. J.   Determination of the Total X-Ray Dose in the SiO_2 Layer of Backside Electron Irradiated ICCD Imagers, *IEEE Trans. Nucl. Sci., December 1979*, Vol. NS-26, pp. 4885–4891.

Infrared Detectors and Associated Cryogenics

BY WAYNE T. GRANT AND RANDOLPH E. LONGSHORE

DETECTORS

159. Infrared detectors[1-5]* provide an electrical output which is a useful measure of the incident infrared radiation. It is usually necessary to cool detectors to cryogenic temperatures to reduce the thermal noise inherent in an electrical transducer. Infrared detectors can be divided into two categories, *thermal detectors* and *quantum detectors.*

160. Thermal detectors (Table 11-25, Par. **11-164**) are of three types. The *bolometer* varies its electrical resistance as a result of temperature changes produced by absorbed infrared radiation.

The *thermocouple* is a junction of two dissimilar metals. When the junction is heated, it produces a voltage across the two open leads. A *thermopile* consists of several thermocouples combined in a single responsive element.

The *pyroelectric detector* produces an observable external electric field when heated by infrared radiation.

161. Quantum detectors (Table 11-26, Par. **11-164**) are of three types. A *photovoltaic detector* is a *pn* semiconductor junction. Absorbed infrared photons produce free charge carriers which, if near the junction, are separated by it, producing an external voltage (open circuit) or an electric current (closed circuit).

A *photoconductor* is a semiconductor in which absorbed infrared photons produce free charge carriers, resulting in a change in the electric conductivity.

A *photoemissive detector* is one in which incident photons impart sufficient energy to surface electrons to free them from the detector surface. The topic of photoemissive surface is treated in Par. **11-121**.

162. Detector Parameters. The parameters most often used in the description of infrared detectors are as follows.

Responsivity R is the ratio of the rms signal voltage (or current) to the rms incident signal power, referred to an infinite load impedance or to the terminals of the detector, $R = V_{s,rms}/P_{s,rms}$. Spectral responsivity R_λ refers to a monochromatic input signal, and blackbody responsivity R_{BB} refers to an input signal having a blackbody spectrum. The units of responsivity are volts per watt (or amperes per watt). Responsivity is a function of wavelength λ, signal frequency f, operating temperature T, and bias voltage V_B.

Impedance times area product ZA is a parameter used to characterize a photovoltaic detector. Here A is the area of the detector, and Z is the dynamic impedance taken from the *I-V* curve of the diode at some operating point; that is, $Z = \delta V/\delta I$. Usually the zero-bias impedance Z_0 is measured to determine the thermal noise and the proper matching to external electric circuits. The unit of ZA is ohm-centimeters^2.

Chopping frequency f_c is the rate at which the blackbody radiation source is mechanically interrupted to provide a strong periodic signal for separating ac components of the instantaneous power from the dc component.

Noise-equivalent power (NEP) is that value of incident rms signal power required to produce an rms signal-to-noise ratio of unity NEP $= V_{s,rms}/R$. Spectral noise-equivalent power (NEP)$_\lambda$ refers to a monochromatic input signal, and blackbody noise-equivalent power (NEP)$_{BB}$ refers to an input signal having a blackbody spectrum. The unit of NEP is the watt. NEP is a function of wavelength λ, detector area A, chopping frequency f_c, electric bandwidth Δf, temperature of blackbody T_{BB}, field of view Ω, and background temperature T_B.

---

*Superior numbers correspond to references in Par. **11-173**.

Detectivity D is the reciprocal of the noise-equivalent power. It can also be expressed as the rms signal-to-noise ratio per unit of rms power incident on the detector. Spectral detectivity D_λ and blackbody detectivity D_{BB} are the reciprocals of NEP and $(NEP)_{BB}$, respectively.

D star $(D)^*$ is a normalization of the reciprocal of the noise-equivalent power to take into account the area and electric-bandwidth dependence, $D^* = \sqrt{A\,\Delta f}/\,NEP = \sqrt{A\,\Delta f}\,D$. By using $(NEP)_{BB}$, D^*_{BB} can be determined, and from the definitions of $(NEP)_{BB}$ and R, $D^*_{BB} = \sqrt{A\,\Delta f}\,V_{s,rms}/V_{n,rms}P_{BB,rms}$, where $P_{BB,rms}$ is the rms power incident on the detector from a black-body at temperature T_{BB}. Since D^*_{BB} depends on T_{BB}, f_c, and Δf, it is sometimes written as $D^*(T_{BB}, f_c, \Delta f)$. Similarly, D^*_λ depends on λ, f_c, and Δf and is sometimes written as $D^*(\lambda, f_c, \Delta f)$. It can be determined from $(NEP)_\lambda$, $D^*_\lambda = \sqrt{A\,\Delta f}\,V_{s,rms}/V_{n,rms}P_{\lambda,rms}$ where $P_{\lambda,rms}$ is the rms power incident on the detector from a monochromatic source.

These figures of merit can be used to compare detectors of different types since the detector with the greater value of D^* is a better detector when the terms in the parentheses are identical.

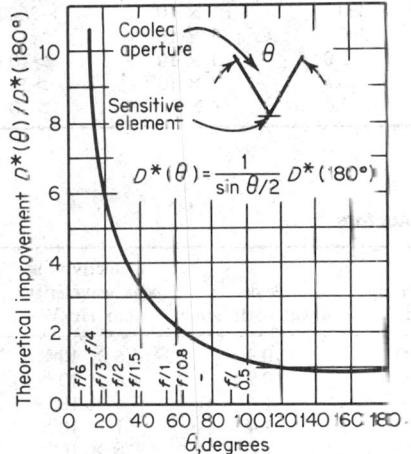

Fig. 11-66. Relative increase in D^* for BLIP detectors obtained by using cold shielding.

This does not hold for detectors limited by noise due to fluctuations in background photon flux. For such detectors, field of view and background temperature must be specified. The units of D^* are $cm \cdot Hz^{1/2}/W$.

Quantum efficiency (QE or η) is the ratio of countable output events to the number of incident photons.

Time constant τ is a measure of the speed of response of a detector. It is usually defined as $\tau = (2\pi f_c)^{-1}$, where f_c is that signal frequency at which the responsivity has fallen to 0.707 times its maximum value.

163. Noise mechanisms in infrared detectors are usually of five types.

Photon Noise (Background Noise). Random fluctuations in the arrival rate of background photons incident on the detector produce random fluctuations in its output signal. Photodetectors whose D^* is limited only by this type of noise are called *background-limited infrared photo-detectors* (BLIP detectors). This value of D^* is the theoretical limit for a photon detector.

Figure 11-66 is a plot of relative improvement in D^* vs. angular field of view for a BLIP detector. Cooled spectral filters also improve the D^*_λ of a BLIP or near-BLIP detector by attenuating radiation at wavelengths which are not of interest.

Johnson Noise (Nyquist or Thermal Noise). The random motion of charge carriers in a resistive element at thermal equilibrium generates a random electric voltage across the element.

Generation-Recombination Noise (GR). Variations in the rate of generation and recombination of charge carriers in the detector create electric noise.

Shot Noise. Since the electric charge is discrete, there is a noise current flowing through the detector as a result of current pulses produced by individual charge carriers

1/f Noise. The mechanism involved in this type of noise is not well understood. It is characterized by a $1/f^n$ noise-power spectrum, where n varies from 0.8 to 2.

164. Detector Data.[6] Tables 11-25 and 11-26 list commercially available detectors. They include the operating temperature, cutoff wavelength, and peak D^*.

Figures 11-67 and 11-68 are plots of D_λ^* vs. wavelength for selected detectors. Since detector noise is a function of operating temperature, D^* will also vary with temperature. Figure 11-69 is a plot of peak D^* vs. temperature for selected detectors. All values tabulated are data from above-average single-element detectors.

165. Charge Transfer Devices for Infrared Detection. In applications requiring many infrared detectors, e.g., thermal imaging, the detectors and signal processors have been

TABLE 11-25   Thermal Detectors

Type	Operating temp, K	Detectivity D^*, $cm \cdot Hz^{1/2}/W$	Wavelength region, μm	Response time, ms	Resistance
Pyroelectric	300	1×10^9	>1	1	10 TΩ
Thermistor	300	5×10^8	1–40	0.1	2 MΩ
Golay cell	300	1.5×10^9	1–2000	15	
Thermocouple	300	1×10^9	>1	10	10 Ω
Germanium bolometer	2.0	1×10^{12}	>10	10	100 kΩ
Tin bolometer	3.7	7×10^{10}	>10	10	100 Ω
Carbon bolometer	2.0	4×10^{10}	>10	1	100 kΩ

TABLE 11-26   Quantum Detectors

Material	Type†	Operating temp, K	Peak wavelength, μm	Detectivity at peak wavelength, $cm \cdot Hz/W$	Response time, μs	Resistance (size-dependent)
Si	PV‡	300	0.9	5.6×10^{13}	10^{-2}	2 kΩ
Si	PV	300	0.9	6×10^{12}	0.2	5 kΩ
Ge	PV	300	1.6	4×10^{11}	1,000	
PbS	PC	300	2.4	1×10^{11}	300	1 mΩ
PbS	PC	193	2.8	5×10^{11}	3,000	5 MΩ
PbSe	PC	300	3.8	3×10^{10}	2	5 MΩ
PbSe	PC	193	4.8	2×10^{10}	30	50 MΩ
PbSe	PC	77	5.0	2×10^{10}	30	5 MΩ
InAs	PV	300	3.4	7×10^9	1	20 Ω
InAs	PV	195	3.2	1×10^{11}	1	5 kΩ
InAs	PV	77	3.0	7×10^{11}	1	100 kΩ
InSb	PV	77	5.0	1×10^{11}	0.1	1 MΩ
HgCdTe	PC	200	5	5×10^{10}	5	500 Ω
HgCdTe	PC	77	11.5	2×10^{10}	1	40 Ω
HgCdTe	PV	200	4.2	1×10^{11}	0.1	100 kΩ
HgCdTe	PV	77	10.6	2×10^{10}	0.1	1 kΩ
PbSnTe	PV	77	11.5	2×10^{10}	0.1	10 kΩ
Ge:Au	PC	77	5	7×10^9	0.1	300 kΩ
Ge:Hg	PC	28	11	2×10^{10}	0.1	100 kΩ
Ge:Cu	PC	5	25	3×10^{10}	0.1	300 kΩ
Si:Ga		30	15	4×10^{10}	0.1	100 kΩ
Si:In		50	5.6	3×10^{11}	0.1	300 kΩ

†PV = photovoltaic, PC = photoconductor.
‡Avalanche.

integrated into a single structure.[7-9] The signal processor provides rapid and efficient readout of the many detected signals by using charge-coupled devices (CCDs) or charge-injection devices (CIDs) (Pars. **11-152** to **11-158**). These structures have been fabricated with InSb,[10,11] extrinsic silicon,[12,13] and HgCdTe.[14,15]

166. Detector Selection. Though no absolute guidelines for the choice of an infrared detector for a specific application are possible, certain criteria can be used to eliminate many of the available detectors. Among them are spectral region of interest, maximum signal frequency, required sensitivity, and available cooling.

COOLERS

167. Cryogenic Cooling. For BLIP performance, temperatures of 200 K or lower are required of intrinsic infrared detectors, and extrinsic detectors require temperatures of 80 K or less (Table 11-26).

The four basic types of cooling systems (see Table 11-27) are:

1. *Open-cycle, expendable systems,* which operate by the transfer of stored cryogens
2. *Mechanical coolers,* which are closed-cycle, expansion engines providing cooling at low temperatures and rejecting heat at high temperatures

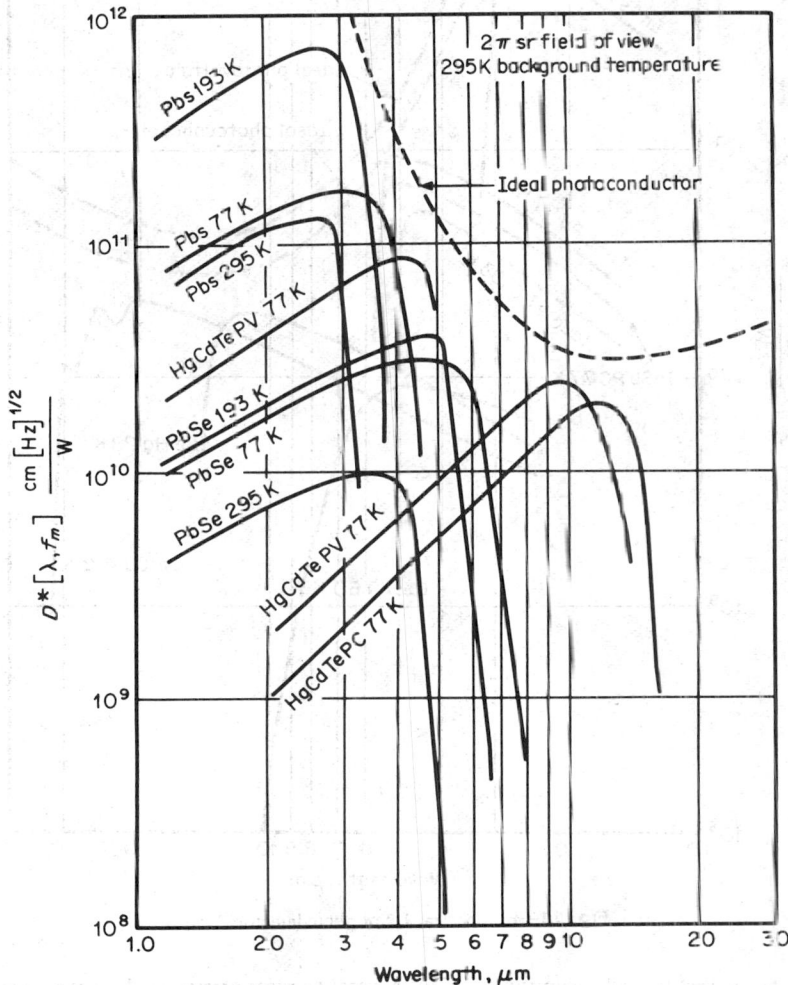

Fig. 11-67. Spectral $D*$ of photodetectors [6]

3. *Thermoelectric coolers,* which operate by using the Peltier cooling effect
4. *Radiation coolers,* which are passive and cool detectors by radiating heat to the low-temperature deep-space environment

168. Cooling Specifications. The particular application and detector type determine the best-suited cooling system. Variables used for cooler specifications include cooling capacity, cooling temperature, cooldown time, temperature stability, reliability, duty cycle, environment, weight, configuration, noise (acoustical, electromagnetic, mechanical), and power (ac, dc, limits). Table 11-27 summarizes cooling features and limitations.

169. Open-Cycle Expendable Systems. Open-cycle systems include the use of high-pressure gas combined with a Joule-Thomson (J-T) expansion valve, cryogenic liquids, or solid cryogenics.

Joule-Thomson.[17,18] The J-T cooling process exploits the cooling effect obtained by the adiabatic expansion of a high-pressure (1,000 to 6,000 lb/in²) gas through an orifice. The cooled,

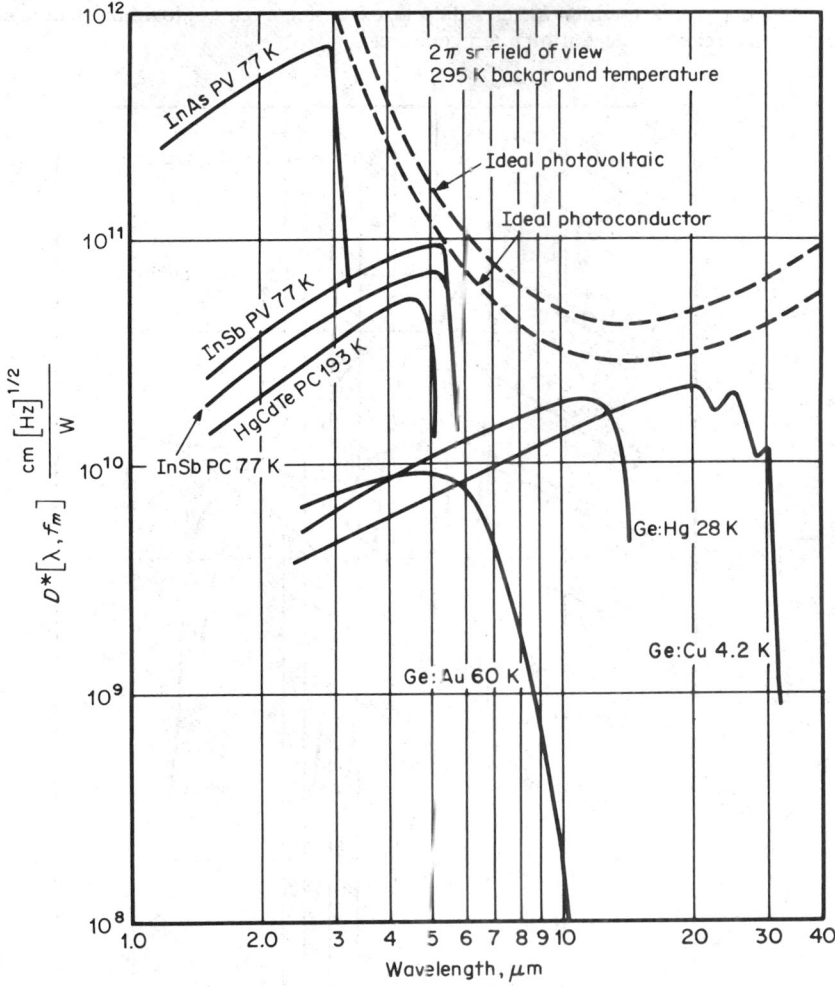

Fig. 11-68. Spectral $D*$ of photodetectors.[6]

expanded gas is used to cool incoming gas. This process of regenerative cooling continues until liquid forms at the orifice. Figure 11-70 shows a typical J-T cooler. Self-regulating throttle valves can be placed in the orifice of the J-T cooler to maintain constant temperature, reduce clogging, and improve gas economy. Some gases, e.g., helium, hydrogen, and neon, require precooling before the J-T cooling effect can occur.

Liquid-Cryogen Storage. Cryogens can be stored as liquids in equilibrium with their vapors (subcritical) or as homogeneous fluids at high pressure and temperatures (supercritical). There are two basic types of storage, direct-contact and liquid-feed. In the direct-contact (or integral) system, the detector is built into a cryogenic dewar which stores the liquid cryogen. In the liquid-feed method, the cryogen is fed to the detector from a remote storage tank. The liquid is transferred by gravity or by gas pressure resulting from the natural pressure buildup in the storage tank.

Solid-Cryogen Storage.[19]   Stored solid cryogen sublimes, causing an increase in pressure and temperature. The detector temperature and the dewar pressure are maintained at constant levels by venting the gas through specially designed ducts. Such coolers have operated successfully in space for a year or more.

170. Mechanical Coolers. Closed-cycle, expansion-engine refrigerators (described below) are used to cool infrared detectors. Cooling is produced by the expansion of a gas from a high pressure to a low pressure, with consequent reduction of working gas temperature. The figure of merit for coolers is the coefficient of performance (COP), the ratio of the produced cooling power to the power supplied. The Carnot cycle[20] is used as a standard of comparison because for given temperature limits its COP is maximum. For a Carnot engine operating

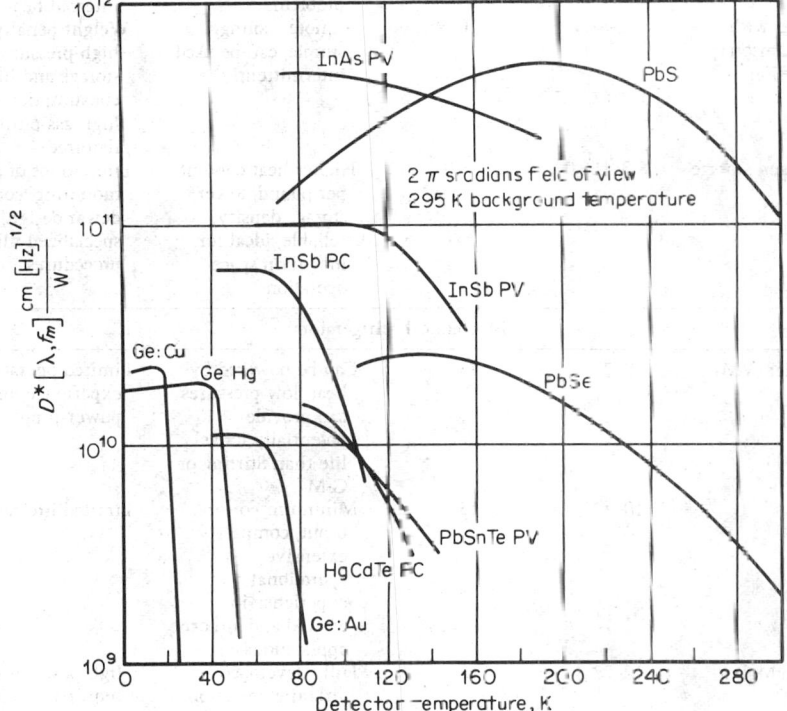

Fig. 11-69. Peak D^* vs. temperature for selected detectors. PC = photoconductive mode, PV = photovoltaic mode.

between the temperatures T_a and T_c, COP = $T_c/(T_a - T_c)$, where heat is absorbed at T_c and heat is rejected at a higher temperature T_a.

Stirling.[21]   In the Stirling cycle, cooling is obtained by cyclic out-of-phase motion of a compression piston and a displacer-regenerator (Fig. 11-71). The working gas is compressed while occupying the ambient space, at temperature T_a, by an upward motion of the compression piston, reducing the gas volume. Heat of compression is rejected to ambient. The COP for an ideal working gas is equal to that of the Carnot engine. This refrigeration cycle is well developed. Stirling cycle refrigerators have the best COP in practice (10^{-3} to 5×10^{-2}) and the best ratio of total weight per watt refrigeration compared with other refrigerators.[22]

Vuilleumier (VM).[23]   This is a heat-driven cycle which is exploited for its long life and low vibration, due in part to inherently very low dynamic forces on moving parts. Coolers have been built that provide refrigeration at 10 K. Cooling is obtained by cyclic out-of-phase motion of two displacer-regenerators (Fig. 11-72). The working gas throughout the entire cooler is compressed by downward motion of the hot displacer-regenerator as it transfers part of the gas at the ambient end into the heated end. COP is equal to that for two Carnot heat engines in series: COP = $(T_c/T_h)(T_h - T_a)/(T_a - T_c)$. Typical COP values of 3×10^{-4} to 2×10^{-2} are obtained.

TABLE 11-27 Cooling-System Characteristics[16]

Cooler type	Temp, K	Cooling capacity, W	Advantages or features	Disadvantages or limitations
Open-cycle expendable systems				
Liquid storage (subcritical)	4.2–77.0	Unlimited	Extensive experience, reliable, remote cooling, lightweight	Limited operating time, complex dewar-design logistics, phase separation, problems in space
Single-phase storage (supercritical)	5.2–126.0	Unlimited	Homogeneous fluid provides increased design flexibility, eliminates two-phase problems	Higher pressure and temperature, increased dewar weight, may require internal heater
Gas storage with Joule-Thomson (J-T) expansion	4.2–87.4	20	Remote cooling, simple, can be used intermittently	Weight penalty for high-pressure gas storage and high consumption rates, high gas purity required
Solid cryogen storage	8.3–150.0	1.0	Higher heat content per pound, lower storage density, reliable, ideal for long-term space operation	Limitations of detector mounting, complex dewar design, specialized filling procedures
Mechanical refrigerators				
Vuilleumier (VM)	10–77	15	Can be powered by heat, low pressures can provide potentially longer life than Stirling or G-M	Limited operational experience, moderate power input
Stirling	10–77	15	Minimum power input, compact, extensive operational experience in ground and airborne applications	Limited life capability
Gifford-McMahon (G-M)	10–77	15	Fully developed for airborne operations, split cycle permits remote cooling	High power input required, limited life capability
Closed-cycle J-T	77	5.0	Eliminates logistics associated with open-cycle system	High input power, limited-life compressors
Thermoelectric				
Single-stage	230–300	100.0	Lightweight, compact, high reliability, low cost	Limited minimum temperature attainable
Multiple-stage, cascaded	145–230	1.0	Lightweight, compact, high reliability, low cost	High power input required at low temperature
Radiation				
Small radiator	80–100	0.10	Completely passive, reliable, long life capabilities, low cost	Orientation may be limited, parasitic heat leak and area become prohibitive as temperature decreases and cooling load increases
Large, external radiators	100–200	10		

Gifford-McMahon[24] (GM) and *Solvay*.[25] By separating the expander from the compressor, a refrigeration system can be constructed that consists of a simple, lightweight cooling unit and a compressor which can be located remotely, connected to the expander with pressure lines. A piston displacer-regenerator is pneumatically moved up and down by timed valving of the high and low working-gas pressure. Cyclic charging and discharging of the expander working-gas pressures with timed piston motion will pump heat from the cold to the ambient end. Heat pumped from the cold end using the GM cycle is rejected at the ambient end of the expander. Heat pumped from the cold end using the modified Solvay cycle is rejected along pressure lines and at the compressor. For these coolers COP values of 2×10^{-4} to 10^{-2} are measured.

J-T Closed Cycle.[26] In this system a compressor is used to supply high-pressure gas to the J-T throttling valve. After expansion, the gas is recycled through the compressor. COP values of 3×10^{-3} to 10^{-2} are obtained for this system.

171. Thermoelectric Cooler (TE).[27] The basic operating principle of the thermoelectric cooler, the Peltier cooling effect, is the absorption or generation of heat as a current passes through a junction of two dissimilar materials (Fig. 11-73). Electrons passing across the junction absorb or give up an amount of energy equal to the transport energy and the energy difference between the disimilar-materials conduction bands. Cryogenic temperatures are reached using heat rejected from one thermoelectric cooler stage to supply thermal input to the stage below. The maximum temperature difference attainable on a practical basis is about 150°C, which implies a minimum attainable temperature of approximately 150 K. Typical performance characteristics are shown in Table 11-28 for TE coolers.

172. Radiation Coolers.[28] Radiation coolers are used in spaceborne applications. These systems are passive and cool by radiating heat from the detector into the low-temperature (4-K) sink of deep space.

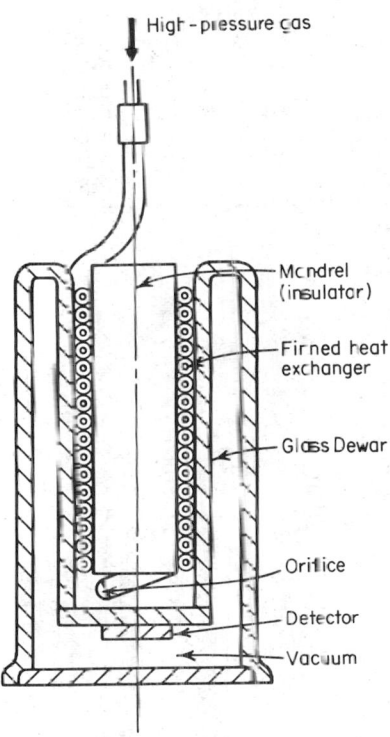

Fig. 11-70. Single-stage, open-cycle Joule-Thomson cooling system.

The radiators consist of a suitably sized cold plate of high emissivity connected to the detectors. The high-vacuum in-space environment minimizes convective heating, but the radiator must be shielded from sunlight and (for near-earth orbits) from thermal emission and reflected sunlight from the earth and its atmosphere. Radiation coolers have been designed for cooling milliwatt-level loads at 85 K and 5-W loads at 135 K.

173. References on Infrared Detectors and Cooling

1. Kruse, P. W., L. D. McGlauchlin, and R.B. McQuistan "Elements of Infrared Technology," Wiley, New York, 1962.

2. Willardson, R. K., and A. C. Beer "Semiconductors and Semimetals" Vol. 5, "Infrared Detectors," Academic, New York, 1970.

3. Willardson, R. K., and A. C. Beer "Semiconductors and Semimetals," Vol. 12, "Infrared Detectors II," Academic, New York, 1977.

4. Limpens, T., and J. Mudar Detectors, Chap. II in W. L. Wolfe and G. J. Zissis (eds.), "The Infrared Handbook," GPO, Washington, 1978.

5. Hudson, R. D., and J. W. Hudson "Infrared Detectors," Dowden, Hutchinson, and Ross, Stroudsburg, Penn., 1975.

6. These data are from Refs. 2 to 5 above and from "The SBRC Brochure," new ed., Santa Barbara Research Center, Goleta, Calif., 1979.

7. Steckl, A. J. Charge-Couple Devices, Chap. 12 in W. L. Wolfe and G. J. Zissis, "The Infrared Handbook," GPO, Washington, 1978.

8. McCaughan, D. V., and B. R. Holeman Applications of CCDs to Imaging, Chap. 5 in M. J. Howes and D. V. Morgan, "Charge-Coupled Devices and Systems," Wiley, New York, 1979.

9. Milton, A. F. Charge Transfer Devices for Infrared Imaging, Chap. 6 in R. J. Keyes (ed.), "Topics in Applied Physics," Vol. 19, Springer-Verlag, Berlin, 1977.

10. Thom, R. D., et al. *Int. Conf. CCD*, San Diego, 1975, pp. 31–41.

Labels in figure: High-pressure gas; Mandrel (insulator); Finned heat exchanger; Glass Dewar; Orifice; Detector; Vacuum

TABLE 11-28   Typical Thermoelectric Cooler Performance[16]

Hot-junction temperature = 300 K

No. of stages	Cooling load, W	Load T, K	Input power, W	ΔT, K	COP
1	5.5	250	27.2	50	0.202
2	0.2	223	4.4	87	0.05
3	0.15	203	3.5	107	0.043
4	0.08	196	7.0	104	0.011
	0.02	186	6.6	105	0.014
	0.01	178	6.0	122	0.0016
6	0.02	180	10.2	120	0.0019
	0.02	170	9.3	130	0.002
8	0.01	145	40	180	0.00025

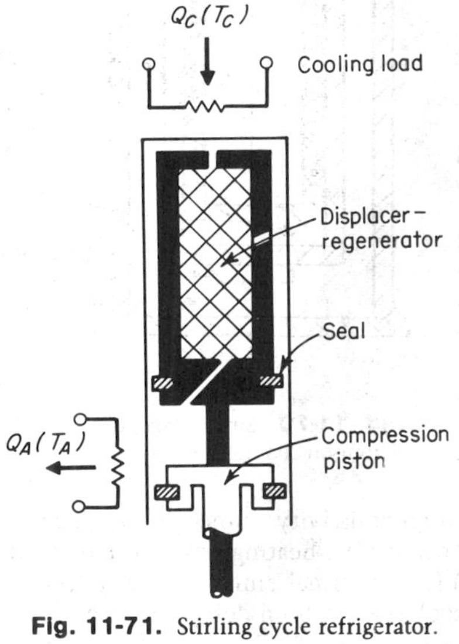

Fig. 11-71. Stirling cycle refrigerator.

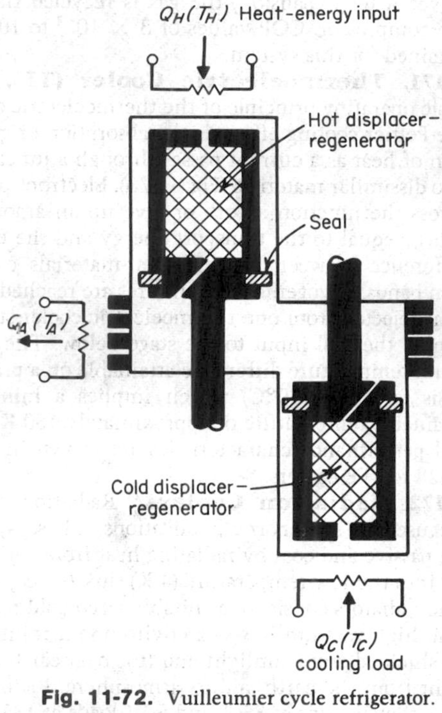

Fig. 11-72. Vuilleumier cycle refrigerator.

11. Kim, J. C., et al.   *Proc. IEDM,* **76**:550–554 (1976).

12. Pines, M. Y., et al.   *Proc. IEDM,* **74**:446–450 (1974).

13. Loh, K. W., et al.   *IEEE Trans. Elec. Dev.,* **ED-24**: 1041–1048 (1977).

14. Buss, D. D., et al.   *Proc. IEDM,* **78**:476–500 (1978).

15. Broudy, R., et al.   *Proc. Soc. Photo-Opt. Instrum. Eng.,* **132**:10–26 (1978).

16. Donabedian, M.   Cooling Systems, Chap. 15 in W. L. Wolfe and G. J. Zissis (eds.), "The Infrared Handbook," GPO, Washington, 1978.

17. Stephens, S.W.   *Infrared Phys.,* **8**(1):25–35 (1968).

18. Buller, J. S.   "Miniature Self-Regulating Rapid Cooling Joule-Thomson Cryostat," Santa Barbara Research Center, Goleta, Calif., May 1969, p. 12; "Infrared Brochure," Santa Barbara Research Center, Goleta, Calif., 1979, pp. 37–39.

19. Fowle, A. A.   *Adv. Cryogenic Eng.,* **2**:198–201 (1965).

20. Zemansky, M. W.   "Heat and Thermodynamics," pp. 157–166, McGraw-Hill, New York, 1957.

21. Kohler, J. W. L.   "Progress in Cryogenics," Vol. 2, p. 41, Heywood, London, 1960; *Sci. Am.,* **212**:119 (April 1965).

22. Jensen, H. L., et al.   Investigation of External Refrigeration Systems for Long Term Cryogenic Storage, *Lockheed Missiles and Space Co. Rep.* LMSC-A903162, May 28, 1970.

23. Timmerhaus, K. D.   *Adv. Cryogen. Eng.,* **14**:332–377 (1968).

24. Ackermann, R. A.   An Investigation of Gifford-McMahon Cycle and Pulse-Tube Refrigerators, *U.S. Army ECOM Tech. Rep.* 3245, March 1970.

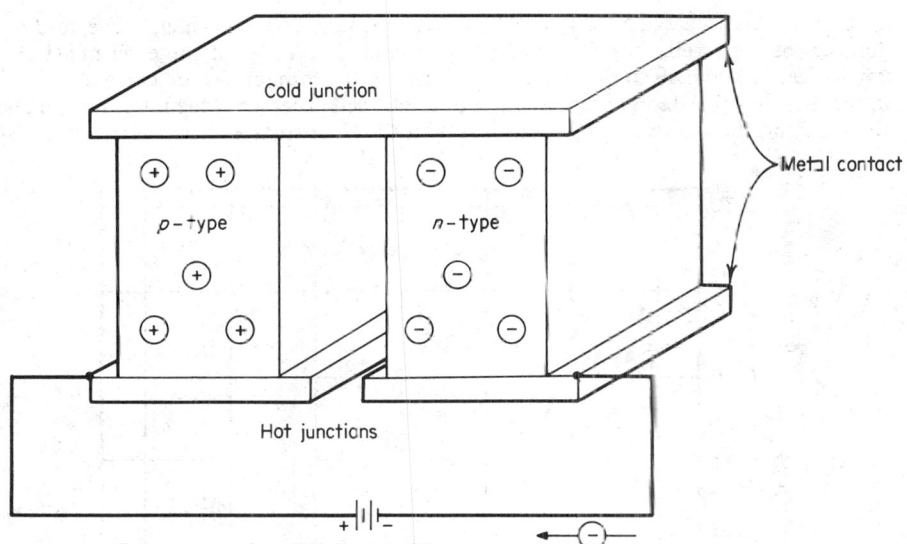

Fig. 11-73. Single-stage thermoelectric refrigerator.

25. Longsworth, R. C.   A Modified Solvay Cycle Cryogenic Refrigerator, 1970 Cryogen. Conf., Boulder, Colo., Pap. K-5.

26. Daunt, J. G.   Miniature Cryogenic Refrigerators, AD 697 972, Stevens Institute of Technology, Hoboken, N.J., August 1969.

27. Ioffe, A. F.   "Semiconductor Thermoelements and Thermoelectric Cooling," Tri-Litho Offset Ltd., London, 1957.

28. Annable, R. V.   Appl. Opt., **9**(1):185–193 (1970).

Solar Cells*

BY KIM W. MITCHELL

174. Description. A solar cell is a semiconductor electric-junction device which absorbs the radiant energy of sunlight and converts it directly and efficiently into electric energy. Solar cells may be used individually as light detectors, e.g., in cameras, or connected in series and parallel to obtain the required values of current and voltage for electric-power generation.

Most solar cells are made from single-crystal silicon and have been too expensive for generating electricity, except for space satellites and remote areas where low-cost conventional power sources are unavailable. Recent research has emphasized lowering solar-cell cost by improving performance and by reducing materials and manufacturing costs. One approach is to use optical concentrators such as mirrors or fresnel lenses to focus the sunlight onto solar cells of smaller area. Other approaches replace the high-cost single-crystal silicon with thin films of amorphous or polycrystalline silicon, gallium arsenide, cadmium sulfide, or other compounds.

175. Solar Radiation. The intensity and quality of sunlight is dramatically different outside the earth's atmosphere and on the surface of the earth, as shown in Fig. 11-74. The number of photons at each energy is reduced upon entering the earth's atmosphere due to reflection, scattering, or absorption by water vapor and other gases. Thus, while the solar energy at normal incidence outside the earth's atmosphere is 1.36 kW/m², on the surface of the earth at noontime on a clear day the intensity is about 1 kW/m².

176. Principles of Operation. The conversion of sunlight into electric energy in a solar cell involves three major processes: (1) absorption of the sunlight in the semiconductor material;

(2) generation and separation of free positive and negative charges which move to different regions of the solar cell, creating a voltage in it; and (3) transfer of these separated charges through electric terminals to the outside application in the form of electric current.

In the first step, the absorption of sunlight by a solar cell depends on the intensity and quality of the sunlight, the amount of light reflected from the front surface of the solar cell, the semi-

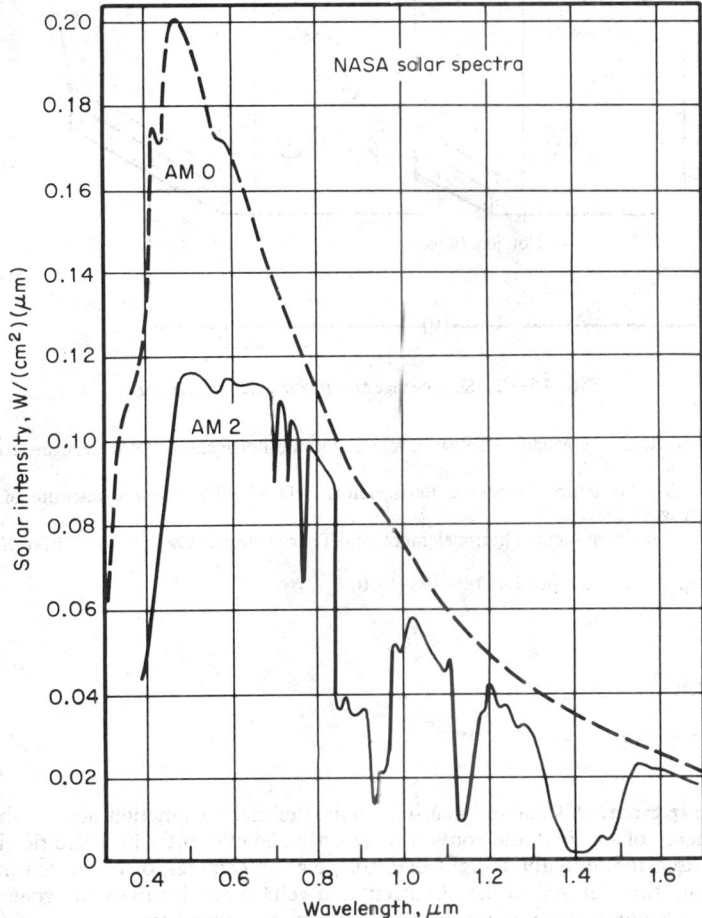

Fig. 11-74. Variation of solar intensity with wavelength of photons for air mass O (AMO) outside the earth's atmosphere and for AM2, a typical spectrum on the surface of the earth. *(From McGraw-Hill Encyclopedia of Science and Technology, 5th ed., 1982.)*

conductor bandgap energy, which is the minimum light (photon) energy the material absorbs, and the layer thickness. Some materials, e.g., silicon require tens of micrometers thickness to absorb most of the sunlight, while others, e.g., gallium arsenide, cadmium telluride, and copper sulfide, require only a few micrometers.

When light is absorbed in the semiconductor, a negatively charged electron and positively charged hole are created. The heart of the solar cell is the electric junction which separates these electrons and holes from each other after they are created by the light. An electric junction can be formed by the contact of (1) a metal to a semiconductor (a Schottky barrier); (2) a liquid to a semiconductor (a photoelectrochemical cell); or (3) two semiconductor regions (a pn junction).

The fundamental principles of the electric junction can be illustrated with the silicon pn junction. Pure silicon, to which a trace amount of a column V element such as phosphorus has been added, is an n-type semiconductor where electric current is carried by free electrons. Each phosphorus atom contributes one free electron, leaving behind the phosphorus atom bound to the crystal structure with a unit positive charge. Similarly, pure silicon to which a trace amount of a column III element such as boron has been added is a p-type semiconductor, where the electric current is carried by free holes. Each boron atom contributes one hole, leaving behind the boron atom with a unit negative charge. The interface between the p- and n-type silicon is called the pn junction. The fixed charges at the interface due to the bound boron and phosphorus atoms create a permanent-dipole charge layer with a high electric field. When photons of light energy from the sun produce electron-hole pairs near the junction, the built-in electric field forces the holes to the p side and the electrons to the n side, as illustrated in Fig. 11-75. This displacement of free charges results in a voltage difference between the two regions of the crystal, the p region being plus and the n region minus. When a load is connected at the terminals, an electron current flows in the direction shown by the arrow and useful electric power is available at the load.

177. Solar-Cell Characteristics. The electrical characteristics of a typical silicon pn-junction solar cell are shown in Fig. 11-76. Figure 11-76a shows open-circuit voltage and short-circuit current as a function of light intensity from total darkness to full sunlight (1,000 W/m²).

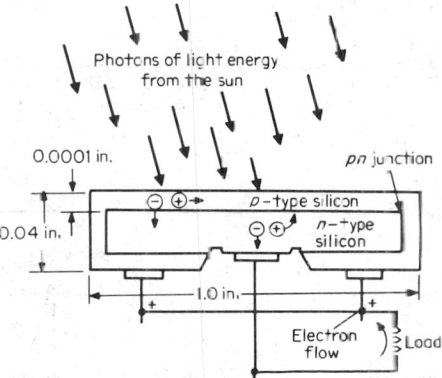

Fig. 11-75. Cross-sectional view of a silicon pn-junction solar cell, illustrating the creation of electron pairs by photons of energy from the sun. *(From McGraw-Hill Encyclopedia of Science and Technology, 5th ed., 1982.)*

The short-circuit current is directly proportional to light intensity and amounts to 28 mA/cm² at full sunlight. The open-circuit voltage rises sharply under weak light and saturates at about 0.6 V for radiation between 200 and 1,000 W/cm². The variation in power output from the solar cell irradiated by full sunlight as its load is varied from short circuit to open circuit is shown in Fig. 11-76b. The maximum power output is about 11 mW/cm² at an output voltage of 0.45 V.

Under these operating conditions, the overall conversion efficiency from solar to electric energy is 11%. The output power as well as the output current is proportional to the irradiated surface area, whereas the output voltage can be increased by connecting cells in series, as in a chemical storage battery. Experimental samples of silicon solar cells have been produced which operate at efficiencies up to 18%, but commercial cell efficiency is around 10 to 12% under normal operating conditions.

Using optical concentration to intensify the light incident on the solar cell, efficiencies above 20% have been achieved with silicon cells and above 25% with gallium arsenide cells. New concepts to split the solar spectrum and illuminate two optimized solar cells of different band-gaps have achieved efficiencies above 28% with expected efficiencies of 35%. Thin-film solar cells currently between 4 and 9% efficiency are expected in low-cost arrays to be above 10%.

178. References on Solar Cells and Arrays

1. Angrist, S.   "Direct Energy Conversion," 3d ed., Allyn and Bacon, Boston, 1976.

2. *IEEE Photovoltaics Spec. Conf.* 10th, Nov. 13–15, 1973; 11th, May 6–8, 1975; 12th, Nov. 15–18, 1976; 13th, June 5–8, 1978; 14th, Jan. 7–10, 1980, Institute of Electrical and Electronics Engineers, New York

3. "Solar Cells and Photocells," 2d ed., International Rectifier, Semiconductor Division, El Segundo, Calif., 1973.

4. Backus, C. E. (ed.)   "Solar Cells," Institute of Electrical and Electronics Engineers, New York, 1976.

5. Chalmers, B.   Photovoltaic Generation of Electricity, *Sci. Am.*, October 1976, Vol. 235, pp. 34–53.

6. Hovel, H. J.   "Solar Cells," Vol. 11 of R. K. Willardson and A. C. Beer (eds), "Semiconductors and Semimetals," Academic, New York, 1975.

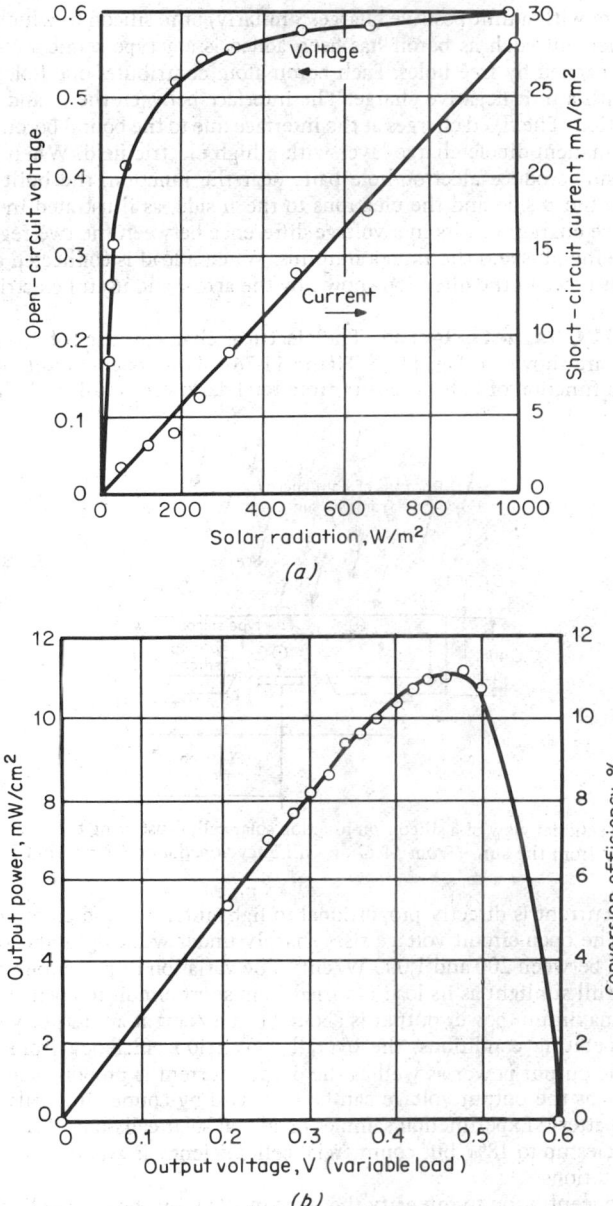

Fig. 11-76. Electrical characteristics of silicon *pn*-junction solar cell at operating temperature of 17°C: (*a*) variation of open-circuit voltage and short-circuit current with light intensity; (*b*) variation in power output as load is varied from short to open circuit. *(From McGraw-Hill Encyclopedia of Science and Technology, 5th ed., 1982.)*

Optical Accessories

179.Index to Literature. A discussion of conventional optics and a listing of standard optics textbooks, handbooks, or brochures is considered to be beyond the scope of this edition of the Handbook. However, in the past few years new types of optical components have been improved and have become increasingly available. Of these, liquid crystals and plastic lenses may be of special interest to the electronics engineer and may not yet be adequately cited in standard optics handbooks. A short list of such references is therefore provided here.

180. References on Optical Accessories

REFERENCES ON LIQUID CRYSTALS

1. Kelker, H., and R. Hatz "Handbook of Liquid Crystals," Verlag Chemie Weinheim, Germany, 1979.
2. Saeva, F. D. "Liquid Crystals: The Fourth State of Matter," Dekker, New York, 1979 (of special interest is the chapter by J. Castellano on applications).
3. Meier, G., E. Sackmann, and J. G. Grabmaier "Applications of Liquid Crystals," Springer-Verlag, Berlin, 1975.
4. *Kodak Publ.* **JJ**-14, Eastman Liquid Crystal Products, Eastman Kodak, Rochester, N.Y.
5. "Licristal," E M Laboratories, Elmsford, N.Y.
6. BDH (British Drughouse) Brochure on Liquid Crystals. Distributed by E M Laboratories, Elmford, N.Y.

REFERENCES ON PLASTIC OPTICS

7. Plastics Design & Processing, Optical Division, Bell & Howell, Chicago, 1979.
8. "The Handbook of Plastic Optics," U.S. Precision Lens, Cincinnati. 1973.

Section 12

Filters and Attenuators*

EDWIN C. JONES, JR. *Professor of Electrical Engineering, Iowa State University; Senior Member, IEEE*

HARRY W. HALE *Professor of Electrical Engineering, Iowa State University; Senior Member, IEEE*

CONTENTS

Numbers refer to paragraphs

*Parts of this section are adapted from D. G. Fink and H. W. Beaty (eds.), "Standard Handbook for Electrical Engineers," 11th ed., McGraw-Hill. New York, 1978, reprinted by permission.

FILTERS

1. Historical Note. The basic concept of the electric wave filter was originated by Campbell and Wagner, working independently, in 1915. The continuation of this work has proceeded along two paths, *image-parameter filter design* and *insertion-loss filter design*.

The design of *image-parameter filters* is based on the concept of a cascade of fundamental sections with matched image impedances. This type of filter design was given a marked impetus by the work of Zobel in 1923 on *m-derived filters* and was dominant for more than 30 years.

Insertion-loss filter design is based on the specification of an appropriate network response in the form of a network function part such as transfer-function magnitude or phase and the synthesis of a lumped network to achieve the specified response. This type of filter design had its origins in the work of Norton, Foster, Cauer, Bode, Darlington, and others and was well developed by 1940. It did not gain immediate acceptance, however, because of the requirement for precise and lengthy computations in the design process. When the advent of digital computers in the 1950s eliminated the computational burden, this type of filter design gained rapid acceptance and has largely supplanted image-parameter design. Today, design of this type of filter is largely a matter of looking up element values in computer-determined tables for various classes of filters and modifying them in a routine fashion.

Image-parameter filters will not be discussed here. Detailed descriptions are available elsewhere.[1]*

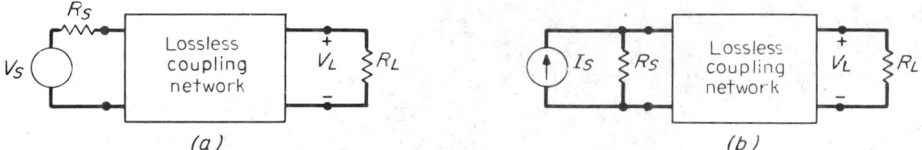

Fig. 12-1. General form of network: (a) with voltage source; (b) with current source.

2. Insertion-Loss Filter Design. The design process is carried out in terms of one of the two general networks shown in Fig. 12-1 according to the type of source. The procedure consists of the following general steps:

1. The filter characteristics are specified.

2. The filter characteristics are approximated in some sense by a realizable network function part. This is often the magnitude of the network transfer function

$$|G_{SL}(j\omega)| = \left| \frac{V_L}{V_S}(j\omega) \right| \qquad (12\text{-}1)$$

$$|Z_{SL}(j\omega)| = \left| \frac{V_L}{i_S}(j\omega) \right| \qquad (12\text{-}2)$$

for the networks of Fig. 12-1a and b, respectively. (For convenience, subsequent references to G_{SL} will omit subscripts and also apply to Z_{SL} except where a specific distinction is made.) The phase function

$$\theta(\omega) = \arg G(j\omega) \qquad (12\text{-}3)$$

is also used in some applications.

3. The coupling network of Fig. 12-1a or b is synthesized to realize the network function part of step 2. In 1939, Darlington[2] showed that this coupling network can be lossless. In addition, it is a ladder network for many common cases. This is an attractive feature.

The emphasis in the design procedure is on steps 1 and 2, their result being parameters that can be used to enter computer-determined tables of element values for normalized prototypes to which scaling and frequency transformation techniques can then be applied.

3. The Classification of Ideal Filters. An ideal filter is said to pass (with no attenuation) all frequencies within its passbands and to stop (with infinite attenuation) all frequencies in its stop bands. The following five basic types of filters are commonly used:

1. The low-pass filter passes frequencies from zero to its cutoff frequency and stops all frequencies higher than the cutoff frequency.

*Superior numbers refer to bibliography, Par. **12-43**.

2. The high-pass filter stops all frequencies below its cutoff frequency and passes all frequencies above it.

3. The bandpass filter passes all frequencies between its lower and upper cutoff frequencies and stops all frequencies outside this range.

4. The band-elimination filter stops all frequencies between its lower and upper cutoff frequencies and passes all frequencies outside this range.

5. The all-pass filter passes all frequencies. Its purpose is to produce a predictable phase shift, and it is used to produce a constant time delay for all frequencies.

These ideal filter characteristics are shown in Fig. 12-2.

4. The Low-Pass Prototype. The low-pass, high-pass, bandpass, and band-elimination filters can be designed by a direct process An alternative procedure consists of first determining a low-pass prototype normalized to make $R_t = 1$ and $\omega_c = 1$ and then using scaling and frequency transformations to arrive at the desired filter. The latter procedure, used here, makes it possible to use a reasonable number of computer-determined tables of element values for low-pass prototypes with various characteristics. This is particularly significant since this type of filter design requires precision in prototype computation to achieve satisfactory results. The following paragraphs are organized around this concept.

5. The Approximation Problem. The ideal low-pass filter characteristics can only be

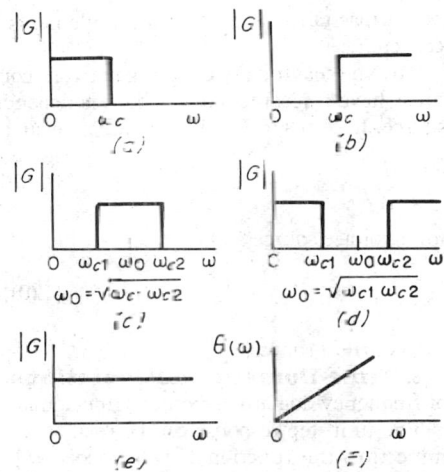

Fig. 12-2. Ideal-filter characteristics: (a) low-pass; (b) high-pass (c) bandpass; (d) band-elimination; (e) all-pass (magnitude), (f) all-pass (phase).

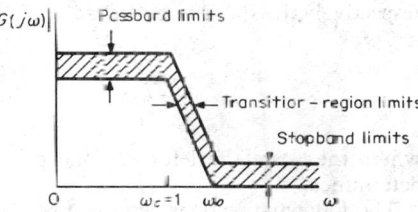

Fig. 12-3. Limits on frequency response.

approximated. A basic problem is to select a transfer-function magnitude $|G(j\omega)|$ that approximates the ideal characteristic *and* results in a realizable coupling network. This is indicated in Fig. 12-3, where limits on the frequency response are suggested. The approximation problem consists of finding an appropriate $|G(j\omega)|$ that will lie within the shaded region.

Generally, the magnitude-squared transfer function can be of the form

$$|G(j\omega)|^2 = K^2 \frac{A(\omega^2)}{D(\omega^2)} \tag{12-4}$$

It is noted that the numerator and denominator must be even, nonnegative polynomials There are, of course, other restrictions. Equation 12-4 can be rewritten as

$$|G(j\omega)|^2 = K^2 \frac{N(\omega^2)}{N(\omega^2) + M(\omega^2)} = K^2 \frac{1}{1 + M(\omega^2)/N(\omega^2)} \tag{12-5}$$

One approach to the approximation problem is to let $N(\omega^2) = 1$, in which case the magnitude-squared transfer function becomes

$$|G(j\omega)|^2 = \frac{K^2}{1 + M(\omega^2)} \tag{12-6}$$

and the problem is to select a polynomial $M(\omega^2)$ such that

$$|M(\omega^2)| \begin{cases} \ll 1 & \omega \ll 1 \\ \gg 1 & \omega \gg 1 \end{cases}$$

and which will lead to a realizable network.

Low-pass filters based on Eq. (12-6) are ladder structures with inductors as the series elements and capacitors as the shunt elements. If the more general form of Eq. (12-5) is used, the resulting filter will again be a ladder but the series elements may be parallel combinations of inductance and capacitance and the shunt elements may be series combinations of inductance and capacitance.

To avoid having the coupling network contain ideal transformers, the constant K of Eq. (12-4) must have a specific value which is dependent on the terminating resistances and the type of source. In terms of Fig. 12-1, K must be such that

$$|G(0)| = \frac{R_L}{R_S + R_L} \tag{12-7}$$

for a voltage source and

$$|Z(0)| = \frac{R_S R_L}{R_S + R_L} \tag{12-8}$$

for a current source.

6. Time-Domain Considerations. Filter specifications are ordinarily given in terms of frequency-domain response. Time-domain response is often an important factor also.

The unit-step response can be used as a measure of time-domain response. In order to determine this, the function $G(s)$ is needed. This can be determined from the magnitude-squared function of Eq. (12-6) by use of analytic continuation to obtain

$$|G(j\omega)|^2_{\omega^2 = -s^2} = G(s)G(-s) = K^2 \frac{1}{1 + M(-s^2)} \tag{12-9}$$

The left-half-plane and right-half-plane poles of $G(s)G(-s)$ must be assigned to $G(s)$ and $G(-s)$, respectively, if $G(s)$ is to be realizable. The result is

$$G(s) = \frac{k}{\prod\limits_{j=1}^{n} (s - s_j)} \tag{12-10}$$

where the s_j are the left-half-plane poles. With $G(s)$ known, the unit-step response can be determined.

The following sections describe five different approximations, four to the ideal low-pass frequency response and one to the ideal all-pass (time-delay) response. In each case the pole locations are tabulated as linear and quadratic factors. The unit-step response for each of the five approximations is described in Par. **12-13** in terms of tabulated figures of merit.

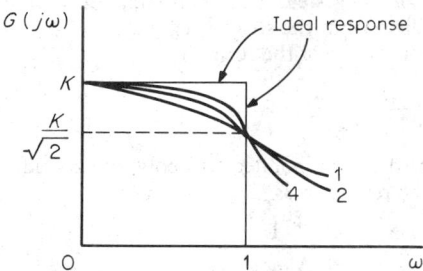

Fig. 12-4. The Butterworth approximation.

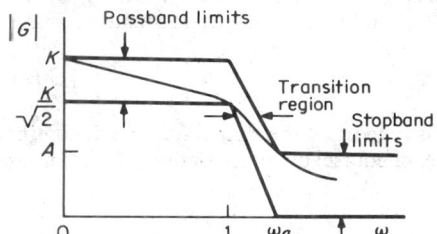

Fig. 12-5. Determination of the minimum value of n for the Butterworth approximation.

7. The Butterworth Approximation. The transfer-function magnitude

$$|G(j\omega)| = K/(1 + \omega^{2n})^{1/2} \tag{12-11}$$

is known as the nth-order Butterworth low-pass approximation. The general form of the approximation is shown in Fig. 12-4.

It is clear that increasing n improves the approximation in both passband and stop band. This also has adverse effects on the time-domain response and results in a larger number of elements in the filter. Thus the choice of n is the smallest value that will satisfy some specific specification on the frequency-domain response.

The typical problem of determining n is illustrated in Fig. 12-5. If the specifications require that

$$|G(j\omega)| \leq A \qquad \omega_c \leq \omega < \infty \tag{12-12}$$

then the value of n required is the smallest integer value satisfying the inequality

$$A \geq K/(1 + \omega_a^{2n})^{1/2} \tag{12-13}$$

It can be shown that the first $2n - 1$ derivatives of the $|G j\omega)|$ of Eq. (12-11) are zero at $\omega = 0$. This approximation is often called a maximally flat approximation for this reason. It is also noted that the derivative at the cutoff frequency is

$$\frac{d}{d\omega} |G(j\omega)|_{\omega=1} = \frac{-nK}{2^{3/2}} \tag{12-14}$$

The frequency response is often described in terms of the attenuation

$$\alpha = 20 \log \frac{|G(j\omega)|}{|G(0)|} = 10 \log (1 + \omega^{2n}) \qquad dB \tag{12-15}$$

It is noted that at the cutoff frequency

$$\alpha_{\omega=1} = 10 \log 2 = 3.0103 \; dB \tag{12-16}$$

This is the usual interpretation of a cutoff frequency, also referred to as the half-power or "3-dB" frequency. It is also noted that for $\omega \gg 1$

$$\alpha \approx 20 \; n \log \omega \qquad dB \tag{12-17}$$

A specification of minimum attenuation α_{min} for $\omega \geq \omega_a$ results in

$$\alpha_{min} \geq 10 \log (1 + \omega^{2n}) \tag{12-18}$$

and a solution for n with the equality sign gives

$$n = \frac{1}{2} \frac{\log (10^{\alpha_{min}/10} - 1)}{\log \omega_a} \tag{12-19}$$

which, for large attenuation, is reasonably approximated by

$$n \approx \frac{\alpha_{min}}{20 \log \omega_a} \tag{12-20}$$

with the next larger integer value being used.

As indicated in Par. **12-6**, the poles of $G(s)$ can be determined from the magnitude-squared function

$$|G(j\omega)|^2 = K^2/(1 + \omega^{2n}) \tag{12-21}$$

by the use of analytic continuation. The specific pole locations are given in Table 12-1 for n ranging from 1 to 10, where they are given as quadratic and linear factors.

8. The Chebyshev Approximation. The Chebyshev approximation is

$$|G(j\omega)| = K/[1 + \epsilon^2 C_n^2(\omega)]^{1/2} \tag{12-22}$$

where $C_n(\omega)$ is the nth-order Chebyshev polynomial and ϵ is a real constant less than 1. Specifically,

$$C_n(\omega) = \begin{cases} \cos (n \cos^{-1} \omega) & \text{for } 0 < \omega \leq 1 \tag{12-23} \\ \cosh (n \cosh^{-1} \omega) & \text{for } \omega \geq 1 \tag{12-24} \end{cases}$$

It is apparent that

$$C_0(\omega) = 1 \tag{12-25}$$

and

$$C_1(\omega) = \omega \tag{12-26}$$

These can be used with the recursion formula

$$C_{n+1}(\omega) = 2\omega C_n(\omega) - C_{n-1}(\omega) \tag{12-27}$$

to develop higher-order polynomials. They are given in Table 12-2.

Each Chebyshev polynomial has real zeros that are within the interval $-1 \le \omega \le 1$ and its maximum and minimum values are $+1$ and -1, respectively, within the interval. Thus

$$C_n^2(\omega) \le 1 \qquad \text{for } 0 \le \omega \le 1 \tag{12-28}$$

$$C_n^2(0) = \begin{cases} 0 & \text{for } n \text{ odd} \\ 1 & \text{for } n \text{ even} \end{cases} \tag{12-29} \tag{12-30}$$

$$C_n^2(1) = 1 \qquad \text{for all } n \tag{12-31}$$

$$C_n^2(\omega) > 1 \qquad \text{for } \omega > 1 \tag{12-32}$$

Figure 12-6 shows the frequency response for the Chebyshev approximation for n even and odd. In either case there are n half cycles (from maximum to minimum and the reverse) in the interval $0 \le \omega \le 1$.

TABLE 12-1   Linear and Quadratic Factors of Butterworth Polynomials — Poles of Butterworth Transfer Functions

n	$[G(s)]^{-1}$	n	$[G(s)]^{-1}$
2	$s^2 + 1.4142136s + 1$	8	$s^2 + 0.3901806s + 1$
3	$s + 1$		$s^2 + 1.1111405s + 1$
	$s^2 + s + 1$		$s^2 + 1.6629392s + 1$
			$s^2 + 1.9615706s + 1$
4	$s^2 + 0.7653669s + 1$		
	$s^2 + 1.8477591s + 1$	9	$s + 1$
5	$s + 1$		$s^2 + 0.3472964s + 1$
	$s^2 + 0.6180340s + 1$		$s^2 + s + 1$
	$s^2 + 1.6180340s + 1$		$s^2 + 1.5320889s + 1$
			$s^2 + 1.8793852s + 1$
6	$s^2 + 0.5176381s + 1$		
	$s^2 + 1.4142136s + 1$		
	$s^2 + 1.9318517s + 1$	10	$s^2 + 0.3128689s + 1$
7	$s + 1$		$s^2 + 0.9079810s + 1$
	$s^2 + 0.4450419s + 1$		$s^2 + 1.4142136s + 1$
	$s^2 + 1.2469796s + 1$		$s^2 + 1.7820130s + 1$
	$s^2 + 1.8019377s + 1$		$s^2 + 1.9753767s + 1$

It is also observed that for the special case of $\epsilon = 1$

$$\left. \frac{d|G(j\omega)|}{d\omega} \right|_{\omega=1} = \frac{-Kn^2}{(2)^{3/2}} \tag{12-33}$$

$$|G(j\omega)| \le K = G(0) \qquad \text{for all } \omega \tag{12-34}$$

and for n even $\qquad |G(j\omega)| \le K = G(0)\sqrt{1 + \epsilon^2} \qquad \text{for all } \omega \tag{12-35}$

The consequence of these conditions is that there are combinations of values of R_S, R_L, and ϵ that cannot be realized when n is even. In particular, the even-order Chebyshev approximation cannot be realized for any value of ϵ when $R_S = R_L$. It will be shown subsequently that the even-order Chebyshev polynomial can be modified to result in a realizable approximation with some sacrifice in stop-band performance.

Two parameters, ϵ and n, are necessary to define a particular Chebyshev approximation. Figure 12-7 illustrates the problem for n odd (the results also apply to n even). It is apparent that ϵ controls the passband limits. The ratio of upper to lower passband limit is $(1 + \epsilon^2)^{1/2}$, and the value of this ratio is sufficient to determine ϵ. The logarithmic function

$$10 \log (1 + \epsilon^2) \tag{12-36}$$

is often used to describe this ratio. Thus a Chebyshev approximation with $\epsilon = 0.7648$ is said to have a 2-dB ripple. Both ϵ and n affect the stop-band response. With ϵ known and with the specifications requiring that

$$G(j\omega) \le A \qquad \text{for } \omega_a \le \omega < \infty \tag{12-37}$$

the inequality

$$A \geq \frac{K}{[1 + \epsilon^2 C_n^2(\omega_a)]^{1/2}}$$ (12-38)

must be satisfied. Equation (12-38) can be used with the known value of ϵ to determine that

$$C_n(\omega_a) \geq \frac{(K^2 - A^2)^{1/2}}{A\epsilon}$$ (12-39)

Thus n can be determined by using $C_n(\omega_a)$ resulting from the equality sign in Eq. (12-39) in conjunction with Eq. (12-24), with the result that

$$n = \frac{\cosh^{-1} C_n(\omega_a)}{\cosh^{-1} \omega_a}$$ (12-40)

with the next larger integer value being used.

TABLE 12-2 Chebyshev Polynomials for $n = 0$ to 10

n	Chebyshev polynomial
0	1
1	ω
2	$2\omega^2 - 1$
3	$4\omega^3 - 3\omega$
4	$8\omega^4 - 8\omega^2 + 1$
5	$16\omega^5 - 20\omega^3 + 5\omega$
6	$32\omega^6 - 48\omega^4 + 18\omega^2 - 1$
7	$64\omega^7 - 112\omega^5 + 56\omega^3 - 7\omega$
8	$128\omega^8 - 256\omega^6 + 160\omega^4 - 32\omega^2 - 1$
9	$256\omega^9 - 576\omega^7 + 432\omega^5 - 120\omega^3 + 9\omega$
10	$512\omega^{10} - 1280\omega^8 + 1120\omega^6 - 400\omega^4 + 50\omega^2 - 1$

It is apparent that decreasing ϵ to improve passband response either degrades the stop-band response or requires a larger n.

The attenuation for the Chebyshev approximation is

$$\alpha = 10 \log [1 - \epsilon^2 C_n^2(\omega)] \quad \text{dB}$$ (12-41)

which is approximated by

$$\alpha \approx 10 \log \epsilon^2 C_n^2(\omega) \quad \text{dB}$$ (12-42)

for $\omega \gg 1$. When we recognize that the highest-powered term in $C_n(\omega)$ is $2^{n-1}\omega^n$, Eq. (12-42) becomes

$$\alpha \approx 20 \log \epsilon + 6(n - 1) + 20n \log \omega$$ (12-43)

as a further approximation.

It was observed for the Butterworth approximation that $\omega = 1$ is the half-power frequency. This is not the case with the Chebyshev approximation except for the special case of $\epsilon = 1$. In

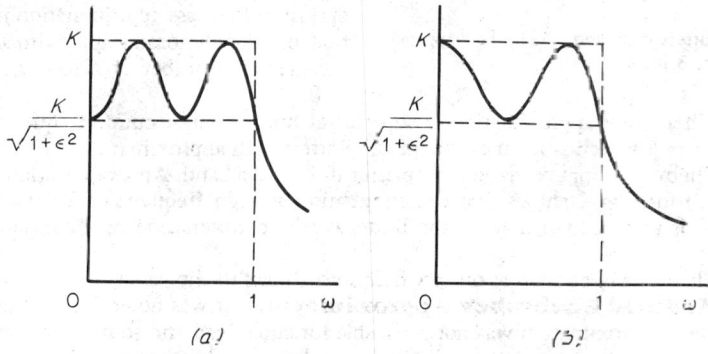

Fig. 12-6. The Chebyshev approximation: (a) n even; (b) n odd.

some applications it is desirable to normalize the Chebyshev approximation so that $\omega = 1$ is the half-power frequency. In this case, the frequency response appears as in Fig. 12-8. This can be done by determining the frequency at which the magnitude of $G(j\omega)$ becomes $K/(2)^{1/2}$, implying that

$$\epsilon^2 C_n^2(\omega_3) = 1 \tag{12-44}$$
or
$$C_n^2(\omega_3) = 1/\epsilon \tag{12-45}$$

There is an explicit expression for the resulting value of ω

$$\omega_{hp} = \cosh\left(\frac{1}{n}\cosh^{-1}\frac{1}{\epsilon}\right) \tag{12-46}$$

The poles of $G(s)$ can be determined from the magnitude-squared transfer function. Table 12-3 gives the quadratic factors for n ranging from 1 to 10 and ripples of 3, 1, 0.5, 0.1, and 0.01 dB rather than the explicit complex pole locations.

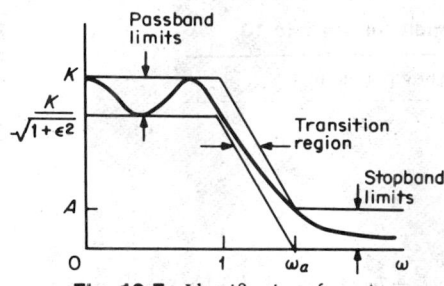

Fig. 12-7. Identification of ϵ and n.

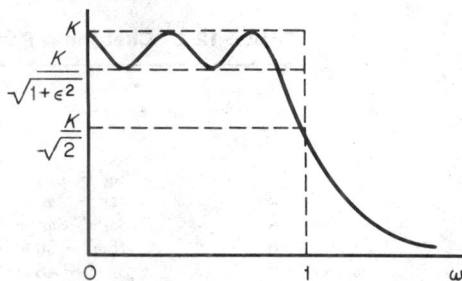

Fig. 12-8. The Chebyshev approximation (n odd) normalized to make $\omega = 1$ the half-power frequency.

The pole locations and the element values in Table 12-13 are based on $\omega = 1$ being the end of the ripple, as in Fig. 12-7. The half-power frequencies ω_{hp} of Eq. (12-46) are given in Table 12-14 for each value of n and ϵ for ready reference and use in scaling if desired.

9. Comparison of Butterworth and Chebyshev Approximations. An equitable comparison of the Butterworth and Chebyshev approximations can be made only when they have the same half-power frequency. In general, this involves normalizing the Chebyshev approximation to a 3-dB bandwidth before comparison. This occurs naturally when $\epsilon = 1$, and considerable insight into the comparative merits of the two approximations can be obtained from this special situation.

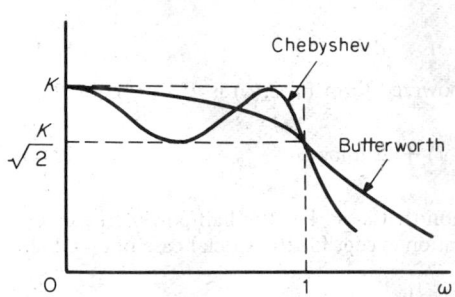

Fig. 12-9. Butterworth and Chebyshev approximations for $n = 3$ and $\epsilon = 1$.

Figure 12-9 shows the Butterworth and Chebyshev ($\epsilon = 1$) frequency responses for $n = 3$. The general relationships shown hold for all values of n, although the details differ. The two approximations are compared as follows:

1. The Butterworth approximation is superior at and near $\omega = 0$. In fact, it is the optimum low-pass approximation in the sense that of all polynomial approximations it has the greatest number of zero derivatives at $\omega = 0$.

2. The Chebyshev approximation is superior at and near the cutoff frequency. It has a derivative at $\omega = 1$ which is n times that of the Butterworth approximation.

3. The Chebyshev approximation is superior in the stop band. An examination of Eqs. (12-17) and (12-43) for $\epsilon = 1$ shows that the attenuation at high frequencies for the Chebyshev approximation is greater than that for the Butterworth approximation by a constant $6(n - 1)$ dB.

4. The Chebyshev approximation sacrifices smoothness in the passband.

10. The Modified Chebyshev Approximation. It was noted earlier that the even-order Chebyshev approximation was not realizable for equal load and source terminations. The reason for this is that the even-ordered Chebyshev polynomial has a maximum at $\omega = 0$ rather than a zero, and thus the transfer function has a relative minimum at $\omega = 0$. Saal[3] has suggested

TABLE 12-3   Linear and Quadratic Factors of Poles of Chebyshev Transfer Functions

n	Ripple, dB	$[G(s)]^{-1}$
2	0.01	$s^2 + 4.4555282s + 10.4258656$
	0.10	$s^2 + 2.3723563s + 3.3140371$
	0.50	$s^2 + 1.4256245s + 1.5162026$
	1.00	$s^2 + 1.0977343s + 1.1025103$
	3.00	$s^2 + 0.6448997s + 0.7079478$
3	0.01	$s + 1.5893705$
		$s^2 + 1.5893705s + 3.2760986$
	0.10	$s + 0.9694057$
		$s^2 + 0.9694057s + 1.3897474$
	0.50	$s + 0.6264565$
		$s^2 + 0.6264565s - 1.1424477$
	1.00	$s + 0.4941706$
		$s^2 + 0.4941706s + 0.9942046$
	3.00	$s + 0.2986202$
		$s^2 + 0.2986202s + 0.8391740$
4	0.01	$s^2 + 0.8217328s + 2.0062632$
		$s^2 + 1.9838384s + 1.2991615$
	0.10	$s^2 + 0.5283127s + 1.3300314$
		$s^2 + 1.2754598s + 0.6229246$
	0.50	$s^2 + 0.3507061s + 1.0635186$
		$s^2 + 0.8466795s + 0.3564119$
	1.00	$s^2 + 0.2790720s + 0.9865049$
		$s^2 + 0.6737394s + 0.2793981$
	3.00	$s^2 + 0.1703405s + 0.9030868$
		$s^2 + 0.4112391s + 0.1959800$
5	0.01	$s + 0.8171468$
		$s^2 + 1.3221715s + 1.0132204$
		$s^2 + 0.5050245s + 1.5722374$
	0.10	$s + 0.5389143$
		$s^2 + 0.8719817s + 0.6359202$
		$s^2 - 0.3330674s + 1.1949371$
	0.50	$s + 0.3623196$
		$s^2 + 0.5862455s - 0.4767670$
		$s^2 + 0.2239258s + 1.0357840$
	1.00	$s + 0.2894933$
		$s^2 + 0.4684101s + 0.4292979$
		$s^2 + 0.1789167s + 0.9833149$
	3.00	$s + 0.1775303$
		$s^2 + 0.2872500s + 0.3770085$
		$s^2 + 0.1097197s + 0.9360255$
6	0.01	$s^2 + 0.3429310s + 1.3719082$
		$s^2 + 0.9369050s + 0.9388954$
		$s^2 + 1.2798360s + 0.5058827$
	0.10	$s^2 + 0.2293867s - 1.1293868$
		$s^2 + 0.6266962s - 0.6963741$
		$s^2 + 0.8560830s - 0.2633614$

TABLE 12-3 Linear and Quadratic Factors of Poles of Chebyshev Transfer Functions (*Continued*)

n	Ripple, dB	$[G(s)]^{-1}$
6	0.50	$s^2 + 0.1553002s + 1.0230228$ $s^2 + 0.4242879s + 0.5900101$ $s^2 + 0.5795881s + 0.1569974$
	1.00	$s^2 + 0.1243620s + 0.9907323$ $s^2 + 0.3397634s + 0.5577196$ $s^2 + 0.4641255s + 0.1247069$
	3.00	$s^2 + 0.0764590s + 0.9548302$ $s^2 + 0.2088899s + 0.5218175$ $s^2 + 0.2853490s + 0.0888048$
7	0.01	$s + 0.5584353$ $s^2 + 1.0062656s + 0.5001050$ $s^2 + 0.6963574s + 0.9231104$ $s^2 + 0.2485271s + 1.2623344$
	0.10	$s + 0.3767779$ $s^2 + 0.6789303s + 0.3302167$ $s^2 + 0.4698343s + 0.7532220$ $s^2 + 0.1676819s + 1.0924460$
	0.50	$s + 0.2561700$ $s^2 + 0.4616024s + 0.2538782$ $s^2 + 0.3194388s + 0.6768835$ $s^2 + 0.1140064s + 1.0161075$
	1.00	$s + 0.2054143$ $s^2 + 0.3701438s + 0.2304501$ $s^2 + 0.2561474s + 0.6534555$ $s^2 + 0.0914180s + 0.9926795$
	3.00	$s + 0.1264854$ $s^2 + 0.2279188s + 0.2042536$ $s^2 + 0.1577247s + 0.6272590$ $s^2 + 0.0562913s + 0.9664830$
8	0.01	$s^2 + 0.9480834s + 0.2716669$ $s^2 + 0.8037463s + 0.5422649$ $s^2 + 0.5370461s + 0.9249484$ $s^2 + 0.1885855s + 1.1955464$
	0.10	$s^2 + 0.6432996s + 0.1456123$ $s^2 + 0.5453631s + 0.4162103$ $s^2 + 0.3644000s + 0.7988938$ $s^2 + 0.1279602s + 1.0694918$
	0.50	$s^2 + 0.4385859s + 0.0880523$ $s^2 + 0.3718151s + 0.3586504$ $s^2 + 0.2484389s + 0.7413338$ $s^2 + 0.0872402s + 1.0119319$
	1.00	$s^2 + 0.3519965s + 0.0702612$ $s^2 + 0.2984083s + 0.3408593$ $s^2 + 0.1993900s + 0.7235427$ $s^2 + 0.0700165s + 0.9941407$
	3.00	$s^2 + 0.2169614s + 0.0502939$ $s^2 + 0.1839310s + 0.3208920$ $s^2 + 0.1228988s + 0.7035754$ $s^2 + 0.0431563s + 0.9741735$

TABLE 12-3   Linear and Quadratic Factors of Poles of Chebyshev Transfer Functions (Continued)

n	Ripple, dB	$|G(s)|^{-1}$
9	0.01	$s + 0.4264123$ $s^2 + 0.8013929s + 0.2988052$ $s^2 + 0.6533015s + 0.5950033$ $s^2 + 0.4264123s + 0.9318274$ $s^2 + 0.1480914s + 1.1516737$
	0.10	$s + 0.2904612$ $s^2 + 0.5453855s - 0.2013455$ $s^2 + 0.4450123s - 0.4975436$ $s^2 + 0.2904623s + 0.8343677$ $s^2 + 0.1003761s + 1.0542140$
	0.50	$s + 0.1984053$ $s^2 + 0.3728800s + 0.1563424$ $s^2 + 0.3039745s + 0.4525406$ $s^2 + 0.1984053s + 0.7893647$ $s^2 + 0.0689054s + 1.0092110$
	1.00	$s + 0.1593305$ $s^2 + 0.2994433s + 0.1423640$ $s^2 + 0.2441084s + 0.4385621$ $s^2 + 0.1593305s - 0.7753862$ $s^2 + 0.0553349s - 0.9952325$
	3.00	$s + 0.0982746$ $s^2 + 0.1846953s - 0.1266357$ $s^2 + 0.1505854s + 0.4228358$ $s^2 + 0.0982745s + 0.7596579$ $s^2 + 0.0341304s + 0.9795042$
10	0.01	$s^2 + 0.7540220s + 0.1701746$ $s^2 + 0.6802131s + 0.3518103$ $s^2 + 0.5398201s + 0.6457029$ $s^2 + 0.3465859s + 0.9395955$ $s^2 + 0.1194254s + 1.1212312$
	0.10	$s^2 + 0.5150591s - 0.0924569$ $s^2 + 0.4646415s - 0.2740926$ $s^2 + 0.3687416s - 0.5679852$ $s^2 + 0.2367467s - 0.8618778$ $s^2 + 0.0815773s + 1.0435134$
	0.50	$s^2 + 0.3522999s + 0.0562789$ $s^2 + 0.3178143s + 0.2379146$ $s^2 + 0.2522139s + 0.5318072$ $s^2 + 0.1619345s + 0.8256998$ $s^2 + 0.0557938s + 1.0073354$
	1.00	$s^2 + 0.2830386s + 0.0450019$ $s^2 + 0.2553328s + 0.2266375$ $s^2 + 0.2026332s - 0.5205301$ $s^2 + 0.1300985s - 0.8144227$ $s^2 + 0.0448289s - 0.9960584$
	3.00	$s^2 + 0.1746691s + 0.0322899$ $s^2 + 0.1575659s + 0.2139255$ $s^2 + 0.1250450s + 0.5073181$ $s^2 + 0.0802838s + 0.8017108$ $s^2 + 0.0276639s + 0.9833464$

a modification of the even-ordered Chebyshev polynomial to obtain a new even-ordered polynomial that has a zero at $\omega = 0$, still has maxima and minima of $+1$ and -1 within the interval $0 < \omega \leq 1$, and increases monotonically for $\omega > 1$. This procedure yields a set of polynomials $\bar{C}_n(\omega)$ which are given in Table 12-4.

TABLE 12-4   Modified Chebyshev Polynomials of Even Order

n	$\bar{C}_n(\omega)$
2	ω^2
4	$5.82842713\omega^4 - 4.82842713\omega^2$
6	$25.9903811\omega^6 - 36.1865335\omega^4 + 11.1961524\omega^2$
8	$109.597711\omega^8 - 210.522708\omega^6$
	$+ 122.034355\omega^4 - 20.109358\omega^2$
10	$452.344415\omega^{10} - 1102.492675\omega^8 + 926.297276\omega^6$
	$- 306.717773\omega^4 + 31.568758\omega^2$

For a given value of n, the modified Chebyshev polynomials have the property that

$$C_n(\omega) > \bar{C}_n(\omega) > C_{n-1}(\omega) \qquad (12\text{-}47)$$

for $\omega > 1$. Thus a low-pass approximation based on these polynomials for an even value of n will be better in the stopband than that using the Chebyshev polynomial of order $n - 1$ but will not be as good as that using the regular Chebyshev polynomial of order n. Thus this modified Chebyshev approximation is a possible alternative in some situations to the increase of n to the next higher odd value to achieve realizability, especially when equal load and source resistances are required.

The extent of the difference between $\bar{C}_n(\omega)$ and $C_n(\omega)$ is shown in Fig. 12-10, where the ratio $\bar{C}_n(\omega)/C_n(\omega)$ for $\omega \geq 1$ is shown for even values of n.

Table 12-5 gives the poles of $G(s)$, in terms of quadratic factors, for $n = 2, 4, 6, 8, 10$ and ripples of 0.01, 0.1, 0.5, 1.0, and 3.0 dB. As for the regular Chebyshev approximation, the pole locations and the element values in Table 12-13 (Par. **12-19**) are based on $\omega = 1$ being the end of the ripple. The half-power frequencies are given in Table 12-14 (Par. **12-19**).

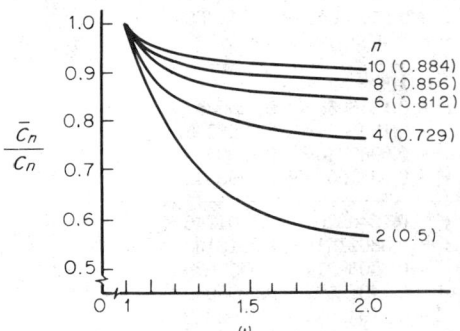

Fig. 12-10. Ratio of modified Chebyshev polynomial to Chebyshev polynomial. Limiting value as ω approaches infinity is in parentheses for each value of n.

TABLE 12-5   Linear and Quadratic Factors Giving Poles of Modified Chebyshev Transfer Functions

n	Ripple, dB	$[G(s)]^{-1}$
2	0.01	$s^2 + 6.4541054s + 20.8277385$
	0.10	$s^2 + 3.6200009s + 6.5522033$
	0.50	$s^2 + 2.3928122s + 2.8627752$
	1.00	$s^2 + 1.9825371s + 1.9652267$
	3.00	$s^2 + 1.4158936s + 1.0023773$
4	0.01	$s^2 + 2.2680440s + 1.6170715$
		$s^2 + 0.9235542s + 2.2098435$
	0.10	$s^2 + 1.5452285s + 0.7991383$
		$s^2 + 0.6059475s + 1.4067406$
	0.50	$s^2 + 1.1191759s + 0.4523364$
		$s^2 + 0.4086395s + 1.0858612$
	1.00	$s^2 + 0.9486848s + 0.3398608$
		$s^2 + 0.3272401s + 0.9921108$
	3.00	$s^2 + 0.6845838\ \ + 0.1932927$
		$s^2 + 0.2013994s + 0.8897428$

TABLE 12-5 Linear and Quadratic Factors Giving Poles of Modified Chebyshev Transfer Functions (*Continued*)

n	Ripple, dB	$[G(s)]^{-1}$
6	0.01	$s^2 + 1.4118389s + 0.5893450$
		$s^2 + 1.0058571s + 0.9699759$
		$s^2 + 0.3640166s + 1.4018414$
	0.10	$s^2 + 0.9995998s + 0.3173248$
		$s^2 + 0.6819269s + 0.6966158$
		$s^2 + 0.2448454s + 1.1404529$
	0.50	$s^2 + 0.7362729s + 0.1875012$
		$s^2 + 0.4662515s + 0.5727968$
		$s^2 + 0.1663135s + 1.0255810$
	1.00	$s^2 + 0.6267633s + 0.1428406$
		$s^2 + 0.3748110s + 0.5343433$
		$s^2 + 0.1333343s + 0.9906678$
	3.00	$s^2 + 0.4554072s + 0.0825380$
		$s^2 + 0.2315851s + 0.4909174$
		$s^2 + 0.0820877s + 0.9518235$
8	0.01	$s^2 + 1.0335548s + 0.3097540$
		$s^2 + 0.8491714s + 0.5489161$
		$s^2 + 0.5591534s + 0.9283901$
		$s^2 + 0.1954150s + 1.2038909$
	0.10	$s^2 + 0.7417790s + 0.1718214$
		$s^2 + 0.5828511s + 0.4083754$
		$s^2 + 0.3806989s + 0.7943820$
		$s^2 + 0.1328521s + 1.0725554$
	0.50	$s^2 + 0.5496051s + 0.1030188$
		$s^2 + 0.4007069s + 0.3416991$
		$s^2 + 0.2600622s + 0.7328336$
		$s^2 + 0.0906706s + 1.0125586$
	1.00	$s^2 + 0.4685593s + 0.0788470$
		$s^2 + 0.3226465s + 0.3205446$
		$s^2 + 0.2088395s + 0.7137508$
		$s^2 + 0.0727952s + 0.9940103$
	3.00	$s^2 + 0.3392712s + 0.0458028$
		$s^2 + 0.1997071s + 0.2963792$
		$s^2 + 0.1258087s + 0.6922959$
		$s^2 + 0.0448874s + 0.9731910$
10	0.01	$s^2 + 0.8175750s - 0.1921666$
		$s^2 + 0.7133475s - 0.3528340$
		$s^2 + 0.5571617s + 0.6426718$
		$s^2 + 0.3555630s + 0.9397252$
		$s^2 + 0.1222554s + 1.1244354$
	0.10	$s^2 + 0.5905036s + 0.1080675$
		$s^2 + 0.4925060s + 0.2665100$
		$s^2 + 0.3816272s + 0.5602769$
		$s^2 + 0.2431732s + 0.8592520$
		$s^2 + 0.0835800s - 1.0445869$
	0.50	$s^2 + 0.4387194s - 0.0652281$
		$s^2 + 0.3394908s + 0.2246543$
		$s^2 + 0.2614400s + 0.5216326$
		$s^2 + 0.1664374s + 0.8217378$
		$s^2 + 0.0571933s + 1.0075592$
	1.00	$s^2 + 0.3742846s + 0.0500294$
		$s^2 + 0.2735736s + 0.2112433$
		$s^2 + 0.2101536s + 0.5095395$
		$s^2 + 0.1337446s + 0.8100361$
		$s^2 + 0.0459557s + 0.9959854$
	3.00	$s^2 + 0.2711265s + 0.0291328$
		$s^2 + 0.1694797s + 0.1958431$
		$s^2 + 0.1297678s + 0.4958770$
		$s^2 + 0.0825542s - 0.7968406$
		$s^2 + 0.0283639s - 0.9829386$

11. The Legendre-Papoulis Approximation. Papoulis[4] has described an approximation

$$|G(j\omega)| = K/[1 + L_n(\omega^2)]^{1/2} \tag{12-48}$$

where the polynomial $L_n(\omega^2)$ is derived, using Legendre polynomials of the first kind, to satisfy the following requirements:

1. $L_n(\omega^2)$ increases monotonically; thus $|G(j\omega)|$ decreases monotonically.
2. $L_n(0) = 0$ and $L_n(1) = 1$.
3. Of all the polynomials satisfying 1 and 2, $L_n(\omega^2)$ has the largest derivative at $\omega = 1$.

The resulting polynomials are tabulated in Table 12-6 for $n = 2$ to 10. The pole locations for $G(s)$ are described in Table 12-7 in terms of the quadratic factors.

TABLE 12-6   The Polynomials $L_n(\omega^2)$

n	
2	ω^4
3	$3\omega^6 - 3\omega^4 + \omega^2$
4	$6\omega^8 - 8\omega^6 + 3\omega^4$
5	$20\omega^{10} - 40\omega^8 + 28\omega^6 - 8\omega^4 + \omega^2$
6	$50\omega^{12} - 120\omega^{10} + 105\omega^8 - 40\omega^6 + 6\omega^4$
7	$175\omega^{14} - 525\omega^{12} + 615\omega^{10} - 355\omega^8 + 105\omega^6 - 15\omega^4 + \omega^2$
8	$490\omega^{16} - 1680\omega^{14} + 2310\omega^{12} - 1624\omega^{10} + 615\omega^8 - 120\omega^6 + 10\omega^4$
9	$1764\omega^{18} - 7056\omega^{16} + 11704\omega^{14} - 10416\omega^{12} + 5376\omega^{10} - 1624\omega^8 + 276\omega^6 - 24\omega^4 + \omega^2$
10	$5292\omega^{20} - 23520\omega^{18} + 44100\omega^{16} - 45360\omega^{14} + 27860\omega^{12} - 10416\omega^{10} + 2310\omega^8 - 280\omega^6 + 15\omega^4$

TABLE 12-7   Linear and Quadratic Factors for Poles of Legendre-Papoulis Approximation

n	$[G(s)]^{-1}$	n	$[G(s)]^{-1}$
2	$s^2 + 1.4142136s + 1.0000000$	8	$s^2 + 0.1378844s + 0.9808397$
			$s^2 + 0.3885518s + 0.7179832$
3	$s + 0.6203318$		$s^2 + 0.6005680s + 0.3828971$
	$s^2 + 0.6903712s + 0.9307119$		$s^2 + 0.7343526s + 0.1675357$
4	$s^2 + 0.4633774s + 0.9476701$		
	$s^2 + 1.0994868s + 0.4307915$	9	$s + 0.3256878$
			$s^2 + 0.1101944s + 0.9844435$
5	$s + 0.4680899$		$s^2 + 0.3145676s + 0.7666498$
	$s^2 + 0.3071734s + 0.9608963$		$s^2 + 0.4971058s + 0.4635058$
	$s^2 + 0.7762796s + 0.4971406$		$s^2 + 0.6187708s + 0.2089807$
6	$s^2 + 0.2303854s + 0.9696012$		
	$s^2 + 0.6179218s + 0.5828947$	10	$s^2 + 0.0918020s + 0.9869313$
	$s^2 + 0.8778030s + 0.2502256$		$s^2 + 0.2650376s + 0.8012497$
7	$s + 0.3821033$		$s^2 + 0.4283460s + 0.5282527$
	$s^2 + 0.1724170s + 0.9764158$		$s^2 + 0.5548108s + 0.2702425$
	$s^2 + 0.4748794s + 0.6621299$		$s^2 + 0.6344130s + 0.1217699$
	$s^2 + 0.6984636s + 0.3060005$		

Figure 12-11 compares the Legendre-Papoulis approximation with the Butterworth and Chebyshev ($\epsilon = 1$) approximations for $n = 3$. The Legendre-Papoulis approximation is smoother than the Chebyshev approximation but is not as smooth as the Butterworth approximation in the passband. The stop-band characteristics of the Legendre-Papoulis approximation are better than those of the Butterworth approximation but are not as good as those of the Chebyshev approximation. It is apparent that the Legendre-Papoulis approximation is in many respects a good compromise between the Butterworth and Chebyshev approximations.

The selection of n follows the same general procedure as for the Butterworth approximation. If the specification

$$A \geq |G(j\omega_a)| = K/[1 + L_n(\omega_a^2)] \tag{12-49}$$

is to be satisfied, then

$$L_n(\omega_a^2) \geq (K^2/A^2) - 1 \tag{12-50}$$

and successively higher-ordered polynomials of Table 12-6 can be evaluated at ω_c until Eq. (12-50) is satisfied.

12. The Bessel Approximation. The ideal all-pass characteristic of Fig. 12-2

$$|G(j\omega)| = K \tag{12-51}$$

and
$$\theta(j\omega) = T\omega \tag{12-52}$$

implies that the output is a replica (scaled by the constant K) of the input but delayed in time by T seconds. The negative derivative of the phase function

$$-\frac{d}{d\omega}\theta(j\omega) = \tau_d(\omega) \tag{12-53}$$

is called the group delay and in the ideal situation is a constant T.

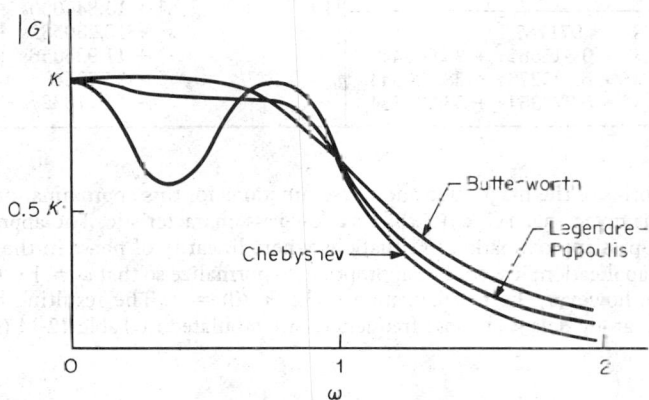

Fig. 12-11. Comparison of Butterworth, Legendre-Papoulis, and Chebyshev ($\epsilon = 1$) approximations for $n = 3$.

One approach to the approximation of the ideal delay characteristic, first described by Thomson,[5] is to make the group delay maximally flat at $\omega = 0$. This leads to a transfer function of the form

$$G(s) = \frac{b_0 K}{b_0 + b_1 s + \cdots + b_{n-1}s^{n-1} + s^n} \tag{12-54}$$

where the denominator polynomial is related to a class of Bessel polynomials. The coefficients for these polynomials are given in Table 12-8 for the approximation to a group delay of 1 s at $\omega = 0$. These polynomials are related by the recursion formula

$$B_n = (2n - 1)B_{n-1} + s^2 B_{n-2} \tag{12-55}$$

The pole locations for $G(s)$ are described in Table 12-9 in terms of the quadratic factors.

TABLE 12-8   The Coefficients of the Bessel Polynomials

n	b_0	b_1	b_2	b_3	b_4	b_5	b_6	b_7
1	1							
2	3	3						
3	15	15	6					
4	105	105	45	10				
5	945	945	420	105	15			
6	10,395	10,395	4,725	1,260	210	21		
7	135,135	135,135	62,370	17,325	3,150	378	28	
8	2,027,025	2,027,025	945,945	270,270	51,975	6,930	630	36

TABLE 12-9   Linear and Quadratic Factors for Poles of Bessel Approximation

n	$[G(s)]^{-1}$	n	$[G(s)]^{-1}$
2	$s^2 + 3s + 3$		
		8	$s^2 + 11.175772s + 31.977224$
3	$s + 2.322185$		$s^2 + 10.409682s + 33.934741$
	$s^2 + 3.677814s + 6.459432$		$s^2 + 8.736578s + 38.569256$
			$s^2 + 5.677968s + 48.432015$
4	$s^2 + 5.792422s + 9.140133$		
	$s^2 + 4.207578 + 11.487799$		
		9	$s + 6.297019$
5	$s + 3.646739$		$s^2 + 12.258736s + 40.589268$
	$s^2 + 6.703912s + 14.272476$		$s^2 + 11.208844s + 43.646648$
	$s^2 + 4.649348s + 18.156314$		$s^2 + 9.276880s + 49.788507$
			$s^2 + 5.958522s + 62.041443$
6	$s^2 + 8.496718s + 18.801128$		
	$s^2 + 7.471416s + 20.852819$		
	$s^2 + 5.031864s + 26.514025$		
		10	$s^2 + 13.844090s + 48.667550$
7	$s + 4.971787$		$s^2 + 13.230582s + 50.582362$
	$s^2 + 9.516582s + 25.666449$		$s^2 + 11.935056s + 54.839151$
	$s^2 + 8.140278s + 28.936544$		$s^2 + 9.772440s + 62.625584$
	$s^2 + 5.371354s + 36.596784$		$s^2 + 6.217832s + 77.442692$

The general forms of the magnitude and phase functions for this approximation are indicated in Fig. 12-12. It is noted that $|G(j\omega)|$ exhibits a low-pass characteristic. The approximation can be used as a low-pass approximation in situations where linearity of phase in the passband is of concern. In this application, it would be appropriate to normalize so that $\omega = 1$ is the half-power frequency. Here, however, the normalization makes $\tau_d(0) = 1$. The resulting half-power frequencies and the group delays at those frequencies are tabulated in Table 12-14 (Par. **12-19**).

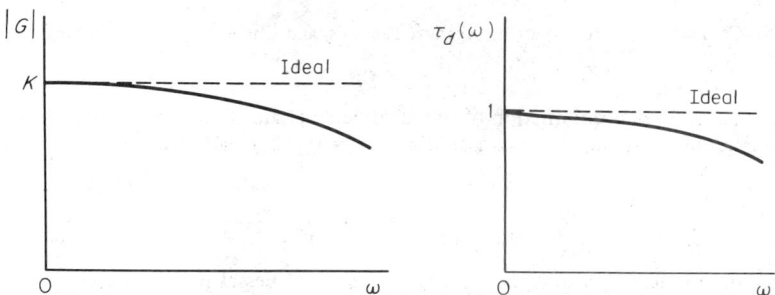

Fig. 12-12.   Magnitude and group delay for maximally flat group-delay approximation.

The parameter n is selected to satisfy either a minimum group-delay requirement or a minimum magnitude requirement at some frequency ω_a, that is,

$$\tau_d(\omega_a) \geq T_a \tag{12-56}$$

or
$$|G(j\omega_a)| \geq A \tag{12-57}$$

The condition of Eq. (12-57) can be stated equivalently in terms of the maximum attenuation at ω_a. Figure 12-13 gives the attenuation vs. ω and the group delay vs. ω for $n = 1$ to 8 and can be used for the selection of n.

13. The Step Response.   The step response can be described in terms of various figures of merit. Figure 12-14 is used to define three (overshoot, rise time, and delay time) used here. Tables 12-10, 12-11, and 12-12 give the percent overshoot, rise time, and delay time, respectively, for each of the five approximations described.

The general form of response illustrated in Fig. 12-14 is typical except for the odd-order Chebyshev approximation and the modified Chebyshev approximations when the ripple is large. Figure 12-15 shows the response for the third-order Chebyshev approximation with a 3-dB ripple. Attention is drawn to the fact that the first relative maximum is less than the final value. This characteristic becomes even more pronounced with increasingly large values of n, several relative maxima occurring before the peak value, as illustrated in Fig. 12-16 for $n = 5$.

14. The Prototype Filter Networks. The all-pole approximations that have been described can be realized by lossless ladder networks terminated in resistance at both ends. The applicable synthesis techniques are described in the literature.[8-10] and only the results are given here. The resulting ladder networks are shown in Figs. 12-17 and 12-18. Figure 12-17 shows four networks classified according to the type of transfer function, Z_{SL} or G_{SL}, being realized and whether n is even or odd. It is noted that the networks of Fig. 12-17c and d are duals of those in Fig. 12-17a and b, respectively. The four networks of Fig. 12-18 can be regarded as being derived from those of Fig. 12-17 by source transformation.

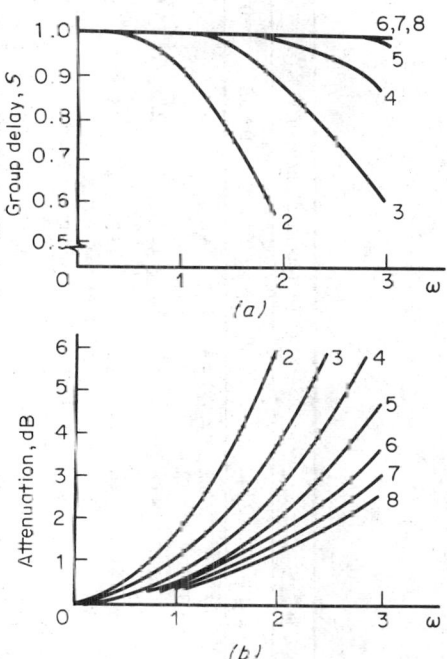

Fig. 12-13. Group delay and attenuation for the Bessel approximation

For a given type of source and value of n there are thus two networks that can be used; e.g., with a current source input and with n odd, either the network of Fig. 12-17b or that of Fig. 12-18d can be used. A significant difference is that the first and last lossless elements in the first case are shunt capacitors and in the second case are series inductors

The notation and organization in Figs. 12-17 and 12-18 are based on the fact that for a given type of approximation and value of n the same numerical element values apply to four different networks. For example, a Butterworth approximation with $n = 2$ and R_s or G_s, according to the

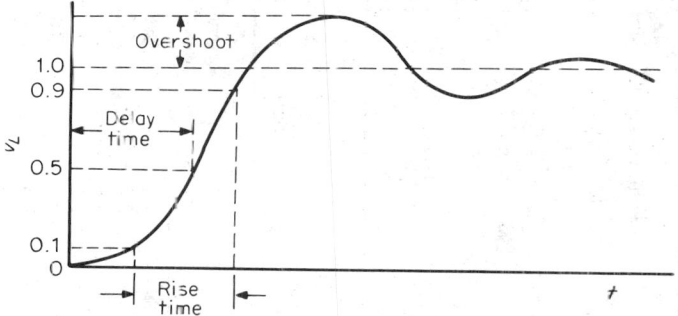

Fig. 12-14. General form of the step response and definitions of figures of merit.

specific network being used, equal to ½ results in the four networks of Fig. 12-19. Here, the numerical values for the lossless elements are the same when they are taken in order from the source end. This fact makes it possible to construct a table of element values which uses as parameters for entry (1) the type of approximation, including the ripple width in decibels, Eq. (12-36), in the case of the Chebyshev approximations; (2) the value of n; and (3) the value of R_s or G_s, according to the particular network in Figs. 12-17 and 12-18 that is being used. Such a table is Table 12-13.

TABLE 12-10 Percent Overshoot

Order	Bessel, %	Butterworth, %	Legendre-Papoulis, %	Chebyshev, %				Modified Chebyshev, %			
				3.0 dB	1.0 dB	0.5 dB	0.1 dB	3.0 dB	1.0 dB	0.5 dB	0.1 dB
2	0.4	4.3	4.3	27.2	14.6	10.7	6.7	4.3	4.3	4.3	4.3
3	0.7	8.1	7.5	2.7	6.4	8.9	10.2				
4	0.8	10.8	11.2	35.7	21.9	18.1	14.5	8.0	10.5	12.1	12.9
5	0.8	12.8	13.3	1.7	10.2	13.2	15.2				
6	0.6	14.3	15.2	38.7	24.9	21.2	18.0	8.5	12.8	15.1	16.5
7	0.5	15.4	16.4	2.0	12.1	15.3	17.7				
8	0.3	16.3	17.6	40.4	26.5	23.0	19.8	8.7	14.0	16.6	18.5
9	0.2	17.1	18.3	3.1	13.3	16.7	19.1				
10	0.1	17.8	19.1	41.5	27.6	24.1	21.0	9.5	14.8	17.6	19.7

TABLE 12-11 Rise Time for Polynomial Filters

Order	Bessel	Butterworth	Legendre-Papoulis	Chebyshev				Modified Chebyshev			
				3.0 dB	1.0 dB	0.5 dB	0.1 dB	3.0 dB	1.0 dB	0.5 dB	0.1 dB
2	1.58	2.2	2.2	1.7	1.6	1.5	1.1	2.2	1.5	1.3	0.8
3	1.25	2.3	2.5	3.2	2.4	2.2	1.7				
4	1.04	2.4	2.7	2.5	2.5	2.4	2.2	3.2	2.6	2.4	2.1
5	0.91	2.6	2.9	3.7	3.0	2.8	2.5				
6	0.82	2.7	3.0	2.9	3.0	2.9	2.8	3.7	3.2	3.0	2.7
7	0.75	2.8	3.2	4.0	3.4	3.3	3.0				
8	0.69	2.9	3.4	3.2	3.3	3.3	3.2	4.1	3.6	3.4	3.2
9	0.64	3.0	3.5	4.3	3.8	3.6	3.4				
10	0.60	3.1	3.6	3.5	3.6	3.6	3.5	4.4	3.9	3.7	3.5

TABLE 12-12 Delay Times for Polynomial Filters

Order	Bessel	Butterworth	Legendre-Papoulis	Chebyshev 3.0 dB	Chebyshev 1.0 dB	Chebyshev 0.5 dB	Chebyshev 0.1 dB	Modified Chebyshev 3.0 dB	Modified Chebyshev 1.0 dB	Modified Chebyshev 0.5 dB	Modified Chebyshev 0.1 dB
2	0.90	1.4	1.4	1.5	1.2	1.1	0.8	1.4	1.0	0.8	0.6
3	0.96	2.1	2.4	3.0	2.4	2.2	1.7				
4	0.98	2.8	3.3	3.6	3.3	3.2	2.7	3.9	3.3	3.1	2.6
5	0.99	3.5	4.2	5.2	4.6	4.3	3.8				
6	0.99	4.2	5.2	5.6	5.4	5.2	4.8	6.1	5.5	5.2	4.7
7	1.00	4.8	6.2	7.3	6.7	6.4	5.8				
8	1.00	5.5	7.1	7.7	7.5	7.3	6.8	8.2	7.6	7.3	6.8
9	1.00	6.2	8.1	9.4	8.8	8.5	7.9				
10	1.00	6.8	9.0	9.7	9.6	9.4	8.9	10.3	9.7	9.4	6.9

15. The Table of Element Values. Table 12-13 gives element values, numbered in order from the source end in the same manner as in Figs. 12-17 and 12-18, for the five all-pole approximations (including five ripple widths in the case of the Chebyshev approximations), n = 2 to 10, and for R_S or G_S equal to 0 and 1. Reference 6 is an excellent source of other tables.

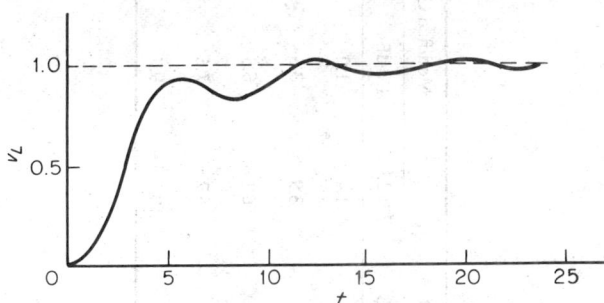

Fig. 12-15. Step response of third-order Chebyshev approximation with a 3-dB ripple.

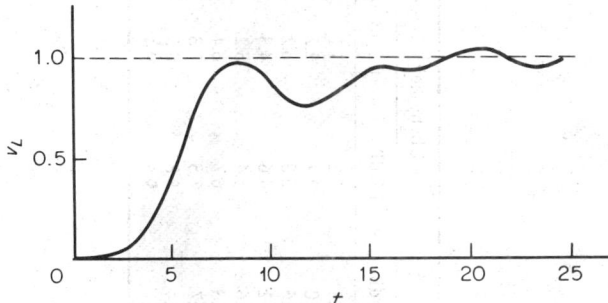

Fig. 12-16. Step response for fifth-order Chebyshev approximation with a 3-dB ripple.

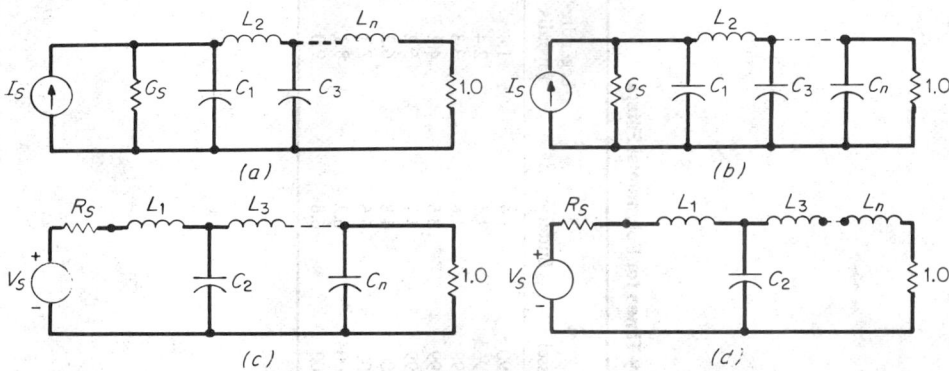

Fig. 12-17. Four filter networks classified according to type of transfer function and n even or odd: (a) Z_{SL}, n even; (b) Z_{SL}, n odd; (c) G_{SL}, n even; (d) G_{SL}, n odd.

The synthesis process for determining the element values in Table 12-13 often involves a choice of location for the zeros of the reflection coefficient. All values in the table where such a choice was made are based on placing those zeros in the left half plane. Other choices for their location would lead to different element values.

Table 12-13 is based on the following normalizations:
1. $R_L = 1$.
2. $\omega_{hp} = 1$ for the Butterworth and Legendre-Papoulis approximations.
3. $\omega = 1$ is the end of the ripple band for the Chebyshev approximations.
4. The group delay at $\omega = 0$ is 1 s for the Bessel approximation.

The half-power frequencies for the Chebyshev and Bessel approximations and the group delay of the half-power frequency for the Bessel approximation have been calculated and are tabulated in Table 12-14.

Only two values of G_s or R_s, 0 and 1, are used in Table 12-13, since these represent the two most commonly encountered situations. Tables for other values of G_s and R_s can be found in Zverev[6] and Weinberg.[10]

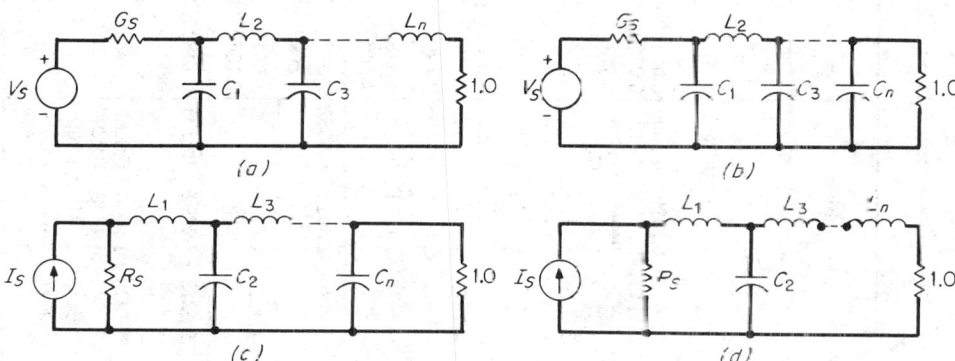

Fig. 12-18. Filter networks derived from those in Fig. 12-17 by source transformations: (a) G_{SL}, n even; (b) G_{SL}, n odd; (c) Z_{SL}, n even; (d) Z_{SL}, n odd.

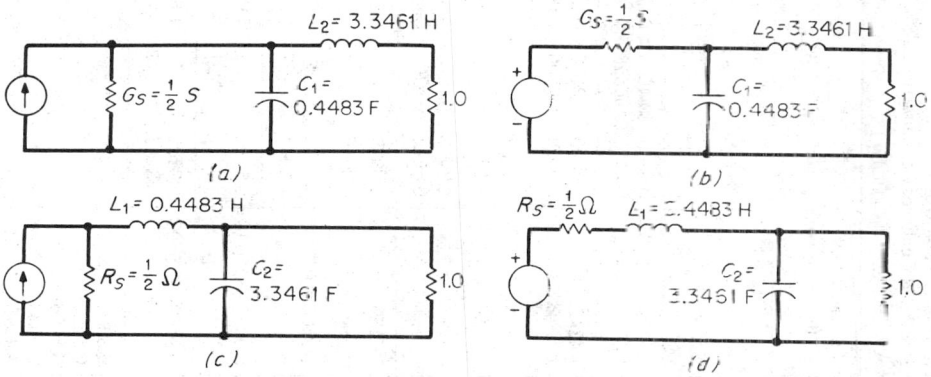

Fig. 12-19. Four networks resulting from a Butterworth approximation with $n = 2$ and R_s or G_s (as appropriate to the type of network) equal to ½.

In the case of n odd and $G_s = R_s = 1$ for the Butterworth and regular Chebyshev approximations, the resulting networks are symmetrical; that is, $C_1 = C_n$, $L_2 = L_{n-1}$, etc. In these particular cases, the element values can be modified easily to realize other values of G_s or R_s. Consider the case of Fig. 12-17b with $n = 5$, redrawn as Fig. 12-20a to emphasize the resulting symmetry. G_s can be changed to some other value by magnitude scaling the element values in the left half of the network. Thus making $G_s = $ ½ would result in the network of Fig. 12-20b, where the two capacitors of value $C_3/2$ and $C_3/4$ can be recombined into a single capacitor. Similar operations can be applied to other symmetrical networks. After the application of this procedure the zeros of the reflection coefficient do not remain in the left half plane.

TABLE 12-13 Low-Pass Filter Circuit Element Values*

Table 12-14 gives half-power frequencies for Bessel and Chebyshev circuits; the half-power frequency is 1.000 rad/s for Butterworth and Legendre-Papoulis; empty spaces indicate unrealizable networks

Filter type	R_s or G_s Element number	Second-order network $n=2$		Third-order network $n=3$			Fourth-order network $n=4$				Fifth-order network $n=5$				
		1	2	1	2	3	1	2	3	4	1	2	3	4	5
Bessel	0	1.0000	0.3333	0.8333	0.4800	0.1667	0.7101	0.4627	0.2899	0.1000	0.6231	0.4215	0.3103	0.1948	0.0667
	1	1.5774	0.4227	1.2550	0.5528	0.1922	1.0598	0.5116	0.3181	0.1104	0.9303	0.4577	0.3312	0.2089	0.0718
Butterworth	0	1.4142	0.7071	1.5000	1.3333	0.5000	1.5307	1.5772	1.0824	0.3827	1.5451	1.6944	1.3820	0.8944	0.3090
	1	1.4142	1.4142	1.0000	2.0000	1.0000	0.7654	1.8478	1.8478	0.7654	0.6180	1.6180	2.0000	1.6180	0.6180
Legendre-Papoulis	0	1.4142	0.7071	1.5909	1.4270	0.7629	1.6120	1.6616	1.4292	0.6399	1.6372	1.7509	1.7358	1.3945	0.6445
	1	1.4142	1.4142	2.1801	1.3538	1.1737	1.5645	1.9584	1.4769	1.0826	1.9990	1.5395	2.0673	1.4780	0.9512
Chebyshev 0.01 dB	0	0.4274	0.2244	0.7907	0.7634	0.3146	1.0421	1.1547	0.8945	0.3564	1.1978	1.3902	1.2739	0.9576	0.3782
	1			0.6292	0.9703	0.6292					0.7563	1.3049	1.5773	1.3049	0.7563
Chebyshev 0.10 dB	0	0.7159	0.4215	1.0895	1.0864	0.5158	1.2453	1.4576	1.1994	0.5544	1.3759	1.5924	1.5562	1.2490	0.5734
	1			1.0316	1.1474	1.0316					1.1468	1.3712	1.9750	1.3712	1.1468
Chebyshev 0.50 dB	0	0.9403	0.7014	1.3465	1.3001	0.7981	1.3138	1.7279	1.3916	0.8352	1.5388	1.6426	1.8142	1.4291	0.8529
	1			1.5963	1.0967	1.5963					1.7058	1.2296	2.5408	1.2296	1.7058
Chebyshev 1.0 dB	0	0.9957	0.9110	1.5088	1.3332	1.0118	1.2817	1.9093	1.4126	1.0495	1.6652	1.5908	1.9938	1.4441	1.0674
	1			2.0236	0.9941	2.0236					2.1349	1.0911	3.0009	1.0911	2.1349
Chebyshev 3.0 dB	0	0.9109	1.5506	2.0302	1.1739	1.6744	1.0578	2.5272	1.2292	1.7195	2.1489	1.3016	2.6224	1.2502	1.7406
	1			3.3487	0.7117	3.3487					3.4813	0.7619	4.5375	0.7619	3.4813
Modified Chebyshev 0.01 dB	1	0.3099	0.3099				0.6266	1.1938	1.1938	0.6266					
0.10 dB	1	0.5525	0.5525				0.9297	1.4346	1.4346	0.9297					
0.50 dB	1	0.8358	0.8358				1.3091	1.5415	1.5415	1.3091					
1.0 dB	1	1.0088	1.0088				1.5675	1.5537	1.5537	1.5675					
3.0 dB	1	1.4125	1.4125				2.2574	1.5107	1.5107	2.2574					

Filter type	Element number R_s or G_x	Sixth-order network n = 6						Seventh-order network n = 7						
		1	2	3	4	5	6	1	2	3	4	5	6	7
Bessel	0	0.5595	0.3821	0.3005	0.2246	0.1400	0.0476	0.5111	0.3487	0.2827	0.2288	0.1704	0.1055	0.0357
	1	0.8376	0.4116	0.3158	0.2364	0.1480	0.0505	0.7677	0.3744	0.2944	0.2378	0.1778	0.1104	0.0375
Butterworth	0	1.5529	1.7593	1.5529	1.2016	0.7579	0.2588	1.5576	1.7988	1.6588	1.3972	1.0550	0.6560	0.2225
	1	0.5176	1.4142	1.9319	1.9319	1.4142	0.5176	0.4450	1.2470	1.8019	2.0000	1.8019	1.2470	0.4450
Legendre-Papoulis	0	1.6348	1.8088	1.8223	1.6795	1.3486	0.5793	1.6391	1.8312	1.8911	1.7988	1.6845	1.3290	0.5787
	1	1.5763	1.9040	1.7442	1.9857	1.4852	0.9160	1.8640	1.5895	2.1506	1.7270	1.9394	1.4770	0.8304
Chebyshev 0.01 dB	0	1.2931	1.5400	1.4922	1.3319	0.9925	0.3907	1.3615	1.6303	1.6290	1.5412	1.3650	1.0137	0.3985
	1							0.7969	1.3924	1.7481	1.6331	1.7481	1.3924	0.7969
Chebyshev 0.10 dB	0	1.4035	1.7236	1.6749	1.5999	1.2752	0.5841	1.4745	1.7395	1.7087	1.7107	1.6236	1.2908	0.5906
	1							1.1812	1.4228	2.0967	1.5734	2.0967	1.4228	1.1812
Chebyshev 0.50 dB	0	1.4042	1.9018	1.7101	1.8494	1.4483	0.8627	1.5983	1.7252	1.9713	1.7369	1.8677	1.4595	0.8686
	1							1.7373	1.2582	2.6383	1.3443	2.6383	1.2582	1.7373
Chebyshev 1.0 dB	0	1.3457	2.0491	1.6507	2.0270	1.4601	1.0773	1.7120	1.6488	2.1194	1.6735	2.0438	1.4692	1.0833
	1							2.1666	1.1115	3.0936	1.1735	3.0036	1.1115	2.1666
Chebyshev 3.0 dB	0	1.0876	2.6309	1.3455	2.6578	1.2606	1.7522	2.1828	1.3281	2.7143	1.3613	2.6752	1.2665	1.7593
	1							3.5185	0.7722	4.6390	0.8038	4.6390	0.7722	3.5185
Modified Chebyshev 0.01 dB	1	0.7190	1.3728	1.6006	1.6006	1.3728	0.7190							
Modified Chebyshev 0.10 dB	1	1.0382	1.5163	1.7892	1.7892	1.5163	1.0382							
0.50 dB	1	1.4611	1.5085	1.9333	1.9333	1.5085	1.4611							
1.0 dB	1	1.7623	1.4528	2.0089	2.0089	1.4528	1.7623							
3.0 dB	1	2.6073	1.2711	2.1729	2.1729	1.2711	2.6073							

*Values for the Legendre-Papoulis approximation by permission from A. Zverev, "Handbook of Filter Synthesis," John Wiley & Sons, Inc., New York, 1967.

TABLE 12-13 Low-Pass Filter Circuit Element Values*

Table 12-14 gives half-power frequencies for Bessel and Chebyshev circuits; the half-power frequency is 1.000 rad/s for Butterworth and Legendre-Papoulis circuits; empty spaces indicate unrealizable networks

Filter type	Element number R_x or G_x	Eighth-order network n = 8								Ninth-order network n = 9								
		1	2	3	4	5	6	7	8	1	2	3	4	5	6	7	8	9
Bessel	0	0.4732	0.3212	0.2639	0.2227	0.1806	0.1338	0.0823	0.0278	0.4424	0.2986	0.2465	0.2129	0.1811	0.1463	0.1077	0.0660	0.0222
	1	0.7125	0.3446	0.2735	0.2297	0.1867	0.1387	0.0855	0.0289	0.6678	0.3203	0.2547	0.2184	0.1859	0.1506	0.1112	0.0682	0.0230
Butterworth	0	1.5607	1.8246	1.7287	1.5283	1.2588	0.9371	0.5776	0.1951	1.5628	1.8424	1.7772	1.6202	1.4037	1.1408	0.8414	0.5155	0.1736
	1	0.3902	1.1111	1.6629	1.9616	1.9616	1.6629	1.1111	0.3902	0.3473	1.0000	1.5321	1.8794	2.0000	1.8794	1.5321	1.0000	0.3473
Legendre-Papoulis	0	1.6345	1.8542	1.9102	1.8673	1.8019	1.6437	1.2869	0.5372	1.6341	1.8625	1.9349	1.8961	1.8815	1.7860	1.6397	1.2740	0.5358
	1	1.5564	1.8501	1.8411	2.0515	1.7872	1.0115	1.4688	0.8205	1.7645	1.6134	2.1585	1.7816	2.0662	1.7755	1.8674	1.4555	0.7695
Chebyshev 0.01 dB	0	1.4036	1.6971	1.7097	1.6708	1.5697	1.3858	1.0275	0.4036	1.4392	1.7371	1.7701	1.7457	1.6949	1.5878	1.3990	1.0370	0.4072
	1									0.8145	1.4271	1.8044	1.7125	1.9058	1.7125	1.8044	1.4271	0.8145
Chebyshev 0.10 dB	0	1.4660	1.8163	1.8070	1.8302	1.7302	1.6380	1.3008	0.5949	1.5182	1.7991	1.8814	1.8343	1.8473	1.7423	1.6476	1.3076	0.5978
	1									1.1957	1.4426	2.1346	1.6167	2.2054	1.6167	2.1346	1.4426	1.1957
Chebyshev 0.50 dB	0	1.4379	1.9571	1.7838	1.9980	1.7508	1.8786	1.4666	0.8725	1.6238	1.7571	2.0203	1.8055	2.0116	1.7591	1.8856	1.4714	0.8752
	1									1.7504	1.2690	2.6678	1.3673	2.7239	1.3673	2.6678	1.2690	1.7504
Chebyshev 1.0 dB	0	1.3691	2.0922	1.7021	2.1453	1.6850	2.0537	1.4751	1.0872	1.7317	1.6707	2.1574	1.7213	2.1582	1.6918	2.0601	1.4790	1.0899
	1									2.1797	1.1192	3.1214	1.1897	3.1746	1.1897	3.1214	1.1192	2.1797
Chebyshev 3.0 dB	0	1.0982	2.6618	1.3687	2.7436	1.3690	2.6852	1.2701	1.7638	2.1970	1.3380	2.7413	1.3827	2.7576	1.3733	2.6915	1.2726	1.7670
	1									3.5339	0.7760	4.6691	0.8118	4.7270	0.8118	4.6691	0.7760	3.5339
Modified Chebyshev 0.01 dB	1	0.7584	1.4274	1.7122	1.7503	1.7503	1.7122	1.4274	0.7584									
Modified Chebyshev 0.10 dB	1	1.0880	1.5265	1.9029	1.8301	1.8301	1.9029	1.5265	1.0880									
Modified Chebyshev 0.50 dB	1	1.5372	1.4679	2.1033	1.8436	1.8436	2.1033	1.4679	1.5372									
Modified Chebyshev 1.0 dB	1	1.8642	1.3833	2.2357	1.8319	1.8319	2.2357	1.3833	1.8642									
Modified Chebyshev 3.0 dB	1	2.8062	1.1473	2.5784	1.7813	1.7813	2.5784	1.1473	2.8062									

Filter type	R_s or G_s / Element number	1	2	3	4	Tenth-order network $n = 10$ 5	6	7	8	9	10
Bessel	0	0.4170	0.2797	0.2311	0.2021	0.1770	0.1504	0.1209	0.0886	0.0541	0.0182
	1	0.6305	0.3002	0.2384	0.2066	0.1808	0.1539	0.1240	0.0911	0.0556	0.0187
Butterworth	0	1.5643	1.8552	1.8121	1.6869	1.5100	1.2921	1.0406	0.7626	0.4654	0.1564
	1	0.3129	0.9080	1.4142	1.7820	1.9754	1.9754	1.7820	1.4142	0.9080	0.3129
Legendre-Papoulis	0	1.0298	1.8741	1.9405	1.9223	1.9102	1.8629	1.7785	1.6082	1.2386	0.5065
	1	1.5286	1.8122	1.8953	2.0409	1.8453	2.0327	1.7839	1.8537	1.4454	0.7575
Chebyshev 0.01 dB	0 / 1	1.4598	1.7731	1.8054	1.8020	1.7664	1.7106	1.6003	1.4096	1.0439	0.4098
Chebyshev 0.10 dB	0 / 1	1.4964	1.8585	1.8600	1.9068	1.8489	1.8579	1.7503	1.0542	1.3124	0.6000
Chebyshev 0.50 dB	0 / 1	1.4539	1.9816	1.8119	2.0432	1.8165	2.0197	1.7645	1.8905	1.4748	0.8771
Chebyshev 1.0 dB	0 / 1	1.3801	2.1111	1.7215	2.1803	1.7307	2.1658	1.6962	2.0645	1.4817	1.0918
Chebyshev 3.0 dB	0 / 1	1.1032	2.6753	1.3771	2.7082	1.3893	2.7655	1.3761	2.6958	1.2744	1.7692
Modified Chebyshev 0.01 dB	1	0.7795	1.4503	1.7608	1.7900	1.8405	1.0495	1.7903	1.7608	1.4503	0.7795
Modified Chebyshev 0.10 dB	1	1.1165	1.5249	1.9621	1.8183	1.9346	1.9346	1.8183	1.9621	1.5249	1.1165
Modified Chebyshev 0.50 dB	1	1.5832	1.4380	2.2072	1.7663	2.0029	2.0029	1.7663	2.2072	1.4380	1.5832
Modified Chebyshev 1.0 dB	1	1.9273	1.3374	2.3826	1.7118	2.0410	2.0410	1.7118	2.3826	1.3374	1.9273
Modified Chebyshev 3.0 dB	1	2.0950	1.0731	2.0001	1.5644	2.1269	2.1209	1.5644	2.8661	1.0731	2.9356

*Values for the Legendre-Papoulis approximation by permission from A. Zverev, "Handbook of Filter Synthesis," John Wiley & Sons, Inc., New York, 1967.

TABLE 12-14   Half-Power Frequencies for Various Chebyshev and Bessel Filters*

	Order of filter								
Type of filter	$n = 2$	$n = 3$	$n = 4$	$n = 5$	$n = 6$	$n = 7$	$n = 8$	$n = 9$	$n = 10$
Chebyshev 0.01 dB	3.303615	1.877180	1.466904	1.291217	1.199412	1.145268	1.110609	1.087064	1.070331
Chebyshev 0.10 dB	1.943219	1.388995	1.213099	1.134718	1.092931	1.068001	1.051927	1.040955	1.033131
Chebyshev 0.50 dB	1.389744	1.167485	1.093102	1.059259	1.041030	1.030090	1.023011	1.018167	1.014707
Chebyshev 1.00 dB	1.217626	1.094868	1.053002	1.033815	1.023442	1.017205	1.013164	1.010396	1.008418
Chebyshev 3.00 dB	1.000594	1.000264	1.000149	1.000095	1.000066	1.000048	1.000037	1.000029	1.000024
Modified Chebyshev 0.01 dB	4.563742		1.532784		1.212468		1.114760		1.072036
Modified Chebyshev 0.10 dB	2.559727		1.246004		1.099300		1.053929		1.033948
Modified Chebyshev 0.50 dB	1.691974		1.108290		1.043913		1.023911		1.015073
Modified Chebyshev 1.00 dB	1.401865		1.061830		1.025105		1.013681		1.008628
Modified Chebyshev 3.00 dB	1.001188		1.000174		1.000071		1.000039		1.000024
Bessel half-power frequency	1.3617	1.7557	2.1140	2.4274	2.7034	2.9517	3.1797	3.3917	3.5910
Delay at half-power frequency	0.8090	0.9349	0.9819	0.9960	0.9993	0.9999	0.9999	1.0000	1.0000

*The Chebyshev filter prototypes have the ripple specified at 1 rad/s. The Bessel filter has a delay of 1 s at very low frequency. The second figure given is the delay at the half-power frequency.

16. Explicit Butterworth and Chebyshev Formulas. Closed-form solutions for doubly terminated Butterworth and Chebyshev filters were derived by Takahasi and restated in Humpherys.[8] They assume that the zeros of the reflection coefficient lie in the left half plane, and so give different networks than the symmetrical network procedure given at the end of Par. **12-15.** This gives the designer a choice in cases where both techniques may be applicable.

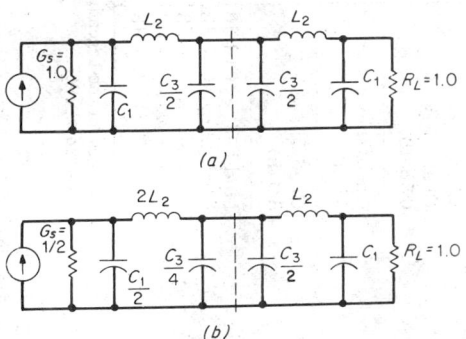

(a)

(b)

Fig. 12-20. The scaling of symmetrical networks to change the value of G_s.

The Butterworth formulas make use of the poles of the transfer functions and require also a factor λ that relates the load and source resistances. The source resistance is assumed to be 1.0, and the formulas are

$$s_i = 2 \sin (\pi i/2n) \qquad (12\text{-}58)$$

$$c_i = 2 \cos (\pi i/2n) \qquad (12\text{-}59)$$

$$\lambda = - \left(\frac{R_L - 1}{R_L + 1} \right)^{1/n} \qquad (12\text{-}60)$$

when the first reactive element is a shunt capacitor, and

$$\lambda = - \left(\frac{G_L - 1}{G_L + 1} \right)^{1/n} \qquad (12\text{-}61)$$

when the first reactive element is a series inductor. Recursive equations give the element values

$$C_1 = \frac{s_1}{1 - \lambda} \tag{12-62}$$

$$C_n = \frac{s_1}{(1 + \lambda)R_L} \quad n \text{ odd} \tag{12-63}$$

or

$$L_n = \frac{s_1 R_L}{1 + \lambda} \quad n \text{ even}$$

$$C_{2m-1}L_{2m} = \frac{s_{4m} - 3s_{4m-1}}{1 - \lambda c_{4m-2} + \lambda^2} \tag{12-64}$$

$$L_{2m}C_{2m+1} = \frac{s_{4m} - 1s_{4m+1}}{1 - \lambda c_{4m} + \lambda^2} \tag{12-65}$$

$$\text{where } m = \begin{cases} 1, 2, \ldots, (n-1)/2 & n \text{ odd} \\ 1, 2, \ldots, n/2 & n \text{ even} \end{cases}$$

When the first element is a series inductor, the roles of L and C are interchanged. An example follows. Develop the prototype network for third-order Butterworth, $R_L = 3$, $R_s = 1$. It follows that

$$\lambda = -0.7937 \tag{12-66}$$
$$s_1 = 1.0000 = s_5 \tag{12-67}$$
$$s_2 = 1.7321 = s_4 \tag{12-68}$$
$$s_3 = 2.0000 \tag{12-69}$$
$$c_1 = 1.7321 \tag{12-70}$$
$$c_2 = 1.0000 \tag{12-71}$$
$$c_3 = 0.0000 \tag{12-72}$$
$$c_4 = -1.0000 \tag{12-73}$$
$$C_1 = \frac{1.0000}{1 - (-0.7937)} = 0.5575 \tag{12-74}$$
$$L_2 = \frac{s_1 s_3}{1 - (-0.7937)(1.0000) + (0.7937)^2 \, C_1} \frac{1}{C_1} = 1.4802 \tag{12-75}$$
$$C_3 = \frac{s_3 s_5}{1 - (-0.7937)(-1.0000) + (0.7937)^2 \, L_2} \frac{1}{L_2} = 1.6158 \tag{12-76}$$

The prototype network thus becomes that of Fig. 12-21.

Similar results are obtained if the equation for C_n is used first and the network is developed from the load end; this serves as a check on the work.

The Chebyshev equations are similar. Define the following terms:

$$\epsilon = \sqrt{10^{r/10} - 1} \qquad \begin{array}{l} \epsilon = \text{ripple factor} \\ r = \text{ripple, dB} \end{array} \tag{12-77}$$

$$A = \begin{cases} 4R_L R_s/(R_L + R_s)^2 & n \text{ odd} \\ 4(1 + \epsilon^2)R_L R_s/(R_L - R_s)^2 \leq 1 & n \text{ even} \end{cases} \begin{array}{l} (12\text{-}78a) \\ (12\text{-}78b) \end{array}$$

$$s_i = 2 \sin (\pi i/2n) \tag{12-79}$$
$$c_i = 2 \cos (\pi i/2n) \tag{12-80}$$

$$k = \left(\frac{1}{\epsilon} + \sqrt{\frac{1}{\epsilon^2} + 1} \right)^{1/n} \tag{12-81}$$

$$h = -\left(\sqrt{\frac{1 - A}{\epsilon^2}} + \sqrt{\frac{1 - A}{\epsilon^2} + 1} \right)^{1/n} \tag{12-82}$$

$$k' = k - 1/k \tag{12-83}$$
$$h' = h - 1/h \tag{12-84}$$

Recursive equations give the element values

$$C_1 = \frac{2s_1/R_s}{k' - h'} \tag{12-85}$$

$$C_n = \frac{2s_1/R_L}{k' + h'} \quad n \text{ odd} \tag{12-86}$$

$$L_n = \frac{2s_1R_L}{k' + h'} \qquad n \text{ even} \tag{12-87}$$

$$C_{2m-1}L_{2m} = \frac{4s_{4m-3}s_{4m-1}}{k'^2 - c_{2i}k'h' + h'^2 + s_{2i}^2} \tag{12-88}$$

$$L_{2m}C_{2m+1} = \frac{4s_{4m-1}s_{4m+1}}{k'^2 - c_{2i}k'h' + h'^2 + s_{2i}^2} \tag{12-89}$$

$$\text{where } m = \begin{cases} 1, 2, \ldots, (n-1)/2 & n \text{ odd} \\ 1, 2, \ldots, n/2 & n \text{ even} \end{cases}$$

When the first element is a series inductor, the roles of L and C are interchanged, while G_s and G_L are substituted for R_s and R_L. As an example, develop the prototype when $R_s = 3$, $R_L = 1$, 0.5 dB ripple, and a fourth-order network is needed. It follows that

$$\epsilon = \sqrt{10^{0.05} - 1} = 0.3493 \tag{12-90}$$
$$A = 0.75[1 + (0.3493)^2] = 0.8415 \tag{12-91}$$
$$s_1 = 0.7654 = s_7 \tag{12-92}$$
$$s_2 = 1.4142 = s_6 \tag{12-93}$$
$$s_3 = 1.8478 = s_5 \tag{12-94}$$
$$s_4 = 2.0000 \tag{12-95}$$
$$c_0 = 2.0000 \tag{12-96}$$
$$c_2 = 1.4142 \tag{12-97}$$
$$c_4 = 0.0000 \tag{12-98}$$
$$c_6 = -1.4142 \tag{12-99}$$
$$k = 1.5582 \tag{12-100}$$
$$h = -1.2766 \tag{12-101}$$
$$k' = 0.9164 \tag{12-102}$$
$$h' = -0.4933 \tag{12-103}$$
$$C_1 = \frac{2(0.7654)}{3[0.9164 - (-0.4933)]} = 0.3620 \tag{12-104}$$
$$L_2 = \frac{4s_1s_3}{C_1(k'^2 - c_2k'h' - h'^2 + s_2^2)} = 4.1985 \tag{12-105}$$
$$C_3 = \frac{4s_3s_5}{L_2(k'^2 - c_4k'h' - h'^2 + s_4^2)} = 0.6399 \tag{12-106}$$
$$L_4 = \frac{4s_5s_7}{C_3(h'^2 - c_6k'h' + h'^2 + s_6^2)} = 3.6172 \tag{12-107}$$

The prototype network is shown in Fig. 12-22.

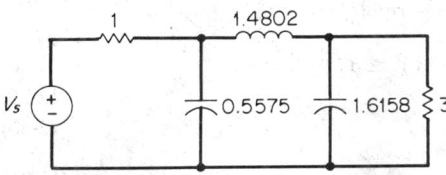

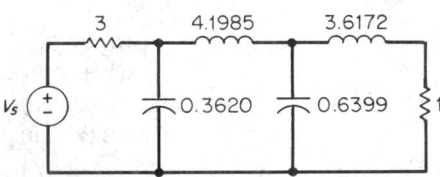

Fig. 12-21. Prototype Butterworth network, $n = 3$, $R_s = 1$, $R_L = 3$.

Fig. 12-22. Prototype Chebyshev network, $R_s = 3$, $R_L = 1$, $n = 4$, 0.5 dB ripple.

17. The Elliptic Filter. The five approximations thus far described are of a class known as *all-pole approximations* because the magnitude-squared function of Eq. (12-6) has no finite zeros. This concentration of the zeros at infinity usually gives more attenuation than required at the higher frequencies and less in the vicinity of the cutoff frequency. If the more general form of Eq. (12-5) is used for the magnitude-squared function, some of the zeros can be placed close to the stop-band edge frequency, with an improvement in the cutoff rate. Specifically, if

$$|G(j\omega)|^2 = K^2 \frac{1}{1 + \epsilon^2 R_n^2(\omega)} \tag{12-108}$$

$$
R_n(\omega) = \begin{cases}
r \dfrac{\omega \displaystyle\prod_{i=1}^{(n-1)/2} (\omega^2 - \omega_{pi}^2)}{\displaystyle\prod_{i=1}^{(n-1)/2} (\omega^2 - \omega_{si}^2)} & n \text{ odd} \qquad\qquad (12\text{-}109) \\[4ex]
r \dfrac{\displaystyle\prod_{i=1}^{n/2} (\omega^2 - \omega_{pi}^2)}{\displaystyle\prod_{i=1}^{n/2} (\omega^2 - \omega_{si}^2)} & n \text{ even} \qquad\qquad (12\text{-}110)
\end{cases}
$$

where

r being a multiplicative constant, the approximation can be made equiripple in both the pass-band and the stop band. Such an approximation is called an *elliptic approximation* because elliptic functions are used in its determination. Figure 12-23 shows such an approximation in a plot of attenuation vs. frequency for $n = 3$ and $n = 4$, illustrative of the general characteristics for even and odd n, respectively.

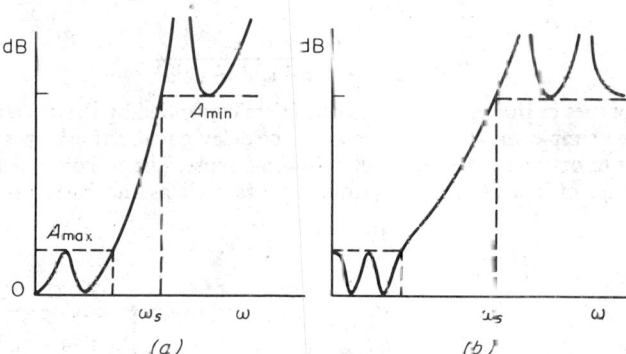

Fig. 12-23. The elliptic approximation for (*a*) $n = 3$ and (*b*) $n = 4$.

Since the even-order elliptic approximation does not have infinite attenuation at infinite frequency, it cannot be realized by an *LC* ladder. A modified even-order elliptic approximation can be obtained by a frequency transformation. This transformation sacrifices some stop-band attenuation to shift a pole of attenuation to ∞. A further modification of the even-order elliptic approximation can be used to make the attenuation zero at zero frequency. This permits an *LC* ladder realization to be equally terminated.

The elliptic approximation is characterized by four parameters.

1. The passband ripple A_{max} or, equivalently, the reflection coefficient ρ, related to A_{max} by $A_{max} = -10 \log (1 - \rho^2)$

2. The order n

3. The minimum stop-band attenuation A_{min}

4. The stop-band edge frequency ω_s or, equivalently, the modular angle θ[6]

Any three of these four parameters can be independently specified.

As for the all-pole approximations, tabulations[7] of pole-zero locations and tables of element values for normalized low-pass filters based on the elliptic approximation are possible. Since three parameters are required to specify an elliptic approximation, these tabulations are voluminous.

18. Delay Equalization. The ideal low-pass filter characteristic would have a constant group delay as well as a constant amplitude throughout the passband so that the various frequency components of a signal would arrive at the output unattenuated and with the proper phase relationship. Deviations from constant amplitude and group delay produce amplitude and phase distortion, respectively. While phase distortion is not a problem in many applications, it is significant when pulse transmission is involved.

With the exception of the Bessel approximation, the various approximations have focused on the amplitude characteristic. While the Bessel approximation has a good group-delay characteristic, it has a significantly poorer amplitude characteristic. One approach to the problem of obtaining filters with good amplitude *and* group-delay characteristics is through the use of delay equalizers. This consists of using one of the approximations with desirable amplitude character-istics in conjunction with an all-pass function that does not affect the amplitude characteristic but modifies the group delay beneficially.

The first-order all-pass function is

$$G(s) = \frac{s - \delta}{s + \delta} \tag{12-111}$$

This function has an amplitude of 1 for all frequencies and a group-delay function

$$\tau_d(\omega) = \frac{2}{\delta} \frac{1}{1 + (\omega/\delta)^2} \tag{12-112}$$

The specific properties of the group-delay function are controlled by the selection of δ.

The second-order all-pass function is

$$G(s) = \frac{(s - \delta)^2 + \beta^2}{(s + \delta)^2 + \beta^2} \tag{12-113}$$

with

$$|G(j\omega)| = 1 \tag{12-114}$$

and

$$\tau_d(\omega) = \frac{4\delta(\omega^2 + \delta^2 + \beta^2)}{(\delta^2 + \beta^2 - \omega^2)^2 + 4\delta^2\omega^2} \tag{12-115}$$

The specific properties of this group-delay function are controlled by the selection of δ and β.

Except for some simple situations, the problem of delay equalization is best approached by using a computer to optimize the group delay in some sense. Blinchikoff and Zverev[21] give an excellent discussion of a least-squares optimization as well as the basic principles of delay equalization.

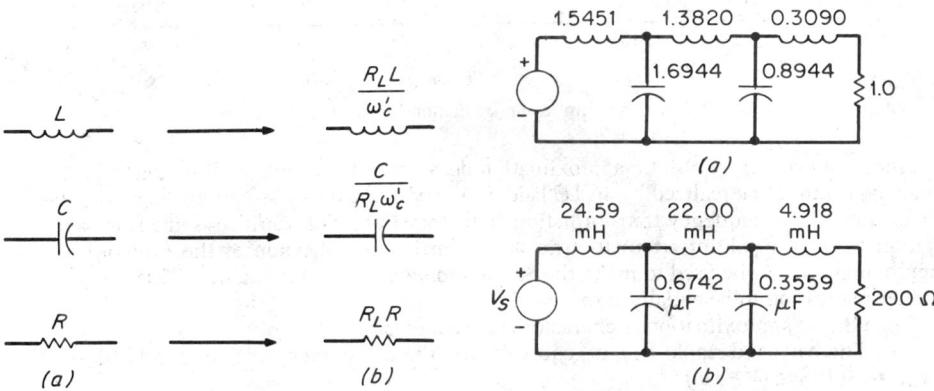

Fig. 12-24. Relations between elements of (a) a normalized prototype and those of (b) a practical low-pass filter.

Fig. 12-25. (a) A normalized prototype and (b) resulting low-pass filter.

19. Practical Filter Design. Table 12-13 gives element values for normalized networks. Practical filters can be designed from these normalized networks by a process consisting of three general steps.

1. Statement of the specifications for the filter
2. Translation of the specifications to equivalent statements for the normalized prototype and their use to determine the parameters necessary to enter Table 12-13
3. Application of frequency transformations and impedance-level scaling to the normalized prototype to determine the filter network

This process is described in the following paragraphs for the various types of filters, accompanied by examples.

20. The Low-Pass Filter. The low-pass filter is related to the normalized prototype by frequency and impedance-level scaling. The frequency in the prototype is related to the frequency in the low-pass filter by

$$\omega = \omega'/\omega_c' \tag{12-116}$$

where the primed quantities refer to the practical low-pass filter, with ω_c' being its cutoff frequency. The application of the frequency scaling implied by Eq. (12-116) and the impedance-level scaling required to change the load resistance from 1.0 to R_L results in the elements of the low-pass filter becoming those indicated in Fig. 12-24b.

Example. A low-pass filter is to be designed to have (a) a maximally flat amplitude-vs.-frequency characteristic at $\omega' = 0$, (b) a cutoff (half-power) frequency of 2 kHz, (c) a load resistance of 200 Ω, (d) a voltage source input with zero resistance, and (e) an attenuation of not less than 20 dB at frequencies greater than 3.2 kHz. A Butterworth approximation is indicated. The frequency in the prototype equivalent to 3.2 kHz is

$$\omega = \frac{(3.2 \times 10^3)(2\pi)}{(2 \times 10^3)(2\pi)} = 1.6 \tag{12-117}$$

and the equivalent design specifications for the prototype require that the attenuation be no less than 20 dB for $\omega \geq 1.6$. This, using Eq. (12-20), results in

$$n \geq \frac{20}{20 \log 1.6} = 4.899 \tag{12-118}$$

and the next larger integer value of 5 is used. The appropriate prototype appears in Fig. 12-25a with element values taken from Table 12-13 for the Butterworth approximation with $n = 5$ and $R_s = 0$. The application of frequency and magnitude scaling by means of the relations in Fig. 12-24 results in the low-pass filter of Fig. 12-25b.

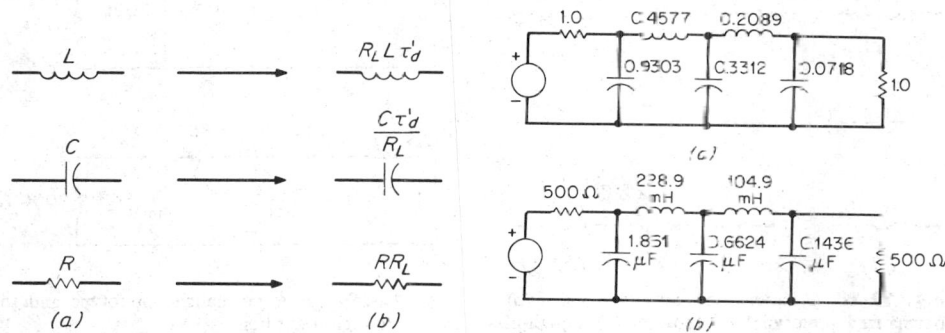

Fig. 12-26. Relations between elements for (a) a normalized prototype and those of (b) a time-delay network.

Fig. 12-27. (a) A normalized prototype and (b) resulting time-delay network.

21. A Time-Delay Network. The frequencies in the time-delay network and its normalized prototype are related by

$$\omega = \tau_d' \omega' \tag{12-119}$$

where the primed quantities again refer to the practical network, τ_d being the group delay evaluated at $\omega' = 0$. The element values of the practical network are related to those of the normalized prototype by frequency and impedance-level scaling. These relations are shown in Fig. 12-26.

Example. A time-delay network is to be designed to have (a) a time delay of 1 ms, the error being no greater than 5% at 500 Hz; (b) a voltage input with a resistance of 500 Ω; and (c) a load resistance of 500 Ω. The prototype frequency equivalent to 500 Hz is

$$\omega = 10^{-3}(2\pi \times 500) = 3.142 \tag{12-120}$$

and the equivalent specifications for the prototype require a time delay of 1 s, the error being no greater than 5% at $\omega = 3.142$. Figure 12-13b is used to determine that $n = 5$ is required to satisfy this requirement. The prototype can be of the form of either Fig. 12-17d or Fig. 12-18b. The latter is chosen with the element values taken from Table 12-13 for the Bessel approximation with $n = 5$ and $G_s = 1$. The prototype is shown in Fig. 12-27a, and the application of the relations in Fig. 12-26 results in the time-delay network of Fig. 12-27b.

22. The High-Pass Filter. The frequency in the high-pass filter is related to that in the normalized prototype by

$$\omega = \omega_c'/\omega' \tag{12-121}$$

where the primed quantities refer to the high-pass filter, ω'_c being the cutoff frequency. The relations between the elements of the normalized prototype and those of the high-pass filter are given in Fig. 12-28.

Example.   A high-pass filter is to be designed to have (a) a cutoff frequency of 7.5 kHz, (b) an equiripple characteristic in the passband with a 0.5-dB ripple ($\epsilon = 0.3493$), (c) an attenuation of at least 35 dB at frequencies below 3 kHz, (d) equal load and source resistances of 2000 Ω, and (e) a voltage source input. The prototype frequency equivalent to 3 kHz is

$$\omega = \frac{2\pi(7.5 \times 10^3)}{2\pi(3 \times 10^3)} = 2.5 \tag{12-122}$$

Thus the prototype is based on a Chebyshev approximation with a 0.5-dB ripple, equal load and source terminations, voltage source input, and an attenuation of at least 35 dB for frequencies greater than $\omega = 2.5$. The last requirement, used with Eq. (12-42), results in $C_n(2.5) \approx 161.0$. This result, used in Eq. (12-40), yields $n = 3.69$, and $n = 4$ is used. Since the regular Chebyshev approximation is not available, the modified Chebyshev approximation is used. The networks of Figs. 12-17c and 12-18a are possible. The former is chosen, and the resulting prototype is shown in Fig. 12-29a. The relations of Fig. 12-28 are then used to determine the element values of the high-pass filter of Fig. 12-29b.

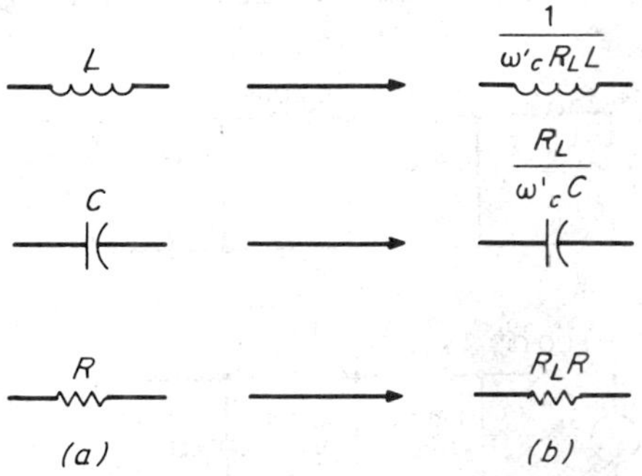

Fig. 12-28. Relations between elements of (a) a normalized prototype and those of (b) a high-pass filter.

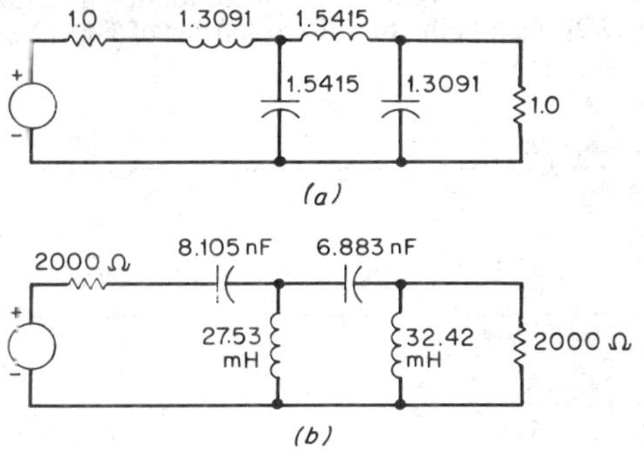

Fig. 12-29. (a) A normalized prototype and (b) resulting high-pass filter.

23. The Bandpass Filter. The frequency in the bandpass filter is related to that in the normalized prototype by

$$\omega = \frac{\omega'_0}{\beta'}\left(\frac{\omega'}{\omega'_0} - \frac{\omega'_0}{\omega'}\right) \tag{12-123}$$

where the primed quantities refer to the bandpass filter, ω'_0 being the center frequency and β' the bandwidth, $\omega'_{c2} - \omega'_{c1}$, as defined in Fig. 12-2c. The elements of the normalized prototype and those of the bandpass filter are related as shown in Fig. 12-30.

Example.   A bandpass filter is to be designed to have (a) a center frequency of 4.0 kHz, (b) a bandwidth of 900 Hz, (c) an equiripple characteristic in the passband with a 1-dB ripple ($\epsilon = 0.5088$), (d) an attenuation of at least 18 dB at 4.9 kHz, (e) equal load and source resistances of 200 Ω, and (f) a current source input. The prototype frequency equivalent to 4.9 kHz is, from Eq. (12-123), 1.816 and the prototype is based on a Chebyshev approximation with a 1-dB ripple, equal load and source terminations, current source input, and an attenuation of at least 18 dB at $\omega = 1.816$. The last requirement, used with Eq. (12-42), results in $C_n(1.8163) \approx 19.65$. This result, used in Eq. (12-40), yields $n = 2.86$, and $n = 3$ is used. Two networks, those of Fig. 12-17a and Fig. 12-18d, are possible. The former is chosen, and the resulting prototype is shown in Fig. 12-31a. The relations of Fig. 12-30 are then used to determine the element values of the bandpass filter of Fig. 12-31b.

The results of the preceding example illustrate a problem in the transformation from prototype to bandpass filter. For even the moderate ratio of center frequency to bandwidth of the example, the ratio of series-arm to shunt-arm inductances in the bandpass filter is large. This creates some practical problems arising from the stray capacitances associated with large inductances. The reader is referred to Humpherys[8] for an excellent discussion of this problem and possible ways of overcoming the difficulty. A similarly large ratio of shunt-arm to series-arm capacitances is also present, although the practical problems are not so severe.

24. The Band-Elimination Filter. The frequency in the band-elimination filter is related to that in the normalized prototype by

$$\omega = \frac{1}{(\omega_0'/\beta')[(\omega_0'/\omega') - \omega'/\omega_0']} \tag{12-124}$$

where the primed quantities refer to the band-elimination filter, ω_0' being the center frequency and β' the bandwidth, $\omega_{c2}' - \omega_{c1}'$, as defined in Fig. 12-2d. The elements of the normalized prototype and those of the band-elimination filter are related as shown in Fig. 12-32.

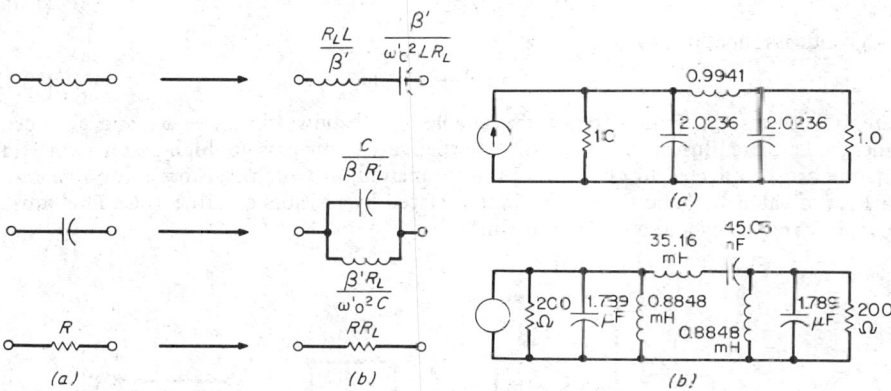

Fig. 12-30. Relations between elements of (a) a normalized prototype and those of (b) a bandpass filter.

Fig. 12-31. (a) A normalized prototype and (b) resulting bandpass filter.

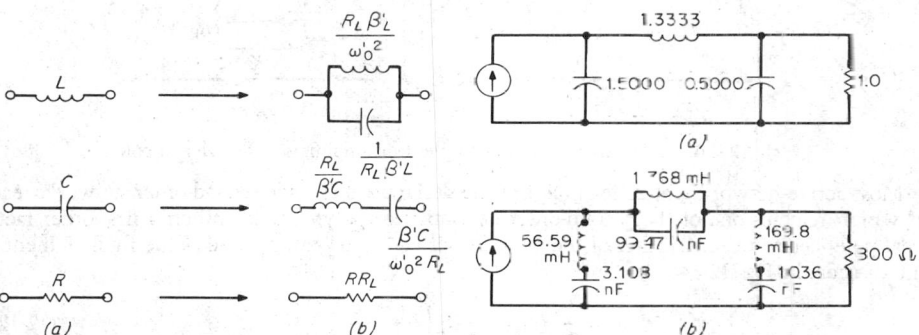

Fig. 12-32. Relations between elements of (a) a normalized prototype and those of (b) a band-elimination filter.

Fig. 12-33. (a) A normalized prototype and (b) resulting band-elimination filter.

Example. A band-elimination filter is to be designed to have (a) a center frequency of 12 kHz, (b) a bandwidth (half-power) of 1.5 kHz, (c) a maximally flat characteristic at $\omega' = 0$, (d) an attenuation of at least 16 dB at 11.6 kHz, (e) a current source input, and (f) $R_L = 800\ \Omega$ and $G_s = 0$. Since the prototype frequency equivalent to 11.6 kHz is, from Eq. (12-124), 1.8432, the prototype is based on a Butterworth approximation with an attenuation of at least 16 dB at this frequency. This requirement, used with Eq. (12-19), results in $n = 2.99$, and $n = 3$ is used. Since $G_s = 0$, the only network available is that of Fig. 12-17b, and the resulting prototype is shown in Fig. 12-33a. The relations of Fig. 12-32 are then used to determine the element values of the band-elimination filter of Fig. 12-33b.

This example illustrates the large ratios of shunt-arm to series-arm inductances and of series-arm to shunt-arm capacitances that result with even a moderate ratio of center frequency to bandwidth. The reader is again referred to Humpherys[8] for a discussion of this difficulty and steps that can be taken to overcome it.

25. Active Network Filters. Advances in recent years in the development of low-cost integrated circuits have made it possible to incorporate these modules into filters. They are attractive in the lower frequency ranges, though many practical circuits now operate to above 100 kHz. A primary reason for choice of an active filter is to eliminate the inductors; a second advantage is that a voltage or current gain is often possible. Huelsman[11,12] describes three general techniques, controlled-source realizations, infinite-gain realizations, and synthetic-inductance networks. The operational amplifier is the principal active element employed, and special-purpose modules are available for many applications.

The quadratic factors given in Tables 12-1, 12-3, 12-5, 12-7, and 12-9 are all for low-pass filters. In every case these are denominator polynomials, while the numerator is a constant. These transfer functions can be changed to high-pass functions by the change of variable

$$s = 1/p \tag{12-125}$$

and to bandpass functions by

$$s = (p^2 + \omega_0^2)/Bp \tag{12-126}$$

where, in both cases, p = new frequency variable, B = bandwidth, $\omega_{c2} - \omega_{c1}$, and ω_0 = center frequency. Both are illustrated in Fig. 12-2. Alternatively, a low-pass-to-high-pass network transformation can be effected by the RC-CR transformation, in which resistors R are replaced by capacitors of value $1/R$ and capacitors C are replaced by resistors of value $1/C$. This must be followed by frequency and impedance scaling.

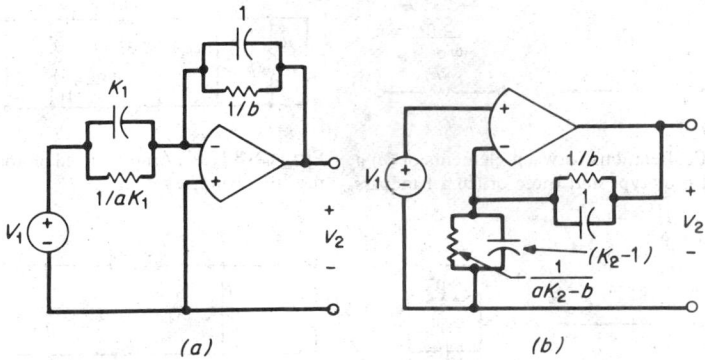

Fig. 12-34. (a) Inverting and (b) noninverting amplifiers to realize real poles.

Most active-network realizations require the designer to cascade second-order networks, each of which realizes one of the second-order denominator polynomials. When a first-order factor (real pole) is also needed, either of the networks of Fig. 12-34 can be used as the final element in the cascade. In Fig. 12-34a

$$\frac{V_2}{V_1}(s) = -K_1 \left(\frac{s + a}{s + b}\right) \tag{12-127}$$

and in Fig. 12-34b,

$$\frac{V_2}{V_1}(s) = K_2 \left(\frac{s + a}{s + b}\right) \tag{12-128}$$

In Fig. 12-34b, $K_2 \geq 1$, and $aK_2 \geq b$.

26. Synthetic Inductance Filters—Gyrators. It is possible to build filters with active elements that replace inductors. These realizations use the published tables for passive low-pass networks and the networks that are transformed from them. They also have the low-sensitivity properties of passive networks. One technique is based on the gyrator,[14] a two-port with a z-matrix representation

$$\begin{bmatrix} V_1 \\ V_2 \end{bmatrix} = \begin{bmatrix} 0 & -a \\ a & 0 \end{bmatrix} \begin{bmatrix} I_1 \\ I_2 \end{bmatrix} \tag{12-129}$$

where a is called the *gyration resistance*. A gyrator terminated in an impedance Z_L is shown in Fig. 12-35, and it is possible to show that

$$V_1/I_1 = Z_{IN} = a^2/Z_L \tag{12-130}$$

when $Z_L = 1/sC$,

$$Z_{IN} = a^2 sC \qquad (12\text{-}131)$$

which is equivalent to an inductance of a^2C. The problem then becomes one of choosing a and C.

Gyrators are available in integrated-circuit form from several electronics manufacturers; the user simply adds the appropriate capacitor. Gyrators can also be built with operational amplifiers. One useful circuit is the Riordan gyrator,[15] shown in Fig. 12-36 with Z_L as one of the five impedances it requires.

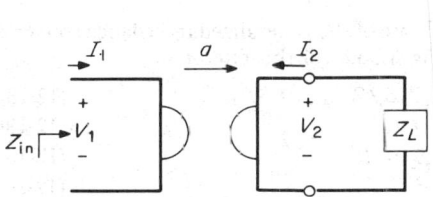

Fig. 12-35. Gyrator terminated with Z_L.

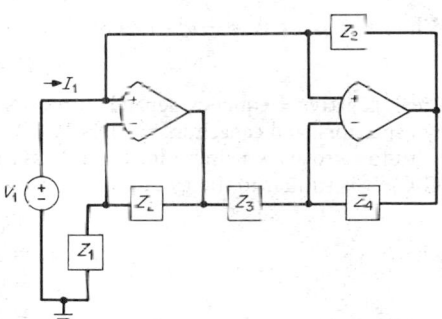

Fig. 12-36. Riordan gyrator.

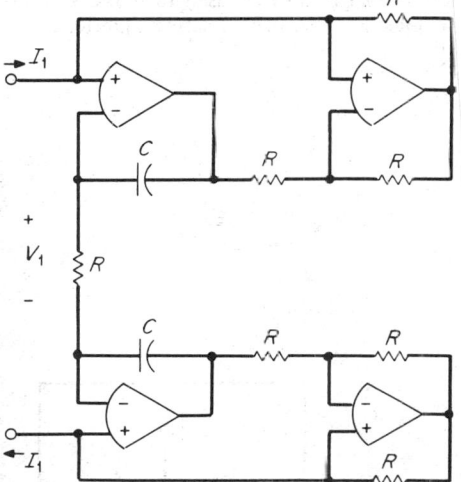

Fig. 12-37. Riordan-type back-to-back gyrator for floating inductors.

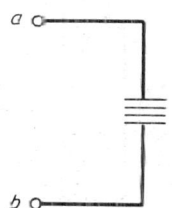

Fig. 12-38. FDNR representation.

It can be shown that

$$\frac{V_1}{I_1} = Z_{IN} = \frac{Z_1 Z_2 Z_3}{Z_4} \frac{1}{Z_L} \qquad (12\text{-}132)$$

When all Z's are resistors R, and $Z_L = 1/sC$,

$$Z_{IN} = R^2 sC \qquad (12\text{-}133)$$

This circuit must be used to replace a grounded inductor, e.g., in a high-pass network. If an ungrounded inductor is needed, the modification shown in Fig. 12-37 can be used, for which

$$Z_{IN} = R^2 sC \qquad (12\text{-}134)$$

Other circuits, for which one side of the added capacitor C may be grounded, can also be used though they may be more difficult to align.

27. Synthetic Inductance Filters—Frequency-Dependent Negative Resistors. Bruton[16] introduced a new type of circuit, called the *frequency-dependent negative resistor* (FDNR); the symbol is shown in Fig. 12-38. To use this idea, consider a new type of impedance scaling, in which passive elements become scaled by A/s; this does not affect the transfer function:

Passive impedance   Scaled impedance

$$\text{---} sL \quad \rightarrow \quad \text{---} AL \tag{12-135a}$$

$$\text{---} R \quad \rightarrow \quad \text{---} AR/s \tag{12-135b}$$

$$\text{---} 1/sC \quad \rightarrow \quad \text{---} A/Cs^2 \tag{12-135c}$$

When $s = j\omega$ or $s^2 = -\omega^2$, the last element becomes

$$Z(j\omega) = -A/D\omega^2 \tag{12-136}$$

a real, negative, frequency-dependent resistor. Thus, inductors are replaced by resistors, resistors by capacitors, and capacitors by FDNRs.

Bruton also gives a circuit for a FDNR, as a special case of the generalized impedance converter (GIC). It is similar to the gyrator and is shown in Fig. 12-39. For this circuit,

$$V_1/I_1 = Z_{IN} = Z_1 Z_3 Z_5 / Z_2 Z_4 \tag{12-137}$$

when

$$Z_1 = Z_3 = 1/sC \tag{12-138}$$

$$Z_2 = Z_4 = Z_5 = R \tag{12-139}$$

$$Z_{IN} = 1/s^2 R C^2 \tag{12-140}$$

which is a FDNR. GICs are available from integrated-electronic-circuit manufacturers, and the user adds resistors and capacitors to make FDNRs. They can also be used to make gyrators. FDNRs are grounded when replacing capacitors in low-pass networks; floating FDNRs can be achieved by back-to-back GICs, as with gyrators.

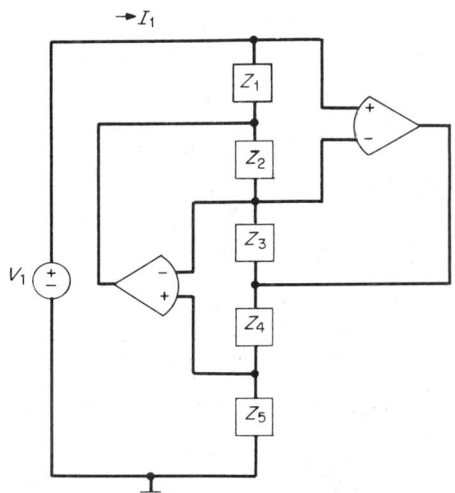

Fig. 12-39. Generalized impedance converter.

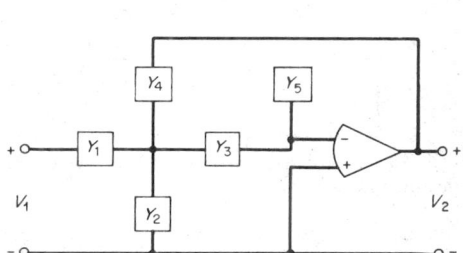

Fig. 12-40. A multiple-feedback, infinite-gain active realization for low-pass, high-pass, and band-pass filters. Table 12-15 indicates the choice of the five active elements for each case.

28. Infinite-Gain, Multiple-Feedback Realization.[12,13] The circuit of Fig. 12-40 shows an operational amplifier with five passive elements, which are either resistors or capacitors. The general voltage transfer function is

$$\frac{V_2}{V_1} = \frac{-Y_1 Y_3}{Y_5(Y_1 + Y_2 + Y_3 + Y_4) + Y_3 Y_4} \tag{12-141}$$

Table 12-15 describes how the five passive elements can be chosen to implement a low-pass, high-pass, or bandpass network.

TABLE 12-15   Element Choice for Active Filter Circuit of Fig. 12-40

Filter desired	Y_1	Y_2	Y_3	Y_4	Y_5
Low-pass	Resistor	Capacitor	Resistor	Resistor	Capacitor
High-pass	Capacitor	Resistor	Capacitor	Capacitor	Resistor
Bandpass	Resistor	Resistor	Capacitor	Capacitor	Resistor

29. State-Variable Realization.[12] This network is a special but important type of infinite-gain realization. It has the advantage that low-pass, high-pass, and bandpass configurations can be realized simultaneously, and it is also easy to adjust and to produce in quantity. The network is shown in Fig. 12-41, and the three possible transfer functions are

Low-pass:

$$\frac{V_{lp}}{V_1} = \frac{\dfrac{R_4(R_5 + R_6)}{R_1 R_2 R_5 C_1 C_2(R_3 + R_4)}}{s^2 + s\left[\dfrac{R_3(R_5 + R_6)}{R_1 C_1(R_3 + R_4)}\right] + \dfrac{R_5}{R_1 R_2 R_5 C_1 C_2}} \tag{12-142}$$

High-pass:

$$\frac{V_{hp}}{V_1} = \frac{\dfrac{s^2 R_4(R_5 + R_6)}{R_5(R_3 + R_4)}}{s^2 + s\left[\dfrac{R_3(R_5 + R_6)}{R_1 C_1(R_3 + R_4)}\right] + \dfrac{R_5}{R_1 R_2 R_5 C_1 C_2}} \tag{12-143}$$

Bandpass:

$$\frac{V_{bp}}{V_1} = \frac{\dfrac{-s R_4(R_5 + R_6)}{R_1 R_5 C_1(R_3 + R_4)}}{s^2 + s\left[\dfrac{R_3(R_5 + R_6)}{R_1 C_1(R_5 + R_4)}\right] + \dfrac{R_6}{R_1 R_2 R_5 C_1 C_2}} \tag{12-144}$$

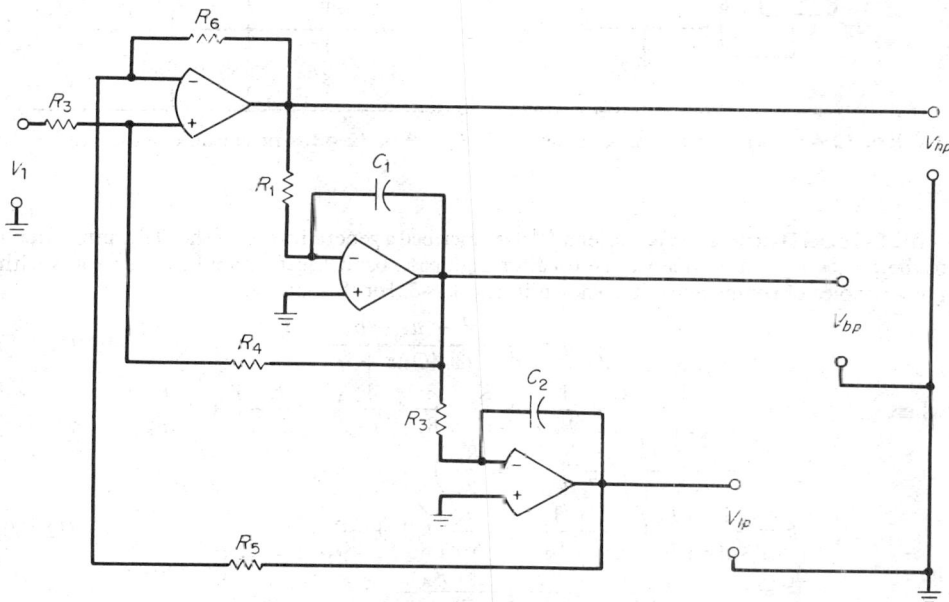

Fig. 12-41. State-variable realization of second-order active filters.

The network has a low output impedance at each terminal, so that RC sections can be added for odd-ordered networks. These sections can be cascaded so that networks of order 4 and higher can be built. They have the property that performance variations with parameter changes are comparable with those of strictly passive networks, but the disadvantage of requiring three operational amplifiers.

As with the other networks, design is a matter of choosing the appropriate quadratic factors from Tables 12-1, 12-3, 12-5, 12-7, or 12-9, matching coefficients with other constraints that arise from technological and economic considerations, and finally, scaling impedance and frequency.

30. Delyiannis Bandpass Circuit. Daryanani[17] describes a useful bandpass circuit shown in Fig. 12-42. For this circuit

$$\frac{V_2}{V_1}(s) = \frac{-s/[R_1 C_2(1 - 1/k)]}{s^2 + (\omega_P/Q_P)s + \omega_P^2} \tag{12-145}$$

where

$$k = 1 + \frac{R_B}{R_A} \tag{12-146}$$

$$\omega_P^2 = \frac{1}{R_1 R_2 C_1 C_2} \tag{12-147}$$

$$\frac{\omega_P}{Q_P} = \frac{1}{R_2 C_1} + \frac{1}{R_2 C_2} - \frac{1}{k-1}\frac{1}{R_1 C_2} \tag{12-148}$$

The circuit can be used with $R_A = 0$; in this case $k \to \infty$, and Eq. (12-145) is readily simplified. Here it is necessary to limit Q_P to about 5.

When Q_P greater than 5 is desired, R_A is included and the difference term in the denominator permits a greater Q_P.

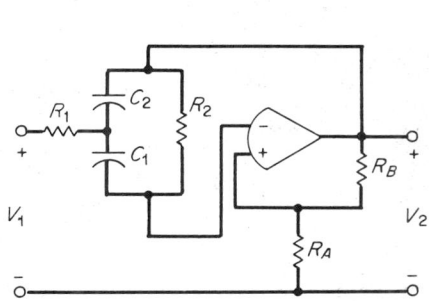

Fig. 12-42. Delyiannis bandpass circuit.

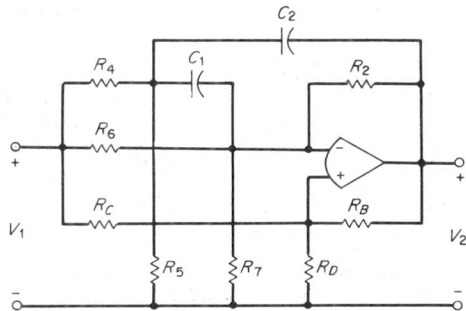

Fig. 12-43. Friend biquadratic circuit.

31. Friend Biquadratic. Friend[17,18] has described a generalization of the Delyiannis circuit of the preceding section that can be used for high-pass, band-reject, and all-pass networks with proper choice of components. It is shown in Fig. 12-43. For this circuit

$$\frac{V_2}{V_1}(s) = \frac{K_2 s^2 + as + b}{s^2 + (\omega_P/Q_P)s + \omega_P^2} \tag{12-149}$$

where

$$a = \frac{K_2}{C_2}\left(\frac{1}{R_1} + \frac{1}{R_2} + \frac{1}{R_3}\right) + \frac{K_2}{C_1}\left(\frac{1}{R_2} + \frac{1}{R_3}\right) - \frac{K_1}{R_1 C_2}\left(1 + \frac{R_A}{R_B}\right)$$
$$- \frac{K_3}{R_3}\left(\frac{1}{C_1} + \frac{1}{C_2}\right)\left(1 + \frac{R_A}{R_B}\right)$$

$$b = \frac{1}{C_1 C_2}\left[\frac{K_2}{R_1}\left(\frac{1}{R_2} + \frac{1}{R_3}\right) - \frac{K_3}{R_1 R_3}\left(1 + \frac{R_A}{R_B}\right)\right] \tag{12-150}$$

$$\frac{\omega_P}{Q_P} = \frac{C_1 + C_2}{C_1 C_2}\left(\frac{1}{R_2} - \frac{R_A}{R_B R_3}\right) - \frac{R_A}{R_B R_1 C_2}$$

$$\omega_P^2 = \frac{1}{R_1 C_1 C_2}\left(\frac{1}{R_2} - \frac{R_A}{R_B R_3}\right)$$

and

$$K_1 = \frac{R_5}{R_4 + R_5} \qquad K_2 = \frac{R_D}{R_C + R_D} \qquad K_3 = \frac{R_7}{R_6 + R_7}$$

$$R_A = \frac{R_C R_D}{R_C + R_D} \qquad R_1 = \frac{R_4 R_5}{R_4 + R_5} \qquad R_3 = \frac{R_6 R_7}{R_6 + R_7} \tag{12-151}$$

With this circuit, normal practice is to make $C_1 = C_2$. If a high-pass network is needed, $a = b = 0$; this leads to constraints on K_1, K_3, and the resistors. For band-reject or notch filters, $a = 0$, other constraints follow from Eq. (12-150), but it is usually possible to adjust them so that all element values are positive.

32. Biquadratics with Generalized Impedance Converters. Temes[19] describes a general method for designing any biquadratic transfer function using two operational amplifiers and eight impedances. The generalized impedance converter (GIC), introduced in Par. **12-27**, can be used. A complete circuit is shown in Fig. 12-44. For this circuit,

$$\frac{V_2}{V_1}(s) = \frac{Y_A(Y_1 Y_3 - Y_0 Y_2) + Y_B Y_2(Y_4 + Y_5)}{Y_1 Y_3(Y_A + Y_5) + Y_2 Y_4(Y_B + Y_0)}$$

(12-152)

Table 12-16 shows how to choose the elements for various types of transfer function. In version 1, the notch frequency is greater than the resonant frequency of the circuit, while in version 2, the notch frequency is less than the resonant

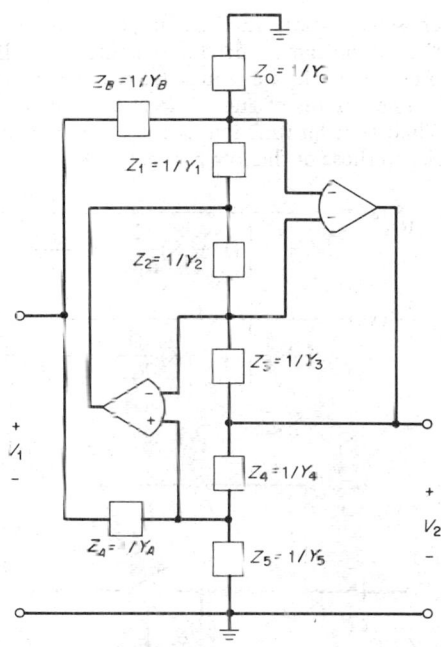

Fig. 12-44. GIC biquadratic filter.

TABLE 12-16 Element Choices for GIC Biquadratic Filters

Filter type	Y_0	Y_1	Y_2	Y_3	Y_4	Y_5	Y_A	Y_B
Low-pass	0	$G_1 + sC_1$	G_2	sC_3	G_-	G_5	0	G_B
High-pass	G_0	G_1	sC_2	G_3	G_4	G_3	0	sC_B
Bandpass	sC_0	G_1	sC_2	G_3	G_4	G_5	0	G_B
Band-stop, version 1	0	sC_1	G_2	G_3	G_4	G_5	sC_A	G_B
Version 2	G_0	G_1	G_2	G_3	sC_4	0	G_A	sC_B
All-pass	G_0	G_1	sC_2	G_3	G_4	C	$G_A = G_4$	sC_B

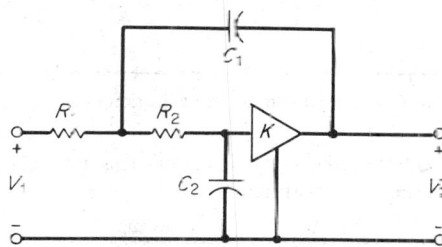

Fig. 12-45. A low-pass active filter network with gain K greater than 0.

frequency. In addition, version 2 requires $G_1 G_3 > G_0 G_2$. These circuits have the property of low sensitivity at the expense of two amplifiers.

33. Controlled Source Realizations. The circuit of Fig. 12-45 is a second-order, low-pass active filter having a gain K that is greater than zero. For this network,

$$\frac{V_2}{V_1}(s) = \frac{K(1/R_1 R_2 C_1 C_2)}{s^2 + \left[(1 - K)\dfrac{1}{R_2 C_2} + \dfrac{1}{R_1 C_1} + \dfrac{1}{R_2 C_1}\right]s + \dfrac{1}{R_1 R_2 C_1 C_2}}$$

(12-153)

Design of this circuit requires choice of a suitable quadratic factor from Tables 12-1, 12-3, 12-5, 12-7, or 12-9, matching coefficients so that the four elements are chosen (two of these may be chosen arbitrarily) and, finally, frequency scaling. When an odd-order circuit is required, a single RC section can be added to the output terminals, with care to avoid changes in K. When higher-order networks are required, a cascade of these sections is possible.

The circuits of Figs. 12-46 and 12-47 give corresponding high-pass and bandpass realizations. Their transfer functions are Eqs. (12-154) and (12-155), respectively. Design techniques are similar to those of the low-pass network.

$$\frac{V_2}{V_1}(s) = \frac{Ks^2}{s^2 + s\left[(1-K)\dfrac{1}{R_1 C_1} + \dfrac{1}{R_2 C_2} + \dfrac{1}{R_2 C_1}\right] + \dfrac{1}{R_1 R_2 C_1 C_2}} \tag{12-154}$$

$$\frac{V_2}{V_1}(s) = \frac{K(1/R_1 C_2)s}{s^2 + s\left[(1-K)\dfrac{1}{R_2 C_2} + \dfrac{1}{R_3 C_2} + \dfrac{1}{R_1 C_1} + \dfrac{1}{R_2 C_1} + \dfrac{1}{R_1 C_2}\right] + \dfrac{1}{R_3 C_1 C_2}\left(\dfrac{1}{R_1} + \dfrac{1}{R_2}\right)} \tag{12-155}$$

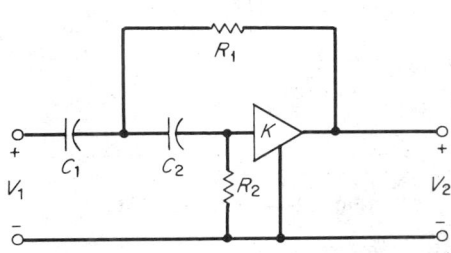

Fig. 12-46. A high-pass active filter network with gain K greater than 0.

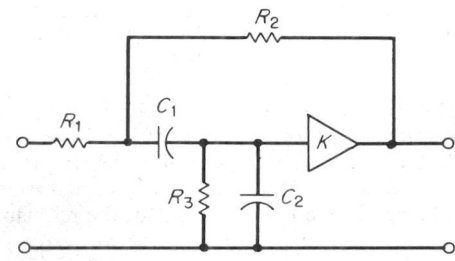

Fig. 12-47. A bandpass active filter network with gain K greater than 0.

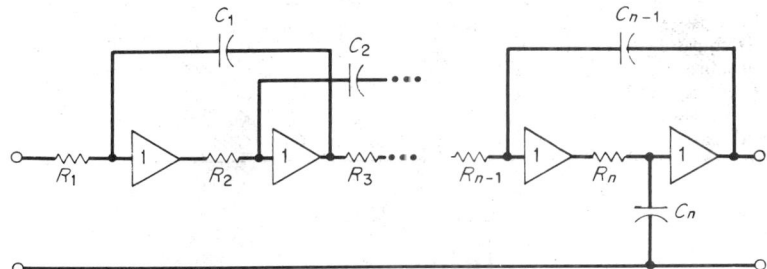

Fig. 12-48. An RC-unity-gain amplifier realization of an active low-pass filter.

Mitra[13] and Huelsman[11] both describe a chain network that may be used for low-pass circuits. It is given in Fig. 12-48, and the transfer function is

$$\frac{V_2}{V_1}(s) = \frac{\omega_1 \omega_2 \omega_3 \cdots \omega_n}{s^n + \omega_1 s^{n-1} + \omega_1 \omega_2 s^{n-2} + \cdots + \omega_1 \omega_2 \omega_3 \cdots \omega_n} \tag{12-156}$$

where

$$\omega_i = 1/R_i C_i \tag{12-157}$$

As an example, consider a third-order Bessel filter, for which

$$\frac{V_2}{V_1}(s) = \frac{15}{s^3 + 6s^2 + 15s + 15} \tag{12-158}$$

choose $1/R_1 C_1 = 6$, $6(1/R_2 C_2) = 15$, and $15(1/R_3 C_3) = 15$. If $C_1 = C_2 = C_3 = 1.0$, then

$$R_1 = \tfrac{1}{6} \qquad R_2 = \tfrac{2}{5} \qquad R_3 = 1$$

Frequency and impedance scaling can be applied as required.

34. Crystal, Mechanical, and Acoustic Coupled-Resonator Filters.[6,8,22,23] In applications such as single-sideband communications it is often necessary to have a bandpass filter with a bandwidth that is a fractional percentage of the center frequency and in which one or both transition regions are very short. Meeting such requirements usually requires a filter in which the resonators are not electrical. Two types of resonator are quartz crystals and mechanical elements, such as disks or rods. Transducers from the electric signal to the mechanical device, output transducers, and resonator-coupling elements are needed.

Crystal filters include resonators made from piezoelectric quartz crystals. The transducers are plates of a conductor deposited on the appropriate surfaces of the crystal, and coupling from one crystal to the next is electrical. The center frequency depends on the size of the crystal, its manner of cutting, and the choice of frequency determining modes of oscillation. It can vary from about 1.0 kHz to 100 MHz. If extreme care is taken, equivalent quality factors (Q's) can be greater than 100,000. These filters can also be very stable with regard to temperature and age.

Mechanical filters use rods or disks as resonating elements, which are coupled together mechanically, usually with wires welded to the resonators. The transducers are magnetostrictive. The frequency range varies from as low as 100 Hz to above 500 kHz. Quality factors above 20,000 are possible and, with proper choice of alloys, temperature coefficients of as low as 1.0 ppm/°C are possible.

Acoustic filters use a combination of crystal and mechanical filter principles. The resonators are monolithic quartz crystals; the transducers are similar to those of crystal filters, but the coupling is mechanical (referred to as acoustic coupling). These filters have many of the properties of crystal filters, but the design techniques have much in common with those of mechanical filters.

Coupled-resonator filters are usually described in terms of an electric equivalent circuit. The direct or mobility analogy (mass to capacitance, friction to conductance, and springs to inductance) is more useful, because the "across" variables of velocity and voltage are analogous, as are the "through" variables of force and current. Equivalent capacitances or inductances and center frequencies are among the common parameters specified for filter elements.

The following paragraphs discuss, in general terms, the design procedure used for coupled-resonator filters, the equivalent circuits used, and some network transformations that enable the designer to implement the design procedure. References 6, 8, 22, and 23 give much additional information, and in particular, Ref. 23 contains an extensive discussion and bibliography. Manufacturer's catalogs are a good source of current data.

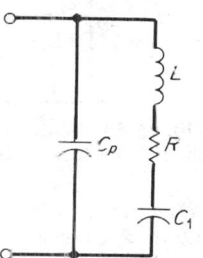

35. Coupled-Resonator Design Procedure. The insertion-loss low-pass prototype filters tabulated in Par. **12-15** can be used to design coupled-resonator bandpass filters. Five steps can be identified in the process, though in some cases the dividing lines become indistinct.

Fig. 12-49. Equivalent circuit for piezoelectric crystal. Because coupling is electrical, a one-port representation is sufficient.

1. Transform the bandpass specifications to a low-pass prototype, using Eq. (12-123). This will take the center frequency to $\omega = 0$ and, usually, the band edge to $\omega = 1$.

2. Choose the appropriate low-pass response e.g., Chebyshev, elliptic, or Butterworth, that meets the transformed specifications. Zeros of transmission are fixed at this time. From this characteristic function determine the transfer function that is needed. The tables presented earlier may be useful, as may other tables cited in the bibliography.

3. Determine the short-circuit y or open-circuit z parameters from the transfer function.

4. If possible, look up or synthesize the appropriate ladder or lattice network needed. At this point, it is still a low-pass prototype. The technique chosen may depend on the expected form of the final network.

5. Use Fig. 12-30 to transform the network into a bandpass network and then use network theorems to adjust the network to a configuration and a set of element values that is practical, i.e., one that matches the resonators.

This process is not one in which success is assured. It may require a variety of attempts before a suitable design is achieved. Equivalent circuits and network theorems are summarized in the following paragraphs.

36. Equivalent Circuits. The most common equivalent circuit for a piezoelectric crystal shows a series-resonant RLC circuit in parallel with a second capacitor, as shown in Fig. 12-49. The parallel capacitor C_p is composed of the mounting hardware and electric plates or the crystal. In practice, the ratio C_p/C cannot be reduced below about 125, but it may be increased if needed. When a filter contains more than one crystal, the coupling is electrical, usually with capacitors.

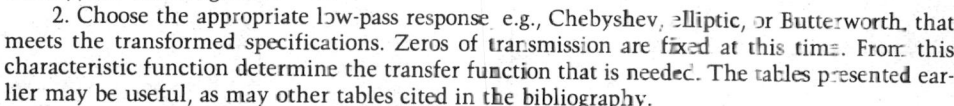

Mechanical filters have an equivalent circuit, as indicated in Fig. 12-50. The resonant circuits L_0, C_R represent the transducer magnetostrictive coils and their tuning capacitances. (In cases of small R_L, it may be more accurate to place C_R in series with L_0.) The resonant circuits L_1, C_1, R_1 and L_n, C_n, R_n include the motional parameters of the transducers. Elements L_2, C_2, R_2, ..., L_{n-1}, C_{n-1}, R_{n-1} represent the motional parameters of the resonant elements, and L_{12}, ..., $L_{n-1,n}$ represent the compliances of the coupling wires.

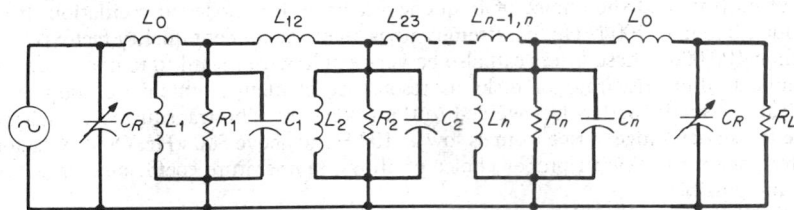

Fig. 12-50. Equivalent circuit for a mechanical filter. A two-port representation allows an electric equivalent circuit for the entire filter.

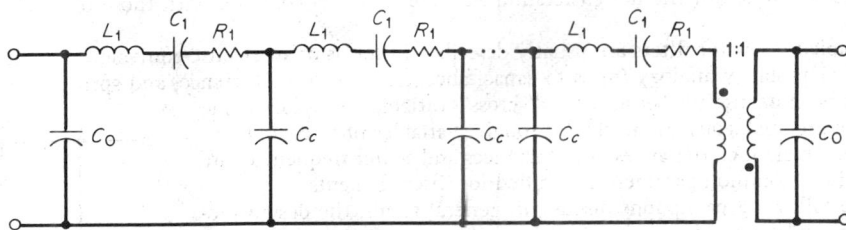

Fig. 12-51. Equivalent circuit for a monolithic crystal or acoustic filter. The 1-to-1 ideal transformer models the 180° phase shift observed in these filters.

The acoustic filter is represented, after substantial development, by the circuit shown in Fig. 12-51. The development has made the circuit easy to use, but the association between the electrical elements and the filter elements is less apparent than in the previous circuits. The ideal transformer at the output accounts for the 180° phase shift observed in these filters. In some analyses, it may be omitted.

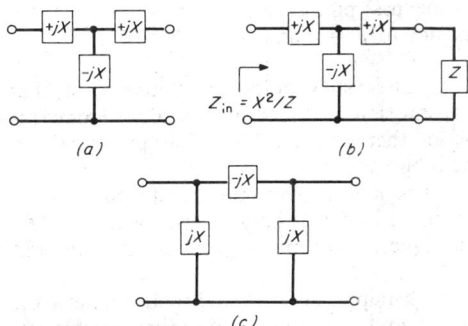

Fig. 12-52. Reactive impedance inverters: (a) T inverter; (b) T inverter with load Z; $Z_{in} = X^2/Z$; (c) π inverter.

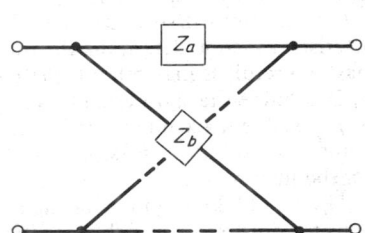

Fig. 12-53. Symmetrical lattice. The dotted diagonal line indicates a second Z_b; the dotted horizontal line, a second Z_a.

37. Network Transformations. In the process of changing a bandpass circuit to meet the configuration of the equivalent circuit of a coupled resonator a variety of equivalent networks may be useful. At one step negative elements may appear. These can be absorbed later in series or parallel with positive elements so that the overall result is positive.

The impedance inverters of Fig. 12-52 can be used to invert an impedance, as indicated. Over a very narrow frequency range they can often be approximated with three capacitors, two of which are negative. An inverter can be used to convert an inductance into a capacitance provided the negative elements can then be absorbed. Other similar reactive configurations can also be used.

Lattice networks (Fig. 12-53) are often used in crystal filters. If the condition prevailing in Fig. 12-54 exists, the equivalent can be used in either direction to effect a change. In particular, the ladder can be transformed into a lattice, which then has the crystal equivalent circuit.

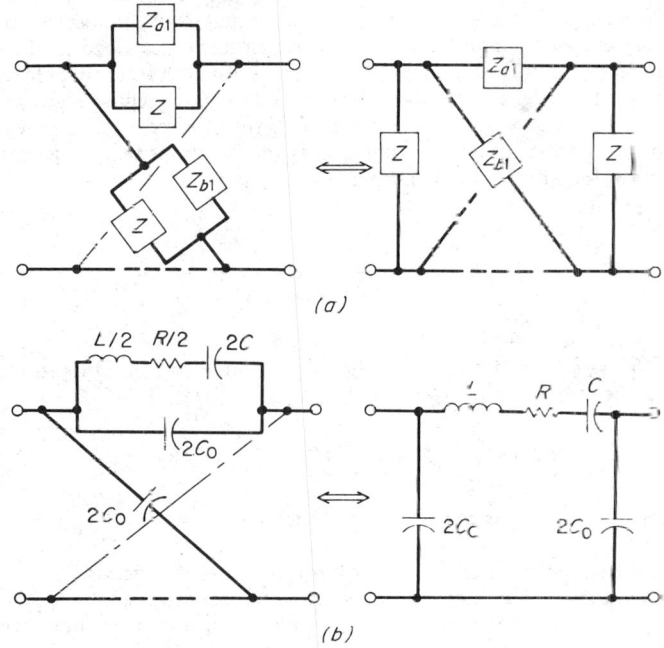

(a)

(b)

Fig. 12-54. Lattice and ladder: (a) general lattice and equivalent circuit; (b) application to crystal filters.

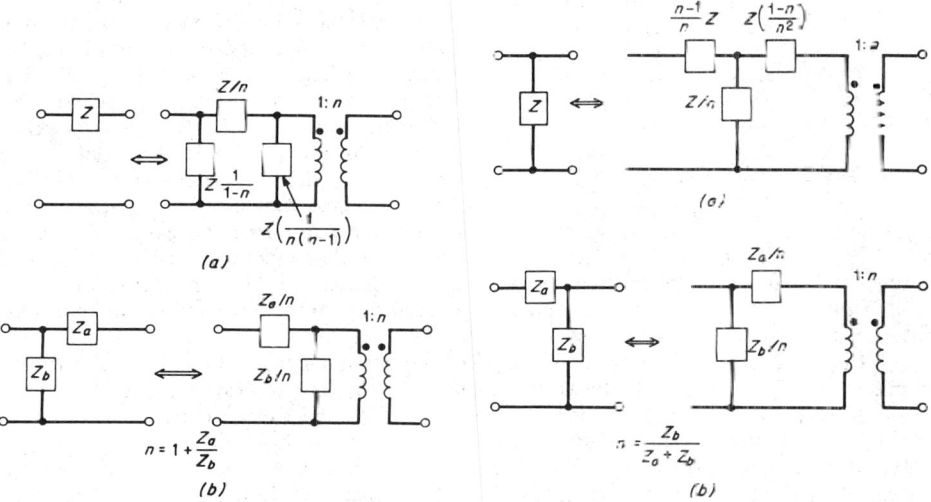

(a)

(b)

Fig. 12-55. Norton's first transformation and a derived network.

(a)

(b)

Fig. 12-56. Norton's second transformation and a derived network.

Two Norton transformations and networks derived from them are shown in Figs. 12-55 and 12-56. They lead to negative elements, and it is expected that they will later be absorbed into positive elements. Humpherys[6] gives another derived Norton transformation that can be used to reduce inductance values. It changes the impedance level on one side of the network. When this is applied to a symmetrical network, the new impedance levels will eventually become directly connected, so that no transformer is needed.

38. Digital Filters.[22,24,25,26,27] A *digital filter* is a circuit or computer program that is linear and time-invariant and operates on discrete time signals. For this section, the term is used to introduce the process of adapting the previous material on continuous time signals to digital operations. Digital filters can be built with conventional digital hardware, such as adders, multipliers, and shift registers for delay. Digital filters can also be implemented as computer programs on suitable general-purpose or special-purpose computers or microcomputers.

The basic analysis tool for digital filters is the z transform, described in the references and almost all system and circuit-theory books. Its role is quite analogous to that of the Laplace transform for continuous time signals. In this development, a working knowledge of the z transform is assumed. It will be used to present the two basic types of filter, nonrecursive and recursive. Four types of mapping from continuous to discrete time functions will be described.

The general form for a digital-filter transfer function is

$$H(z) = \frac{V_2(nT)}{V_1(nT)} = \frac{\displaystyle\sum_{i=0}^{M} a_i z^{-i}}{1 + \displaystyle\sum_{i=1}^{N} b_i z^{-i}} \tag{12-159}$$

where, $M \le N$, V_2 = output signal, V_1 = input signal, and T = sampling interval. Solution of this equation for the current value of the output, $V_2(kT)$, gives

$$V_2(kT) = \sum_{i=0}^{M} a_i V_1(kT - iT) - \sum_{k=1}^{N} b_i V_2(kT - iT) \tag{12-160}$$

Thus, the current value of V_2 is computed from the $N - 1$ preceding values of V_2 and the M values of V_1.

Two important cases arise. If all the b_i are zero, the output depends only on values of the input. This is the nonrecursive filter (FIR, for *finite-impulse-response filter*). Filters in this form usually require many input samples and have good phase properties, finite memory, and no stability problems. Figure 12-57 shows a configuration for a three-stage FIR filter.

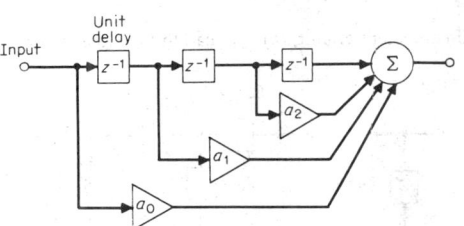

Fig. 12-57. Nonrecursive filter of order 3.

39. Recursive Filters. When the b_i are not all zero, a recursive, or *infinite-impulse-response* (IIR), filter is the result. Such filters tend to have fewer terms and poorer phase properties than the FIR. They also exhibit infinite memory. The IIR function can be expressed either in factored form, typically second-degree terms, or in a partial-fraction expansion, again of second-degree terms. A partial-fraction expansion can be realized as a parallel combination of realizations of the second-degree terms, while a product is realized as a cascade connection. Two possible realizations of a second-degree factor are shown in Fig. 12-58.

40. Design Process. Infinite-impulse-response filters can be developed from continuous time functions in the following steps:

1. Choose an analog filter by obtaining an appropriate transfer function $H(s)$ that meets the specifications. This may be a Butterworth, Legendre, Chebyshev, or Bessel function.

2. Map the result into $H(z)$. Choose a suitable mapping and check the results carefully.

3. Realize, either in hardware or software.

41. Mapping Functions. Lam[25] describes a numerical integration technique in which the various derivatives of a continuous signal are approximated by finite differences. The process leads to the substitution

$$s = (1 - z^{-1})/T \tag{12-161}$$

where T is the sampling interval. This substitution works well for low-pass functions in which T is small. Table 12-17 shows an example.

A second substitution comes from the impulse invariant transformation.[22,25,27] This mapping

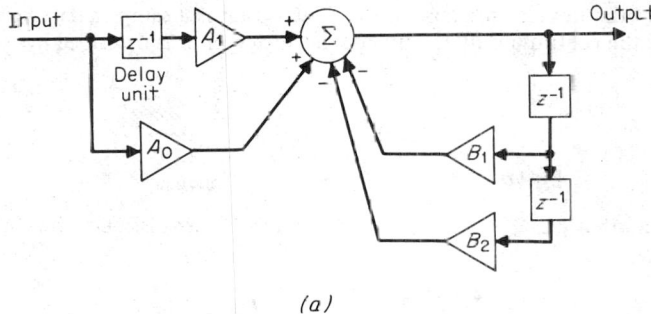

(a)

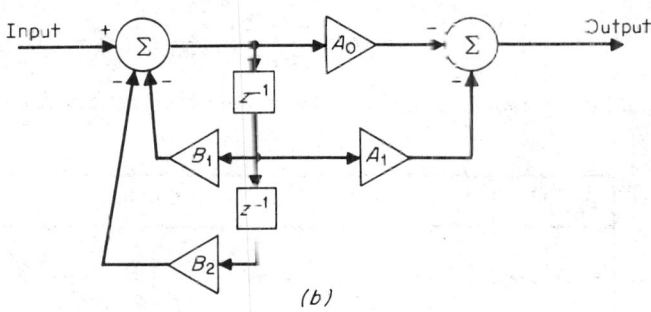

(b)

Fig. 12-58. Two realizations of second-order digital filters

$$H(z) = \frac{A_0 + A_1 z^{-1}}{1 + B_1 z^{-1} + B_2 z^{-2}}$$

For first-order filters blocks A_1 and B_2 are eliminated.

TABLE 12-17 Four Digital-Filter Transformations
Second-order Bessel continuous-time approximation, $H(s) = 1/(s^2 + 3s + 3)$

Transformation	Substitution or definition	Application to $F(s) = \dfrac{1}{s^2 + 3s + 3}$
Numerical, derivative or Euler	$s = \dfrac{1 - z^{-1}}{T}$ $z^{-1} = 1 - sT$	$H(z) = \dfrac{T^2}{z^{-2} - (\varepsilon T + 2)z^{-1} + (1 + 3T + T^2)}$
Impulse invariant	$H(s) = \sum\limits_{i=1}^{N} \dfrac{\alpha_i}{s - p_i}$ $H(z) = \sum\limits_{i=1}^{N} \dfrac{\alpha_i}{1 - z^{-1}e^{p_i T}}$	$p_{1,2} = \dfrac{-3}{2} \pm j \sqrt{\dfrac{3}{2}}$ $\alpha_{1,2} = j\sqrt{3},\ -j\sqrt{3}$ After simplification $H(z) = \dfrac{2\sqrt{3}\,\varepsilon^{(-3/2)}T \sin(\sqrt{3}T/2)}{z^{-2}e^{-3T} - 2z^{-1}e^{-3/2}T \cos(\sqrt{3}T/2) + 1}$
Bilinear	$s = \dfrac{2}{T}\dfrac{1 - z^{-1}}{1 + z^{-1}}$	$H(z) = \dfrac{3T^2 z^{-1} + 6T^2 z^{-1} + 3T^2)}{(1 - 6T + 3T^2)z^{-2} - (8 - 6T^2)z^{-1} + (4 + 6T + 3T^2)}$
Matched	$s \rightarrow e^{sT} = z$	$H(z) = \dfrac{3}{1 - 2z^{-1}e^{-(3/2)}T \cos(\sqrt{3}T/3) + z^{-2}e^{-T}}$

ensures that the impulse response of the digital filter is a sampled-data version of the impulse response of the continuous filter. Suppose the analog filter is described by

$$H(s) = \frac{\sum\limits_{i=1}^{M} a_i s^i}{\sum\limits_{i=1}^{N} b_i s^i} = \sum\limits_{i=1}^{N} \frac{\alpha_i}{s - p_i} \tag{12-162}$$

This assumes that $M < N$, $b_N \neq 0$, $b_0 \neq 0$, and the p_i are distinct. The corresponding impulse response is

$$h(kT) = \sum\limits_{i=1}^{N} \alpha_i e^{p_i kT} U(kT) \tag{12-163}$$

where $U(kT)$ is the unit-step sequence. From this,

$$H(z) = \sum\limits_{i=1}^{N} \frac{\alpha_i}{1 - z^{-1} e^{p_i T}} \tag{12-164}$$

This is useful for low-pass and some band-limited bandpass functions. An example is shown in Table 12-17.

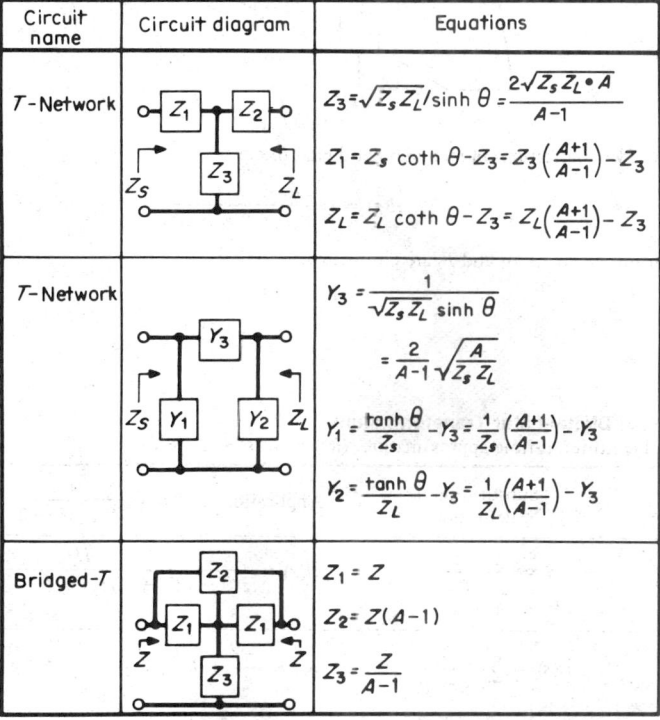

Fig. 12-59. Attenuator networks and equations.

A third transformation is the bilinear transformation[25,26]

$$s = \frac{2}{T} \frac{1 - z^{-1}}{1 + z^{-1}} \tag{12-165}$$

which is useful when the passband or stop-band characteristics are relatively constant in magnitude and when the impulse or step response characteristics need not be preserved. This equation maps the left half s plane into the exterior of the unit circle in the z plane. Since this warps the frequency axis by the equation

$$f_s T = (1/\pi) \tan \pi f_z T \tag{12-166}$$

it may be necessary to "predistort" the frequencies in the analog filter before this substitution is made. This is easy to implement in a partial-fraction expansion of the original analog function. Table 12-17 also shows an example.

A fourth transformation is given by Golden[22,27] and is called the *matched z transform*. In it,

$$s \to e^{sT} = z \tag{12-167}$$

so that a pole at $s = -\alpha$ becomes a pole in the digital filter at $e^{-\alpha T}$. This is identical to the impulse invariant transformation for denominator factors but can lead to different numerator factors than that one. It can be used for wide bandpass and high-pass filters. In general, it leads to a cascade realization, but a partial-fraction expansion can be used to get a parallel realization. The substitution is simply written

$$(s + \alpha)^2 + \beta^2 \Leftrightarrow 1 - 2z^{-1}e^{-\alpha T} \cos \beta T + z^{-2}e^{-2\alpha T} \tag{12-168}$$

Table 12-17 shows an example. Some tables of coefficients have been published. An extensive tabulation is given by Genesio.[28]

ATTENUATORS

42. Attenuator Network Design. Attenuators are passive circuits designed to introduce a fixed power loss between a source and a load while matching the impedances. The power loss is independent of the direction of power flow. Circuits and design equations are given for T networks, π networks, and bridged-T networks. The equations are suitable for use on most calculators, though some compromise to achieve practical values may be needed. The symbols used are

Z_s = source impedance
Z_L = load impedance
A = ratio of available power to desired load power = $10^{B/10}$ (12-169)
B = 10 log A
θ = ½ ln A = ½ ln $10^{B/10}$

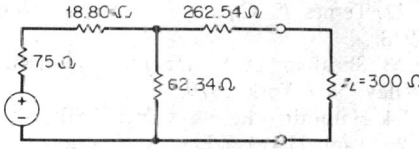

Fig. 12-60. A 14.0-dB attenuator between a 75-Ω source and a 300-Ω load.

The basic equations are found in Cauer,[20] but the notation is different. The first two networks given in Fig. 12-59 are for unbalanced, unsymmetrical configurations. For symmetrical networks, $Z_S = Z_L$, which leads to $Z_1 = Z_2$. For balanced T networks, Z_1 and Z_2 are divided by 2, with half of each element in each series arms. The last network is given only in symmetrical form. The equations are valid for resistive and complex impedances.

As an example, let it be desired to design an attenuator to match a 75-Ω source to a 300-Ω load and to introduce a 14.0-dB loss. A T section is desired. In the terms of Eq. (12-169),

$$Z_S = 75\ \Omega \quad Z_L = 300\ \Omega \quad B = 14.0\ \text{dB} \quad A = 25.12$$
$$\theta = 1.612 \quad Z_3 = 62.34\ \Omega \quad Z_1 = 18.88\ \Omega \quad Z_2 = 262.54\ \Omega \tag{12-170}$$

The computed network is shown in Fig. 12-60.

43. Bibliography on Filters

1. Ruston, H., and J. Bordogna "Electric Networks: Functions, Filters, Analysis," McGraw-Hill, New York, 1966.

2. Darlington, S. Synthesis of Reactance 4-Poles Which Produce Prescribed Insertion Loss Characteristics, Including Special Applications to Filter Design, J. Math. Phys., 1939, Vol 18, pp. 257–353.

3. Saal, R., and E. Ulbrich On the Design of Filters by Synthesis, IRE Trans. Circuit Theory, 1958, Vol. CT-5, pp. 284–327.

4. Papoulis, A. A New Class of Filters, Proc. IRE, 1959, Vol. CT-6, pp. 277–281.

5. Thomson, W. E. Delay Networks Having Maximally Flat Frequency Characteristics, Proc. IEE, Pt. 3, November 1949, Vol. 96, pp. 487–490.

6. Zverev, A. I. "Handbook of Filter Synthesis," Wiley, New York, 1967.

7. Hansell, G. E. "Filter Design and Evaluation," Van Nostrand Reinhold, New York, 1969.

8. Humpherys, D. S. "The Analysis, Design, and Synthesis of Electric Filters," Prentice-Hall, Englewood Cliffs, N.J., 1970.

9. Van Valkenburg, M. E. "Introduction to Modern Network Synthesis," Wiley, New York, 1960.

10. Weinberg, L. "Network Analysis and Synthesis," McGraw-Hill, New York, 1962.

11. Huelsman, L. P. Modern Techniques of Active Filter Design, *Proc. Natl. Electron. Conf.*, 1974, Vol. 29, pp. 449–453.

12. Huelsman, L. P. "Theory and Design of Active RC Circuits," McGraw-Hill, New York, 1968.

13. Mitra, S. K. "Analysis and Synthesis of Linear Active Networks," Wiley, New York, 1969.

14. Tellegen, B. D. H. The Gyrator, A New Electric Network Element, *Philips Res. Rep.*, April 1948, Vol. 3, pp. 81–101.

15. Riordan, R. H. S. Simulated Inductors Using Differential Amplifiers, *Electron. Lett.*, 1967, Vol. 3, pp. 50–51.

16. Bruton, L. T. Network Transfer Functions Using the Concept of Frequency-Dependent Negative Resistance, *IEEE Trans. Circuit Theory*, August 1969, Vol. CT-16, No. 3, pp. 406–408.

17. Daryanani, Gobind "Principles of Active Network Synthesis and Design," Wiley, New York, 1976.

18. Friend, J. J., C. A. Harris, and D. Hilberman STAR: An Active Biquadratic Filter Section, *IEEE Trans. Circuits Syst.*, February 1975, Vol. CAS-22, No. 2, pp. 115–121.

19. Temes, G. C., and J. W. La Patra "Introduction to Circuit Synthesis and Design," McGraw-Hill, New York, 1977.

20. Cauer, W. "Synthesis of Linear Communication Networks," McGraw-Hill, New York, 1958.

21. Blinchikoff, H. J., and A. I. Zverev "Filtering in the Time and Frequency Domains," Wiley, New York, 1976.

22. Temes, G. C., and S. K. Mitra "Modern Filter Theory and Design," Wiley, New York, 1973.

23. Sheahan, D. F., and R. A. Johnson "Modern Crystal and Mechanical Filters," IEEE–Wiley, New York, 1979.

24. Hamming, R. W. "Digital Filters," Prentice-Hall, Englewood Cliffs, N.J., 1977.

25. Lam, Harry Y-F. "Analog and Digital Filters: Design and Realization," Prentice-Hall, Englewood Cliffs, N.J., 1977.

26. Kuo, F. F., and J. F. Kaiser (eds.) "System Analysis by Digital Computer," Wiley, New York, 1966.

27. Special Issue on Digital Filters *IEEE Trans. Audio Electroacoust.*, September 1968, Vol. AU-26, No. 3.

28. Genesio, R., A. Laurentini, V. Mauro, and A. R. Meo "Butterworth and Chebyshev Digital Filters," Elsevier, Amsterdam, 1973.

GENERAL REFERENCES

29. Balabanian, N. "Network Synthesis," Prentice-Hall, Englewood Cliffs, N.J., 1958.

30. Belevitch, V. Summary of the History of Circuit Theory, *Proc. IEEE*, May 1962, Vol. 50, No. 5, pp. 848–855.

31. Bode, H. W. "Network Analysis and Feedback Design," Van Nostrand, Princeton, N.J., 1945.

32. Budak, A. "Passive and Active Network Analysis and Synthesis," Houghton Mifflin, Boston, 1974.

33. Cauer, W. New Theory and Design of Wave Filters, *Physics*, 1932, Vol. 2, pp. 242–268.

34. Craig, J. W. "Design of Lossy Filters," M.I.T. Press, Cambridge, Mass., 1970.

35. Guillemin, E. A. "Communication Networks," Vol. II, Wiley, New York, 1935.

36. Huelsman, L. P. "Active Filters: Lumped, Distributed, Digital, and Parametric," McGraw-Hill, New York, 1970.

37. Javid, M., and E. Brenner "Analysis, Transmission and Filtering of Signals," McGraw-Hill, New York, 1963.

38. Mitra, S. K. (ed.) "Active Inductorless Filters," IEEE Press, New York, 1971.

39. Orchard, H. J. The Roots of the Maximally Flat-Delay Polynomials, *IEEE Trans. Circuit Theory*, 1965, Vol. CT-12, pp. 452–454.

40. Szentirmai, G. (ed.) "Computer-Aided Filter Design," IEEE Press, New York, 1973.

41. Van Valkenburg, M. E. (ed.) "Circuit Theory: Foundations and Classical Contributions," Dowden, Hutchinson, and Ross, Stroudsburg, Pa., 1974.

42. Antoniou, A. "Digital Filters," McGraw-Hill, New York, 1979.

43. Johnson, D. E., J. R. Johnson, and H. P. Moore   "A Handbook of Active Filters," Prentice-Hall, Englewood Cliffs, N.J., 1980.

44. Johnson, D. E.   "Introduction to Filter Theory," Prentice-Hall, Englewood Cliffs, N.J., 1976.

45. Lindquist, C. S.   "Active Network Design," Steward, Long Beach, Calif., 1977.

46. Oppenheim, A. V. (ed.)   "Application of Digital Signal Processing," Prentice-Hall, Englewood Cliffs, N.J., 1978.

47. Rhodes, J. D.   "Theory of Electrical Filters," Wiley, London, 1976.

48. Warner, F. L.   "Microwave Attenuation Measurement," Institution of Electrical Engineers, London, 1977.

49. Huelsman, L. P., and P. E. Allen   "Introduction to the Theory and Design of Active Filters," McGraw-Hill, New York, 1980.

Amplifiers and Oscillators

G. BURTON HARROLD *Senior Engineer, Electronics Laboratory, General Electric Company, Member IEEE*

SAMUEL M. KORZEKWA *Senior Engineer, Electronics Laboratory, General Electric Company*

ROBERT J. McFAYDEN *Consulting Engineer, Electronics Laboratory, General Electric Company, Member IEEE*

STEPHEN W. TEHON *Consulting Engineer, Electronics Laboratory, General Electric Company, Fellow, IEEE*

RICHARD W. FRENCH *ELEMEK Inc., Member IEEE*

JOSEPH P. HESLER *President, AKF Design Ltd.*

CHANG S. KIM *President, Central Research Lab., Taihan Electric Wire Co., Ltd., Member, IEEE*

HAROLD W. LORD *Consulting Engineer, Fellow IEEE*

JOHN W. LUNDEN *Manager, Optoelectronic Systems Operation, General Electric Company, Member IEEE*

CONRAD E. NELSON *Senior Consulting Engineer, Heavy Military Electronic Systems, General Electric Company, Senior Member IEEE*

GUNTER K. WESSEL *Professor, Physics Department, Syracuse University*

CONTENTS

Numbers refer to paragraphs

Principles of Operation—Amplifiers

BY G. B. HARROLD

1. Gain. In most amplifier applications the prime concern is gain. A generalized amplifier is shown in Fig. 13-1; the most widely applied definitions of gain using the quantities defined there are

$$\text{Voltage gain } A_v = e_{22}/e_{11} \qquad \text{Current gain } A_i = i_2/i$$

$$\text{Available power from source } P_{avs} = \frac{|e_s|^2}{4 \text{ Re } Z_s}$$

where Re = real part of complex impedance

$$\text{Output load power } P_L = \frac{|e_{22}|^2}{\text{Re } Z_L} \qquad \text{Input power } P_I = \frac{|e_{11}|^2}{\text{Re } Z_{in}}$$

$$\text{Available power at output } P_{avo} = \frac{|e_{22}|^2}{4 \text{ Re } Z_{out}} \qquad \text{Transducer gain } G_T = P_L/P_{avs}$$

$$\text{Available power gain } G_A = P_{avo}/P_{avs} \qquad \text{Power gain } G = P_L/P_I$$

$$\text{Insertion power gain } G_I = \frac{\text{power into load with network inserted}}{\text{power into load with source connect to load}}$$

2. Bandwidth and Gain-Bandwidth Product. Bandwidth is a measure of the range of frequencies within which an amplifier will respond. The frequency range (passband) is usually measured between the half-power (3-dB) points on the output-response-vs.-frequency curve, for constant input. In some cases it is defined at the quarter-power points (6 dB). See Fig. 13-2.

The gain-bandwidth product of a device is a commonly used figure of merit. It is defined for a bandpass amplifier as

$$F_a = A_r B$$

where F_a = figure of merit (rad/s), A_r = reference gain, either the maximum gain or the gain at the frequency where the gain is purely real or purely imaginary, and B = 3-dB bandwidth (rad/s).

For low-pass amplifiers

$$F_a = A_r W_H$$

where F_a = figure of merit (rad/s), A_r = reference gain, and W_H = upper cutoff frequency (rad/s).

In the case of vacuum tubes and certain other active devices this definition is reduced to

$$F_a = g_m / C_T$$

where F_a = figure of merit (rad/s), g_m = transconductance of active device, and C_T = total output capacitance, plus input capacitance of subsequent stage.

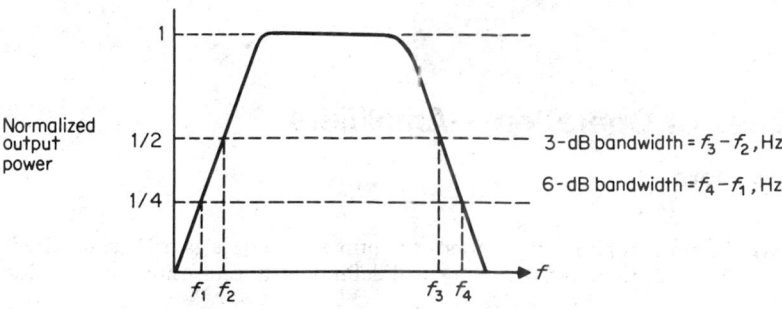

Fig. 13-1. Input and output quantities of generalized amplifier.

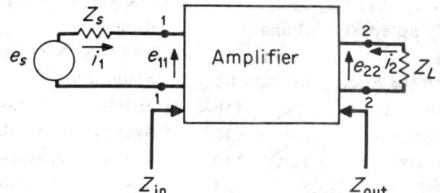

Fig. 13-2. Amplifier response and bandwidth.

3. Noise. The major types of noise are illustrated in Fig. 13-3. Important relations and definitions in noise computations are:

Noise factor
$$F = \frac{S_i/N_i}{S_o/N_o}$$

where S_i = signal power available at input, S_o = signal power available at output, N_i = noise power available at input at T = 290 K, and N_o = noise power available at output,

Available noise power $$P_{n,av} = \frac{e_n^2}{4R} = KTB \qquad \text{for thermal noise}$$

where the quantities are as defined in Fig. 13-3,

Excess noise factor $$F - 1 = N_e/N_i$$

where $F - 1$ = excess noise factor, N_e = total equivalent device noise referred to input, and N_i = thermal noise of source at standard temperature.

Noise temperature $\qquad\qquad\qquad T = P_{n,av}/KB$

where $P_{n,av}$ = average noise power available.

At a single input-output frequency in a two-port,

Effective input noise temperature $\qquad T_e = 290(F - 1)$

Noise Factor of Transmission Lines and Attenuators. The noise factor of two ports composed entirely of resistive elements at room temperature (290 K) and an impedance matched loss of L = $1/G_A$ is $F = L$.

Cascaded noise factor $\qquad\qquad\qquad F_T = F_1 + (F_2 - 1)/G_r$

where F_T = overall noise factor, F_1 = noise factor of first stage, F_2 = noise factor of second stage, and G_A = available gain of first stage.

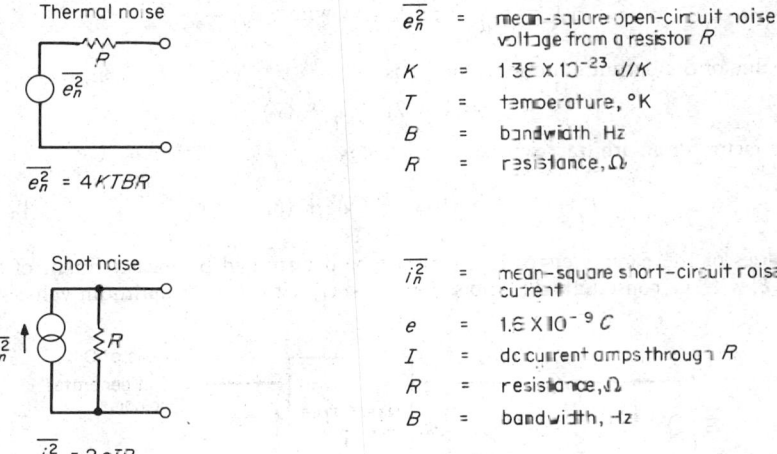

Thermal noise

$e_n^2 = 4KTBR$

e_n^2	=	mean-square open-circuit noise voltage from a resistor R
K	=	1.38×10^{-23} J/K
T	=	temperature, °K
B	=	bandwidth Hz
R	=	resistance, Ω

Shot noise

$i_n^2 = 2eIB$

i_n^2	=	mean-square short-circuit noise current
e	=	1.6×10^{-9} C
I	=	dc current amps through R
R	=	resistance, Ω
B	=	bandwidth, Hz

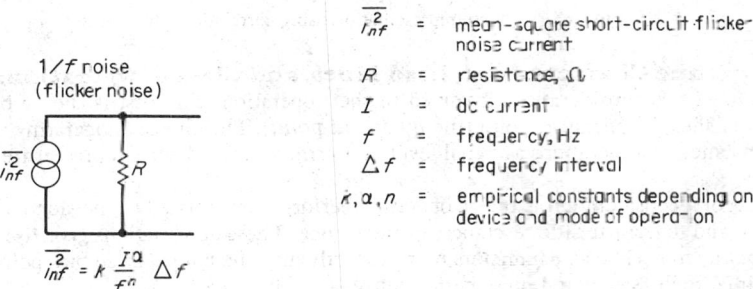

1/f noise
(flicker noise)

$i_{nf}^2 = k \dfrac{I^a}{f^n} \Delta f$

i_{nf}^2	=	mean-square short-circuit flicker noise current
R	=	resistance, Ω
I	=	dc current
f	=	frequency, Hz
Δf	=	frequency interval
k, a, n	=	empirical constants depending on device and mode of operation

Fig. 13-3. Noise-equivalent circuits.

System Noise Temperature. Space probes and satellite communication systems using low-noise amplifiers and antennas directed toward outer space make use of system noise temperatures. When we define T_A = antenna temperature, L = waveguide numeric loss (greater than 1), T_{E1} = amplifier noise temperature, G_A = amplifier available gain, F = postamplifier noise factor, and B = postamplifier bandwidth, this temperature can be calculated as

$$T_{sys} = T_A - |L - 1|290° + LT_{E1} + \frac{(F - 1)290L}{G_A}$$

The quantity of interest is the output signal-to-noise ratio where S_A is available signal power at the antenna (assuming the antenna is matched to free space)

$$S/N = S_A/KT_{sys}B \qquad K = 1.38 \times 10^{-23}$$

Generalized Noise Factor. It can be shown[1]* that a general representation of noise performances can be expressed in terms of Fig. 13-4. This is the representation of a noisy two-port in terms of external voltage and current noise sources with a correlation admittance. In this case the noise factor becomes

$$F = 1 + \frac{G_U}{G_s} + \frac{R_N}{G_s}[(G_s + G_\gamma)^2 + (B_s + B_\gamma)^2]$$

where F = noise factor, G_s = real part of Y_s, B_s = imaginary part of Y_s, G_u = conductance due to the uncorrelated part of the noise current, Y_γ = correlation admittance between cross product of current and voltage noise sources, G_γ = real part of Y_γ, B_y = imaginary part of Y_γ, and R_N = equivalent noise resistance of the noise voltage.

The optimum source admittance is $Y_{opt} = G_{opt} + jB_{opt}$

$$G_{opt} = \left(\frac{G_u + R_N G_\gamma^2}{R_N} \right)^{1/2} \qquad \text{where } B_{opt} = -B_{gg}$$

and the value of the optimum noise factor F_{opt} is

$$F_{opt} = 1 + 2R_N(G_\gamma + G_0)$$

The noise factor for an arbitrary source impedance is

$$F = F_{opt} + \frac{R_N}{G_s}[(G_s - G_0)^2 + (B_s - B_0)^2]$$

The values of the parameters of Fig. 13-4 can be determined by measurement of (1) noise figure vs. B_s with G_s constant and (2) noise figure vs. G_s with B_s at its optimum value.

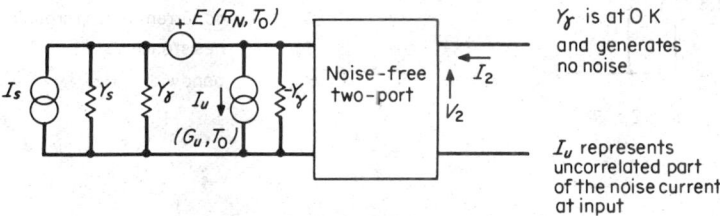

Fig. 13-4. Noise representation using correlation admittance.

4. Dynamic Characteristic, Load Lines, and Class of Operation. Most active devices have two considerations involved in their operation. The first is the dc bias condition that establishes the operating point (the *quiescent point*). The choice of operating point is determined by such considerations as signal level, uniformity of the device, temperature of operation, etc.

The second consideration is the ac operating performance, related to the slope of the dc characteristic and to the parasitic reactances of the device. These ac variations give rise to the *small-signal parameters*. The ac parameters may also influence the choice of dc bias point when basic constraints, such as gain and noise performance, are considered.

For frequencies of operation where these parasites are not significant, the use of a load line is valuable. The class of amplifier operation is dependent upon its quiescent point, its load line, and input signal level. The types of operation are shown in Fig. 13-5.

5. Distortion. Distortion takes many forms, most of them undesirable. The basic causes of distortion are nonlinearity in amplitude response and nonuniformity of phase response. The most commonly encountered types of distortion are as follows:

Harmonic distortion is due to nonlinearities in the amplitude transfer characteristics. The typical output contains not only the fundamental frequency but integer multiples of it.

Crossover distortion is due to the nonlinear characteristics of a device when changing operating modes (e.g., in a push-pull amplifier). It occurs when one device is cut off and the second turned on if the crossover is not smooth between the two modes.

Intermodulation distortion is a spurious output resulting from the mixing of two or more

*Superior numbers correspond to numbered references, Par. **13-17**.

signals of different frequencies. The spurious output occurs at the sum or difference of integer multiples of the original frequencies.

Cross-modulation distortion occurs when two signals pass through an amplifier and the modulation of one is transferred to the other.

Phase distortion results from the deviation from a constant slope of the output-phase-vs.-frequency response of an amplifier. This deviation gives rise to echo responses in the output that precede and follow the main response, and a distortion of the output signal when an input signal having a large number of frequency components is applied.

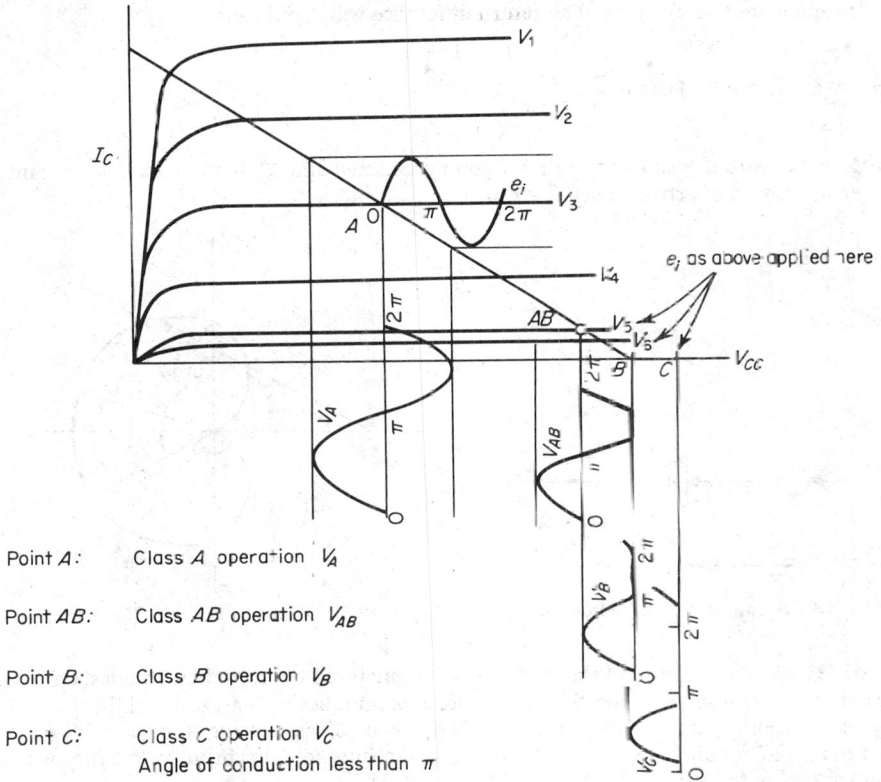

Point *A:* Class *A* operation V_A

Point *AB:* Class *AB* operation V_{AB}

Point *B:* Class *B* operation V_B

Point *C:* Class *C* operation V_C
 Angle of conduction less than π

Fig. 13-5. Classes of amplifier operation. Class S operation is a switching mode in which a square-wave output is produced by a sine-wave input.

6. Feedback Amplifiers. Feedback amplifiers fall into two categories: those having positive feedback (usually oscillators) and those having negative feedback. The positive-feedback case is discussed under oscillators (Par. **13-12**). The following discussion is concerned with negative-feedback amplifiers.

7. Negative Feedback. A simple representation of a feedback network is shown in Fig. 13-6. The closed-loop gain is given by

$$e_2/e_1 = A/(1 - BA)$$

where A = forward gain with feedback removed and B = fraction of output returned to input.
 For negative feedback, A provides a 180° phase shift in midband, so that

$$1 - AB > 1 \quad \text{in this frequency range}$$

The quantity $1 - AB$ is called the *feedback factor*, and if the circuit is cut at any X point in Fig. 13-6, the open-loop gain is AB.

It can be shown that for large loop gain AB the closed-loop transfer function reduces to

$$e_2/e_1 \approx 1/B$$

The gain then becomes essentially independent of variations in A. In particular, if B is passive, the closed-loop gain is controlled only by passive components. It can also be shown[2] that feedback has no beneficial effect in reducing unwanted signals at the input of the amplifier, e.g., input noise, but does reduce unwanted signals generated in the amplifier chain (e.g., output distortion).

The return ratio can be found if the circuit is opened at any point X (Fig. 13-6) and a unit signal P is injected at that X point. The return signal P' is equal to the return ratio, since the input P is unity. In this case the return ratio T is the same at any point X and is

$$T = -AB$$

The minus sign is chosen because the typical amplifier has an odd number of phase reversals and T is then a positive quantity. The return difference is by definition

$$F = 1 + T$$

It has been shown by Bode that

$$F = \Delta/\Delta^0$$

where Δ = network determinant with XX point connected and Δ^0 = network determinant of amplifier when gain of active device is set to zero.

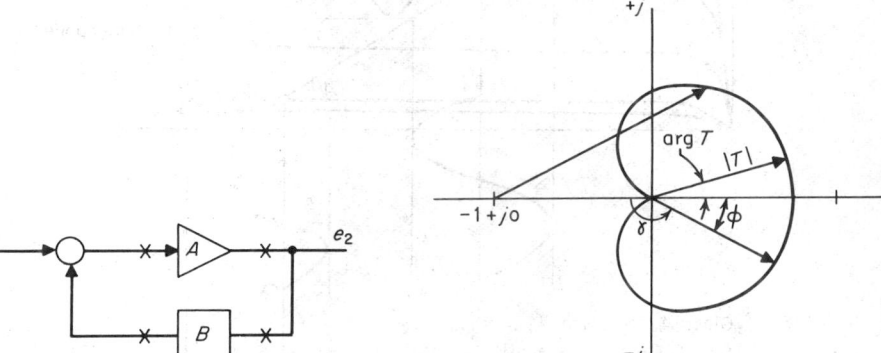

Fig. 13-6. Amplifier with feedback loop.   **Fig. 13-7.** Nyquist diagram for determining stability.

8. Stability. The stability of the network can be analyzed by several techniques. Of prime interest are the Nyquist, Bode, Routh, and root-locus techniques of analyzing stability.

Nyquist Method. The basic technique of Nyquist involves plotting T on a polar plot as shown in Fig. 13-7 for all values $s = j\omega$ for ω between minus and plus infinity. Stability is then determined by the following method:

1. Draw a vector from the $-1 + j0$ point to the plotted curve and observe the rotation of this vector as ω varies from $-\infty$ to $+\infty$. Let R be the net number of counterclockwise revolutions of this vector.

2. Determine the number of roots of the denominator of $T = -AB$ which have positive real parts. Call this number P.

3. The system is stable if and only if $P = R$. Note that in many systems A and B are stable by themselves, so that P becomes zero and the net counterclockwise revolution N becomes zero for stability.[3]

Bode's Technique. A technique that has found wide use in determining stability and performance, especially in control systems, is the Bode diagram. The assumptions used here for this method are that $T = -AB$, where A and B are stable when the system is open-circuited and consists of minimum-phase networks. It is also necessary to define a phase margin γ such that $\gamma = 180 + \phi$, where ϕ is the phase angle of T and is positive when measured counterclockwise from zero, and γ, the phase margin, is positive when measured counterclockwise from the 180° line (Fig. 13-7). The stability criterion under these conditions reads: Systems having a positive phase margin when their return ratio equal to 20 log $|T|$ goes through 0 dB (i.e., where $|T|$ crosses the unit circle in the Nyquist plot) are stable; if a negative γ exists at 0 dB, the system is unstable.

The usefulness of this technique lies in the fact that Bode's theorems show that the phase angle of a system is related to the attenuation or gain characteristic as a function of frequency. It is shown in Ref. 3 that the use of Bode's technique relies heavily on straight-line approximation. Figure 13-8 shows the use of these straight-line approximations applied to the factored form

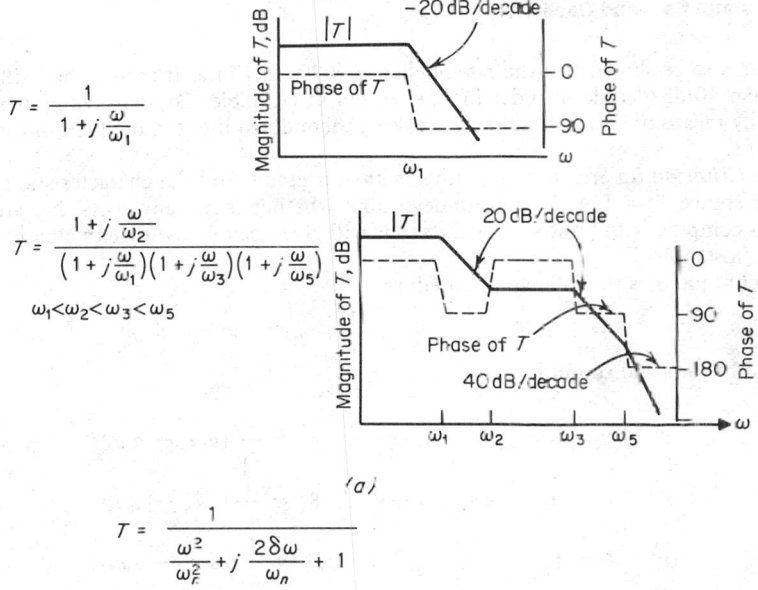

$$T = \frac{1}{1 + j\frac{\omega}{\omega_1}}$$

$$T = \frac{1 + j\frac{\omega}{\omega_2}}{\left(1 + j\frac{\omega}{\omega_1}\right)\left(1 + j\frac{\omega}{\omega_3}\right)\left(1 + j\frac{\omega}{\omega_5}\right)}$$

$$\omega_1 < \omega_2 < \omega_3 < \omega_5$$

(a)

$$T = \frac{1}{\frac{\omega^2}{\omega_r^2} + j\frac{2\delta\omega}{\omega_n} + 1}$$

(Second–order function with damping)

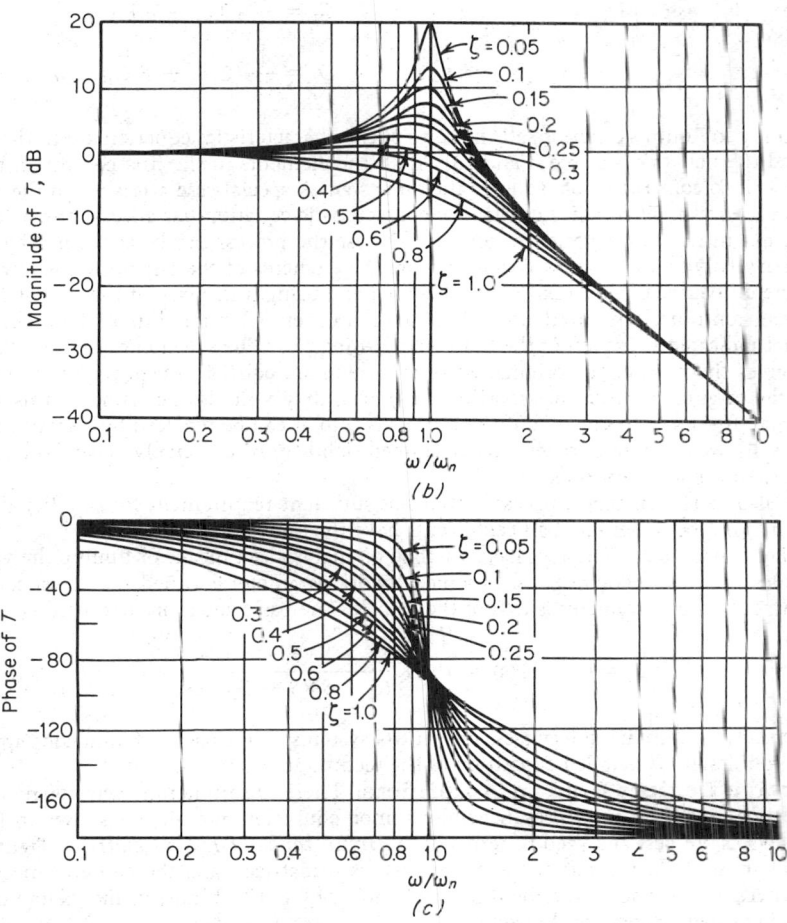

Fig. 13-8. Bode method of determining stability: (a) typical straight-line approximation; (b) magnitude of T vs. frequency; (c) phase of T vs. frequency.

of T. It can also be shown that the rate of change of 20 log $|T|$ as it crosses the 0-dB axis must be less than 40 dB/decade in order for the system to be stable. Complicated systems may be analyzed by means of this technique by breaking them down into the products of their various components.

Routh's Criterion for Stability. Routh's method is used to test the characteristic equation or return difference $F = 1 + T = 0$, to determine whether it has any roots that are real and positive or complex with positive real parts that will give rise to growing exponential responses and hence instability.

This technique uses the following procedure:

$$F = 1 + T = a_0 s^n + a_1 s^{n-1} + a_2 s^{n-2} + \cdots + a_n$$

is arranged in rows and columns as

s^n	a_0	a_2	a_4	a_6 $\cdots$	$B_1 = \dfrac{1}{a_1}(a_1 a_2 - a_0 a_3)$
s^{n+1}	a_1	a_3	a_5	a_7 $\cdots$	$B_3 = \dfrac{1}{a_1}(a_1 a_4 - a_0 a_5)$
s^{n-2}	B_1	B_3	B_5		$B_5 = \dfrac{1}{a_1}(a_1 a_6 - a_0 a_7)$
s^{n-3}	C_1	C_3			$C_1 = \dfrac{1}{B_1} B_1 a_3 - a_1 B_3)$
s^{n-4}	D_1	$\cdots$			$C_3 = \dfrac{1}{B_1}(B_1 a_5 - a_1 B_5)$
					$D_1 = \dfrac{1}{C_1}(C_1 B_3 - B_1 C_3)$

According to Routh's criterion, all the roots of the characteristics equation are in the left half plane and the network is stable if and only if all the elements in the first column of the array a_0, a_1, B_1, ... so constructed above have the same sign. A special case arises when the elements of a row all vanish. This indicates that the characteristic equation has at least one pair of roots equal in magnitude but opposite in sign. In this case the process can be continued by forming an auxiliary polynominal whose coefficients are the elements of the last nonvanishing row. By taking the derivative of this auxiliary polynomial and using it in place of the row of zeros, the process can continue. This auxiliary polynomial is an even polynomial in s^2, the highest power being that indicated at the left of the last nonvanishing row. The solution of this auxiliary polynomial gives the roots of the original equation which are equal in magnitude but opposite in sign. If the original equation has roots on the j axis, they will also be found in this auxiliary polynomial. If the lead coefficient of the process is zero, it can be replaced by a small number δ. Then δ is allowed to approach zero to determined stability. Alternatively, s can be replaced by $1/P$ and then the test performed.

It can also be shown that a necessary but not sufficient requirement for stability is that all powers of s in F be present and all coefficients have the same sign.

Root-Locus Method. The root-locus method of analysis is a means of finding the variations of the poles of a closed-loop response as some network parameter is varied. The most convenient and commonly used parameter is that of the gain K. The basic equation then used is

$$F = 1 + KT(s) = 1 - K \frac{(S - S_2)(S - S_4) \cdots}{(S - S_1)(S - S_3) \cdots} = 0$$

This is a useful technique in feedback and control systems, but it has not found wide application in amplifier design. A detailed exposition of the technique is found in Ref. 4.

9. Active Devices Used in Amplifiers. There are numerous ways of representing active devices and their properties. Several common equivalent circuits are shown in Fig. 13-9. Active devices are best analyzed in terms of the *immittance* or *hybrid matrices*. Figures 13-10 and 13-11 show the definition of the commonly used matrices, and their interconnections are shown in Fig. 13-12. The requirements at the bottom of Fig. 13-12 must be met before the interconnection of two matrices is allowed.

The matrix that is becoming increasingly important at higher frequencies is the S matrix.

Here the network is embedded in a transmission-line structure, and the incident and reflected powers are measured and reflected coefficients and transmission coefficients are defined.

10. Cascaded and Distributed Amplifiers. Most amplifiers are cascaded (i.e., connected to a second amplifier). The two techniques commonly used are shown in Fig. 13-13. In the cascade structure the overall response is the product of the individual responses; in the dis-

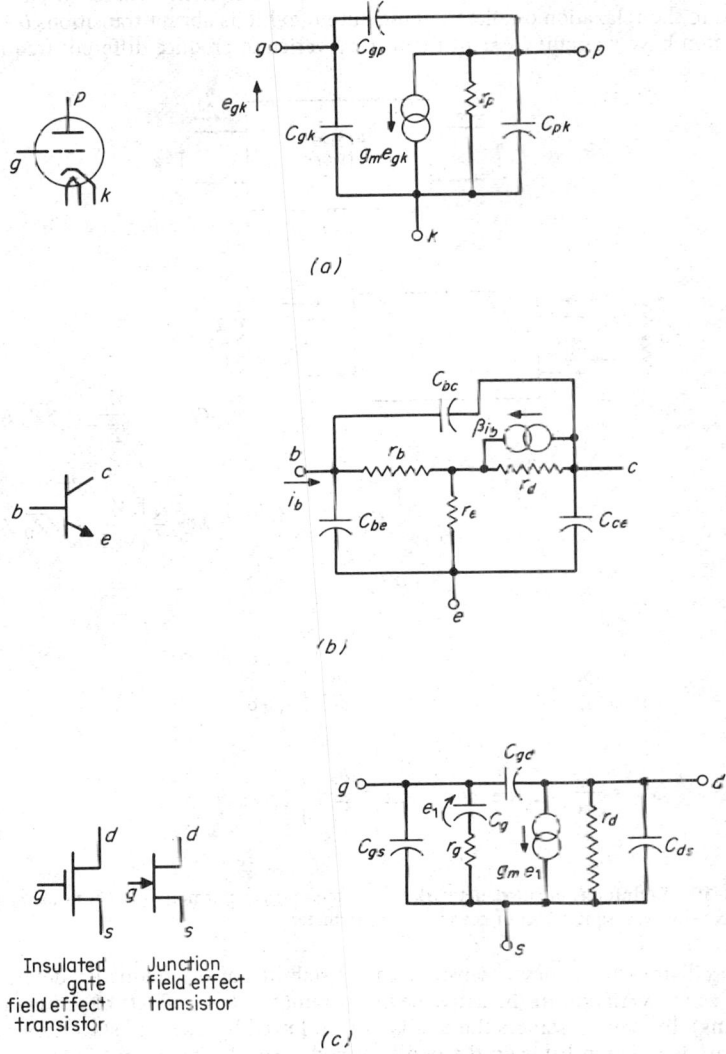

Fig. 13-9. Equivalent circuits of active devices: (a) vacuum tube; (b) bipolar transistor; (c) field-effect transistor (FET).

tributed structure the response is one-half the sum of the individual responses, since each stage's output is propagated in both directions. In cascaded amplifiers the frequency response and gain are determined by the active device as well as the interstage networks. In simple audio amplifiers these interstage networks may become simple RC combinations, while in rf amplifiers they may become critically coupled double-tuned circuits. Interstage coupling networks are discussed in subsequent sections.

In distributed structures (Fig. 13-13b), actual transmission lines are used for the input to the amplifier, while the output is taken at one end of the upper transmission line. The propagation time along the input line must be the same as that along the output line, or distortion will result. This type of amplifier, noted for its wide frequency response, is discussed in Par. **13-63**.

Principles of Operation—Oscillators

BY G. B. HARROLD

11. Introduction. An oscillator can be considered as a circuit that converts a dc input to a time-varying output. This discussion deals with oscillators whose output is sinusoidal, as opposed to the relaxation oscillator whose output exhibits abrupt transitions (see Sec. **16**). Oscillators often have a circuit element that can be varied to produce different frequencies.

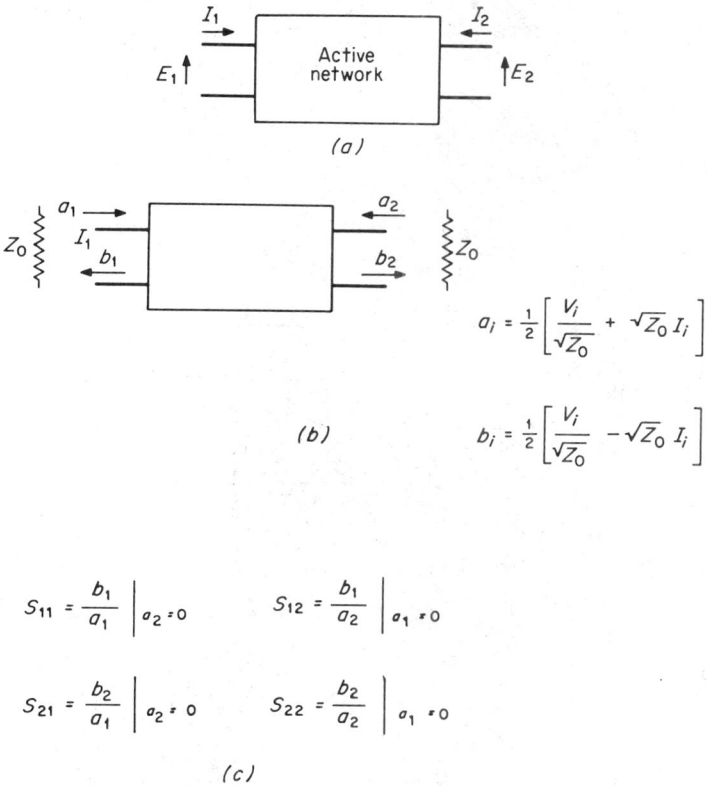

Fig. 13-10. Definitions of active-network parameters: (*a*) general network; (*b*) ratios a_i and b_i of incident and reflected waves (square root of power); (*c*) *s* parameters

An oscillator's frequency is sensitive to the stability of the frequency-determining elements as well as the variation in the active-device parameters (e.g., effects of temperature, bias point, and aging). In many instances the oscillator is followed by a second stage serving as a buffer, so that there is isolation between the oscillator and its load. The amplitude of the oscillation can be controlled by automatic gain control (AGC) circuits, but the nonlinearity of the active element usually determines the amplitude. Variations in bias, temperature, and component aging have a direct effect on amplitude stability.

12. Requirements for Oscillation. Oscillators can be considered from two viewpoints: as using positive feedback around an amplifier or as a one-port network in which the real component of the input immittance is negative. An oscillator must have frequency-determining elements (generally passive components), an amplitude-limiting mechanism, and sufficient closed-loop gain to make up for the losses in the circuit. It is possible to predict the operating frequency and conditions needed to produce oscillation from a Nyquist or Bode analysis. The prediction of output amplitude requires the use of nonlinear analysis, commonly in the form of graphical techniques.

13. Oscillator Circuits. Typical oscillator circuits applicable up to UHF frequencies are shown in Fig. 13-14. These are discussed in detail in the following subsections. Also of interest are crystal oscillators. In this case the crystal is used as the passive frequency-determining element. The frequency range of crystal oscillators extends from a few hundred hertz to over 200 MHz by use of overtone crystals. The analysis of crystal oscillators is best done using the equivalent circuit of the crystal (see also Sec. 7).

14. Synchronization. Synchronization of oscillators is accomplished by using phase-locked loops or by direct low-level injection of a reference frequency into the main oscillator. The diagram of a phase-locked loop is shown in Fig. 13-15 and that of an injection-locked oscillator in Fig. 13-16. Detailed discussions are contained in Refs. 5 to 7, Par. 13-17.

$$\begin{bmatrix} E_1 \\ E_2 \end{bmatrix} = \begin{bmatrix} z_{11} & z_{12} \\ z_{21} & z_{22} \end{bmatrix} \times \begin{bmatrix} I_1 \\ I_2 \end{bmatrix}$$

$$z_{11} = \left(\frac{E_1}{I_1}\right)_{I_2=0} \qquad z_{12} = \left(\frac{E_1}{I_2}\right)_{I_1=0}$$

$$z_{21} = \left(\frac{E_2}{I_1}\right)_{I_2=0} \qquad z_{22} = \left(\frac{E_2}{I_2}\right)_{I_1=0}$$

$$\begin{bmatrix} I_1 \\ I_2 \end{bmatrix} = \begin{bmatrix} y_{11} & y_{12} \\ y_{21} & y_{22} \end{bmatrix} \times \begin{bmatrix} E_1 \\ E_2 \end{bmatrix}$$

$$y_{11} = \left(\frac{I_1}{E_1}\right)_{E_2=0} \qquad y_{12} = \left(\frac{I_1}{E_2}\right)_{E_1=0}$$

$$y_{21} = \left(\frac{I_2}{E_1}\right)_{E_2=0} \qquad y_{22} = \left(\frac{I_2}{E_2}\right)_{E_1=0}$$

$$\begin{bmatrix} E_1 \\ I_2 \end{bmatrix} = \begin{bmatrix} h_{11} & h_{12} \\ h_{21} & h_{22} \end{bmatrix} \times \begin{bmatrix} I_1 \\ E_2 \end{bmatrix}$$

$$h_{11} = \frac{1}{y_{11}} \qquad h_{12} = \left(\frac{E_1}{E_2}\right)_{I_1=0}$$

$$h_{21} = \left(\frac{I_2}{I_1}\right)_{E_2=0} \qquad h_{22} = \frac{1}{z_{22}}$$

$$\begin{bmatrix} I_1 \\ E_2 \end{bmatrix} = \begin{bmatrix} l_{11} & l_{12} \\ l_{21} & l_{22} \end{bmatrix} \times \begin{bmatrix} E_1 \\ I_2 \end{bmatrix}$$

$$l_{11} = \frac{1}{z_{11}} \qquad l_{12} = \left(\frac{I_1}{I_2}\right)_{E_1=0}$$

$$l_{21} = \left(\frac{E_2}{E_1}\right)_{I_1=0} \qquad l_{22} = \frac{1}{y_{22}}$$

$$\begin{bmatrix} E_1 \\ I_1 \end{bmatrix} = \begin{bmatrix} a_{11} & a_{12} \\ a_{21} & a_{22} \end{bmatrix} \times \begin{bmatrix} E_2 \\ -I_2 \end{bmatrix}$$

$$a_{11} = \frac{1}{l_{21}} \qquad a_{12} = -\frac{1}{y_{21}}$$

$$a_{21} = \frac{1}{z_{21}} \qquad a_{22} = -\frac{1}{h_{21}}$$

$$\begin{bmatrix} E_2 \\ I_2 \end{bmatrix} = \begin{bmatrix} b_{11} & b_{12} \\ b_{21} & b_{22} \end{bmatrix} \times \begin{bmatrix} E_1 \\ -I_1 \end{bmatrix}$$

$$b_{11} = \frac{1}{h_{12}} \qquad b_{12} = -\frac{1}{y_{12}}$$

$$b_{21} = \frac{1}{z_{12}} \qquad b_{22} = -\frac{1}{l_{12}}$$

$$\begin{bmatrix} b_1 \\ b_2 \end{bmatrix} = \begin{bmatrix} S_{11} & S_{12} \\ S_{21} & S_{22} \end{bmatrix} \times \begin{bmatrix} a_1 \\ a_2 \end{bmatrix}$$

$$S_{11} = \left(\frac{b_1}{a_1}\right)_{a_2=0} \qquad S_{12} = \left(\frac{b_1}{a_2}\right)_{a_1=0}$$

$$S_{21} = \left(\frac{b_2}{a_1}\right)_{a_2=0} \qquad S_{22} = \left(\frac{b_2}{a_2}\right)_{a_1=0}$$

Fig. 13-11. Network matrix terms.

15. Harmonic Content. The harmonic content of the oscillator output is related to the amount of oscillator output power at frequencies other than the fundamental. From the viewpoint of a negative-conductance (resistance) oscillator,[8] better results are obtained if the curve of the negative conductance (or resistance) vs. amplitude of oscillation is smooth and without an inflection point over the operating range. Harmonic content is also reduced if the oscillator's operating point Q is chosen so that the range of negative conductance is symmetrical about Q on the negative conductance-vs.-amplitude curve. This can be done by adjusting the oscillator's bias point within the requirement of $|G_C| = |G_D|$ for sustained oscillation (see Fig. 13-17).

16. Stability. The stability of the oscillator's output amplitude and frequency from a negative-conductance viewpoint depends on the variation of its negative conductance with operating point and the amount of fixed positive conductance in the oscillator's associated circuit. In particular, if the change of bias results in vertical translation of the conductance-(resistance)-vs.-amplitude curve, the oscillator's stability is related to the change of slope at the point where the circuit's fixed conductance intersects this curve (point Q in Fig. 13-17). If the $|G_D|$ curve is of the shape of $|G_D|_2$, the oscillation can stop when a large enough change in bias point occurs for

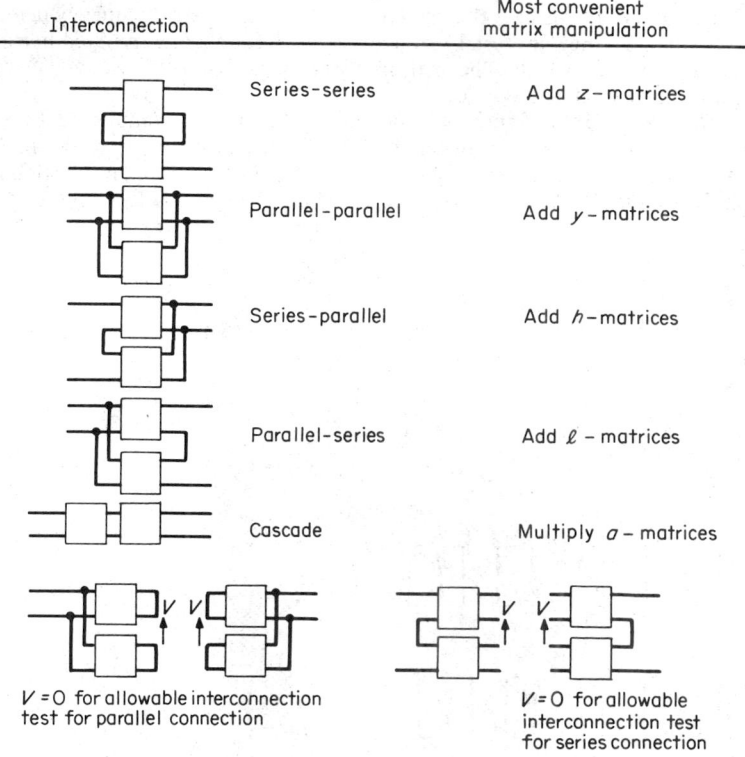

Interconnection	Most convenient matrix manipulation

Series–series — Add z–matrices

Parallel–parallel — Add y–matrices

Series–parallel — Add h–matrices

Parallel–series — Add ℓ – matrices

Cascade — Multiply a – matrices

$V = 0$ for allowable interconnection test for parallel connection

$V = 0$ for allowable interconnection test for series connection

Fig. 13-12. Matrix equivalents of network interconnections.

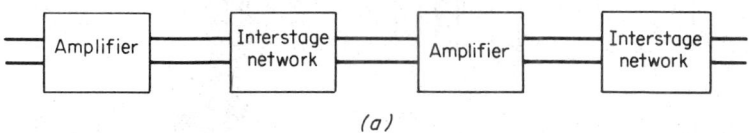

Amplifier — Interstage network — Amplifier — Interstage network

(a)

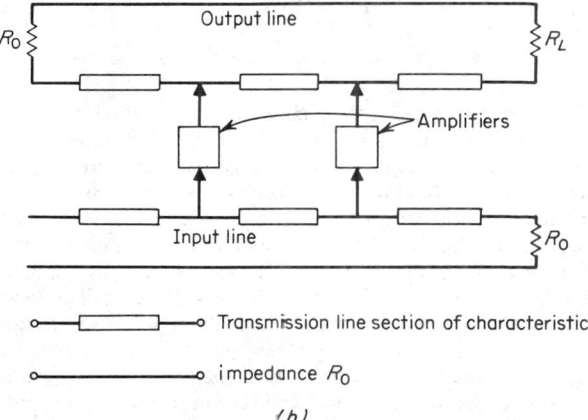

Output line

R_O R_L

Amplifiers

Input line R_O

Transmission line section of characteristic

impedance R_O

(b)

Fig. 13-13. Multiamplifier structures: (a) cascade; (b) distributed.

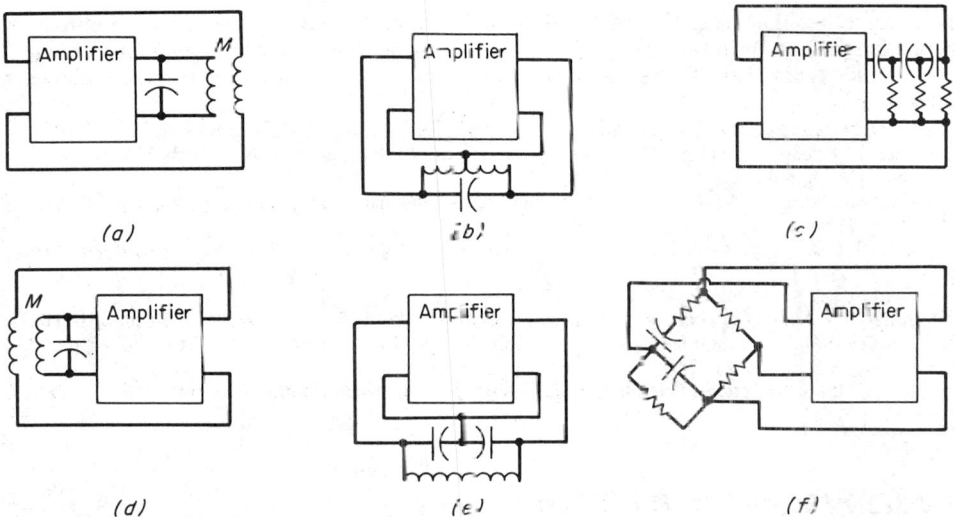

Fig. 13-14. Types of oscillators: (a) tuned-output; (b) Hartley; (c) phase-shift; (d) tuned-input; (e) Colpitts; (f) Wien bridge.

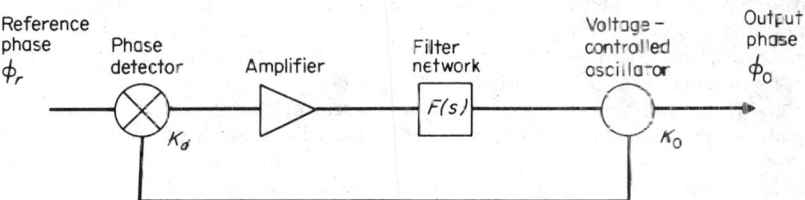

Fig. 13-15. Phase-locked-loop oscillator.

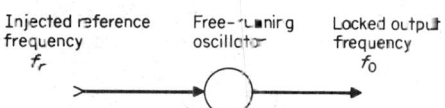

Fig. 13-16. Injection-locked oscillator.

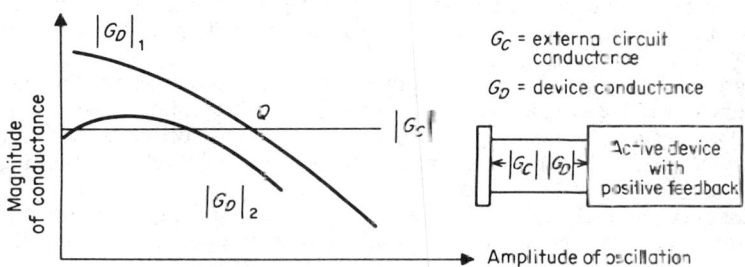

Fig. 13-17. Device conductance vs. amplitude of oscillator.

$|G_D|$ to be less than $|G_C|$ for all amplitudes of oscillation. Stabilization of the amplitude of oscillation may occur in the form of modifying G_C, G_D or both to compensate for bias changes.

Particular types of oscillators and their parameters are discussed in detail in the following subsections.

17. References on Principles of Amplifiers and Oscillators

1. Shea, R. F. (ed.)   "Amplifier Handbook," pp. 7-11 and 7-19, McGraw-Hill, New York, 1968.
2. Ibid., pp. 6-4 and 6-5.
3. Chestnut, H., and R. W. Mayer   "Servomechanism and Regulating System Design," Vol. I, p. 145, Wiley, New York, 1959.
4. Truxal, J. C.   "Automatic Feedback Control System Synthesis," pp. 221–227, McGraw-Hill, New York, 1955.
5. Gardiner, F. M.   "Phaselock Techniques," Wiley, New York, 1966.
6. Adler, R.   A Study of Locking Phenomena in Oscillators, *Proc. IRE*, June 1946, Vol. 34, pp. 351–357.
7. Mackey, R. C.   Injection Locking of Klystron Oscillators, *IRE Trans. MTT*, July 1962, Vol. 10, pp. 228–235.
8. Reich, H. J.   "Functional Circuits and Oscillators," Sec. 74, Van Nostrand, New York, 1961.

Audio-Frequency Amplifiers

BY S. M. KORZEKWA

18. Preamplifiers. *General Considerations.*   The function of a preamplifier is to amplify a low-level signal to a higher level before further processing or transmission to another location. The required amplification is achieved by increased signal voltage and/or impedance reduction. The amount of power amplification required varies with the particular application. A general guideline is to provide sufficient preamplification to ensure that further signal handling adds minimal (or acceptable) signal-to-noise degradation.

Signal-to-Noise Considerations.   The design of a preamplifier must consider all potential signal degradation from sources of noise, whether generated externally or within the preamplifier itself.

Examples of externally generated noise are hum and pickup, which may be introduced by the input-signal lines or the power-supply lines. Shielding of the input-signal lines often proves to be an acceptable solution. The preamplifier should be located close to the transmitting source, and the preamplifier power gain must be sufficient to override interference that remains after these steps are taken.

A second major source of noise is that internally generated in the preamplifier itself. The noise figure specified in decibels for a preamplifier, which serves as a figure of merit, is defined as the ratio of the available input-to-output signal-to-noise power ratios:

$$F = \frac{S_i/N_i}{S_o/N_o}$$

where F = noise figure of preamplifier, S_i = available signal input power, N_i = available noise input power, S_o = available signal output power, and N_o = available noise output power.

Design precautions to realize the lowest possible noise figure include the proper selection of the active device, optimum input and output impedances, correct voltage and current biasing conditions, and pertinent design parameters of devices. Figure 13-18 illustrates the effects on the noise figure of specific transistors due to changes in source resistance, frequency, and current bias.[5][*]

Sensor System Characteristics.   A preamplifier can be used to compensate the sensor's signal characteristics to realize a specific effect, as in phonograph pickups. In a magnetic sensor, when the stylus motion causes an output signal proportional to stylus velocity, the output signal level vs. frequency increases at 20 dB/decade for a constant input-signal level. If the loading is varied, the signal output will be further modified.

In contrast, a ceramic phonograph pickup is basically capacitive; i.e., its high impedance is susceptible to loading and the output signals are proportional to the amplitude of the stylus

---

[*]Superior numbers correspond to references in Par. **13-33**.

movement. The signals derived from magnetic tapes are often processed via the preamplifier used for phonograph records (see Sec. **19**). The standard RIAA recording characteristic (Fig. **19-139**) indicates the proper system frequency response when using an inductive pickup. The specified frequency shaping takes into account both the low-frequency limitations of the original recording and also the higher-frequency performance of the sensor. The resultant frequency compensation is ultimately obtained via preamplifier design, pickup loading, or both. Note that

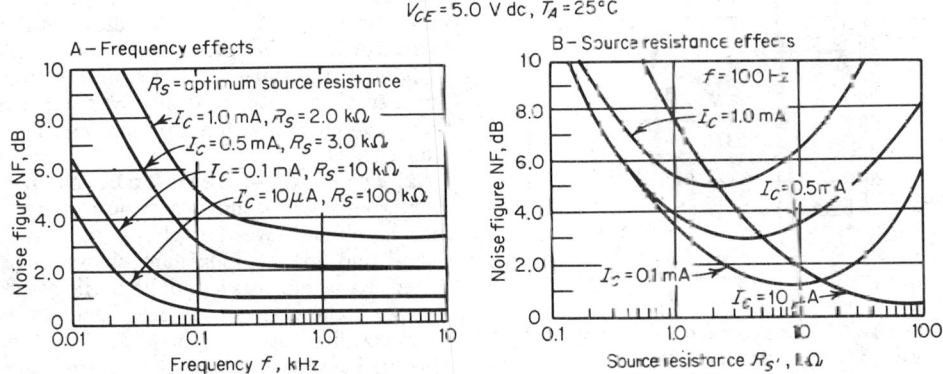

$$V_{CE} = 5.0 \text{ V dc}, \ T_A = 25°C$$

Fig. 13-18. Noise-figure data for the 2N5088 and 2N5089 silicon npn transistors.[3]

what is being specified via these standards is essentially the system transfer function, which takes into account specific processing and sensor limitations and/or attributes.

Figure 13-19 represents a phonograph preamplifier designed for a 5,000- to 10,000-pF-capacitance ceramic cartridge. With an Astatic model 137 cartridge, the output reference level of 1 V is 13 dB below maximum output and 69 dB above the unweighted noise level. The total harmonic distortion is less than 0.6% at the reference level. This preamplifier is properly equalized in accordance with the RIAA standard.[3]

19. Low-Level Amplifiers. The low-level designation applies to amplifiers operated below maximum permissible power-dissipation, current, and voltage limits. Thus many low-level amplifiers are purposely designed to realize specific attributes other than delivering the maximum attainable power to the load, such as gain stability, bandwidth, optimum noise figure, low cost, etc.

In an amplifier designed to be operated with a 24-V power supply and a specified load termination, for example, the operating conditions may be such that the active devices are just within their allowable limits. If operated at these maximum limits, this is not a low-level amplifier. However, if this amplifier also fulfills its performance requirements at a reduced power-supply voltage of 6 V, with resulting much lower internal dissipation levels, it becomes a low-level amplifier.

20. Medium-Level and Power Amplifiers. The medium-power designation for an amplifier implies that some active devices are operated near their maximum dissipation limits, and precautions must be taken to protect these devices. If power-handling capability is taken as the criterion, the 5- to 100-W power range is a current demarcation line. As higher-power-handling devices come into use, this range will tend to shift to higher power levels.

The amount of power that can safely be handled by an amplifier is usually dictated by the dissipation limits of the active devices in the output stages, the efficiency of the circuit, and the means used to extract heat to maintain devices within their maximum permissible temperature limits. The classes of operation (A, B, AB, C) are discussed in Par. **13-29**. When single active devices

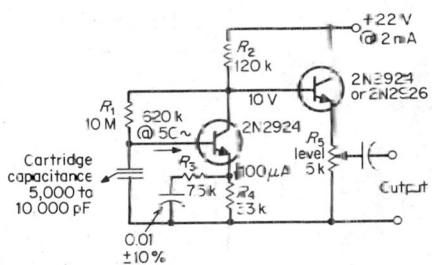

Fig. 13-19. Phonograph preamplifier for ceramic cartridge, with RIAA equalization.[3]

do not suffice, multiple series or parallel configurations can be used to achieve higher voltage or power operation. Figure 13-20 illustrates an audio output power stage capable of providing 70 W to a load without using an output transformer.[6]

21. Multistage Amplifiers. An amplifier may take the form of a single stage or a complex single stage, or it may employ an interconnection of several stages. Various biasing, coupling, feedback, and other design alternatives influence the topology of the amplifier. For a multistage amplifier, the individual stages may be essentially identical or radically different. Feedback techniques may be used at the individual stage level, at the amplifier functional level, or both, to realize bias stabilization, gain stabilization, output-impedance reduction, etc.

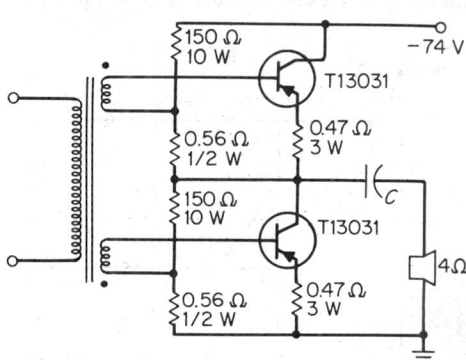

Fig. 13-20. Audio output power amplifier of 70 W power. *(Texas Instruments, Inc.)*

22. Typical Electron-Tube Amplifier. Figure 13-21 shows a typical electron-tube amplifier stage. For clarity the signal-source and load sections are shown partitioned. For a multistage amplifier the source represents the equivalent signal generator of the preceding stage. Similarly, the load indicated includes the loading effect of the subsequent stage, if any.

Figure 13-22 is a simplified small-signal ac equivalent of the amplifier shown in Fig. 13-21. The voltage gain from the grid of the tube to the output can be calculated to be

$$A_{v1} = -\frac{\mu R_l}{r_p + R_l}$$

Similarly, the voltage gain from the source to the tube grid is

$$A_{v2} = \frac{R_1}{(R_1 + R_g) + 1/j\omega C}$$

Combining the above equations gives the composite amplifier voltage gain

$$A_v = -\frac{\mu R_1 R_l}{(r_p + R_l)[(R_1 + R_g) + 1/j\omega C]}$$

This example illustrates the fundamentals of an electron-tube amplifier stage. Many excellent references treat this subject in detail.[7,8]

23. Typical Transistor Amplifier. The analysis techniques used for electron-tube amplifier stages generally apply to transistorized amplifier stages. The principal difference is that different active-device models are used.

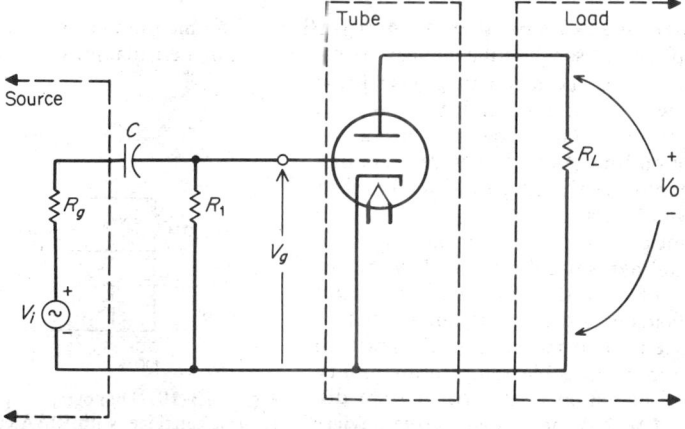

Fig. 13-21. Typical triode electron-tube amplifier stage (biasing not shown).

The typical transistor stage shown in Fig. 13-23 illustrates a possible form of biasing and coupling. The source section is partitioned and includes the preceding-stage equivalent generator, and the load includes subsequent stage-loading effects. Figure 13-24 shows the generalized h-equivalent circuit representation for transistors. Table 13-1 lists the h-parameter transformations for the common-base, common-emitter, and common-collector configurations.

Table 13-1. h **Parameters of the Three Configurations**[2]

	Common-base	Common-emitter	Common-collector
h_{11}	h_{ib}	$h_{ib}(h_{fe} + 1)$	$h_{ib}(h_{fe} + 1)$
h_{12}	h_{rb}	$h_{ib}h_{ob}(h_{fe} + 1) - h_{rb}$	1
h_{21}	h_{fb}	h_{fe}	$-(h_{fe} + 1)$
h_{22}	h_{ob}	$h_{ob}(h_{fe} + 1)$	$h_{cb}(h_{fe} + 1)$

While these parameters are complex and frequency-dependent, it is often feasible to use simplifications. Most transistors have their parameters specified by their manufacturers, but it may be necessary to determine additional parameters by test.

Figure 13-25 illustrates a simplified model of the transistor amplifier stage of Fig. 13-23. The common-emitter h parameters are used to represent the equivalent transistor. The voltage gain for this stage is

$$A_v = \frac{V_o}{V_i} = -\frac{h_{fe}R_l}{R_g + h_{ie}}$$

The complexity of analysis depends on the accuracy needed. Currently, most of the more complex analysis is performed with the aid of computers. Several transistor-amplifier-analysis references treat this subject in detail.[2-4]

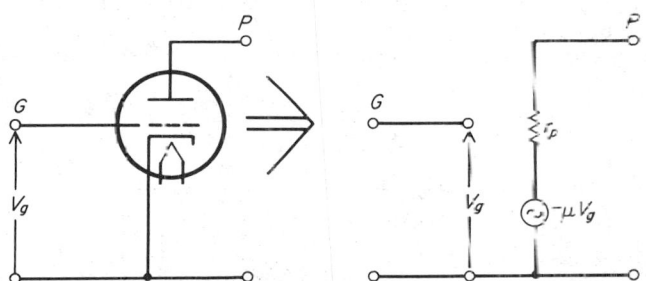

Fig. 13-22. Equivalent circuit (right) of electron tube (left).

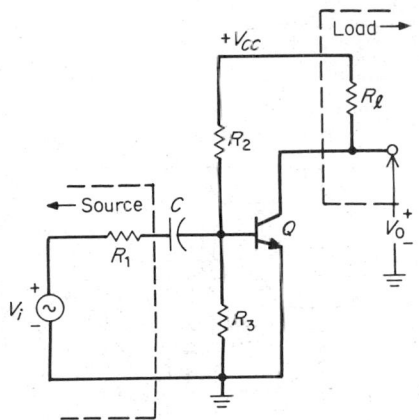

Fig. 13-23. Typical bipolar transistor-amplifier stage.

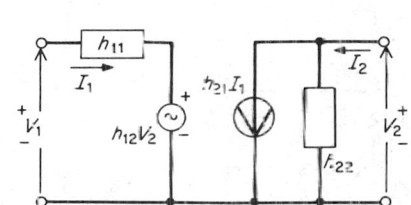

Fig. 13-24. Equivalent circuit of transistor, based on h parameters.

24. Typical Multistage Transistor Amplifier. Figure 13-26 is an example of a capacitively coupled three-stage transistor amplifier. It has a broad frequency response, illustrating the fact that an audio amplifier can be useful in other applications. The component values marked on the diagram are

$$R_1 = 16,000 \ \Omega \qquad R_2 = 6,200 \ \Omega \qquad R_3 = 1,600 \ \Omega \qquad R_4 = 1,000 \ \Omega$$
$$R_L = 560 \ \Omega \qquad Q_1, Q_2, Q_3, = 2N1565 \qquad C_1 = 10 \ \mu F \qquad C_2 = 100 \ \mu F$$

This amplifier is designed to operate over a range of -55 to $+125°C$, with an output voltage swing of 2 V peak to peak and frequency response down 3 dB at approximately 200 Hz and 2 MHz. The overall gain at 1,000 Hz is

Temperature, °C	Gain, dB
-55	83
$+25$	88
$+125$	91

25. Biasing Methods. The biasing scheme used in an amplifier determines the ultimate performance that can be realized. Conversely, an amplifier with poorly implemented biasing may suffer in performance, and be susceptible to catastrophic circuit failure due to high stresses within the active devices. In view of the variation of parameters within the active devices, it is important that the amplifier function properly even when the initial and/or end-of-life parameters of the devices vary.

26. Electron-Tube Biasing. Biasing is intended to maintain the quiescent currents and voltages of the electron tube at the prescribed levels. The tube-plate characteristics represent the biasing relations between the tube parameters.

The single-stage amplifier shown in Fig. 13-27 illustrates self-biasing (cathode biasing) via the cathode feedback components R_k and C_k. The cathode current tends to increase until the correct quiescent positive cathode bias results.

The principal bias parameters (steady-state plate and grid voltages) can be readily identified by

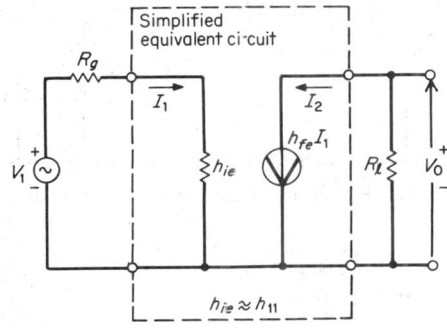

Fig. 13-25. Simplified equivalent circuit of transistor.

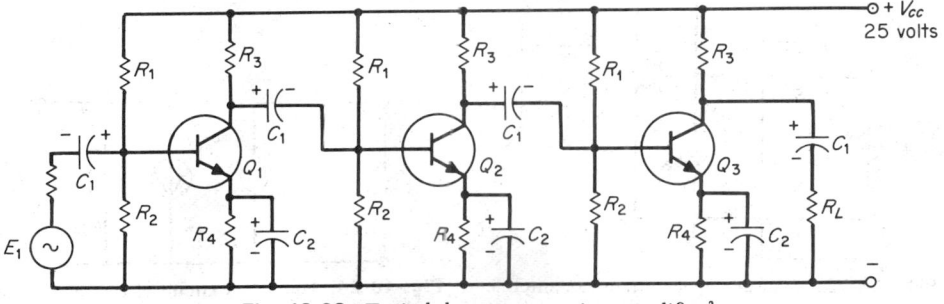

Fig. 13-26. Typical three-stage transistor amplifier.[2]

the construction of a load line on the plate characteristic, as illustrated in Fig. 13-28. The operating point Q is located at the intersection of the selected plate characteristic with the load line. The load-line endpoints are determined by

$$i_b = \begin{cases} 0 & E_b = E_{bb} \\ E_{bb}/R_L & e_b = 0 \end{cases}$$

It is evident that there are many possible grid-voltage–anode-current combinations, each resulting in a different quiescent I_b current. When a particular grid bias voltage is selected, the resulting I_b current directly determines the value of cathode resistor R_k. The capacitor C_k is selected to give the required frequency performance.

27. Transistor Biasing. Although the methods of biasing a transistor-amplifier stage are in many respects similar to those of an electron-tube amplifier, there are many different types of transistors, each characterized by different curves. Bipolar transistors are generally characterized by their collector and emitter families, while field-effect transistors have different characterizations. The npn transistor requires a positive base bias voltage and current (with respect to its emitter) for proper operation; the converse is true for a pnp transistor.

Figure 13-29 illustrates a common biasing technique. A single power supply is used, and the transistor is self-biased with the unbypassed emitter resistor R_e. Although a graphical solution

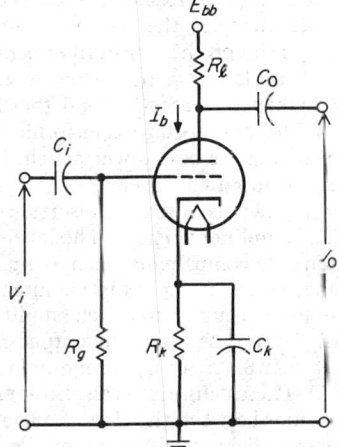

Fig. 13-27. Self-biased electron-tube stage.

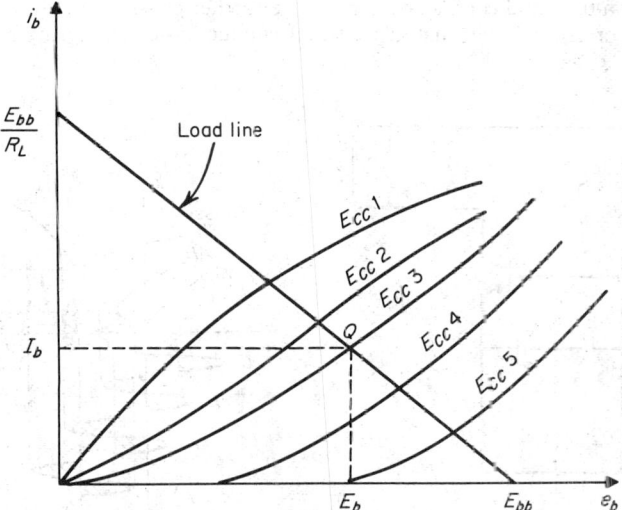

Fig. 13-28. Load-line determination of operating point Q.

of the value of R_e could be found by referring to the collector-emitter curves, an iterative solution, described below, is also commonly used.

Because the performance of the transistors depends on the collector current and collector-to-emitter voltage, they are often selected as starting conditions for biasing design. The unbypassed emitter resistor R_e and collector resistor R_c, the primary voltage-gain-determining components, are determined next, taking into account other considerations such as the anticipated maximum signal level, available power supply V_{cc}, etc. The last step is to determine the R_1 and R_2 values.

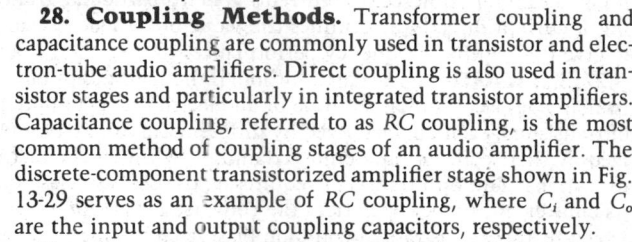

28. Coupling Methods. Transformer coupling and capacitance coupling are commonly used in transistor and electron-tube audio amplifiers. Direct coupling is also used in transistor stages and particularly in integrated transistor amplifiers. Capacitance coupling, referred to as RC coupling, is the most common method of coupling stages of an audio amplifier. The discrete-component transistorized amplifier stage shown in Fig. 13-29 serves as an example of RC coupling, where C_i and C_o are the input and output coupling capacitors, respectively.

Transformer coupling is commonly used to match the input and output impedances of electron-tube amplifier stages. Since the input impedance of an electron tube is very high at audio frequencies, the design of an electron-tube stage depends primarily on the transformer parameters. The much lower input impedances of transistors demand that many other factors be taken into account, and the design becomes more complex.

Fig. 13-29. Capacitively coupled *npn* transistor-amplifier stage.

The output-stage transformer coupling to a specific load is often the optimum method of realizing the best power match. Figure 13-30 illustrates a typical transformer-coupled transistor audio-amplifier stage.

The direct-coupling approach is now also used for discrete-component transistorized amplifiers, and particularly in integrated amplifier versions. The level-shifting requirement is realized by selection from the host of available components, such as *npn* and *pnp* transistors and zener diodes. Since it is difficult to realize large-size capacitors via integrated-circuit techniques, special methods have been developed to direct-couple integrated amplifiers.

29. Classes A, B, AB, and C Operation. The output or power stage of an amplifier is usually classified as operating class A, B, AB, or C, depending on the conduction characteristics of the active devices (see Fig. 13-5). These definitions can also apply to any intermediate amplifier stage. Figure 13-31 illustrates relations between the class of operation and conduction using transistor parameters.[2] This figure would be essentially the same for an electron-tube amplifier with the tube plate current and grid voltage as the equivalent device parameters.

Subscripts may be used to denote additional conduction characteristics of the device. For example, the electron-tube grid conduction can also be further classified as A_1, to show that no grid current flows, or A_2, to show that grid-current conduction exists during some portion of the cycle.

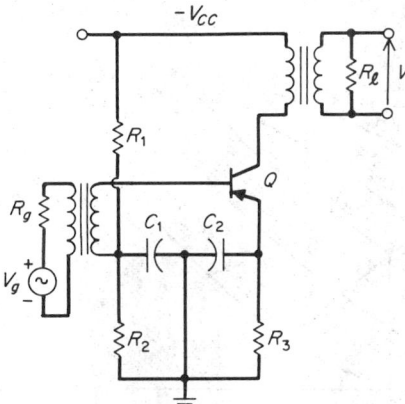

Fig. 13-30. Transformer-coupled *pnp* transistor-amplifier stage.

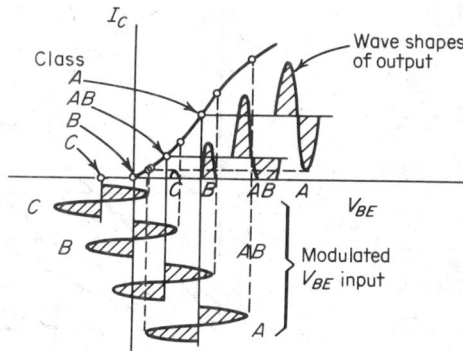

Fig. 13-31. Classes of amplifier operation, based on transistor characteristics.

30. Push-Pull Amplifiers. In a single-ended amplifier the active devices conduct continuously. The single-ended configuration is generally used in low-power applications, operated in class A. For example, preamplifiers and low-level amplifiers are generally operated single-ended, unless the output power levels necessitate the more efficient power handling of the push-pull circuit.

In a push-pull configuration there are at least two active devices that alternately amplify the negative and positive cycles of the input waveform. The output connection to the load is most

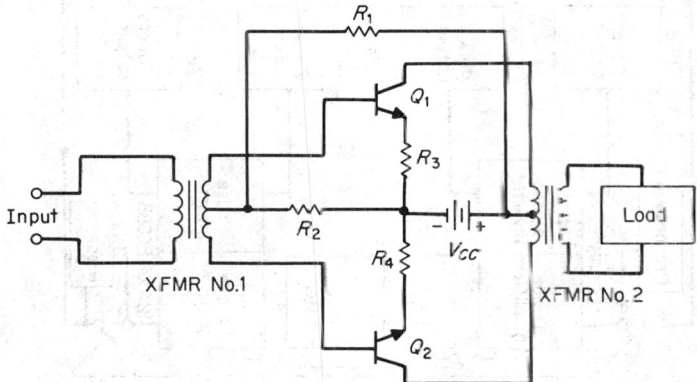

Fig. 13-32. Transformer-coupled push-pull transistor stage.

often transformer-coupled. An example of a transformer input and output in a push-pull amplifier is illustrated in Fig. 13-32. Direct-coupled push-pull amplifiers and capacitively coupled push-pull amplifiers are also feasible, as illustrated in Fig. 13-33.

The active devices in push-pull are usually operated either in class B or AB because of the high power-conversion efficiency. Feedback techniques can be used to stabilize gain, stabilize biasing or operating points, minimize distortion, etc.

31. Output Amplifiers. The function of an audio output amplifier is to interface with the preceding amplifier stages and to provide the necessary drive to the load. Thus the output-amplifier designation does not uniquely identify a particular amplifier class. When several different types of amplifiers are cascaded between the signal source and its load, e.g., a high-power speaker, the last-stage amplifier is designated as the output amplifier. Because of the high power requirements, this amplifier is usually a push-pull type operating either in class B or AB.

32. Stereo Amplifiers. A stereo amplifier provides two separate audio channels properly phased with respect to each other. The objective of this two-channel technique is to enhance the audio reproduction process, making it more realistic and lifelike. It is also feasible to extend the system to contain more than two channels of information (e.g., four-channel, see Sec. 19). A

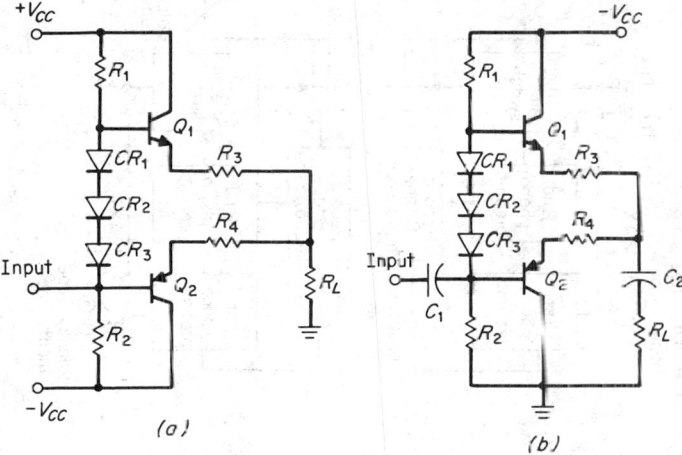

Fig. 13-33. (a) Direct- and (b) capacitively coupled push-pull stages.

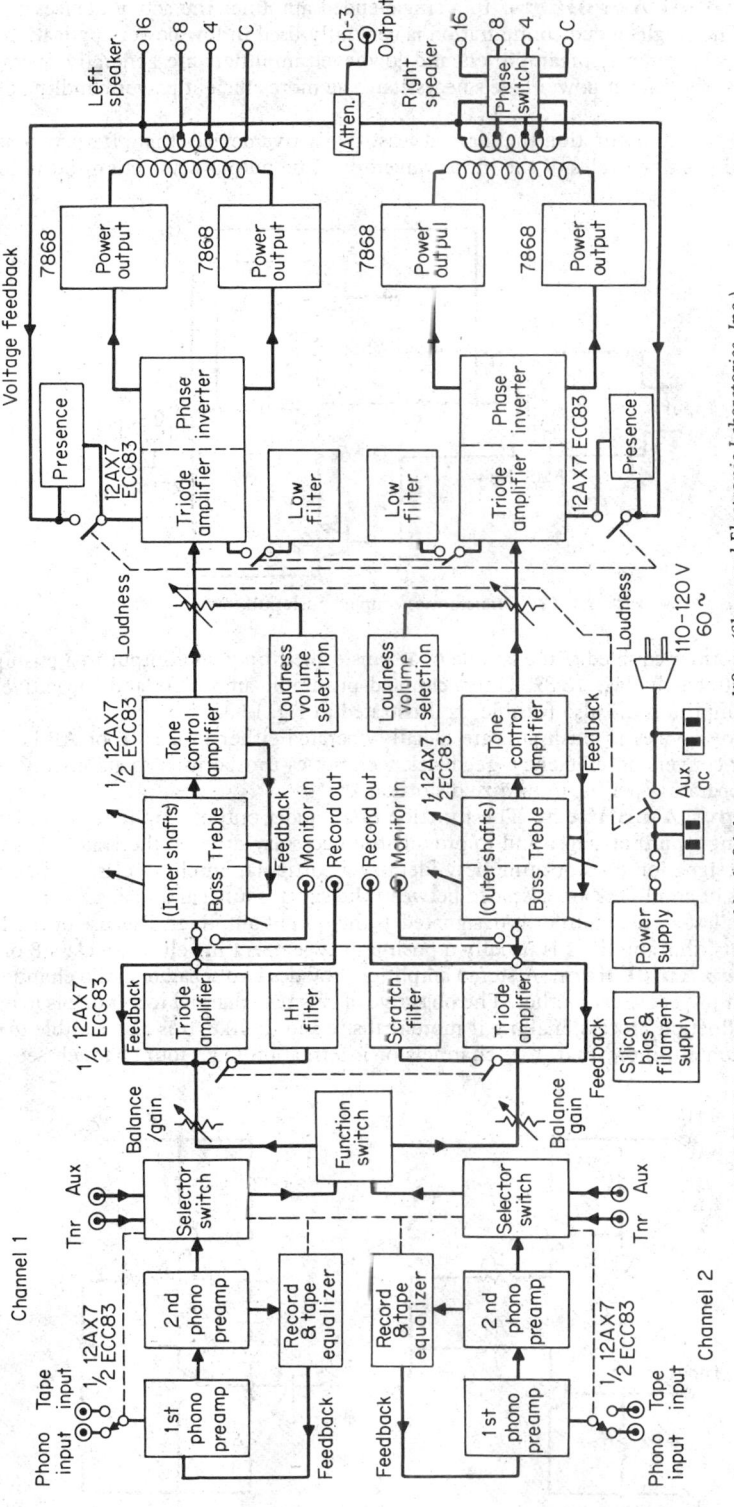

Fig. 13-34. Block diagram of high-fidelity stereo amplifier. (*Sherwood Electronic Laboratories, Inc.*)

stereo amplifier is a complete system that contains its power supply and other commonly required control functions.

Each channel has its own preamplifier, medium-level stages, and output power stage, with different gain and frequency responses for each mode of operation, e.g., for tape, phonograph, etc. The input signal is selected from the phonograph input connection, tape input or a tuner output. Special-purpose trims and controls are also used to optimize performance on each mode. The bandwidth of the amplifier extends to 20 kHz or higher.

The block-diagram representation of a high-quality stereo amplifier is shown in Fig. 13-34. The performance specifications[2] for this amplifier are as follows:

Power output. Stereophonic, each channel 40 W music power (36 W continuous, 72 W peak). Monophonic, 80 W music power (72 W continuous, 144 W peak), at 1½% intermodulation distortion (60 Hz to 7 kHz 4:1)

Outputs. 16-, 8-, and 4-Ω left and right speakers; two recording, third channel

Inverse feedback. 16 dB

Damping factor. 5:1

Frequency response (36 W). 20 Hz to 20 kHz ± ½ dB

Tone-control range. 15 kHz, 17-dB boost or cut; 40 Hz, 16-dB boost, 19-dB cut

Rumble filter. 27 Hz, 17-dB rejection; 70 Hz less than 1 dB down

Sensitivity. Radio 0.25 V, tape 1.4 mV, phonograph 1.2 mV; all inputs adjustable with level control

Maximum input. Phonograph, 200 mV for less than 1% distortion; radio, adjustable with level control

Maximum hum and noise. Volume control, minimum 100 dB (weighted) below rated output; radio input (controls maximum), 90 dB (weighted) below rated output; phonograph input (controls flat), 60 dB below rated output, 72 dB below 10 mV (equivalent to ½ μV referred to input grid)

Interchannel crosstalk. Less than −50 dB at 1 kHz

Power consumption. 110 to 120 V, 60 Hz, 150 W, 1.3 A

Tube complement. Four 7868, six 12AX7/ECC83, four silicon rectifiers

33. References on Audio Amplifiers

1. IEEE Standard on Definitions of Terms for Audio and Electroacoustics, *IEEE Trans. Audio Electroacoust.*, June 1966, Vol. AU-14, No. 2.

2. Shea, R. F. "Amplifier Handbook," McGraw-Hill, New York, 1968.

3. "Transistor Manual," 7th ed., General Electric Co., 1964.

4. Texas Instruments, Inc. "Transistor Circuit Design," McGraw-Hill, New York, 1963.

5. Brubaker, R. Semiconductor Noise Figure Considerations, Motorola Semiconductor Products, Inc., *Appl. Note*, AN-421, Phoenix, Ariz., 1968.

6. Markus, J. "Sourcebook of Electronic Circuits," McGraw-Hill, New York, 1968.

7. Landee, R., D. Davis, and A. Albrecht "Electronic Designers' Handbook," McGraw-Hill, New York, 1957.

8. "Reference Data for Radio Engineers," 5th ed., International Telephone and Telegraph Corp., 1968.

Audio Oscillators

BY R. J. McFADYEN

34. General Considerations. In the strict sense, an audio oscillator is limited to frequencies from about 15 to 20,000 Hz,[1]* but a much wider frequency range is included in most oscillators used in audio measurements since knowledge of amplifier characteristics in the region above audibility is often required.

For the production of sinusoidal waves, audio oscillators consist of an amplifier having a non-linear power gain characteristic, with a path for regenerative feedback. Single- and multistage transistor amplifiers with *LC* or *RC* feedback networks are most often used. The term *harmonic oscillator* is used for these types. Relaxation *oscillators*, which may be designed to oscillate in the audio range, exhibit sharp transitions in the output voltages and currents. Relaxation oscillators are treated in Sec. **16**.

*Superior numbers correspond to numbered references, Par. **13-42**.

The instantaneous excursion of the operating point in a harmonic oscillator is restricted to the range where the circuit exhibits an impedance with a negative real part. The amplifier supplies the power, which is dissipated in the feedback path and the load. The regenerative feedback would cause the amplitude of oscillation to grow without bound were it not for the fact that the dynamic range of the amplifier is limited by circuit nonlinearities. Thus, in most sine-wave audio oscillators, the operating frequency is determined by passive-feedback elements, whereas the amplitude is controlled by the active-circuit design.

Analytical expressions predicting the frequency and required starting conditions for oscillation can be derived using Bode's amplifier feedback theory and the stability theorem of Nyquist[2] (see Par. **13-8**). Since this analytical approach is based on a linear-circuit model, the results are approximate but usually suitable for design of sinusoidal oscillators. No prediction on waveform amplitude results, since this is determined by nonlinear-circuit characteristics. Estimates of the waveform amplitude can be made from the bias and limiting levels of the active circuits. Separate

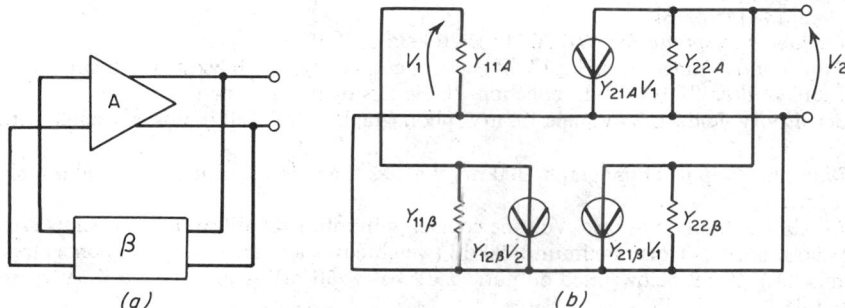

Fig. 13-35. Oscillator representations: (a) generalized feedback circuit; (b) equivalent-y-parameter circuit.

limiters and AGC techniques are also useful for controlling the amplitude to a prescribed level. Graphical and nonlinear analysis methods[3] can also be used for obtaining a prediction of the amplitude of oscillation.

A general formulation suitable for a linear analysis of almost all audio oscillators can be derived from the feedback diagram[4] in Fig. 13-35. Note that the amplifier internal feedback generator has been neglected; that is, y_{12A} is assumed to be zero. This assumption of unilateral amplification is almost always valid in the audio range even for single-stage transistor amplifiers.

The stability requirements for the circuit are derived from the closed-loop-gain expression

$$A_c = A/(1 - A\beta) \tag{13-1}$$

where the gain A is treated as a negative quantity for an inverting amplifier. Infinite closed-loop gain occurs when $A\beta$ is equal to unity, and this defines the oscillatory condition. In terms of the equivalent circuit parameters used in Fig. 13-35,

$$1 - A\beta = 1 - y_{21A} \frac{y_{12\beta}}{(y_{11A} + y_{11\beta})(y_{22A} + y_{22\beta}) - y_{12\beta}y_{21\beta}} \tag{13-2}$$

In the audio range, y_{21A} remains real, but the fractional portion of the function is complex because β is frequency-sensitive. Therefore, the open-loop gain $A\beta$ can be expressed in the general form

$$A\beta = y_{21A} \frac{A_r + jA_i}{B_r + jB_i} \tag{13-3}$$

It follows from Nyquist's stability theorem that this feedback system will be unstable if, first, the phase shift of $A\beta$ is zero and, second, the magnitude is equal to or greater than unity. Applying this criterion to Eq. (13-3) yields the following two conditions for oscillation:

$$A_i B_r - A_r B_i = 0 \tag{13-4}$$

$$y_{21}^2 \geq \frac{B_r^2 + B_i^2}{A_r^2 + A_i^2} \tag{13-5}$$

Equation (13-4) results from the phase condition and determines the frequency of oscillation. The inequality in Eq. (13-5) is the consequence of the magnitude constraint and defines the necessary condition for sustained oscillation. Equation (13-5) is evaluated at the oscillation frequency determined from Eq. (13-4).

A large number of single-stage oscillators have been developed in both vacuum-tube and transistor versions. The transistor circuits followed by direct analogy from the earlier vacuum-tube circuits. In the following examples, transistor versions are illustrated, but the y-parameter equations apply to other devices as well.

35. LC Oscillators.[4-6] The *Hartley oscillator* circuit is one of the oldest forms; the transistor version is shown in Fig. 13-36. With the collector and base at opposite ends of the tuned

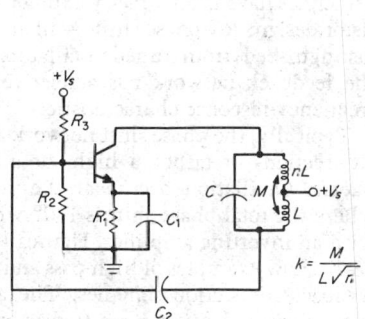

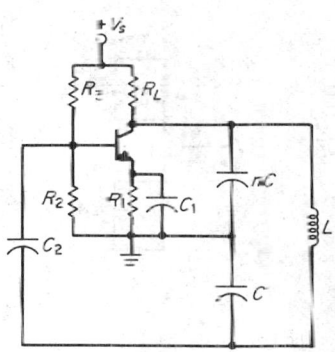

Frequency of oscillation:

$$\omega^2 = \frac{1}{LC\,(1+2k\sqrt{n}+n)+nL^2\,(1-k^2)(y_{11A}\,y_{22A})}$$

Condition for oscillation:

$$y_{21A} \geq \frac{y_{11A}+ny_{22A}+n\omega^2\,LC\,(1-k^2)(y_{11A}\,y_{22A})}{k\,\sqrt{n}+n\omega^2 LC\,(1-r^2)}$$

Frequency of oscillation:

$$\omega^2 = \frac{1}{LC}\left(1+\frac{1}{n}\right)+\frac{1}{n^2C^2}\,(y_{11A}\,y_{22A})$$

Condition for oscillation:

$$y_{21A} \geq \omega^2\,LC\left(ny_{11A}+y_{22A}\right)-(y_{11A}+y_{22A})$$

Fig. 13-36. Hartley oscillator circuit.

Fig. 13-37. Colpitts oscillator circuit.

circuit, the 180° phase relation is secured, and feedback occurs through mutual inductance between the two parts of the coil. In the figure, the frequency and condition for oscillation are expressed in terms of the transistor y parameters and feedback inductance L, inductor coupling coefficient k, inductance ratio n, and tuning capacitance C. The admittance parameters of the bias network R_1, R_2, and R_3, as well as the reactance of bypass capacitor C and coupling capacitor C_2, have been neglected. These admittances could be included in the amplifier y parameters in cases where their effect is not negligible.

If

$$\frac{C}{L} \gg \frac{n(1-k^2)(y_{11A}y_{22A})}{1+2k\sqrt{n}+n} \tag{13-6}$$

the frequency of oscillation will be essentially independent of transistor parameters.

The transistor version of the *Colpitts oscillator* is shown in Fig. 13-37. Capacitors C and nC in the combination with the inductance L determine the resonant frequency of the circuit. A fraction of the current flowing in the tank circuit is regeneratively fed back to the base through the coupling capacitor C_2. Bias resistors R_1, R_2, R_3, and R_L, as well as capacitors C_1 and C_2, are chosen so as not to affect the frequency or conditions for oscillation. Alternatively, the bias element admittances may be included in the amplifier y parameters.

In the Colpitts circuit, if the ratio of C/L is chosen so that

$$\frac{C}{L} \gg \frac{y_{11A}y_{22A}}{1+n} \tag{13-7}$$

the frequency of oscillation is essentially determined by the tuned-circuit parameters.

Another oscillator configuration useful in the audio-frequency range is the tuned-collector circuit shown in Fig. 13-38. Here regenerative feedback is furnished via the transformer turns ratio n from the collector to base. If the ratio of C/L is such that

$$\frac{C}{L} \gg N^2 y_{11A} y_{22A}(1 - k^2) \tag{13-8}$$

the frequency of oscillation is specified by $\omega^2 = 1/LC$. This circuit can be tuned over a wide range by varying the capacitor C and is compatible with simple biasing techniques.

36. RC Oscillators.[4,5] Audio sinusoidal oscillators can be designed using an RC ladder network (of three or more sections) as a feedback path in an amplifier. This scheme originally appeared in vacuum-tube circuits, but the principles have been directly extended to transistor design. RC phase-shift oscillators can be distinguished from tuned oscillators in that the feedback network has a relatively broad frequency-response characteristic.

Typically, the phase shift network has three RC sections of either a high- or a low-pass nature. Oscillation occurs at the frequency where the total phase shift is 180° when used with an inverting amplifier. Figures 13-39 and 13-40 show examples of high-pass and low-pass feedback-connection schemes. The amplifier is a differential pair with a transistor current source, a configuration which is common in integrated-circuit amplifiers. The output is obtained at the opposite collector from the feedback connection, since this minimizes external loading on the phase shift network. The conditions for, and the frequency of, oscillation are derived, assuming that the input resistance of the amplifier, which loads the phase shift network, has been adjusted to equal the resistance R. The load resistor R_L is considered to be part of the amplifier output resistance, and it is included in y_{22A}.

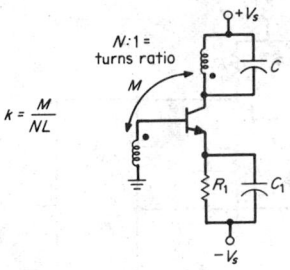

Frequency of oscillation:

$$\omega^2 = \frac{1}{LC + N^2 L^2 \, y_{11A} \, y_{22A} \, (1 - k^2)}$$

Condition for oscillation:

$$y_{21A} \geq \frac{1}{Nk} (N^2 Y_{11A} + Y_{22A}) - \frac{\omega^2 NLCY_{11A}}{k} (1 - k^2)$$

Fig. 13-38. Tuned-collector oscillator.

37. Null-Network Oscillators.[4,5,7] In almost all respects null-network oscillators are superior to the RC phase-shift circuits described in Par. **13-36**. While many null-network configurations are useful (including the bridged-T and twin-T), the Wien bridge design predominates.

The general form for the Wien bridge oscillator is shown in Fig. 13-41. In the figure, an ideal differential voltage amplifier is assumed, i.e., one with infinite input impedance and zero output impedance. An integrated-circuit operational amplifier that has a differential input stage is a practical approximation to this type of amplifier and is often used in bridge-oscillator designs.

The Wien bridge is used as the feedback network, with positive feedback provided through

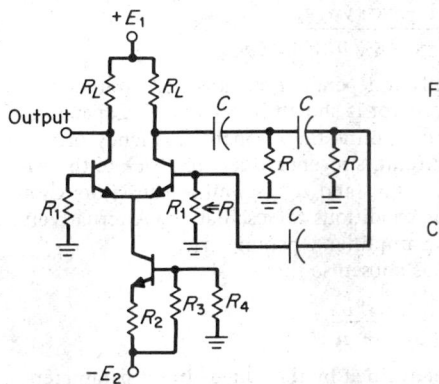

Frequency of oscillation:

$$\omega^2 = \frac{y_{22A}}{2C^2 R(2 + 3 Ry_{22A})}$$

Condition for oscillation:

$$y_{21A} \geq \frac{1}{R} \left(\frac{1 + 5 R/R_L}{\omega^2 R^2 C^2} - \frac{R}{R_L} - 3 \right)$$

Fig. 13-39. RC oscillator with high-pass feedback network.

the RC branches for regeneration and negative feedback through the resistor divider. Usually the resistor-divider network includes an amplitude-sensitive device in one or both arms which provides automatic correction for variation of the amplifier gain. Circuit elements such as a tungsten lamp, thermistor, and field-effect transistor used as the voltage-sensitive resistance element maintain a constant output level with a high degree of stability. Amplitude variations of less than $\pm 1\%$ over the band from 10 to 100,000 Hz are realizable.[8] In addition, since the amplifier is never driven into the nonlinear region, harmonic distortion in the output waveform is minimized. For the connection shown in Fig. 13-41, an increase in V will cause a decrease in R_2, restoring V to the original level.

The lamp or thermistor have thermal time constants that set at a lower frequency limit on this method of amplitude control. When the period is comparable with the thermal time con-

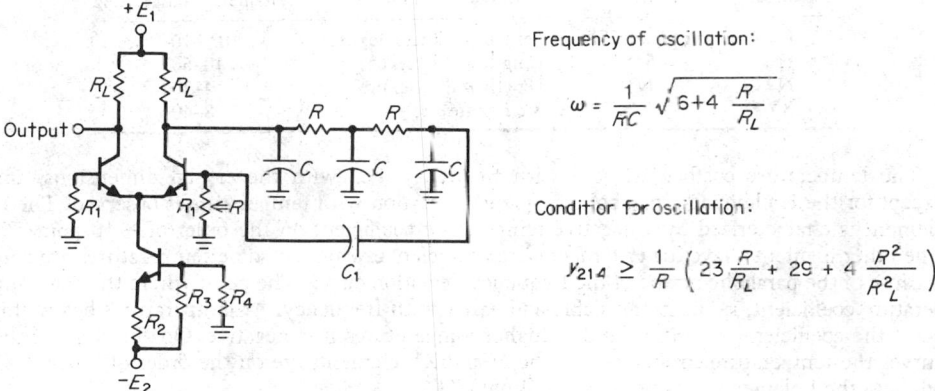

Frequency of oscillation:

$$\omega = \frac{1}{RC} \sqrt{6 + 4\frac{R}{R_L}}$$

Condition for oscillation:

$$y_{214} \geq \frac{1}{R}\left(23\frac{P}{R_L} + 29 + 4\frac{R^2}{R_L^2}\right)$$

Fig. 13-40. RC oscillator with low-pass feedback network.

stant, the change in resistance over an individual cycle distorts the output waveform. There is an additional degree of freedom with the field-effect transistor, since the control voltage must be derived by a separate detector from the amplifier output. The time constant of the detector, and hence the resistor, are set by a capacitor, which can be chosen commensurate with the lowest oscillation frequency desired.

At ω_0 the positive feedback predominates, but at harmonics of ω_0 the net negative feedback reduces the distortion components. Typically, the output waveform exhibits less than 1% total harmonic distortion. Distortion components well below 0.1% in the mid-audio-frequency range are also achieved.[8]

Unlike LC oscillators, in which the frequency is inversely proportional to the square root of L and C, in the Wien bridge ω_0 varies as $1/RC$. Thus, a tuning range in excess of 10:1 is easily achieved. Continuous tuning within one decade is usually accomplished by varying both capacitors in the reactive feedback branch. Decade changes are normally accomplished by switching both resistors in the resistive arm. Component tracking problems are eased when the resistors and capacitors are chosen to be equal.

Almost any three-terminal null network can be used for the reactive branch in the bridge; the resistor divider network adjusts the degree of unbalance in the manner described. Many of these networks lack the simplicity of the Wien bridge since they may require the tracking of three components for frequency tuning. For this reason networks such as the bridged-T and twin-T are usually restricted to fixed-tuned applications.

38. Low-Frequency Crystal Oscillators.[9-13] Quartz-crystal resonators are

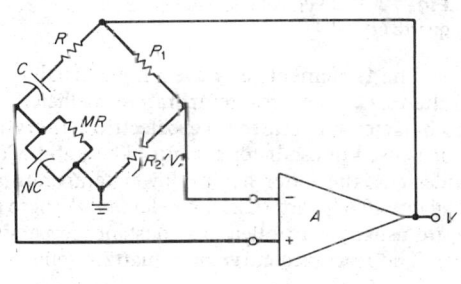

Frequency of oscillation $(M = N = 1)$:

$$\omega_0 = \frac{1}{RC}$$

Condition for oscillation:

$$A \geq \delta = \frac{3(R_1 - R_2)}{R_1 - 2R_2}$$

Fig. 13-41. Wien bridge oscillator circuit.

used where frequency stability is a primary concern. The frequency variations with both time and temperature are several orders of magnitude lower than obtainable in LC or RC oscillator circuits. The very high stiffness and elasticity of piezoelectric quartz make it possible to produce resonators extending from approximately 1 kHz to 200 MHz. The performance characteristics of crystal depend on both the particular cut and the mode of vibration (see Sec. 7). For convenience, each "cut-mode" combination is considered as a separate piezoelectric element, and the more commonly used elements have been designated with letter symbols. The audio-frequency range (above 1 kHz) is covered by elements J, H, N, and XY, as shown in Table 13-2.

TABLE 13-2   Low-Frequency Crystal Elements

Symbol	Cut	Mode of vibration	Frequency range, kHz
J	Duplex 5°X	Length-thickness flexure	0.9–10
H	5°X	Length-width flexure	10–50
N	NT	Length-width flexure	4–200
XY	XY	XY flexure	8–40

The temperature coefficients vary with frequency, i.e., with the crystal dimensions, and except for the H element, a parabolic frequency variation with temperature is observed. The H element is characterized by a negative temperature coefficient on the order of -10 ppm/°C. The other elements have lower temperature coefficients, which at some temperatures are zero because of the parabolic nature of the frequency-deviation curve. The point where the zero temperature coefficient occurs is adjustable and varies with frequency. At temperatures below this point the coefficient is positive, and at higher temperatures it is negative. On the slope of the curves the temperature coefficients for the N and XY elements are on the order of 2 ppm/°C, whereas the J element is about double at 4 ppm/°C.

Although the various elements differ in both cut and mode of vibration, the electric equivalent circuit remains invariant. The schematic representation and the lumped constant equivalent circuit are shown in Fig. 13-42. As is characteristic of most mechanical resonators, the motional inductance L resulting from the mechanical mass in motion is large relative to that obtainable from coils. The extreme stiffness of quartz makes for very small values of the motional capacitance C, and the very high order of elasticity allows the motional resistance R to be relatively low. The shunt capacitance C_0 is the electrostatic capacitance existing between crystal electrodes with the quartz plate as the dielectric and is present whether or not the crystal is in mechanical motion. Some typical values for these equivalent-circuit parameters are shown in Table 13-3.

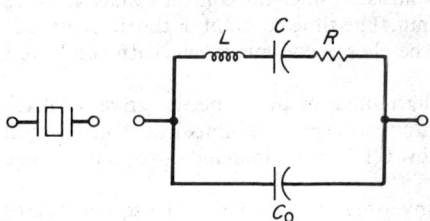

Fig. 13-42. Symbol and equivalent circuit of a quartz crystal.

The H element can have a high Q value when mounted in a vacuum enclosure; however, it then has the poorest temperature coefficient. The N element exhibits an excellent temperature characteristic, but the piezoelectric activity is rather low, so that special care is required when it is used in oscillator circuits. The J and XY elements operate well in low-frequency oscillator designs, the latter having lower temperature drift. For the same frequency the XY crystal is about 40% longer than the J element. Where extreme frequency stability is required, the crystals are usually controlled to a constant temperature

The reactance curve of a quartz resonator is shown in Fig. 13-43. The zero occurs at the fre-

TABLE 13-3   Typical Crystal Parameter Values

Element	Frequency, kHz	L, H	C, pF	R, kΩ	C_0, pF	Q, approx
J	10	8,000	0.03	50	6	20,000
H	10	2,500	0.1	10	75	20,000
N	10	8,000	0.03	75	30	10,000
XY	10	12,000	0.02	30	20	30,000

quency f_s, which corresponds to series resonance of the mechanical L and C equivalences. The antiresonant frequency f_p is dependent on the interelectrode capacitance C_0. Between f_s and f_p the crystal is inductive, and this frequency range is normally referred to as the *crystal bandwidth*

$$BW = f_s/(2C_0/C) \tag{13-9}$$

In oscillator circuits the crystal can be used as either a series or a parallel resonator. At series resonance the crystal impedance is purely resistive, but in the parallel mode the crystal is operated between f_s and f_p and is therefore inductive. For oscillator applications the circuit capacitance shunting the crystal must also be included when specifying the crystal since it is part of the resonant circuit. If a capacitor C_L, that is, a negative reactance, is placed in series with a crystal, the combination will series-resonate at the frequency f_R of zero reactance for the combination.

$$f_R = f_s \left[1 - \frac{1}{(2C_0/C)(1 + C_L/C_0)}\right] \tag{13-10}$$

The operating frequency can vary in value due to changes in the load capacitance, and this variation is prescribed by

$$\Delta f_R = \frac{f_s}{2C_0/C} \frac{\Delta C_L/C_0}{(1 + C_L/C_0)^2} \tag{13-11}$$

This effect can be used to "pull" the crystal for initial alignment, or if the external capacitor is a voltage-controllable device, a VCO with a range of about $\pm 0.01\%$ can be constructed. Phase changes in the amplifier will also give rise to frequency shifts since the total phase around the loop must remain at $0°$ to maintain oscillation.

Although single-stage transistor designs are possible, more flexibility is available in the circuit of Fig. 13-44, which uses an integrated-circuit operational amplifier for the gain element. The crystal is operated in the series mode, and the amplifier gain is precisely controlled by the negative-feedback divider R_2 and R_3. The output will be sinusoidal if

$$\frac{V_D R_1}{R_1 + R} \left(1 + \frac{R_3}{R_2}\right) < V_{\text{lim}} \tag{13-12}$$

where V_D = limiting diode forward voltage drop and V_{lim} = limiting level of amplifier output.

Low-cost electronic wristwatches use quartz crystals for establishing a high degree of timekeeping accuracy. A high-quality mechanical watch may have a yearly accuracy on the order of 20 min, whereas many quartz watches are guaranteed to vary only 1 min/year.[15]

Generally the XY crystal is used, but other types are continually being developed to improve

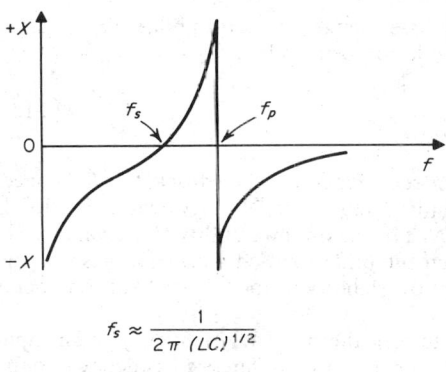

$$f_s \approx \frac{1}{2\pi (LC)^{1/2}}$$

$$f_p \approx \frac{1}{2\pi \left(\dfrac{LC}{1 + C/C_0}\right)}$$

Fig. 13-43. Quartz-crystal reactance curve.

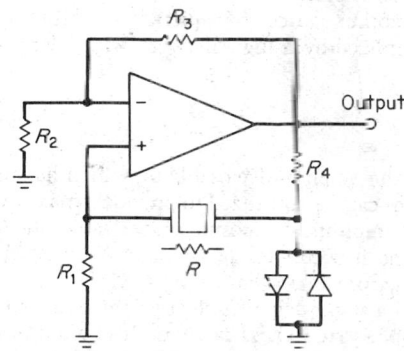

Condition for oscillation:

$$\left(\frac{R}{R_1 + R_4 + R}\right)\left(1 - \frac{R_3}{R_2}\right) \geq 1$$

Fig. 13-44. Crystal oscillator using an integrated-circuit operational amplifier.

accuracy, reduce size, and lower manufacturing cost. The active gain elements for the oscillator are part of the integrated circuit which contains the electronics for the watch functions. The flexure or tuning-fork frequency is set generally to 32,768 Hz, which is 2^{15} Hz. This frequency reference is divided down on the integrated circuit to provide seconds, minutes, hours, day of the week, date, month, etc.

A logic gate or inverter is often used as the gain element in the oscillator circuit.[17] A typical configuration is shown in Fig. 13-45. The resistor R_1 is used to bias the logic inverter for class A amplifier operation. The resistor R_2 helps reduce both voltage sensitivity of the network and crystal power dissipation. The combination of R_2 and C_2 provides added phase shift for good oscillator start-up. The series combination of capacitors C_1 and C_2 provides the parallel load for the crystal. C_1 can be made tunable for precise setting of the crystal oscillation frequency. The inverter provides the necessary gain and 180° phase shift. The π network consisting of the capacitors and the crystal provides the additional 180° phase shift needed to satisfy the conditions for oscillation.

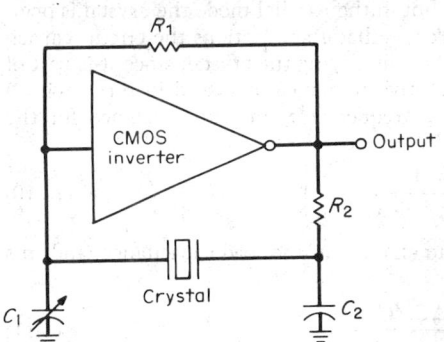

Fig. 13-45. Crystal oscillator using a logic gate for the gain element.

39. Frequency Stability. Many factors contribute to the ability of an oscillator to hold a constant output frequency over a period of time and range from short-term effects, caused by random noise, to longer-term variations, caused by circuit parameter dependence on temperature, bias voltage, and the like. In addition to the temperature and aging effects of the frequency-determining elements, nonlinearities, impedance loading, and amplifier phase variations also contribute to instability.

Harmonics generated by circuit nonlinearities are passed through the feedback network, with various phase shifts, to the input of the amplifier.[14] Intermodulation of the harmonic frequencies produces a fundamental frequency component that differs in phase from the amplifier output. Since the condition $A\beta = 1$ must be satisfied, the frequency of oscillation will shift so that the network phase shift cancels the phase perturbation caused by the nonlinearity. Therefore, the frequency of oscillation is influenced by an unpredictable amplifier characteristic, namely, the saturation nonlinearity. This effect is negligible in the Wien bridge oscillator, where automatic level control keeps harmonic distortion to a minimum.

The relationships shown in Fig. 13-41 were derived assuming that the amplifier does not load the bridge circuit on either the input or output sides. In the practical sense this is never true, and changes in the input and output impedances will load the bridge and cause frequency variations to occur.

Another source of frequency instability is small phase changes in the amplifier. The effect is minimized by using a network with a large stability factor, defined by

$$S = \frac{d\phi}{d\omega/\omega_0}\bigg|_{\omega=\omega_0} \tag{13-13}$$

For the Wien bridge oscillator, which has amplitude-sensitive resistive feedback, the RC impedances can be optimized to provide a maximum stability factor value.[15,16] As shown in Fig. 13-41, this amounts to choosing proper values for M and N. The maximum stability-factor value is $A/4$, and it occurs for $N = \frac{1}{2}$ and $M = 2$. Most often the bridge is used with equal resistor and capacitor values; that is, $M = N = 1$, in which case the stability factor is $2A/9$. This represents only a slight degradation from the optimum.

40. Synchronization. It is often desirable to lock the oscillator frequency to an input reference. Usually this is done by injecting sufficient energy at the reference frequency into the oscillator circuit. When the oscillator is tuned sufficiently close to the reference, natural oscillations cease and the synchronization signal is amplified to the output. Thus the circuit appears to oscillate at the injected signal frequency. The injected reference is amplitude-stabilized by the AGC or limiting circuit in the same manner as the natural oscillation. The frequency range over which locking can occur is a linear function of the amplitude of the injected signal. Thus, as the synchronization frequency is moved away from the natural oscillator frequency, the amplitude threshold to maintain lock increases. The phase error between the input reference and the oscillator output will also deviate as the input frequency varies from the natural frequency.

Methods for injecting the lock signal vary and depend on the type of oscillator under consideration. For example, LC oscillators may have signals coupled directly to the tank circuit, whereas the lock signal for the Wien network is usually coupled into the center of the resistive side of the bridge, i.e., the junction of R_1 and R_2 in Fig. 13-41.

If the natural frequency of oscillation can be voltage-controlled, synchronization can be accomplished with a phase-locked loop.[17] Replacing both R's with field-effect transistors, or alternatively shunting both C's with varicaps, provides an effective means for voltage controlling the frequency of the Wien bridge oscillator. Although more complicated in structure, the phase-locked loop is more versatile and has many diverse applications (see Pars. **13-14** and **13-17**).

41. Piezoelectric Annunciators. Another important class of audio oscillators uses piezoelectric elements for both frequency control and audible-sound generation. Because of their low cost and high efficiency these devices are finding increasing use in smoke detectors, burglar alarms, and other warning devices. Annunciators using these elements typically produce a sound level in excess of 85 dB measured at a distance of 10 ft.

Usually the element consists of a thin brass disk to which a piezoceramic material has been attached. When an electric signal is applied across its surfaces, the piezoceramic disk attempts to change diameter. The brass disk to which it is bonded acts as a spring restraining force on one surface of the ceramic. The brass plate serves as one electrode for applying the electric signal to the ceramic. On the other surface a fired-on silver paste is used as an electrode The restraining action of the brass disk causes the assembly to change from a flat to a convex shape. When the polarity of the electric signal reverses, the assembly flexes in the other direction to a concave shape. When the device is properly mounted in a suitable horn structure, this motion is used to produce high-level sound waves. One useful method is to clamp the disk at nodal points, i.e., at a distance from the center of the disk where mechanical motion is at a vibrational null.

The piezoelectric assembly will produce sound levels more efficiently when excited near the series-resonant frequency. The simple equivalent circuit used for the quartz crystal (Fig. 13-42) also applies to the piezoceramic assembly for frequencies near resonance. Generally the piezoelectric element is used as the frequency-determining element in an audio oscillator. The advantage of this method is that the excitation frequency is inherently near the optimum value, since it is self-excited. A typical piezoceramic 1 in in diameter mounted on a 1¾-in brass disk would have the following equivalent values: $C_0 = .02\ \mu F$, $C = 0.0015\ \mu F$, $L = 2\ H$, $R = 500\ \Omega$, $Q = 75$, $f_s = 2.9\ kHz$, and $f_p = 3.0\ kHz$.

A basic oscillator, capable of producing high-level sound, is shown in Fig. 13-46. The inductor L_1 provides a dc path to the transistor and broadly tunes the parallel input capacitance of the piezoelectric element. C_1 is an optional capacitor which adds to the input shunt capacitance for optimizing the drive impedance of the element. Resistor R_1 provides base-current bias to the transistor so that oscillations can start.

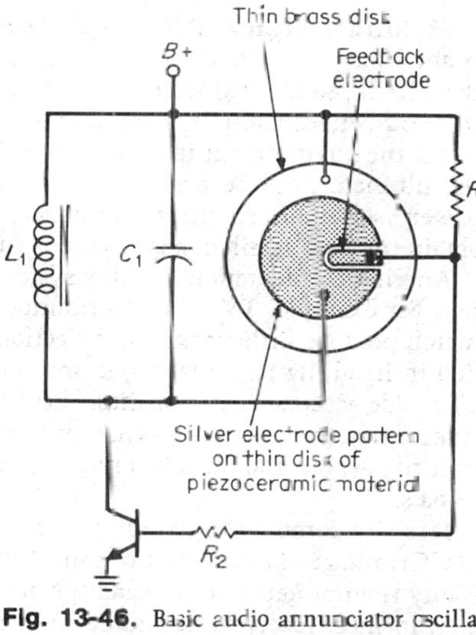

The element has a third small electrode etched in the silver pattern. It is used to derive a feedback signal which, when resistively loaded by R_1, pro-

Fig. 13-46. Basic audio annunciator oscillator circuit using a thin-disk piezoelectric transducer.

vides an in-phase signal to the base for sustaining circuit oscillation. The circuit operates like a blocking oscillator in that the transistor is switched on and off and the collector voltage can fly above B-plus because of the inductor L_1.

The collector load consisting of L_1 and C_1 can be replaced with a resistor, in which case the audio output will be less.

42. References on Audio Oscillators

1. IEEE Standard on Definitions of Terms for Audio and Electroacoustics, *IEEE Trans. Audio Electroacoust.*, June 1966, Vol. AU-14.

2. Bode, H. W. "Network Analysis and Feedback Amplifier Design," Van Nostrand, New York, 1945.

3. Cunningham, W. J.   "Nonlinear Analysis," McGraw-Hill, New York, 1958.

4. Hakim, S. S.   "Juncion Transistor Circuit Analysis," Wiley, New York, 1962.

5. Reich, H. J.   "Functional Circuits and Oscillators," Van Nostrand, New York, 1961.

6. Millman, J.   "Vacuum Tube and Semiconductor Electronics," McGraw-Hill, New York, 1958.

7. Strauss, L.   "Wave Generation and Shaping," 2d ed., McGraw-Hill, New York, 1970.

8. Owen, R. E.   Solid State *RC* Oscillator Design for Audio Use, *J. Audio Eng. Soc.*, January 1966, Vol. 14, No. 1.

9. Silver, J. F.   "Design Notes for Quartz Crystals in Oscillators and Filter Applications," CTS Knight Co., Sandwich, Ill., 1962.

10. Buchanan, J. P.   "Handbook of Piezoelectric Crystals for Radio Equipment Designers," *WADC Tech. Rep.* 54-248, Wright Air Development Center, 1954.

11. Firth, D.   "Quartz Crystal Oscillator Circuits Design Handbook," Magnavox Co., Fort Wayne, Ind., 1965.

12. Von Willisen, F. K.   The Quartz Watch: Its Life and Times, *IEEE Spectrum*, June 1979, pp. 18–23.

13. Frerking, M. E.   "Crystal Oscillator Design and Temperature Compensation," Van Nostrand Reinhold, New York, 1978.

14. Groszkowski, J.   The Interdependence of Frequency Variation and Harmonic Content and the Problem of Constant Frequency Oscillators, *Proc. IRE*, July 1933, Vol. 21, pp. 958–981.

15. Stevens, B. L., and R. P. Manning   Improvements in the Theory and Design of *RC* Oscillators, *IEEE Trans. Circuit Theory*, November 1971, Vol. CT-18, No. 6, pp. 636–643.

16. Mehta, V. B.   Comparison of *RC* Networks for Frequency Stability in Oscillators, *Proc. IEEE*, February 1965, Vol. 112, pp. 296–300.

17. Gardner, F. M.   "Phaselock Techniques," Wiley, New York, 1967.

Radio-Frequency Amplifiers

BY G. B. HARROLD

43. Small-Signal RF Amplifiers. The prime considerations in the design of first-stage rf amplifiers are gain and noise figure. As a rule, the gain of the first rf stage should be greater than 10 dB, so that subsequent stages contribute little to the overall amplifier noise figure. The trade-off between amplifier cost and noise figure is an important design consideration. For example, if the environment in which the rf amplifier operates is noisy, it is uneconomic to demand the ultimate in noise performance. Conversely, where a direct trade-off exists in transmitter power vs. amplifier noise performance, as it does in many space applications, money spent to obtain the best possible noise figure is fully justified.

Another consideration in many systems is the input-output impedance match of the rf amplifier. For example, TV cable distribution systems require an amplifier whose input and output match produce little or no line reflections. The performance of many rf amplifiers is also specified in handling large signals, to minimize cross- and intermodulation products in the output. The wide acceptance of transistors has placed an additional constraint on first-stage rf amplifiers, since many rf transistors having low noise, high gain, and high frequency response are susceptible to burnout and must be protected to prevent destruction in the presence of high-level input signals.

It is also common to require that first rf stages be gain-controlled by automatic gain control (AGC) voltage. The amount of gain control and the linearity of control are system parameters. Many rf amplifiers have the additional requirement that they be tuned over a range of frequencies. In most receivers, regardless of configuration, local-oscillator leakage back to the input is strictly controlled by government regulation. Finally, the rf amplifier must be stable under all conditions of operation.

44. Device Evaluation for RF Amplifiers. An important consideration in an rf amplifier is the choice of active device. This information on device parameters can often be found in published data sheets. If parameter data are not available or not a suitable operating point, the following characterization techniques can be used.

*Rx Meter.** This measurement technique is usually employed at frequencies below 200 MHz for active devices that have high input and output impedances. The technique[1,2]† is summarized in Fig. 13-47 with assumptions tacit in these measurements. The biasing techniques are

---

*Trademark of the Hewlett Packard Co., Palo Alto, Calif.

†Superior numbers correspond to numbered references, Par. **13-57**.

shown. In particular, the measurement of h_{22b} requires a very large resistor R_e to be inserted in the emitter, and this may cause difficulty in achieving the proper biasing. Care should be taken to prevent burnout of the bridge when a large dc bias is applied. The bridge's drive to the active device may be reduced for more accurate measurement by varying the B-plus voltage applied to the internal oscillator.

*Vector Voltmeter.** This characterization technique measures the S parameters defined in Par. **13-9** and Fig. 13-10. The measurement consists in inserting the device in a transmission

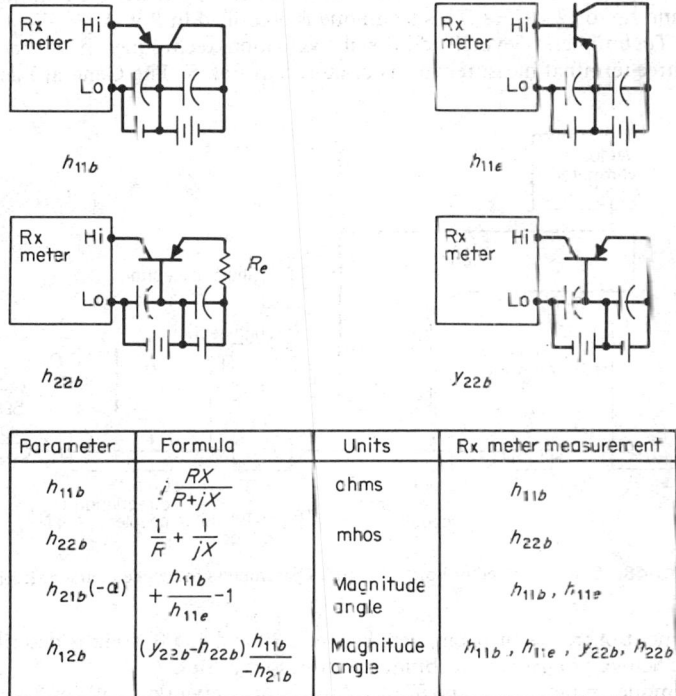

Parameter	Formula	Units	Rx meter measurement
h_{11b}	$j\dfrac{RX}{R+jX}$	ohms	h_{11b}
h_{22b}	$\dfrac{1}{R}+\dfrac{1}{jX}$	mhos	h_{22b}
$h_{21b}(-\alpha)$	$+\dfrac{h_{11b}}{h_{11e}}-1$	Magnitude angle	$h_{11b},\ h_{11e}$
h_{12b}	$(y_{22b}-h_{22b})\dfrac{h_{11b}}{-h_{21b}}$	Magnitude angle	$h_{11b},\ h_{11e},\ y_{22b},\ h_{22b}$

Assumes: Determinate of $|h| \ll h_{21}$ and $h_{12} \ll 1$

R and X are Rx meter's reading of parallel resistance and reactance

Fig. 13-47. Use of the Rx meter in device characterization.

line, usually 50 Ω characteristic impedance, and measuring the incident and reflected voltages at the two ports of the device. The setup is shown in Fig. 13-48 and discussed in Ref. 3.

This approach is used between frequencies of 100 and 1,000 MHz, on devices whose impedance levels are compatible with 50 Ω. The vector voltmeter measures the voltages present at A and B and gives the phase of voltage B referenced to that of voltage A. The initial calibration is performed by placing a short-circuited section in place of the transistor jig, the probe B at B_1, and adjusting the line stretcher to give zero-degree phase shift between inputs A and B. This calibration can usually be held across a reasonable frequency band. The short-circuited section is then replaced by the transistor in its jig, and bias is applied. S_{11} is then found by taking the ratio of the vector voltages V_A and V_{B_1}.

$$S_{11} = \frac{V_{B_1}}{V_A} \bigg/ \arg V_{B_1} - \arg V_A$$

The calculation is simplified if V_A is taken as one unit of voltage, with argument zero. Care must be taken not to drive the transistor into saturation, or the measurement of the small-signal parameters will not be valid.

Next, with a through-section replacing the transistor jig and the B probe in the B_1 position, the phase difference between V_A and V_B is again adjusted to zero by means of the line stretcher.

*Trademark of the Hewlett Packard Co., Palo Alto, Calif.

By replacing the through-section with the transistor in its jig and applying bias, S_{21} can be measured as the ratio between the vector voltages V_A and V_B

$$S_{21} = \frac{V_{B_2}}{V_A} \bigg/ \text{arg } V_{B_2} - \text{arg } V_A$$

The measurements of S_{12} and S_{22} are achieved by using the above calibration technique but reversing the transistor jig.

*Use of 8743 Reflectometer.** This modification of the S-parameter technique can be used above 1 GHz and up to 12.4 GHz. This technique is described in Ref. 4.

Additional Techniques. Several additional evaluation techniques have been used. The Wayne-Kerr three-terminal measurement is discussed in Ref. 5. The General Radio Bridge GR

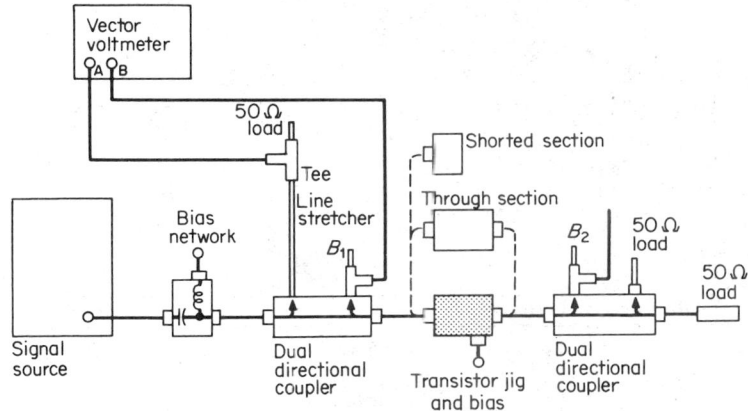

Fig. 13-48. Use of the vector voltmeter and S parameters in device characterization.

1607 measurement of transistor parameters between 30 and 1,000 MHz is described in Ref. 6, and the Rhode-Schwartz Diagraph technique is discussed in Ref. 7.

A new technique employs the type 8750A Automatic Network Analyzer,* which has computer control to measure device parameters automatically from 100 MHz to 12 GHz, with CRT display and teletypewriter printout of the desired parameters.

45. Noise in RF Amplifiers. A common technique employing a noise source to measure the noise performance of an rf amplifier is shown in Fig. 13-49. Initially the external noise source (a temperature-limited diode) is turned off, the 3-dB pad short-circuited, and the reading on the output power meter recorded. The 3-dB pad is then inserted, the noise source is turned on, and its output increased until a reading equal to the previous one is obtained. The noise figure can then be read directly from the noise source, or calculated from 1 plus the added noise per unit bandwidth divided by the standard noise power available[8] KT_0, where $T_0 = 290$ K and $K =$ Boltzmann's constant $= 1.38 \times 10^{-23}$ J/K.

At higher frequencies, the use of a temperature-limited diode is not practical, and a gas-discharge tube or a hot-cold noise source is employed. The Y-factor technique of measurement is used.[9] The output from the device to be measured is put into a mixer, and the noise output converted to a 30- or 60-MHz center-frequency (i.f.) output. A precision attenuator is then inserted between this i.f. output and the power-measuring device. The attenuator is adjusted to give the same power reading for two different conditions of noise power output represented by effective temperatures T_1 and T_2. The Y factor is the difference in decibels between the two precision attenuator values needed to maintain the same power-meter reading.[8,9] The noise factor is

$$F = \frac{(T_2/290) - T_1 Y/290}{Y - 1} + 1$$

where $T_1 =$ effective temperature at reference condition 1, $T_2 =$ effective temperature of reference condition 2, and $Y =$ decibel reading defined in the text, converted to a numerical ratio.

*Manufactured by the Hewlett-Packard Co.

In applying this technique it is often necessary to correct for the second-stage noise. This is done by use of the cascade formula

$$F_1 = F_T - (F_2 - 1)/G_1$$

where F_1 = noise factor of first stage, F_T = overall noise factor measured, F_2 = noise factor of second-stage mixer and i.f. amplifier, and G_1 = available gain of first stage.

It is often necessary to determine the optimum source impedance and best noise figure for a particular device. Since many devices have essentially the same first-order noise representation,

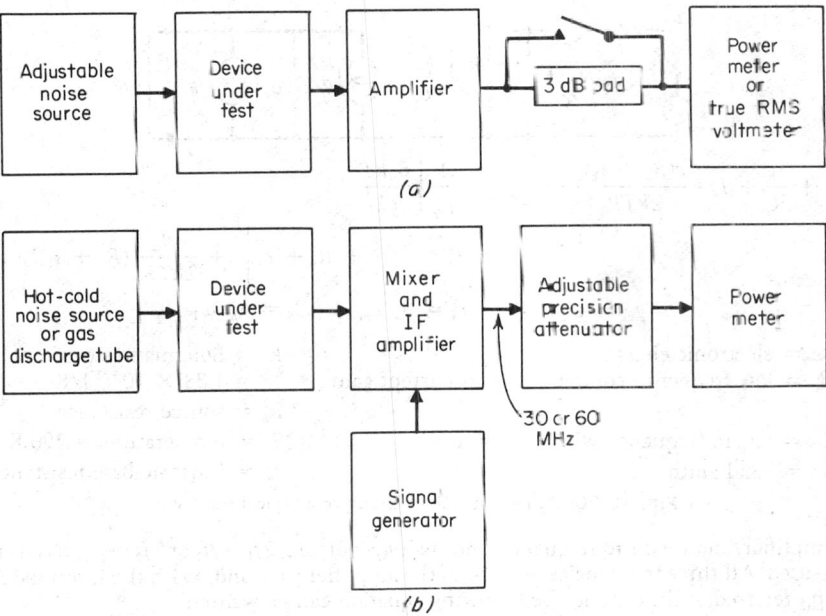

(a)

(b)

Fig. 13-49. Noise-measurement techniques: (a) at low frequencies; (b) at high frequencies.

a generalized analysis can be used. For example, an analysis based on Ref. 10 shows that, for a typical transistor, a first-order assessment of its noise performance, neglecting $1/f$ noise, can be made using the equivalent circuit of Fig. 13-50. The analysis technique is to determine noise-power output from all sources, internal as well as external, and then find the output noise power due to the external source. The noise figure is the ratio of the first divided by the second.

By applying the usual minimization procedures, the optimum source resistance $R_{s,opt}$ becomes

$$R_{s,opt} = \frac{1}{g_m} \sqrt{h_{FE}} \sqrt{1 + 2g_m r_b}$$

and the minimum noise figure $F_{0,min}$ becomes

$$F_{0,min} = 1 + \frac{1}{\sqrt{h_{FE}}} \sqrt{1 + 2g_m r_b}$$

where h_{FE} is the dc beta, and the other terms are as defined in Fig. 13-50. The above analysis is not valid at the high-frequency end of the transistor's performance, and the optimum source impedance for best noise performance is not necessarily that source impedance for optimum power gain.

The typical noise-figure-frequency characteristic of an active device is relatively flat in the middle of its frequency range, and the optimum source impedance in this region can be varied within limits without seriously affecting the noise performance.

The question of input match is determined by the various required performance characteristics. If best noise performance is required, the input matching network must transform the source impedance to the optimum-noise source impedance for the active device. On the other hand, the matching network may be designed for optimum power gain or to present an input impedance that is constant across a wide band of frequencies required by coaxial signal distribution systems.[11]

46. Large-Signal Performance of RF Amplifiers. The large-signal performance of an rf amplifier can be specified in many ways. A common technique is to specify the input where the departure from a straight-line input-output characteristic is 1 dB. This point is commonly called the *1-dB compression point*. The greater the input before this compression point is reached the better the large-signal performance.

Another method of rating an rf amplifier is in terms of its third-order intermodulation performance. Here two different frequencies, f_1 and f_2, of equal powers, p_1 and p_2, are inserted into

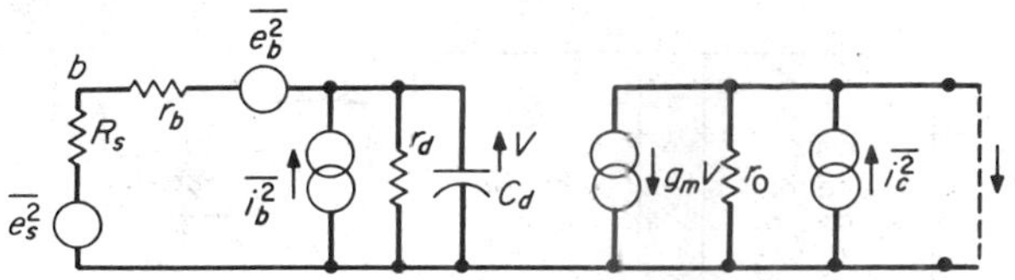

$$F = 1 + \frac{r_b}{R_s} + I_B \cdot \frac{e(R_s + r_b)^2}{2kTR_s} + \frac{eI_C}{2kTR_s} \cdot \frac{1}{\beta_o^2}\left[\frac{\beta_okT}{eI_C}\right.$$

$$\left. + R_s + r_b \right]^2 + \frac{eI_c}{2kTR_s}(R_s + r_b)^2 \cdot (\omega/\omega_T)^2$$

where $\qquad g_m = eI_C/kT \qquad r_d = \beta_o/g_m \qquad C_d = g_m/\omega_T$

and $\quad$ e = electronic charge $\qquad\qquad\qquad\qquad$ k = Boltzmann's constant
$\quad\beta_o$ = low-frequency common-emitter current gain $\qquad$ = 1.38×10^{-23} J/K
$\qquad\qquad\qquad\qquad\qquad\qquad\qquad\qquad\qquad\quad R_s$ = source resistance
$\quad\omega_T$ = radian frequency where $h_{fe} = 1$ $\qquad\qquad\quad$ T = temperature = 290 K
$\quad\Delta f$ = bandwidth $\qquad\qquad\qquad\qquad\qquad\qquad\qquad\quad r_b$ = intrinsic base resistance

Fig. 13-50. Noise equivalent circuit of a typical transistor.

the rf amplifier, and the third frequency, internally generated, $2f_1 - f_2$ or $2f_2 - f_1$, has its power p_{12} measured. All three frequencies must be in the amplifier passband. With the intermodulation power p_{12} referred to the output, the following equation can be written:

$$P_{12} = 2P_1 + P_2 + K_{12}$$

where P_{12} = intermodulation output power at $2f_1 - f_2$ or $2f_2 - f_1$, P_1 = output power at input frequency f_1, P_2 = output power at input frequency f_2, all in decibels referred to (0 dBm) and K_{12} = constant associated with the particular device.

The value of K_{12} in the above formula can be used to rate the performance of various device choices. Higher orders of intermodulation products can also be used.[12]

A third measure of large-signal performance commonly used is that of cross-modulation. In this instance, a carrier at f_D with no modulation is inserted into the amplifier. A receiver is then placed at the output and tuned to this unmodulated carrier. A second carrier at f_1 with amplitude-modulation index M_l is then added. The power P_l of f_l is increased, and its modulation is partially transferred to f_D. The equation becomes

$$10 \log (M_K/M_l) = P_l + K$$

where M_K = cross-modulation index of originally unmodulated signal at f_D, M_l = modulation index of signal F_l, P_l = output power of signal at f_l, all in decibels referred to 1 mW (dBm), and K = cross-modulation constant.

47. Maximum Input Power. In addition to the large-signal performance, the maximum power of voltage input into an rf amplifier is specified, with a requirement that device burnout must not occur at this input. There are two ways of specifying this input: by a stated pulse of energy or by a requirement to withstand a continuously applied large signal. It is also common to specify the time required to unblock the amplifier after removal of the large input. With the increased use of transistors (FETs especially) having good noise performance, these overload characteristics have become a severe problem. In many cases, conventional or zener diodes, in a back-to-back configuration shunting the input, are used to reduce the amount of power the input of the active devices must dissipate.

48. RF Amplifiers in Receivers. RF amplifiers intended for the first stages of receivers have additional restrictions placed upon them. In most cases, such amplifiers are tunable across a band of frequencies with one or more tuned circuits. The tuned circuits must track across the frequency band, and in the case of the superheterodyne, tracking of the local oscillator is necessary so that a constant frequency difference (i.f.) is maintained. The receiver's rf section can be tracked with the local oscillator by the two- or the three-point method, i.e., with zero error in the tracking at either two or three points. The design technique is discussed in Ref. 13.

A second consideration peculiar to rf amplifiers used for receivers is the automatic gain control (AGC). This requirement is often stated by specifying a low-level rf input to the receiver and

TABLE 13-4 Allowable Spurious Radiation from Receivers

Frequency of radiation, MHz	Signal strength at 100 ft or more from receiver, $\mu V/m$
0.45–25*†	
For TV receivers 0.45–25	<100
All other receivers:	
Up to 9	<100
9–10	100–1,000 linear increase
10–25	<1,000
Over 25–70†	<32
70–130	<50
130–174	<50–<150
174–260	<150
260–1,000	<150–<500 linear interpolation
470–1,000	<500

*Measurement in this frequency band of rf voltage is between each power line and ground for receivers so powered.
†Inclusive.

noting the power out. The rf signal input is then increased with the AGC applied until the output power has increased a predetermined amount. This becomes a measure of the AGC effectiveness. The AGC performance can also be measured by plotting a curve of rf input vs. AGC voltage needed to maintain constant output, compared with the desired performance.

A third consideration in superheterodynes is the leakage of the local oscillator in the receiver to the outside. This spurious radiation is carefully specified[14] by the Federal Communications Commission (FCC) in the United States, to be according to Table 13-4. Reference 14 states the various radiation levels for community antenna television systems and suggests techniques of measurement.

49. Design Using Immittance and Hybrid Parameters. The general gain and input-output impedance of an amplifier can be formulated, in terms of the Z or Y parameters, to be

$$Y_{in} = y_{11} - \frac{y_{12}y_{21}}{y_{22} + y_L} \qquad Y_{out} = y_{22} - \frac{y_{12}y_{21}}{y_{11} + y_s}$$

where y_L = load admittance, y_s = source admittance, Y_{in} = input admittance, Y_{out} = output admittance, G_T = transducer gain, and the transducer gain is

$$G_T = \frac{4 \operatorname{Re} y_s \operatorname{Re} y_L |y_{21}|^2}{|(y_{11} + y_s)(y_{22} + y_L) - y_{12}y_{21}|^2}$$

for the y parameters, and interchange of z for y is allowed.

The stability of the circuit can be determined by either Linvill's C or Stern's k factor as defined below. Using the y parameters, $y_{ij} = g_{ik} + jB_{ik}$, these are

Linvill:
$$C = \frac{|y_{12}y_{21}|}{2g_{11}g_{22} - \operatorname{Re} y_{12}y_{21}}$$

where $C < 1$ for stability does not include effects of load and source admittance.

Stern:
$$k = \frac{2(g_{11} + g_s)(g_{22} + g_L)}{|y_{12}y_{21}| + \operatorname{Re} y_{12}y_{21}}$$

where $k > 1$ for stability, g_L = load conductance, and g_s = source conductance.

The preceding C factor defines only unconditional stability; i.e., no combination of load and source impedance will give instability. Rollett[15] has shown that there is an invariant quantity K defined as

$$K = \frac{2 \, \mathrm{Re} \, \gamma_{11} \, \mathrm{Re} \, \gamma_{22} - \mathrm{Re} \, \gamma_{12}\gamma_{21}}{|\gamma_{21}\gamma_{12}|} \qquad \begin{array}{l} \mathrm{Re} \, \gamma_{11} > 0 \\ \mathrm{Re} \, \gamma_{22} > 0 \end{array}$$

where γ represents either the y, z, g, or h parameters, and $K > 1$ denotes stability.

This quantity K has then been used to define maximum available power gain G_{max} (only if $K > 1$)

$$G_{max} = |\gamma_{21}/\gamma_{12}|(K - \sqrt{K^2 - 1})$$

To obtain this gain, the source and load immittance are found to be $(K > 1)$

$$\gamma_s = \frac{\gamma_{12}\gamma_{21} + |\gamma_{12}\gamma_{21}|(K + \sqrt{K^2 - 1})}{2 \, \mathrm{Re} \, \gamma_{22}} - \gamma_{11} \qquad \gamma_s = \text{source immittance}$$

$$\gamma_L = \frac{\gamma_{12}\gamma_{21} + |\gamma_{12}\gamma_{21}|(K + \sqrt{K^2 - 1})}{2 \, \mathrm{Re} \, \gamma_{11}} - \gamma_{22} \qquad \gamma_L = \text{load immittance}$$

The procedure is to calculate the K factor, and if $K > 1$, calculate G_{max}, γ_s, and γ_L. If $K < 1$, the circuit can be modified either by use of feedback or by adding immittances to the input-output.

50. Design Using S Parameters. The advent of automatic test equipment and the extension of vacuum tubes and transistors to the gigahertz frequency range have led to design procedures using the S parameters. These parameters were described in Par. **13-44**. Following the previous discussion, the input and output reflection coefficient can be defined as

$$p_{in} = S_{11} + p_L \frac{S_{12}S_{21}}{1 - p_L S_{22}} \qquad p_L = \frac{Z_L - Z_0}{Z_L + Z_0}$$

$$p_{out} = S_{22} + p \frac{S_{12}S_{21}}{1 - pS_{11}} \qquad p_s = \frac{Z_s - Z_0}{Z_s + Z_0}$$

$$Z_0 = \text{characteristic impedance}$$

where p_{in} = input reflection coefficient and p_{out} = output reflection coefficient. The transducer gain can be written

$$G_{transducer} = \frac{|S_{21}|^2(1 - |p_s|^2)(1 - |p_L|^2)}{|(1 - S_{11}p_s)(1 - S_{22}p_L) - S_{21}S_{12}p_s p_L|^2}$$

The unconditional stability of the amplifier can be defined by requiring the input (output) impedance to have a positive real part for any load (source) impedance having a positive real part.[16]

This requirement gives the following criterion:

$$|S_{11}|^2 + |S_{12}S_{21}| < 1 \qquad |S_{22}|^2 + |S_{12}S_{11}| < 1$$

and

$$\eta = \frac{1 + |\Delta_s|^2 - |S_{11}|^2 - |S_{22}|^2}{2|S_{12}S_{21}|} > 1 \qquad \Delta_s = S_{11}S_{22} - S_{12}S_{21}$$

Similarly, the maximum transducer gain, for $\eta > 1$, becomes

$$G_{max \, transducer} = |S_{21}/S_{12}|(\eta \pm \sqrt{\eta^2 - 1})$$

(positive sign when $|S_{22}|^2 - |S_{11}|^2 - 1 + |\Delta_s|^2 > 0$) for conditions listed above.

The source and load to provide conjugate match to the amplifier when $\eta > 1$ are the solutions of the following equations, which give $|p_s|$, and $|p_L|$ less than 1

$$p_{ms} = C_1^* \frac{B_1 \pm \sqrt{B_1^2 - 4|C_1|^2}}{2|C_1|^2} \qquad p_{mL} = C_2^* \frac{B_2 \pm \sqrt{B_2^2 - 4|C_2|^2}}{2|C_2|^2}$$

where

$$B_1 = 1 + |S_{11}|^2 - |S_{22}|^2 - |\Delta_s|^2 \qquad B_2 = 1 + |S_{22}|^2 - |S_{11}|^2 - |\Delta_s|^2$$
$$C_1 = S_{11} - \Delta_s S_{22}^* \qquad C_2 = S_{22} - \Delta_s S_{11}^*$$

the star (*) denoting conjugate.

If $|\eta| > 1$ but η is negative or $|\eta| < 1$, it is not possible to match simultaneously the two-port with real source and load admittances. Graphical techniques have been found to allow design of rf amplifiers.[17,18] In addition, numerous computer programs are available to aid in the design.

51. Intermediate-Frequency Amplifiers. Intermediate-frequency amplifiers consist of a cascade of a number of stages whose frequency response is determined either by a filter or by tuned interstages. The design of the individual active stages follows the techniques discussed earlier, but the interstages become important for frequency shaping. There are various forms of interstage networks; several important cases are discussed below.

Synchronous-Tuned Interstages.   The simplest forms of tuned interstages are synchronously tuned circuits. The two common types are the single- and double-tuned interstage.

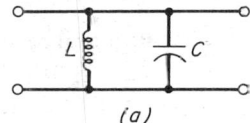

(a)

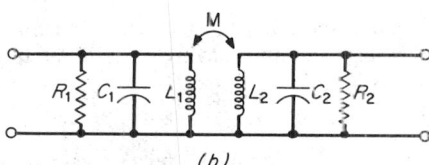

(b)

Fig. 13-51. Interstage coupling circuits: (*a*) single-tuned; (*b*) double-tuned.

The governing equations are:

Single-tuned interstage (Fig. 13-51*a*):

$$A(j\omega) = -A_r \frac{1}{1 + jQ_L(\omega/\omega_0 - \omega_0/\omega)}$$

where Q_L = loaded Q of the tuned circuit greater than 10, ω_0 = resonance frequency of the tuned circuit = $1\sqrt{LC}$, ω = frequency variable, A_r = midband gain equal to g_m times the midband impedance level. For an *n*-stage amplifier with *n* interstages.

$$A_T = A^n(j\omega) = A_r^n \left[1 + \left(\frac{\omega^2 - \omega_0^2}{B\omega}\right)^2\right]^{-n/2}$$

where $B = \omega_0/Q_L$ = single-stage bandwidth, n = number of stages, ω_0 = center frequencies, and Q_L = loaded Q. $B_n = B\sqrt{2^{1/n} - 1}$ is the overall bandwidth reduction due to n cascades.

Double-tuned interstage (Fig. 13-51*b*):

$$A(j\omega) = \frac{g_m k}{C_1 C_2(1 - k^2)\sqrt{L_1 L_2}} \frac{j\omega}{\omega^4 - ja_1\omega^3 - a_2\omega^2 + ja_3\omega + a_4}$$

(for a single double-tuned stage), where

$$a_1 = \omega_r\left(\frac{1}{Q_1} + \frac{1}{Q_2}\right) \qquad a_2 = \frac{\omega_r^2}{Q_1 Q_2} + \frac{1}{1 - k^2}(\omega_1^2 + \omega_2^2)$$

$$a_3 = \frac{\omega_r}{1 - k^2}\left(\frac{\omega_2^2}{Q_1} + \frac{\omega_1^2}{Q_2}\right) \qquad a_4 = \frac{\omega_1^2 \omega_2^2}{1 - k^2}$$

The circuit parameters are

R_1 = total resistance primary side
C_1 = total capacitance primary side
L_1 = total inductance primary side
R_2 = total resistance secondary side
C_2 = total capacitance secondary side
L_2 = total inductance secondary side
M = mutual inductance = $k\sqrt{L_1 L_2}$
k = coefficient of coupling

ω_r = resonant frequency of amplifier
$\omega_1 = 1/\sqrt{L_1 C_1}$
$\omega_2 = 1/\sqrt{L_2 C_2}$
Q_1 = primary Q at $\omega_r = \omega_r C_1 R_1$
Q_2 = secondary Q at $\omega_r = \omega_r C_2 R_2$
g_m = transconductance of active device at midband frequency

Simplification. If $\omega_1 = \omega_2 = \omega_0$, that is, primary and secondary tuned to the same frequency, then

$$\omega_r = \omega_0/\sqrt{1 - k^2}$$

is the resonant frequency of the amplifier and

$$A(j\omega_r) = \frac{+jkg_m\sqrt{R_1 R_2}}{\sqrt{Q_1 Q_2}(k^2 + 1/Q_1 Q_2)}$$

is the gain at this resonant frequency.

For maximum gain,

$$k_c = \frac{1}{\sqrt{Q_1 Q_2}} = \text{critical coupling}$$

and for maximum flatness,

$$k_T = \sqrt{\frac{1}{2}\left(\frac{1}{Q_1^2} + \frac{1}{Q_2^2}\right)} = \text{transitional coupling}$$

If k is increased beyond k_T, a double-humped response is obtained.

Overall bandwidth of an n-stage amplifier having equal Q circuits with transitional coupled interstages whose bandwidth is B is

$$B_n = B(2^{1/n} - 1)^{1/4}$$

The governing equations for the double-tuned-interstage case are shown above. The response for various degrees of coupling related to $k_T = k_C$ in the equal-coil-Q case is shown in Fig. 13-52.

52. Maximally Flat Staggered Interstage Coupling. This type of coupling consists of n single-tuned interstages that are cascaded and adjusted so that the overall gain function is maximally flat. The overall cascade bandwidth is B_n, the center frequency of the cascade is ω_c, and each stage is a single-tuned circuit whose bandwidth B and center frequency are determined from Table 13-5. The gain of each stage at cascade center frequency is $A(j\omega_c) = -g_m/$

Table 13-5.   Design Data for Maximally Flat Staggered n-tuples

n	Name of circuit	No. of stages	Center frequency of stage	Stage bandwidth
2	Staggered pair	2	$\omega_c \pm 0.35B_n$	$0.71B_n$
3	Staggered triple	2	$\omega_c \pm 0.43B_n$	$0.50B_n$
		1	ω_c	$1.00B_n$
4	Staggered quadruple	2	$\omega_c \pm 0.46B_n$	$0.38B_n$
		2	$\omega_c \pm 0.19B_n$	$0.92B_n$
5	Staggered quintuple	2	$\omega_c \pm 0.29B_n$	$0.81B_n$
		2	$\omega_c \pm 0.48B_n$	$0.26B_n$
		1	ω_c	$1.00B_n$
6	Staggered sextuple	2	$\omega_c \pm 0.48B_n$	$0.26B_n$
		2	$\omega_c \pm 0.35B_n$	$0.71B_n$
		2	$\omega_c \pm 0.13B_n$	$0.97B_n$
7	Staggered septuple	2	$\omega_c \pm 0.49B_n$	$0.22B_n$
		2	$\omega_c \pm 0.39B_n$	$0.62B_n$
		2	$\omega_c \pm 0.22B_n$	$0.90B_n$
		1	ω_c	$1.00B_n$

For $Q_L > 20$

$C_T[B + j(\omega_c^2 - \omega_0^2)/\omega_c]$, where C_T = sum of output capacitance and input capacitance to next stage and wiring capacitance of cascade, B = stage bandwidth, ω_0 = center frequency of stage, and ω_c = center frequency of cascade.[20]

53. Other I.F. Interstage Coupling Systems. There are several other methods of obtaining the desired passband response of an i.f. amplifier. The design of cascaded single-tuned

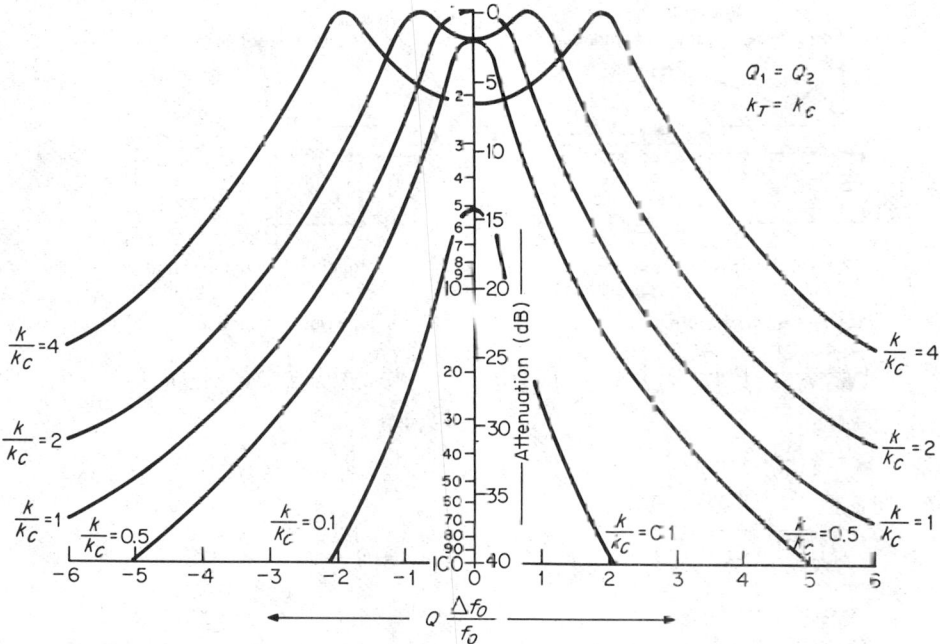

Fig. 13-52. Selective curves for two identical circuits in a double-tuned interstage circuit, at various values of k/k_c.

circuits with active devices in between can also achieve a Chebyshev ripple passband response.[20] Also coming into vogue, with the use of integrated circuits, is the use of lumped gain followed by a passive filter and additional gain. The amount of gain before and after the filter is a system compromise between noise performance and the overloading of the wide-band stages that are used.

Radio-Frequency Oscillators

BY G. B. HARROLD

54. General Considerations. Oscillators at rf frequencies are usually of the class A sine-wave-output type.

RF oscillators (in common with audio oscillators, see Pars. **13-34** to **13-42**) may be considered either as one-port networks that exhibit a negative real component at the input or as two-port-type networks consisting of an amplifier and a frequency-sensitive passive network that couples back to the input port of the amplifier. It can be shown that the latter type of feedback oscillator also has a negative resistance at one port. This negative resistance is of a dynamic nature and is best defined as the ratio between the fundamental components of voltage and current.

The sensitivity of the oscillator's frequency is directly dependent upon the effective Q of the frequency-determining element and the sensitivity of the amplifier to variations in temperature, voltage variation, and aging. For example, the effective Q of the frequency-determining element is important because the percentage change in frequency required to produce the compensating phase shift in a feedback oscillator is inversely proportional to the circuit Q, thus the larger the effective Q the greater the frequency stability. The load on an oscillator is also critical to the frequency stability since it affects the effective Q and in many cases the oscillator is followed by a buffer stage for isolation.

It is also desirable to provide some means of stabilizing the oscillator's operating point, either by a regulated supply, dc feedback for bias stabilization, or oscillator self-biasing schemes such

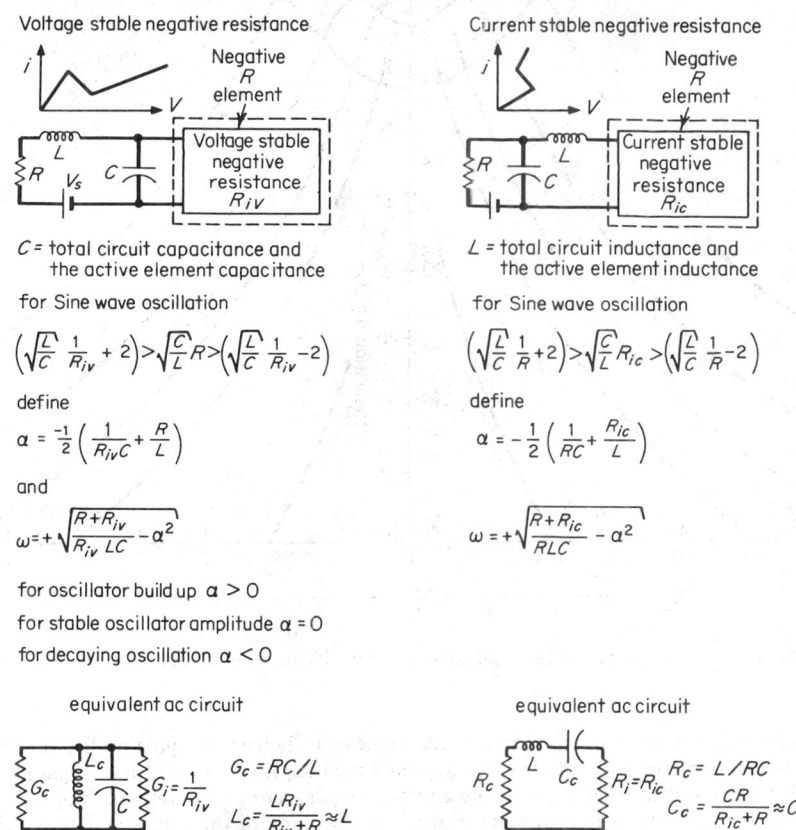

Fig. 13-53. General analysis of negative-resistance oscillators.

as grid-leak bias. This stabilizes not only the frequency but also the output amplitude, by tending to compensate any drift in the active device's parameters. It is also necessary to eliminate the harmonics in the output since they give rise to cross-modulation products producing currents at the fundamental frequency that are not necessarily in phase with the dominant oscillating mode. The use of high-Q circuits and the control of the nonlinearity helps in controlling harmonic output. The basic block diagrams for feedback oscillators are shown in Fig. 13-16.

55. Negative-Resistance Oscillators. The analysis of the negative-impedance oscillator is shown in Fig. 13-53. The frequency of oscillation at buildup is not completely determined by the LC circuit but has a component that is dependent upon the circuit resistance. At steady state, the frequency of oscillation is a function of $1 + R/R_{iv}$ or $1 + R_{ic}/R$, depending on the particular circuit where the ratios R/R_{iv}, R_{ic}/R are usually chosen to be small. While R is a fixed function of the loading, R_{ic} or R_{iv}/R must change with amplitude during oscillator buildup, so that the condition of $\alpha = 0$ can be reached. Thus R_{iv}, R_{ic} cannot be constant but are dynamic impedances defined as the ratio of the fundamental voltage across the element to the fundamental current into the element.

The type of dc load for biasing and the resonant circuit required for the proper operation of a negative-resistance oscillator depend on the type of active element. It can be shown[21] that R must be less than $|R_{iv}|$ or that R must be greater than $|R_{ic}|$ in order for oscillation to build up and be sustained.

The detailed analysis of the steady-state oscillator amplitude and frequency can be undertaken by graphical techniques. The magnitude of G_i or R_i is expressed in terms of its voltage dependence. Care must be taken with this representation, since the shape of the G_i or R_i curve depends upon the initial bias point.

The analysis of negative-resistance oscillators can now be performed by means of admittance diagrams. The assumption for oscillation to be sustaining is that the negative-resistance element, having admittance y_i, must equal $-y_e$, the external circuit admittance. This can be summarized by $G_i = -G_e$ and $B_i = -B_e$. A typical set of admittance curves is shown in Fig. 13-54. In this construction, it is assumed that $B_i = -B_e$, even during the oscillator buildup. Also shown is the fact that G_i at zero amplitude must be larger than G_e so that the oscillator can be started, that is, $\alpha > 0$, and that it may be possible to have two or more stable modes of oscillation.

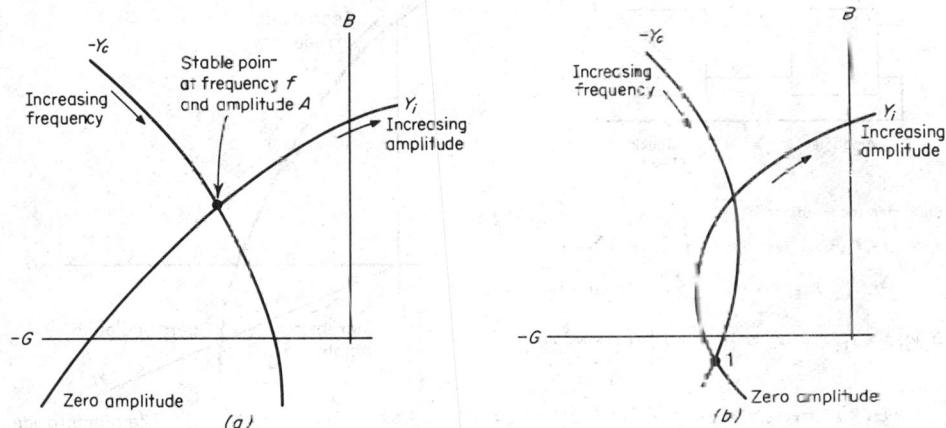

Fig. 13-54. Admittance diagrams of voltage-stable negative-resistance oscillators. (a) self-starting case, $\alpha > 0$; (b) circuit starts oscillating only if externally excited beyond point 1.

56. Feedback Oscillators. Several techniques exist for the analysis of feedback oscillators.[21-23] In the generalized treatment, the active element is represented by its y parameters whose element values are at the frequency of interest, having magnitudes defined by the ratio of the fundamental current divided by fundamental voltage. The general block diagram and equations are shown in Fig. 13-55. Solution of the equations given yields information on the oscillator's performance. In particular, equating the real and imaginary parts of the characteristic equation gives information on amplitude and frequency of oscillation.

In many instances, many simplifications to these equations can be made. For example, if y_{11} and y_{12} are made small (as in vacuum-tube amplifiers), then

$$y_{21} = -(1/z_{21})(y_{22}z_{11} + 1) = -1/Z$$

This equation can be solved by equating the real and imaginary terms to zero to find the frequency and the criterion for oscillation of constant amplitude. This equation can also be used to draw an admittance diagram for oscillator analysis.

These admittance diagrams are similar to those discussed under negative-resistance oscillators (Par. **13-55**). The technique is illustrated in Fig. 13-56.

At higher frequencies, the S parameters can also be used to design oscillators (Fig. 13-57). The basis for the oscillator is that the magnitude of the input reflection coefficient must be greater than unity, causing the circuit to be potentially unstable (in other words, it has a negative real part for the input impedance). The input reflection coefficient with a Γ_L output termination is

$$S'_{11} = S_{11} + \frac{S_{12}S_{21}\Gamma_L}{1 - S_{22}\Gamma_L}$$

Either by additional feedback or adjustment of Γ_L it is possible to make $|S'_{11}| > 1$. Next, establishing a load Γ_s such that it reflects all the energy incident upon it will cause the circuit to oscillate. This criterion is stated as

$$\Gamma_s S'_{11} = 1$$

at the frequency of oscillation.

This technique can be applied graphically, using a Smith chart as before. Here the reciprocal of S'_{11} is plotted as a function of frequency since $S'_{11} > 1$. Now choose either a parallel or a series-tuned circuit and plot its Γ_s. If f_1 is the frequency common to $1/S'_{11}$ and Γ_s and satifies the above criterion, the circuit will oscillate at this point.

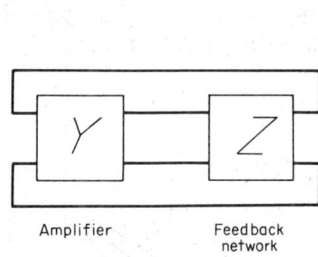

Characteristic equation:

$$y_{21} z_{21} + y_{11} z_{22} + y_{22} z_{11} + y_{12} z_{12} + \Delta_y \Delta_z + 1 = 0$$

$$\Delta_y = y_{11} y_{22} - y_{12} y_{21} \qquad \Delta_z = z_{11} z_{22} - z_{12} z_{21}$$

If $y_{21} \gg y_{12}$ $y_{12} \approx 0$ and $[z]$ passive $z_{12} = z_{21}$

Then:

$$y_{21} z_{12} + y_{11} z_{22} + y_{22} z_{11} + \Delta_y \Delta_z + 1 = 0$$

Fig. 13-55. General analysis of feedback oscillators.

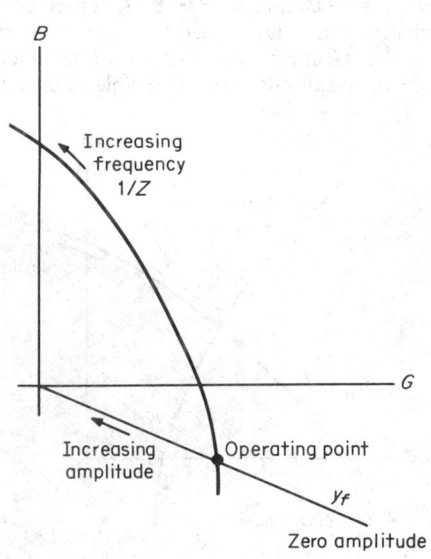

Fig. 13-56. Admittance diagram of feedback oscillator.

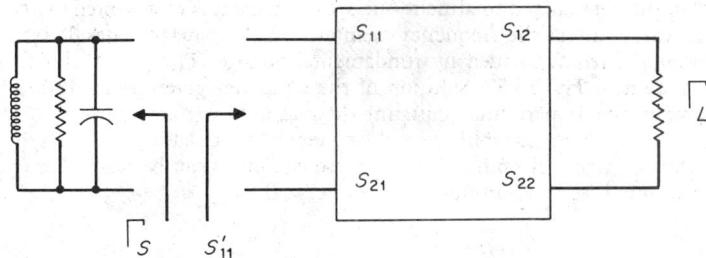

Fig. 13-57. S-parameter analysis of oscillators.

57. References on RF Amplifiers and Oscillators

1. McCasland, G. P.   Transistor Measurements with the HF-VHF Bridge, *Boonton Radio Notebook* 19, pp. 1–6, 1958.

2. McCasland, G. P.   More about Transistor Measurements, *Boonton Radio Notebook* 20, 1959.

3. Transistor Parameter Measurements, *Hewlett Packard Appl.*, Note 77-1, Palo Alto, Calif.

4. An Advanced New Network Analyzer for Sweep Measuring Application and Phase from 0.1 to 12.4 GHz, *Hewlett Packard J.*, February 1967, Vol. 18, No. 6.

5. Scarlett, R. M.   Measuring Transistor Parameters with Wayne Kerr R. F. Bridges, *Electron Des.*, July 22, 1959, Vol. 15, No. 7, pp. 38–41.

6. Thurston, W. R., and R. A. Soderman   Type 1607-A Transfer Function and Admittance Bridge, *Gen. Radio Exp.*, May 1959, Vol. 33, No. 5, pp. 3–11.

7. Abraham, R. P., and R. J. Kirkpatrick   Transistor Characterization at VHF, *Bell Telephone Syst. Monogr.* 3007, 1957.

8. "Instruction Book for VHF-UHF Noise Generator PRD Type 904-A," PRD Electronics (Div. of Harris-Intertype Corp.), Westbury, N.J., 1967.

9. "Instruction Book for AIL Type 70 Hot-Cold Body Standard Noise Generator," Cutler-Hammer, Deer Park, N.Y., 1968.

10. Chenette, E. R.   Low Noise Transistor Amplifiers, *Solid State Des.*, February 1964, pp. 27–30.

11. Lo, A. W., R. O. Endres, et al.   "Transistor Electronics," pp. 318–326, Prentice-Hall, Englewood Cliffs, N.J., 1955.

12. Interference Analysis of New Components and Circuits, Rome Air Dev. Cen. *Tech. Doc. Rep.* RADC-TDR-64-161, AD 601850, May 1964.

13. "Radiotron Designer's Handbook," pp. 1002–1017, Amalgamated Wireless Co. Sydney, Australia, 1953.

14. FCC Rules and Regulations, Vol. II, Sec. 15.61, August 1969.

15. Rollett, J. M.   Stability and Power Gain Invariants of Linear Two Ports, *IRE Trans. Circuit Theory*, March 1962, Vol. CT-9, pp. 29–32.

16. Haykin, S. S.   "Active Network Theory," pp. 272–276, Addison-Wesley, Reading, Mass., 1970.

17. Bodway, G. E.   Two Port Power Flow Analysis Using Generalized Scattering Parameters, *Microwave J.*, May 1967, Vol. 10, No. 6.

18. Linvill, J. G., and J. F. Gibbons   "Transistor and Active Circuits," Chaps. 10 and 11, McGraw-Hill, New York, 1961.

19. Anderson, R. W.   S-Parameter Techniques for Faster, More Accurate Network Design, *Hewlett Packard J.*, February 1967, Vol. 18, No. 6.

20. Martin, T. L., Jr.   "Electronic Circuits," Chaps. 4 and 5, Prentice-Hall, Englewood Cliffs, N.J., 1955.

21. Reich, H. J.   "Functional Circuits and Oscillators," p. 317, Van Nostrand, New York, 1961.

22. Reference 20, Chap. 10.

23. Seely, S.   "Electronic Tube Circuits," Chap. 12, McGraw-Hill, New York, 1950.

Broadband Amplifiers

BY J. W. LUNDEN

58. Introduction. In broadband amplifiers signals are amplified so as to preserve over a wide band of frequencies such characteristics as signal amplitude, gain response, phase shift, delay, distortion, and efficiency. The width of the band depends upon the active device used, the frequency range, and power level in the current state of the art. As a general rule, above 100 MHz, a 20% or greater bandwidth is considered broadband, whereas an octave or more is typical below 100 MHz. As the state of the art advances, it is becoming more common to achieve octave-bandwidth amplifiers above 100 MHz and well into the microwave region using new bipolar and FET active devices.

It is becoming increasingly uncommon to use tube devices for new amplifier designs. In the following discussion both tube and bipolar transistor notations appear for generality. The concepts are the same, and only the active-device equivalent circuits need be changed to incorporate FET devices as the active elements

$$[g_m \text{ (tube)} = g_m \text{ (FET)}, \ r_p \text{ (tube)} = r_{ds} \text{ (FET), etc.}]$$

59. Low-, Mid-, and High-Frequency Performance. Consider the basic common-cathode and common-emitter-broadband RC coupled configurations shown in Fig. 13-58. Simplified low-frequency equivalent circuits are shown in Fig. 13-59.

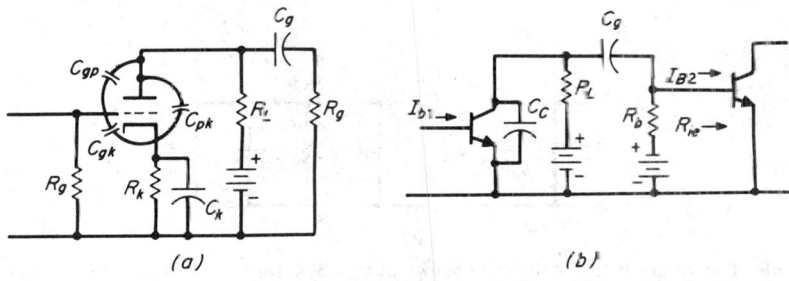

(a)   *(b)*

Fig. 13-58. *RC*-coupled stages: (a) electron-tube form; (b) transistor form.

The voltage gain of the tube amplifier stage under the condition that all reactances are negligibly small is the midband value (at frequency f)

$$(A_{mid})_{tube} = \frac{-g_m}{1/r_p + 1/R_L + 1/R_g} = \frac{-\mu[R\,R_g/(R_L + R_g)]}{r_p + R_L R_g/(R_L + R_g)} \approx -g_m R_L$$

If the low-frequency effects are included, this becomes

$$(A_{low})_{tube} = \frac{g_m R_L}{1 + 1/j\omega R_g C_g} \frac{1 + 1/j\omega R_K C_K}{1 + (1 + g_m R_K)/j\omega C_K R_K}$$

The low-frequency cutoff is due principally to two basic time constants, $R_g C_g$ and $R_K C_K$. For C_K values large enough for the time constant to be much longer than that associated with C_g, a low-frequency cutoff or half-power point can be determined as

$$(f_1)_{tube} = \frac{1}{2\pi C_g[R_g + r_p R_L(r_p + R_L)]}$$

If the coupling capacitor is very large, the low-frequency cutoff is due to C_K. The slope of the actual rolloff is a function of the relative effect of these two time constants. Therefore, the design of coupling and bypass circuits to achieve very-low-frequency response requires very large values of capacitance.

Similarly, for a transistor stage, the midband current gain can be determined as

$$(A_{mid})_{transistor} = \frac{-\alpha r_c R_L}{[R_L + r_c(1 - \alpha)]\left[R_{ie} + \dfrac{R_L r_c(1 - \alpha)}{R_L + r_c(1 - \alpha)}\right]} \approx \frac{-\alpha}{1 - \alpha} \frac{R_L}{R_L + R_{ie}}$$

where

$$R_{ie} = r_b + \frac{r_e}{1 - \alpha}$$

When low-frequency effects are included, this becomes

$$(A_{low})_{transistor} \approx \frac{-\alpha}{1 - \alpha} \frac{R_L}{R_L + R_{ie} - j/\omega C_g} \qquad \text{for } R_L \ll r_c(1 - \alpha)$$

and

$$(f_1)_{transistor} = \frac{1}{2\pi C_g} \frac{1}{R_{ie} + \dfrac{R_L r_c(1 - \alpha)}{R_L + r_c(1 - \alpha)}} \approx \frac{1}{2\pi C_g} \frac{1}{R_{ie} + R_L}$$

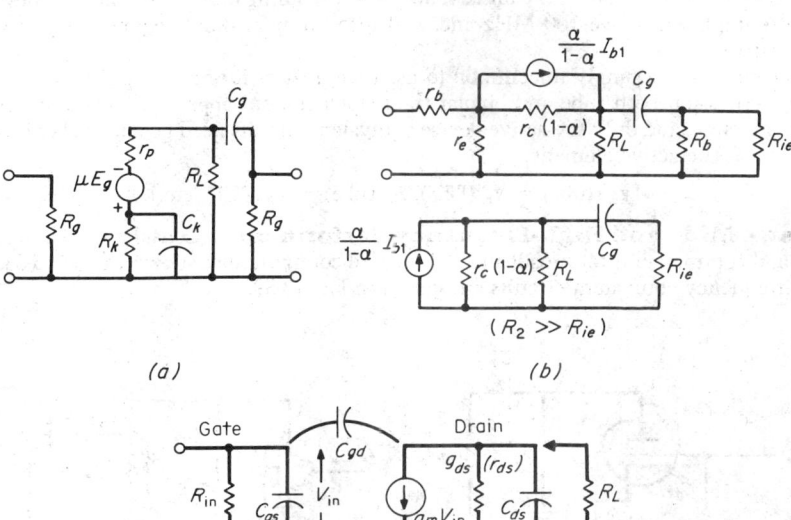

(a)   (b)   (c)

Fig. 13-59. Equivalent circuits of the stages shown in Fig. 13-58: (a) electron-tube form; (b) transistor forms; (c) FET form.

If the ratio of low- to midfrequency voltage or current gain is taken, its reactive term goes to unity at $f = f_1$, that is, the cutoff frequency

$$\frac{A_{\text{low}}}{A_{\text{mid}}} = \frac{1}{1 - jf_1/f)} \qquad \phi_{\text{low}} = \tan^{-1}\frac{f_1}{f}$$

These quantities are plotted in Fig. 13-60 for a single time-constant rolloff.

Caution should be exercised in assuming that interelectrode reactances are negligible. Although this is generally the case, gain multiplicative effects can result in input or output values greater than the values assumed above, e.g., by the Miller effect:

$$C_{\text{in}} = C_{gk} + C_{gp}(1 + g_m R_p)$$

Typically, the midfrequency gain equation can be used for frequencies above that at which $X_c = R_g/10$ and below that at which $X_{cg} = 10R_gR_L/(R_g + R_L)$ (for the tube circuit).

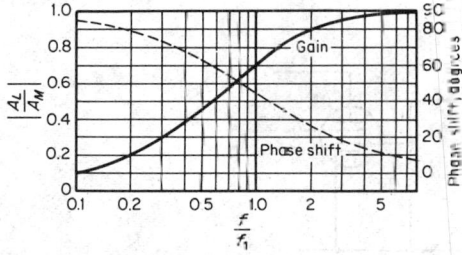

Fig. 13-60. Gain and phase-shift curves at low frequencies.

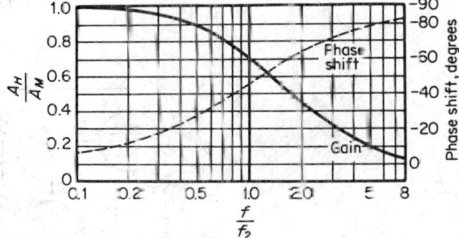

Fig. 13-61. Gain and phase-shift curves at high frequencies.

If the frequency is increased further, a point is reached where the shunt reactances are no longer high with respect to the circuit resistances. At this point the coupling and bypass capacitors can be neglected. The high-frequency gain can be determined as

$$(A_{\text{high}})_{\text{tube}} = \frac{-g_m}{1/r_p + 1/R_L + j\omega C_L}$$

where C_L is the effective total interstage shunt capacitance

$$(A_{\text{high}})_{\text{transistor}} \approx \frac{-\alpha}{1 - \alpha} \frac{1}{1 + R_{ie}\left(\frac{1}{R_L} - \frac{j\omega C_c}{1 - \alpha}\right)} \qquad \text{for } R_L \ll r_c(1 - \alpha)$$

The ratio of high- to midfrequency gains can be taken and upper cutoff frequencies determined

$$\left(\frac{A_{\text{high}}}{A_{\text{mid}}}\right)_{\text{tube}} = \frac{1}{1 + j\omega C_g \dfrac{1}{(1/r_p) + (1/R_L) + 1/R_g}}$$

$$(f_2)_{\text{tube}} = \frac{1}{2\pi C_g}\left(\frac{1}{r_p} + \frac{1}{R_L} + \frac{1}{R_g}\right)$$

$$\left(\frac{A_{\text{high}}}{A_{\text{mid}}}\right)_{\text{transistor}} = \frac{1}{1 + \dfrac{j\omega C_c r_c R_L R_{ie}}{R_{ie}[R_L + r_c(1 - \alpha)] + R_L r_c(1 - \alpha)}}$$

$$(f_2)_{\text{transistor}} \approx \frac{1 - \alpha}{2\pi C_c}\left(\frac{1}{R_L} + \frac{1}{R_{ie}}\right) \qquad \text{and} \qquad \phi_{\text{high}} = -\tan^{-1}(f/f_2)$$

Dimensionless curves for these gain ratios and phase responses are plotted in Fig. 13-61.

60. Compensation Techniques. To extend the cutoff frequencies f_1 and f_2 to lower or higher values, respectively, compensation techniques can be used.

Figure 13-62 illustrates two techniques for low-frequency compensation. If the condition $R_x C_x = C_x R_x R_L/(R_x + R_L)$ is fulfilled (in circuit a or b), the gain relative to the midband gain is

$$\frac{A_{\text{low}}}{A_{\text{mid}}} = \frac{1}{1 - j(1/\omega R_g C_g)[R_L/(R_L + R_x)]} \qquad \text{and} \qquad f_1 = \frac{1}{2\pi R_g C_g}\frac{R_L}{R_L + R_x}$$

Hence, improved low-frequency response is obtained with increased values of R_x. This value is related to R_L and restricted by active-device operating considerations. Also, R_L is dependent on the desired high-frequency response. It can be shown that equality of time constants $R_L C_x = R_g C_g$ will produce zero phase shift in the coupling circuit (for $R_x > 1/\omega C_x$). The circuit shown in Fig. 13-62c is more critical. It is used with element ratios set to $R_L/R_x = R_g/R_c$ and $C_x/C_g = R_c/R_x$.

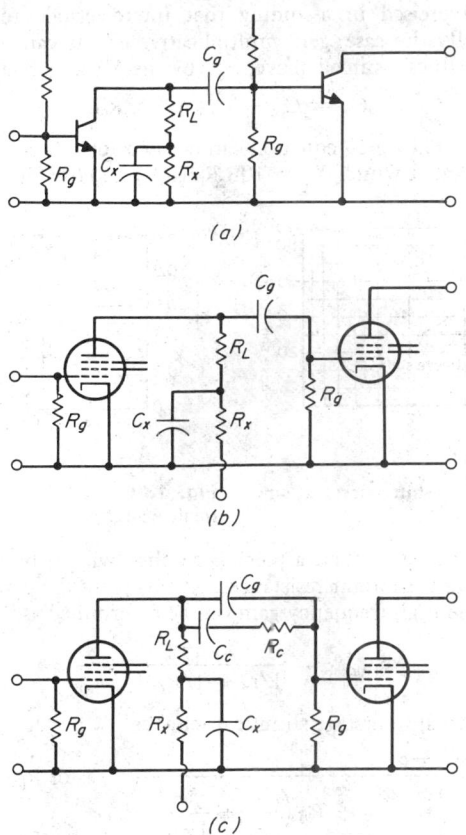

Fig. 13-62. Low-frequency compensation networks: (a) transistor version; (b),(c) tube versions.

Various compensation circuits are also available for high-frequency-response extension. Two of the most common, the series- and shunt-compensation cases, are shown in Fig. 13-63. The high-frequency-gain expressions of these configurations can be written

$$\left|\frac{A_{high}}{A_{mid}}\right| = \sqrt{\frac{1 + a_1(f/f_2)^2 + a_2(f/f_2)^4 + \cdots}{1 + b_1(f/f_2)^2 + b_2(f/f_2)^4 + b_3(f/f_2)^6 + \cdots}}$$

The coefficients of the terms decrease rapidly for the higher-order terms, so that if $a_1 = b_1$, $a_2 = b_2$, etc., to as high an order of the f/f_2 ratio as possible, a maximally flat response curve is obtained.

For the phase response, $d\phi/d\omega$ can also be expressed as a ratio of two polynomials in f/f_2, and a similar procedure can be followed. A flat time-delay curve results. Unfortunately, the sets of conditions for flat gain and linear phase are different, and compromise values must be used.

Shunt Compensation. The high-frequency gain and time delay for the shunt-compensated stage are

$$\left|\frac{A_{high}}{A_{mid}}\right| = \sqrt{\frac{1 + \alpha^2(f/f_2)^2}{1 + (1 - 2\alpha)(f/f_2)^2 + \alpha^2(f/f_2)^4}} \qquad \phi = -\tan^{-1}\frac{f}{f_2}\left[1 - \alpha + \left(\frac{f}{f_2}\right)^2 \alpha^2\right]$$

where $\qquad\qquad \alpha = L/C_g R_L^2 \qquad$ and $\qquad R_g \gg R_L$

Generalized amplitude and phase (delay) response curves for the shunt-compensated amplifiers are shown in Fig. 13-64 for several values of α.

A case when R_g cannot be assumed to be high, such as the input of a following transistor stage, is considerably more complex, depending on the transistor equivalent circuit used. This is particularly true when operating near the transistor f_T and/or above the VHF band.

Series Compensation. In the series-compensated circuit the ratio of C_s to C_g is an additional parameter. If this can be optimized, the circuit performance is better than in the shunt-compen-

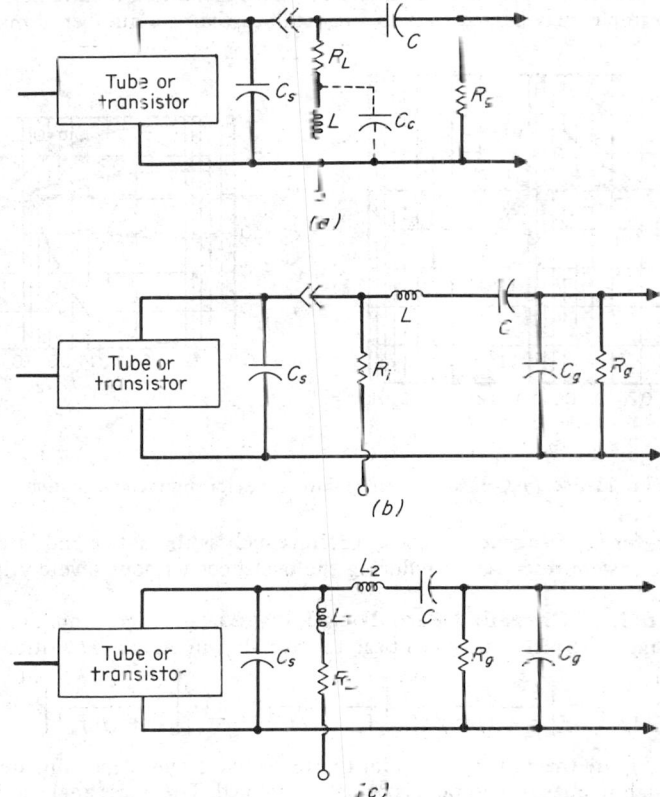

Fig. 13-63. High-frequency compensation schemes: (a) shunt; (b) series; (c) shunt-series.

sated case. Typically, however, control of this parameter is not available due to physical and active-device constraints.

The gain and phase response for series compensation are

$$\left| \frac{A_{high}}{A_{mid}} \right| = \left\{ \frac{1}{1 + \left(\frac{f}{f_2}\right)^2 \left[1 - \frac{2LC_g}{R_L^2(C_s + C_g)^2}\right]} + \left(\frac{f}{f_2}\right)^4 \left[\frac{L^2C_g^2}{R_L^4(C_s + C_g)^4} - \frac{2LC_sC_g}{R_L^2(C_s + C_g)^3}\right] \right.$$
$$\left. + \left(\frac{f}{f_2}\right)^6 \frac{L^2C_s^2C_g^2}{R_L^4(C_s + C_g)^6} \right\}^{1/2}$$

$$\phi = -\tan^{-1} \frac{\dfrac{f}{f_2} \left[\left(\dfrac{f}{f_2}\right)^2 \dfrac{LC_sC_g}{R_L^2(C_s + C_g)^3} - 1\right]}{1 - \left(\dfrac{f}{f_2}\right)^2 \dfrac{LC_g}{R_L^2(C_s + C_g)^2}}$$

For maximal flatness, $C_2/(C_1 + C_2) = 0.75 = K_2$ and $L = mR_L^2(C_s + C_g)$, with $m = 0.667$. The maximum improvement in rise time or bandwidth with this technique is about a factor of 4. The response curves for series compensation are given in Fig. 13-65 for several values of m

These two basic techniques can be combined to improve the response at the expense of complexity. The shunt-series-compensation case (Fig. 13-66) and the so-called "modified" case are examples. The latter involves a capacitance added in shunt with the inductance L or placing L between C_s and R_L.

Response curves for the latter case are given in Fig 13-67. For the modified-shunt case, the added capacitance C_e permits an additional degree of freedom and associated parameter. $K_1 = C_e/C_s$.

Other circuit variations exist for specific broadband compensation requirements. Phase compensation, for example, may be necessary as a result of cascading a number of minimum-phase

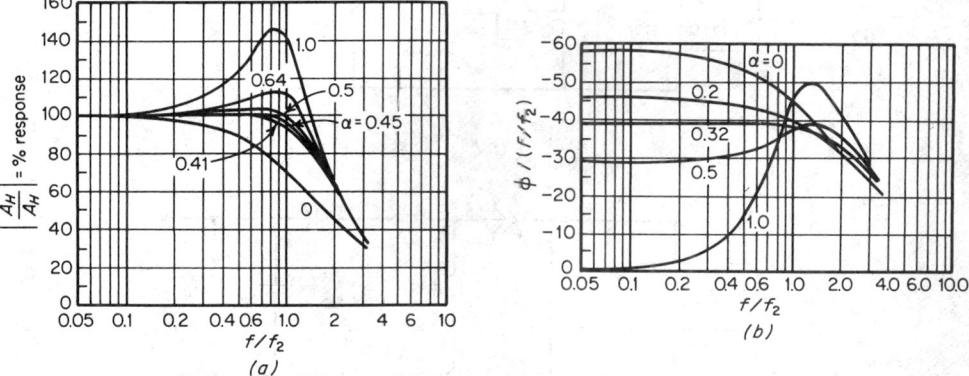

Fig. 13-64. (a) Gain and (b) phase-shift curves for shunt compensation.

circuits designed for flat frequency response. Circuits such as the lattice and bridged-T can be used to alter the system response by reducing the overshoot without severely increasing the overall rise time.

61. Bandwidth of Cascaded Broadband Stages. When an amplifier is made up of n cascaded RC stages, not necessarily identical, the overall gain A_n can be written

$$\left|\frac{A_n}{A_{mid}}\right| = \left[\frac{1}{1 + (f/f_a)^2}\right]^{1/2}\left[\frac{1}{1 + (f/f_b)^2}\right]^{1/2}\cdots\left[\frac{1}{1 + (f/f_n)^2}\right]^{1/2}$$

where f_a, f_b, ..., f_n are the f_1 or f_2 values for the respective stages, depending on whether the overall low- or high-frequency gain ratio is being determined. The phase angle is the sum of the individual phase angles. If the stages are identical, $f_a = f_b = f_x$ for all, and

$$\left|\frac{A_n}{A_{mid}}\right| = \left[\frac{1}{1 + (f/f_x)^2}\right]^{n/2}$$

Stagger Peaking. The principle of stagger tuning has been well known for some time. A number of individual bandpass amplifier stages are cascaded with frequencies skewed according

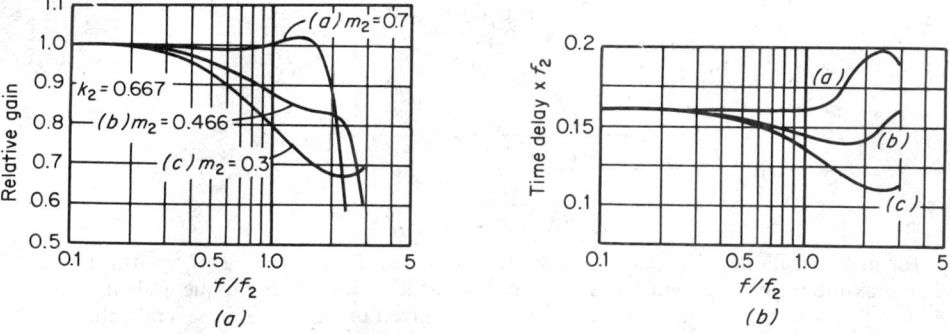

Fig. 13-65. (a) Gain and (b) time-delay curves for series compensation.[*]

to some predetermined criteria. The most straightforward is with the center frequencies adjusted so that the f_2 of one stage coincides with the f_1 of the succeeding stage, etc. The overall gain bandwidth then becomes

$$(GBW)_n = \sum_{n=1}^{N} (GBW)_1$$

A significant simplifying criterion of this technique is stage isolation. Isolation, in transistor stages particularly, is not generally high, except at low frequencies. Hence the overall design

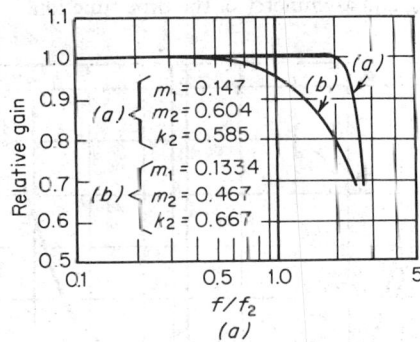

$$k_2 = \frac{C_g}{C_s + C_g}$$

$$m_1 = \frac{L_1}{(C_s + C_g) R_L^2}$$

$$m_2 = \frac{L_2}{(C_s + C_g) R_L^2}$$

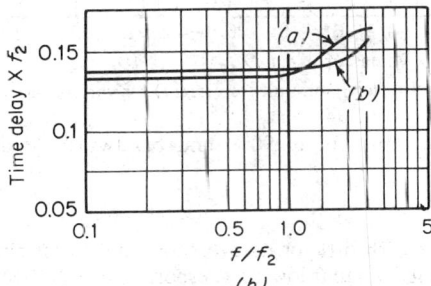

Fig. 13-66. (a) Gain and (b) time delay of shunt-series compensation.[4]

equations and subsequent overall alignment can be significantly complicated due to the interactions. Complex device models and computer-aided design greatly facilitate the implementation of this type of compensation. The simple shunt-compensated stage has found extensive use in stagger-tuned pulse-amplifier applications.

62. Transient Response. Time-domain analysis is particularly useful for broadband applications. Extensive theoretical studies have been made of the separate effects of nonlineari-

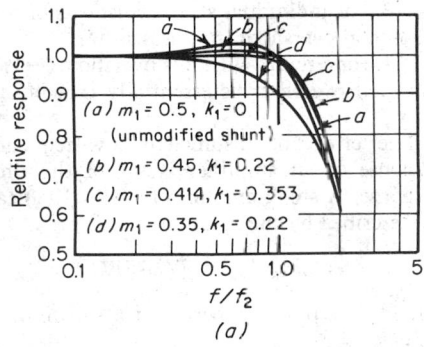

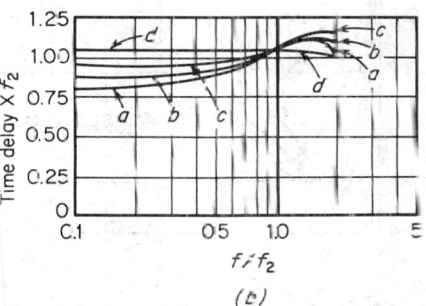

Fig. 13-67. (a) Gain and (b) time delay of modified shunt compensation.[4]

ties of amplitude and phase response.[1-3] These effects can be investigated starting with a normalized low-pass response function.

$$A(jw)/A(0) = \exp(a^m w^\pi - jb^n w^n)$$

where a and m = constants describing amplitude-frequency response and b and n = constants describing phase-frequency response. Figure 13-68 illustrates the time response to an impulse and a unit-step forcing function for various values of m, with $n = 0$. Rapid change of amplitude with frequency (large m) results in overshoot. Nonzero, but linear, phase-frequency characteristics ($n = 1$) result in a delay of these responses, without introducing distortion. Further increase in n results in increased ringing and asymmetry of the time function.

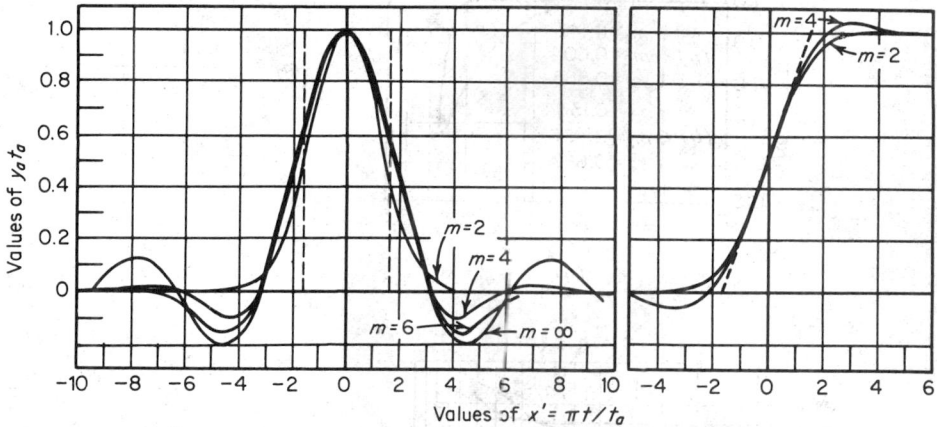

Fig. 13-68. Transient responses to unit impulse *(left)* and unit step *(right)* for various values of m.[2]

An empirical relationship between rise time (10 to 90%) and bandwidth (3 dB) can be expressed as

$$t_r \cdot BW = K$$

where K varies from about 0.35 for circuits with little or no overshoot to 0.45 for circuits with about 5% overshoot. K is 0.51 for the ideal rectangular low-pass response with 9% overshoot; for the gaussian amplitude response with no overshoot, $K = 0.41$.

The effect on rise time of cascading a number of networks n depends on the individual network pole-zero configurations. Some general rules follow.

1. For individual circuits having little or no overshoot, the overall rise time is

$$t_{rt} = (t_{r1}^2 + t_{r2}^2 + t_{r3}^2 + \cdots)^{1/2}$$

2. If $t_{r1} = t_{r2} = t_{rm}$,

$$t_{rt} = 1.1\sqrt{n}t_{r1}$$

3. For individual stage overshoots of 5 to 10%, total overshoot increases as $\sqrt{n}$.

4. For circuits with low overshoot ($\sim$1%), the total overshoot is essentially that of one stage.

The effect of insufficient low-frequency response of an amplifier is sag of the time response. A small amount of sag ($<$10%) can be described by the formula

$$E_{sag}/E_{tot} = T/2.75RC$$

where T is one-half period of a square-wave voltage.

Figure 13-69 illustrates the response to a step

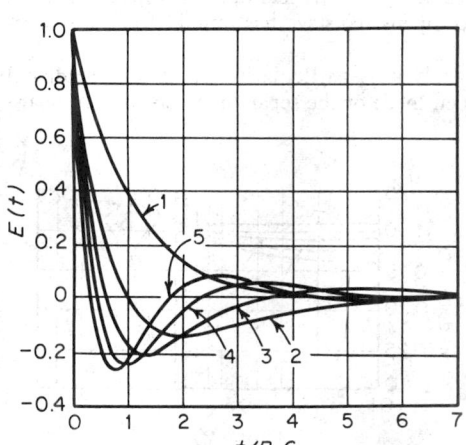

Fig. 13-69. Response to a unit step of n capacitively coupled stages of the same time constant.[5]

function of n capacitor-coupled stages with the same time constant. The effect of arithmetic addition of individual stage initial slopes determining the net total initial transition slope can be seen.

Figure 13-70 shows the effect of high-frequency shunt compensation on the time response to a step function. The series peaked time response is shown in Fig. 13-71. Similar curves for the many other compensation circuits and variations are given in the literature.[4-6]

63. Distributed Amplifiers. This technique is useful for operation of devices at or near the high-frequency limitation of gain band width. While the individual stage gain is low, the stages are cascaded so that the gain response is additive instead of multiplicative. The basic principle is to allow the input and output capacitances to form the shunt elements of two delay lines. This is shown in Fig. 13-72.

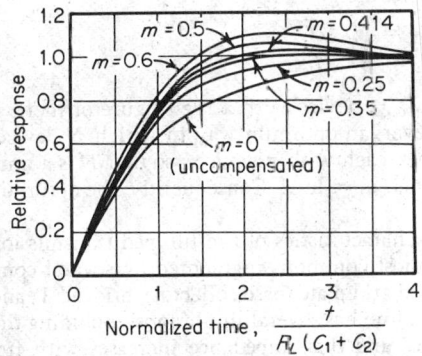

Fig. 13-70. Response to a unit step of a shunt-compensated stage.[4]

Fig. 13-71. Response to a unit step of a series-compensated stage.[4]

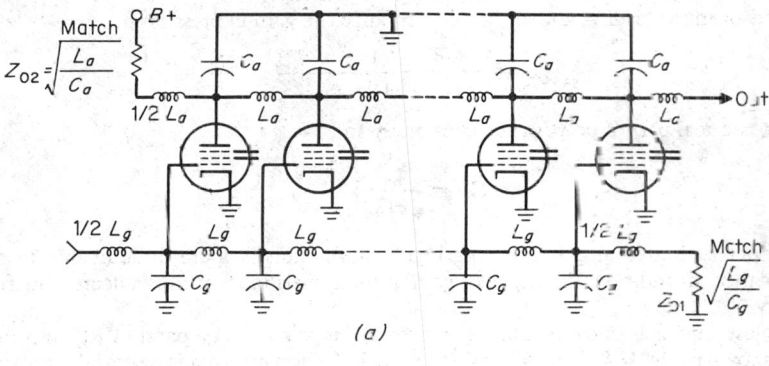

(a)

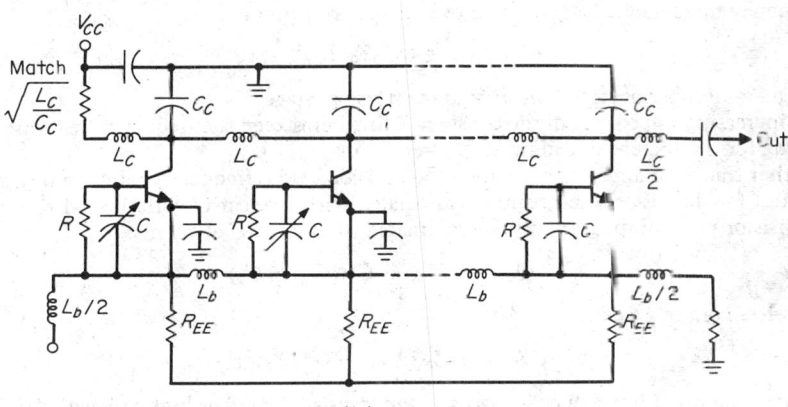

(b)

Fig. 13-72. Distributed amplifier circuits: (a) tube version; (b) transistor version.

If the delay times per section in the two lines are the same, the output signals traveling forward add together without relative delay. Care must be taken to ensure proper terminating conditions. The gain produced by n devices, each of transconductance g_m, has a value of $G = ng_m Z_{02}/2$. Performance to very low frequencies can be achieved. The high-frequency limit is determined by the cutoff frequencies of the input and output lines or effects within the active devices themselves other than parasitic shunt capacities (e.g., transit time or alpha-fall-off effects).

For input and output lines of different characteristic impedances, the overall gain is given by

$$G = \frac{2Z_{01}}{Z_{01} + Z_{02}} \frac{ng_m Z_{02}}{2}$$

The characteristic impedances and cutoff frequencies are given as

$$Z_{01} = L_g/\sqrt{C_g} \qquad Z_{02} = L_a/\sqrt{C_a}$$
$$f_{c1} = \pi/\sqrt{L_g C_g} \qquad f_{c2} = \pi/\sqrt{L_a C_a}$$

There does not seem to be an optimum choice for Z_{02}; a device with a high figure of merit is simply chosen, and the rest follows. There is, however, an optimum way in which N devices can be grouped, namely in m identical cascaded stages, each with n devices, so that N is a minimum for a given overall gain A. N is a minimum when $m = \ln A$. Consequently, the optimum gain per stage $G = e = 2.72$.

Various techniques are utilized to determine the characteristics of the lumped transmission lines. Constant-K and m-derived filter sections are most common, augmented by several compensation circuit variations.[7,8] The latter include paired grid-plate (base-collector), bridged-T, and resistive loading connections. The constant-K lumped line has several limitations, including the fact that the termination is not a constant resistance and that impedance increases with frequency and time delay also changes with frequency.

m-derived terminating half sections and time-delay equalizing sections[9] result in a frequency-amplitude response that is quite flat at very high frequencies.

The effects of input loading and/or line loss modify the gain expression to

$$G = \frac{2(GB)}{f_c} \frac{e^{-a}(1 - e^{-n\alpha})}{1 - e^{-\alpha}}$$

where α = the real part of propagation constant γ and

$$GB = \frac{g_m}{2\pi C_{eff}}\bigg|_{tube}$$

The design of distributed amplifiers using bipolar transistors is more difficult due to the low input impedance. In addition, the intrinsic gain of the transistor (β or α) is a decreasing function of frequency.

The simplest approach in overcoming this problem is to connect a parallel RC in series with the base, as shown in Fig. 13-72b. By setting $RC = \beta_0/2\pi f_T$ the gain is made essentially independent of frequency up to about $f_T/2$, with an overall voltage gain of

$$A_v = (n g_0 Z_0/2R)^m$$

where m = number of stages and n = transistors per stage.

The increasing impedance of the constant-K line helps keep the frequency response flat, compensating for the increased loading of the transistor.

Another transistor approach[10] involves the division of the frequency range into three regions to account for the losses. Each region is associated with a linear-Q variation with frequency of the transistor input-output circuits, approximated by shunt RC elements.

$$Q_{in} = w(C_{in} + C)/(G_{in} + G_k)$$

The lossless voltage gain is

$$A_L = f_T Z_0/[2f(1 - (f/f_c)^2)^{1/2}]$$

In region I, the input impedance is high and the transistor supplies high voltage gain. In regions II and III, the voltage gains are given by

$$A_{\rm II} = A_L n \frac{1 - nK_{11}Z_0(f/f_c)^2}{4r_b'[1 - (f/f_c)^2]^{1/2}} \qquad A_{\rm III} = A_L n \frac{1 - nK_{11}Z_0}{2r_b'[1 - (f/f_c)^2]^{1/2}}$$

where $K_{11} = (f_c/f_Q)^2$ $K_{111} = 1$ $f_Q = $ frequency of min Q

64. Broadband Matching Circuits. Broadband impedance transformation interstage-coupling matching can be achieved with balun transformers,[11] quarter-wave transmission-line sections,[12] lumped reactances in configurations other than discussed above, and short lengths of transmission lines.[13]

Balun Transformers. In conventional coupling transformers, the interwinding capacitance resonates with the leakage inductance, producing a loss peak. This limits the high-frequency response. A solution is to use transmission-line transformers in which the turns are arranged physically to include the interwinding capacitance as a component of the characteristic impedance of a transmission line. With this technique, bandwidths of hundreds of megahertz can be achieved. Good coupling can be realized without resonances, leading to the use of these transformers in power dividers, couplers, hybrids, etc.

Typically, the lines take the form of twisted wire pairs, although coaxial lines can also be used. In some configurations, the length of the line determines the upper cutoff frequency. The low-frequency limit is determined by the primary inductance. The larger the core permeability the fewer the turns required for a given low-frequency response. Ferrite toroids have been found to be satisfactory with widely varying core-material characteristics. The decreasing permeability with increasing frequency is offset by the increasing reactance of the wire itself, causing a wideband, flat-frequency response.

Quarter-Wave Transformers. The quarter-wavelength line transformer is another well-known element. It is simply a transmission line one-quarter wavelength long, with a characteristic impedance

$$Z_{\rm line} = \sqrt{Z_{\rm in}Z_{\rm out}}$$

where $Z_{\rm in}$ and $Z_{\rm out}$ are the terminating impedances. The insertion loss of this line section is

$$10 \log \left[1 + \frac{(r-1)^2}{4r} \cos^2 \theta \right] \qquad \text{(dB)}$$

where $r = Z_{\rm in}/Z_{\rm out}$ and $\theta = 2\pi L/\lambda = 90°$ at f_0

Figure 13-73 shows the bandwidth performance of the quarter-wave line for several matching ratios. Such lines may be cascaded to achieve broader-band matching by reducing the matching ratio required of each individual section.

Lumped and Pseudo-Lumped Transformations. The familiar lumped-element ladder network, depending on the element values and realization, approximates a short step transmission-line transformer or a tapered transmission line.[14] Convenient tables of element values as a function of bandwidth, ripple, transformation ratio, and number of elements are available.[15,16]

65. Feedback and Gain Compensation. *Feedback.* The bandwidth of amplifiers can be increased by the application of inverse feedback. A multiplicity of feedback combinations and analysis techniques are available and extensively treated in the literature.[18-20] In addition, the related concept of feedforward has been investigated and successfully implemented.[21] Figure 13-74 shows four feedback arrangements and formulas describing their performance. A major consideration, particularly for rf applications, is the control of impedance and loop-delay characteristics to ensure stability.

Gain Compensation. The power gain of a transistor amplifier typically falls 6 dB/octave with increasing frequency above the f_s value. The gain can be leveled (bandwidth-widened) by exact matching of the source impedance to the device at the upper frequency only, causing increasing mismatch with decreasing fre-

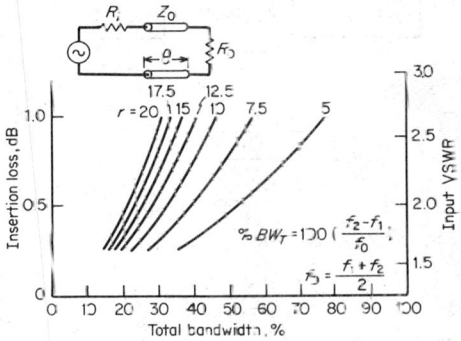

Fig. 13-73. Bandwidth of a quarter-wave matching transformer.

quency and the associated gain loss. The overall flat gain is the value at the high best-match frequency. Sufficient resistive loss is usually required in this interstage to prevent instabilities in either driver or driven stage due to the mismatch unloading.

66. Power-Combining Amplifiers. Many circuit techniques have been developed to obtain relatively high output powers with given modest-power devices. Two approaches are the direct-paralleling and hybrid splitting-combining techniques. The direct-paralleling approach is limited by device-to-device variations and the difficulties in providing balanced conditions due to the physical wavelength restrictions. A technique commonly used at UHF and microwave frequencies to obtain multioctave response incorporates matched stages driven from hybrid couplers in a balanced configuration. The coupler offers a constant-source impedance (equal to the driving-port impedance) to the two stages connected to its output ports. Power reflected due to the identical mismatches is coupled to the difference port of the hybrid and dissipated in the

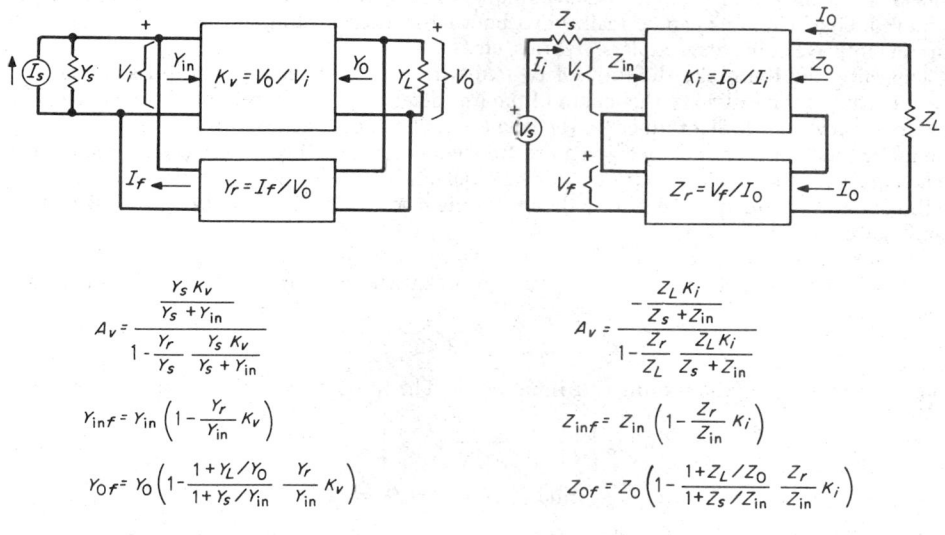

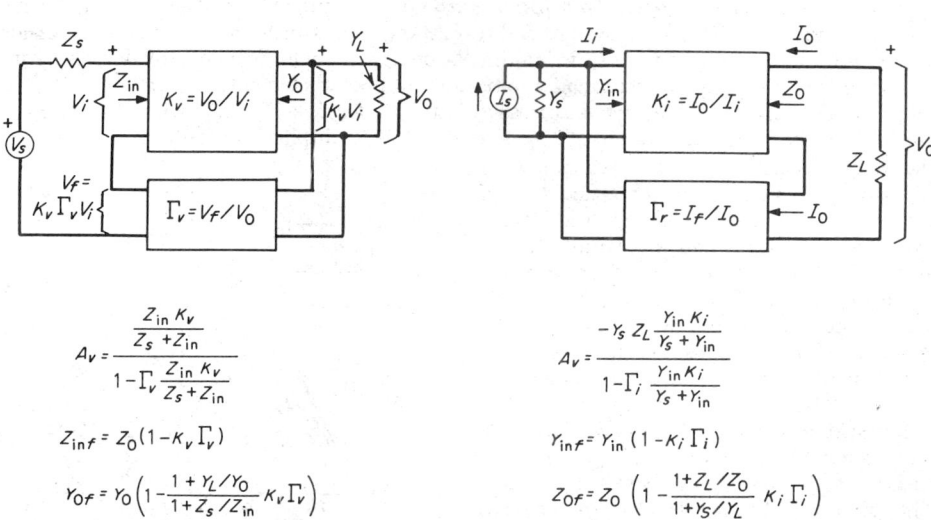

Fig. 13-74. Four methods of applying feedback, showing influence on gain and input and output immittances. *(From Hakim "Junction Transistor Circuit Analysis," by permission of John Wiley & Sons, Inc., New York.)*

idler load. The hybrid splitting-combining approach has proved quite effective in implementing high-output-level requirements (e.g., kilowatts at 1.4 GHz with transistors).

The hybrid splitting-combining approach enhances circuit operation. In particular, quadrature hybrids effect a VSWR-canceling phenomenon which results in extremely well-matched power-amplifier inputs and outputs that can be broadbanded upon proper selection of particular hybrid trees. Also, the excellent isolation between devices makes reliable power-amplifier service possible.

Several hybrid-directional-coupler configurations are possible, including the split-T, branch-line, magic-T, backward-wave, lumped, etc. Important factors in the choice of hybrid for this

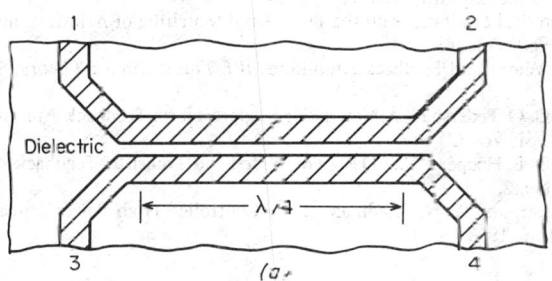

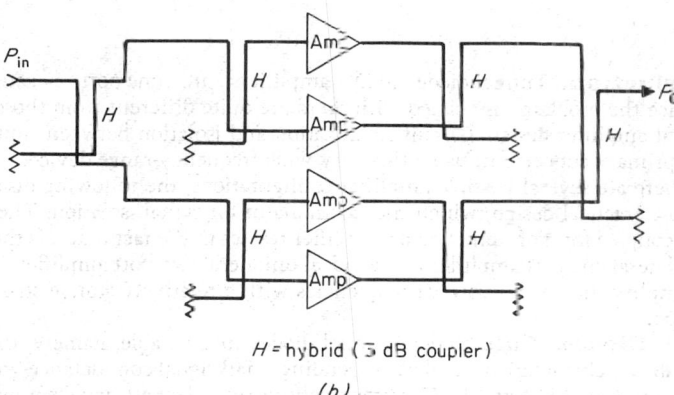

H = hybrid (3 dB coupler)

(b)

Fig. 13-75. Wide-band combining techniques: (*a*) wide-band quarter-wave coupler; (*b*) balanced combining amplifier configuration.

application are coupling bandwidth, isolation, fabrication ease, and form. The equiamplitude, quadrature-phase, reverse-coupled TEM $\lambda/4$ coupler is a particularly attractive implementation due to its bandwidth and amenability to strip-transmission-line circuits. Figure 13-75a illustrates this coupler type.

67. References on Broadband Amplifiers

1. Wheeler, H. A.   The Interpretation of Amplitude and Phase Distortion in Terms of Paired Echoes, *Proc. IRE*, June 1939, Vol. 27.
2. DiToro, M. J.   Phase and Amplitude Distortion in Linear Networks, *Proc. IRE*, January 1948, Vol. 36.
3. Bangert, J. T.   Practical Applications of Time Domain Theory, *IRE WESCON Conv. Rec.*, 1959, Pt. 3.
4. Glasford, G. M.   "Fundamentals of Television Engineering," McGraw-Hill, New York, 1955.
5. Valley, G. E., Jr., and H. Wallman   "Vacuum Tube Amplifiers," McGraw-Hill New York. 1948.
6. Muller, F. A.   High Frequency Compensation of *RC* Amplifiers, *Proc. IRE*. August 1954.
7. Lewis, I. A. D., and F. H. Wells   "Millimicrosecond Pulse Techniques," Pergamon, New York, 1959.
8. Moore, A. D.   Synthesis of Distributed Amplifiers for Prescribed Amplitude Response, *Stanford Univ. Tech. Rep.* 53, August 1952.
9. Ginzton, E. L., W. R. Hewlett, J. H. Jasberg, and J. D. Noe   Distributed Amplification, *Proc. IRE*, August 1948.
10. Roeshot, L. F.   UHF Broadband Transistor Amplifiers, *Elec. Des. News*, January–March, 1963.

11. Ruthroff, C. L.   Some Broad-Band Transformers, *Proc. IRE*, August 1959, pp. 1337–1342.

12. Young, L.   Tables of Cascaded Homogeneous Quarter Wave Transformers, *IRE Trans GMTT*, April 1959, Vol. MTT-7.

13. Cohn, S. B.   Optimum Design of Stepped Transmission Line Transformers, *IRE Trans. GMTT*, April 1955, Vol. MTT-3.

14. Womack, C. P.   The Use of Exponential Transmission Lines in Microwave Components, *IRE Trans.*, March 1962, Vol. MTT-10.

15. Matthaei, G. L.   Tables of Chebyshev Impedance Transforming Networks of Low Pass Filter Form, *Proc. IEEE*, August 1964.

16. Matthaei, G. L., L. Young, and E. M. T. Jones   "Microwave Filters, Impedance Matching Networks, and Coupling Structures," McGraw-Hill, New York, 1964.

17. Fano, R. M.   Theoretical Limitations on the Broad-Band Matching of Arbitrary Impedance, *J. Franklin Inst.*, January–February 1950, Vol. 249.

18. Waldhauer, F. D.   Wide Band Feedback Amplifiers, *IRE Trans. Circuit Theory*, September 1957, Vol. CT-4.

19. Ghausi, M. S., and D. O. Pederson   A New Design Approach for Feedback Amplifiers, *IRE Trans. Circuit Theory*, September 1961, Vol. CT-9.

20. Cherry, E. M., and D. E. Hooper   The Design of Wide-Band Transistor Feedback Amplifiers, *Proc. IEE*, February 1963, Vol. 110, No. 2.

21. Seidel, H., H. Beurrier, and A. N. Friedman   Error-Controlled High Power Linear Amplifier at VHF, *Bell Syst. Tech. J.*, May–June 1968.

Tunnel-Diode Amplifiers

BY C. S. KIM

68. Introduction. Tunnel-diode (TD) amplifiers are one-port negative-conductive devices.[1]* Hence the problems associated with them are quite different from those encountered in conventional amplifier design. Circuit stabilization and isolation between input and output terminals are primary concerns in using this very wide-frequency-range device.

Although there are several possible amplifier configurations, the following discussion is limited to the most practical design, which uses a circulator for signal isolation. The advantage of the circulator-coupled form of tunnel-diode amplifier resides in the fact that it is thereby possible to convert a bilateral one-port amplifier into an ideal unilateral two-port amplifier. Tunnel diodes can provide amplification at microwave frequencies with a relatively simple structure and at a low noise figure.

69. Tunnel Diodes. Three kinds of tunnel diodes are available, namely, those using Ge, GaAs, and GaSb. *V-I* characteristics and corresponding small-signal conductance-voltage relationships are shown in Figs. 13-76 and 13-77, respectively. A typical small-signal tunnel-diode equiv-

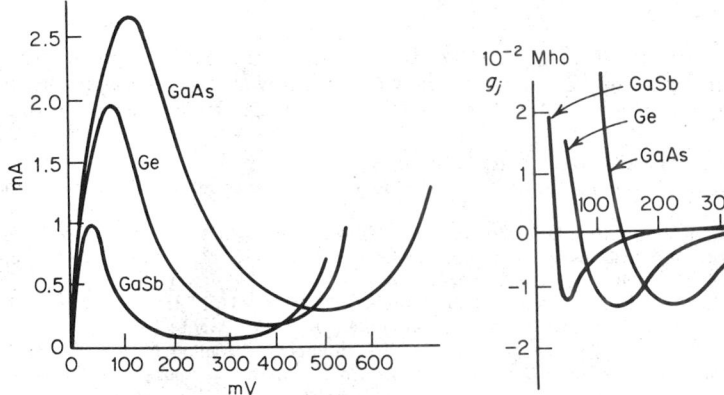

Fig. 13-76. Characteristics of tunnel diodes.

Fig. 13-77. Voltage-vs.-g_j characteristics of tunnel diodes of Fig. 13-76.

*Superior numbers correspond to numbered references, Par. **13-73**.

alent circuit[2] is shown in Fig. 13-78. Here g_j, C_j, r_s, L, and C_p are the small-signal conductance, junction capacitance, series resistance, series inductance, and shunt capacitance. Noise generators e_s and i_j are included for subsequent discussion.

70. Stability. Several studies of the stability conditions of tunnel-diode amplifiers have been reported.[3,4] The stability criteria are derived from the immittance expression across the diode terminals and are quite complicated.

Using the short-circuit stable condition,[5] the stability criteria can be simplified considerably. With reference to Fig. 13-78, with the external circuit connected, the total admittance Y_T across g_j can be expressed as

$$Y_T(s) = Y_i(s) - g_j = p(s)/q(s)$$

where $Y_i(s)$, the admittance facing g_j, is a positive real function.[6] Since Y_i is connected in parallel with a short-circuit stable device with negative conductance g_j, Y_T is short-circuit-stable.[5,7] This

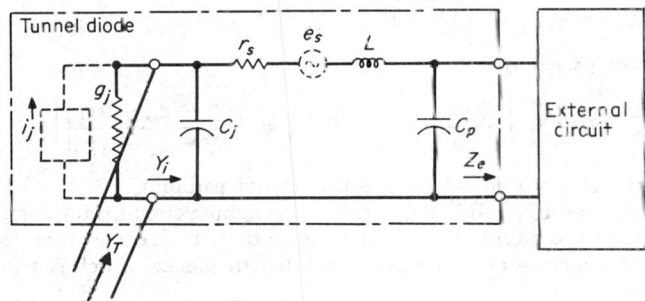

Fig. 13-78. Small-signal equivalent circuit of tunnel-diode amplifier.

implies that, since $q(s)$ is always a Hurwitz polynomial, the stability condition is that $p(s)$ must also be a Hurwitz polynomial.[6]

A simple graphical interpretation of this stability condition is as follows. The plot* of $Y_T(\omega)$ can be obtained from the plot of $Y_i(\omega)$ by shifting the imaginary axis by $|g_j|$ along the real axis, as shown in Fig. 13-79. Since $q(s)$ has no roots in the right half plane, any encirclement of the origin of $Y_T(\omega)$ must come from the right-half-plane roots of $p(s)$ only. Therefore, the circuit will be stable if and only if $Y_T(\omega)$ does not encircle the origin (Fig. 13-79a). If g_j becomes large so that the origin is encircled by $Y_T(\omega)$ (Fig. 13-79b), the circuit is unstable.

71. Tunnel-Diode-Amplifier Design. A simplified block diagram of a circulator coupled TD amplifier is shown in Fig. 13-80. The three basic circuit parts are the tunnel diode the stabilizing circuit, and the tuning circuit, which includes a four-port circulator. The following conditions are imposed on the amplifier design:

1. In the band
   a. $G_i = \text{Re } Y_i$ is slightly larger than $|g_j|$.
   b. G_i is contributed by the tuning circuit only.
   c. $B_i = \text{Im } Y_i = 0$ at the center frequency f_0 and small in the band.

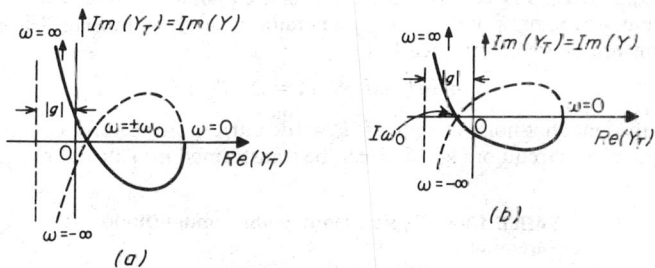

Fig. 13-79. Representative plots of real and imaginary parts of Y_T: (a) stable condition; (b) unstable condition.

*Here, the plot of $YT_r(\omega)$ represents the case of $Z_e(\omega)$ short-circuited. However, similar plots can be obtained for more general cases.

2. Outside the band

    a. $G_i = \mathrm{Re}\ Y_i$ is larger than $|g_j|$

    b. If $G_i \leq |g_j|$, B_j should not be zero.

To satisfy these conditions, a stabilizing circuit, shown in Fig. 13-80, is required. This circuit is designed so that the following relationships are satisfied:

$$Y_1(f_0) = Y_1(3f_0) = Y_2(f_0) = 0 \qquad Y_1(2f_0) = Y_1(4f_0) = Y_2(3f_0) = 1/R \qquad Y_s = Y_1 + Y_2$$

where f_0 = center frequency.

The equivalent circuit of Fig. 13-78 can be transformed into the parallel equivalent circuit of Fig. 13-81. The following identities relate the parameters of Figs. 13-78 and 13-81:

$$g_{jp} = 1/|Z_s|^2 g_j X \qquad g_{sp} = r_s/|Z_s|^2$$

$$B = \omega C_p - \frac{\omega L - (1 - 1/X)1/\omega C_j}{|Z_s|^2} \qquad \text{where } X = 1 + \omega^2 c_j^2/g_j^2$$

$$|Z_s|^2 = \gamma_s - 1/|g_j| X + [\omega L - (1 - 1/X)1/\omega C_j]^2$$

The gain can be expressed by

$$|\Gamma|^2 = \left| \frac{Y_e - Y_d}{Y_e + Y_d} \right|^2 \qquad |\Gamma|^2_{f=f_0} = \left| \frac{G_e - g_{sp} + |g_{jp}|}{G_e + g_{sp} - |g_{jp}|} \right|^2$$

where $B_e - B = 0$ at $f = f_0$, in which B_e is the reactive part of Y_e.

In this case, the tuning circuit is a more general matching circuit using combinations of parallel and series transmission lines. It should be noted that $|\Gamma|$ becomes 1, or $g_{sp} = |g_{jp}|$, as the operating frequency increases to f_r, the resistive cutoff frequency, which is defined by

$$f_r = \frac{|g_j|}{2\pi C_j} \sqrt{1 - \frac{1}{r_s|g_j|}}$$

It is desirable to have a device with f_r several times (at least 3) larger than f_0. Furthermore, it is desirable to make the self-resonance frequency f_x

$$f_x = \frac{1}{2\pi} \sqrt{\frac{1}{LC_j} - \left(\frac{g_j}{C_j} \right)^2}$$

as high as possible (higher than f_c) to improve the stability margin.

The gain expression can be modified to include the imput amplifier admittance Y_a of Fig. 13-80 as follows:

$$|\Gamma|^2 = \left| \frac{Y_0 - Y_a}{Y_0 + Y_a} \right|^2$$

Similarly, the bandwidth can be determined from the expression for Γ^2 and f_x, above.

Typical germanium tunnel diode parameters pertinent to S-, C-, and X-band amplifiers with approximate gains of 10 dB are shown in Table 13-6.

72. Noise Figure in TD Amplifiers. A noise equivalent circuit can be completed[9] by inserting a current generator, i_j^2 and a voltage generator e_s^2, as shown in Fig. 13-78 (dotted lines). The mean-square values are determined from

$$\overline{i_j^2} = 2eI_{eq}\,\Delta f \qquad \overline{e_s^2} = 2KTr_s\,\Delta f$$

where I_{eq} = equivalent shot-noise current = I_{dc} = dc current in negative-conductance region.

The noise equivalent circuit of Fig. 13-78 can be transformed into a parallel-equivalent circuit

TABLE 13-6 Typical Germanium Tunnel-Diode Parameters

	S band	C band	X band
$|r_j| = 1/g_j$, Ω	70	70	70
C_j, pF	<1	<0.5	<0.2
L, H	<0.1	<0.1	<0.2
r_s, Ω	<0.1r_j	<0.1r_j	<0.1r_j

of Fig. 13-81 having current generators $\overline{i_{jp}^2}$ and $\overline{i_{sp}^2}$ (dotted lines). $\overline{i_{jp}^2}$ and $\overline{i_{sp}^2}$ can be derived from the two equivalent circuits to be

$$\overline{i_{jp}^2} = 4KT(G_{eq}/|g_j|)|g_{jp}|\,\Delta f \qquad \overline{i_{sp}^2} = 4KTg_{sp}\,\Delta f$$

where

$$G_{eq} = eI_{eq}/2KT = 20I_{eq} \text{ at room temperature}$$

In Fig. 13-82, the noise figure F of a circulator-coupled TD amplifier is derived, assuming that the stability circuit is opened at the operating frequency, as follows:

$$F = \frac{\text{total mean-square noise currents appearing at port 3}}{\text{mean-square noise current appearing at port 3 contributed by source}}$$

$$= \frac{\overline{i_s^2}|\Gamma|^2 + \overline{i_2^2}|\Gamma - 1|^2}{\overline{i_s^2}|\Gamma|^2}\left(1 + \frac{\overline{i_2^2}|\Gamma - 1|^2}{\overline{i_s^2}|\Gamma|^2}\right)$$

where

$$\overline{i_s^2} = 4KT_0G_0\,\Delta f \qquad \overline{i_2^2} = \overline{i_{sp}^2} + \overline{i_{jp}^2} \qquad \Gamma = \frac{Y_0 - Y_a}{Y_0 + Y_a}$$

If $|\Gamma|$ approaches infinity, F becomes

$$F = 1 + \frac{\overline{i_2^2}}{\overline{i_s^2}} = 1 + \frac{(g_{sp} + G_{eq}|g_{jp}|/|g_j|)T}{G_sT_0} = \frac{1 + G_{eq}/|g_j|}{(1 - r_s|g_j|)[1 - (f/f_r)^2]}$$

For low F_0, $r_s|g_j|$, f/f_r, and $G_{eq}/|g_j|$ should be made small. If I_{eq} could be approximated by I_{dc} in the negative conductive region, $I_{dc}/|g_j|$ would be determined by the material used. $I_{dc}/|g_j|$ is minimum at the voltage where $|g_j|$ becomes maximum.

A typical $r_s|g_j|$ value is 0.1. Therefore, if $f/f_r = \frac{1}{3}$, then for $I_{eq}/|g_j| = 0.06$ (Ge) and $I_{eq}/|g_j| = 0.04$ (GaSb), F_0 becomes, in each case,

$$F_0 = \begin{cases} \dfrac{1 + 20(0.06)}{(1 - 0.1)(1 - 1/3^2)} = 4.4 \text{ dB} & \text{for Ge} \\[2mm] \dfrac{1 + 20(0.04)}{(1 - 0.1)(1 - 1/3^2)} = 3.5 \text{ dB} & \text{for GaSb} \end{cases}$$

Tunnel diodes provide good noise performance with a relatively simple amplifier structure

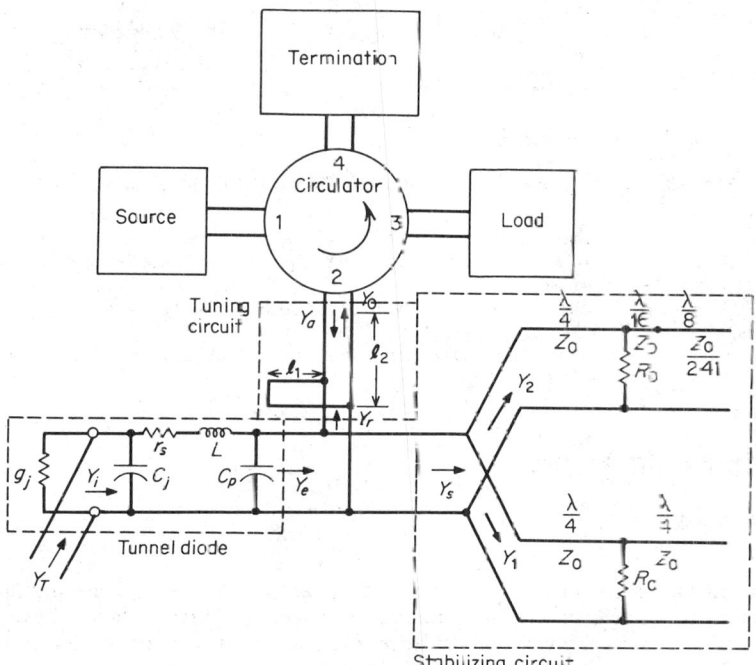

Termination

4
Circulator

Source 1 3 Load

2

Tuning circuit Y_a ℓ_2

ℓ_1 Y_r

g_j Y_i C_j C_p Y_e

Tunnel diode

Y_T

$\frac{\lambda}{4}$ $\frac{\lambda}{16}$ $\frac{\lambda}{8}$
Z_0 Z_0 Z_0
R_0 241

Y_2

Y_s

$\frac{\lambda}{4}$ $\frac{\lambda}{4}$
Y_1 Z_0 Z_0
R_C

Stabilizing circuit

Fig. 13-80. Tunnel-diode amplifier using circulator.

requiring only a dc source for bias. It is therefore a useful low-noise, small-signal amplifier for microwave applications.

73. References on Tunnel-Diode Amplifiers

1. Shea, R. F.   "Amplifier Handbook," Chap. 12, McGraw-Hill, New York, 1968.

2. Kim, C. S., and C. W. Lee   Microwave Measurements on Tunnel Diode Parameters, *Microwave*, November 1964.

3. Sniden, L. I., and D. C. Yonta   Stability Criteria for Tunnel Diodes, *Proc. IRE*, July 1961, Vol. 49, pp. 1206–1207.

4. Hines, M. E.   High-Frequency Negative-Resistance Circuit Principles for Esaki Diodes Application, *Bell Syst. Tech. J.*, May 1960, Vol. 39.

5. Bode, H. W.   "Network Analysis and Feedback Amplifier Design," Van Nostrand, Princeton, N.J., 1945.

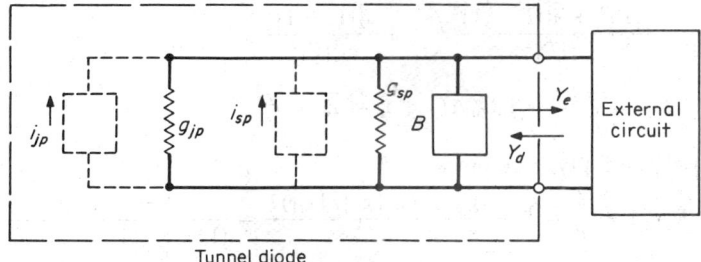

Fig. 13-81.   Parallel version of equivalent of Fig. 13-78.

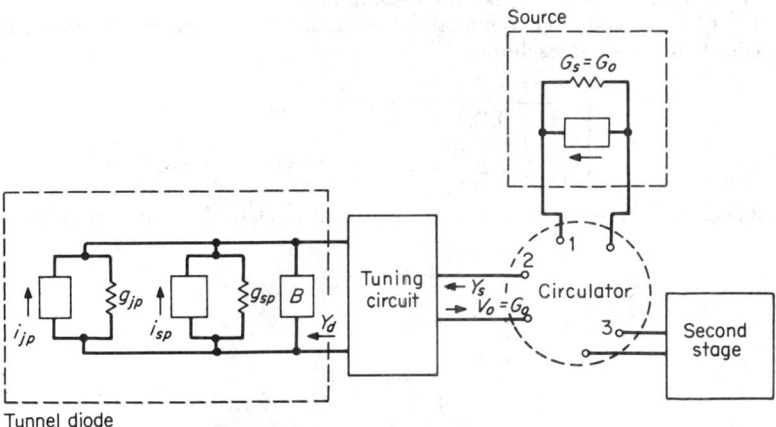

Fig. 13-82.   Noise equivalent circuit of circulator-coupled tunnel-diode amplifier.

6. Guillemin, E. A.   "Synthesis of Passive Networks," Wiley, New York, 1957.

7. Kim, C. S., and A. Brandli   High Frequency High Power Operation of Tunnel Diodes, *IRE Trans. Circuit Theory*, December 1962, pp. 416–426.

8. Kuh, E. S., and J. D. Patterson   Design Theory of Optimum Negative Resistance Amplifier, University of California, Electronics Research Laboratory, Dec. 6, 1960.

9. Tiemann, J. J.   Shot Noise in Tunnel Diode Amplifier, *Proc. IRE*, August 1960, Vol. 48, No. 8, p. 1418.

Parametric Amplifiers

BY C. E. NELSON

74. Introduction.   The term *parametric amplifier* (paramp) refers to an amplifier (with or without frequency conversion) using a nonlinear or time-varying reactance. Development of low-loss variable-capacitance (varactor) diodes resulted in the development of varactor-diode parametric amplifiers with low noise figure in the microwave frequency region. Types of paramps include one-port, two-port (traveling-wave), degenerate (pump frequency twice the signal frequency), nondegenerate, multiple pumps, and multiple idlers. The most widely used amplifier is the nondegenerate one-port paramp with a circulator, because it achieves very good noise figures without undue circuit complexity.

75. One-Port Paramp with Circulator. The one-port paramp with circulator is illustrated in Fig. 13-83, and the simplified circuit diagram is shown in Fig. 13-84. The input signal and the amplified output signal (at the same frequency) are separated by the circulator. The cw pump source is coupled to a back-biased varactor to drive the nonlinear junction capacitance at the pump frequency. The signal and pump currents mix in the nonlinear varactor to produce voltages at many frequencies. The additional (idler) filter allows only the difference or idler current to flow at the idler frequency; i.e.,

$$f_i = f_p - f_s$$

The idler current remixes with the pump current to produce the signal frequency again. The phasing of this signal due to nonlinear reactance mixing is such that the original incident signal is reenforced (i.e., amplified) and reflected back to the circulator. The one-port paramp at band center is essentially a negative-resistance device at the signal frequency.

76. Power Gain and Impedance Effects. The one-port paramp power gain at the signal frequency is

$$G = \left| \frac{Z_{in} - R_g}{Z_{in} + R_g} \right|^2$$

where R_g = signal-circuit equivalent generator resistance, and Z_{in} = input impedance at the signal frequency.

For a single-tuned signal circuit and a single-tuned idler circuit the input impedance at the signal frequency is

$$Z_{in} = \beta_s R_d + jX_s - \frac{\sigma \beta_s \beta_i R_d^2}{\beta_i R_d - jX_i} \qquad \text{where } \sigma = \frac{m_i^2 f_e^2}{\beta_s \beta_i f_s f_i}$$

The diode loss resistance R_d has been modified at the signal and idler frequencies to include circuit losses (that is, $\beta \geq 1$). The signal and idler reactances are X_s and X_i, respectively, and must include the varactor junction capacitance. (In the simplified circuit the diode package capacitance has been neglected.) For a fully pumped varactor the cutoff frequency is

$$f_c = 1/2\pi R_d C_R$$

where C_R = junction capacitance at reverse breakdown. The modulation ratio m_1 for an abrupt-junction diode is 0.25.[3]*

77. Bandwidth and Noise. Factors that determine the overall paramp bandwidth include the varactor characteristics (cutoff frequency, junction and package capacitance, and lead

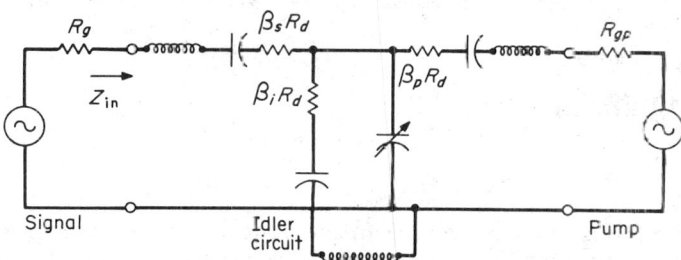

Fig. 13-84. Circuit of one-port paramp using single-tuned resonators.

*For references, see Par. **13-80**.

inductance), choice of idler (pump) frequency, the nature of the signal and idler resonant circuits and the choice of band-center gain. Multiple-tuned signal and idler circuits are often used to increase the overall paramp bandwidth.

At band center the effective noise temperature of the one-port paramp (due to the diode and circuit loss resistances at the signal and idler frequencies) is

$$T_e = T_d \frac{G-1}{G} \left(\frac{f_p}{f_i} \frac{\sigma}{\sigma - 1} - 1 \right)$$

where T_d = temperature of varactor junction and loss resistances. The effective noise temperature can be reduced by cooling the paramp below room temperature and/or by proper choice of pump frequency. Circulator losses must be included when determining the overall paramp noise figure.

78. Pump Power. The pump power required at the diode to fully drive a single varactor is

$$P_p = k_1 \beta_p R_d [\omega_p C_R (\phi - V_R)]^2 \quad \text{(watts)}$$

where the pump circuit loss resistance and diode resistance = $\beta_p R_d$, and k_1 = 0.5 for an abrupt-junction varactor.

79. Gain and Phase Stability. The stability of the cw pump source (amplitude and frequency) is a significant factor in the overall paramp gain and phase stability. For small pump-power changes an approximate expression for the paramp power-gain variation is

$$\frac{\Delta G}{G} \approx \frac{G-1}{\sqrt{G}} \frac{\Delta P_p}{P_p}$$

Environmental temperature changes often require a temperature-regulated enclosure for the paramp and pump source.

At low levels of nonlinear distortion, the amplifier third-order relative power intermodulation product at band center is a function of the one-port amplifier gain and the incident signal power level; i.e.,

$$\Delta(\text{IMP}) \propto \frac{(G-1)^4 P_s^2}{G} \approx G^3 P_s^2 \approx G P_{\text{out}}^2$$

Thus low-band-center amplifier gains reduce this nonlinear distortion.

80. References on Parametric Amplifiers

1. Manley, J. M., and H. E. Rowe   Some General Properties of Nonlinear Elements, Pt. 1, General Energy Relations, *Proc. IRE*, July 1956, Vol. 44, pp. 904–913.

2. Blackwell, L. A., and K. L. Kotzebue   "Semiconductor-Diode Parametric Amplifiers," Prentice-Hall, Englewood Cliffs, N.J., 1961.

3. Penfield, P., Jr., and R. P. Rafuse   "Varactor Applications," The M.I.T. Press, Cambridge, Mass., 1962.

4. Chang, K. K. N.   "Parametric and Tunnel Diodes," Prentice-Hall, Englewood Cliffs, N.J., 1964.

5. Aitchison, C. S., R. Davies, and P. J. Gibson   A Simple Diode Parametric Amplifier Design for the Use at S, C. and X Band, *IEEE Trans.*, January 1967, Vol. MTT-15, pp. 22–31.

6. Porra, V., and P. Somervuo   Broadband Matching of a Parametric Amplifier by Using Fano's Method, *IEEE Trans.*, October 1968, Vol. MTT-16, pp. 880–882.

7. Takahashi, S., M. Nojima, T. Fukuda, and A. Yamada   K-Band Cryogenically Cooled Wide-Band Non-degenerate Parametric Amplifier, *IEEE Trans.*, December 1970, Vol. MTT-18, pp. 1176–1178.

8. Okean, H. C., C. M. Allen, E. W. Sard, and H. Weingart   Integrated Parametric Amplifier Module with Self-Contained Solid-State Pump Source, *IEEE Trans.*, May 1971, Vol. MTT-19, pp. 491–493.

Maser Amplifiers

BY G. K. WESSEL

81. Introduction. A *maser* is a microwave active device whose name is an acronym derived from microwave amplification by stimulated emission of radiation. A *laser* is an amplifier or oscillator also based on stimulated emission but operating in the optical part of the spectrum. The expressions microwave, millimeter, submillimeter, infrared, optical, and ultraviolet maser are also in common use. Here the word maser is used for the whole frequency range of devices, the *m* standing for molecular.

A description of the laser principle, some of its properties and means of achieving laser operation, is given in Sec. **11**. A complete treatment of the subject is available in monographs[1-4]* and review articles.[5,6]

In these devices, no tubes or transistors are employed in the radiation process, but the properties of atoms or molecules—as gases, liquids, or solids—serve for the amplification of the signals. Oscillators require the addition of a feedback mechanism. The interaction of the electromagnetic radiation with the maser material occurs in a suitable cavity or resonance structure. This structure often serves the additional purpose of generating a desired phase relationship between different spatial parts of the signal. The application of external energy, required for the amplification or oscillation process, is referred to as *pumping*. The pumping power consists, in many cases, of external electromagnetic radiation of a different frequency (usually higher) from that for the signal.

Microwave masers are used as low-noise preamplifiers and as time and frequency standards. The stimulated emission properties of atoms and molecules at optical frequencies, with their relatively high noise content, make the laser more useful for high-power light amplification and oscillation.

82. Maser Principles. In the processes of emission and absorption, atoms or molecules interact with electromagnetic radiation. It is assumed that the atoms possess either sharp internal-energy states or broader energy bands. A change of internal energy from one state to another is accompanied by absorption or emission of electromagnetic radiation, depending on the direction of the process. The difference in energy from the original to the final state is proportional to the frequency of the radiation, the proportionality constant being Planck's constant h.

The processes of the spontaneous emission and absorption were defined by Einstein:[7]

Spontaneous emission:
$$-\frac{dN_2}{dt} = A_{21}N_2 \tag{13-14}$$

Absorption:
$$-\frac{dN_1}{dt} = B_{12}\mu(\nu_{12}, T)N_1 \tag{13-15}$$

where N_1 and N_2 = population densities of atoms or molecules of material having energy states E_1 and E_2 and an energy separation $E_2 - E_1 = h\nu_{12}$ between them, and A_2 and B_{12} are the Einstein coefficients.

At high temperature, an inconsistency in the equilibrium system arises unless a second emission process, stimulated emission, takes place. The rate of stimulated emission is defined very similarly to that of absorption (the stimulated emission is sometimes called *negative absorption*):

$$-\frac{dN_2}{dt} = B_{21}\mu(\nu_{12}, T)N_2 \tag{13-16}$$

In temperature equilibrium, the rates of emission must be equal to the rate of absorption

$$A_{21}N_2 + B_{21}\mu N_2 = B_{12}\mu N_1 \tag{13-17}$$

Assuming that the ratio of the population densities is equal to the Boltzmann factor,†

$$N_1/N_2 = \exp(h\nu_{12}/kT) \tag{13-18}$$

Equation (13-17) yields, for the radiation density,

$$\mu(\nu_{12}, T) = \frac{A_{21}/B_{21}}{B_{12}/B_{21} \exp(h\nu_{12}/kT) - 1} \tag{13-19}$$

Equation (13-19) is identical with Planck's radiation law if we set

$$B_{12} = B_{21} \tag{13-20a}$$

and
$$A_{21}/B_{21} = 8\pi h\nu_{12}^3/c^3 \tag{13-20b}$$

where c = velocity of light.

Equation (13-20a) shows the close relationship between stimulated emission and absorption. The rates of population decrease, for these two processes [Eqs. (13-15) and (13-16)] depend only on their respective population densities. Neglecting the spontaneous emission, the net absorption

---

*Superior numbers correspond to numbered references, Par. **13-92**.
†It is assumed that the statistical weights are equal to 1.

or the net stimulated emission of an incoming radiation depends, therefore, only on the difference in population density: for absorption, if $(N_1 - N_2) > 0$; for amplification, if $(N_1 - N_2) < 0$. In particular, if a system is in quasi equilibrium such that the upper energy state, E_2, is more populated than the lower one, E_1, it is capable of amplifying electromagnetic radiation.

To create and maintain the quasi equilibrium requires application of external energy since the natural equilibrium has the opposite population excess $(N_1 > N_2)$. Different masers differ widely in the methods of how to accomplish the reverse in population density.

83. Properties of Masers. The properties of masers can be understood by analyzing Eqs. (13-14) to (13-20b), as follows.

Signal-to-Noise Ratio. From Eqs. (13-14) and (13-16) it follows that the Einstein coefficients A and B have different dimensions but the ratio $B_{21}\mu/A_{21}$ is dimensionless. Here μ is proportional to the strength of the incoming signal and $B_{21}\mu$ is proportional to the amplified signal strength. A_{21} is proportional to the noise contribution due to spontaneous emission. After rewriting Eq. (13-20b),

$$B_{21}\mu/A_{21} = c^3\mu/8\pi h\nu_{12}^3 \qquad (13\text{-}21)$$

the ratio is found to be proportional to the signal-to-noise ratio of the amplifier. It thus becomes clear that the noise contribution (thermal or Johnson noise) is very small at microwave frequencies, whereas it becomes very large at optical frequencies (an increase of 15 orders of magnitude for an increase of 5 orders for the frequency). Microwave masers are therefore very useful as low-noise microwave preamplifiers.

Infrared, optical, and ultraviolet masers (lasers), on the other hand, are commonly used for power amplification and as powerful light sources (oscillators). The high content of spontaneous-emission noise does not make them easily applicable for low-noise amplification or sensitive detection of light, except in a few special cases.

84. Linearity and Line Width. The proportionality between the incoming signal strength, the radiation density μ, and the amplified signal strength means that a maser is a linear amplifier as long as it is not driven into saturation. The latter occurs when the excess population becomes $N_2 - N_1 \approx N_2$.

Line Width. In the absence of a strong interaction of the individual atoms of the material, either with the environment or with each other, the line width is determined by the average length of time of interaction of an atom with the radiation field. However, if regeneration due to feedback is taking place, the line width may be much narrower because of the following effect. It is assumed that the incoming radiation has a frequency distribution as shown in Fig. 13-85. The amplified signal is proportional to the rate of stimulated emission dN_2/dt. According to Eq. (13-16), the latter is proportional to $\mu(\nu_{12})$. Thus the center portions of the line will be more amplified than the wings, leading to increasingly narrower line widths. In an oscillator the final width is ultimately limited by statistical fluctuations due to the quantum nature of the radiation.

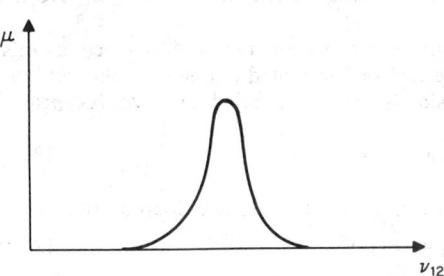

Fig. 13-85. Frequency distribution of radiation incoming to maser.

The narrow line width coupled with the low-noise property of a microwave maser makes the latter useful as a time and frequency standard device. To reduce interaction, gases are used as microwave maser materials.

In low-noise microwave amplifiers one is usually interested in a relatively large bandwidth (typically 50 MHz or more). For such amplifiers solid-state materials are used, with strong spin-spin interactions of atoms (for example, Cr^{3+} in ruby).

85. Tuning. Solid-state microwave masers can be tuned over a wide range of frequencies by adjustment of an external magnetic field, while the relative tuning range by a magnetic field is smaller by a factor of 10^{-4} to 10^{-5} in optical masers. Generally, in lasers, one has to depend on fixed frequencies wherever a suitable spectral line occurs in the material. However, modern liquid-dye lasers can be tuned over a relatively wide range in the visible spectrum.

86. Power Output. The power output due to stimulated emission depends on the number of excess atoms $(N_2 - N_1)$ available for the transition. Assuming p to be the probability of a transition per second, the power output is given by

$$P = (N_2 - N_1)h\nu_{12}p \qquad (13\text{-}22)$$

Equation (13-22) shows that the power output of a microwave gas maser must be very low (for example, 100 to 1 pW) because the population density and the frequency are small. On the other hand, solid-state lasers and some gas lasers with very high efficiencies may have cw power outputs of 1,000 W and more.

87. Coherence. The similarity between absorption and stimulated emission [as expressed in Eqs. (13-15), (13-16), and (13-20a)] means that the stimulated emission is a coherent process. In practice, this means that a beam of radiation will be amplified only in the direction of propagation of the incoming beam and when all atoms participate coherently in the amplification. This is very different from the noise-producing spontaneous emission, where the radiation is emitted isotropically and no phase relationship exists between the radiation coming from different atoms.

The coherence of the stimulated emission enables one to produce coherent light and radiation patterns similar to those which are well known and applied in the radio and microwave parts of

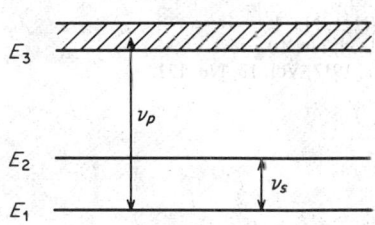

Fig. 13-86. Three-level maser.

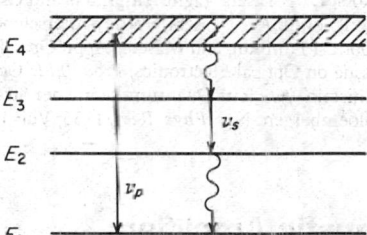

Fig. 13-87. Four-level maser.

the spectrum. In particular, it is possible by a suitable geometry of the device to produce a plane-parallel light beam whose divergence angle is limited only by diffraction. The divergence angle is given by

$$\alpha \approx \lambda/d \tag{13-23}$$

88. Time and Frequency Standards. The ammonia and hydrogen atomic-beam masers are used as time and frequency standards with operating frequencies of about 24 GHz and about 1.4 GHz, respectively. The reverse of population difference is achieved by focusing atoms in the excited state into a suitable microwave cavity, whereas the atoms in the lower state are defocused by a special focuser and do not reach the interior of the cavity. The microwave field in the cavity causes stimulated emission and amplification. If the cavity losses can be overcome, oscillations will set in. Time and frequency accuracy of 1 part in 10^{12} or higher can be obtained with modern maser standard devices.

89. Three- and Four-Level Devices. In many of the masers of all frequency regions, population reverse is obtained by a *three-level-maser* method first described by Bloembergen.[8] Radiation from an external power source at the pumping frequency ν_p (Fig. 13-86) is applied to the material. E_3 may be a band to make the pumping more economical. The material is contained in a suitable cavity, a slow-wave structure, or other resonance structure which is resonant at the signal frequency ν_s. Under favorable conditions of the pumping power, the frequency ratio ν_s/ν_p, and the involved relaxation times, an excess population in the excited state E_2 over the ground state E_1 can be obtained. The interaction with the electromagnetic field of frequency ν_s causes stimulated emission. Thus amplification (and if required, oscillation) is produced.

The four-level maser (Fig. 13-87) has the additional advantage that the population of the ground state does not have to be reduced to less than 50% of its original equilibrium value and the level E_4 can remain a relatively wide band. The transfer of energy from E_4 to E_3, and E_2 to E_1, is by spontaneous decay or some other means like spin-lattice relaxation. Most solid-state microwave and optical masers are of the three- or four-level variety.

90. Semiconductor Lasers. Stimulated light emission can be obtained by carrier injection in certain semiconductor materials. Upon application of an external dc potential, recombination light emission can take place at the interface of an n- and p-type semiconductor (for example Zn and Te in GaAs). The semiconductor laser requires only a very simple power source (a few volts of direct current). However, the laser material has to be cooled to liquid-nitrogen temperature and the power output is relatively low.

91. Applications. The particular properties of devices based on stimulated emission have resulted in numerous applications which often surpass the capabilities of standard devices. In some instances, e.g., holography, new fields have opened up through the emergence of lasers and masers.

Solid-state microwave preamplifiers for communication, radar, and radio astronomy have by far the best signal-to-noise ratio of all microwave amplifiers. The hydrogen-beam maser is the most accurate of all frequency and time standards.

Lasers have revolutionized the field of optical instrumentation and spectroscopy. They have made possible the new field of nonlinear optics. The laser is or may be used in many applications where large energy densities are required, as in microwelding, medical surgery, or even in cracking rocks in tunnel building. Lasers can be used as communication media where extremely large bandwidths are required. Optical radar, with its high precision due to the short wavelength, uses lasers as powerful, well-collimated light sources. Their range extends as far as the moon.

92. References on Maser Amplifiers

1. Singer, J. R. "Masers," Wiley, New York, 1959.
2. Unger, H. G. "Introduction to Quantum Electronics," Pergamon, New York, 1970.
3. Ross, D. "Lasers, Light Amplifiers and Oscillators," Academic, New York, 1969.
4. Lengyel, B. A. "Lasers," Wiley-Interscience, New York, 1971.
5. Optical Pumping and Masers, *Appl. Opt.*, January 1962, Vol. 1, No. 1.
6. Issue on Optical Electronics, *Proc. IEEE*, October 1966, Vol. 54, No. 10.
7. Einstein, A. Zur Quantumtheorie der Strahlung, *Z. Fhys.*, 1917, Vol. 18, No. 121.
8. Bloembergen, N. *Phys. Rev.*, 1956, Vol. 104, No. 324.

Acoustic Amplifiers

BY S. W. TEHON

93. Acoustoelectric Interaction. The acoustic amplifier stems from the announcement by Hutson, McFee, and White,[1]* in 1961, that they had observed a sizable influence on acoustic waves in single crystals of CdS caused by a bias current of charge carriers. CdS is both a semiconductor and piezoelectric crystal, and the interaction was found to involve an energy transfer via traveling electric fields generated by acoustic waves, producing, in turn, bunching of the drift carriers. Quantitative analyses showing that either loss or gain in wave propagation could be selected by controlling the drift field were published by White[2] and Kino.[3]

Figure 13-88 illustrates the nature of the interaction. As an acoustic wave propagates in the

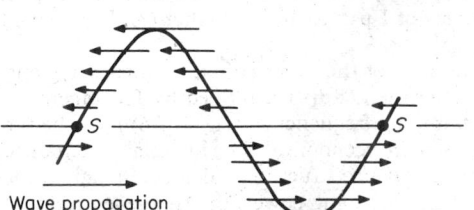
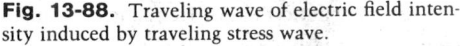

Wave propagation

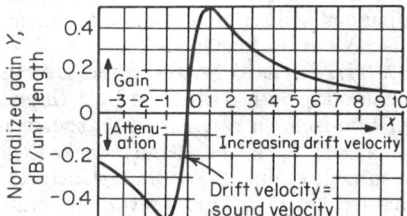

Fig. 13-88. Traveling wave of electric field intensity induced by traveling stress wave.

Fig. 13-89. Normalized gain as a function of bias field. [*After R. F. Shea (ed.), "Amplifier Handbook," Chap. 30, McGraw-Hill, New York, 1968.*]

crystal its stresses induce a similar pattern of electric field through piezoelectric coupling. Since the coupling is linear, compressive stresses in half the wave induce a forward-directed field, and tensile stresses in the remainder of the wave induce a backward-directed field. Drifting charge carriers tend to bunch at points of zero field to which they are forced by surrounding fields. When a charge carrier loses velocity in bunching, it gives its excess kinetic energy to the acoustic wave; when it gains velocity, it extracts energy from the wave. Therefore the drift field is effective for determining attenuation or amplification in the range near which carriers move at the speed of sound. Figure 13-89 illustrates the gain characteristic as a function of drift velocity. At the zero-crossover point, the carrier velocity is just equal to the velocity of propagation for the

*Superior numbers correspond to numbered references, see Par. **13-98**.

acoustic wave; the gain ranges from maximum attenuation up to maximum amplification over a relatively small change in bias. Beyond this range of drift velocity, bunching becomes decreasingly effective, and the interaction has less effect on the acoustic wave.

94. Piezoelectric Materials. Piezoelectricity is the linear, reversible coupling between mechanical and electric energy due to displacement of charges bound in molecular structure. Pressure applied to a piezoelectric material produces a change in observed surface density of charge, and conversely, charge applied over the surfaces produces internal stress and strain. If S is the strain, T the stress, E the electric field intensity, and D the dielectric displacement, the piezoelectric effect at a point in a medium is described by the pair of linear equations

$$S = sT + dE \qquad D = dT + \epsilon E$$

where constant s = elastic compliance, ϵ = permittivity, and d = piezoelectric constant. The ratio $d_2/s\epsilon$, calculated from these constants for a particular piezoelectric material, is defined as k^2, where k is the coefficient of electromechanical coupling. As a consequence of conservation of energy, it can be shown that k is a number less than unity and that a fraction k^2 of applied energy (mechanical or electric) is stored in the other form (electric or mechanical.) Since D and E are vectors and S and T are tensors, the piezoelectric equations are tensor equations and are equivalent to nine algebraic equations, describing three vector and six tensor components.

Materials that are appreciably piezoelectric are either crystals with anisotropic properties or ceramics with ferroelectric properties which can be given permanent charge polarization through dielectric hysteresis. Ferroelectric ceramics, principally barium titanate and lead zirconate titanate, are characterized by relative dielectric constants ranging from several hundred to several thousand, by coupling coefficients as high as 0.7, and by polycrystalline grain structure which will propagate acoustic waves with moderate attenuation at frequencies extending up to the low-megahertz range.

Single crystals are generally suited for much higher acoustic frequencies, and in quartz acoustic-wave propagation has been observed at 125 GHz. Quartz has low loss but a low coupling coefficient. Lithium niobate is a synthetic crystal, ferroelectric and highly piezoelectric; lithium tantalate is somewhat similar. The semiconductors cadmium sulfide and zinc oxide are moderately low in loss and show appreciable coupling.

95. Stress Waves in Solids. Acoustic waves propagate with low loss and high velocity (5,000 m/s) in solids. Solids also have shear strength, whereas sound waves in gases and fluids are simple pressure waves, manifested as traveling disturbances measurable by pressure and longitudinal motion.

Sound waves in solids involve either longitudinal or transverse particle motion. The transverse waves may be propagation of simple shear strains or may involve bending in flexural waves. Since different modes of waves travel with different velocities, and since both reflections and mode changes can occur at material discontinuities, the general pattern of wave propagation in a bounded solid medium is quite complicated.

Bulk waves are longitudinal or transverse waves traveling through solids essentially without boundaries; e.g., the wavefronts extend over many wavelengths in all directions. A solid body supporting bulk waves undergoes motion and stress throughout its volume. A *surface wave* follows a smooth boundary plane, with elliptical particle motion which is greatest at the surface and drops off so rapidly with depth that almost all the energy is carried in a one-wavelength layer at the surface.

Ideally, the wave medium for surface-wave propagation is regarded as infinitely deep; practically if it is many wavelengths deep, its properties are equivalent. A surface wave following a surface free from forces is a Rayleigh wave; if an upper material with different elastic properties bounds the surface, the motion may be a Stonely wave.

96. Bulk-Wave Devices. Most acoustic amplifiers utilizing bulk waves have the components shown in Fig. 13-90. The input and output transducers are piezoelectric crystals or

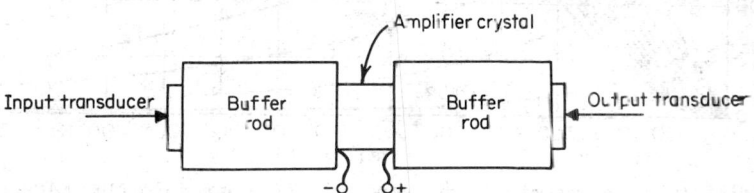

Fig. 13-90. Typical structure for acoustic amplification measurements. [*After R. F. Shea (ed.), "Amplifier Handbook," Chap. 30, McGraw-Hill, New York, 1968.*]

deposited thin layers of piezoelectric material, used for energy conversion at the terminals. The amplifier crystal is generally CdS, which is not only piezoelectric but also an n-type semiconductor. Electrodes are attached at the input and output surfaces of the CdS crystal, for bias current. Since the mobility of negative-charge carriers in CdS is only about 250 cm^2/V·s, large bias fields are required to provide a drift velocity equal to acoustic velocity: in CdS, 4,500 m/s, for longitudinal waves and 1,800 m/s for shear waves. The buffer rods shown in Fig. 13-90 are therefore added for electrical isolation of the bias supply. Furthermore, the transducers and amplifying crystal are cut in the desired orientation, to couple to either longitudinal or shear waves.

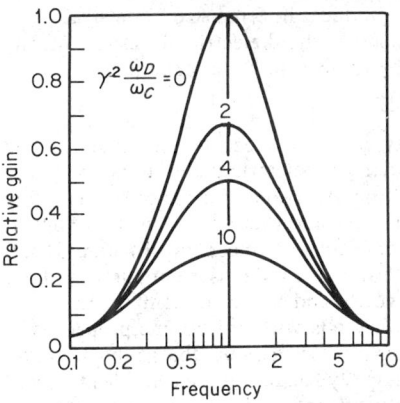

Fig. 13-91. Normalized plots of gain vs. frequency. Note symmetry of frequency response and effects of bias and critical frequency values. [*After R. F. Shea (ed.), "Amplifier Handbook," Chap. 30, McGraw-Hill, New York, 1968.*]

In the analysis of bulk-wave amplification[2,3] the crystal properties are characterized by f_D = diffusion frequency = $\omega_D/2\pi$; f_C = dielectric relaxation, or conductivity frequency = $\omega_C/2\pi$; $\gamma = (1 - f)$ times the drift velocity divided by the acoustic phase velocity, where f is the fraction of space charge removed from the conduction band by trapping; k = coefficient of electromechanical coupling; and $\bar{\omega}$ = radian frequency of maximum gain = $\sqrt{\omega_C\omega_D}$. Figure 13-91 shows the computed gain curves, using these constants, for four values of drift velocity. The frequency of maximum gain can be selected by control of crystal conductivity, which is easily accomplished in CdS by application of light. The amount of gain, and to some extent the bandwidth, are controlled by the bias field.

Figure 13-92 shows characteristics specifically calculated for CdS with shear acoustic-wave amplification. Large bias voltages (800 to 1,300 V/cm) are required, and extremely high gains are possible. Generally, heat dissipation due to the power supplied by bias is so high (up to 13 W/cm^3 at γ = 0.76) that only pulsed operation is feasible.

97. Surface-Wave Devices. The disadvantages in bulk-wave amplifiers, evident as excessive bias requirements, could be alleviated if materials with high mobility were available. However, the amplifier crystal must show low acoustic loss, high piezoelectric coupling, and high mobility; a suitable material combining these properties has not been found. Figure 13-93 shows an acoustic amplifier structure which operates with acoustic surface waves and provides

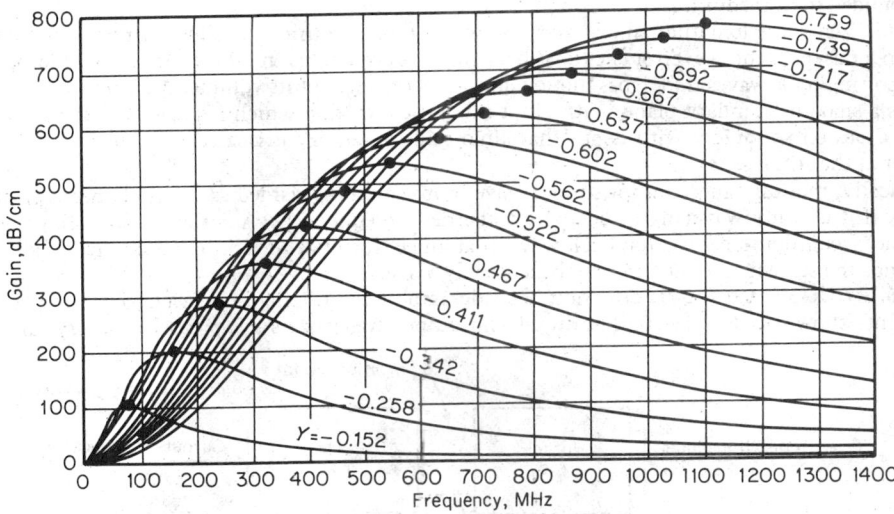

Fig. 13-92. Gain vs. frequency for CdS shear-wave amplification at f_D = 796 MHz. [*After R. F. Shea (ed.), "Amplifier Handbook," Chap. 30, McGraw-Hill, New York, 1968.*]

charge carriers in a thin film of silicon placed adjacent to the insulating piezoelectric crystal. Coupling for the interaction takes place in the electric field across the very small gap between piezoelectric and semiconductor surfaces. Transducers are formed by metallic fingers, interlaced to provide field for piezoelectric coupling in the regions between parallel fingers. This interdigital-array technique is flexible, providing means for complicated transducer designs.

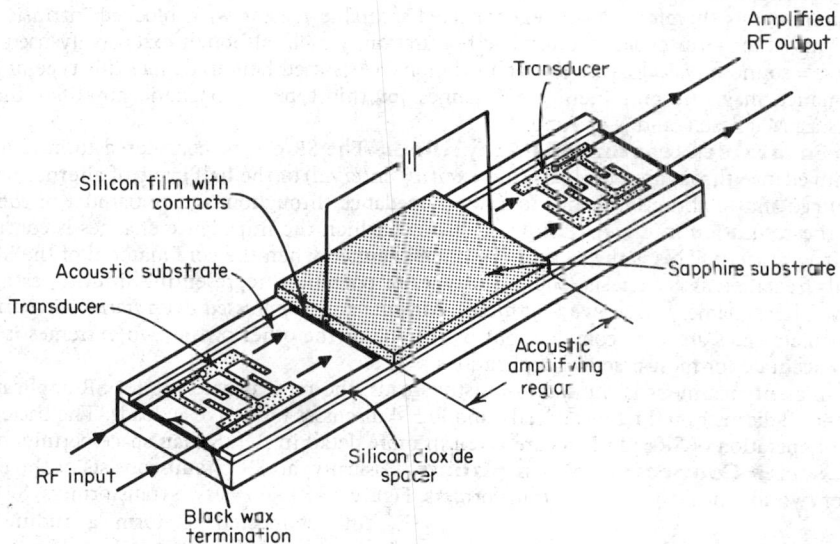

Fig. 13-93. Structure of a surface-wave acoustic amplifier. *(After J. H. Collins and P. J. Hagon, Electronics, Dec. 8, 1969.)*

Typically, gain in an amplifier of this type can be as much as 100 dB in a crystal less than 1 cm long, operating with as much as 50% bandwidth at frequencies up to several hundred megahertz. The amount and direction of gain are controlled by the bias field, and the bandwidth is determined by geometry of the interdigital arrays. The silicon film operates at much lower bias field than required in bulk-wave amplifiers and can be deposited on a sapphire substrate in narrow strips to permit parallel excitation at low voltage. The upper frequency limits are set by resolution in the photolithographic processes used to form the fingers and have been extended to more than 3 GHz by using electron-beam processing to secure high resolution.

98. References on Acoustic Amplifiers

1. Hutson, A. R., J. H. McFee, and D. L. White   Ultrasonic Amplification in CdS, *Phys. Rev. Lett.*, Sept. 15, 1961, Vol. 7, No. 6, pp. 237–239.

2. White, D. L.   Amplification of Ultrasonic Waves in Piezoelectric Semiconductors, *J. Appl. Phys.*, August 1962, Vol. 33, No. 8, pp. 2547–2554.

3. Kino, G. S.   Acoustoelectric Interactions in Acoustic-Surface-Wave Devices, *Proc. IEEE*, May 1976, Vol. 64, No. 5, pp. 724–748.

Magnetic Amplifiers

BY H. W. LORD

99. Static magnetic amplifiers can be divided into two classes, identified by the terms saturable reactor and self-saturating magnetic amplifier. A *saturable reactor* is defined as "an adjustable inductor in which the current-vs.-voltage relationship is adjusted by control magnetomotive forces applied to the core."[1]* A *magnetic amplifier* is defined as "a device using saturable reactors either alone or in combination with other circuit elements to secure amplification or control."[1] A *simple magnetic amplifier* is defined as "a magnetic amplifier consisting only of

---

*Superior numbers correspond to numbered references, Par **13-110**.

saturable reactors."[1] The abbreviation SR is used in this section to denote a *saturable reactor* and/or simple magnetic amplifier.

A self-saturating magnetic amplifier is a magnetic amplifier in which half-wave rectifying circuit elements are connected in series with the output windings of saturable reactors. It has been shown[2] that saturable reactors can be considered to have negative feedback. Half-wave rectifiers in series with the load windings will block this intrinsic feedback. A self-saturating magnetic amplifier is therefore a parallel-connected saturable reactor with blocked intrinsic feedback. This latter term avoids the term self-saturation, which, although extensively used, does not have a sound physical basis. The abbreviation MA is used here to denote this type of high-performance magnetic amplifier. Trade names for this type of magnetic amplifier include Amplistat, Magnestat, and Mag-Amp.

100. Saturable-Reactor (SR) Amplifiers. The SR can be considered to have a very high impedance throughout one part (the *exciting interval*) of the half cycle of alternating supply voltage and to abruptly change to a low impedance throughout the remainder of the half cycle (the *saturation interval*). The phase angle at which the impedance changes is controlled by a direct current. This is the type of operation obtained when the core material of the SR has a highly rectangular hysteresis loop. Two types of operation, representing limiting cases, are discussed here, namely, *free even-harmonic currents* and *suppressed even-harmonic currents*. Intermediate cases are very complex, but use of one or the other of the two extremes is sufficiently accurate for most practical applications.

The present treatment is limited to resistive loads, the most usual type for SR applications. Reference 3 discusses inductive dc loads, and Ref. 4 discusses inductive ac loads. The basic principles of operation of SRs and MAs are given in more detail in Ref. 5 than space permits here.

101. Series-Connected SR Amplifiers. Basically, an SR circuit consists of the equivalent of two identical single-phase transformers. Figure 13-94 shows two transformers, SR$_A$ and SR$_B$, interconnected to form a rudimentary series-connected SR circuit.

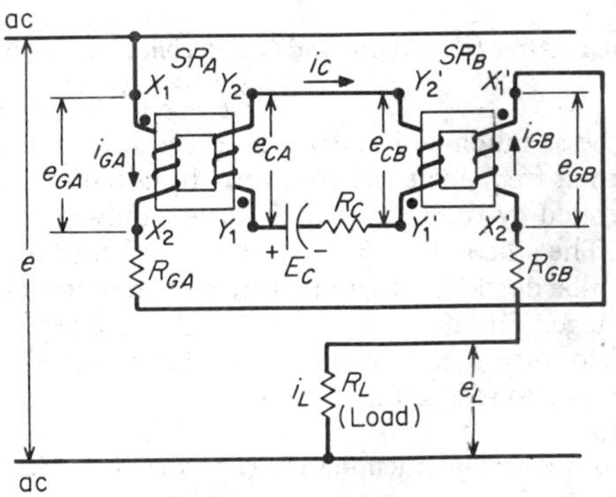

Fig. 13-94. Series-connected SR amplifier.

The two series-connected SR windings in series with the load are called *gate windings*, and the other two series-connected windings are called *control windings*. Note, from the dots that indicate relative polarity, that the gate windings are connected in series additive and the control windings are connected in series subtractive. By reason of these connections, the fundamental power frequency and all *odd* harmonics thereof will not appear across the total of the two control windings but any *even* harmonic induced in one control winding will be additive with respect to a corresponding even harmonic induced in the other control winding.

For this connection, SR$_A$ and SR$_B$ are normally so designed that each gate winding will accommodate one-half the alternating voltage of the supply without producing a peak flux density in the core that exceeds the knee of the magnetization curve, assuming no direct current is flowing in the control winding. Under these conditions, if SR$_A$ is identical with SR$_B$ one-half of the supply voltage appears across each gate winding, and the net voltage induced in the control circuit is zero. The two SRs operate as transformers over the entire portion of each half cycle.

When a direct current is supplied to the control circuit, each SR will have a saturation interval during a part of each cycle, SR$_A$ during half cycles of one polarity and SR$_B$ during half cycles of the opposite polarity. The ratio of the saturation interval to the exciting interval can be controlled by varying the direct current in the control winding.

When an SR core has a saturation interval during part of half cycles of one polarity, and there is a load or other impedance in series with the gate winding, even-harmonic voltages are induced in all other windings on that core. In the series connection, one SR gate winding can be the series impedance for the SR, and large even-harmonic voltages will appear across the individual gate windings and control windings when direct current flows in the control windings. The amount of even-harmonic current which flows in the control circuit of Fig. 13-94 will depend upon the impedance of the control circuit to the harmonic voltages. If the control-circuit impedance between terminals Y_1 and Y_1' is low with respect to the induced harmonic voltages of the control windings (usually referred to as a *relatively low* control-circuit impedance), the harmonic currents can flow freely in this circuit, and the SR circuit is identified by the term *free even-harmonic currents*. If the control-circuit impedance is high with respect to the induced harmonic voltages (a *relatively high* control-circuit impedance), the harmonic-current flow is suppressed and the SR circuit is identified by the term *suppressed even-harmonic currents*.

If R_C in Fig. 13-94 is relatively low and the source impedance of the dc control current is low, the circuit is of the free even-harmonic-currents type and the two SRs are tightly coupled together by the control circuit. If one SR is in its saturation interval, it will reflect a low impedance to the gate winding of the other SR, even though it is then operating in its exciting interval. Thus, during the saturation interval of any core, transformer action causes both gate windings to have low impedances and current can flow from the ac source to the load, with correspond-

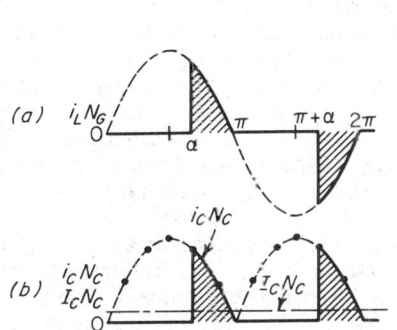

Fig. 13-95. Currents in circuit of Fig. 13-94, for free even-harmonic-current conditions: (*a*) gate current; (*b*) control-circuit current.

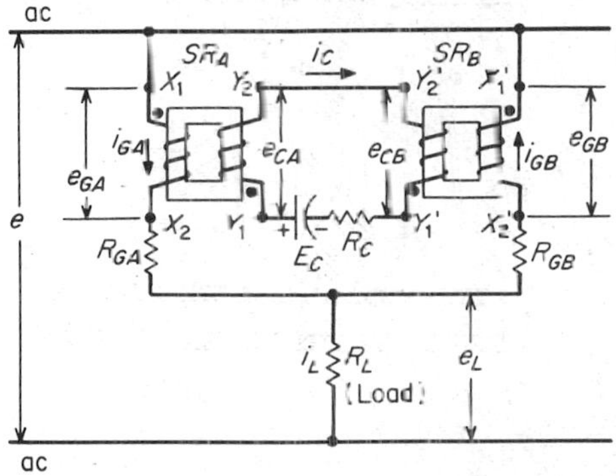

Fig. 13-96. Parallel-connected SR amplifier.

ingly high harmonic circulating currents in the control circuit. When both cores are in their exciting intervals, only the low core-exciting current can flow and the current circulating in the control circuit is substantially zero.

If the exciting current is so low as to be negligible, and if N_G are the turns in each gate winding, N_C are the turns in each control winding, I_G is the rectified average of the load current, and I_C is the average value of the control current, then applying the law of equal ampere-turns for transformers provides the following expression for the circuit in Fig. 13-94: $I_C N_C = I_G N_G$. This is the law of equal ampere-turns for the series-connected SR with resistive load. It applies to operation in the so-called *proportional region*, the upper limit of this region being that point in the control characteristic where the load current is limited solely by the load circuit resistance. Figure 13-95 shows the gate-current circuit and control-circuit current at one operating point in the proportional region. An increase in control current causes a decrease in the angle α, thereby increasing I_L, the rectified average of the current to the load.

102. Parallel-Connected SR Amplifiers. Figure 13-96 shows the circuit diagram for the parallel-connected SR. Each gate winding is connected directly between the ac supply and the load resistance, thus providing two parallel paths through the SR. There is therefore a low-impedance path for the free flow of even-harmonic currents, even though the impedance of the control circuit happens to be relatively high. As a result, the parallel-connected SR is always of the *free flow of even-harmonic-currents* type, the cores operate in the same manner as described in Par. **13-101**, and the waveshapes of currents to the load are as shown in Fig. 13-95. However, the gate winding currents are different, as shown in Fig. 13-97.

It is obvious from Fig. 13-97 that $I'_L = -I_L/2$. Since, except for the low-excitation requirements for the core, the net ampere-turns acting upon the core must be zero during the exciting interval, then $I'_L N_G + I_C N_C = 0$. Using the above equation to eliminate I'_L, the result is $I_C N_C = I_L N_G/2$. Figure 13-98 shows the control characteristic of saturable core amplifiers applicable both to free and suppressed even-harmonic modes of operation.

103. Series-Connected SR Circuit with Suppressed Even Harmonics. This circuit is usually analyzed by assuming operation into a short circuit and the control current from a current source. Figure 13-99 shows idealized operating minor hysteresis loops for this circuit, and Fig. 13-100 shows pertinent current and voltage waveshapes lettered to correspond to Fig. 13-99. Note that this circuit supplies a square wave of current to the load so long as it is operating in the proportional region. Energy is interchanged between the power-supply circuit and the control circuit so as to accomplish this type of operation. If a large inductor is used to maintain substantially ripple-free current in the control circuit, the energy interchanged between the supply voltage and control circuit is alternately stored in, and given up by, the control-circuit inductor.

104. Gain and Speed of Response of SRs. It is obvious from the generalized control characteristic that the current gain is directly proportional to the ratio of the control winding turns to the gate winding turns. Thus $I_L/I_C = N_C/N_G$. If R_C is the control-circuit resistance and R_L is the load resistance, the power gain GP is $N_C^2 R_L^2 / N_G^2 R_C^2$.

The response time of saturable reactors is the combination of the time constants of the control circuit and of the gate-winding circuit. Expressed in terms of the number of cycles of the supply frequency, the time constant of the control winding of a series-connected SR is[6] $\tau_C = R_0 N_C^2 / 4 R_C N_C^2$ cycles. If the SR is operating underexcited, a transportation lag may cause an additional delay in response.[7]

105. High-Performance Magnetic Amplifier (MA). If a rectifier is placed in series with each gate winding of a parallel-connected SR and the rectifiers are poled to provide an ac output, it becomes a type of high-gain MA circuit called the *doubler circuit*. When this is done, the law of equal ampere-turns no longer applies, and the transfer characteristic is mainly determined by the magnetic properties of the SR cores. The design of an MA therefore requires more core-materials data than are required for SRs in simple magnetic amplifier circuits. The text of this section assumes that the designer has the required magnetic-core-materials data on hand, since such data are readily available from the manufacturers of cores for use in high-performance MAs.

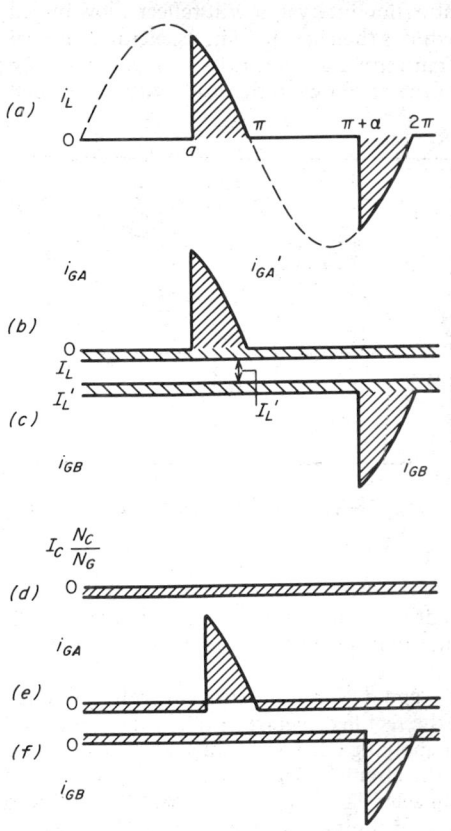

Fig. 13-97. Currents in the circuit of Fig. 13-96 for free even-harmonic-current conditions: (a) load current; (b) components of load current in SR_A gate winding; (c) same in SR_B; (d) control-circuit current; (e) gate current in SR_A; (f) same in SR_B.

Descriptions of the operation of MAs require several terms which are defined here for reference.[8]

Firing. In a magnetic amplifier, the transition from the unsaturated to the saturated state of the saturable reactor during the conducting or gating alternation. Firing is also used as an adjective modifying phase or time to designate when the firing occurs.

Gate angle (firing angle). The angle at which the gate impedance changes from a high to a low value.

Gating. The function or operation of a saturable reactor or magnetic amplifier that causes it, during the first portion of the conducting alternation of the ac supply voltage, to block substantially all the supply voltage from the load and during a later portion allows substantially all the supply voltage to appear across the load. The "gate" is said to be virtually closed before firing and substantially open after firing.

Reset, degree of. The reset flux level expressed as a percentage or fraction of the reset flux level required to just prevent firing of the reactor in the subsequent gating alternation under given conditions.

Reset flux level. The difference in saturable-reactor core flux level between the saturation level and the level attained at the end of the resetting alternation.

Resetting (presetting). The action of changing saturable-reactor core flux level to a controlled ultimate reset level, which determines the gating action of the reactor during the subsequent gating alternation. The terms resetting and presetting are synonymous in common usage.

Resetting half cycle. The half cycle of the MA ac supply voltage at which resetting of the saturable reactor may take place.

106. Half-Wave MA Circuits. Figure 13-101 shows a half-wave MA circuit with control from a source of controllable direct *current.* Figure 13-102 shows a half-wave MA circuit with control from a controllable source of direct *voltage.* The following sequence of operation is the same for both circuits and is described in connection with Fig. 13-103.

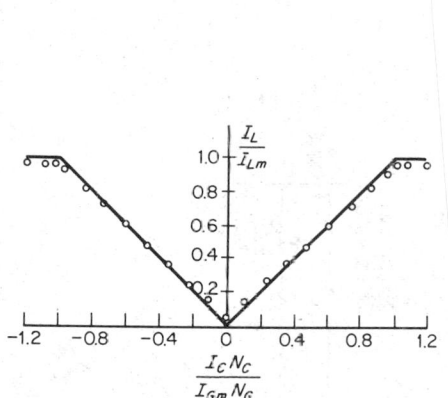

Fig. 13-98. Control characteristic of SR amplifier, applying both to free and suppressed even-harmonic conditions.

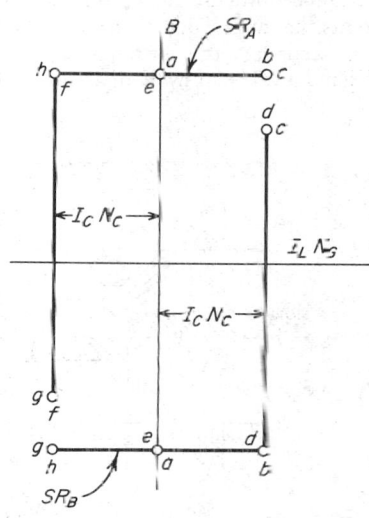

Fig. 13-99. Idealized operating hysteresis loops for suppressed even-harmonic-current condition.

1. During the gating half cycle (diode REC or REC_1 conducting) the core flux increases toward a saturation level (3') from some reset flux level (0) and the current to the load R_t is very low.

2. When saturation of the SR occurs at 2', firing occurs and current flows to the load for the rest of the gating half cycle, leaving the core flux at 4'.

3. During the resetting half cycle (diode REC or REC_1 blocking) the SR core is reset from 4' through 5' to a value of reset flux level corresponding to B' and 0.

The waveshape of the current to the load is the same for both circuits, being a portion of one polarity of sine wave (phase-controlled half-wave). The two types of control circuits differ as follows:

1. The curve on the reset portion of the hysteresis loop between 4' and 0 is as shown in Fig. 13-103 for the current control type of control circuit (Fig. 13-101), but for the circuit of Fig. 13-102, the resetting portion of the curve in the region 5' is not a vertical line and may coincide with the outer major hysteresis loop for a substantial portion of the resetting period.

2. The output of the circuit of Fig. 13-101 is a maximum for zero dc control current, but for Fig. 13-102 the output is a minimum for zero dc control voltage.

The operation of a half-wave MA can be summarized as follows: (1) During the gating half cycle, the core acts to withhold current from the load for a portion of each half cycle, the length of the withholding period being determined by the degree of reset of core flux provided during the immediately preceding resetting half cycle. (2) During the resetting half cycle, the degree of reset can be controlled by varying the amount of direct current of proper polarity in the control winding, or by varying the amount of average voltage of proper polarity applied to the control winding. (3) The amount of current and power required to reset the core is primarily a function of the excitation requirements of the core and bears no direct relation to the power delivered to the load.

107. Full-Wave MA Circuits. Two half-wave MA circuits can be combined to provide either full-wave ac output (Fig. 13-104) or full-wave dc output (Fig. 13-105). During the gating half cycle, all gate windings act in the same manner as described for the half-wave MA circuit. Because of an interaction of the common load resistor voltage with the availability of inverse voltage across a rectifier during its resetting half cycle, each half-wave part of Fig. 13-104 cannot be assumed to operate independently of its mate. The differences in control characteristics are shown by Fig. 13-106, where curve A applies to the circuit of Fig. 13-101 and curve B applies to the circuit of Fig. 13-104.

108. Equations for MA Design. For SRs in simple MA circuits, since the ampere-turn balance rules apply, their design is so similar to transformer design that ordinary transformer design procedures are applicable. For SRs in MA circuits, the gate windings must carry rms load currents that are related to the ac and dc load currents in the same manner that rectifier transformer secondary rms currents are related to the load current. Gate windings must be able to withstand a full half cycle of ac supply voltage, without saturation, for a total flux swing from

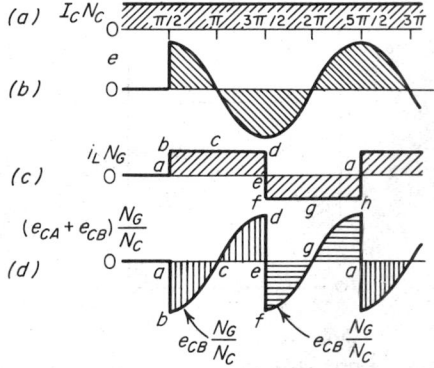

Fig. **13-100.** Series-connected suppressed even-harmonic condition: (a) control-circuit current; (b) applied input voltage; (c) gate winding and load current; (d) voltage across series-connected control coils.

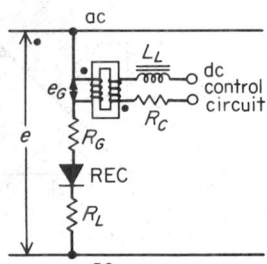

Fig. **13-101.** Half-wave MA circuit.

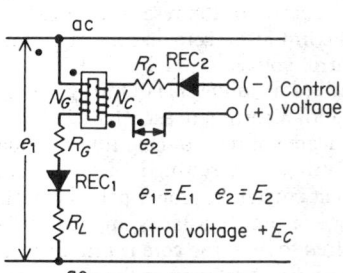

Fig. **13-102.** Voltage-controlled MA circuit.

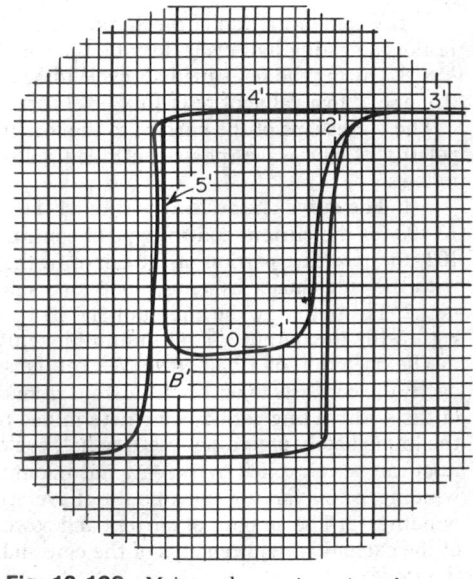

Fig. **13-103.** Major and operating minor dynamic hysteresis loops for circuit of Fig. 13-101.

$-B\mu_m$ to $+B\mu_m$, where $B\mu_m$ is the flux density at which the permeability of the core material is a maximum. In the cgs system of units:

Maximum ac supply voltage, rms: $E_e = 4.44 N_G B_{\mu m} A f \times 10^{-8}$, where N_G is gate winding turns, A is core area, and f is supply frequency.

Maximum load current for firing angle $\alpha = 0°$ (rectified average): $I_{LM} = E_e/1.11(R_G + R_L)$, where R_G is total gate winding resistance, including diode drops, and R_L is the effective load resistance.

For current-controlled MA circuits the following equations apply:

Minimum load current ($\alpha = 180°$); $I_{LX} = 2I_X = 2H_{af}l/0.4\pi N_G$, where $I_X =$ exciting current (A) of one SR core, $H_{af} =$ sine flux coercive force of the core material (Oe) and $l =$ mean length of magnetic circuit of one core.

Control current for upper end of control current: $I_C = H_c l/0.4\pi N_C$, where $H_C =$ dc coercive force of the core material.

Control current for minimum load-current point: $I_C = H_{af}l/0.4\pi N_C$. Ampere-turn gain $G_{AT} = (I_{LM}/2I_X)\pi$.

Time constant $\tau_C = 0.9\pi E_e N_C^2/4fl_X R_C N_G^2$ seconds, where R_C is the control circuit resistance

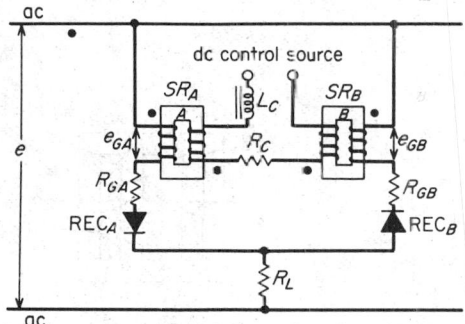

Fig. 13-104. Full-wave ac-output MA circuit with high-impedance control circuit.

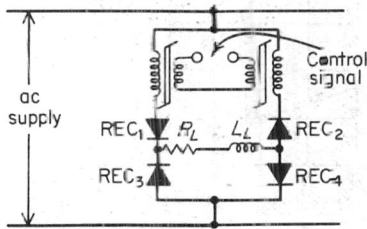

Fig. 13-105. Full-wave bridge MA circuit with inductive load.

and assuming $\alpha = 90°$. This equation does not take into account a transportation or any under-excited effects which increase the delay time.

Power gain $G_p = G_{AT}^2(N_C/N_G)^2(R_L/R_C)$, where $k_f =$ form factor (ratio of rms to average values). If only average values are used, k_f is unity. At $\alpha = 90°$, dynamic power gain $G_D = G_p/\tau_C = \pi f I_{LM} R_L/I_X(R_G + R_L)$ per second.

For voltage-controlled MA circuits, the following equations apply to Fig. 13-102:

AC bias voltage $E_2 > E_1 N_C/N_G$ and of indicated relative polarities. A value of $E_2 = 1.2 E_1 N_C/N_G$ is usually adequate. Since the dc control voltage E_C acts to inhibit the resetting effect of E_2, a firing angle $\alpha = 0°$ occurs when $E_C \geq \sqrt{2}E_2$.

A single-stage amplifier has a half-cycle transport lag and no other time delay unless the resetting is inhibited by common-load interactions with other half-wave-amplifier elements.

Maximum average load current $I_{LM} = 0.9E_1/(R_G + R_L)$.

Minimum average load current $I_{LX} = I_X = H_{af}l/0.4\pi N_G$.

The calculation of the power gain of voltage-controlled MA circuits largely depends upon the control circuit. The dc control voltage of Fig. 13-102 actually must absorb power from the ac bias circuit during much of its control range. Assuming that $E_2 = 1.2 E_1 N_C/N_G$ and a resistor R_C' is connected across the control source, which has a drop of $0.2E_2$ at cutoff, then $R_C' = 0.24 E_1 N_C^2/N_{Gx}^2$. The maximum dc control voltage is then $1.2\sqrt{2}E_1 N_2/N_G$.

When the dc control voltage $E_C > \sqrt{2}E_2$, rectifier REC$_2$ is blocking all the time, and so the only load on the control source is R_C'. Therefore the maximum power from the control source $P_C = E_C^2/R_C' = 5\sqrt{2}E_1 I_X$.

The maximum power output is $P_0 = I_{LM}^2 R_L(0.9E_1)^2/(R_G + R_L)^2$.

The power gain is then $G_p = P_0/P_C$ if the gain is assumed to be linear over most of the control characteristic. This can also be written $G_P = I_{LM}^2 G_p - I_{LM}^2 R_L/5\sqrt{2}E_1 I_X$.

109. Core Configurations. Figures 13-107 and 13-108 show two coil-and-core configurations commonly used for SRs. The best core-and-coil geometry for the SRs and MAs is the toroid-shaped core with the gate winding uniformly wound over the full 360° of the core. Full-wave operation requires two such cores. After winding a gate winding on each core of a matched pair, the two cores and coils can be stacked together coaxially and the required control coils wound over the stack. The gate coils are so connected into the load circuit that no fundamental or odd-harmonic voltage is induced in the control windings.

110. References on Magnetic Amplifiers

1. "The International Dictionary of Physics and Electronics," Van Nostrand, Princeton, N.J., 1956.
2. Storm, H. F.   "Magnetic Amplifiers," Chap. 5, Wiley, New York, 1955.
3. Reference 2, Chaps. 11 and 12.
4. Wilson, T. G.   Series-Connected Magnetic Amplifier with Inductive Loading, *Trans. AIEE*, 1952, Vol. 71, Pt. I, pp. 101–110.
5. Shea, R. F.   "Amplifier Handbook," Chaps. 8 and 21, McGraw-Hill, New York, 1968.
6. Reference 2, p. 148.
7. Reference 2, p. 150.
8. Terms and Definitions for Magnetic Amplifiers, AIEE Committee Report, *Trans. AIEE*, 1954, Vol. 73, Pt. I, pp. 265–270.

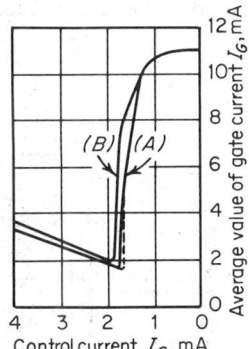

Fig. 13-106. MA circuit control characteristics: (A) half-wave circuit; (B) full-wave ac-output circuit.

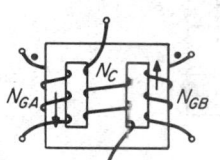

Fig. 13-107. Three-legged SR core-and-coil configuration.

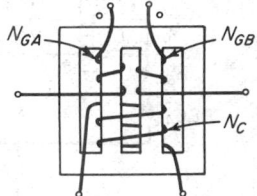

Fig. 13-108. Four-legged SR core-and-coil configuration.

Direct-Coupled, Operational, and Servo Amplifiers

BY S. M. KORZEKWA

111. Direct-Coupled Amplifiers. A direct-coupled amplifier has frequency response that starts at zero frequency (dc) and extends to some specified upper limit. To obtain the zero-frequency capability, such amplifiers are normally direct-coupled throughout; i.e., they do not use capacitive or transformer coupling (except for auxiliary higher-frequency compensation or signal transmission).

The primary sources of error of a direct-coupled amplifier are initial offset, drift, and gain variations, errors usually dependent on temperature, aging, etc. The gain-variation problem can be minimized by feedback gain-stabilization techniques, but offset and drift errors are usually not handled so effectively by feedback techniques. A bias shift or drift error cannot be distinguished from a signal response because their output responses are identical. Various techniques are available to minimize this drift problem, and several different methods may be used in an amplifier.

One method of minimizing drift is to use a balanced topology, as in differential amplifiers, whereby the drift errors tend to cancel out. A more complicated but effective method is the modulated-carrier-amplifier approach: the signal is first converted into a carrier signal, amplified using an ac-coupled amplifier, and then demodulated to a baseband signal. The function performed by the modulated-carrier amplifier is identical with that of a direct-coupled amplifier, but the signal processing technique is different.[1,2]*

An example of a direct-coupled transistor amplifier is illustrated in Fig. 13-109. The primary signal path, shown in heavy lines, directly couples the input signal to the amplifier output. This configuration also provides gain and bias stabilization via the direct-coupled R_3 feedback path. Usually the low-frequency amplifier voltage gain is primarily determined via the R_1 and R_3

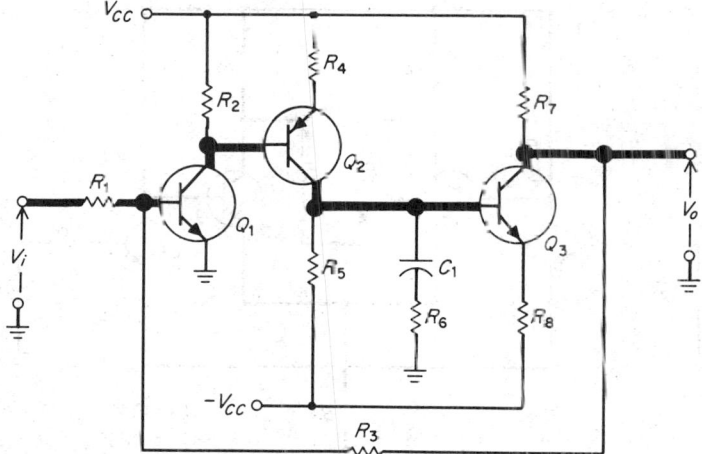

Fig. 13-109. Direct-coupled transistor amplifier.

components, while the function of the C_1 capacitor and R_6 resistor is to provide high-frequency compensation or stabilization. Sources of error for this topology include the initial V_{BE} offset of the Q_1 input transistor and the subsequent drift caused by the temperature dependence of this offset.

112. Differential Amplifiers. A differential amplifier is a dual-input amplifier that amplifies the difference between its two signal inputs. This amplifier may have an output that is single-ended (one output) or it may have a differential output.

The differential amplifier eliminates or greatly minimizes many common sources of error. The drift problem encountered in direct-coupled amplifiers can be handled more effectively by the differential approach. A second major advantage of a differential amplifier is its ability to reject common-mode signals, i.e., unwanted signals present at both of the amplifier inputs or other common points. Common-mode performance is usually a critical requirement in instrumentation amplifiers.

The basic differential active-device circuits commonly used in differential amplifiers are shown in Fig. 13-110. Such differential pairs can be constructed using separate devices or in integrated-circuit form. The integrated package yields additional advantages since the parameter

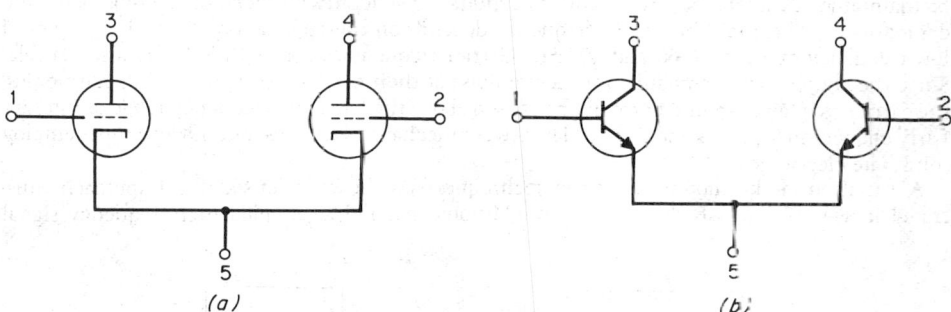

Fig. 13-110. Differential amplifier pairs: (a) tube version; (b) transistor version.

*Superior numbers correspond to numbered references, Par. **13-117.**

differences between the units of the integral differential pair are usually much less than if separate devices are used. Thus the units of such integral pairs tend to track differentially more closely, even though their individual parameters may vary in absolute value. Also, many of the passive components in the integrated amplifiers track better. Figure 13-111 shows a typical differential transistor amplifier.[3] For further detailed information, such as analysis procedures, design techniques, and application data, refer to the literature.[4,5]

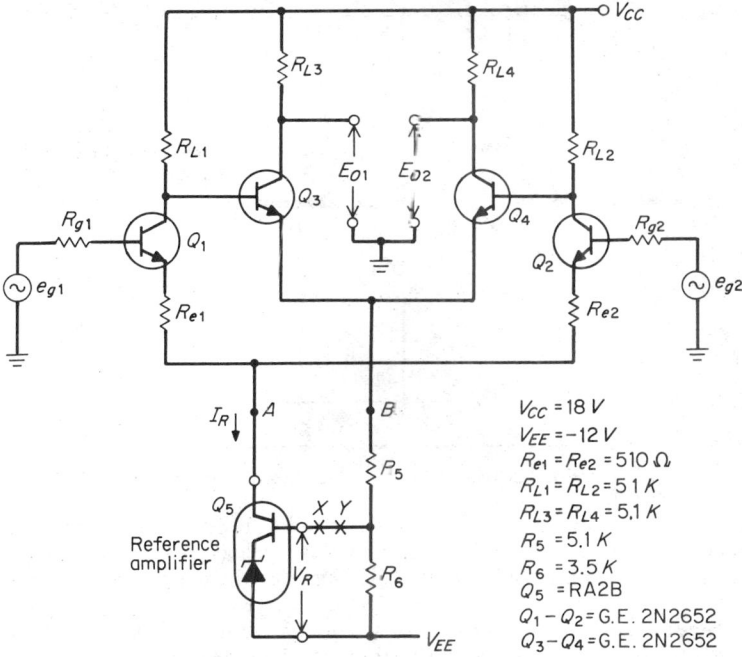

$V_{CC} = 18\ V$
$V_{EE} = -12\ V$
$R_{e1} = R_{e2} = 510\ \Omega$
$R_{L1} = R_{L2} = 51\ K$
$R_{L3} = R_{L4} = 5.1\ K$
$R_5 = 5.1\ K$
$R_6 = 3.5\ K$
$Q_5 = RA2B$
$Q_1 - Q_2 = G.E.\ 2N2652$
$Q_3 - Q_4 = G.E.\ 2N2652$

Fig. 13-111. Two-stage differential amplifier with common-mode feedback. *(General Electric Co.)*

113. Chopper (Modulated-Carrier) Amplifiers. The chopper amplifier performs the function of a direct-coupled amplifier by using a carrier frequency and an ac-coupled amplifier. The modulated-carrier approach is used specifically to minimize drift and offset types of errors. Although this approach is often more complex to implement, the performance improvement makes this technique desirable for applications demanding low-drift performance.

Figure 13-112 shows the block diagram of a modulated-carrier amplifier. While the input and output signals are dc-coupled, the interstage coupling between the modulator, amplifier, and demodulator may be either a capacitor or a transformer.

Since modulator and demodulator are usually operated synchronously, the carrier delay must be maintained at acceptable levels. Low-pass filters are generally included in the modulator and demodulator. The choice of carrier frequency depends on the application. Typical examples of low-frequency carriers are 60 and 400 Hz. Carrier frequencies above 10 kHz are also feasible. Since the chopper and demodulator generate noise at their carrier frequency and its harmonics, the carrier frequency should normally be chosen above the baseband frequency range of interest. Early chopper modulators and demodulators were mechanical devices, but modern types employ solid-state electronics.

A variation of the modulated-carrier technique uses the chopper-stabilized approach illustrated in Fig. 13-113, which includes an additional parallel ac-coupled high-frequency signal

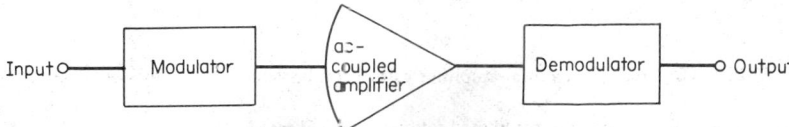

Fig. 13-112. Modulated-carrier (chopper) amplifier.

path. The two signal paths have gains and bandwidths tailored to a crossover frequency; e.g., when the high-frequency signal path becomes dominant, the modulated-carrier signal path ceases to contribute to the sum total. This variation offers the low-drift advantages of the carrier-modulated system with much higher bandwidth.

In a typical example of a commercially available chopper-stabilized amplifier (Analog Devices model 210/211 chopper-stabilized operational amplifier) the following characteristics are obtained:

Voltage drift	$0.5\ \mu V/^\circ C$	Overload recovery	$0.2\ \mu s$
Current drift	1 pA/°C	Long-term stability	$1\ \mu V/d$
Slewing rate	100 V/μs	Voltage gain	10^8
Bandwidth	20 MHz	Noise	$5\ \mu V$ (p-p)

114. Operational Amplifiers. An operational amplifier is intended to realize specific signal processing functions. For example, the same operational amplifier (depending on the externally added components) can be used as an integrating amplifier, a differentiating amplifier, an active filter, or an oscillator, among others. Applications of operational amplifiers also include such functions as impedance transformers, regulators, and signal conditioning. They are versatile building blocks that can also be used in nonlinear applications to realize functions such as log-arithmic amplifiers, comparators, ideal rectifiers, etc.

The early application of operational amplifiers was largely in the area of analog computations. The requirements placed upon the amplifiers were severe, and the cost was high. Currently, however, these high-performance functions are available at low cost and are widely used.

An operational amplifier can be either direct-coupled or ac-coupled. Most operational amplifiers have differential inputs and consequently realize common-mode rejection; however, many operational amplifiers are single-ended. Their power capability covers a wide range, and they can function as a power driver. The open-loop bandwidth can range from below 1 kHz up to the megahertz range. Voltage gain can be designed from unity to above 100 000.

To design or use the many excellent models now available commercially, it is necessary to understand the attributes and limitations of these versatile building blocks. There are many excellent references[1,4] that include design approaches, analysis techniques, and application data.

An example of a currently available and widely used integrated transistorized operational amplifier is shown in Fig. 13-114. Its electrical characteristics are listed in Table 13-7.

115. Application of Operational Amplifiers. Figure 13-115 shows the block dia-gram for an operational amplifier with associated external elements. A_1 represents the transfer function of the amplifier, either current gain or voltage gain, which is generally frequency-dependent. Z_i and Z_f are the primary elements that normally determine the closed-loop transfer function for this operational amplifier. The indicated Z_{in} and Z_l are equivalent summing node and load impedances, respectively, which are usually factored into the error terms of the com-posite transfer function. All the elements are general impedances that may be real or complex. Thus the simplified transfer function of this operational amplifier can be written

$$e_o/e_i \approx -Z_f/Z_i \qquad (13\text{-}24)$$

The error terms of the complete transfer function equation are not shown in Eq. (13-24); how-ever, the complete equation with error terms can be readily derived or obtained from the liter-

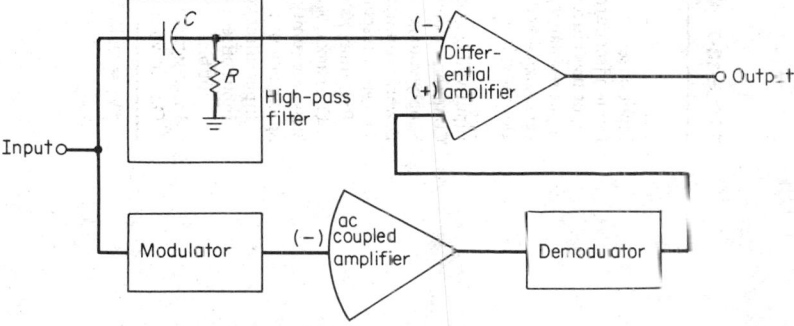

Fig. 13-113. Chopper-stabilized amplifier.

TABLE 13-7 Characteristics of Operational Amplifier in Fig. 13-114*

Parameter	Conditions	Min	Typical	Max
Input offset voltage, mV	$R_s < 10$ kΩ	...	1.0	5.0
Input offset current, nA		...	30	200
Input bias current, nA		...	200	500
Input resistance, MΩ		0.3	1.0	...
Large-signal voltage gain	$R_L \geq 2$ kΩ, $V_{out} = \pm 10$ V	50,000	200,000	...
Output voltage swing, V	$R_L \geq 10$ kΩ	±12	±14	
	$R_L \geq 2$ kΩ	±10	±13	
Input voltage range, V		±12	±13	
Common-mode rejection ratio, dB	$R_s \leq 10$ kΩ	70	90	
Supply-voltage rejection ratio, µV/V	$R_s \leq 10$ kΩ	...	30	150
Power consumption, mW		...	50	85
Transient response (unity gain):	$V_{in} = 20$ mV, $R_L = 2$ kΩ, $C_L \leq 100$ pF			
Rise time, µs		...	0.3	
Overshoot, %		...	5.0	
Slew rate (unity gain), V/µs	$R_L \geq 2$ kΩ	...	0.5	
Specifications for $-55° \leq T_A \leq +125°C$:				
Input offset voltage, mV	$R_s \leq 10$ kΩ	...		5.0
Input offset current, nA		...		500
Input bias current, µA		...		1.5
Large-signal voltage gain	$R_L \geq 2$ kΩ, $V_{out} = \pm 10$ V	25,000		
Output voltage swing, V	$R_L \geq 2$ kΩ	±10		

*$V_s = \pm 15$ V, $T_A = 25°C$ unless otherwise specified.

ature.[1,2,5] Thus, if the required constraints are adhered to, Eq. (13-24) can be used to generate various transfer functions, as illustrated in the examples given below.

Integrating Amplifier. An example of an integrating amplifier is shown in Fig. 13-116. Assume that the external-component values are as indicated and that the A_1 amplifier is that previously described in Fig. 13-114. Using Laplace transform notation, the integrator transfer function becomes

$$(e_o/e_i)S = (-)1/RCS \tag{13-25}$$

Using the component values specified in the figure, the integrator transfer function can then be written

$$(e_o/e_i)S = (-)5,000/S \tag{13-26}$$

An alternative frequency-domain representation of the above integrator transfer function can also be used.

$$(e_o/e_i)f = (-)800/jf \tag{13-27}$$

The closed- and open-loop frequency responses of this operational integrator amplifier are illustrated in Fig. 13-116b.

The frequency range of application depends primarily on the accuracy desired. In this example, 1-Hz to 10-kHz frequency is considered a realistic range of operation. For very low frequen-

Schematic diagram

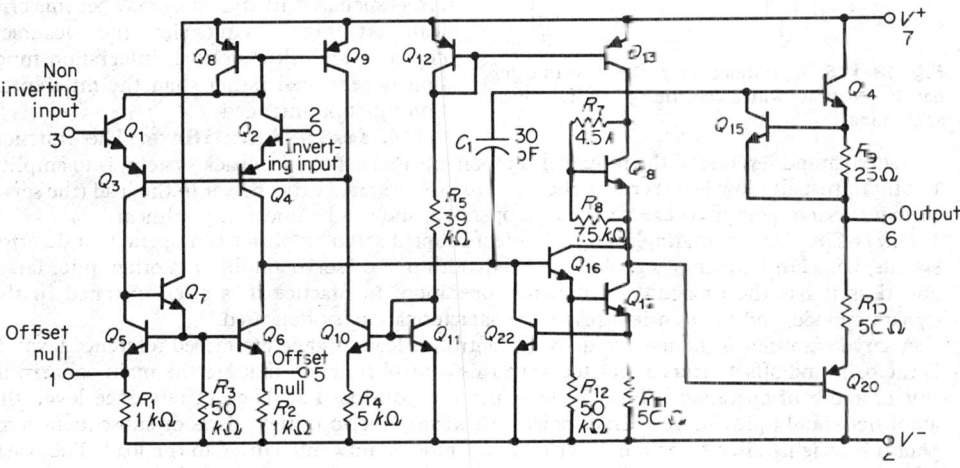

Connection diagram
(top view)

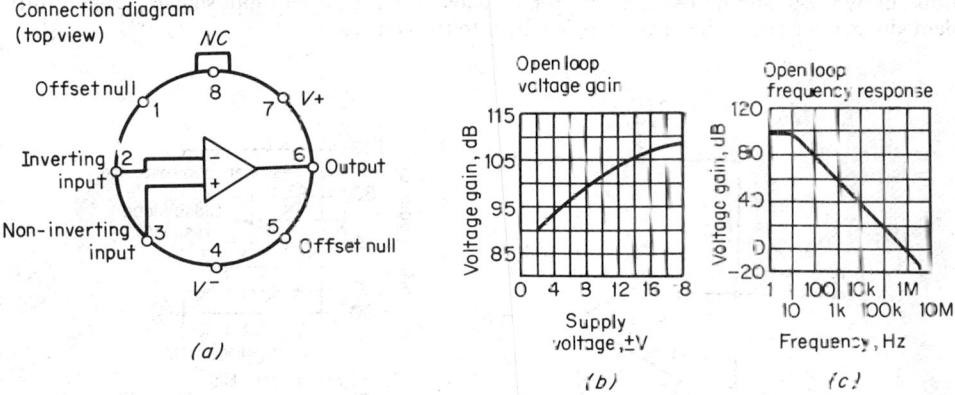

(a)

(b)

(c)

Fig. 13-114. Integrated-circuit version of an operational amplifier: (a) schematic and connection diagrams; (b) open-loop voltage gain; (c) open-loop frequency response. (Fairchild model μA 741). *(Fairchild Semiconductor Corporation.)*

cies the error tends to increase due to inadequate excess gain within the operational amplifier, whereas for high frequencies the closed-loop gain becomes low and drift and offset errors may then become important.

Differentiating Amplifier. An example of a differentiating amplifier is shown in Fig. 13-117. The following simplified closed-loop equations are applicable:

$$(e_o/e_i)S \approx (-)RCS \tag{13-28}$$
$$(e_o/e_i)S \approx (-)0.165S \tag{13-29}$$
$$(e_o/e_i)f \approx (-)jf \tag{13-30}$$

The frequency responses of this operational amplifier differentiator are shown in Fig. 13-117b. The open-loop response of the operational amplifier of Fig. 13-114 is used, and the differentiator closed-loop response is superimposed onto it.

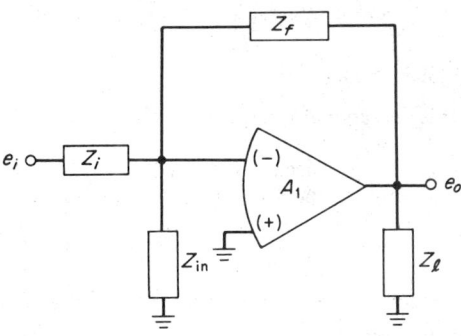

Fig. 13-115. Operational amplifier A_1 with external impedances which determine its functional application.

The frequency range of this operational amplifier differentiator depends primarily on the closed-loop differentiation accuracy required. The difference between open- and closed-loop transfer functions (usually referred to as *excess gain*) can be used to predict the accuracy of the function being generated. A realistic frequency range is from 0.01 to 100 Hz. At very low frequencies the closed-loop gain becomes very small, and consequently errors such as drift and offset may become critical. At high frequencies the accuracy degrades, and ultimately an integration function is generated rather than the differentiation function intended.

116. Servo Amplifiers. The function of a servo amplifier, one of the principal components in a control feedback system, is to amplify the input (usually low-level) error signals and to provide rated drive power to the load (the servo actuator). Servo amplifiers can be dc- or ac-operated, and can be linear or nonlinear.

Direct-Coupled Servo Amplifiers. A direct-coupled servo amplifier can operate on dc error signals, i.e., zero frequency signals. The bandwidth of a dc servo amplifier is often quite large, and thus it has the capability of dynamic operation. In practice it is often operated in the dynamic mode, and the transient-response characterization is often used.

A servo amplifier is intended to drive or control its load to some prescribed reference level. It is the drift and offset errors associated with this control function that are the main concern in this dc mode of operation. Thus, if the actuator is positioned at its exact reference level, the amplifier should provide zero drive power. However, due to initial offsets or subsequent temperature or aging effects, the amplifier may still provide unwanted drive to the load. The usual solution is to minimize these problems by circuit design, e.g., tracking or balanced configurations, or by use of an intermediate-carrier modulated by the dc error input signals, subsequently demodulated for use as the direct-coupled drive to the actuator.

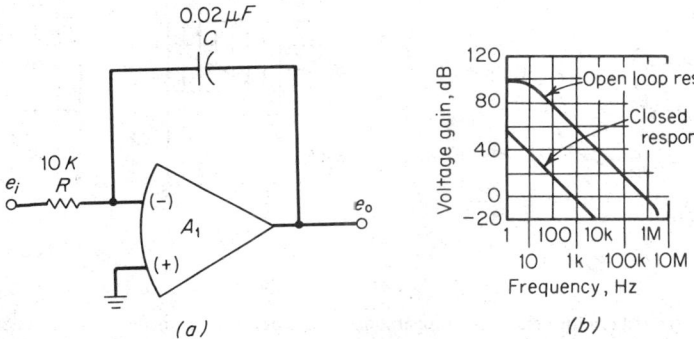

(a)   *(b)*

Fig. 13-116. Operational amplifier of Fig. 13-114 used for integration: (a) basic schematic; (b) frequency response.

AC Servo Amplifiers. An ac servo amplifier operates at a selected fixed frequency and is consequently a carrier-system amplifier. The most commonly used carrier frequencies are 60 and 400 Hz, but ac servo systems can be designed to operate at almost any frequency compatible with the servo actuator used.

The principal advantage of ac servo amplifiers is that the previously discussed drift and offset problems present in dc servo amplifiers are virtually eliminated. Another advantage is that the prime power from the actuator now can be supplied directly from the line, for example, 60-Hz 115-V source.

For applications wherein the servo actuator or load is a servomotor, the load is often tuned to the carrier frequency via the addition of a capacitor in parallel with the motor. Note that the

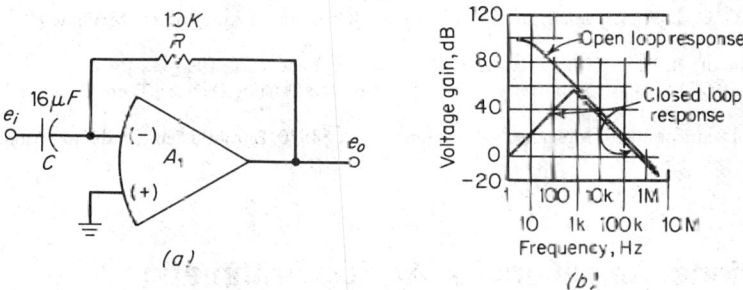

Fig. 13-117. Operational amplifier of Fig. 13-114 used for differentiation. (*a*) basic schematic; (*b*) frequency response.

servomotor impedance is a function of the motor speed; however, the losses (usually resistive) change, whereas the parallel inductance remains virtually constant. Thus tuning the load causes it to become real (resistive), and consequently the efficiency of the amplifier output stage improves significantly.

The output stage of an ac servo amplifier is usually operated class B or AB and in a push-pull configuration to obtain the best possible drive efficiency. Figure 13-118 shows an example of a push-pull tuned-load ac servo amplifier.[3]

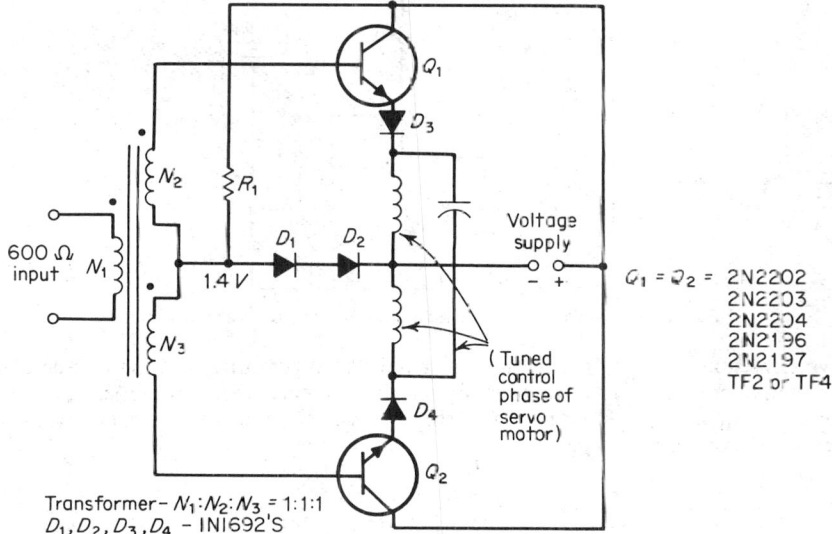

Fig. 13-118. Servo amplifier used as a motor drive, in the power range of 1 to 4 W. (*General Electric Company.*)

Nonlinear Servo Amplifiers. This class of servo amplifier uses the load to filter the highly nonlinear drive signals resulting in improved drive efficiencies and higher realizable driver-power capabilities. The drivers essentially act as switches wherein the dissipation losses are low when the switches are either on or off. Since the loads or servo actuators are usually highly reactive, they can be advantageously used in this manner. The drive power can be readily derived from a dc source via a pulse-width-modulated scheme. In addition, an ac source of power can also be used via a phase-modulation technique. Some examples of high-power switching devices used for ac nonlinear servo amplifiers are thyratrons and silicon controlled rectifiers (SCRs).

117. References on Direct-Coupled, Operational, and Servo Amplifiers
1. Shea, R. F. "Amplifier Handbook," McGraw-Hill, New York, 1968.
2. Hunter, L. P. "Handbook of Semiconductor Electronics " McGraw-Hill, New York, 1970.
3. "Transistor Manual," 7th ed., General Electric Co., 1964.
4. Tobey, G. E., L. P. Huelsman, and J. G. Graeme "Operational Amplifiers," McGraw-Hill, New York, 1971.
5. Middlebrook, R. D. "Differential Amplifiers," Wiley, New York, 1963.
6. Model 210/211 Chopper Stabilized Operational Amplifiers, Analog Devices, Tech. Data Pub., Cambridge, Mass.
7. Fairchild Semiconductor Integrated Circuit Data Catalog 1970, Fairchild Semiconductor, Mountain View, Calif., 1969.

Operational Amplifiers for Analog Arithmetic

BY J. P. HESLER

118. Analog Multiplier Circuits. Circuits used for analog multiplication fall into three categories: transconductance multipliers, averaging-type multipliers, and exponential multipliers.

119. Transconductance Multipliers. The most prominent type of transconductance multiplier uses the property of the bipolar transistor; i.e., its collector current and transconductance are linearly related. The balanced transistor differential amplifier used in these circuits offers good accuracy, wide bandwidth, and low cost.

In the differential amplifier as shown in Fig. 13-119, the collector currents I_{c_1} and I_{c_2} are functions of the differential input voltage $V_{b_1} - V_{b_2}$. It is assumed that the current transfer ratios of

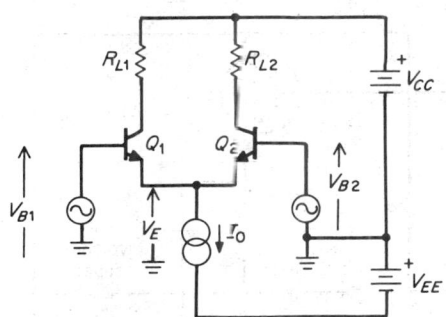

Fig. 13-119. Basic differential amplifier for analog multiplication.

the two transistors (Q_1 and Q_2), the junction ambient temperatures, and the base-emitter junction saturation currents are the same, as is probable in integrated-circuit fabrication.

The diode equation governs the relationship between the transistor emitter currents I_{E_i} and the base-to-emitter voltages V_{BE_i}

$$I_E = I_S \exp\left(\frac{V_{BE}q}{kT} - 1\right)$$

where I_S = junction saturation current, q = electron charge = 1.6×10^{19} C, k = Boltzmann's constant = 1.38×10^{-23} W/s·K, and T = junction temperature (K).

The relationships of collector currents, shown graphically in Fig. 13-120 are

$$I_{c_1} = \frac{\alpha_1 I_0}{1 + \exp\left[(V_{BE_2} - V_{BE_1})q/kT\right]} \qquad I_{c_2} = \frac{\alpha_2 I_0}{1 + \exp\left[(V_{BE_1} - V_{BE_2})q/kT\right]}$$

For zero differential input $V_{BE_1} = V_{BE_2}$ and $\alpha_1 = \alpha_2$, and so

$$I_{c_1} = I_{c_2} = \alpha I_0/2$$

At zero input, $\Delta V_{BE} = 0$, the maximum single-ended transconductance occurs at

$$gm_{max} = \alpha I_0/(4kt/q)$$

For a differential output connection the maximum transconductance is twice this value. The linear range of transconductance extends over a differential input signal range of about 50 mV peak to peak at room temperature.

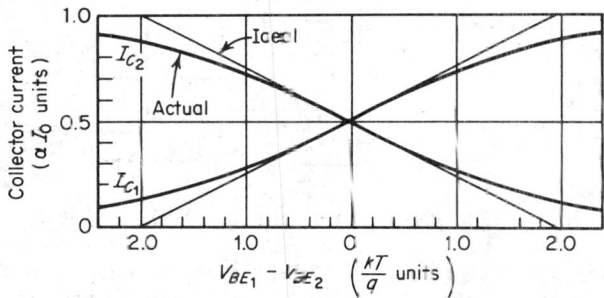

Fig. 13-120. Transfer curves on which analog multiplication is based.

Linear multiplication occurs when one input varies the I_0 term and the other input is used to vary ΔV_{BE}.

$$\Delta I_c = gm \, \Delta V_{BE} = \frac{\alpha I_0}{2kT/q} \Delta V_{BE}$$

The linear range of input voltage can be extended by inserting resistance in series with each emitter. This increase in allowable signal is accompanied by a corresponding reduction in transconductance. For I_0 equal to 2 mA, the addition of 50 Ω in series with each emitter increases the linear input swing by a factor of 3 and reduces the transconductance to one-third. Optimization of the transconductance multiplier operation requires a trade-off between linearity and error sources due to offset voltages and thermal noise.

A typical four-quadrant multiplier using the differential amplifier as building blocks is shown in Fig. 13-121. This circuit is generally used with differential inputs to obtain maximum linearity and common-mode signal rejection.

120. Averaging-Type Multipliers. Two types of averaging multipliers have found extensive use. The first type, the pulse height and width multiplier, uses one of the input variables to modulate the height or amplitude of a pulse train, while the pulse width is modulated with the other input variable. The pulse area, averaged over a suitable period, is proportional to the product of the input variables. High pulse rates permit short averaging intervals and fast response. These multipliers offer very good accuracy and stability but are more expensive than the transconductance multipliers. A block diagram of a typical pulse height and width averaging multiplier is shown in Fig. 13-122.

A second version of the averaging multiplier is the triangle averaging multiplier. In this type a high-frequency triangular waveform is generated and combined with the input variables to form an averaged output proportional to the product of the input signals. These multipliers are less accurate than the pulse height and width multipliers and are being displaced by transconductance multipliers.

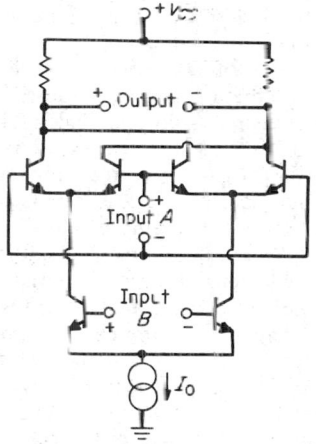

Fig. 13-121. Four-quadrant transconductance multiplier.

A third type of averaging multiplier is the time-base multiplier. This approach uses a comparator to sense the time interval required for the integral of a reference input voltage to equal the amplitude of a sample of one input variable. During the same interval, 0 to T, the other input variable is integrated to produce an output V. This type of circuit can be built with a few components to provide moderate accuracy, 1 to 5%, at low cost.

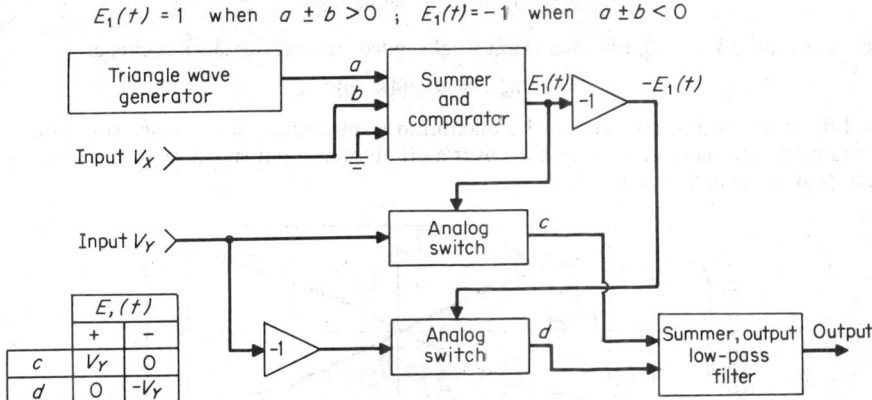

$E_1(t) = 1$ when $a \pm b > 0$; $E_1(t) = -1$ when $a \pm b < 0$

Fig. 13-122. Block diagram of an averaging-type multiplier.

121. Exponential Multipliers. The first electronic multipliers were of the exponential type. In these circuits, resistor-diode networks are designed to provide a current or voltage output approximating the square of the input. These multipliers are also called *quarter-square* multipliers based on the identity

$$XY = \tfrac{1}{4}[(X + Y)^2 - (X - Y)^2]$$

The piecewise-linear approximations to the squared response result in "lumpy" error characteristics. Also, the amount of circuitry required to compute the quarter-square algorithm is expensive. Although these multipliers are capable of good accuracy and bandwidth, they are becoming obsolete.

A second type of exponential multiplier uses logarithmic amplifiers, a summer, and an antilog amplifier to implement the relationship

$$XY = \text{antilog}_a (\log_a X + \log_a Y)$$

Semiconductor diodes are available with excellent logging characteristics over many decades of bias current. These diodes are used with operational amplifiers to realize the necessary functions. The circuits can provide moderate to good accuracy (with thermal compensation) and good bandwidth. Applications are restricted by unipolar input requirements and differential drift due to thermal effects.

122. Multiplier Error Sources. *Offset Error.* There are two subclasses of offset error. The first is static offset caused by variances in component parameters such as saturation current I_s, in transistors and diodes. The second error is due to drift in component parameters. While the initial static offset can be trimmed with external adjustments, the drift components must be compensated by introducing complementary temperature coefficients within the circuit. A common source of drift is local heating of diode junctions and resistive components having nonzero temperature coefficients.

Feedthrough Error. Feedthrough errors result from nonideal transfer characteristics. Two feedthrough error contributions can exist. The first, E_{FY}, is defined as the output due to the input variable Y when the X input is zero. The second, E_{FX}, is the complementary function due to an X input. Feedthrough errors may result in dc, fundamental, and harmonic components of the contributing input signal.

Gain and Nonlinearity Errors. Gain errors produce output deviations from the expected scale factor of the multipliers. Nonlinearity of the transfer functions of the multiplier can produce additional error contributions, as previously discussed. Gain is most apt to vary as a function of the combined input-signal values because the internal components are operated over a range of bias conditions.

In certain multiplier applications, required transient responses may overtax the bandwidth capabilities of the circuits. In these instances additional error terms may appear as a result of limited slew rate of the circuits and differential phase response between the input channels.

123. Squaring Circuits. Squaring circuits are readily implemented by introducing the variable to both inputs of a multiplier circuit. Alternatively, the resistor-diode squaring circuits used in exponential multipliers may be applied directly.

124. Dividing and Square-Root Circuits. Division and square-root functions can be implemented with basic multiplier circuits by altering the interconnections. A typical dividing-

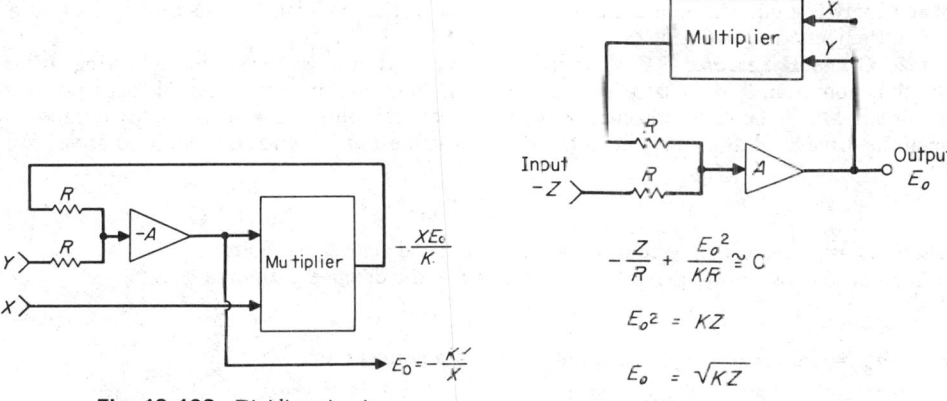

Fig. 13-123. Dividing circuit.

$$-\frac{Z}{R} + \frac{E_o^2}{KR} \cong C$$

$$E_o^2 = KZ$$

$$E_o = \sqrt{KZ}$$

Fig. 13-124. Square-root circuit.

circuit connection is shown in Fig. 13-123. The multiplier output is fed back through a summing amplifier to one of the multiplier inputs. The summing amplifier maintains an equivalence between the numerator Y and the multiplier output XE_0/K.

$$Y/R = -XE_0/KR \qquad \text{or} \qquad E_0 = -KY/X$$

Square-root circuits can be implemented in a similar method with another feedback connection involving a multiplier. In this instance, the output is used as each of the two multiplier-input variables, and the multiplier output is summed with the variable whose square root is desired. The square-root connection is shown in Fig. 13-124. The nonnegative input limitations should be stated for division and square-root connections.

125. References on Analog Multiplier Circuits

1. Evaluating, Selecting, and Using Multiplier Circuit Modules for Signal Manipulation and Function Generation, Analog Devices, Cambridge, Mass., 1970.

2. Programmable Multiplier/Divider IC Includes a Variable Scale Factor, *Elect. Des. News*, Nov. 15, 1972, p. 72.

3. Cate, T. Modern Techniques of Analog Multiplication, *Electron. Eng.*, April 1970, pp. 75–79.

4. Abbott, H. W., and V. P. Mathis Elapsed Time Computation, *Proc. Nat. Electron. Conf.*, Chicago, Vol. 25, Oct. 12–14, 1959.

High-Power Amplifiers

BY R. W. FRENCH

126. Thermal Considerations. A problem common to all high-power amplifier and oscillator equipment is removal of the excess thermal energy produced in the active devices and other circuit components so that operating temperatures consistent with reliable performance can be maintained. Available cooling methods are radiation, natural convection, forced convection, liquid, evaporative, and conduction. Radiation and evaporative cooling depend for suitable operation upon a rather high temperature for the device being cooled and are thus generally restricted to use with vacuum tubes. The remaining methods are suitable for use in both solid-state and vacuum-tube systems.

127. High-Power Broadcast-Service Amplifiers. Transmitters for amplitude modulation broadcast service may employ high-level (plate) modulation or low-level (grid or screen) modulation of the output stage or modulation of an intermediate stage, followed by linear amplification in the following stage(s). Generally, the last approach is used in very-high-power transmitters, because of the difficulty and expense in designing and building a modulation transformer to handle the high audio modulating power required. High-level modulation has been used in AM transmitters up to at least 250 kW carrier power with a modulator output power requirement of 125 kW. By a unique design in which the positive terminal of the modulator plate supply is grounded, an autotransformer is used as the modulation transformer, with a significant reduction in size and cost.

128. Class B Linear RF Amplifiers. The conventional means for achieving linear amplification of an AM signal is the class B rf amplifier circuit, often referred to simply as a *linear amplifier*. The plate efficiency, plate dissipation, and output power are highly dependent upon the drive level. It is convenient, therefore, to define a *drive ratio* or normalized drive level k.

$$k = E_{pm}/E_{bb} \tag{13-31}$$

where E_{bb} = dc plate supply voltage and E_{pm} = peak ac plate signal voltage.

In an ideal class B amplifier with sinusoidal drive the dc plate current is

$$I_b = k i_{bm}/\pi \tag{13-32}$$

where i_{bm} = peak value of plate current at full output power level.

Thus

DC plate input power $\qquad P_{DC} = E_{bb}I_b = kE_{bb}i_{bm}/\pi \tag{13-33a}$

Output power $\qquad P_o = (E_{pm})(ki_{bm})/4 = k^2E_{bb}i_{bm}/4 \tag{13-33b}$

Plate efficiency $\qquad \eta_P = \dfrac{P_o}{P_{DC}} = \dfrac{k^2 E_{bb}i_{bm}/4}{kE_{bb}i_{bm}/\pi} = k\dfrac{\pi}{4} \tag{13-34}$

Plate dissipation $\qquad P_D = P_{DC}(1 - \eta_P) = \dfrac{kE_{bb}i_{bm}}{\pi}\left(1 - \dfrac{k\pi}{4}\right) \tag{13-35}$

The maximum output power level occurs at full drive, i.e., for $k = 1$. Substituting $k = 1$ into Eq. (13-33b) gives

$$P_{o,max} = E_{bb}i_{bm}/4 \tag{13-36}$$

Normalizing P_{DC}, P_o, and P_D given by Eqs. (13-33a), (13-33b), and (13-35), respectively, by the maximum output power $P_{o,max}$ from Eq. (13-26) gives

$$P_{DC}/P_{o,max} = k4/\pi \tag{13-37}$$
$$P_o/P_{o,max} = k^2 \tag{13-38}$$
$$P_D/P_{o,max} = (4k/\pi) - k^2 \tag{13-39}$$

Equations (13-37) to (13-39) and (13-34) are plotted in Fig. 13-125 vs. the normalized drive level k defined in Eq. (13-31). The necessity for a high drive level to obtain a high efficiency is evident from the curves in this figure.

The greatest single limitation on the plate efficiency obtainable from a practical class B amplifier is the inability to achieve a value of unity for the normalized drive level k, that is, the inability to achieve a peak ac plate voltage swing equal to the dc plate voltage. The minimum instantaneous plate voltage must never fall below the instantaneous peak positive grid voltage in a triode and should preferable be 2 to 3 times the latter; hence k must always be less than 1. A reasonable value for k in the best presently available triodes is about 0.9. Thus from Fig. 13-125, a maximum plate efficiency of 70.6% could be realized. Assuming an efficiency of 0.9 for the output-coupling circuitry, an overall plate circuit efficiency of 63.5% could be expected.

Class B amplifiers using transistors can approach the theoretical maximum collector efficiency of $\pi/4$, or 78.54%, quite closely, as a consequence of the very low collector saturation voltage of transistors, which gives a correspondingly high value for k.

If an ideal class B amplifier is used to amplify an AM signal, the efficiency is less than that attainable with a cw signal because the average value of the drive ratio k is less for the AM signal. With no modulation (carrier condition), the amplifier drive is adjusted to produce a plate voltage swing equal to one-half the maximum, so that, with 100% modulation, the full maximum swing is used on positive-modulation peaks. The carrier condition corresponds to a k of 0.5 in the ideal amplifier, and from Fig. 13-125, it is seen that the theoretical plate efficiency for this

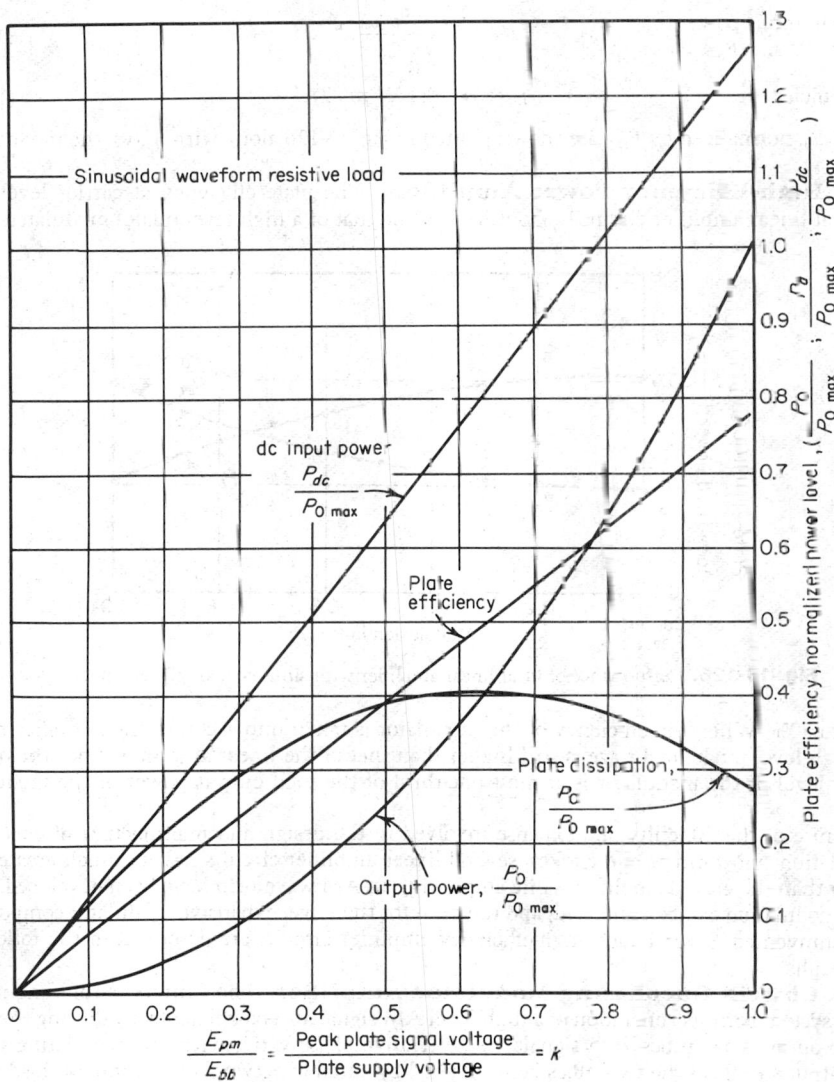

Fig. 13-125. Ideal class B amplifier characteristics

condition is 39.3%, or half the theoretical maximum. It is evident from the curve for dc plate input power in Fig. 13-125 that there will be no change in average input power with modulation, assuming a modulating waveform having zero average value. However since output power varies as k^2, the output power, and thus the efficiency, should increase with modulation depth.

By making use of the relationship between the modulation index m of the AM signal and the normalized drive ratio k, expressions for efficiency, plate output power, and dc plate input power can be derived for the ideal linear amplifier. The required equation relating m and k is

$$k = \tfrac{1}{2}(1 + m \cos \omega_a t) \qquad (13\text{-}40)$$

where ω_z = modulating frequency. The derived equations are

Plate dissipation:
$$P_D = \frac{E_{bb}i_{bm}}{\pi}\left[\frac{1}{2} - \frac{\pi}{16}\left(1 + \frac{m^2}{2}\right)\right]$$
(13-41)

Output power:
$$P_o = \frac{E_{bb}i_{bm}}{16}\left(1 + \frac{m^2}{2}\right)$$
(13-42)

DC plate input power
(sum of P_D and P_o):
$$P_{DC} = \frac{E_{bb}i_{bm}}{2\pi}$$
(13-43)

Plate efficiency:
$$\eta_P = (\pi/8)(1 + m^2/2)$$
(13-44)

P_D and P_o, normalized by P_{DC}, are shown plotted in Fig. 13-126 along with η_p, vs. the modulation index.

129. High-Efficiency Power Amplifiers. The plate efficiency at carrier level in a practical linear amplifier circuit is about 33%, while that of a high-level (plate) modulated stage

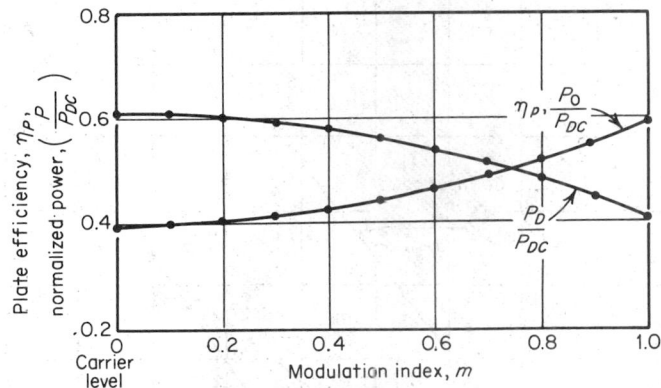

Fig. 13-126. Performance of ideal linear amplifier with sinusoidal amplitude modulation.

is 65 to 80%. When the efficiency of the modulator is taken into account, the net efficiency of the high-level modulated stage is still higher than that of the linear amplifier, since the output power level of the modulator is at most one-third of the total output power of the modulated stage.

Because of the difficulty and expense involved in the design and manufacture of very large modulation transformers and chokes, several linear amplifier circuits having much greater efficiency than the class B amplifier while amplifying an AM waveform have been developed. Low-level modulation can be employed, and the need for the large, expensive modulator components is circumvented. Several such high-efficiency amplifier circuits are described in the following paragraphs.

130. Chireix Outphasing Modulated Amplifier. The Chireix outphasing modulation system permits generation of a high-level AM signal at good efficiency by driving the grids of two output-stage tubes with signals whose relative phase varies with the modulating signal, the output signals of the two tubes being applied by suitable networks to a common load.

Figure 13-127 shows the basic circuit arrangement of the output stage. If the control grids of the two tubes are driven with signals which are 180° out of phase, the output power will be zero, which would correspond to a negative-modulation peak of a 100% modulated AM signal. By causing the angle θ to vary with modulating signal amplitude, the output power from the stage is made to vary.

Because of the outphasing method employed to produce the modulated output signal, the power factor associated with the output stage is of special significance. The basic modulation scheme would produce a unity power factor at zero output level which would decrease with increasing output level. To modify this undesirable characteristic the output plate circuits are detuned, one above and one below resonance, to produce an offset or dephasing of $\theta = \theta_0$. For $\theta_0 = 18°$, the power factor is zero for zero output ($\theta = 0$), rises rapidly to unity at $\theta = \theta_0$, and remains at 0.9 or higher for $\theta \geq \theta_0$. The low power factor at low output level does not have a significant effect on overall efficiency. The angle θ will have a value of approximately 25° at carrier level and 45 to 50° at a 100% positive-modulation peak. The linearity of the modulation characteristic is good, and efficiency is reasonably high, 60% being typical at carrier level, decreasing slightly with modulation.

The phase-modulated carrier-frequency driving signals for the output stage are generated at low level and amplified to the level required. A suppressed-carrier AM signal is produced by combining the carrier and modulating signals in a balanced modulator. To this suppressed-carrier signal is added a carrier having a quadrature phase relative to the original (suppressed) carrier. Two such composite signals are generated having opposite sense of phase rotation. The amplitude variation of the composite signal is only about 10% for a phase variation of 50°

An important difficulty in implementing the Chireix outphasing modulation circuit is main-

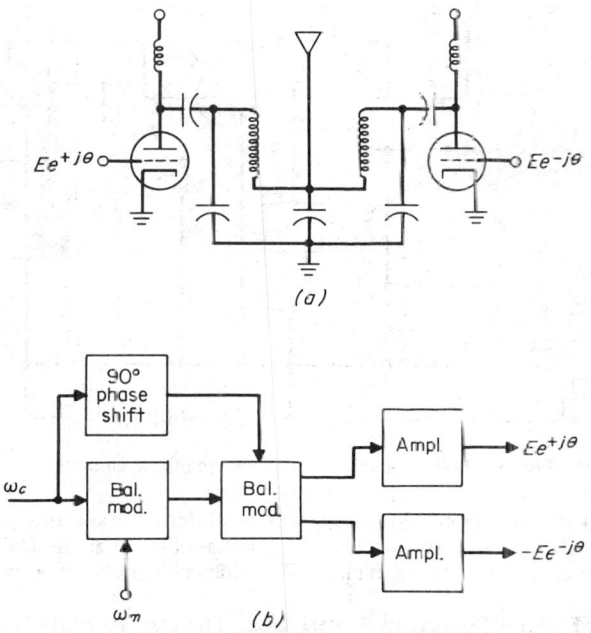

Fig. 13-127. Chireix outphasing modulated amplifier: (a) output circuit, (b) drive signal generation.

taining the very stable phase characteristics of the circuits producing the two driving signals for the output stage, since a relative phase of ±45° between the two signals produces 100% modulation.

131. Dome High-Efficiency Modulated Amplifier. In the Dome high-efficiency modulated circuit, shown schematically in Fig. 13-128, modulation is achieved by load-line modification during positive-modulation swings and by linear amplification during negative-modulation swings. Load-line modification is achieved by absorption of a portion of the generated rf power; however, a major portion of the absorbed power is returned to the plate power supply, rather than being dissipated, and high-power efficiency results.

The operation of the circuit shown in Fig. 13-128 is as follows. Tube V1 is used in a plate-modulated driver stage supplying power to the grid circuit of the power amplifier tube V2. The total load impedance in the plate circuit of V2 is that load impedance reflected into the primary of rf transformer T1 (from the antenna) in series with the impedance appearing across the C8 terminals of the 90° phase-shift network consisting of C8, C9, and L4. The impedance appearing across the C8 terminals of the 90° network is inversely related to the impedance across the other terminals of the network, i.e., the effective ac impedance of tube V3. Thus, with tube V3 cut off, a short circuit is reflected across the C8 terminals of the network. Tube V3 is a modulated rectifier, the audio modulating signal voltage being applied to its control grid. Dome calls this tube a *modifier*.

With no modulation (carrier condition), tube V3 is biased to cut off and the drive to V2 is adjusted until the output power is equal to 4 times carrier power (corresponding to a positive-modulation peak). Note that all the plate signal voltage of V2 appears across the primary of the transformer T1 for this condition. The bias on tube V3 is then reduced, lowering the ac impedance of the tube, and therefore reflecting an increasing impedance at the C8 terminals of the 90° network. The bias on V3 is reduced until carrier power level is being delivered to the antenna. An amount of power equal to the carrier power is then being rectified by V3 and returned to the plate power supply, except for that portion dissipated on the V3 plate. The drive level to V2 is now adjusted so that the tube is just out of saturation.

For positive-modulation swings, tube V3 grid voltage is driven negative, reaching cutoff for a positive-modulation peak (100%). For negative-modulation swings, tube V2 acts as a linear amplifier. V3 does not conduct during the negative-modulation swing since the peak rf voltage on its plate is less than the dc supply voltage on its cathode.

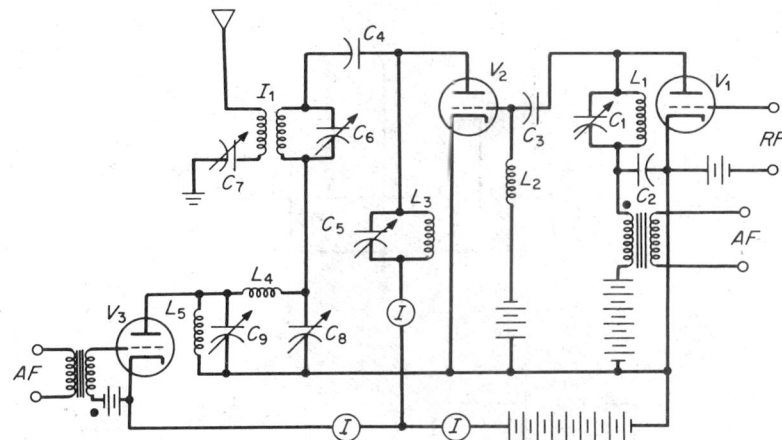

Fig. 13-128. Dome high-efficiency modulated amplifier.

Because of the serial loss incurred in the power amplifier tube and in the modifier tube for energy returned to the power supply, the circuit is not as efficient as the Doherty circuit, 55 to 60% efficiency at carrier level being typical. The efficiency does not vary appreciably with modulation.

132. Doherty High-Efficiency Amplifier. The Doherty amplifier circuit is perhaps the most widely applied of the several types of high-efficiency circuits. This circuit is a high-efficiency linear amplifier circuit as opposed to high-efficiency circuits that achieve modulation as well as amplification. Doherty was the first to employ the 90° network as an impedance inverter to achieve load-line modification as a function of the power level in an amplifier stage.

In the circuit shown in Fig. 13-129a, the carrier tube V1 is biased class B, and its loading and drive are such that it is operating at maximum linear voltage swing at carrier level. Tube V2 is biased class C such that at carrier level it is just beginning to conduct plate current. Each tube delivers an output power equal to twice carrier power when working into a load impedance of $R\ \Omega$. At carrier level the reflected load impedance at the plate of V1 is $2R\ \Omega$, which is the correct value of load impedance for carrier-level output power at full plate voltage swing. The impedance-inverting property of the 90° network is like that of a quarter-wave transmission line; i.e.,

$$Z_{in} = Z_0^2 / Z_{load}$$

where Z_0 = the characteristic impedance of the line. If the three reactances in the 90° network are each equal to $R\ \Omega$, the input impedance is

$$R_{in} = R^2 / R_{load}$$

Thus, for $R_{load} = R/2$,

$$R_{in} = \frac{R^2}{R/2} = 2R$$

The 90° phase lead network in the grid circuit of V1 is required to compensate for the 90° phase lag produced by the 90° network (impedance-inverting network) in the plate circuit. For negative-modulation swings the carrier tube performs as a linear amplifier with load impedance of $2R$ Ω, and the peak tube is inoperative.

On positive-modulation swings the peak tube conducts and contributes power to the $R/2$ load resistance. This is equivalent to connecting a negative resistance in shunt with the $R/2$ load resistance, so that the load resistance seen at the load end of the 90° network increases. This increase is reflected through the network as a decrease in load resistance at the plate of V1, causing an increase in its output current, and hence in its output power. The drive levels on the tubes are adjusted so that each contributes the same power (equal to twice carrier power) at a positive-modulation peak. For this condition a load of R Ω is presented to each tube.

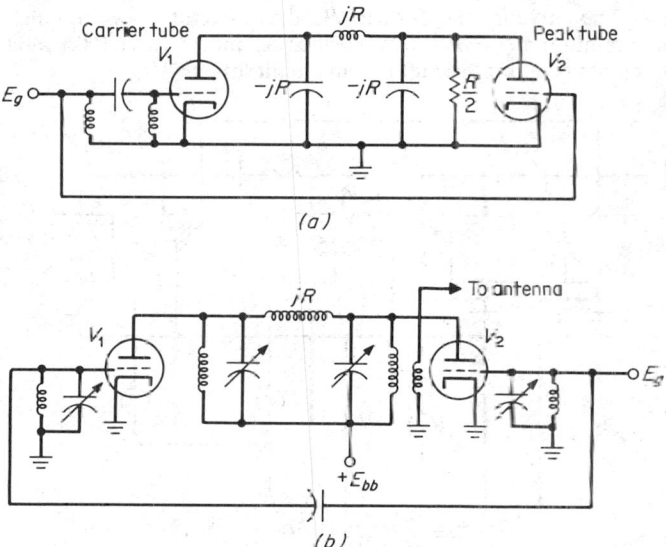

Fig. 13-129. Doherty high-efficiency amplifier.

The important aspect of the circuit operation is the change in load impedance on tube V1 with modulation which enables it to deliver increased output power at constant plate voltage swing, and thus at high efficiency and good linearity. Some distortion of the modulation envelope will occur at carrier level as the peak tube comes into operation. The distortion can be reduced by a reduction in bias on the peak tube, but this causes a reduction in efficiency. Envelope feedback is normally used to improve the linearity. The efficiency of the Doherty circuit is 60 to 65%, essentially independent of modulation depth.

A more practical circuit for the Doherty amplifier is shown in Fig. 13-129b, where the shunt reactances of the phase-shift networks are supplied by detuning the related tuned circuits, the tuned circuits in the grid circuits being tuned above the operating frequency, while those in the plate circuits are tuned below the operating frequency.

133. Terman-Woodyard High-Efficiency Modulated Amplifier. The Terman-Woodyard high-efficiency grid-modulated amplifier uses the basic scheme of Doherty for achieving high efficiency, i.e., the impedance-inverting property of a 90° phase-shift network. However, the Terman-Woodyard circuit employs grid modulation of both the carrier tube and the peak tube, which allows both tubes to be operated class C, with the result that an increase in efficiency over the Doherty circuit is obtained. It has the highest efficiency of the systems currently available.

Referring to Fig. 13-130, V1 is the carrier tube and V2 is the peak tube. In the absence of modulation, V1 is operating as a class C amplifier, supplying the carrier power, and V2 is biased so that it is just starting conduction. The efficiency at carrier level is thus essentially that of a class C amplifier.

The shunt reactances of the 90° phase-lead network in the grid circuit of V1 and the 90° phase-lag network (impedance-inverting network) in the plate circuit of V1 are provided by detuning the tuned circuits L1-C3, L2-C5, L3-C7, and C10 and the primary inductance of T1. C2 is the series element of the phase-lead network, and L4 is the series element of the phase-lag network. Capacitor C9 enables the metering of the individual plate currents of V1 and V2.

Resistor $R1$ is included to prevent overdriving of the grid of $V1$ on positive-modulation swings. The relative modulating voltage applied to the two tubes is controlled by adjustment of $R2$.

During a positive-modulation swing, tube $V2$ conducts, and at a (100%) modulation peak the two tubes are supplying equal amounts of power to the load, similar to the Doherty amplifier operation. During a negative-modulation swing, $V2$ is cut off and $V1$ performs as a standard grid-modulated amplifier.

Terman and Woodyard show curves of plate efficiency as a function of modulation percentage (sinusoidal modulation) for both circuits based upon a peak plate voltage swing equal to 80% of the dc plate supply voltage. The curves indicate an efficiency for the Doherty amplifier of 62.8% at carrier level, dropping to a minimum of 57% at 50% modulation and rising back to 62.8% at 100% modulation. The curve for the Terman-Woodyard circuit shows an efficiency of 80% at carrier level and a minimum of 68% at 50% modulation and 73.4% at 100% modulation. These efficiency values do not take into account output-circuit losses.

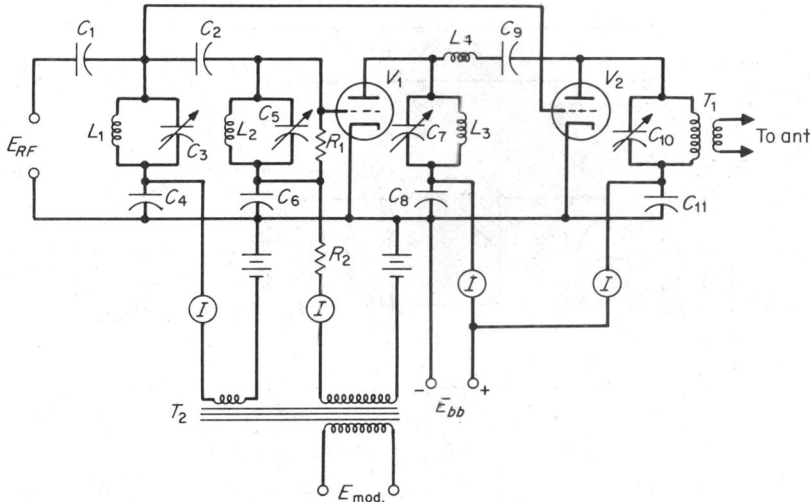

Fig. 13-130. Terman-Woodyard high-efficiency modulated amplifier.

The adjustments in rf drive level, modulation voltage amplitude, and tuning and loading to obtain proper operation of the Terman-Woodyard circuit are tedious. The application of envelope feedback is very desirable in order to reduce distortion and minimize the effects of misadjustment of the operating conditions on performance. Large amounts of feedback can normally be applied in this circuit because the feedback loop includes only the one rf stage.

134. Induction Heating Circuits. Induction heating is achieved by placing a coil carrying alternating current adjacent to a metal workpiece so that the magnetic flux produced by the current in the coil induces a voltage in the workpiece, which produces the necessary current flow.

Power sources for induction heating, in addition to direct use of commercial power, include spark-gap converters, motor-generator sets, vacuum-tube oscillators, and inverters. Motor-generator sets generally provide outputs from 1 kW to more than 1 MW and from 1 to 10 kHz. Spark-gap converters are generally used for the frequency range from 20 to 400 kHz and provide output power levels up to 20 kW. Vacuum-tube oscillators operate at frequencies from 3 kHz to several MHz and provide output levels from less than 1 kW to hundreds of kilowatts. Inverters using mercury-arc tubes have been used up to about 3 kHz. Solid-state inverters have been developed in recent years which operate at frequencies up to about 10 kHz and at power levels of several megawatts. These solid-state inverters generally employ thyristors (silicon controlled rectifiers) and are replacing motor-generator sets and mercury-arc inverters.

Induction heaters using vacuum tubes generally operate at frequencies of 300 kHz and higher and are available for output power levels from about 5 to 200 kW in single equipments. A single power tube is usually employed in an oscillator circuit. A simplified circuit diagram of a typical 20-kW induction heater is shown in Fig. 13-131.

135. Dielectric Heating. Whereas induction heating is used to heat materials which are electrical conductors, dielectric heating is used to heat nonconductors or dielectric materials.

The basic arrangement for a dielectric heating system is that of a capacitor in which the material to be heated forms the dielectric or insulator. The heat generated in the material is proportional to the loss factor, which is the product of the dielectric constant and the power factor. Because the power factor of most dielectric materials is quite low at low frequencies, the range of frequencies employed for dielectric heating is higher than for induction heating, extending from a few megahertz to a few gigahertz.

The power generated in a material is given by

$$P = 141\,AV^2 f \frac{K \cos \phi}{t} \times 10^{-6} \quad \text{(watts)}$$

where V = voltage across material (V), f = frequency of power source (MHz), A = area of material (in²), K = dielectric constant, t = material thickness (in), and $\cos \phi$ = power factor of dielectric material.

The voltage that can be applied to a particular material is limited by the insulation properties

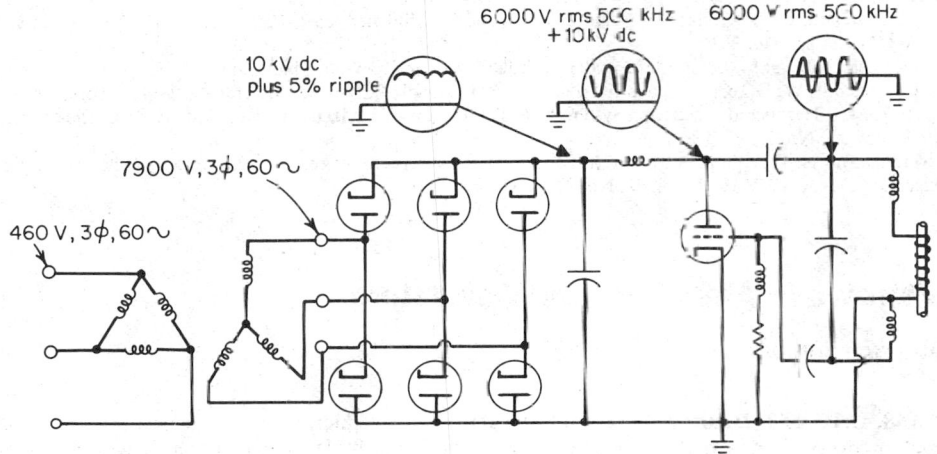

Fig. 13-131. Induction heating circuit of 20 kW power.

of the material at the required process temperature. The frequency that can be used is limited by voltage standing waves on the electrodes, which will be appreciable when the electrode dimensions are comparable with one-eighth wavelength (10% voltage variation).

136. Transistors in High-Power Amplifiers. Many high-power amplifier circuits, including those described in this paragraph, can be implemented by using solid-state power devices. Bipolar transistors have been successfully applied in induction heaters, sonar transmitters, and broadcast transmitters. The sonar transmitters have employed class B, class C, and the so-called class D power amplifier circuits. In the class D circuit the active device is used as an on-off switch, and output power variations are achieved by pulse-width modulation or supply-voltage control. A tuned load or one incorporating a low-pass filter is used with the class D amplifier.

Recent significant improvements in the characteristics of power MOSFET transistors make them very attractive as high-power amplifier devices in such applications as induction heating, broadcast transmitters, sonar transmitters, and dielectric heating. The device characteristics of low drive power, fast switching, absence of secondary breakdown, and adaptability to paralleling make them suitable for use in many high-power amplifiers.

137. References on High-Power Amplifiers

1. Chireix, H.   High Power Outphasing Modulation, *Proc. IRE*, November 1935, Vol. 23, pp 1370-1392.

2. Dome, R. B.   High-Efficiency Modulation System, *Proc. IRE*, August 1938, Vol. 26, pp. 963-982.

3. Doherty, W. H.   A New High Efficiency Power Amplifier for Modulated Waves, *Proc. IRE*, September 1936, Vol. 24, pp. 1163-1182.

4. Terman, F. E., and J. R. Woodyard   A High-Efficiency Grid-Modulated Amplifier, *Proc. IRE*, August 1938, Vol. 26, pp. 929-945.

5. Fisher, S. T.   A New Method of Amplifying with High Efficiency a Carrier Wave Modulated in Amplitude by a Voice Wave, *Proc. IRE*, January 1946, Vol. 34, pp. 3p-13p.

6. Sainton, J. B.   A 500 Kilowatt Medium Frequency Broadcast Transmitter, *Machlett Cathode Press*, 1965, Vol. 22, No. 4.

7. Greenville, U.S.A., the World's Largest Broadcast Facility 4.8 MW, *Machlett Cathode Press*, 1965, Vol. 22, No. 4.

8. Curtis, F. W.   "High-Frequency Induction Heating," McGraw-Hill, New York, 1964.

9. Brown, G. H., C. N. Hoyler, and R. A. Bierwirth   "Theory and Application of Radio-Frequency Heating," Van Nostrand, New York, 1947.

10. Stansel, N. R.   "Induction Heating," McGraw-Hill, New York, 1949.

11. Spash, D. I.   "High Frequency Heating," Pt. 3, Vol. 3 of A. H. Beck (ed.), "Handbook of Vacuum Physics," Macmillan-Pergamon, New York, 1964.

12. Cable, J.   "Induction and Dielectric Heating," Reinhold, New York, 1954.

13. Hughes, L. E. G., and F. W. Holland (eds.)   "Electronic Engineer's Reference Book," 3d ed., Heywood, London, 1967.

14. Frank, W. E.   New Developments in High-Frequency Power Sources, *IEEE Trans. Ind. Gen. Appl.*, Vol. IGA-6, No. 1, pp. 29–35.

15. Ross, N. V.   A System for Induction Heating of Large Steel Slabs, *IEEE Trans. Ind. Gen. Appl.*, Vol. IGA-6, No. 5, pp. 449–454.

16. Dewan, S. B., and G. Havas   A Solid-State Supply for Induction Heating and Melting, *IEEE Trans. Ind. Gen. Appl.*, Vol. IGA-5, No. 6, pp. 686–692.

17. Hatchard, D. G.   Induction Heating of Bars and Semifinished Steel, *IEEE Trans. Ind. Gen. Appl.*, Vol. IGA-2, No. 5, pp. 346–352.

18. Shea, R. F. (ed.)   "Amplifier Handbook," Chap. 20, Mc-Graw-Hill, New York, 1968.

19. Cockrell, W. D. (ed.)   "Industrial Electronics Handbook," Chap. 5c, McGraw-Hill, New York, 1958.

20. A New Hardboard Production System Uses a 600 Kilowatt Dielectric Heater, *Machelett Cathode Press*, 1968, Vol. 25, No. 2.

21. Collins, H. W., and B. Pelly   HEXFET, A New Power Technology, Cuts On-Resistance, Boosts Ratings, *Electron. Des.*, June 7, 1979, Vol. 27, No. 12, p. 36.

Microwave Amplifiers and Oscillators

BY J. W. LUNDEN

138. IMPATT Diode Circuits. The generation of microwave power in a reverse-biased *pn* junction was originally suggested in 1958 by Read.[1]* He proposed that the finite delay between applied rf voltage and the current generated by avalanche breakdown, with the subsequent drift of the generated carriers through the depletion layer of the junction, would lead to negative resistance at microwave frequencies. The diode is biased in the avalanche-breakdown region. As the rf voltage rises above the dc breakdown voltage during the positive half cycle, excess charge builds up in the avalanche region, reaching a peak when the rf voltage is zero. Hence this charge waveform lags the rf voltage by 90°. Subsequently, the direction of the field in the diode causes the multiplied carriers to drift across the depletion region. This, in turn, induces a positive current in the external circuit while the diode rf voltage is going through its negative half cycle. This is equivalent to negative resistance, which is maximum when the transit angle is approximately 0.74π.

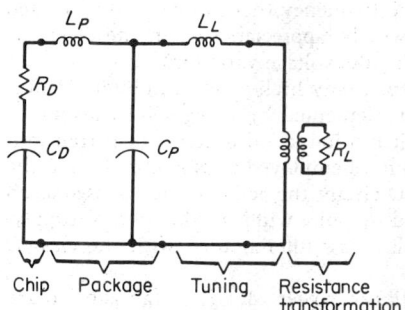

Fig. 13-132. Approximate rf equivalent circuit of IMPATT oscillator.

A simplified equivalent circuit of the IMPATT diode circuit is shown in Fig. 13-132. The resistance R_D includes both the parasitic positive resistance due to contacts, bulk material, etc., and the dynamic negative resistance. The net magnitude is typically in the range −0.5 to −4.0 Ω and varies with current. The capacitance C_D is the voltage-sensitive depletion-layer capacitance and can be approximated sufficiently accurately with the value at breakdown. The diode resistance variation results in a stable operating point for any positive load resistance equal to or less than the diode's peak negative value.

Oscillations will build up and be maintained at the frequency for which the net inductive

---

*Superior numbers correspond to numbered references, Par. **13-149**.

reactance of the package parasitics and the external load equals the capacitive reactance of C_D. The values for L_P and C_P vary with package or mounting style. Typical values range from 0.3 to 0.6 nH and 0.2 to 0.4 pF, respectively. It is important to minimize these parasitics since they limit the operating frequency and bandwidth.

IMPATT diodes can be used in several mounting configurations, including coaxial, waveguide, strip-line, or microstrip. It is important, to ensure that a good heat flow path is provided[2] (due to the typically low efficiency) and that low-electrical-resistance contacts be made to both anode and cathode.

In avalanche breakdown, the diode tends to look like a voltage source. Hence a current source is desirable for dc bias. Several circuits are possible. The RC bias circuit (Fig. 13-133a) is the

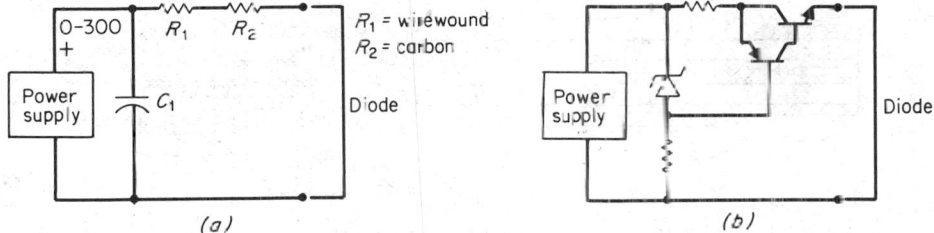

Fig. 13-133. Bias circuits for IMPATT diodes: (a) RC type; (b) transistor-regulated type.

simplest but is inefficient, and the transistor current regulator (Fig. 13-133b) may be more desirable. In either case, the loading of the diode with shunt capacitance or a resonance path to ground (at some frequency) must be avoided to prevent instabilities (noise and/or spurious frequencies).

Two broadly tunable diode loading circuits are the multiple-slug cavity[3] and the variable-package-inductance types. Coaxial implementations of these two techniques are shown in the cross sections of Fig. 13-134. In Fig. 13-134a slugs $\lambda/8$ and/or $\lambda/4$ long at the desired center frequency of operation with characteristic impedance of between 10 and 20 Ω are adjusted in position along the centerlines to provide a load reactance equal to the negative of the diode reactance and a load resistance equal to or less than the magnitude of the diode net negative resistance. Circuit b tunes the diode by recessing it into the holder, effectively decreasing the net series inductance, which is resonant with the diode capacitance. Single- or multisection transformers can also be included for resistive matching to the load R_L. Similar waveguide-circuit implementations can be used with typically narrower bandwidths, better frequency stability, and lower FM noise.

The IMPATT diode can function as an amplifier if the load resistance presented to it is larger in magnitude than the negative resistance of the diode. Typically, a circulator is used in conjunction with the tuned diode circuit to separate input and output signals. At the center frequency, the power gain of the amplifier is given by

$$G_0 = \left(\frac{R_D - R_L}{R_D + R_L} \right)^2$$

where R_D and R_L are as shown in Fig. 13-132.

In general, amplifier operation for a given diode requires a smaller shunt tuning capacitor or a larger transformer characteristic impedance. An estimate of the diode rf current and voltage can be obtained[3] from

$$V_D \approx \frac{\sqrt{2 P_o R_L}}{\omega C_j(V_b)} \left(1 + \frac{1}{\sqrt{G_0}} \right) \qquad I_D \approx \omega C_j(V_e) V_D$$

where V_b = diode breakdown voltage, C_j = junction capacitance, P_o = power output, and G_0 = gain. Higher output powers have been achieved by multiple-diode series-parallel configurations and/or hybrid combining techniques.

In general, properly designed IMPATT oscillator circuits can have noise performance comparable with reflex klystron or Gunn oscillators. The noise performance of both Si and GaAs IMPATT diode amplifiers and oscillators has been extensively treated in the literature.[4-7]

139. TRAPATT Diode Circuits. The TRAPATT (trapped plasma avalanche triggered transit) mode of operation is characterized by high efficiency, operation at frequencies well below the transit-time frequency, and a significant change in the dc operating point when the diode switches into the mode. The basic understanding of this high-efficiency mode of operation

has proceeded from consideration of IMPATT diode behavior with large signals.[9,10] Two saturation mechanisms in the IMPATT diode, space-charge suppression of the avalanche and carrier
trapping, reduce the power generated at the transit-time frequency, but play an important role
in establishing the "trapped-plasma" states for high-efficiency operation.

To manifest the high efficiency of the TRAPATT mode, four important circuit conditions
must be met: large IMPATT-generated voltage swings must be obtained by trapping the
IMPATT oscillation in a high-Q-cavity circuit; selective reflection and/or rejection of all sub-
harmonics of the IMPATT frequency except the desired subharmonic must be realized (typi-

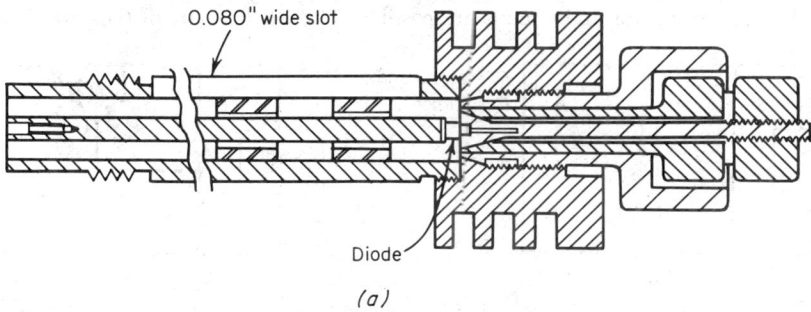

0.080" wide slot

Diode

(a)

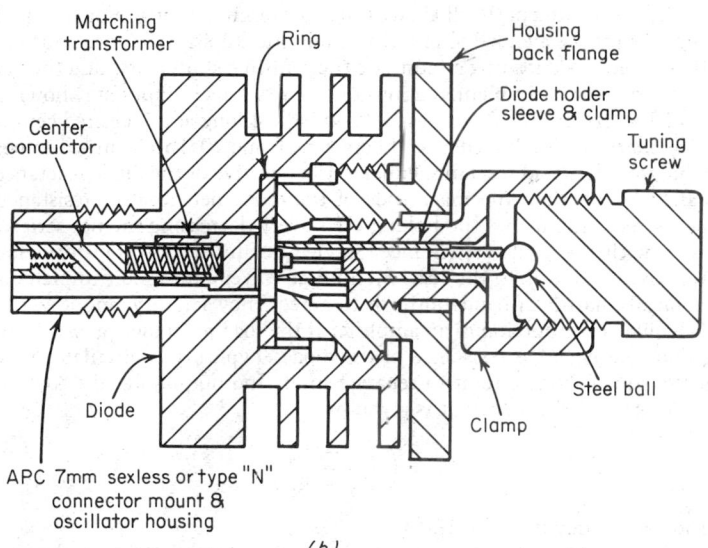

Matching transformer

Ring

Housing back flange

Diode holder sleeve & clamp

Center conductor

Tuning screw

Diode

Steel ball

Clamp

APC 7mm sexless or type "N"
connector mount &
oscillator housing

(b)

Fig. 13-134. Mechanical tuning of IMPATT circuit: (*a*) multiple-slug type; (*b*) variable-inductance (diode-recess) type. (*Hewlett-Packard Company.*)

cally with a low-pass filter); sufficient capacitance must be provided near the diode to sustain the
high-current state; and tuning or matching to the load must be provided at the TRAPATT frequency. Figure 13-135 illustrates a widely used circuit that achieves the foregoing conditions.

The TRAPATT diode is typically represented by a current pulse generator and the diode's
depletion-layer capacitance, as shown in the simplified schematic of Fig. 13-136. This permits a
simple interpretation of operation. Initially, a large voltage pulse reaches the diode, triggering
the traveling avalanche zone in the diode. This initiates a high-current low-voltage state. The
drop in voltage propagates down the line l, is inverted (due to the -1 reflection coefficient of
the low-pass filter), and travels back toward the diode. The process then repeats. Consequently,
the frequency of operation is inversely proportional to the length of line l. The period of oscillation is slightly modified by the finite time required for the diode voltage to drop to zero. The
inductance L is due to the bond lead and is helpful in driving the diode voltage high enough to
initiate the TRAPATT plasma state. The capacitance C is provided to supply the current

required by the diode to the extent that the transmission line is insufficient. The total capacitance which can be discharged during the short interval of high conduction current in the diode is

$$C_T = C + \tau_P/Z_0$$

where τ_P = time high-current or trapped plasma state exists. τ_P must be at least as large as the transit time of carriers through the diode, thus putting an upper limit on C. This is contrary to the high-current requirements to initially drive the avalanche zone through the diode (10 to 20 kA/cm²) in large-area devices.

Since the TRAPATT frequency is generally an integral submultiple of the diode transit-time frequency, the line length l ($l = \lambda/2$ at f_{TRAPATT}) presents the low-pass-filter short circuit to the

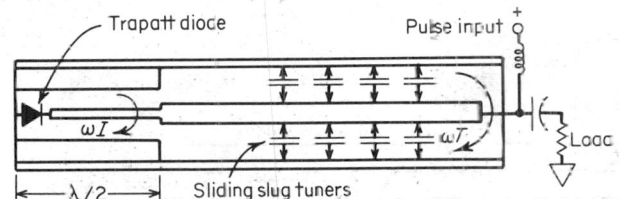

Fig. 13-135. Typical TRAPATT diode circuit configuration.

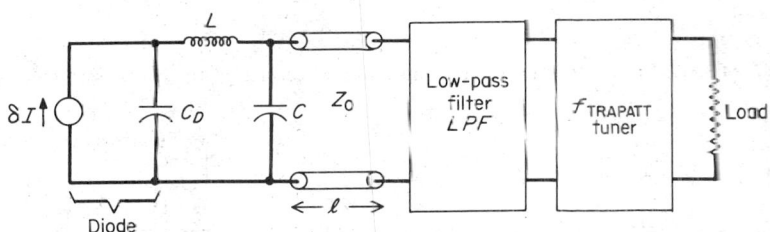

Fig. 13-136. Simplified schematic diagram of TRAPATT circuit.

diode as a series resonance at the IMPATT frequency. This net series resonance, however, includes the diode series-reactive elements. Further, this circuit should have a high Q at the transit-time frequency to reduce the buildup time to TRAPATT initiation.

Several TRAPATT circuit configurations are shown in Fig. 13-137. The coaxial cavity circuit in Fig. 13-137a places the diode in the reentrant gap of the half-wave cavity resonator. The output coupling loop passes only the fundamental to a triple stub tuner to match to the load. Proper dc biasing and bypassing are included. This circuit is good into the lower L band. The lumped circuit (Fig. 13-137b) is compact and very useful for VHF and UHF. The series capacitor is resonated with the inductance of a copper bar and the self-inductance of the diode. The trimmer controls the resonance of the third and fifth harmonics. The circuit of Fig. 13-137c is a variation on the circuit of Fig. 13-135. The use of additional filter sections and/or lumped elements provides better harmonic filtering and results in higher efficiencies. Circuits analogous to those of Fig. 13-135 can be implemented in waveguide for the higher frequencies. In all these circuits, the presence of the third and fifth harmonics has been found essential for stable and high-efficiency performance. Higher power levels can be achieved by operating multiple diodes in series and/or parallel configurations.[12]

Another useful technique for extending both power and frequency is the antiparallel diode configuration. The circuit consists of two diodes, placed with opposite polarity approximately one-half fundamental wavelength apart in a transmission line.[13] Operation is similar to a free-running multivibrator. Output may be extracted with a transmission line connected to the midpoint of the diodes, followed by the usual low-pass filter. The position of this filter should be adjusted so that the round-trip delays from midpoint to diodes and the filter are equal. A micro-strip circuit realization for antiparallel operation is given in Fig. 13-138.

Operation of the anomalous avalanche diode for microwave amplification has also been established.[15] A 10-dB dynamic range is typical, with a low-level threshold gain decreasing with increasing power level to a saturated condition. The pulsed bias can be replaced by a simple dc source and storage capacitor. Unlike the *locked-oscillator* mode of operation, only a small resid-

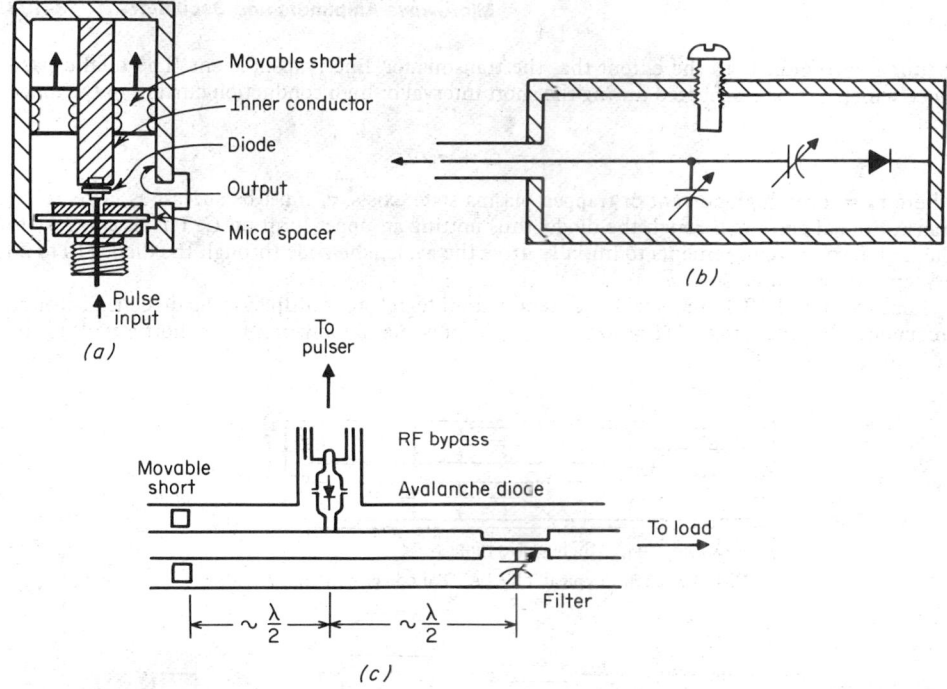

Fig. 13-137. TRAPATT diode circuit arrangements: (*a*) tuned coaxial cavity; (*b*) lumped circuit; (*c*) coaxial circuit.

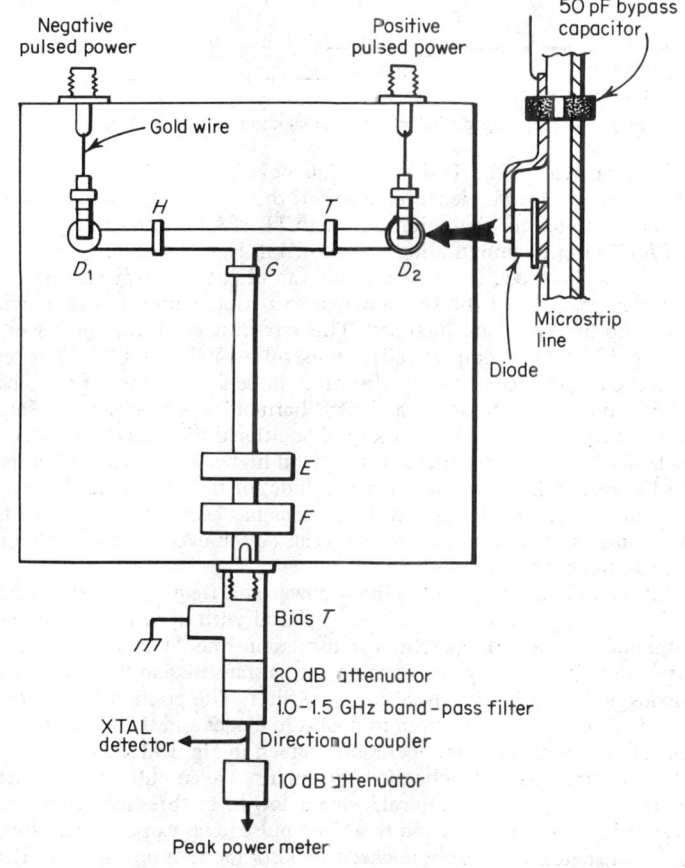

Fig. 13-138. Microstrip version of antiparallel TRAPATT circuit.

ual output is present without the input signal. The locked oscillator will typically display only a 3:1 power change between locked and unlocked cases.

TRAPATT operation has yielded output power levels of 10 to 500 W with efficiencies of 20 to 75% in the frequency range from 0.5 to 10 GHz.

140. Transferred Electron Effect Device (TED) Circuits. This class of circuit, using both the Gunn and LSA (limited-space-charge accumulation) devices depends on the internal negative resistance due to carrier motion in the semiconductor at high electric fields.[16,17] When the material is biased above the critical threshold field, a negative dielectric relaxation time is exhibited, which results in amplification of any carrier concentration fluctuations, causing a deviation from space-charge neutrality. The resultant domain drifts toward the anode and is extinguished, and a new domain is formed at the cathode. The current through the device consists of a series of narrow spikes with a period equal to the transit time of the domain.

When an rf voltage is superimposed on the bias, in a given period of time, the terminal voltage can be below both the threshold voltage V_{th} and the domain-sustaining voltage V_s. The domain is quenched at any place in the device when the latter occurs, and the nucleation of a new domain is delayed until the voltage again exceeds V_{th}. Therefore the frequency of oscillation is determined by the resonant circuit, including the impedance of the device. Experimental results and computer modeling have shown that the device can be tuned over greater than an octave bandwidth by the external circuit cavity. Although other modes of operation are possible, depending on the characteristics of the external circuit, the LSA mode appears to be the most important.

An approximate equivalent circuit is given in Fig. 13-139, with values dependent on frequency, bias, and power level. The capacitance includes the diode static capacitance, in addition to that due to traveling high-field domains.

One of the simplest tuned circuits is the coaxial-line cavity, as shown in Fig. 13-140a. The diode is mounted concentric with the line, at one end to facilitate heat sinking. The frequency of oscillation is determined primarily by the length of the cavity. The position of the output coupling loop (or plate) determines the load impedance.

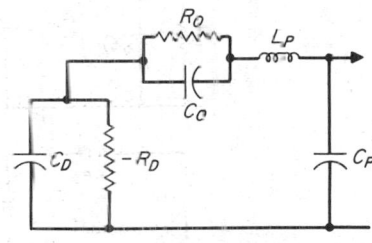

C_D = Domain capacitance

$-R_D$ = Negative differential resistance

P_0, C_C = Due to bulk material

C_P, L_P = Packaging parasitics

Fig. 13-139. Approximate equivalent circuit of a TED device and its package.

A rectangular waveguide cavity configuration (Fig. 13-140b) is more widely used due to its higher Q and better performance at X band and higher frequencies. The diode post acts as a large inductive susceptance, which, with the inductive iris, produces the resonant frequency for which the length l is $\lambda/2$. The tuning rod lowers the frequency as its insertion length increases.

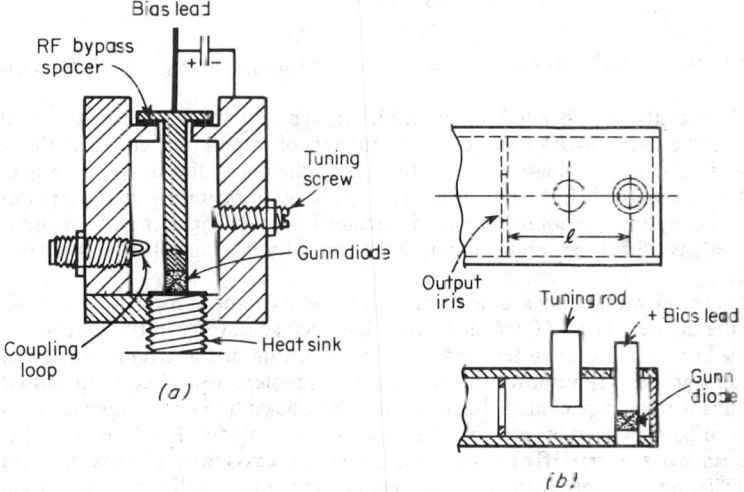

Fig. 13-140. Gunn diode cavity circuits: (a) coaxial form; (b) waveguide-cavity form.

In addition to mechanical tuning, both YIG[18] and varactor[19] tuning techniques are applicable. YIG tuning has the potential for the widest tuning range but is limited in tuning speed, hysteresis, and physical bulk. Figure 13-141 illustrates a typical varactor-tuned Gunn oscillator equivalent circuit and characteristics. The varactor diode Q is highest at the maximum reverse voltage and decreases with decreasing voltage due to an increase in R_{vs}. Figure 13-142 illustrates two varactor-tuned implementation techniques.

The noise characteristic of a GaAs TED is comparable with that of a klystron. Various noise-source models and measuring equipments are discussed in the literature. Several methods can be employed to reduce the AM and FM noise of a Gunn oscillator, including increasing the loaded Q of the cavity circuit; biasing at or near the frequency and/or power turnover points (i.e., bias at which $df/dv = 0$ and $dP_o/dv = 0$); minimizing power-supply ripple; and diode selection.

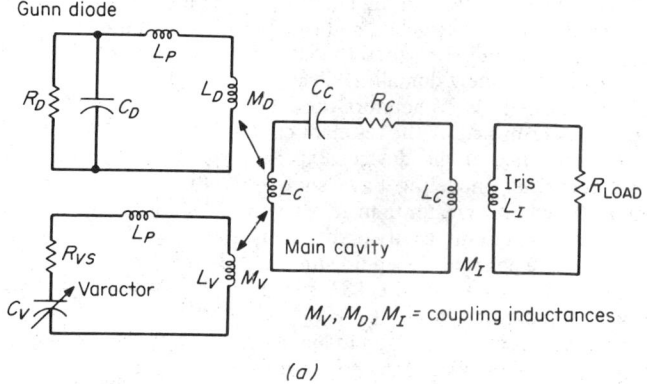

(a)

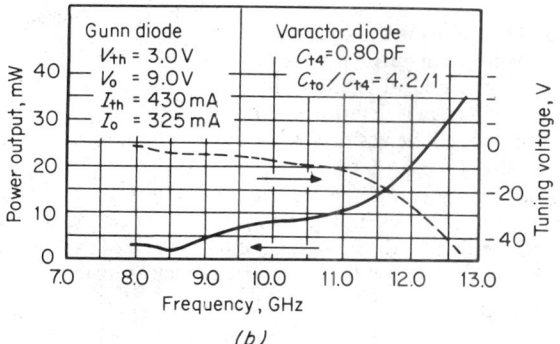

(b)

Fig. 13-141. Varactor tuning of Gunn oscillator: (a) simplified equivalent circuit; (b) tuning characteristic.

The TED can also be operated as an amplifier, typically using circulator or hybrid techniques[20,21] similar to the IMPATT circuits. Parameters of importance are saturation characteristics, bandwidth, gain and phase tracking, linearity, efficiency, and dynamic range. The block diagram of a four-stage chain is shown in Fig. 13-143a, with operating characteristics in Fig. 13-143b. Hybrid coupling was found to provide greater linear power output. Gains in excess of 20 dB can be realized. Efficiency and bandwidth are typically less than 10% and greater than 35%, respectively.

TEDs can be operated with a pulsed bias, allowing extremely high power density without damage to the device. Over 1,000 W at 10 GHz has been achieved[22] in the LSA mode with short low-duty cycle pulses. Reactive termination and unloading of the circuit at the harmonic frequencies can significantly improve the efficiency. A problem associated with pulsed oscillators is the significant frequency change during each pulse caused by rapid temperature rise. Starting-time jitter can be alleviated by *priming*, i.e., injecting a weak cw signal into the circuit.

141. Transistor Amplifier and Oscillator Microwave Circuits. Silicon bipolar and GaAs FETs are available with cutoff frequencies extending well into X and K bands. Equivalent circuits for these transistor types are given in Fig. 13-144. The intrinsic chip element values, with variations considered at low frequencies, are further modified by high-frequency effects. A small-signal figure of merit has been defined in terms of the contributing time constants.

$$K = (\text{power gain})^{1/2}(\text{bandwidth}) = 1/4\pi(r_b'C_c\tau_{ec})^{1/2}$$

where $\tau_{ec} = \tau_e + \tau_b + \tau_1 + \tau_c$, τ_e = emitter barrier charging time, τ_b = base transit time, τ_1 = collector transit time, and τ_c = collector depletion-layer charging time. The maximum frequency of oscillation is defined as the frequency for which the power gain is unity

$$f_{max} \approx 1/4\pi(r_b'C_c\tau_{ec})^{1/2}$$

The application of transistors at microwavelengths demands that considerable attention be given to packaging, fixturing, and impedance characterization. Historically, characterization has taken the form of f_T, $r_b'C_c$ specification and/or h-y-parameter techniques. A more desirable method is by scattering parameters. Scattering parameters describe the relationship between the incident and reflected power waves in any N-port network.[23,24] As such, this technique offers

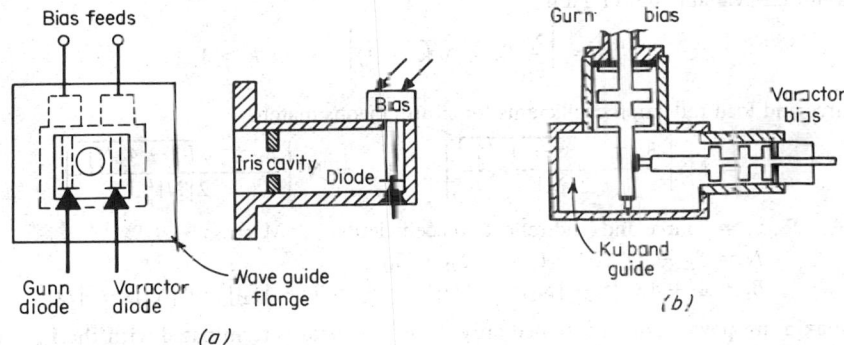

Fig. 13-142. Varactor tuning techniques for Gunn diodes: (a) front and side views of double-port waveguide-type; (b) type used at K_u band.

substantial advantages, including remote measurement, broadband (no tuning), stability (no short-circuited or open terminations), accuracy, and ease of measurement.

From Fig. 13-145, the scattering equations describing the two-port network can be written.

$$b_1 = S_{11}a_1 + S_{12}a_2 \qquad b_2 = S_{21}a_1 + S_{22}a_2$$

Solving for the S parameters yields

$$S_{11} = \left.\frac{b_1}{a_1}\right|_{a_2=0} = \text{input reflection coefficient with } Z_L = Z_0$$

$$S_{12} = \left.\frac{b_1}{a_2}\right|_{a_1=0} = \text{reverse transmission gain with } Z_L = Z_0$$

$$S_{21} = \left.\frac{b_2}{a_1}\right|_{a_2=0} = \text{forward transmission gain with } Z_L = Z_0$$

$$S_{22} = \left.\frac{b_2}{a_2}\right|_{a_1=0} = \text{output reflection coefficient with } Z_L = Z_0$$

Other linear two-port parameters can be calculated from S parameters (for example, y parameters for feedback analysis). Either manual or complex automatic measurement techniques can be used. In either case, the transistor chip package is embedded in a system with a given reference impedance Z_0 and well-defined reference wave planes.

A number of useful relationships can be calculated from the S parameters and used in amplifier-oscillator design.

Reflection coefficient-impedance relationship:

$$S_{11} = \frac{Z - Z_0}{Z + Z_0} \qquad \text{where } Z_0 = \text{reference impedance}$$

Input reflection coefficient with arbitrary Z_L:

$$S_{11}' = S_{11} + \frac{S_{12}S_{21}\Gamma_L}{1 - S_{22}\Gamma_L}$$

Output reflection coefficient with arbitrary Z_s:

$$S_{22}' = S_{22} + \frac{S_{12}S_{21}\Gamma_S}{1 - S_{11}\Gamma_S}$$

Stability factor:

$$K = \frac{1 + |D|^2 - |S_{11}|^2 - |S_{22}|^2}{2(S_{12}S_{21})}$$

Transducer power gain:

$$G_T = \frac{|S_{21}|2(1 - |\Gamma_S|^2)(1 - |\Gamma_L|^2)}{|(1 - S_{11}\Gamma_S)(1 - S_{22}\Gamma_L) - S_{12}S_{21}\Gamma_L\Gamma_S|^2}$$

Maximum available power gain:

$$G_{max} = \left| \frac{S_{21}}{S_{12}} (k \pm \sqrt{k^2 - 1}) \right| \qquad \text{for } k > 1$$

Source and load reflection coefficients for simultaneous match:

$$\Gamma_{ms} = M^* \left[\frac{B_1 \pm \sqrt{B_1^2 - 4|M|^2}}{2|M|^2} \right] \qquad \Gamma_{mL} = N^* \left[\frac{B_2 \pm \sqrt{B_2^2 - 4|N|^2}}{2|N|^2} \right]$$

where Γ_S, Γ_L = source and load reflection coefficients $M = S_{11} - DS_{22}^*$
      $N = S_{22} - DS_{11}^*$ $D = S_{11}S_{22} - S_{12}S_{21}$
      $B_1 = 1 + |S_{11}|^2 - |S_{22}|^2 - |D|^2$ $B_2 = 1 + |S_{22}|^2 - |S_{11}|^2 - |D|^2$

The maximum power gain is obtained only if the transistor is terminated with the Γ_{ms} and Γ_{mL} resultant impedances. A lossless transforming network is placed between the source and load to realize this transformation.

Generally, the embedding circuits used take the form of simple ladder networks; series-shunt combinations of L's and C's or their transmission-line equivalents. These elements can be determined by moving on the Smith chart from the value of the terminating impedance to the center of the chart along constant resistance-conductance, reactance-susceptance contours for series-shunt elements, respectively. The circuits are typically implemented in strip-line or lumped form, as shown in Fig. 13-146.

Optimum design at more than one frequency requires plotting of gain circles at each frequency, with subsequent terminating impedance iteration to obtain the best compromise across

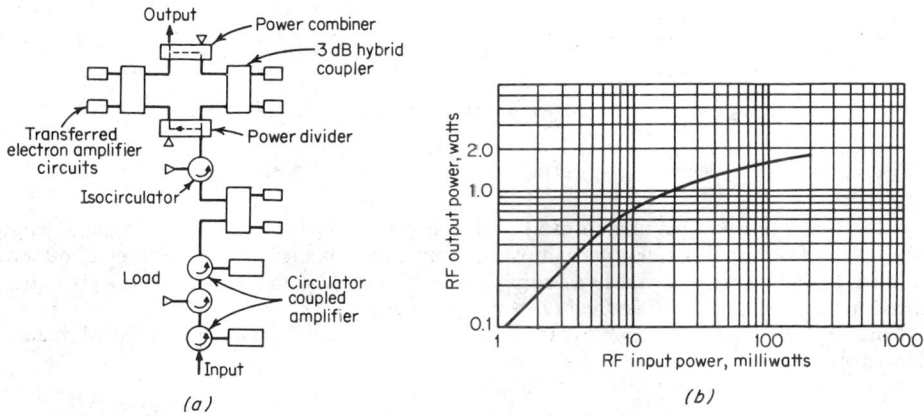

(a)   (b)

Fig. 13-143. Four-stage ganged TED amplifier chain for 5.7 GHz: (a) block diagram; (b) power-transfer curve.

the band. Several computer-aided optimization programs[25] are available to simplify this routine. A plot of gain circles and impedance loci for a typical S-band design is shown in Fig. 13-147, which corresponds to the collector circuit of Fig. 13-146a.

142. Noise Performance of Microwave Transistor Circuits. The noise performance of a well-designed amplifier depends almost entirely upon the noise figure of the first

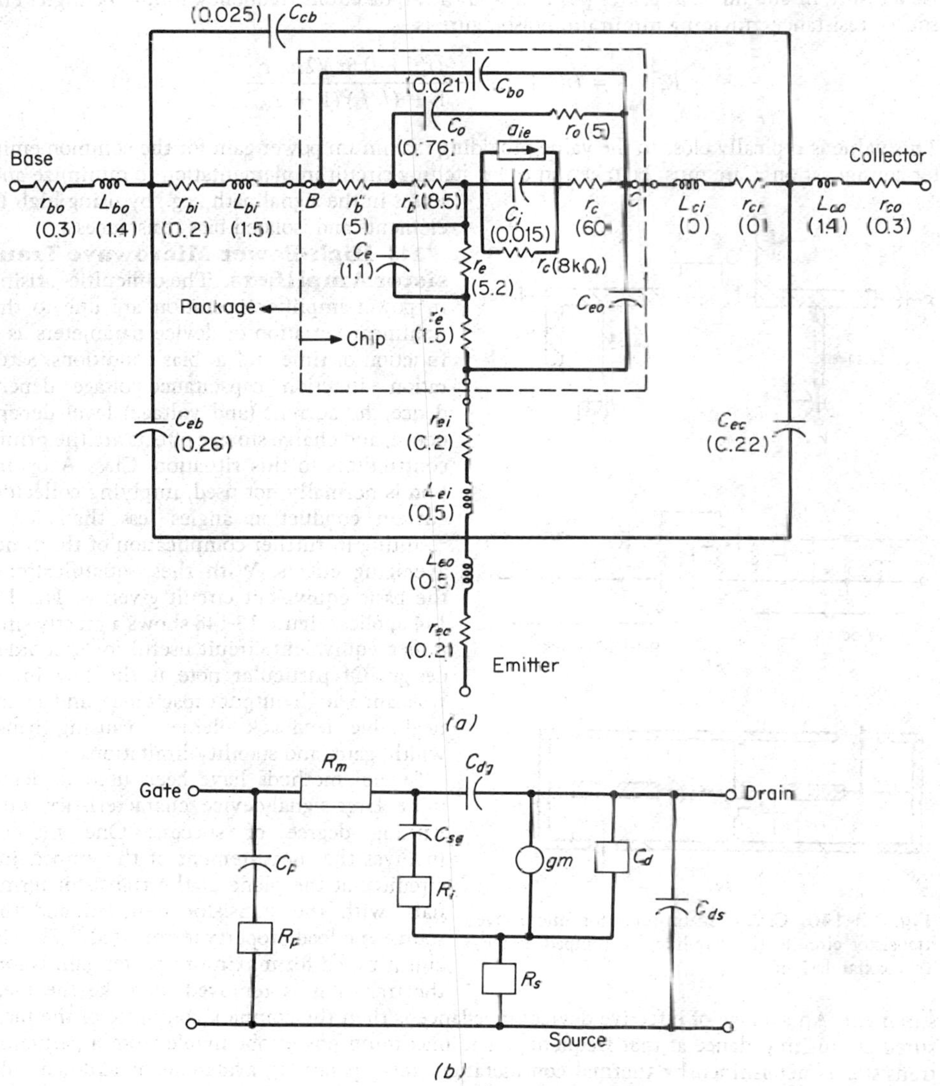

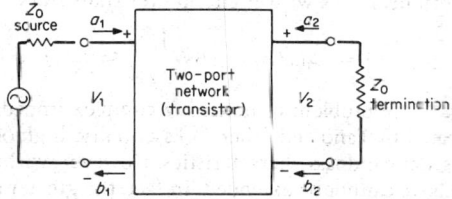

Fig. 13-144. Equivalent circuits of microwave solid-stage devices: (a) bipolar transistor; (b) Schottky barrier FET (chip only).

Fig. 13-145. Basic two-port configuration for a microwave transistor.

transistor. It has been shown[26] that the noise factor of a transistor amplifier is related to its equivalent circuit parameters and external circuit by

$$NF = 1 + \frac{r_{bb}'}{R_g} + \frac{r_e}{2R_g} + \frac{(r_{bb}' + r_e + R_g)^2}{2r_e R_g h_{feo}}\left[1 + \left(\frac{f}{f_a}\right)^2 (1 + h_{feo})\right]$$

As a result, r_{bb}' should be as low as possible, and the alpha cutoff frequency should be high. The source resistance providing minimum noise figure is

$$R_g\bigg|_{F_{min}} = (r_{bb}' + r_e)^2 + \frac{(r_{bb}' + 0.5r_e)(2h_{feo}r_e)}{1 + (f/f_a)^2(1 + h_{feo})}$$

This value is typically close to the value providing maximum power gain for the common-emitter configuration. Care must be taken in the matching-circuit implementation to minimize any losses in the signal path, e.g., by using high-Q elements and isolated bias resistances.

143. High-Power Microwave Transistor Amplifiers. The difficulties arising in power-amplifier operation are due to the nonlinear variation of device parameters as a function of time and to bias conditions. Saturation, junction capacitance-voltage dependence, h_{fe} current (and voltage) level dependence, and charge-storage effects are the prime contributors to this situation. Class A operation is normally not used, implying collector-current conduction angles less than 360°, resulting in further complication of the time-averaging effects. With these qualifications, the basic equivalent circuit given in Fig. 13-144 applies. Figure 13-148 shows a greatly simplified equivalent circuit useful for first-order design. Of particular note is the low input resistance, high output capacitance, and a non-negligible feedback element causing bandwidth, gain, and stability limitations.

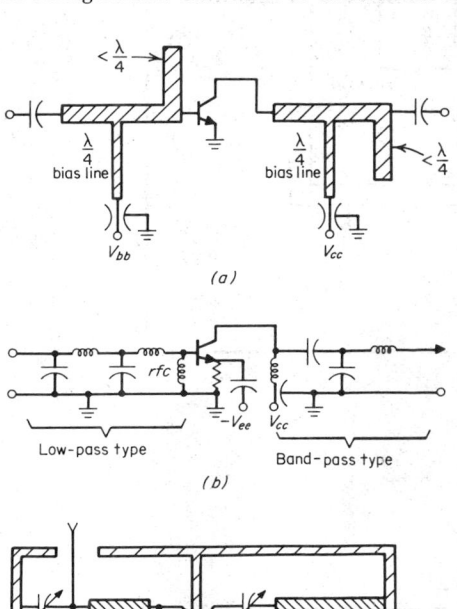

Fig. 13-146. Circuit realizations for microwave transistor circuits: (a) strip-line; (b) lumped design; (c) coaxial design.

Several methods have been used to determine large-signal-device characteristics with varying degree of success. One method involves the measurement of the embedding circuitry at the plane of the transistor terminals, with the transistor removed and the source and load property terminated.[27] The circuit is tuned for maximum power gain before the transistor is removed to make the measurement. An average or effective device impedance is then the complex conjugate of the measured circuit impedance at that frequency. The maximum power obtainable from a particular transistor is determined by thermal considerations (cw operation), avalanche-breakdown voltages, current-gain falloff at high current levels, and second-breakdown effects.

Bandwidth. The input impedance has been found to be the primary bandwidth-limiting element. R_S varies inversely with the area of the transistor. Hence, for a given package L_S, the Q increases and bandwidth decreases with higher-power transistors

$$Q = \omega_0 L_S / R_S \qquad BW\bigg|_{3\ dB} \approx f_0/Q$$

Impedance Matching. The problem of matching complex impedances over a wide band of frequencies has been treated by Fano and others.[17] Essentially, high-order networks can be used to achieve nearly rectangular bandpass characteristics. However, without mismatching, the 3-dB bandwidth determined above cannot be extended. In fact, the greater the ratio of generator resistance R_g to transistor input resistance R_s the greater will be the reduction of the intrinsic band-width, for a given ripple and number of circuit elements. Hence the external circuit design must consider both the transistor-input-circuit Q value and the impedance level relative to the driving source for a given bandwidth.

Either lumped- or transmission-line-element networks of relative simplicity are typically used to achieve the necessary input-output matching. Although the bandpass type yields somewhat

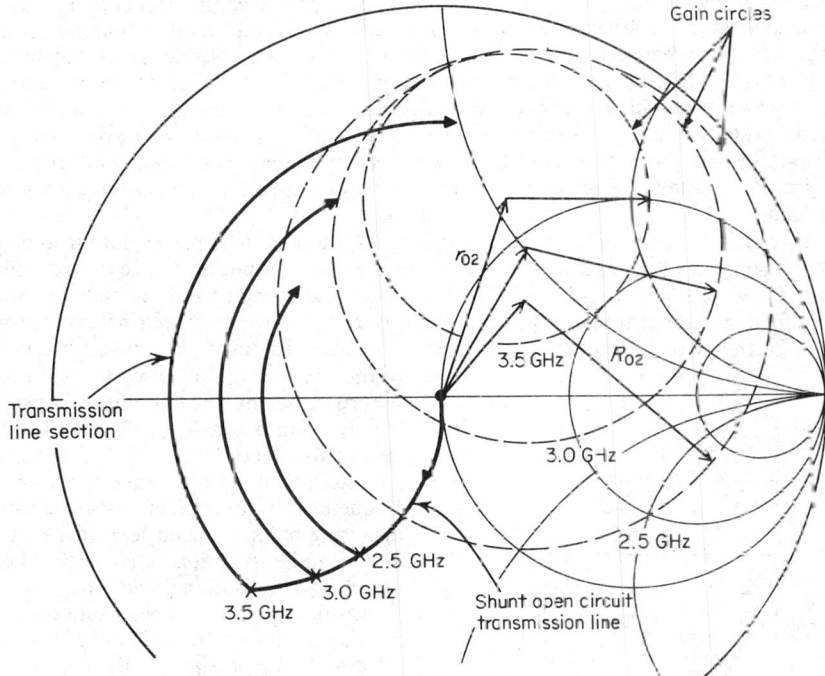

Fig. 13-147. Use of Smith chart in microwave transistor circuit design.

better performance, the low-pass configuration is more convenient to realize physically. Quarter-wave line sections are particularly useful for broadband impedance transformations and bias feed-bypassing functions. Eighth-wave transformers are useful to match the small complex impedances directly without tuning-out mechanisms. The input impedance to a $\lambda/8$ section is real if it is terminated in an impedance with magnitude equal to the Z_0 of the line.

Load Resistance. The desired load resistance can be determined to first order by

$$R_L \approx (V_{CC} - V_{CE,sat})^2 / 2P_o$$

where P_o = desired fundamental power output and V_{CC} = collector supply voltage. This expression is altered by several factors, including circuit Q, harmonic frequencies, leakage, current conduction angle, etc. Assuming only the presence of the fundamental frequency, V_{CC} is limited to $\frac{1}{2}BV_{CBO}$ by resonant effects. Recently, it has been shown that BV_{CBO} can be exceeded for short pulses without causing avalanche.

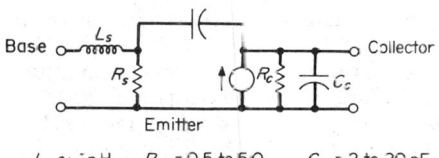

$L_s \approx 1\,\mathrm{nH}$; $R_s = 0.5\,\mathrm{to}\,5\,\Omega$, $C_c = 2\,\mathrm{to}\,20\,\mathrm{pF}$

Fig. 13-148. Equivalent circuit of microwave power transistor.

Power Gain. The power gain depends on the dynamic f_T or large-signal current gain-bandwidth capability, the dynamic input impedance, and the collector load impedance. A simple expression for power gain is

$$PG = \frac{(f_T/f)^2 R_L}{4R_e(Z_{in})}$$

The current gain at f_T is of particular importance. The effect of parasitic common-terminal inductance is to reduce this gain in the common-emitter and to cause regeneration in the common-base connection. The latter configuration is more commonly used at frequencies near or above the f_T value, whereas the former generally results in a more stable circuit below f_T. This situation is highly dependent on the parasitic-element situation with respect to the specific frequency. Various forms of instabilities, such as hysteresis (jump modes), parametric, low-frequency, and thermal, can occur due to the parameter values changing with time-varying high-signal levels. Usually, most of these difficulties can be eliminated or minimized by careful design of bias circuits (including fewer elements), ground returns, parasitics, and out-of-band terminating impedances.

The *collector efficiency* of a transistor amplifier is the ratio of rf power output to dc power input. High efficiency implies low circuit losses, high ratio of output resistance to load resistance, high f_T and low collector saturation voltage. In addition, experiments and calculations show that for high efficiency, the impedance presented to the collector by the output network should be inductive for the favored generation of second harmonic. If the phase is correct, the amplitude of the fundamental is raised beyond the limit otherwise set by the difference between the supply voltage and $V_{C,sat}$. Figure 13-149 illustrates this effect.

A high value of f_T relative to the operating frequency improves efficiency by causing the operating point to spend less time (per cycle) in the high-dissipation active region between cutoff and saturation. The integrity of the transistor die bond has been found to have a substantial effect on efficiency due to the effects of low intrinsic bulk collector resistance and thermal gradient.

144. Transistor Oscillators. Transistor oscillators can be designed by choosing the source (or load) terminating impedances such that $S'_{11}\Gamma_s \geq 1 (S'_{22}\Gamma_L \geq 1)$. The design is facilitated by plotting stability circles[28] on the Smith chart.

A number of circuit configurations are appropriate, including the standard Colpitts, Hartley, Clapp, coupled-hybrid, etc. The major difference, however, is the proper inclusion of device parasitic reactances into the intended configuration. The frequency-determining element(s) may be in the input, output, or feedback circuit but should have a high Q for good frequency stability. Care must be taken to decouple or resistively load the circuit at frequencies outside the band where oscillatory conditions are satisfied. This is particularly true for lower frequencies where the current gain is much better.

The common-base and common-collector transistor connections are generally more unstable than common-emitter and most commonly used, depending on the power requirements and frequency of oscillation.

Several methods are utilized to vary the frequency dynamically. Bias variation is effective only over a narrow band and results in substantial power variation. The YIG sphere provides very high Q but is somewhat bulky and of limited tuning speed. The varactor diode provides very low power, octave-tuning bandwidth, and fast tuning in a form compatible with hybrid integration. Figure 13-150 shows the schematic of an S-band varactor-tuned wide-band oscillator implemented in hybrid thin-film form. The frequency-determining elements are connected in

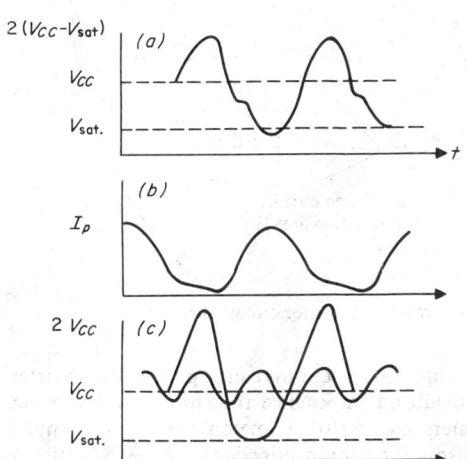

Fig. 13-149. Typical voltage-current waveforms in power applications: (*a*) collector voltage; (*b*) collector current; (*c*) second-harmonic enhancement.

the high-Q common-base input circuit. This also allows maximum isolation from load mismatch effects (pushing) and permits the balance of the circuitry to be low-Q (broadband).

Both high- and low-power oscillators are large-signal; i.e., the output power is limited primarily by beta falloff, with increasing current at a given frequency and collector voltage.

145. Traveling-Wave-Tube Circuits. The traveling-wave tube[29] is a unique structure capable of providing high amplification of rf signals varying in frequency over several octaves without the need for any tuning or voltage adjustment (see Sec. **9**). Figure 13-151 shows the principal components of an amplifier using this tube. Electrons emitted by the electron-gun assembly are sharply focused, drawn through the length of the slow-wave rf structure, and eventually dissipated in the collector. Synchronism between the rf electromagnetic wave and the

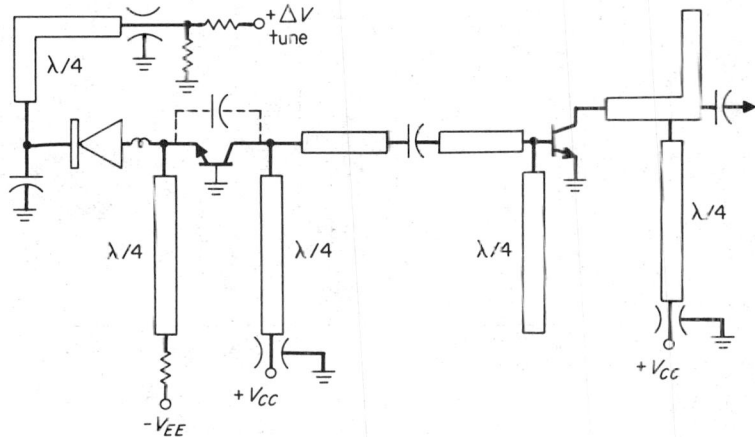

Fig. 13-150. Varactor-tuned microstrip oscillator-buffer circuit.

beam electrons results in a cumulative interaction which transfers energy from the dc beam to the rf wave. For details on this type of amplifier see Pars. **9-34** to **9-39**.

Figure 13-152 shows the gain characteristics of a typical broadband TWT amplifier. As more energy is extracted from the electron beam, it slows down. This loss of synchronism results in lower gains at higher power levels. One advantage of this overdrive characteristic is the protection of following stages against strong signals.

The backward-wave oscillator[30] is essentially a TWT device making use of the interaction of the electron stream with an electromagnetic wave whose phase and group velocities are 180° apart. At a sufficient beam current, oscillations are produced as discussed above, without a reverse-wave attenuator. These devices are voltage-tunable. Frequency is proportional to the ½ power of the cathode-helix voltage and the dimensions of the structure. Multioctave tuning is possible, depending on output-power-variation requirements. These oscillators have low pulling

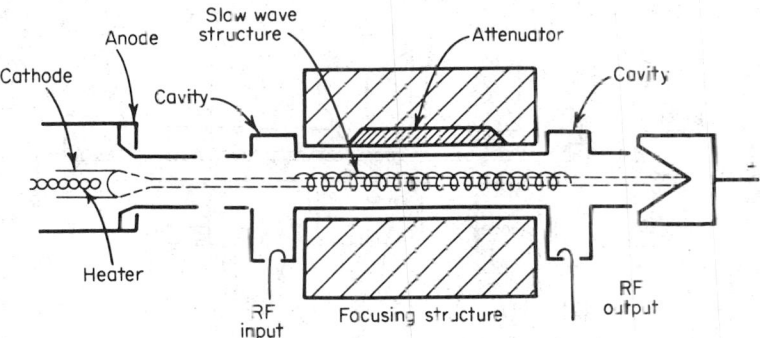

Fig. 13-151. Traveling-wave amplifier circuit.

figures and, typically, high pushing figures. Frequency stability is excellent, usually dependent on power-supply variations. These devices are typically low-power (<1 W).

146. Klystron Oscillators and Amplifiers. The chief advantages of the klystron amplifier oscillator are that it is capable of large stable output power (10 MW) with good efficiency (40%) and high gain (70 dB). Basically, the mechanism involves modulation of the velocity of electrons in a beam by an input rf signal. This is converted into a density-modulated

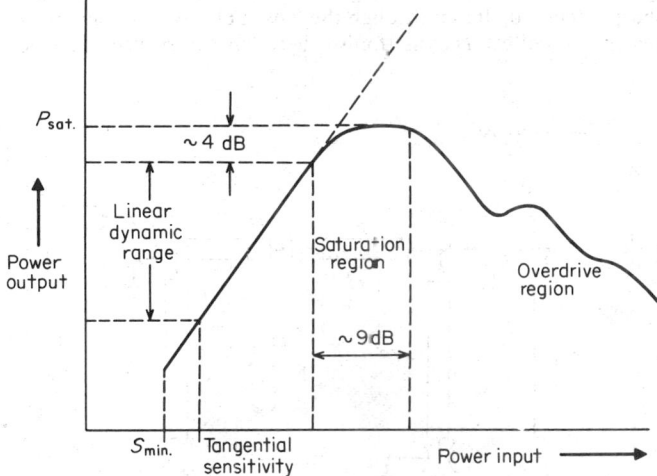

Fig. 13-152. Typical gain characteristic of a broadband TWT amplifier.

(bunching) beam from which a resonant cavity extracts the rf energy and transforms it to a useful load.

Klystron amplifiers (see Pars. **9-24** to **9-33**) can be conveniently divided into three categories: (1) two-resonator single-stage high- and low-noise voltage amplifiers, (2) two-resonator single stage *(optimum bunching)* power amplifiers, and (3) multiresonator cascade-stage voltage and power amplifiers. The power gain of a two-cavity voltage amplifier can be given by[8]

$$G = \frac{M_1^2 M_2^2}{240\beta} \left(\frac{\pi a}{\lambda} \right)^2 \frac{G_0}{(G_{BR})^2}$$

where M_1, M_2 = beam coupling factors, a = beam radius, G_0 = beam conductance, G_{BR} = cavity shunt conductance contributions due to beam loading and ohmic losses, and β = (electron velocity)/(velocity of light).

A simplified schematic representation of a klystron amplifier is shown in Fig. 13-153. Multi-

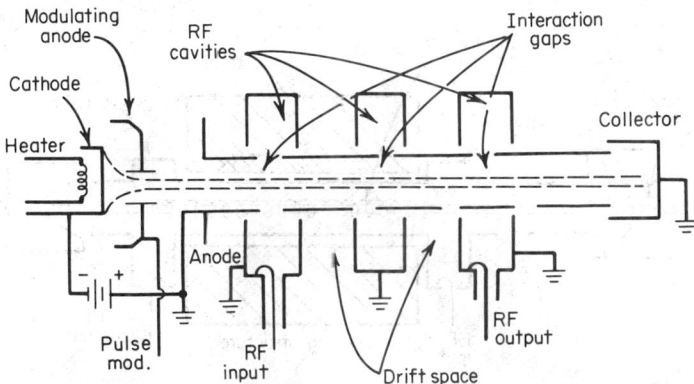

Fig. 13-153. Klystron amplifier structure.

cavity tubes are typically used for high pulse power and cw applications. The intermediate cavities serve to remodulate the beam, causing additional bunching and higher gain-power output. Optimum power output is obtained with the second cavity slightly detuned. Further, loading of this cavity serves to increase the bandwidth (at the expense of gain). These tubes typically use magnetically focused high-perveance beams.

The broadbanding of a multicavity klystron is accomplished in a manner analogous to that of multistage i.f. amplifiers. A common technique is stagger tuning, which is modified somewhat,

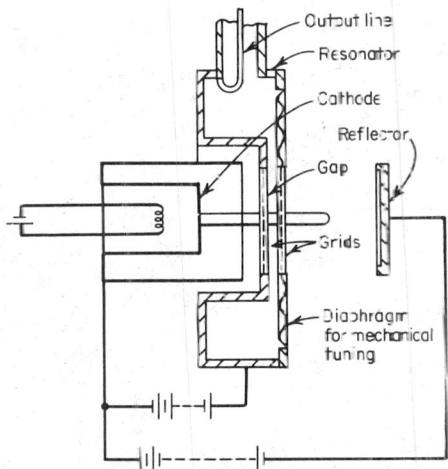

Fig. 13-154. Reflex klystron with coaxial-line loop output.

due to nonadjacent cavity interactions. The gain and bandwidth of multicavity klystrons have been calculated by Krèuchen.[32]

The two-cavity amplifier can be made to oscillate by providing a feedback loop from output to input with proper phase relationship. Klystrons can also be used for frequency multiplication, using the high harmonic content of the bunched-beam current waveforms.

Reflex Klystron. A simple klystron oscillator results if the electron-beam direction is reversed by a negative electrode, termed the *reflector.* A schematic diagram of such a structure is given in Fig. 13-154. Performance data for a reflex klystron are usually given in the form of a reflector-characteristic chart. This chart displays power output and frequency deviation as a function of reflector voltage. Two distinct classes of reflex klystrons are low-power tubes for oscillator, pump, and test applications and higher-power tubes (10 W) for frequency-modulator applications. Operating voltage varies from 300 to 2,000 V with bandwidths up to ~200 MHz.

147. Crossed-Field-Tube Circuits.[34] Practically all crossed-field tubes have integrally attached distributed constant circuits. Operation within a critical range of beam current is necessary to maintain the proper bandpass characteristics.

Magnetron.[35] The original microwave tube was a magnetron diode switch with oscillations due to the cyclotron resonance frequency. Several oscillator circuits were used until the standard cavity-resonator magnetron was introduced in 1940 (see Pars. **9-40 to 9-50**).

The relations between power, frequency, and voltage vs. the load admittance are shown in the Rieke diagram (Fig. 13-155). Such charts illustrate the compromises necessary to obtain desired operating conditions. In general, good efficiency results from increasing anode current and magnetic field strength.

Various load effects can be considered with the Rieke diagram. The pulling figure [measure of frequency change for a defined load mismatch −(SWR) = 1.5 at all phase angles] and long-line effect are examples.

The area of highest power on the Rieke diagram is called the *sink* and represents the tightest load coupling to the tube. The highest efficiency results in this region. However, a poor spectrum or instability typically results. The buildup of oscillations in the antisink region is closer to ideal. However, this lightly loaded condition may also result in instability. Stability is a measure of the percentage of missing pulses, usually defined at 30% energy loss.

Tuning. Methods used to tune a magnetron are classified as mechanical, electronic, and voltage tuning. In the mechanical methods, the frequency of oscillation is changed by the motion

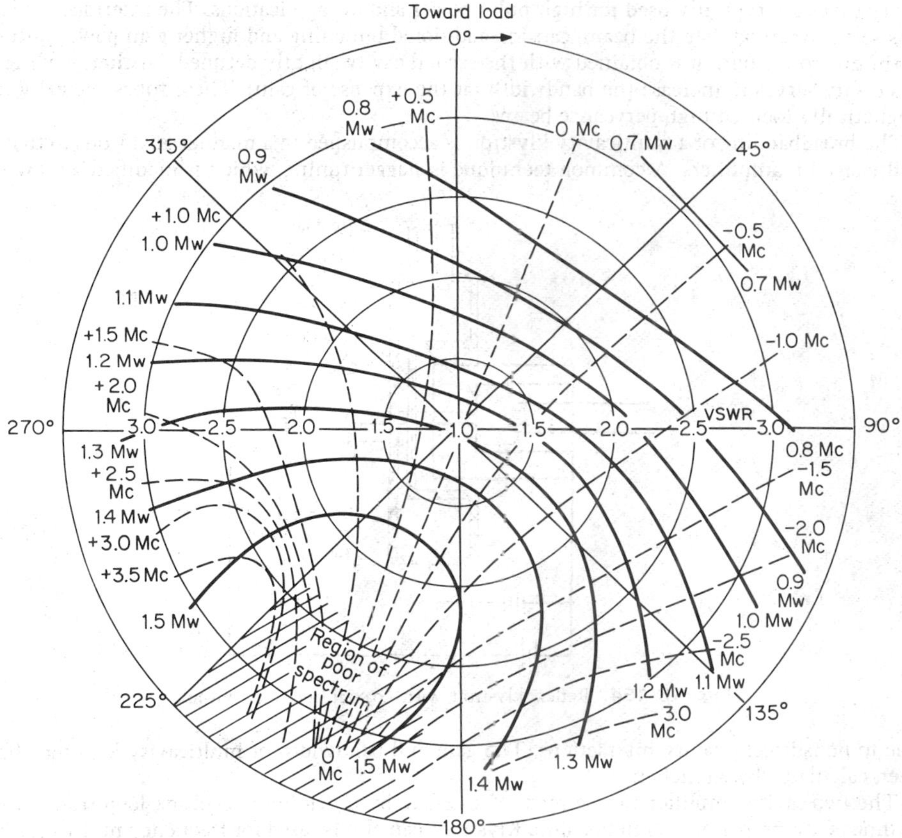

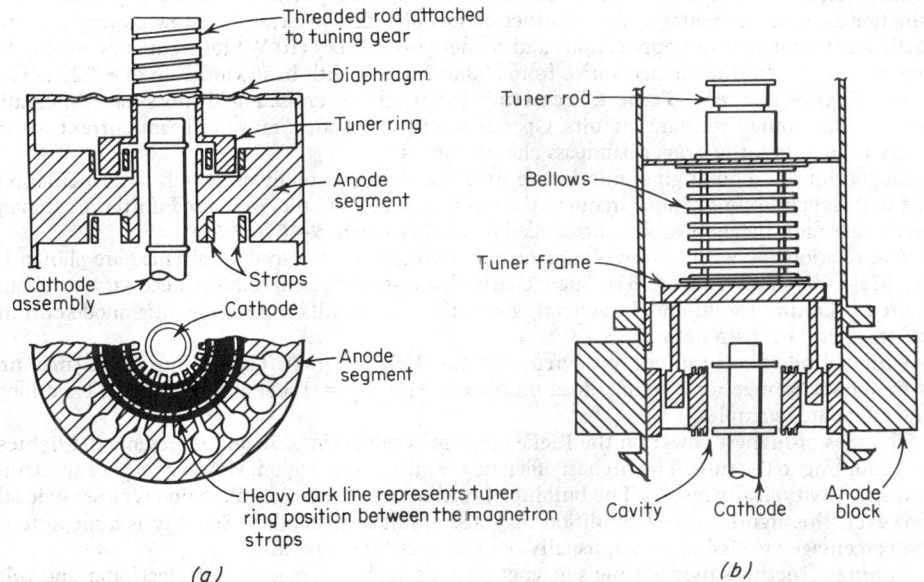

Fig. 13-155. Rieke diagram for an L-band magnetron at 1250 MHz. Solid curves are contours of constant power; dashed curves of constant frequency. *(Raytheon Company.)*

Fig. 13-156. Mechanical magnetron tuning mechanisms: *(a)* capacitance type; *(b)* inductance type.

of an element in the resonant circuit. Two types are shown in Fig. 13-156. An electron beam injected into the cavities will change the effective dielectric constant and hence the frequency. By using the frequency pushing effect it has been possible to tune the magnetron over a frequency range of 4 to 1 (typically, 0.1 to 2 MHz/V) using voltage tuning. The frequency change is usually linear, but the power output is not constant over the tuning range. Magnetrons with power output greater than 5 MW and efficiencies 50% or more are available.

Amplitron. The Amplitron is essentially a magnetron with two external couplings, enabling amplifier operation. It is characterized by high power, broad bandwidth, very high efficiency, and low gain. The output is independent of of input but dependent on dc input. It acts as a low-loss passive transmission line in the absence of high voltage. A typical plot of power for an Amplitron is shown in Fig. 13-157. Conversion efficiency of an Amplitron can be as high as 85%. The gain can be increased by inserting mismatches into the input and output transmission lines.

148. GaAs Field-Effect Transistor Circuits. The state of the art of gallium arsenide (GaAs) field-effect transistors for both small-signal and power applications is rapidly improving, thanks to the intrinsically higher carrier mobility in GaAs, coupled with significantly improved fabrication techniques, e.g., in the area of photolithography. Many devices fabricated today use feature sizes of 1μm or less. GaAs FETs are particularly attractive since the parameter values are less variable as a function of operating conditions and their imped-

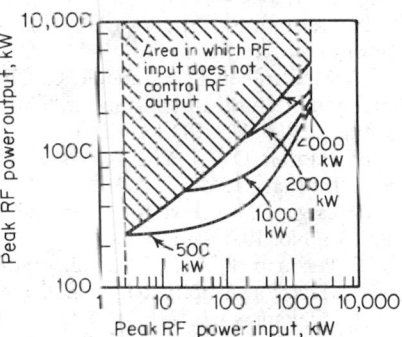

Fig. 13-157. Gain characteristic of model QK520 L-band amplifier.

ances are generally higher than in comparable bipolar devices. As a result, GaAs FET circuits are easier to design and typically achieve wider bandwidth, lower noise, and higher linearity.

Power devices are usually composed of a number of individual cells, combined in parallel on the chip. Gate widths of 5 to 10 mm are typical. As with bipolar power devices, package parasitics are likely to dominate. Hence, internally matched chip carrier structures are generally used. The input circuit comprises a small SiO_2 chip capacitor connected to the FET gate with a bond wire, thus forming a π-matching section with the FET input capacitance. Input and output circuits are typically LC ladder networks implemented in both discrete and semidistributed circuit forms. The common-source configuration has been found to work well for both amplifiers and oscillators.

With these devices, the following results have been achieved: multioctave small-signal amplification; noise figures of 1 to 2 dB (S band); output power of 10 W at X band; and 5-W power amplification across the band of 4 to 8 GHz.[36]

149. References on Microwave Amplifiers and Oscillators

1. Read, W. T. A Proposed High-Frequency Negative Resistance Diode, *Bell Syst. Tech. J.*, 1958, Vol. 37.

2. Haitz, R. H., et al. A Method for Heat Flow Resistance Measurements in Avalanche Diodes, *IEEE Trans. Electron Dev.*, May 1969, Vol ED-16, p. 438.

3. Microwave Power Generation and Amplification Using IMPATT Diodes, *Hewlett Packard Appl. Note 935*, June 1971.

4. Gummel, H. K., and J. L. Blue A Small Signal Theory of Avalanche Noise in IMPATT Diodes, *IEEE Trans. Electron Dev.* 1967, Vol. ED-14.

5. Hines, M. E. Noise Theory for Read Type Avalanche Diodes, *IEEE Trans. Electron Devices*, 1966, Vol. ED-13.

6. Scherer, E. F. Investigations of the Noise Spectra of Avalanche Oscillators, *IEEE Trans.*, September 1966, Vol. GMTT-16.

7. Kuvas, R. L. Noise in IMPATT Diode Oscillators and Amplifiers, *Proc. 3d Bienn. Cornell Elec. Eng. Conf.*, Ithaca, N.Y., 1971.

8. Chan, V. W., and P. A. A. Levine Comparative Study of IMPATT Diode Noise Properties, *Proc. 3d Bienn. Cornell Elec. Eng. Conf.*, Ithaca, N.Y., 1971.

9. Prager, H. J., K. K. N. Chang, and S. Weisbrod High Power, High Efficiency Silicon Avalanche Diodes at Ultra High Frequencies, *Proc. IEEE Lett.*, April 1967, Vol. 55.

10. Evans, W. J. Circuits for High Efficiency Avalanche-Diode Oscillators, *IEEE Trans.*, December 1969, Vol. MTT-17, No. 12.

11. Ibid.

12. Liu, S. G., and J. J. Risko   Fabrication and Performance of Kilowatt L-Band Avalanche Diodes, *RCA Rev.*, March 1970, Vol. 31.

13. Kawamotto, H.   Anti-parallel Operation of Four High Efficiency Avalanche Diodes. *IEEE ISSCC Dig.*, February 1971.

14. Deloach, B. C., Jr., and D. L. Scharfetter   Device Physics of TRAPATT Oscillators, *IEEE Trans. Electron Dev.*, January 1970, Vol. ED-17.

15. Liu, S. G., H. J. Prager, K. K. N. Chang, J. J. Risko and S. Weisbrod   High Power Harmonic Extraction and Triggered Amplification with High Efficiency Avalanche Diodes, *IEEE ISSCC Dig.*, 1971.

16. Gunn, J. B.   Microwave Oscillators of Current in III-IV Semiconductors, *Solid State Commun.*, 1963, Vol. 1.

17. McCumber, D. E., and A. G. Chynoweth   Theory of Negative-Conductance Amplification and of Gunn Instabilities in "Two-Valley" Semiconductors, *IEEE Trans. Electron Devices*, January 1966, Vol. ED-13, No. 1.

18. Hanson, D. C.   YIG Tuned TED Using Thin Film Microcircuits, *IEEE ISSCC Digest Tech. Pap.*, February 1969.

19. Large, D.   Octave Band Varactor-Tuned Gunn Diode Sources, *Microwave J.*, October 1970, Vol. 13, No. 10.

20. Perlman, B. S., L. C. Upadhyayula, and R. E. Marx   Wide-Band Reflection Type Transferred Electron Amplification, *IEEE Trans.*, November 1970, Vol. MTT-8, No. 11.

21. Siekanowicz, N. W., B. S. Perlman, B. E. Berson, R. E. Marx, and W. E. Klatskin   Performance of Medium Power, High Gain, C. W. Transferred Electron Amplifiers at C-Band, *Proc. 3d Bienn. Cornell Elect. Eng. Conf., Ithaca, N.Y., 1971.*

22. Camp, W. O., J. S. Bravman, and D. W. Woodard   The Operation of Very High Power LSA Transmitters, *Proc. 3rd Bienn. Cornell Elect. Eng. Conf., Ithaca, N.Y., 1971.*

23. Kurokawa, K.   Power Waves and the Scattering Matrix, *IEEE Trans. GMTT*, March 1965, Vol. MTT-13, No. 2.

24. Bodway, G.   Two Part Power Flow Analysis Using Generalized Scattering Parameters, *Microwave J.*, May 1967, Vol. 10, No. 6.

25. Gelnovatch, V. G., I. L. Chase, and T. Arell   A 2–4 GHz Integrated Transistor Amplifier Designed by an Optimal Seeking Computer Program, *IEEE ISSCC Dig.*, February 1970.

26. Hunter, L. P.   "Handbook of Semiconductor Electronics" 3d ed., McGraw-Hill, New York, 1969.

27. Lee, H. C.   Microwave Power Transistors, *Microwave J.*, February 1969.

28. Froehner, W. H.   Quick Amplifier Design Using Scattering Parameters, *Electronics*, 1967.

29. Pierce, J. R.   "Traveling Wave Tubes," Van Nostrand, New York, 1950.

30. Heffner, H.   Analysis of the Backward Wave Traveling Wave Tube, *Proc. IRE*, June 1954, Vol. 42.

31. Hamilton, D. R., J. K. Knipp, and J. B. H. Kuper   "Klystrons and Microwave Triodes," McGraw-Hill, New York, 1948.

32. Kreuchen, K. H., B. A. Auld, and N. E. Dixon   A Study of the Broad Band Frequency Response of the Multi-Cavity Klystron Amplifier, *J. Electron.*, May 1957, Vol. 2.

33. Dain, J.   Ultra High Frequency Power Amplifiers, *Proc. IEEE*, November 1958, Vol. 105, Pt. B.

34. Okress, E.   "Crossed Field Microwave Devices," Vols. 1 and 2, Academic, New York, 1961.

35. Collins, G. B.   "Microwave Magnetrons," MIT Radiation Laboratory Series, Vol. 6, McGraw-Hill, New York, 1948.

36. Ohta, K., S. Jodai, N. Fukuden, Y. Hirano, and M. Itoh   A 5 Watt 4–8 GHz GaAs FET Amplifier, *Mircowave J.*, November 1979.

37. Takayama, Y., K. Honjo, A. Higahisaka, and F. Hasegawa,   Internally Matched Microwave Broadband Linear Power FET, *ISSCC Dig.*, February 1977, Vol. 20.

Section 14

Modulators, Demodulators, and Converters

JOSEPH L. CHOVAN *Senior Engineer, Electronics Laboratory General Electric Company*

MYRON D. EGTVEDT *Senior Engineer, Electronics Laboratory, General Electric Company, Member, IEEE*

GLENN B. GAWLER *Senior Engineer, Electronics Laboratory, General Electric Company, Member, IEEE*

NOBLE POWELL *Consulting Engineer, Electronics Laboratory, General Electric Company, Member, IEEE*

JOSEPH P. HESLER *President, AKF Design Ltd.*

GEORGE F. PFEIFER *Senior Engineer, Solid State Applications Operation, General Electric Company, Member, IEEE*

CONTENTS

Numbers refer to paragraphs

Amplitude Modulators and Demodulators

BY JOSEPH P. HESLER

1. Amplitude Modulation. Frequency translation is the key to radio communications in that it produces signal energy, in proportion to variations of an information source, at frequencies that have desirable transmission characteristics, such as antenna size, freedom of interference from similar information sources, line-of-sight to long-range propagation, and freedom of interference from particular noise sources. Frequency translation permits the efficient utilization of open and closed propagation media by many simultaneous users and/or signals.

One of the most used forms of frequency translation is linear modulation, the most common of which is amplitude modulation. In general, amplitude modulation consists in varying the magnitude of a carrier signal in direct correspondence to the instantaneous fluctuations of a modulating signal source, as illustrated in Fig. 14-1.

Variations of the basic amplitude-modulation process have been developed to achieve more efficient spectrum utilization and to reduce transmitter power requirements. These include suppressed-carrier systems such as vestigial-sideband, single-sideband, and double-sideband modulation systems. The companion form of frequency translation used in amplitude-modulation systems is *detection*. This is the process whereby the originally translated information is recovered as a baseband signal. Linear amplitude modulation has been the most widely used form of frequency translation in general communications for three reasons: relative ease of implementation, efficient utilization of bandwidth, and availability of devices to implement a simple detection procedure.

It can be argued that amplitude modulation includes such modulation methods as pulse-code keying, pulse-amplitude modulation, frequency-shift keying (the sequential keying of multiple carrier signals), and variations of them, such as pulse-position modulation and pulse-width modulation. This subsection (Pars. **14-1** to **14-15**) is restricted to the amplitude modulation by signal sources whose outputs are continuous time functions. The special types of modulation mentioned above are discussed in Pars. **14-26** to **14-41**.

The general expression for the output of a linear amplitude modulator with a sinusoidal modulation input is

$$E = E_0(1 + m \sin \omega_m t) \sin (\omega_c t + \phi) \qquad (14\text{-}1)$$

where E_0 = peak amplitude of carrier signal, ω_m = modulating signal frequency (rad/s), ω_c = carrier frequency (rad/s), m = modulation index, ϕ = arbitrary carrier phase angle (rad), and t = time (s).

Expansion of Eq. (14-1) provides

$$E = E_0 \sin (\omega_c t + \phi) + \frac{mE_0}{2} \cos [(\omega_c - \omega_m)t + \phi] - \frac{mE_0}{2} \cos [(\omega_c + \omega_m)t + \phi] \qquad (14\text{-}2)$$

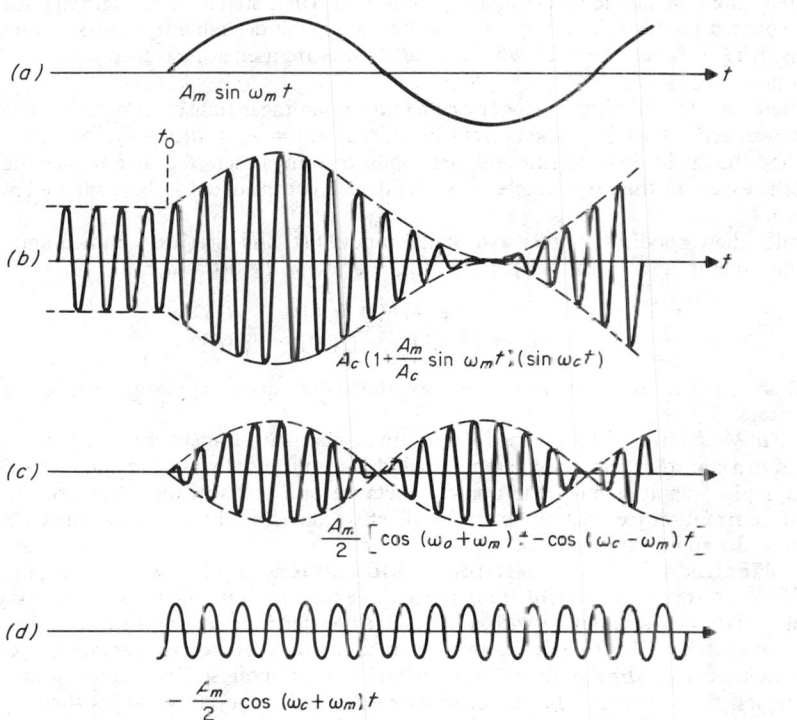

Fig. 14-1. Amplitude modulation: (a) modulating signal; (b) double-sideband amplitude-modulated signal; (c) double-sideband suppressed-carrier amplitude-modulated signal; (d) single-sideband suppressed-carrier amplitude-modulated signal.

Note that the carrier signal is reproduced exactly as if it carried no modulation. The carrier in itself does not carry any information. The second and third terms in Eq. (14-2) represent sideband signals produced in the modulation process. These signals are displaced from the carrier signal in the frequency spectrum, on each side of the carrier, by a frequency difference equal to the modulating-signal frequency. The magnitudes of the sideband signals are equal and are proportional to the modulation index m.

Both positive and negative amplitude modulation can be produced in an unsymmetrical manner.

For each case the amplitude-modulation index m is defined as

$$m = \begin{cases} \dfrac{E_{max} - E_0}{E_0} & \text{positive modulation} \\[2ex] \dfrac{E_0 - E_{min}}{E_0} & \text{negative modulation} \end{cases} \qquad (14\text{-}3)$$

where E_{max} = peak amplitude of modulated carrier and E_0 = peak amplitude of unmodulated carrier.

The maximum negative-modulation index of unity results from the reduction of the instantaneous carrier envelope to zero. The positive-modulation index is not limited. The maximum symmetrical amplitude modulation that can be produced corresponds to a modulation index of unity.

For a complex modulating signal $G(t)$ the modulated carrier spectrum is

$$F(\omega) = \mathscr{F}\{E_0[1 + mG(t)] \sin (\omega_c t + \phi)\} = \frac{E_0}{2\pi} \int_{-\infty}^{\infty} [1 + mG(t) \sin (\omega_c t + \phi)e^{-j\omega t} dt \quad (14\text{-}4)$$

where $F(\omega)$ = Fourier transform of the time function $\mathscr{E}[f(t)]$.

2. Types of Amplitude Modulation. Generation of an amplitude-modulated waveform requires the multiplication of signals, $f_1(t)f_2(t)$. Both signals need not be time-variant; e.g., a carbon microphone modulates a dc potential with voice signals. The main applications, however, concern the modulation of *carrier* signals to exploit desirable transmission characteristics, that is, $f_1(t)[A \sin (\omega_c t + \phi)]$. Two classes of circuits are used as modulators, *square-law* devices and *linear* modulators.[16-18]*

Square-Law Modulation.   Any device having a nonlinear transfer function can be expressed in a power series form, e.g., the current in a diode, $i(e) = a_0 + a_1 e + a_2 e^2 + a_3 e^3 + \cdots$.

When the diode characteristic and bias conditions are chosen so as to enhance the coefficient a_2 with respect to the other coefficients, the device is considered to be a square-law nonlinear element.[2]

Under these conditions, when two signal inputs $f_1(t)$ and $f_2(t)$ are summed and used as the driving function e, a significant portion of the output will exist as $a_2 e^2$

$$e = f_1(t) + f_2(t) \quad (14\text{-}5)$$
$$e^2 = 2f_1(t)f_2(t) + [f_1(t)]^2 + [f_2(t)]^2 \quad (14\text{-}6)$$

Suitable nonlinear characteristics are exhibited by various types of rectifiers, triodes, and transistors.

Linear Modulation.   Linear modulators are devices with transfer functions that are linearly related to a control parameter. Examples include the outputs of class C rf amplifiers as a function of the B-plus supply voltage, the transconductance gain of transistor differential amplifiers vs. emitter current-source magnitude, and Hall effect devices whose transconductance is proportional to the applied magnetic field.[6]

3. Methods of Amplitude Modulation. *Square-Law Amplitude Modulators.*[2,17-19]   A square-law modulator requires three features: a method of summing the two input signals $f_1(t)$ and $f_2(t)$, a device with a nonlinear transfer function, and a tuned circuit and coupling network for extracting the desired modulation products. Voltage summing or current summing are used depending on the transfer characteristic of interest. The nonlinear device is biased in a region that enhances the second-order coefficient of the power series that represents the nonlinear transfer function. The most common devices used for this type of modulation are semiconductor diodes and vacuum-tube triodes. An example of a typical circuit is shown in Fig. 14-2.

The efficiency of this type of amplitude modulation is generally low, and all the output energy is supplied by the driving functions.

Consider two input signals:

$$e_m = E_m \cos \omega_m t \qquad \text{modulating signal} \quad (14\text{-}7)$$
$$e_c = E_c \cos \omega_c t \qquad \text{carrier signal} \quad (14\text{-}8)$$

The input applied to the nonlinear device is

$$e_s = E_m \cos \omega_m t + E_c \cos \omega_c t \quad (14\text{-}9)$$

If the transfer function is represented by the two terms of interest from a Taylor series,

$$e_0 = a_1 e_s + a_2 e_s^2 \quad (14\text{-}10)$$

---

*Superior numbers correspond to numbered references, Par. **14-15**.

The output components resulting are

$$e_0 = (a_2/2)(E_m^2 + e_c^2) \qquad \text{dc rectified component}$$
$$+ a_1 E_m \cos \omega_m t \qquad \text{modulating signal}$$
$$+ a_1 E_c \cos \omega_c t \qquad \text{carrier}$$
$$+ (a_2/2)E_m^2 \cos^2 2\omega_m t \qquad \text{second harmonic of modulation} \qquad (14\text{-}11)$$
$$+ (a_2/2)E_c^2 \cos^2 2\omega_c t \qquad \text{second harmonic of carrier}$$
$$+ a_2 E_c E_m \cos (\omega_c - \omega_m)t \qquad \text{lower sideband}$$
$$+ a_2 E_c E_m \cos (\omega_c + \omega_m)t \qquad \text{upper sideband}$$

There would be other terms, also, from the higher-order coefficients of the Taylor series. The degree of modulation is expressed as

$$\text{Modulation index} = 2(a_2/a_1)E_m \qquad (14\text{-}12)$$

The desired outputs for double-sideband amplitude modulation are

$$e_5 = a_1 E_c \cos \omega_c t + a_2 E_c E_m \cos (\omega_c \pm \omega_m)t \qquad (14\text{-}13)$$

The square-law devices are reciprocal in that the modulating frequency will appear as an output if a modulated signal is applied as the input. Thus the square-law device can also be used as a demodulator or detector.

4. Low- and Medium-Power Linear Modulators.[1,3,6-8,19,21,23,25] Many applications exist in mobile equipment for amplitude modulators with output powers from milliwatts to tens of watts. Transistor circuits are used almost exclusively for these circuits. Carrier frequencies above 1 GHz can be used in the lower-power transistor circuits. The most common methods of amplitude modulation used are class C collector-modulated stages with the rf applied in the common-emitter or common-base configuration, as shown in Fig. 14-3. The common-emitter configuration provides the maximum power gain and excellent efficiency. Common-base stages are used to increase the upward modulation capabilities, where the maximum modulation indices are important. Increased linearity can be achieved at the expense of efficiency by biasing the amplifier class B so that a nominal collector current flows under no-modulation conditions.

For class C operation the transistor transfer characteristics of the modulated amplifier are determined from the large-signal input and output parallel equivalent-impedance data.[12,15] These are determined experimentally or provided on device specification sheets. The transistors are operated in a very nonlinear manner as class C amplifiers; therefore the experimental data should be representative of the expected operating point, because the small-signal transistor parameters are not adequate.

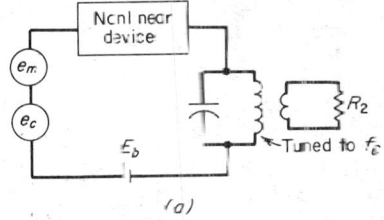

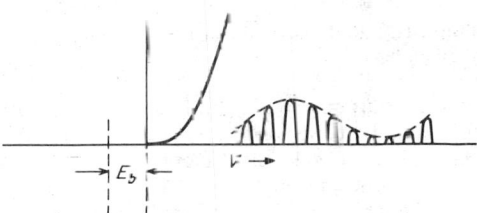

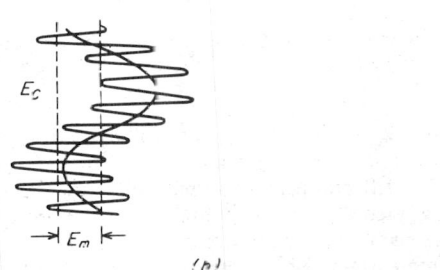

Fig. 14-2. Square-law modulators.

5. Power Relationships. The instantaneous rms output voltage E varies about the unmodulated carrier rms level E_0. The maximum rms output is $E_0(1 + m)$, where m is the modulation index which can vary from zero to unity for symmetrical modulation. The minimum rms output is $E_0(1 - m)$. The unmodulated power into the load R is

$$P_o = E_0^2/R \tag{14-14}$$

The peak power into the load is

$$P_{max} = [E_0(1 + m)]^2/R \tag{14-15}$$

The minimum is

$$P_{min} = [E_0(1 - m)]^2/R \tag{14-16}$$

The average power into the load for sinusoidal modulation is

$$P_{av} = P_o(1 + m^2/2) = (E_0^2/R)(1 + m^2/2) \tag{14-17}$$

The unmodulated output power E_0^2/R is supplied by the class C amplifier, and the sideband energy $(E_0^2/R)(m^2/2)$ is supplied by the modulator. The class C amplifier can be biased very close to the peak modulated-output envelope swing, $V_{CC} \approx \sqrt{2}mE_0$. For 100% modulation, $V_{CC} \approx \sqrt{2}E_0$. The dissipation in the output voltage stage is the difference between the total input power, consisting of the dc collector bias and the input rf drive, and the output rf power. The input rf drive for an amplifier with power gain A is

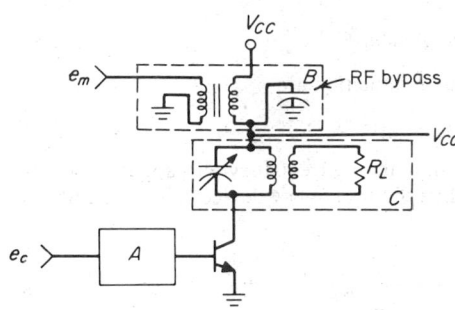

$$P_{drive} = E_0^2/RA \tag{14-18}$$

The input dc bias, unmodulated, is slightly greater than

$$P_{dc} = \sqrt{2}E_0\bar{I}_C \tag{14-19}$$

where $\bar{I}_C$ = average dc collector current.

The output transistor dissipation unmodulated is

$$P_{TR_o} \approx \sqrt{2}E_0\bar{I}_C + P_{drive} - E_0^2/R \tag{14-20}$$

$$P_{TR_o} \approx \sqrt{2}E_0\bar{I}_C - \frac{E_0^2}{R}\frac{A-1}{A} \tag{14-21}$$

A = input impedance matching network

B = modulator circuit for producing V_{CC} which follows the modulating signal, e_m, and provides a low source impedance at f_C

C = Frequency selective network, $f_c \pm f_m$, and output impedance matching network

Fig. 14-3. Collector-modulated transistor (class C) rf amplifier.

In higher-power systems the output transistors can be paralleled. At higher frequencies the gain of the output transistors may be such that the output power is limited by the dissipation in the driver. In these instances the driver may also be collector-modulated to achieve adequate upward modulation and to reduce the power dissipation in the driver.[9]

The input and output impedances of the transistor class C amplifiers are characteristically low. Typical parallel equivalent input impedances are

$$2 < R_{in} < 50\ \Omega \qquad 30 < C_{in} < 5,000\text{ pF}$$

The collector load impedance-resistance component is determined from the bias voltage and output power

$$R_L' = (V_{CC})^2/2P_o \tag{14-22}$$

where V_{CC} = dc collector bias voltage and the output voltage swing is $2V_{CC}$ peak to peak and P_o = unmodulated output power.

The collector parallel output capacitance depends on the device geometry and may range from a few picofarads for very-high-frequency lower-power devices to a few hundred picofarads for large-geometry high-power devices.

Interstage and output networks are used to obtain conjugate matches for maximum power gain.

Care must be exercised in the selection of components, especially capacitors, used in the input matching networks. The low input impedance and high ratio of reactive to resistive components of the input impedance can result in very high circulating currents in these networks.

The class C transistor amplifiers have the advantage over tube circuits that a zero bias condition at the base-to-emitter junction will reduce the collector current to the value of collector leakage current; hence loss of drive will not result in destructive device dissipation, as may be the case with grid-leak biased tube circuits.

Other types of linear transistor modulators can be used where efficiency is less critical or where very-wide-band operation precludes effective output filtering for the elimination of harmonics. A differential amplifier circuit, as shown in Fig. 14-4, will have a transconductance gain that is very nearly proportional to the emitter current-source magnitude. Modulation of the cur-

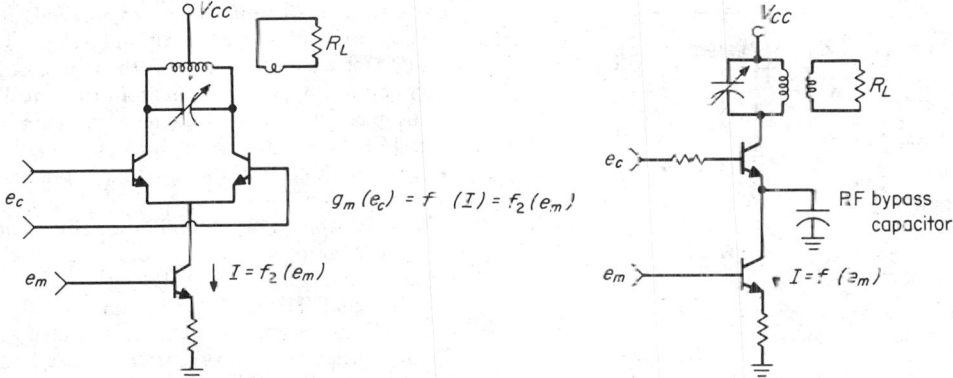

$$g_m (e_c) = f (I) = f_2 (e_m)$$

$$I = f_2 (e_m)$$

$$I = f (e_m)$$

Fig. 14-4. Differential amplifier amplitude modulator (without dc bias details).

Fig. 14-5. Two-transistor transconductance modulator.

rent source will produce a nearly ideal multiplication of the rf input signal and the modulation signal for a wide range of low current levels. The differential amplifier must be biased class A with minimum dc offset at the base inputs. A single transistor multiplier can also be used, as shown in Fig. 14-5 for very-low-level outputs. An emitter bypass capacitor for the rf signal is used in place of the additional transistor for the rf return.

6. High-Power Linear Modulators. High-power linear modulators, 50 W and up, are generally constructed using class C plate-modulation vacuum-tube circuits (Fig. 14-6). Some of

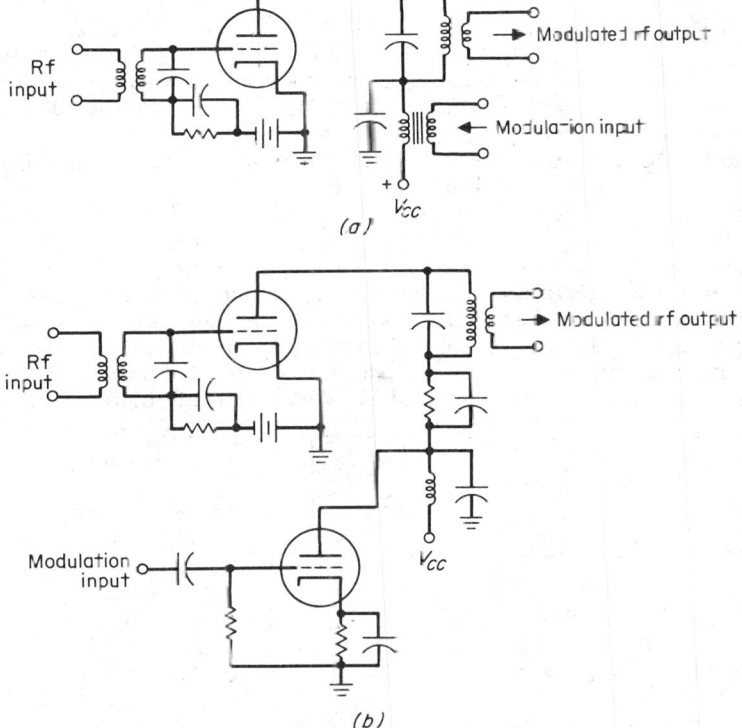

Fig. 14-6. Typical class C plate modulators: (a) modulation transformer; (b) class A modulating driver.

the intermediate power and frequency applications use paralleled transistor configurations with class C collector modulation.[14]

Triodes, tetrodes, and pentodes are used in the vacuum-tube circuits. Multigrid tubes require screen-grid modulation in conjunction with the control-grid modulation to achieve space-charge modulation and to minimize screen current and screen dissipation. The two methods of screen modulation commonly used are: (1) self-bias of the screen grid with a bypassed dropping resistor or inductor from the screen to the plate supply or the screen-grid supply (Fig. 14-7*a*) and (2) screen modulation via a separate winding on the modulation transformer (Fig. 14-7*b*).

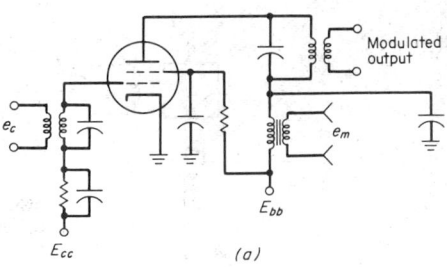

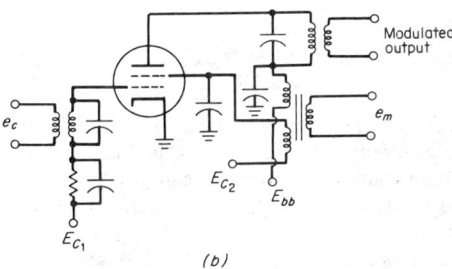

Fig. 14-7. Multigrid class C modulation circuits: (*a*) screen modulation via *RC* from unmodulated plate supply; (*b*) screen modulation via separate winding on modulation transformer.

7. Grid Modulation. The amplitude of a class C rf amplifier output can also be modulated by changing the grid bias with the modulating signal. The modulating signal is added to the rf input signal. The effect is to change the magnitude of the plate-current pulses, and hence the fundamental component of the plate current (Fig. 14-8).

The disadvantage associated with grid modulation is that the fixed-plate supply voltage must be twice the peak rf voltage without modulation. This causes high plate dissipation and lowers the plate efficiency to the range of 35 to 45% when unmodulated.

The carrier power obtained from a plate-modulated class C amplifier is about 3 times that available from a grid-modulated circuit using the same tube.

The principal advantage of grid modulation is the reduction in modulator voltage and power. Grid modulation is used in systems where plate-modulation transformers cannot provide adequate bandwidth and a class A modulator is required. Linearity in grid-modulated class C amplifiers is more difficult to obtain at a high modulation index while maintaining maximum efficiency.

8. Cathode Modulation of Class C RF Amplifiers. Cathode modulation can be used with a class C rf amplifier, as shown in Fig. 14-9. The modulation transformer output varies the grid-cathode as well as the plate-cathode voltages. The ratio of grid and plate modulation can be selected by varying the tap; thus the circuit provides a means of producing varying combinations of grid and plate modulation. Some grid-leak bias is normally used to improve linearity.

9. Modified AM Methods. The information transmitted by an amplitude-modulated carrier is contained wholly in the modulation sidebands. The transmission of the carrier energy simplifies the receiver-detector implementation but adds no information. In addition, each sideband contains the same information, and only one is required to transmit the intelligence. Elimination of the carrier and/or one sideband can effect a substantial transmitter power saving.

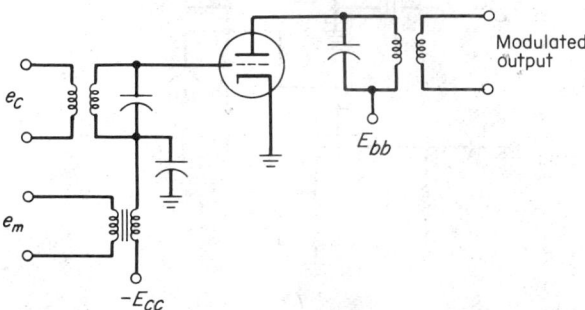

Fig. 14-8. Grid-modulated class C rf amplifier.

For 100% amplitude modulation the carrier power is two-thirds of the transmitter power and each sideband one-sixth. Elimination of the carrier will result in *double-sideband suppressed-carrier modulation*. Elimination of one sideband while retaining the carrier or a substantial portion of the carrier results in *vestigial-sideband transmission*. Elimination of the carrier and one sideband is called *single-sideband suppressed-carrier modulation*.

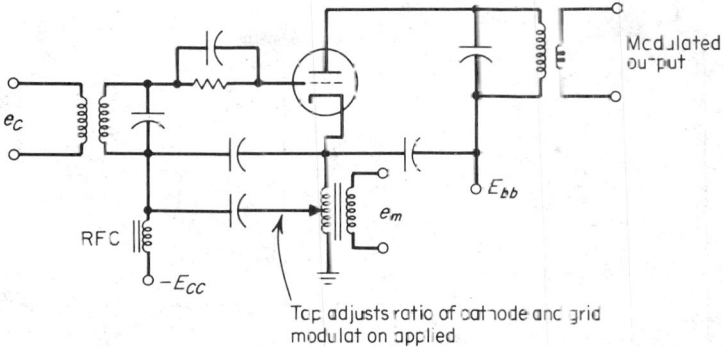

Fig. 14-9. Cathode-modulated class C rf amplifier.

The easiest of these modulation schemes to implement is the vestigial-sideband transmission, both from a transmitter and a receiver viewpoint. The unwanted sideband is generally filtered out at low levels in the transmitter chain and is known as *transmitter attenuation* (TA). One sideband can also be eliminated in the receiver by selective filtering and is called *receiver attenuation* (RA). The latter scheme is not practical from a spectrum-utilization sense; hence it is normally used only as in television broadcast to complete unwanted sideband rejection performed primarily at the transmitter.

A vector notation can be used to illustrate the phenomena of various types of amplitude modulation. The vector is a complex function. The sinusoidal function of time that is of interest is the real part or the vector projection on the real axis of the complex plane. Thus the real part of $Ae^{j\omega t}$ is $A \cos \omega t$, as shown in Fig. 14-10a. The projection on the real axis of the vector $Ae^{j\omega t}$ can be considered as the carrier signal in subsequent amplitude-modulation discussions. Since the unmodulated carrier signal in amplitude-modulation processes is a fixed peak amplitude and fixed frequency function of time, a modified vector representation can be used to describe the envelope of the modulated waveform. The modified vector diagram maintains the carrier vector as a fixed, nonrotating vector. By this means subsequent illustrations are referenced to a complex plane rotating at the carrier angular rate, and the projection on the real axis corresponds to the *modulation envelope variations with time*.

For an amplitude-modulation system using a sinusoidal modulating function at 100% modulation, $m = 1$, the addition of two vectors to the basic vector diagram is required. The two additional vectors represent the sideband signals produced in the modulation process. The two sideband signals are displaced on either side of the carrier signal in the modulated signal spectrum by a frequency equal to the modulating frequency. Therefore, with respect to the modified reference system, the lower sideband vector will rotate clockwise at the modulation-signal angular rate, and the upper sideband will rotate counterclockwise at the same rate.

The relative phases of the two sideband vectors are such that no change in carrier phase angle is introduced when they are both present; i.e., the imaginary parts of the sideband contributions cancel. The initial angles are set by the phase of the modulating function. As the sideband vectors rotate in the modified complex plane, the vector sum of the carrier plus sidebands describes sinusoidal projection on the real axis at the modulating frequency rate.

Note that the unmodulated carrier projection is constant and equal to the peak carrier magnitude. The 100% modulated carrier projection is nonnegative. If the original complex plane were used, the entire vector system would rotate at the carrier angular rate, and the projection on the real axis would be the actual time function produced by the modulator. This function would have negative portions and would fill the envelope function and its image with amplitude-modified sinusoids at the carrier frequency, as shown in Fig. 14-10b with the dashed lines within the envelope.

10. Vestigial-Sideband Systems. Vestigial-sideband transmission introduces angle modulation on the carrier because the symmetry of the contrarotating sideband vectors is lost, as illustrated in Fig. 14-10*d*. A standard envelope detector can still be used to recover the modulating signal, however. The primary objective in the application of vestigial-sideband transmission is to conserve spectrum in the transmission medium.

11. Suppressed-Carrier Systems. Suppressed-carrier systems for AM transmission and reception require modifications to the receiver. The double-sideband suppressed-carrier signal cannot be envelope-detected without the reinsertion of a carrier signal. The frequency and the

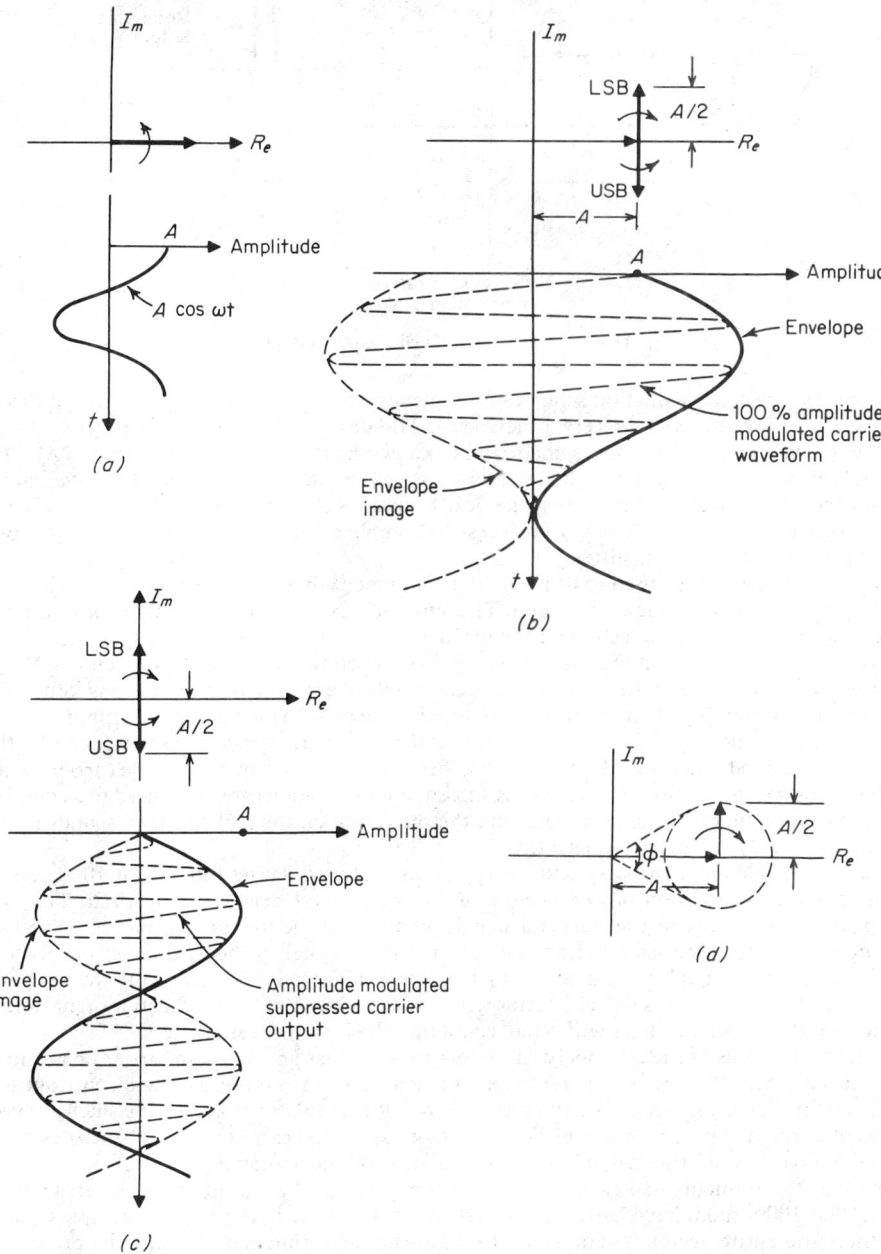

Fig. 14-10. Vector representation of amplitude modulation. (*a*) Rotating carrier vector and its real projection. (*b*) Vector representation with sideband vectors resulting from 100% amplitude modulation. Envelope function is locus of projection on the real axis with a nonrotating carrier vector. (*c*) Double-sideband suppressed-carrier vector representation. Envelope function is locus of projection on the real axis with coordinate rotation as in (*b*). (*d*) Vector representation of a single-sideband amplitude-modulated waveform showing peak-to-peak carrier phase modulation ϕ resulting from the absence of the other sideband signal.

phase of the reinserted carrier are critical. This type of transmission is used with special phase-locked receivers or with the transmission of low-level carrier to permit the reconstitution of the carrier frequency and phase at the receiver.

A double-sideband suppressed-carrier signal can be generated with a balanced modulator, as shown in Fig. 14-11a.

For a sinusoidal modulating signal the outputs of the two modulators are

$$e_1 = E_c \cos \omega_c t + \frac{E_m}{2} \cos (\omega_c + \omega_m)t + \frac{E_m}{2} \cos (\omega_c - \omega_m)t \qquad (14\text{-}23)$$

$$e_2 = E_c \cos \omega_c t \frac{E_m}{2} \cos [(\omega_c + \omega_m)t + \pi] + \frac{E_m}{2} \cos [(\omega_c - \omega_m)t + \pi] \qquad (14\text{-}24)$$

When these two signals are combined in push-pull, the output becomes

$$e_0 = e_1 + e_2 = E_m \cos (\omega_c + \omega_m)t + E_m \cos (\omega_c - \omega_m)t \qquad (14\text{-}25)$$

The nonlinear devices can be diodes or modulated class C rf amplifiers. The balanced modulator simplifies the tuned-output-circuit design because all even harmonics of the modulating process tend to be canceled along with the carrier.

Single-sideband suppressed-carrier modulation simplifies the receiver design in some applications. The phase of the reinserted carrier at the receiver is not critical as in double-sideband suppressed-carrier systems. Also, a frequency error in the reconstituted carrier will result only in a corresponding shift in the demodulated signal frequencies, which may be tolerable. If accurate modulation-frequency preservation is required, a low-level carrier can be transmitted to aid in the reconstruction of the carrier signal at the receiver. The reinserted carrier amplitude with respect to the received sideband signal is generally made large to minimize angle modulation of the carrier at the detector.

Two methods are usually employed to generate suppressed-carrier signal-s deband transmissions. The most direct method is to filter out the undesired sideband and carrier at low levels in the transmitter chain. This filtering problem is difficult when the modulation sidebands of interest are close to the carrier and the carrier frequency is high. This problem can be eased by using successive modulators where the first carrier frequency is low. The single-sideband signal is filtered from the output of a low-frequency-carrier balanced modulator to eliminate the lower sideband. This signal is then used to modulate a second balanced modulator. The

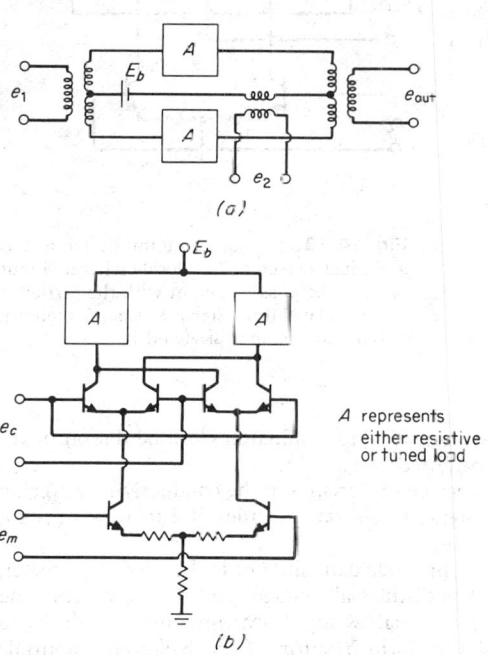

Fig. 14-11. Double-sideband suppressed-carrier modulators: (a) general form of a balanced modulator for suppressed-carrier output using nonlinear elements A and bias voltage E_b; (b) four-quadrant multiplier.

A represents either resistive or tuned load

balanced modulators eliminate the carriers and cause the second-modulation sidebands to be separated by twice the first carrier frequency to ease the second filtering.

An alternative method of suppressed-carrier single-sideband modulation is by unwanted sideband cancellation. Two balanced modulators are used, with their outputs combined in push-pull. One modulator is driven with carrier and modulating signals that have been shifted 90° with respect to the inputs to the other modulator. Special wide-band phase-shift networks are required to handle the modulation input because orthogonality must be maintained across the bandwidth of the modulating signal.

12. Modulated Oscillators. A direct modulated class C oscillator can be used as an AM transmitter. The linearity of such circuits is generally as good as or better than the plate-modulated class C amplifiers. The main disadvantage is the tendency for carrier-frequency pulling which results from the changes in the oscillator operating point as the modulation signal varies.

A very useful circuit for the generation of low-level double-sideband suppressed-carrier signals is the four-quadrant multiplier. The same circuit will also operate as a double-sideband modulator with carrier and as low-level demodulator in phase-locked receiver systems. Multipliers are available in integrated-circuit form, and they can be used at frequencies from dc to beyond 100 MHz (Fig. 14-11b).

AMPLITUDE DEMODULATORS

13. Detectors. The most common AM detector or demodulator is a diode rectifier. The ideal diode detector passes current in only one direction and will essentially follow the envelope of an amplitude-modulated waveform when used in a circuit as shown in Fig. 14-12 (see also Pars. **14-60** to **14-68**).

The charging time constant R_sC must be short, so that the capacitor voltage follows the input signal E_{rf} when the diode is forward-biased or conducting. Conversely, the discharge time constant R_LC must be long enough to retain most of the rectified voltage between cycles of the carrier signal but not so large that the capacitor voltage will not discharge at the maximum rate of change of the input signal

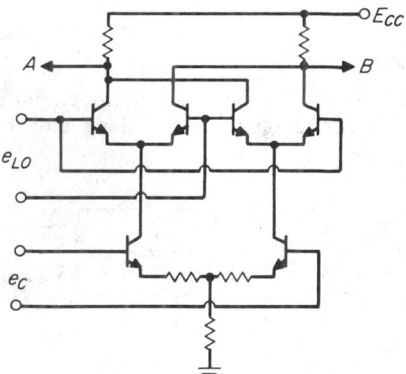

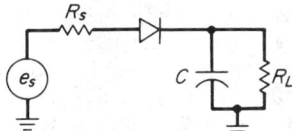

Fig. 14-12. Diode envelope detector.

Fig. 14-13. Four-quadrant multiplier used as a product detector. The local-oscillator input e_{LO} must be phase-coherent with the carrier of the modulated input signal e_c. Complementary outputs are obtained at A and B.

envelope. The envelope detector is essentially insensitive to residual angle modulation of the carrier, and it is therefore usable in single-sideband receivers.

Practical diode rectifiers have nonlinear resistance characteristics in the conduction bias region and are therefore operated at fairly high input signal levels, on the order of 2 to 10 V peak for semiconductor diodes.

14. Product Detectors. Another type of amplitude demodulator is the product detector, or multiplier circuit (Fig. 14-13). This circuit has distinct advantages and disadvantages. The advantages include the ability to detect lower-level signals with a linear response; the ability to differentiate phase reversals in the modulated waveform, resulting from balanced amplitude modulation with suppressed carrier; and the ability to produce, in some designs, error signals for automatic-frequency-control systems in receivers.[20]

The analytical expression for the output of a product detector is

$$e(t) = \underbrace{E[1 + m \sin (\omega_m t + \phi_m)] \cos (\omega_c t + \phi_c)}_{\text{amplitude-modulated signal}} \underbrace{\cos (\omega_c t + \phi_d)}_{\text{local-oscillator signal}} \qquad (14\text{-}26)$$

Expansion of Eq. (14-26) shows that the product of the incoming carrier signal, $\cos (\omega_c t + \phi_c)$, and the local-oscillator signal, $\cos (\omega_c t + \phi_c)$, produces a dc term except when these two inputs are in quadrature phase. The output dc term is proportional to the cosine of the relative phase angle of the carrier and local-oscillator signals. A four-quadrant multiplier circuit capable of performing this type of detection is shown in Fig. 14-13.

The details of the dc bias network are not included in the elementary schematic of the multiplier circuit. The input signals, rf and local-oscillator, are applied to either input port. Balanced-differential or single-ended inputs can be used, although the balanced inputs give the added performance of increased linearity and common-mode signal rejection. Two outputs of opposite polarity are available at the collectors of the upper-rank transistors. Design options are available to increase the efficiency and linearity of the circuit. Normally, an overdrive is applied to the local-oscillator port to make that section of the multiplier operate in a switching mode. The effect is to multiply the rf signal with a square wave at the carrier frequency instead of a sine wave. This type of operation also produces additional outputs at the higher harmonics of the carrier frequency.

For balanced linear in-phase inputs at both ports, the outputs consist of

$$e_0 = \frac{1}{2}(A_{LO} + A_{LO} \cos 2\omega_c t)E[1 + m \sin (\omega_m t + \phi_m)] \tag{14-27}$$
$$= \frac{1}{2}A_{LO}E[1 + m \sin (\omega_m t + \phi_m)] + \frac{1}{2}A_{LO}E \cos 2\omega_c t$$

The linearity versus dc offset of the multiplier can be improved by using emitter degeneration at the rf signal port. Without degeneration the maximum peak-to-peak differential drive that will not cause distortion is on the order of 50 mV. At this signal input level the dc offsets in the circuit may cause undesirable output voltage shifts. A compromise can be made to minimize the ratio of output dc offset to peak signal output by degenerating the rf port input and applying a larger input signal level.

The disadvantages of the product detector are relative circuit complexity and the need for a phase-coherent local-oscillator signal. The circuit complexity can be circumvented for input signal frequencies up to 100 MHz by using integrated-circuit multipliers. This approach also minimizes the possibility of serious dc offset problems due to device mismatch. The generation of the coherent local-oscillator signal can be achieved in two ways. For simple DSB amplitude-modulated signals, the carrier can be stripped from the input rf signal, with a parallel limiting amplifier with narrow bandwidth. This approach cannot be used in suppressed-carrier AM systems, where the carrier phase reversals are introduced in the modulation process.

The more general approach is to use an additional multiplier circuit as a phase detector. When the rf and local-oscillator signals are equal in frequency and in phase quadrature, the multiplier output has no dc component. Any relative phase shift from quadrature will produce an odd-function-error signal at dc. This signal can be used with a voltage-controlled oscillator to correct the phase of the local oscillator or input rf signal. The system described is a phase-locked loop. The bandwidth of this loop can be controlled independently from the rf bandwidth for optimum acquisition and noise-suppression characteristics. This type of system is capable of producing a stable and noise-free local-oscillator signal that tracks any variations in the frequency and phase of the input rf signal.

With additional modifications the phase-locked receiver system is capable of reinserting the desired carrier in suppressed-carrier double- and vestigial-sideband systems.

15. References on Amplitude Modulators and Demodulators

1. Wilson, J. P.   A Simple High-Speed Analogue Multiplier, *Electron. Eng.* (Great Britain), January 1967, Vol. 39, pp. 11–14.

2. Leenov, D.   PIN Diode Microwave Switches and Modulators, *Solid State Des.*, April 1965, Vol. 6, pp. 37–40.

3. DeKold, R.   Amplitude Modular Is Highly Linear, *Electronics*, June 5, 1972, pp. 101–102.

4. Minton, R.   Design Trade-offs for RF Transistor Power Amplifiers, *Electron. Eng.*, March 1967, pp. 68–73.

5. Hejhall, R.   For High-Frequency Communications Equipment Use Balanced Modulators *EDN/EEE*, Feb. 15, 1972, pp. 28–32.

6. Oppenheimer, M.   In IC Form, Hall-Effect Devices Can Take On Many New Applications, *Electronics*, Aug. 2, 1971, pp. 46–49.

7. Cote, T.   Modern Techniques of Analog Multiplication, *Electron. Eng.*, April 1970, pp. 75–79.

8. Gilbert, B.   A Precise Four Quadrant Multiplier with Subnanosecond Response, *IEEE J. Solid State Circuits*, December 1968, Vol. SC-3, No. 4, pp. 365–373.

9. Rheinfelder, W. A.   Modulation of Driver Stage to Increase Power Output of AM Transmitter, *Motorola Semiconductor Appl. Note* AN-114. reprint from *Semiconductor Prod. Mag.* March 1962.

10. Hejhall, R.   Getting Transistors into Single-Sideband Amplifiers, *Motorola Semiconductor Prod. Appl. Note* AN-150, reprinted from *Electronics*.

11. Principles and Techniques of Single-Sideband Modulation, *Electro-Technology*, July 1962.

12. Brubaker, R. J.   An All-Solid-State Marine Band Transmitter, *Motorola Semiconductor Prod. Appl. Note* AN-156, reprinted from *SSD/CDE*.

13. Hejhall, R. C.   A 50 Watt 50 MHz Solid State Transmitter, *Motorola Semiconductor Prod. Appl. Note* AN-246.

14. Brubaker, R.   A Broadband 4-Watt Aircraft Transmitter, *Motorola Semiconductor Prod. Appl. Note* AN-481.

15. Martens, C.   A 40-W, 50-MHz Transmitter for 12.5 Volt Operation, *Motorola Semiconductor Prod. Appl. Note* AN-502.

16. Terman, F. E.   "Electronic and Radio Engineering," Chap. 15, McGraw-Hill, New York, 1955.

17. Landee, R. W., D. C. Davis, and A. P. Albrecht   "Electronic Designers' Handbook," Secs. 5.1–5.4, McGraw-Hill, New York, 1957.

18. Gray, T. S.   "Applied Electronics," 2d ed., Chap. 12, Arts. 1–14, Wiley, New York, 1955.

19. Bilotti, A.   Application of a Monolithic Analog Multiplier, *IEEE J. Solid State Circ.* December 1968, Vol. SC-3, No. 4.

20. Costas, J. P.   Synchronous Communications, *Proc. IRE*, December 1956, Vol. 44, pp. 1713–1718.

21. Shapiro, G. R.   Analog Multipliers Offer Solutions to Video Modulation Problems, *EDN*, Sept. 1, 1972, pp. 40–41.

22. Microwave Power Transistor Brochure MPT-700, RCA Solid State Division.

23. Analog Multiplier Principles, *Analog Dev. Tech. Bull.* AD530.

24. A New Linear Power Transistor for SSB Equipment, *RCA Commercial Eng. Appl. Note* AN-4591, Harrison, N.J.

25. Wideband Analog Multipliers 424/5, Analog Devices, Cambridge, Mass.

26. Carson, J. R.   The Equivalent Circuit of the Vacuum-Tube Modulator, *IRE Proc.*, 1921, Vol. 9, pp. 243–249.

27. Neilsen, J. R.   Transformerless Ring Modulator, *EEE*, February 1970, p. 116.

28. Sonde, B. S.   (Correspondence), Micropower Amplitude Modulator, *Proc. IEEE*, July 1971, pp. 114–116.

29. Kelly, R. G.   Linear Modulator Has Excellent Temperature Stability, *EEE*, July 1968, p. 102.

30. Pichard, A.   100% Amplitude Modulation with Two Transistors, *Electronics*, June 12, 1967, pp. 104–105.

31. McDermott, C.   Suppressed Carrier Modulator with Noncritical Components, *Electronics*, Oct. 31, 1966, p. 70.

32. Rockwell, R. J.   Cathanode Modulation System, *IEEE Trans. Broadcast.*, January 1967, Vol. BC-13, p. 19.

Frequency and Phase (Angle) Modulators

BY N. R. POWELL

16. Angle Modulation. The representation of angle modulation is conveniently made in terms of the notion of the analytic function.[1]* For real continuous functions of time $x(t)$, consider

$$m(t) = x(t) + jHx(t)$$

where $x(t)$ = real continuous function, $j = (-1)^{1/2}$, Hx = Hilbert transformation of x, and

$$Hx(t) = \pi^{-1} \int_{-\infty}^{\infty} x(\tau)(t - \tau)^{-1}\, d\tau$$

The angle of $m(t)$ is said to be the angle θ, defined by

$$\theta = \tan^{-1}(Hx/x)$$

Angle modulation may be considered as the change in θ with time or as that portion of the total change in θ which can be associated with the phenomenon of interest. If the relationship between the changes in θ and the effect of interest is direct, the modulation is called *phase modulation* (PM) and the devices producing this relationship *phase modulators*. If the relationship between the changes in the derivative $d\theta/dt$ and the effect of interest is direct, the modulation is called *frequency modulation* (FM), and the devices producing this relationship are called *frequency modulators*.

The derivative $d\theta/dt = \dot{\theta}$ is related to the components of the analytic function $m(t)$ by

$$\dot{\theta} = \frac{\begin{vmatrix} x & Hx \\ \dot{x} & H\dot{x} \end{vmatrix}}{x^2 + (Hx)^2} \tag{14-28}$$

for functions $x(t)$ for which the differential and Hilbert operators commute.

Thus in angle modulators, whether implemented functionally in the general form (as with a general-purpose computer) suggested by the representations of θ and $\dot{\theta}$, or as some special combination of electronic networks, a direct or proportional relationship is established within the device between either of these two functions and an effect, call it the input $v(t)$, of interest.

---

*Superior numbers correspond to numbered references, Par. **14-25**.

Diagrammatically, a linear angle modulator can be considered to be a device that transforms $v(t)$, as shown in Fig. 14-14.

Demodulators simply perform the inverse of this operation, providing an output function proportional to $\theta(t)$ from a function proportional to the angle of $m(t)$ as an input.

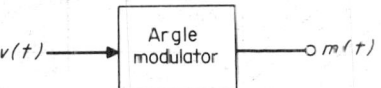

Fig. 14-14. Linear angle modulator.

17. Angle-Modulation Spectra. The spectral distribution of power for angle-modulated waveforms varies widely with the nature of the input function $v(t)$.

Random Modulation. For a random process having sample functions

$$x(t) = b \sin (2\pi ft + \phi) \qquad b\text{-const}$$

where f and ϕ = independent random variables, ϕ being uniformly distributed over $-\pi \leq \phi \leq \pi$ and f having a symmetric probability density $p(f)$, this stationary random process has a spectral density function

$$S_{xx}(f) = (b^2/2)p(f)$$

Note that if $p(f)$ is not a discrete distribution, the process is not in general periodic and $S_{xx}(f)$ must be considered continuous.

Periodic Modulation. For a random process having sample functions

$$x(t) = b \cos [\omega_c t - \phi(t) + \theta] \qquad b, \omega_c\text{-const}$$

θ uniformly distributed over $-\pi \leq \theta \leq \pi$, $\phi(t)$ is a stationary process independent of θ; that is,

$$\phi(t) = d \cos (\omega_m t + \theta') \qquad d, \omega_m\text{-const}$$

For θ' uniformly distributed over $-\pi \leq \theta' \leq \pi$, the spectral density function is

$$S_{xx}(f) = (b^2/4) \left(J_0^2(d)[\delta(f - f_c) + \delta(f + f_c)] \right.$$

$$\left. + \sum_{n=1}^{\infty} J_n^2(d)\{\delta[f - (f_c \pm nf_m)] + \delta[j + (f_c \pm nf_m)]\} \right) \qquad (14\text{-}29)$$

where $J_n(d)$ = nth-order Bessel function of first kind evaluated at d, δ = Kronecker delta function, and $S_{xx}(f)$ = Fourier transformation of autocorrelation function of x.

Deterministic Modulation. For a function of a completely specified type, e.g.,

$$x(t) = b \sin (\omega_c t + d \sin \omega_m t)$$

frequently it is possible to reexpress $x(t)$ in terms of $J_n(d)$ as

$$x(t) = \sum_{n=-\infty}^{\infty} J_n^2(d) \sin (\omega_c t + n\omega_m t) \qquad (14\text{-}30)$$

Examination of Eqs. (14-28) and (14-29) and tables of Bessel functions[2] permits the construction of Fig. 14-15, showing[3] the implied increase in bandwidth vs. modulation index for FM. Modulation index for sinusoidal modulation can be defined as

$$d = |\Delta f|/f_m$$

where $|\Delta f| = df_m$ = amount of instantaneous frequency change θ to be associated with the input $v(t)$.

Such a set of curves can be readily constructed for most deterministic modulation waveforms, since $J_n(d)$ decreases monotonically and rapidly with n for $n > d > 1$. Using the criteria for n indicated for curves A, B, and C, the frequency range (or bandwidth) centered about ω_c, which contains all such spectral components, is indicated. These are the components that are found to be below the value of nf_m for which $J_n(d)$ is monotonically decreasing and equal to the value for each case. Thus the bandwidth (BW) required at the frequency ω_c is

$$BW = 2df_n (1 + l) \qquad (14\text{-}31)$$

Such criteria should be used with care, since the relationship which these measures bear to distortion of the input $v(t)$ when carried at frequency ω_c through linear-tuned circuits as angle

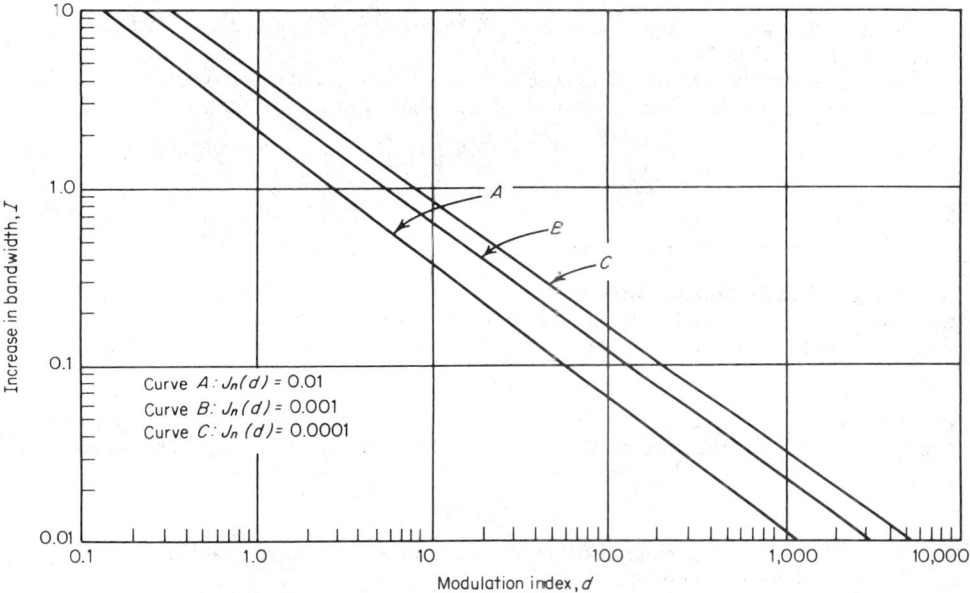

Fig. 14-15. Bandwidth increase vs. modulation index.

modulation is rather indirect. An approximation for the signal-to-distortion ratio (SDR) for gaussian baseband modulation, uniform in $(-B, B)$ of modulation index d, which is passed through a single-pole bandpass filter with half-bandwidth f_c, is[9]

$$\text{SDR} \approx 15/2B^2 d^4 \qquad B/f_c < 0.3 \qquad (14\text{-}32)$$

18. Angle-Modulation Signal-to-Noise Improvement. One of the principal reasons for using frequency modulation in communications and telemetry systems is that it provides a convenient and power-efficient method of trading power for bandwidth while providing high-quality transmission of the input. This is expressed in phase modulation by the relationship

$$\left.\frac{S}{N}\right|_{f_m} = ad^2 \left.\frac{C}{N}\right|_{f_m} \qquad (14\text{-}33)$$

where $(S/N)|_{f_m}$ = demodulated output signal-to-noise ratio measured in a bandwidth f_m, $(C/N)|_{f_m}$ = demodulator input signal-to-noise ratio measured in a bandwidth f_m, d = modulation index, a = a constant of proportionality. The constant of proportionality a is unity for sinusoidal phase modulation of constant-amplitude sinusoidal carrier. The constant varies between 0.5 and 3.0 with class of modulation waveforms, type of network compensation, and noise spectrum; however, Eq. (14-33) can be conservatively applied to FM single-channel voice systems for a = ⅔ and preemphasis and deemphasis networks which preshape the spectrum of the modulation to match the sloped noise spectrum and to restore the original spectrum after demodulation.

The trade-off between the signal-to-noise improvement and the required bandwidth corresponding to curve A in Fig. 14-15 is shown for FM in Fig. 14-16. These curves have been prepared for a conventional demodulator operating at an input carrier-to-noise ratio 1 dB above the threshold (13 dB) measured in the input noise bandwidth to the demodulator, with a constant of proportionality a equal to 0.5. Output signal-to-noise ratio referred to the output information bandwidth $(S/N)_{f_m}$ is

$$(S/N)|_{f_m} = 10 + G \text{ (dB)} \qquad (14\text{-}34)$$

where G is obtained from the figure along with r, the ratio of premodulator bandwidth to output information bandwidth. The carrier-to-noise ratio in a noise bandwidth equal to the output information bandwidth is

$$(C/N)|_{f_m} = 13 + 10 \log r \text{ (dB)} \qquad (14\text{-}35)$$

19. Noise-Threshold Properties of Angle Modulation. The signal-to-noise improvement represented by Eq. (14-33) is achievable only when the input carrier-to-noise ratio is above certain minimum levels. These levels depend upon the type of modulating waveforms, the type of noise interference prevalent, and the type of demodulator. As the foregoing discussion indicates, whenever a demodulator without phase or frequency feedback is used, an input carrier-to-noise ratio (measured in the premodulator bandwidth) of roughly 12 dB is required. Unless this condition is met, a small decrease in carrier-to-noise ratio will result in a sharp decrease in output signal-to-noise ratio, accompanied by undesirable noise effects, such as loud clicking sounds in the case of voice modulation.

For properly designed feedback demodulators, the noise threshold is not determined by the condition prevalent in the premodulator bandwidth as much as by the closed-loop noise bandwidth of the demodulator and the match between internal loop filtering, modulation waveforms, and noise properties.

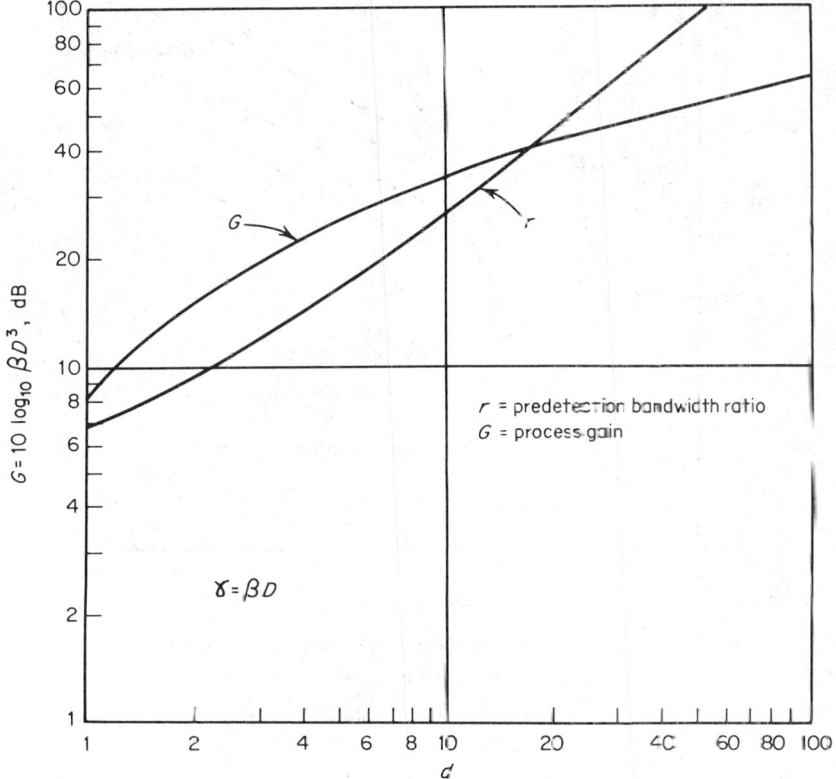

Fig. 14-16. Process gain and rf bandwidth vs. deviation ratio.

A set of curves illustrating the effect of thresholding in properly designed feedback demodulators is shown in Fig. 14-17. Curve 1 is an information-theoretic limit based upon gaussian modulation, infinite predemodulator bandwidth, and the Shannon upper bound on information flow. The Wiener-Hopf filter limit corresponds to the use of optimal feedback-demodulator filter design and is shown as curve 2. Curve 3 corresponds to the use of a phase-locked demodulator employing frequency feedback,[4] with a simple proportional plus integral filter, the time constants for which are selected according to the prevalent signal-to-noise ratio and the modulation parameters. Curves 4 and 5 represent threshold performance achievable with a fixed-parameter design using the same linear filter approximation to the optimum filter. The noise threshold for curve 4 is, approximately,

$$(S/N)|_{f_m} = 0.48 \times 10^{-2}(C/N)^5|_{f_m} \tag{14-36}$$

The modulation indices shown for operation above threshold are for single-channel voice and single-channel gaussian[5] modulation.

20. Angle Modulators. Angle modulators for communications and telemetry purposes generally fall into the category of what may be termed "hard" oscillators having relatively high-Q frequency-determining networks; or they fall into the category of "soft" oscillators having supply and bias sources as the frequency-determining networks. Examples of each are shown in Fig. 14-18.

Control of the hard oscillator is executed by symmetrical incremental variation of the reactive components. For the case of a Hartley oscillator, Z_1 and Z_2 are inductors and Z_3 is a capacitor, allowing the use of varicaps paralleling Z_3 as the voltage-controllable reactance. In such a case

$$f_0 = (2\pi)^{-1}[C_3(L_1 + L_2)]^{-1/2}$$

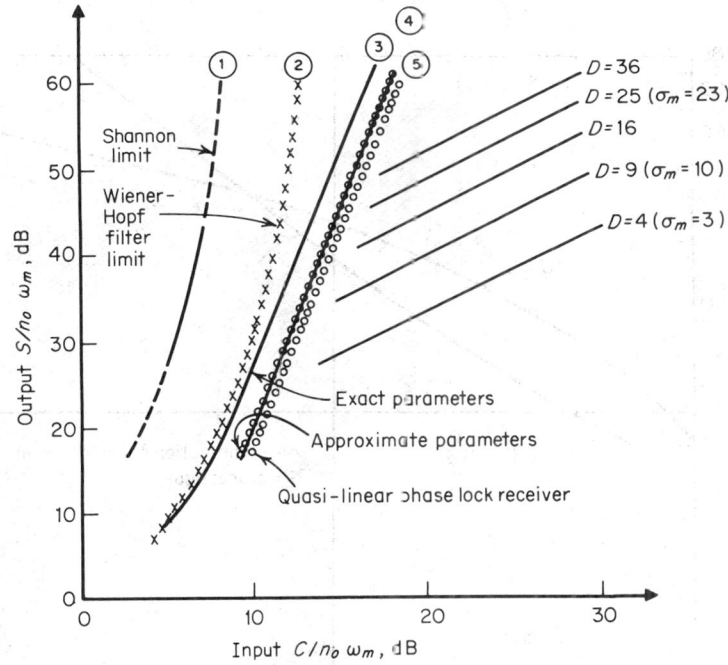

Note: Curves 1,2,5 correspond to gaussian modulation, σ_m

Curves 3,4 correspond to voice modulation, D

Fig. 14-17. Threshold properties of FM.

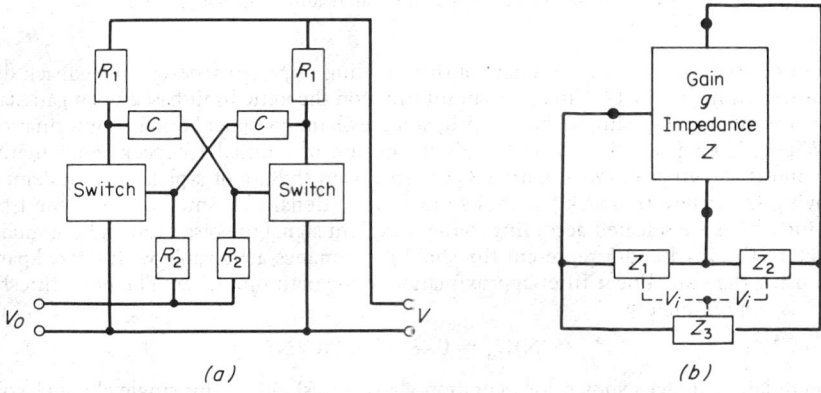

Fig. 14-18. Voltage-controlled oscillators: (*a*) soft oscillator; (*b*) hard oscillator.

where C_3 = total capacitance of varicaps and fixed capacitor of Z_3. Note that the bandwidth of this modulator is determined by the frequency-determining impedances of the network, i.e., the overall Q and center frequency. In view of the need for certain minimum bandwidth requirements from Fig. 14-15 and the need for good oscillator stability, the total frequency deviation required is sometimes obtained by following the oscillator with a series of frequency multipliers and frequency translators, as shown in Fig. 14-19. This configuration permits the attainment of good oscillator stability, constant proportionality between output frequency change and input voltage change, and the necessary modulator bandwidth to achieve wide-band FM.

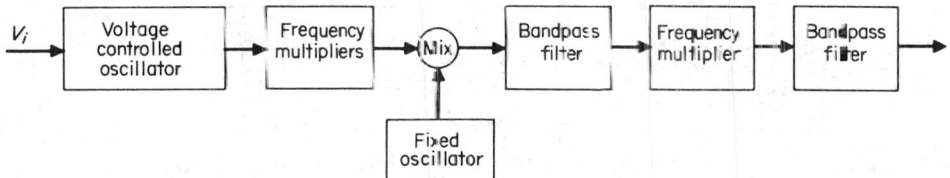

Fig. 14-19. Frequency-modulator configuration.

Control for the soft oscillator is introduced as a change in the switching level of the active-device switches. The frequency of a transistor version of the oscillator is, roughly,

$$f = [2R_1 C \ln (1 + V/V_i)]^{-1}$$

Since this type of oscillator is a relaxation oscillator, the rate at which the frequency of oscillation can be changed is limited only by the rate at which the switching points can be altered by voltage control. Modulators of this kind can be designed with bandwidths greater than the frequency of oscillation. The disadvantage of such networks is the relatively poor frequency stability compared with the high-Q hard oscillators.

Angle Demodulators

BY N. R. POWELL

21. Discriminators. Basic angle demodulators can be designed using balanced tuned networks with suitably connected nonlinearities (such as diode switches); or angle demodulators can be designed using voltage-controlled oscillators (VCO), multipliers, and appropriate feedback networks.

The former are simpler to implement than the latter but require much higher input signal-to-noise ratios for operation above the noise threshold at which the angle modulation produces signal process gain. Two popular versions of the discriminator type of frequency demodulator are shown in Fig. 14-20. The diode discriminator is designed so that one diode conducts more with increases, the other with decreases, in frequency. For greatest linearity, the mutual coupling between the tuned circuits is generally greater than unity. The other basic type of conventional demodulator simply implements a version of the definition [Eq. (14-28)] of FM for the case of a constant-amplitude sinusoidal waveform. The mixing operation can be replaced by a simple phase shifter that provides a 90° phase relationship between x and Hx. Frequently, the reference oscillator will be phase-locked to the average value of the $J_0(d)$ component indicated by Eqs. (14-29) and (14-30).

22. Feedback Demodulators (FMFB). These angle demodulators have the advantage of lower distortion, lower noise threshold, and little or no drift in center frequency of operation and can be designed to be less sensitive to interference. No limiter is required for good performance, and there is no requirement to maintain a minimum predemodulator input carrier-to-noise ratio to avoid noise threshold.

A block diagram of a basic synchronous-filtering demodulator is shown in Fig. 14-21. The synchronous filter is indicated by the dashed lines which enclose a phase-locked loop designed to follow the instantaneous excursions in the phase of the angle-modulated signal ϕ_2. The mixer, i.f. amplifier, discriminator, filter 2, and the voltage-controlled oscillator VCO-2 form a frequency feedback loop for the demodulator.

Aside from the synchronous filter, the basic configuration can be considered that of a conventional FMFB demodulator[6] which compresses the wide-band FM input signal so that it can be passed through a relatively narrow bandwidth fixed-tuned filter and to a discriminator for detection. It should be noted that the configuration shown here can also be considered simply as a phase-locked loop (PLL) with frequency feedback around it.

The significant advantages of each of these techniques (FMFB and PLL) can be combined. It is important to consider the design from both the synchronous and frequency-feedback viewpoints.

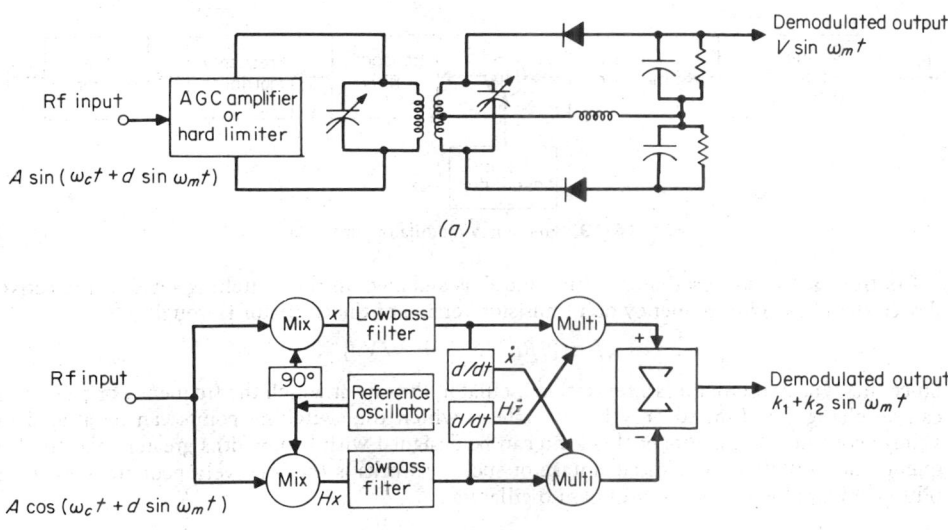

(a)

(b)

Fig. 14-20. Conventional demodulators: (a) diode discriminator; (b) phase-shift discriminator.

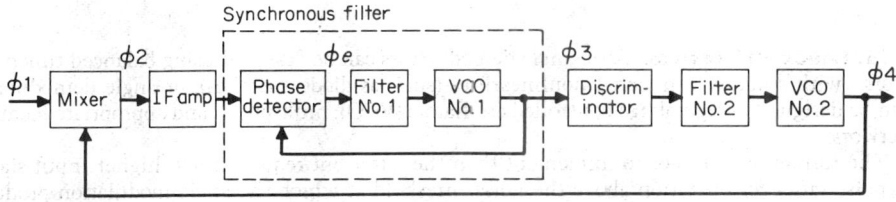

Fig. 14-21. Block diagram of basic feedback demodulator.

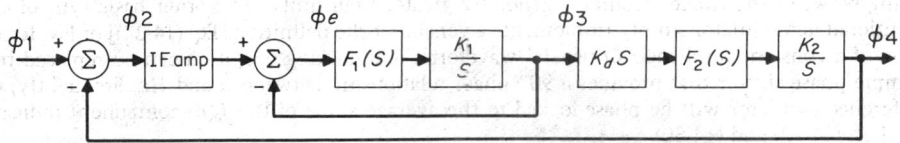

Fig. 14-22. Equivalent network of basic demodulator shown in Fig. 14-21.

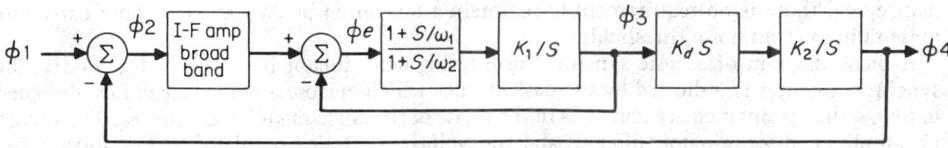

Fig. 14-23. Direct linear equivalent form of basic demodulator.

Note in this regard Figs. 14-22 and 14-23, in which $F_1(S)$ and $F_2(S)$ have been assigned. Observe that Fig. 14-24 is an FMFB equivalent linear form obtained by substituting the closed-loop transfer function for the synchronous filter; however, by retaining the inner loop and combining phase comparators, we obtain the synchronous phase-locked-loop form, as shown in Fig. 14-25. To the extent that linear analysis and quasi-linear substitutions can be made in these block diagrams, the remarks which follow are thus relevant to both PLL and FMFB forms implemented with synchronous filters or with broadband amplifiers in cascade with single-tuned filters.

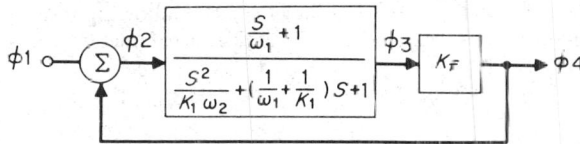

Fig. 14-24. FMFB demodulator.

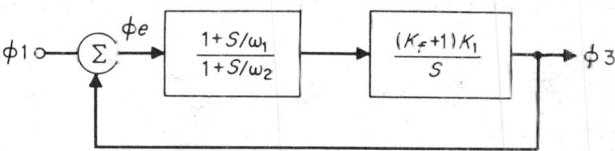

Fig. 14-25. PLL demodulator.

23. Phase-Locked-Loop Demodulators. The introduction of the phase-locked loop between the i.f. amplifier and discriminator may be viewed simply as a means by which the i.f. signal can be tracked, limited, and filtered regardless of Doppler shift or oscillator drift. The synchronous filter is employed in conjunction with the relatively wide-bandwidth i.f. amplifier shown on the block diagram to perform this critical filtering function, as well as to take advantage of the phase coherence between the FM signal sidebands.

The operation of the demodulator can be understood from an examination of the equivalent network shown in Fig. 14-22, in which the transfer functions relate to phase as the input and output variables. If the bandwidth of the i.f. amplifier is broad by comparison with both the synchronous-filter bandwidth and significant modulation sidebands, it can be ignored in the equivalent linear representation of the demodulator. Since the synchronous-filter function is

$$\frac{\phi_3(s)}{\phi_2(s)} = \frac{(K_1/s)F_1(s)}{1 + (K_1/s)F_1(s)} \tag{14-37}$$

the signal ϕ_2 is related to the input by

$$\frac{\phi_2(s)}{\phi_1(s)} = \frac{1}{1 + K_1 K_2 K_d F_1(s)F_2(s)/\{s[1 + (K_1/s)F_1(s)]\}} \tag{14-38}$$

For frequency components of $\phi_1(s)$ lying well within the bandwidths of $F_2(s)$ and the synchronous filter, this transfer function reduces to the familiar form of a type-zero feedback network; i.e.,

$$\frac{\phi_2(s)}{\phi_1(s)} = \frac{1}{1 + K_2 K_d} = \frac{s\phi_2(s)}{s\phi_1(s)} \tag{14-39}$$

Since this is also the transfer function with respect to frequency variations, the compression of frequency excursions is evident. If

$$\phi_1(t) = (\Delta\omega/\omega_m) \sin \omega_m t = D \sin \omega_m t \tag{14-40}$$

where $\phi(t)$ represents the instantaneous variation of the phase of the input signal relative to some reference carrier phase, and if the synchronous filter follows this instantaneous variation, the effective phase excursion to the discriminator is reduced to

$$\phi_3(t) = \frac{D}{1 + K_f} \sin \omega_m t \qquad K_f = K_2 K_d \tag{14-41}$$

if ω_m is well within the passband of $F_2(s)$ and the phase-locked loop. This reduction in deviation ratio by the use of frequency feedback gain K_f suggests that an optimum gain and synchronous-filter bandwidth combination should be sought for given modulation and noise characteristics, just as the proper i.f. amplifier and K_f must be chosen in a conventional FMFB demodulator.

Figure 14-24 shows the closed-loop transfer function of the synchronous filter along with the frequency-feedback loop. It has been assumed that the i.f. amplifier bandwidth is very broad compared with the bandwidth occupied by the significant portions of the signal spectrum as it appears at i.f. frequencies. This amounts to assuming that the dispersive effect produced by the i.f. amplifier is negligible. If the phase-locked-loop gain can be considered large compared with the filter zero ($K_1 \gg \omega_1$ is usually satisfied in practice) the following relationships between the synchronous-filter and the complete feedback-demodulator parameters can be written

$$\zeta_f^2 = (K_f + 1)\zeta_\phi^2 \qquad \omega_{nf}^2 = (K_f + 1)\omega_{n_\phi}^2 \qquad B_{nf} = B_{n_\phi}\left(1 + K_f\frac{1}{1 + 1/4\zeta_\phi^2}\right) \qquad (14\text{-}42)$$

where ζ_f = damping of demodulator, ω_{nf} = natural frequency of demodulator, ζ_ϕ = damping of synchronous filter, ω_{n_ϕ} = natural frequency of synchronous filter, B_{nf} = noise bandwidth of demodulator, B_{n_ϕ} = noise bandwidth of synchronous filter, and in terms of the actual network parameters

$$\zeta_\phi^2 = \frac{K_1\omega_2}{4\omega_1^2} \qquad \omega_{n_\phi}^2 = K_1\omega_2 \qquad B_{n_\phi} = \frac{1}{2}\left(\frac{K_1\omega_2}{\omega_1} + \omega_1\right) \quad (\text{Hz}) \qquad (14\text{-}43)$$

Note that all the demodulator response variables can be controlled by simple RC adjustments. If, as previously, $\phi_1(t) = D \sin \omega_m t$,

$$\phi_2(t) = \frac{D}{K_f + 1}\left\{\sin \omega_m t + \frac{K_f\omega_m^2/\omega_{nf}^2}{[(1 - \omega_m^2/\omega_{nf}^2)^2 + (\omega_m/\omega_1)^2]^{1/2}}\sin(\omega_m t + \psi_1)\right\} \qquad (14\text{-}44)$$

where $\psi_1 = \pi - \tan^{-1}\dfrac{K_1}{\omega_1}\dfrac{1}{1 - (\omega_m/\omega_{nf})^2} \approx 90°$ $\qquad$ for $K_1/\omega_1 \gg 1$ and $(\omega_m/\omega_{nf})^2 \ll 1$

$$\phi_4(t) = \frac{K_f D}{K_f + 1}\left\{\sin \omega_m t - \frac{(\omega_m/\omega_{nf})^2 \sin(\omega_m t + \psi_1)}{[(1 - \omega_m^2/\omega_{nf}^2)^2 + (\omega_m/\omega_1)^2]^{1/2}}\right\} \qquad (14\text{-}45)$$

$$\phi_3(t) = \frac{D}{K_f + 1}\left\{\sin \omega_m t - \frac{(\omega_m/\omega_{nf})^2 \sin(\omega_m t + \psi_1)}{[(1 - \omega_m^2/\omega_{nf}^2)^2 + (\omega_m/\omega_1)^2]^{1/2}}\right\} \qquad (14\text{-}46)$$

$$\phi_e(t) = \frac{(D\omega_m^2/\omega_{nf}^2)\sin(\omega_m t + \psi_1)}{[(1 - \omega_m^2/\omega_{nf}^2)^2 + (\omega_m/\omega_1)^2]^{1/2}} \qquad (14\text{-}47)$$

These four equations describe the instantaneous time relationships which exist throughout the demodulator. Examination of them along with relationships (14-42) indicates how the demodulator parameters can be used to improve the response characteristics of the demodulator and to keep the phase error in Eq. (14-47) within the proper bounds.

24. FM Feedback-Demodulator Design Formulas. *Acquisition.* The rate at which the VCO in a phase-locked demodulator can be swept through the frequency and phase at which pull-in and phase lock will occur is indicated by the representative curves[7]* of Fig. 14-26.

Pull-In. The frequency difference $\Delta\omega$ (between an unmodulated carrier and the VCO) within which a phase-locked loop will pull into phase synchronism is, roughly,[8]

$$|\Delta\omega|_p = (2pK\omega_n)^{1/2} \qquad \text{for } K/\omega_n \gg 1$$

where K = total open-loop gain (rad/s), ω_n = loop natural frequency (rad/s), and p = dimensionless constant ≈ 1.

The time required for pull-in is given by[9]

$$T = 4(\Delta f)^2/E_n^3 \qquad (\text{seconds})$$

for loop damping of 0.5 and $|\Delta f| < 0.8|\Delta f|_p$, where Δf = difference frequency (Hz) and B_n = closed-loop noise bandwidth (Hz).

*Superior numbers correspond to numbered references, Par. **14-25**.

Stability.   The condition of sustained beat-note stability without the loop capacity to be swept into lock, which is exhibited by high-gain narrow-bandwidth phase-locked demodulators, is a condition of loop oscillation arising from the presence of the phase detector nonlinearity and extraneous memory, such as that in the tuned circuits in phase detectors and in voltage-controlled oscillators. This condition can be predicted from the approximate condition

$$\frac{K^2 G(\omega_0)|}{2|\omega_0|^2} [\cos \phi(\omega_0)] + 1 = 0$$

where K = total open-loop gain, G = complete transfer function between phase detector and VCO, and ϕ = angle of G. This condition can be used to predict the nonlinear network oscillations preventing lockup.

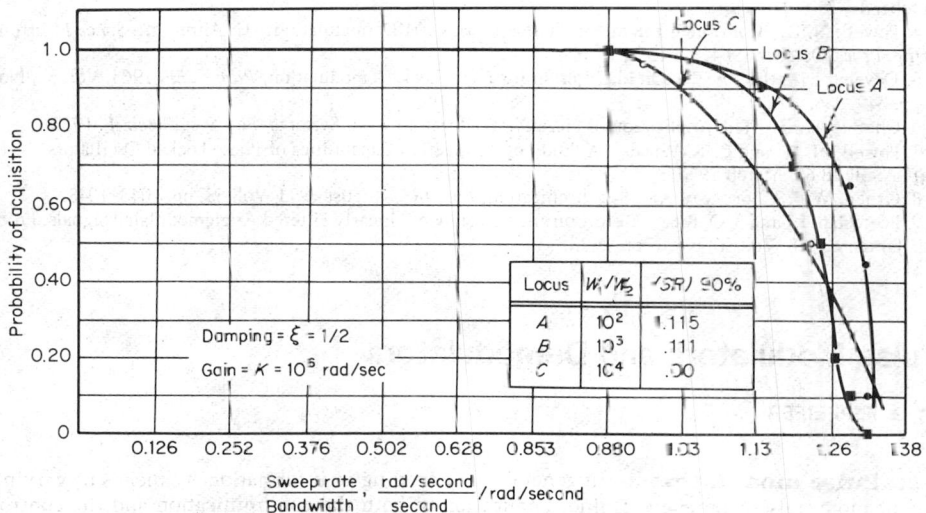

Fig. 14-26. Probability of acquisition vs. sweep-rate per unit bandwidth.

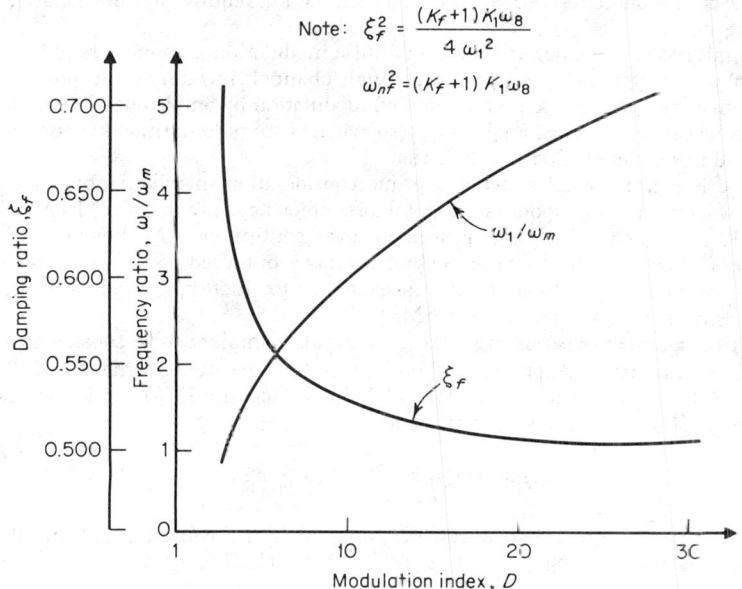

Fig. 14-27. Parametric relationships for minimum threshold (voice).

Parameter Variations.   The rate at which second-order demodulators damped for minimum-noise bandwidth can have the loop natural frequency changed at constant damping is given by[4]

$$\frac{d\omega_n/dt}{\omega_n^2} \leq 0.1$$

for peak phase errors less than $5°$, where ω_n is the natural frequency of the loop.

Minimum-Threshold Parameters.   The parametric relationships for minimum-noise threshold for the demodulator in Fig. 14-23 are shown in Fig. 14-27.

25. References on Angle Modulators and Demodulators

1. Bedrosian, E.   The Analytic Signal Representation of Modulated Waveforms, *Proc. IRE*, October 1962, Vol. 50, p. 2071.

2. Janhke, E., and F. Emde   "Tables of Functions with Formulae and Curves," Dover, New York.

3. Corrington, M. S.   Variation of Bandwidth with Modulation Index in Frequency Modulation, *Proc. IRE*, October 1947.

4. Powell, N. R.   Controlled Parameter Phase-Feedback FM Demodulation, *7th Annu. Int. Space Electron. Symp. Proc.*, October 1964, 6-b-1.

5. Develet, J. A., Jr.   A Threshold Criterion for Phase-Lock Demodulation, *Proc. IEEE*, 1963, Vol. 51, No. 2, pp. 349–356.

6. Chaffee, H. C.   The Application of Negative Feedback to F. M. Systems, *Bell Syst. Tech. J.*, 1939.

7. Powell, N. R., and C. R. Woods   A Study of Acquisition Capabilities of Phase-Locked Oscillators, G. E. *Rep. ASER 28-60*, March 1960.

8. Gruen, W. J.   Theory of AFC Synchronization, *Proc IRE*, August 1953, Vol. 53, pp. 1043–1048.

9. Bedrosian, E., and S. O. Rice   Distortion and Crosstalk of Linearly Filtered Anglemodulated Signals, *Proc. IEE*, January 1968, Vol. 56, No. 1, pp. 2–13.

Pulse Modulators and Demodulators

BY G. F. PFEIFER

26. Pulse modulation is, in general, the encoding of information by means of varying one or more pulse parameters. It finds application in both the communication and the control fields. The control applications are usually confined to the use of pulse-time modulation (PTM) and pulse-frequency modulation (PFM), where on-off control power can be used to minimize device dissipation. All pulse modulation schemes require sampling analog signals, and some, such as pulse-code modulation (PCM) and delta modulation, require the additional quantization of the analog signals.

In communications, the chief application of pulse modulation is found where it is desired to time-multiplex by interleaving a number of single-channel, low-duty-cycle pulse trains. The pulse trains may, in turn, be used for compound modulation by amplitude or angle modulation of a continuous carrier. In usual applications, subcarriers are pulsed, time-division-multiplexed, and then used to frequency-modulate a carrier.

Since noise is present in all systems, a prime consideration in modulation selection is the choice of a waveform based upon its signal-to-noise efficiency. For instance, PTM is more efficient than PAM, which offers no improvement over continuous AM; however, PTM is less efficient than PCM or delta modulation. A chief advantage of pulsed systems such as PTM, PCM, and delta is improved signal-to-noise ratio in exchange for increased bandwidth, in the same manner as continuous FM improves over AM.

27. Sampling and Smoothing. An ideal impulse sampler can be considered as the multiplication of an impulse train, period T seconds, with the continuous signal $f(t)$. Following the notation[1] of Ref. 6,† this is shown in Fig. 14-28a for the impulse train defined as $p_T(t) = \text{rep}_T[\delta(t)]$ where $\delta(t)$ is an impulse at $t = 0$ and

$$\text{rep}_T[u(t)] = \sum_{n=-\infty}^{\infty} u(t - nT) \tag{14-48}$$

The output spectrum function is the convolution of $F(f)$, the Fourier transform of the input signal, and the transform of $p_T(t)$, which is $(1/T)\,\text{comb}_{1/T}(1)$, defined as

†Numbered references are listed in Par. **14-41**.

$$\text{comb}_{1/T}\,[U(f)] = \sum_{n=-\infty}^{\infty} U\left(\frac{n}{T}\right) \delta\left(f - \frac{n}{T}\right) \qquad (14\text{-}49)$$

Thus the transform $R(f) = F(f) * (1/T)\,\text{comb}_{1/T}\,(1)$ with spectrum is as shown in Fig. 14-28b. The result of ideal impulse sampling has been to repeat the original signal spectrum, assumed to be band-limited, each $1/T$ Hz and multiply each by a $1/T$ scale factor. Since all the signal information is present in each lobe of Fig. 14-28b, it is only necessary to recover a single lobe through filtering in order to recover the signal function reduced by a scale factor.

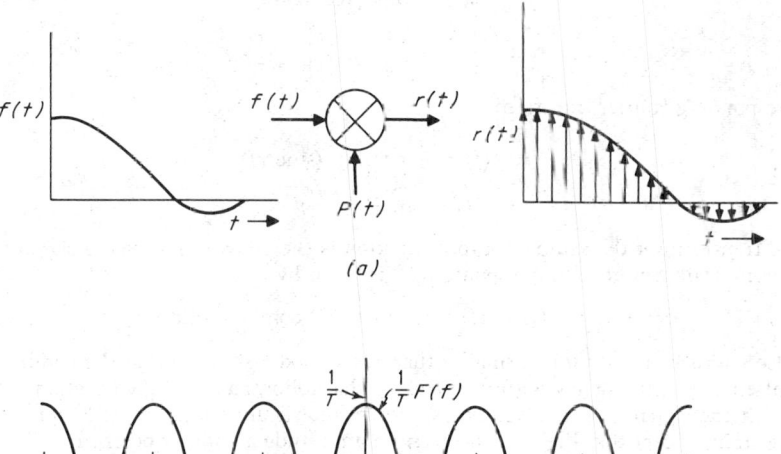

Fig. 14-28. Pulse modulation: (a) output spectrum; (b) sampling configuration.

Consider an ideal low-pass rectangular filter of bandwidth f_f Hz defined as $T(jf) = A(f)\exp[-j^\theta(jf)]$, where

$$A(f) = \begin{cases} 1 & |f| < f_f \\ 0 & |f| > f_f \end{cases} \quad \text{and} \quad \theta(jf) = 2\pi\alpha f \quad \text{for all } f$$

The cutoff frequency f_f is adjusted to select the output spectral lobe about zero $f_c < f_f < 1/T - f_c$ and will fall in the guard band between lobes. That portion of filter output $R(f)$ selected is

$$R_0(f) = (1/T)F(f)\exp[-j2\pi\alpha f] \qquad (14\text{-}50)$$

with inverse transform

$$r_0(t) = (1/T)f(t - \alpha) \qquad (14\text{-}51)$$

which is identical with the signal function, with the amplitude reduced by a scale factor and function shifted by α seconds. If $\alpha = 0$, signifying no delay, the filter is termed a "cardinal data hold"; otherwise, it is an "ideal low-pass filter." Unfortunately, these filters cannot be realized in practice, since they are required to respond before they are excited.[2] Examination of Fig. 14-28b gives rise to the sampling theorem accredited to Shannon and/or Nyquist which states that, when a continuous time function with band-limited spectrum $-f_c < f < f_c$ is sampled at twice the highest frequency, $f_s = 2f_c$, the original time function can be recovered. This corresponds to the point where the sampling frequency $f_s = 1/T$ is decreased so that the spectral lobes of Fig. 14-28b are just touching. To decrease f_s beyond the value of $2f_c$ would cause spectral overlap and make recovery with an ideal filter impossible. A more general form of the sampling theorem states that any $2f_c$ independent samples per second will completely describe a band-limited signal, thus removing the restriction of uniform sampling, as long as independent samples are used.[2] In general, for a time-limited signal of T seconds band-limited to f_c Hz, only $2f_cT$ samples are needed to specify the signal completely.

In practice, the signal is not completely band-limited, so that it is common to allow for a greater separation of spectral lobes, called the *guard band*. This guard band is generated simply by sampling at greater than $2f_c$, as in the case for Fig. 14-28b. Although the actual tolerable overlap depends on the signal spectral slope, setting the sampling rate at about $3f_c = f_s$ is usually adequate to recover the signal.

In practice, narrow but finite-width pulse trains are used in place of the idealized impulse sampling train. To determine the effect of finite-width pulses, consider the pulse train made up of τ duration pulses repeating at a $1/T$ rate represented by

$$s(t) = \text{rep}_T [\text{rect } (t/\tau)] \tag{14-52}$$

where

$$\text{rect} \left(\frac{t}{\tau} \right) = \begin{cases} 1 & |t| < \tau/2 \\ 0 & |t| > \tau/2 \end{cases}$$

The corresponding Fourier transform is

$$S(f) = \frac{\tau}{T} \text{comb}_{1/\tau} (\text{sinc } \tau f) \tag{14-53}$$

where

$$\text{sinc } \tau f = (\sin \pi \tau f) / \pi \tau f$$

Since the transform of the sampler output function is the convolution of the signal transform and the pulse train, the resulting response $R_s(f)$ is given by

$$R_s(f) = F(f) * S(f) = F(f) * (\tau/T) \text{comb}_{1/T} (\text{sinc } \tau f) \tag{14-54}$$

The pulse width is usually much smaller than the period $\tau \ll T$, so that the comb function is an impulse train in frequency with an envelope that follows a $(\sin x)/x$ function. The convolution with the signal $F(f)$ thus yields a spectrum that is almost the same as for the impulse sampling train, except that the lobes decrease in amplitude as sinc τf determined by the pulse width τ. The ideal reconstruction filter can be approximated by commonly used hold circuits, the characteristics of which are shown[10] in Fig. 14-29.

28. Pulse-Amplitude Modulation (PAM). Pulse-amplitude modulation is essentially a sampled-data type of encoding where the information is encoded into the amplitude of a train of finite-width pulses. The pulse train can be looked upon as the carrier in much the same way as the sine wave is for continuous-amplitude modulation. There is no improvement in signal-to-noise when using PAM, and furthermore, PAM is not considered wide-band in the sense of FM or pulse-time modulation (PTM). Thus PAM would correspond to continuous AM, while PTM corresponds to FM. Generally, PAM is used chiefly for time-multiplex systems employing a number of channels sampled, consistent with the sampling theorem.

There are a number of ways of encoding information as the amplitude of a pulse train. They include both bipolar and unipolar pulse trains for both instantaneous or square-topped sampling and for exact or top sampling. In top sampling, the magnitude of the individual pulses follows the modulating signal during the pulse duration, while for square-topped sampling, the individual pulses assume a constant value, depending on the particular exact sampling point that occurs somewhere during the pulse time. These various waveforms are shown in Fig. 14-30.

The spectrum for the top-modulation bipolar sampling case is given by Eq. (14-54), since this type of modulation, shown in Fig. 14-30c, is simply sampling with a finite-pulse-width train. Carrying out the convolution indicated yields

$$R_{\text{STB}}(f) = \frac{\tau}{T} \sum_{n=-\infty}^{\infty} \left(\text{sinc } \frac{\tau n}{T} \right) F \left(f - \frac{n}{T} \right) \tag{14-55}$$

The spectrum for top-modulation bipolar sampling, using a square-topped rectangular spectrum for the original signal spectrum, is shown in Fig. 14-31a. The signal spectrum repeats with a $(\sin x)/x$ scale factor determined by the sampling pulse width, with each repetition a replica of $F(f)$.

Unipolar sampling can be implemented by adding a constant bias A to $f(t)$, the signal, to produce $f(t) + A$, where A is large enough to keep the sum positive; that is, $A > |f(t)|$. Sampling the new sum signal by multiplication with the pulse train results in the unipolar top-modulated waveform of Fig. 14-30e. The spectrum can be found by substituting the new signal into Eq. (14-54) and results in

$$R_{\text{STU}}(f) = \frac{\tau}{T} \sum_{n=-\infty}^{\infty} \left(\text{sinc } \frac{\tau n}{T} \right) \left[F \left(f - \frac{n}{T} \right) + A\delta \left(f - \frac{n}{T} \right) \right] \tag{14-56}$$

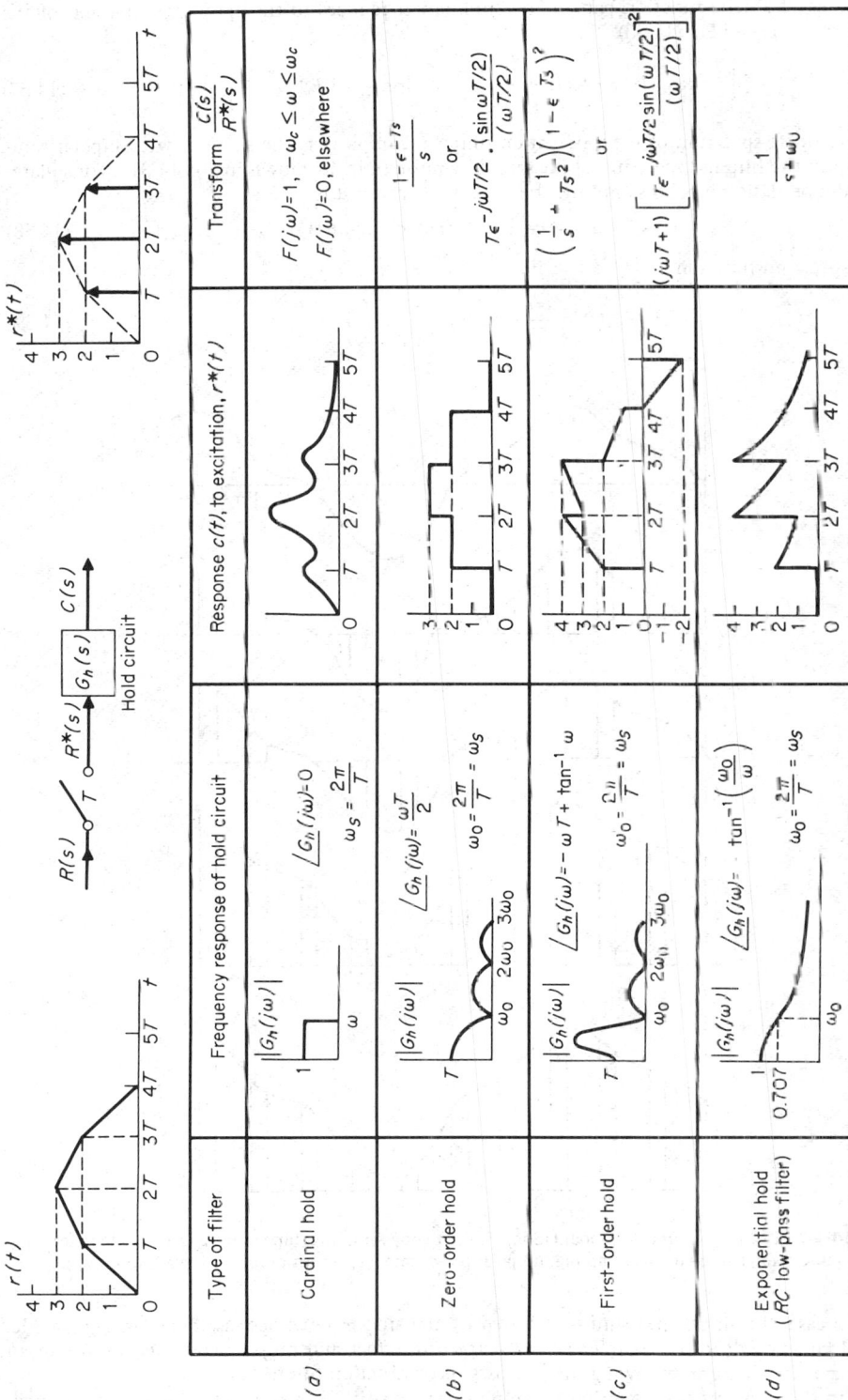

Fig. 14-29. Characteristics of various filter circuits.[10]

However, the delta-function part of the summation reduces to the spectrum function of the pulse train $S(f)$ by Eq. (14-53). Thus

$$R_{STU}(f) = AS(f) + \frac{\tau}{T} \sum_{n=-\infty}^{\infty} \left(\text{sinc} \frac{\tau n}{T} \right) F \left(f - \frac{n}{T} \right)$$ (14-57)

The resulting spectrum of top-modulation unipolar sampling is the same as with bipolar sampling plus the impulse spectrum of the sampling pulse train, as shown in Fig. 14-31b. For square-topped-modulation bipolar sampling, the time-domain result is

$$r_{SSB}(t) = \text{rect} \, (t/\tau) * \text{comb}_T \, f(t)$$ (14-58)

with spectrum function

$$R_{SSB}(f) = \frac{\tau}{T} (\text{sinc} \, f\tau) \sum_{n=-\infty}^{\infty} F \left(f - \frac{n}{T} \right)$$ (14-59)

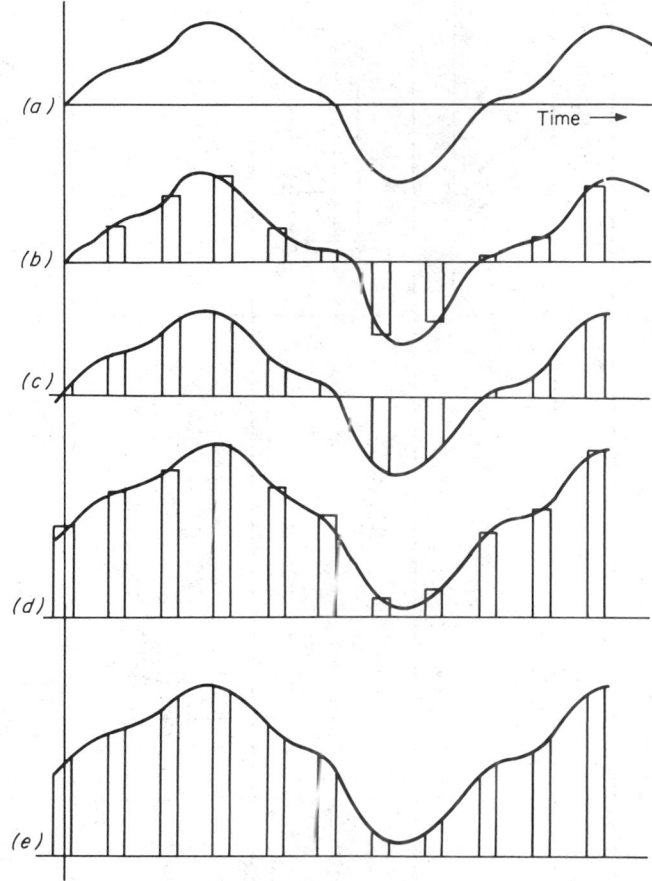

Fig. 14-30. PAM waveforms: (a) modulation; (b) square-top sampling, bipole pulse train; (c) top sampling, bipole pulse train; (d) square-top sampling, unipolar pulse train; (e) top sampling, unipolar pulse train.

In this case, the signal spectrum is distorted by the sinc $f\tau$ envelope, as shown in Fig. 14-30c. This frequency distortion is referred to as *aperture effect* and may be corrected by use of an equalizer sinc $f\tau$ form, following the low-pass reconstruction filter.

As in the previous case of unipolar sampling, the resulting spectrum for square-topped modulation will contain the pulse-train spectrum, as shown in Fig. 14-30d. The expression is

$$R_{SSU}(f) = AS(f) + \frac{\tau}{T} (\text{sinc} \, f\tau) \sum_{n=-\infty}^{\infty} F \left(f - \frac{n}{T} \right)$$ (14-60)

The signal information is generally recovered, in PAM systems, by use of a low-pass filter which acts on the reduced signal energy around zero frequency, as shown in Fig. 14-31.

29. Pulse-Time (PTM), Pulse-Position (PPM), and Pulse-Width (PWM) Modulation. In PTM the information is encoded into the time parameter instead of, for instance, the amplitude, as in PAM. There are two basic types of PTM: pulse-position modulation (PPM) and pulse-width modualtion (PWM), also known as pulse-duration (PDM) or pulse-length (PLM) modulation. The PTM allows the power-driver circuitry to operate at saturation

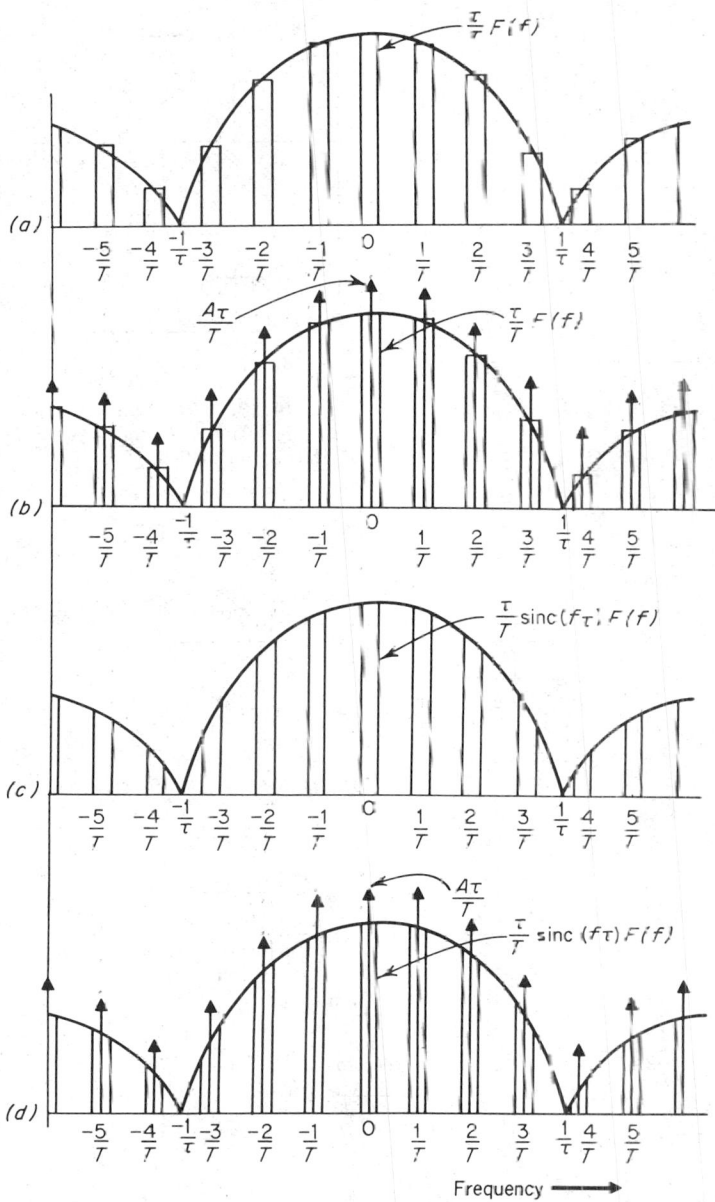

Fig. 14-31. PAM spectra: (*a*) top modulation, bipolar sampling; (*b*) top modulation, unipolar sampling; (*c*) square-top modulation, bipolar sampling (*d*) square-top modulation, unipolar sampling.

level, thus conserving power loss. Operating driver circuitry full on, full off, is especially important for heavy-duty high-load control applications, as well as for communication applications.

In PPM the information is encoded into the time position of a narrow pulse, generally with respect to a reference pulse. The basic pulse width and amplitude are kept constant, while only

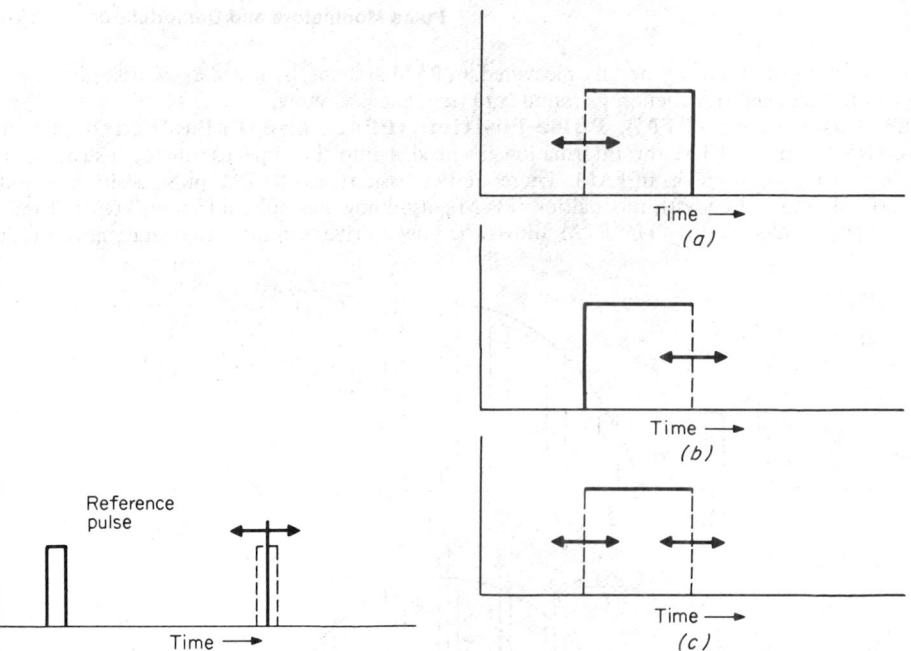

Fig. 14-32. PPM time waveform.

Fig. 14-33. PWM time waveforms: (*a*) leading-edge modulation; (*b*) trailing-edge modulation; (*c*) both-edge modulation.

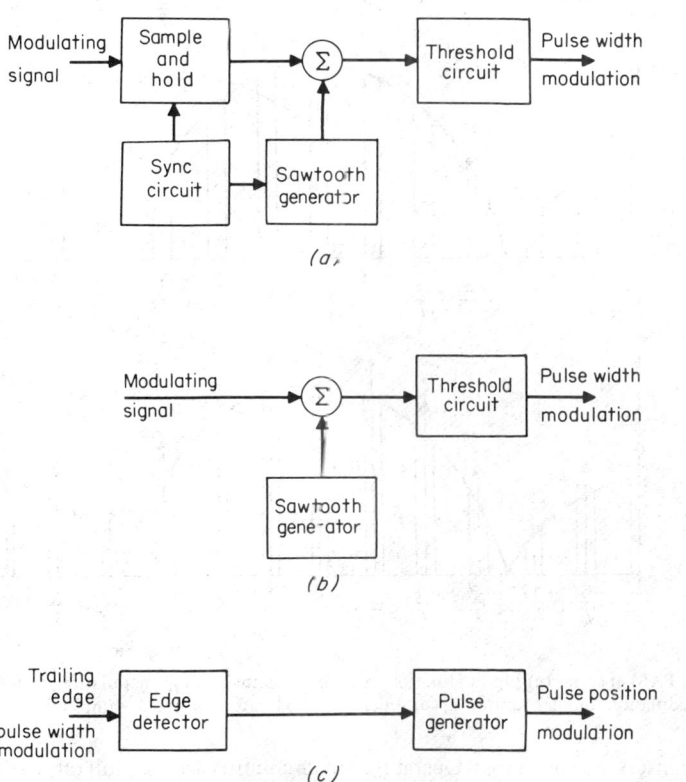

Fig. 14-34. PTM generation: (*a*) pulse-width-modulation generation, uniform sampling; (*b* pulse-width-modulation generation, nonuniform sampling; (*c*) pulse-position-modulation generation.

the pulse position is changed, as shown in Fig. 14-32. There are three cases of PWM which are the modulation of the leading edge, trailing edge, or both edges, as displayed in Fig. 14-33. In this case the information is encoded into the width of the pulse, with the pulse amplitude and period held constant. The derivative relationship existing between PPM and PWM can be illustrated by consideration of trailing-edge PWM modulation. The pulses of PPM can be derived from the edges of trailing-edge PWM (Fig 14-33b) by differentiation of the PWM signal and a sign change of the trailing-edge pulse. Pulse-position modulation is essentially the same as PWM, with the information-carrying variable edge replaced by a pulse. Thus, when that part of the signal power of PWM that carries no information is deleted, the result is PPM.

Generally, in PTM systems a guard interval is necessary due to the pulse rise times and system responses. Thus 100% of the interpulse period cannot be used without considerable channel crosstalk due to pulse overlap. It is necessary to trade off crosstalk vs. channel utilization at the system design level. Another consideration is that the information sampling rate cannot exceed the pulse repetition frequency and would be less for a single channel of a multiplexed system where channels are interwoven in time.

Generation of PTM. There are two basic methods of pulse-time modulation: (1) based on uniform sampling in which the pulse-time parameter is directly proportional to the modulating signal at uniformly sampled points and (2) in which there is some distortion of the pulse-time parameter due to the modulation process. Both methods of modulation are illustrated in Fig. 14-34 for PWM. Basically, PPM can be derived from trailing-edge PWM, as shown in Fig. 14-34c, by use of an edge detector or differentiator and a standard narrow-pulse generator.

In the uniform sampling case for PWM of Fig. 14-34a, the modulating signal is sampled uniformly in time and the special PAM derived by a sample-and-hold circuit as shown in Fig. 14-35a. This PAM signal provides a pedestal for each of the three types of sawtooth waveforms producing leading, trailing, or double-edge PWM, as shown in Fig. 14-35c, e, and g, respectively. The uniform sampled PPM is shown in Fig. 14-35h as derived from the trailing-edge modulation of g.

Nonuniformly sampled modulation, termed *natural sampling* by some authors, is shown in Fig. 14-36, and results from the method of Fig. 14-34b, where the sawtooth is added directly to the modulating signal. In this case the modulating waveform influences the time when the samples are actually taken. This distortion is small when the modulating-amplitude change is small during the interpulse period T. The distortion is caused by the modulating signal distorting the sawtooth waveform when they are added, as indicated in Fig. 14-34b. The information in the PPM waveform is similarly distorted because it is derived from the PWM waveform, as shown in Fig. 14-35h.

30. Pulse-Time Modulation Spectra. The spectra are smeared in general, for most modulating signals, and are difficult to derive; however, some idea of what happens to the spectra with modulation is possible by considering a sinusoidal modulation of form

$$A \cos 2\pi f_s t \tag{14-61}$$

The amplitude $A < T/2$, where T is the interpulse period, assuming no guard band.

For PPM with uniform sampling and unity pulse amplitude, the spectrum is given by

$$
\begin{aligned}
x(t) = \frac{\tau}{T} &+ \frac{2\tau}{T} \sum_{m=1}^{\infty} (\text{sinc } mf_0) J_c(2\pi A m f_0) \cos 2\pi m f_c t \\
&+ \frac{2\tau}{T} \sum_{n=1}^{\infty} \text{sinc } (nf_s) J_n(2\pi A n f_s) \cos \left(2\pi n f_s t - \frac{n\pi}{2} \right) \\
&+ \frac{2\tau}{T} \sum_{m=1}^{\infty} \sum_{n=1}^{\infty} \left\{ \text{sinc } (mf_0 + nf_s) J_n[2\pi A(mf_0 - nf_s)] \cos \left[2\pi(mf_0 + nf_s)t - \frac{n\pi}{2} \right] \right. \\
&+ \left. \text{sinc } (nf_s - mf_c) J_n [2\pi A(nf_s - mf_0)] \cos \left[2\pi(nf_s - mf_0)t - \frac{n\pi}{2} \right] \right\} \tag{14-62}
\end{aligned}
$$

where τ = pulse width, T = pulse period, f_s = modulation frequency, J_n = Bessel function of first kind, nth order, and $f_0 = 1/T$.

As is apparent, all the harmonics of the pulse-repetition frequency and the modulation frequency are present, as well as all possible sums and differences. The dc level is τ/T, with the harmonics carrying the modulation. The pulse shape affects the line amplitudes as a sinc function, reducing the spectra for higher frequencies.

The spectrum for PWM is similar to that of PPM, and for uniformly sampled trailing-edge sinusoidal modulation is given by

$$x(t) = \frac{1}{2} + \frac{1}{\pi T} \sum_{m=1}^{\infty} \frac{1}{mf_0} \cos \left[2\pi mf_0 t + \frac{\pi}{2}(2m-1) \right]$$

$$+ \frac{1}{\pi T} \sum_{m=1}^{\infty} \frac{1}{mf_0} J_0(2\pi A m f_0) \cos \left(2\pi m f_0 t - \frac{\pi}{2} \right)$$

$$+ \frac{1}{\pi T} \sum_{n=1}^{\infty} \frac{1}{nf_s} J_n(2\pi A n f_s) \cos \left[2\pi n f_s t - (n+1)\frac{\pi}{2} \right]$$

$$+ \frac{1}{\pi T} \sum_{\substack{m=1 \\ n=1}}^{\infty} \left\{ \frac{1}{mf_0 + nf_s} J_n[2\pi A(mf_0 + nf_s)] \cos \left[2\pi(mf_0 + nf_s)t - (n+1)\frac{\pi}{2} \right] \right.$$

$$\left. + \frac{1}{nf_s - mf_0} J_n[2\pi A(nf_s - mf_0)] \cos \left[2\pi(nf_s - mf_0)t - (n+1)\frac{\pi}{2} \right] \right\} \qquad (14\text{-}63)$$

The same comments apply for PWM as for PPM.

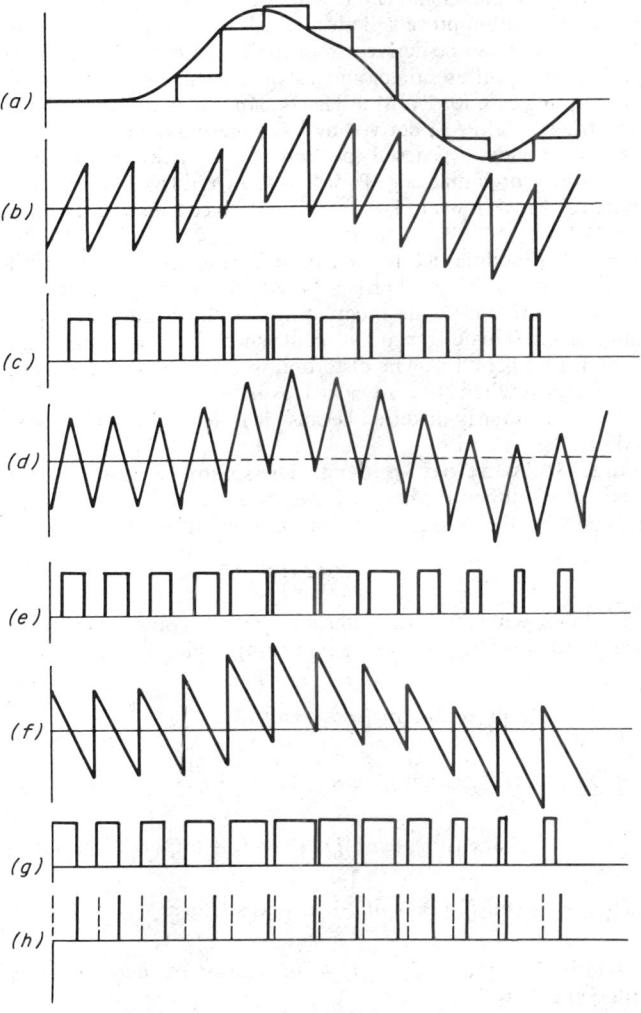

Fig. 14-35. Pulse-time modulation, uniform sampling: (a) modulating signal and sample-and-hold waveform; (b) sawtooth added to sample-and-hold waveform; (c) leading-edge modulation; (d) sawtooth added to sample-and-hold waveform; (e) double-edge modulation; (f) sawtooth added to sample-and-hold waveform; (g) trailing-edge modulation; (h) pulse-position modulation (reference pulse dotted).

A more compact form is given in Ref. 7 (Par. **14-41**) for PPM and PWM, respectively, as

$$x(t) = \frac{1}{T} \sum_{\substack{m=\infty \\ n=\infty}} (-j)^n J_n[2\pi A(mf_0 + nf_s)]P(mf_0 + nf_s)\exp[j2\pi(mf_0 + nf_s)t] \quad (14\text{-}64)$$

where $P(f)$ is Fourier transform of the pulse shape $p(t)$, and

$$x(t) = \frac{1}{2} + \frac{1}{T} \sum_{\substack{x=-\infty \\ m \neq 0}}^{\infty} j^{2m-1} \frac{e^{j2\pi mf_0 t}}{2\pi mf_0} - \frac{1}{T} \sum_{\substack{n=-\infty \\ n=-\infty \\ |m| + |n| \neq 0}}^{\infty}$$

$$(-j)^{n+1} \frac{J_n[2\pi A(mf_0 + nf_s)]}{2\pi(mf_0 + nf_s)}\exp[j2\pi(mf_0 + nf_s t)] \quad (14\text{-}65)$$

31. Demodulation of PTM. Demodulation of PWM or PPM can be accomplished by low-pass filtering if the modulation is small compared with the interpulse period. However, in general, it is best to demodulate on a pulse-to-pulse basis that usually requires some form of synchronization with the pulses. The distortion introduced by nonuniform sampling cannot be eliminated and will be present in the demodulated waveform. However, if the modulation is small compared with the interpulse period T, the distortion will be minimized.

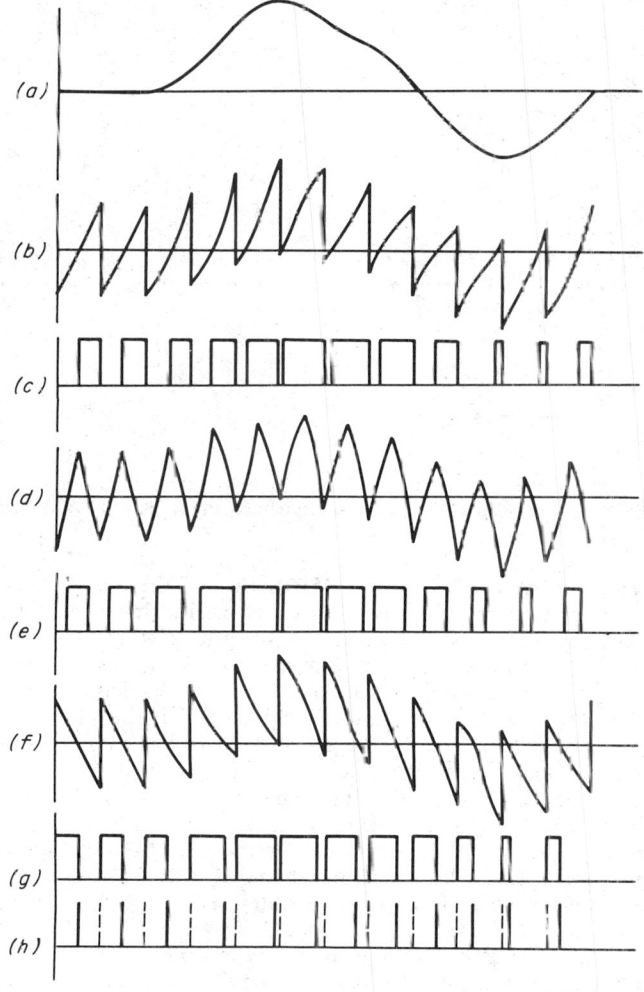

Fig. 14-36. Pulse-time modulation, nonuniform sampling. (a) modulating signal; (b) sawtooth added to modulation; (c) leading-edge modulation; (d) sawtooth added to modulation; (e) double-edge modulation; (f) sawtooth added to modulation; (g) trailing-edge modulation; (h) pulse-position modulation.

To demodulate PWM each pulse can be integrated and the maximum value sampled and held and low-pass-filtered, as shown in Fig. 14-37a. To sample and reset the integrator, it is necessary to derive sync from the PWM waveform, in this case trailing-edge-modulated.

Generally, PPM is demodulated by conversion to PWM and then demodulated as PWM. Although in some demodulation schemes the actual PWM waveform may not exist as such, the general demodulation scheme is the same. PPM can be converted to PWM by the configuration of Fig. 14-37b. The PPM signal is applied to an amplitude threshold, usually termed a *slicer*, that rejects noise except near the pulses. The pulses are applied to a flip-flop synchronized to one particular state by the reference pulse, and it generates the PWM as its output. More detailed information on PTM is available.[1–5,7,8]

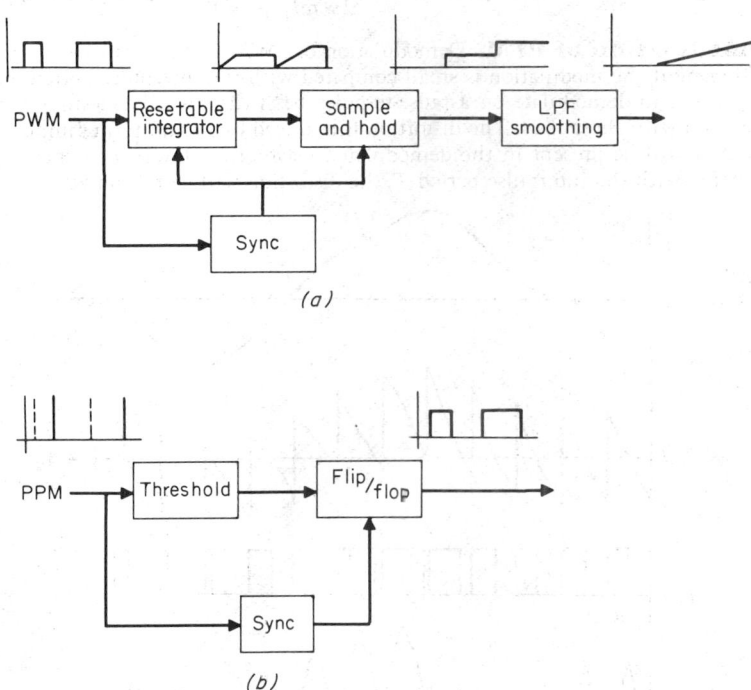

Fig. 14-37. Pulse-time demodulation: (a) PWM demodulation; (b) PPM to PWM for demodulation.

32. Pulse Frequency Modulation (PFM). In PFM the information is contained in the frequency of the pulse train, which is composed of narrow pulses. The highest frequency possible ideally occurs when there is no more interpulse spacing left for finite-width pulses. This frequency, given by $1/\tau$, where τ is the pulse width, will not be achieved in practice, due to the pulse rise time. The lowest frequency is determined by the modulator, usually a voltage-controlled oscillator (VCO), in which in practice a 100:1 ratio of high to low frequency is easily achievable. Examination of Fig. 14-38 indicates why PFM is used mostly for control purposes rather than communications. The wide variation and uncertainty of pulse position do not lend themselves to time multiplexing, which requires the interweaving of channels in time. Since one of the chief motivations of pulse modulation in communication systems is to be able to time-multiplex a number of channels, PFM is not used. On the other hand, PFM is a good choice for on-off control applications, especially where fine control is required. A classic example of PFM control is for the attitude control of near-earth satellites that have on-off gas thrusters where a very close approximation to a linear system response is achievable.

Generation of PFM. Basically, PFM is generated by modulation of a VCO as shown in Fig. 14-39a. A constant reference voltage is added to the modulation so that the frequency can swing above and below the reference-determined value. For control applications it is usually required that the frequency follow the magnitude of the modulation, its sign determining which actuators are to be turned on, as shown in Fig. 14-39b.

Spectrum for PFM. The spectrum can be determined from consideration of the following expression,[1] for the case of sinusoidal modulation of the form $\beta \sin(\omega_m t + \phi)$, where $\beta = \Delta\omega/\omega_m$ and the pulse train is of amplitude A with τ width pulses and spacing T:

$$x(t) = \frac{A\tau}{T} + \sum_{k=1}^{\infty}\left(\text{sinc}\,\frac{\kappa\tau}{T}\right)\left(J_0(k\beta)\cos\frac{2\pi kt}{T} + \sum_{n=1}^{\infty}J_n(k\beta)\left|\cos\left[\left(\frac{2\pi k}{T} + 2\pi nf_m\right)t\right.\right.\right.$$

$$\left.\left.\left. + n\phi\right] + (-1)^n\cos\left[\left(\frac{2\pi k}{T} - 2\pi nf_m\right)t - n\phi\right]\right|\right) \qquad (14\text{-}66)$$

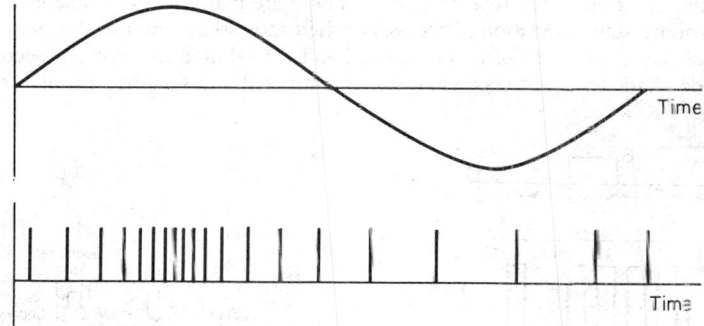

Fig. 14-38. PFM modulation.

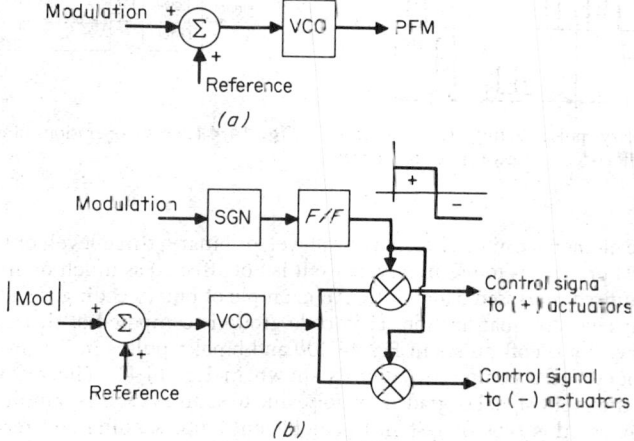

Fig. 14-39. Generation of PFM: (*a*) PFM modulation; (*b*) PFM for control.

The kth harmonic of the pulse repetition frequency is frequency-modulated with modulation index $k\beta$. Hence it could be demodulated with no harmonic distortion by a bandpass filter set to extract one of the harmonics and a frequency discriminator. Also note that the spectral amplitude decreases as a sinc function due to its relationship with a rectangular pulse.

33. Pulse-Code Modulation (PCM). In PCM the signal is encoded into a stream of digits. This differs from the other forms of pulse modulation by requiring that the sample values of the signal be quantized into a number of levels and subsequently coded as a series of pulses for transmission. By selecting enough levels, the quantized signal can be made to approximate closely the original continuous signal at the expense of transmitting more bits per sample. The PCM scheme lends itself readily to time multiplexing of channels and will allow widely different types of signals; however, synchronization is strictly required. This synchronization of the system can be on a single-sample or code-group basis. The synchronizing signal is most likely inserted with a group of samples from different channels, on a frame or subframe basis to conserve space.

The motivation behind modern PCM is that improved implementation techniques of solid-state circuitry allow extremely fast quantization of samples and translation to complex codes with reasonable equipment constraints. PCM is an attractive way to trade bandwidth for signal-to-noise and has the additional advantage of transmission through regenerative repeaters with a signal-to-noise ratio that is substantially independent of the number of repeaters. The only requirement is that the noise, interference, and other disturbances be less than one-half a quantum step at each repeater. Also, systems can be designed that have error-detecting and error-correcting features.

34. PCM Coding and Decoding. Coding is the generation of a PCM waveform from an input signal, and decoding is the reverse process. There are many ways to code and many code groups to use; hence standardization is necessary when more than one user is considered. Each sample value of the signal waveform is quantized and represented to sufficient accuracy by an appropriate code character. Each code character is composed of a specified number of code ele-

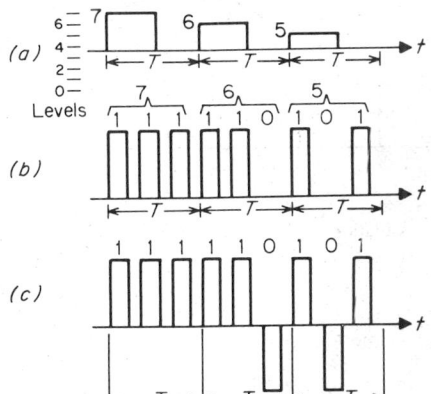

Fig. 14-40. Binary pulse coding: (a) quantized samples; (b) on-off coded pulses; (c) bipolar coded pulses.[3]

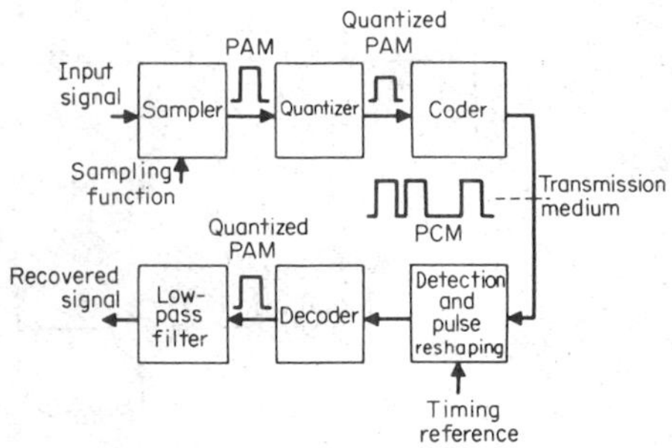

Fig. 14-41. Basic operations of a PCM system.[3]

ments. The code elements can be chosen as two-level, or binary; three-level, or ternary; or n-ary. However, general practice is to use binary, since it is not affected as much by interference introduced by the required increased bandwidth. An example of binary coding is shown in Fig. 14-40 for 3-bit or eight levels of quantization. Each code group is composed of three pulses, with the pulse trains shown for on-off pulses in Fig. 14-40b and bipolar pulses in Fig. 14-40c.

A generic diagram of a complete system is shown in Fig. 14-41. The recovered signal is a delayed copy of the input signal degraded by noise due to sources such as sampling, quantization, and interference. For this type of system to be efficient, both sending and receiving terminals must be synchronized. This synchronism is required to be monitored continuously and be capable of establishing initial synchronism when the system is out of frame. The synchronization is usually accomplished by use of special sync pulses that establish frame, subframe, or word sync.

There are three basic ways to code, namely, feedback and subtraction, pulse counting, and parallel comparison.[5] In *feedback subtraction* the sample value is compared with the most significant code-element value and that value subtracted from the sample value if the element value is less. This process of comparison and subtraction is repeated for each code-element value down to the least significant bit. At each subtraction the appropriate code element or bit is selected to complete the coding. In *pulse counting* a gate is established by using the PWM pulse corresponding to a sample value. Clock pulses are gated using the PWM gate and are counted in a counter. The output of a decoding network attached to the counter is read out as the PCM. *Parallel comparison* is the fastest method since the sampled value is applied to a number of different threshold values. The thresholders are read out as the PCM.

35. System Considerations for PCM. Quantization introduces an irremovable error into the system, referred to as *quantization noise*. This kind of noise is characterized by the fact that its magnitude is always less than one-half a quantum step, and it can be treated as uniformly distributed additive noise with zero mean value and rms value equal to $1/\sqrt{12}$ times the total height of a quantum step.[3] When the ratio of signal power to quantization noise power at the quantizer output is used as a measure of fidelity the improvement with quantizer levels is as shown in Fig. 14-42 for different kinds of signals.

In general, using an n-ary code with m pulses allows transmission of n^a values. For the binary code this reduces the 2^m values which approximate the signal to 1 part in $2^m - 1$ levels. Encoding into plus and minus pulses, assuming either pulse is equally likely, results in an average power of $A^2/4$, which is half the on-off power of $A^2/2$, where the total pulse amplitude, peak to peak,

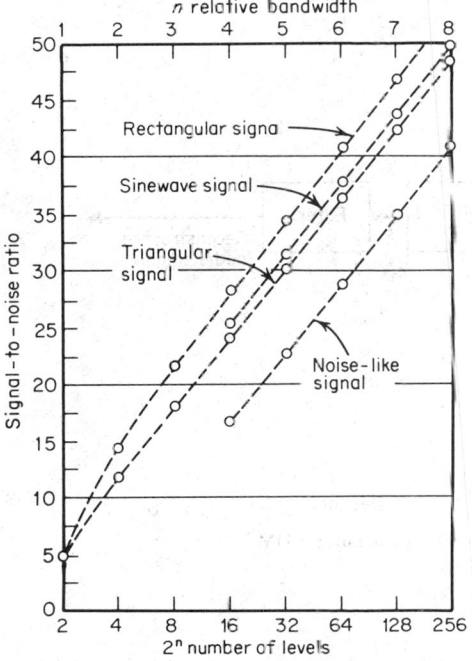

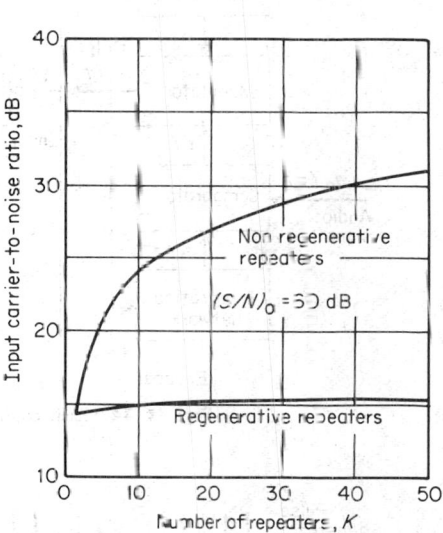

Fig. 14-42. PCM signal-to-noise improvement with number of quantization levels.[3]

Fig. 14-43. Input carrier-to-noise ratio vs. number of repeaters for constant-output signal-to-noise ratio.[1]

is A. The channel capacity for a system sampled at the Nyquist rate of $2f_m$ and quantized into s levels is

$$C = 2f_m \log_2 s \quad \text{(bits/s)} \tag{14-67}$$

or for m pulses of n values each

$$C = mf_m \log_2 n^* \quad \text{(bits/s)} \tag{14-68}$$

Since the encoding process squeezes one sample into m pulses, the pulse widths are effectively reduced by $1/m$; thus the transmission bandwidth is increased by a factor of m, or $B = mf_m$.

The maximum possible ideal rate of transmission of binary bits is

$$C = B \log_2 (1 + S/N) \quad \text{(bits/s)} \tag{14-69}$$

according to Shannon.[2] For a system sampled at the Nyquist rate, quantized to $K\sigma$ per level and using the plus and minus pulses, the channel capacity is

$$C = B \log_2 (1 + 12S/K^2N) \quad \text{(bits/s)} \quad N = \sigma^2 \tag{14-70}$$

where S = average power over large time interval and σ = rms noise voltage at decoder input.

There exists in PCM a fairly definite threshold, as indicated in Table 14-1, at about 20 dB (ratio of 9.2) peak signal pulse to rms noise voltage, where the error rate decreases quite rapidly.

TABLE 14-1 Probability of Error

S/N, dB	13	17	20	21	22	23
Probability of error	10^{-2}	10^{-4}	10^{-6}	10^{-8}	10^{-10}	10^{-12}

Using the threshold ratio of 9.2 for an average error rate of about 10^{-6} yields a PCM binary system that requires 7 times (8.5 dB) the power of the ideal binary system. Although PCM is much more efficient than uncoded systems such as FM and PPM, it is still 8.5 dB less efficient than the ideal system.

The effect of regenerative repeaters is illustrated in Fig. 14-43, where the input carrier signal-to-rms noise ratio is expressed as a function of the number of repeaters for a 60-dB output signal-to-noise. Note that for regenerative repeaters the 60-dB output is achievable relatively independently of the number of repeaters used.

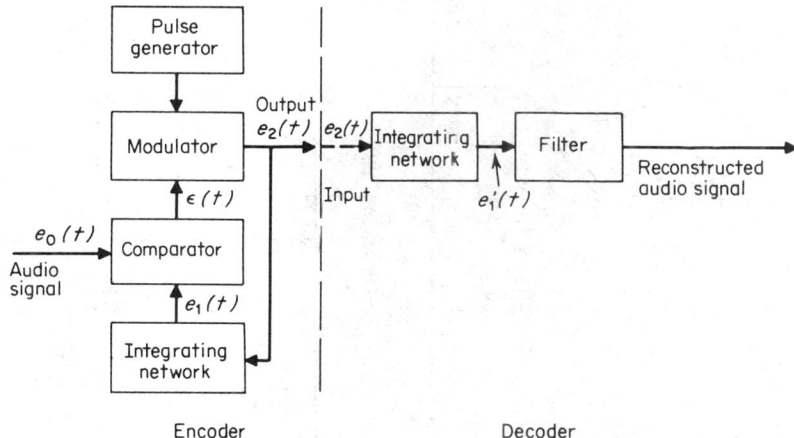

Fig. 14-44. Basic coding-decoding diagram for DM.[1]

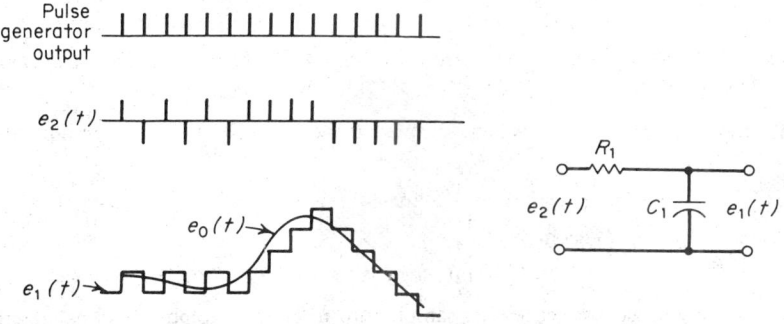

Fig. 14-45. Delta-modulation waveforms using single integration.[1]

36. Delta Modulation (DM). Delta modulation is basically a one-digit PCM system where the analog waveform has been encoded in a differential form. In contrast to the use of n digits in PCM, simple DM uses only one digit to indicate the changes in the sample values. This is equivalent to sending an approximation to the signal derivative. At the receiver the pulses are integrated to obtain the original signal. Although DM can be simply implemented in circuitry, it requires a sampling rate much higher than the Nyquist rate of $2f_m$ and a wider bandwidth than a comparable PCM system. Most of the other characteristics of PCM apply to DM.

Delta modulation differs from differential PCM in which the difference in successive signal samples is transmitted.[11] In DM only 1 bit is used to express and transmit the difference. Thus DM transmits the sign of successive slopes.

Coding and Decoding DM. There are a number of coding and decoding variations in DM, such as single-integration, double-integration, mixed-integration, delta-sigma, and high-information DM (HIDM). In addition, companding the signal which is compressing the signal at transmission and expanding it at reception is also used to extend the limited dynamic range. The simple single-integration DM of the coding-decoding scheme is shown in Fig. 14-44. In the encoder the modulator produces positive pulses when the sign of the difference signal $\epsilon(t)$ is positive and negative pulses otherwise; and the output pulse train is integrated and compared with the input signal to provide an error signal $\epsilon(t)$, thus closing the encoder feedback loop. At the receiver the pulse train is integrated and filtered to produce a delayed approximation to the signal, as shown in Fig. 14-45. The actual circuit implementation with operational amplifiers and logic circuits is very simple.

By changing to a double-integration network in the decoder, a smoother replica of the signal is provided. This decoder has the disadvantage, however, of not recognizing changes in the slope of the signal. This gave rise to a scheme to encode differences in slope instead of amplitude,

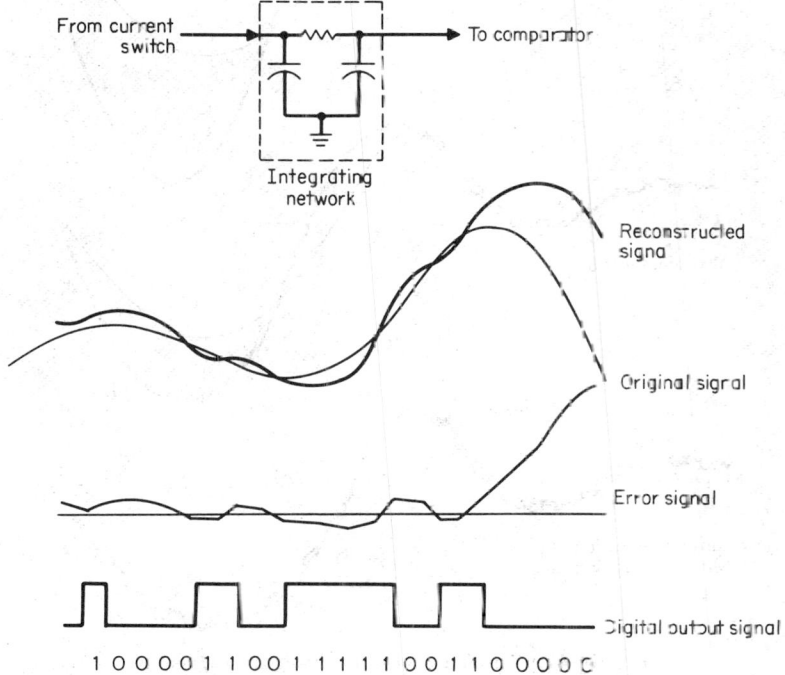

Fig. 14-46. Waveforms for delta coder with double integration.

leading to coders with double integration; however, systems of this type are marginally stable and can oscillate under certain conditions.[11] Waveforms of a double-integrating delta coder are shown in Fig. 14-46. Single and double integration can be combined to give improved performance while avoiding the stability problem. These mixed systems are often referred to in the literature as *delta modulators* with double integration.[11] A comparison of waveforms is shown in Fig. 14-47.

System Considerations for DM.   The synthesized waveform can change only one level each clock pulse; thus DM overloads when the slope of the signal is large. The maximum signal power will depend on the type of signal, since the greatest slope that can be reproduced is the integration of one level in one pulse period. For a sine wave of frequency f, the maximum-amplitude signal is

$$A_{max} = f_s \sigma / 2\pi \tag{14-71}$$

where f_s = sampling frequency and σ = one quantum step.

It has been observed that a DM system will transmit a speech signal without overloading if the amplitude of the signal does not exceed the maximum permissible amplitude of an 800-Hz sine wave. The DM coder overload characteristic is shown in Fig. 14-48 along with the spectrum

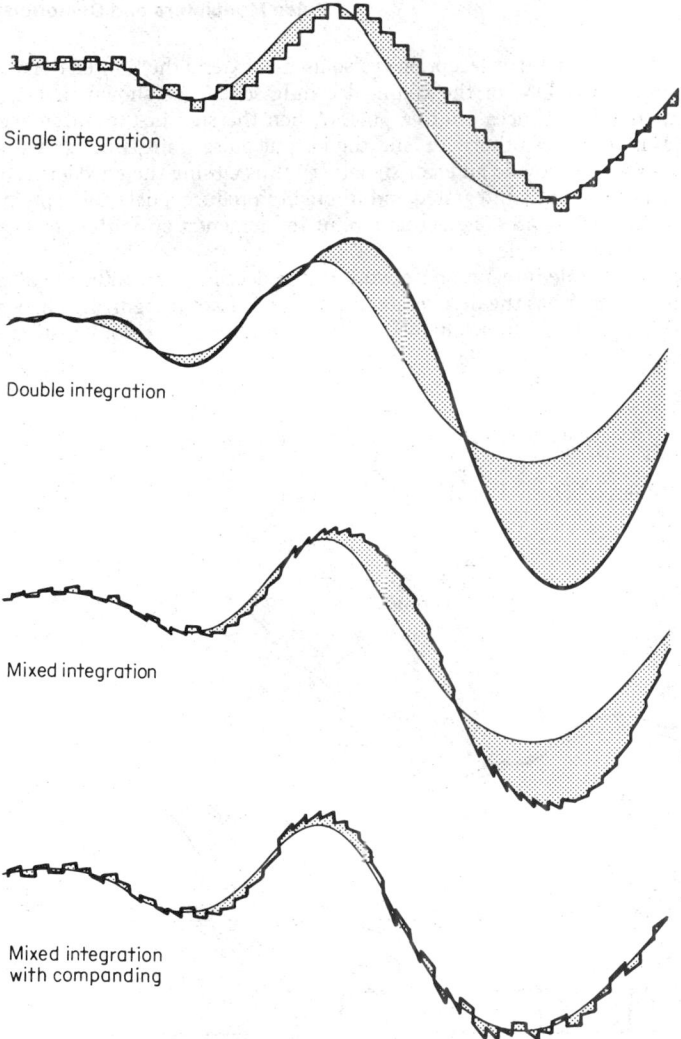

Single integration

Double integration

Mixed integration

Mixed integration
with companding

Fig. 14-47. Waveforms for various integrating systems.[11]

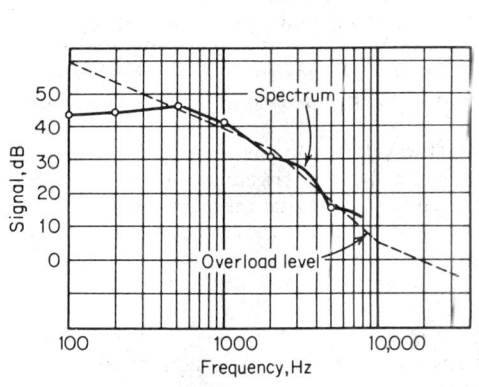

Fig. 14-48. Spectrum of the human voice compared with delta-coder overload level.[11]

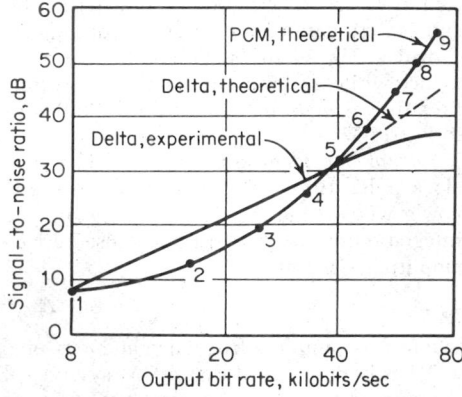

Fig. 14-49. Signal-to-noise ratio for delta modulation and PCM.[1]

of a human voice. Notice that they decrease in frequency together, indicating that DM can be used effectively with speech transmission. Generally speaking, transmission of speech is the chief application of DM, although various modifications and improvements are being studied to extend DM to higher frequencies and transmission of the lost dc component.

Among these techniques is delta-sigma modulation, where the signal is integrated and compared with an integrated approximation to form the error signal similar to $\epsilon(t)$ of Fig. 14-44. The decoding is accomplished with a low-pass filter and requires no integration.

The signal-to-quantization noise ratio for single-integration DM is given by[11]

$$S/N = 0.2 f_s^{3/2}/f f_0^{1/2} \qquad (14\text{-}72)$$

where f_s = sampling frequency, f = signal frequency, f_0 = signal bandwidth. For double or mixed DM[11]

$$S/N = 0.026 f_s^{5/2}/f f_0^{3/2} \qquad (14\text{-}73)$$

A comparison of signal-to-noise ratio for DM and PCM is shown in Fig. 14-49, along with an experimental DM system for voice application. Note that DM at 40 kilobits/s sampling rate is equal in performance with a 5-bit PCM system.

Extended-Range DM. A system termed high-information DM (HIDM, developed by M. R. Winkler in 1963) falls in the category of companded systems and encodes more information in the binary sequence than normal DM. Basically, the method doubles the size of the quantization step when two identical, consecutive binary values appear and takes one-half of the step after each transition of the binary train. The HIDM system is capable of reproducing the signal with smaller quantization and overload errors. This technique also increases the dynamic range. The response of HIDM compared with that of DM is shown in Fig. 14-50. For a more extensive discussion of recent companding schemes see Ref. 11.

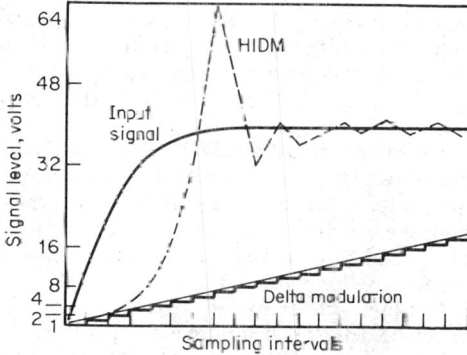

Fig. 14-50. Step response for a high-information delta modulation.[11]

Implementation of HIDM is similar to that of DM, as shown in Figs. 14-51 and 14-52, with the difference only in the demodulator. The flip-flop of Fig. 14-52 changes state on the polarity of the input pulses. While the impulse generator initializes the experimental generators each pulse time, the flip-flop selects either the positive or negative one. The integrator adds and smooths the exponential waveforms to form the output signal. The scheme has a dynamic range with slope limiting of 11.1 levels per pulse period, which is much greater than DM and is equivalent to a 7-bit linear-quantized FDM system.[1]

37. Digital Modulation. Digital modulation is concerned with the transmission of a binary pulse train over some medium. The output of, say, a PCM coder would be used to modulate a carrier for transmission. This modulation is treated here; for typical and more extensive

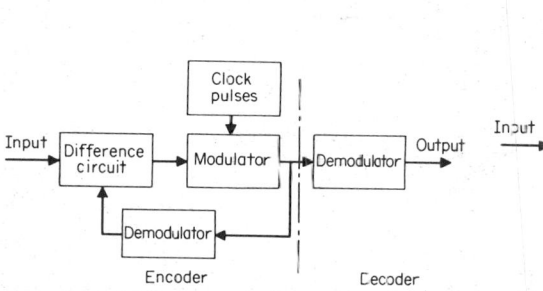

Fig. 14-51. Block diagram of HIDM system.[1]

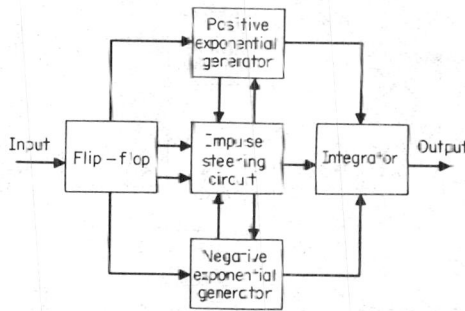

Fig. 14-52. Block diagram of HIDM demodulator.

system detail, see Ref. 12, Par. **14-41**. In PCM systems, for instance, the high-quality reproduction of the analog signal is a function only of the probability of correct reception of the pulse sequences. Thus the measure of digital modulation is the probability of error resulting from the digital modulation. The three basic types of digital modulation, amplitude-shift keying (ASK), frequency-shift keying (FSK), and phase-shift keying (PSK), are treated below.

38. Amplitude-Shift Keying (ASK). In ASK the carrier amplitude is turned on or off, generating the waveform of Fig. 14-53 for rectangular pulses. Pulse shaping such as raised cosine, etc., is sometimes used to conserve bandwidth. The elements of a binary ASK receiver are shown in Fig. 14-54. The detection can be either coherent or noncoherent. However, if the added complexity of coherent methods is to be applied, a higher performance can be achieved by using one of the other methods of digital modulation.

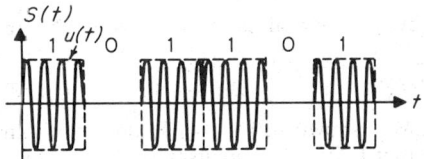

The error rate of ASK with noncoherent detection is given in Fig. 14-55. Note that the curves approach constant values of error for high signal-to-noise ratios.

Fig. 14-53. ASK modulation.[3]

The probability of error for the coherent detection scheme of Fig. 14-54c is shown in Fig. 14-56. The coherent-detection operation is equivalent to bandpass filtering of the received signal plus noise, followed by synchronous detection, as shown. At the optimum threshold shown in Fig. 14-56, the probability of error of marks and spaces is the same. The curves also tend toward a constant false-alarm rate, as in the noncoherent case.

39. Frequency-Shift Keying (FSK). In FSK the frequency is shifted rapidly between one of two frequencies. Generally, two filters are used in favor of a conventional FM detector to discriminate between the marks and spaces, as illustrated in Fig. 14-57. As with ASK, either noncoherent or coherent detection can be used, although in practice coherent detection is not often used. This is because it is just as easy to use PSK with coherent detection and achieve superior performance.

In the noncoherent FSK system shown in Fig. 14-58a, the largest of the output of the two envelope detectors determines the mark-space decision. Using this system results in the curve for noncoherent FSK in Fig. 14-59. Comparison of the noncoherent FSK error with that of the noncoherent ASK results in the conclusion that both achieve an equivalent error rate at the same average SNR at low error rates.[3] FSK requires twice the bandwidth of ASK because of the use of two tones. In ASK, in order to achieve this performance, it is required to optimize the detection

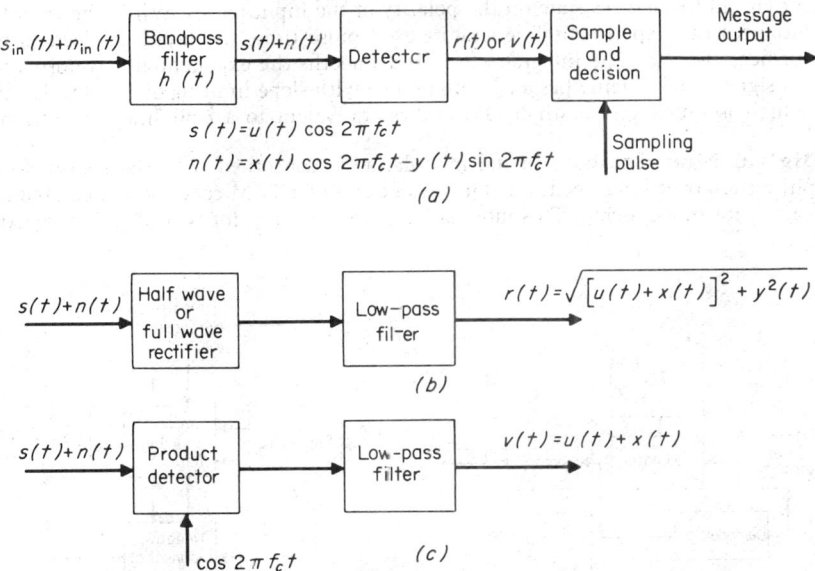

$$s(t) = u(t)\cos 2\pi f_c t$$
$$n(t) = x(t)\cos 2\pi f_c t - y(t)\sin 2\pi f_c t$$

(a)

$$r(t) = \sqrt{[u(t) + x(t)]^2 + y^2(t)}$$

(b)

$$v(t) = u(t) + x(t)$$

$\cos 2\pi f_c t$ (c)

Fig. 14-54. Elements of a binary digital receiver: (a) elements of a simple receiver; (b) noncoherent (envelope) detector; (c) coherent (synchronous) detector.[3]

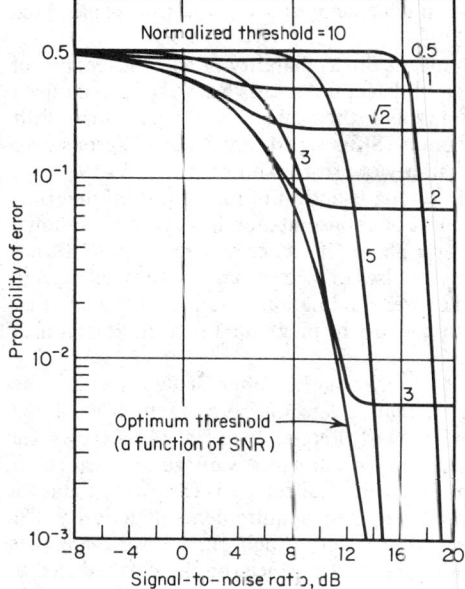

Fig. 14-55. Error rate for on-off keying, noncoherent detection.[3]

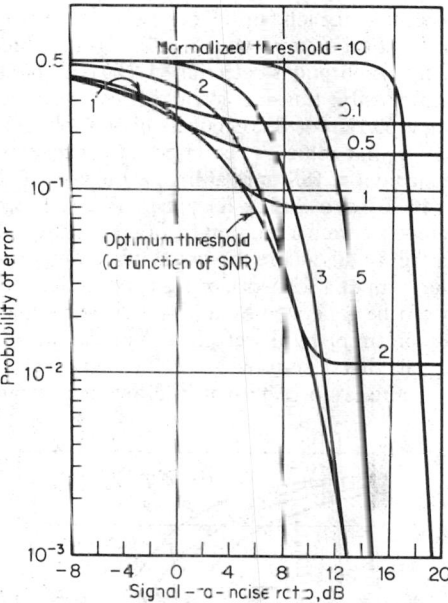

Fig. 14-56. Error rate for on-off keying, coherent detection.[3]

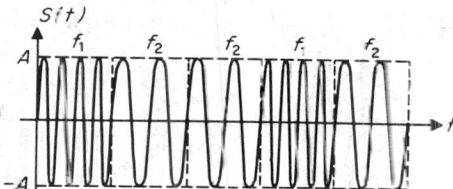

Fig. 14-57. FSK waveform, rectangular pulses.[3]

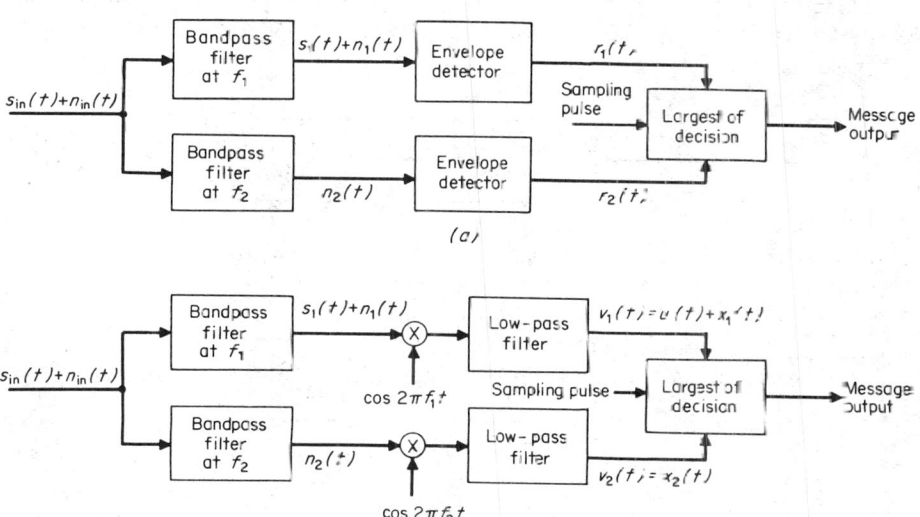

Fig. 14-58. Dual-filter detection of binary FSK signals: (a) noncoherent detection tone f_1 signaled. (b) coherent detection tone f_1 signaled.[3]

threshold at each SNR. The FSK system threshold is independent of SNR, and thus is preferred in practical systems where fading is encountered.

By synchronous detection of FSK (Fig. 14-58b) is meant the availability of an exact replica of each possible transmission at the receiver. The coherent detection process has the effect of rejecting a portion of the bandpass noise. Coherent FSK involves the same difficulties as phase-shift keying but achieves poorer performance. Also, coherent FSK is significantly advantageous over noncoherent FSK only at high error rates. The probability of error is shown in Fig. 14-59.

40. Phase-Shift Keying (PSK). Phase-shift keying is optimum in the minimum-error-rate sense from a decision-theory point of view. The PSK of a constant-amplitude carrier is shown in Fig. 14-60, where the two states are represented by a phase difference of π rad. Thus PSK has the form of a sequence of plus and minus rectangular pulses of a continuous sinusoidal carrier. It can be generated by double-sideband suppressed-carrier modulation by a bipolar rectangular waveform or by direct phase modulation. It is also possible to phase-modulate more complex signals than a sinusoid.

There is no performance difference in binary PSK between the coherent detector and the normal phase detector, both of which are shown in Fig. 14-61. Reference to Fig. 14-59 shows that there is a 3-dB design advantage for ideal coherent PSK over ideal coherent FSK, with about the same equipment requirements. Practically, PSK can suffer if very much phase error $\Delta\phi$ is present in the system, since the signal is reduced by $\cos\Delta\phi$. This phase error can be introduced by relative drifts in the master oscillators at transmitter or receiver or be due to phase drift or fluctuation in the propagation path. In most cases this phase error can be compensated at the expense of requiring long-term smoothing.

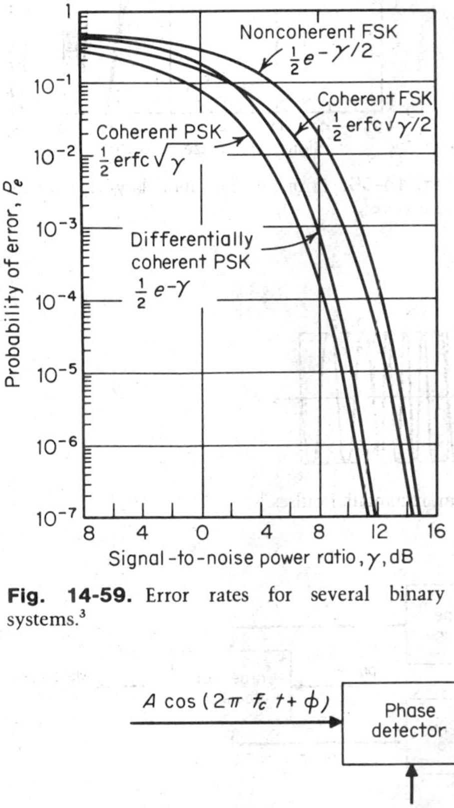

Fig. 14-59. Error rates for several binary systems.[3]

Fig. 14-60. PSK signal, rectangular pulses.[3]

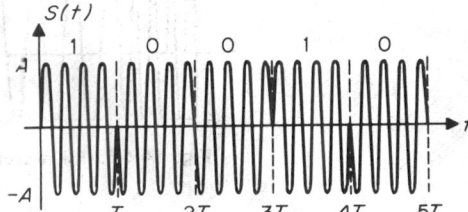

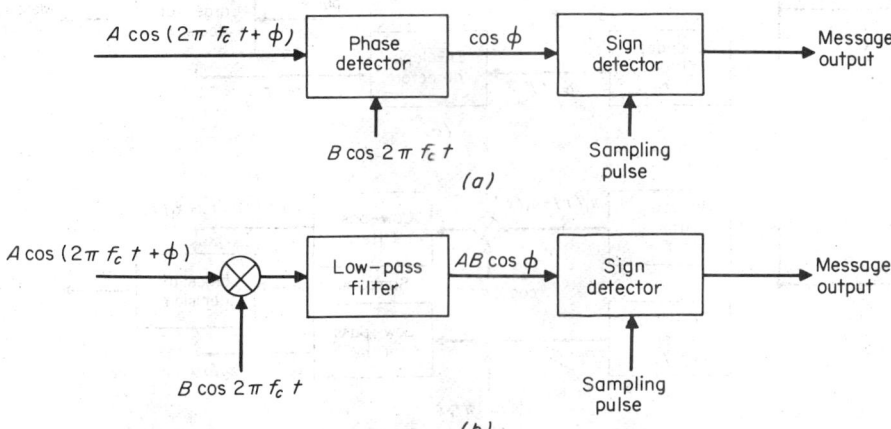

Fig. 14-61. Two detection schemes for ideal coherent PSK: (a) phase detection; (b) coherent detection.[3]

An alternative to PSK is differential phase-shift keying (DPSK), where it is required that there be enough stability in the oscillators and transmission path to a low negligible phase change from one information pulse to the next. Information is encoded differentially in terms of phase change between two successive pulses. For instance, if the phase remains the same from one pulse to the next (0° phase shift), a mark would be indicated. However, a phase shift of π from the previous pulse to the next would indicate a space. A coherent detector is still required where one input is the current pulse with the other input the previous pulse.

The probability of error is shown in Fig. 14-59. At all error rates DPSK requires 3 dB less SNR than noncoherent FSK for the same error rate. Also, at high SNR, DPSK performs almost as well as ideal coherent PSK at the same keying rate and power level.

41. References on Pulse Modulators and Demodulators

1. Panter, P. F.   "Modulation, Noise, and Spectral Analysis," McGraw-Hill, New York, 1965.
2. Schwartz, M.   "Information Transmission, Modulation and Noise," 2d ed., McGraw-Hill, New York, 1970.
3. Stein, S., and J. J. Jones   "Modern Communication Principles," McGraw-Hill, New York, 1968.
4. Landee, R., D. Davis, and A. Albrecht   "Electronic Designer's Handbook," McGraw-Hill, New York, 1957.
5. Black, H. S.   "Modulation Theory," Van Nostrand, New York, 1953.
6. Woodward, P. M.   "Probability and Information Theory, with Applications to Radar," Pergamon, New York, 1953.
7. Rowe, H. E.   "Signals and Noise in Communication Systems," Van Nostrand, New York, 1965.
8. Nichols, M. H., and L. L. Rauch   "Radio Telemetry," Wiley, New York, 1954.
9. Truxal, J. G.   "Automatic Feedback Control System Synthesis," McGraw-Hill, New York, 1955.
10. Mishkin, E., and L. Braun   "Adaptive Control Systems," McGraw-Hill, New York, 1961.
11. Schindler, H. R.   Delta Modulation, *IEEE Spectrum*, October 1970, p. 69.
12. Handbook of Digital Communications, special issue of *Microwave Syst. News*, October 1979, Vol. 9, No. 11, pp. 13–24, 30–134.

Spread-Spectrum Modulation

BY M. D. EGTVEDT

42. Spread-Signal Modulation.

In a receiver designed exactly for a specified set of possible transmitted waveforms (in the presence of white noise and in the absence of such propagation defects as multipath and dispersion), the performance of a matched filter or cross-correlation detector depends only on the ratio of signal energy to noise power density E/n_0, where E is the received energy in one information symbol and $n_0/2$ is the rf noise density at the receiver input. Since signal bandwidth has no effect on performance in white noise, it is interesting to examine the effect of spreading the signal bandwidth in situations involving jamming, message and traffic-density security, and transmission security. Other applications include random-multiple-access communication channels,[11]* multipath propagation analysis,[4] and ranging. Newer references [20,21] provide extensive treatment.

The information-symbol waveform can be characterized by its time-bandwidth (TW) product. Consider a binary system with the information symbol defined as a *bit* (of time duration T), while the fundamental component of the binary waveform is called a *chip*. For this *direct-sequence* system, the ratio (chips per bit) is equal to the TW product. An additional requirement on the symbol waveforms is that their cross-correlation with each other and the noise or extraneous signals be minimal.

Spread-spectrum systems occupy a signal bandwidth much larger (> 10) than the information bandwidth, while the conventional systems have a TW of well under 10. FM with a high modulation index might slightly exceed 10 but is not optimally detectable and has a processing gain only above a predetection signal-to-noise threshold.

43. Nomenclature of Secure Systems.

While terminology is not subject to rigorous definition the following terms apply to the following material:

Security and Privacy. Relate to the protection of the signal from an unauthorized receiver. They are differentiated by the sophistication required. Privacy protects against a casual listener with little or no analytical equipment, while security implies an interceptor familiar with the

---

*Superior numbers correspond to numbered references, Par. **14-49**.

principles and using an analytical approach to learn the *key*. Protection requirements must be defined in terms of the interceptor's applied capability and the time value of the message. Various forms of *protection* include:

Crypto security. Protects the information content, generally without increasing the TW product.

Antijamming (AJ) security. Spreads the signal spectrum to provide discrimination against energy-limited interference by using cross-correlation or matched-filter detectors. The interfer-

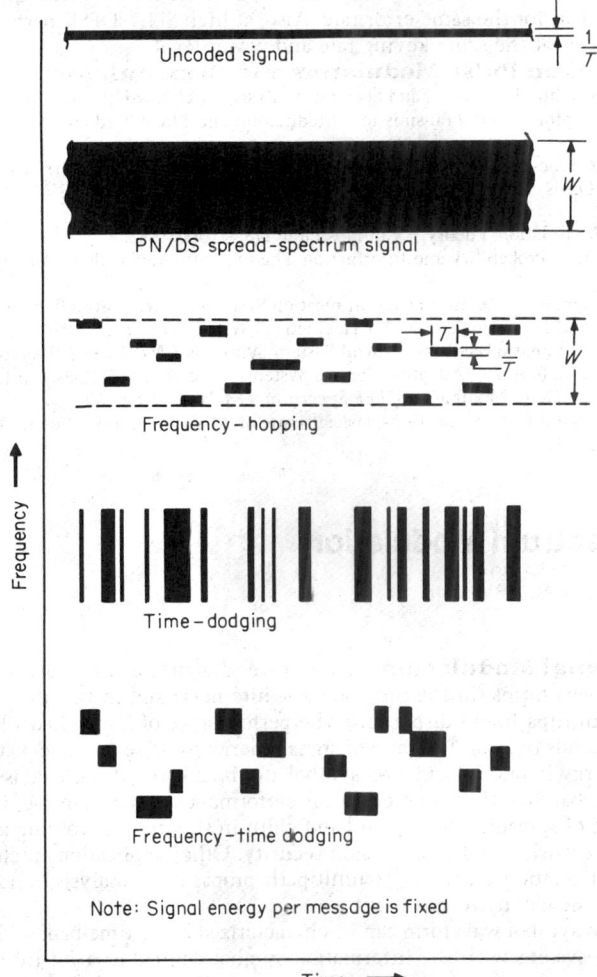

Fig. 14-62. Spectral occupancy vs. time characteristics of spread-spectrum signals.

ence may be natural (impulse noise), inadvertent (as in amateur radio or aircraft channels), or deliberate (where the jammer may transmit continuous or burst cw, swept cw, narrow-band noise, wide-band noise, or replica or deception waveforms).

Traffic-density security. Involves capability to switch data rates without altering the apparent characteristics of the spread-spectrum waveform. The TW product (processing gain) is varied inversely with the data rates.

Transmission security. Involves spreading the bandwidth so that, beyond some range from the transmitter, the transmitted signal is buried in the natural background noise. The process gain (TW) controls the reduction in detectable range vis-à-vis a "clear" signal.

Use in Radar.   It is usual to view radar applications as a variation on communication; i.e., the return waveforms are known except with respect to noise, Doppler shift, and delay. Spectrum spreading is applicable to both cw and pulse[2] radars. The major differentiation is in the choice of cross-correlation or matched-filter detector. The TW product is the key performance parameter, but the covariance function properties must frequently be determined to resolve Doppler shifts as well as range delays.

44. Classification of Spread-Spectrum Signals. Spread-spectrum signals can be classified on the basis of their spectral occupancy vs. time characteristics, as sketched in Fig. 14-62. Direct-sequence (DS) and pseudo-noise (PN) waveforms provide continuous full coverage, while frequency-hopping (FH), time-dodging, and frequency-time dodging (F-TD), fill the frequency-time plane only in a long-term averaging sense.

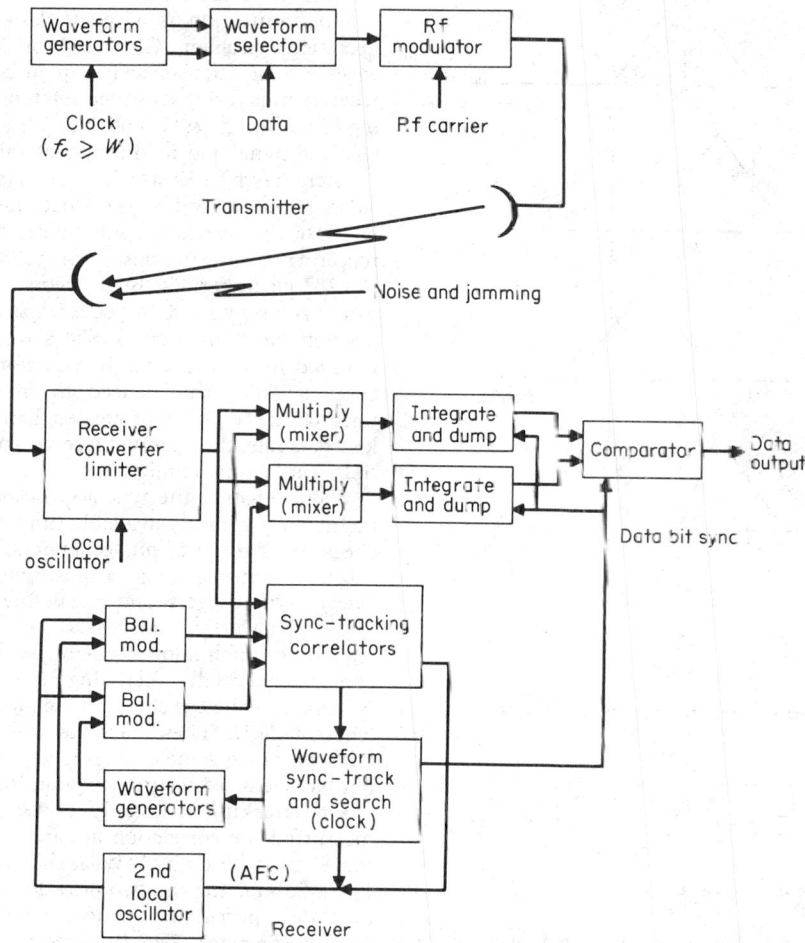

Fig. 14-63. Direct-sequence link for spread-spectrum system.

DS waveforms are pseudo-random digital streams generated by digital techniques and transmitted without significant spectral filtering. If heavy filtering is used, the signal amplitude statistics become quite noiselike, and this is called a *PN waveform*. In either case correlation detection is generally used because the waveform is dimensionally too large to implement a practical matched filter, and the sequence generator is relatively simple and capable of changing codes.

In FM schemes the spectrum is divided into subchannels spaced orthogonally at $1/T$ separations. One or more (e.g., two for FSK) are selected by pseudo-random techniques for each data bit. In time-dodging schemes the signal burst time is controlled by pulse repetition methods, while F-TD combines both selections. In each case a jammer must either jam the total spectrum continuously or accept a much lower effectiveness (approaching $1/TW$). Frequency-hopped signals can be generated using SAW chirp devices.[22]

45. Correlation-Detection Systems. The basic components of a typical direct-sequence (DS) type of link are shown in Fig. 14-63. The data are used to select the appropriate waveform, which is shifted to the desired rf spectrum by suppressed-carrier frequency-conversion techniques, and transmitted. At the receiver identical locally generated waveforms multiply with the incoming signal. The stored reference signals are often modulated onto a local oscillator, and the incoming rf may be converted to an intermediate frequency, usually with rf or i.f. limiters.

The mixing detectors are followed by linear integrate-and-dump filters, with a "greatest of" decision at the end of each period. The integrator is either a low-pass or bandpass quenchable narrow-band filter. Digital techniques are increasingly being used.

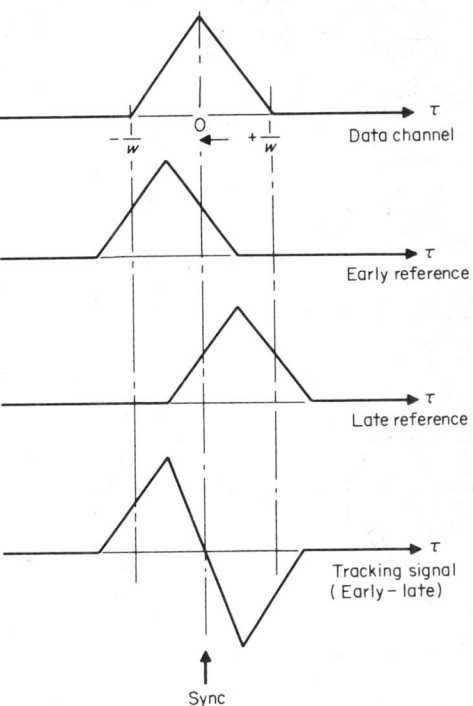

Sync

Fig. 14-64. Sync tracking by early-late correlators.

Data channel

Early reference

Late reference

Tracking signal
(Early – late)

Synchronization is a major design and operational problem. Given a priori knowledge of the transmitted sequences, the receiver must bring its stored reference timing to within $\pm 1/(2W)$ of the width of the received signal and hold it at that value. In a system having a 19-stage *pn* generator, a 1-MHz *pn* clock, and a 1-kHz data rate, the width of the correlation function is $\pm \frac{1}{2} \mu s$, repeating $\frac{1}{2}$ s separations, corresponding to 524,287 clock periods. In the worst case, it would be necessary to inspect each sequence position for 1 ms; that is, 524 s would be required to acquire sync. If oscillator tolerances and/or Doppler lead to frequency uncertainties equal to or greater than the 1-kHz data rate, then parallel receivers or multiple searches are required.

Ways to reduce the sync acquisition time include using jointly available timing references to start the *pn* generators, using shorter sequences for acquisition only; "clear" sync triggers; and paralleling detectors. Titsworth[3] discusses composite sequences which allow acquiring each component sequentially, searching $N_1 + N_2 + N_3$ delays, while the composite sequence has length $N_1 N_2 N_3$. These methods have advantages for space-vehicle ranging applications but have reduced security to jamming.

Sync tracking is usually performed by measuring the correlation at early and late times, $\pm \tau$, where $\tau \leq 1/W$, as shown in Fig. 14-64. Subtracting the two provides a useful time discrimination function, which controls the *pn* clock. The displaced values can be obtained by two tracking-loop correlators or by time-sharing a single unit. "Dithering"

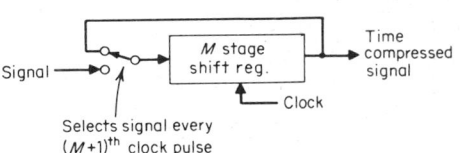

Fig. 14-65. Delay-line time compression (deltic) configuration.

Signal → M stage shift reg. → Time compressed signal

Clock

Selects signal every $(M+1)^{th}$ clock pulse

the reference signal to the signal correlator can also be used, but with performance compromises.

The tracking function can also be obtained by using the time derivative of one of the inputs

$$\frac{d}{d\tau} \varphi_{XY}(\tau) = \overline{\frac{dX(t)}{dt} \cdot Y(t + \tau)}$$

A third approach has been to add by mod 2 methods the clock to the transmitted *pn* waveform. The spectral envelope is altered, but very accurate peak tracking can be accomplished by phase locking to the recovered clock.

46. Limiters in Spread-Spectrum Receivers. Limiters are frequently used in spread-spectrum receivers to avoid overload saturation effects, such as circuit recovery time, and incidental phase modulation. In the usual low-input signal-noise range, the limiter tends to normalize the output noise level, which simplifies the decision circuit design. In repeater applications (e.g., satellite), a limiter is desirable to allow the transmitter to be fully modulated regardless of the input-signal strength. When AGC is used, the receiver is highly vulnerable to pulse jamming, while the limiter causes a slight reduction of the instantaneous signal-to-jamming ratio and a proportional reduction of transmitter power allocated to the desired signal.

The signal-to-jamming ratios (SJRs) in and out of a limiter can be expressed by

$$SJR_{out} = \alpha \cdot SJR_{in}$$

where α is a function of the signal and jamming and the time and spectral characteristics of each. This problem has been analyzed[5,6,7,14,16,17] for various cases. Jones[5] covers the case of cw signal against a cw jammer plus noise or noise alone (including noise jamming). The problem of gaussian signals has been treated also, for gaussian noise, by Price[13] and others. As an approximation, α can be taken as 1 to 2 dB loss for noise jamming and up to 6 dB loss for cw jamming.

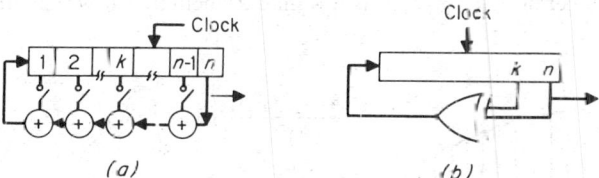

Fig. 14-66. Maximal-length-sequence (MLS) system.

Additional discrimination against narrow-band jamming can be obtained by fixed or adaptive notch filters or by the Kirbar fix,[18] in which the spectrum is divided into several contiguous bands, each of which is limited, and the outputs combined. This procedure limits the cw energy to $1/n$ times the total, where n is the number of contiguous bands.

47. Deltic-Aided Search. The sync search can be accelerated by use of deltic-aided (delay-line time compression) circuits if logic speeds permit.[19] The basic deltic consists of a recirculating shift register (or a delay line) which stores M samples, as shown in Fig. 14-65. The incoming spread-spectrum signal must be sampled at a rate above W (W = bandwidth). During each intersample period the shift register is clocked through $M + 1$ shifts before accepting the next sample. If $M \geq 2W$, a signal period at least equal to the data integration period is stored and is read out at M different delays during each period T, permitting many high-speed correlations against a similarly accelerated (but not time-advancing) reference.

For a serial-deltic and shift-register delay line the clock rate is at least $4TW^2$. Using a deltic with K parallel interleaved delay lines, the internal delay lines are clocked at $4TW^2/K^2$ and the demultiplexed output has a bit rate of $4TW^2/K$, providing only M/K discrete delays. This technique is device-limited to moderate signal bandwidths, primarily in the acoustic range up to about 10 kHz.

48. Waveforms. The desired properties of a spread-spectrum signal include:

An autocorrelation function which is unity at $\tau = 0$ and zero elsewhere

A zero cross-correlation coefficient with noise and other signals

A large library of orthogonal waveforms

Maximal-Length Linear Sequences. A widely used class of waveforms is the maximal-length sequence (MLS) generated by a tapped re-fed shift register,[9,12,14] as shown in Fig. 14-66a and as one-tap unit in Fig. 14-66b. The mod 2 half adder ($\oplus$) and EXCLUSIVE-OR logic gate are identical for 1-bit binary signals. Analyses of this mode of operation are given by Birdsall and Ristenbatt[1] and by Golomb.[10]

If analog levels $+1$ and -1 are substituted, respectively, for 0 and 1 logic levels, the circuit is observed to function as a 1-bit multiplier.

Pertinent properties of the MLS are as follows. Its length, for an n-stage shift register is $2^n - 1$ bits. During $2^n - 1$ successive clock pulses, all n-bit binary numbers (except all zeros) will have been present. The autocorrelation function is unity at $\tau = 0$, and at each $2^n - 1$ clock pulses displacement, and $1/(2^n - 1)$ at all other displacements. This assumes that the sequences repeat cyclically; i.e., the last bit is closed onto the first. The autocorrelation function of a single (noncyclic) MLS shows significant time side lobes (Frank[8]). Titsworth[3] has analyzed the self-noise of incomplete integration over p chips, obtaining for MLSs,

$$\sigma^2(t) = (p - t)(p^2 - 1)/p^3 t$$

which approaches $1/t$ for the usual case of $p \gg t$. Since $t \approx TW$, the self-noise component is usually negligible.

Another self-noise component is frequently present due to amplitude and dispersion differences, caused by filtering, propagation effects, and circuit nonlinearities. In addition to intentional clipping, the correlation multiplier is frequently a balanced modulator, which is linear only to the smaller signal, unless deliberately operated in a bilinear range.[13,15] The power spectrum is shown in Fig. 14-67. The envelope has a $(\sin^2 X)/X^2$ shape ($X = \pi\omega/\omega_{clock}$), while the individual lines are separated by $\omega_{clock}/(2^n - 1)$.

An upper bound on the number of MLS for an n-stage shift register is given in terms of the Euler ϕ function:

$$N_u = \phi(2^n - 1)/n \le 2^{(n - \log_2 n)}$$

where $\phi(k)$ = number of positive integers less than k, including 1, which are relatively prime to k.

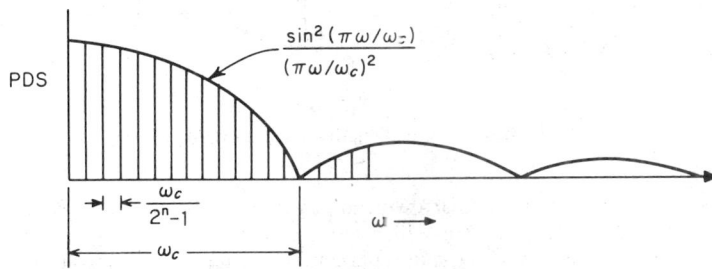

Fig. 14-67. Spectrum of MLS system.

49. References on Spread-Spectrum Modulation

1. Birdsall, T. G., and M. P. Ristenbatt Introduction to Linear Shift-Register Generated Sequences, *Univ. Mich. Res. Inst. Tech. Rep. 90*, Ann Arbor, Mich., October 1958.

2. Cook, C. E. Pulse Compression: Key to More Efficient Radar Transmission, *Proc. IRE*, March 1960, p. 310.

3. Titsworth, R. C. Correlation Properties of Cyclic Sequences, *Calif. Inst. Technol. Jet Propulsion Lab. Tech Rep. 32-388*, Pasadena, Calif., July 1963.

4. Price, R., and P. E. Green, Jr. A Communication Technique for Multipath Channels, *Proc. IRE*, March 1958, pp. 555–570.

5. Jones, J. J. Hard Limiting of Two Signals in Random Noise, *IEEE Trans. Inf. Theory*, January 1963, Vol. IT-9.

6. Manasse, R., R. Price, and R. Lerner Loss of Signal Detectability in Bandpass Limiters, *IRE Trans. Inf. Theory*, May 1958, Vol. IT-4.

7. Davenport, W. B., Jr. Signal-to-Noise Ratios in Bandpass Limiters, *J. Appl. Phys.*, June 1953, Vol. 24, pp. 720–727.

8. Frank, R. L. Polyphase Codes with Good Nonperiodic Correlation Properties, *IEEE Trans. Inf. Theory*, January 1963, Vol. IT-9.

9. Marsh, R. W. "Table of Irreducible Polynomials over GF(2) through Degree 19," NSA, distributed by Commerce Dept. Office of Technical Services, Washington, October 1957.

10. Golomb, S. W. "Shift Register Sequences," Holden-Day, San Francisco, Calif., 1967.

11. Costas, J. P. Poisson, Shannon, and the Radio Amateur, *Proc. IRE*, December 1959, p. 2058.

12. Peterson, W. W. "Error Correcting Codes," Wiley-M.I.T. Press, New York, 1961.

13. Price, R. A Useful Theorem for Nonlinear Devices Having Gaussian Inputs, *IRE Trans. Inf. Theory*, June 1958.

14. Nikiforuk, P. N., and M. M. Gupta A Bibliography on the Properties, Generation and Control System Applications of Shift Register Sequences, *Int. J. Control*, 1969, Vol. 9, No. 2, pp. 217–234.

15. Green, P. E., Jr. The Output Signal-to-Noise Ratio of Correlation Detectors, *IRE Trans. Inf. Theory*, March 1957.

16. Bussgang, J. J. Cross-Correlation Functions of Amplitude-Distorted Gaussian Signals, *M.I.T. Res. Lab. Elec. Tech. Rep. 216*, March 1952.

17. Baum, R. F. The Correlation Function of Smoothly Limited Gaussian Noise, *IRE Trans. Inf. Theory*, September 1957.

18. Kirkpatrick, G. M. "Signal Processing Arrangement with Filters in Plural Channels Minimizing Underdesirable Interference to Narrow and Wide Pass Bands," U.S. Pat. 3,112,452, November 1963.

19. Allen, W. B., and E. C. Westerfield Digital Compressed-Time Correlators and Matched Filters for Active Sonars, *J. Acoust Soc Am.*, January 1964, pp. 121–139.

20. Dixon, R. C. "Spread Spectrum Systems," Wiley-Interscience, New York, 1976.

21. Dixon. R. C. (ed.) "Spread Spectrum Techniques," IEEE Press, New York, 1976.

22. Patterson, E. W., et al. Frequency-Hopped Waveform Synthesis by Using S.A.W. Chirp Filters, *Electro. Lett.*, Oct. 13, 1977, Vol. 13, No. 21, pp. 633–635.

Optical Modulators and Demodulators

BY J. L. CHOVAN

50. Modulation of Beams of Radiation. This discussion of optical modulators is restricted to devices that operate on a directed beam of optical energy to control its intensity, phase, or frequency, according to some time-varying modulating signal. Devices that deflect a light beam or spatially modulate a light beam, such as light-valve projectors, are treated in Sec. **20**.

Phase or frequency modulation requires a coherent light source, such as a laser. Optical heterodyning is then used to shift the received signal to lower frequencies, where conventional FM demodulation techniques can be applied.

Intensity modulation can be used on incoherent as well as coherent light sources. However, the properties of some types of intensity modulators are wavelength-dependent. Such modulators are restricted to monochromatic operation but not limited to the extremely narrow laser line widths required for frequency modulation.

Optical modulation depends on either perturbing the optical properties of some material with a modulating signal or mechanical motion to interact with the light beam. Modulation bandwidths of mechanical modulators are limited by the inertia of the moving masses. Optical-index modulators generally have a greater modulation bandwidth but typically require critical and expensive optical materials.

Optical-index modulation can be achieved with electric or magnetic fields or by mechanical stress. Typical modulator configurations are presented below, as in heterodyning, which is often useful in demodulation. Optical modulation can also be achieved using semiconductor junctions. This approach, which is comparatively new and presently under development is discussed in Ref. 4.*

51. Optical-Index Modulation: Electric Field Modulation. *Pockels and Kerr Effects.* In some materials, an electric field vector **E** can produce a displacement vector **D** whose direction and magnitude depend on the orientation of the material. Reference 1 shows that such a material can be completely characterized in terms of three independent dielectric constants associated with three mutually perpendicular natural directions of the material. If all three dielectric constants are equal, the material is *isotropic*. If two are equal and one is not, the material is *uniaxial*. If all three are unequal, the material is *biaxial*.

The optical properties of such a material can be described in terms of the *ellipsoid of wave normals* (Fig. 14-68). This is an ellipsoid whose semiaxes are the square roots of the associated dielectric constants. The behavior of any plane monochromatic wave through the medium can be determined from the ellipse formed by the intersection of the ellipsoid with a plane through the center of the ellipsoid and perpendicular to the direction of wave travel. The instantaneous electric field vector **E** associated with the optical wave has components along the two axes of this ellipse. Each of these components travels with a phase velocity that is inversely proportional to the length of the associated ellipse axis.

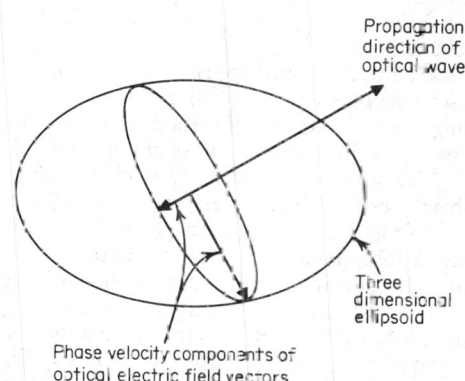

Propagation direction of optical wave

Three dimensional ellipsoid

Phase velocity components of optical electric field vectors

Fig. 14-68. Ellipsoid of wave normals.

*Numbered references are listed in Par. **14-59**.

Thus there is a differential phase shift between the two orthogonal components of the electric field vector after it has traveled some distance through such a birefringent medium. The two orthogonal components of the vector vary sinusoidally with time but have a phase difference between them, which results in a vector whose magnitude and direction vary to trace out an ellipse once during each optical cycle. Thus linear polarization is converted into elliptical polarization in a birefringent medium.

In some materials it is possible to induce a perturbation in one or more of the ellipsoid axes by applying an external electric field. This is the electrooptical effect (Par. **14-54**). The electrooptical effect is most commonly used in optical modulators presently available. More detailed configurations using these effects are discussed in Par. **14-54**. Reference 12 presents design considerations for various configurations and tabulates material properties.

Stark Effect.   Materials absorb and emit optical energy at frequencies which depend on molecular or atomic resonances characteristic of the material. In some materials an externally applied electric field can perturb these natural resonances. This is known as the Stark effect.

Reference 3 discusses a modulator for the CO_2 laser in the 3- to 22-μm region. The laser output is passed through an absorption cell whose natural absorption frequency is varied by the modulating signal, using the Stark effect. Since the laser frequency remains fixed, the amount of absorption depends on how closely the absorption cell is tuned to the laser frequency. Intensity modulation results.

52. Magnetic Field Modulation. *Faraday Effect.*   Two equal-length vectors circularly rotating at equal rates in opposite directions in space combine to give a nonrotating resultant whose direction in space depends on the relative phase between the counterrotating components. Thus any linearly polarized light wave can be considered to consist of equal right and left circularly polarized waves.

In a material which exhibits the Faraday effect, an externally applied magnetic field causes a difference in the phase velocities of right and left circularly polarized waves traveling along the direction of the applied magnetic field. This results in a rotation of the electric field vector of the optical wave as it travels through the material. The amount of the rotation is controlled by the strength of a modulating current producing the magnetic field. Reference 4 discusses an infrared modulator (1.2 to 4.5 μm) that uses the Faraday effect in yttrium-iron-garnet (YIG).

Zeeman Effect.   In some materials the natural resonance frequencies at which the material emits or absorbs optical energy can be perturbed by an externally applied magnetic field. This is known as the Zeeman effect.

Intensity modulation can be achieved using an absorption cell modulated by a magnetizing current in much the same manner as the Stark effect absorption cell is used. The Zeeman effect has also been used to tune the frequency at which the active material in a laser emits.

53. Mechanical-Stress Modulation. In some materials the ellipsoid of optical-wave normals can be perturbed by mechanical stress. An acoustic wave traveling through such a medium is a propagating stress wave which produces a propagating wave of perturbation in the optical index.

When a sinusoidal acoustic wave produces a sinusoidal variation in the optical index of a thin isotropic medium, the medium can be considered, at any instant of time, as a simple phase grating. Such a grating diffracts a collimated beam of coherent light into discrete angles whose separation is inversely proportional to the spatial period of the grating.

This situation is analogous to an rf carrier phase-modulated by a sine wave. A series of sidebands results which correspond to the various orders of diffracted light. The amplitude of the mth order is given by an mth-order Bessel function whose argument depends on the peak phase deviation produced by the modulating signal. The phases of the sidebands are the appropriate integral multiples of the phase of the modulating signal.

The mth order of diffracted light has its optical frequency shifted by m times the acoustic frequency. The frequency is increased for positive orders and decreased for negative orders.

Similarly, a thick acoustic grating refracts light mainly at discrete input angles. This condition is known as *Bragg reflection* and is the basis for the *Bragg modulator* (Fig 14-69). In the Bragg modulator, essentially all the incident light can be refracted into the desired order, and the optical frequency is shifted by the appropriate integral multiple of the acoustic frequency.

Figure 14-69 shows the geometry of a typical Bragg modulator. The input angles for which Bragg modulation occurs are given by

$$\sin \theta = m\lambda/2\Lambda$$

where θ = angle between propagation of input optical beam and planar acoustic wavefronts, λ = optical wavelength in medium, Λ = acoustic wavelength in medium, $m = \pm 1, \pm 2, \pm 3,$

..., and $m\theta$ = angle between propagation direction of output optical beam and planar acoustic wavefronts.

The ratio of optical to acoustic wavelength is typically quite small, and m is a low integer, so that the angle θ is very small. Critical alignment is thus required between the acoustic wavefronts and the input light beam.

If the modulation bandwidth of the acoustic signal is broad, the acoustic wavelength varies, so that there is a corresponding variation in the angle θ for which Bragg reflection occurs. To

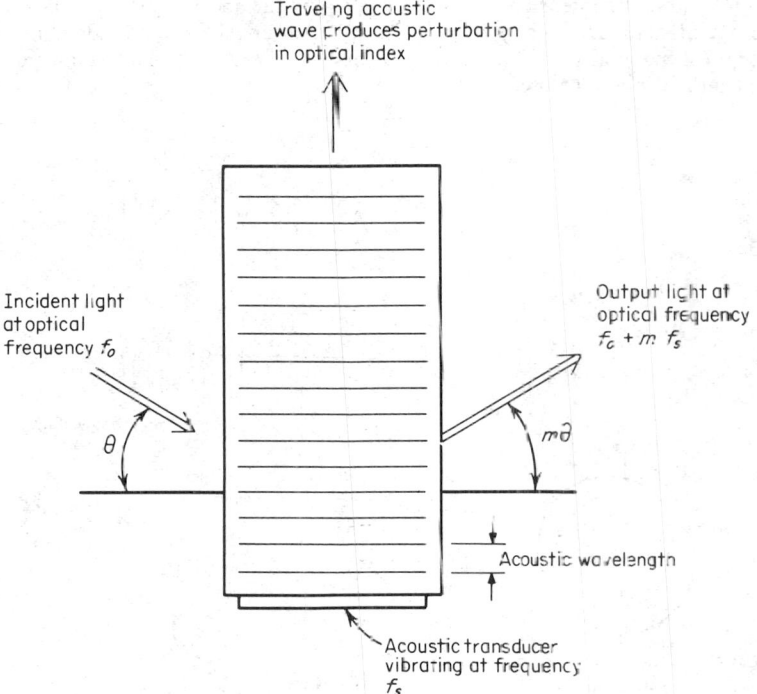

FIG. 14-69. The Bragg modulator.

overcome this problem, a phased array of acoustic transducers is often used to steer the angle of the acoustic wave as a function of frequency in the desired manner.

A limitation on bandwidth is the acoustic transit time across the optical beam. Since the phase grating in the optical beam at any instant of time must be essentially constant frequency if all the light is to be diffracted at the same angle, the bandwidth is limited so that only small changes can occur in this time interval. Reference 5 presents a figure of merit from Bragg modulator materials. References 6 and 7 review acoustooptical devices and compare several materials used in such devices. Lithium niobate is a material commonly used in commercially available Bragg modulators.

54. Modulator Configurations: Intensity Modulation. *Polarization Changes.* Linearly polarized light can be passed through a medium exhibiting an electrooptical effect and the output beam passed through another polarizer. The modulating electric field controls the eccentricity and orientation of the elliptical polarization and hence the magnitude of the component in the direction of the output polarizer. Typically, the input linear polarization is oriented to have equal components along the fast and slow axes of the birefringent medium, and the output polarizer is orthogonal to the input polarizer. The modulating field causes a phase differential varying from 0 to π rad. This causes the polarization to change from linear (at 0) to circular (at $\pi/2$) to linear normal to the input polarization (at π). Thus the intensity passing through the output polarizer varies from 0 to 100% as the phase differential varies from 0 to π rad.

Figure 14-70 shows this typical configuration. The following equations relate the optical intensity transmission of this configuration to the modulation.

$$I_o/I_i = \tfrac{1}{2}(1 - \cos \phi)$$

where I_o = output optical intensity, I_i input optical intensity, and ϕ = differential phase shift between fast and slow axes.

In the Pockels effect the differential phase shift is linearly related to applied voltage; in the Kerr effect it is related to the voltage squared.

$$\phi = \begin{cases} \pi v/V & \text{Pockels effect} \\ \pi (v/V)^2 & \text{Kerr effect} \end{cases}$$

where v = modulation voltage and V = voltage to produce π rad differential phase shift.

Figure 14-71 shows the intensity transmission given by the above expression. The most linear part of the modulation curve is at $\phi = \pi/2$. Often a quarter-wave plate is added in series with the electrooptical material to provide this fixed bias at $\pi/2$. A fixed-bias voltage on the electrooptical material can also be used.

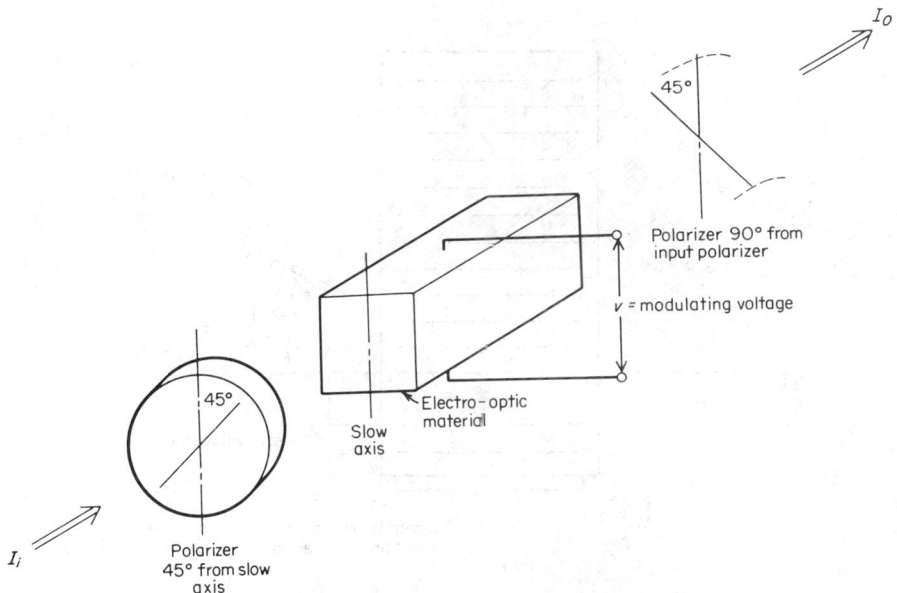

Fig. 14-70. Electrooptical intensity modulator.

This arrangement is probably the most commonly used broadband intensity modulator. Early modulators of this type used a uniaxial Pockels cell with the electric field in the direction of optical propagation. In this arrangement, the induced phase differential is directly proportional to the optical-path length, but the electric field is inversely proportional to this path length (at fixed voltage). Thus the phase differential is independent of the path length and depends only on applied voltage. Typical materials require several kilovolts for a differential phase shift of π in the visible-light region.

Since the Pockels cell is essentially a capacitor, the energy stored in it is $\frac{1}{2}CV^2$ where C is the capacitance and V is the voltage. This capacitor must be discharged and charged during each modulation cycle. Discharge is typically done through a load resistor, where this energy is dissipated. The high voltages involved mean that the dissipated power at high modulation rates is appreciable.

The high-voltage problem can be overcome by passing light through the medium in a direction normal to the applied electric field. This permits a short distance between the electrodes (so that a high-E field is obtained from a low voltage) and a long optical path in the orthogonal direction (so that the cumulative phase differential is experienced).

Unfortunately, materials available are typically uniaxial, having a high eccentricity in the absence of electric fields. When oriented in a direction that permits the modulating electric field to be orthogonal to the propagation direction the material has an inherent phase differential which is orders of magnitude greater than that induced by the modulating field. Furthermore, minor temperature variations cause perturbations in this phase differential which are large compared with those caused by modulation.

This difficulty is overcome by cascading two crystals which are carefully oriented so that temperature effects in one are compensated for by temperature effects in the other. The modulation electrodes are then connected so that their effects add. This approach is discussed in Ref. 8. Commercially available electrooptical modulators are of this type.

The Kerr effect is often used in a similar arrangement. Kerr cells containing nitrobenzene are commonly used as high-speed optical shutters.

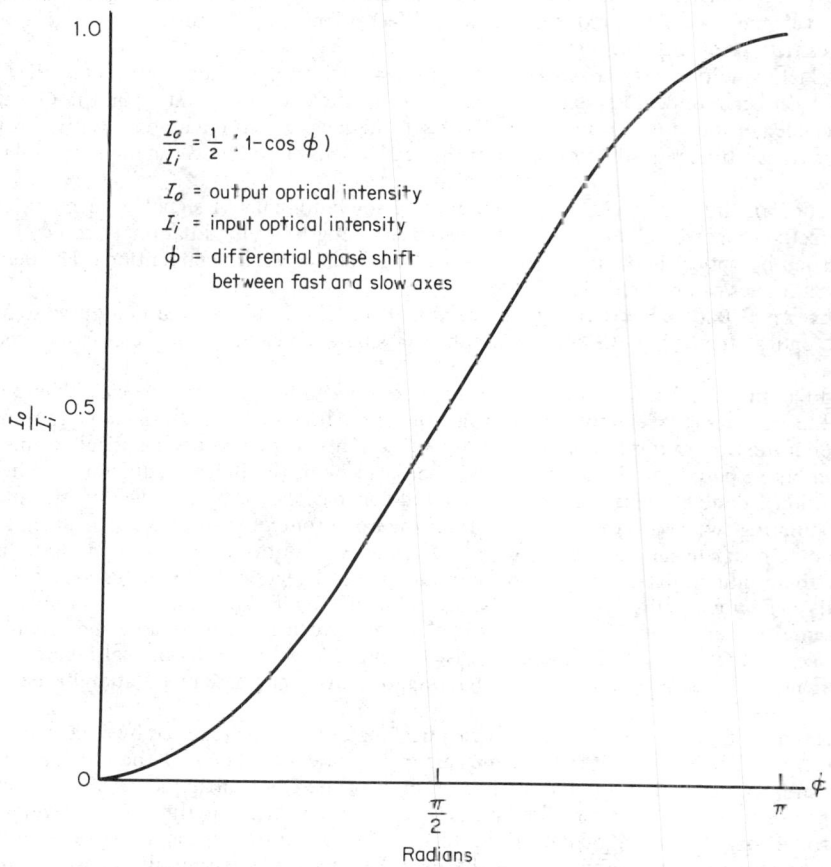

$$\frac{I_o}{I_i} = \frac{1}{2}(1-\cos\phi)$$

I_o = output optical intensity
I_i = input optical intensity
ϕ = differential phase shift between fast and slow axes

Fig. 14-71. Transmission of electrooptical intensity modulator.

Polarization rotation produced by the Faraday effect is also used in intensity modulation by passing through an output polarizer in a manner similar to that discussed above. The Faraday effect is more commonly used at wavelengths where materials exhibiting the electrooptical effect are not readily available.

Controlled Absorption. As noted above, the frequency at which a material absorbs energy due to molecular or atomic resonances can be tuned over some small range in materials exhibiting the Stark or Zeeman effect. Laser spectral widths are typically narrow compared with such an absorption line width. Thus the absorption of the narrow laser line can be modulated by tuning the absorption frequency over a range near the laser frequency. Although such modulators have been used, they are not as common as the electrooptical modulators discussed above.

55. Phase and Frequency Modulation of Beams. *Laser-Cavity Modulation.* The distance between mirrors in a laser cavity must be an integral number of wavelengths. If this distance is changed by a slight amount, the laser frequency changes to maintain an integral number. The following equation relates the change in cavity length to the change in frequency:

$$\Delta f = \frac{C}{L}\frac{\Delta L}{\lambda}$$

where Δf = change in optical frequency, ΔL = change in laser-cavity length, L = laser-cavity length, λ = optical wavelength of laser output, and C = velocity of light in laser cavity.

In a cavity 1 m long, a change in mirror position of one optical wavelength produces about 300 MHz frequency shift. Thus a laser can be frequency-modulated by moving one of its mirrors with an acoustic transducer, but the mass of the transducer and mirror limit the modulation bandwidths that can be achieved.

An electrooptical cell can be used in a laser optical cavity to provide changes in the optical-path length. The polarization is oriented so that it lies entirely along the axis of the modulated electrooptical material. This produces the same effect as moving the mirror but without the inertial restrictions of the mirror's mass.

Under such conditions, the ultimate modulation bandwidth is limited by the Q of the laser cavity. A light beam undergoes several reflections across the cavity, depending on the Q, before an appreciable portion of it is coupled out. The laser frequency must remain essentially constant during the transit time required for these multiple reflections. This limits the upper modulation frequency.

Modulation of the laser-cavity length produces a conventional FM signal with modulating signal directly proportional to change in laser-cavity length. Demodulation is conveniently accomplished by optical heterodyning to lower rf frequencies where conventional FM demodulation techniques can be used.

56. External Modulation. The Bragg modulator (Fig. 14-69) is commonly used to modulate the optical frequency. As such it produces a single-sideband suppressed-carrier type of modulation.

Demodulation can be achieved by optical heterodyning to lower rf frequencies, where conventional techniques can be employed for this type of modulation. It is also possible to reinsert the carrier at the transmitter for a frequency reference. This is done by using optical-beam splitters to combine a portion of the unmodulated laser beam with the Bragg modulator output.

Conventional double-sideband amplitude modulation has also been achieved by simultaneously modulating two laser beams (derived from the same source) with a common Bragg modulator to obtain signals shifted up and down. Optical-beam splitters are used to combine both signals with an unmodulated carrier.[9] Conventional power detection demodulates such a signal.

Optical phase modulation is commonly accomplished by passing the laser output beam through an electrooptical material, with the polarization vector oriented along the modulated ellipsoid axis of the material. Demodulation is conveniently achieved by optical heterodyning to rf frequencies, FM demodulation, and integrating to recover the phase modulation in the usual manner.

For low modulation bandwidths, the electrooptical material can be replaced by a mechanically driven mirror. The light reflected from the mirror is phase-modulated by the changes in the mirror position. This effect is often described in terms of the Doppler frequency shift, which is directly proportional to the mirror velocity and inversely proportional to the optical wavelength.

57. Traveling-Wave Modulation. In the electrooptical and magnetooptical modulators described thus far, it is assumed that the modulating signal is essentially constant during the optical transit time through the material. This sets a basic limit on the highest modulating frequency that can be used in a lumped modulator.

This problem is overcome in a traveling-wave modulator. The optical wave and the modulation signal propagate with equal phase velocities through the modulating medium, allowing the modulating fields to act on the optical wave over a long path, regardless of how rapidly the modulating fields are changing. The degree to which the two phase velocities can be matched determines the maximum interaction length possible.

Reference 10 describes such a traveling-wave optical modulator using microwave modulation frequencies in a carbon disulfide Kerr cell.

58. Optical Heterodyning. Two collimated optical beams, derived from the same laser source and illuminating a common surface, produce straight-line interference fringes. The distance between fringes is inversely proportional to the angle between the beams. Shifting the phase of one of the beams results in a translation of the interference pattern, such that a 2π-rad phase shift translates the pattern by a complete cycle. An optical detector having a sensing area small compared with the fringe spacing has a sinusoidal output as the sinusoidal intensity of the interference pattern translates across the detector.

A frequency difference between the two optical beams produces a phase difference between the beams that changes at a constant rate with time. This causes the fringe pattern to translate across the detector at a constant rate, producing an output at the difference frequency. This technique is known as *optical heterodyning* in which one of the beams is the signal beam, the other the local oscillator.

The effect of the optical alignment between the beams is evident. As the angle between the two collimated beams is reduced, the spacing between the interference fringes increases, until the spacing becomes large compared with the overall beam size. This permits a large detector which uses all the light in the beam. If converging or diverging beams are used instead of collimated beams, the situation is similar, except that the interference fringes are curved instead of straight. Making the image of the local-oscillator point coincide with the image of the signal-beam point causes the desired infinite fringe spacing.

Optical heterodyning provides a convenient solution to several possible problems in optical demodulation. In systems where a technique other than simple amplitude modulation has been used (e.g., single-sideband, frequency, or phase modulation) optical heterodyning permits shifting to frequencies where established demodulation techniques are readily available.

In systems where background radiation, such as from the sun, is a problem, heterodyning permits shifting to lower frequencies, so that filtering to the modulation bandwidth removes most of the broadband background radiation. The required phase front alignment also eliminates background radiation from spatial positions other than that of the signal source.

Many systems are limited by thermal noise in the detector and/or front-end amplifier. Cooled detectors and elaborate amplifiers are often used to reduce this noise to the point that photon noise in the signal itself dominates. This limit also can be achieved in an optical heterodyne system with noncooled detector and normal amplifiers by increasing the local-oscillator power to the point where photon noise in the local oscillator is the dominant noise source.[1] Under these conditions, the signal-to-noise power ratio is given by the following equation:

$$S/N = \eta\lambda P/2hBC$$

where S/N = signal-power–noise-power ratio, η = quantum efficiency of photo detector, λ = optical wavelength, h = Planck's constant, C = velocity of light, B = bandwidth over which S/N is evaluated, and P = optical signal power received by detector.

59. References on Optical Modulators and Demodulators

1. Born, M., and E. Wolf "Principles of Optics," Chap 14, Pergamon, New York, 1959.
2. Ibid., Chap. 12.
3. Fandman, A., H. Marantz, and V. Early Light Modulation by Means of Stark Effect in Molecular Gases-Application to CO_2 Lasers, *Appl. Phys. Lett.*, Dec. 1, 1969, pp. 357–360.
4. Nelson, D. F. The Modulation of Laser Light, *Sci. Am.*, June 1968.
5. Gordon, E. I. Figure of Merit for Acoustic-Optical Deflection and Modulation Devices, *IEEE J. Quantum Electron.* (correspondence), May 1966, pp. 104–105.
6. Gordon, E. I. A Review of Acousto-optical Deflection and Modulation Devices, *Proc. IEEE*, October 1966, pp. 1391–1401.
7. Adler, R. Interaction between Light and Sound, *IEEE Spectrum*, May 1967, pp. 42–54.
8. Fey, J. M., and R. J. Webb Low Voltage Light-Amplitude Modulation, *Electron. Lett.*, May 31, 1968, pp. 213–215.
9. Dixon, R. W., and E. I. Gordon Acoustic Light Modulator Using Optical Heterodyne Mixing, *Bell Syst. Tech. J.*, 1967, Vol. 46, p. 367.
10. Chenaweth, A. J., O. L. Gadely, and D. F. Holshouser Carbon Disulfide Traveling Wave Kerr Cells, *Proc. IEEE*, October 1966, pp. 1414–1418.
11. Pratt, W. K. "Laser Communication System," Chap. 10, Wiley, New York, 1969.
12. Kaminow, I. P., and E. H. Turner Electro-Optic Light Modulators, *Proc. IEEE*, October 1966, p. 1374.

Frequency Converters and Detectors

BY G. B. GAWLER

60. General Considerations of Frequency Converters. A frequency converter usually consists of an oscillator (called a *local oscillator* or LO) and a device used as a mixer. The mixing device is either nonlinear or its transfer parameter can be made to vary in synchronism with the local oscillator. A signal voltage with information in a frequency band centered at frequency f_s enters the frequency converter, and the information is reproduced in the intermediate-frequency (i.f.) voltage leaving the converter. If the local-oscillator frequency is designated f_{LO}, the i.f. voltage information is centered about a frequency $f_{if} = f_{LO} \pm f_s$. The situation is shown pictorially in Fig. 14-72. Characteristics of interest for design in systems using frequency converters are gain, noise figure, image rejection, spurious responses, intermodulation and cross-modulation capability, desensitization, local-oscillator to rf and to i.f. isolation. These characteristics will be discussed at length in the descriptions of different types of frequency-converter mixers and their uses in various systems. First, explanations are in order for the above terms.

Frequency-Converter Gain.   The available power gain of a frequency converter is the ratio of power available from the i.f. port to the power available at the signal port. Similar definitions apply for transducer gain and power gain.

Noise Figure of Frequency Converter.   The noise factor is the ratio of noise power available at the i.f. port to the noise power available at the i.f. port due to the source alone at the signal port.

Image Rejection.   For difference mixing $f_{if} = f_{LO} - f_s$, and the image is $2f_{LO} - f_s$. For sum mixing $f_{if} = f_{LO} + f_s$, and the image is $2f_{LO} + f_s$. An undesired signal at the difference mixing

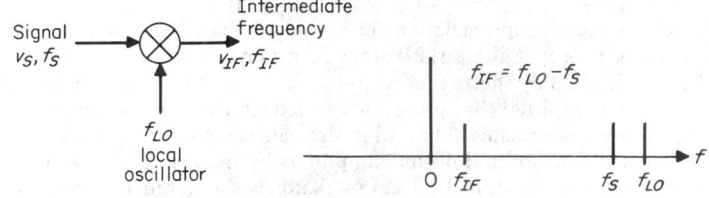

Fig. 14-72. Frequency-converter terminals and spectrum.

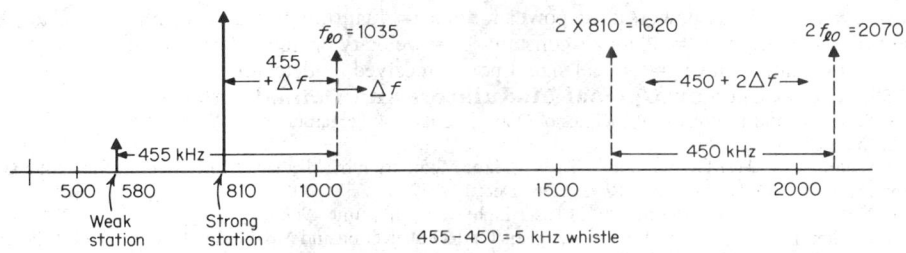

Fig. 14-73. Spurious response in AM receiver.

frequency $2f_{LO} - f_s$ results in energy at the i.f. port. This condition is called *image response* and attenuation of the image response is image rejection, measured in decibels.

Spurious Responses.   Spurious external signals reach the mixer and result in generation of undesired frequencies that may fall into the intermediate-frequency band.[26]* The condition for an interference in the i.f. band is

$$mf'_s \pm nf_1 = \pm f_{if}$$

where m and n are integers and f'_s represents spurious frequencies at the signal port of the mixer.

Example.   There is a strong local station in the broadcast band at 810 kHz and a weak distant station at 580 kHz. A receiver is tuned to the distant station, and a whistle, or beat, at 5 kHz is heard on the receiver (refer to Fig. 14-73).

An analysis shows that the second harmonic of the local oscillator interacts with the second harmonic of the 810-kHz signal to produce a mixer output at 450 kHz in the i.f. band of the receiver:

$$580 + 455 = 1,035 \text{ kHz} = \text{LO frequency}$$
$$2 \times 1,035 - 2 \times 810 = 450 \text{ kHz} = \text{i.f. interference frequency}$$

The interference at 450 kHz then mixes with the 455-kHz desired signal in the second detector to produce the 5-kHz whistle. Notice that if the receiver is slightly detuned upward by 5 kHz, the whistle will zero-beat. Further upward detuning will create a whistle of increasing frequency.

*Superior numbers correspond to numbered references, Par. **14-69.**

Karpen and Mohr[26] have shown that a mixer spurious response of signal order m is of the form

$$E(m, n) = k(m, n)E_S^m$$

or in terms of decibels,

$$P(m, n) = mP_s' + K(m, n)$$

where n is the order of the LO and

$$mf_s' \pm nf_{LO} = \pm f_{if}$$

and P denotes power (dBm or dBw).

Significance of the Formulation of $P(m, n)$.   Extensive measurements show that larger values of numbers m and n give smaller and smaller values of $P(m, n)$. Therefore, for practical purposes, a limited range of m and n can be selected for any mixer evaluation, and the numbers $K(m, n)$ can be found from a set of measurements on a mixer. Figure 14-74 shows some sample measurements from Karpen and Mohr.[26] The lines are labeled $n \times m$. Notice that higher orders give less output at a given input level.

As an example, take a double conversion system. A spur-chart analysis such as that in the ITT Handbook[27] or Hoigaard[24] can be performed to select first and second i.f. frequencies to avoid spurious responses for the first few orders of m, n. Then the numbers $K(m, n)$ above can be used to determine filtering required to keep spurious responses $P(m, n)$ below system threshold sensitivity. Charts of the $K(m, n)$ values for Schottky diode doubly balanced mixers are available.

For a simple numerical example (refer to Fig. 14-75), suppose the first i.f. is 100 MHz and a spurious signal is due to the first LO at $f_s' = 200$ MHz. Let $K(3, 1) = 35$, and require that spurious responses be below -110 dBm. What is the required filtering at 200 MHz in the 100-MHz i.f. strip?

$$-110 \text{ dBm} = P(3, 1) = 3P_s' + K(3, 1) = 3P_s' - 35$$

or

$$P_s' = \frac{1}{3}(-110 + 35) = -25 \text{ dBm}$$

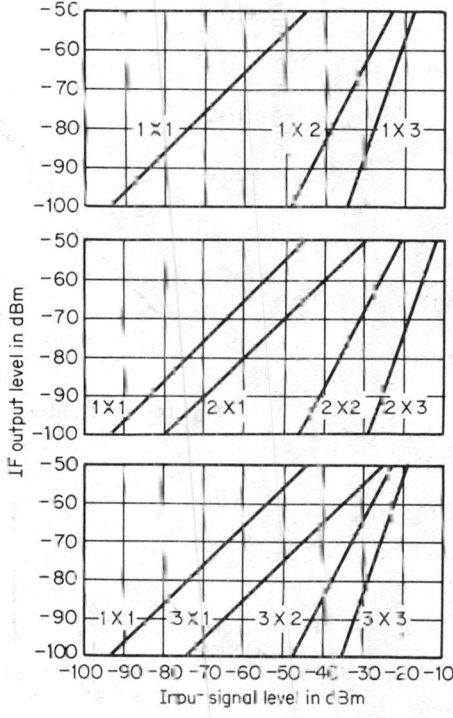

Fig. 14-74. Spurious-response behavior of a balanced mixer.[26]

That is, the level of the 220-MHz leakage from the first LO at the second mixer input must be no higher than -25 dBm.

61. Intermodulation. Intermodulation is particularly troublesome because a pair of strong signals that pass through a receiver preselector can cause interference in the i.f. passband, even though the strong signals themselves do not enter the passband. Epstein et al.[29] show typical generating mechanisms for intermodulation and cross-modulation in mixers.

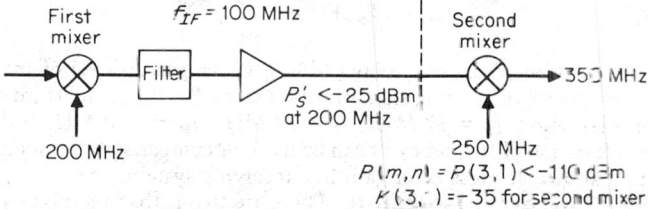

Fig. 14-75. Spurious-response analysis.

Consider two undesired signals at 97 MHz passing through a superheterodyne receiver tuned to 100 MHz. Suppose, further, that the i.f. is so selective that a perfect mixer allows no response to the signals (see Fig. 14-77). Third-order intermodulation in a physically realizable mixer will result in interfering signals at the i.f. frequency and 9 MHz away (corresponding to 100 and 91 MHz rf frequencies, respectively). Fifth-order intermodulation will produce interferences 3 and 12 MHz from the intermediate frequency (103 and 88 MHz rf frequencies).

There is a formula for variation of intermodulation products that is quite useful. Figure 14-76 shows typical variations of desired output and intermodulation with input power level. Desired

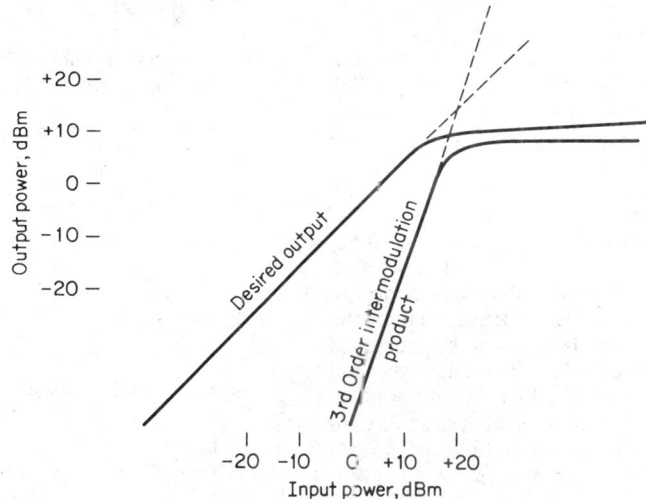

Fig. 14-76. Third-order intermodulation intercept power.

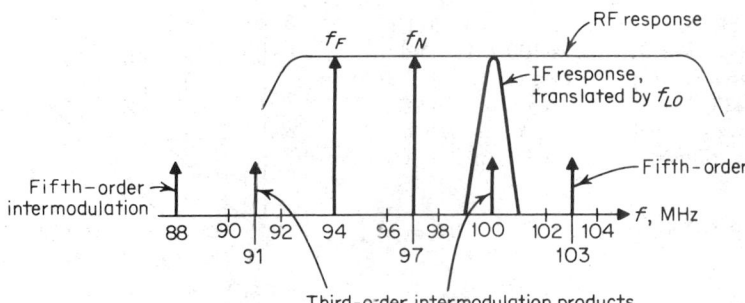

Fig. 14-77. Intermodulation in a superheterodyne receiver.

output increases 1 dB for each 1-dB increase of input level, whereas third-order intermodulation increases 3 dB for each 1-dB increase of input level. At some point the mixer saturates and the above behavior no longer obtains. Since the interference of the intermodulation product is primarily of interest near the system sensitivity limit (usually somewhere below -20 dBm), the 1 dB per 1 dB and 3 dB per 1 dB patterns hold. The formula can be written

$$P_{21} = 2P_N + P_F - 2P_{I_{21}}$$

where P_{21} = level of intermodulation product (dBm), P_N = power level of interfering signal nearest P_{21}, and P_F = power interfering signal farthest from P_{21}. $P_{I_{21}}$ is the third-order intercept power. For proper orientation, f_N = 97 MHz, f_F = 94 MHz, f_{21} = 100 MHz in Fig. 14-77. The intercept power is a function of frequency. It can be used for comparisons between mixer designs and for determining allowable preselector gain in a receiving system.

62. Frequency-Converter Isolation. There are two paths in a mixer where isolation is important. The so-called *balanced mixers* give some isolation of the local-oscillator energy at the rf port. This keeps the superheterodyne receiver from radiating excessively. The doubly balanced mixers also give rf-to-i.f. isolation. This keeps interference in the receiver rf environment from penetrating the mixer directly at the i.f. frequency. Less important, but still significant, is the LO-to-i.f. isolation. This keeps LO energy from overloading the i.f. amplifier. Also, in multiple-conversion receivers low LO-to-i.f. leakage minimizes spurious responses in subsequent frequency converters.

Desensitization. A strong signal in the rf bandwidth not directly converted to i.f., drives the operating point of the mixer into a nonlinear region. The mixer gain is then either decreased or increased. In radar, the characteristic of concern is pulse desensitization. In television receivers the characteristic is called cross-modulation. Here the strong undesired adjacent TV station modulates the mixer gain, especially during synchronization intervals, where the signal is strongest. The result appears in the desired signal as a contrast modulation of picture with the pattern of the undesired sync periods, corresponding to mixer gain pumping by the strong adjacent channel.

63. Schottky Diode Mixers. The Schottky barrier diode is an improvement over the point-contact diode. The Schottky diode has two features that make it very valuable in high-frequency mixers: (1) it has low series resistance and virtually no charge storage, which results in low conversion loss; (2) it has a noise-temperature ratio very close to unity. The noise factor of a mixed-i.f. amplifier cascade is[1]

$$F = L_M(t_D + F_{if} - 1)$$

where L_M = mixer loss, t_D = diode noise-temperature ratio, and F_{if} = i.f. noise factor. Since t_D is near unity and L_M is in the range[2] of 2.4 to 6 dB, overall noise factor is quite good, with F_{if} near 1.5 dB in well-designed systems.

Basic theory for diode mixer operation is given in Refs. 5 to 7 (Par. **14-69**). Torrey and Whitmer[5] show that the complete conversion matrix involves LO harmonic sums and differences, as well as signal, i.f., and image frequencies. They restrict their treatment of crystal rectifiers to the third-order matrix

$$\begin{bmatrix} I_1 \\ I_2 \\ I_3^* \end{bmatrix} = [Y] \begin{bmatrix} V_1 \\ V_2 \\ V_3^* \end{bmatrix} \quad Y = \begin{bmatrix} y_{11} & y_{12} & y_{13} \\ y_{21} & y_{22} & y_{23} \\ y_{31} & y_{32} & y_{33} \end{bmatrix}$$

where 1 denotes signal port; 2, i.f. port; and 3, image port.

With point-contact diodes, the series resistance is so large that not much improvement is realized by terminating the image frequency, and terminating the other frequencies involved is less significant.

With the advent of Schottky barrier diodes, which have much smaller series resistances, proper termination of pertinent frequencies, other than signal and i.f. frequencies, results in a minimizing of conversion loss. This, in turn, leads to a minimizing of noise figure.

An outline of the mathematical approach to conversion-loss minimization involves the maximum available gain concept discussed by Fukui.[30] First, let the conversion-loss matrix be reduced to a two-port matrix for signal and i.f. frequencies by termination of all other pertinent frequencies:

$$\begin{bmatrix} I_1 \\ I_2 \end{bmatrix} = \begin{bmatrix} y_{11} & y_{12} \\ y_{21} & y_{22} \end{bmatrix} \begin{bmatrix} V_1 \\ V_2 \end{bmatrix}$$

where 1 = signal port and 2 = i.f. port.

Then the maximum available gain is

$$\left| \frac{y_{21}}{y_{12}} \right| \frac{1}{k + \sqrt{k^2 - 1}} \qquad \text{where } k = \frac{2g_{11}g_{22} - \text{Re } y_{12}y_{21}}{|y_{12}y_{21}|}$$

Although maximum available gain corresponds to minimum available loss, the process is not so simple. There are constraints on how pertinent frequencies (such as the image) are terminated, because a diode has only one physical port and it is difficult to control the impedances presented to all these frequencies. Nevertheless, conceptually the effort should be made to terminate the pertinent frequencies in such a manner that signal-i.f. two-port having the largest value of maximum available gain is obtained.

Herold et al.[6] discuss termination of the image frequency, while Johnson[4] includes termination of the sum frequency as well as termination at harmonic sums and differences. Konishi[2] has measured a 4-dB noise figure for an image-terminated mixer at 12 GHz. Cong et al[31] have reported a mixer noise temperature of 860 K at 115 GHz. This corresponds to a 6-dB noise figure. Appropriate terminations of LO harmonics are a key factor.

Several different configurations are used with Schottky mixers. Figure 14-78 shows an image-rejection mixer, which is used for low i.f. frequency systems where rf filtering of the image is impractical. Kurpis and Taub[3] report 20-dB image rejection, 11-dB noise figure (including i.f.), and 8- to 12-GHz frequency range. Their image-rejection mixer is balanced and has a 200-Mhz i.f.

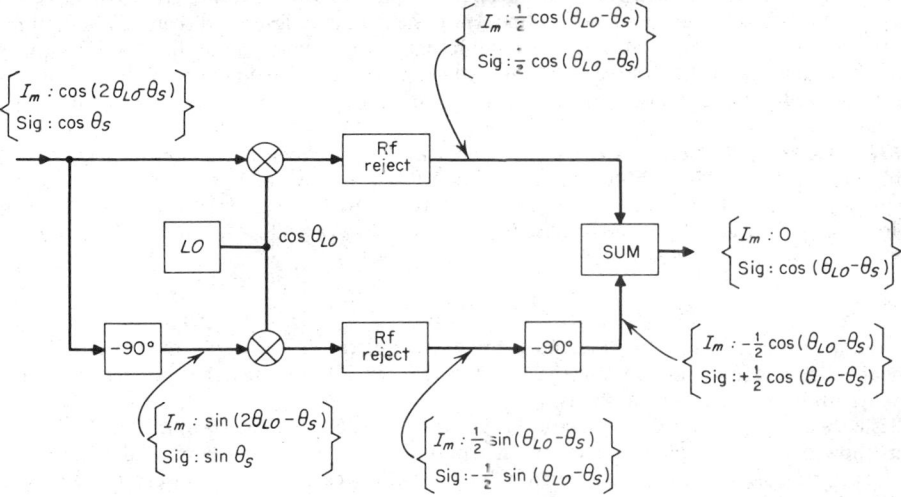

Fig. 14-78. Mixer designed for image rejection.

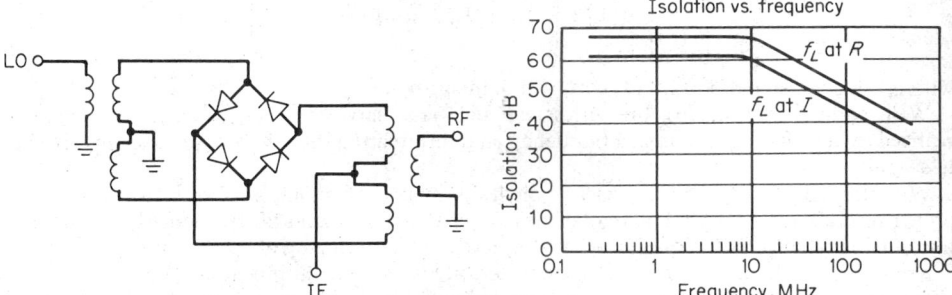

Fig. 14-79. Doubly balanced mixer.

Fig. 14-80. Rf-to-i.f. isolation in doubly balanced mixer. (*Model MI, RELCOM Division of Watkins-Johnson.*)

There is a general rule of thumb for obtaining good intermodulation, cross-modulation, and desensitizable performance in mixers. It has been found experimentally that pumping a mixer harder extends its range of linear operation. The point-contact diode had a rapidly increasing noise figure with high LO power level and could easily burn out with too much power. The Schottky diode, however, degrades in noise figure relatively slowly with increasing LO power, and it can tolerate quite large amounts of power without burnout. There is a limit to this process of increasing LO power: the Schottky diode series resistance begins to appear nonlinear. This leads to another rule of thumb: pump the diode between two linear regions, and spend as little time as possible in the transition region. Application of these two rules leads to the doubly balanced Schottky mixer. The reason for this is that one pair of diodes conducts hard and holds the other pair off. Hence large LO power is required, and one diode pair is conducting well into its linear region while the other diode pair is held in its linear nonconducting region.

64. Doubly Balanced Mixers. The diode doubly balanced mixer, or ring modulator, is shown in Fig. 14-79. This configuration has been analyzed by Caruthers,[8] Tucker,[9] and Belevitch.[10] The doubly balanced mixer is used up to and beyond 1 GHz in this configuration. The noise-figure optimization process previously discussed applies to this type of mixer. It exhibits good LO-to-rf and LO-to-i.f. isolation, as shown in Fig. 14-80. Typical published data on mixers quote 30 dB rf-to-i.f. isolation below 50 MHz and 20 dB isolation from 50 to 500 MHz.

Figure 14-81 gives a list of spurious responses for 50-MHz operation of an available mixer. With reference to the spurious-response formula (Par. **14-50**) relating $P(m, n)$, P'_s, and $K(m, n)$, the numbers shown in Fig. 14-81 are $P'_s - P(m, n)$ in decibels for $P'_s = 10$ dBm. From these

Harmonic intermodulation signals

Harmonics of f_R	0	f_L	$2f_L$	$3f_L$	$4f_L$	$5f_L$	$6f_L$	$7f_L$	$8f_L$
$7f_R$	98	93	>100	90	94	84	>100	83	95
$6f_R$	97	89	96	90	97	95	>100	92	>100
$5f_R$	83	73	86	70	92	67	92	67	82
$4f_R$	85	82	96	85	95	84	96	85	91
$3f_R$	66	61	64	55	65	51	69	44	72
$2f_R$	70	65	79	65	82	66	80	67	77
f_R	24	0	37	13	40	23	43	32	45
0		27	37	26	53	34	54	38	55

Values of $P'_s - P(m,n)$ for $P'_s = -10$ dBm

Fig. 14-81. Spurious-response chart of mixer shown in Fig. 14-80. (*RELCOM Division of Watkins-Johnson.*)

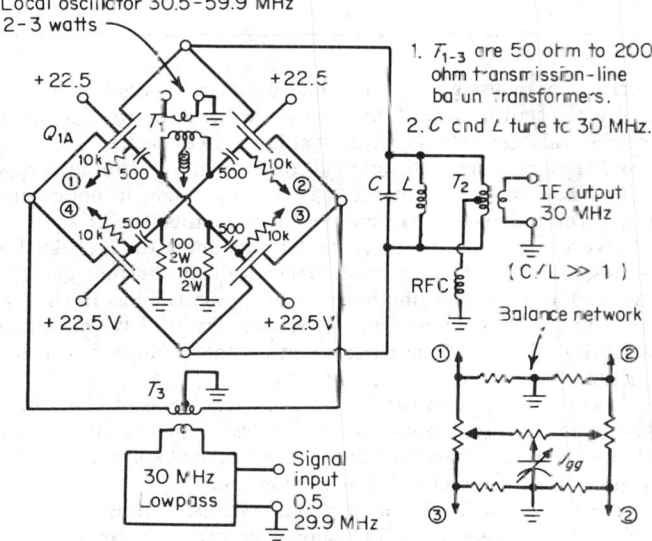

Fig. 14-82. Broadband MOSFET mixer of high dynamic range.

values an array of $K(m, n)$ can be calculated. Then other sets of $K(m, n)$ values can be found by measurement for other frequencies. The information for this mixer can be compared with similar data for other mixers to determine optimum mixer designs for spurious-response performance. Of course, the data can also be used to determine filtering requirements in design of superheterodyne receiving systems. Finally, an intermodulation ratio of -60 dB for -10 dBm applied is quoted. This corresponds to a third-order intercept power of $+20$ dBm.

References 32 and 33 describe the Volterra series techniques for calculating intermodulation characteristics of mixer divices.

Another feature of balanced mixers is their LO noise-suppression capability. Reference 15 gives a formula for LO noise suppression in point-contact diode mixers. Modern mixers using Schottky diodes in a ring modulator provide somewhat better LO noise suppression. The ring modulator provides suppression only to AM LO noise, not FM noise.

Balanced mixers can take forms other than the ring modulator. Rafuse[18] has described a doubly balanced mixer using MOS field-effect transistors to achieve high dynamic range. With a local-oscillator power of 3 W at 40 MHz, he measures a third-order intercept power of $+60$ dBm. The MOSFET mixer is shown in Fig. 14-82.

Another balanced mixer uses a pair of high-gain junction FETs at 250 MHz. The reported third-order intercept is $+32$ dBm, with noise figure at 6.5 dB and gain at 3 dB.

65. Parametric Converters. Parametric converters make use of time-varying energy-storage elements. Their operation is in many ways similar to that of parametric amplifiers (see Sec. **13**). The difference is that output and input frequencies are the same in parametric amplifiers, while the frequencies differ in parametric converters. The device most widely used for microwave parametric converters today is the varactor diode, which has a voltage-dependent junction capacitance. The time variation of varactor capacitance is provided by a local oscillator, usually called the *pump*.

Attainable gain of a parametric converter is limited by the ratio of output to input frequencies. Therefore up conversion is generally used to achieve some gain. Because lower-sideband up conversion results in negative resistance,[12] the upper sideband is generally used. This results in simpler circuit elements to achieve stability.

Gemulla[13] and Maninger[14] mention achievable gain of 2 to 3 dB and noise figure of about 3 dB in the X-band region. Maninger quotes a third-order intermodulation intercept of $+10$ dBm, while Gemulla gives $+5$ dBm at 100 mW pump power. Maninger reports these spurious responses, $mf_2 - nf_{LO} = \pm f_{if}$:

m	n	Interfering input level, dBm	Distortion product level referred to input, dBm
2	1	-4	-95
1	2	-75	-90

There is a distinct advantage to up conversion; image rejection is easily achievable by a simple low-pass filter. Finally, the above values of intercept power are quite good, compared with values achievable with doubly balanced Schottky mixers at lower frequencies.

66. Transistor Mixers. One of the original concerns in transistor mixers was their noise performance.[19,20] The base spreading resistance r_b is very important in noise performance. The reason is that mixing occurs across the base-emitter junction; then the i.f. signal is amplified by transistor action. However, r_b is a lossy part of the termination at the i.f., signal, image, and all other frequencies present in the mixing process. Hence r_b dissipates some energy at each of the frequencies present, and all these contributions add to appear as a loss in the signal-to-i.f. conversion. This loss, in turn, degrades noise figure. Vogel and Strutt[20] have measured as low as a 3-dB noise figure at 1 MHz for an OC45 germanium unit. Other values range as high as 15 dB. A 6-dB noise figure at 200 MHz has been reported.[34]

Manufacturers do not promote transistors used as mixers, probably because of their inter-modulation and spurious-response performance. Estimates of intermodulation intercept power go as high as $+12$ dBm, while one measurement gave $+5$ dBm at 200 MHz. However, a cascode transistor mixer is used in a commercial VHF television tuner.

Characterization of Linearity in TV Tuners   The television tuner is a good example for discussing nonlinearities in mixers. Table 14-2 shows the various types of interferences generated, and they all result from the third-order, or cubic, nonlinearity. This explains the appearance of P_{h1} in each formula. Cross-modulation is the nonlinearity observed for characterizing the tuner, while the other three types actually cause interferences observed on the picture tube. Figure 14-83 shows the frequencies involved in the 920-kHz beat generation.

Investigators who measure cross-modulation performance in TV tuners usually apply the

TABLE 14-2   Interference Mechanisms in TV Tuners*

Interference	Mechanism	Formula
Intermodulation	$(v_N + v_F)^3 \rightarrow 3v_N^2 v_F$	$P_{21} = 2P_N + P_F - 2P_{h1}$
Cross-modulation	$[v_U(1 + m_U \cos \theta_M) + v_D]^3 \rightarrow 6m_U \cos \theta_M v_U^2 v_D$	$20 \log (m_X/m_U) = 2(P_U - P_{h1}) + 12$ dB
Desensitization	$(v_U + v_D)^3 \rightarrow 3v_U^2 v_D$	$20 \log |\Delta G/G| = 2(P_U - P_{h1}) + 6$ dB
920-kHz beat	$(v_S + v_P + v_C)^3 \rightarrow 5v_S v_P v_C$	$P_{beat} = P_S + P_P + P_C - 2P_{h1} + 6$ dB

*All power in decibels referred to 1 mW; U = undesired, D = desired, S = sound carrier, P = picture carrier, C = chroma, m_X = cross-modulation index, m_U = modulation index on undesired signal.

undesired signal at 6 MHz or so from band center of the desired signal. This allows the rf-mixer interstage filter to provide attenuation and does not give a true measure of mixer linearity. The measurement does provide information on performance in the presence of a strong adjacent channel, because there is a correspondence between cross-modulation and the densensitization due to the sync tips of the adjacent channel. However, the 920-kHz beat that occurs due to the mixer is not controlled by results of the above measurement. The latter phenomenon results

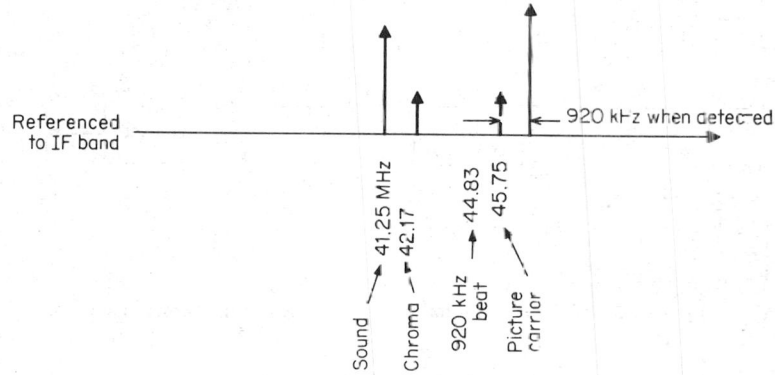

Fig. 14-83. Generation of 920-kHz beat in TV tuners.

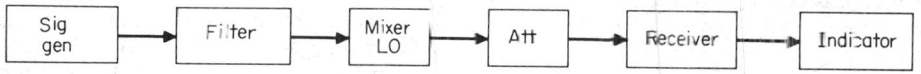

Fig. 14-84. Test equipment for measuring mixer spurious responses.

from mixing of the picture chrominance and sound carriers to produce a distortion product 920 kHz away from the picture carrier. The three carriers and the distortion product are all within the desired passband. Hence the cross-modulation measurement taken with the undesired signal carrier 6 MHz away is not adequate. Nevertheless, the 920-kHz beat results from a third-order phenomenon and is therefore related to both the cross-modulation and intermodulation phenomena. An extra set of measurements taken near band center would suffice to predict performance for the 920-kHz beat.

67. Measurement of Spurious Responses. Figure 14-84 shows an arrangement for measuring mixer spurious responses. The filter following the signal generator implies that generator harmonics are down, say 40 dB. This ensures that frequency-multiplying action is due only to the mixer under test. The attenuator following the mixer can be used to be sure that a spurious response of the receiver is not being measured. That is, a 6-dB change in attenuator setting should be accompanied by a 6-dB change on the indicator.

Generally the most convenient way of performing the spurious-response test is first to obtain an indication of the indicator. Then tune the signal generator to the desired frequency and record the level required to obtain the original response. This should be repeated at one or two more levels of the undesired signal to ensure that the spur follows the appropriate laws. For example, if the response is fourth-order (4 times the signal frequency ± n times the LO frequency), the measured value should change 4 db for a 1-dB change in undesired frequency level. The order of the spurious response can be determined by either of two methods. The first method is simply by knowing with some accuracy the undesired signal frequency and the LO frequency and then determining the harmonic numbers required to obtain the i.f. frequency. The other technique entails observing the incremental changes of the i.f. frequency with known changes in the undesired signal frequency and the LO signal frequency.

This completes the measurement for one spurious response. The procedure should be repeated for each of the spurious responses to be measured.

The intermodulation test setup is shown in Fig. 14-85. In general, a diplexer is preferable to a directional coupler for keeping generator 1 signal out of generator 2. This is necessary so that the measurement is not limited by the test setup. A good idea would be to establish that no third-order intermodulation occurs due to the setup alone. To do this, initially remove the mixer-LO circuit. Then tune generator 1 off from center frequency to about 10 or 20 dB down on the skirt of the receiver preselector. Tune generator 2 twice this amount from the receiver center frequency. Set generator levels equal and at some initial value, say −30 dBm. Then vary one generator frequency slightly and look for a response peak on the indicator. If none is noticed,

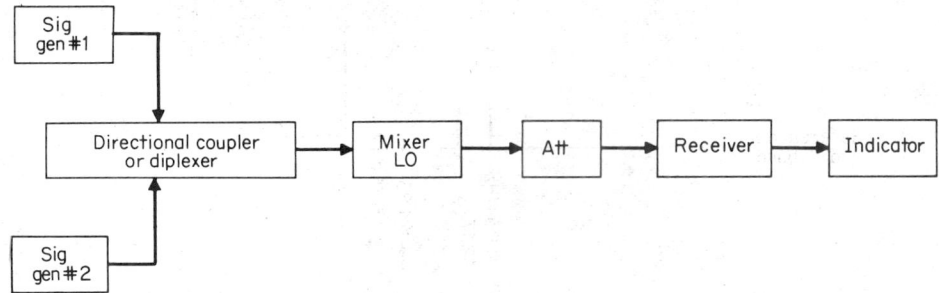

Fig. 14-85. Test equipment for measurement of mixer intermodulation.

increase the generator level to −20 dBm and repeat the procedure. Usually, except for very good receivers, the third-order intermodulation response is found. Vary the attenuator by 6 dB, and look for a 6-dB variation in the indicator reading. If the latter is not 6 dB but 18 dB, intermodulation is occurring in the receiver. If the indicator variation is between 6 and 18 dB, intermodulation is occurring in the circuitry preceding the attenuator and in the receiver. To obtain trustworthy measurements with a mixer in the test position, the indicator should read at least 20 dB greater than without the mixer, while the generator levels should be lower by mixer gain +10 dB than they were without the mixer. This ensures that the test setup is contributing an insignificant amount to the intermodulation measurement.

With the mixer in test position and the above conditions satisfied, obtain a reading on the indicator and let the power referred to the mixer input be denoted by P (dBm). Turn down both generator levels, and retune generator 1 to center frequency. Adjust generator 1 level to obtain the previous indicator reading. This essentially calibrates the measurement setup. Denote generator 1 level referred to the mixer input by P_{21} dBm. Then the intermodulation intercept power is given by

$$P_{I_{21}} \text{ dBm} = (3P - P_{21})/2$$

The subscripts on intercept power $P_{I_{21}}$ refer to second order for the near frequency and first order for the far frequency (see Fig. 14-76).

The procedure should be repeated for one or two lower values of P. The corresponding values of $P_{I_{21}}$ should asymptotically approach a constant value. The constant value of $P_{I_{21}}$ so obtained is then a valid number for predicting behavior of the mixer near its sensitivity limit.

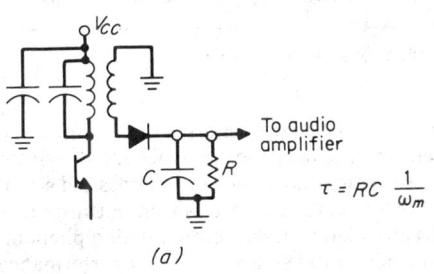

$$\tau = RC \; \frac{1}{\omega_m}$$

(a)

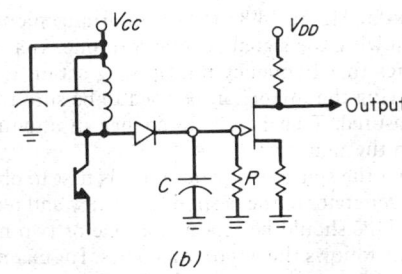

(b)

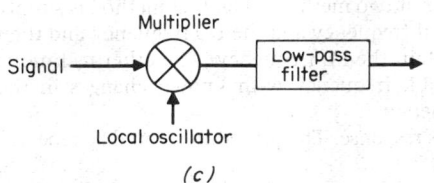

(c)

Fig. 14-86. AM detectors: *(a)* AM envelope detector; *(b)* peak detector; *(c)* product detector.

68. Detectors (Frequency Deconverters). Detectors have become more complex and versatile since the advent of integrated circuits. Up to the mid-1950s most radio receivers used the standard single-diode envelope detector for AM and a Foster-Seeley discriminator or ratio detector for FM. Today, integrated circuits are available with i.f. amplifier, detector, and audio-amplifier functions in a single package.

Figure 14-86 shows three conventional AM detectors. In Fig. 14-86a an envelope detector is shown. In order for the detected output to follow the modulation envelope faithfully, the RC time constant must be chosen so that $RC < 1/\omega_m$, where ω_m is the maximum angular modulation frequency in the envelope. Figure 14-86b shows a peak detector. Here the RC time constant

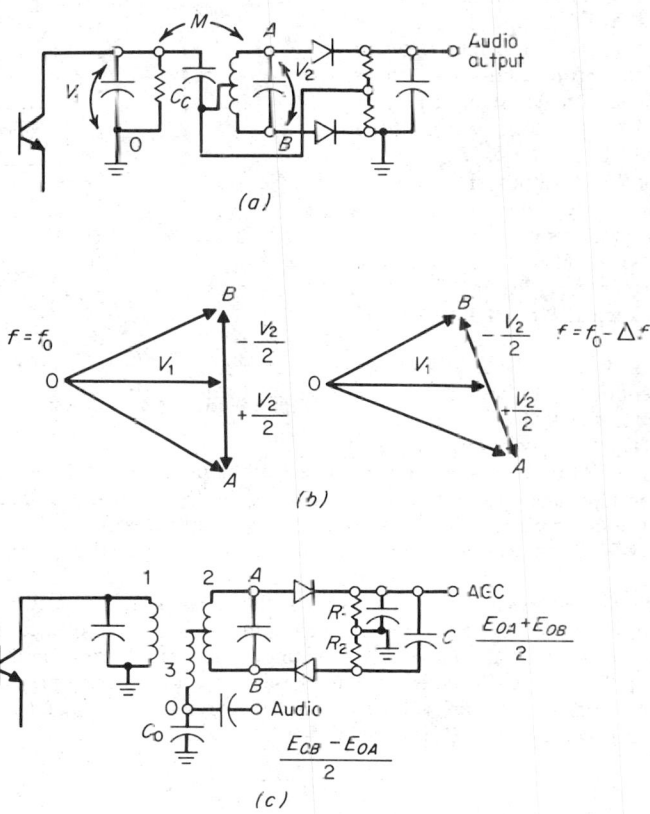

Fig. 14-87. FM detectors: (a) Foster-Seeley FM discriminator; (b) phasor diagrams; (c) ratio detector.

is chosen large, so that C stays charged to the peak voltage. Usually, the time constant depends on the application. In a television field-strength meter, the charge on C should not decay significantly between horizontal sync pulses separated by 62.5 μs. Hence a time constant of 1 to 6 ms should suffice. On the other hand, an AGC detector for single-sideband use should have a time constant of 1 s or longer.

Figure 14-86c shows a product (synchronous) detector. This type of detector has been used since the advent of single-sideband transmission. The product detector multiplies the signal with the LO, or beat frequency oscillator (BFO), to produce outputs at sum and difference frequencies. Then the low-pass filter passes only the difference frequency. The result is a clean demodulation with a minimum of distortion for single-sideband signals.

The two classical FM detectors widely used up to the present are the Foster-Seeley discriminator and the ratio detector. Figure 14-87 shows the Foster-Seeley discriminator and its phasor diagrams. The circuit consists of a double-tuned transformer, with primary and secondary voltages series-connected. The diode connected to point A detects the peak value of $V_1 + V_2/2$, and the diode at B detects the peak value of $V_1 - V_2/2$. The audio output is then the difference between the detected voltages. When the incoming frequency is in the center of the passband,

V_2 is in quadrature with V_1, the detected voltages are equal, and audio output is zero. Below the center frequency the detected voltage from B decreases, while that from A increases, and the audio output is positive. By similar reasoning, an incoming frequency above band center produces a negative audio output. Optimum linearity requires that $KQ = 2$, where K is the transformer coupling and Q is the primary and secondary quality factor.

Figure 14-87c shows a ratio detector, which has an advantage over the Foster-Seeley discriminator in being relatively insensitive to AM. The ratio detector uses a tertiary winding (winding 3) instead of the primary voltage, and one diode is reversed. However, the phasor diagrams also apply to the ratio detector. The AM rejection feature results from choosing the $(R_1 + R_2)C$ time constant large compared with the lowest frequency to be faithfully reproduced. The voltages E_{OA} and E_{OB} represent the detected values of rf voltages across OA and OB, respectively. With the large time constant above, voltage on C changes slowly with AM and the conduction angles of the diodes vary, loading the tuned circuit so as to keep the rf amplitudes relatively constant.

Capacitor C_0 is chosen to be an rf short circuit but small enough to follow the required audio variations. In the AM rejection process, AF voltage on C_0 does not follow the AM

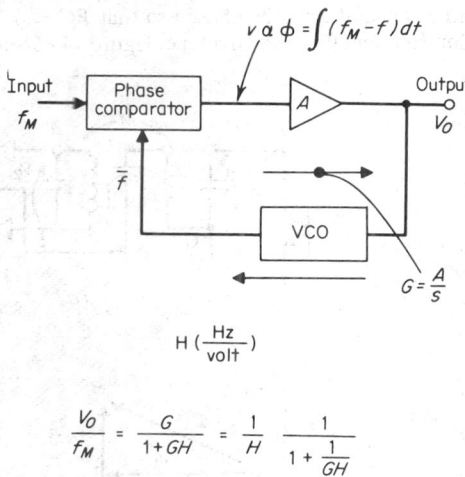

$$v \alpha \phi = \int (f_M - f) dt$$

$$H \left(\frac{Hz}{volt} \right)$$

$$\frac{V_0}{f_M} = \frac{G}{1 + GH} = \frac{1}{H} \frac{1}{1 + \frac{1}{GH}}$$

Fig. 14-88. FM detector using phase-locked loop.

because the charge put on by one diode is removed by the other diode. With FM variations on the rf, voltage on C_0 changes to reach the condition, again, that charge put on C_0 by one diode is removed by the other diode. The ratio detector is generally used with little or no previous limiting of the rf, while the Foster-Seeley discriminator must be preceded by limiters to provide AM rejection.

With the recent trend toward integrated circuits, there has been increased interest in using phase-locked loops and product detectors. These techniques have been selected because they do not require inductors, which are not readily available in integrated form. Figure 14-88 shows a phase-locked loop (PLL) as an FM detector. The phase comparator merely provides a dc voltage proportional to the difference in phase between signals represented by f_M and f. Initially, f and f_M are unequal, but because of high loop gain, $GH \gg 1$, f and f_M quickly become locked and

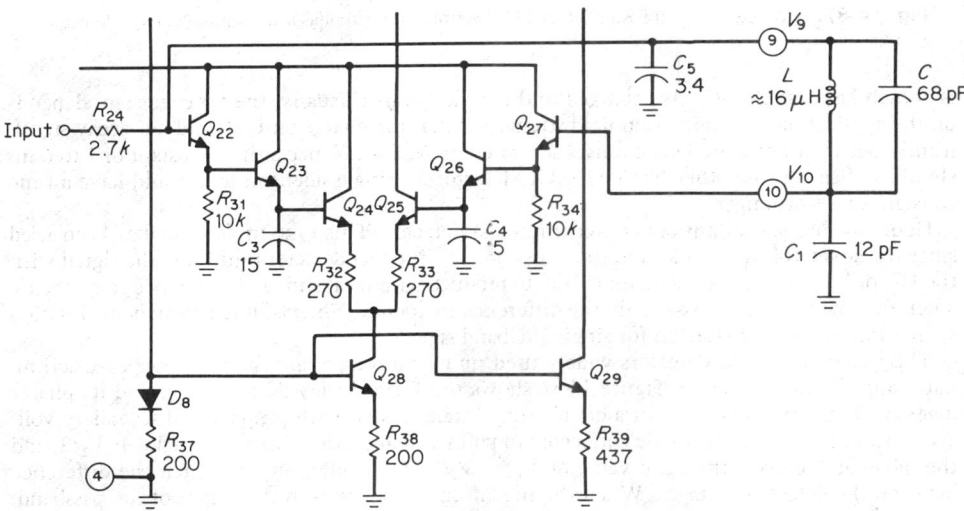

Fig. 14-89. Differential peak detector for TV FM sound, a portion of the CA3065 integrated circuit. (*RCA Corporation.*)

stay locked. Then as f_M varies, f follows exactly. But because of the high loop gain, response is essentially $1/H$, which is the voltage-controlled oscillator (VCO) characteristic. Hence the PLL serves as an FM detector. Grebene and Camenzind[24,25] give more detail on integrated-circuit implementation of the PLL technique.

AM product detectors also make use of the PLL to provide a carrier locked to the incoming signal carrier. The output of the VCO is used to drive the product detector. Grebene[25] describes such an arrangement in integrated-circuit form. Probably one of the most stringent uses of the product detector is in an FM stereo decoder. The *left minus right* (L − R) subcarrier is located at 38 kHz with sidebands from 23 to 53 kHz. There may also be an SCA signal centered about 67 kHz which is used to provide a music service for restaurants and commercial offices. The L − R product detector is driven by a 38-kHz VCO, the output of which also goes to a 2-to-1 counter. The counter output is compared with the 19-kHz pilot signal in a phase comparator, and the phase-comparator output then controls the VCO. Because of the relatively small pilot signal and the presence of L + R, L − R, and SCA information, the requirement for phase locking is stringent.

An interesting new FM detector is the *differential peak detector*. It is included in the RCA

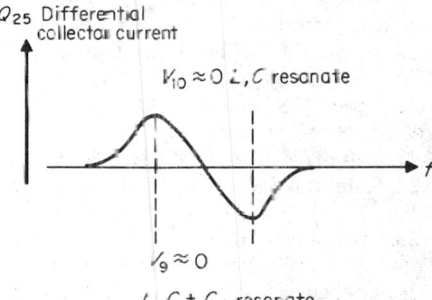

$$V_9 = F(\omega)\left[1-\omega^2 L(C+C_1)\right] \qquad C_1 = 12 \text{ pF}$$
$$V_{10} = F(\omega)(1-\omega^2 LC) \qquad C = 68 \text{ pF}$$
$$F(\omega) = \text{three pole transfer function}$$

Q_{25} Differential collector current

$V_{10} \approx 0$ L,C resonate

$V_9 \approx 0$

$L,C + C_1$ resonate

Fig. 14-90. Characteristic of differential peak FM detector.

CA3065 integrated circuit, which serves as a complete TV sound system (excluding audio output stage). The FM detector portion is shown in Fig. 14-89. Transistors Q23 and Q26 are peak detectors, and the difference of their detected voltages divided by R32 + R33 is the dc differential current at the Q25 collector. Transistors Q22 and Q27 are buffers for rf voltages. Figure 14-90 gives an idea of detector operation. Starting from low frequencies, voltages at Q23 and Q26 detectors are near equal, and differential output current is near zero. As frequency increases, the LC parallel circuit series-resonates with C_1, so that a short circuit appears at V_9 (pin 9). This is a maximum at V_{10} (pin 10). At higher frequencies LC approaches parallel resonance, where $V_{10} \approx 0$ and V_9 reaches maximum. This, along with the differential operation, results in the S shape required for FM detection.

69. References on Frequency Converters and Detectors

1. Osborne, T. L., L. U. Kibler, and W. W. Snell   Low-Noise Receiving Down Converter, *Bell Syst. Tech. J.*, July–August 1969, Vol. 48, No. 6, pp. 1651–1663.

2. Konishi, Y.   12-GHz-Band FM Receiver for Satellite Broadcasting, *IEEE Trans.*, 1978. Vol. MTT-26, No. 10, pp. 720–725.

3. Kurpis, G. P., and J. J. Taub   Wideband X-Band Microstrip Image Rejection Balanced Mixer, *IEEE Trans.*, 1970, Vol. MTT-18, No. 12, pp. 1181–1182.

4. Johnson, K. M.   X-Band Integrated Circuit Mixer with Reactively Terminated Image, *IEEE Trans.*, 1968, Vol. ED-15, No. 7, pp. 450–459.

5. Torrey, H. C., and A. C. Whitmer   "Crystal Rectifiers," M.I.T. Radiation Laboratory Series, Vol. 15, pp. 111–178, McGraw-Hill, New York, 1948.

6. Herold, E. W., R. R. Bush, and W. R. Ferris   Conversion Loss of Diode Mixers Having Image-Frequency Impedance, *Proc. IRE*, September 1945, Vol. 33, pp. 603–609.

7. Strum, P. D.   Some Aspects of Crystal Performance, *Proc. IRE*, July 1953, Vol. 41, pp. 875–889.

8. Caruthers, R. S.   Copper Oxide Modulations in Carrier Telephone Systems, *Bell Syst. Tech. J.*, April 1939, Vol. 18, pp. 315–337.

9. Tucker, D. G.   Intermodulation Distortion in Rectifier Modulators, *Wireless Eng.*, June 1954, Vol. 31, pp. 145–152.

10. Belevitch, V.   Non-linear Effects in Rectifier Modulators, *Wireless Eng.*, 1950, Vol. 27, p. 130.

11. Karpen, E. W., and R. J. Mohr   Graphical Presentation of Spurious Responses in Tunable Superheterodyne Receivers, *IEEE Trans.*, December 1966, Vol. EMC-8, No. 4, pp. 192–196.

12. Manley, J. M., and H. E. Rowe   Some General Properties of Non-linear Elements. I: General Energy Relations, *Proc. IRE*, July 1956, Vol. 44, pp. 904–913; II: Small Signal Theory, May 1958, Vol. 46, pp. 850–860.

13. Gemulla, W. J.   Parametric Up-Converters for Low-Noise Broadband Microwave Receivers, *WESCON Tech. Pap.*, 1970, Vol. 14, Sess. 9, Pap. 3.

14. Maninger, L.   Wideband Receivers Using Varacter Diode Frequency Upconverters, *Microwave J.*, August 1968, Vol. 11, pp. 49–55.

15. Pound, R. V.   "Microwave Mixers," M.I.T. Radiation Laboratory Series, Vol. 16, Chap. 6, McGraw-Hill, New York, 1948.

16. Ohtomo, M.   Experimental Evaluation of Noise Parameters in Gunn and Avalanche Oscillators, *IEEE Trans.*, July 1972, Vol. MTT-20, No. 7, pp. 425–437.

17. Taub, J. J., and P. J. Giordano   Use of Crystals in Balanced Mixers, *IRE Trans.*, July 1954, Vol. MTT-2, pp. 26–38.

18. Rafuse, R. P.   Symmetric MOSFET Mixers of High Dynamic Range, *ISSCC Dig. Tech. Pap.*, 1968, pp. 122–123.

19. Webster, R.   The Noise Figure of Transistor Converters, *IRE Trans.*, November 1961, Vol. BTR-7, pp. 50–65.

20. Vogel, J. S., and M. J. O. Strutt   Noise in Transistor Mixers, *Proc. IEEE*, February 1963, Vol. 51, No. 2, pp. 340–349.

21. Van der Ziel, A.   "Noise," Prentice-Hall, Englewood Cliffs, N.J., 1954.

22. Knight, M. B.   A New Miniature Beam-Deflection Tube, *RCA Rev.*, June 1960, pp. 266–289.

23. Schlesinger, K.   The Synchrotector: A Sampling Detector for Television Sound, *IRE Trans.*, July 1956, Vol. BTR-2, pp. 34–42.

24. Grebene, A. B., and H. R. Camenzind   Frequency-Selective Integrated Circuits Using Phase-Lock Techniques, *IEEE J. Solid State Circ.*, August 1969, Vol. SC-4, No. 4, pp. 216–225.

25. Grebene, A. B.   An Integrated Frequency-Selective AM/FM Demodulator, *IEEE Trans. Broadcast TV Receivers*, May 1971, Vol. BTR-17, pp. 71–80.

26. Karpen, E. W., and R. J. Mohr   Graphical Presentation of Spurious Responses in Tunable Superheterodyne Receivers, *IEEE Trans. Electromagn. Compatibility*, December 1966, Vol. EMC-8, No. 4, pp. 192–196.

27. "Reference Data for Radio Engineers," 4th ed., p. 774, International Telephone and Telegraph Corp., 1962.

28. Hoigaard, J. C.   Spurious Frequency Generation in Frequency Converters, *Microwave J.*, July 1967, pp. 61–64; August 1967, pp. 78–82.

29. Ebstein, B., R. Huenemann, and R. Sea   The Correspondence of Intermodulation and Cross Modulation in Amplifiers and Mixers, *Proc. IEEE*, August 1967, Vol. 55, No. 8, pp. 1514–1516.

30. Fukui, H.   Available Power Gain, Noise Figure, and Noise Measure of Two-Ports and Their Graphical Representations, *IEEE Trans. Circ. Theory*, June 1966, Vol. CT-13, No. 2, pp. 137–142.

31. Cong, H., A. R. Kerr, and R. J. Mattauch   The Low-Noise 115-GHz Receiver on the Columbia-GISS 4-Ft Radio Telescope, *IEEE Trans.*, 1979, Vol. MTT-27, No. 3, pp. 245–248.

32. Swerdlow, R. B.   Analysis of Intermodulation Noise in Frequency Converters by Volterra Series, *IEEE Trans.*, 1978, Vol. MTT-26, No. 4, pp. 305–313.

33. Narayanan, S., and H. C. Poon   An Analysis of Distortion in Bipolar Transistors Using Integral Charge Control Model and Volterra Series, *IEEE Trans.*, 1973, Vol. CT-20, No. 4, pp. 341–351.

34. "Texas Instruments Transistor and Diode Data Book," 1st Ed., 1973. See Device Type TIS126, pp. 4-538 to 4-540.

Power Electronics

W. E. NEWELL *(Deceased) Technical Staff, Westinghouse Research Laboratories; Senior Member, IEEE*

P. F. PITTMAN *Technical Staff, Westinghouse Research Laboratories; Senior Member, IEEE*

J. C. ENGEL *Technical Staff, Westinghouse Research Laboratories*

PETER WOOD *Technical Staff, Westinghouse Research Laboratories*

T. M. HEINRICH *Technical Staff, Westinghouse Research Laboratories; Senior Member, IEEE*

R. M. OATES *Technical Staff, Westinghouse Research Laboratories; Member, IEEE*

L. GYUGYI *Technical Staff, Westinghouse Research Laboratories*

J. W. MOTTO *Formerly staff member, Westinghouse Research Laboratories*

B. R. PELLY *Formerly staff member, Westinghouse Research Laboratories*

CONTENTS

Numbers refer to paragraphs

Introduction*

BY W. E. NEWELL

1. Power Electronics Defined. Power-electronics technology deals with the application of electronic power devices and associated components to the conversion, control, and conditioning of electric power. The primary characteristics of electric power which are subject to control include its basic form (ac or dc), its effective voltage or current (including the limiting cases of initiation and interruption of conduction), its frequency, and power factor (if ac). The control of electric power is frequently desired as a means for achieving control or regulation of one or more nonelectrical parameters, e.g., the speed of a motor, the temperature of an oven, the rate of an electrochemical process, or the intensity of lighting.

2. Efficiency Requirements. Aside from the obvious difference in function, power-electronics technology differs markedly from the technology of low-level electronics for information processing in that much greater emphasis is required on achieving high power efficiency. Few low-level circuits exceed a power efficiency of 15%, but few power circuits can tolerate a power efficiency less than 85%. High efficiency is vital, first, because of the economic value of the wasted power and, second, because of the cost of dissipating the heat it generates. This high efficiency cannot be achieved by simply scaling up low-level circuits; a different approach must be adopted.

Variable-Resistance Approach. In theory, the desired control of power could be achieved by means of a high-speed rheostat connected in series with the load. Variation of the rheostat's resistance from zero to infinity permits continuous and smooth control of the load voltage and power from their maximum values to zero. However, the power efficiency is proportional to the load voltage and is only 50% when the load voltage is equal to half of the source voltage. At its intermediate settings, the rheostat must be capable of dissipating up to 25% of the maximum load power. The consequences of this dissipation are trivial in the milliwatt power region. They can be tolerated in the power region of watts, where other approaches may cause more severe problems for certain applications, but in the kilowatt power region and above, this approach becomes completely impracticable.

On-Off Approach. This alternative is based on the fact that the rheostat dissipates very little power in its two extreme positions, corresponding to a closed or open switch. In this case, control is achieved by means of the timing of repetitive switching action. Because of wear and limited switching speed, mechanical switches are ordinarily not suitable, but electronic switches have made this approach feasible into the megawatt power region while maintaining high power efficiencies over wide ranges of control. However, the inherent nonlinearity of the switching action leads to the generation of distortion and other unwanted effects that must be considered in the design process (Wood 1981).†

---

*Overall coordination of Sec. 15 in the first edition was the work of the late W. E. Newell. The section was revised for the second edition by John Rosa, of the Westinghouse Research and Development Center.

†Author and date references in parentheses correspond to entries in the Bibliography in Par. **15-52**.

3. Types of Switching Devices. The origins of modern power electronics can be traced to the technology of rectifiers and inverters developed many years ago (Rissik, 1939) to utilize mercury-arc devices. Today solid-state power-switching devices have won nearly universal acceptance in these applications and are making many new applications feasible because of their greater reliability, faster speed, higher efficiency, smaller size, and lower cost. Much of the technology of power electronics is devoted to the problems of utilizing the full capability of state-of-the-art solid-state power devices while minimizing the effects of their inherent limitations.

For example, most power converters depend on the ability to connect any of two or more input lines to any of two or more output lines, a function served admirably by a multipole selector switch. The lack of electronic switching devices capable of fulfilling this function directly leads to a large (and to the beginner, often bewildering) variety of device and circuit configurations for performing the function indirectly by means of unilateral, single-pole single-throw switches, which can be turned on but some of which cannot be turned off by a control signal.

Because of the key role played by solid-state switching devices, it is advisable to summarize their main characteristics and distinguish between their main types before proceeding further. Most solid-state devices are inherently *unilateral;* i.e., they are designed to carry current in only one direction, known as the *forward direction.**

The first type of solid-state device which should be distinguished is a *two-terminal switch,* which "closes" automatically when forward-polarity voltage is applied. Such devices are commonly known as *rectifier diodes* or simply diodes. The intensity and shape of current through the diode are determined by the supply voltage, the load, and the circuit configuration. The diode turns off automatically when the circuit conditions mandate current flow in the reverse direction and blocks voltage in the reverse direction.

The second type of solid-state switch has an additional control terminal, the *gate.* The device will block forward voltage until an appropriate current pulse is applied to the gate. Thereafter the device continues to conduct as long as forward current flows. This type of device is known officially as a *reverse-blocking triode thyristor* but is commonly called thyristor or SCR (for *silicon-controlled rectifier*). As in the diode, the intensity and shape of current through the thyristor are determined by the supply voltage, the load, and the circuit configuration. The gate cannot interrupt forward current. The thyristor turns off automatically when the circuit conditions mandate current flow in the reverse direction and blocks voltage in both reverse and (in the absence of gate signal) in the forward direction.

The third type of device is similar to the thyristor except that it also has *turnoff* ability. Forward current can be terminated by applying an appropriate negative current pulse to the gate. Devices of this type are known as *turnoff thyristors* or *gate-controlled switches* (GTO or GCS).

The fourth type of device has an additional control terminal, the *base.* The intensity of current applied to the base determines the maximum value of forward current the device will conduct with negligible voltage drop. This device is called a *transistor.* The transistor is a solid-state switching device with a forward current limit determined by the base signal. Removal of the base signal terminates conduction. The transistor is not suitable for blocking reverse voltage.

4. Commutation. The functioning of power-electronic circuits is based on the operation of solid-state switching devices. The "opening" of solid-state switches is frequently termed *commutation.* In a broad sense, commutation refers to the process by which forward current is interrupted or transferred from one switching device to another. In most circuits operating from an ac source, only turn-on control is applied and turnoff occurs naturally, either when the cyclic alternation of the ac voltage causes the polarity of voltage across a given device to reverse† or when turning on a switching device results in turning off the device in conduction by applying to it the ac line voltage in reverse direction. Both phenomena are called *natural commutation;* the second is usually termed *line commutation.*

Most circuits supplied from a dc source and some ac circuits must be able to interrupt current. Because of the limited power capacity of presently available devices with inherent turnoff capability, it is often necessary to achieve turnoff by indirect means. These include auxiliary components which momentarily reverse bias and divert the current from a turn-on thyristor, lacking inherent turnoff ability, until it reestablishes its blocking state. This artificial turnoff of devices which are capable only of turn-on is known as *forced commutation.*

---

*One exception is the *triac,* which is the integration of two unilateral thyristors in parallel opposition.

†Care should be taken to distinguish between the reversal of device voltage and the reversal of source voltage. With transient effects neglected, the device voltage reverses as its current passes through zero. This may or may not occur in synchronism with the reversal of source voltage.

Device name	Structure	Symbol	V-I characteristic	Principal use
Conventional diode				Rectifier
Avalanche diode				Rectifier
Thyristor				Controlled turn-on switch
Triac				Bidirectional controlled turn-on switch
Transistor				Controlled turn-on and turn-off switch

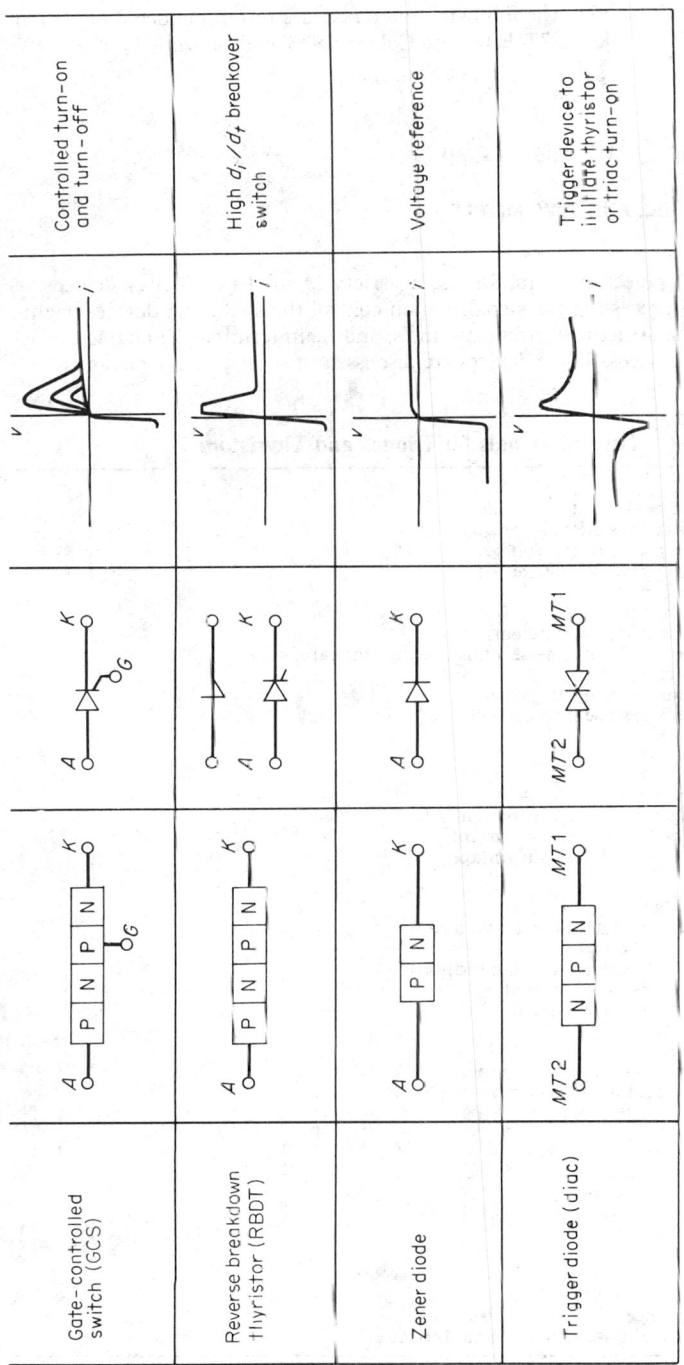

Fig. 15-1. Devices commonly used in power circuits.

For an overview of power electronics, the following references are recommended: Gutzwiller, 1967; Storm, 1969; Hoft, 1972; Kusko, 1972; Bates and Colyer, 1973; and Newell, 1974.

Solid-State Power Devices

BY P. F. PITTMAN, J. C. ENGEL, AND J. W. MOTTO

5. Introduction. Power-electronic circuits use a variety of solid-state power devices, as well as low-level devices for processing the signals which control the switching devices. Figure 15-1 summarizes the junction structures, circuit symbols, and main-terminal VI characteristics of the most common power devices. These devices are discussed in subsequent paragraphs.

Table 15-1. Letter Symbols for Diodes and Thyristors

Ratings:

V_{DSM}	Nonrepetitive peak off-state voltage
V_{DRM}	Repetitive peak off-state voltage
V_{RSM}	Nonrepetitive peak reverse voltage
V_{RRM}	Repetitive peak reverse voltage
$I_{T.rms}$	Rms on-state current
$I_{T,av}$	Average on-state current
I_{TSM}	Surge (nonrepetitive) on-state current
I^2t	Nonrepetitive (rms) ampere^2-seconds overcurrent capability
I_{GM}	Peak gate current
V_{GRM}	Repetitive peak reverse gate voltage
I_{GRM}	Repetitive peak reverse gate current
P_{GM}	Peak gate power
$P_{G,av}$	Average gate power
T_J	Virtual junction temperature
T_{stg}	Storage temperature
$P_{av,max}$	Maximum average device dissipation
di/dt	Critical rate of rise of on-state current
dv/dt	Critical rate of rise of off-state voltage

Characteristics:

v_{BO}	Breakover voltage
V_{TM}	Maximum (steady-state) on-state voltage
$V_{T,av}$	Average on-state voltage
V_{TO}	Dynamic on-state voltage (during turn-on)
I_{DRM}	Repetitive peak off-state current
I_{RRM}	Repetitive peak reverse current
I_L	Latching current
I_H	Holding current
I_{DSM}	Nonrepetitive peak off-state current
I_{RSM}	Nonrepetitive peak reverse current
I_{RQM}	Peak reverse recovery current
V_{GT}	Gate trigger voltage
I_{GT}	Gate trigger current
V_{GD}	Gate nontrigger voltage
I_{GD}	Gate nontrigger current
t_d	Gate-controlled delay time
t_r	Gate-controlled rise time
t_f	Fall time
t_{rr}	Reverse recovery time
t_q	Circuit-commutated turn-off time
$R_{\theta JC}$	Thermal resistance, junction to case
$Z_{\theta JC(t)}$	Transient thermal impedance, junction to case

Tables 15-1 and 15-2 list the standard letter symbols used to designate the more important ratings and characteristics of power devices.

Further information on solid-state devices is available in manufacturers' handbooks, in textbooks (Gentry et al., 1964; Hibberd, 1968; Delhom, 1968), and in several IEEE publications on power devices (IEEE, 1967, 1970). Pertinent standards include USAS C34.2-1968 and EIA-1972.

RECTIFIER DIODES

6. Conventional Diodes. A conventional semiconductor diode employs a single *pn* junction, as shown in Fig 15-1. The diode *Vi* characteristic is such that current can easily flow in one direction while its flow in the other direction is blocked by the junction. The physical theory of the action of a *pn* junction under conditions of forward and reverse bias is discussed in Sec. **7**, Pars. **7-68** and **7-70**.

Silicon diodes have increased in power-handling capability at least as rapidly as triode power-control devices. Diodes with current and voltage ratings of 5,000 A or 5,000 V repetitive reverse voltage are readily available, although not in the same device. Larger devices are in development, and as a result, diode power-handling capabilities will continue to increase.

Silicon-diode ratings include voltage, current, and junction temperature. The device current rating I_T is primarily determined by the area of the silicon slice, power dissipation, and the method of heat sinking, while the spread of voltage ratings V_{RRM} is determined by silicon resistivity and slice thickness.

Reverse voltage ratings are designated as repetitive V_{RRM} and nonrepetitive V_{RSM}. The repetitive value pertains to steady-state operating conditions, while the nonrepetitive peak value applies to occasional transient or fault conditions. Care must be exercised when applying a device to ensure that the voltage rating is never exceeded, even momentarily. When the blocking capability of a conventional diode is exceeded, leakage currents flow through the localized areas at the edge of the crystal. The resulting localized heating can easily cause instant device failure.

Table 15-2.   Letter Symbols for Transistors

Ratings:	
V_{EBO}	Emitter-to-base dc voltage (collector open)
$V_{CEO,sus}$	Collector-to-emitter sustaining voltage (base open)
V_{CEO}	Collector-to-emitter voltage (base open)
V_{CEV}	Collector-to-emitter voltage (base at specified voltage)
I_B	Base current (dc)
I_C	Collector current (dc)
P_T	Total power dissipation
Characteristics:	
h_{FE}	Dc current gain (common emitter)
$V_{CE,sat}$	Collector-to-emitter saturation voltage
$V_{BE,sat}$	Base-to-emitter saturation voltage
I_{EBO}	Emitter-to-base cutoff current (collector open)
t_s	Storage time
C_{ob}	Output capacitance
f_T	Gain bandwidth
SOA	Safe operating area

Although even low-energy reverse overvoltage transients are likely to be destructive, the silicon diode is remarkably rugged with respect to *forward current transients*. This property is demonstrated by the I_{TSM} rating which permits one-half-cycle peak surge current of nearly 10 times the I_T rating. For shorter current pulses, less than 4 ms, the surge current is specified by an I^2t rating similar to that of a fuse.

Proper circuit design must ensure that the *maximum average junction temperature* will never exceed its design limit of typically 200°C. Good design practice for high reliability, however, limits the maximum junction temperature to a lower value. The average junction-temperature rise above ambient is calculated by multiplying the average power dissipation, given approximately by the product of V_T and I_T, by the thermal resistance. Transient junction temperatures can be computed from the transient thermal-impedance curve.

Device ratings are normally specified at a given *case temperature and operating frequency*. The proper use of a device at other operating conditions requires an appreciation of certain basic device characteristics. This is especially true in applications where the operating conditions of a number of devices are interdependent, as in series and parallel operation. For example, the forward voltage drop of a silicon diode has a negative temperature coefficient of 2 mV/°C for currents below the rated value. This variation in forward drop must be considered when devices are to be operated in parallel.

The reverse blocking voltage of a diode, at a specified reverse current, can either decrease or increase with temperature. The tendency to decrease comes from the fact that the reverse leakage current of a junction increases with temperature, thereby decreasing the voltage attained at a given measuring-current level. If the leakage current is very low, the maximum reverse voltage will be determined by avalanche breakdown in the silicon, which has a coefficient of approximately 0.1%/°C. It should be noted that the reverse blocking voltage of a conventional diode is usually determined by imperfections at the edge of the die, and thus an ideal avalanche breakdown is usually not observed.

The *reverse recovery time* of a diode causes its performance to degrade with increasing frequency. Because of this effect, the rectification efficiency of a conventional diode used in a power circuit at high frequency is poor. In order to serve this application, a family of fast-recovery diodes has been developed and is presently available at a premium price. The stored charge of these devices is low, with the result that the amplitude and duration of the sweep-out current are greatly reduced compared with those of a conventional diode. However, improved turnoff characteristics of the fast-recovery diodes are obtained at some sacrifice in blocking voltage and forward drop compared with a conventional diode.

7. Avalanche Diodes. An avalanche diode differs from a conventional diode in its ability to repeatedly absorb large reverse energy pulses of a joule or more. This capability results from the fact that avalanche breakdown is a bulk crystal phenomenon, whereas the edge breakdown of the conventional diode occurs due to defects in the crystal lattice. To ensure that bulk breakdown occurs first, the edges of an avalanche diode are tapered or beveled to reduce the local field intensity. The positive temperature coefficient of the avalanche voltage (approximately 0.1%/°C), coupled with the heating that results from localized reverse currents, causes the area of reverse conduction to spread throughout the entire diode. The reverse energy is thus absorbed by the complete wafer, not a small localized spot.

The ratings of current, forward voltage drop, reverse blocking voltage, and operating and storage temperatures for an avalanche diode are similar to those of a conventional diode of the same wafer size. The difference in reverse characteristics results in the repetitive and nonrepetitive peak-voltage ratings being replaced by reverse-pulse power ratings. These are usually given in curve form, as shown in Fig. 15-2 for a 250-A device. As can be seen, the device can absorb nearly 100 kW for 10 μs (1 J), which makes it relatively insensitive to ordinary transient or fault conditions in a circuit.

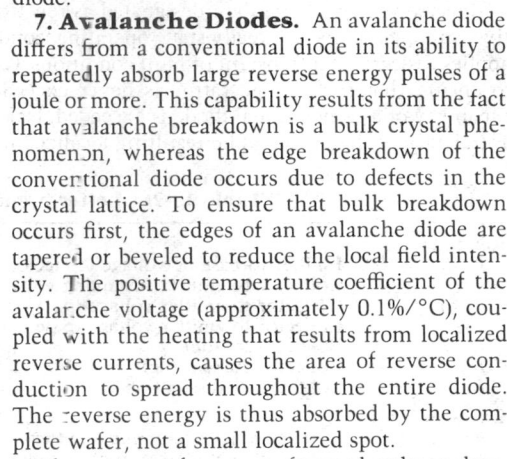
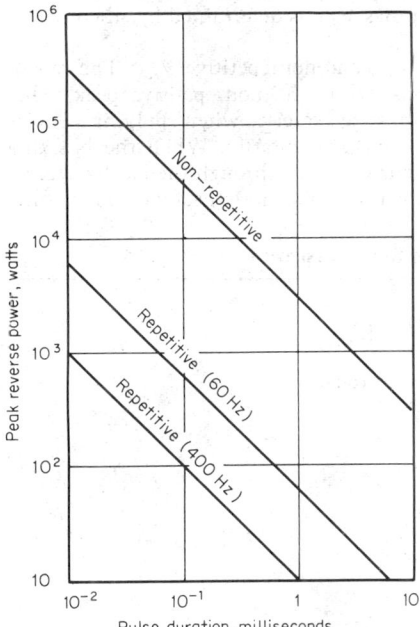
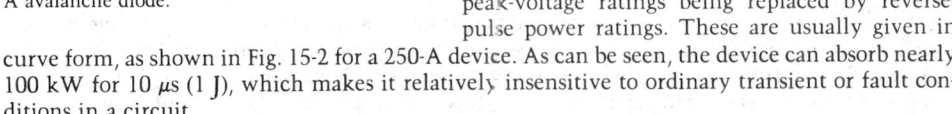

Fig. 15-2. Power-surge rating for a typical 250-A avalanche diode.

Following forward conduction, several microseconds are required for an avalanche diode to regain its reverse blocking voltage. If several diodes are connected in series, the variation in the reverse recovery times of individual diodes forces the first diodes that recover to pass the high reverse recovery current of the remaining diodes while supporting their full avalanche voltage. Such a condition would be instantly fatal to an ordinary diode but is not harmful to an avalanche diode. This characteristic can therefore greatly simplify high-voltage rectifier circuitry where series operation is required. Because of the extra steps involved in its manufacture, an avalanche diode is ordinarily more expensive than a conventional diode and thus is usually not employed unless its special features are required.

TURN-ON DEVICES

8. Three-Terminal Devices. Power at a high level can be controlled by a signal at a much lower level when a three-terminal, or triode, device is used. In general, load power flows through two of the device terminals when a control signal, referenced to one of the power terminals, is applied to the third, or control, terminal. The control signal may bear a linear relationship to the load current, as in the transistor; it may serve only to initiate conduction, as in the thyristor and triac; or it may serve both to turn on and turn off load current, as in the gate-controlled switch.

9. Thyristors (SCRs). The thyristor is a triode semiconductor device composed of four layers of silicon, as discussed briefly in Sec. 7, Par. 7-72. In contrast to the linear relation which exists between load and control currents in a transistor, the thyristor is bistable.

The four-layer structure of the thyristor is shown in Fig. 15-1. The anode and cathode terminals, whose names were carried over from vapor-rectifier-tube terminology, are connected in series with the load to which power is to be controlled. The thyristor is turned on by application of a low-power control signal between the third terminal, or gate, and the cathode.

The anode-to-cathode V_I characteristic of a thyristor is also shown in Fig. 15-1 for several fixed values of gate current. The reverse characteristic is determined by the outer two junctions, which are reverse-biased in this region. With zero gate current, the forward characteristic in the off, or blocking, state is determined by the center junction, which is reverse-biased. However, if the applied voltage exceeds the forward blocking voltage, the thyristor switches to its on, or conducting, state. The effect of gate current is to lower the blocking voltage at which switching takes place.

This behavior can be explained in terms of the two-transistor analog shown in Fig. 15-3. The two transistors are regeneratively coupled so that if the sum of their current gains (α's) exceeds unity, each drives the other into saturation. In the forward blocking state, the leakage current is small, both α's are small, and their sum is less than unity. Gate current increases the current in both transistors, increasing their α's When the sum of the two α's equals 1, the thyristor switches to its on state.

The form of the *gate-to-cathode VI characteristic* of the thyristor is similar to that of a diode. With positive gate bias, the gate-cathode junction is forward-biased and permits the flow of a large current in the presence of a low voltage drop. When negative gate voltage is applied, the gate-cathode junction is reverse-biased and prevents the flow of current until the avalanche breakdown voltage is reached.

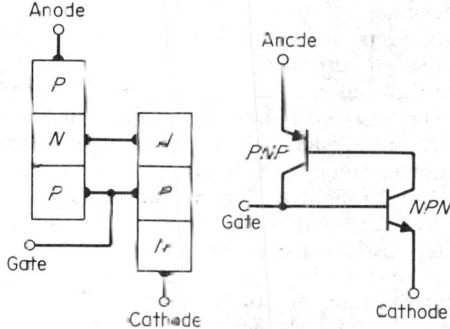

Fig. 15-3. Two-transistor analog of a thyristor.

Recently, minute gate-cathode shorts have been intentionally introduced into the thyristor during fabrication to reduce the sensitivity of the device to unwanted dv/dt triggering. One result of the presence of the shorts is increased reverse gate-leakage current.

A summary is provided in Table 15-1 of some of the ratings which must be considered when choosing a thyristor for a given application. Both forward and reverse repetitive and nonrepetitive voltage ratings must be considered, and a properly rated device must be chosen so that the maximum voltage ratings are never exceeded. In most cases, either forward or reverse voltage transients in excess of the nonrepetitive maximum ratings result in destruction of the device.

The *maximum rms* or *average current ratings* given are usually those which cause the junction to reach its maximum rated temperature. Because the maximum current will depend upon the current waveform and upon thermal conditions external to the device, the rating is usually shown as a function of case temperature and conduction angle. The peak single half-cycle surge-current rating must be considered, and in applications where the thyristor must be protected from damage by overloads, a fuse with an I^2t rating smaller than the maximum rated value for the device must be used. Maximum ratings for both forward and reverse gate voltage, current, and power also must not be exceeded.

The maximum rated *operating junction temperature* must not be exceeded, since device performance, in particular voltage-blocking capability will be degraded. Junction temperature cannot be measured directly but must be calculated from a knowledge of steady-state thermal resistance and the average power dissipation. For transients or surges, the transient thermal impedance must be used.

The *maximum average power dissipation* is related to the maximum rated operating junction temperature and the case temperature by the steady-state thermal resistance. In general, both the maximum dissipation and its derating with increasing case temperature are provided.

The number of thyristor characteristics specified varies widely from one manufacturer to another — whether only typical values are given of minima and maxima are also provided and whether charts or graphs showing their variations are included. Table 15-1 summarizes some of the characteristics provided.

Included are the values of conducting drop under both steady-state and turn-on transient conditions. Junction leakage and saturation currents are usually given for the maximum rated operating junction temperature. Gate conditions of both voltage and current to ensure either non-triggered or triggered device operation are included.

The *turn-on* and *turnoff transients* of the thyristor are characterized by switching times listed in Table 15-1. The turn-on transient can be divided into three intervals: gate-delay interval, turn-on of initial area, and spreading interval.

The *gate-delay interval* is simply the time between application of a turn-on pulse at the gate and the time the initial area turns on. This delay decreases with increasing gate drive current and is of the order of a few microseconds.

The second interval, the time required for *turn-on of the initial area*, is quite short, less than 1 μs. In general, the initial area turned on is a small percentage of the total useful device area. After the initial area turns on, conduction spreads throughout the device by diffusion and in general fills the entire area of the thyristor in about 50 μs.

During the third, or *spreading*, *interval*, load current flows, but the forward drop is abnormally high until spreading has been completed. The presence of this spreading interval has given rise to a load current rate-of-rise, or *di/dt*, rating for a thyristor. The *di/dt* rating for a given device operating at a given set of conditions is governed by the maximum local hot-spot temperature generated within the device at the restricted initial area turned on.

Localized heating due to excessive load current rate of rise can be a problem at 60 Hz if the load current is large enough, and it becomes increasingly difficult as the frequency of operation is increased. Localized heating from high *di/dt* applications has led to a whole new family of thyristors designed to minimize this problem. These new devices, which have special gate structures, turn on a much larger initial area than conventional thyristors, and, as a result, greater *di/dt* can be tolerated without increasing the localized hot-spot temperature. Rated values of *di/dt* for this new family of thyristors are of the order of 500 A/μs, compared with about 50 A/μs for conventional thyristors.

Thyristor turnoff times range from 3 to 250 μs. Turnoff time is generally important when force-commutated turnoff is used. A thyristor which has been carrying current is force-commutated off by suddenly reducing the current to zero at a rapid rate and then applying a reverse bias. Before the device will block reverse current, a *reverse sweep-out current* will flow until all stored charge is removed. When the flow of sweep-out current stops, the device will block reverse voltage. A short time later, the reverse voltage can be reduced to zero and forward voltage reapplied at a fixed rate, or *dv/dt*. If forward voltage is reapplied too soon or at too rapid a rate, the thyristor will turn on immediately, causing the circuit to malfunction. *dv/dt* values of the presently available thyristors range from 50 to 300 V/μs.

Both *steady-state thermal resistance* and *transient thermal impedance* are specified. In general $R_{\theta JC}$ is given as a number, while $Z_{\theta JC}$ is presented in the form of a graph. The double-sided flat package which makes heat removal possible from both sides of the device has become widely accepted and is replacing the stud-mounted package in many applications. On some data sheets, vendors provide data for both single- and double-sided cooling. Care must be taken to use the appropriate data which apply to the intended method of use.

10. Triacs. The triac is equivalent to two thyristors connected together in inverse parallel. When a single thyristor is used to control power in an ac circuit, either by off-on switching or phase control, other devices, such as diodes in a bridge arrangement, must be connected around the thyristor to accommodate both directions of current flow. On the other hand, the diodes can be eliminated if a second thyristor is connected in inverse parallel across the first one.

The triac was developed to satisfy the need for ac power control with a single device. The structure of the device is shown in Fig. 15-1. The triac is basically a five-layer silicon device of the form *npnpn*, as shown. The n regions at the extreme ends are shorted to the adjacent p regions by the contact metallization. The overall result of these shorts is to minimize the effect of the n region, which is positively biased. As a result, the triac looks like a *pnpn* configuration independent of the polarity of the bias voltage and thus performs functionally as though two thyristors were connected in inverse parallel.

The symbol used for the triac (Fig. 15-1) is a combination of the symbols for two thyristors. The terminology used to designate the device terminals differs from one vendor to another.

A family of *triac characteristics* is shown in Fig. 15-1 with gate current as a variable. Load current may be triggered on by any combination of MT2 voltage and gate-current polarities, but triggering levels differ depending upon the combination used. If unidirectional gate-current pulses are used for applications where the MT2 voltage alternates in polarity, a preferred gate-current polarity is recommended by the vendor and should be used.

The types of ratings specified for a triac are quite similar to those used for a thyristor with a few exceptions. A list of parameters used in the rating of a thyristor is shown in Table 15-1. In the case of the triac, the rated blocking voltage V_{DRM} applies to either polarity of MT2 voltage. In addition, because the gate characteristic is symmetrical, the reverse gate-voltage ratings do not apply.

The thyristor characteristics listed in Table 15-1 apply to the triac as well, with the following exceptions. Breakover-voltage and conducting-voltage drop apply to both polarities of MT2 voltage and current instead of only one polarity, as for the thyristor. All load-current ratings must be changed to accommodate the flow of bidirectional load current. In addition, gate characteristics for both triggering and nontriggering must be changed to accommodate both polarities of gate and MT2 voltage.

The *turn-on transient characteristics* must be enlarged to include turn-on with both polarities of MT2 voltage. The turnoff transient must be characterized differently because circuit-commutated turnoff does not apply to the triac. During the turnoff transient of a thyristor, reverse bias is applied to the device following sweep-out. Such a condition is impossible to establish in a triac because it will turn on with potential of either polarity. In general, values for fall time, reverse-recovery time, and turnoff time are not specified because they are of little value in most triac applications. Owing to the special structure of the triac, the rate of reapplication of blocking voltage following termination of load current is much lower than that allowable with the thyristor. In general, the reapplied voltage of the triac must be held to 5 V/μs.

TURNOFF DEVICES

11. Power Transistors. The thyristor and triac discussed in the previous section can initiate the flow of load current by turning on as the result of the application of a low-level gating signal. Once load current flows, however, the switching device loses control and current must be made to go to zero by external means. The transistor and gate-controlled switch are capable not only of initiating the flow of current but of interrupting it also under the control of a low-level gating signal.

The transistor is a triode device composed of three layers of semiconductor material, either *npn* or *pnp*, as shown in Fig. 15-1. Transistor operation and fabrication are discussed in Sec. 7.

Most power transistor circuits, other than linear regulators, do not use the linear relationship between collector and base currents. The transistor is driven from cutoff to saturation to operate as an off-on switch, avoiding the high dissipation encountered with sustained operation in the linear region.

The rating which defines the collector-to-emitter voltage blocking capability is the *sustaining voltage*, $V_{CEO,sus}$. This rating should never be exceeded even on a transient basis, except in very special circumstances, if reliable operation is desired. A similar maximum rating applies to reverse base-to-emitter voltage V_{BEO}, which, if exceeded, will cause the emitter to fail.

Ratings for maximum peak and continuous *collector currents* are given for the most favorable conditions of ambient temperature. A rating for maximum power dissipation at 25°C case temperature is also provided, as is its variation with increasing case temperature. One of these ratings will determine the maximum allowable collector current for a given application. A similar maximum rating for *base current* is specified and must be observed.

The number and types of characteristics specified on data sheets, and whether only typical values or minima or maxima are given, vary widely between vendors. Values for reverse-biased junction *leakage* currents (collector cutoff currents) are sometimes given at room and at elevated temperatures (I_{CEX}, I_{CEO}, and I_{EBO}). In general, typical values for switching times comprising delay t_d, rise t_r, storage t_s, and fall times t_f are provided although some data sheets provide maximum limits also. Other characteristics include thermal resistance $R_{\theta JC}$, collector-to-base saturation voltage $V_{CE,sat}$, base-to-emitter saturation voltage $V_{BE,sat}$, and small-signal current gain h_{FE}.

Of special importance in switching circuits is the *safe operating area* (SOA) *characteristic* usually supplied in graphical form. If the device is not applied within the limits of the SOA curves, it may latch up during a switching transient, causing the circuit to malfunction and possibly causing device failure. For a more complete discussion of transistor SOA characteristics, see Sec. **7**, Par. **7-88**, and RCA, 1971, pp. 134–145, referred to in Par. **15-52**. See also Sec. **28**.

12. Gate-Controlled Switch (GCS). A gate-controlled switch or *turnoff thyristor* is a four-layer triode silicon semiconductor device which resembles the thyristor in performance but has the additional attribute that anode current can be turned off by control of the gate. It is effectively a semiconductor analog to an electromechanical toggle switch.

The basic four-layer structure of the GCS, shown in Fig. 15-1, is identical to that of the thyristor. The difference in performance arises from the differences in geometry of the cathode and gate designs on the upper surface of the device. In the design of the GCS, the gate structure is distributed throughout the cathode to permit the gate to exert maximum influence on all parts of it. This is not true of the thyristor, where only a small area of the cathode is in intimate contact with the gate.

The form of the *VI* characteristic of the GCS is the same as that of the thyristor, as shown in Fig. 15-1. The parameters used to rate a GCS are similar to those used for the thyristor with several exceptions. All those shown in Table 15-1 apply to the GCS also. For the GCS, however, it is also necessary to rate the maximum peak anode current which can be interrupted by a signal applied to the gate. When this maximum anode current is exceeded, gate current no longer can cause anode current to cease and the device may be destroyed.

All the characteristics shown in Table 15-1 for thyristors apply to the GCS except reverse recovery time and circuit-commutated turnoff time. For the GCS, turnoff is initiated by the device itself, and, in general, anode voltage does not reverse but builds up in the forward direction as anode current goes to zero.

To the characteristics listed, a *turnoff gain* must be added. This characteristic is the ratio of anode current to required gate current when the turnoff transient begins. Because the value of this characteristic depends upon many factors, including the level of anode current and the length of time reverse gate current has been applied, the conditions under which it is measured must be specified.

Naturally Commutated Circuits

AC-DC CONVERTERS

BY B. R. PELLY

13. General Considerations. The basic feature common to all the ac-dc converters described in the following paragraphs is that they are connected to a source of ac voltage which causes natural commutation. In most cases, power flow is from the ac terminals to a dc load, and the process is known as *rectification.* However, some members of this general family of converters can be controlled so that power flow occurs from the dc terminals back into the ac line or into a synchronous motor operated at variable frequency and thus at variable speed (Rosa, 1979). This process is known as *synchronous inversion* to distinguish it from inversion into a passive ac load. In the latter case, forced commutation is usually required.

The simplest member of this family of converters is the well-known *half-wave single-phase rectifier.* Although widely used for low-current dc power supplies, this type of rectifier is not used for higher power because of the large ripple voltage in its output and because the unidirectional current causes dc magnetization of transformer cores.

The number of different converter circuits is very large. Two basic configurations should first be distinguished, the *bridge circuit* (also known as double-way circuit) and the *midpoint circuit* (also known by the names center tap and single way).

Converter circuits are also distinguished according to whether the ac line is *single-phase* or *3-phase.* The effective number of phases of a 3-phase line can be further increased by connecting transformer windings to give intermediate phase shifts. Increasing the number of phases increases the *ripple frequency* of both the dc output voltage and the ac line current, making filtering easier. The ratio of the fundamental ripple frequency of the dc voltage to the ac line frequency is known as the *pulse number.*

If all the switching devices are diodes, the converter can operate only as an *uncontrolled rectifier* with the average dc output voltage fixed by the input ac voltage and by the circuit configuration. If half of the diodes in a bridge are replaced by thyristors, the average dc output voltage can be controlled by changing the phase angle at which the thyristors are fired, but the circuit is still capable only of rectifier action with power flow from the ac terminals to the dc terminals. Such circuits, known as *half-controlled converters* or *semiconverters*, belong to the category of *one-quadrant converters* since only one polarity of dc voltage and one polarity of dc current are possible.

Replacement of all the diodes in a rectifier by thyristors produces a fully controlled converter, or *two-quadrant converter*. This type permits dc current to flow in only one direction, but the dc voltage may have either polarity. With one polarity, power flows from the ac to the dc terminals, and the converter acts as a *rectifier* With the opposite polarity of dc voltage, net power flows from the dc terminals to the ac terminals, and the circuit acts as a *synchronous inverter*.

In some applications both polarities of both dc current and dc voltage must be permitted. This *four-quadrant action* can be achieved by interconnecting two similar two-quadrant converters, the combination being known as a *dual converter*.

The applications for this family of naturally commutated converters embrace a very wide range, including dc power supplies for electronic equipment, battery chargers, and speed controllers for fractional-horsepower motors, as well as dc supplies delivering many thousands of amperes for electrochemical and other industrial processes, high-performance reversing drives for dc machines rated at thousands of horsepower, and high-voltage dc transmission in the megawatt power region.

Throughout the discussion which follows, unless otherwise stated, the following simplifying assumptions are made:

1. The voltage drop across switching devices is neglected while they are conducting, and the leakage current is neglected while they are blocking. Stray resistances are neglected.

2. Device turn-on and turnoff occur instantaneously.

3. The dc terminals are connected to an ideal filter (an infinite inductance), which suppresses all ripple current.

AC-DC converters are treated in greater detail in Pelly (1971) and Schaefer (1965).

14. Two-Quadrant Converters. The circuit configurations, waveforms, and design relationships for various one- and two-quadrant converters are tabulated in subsequent paragraphs. In this paragraph, the operation of several typical circuits is discussed.

Two-Pulse Midpoint Circuit. Figure 15-4 shows a single-phase two-pulse midpoint converter and the associated source and load waveforms for various values of the firing delay angle α. For $\alpha = 0$, the converter is equivalent to an uncontrolled rectifier, and the thyristors could be replaced by diodes. During the positive half cycle of the supply voltage, thyristor Th1 and transformer secondary S1 carry the load current, the voltage across the load is v_{s1}, and thyristor Th2 is reverse-biased. During the negative half cycle, Th2 and S2 carry the load current, the load voltage is v_{s2}, and Th1 is reverse-biased. The load-voltage waveform consists of a direct component V_{do} plus a superimposed ac ripple having a fundamental frequency which is twice the supply frequency (hence the name, two-pulse). The fundamental component of the supply current is in phase with the supply voltage.

As α is increased in the range $0 < \alpha < 90°$, the delay in firing causes the average load voltage to decrease, as shown in Fig. 15-4b. Note that the assumption of smooth load current, i.e., constant throughout the cycle, implies a highly inductive load.* When the instantaneous load voltage goes negative, the reactive emf of the inductance forces power back into the source in order to maintain the current constant. However, over a half cycle, the net power flow is from the ac source to the dc load, and the supply current has a lagging power factor.

When α becomes equal to $90°$, the instantaneous load voltage is negative for as long as it is positive, so that the average dc component of load voltage is zero (see Fig. 15-4c), and the supply current lags the supply voltage by $90°$.

If α is increased beyond $90°$, the continuous current flow can be maintained only if an external negative dc source is connected to the dc terminals. Net power flow is from the dc terminals to the ac terminals, and the converter is performing synchronous inversion (see Fig. 15-4d). Since the polarity of the average dc voltage has reversed, operation has shifted from quadrant I to

---

*Under this assumption, Th1 cannot cease conduction until Th2 is fired. Therefore each thyristor conducts for 180°. If, however, the load is purely resistive, each thyristor will cease conduction when its half of the supply voltage goes negative, and the current will pulsate

quadrant IV. But because the current cannot reverse, quadrants II and III are forbidden. Hence this is a two-quadrant converter.

Finally, in Fig. 15-4e, α is nearly equal to 180°, and the dc voltage approaches its maximum negative value. In practice, α must be limited to about 160° or less to permit the thyristor which is being commutated off to regain its blocking ability before forward voltage is reapplied to it. Otherwise a commutation failure occurs. Operation in the inversion region is frequently

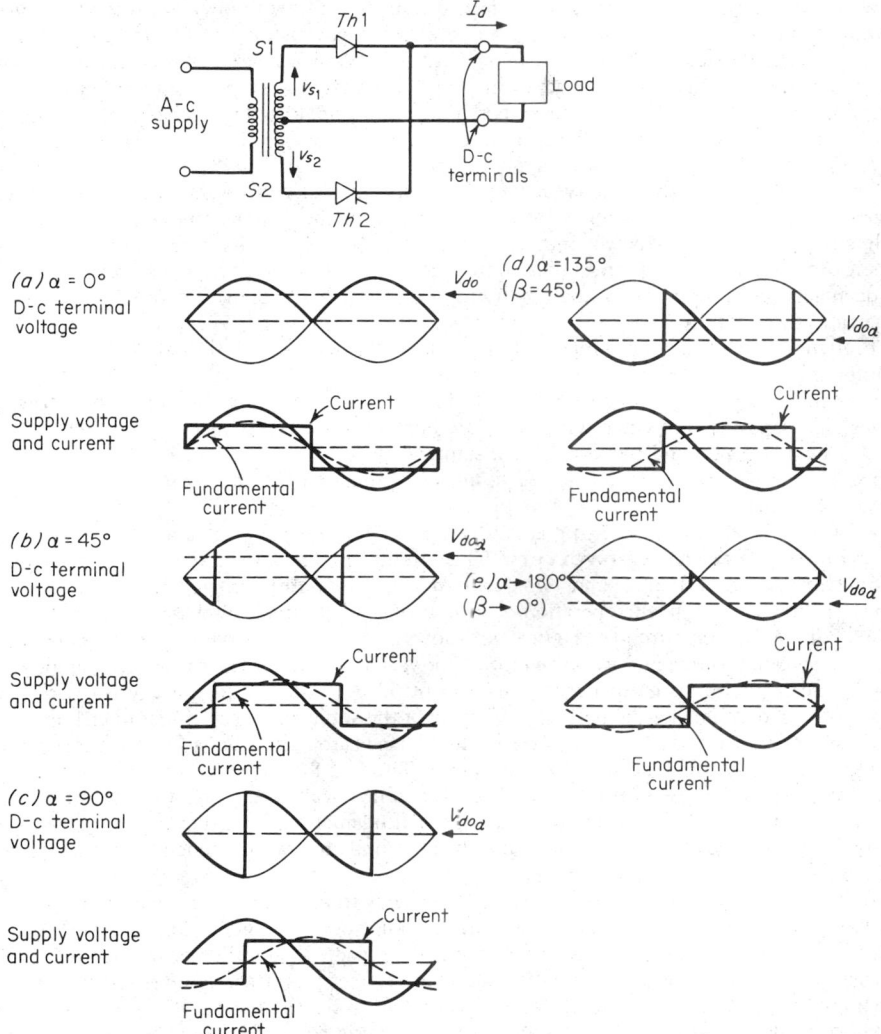

Fig. 15-4. Two-pulse midpoint converter circuit and associated waveforms (smooth direct current assumed).

described in terms of the advance angle $\beta = 180° - \alpha$. The margin of safety from commutation failure is described by the recovery angle δ between the completion of commutation and the next zero crossing at which forward voltage is reapplied.

Three-Pulse Midpoint Circuit.   The simplest type of phase-controlled converter which operates from a 3-phase supply is the three-pulse midpoint circuit, shown in Fig. 15-5 with idealized waveforms. The zigzag connection of the transformer secondary windings prevents dc magnetization of the transformer core by permitting equal and opposite currents to flow in the two secondary windings in each phase.

The waveforms illustrate that this circuit has the same basic operating characteristics as the two-pulse circuit of Fig. 15-4. That is, continuous control of the mean dc terminal voltage from maximum positive to maximum negative is achieved by controlling the phase of the thyristor firing pulses through a theoretical range of 180°. This is accompanied by a continuously increasing shift in the phase of the input current from 0 to 180° lagging. In fact, these characteristics are common to all two-quadrant converters.

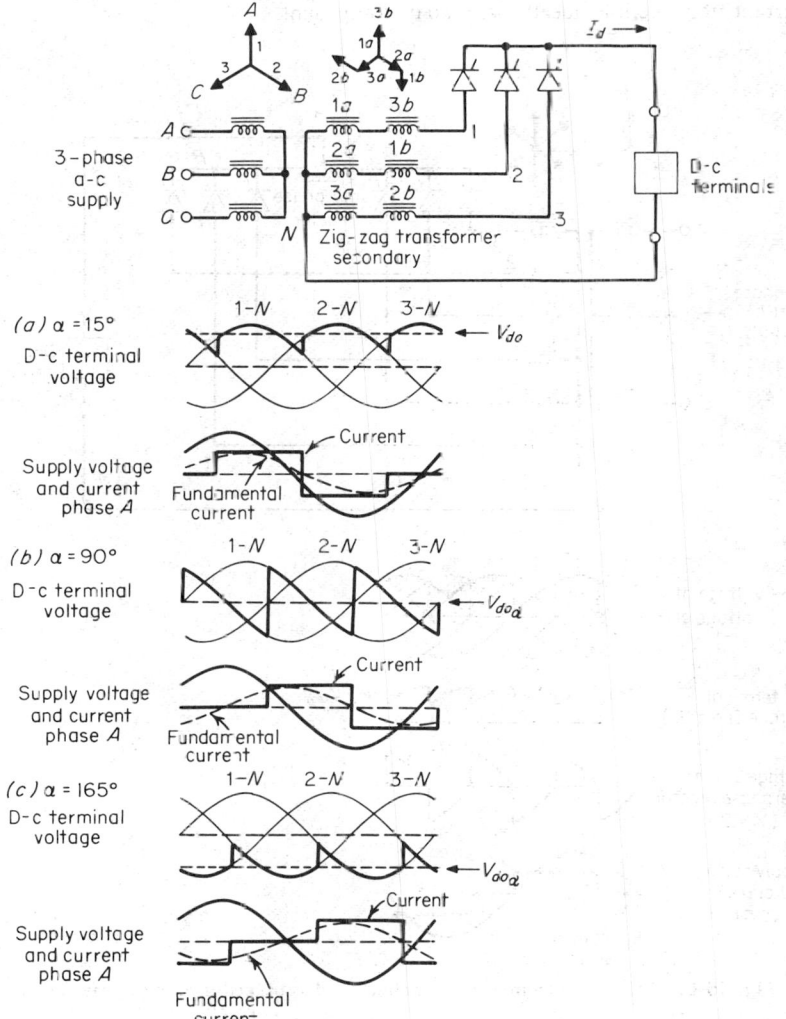

Fig. 15-5. Three-pulse midpoint converter circuit and associated theoretical waveforms.

Six-Pulse Midpoint Circuit. The outputs of two three-pulse converters having mutually displaced input voltages can be combined in parallel through an interphase reactor as shown in Fig. 15-6. Each three-pulse converter operates independently of the other. In the ideal case, the load current is shared equally between the two groups, and there is no dc magnetization of the core of the interphase reactor. In practice, some relatively small unbalance of currents may occur.

Because of the phase displacement between the ac ripple voltages at the dc terminals of the individual groups, the fundamental frequency of the ripple in the output voltage is 6 times the input frequency. The fundamental frequency of the ripple voltage across the interphase reactor is 3 times the input frequency.

If the interphase reactor is eliminated by making a solid connection between the dc terminals of the three-pulse groups, a six-pulse voltage waveform is still obtained at the output. However, the utilization factor of the circuit decreases because each thyristor then conducts for only 60° instead of the previous 120°.

Six-Pulse Bridge Circuit. The dc terminals of two three-pulse groups can be connected in series to give an overall six-pulse operation. The resulting bridge, one of the most commonly used converter circuits, is shown in Fig. 15-7. So far as the ac lines are concerned, the bridge circuit contains two similar oppositely poled groups of rectifying devices; thus, it draws a balanced current from the line, ideally with no dc component.

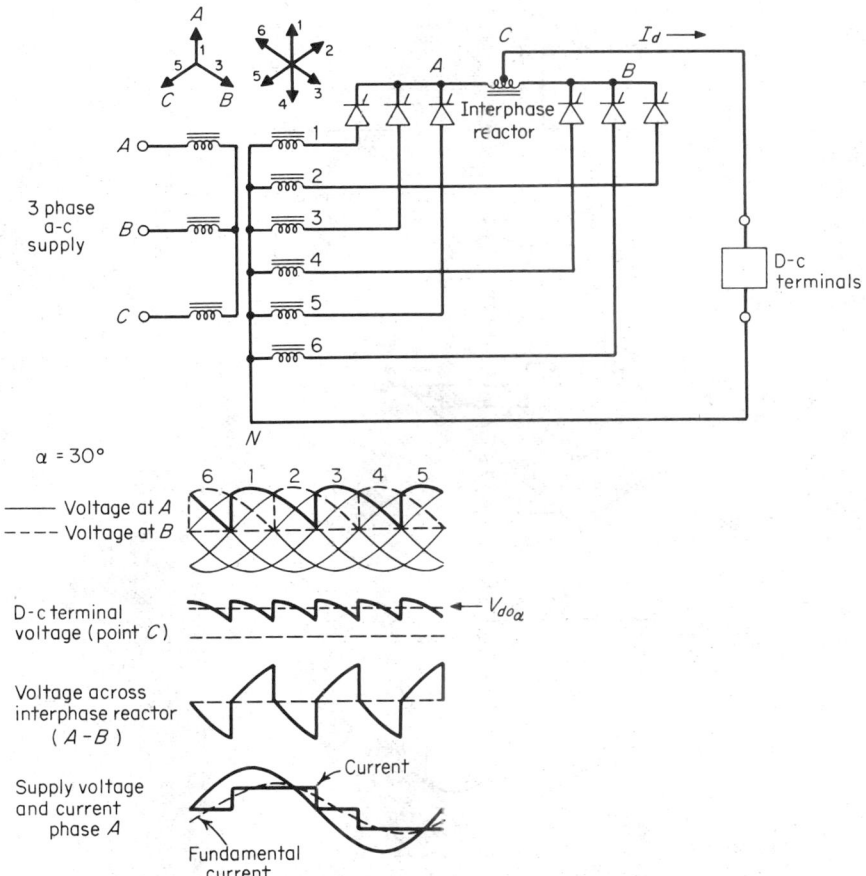

Fig. 15-6. Six-pulse midpoint converter circuit and associated theoretical waveforms.

Higher Pulse Numbers. Other converter-circuit configurations having higher pulse numbers can be constructed by connecting the dc terminals of individual groups, with suitably displaced ac voltages, in series or parallel, or by combining series and parallel connections into one system.

In practice, a thyristor conduction angle of 120° is greatly preferred. Thus, almost all practical multipulse converter circuits comprise combinations of the basic three-pulse commutating groups. Each group within the system operates essentially independently of all the other groups. When the dc terminals of individual groups must be connected in parallel, the connections are made through interphase reactors to maintain independent operation of the groups. On the other hand, groups can be connected in series with "solid" connections at the dc terminals. Series connections of bridges, however, require isolation between the transformer secondaries connected to the individual bridges.

15. One-Quadrant Converters. Many applications require operation with only one polarity of dc output voltage; i.e., they operate only in the rectifying mode. In this case, it is generally advantageous to connect uncontrolled diodes into certain parts of the circuit.

In bridge-connected circuits (but not midpoint circuits) uncontrolled diodes can be used in place of half of the thyristors. With this *half-controlled converter*, it is possible to control the mean dc terminal voltage continuously from maximum to virtually zero, but reversal of the mean voltage is not possible.

The half-controlled bridge has economic advantages over the fully controlled circuit because diodes are less expensive than equivalent thyristors. In addition, the input-power factor at relatively low levels of output voltage is improved over that of a fully controlled converter. Except for a single-phase bridge circuit, however, this advantage is obtained at the expense of a 2:1 reduction in ripple frequency at the dc terminals.

Either bridge or midpoint two-quadrant converters can be limited to one-quadrant operation by connecting a *freewheeling diode* across the dc terminals to conduct when the terminal voltage instantaneously tends to go negative. The diode reduces the ripple and improves the input-power factor for low dc output voltages. A further feature of the freewheel diode is that it provides a bypass path for inductive load currents if the supply lines become disconnected, thereby preventing reverse-voltage surges.

Both the half-controlled converter and the fully controlled converter with freewheeling diode have the advantage that the ratio of input current to dc output current decreases as the output voltage is reduced toward zero. In an ideal two-quadrant converter this ratio remains constant.

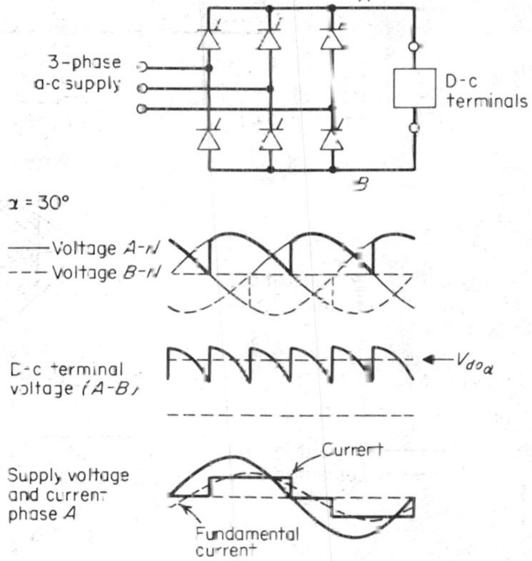

16. Commutation Overlap. The preceding discussion and the idealized waveforms that have been shown assume instantaneous commutation of current from one thyristor which is turning off to the next which is being turned on. In practice, circuit inductance causes conduction in the two devices to overlap for a time that is usually not negligible. The process is known as commutation overlap, and its duration relative to the period of a cycle is expressed in terms of the overlap angle u.

The physical explanation of commutation

Fig. 15-7. Six-pulse bridge converter and associated theoretical waveforms.

overlap depends upon the fundamental voltage-current relationship of an inductor, $\Delta i = (1/L)\int v\,dt$, which states that the change in current is equal to the voltage-time area, i.e., integral, divided by the inductance.* Hence transformer-leakage inductance and inductance in the ac line introduce a delay until the voltage-time area is sufficient to bring about the necessary redistribution of currents. During this delay, the current in the thyristor being turned on increases, and that in the thyristor turning off decreases at the same rate, since the total current is constant. If each thyristor has an equal series inductance, during the overlap the dc terminal voltage will be the average of the two source voltages to which the thyristors are connected.

The effects of commutation overlap are illustrated in Fig. 15-8 for a three-pulse group having inductance L in series with each thyristor. The voltage at the dc terminal is shown as the current commutates from Th1 to Th2 after a phase delay α. The voltage-time area required to change the current in L_B from zero to I_d is shown shaded and is subtracted from the ideal output-voltage waveform. The average value of voltage withheld from the dc terminals is directly proportional to the product of the direct current and the inductance, and it is independent of the firing-delay angle. Thus during the overlap, the output voltage follows the curve $(v_A + v_B)/2$.

---

*This relationship is widely useful in analyzing the smoothing action of an interphase or filter reactor and other aspects of converter operation.

In general, the relationship between the firing-delay angle α and the overlap angle u for a three-pulse commutating group is

$$\cos \alpha - \cos (u + \alpha) = \sqrt{\frac{2}{3}} \frac{X_c I_d}{V_s}$$

For inverter operation, the corresponding relationship between the advance angle β and the recovery angle $\delta = \beta - u$ is

$$\cos \delta - \cos \beta = \sqrt{\frac{2}{3}} \frac{X_c I_d}{V_s}$$

For definitions of X_c, I_d, and V_s, see Table 15-3. These relationships are also valid for multipulse converters consisting of noninteracting three-pulse groups.

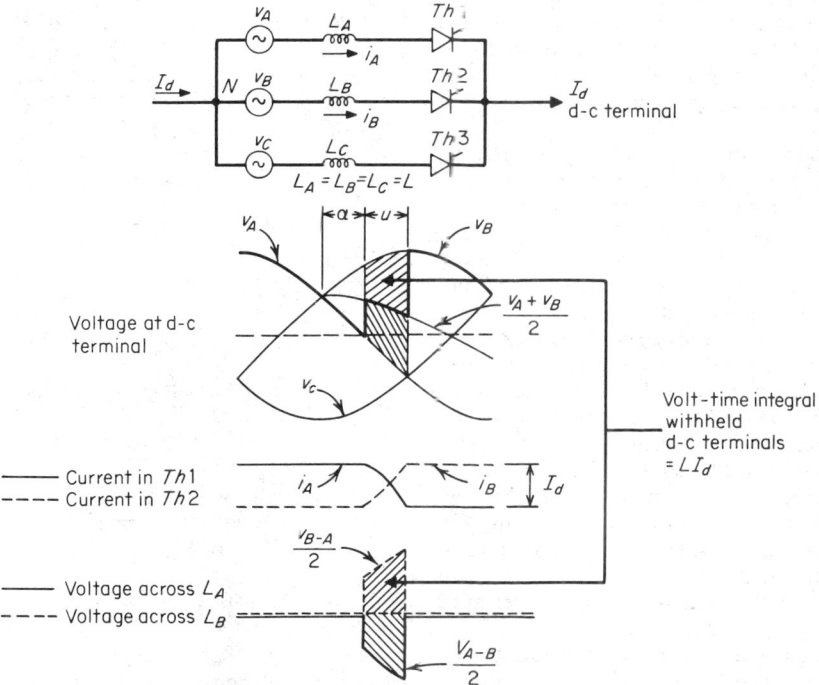

Fig. 15-8. The commutation process for a three-pulse group.

At the input side of a converter, the effect of commutation overlap is to cause rounding of the edges of the waveforms of line current. This means that the amplitudes of the higher-order harmonic terms are progressively reduced, compared with the theoretial amplitudes of these components with no overlap. In addition, the duration of each segment of the waveform of the input current is stretched by the overlap angle, resulting in a slight additional lagging phase shift of the fundamental component of current.

17. Waveforms and Data for Converter Circuits. Table 15-3 lists the letter symbols most frequently used in the analysis of converter circuits. Tables 15-4 to 15-6 summarize the idealized waveforms and design relationships for the more common single- and 3-phase, one- and two-quadrant converters.

The relationships between firing angle and the principal harmonic components in the dc terminal voltage of these converters are shown in Figs. 15-9 and 15-10.

For all two-quadrant converters, the input-displacement factor is equal to the dc voltage ratio; and for all half-controlled converters, the input-displacement factor is equal to the square root of the dc voltage ratio. These relationships are indicated in the tables. The corresponding relationships for the converter circuits with freewheel diodes are illustrated in Fig. 15-11.

Table 15-3.   Letter Symbols Used in the Analysis of Converter Circuits

V_S	Rms value of phase-to-neutral voltage at converter input terminals
$\hat{V}_s$	Peak value of V_s
V_n	Rms value of phase-to-neutral voltage at primary converter transformer
h	Ratio of V_n to V_S
V_d	Average value of voltage at dc terminals of converter under load, at any firing angle
$V_{d_{0\alpha}}$	Average value of voltage at dc terminals of converter at firing angle α, with no commutation overlap
$V_{d_{max}}$	Maximum possible average value of voltage at dc terminals of converter, obtained at $\alpha = 0°$, with no commutation overlap
V_{FB}	Maximum instantaneous value of forward blocking voltage applied across thyristor
V_{RB}	Maximum instantaneous value of reverse blocking voltage applied across thyristor
V_D	Maximum instantaneous value of reverse blocking voltage applied across diode
r	Ratio of V_e to $V_{d.max}$
I_d	Direct current at output of converter
I_1	Rms value of the fundamental component of converter input line current
I_{1P}	Rms value of the "in-phase" or "power" component of I_1
I_{1Q}	Rms value of the "quadrature" or "reactive" component of I_1
$I_{av,Th}$ $(I_{av,D})$	Average value of thyristor (diode) current
$I_{rms,Th}$ $(I_{rms,D})$	Rms value of thyristor (diode) current
P_d	Average power at output of converter
P_{d_0}	Theoretical average power at output of converter, at $\alpha = 0°$ with no commutation overlap
VA_0	Theoretical rms volt-amperes of transformer windings at $\alpha = 0°$ with no commutation overlap
L_s	Line-to-neutral commutating inductance at transformer secondary
L_p	Line-to-neutral commutating inductance at transformer primary
X_c	Commutating reactance at input frequency, referred to transformer secondary
α	Converter firing-delay angle, measured from the point at which the converter operates as if it were an uncontrolled rectifier circuit
β	Inverter advance angle; angle in advance of the zero crossing of the line-to-line commutating voltage at which the commutation is initiated: $\beta = 180° - \alpha$
δ	Inverter recovery angle; angle in advance of the zero crossing of the line-to-line commutating voltage at which the commutation is completed: $\delta = 180° - (\alpha + u)$
u	Commutation overlap angle
u^*	Overlap angle for commutation of current into freewheeling path
ϕ	Displacement angle between fundamental component of converter input current and associated line-to-neutral voltage
$\cos \phi$	Displacement factor of fundamental component of converter input current: $\cos \phi = I_{1P} / \sqrt{I_{1P}^2 + I_{1Q}^2}$
λ	Power factor at a given point in the converter input circuit; ratio of the average power to the rms volt-amperes
μ	Distortion factor of the current at a given point in the converter input circuit; ratio of the rms value of the fundamental component to the total rms value $\mu = \lambda /(\cos \phi)$
p	Pulse number of converter = ratio of fundamental output ripple frequency to ac supply frequency (with steady delay angle)
ω	Angular frequency of input supply

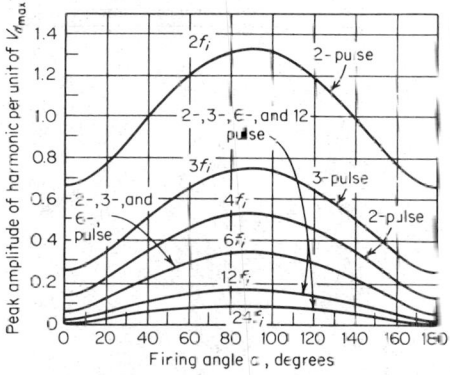

Fig. 15-9. Variation with firing angle of the principal harmonic components present in the dc terminal voltage of various two-quadrant converters with continuous conduction and no commutation overlap. *(From B. R. Pelly, "Thyristor Phase-Controlled Converters and Cycloconverters," Wiley-Interscience, New York, 1971; used by permission.)*

Table 15-4. Waveforms and Data for Various Rectifier and Converter Circuits

	2-pulse midpoint rectifier	2-pulse midpoint converter (2-quadrant)	2-pulse midpoint converter with freewheel diode (1-quadrant)	2-pulse bridge rectifier	2-pulse bridge converter (2-quadrant)
Circuit	Circuit same as converter, with diodes instead of thyristors	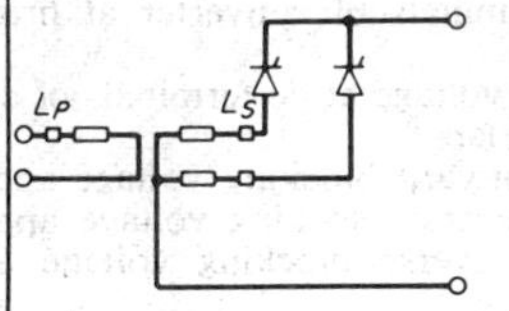	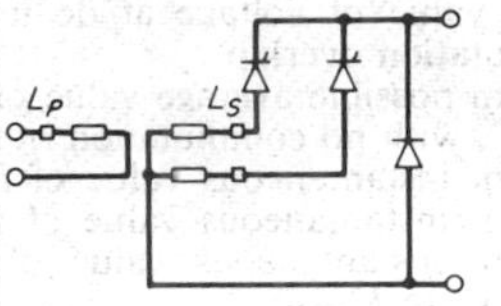	Circuit same as converter, with diodes instead of thyristors	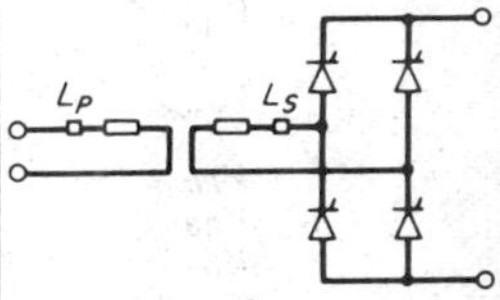
D-c terminal voltage		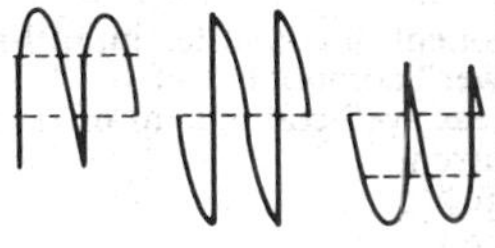 $$V_d = \frac{2}{\pi}\hat{V}_s \cos\alpha - I_d\frac{X_c}{\pi}$$ $$X_c = \omega\left(\frac{2L_p}{h^2}+L_s\right)$$ For harmonic distortion, see Fig. 9	 $$V_d = \frac{2}{\pi}\hat{V}_s\left(\frac{1+\cos\alpha}{2}\right) - I_d\frac{X_c}{\pi}$$ $$X_c = \omega\left(\frac{2L_p}{h^2}+L_s\right)$$ For harmonic distortion, see Fig. 10		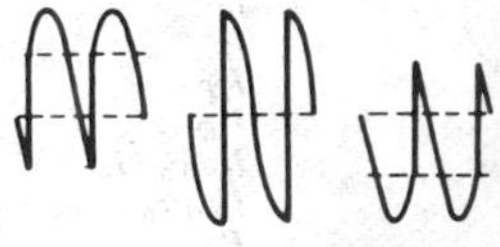 $$V_d = \frac{2}{\pi}\hat{V}_s\cos\alpha - I_d\frac{2X_c}{\pi}$$ $$X_c = \omega\left(\frac{L_p}{h^2}+L_s\right)$$ For harmonic distortion, see Fig. 9
Device voltage and current		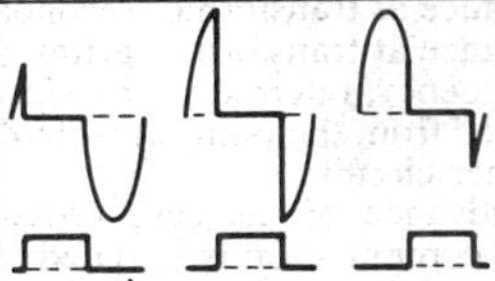 $V_{FB}=2.0\hat{V}_s$ $I_{AVTh}=0.5I_d$ $V_{RB}=2.0\hat{V}_s$ $I_{RMSTh}=0.707I_d$	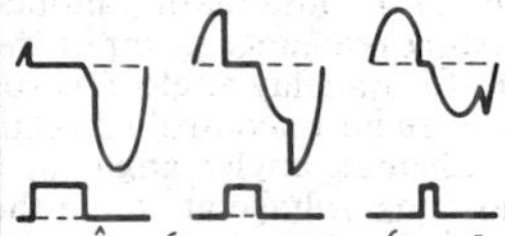 $V_{FB}=\hat{V}_s$ $V_{RB}=2\hat{V}_s$ $V_D=2\hat{V}_s$ $\begin{pmatrix}I_{AVTh}=0.5I_d\\I_{RMSTh}=0.707I_d\end{pmatrix}(\alpha=0)$ $\begin{pmatrix}I_{AVD}=I_d\\I_{RMSD}=I_d\end{pmatrix}(\alpha=\pi)$		 $V_{FB}=\hat{V}_s$ $I_{AVTh}=0.5I_d$ $V_{RB}=\hat{V}_s$ $I_{RMSTh}=0.707I_d$
Transformer secondary voltage and current		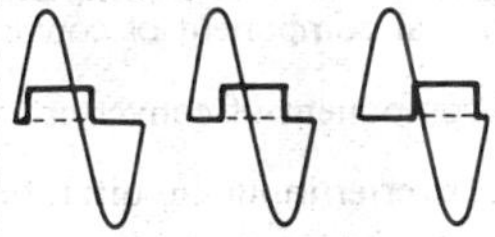 $I_{RMS}=0.707I_d$ $\cos\phi=\cos\alpha=r$ $VA_o=1.57P_{do}$ $\lambda=0.637r$	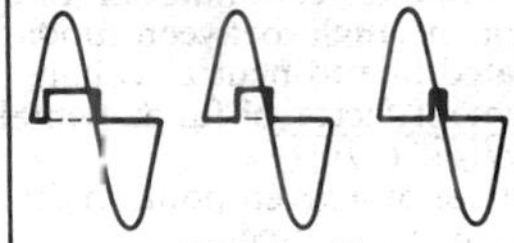 $I_{RMS}=0.707I_d\,(\alpha=0)$ $VA_o=1.57P_{do}$ $\cos\phi=\cos\frac{\alpha}{2}=\sqrt{r}$ λ see Fig.12		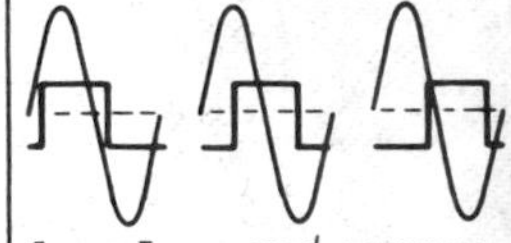 $I_{RMS}=I_d$ $\cos\phi=\cos\alpha=r$ $VA_o=1.11P_{do}$ $\lambda=0.9r$
Transformer primary voltage and current		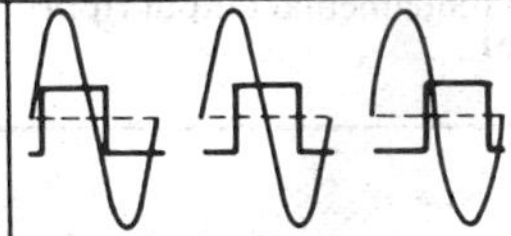 $I_{RMS}=I_d/h$ $\cos\phi=\cos\alpha=r$ $VA_o=1.11P_{do}$ $\lambda=0.9r$ For harmonic distortion, see Fig. 13	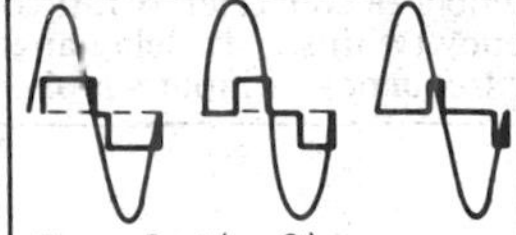 $I_{RMS}=I_d/h\,(\alpha=0)$ $V_{L_o}=1.11P_{do}$ $\cos\phi=\cos\frac{\alpha}{2}=\sqrt{r}$ λ—see Fig.12 For harmonic distortion, see Fig. 13		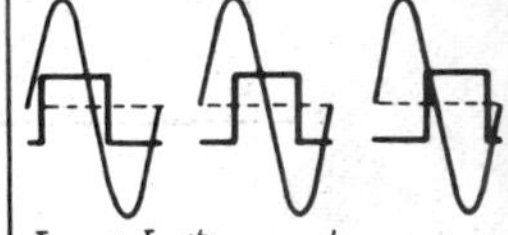 $I_{RMS}=I_d/h$ $\cos\phi=\cos\alpha=r$ $VA_o=1.11P_{do}$ $\lambda=0.9r$ For harmonic distortion, see Fig. 13

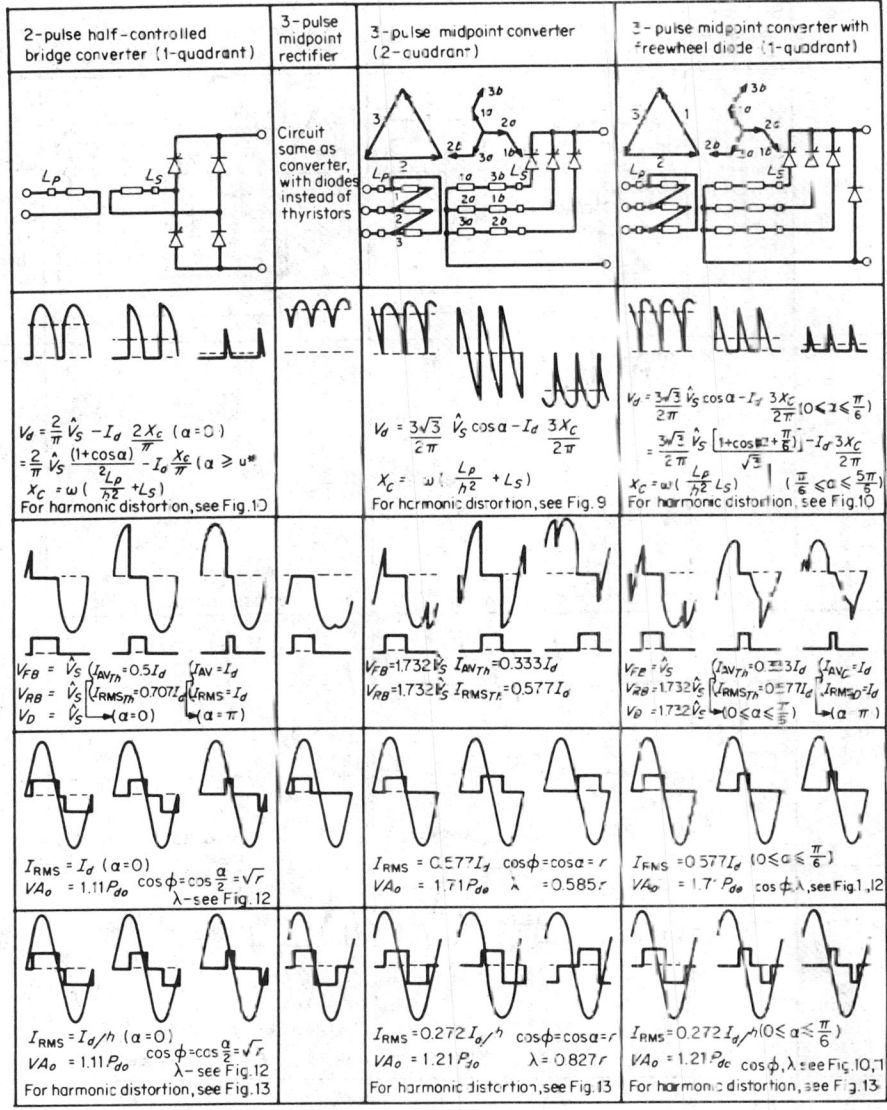

2-pulse half-controlled bridge converter (1-quadrant)	3-pulse midpoint rectifier	3-pulse midpoint converter (2-quadrant)	3-pulse midpoint converter with freewheel diode (1-quadrant)
	Circuit same as converter, with diodes instead of thyristors		
$V_d = \frac{2}{\pi}\hat{V}_S - I_d\,\frac{2X_c}{\pi}$ $(\alpha=0)$ $= \frac{2}{\pi}\hat{V}_S\frac{(1+\cos\alpha)}{2} - I_a\frac{X_c}{\pi}$ $(\alpha \geqslant u^*)$ $X_c = \omega(\frac{2L_p}{h^2}+L_S)$ For harmonic distortion, see Fig.10		$V_d = \frac{3\sqrt{3}}{2\pi}\hat{V}_S\cos\alpha - I_d\frac{3X_c}{2\pi}$ $X_c = \omega(\frac{L_p}{h^2}+L_S)$ For harmonic distortion, see Fig. 9	$V_d = \frac{3\sqrt{3}}{2\pi}\hat{V}_S\cos\alpha - I_d\frac{3X_c}{2\pi}$ $(0 \leqslant \alpha \leqslant \frac{\pi}{6})$ $= \frac{3\sqrt{3}}{2\pi}\hat{V}_S\left[1+\cos(\alpha+\frac{\pi}{6})\right] - I_d\frac{3X_c}{2\pi}$ $(\frac{\pi}{6} \leqslant \alpha \leqslant \frac{5\pi}{6})$ $X_c = \omega(\frac{L_p}{h^2}L_S)$ For harmonic distortion, see Fig.10
$V_{FB} = \hat{V}_S$ $(I_{AVTh}=0.5I_d$ $I_{AV}=I_d$ $V_{RB} = \hat{V}_S$ $I_{RMSTh}=0.707I_d$ $I_{RMS}=I_d$ $V_D = \hat{V}_S$ $(\alpha=0)$ $(\alpha=\pi)$		$V_{FB}=1.732\hat{V}_S$ $I_{AVTh}=0.333I_d$ $V_{RB}=1.732\hat{V}_S$ $I_{RMSTh}=0.577I_d$	$V_{FB}=\hat{V}_S$ $(I_{AVTh}=0.33I_d$ $I_{AVC}=I_d$ $V_{RB}=1.732\hat{V}_S$ $I_{RMS}=0.577I_d$ $I_{RMS}=I_d$ $V_D=1.732\hat{V}_S$ $(0\leqslant\alpha\leqslant\frac{\pi}{6})$ $(\alpha=\pi)$
$I_{RMS}=I_d$ $(\alpha=0)$ $VA_o = 1.11P_{do}$ $\cos\phi=\cos\frac{\alpha}{2}=\sqrt{r}$ λ—see Fig.12		$I_{RMS}=0.577I_d$ $\cos\phi=\cos\alpha=r$ $VA_o = 1.71P_{do}$ $\lambda = 0.585r$	$I_{RMS}=0.577I_d$ $(0\leqslant\alpha\leqslant\frac{\pi}{6})$ $VA_o = 1.71P_{do}$ $\cos\phi$, see Fig.1,12
$I_{RMS}=I_d/h$ $(\alpha=0)$ $VA_o = 1.11P_{do}$ $\cos\phi=\cos\frac{\alpha}{2}=\sqrt{r}$ λ—see Fig.12 For harmonic distortion, see Fig.13		$I_{RMS}=0.272I_d/h$ $\cos\phi=\cos\alpha=r$ $VA_o = 1.21P_{do}$ $\lambda=0.827r$ For harmonic distortion, see Fig.13	$I_{RMS}=0.272I_d/h$ $(0\leqslant\alpha\leqslant\frac{\pi}{6})$ $VA_o = 1.21P_{dc}$ $\cos\phi,\lambda$ see Fig.10,1 For harmonic distortion, see Fig.13

Table 15-5. Waveforms and Data for Various Rectifier and Converter Circuits

	6-pulse midpoint rectifier	6-pulse midpoint converter (2-quadrant)	6-pulse midpoint converter with freewheel diode(1-quadrant)	6-pulse bridge rectifier
Circuit	Circuit same as converter, with diodes instead of thyristors	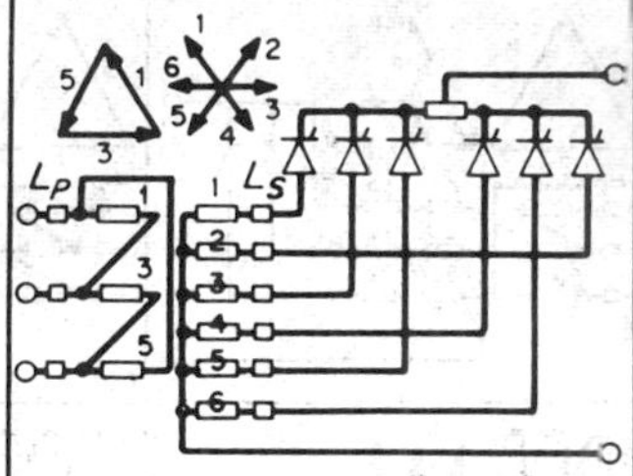	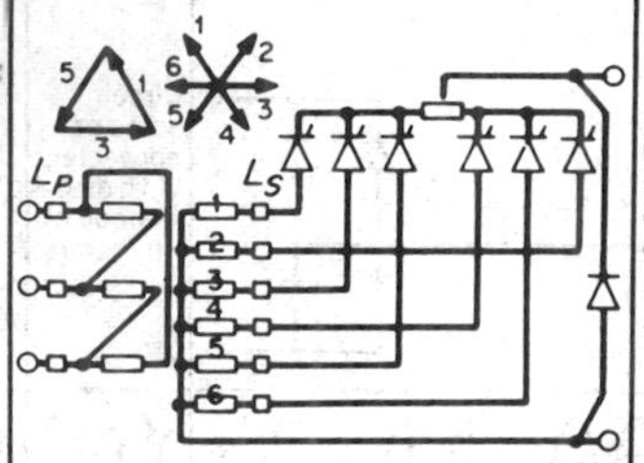	Circuit same as converter, with diodes instead of thyristors
D-c terminal voltage	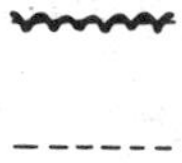	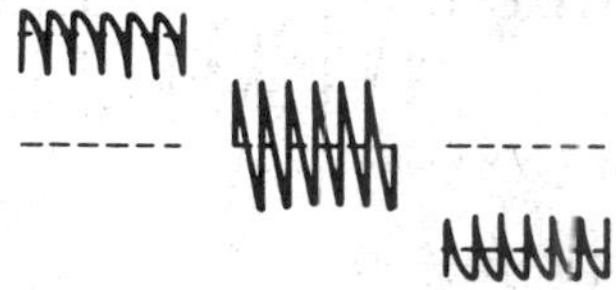 $$V_d = \frac{3\sqrt{3}}{2\pi}\hat{V}_s\cos\alpha - I_d\frac{3X_c}{4\pi}$$ $$X_c = \omega\left(\frac{L_p}{h^2}+L_s\right)$$ For harmonic distortion, see Fig 9	 $$V_d = \frac{3\sqrt{3}}{2\pi}\hat{V}_s\cos\alpha - I_d\frac{3X_c}{4\pi}\ (0\leqslant\alpha\leqslant\frac{\pi}{3})$$ $$= \frac{3\sqrt{3}}{2\pi}\hat{V}_s[1+\cos(\alpha+\frac{\pi}{3})]-I_d\frac{3X_c}{2\pi}$$ $$X_c=\omega\left(\frac{L_p}{h^2}+L_s\right)\ |\ (u^*+\frac{\pi}{3}\leqslant\alpha\leqslant2\frac{\pi}{3})$$ For harmonic distortion, see Fig.10	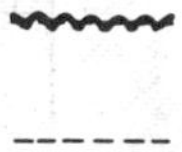
Device voltage and current		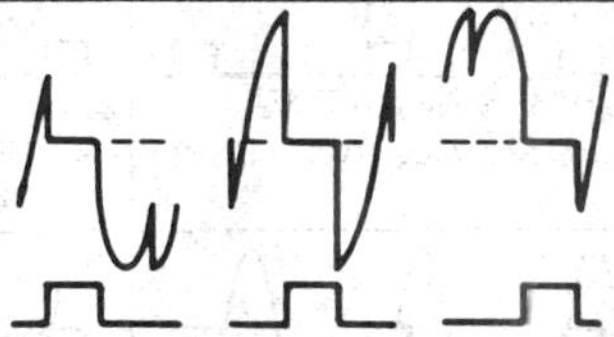 $$V_{FB}=1.732\,\hat{V}_s \qquad I_{AVTh}=0.167\,I_d$$ $$V_{RB}=1.732\,\hat{V}_s \qquad I_{RMSTh}=0.288\,I_d$$	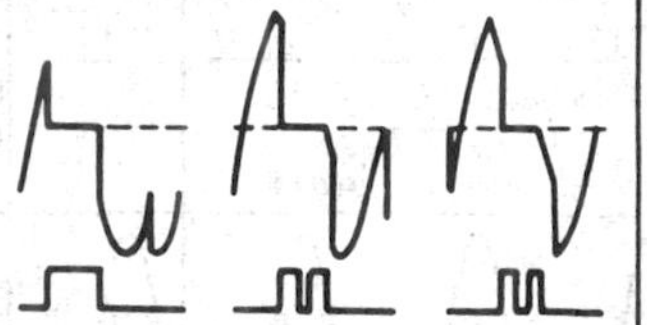 $$V_{FB}=1.5\,\hat{V}_s \quad \begin{pmatrix} I_{AVTh}=0.167\,I_d\\ I_{RMSTh}=0.288\,I_d\\ (0\leqslant\alpha\leqslant\frac{\pi}{3}) \end{pmatrix} \begin{pmatrix} I_{AVD}=I_d\\ I_{RMSD}=I_d\\ (\alpha=\pi) \end{pmatrix}$$ $$V_{RB}=1.732\,\hat{V}_s$$	
Transformer secondary voltage and current		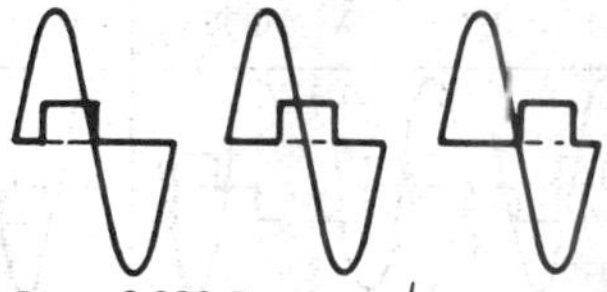 $$I_{RMS}=0.288\,I_d \qquad \cos\phi=\cos\alpha=r$$ $$VA_o=1.48\,P_{do} \qquad \lambda=0.675\,r$$	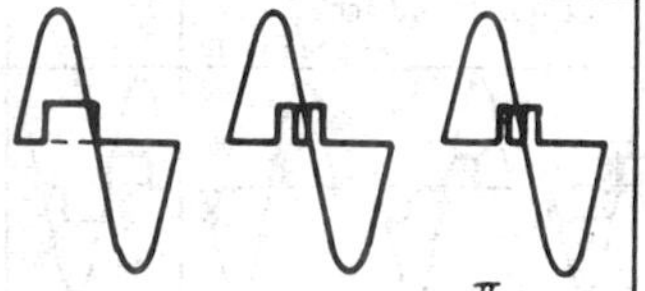 $$I_{RMS}=0.288\,I_d \quad (0\leqslant\alpha\leqslant\frac{\pi}{3})$$ $$VA_o=1.48\,P_{do} \qquad \cos\phi,\lambda,\text{see Fig.11,12}$$	
Transformer primary voltage and current		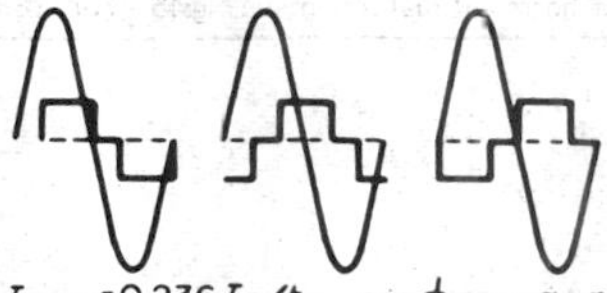 $$I_{RMS}=0.236\,I_d/h \qquad \cos\phi=\cos\alpha=r$$ $$VA_o=1.05\,P_{do} \qquad \lambda=0.955\,r$$ For harmonic distortion, see Fig.13	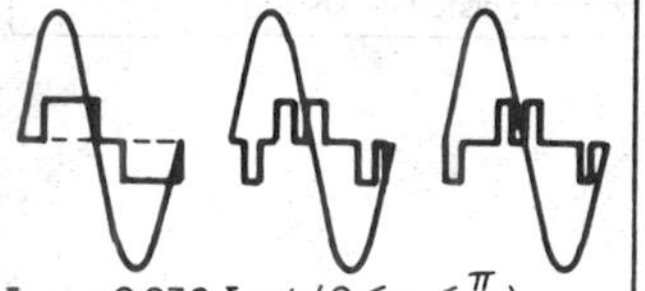 $$I_{RMS}=0.236\,I_d/h \quad (0\leqslant\alpha\leqslant\frac{\pi}{3})$$ $$VA_o=1.05\,P_{do} \qquad \cos\phi,\lambda,\text{see Fig. 11,12}$$ For harmonic distortion, see Fig 13.	

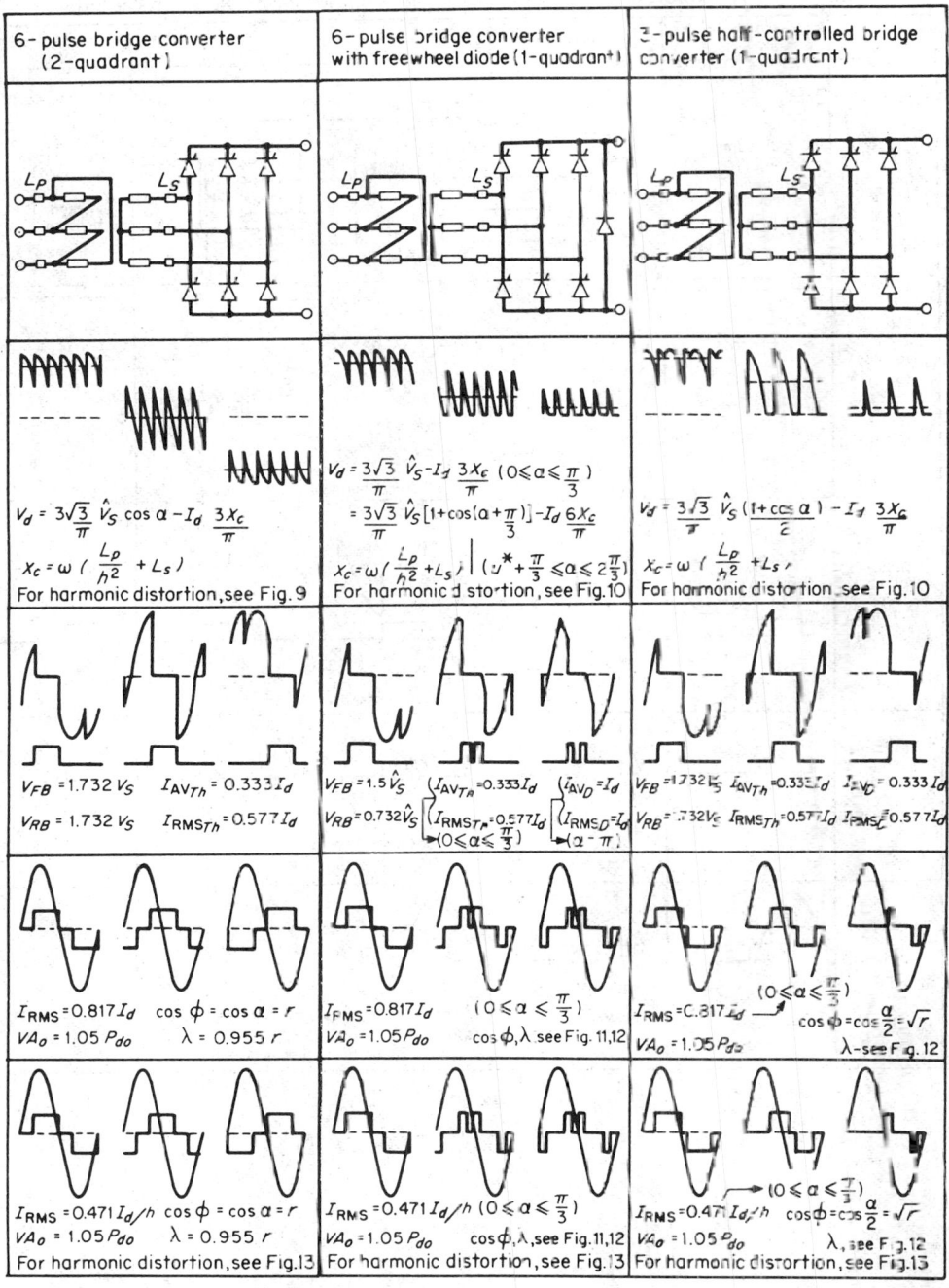

6-pulse bridge converter (2-quadrant)	6-pulse bridge converter with freewheel diode (1-quadrant)	3-pulse half-controlled bridge converter (1-quadrant)
$V_d = 3\frac{\sqrt{3}}{\pi} \hat{V}_S \cos\alpha - I_d \frac{3X_c}{\pi}$ $X_c = \omega\left(\frac{L_p}{h^2} + L_s\right)$ For harmonic distortion, see Fig. 9	$V_d = \frac{3\sqrt{3}}{\pi} \hat{V}_S - I_d \frac{3X_c}{\pi}$ $(0 \leqslant \alpha \leqslant \frac{\pi}{3})$ $= \frac{3\sqrt{3}}{\pi}\hat{V}_S[1+\cos(\alpha+\frac{\pi}{3})] - I_d\frac{6X_c}{\pi}$ $X_c = \omega\left(\frac{L_p}{h^2}+L_s\right)$ $(u^* + \frac{\pi}{3} \leqslant \alpha \leqslant 2\frac{\pi}{3})$ For harmonic distortion, see Fig. 10	$V_d = \frac{3\sqrt{3}}{\pi}\hat{V}_S(1+\cos\alpha) - I_d\frac{3X_c}{\pi}$ $X_c = \omega\left(\frac{L_p}{h^2}+L_s\right)$ For harmonic distortion, see Fig. 10
$V_{FB} = 1.732\,V_S$ $I_{AV_{Th}} = 0.333\,I_d$ $V_{RB} = 1.732\,V_S$ $I_{RMS_{Th}} = 0.577\,I_d$	$V_{FB} = 1.5\,\hat{V}_S$ $V_{RB} = 0.732\,\hat{V}_S$ $\begin{cases} I_{AV_{Th}} = 0.333\,I_d \\ I_{RMS_{Th}} = 0.577\,I_d \\ (0 \leqslant \alpha \leqslant \frac{\pi}{3}) \end{cases}$ $\begin{cases} I_{AV_D} = I_d \\ I_{RMS_D} = I_d \\ (\alpha - \pi) \end{cases}$	$V_{FB} = 1.732\,V_S$ $I_{AV_{Th}} = 0.333\,I_d$ $I_{AV_D} = 0.333\,I_d$ $V_{RB} = 1.732\,V_S$ $I_{RMS_{Th}} = 0.577\,I_d$ $I_{RMS_D} = 0.577\,I_d$
$I_{RMS} = 0.817\,I_d$ $\cos\phi = \cos\alpha = r$ $VA_o = 1.05\,P_{do}$ $\lambda = 0.955\,r$	$I_{FMS} = 0.817\,I_d$ $(0 \leqslant \alpha \leqslant \frac{\pi}{3})$ $VA_o = 1.05\,P_{do}$ $\cos\phi, \lambda$ see Fig. 11,12	$(0 \leqslant \alpha \leqslant \frac{\pi}{3})$ $I_{RMS} = 0.817\,I_d$ $\cos\phi = \cos\frac{\alpha}{2} = \sqrt{r}$ $VA_o = 1.05\,P_{do}$ λ-see Fig. 12
$I_{RMS} = 0.471\,I_d/h$ $\cos\phi = \cos\alpha = r$ $VA_o = 1.05\,P_{do}$ $\lambda = 0.955\,r$ For harmonic distortion, see Fig. 13	$I_{RMS} = 0.471\,I_d/h$ $(0 \leqslant \alpha \leqslant \frac{\pi}{3})$ $VA_o = 1.05\,P_{do}$ $\cos\phi, \lambda,$ see Fig. 11,12 For harmonic distortion, see Fig. 13	$(0 \leqslant \alpha \leqslant \frac{\pi}{3})$ $I_{RMS} = 0.471\,I_d/h$ $\cos\phi = \cos\frac{\alpha}{2} = \sqrt{r}$ $VA_o = 1.05\,P_{do}$ $\lambda,$ see Fig. 12 For harmonic distortion, see Fig. 13

Table 15-6. Waveforms and Data for Various Rectifier and Converter Circuits

	12-pulse midpoint rectifier	12-pulse midpoint converter (2-quadrant)		12-pulse bridge rectifier
Circuit	Circuit same as converter, with diodes instead of thyristors			Circuit same as converter, with diodes instead of thyristors
D-c terminal voltage		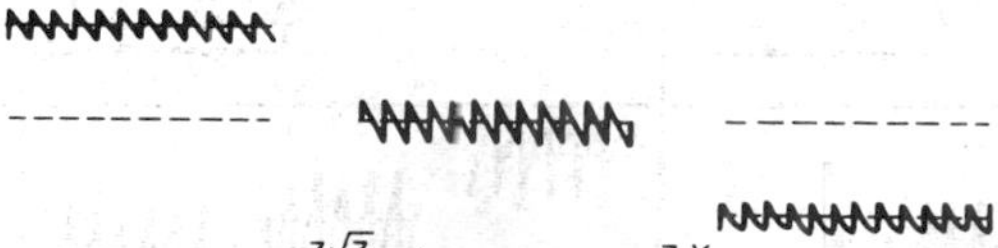 $$V_d = \frac{3\sqrt{3}}{2\pi}\,\hat{V}_S\,\cos\alpha - I_d\,\frac{3X_C}{8\pi}$$ $$X_C = \omega\left(\frac{L_P}{h^2} + L_S\right)$$ For harmonic distortion, see Fig. 9		
Device voltage and current	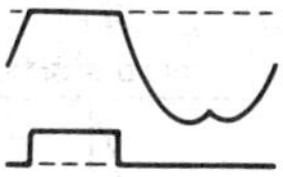	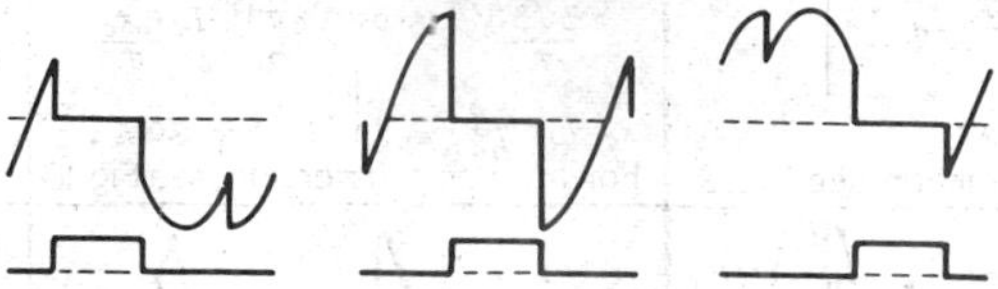 $V_{FB} = 1.732\,\hat{V}_S$ $\quad I_{AV\,Th} = 0.083\,I_d$ $V_{RB} = 1.732\,\hat{V}_S$ $\quad I_{RMS\,Th} = 0.144\,I_d$		
Transformer secondary voltage and current	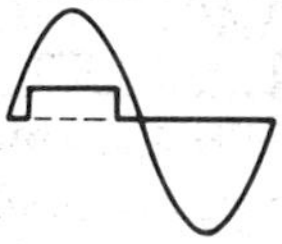	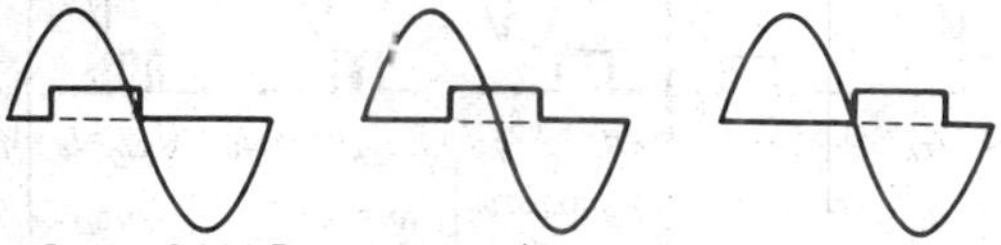 $I_{RMS} = 0.144\,I_d$ $\quad \cos\phi = \cos\alpha = r$ $VA_o = 1.48\,P_{do}$ $\quad \lambda = 0.675\,r$		
Transformer primary voltage and current	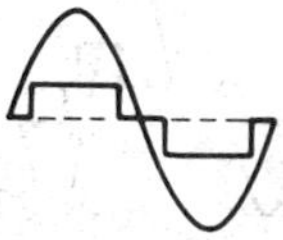	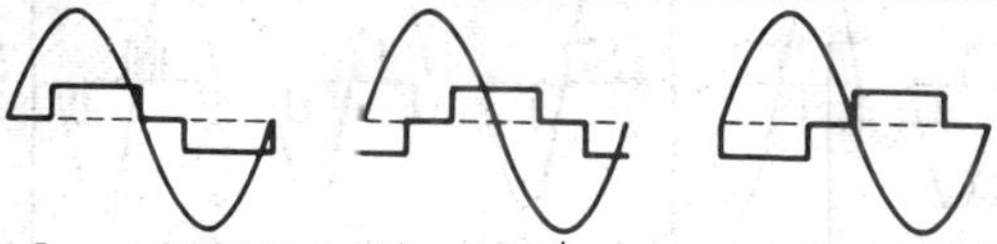 $I_{RMS} = 0.204\,I_d/h\,.T1)$ $\quad \cos\phi = \cos\alpha = r$ $\quad VA = 1.05\,P_{do}$ $= 0.118\,I_d/h\,(T2)$ $\quad \lambda = 0.955 = r$ For harmonic distortion, see Fig. 13		

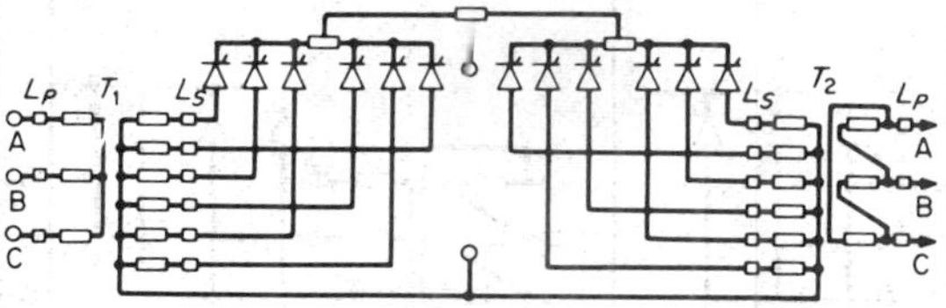

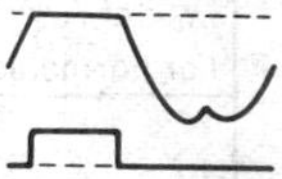

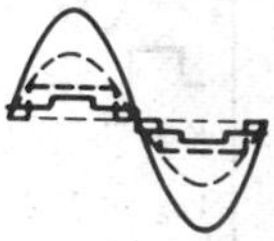

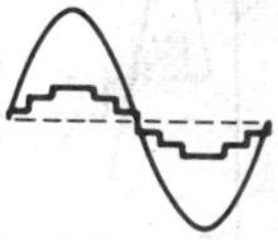

12 - pulse bridge converter (2 - quadrant)

6 - pulse half - controlled bridge converter
(shifted input voltages   (1 - quadrant)

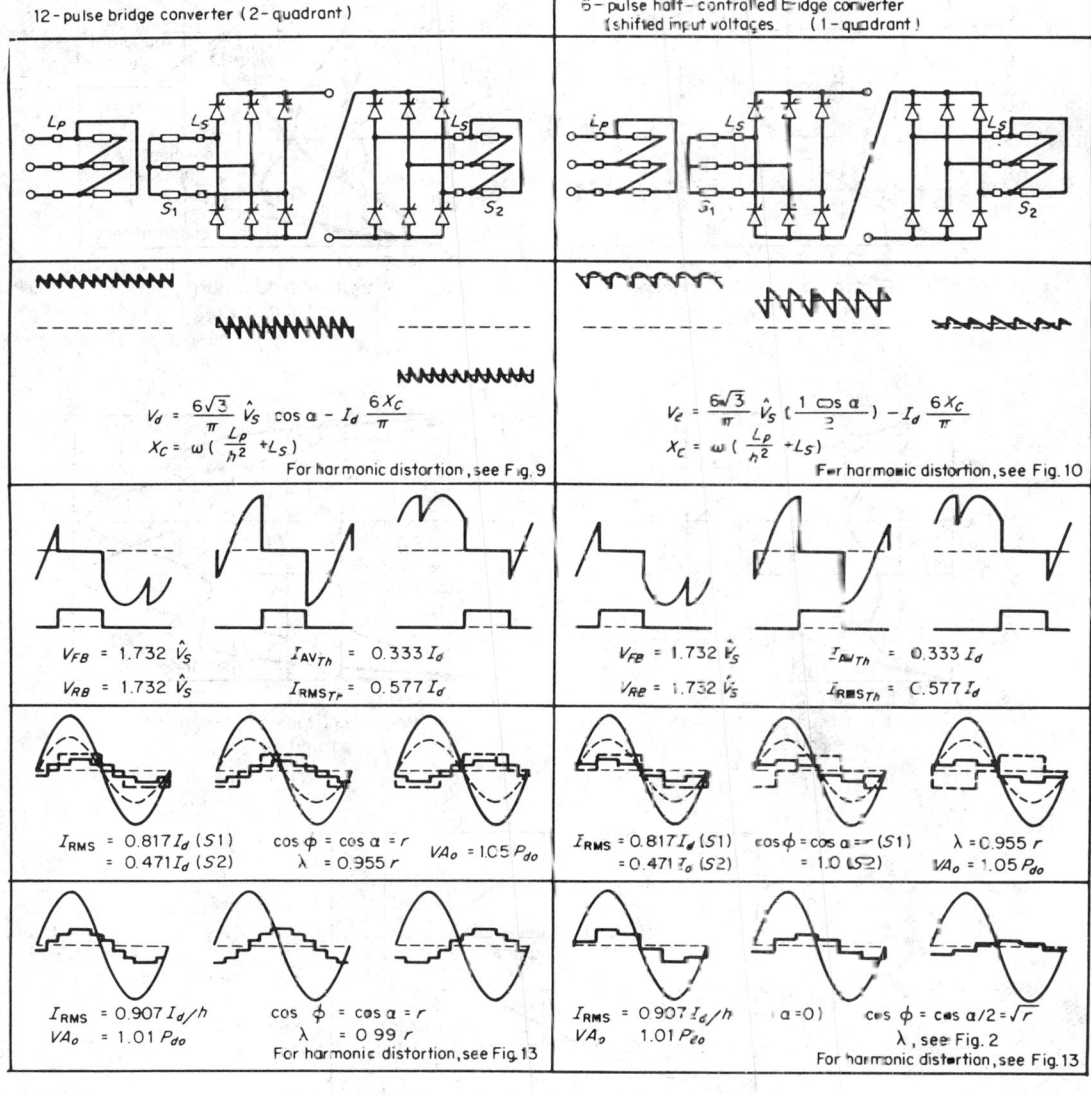

$$V_d = \frac{6\sqrt{3}}{\pi}\, \hat{V_S}\, \cos\alpha - I_d\, \frac{6X_C}{\pi}$$

$$X_C = \omega\left(\frac{L_P}{h^2} + L_S\right)$$

For harmonic distortion, see Fig. 9

$$V_d = \frac{6\sqrt{3}}{\pi}\, \hat{V_S}\left(\frac{1+\cos\alpha}{2}\right) - I_d\, \frac{6X_C}{\pi}$$

$$X_C = \omega\left(\frac{L_P}{h^2} + L_S\right)$$

For harmonic distortion, see Fig. 10

$V_{FB} = 1.732\,\hat{V_S}$

$V_{RB} = 1.732\,\hat{V_S}$

$I_{AV_{Th}} = 0.333\, I_d$

$I_{RMS_{Tr}} = 0.577\, I_d$

$V_{FB} = 1.732\,\hat{V_S}$

$V_{RB} = 1.732\,\hat{V_S}$

$I_{AV_{Th}} = 0.333\, I_d$

$I_{RMS_{Th}} = 0.577\, I_d$

$I_{RMS} = 0.817\, I_d$ (S1)

$\quad\ = 0.471\, I_d$ (S2)

$\cos\phi = \cos\alpha = r$

$\lambda = 0.955\, r$

$VA_o = 1.05\, P_{do}$

$I_{RMS} = 0.817\, I_d$ (S1)

$\quad\ = 0.471\, I_d$ (S2)

$\cos\phi = \cos\alpha = r$ (S1)

$\quad\quad\quad = 1.0$ (S2)

$\lambda = 0.955\, r$

$VA_o = 1.05\, P_{do}$

$I_{RMS} = 0.907\, I_d/h$

$VA_o = 1.01\, P_{do}$

$\cos\phi = \cos\alpha = r$

$\lambda = 0.99\, r$

For harmonic distortion, see Fig. 13

$I_{RMS} = 0.907\, I_d/h$

$VA_o = 1.01\, P_{do}$

$(\alpha = 0)\qquad \cos\phi = \cos\alpha/2 = \sqrt{r}$

λ, see Fig. 2

For harmonic distortion, see Fig. 13

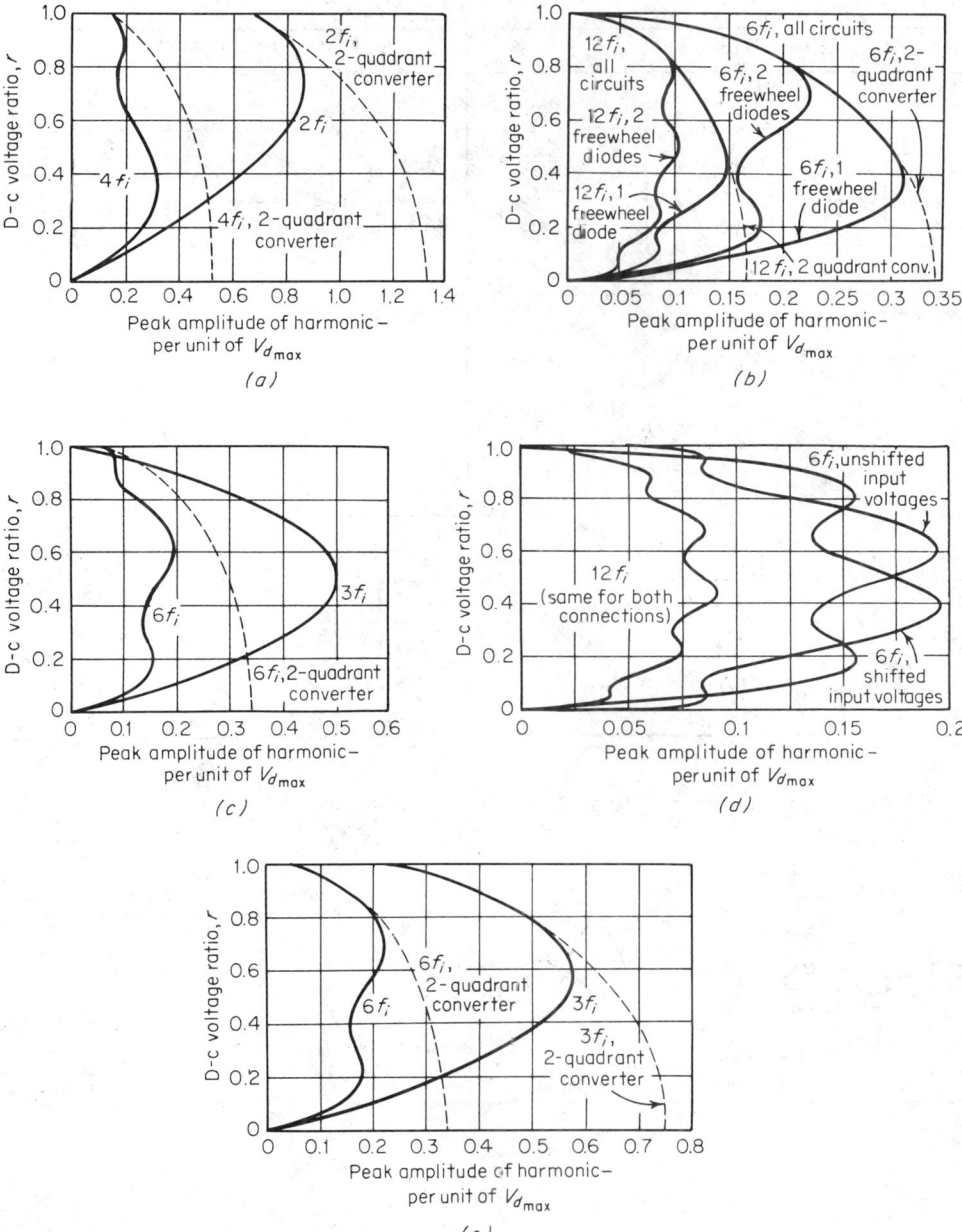

Fig. 15-10. Principal harmonic components in the dc terminal voltage of various one-quadrant converters with no commutation overlap. Curves for corresponding two-quadrant converters are shown for comparison. (a) Two-pulse half-controlled bridge circuit and two-pulse midpoint circuit with freewheel diode; (b) six-pulse circuit with one and two freewheel diodes; (c) three-pulse half-controlled bridge circuit; (d) six-pulse half-controlled bridge circuit, with 30° "shifted" and "unshifted" input voltages for the two bridges; (e) three-pulse circuit with freewheel diode. *(From B. R. Pelly, "Thyristor Phase-Controlled Converters and Cycloconverters," Wiley-Interscience, New York, 1971; used by permission.)*

For all two-quadrant converters, the input-power factor and the dc voltage ratio are also directly proportional to each other, as indicated in the tables. The corresponding relationships for the one-quadrant converters are illustrated in Fig. 15-12.

Figure 15-13 shows the principal harmonic components present in the input line current of each of the converters shown in Tables 15-4 to 15-6.

All of the above theoretical relationships assume ripple-free current at the dc terminals of the converter, with no commutation overlap.

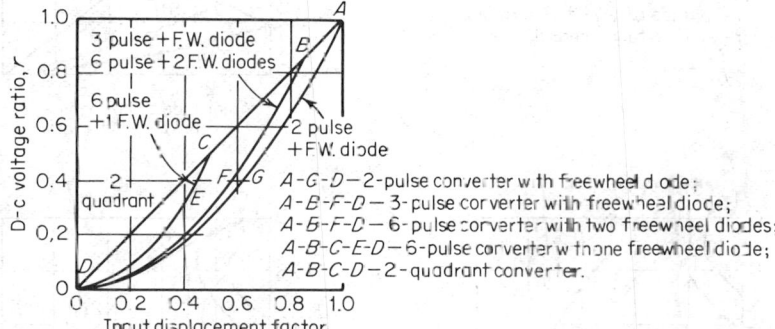

Fig. 15-11. Relationships between the dc terminal voltage ratio and the input-displacement factor for two-, three-, and six-pulse converters with freewheel diode and no commutation overlap. *A-G-D*, two-pulse with freewheel diode; *A-B-F-D*, three-pulse with freewheel diode; *A-B-F-D*, six-pulse with two freewheel diodes; *A-B-C-E-D*, six-pulse with one freewheel diode; *A-B-C-D*, two-quadrant converter.

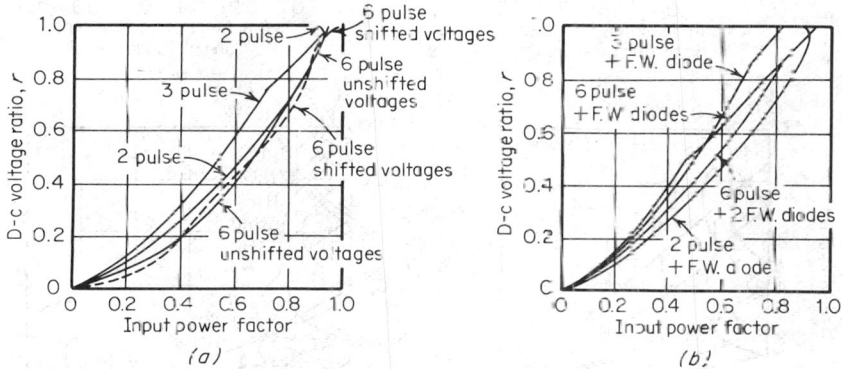

Fig. 15-12. Relationships between the dc voltage ratio and the input-power factor for various one-quadrant converters with no commutation overlap: (*a*) two-, three-, and six-pulse half-controlled circuits; (*b*) two-, three-, and six-pulse circuits with freewheel diodes. These curves apply to transformer primary and secondary for bridge circuits and to primary for midpoint circuits. For transformer-secondary power factor of midpoint circuits, multiply by 0.707.

18. Four-Quadrant Converters. A four-quadrant converter, or *dual converter*, can operate with both polarities of both voltage and current at the dc terminals. Such converters permit, for example, dc motors to be driven and regeneratively braked in both forward and reverse directions. A six-pulse bridge four-quadrant converter formed by paralleling two oppositely polarized two-quadrant converters is shown in Fig. 15-14.

If both converters are active simultaneously, in principle one operates as a rectifier while the other operates as an inverter with the same average voltage. In practice, the instantaneous difference between the voltages of the two converters tends to cause a large circulating current. One solution is to parallel the two converters through a circulating-current reactor, as shown in Fig. 15-14. Another solution, which is usually preferable, involves deactivating the idle converter either by removing its firing pulses or by appropriately adjusting its relative delay angle. This control can be achieved automatically in various ways (Pelly, 1971).

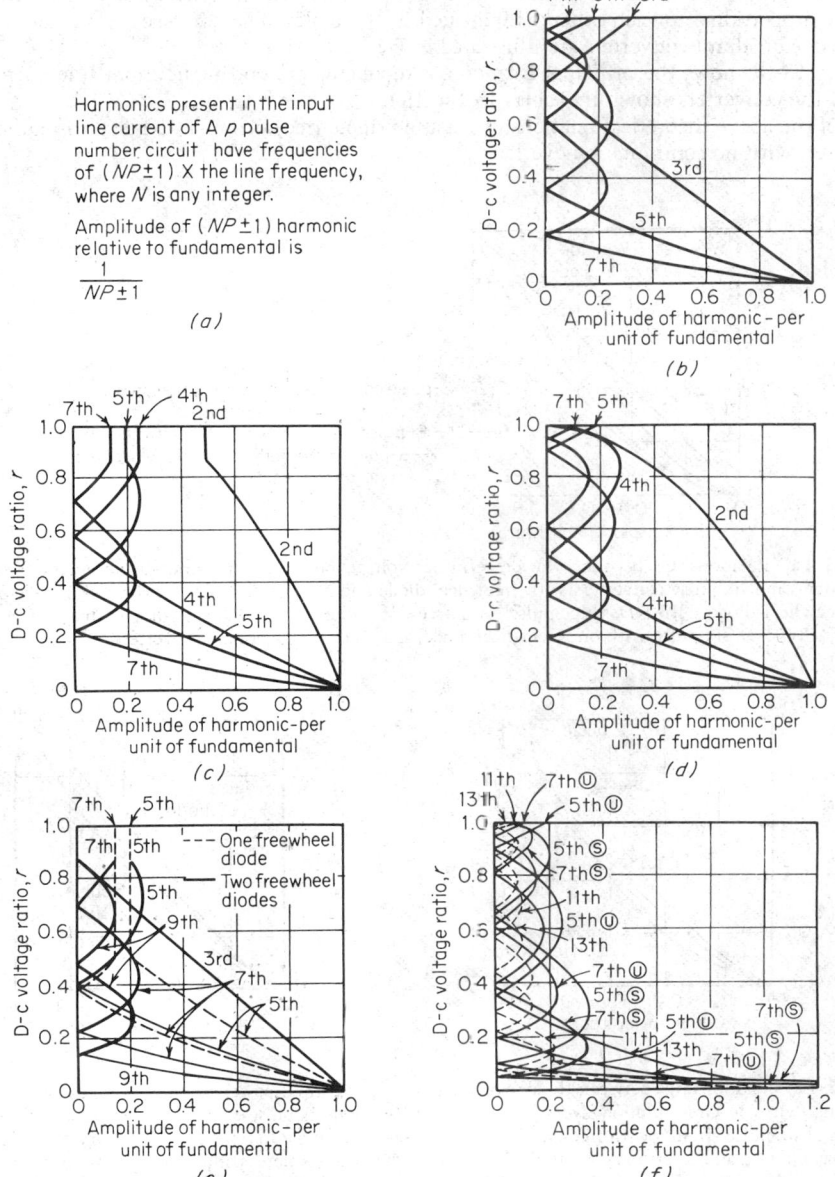

Harmonics present in the input
line current of A p pulse
number circuit have frequencies
of $(NP\pm1) \times$ the line frequency,
where N is any integer.

Amplitude of $(NP\pm1)$ harmonic
relative to fundamental is

$$\frac{1}{NP\pm1}$$

(a)

Fig. 15-13. Principal harmonic components in the input line current of various converter circuits (with smooth direct current and no commutation overlap).

AC SWITCHES AND REGULATORS

BY P. WOOD

19. AC Switches. Various applications require the power of an ac source to be regulated or switched on and off. Naturally commutated solid-state switches can perform these functions. Such switches and regulators require bilateral conduction. An obvious way to provide for bilateral conduction is by applying two thyristors in the antiparallel connection, shown in Fig. 15-15a. The triac, whose symbol is shown in Fig. 15-15b, integrates both thyristors into a single device structure and is widely used where its power and frequency ratings suffice. Two other configurations for a bilateral switch are shown in Fig. 15-15c and d.

20. On-Off Control and Protection. Solid-state ac switches can be used to replace mechanical relays and circuit breakers for on-off control and protection of electric circuits. Since there are no moving contacts to wear or arc, useful life and reliability can be very long. Electromagnetic noise and transient surges can be minimized by delaying turn-on until a voltage zero.

The fast response of solid-state switches can be used in circuit breakers. Normally, ac switches are switched off by removing the gate firing signals, allowing the conducting devices to commutate naturally at the next current zero. Thyristors in series with the power line can be used as a current-limiting circuit breaker. However, subcycle interruption of a fault current before it reaches its peak requires the use of forced commutation.

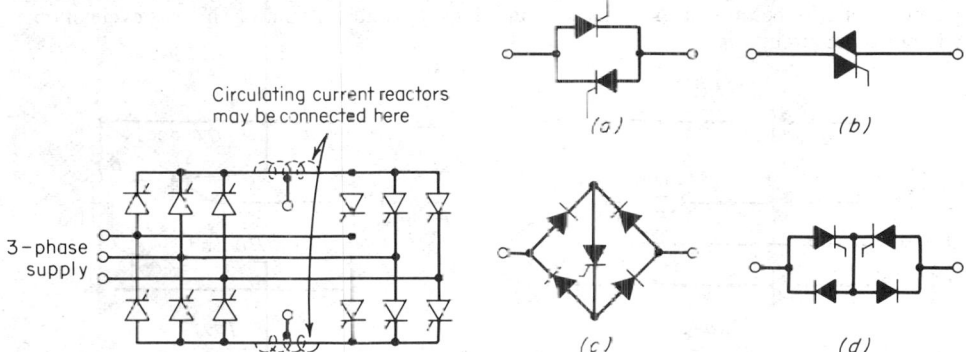

Circulating current reactors may be connected here

3-phase supply

Fig. 15-14. Six-pulse bridge dual-converter circuit.

Fig. 15-15. Single-phase ac switches.

In some applications, a thyristor across the power line is fired when a fault is detected, thereby blowing a fuse in series with the line. The fuse clears the circuit, and the thyristor "crowbar" limits the voltage surge across the protected circuit. In certain applications, a hybrid combination of solid-state and mechanical switches is advantageous. The transiently rated solid-state switch eliminates arcing and allows precise control of the instant of switching, whereas the parallel mechanical contacts carry the continuous load current and prevent heat losses in the solid-state devices.

21. AC Power Regulators. Since the instant of turn-on of the ac switches of Fig. 15-15 can be controlled by proper timing of the gate signal and they turn off automatically when the current flow therein attempts to reverse, they can be used for the continuous control of an ac quantity. The control may be merely *transitory*, as in soft starting an induction motor or limiting the inrush current to a transformer, or *perpetual*, as in the control of resistive heating elements, incandescent lamps, and the reactors of a static VAR (reactive volt-ampere) generator.

The basic single-phase ac regulator is depicted in Fig. 15-16, using a triac as the ac switch (but any of the ac switch combinations shown in Fig. 15-15 is applicable). The various three-phase arrangements possible are shown in Fig. 15-17. The first of these, the wye-connected regulator with a neutral connection (Fig. 15-17a), exhibits behavior identical to that of the single-phase regulator, since it is merely a threefold replica of the single-phase version.

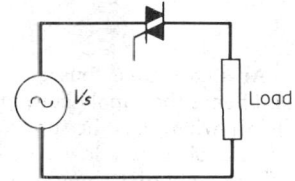

Fig. 15-16. Single-phase ac regulator.

The delta-connected regulator arrangement of Fig. 15-17b is also essentially similar in behavior to the single-phase regulator insofar as load voltages and currents are concerned. Because of the delta connection, however, any symmetrical zero sequence components of the load currents will not flow in the supply lines but will only circulate in the delta-connected loads.

The three-phase three-wire wye-switched regulator of Fig. 15-17c behaves differently because two switches must be closed for current to flow in any load. Shown delta-loaded, it may also have the loads wye-connected without a neutral return. In this connection, each ac switch may consist of the antiparallel combination of a thyristor and a diode. The normal wye-delta transformations apply to load voltages and currents.

The "British delta" circuit of Fig. 15-17d behaves in the same way as a wye-switched regulator in which thyristors with inverse parallel connected diodes are used as the switches and is unique in that only unidirectional current capability is required of its switches.

When the loads are essentially resistive, two methods of control are currently employed. The technique known as *integral-cycle control* operates the regulator by keeping the switch(es) closed for some number m of complete cycles of the supply and then keeping the switch(es) open for some number n of cycles. The power delivered to the load(s) is then simply $m/(m + n)$ times the power delivered if the switch(es) are kept permanently closed, for the single-phase, wye with neutral, and delta-connected regulators.

For the wye-switched (without neutral) and British delta regulators, the power delivered is slightly greater than $m/(m + n)$ times the power at full switch conduction, and dc and unbalanced ac components develop in the supply unless special control techniques are used. These phenomena arise because of the transient conditions inevitably attending the first cycle of operation of these circuits.

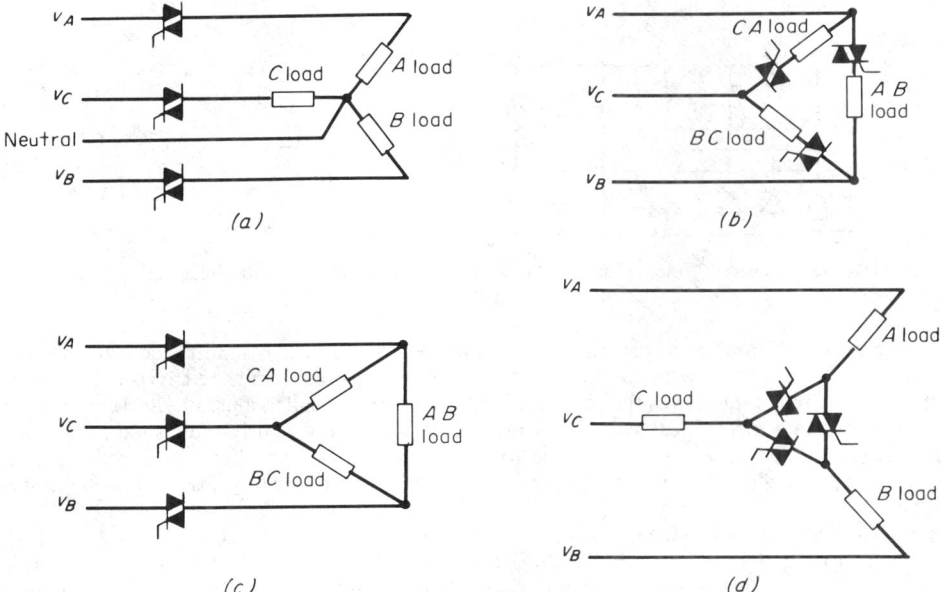

Fig. 15-17. Three-phase ac regulators.

An undesirable consequence of integral-cycle control is that the load voltages and currents, and hence the supply currents, contain sideband components having frequencies $f_s[1 \pm pm/(m + n)]$, where f_s is the supply frequency and p any integer, 1 to infinity. Many of these frequencies are obviously lower than the supply frequency and may create problems for the supply system and other connected loads. The existence of this type of unwanted components in the voltage and current spectra makes integral-cycle control unsuitable for inductive loads (including loads fed by transformers). Since none of the sidebands are zero sequence, the line currents of an integral-cycle-controlled delta regulator are identical to the properly transposed load currents.

Integral-cycle control results in unity displacement factor (the cosine of the angle between the fundamental component of supply current and the supply voltage). The power factor of the burden they impose on the supply with pure resistive loads is $[m/(m + n)]^{0.5}$. This is true because any regulator which forces the load current to flow in the supply while reducing the rms voltage applied to a resistive load has a power factor equal to the rms load voltage divided by the rms supply voltage.

The other method of control commonly used is termed *phase-delay control*. It is implemented by delaying the closing of the switch(es) by an angle α (called the firing angle) from each zero crossing of the supply voltage(s) and allowing the switch(es) to open again on each succeeding current zero. The load voltages and currents in this case contain only harmonics of the supply frequency as unwanted components, and except for the regulator shown in Fig. 15-17c, using

thyristor-inverse diode switches, only odd-order harmonics are present. Thus this control technique can be used with inductive loads.

The general expressions for the load voltages and currents produced by the single-phase regulator are very cumbersome but simplify considerably for the two cases of greatest practical
importance, pure resistive and pure inductive loads. For a pure resistive load with a supply voltage $V \cos \omega_s t$, the fundamental component of load voltage is given by

$$V_{DIR} = \left(1 - \frac{\alpha}{\pi} + \frac{\sin 2\alpha}{2\pi}\right) V \cos \omega_s t + \frac{\sin^2 \alpha}{\pi} V \sin \omega_s t$$

and the total rms load voltage by

$$V_{RMSR} = \frac{V}{\sqrt{2}} \left(1 - \frac{\alpha}{\pi} + \frac{\sin 2\alpha}{2\pi}\right)^{1/2}$$

where α is the firing angle measured from the supply-voltage zero crossings. For pure inductive
load it is convenient to define the firing angle $\alpha' = \alpha - \pi/2$, so that at full output $\alpha' = 0$. The
fundamental voltage component is then given by

$$V_{DIL} = \left(1 - \frac{2\alpha'}{\pi} - \frac{\sin 2\alpha'}{\pi}\right) V \cos \omega_s t$$

and the total rms voltage by

$$V_{RMSL} = \frac{V}{\sqrt{2}} \left(1 - \frac{2\alpha'}{\pi} - \frac{\sin 2\alpha'}{\pi}\right)^{1/2}$$

The same relationships apply to the three-phase circuits, which are in effect triplicates of the
single-phase circuit (Fig. 15-17a and b); more complex relationships exist for the remaining
three-phase circuits.

The use of phase-delay control results in decreasing lagging displacement factor with increasing firing angle. Maximum displacement factor is obtained at full output, equaling the power
factor of the given load. At a reduced power setting the power factor is less than the displacement
factor; the ratio of the two equals the ratio of the fundamental line currents vs. the total rms
line currents. This ratio is less than unity due to the presence of harmonic currents.

The load voltages and currents and, more importantly, the line currents of phase-delay-controlled regulators have lower total rms distortion than those of integral-cycle-controlled regulators. Among the circuits shown, the delta regulator of Fig. 15-17b is most beneficial; since the
triple n harmonics (those of orders which are integer multiples of 3) in its load currents are zero
sequence, they do not flow in the supply lines and the circuit has both a better power factor and
lower total line-current distortion than integral-cycle regulators or the phase-delay-controlled
wye regulators with neutral.

For the wye regulator without neutral, the range of α is 0 to $7\pi/6$ rad, provided fully bilateral
switches are used; for the British delta regulator and the wye regulator without neutral using
thyristor-inverse diode switches, the range is 0 to $5\pi/6$ rad. When phase-delay regulators are
used with inductive loads, the range of α used for control is reduced because current-zero crossings lag voltage-zero crossings and thus abrogate part of the delay obtained with resistive loads.
The regulators most commonly used with inductive loads are the single-phase, wye with neutral
and the delta, for which the range of α becomes ϕ to π, where ϕ is the load phase angle.

22. Static VAR Generators. The delta regulator with purely inductive loading finds
extensive use in the *static VAR generator* (SVG) (Gyugyi et al., 1978, 1980). A basic SVG consists
of three delta-connected inductors with phase-controlled switches ($\pi/2 \leq \alpha \leq \pi$) and three
fixed capacitive branches which may be delta- or wye-connected. The capacitive branches draw
a fixed current from the supply, leading the voltage by $\pi/2$ rad. The fundamental current in the
inductors is lagging the voltage by $\pi/2$ rad. Its amplitude can be varied, by phase controlling the
switches, from the full inductor current to zero.

Hence the net reactive volt-ampere burden on the supply can be continuously controlled from
the full capacitive VAR, when $\alpha = \pi$ and the inductor currents are zero, to the difference
between the capacitive- and inductive-branch VARs when $\alpha = \pi/2$ and full inductor currents
flow. This difference will be zero if inductive-branch VARs are made equal to capacitive-branch
VARs and become an inductive burden if inductive VARs at full conduction exceed the capacitive VARs. Since the firing angle α can be varied on a half-cycle–to–half-cycle basis, extremely
rapid changes in VAR supply (capacitive burden) or demand (inductive burden) can be
accomplished.

SVGs can be used to supply shunt-reactive compensation on ac transmission and distribution systems, helping system stability and voltage regulation. They can also be used to provide damping of the subsynchronous resonances, which often prove troublesome during transient disturbances on series capacitor-compensated transmission systems, and to reduce the voltage fluctuations (flicker) produced by arc-furnace loads. In the latter application, their ability to accomplish dynamic load balancing is especially valuable.

An SVG which can provide control of reactive power supply or demand can obviously compensate for an unbalanced reactive load. It can also act as a Steinmetz balancer, providing the reactive power exchange between phases necessary to transform an unbalanced resistive load into a perfectly balanced and totally active (real) power load on the supply system.

This action can be explained as follows. Suppose a single-phase resistive load is connected between lines A and B of a three-phase system. Then the current it draws will be in phase with the AB voltage, and thus the A-line current created will lead the A-phase (line-to-neutral) voltage by $\pi/6$ rad and the B-line current will lag the B-phase voltage by $\pi/6$ rad.

If equal-impedance purely reactive loads are now connected to the BC and CA line pairs, capacitance on BC and inductive on CB, they create currents with the following phase relationships to the phase voltages:

In the A line, lagging by $2\pi/3$ rad.
In the B line, leading by $2\pi/3$ rad.
In the C line, one leading by $\pi/3$ rad and the other lagging by $\pi/3$ rad.

The result in the C line is clearly an in-phase, wholly real current. If the impedances are of appropriate magnitude, their lagging and leading quadrature contributions in the A and B lines, respectively, can be made to cancel the lagging and leading quadrature currents created therein by the single-phase resistive load. The impedance required is $\sqrt{3}$ times the resistance. Obviously an SVG capable of providing either leading or lagging line-to-line loading on any of the line pairs can be used to balance a single-phase resistive load on any one line pair; by extension, it can be used to balance any unbalanced load. It can respond rapidly to changes in the degree of unbalance existing and thus dynamically balance the load despite the fluctuating unbalance typically created by an arc furnace.

In addition to a varying reactive fundamental current, an SVG operating other than at full or zero conduction in its reactive branches generates harmonic currents. Thus at least part of the capacitive branch is usually realized in the form of tuned harmonic filters to limit harmonic injection to the ac supply system. Maximum harmonic amplitudes relative to maximum fundamental are:

Harmonic order	3d	5th	7th	9th	11th	13th
Maximum amplitude, %	13.8	5.05	2.59	1.57	1.05	0.752

with diminishing amplitudes of the higher-order components.

When the SVG is in balanced operation, the triple n harmonics (3d and 9th in the table above) do not flow in the supply, being zero sequence. When operation is unbalanced in order to balance an unbalanced real load, positive and negative sequence components of the triple n harmonics develop and of course do flow in the supply unless filtering is provided for them.

Circuits Using Forced Commutation

23. General Considerations. *Forced commutation* is the general name given to the use of power semiconductor devices which have turnoff ability and to a variety of techniques and circuits which permit current to be interrupted by power semiconductor devices without inherent turnoff ability (see also Par. **15-4**).

One type of forced commutation uses *resonance* to generate an alternation which brings the current in a conducting thyristor to zero. Such circuits typically require large reactive elements with considerable energy storage, but the functions of commutation and filtering for waveform improvement can often both be performed by the same reactive elements (Rice, 1970, pp. 8-47 to 8-54).

Another type of forced commutation, known as *impulse commutation*, momentarily diverts the load current while reverse-biasing the thyristor until it regains its blocking state. Some circuit configurations include a capacitor which achieves an automatic transfer of current when the next load-carrying thyristor is fired. These circuits are said to be self-commutated. Other circuits use an auxiliary thyristor in series with a capacitor to divert load current briefly and then cease conduction.

The *energy storage* required to achieve impulse commutation is considerably less than that required for resonant commutation. Because of the crucial role played by stored energy in achieving forced commutation, the fault current which can be safely interrupted is limited by the commutation circuit rather than by the thyristors.

The *power capacity* required by many applications of the circuits to be described can be obtained only by using conventional thyristors. Achieving reliable and economical forced commutation is then a major factor in the design process For applications within the available ratings of power transistors or gate-controlled switches, these devices can be used to avoid the costs of the additional components needed for forced commutation.

In its on state, a transistor is generally operated in or near saturation, where its dissipation is minimized. If the collector current appreciably exceeds the product of the base current and the gain, the transistor comes out of saturation and is subject to very high dissipation, which can cause permanent damage very quickly. The transistor should never be exposed to voltage and current outside its safe operating area (SOA) (see Sec. 7, Par. **7-88**). See also Sec. **28**.

All circuits which are required to interrupt current in an inductive load must provide a safe path for absorbing the energy stored in the inductance. Otherwise the inductive voltage spikes are likely to destroy the long-term reliability, if not the immediate operation of the circuit. A freewheeling diode across the load is a common technique which permits the energy to be dissipated in the load itself. In circuits where the load is resonated, this energy is transferred to a capacitor, where it is available for reuse.

24. DC Regulators (Choppers). The power supplied from a dc source to a dc load can be regulated with high efficiency by a series switch that repetitively opens and closes, thereby "chopping" the current flowing between the two. Hence dc regulators are frequently called choppers. Control is achieved by varying the relative on time or duty cycle. Obviously the duty cycle can be changed by changing the on time, the off time, or both, and all three control modes are used in practical choppers (Becford and Hoft, 1964, Chap. 10).

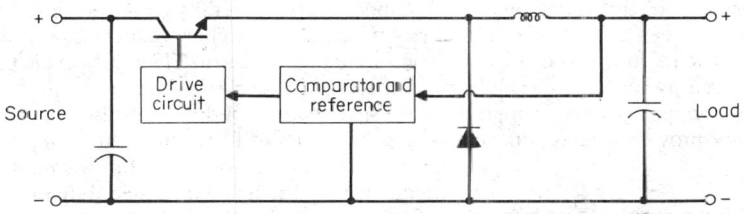

Fig. 15-18. Chopper using a transistor switch.

25. Transistor Choppers. A typical transistor chopper is shown in Fig. 15-18. The transistor is switched by a drive circuit connected to its base. The duty cycle of the drive circuit is determined by a circuit which compares the load voltage with a reference voltage. The capacitor across the source protects the transistor from inductive spikes when the source current is interrupted, and a freewheeling diode shunts inductive spikes from the load. An *LC* filter between the chopper and the load reduces the ripple in the output voltage.

The frequency at which the switch is operated is determined by the desired efficiency and size. As the frequency is increased, the size of the filter decreases, but the switching losses in the transistor and diode increase, reducing efficiency and requiring a larger heat sink. Transistor choppers are typically operated at frequencies in the range of 1 to 4 kHz. However, in some specialized applications where the power requirements are low and minimum size and weight are important, frequencies above 20 kHz can be used.

26. Thyristor Choppers. The principles of operation of thyristor choppers are basically the same as for transistor choppers except that forced commutation is required with thyristors. Of the variety of commutation circuits in use, two typical circuits are shown enclosed in dashes in Fig. 15-19. In the first circuit (*a*), capacitor *C* is first charged by switching on thyristor Q1. Resonant charging through *L* causes the voltage across *C* to rise to a peak value which is greater than the supply voltage, after which Q1 ceases to conduct. After *C* is charged, main thyristor Q2 is switched on and current is delivered to the load. To turn Q2 off, Q3 is fired. The voltage on *C* reverse-biases Q2 until it is able to block forward voltage. When *C* is discharged, Q3 ceases to conduct.

In the second circuit (b), capacitor C is charged to the supply voltage by switching thyristor Q1 on. After C has charged, Q2 is switched on to deliver a load pulse and to allow the voltage on C to reverse by resonantly discharging through the path provided by diode D1 and L. After the voltage on C has reversed, Q1 may be switched on to commutate Q2 off.

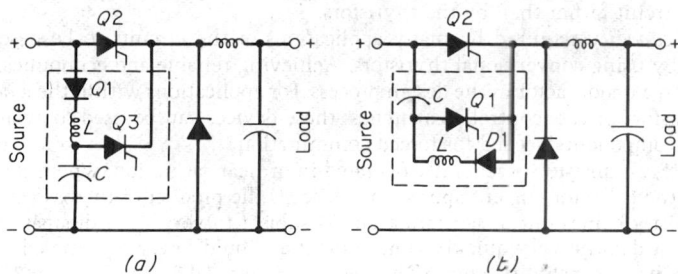

(a)   (b)

Fig. 15-19. Two typical thyristor choppers showing circuits for forced commutation.

The maximum current which can be reliably controlled in a thyristor chopper depends upon the commutating capability designed into the circuit. In many applications, if the current exceeds that which the commutating circuit can switch off, the main thyristor will remain on and the full supply voltage will be delivered to the load. Normal operation can then be restored only by opening the source or load circuit. For this reason, many thyristor choppers have automatic current limiting built into their control circuits. When the load current approaches a preset limit (determined either by the commutating circuit or by the thyristor rating), the control signal from the limiter overrides the signal from the output comparator and the chopper operates in a current-limited mode.

The *frequency of operation* of thyristor choppers is lower than that obtained with transistors because of the combined effects of device switching and recovery times and the di/dt and dv/dt limitations of thyristors. Frequencies under 2 kHz generally provide the most favorable compromise between efficiency, size, and cost. If necessary, higher frequencies can be achieved by replacing the main thyristor by a number of parallel thyristors. The duty cycle of each thyristor is reduced by time-shared conduction on sequential cycles.

Thyristor choppers find wide application for speed control of dc motors used in traction applications. They provide smooth, efficient control from standstill to full vehicle speed. Choppers are used both with series- and shunt-wound motors and frequently incorporate controlled dynamic or regenerative braking of the wheels. The range of traction applications varies from battery-powered forklift trucks, using controllers operating at 36 V and several hundred amperes, to high-speed train drives using series thyristors operating at several thousand volts and several hundred amperes.

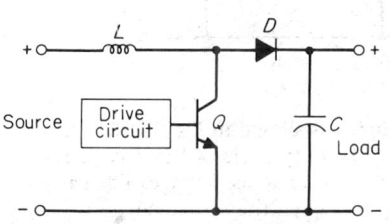

Fig. 15-20. Voltage step-up chopper.

27. Voltage Step-Up Choppers. Choppers can also be used to produce a dc output voltage which is higher than the supply voltage. This is accomplished by means of the circuit shown in Fig. 15-20. Initially transistor Q is turned on, and the source stores energy in inductor L. When Q is turned off, the energy in L discharges into the load through diode D. During this discharge period, the voltage across L adds to the source voltage. Capacitor C across the load smooths the voltage pulsations. This circuit is equivalent to a dc step-up transformer. With the source and load terminals interchanged, this circuit is suitable for regenerating power from the dc circuit into the source and is used for implementing regenerative braking of a dc motor.

INVERTERS

BY T. M. HEINRICH AND R. M. OATES

28. General Considerations. An inverter is a power converter in which the normal direction of power flow is from a dc source to an ac load. In contrast to naturally commutated converters, which may operate as synchronous inverters as described in Pars. **15-13** to **15-18**, the thyristor inverters to be discussed here must be force-commutated unless the load happens to have a leading power factor. However, the power flow is still reversible. By properly phasing the control signals, the dc source can be made to absorb power from an active ac load, such as a motor which is being dynamically braked.

Table 15-7 shows the four basic inverter circuits, corresponding to a center-tapped load, a center-tapped source, and single- and 3-phase bridges. The relationships given were derived with the aid of the following simplifying assumptions: the switches operate instantaneously and have no voltage drop when closed or leakage current when open; the load acts as an ideal filter which blocks all current harmonics regardless of the shape of the ac voltage; the load impedance has phase angle φ at fundamental frequency. As illustrated, the switches must block only one polarity of voltage, but they must be capable of conducting both polarities of current. In practice these switches are implemented by shunting a transistor or thyristor by a diode which carries the reverse current.

29. Self-Commutated Inverters. The half-bridge thyristor inverter shown in Fig. 15-21 illustrates the principles of self-commutation. R_0 is the load resistor, L_c and C_c are the commutating components, and L_f and C_f constitute a simple filter to smooth the load-voltage waveform. When Th1 is turned on to initiate the first positive half cycle, C_c charges to a voltage of $+E/2$. When the first negative half cycle is to be initiated, Th2 is turned on, causing the lower half of the source voltage to add to the voltage stored on C_c across the lower half of L_c.

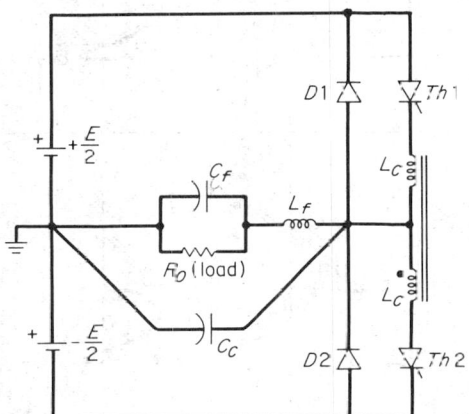

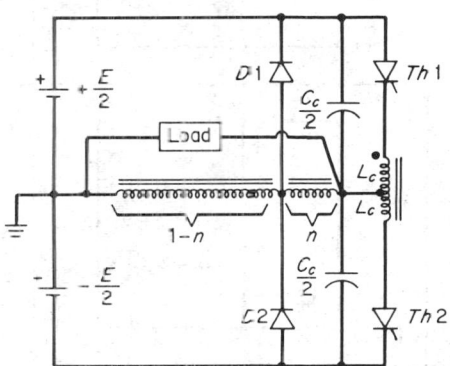

Fig. 15-21. Half-bridge thyristor inverter with forced commutation.

Fig. 15-22. Half-bridge thyristor inverter with forced commutation and energy-recovery transformer.

Because of the mutual coupling between the two halves of the commutating inductor, a voltage equal to E is induced in the upper half. This voltage causes Th1 to become reverse-biased so that it begins to turn off. C_c and L_c are designed to hold a reverse bias on Th1 long enough for it to recover its blocking ability. Thus the firing of Th2 automatically transfers current from Th1 to Th2, and, following the commutation transient, the polarity of the load voltage reverses, causing C_c to charge to $-E/2$. When Th1 is again fired, Th2 is turned off in a complementary manner, and the first full cycle of operation is complete.

During the time that C_c is recharging toward $-E/2$, the current in the lower half of L_c approximately doubles. Because of diode D2, the voltage across inductor L_c cannot reverse by more than the combined forward drops of D2 and Th2. The excess current so trapped continues to circulate through D2 and Th2 until all the trapped inductor energy is dissipated. Aside from the extra dissipation in the devices, the trapped energy is bothersome in that it hinders the commutation process. If it is not removed, the inductor current for subsequent commutations will become progressively greater until C_c can no longer supply sufficient commutating energy.

A resistor may be inserted in series with diodes D1 and D2 to dissipate the trapped energy, but for frequencies up to about 400 Hz, most of this energy can be recovered by employing a tapped transformer primary (McMurray and Shattuck, 1961). A practical circuit with trapped-energy-recovery transformer is shown in Fig. 15-22. The tap at n provides an additional voltage in the discharge loop to absorb energy from L_c. The energy absorbed by the transformer is passed along to the load or returned to the dc source if the load is unreceptive. The tap n is generally

TABLE 15-7 Basic Inverter Circuits

	Center-tap	Half-bridge	Full-bridge	Three-phase bridge
Circuit diagram	$\eta:1:1$ Turns ratio			
Circuit name	Center-tap	Half-bridge	Full-bridge	Three-phase bridge
Output voltage V_{out} — Voltage waveform	ηE_{DC}, $-\eta E_{DC}$ (π, 2π)	$\frac{1}{2}E_{DC}$, $-\frac{1}{2}E_{DC}$ (π, 2π)	E_{DC}, $-E_{DC}$ (π, 2π)	E_{DC}, $-E_{DC}$ ($\frac{2}{3}\pi$, π, 2π, $\frac{5}{3}\pi$) — Contains no third harmonic
Output voltage V_{out} — RMS value of V_{out} (fundamental component only)	$\frac{2\sqrt{2}}{\pi}\,\eta\,E_{DC}$	$\frac{\sqrt{2}}{\pi}E_{DC}$	$\frac{2\sqrt{2}}{\pi}E_{DC}$	$\frac{\sqrt{6}}{\pi}E_{DC}$
Input current I_{DC} — Waveform	$I_L\sqrt{2}$ (φ, π, 2π)	$I_L\sqrt{2}$ (φ, π, 2π)	$I_L\sqrt{2}$ (φ, π, 2π)	(π, 2π)

Input current I_{DC}				
I_{DC} (avg value)	$\frac{2\sqrt{2}}{\pi}\,\eta I_L\cos\varphi$	$\frac{\sqrt{2}}{\pi}\,I_L\cos\varphi$	$\frac{2\sqrt{2}}{\pi}\,I_L\cos\varphi$	$\frac{3\sqrt{2}}{\pi}\,I_L\cos\varphi$
$\dfrac{I_{PK}}{I_{DC}\text{ (avg)}}$	$\dfrac{\pi}{2\cos\varphi}$	$\dfrac{\pi}{\cos\varphi}$	$\dfrac{\pi}{2\cos\varphi}$	$\dfrac{\pi}{3\cos\varphi}$ $\left(0\le\omega\le\frac{\pi}{6}\right)$
$\dfrac{f_{ripple}}{f_{inverter}}$	2	1	2	6
Switch stress				
Voltage waveform	($2E_{DC}$)	(E_{DC})	(E_{DC})	(E_{DC})
Current waveform	($\eta I_L\sqrt{2}$ / $\eta I_L\sqrt{2}\cos\varphi$)	($I_L\sqrt{2}$ / $I_L\sqrt{2}\cos\psi$)	($I_L\sqrt{2}$ / $I_L\sqrt{2}\cos\varphi$)	($I_L\sqrt{2}$ / $I_L\sqrt{2}\cos\varphi$)
RMS value of reverse current I_{REV}	$\frac{1}{2}\eta I_L\sqrt{\dfrac{2\varphi-\sin 2\varphi}{\pi}}$	$\frac{1}{2}I_L\sqrt{\dfrac{2\varphi-\sin 2\varphi}{\pi}}$	$\frac{1}{2}I_L\sqrt{\dfrac{2\varphi-\sin 2\varphi}{\pi}}$	$\frac{1}{2}I_L\sqrt{\dfrac{2\varphi-\sin 2\varphi}{\pi}}$
RMS value of forward current as a function of I_{REV}	$\sqrt{\dfrac{\eta^2 I_L^2}{2}-(I_{REV})^2}$	$\sqrt{\dfrac{I_L^2}{2}-(I_{REV})^2}$	$\sqrt{\dfrac{I_L^2}{2}-(I_{REV})^2}$	$\sqrt{\dfrac{I_L^2}{2}-(I_{REV})^2}$

placed at 10 to 20% of the primary turns, tending toward 20% if the dc input voltage is low and inverter frequency is high. Note that in Fig. 15-22 the commutating capacitor C_c has been split between the +dc and −dc supplies. With this arrangement a center-tapped dc supply is unnecessary for commutation, and two half bridges can be combined into a full bridge, or three half bridges can be combined into a 3-phase bridge, as shown in Table 15-7.

The values of the commutating capacitor and commutating inductor are given by

$$C_c = \frac{t_r \hat{I}}{0.425E} \quad \text{(farads)} \quad \text{and} \quad L_c = \frac{t_r E}{0.425 \hat{I}} \quad \text{(henrys)}$$

where t_r = turnoff time required by thyristor, E = total dc supply voltage, and $\hat{I}$ = maximum thyristor anode current to be commutated.

30. Auxiliary Commutation. To circumvent the trapped-energy problem, auxiliary commutation can be used. A half-bridge circuit which uses auxiliary thyristors for commutation is shown in Fig. 15-23. This circuit was suggested by W. McMurray. It has better voltage-regulation characteristics than the self-commutated circuit and can be used at frequencies up to about 5 kHz.

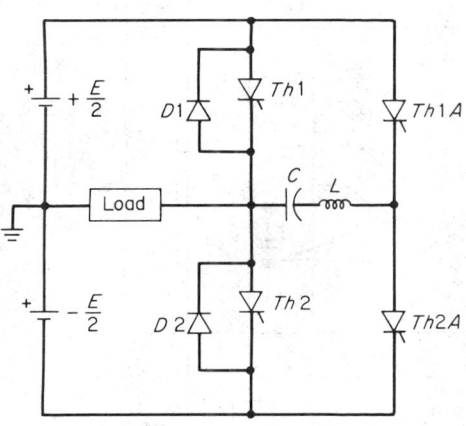

Operation of the circuit is initiated by firing Th1 and Th2A, thereby applying +E/2 to the load and charging the commutating capacitor C. When C is fully charged, the current in Th2A goes to zero and it ceases conduction. To end the first half cycle, auxiliary thyristor Th1A is fired. Inductor L limits the rate of current increase in D1 and Th1A. As the current in L increases, the load current is diverted from Th1 to Th1A and C. After a delay of about $2.4\sqrt{LC}$ s, the forward drop across D1 reverse biases Th1 and turns it off. Then Th2 is fired to begin the negative half cycle. In the

Fig.   15-23. Auxiliary   impulse-commutated inverter.

meantime, C charges to the opposite polarity for the next commutation before Th1A ceases conduction. Th2 is turned off by Th2A in the same way that Th1 was turned off by Th1A.

The values of the commutating components are given by (Bedford and Hoft, 1964, p. 180):

$$C = 0.893 \frac{\hat{I} t_r}{E} \quad \text{(farads)} \quad \text{and} \quad L = 0.397 \frac{E t_r}{\hat{I}} \quad \text{(henrys)}$$

where E, $\hat{I}$, and t_r are as defined previously.

The circuit as shown generates severe dv/dt transients on all the thyristors, which require snubber circuits for protection (see Par. **15-46** on the protection of thyristors).

31. Output-Voltage Waveform. For some applications, such as motor drives and dc-to-dc converters, a square-wave output from an inverter may be acceptable. In many applications, however, sinusoidal voltage waveforms with limited total harmonic distortion are desired. A typical limit in equipment specifications would be 5% total harmonics relative to the magnitude of the fundamental frequency.

Various second- and third-order filter networks are commonly used to eliminate undesirable harmonics from the inverter output, but all tend to be large, heavy, costly, and, in general, highly load-dependent. For this reason, it is desirable to provide an inverter waveform which is inherently devoid of low-order harmonics. Higher-order harmonics can then be filtered with a relatively small network, producing an output waveform which is nearly sinusoidal. Common methods for producing such waveforms from square-wave inverters can be placed in two main categories: harmonic neutralization and pulse-width modulation. Harmonic neutralization involves a combination of several phase-shifted square-wave inverters, each switching at the fundamental frequency, whereas pulse-width modulation involves switching a single inverter at a frequency higher than the fundamental (Kernick et al., 1962). Both schemes give satisfactory results, and actual selection of a method would depend on many factors such as the output-power level, the fundamental frequency, the speed of the switching devices, and the type of commutation circuit. Harmonic neutralization is especially suited for 3-phase outputs.

32. Harmonic Neutralization. A harmonic-neutralized inverter consists of N square-wave inverter stages which are sequentially phase shifted by $180/N$ electrical degrees (Kernick and Heinrich, 1964; Heinrich, 1967). In general, for a polyphase ac output, each inverter stage contributes to the output of each phase by means of a process of phasor addition performed by transformer windings. In place of an overall square-wave output containing all odd harmonics of the fundamental frequency, the output voltage is a stepped approximation to a sine wave in which most of the harmonics have been neutralized. The remaining harmonics occur in pairs and have frequencies of $2kN \pm 1$, where $k = 1, 2, 3, \dots$. The amplitudes of the harmonics which remain, relative to the fundamental, are inversely proportional to their frequencies, as in a square wave.

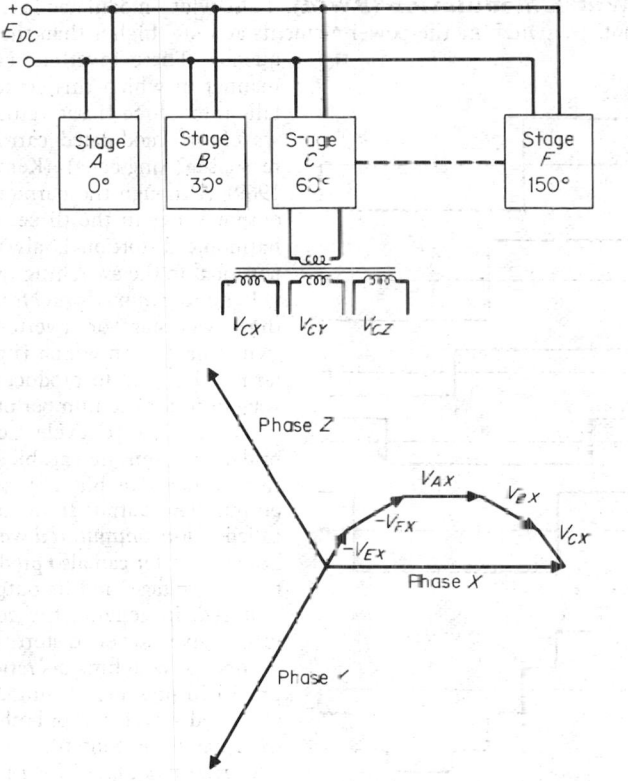

Fig. 15-24. Six-stage harmonic-neutralized inverter.

Each stage of the inverter will share the total output power equally if the load is balanced, but the voltage contributed to each phase by each stage must be properly adjusted. In general, these voltages will not be equal but are given by $(\pi V_{rms} \cos \Psi_{MW})/\sqrt{2}N$, where V_{rms} is the desired line-to-neutral output voltage and Ψ_{MW} is the phase angle between stage M and phase W.

As an example, consider the 3-phase six-stage inverter shown in Fig. 15-24. The firing angles of the respective stages are separated by $180°/6 = 30°$, giving the phasor diagrams shown for the fundamental components of the individual stages and phases. It is assumed that stage A and phase X are each arbitrarily assigned a $0°$ phase angle and the line-to-neutral voltage is to be 120 V. Using the relationship above, the transformer-turns ratios in the various stages should be chosen as follows for phase X:

$$V_{AX} = (\pi \times 120 \cos 0°)/(\sqrt{2} \times 6) = 44 \text{ V}$$
$$V_{BX} = (\pi \times 120 \cos 30°)/(\sqrt{2} \times 6) = 38 \text{ V}$$

Similarly, $V_{CX} = 22$ V, $V_{DX} = 0$, $V_{EX} = -22$ V, $V_{FX} = -33$ V. Since $V_{DX} = 0$, no winding is needed, and phase X is formed by the series connection of the other five windings.

The individual square waves and the corresponding output-voltage waveform for phase X are shown in Fig. 15-25. In a similar way, the contribution of each stage to the other two phases can be calculated. The only harmonics present in the output waveform and their amplitudes relative to the fundamental are the eleventh ($\frac{1}{11}$) and thirteenth ($\frac{1}{13}$), twenty-third ($\frac{1}{23}$) and twenty-fifth ($\frac{1}{25}$), etc.

The ripple frequency of the current into the inverter is $2N$, or 12 times the line frequency, thereby reducing the size of the input filter. The combined rating of the transformers is about 1.4 times the rating of the inverter.

Although the inverter described synthesizes the output from isolated single-phase stages, many variations are possible depending on the particular application. For instance, the same result could be achieved using a pair of 3-phase bridge inverters and 3-phase transformers (Oates, 1970).

33. Pulse-Width Modulation (PWM).

Pulse-width-modulated inverters approximate sine-wave outputs by switching the power elements at a rate higher than the fundamental frequency. These inverters, categorized by the manner in which this switching takes place, fall into three basic groups: programmed waveform, modulated carrier, and optimum response (bang-bang) (Kernick and Haque, 1969). Although the harmonic content of the output varies in the three methods, the total harmonic distortion is always inversely proportional to the switching rate.

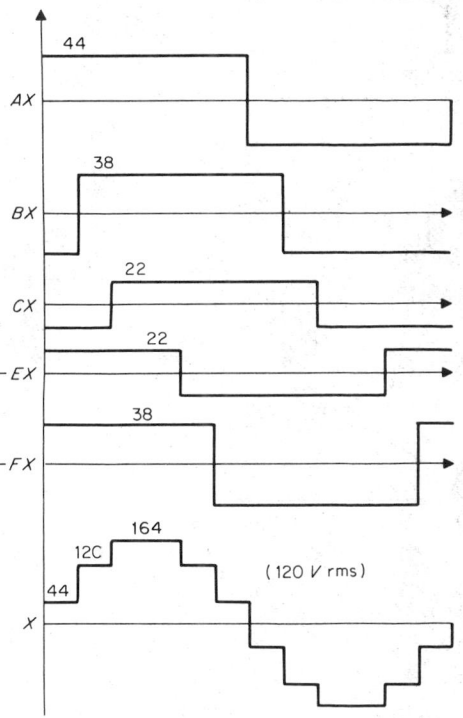

Fig. 15-25. Waveform from individual inverter stages summed to form phase X.

In a *programmed-waveform* PWM inverter, the power stage or inverter is given a fixed switching pattern which is periodic. This pattern is designed to produce the best possible waveform for the number of switching operations permitted per cycle. Center-tap and half-bridge inverters are capable of providing positive or negative but not zero instantaneous output. The output from such an inverter is called a *noncommutated waveform*. The full-bridge inverter can also produce a zero instantaneous voltage, and its output is called *commutated*. In general, the commutated waveforms give lower distortion for the same number of switching operations per cycle. Figure 15-26 presents a summary of useful programmed waveforms of both types along with their harmonic content.

Carrier-modulated PWM is usually implemented by comparing a reference sine wave at the fundamental frequency to a triangular wave signal having a fixed frequency higher than the fundamental (Ravas et al., 1967). The power elements are switched at the intersection points of these two signals, as shown in Fig. 15-27. Distortion of the output waveform occurs at the carrier frequency and its sidebands and at multiples of the carrier frequency and their associated sidebands. This distortion may or may not be harmonic, depending on whether or not the carrier frequency is synchronized with the fundamental reference. The magnitude of the distortion depends on the degree of modulation (relative magnitude of the sine-wave peak compared to the carrier peak) and is lowest at 100% modulation.

Another type of PWM, known as *optimum-response switching*, is shown in Fig. 15-28 (Geyer and Kernick, 1971). This scheme, unlike the others, must operate with an output filter, and it must have closed-loop control. Hysteresis in the feedback path sets the allowable deviation of the output from a sinusoidal reference. The switching rate varies throughout the cycle and is determined by the amount of hysteresis and by the characteristics of the load and filter. Very high switching rates are generally required to keep the error small.

The control for such an inverter is very simple, and voltage regulation is automatically accomplished. However, many applications will not permit optimum-response PWM because of the inherent voltage ripple and the asynchronous output waveform.

34. Voltage Control. Most inverter applications require direct control of the output voltage. For motor-drive inverters, it must be continuously variable from zero to full value, depending upon torque and speed requirements. For ac power-supply inverters, the voltage must be held nearly constant over a certain load and input range. In addition, many inverters are required to provide a specified amount of current into a short circuit, making it necessary to cut back the output voltage to nearly zero. A typical load profile is shown in Fig. 15-29.

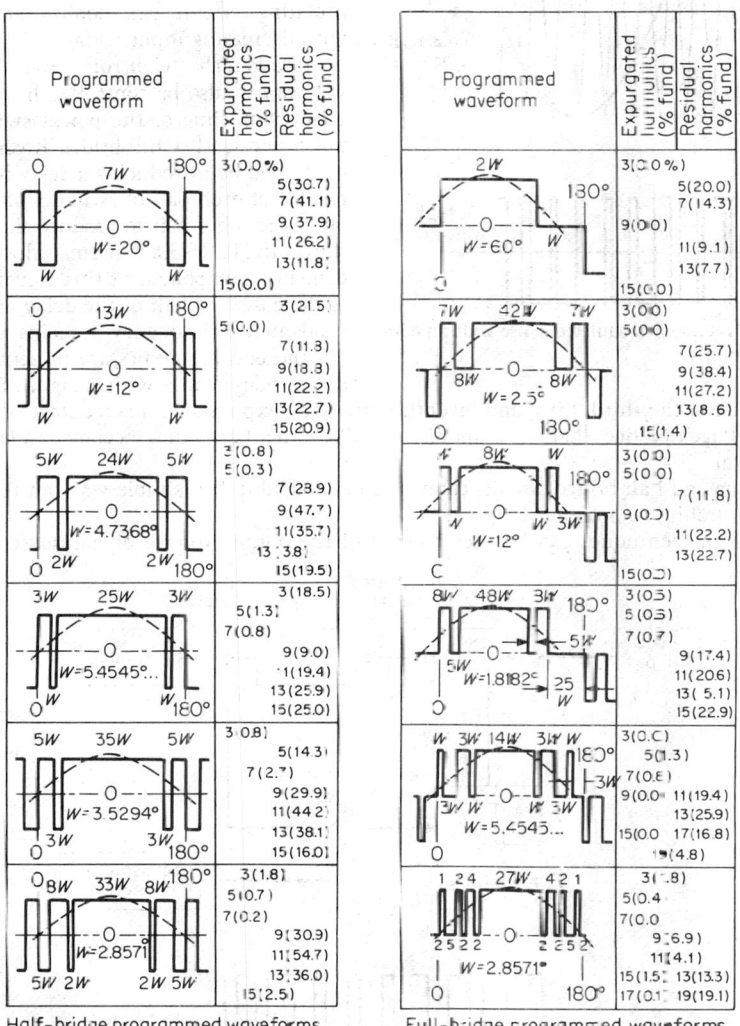

Half-bridge programmed waveforms   Full-bridge programmed waveforms

Fig. 15-26. Summary of programmed waveforms and their harmonic content. W is the unit increment of time for each waveform in degrees.

Varying the dc input voltage and internal pulse-width control are the most common methods of controlling the output voltage where this control is not inherent, as it is in carrier-modulated PWM and optimum-response inverters.

DC-Input Control. Control of the dc input is the most straightforward method. If an inverter's switching pattern remains invariant, the output voltage is directly proportional to the dc-input voltage for all types of inverters. If the power source is ac, a phase-controlled rectifier can be used to control the dc input to the inverter. If the power source is dc, it is necessary to use a dc regulator, i.e., chopper.

The main advantages of using dc-input control are that the switching requirements and control complexity are not increased. In addition, the harmonic content of the output-voltage waveform does not vary with the input voltage. However, it has the disadvantage that the power must often pass through an extra stage of conditioning, thus reducing overall efficiency. Also, it is often impractical to use input-voltage control when the control range is large because the amplitude of current a given commutating circuit can commutate decreases with decreasing input voltage.

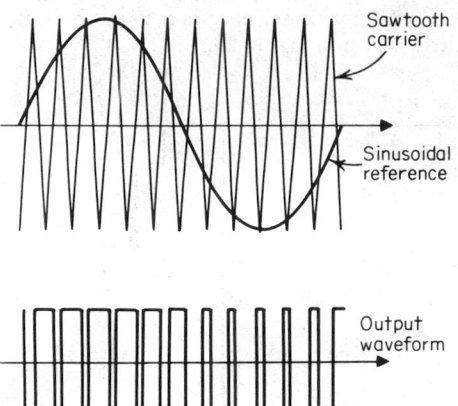

Fig. 15-27. Carrier-modulated pulse-width-modulation waveforms.

Pulse-Width Control. Inverter output voltage can also be controlled by varying the conduction time of the power switches. The pulse width of a full-bridge inverter can be controlled by introducing a delay between the turnoff of each pair of switches and the turn-on of the other pair to produce the waveform shown in Fig. 15-30. The rms value of the fundamental component of this waveform varies as the cosine of half of the delay angle θ. The fundamental frequency remains unchanged. All the odd harmonics are present, but their magnitudes change with θ. Figure 15-31 shows the variation of the third, fifth, and seventh harmonics, expressed as a percentage of the fundamental voltage. Notice that as θ approaches 180°, the harmonics become as large as the fundamental.

Center-tap and half-bridge circuits cannot be modulated in this simple way but require more complex switching at a higher frequency.

Pulse-width techniques can be used to control the output voltage of harmonic-neutralized

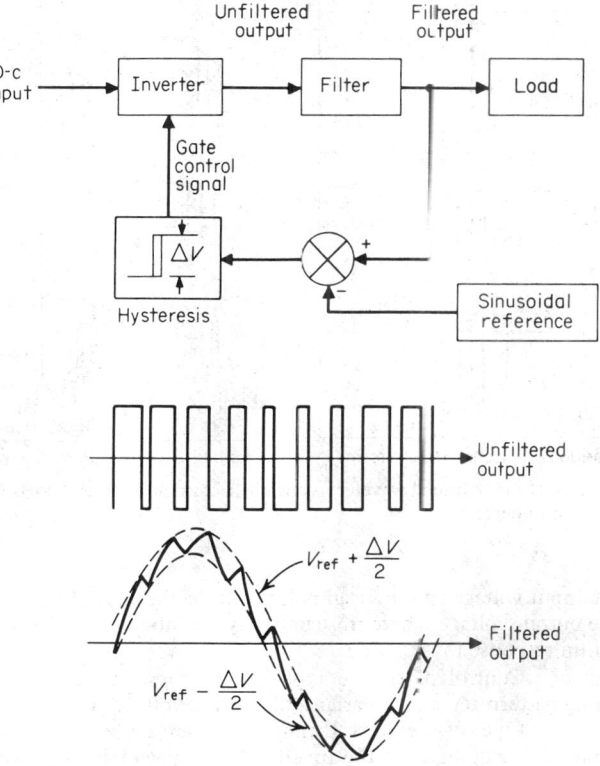

Fig. 15-28. Optimum-response pulse-width modulation.

inverters by controlling each individual inverter stage. The output continues to obey the cosine dependence on θ, and all neutralized harmonics remain neutralized. The remaining harmonics vary with θ.

Pulse-width control of programmed-waveform inverters is more difficult. To preserve the harmonic cancellation, each conduction period must be reduced by the same proportion while maintaining its relative position within the cycle constant. The width of each pulse *(picket)* must be reduced from both directions about the center of that picket.

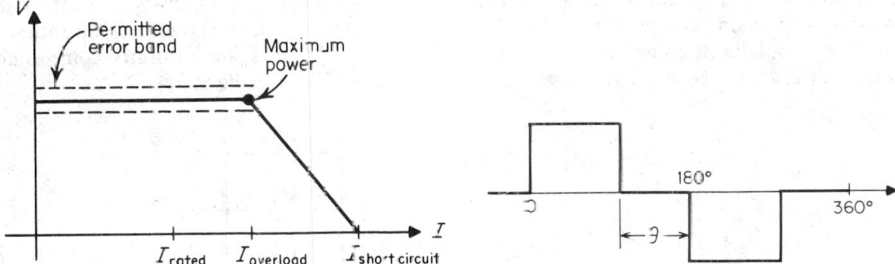

Fig. 15-29. Typical inverter load profile.

Fig. 15-30. Pulse-width voltage control of a full-bridge inverter.

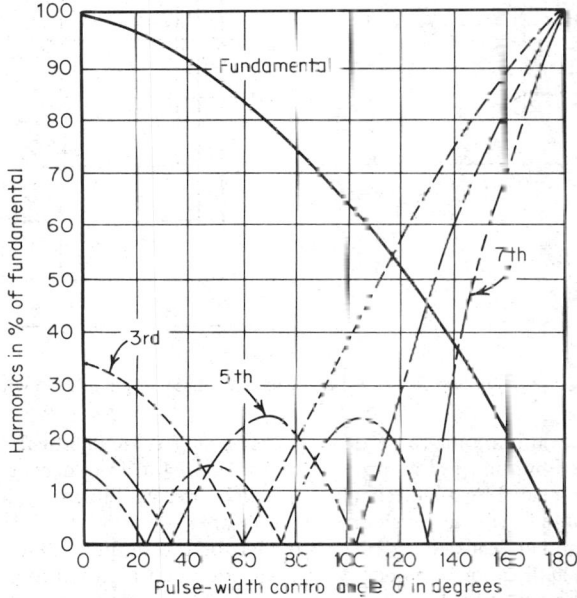

Fig. 15-31. Variation of harmonics for a pulse-width-controlled square wave.

For carrier-modulated and optimum-response PWM inverters, pulse-width control is accomplished by simply reducing the width of each pulse. This reduction occurs automatically as the amplitude of the sinusoidal reference is decreased.

POWER FREQUENCY CHANGERS

BY L. GYUGYI

35. Basic Principles and Circuits. Power frequency changers are static systems usually employing solid-state switching devices, capable of directly, i.e., without an intermediate dc link, converting single or polyphase ac power of a given frequency to single or polyphase power of a chosen frequency. They may be used to link two ac power systems of different frequencies, to provide power at controllable frequency for variable-speed ac motor drives, or to convert the output of variable-speed ac generators to constant frequency.

Functionally, frequency changers are wave synthesizers. They fabricate the output-voltage wave(s) of desired amplitude and frequency by sequentially applying appropriate segments of the input-voltage wave(s) to the output. This is accomplished by arrays of static switches arranged to make *bilateral connections*, for controlled time intervals, between the input and output terminals, i.e., between the supply voltages and loads.

Frequency changers generally require controllable power switches with intrinsic turn-on and turnoff ability (such as transistors and gate-controlled switches) or switches with controllable turn-on ability (such as thyristors and triacs) complemented by auxiliary forced-commutation circuitry to implement controllable turnoff. A notable exception is the naturally commutated cycloconverter (Par. **15-40**), which uses conventional controlled rectifiers.

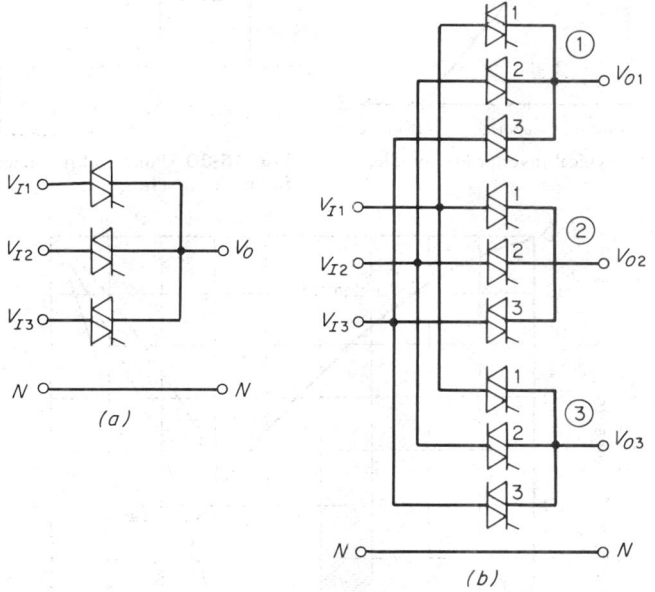

Fig. 15-32. Three-pulse frequency changers: (*a*) with single-phase output and (*b*) with three-phase output.

The basic circuit configurations of static frequency changers are identical with polyphase converters characterized by their pulse number (see Tables 15-4 to 15-6) except that each unidirectional thyristor is replaced by a bidirectional ac switch. Typical bilateral solid-state switch configurations, applicable in frequency-changer circuits, are shown in Fig. 15-15. As in converters, increased pulse number leads to reduced distortion of the output-voltage and input-current waves. In practical applications, frequency changers are often required to produce 3-phase output; in this case three identical converter circuits, one for each output phase, are employed. Three-pulse frequency changers with single-phase and 3-phase output are shown in Fig. 15-32a and b, respectively. The bilateral-switch symbols represent any one of the previously described bidirectional solid-state switch arrangements.

Frequency changers fabricate the output-voltage wave with the desired (or "wanted") frequency and amplitude by sequentially connecting the input voltages to the output(s) for appropriate time intervals. The output-voltage wave(s) are thus composed of segments of the input-voltage waves. The length of each segment is determined by the duration of closure of the corresponding switch. However, an output-voltage wave of given frequency and amplitude can be obtained in several distinctly different ways (Gyugyi and Pelly, 1976) characterized by the control (modulation) of the repetition rate and/or duration of switch closures. The method of output-waveform fabrication uniquely determines the external performance characteristics of the frequency changer, the most important of which are the distortion of the output-voltage and input-current waves and the input-displacement and power factors.

36. Fundamental Principles. Consider the simple three-pulse frequency-changer circuits shown in Fig. 15-32. These circuits convert 3-phase input power of frequency f_i into a single- or 3-phase output power of frequency f_o. The relationship between the input and the generated output *voltage waves* can be described by the matrix equation

$$[v_o(t)] = [H(t)][v(t)]$$

or

$$
\begin{bmatrix} v_{O1}(t) \\ v_{O2}(t) \\ v_{O3}(t) \end{bmatrix}
=
\begin{bmatrix} h_{11}(t) & h_{12}(t) & h_{13}(t) \\ h_{21}(t) & h_{22}(t) & h_{23}(t) \\ h_{31}(t) & h_{32}(t) & h_{33}(t) \end{bmatrix}
\begin{bmatrix} v_{I1}(t) \\ v_{I2}(t) \\ v_{I3}(t) \end{bmatrix}
\tag{15-1}
$$

where v_{O1}, v_{O2}, v_{O3} are the time functions of generated voltage waves; v_{I1}, v_{I2}, v_{I3} are the three input-voltage waves, which are usually sinusoids, i.e.,

$$v_{I1} = V_I \sin \omega_I t \qquad v_{I2} = V_I \sin (\omega_I t - 2\pi/3) \qquad v_{I3} = V_I \sin (\omega_I t - 4\pi/3)$$

and each h_{ij} ($i = 1, 2, 3; j = 1, 2, 3$) is a time-varying existence function which defines whether a given switch h_{ij}, connecting output terminal i to input terminal j, is open ($h_{ij} = 0$) or closed ($h_{ij} = 1$) at a given time t.

The input *current waves* drawn from the supply by a three-pulse frequency changer can be similarly expressed, by transposing matrix $[H(t)]$, in terms of the output (load) currents:

$$[i(t)] = [H(t)]^T[i_o(t)]$$

or

$$
\begin{bmatrix} i_{I1}(t) \\ i_{I2}(t) \\ i_{I3}(t) \end{bmatrix}
=
\begin{bmatrix} h_{11}(t) & h_{21}(t) & h_{31}(t) \\ h_{12}(t) & h_{22}(t) & h_{32}(t) \\ h_{13}(t) & h_{23}(t) & h_{33}(t) \end{bmatrix}
\begin{bmatrix} i_{O1}(t) \\ i_{O2}(t) \\ i_{O3}(t) \end{bmatrix}
\tag{15-2}
$$

where i_{I1}, i_{I2}, i_{I3} are the three input current waves, i_{O1}, i_{O2}, i_{O3} are the three output current waves, which for computations are usually assumed to be symmetrically displaced sinusoids, and each h_{ij} is an appropriate existence function introduced in Eq. (15-1).

The basic sets of the three existence functions and related input- and output-voltage waveforms of the three-pulse power circuit shown in Fig. 15-32 are illustrated in Fig. 15-33a and b for the trivial case of zero output frequency and zero output voltage. Note that the output-voltage wave of Fig. 15-33a is identical with the output of a unidirectional naturally commutated ac-dc converter conducting continuous positive load current when the delay angle α is 90° (see Fig. 15-5). Similarly that of Fig. 15-33b is obtained from a converter conducting negative load current at $\alpha = 90°$. A bidirectional converter employing bilateral turnoff switches can produce either of these two waveforms, depending on which of the two sets of complementary existence functions (h_{ij} or h_{ijx}) describes its operation. This free option is utilized in devising methods of output-waveform fabrication which provide desired operating and performance characteristics for frequency changers.

The unmodulated existence functions represent rectangular pulses with repetition period $1/f_I$, pulse duration $\frac{1}{3}f_I$, and amplitude unity. To obtain the steady-state output voltages of the frequency changer in explicit mathematical form, the existence functions h_{ij} and h_{ijx} shown in Fig. 15-33a and b can be expanded into the Fourier series

$$h_{ij}(\omega_I t) = \frac{1}{3} - \frac{2}{\pi} \sum_{n=0}^{\infty} \frac{\sin (n2\pi/3)}{n} \cos \left\{ n \left[\omega_I t - (j - 1)\frac{2\pi}{3} \right] \right\} \tag{15-3}$$

and

$$h_{ijx}(\omega_I t) = \frac{1}{3} + \frac{2}{\pi} \sum_{n=0}^{\infty} \frac{\sin (n\pi/3)}{n} \cos \left\{ n \left[\omega_I t - (j - 1)\frac{2\pi}{3} \right] \right\} \tag{15-4}$$

where $i = 1, 2, 3$, $j = 1, 2, 3$, and subscript π indicates that the second set is displaced by π rad with respect to the first.

The existence functions defined by Eqs. (15-3) and (15-4) describe two complementary but otherwise equivalent modes of operation of the switches in the power converter which result in zero desired output frequency and voltage. (The output waveform consists entirely of "unwanted" components.)

A desired output frequency differing from zero is obtained by appropriately modulating the repetition frequency of the basic existence functions, which is equivalent to varying the commencement and/or duration of the conduction intervals of the corresponding switches. Mathematically this means that a modulating function $M(\omega_0 t)$ ($\omega_0 = 2\pi f_0$; f_0 is the "wanted" output frequency), is added to the arguments of the basic existence functions given by Eqs. (15-3) and (15-4). To keep the wanted (fundamental) component identical in the two output-voltage waves obtainable by the use of the two complementary sets of existence functions, the modulating functions must also be mutually displaced by π. Similarly, for balanced 3-phase output, the modulating functions used to generate the three output voltage waves must also be mutually displaced by $2\pi/3$. Thus

$$h_{ij}(\omega_i t, \omega_0 t) = \frac{1}{3} - \frac{2}{\pi} \sum_{n=0}^{\infty} \frac{\sin(n2\pi/3)}{n} \cos\left(n\left\{\omega_i t + M\left[\omega_0 t - (i-1)\frac{2\pi}{3}\right] - (j-1)\frac{2\pi}{3}\right\}\right)$$

(15-5)

and

$$h_{ijr}(\omega_i t, \omega_0 t) = \frac{1}{3} + \frac{2}{\pi} \sum_{n=0}^{\infty} \frac{\sin(n\pi/3)}{n} \cos\left(n\left\{\omega_i t + M\left[\omega_0 t - (i-1)\frac{2\pi}{3} - \pi\right] - (j-1)\frac{2\pi}{3}\right\}\right)$$

(15-6)

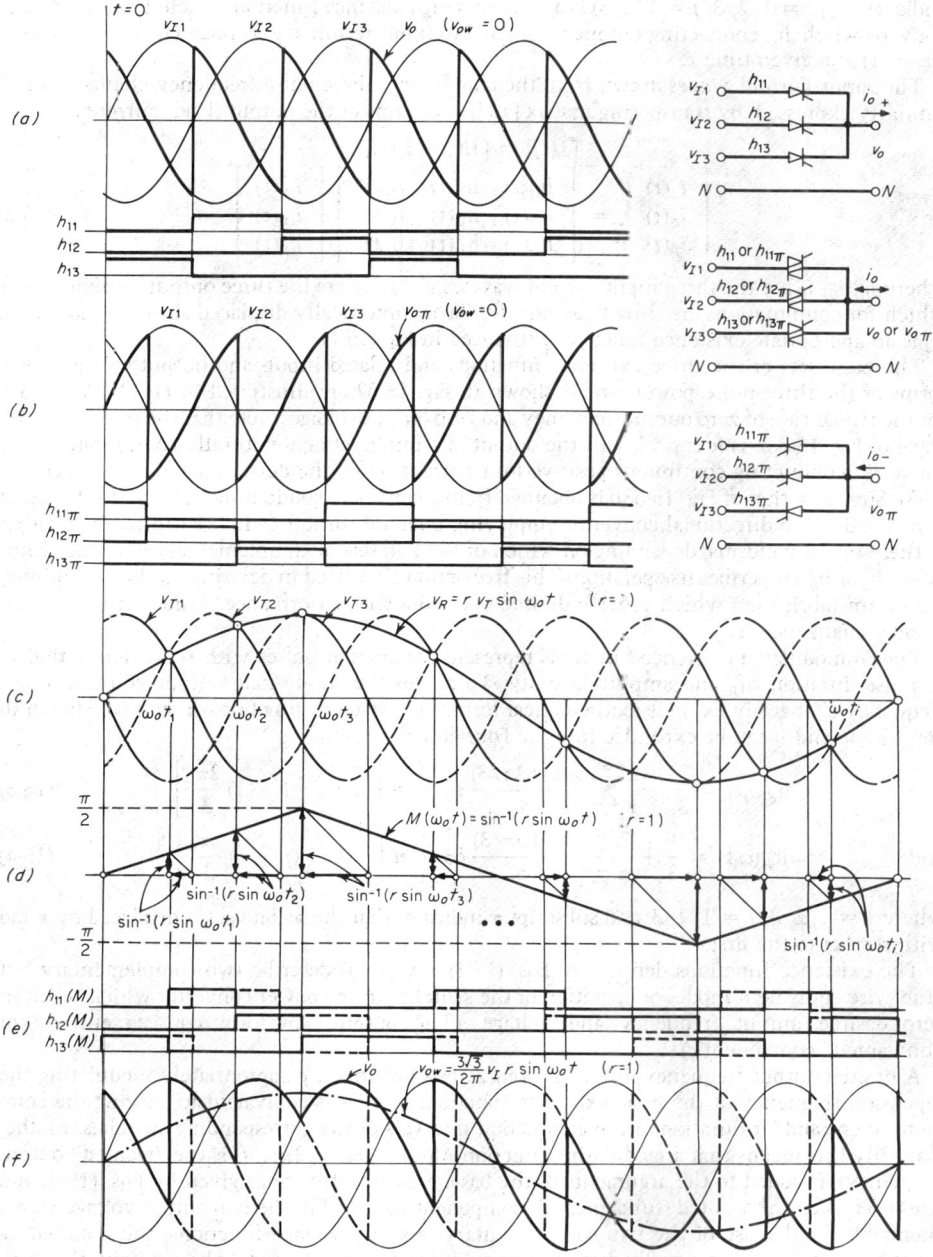

Fig. 15-33. Waveforms illustrating the degeneration of the two complementary waveforms V_O and $V_{O\pi}$.

The modulating function in Eqs. (15-5) and (15-6) can be visualized as a time-varying angle which effectively changes (modulates) the repetition frequency of the existence functions from or about their quiescent frequency f_I.

Mathematical expressions for the resulting output-voltage waveforms of a general P-pulse frequency changer can be obtained from Eqs. (15-1) and (15-5) and (15-1) and (15-6) in terms of the modulating functions $M(\omega_0 t)$ and $M(\omega_0 t - \pi)$, respectively (Gyugyi and Pelly, 1976):

$$v_{Oi} = \frac{3\sqrt{3}}{2\pi} V_I \sin\left\{ M\left[\omega_0 t - (i-1)\frac{2\pi}{3} \right] \right\}$$

$$+ \frac{3\sqrt{3}}{2\pi} V_I \sum_{k=1}^{\infty} \left(\frac{\sin\left\{ Pk\omega_i t + (Pk-1)M\left[\omega_0 t - (i-1)2\pi/3 \right] \right\}}{Pk-1} \right.$$

$$\left. + \frac{\sin \cdot Pk\omega_i t + (Pk+1)M\left[\omega_0 t - (i-1)2\pi/3 \right]}{Pk+1} \right] \quad (15\text{-}7)$$

and

$$v_{Oi\pi} = -\frac{3\sqrt{3}}{2\pi} V_I \sin\left\{ M\left[\omega_0 t - (i-1)\frac{2\pi}{3} - \pi \right] \right\}$$

$$- \frac{3\sqrt{3}}{2\pi} V_I \sum_{k=1}^{\infty} (-1)^k \left(\frac{\sin\left\{ Pk\omega_i t + (Pk-1)M\left[\omega_0 t - (i-1)(2\pi/3) - \pi \right] \right\}}{Pk-1} \right.$$

$$\left. + \frac{\sin\left\{ Pk\omega_i t + (Pk+1)M\left[\omega_0 t - (i-1)(2\pi/3) - \pi \right] \right\}}{Pk+1} \right) \quad (15\text{-}8)$$

where $i = 1, 2, 3, P$ is the pulse number and $M(\omega_c t)$ specifies the modulation of the basic repetition frequency f_I of the existence functions.

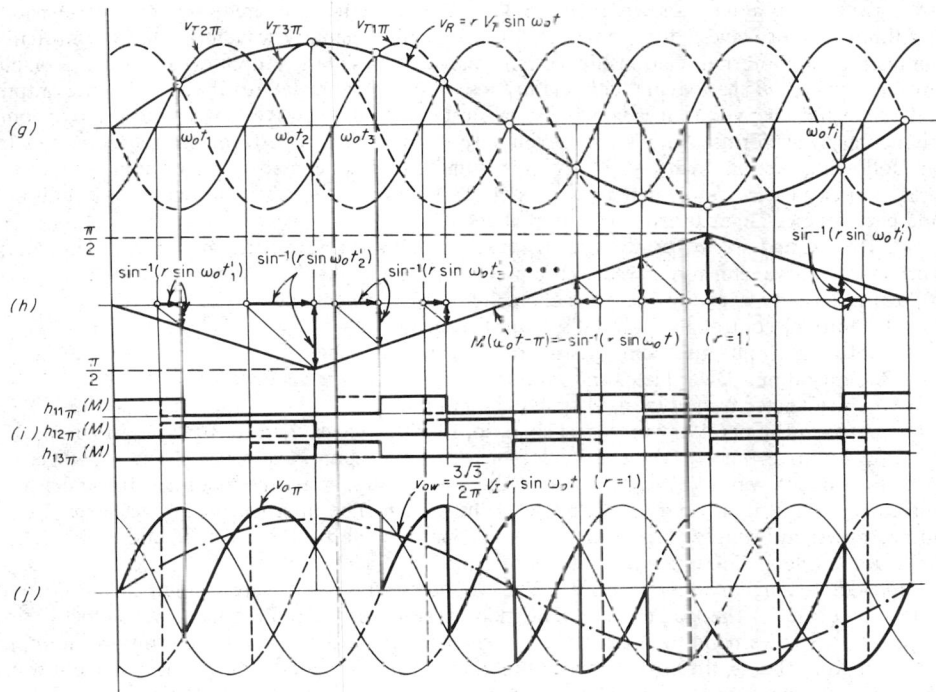

Fig. 15-33 (continued).

Similar equations can be written for the input-current wave, and after laborious computation the performance characteristics can be numerically obtained.

Equations (15-7) and (15-8) indicate that the modulating function entirely determines the operation and performance characteristics of a frequency-changer circuit defined by its pulse number P. The modulating function is actually a mathematical description of the control defining the operation of the power switches and thereby the method of output-waveform fabrication. Various control methods (modulating functions) can be applied to the same power circuit to generate output-voltage (or input-current) waveforms of widely differing characteristics to meet practical requirements. In the following, the five most important operation modes, resulting in practically desirable output waveforms and performance characteristics, will be summarized and illustrated for the case of the single-phase output, three-pulse circuit shown in Fig. 15-32a. All the output waveforms considered can be derived from the two complementary output-voltage waveforms obtained from Eqs. (15-7) and (15-8) by the substitution of the modulating function

$$M(\omega_0 t) = \sin^{-1}(r \sin \omega_0 t) \qquad (15-9)$$

where r is the output-voltage ratio, that is, $r = V_O/V_{O\max}$ $(r \leq 1)$. The practical derivation of this modulating function and the subsequent generation of the two complementary waveforms v_O and $v_{O\bar{r}}$ are graphically illustrated for $r = 1$ in Fig. 15-33c to j.

Figure 15-33c and d illustrates that the modulating function $M(\omega_0 t)$ of Eq. (15-9) is a mathematical expression for the well-known sine-wave crossing technique widely used to control the firing angle of thyristor converters (Pelly, 1971, Chap. 9). Using this technique, the magnitude of $M(\omega_0 t)$, and thus the modulation of each existence function h_{ij} about its quiescent point (zero output), is determined by the crossing point of a corresponding timing wave v_T with a reference sinusoid v_R. Derivation of the complementary sets of $M(\omega_0 t)_r$ and $h_{ij\bar{r}}$ are shown in Fig. 15-33g to i. The timing waves v_{T1}, v_{T2}, v_{T3}, $v_{T1\bar{r}}$, $v_{T2\bar{r}}$, and $v_{T3\bar{r}}$ are opposite half-period sections of sine waves synchronized to the source voltages with a phase relationship such that at zero reference the mean of the output voltages v_O and $v_{O\bar{r}}$ is zero.

Figure 15-33e, f, i, and j shows that the output waveforms v_O and $v_{O\bar{r}}$, at $r = 1.0$, are generated by the act of periodically stepping up and down (v_O) and down and up $(v_{O\bar{r}})$ the original f_I repetition frequency of the existence functions, and thus that of the power switches, to f_I and f_O and $f_I - f_O$, respectively.

The spectral characteristics (frequency and amplitude) of the two complementary output-voltage waveforms v_O and $v_{O\bar{r}}$ shown in Fig. 15-33f and j are identical, as are those of the corresponding input-current waves. Since, however, the two complementary waveforms have a mutually complementing internal relationship (certain characteristics of the output cycle of v_O, observable during a given *half* period, are identical to those of $v_{O\bar{r}}$, observable during the following output half cycle and vice versa), it is possible to fabricate new output waveshapes from the two complementary waveforms which satisfy given output- and/or input-performance requirements. In the following section, synthesis of the four important frequency-changer output waveforms, using the two basic complementary waves, is described, and the pertinent operating conditions and performance characteristics are summarized.

37. Practical Frequency Changers. Utilizing the properties of the complementary output-voltage waveforms derived in the preceding paragraphs, frequency changers having the following special, *mutually exclusive* characteristics can be devised:

1. Unity or controllable-input displacement factor
2. Natural (input-line) commutation of the power switches
3. Unity input-power factor and minimum output-voltage distortion
4. Unrestricted output-to-input frequency ratio

To establish the necessary operating conditions for the above characteristics, consider Fig. 15-34a and b where the two three-pulse complementary output waveforms v_O and $v_{O\bar{r}}$, together with the voltage waves $(v_{I1}, v_{I2}, \text{and } v_{I3})$ of the 3-phase supply and an assumed sinusoidal load current i_0 having an arbitrary phase angle ϕ_0, are shown. The input-current waveshapes, i_{I1}, i_{I2}, and i_{I3}, flowing in supply phases 1, 2, and 3, are shaded by vertical, horizontal, and crosshatched lines, respectively. The following observations may be made.

Input-Phase-Angle Characteristics. The input-current waves are composed of current "blocks" cut out of the load current. The phase position of the individual current blocks with respect to the corresponding input voltages over a complete output cycle determines the input phase angle ϕ_I, that is, the angle between the fundamental component of the input-current wave and the corresponding phase voltage, and the displacement factor (cos ϕ_I). For v_O (Fig. 15-34a), the phase angles of the input-current blocks lag the corresponding phase voltages during the positive output-*current* half cycle, and then they lead those during the negative half cycle. Conversely, for v_{O_r} (Fig. 15-34b), the phase position of the input-current blocks is opposite; i.e., they lead the corresponding phase voltages during the positive output-current half cycle, and they lag those during the negative half cycle. The net input phase angle averaged over a complete output cycle is therefore zero, and the input displacement is unity for both v_O and v_{O_r}, regardless of the load phase angle.

Commutation Characteristics. At each switching point of waveform v_O the "incoming" input voltage is more positive than the "outgoing" one. The opposite is true for waveform v_{O_r}. Consequently, v_O satisfies the conditions of natural commutation for a *positive* unidirectional converter, and v_{O_r} satisfies those for a *negative* converter. Therefore, only during the positive (negative) half cycle of the output *current* can v_C (v_{O_r}) be produced by a naturally commutated converter.

Spectral Characteristics. The switches of the converter generating waveform v_O are operated at a constant rate of $f_i + f_O$ during the half-cycle interval when the slope of the wanted (fundamental) component is positive, and they are operated at a constant rate of $f_i - f_O$ when the slope of the wanted component is negative. Conversely, the switches of the converter generating v_{O_r} are operated at the "slow" rate of $f_i - f_O$ to produce the half-cycle sections of the output waveform with positive slope, and they are operated at the 'fast" rate of $f_i + f_O$ to produce the other half-cycle sections with negative slope. The half-cycle waveform intervals with "fast" switching rates can generally be characterized by unwanted components having frequencies consisting of *sums* of multiples of f_i and f_O, whereas the intervals with 'slow" switching rates can be characterized by unwanted components having frequencies consisting of *differences* of multiples of f_i and f_O. The total waveform, v_C as well as v_{O_r}, therefore can be characterized by a frequency spectrum consisting of both sums and differences of multiples of the input and output frequencies.

When these complementary characteristics of waveforms v_O and v_{O_r} are used, the operating and performance characteristics of the following practically important frequency changers can be established.

38. Unity-Displacement-Factor Frequency Changer (UDFFC).
As was established under *input-phase-angle characteristics* in Par. **15-37**, the input-displacement factor of a bilateral converter generating either complementary output-voltage waveform v_O or v_{O_r} is unity independently of the load-power factor. For this reason, a frequency changer controlled to fabricate either v_O or v_{O_r} is termed a unity-displacement-factor frequency changer (UDFFC).

The pertinent characteristics of the UDFFC are summarized in the table adjoining the waveforms of Fig. 15-34a and b.

39. Controllable-Displacement-Factor Frequency Changer (CDFFC).
The controllable-displacement-factor frequency changer uses power switches with intrinsic turnoff ability (or with external forced commutation). The bidirectional converter is commutated so as to generate alternating half-period intervals of the complementary waveforms v_C and v_{O_r}. The phase position of the input current during successive input cycles used to fabricate v_O (v_{O_r}) varies from lagging (leading) to leading (lagging) as the ac output current goes through a full cycle. Thus, by appropriately choosing alternate half-period sections of v_O and v_{O_r} to fabricate the final output waveform, the input-displacement factor may be made lagging, leading, or unity regardless of the phase angle of the output (load) current. The half-period sections constituting the output waveform can conveniently be defined for a given input phase angle (displacement factor) by an angle σ (measured at $\omega_0 = 2\pi f_O$ angular frequency) specifying the point of changeover between v_O and v_{O_r} with respect to the zero crossing point of the output current, as illustrated in Fig. 15-34c. The relationship between the input phase angle and angle σ is shown in Fig. 15-35 for various ϕ_O load phase angles.

The intervals during which the converter switches are operated at high and low rates ($f_i + f_O$ and $f_i - f_O$) depend upon the angle σ, that is, the output-to-input displacement-factor transfer. The frequencies of the dominant unwanted components may thus be either the sums or differences, or both, of multiples of f_i and f_C. The frequency spectrum is therefore a function of σ.

The characteristics of the CDFFC are summarized in the table adjoining Fig. 15-34c. The unique input characteristic of the CDFFC offers a number of intriguing application possibilities. One of them is a power-generating system which utilizes a squirrel-cage induction machine and a CDFFC. In this system the reactive-excitation requirement of the induction generator and the reactive kilovolt-ampere demand of the load are provided by the static frequency changer itself. The frequency changer thus has two basic functions: (1) it converts the generally variable generator frequency to precisely regulated output frequency, and (2) it provides a controllable excitation for the induction machine.

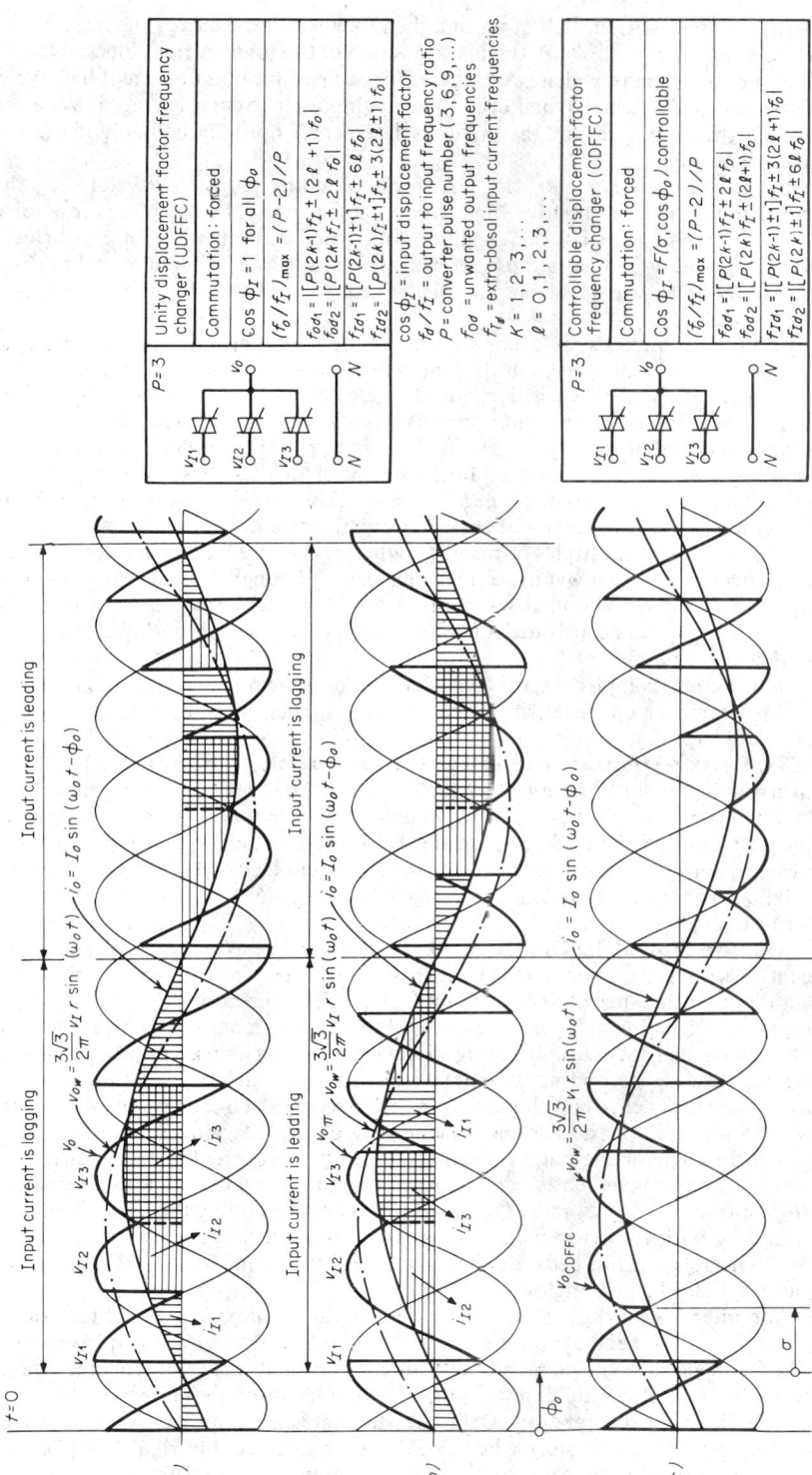

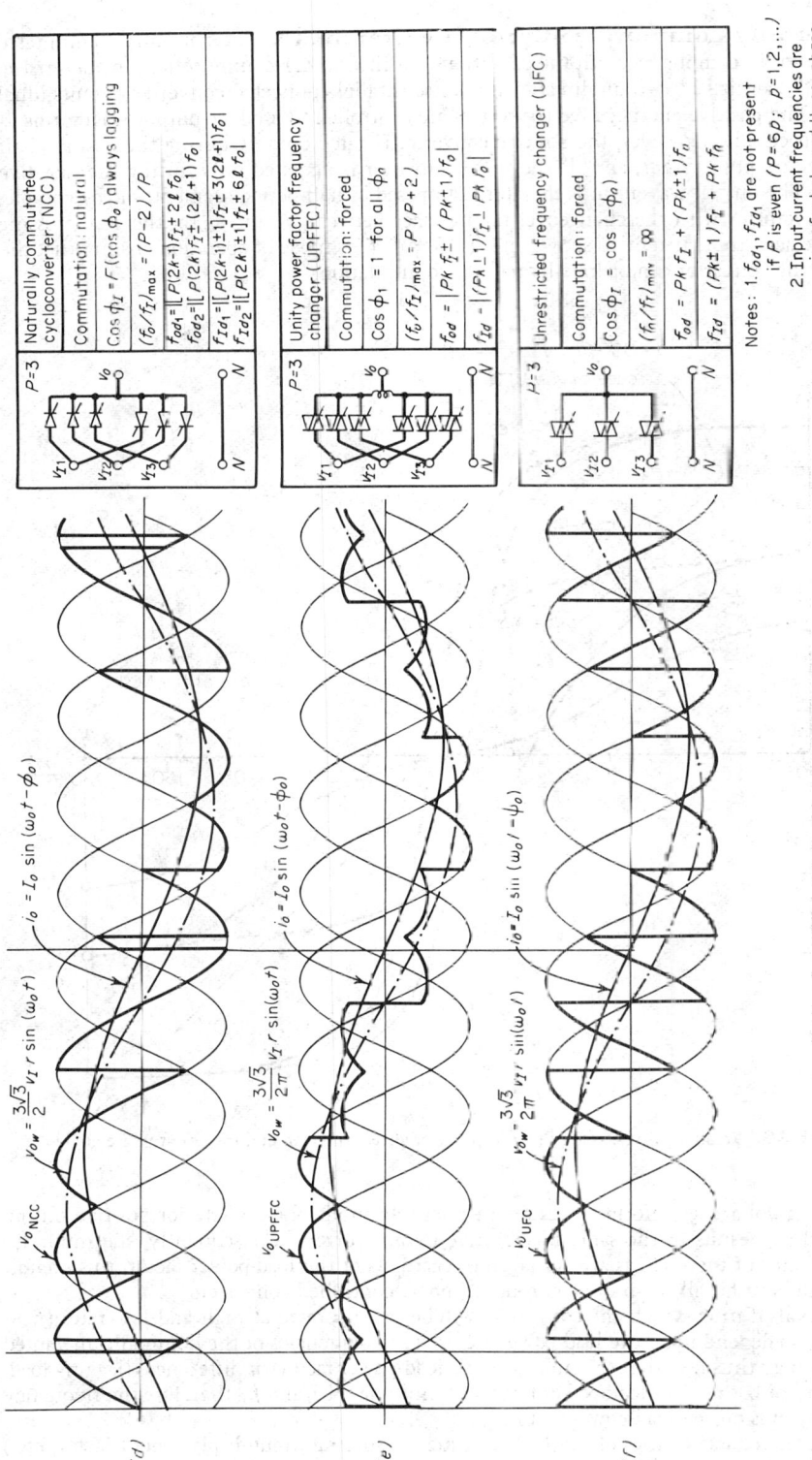

Fig. 15-34. Waveforms illustrating the derivation of practical frequency-changer output waveforms from the two complementary waves.

40. Naturally Commutated Cycloconverter (NCC). The naturally commutated cycloconverter, in compliance with the conditions outlined under *commutation characteristics* in Par. **15-37**, consists of two unidirectional inverse-parallel connected converter circuits (dual converter). The positive and negative converters are controlled to produce output waveforms v_O and v_{O_r}, respectively. However, the positive converter is gated on only during the positive half cycles of the output *current*, and the negative converter is operated only during the negative half cycles. The output waveform is therefore composed of half-period segments of the complementary waveforms with the changeover taking place between v_O and v_{O_r} at the zero crossing points of the ac output current, as shown in Fig. 15-34d. This mode of operation ensures that the switches of the converter can operate by natural commutation.

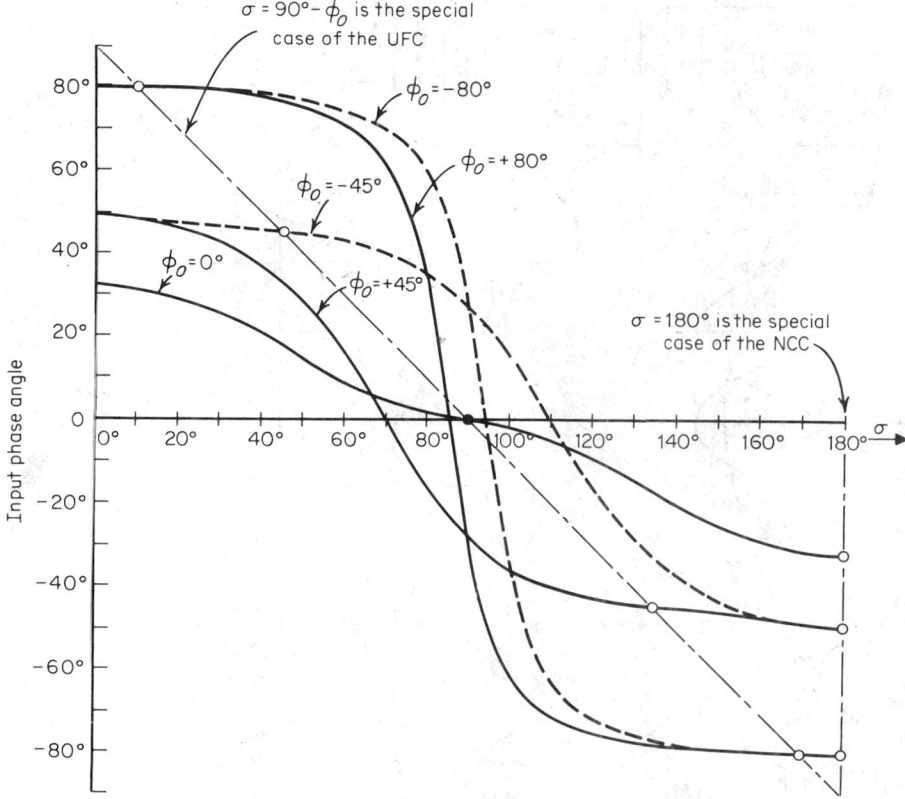

Fig. 15-35. Relationship between the load phase angle ϕ_O, angle σ, and the input phase angle.

The output voltage waveform v_O results in a lagging input phase angle for positive output current, and v_{O_r} results in the same for negative output current. Consequently, the input-displacement factor of the NCC is always lagging regardless of the load-power factor; this characteristic is inherent for all naturally commutated phase-controlled converters.

The intervals during which the converter switches are operated at high and low rates ($f_1 + f_O$ and $f_1 - f_O$) depend upon the load-power factor. The frequencies of the dominant unwanted components may thus be either the sums (leading load-power factor) or differences (lagging load-power factor), or both (unity load-power factor), of multiples of f_1 and f_O; therefore the frequency spectrum depends on the load-power factor.

Note that the characteristics of the NCC (frequency spectra, input-displacement factor, etc.) are identical to those of the CDFFC operated at fixed $\sigma \equiv 180°$ (see Fig. 15-35 and tables adjoining Fig. 15-34c and d).

The practical significance of the NCC is due to the fact that presently the voltage and current ratings of thyristors are considerably higher than those of other semiconductor switches having internal turnoff ability. For this reason, the NCC currently offers the most economical if not the only feasible solution to very high power frequency-changer applications.

41. Unity-Power-Factor Frequency Changer (UPFFC). The unity-power-factor frequency changer is the combination of two bilateral converter circuits operated from a common ac source and supplying the same load. One converter is controlled to generate v_O, the other v_{Or}. The final output waveform is produced by summing or generating the arithmetic mean of the two complementary waveforms (see Fig. 15-34e). Note that the output rating of the combined system is the sum of the ratings of the constituent converters

The input-displacement factor of each constituent converter, and thus that of the combined system, is unity. The advantage of this arrangement is that in addition to unity input-displacement factor (regardless of load-power factor), certain groups of unwanted components present in the output-voltage (input-current) waves of the constituent converters cancel out, resulting in greatly improved frequency spectra, increased f_O/f_i ratio, and rms distortion decreased by a factor of $\sqrt{2}$ (see table adjoining Fig. 15-34e). The reduction in the distortion of the input-current wave results in a "near unity" input-power factor λ, which is the product of the input-displacement and current-distortion factors ($\lambda = 0.9$ for a three-pulse, $\lambda = 0.977$ for a six-pulse, and $\lambda = 0.995$ for a twelve-pulse system).

The UPFFC is particularly advantageous in applications where the required output power is higher than that obtainable from a single converter, in which case multiple converters can be used advantageously to increase the power rating as well as to improve the performance of the system.

42. Unrestricted Frequency Changer (UFC). The unrestricted frequency changer utilizes a single bilateral converter whose switches are operated at the constant "fast" rate of $f_i + f_O$. Therefore, as discussed under *spectral characteristics* in Par. **15-37**, the output waveform of the UFC can be synthesized from half-period sections of the two complementary waveforms, v_O providing the output when the slope of the wanted component is positive and v_{Or} when it is negative. The changeover points between v_O and v_{Or} thus coincide with the peaks of the wanted voltage component (see Fig. 15-34f). Because of the constant switching rate of $f_i + f_O$, the frequencies of the unwanted components are only sums of multiples of f_i and f_O and therefore are always higher than f_O, regardless of the f_O/f_i ratio (see table adjoining Fig. 15-34f). Consequently, the UFC can generate a high-quality output waveform having a frequency which may be lower or higher than the ac supply frequency or equal to it. The described operation of the converter switches also results in a unique output- to input-power-factor transfer characteristic; i.e., the UFC reflects the negative of the load phase angle back to the source (Fig. 15-35) and therefore the input- and output-displacement factors are mirror images of each other (an inductive load is seen capacitive and vice versa).

The UFC is an ideal system to provide ac output power over a wide frequency range (which may extend from zero to well above the ac supply frequency) to control the speed of ac motors.

43. Control of the Output Voltage. In the previously described operation modes, frequency changers supply ac power at the maximum output voltage obtainable from the given supply voltages. In many practical applications, e.g., speed control of ac motors and regulated ac supplies, the effective value of the output-voltage waveform, i.e., the amplitude of the wanted component, has to be controllable independently of the input source. This can be achieved by varying either the depth of modulation used to generate the output waveform or the conduction intervals of the switches while maintaining their repetition rate as required for maximum output voltage (pulse-width modulation). The first type of voltage control can be accomplished simply through sine-wave crossing control by varying the amplitude of the reference wave. This method is compatible, i.e., does not significantly affect the output and input performance characteristics obtained at maximum voltage, with the UDFFC, NCC, CDFFC, and UPFFC (Gyugyi and Pelly, 1976). Typical three-pulse NCC and UFFFC output waveforms with a relative amplitude of 0.7 ($v_O/v_{O,\max} = 0.7$) are shown in Fig. 15-36a and b, respectively.

Pulse-width-modulation voltage control is the only type which is completely compatible with the UFC (Gyugyi and Pelly, 1976). Its essence is to subdivide the conduction periods into *active* and *passive* intervals. During the active interval, the switches are operated in the usual manner. During the passive interval, the load is reconnected to the input phase used for the preceding active interval, as illustrated by the three-pulse UFC waveform in Fig. 15-36c. [Note that for even-pulse-number converters (six, twelve, etc.), the output voltage is zero and the load is actually shorted during the passive intervals.] By controlling the relative duration of the active and passive intervals within the original conduction period, the mean output voltage can be continuously varied from maximum to zero at any given output frequency.

The amplitudes of the dominant unwanted components present in the output voltage waves of three-, six-, and twelve-pulse NCC, CDFFC, UPFFC, and UFC are given as per unit values of $v_{O,max}$ in Table 15-8 for five discrete values of the output-voltage ratio $r = v_O/v_{O,max}$. The data presented are pertinent to the sine-wave crossing control for the NCC (at unity load-power factor, that is, $\phi_0 = 0$ or $180°$), the CDFFC (at $\sigma + \phi_0 = 0$ or $180°$), and the UPFFC. For the UFC, the data are relevant to PWM control.

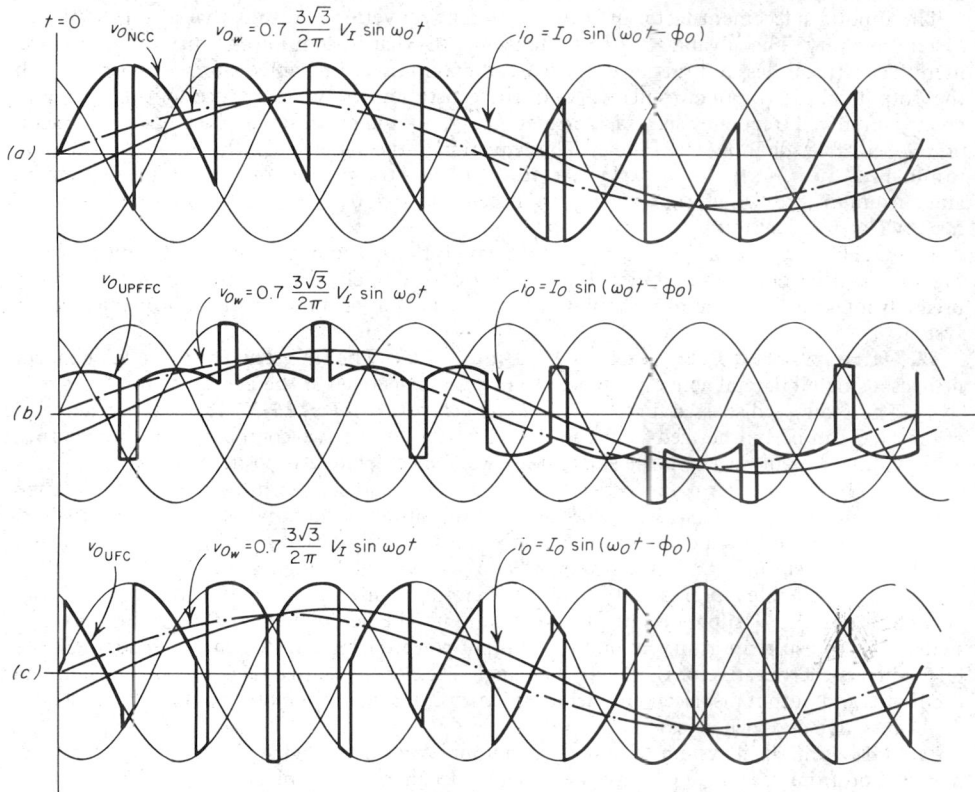

Fig. 15-36. Waveforms illustrating amplitude control of the output voltage by reducing the depth of modulation (*a* and *b*, NCC and UPFFC) and employing pulse-width modulation (*c*, UFC).

Thermal Considerations

44. Thermal Resistances. The internal or junction temperature of a solid-state device has considerable influence on the terminal characteristics of the device. The reverse leakage current of a *pn* junction, for example, increases rapidly with temperature. In a thyristor, this leakage current in the forward blocking state plays a role which is similar to gate current. Hence if the junction temperature of a thyristor exceeds its rated maximum (usually 125°C) even instantaneously, the thyristor cannot be depended upon to block forward voltage until the temperature drops. As a result, the circuit may temporarily malfunction or be permanently faulted.

Consequently, precautions must be taken to ensure that both average and instantaneous peak junction temperatures are within the limits of good design. The average junction-temperature rise can be calculated by multiplying the dissipation (averaged over a cycle) by the total steady-state thermal resistance from the junction to the cooling medium, which is assumed to have a known constant temperature. The total thermal resistance is the sum of several components, including the junction-to-case thermal resistance of the device itself, the thermal resistance of the heat sink to the coolant, and the thermal resistances of any interfaces through which the heat must flow.

In normal 60-Hz operation, the average dissipation will be determined primarily by the for-

Table 15-8. Amplitudes of the Dominant Unwanted Frequency Components of the Output Voltage for Various Voltage Ratios

Number of Pulses		l	ε				
			1.0	0.8	0.6	0.4	0.2
NCC at $\phi_0 = 0$ and CDFFC at $\sigma + \phi_0 = \begin{cases} 0 \\ 180° \end{cases}$							
3	$3f_1 \pm 2lf_0$	0	0.318	0.470	0.596	0.682	0.734
		1	0.272	0.276	0.230	0.163	0.084
		2	0.164	0.112	0.071	0.040	0.018
		3	0.064	0.038	0.025	0.015	0.007
		4	0.021	0.018	0.013	0.008	0.004
6	$6f_1 \pm (2l+1)f_0$	0	0.006	0.069	0.130	0.178	0.208
		1	0.055	0.123	0.138	0.120	0.090
		2	0.105	0.111	0.084	0.060	0.047
		3	0.100	0.065	0.045	0.036	0.032
		4	0.055	0.035	0.029	0.026	0.025
		5	0.026	0.023	0.022	0.021	0.020
		6	0.022	0.021	0.018	0.017	0.017
12	$12f_1 \pm (2l+1)f_0$	0	0.002	0.022	0.035	0.057	0.091
		1	0.001	0.021	0.045	0.071	0.059
		2	0.007	0.029	0.061	0.056	0.029
		3	0.005	0.052	0.051	0.030	0.018
		4	0.027	0.053	0.032	0.017	0.013
		5	0.050	0.036	0.018	0.012	0.010
		6	0.049	0.020	0.012	0.010	0.009
		7	0.027	0.013	0.009	0.008	0.007
		8	0.011	0.009	0.008	0.007	0.006
UPFCC							
3	$3f_1$		0.000	0.099	0.310	0.530	0.690
	$3f_1 \pm 2f_0$		0.250	0.275	0.210	0.110	0.028
	$3f_1 \pm 4f_0$		0.125	0.050	0.018	0.004	0.000
6	$6f_1 \pm f_0$		0.000	0.061	0.025	0.169	0.163
	$6f_1 \pm 3f_0$		0.000	0.098	0.133	0.068	0.012
	$6f_1 \pm 5f_0$		0.100	0.089	0.031	0.006	0.000
	$6f_1 \pm 7f_0$		0.071	0.013	0.003	0.000	0.000
12	$12f_1 \pm f_0$		0.000	0.014	0.020	0.047	0.086
	$12f_1 \pm 3f_0$		0.000	0.019	0.035	0.063	0.033
	$12f_1 \pm 5f_0$		0.000	0.027	0.047	0.040	0.003
	$12f_1 \pm 7f_0$		0.000	0.026	0.046	0.007	0.000
	$12f_1 \pm 9f_0$		0.000	0.052	0.012	0.001	0.000
	$12f_1 \pm 11f_0$		0.045	0.019	0.001	0.000	0.000
	$12f_1 \pm 13f_0$		0.038	0.002	0.000	0.000	0.000
UFC							
3	$3f_1 + 2f_0$		0.500	0.678	0.822	0.922	0.981
	$3f_1 + 4f_0$		0.250	0.038	0.171	0.346	0.464
6	$6f_1 + 5f_0$		0.200	0.350	0.399	0.335	0.196
	$6f_1 + 7f_0$		0.143	0.071	0.235	0.229	0.179
12	$12f_1 + 11f_0$		0.091	0.176	0.037	0.141	0.162
	$12f_1 + 13f_0$		0.077	0.121	0.111	0.075	0.148

ward current and voltage drop and by the proportion of the cycle during which conduction takes place. Manufacturers' data sheets usually show this dissipation plotted against average current for various conduction angles. For operation at higher frequencies, the switching dissipation during turnoff and turn-on must be included and may become predominant (Golden, 1971).

The junction-to-case thermal resistance is determined by the internal construction of the device and is specified on the manufacturer's data sheet. Typical values range from about 1°C/W for a device with a ¼-in stud to 0.06°C/W for double-sided cooling of a flat device. To minimize the interface resistance, the contacting surfaces must be smooth and flat and should be properly covered by a thermal lubricant. Stud nuts should be tightened to the recommended torque to minimize thermal resistance without causing internal damage from excessive mechanical stresses. Typical case-to-sink thermal resistances range from 0.14 to 0.05°C/W.

The heat sink-to-ambient thermal resistance varies widely, depending on the type of heat sink and on the type and flow rate of the coolant. The thermal resistance of several typical air-cooled heat sinks is plotted in Fig. 15-37 as a function of air velocity. Water-cooled heat sinks are usually designed for a thermal resistance of 0.1°C/W or less.

45. Transient Thermal Impedance. The calculation of instantaneous junction temperature involves thermal capacity (specific heat and mass) as well as thermal resistance. Because the masses of solid-state power devices are very small compared with those of other electrical components, e.g., transformers and motors, with comparable power ratings, their thermal capacities are also much smaller. Consequently, the thermal response of a solid-state device to transient, overload, and fault conditions requires close attention if long life and reliable operation are to be achieved.

Manufacturers' data sheets for power devices normally contain a graph of transient thermal impedance plotted as a function of time. Physically this plot shows how the junction temperature would rise as a function of time following the application of a 1-W step function of dissipation. In practice, the transient temperature rise resulting from a rectangular pulse of power of duration t is obtained by multiplying the transient thermal impedance at t by the amplitude of the power pulse. If the power during the pulse is not constant, a stepped approximation and linear superposition can be used to calculate the temperature response because heat conduction is a linear process (Gutzwiller and Sylvan, 1961). The same curve of transient thermal impedance can also be used to calculate the rate of cooling of the junction following a power pulse.

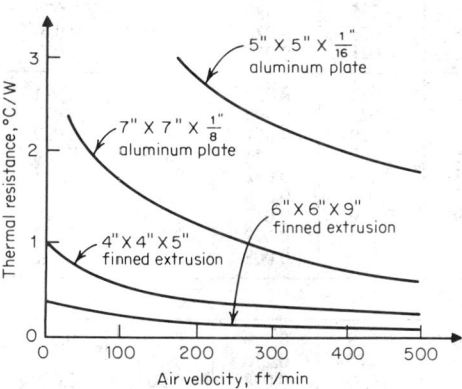

Fig. 15-37. Steady-state thermal resistance of typical air-cooled heat sinks.

This method of finding the instantaneous junction temperature is valid once the total area of the device is turned on. During the first 20 μs or so of the turn-on transient, the conducting area is increasing rapidly, and the hot-spot temperature is held to a safe value by observing the turn-on di/dt rating (Mapham, 1964).

For long pulse durations, internal heat flow approaches equilibrium, and the transient thermal impedance approaches the steady-state thermal resistance as an upper limit. The time at which the transient thermal impedance becomes nearly equal to its limit is the thermal time constant of the device and its package. If the duration of a transient overload exceeds this time, the transient thermal impedance of the heat sink must be added to that of the device to obtain the thermal response. For shorter times, the transient temperature rise is determined solely by the internal construction of the device.

46. Protection of Thyristors. In many applications, if a thyristor or other solid-state device is specified only on the basis of its steady-state ratings, it is likely to fail immediately or after very short service. To achieve the reliability and long life for which these devices are well known, the surrounding circuit must be carefully designed to protect them from transients and overloads which exceed their ratings. Of particular importance are both the maximum dv/dt and peak voltage of transients when a device is in its off state, the maximum di/dt when it is being turned on, and the peak current once it is fully on.

A capacitor (typically 0.1 to 1 μF) is frequently connected in parallel with a thyristor to act as a "snubber" which limits dv/dt to prevent unintentional firing and also absorbs energy from voltage spikes. A resistor of 10 to 50 Ω is usually required in series with this capacitor to prevent excessive di/dt when the thyristor is turned on (McMurray, 1972; Rice and Nickels, 1968; von Zastrow and Galloway, 1965).

Excessive voltage transients may also be limited by nonlinear voltage suppressors, which are matched to the maximum voltage rating of the thyristor and have sufficient energy capacity to dissipate the transient (Gutzwiller, 1967; IEEE Comm. Rep., 1970; Harnden, 1972; Lawatsch and Weisshaar, 1972).

The di/dt at turn-on is usually limited by the inductance inherent in the circuit or by an inductor added for this purpose (Paice and Wood, 1967). High-frequency inverters and other applications requiring high di/dt can utilize fast turn-on thyristors developed especially for this purpose. The gate construction of these devices causes conduction to be initiated over a larger area than in conventional thyristors, thereby increasing the di/dt rating. The magnitude and rise time of the signal applied to the gate also influence the di/dt capability of a device. Manufacturers' recommendations should be followed for dependable performance (Dyer, 1966).

Currents which exceed the steady-state rating of a device fall into two categories. Abnormal or fault conditions presumably occur infrequently. Under these conditions the equipment is usually made temporarily inoperative if necessary to prevent the destruction of the solid-state devices. The surge rating of a device determines when this must be done. Although the terminal characteristics of a device should not be noticeably different after it has been stressed to its surge rating, this is a "nonrepetitive" rating. A device should not be exposed to more than 100 surges of this intensity during the entire expected lifetime.

The surge current through a device is normally limited by a fast-acting series fuse. In the past, I^2t ratings have been used to achieve proper coordination between the fuse and the device which it protects (Gutzwiller, 1958). However, this rating is not constant but varies with the peak current and the clearing time of the fuse. Instead $I^2\sqrt{t}$ has been proposed as a rating which is more nearly constant and hence more valid (Motto, 1971). Another approach to protection which may sometimes be used in place of fuses involves modification of the gate circuit such that subsequent firing pulses are disabled when a fault is detected.

In some applications, such as motor controls, the solid-state devices are repetitively loaded beyond their steady-state ratings for short periods of time. This type of overload is "normal" in these applications and obviously must not blow a fuse each time. The capacity of a device to withstand this type of overload is determined by the allowable peak junction temperature and the transient thermal impedance. Applications of this type may require that oversize solid-state devices be chosen on the basis of their overload capacity rather than their steady-state ratings.

High-voltage or high-current applications which exceed the available ratings of individual solid-state devices require that multiple devices be combined in series or parallel arrays. In theory, matched devices could be used in an array, but in practice typical spreads of device characteristics frequently necessitate equalizing networks to avoid stress concentrations under both steady-state and transient conditions (Rice, 1970; Grafham and Hey, 1972).

Parallel devices tend to share current unequally, with the highest current flowing in the device with the lowest forward drop. An impedance in series with each device helps to equalize the steady-state current sharing. Fortunately at high current levels the temperature coefficient of forward voltage becomes positive, so that surge currents tend to be self-equalizing.

Because of the variation in leakage currents, series devices share blocking voltages unequally in steady state. Also, the differences in junction capacitance and switching times cause unequal voltage distributions during transients, turn-on, or turnoff. An RC network in parallel with each device can be designed to equalize both the steady-state and transient voltage distribution, and turn-on times and stored charges are kept to close tolerances.

POWER FILTERS

47. General Considerations. Of necessity, high-power controls and converters use switching devices because they permit high power efficiency to be achieved. But the switching action generates transients and spurious frequencies (usually harmonics of the fundamental frequency) which may have intolerable effects on the power source, the load, and/or other nearby equipment, by way of electromagnetic interference (EMI) radiated or conducted through the supply line. Frequently the internal design of a power converter, such as a harmonic-neutralized inverter, is chosen to minimize the most troublesome spurious frequencies. In addition it is often necessary to filter the input, the output, or both.

Because all power filters must handle substantial volt-amperes, their component dissipation, cost, size, and weight are all important design factors. Consequently, they are usually not optimized by conventional small-signal filter-synthesis procedures. Basically, a low-loss filter operates by storing energy during an unwanted peak in a waveform and then discharging the energy back into the circuit during an unwanted trough. The cost of the filter obviously increases as the required energy storage increases.

Power filters can be classified according to whether their main purpose is to improve the power waveform or to remove EMI. Filters for waveform improvement usually deal with frequencies in the audio range. EMI filters are usually concerned with frequencies of 455 kHz or higher, although coupling to telephone lines or interference with low and very low frequency communications can be a problem at much lower frequencies.

48. Input Filters. Input filters for waveform improvement normally consist of three to five series-resonant traps across the input power lines. These traps provide a low-impedance path in which the dominant low-order harmonic currents required by the converter can circulate. Without a filter, these currents would have to circulate through the source impedance of the supply line, thereby deteriorating the voltage waveform.

$L_1 C_1 R_1$ in Fig. 15-38 is a single-tuned trap of this type. The Q of these traps is relatively high, with only enough damping to accommodate variations in line frequency and changes in com-

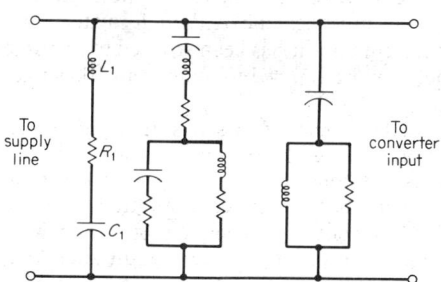

Fig. 15-38. Typical input filter for waveform improvement.

ponent values due to initial tolerances, aging, and temperature variation. Although the ease of adjustment is decreased, the cost of filter components can be reduced by combining two single-tuned traps into a double-tuned trap, as shown in the middle of Fig. 15-38.

In contrast to the low-order harmonics which must be individually suppressed by high-Q traps, higher-order harmonics can usually be adequately suppressed by a single damped filter section like that shown at the right of Fig. 15-38.

It should be noted that all these input filters draw leading current at the fundamental frequency, a property which is often useful for power-factor correction. It should also be observed that poles of impedance interleave the zeros which suppress the dominant harmonics. Care must be taken to assure that residual harmonics at other frequencies do not excite undesired resonance at these poles.

If the output frequency is variable, either intentionally (as in a variable-frequency cycloconverter) or because of appreciable variation in the supply frequency, high-Q traps are unsatisfactory and broadband filters must be used.

Design procedures for input-power filters are given in Cory (1965, Chap. 7) and Kimbark (1971, Chap. 8).

Input filters for EMI suppression usually consist of one or more low-pass LC L sections between the converter and the supply lines. The input and output terminals of the filter are transposed from the usual low-pass section, as shown in Fig. 15-39, because it is desired to minimize the current-transfer ratio rather than the voltage-transfer ratio. With this transposition, treating the converter as a current source and the supply line as a zero-impedance load, conventional design tables can be used (Geffe, 1964, App. 3). Second-order filters are usually critically damped, i.e., Butterworth response, while filters of higher order can be made to have a steeper edge to the stop band by using the Chebyshev design criterion.

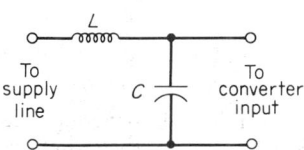

Fig. 15-39. Transposed low-pass input filter to prevent EMI from entering the supply lines.

49. Output Filters. Output filters for waveform improvement can be divided into two categories: dc filters for rectifiers and choppers and ac filters for inverters and cycloconverters.

Conventional single-section low-pass LC filters are used almost universally for dc applications. The inductor is usually chosen to be larger than the critical value which will maintain continuous current for the worst-case ripple-load combination (Distler and Munshi, 1965). The capacitor is then chosen to obtain the desired reduction in ripple voltage. However, the resonant frequency must also be chosen so that it does not coincide with a residual harmonic below the fundamental ripple frequency. Although in theory these harmonics of the supply frequency are cancelled, the cancellation is never perfect in practice.

The amplitudes of the unfiltered harmonics are summarized in Figs. 15-9 and 15-10, and design data for rectifier filters are presented in Terman (1943, Sec. 8) and Langford-Smith (1953, Chap. 31).

Simple second-order low-pass sections are ordinarily not used to filter the ac output of inverters and cycloconverters because of their insertion loss at the fundamental frequency. However, the series arm can be resonated to minimize this loss in fixed-frequency equipment, as in the Ott filter, which also supplies commutating capacitance (Ott, 1963; Rice, 1970, pp. 8-47 to 8-54). The shunt arms can be series-resonated at the dominant low-order harmonics and/or parallel-resonated at the fundamental frequency.

Passive filters are not suitable for use with variable-frequency converters unless the frequency range is restricted so that the lowest harmonic of the lowest fundamental frequency is considerably higher than the highest fundamental frequency.

50. Filter Components. The capacitors used in power filters are required to pass high currents. Therefore they should always be of extended-foil construction with a dielectric having a low loss over the required frequency range. Paper-oil, plastic-film, polycarbonate, or mica capacitors are generally suitable but must be chosen to have adequate transient ratings as well as steady-state ratings. Filters for dc applications nearly always use electrolytic capacitors because of their lower cost and smaller volume, even if their limited ripple-current capacity necessitates overdesign of the filter to stay within their ratings. The filter designer must also remember that in certain applications, e.g., a rectifier feeding dc power to an inverter, the load may cause significant additional ripple currents.

The design of inductors for dc filters is well established (Langford-Smith, 1953, Chap. 5.6; Terman, 1943, Sec. 2). Chokes for ac filters may be air- or iron-cored, depending on the required inductance, kilovolt-ampere rating, and frequency range. Even iron-cored inductors are designed with an air gap which determines the inductance. The design of low-loss inductors is complicated by skin effect and winding capacitance, which often dictate the use of strip, tubular, or Litz conductors, and by fringing flux at the gap, which may require using powdered iron to distribute the gap.

51. Active Filters. Conventional passive filters frequently represent a substantial part of the total cost, weight, and size of power electronics equipment. For this reason, improved types of filters are constantly being sought. Active filters represent a new approach still in its infancy about which little has been published. The approach is related to the theory of low-level active-feedback filters but requires considerable adaptation to achieve power efficiency within the capability of available devices. In effect, a power operational amplifier in a feedback loop with a single energy-storage element (an inductor or capacitor) serves to minimize the difference between the actual waveform and the desired waveform. The "filter" may become an integral part of the converter itself (see the optimum-response inverter described in Fig. 15-28). Although the cost of active filters tends to be high at present, their adaptability permits problems to be solved that could not be solved otherwise, as in a variable-frequency inverter, for instance.

52. Bibliography

Bates, J. J., and R. E. Colyer (1973) The Impact of Semiconductor Devices on Electrical Power Engineering, *Radio Electron. Eng.*, Vol. 43, No. 1/2, pp. 115–124.

Bedford, B. D., and R. G. Hoft (1964) "Principles of Inverter Circuits," Wiley, New York.

Cory, B. J. (ed.) (1965) "High Voltage Direct Current Convertors and Systems," Macdonald, London.

Davis, R. M. (1971) "Power Diode and Thyristor Circuits," Cambridge University Press, London.

Delhom, L. A. (1968) "Design and Application of Transistor Switching Circuits," McGraw-Hill, New York.

Distler, R. J., and S. G. Munshi (1965) Critical Inductance and Controlled Rectifiers, *IEEE Trans.*, Vol. IECI-12, pp. 34–37.

Dyer, R. F. (1966) The Rating and Application of SCR's Designed for Power Switching at High Frequencies, *IEEE Trans.*, Vol. IGA-2, pp, 5–15.

EIA (1971) Recommended Standards for Thyristors, EIA Stand. RS-397.

Geffe, P. (1964) 'Simplified Modern Filter Design," Rider, New York.

Gentry, F. E., et al. (1964) "Semiconductor Controlled Rectifiers: Principles and Applications of PNPN Devices," Prentice-Hall, Englewood Cliffs, N J.

Geyer, M. A., and A. Kernick (1971) Time Optimal Response Control of a Two-Pole Single-Phase Inverter, *Power Cond. Spec. Conf. Rec.*, pp. 101–109, IEEE Pub. 71C15-AES.

Golden, F. B. (1971) Thyristor Switching Losses and Their Measurement, *Direct Curr. Power Electron.*, Vol. 2, pp. 112–120.

Grafham, D. R., and J. C. Hey (eds.) (1972) "SCR Manual," 5th ed., General Electric Semiconductor Products Dept., Syracuse, N.Y.

Gutzwiller, F. W. (1958) The Current-Limiting Fuse as Fault Protection for Semiconductor Rectifiers, *Trans. AIEE*, Vol. 77, pt. I, pp. 751-754.

—— (1967) Thyristors and Rectifier Diodes — The Semiconductor Workhorses, *IEEE Spectrum*, August 1967, pp. 102–111.

—— (ed.) (1967) "General Electric SCR Manual," 4th ed., General Electric Semiconductor Products Department, Syracuse, N.Y.

——— and T. P. Sylvan (1961)   Power Semiconductor Ratings under Transient and Intermittent Loads, *Trans. AIEE*, Vol. 79, Pt. 1, pp. 699–705.

Gyugyi, L., R. A. Otto, and T. H. Putman (1978)   Principles and Applications of Static, Thyristor Controlled Shunt Compensators, *IEEE Trans. Power Apparat. Syst.*, Vol. PAS-97, No. 5.

——— and B. R. Pelly (1976)   "Static Power Frequency Changers," Wiley, New York.

——— and E. R. Taylor, Jr. (1980)   Characteristics of Static, Thyristor-Controlled Shunt Compensators for Power Transmission System Applications, *IEEE Trans. Power Eng. Soc.*, Vol. F 80, p. 236.

Harm, R. S., H. R. Emerick, and W. J. Savage (1980)   "Power Semiconductor User's Manual," Westinghouse, Youngwood, Pa.

Harnden, J. D., et al. (1972)   Metal-Oxide Varistor: A New Way to Suppress Transients, *Electronics*, Oct. 9, pp. 91–95.

Heinrich, T. M. (1967)   Static Inverter with Neutralization of Harmonics, M.S. dissertation, University of Pittsburgh, Pa.

Hibberd, R. G. (1968)   "Solid State Electronics," McGraw-Hill, New York.

Hnatek, E. R. (1971)   "Design of Solid-State Power Supplies," Van Nostrand Reinhold, New York.

Hoft, R. G. (1972)   "Static Power Converters in the USA," *Conf. Rec. IEEE Int. Semiconductor Power Converter Conf.*, pp. 2-8-1 to 2-8-8.

IEEE (1967)   Special Issue on High-Power Semiconductor Devices, *Proc. IEEE*, Vol. 55, August.

IEEE (1970)   Special Issue on High-Power Semiconductor Devices, *IEEE Trans.*, Vol. ED-17, September.

IEEE (1972)   "Applications of High Power Semiconductor Devices," IEEE Press, New York.

IEEE Comm. Rep. (1970)   Bibliography on Surge Voltages in AC Power Circuits Rated 600 Volts and Less, *IEEE Trans.*, Vol. PAS-89, pp. 1056–1061.

Kernick, A., et al. (1962)   Static Inverter with Neutralization of Harmonics, *Trans. AIEE*, Vol. 81, Pt. 2, pp. 59–68.

——— and I. U. Haque (1969)   Programmed Waveform Static Inverter, *Proc. 23d Annu. Power Sources Conf.*, pp. 59–63, sponsored by U.S. Army Electronics Command, Fort Monmouth, N.J.

——— and T. M. Heinrich (1964)   Controlled Current Feedback in a Static Inverter with Neutralization of Harmonics, *IEEE Trans. Aerosp.*, Vol. 2, pp. 985–992.

Kimbark, E. W. (1963)   A Chart Showing the Relations between Electrical Quantities on the AC and DC Sides of a Converter, *IEEE Trans.*, Vol, PAS-82, pp. 1050–1054.

——— (1971)   "Direct Current Transmission," Vol. I, Wiley-Interscience, New York.

Kusko, A. (1969)   "Solid-State DC Motor Drives," M.I.T. Press, Cambridge, Mass.

——— (1972)   Solid-State Motor-Speed Controls, *IEEE Spectrum*, October, pp. 50–55.

Langford-Smith, F. (ed.) (1953)   "Radiotron Designer's Handbook," RCA, Harrison, N.J.

Lawatsch, H., and E. Weisshaar (1972)   A Silicon Voltage Limiter for Power Thyristor Circuits, *Brown Boveri Rev.*, pp. 476–482.

Mapham, N. (1964)   The Rating of Silicon Controlled Rectifiers when Switching into High Currents, *Trans. AIEE*, Vol. 83, Pt. 1, pp. 515–519.

McMurray, W. (1972)   Optimum Snubbers for Power Semiconductors, *IEEE Trans.*, Vol. IA-8, pp. 593–600.

——— and D. P. Shattuck (1961)   A Silicon Controlled Rectifier Inverter with Improved Commutation, *Trans. AIEE*, Vol. 80, Pt. 1, pp. 511–542.

Motto, J. W. (1971)   A New Quantity to Describe Power Semiconductor Subcycle Current Ratings, *IEEE Trans.*, Vol. IGA-7, pp. 510–517.

Newell, W. E. (1974)   Power Electronics—Emerging from Limbo, *IEEE Trans.*, Vol. IA-10, January-February.

Oates, R. M. (1970)   Inverter Harmonic Neutralization Using Interphase Reactors, M.S. dissertation, University of Pittsburgh, Pa.

Ott, R. R. (1963)   A Filter for SCR Commutation and Harmonic Attenuation in High-Power Inverters, *IEEE Trans. Commun. Electron.*, Vol. 82, pp. 259–262.

Paice, D. A., and P. Wood (1967)   Nonlinear Reactors as Protective Elements for Thyristor Circuits, *IEEE Trans.*, Vol. MAG-3, pp. 228–232.

Pelly, B. R. (1971)   "Thyristor Phase-Controlled Converters and Cycloconverters," Wiley-Interscience, New York.

Ravas, R. J., et al. (1967)   Staggered Phase Carrier Cancellation: A New Circuit Concept for Lightweight Static Inverters, *EASTCON Tech. Conv. Rec., Suppl. IEEE Trans.*, Vol. AES-3, No. 6, pp. 432–444; *IEEE Pub.* 10-C-57.

RCA (1971) "Solid-State Power Circuits (SP-52)," RCA Solid State Division, Somerville, N.J.

Rice, J. B., and L. E. Nickels (1968) Commutation dv/dt Effects in Thyristor Three-Phase Bridge Converters, *IEEE Trans.,* Vol. IGA-4, pp. 565–572.

Rice, L. R. (ed.) (1970) "Westinghouse Silicon Controlled Rectifier Designers Handbook," 2d ed., Westinghouse Semiconductor Division, Youngwood, Pa.

Rissik, H. (1939) "The Fundamental Theory of Arc Converters," Chapman & Hall, London (contains extensive bibliography).

Rosa, J. (1979) Utilization and Rating of Machine Commutated Inverter Synchronous Motor Drives. *IEEE Trans. on Industr. Appl.,* Vol. IA-15, No. 2, pp. 155–164.

Schaeffer, J. (1965) "Rectifier Circuits: Theory and Design," Wiley, New York.

Storm, H. F. (1969) Solid-State Power Electronics in the USA, *IEEE Spectrum,* October, pp. 49–59.

Terman, F. E. (1943) "Radio Engineers' Handbook," McGraw-Hill, New York.

USAS C34.2-1968 "USA Standard Practices and Requirements for Semiconductor Power Rectifiers," American National Standards Institute, New York.

Wood, P. (1981) "Switching Power Converters," Van Nostrand Reinhold, New York.

Zastrow, E. E. von, and J. H. Galloway (1965) Commutation Behavior of Diffused High Current Rectifier Diodes, *IEEE Trans.,* Vol. IGA-1, pp. 157–166.

Pulsed Circuits, Logic Circuits, and Waveform Generators

PAUL G. A. JESPERS *Professor of Electrical Engineering, Bâtiment Maxwell, Catholic University of Louvain, Louvain-la-Neuve, Belgium; Senior Member, IEEE*

CONTENTS

Numbers refer to paragraphs

PASSIVE WAVEFORM SHAPING

1. Linear Passive Networks. Waveform generation is customarily performed in active nonlinear circuits. Since passive networks, linear as well as nonlinear, enter into the design of pulse-forming circuits, this survey starts with the study of the transient behavior of passive circuits.

Among linear passive networks, the single-pole RC and RL networks are the most widely used. Their transient behavior in fact has a broad field of applications since the responses of many complex higher-order networks are dominated by a single pole; i.e., their response to a step function is very similar to that of a first-order system.

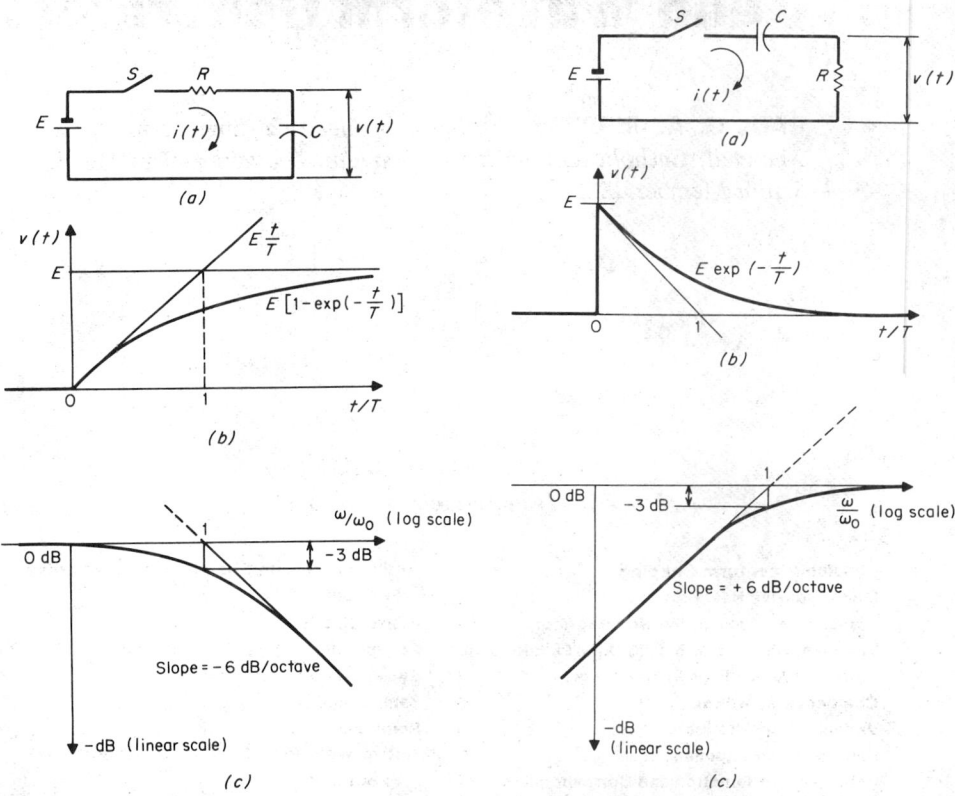

Fig. 16-1. (a) RC integrator circuit; (b) voltage vs. time across capacitor; (c) attenuation vs. angular frequency.

Fig. 16-2. (a) RC differentiator circuit; (b) voltage across resistor vs. time; (c) attenuation vs. angular frequency.

2. Transient Analysis of the RC Integrator. The step-function response of the RC circuit shown in Fig. 16-1a, after closing of the switch S, is given by

$$V(t) = E[1 - \exp(-t/T)] \tag{16-1}$$

where T = time constant = RC. The inverse of T is called the cutoff pulsation ω_0 of the circuit.

The Taylor-series expansion of Eq. (16-1) yields

$$V(t) = E\frac{t}{T}\left(1 - \frac{t}{2!T} + \frac{t^2}{3!T^2} - \cdots\right) \tag{16-2}$$

When the values of t are small compared with T, a first-order approximation of Eq. (16-2) is

$$V(t) \approx Et/T \tag{16-3}$$

In other words, the RC circuit of Fig. 16-1 behaves like an imperfect integrator. The relative error ϵ with respect to the true integral response is given by

$$\epsilon = -\frac{t}{2!T} + \frac{t^2}{3!T^2} - \frac{t^3}{4!T^3} + \cdots$$

The theoretical step-function response of (16-1) and the ideal-integrator output of (16-3) are represented in Fig. 16-1b.

Small values of t with respect to T correspond in the frequency domain (Fig. 16-1c) to frequency components situated above ω_0, that is, the transient signal whose spectrum lies to the right of ω_0 in the figure. In that case, the difference is small between the response curve of the RC filter and that of an ideal integrator (represented by the -6 dB/octave line in the figure). The circuit shown in Fig. 16-1a thus approximates an integrator, provided either of the following conditions is satisfied: (1) the time under consideration is much smaller than T or (2) the spectrum of the signal lies almost entirely above ω_0.

3. Transient Analysis of the RC Differentiator. When the resistor and the capacitor of the integrator are interchanged, the circuit (Fig. 16-2a) is able to differentiate signals. The step-function response (Fig. 16-2b) of the RC differentiator is given by

$$v(t) = E \exp(-t/T) \qquad (16\text{-}4)$$

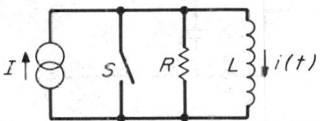

Fig. 16-3. RL current-integrator circuit, the dual of the circuit in Fig. 16-1a.

Fig. 16-4. RL current-differentiator circuit, the dual of the circuit in Fig. 16-2a.

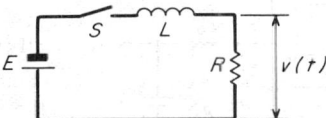

Fig. 16-5. RL voltage integrator.

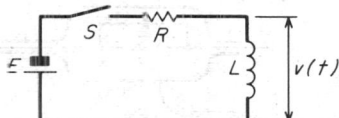

Fig. 16-6. RL voltage differentiator.

The time constant T is equal to the product RC, and its inverse ω_0 represents the cutoff of the frequency response of the circuit. As the values of t become large compared with T, the step-function response becomes more like a sharp spike; i.e., it increasingly resembles the delta function (Par. **16-8**).

The response differs from the ideal delta function, however, because both its amplitude and its duration are always finite quantities. The area under the exponential pulse, equal to ET, is the important quantity in applications where such a signal is generated to simulate a delta function, as in the measurement of the impulse response of a system. These considerations may be transported in the frequency domain (Fig. 16-2c).

4. Transient Analysis of RL Networks. Circuits involving a resistor and an inductor are also often used in pulse formation. Since integration and differentiation are related to the functional properties of first-order systems rather than to the topology of actual circuits, RL networks may perform the same functions as RC networks. The duals of the circuits represented in Figs. 16-1 and 16-2, respectively, are shown in Figs. 16-3 and 16-4 and exhibit identical functional properties. In the first case, the current in the inductor increases exponentially from zero to I with a time constant equal to L/R, while in the second case it drops exponentially from the initial value I to zero, with the same time constant. Similar behavior can be obtained regarding voltage instead of current by changing the circuit from Fig. 16-3 to that of Fig. 16-5 and from Fig. 16-4 to Fig. 16-6, respectively. This duality applies also to the RC case.

5. Compensated Attenuator. The compensated attenuator is a widely used network, e.g., as an attenuator probe used in conjunction with cathode-ray oscilloscopes. The compensated attenuator (Fig. 16-7) is designed to perform the following functions:

1. To provide remote sensing with a very high input impedance, thus producing a minimum perturbation to the circuit under test.

2. To deliver a signal to the receiving end (usually the input of a wide-band oscilloscope) which is an accurate replica of the signal at the input of the attenuator probe. These conditions

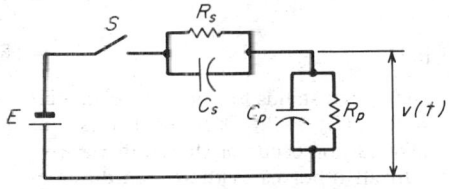

Fig. 16-7. Compensated attenuator circuit.

can be met only by introducing substantial attenuation to the signal being measured, but this is a minor drawback since adequate gain to compensate the loss is usually available.

Diagrams of two types of cathode-ray oscilloscope attenuator probes are given in Fig. 16-8, similar to the circuit of Fig. 16-7. In both cases, the coaxial-cable capacitance parallels the input capacitance of the receiver end; C_p represents the sum of both capacitances.

The shunt resistor R_p has a high value, usually 1 MΩ, while the series resistor R_s is typically 9 MΩ. The dc attenuation ratio of the attenuator probe therefore is 1:10, while the input imped-ance of the probe is 10 times that of the receiver.

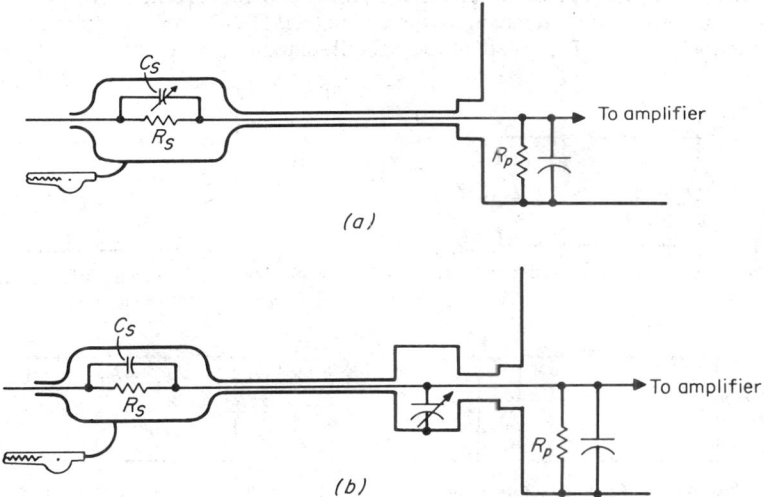

Fig. 16-8. Coaxial-cable type of attenuator circuit: (*a*) series adjustment; (*b*) shunt adjustment.

At high frequencies the parallel and series capacitors C_p and C_s play the same role as the resistive attenuator. Ideally these capacitors should be kept as low as possible to achieve a high input impedance even at high frequencies. Since it is impossible to reduce C_p below the capaci-tance of the coaxial cable, there is no alternative other than to insert the appropriate value of C_s to achieve a constant attenuation ratio over the required frequency band. In consequence, as the frequency increases, the nature of the attenuator changes from resistive to capacitive. However, the attenuation ratio remains unaffected, and no signal distortion is produced. The condition that ensures constant attenuation ratio is given by

$$R_p C_p = R_s C_s \qquad (16\text{-}5)$$

The step-function response of the compensated attenuator illustrates clearly how distortion may occur when the above condition is not met. The output voltage $V(t)$ of the attenuator is given by

$$V(t) = \frac{C_s}{C_s + C_p} \left\{ 1 - (1 - K) \left[1 - \exp\left(-\frac{t}{T}\right) \right] \right\} E \qquad (16\text{-}6)$$

where K represents the ratio of the resistive attenuation factor to that of the capacitive atten-uation factor

$$K = \frac{R_p}{R_p + R_s} \frac{C_s}{C_p + C_s}$$

and

$$T = (R_p \,\|\, R_s)(C_s + C_p) \qquad (16\text{-}7)$$

The $\|$ sign stands for the parallel combination of two elements, e.g., in the present case $R_p \,\|\, R_s = R_p R_s/(R_p + R_s)$. Only when K is equal to 1, in other words when Eq. (16-5) is satisfied, will no distortion occur, as shown in Fig. 16-9.

In all other cases there is a difference between the initial amplitude of the step-function response (which is controlled by the attenuation ratio of the capacitive divider) and the steady-state response (which depends on the resistive divider only).

A simple adjustment to compensate an attenuator consists in trimming one capacitor, either C_p and C_s, to obtain the proper step-function response. Adjustments of this kind are provided in attenuators like those shown in Fig. 16-8.

6. Variable Step Attenuators. A variable step attenuator is a set of fixed compensated attenuators that may either be interchanged or placed in cascade to achieve desired levels of

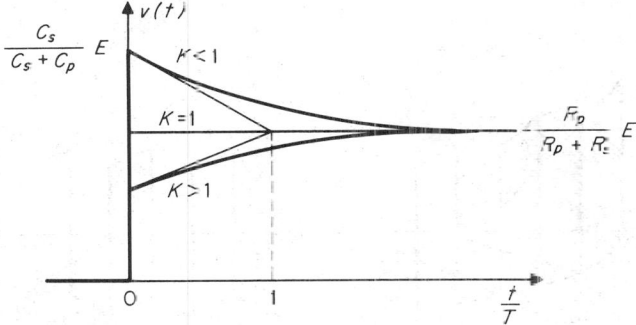

Fig. 16-9. Voltage vs. time responses of attenuator, showing correctly compensated condition at $K = 1$.

attenuation. The conditions imposed on a variable step attenuator are like those enumerated in Par. **16-5**, but an additional requirement is introduced, namely, the requirement for constant input impedance when cells are cascaded. This introduces a different structure compared with the compensated attenuator, as shown in Fig. 16-10. The resistances R_p and R_s must be chosen so

that the impedance R is kept constant. The capacitor C_s is adjusted to compensate the attenuator, while C_p provides the required additional capacitance to make the input susceptance equal to that of the load.

7. Periodic Input Signals. Repetitive transients are typical input signals to the majority of pulsed circuits. In linear networks there is no difficulty in predicting the response of circuits to a succession of periodic step functions, alternatively positive and negative, since in linear circuits the principle of superposition holds. There is no need for transient solution of linear circuits when the period of the excitation lasts

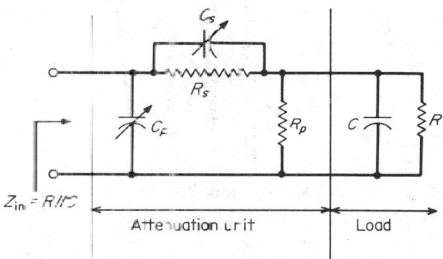

Fig. 16-10. Compensated attenuator suitable for use in cascaded circuits.

for a time longer than that needed to attain the steady state. Hence we restrict our attention here to two simple cases, the square-wave response of an RC integrator and an RC differentiator.

Figure 16-11 represents, at the left, the buildup of the response of the RC integrator, assuming that the period τ of the input square wave is smaller than the time constant of the circuit T. On the right in the figure the steady-state response is shown. The triangular waveshape represents a fair approximation to the integral of the input square wave. The triangular wave is superimposed on a dc pedestal of amplitude $E/2$. Higher repetition rates of the input reduce the amplitude of the triangular wave without affecting the dc pedestal. When the frequency of the input

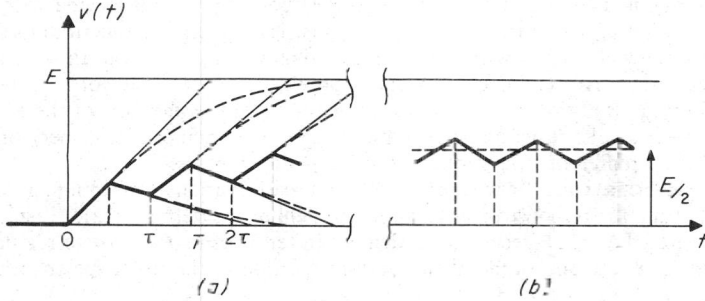

(a)   (b)

Fig. 16-11. RC integrator with square-wave input of period smaller than RC. (a) initial buildup; (b) steady state.

square wave is high enough, the dc component is the only remaining signal; i.e., the RC integrator then acts like an ideal low-pass filter.

A similar presentation of the behavior of the RC differentiator is shown in Fig. 16-12a and b. The steady-state output in this case is symmetrical with respect to the zero axis because no dc component can flow through the series capacitor. When, as shown in Fig. 16-12b, no overlapping of the pulses occurs, the steady-state solution is obtained from the first step.

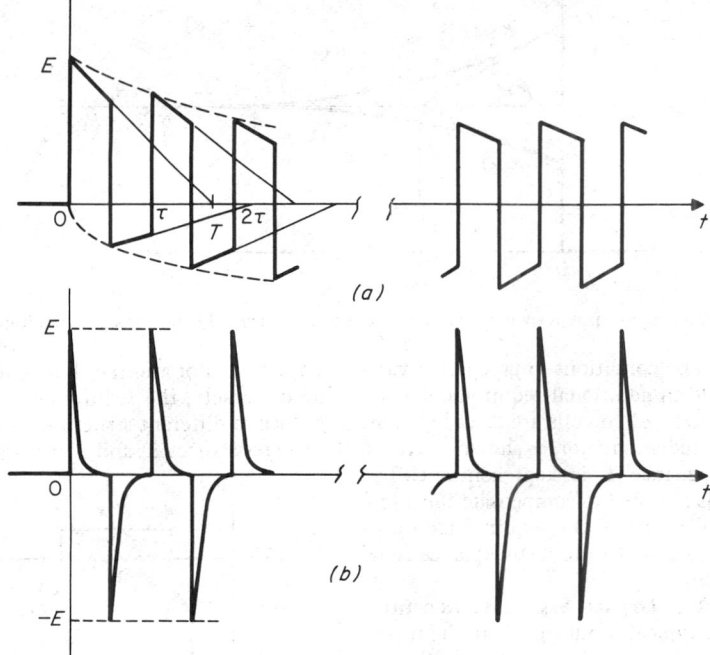

Fig. 16-12. RC differentiator with square-wave input: (a) period of input signal smaller than RC; (b) input period longer than RC.

8. Delta (Dirac) Function and Corresponding Physical Impulse. The step function and the delta function (Dirac function) are widely used to determine the dynamic behavior of physical systems. Theoretically the delta function is a pulse of infinite amplitude and infinitesimal duration but having a finite area (product of amplitude and time). In practice the question of the equivalent physical impulse arises. The answer to this question involves the system under consideration as well as the impulse itself.

The delta function has a constant amplitude over the whole frequency spectrum from zero to infinity. Other transient signals of finite area (amplitude $\times$ time) have different spectral distributions. On a logarithmic scale of frequency, the spectrum of any finite-area transient signal tends to be constant between zero and a cutoff frequency which depends on the shape of the signal. The shorter the duration of the signal, the wider the constant-amplitude portion of the spectrum.

If such a signal is used in a system whose useful frequency band is located within the cutoff frequency of the signal spectrum, the system response is indistinguishable from its delta impulse response. Any transient signal with a finite area, whatever its shape, can thus be considered as a delta function relative to the given system, provided that the flat portion of its spectrum embraces all or a part of the system's useful frequency range. A measure of the effectiveness of a pulse to serve as a delta function is given by the approximation of useful spectrum bandwidth $B = 1/\tau$, where τ represents the midheight duration of the pulse.

9. Pulse Generators. Very short pulses are used in various applications in order to measure the delta-function response of systems. In the field of rf interference, for instance, the basic response curve of the CISPR receiver[1]* is defined in terms of its response to regularly repeated pulses. In this case, the amplitude of the uniform portion of the pulse spectrum must be cali-

*Superior numbers correspond to numbered references, Par. **16-46**.

brated, i.e., the area under the pulse must be a known constant which is a function of a limited number of circuit parameters.

The step-function response of an RC differentiator provides such a convenient signal. Its area is given by the amplitude of the input step multiplied by the time constant RC of the circuit. Moreover, since the signal is exponential in shape, its -3-dB spectrum bandwidth is equal to $1/RC$. In the circuit of Fig. 16-13, R_1 is much larger than R; when the switch S is open, the capacitor charges to the voltage E of the dc source. When the switch is suddenly closed, the capacitor discharges through R, producing an exponential signal of known amplitude and duration (known area).

A circuit based on the same principle is shown in Fig. 16-14. Here the coaxial line plays the role of energy storage source. If the line is lossless, its characteristics impedance is given by R_0, the propagation delay is equal to τ, and the Laplace transform of the voltage drop across R is

$$V(p) = (1/p)E[1 - (R_0/R)\coth p\tau]^{-1} \tag{16-8}$$

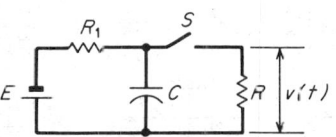

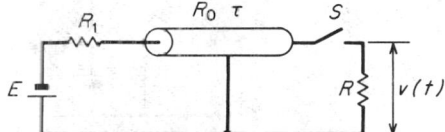

Fig. 16-13. RC pulse-generator circuit with large series resistance R.

Fig. 16-14. Coaxial-cable version of RC pulse generator.

When the line is matched to the load, Eq. (16-8) reduces to

$$V(p) = (1/2p)E(1 - e^{-p\cdot\tau}) \tag{16-9}$$

which indicates that $V(t)$ is a square wave of amplitude $E/2$ and duration 2τ. The area of the pulse is equal to the product of E and the time constant τ; both quantities can be kept reasonably constant. Since the bandwidth of this circuit is larger than that of an exponential pulse of the same area (Fig. 16-13) by the factor π, this generator is particularly suitable for very high frequency applications.

Very wide bandwidth pulse generators based on this principle use a coaxial mercury-wetted switch built into the line (Fig. 16-15) to achieve low standing-wave ratios. A bandwidth of several gigahertz can be obtained in this manner.

In the coaxial circuits, any impedance mismatch causes reflections to occur at both ends of the line, replacing the desired square-wave signal by a succession of steps of decreasing amplitude. The cutoff frequency of the spectrum is lowered thereby, and its shape above the uniform part is drastically changed. Below cutoff frequency, however, the spectrum amplitude is given by $E\tau R/R_0$.

When the finite closing time of the switch is taken into account, it can be shown that only the width of the spectrum is reduced without affecting its absolute value below the cutoff frequency. Stable calibrated pulse generators can also be built using electronic instead of mechanical switches.

10. Nonlinear-Passive-Network Waveshaping. Nonlinear passive networks offer wider possibilities for waveshaping than linear networks, especially when energy-storage elements such as capacitors or inductors are used with nonlinear devices. Since the analysis of the behavior of such circuits is difficult, we first consider purely resistive nonlinear circuits.

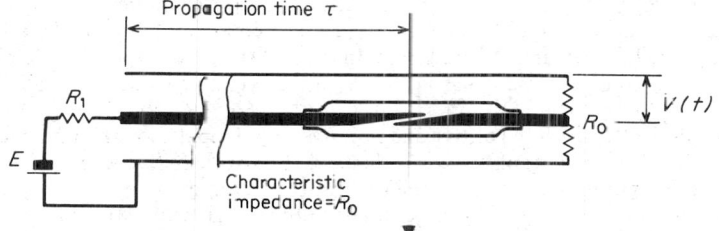

Fig. 16-15. Use of mercury-wetted switch contacts in coaxial pulse generator.

11. Resistive Clamping Circuits. Diodes provide a simple means for clamping a voltage to a constant value. Both forward conduction and avalanche (zener) breakdown are well suited for this purpose. Avalanche breakdown usually offers sharper nonlinearity than forward biasing, but the latter consumes less power.

Clamping action can be obtained in many different ways. The distinction between series and parallel clamping is shown in Fig. 16-16. Clamping occurs in the first case when the diode conducts; in the second it is blocked.

Since the diode is not an ideal device, it is useful to introduce an equivalent network that takes into account some of its imperfections. The complexity of the equivalent network is a trade-off between accuracy and ease of manipulation.

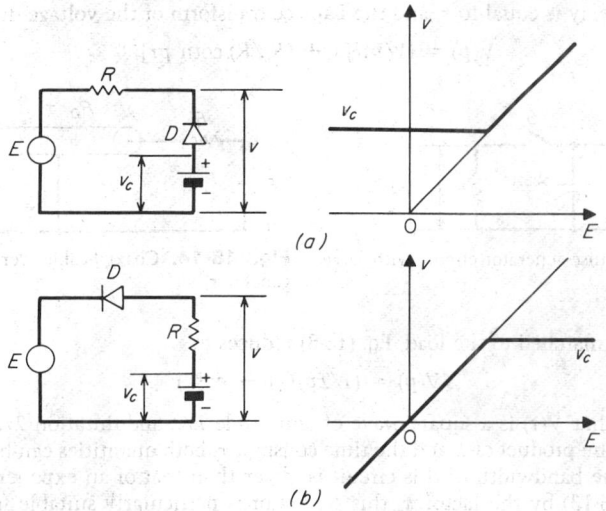

Fig. 16-16. Diode clamping circuit and voltage vs. time responses: (a) shunt diode; (b) series diode.

The physical diode is characterized by

$$I = I_s[\exp(V/V_T) - 1] \qquad (16\text{-}10)$$

where I_s = leakage current and $V_T = kT/q$, typically 26 mV at room temperature.

The leakage current is usually quite small, typically 100 pA or less. Therefore, V must be at least several hundred millivolts, typically 600 mV or more, to attain values of forward current I in the range of milliamperes. A first approximation of the forward-biased real diode consists therefore of a series combination of the ideal diode and a small emf (Fig. 16-17). Moreover, to take into account the finite slope of the forward characteristic, a better approximation is obtained by inserting a small resistance in series. If the leakage current under reverse-bias conditions also is important, an additional improvement consists in introducing a current source in parallel.

12. Nonlinear Passive Networks with Storage Elements. There is no simple theory of the behavior of nonlinear circuits with storage elements, such as capacitors or inductances. Acceptable solutions can be found, however, by breaking the analysis of the circuit under investigation into a series of linear solutions. A typical example is the dc restorer circuit.

13. DC Restorer Circuit. The circuit shown in Fig. 16-18 resembles the RC differentiator but exhibits properties which differ substantially from those examined previously. The diode D is assumed to be ideal to simplify the analysis of the circuit, which is carried out in two steps, i.e., with the diode forward- and reverse-biased. In the first step, the output of the circuit is short-circuited; in the second, the diode has no effect, and the circuit is identical to the linear RC differentiator.

When a series of alternatively positive and negative steps is applied at the input, after the first positive step is applied, no output voltage is obtained. The first positive step causes a large transient current to flow through the diode and charges the capacitor. Since D is assumed to be an ideal short circuit, the current will be close to a delta function (Par. **16-8**) as long as the internal impedance of the generator connected at the input is small.

In practice, the finite series resistance of the diode must be added to the generator internal impedance, but this does not affect the load time constant significantly, since it is smaller than the time between the first positive step and the following negative step. This allows the circuit to attain the steady-state conditions between steps. When the input voltage suddenly returns to zero, the output voltage undergoes a large negative swing whose amplitude is equal to that of the input step. The diode is then blocked, and the capacitor discharges slowly through the resistor. If the time constant is assumed to be much larger than the period of the input wave, when the second positive voltage step is applied, the output voltage swings back to zero and only a small current flows through the forward-biased diode to restore the charge lost when the diode was under the reverse-bias condition.

If the finite resistance of the diode is taken into consideration, a series of short positive exponential pulses must be added to the output signal, as shown in the lower part of Fig. 16-13. The first pulse, which corresponds to the initial full charge on the capacitor, is substantially higher than the following pulse, but this is of little importance in the operation of the circuit.

An interesting feature of the dc restorer circuit lies in the fact that although no dc component can flow from input to output, the output signal has a well-defined level, although determined only by the amplitude of the negative steps (assuming of course that the lost charge between two steps is negligible). This circuit is used extensively in video systems to prevent the average brightness level of the image from being affected by its varying video content. In this case, the reference steps are the line-synchronizing pulses (see Par. **20-11**).

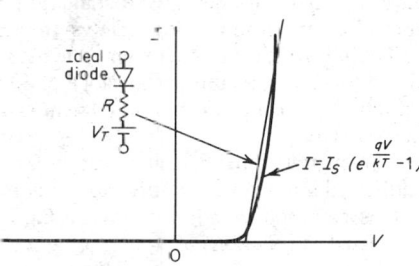

Fig. 16-17. Actual and approximate current-voltage characteristics of ideal and real diodes.

PASSIVE AND ACTIVE ELEMENTS USED AS SWITCHES

14. The Ideal Switch. An ideal switch is a two-pole device that satisfies the following conditions:

Closed-switch condition. The voltage drop across the switch is zero whatever the current flowing through the switch may be.

Open-switch condition. The current through the switch is zero whatever the voltage across the switch may be.

Mechanical switches are usually electrically

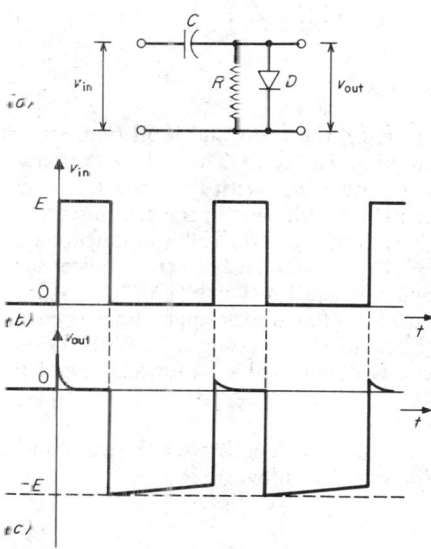

Fig. 16-18. (*a*) DC restorer circuit; (*b*) input signal; (*c*) output signal.

ideal, but they suffer from other drawbacks; e.g., their switching rate is low, and they exhibit jitter. Moreover bouncing of the contacts may be experienced after closing, unless mercury-wetted contacts are used. Electronic switches do not exhibit these effects, but they are less ideal in their electrical characteristics.

15. Bipolar-Transistor Switches (Static Characteristics). The bipolar transistor approximates an open switch between emitter and collector when its base terminal is open or when both junctions are reverse-biased or even only slightly forward-biased. Inversely, under saturated conditions, the transistor resembles a closed switch with a small voltage drop in series, typically 50 to 200 mV. This drop may be considered negligible in many applications.

A rigorous approach to the transistor static switching characteristics is based on the Ebers and Moll transport equations

$$\begin{bmatrix} I_E \\ \\ I_C \end{bmatrix} = I_s \begin{bmatrix} -\dfrac{1}{\beta_F} - 1 & 1 \\ \\ 1 & -1-\dfrac{1}{\beta_R} \end{bmatrix} \begin{bmatrix} \exp(V_E/V_T) - 1 \\ \\ \exp(V_C/V_T) - 1 \end{bmatrix} \qquad (16\text{-}11)$$

where V_E, V_C = voltage drops across emitter and collector junctions, respectively (positive voltages stand for forward bias, negative for reverse bias); $V_T = kT/q$ (typically 26 mV at room temperature); I_S = saturation current; and β_F, β_R represent forward (I_C/I_B) and reverse (I_E/I_B) current gains, respectively, with $V_E > 0$, $V_C < 0$ in the first case and $V_E < 0$, $V_C > 0$ in the second.

The saturation current I_S governs the leakage current flowing through the transistor under blocked conditions. It is always exceedingly small, and since it usually amounts to 10^{-14} or 10^{-15} A. it is difficult to measure. A standard procedure is to determine I_S from the plot representing the collector current in log scale vs. the emitter forward bias V_E in linear scale.

To find the saturation current, one must extrapolate the part of the curve which can be assimilated to a straight line with a slope of 60 mV/decade to the intercept with the vertical axis for which $V_E = 0$. The saturation current can also be obtained with emitter and collector terminals permutated.

The current gains β_F and β_R can be evaluated by means of the same experimental setup. An additional ammeter is required to measure I_B.

It is common practice to rewrite Eq. (16-11) so that the emitter and collector currents are expressed as functions of

$$I_F = I_S[\exp(V_E/V_T) - 1] \tag{16-12}$$
and
$$I_R = I_S[\exp(V_C/V_T) - 1] \tag{16-13}$$

With these definitions, Eq. (16-11) can be expressed as

$$I_C = I_F - I_R - I_R/\beta_R \tag{16-14}$$
$$I_E = -I_F - (I_F/\beta_F) + I_R \tag{16-15}$$

Hence, the Ebers and Moll transport model is found. This is illustrated by the equivalent circuit of Fig. 16-19. The leakage currents of the two diodes D_1 and D_2 are given respectively by I_S/β_F and I_S/β_R. With this model, it is possible to compute the currents flowing through the transistor under any of the circumstances considered in the beginning of Par. **11-16**.

For instance, if the collector junction is reverse-biased and a small positive bias of, for example, $+100$ mV is established across the emitterjunction, the reverse current I_R is almost equal to $-I_S$ and I_F is equal to $I_S \exp(100/26)$ or $46.8 I_S$. Hence, from Eqs. (16-14) and (16-15), the collector current is found to be approximately equal to $48 I_S$, and the emitter current is approximately the same with opposite sign. With the assumption that I_S is equal to 10 pA, both I_C and I_E are essentially negligible. A fortiori, I_B as derived from Eqs. (16-14) and (16-15) is also small:

$$I_B = (I_F/\beta_F) + I_R/\beta_R \tag{16-16}$$

To drive current through the transistor, the voltage across one of the two junctions must reach the value given by

$$V_E = V_T \ln(I_F/I_S) \tag{16-17}$$
or
$$V_C = V_T \ln(I_R/I_S) \tag{16-18}$$

derived from Eqs. (16-12) and (16-13). Thus V_E, V_C, or both must reach at least 0.5 V, before current starts flowing through the transistor. Hence, in most applications, the voltage drop across a forward-biased junction amounts to 0.6 or 0.7 V.

Usually, the transistor operates in the saturation region when it must approximate a closed switch. The voltage drop between the emitter and collector terminals is then given by

$$V_{CE,sat} = V_T \ln \frac{n + (n + \beta_F)/\beta_R}{n - 1} \tag{16-19}$$

where n represents the ratio $\beta_F I_B/I_C$, assumed larger than 1. For most transistors, this voltage drop falls between 50 and 200 mV. The inevitable resistance in series with the collector increases this voltage by a few tens of millivolts.

A common situation arises when I_C is almost equal to zero; e.g., when a bipolar transistor is used to set the potential across a capacitor. In this case, Eq. (16-19) becomes

$$V_{CE,sat} = V_T \ln(1 + 1/\beta_R) \tag{16-20}$$

Similarly, with interchanged emitter and collector terminals, the voltage drop is given by

$$V_{EC,sat} = V_T \ln(1 + 1/\beta_F) \tag{16-21}$$

Since β_F usually is much larger than β_R, $V_{EC/sat}$ may be smaller than 1 mV provided β_F is at least equal to 25. Consequently, inverted bipolar transistors can be considered as switches with a very small series voltage drop, provided that the current flowing through the transistor is kept small.

The two situations examined so far (open or closed switch) correspond in Fig. 16-20 respectively to $I_B = 0$ and to the part of the curves superimposed and closely parallel to the collector-current axis. The fact that all the curves have nearly the same vertical shape means that the series resistance of the saturated transistor is quite small. Since the characteristics do not coincide with the vertical coordinate axis, a small series emf must be considered, as previously stated.

A third region of interest exists where the transistor plays the role of a current switch instead of a voltage switch. It concerns the switching from blocked conditions to any point within the active region or vice versa. Conceptually, the transistor can be compared to a controlled current

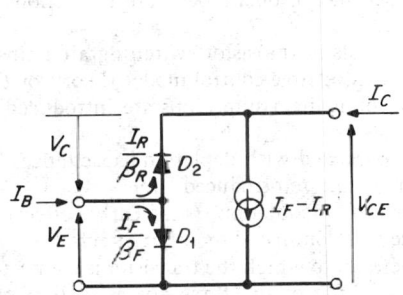

Fig. 16-19. Equivalent circuit of the Ebers and Moll transport model of the bipolar transistor.

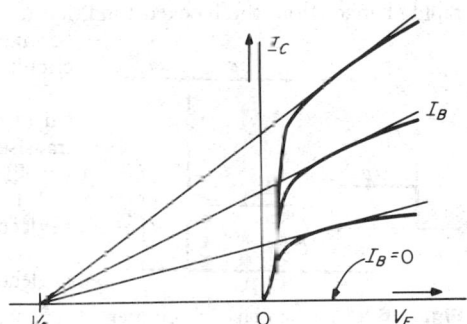

Fig. 16-20. Typical common-emitter characteristics of the bipolar transistor. V_A is called the Early voltage.

source which is switched on or off. However, because of the Early effect, the current cannot be constant when the transistor is on. The Ebers and Moll model is inappropriate to describe this effect. A better expression of I_C is

$$I_C = I_s \exp (V_E/V_T)(1 + V_{CE}/V_A) \tag{16-22}$$

where V_A is called the *Early voltage*. Equation (16-22) is illustrated in Fig. 16-20. The finite output conductance of the transistor is given by I_C/V_A.

16. Field-Effect-Transistor Switches (Static Characteristics). Junction field-effect transistors and insulated-gate field-effect transistors are often used as switches. Their static behavior is simpler to analyze than that of bipolar transistors. When the transistor is blocked, the residual current flowing through the input-output terminals is the sum of the leakage currents between junctions and the substrate plus a very small source-to-drain current due to the weak inversion under the gate region. These leakage currents typically range from picoamperes to nanoamperes.

When a field-effect transistor is turned on, its characteristics differ substantially from those of a bipolar switch. Since there is no residual emf in series, the transistor is comparable to a resistor whose conductance G is

$$G = \mu C_{ox}(W/L)(V_G - V_{T0} - \lambda V) \tag{16-23}$$

where μ = mobility in inversion layer, C_{ox} = gate inversion layer capacitance per unit area, W = width of channel, L = length of channel, V_G = gate-to-substrate voltage, V_{T0} = gate-threshold voltage under zero back bias, V = source voltage (assumed identical to drain voltage in order to verify closed-switch conditions), and λ = dimensionless coefficient which takes substrate effect into account (a satisfactory value of λ is 1.1). The mobility μ is not a constant since it is influenced by the back bias. A good approximation of μ is given by

$$\mu = \mu_0/[1 + \theta(V_G - V_{T0} - \lambda V)] \tag{16-24}$$

Here θ is the substrate sensitivity factor, which usually is around 0.05 to 0.04 V^{-1}. The remaining constant μ_0 represents the zero-back bias mobility. Usually the product $\mu_c C_{ox}$ is of interest rather than both coefficients separately, since this is a technology-dependent constant. For instance, if we consider an n-channel transistor with a 100-nm-thick oxide, the factor $\mu_0 C_{ox}$ is approximately equal to 20 μA/V^2.

The conductance of the MOS switch thus varies more or less linearly with the voltage applied to the gate. For instance, an n-channel transistor with a W/L ratio of 10 and a threshold voltage of 1 V, exhibits a conductance of 0.7 mS for $V_G = 5$ V, with source and drain at the substrate potential. For $V_G = 3$ V, the conductance amounts to 0.37 mS.

17. Dynamic Behavior of Bipolar Switches.[2-4] The imperfections of transistors with respect to their dynamic behavior have a more limiting effect than those for static behavior. This is due to the inevitable reactive contributions associated with the diffusion mechanism of minority carriers in the base region and with the junction depletion capacitances. These reactive contributions are nonlinear in essence, the first being a function of the current flowing through the base, the second depending on the voltage across the junctions. Since switching involves rapid changes from the blocked condition to saturation and vice versa, the nonlinearities must be taken into account to obtain accurate predictions of circuit behavior.

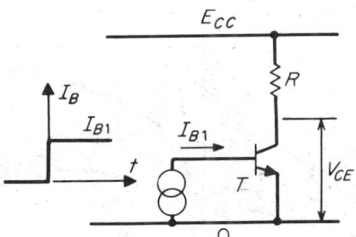

Fig. 16-21. Basic bipolar inverter stage.

The fundamentals of transistor switching are embodied in the so-called charge-control-model theory of the transistor. The following assumptions are introduced to simplify the analysis:

1. Effects associated with depletion capacitances are neglected initially and reintroduced later.

2. Base-width modulation (the Early effect) is neglected since it is dominated by other effects.

Because the circuit to which the transistor is connected controls the switching of the transistor as well as the physical characteristics of the device itself, a classical inverter circuit is considered (Fig. 16-21). The transistor is assumed to be blocked initially. If a current step of amplitude I_{B1} is suddenly applied to the base terminal, the transistor is driven in the active region first and in saturation thereafter, provided the base current I_{B1} is larger than

$$I_{B0} = \frac{1}{\beta_F} \frac{E_{cc}}{R} = \frac{I_{C,max}}{\beta_F} \tag{16-25}$$

This represents the amplitude of the base-current step which drives the transistor at the boundary between the two regions. As long as the transistor operates in the active region, the forward current I_F increases and the reverse current I_R remains at zero. Once saturation is reached, I_R starts increasing also.

In the charge-control model of the transistor, the important quantities to consider are the forward Q_F and reverse Q_R charges accumulated in the base. The assumption that Q_F and Q_R are proportional respectively to I_F and I_R is made:

$$Q_F = \tau_F I_F \tag{16-26}$$

$$Q_R = \tau_R I_R \tag{16-27}$$

where τ_F = forward transit time and τ_R = reverse transit time.

These time constants are usually quite different, since the charges in the base associated with forward and reverse injections differ substantially because of the asymmetry of planar transistors. The emitter-junction area is always much smaller than that of the collector junction; therefore Q_F is much smaller than Q_R. The proportionality between Q_F and I_F furthermore implies that Q_F will be proportional to the base current under steady-state conditions in the active region. The actual proportionality coefficient $\tau_F \beta_F$ is written τ_{BF} in this discussion. Similarly, under reverse bias, τ_{BR} is equal to $\beta_R \tau_R$.

Under transient conditions, the forward current I_F is controlled by Q_F. Hence, the forward charge Q_F increases or decreases by some charge whose derivative represents the transient current supplied by the base terminal. As long as the transistor is in the active region, one can write:

$$I_{B1} = \frac{Q_F}{\tau_{BF}} + \frac{dQ_F}{dt} \tag{16-28}$$

This first-order equation in fact considers the base to be a lumped capacitor. In reality, the base acts like a distributed line, but the effect can be neglected since the width of the base is always very small compared with the diffusion distance of minority carriers.

Solving Eq. (16-28) leads to

$$Q_c = \tau_{BF} I_{B1}[1 - \exp(-t/\tau_{BF})] \tag{16-29}$$

The rise time t_r of the collector current, defined as the time between the 10 and 90% levels of the steady-state collector current, is given by

$$t_r = 2.2\tau_{BF} \tag{16-30}$$

However, with the assumption that $I_{B1} > I_{B0}$, the transistor enters into saturation as soon as I_F is equal to $I_{C,\max}$ or E_{cc}/R. Hence, the collector current suddenly stabilizes before steady-state conditions are actually reached. The response time t_r is thus smaller than given by Eq. (16-30) and must be expressed as

$$t_r = \tau_{BF} \ln \frac{1 - 0.1/n}{1 - 0.9/n} \tag{16-31}$$

with n, the overdrive factor, defined as the ratio of I_{B1} to I_{B0}. When $n \gg 1$, the expression can be simplified

$$t_r \approx \tau_{BT} (0.8/n) \tag{16-32}$$

Once the transistor enters saturation, the collector junction also injects minority carriers in the base, since it is forward-biased as well as the emitter junction. Hence Eqs. (16-26) and (16-27) both hold simultaneously. The base current, I_{B1} no longer satisfies Eq. (16-28) because two additional terms must be added to take care of the inverted functioning mode:

$$I_{B1} = \frac{Q_F}{\tau_{BF}} + \frac{dQ_F}{dt} + \frac{Q_R}{\tau_{BR}} + \frac{dQ_R}{dt} \tag{16-33}$$

To solve this equation, another relation between Q_F and Q_R is needed which characterizes the saturated mode of operation. It suffices therefore to rewrite Eq. (16-14) with the assumption that I_C is constant and equal to $I_{C,\max}$ and to replace I_F and I_R by Q_F/τ_F and Q_R/τ_R according to Eqs. (16-26) and (16-27). The result is

$$I_{C,\max} = \frac{Q_F}{\tau_F} - \frac{Q_R}{\tau_R}\left(1 + \frac{1}{\beta_R}\right) \tag{16-34}$$

After derivation of the above expression and elimination of dQ_F/dt between Eqs. (16-33) and (16-34), the differential equation describing the behavior of Q_R is found:

$$\frac{I_{B1} - I_{C,\max}/\beta_F}{1 + (\tau_F/\tau_R)(1 + 1/\beta_R)} = \frac{Q_R}{\tau_{3s}} + \frac{dQ_R}{dt} \tag{16-35}$$

with

$$\tau_{ES} = \frac{(\tau_F + \tau_R)\beta_F\beta_R + \tau_R\beta_F}{1 + \beta_F + \beta_F} \tag{16-36}$$

The time constant τ_{ES} also governs Q_F since Q_F and Q_R are related by Eq. (16-34). Thus τ_{BS} is the only representative time constant of the saturated transistor. In this mode of operation neither τ_{BF} nor τ_{BR} can be considered separately because of the simultaneous injection in the base from the two junctions. The time constant τ_{BS} averages in some manner τ_{BF} and τ_{BR}, as is evident from Eq. (16-36), especially when $\beta_R > 1$, for which

$$\tau_{BS} \approx \frac{\tau_F + \tau_R}{(1/\beta_F) + 1/\beta_R} \tag{16-37}$$

To summarize, Eqs. (16-28) and (16-35) describe the transient behavior of the transistor in the active region first and then in saturation. The fact that the collector current I_C as well as V_{CE}, remains constant once saturation is reached does not mean that steady-state conditions are met, inasmuch the internal charge distribution within the base increases exponentially with the time constant τ_{BS}.

For example, consider a typical switching transistor with $\beta_F = 80$; $\beta_R = 2$; $\tau_F = 0.1$ ns; and $\tau_R = 10$ ns. For this transistor $\tau_{BF} = 8$ ns, $\tau_{BR} = 20$ ns, and $\tau_{BS} = 19.6$ ns. Applying an overdrive factor n of 10 yields a rise time t_r as short as 0.67 ns, which is negligible compared with the time needed to read steady-state conditions in the saturated mode of operation. This time is equal to 2.2 τ_{BS}, or 43 ns in the present case, but the value of t_r obtained is unrealistic because the transition capacitances have not yet been taken into account. We now introduce them in an attempt to predict the actual rise time.

Of the two transition capacitances associated with the emitter and collector depletion layers, respectively called C_{TE} and C_{TC}, only the latter has a substantial effect on the switching speed of the inverter stage. The charge accumulated in the emitter transition capacitance, in fact, does not vary substantially because the emitter-junction bias is almost constant in the active region as well as in saturation. This is not true, however, when the transistor is blocked before turning on, for the emitter junction is reverse-biased before switching. When this is the case, a small additional delay is introduced, during which the emitter-junction voltage increases until it reaches 0.6 to 0.7 V in order to enter in the active region.

Once the transistor enters this region, the collector voltage drops rapidly from E_{cc} to almost 0 V. Hence, a large current flows through the collector transition capacitance (also called the Miller capacitance). This current represents an appreciable part of I_{B1}, and the current available to turn the transistor on is substantially reduced. The actual base current I_B of the intrinsic transistor can easily be computed from

$$I_B = I_{B1} - RC_{TC}\frac{dI_F}{dt} \tag{16-38}$$

or

$$I_B = I_{B1} - \frac{RC_{TC}}{\tau_F}\frac{dQ_F}{dt} \tag{16-39}$$

In order to determine the correct rise time t_r, it is sufficient to replace I_{B1} in Eq. (16-28) by I_B as given by Eq. (16-39). Hence, the following differential equation is found for Q_F:

$$I_{B1} = \frac{Q_F}{\tau_{BF}} + \left(1 + \frac{RC_{TC}}{\tau_F}\right)\frac{dQ_F}{dt} \tag{16-40}$$

Thus, instead of Eq. (16-29), one gets

$$Q_F = \tau_{BF}I_{B1}[1 - \exp(-t/\tau'_{BF})] \tag{16-41}$$

where

$$\tau'_{BF} = \tau_{BF}(1 + R_C C_{TC}/\tau_F) \tag{16-42}$$

Hence, Eqs. (16-30) and (16-31) must be corrected, in the same manner, to find the correct rise time t_r.

The drastic effect of the Miller capacitance can be made clear with the example considered above, considering a collector transition capacitance C_{TC} of 0.5 pF, typical of an integrated switching transistor, and a collector resistance R of 1 kΩ. From Eq. (16-42) the time constant τ'_{BF} is found to be 6 times larger than τ_{BF}. Hence, t_r is also 6 times larger and reaches the more realistic value of 4.0 ns.

No correction is needed with respect to the saturation time because the voltages across both junctions then remain constant.

Having examined the turn-on mechanism of the bipolar inverter, we now discuss the turnoff, assuming that enough time has elapsed to allow steady-state conditions to be reached in the base. The base current which controls the turnoff of the transistor is assumed to be a step function of amplitude I_{B2} whose polarity is opposed to that of I_{B1}. The analysis of the turnoff mechanism is divided into two parts: the desaturation of the transistor and the active region. During the first part, the base charges Q_F and Q_R decay simultaneously, and the transistor enters into the active region at the precise moment when Q_R is equal to zero. Hence, Eq. (16-35) can be used to evaluate the time t_s needed to make Q_R equal to zero. Straightforward calculations lead to

$$t_s = \tau_{BS}\ln\frac{m - n}{m - 1} \tag{16-43}$$

where m is the underdrive factor, defined as the ratio of I_{B2} to I_{B0} (notice that m and n have opposite signs).

Once desaturation is achieved, the transistor enters the active region described by Eq. (16-40). Hence, the decay time t_d is given by

$$t_d = \tau_{BF}\ln\frac{1 - 0.9/m}{1 - 0.1/m} \tag{16-44}$$

To cast light on the magnitude of the delay t_s, the above example is used again, assuming that a current I_{B2} equal to $-5I_{B0}$ is extracted from the base to desaturate the transistor. Hence, with $m = -5$ and $n = +10$, t_s is found to be 18 ns and t_d to be 7.0 ns.

The actual switching of the transistor is illustrated in Fig. 16-22 with the corresponding control base current. The fact that I_B decays to zero after V_{CE} has reached 5 V is not consistent with the assumption that I_B is delivered by a current source. Usually, this is not the case, however, for I_B is normally supplied by a voltage generator with a large series resistance. Once the transistor is blocked, no current flows in the base terminal. The generator can no longer be considered then as a current source. This explains why I_B decays to zero with a time constant approximately equal to the product of the base generator resistance and the sum of the transition capacitances.

To determine experimentally the parameters to predict the switching times of a bipolar inverter, only two time constants are required, for instance, τ_{BF} and τ_{BS}. The time constant τ_{BR} is not needed if τ_{BS} is known; moreover, its correct measurement is difficult. The measurement of the forward time constant τ_{BF} is simple; e.g., it can be determined directly from the response to a small current step in order to stay within the active region. However, this measurement does include the Miller capacitance, and thus it is τ_{BF} which is actually determined. To obtain τ_{BF} it is necessary to measure the average value of the collector transition capacitance C_{TC} with a high-frequency bridge.

Last, the saturation time constant must be evaluated. One method consists in measuring several desaturation times t_s vs. various factors n and m and to plot t_s vs. $\log[(m-n)/(m-1)]$. A straight line crossing the origin should be obtained according to Eq. (16-43). Its slope represents τ_{BS}.

18. Dynamic Switching of Diodes. The inverse base current, due to the mechanism of minority-carrier storage, also exists in diodes. This effect causes the reverse-biased conventional diode to act like a short circuit for a very short time after forward conduction. This time is a monotonic function of the amplitude of the forward current. Desaturation times of junction diodes usually range in nanoseconds.

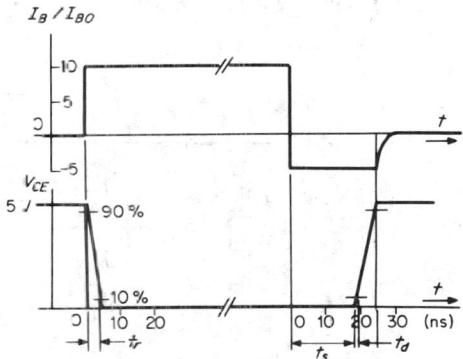

Fig. 16-22. Dynamic behavior of the bipolar inverter. The upper trace represents the base current, the lower trace the collector-to-emitter voltage. The desaturation time t_s is clearly visible.

Schottky barrier diodes, on the other hand, operate by field-emission effects and their storage times are exceedingly small. An interesting application of Schottky barrier diodes is shown in Fig. 16-23. In this circuit, the Schottky barrier diode is reverse-biased at all times when the transistor approaches the saturation region. Once this happens, the collector-to-emitter voltage drops below the base-to-emitter voltage by about 0.4 V. This is sufficient to make the Schottky diode conduct. The corresponding collector-to-emitter voltage of the transistor is still around 0.3 V, and saturation is avoided. In fact, the Schottky diode and the collector-junction diode D_2 of Fig. 16-19 are put in parallel.

Since the voltage across the Schottky diode is small, D_2 can never conduct. Hence, the reverse charge Q_R is always negligible, and the desaturation time t_s is almost zero whatever the base overdrive factor may be. The base current above I_{BC}, in fact, is derived through the Schottky barrier diode to ground by means of the transistor itself. Figure 16-24 shows that the Schottky diode can be implemented, for a planar npn transistor, by extending the base contact over the collector region to take advantage of the Schottky diode formed by the aluminum to the lightly doped n-type collector.

19. Dynamic Behavior of Field-Effect Switches.[5,6] Large-scale integration technology takes advantage of the possibilities offered by insulated-gate field-effect transistors

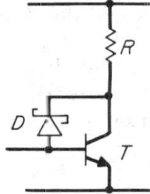

Fig. 16-23. A Schottky diode D prevents T from going into saturation.

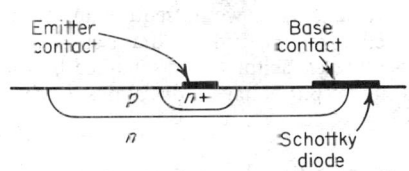

Fig. 16-24. Planar npn transistor and Schottky diode in integrated-circuit form.

(IGFET), such as MOS aluminum and silicon gate transistors (see Sec. **8**). Their small size and ease of fabrication make them attractive for producing complex digital systems on a single chip. However, their rather low speed, compared with bipolar transistors, is a drawback.

In examining the dynamics of FET switches, we again consider an inverter stage. The actual inverting transistor is always an enhancement-mode transistor to avoid the second power supply which would be necessary if the inverter were a depletion-mode transistor. The output load is formed by another FET, like that shown in Fig. 16-25, since the silicon area ("real estate") needed for such a transistor is far less than that needed by an equivalent resistor. A 20-kΩ resistor, for instance, made from a typical 200 Ω/□ diffusion requires 100 squares; i.e., if the section of the diffused ribbon is 6 μm, a total length of 0.6 mm is required, whereas a 20-μm square is sufficient for a FET having the same resistance.

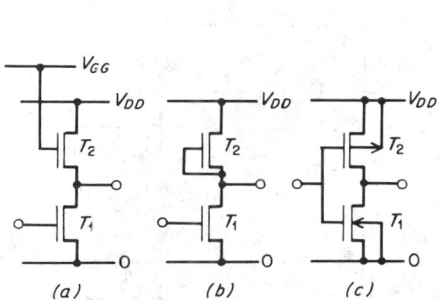

Fig. 16-25. Three types of MOS inverters: (a) T_1 and T_2 are two enhancement transistors, p- or n-channel; (b) T_1 is an enhancement transistor and T_2 a depletion load, p- or n-channel; (c) CMOS inverter.

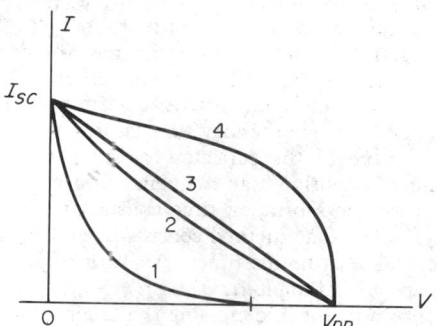

Fig. 16-26 Load characteristics of the MOS inverter; curve 1 = enhancement-mode transistor with single power supply; curve 2 = enhancement-mode transistor with two power supplies; curve 3 = resistive load; curve 4 = depletion-mode transistor.

The load transistor may be either enhancement-mode or depletion-mode (Fig. 16-25a and b). In Fig. 16-25a the gate is connected to V_{GG} or to the drain supply V_{DD}, the latter resulting in substantially longer switching times, however. Figure 16-25b represents a good trade-off between switching time and power consumption. However, since both enhancement and depletion load transistors must be realized on the same chip, ion implantation must be used. Both circuits in Fig. 16-25a and b can be implemented in either p- or n-channel technology. Figure 16-25c represents the complementary MOS inverter (CMOS); the two transistors T_1 and T_2 are enhancement-mode transistors with n and p channels, respectively.

In analyzing the switching of the MOS inverter, parasitic capacitances must be taken into consideration. The source and drain are in fact reverse-biased junctions exhibiting a capacitive coupling with respect to the substrate. The gate also is capacitively linked to the substrate and to the source. The gate-drain capacitance (the Miller capacitance) is particularly important in respect to the switching speed. All these capacitances are transition capacitances. Their effect upon the switching of the inverter stage completely overrules the effects associated with the dynamic behavior of the inversion layer. Indeed the propagation time in the channel is always one or two orders of magnitude smaller than the actual switching time of the inverter stage and can thus be neglected without difficulty.

We will divide our analysis into two parts: switching the inverter transistor starting from blocked conditions and the opposite situation. We assume that a voltage step V_G is applied to the gate of T_1. No sudden change of output voltage is experienced after the input step has been applied to the blocked transistor T_1, since instantaneous unloading of the lumped output parasitic capacitance C would require an infinite current pulse. The inverter T_1 thus is driven into the active region, and the drain current is almost independent of the drain-to-source voltage. Hence the capacitor C is discharged by the more or less constant current flowing through T_1, and the output voltage drops almost linearly. With the drain current given by

$$I_{DI} = \mu C_{ox}(W/L)(V_G - V_{TE})^2/2\lambda \qquad (16\text{-}45)$$

the response time t_r is approximately given by

$$t_r = (C/I_{DI})V \qquad (16\text{-}46)$$

where the subscript I is relative to the enhancement-mode transistor T_1, V_{TE} is its threshold voltage, and V is the output voltage of the inverter stage before switching. Equation (16-46) is written on the assumption that the current flowing through the load transistor is small compared with the actual drain current of the inverter T_1. Furthermore, the so-called "linear region" of T_1, which corresponds to small drain-to-source voltages, is not considered since it does not influence t_r substantially.

The switching time t_r usually is small, typically 10 to 50 ns, depending on the actual value of the parasitic capacitance C and on the characteristics of the inverter transistor T_1. In fact, most MOS inverters, except CMOS, suffer more from long switching-off times than from exaggerated t_r times. We therefore focus our analysis on the switching off of the inverter stage.

When T_1 is turned off, we assume that the current I_D provided by T_1 instantaneously drops to zero. Now the output parasitic capacitance C must be loaded by means of the load transistor only. The characteristics of the load transistor are illustrated in Fig. 16-26, in relation to circuits in Fig. 16-25a and b. To compare their performances, load transistors having the same short-circuit I_{SC} current are considered. The short-circuit current is defined as the current which flows through the transistor when the output voltage V_e is made equal to zero.

It is obvious that the shortest loading time of C will be obtained with the device delivering a current that stays more or less constant with increasing voltage V_o. This is achieved when the load transistor is a depletion-mode transistor with the gate connected to the source, as shown in Fig. 16-25b. In this case, in fact, the current I_{D2} flowing through the load transistor is not strongly influenced by the drain voltage as long as the transistor is saturated (saturation of MOS transistors corresponds to the part of the characteristics where the drain current is almost horizontal in the diagram of I_D vs. drain-to-source voltage; confusion with the saturated region of bipolar transistors must be avoided). The expression of the drain current I_{D2} for Fig. 16-25b is

$$I_{D2} = \mu C_{ox}(W/L)_2 \, [V_o(1 - \lambda) - V_{TD}]^2/2\lambda \tag{16-47}$$

where the subscript L relates to the depletion-load transistor T_2 and V_{TD} represents its threshold voltage. The dimensionless factor λ, defined in Par. **16-16**, simulates the substrate effect; it is influenced by the device technology, mainly by the substrate impurity concentration. It was mentioned earlier that λ falls generally between 1.1 and 1.2. The difference $1 - \lambda$ is thus not very well known. This is always the problem with depletion-mode transistors operating as loads. Unfortunately no simple model of depletion transistors is available.

The enhancement-mode load transistor (Fig. 16-25a) can be analyzed along the same lines. The much greater sensitivity of the drain current I_{D2} than V_o is due to the fact that the gate is not connected to the source. The actual gate-to-source voltage thus decreases while V_o increases. Unfortunately it is not possible to connect the gate as in Fig. 16-25b, for T_2 is an enhancement-mode transistor; its drain current would be zero.

The collector current I_{D2} of T_2 in Fig. 16-25a is given by

$$I_{D2} = \mu C_{ox} \left(\frac{W}{L} \right) \, [V_{GG} - V_{TE} - \frac{\lambda}{2}(V_{DD} + V_o)](V_{DD} - V_o) \tag{16-48}$$

where V_{TE} represents the threshold voltage of the enhancement-load transistor. It is assumed that T_2 is in the linear, or so-called unsaturated, region. The situation where T_2 is saturated also exists, especially when the gate of T_2 is connected to V_{DD}. Under these circumstances, the current I_{D2} is given by

$$I_{D2} = \mu C_{ox}(W/L)_L(V_G - V_{TE} - V_o)^2/2\lambda \tag{16-49}$$

Notice that in this case the output voltage V_o cannot reach V_{DD} but levels off at $(V_{DD} - V_{TE})/\lambda$. The curves shown in Fig. 16-26 represent Eqs. (16-47) to (16-49). The straight line corresponds to a purely resistive load.

Evaluation of the actual switching-off time, can be achieved by solving

$$dt = C \, dV_o/I_{D2}(V_o) \tag{16-50}$$

The expression $I_{D2}(V_o)$ must be chosen according to the circuit considered. One way to proceed is first to define a time constant τ which is equal to the product of C times a pseudo resistor R. This resistor represents the ratio of the maximum output voltage V_o divided by the short-circuit current I_{SC}. The maximum of V_o is always given by V_{DD} except when the gate of the enhancement-mode load transistor is connected to V_{DD} [case (a) without V_{GG}]. In the latter situation, the source voltage of T_2 is kept below V_{DD} to allow the very small leakage current of the junctions connected to the output node to flow through the transistor.

With the assumption made above that all load transistors have the same short-circuit current, the resistor R thus is the same, except for the last case. The time constant τ is the same, and a comparison of the switching performances of the circuits in Fig. 16-25a and b is easy. For Fig. 16-25b Eq. (16-50) yields

$$t = \tau \frac{1}{1 - (1 - \lambda)V_o/V_{TD}} \qquad (16\text{-}51)$$

and the switching time can be computed by making V_o equal to V_{DD}. The circuit in Fig. 16-25a leads to

$$t = \tau \frac{1}{1 - m} \ln \frac{1 - mV_o/V_{DD}}{1 - V_o/V_{DD}} \qquad (16\text{-}52)$$

where

$$m = \frac{\lambda V_{DD}}{2(V_{GG} - V_{TE}) - \lambda V_{DD}} \qquad (16\text{-}53)$$

The time needed for V_o to go from $0.1V_{DD}$ to $0.9V_{DD}$ can be determined from the above expressions. Finally, when the gate of T_2 is connected to V_{DD}, one obtains

$$t = \tau \frac{V_o/V_{o,\text{max}}}{1 - V_o/V_{o,\text{max}}} \qquad (16\text{-}54)$$

where $V_{o,\text{max}}$ is given by $(V_{DD} - V_{TE})/\lambda$.

In a typical example, an MOS load transistor is considered with a short-circuit current of 50 μA. If the supply voltage is 5 V, as it usually is, resistor R is equal to 100 kΩ. With an average value of 0.5 pF for the lumped parasitic capacitor C, a time constant τ of 50 ns is obtained. We consider a threshold voltage of $+ 1$ V for the enhancement-mode and -2 V for the depletion-mode transistor and a typical λ of 1.1. The switching-off time computed by means of Eq. (16-51) is 1.3τ or 65 ns, for the inverter with a depletion load (case b). With Eqs. (16-52) and (16-53), the switching time amounts to 2.8τ or 140 ns between 10 and 90% levels of the steady-state output voltage. This corresponds to the enhancement-mode load (case a) considering $V_{GG} = 12$ V. Finally, for the enhancement-mode load with a single supply of 12 V, Eq. (16-54) yields a switching time of 3.9τ, or 200 ns. Assuming that V_o changes from 10 to 80% levels of the steady-state value $V_{o,\text{max}}$, this is equal to 10 V in the present case.

It is obvious that the maximum switching frequencies of the various inverters are quite different. If we define the maximum switching frequency as the reciprocal of 4 times the switching time, a value of 4 MHz is found for the n-channel inverter with the depletion load, 2 MHz for the inverter with the enhancement-mode load and a double power supply, and only 1.3 MHz if a single power supply is used. The performances of the last circuit are thus poor. Not only is the speed poor but the power consumption is also higher than in the two first cases. The most satisfactory performance is obtained with the circuit in Fig. 16-25b, since it does not require two power supplies like the circuit in Fig. 16-25a.

It was stated earlier that the switching-on time is not a limiting factor. In order to illustrate this, let us evaluate the current I_{DI} flowing through the inverting transistor T_1, assuming that V_G reaches 5 V and that the aspect ratio W/L of the inverter is equal to 1. Hence, I_{DI} determined from Eq. (16-45) is equal to 150 μA with μC_{ox} given by the same value as in Par. **16-16**. The switching-on time computed from Eq. (16-46) is equal to approximately 17 ns. This is smaller than the shortest time found for switching off. Consequently, in a chain of MOS inverters, only the switching times given by Eqs. (16-51), (16-52), and (16-54) should be considered.

A different conclusion must be drawn with respect to the CMOS inverter of Fig. 16-25c. Here, both transistors act like inverting enhancement-mode transistors. Thus the switching times are governed by the turn-on characteristics since the CMOS inverter behaves like two complementary switches, one always off, the other on. The switching times are given by Eqs. (16-45) and (16-46). Since the mobility of holes is approximately 3 times smaller than that of electrons, the switching time of n-channel transistors is approximately 3 times shorter than that of p-channel transistors. Unequal ratios (W/L) can therefore be chosen.

Considering the previous example, a maximum switching frequency of 10 to 15 MHz is found for the CMOS inverter. This excellent figure is obtained at the price of increased silicon real estate (p wells must be implemented in the n-type substrate) and a more elaborate technology. An important property of CMOS inverters is the absence of dc consumption, a feature which is not shared by the circuit in Fig. 16-25a or b. This makes CMOS very attractive for micropower applications such as solid-state wristwatches.

20. Speed vs. Power.[7] The power consumption of inverter stages results both from the direct current delivered by the power supply and from the repetitive discharging process of the unavoidable parasitic capacitor C, which is present at each input-output node. If the switching

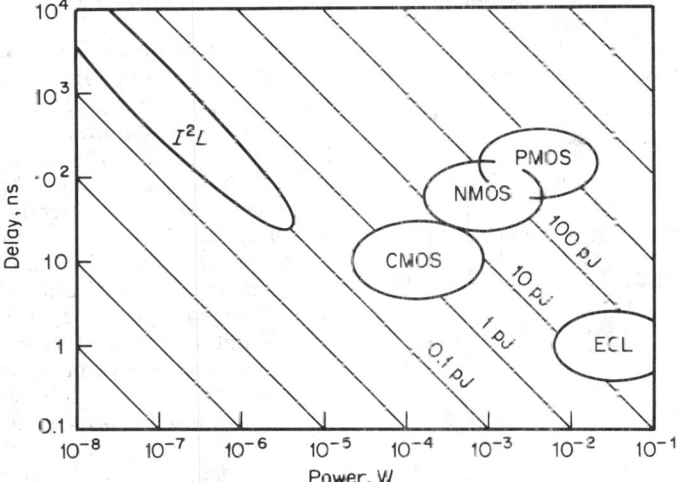

Fig. 16-27. Propagation delay vs. power dissipation for various inverter stages.

frequency is represented by f and a mark-space ratio of 1 is considered, the average power consumption of the inverter stage is given by

$$P = (\tfrac{1}{2}I_{cc} + CV_{DD}f)V_{DD} \tag{16-55}$$

The factor in parentheses represents the average current flowing through the inverter stage.

In the inverter stages considered in Par. **16-19**, the static power consumption $\tfrac{1}{2}I_{cc}V_{DD}$ is equal to 125 μW, and the dynamic power consumption CV_{DD}^2f equals 62.5 μW, for a switching frequency of 5 MHz. Since CMOS logic has virtually no static consumption, it has a unique advantage at low or medium frequencies. When the switching frequency approaches several megahertz, however, the power consumption of all MOS inverter families tends to be dominated by the dynamic power. Hence, they are approximately the same. The only way to get the power consumption down is to use smaller power-supply voltages and/or scale down the MOS transistor sizes.

The propagation delay vs. power provides a good means of comparing the performances of various inverter stages. The propagation delay is defined as the minimum time between transitions in a chain of inverters. Generally, one considers an uneven number of inverters in a closed-loop chain to obtain a self-oscillatory system. It is easy to determine both the propagation delay and the power consumption.

A good figure of merit for an inverter is its energy consumption per cycle, obtained by dividing P by f and generally expressed in picojoules.

In the inverter stages considered earlier, the minimum energy per cycle corresponds to repetition rates for which the dynamic power is larger than the static-power consumption. Under these circumstances, the minimum energy per cycle for the MOS inverter that was considered above is equal to CV_{DD}^2, or 12.5 pJ. Constant-energy-per-cycle loci are straight lines in the delay-vs.-power diagram. An illustration is given in Fig. 16-27, in which several logic families (inverters) are compared. It is obvious that bipolar inverters are faster than MOS inverters, but their power consumption is always higher too. Among the members of the MOS family, n-channel inverters are faster than p-channel because of the superior mobility of electrons over holes.

21. Logic Gates (Bipolar).[8,9] The most important application of inverters is logic gates, and in this field, integrated circuits have almost completely eliminated the discrete-component-circuit approach. We restrict our discussion to integrated-circuit logic gates. Among the main families of monolithic bipolar integrated-circuit logic gates, we will examine the resistor-transistor logic (RTL), emitter-coupled logic (ECL), and transistor-transistor logic (TTL or T²L).

RTL Gates.   Resistor-transistor logic is an extension of the classical inverter stage, using several parallel-input transistors instead of one (Fig. 16-28). The voltage at terminal F can be high, e.g., logic 1, only if and when all input transistors are blocked simultaneously. Any other combination, with one or more transistors saturated, leads necessarily to a low output voltage (logic 0). The logic function this circuit performs therefore is the NOR function ($F = \overline{A + B + C}$) for three-input gate considered in Fig. 16-28.

The single-input inverter stage also provides a logical function, since it negates its input signal. For instance, if we consider the second inverter in Fig. 16-28, the output signal this stage delivers negates, or complements, the previous NOR function, providing an OR output ($F = A + B + C$). Logic gates are interconnected to provide a variety of desired logic functions. Appropriate synthesis procedures are currently available (see Secs. **8** and **23**).

A fundamental requirement on all logic circuits is that any output terminal be able to feed several gates of the same logic family. The upper limit to the number of loading gates allowable for a single output terminal is known as the *fan-out* capability of the circuit. We have noted the connection between the NOR and the OR terminals of the RTL logic circuit represented in Fig. 16-28, a typical example of cascading of two identical stages. When the NOR output signal is low, V_F is of the order of magnitude of $V_{CE,sat}$, blocking the second inverter stage and preventing current from flowing through R_B. When the same output is high, V_F is fixed by the attenuator formed by R_C and R_B, and by E_{cc} and the emitter voltage drop V_{BE} of the second inverter

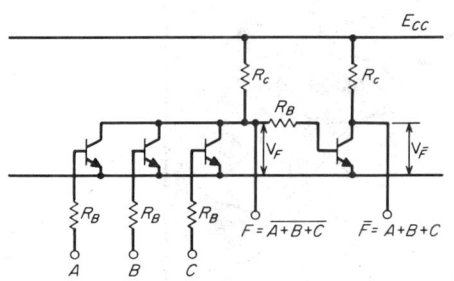

Fig. 16-28. RTL gates with NOR and OR output.

$$V_F = V_{BE} + \frac{R_B}{R_B + R_C}(E_{cc} - V_{BE}) \tag{16-56}$$

If R_B were equal to zero, the output voltage V_F would be equal to V_{BE} and the logical swing, i.e., the difference between the logic 1 and logic 0, would be very small, namely, the difference between V_{BE} and $V_{CE,sat}$ (typically 400 mV).

This situation leads to increased noise susceptibility and reduces the acceptable tolerance on the logic 1, since V_{BE} may vary by a few tens of millivolts from junction to junction. Hence, taking R_B equal to a few hundred ohms (typically 200 Ω to 1 kΩ) yields a larger logic swing, better noise immunity, and smaller tolerance on the logic 1.

The values of R_L and R_B are determined by the following considerations. When an inverter stage is saturated, the collector current is given by

$$I_{C,sat} \approx E_{cc}/R_L \tag{16-57}$$

Therefore the corresponding base current I_B must be

$$I_B \geq I_{C,sat}/h_{FE} = E_{cc}/h_{FE}R_L \tag{16-58}$$

This inequality must be satisfied with the largest possible number of logic gates connected in parallel (Fig. 16-29). The actual value of base current will be

$$I_B = \frac{1}{N}\frac{E_{cc}}{R_L + R_B/N} \tag{16-59}$$

Combining Eqs. (16-58) and (16-59) yields the maximum number of gates that can be connected to the same output terminal

$$N \leq h_{FE} - R_B/R_L \tag{16-60}$$

This expression characterizes the fan-out capability of the circuit under consideration. Since the smaller the value of N, the more gates are overdriven, resulting in faster switching times, it is a good practice to chose N not to exceed one-fifth of the right-hand term of the inequality in Eq. (16-60).

N is determined mainly by h_{FE}. High fan-out capabilities thus require high current gain, which may be achieved by using an additional buffer output stage. A typical example is provided in Fig. 16-30. Fan-out figures of 10 to 20 are readily available in this manner, with switching times not longer than 10 ns.

RTL circuits are normally saturated when the output signal is at low level, and their switching behavior is controlled by the mechanism of minority-carrier storage in the base. Hence RTLs, like other saturated logic, are slower than nonsaturated-logic integrated-circuit gates.

ECL Gates. Emitter-coupled logic gates are typical of nonsaturated logic circuits. The basic ECL circuit (Fig. 16-31) resembles the differential amplifier, and one side comprises a series of parallel transistors as in RTL circuits. Better understanding is obtained by considering the ECL circuit as a parallel combination of emitter followers comprising T_A, T_B, T_C, and T_D, with a common load R_E. The voltage across R_E always is equal to the highest of the input voltages V_A, V_B, V_C, or V_D. Thus if V_A, V_B, V_C are all simultaneously more negative than V_D, none of the three parallel transistors conducts and the output voltage V_F is high. When one of the input voltages at A, B, or C is more positive than V_D, V_F drops to a minimum voltage. The circuit thus realizes the NOR function, as in the RTL case. The second output negates the NOR output and provides an additional OR gate, also as in the RTL circuit, although the interaction between the NOR and OR sections is different, i.e., through emitter coupling.

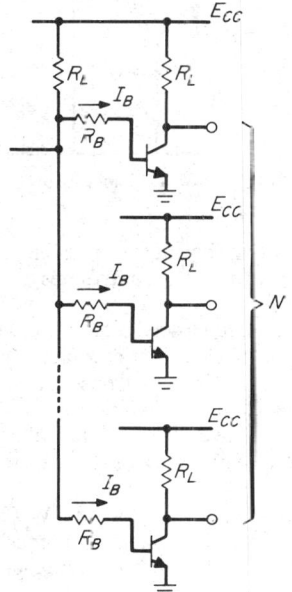

Fig. 16-29. Fan-out of logic gates.

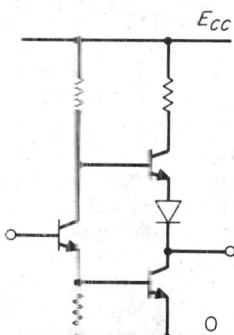

Fig. 16-30. Buffer stage used to improve fan-out capability of RTL gates.

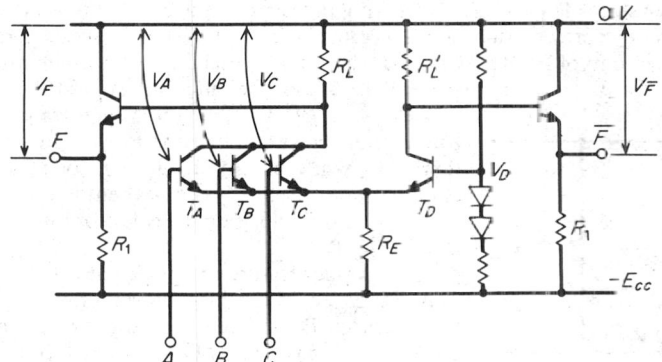

Fig. 16-31. Typical ECL logic-gate circuit. Typical values are $R_L = 270\ \Omega$, $R_L' = 300\ \Omega$, $R_E = 1.25\ k\Omega$, $R_1 = 2\ k\Omega$, $V_D = -1.15\ V$, $V_{EE} = -5.2\ V$.

ECL logic circuits require input-signal swings whose range exceeds the dynamic range of the typical differential amplifier, e.g., of the order of 1 V. A typical input-output curve is shown in Fig. 16-32. Both output signals are always negative (opposite to the RTL case). Hence RTL and ECL circuits cannot be cascaded unless special interface circuits are provided. ECL circuits may be used in series without any problem, since the two output emitter-follower circuits represented in Fig. 16-31 shift the signals downward at the common collector of the triple $T_A T_B T_C$ combination and at the collector of T_D by an amount equal to V_{BE}. Furthermore, these two output stages provide low output impedances.

An important difference between ECL logic gates and other integrated-circuit logic gates lies in the fact that all transistors in an ECL circuit are operated as emitter followers, or are blocked.

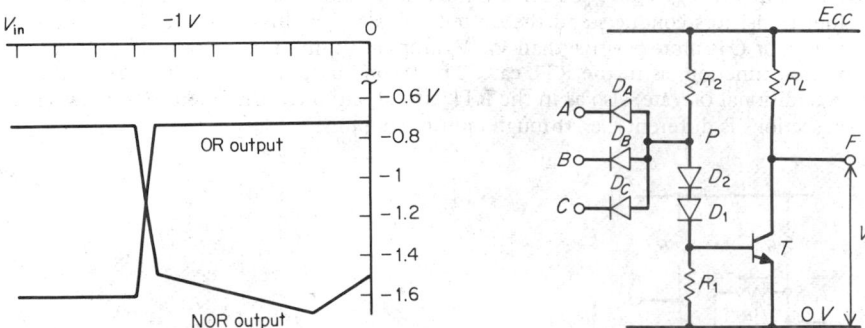

Fig. 16-32. OR and NOR outputs of ECL gates, showing voltage range between logic 1 and 0.

Fig. 16-33. DTL logic-gate circuit.

Hence saturation never occurs, and switching times are among the shortest possible, typically 3 to 5 ns or less. The presence of the emitter-follower buffers also contributes to the rapid switching.

T^2L *Gates.*[8,9] Transistor-transistor logic uses integrated logic closely related to DTL (diode-transistor logic). The basic DTL circuit is shown in Fig. 16-33. We first consider the situation where no current is flowing through the input diodes D_A, D_B, and D_C. Consequently, current flows through R_2, D_1, and D_2, dividing itself between R_1 and the base terminal of T. The latter current is intended to drive the transistor into saturation or to overdrive it. The voltage at point P is easy to compute, since it is equal to the sum of the voltage drops across three forward-biased junctions. Thus it amounts to approximately 2.2 V.

Next consider the case when one of the input voltages at A, B, or C is close to 0 V. The current through R_2 then passes through one of the input diodes, and the voltage at point P drops to approximately V_{BE}, bringing T to cutoff because of the presence of D_1 and D_2. The logic function is thus not the NOR of the RTL and ECL cases but a NAND function.

One of the drawbacks of this circuit is the relatively slow operation. This is due partly to the fact that T saturates but also to the fact that integrated diodes are passive elements embedded with parasitic capacitances. No opportunity is available to regenerate the signal except in the output inverter stage. Typical DTL switching times are of the order of tens of nanoseconds.

The schematic representation of a T^2L gate shown in Fig. 16-34 reveals certain similarities with the DTL circuit of Fig. 16-33 and also some substantial differences. The multiemitter transistor T_2 is used in the same manner as the diodes D_A, D_B, D_C and D_1, D_2 in the DTL circuit. When transistor T_1 is on, its base current flows through the forward-biased collector junction of T_2 while no current flows through any of the input emitters. Conversely, T_1 is off when one or more emitters of T_2 are grounded, blocking the collector junction in series with the base of T_1.

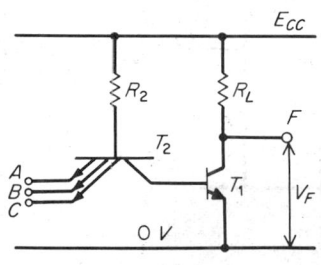

Fig. 16-34. TTL logic gate circuit.

An important feature of this circuit is that the passage of T_1 from saturation to the blocked condition is greatly aided by the transistor T_2. To show this, suppose that all input signals are initially high; that is, T_1 is saturated. If one or more of the input signals drops suddenly to zero, T_1 must switch off, but this requires first that the base be emptied of excess minority-carrier charges. Consequently, reversing the base current is required to achieve fast recovery. This happens because T_2 is then in the active region and its collector current can be very high. Once the charge in the base of T_1 is emptied, no further current flows through the collector of T_2 and this transistor goes into saturation. Hence, very fast switching is possible in spite of the fact that T²L circuits fall in the saturated-logic category. Typical T²L switching times are of the order of 6 ns.

The output-signal swing of T²L circuits is easily determined since the smallest output voltage is $V_{CE,sat}$ and the highest corresponds to no current in the load, making V_F dependent on the actual load connected to the output terminal.

Wired Logic. In the families of logic circuits just discussed the output terminal of a logic gate feeds several other gates of the same family. Paralleling two or more output terminals is not permitted, since this would cause improper functioning of the gates. An exception, called *wired logic*, exists whereby output terminals of two or more gates are connected in parallel to a common-load resistor.

RTL circuitry provides a good example of wired logic, since any number of collectors of transistors belonging to other RTL gates can be connected to the same output load R_L without trouble. Wired logic may result in saving components, as in the example shown in Fig. 16-35, representing two slightly modified T²L gates connected to a common-load resistor R and to a common-emitter resistor R_E The voltage drops across R_L and R_E are used to drive the output buffer (totem-pole buffer amplifier). In this way, the NOR function of two AND gates is provided in a single circuit.

22. Logic Gates (Integrated Injection Logic).[10,11] In an attempt to decrease the power consumption of bipolar inverters, the value of the collector resistance R must be increased. This requirement conflicts with the trend toward smaller geometries. The collector resistance can be substantially increased, however, without sacrificing silicon real estate if a common-base *pnp* transistor is used instead of a resistor, as shown in Fig. 16-36a. The load curve of the inverter stage is shown in Fig. 16-36b, the well-known common-base characteristic of the *pnp* transistor. When the inverter is in the on condition, all the current supplied by the common-base *pnp* transistor T_2 is short-circuited to ground through the saturated transistor T_1. In the opposite situation, the collector current of T_2 is derived at the output terminal.

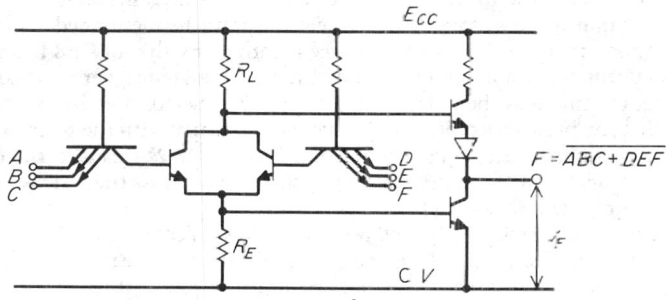

Fig. 16-35. Wired-logic circuit.

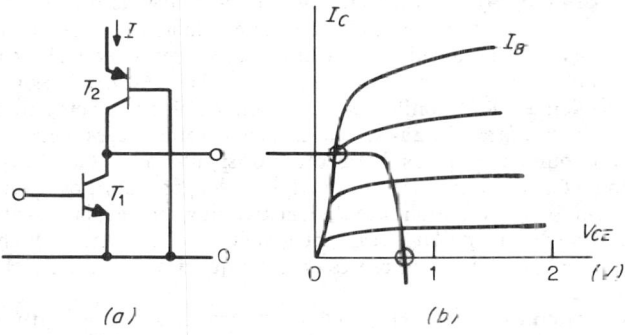

(a)   *(b)*

Fig. 16-36. The common-base *pnp* transistor can be used as load for bipolar inverters. The power supply is provided by the emitter current I.

Wired logic can be used as in other RTL circuits. An illustration of a wired-logic configuration is given in Fig. 16-37. Here the load transistor T_2 supplies the base current of another inverter T_3. In integrated injection logic (I^2L), this configuration is implemented in a very attractive manner. Figure 16-38 shows the integrated counterpart of transistors T_2 and T_3. Transistor T_2 in fact is formed by the lateral pnp transistor whose collector is merged with the base of the npn transistor T_3 (this is the origin of the name MTL merged transistor logic sometimes given this device). The npn transistor T_3 is used upside down with respect to the connection of the classical diffused bipolar transistor. Its emitter is always connected to ground. Several collectors are shown in Fig. 16-38, which reminds one of the multiemitter structure of T^2L.

The outstanding feature of I^2L logic is its small size, which results from the fact that many connections preexist in this device. All emitters are grounded, and a single base terminal is used whenever transistor inputs must be parallel. Moreover, the fact that the collectors are the smallest geometries available and that they are naturally isolated from each other helps reduce the

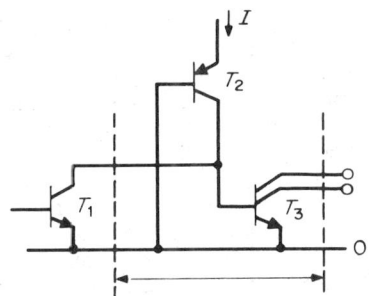

Fig. 16-37. Basic I^2L cell.

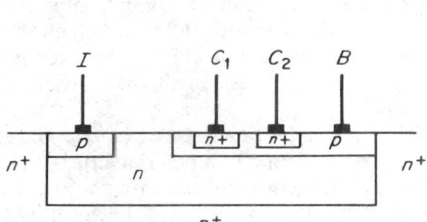

Fig. 16-38. The integrated I^2L cell.

size. Yet the upside-down connection of the inverting transistor leads to some drawbacks. For instance, the gain of the inverted transistor is poor, but this is not a fundamental problem in logic circuits. More important is the slow speed of operation which results from the very poor transition frequency f_T of the inverted transistor (only a few megahertz).

From a circuit point of view, some unusual features must be mentioned. I^2L gates are single-input, multioutput structures. The supply source is formed by the forward-biased emitter junction of the lateral pnp transistor, which is called the *injector*. Hence, very low power dissipation can be achieved, inasmuch as the current supplied by the injector may be exceedingly small. It can be shown that the propagation delay of I^2L increases linearly with the reciprocal of the power consumption. This unique feature of I^2L, illustrated in Fig. 16-27, leads to the conclusion that I^2L, like CMOS, is ideally suited to micropower applications such as the wristwatch. The energy per cycle of I^2L can be as low as 0.1 pJ.

The power-delay product is a constant because at low injector current only the transition capacitances play a significant role. When the injector current increases to the point where the diffusion capacitance must also be taken into account, the propagation delay tends to become constant while the power continues to increase. The value of the propagation delay is then a minimum. It can hardly go below 20 to 10 ns unless special technologies are used.

23. Logic Gates (MOS).[5,6] MOS logic gates are currently used in medium- or large-scale integrated circuits for their small size and relative ease of fabrication. While no specific families of logic gates exist using FETs, a wide variety of manufacturer-wired logic is available.

The basic NOR and NAND FET structures are shown in Fig. 16-39. The NOR gate simply reproduces the discrete-element RTL configuration, the only difference being that the load resistor has been replaced by a saturated transistor, as in FET inverter stages. Less power is consumed, and higher circuit complexity is possible because of smaller size. The drawback, however, is speed of operation, i.e., switching times 10 to 20 times longer than those available with bipolar transistors. One should not overemphasize this relative sluggishness, because the increased complexity possible in medium- and large-scale integrated circuits permits a high degree of decentralization and redundancy. The multiprocessor technique used in the field of minicomputers is typical of this trend.

The basic NAND circuit shown in Fig. 16-39b is not commonly used with bipolar transistors, since the latter introduce unwanted pedestals in series with each saturated transistor. The FET transistors behave more like small resistors in series, and the total voltage drop across them is negligible compared with the normal input dynamic range.

An illustration of wired-logic FET circuitry is shown in Fig. 16-40, which represents an EXCLUSIVE-OR gate using four switching transistors with three load transistors and one inverter stage. The same circuit using only NOR or NAND gates would be more complex.

24. Transistor Switches Other than Logic Gates. Transistor switches are extensively used in applications other than logic gates covering a wide variety of both digital and analog applications. A typical illustration is the circuit converting the frequency of a signal into a proportional current, the so-called *diode pump*. This circuit (Fig. 16-41) comprises a capacitor C, two diodes D_1 and D_2, and a switch formed by a transistor T and a resistor R. The transistor is assumed to be driven periodically by a square-wave source, alternatively on and off. When T is blocked, the capacitor C charges through the diode D_1, while D_2 has no effect. As soon as the voltage across C has reached its steady-state value E_c, T may be turned on abruptly. The voltage

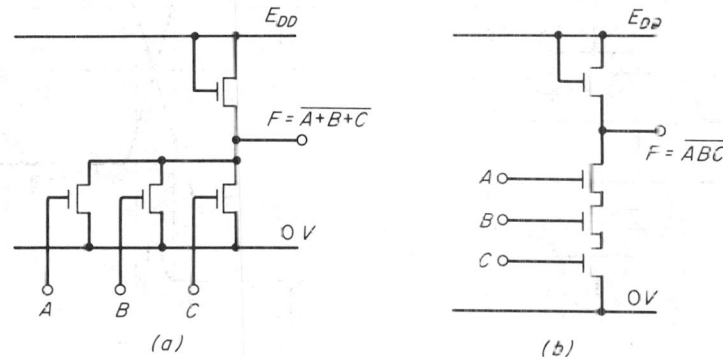

Fig. 16-39. MOS gates (a) NOR; (b) NAND.

with respect to ground at point A becomes negative, and D_1 is blocked while D_2 is forward-biased. The capacitor thus discharges itself in the load (in Fig. 16-41 an ammeter, but it could be any other circuit element that does not exhibit storage), allowing V_A to reach 0 V before T is again turned off. The charge fed to the load thus amounts to CE_{cc} coulombs. If we suppose that this process is repeated periodically, the average current in the load is given by

$$I = fCE_{cc} \qquad (16\text{-}61)$$

where f represents the switching repetition rate.

The diode-pump circuit provides a pulsed current whose average value is proportional to the frequency of the square-wave generator controlling the switching transistor T. The proportionality would of course be lost if the load exhibited storage, e.g., if the load were a parallel combination of a resistor and a capacitor in order to obtain a voltage drop proportional to the average current. Using an operational amplifier, as shown in the right side of Fig. 16-41, circumvents the problem.

The requirements on the switching transistor in this application are different and in many respects more stringent than for logic gates.

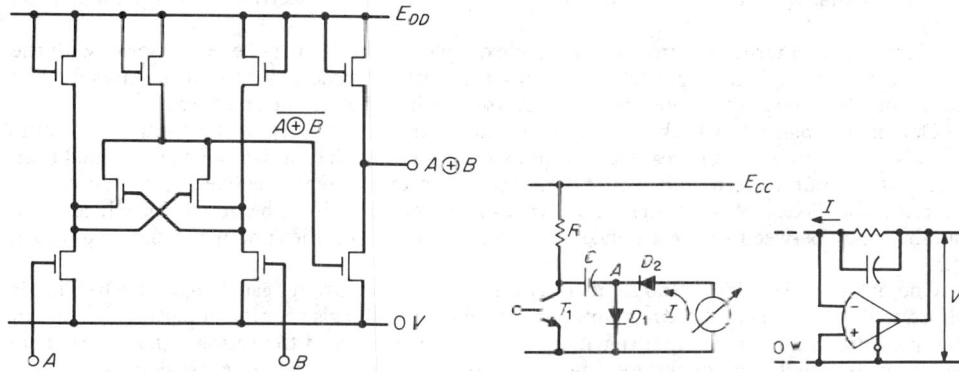

Fig. 16-40. Wired-logic EXCLUSIVE-OR circuit.   **Fig. 16-41.** Diode-pump circuit.

The transistor in a logic circuit provides a way of defining two well-distinguished states, logic 1 and 0. Nothing further is required whether these two states approach an actual short circuit or an open circuit. In the diode-pump circuit, however, the total switching characteristics are important, since the residual voltage drop across the saturated transistor of Fig. 16-41 influences the charge transfer from C to the load, thereby also introducing unwanted temperature sensitivity. The main difference lies in the fact that while T is operated as a logical element, the purpose of the circuit actually is to deliver an analog signal.

There are many other examples where the characteristics of switching transistors influence the accuracy of given circuits or instruments, as in the digital voltmeter based on the weighting

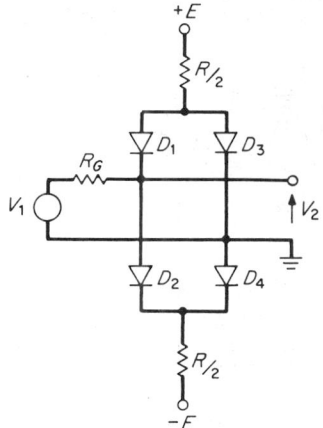

Fig. 16-42. Amplitude-gating circuit.

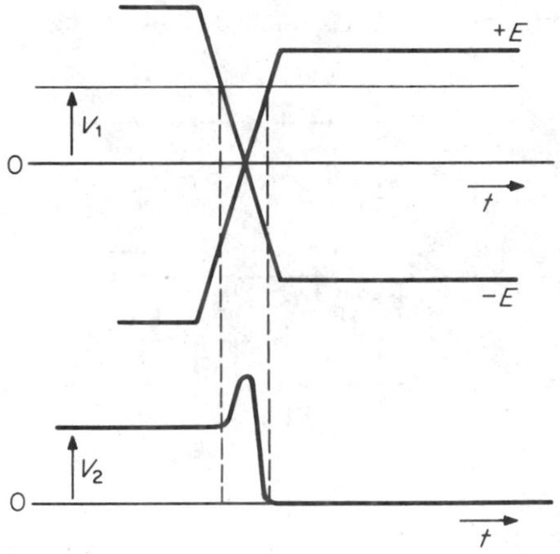

Fig. 16-43. Transient signal *(bottom)* caused by finite rise times of gating signals.

principle. An even more critical problem pertains to amplitude gating, since this class of applications requires switches which correctly transfer analog signals without introducing distortion. Furthermore, positive and negative signals must be transmitted equally well, and noise introduced by the gating signals must be minimized.

Amplitude Gating.   One of the simplest and most effective approaches is to use balanced diode bridges. The pedestals introduced by the diodes then cancel out to the extent that symmetry is achieved.

A typical amplitude-gating circuit is shown in Fig. 16-42. We assume first that balanced gating signals are applied to the two equal resistors symmetrically connected to the bridge. With the polarities indicated in Fig. 16-42, all four diodes conduct, and the output voltage V_2 is zero, assuming that the voltage drops across the diodes are identical. The existence of a small input voltage V_1 does not alter this condition if the resulting current unbalance is kept small. Since diodes are insensitive to current as far as forward-voltage drops are concerned, the error at the output can be neglected.

In the opposite situation, when the polarities of the gating sources are reversed and the diodes are all back-biased, the output signal V_2 becomes equal to V_1 since the input generator is no longer loaded.

Many circuit alternatives are available. For example, the bridge may be put in series with the input generator and an output load, replacing the shunt approach by series transmission. The open and close control functions must be reversed with respect to the shunt case.

One of the major drawbacks of electronic gates is their sensitivity to the controlling gating signals. Fast switching requires steep control-voltage steps, which may produce unwanted transients at the output terminals, particularly those due to differences between junction capacitances of the diodes. Minority-carrier storage causes another problem because it introduces measurable delay between the appearance of the control signal and the time when the diodes turn off.

One of the most difficult transients to eliminate is caused by the gated signal itself. This is illustrated in Fig. 16-43, showing two symmetrical control signals opening or closing the gate in the presence of a nonzero input signal V_1. The finite rise times of the gating signals cause nonsimultaneous switching of the four diodes. A strong unbalance therefore exists during a very short interval of time which produces a sharp spike superimposed on the gated output signal. The amplitude and the polarity of this spurious transient depend on the gated signal, which vanishes when V_1 is equal to zero.

Field-effect transistors can also be used to perform amplitude gating. A typical application is switched-capacitor filters.[12] Figure 16-44a illustrates an elementary switched-capacitor network. In this circuit, the capacitor C is connected back and forth between the two terminals so that a charge $C(V_1 - V_2)$ is transferred at each cycle. Hence, if the repetition rate is f, the switched-capacitor network allows an average direct-current $C(V_1 - V_2)f$ to flow from one terminal to the other. It is thus equivalent to a resistor whose value is $1/Cf$.

If another capacitor C_o is connected at the output port, an elementary sampled-data RC circuit is built with a time constant equal to C_o/Cf. This time constant depends only on the clock

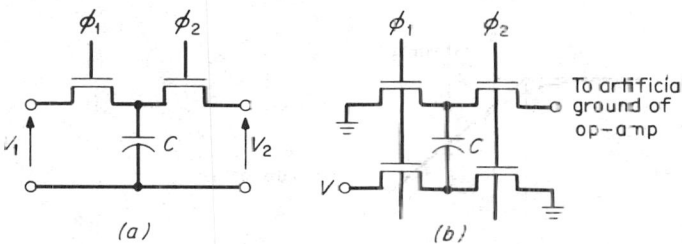

Fig. 16-44. Switched-capacitor networks. Circuit (a) is more sensitive to stray capacitance than circuit (b).

frequency f and on the ratio of two capacitors. Hence, relatively large time constants can be achieved with good accuracy using very small capacitors and MOS transistor switches. In practice, MOS capacitors of a few picofarads can be made with ratios better than 0.1%. In addition, a slight modification (see Fig. 16-44b) of the circuit avoids the stray capacitance which would otherwise affect the accuracy adversely. Hence, fully integrated switched-capacitor RC active filters can be designed to tight specifications, e.g., for telephone applications.

According to Eq. (16-23), a drawback of amplitude-gating MOS switches may be the large variations of the conductance in the on state which are experienced when the actual channel potential approaches the value of the gate voltage V_G. This problem is virtually eliminated in the circuit of Fig. 16-45, representing a CMOS amplitude-gating switch. In this circuit, the p- and n-channel MOS transistors are in the on state simultaneously since complementary control signals are provided to the gates. Their conductances are influenced by the level of the signal to be transferred, as in the simple transistor MOS switch, but since the conductances vary in opposite directions and must be added because of the paralleling of the two transistors, the overall conductance variation of the switch is substantially reduced.

ACTIVE WAVEFORM SHAPING USING NEGATIVE FEEDBACK

25. Active Circuits. Linear active networks used for waveshaping may employ negative feedback to improve the performance of the waveshaping circuits. Of the linear active waveshaping circuits, the operational amplifier-integrator is most widely used.

26. RC Operational Amplifier-Integrator. In Fig. 16-46 it is assumed that the operational amplifier has infinite input impedance, zero output impedance, and a high negative gain A. The overall transfer function is

$$\frac{A}{1 + p(1 - A)T} \quad \text{where } T = RC \quad (16\text{-}62)$$

This function represents a first-order system with gain A and a cutoff frequency which is approximately $|A|$ times lower than the inverse of the time constant T of the RC circuit. In Fig. 16-46b the fre-

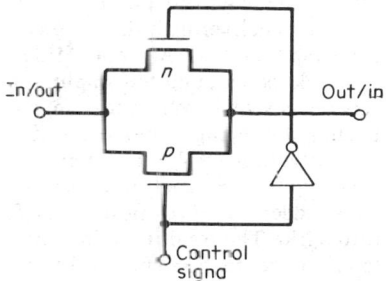

Fig. 16-45. CMOS analog switch.

quency response of the active circuit is compared with that of the passive RC integrator. The widening of the spectrum useful for integration is clearly visible. For instance, an integrator using an operational amplifier with a gain of 10^4 and a RC network having a 0.1-s time constant has a cutoff frequency as low as 1.6 mHz.

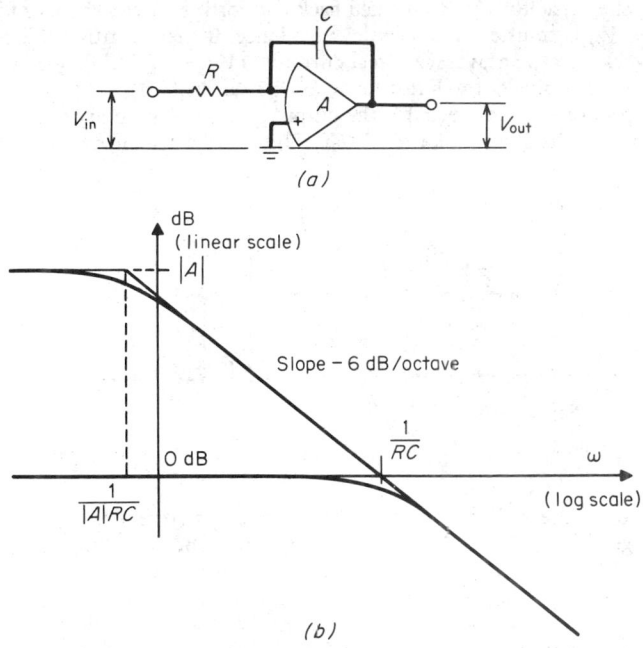

(a)

(b)

Fig. 16-46. (a) Operational amplifier-integrator; (b) gain vs. angular frequency.

In the time domain, the Taylor expansion of the amplifier-integrator response to the step function is

$$V(t) = E\frac{t}{T}\left[1 - \frac{t}{2!\,|A|\,T} + \frac{t^2}{3!(|A|\,T)^2} - \cdots\right] \tag{16-63}$$

This shows that any desired degree of linearity of $V(t)$ can be achieved by providing sufficient gain in the amplifier.

27. Sweep Generators.[13] Sweep generators (also called time-base circuits) produce linear voltage or current ramps vs. time. They are widely used in applications such as cathode-ray displays, digital voltmeters, and television. In almost all such circuits the linearity of the ramp results from charging or discharging a capacitor through a constant-current source. The difference between circuits used in practice rests in the manner of realizing the constant-current property of the source. Sweep generators may also be looked upon as integrators with a constant-amplitude input signal. The latter point of view shows that RC operational amplifier-integrators provide the basic structure for sweep generation.

Circuits delivering a linear voltage sweep fall into two categories, the Miller time base and bootstrap time base. A simple Miller circuit (Fig. 16-47) comprises a capacitor C connected in a feedback loop around the amplifier formed by T_1 and its output-load resistor R_L. Transistor T_2 acts like a switch. When it is on, all the current flowing through the base resistor R_B is driven to ground, keeping T_1 blocked, since the voltage drop across T_2 is lower than the normal base-to-emitter voltage of T_1. The output signal V_{CE} of T_1 is thereby clamped at the level of the power-supply voltage E_{cc} and the voltage drop across the capacitor C is approximately the same. When T_2 is suddenly turned off, it drives T_1 into the active region and causes collector current to flow through R_L. The resulting voltage drop across R_L is coupled capacitively to the base of T_1, tending to minimize the base current; i.e., the negative-feedback loop is closed. The collector-to-emitter voltage V_{CE} of T_1 subsequently undergoes a linear voltage sweep downward, as illustrated in Fig. 16-47b.

The circuit behaves in the same manner as the RC operational amplifier described in Par. **16-26**. Almost all the current flowing through R_B is derived through the feedback capacitor, and only a very small part is used for controlling the base of T_1. The feedback loop opens when T_1 enters into saturation, and the voltage gain of the amplifier becomes negligible.

When T_2 is subsequently turned on again, blocking T_1 and recharging C through R_L and the saturated switch, the output voltage V_{CE} rises again according to an exponential curve with time constant $R_L C$.

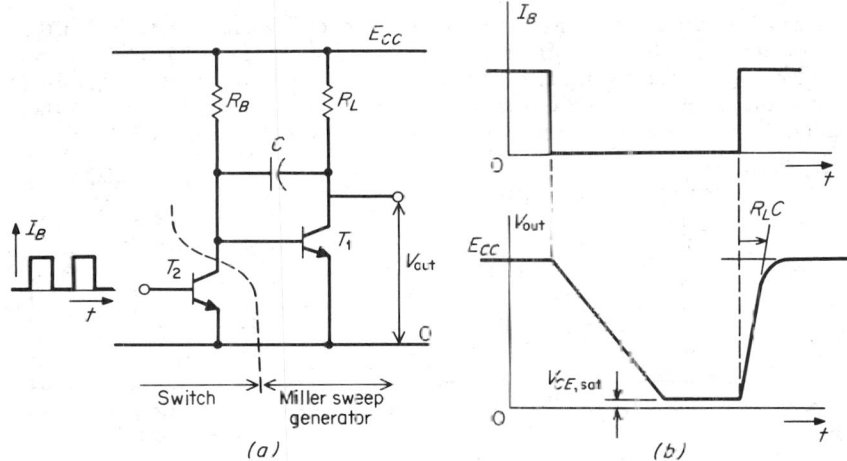

Fig. 16-47. Miller sweep generator:(a) circuit (b) input and output vs. time.

Figure 16-48 shows a typical bootstrap time-base circuit. It differs from the Miller circuit in that the capacitor C is not a part of the feedback loop. Instead the amplifier is replaced by an emitter-follower circuit delivering an output signal V_{out} which reproduces the voltage drop across the capacitor. C is charged through resistor R_B from a floating voltage source formed by the capacitor C_0 (C_0 is large compared with C).

First, we consider that the switch T_2 is on. Direct current then flows through the series combination formed by the diode D, the resistor R_B, and the saturated transistor T_2. The emitter follower T_1 is blocked since T_2 is saturated. Moreover, the capacitor C_0 can charge through the path formed by the diode D and the emitter resistor R_E, and the voltage drop across its terminals is equal to E_{CC}. When T_2 is cut off, the current through R_B flows into the capacitor C, causing the voltage drop across its terminals to rise gradually, driving T_1 into the active region. Because T_1 is a unity-gain amplifier, V_{out} is a replica of the voltage drop across C.

Since C_0 acts as a floating dc voltage source, diode D is reverse-biased almost immediately. The current flowing through R_B is supplied exclusively by C_0, and the voltage across it stays constant at the voltage drop across C_0 minus the base-to-emitter voltage of T_1.

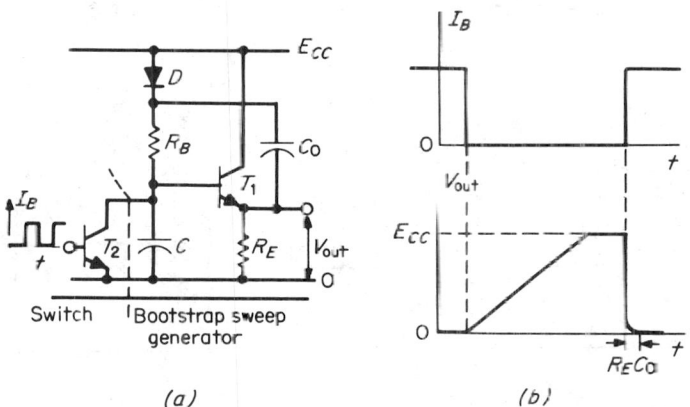

Fig. 16-48. Bootstrap sweep generator:(a) circuit (b) input and output vs time.

Considering that the base current of T_1 represents only a small fraction of the total current flowing through R_B, it is evident that the charging of capacitor C occurs through a constant-current source and that therefore a linear voltage ramp is obtained as long as the output voltage of T_1 is not clamped to the level of the power-supply voltage E_{CC}.

The corresponding output waveforms are shown in Fig. 16-48b. After T_2 is switched on again, C discharges rapidly, causing V_{out} to drop, while the diode D again is forward-biased and the small charge lost by C_0 is restored. In practice, C_0 should be at least 100 times larger than C to ensure a quasi-constant voltage source.

More detailed analysis of the Miller and bootstrap sweep generators reveals that they are in fact equivalent. We redraw the Miller circuit as shown at the left of Fig. 16-49. Remembering that the operation of the sweep generator is independent of which output terminal is grounded, we ground the collector of T_1 and redraw the corresponding circuit. As shown at the right in the figure, this is a bootstrap circuit, and the two circuits thus are equivalent.

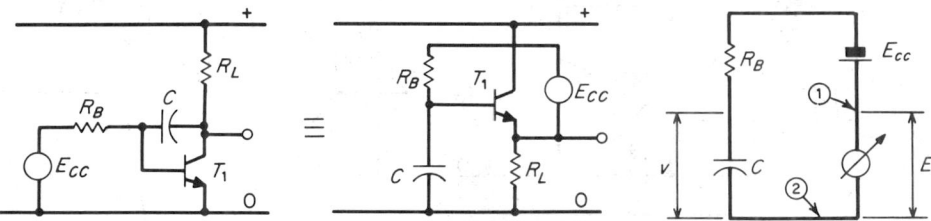

Fig. 16-49. Equivalency of the Miller and bootstrap sweep generators.

Fig. 16-50. Basic loop of sweep-generator circuits.

Any sweep generator can be regarded as a simple loop (Fig. 16-50) comprising the capacitor C delivering a voltage ramp, the loading resistor R_B, and the series combination of two sources: a constant voltage source E_{CC} and a variable source whose emf E reproduces the voltage drop V across the capacitor. The voltage drop across R_B consequently remains constant at E_{CC}, and so the loop current is also constant. The voltage ramp consequently is given by

$$E = V = (E_{CC}/R_B C)t \qquad (16\text{-}64)$$

Grounding terminal 1 yields the Miller network, while grounding terminal 2 leads to the bootstrap circuit.

Since the degree of linearity is one of the essential features of sweep generators, we consider the equivalent networks represented in Fig. 16-51. Starting with the Miller circuit, we determine the impedance in parallel with C

$$|A|(R_B \parallel h_{11}) \qquad (16\text{-}65)$$

where $|A|$ is the absolute value of the voltage gain of the amplifier

$$|A| = (h_{21}/h_{11})R_L$$

Next, considering the bootstrap circuit, we calculate the input impedance of the unity-gain amplifier to determine the loading impedance acting on C. This impedance is

$$R_L h_{21} R_B/(R_B + h_{11}) \qquad (16\text{-}66)$$

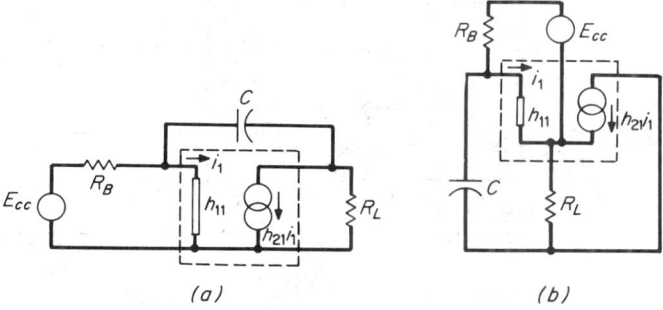

(a)   *(b)*

Fig. 16-51. Equivalent forms of (a) Miller and (b) bootstrap sweep circuits.

which turns out to be the same as that given in Eq. (16-65); i.e, the two circuits are equivalent. To determine the degree of linearity it is sufficient to consider the common equivalent circuit of Fig. 16-52 and to calculate the Taylor expansion of the voltage V

$$V = \frac{E_{CC}}{R_B C} t \left[1 - \frac{t}{2! |A| (R_E \parallel h_{11}) C} + \frac{t^2}{3! [|A| (R_B \parallel h_{11}) C]^2} - \cdots \right] \tag{16-67}$$

The higher the voltage gain $|A|$, the better the linearity. Thus, an integrated operational amplifier in place of T_1 leads to excellent performance in both the Miller and the bootstrap circuit. Voltage gains as high as 10,000 are readily obtained for this purpose.

28. Sample-and-Hold Circuits.[14] Sample-and-hold circuits are widely used to store an analog voltage accurately over a time ranging from microseconds to minutes. They are basically switched-capacitor networks like those considered in Par. **16-24**, but since the analog voltage

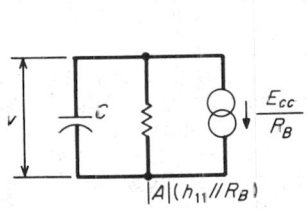

Fig. 16-52. Common equivalent circuit of sweep generators.

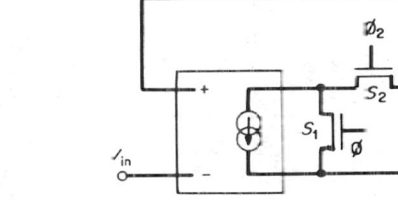

Fig. 16-53. Typical sample-and-hold circuit.

across the storage capacitor in the hold mode must be available at the output terminal of the sample-and-hold circuit under low impedance, a buffer amplifier is needed. Op-amps with FET input are commonly used for this purpose to minimize the hold-mode droop. The schematic of a widely used integrated circuit is shown in Fig. 16-53.

Storage and readout are achieved by an FET input op amp in the hold mode. During the acquisition time, the positions of the two switches are changed simultaneously. Current is supplied by the voltage-dependent current source to minimize the voltage difference between input and output terminals. As soon as S_1 and S_2 return to their initial positions, V_{out} ceases to follow V_{in} and remains unchanged.

The main requirements for sample-and-hold circuits are low hold-mode droop, short settling time in the acquisition mode, low offset voltage, and hold-mode feedthrough. The hold-mode droop is dependent on the leakage current of the op-amp inverting node. Short settling times require high-slew-rate op-amps and large current-handling capabilities for both the current source and the op-amp. The offset voltage is determined by the differential amplifier which controls the current source. Finally, feedthrough is due to imperfect isolation between the current source and the op-amp. For this reason, a double switch is preferred to a single series switch.

Another important feedthrough problem is related to the unavoidable gate-to-source or drain-overlap capacitance of the MOS switch S_2. When the gate-control signal is switched off, some small charge is always transferred capacitively to the storage capacitor and a small voltage step is superimposed on the output terminal when the circuit enters the hold state. Minimization of this effect can be achieved by increasing the ratio of the storage capacitance to the switch-overlap capacitance. Since the latter cannot be made equal to zero, the storage capacitance must be chosen sufficiently large, but this inevitably lengthens the settling time. One means of alleviating the problem is to compensate the switching charge due to the control signal by injection of an equal and opposite charge on the inverting input node of the op-amp. This can be achieved by means of a dummy transistor controlled by the inverted signal.

29. Nonlinear Active Networks. The use of nonlinear devices with negative feedback accentuates the character of waveshaping networks. In many circumstances, this leads to an idealization of the nonlinear character of the devices considered. A good example of this is given by the ideal rectifier circuit.

Ideal Rectification with Nonlinear Feedback   The negative-feedback loop in the circuit shown in Fig. 16-54 is formed by two parallel branches. Each comprises a diode connected in such manner that if V_1 is positive, the current injected by resistor R_1 flows through D_1, and if V_1 is negative, it flows through D_2. Furthermore, a resistor R_2 is placed in series with D_1, and the output voltage V_2 is taken at the node between R_2 and D_1. Hence, V_2 is given by $-(R_2/R_1)V_1$ when V_1 is positive, independently of the forward voltage drop across D_1, and it is zero when V_1 is negative.

When D_1 is forward-biased, the voltage V at the output of the op-amp adjusts itself to force the current flowing through D_1 and R_1 to be exactly the same as through R_1. This means that V, in some circumstances, may be much larger than V_2, especially when V_2 (and thus also V_1) is of the order of millivolts. In fact, V exhibits approximately the same shape as V_2 plus an additional pedestal of approximately 0.6 to 0.7 V. Typical waveforms obtained with a sinusoidal voltage of a few tens of millivolts are shown in Fig. 16-55.

The quasi-ideal rectification characteristic of this circuit is readily understood by considering the Norton equivalent network seen from R_2 and D_1 in series. It consists of a current source

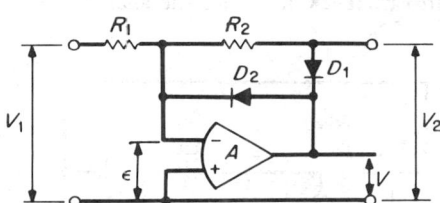

delivering the current V_1/R_1 in parallel with an almost infinite resistor $|A|R$, where A represents the voltage gain of the op-amp. Hence, the current flowing through the branch formed by R_2 and D_1 is delivered by a quasi-ideal current source, and the voltage drop across R_2 is unaffected by the series diode D_1. As for D_2, it is required to prevent the feedback loop from being opened when V_1 is negative. If this should happen, the artificial ground at the input of the op-amp would be lost and V_2 would not be zero.

Fig. 16-54. The precision rectifier using negative feedback is almost an ideal rectifier.

Other negative-feedback configurations leading to very high output impedances are equally powerful in achieving ideal rectification characterics. For instance, the unity-gain amplifier used in instrumentation has wide linear ac measurement capabilities (Fig. 16-56).

ACTIVE WAVEFORM SHAPING USING POSITIVE FEEDBACK

30. Positive Feedback. Positive feedback can be used in stable as well as unstable (free-running) circuits. Unstable networks include harmonic oscillators and free-running relaxation circuits. The latter may be self-excited or synchronized by external trigger pulses. Trigger pulses can also be used to execute transitions between stable states. A free-running relaxation circuit is called an *astable multivibrator* network. Monostable and bistable multivibrator circuits also occur, with one and two distinct stable states, respectively.

The degree to which positive feedback is used in harmonic oscillators differs substantially from that of astable, monostable, and bistable pulse circuits. In the oscillator the total loop gain

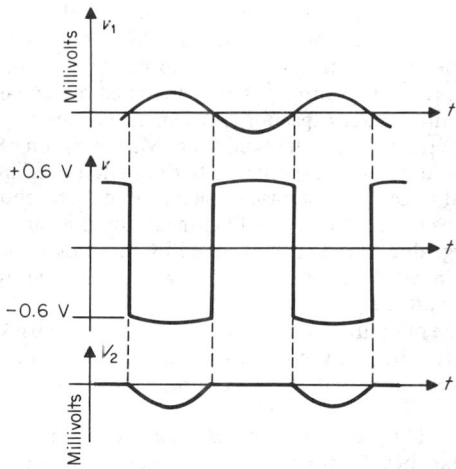

Fig. 16-55. Waveforms of circuit in Fig. 16-54.

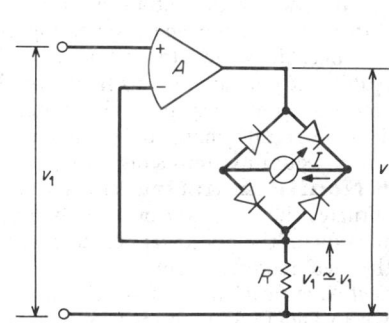

Fig. 16-56. Feedback rectification circuit used in precision measurements.

must be kept close to 1, and it need compensate only for the small losses in the resonating tank circuit. In the pulsed circuits, on the other hand, positive feedback permits fast switching from one state to another, e.g., from cutoff to saturation and vice versa. Before and after these transitions occur, the circuit usually is passive. Switching occurs in an extremely short time, typically less than 10 ns. After switching, the circuit operates more slowly, approaching steady-state conditions.

It is common practice to call the switching time the *regeneration time* and the time needed to reach final steady-state conditions the *resolution time*. The resolution time may range from tens of nanoseconds to several seconds or more, depending on the circuit design.

An important feature of regenerative circuits is that their switching times are essentially independent of the steepness of the trigger-signal waveshape. Once the level of instability is reached, the transition occurs at a rate fixed by the total loop gain and by the reactive parasitics

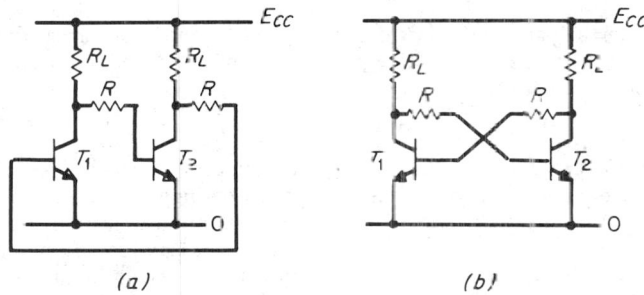

Fig. 16-57. The Eccles-Jordan bistable circuit (flip-flop): (a) in the form of two cascaded amplifiers with output connected to input; (b) as customarily drawn, showing symmetry of connections.

of the circuit itself but independent of the rate of change of the trigger signal. Regenerative circuits thus provide means of restoring short rise times to pulses broadened by transmission through large systems.

Positive-feedback pulse circuits are necessarily nonlinear. The most conventional way to study their behavior is to use a piecewise-linear analysis technique.

31. Collector-Coupled Bistable Circuits.[15] Two cascaded common-emitter transistor stages realize an amplifier with a high positive gain. Connecting the output to the input (Fig. 16-57) produces an unstable network known as the *Eccles-Jordan bistable circuit* or *flip-flop*. Under steady-state conditions one transistor is saturated and the other is blocked.

In a typical case, the circuit of Fig. 16-57 has the value $R_L = 1$ kΩ, $R = 2.2$ kΩ, and $E_{cc} = 6$ V. Suppose T_1 is at cutoff, and consider the Thevenin equivalent network connected to the base of T_2. It can be viewed as an emf of $+6$ V and series resistances of 3.2 kΩ. The base current of T_2 is given by

$$I_{B2} = (6 - 0.7)/3.2 = 1.66 \text{ mA}$$

If T_2 is saturated, the collector current is equal to E_{cc}/R_2 or 6 mA. One can conclude that a current gain of 4 is sufficient to ensure saturation of T_2. Hence the collector-to-emitter voltage across T_2 will be very small ($V_{CE,sat}$), and consequently T_1 is blocked, as stated initially. The reverse situation, with T_1 saturated and T_2 cut off, is governed by the same considerations for reasons of symmetry. Two distinct stable states thus are possible.

When one of the transistors is suddenly switched from one state to the opposite, the other transistor automatically undergoes an opposite transition. At a given time both transistors conduct simultaneously, increasing the loop gain suddenly from zero to a high positive value. This corresponds to the regenerative phase, during which the circuit becomes active.

It is difficult to compute the regeneration time since the operating points of both transistors move through the entire active region, causing large variations of the small-signal parameters of both transistors. Although determination of the regeneration time on the basis of a linear model is unrealistic and leads only to a rough approximation, we briefly examine this problem since it illustrates how unstable networks can be analyzed.

First, we introduce two capacitors in parallel with the two resistors R. These capacitors provide a direct connection from collector to base under transient conditions and hence increase the high-frequency loop again. The circuit can now be described by the network of Fig. 16-58, which consists of a parallel combination of two reversed transistor hybrid equivalents without extrinsic base resistances (for calculation convenience) and with two load admittances Y which combine the load and resistive part of each transistor's input. Starting from the admittance matrix of one of the transistors with its load, we equate the determinant of the parallel combination

$$\begin{vmatrix} p(C_\pi + C_{TC}) & -pC_{TC} \\ \dfrac{I}{V_T} - pC_{TC} & pC_{TC} + Y \end{vmatrix}$$

$$(16\text{-}68)$$

to zero to find the natural frequencies of the circuit. This leads to

$$pC_\pi + (I/V_T) + Y = 0 \tag{16-69}$$

and

$$p(C_\pi + 4C_{TC}) + Y - I/V_T = 0 \tag{16-70}$$

where C_π stands for the parallel combination of C_{TE} and the diffusion capacitance $\tau_F I/V_T$. Only Eq. (16-70) has a zero with a real positive part, producing an increasing exponential function with time constant approximately equal to

$$\tau = (C_\pi + 4C_{TC})V_T/I \tag{16-71}$$

Since the diffusion capacitance overrules the transition capacitances at high current, Eq. (16-71) reduces finally to τ_F. This yields extremely short switching times. For instance, a transistor with a maximum transition frequency f_F of 300 MHz has a τ_F equal to 0.53 ns, and the regeneration time (defined as the time elapsing between 10 and 90% of the total voltage excursion from cutoff to saturation or vice versa) is equal to $2.2\tau_F$, or 1.2 ns. This may be about one order of magnitude below the actual switching time.

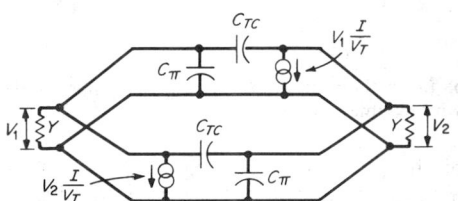

Fig. 16-58. Flip-flop circuit showing capacitances that determine time constants.

A more accurate but much more elaborate analysis, taking into account the influence of the extrinsic base resistance and the transition capacitances in the region of small-collector current, would require a computer simulation based on the dynamic large-signal model of the bipolar transistor of Par. **16-17**. Nevertheless, Eq. (16-71) clearly pinpoints the factors controlling the regeneration time: the transconductance and the unavoidable parasitic capacitances (in other words, the merit factor of the transistor). This is a quite general statement, which is verified in many other positive-feedback switching circuits.

We next consider the principal time constants controlling the resolution time of this circuit, using the same numerical values. We suppose T_1 initially nonconducting and T_2 saturated. The sudden turnoff of T_2 is simulated by opening the short-circuit switch S_2 in Fig. 16-59a. Immediately, V_{CE2} starts increasing toward E_{cc}. The base-to-ground voltage V_{BE1} of T_1 consequently rises above its 0-V steady-state value with a time constant fixed only by the total parasitic capacitance C_T at the collector of T_2 and base of T_1 times the resistor R_L. Hence the time constant is

$$\tau_1 = R_L C_T \tag{16-72}$$

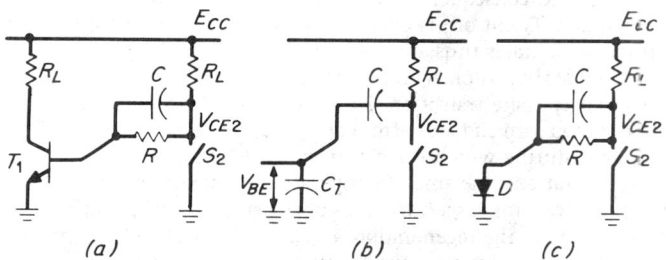

(a)   (b)   (c)

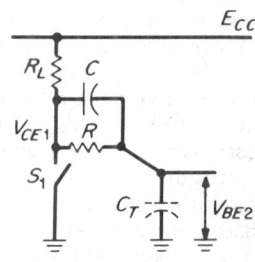

Fig. 16-59. Piecewise analysis of flip-flop switching behavior. Transistor T_2 is assumed to be on before (a). The opening of S_2 simulates cutoff. The new steady-state conditions are reached in (c).

Fig. 16-60. The longest time constant is experienced when T_1 is turned on. This is simulated by the closure of switch S_1.

This time is normally extremely short; e.g., a parasitic capacitance of 1 pF yields a time constant of 1 ns. The charge accumulated across C evidently cannot change, for C is larger than C_T. So V_{BE1} and V_{CE2} increase at the same rate. This situation is illustrated in Fig. 16-59b. When V_{BE1} reaches approximately 0.7 V, T_1 starts conducting and a new situation arises, illustrated in Fig. 16-59 by the passage from (b) to (c). This is when regeneration actually takes place, forcing T_1 T_1 to go into saturation very rapidly. With the regeneration period neglected, case (c) is characterized by the time constant

$$\tau_2 = (R_L/R)C$$

For instance, if C is equal to 10 pF, τ_2 yields 7 ns. Although this time constant is much longer than τ_1, it is still not the longest, for we have not yet considered the evolution of V_{BE2}.

It is considered in Fig. 16-60, where the saturated transistor T_1 is replaced by the closing of S_1. The problem is the same as for the compensated attenuator (Par. 16-5). Since overcompensation is achieved, V_{BE2} undergoes a large negative-voltage swing almost equal to E_{cc} before climbing toward its steady-state value 0 V. The time constant of this third phase is given by

$$\tau_3 = RC \qquad (16\text{-}74)$$

In the present case, τ_3 equal 22 ns.

The voltage variations vs. time of the flip-flop circuit thus far analyzed are reviewed in Fig. 16-61 with the assumption that the regeneration time is negligible. It is evident that C plays a dual

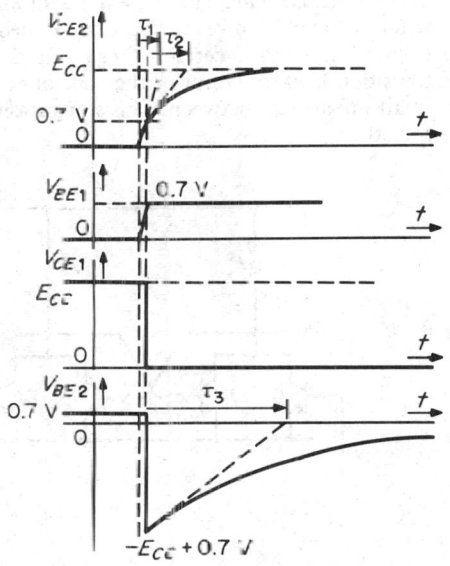

Fig. 16-61. Voltage variations vs. time of flip-flop circuit.

role. The first is favorable since it ensures fast regeneration and efficiently removes excess charges from the base of the saturated transistor, but the second is unfavorable since it increases the resolution time and sets an upper limit to the maximum repetition rate at which the flip-flop can be switched. The proper choice of C as well as of R_L and R must take this fact into consideration. Small values of the resistances make high repetition rates possible at the price of increased dc power consumption.

32. Integrated-Circuit Flip-Flops.[8] The Eccles-Jordan circuit (Fig. 16-57) considered in Par. 16-31 is the basic structure of integrated bistable circuits. Of course the capacitor C is not present, and this causes increased sluggishness. However, as mentioned earlier, speed is obtained at the price of increased power consumption. Another way to alleviate the problem is to avoid saturation by means of Schottky clamps (see Par. 16-18). In this manner both speed and low power can be achieved simultaneously This is known as *low-power Schottky logic*.

The integrated flip-flop can be viewed as two cross-coupled single-input NOR or NAND circuits. In fact, integrated flip-flops vary only in the way external signals act upon them for control purposes. A typical example is given in Fig. 16-62 with the corresponding logic-symbol representation. The triggering inputs are called *set* S and *reset* R terminals. Additional transistors are necessary to provide the actual triggering, but this is not a problem in integrated circuits because transistors are the smallest devices available on the chip.

The truth tables for NOR and NAND bistable circuits are

		NOR bistable			NAND bistable		
R	S	Q_1	Q_2	Line	Q_1	Q_2	Line
0	0	Q	$\bar{Q}$	1	1	1	5
0	1	1	0	2	1	0	6
1	0	0	1	3	0	1	7
1	1	0	0	4	Q	$\bar{Q}$	8

Lines 1 and 8 correspond to situations where the S and R inputs are both inactive, leaving the bistable circuit in one of the two possible states indicated in the tables above by the letters Q

and $\overline{Q}$ (Q may be either 1 or 0). If a specified output state is required, a pair of adequate complementary dc trigger signals can be applied to the S and R inputs simultaneously.

For instance, if the output pair is to be characterized by $Q_1 = 1$ and $Q_2 = 0$, the necessary input combination, for NOR and NAND bistable circuits, is $S = 1$ and $R = 0$. Changing S back from 1 to 0 does not change anything in the output state in the NOR bistable. The same is true if S is made equal to 1 in the NAND bistable. In both cases, the flip-flop exhibits infinite memory of the imposed state. The name *sequential circuit* is given to this class of networks.

Lines 4 and 5, however, must be avoided, for the passage from line 4 to line 1 or from line 5 to line 8 leads to uncertainty regarding the final state of the bistable circuit. In fact, the final transition is entirely out of the control of the input, since in both cases it results solely from small imbalances between transistor parasitics that allow faster switching of one inverter than the other.

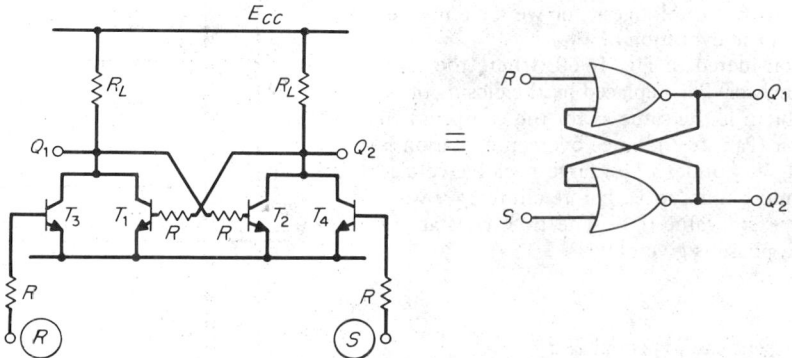

Fig. 16-62. DC-coupled version of flip-flop, customarily used in integrated-circuit versions of this circuit.

33. Synchronous Bistable Circuits. Sequential networks may be either asynchronous or synchronous. The asynchronous class describes circuits in which the application of an input control signal triggers the bistable circuit immediately. This is true of the circuits thus far considered. In the synchronous class, changes of state occur only at selected times, after a clock signal has occurred.

Synchronous circuits are insensitive to hazard conditions, whereas asynchronous circuits are severely troubled by this effect, originating in the propagation delay for which each bistable network or gate is responsible. These delays, although individually very small (typically of the order of a few nanoseconds), are responsible for introducing differential delays between signals that must travel through different numbers of logic circuits. Unwanted signal combinations may therefore appear for short periods and may be interpreted erroneously.

Synchronous circuits do not suffer from this limitation because they conform to the control signals only when the clock pulse is present, usually after the transient spurious combinations are over. A simple synchronous circuit is shown in Fig. 16-63. The inhibition action provided by the absence of the clock signal is provided by a pair of input AND circuits. Otherwise nothing is changed with respect to the classical bistable network.

A difficulty occurs in a cascade of bistable circuits, used to realize a time-division network or a shift register. Instead of each circuit controlling its closest neighbor, when a clock signal is applied, the set and reset signals of the first bistable jump from one circuit to the next, traveling throughout the entire chain in a time which may be shorter than the duration of the clock pulse. To prevent this, a time delay must be introduced between the gating NAND circuits (Fig. 16-63) and the bistable network, so that changes of state can occur only after the clock signal has disappeared.

One approach is to take advantage of storage effects in bipolar transistors, but the so-called *master-slave* association, shown in Fig. 16-64, is preferred. In this circuit, intermediate storage is realized by an auxiliary clocked bistable network controlled by the complement of the

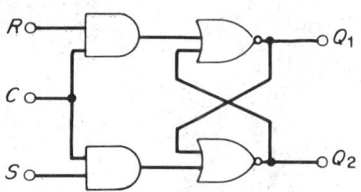

Fig. 16-63. Synchronous flip-flop.

clock signal. The additional circuit complexity is appreciable, but this approach is practical in integrated-circuit technology.

The master-slave bistable truth table can be found from that of the synchronous circuit in Fig. 16-63, which in turn can be deduced from the truth table in Par. **16-32**. One problem remains, however, the forbidden 1,1 input pair which is responsible for ambiguous states each time the clock goes to zero. To solve this problem, the JK bistable is introduced (see Fig. 16-65). The main

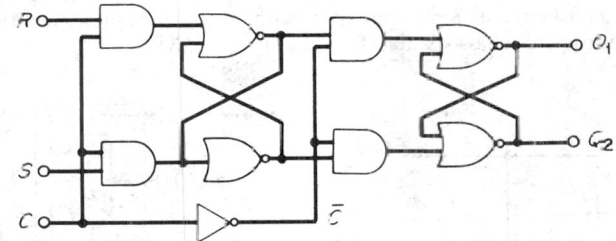

Fig. 16-64. Master-slave synchronous flip-flop.

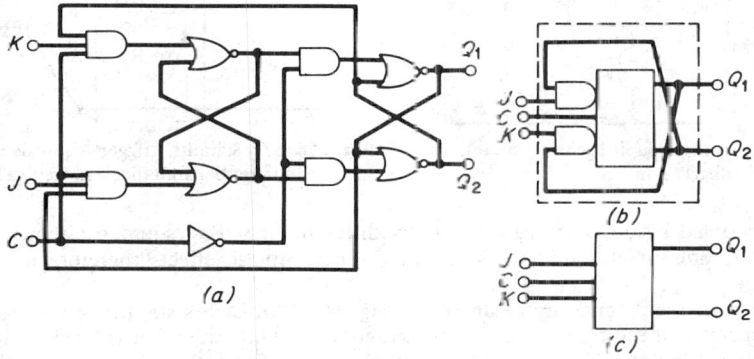

Fig. 16-65. JK flip-flop.

difference from the circuit of Fig. 16-64 is the introduction of a double feedback loop. Hence the S and R inputs become respectively JQ and $K\overline{Q}$. As long as the J and K inputs are not simultaneously equal to 1, nothing in fact is changed with respect to the behavior of the SR synchronous circuit.

When J and K are both high, the cross-coupled output signals fed back to the input gates cause the flip-flop to toggle under control of the clock signal. The truth table then becomes

J	K	Q_{n+1}	Line
0	0	Q_n	1
1	0	1	2
0	1	0	3
1	1	$\overline{Q}_n$	4

Q_{n+1} stands for Q at the clock time $n+1$ and Q_n for Q at clock time n. Lines 1 to 3 match the corresponding lines of the NOR bistable truth table in Par. **16-32**. Line 4 indicates that state transitions occur each time the clock signal goes from high to low. The corresponding logic equation of the JK flip-flop therefore is

$$Q_{n+1} = J\overline{Q}_n + \overline{K}Q_n \tag{16-75}$$

When only one control signal is used, J for instance, and the other, K, is obtained by negation of the J signal, a new type of bistable is found which is called the D flip-flop. The name given to the J input is D. Since K is equal to $\overline{J}$, Eq. (16-75) reduces to

$$Q_{n+1} = D \tag{16-76}$$

Hence, in this circuit the output is set by the input D after a clock cycle has elapsed. Notice that the flip-flop is insensitive to changes of D occurring while the clock is high.

D flip-flops without the master-slave configuration also exist, but their output state follows the D signal if changes occur while the clock is high. These bistables can be used to latch data. Several D flip-flops controlled by the same clock form a register for data storage. The clock signal then is an enable signal.

34. Emitter-Coupled Bistables (Schmitt Circuits).[16] In the basic Schmitt circuit represented in Fig. 16-66 bistable operation is obtained by the positive-feedback loop formed by the common-base and common-collector transistor pair (respectively T_1 and T_2). The Schmitt circuit can also be considered as a differential amplifier with a positive-feedback loop, which turns out to be a series-parallel association (Fig. 16-67).

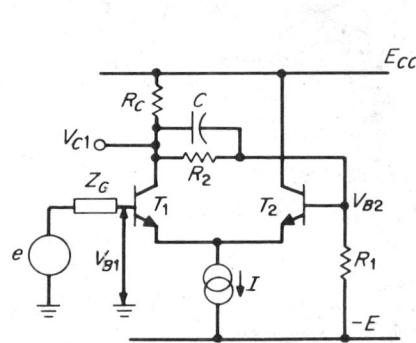

Fig. 16-66. Emitter-coupled Schmitt circuit, showing positive-feedback loop.

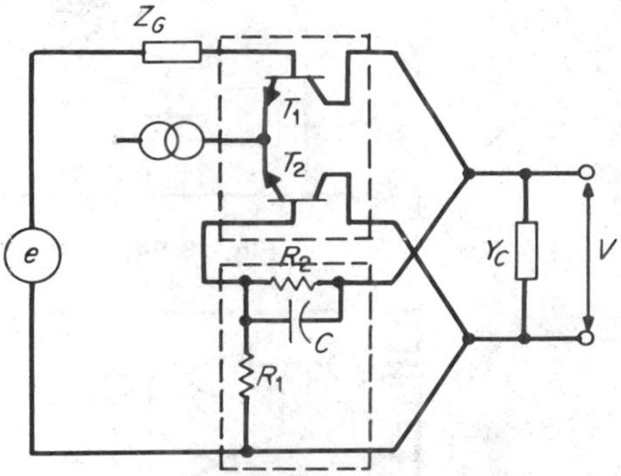

Fig. 16-67. Schmitt trigger equivalent circuit, showing differential amplifier and feedback loop.

Emitter-coupled bistables are fundamentally different from Eccles-Jordan circuits, since no transistor saturates in either of the two stable states. Storage effects therefore need not be considered.

The two permanent states are examined in Fig. 16-68. In each state, (a) as well as (b), one transistor operates in the common-collector configuration while the other is blocked. In Fig. 16-68a, the collector voltage V_{C1} of T_1 and base voltage V_{B2} of T_2 are given by

$$V_{C1} = E_{cc} \frac{R_1 + R_2}{R_1 + R_2 + R_c}$$

$$V_{B2} = E_{cc} \frac{R_1}{R_1 + R_2 + R_c} = V_h \qquad (16\text{-}77)$$

When the other stable state (b) is considered,

$$V_{C1} = (E_{cc} - R_c I) \frac{R_1 + R_2}{R_1 + R_2 + R_c}$$

$$V_{B2} = (E_{cc} - R_c I) \frac{R_1}{R_1 + R_2 + R_c} = V_l \qquad (16\text{-}78)$$

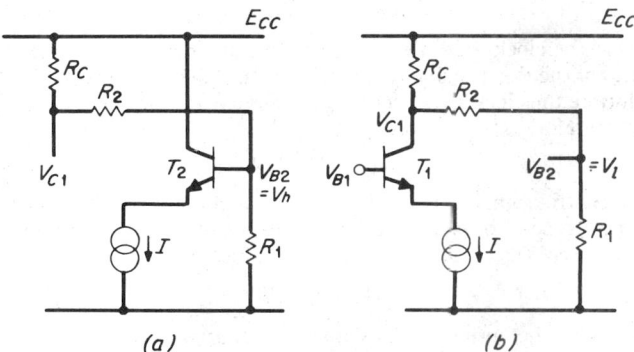

Fig. 16-68. Execution of transfer in Schmitt circuit: (a) with T_1 blocked; (b) with T_1 conducting.

The situation depicted in Fig. 16-68a remains unchanged as long as the input voltage V_{B1} applied to T_1 is kept below the actual value V_h. In the other state (b), T_2 will be off as long as V_{B1} is larger than V_l. A range of input voltages between V_h and V_l thus exists where either of the two states is possible.

To alleviate the ambiguity, let us consider an input voltage below the smallest of the two possible values of V_{B2} so that the transistor T_1 necessarily is blocked. This corresponds to the situation of Fig. 16-68a. Now let the input volt-
age be gradually increased. Nothing will happen until V_{B1} approaches V_h. When the difference between the two base voltages is reduced to 100 mV or less, T_1 will start conducting and the volt-age drop across R_c will lower V_{32}. The emitter current of T_2 will consequently be reduced, and more current will be fed back to T_1. Hence, an unstable situation is created which ends when T_1 takes over all the current delivered by the current source and T_2 is blocked.

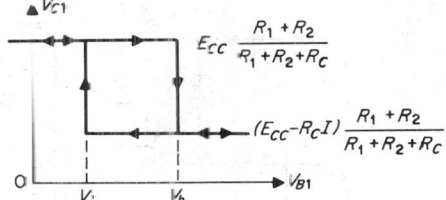

Fig. 16-69. Input-output characteristic of Schmitt circuit, showing rectangular hysteresis.

Now the situation depicted in Fig. 16-68b is reached. The base voltage of T_2 becomes V_l, and the input voltage may either continue to increase or decrease without anything else happening as long as V_{B1} has not reached V_l. When V_{B1} approaches V_l, another unstable situation is created causing the switching from (b) to (a). Hence, the input-output characteristic of the Schmitt trigger is as shown in Fig. 16-69 with a rectangular hysteresis loop.

Schmitt triggers are suitable for detecting the moment when an analog signal crosses a given dc level. They are widely used in oscilloscopes to provide time-base synchronization. This is illustrated in Fig. 16-70, which shows a periodic signal triggering a Schmitt circuit and the cor-responding output waves. It is possible to modify the switching levels by changing the operating points of the transistors electrically, e.g., by modifying the current delivered by the current source.

In many applications, the width of the hysteresis does not play a significant role. The width can be decreased, however, by increasing the attenuation of the resistive divider formed by R_1 and R_2, but one should not go below 1 V because sensitivity to variations in circuit components or supply voltage may be experienced. Furthermore, the increased attenuation in the feedback loop must be compensated for by a corresponding increase in the differential amplifier gain. Otherwise the loop gain may fall below 1, preventing the Schmitt circuit from triggering. A hysteresis of a few millivolts is therefore difficult to achieve.

A much better solution is to use *comparators* instead of Schmitt triggers. A typical integrated comparator is shown in Fig. 16-71. It is mainly a medium-gain amplifier (10^3 to 10^4) with a very fast response (a few tens of nanoseconds) and an excellent slew rate. Comparators are not designed to be used as linear amplifiers like op-amps. Their large gain-bandwidth product makes them inap-propriate for feedback configurations. They inev-itably oscillate in any type of closed loop. In an open-loop configuration, however, they behave like clipping circuits with an exceedingly small input hysteresis which is equal to the output-volt-age swing, usually 5 V, divided by the gain. The main difference compared with Schmitt triggers is their high sensitivity. Another important differ-ence is the fact that comparators do not actually switch like Schmitt triggers. This is not a signifi-cant problem, however, except for very slowly varying input signals.

In the circuit of Fig. 16-66, the common-emitter current source can be replaced by a resistor. This is a satisfactory and economical solution, but it introduces some common-mode sensitivity. The

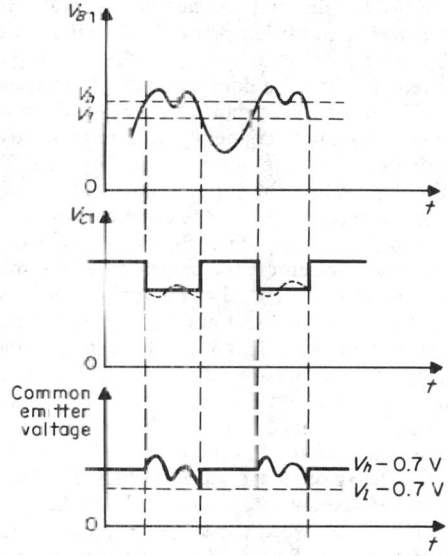

Fig. 16-70. Trigger input and output voltage of Schmitt circuit; solid line delivered by a cur-rent source; broken line delivered by a resistor.

output signal does not look square, as shown in Fig. 16-70 by dashed lines. This unwanted effect can be avoided by taking the output signal at the collector of T_2 through an additional resistor, since the current flowing through T_2 is constant. An additional advantage of the latter circuit is that the output load does not interfere with the feedback loop.

35. Integrated-Circuit Schmitt Triggers.[17] Basically a Schmitt trigger can always be implemented by means of an integrated differential amplifier and an external positive-feedback loop. If the amplifier is an op-amp, poor switching characteristics are obtained unless an

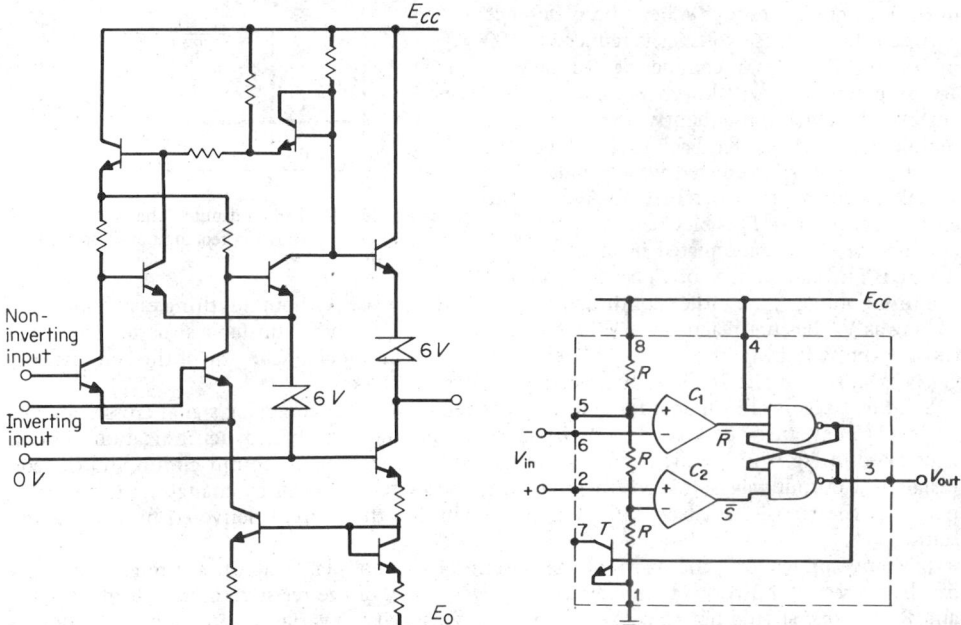

Fig. 16-71. Bipolar integrated version of a comparator.

Fig. 16-72. Precision integrated Schmitt trigger.

amplifier with a very high slewing rate is chosen. If a comparator is considered instead of an op-amp, switching will be fast but generally the output signal will exhibit spurious oscillations during the transition period. The oscillatory character of the output signal is due to the trade-off between speed and stability which is typical of comparators compared with op-amps. Any attempt to create a dominant pole, in fact, inevitably would ruin their speed performance.

The integrated-circuit counterpart of the Schmitt trigger is shown in Fig. 16-72. It consists of two comparators connected to a resistive divider formed by three equal resistors R. Input terminals 2 and 6 are normally tied together. The output signals of the two comparators control a flip-flop. When the input voltage is below $E_{cc}/3$, the flip-flop is set. Similarly when the input voltage exceeds $2E_{cc}/3$, the circuit is reset.

The actual state of the flip-flop in the range between $E_{cc}/3$ and $2E_{cc}/3$ will depend on how the input voltage enters the critical zone. For instance, if the input voltage starts below $E_{cc}/3$ and is increased so that it changes the state of comparator C_2 but not that of comparator C_1, both $\bar{S}$ and $\bar{R}$ are equal to 1 and the flip-flop remains set. The state changes only if the input voltage exceeds the limit $2E_{cc}/3$. Similarly, if the input voltage is lowered, setting the flip-flop will occur only when $E_{cc}/3$ is reached. Hence the circuit of Fig. 16-72 behaves like a Schmitt trigger with a hysteresis width $E_{cc}/3$ depending only on the resistive divider $3R$ and the offset voltages of the two comparators C_1 and C_2. This circuit can therefore be considered as a precision Schmitt trigger and is widely used as such.

36. Monostable and Astable Circuits (Discrete Components).[13] Figures 16-73 and 16-74 show monostable and astable collector-coupled pairs, respectively. The fundamental difference between these circuits and the bistable networks of Par. **16-31** lies in how dc biasing is achieved. In Fig. 16-73a, T_2 is normally conducting except when a negative trigger pulse drives this transistor into the cutoff region. T_1 necessarily undergoes the inverse transition, suddenly producing a large negative voltage step at the base of T_2 shown in Fig. 16-73b. V_{BE2}, however, cannot remain negative since the base is connected to the positive-voltage supply through the resistor R_1. The base voltage must thus rise toward E_x, with a time constant R_1C. As soon as the emitter junction of T_2 becomes forward-biased, the monostable circuit will change its state again and the circuit remains in that state until another trigger signal is applied.

The time T between the application of a trigger pulse and the instant T_2 saturates again is given approximately by

$$T = \tau \log_n 2 = 0.693\tau \qquad (16\text{-}79)$$

where $\tau = R_1C$. The supply voltage E_{CC} must be large compared with the forward-voltage drop of the emitter junction of T_2 for this expression to apply.

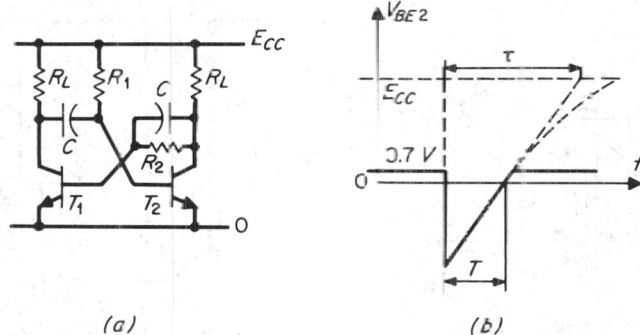

(a)   (b)

Fig. 16-73. Monostable collector-coupled pair (a) circuit; (b) output vs. time characteristics.

The astable collector-coupled pair, or free-running multivibrator (Fig. 16-74), operates according to the same scheme except that steady-state conditions are never reached. The base-bias networks of both transistors are connected to the same positive power supply. The period of the multivibrator thus equals $2T$ if the circuit is symmetrical, and the repetition rate F_r is given by

$$F_r = \frac{1}{2\tau \log_n 2} = \frac{0.721}{RC} \qquad (16\text{-}80)$$

An astable network need not be triggered, of course, but pulses fed to a multivibrator from an external generator may synchronize it. The mechanism can be understood on the basis of Fig. 16-75, where we assume that regularly spaced negative pulses are applied to the collectors of the multivibrator or to the bases through small capacitances. Figure 16-75 shows typical base-voltage curves obtained when the multivibrator is synchronized by means of pulses running 6 times faster than the period of the multivibrator itself. Starting arbitrarily from a state transition, at the time t_0, we see that the pulse appearing at t_1 is not sufficient to drive the blocked transistor into the active region because its base is still heavily reverse-biased. The second pulse does not cause switching either, but the third is successful.

If the multivibrator is symmetrical, similar situations will occur respectively at t_4, t_5, and t_6. State transitions consequently occur every third trigger pulse, and the oscillation frequency of

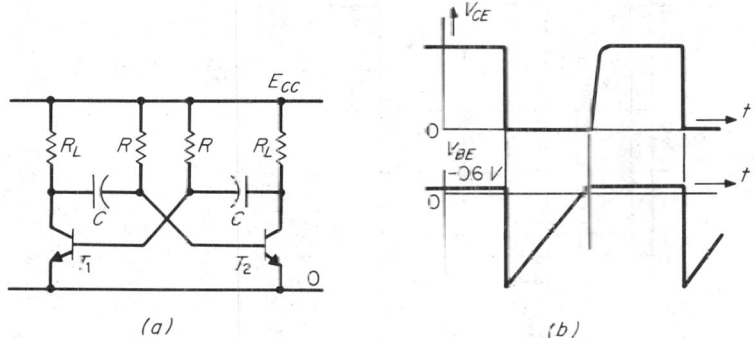

(a)   (b)

Fig. 16-74. Astable (free-running) flip-flop: (a) circuit; (b) output vs. time characteristics.

the multivibrator is 6 times lower than that of the pulse generator. Symmetry is not absolutely necessary. In fact, a multivibrator may run at uneven fractions of the pulse-repetition rate.

Synchronization of astable networks provides an easy method of time division, but this method suffers from its high sensitivity to variations of the circuit elements and power supply. The larger the time division, the more critical the adjustments. Time division by a factor of 10, for instance, is completely unreliable. The availability of integrated counters has completely eliminated the synchronized-multivibrator method of time division.

An interesting discrete-component astable circuit derived from the Schmitt trigger is shown in Fig. 16-76. The capacitor C provides a short circuit between the emitters of the two transistors, closing the positive-feedback loop during the regeneration time. As long as one or the other of

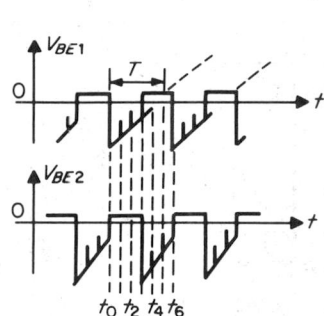

Fig. 6-75. External synchronization of a flip-flop.

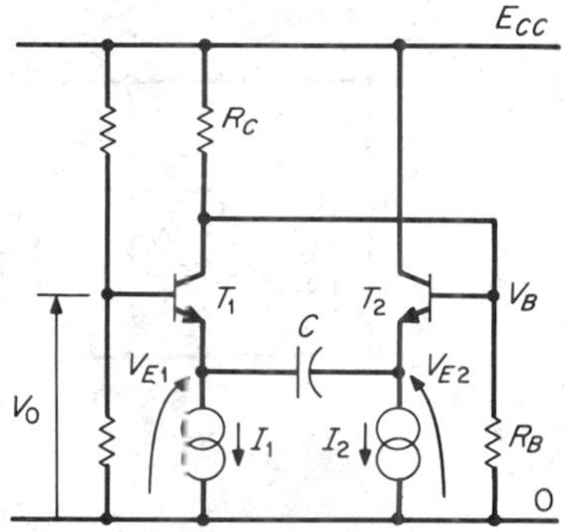

Fig. 16-76. Discrete-component emitter-coupled astable circuit.

the two transistors is cut off, C offers a current sink to the current source connected to the emitter of the blocked transistor. The capacitor thus is periodically charged and discharged by the two current sources, and the voltage across its terminal exhibits a triangular waveform. The collector current of T_1 is either zero or $I_1 + I_2$, so that the resulting voltage step across R_C is ($R_B \parallel R_C)(I_1 + I_2)$. Since the base of T_2 is directly connected to the collector of T_1, the same voltage step controls T_2 and determines the width of the input hysteresis, i.e., the maximum amplitude of the voltage sweep across C. The period of oscillation is computed from

$$T = C\left(\frac{1}{I_1} + \frac{1}{I_2}\right)(R_B \parallel R_C)(I_1 + I_2) \tag{16-81}$$

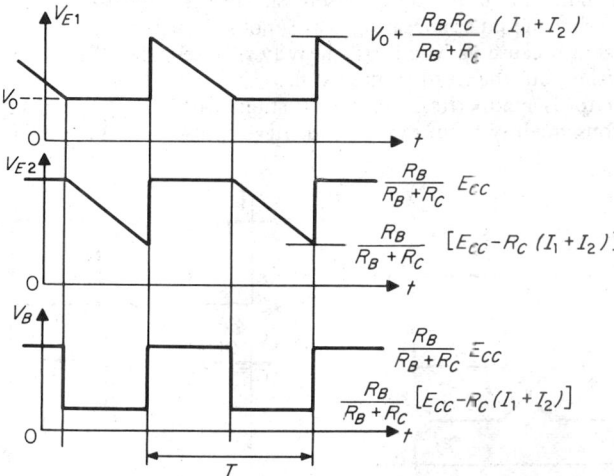

Fig. 16-77. Waveforms of circuit in Fig. 16-76.

when, as is usual, both current sources deliver equal currents, the expression for T reduces to

$$T = 4(R_8 \parallel R_9)C \qquad (16\text{-}82)$$

T does not depend, in this case, on the amplitude of the current because changes in current in fact modify the amplitude and the slope of the voltage sweep across C in the same manner. A review of the waveforms obtained at various points of the circuit is given in Fig. 16-77.

37. Integrated Monostable and Astable Circuits.[17] Integrated monostable and astable circuits can be derived from the precision Schmitt trigger circuit shown in Fig. 16-72. The monostable configuration is illustrated in Fig. 16-78a.

Under steady-state conditions, the flip-flop is set and the transistor T saturated. The input voltage V_{in} is kept somewhere between the two triggering levels $E_{cc}/3$ and $2E_{cc}/3$. To initiate the monostable condition, it is sufficient that V_{in} drops, even for a very short time, below $E_{cc}/3$ in order to reset the flip-flop and prevent T from conducting. The current flowing through R_1 then charges C until the voltage V_C reaches the triggering level $2E_{cc}/3$. Immediately thereafter, the circuit switches to the opposite state and transistor T discharges C. The monostable circuit remains in that state until a new triggering pulse V_{in} is fed to the comparator C_2. The waveforms V_{in}, V_C, and V_{out} are shown in Fig. 16-78b. This circuit is also called a *timer* because it provides constant-duration pulses, triggered by a short input pulse. A slight modification of the external control circuitry may change this monostable into a retriggerable timer.

The astable version of the precision Schmitt trigger is shown in Fig. 16-79. Its operation is easily understood from the preceding discussion. The capacitor C is repetitively discharged

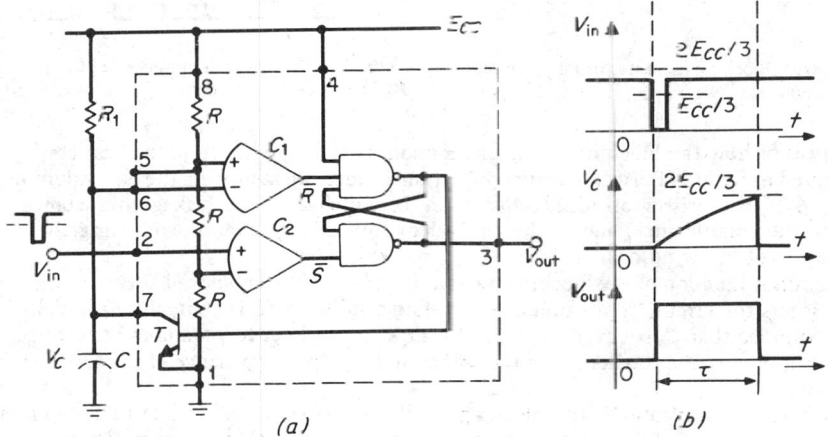

Fig. 16-78. Monostable precision Schmitt trigger: (a) circuit; (b) waveforms.

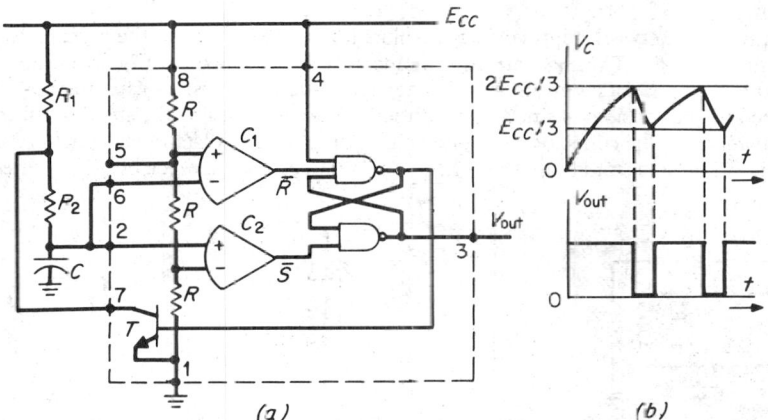

Fig. 16-79. Astable precision Schmitt trigger: (a) circuit; (b) waveforms.

through R_2 in series with the saturated transistor T and recharged through $R_1 + R_2$. The voltage V_C therefore consists of two distinct exponentials clamped between the triggering levels $E_{cc}/3$ and $2E_{cc}/3$. The frequency is equal to $1.44/(R_1 + 2R_2)C$. Because of the precision of the triggering levels, excellent short-term frequency stability can be achieved with this circuit.

38. The Blocking Oscillator.[18] The regenerative circuits examined thus far have a positive-feedback loop comprising at least two active elements. The blocking oscillator, on the other hand, needs only one transistor, used with a pulse transformer. The circuit derives from the harmonic oscillator, but its very close coupling makes it an extremely nonlinear network that can operate in a monostable mode or free-running mode. The resolution time, as in other pulse-generating circuits, depends on reactive components, usually the magnetizing inductance of the pulse transformer, but saturation of the transformer or other elements may dominate if the circuit is improperly designed.

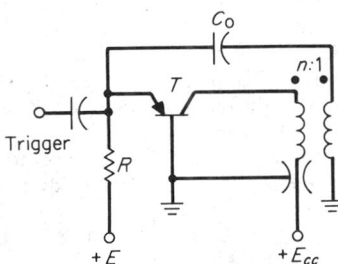

Fig. 16-80. Blocking oscillator using common-base configuration.

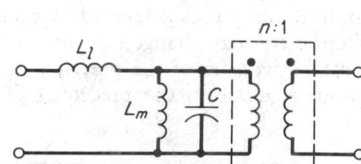

Fig. 16-81. Equivalent network of the pulse transformer used in Fig. 16-80.

Analysis of how the blocking oscillator functions is based on the common-base configuration represented in Fig. 16-80. For the actual pulse transformer, we substitute the equivalent network of Fig. 16-81, comprising an ideal transformer, of which L_l is the leakage inductance, L_m the magnetization inductance, and C the equivalent lumped capacitance, including the parasitic capacitances of the windings.

The emitter junction of the blocking oscillator is assumed to be blocked at the time a negative pulse triggers the circuit. Thus under steady-state conditions C_0 is charged to the voltage E. It will be assumed that C_0 behaves like a floating dc source. Moreover, R and the trigger generator need not be considered further, since the common-base input impedance of the active transistor is so small.

With these simplifications, the blocking oscillator can be considered to be a common-base unity-gain current amplifier followed by a current transformer that injects in the emitter a current n times larger than the collector current. The resulting regeneration time is very short, a few nanoseconds, depending primarily on the transistor switching characteristics and the series leakage current.

The equivalent network representing the blocking oscillator during the regeneration time is depicted in Fig. 16-82. The magnetization inductance is neglected because switching occurs so fast that the magnetization current remains negligibly small. The stray capacitance of the transformer need not be considered since it is essentially short-circuited by the small input impedance of the transformer. The collector-to-ground voltage drops almost instantaneously to zero, causing saturation of the transistor before the collector current becomes appreciable. The collector junc-

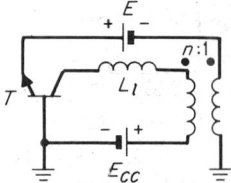

Fig. 16-82. Equivalent network of blocking oscillator during regeneration time.

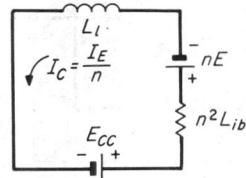

Fig. 16-83. Collector loop of circuit in Fig. 16-82.

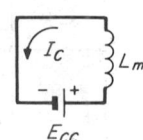

Fig. 16-84. Path of collector current through magnetization inductance.

tion becomes essentially a short circuit. Hence the collector loop can be represented as in Fig. 16-83, with the emitter load seen through the ideal transformer.

The current in the loop increases very rapidly until a maximum is reached:

$$(1/n)I_{E,max} = (E_{CC} - nE)/n^2 h_{ib} \tag{16-83}$$

The rise time

$$t_r = 2.2L_f/n^2 h_{ib}$$

is of the order of 10 ns or less. During this period, emitter and collector current vary in the same manner, but the emitter current is n times larger than the collector current and the transistor is thus heavily saturated.

This situation could persist indefinitely if the magnetization inductance were infinite. This not being the case, additional collector current starts flowing in the magnetization inductance according to Fig. 16-84. The total collector current is given by the sum

$$I_c = \frac{1}{n}\left(\frac{E_{cc}}{n} - E\right)\frac{1}{h_{ib}} + \frac{E_{cc}}{L_m}t \tag{16-84}$$

This expression remains valid as long as no saturation is experienced; otherwise I_c departs suddenly from its linear increase, shown in Fig. 16-85, and increases much faster. Whether or not this happens, I_c at some point becomes equal to I_E and the transistor enters the active region. A new regeneration phase then takes place, tending to cut the transistor off as fast as occurred during the first regeneration period. The collector current vanishes almost instantaneously, and the energy accumulated in the magnetization inductance causes a rapid increase of the collector voltage, substantially above E_{CC}. The combination of L_m and C produces ringing, unless an external damping circuit is used. Avalanche breakdown is likely to occur unless steps are taken to prevent it.

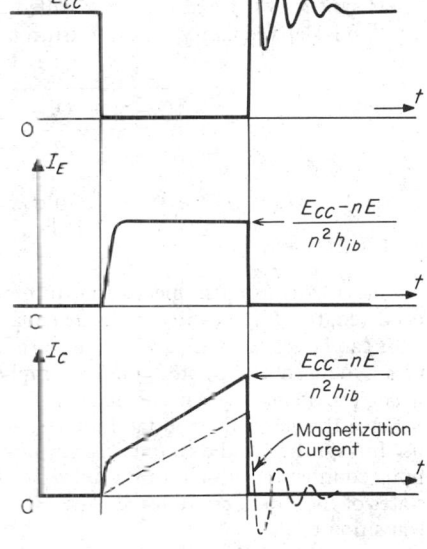

Fig. 16-85. Waveforms of blocking oscillator.

Blocking oscillators have been used extensively in television applications. They are less used now for the obvious reason that there is no way to put them into integrated-circuit form. Moreover, their very sharp rise times combined with the risk of magnetic coupling causes them to become radio-interference generators.

DIGITAL AND ANALOG SYSTEMS

39. Integrated Systems. With the trend toward higher integration levels, an increasing number of ICs is being designed which combine some of the circuits seen before in order to build large systems, digital as well as analog, or even mixed, such as wave generators, A/D and D/A converters, etc. Some of them are reviewed below.

40. Counters.[8,19] To count any number N of events, at least K flip-flops are required, such that

$$2^K \geq N \tag{16-85}$$

Ripple Counters. JK flip-flops with J and K inputs equal to 1 are divide-by-2 circuits. Hence, a cascade of K flip-flops with each output Q driving the clock of the next circuit forms a divide-by-2^K chain, or a binary counter (see Fig. 16-86). The main drawback of this circuit is its increasing propagation delay with K. When all the flip-flops switch, the clock signal must in fact ripple through the entire counter. Hence, enough time must be allowed to obtain the correct count. Furthermore, the delays between the various stages of the counter may produce glitches, e.g., when decoding is achieved.

Synchronous Binary Counters. Minimization of delay and glitches can be achieved by designing synchronous instead of asynchronous counters. In a synchronous counter all the clock inputs of the *JK* flip-flops are driven in parallel by a single clock signal. The control of the counter is achieved by driving the *J* input by means of the AND combination of all the preceding Q outputs, as shown in Fig. 16-87. In this manner, all state changes occur on the same trailing edge of the clock signal. The only remaining requirement is to allow enough time between clock pulses for the propagation through a flip-flop and one AND gate. The drawback of course is increased complexity with the order K.

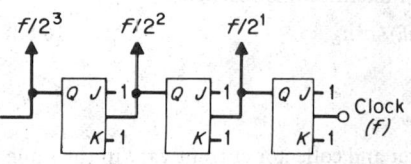

Fig. 16-86. Ripple counter formed by cascading flip-flops.

Divide-by-N Synchronous Counters. Whenever the number N of counts cannot be expressed in binary form, auxiliary decoding circuitry is required. This is true also for counters that provide truncated and irregular count sequences. Their synthesis is based on the so-called transition tables. The basic transition table of the *JK* flip-flop is derived easily from the truth table of Par. **16-33**:

Q_n	Q_{n+1}	*J*	K	Line
0	0	0	$\times$	1
0	1	1	$\times$	2
1	0	$\times$	1	3
1	1	$\times$	0	4

Line 1, for instance, means that in order to maintain Q equal to 0 after a clock signal has occurred, the *J* input must be made equal to 0 whatever K may be ($\times$ stands for "don't care"). This can be easily verified with the truth table (lines 1 and 3) of Par. **16-33**. Hence, the synthesis of a synchronous counter consists simply of determining the *J* and K inputs of all flip-flops needed to obtain a given sequence of states.

Once the *J* and K truth tables have been obtained, classical minimization procedures can be used to synthesize the counter. For instance, let us consider the design of a divide-by-5 synchronous counter for which a minimum of three flip-flops is required. First, the present and next states of the flip-flops are listed. Then the required *J* and K inputs are found by means of the *JK* transition table:

	Present state			Next state			JK inputs					
	Q_3	Q_2	Q_1		Q_3	Q_2	Q_1	J_3	K_3	J_2	K_2	J_1	K_1
0	0	0	0	1	0	0	1	0	$\times$	0	$\times$	1	$\times$
1	0	0	1	2	0	1	0	0	$\times$	1	$\times$	$\times$	1
2	0	1	0	3	0	1	1	0	$\times$	$\times$	0	1	$\times$
3	0	1	1	4	1	0	0	1	$\times$	$\times$	1	$\times$	1
4	1	0	0	0	0	0	0	$\times$	1	0	$\times$	0	$\times$

Using Karnaugh minimization techniques, one finds

$$J_3 = Q_1 Q_2 \qquad K_3 = 1$$
$$J_2 = Q_1 \qquad K_2 = Q_1$$
$$J_1 = \overline{Q_3} \qquad K_1 = 1$$

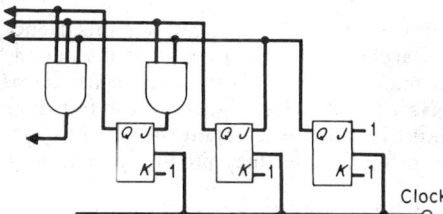

Fig. 16-87. Synchronous counter.

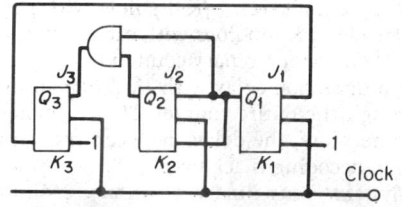

Fig. 16-88. Synchronous divide-by-5 circuit.

The corresponding counter is shown in Fig. 16-88. With this procedure it is quite simple to synthesize a decimal counter with four flip-flops. The same method also applies to the synthesis of D flip-flop counters.

Up-Down Counters. Upward and downward counters differ from the preceding ones only by the fact that an additional bit is provided to select the proper J and K controls for up or down count. The same synthesis methods are applicable.

Presettable Counters. Since a counter is basically a chain of flip-flops, parallel loading by any number within the count sequence is readily possible. This can be achieved by means of the set terminals, for instance. Hence, the actual count sequence can be initiated from any arbitrary number.

41. Shift Registers.[8] Shift registers are chains of flip-flops connected so that the state of each can be transferred to its next left or right neighbor under control of the clock signal. Shift registers can be built with JK as well as D flip-flops. An example of a typical bidirectional shift register is shown in Fig. 16-89. Shift registers, like counters, may be loaded in parallel or serial mode. This is also true for reading out.

In MOS technology, dynamic shift registers can be implemented in a very simple manner. Memorization of the state occurs electrostatically rather than by means of a flip-flop. The information is stored in the form of the charge on the gate of an MOS transistor. Retention times of 1 ms are readily available and reliable although the input capacitance of MOS transistors is usually less than 1 pF. The main advantage of MOS dynamic shift registers is the area saving resulting from the replacement of a flip-flop by a single MOS transistor. A typical dynamic 2-phase shift register (Fig. 16-90) consists of two cascaded MOS inverters connected by series switches.

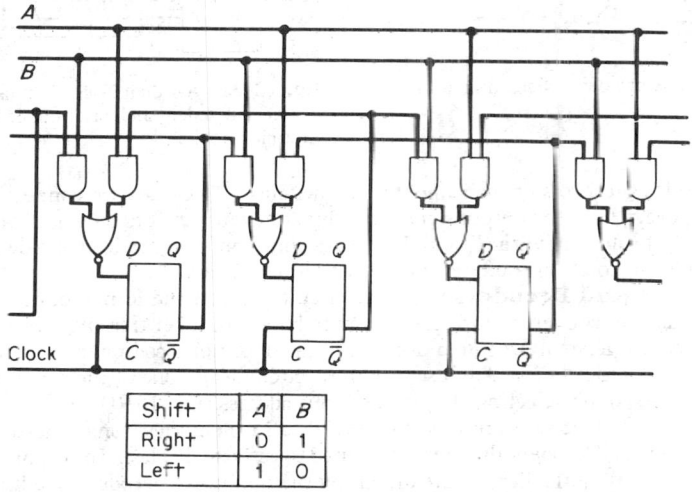

Shift	A	B
Right	0	1
Left	1	0

Fig. 16-89. Bidirectional shift register.

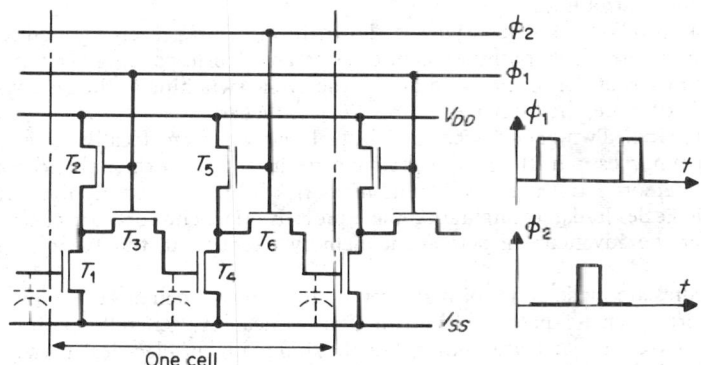

Fig. 16-90. A 2-phase MOS dynamic shift register.

These switches, T_3 and T_6, are divided into two classes: *odd* switches are controlled by the signal ϕ_1, and *even* switches are controlled by ϕ_2.

The control signals determine two phases which are interlaced. When ϕ_1 turns on the odd switches, the output signal of the first inverter controls the next inverter but T_6 prevents the information from going further. The interrupters are on during a very short time, of the order of 1 or 2 μs. The rest of the time, retention of the information occurs. The signal jumps from one inverter to the next until it reaches the last stage. Feeding the output signal back to the input yields memory of indefinite duration, provided the control signals continue indefinitely.

The power consumption of dynamic shift registers is very small compared with that of bipolar circuits. The static power consumption of 2-phase dynamic MOS memories can be cut even

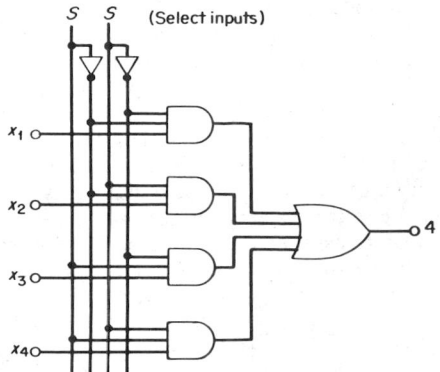

Fig. 16-91. A one-out-of-4 decoding network.

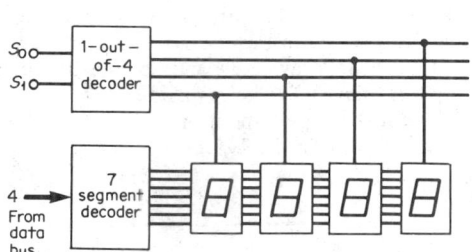

Fig. 16-92. A 4-digit 7-segment display using a 7-segment decoder and a 1-out-of-4 decoder for multiplexing.

further if the load resistors of each inverter are switched. The loads are connected only when required, i.e., every time the output signal of an inverter must shift to its neighbor. Figure 16-90 shows a practical solution, with T_2 and T_3 serving simultaneously as load and switch. Another even more efficient solution is offered by 4-phase shift registers.

42. Encoders and Decoders.[19] Coding circuits convert the format of digital data. They are currently used to convert numbers, e.g., a pure binary number that must be changed into a BCD (binary-coded decimal) or into a decimal number. Among coding circuits, the 1-out-of-n codes or address decoders are widely used. In these circuits, any one of the n output signals can be given the value 1 by selecting the proper input address (Fig. 16-91). All the other outputs remain at 0. Address decoders are used to select data in memories, sometimes in combination with demultiplexers. Decoders also are widely used to activate displays. For instance, the control lines of 7-segment numeric displays are driven by the outputs of decoders, the inputs of which are fed by the 4 bits representing the binary code of the numbers to be displayed. With multiple-digit displays, 7-segment decoders and 1-out-of-n decoders are used in conjunction for the minimization of the control lines.

A typical multiplexed 4-digit display is shown in Fig. 16-92. Each light-emitting diode is selected by one of the 7 lines of the segment decoder, and a corresponding 1-out-of-4 line display is selected by means of the binary word S_0S_1. If the cyclic switching of the displays is done at a speed of 60 Hz or more, the human eye cannot detect flicker.

43. Memories. Two-state devices and bistable circuits[20] are ideally suited for memory (binary-storage) purposes. In the past, magnetic cores have offered the only economic solution for large-size memories, but use of solid-state memories (which rely on bistable circuits derived from the basic Eccles-Jordan configuration) now prevails. Magnetic cores are cheap and reliable, and they offer the advantage of permanent memory retention in the event of power-supply failure.

Magnetic cores are made either of high-permeability alloys, such as Permalloy, or of ferrites (see Sec. **6**). Ferromagnetic materials having a high permeability generally exhibit low coercive force. Ferrites have smaller permeability, but they offer higher electrical resistance and thus smaller losses. The mechanism of magnetization can be understood on the basis of microscopic theory. Within the material small domains exist where the molecular magnetic moments have the same direction. The domains are separated by walls, and their magnetic moments are distributed randomly on a macroscopic scale. An external field aligns the magnetic moments of those domains already close to the imposed direction. The domain walls then become unstable, and some of the domains grow rapidly until they encompass almost the entire material, causing a large flux-density variation followed by saturation. Further increasing the magnetization force causes only a few remaining domains, e.g., those which were oriented in the reverse direction, to rotate. After the external field is removed, the magnetic domains relax somewhat and adopt "easy" magnetization directions, which depart little from the imposed one. A reverse magnetization is required to return to the initial random situation.

When a step-function magnetizing force is applied to a magnetic core, the resulting emf developed across a test winding exhibits the shape indicated in Fig. 16-93. The initial spike is the

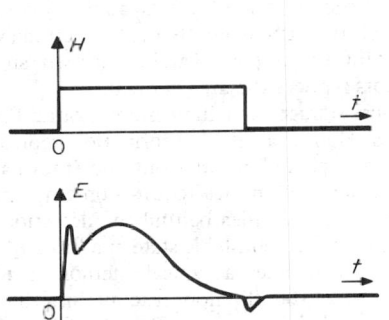

Fig. 16-93. Change in magnetizing force of core-memory cell and resulting induced voltage.

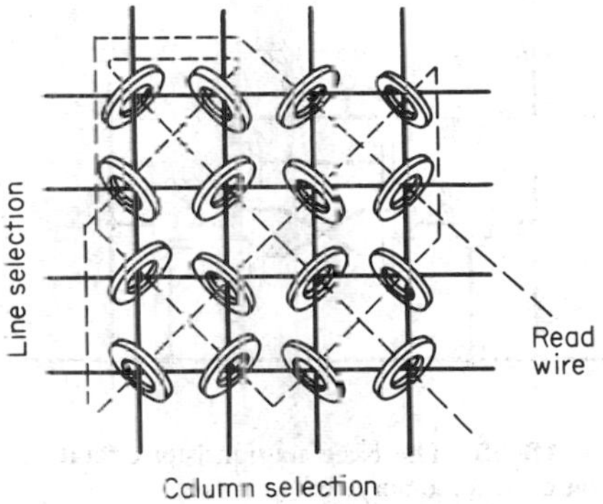

Fig. 16-94. Two-dimensional memory-core array.

result of rapid rotation of some domains that are already oriented close to the imposed direction. The slower and more important flux variation that follows is due to propagation of changes in the domain walls. The last negative spike corresponds to a small flux-density variation after the excitation is removed, and the magnetic moments relax toward their nearest axis of easy magnetization.

The longest time constant involved, which is associated with the wall-propagation mechanism is usually of the order of 100 ns. When a core has been magnetized previously in some direction and a new magnetizing pulse with the same direction is imposed, propagation of walls is almost nonexistent. This means that the long pulse in Fig. 16-93 vanishes but rotation is still experienced, so that the two short spikes still remain. In applications where cores are used as memory media, some care must be taken not to interpret rotation spikes as useful bits. The use of the strobing-pulse technique or slower rise and decay times of the magnetizing force may help cope with this problem.

Spurious signals become more important as the number of interrogated magnetic cores increases. In computer memories it is usual practice to wind the reading wire of a core matrix in such a way that unwanted effects cancel out to the maximum possible extent. A practical solution used in an elementary two-dimensional matrix is shown in Fig. 16-94. Selecting the information contained in a given core is performed by the so-called *coincidence-current technique*. In such a matrix only the core situated at the intersection of a selected row and column can switch, whereas those on the same row but a different column (or vice versa) are subjected to only half the requisite magnetizing force and hence do not switch.

Reading the content of a magnetic-core matrix destroys the available information. It is thus necessary to restore the information after each reading operation if further storage of the data is required.

Semiconductor memories are now successful competitors of core matrices, since large-scale integration, combined with mass production, has lowered their cost considerably. In the early 1970s solid-state memories became cheaper than core memories for capacities of the order of 1,024 to 4,096 bits.

Solid-State Memories.[21,22]   The static memory flip-flop is the solid-state counterpart of core memories. It is basically an Eccles-Jordan bistable pair, using either bipolar or field-effect transistors. An illustration of a six-transistor static MOS memory cell is given in Fig. 16-95. The state of the cell can be read nondestructively by means of the word line (WL). When the WL is in the high state, the bit lines (BL and $\overline{BL}$) can be used for readout as well as for write-in purposes. Storage of the information is obtained by lowering the voltage of the WL. Retention is lost if the power supply fails (unlike core memories).

The efforts of semiconductor memory manufacturers during the 1970s have been aimed at decreasing the power consumption, area per cell, and access time of semiconductor memories. Less power per cell is consumed if the load transistors are eliminated at the expense of refresh circuitry. The static memory cell then becomes a dynamic memory cell.

The smallest dynamic memory cell (Fig. 16-96) consists of a single MOS transistor.

The word line (WL) controls the gate of a MOS switch which connects the diffused bit line to a storage capacitor. Charge is pumped from (or fed back to) the bit line by changing the bit-line voltage. Storage of the charge is done by means of the inversion layer capacitively coupled to the substrate and to another gate electrode with a constant dc voltage. Paralleling two storage capacitors reduces the area per cell.

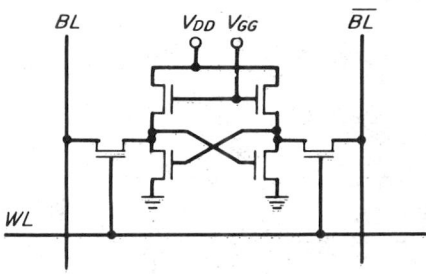

Fig. 16-95. The basic six-transistor circuit of a static MOS memory cell.

Typical storage capacitances are around 0.1 pF and less. Hence, a bit of information represents less than 1 pJ and reading out the information stored in large memories requires on-chip amplification. The basic idea behind the detection circuit is to use the metastable state of a flip-flop. Figure 16-97 illustrates a typical memory readout circuit. First, the flip-flop readout amplifier is activated and the transistor T is turned on. The circuit resumes the cascade connection of two MOS inverters, with a single short-circuit transistor connected between gates and drains. Hence, local negative-feedback loops overrule the effect of the global positive-feedback loop, and a stable state is obtained.

The drain-to-ground voltages of T_1 and T_2 necessarily are the same as the gate-to-ground voltages. For the same reason, bit lines BL_1 and BL_2 are precharged and their voltages with respect to ground are approximately half of the power-supply voltage.

Second, T is opened, leaving the flip-flop in its metastable state while the word line WL_1 selects the cell to be read. The voltage across the cell before readout is zero or equal to the power-supply voltage, depending on the binary information previously stored in the cell. When the cell is connected to the bit line, it modifies the bit-line voltage by some small amount, for instance ± 0.1 V. This creates a small unbalance between the two bit lines, causing the flip-flop to switch to one of its two possible stable states. Which state is determined by the sign of the unbalance.

Thus the final state is a replica of the selected cell content. In fact, more than one cell is selected, since a dummy cell (selected by WL_2) is interrogated at the same time. This trick ensures a better compensation of the inevitable parasitic clock feedthrough injected from word lines into the bit lines through the gate to diffusion overlap capacitances. As a result smaller useful signals can be detected and larger memories built. For instance, 128 memory sites can

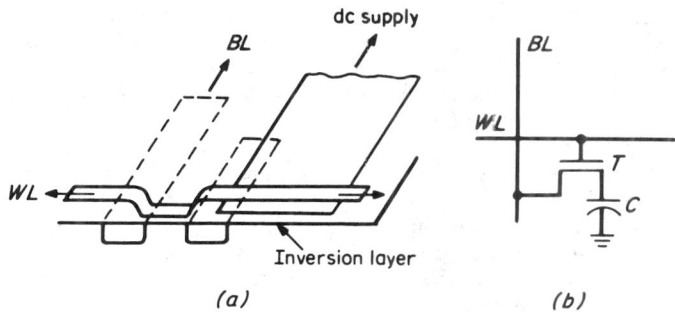

(a)   (b)

Fig. 16-96. The one-transistor MOS memory cell: (a) IC implementation; (b) equivalent circuit.

share the same bit line without too much trouble, although the ratio of bit-line capacitance to cell capacitance is unfavorable.

To access a memory word, e.g., 8 bits selected by the same word line, an address decoder is required with K bits to select 2^K word lines. This approach may lead to unfavorable geometry. For instance, an 11-bit address suffices to address 2,048 lines with 8-bit words in a 16-kilobit memory.

A much better use of silicon real estate can be made, for instance, if the memory is accessed as shown in Fig. 16-98. Here the decode-X network controls 128 word lines by means of 7-bit

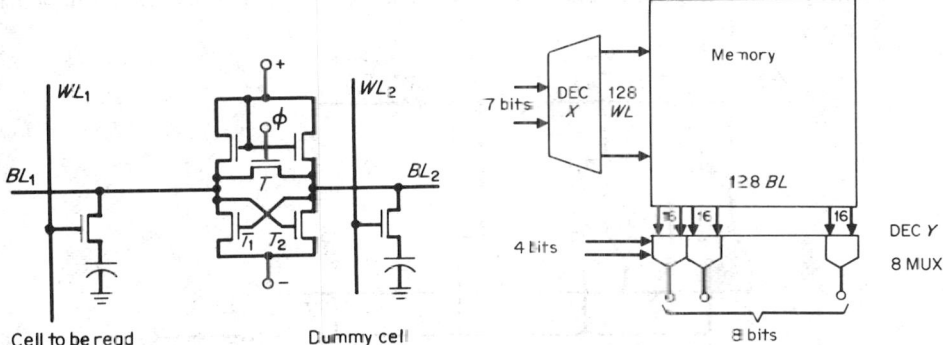

Fig. 16-97. The readout amplifier of dynamic one-transistor-per-cell MOS memories consists of a flip-flop in the metastable state in order to enhance sensitivity and speed and to minimize silicon area.

Fig. 16-98. On-chip architecture of a 16-kilobit memory with 8 parallel outputs.

words. Since the memory array is a square, 128 output signals are available. These 128 bits are divided into blocks connected to 8 multiplexers with 16 inputs each, and 4-bit words control the 16 multiplexers in parallel. This is called the decode-Y network. At the output, 8 parallel bits are available. The total number of bits required to select a word thus amounts to $7 + 4 = 11$. Hence, 2,048 words, each 8 bits long, are readily accessed in an approximately square matrix.

RAM. Random-access read-write memories (RAMs) are the counterparts of magnetic-core memories. They are used to store binary information and to read it out. Dynamic RAMs must be refreshed regularly, e.g., each millisecond.

ROM. In read-only memories (ROMs) the information is frozen; i.e., these memories are nonvolatile. Information can only be read out. An ROM memory usually is a square matrix of MOS transistors, some conducting, others not. Access to the information is the same as in RAMs. Whether a transistor is conducting or not has been determined during the fabrication process, e.g., some transistors have a thin oxide layer under the gate; others have a thick oxide layer.

PROM. The requirement for committing to a particular code, inherent in the fixed structure of ROMs, is a serious disadvantage in many applications. PROMs (programmable ROMs) allow the user to program ROMs himself. Various principles for programming exist, e.g., fuse links and irreversible shifts of the threshold voltages of the memory transistors.

EPROM. Erasable programmable ROMs offer a better trade-off than PROMs. The user can not only program the memory but also erase the contents. The principle of an EPROM memory cell is to store charge on the floating gate of an MOS transistor through the insulating layer by avalanching the neighboring source or drain. This procedure injects hot electrons onto the gate, where they are trapped for very long periods, i.e., years. To erase the information stored in an EPROM, x-ray or ultraviolet irradiation is needed to provide the energy required to unload the gate.

44. D/A and A/D Converters. Digital signal processing is often preferred to analog signal processing, but the complexity (and the signal bandwidth) of digital circuits is generally much greater than that of analog circuits. The latter unfortunately lack accuracy and are more subject to signal degradation compared with digital circuits. Whatever the reasons may be to prefer one solution to another, in many applications (such as audio, video, and telecommunications) both analog and digital processing may be combined. Moreover, even when digital processing is considered alone, the input signals generally are analog and an initial conversion is then necessary.

The Resistive-Ladder D/A Converter.[13]   Conversion from digital data to analog signals can be performed with the circuit represented in Fig. 16-99. In this ladder network the input data are represented by the K bits of a binary number to be converted. Each bit controls a double-throw switch. In Fig. 16-99 the MSB (most significant bit) controls the switch S_3 and the LSB (least significant bit) controls S_0. If a bit is 0, the position of the switch is as shown in the figure. If it is 1, it is the opposite. Whatever the actual positions of the switches may be, all resistors $2R$ have their lower end connected to ground, either directly or by means of the virtual ground of the op-amp. Thus the currents in the ladder network are completely independent of the actual

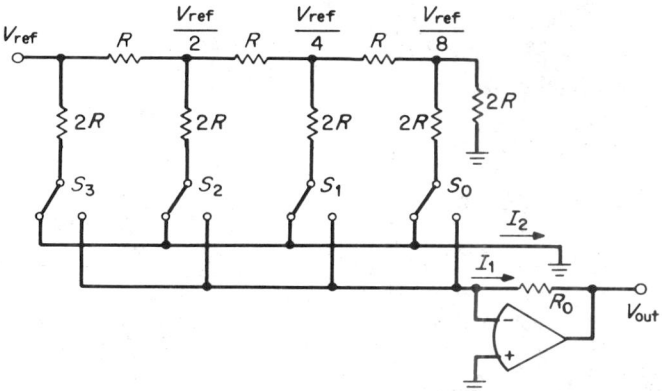

Fig. 16-99. Resistive-ladder D/A converter.

state of the switches. Hence, the voltages at the upper nodes of the ladder network can easily be determined.

In fact, if we consider the LSB side of the ladder network first, we notice two parallel resistors having the same value $2R$. The voltage at the upper node facing the switch S_0 thus necessarily is half of the voltage at the node facing S_1. Repeating the observation from node to node, one comes to the conclusion that the voltages are respectively $V_{ref}/8$, $V_{ref}/4$, $V_{ref}/2$, and V_{ref}. Hence, the currents in the resistors $2R$ are monotonic binary fractions of the reference current $V_{ref}/2R$. Thus the output voltage V_{out} of the op-amp is a linear image of the binary input number controlling the switches. Notice that this result is obtained by means of a ladder network having only two different values of resistors.

A/D Converters: Successive-Approximation Method.[13]   The D/A converter described above can be put in a feedback loop, as shown in Fig. 16-100, to create an A/D converter. The output voltage of the resistive-ladder network is compared with the unknown voltage V_x. The amplified error signal controls the switches of the D/A converter following the so-called successive-weights algorithm. In this method, the MSB of the D/A converter is sensed first, and the output of the converter is compared with V_x. If V_x is larger than $V_{ref}/2$, the next MSB is sensed and V_x is compared with $(V_{ref}/2 + V_{ref}/4)$.

The procedure follows unaltered until the output voltage of the D/A converter becomes larger than V_x. When this happens, the last switch that was activated is reset and the next one is set.

The same procedure is carried out until the LSB is finally tested. The positions of the switches then express the binary number representing the closest approximation of the unknown analog signal V_x.

The most significant feature of the successive-approximation A/D converter is its speed. Only K tests are required to convert an analog signal with a resolution of 1 out of $2K$. To obtain this result, however, the unknown signal must be constant while the conversion

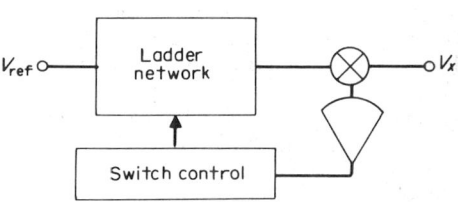

Fig. 16-100. A D/A converter can be changed into an A/D converter by means of a feedback loop.

procedure is in progress. Otherwise the algorithm may fail. To avoid this, it may be necessary to place a sample-and-hold circuit in front of the converter.

A/D Converters: Charge-Redistribution Method.[23] The ladder networks used in the converters described above generally are made with thin-film resistors in order to avoid the nonlinearities and tracking errors of integrated resistors. To design a fully integrated converter one must take advantage of the fact that integrated capacitors offer both better linearity and smaller ratio tolerances than resistors. The integrated counterpart of the successive-approximation A/D converter is shown in Fig. 16-101. In this circuit an array of binary weighted capacitors is used.

The conversion is accomplished by a sequence of three operations. In the first, the unknown voltage V_x is sampled. The corresponding positions of the switches are as illustrated in the figure. The total stored charge is proportional to V_x. In the second step switch S_A is opened, and the positions of all the S_i switches are changed. The bottom plates of all capacitors are thus grounded,

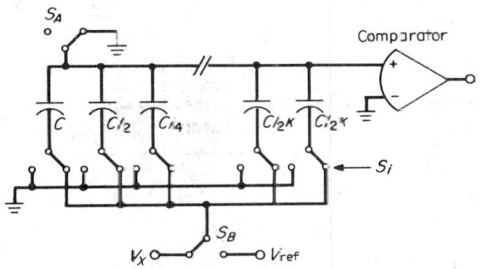

Fig. 16-101. A switched-capacitor A/D converter which can be fully integrated.

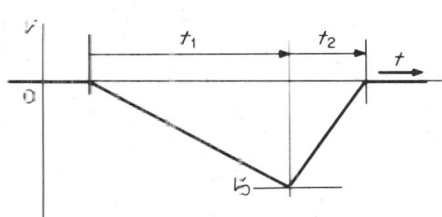

Fig. 16-102. The dual-slope A/D converter principle.

and consequently the voltage at the input node of the comparator equals $-V_x$. The third step is initiated by raising the bottom plate of the MSB capacitor from ground to the reference voltage V_{ref}. This is done by again changing the position of the MSB switch and connecting S_B to V_{ref} instead of to V_x. The voltage at the input node of the comparator is thus increased by $+V_{ref}/2$, so that the comparator's output is a logic 1 or 0, according to the sign of $(-V_x + V_{ref}/2)$.

The same procedure then is initiated as for the ladder-network converter. That is, V_x is compared with $V_{ref}/2 + V_{ref}/4$ if the result of the previous operation is negative; otherwise the MSB switch returns to its initial position, and a new test is carried out with the next bit.

Charge-redistribution converters have been integrated with 3-bit accuracies and are now successfully used in LSI applications where analog-signal processing is done internally by digital hardware.

A/D Converter: Dual-Ramp Method. Another integrated converter widely accepted in the field of electrical measurements is the dual-ramp A/D converter. The principle of this device is first to integrate the unknown voltage V_x during a fixed duration τ_1 (see Fig. 16-102). Then the input signal V_x is replaced by a reference voltage V_{ref}. Since the polarities of both signals must be opposite, the integrator provides an output ramp with opposite slope. It takes a time t_2 for the integrator output voltage to return from V_0 to zero. Hence

$$V_0 = -\frac{1}{RC} \int_0^{t_1} V_x \, dt \tag{16-86}$$

and
$$V_0 = (V_{ref}/RC)t_2 \tag{16-87}$$

Consequently, if V_x is a constant

$$V_x = -V_{ref}t_2/t_1 \tag{16-88}$$

the time t_1 is determined by counting 2^K clock signals by means of a counter with K stages. At the moment the counter overflows, switching of the input signal from V_x to V_{ref} occurs. The counter is automatically reset, and a new counting sequence initiated. When the integrator output voltage has returned to zero, the comparator stops the counting sequence. Thus the actual count is a direct measure of t_2.

The dual-ramp converter has a number of interesting features. Its accuracy is not influenced by the value of the integration time constant RC; neither is it sensitive to the long-term stability of the clock-frequency generator. The comparator offset can easily be compensated by autozeroing techniques. The only signal which actually controls the accuracy of the A/D converter is the reference voltage V_{ref}. Excellent thermal stability of V_{ref} can be achieved by means of integrated band-gap reference sources.

Another interesting feature of the dual-ramp A/D converter is the fact that since V_x is integrated during a constant time t_1, any spurious periodic signal with zero mean whose period is a submultiple of t_1 is automatically canceled. Hence, by making t_1 equal to an entire number of periods of the power supply one obtains excellent hum rejection. The drawback with respect to the successive-approximation A/D converter is the much longer conversion time required.

45. Function Generators. Integrated function generators consist generally of a free-running relaxation oscillator controlling a nonlinear shaping circuit. A typical block diagram of a

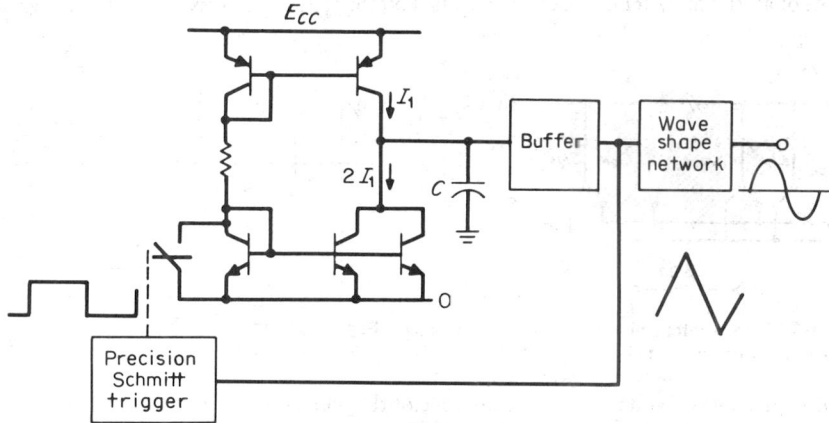

Fig. 16-103. A typical integrated wave generator that can deliver a square wave, a triangular wave, and a sine wave.

function generator is shown in Fig. 16-103. The relaxation oscillator is a combination of a time-base generator and a Schmitt trigger. The time base in the present case is obtained by successive loading and discharging of a capacitor using two current sources. One is a constant current source I_1 and the other a controlled current source delivering a current step equal to $-2I_1$ or zero. Hence, the voltage across the capacitor C is a sawtooth.

The switching of the controlled current source is monitored by the logical output signal of the Schmitt trigger. This last circuit is in fact a precision Schmitt trigger like the one of Par. **16-35**. Thus the oscillating voltage across C is obtained in the same manner. The output sawtooth signal is buffered and drives a network consisting of resistors and diodes which changes the sawtooth into a more or less sinusoidal voltage.

The advantage of function generators over RC or op-amp oscillators is their excellent amplitude stability vs. frequency. Also frequency modulation can easily be achieved by changing the current delivered by the two current sources. This type of function generator can be frequency-swept very rapidly over a wide dynamic range without spurious amplitude transients since no selective network is involved.

46. References

1. International Electrotechnical Commission (IEC)   "Spécifications de l'appareillage de mesure CISPR pour les fréquences comprises entre 25 et 300 MHz," 1961.

2. Ebers, J. J., and J. L. Moll   Large-Signal Behavior of Junction Transistors, *Proc. IRE*, December 1954, Vol. 42, pp. 1761–1772.

3. Moll, J. L.   Large-Signal Transient Response of Junction Transistors, *Proc. IRE*, December 1954, Vol. 42, pp. 1773–1784.

4. Getreu, I.   "Modeling the Bipolar Transistor," Tektronix Inc., Beaverton, Oreg., 1976.

5. Carr, W. N., and J. P. Mize   "MOS/LSI Design and Applications," Texas Instruments Electronics Series, McGraw-Hill, New York, 1972.

6. Penney, W. M. (ed.)   "MOS Integrated Circuits," Van Nostrand Reinhold, New York, 1972.

7. Glaser A., and G. E. Subak-Sharpe   "Integrated Circuit Engineering" Addison-Wesley, Menlo Park, Calif., 1977.

8. Taub, H., and D. Schilling "Digital Integrated Electronics," McGraw-Hill, New York, 1977.

9. Hamilton, D., and W. Howard "Basic Integrated Circuits," McGraw Hill, New York, 1975.

10. Hart K., and A. Slob Integrated Injection Logic: A New Approach to LSI, *IEEE J.S.S.C.*, Oct. 1972, Vol. SC-7, no. 5, pp. 346–351.

11. Berger, H., and S. Wiedmann Merged-Transistor Logic (MTL): A Low-Cost Bipolar Logic Concept, *IEEE J. Solid-State Circuits*, October 1972, Vol. SC-7, no. 5, pp. 340–346.

12. Hosticka, B. J., R. W. Brodersen, and P. R. Gray MOS Sampled Data Recursive Filters Using State Variable Techniques, *IEEE J. Solid State Circuits* 1977, Vol. SC-12, pp. 600–608; R. W. Brodersen, P. R. Gray, and D. A. Hodges, MOS Switched-Capacitor Filters, *Proc. IEEE*, January 1979, Vol. 67, No. 1, pp. 61–75.

13. Millman, J. "Microelectronics Digital and Analog Circuits and Systems," McGraw Hill, New York, 1979.

14. Zuch, E. L. Sample and Hold Circuits, *Electron. Des.* Nov. 8, 1973, No. 23, pp. 84–89; Dec. 6, 1978, No. 25, pp. 80–87; Dec. 20, 1978, No. 25, pp. 84–90.

15. Eccles, W. H., and F. W. Jordan A Trigger Relay Utilizing Three Electrode Thermionic Vacuum Tubes, *Radio Rev.*, 1919, Vol. 1, No. 3, pp. 143–146.

16. Schmitt, O. H. A Thermionic Trigger, *J. Sci. Instrum*, 1938, Vol. 15, p. 24.

17. Tietze, U., and C. Schenk "Advanced Electronic Circuits," Springer, Berlin, 1978.

18. Linvill, J. G., and R. H. Mattson Junction Transistor Blocking Oscillator, *Proc. IRE*, 1955, Vol. 43, No. 11, pp. 1632–1639.

19. Greenfield, J. D. "Practical Digital Design Using IC's," Wiley, New York, 1977.

20. Chen, T. C., and A. Papoulis Domain Theory in Core Switching, *Proc. IRE*, 1958, Vol. 46, No. 5, pp. 839–849.

21. Stein, K. U., and H. Friedrich A 1-mil^2 Single-Transistor Cell in n-Silicon Gate Technology, *IEEE J. Solid-State Circuits*, 1973, No. 8, pp. 319–323.

22. Hodges, D. A. (ed.) "Semiconductor Memories," IEEE Press, New York, 1972.

23. McCreary J., and P. R. Gray All-MOS Charge Redistribution Analog-to-Digital Conversion Techniques, *IEEE J. Solid-State Circuits*, December 1975, Vol. 10, pp. 371–379.

Measurement and Control Circuits

FRANCIS T. THOMPSON *Manager, Electronics Technology Division, Research and Development Center, Westinghouse Electric Corporation; Fellow, IEEE; Member, Instrument Society of America*

CONTENTS

Numbers refer to paragraphs

Principles of Measurement Circuits*

DEFINITIONS AND PRINCIPLES OF MEASUREMENT

1. Precision is a measure of the spread of repeated determinations of a particular quantity. Precision depends on the resolution of the measurement means and variations in the measured

*The author is indebted to I. A. Whyte, L. C. Vercellotti, T. H. Putman, T. M. Heinrich, T. I. Pattantyus, and R. A. Mathias for their suggestions and constructive criticisms. The author wishes to thank Mrs. Sandra Mahan and Mrs. Leslie Arthrell for their fine work in typing the manuscript.

value caused by instabilities in the measurement system. A measurement system may provide precise readings, all of which are inaccurate because of an error in calibration or a defect in the system.

2. Accuracy is a statement of the limits which bound the departure of a measured value from the true value. Accuracy includes the imprecision of the measurement along with all the accumulated errors in the measurement chain extending from the basic reference standards to the measurement in question.

3. Errors may be classified into two categories, systematic and random. *Systematic errors* are those which consistently recur when a number of measurements are taken. Systematic errors may be caused by deterioration of the measurement system (weakened magnetic field, change in a reference resistance value), alteration of the measured value by the addition or extraction of energy from the element being measured, response-time effects, and attenuation or distortion of the measurement signal. *Random errors* are accidental, tend to follow the laws of chance, and do not exhibit a consistent magnitude or sign. Noise and environmental factors normally produce random errors but may also contribute to systematic errors.

Table 17-1.   Factors for Establishing Confidence Interval*

Number of observations	Confidence level			
	0.50	0.90	0.95	0.99
	Confidence interval			
2	$X \pm 1.00s$	$X \pm 6.31s$	$X \pm 12.71s$	$X \pm 63.66s$
3	$X \pm 0.82s$	$X \pm 2.92s$	$X \pm 4.30s$	$X \pm 9.92s$
4	$X \pm 0.77s$	$X \pm 2.35s$	$X \pm 3.18s$	$X \pm 5.84s$
5	$X \pm 0.74s$	$X \pm 2.13s$	$X \pm 2.78s$	$X \pm 4.60s$
6	$X \pm 0.73s$	$X \pm 2.02s$	$X \pm 2.57s$	$X \pm 4.03s$
7	$X \pm 0.72s$	$X \pm 1.94s$	$X \pm 2.45s$	$X \pm 3.71s$
8	$X \pm 0.71s$	$X \pm 1.90s$	$X \pm 2.37s$	$X \pm 3.50s$
9	$X \pm 0.71s$	$X \pm 1.86s$	$X \pm 2.31s$	$X \pm 3.36s$
10	$X \pm 0.70s$	$X \pm 1.83s$	$X \pm 2.26s$	$X \pm 3.25s$
11	$X \pm 0.70s$	$X \pm 1.81s$	$X \pm 2.23s$	$X \pm 3.17s$
16	$X \pm 0.69s$	$X \pm 1.75s$	$X \pm 2.13s$	$X \pm 2.95s$
∞	$X \pm 0.67s$	$X \pm 1.64s$	$X \pm 1.96s$	$X \pm 2.58s$

* Modified and abridged from Table IV of R. A. Fisher and F. Yates, "Statistical Tables for Biological, Agricultural and Medical Research," Oliver & Boyd, Edinburgh, 1963. By permission of the authors and publishers.

The arithmetic average of a number of observations should be used to minimize the effect of random errors. The arithmetic average or mean X of a set of n readings $X_1, X_2, \ldots, X_n$ is

$$X = \Sigma X_i / n$$

The dispersion of these readings about the mean is generally described in terms of the standard deviation σ, which can be estimated for n observations by

$$s = \sqrt{\frac{\Sigma(X_i - X)^2}{n - 1}}$$

where s approaches σ as n becomes large.

A *confidence interval* can be determined within which a specified fraction of all observed values may be expected to lie. The *confidence level* is the probability of a randomly selected reading falling within this interval. Confidence intervals are given in Table 17-1 as a function of the number of observations and the required confidence level. Detailed information on measurement errors is given in Ref. 1, Par. **17-158**.

4. Standardization and calibration involve the comparison of a physical measurement with a reference standard. Calibration normally refers to the determination of the accuracy and linearity of a measuring system at a number of points, while standardization involves the adjustment of a parameter of the measurement system so that the reading at one specific value is in correspondence with a reference standard. The numerical value of any reference standard should be capable of being traced through a chain of measurements to a National Reference Standard maintained by the National Bureau of Standards.

5. The range of a measurement system refers to the values of the input variable over which the system is designed to provide satisfactory measurements. The range of an instrument used for a measurement should be chosen so that the reading is large enough to provide the desired precision. An instrument having a linear scale which can be read within 1% at full scale can be read only within 2% at half scale.

6. The resolution of a measuring system is defined as the smallest increment of the measured quantity which can be distinguished. The resolution of an indicating instrument depends on the deflection per unit input. Instruments having a square-law scale provide twice the resolution at full scale as linear-scale instruments. Amplification and zero suppression can be used to expand the deflection in the region of interest and thereby increase the resolution. The resolution is ultimately limited by the magnitude of the signal that can be discriminated from the noise background.

7. Noise may be defined as any signal which does not convey useful information. Noise is introduced in measurement systems by mechanical coupling, electrostatic fields, and magnetic fields. The coupling of external noise can be reduced by vibration isolation, electrostatic shielding, and electromagnetic shielding. Electrical noise is often present at the power-line frequency and its harmonics, as well as at radio frequencies.

In systems containing amplification, the noise introduced in low-level stages is most detrimental because the noise components within the amplifier passband will be amplified along with the signal. The noise in the output determines the lower limit of the signal that can be observed.

Even if external noise is minimized by shielding, filtering, and isolation noise will be introduced by random disturbances within the system caused by such mechanisms as the Brownian motion in mechanical systems, Johnson noise in electrical resistance and the Barkhausen effect in magnetic elements. Johnson noise is generated by electron thermal agitation in the resistance of a circuit. The equivalent rms noise voltage developed across a resistor R at an absolute temperature T is equal to $\sqrt{4kTR\,\Delta f}$, where k is Boltzmann's constant (1.38×10^{-23} J/K) and Δf is the bandwidth in hertz over which the noise is observed.

8. The bandwidth Δf of a system is the difference between the upper and lower frequencies passed by the system (see Par. **17-43**). The bandwidth determines the ability of the system to follow variations in the quantity being measured. The lower frequency is zero for dc systems, and their response time is approximately equal to $1/(3\Delta f)$. Although a wider bandwidth improves the response time, it makes the system more susceptible to interference from noise.

9. Environmental factors which influence the accuracy of a measurement system include temperature, humidity, magnetic and electrostatic influences, mechanical stability, shock, vibration, and position. Temperature changes can alter the value of resistance and capacitance, produce thermally generated emfs, cause variations in the dimensions of mechanical members, and alter the properties of matter. Humidity affects resistance values and the dimensions of some organic materials. DC magnetic and electrostatic fields can produce an offset in instruments which are sensitive to these fields, while ac fields can introduce noise. The lack of mechanical stability can alter instrument reference values and produce spurious responses. Mechanical energy imparted to the system in the form of shock or vibration can cause measurement errors and, if severe enough, can result in permanent damage. The position of an instrument can affect the measurements because of the influence of magnetic, electrostatic, or gravitational fields.

TRANSDUCERS, INSTRUMENTS, AND INDICATORS

10. Transducers are used to respond to the state of a quantity to be measured and to convert this state into a convenient electrical or mechanical quantity. Transducers can be classified according to the variable to be measured. Variable classifications include mechanical, thermal, physical, chemical, nuclear-radiation, electromagnetic-radiation, electrical, and magnetic, as detailed in Sec. **10**.

11. Instruments can be classified according to whether their output means is analog or digital. Analog instruments include the d'Arsonval (moving-coil) galvanometer, dynamometer instrument, moving-iron instrument, electrostatic voltmeter, galvanometer oscillograph, cathode-ray oscilloscope, and potentiometric recorders. Digital-indicator instruments provide a numerical readout of the quantity being measured and have the advantage of allowing unskilled people to make rapid and accurate readings.

12. Indicators are used to communicate output information from the measurement system to the observer.

MEASUREMENT CIRCUITS

13. Substitution circuits are used in the comparison of the value of an unknown electrical quantity with a reference voltage, current, resistance, inductance, or capacitance. Various potentiometer circuits are used for voltage substitution, and divider circuits are used for voltage, current, and impedance comparison. A number of these circuits and the reference components used in them are described in Pars. **17-18** to **17-25**.

14. Analog circuits are used to embody mathematical relationships which permit the value of an unknown electrical quantity to be determined by measuring related electrical quantities. Analog-measurement techniques are discussed in Par. **17-38**, and a number of special-purpose measurement circuits are described in Par. **17-94**.

15. Digital processing of analog quantities which have been converted into digital values by an analog-to-digital (A/D) converter is frequently used. The microprocessor offers a number of advantages for digital processing because of its computational power and programmable flexibility, as discussed in Pars. **17-55** and **17-56**.

16. Bridge circuits provide a convenient and accurate method of determining the value of an unknown impedance in terms of other impedances of known value. The circuits of a number of impedance bridges and the amplifiers and detectors used for bridge measurements are described in Pars. **17-60** to **17-93**. The advent of microprocessors has made possible a new generation of automatic impedance bridges (see Par. **17-83**).

17. Transducer amplifying and stabilizing circuits are used in conjunction with measurement transducers to provide an electric signal of adequate amplitude which is suitable for use in measurement and control systems. These circuits, which often have severe linearity, drift, and gain-stability requirements, are described in Pars. **17-46** to **17-59**.

Substitution and Analog Measurements

VOLTAGE SUBSTITUTION

18. The constant-current potentiometer, which is used for the precise measurement of unknown voltages below 1.5 V, is shown schematically in Fig. 17-1. For a constant current, the output voltage V_o is proportional to the resistance included between the sliding contacts. In this circuit all the current-carrying connections can be soldered, thereby minimizing contact-resistance errors. When the sliding contacts are adjusted to produce a null, V_o is equal to the unknown emf and no current flows in the sliding contacts. At null, no current is drawn from the unknown emf, and therefore the measured voltage is independent of the internal resistance of the source.

The circuit of a multirange commercial potentiometer is shown in Fig. 17-2. The instrument is standardized with the range switch in the highest range position as shown and switch S connected to the standard cell. The calibrated standard-cell dial is adjusted to correspond to the known voltage of the standard cell, and the standardizing resistance is adjusted to obtain a null on the galvanometer. This procedure establishes a constant current of 20 mA through the potentiometer. The unknown emf is connected to the emf terminals, and switch S is thrown to the emf position. The unknown emf can be read to at least five significant figures by adjusting the tap slider and the 11-turn 5.5-Ω potentiometer for a null on the galvanometer. The range switch reduces the potentiometer current to 2 or 0.2 mA for the 0.1 and the 0.01 ranges, respectively, thereby permitting lower voltages to be measured accurately. Since the range switch does not alter the battery current (22 mA), the instrument remains standardized on the lower ranges. When making measurements, the current should be checked using the standard cell to ensure that the current has not drifted from the standardized value.

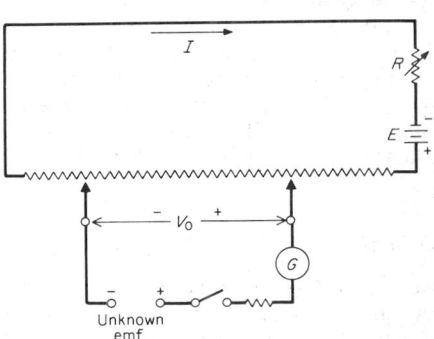

Fig. 17-1. Constant-current potentiometer.

19. The constant-resistance potentiometer of Fig. 17-3 uses a variable current through a fixed resistance to generate a voltage for obtaining a null with the unknown emf. The constant-resistance potentiometer is used primarily for measurements in the millivolt and microvolt range.

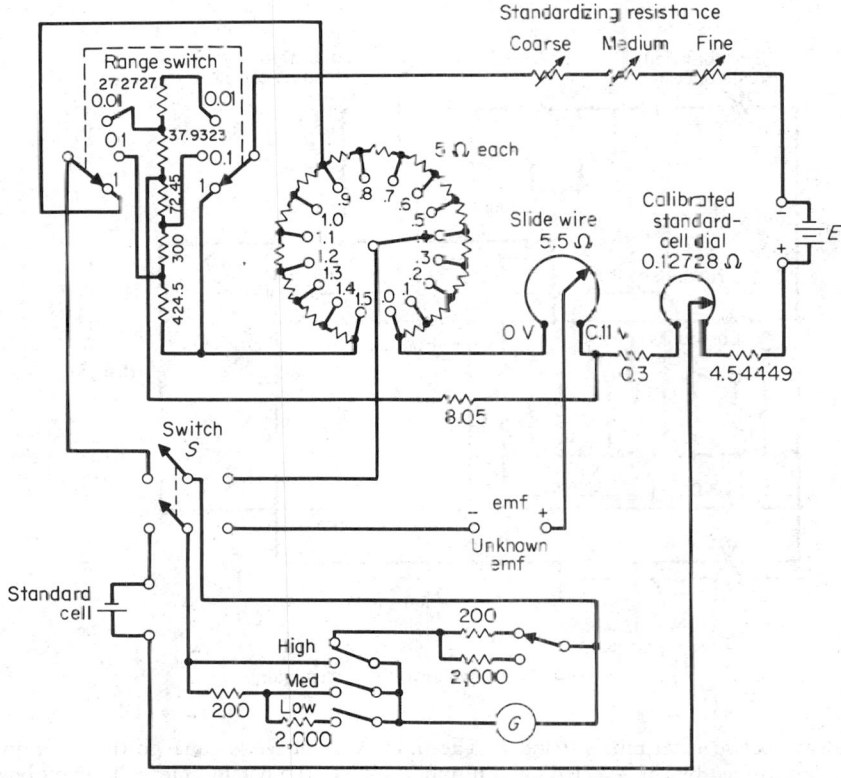

Fig. 17-2. K2 potentiometer *(Leeds and Northrup.)*

20. The microvolt potentiometer, or low-range potentiometer, is designed to minimize the effect of contact resistance and thermal emfs. Thermal shielding is used to minimize temperature differences. The galvanometer is connected to the circuit through a special Wenner thermo-free reversing key of copper and gold construction to eliminate thermal effects in the galvanometer circuit.

A typical microvolt potentiometer circuit consisting of two constant-current decades and a constant-resistance element is shown in Fig. 17-4. The constant-current decades use Diesselhorst rings, in which the constant current entering and leaving the ring divides between two paths. The *IR* drop across the resistance in the isothermal shield increases in 10 equal increments as the dial switch is rotated. The switch contacts are in the constant-current supply circuit, and therefore the effects of their *IR* drops and thermal emfs are minimized.

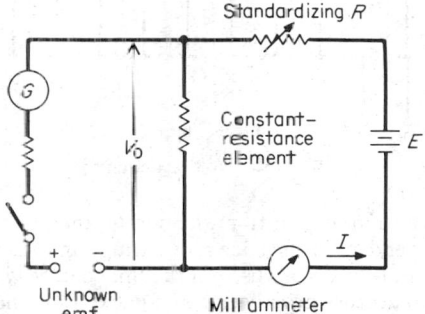

Fig. 17-3. Constant-resistance potentiometer.

A 100-division milliammeter associated with the constant-resistance element provides nearly 3 additional decades of resolution. Readings to 10 nV are possible with this type of potentiometer.

DIVIDER CIRCUITS

21. The volt box (Fig. 17-5) is used to extend the voltage range of a potentiometer. The unknown voltage is connected between 0 and an appropriate terminal, for example, ×100. The potentiometer is connected between the 0 and P output terminals. When the potentiometer is balanced, it draws no current, and therefore the current drawn from the source flows through

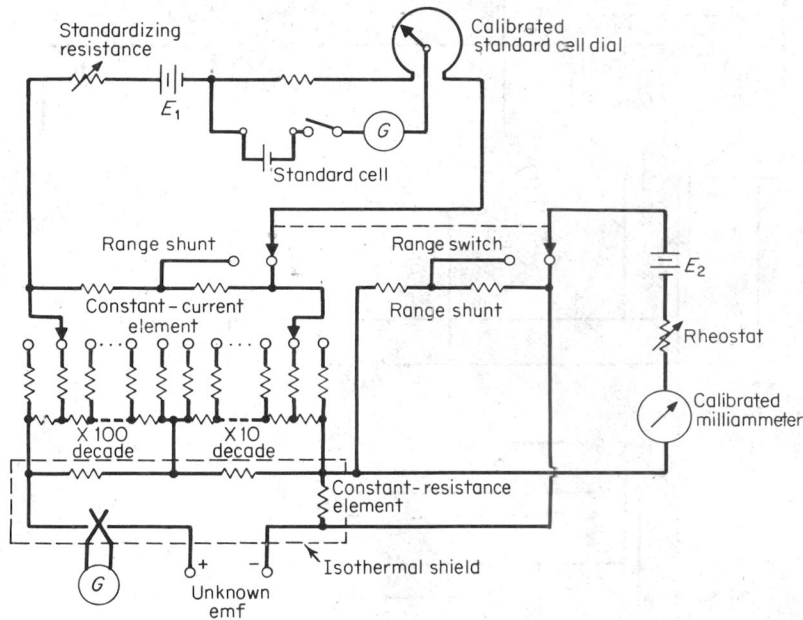

Fig. 17-4. Microvolt potentiometer.[2]

the resistor between terminals 0 and P. The unknown voltage is equal to the potentiometer reading multiplied by the selected tap multiplier. Unlike the potentiometer, the volt box does load the voltage source. Typical resistances range from about 200 to 1000 Ω/V. The higher resistance values minimize self-heating and do not load the source as heavily. Errors due to leakage currents which could flow through the insulators supporting the resistors are minimized by using a guard circuit (see Par. **17-87**).

22. Decade voltage dividers provide a wide range of precisely defined and very accurate voltage ratios. The Kelvin-Varley vernier decade circuit is shown in Fig. 17-6. The slide arms in the first 3 decades are arranged so that they always span two contacts. The shunting effect of the second gang resistance across the slide arms of the first decade is equal to $2R$, thereby giving a net resistance of R between the slide-arm contacts. With no current drawn from the output, the resistance loading on the input is equal to $10R$ and is independent of the slide-arm settings. In each of the first 3 decades, 11 resistors are used, while only 10 resistors are used in the final decade, which has a single sliding contact. Potentiometers with 6 decades have been constructed using the Kelvin-Varley circuit.

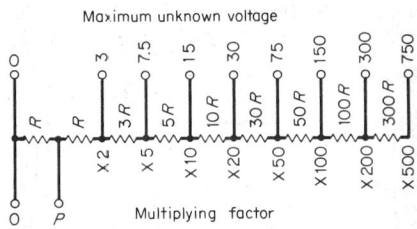

Fig. 17-5. Volt-box circuit.[2]

DECADE BOXES

23. Decade resistor boxes contain an assembly of resistances and switches, as shown in Fig. 17-7. The power rating of each resistance step is approximately constant; therefore, each decade has a different maximum current rating which should not be exceeded. Boxes having 4 to 7 decades are available with accuracies of 0.02%. Two typical 7-decade boxes provide resistance values from 0 to 1,111,111 Ω in 0.1-Ω steps and values from 0 to 11,111,110 Ω in 1-Ω steps. The accuracy at higher frequencies is affected by skin effect, series inductance, and shunt capacitance. The equivalent circuit of a resistance decade is shown in Fig. 17-8, where ΔL is the undesired incremental inductance added with each resistance step ΔR. Silver contacts are used to

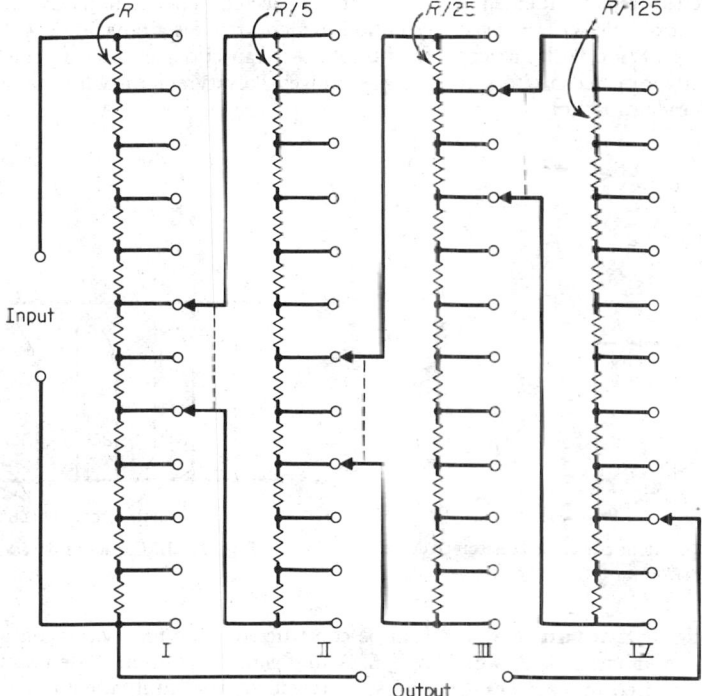

Fig. 17-6. Decade voltage divider.[2]

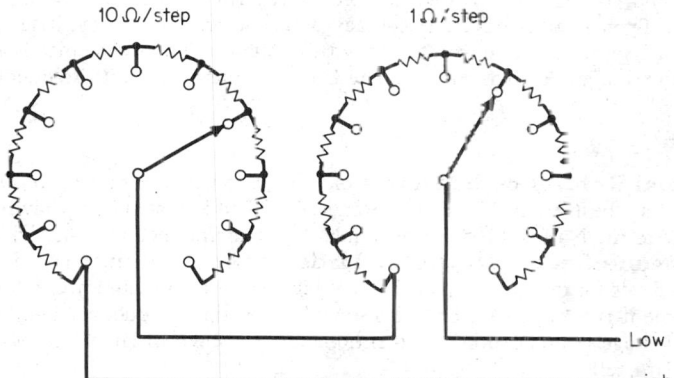

Fig. 17-7. Decade resistance box.

obtain a zero resistance R_0, as low as 1 mΩ/decade at dc. Zero inductance values L_0 as low as 0.1 μH/decade are obtainable. The shunt capacitance for the configuration of Fig. 17-7 is a function of the highest decade in use, i.e., not set at zero. The shunt capacitance with the low terminal connected to the shield is typically 10 to 15 pF for the highest decade in use plus an equal value for each higher decade not in use.

Some applications, e.g., the determination of small inductances at audio frequency and the determination of resistance at radio frequency by the substitution method, require that the equivalent series inductance of the resistance box remain constant, independent of the resistance setting.[3]* In the inductively compensated decade resistance box small copper-wound coils each having an inductance equal to the inductance of an individual resistance unit are selected by the decade switch so as to maintain a constant total inductance.

24. Decade capacitor units generally consist of four capacitors which are selectively connected in parallel by a four-gang 11-position switch (Fig. 17-9). The individual capacitors and their associated switch are shielded to ensure that the selected steps add properly.[4] *Decade capacitor boxes* are available with 6-decade resolution, which provides a range of 0 to 1.11111 μF in increments of 1 pF and with an accuracy of 0.05%. Air capacitors are used in the 1- and 10-pF decades, and silver-mica capacitors in the higher ranges. Polystyrene capacitors are used in some less precise decade capacitors.

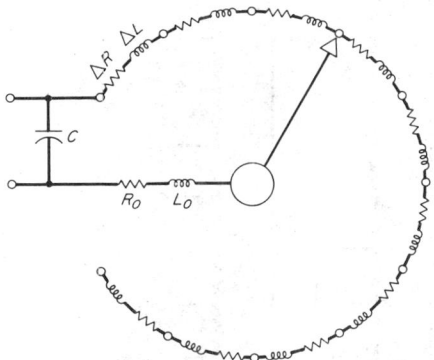

Fig. 17-8. Equivalent circuit of a resistance decade. (*General Radio Co.*)

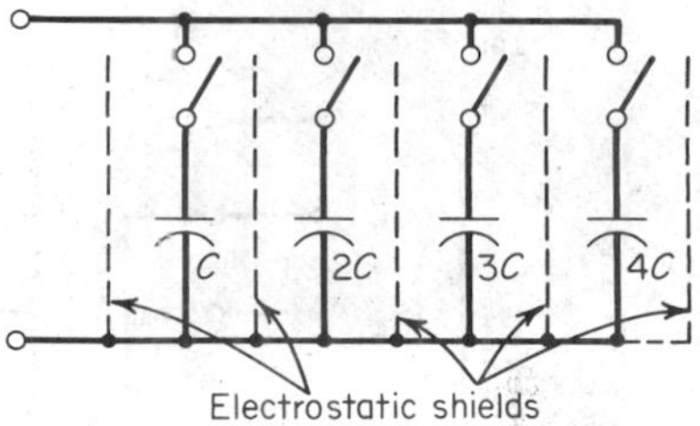

Fig. 17-9. Capacitor decade.[3]

25. Decade inductance units can be constructed using four series-connected inductances of relative values 1, 2, 3, 4 or 1, 2, 2, 5. A four-gang 11-position switch is used to short-circuit the undesired inductances. Care must be taken to avoid mutual coupling between the inductances. *Decade inductance boxes* are available with individual decades ranging from 1 mH to 10 H total inductance. A commercial single-decade unit consists of an assembly of four inductors wound on molybdenum-Permalloy dust cores and a switch which enables consecutive values to be selected. Typical units have an accuracy of 1% at zero frequency. The effective series inductance of a typical decade unit increases with frequency. The inductance is also a function of the ac current and any dc bias current. The Q of the coils varies with frequency.

STANDARDS

26. National Reference Standards of voltage, resistance, inductance, and capacitance are maintained at the National Bureau of Standards. Secondary standards, having a numerical value traceable to the National Reference Standards, are maintained in many laboratories.

27. Josephson-effect voltage standards have been used since 1972 by the National Bureau of Standards for maintenance of the U.S. legal volt[5] to an accuracy of a few parts in 10^8. This measurement is based on the ac Josephson effect in superconductors. The Josephson junction acts as a voltage reference for the potentiometer, as shown in Fig. 17-10, which is used to calibrate standard cells.

Resonant thin-film tunnel junctions of Pb films with a lead oxide insulation barrier on a glass substrate are mounted in a probe in a superinsulated helium Dewar system. The junction is irradiated at approximately 9 GHz using a microwave source which is phase locked to a high-stability 100-MHz temperature-controlled crystal oscillator. The Josephson junction produces discrete voltages given by the relationship $V = nhf/2e$, where f = microwave frequency, h = Planck's constant, e = electron charge, and n = integer. A junction operated on the 240th step (n = 240) produces a voltage of 4.5 mV. The desired step n is selected by adjusting the bias supply current and the IV characteristic can be viewed using the oscilloscope. The potentiometer

*Superior numbers correspond to References, Par. **17-158**.

consisting of a Hamon network (series-parallel reconnectable resistors) and a Kelvin-Varley divider is used to compare the standard cell voltage with the junction voltage.

A portable Josephson system having an accuracy of 0.4 ppm has been developed for use in Standards Laboratories.[6] This instrument is designed to calibrate cadmium-sulfate standard cells against a time-invariant Josephson junction reference, thereby replacing the large groups of cells typically used as a voltage reference.

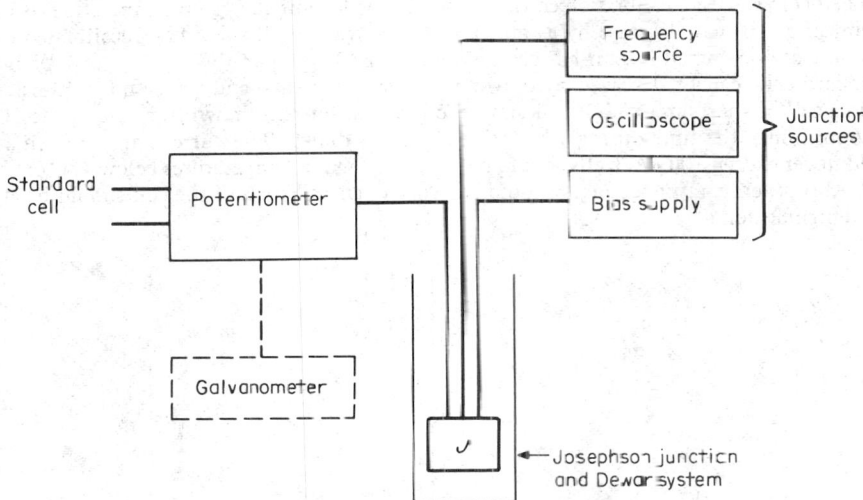

Fig. 17-10. Josephson-effect voltage standard.

28. The standard cell is used as a reference standard in electrical measurements and therefore must possess stability, long life, low temperature coefficient, and reproducibility. Two types of standard cells are used: the Weston saturated cell is used as an accurate reference standard in large laboratories; the Weston unsaturated cell, although less accurate (0.005%), has a lower temperature coefficient and is used as a reference in potentiometers and recording meters.

The Weston saturated cell (Fig. 17-11) has a positive electrode of metallic mercury, with mercurous sulfate as a depolarizer. The negative electrode is a cadmium-mercury amalgam (10% Cd). The electrolyte consists of a saturated solution of cadmium sulfate with an excess of cadmium sulfate crystals. This normal solution is usually acidified with sulfuric acid (0.04 to 0.08 N). The emf of acid cells is lower than that of the normal cell, $E_{20} = 1.018636 - 0.00060 N - 0.00005 N^2$, where E_{20} is the cell voltage at 20°C and N is the normality of the sulfuric acid.[7] The saturated cell is reproducible within a few microvolts but has a temperature coeffi-

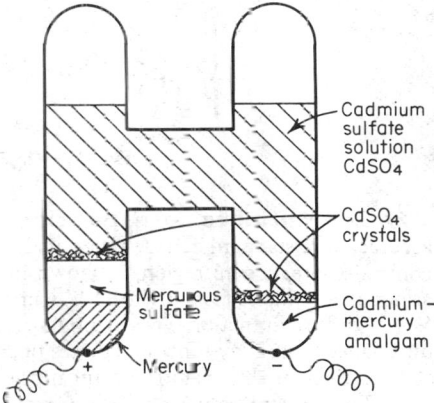

Fig. 17-11. Saturated standard cell.[25]

cient of -39.4 $\mu V/°C$ at 20°C ($- 52.9$ $\mu V/°C$ at 28°C). The formula of Vigoureux and Watts[8] is applicable to 10% amalgam cells:

$$E_t = E_{20} - 39.39 \times 10^{-6}(t - 20) - 0.903 \times 10^{-6}(t - 20)^2$$
$$+ 6.50 \times 10^{-9}(t - 20)^3 - 0.15 \times 10^{-9}(t - 20)^4$$

where t is the temperature in degrees Celsius. Large temperature gradients must be avoided since the temperature coefficient of the positive limb of the cell is about 310 $\mu V/°C$ while that of the negative limb is -350 $\mu V/°C$ at 20°C. The reference group of cells at the National Bureau of Standards is maintained at about 30°C in an air bath in which the temperature is held constant within 20 × 10^{-6}°C. These cells are calibrated using the Josephson voltage standard (Par. **17-27**).

The Weston unsaturated cell uses the same electrode structure, but the concentration of the electrolyte solution is chosen for saturation at 4°C and therefore is unsaturated at room temperature. Since it has a temperature coefficient of less than $-10\ \mu\text{V/}°\text{C}$, it can be used without temperature correction if the operating temperature is within a few degrees of the calibration temperature. New unsaturated cells ordinarily range between 1.0190 and 1.0194 absolute volts, and a cell comes with a calibration certificate specifying its emf at a given temperature. Because the emf of a cell may decrease by as much as 0.01%/year, a cell should be recalibrated every year and discarded when its emf has decreased to 1.0183 V.

Standard cells can be damaged by drawing excessive currents, and under no circumstances should a cell be short-circuited. It is desirable to limit the current drawn from the cell to 10 μA and to minimize the time during which this current is drawn. The current drawn from a cell should never exceed 100 μA. Cells should never be exposed to temperatures below 4°C or above 40°C. Abrupt temperature changes should be avoided, and all parts of the cell should be at the same temperature.

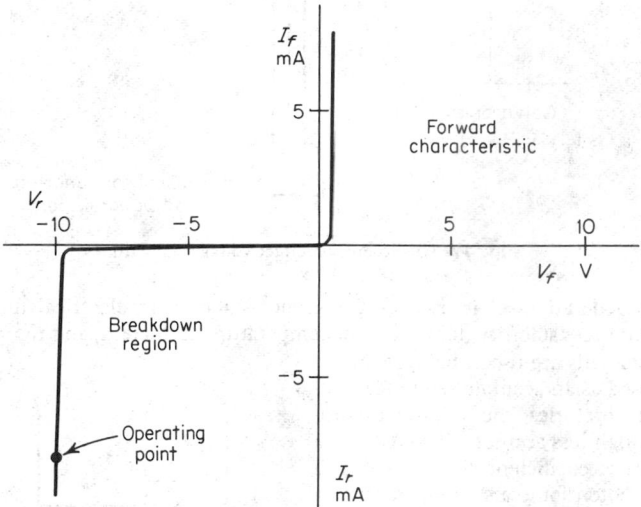

Fig. 17-12. Zener-diode characteristic.

29. Zener diodes are used as voltage references in instruments which do not require the accuracy of a standard cell. A zener diode is a reverse-biased silicon *pn* junction having a well-controlled breakdown region, as shown in Fig. 17-12. Alloy-junction zener diodes are available at nominal voltages from 2.4 to 12 V, while diffused-junction types are available from 6.8 to 200 V. The diffused-junction types exhibit a well-defined breakdown knee and have a lower dynamic impedance than alloy-junction devices under comparable operating conditions. The slope of the characteristic at the operating point in the breakdown region determines the dynamic impedance. The dynamic impedance is a function of the nominal voltage and the operating current. The temperature coefficient is a function of the zener voltage, as shown in Fig. 17-13.

30. The temperature-compensated zener diode consists of a positive-temperature-coefficient reverse-biased zener diode connected in series with one or more negative-temperature-coefficient forward-biased diodes within a single package. Specified temperature coefficients as low as 0.0002%/°C are available. The zener diode must be operated at a specified current to obtain the rated temperature coefficient. A typical circuit is shown in Fig. 17-14. For small changes in the supply voltage E, the zener diode may be replaced by its dynamic resistance r_d, and the change in reference voltage is

$$\Delta E_r = \Delta E \frac{1}{1 + R_1/r_d + R_1/R_2}$$

For the values given $\Delta E_r = \Delta E/74.88$. The higher the value of R_2, the lower the reference-voltage change since R_1 can be increased for a given zener-diode current. If R_2 is infinite and R_1

is selected as 1,560 Ω to maintain the specified zener-diode current, then $\Delta E_r = \Delta E/105$. Resistor R_2 can be made up of a pair of series-connected resistors, permitting a portion of E_r to be obtained at their junction. If E_r is to be held within 0.001%, the voltage E of Fig. 17-14 must be regulated to better than 0.1%. Stable low-temperature-coefficient resistors must be used to achieve this degree of stability. Precision integrated-circuit voltage references are available with temperature coefficients as low as 5 ppm/°C, long-term stability of 25 ppm per 1,000 h, and load regulation of 200 μV/mA (Analog Devices AD581).

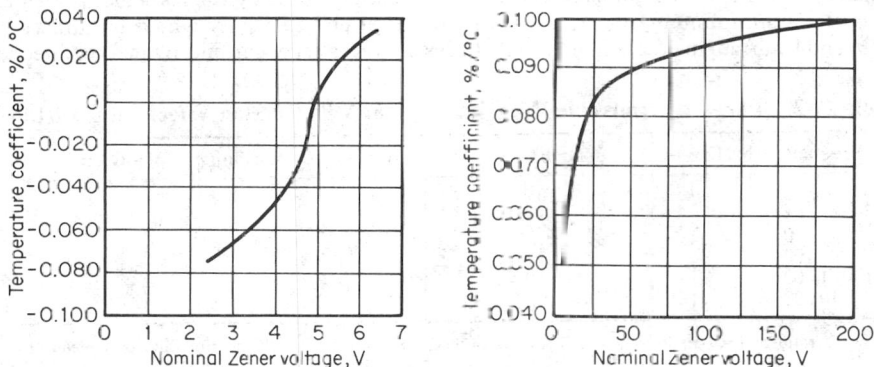

Fig. 17-13. Zener-diode temperature coefficient: (a) alloy-junction type; (b) diffused-junction type. *(Motorola, Inc.)*

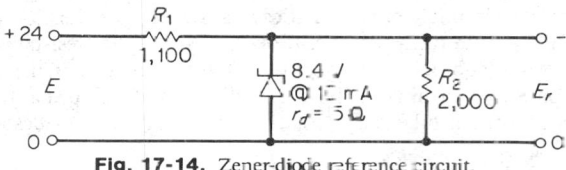

Fig. 17-14. Zener-diode reference circuit.

31. Current standards are precision four-terminal resistors having a pair of current terminals through which the current is passed. The voltage drop across the resistor is measured by a precision potentiometer connected to the voltage terminals of the resistor. A standard current is therefore a quantity that is derived from the voltage of a standard cell and a known resistance.

32. The National Reference Standard of resistance consists of a group of 10 specially constructed 1-Ω resistors maintained at the National Bureau of Standards to an accuracy of 0.08 ppm. They are intercompared regularly and are also compared with the reference standards of other countries to ensure that their values are constant.

33. Resistance standards are made using high resistivity metal in the form of wire or strip which is fully annealed to remove residual strains and sealed from contact with the air or other oxidizing agents. Manganin is widely used because it has a low temperature coefficient and a low thermoelectric emf at junctions with copper (see Table 17-2). Evanohm is used for high-

Table 17-2.   Properties of Resistance Wire[3]

| Material | Typical composition | Resistivity | | Thermal emf against copper, μV/°C |
		$\mu\Omega$·cm	Ω/mil·ft	Temperature coefficient, ppm/°C	
Nichrome, nichrome I to V, chromel A, tophet, etc.	Ni 80%, Cr 20%	08	650	150	22
Advance, ideal, cupron, copel, constantan, etc.	Ni 45%, Cu 55%	45	290	±20	43
Manganin	Ni 4%, Cu 84%, Mn 12%	45	290	±15	<3.0
Evanohm, Karma, 331 alloy	Ni 74.5%, Cr 20%, balance Al, and Fe or Cu	133	800	±20	<2.5
Copper	Cu 99.9+%	1.724	10.37	3,950	

resistance standards because it exhibits similar characteristics and has a much higher resistivity. Oil baths are often used to dissipate heat and maintain a constant temperature.

The increase in resistance due to skin effect must be taken into account at high frequency. The ratio of ac to dc resistance of an isolated wire is a function of the factor $d\sqrt{f/p}$, where d is the wire diameter, f the frequency, and p the resistivity. The largest wire diameter that can be used at various frequencies and still keep the ac resistance within 1% of the dc resistance is given in Table 17-3. Wire-wound resistors have undesired series inductance and shunt capacitance associated with them. The inductance can be minimized by winding the resistance so that each turn encloses a minimum area, by selecting a diameter and resistivity which minimizes wire length, and by arranging the winding so that adjacent turns carry current in opposite directions.

Table 17-3.   Largest Permissible Wire Diameter in Mils for Skin-Effect Ratio of 1.01[3]*

Frequency, MHz	Nichrome	Advance and Manganin	Copper
0.1	110	74	14
1	35	23	4.4
10	11	7.4	1.4
100	3.5	2.3	0.44
1,000	1.1	0.74	0.14
10,000	0.35	0.23	0.04

* For a ratio of 1.001 multiply above diameters by 0.55. For a ratio of 1.10 multiply above diameters by 1.78.

Low shunt capacitance is obtained by arranging the winding so that adjacent turns are widely spaced and have a minimum potential difference between them. The Ayrton-Perry winding (Fig. 17-15b) incorporates these desirable characteristics and is used for precision resistors up through 200 Ω. The more easily wound mica card resistor is used for higher resistance values.

Standard resistors of 1 Ω and below are constructed with four terminals, while some higher-value resistors have only two terminals. The standard resistor developed at the National Bureau of Standards is a four-terminal oil-cooled resistor having current lugs designed to be used in mercury cups.

(a)   (b)

Fig. 17-15. Resistor windings: (a) mica card; (b) Ayrton-Perry.[9]

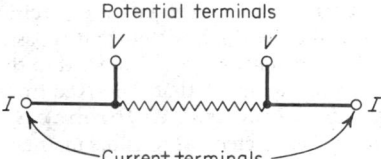

Potential terminals

Current terminals

Fig. 17-16. Four-terminal resistor.

34. Four-terminal resistors are used because a two-terminal resistor is subject to errors due to contact resistance. Assuming an unpredictable series resistance component of 1 mΩ, a 1-Ω two-terminal resistor has an uncertainty of 0.1% while a 0.1-Ω two-terminal resistor has an uncertainty of 1%. This source of error is overcome by the use of four terminals, as shown in Fig. 17-16. The current enters and leaves through the outermost terminals, which are called the *current terminals*. The voltage terminals are arranged so that the voltage drop occurs between them when the current passes through the resistor. The current drawn by the potential measuring device should be negligible. Potentiometers, which draw no current at null, are used to make precise potential measurements.

35. Standard capacitors with values known to 2 parts in 10^8, used in the measurement of the absolute farad and absolute ohm, have been constructed at the National Bureau of Standards using four equal, closely spaced cylindrical rods in a geometry prescribed by Thompson and by Lampard.[10,11] The usual standard capacitors up to 1,000 pF use multiple parallel plates with a dry-air or nitrogen dielectric. Invar, a low-expansion alloy, is used to secure a low temperature coefficient, and good stability is achieved by mounting fully annealed components in a strain-free manner.

Three-terminal shielded construction is used to provide a definite value of capacitance which is independent of external objects and fields. With two-terminal construction, the capacitance from each electrode to surrounding objects forms a second capacitance in parallel with the desired capacitor. The equivalent circuit of a commercial three-terminal reference capacitor is given in Fig. 17-17. This capacitor which has a drift of less than 20 ppm/year, is available in 10-, 100-, and 1,000-pF values calibrated against working standards whose absolute value is known to ±5 ppm.

Precision variable capacitors, which can be set to 40 ppm, are commercially available in capacitance ranges up to 1,100 pF. A four-terminal ac-only decade capacitor is available that consists of a 1-μF polystyrene capacitor and a transformer that multiplies the effective capacitance in decade steps up to 1 F with accuracies better than 1% at 120 Hz.

36. Inductance standards consist of single-layer solenoids wound on dimensionally stable forms of fused silica. The windings are laid in a lapped groove to ensure uniform winding pitch. These solenoids are large and bulky since their dimensions must be large enough to permit their values to be computed from measured dimensions.

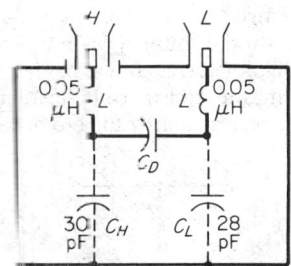

Fig. 17-17. Equivalent circuit showing direct capacitance C_D and average values of residual inductance L and terminal capacitances C_H and C_L. (*General Radio Co.*)

Multilayer coils wound on ceramic, marble, or Bakelite forms are usually used as working standards. Toroidal cores are often used because they are nearly immune to external magnetic fields. The effective inductance is a function of frequency because of the capacitance associated with the winding. The frequency with which the calibrated value is associated should be given. Standard inductors, which are stable within ±0.01%/year, are commercially available from 50 μH to 10 H. They are calibrated at 100, 200, 400, and 1,000 Hz against standards certified to ±0.02%.

37. Mutual inductance standards, which are computable from measured dimensions, are constructed using single-layer primaries having two or three sections spaced so that there is a region in which the field gradients are very small.[7] The multilayer secondary winding is placed in this region to minimize the effect of positional errors.

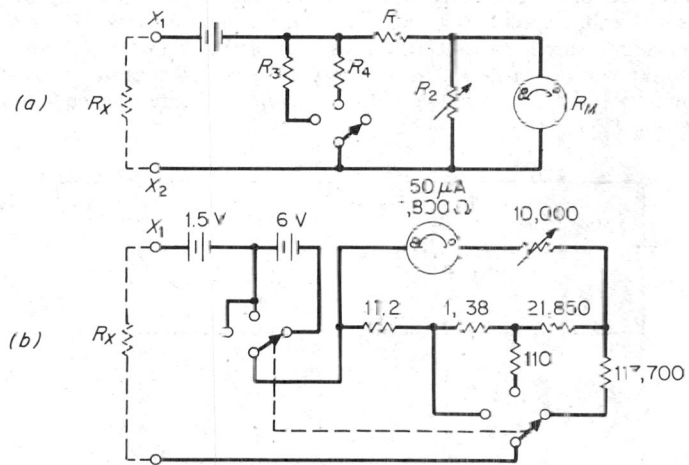

Fig. 17-18. Series-type ohmmeters. (*b Simpson Electric Co.*)

ANALOG MEASUREMENTS

38. Ohmmeter circuits provide a convenient means of obtaining an approximate measurement of resistance.

The basic series-type ohmmeter circuit of Fig. 17-18a consists of an emf source, series resistor R_1, and d'Arsonval milliammeter. Resistor R_2 is used to compensate for changes in battery emf and is adjusted to provide full-scale meter deflection (0-Ω indication) with terminals X_1 and X_2 short-circuited. No deflection (infinite resistance indication) is obtained with X_1 and X_2 open-circuited. When an unknown resistor R_x is connected across the terminals, the meter deflection varies inversely with the unknown resistance. With the range switch in the position shown, half-scale deflection is obtained when the external resistance is equal to $R_1 + R_2R_M/(R_2 + R_M)$. A multirange meter can be obtained using current-shunting resistors R_3 and R_4. A typical commercial ohmmeter circuit (Fig. 17-18b) having midscale readings of 12 Ω, 1,200 Ω, and 120 kΩ uses an Ayrton shunt for range selection and a higher battery voltage for the highest resistance range.

In the shunt-type ohmmeter the unknown resistor R_x is connected across the d'Arsonval milliammeter, as shown in Fig. 17-19a. The variable resistance R_1 is adjusted for full-scale deflection (infinite-resistance indication) with terminals X_1 and X_2 open-circuited. The ohm scale, with 0 Ω corresponding to zero deflection, is the reverse of the series-type ohmmeter scale. The resistance range can be lowered by switching a shunt resistor across the meter. With the range switch selecting shunt resistor R_2, half-scale deflection occurs when R_x is equal to the parallel combination of R_1, R_M, and R_2. The shunt-type ohmmeter is therefore most suited to low-resistance measurements.

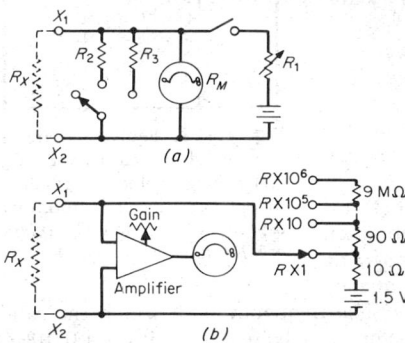

The use of a high-input impedance amplifier between the circuit and the d'Arsonval meter permits the shunt-type ohmmeter to be used for high- as well as low-resistance measurements. A commercial ohmmeter (Fig. 17-19b) uses a field-effect-amplifier input stage which draws negligible current. The amplifier gain is adjusted to provide full-scale deflection with terminals X_1 and X_2 open-circuited. Half-scale deflection occurs when R_x is equal to the total selected tap resistance.

Fig. 17-19. Shunt-type ohmmeter. *(Triplett Electrical Instrument Co.)*

39. Voltage-drop (or fall-of-potential) methods for determining resistance involve measuring the current flowing through the resistor with an ammeter, measuring the voltage drop across the resistor with a voltmeter, and calculating the resistance using Ohm's law. The circuit of Fig. 17-20a should be used for low-resistance measurements since the current drawn by the voltmeter V/R_v will be small with respect to the total current I. The circuit of Fig. 17-20b should be used for high-resistance measurements since the resistance of the ammeter R_A will be small with respect to the unknown resistance R_x. An accuracy of 1% or better can be obtained using 0.5% accurate instruments if the voltage source and instrument ranges are selected to provide readings near full scale.

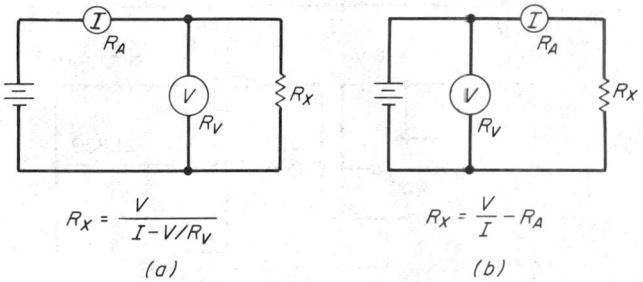

$$R_x = \frac{V}{I - V/R_v}$$

$$R_x = \frac{V}{I} - R_A$$

(a)   (b)

Fig. 17-20. Fall-of-potential method.

40. The loss-of-charge method is useful for measuring insulation resistance and other very high leakage resistances. The unknown resistor R is connected in parallel with a known capacitor C having a leakage resistance r, as shown in Fig. 17-21. Key A is momentarily depressed to charge the capacitor. Key B is closed immediately after key A is opened, and the ballistic throw d_1 of the galvanometer recorded. The process is repeated, but now a time of t seconds is allowed to pass from the time key A is opened until key B is closed and a deflection d_2 observed. The time constant of the circuit in seconds is[2]

$$\tau = t/[2.303 \log (d_1/d_2)]$$

The time constant τ_1 is experimentally determined for the configuration of Fig. 17-21 where $\tau_1 = CRr/(R + r)$. The resistor R is disconnected, and the time constant τ_2 of the capacitor and its leakage resistance are determined by repeating the measurements, where $\tau_2 = Cr$. The resistance R is

$$R = \tau_1\tau_2/[C\tau_2 - \tau_1]$$

The *loss-of-charge method* can also be used to determine the value of an unknown capacitance C having an unknown leakage resistor r if the value of R is known. The above measurements are performed to determine τ_1 and τ_2. The values of C and r are

$$C = \frac{\tau_1\tau_2}{R(\tau_2 - \tau_1)} \qquad r = R\frac{\tau_2 - \tau_1}{\tau_1}$$

41. Resonance methods can be used to measure the inductance, capacitance and Q factor of components at radio frequencies. In Fig. 17-22 resistors R_1 and R_2 couple the oscillator voltage e to a series-connected known capacitance and an unknown inductance represented by effective inductance L' and effective series resistance r'. Resistor R_2 is chosen to be small with respect to resistance r', thereby minimizing the effect of source resistance of the injected voltage.

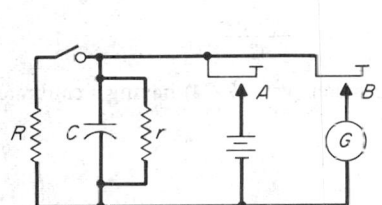

Fig. 17-21. Loss-of-charge method.[2]

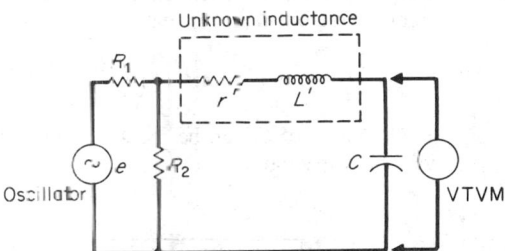

Fig. 17-22. Inductance measurement.

A circuit containing reactive components is in resonance when the supply current is in phase with the applied voltage. The series circuit of Fig. 17-22 is in resonance when the inductive reactance $X_{L'}$ is equal to the capacitive X_C, which occurs when

$$\omega^2 = \omega_0^2 = 1/L'C$$

where $X_{L'} = \omega L'$, $X_C = 1/\omega C$, $\omega = 2\pi f$, $\omega_0 = 2\pi f_0$, f_0 = resonant frequency (Hz), L' = effective inductance (H), and C = capacitance (F).

If L', r', C, and e are constant and the oscillator frequency ω is adjusted until the voltage read across C by the vacuum-tube voltmeter is maximum, the frequency ω will be slightly less than the resonant frequency ω_0:

$$\omega^2 = \frac{1}{L'C} - \frac{(r' + R_s)^2}{2L'^2} = \omega_0^2\left(1 - \frac{1}{2Q^{*2}}\right)$$

where $R_s = R_1R_2/(R_1 + R_2)$ and $Q^* = \omega_0 L'/(r' + R_s)$. If $Q^* \geq 10$, ω and ω_0 will differ by less than 0.3%. The ratio m of the voltage across C to the voltage across R_2 can be measured while operating at ω. If $R_s \ll r'$, the value of the effective Q of the unknown inductance is related to m by

$$m = \frac{2Q'^2}{\sqrt{4Q'^2 - 1}} \qquad Q' = \sqrt{\frac{m}{2(m - \sqrt{m_2 - 1})}}$$

where $Q' = \omega_0 L'/r'$ and $m = V_C/V_{R_2}$ measured at ω. The values of m and Q' are nearly equal for high values of Q'; the difference is less than 0.4% for $Q' > 10$. If R_s is not small with respect to r', its value affects the determination of Q' only indirectly through its effect on ω. If $R_s = r$ and $Q' \geq 10$, the determination of Q' by the above equation is in error by less than 1%.

If ω, L', r', and e are constant and the capacitance C is adjusted until the voltage across it is maximum, the capacitance value C will be slightly less than the capacitance value C_R needed for resonance at the frequency ω:

$$C = \frac{1}{\omega^2L'}\left(\frac{1}{1 + 1/Q^{*2}}\right) = C_R\frac{1}{1 + 1/Q^{*2}}$$

where $\omega = \omega_0$, $Q^* = \omega_0 L'/(r' + R_s)$, and $R_s = R_1 R_2/(R_1 + R_2)$. For $Q^* \geq 10$, C differs from C_R by less than 1%. If $R_s \ll r'$, the value of the effective Q of the unknown inductance is related to m by

$$m = \sqrt{Q'^2 + 1} \qquad Q' = \sqrt{m^2 - 1}$$

where $Q' = \omega L'/r'$, $m = V_C/V_{R_2}$, $\omega = \omega_0 = 2\pi f_0$, and f_0 is the resonant frequency in hertz for L and C_R.

The circuit of Fig. 17-22 can be used to find the value of an unknown capacitance if a stable inductance of known value is available. Similar circuits are used in Q meters (see Par. **17-94**). Discussions of series resonance, parallel resonance, and Q factor are given in Refs. 3, 12, and 13.

Lead lengths should be kept short when making measurements at high frequencies, and the capacitance of the vacuum-tube voltmeter must be added to capacitor C if it is not negligible.

42. The self-capacitance C_0 of an inductance is generally not negligible and can cause discrepancies in measurements. An inductance can be represented by the equivalent circuit of Fig. 17-23, where C_0 is the self-capacitance, L is the true inductance, and r is the true series resistance. The self-capacitance causes the effective inductance L' and the effective series resistance r' of the equivalent unknown inductance as measured in the circuit of Fig. 17-22 to differ somewhat from L and r, that is,

$$r' = r\left(\frac{C + C_0}{C}\right)^2 \qquad \text{and} \qquad L' = L\frac{C + C_0}{C}$$

The value of C_0 can be measured with the aid of a Q meter (Par. **17-94**) having a calibrated variable capacitance.

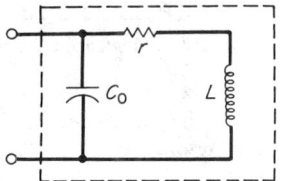
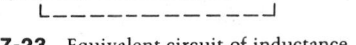

Fig. 17-23. Equivalent circuit of inductance.

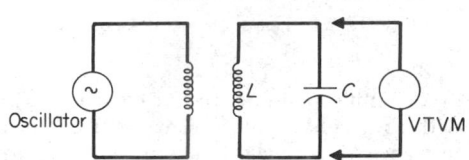

Fig. 17-24. Loosely coupled tuned circuit.

43. The bandwidth of a tuned circuit can be determined by measuring the frequencies f_1 and f_2 at which the capacitor voltage in Fig. 17-24 is 0.707 times the maximum capacitor voltage. The oscillator, which maintains a constant voltage, is loosely coupled to the tuned circuit. The bandwidth Δf is equal to the difference $f_2 - f_1$ and is a function of the resonant frequency f_0 and the Q of the tuned circuit: $\Delta f = f_0/Q$.

44. The circuit Q of a tuned circuit for sinusoidal excitation is defined as 2π(maximum stored energy)/(energy dissipated per cycle). Circuit Q can be measured directly using the circuit of Fig. 17-24. In this circuit the self-capacitance C_0 appears directly in parallel with the tuning capacitor C. The bandwidth is determined as explained above, and the Q is equal to $f_0/(f_2 - f_1)$. If the circuit of Fig. 17-22 is used, the self-capacitance no longer appears in parallel with the tuning capacitor C and the effective Q is measured. The real Q can be calculated if the self-capacitance C_0 is known: $Q = Q'(C + C_0)/C$, where Q' is the effective Q.

Transducer-Input Measurement Systems

45. Transducers are used to convert the quantity to be measured into an electric signal. Transducer types and their input and output quantities are discussed in Sec. **10**.

TRANSDUCER SIGNAL CIRCUITS

46. Amplifiers are often required to increase the voltage and power levels of the transducer output and to prevent excessive loading of the transducer by the measurement system. The design of the amplifier is a function of the performance specifications, which include required amplification in terms of voltage gain or power gain, frequency response, distortion permissible at a given maximum signal level, dynamic range, residual noise permissible at the output, gain stability, permissible drift (for dc amplifiers), operating-temperature range, available supply voltage, permissible power consumption and dissipation, reliability, size, weight, and cost.

47. Capacitive-coupled amplifiers (ac amplifiers) are used when it is not necessary to preserve the dc component of the signal. AC amplifiers are used with transducers that produce a modulated carrier signal. Low-level amplifiers increase the signal from millivolts to several volts. The two-stage class A capacitor-coupled transistor amplifier of Fig. 17-25 has a power gain of 64 dB and a voltage gain of approximately 1,000. Design information, an explanation of biasing, and equations for calculating the input impedance and various gain values are given in Ref. 14. An excellent ac amplifier can be obtained by connecting a coupling capacitor in series with resistor R_1 of the operational amplifier of Fig. 17-29. The capacitor should be selected so that $C > 1/2\pi fR_1$, where f is the lowest signal frequency to be amplified. Class E transformer-coupled amplifiers, which are often used for higher power-output stages, are also discussed in Ref. 14.

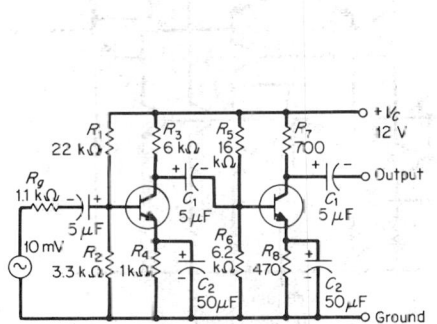

Fig. 17-25. Two-stage cascaded common-emitter capacitive-coupled audio amplifier.[14]

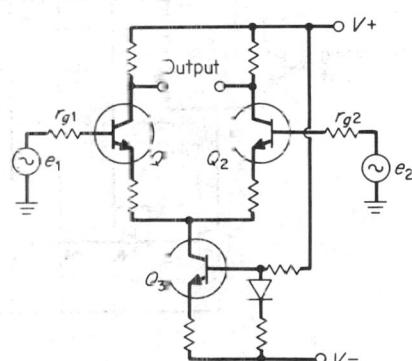

Fig. 17-26. Differential amplifier.

48. Direct-coupled amplifiers are used when the dc component of the signal must be preserved. These designs are more difficult than those of capacitive-coupled amplifiers because changes in transistor leakage currents, gain, and base-emitter voltage drops can cause the output voltage to change for a fixed input voltage, i.e., cause a dc-stability problem. The dc stability of an amplifier is determined primarily by the input stage since the equivalent input drift introduced by subsequent stages is equal to their drift divided by the preceding gain. Balanced input stages, such as the differential amplifier of Fig. 17-26, are widely used because drift components tend to cancel. By selecting a pair of transistors, Q_1 and Q_2, which are matched for current gain within 10% and base-to-emitter voltage within 3 mV, the temperature drift referred to the input can be held to within 10 μV/°C. Transistor Q_3 acts as a constant-current source and thereby increases the ability of the amplifier to reject common-mode input voltages. For applications where the generator resistance r_g is greater than 50 kΩ, current offset becomes dominant, and lower overall drift can be obtained by using field-effect transistors in place of the bipolar transistors Q_1 and Q_2. Voltage drifts as low as 0.6 μV/°C can be obtained using integrated-circuit operational amplifiers (see Fig. 17-27).

49. Operational amplifiers are widely used for amplifying low-level ac and dc signals. They usually consist of a balanced input stage, a number of direct-coupled intermediate stages, and a low-impedance output stage. They provide high open-loop gain which permits the use of a large amount of gain-stabilizing negative feedback (see Par. **17-127**). The schematic of an integrated-circuit operational amplifier which is intended for use in instrumentation applications along with some performance specifications is given in Fig. 17-27. With input voltages e_1, e_2, and e_{cm} equal to zero, the output of the amplifier of Fig. 17-28 will have an offset voltage E_{os} defined by[15]

$$E_{os} = V_{os}\frac{R_1 + R_2}{R_4} + I_{b1}R_2 - I_{b2}\frac{R_3R_4(R_1 + R_2)}{R_1(R_3 + R_4)}$$

where I_{b1} and I_{b2} are bias currents that flow into the amplifier when the output is zero and V_{os} is the input offset voltage that must be applied across the input terminals to achieve zero output. The input bias current specified for an operational amplifier is the average of I_{b1} and I_{b2}.

Since the bias currents are approximately equal, it is desirable to choose the parallel combination of R_3 and R_4 equal to the parallel combination of R_1 and R_2. For this case, $E_{os} = V_{os}(R_1 + R_2)/R_1 + I_{os}R_2$, where offset current $I_{os} = I_{b1} - I_{b2}$.

In the ideal case, where V_{os} and I_{os} are zero, the output voltage E_0 as a function of signal voltage e_1 and e_2 and common-mode voltage e_{cm} is

$$E_0 = -e_1\frac{R_2}{R_1} + e_2\frac{R_4(R_1 + R_2)}{R_1(R_3 + R_4)} + e_{cm}\frac{R_4(R_1 + R_2) - R_2(R_3 + R_4)}{R_1(R_3 + R_4)}$$

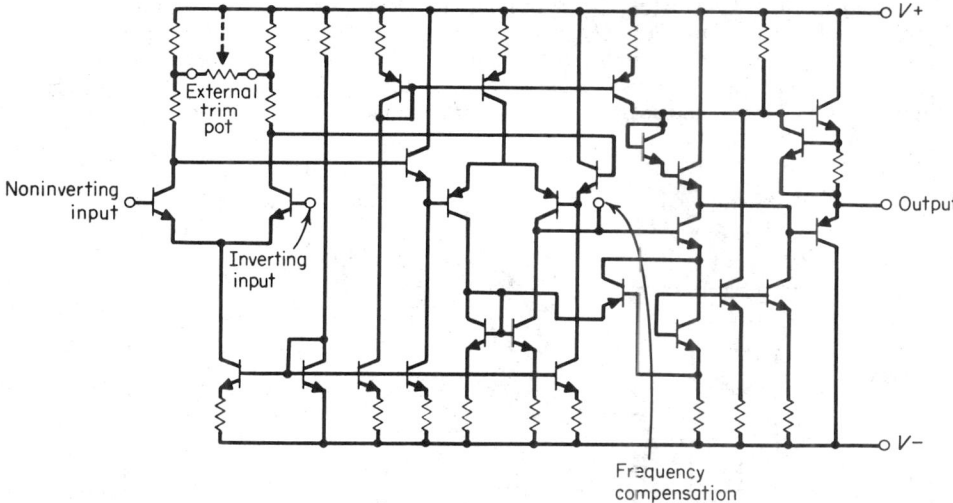

Fig. 17-27. 725 = μA instrumentation operational amplifier. *(Fairchild Semiconductor Co.)*
Typical specifications are:

Input offset voltage		Input resistance	1.5 mΩ	Average input offset	
(no external trim)	0.5 mV	Open-loop gain	3 × 10^6	drift, external trim	0.6 μV/°C
Input offset current	2 nA	Common-mode		No external trim	2 μV/°C
Input bias current	50 nA	rejection ratio	120 dB		

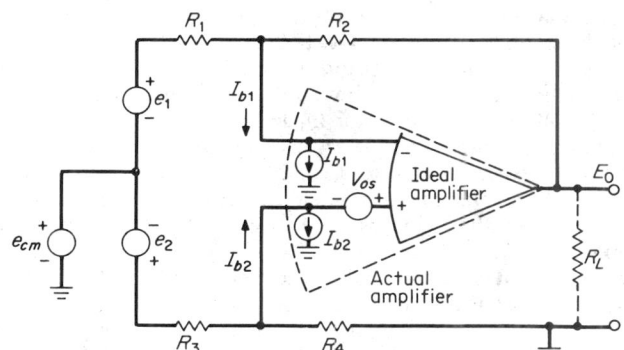

Fig. 17-28. Operational amplifier.[15]

Maximum common-mode rejection can be obtained by choosing $R_4/R_3 = R_2/R_1$, which reduces the above equation to $E_0 = R_2(e_2 - e_1)/R_1$. The common-mode signal is not entirely rejected in an actual amplifier but will be reduced relative to a differential signal by the common-mode rejection ratio of the amplifier. Minimum drift and maximum common-mode rejection, which are important when terminating the wires from a remote transducer, can be obtained by selecting $R_3 = R_1$ and $R_4 = R_2$.

Where common-mode voltages are not a problem, the simple inverting amplifier (Fig. 17-29) is obtained by replacing e_{cm} and e_2 with short circuits and combining R_3 and R_4 into one resistor, which is equal to the parallel equivalent of R_1 and R_2. The input impedance of this circuit is equal to R_1.

Similarly, the simple noninverting amplifier (Fig. 17-30) is obtained by replacing e_{cm} and e_1 with short circuits. The voltage follower is a special case of the noninverting amplifier where $R_1 = \infty$ and $R_2 = 0$. The input impedance of the circuit of Fig. 17-30 is equal to the parallel combination of the common-mode input impedance of the amplifier and impedance Z_{id} [1 + $(AR_1)/(R_1 + R_2)$], where Z_{id} is the differential-mode amplifier input impedance and A is the amplifier gain. Where very high input impedances are required, as in electrometer circuits, operational amplifiers having field-effect input transistors are used to provide input resistances up to $10^{12}\ \Omega$.

An ac-coupled amplifier can be obtained by connecting a coupling capacitor in series with the input resistor of Fig. 17-29. The capacitor value should be selected so that the capacitive reactance at the lowest frequency of interest is lower than the amplifier input impedance R_1.

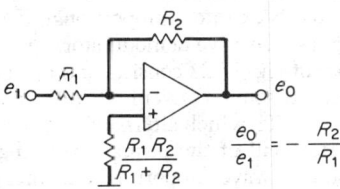

Fig. 17-29. Inverting amplifier.

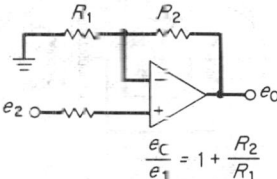

Fig. 17-30. Noninverting amplifier.

Operational amplifiers are useful for realizing filter networks and integrators (see Par. **17-54**). Other applications include absolute-value circuits (see Par. **17-145**), logarithmic converters, non-linear amplification, voltage-level detection, function generation, and analog multiplication and division.[15] Care should be taken not to exceed the maximum supply voltage and maximum common-mode voltage ratings and also to be sure that the load resistance R_L is not smaller than that permitted by the rated output.

50. The charge amplifier is used to amplify the ac signals from variable-capacitance transducers and transducers having a capacitive impedance such as piezoelectric transducers. In the simplified circuit of Fig. 17-31a, the current through C_s is equal to the current through C_1, and therefore

$$C_s \frac{\partial e_s}{\partial t} + e_s \frac{\partial C_s}{\partial t} = -C_1 \frac{de_0}{dt}$$

For the piezoelectric transducer, C_s is assumed constant, and the gain $\delta e_0/\delta e_s = -C_s/C$. For the variable-capacitance transducer, e_s is constant, and the gain $de_0/dC_s = -e_s/C_1$. A practical circuit requires a resistance across C_1 to limit output drift. The value of this resistance must be greater than the impedance of C_1 at the lowest frequency of interest. A typical operational amplifier having field-effect input transistors has a specified maximum input current of 2 nA, which will result in an output offset of only 0.2 V if a 100-MΩ resistance is used across C_1. It is preferable to provide a high effective resistance by using a network of resistors, each of which has a value of 1 MΩ or less.

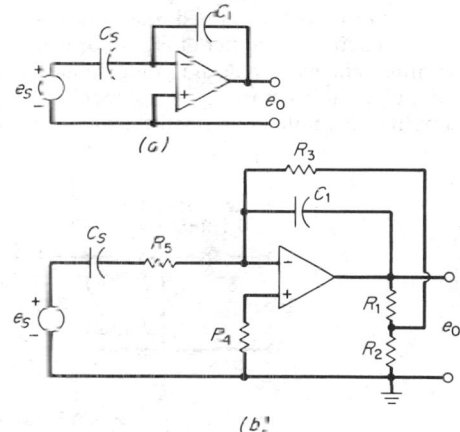

Fig. 17-31. Charge amplifier.

The effective feedback resistance R' in the practical circuit of Fig. 17-31b is given by $R' = R_3(R_1 + R_2)/R_2$, assuming that $R_3 > 10R_1R_2/(R_1 + R_2)$. Output drift is further reduced by selecting $R_4 = R_3 + R_1R_2/(R_1 + R_2)$. Resistor R_5 is used to provide an upper frequency rolloff at $f = 1/2\pi R_5C_s$, which improves the signal-to-noise ratio.

51. Amplifier-gain stability is enhanced by the use of feedback (see Par. **17-127**) since the gain of the amplifier with feedback is relatively insensitive to changes in the open-loop amplifier gain G provided that the loop gain GH is high. For example, if the open-loop gain G changes by 10% from 100,000 to 90,000 and the feedback divider gain H remains constant at 0.01, the closed-loop gain $G/(1 + GH) \approx 99.9$ changes only 0.011%.

If a high closed-loop gain is required, simply decreasing the value of H will reduce the value of GH and thereby reduce the accuracy. The desired accuracy can be maintained by cascading two or more amplifiers, thereby reducing the closed-loop gain required from each amplifier. Each amplifier has its own individual feedback, and no feedback is applied around the cascaded amplifiers. In this case, it is unwise to cascade more stages than needed to achieve a reasonable value of GH in each individual amplifier, since excessive loop gain will make the individual stages more prone to oscillation and the overall system will exhibit increased sensitivity to noise transients.

52. Chopper amplifiers are used for amplifying dc signals in applications requiring very low drift. The dc input signal is converted by a chopper to a switched ac signal having an amplitude proportional to the input signal and a phase of 0 or 180° with respect to the chopper reference frequency, depending on the polarity of the input signal. This ac signal is amplified by an ac amplifier, which eliminates the drift problem, and then is converted back into a proportional dc output voltage by a phase-sensitive demodulator.

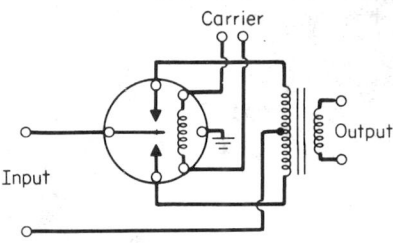

Fig. 17-32. Full-wave chopper.

The chopper of Fig. 17-32 consists of a mechanical switch driven at some convenient reference frequency such as 60 Hz, which alternately connects the input across each half of the primary winding. The output is a square wave proportional to the input, which meets the requirement of having zero amplitude for a zero input signal. Different configurations can be used with the mechanical chopper, and other types of elements such as field-effect transistors, photoconductors, junction transistors, and diodes operated in a switching mode can be used in place of mechanical switches.

The frequency response of a chopper amplifier is theoretically limited to one-half the carrier frequency. In practice, however, the frequency response is much lower than the theoretical limit. High-frequency components in the input signal exceeding the theoretical limit are removed to avoid unwanted beat signals with the chopper frequency.

The chopper amplifier of Fig. 17-33 consists of a low-pass filter to attenuate high frequencies, an input chopper, an ac amplifier, a phase-sensitive demodulator, and a low-pass output filter to attenuate the chopper ripple component in the output signal. The frequency response of this amplifier is limited to a small fraction of the chopper frequency.

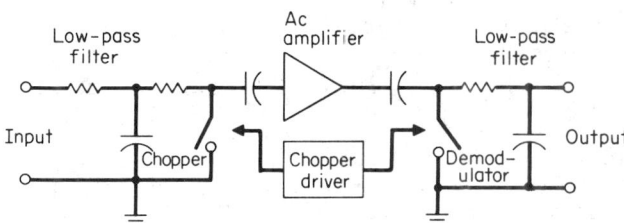

Fig. 17-33. Chopper amplifier.[15]

The frequency-response limitation of the chopper amplifier can be overcome by using the chopper amplifier for the dc and low-frequency signals and a separate ac amplifier for the higher-frequency signals, as shown in Fig. 17-34. Simple shunt field-effect-transistor choppers Q_1 and Q_2 are used for modulation and detection, respectively. Capacitor C_T is used to minimize spikes at the input to the ac amplifier.

53. Modulator-demodulator systems avoid the drift problems of dc amplifiers by using a modulated carrier which can be amplified by ac amplifiers (Fig. 17-35). Inputs and outputs may be either electrical or mechanical.

The varactor modulator (Fig. 17-36) takes advantage of the variation of diode-junction capacitance with voltage to modulate a sinusoidal carrier. The carrier and signal voltages applied to the diodes are small, and the diodes never reach a low-resistance condition. Input bias currents of the order of 0.01 pA are possible. For zero signal input, the capacitance values of the diodes are equal, and the carrier signals coupled by the diodes cancel. A dc-input signal will increase the capacitance of one diode while decreasing the capacitance of the other and thereby produce a carrier unbalance signal which is coupled to the ac amplifier by capacitor C_2. A phase-sensitive demodulator, such as field-effect transistor Q_2 of Fig. 17-34, may be used to recover the dc signal.

The magnetic amplifier and second-harmonic modulator can also be used to convert dc-input signals to modulation on a carrier, which is amplified and later demodulated. Mechanical-input modulators include ac-driven potentiometers (Par. **17-113**), linear variable differential transformers (Par. **17-146**), and synchros (Par. **17-113**). The amplified ac carrier can be converted directly into a mechanical output by a two-phase induction servomotor (Par. **17-154**).

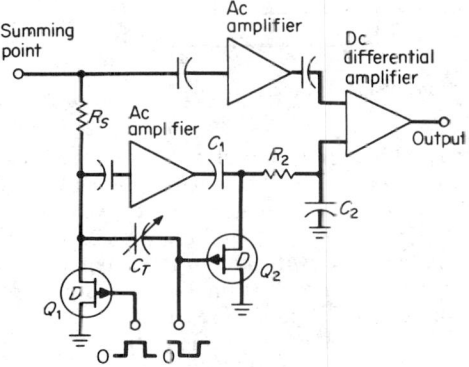

Fig. 17-34. Chopper-stabilized dc amplifier.[16]

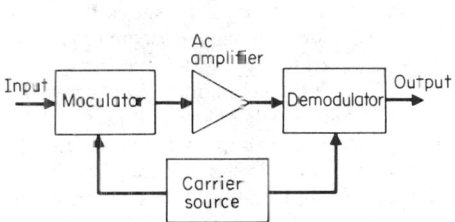

Fig. 17-35. Modulator-demodulator system.

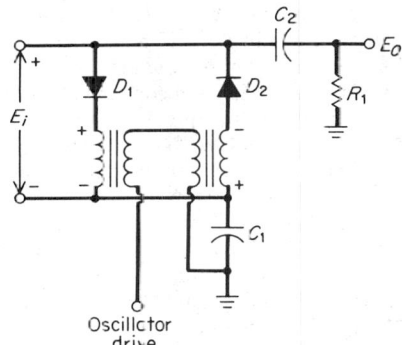

Fig. 17-36. Basic varactor modulator.[15]

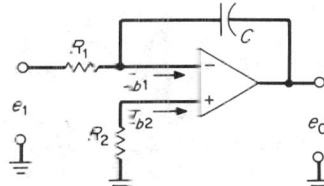

Fig. 17-37. Analog integrator.

54. Integrators are often required in systems where the transducer signal is a derivative of the desired output, e.g., when an accelerometer is used to measure the velocity of a vibrating object. The output of the analog integrator of Fig. 17-37 consists of an integrated signal term plus error terms caused by the offset voltage V_{os} and the bias currents I_{b1} and I_{b2} (see Par. **17-49**)

$$e_0 = -\frac{1}{R_1 C} \int e \; dt + \frac{1}{R_1 C} \int (V_{os} + I_{B_1} R_1 - I_{b_2} R_2) \; dt$$

These error terms will cause the integrator to saturate unless the integrator is reset periodically or a feedback path exists which tends to drive the output toward a given level within the linear range. In the accelerometer integrator, accurate integration may not be required below a given frequency, and the desired stabilizing feedback path can be introduced by incorporating a large effective resistance across the capacitor using the technique shown in Fig. 17-31b. In this case, the integrator response is approximated by the low-pass network characteristic of Table 17-7, Par. **17-139**.

55. Digital processing of analog quantities is frequently used because of the availability of high-performance analog-to-digital converters (A/D), digital-to-analog converters (D/A), microprocessors, and other special digital processors. The analog input signal (Fig. 17-38) is converted into a sequence of digital values by the A/D converter. The digital values are processed

by the microprocessor (Par. **17-56**) or other digital processor, which can be programmed to provide a wide variety of functions including linear or nonlinear gain characteristics, digital filtering, integration, differentiation, modulation, linearization of signals from nonlinear transducers,
computation, and self-calibration.

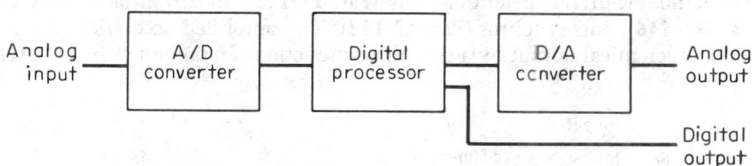

Fig. 17-38. Digital processing of analog signals.

The digital output may be used directly or converted back into an analog signal by the D/A
converter. Commercial A/D converter modules are available with 14-bit resolution and 12-μs
conversion times using the successive-approximation technique (Analog Devices ADC1131).
Monolithic integrated-circuit A/D converters are available with 10-bit resolution and 25-μs conversion times (Analog Devices AD571). An 18-bit D/A converter with a full-scale current output
of 100 mA and a compliance voltage range of ±12 V has been built at the National Bureau of
Standards.[17] It exhibits a settling time to ½ least significant bit (2 ppm) of less than 20 μs. Commercial 16- and 18-bit D/A converters are available with 2-mA full-scale outputs (Analog
Devices DAC1136/1138, Burr Brown DAC 70).

The application of A/D converters, D/A converters, and sample-and-hold (S/H) circuits
requires careful consideration of their static and dynamic characteristics. Discussions of performance terminology and measurement tech-
niques are given in Refs. 18 and 19.

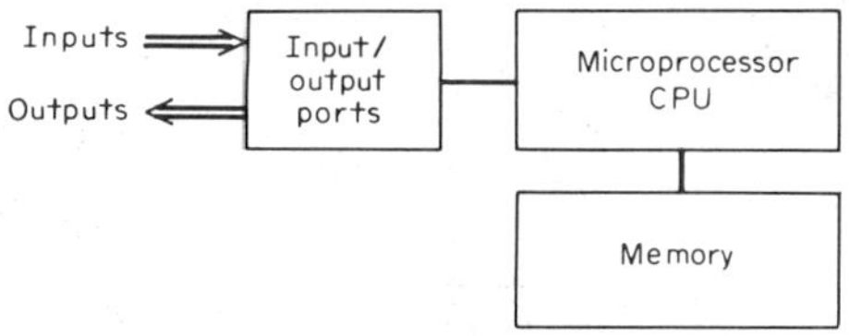

56. The microprocessor is revolution-
izing instrumentation and control. The digital
processing power that previously cost many
thousands of dollars has been made available
on a silicon integrated-circuit chip for only a
few dollars. A microcomputer system (Fig. 17-
39) consists of a microprocessor central processing unit (CPU), memory, and input-output
ports. A typical microprocessor system consists of the 8085 microprocessor, an 8155 RAM/I0/
timer chip, which provides 256 bytes of random-access memory and 22 input-output lines, and
a number of 2,048- or 4,096-byte read-only-memory (ROM) chips. For applications requiring only
a limited memory capability the 3870 microcomputer combines the CPU, a timer, 32 input-
output lines, 64 bytes of RAM and 2 kilobytes of ROM on a single chip. The microprocessor can
provide a number of features at little incremental cost by means of software modification. Typical features include automatic ranging, self-calibration, self-diagnosis, data conversion, data processing, linearization, regulation, process monitoring, and control. Instruments using microprocessors are described in Pars. **17-83** and **17-110**.

Fig. 17-39. Microcomputer system.

57. Voltage comparators are useful in a number of applications, including signal comparison with a reference, positive and negative signal-peak detection, zero-crossing detection, A/
D successive-approximation converter systems, crystal oscillators, voltage-controlled oscillators,
magnetic-transducer voltage detection, pulse generation, and square-wave generation. Compara-

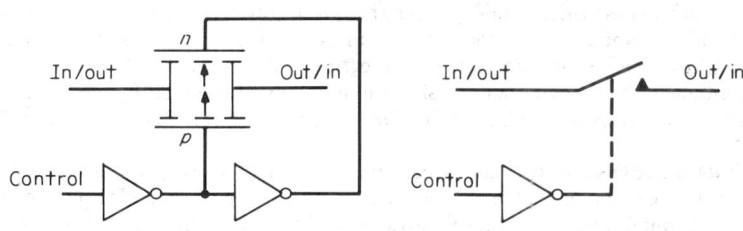

Fig. 17-40. Bilateral analog switch.

tors with input offset voltages of 2 mV and input offset currents of 5 nA are commercially available.

58. Analog switches using field-effect-transistor (FET) technology are used in general-purpose high-level switching (± 10 V), multiplexing, A/D conversion, chopper applications, set-point adjustment, and bridge circuits. Logic and schematic diagrams of a typical bilateral switch are shown in Fig. 17-40. The switch provides the required isolation between the data signal and the control signal. Typical switches provide zero offset voltage, on resistance of 60 Ω, and leakage current of 0.1 nA. Multiple-channel switches are commercially available in a single package.

59. Output Indicators. A variety of analog and digital output indicators can be used to display and record the output from the signal-processing circuitry.

Bridge Circuits, Detectors, and Amplifiers

PRINCIPLES OF BRIDGE MEASUREMENTS

60. Bridge circuits are used to determine the value of an unknown impedance in terms of other impedances of known value. Highly accurate measurements are possible because a null condition is used to compare ratios of impedances.

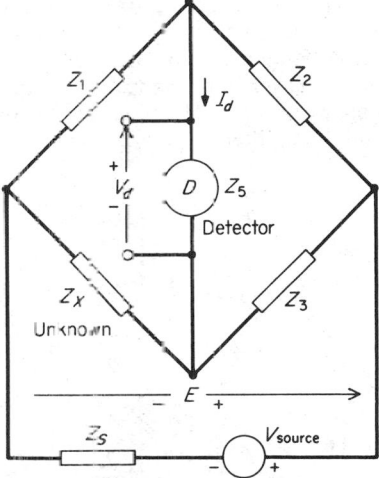

Fig. 17-41. Basic impedance bridge.

The most common bridge arrangement (Fig. 17-41) contains four branch impedances, a voltage source, and a null detector. Galvanometers, alone or with chopper amplifiers, are used as null detectors for dc bridges; while telephone receivers, vibration galvanometers, and tuned amplifiers with suitable detectors and indicators are used for null detection in ac bridges (see Pars. **17-89** to **17-93** and Ref. 25). The voltage across an infinite-impedance detector is

$$V_d = \frac{(Z_1 Z_3 - Z_2 Z_x)E}{(Z_1 + Z_2)(Z_3 + Z_x)}$$

If the detector has a finite impedance Z_5, the current in the detector is

$$I_d = \frac{(Z_1 Z_3 - Z_2 Z_x)E}{Z_5(Z_1 + Z_2)(Z_3 + Z_x) + Z_1 Z_2(Z_3 + Z_x) + Z_3 Z_x(Z_1 + Z_2)}$$

where E is the potential applied across the bridge terminals.

61. A null or balance condition exists when there is no potential across the detector. This condition is satisfied, independent of the detector impedance, when $Z_1 Z_3 = Z_2 Z_x$. Therefore, at balance the value of the unknown impedance Z_x can be determined in terms of the known impedances Z_1, Z_2, and Z_3:

$$Z_x = Z_1 Z_3 / Z_2$$

Since the impedances are complex quantities, balance requires that both magnitude and phase angle conditions be met: $|Z_x| = |Z_1| \cdot |Z_3| / |Z_2|$ and $\theta_x = \theta_1 + \theta_3 - \theta_2$. Two of the known impedances are usually fixed impedances, while the third impedance is adjusted in resistance and reactance until balance is attained.

62. The sensitivity of the bridge can be expressed in terms of the incremental detector current ΔI_d for a given small per-unit deviation δ of the adjustable impedance from the balance value. If Z_1 is adjusted, $\delta = \Delta Z_1 / Z_1$ and

$$\Delta I_d = \frac{Z_3 Z_x E \delta}{(Z_3 + Z_x)^2 [Z_5 + Z_1 Z_2 / (Z_1 + Z_2)] - Z_3 Z_x^2 (Z_3 + Z_x)]}$$

where Z_5 is the detector impedance.

If a high-input-impedance amplifier is used for the detector and impedance Z_5 can be considered infinite, the sensitivity can be expressed in terms of the incremental input voltage to the detector ΔV_d for a small deviation from balance

$$\Delta V_d = Z_3 Z_x E\delta / (Z_3 + Z_x)^2 = Z_1 Z_2 E\delta / (Z_1 + Z_2)^2$$

where $\delta = \Delta Z_1 / Z_1$ and ΔZ_1 is the deviation of impedance Z_1 from its balance value Z_1. Maximum sensitivity occurs when the magnitudes of Z_3 and Z_x are equal (which for balance implies that the magnitudes of Z_1 and Z_2 are equal). Under this condition, $\Delta V_d = E\delta/4$ when the phase angles θ_3 and θ_x are equal; $\Delta V_d = E\delta/2$ when the phase angles θ_3 and θ_x are in quadrature; and ΔV_d is infinite when $\theta_3 = -\theta_x$, as is the case with lossless reactive components of opposite sign. In practice, the value of the adjustable impedance must be sufficiently large to ensure that the resolution provided by the finest adjusting step permits the desired precision to be obtained. This value may not be compatible with the highest sensitivity, but adequate sensitivity can be obtained for an order-of-magnitude difference between Z_3 and Z_x or Z_1 and Z_2, especially if a tuned-amplifier detector is used.

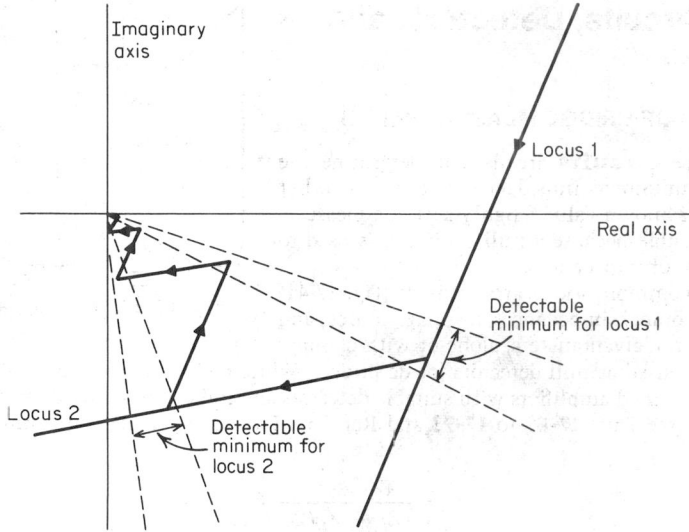

Fig. 17-42. Linearized convergence locus.[20]

63. Interchanging the source and detector can be shown to be equivalent to interchanging impedances Z_1 and Z_3. This interchange does not change the equation for balance but does change the sensitivity of the bridge. For a fixed applied voltage E higher sensitivity is obtained with the detector connected from the junction of the two high-impedance arms to the junction of the two low-impedance arms.

64. The source voltage must be carefully selected to ensure that the allowable power dissipation and voltage ratings of the known and unknown impedances of the bridge are not exceeded. If the bridge impedances are low with respect to the source impedance Z_s, the bridge-terminal voltage E will be lowered. This can adversely affect the sensitivity, which is proportional to E. The source for an ac bridge should provide a pure sinusoidal voltage since the harmonic voltages will usually not be nulled when the balance is achieved at the fundamental frequency. A tuned detector is helpful in achieving an accurate balance.

65. Balance Convergence.[20] The process of balancing an ac bridge consists of making successive adjustments of two parameters until a null is obtained at the detector. It is desirable that these parameters not interact and that convergence be rapid.

The equation for balance can be written in terms of resistances and reactances as

$$R_x + jX_x = (R_1 + jX_1)(R_3 + jX_3)/(R_2 + jX_2)$$

Balance can be achieved by adjusting any or all of the six known parameters, but only two of them need be adjusted to achieve the required equality of both magnitude and phase (or real and imaginary components). In ratio-type bridge, one of the arms adjacent to the unknown, either Z_1 or Z_3, is adjusted. Assuming that Z_1 is adjusted, then to make the resistance adjustment independent of the change in the corresponding reactance, the ratio $(R_3 + jX_3)/(R_2 + jX_2)$ must be either real or imaginary but not complex. If this ratio is equal to the real number k, then for balance $R_x = kR_1$ and $X_x = kX_1$. In a product-type bridge, the arm opposite the unknown, Z_2, is adjusted for balance, and the product Z_1Z_3 must be either real or imaginary to make the resistance adjustment independent of the reactance adjustment.

Near balance, the denominator of the equation giving the detector voltage (or current) changes little with the varied parameter, while the numerator changes considerably. The usual convergence loci, which consist of circular segments, can be simplified to obtain linear convergence loci by assuming that the detector voltage near balance is proportional to the numerator, $Z_1Z_3 - Z_2Z_x$. Values of this quantity can be plotted on the complex plane. When only a single adjustable parameter is varied, a straight-line locus will be produced as shown in Fig. 17-42. Varying the other adjustable parameter will produce a different straight-line locus. The rate of convergence to the origin (balance condition) will be most rapid if these two loci are perpendicular, slow if they intersect at a small angle, and zero if they are parallel. The cases of independent resistance and reactance adjustments described above correspond to perpendicular loci.

RESISTANCE BRIDGES

66. The Wheatstone bridge is used for the precise measurement of two-terminal resistances. The lower limit for accurate measurement is about 1 Ω because contact resistance is likely to several milliohms. For simple galvanometer detectors, the upper limit is about 1 MΩ, which can be extended to 10^{12} Ω by using a high-impedance high-sensitivity detector and a guard terminal to substantially eliminate the effects of stray leakage resistance to ground.

The Wheatstone bridge (Fig. 17-43) although historically older, may be considered as a resistance version of the impedance bridge of Par. **17-60**, and therefore the sensitivity equations are applicable. At balance

$$R_x = R_1R_3/R_2$$

Known fixed resistors, having values of 1, 10, 100, or 1,000 Ω, are generally used for two arms of the bridge, for example, R_2 and R_3. These arms provide a ratio R_3/R_2 which can be selected from 10^{-3} to 10^3. Resistor R_1,

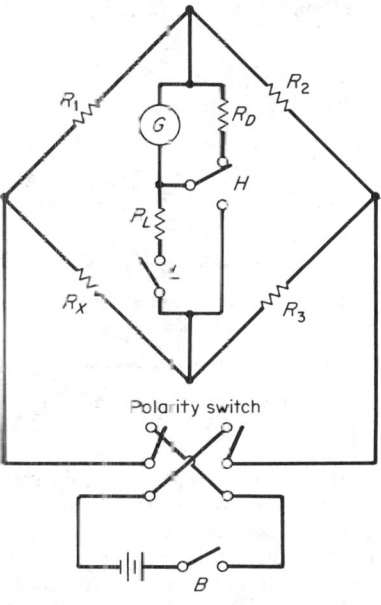

Fig. 17-43. Wheatstone bridge.

typically adjustable to 10,000 Ω in 1- or 0.1-Ω steps, is adjusted to achieve balance. The ratio R_3/R_2 should be chosen so that R_1 can be read to its full precision. The magnitudes of R_2 and R_3 should be chosen to maximize the sensitivity while taking care not to draw excessive current.

An alternate arrangement using R_1 and R_2 for the ratio resistors and adjusting R_3 for balance will generally provide a different sensitivity (see Pars. **17-62** and **17-63**).

The battery key B should be depressed first to allow any reactive transients to decay before the galvanometer key is depressed. The low-galvanometer-sensitivity key L should be used until the bridge is close to balance. The high-sensitivity key H is then used to achieve final balance. Resistance R_D provides critical damping between galvanometer measurements. The battery connections to the bridge may be reversed and two separate resistance determinations made to eliminate any thermoelectric errors.

67. The Kelvin double bridge (Fig. 17-44) is used for the precise measurement of low-value four-terminal resistors in the range 1 $\mu\Omega$ to 10 Ω. The resistance to be measured X and a standard resistance S are connected by means of their current terminals in a series loop containing a battery, an ammeter, an adjustable resistor, and a low-resistance link l. Ratio-arm resistances A and B and α and β are connected to the potential terminals of resistors X and S as shown. The equation for balance is

$$X = S\frac{A}{B} + \frac{\beta l}{\alpha + \beta + 1}\left(\frac{A}{B} - \frac{\alpha}{\beta}\right)$$

If the ratio α/β is made equal to the ratio A/B, the equation reduces to $X = S(A/B)$.

The equality of the ratios should be verified after the bridge is balanced by removing the link. If $\alpha/\beta = A/B$, the bridge will remain balanced. Lead resistances r_1, r_2, r_3, and r_4 between the bridge and the potential terminals of the resistors may contribute to ratio unbalance unless they have the same ratio as the arms to which they are connected. Ratio unbalance caused by lead resistance can be compensated by shunting α or β with a high resistance until balance is obtained with the link removed.

In some bridges a fixed standard resistor S having a value of the same order of magnitude as resistor X is used. Fixed resistors of 10, 100, or 1,000 Ω are used for two arms, for example, B and β, with B and β having equal values. Bridge balance is obtained by adjusting tap switches to select equal resistances for the other two arms, for example, A and α, from values adjustable up to 1,000 Ω in 0.1-Ω steps. In other bridges, only decimal ratio resistors are provided for A, B, α, and β, and balance is obtained by means of an adjustable standard having nine steps of 0.001 Ω each and a Manganin slide bar of 0.0011 Ω.

The battery connection should be reversed and two separate resistance determinations made to eliminate thermoelectric errors.

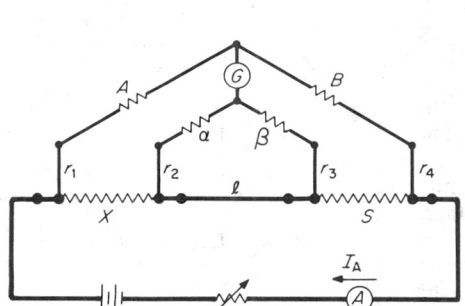

Fig. 17-44. Kelvin double bridge.[2] $A + B$ is typically 1,000 Ω and $\alpha + \beta$ is typically 1,000 Ω.

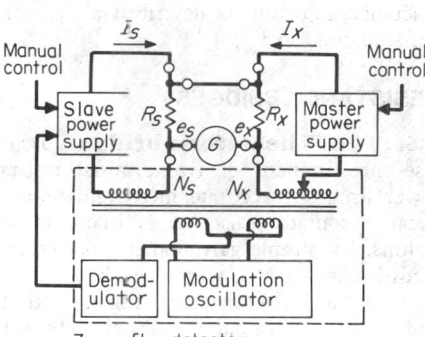

Fig. 17-45. Comparator ratio bridge.[21]

68. The dc-comparator ratio bridge (Fig. 17-45) is used for very precise measurement of four-terminal resistors. Its accuracy and stability depend mainly on the turns ratio of a precision transformer. The master current supply is set at a convenient fixed value I_x. The zero-flux detector maintains an ampere-turn balance, $I_x N_x = I_s N_s$, by automatically adjusting the current I_s from the slave supply as N_x is manually adjusted. A null reading on the galvanometer is obtained when $I_s R_s = I_x R_x$. Since the current ratio is precisely related to the turns ratio, the unknown resistance $R_x = N_x R_s / N_s$. Fractional turn resolution for N_x can be obtained by diverting a fraction of the current I_x as obtained from a decade current divider through an additional winding on the transformer. Turn ratios have been achieved with an accuracy of better than 1 part in 10^7. The zero-flux detector operates by superimposing a modulating mmf on the core using modulation and detector windings in a second-harmonic modulator configuration. The limit sensitivity of the bridge is set by noise and is about 3 μA·turns.

69. Murray and Varley bridge circuits are used for locating faults in wire lines and cables. The faulted line is connected to a good line at one end by means of a jumper to form a loop. The resistance r of the loop is measured using a Wheatstone bridge. The loop is then connected as shown in Fig. 17-46 to form a bridge in which one arm contains the resistance R_x between the test set and the fault and the adjacent arm contains the remainder of the loop resistance. The galvanometer detector is connected across the open terminals of the loop, while the voltage supply is connected between the fault and the junction of fixed resistor R_2 and variable resistor R_3. When balance is attained

$$R_x = rR_3/(R_2 + R_3)$$

where r is the resistance of the loop. Resistance R_x is proportional to the distance to the fault. In the Varley loop of Fig. 17-47, variable resistor R_1 is adjusted to achieve balance and

$$R_x = \frac{rR_3 - R_1 R_2}{R_2 + R_3}$$

where r is the resistance of the loop.

INDUCTANCE BRIDGES

70. General. Many bridge types are possible since the impedance of each arm may be a combination of resistances, inductances, and capacitances. A number of popular inductance bridges are shown in Fig. 17-48. In the balance equations L and M are given in henrys, C in farads, and R in ohms; ω is 2π times the frequency in hertz. The Q of an inductance is equal to $\omega L/R$, where R is the series resistance of the inductance.

71. The symmetrical inductance bridge (Fig. 17-48a) is useful for comparing the impedance of an unknown inductance with that of a known inductance. An adjustable resistance is connected in series with the inductance having the higher Q, and the inductance and resistance values of this resistance are added to those of the associated inductance to obtain the impedance of that arm. If this series resistance is adjusted along with the known inductance to obtain balance, the resistance and reactance balances are independent and balance convergence is rapid.

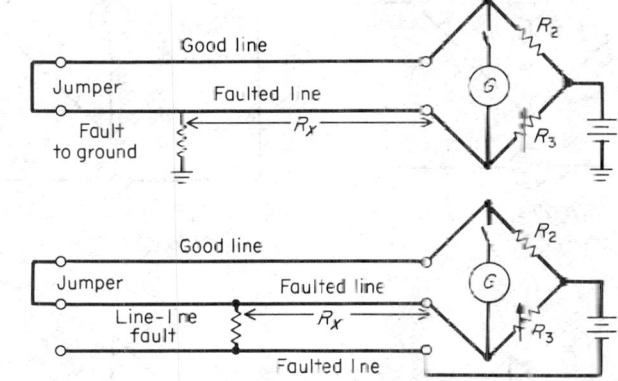

Fig. 17-46. Murray loop-bridge circuits.

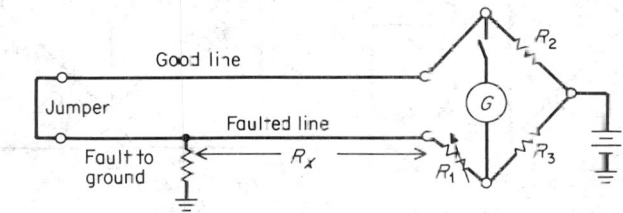

Fig. 17-47. Varley loop circuit.

If only a fixed inductance is available, the series resistance is adjusted along with the ratio R_3/R_2 until balance is obtained. These adjustments are interacting, and the rate of convergence will be proportional to the Q of the unknown inductance. Care must be taken to avoid inductive coupling between the known and unknown inductances since it will cause a measurement error.

72. The Maxwell-Wien bridge (Fig. 17-48b) is widely used for accurate inductance measurements. It has the advantage of using a capacitance standard which is more accurate and easier to shield and produces practically no external field. R_2 and C_2 are usually adjusted since they provide a noninteracting resistance and inductance balance. If C_1 is fixed and R_2 and R_1 or R_3 are adjusted, the balance adjustments interact and balancing may be tedious.

73. Anderson's bridge (Fig. 17-48c) is useful for measuring a wide range of inductances with reasonable values of fixed capacitance. The bridge is usually balanced by adjusting r and a resistance in series with the unknown inductance. Preferred values for good sensitivity are $R_1 = R_2 = R_3/2 = R_x/2$ and $L/C = 2R_x^2$. This bridge is also used to measure the residuals of resistors using a substitution method to eliminate the effects of residuals in the bridge elements.

74. Owen's bridge (Fig. 17-48d) is used to measure a wide range of inductance values in terms of resistance and capacitance. The inductance and resistance balances are independent if R_3 and C_3 are adjusted. The bridge can also be balanced by adjusting R_1 and R_3.

This bridge is useful for finding the incremental inductance of iron-cored inductors to alternating current superimposed on a direct current. The direct current can be introduced by connecting a dc-voltage source with a large series inductance across the detector branch. Low-impedance blocking capacitors are placed in series with the detector and the ac source.

75. Hay's bridge (Fig. 17-48e) is similar to the Maxwell-Wien bridge and is used for measuring inductances having large values of Q. The series $R_2 C_2$ arrangement permits the use of smaller resistance values than the parallel arrangement. The frequency-dependent $1/Q_x^2$ term in the inductance equation is inconvenient since the dials cannot be calibrated to indicate inductance directly unless the term is neglected, which causes a 1% error for $Q_x = 10$.

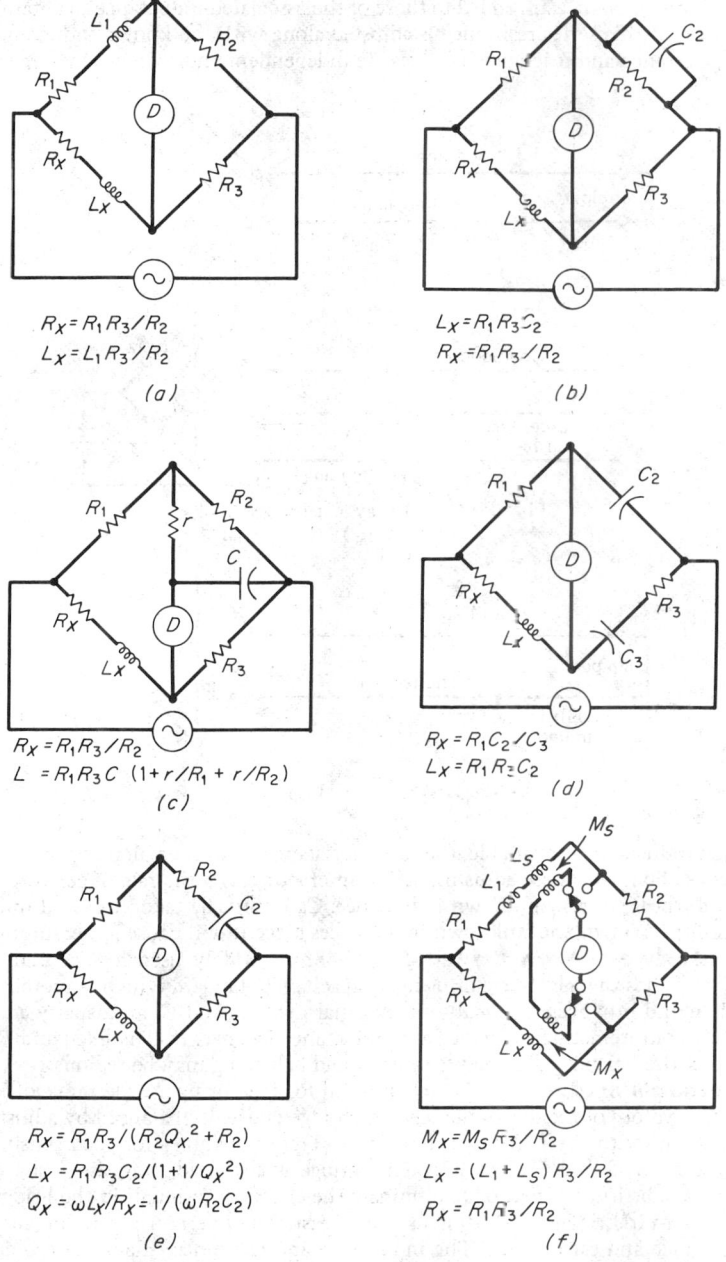

$$Rx = R_1 R_3 / R_2$$
$$Lx = L_1 R_3 / R_2$$

(a)

$$Lx = R_1 R_3 C_2$$
$$Rx = R_1 R_3 / R_2$$

(b)

$$Rx = R_1 R_3 / R_2$$
$$L = R_1 R_3 C (1 + r/R_1 + r/R_2)$$

(c)

$$Rx = R_1 C_2 / C_3$$
$$Lx = R_1 R_2 C_2$$

(d)

$$Rx = R_1 R_3 / (R_2 Q_x^2 + R_2)$$
$$Lx = R_1 R_3 C_2 / (1 + 1/Q_x^2)$$
$$Q_x = \omega Lx/Rx = 1/(\omega R_2 C_2)$$

(e)

$$Mx = Ms R_3 / R_2$$
$$Lx = (L_1 + Ls) R_3 / R_2$$
$$Rx = R_1 R_3 / R_2$$

(f)

Fig. 17-48. Inductance bridges: (a) symmetrical inductance bridge;[3] (b) Maxwell-Wien bridge;[3] (c) Anderson's bridge;[2] (d) Owen's bridge;[2] (e) Hay's bridge;[3] (f) Campbell's bridge.[2]

This bridge is also used for determining the incremental inductance of iron-cored reactors, as discussed for Owen's bridge.

76. Campbell's bridge (Fig. 17-48f) for measuring mutual inductance makes possible the comparison of unknown and standard mutual inductances having different values. The resistances and self-inductances of the primaries are balanced with the detector switches to the right by adjusting L_1 and R_1. The switches are thrown to the left, and the mutual-inductance balance is made by adjusting M_s. Care must be taken to avoid coupling between the standard and unknown inductances.

CAPACITANCE BRIDGES

77. Capacitance bridges are used to make precise measurements of capacitance and the associated loss resistance in terms of known capacitance and resistance values. Several different bridge circuits are shown in Fig. 17-49. In the balance equations R is given in ohms and C in

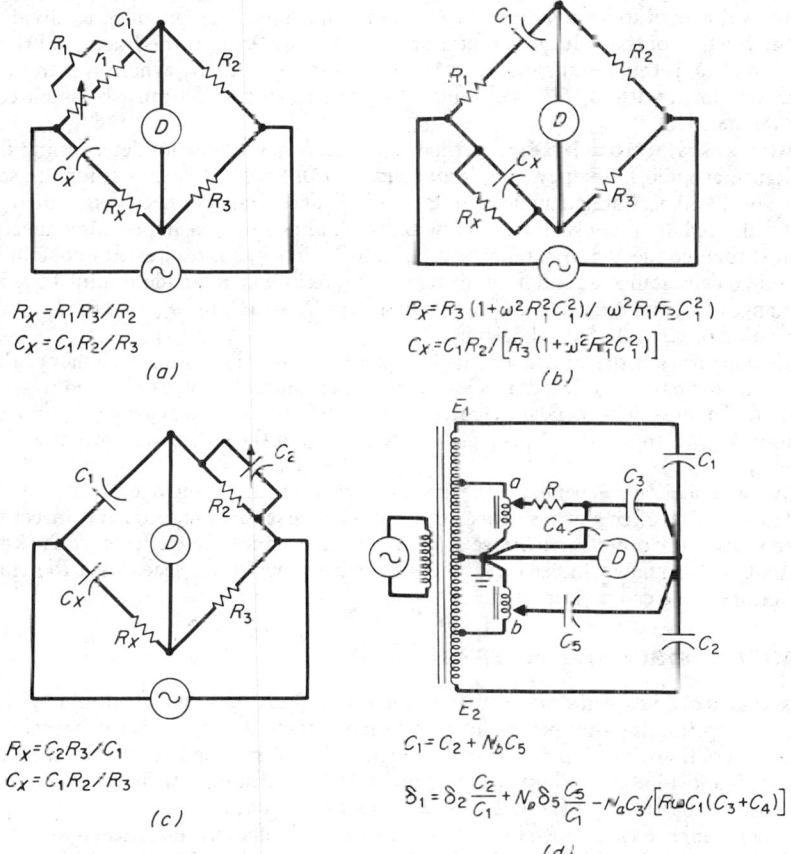

$$R_x = R_1 R_3 / R_2$$
$$C_x = C_1 R_2 / R_3$$
(a)

$$R_x = R_3(1+\omega^2 R_1^2 C_1^2)/ \omega^2 R_1 R_2 C_1^2$$
$$C_x = C_1 R_2/[R_3(1+\omega^2 R_1^2 C_1^2)]$$
(b)

$$R_x = C_2 R_3 / C_1$$
$$C_x = C_1 R_2 / R_3$$
(c)

$$C_1 = C_2 + N_b C_5$$
$$\delta_1 = \delta_2 \frac{C_2}{C_1} + N_a \delta_5 \frac{C_5}{C_1} - N_a C_3/[R\omega C_1(C_3+C_4)]$$
(d)

Fig. 17-49. Capacitance bridges: (a) series-resistance-capacitance bridge [3] (b) Wien bridge;[3] (c) Schering's bridge;[3] (d) transformer bridge.[2]

farads, and ω is 2π times the frequency in hertz. The loss angle δ of a capacitor may be expressed either in terms of its series loss resistance r_s, which gives $\tan \delta = \omega C r_s$, or in terms of the parallel loss resistance r_p, in which case, $\tan \delta = 1/\omega C r_p$.

78. The series RC bridge (Fig. 17-49a) is a resistance-ratio bridge used to compare a known capacitance with an unknown capacitance. The adjustable series resistance is added to the arm containing the capacitor having the smaller loss angle δ.

79. The Wien bridge (Fig 17-49b) is useful for determining the equivalent capacitance C_x and parallel loss resistance R_x of an imperfect capacitor, e g., a sample of insulation or a length of cable.

An important application of the Wien bridge network is its use as the frequency-determining network in RC oscillators (see Par. **17-102**).

80. Schering's bridge (Fig. 17-49c) is widely used for measuring capacitance and dissipation factors. The unknown capacitance is directly proportional to known capacitance C_1. The dissipation factor $\omega C_x R_x$ can be measured with good accuracy using this bridge. The bridge is also used for measuring the loss angles of high-voltage power cables and insulators. In this application, the bridge is grounded at the R_2/R_3 node, thereby keeping the adjustable elements R_2, R_3, and C_2 at ground potential.

81. The transformer bridge is used for the precise comparison of capacitors, especially for three-terminal shielded capacitors.[22,23] A three-winding toroidal transformer having low leakage reactance[24] is used to provide a stable ratio, known to better than 1 part in 10^7. In Fig. 17-49d, capacitors C_1 and C_2 are being compared, and a balancing scheme using inductive-voltage dividers a and b is shown. It is assumed that $C_1 > C_2$ and loss angle $\delta_2 > \delta_1$. In-phase current to balance any inequality in magnitude between C_1 and C_2 is injected through C_5 while quadrature current is supplied by means of resistor R and current divider $C_3/(C_3 + C_4)$. The current divider permits the value of R to be kept below 1 MΩ. Fine adjustments are provided by dividers a and b. N_a is the fraction of the voltage E_1 that is applied to R, while N_b is the fraction of the voltage E_2 applied to C_5. δ_1 is the loss angle of capacitor C_1 and $\tan \delta_1 = \omega C_1 r_1$, where r_1 is the series loss resistance associated with C_1. The reactance of C_3 and C_4 in parallel must be small compared with the resistance of R.

82. The substitution-bridge method is particularly valuable for determining the value of capacitance at radio frequency. *The shunt-substitution method* is shown for the series RC bridge in Fig. 17-50. Calibrated adjustable standards R_s and C_s are connected as shown, and the bridge is balanced in the usual manner with the unknown capacitance disconnected. The unknown is then connected in parallel with C_s, and C_s and R_s are readjusted to obtain balance. The unknown capacitance C_x and its equivalent series resistance R_x are determined by the rebalancing changes ΔC_s and ΔR_s in C_s and R_s, respectively: $C_x = \Delta C_s$ and $R_x = \Delta R_s (C_{s1}/C_x)^2$, where C_{s1} is the value of C_s in the initial balance.

In *series substitution* the bridge arm is first balanced with the standard elements alone, the standard elements having an impedance of Z_{s1}, and then the unknown is inserted in series with the standard elements. The standard elements are readjusted to an impedance Z_{s2} to restore balance. The unknown impedance Z_x is equal to the change in the standard impedance, that is, $Z_x = Z_{s1} - Z_{s2}$.

Measurement accuracy depends on the accuracy with which the changes in the standard values are known. The effects of residuals, stray capacitance, stray coupling, and inaccuracies in the impedances of the other three bridge arms are minimal, since these effects are the same with and without the unknown impedance. The proper handling of the leads used to connect the unknown impedance can be important.[4]

IMPEDANCE BRIDGES AND METERS

83. Automatic impedance bridges incorporate microprocessor control of ranging and decimal-point positioning that permit automated measurement of inductance, capacitance, and resistance in less than 1 s. Typical of the new generation of microprocessor instrumentation is the General Radio 1658 Digibridge, which automatically measures a wide range of L, C, R, D, and Q values at 100 Hz to 1 kHz with a basic accuracy of 0.1%.

Similar performance is provided by the Hewlett-Packard microprocessor-based 4274A (100 Hz to 100 kHz) and 4275A (10 kHz to 10 MHz) Multifrequency LCR Meters, which measure the impedance of the device under test at a selected frequency and compute the values of L, C, R, D, and Q as well as the impedance, reactance, conductance, susceptance, and phase angle.

FACTORS AFFECTING ACCURACY

84. Stray Capacitance and Residuals. The bridge circuits of Figs. 17-48 and 17-49 are idealized since stray capacitances which are inevitably present and the residual inductances associated with resistances and connecting leads have been neglected. These spurious circuit elements can disturb the balance conditions and result in serious measurement errors. Detailed discussions of the residuals associated with the various bridges are given in Ref. 20.

85. Shielding and grounding can be used to control errors caused by stray capacitance. Stray capacitances in an ungrounded, unshielded series RC bridge are shown schematically by

C_1 through C_2 in Fig. 17-51. The elements of the bridge may be enclosed in the grounded metal shield, as shown schematically in Fig. 17-52. Shielding and grounding eliminate some capacitances and make the others definite localized capacitances which act in a known way as illustrated in Fig. 17-53. The capacitances associated with terminal D shunt the oscillator and have no adverse effect. The possible adverse effects of the capacitance associated with the output diagonal EF are overcome by using a shielded output transformer. If the shields are adjusted so that $C_{22}/C_{21} = R_a/R_b$, the ratio of the bridge is independent of frequency. Capacitance C_{24} can be taken into account in the calibration of C_s, and capacitance C_{23} can be measured and its shunting effect across the unknown impedance can be calculated. Shielding, which is used at audio frequencies, becomes more necessary as the frequency and impedance levels are increased.

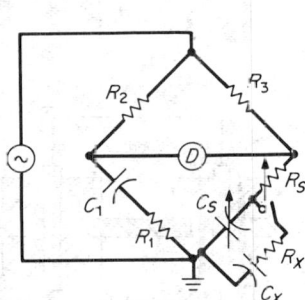

Fig. 17-50. Substitution measurement.[3]

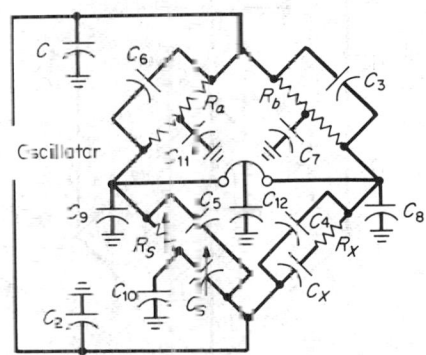

Fig. 17-51. Stray capacitances in unshielded and ungrounded bridge.[3]

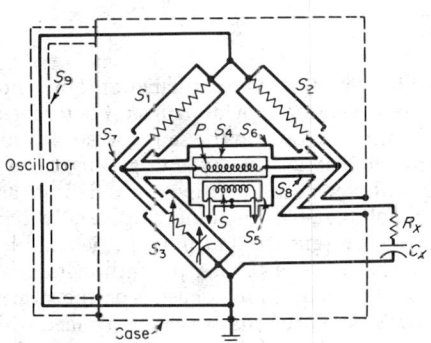

Fig. 17-52. Bridge with shields and ground.[3]

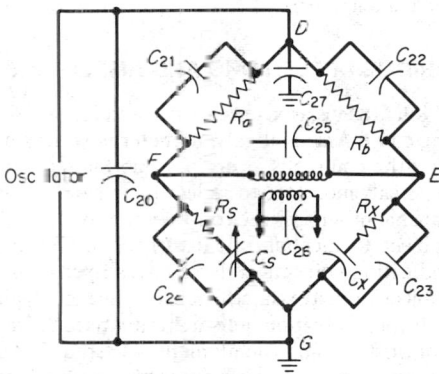

Fig. 17-53. Schematic circuit of shielded and grounded bridge.[3]

86. The Wagner ground connection (Fig. 17-54) can be used in place of shielding at lower frequencies if the utmost in precision is not required. The goal of the Wagner ground is to establish a ground connection on the oscillator in a manner that will bring the detector diagonal to ground potential. The measurement procedure is to balance the bridge as well as possible while ignoring the Wagner system. One end of the detector is then grounded as shown, and the potentiometer P and the balancing capacitor C (if present) are adjusted for a null. The detector is reconnected across the bridge, and the bridge is adjusted for a more accurate null. This process may be repeated as necessary to achieve better accuracy.

87. Guard circuits (Fig. 17-55) are often used at critical circuit points to prevent leakage currents from causing measurement errors. In an unguarded circuit surface leakage current may bypass the resistor R and flow through the detector G, thereby giving an erroneous reading. If a guard ring surrounds the positive terminal post (as in the circuit of Fig. 17-55), the surface leakage current flows through the guard ring and a noncritical return path to the voltage source. A true reading is obtained since only the resistor current flows through the detector.

88. Coaxial leads and twisted-wire pairs may be used in connecting impedances to a bridge arm in order to minimize spurious-signal pickup from electrostatic and electromagnetic fields. It is important to keep lead lengths short, especially at high frequencies.

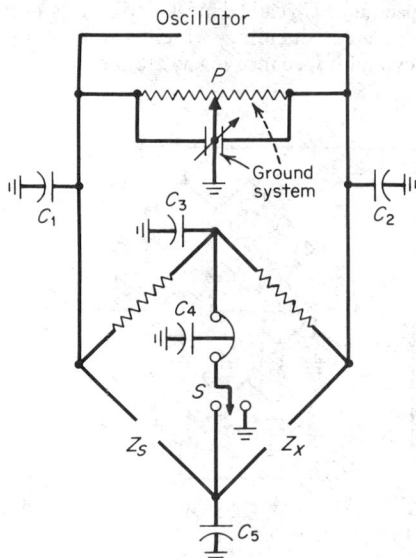

Fig. 17-54. Resistance-ratio bridge with Wagner ground connection.[3]

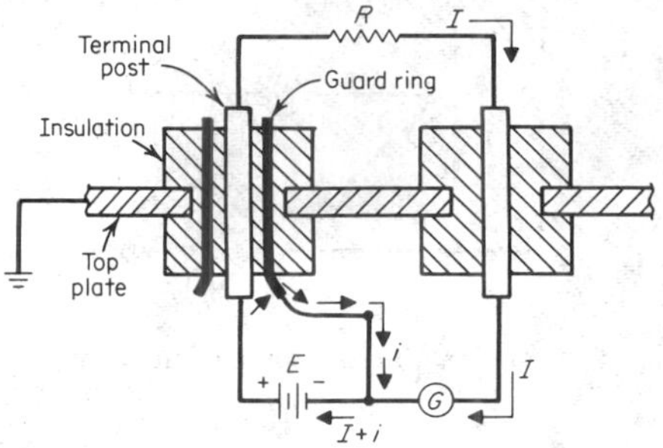

Fig. 17-55. Leakage current in guarded circuit. *(Leeds and Northrup.)*

BRIDGE DETECTORS AND AMPLIFIERS

89. Galvanometers are used for null detection in dc bridges. The permanent-magnet moving-coil d'Arsonval galvanometer is widely used. The suspension provides a restoring torque so that the coil seeks a zero position for zero current. A mirror is used in the sensitive suspension-type galvanometer to reflect light from a fixed source to a scale. This type of galvanometer is capable of sensitivities on the order of 0.001 μA per millimeter scale division but is delicate and subject to mechanical disturbances. Galvanometers for portable instruments generally have indicating pointers and use taut suspensions which are less sensitive but more rugged and less subject to disturbances. Sensitivities are typically in the range of 0.5 μA per millimeter scale division. Galvanometers exhibit a natural mechanical frequency which depends on the suspension stiffness and the moment of inertia. Overshoot and oscillatory behavior can be avoided without an excessive increase in response time if an external resistance of the proper value to produce critical damping is connected across the galvanometer terminals.

90. Null-detector amplifiers, incorporating choppers or modulators, are used to amplify the null output signal from dc bridges to provide higher sensitivity and permit the use of rugged, less sensitive microammeter indicators. Null-detector systems are available with sensitivities of 10 nV per division for a 300-Ω input impedance. Nonlinear responses can be provided so that large unbalance signals do not cause off-scale deflections. Proper design is required to avoid problems caused by zero drift, amplifier noise, thermal emfs, and stray electric and magnetic fields. Typical chopper and modulator amplifiers are described in Pars. **17-52** and **17-53**.

91. Telephone receivers are often used as null detectors in ac bridges operating in the range of 200 Hz to 10 kHz. Maximum sensitivity, which depends on the receiver characteristics and the acuteness of hearing of the observer, usually occurs between 1,000 and 2,000 Hz. At maximum sensitivity, signals produced by currents of 10 nA are audible in a quiet room. The effective impedance of the receiver should match the output impedance of the bridge for maximum power transfer. *Transformers* can be used for impedance matching and for overcoming the effect of stray capacitances associated with the output branch (see Par. **17-85**).

92. Vibration galvanometers provide better sensitivity than telephone receivers for frequencies below 300 Hz. The moving-coil type consists of a small coil and mirror mounted on a suspension that can be tuned by adjusting its length and tension. The coil, which is in a permanent magnet field, vibrates when alternating current flows in it. High sensitivity and frequency selectivity are obtained by tuning the lightly damped coil-suspension system to resonance at the applied frequency. The amplitude of vibration is proportional to the current.

93. Frequency-selective amplifiers are extensively used to increase the sensitivity of ac bridges. An ac amplifier with a twin-T network in the feedback loop provides full amplification at the selected frequency but falls off rapidly as the frequency is changed. Rectifiers or phase-sensitive detectors are used to convert the amplified ac signal into a direct current to drive a dc microammeter indicator. Cathode-ray-tube displays are also used to indicate the deviation from null conditions. Amplifier and detector circuits are described in Pars. **17-46** to **17-59**.

MISCELLANEOUS MEASUREMENT CIRCUITS

94. The Q meter is used to measure the quality factor Q of coils and the dissipation factor of capacitors; the dissipation factor is the reciprocal of Q. The Q meter provides a convenient method of measuring the effective values of inductors and capacitors at the frequency of interest.

The simplified circuit of a Q meter is shown in Fig. 17-56, where an unknown impedance of effective inductance L' and effective resistance r' is being measured. A sinusoidal voltage e, typically 0.01 V, is injected by L-106, in series with the circuit containing the unknown impedance and the tuning capacitor C. L-106 is selected to have a small impedance in comparison with the unknown impedance.

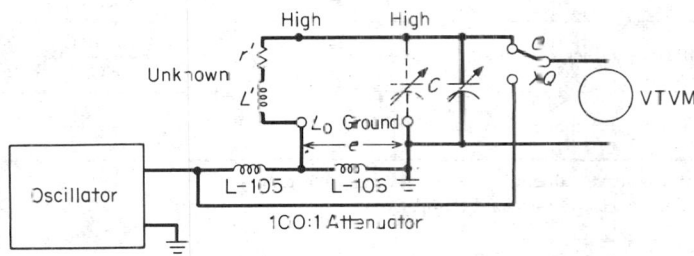

Fig. 17-56. Q meter. *(Hewlett-Packard Co.)*

Either the oscillator frequency or the tuning-capacitor value is adjusted to bring the circuit to approximate resonance, as indicated by a maximum voltage across capacitor C. At resonance $X_L = X_C$, where $X_L = 2\pi f L'$, $X_C = 1/(2\pi f C)$, L' is the effective inductance in henrys, C is the tuning capacitance in farads, and f is the frequency in hertz. The current at resonance is $I = e/R$, where R is the sum of the resistances of the unknown and the internal circuit. The voltage across capacitor C is $V_C = IX_C = eX_C/R$, and the indicated circuit Q is equal to V_C/e. In practice, the oscillator output voltage is adjusted to 1 V using a high-impedance voltmeter, typically a vacuum-tube voltmeter, and the indicated circuit Q is read with the voltmeter connected across C. The Q scale of the voltmeter is calibrated to read 100 for 1-V signal. Frequency-response errors of the voltmeter tend to cancel since the ratio of V_C/e is determined. Corrections for residual resistances and reactances in the internal circuit become increasingly important at higher frequencies. It should be noted that the Q meter increases the effective Q and effective inductance of the unknown impedance (see Pars. **17-41** to **17-44**). For low values of Q, neglecting the difference between the resonance and the approximate resonance achieved by maximizing the capacitor voltage may result in an unacceptable error. Exact equations are given in Par. **17-41**.

95. Substitution-measurement methods are used with the Q meter to cancel the effects of residual impedances (Figure 17-57). The parallel connection is used for high-impedance components while the series connection is used for low-impedance components. The circuit is resonated with the parallel unknown impedance removed (or series unknown impedance short-circuited), and values C_1 and Q_1 are recorded. The unknown impedance is introduced, the circuit resonated by adjusting C, and the new values C_2 and Q_2 recorded. The parameters of the unknown are given in Table 17-4.

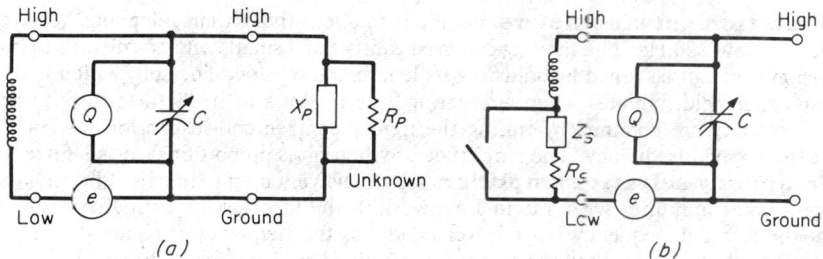

Fig. 17-57. Q-meter substitution measurements: (a) parallel; (b) series. (Hewlett-Packard Co.)

TABLE 17-4   Impedance Parameters from Parallel and Series Substitution Measurements

Parameter of unknown	From parallel measurements	From series measurements
Effective Q	$Q = \dfrac{Q_1 Q_2 |C_1 - C_2|}{\Delta Q C_1}$	$Q = \dfrac{Q_1 Q_2 |C_1 - C_2|}{C_1 Q_1 - C_2 Q_2}$
Effective resistance	$R_p = \dfrac{Q_1 Q_2}{\omega C_1 \Delta Q}$	$R_s = \dfrac{(C_1/C_2)Q_1 - Q_2}{\omega C_1 Q_1 Q_2}$
Effective reactance*	$X_p = \dfrac{1}{\omega(C_2 - C_1)}$	$X_s = \dfrac{C_1 - C_2}{\omega C_1 C_2}$
Effective inductance	$L_p = \dfrac{1}{\omega^2(C_2 - C_1)}$	$L_s = \dfrac{C_1 - C_2}{\omega^2 C_1 C_2}$
Effective capacitance	$C_p = C_1 - C_2$	$C_s = \dfrac{C_1 C_2}{C_2 - C_1}$

*A positive value indicates an inductive reactance.
Source: Hewlett-Packard Company.

96. The twin-T measuring circuit of Fig. 17-58 is used for admittance measurements at radio frequencies. This circuit operates on a null principle similar to a bridge circuit, but it has an advantage in that one side of the oscillator and detector are common and therefore can be grounded. The substitution method is used with this circuit, and therefore the effect of stray capacitances is minimized. The circuit is first balanced to a null condition with the unknown admittance $G_x + jB_x$ unconnected.

$$G_L = \omega^2 R C_1 C_2(1 + C_a/C_3)$$
$$L = 1/[\omega^2(C_b + C_1 + C_2 + C_1 C_2/C_3)]$$

The unknown admittance is connected to terminals a and b, and a null condition is obtained by readjusting the variable capacitors to values C'_a and C'_b. The conductance G_x and the susceptance B_x of the unknown are proportional to the changes in the capacitance settings:

$$G_x = \omega^2 R C_1 C_2(C'_a - C_a)/C_3$$
$$B_x = \omega(C_b - C'_b)$$

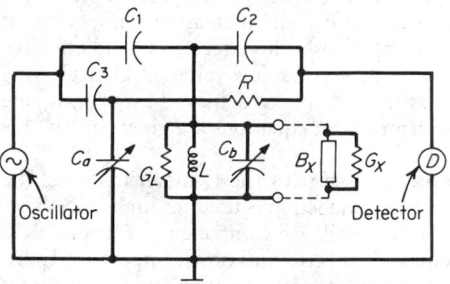

Fig. 17-58. Twin-T measuring circuit. (*General Radio Co.*)

97. Measurement of Coefficient of Coupling.[9] Two coils are inductively coupled when their relative positions are such that lines of flux from each coil link with turns of the other coil. The mutual inductance M in henrys can be measured in terms of the voltage e induced in one coil by a rate of change of current di/dt in the other coil; $M = -e_1/(di_2/dt) = -e_2/(di_1/dt)$. The maximum coupling between two coils of self-inductance L_1 and L_2 exists when all the flux from each of the coils links all the turns of the other coil; this condition produces the maximum value of mutual inductance, $M_{max} = \sqrt{L_1 L_2}$. The *coefficient of coupling* k is defined as the ratio of the actual mutual inductance to its maximum value; $k = M/\sqrt{L_1 L_2}$.

The value of mutual inductance can be measured using Campbell's mutual-inductance bridge (Par. **17-76**). Alternately, the mutual inductance can be measured using a self-inductance bridge.

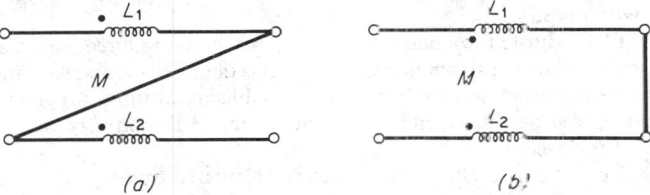

Fig. 17-59. Mutual inductance connected for self-inductance measurement.[2]

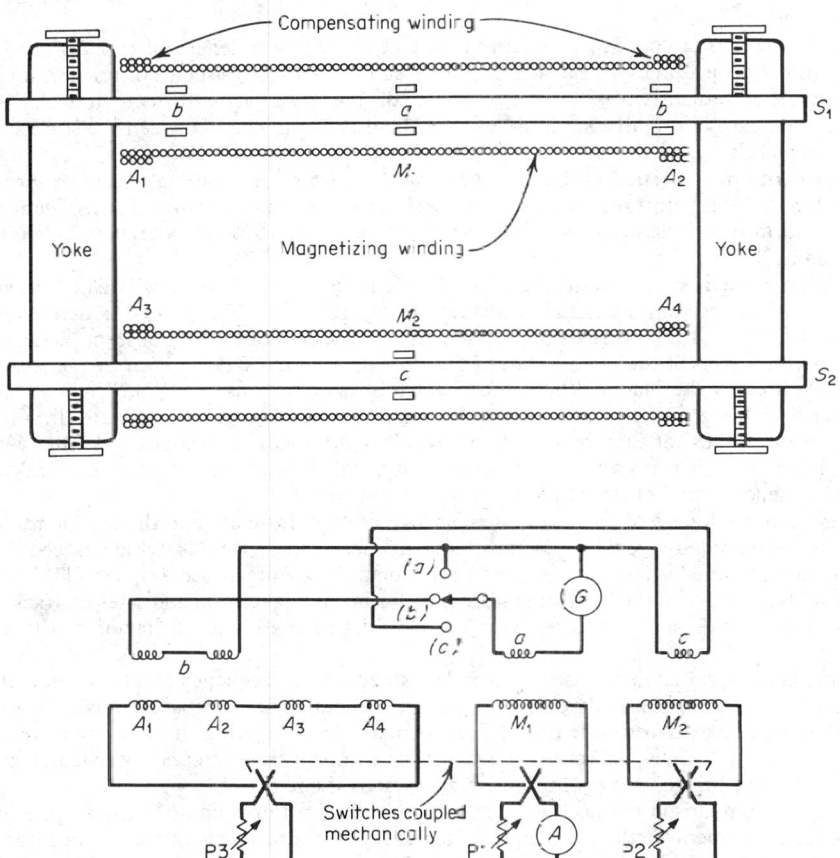

Fig. 17-60. Burrows permeameter.[27]

When the coils are connected in series with the mutual-inductance emf aiding the self-inductance emf (Fig. 17-59d), the total inductance $L_a = L_1 + L_2 + 2M$ is measured. With the coils connected with the mutual-inductance emf opposing the self-inductance emf (Fig. 17-59b), inductance $L_b = L_1 + L_2 - 2M$ is measured. The mutual inductance is $M = (L_a - L_b)/4$.

98. Permeameters are used to test magnetic materials. By simulating the conditions of an infinite solenoid the magnetizing force H can be computed from the ampere-turns per unit length. When H is reversed, the change in flux linkages in a test coil induces an emf whose time integral can be measured by a ballistic galvanometer. The *Burrows permeameter* (Fig. 17-60) uses two magnetic specimen bars S_1 and S_2, usually 1 cm in diameter and 30 cm long, joined by soft-iron yokes. High precision is obtainable for magnetizing forces up to 300 Oe. The currents in magnetizing windings M_1 and M_2 and in compensating windings A_1, A_2, A_3, and A_4 are adjusted independently to obtain uniform induction over the entire magnetic circuit. Windings A_1, A_2, A_3, and A_4 compensate for the reluctance of the joints. The reversing switches are mechanically coupled and operate simultaneously. Test coils a and c each have n turns, while each half of the test coil b has $n/2$ turns. Coils a and b are connected in opposing polarity to the galvanometer when the switch is in position b, while coils a and c are opposed across the galvanometer for switch position c.

Potentiometer P1 is adjusted to obtain the desired magnetizing force, and potentiometers P2 and P3 are adjusted so that no galvanometer deflection is obtained on magnetizing current reversal with the switches in either position b or c. This establishes uniform flux density at each coil. The switch is now set at position a and the galvanometer deflection d is noted when the magnetizing current is reversed.

The values of B in gauss and H in oersteds can be calculated from

$$H = \frac{0.4\pi NI}{l} \qquad B = 10^8 \frac{dkR}{2an} - \frac{A - a}{a} H$$

where N = turns of coil M_1, I = current in coil M_1 (A), l = length of coil M_1 (cm), d = galvanometer deflection, k = galvanometer constant, R = total resistance of test coil a circuit, a = area of specimen (cm²), A = area of test coil (cm²), and n = turns in test coil a. The term $(A - a)H/a$ is a small correction term for the flux in the space between the surface of the specimen and the test coil.

Other permeameters such as the Fahy permeameter, which requires only a single specimen, the Sanford-Winter permeameter, which uses a single specimen of rectangular cross section, and Ewing's isthmus permeameter, which is useful for magnetizing forces as high as 24,000 G, are discussed in Ref. 7.

99. The frequency standard of the National Bureau of Standards is based on atomic resonance of the cesium atom and is accurate to 1 part in 10^{13}. The second is defined as the duration of 9,192,631,770 periods of the radiation corresponding to the transition between the two hyperfine levels of the ground state of the atom of cesium 133. Reference frequency signals are transmitted by the National Bureau of Standards' radio stations WWV and WWVH at 2.5, 5, 10, and 15 MHz. Pulses are transmitted to mark the seconds of each minute. In alternate minutes during most of each hour, 500- or 600-Hz audio tones are broadcast. A binary-coded-decimal time code is transmitted continuously on a 100-Hz subcarrier. The carrier and modulation frequencies are accurate to better than 1 part in 10^{11}.

These frequencies are offset by a known and stable amount relative to the atomic-resonance frequency standard to provide "Coordinated Universal Time" (UTC), which is coordinated through international agreements by the International Time Bureau (see Ref. 26). UTC is maintained within ± 0.9 s of the UT1 time scale used for precise navigation and satellite tracking by adding leap seconds about once per year to UTC, depending on the behavior of the earth's rotation.

Quartz-crystal oscillators are used as secondary standards for frequency and time-interval measurement. They are periodically calibrated using the standard radio transmissions.

100. Frequency measurements can be made by comparing the unknown frequency with a known frequency, by counting cycles over a known time interval, by balancing a frequency-sensitive bridge, or by using a calibrated resonant circuit.

Frequency-comparison methods include using Lissajous patterns on an oscilloscope and heterodyne measurement methods. In Fig. 17-61, the frequency to be measured is compared with a harmonic of the 100-kHz reference oscillator. The difference frequency lying between 0 and 50 kHz is selected by the low-pass filter and compared with the output of a calibrated audio oscillator using Lissajous patterns. Alternately, the difference frequency and the audio oscillator frequency may be applied to another detector capable of providing a zero output frequency.

101. Digital frequency meters provide a convenient and accurate means for measuring frequency. The unknown frequency is counted for a known time interval, usually 1 or 10 s, and displayed in digital form. The time interval is derived by counting pulses from a quartz-crystal oscillator reference. Frequencies as high as 500 MHz can be measured by using scalers (frequency dividers). At low frequencies, for example 60 Hz, better resolution is obtained by measuring the period $T = 1/f$. A counter with a built-in computer is available which measures the period at low frequencies and automatically calculates and displays the frequency.

102. A frequency-sensitive bridge can be used to measure frequency to an accuracy of about 0.5% if the impedance elements are known. The Wien bridge of Fig. 17-62 is commonly used, R_3 and R_4 being identical slide-wire resistors mounted on a common shaft. The equations for balance[7] are $f = 1/(2\pi\sqrt{R_3R_4C_3C_4})$ and $R_1/R_2 = R_4/R_3 + C_3/C_4$.

In practice, the values are selected so that $R_3 = R_4$, $C_3 = C_4$, and $R_1 = 2R_2$. Slide wire r, which has a total resistance of $R_1/100$, is used to correct any slight tracking errors in R_3 and R_4. Under these conditions $f = 1/2\pi R_4C_4$. A filter is needed to reject harmonics if a null indicator is used since the bridge is not balanced at harmonic frequencies.

103. Time intervals can be measured accurately and conveniently by gating a reference frequency derived from a quartz-crystal oscillator standard to a counter during the time interval to be measured. Reference frequencies of 10, 1, and 0 1 MHz, derived from a 10-MHz oscillator, are commonly used.

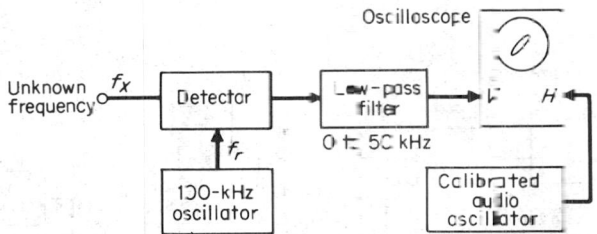

Fig. 17-61. Heterodyne frequency-comparison method.

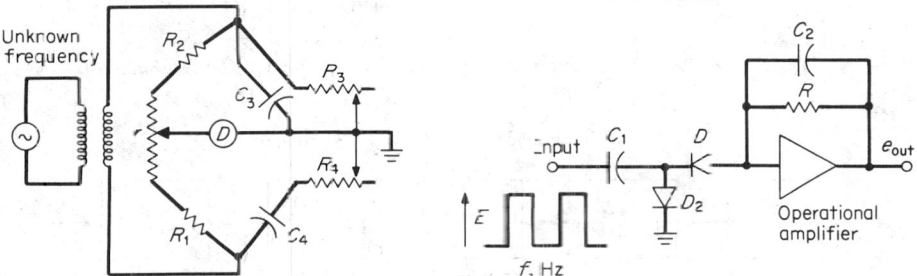

Fig. 17-62. Wien frequency bridge.[7]   **Fig. 17-63.** Frequency-to-voltage converter.

104. Analog frequency circuits that produce an analog output proportional to frequency are used in control systems and to drive frequency-indicating meters. In Fig. 17-63, a fixed amount of charge proportional to $C_1(E - 2d)$, where d is the diode-voltage drop, is withdrawn through diode D_1 during each cycle of the input. The current through diode D_1, which is proportional to frequency, is balanced by the current through resistor R, which is proportional to e_{out}. Therefore, $e_{out} = fRC_1(E - 2d)$. Temperature compensation is achieved by adjusting the voltage E with temperature so that the quantity $E - 2d$ is constant.

105. Frequency analyzers are used for measuring the frequency components and analyzing the spectra of acoustic noise, mechanical vibrations, and complex electric signals. They permit harmonic and intermodulation distortion components to be separated and measured. A simple analyzer consists of a narrow-bandwidth filter which can be adjusted in frequency or swept over the frequency range of interest. The output amplitude in decibels is generally plotted as a function of frequency using a logarithmic frequency scale. Desirable characteristics include wide dynamic range, low distortion, and high stop-band attenuation. Analog filters which operate at the frequency of interest exhibit a constant-percentage bandwidth, for example, 1%, while those using heterodyne techniques provide a constant bandwidth, for example, 10 Hz. The signal must be averaged over a period inversely proportional to the filter bandwidth if the reading is to be within given confidence limits of the long-time average value.[27]

106. Real-time frequency analyzers are available which perform ⅓-octave spectrum analysis on a continuous real-time basis. The analyzer of Fig. 17-64 uses 30 separate filters each having a bandwidth of ⅓ octave to achieve the required speed of response. The multiplexer sequentially samples the filter output of each channel at a high rate. These samples are converted into a binary number by the A/D converter. The true rms values for each channel are computed from these numbers during an integration period adjustable from ⅛ to 32 s and stored in the memory. The rms value for each channel is computed from 1,024 samples for integration periods of 1 to 32 s.

Real-time analyzers are also available for analyzing narrow-bandwidth frequency components in real time. The required rapid response time is obtained by sampling the input waveform at 3 times the highest frequency of interest using an A/D converter and storing the values of a large number of samples in a digital memory. The frequency components can be calculated in real time by a microprocessor using fast Fourier transforms.

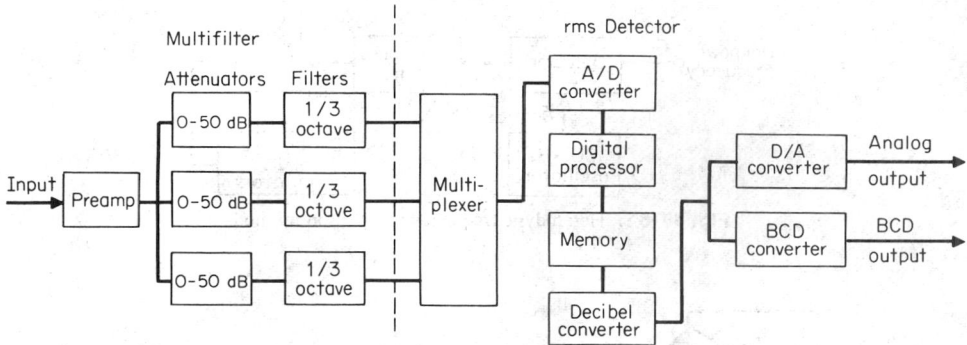

Fig. 17-64. Real-time analyzer using 30 attenuators and filters. *(General Radio Co.)*

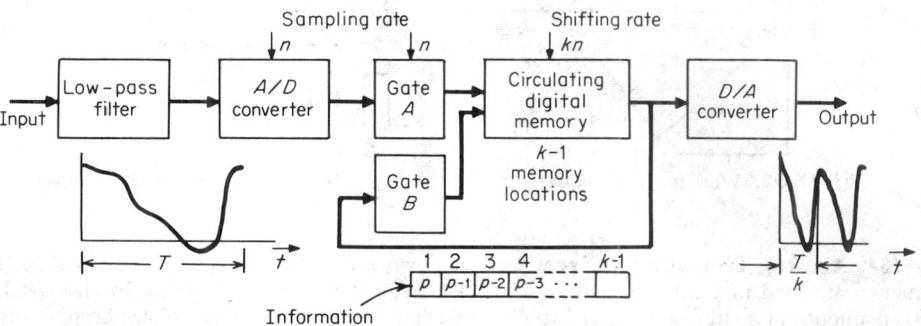

Fig. 17-65. Time-compression system.

107. Time-compression systems can be used to preprocess the input signal so that analog filters can be used to analyze narrow-bandwidth frequency components in real time. The time-compression system of Fig. 17-65 uses a recirculating digital memory and a D/A converter to provide an output signal having the same waveform as the input with a repetition rate which is k times faster. This multiplies the output-frequency spectrum by a factor of k and reduces the time required to analyze the signal by the same factor. The system operates as follows. A new sample is entered into the circulating memory through gate A during one of each k shifting periods. Information from the output of the memory recirculates through gate B during the remaining $k - 1$ periods. Since information experiences k shifts between the addition of new samples in a memory of length $k - 1$, each new sample p is entered directly behind the previous sample $p - 1$, and therefore the correct order is preserved. $(k - 1)/n$ seconds is required to fill an empty memory, and thereafter the oldest sample is discarded when a new sample is entered.

108. Frequency synthesizers provide a sinusoidal output voltage which is tunable with high resolution over a wide frequency range and yet have the stability and accuracy of a crystal oscillator reference. They are useful for providing accurate reference frequencies and for making measurements on filter networks, tuned circuits, and communication equipment. High-precision units feature 7-decade digital frequency selection plus a continuously adjustable decade that can be manually adjusted or electrically swept over a selectable frequency range. In the simplified block diagram of Fig. 17-66, the signal from the 5-MHz reference oscillator is sequentially processed by the digit insertion units, beginning with the least significant digit. The frequencies in parentheses are given for digit settings 8, 3, 5, 7, 2, 4, 6. Phase-locked 42-MHz and 3.0-, 3.1-, ... , 3.9-MHz signals derived from the reference oscillator are fed to all the digit-insertion units for synchronization. All digit-insertion units are identical and process a signal near 5 MHz, as shown in Fig. 17-67. The signal from the most significant digit-insertion unit is mixed with the 5-MHz reference to provide an output frequency between 0 and 100 kHz. A continuously

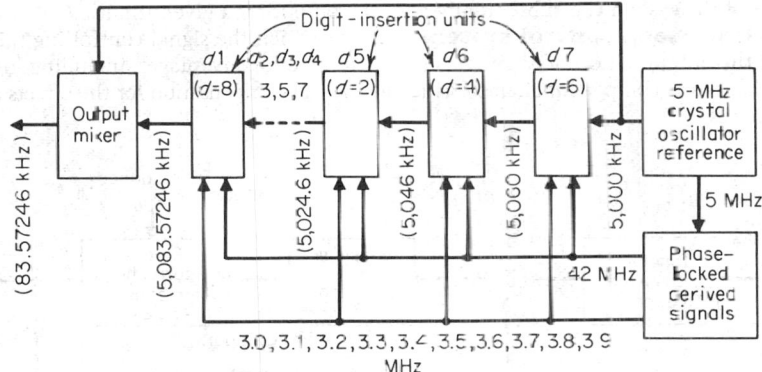

Fig. 17-66. Coherent decade frequency synthesizer. (General Radio Co.)

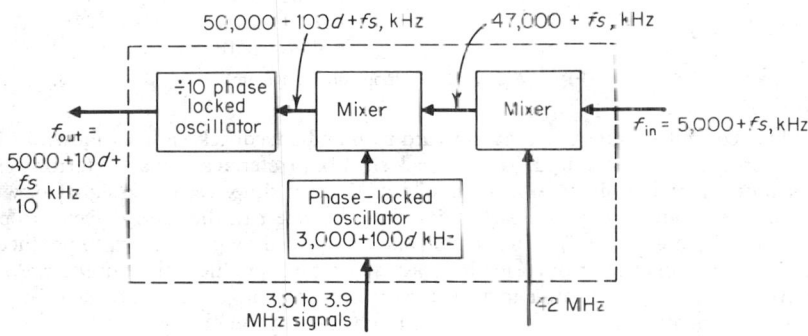

Fig. 17-67. Digit-insertion unit. (General Radio Co.)

adjustable decade unit can be substituted for any digit-insertion unit if a continuously adjustable frequency is desired. An output-frequency range of 0 to 1 MHz can be obtained by frequency-multiplying the output from the most significant decade by 10 and mixing it with a 50-MHz signal in the output mixer.

109. Time-domain reflectometry is used to identify and locate cable faults. The cable-testing equipment is connected to a line in the cable and sends an electric pulse that is reflected back to the equipment by a fault in the cable. The original and reflected signals are displayed on an oscilloscope. The type of fault is identified by the shape of the reflected pulse, and the distance is determined by the interval between the original and reflected pulses. Accuracies of 2% are typical.

110. A low-frequency voltmeter using a microprocessor has been developed that is capable of measuring the true rms voltage of approximately sinusoidal inputs at voltages from 2 mV to 10 V and frequencies from 0.1 to 120 Hz.[28] A combination of computer algorithms is used to implement the voltage- and harmonic-analysis functions. Harmonic distortion is calculated using a fast Fourier transform algorithm. The total autoranging, settling, and measurement time is only two signal periods for frequencies below 10 Hz.

Principles of Control Systems

TYPES OF CONTROL SYSTEMS

111. An open-loop control system is one in which the signal controlling the output is independent of the output. The d'Arsonval meter is an example of an open-loop system. The accuracy of an open-loop system depends on its calibration. Changes in the characteristics of the components of the system can substantially alter the output for a given input.

112. A closed-loop control system is one in which the signal controlling the output depends on the output. Closed-loop systems have a number of advantages, including lower sensitivity to changes in component characteristics and partial compensation for the effects of external disturbances.

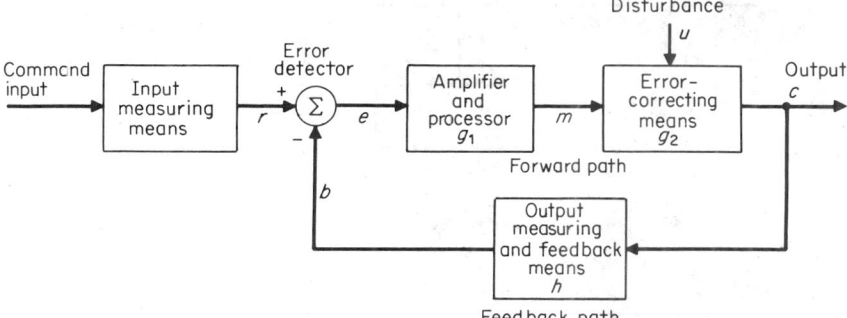

Fig. 17-68. Closed-loop control system.

Closed-loop control systems contain a forward path and a feedback path, as shown in Fig. 17-68. The lower-case signals c (output), b (feedback signal), r (reference signal), e (error signal), m (processor output), and u (disturbance) are all functions of time. Feedback signal b, which is proportional to the output c, is compared with the reference r in the error detector to produce error signal e, where $e = r - b$. The error signal is amplified by processor g_1 to produce signal m. This signal controls the error-correcting means g_2 which produces the output c. Although the output c is relatively insensitive to disturbance u and to changes in the characteristics of the forward path elements g_1 and g_2, it is sensitive to changes in the characteristics of the output-measuring and feedback means h, the error detector, and the input-measuring means. These elements must be selected carefully if accuracy is to be achieved. Fortunately, they usually operate at low power and are usually considerably simpler and more stable than elements g_1 and g_2.

The *output variable* may be any controllable quantity such as voltage, current, position, speed, torque, or temperature. When the controlled variable is a mechanical position or a time derivative of position such as velocity or acceleration, the feedback control system is called a *servomechanism.*

CLOSED-LOOP CONTROL-SYSTEM ELEMENTS

113. The error-detection subsystem consists of the input-measuring means, the output-measuring and feedback means, and the error detector. This subsystem is very important since it directly affects the accuracy of the closed-loop system. Subsystem characteristics of interest include the energy required to measure the command input, accuracy, size, reliability, linearity, noise, signal level, and resolution. The *output-measuring means* measures the output variable *c* or a function of it and provides an output signal which may be further processed by the feedback means before being applied to the error detector. The *input-measuring means* converts the *command input*, which can be any convenient quantity, into a suitable *reference signal* for the error detector. The *error detector* produces an error signal *e* proportional to the difference between the output of the input-measuring means and the feedback means. A number of error-detection subsystems are given in Table 17-5 and Fig. 17-69.

Some error-detection subsystems incorporate thermistor bridges, photoconductors, linear variable differential transformers, Hall effect devices, proximity detectors, gyroscopes, accelerometers, digital angle transducers, and heterodyne frequency-error detectors. In other error-detection subsystems, A/D converters are used to provide digital signals to error detectors which employ digital subtraction. Microprocessors can be used to perform digital subtraction and error-signal processing (see Par **17-116**).

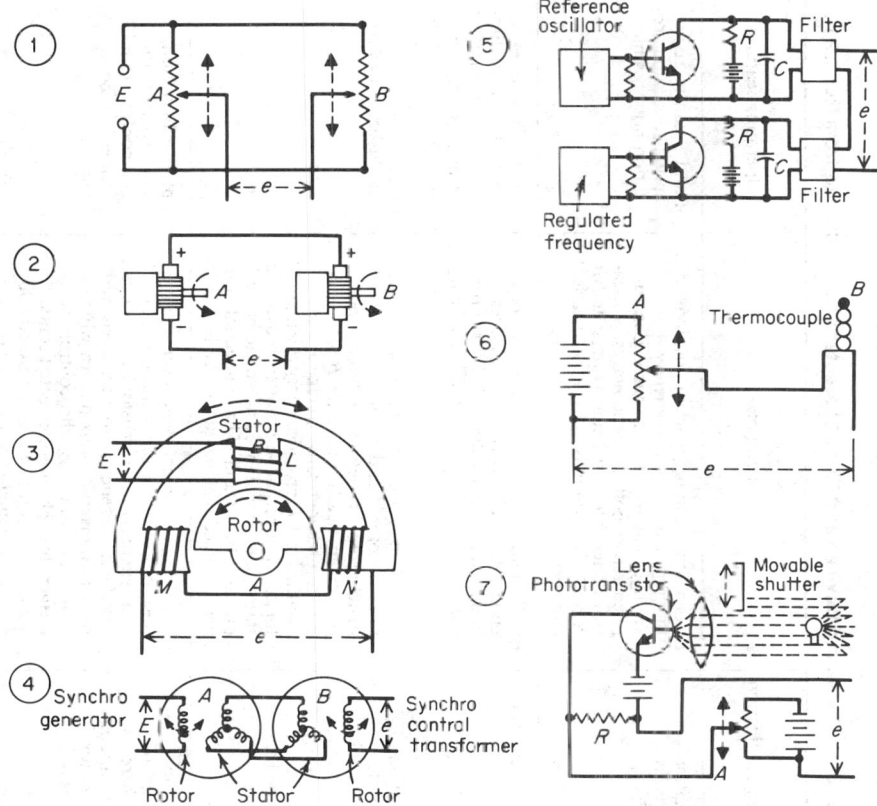

Fig. 17-69. Typical error-detection subsystems[25] (see Table 17-5).

114. The error-signal amplifier and processor increases the power level of the error signal so that it is sufficiently powerful to control the error-correcting means. The peak control-power requirement is often determined by the speed with which a mechanical device such as a throttle or valve must be operated or by the rate at which a control current must be changed. The frequency-response characteristic of the amplifier usually requires careful design to obtain good transient response and adequate system stability (see Par. **17-139**). Typical amplifying devices include transistors, operational amplifiers, silicon-controlled rectifiers, relays, generators, and valves (see Fig. 17-70 and Pars. **17-49** and **17-150**). Amplifiers may be cascaded where high gains are required (see Table 17-6).

115. The error-correcting means is a device capable of supplying the power required to change the output and thereby reduce the error. Typical devices include dc and ac motors, solenoids, stepping motors, hydraulic motors and pistons, prime movers, and fuel burners.

TABLE 17-5 Typical Electrical Error-Detecting Devices and Their Characteristics (Adapted from Ref. 29)

Number in Fig. 17-69	Type	Main application	Operation	Operating features	Accuracy limited by	Features determining energy required to vary reference quantity measurement r
1	Dc or ac resistance bridge	Position control	Error voltage e appears when positions of moving arms of potentiometers A and B are not matched: power source E is applied across both potentiometers; A measures reference position as voltage and B regulated position as voltage, their difference being e	A and B can be remote; continuous rotation not possible	Potentiometer winding	Contact arm and bushing friction
2	Dc tachometer bridge	Speed control	Error voltage e appears when speeds of tachometers A and B vary: A measures reference speed as a voltage and B regulated speed as a voltage, their difference being e	A and B can be remote; top speed limited by commutator; A may be replaced by a potentiometer	Tachometer accuracy; commutator resistance	Brush and bearing friction
3	Ac magnetic bridge	Position control, particularly for gyro pickups, where very small forces prevail	Error voltage e appears when relative positions of rotor A and stator B do not match: rotor A measures reference position magnetically and stator B regulated position magnetically; voltage E across exciting coil L provides energy; when rotor covers unequal areas of each exposed stator pole (unbalanced magnetic bridge), pickup coils M and N have unequal voltages induced; voltage difference is e	Limited rotation; air gap usually small	Machining tolerance, magnetic fringing, and voltage-phase shift	Load taken from e; bearing friction
4	Ac synchro-system	Position control where continuous rotation is desired	Error voltage e appears whenever relative positions of rotors of synchrogenerator A and synchrocontrol transformer B are not matched: reference position is measured by A as a magnetic-flux pattern which is transmitted to the synchrocontrol transformer through interconnected stator windings; if rotor of B is not exactly 90° from transmitted flux pattern, e is produced	Unlimited rotation; synchrogenerator and control transformer can be remote	Machining tolerance, accuracy of winding distribution	Distributed or nondistributed winding of control transformer rotor; load taken from e; bearing and slipping friction

						Input impedance
5	Frequency bridge	Frequency control	Error voltage e appears when reference and regulated frequencies differ; transistor channel A produces a filtered sawtooth wave which gives a dc voltage inversely proportional to the reference frequency; transistor channel B produces a similar voltage as a measure of regulated frequency; difference of these dc voltages is e	A and B can be remote; transistors operate in the switching mode; wide range of frequencies can be covered	Temperature and aging effects on transistors and circuit elements	
6	Millivolt bridge	Temperature control	Error voltage e appears whenever regulated temperature differs from reference temperature; regulated temperature is measured as a voltage by the thermoelectric effect of two dissimilar metals B; reference temperature is represented as a voltage from battery-potentiometer source A; difference in these voltages is e	A and B can be remote; wide range of temperature can be covered	Ability to detect very low-milli-volt signals	Contact arm and bushing friction; if electronic voltage source A is used, input impedance
7	Photo-transistor bridge	Position control by intercepting a light beam	Error voltage e appears when movable shutter is in other than desired position; light reaching phototransistor B measures shutter position; this light is measured as a voltage by the phototransistor-current variation; A reference position of shutter is represented by battery-potentiometer voltage; difference of these voltages is e	A and B can be remote; transparent surfaces through which light travels must be kept clean	Continued accuracy of light source and phototransistor	Contact arm and bushing friction; if electronic voltage source A is used, input impedance

TABLE 17-6　Typical Electrical Amplifiers and Their Characteristics (Adapted from Ref. 29)

Schematic (see Fig. 17-70)	Type	Gate element	Possible input units	Possible output units	Possible power-amplification factor	Devices represented by load L	Power control
(a), (b)	Junction or field-effect transistor	Base or gate	Microwatts	Watts	1×10^5	Relay, motor, generator field, impedance, solenoid	Continuous
(c)	Silicon controlled rectifier	Gate	Milliwatts	Watts or kilowatts	1×10^5	Relay, motor, generator field, impedance, solenoid	Continuous
(d)	Relay	Contact	Watts	Watts or kilowatts	1×10^3	Relay, motor, generator field, impedance, solenoid	On-off
(e)	Generator	Field	Watts	Watts or kilowatts	50	Motor, impedance	Continuous
(f)	Saturable reactor	Dc coil	Milliwatts	Watts	3×10^2	Generator field, impedance	Continuous
(g)	Silverstat	Contacts	Grams	Watts	$1 \times 10^7 \times l$	Generator field, impedance	Stepped

17-44

116. Digital signal processing can be used in implementing the error detector and processor of Fig. 17-68. An A/D converter can be used to convert the output quantity into a digital signal. The reference signal may already exist in digital form e.g., as a number stored in a computer memory. The error is formed by digital subtraction in a microprocessor or special-purpose digital processor. Error signal processing can be performed by a digital algorithm. The processor output signal m may be provided by a D/A converter, or, alternatively, only contact-closure outputs may be provided by the processor.

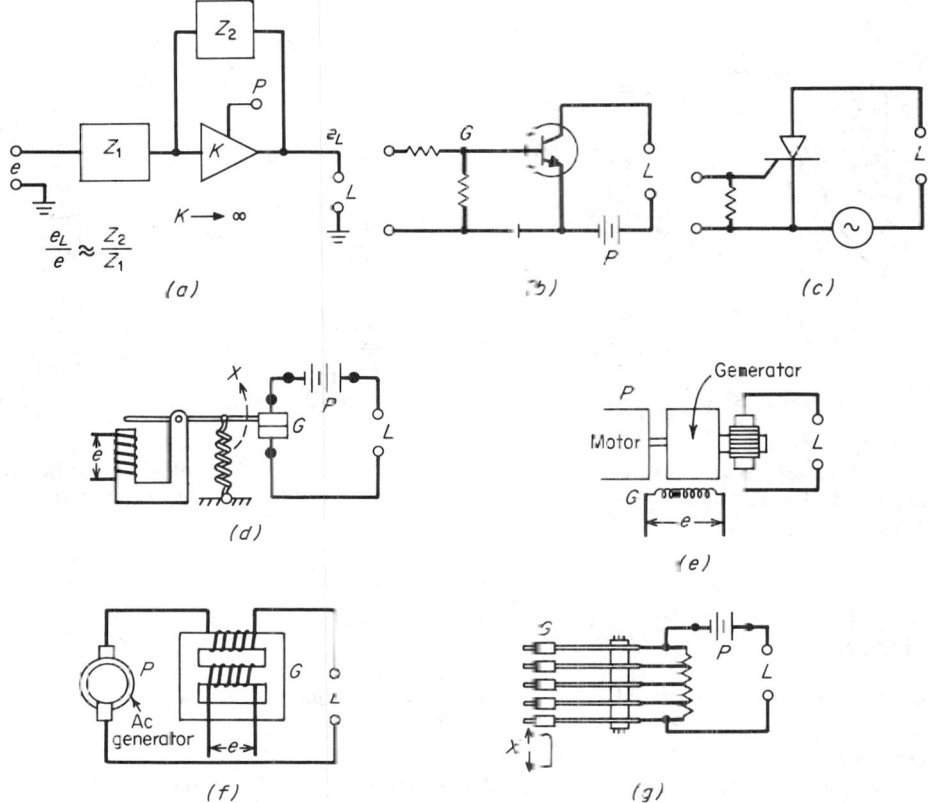

Fig. 17-70. Typical electrical amplifiers: (a) operational amplifier; (b) transistor; (c) silicon-controlled rectifier; (d) relay; (e) generator; (f) saturable reactor; (g) silversat⁹ (see Table 17-5).

TIME RESPONSE AND FREQUENCY RESPONSE

117. The time response of an element or system is often defined in terms of its response to a step input, as shown in Fig. 17-71. The *response time* is the time for the output to reach a specified value. The *rise time* is usually defined as the time required for the output to rise from 10 to 90% of the final value. *Delay time* is often specified as the time to reach 50% of the final value. The *settling time* is the time required for the output to reach and remain within a specified percentage (usually 2 or 5%) of its final value. *Overshoot* is the maximum positive value of the output minus the final output value.

A first-order system (Figs. 17-72 and 17-73) has a single pole (see Par. **17-121**), and the output response can be described in terms of the time constant τ. The *time constant* is defined as the time in seconds for the transient term to decay to $1/e$ of its initial value, where $1/e = 0.368$. A second-order system has two poles, which can be found by solving the quadratic equation. If these poles are complex, the output response may exhibit overshoot and a damped oscillatory response, as shown in Fig. 17-71. The response is a function of the location of the roots (see Par. **17-121**).

118. The frequency response of a component or system for a sinusoidal input is given by (1) the steady-state ratio of the output magnitude to the input magnitude and (2) the output-to-input phase difference for input frequencies over the range of interest. The frequency response of the RC circuit of Fig. 17-72 is given in Fig. 17-73. The magnitude ratio and frequency are usually plotted on logarithmic scales. The magnitude ratio M is often given in decibels M_{dB}, where $M_{dB} = 20 \log M$. The angular frequency ω in radians per second is used as the abscissa, where $\omega = 2\pi f$. The dashed lines of Fig. 17-73 show the approximate straight-line (asymptote) response. The frequency response can be found experimentally by applying a sinusoidal signal and measuring the amplitude and phase characteristics over the frequency range of interest.

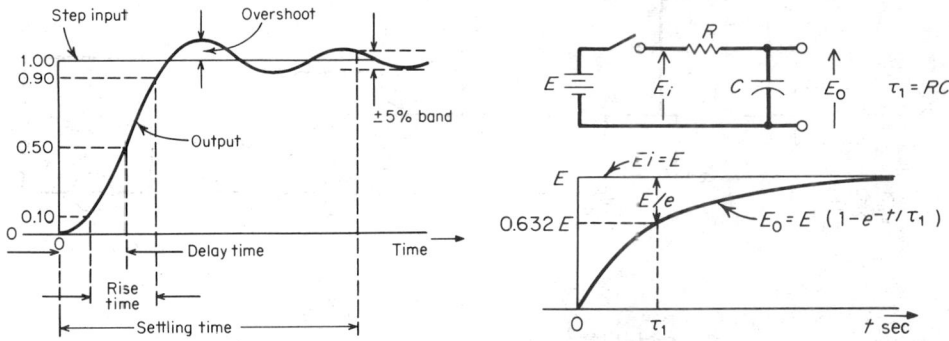

Fig. 17-71. Response to step input.   Fig. 17-72. Response of first-order system.

119. The Laplace transform and its inverse provide a convenient means for finding the transient and steady-state response of a system. The Laplace transform $\mathcal{L}[f(t)]$ is a function $F(s)$ of the complex variable $s = \sigma + j\omega$

$$\mathcal{L}[f(t)] = F(s) = \int_0^\infty f(t)e^{-st}\, dt$$

where $f(t)$ is a function of time t defined for $t \geq 0$. The *inverse Laplace transform* $\mathcal{L}^{-1}[F(s)]$ is defined as

$$\mathcal{L}^{-1}[F(s)] = f(t) = \frac{1}{2\pi j}\int_{c-j\infty}^{c+j\infty} F(s)e^{ts}\, ds$$

where $t \geq 0$ and $c >$ real parts of all singularities of $F(s)$.[32]

It is generally not necessary to use these definitions because of the availability of *transform pairs*.[30] Table **3-3**, Par. **3-18**, Sec. **3** lists a number of Laplace transform pairs. The *time response* of a system is found by (1) converting the differential equations and initial conditions describing the system into a function of the complex variable s using the Laplace transform or the trans-form-pair table; (2) solving for the desired output by algebraically manipulating the functions of s; (3) expressing the input signal as a function of s; (4) combining the results of steps 2 and 3 to form a final $F(s)$; (5) using the inverse Laplace transform or the transform-pair table to find the time response $f(t)$. This last step may involve using techniques (such as the partial-fraction expansion) to put $F(s)$ in a form where the terms of $F(s)$ can be matched with those of the transform-pair table (Table **3-3**, Sec. **3**).

The step-by-step Laplace transform procedure detailed above is illustrated by finding the time response of the circuit of Fig. 17-72 using the transforms from Table **3-3**, Par. **3-18**.

Step 1:
$$e_o = \frac{1}{C}\int i\, dt \rightarrow E_o = \frac{1}{Cs} + \frac{f^{-1}(0^+)}{Cs}$$
$$e_i = iR + \frac{1}{C}\int i\, dt \rightarrow E_i = IR + \frac{I}{Cs} + \frac{f^{-1}(0^+)}{Cs}$$

where $f^{-1}(0^+)$ is the initial capacitor voltage, which for this example is assumed to be zero.

Step 2:
$$E_o = E_i/(1 + RCs)$$

Step 3: $E_i = E/s$ for a step input of magnitude E at $t = 0$

Step 4:
$$E_o = E/[s(1 + RCs)]$$

Step 5: Using a partial-fraction expansion gives

$$E_o = \frac{E}{s} - \frac{E}{s + 1/RC} \rightarrow e_z = E - Ee^{-(t/RC)}$$

where E_i, E_o, I are functions of s and e_i, e_o, i are functions of t.

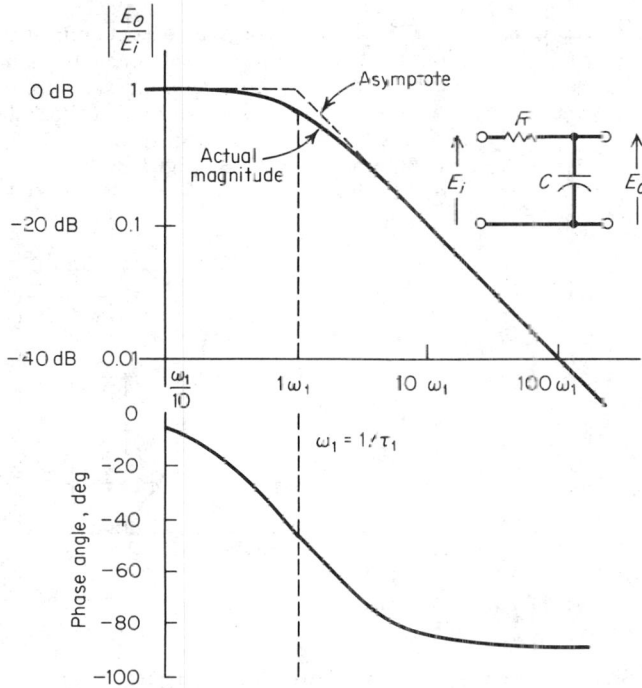

Fig. 17-73. Frequency response of RC circuit.

TRANSFER FUNCTIONS

120. The transfer function $G(s)$ of a linear system is the Laplace transform of the output-to-input ratio for the condition where the initial stored energy is zero. The transfer function for the RC network of Fig. 17-73 is $E_o/E_i = 1/(1 + RCs)$, since $\tau_1 = RC$. The frequency response characteristic can be found from the transfer function by substituting $j\omega$ for s. Transfer functions of a number of networks are given in Table 17-7 of Par. **17-139**.

121. The roots of a transfer function can be plotted on the complex s plane. Values of s which make the denominator of the function equal to zero are called *poles*; values of s which make the numerator of the function equal to zero are called *zeros*. The first-order system of Fig. 17-72, which involves s to the first power, has a single pole on the real axis at $s = -1/\tau$.

A *second-order system* has a transfer function containing a denominator term which can be written as a quadratic in s:

$$s^2 + 2\zeta\omega_0 s + \omega_0^2$$

The poles of the transfer function are $s_1 = -\zeta\omega_0 + \omega_0\sqrt{\zeta^2 - 1}$ and $s_2 = -\zeta\omega_0 - \omega_0\sqrt{\zeta^2 - 1}$. The *undamped natural frequency* of the system is ω_0, and the *damping factor* is ζ. For $\zeta = 1$, the poles are real and equal ($s_1 = s_2 = -\omega_0$); the system is critically damped and will exhibit no overshoot. For values of $\zeta > 1$ the system is overdamped, the poles are real and unequal, and no overshoot will occur. For values of $\zeta < 1$ the poles are complex conjugates (s

$= \zeta\omega_0 \pm j\omega_0 \sqrt{1 - \zeta^2}$), and the system will exhibit a damped oscillatory response with overshoot. The unit-step-function responses for a second-order system having a transfer function $C/R = \omega_0^2/(s^2 + 2\zeta\omega_0 s + \omega_0^2)$ are given in Fig. 17-74. The closed-loop frequency response will exhibit a peak for low values of ζ, as shown in Fig. 17-75. For $\zeta = 0$, the undamped case, the roots lie on the imaginary axis at $\pm\omega_0$, and the response is a constant-amplitude sinusoid.

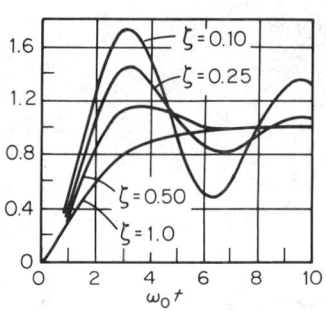

Fig. 17-74. Unit-step-function responses.[31]

BLOCK DIAGRAMS

122. Block diagrams containing the transfer functions of the elements of a system are useful in analyzing the characteristics of the system. The variables are given as functions of the complex operator s. The block diagram of a simple system of the form of that in Fig. 17-68 is given in Fig. 17-76. Since $C = G_1 G_2 E$ and $E = R - CH$, the closed-loop transfer function $C/R = G\ G_2/(1 + G_1 G_2 H)$.

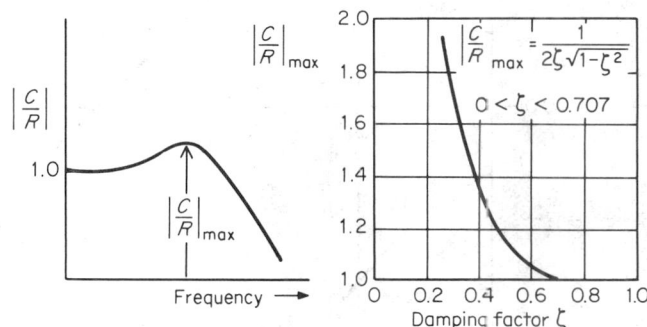

Fig. 17-75. Frequency response corresponding to Fig. 17-74.[33]

123. Simplification of Block Diagrams. Block diagrams can be simplified by combining blocks algebraically. Concentric feedback loops (see Fig. 17-77b) can be removed if the relationship $C/R = G/(1 + GH)$ is applied, starting with the innermost loop. If the loops are interconnected (see Fig. 17-77a), the diagram can usually be reduced to a concentric state by shifting a signal-pickoff point or by shifting the point at which a signal is applied to a summing junction. These rules must be followed: If a signal-pickoff branch is moved from the input to the output of a block G, the function $1/G$ must be inserted in a branch. Conversely, if the branch is moved from the output to the input of a block G, the function G must be inserted in the branch. The pickoff branch must not be moved past a summing point. A branch feeding a summing point may be moved from the input to the output of a block G if the function G is inserted in the branch. If the branch is moved from the output to the input of block G, the function $1/G$ must be inserted in the branch. The summing point must not be moved past a signal-pickoff point. An example of the reduction of a feedback system with interconnected loops is given in Fig. 17-77. Additional examples of block-diagram simplification and a systematic method for signal-flow-graph reduction are given in Refs. 32 and 33.

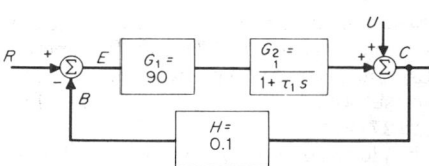

Fig. 17-76. Closed-loop system.

SYSTEM PERFORMANCE SPECIFICATIONS

124. Performance specifications usually include the response of the system to a step input, measured in terms of response time, rise time, delay time, settling time, and overshoot, as described in Par. **17-117**. Other specifications include bandwidth, phase margin, sensitivity to gain changes, sensitivity to load disturbances, and error coefficients.

125. The bandwidth of the system is usually specified in terms of the frequency at which the magnitude of the closed-loop response is down 3 dB. Bandwidth is used as a means of specifying performance related to the speed of response. Excessive bandwidth should be avoided because noise is proportional to bandwidth.

126. Phase margin is used as a method of specifying the relative stability of the system. The phase margin is equal to 180° plus the phase angle of the open-loop transfer function GH for the frequency at which GH has unity magnitude (see Fig. 17-79).

127. The sensitivity of the system to gain changes and load disturbances depends on the loop gain at the frequency of interest. For $U=0$ in the closed-loop system of Fig. 17-76, $C = G_1 G_2 E$ and $E = R - CH$, and therefore $C/R = G_1 G_2/(1 + G_1 G_2 H)$. If the value of the loop gain $G_1 G_2 H$ is much greater than unity, C/R is approximately equal to $1/H$, and therefore the closed-loop system is insensitive to changes in the forward-path elements G_1 and G_2. For the given example, $G_1 G_2 H = 9$ at low frequency, and $C/R = 9/(1 + 0.1 T_2 s)$. The use of feedback has reduced the time constant of the closed-loop system by a factor of $1 + G_1 G_2 H$. The sensitivity to changes in G_1 is reduced by the same factor; a 10% change in G_1 changes C/R by only 1%.

The sensitivity to load disturbance U in Fig. 17-76 is reduced by the factor $1 + G_1 G_2 H$ since the transfer function $C/U = 1/(1 + G_1 G_2 H)$.

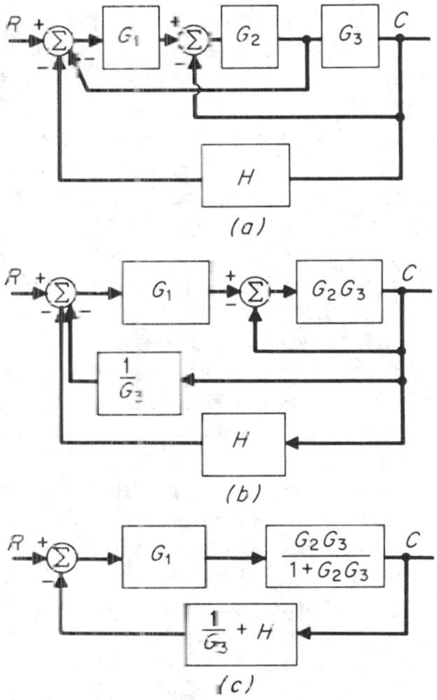

128. Static error coefficients are used as a measure of the effectiveness of closed-loop systems for specified position, velocity, and acceleration input signals, i.e., unit-step, unit-ramp, and unit-parabola inputs, respectively. For the unity-feedback system of Fig. 17-78 these coefficients are defined as follows:

$$K_p = \lim_{s \to 0} G(s) \qquad K_v = \lim_{s \to 0} sG(s) \qquad K_a = \lim_{s \to 0} s^2 G(s)$$

where K_p = position-error coefficient, K_v = velocity-error coefficient, and K_a = acceleration-error coefficient.

For any given system, only one of these coefficients has a finite nonzero value. The type 0 system, which has no net poles at the origin, has a finite nonzero value of K_p, and the steady-state error for a unit-step input is equal to $1/(1 + K_p)$. The type 1 system, which has a simple pole at the origin, has a finite nonzero value of K_v, and the steady-state error for a unit-ramp input is $1/K_v$. The type 2 system, which has two poles at the origin (two integrations), has a finite nonzero value of K_a, and the steady-state error for a unit-parabola input is $1/K_a$ (see Ref. 32).

Fig. 17-77. Simplification of a block diagram

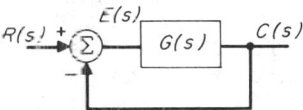

Fig. 17-78. Unity-feedback system.

SYSTEM STABILITY

129. The stability of feedback systems is of great importance since an unstable system will not be effective in maintaining the controlled variable at approximately the desired value. Large oscillations can have a destructive effect on the error-correcting device. Relative stability is also of importance since the performance of a system which exhibits excessive overshoot or an underdamped characteristic may be unsatisfactory in many applications.

The cause of instability can be understood by examining the closed-loop control system of Fig. 17-68. Let us open the feedback loop by disconnecting the input to block g_1 from the error-detector output. *Negative feedback* exists if in response to a positive change in the input to g_1

the error-detector output changes in a negative direction. *Positive feedback* exists if a positive change in the input to g_1 causes the error-detector output to change in the positive direction. In this latter case, if the loop is closed, any change in the input to g_1 will be reinforced by the feedback. If the loop gain G_1G_2H with positive feedback is unity, no input is required to produce an output and the system is unstable.

Phase shifts in the open-loop elements can cause a negative-feedback system to exhibit positive feedback at some frequencies. If the phase shift is $-180°$ at a given frequency, the feedback will reinforce an applied input at this frequency. The system will be unstable if the magnitude of the open-loop gain G_1G_2H is equal to unity at this frequency.

130. Transportation delays occur in distributed systems, e.g., the time delay experienced by a signal traveling along a transmission line. An ideal transportation-delay element will faithfully reproduce an input signal after a delay of T seconds. Transportation delays are detrimental to stability because they provide no attenuation and produce a phase lag that increases linearly with frequency, that is, $\theta = 2\pi fT$, where T is the transportation delay in seconds, f is in hertz, and θ is the phase lag in radians. The Laplace transform of a transportation delay of T seconds is e^{-Ts}.

131. Nonlinearities in open-loop elements will cause their characteristics to be a function of the operating point. Common forms of nonlinearity include saturation, dead band, backlash, static friction, and square-law transfer characteristics. Nonlinear systems are often analyzed by linearizations about a fixed operating point. This permits the application of linear control theory. The reader should consult Refs. 32 and 33 for techniques applicable to nonlinear systems, such as quasi linearization, describing function analysis and the phase-plane representation of system characteristics. Backlash can cause low-level oscillations to be present in systems which would otherwise be stable. Saturation can cause instability in systems which exhibit poor stability at reduced values of open-loop gain, i.e., conditionally stable systems.

132. Noise, which can be defined as any signal that does not convey useful information, must be considered in the design of closed-loop systems. Excessive noise can cause saturation in amplifying stages and thereby cause stability problems. Noise can be controlled by avoiding excessive bandwidth and by providing adequate attenuation for any frequencies at which external unwanted signals are coupled into the system.

MATHEMATICAL ANALYSIS OF LOOP STABILITY

133. The Laplace transform method of determining the stability of a linear system involves solving for the time response of the closed-loop transfer function C/R, where $C/R = G/(1 + GH)$. Simplification of the transfer functions may result in C/R being expressed in polynomial forms, i.e.,

$$\frac{C}{R} = \frac{b_0s^m + b_1s^{m-1} + \cdots + b_m}{a_0s^n + a_1s^{n-1} + \cdots + a_n}$$

The poles of C/R, that is, the values of s which make the denominator zero (which are the same values of s that make $1 + GH = 0$), determine the form of the transient terms. If all the poles lie in the left half of the s plane, i.e., have negative real parts, all exponential transient terms will decay to zero and the system will be stable. The system will be unstable if the poles lie in the right half of the s plane since the transient terms will grow exponentially. Any poles on the imaginary axis will cause a constant-amplitude sinusoidal oscillation, and the system will be unstable. The solution of the time response of the closed-loop system for a step input will provide both transient response and the steady-state response. Digital-computer routines can be used to reduce the labor involved in obtaining the solution.

134. The frequency-response method of determining stability involves plotting the magnitude and phase of the open-loop function GH for $s = j\omega$, where ω assumes all values from $-\infty$ to $+\infty$. If GH has any poles lying on the $j\omega$ axis (including the origin), these points are bypassed by making s traverse a semicircle of near-zero radius to the right of the pole. The magnitude and phase-angle information can be plotted on a polar diagram, as shown in Fig. 17-79. The plot for negative values of ω (dashed lines) can be obtained by reflecting the plot for the positive values of ω about the real axis. The mass, inertia, series inductance, and shunt capacitance of physical systems will cause GH to approach zero, as ω approaches infinity. If desired, a pole at an appropriately high frequency can be added to the GH function to represent these effects.

135. The Nyquist criterion for a linear system which has a stable open-loop function GH can be stated as follows. The closed-loop system will be stable only if a vector from the $-1 + j0$ point to the point on the GH polar plot corresponding to a particular value of ω has a zero net counterclockwise rotation as ω varies from $-\infty$ to $+\infty$. The number of counterclockwise rotations will be equal to the number of right-half-plane poles of the closed-loop system (see Ref. 34).

If the system is open-loop unstable and GH has P poles with positive real parts, then the closed-loop system will be stable only if the vector defined above makes P counterclockwise revolutions as ω varies from $-\infty$ to $+\infty$.

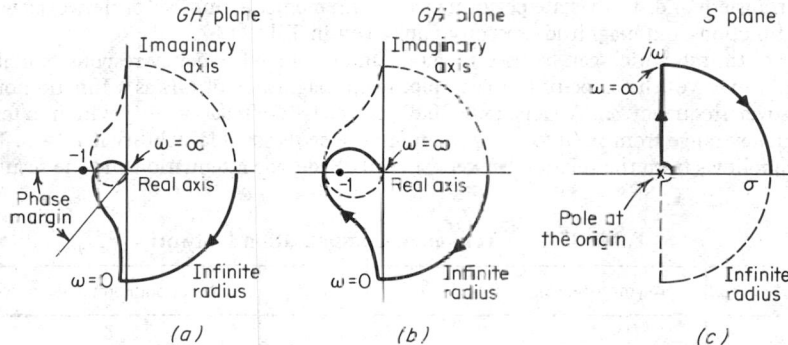

Fig. 17-79. Polar diagrams: (*a*) system stable at low gain; (*b*) system unstable at high gain; (*c*) *s*-plane contour.

136. The Bode representation, which is a direct extension of the Nyquist criterion, can be used to determine the stability of a linear closed-loop negative-feedback system which has a stable open-loop function. The magnitude and phase shift of the open-loop function GH are plotted using the same coordinates as Fig. 17-73. The phase shift should be examined at all crossover points, i.e., points at which the gain magnitude is unity (0 dB). If the magnitude passes through unity only once, the system will be stable if the phase lag at crossover is less than 180°. If the magnitude crosses unity gain at two points, the system will be stable if the phase lead is less than 180° at the low-frequency crossover point and the phase lag is less than 180° at the high-frequency crossover point. For systems having more than two crossover points, it is safer to use the Nyquist criterion.

The Bode representation provides a convenient method of analyzing simple systems and is very useful in designing stabilization networks. Detailed design information is given in Ref. 36.

137. The root-locus method is a powerful tool permitting the designer to maintain control over both the frequency response and the transient response of a feedback system. It permits the location of the closed-loop poles and zeros to be examined as a function of system gain. Various stabilization techniques can

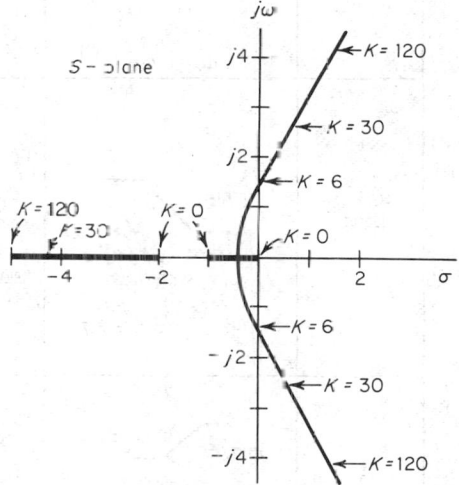

Fig. 17-80. Root-locus plot.[32]

be interpreted in terms of their effect on the locations of these poles and zeros. For the negative-feedback system of Fig. 17-68, the root loci consist of all points in the s plane at which the phase of the open-loop transfer function GH is $180° + n360°$, where n is zero or an integer. The poles of the closed-loop system move along these loci as an open-loop gain is changed. The loci start at the open-loop poles (zero open-loop gain) and terminate at the open-loop zeros (infinite loop gain). The root loci for a simple feedback system are shown in Fig. 17-80. Rules for constructing the loci and for their use in system design are given in Ref. 32.

138. State-space methods provide a systematic approach to the solution of large sets of differential equations describing the dynamics of a complex physical system. The matrix and vector representation is convenient for digital-computer calculations. The reader is referred to Refs. 37 to 39.

METHODS OF STABILIZATION

139. Stabilization networks are often used to modify the open-loop frequency response to meet transient and steady-state performance requirements. A number of networks with their transfer functions and magnitude responses are given in Table 17-7.

A rule of thumb which can be used to obtain an open-loop frequency response suitable as an initial trial involves the slope of the Bode plot (gain magnitude of GH as a function of ω) near the crossover frequency ω_c. A slope of -20dB/decade $[d(GH)/d\omega = -1]$ which extends over the frequency range from ω_c/n to $n\omega_c$ will often provide reasonable stability if $n > 3$. This rule of thumb follows from the relation between a -20 dB/decade attenuation and the resulting $90°$

Table 17-7. Frequency Compensation Networks

Network	Asymptotic response	Series network	Operational amplifier network
Low pass	Gain,dB ω_1 $\omega_1 = 1/R_1 C$	$\dfrac{E_{out}}{E_{in}} = \dfrac{1}{1+R_1 CS}$	$\dfrac{E_{out}}{E_{in}} = -\dfrac{R_1}{R_2(1+R_1 CS)}$
Lag	Gain,dB ω_1 ω_2 $\omega_1 = 1/(R_1+R_2)C$ $\omega_2 = 1/R_2 C$	$\dfrac{E_{out}}{E_{in}} = \dfrac{1+R_2 CS}{1+(R_1+R_2)CS}$	$\dfrac{E_{out}}{E_{in}} = -\dfrac{R_1(1+R_2 CS)}{R_3[1+(R_1+R_2)CS]}$
Lead	Gain,dB ω_1 ω_2 $\omega_1 = 1/R_1 C = 1/R_a C$ $\omega_2 = \dfrac{4R_1+R_2}{4R_1{}^2 C} = \dfrac{R_a+R_b}{R_a R_b C}$	$\dfrac{E_{out}}{E_{in}} = \dfrac{R_b(1+R_a CS)}{(R_a+R_b)(1+\frac{R_a R_b}{R_a+R_b}CS)}$	$\dfrac{E_{out}}{E_{in}} = -\dfrac{4R_1 R_2}{R_3(4R_1+R_2)}\left[\dfrac{1+R_1 CS}{1+\frac{4R_1{}^2}{4R_1+R_2}CS}\right]$
Lag-lead	Gain,dB ω_1 ω_2 ω_3 ω_4 $\omega_1 = 1/(R_1+R_2)C_2 \cong 1/(R_a+R_b+R_c)C_b$ $\omega_2 = 1/R_2 C_2 = 1/R_b C_b$ $\omega_3 = 1/(R_3+R_4)C_4 = 1/R_a C_a$ $\omega_4 = 1/R_4 C_4 \cong (R_a+R_b+R_c)/R_a C_a(R_b+R_c)$	$\dfrac{E_{out}}{E_{in}} = \dfrac{(1+R_a C_a S)(1+R_b C_b S)}{[(R_b+R_c)R_a C_a C_b]S^2}$ $+[R_a C_a+(R_a+R_b$ $\qquad + R_c)C_b]S+1$	$\dfrac{E_{out}}{E_{in}} = -\dfrac{R_1}{R_3}\dfrac{(1+R_2 C_2 S)[1+(R_3+R_4)C_4 S]}{[1+(R_1+R_2)C_2 S](1+R_4 C_4 S)}$

phase lag for a minimum-phase-shift network, i.e., a network containing no right-half-plane zeros or poles. The slope at the crossover frequency ω_c has the greatest influence on the phase lag at crossover, and the slope at frequency ω becomes less important as the ratio ω/ω_c or ω_c/ω becomes large with respect to unity.

The error and disturbance specifications often require a relatively high gain at very low frequency. A *lag network*, as shown in Table 17-7, is useful for providing rapid attenuation between this very low frequency and the frequency at which crossover is desired. A Bode plot illustrating the effect of the lag network on the open-loop frequency response is shown in Fig. 17-81. A *lead network*, as shown in Table 17-7, can be used to reduce the slope of the gain magnitude near crossover and thereby reduce the phase lag of the open-loop function near crossover. Lead networks must be used cautiously because they increase the gain at high frequencies and consequently increase the system susceptibility to noise. The *lead-lag network*, Table 17-7, is useful for providing a combination of the lead and lag network effects.

140. Minor-loop feedback is often used successfully to modify the frequency response of elements within the open loop. A good discussion of this technique is given in Ref. 36.

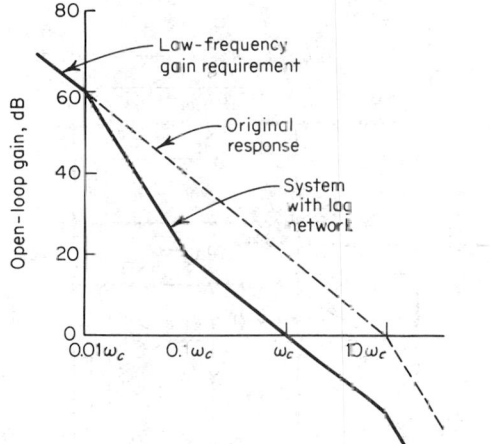

Fig. 17-81. System having lag network with breaks at 0.01 ω_c and 0.1 ω_c.

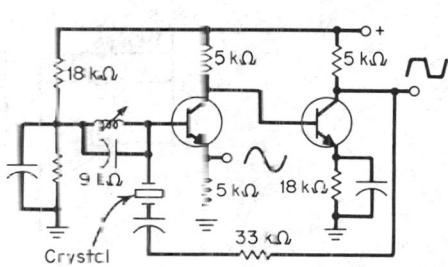

Fig. 17-82. Low-frequency crystal oscillator.

AUTOMATIC CONTROL CIRCUITS

141. General. A typical feedback control system, shown in block-diagram form in Fig. 17-68, consists of an output-measuring means, an input-measuring means, an error detector, an error-signal amplifier and processor, and a final controller (error-correcting means). The components and circuits used to embody the system are discussed in this section.

142. Transducers are used to measure the input, output, and feedback quantities and to convert them into a form suitable for the error-detecting means. The resolution, accuracy, and linearity of these devices are very important because the performance of the control system can be no better than that of the measurement means. Characteristics such as input, energy, size, reliability, output-signal level, and noise output are also important. The characteristics of a number of transducers are detailed in Par. **17-10**, and some typical transducers used to provide an electric error signal are shown in Fig 17-69.

REFERENCE AND MEASURING-ELEMENT CIRCUITS

143. Reference standards are used to provide a fixed input for electrical regulator systems and transducer systems. DC reference standards include standard cells (Par. **17-28**) and zener-diode references (Par. **17-29**). Frequency-reference standards include tuning-fork oscillators for audio frequencies and quartz-crystal oscillators for audio and radio frequencies.

144. The quartz-crystal oscillator of Fig. 17-82 is designed to operate with low-frequency duplex crystals. To ensure oscillator start-up, the transistors must be biased in the linear range under zero signal conditions and provide high loop gain. The tuned circuit provides a sinusoidal output and ensures that the crystal will oscillate only in the desired mode.

While this crystal-oscillator circuit provides an accurate frequency reference, some applications require a well-defined amplitude as well.

145. A constant-amplitude sinusoidal reference is provided by the regulator circuit of Fig. 17-83. The oscillator signal is attenuated by resistance R_1 and r_d, the dynamic impedance of diode D_1, where $r_d = kT/qI$ and k = Boltzmann's constant = 1.38×10^{-23} J/K, T = absolute temperature (K), q = electron charge = 1.6×10^{-19} C, and I = diode current (A). Diode D_1 acts as a variable-gain element since its average dynamic impedance r_d is inversely proportional to the current through resistance R_2. The signal is amplified by amplifier A_1, and unwanted harmonics are attenuated by the tuned LC filter before final amplification in amplifier A_2. The output is measured by the absolute-value circuit of amplifiers A_3 and A_4 (see Ref. 40). If capacitor C_4 were removed, the waveform at point B would correspond to the instantaneous absolute value of the output. With capacitor C_4 present, the voltage at point B is equal to the average of the absolute value of the output. The voltage at point B is compared with the voltage from zener reference diode D_3 to produce an error voltage E which controls the dynamic impedance of variable-gain element D_1.

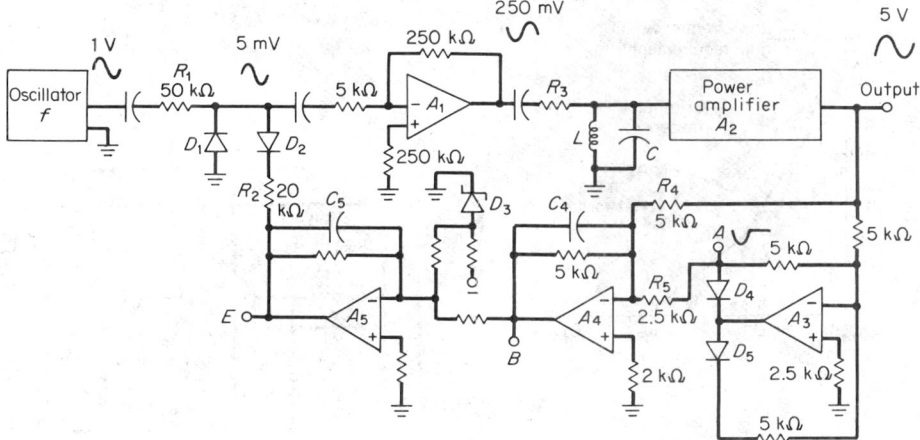

Fig. 17-83. Constant-amplitude sinusoidal reference.

146. Measuring-element circuits in conjunction with transducers perform the function of converting outputs or inputs into electric signals representative of measured quantities. The error-detection subsystems of Fig. 17-69 contain an input-measuring circuit and an output-measuring circuit, which are connected with their outputs in series to produce an error signal. Error signals can also be derived by parallel summing of currents, as in Fig. 17-84.

The linear variable differential transformer of Fig. 17-84 converts the mechanical-output position into sinusoidal signals e_a and e_b. The amplitude of these signals depends on the position of the magnetic core, and they are equal when the core is centered. The measuring-element circuit associated with the transducer overcomes possible summing problems, which could result from a phase difference between signals e_a and e_b, by rectifying each of these signals separately before summing them. Full-wave rectification is achieved in the absolute-value circuit of operational amplifier A_1 (see Ref. 40) by combining the original sinusoidal signal with its half-wave-rectified version. The current through resistor R_2, which is proportional to the inverted half-wave-rectified waveform, has twice the amplitude as the sinusoidal current through resistor R_1. The rectifier circuit associated with signal e_a produces a positive output, while the circuit associated with signal e_b provides a negative output. These signals are applied to the summing junction of operational amplifier A_3 along with the reference-signal current from resistor R_5. Much of the ripple associated with the full-wave-rectified waveforms is removed by the low-pass filtering action of amplifier A_3.

ERROR-SIGNAL PROCESSING CIRCUITS

147. Error-signal amplifying and processing circuits perform the functions of increasing the voltage and power level of the error signal, providing a frequency response which is conducive to dynamic stability, and on occasion providing a specified nonlinear response characteristic. *Offset and drift* must be minimized in direct-coupled systems which operate on a dc error component, since they will cause the output to differ from the desired value. The chopper amplifiers and operational amplifiers discussed in Pars. **17-47** to **17-59** are used for error amplification. A chopper-stabilized amplifier having a parallel path for higher-frequency components (see Fig. 17-34) is useful for amplifying low-level error signals without introducing a dc offset. Higher-level error signals can be processed by operational amplifiers.

148. The frequency-response characteristics of the open-loop system can be modified by the use of frequency-sensitive networks in the error-signal processing circuits. Series networks for use between amplifying stages and frequency-sensitive operational amplifier networks are given in Table 17-7. The analog integrator (Par. **17-54**) is equivalent to the low-pass network of Table 17-7 with an infinite value of RC.

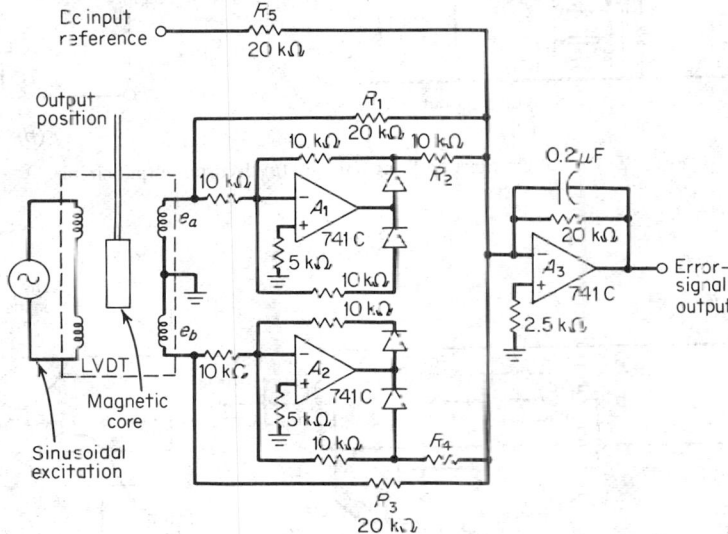

Fig. 17-84. Linear variable differential-transformer circuit.

149. AC carrier systems usually have error signals in the form of suppressed-carrier amplitude-modulated signals. These signals are generated by a number of transducers including ac-energized potentiometers, synchros, and linear variable differential transformers. The error signal can be amplified by the ac amplifiers described in Pars. **17-46** to **17-59**. Frequency compensation is more difficult since the amplitude of the sidebands must be acted on by the frequency-sensitive networks. The design of these systems is discussed in Ref. 41.

POWER-CONTROL CIRCUITS

150. Power-output circuits are used in a wide variety of control systems (see also Sec. 15). AC phase control is used to control the ac-power input to electric heating systems, lighting systems, arc-welding systems, solenoids, actuators, and low-power universal motors. Thyristor systems can supply regulated dc power to actuators, solenoids, battery chargers, and power supplies as well as dc motors. Inverters, which convert dc power to ac power, are used to provide controlled voltage at a controlled frequency in aircraft power systems, land-based power systems, and induction heating, as well as for induction motor drives. Details are given in Sec. 15.

151. Thyristor firing circuits control the power output of thyristor power converters by adjusting the phase angle ϕ of the thyristor-gate firing signal with respect to the ac supply voltage. The firing angle in the circuit of Fig. 17-85 can be controlled by means of a positive input signal which determines the capacitor-charging current provided by transistor Q_2. Near the end of each half cycle the voltage V_1 falls to zero, which causes the capacitor to discharge through unijunction transistor Q_3. When voltage V_1 rises at the beginning of the next half cycle, the capacitor charges at a rate determined by transistor Q_2. When the threshold voltage of the unijunction transistor is reached, it fires and discharges the capacitor rapidly through the pulse transformer, thereby providing a pulse to the thyristor gates. The threshold voltage of the unijunction transistor V_p is equal to $\eta V_{BB} + V_D$, where $\eta = 0.75 \pm 10\%$ for 2N2647, V_{BB} is the voltage between terminals B_1 and B_2 and V_D is a diode-voltage drop, approximately 0.5 V.[42] A simple, manually adjusted circuit can be obtained by replacing transistor Q_2 with a variable resistor.

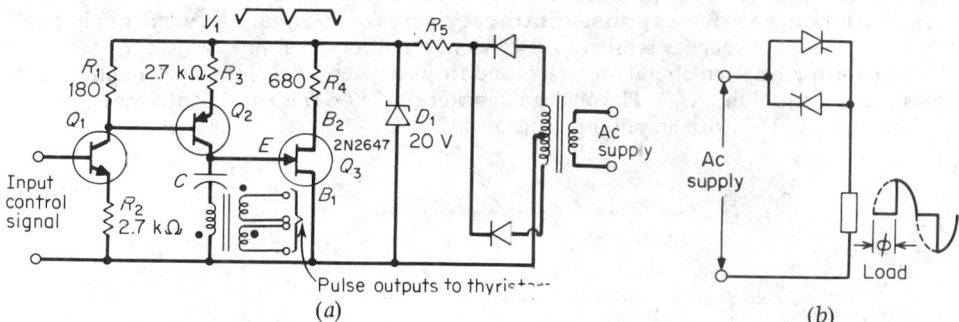

Fig. 17-85. (a) Unijunction firing circuit;[42] (b) thyristor output circuit.

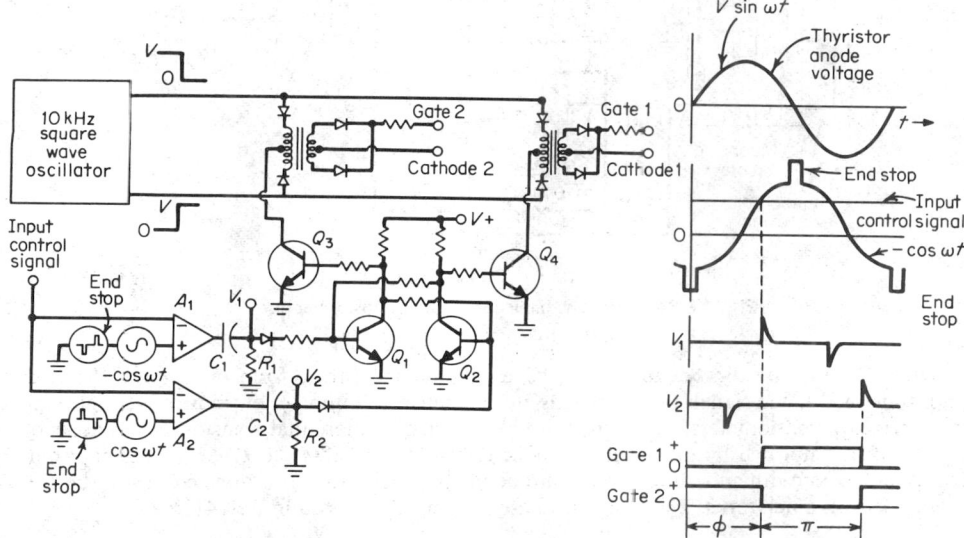

Fig. 17-86. Cosine crossing firing circuit.

When operating with a highly inductive load or a counteremf load, the thyristor may require a longer firing pulse than available from the unijunction circuit. Furthermore, a linear transfer function from the input-control signal to the thyristor converter output can be obtained by using a cosine waveform for phase control, as shown in Fig. 17-86. When the $-\cos \omega t$ voltage to comparator A_1 becomes positive with respect to the input-control signal, differentiator $C_1 R_1$ produces a positive pulse which triggers bistable $Q_1 - Q_2$, causing Q_1 and Q_4 to conduct and provide a positive output to thyristor gate 1. The gate waveforms are maintained within the proper phase range by limiting the voltage excursions of the input-control signal to less than the amplitude of the end stops, which are derived from the zero crossings of the line voltage. A discussion of firing circuits for multiphase applications is given in Refs. 43 and 49.

152. Field control of a dc machine using a pair of thyristors is illustrated in the Ward-Leonard drive of Fig. 17-87. This circuit can supply only unidirectional field current but is capable of positive or negative voltage forcing to achieve rapid changes in the current. The firing pulses from Fig. 17-85 or 17-86 may be used. The RC network across the thyristor prevents excessive rates of change of voltage dV/dt, which could cause spurious firing. The inductance L, which may be supplied by transformer leakage reactance, limits the rate of rise of thyristor current to a safe value and determines the current commutation time. The dc component of the output for this two-quadrant converter, neglecting thyristor losses and commutation losses, is $(2V_{peak} \cos \phi)/\pi$, where ϕ is the firing-angle delay. The field current controls the generator output which in turn drives the dc motor.

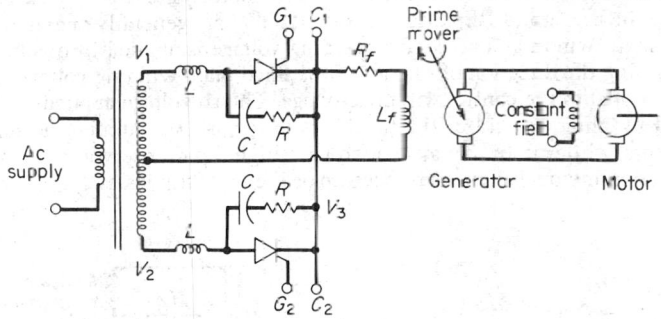

Fig. 17-87. Thyristor control of Ward-Leonard drive.

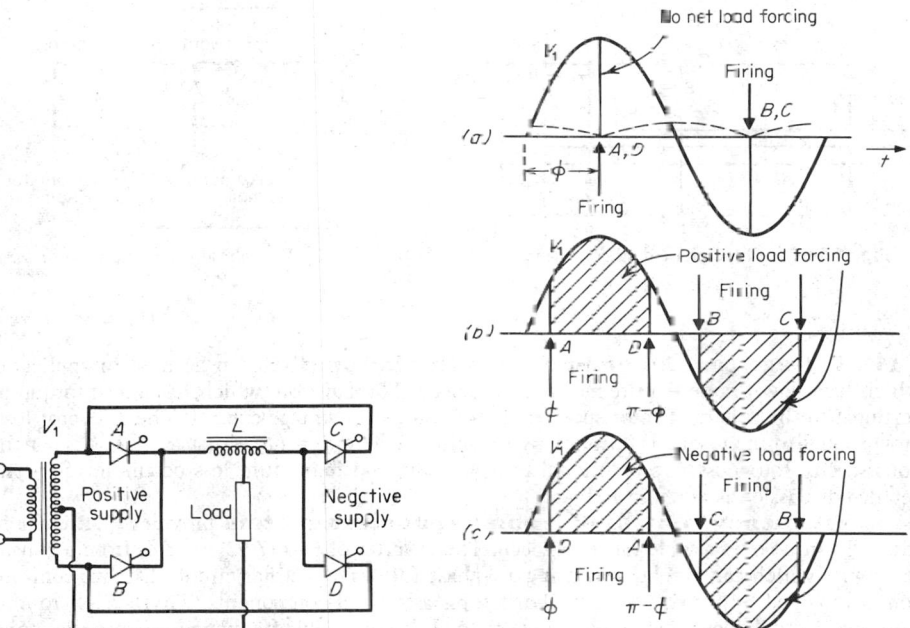

Fig. 17-88. Single-phase dual converter.

Fig. 17-89. Dual converter operation.

If a bidirectional motor drive is required, a circuit capable of supplying bidirectional field current (Fig. 17-88) should be used. The R, L, C transient-suppression networks associated with each thyristor have not been shown for the sake of clarity. The tapped inductance L is required to limit circulating current between the positive and negative current supplies. When zero field current is desired, the thyristors are fired as shown in Fig. 17-89a. For positive current the firing angles of thyristors A and B are advanced while those of thyristors C and D are retarded, as shown in Fig. 17-89b. The magnitude of the dc voltage applied to the load, neglecting commutation losses and losses in the thyristors and the current-limiting reactor, is $(2V_{peak} \cos \phi)/\pi$, where ϕ is the delay angle for firing the phase-advanced thyristor.

153. Armature control of dc motors can be accomplished by using higher-power thyristors. Motors up to 5 hp can be driven from single-phase supplies such as the center-tapped supply of Fig. 17-88 or an equivalent non-center-tapped circuit which uses a four-thyristor positive-current bridge and a four-thyristor negative-current bridge. Both these circuits and the 3-phase circuit of Fig. 17-90 are capable of driving the motor in either direction and will return power from the motor to the supply during deceleration. The circuit of Fig. 17-90 is used for driving high-power dc motors. The operation of these circuits is discussed in Refs. 43 and 44.

154. Two-phase induction motors (ac servomotors) are used as low-power actuators and are available in sizes up to 1 hp. Motors below 200 W are generally operated with a continuously excited main winding. The control-winding voltage is applied in quadrature, with the polarity determining the direction of rotation. For a fixed main-winding voltage, the torque and speed are proportional to the control-winding voltage. The ac voltage applied to the servomotor can be adjusted by phase control of the thyristors in Fig. 17-91. Circuit a is suitable for low-power servomotors, while circuit b may be used with higher-power servomotors. Voltage control of 2-phase induction motors is inefficient because of the high slip losses.

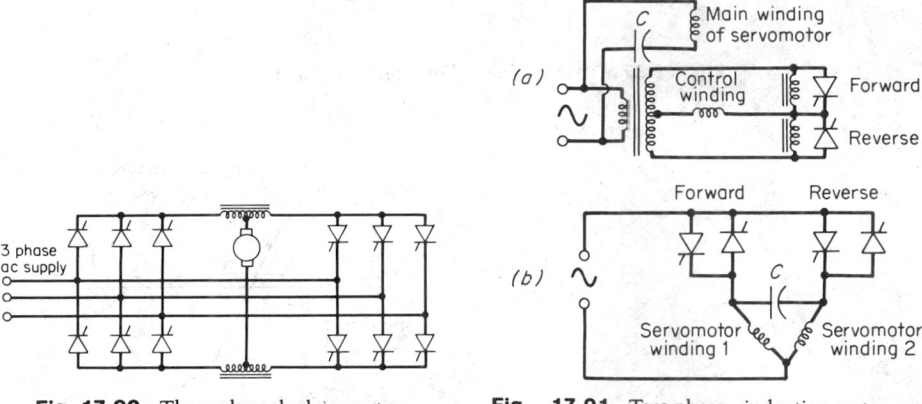

Fig. 17-90. Three-phase dual converter.

Fig. 17-91. Two-phase induction-motor servo circuits.[45]

155. Voltage control of 3-phase induction motors can be used in applications which have low torque requirements at low speed. The fan load, which has a load torque proportional to the square of rotor speed ω_r, is well suited to voltage control. The I^2R rotor loss is proportional to the torque multiplied by the slip S, where $S = (\omega_s - \omega_r)/\omega_s$ and ω_s is synchronous speed. Rotor loss is proportional to $S(1 - S)^2$, and maximum loss occurs for $S = \frac{1}{3}$. A detailed discussion is given in Ref. 46.

156. Frequency control of 3-phase induction motors provides high efficiency and full torque over a wide range of speed. The inverter of Fig. 17-92 operates from a constant dc supply, which can be obtained using a 3-phase full-wave-rectifier circuit. DC bus commutation is provided by thyristors A to D and their associated components. Thyristors 1 to 6 and their associated diodes provide a pulse-width-modulated variable-frequency drive to the 3-phase induction motor. Pulse-width modulation permits voltage adjustment by time-ratio control of the conduction interval so as to maintain constant volts per hertz, which provides nearly constant motor flux. The number of pulses per cycle is increased at lower frequencies to reduce harmonics and allow higher output torque. Detailed discussions of this type of drive are given in Refs. 47 and 48. The speed of a 3-phase induction motor can also be controlled using a cycloconverter connected to the 3-phase 60-Hz supply, eliminating the need for the dc power supply.[51]

157. Feedback signals proportional to motor speed, armature voltage, and armature current are used in torque- and speed-regulating systems. *Tachometers* can provide speed-sensing signals accurate to $\pm 0.1\%$ of full speed.

Armature voltage, from a dc motor under the condition of constant field current, can be used to derive a speed-sensing signal accurate to about $\pm 2\%$. The desired signal V_{emf}, which is proportional to speed and field current, is obtained from the armature voltage V_a by correcting for the IR drop due to armature current I_a:

$$V_{emf} = V_a - I_a R_a$$

where R_a is the resistance of the armature circuit.

A *current-sensing* signal for making a correction in the above equation or for use in torque control or current control can be obtained by using a current shunt or a dc transductor in the armature circuit or by using current transformers in the ac supply to the thyristors.

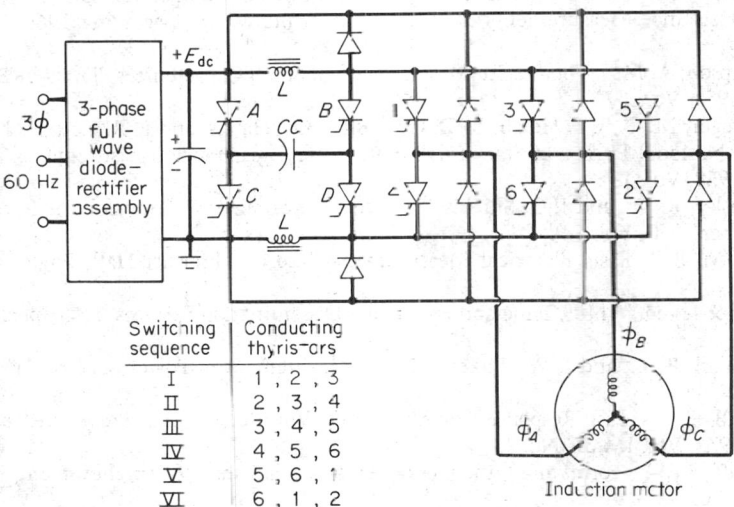

Fig. 17-92. Forced-commutated-inverter induction-motor drive.[7]

158. References

1. Considine, D. M., and S. D. Ross (eds.) "Handbook of Applied Instrumentation," McGraw-Hill, New York, 1964.

2. Fink, D. G., and J. M. Carroll (eds.) "Standard Handbook for Electrical Engineers," 10th ed., McGraw-Hill. New York, 1968.

3. Terman, F. E., and J. M. Pettit "Electronic Measurements," 2d ed., Mc-Graw Hill, New York, 1952.

4. Field, R. F. Connection Errors in Capacitance Measurements, Gen. Radio Exp., May 1947, Vol. 21.

5. Field, B. F., T. F. Finnegan, and J. Toots Volt Maintenance at NBS via 2e/h: A New Definition of the NBS Volt, Metrologia, 1973, Vol. 9, pp. 155–166.

6. Field, B. F., and V. W. Hesterman Laboratory Voltage Standard based on 2 e/h, IEEE Trans. Instrum. Meas., December 1976, Vol. IM-25, No. 4, pp. 509–511.

7. Harris, F. K. "Electrical Measurements," Wiley, New York, 1952.

8. Vigoureux and Watts Proc. Phys. Soc. (Lond.), 1933, Vol. 45, p. 172.

9. Terman, F. E. "Radio Engineers' Handbook," McGraw-Hill, New York, 1943.

10. Cutkosky, R. D. New NBS Measurements of the Absolute Farad and Ohm, IEEE Trans. Instrum. Meas., December 1974, Vol. IM-23, No. 4, pp. 305–309.

11. Lampard, D. G. A New Theorem in Electrostatics with Application to Calculable Standards of Capacitance, Proc. Inst. Elec. Eng. (Lond.), 1957, Vol. 104, Pt. C, pp. 271–280.

12. "The Royal Signals Handbook of Line Communication," H. M. Stationery Office, London, 1947.

13. Resonance Curves, Wireless World, January 1953, pp. 29–33.

14. Walston, J. A., and J. R. Miller (eds.) "Transistor Circuit Design," McGraw-Hill, New York, 1963.

15. Tobey, G. E., J. G. Graeme, and L. P. Huelsman (eds) "Operational Amplifiers: Design and Applications," McGraw-Hill, New York, 1971.

16. Sevin, L. J., Jr.   "Field Effect Transistors," McGraw-Hill, New York, 1965.

17. Schoenwetter, H. K.   A High-Speed Low-Noise 18-Bit Digital-to-Analog Converter, *IEEE Trans. Instrum. Meas.*, December 1978, Vol. IM-27, No. 4, pp. 413–417.

18. Tewksbury, S. K., F. C. Meyer, D. C. Rollenhagen, H. K. Schoenwetter, and T. M. Souders   Terminology Related to the Performance of S/H, A/D, and D/A Circuits, *IEEE Trans. Circ. Syst.*, July 1978, Vol. CAS-25, No.7.

19. Souders, T. M.   A Bridge Circuit for the Dynamic Characterization of Sample/Hold Amplifiers, *IEEE Trans. Instrum. Meas.*, December 1978, Vol. IM-27, No. 4.

20. Hague, B.   "Alternating Current Bridge Methods," 5th ed., Pitman, London, 1957.

21. MacMartin, M. P., and N. L. Kusters   A Direct-Current Comparator Ratio Bridge for Four-Terminal Resistance Measurements, *IEEE Trans. Instrum. Meas.*, December 1966, Vol. IM-15, No. 4.

22. Thompson, A. M.   The Precise Measurement of Small Capacitances, *Trans. IRE, Instrum.*, December 1958, Vol. 1-7.

23. McGregor, M. C., J. F. Hersh, R. D. Cutkosky, F.K. Harris, and F. R. Kotter   New Apparatus at the National Bureau of Standards for Absolute Capacitance Measurement, *Trans. IRE*, December 1958, Vol. 1-7.

24. Cutkosky, R. D., and J. Q. Shields   Precise Measurement of Transformer Ratios, *Trans. IRE*, December 1960, Vol. 1-9.

25. Stout, M. B.   "Basic Electrical Measurements," 2d ed., Prentice-Hall, Englewood Cliffs, N.J., 1960.

26. Howe, S. L. (ed.)   NBS Time and Frequency Dissemination Services, *NBS Spec. Publ. 432*, 1976.

27. Blackman, R. B., and J. W. Tukey   "The Measurement of Power Spectra," Dover, New York, 1958.

28. Field B. F.   A Fast Response Low-Frequency Voltmeter, *IEEE Trans. Instrum. Meas.*, December 1978, Vol. IM-27, No. 4.

29. Herwald, S. W.   Forms and Principles of Servomechanisms, *Westinghouse Eng.*, 1947, Vol. 6, pp. 149–155.

30. Gardner, M. F., and J. L. Barnes   "Transients in Linear Systems," Vol. 1, "Lumped Constant Systems," App. A, Wiley, New York, 1942.

31. James, H. M., N. B. Nichols, and R. S. Phillips   "Theory of Servomechanisms," MIT Rad. Lab. Ser., Vol. 25, p. 143, McGraw-Hill, New York, 1947.

32. Truxal, J. G.   "Automatic Feedback Control System Synthesis," McGraw-Hill, New York, 1955.

33. Grabbe, E. M., S. Ramo, and D. E. Wooldridge (eds.)   "Handbook of Automation, Computation and Control," Vol. 1, "Control Fundamentals," pp. 22–02 and 22–07, Wiley, New York, 1958.

34. Chestnut, H., and R. W. Mayer   "Servomechanisms and Regulating System Design," Vol. 1, Chap. 8, Wiley, New York, 1951.

35. Thaler, G. J., and R. G. Brown   "Servomechanism Analysis," Chap. 7, McGraw-Hill, New York, 1953.

36. Bower, J. L., and P. M. Schultheiss   "Introduction to the Design of Servomechanisms," Wiley, New York, 1958.

37. Gantmacher, F. R.   "The Theory of Matrices," Vols. I and II, Chelsea, New York, 1960.

38. Chen, C. T.   "Introduction to Linear System Theory," Holt, New York, 1970.

39. Enns, M., J. R. Greenwood III, J. E. Matheson, and F. T. Thompson   Practical Aspects of State-Space Methods, Pt. 1, System Formulation and Reduction, *IEEE Trans. Mil. Electron.*, April 1964, Vol. MIL-8, No. 2, pp. 81–93; Pt. 2, System Analysis and Simulation, *IEEE Conf. 1964*, Pap. 5, Sess. 17, pp. 501–520.

40. "Applications Manual for Operational Amplifiers," Philbrick/Nexus Research (Teledyne Philbrick), 1968.

41. Ivey, K. A.   "AC Carrier Control Systems," Wiley, New York, 1964.

42. "G. E. Transistor Manual," 7th ed., 1964.

43. Pelly, B. R.   "Thyristor Phase-Controlled Converters and Cycloconverters," Wiley, New York, 1971.

44. Bedford, B. D., and R. G. Hoft   "Principles of Inverter Circuits," Wiley, New York, 1964.

45. Truxal, J. B.   "Control Engineers' Handbook," McGraw-Hill, New York, 1958.

46. Paice, D. A.   Induction Motor Speed Control by Stator Voltage Control, *IEEE Trans. Power Appar. Syst.*, February 1968, Vol. PAS-87, No. 2.

47. Dewan, S. B., and D. L. Duff Optimum Design of an Input-Commutated Inverter for AC Motor Control, *IEEE Conf. Rec. 1968 IGA Group Meet., October 1968*, pp. 443–455.

48. Dinger, E. H. Digital Application of AC Motor Controls, *IEEE Conf. Rec. 1968 IGA Group Meet., October 1968*, pp. 271–280.

49. "G. E. SCR Manual," 5th ed., 1972.

50. Kusko, A. "Solid-State DC Motor Drives," M.I.T. Press, Cambridge, Mass., 1969.

51. Gyugyi, L., and B. R. Pelly "Static Power Frequency Changers," pp. 357–383, Wiley, New York, 1976.

Section 18

Antennas and Wave Propagation

M. C. BAILEY   *Head, Antenna Research Group Langley Research Center, National Aeronautics and Space Administration; Member IEEE*

WILLIAM F. CROSWELL   *Head, Electromagnetic Research Branch, Langley Research Center, National Aeronautics and Space Administration; Senior Member IEEE*

RICHARD C. KIRBY   *Director, International Radio Consultative Committee (CCIR), Geneva, Switzerland, Fellow IEEE*

CONTENTS

Numbers refer to paragraphs

Antennas

BY M. C. BAILEY AND W. F. CROSWELL

PROPERTIES OF ANTENNAS AND ARRAYS

1. Antenna Principles. The radiation properties of antennas can be obtained from source currents or fields distributed along a line or about an area or volume, depending upon the antenna type. The magnetic field H can be determined from the vector potential as

$$H = \frac{1}{\mu} \nabla \times A \tag{18-1}$$

where μ is the permeability of the source medium. To determine the form of A first consider an infinitesimal dipole of length L and current I aligned with the z axis and placed at the center of the coordinate system given in Fig. 18-1.

The vector potential which satisfies the wave equation in this case is

$$A = z[\mu IL \exp(-jkr)]/4\pi r \tag{18-2}$$

where $k = 2\pi/\lambda$, $j = \sqrt{-1}$, and r = radial distance away from origin in Fig. 18-1. From Eqs. (18-1) and (18-2) and Maxwell's equations, the fields of a short current element are

$$H_\phi = \frac{jkIL \sin\theta}{4\pi r}\left(1 + \frac{1}{jkr}\right)\exp(-jkr)$$

$$E_\theta = \frac{jkLI\eta \sin\theta}{4\pi r}\left(1 + \frac{1}{jkr} - \frac{1}{k^2r^2}\right)\exp(-jkr) \tag{18-3}$$

$$E_r = \frac{IL\eta}{2\pi r}\cos\theta\left(1 + \frac{1}{jkr}\right)\exp(-jkr)$$

where $\eta = \sqrt{\mu/\epsilon}$ and ϵ = permittivity of source medium. With a transformation of coordinates, the vector potential of a short current element at an arbitrary location is

$$A(x, y, z) = [\mu IL \exp(-jkR)]/4\pi R \tag{18-4}$$

$$R = \sqrt{(x - x')^2 + (y - y')^2 + z - z')^2}$$

where I = vector current along dipole axis and R = distance between wire dipole located at (x', y', z') and observation point (x, y, z).

By superposition, these results can be generalized to the vector of an arbitrary oriented volume-current density J given by

$$A(x, y, z) = \frac{\mu}{4\pi}\int_r J(x', y', z')\frac{\exp(-jkR)}{R} dx'\, dy'\, dz' \tag{18-5}$$

For a surface current, the volume-current integral in Eq. (18-5) reduces to a surface integral of $J[\exp(-jkR)]/R$, and for a line current reduces to a line integral of $I[\exp(-jkR)]/R$. The fields of all physical antennas can be obtained from the knowledge of J alone. However, in the synthesis of antenna fields the concept of a magnetic volume current M is useful, even though the magnetic current is physically unrealizable.

In a homogeneous medium the electric field can be determined by

$$E = -\frac{1}{\epsilon}\nabla \times F \qquad F = \frac{\epsilon}{4\pi}\int_r M(x', y', z')\frac{\exp(-jkR)}{R} dx'\, dy'\, dz' \tag{18-6}$$

The potentials for magnetic surface and line currents are determined in a manner similar to that for the electric currents. These electric and magnetic potentials are similar and are in fact duals.

Examples of antennas that have a dual property are the thin dipole in free space and the thin slot in an infinite ground plane. The fields of an electric source J can be determined using Eqs. (18-1) and (18-6) and Maxwell's equations. From the far-field conditions and the relationships between the unit vectors in the rectangular and spherical coordinate systems, the far fields of an electric current source J are

$$\eta H_\phi^J = E_\theta^J = \frac{-j\eta k \exp(-jkr)}{4\pi r} \int_v (J_{x'} \cos\theta \cos\phi + J_{y'} \cos\theta \sin\phi -$$

$$J_{z'} \sin\theta) \exp[jk(x' \sin\theta \cos\phi + y' \sin\theta \sin\phi + z' \cos\theta)] \, dx' \, dy' \, dz' \quad (18\text{-}7)$$

$$-\eta H_\theta^J = E_\phi^J = \frac{j\eta k \exp(-jkr)}{4\pi r} \int_v (J_{x'} \sin\phi - J_{y'} \cos\phi) \exp[jk(x' \sin\theta \cos\phi +$$

$$y' \sin\theta \sin\phi + z' \cos\theta)] \, dx' \, dy' \, dz' \quad (18\text{-}8)$$

In a similar manner the radiated far fields from a magnetic current **M** are

$$\eta H_\phi^M = E_\theta^M = \frac{-jk \exp(-jkr)}{4\pi r} \int_v (M_{y'} \cos\phi - M_{x'} \sin\phi) \exp[jk(x' \sin\theta \cos\phi$$

$$+ y' \sin\theta \sin\phi + z' \cos\theta)] \, dx' \, dy' \, dz' \quad (18\text{-}9)$$

$$\eta H_\theta^M = E_\phi^M = \frac{jk \exp(-jkr)}{4\pi r} \int_v (M_{x'} \cos\phi \cos\theta + M_{y'} \sin\phi \cos\theta$$

$$- M_{z'} \sin\theta) \exp[jk(x' \sin\theta \cos\phi + y' \sin\theta \sin\phi + z' \cos\theta)] \, dx' \, dy' \, dz' \quad (18\text{-}10)$$

2. Currents and Fields in an Aperture. For aperture antennas such as horns, slots, waveguides and reflector antennas, it is sometimes more convenient or analytically simpler to calculate patterns by integrating the currents or fields over a fictitious plane parallel to the physical aperture than to integrate the source currents. Obviously, the fictitious plane can be chosen

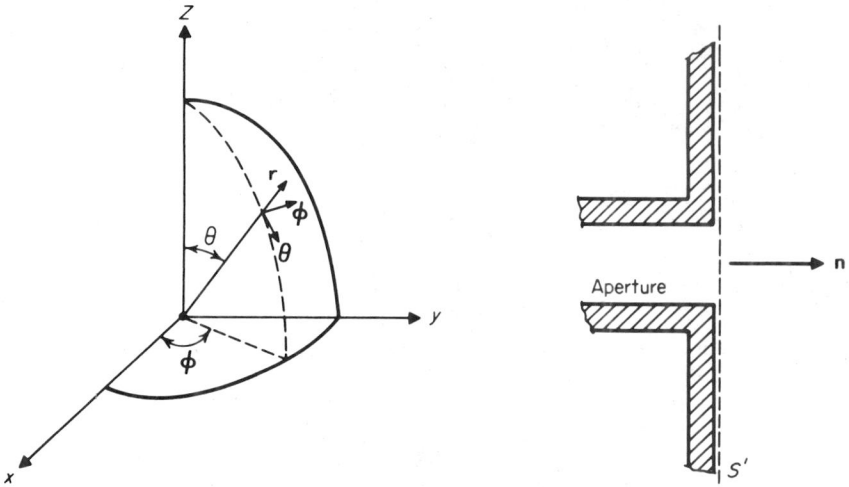

Fig. 18-1. Spherical coordinate system with unit vectors.

Fig. 18-2. Equivalent aperture plane for far-field calculations; $\mathbf{M} = 2\mathbf{E}_s' \times \mathbf{n}$, $\mathbf{J} = 2\mathbf{n} \times \mathbf{H}_s'$, and $\mathbf{J} = \mathbf{n} \times \mathbf{H}_s'$, $\mathbf{M} = \mathbf{E}_s' \times \mathbf{n}$.

to be arbitrarily close to the aperture plane. If the integration is chosen to be an infinitesimal distance away from the aperture plane, the fields to the right of s' in Fig. 18-2 can be found using either of the equivalent currents

$$\mathbf{M}_s = 2\mathbf{E}_s \times \mathbf{n} \qquad (18\text{-}11a)$$
$$\mathbf{J}_s = 2\mathbf{n} \times \mathbf{H}_s \qquad (18\text{-}11b)$$

or $\qquad \mathbf{J}_s = \mathbf{n} \times \mathbf{H}_s \qquad$ and $\qquad \mathbf{M}_s = -\mathbf{n} \times \mathbf{E}_s \qquad (18\text{-}11c)$

If the surface s' is chosen to be away from the ground plane, the form in Eq. (18-11b) is to be used if accurate values are to be computed. The equivalent magnetic current $\mathbf{M}_s$ of Eq. (18-11a) is commonly used for apertures in a very large ground plane (the effect of ground-plane size is discussed later) since the tangential $\mathbf{E}_s$ is zero on the ground plane and a good approximation to the aperture field can be obtained.

It should be noted that tangential $\mathbf{H}_s$ is not zero on a perfectly conducting plane and therefore the equivalent $\mathbf{J}_s$ form in Eq. (18-11b) is seldom used accurately in antenna problems. The combined electric and magnetic current given in Eq. (18-11c) is the general Huygens' source and is useful for aperture problems where the electric and magnetic fields are small outside the aperture; in limited cases the waveguide without a ground plane, a small horn and a large tapered aperture can be approximated this way.

Another form of computing fields is the Stratton-Chu formulation,[1]* which in addition to currents J_s and M_s requires a knowledge of $n \cdot H_s$, as shown by Tai.[2] This form therefore has no practical advantage over other forms in Eq. (18-11) although all equations give identical fields.

3. Far Fields of Particular Antennas. From the field equations stated previously or coordinate transformations of these equations, the far-field pattern of antennas can be determined when the near-field or source currents are known. Approximate forms of these fields or currents can often be estimated, giving good pattern predictions for practical purposes.

4. Electric Line Source. Consider an electric line source (current filament) of length L centered on the z' axis of Fig. 18-1 with a time harmonic-current $I(z')e^{j\omega t}$. The fields of this antenna are, from Eq. (18-7),

$$E_\theta = \frac{j\eta k \sin\theta \exp(-jkr)}{4\pi r} \int_{-L/2}^{L/2} I(z') \exp[-(jkz' \cos\theta)] \, dz' \qquad E_\phi = 0$$

For the short dipole where $kL \ll 1$ and $I(z') = I_0$

$$E_\theta = [j\eta k L I_0 \exp(-jkr) \sin\theta]/4\pi r$$

which agrees with Eq. (18-3). Fields of other current filament antennas are given in Table 18-1.

5. Electric Current Loop. The far fields of an electric current loop of radius a, centered in the xy plane of Fig. 18-1, which has a current flowing on it can be obtained by returning to the vector potential A and deriving the expressions similar to Eqs. (18-7) and (18-8) using a potential $A_\phi = A_r$. The resulting field is

$$E_\theta = \frac{-j\eta \exp(-jkr) \cos\theta}{4\pi r} \int_0^{2\pi} I(\phi') \sin(\phi - \varphi') \exp[jka \cos(\phi - \varphi') \sin\theta] \, d\phi'$$

$$E_\phi = \frac{-jk \exp(-jkr) \cos\theta}{4\pi r} \int_0^{2\pi} I(\phi') \cos(\phi - \phi') \exp[jka \cos(\phi - \phi') \sin\theta] \, d\phi'$$

The fields for the constant current loop, $I(z') = I_0$ with a radius $a \ll \lambda$, are

$$E_\phi = (k^2 a^2 \eta / r) I_0 \exp(-jkr) \sin\theta$$

Note that $E_\theta = 0$. The field of the small loop is similar to the field produced by a constant magnetic current source of length L, where $L \ll \lambda$. Due to the similarity of the pattern shape and the far-field polarization of the electric current loop and the magnetic current source, many authors have designated the loop as a magnetic source. This nomenclature is poor since it is clear that the loop is merely another wire antenna. Further characteristics of wire antennas are given in Pars. **18-22** to **18-33**.

6. Elementary Huygens' Source. Assume that constant electric and magnetic current sources $J_x = J_0$ and $M_y = M_0$ of equal length $L \ll \lambda$ are simultaneously placed at the origin of Fig. 18-1. If the currents are adjusted such that $\eta J_0 L = M_0 L$, the far fields of this source are

$$E_\theta = \frac{-jk \exp(-jkr)}{4\pi r} \cos\phi \, (1 + \cos\theta) J_0 L$$

$$E_\phi = \frac{jk \exp(-jkr)}{4\pi r} \sin\phi \, (1 + \cos\theta) J_0 L$$

The unique feature of this fictitious source compared with the electric or magnetic current element alone is the obliquity factor $(1 + \cos\theta)$, which tends to cancel the far-field radiation pattern in the region $\pi/2 \le \theta \le \pi$ due to its cardioid shape. Aperture antennas have field distributions which can be constructed from Huygens' source elements having the patterns described above.

7. Aperture in Infinite Ground Plane. With the equivalent current $M_s = -2z \times E_s$, the far field of a waveguide aperture opening onto an infinite ground plane can be obtained by integrating the aperture field since the tangential electric field is zero on the ground plane. From the magnetic current and Eqs. (18-7) to (18-10), the far-field patterns of the dominant-mode circular and rectangular waveguides can be derived as given in Table 18-1. Also given in Table 18-1 are the fields of the same antennas neglecting the ground-plane fields.

---

*Superior numbers correspond to numbered references in Par. **18-64**.

Table 18-1. Gains. Patterns. Side-Lobe Level. and Beam Widths of Typical Antenna

Antenna type	Description	Pattern Expressions	Directivity D	First side lobe, dB	Beam width, deg	
					3 dB	First nulls
Short dipole electric or magnetic		Electric: $E_\theta = j_\eta \dfrac{kLI_0}{4\pi r} \exp(-jkr)\sin\theta$	$3\!/\!2$	$\ldots$	90	
		Magnetic: $E_\phi = \dfrac{-jk^2 L}{4\pi r} M_0 \exp(-jkr)\sin\theta$	$3\!/\!2$	$\ldots$	90	
Dipole, $I(z') = I_0 \sin k(L - z')$		$E_\theta = j_\eta \dfrac{I_0 \exp(-jkr)}{2\pi r} \dfrac{\cos(kL\cos\theta) - \cos kL}{\sin\theta}$ for $L = \lambda/4$: $E_\theta = j_\eta \dfrac{I_0 \exp(-jkr)}{2\pi r} \dfrac{\cos(\pi/2 \cos\theta)}{\sin\theta}$	1.64	$\ldots$	78	
Small loop, $I(\phi') = I_0$, $a < \lambda$		$E_\phi = \dfrac{\eta ka}{2r} I_0 \exp(-jkr) J_1(ka \sin\theta)$ or $E_\phi \approx \dfrac{k^2 a^2}{r} \eta I_0 \exp(-jkr)\sin\theta$	$3\!/\!2$	$\ldots$	90	
Annular slot in a ground plane, $V_0 = E_0 b$, $a, b \ll \lambda$		$E_\theta = \dfrac{kV_0 a}{r} \exp(-jkr) J_1(ka \sin\theta)$ or $E_\phi \approx \dfrac{kV_0 a}{r} \exp(-jkr)\dfrac{ka}{2}\sin\theta$	$3\!/\!2$			
Thin half-wave slot in a ground plane, $V_0 = 2aE_0$, $V(x') = V_0 \cos\dfrac{\pi x'}{b}$		$E_\phi = \dfrac{-jV_0}{\pi r} \exp(-jkr)\cos\!\left(\dfrac{\pi}{2}\sin\theta\right)$ yz plane $E_\theta = j\dfrac{V_0}{\pi r}\exp(-jkr)$ xz plane	1.64	$\ldots$	78	

Aperture type	Field components	Directivity	Sidelobe level (dB)	Half-power beamwidth	Beamwidth between first nulls
Rectangular aperture, TE_{01} mode, E, H = 0 outside aperture, $2a, 2b > \lambda$	$E_\theta = \dfrac{jk\,\exp(-jkr)}{r}E_0(1+\cos\theta)4ab\,\dfrac{\sin(ka\sin\theta)}{ka\sin\theta}$ yz plane $E_\phi = \dfrac{jk\,\exp(-jkr)}{2r}E_0(1+\cos\theta)2ab\,\dfrac{\cos(kb\sin\theta)}{\pi^2-(kb\sin\theta)^2}$ xz plane	$10.2\dfrac{ab}{\lambda^2}$	-13.2 -30.0	xz plane: $50\dfrac{\lambda}{b}$ yz plane: $65\dfrac{\lambda}{a}$	$115\sin^{-1}\dfrac{\lambda}{b}$ $115\sin^{-1}\dfrac{\lambda}{a}$
Circular waveguide, TE_{11} mode, E, H = 0 outside aperture, $2a > 2\lambda$	$E_\theta = \dfrac{jk\,\exp(-jkr)}{2r}E_0(1+\cos\theta)kaJ(x'_{\|})J_1(x'_{\|})\dfrac{ka\sin\theta}{ka\sin\theta}$ yz plane $E_\phi = \dfrac{jk\,\exp(-jkr)}{2r}E_0ka(1+\cos\theta)J_1(x'_{\|})\dfrac{J_1(ka\sin\theta)}{1-\left(\dfrac{k\sin\theta}{x'_{\|}}\right)^2}$ xz plane	$10.5\dfrac{(\pi a^2)}{\lambda^2}$	-17.2 -38.0	yz plane: $14.9\dfrac{\lambda}{a}$ xz plane: $25.1\dfrac{\lambda}{a}$	$115\sin^{-1}\dfrac{\lambda}{a}$ $115\sin^{-1}\dfrac{\lambda}{a}$
Uniform rectangular aperture in infinite ground plane, $E_y = E_0$	$E_\phi = \dfrac{jk\,\exp(-jkr)}{4\pi r}2E_0\cos\theta\,4ab\,\dfrac{\sin(kb\sin\theta)}{kb\sin\theta}$ xz plane $E_\theta = \dfrac{-jk\,\exp(-jkr)}{4\pi r}2E_0\,4ab\,\dfrac{\sin(ka\sin\theta)}{ka\sin\theta}$ yz plane	$\dfrac{4\pi}{\lambda^2}\times\text{area}$ $\dfrac{16mab}{\lambda^2}$	-13.2	xz plane: $50.5\dfrac{\lambda}{b}$ yz plane: $50.5\dfrac{\lambda}{a}$	$115\dfrac{\lambda}{b}$ $115\dfrac{\lambda}{a}$
Circular aperture, uniform distribution	$E_\theta = \dfrac{2j\pi a^2}{\lambda r}(1+\cos\theta)\exp(-jkr)\dfrac{J_1(ka\sin\theta)}{ka\sin\theta}$	$\dfrac{4\pi}{\lambda^2}\times\text{area}$ or $\left(\dfrac{2\pi a}{\lambda}\right)^2$	-17.6	$58.5\dfrac{\lambda}{2a}$	$140\dfrac{\lambda}{2a}$
Circular aperture, $1-\left(\dfrac{\rho'}{a}\right)^2$ distribution	$E_\theta = \dfrac{2j\pi a^2}{\lambda r}(1+\cos\theta)\exp(-jkr)\dfrac{2J_2(ka\sin\theta)}{(ka\sin\theta)^2}$	$0.75\left(\dfrac{2\pi a}{\lambda}\right)^2$	-24.6	$72.7\dfrac{\lambda}{2a}$	$189\dfrac{\lambda}{2a}$
Circular aperture $\left[1-\left(\dfrac{\rho'}{a}\right)^2\right]^2$ distribution	$E_\theta = \dfrac{2j\pi a^2}{\lambda r}(1+\cos\theta)\exp(-jkr)\dfrac{8J_3(ka\sin\theta)}{(ka\sin\theta)^3}$	$0.56\left(\dfrac{2\pi a}{\lambda}\right)^2$	-30.6	$84.3\dfrac{\lambda}{2a}$	$232\dfrac{\lambda}{2a}$

Table 18-1. Gains, Patterns, Side-Lobe Level, and Beam Widths of Typical Antenna —*Concluded*

Antenna type	Description	Pattern Expressions	Directivity D	First side lobe, dB	Beam width, deg	
					3 dB	First nulls
Constant electric or magnetic line source, $J_x = J_0$ or $M_x = M_0$		Electric ($x'z'$ plane:) $$E_\theta = \frac{-j\eta\,\exp(-jkr)}{4\pi r}\,\frac{J_0 L}{2}\cos\theta\,\frac{\sin(\pi L/\lambda\,\sin\theta)}{\pi L/\lambda\,\sin\theta}$$ Magnetic ($x'z'$ plane:) $$E_\phi = \frac{jk\,\exp(-jkr)}{4\pi r}\,\frac{M_0 L}{2}\cos\theta\,\frac{\sin(\pi L/\lambda\,\sin\theta)}{\pi L/\lambda\,\sin\theta}$$	1.0 (normalized)	-13.2	$50.8\,\dfrac{\lambda}{L}$	$114.6\,\dfrac{\lambda}{L}$
Electric or magnetic line source, $J_x = J_0\cos\dfrac{\pi x'}{l}$ or $M_x = M_0\cos\dfrac{\pi x'}{L}$		Electric ($x'z'$ plane:) $$E_\theta = \frac{-j\eta\,\exp(-jkr)}{4\pi r}\,\frac{J_0\pi L}{2}\cos\theta\,\frac{\cos(\pi L/\lambda\,\sin\theta)}{(\pi/2)^2 - (\pi L/\lambda\,\sin\theta)^2}$$ Magnetic ($x'z'$ plane:) $$E_\phi = \frac{jk\,\exp(-jkr)}{4\pi r}\,\frac{M_0\pi L}{2}\cos\theta\,\frac{\cos(\pi L/\lambda\,\sin\theta)}{(\pi/2)^2 - (\pi L/\lambda\,\sin\theta)^2}$$	0.810	-23.2	$68.8\,\dfrac{\lambda}{L}$	$171.8\,\dfrac{\lambda}{L}$
Electric or magnetic line source, $J_x = J_0\cos^2\dfrac{\pi x'}{l}$ or $M_x = M_0\cos^2\dfrac{\pi x'}{l}$		$x'z'$ plane: $$E_\theta = \frac{-j\eta\,\exp(-jkr)}{4\pi r}\,J_0\cos\theta$$ $$\left[\frac{L}{2}\frac{\sin(\pi L/\lambda\,\sin\theta)}{\pi L/\lambda\,\sin\theta} - \frac{\pi^2}{\pi^2 - (\pi L/\lambda\,\sin\theta)^2}\right]$$ $$E_\phi = \frac{jk\,\exp(-jkr)}{4\pi r}\,M_0\cos\theta$$ $$\left[\frac{L}{2}\frac{\sin(\pi L/\lambda\,\sin\theta)}{\pi L/\lambda\,\sin\theta} - \frac{\pi^2}{\pi^2 - (\pi L/\lambda\,\sin\theta)^2}\right]$$	0.667	-31.5	$83.2\,\dfrac{\lambda}{L}$	$229.2\,\dfrac{\lambda}{L}$

8. Simple Arrays. Consider a linear array of radiating elements which, for simplicity, are assumed to be equally spaced at a distance d apart as illustrated in Fig. 18-3. The field at a large distance away from the mth element can be written

$$E_m = f_m(\theta, \phi) \frac{\exp(-jkr)}{r} \exp(jk_m d \cos \theta) \tag{18-12}$$

By superposition the field of an array of N elements is given by

$$E_N = \frac{-\exp(-jkr)}{r} \sum_{m=1}^{N} f_m(\theta, \phi) \exp[-j(m - 1)kd \cos \theta] \tag{18-13}$$

if the element at the origin is chosen as a reference. If each element has an identical pattern, $f_m(\theta, \phi) = a_m E(\theta, \phi)$ and

$$E_N(\theta, \phi) = E(\theta, \phi) f(\psi) \qquad f(\psi) = \sum_{m=1}^{N} a_m \exp j\psi \tag{18-14}$$

where a_m is a complex number representing the excitation current (voltage in the case of slots) for the mth element and $\psi = -(m - 1)kd \cos \theta$. The function $f(\psi)$ is commonly called the *array factor* or *array polynomial*, and the factorization process given in Eq. (18-14) is called *pattern multiplication*.

9. Uniform Linear Array. Suppose the array in Fig. 18-3 is fed uniformly in amplitude and has a phase shift δ between adjacent elements. Here $c_m = e^{-jm\delta}$ and $\psi = kd \cos \theta - \delta$, and consequently $|f(\theta, \phi)|^2$ is

General:

$$|f(\theta, \phi)|^2 = \left| \frac{\sin^2[(N/2)(kd \cos \theta - \delta)]}{N^2 \sin^2[(kd/2) \cos \theta - \delta]} \right| \tag{18-15}$$

Broadside $\delta = 0$:

$$|f(\theta, \phi)|^2 = \left| \frac{\sin^2[(N/2)(kd \cos \theta)]}{N^2 \sin^2[(kd/2) \cos \theta]} \right| \tag{18-16}$$

End fire $\delta = kd$:

$$|f(\theta, \phi)|^2 = \left| \frac{\sin^2[(Nkd/2)(\cos \theta - 1)]}{N^2 \sin^2[(kd/2)(\cos \theta - 1)]} \right| \tag{18-17}$$

Due to the ϕ rotational symmetry in Eq. (18-15), the pattern of the line source has rotational symmetry about the z axis. Since the cone angle of the pattern decreases with scan angle from broadside, the pattern directivity remains constant at the value N regardless of scan angle. By choosing the phase shift between elements so that $\delta = kd + 2.94/N$ the sharpest pattern in the end-fire direction ($\theta = 0$) is obtained. In this case

$$|f(\theta, \phi)|^2 = \left| \frac{\sin^2[(Nd/2)(k \cos \theta - k')]}{N^2 \sin^2[(d/2)(k \cos \theta - k')]} \right| \tag{18-18}$$

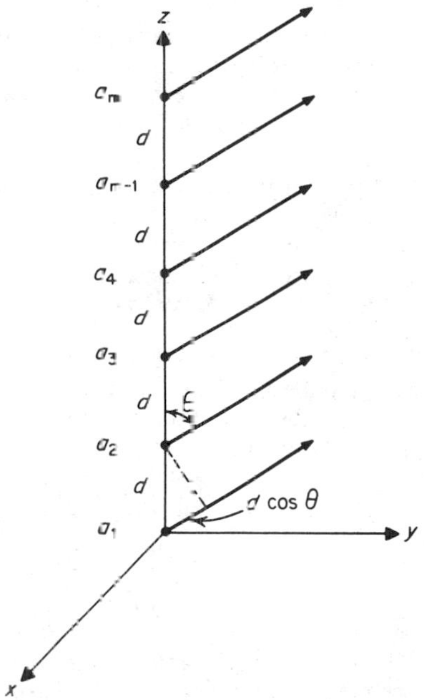

Fig. 18-3. Geometry of a linear array of equally spaced elements.

where $k' = k + 2.94/Nd$. This phase condition, which is determined graphically, is called the *Hansen-Woodward condition* for superdirectivity. This extra directivity may be several decibels in practical antennas.

10. Circular Arrays. Consider an array of N equally spaced elements about a circle of radius a in the xy plane of Fig. 18-1. If the azimuthal location of the mth element is $\phi_m = 2\pi m/N$, then for an element excitation of the form $a_m = A_m e^{j\alpha m}$ the array factor is given by

$$f(\theta, \phi) = \sum_{m=1}^{N} A_m \exp\{j[\alpha_m + ka \sin \theta \cos(\phi - \phi_m)]\} \tag{18-19}$$

For arrays having a large number of elements, $|A_m| = $ const, the array-factor pattern in the xy plane can be approximated by

$$|f(\theta, \phi = \pi/2)| \approx |J_0[2ka \sin \tfrac{1}{2}(\phi - \phi_0)]|$$

which is a directional beam with a maximum of $\phi = \phi_0$. If the pattern of each antenna in the circular array is of the form $F(\phi) = \sum_{m=0}^{\infty} A_m \cos^m \phi$, the pattern of a uniformly excited circular array of N such elements is approximately given by[3]

$$\Phi(\theta, \phi) \approx N \sum_{m=0}^{M} A_m(-i)^m \frac{d^m}{dz^m} [J_0(z) + 2j^N J_m(z) \cos N\phi] \qquad (18\text{-}20)$$

where $M < N$ and $Z = ka \sin \theta$. A design curve for determining the number of sources N required in a circular array of circumference Z to produce an omnidirectional pattern with 0.5 dB ripple or less is given in Fig. 18-4. This design curve works fairly well independent of element pattern except for $F(\phi) = 1$. In this case Eq. (18-20) has infinite ripple for Z, where $J_0(Z) = 0$, independent of the number of elements in the array. This design technique has been applied successfully to several practical arrays.[4,5]

11. Planar Arrays. Now consider a planar array of equally spaced elements in the yz plane of Fig. 18-3, where the elements in the y direction are a distance d from the elements on the z axis. With the origin as a phase reference, the array factor is given by

$$f(\theta, \phi) = \sum_{n=1}^{N} \sum_{m=1}^{M} a_{mn} \exp[jk(nd \cos \theta + md \sin \theta \sin \phi)] \qquad (18\text{-}21)$$

where the excitation coefficient $a_{mn} = A_{mn} \exp[jk(m\delta_y + n\delta_z)]$. If $A_{mn} = 1$ and $\delta_y = \delta_z = 0$, the array has a pattern maximum at broadside and has the array factor

$$|f(\theta, \phi)|^2 = \left| \frac{\sin^2 (Nkd/2 \cos \theta)}{N^2 \sin [(kd/2) \cos \theta]} \right| \left| \frac{\sin^2 [(Mkd/2) \cos \theta \sin \phi]}{M^2 \sin^2 [(kd/2) \cos \theta \sin \phi]} \right| \qquad (18\text{-}22)$$

The pattern in the two principal planes of the uniformly excited array is identical in form to the linear array in the same plane. As the array is scanned from broadside, the pattern broadens and becomes asymmetrical. The directivity of the planar array (unlike that of the linear array) decreases from the broadside value by a factor $\cos \theta$.

12. Gain and Directivity. Antennas that are several wavelengths or larger in dimension have far-field patterns for which most of the radiated energy is restricted to narrow angular regions. Several useful measures of how the pattern is concentrated are gain, directivity, effective area, and beam efficiency. The beam efficiency is important for low-noise communication systems, microwave radiometry, and radio-astronomy antennas. The definitions of directional directivity $D(\theta, \phi)$, directivity D, gain G, and directional gain $G(\theta, \phi)$ are

$$G(\theta, \phi) = N_A(1 - |\Gamma|^2)[D(\theta, \phi)] = N_A(1 - |\Gamma|^2) \left[4\pi \frac{\text{power radiated in direction } (\theta, \phi)}{\text{total power radiated by antenna}} \right]$$

$$G = N_A(1 - |\Gamma|^2)D = N_A(1 - |\Gamma|^2) \left(4\pi \frac{\text{maximum power radiated by antenna}}{\text{total power radiated by antenna}} \right)$$

The term N_A is the antenna efficiency related to I^2R losses, and Γ is the reflection coefficient as seen at the antenna input terminals. In equation form

$$D(\theta, \phi) = \frac{4\pi |E(\theta, \phi)|^2}{\int_0^{2\pi} \int_0^{\pi} |E(\theta, \phi)|^2 \sin \theta \, d\theta \, d\phi} \qquad (18\text{-}23)$$

where $E(\theta, \phi) = $ electric field of the antenna, and

$$D = \frac{4\pi |E(\theta, \phi)|_{\max}^2}{\int_0^{2\pi} \int_0^{\pi} |E(\theta, \phi)|^2 \sin \theta \, d\theta \, d\phi} \qquad (18\text{-}24)$$

13. Effective Area. The effective area of a receiving antenna is

$$A_p(\theta, \phi) = (\lambda^2/4\pi)N_A(1 - |\Gamma|^2)D(\theta, \phi)|\rho_1 \cdot \rho_2|^2 \qquad (18\text{-}25)$$

where $|\rho_1 \cdot \rho_2|^2$ is the polarization loss, which accounts for the difference in polarization of the incoming wave to the polarization of the receiving antenna in direction (θ, ϕ). If the antenna polarization is matched to the incoming polarization, $|\rho_1 \cdot \rho_2|^2 = 1$; the polarization factor is described in detail in Par. **18-18**. The term Γ is the voltage-reflection coefficient observed at the antenna terminals if it is fed as a transmitting antenna. The polarization and reflection coefficient are not always included in the definition of effective area as a matter of standard definition; the standard definition is really the maximum effective area. However, in practice, these additional factors are physically present and must be included.

14. Summary: Pattern, Directivity, and Beam Width. The radiation-pattern expressions, directivity, and beam width of typical antennas are given in Table 18-1. The factor e^{-jkr}/r is suppressed in these expressions, along with the harmonic time dependence. The half-power beam width (3 dB) is defined as the angle between half-power levels in the main lobe of a directional antenna. Another commonly used beam width is the total angle between the first nulls of the main lobe of an antenna pattern. Note that the commonly used beam width λ/D, in radians, is one-half the beam width between nulls of the main lobe of an antenna.

15. Beam Efficiency. In addition to the gain of an antenna, the beam efficiency is a very useful parameter for judging the quality of receiving antennas intended for measurement of noise signals from an extended source. The beam efficiency is a measure of the ability of an antenna to discriminate between the received signal in the main beam and unwanted signals received through the side lobes in other directions. Assuming that the antenna aperture is in the xy plane in Fig. 18-1, the beam efficiency is defined as

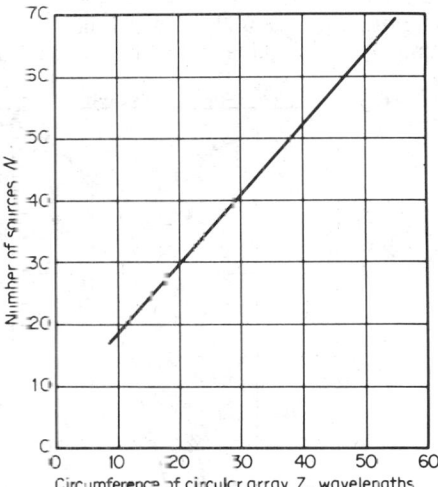

Fig. 18-4. The number of sources required in a circular array to produce an omnidirectional pattern within ±0.5 dB. For elements having circular patterns, nulls occur at $J_0(Z) = 0$ independent of the number of sources N.

The vertical axis of the figure reads "Number of sources N" and the horizontal axis reads "Circumference of circular array Z, wavelengths".

$$BE = \frac{\text{power radiated in a cone angle } \theta}{\text{total power radiated by the antenna}} = \frac{\int_0^{2\pi} \int_0^{\theta_1} |E(\theta, \phi)|^2 \sin \theta \, d\theta \, d\phi}{\int_0^{2\pi} \int_0^{\pi} |E(\theta, \phi)|^2 \sin \theta \, d\theta \, d\phi} \tag{18-26}$$

To gain some insight into how the beam efficiency varies as a function of the side-lobe level and position, calculations for a circular aperture with symmetrical aperture distributions of the form $f(\rho') = [1 - (\rho'/a)^2]^p$ are presented along with similar calculations for the rectangular aperture with $\cos^n \psi$ distribution in Fig. 18-5. It is interesting to note that the beam efficiency does not reach 90% until the angle $2\theta_1$, which is about 2 to 3 times the beamwidth angle.

16. Antenna Temperature. An antenna located on the earth and pointing at an angle (θ, ϕ) to the sky will receive noise from all directions. The amplitude of this noise as seen at the antenna terminals will depend upon the noise source (warm earth, cosmic noise, water vapor, radio stars, etc.), the antenna orientation, and the operating frequency and polarization.[6] The equivalent received noise power in a receiver matched to the antenna terminal impedance by a lossless transmission line is

$$P_n = kTB \tag{18-27}$$

where k = Boltzmann's constant, T = temperature (K) and B = bandwidth (Hz).

Noise power for a fixed bandwidth may be thought of as an equivalent temperature. Consequently, if it is assumed that the various noise sources that make up the antenna noise environment have an equivalent temperature $T(\theta, \phi)$, the apparent antenna temperature is given by

$$T_A = \frac{\int_0^{2\pi} \int_0^{\pi} G(\theta, \phi) T(\theta, \phi) \sin \theta \, d\theta \, d\phi}{\int_0^{2\pi} \int_0^{\pi} G(\theta, \phi) \sin \theta \, d\theta \, d\phi} \tag{18-28}$$

The most important natural emitter of noise at microwave frequencies is the ground at 377 K compared with the sky temperature at a few degrees. Therefore, antennas which have low side and back lobes will have low apparent antenna temperatures T_A. Several low-noise antenna designs with antenna temperatures as low as 2 K have been reported.[7-13]

If the antenna has losses and is not matched to the receiver for maximum power transfer, the antenna temperature will be higher than that predicted by Eq. (18-28). The contribution due to

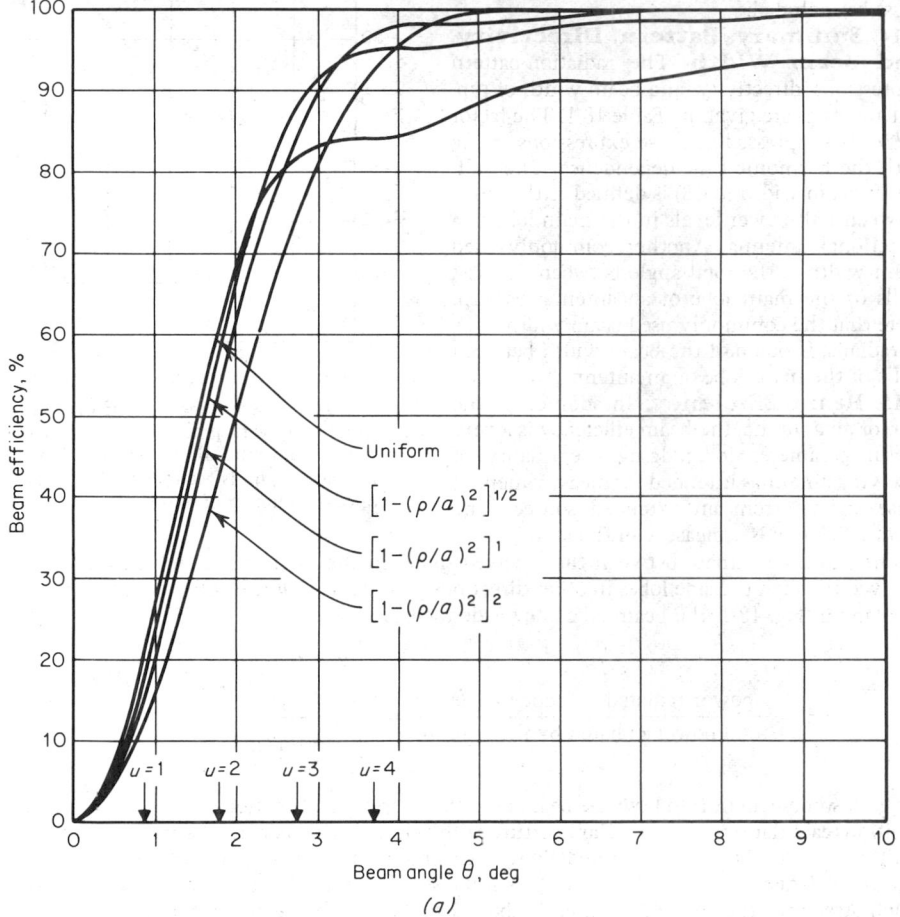

Fig. 18-5. Beam efficiency vs. angle for various distributions of a circular aperture: (a) 20λ in diameter and (b) 20λ on a side. Note that these data can be scaled for other apertures using the parameter $u = 2\pi a/\lambda \sin\theta$, as noted on the beam-angle scale.

the particular mismatch and losses of every component must be analyzed for each particular receiving system in detail. An outline of an excellent method of analysis is available.[15] For a receiving system with no mismatch loss, the apparent temperature T_a is given by

$$T_a = (1 - L)T_A + LT_0 \qquad (18\text{-}29)$$

where T_A = antenna temperature given by (18-28) and T_0 = physical temperature of lossy device. For example, a 10-cm length of precision coaxial cable has a loss of about 0.013 dB (L = 0.003). If T_A = 100 K and T_0 = 300 K, then from (18-29) the apparent temperature T_a = 100.6 K. This increase, while small, is important in remote sensing with radiometric systems, where absolute temperatures are measured to ±0.1 K.

While most reflector and horn antennas have small losses, microwave radiometer systems are being constructed which require antennas having losses known to within 0.01 dB.[14]

17. Friis Transmission Formula. Assume that there is a source antenna and an antenna under test located at a distance r apart such that

$$r \geq 2(d_t)^2/\lambda \qquad (18\text{-}30)$$

where d_t is the maximum aperture dimension of the antenna under test. The distance specified by Eq. (18-30) is the so-called *far-field distance*. The far-field distance is commonly specified as the distance where the phase front of a spherical wave over a planar aperture will not exceed $\pi/8$ rad. For special purposes, such as the measurement of deep nulls or extremely precise side-lobe

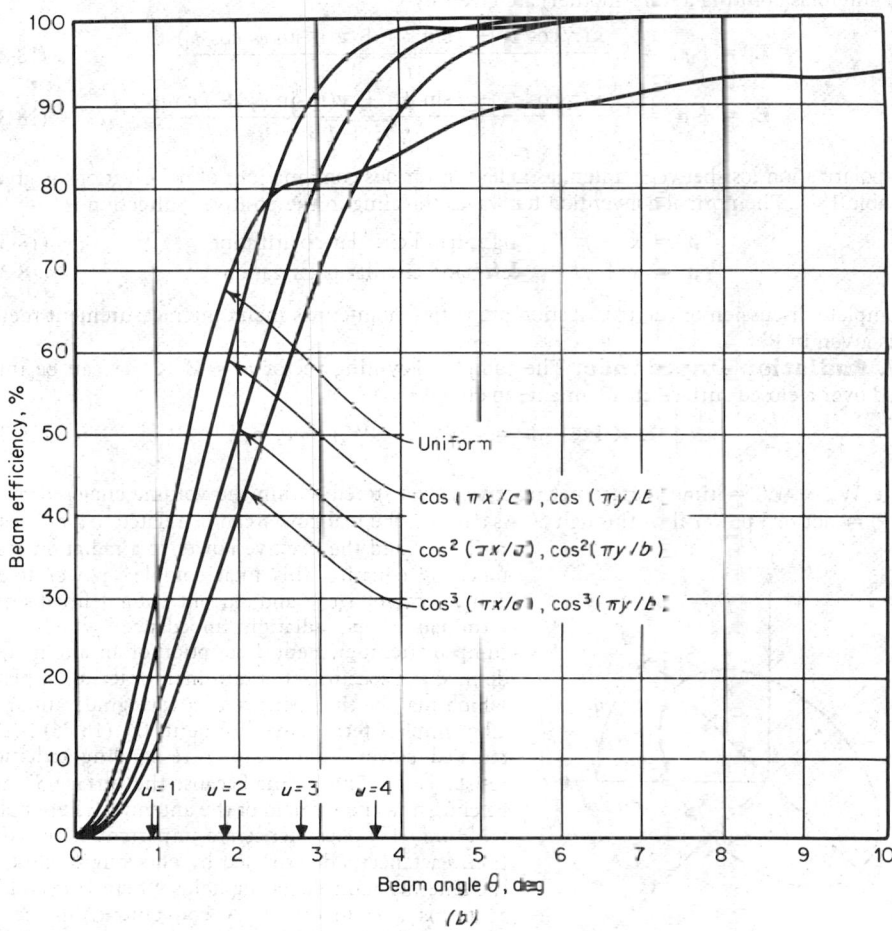

Fig. 18-5. (*Continued*)

levels, the far-field distance may have to be extended further; curves using other criteria are available.[16] The power received at the terminals of one antenna located in the far field of a second antenna can be expressed as a fraction of the transmitted power as

$$P_R = P_T[\lambda/4\pi r]^2 N_{AT} N_{AR} D_T(\theta_T, \phi_T) D_R(\theta_R, \phi_R)(1 - |\Gamma_T|^2)(1 - |\Gamma_R|^2)|\rho_R \cdot \rho_T|^2 \qquad (18\text{-}31)$$

where N_{AT}, N_{AR} = loss efficiencies of antennas and $D_T(\theta_T, \phi_T), D_R(\theta_R, \phi_R)$ = directivities of antennas in the direction one antenna is pointing toward the other. $|\Gamma_T|^2$ and $|\Gamma_R|^2$ are the reflected power due to mismatch of the antenna terminals and $|\rho_R \cdot \rho_T|^2$ is the polarization loss. The term in brackets in Eq. (18-31) is the so-called *free-space loss* which is due to spherical spreading of the energy radiated by an antenna. In the far field all antennas appear as a spherical wave emanating from a point source located at the phase center of the antenna, where the phase center may be a function of the observation angle.

18. Polarization. Consider a plane wave propagating in the z direction which has an arbitrary plane polarization with an axial ratio r_A and a tilt angle as shown in Fig 18-6. The polarization ellipse is the locus of the tip of the electric field vector as the wave propagates in space as a function of time. The axial ratio r_A is defined as the ratio of the major to minor axis of the ellipse referenced to a coordinate system. The field expression for this arbitrary plane-polarized wave is

$$E = C[\mathbf{x}(r_A \cos \phi + j \sin \phi) + \mathbf{y}(r_A \sin \phi - j \cos \phi)] \exp(-jkz)$$

When the z-phase dependence is neglected except for the sign, the normalized fields of two different plane-polarized waves with the wave number 2 propagating in the negative z direction (two antennas pointing at one another) are given by

$$\mathbf{E}_1 = E_1 \rho_1 = E_1 \frac{\mathbf{x}(r_1 \cos \phi_1 + j \sin \phi_1) + \mathbf{y}(r_1 \sin \phi_1 \cos \phi_1)}{\sqrt{r_1^2 + 1}} \tag{18-32}$$

$$\mathbf{E}_2 = E_2 \rho_2 = E_2 \frac{\mathbf{x}(r_2 \cos \phi_2 - j \sin \phi_2) + \mathbf{y}(r_2 \sin \phi_2 + j \cos \phi_2)}{\sqrt{r_2^2 + 1}} \tag{18-33}$$

The polarization loss between antennas radiating various combinations of polarizations is given in Table 18-2. The normal convention for waves traveling in the positive z direction is

$$\rho_1 = x - jy \qquad \text{right-hand circular polarization} \tag{18-34}$$
$$\rho_2 = x + jy \qquad \text{left-hand circular polarization} \tag{18-35}$$

A complete discussion of the polarization properties of antennas including measurement methods is given in Ref. 6.

19. Radiation Impedance. The complex Poynting vector $\rho = \mathbf{E} \times \mathbf{H}^*$ can be integrated over a closed surface about an antenna to give

$$\iint_{s'} \mathbf{E} \times \mathbf{H}^* \cdot ds = 4j\omega(W_m - W_e) + P_s = VI^* \tag{18-36}$$

where $W_m - W_e$ = time-average net reactive power stored within the volume enclosed by S' and P_s = net real power flow through S'. As a result, the real power can be related to a radiation resistance and the reactive power to a radiation reactance by equating this total complex power to an equivalent voltage V and current I at a defined set of terminals. This radiation impedance, which is a lumped-circuit-element description of an antenna, is defined at a specific set of terminals or terminal plane which may be the aperture of a waveguide antenna. The simplest term to determine in Eq. (18-36) is the radiated power P_s and the corresponding radiation resistance R_r. This is true because the surface S' may be chosen in the far field of the antenna, where fields with only $1/r$ dependence are important. The radiation reactance is determined by choosing S' close to the antenna and integrating fields where $1/r^2$ and $1/r^3$ terms are important. A convenient surface to choose for these calculations is a sphere of radius r surrounding the antenna.

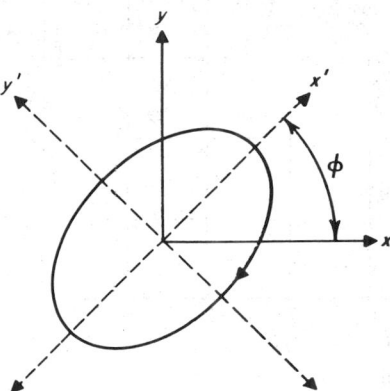

Fig. 18-6. Polarization ellipse.

20. Radiation Resistance of Short Current Filament. The far fields of a current filament from Eq. (18-3) are

$$E_\theta = \frac{jkI_0 \eta L \sin \theta \exp(-jkr)}{4\pi r} \qquad H_\phi = \frac{jkI_0 L \sin \theta \exp(-jkr)}{4\pi r}$$

When these fields are used, P_s from Eq. (18-36) is given by

$$P_s = \int_0^{2\pi} \int_0^\pi E_\theta H_\phi^* \sin \theta \, r^2 \, d\theta \, d\phi = \frac{\eta \pi I_0^2}{3} \left(\frac{L}{\lambda} \right)^2 = I_0^2 R_r$$

21. Array Impedance. The pattern expressions for arrays neglected coupling between elements in the array. Mutual coupling will not only affect the input impedance of each element input terminal as a function of scan angle but will also change the pattern of each element in a manner dependent upon the element location in the array. The determination of the mutual and self-impedance of an element in an array and the effect of these changes upon array patterns

Table 18-2. Polarization Loss between Plane-polarized Waves

Type	ρ_1	ρ_2	$\lvert \rho_1 \cdot \rho_2 \rvert^2$
Linear to linear. $r_1 = r_2 \to \infty$	$x \cos \phi_1 + y \sin \phi_1$	$x \cos \phi_2 + y \sin \phi_2$	$\cos^2(\phi_1 - \phi_2)$
Linear to circular. $(r_1 = 1, \phi = 0 \text{ to } \pi)$	$\dfrac{x \pm jy}{\sqrt{2}}$	$x \cos \phi_2 + y \sin \phi_2$	$\dfrac{1}{2}$
Circular to circular. $r_1 = r_2 = 1$, $\phi_1 = 0 \text{ or } \pi$, $\phi_2 = 0 \text{ or } \pi$	$\dfrac{x + jy}{\sqrt{2}}$	$\dfrac{x - jy}{\sqrt{2}}$	1
	$\dfrac{x + jy}{\sqrt{2}}$	$\dfrac{x + jy}{\sqrt{2}}$	0
General case	$\dfrac{x(r_1 \cos \phi_1 + j \sin \phi_1)}{\sqrt{r_1^2 + 1}}$ $+ \dfrac{y(r_1 \sin \phi_1 - j \sin \phi_1)}{\sqrt{r_1^2 + 1}}$	$\dfrac{x(r_2 \cos \phi_2 - j \sin \phi_2)}{\sqrt{r_2^2 + 1}}$ $+ \dfrac{y(r_2 \cos \phi_2 + j \cos \phi_2)}{\sqrt{r_2^2 + 1}}$	$\dfrac{(1 + r_1^2 + r_2^2 + r_1^2 r_2^2 + 4 r_1 r_2) + (1 + r_1^2 r_2^2 - r_1^2 - r_2^2)\cos^2(\phi_1 - \phi_2)}{2(r_1^2 + 1)(r_2^2 + 1)}$

is a specialized problem. In general, however, arrays of antennas radiating into a linear medium will have to satisfy the same equations as any linear system with n pairs of terminals, or

$$V_1 = Z_{11}I_1 + Z_{12}I_2 + \cdots + Z_{in}I_n \qquad V_n = Z_{n1}I_1 + Z_{n2}I_2 + \cdots Z_{nn}I_n$$

Specific examples of array pattern and impedance properties are given later in this section.

WIRE ANTENNAS

22. Analysis of Wire Antennas. The development of wire antennas has been extensive, since such antennas are simple to analyze and construct. The classical analysis of wire antennas such as dipoles, loops, and loaded-wire antennas has been developed by Hallen and R. W. P. King and his students; a good summary of theoretical and experimental results, including specific impedance curves and design data, is given in Ref. 17. The unfortunate drawback of this analysis is that each new wire-antenna configuration presents another analytical problem which must be solved before design computations can be made. A systematic method of solving wire-antenna problems using computerized matrix methods has been developed by extending the analyses of scattering by wire objects by Richmond[18-21] and Harrington.[22-25] These matrix analysis methods have been applied to wire antennas[26-41] to determine the input impedance, current distribution, and radiation patterns by subdividing any particular wire antenna into segments and determining the mutual coupling between any one segment and all other segments. The method therefore can treat any arbitrary wire configuration, including loading and arrays, the limitation being the storage capacity of available digital computers and the patience of the programmer.

23. Numerical Method. For a single-wire antenna, the antenna is subdivided into N sections. The current on the antenna can be expressed as

$$I = \sum_{n=1}^{N} I_n F_n \tag{18-37}$$

where F_n is a known expansion function such as a pulse, triangle, or piecewise sinusoid and the coefficients I_n are unknowns to be determined. With this current expansion for the segmented antenna a matrix can be written in the form

$$[Z][I] = [V] \tag{18-38}$$

where the column vector $[V]$ is known and the column vector $[I]$ is to be determined. The square impedance matrix $[Z]$ is completely defined by the geometry and choice of the expansion function F_a and a suitable testing function W_n, where if Galerkin's method is used, $W_n = F_n$. By substituting Eq. (18-37) into the vector potential A given by Eq. (18-5), and after considerable algebraic manipulation[25] the impedance-matrix element is given by

$$Z_{mn} = \int dl \int dl' \left(j\omega\mu W_m \cdot F_n + \frac{\partial W_n}{\partial l'} \cdot \frac{\partial F_n}{\partial l} \right) \frac{\exp(-jkR)}{R} \tag{18-39}$$

Once Z_{mn} is known, $[I]$ is defined by Eq. (18-37), and hence the input impedance is known. The basic analytical methods used by Richmond and Harrington are similar except for the form of the current distribution on the fundamental wire segment.[26,27] The use of the piecewise-sinusoidal current distribution on individual wire segments by Richmond may have computational advantages. Computer programs for obtaining the radiation properties of many types of wire antennas are readily available.[26-28,30-33]

24. Finite Gap and Feed Line. The effects of the finite gap in a linear antenna and the connecting balanced-input transmission line used as a feeder are neglected in the classical analysis. These effects have been determined by the matrix method by J. S. Chatterjee. Feeder gaps up to $\frac{1}{10}\lambda$ in a half-wave dipole with a radius $a = 0.000125\lambda$ will produce about a 2% change in the input susceptance of the antenna.

The effect of a feeder transmission line, however, is much more pronounced, as shown in Fig. 18-7. In this figure the input impedance Y_T is plotted; it is computed by transforming the dipole admittance down the transmission line d, assuming that the normal transmission-line equations hold. Y_{in} in this figure is the admittance as seen at the input terminals of the antenna, where the feeder line and dipole arms are all considered as part of the radiating system.

It appears that the best design uses feeder lines that are an odd multiple of $\frac{1}{4}$ wavelength long. Many types of dipole antennas can be fed with an unbalanced transmission line with a balun to minimize feeder-line currents so that the effect of the feeder line is small. An excellent description of baluns is given in Ref. 42.

25. Thin and Fat Dipoles. The impedance and radiation patterns of the thin dipoles depend on the length of the wires and the location of the feed point. Detailed impedance properties and patterns of thin dipoles are available.[43] The bandwidth properties of wire antennas can be improved by using either "thick" wires or "fat" shapes. A thick-dipole design using an open balun design has been used to obtain bandwidths nearly 1.8 to 1,[44] and "fat" dipole designs are available that exhibit 2 to 1 bandwidths.[39,45] Both these designs are primarily useful when mounted over a ground plane, and they can be arranged in crossed pairs to produce circular polarization.

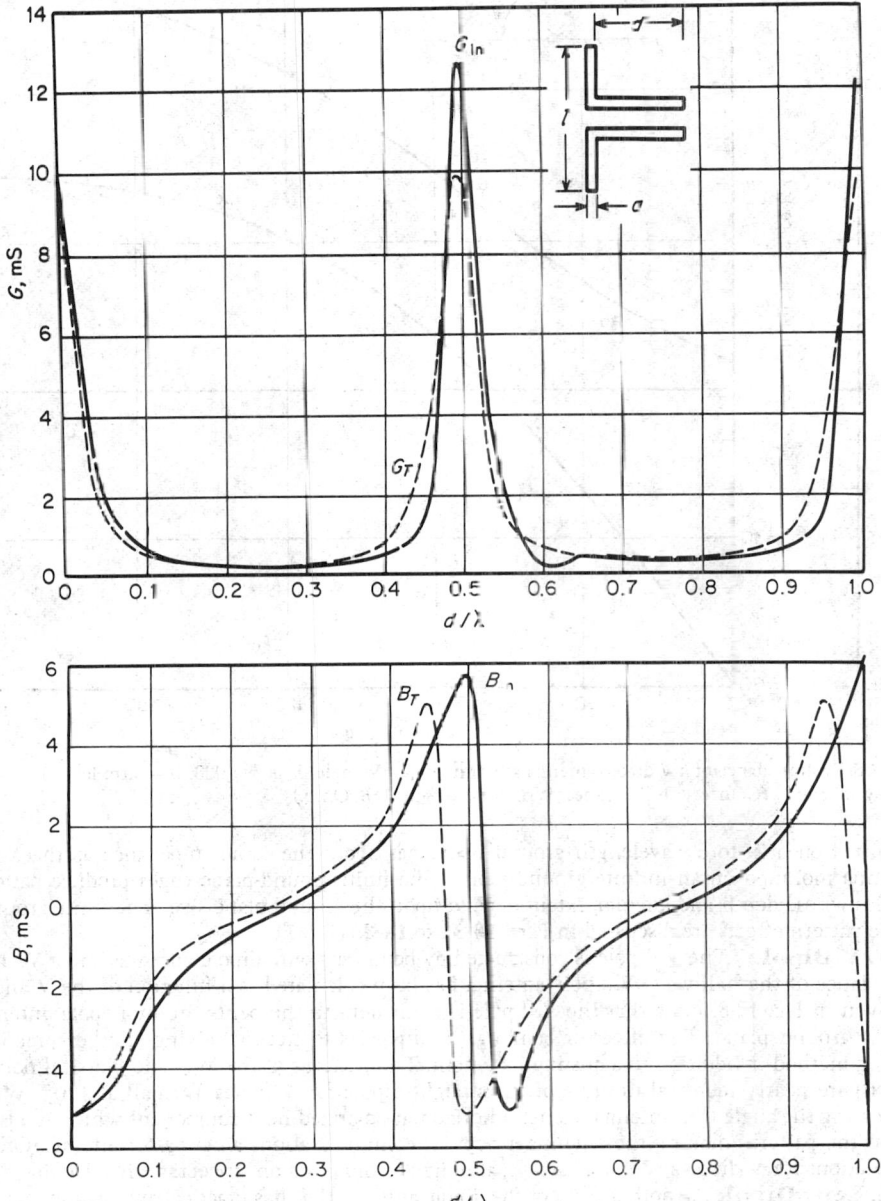

Fig. 18-7. The input admittance of a dipole with a connecting two-wire transmission line, where $Y_T = G_T + jB_T$ is the value using ordinary transmission-line theory and $Y_{in} = G_{in} + B_{in}$ is the value obtained with the integral-equation solution. *(Courtesy of J. S. Chatterjee, Langley Research Center.)*

26. Wire Antennas over Ground Planes. The wire antenna mounted over a ground plane forms an image in the ground plane such that its pattern is that of the real antenna and the image antenna and the impedance is one-half of the impedance of the antenna and its image when fed as a physical antenna in free space. For example, the quarter-wave monopole mounted on an infinite ground plane has an impedance equal to one-half the free-space impedance of the half-wave dipole. The advantage of the ground-plane-mounted wire antenna is that the coaxial feed can be used without disrupting the driving-point impedance. In practice an antenna

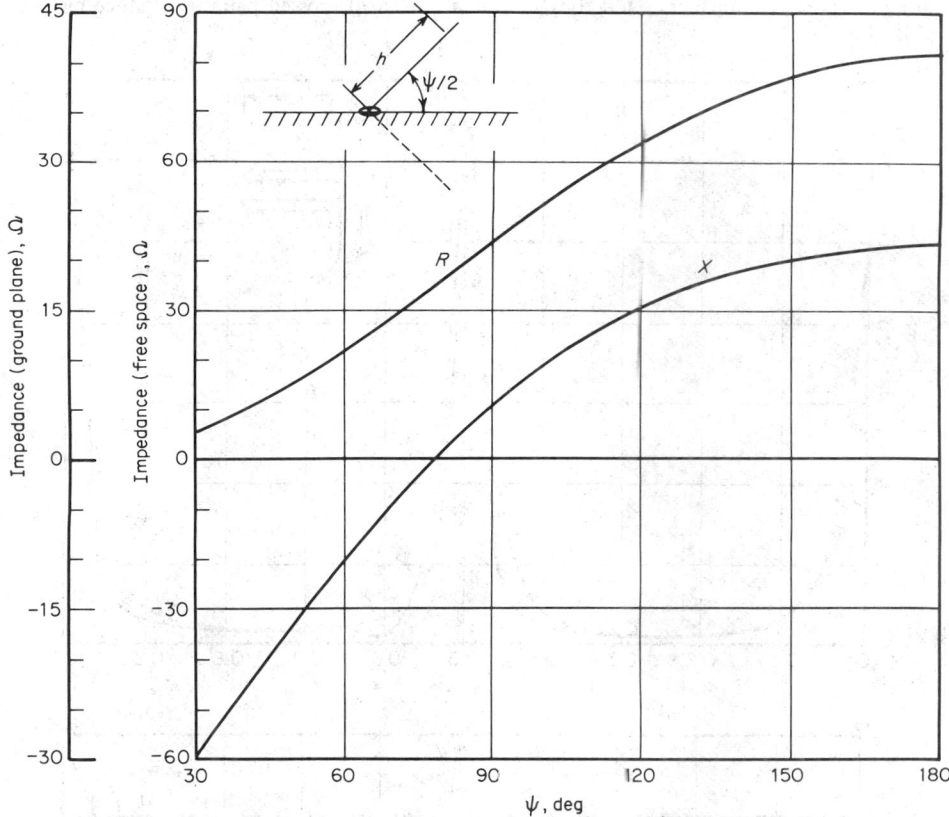

Fig. 18-8. Impedance of a V-dipole antenna as a function of V angle. $h/a = 1,000$, $h = $ arm length $= \lambda/4$, $a = $ wire radius. *(Courtesy of J. E. Jones, Wright Patterson AFB, Ohio.)*

mounted on a 2- to 3-wavelength ground plane has about the same impedance as the same antenna mounted on an infinite ground plane. The finite-ground-plane edges produce pattern ripples whose depth and angular extent depend upon the ground-plane size. The finite-ground-plane pattern effects are discussed in Pars. **18-31** to **18-33**.

27. V Dipole. The V dipole is constructed by bending a wire dipole antenna into a V. The impedance of the half-wave V-dipole antenna has been calculated as a function of the V angle, as given in Fig. 18-8. Note that the V dipole is equivalent to the bent-wire monopole antenna over a ground plane. The effect of bending the dipole is to tune it, giving another practical tuning method in addition to adjusting its length. The patterns of the V dipole (vertical polarization) are nearly identical to those of the straight dipole for ψ angles as small as 120°. With decreasing tilt angle this antenna excites a horizontal polarized field component which tends to fill in the pattern, making the antenna a popular communications antenna for aircraft. Other calculations for V dipoles, V-dipole arrays, and dipoles mounted on spacecraft are available.[26]

28. Bent Dipole. Another form of the dipole antenna that has practical application, particularly for ground-plane or airplane applications, is the bent-wire dipole formed by bending the wire 90° some distance out from the feed point. The impedance of the bent wire is given in Fig. 18-9 for both the free-space and ground-plane case. Note that this antenna can also be tuned by adjusting the lengths perpendicular and parallel to the driving point. The radiation pattern in the plane of this antenna is nearly omnidirectional for values of $H_1 \leq 0.10$, after which the pattern approaches that of the vertical half-wave dipole. Other forms of this antenna can be constructed, including loading to reduce the effective length. Indeed with these computer-analysis methods many other forms can be cheaply designed.

29. Loop Antenna. Another useful classical antenna is the loop antenna. As stated earlier, many investigators have erroneously designated this antenna as a magnetic dipole when indeed it is just another form of the wire antenna. The admittance of the loop antenna can be computed using the matrix method by approximating the loop with a polygon having the same electrical

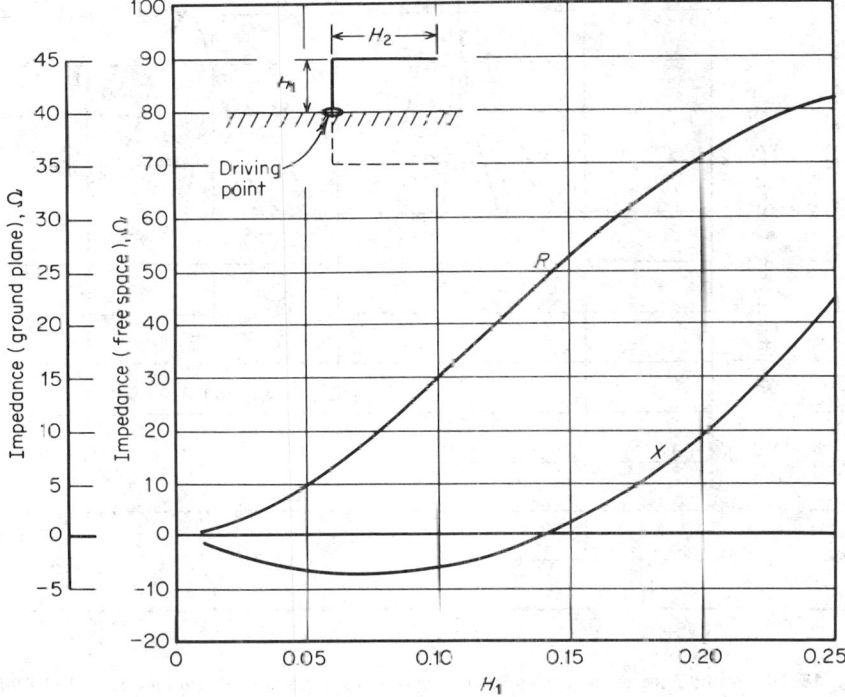

Fig. 18-9. Impedance of half-wave bent-dipole antenna, $H_1 + H_2 = \lambda/4$. (Courtesy of J. E. Jones, Wright Patterson AFB, Ohio.)

length. The admittance of a 12-sided polygon, which is identical to the admittance of a loop of the same length, has been computed and is given in Fig 18-10. These results have been verified in the ground-plane case experimentally.[32] Indeed the square loop or any other multisided loop with the same electrical-perimeter length has approximately the same admittance as the circular loop. Another method of improving the impedance of a loop is to add turns or load the loop with discrete lumped capacitances.[32,46-43] The patterns and impedance of loop antennas mounted on aircraft structures have been studied, and computer programs are available.[33]

30. Wire Antennas near Ground Planes. Although the impedance of wire antennas mounted on ground planes several wavelengths in dimension for practical purposes is similar to the impedance of the same antenna mounted on an infinite ground plane, the patterns of wire antennas on finite ground planes strongly depend upon the ground-plane size. In recent years the geometrical theory of diffraction (GTD) has been successfully applied to such problems.[49-54] These published results and the method of analysis are of great interest to antenna engineers since the geometry is the practical one of interest.

31. Loop above a Finite Ground Plane. The loop above a finite ground plane is an antenna commonly used as an array element in VHF omnirange stations located near all airports as an aircraft landing aid. Since the pattern of the small loop is symmetrical in azimuth, the elevation pattern of a loop over a finite circular ground plane can be computed using a pair of closely spaced line sources fed out of phase and located over a finite-width conducting strip.[52]

This simplified two-dimensional geometry illustrates one aspect of the GTD method and is briefly outlined here.

Consider the geometry of the line-source pair a distance d above a ground plane $2x_0$ in width, as shown in Fig. 18-11. For purposes of analysis the field-pattern space can be broken down into three distinct regions. Region I includes the fields of the source antenna that are reflected by the strip. It contains the incident, reflected, and diffracted fields. Region III is the *shadow* region,

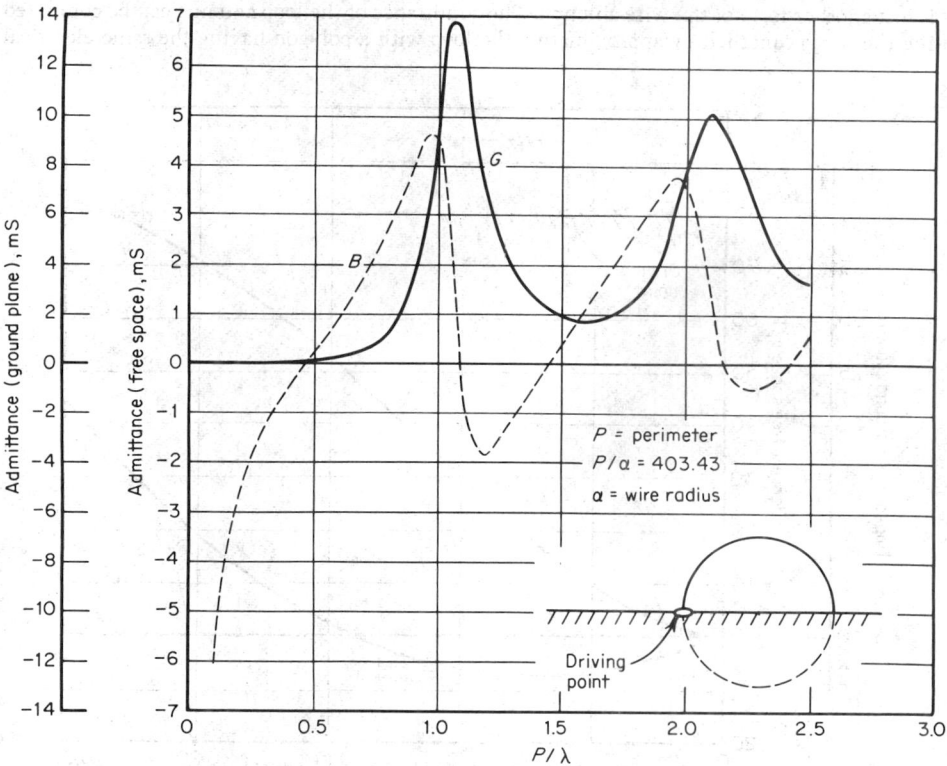

Fig. 18-10. Admittance of the loop antenna. *(Courtesy of J. E. Jones, Wright Patterson AFB, Ohio.)*

where a far-field observer can never see the source-region antenna. Energy reaches region III through diffraction of the line-source-pair fields by the edges of the finite ground plane. Region II is bounded by the incident shadow boundary (ISB) and the reflected shadow boundary (RSB); it contains both incident and diffracted fields, which together provide a smooth transition between fields in the lit and shadow regions. (RSB occurs at $\phi = \alpha$ and $\phi = \pi - \alpha$ from the law of reflection; ISB from geometrical optics occurs at $\phi = \alpha$ and $\phi = \pi + \alpha$.)

To determine the fields in these three regions first assume that the ground plane is infinite. In this case, the incident radiated field E_z^i from the line-source pair will add to its image to give

$$E_z^T = j2 \cos\phi \sin(kd \sin\phi)\, \frac{\exp(-jk\rho)}{\rho^{1/2}} \tag{18-40}$$

where

$$E_z^i = \cos\phi \exp(jkd \sin\phi)\, \frac{\exp(-jk\rho)}{\rho^{1/2}}$$

The fields diffracted from the edges of the ground plane can be determined using reported techniques.[55,51] The fields diffracted by edges 1 and 2 are denoted E_{zD_1} and E_{zD_2} and can be written as

$$E_{zD_1} = \cos\alpha\, [V_B(h, \pi - \phi - \alpha, 2) - V_B(h, \pi - \phi + \alpha, 2)] \exp(jkx_0 \cos\phi)\, \frac{\exp(-jk\rho)}{\rho^{1/2}} \tag{18-41}$$

$$E_{zD_2} = -\cos\alpha\, [V_B(h, \phi + \alpha, z) - V_B(h, \phi + \alpha, z)] \exp(-jkx_0 \cos\phi)\, \frac{\exp(-jk\rho)}{\rho^{1/2}} \tag{18-42}$$

where the function $V_B(r, \beta, 2)$ is the diffraction coefficient of a two-dimensional wedge[56-59] with the wedge angle set to zero for the thin-ground-plane case given by

$$V_B(r, \beta, z) = \pm \frac{1 + j}{2} \exp{(jkr \cos{6})} \cdot [C(z_1) - 0.5] - j[S(z_1) - 0.5]\} \qquad (18\text{-}43)$$

where $z_1 = [(2k\alpha/\pi)(1 + \cos{\beta})]^{1/2}$. From Eqs. (18-41) to (18-43) the fields in the three regions are

$$E_z^I = E_z^T + E_{zD_1} + E_{zD_2} \qquad E_z^{II} = E_z^i + E_{zD_1} + E_{zD_2} \qquad E_z^{III} = E_{zD_1} + E_{zD_2} \qquad (18\text{-}44)$$

The pattern of this antenna has a null at $\phi = 90°$ and a maximum value at some angle above the ground plane, which depends upon the spacing d and the ground-plane size $2x_0$. The value

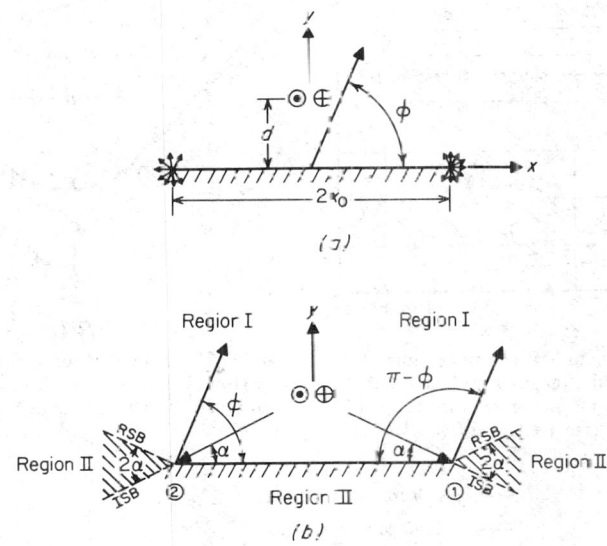

Fig. 18-11. Linear array of line sources and its diffraction mechanism: (a) two-element array above a ground plane; (b) diffraction mechanism. (*Adapted from Balanis.*[52])

of the field along the ground-plane edge ($\phi = 0$) is also of interest to the antenna designer. The variation of these field parameters is given in Fig. 18-12. For a small loop of radius r (or a large loop with the same radius r but a constant current) placed at a height d over a ground plane of dimension $2x_0$, the diffraction field will be small if the following condition is satisfied:

$$[(2x_0 + r)^2 + d^2]^{1/2} - [(2x_0 - r)^2 + d^2]^{1/2} = N\lambda$$

where N is an integer.

32. Horizontal Dipole over a Finite Ground Plane. The GTD method can be applied to the horizontal dipole over a finite ground plane,[53] the geometry of which is shown in Fig. 18-13. Also shown in Fig. 18-13 is the geometry of the same dipole placed over a cylinder. The purpose of this antenna design is to minimize the ripple or field variation in the pattern above the ground plane and simultaneously achieve a low back-lobe level. These field parameters are plotted as a function of ground-plane size and dipole spacing in Fig. 18-14. Also plotted in Fig. 18-14 are similar design curves for the dipole spaced the same parametric distances above a perfectly conducting cylinder having a diameter equal to the finite-ground-plane width. These calculations were made by programming available formulas.[60] The cylinder curvature allows one to obtain a better back-lobe level for a given pattern variation or ripple in the forward region. Experimentally, it has been determined that the rear part of the metal cylinder can be substantially removed with little effect.

33. Dipole or Monopole on a Finite Ground Plane. The radiation patterns of a half-wave dipole over a finite ground plane or the monopole mounted on a finite ground plane can be determined using GTD. These antennas have patterns which can be characterized by a null off the end of the wire, a maximum value at some angle above the ground plane, and energy

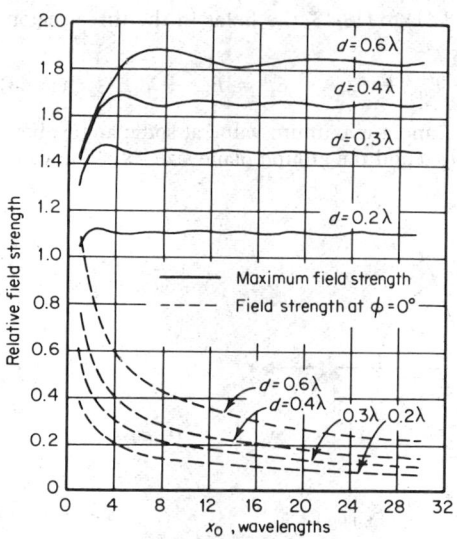

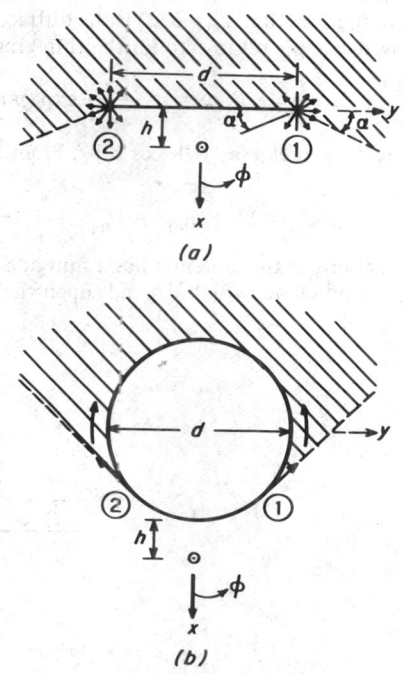

Fig. 18-12. Variations of the maximum field strength and the field strength at the angle of the ground-plane edge as a function of line-source spacing and ground-plane size. (*Adapted from Balanis.*[52])

Fig. 18-13. Radiation mechanism of dipole near finite ground plane and circular conducting cylinder: (*a*) ground plane (*b*) circular cylinder. (*From Balanis and Cockrell.*[53])

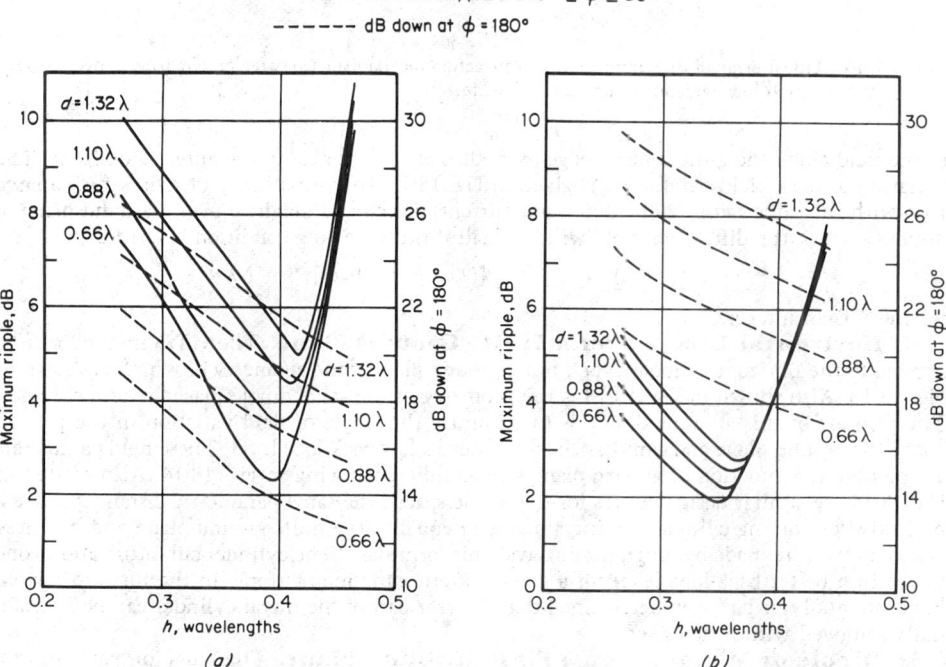

Fig. 18-14. Variations of maximum ripple in $270° \leq \phi \leq 90°$ region and radiation at $\phi = 180°$ as functions of dipole position h near (*a*) the ground plane and (*b*) circular conducting cylinder. (*From Balanis and Cockrell.*[53])

diffracted about the back of the ground plane. All these field properties, except the null, depend upon the ground-plane size and (in the case of the dipole) the spacing above the ground plane. These parameters are plotted in Fig. 18-15. Note that little is to be gained by increasing the ground-plane width beyond 5λ. For circularly symmetric ground planes a caustic point is created which can significantly increase the radiation behind the ground plane.

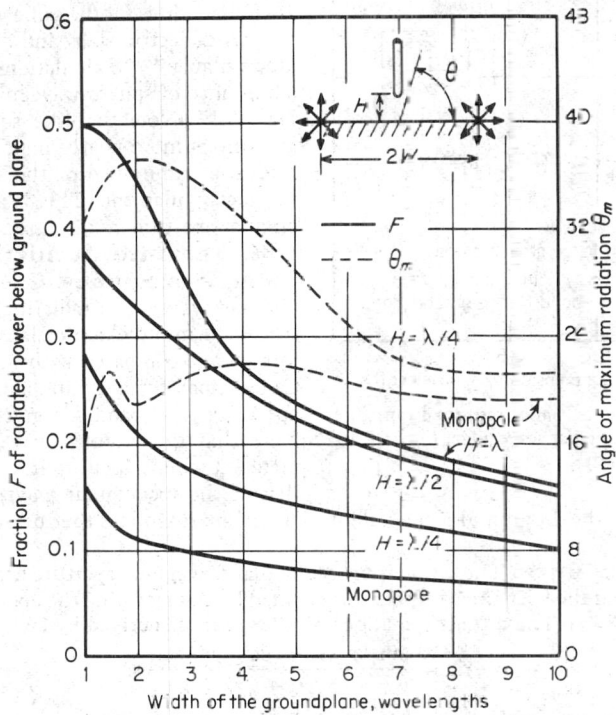

Fig. 18-15. Pattern characteristics of the half-wave dipole and the monopole over a finite ground plane. *(Courtesy of J. S. Chatterjee, Langley Research Center.)*

WAVEGUIDE ANTENNAS

34. General Considerations. The waveguide antenna, which consists of a dominant-mode-fed waveguide opening onto a conducting ground plane, is very useful for many applications such as a feed for reflector antennas or a flush-mounted antenna for aircraft or spacecraft. For flush-mounting purposes it is sometimes desirable or necessary to cover the ground plane with dielectric layers to protect the aperture from the external environment or in some instances to put dielectric plugs in the feed-waveguide section. The impedance properties of waveguide antennas have been studied extensively both theoretically and experimentally, particularly for the rectangular waveguide,[61-68] the circular waveguide,[69-71] and the coaxial waveguide or so-called annular slot.[72-75] For unloaded apertures, the assumption of the single-mode trial field in the impedance variational solution has proved adequate for practical purposes.[64,67,70,73] The impedance of these antennas is, for many practical purposes, relatively independent of ground-plane size so long as the ground plane is 2λ in dimension or greater.[64,70] However, the radiation pattern of the waveguide antenna mounted on finite ground planes is very dependent upon the ground-plane size. The effects of diffraction by ground-plane edges can be treated by the geometrical theory of diffraction (GTD)[76] in a manner similar to that used for wire antennas above a finite ground plane given in Pars. **18-30** to **18-33**.

35. Aperture Admittance of Rectangular Waveguides. Calculations and measurements of the aperture admittance of rectangular waveguides radiating into a half space (including free space) were made in the early 1950s for a variety of width-to-height ratios.[61]

Similar calculations have been made[64] assuming the dominant mode as a trial function. These calculations are compared with measured results in Fig. 18-16, where the waveguide flange was used as a ground plane. (Note that the X-band flange was 1.62 by 1.62 in and the S-band flange was 6.42 by 6.42 in.) Other measurements were made with up to 10λ ground planes with similar results, as given in Fig. 18-16. Calculations for the same waveguide under various dielectric slabs and air-gap tolerances are available.[81–83] Calculations of the aperture admittance of square waveguides are given in Fig. 18-17. Note that the square waveguide aperture is more nearly matched to the characteristic admittance of the waveguide than the rectangular one. This is also true for the circular aperture, as discussed in Par. **18-36**.

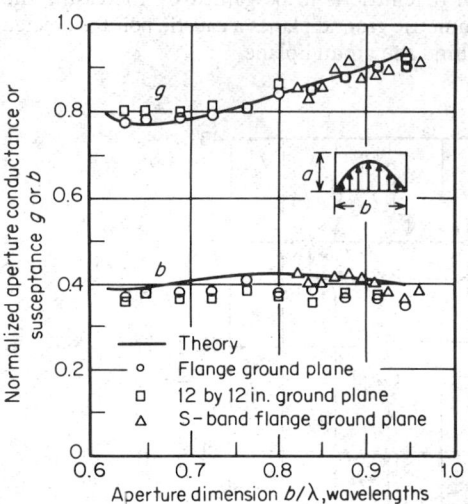

Fig. 18-16. Computed and measured aperture admittance of a rectangular waveguide, $b/a = 2.25$. (*From Croswell et al.*[64])

36. Aperture Admittance of Circular Waveguides. Calculations for the circular waveguide radiating both into free space and into dielectric slabs have been performed and compared with measurements.[69,70] Calculations for a 1.5-in-diameter waveguide operating at C band are given in Fig. 18-18. Note that the circular waveguide is nearly matched when radiating into free space. Like that of the rectangular waveguide, the aperture admittance of the 2λ ground-plane antenna closely approximates the infinite-ground-plane model.

37. Dielectric Plugs. The circular or rectangular waveguide aperture can be sealed with a dielectric plug and the aperture admittance computed by transforming the aperture admittance as given by Eq. (18-45). These transformations, similar to those derived by Swift,[66] are

$$Y_{in} = \frac{\tan kz_{mn}z_0 + jkz_{mn}y/kz'_{mn}}{y \tan kz_{mn}z_0 + jkz_{mn}/kz_{mn}} \tag{18-45}$$

where

$$y = \frac{kz_{mn}}{kz'_{mn}} \frac{Yap + j \tan kz_{mn}z_0}{1 + jYap \tan kz_{mn}z_0} \tag{18-46}$$

where Yap = admittance computed for particular waveguide, $kz_{mn} = k_0 \sqrt{\epsilon_r - (\lambda/\lambda_{cmn})^2}$, $kz'_{m'n} = k_0 \sqrt{1 - (\lambda/\lambda_{cmn})^2}$, λ = operating free-space wavelength, λ_{cmn} = cutoff wavelength for dominant waveguide mode, ϵ_r = dielectric constant of plug, $k_0 = 2\pi/\lambda$, and z_0 = thickness of plug. If the plug thickness z_0 is chosen so that $kz_{mn}z_0 = n\pi$, the input admittance equals the aperture admittance without the plug. The transformations given in Eqs. (18-45) and (18-46) assume that only the dominant waveguide mode propagates in the plug-loaded section of the waveguide. For the rectangular waveguide care should be taken with dielectric plugs since resonances associated with the TE₀₃ mode propagating in the plug-loaded section can occur.[66,84] These resonances can cause serious disruptions in both the admittance and radiation pattern; however, they are very narrow band and can usually be located by a swept-frequency VSWR measurement system. A similar effect has been found in plug-loaded horns used on reentry vehicles and waveguide-fed rectangular-cavity antennas. Such resonances have not been observed in dielectric-plug-loaded circular-waveguide antennas. This effect is similar to that found in dielectric-plug-loaded waveguide arrays.

38. Admittance of Waveguide Antennas on Cylinders. The aperture admittance of waveguide antennas mounted on curved surfaces is of interest for designing flush-mounted missile or aircraft antennas. Extensive calculations to determine the effect of cylinder curvature upon the aperture admittance of a TE₀₁-mode-fed axial slot have been performed.[85] These results clearly show that the axial slot mounted on a cylinder with a circumference $ka \geq 3$ has about the same aperture admittance as the same antenna mounted on an infinite ground plane. If the rectangular waveguide is rotated to be equivalent to the circumferential slot, it has been observed experimentally that this antenna is more sensitive to curvature since the H-plane dimension affects the waveguide wavelength.

39. Patterns of Waveguides on Finite Ground Planes. The problem of waveguide antennas on finite ground planes can also be treated by the geometrical theory of diffrac-

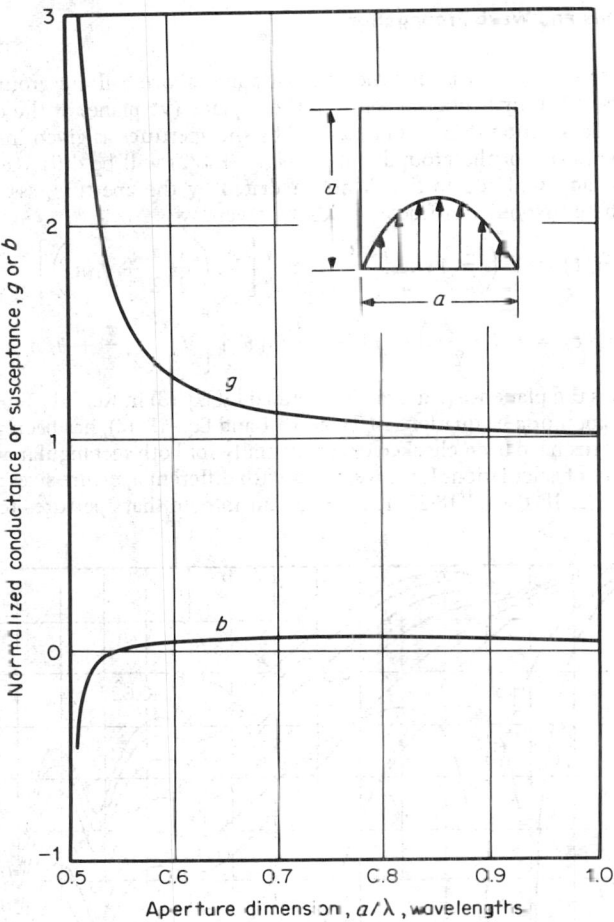

Fig. 18-17. Aperture admittance of a square waveguide.

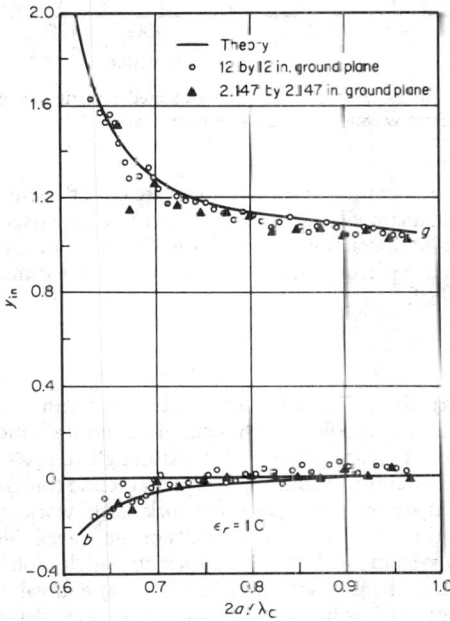

Fig. 18-18. Measured and computed aperture admittance of a circular waveguide. (*From Bailey and Swift.*[70])

tion[86] in a manner similar to that for the wire antennas above a finite ground plane. Edge or diffraction effects will primarily occur only in the E plane (yz plane) of the circular or rectangular waveguide. In addition to the field radiated by the apertures as given in Table 18-1, the E field along the boundary of the ground plane $E_\theta(\theta = +\pi/2)$ will be diffracted by the edges of the ground plane and will add to the E_θ field radiated by the aperture, assuming an infinite ground plane. These first-order diffracted fields are given by

$$E_\theta^{(1)}(\theta) = E_\theta\left(\frac{\pi}{2}\right) \exp\left(\frac{jkl}{2}\sin\theta\right)\left[V_B\left(\frac{1}{2'},\frac{\pi}{2}+\theta, n\right)\right]$$

$$E_\theta^{(2)}(\theta) = E_\theta\left(\frac{-\pi}{2}\right) \exp\left(\frac{-jkl}{2}\sin\theta\right)\left[V_B\left(\frac{1}{2'},\frac{\pi}{2}+\theta, n\right)\right]$$

(18-47)

where $V_B(r, \psi, n)$ is the plane-wave diffraction defined by Eq. (3) in Ref. 51. This solution, which is the sum of the appropriate equations in Table 18-1 and Eq. (18-47), has been programmed, and the resulting patterns have been checked experimentally for both rectangular and circular waveguides. A summary of calculations for waveguides with different aperture sizes and ground-plane sizes is given in Figs. 18-19 and 18-20. It is important to note that apertures fed by waveguides

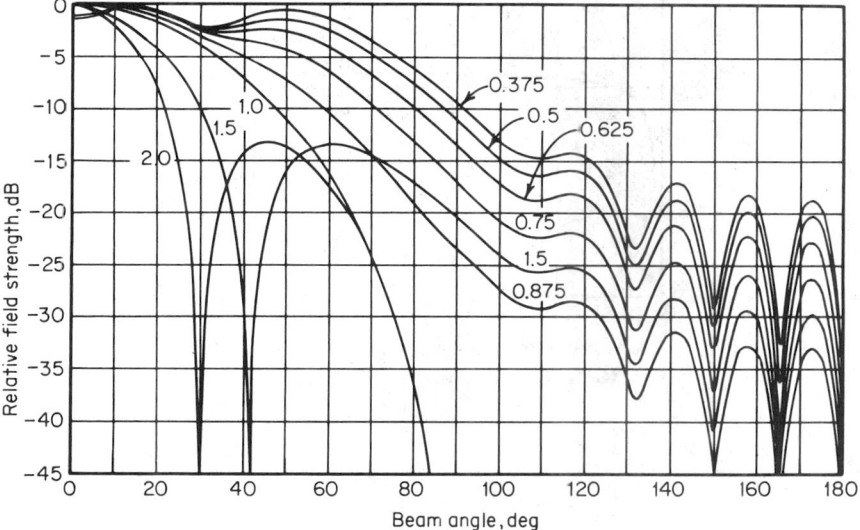

Fig. 18-19. *E*-plane radiation patterns of a TE_{01}-mode-excited rectangular aperture opening onto a finite ground plane vs. aperture size in wavelengths. Ground-plane size = 4λ.

larger than those where the next higher-order modes are cut off were included since small-angle horns preserve the dominant mode in the horn aperture that exists in the feed waveguide. It should also be noted that no diffractions occur in the E plane for a 1λ-wide rectangular aperture or a 1.22λ-diameter circular aperture. Indeed, for these aperture dimensions the E- and H-plane patterns are nearly identical.

SLOT ANTENNAS

40. General Description. The slot antenna, cut in an infinite ground plane, is the complementary antenna to the strip dipole[87] and has pattern and impedance properties which can be related to the linear antenna given in Table 18-1. Although the pattern of the thin slot on an infinite ground plane is similar for various types of slot-antenna configurations, the input impedance is highly dependent upon the type of feed network. Early work concerned the input impedance of waveguides with thin slots cut into the walls of the waveguide.[88,89] Using this work as a basis, Oliner developed a systematic design procedure for single slots in the wall of the rectangular waveguide in the form of equivalent circuits.[90,91] These equivalent circuits have been modified to account for a stratified medium outside the waveguide[92,93] and dielectric loading inside the feed waveguide.[94] Experimental data for the impedance of waveguides with slots are available,[91,95,96] including dielectric-slab covers[97-100] and plasma slabs.[101] Coupling between shunt slots in a waveguide has been considered, and extensive experimental data[102] are available. Simmons[103] has designed crossed slots in the broad wall of a rectangular waveguide to produce circular polarization. This very useful design has been extended to include dielectric covers[104] and has been constructed in arrays.[105] In addition to slotted waveguide antennas other very useful cavity-backed or waveguide-fed slot antennas have been reported, including design equations and experimental data. The dielectric-loaded waveguide-fed cavity antenna has been analyzed,[106] along with a dipole-fed cavity, to produce circular polarization,[107,108] and the T-bar-fed slot.[109] Very low profile slot antennas utilizing shallow cavities fed by coaxial cables have been designed for aircraft use[110] as well as strip-line-fed cavity-backed slots covered by dielectric slabs.[111]

The impedance properties of thin-slot antennas are relatively independent of ground-plane size if the ground plane is larger than about 1 to 2 wavelengths. Confirmation of this fact experimentally has been made.[112] Indeed it has been shown that the slot antenna mounted on a cylinder with $ka \geq 3$ has about the same impedance as a slot mounted on an infinite ground plane[85] and that the circumferential slot on a sphere has about the same impedance as the same slot mounted on an infinite cylinder, even for $ka \approx 1$ to 2.[113] Therefore there appears to be little

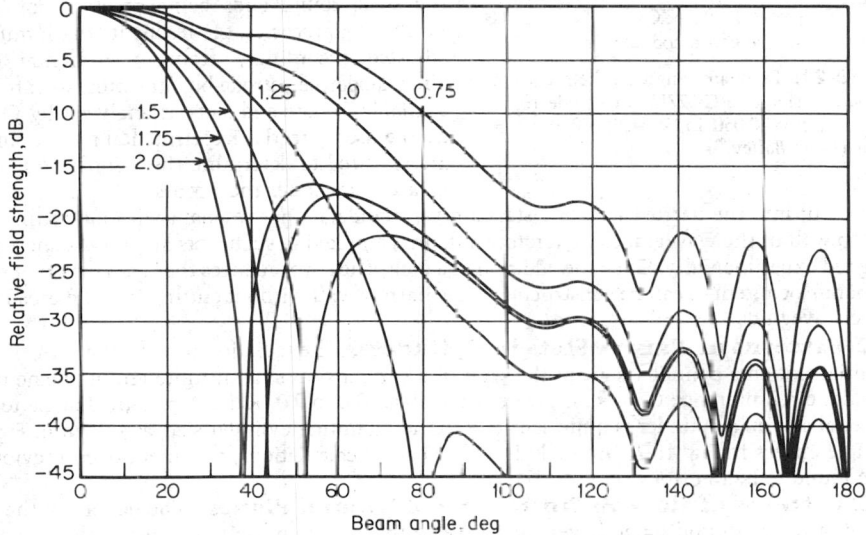

Fig. 18-20. E-plane radiation patterns of a TE_{11}-mode-excited circular aperture opening onto a finite ground plane vs. aperture size in wavelengths. Ground-plane size = 4λ.

practical justification for studying the impedance of waveguide or slot antennas mounted on geometric shapes other than the plane unless the object is smaller than several wavelengths. Thus aperture admittance is sensitive only to the local region about the aperture. This is not true of the radiation patterns of slot antennas, which are very sensitive to both the size and radius of curvature of the mounting ground plane.

The patterns of slot antennas on finite ground planes have been studied both experimentally and theoretically,[55,76,112,114] including the effects of dielectric slabs.[15-17] Some of the early work concerned patterns of slots on cylinders[113-122] with an excellent summary given by Wait[123] and by Compton and Collin.[124] One should be careful using these calculations[123] for precise agreement since some of these results have been found to be in error by several decibels in some angular regions. It is suggested that the formulations are generally correct and only require programming using modern computers. The patterns of slot antennas have been studied when mounted on spherical objects including spacecraft[4,5,125-127] and on cones.[128,129] From the geometrical theory of diffraction (GTD) the patterns of slots on elliptical cylinders[130,131] and on cylinders and three-dimensional objects of arbitrary convex cross section have been determined.[132]

41. Narrow Slots and Covers in Rectangular Waveguides. The impedance properties of the slot in the broad wall of a waveguide can be obtained by modifying Oliner's equations[90,91] to include the addition of a stratified medium outside.[92,93] The power radiated by the slot in the broad wall of the waveguide is related to the orientation of the slot and the slot length. For maximum coupling the slot must be resonant, e.g., the input susceptance is zero. The general effect of adding a dielectric cover to a slot resonant in free space is to lower the resonant frequency. As the thickness of the dielectric is increased, the change in resonant frequency and conductance deviates about a central value. This central value, as it turns out, is the case of a semi-infinite thickness of dielectric material. For dielectric layers thicker than about $0.2\lambda\epsilon$ the admittance and resonant length are within 5% of the value for a semi-infinite medium. The resonant length of a shunt slot as a function of the dielectric constant of an external semi-infinite medium is given in Fig. 18-21 along with approximate values based upon empirical and quasi-static approximation.[97]

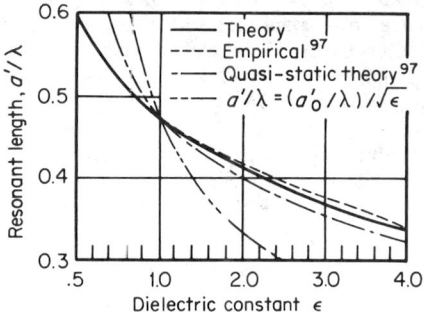

Fig. 18-21. Resonant length of dielectric-covered shunt slot in RG-52/U waveguide (b' = 0.0625 in, t = 0.050 in, a = 0.900 in, b = 0.400). (*From Bailey.*[92])

For design purposes, the slot length should be chosen using Fig. 18-21 or computed for a particular slab thickness using modifications to Oliner's formulas by Bailey.[92] If round ends are used for the slots, round ends being practical for some machining processes, the resonant length must be adjusted accordingly from experimental data. Since modern electrode-burning processes are now available, square-end slots as analyzed by Oliner can be used directly. Experimental measurements are required to determine the resonant length to an accuracy better than $\pm 5\%$.

Slots cut into the narrow wall of a standard waveguide are not resonant without cutting into the top wall of the waveguide.[95] Therefore antennas employing such slots are, for the most part, designed experimentally. Since the addition of a dielectric cover reduces the resonant-slot length, resonant-slot antennas can be constructed in the narrow wall without cutting into the broad wall of the waveguide.

42. Patterns of Narrow Slots in Cylinders. The radiation pattern of a thin circumferential slot on a cylinder is about the same as a similar slot on an infinite ground plane if the cylinder circumference $C = ka$ is greater than about 8.0 to 9.0 (Ref. 60, p. 250). The pattern of the axial slot on a cylinder is quite sensitive to the mounting cylinder size, as shown in Fig. 18-22. The curves in Fig. 18-22 are included since earlier calculations,[123] as mentioned previously, were found to be in error.

43. Patterns of Slots on Finite, Coated Ground Planes. The pattern of the thin slot on a finite ground plane is very sensitive to the ground-plane size and the thickness and dielectric properties of the coating.[115] With the coating, energy is coupled out at the slot onto a surface wave which propagates to the truncated edges and then radiates. This edge radiation interferes with the direct radiation from the slot and produces a ripple structure in the pattern.

HORN ANTENNAS

44. Horn Antennas. The horn antenna may be thought of as a natural extension of the dominant-mode waveguide feeding the horn in a manner similar to the wire antenna, which is a natural extension to the two-wire transmission line. The most common type of horns are the E-plane sectoral, H-plane sectoral, and pyramidal horn, formed by expanding the walls of the TE_{01}-mode-fed rectangular waveguide or the conical horn formed by expanding the wall of the TE_{11}-mode-fed circular waveguide. Early work concerned the determination of the forward radiation patterns, directivity, and approximate impedance of sectoral, pyramidal, and conical horns, including comprehensive experimental studies.[133-142] An excellent summary of this work is given by Compton and Collin.[143] In later work, the input impedance, wide-angle side lobes, and back lobes of certain horn antennas have been determined to a high precision using the GTD.[144-148]

Special horn types, such as the multimode horn,[148-152] diagonal horn,[153] and corrugated horn,[152,154,155] have the primary feature of reducing E-plane side lobes to levels similar to that in the H-plane of the corresponding horn. This improvement in E-plane side-lobe level reduces the on-axis gain but results in a remarkable improvement in the beam efficiency. Horns modified with plates,[156] chokes,[157] grilles,[158] pins,[159] and wires[160] produce pattern shaping, including reduction of E-plane side lobes. Dielectric-loaded waveguides and horns offer improved pattern per-

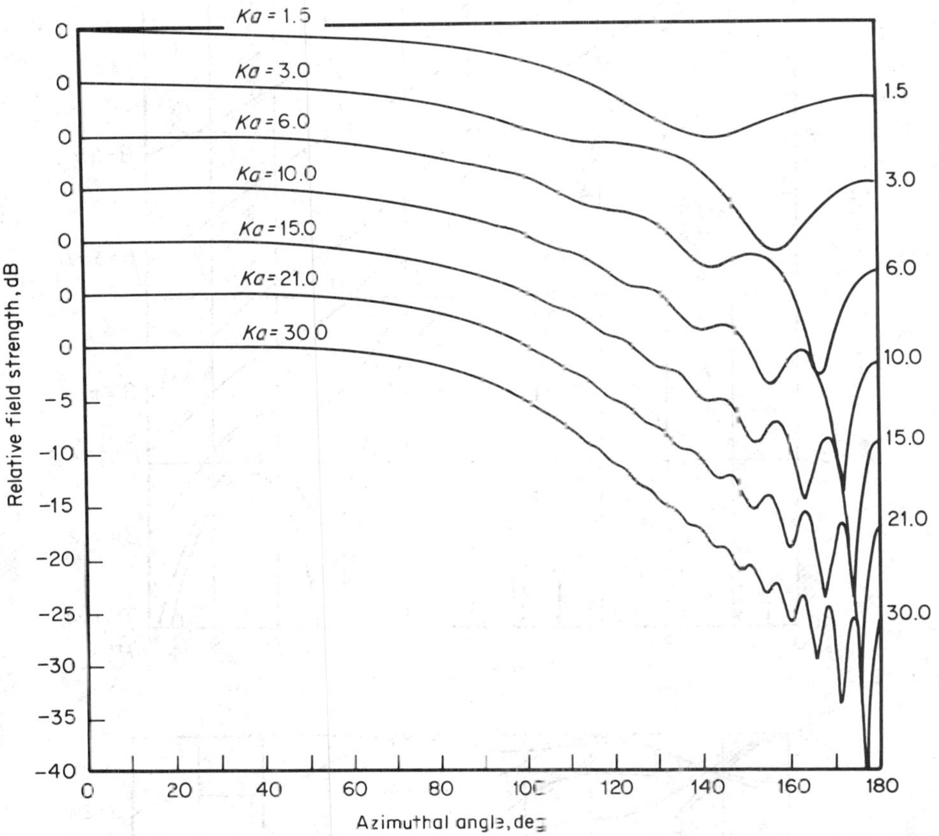

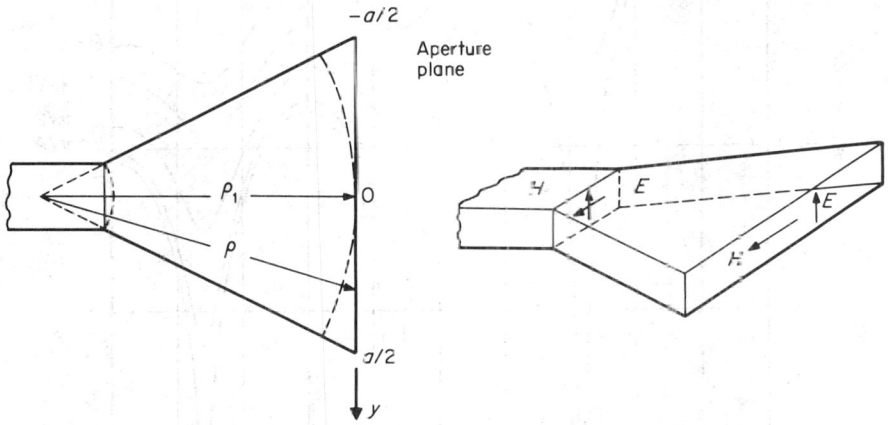

Fig. 18-22. Radiation patterns of an axial infinitesimal slot on a cylinder vs. cylinder circumference Ka in wavelengths. Note that the vertical scale for each pattern is displaced 5 dB for clarity.

Fig. 18-23. Geometry of an H-plane sectoral horn (*From Compton and Collin.*[143])

formance over unloaded horns. Ridged[167] and tapered[168] horn designs improve the bandwidth characteristics, and dual polarized configurations are also available.[169] Other useful horn designs include the horn-reflector antenna.[170] Experimental studies have determined horn-radiation properties to a great precision. The precise phase center has been determined,[171] as have the precise gain,[148,172] beam efficiency, wall losses, and input VSWR.[152]

45. Sectoral and Pyramidal Horns. Consider an H-plane sectoral horn, with the geometry given in Fig. 18-23 fed by a rectangular waveguide supporting the dominant TE_{01} mode.

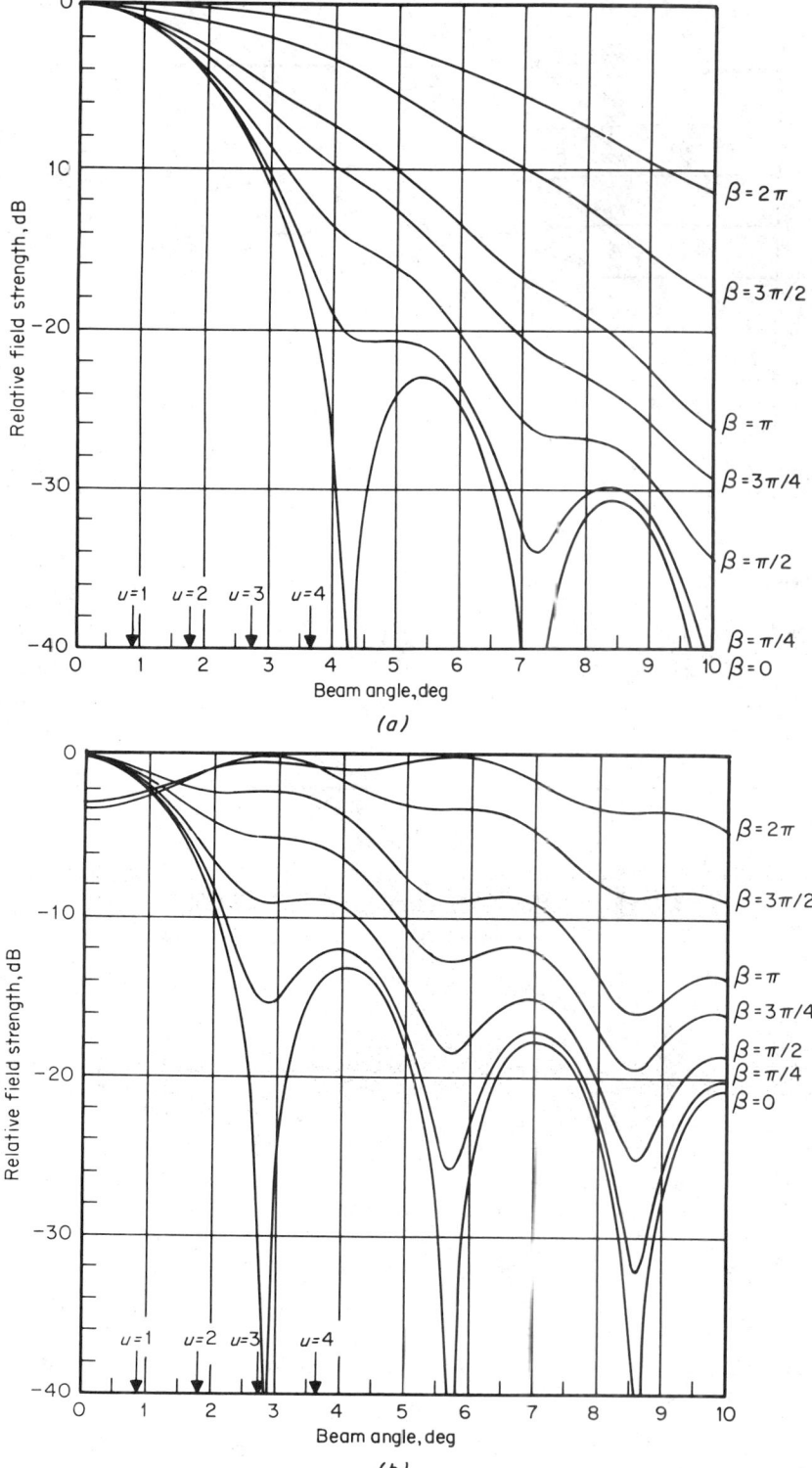

Fig. 18-24. (a) H-plane pattern of a 20λ horn with various amounts of phase taper of the form $\beta x^2/a^2$. These data can be scaled for other apertures using the parameter $u = 2\pi a/(\lambda \sin \theta)$, as noted on the beam-angle scale. (b) E-plane pattern of a 20λ aperture horn with various amounts of aperture phase taper of the form $\beta y^2/b^2$. These data can be scaled for other aperture sizes using the parameter $u = 2\pi b/(\lambda \sin \theta)$, as noted on the beam-angle scale.

The modes that can exist in this horn are the TE_{0m} modes of the form[184]

$$E_x = A \cos \frac{m\pi\phi}{\phi_0} H_{vm}^{(1)}(k_0\rho)$$

$$H_\rho = -jA \frac{m\pi}{k_0\eta_0\phi_0\rho} \sin \frac{m\pi\phi}{\phi_0} H_{vm}^{(1)}(k_0\rho) \qquad H_\phi = -j\frac{A}{\eta_0} \cos \frac{m\pi\phi}{\phi_0} H_{vm}^{(1)\prime}(k_0\rho)$$

where η_0 = free-space wave impedance, $H_{vm}'(\psi) = dH_{vm}(\psi)/d\psi$, $vm = m\pi/\phi_0$, and m = mode number. For the H-plane sectoral horn fed by a rectangular waveguide the throat dimension $\rho_0\phi_0 = a$ will support only the $m = 1$ mode in the throat horn section. If the angle of the horn is less than about 18°, the TE_{01}-dominant-fed waveguide mode will appear at the aperture without appreciable contribution from the higher-order modes. The curvature of the phase fronts in the horn section, however, means that the plane-horn aperture mouth will exhibit a phase error which will vary as $-k_0y^2/2\rho_1$ over the aperture. As a result, for the H-plane sectoral horn the aperture field will be of the form

$$E_x(x, y) = A \cos (\pi y/a)e^{-jk_0y^2/\rho}$$

If the currents on the outside of the horn aperture are neglected, this aperture field can be integrated using Eqs. (18-7) to (18-10) to obtain the far-field radiation pattern. A normalized form of these radiation patterns for the E-plane and the H-plane of sectoral and pyramidal horns greater than 2λ in aperture dimension is given in Fig. 18-24. Note that for small horns, with little phase error, the patterns of large-waveguide antennas given in Fig. 18-19 should be used.

For horns with angles the aperture may have more than one mode, and the radiation patterns given in Fig. 18-24 may be in error. The geometrical theory of diffraction can be used to compute patterns of such horn geometries as well as those with smaller plane angles.[152]

Although the E-plane sectoral horn and H-plane sectoral horn may be considered as extensions of the feed waveguide into a cylindrical region, which can then be analyzed using cylindrical modes, the pyramidal horn cannot be analyzed. For practical purposes, however, the patterns in the E and H planes of the pyramidal horn are nearly identical to those of the E- or H-plane sectoral horn with the same dimensions in corresponding planes.

The impedance of horns which have an angle less than 18° has been determined by Cockrell[173] to be the aperture impedance of a rectangular waveguide excited in the TE_{01} mode and having the same dimensions as the horn aperture. Cockrell performed calculations for H-plane sectoral horns and pyramidal horns using the equations for waveguide apertures and verified the calculations experimentally for horns radiating into free space and dielectric slabs. The input impedance of E-plane sectoral horns can be determined either similarly or using GTD.[148]

The gain of pyramidal and sectoral horns has been determined to accuracies of about ±0.2 dB even for frequencies as high as 38 GHz.[172] The most common gain standard is the pyramidal horn, which has a gain equal to[174]

$$\text{Gain}_{dB} = 10(1.008 + \log a_\lambda b_\lambda) - (L_E + L_H) \qquad (18\text{-}48)$$

where a_λ, b_λ = aperture dimensions in wavelengths and L_E, L_H = loss due to phase error in E and H planes of horn as given in Fig. 18-25. Gain curves for other horns and an excellent summary of horn-design information are given by Jakes[74] and Compton and Collin.[143]

46. Conical Horns. The conical horn formed as an extension of the circular waveguide excited in the TE_{11} mode has been thoroughly studied by King.[14] The radiation patterns of this antenna for horn angles less than about 18° can be obtained by integrating the dominant TE_{11}-mode field, with quadratic phase error, using Eqs. (18-7) to (18-10). As in rectangular-waveguide-fed horns, the fields outside the aperture are neglected, and therefore the wide-angle side lobes and back lobes will be computed incorrectly with such a procedure. The impedance of conical horns with an angle less than 18° can be obtained in a manner similar to that for pyramidal or sectoral horns. In this method the impedance is computed using the equations for circular waveguide apertures assuming that the TE_{11}-mode-excited circular waveguide has a diameter equal to the diameter of the horn aperture. Care must be used in exciting this horn since any feed asymmetry may excite the TM_{01} mode in the feed waveguide.

The gain of the conical horn has been determined to be[176]

$$\text{Gain}_{dB} = 20 \log C_\lambda - L \qquad (18\text{-}49)$$

where C_λ = circumference of horn aperture and L = gain loss due to phase error given by curve in Fig. 18-26. It should be carefully noted that the gain given in Eqs. (18-48) and (18-49) neglects the losses in the conducting walls of the entire horn and the VSWR. For many applications, such as absolute radiometric calibrations, the horn-wall losses must be accounted for and measured. Available measurement techniques for determining low-wall losses include shorting the horn aperture and measuring the input VSWR using a four-probe technique.[152]

47. Low-Side-Lobe Design. In addition to using a corrugated surface or loading of some kind, a simple small-aperture low-side-lobe antenna can be designed over a restricted bandwidth

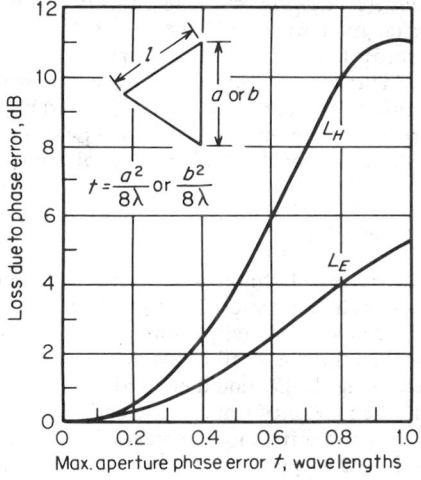

Fig. 18-25. Loss correction for phase error in sectoral and pyramidal horns. (*From Jakes.*[174])

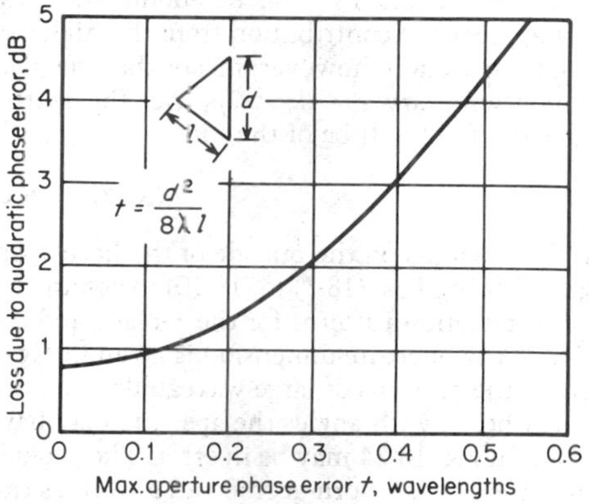

Fig. 18-26. Loss correction for phase error in conical horns. (*From Jakes.*[174])

using a particular choice of horn-aperture dimensions. Notice in Figs. 18-19 and 18-20 that the rectangular-waveguide TE_{01} mode, with a 1λ E-plane dimension, and the circular-waveguide TE_{11} mode, with a 1.22λ diameter, have no E-plane side lobes. The patterns predicted using these dimensions have been experimentally verified using a small-angle square or conical horn to feed the particular aperture. It should be noted that the E- and H-plane patterns of these particular horns are nearly identical. The possible explanation for this performance can be obtained from GTD, in that the E-plane edge diffractions from the two edges of the horn cancel precisely for the dimensions stated. Note, however, that a similar cancellation will not occur for horn dimensions which are multiples of those stated.

REFLECTOR ANTENNAS

48. The reflector antenna is formed by a radiating feed antenna and some sort of reflecting ground plane. Simple forms of reflector antennas, the loop or dipole spaced over a finite ground plane, were described in detail in Pars. **18-30** to **18-33**. A further example of such an antenna is the dipole in the corner reflector, extensive experimental design data for which have been reported.[175,176] The most common type of reflector antenna is one where the reflector is parabolic or spherical. Unusual reflector antennas include the plane-surface reflector array,[177] which can be collimated and/or scanned by adjustable switches in the elements that form the reflecting surface, and the multiplate antenna[178,179] where the reflecting surface is segmented and adjusted mechanically to produce a collimated and scannable secondary-antenna pattern.

The theory commonly used for the direct-fed parabolic antenna is that of Silver,[180] using physical optics. As outlined by Silver, this theory is adequate to predict the gain, radiation pattern, and the level of the first few side lobes of the secondary pattern, neglecting blockage and scattering by feed struts. The effect of feed-strut blockage can be estimated using geometrical optics;[181] however, an analysis of the diffraction of struts using a more rigorous formulation is necessary to improve the quantitative understanding of the problem. The radiation-pattern characteristics of the offset-fed parabola have been determined approximately using an extension of Silver's formulation to include this geometry.[182,183]

The spherical reflector is a good design for a scanning reflector antenna thanks to its geometrical symmetry; however, a point source at the focal region will not produce a set of parallel rays from the secondary reflector. To correct this spherical-aberration error a line-source feed is employed.[184,185] By phasing this line source the beam can be scanned to other positions.

Another way of feeding the parabolic reflector antenna is to use a subreflector in the focal region of the parabola and illuminate the subreflector from the parabolic surface. The principal advantages of the Cassegrain system or other methods of folded optics are the increase in effective f/d ratio and the simple mechanical location of the feed so that cooled receivers used with such antennas can be serviced in a more practical manner.

The chief disadvantage is the aperture blockage of the subreflector, which restricts the application of this principle to large apertures. The Cassegrain antenna has been analyzed extensively, and computer programs are available.[186-195] This design effort resulted in the design and construction of the 210-ft dish antenna[196] used in the worldwide space-receiving network. In order to achieve all-weather operating capability a precision 120-ft Cassegrain dish under a 150-ft radome has been constructed at Millstone Hill, Massachusetts. The radome produces about 1 to 2.8 dB loss in the microwave to millimeter-wavelength region.[197]

Besides ground-based antennas, special reflector types for use as erectable spacecraft antennas have been developed. Analysis of the gored reflector[198] is available, as is the design of a conical-reflector antenna.[199] An excellent review of reflector-antenna theory and design is available.[200]

49. Horn Feeds for Parabolic-Reflector Antennas. One of the most commonly used feeds for parabolic-reflector antennas is the flared-horn antenna fed by rectangular waveguide. As discussed in Par. **18-45**, the rectangular horn with a flare angle less than 18° has the same aperture field as the dominant-mode rectangular waveguide feeding the horn. This section describes a design procedure for determining the dimensions of a TE_0-mode-fed horn which will optimize the resultant aperture efficiency of the parabolic-reflector antenna. The design procedure and the equations for calculating the radiation-field parameters are based upon the work of Rudge and Withers,[201] with modifications.[202] The procedure does not account for tolerance errors, aperture blockage, or scattering by support struts.

The principal components of the focal-plane electric field distribution of a parabolic reflector are related to the reflector-aperture plane-field distribution by the expression

$$E(r, \phi') = \frac{jk}{2\pi} \int_0^{2\pi} \int_0^a \frac{F(u, \phi)}{(1 - u^2)^{1/2}} \exp\left[jktu \cos(\phi - \phi')\right] u \, du \, d\phi'$$

where the coordinate system is given in Fig. 18-27 and $u = \sin \theta$, $\hat{u} = \sin \theta_{max}$, $k = 2\pi/\lambda$. An example of what the focal-plane distribution from an incoming plane wave looks like for several different parabolic reflectors is given in Fig. 18-28. By adjusting the E- and H-plane widths of the rectangular feed horn to correspond to the -10-dB lobe width of the focal-plane distribution the maximum amount of energy will be collected from the parabolic reflector by the TE_{01} mode in the feed horn.

Then by definition the optimum TE_{01}-mode horn will be one whose dimensions are adjusted to fit the -10-dB levels of the focal-plane field. These optimum dimensions are

$$2\hat{x} = 0.95 \, W_u \quad E \text{ plane} \qquad 2\hat{y} = 1.29 W_u \quad H \text{ plane}$$

where W_u is the focal-plane 10-dB lobe width in wavelengths. Since the peak value of the main lobe of the focal-plane field is a function of the reflector f/d ratio due to reflector curvature, the maximum reflector-aperture efficiency, beam width, spillover, and side-lobe level are functions

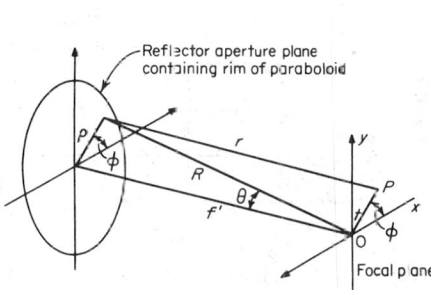

Fig. 18-27. Geometry of parabola, including focal-plane geometry. (*From Rudge and Withers.*[201])

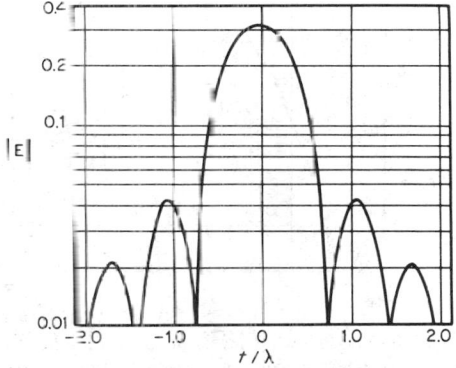

Fig. 18-28. Principal component of the focal-plane electric field distribution. $f/d = 0.50$. (*From Rudge and Withers.*[201])

of the aperture diameter and the f/d ratio. The horn dimensions for the optimized horn will result in a horn pattern whose shape is symmetrical; over the illuminated part of the dish it can be approximated by $\sin [(h \sin \theta)/h]$, where $h = 2\dot{x}\pi/\lambda$. From this approximate expression, the far-field pattern of the reflector antenna fed by this horn for small angles ψ off of boresight can be expressed as

$$E(\alpha, n) = \int_0^{\theta_{max}} \frac{\sin (h \sin \theta)}{h} J_0 \left(4\pi\alpha n \tan \frac{\theta}{2} \right) d\theta$$

where $\alpha = f/d$, $\psi = n\lambda/d$, and $d = $ diameter of parabola in wavelengths. Although this field expression is certainly approximate, it has been found sufficiently accurate for design purposes. Based upon this focal-plane concept the following simple three-step design procedure can be used:

1. Given the f/d ratio, determine $\hat{u} = \sin \theta_{max}$ from Fig. 18-29.
2. Given the f/d ratio, obtain the -10-dB focal-plane lobe width W_u, from Fig. 18-29.
3. From $\hat{u}$ determine the correction T_c due to the curvature of the dish from Fig. 18-30.

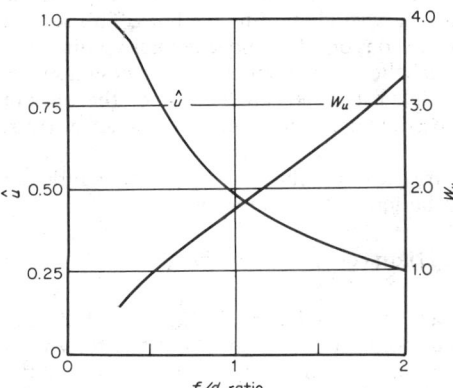

Fig. 18-29. Variation $\hat{u}$ of the angle subtended by the reflector and W_u, the 10-dB lobe-width factor, with the f/d ratio of the reflector. (*Adapted from Rudge and Withers.*[201])

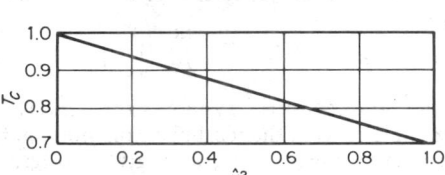

Fig. 18-30. Reflector-curvature correction factor T_c. (*From Rudge and Withers.*[201])

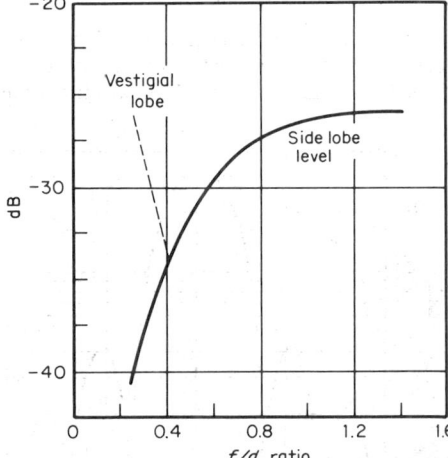

Fig. 18-31. Predicted levels of first side lobes in principal planes or radiation pattern relative to peak of main beam. Circular reflector fed by optimized rectangular horn. (*From Rudge and Withers.*[201])

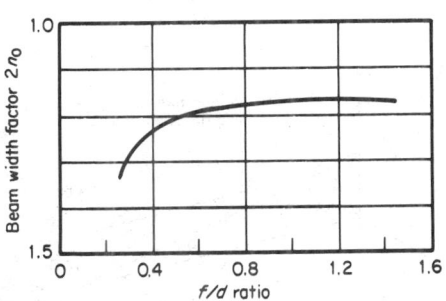

Fig. 18-32. The beam-width factor for circular reflectors. The -3-dB beam width is given by $2n_0\lambda/d$. (*Courtesy of A. W. Rudge.*)

Using the values of $\hat{u}$, W_u, and T_c thus determined. one can obtain the following parameters of interest:

Feed-horn dimensions: $2\hat{x} = 0.95 W_u$, $2\hat{y} = 1.29 W_a$

Reradiated power (not collected by the horn) $= 2.6 T_c \%$

Aperture efficiency: $N_A = 86.5 T_c \%$

Gain $= N_A \tau^2 d^2/\lambda^2 = 0.865 T_c \pi^2 d^2/\lambda^2$

Side-lobe level in decibels in Fig. 18-31

Beam width in Fig. 18-32

Note that this optimum design will not always result in a -10-dB reflector-aperture plane taper, commonly employed as a design practice. Also note that in Fig. 18-31 a distinction is made between the vestigial lobe and the first side-lobe level. For f/d ratios less than 0.4 the vestigial lobe can merge into the main lobe, since it is in phase with the main lobe, and produce a ridged pattern. For $f/d > 0.4$ the vestigial lobe will be no higher than the first side lobe. Using the focal-plane analysis, parabolic antennas with other than "optimum horns" can be evaluated.[201] Also more complex multimode feeds can be designed which collect all the significant focal-plane energy at the expense of the increased blockage; or in the case of off-axis or offset-fed parabolic reflectors the feed can be adjusted to correct for loss of gain due to coma lobes, etc.[203]

50. Random Errors. The major limitation in achieving the maximum aperture efficiency of a reflector antenna is the decrease in directivity caused by random tolerance errors in the surface of the reflector antenna due to constructive errors or thermal distortion. The directivity of a reflector antenna in the presence of random surface errors has been determined by Ruze[204] to be

$$D = \pi^2 \frac{d^2}{\lambda^2} N_A \exp\left[-\left(\frac{4\pi}{\lambda_0}\right)^2 \bar{\Delta}^2\right] \qquad (18\text{-}50)$$

where N_A is the aperture efficiency, d is the parabola diameter, and $\bar{\Delta}^2$ is the mean square mechanical distortion of the surface of the reflector in the same dimensions as the free-space wavelength λ_0.

ARRAYS

51. Radiation Patterns. The simple equations for the radiation patterns of point sources or uncoupled antennas, using pattern multiplication are outlined in Pars. **18-8** to **18-11**. The basic mutual-coupling impedance formulas for wire antennas are outlined in Par. **18-21**, which, with the proper definition of terms, can be applied to arrays of other antenna types. The basic design problem in antenna arrays, including the prediction of array radiation characteristics, is related to main-side-lobe and grating-lobe properties and input impedance. These fundamental characteristics are functions not only of the array spacing and operating wavelength but also of the array geometry and scan angle. The quantitative determination of array radiation properties can be made only by computations using an accurate analysis for a particular element type or by measurement. On a qualitative basis there are some conceptual results which are useful and give a physical insight into array design and performance. A few of these concepts will be discussed briefly here. A good summary of the work in phased arrays, including theory, methods of simulation, and design, is given in a special issue of the *Proceedings of the IEEE*.[205]

52. Grating Lobes in Planar Arrays. Consider a planar array located in the xy-plane in Fig. 18-33 with the spacing and phasing as shown. As derived by von Aulock[206]

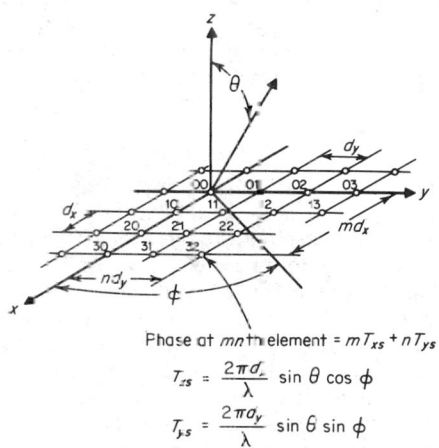

Phase at mnth element $= m T_{xs} + n T_{ys}$

$$T_{xs} = \frac{2\pi d_x}{\lambda} \sin\theta \cos\phi$$

$$T_{ys} = \frac{2\pi d_y}{\lambda} \sin\theta \sin\phi$$

Fig. 18-33. Planar-array element geometry and phasing.

the radiation pattern of a planar array can be written in terms of the direction cosines $\cos \alpha_x$ and $\cos \alpha_y$, where

$$\cos \alpha_x = \sin \theta \cos \phi \qquad \cos \alpha_y = \sin \theta \sin \phi$$

The direction of beam scan is measured by the angle ϕ counterclockwise from the $\cos \alpha_x$ axis as given by the direction cosines $\cos \alpha_{xs}$, $\cos \alpha_{ys}$, where

$$\phi = \tan^{-1}\left[(\cos \alpha_{ys})/(\cos \alpha_{xs})\right]$$

The angle of scan θ can be determined from $\sin \theta$, which is the distance of the point ($\cos \alpha_{xs}$, $\cos \alpha_{ys}$) from the origin. All physically observable grating lobes occur in the region $\cos^2 \alpha_x + \cos^2 \alpha_y \leq 1$. The array factor of a rectangular $M \times N$ array scanned to a direction given by $\cos \alpha_{xs}$, $\cos \alpha_{ys}$ can be expressed as[206]

$$E(\cos \alpha_{xs}, \cos \alpha_{ys}) = \sum_{m-0}^{M-1}\sum_{n=0}^{N-1} |A_{mn}| \exp\left\{j[m(T_x - T_{xs}) + n(T_y - T_{ys})]\right\} \qquad (18\text{-}51)$$

where
$$T_x = (2\pi/\lambda)d_x \cos \alpha_x \qquad T_y = (2\pi/\lambda)d_y \cos \alpha_y$$
$$T_{xs} = (2\pi/\lambda)d_x \cos \alpha_{xs} \qquad T_{ys} = (2\pi/\lambda)d_y \cos \alpha_{ys}$$

and A_{mn} is the excitation amplitude of the mnth element. The location of the grating lobes is highly dependent upon the array configuration. The grating lobes for a rectangular array are located at

$$\begin{aligned} \cos \alpha_{xs} - \cos \alpha_x &= \pm(\lambda/d_x)p \\ \cos \alpha_{ys} - \cos \alpha_y &= \pm(\lambda/d_y)q \end{aligned} \qquad \text{for } p, q = 0, 1, 2, \ldots \qquad (18\text{-}52)$$

where the lobe for $p = q = 0$ represents the main beam. The condition for no grating lobes for a scanned array is

$$\cos \alpha_{xs} + \cos \alpha_x = \pm(\lambda/2d_x)p \qquad \cos \alpha_{ys} - \cos \alpha_y = \pm(\lambda/2d_y)q \qquad (18\text{-}53)$$

where the lobe for $p = q = 0$ represents the main beam. The grating lobes for a triangular lattice with elements of (md_x, nd_y), where $m + n$ is even, are located at

$$\lambda/d_x = \lambda/d_y \leq 1 + \sin \theta_m \qquad (18\text{-}54)$$

where θ_m is the maximum scan angle, and for a triangular lattice array

$$\lambda/d_y = \lambda/\sqrt{3}d_x \leq 1 + \sin \theta_m \qquad (18\text{-}55)$$

Note that for linear arrays the condition given in Eq. (18-54) holds. In general for scanning arrays no grating lobes will ever exist in real space for element spacings less than or equal to $\lambda/2$. It may be observed that the triangular lattice arrays require less element population than the square array to obtain equivalent grating-lobe suppression. Other triangular grids or configurations can be obtained with even less element population[207] for the same grating-lobe performance.

53. Input Impedance as a Function of Scan Angle. As a first approximation a large array can be treated as large continuous aperture. Wheler[208] has shown that the impedance of a large aperture approximating an array may be thought of as an impedance sheet whose normalized impedance and reflection coefficient vary as

$$\eta_{\text{aperture}} = (1 - \sin^2 \theta \cos^2 \phi)/\cos \theta \qquad (18\text{-}56)$$

which results in
$$|\Gamma| = \tan^2(\theta/2) \qquad (18\text{-}57)$$

for $\phi = 0°$ or $\pi/2$. This simple impedance-sheet concept acts as an upper bound on the input-impedance variation of the central element of a large array as a function of scan angle. Perhaps this model inspired the erroneous description of the scan-angle reflection peak as related to a surface wave for uncoated arrays. An indication of the accuracy of this simple bound is given in Fig. 18-34 from dipole calculations by Allen.[209] It should be noted that the reflection peak is the result of a null in the element pattern of each element in the infinite array at the scan angle corresponding to the reflection peak. For waveguide arrays the addition of fences in between waveguides[210] will extend the scan-angle range to wider angles by filling in the reflection null in the element patterns. Placing dielectric plugs and/or sheets in and/or over waveguide arrays can have the same effect as metal fences[211] for particular choices of parameters, since higher-order modes induced in the apertures fill in the element patterns in the same manner as fences. The impedance variation as a function of scan angle, for infinite arrays, can be obtained using waveguide simulators.[212,213]

The coupling effects in finite-sized arrays of antenna elements upon the input impedance of a single element are uniquely dependent upon the detailed antenna configuration, number of elements, and the location of the element in the array. The 100% reflection peak observed for an element in an infinite array[214] at a particular scan angle is modified somewhat for finite arrays.

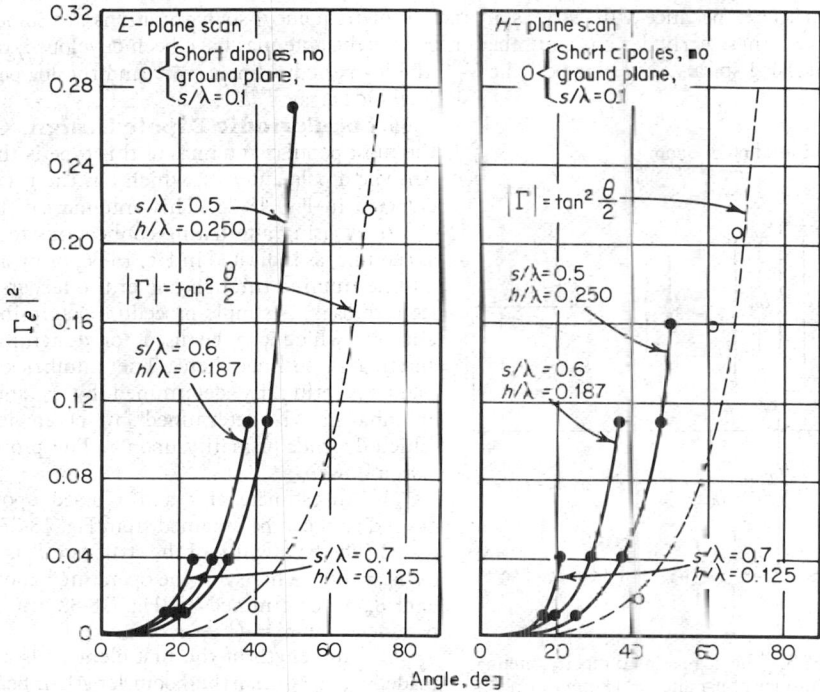

Fig. 18-34. Scanned mismatch variation for different element spacings (h/λ is the dipole spacing above a ground plane). (*Calculations by Allen.[215]*)

54. Coupling between Pairs of Antennas. Many times the antenna designer is concerned with the relatively simple problem of coupling between pairs of antennas mounted on a common ground plane. A typical waveguide-coupling problem of this type is that of two circular waveguides excited in the TE_{11} mode, as depicted in Fig. 18-35. Notice that the E-plane coupling is generally higher than H-plane coupling, a common characteristic for many ground-plane-mounted aperture antennas. The coupling between waveguide apertures is modified for a finite ground plane,[216] especially for well-isolated apertures. A very general study of coupling (isolation) between pairs of various antenna types is available, including theory, nomographs, and extensive experimental data.[217] The usefulness of such flat-ground-plane data has been enhanced since the coupling (isolation) between aperture antennas mounted on a cylinder has been shown to be about the same as that of the identical antennas mounted at the same spacing on a flat ground plane.[218]

LOG-PERIODIC ANTENNAS

55. The frequency-independent antenna is specified only by angles. It was suggested by Rumsey[219] in 1954. The simplest

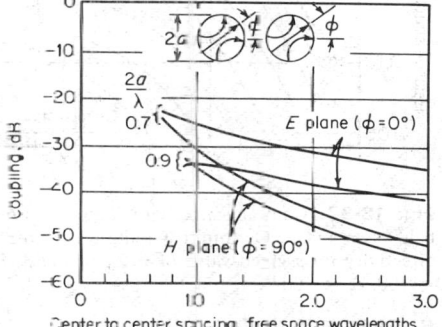

Fig. 18-35. Coupling between TE_{11} modes in two circular apertures radiating into free space.

form of such antennas is the equiangular spiral,[220] although early models with the frequency-independent idea included the tapered helix.[221,222] All antenna shapes which are completely specified by angles must extend to infinity; thus any physically realizable frequency-independent antenna has bandwidth limitations due to end effects. A simple modification of the frequency-independent antenna is the logarithmically periodic antenna,[223] whose properties vary periodically with the logarithm of the frequency. This modification tends to minimize the end effect, although the impedance will vary as a function of frequency; such variations are sometimes small. From these early designs a number of log-periodic antennas have been developed, including conical log spirals,[224] the log-periodic V,[225] the log-periodic dipole,[226-228] and the log-periodic Yagi-Uda array.[229]

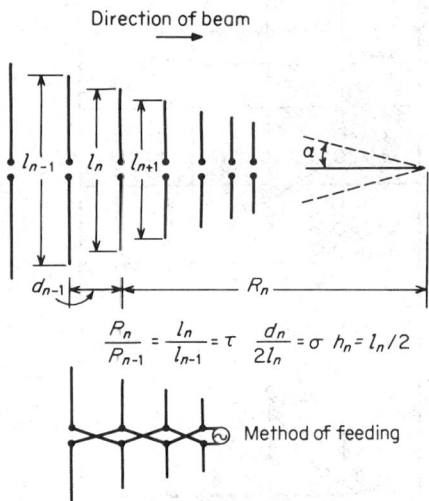

Fig. 18-36. The log-periodic dipole antenna, with definitions of parameters. (*From Carrel.[227]*)

56. Log-Periodic Dipole Design. One of the most popular antennas of this type is the log-periodic dipole antenna, which has the geometry depicted in Fig. 18-36. This antenna can be fed either by using alternating connections to a balanced line, as indicated in Fig. 18-36, or by a coaxial line running through one of the feeders from front to back. A simple procedure determined by Carrel[227] which can be used for designing this antenna is outlined here. The number of elements is primarily determined by τ, and the antenna size is determined by boom length, which depends primarily upon σ. The procedure is as follows.

1. An estimate of τ and σ based upon the desired gain can be obtained from Fig. 18-37.

2. The bandwidth of the structure B_s is given by $B_s = BB_{ar}$, where B is the operating bandwidth and B_{ar} is determined in Fig. 18-38 using the parameter $\tan \alpha = (1 - \tau)/4\sigma$.

3. The length of the first element is always made $\lambda_{max}/2$, so that the boom length L between the largest and smallest elements can be found from $L/\lambda_{max} = \frac{1}{4}(1 - 1/B_s) \cot \alpha$.

4. The number of elements required is given by $N = 1 + [(\log B_s)/\log (1/\tau)]$.

By several iterations of this design procedure a minimum boom length can be obtained. The relative feeder impedance of the design can be found using available data.[227] The design of this antenna, including patterns and impedance, can be treated by the integral-equation methods outlined in Pars. **18-22** to **18-23**. Even shorter designs of the log-periodic dipole antenna have been reported.[228]

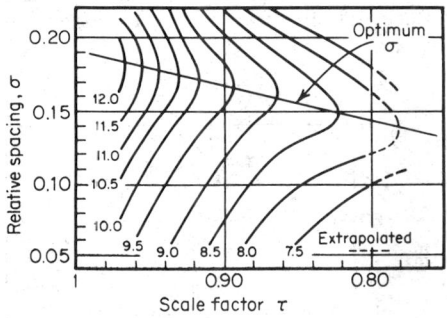

Fig. 18-37. Constant-directivity contours in decibels vs. τ and σ. Optimum σ indicates maximum directivity for a given value of τ. $Z_0 = 100 \ \Omega$, $h/a = 100$, $Z_r = 100 \ \Omega$.

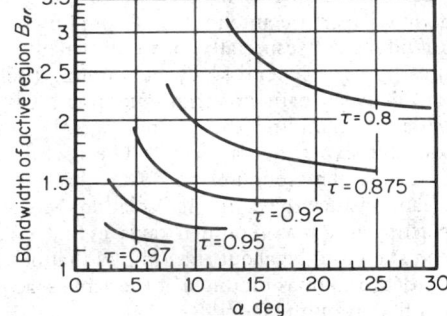

Fig. 18-38. Bandwidth of active region B_{ar} vs. α for several values of τ, for $Z_0 = 100 \ \Omega$, $h/a = 125$. (*From Carrel.[227]*)

SURFACE-WAVE ANTENNAS

57. General Description. A wide class of so-called *surface-wave antennas* has been devised, e.g., the Yagi, backfire, helix, cigar, and polyrod antenna. The surface-wave nomenclature is related to the idea that these antennas, if infinite in length, will support a wave which travels along the structure at a velocity slower than the velocity of light in free space. Data for the phase velocity along such antenna structures are available.[230-232] If the parameters of the antenna structure are chosen so that the resultant phase velocity causes the Hansen-Woodward condition to be met on the finite length of the antenna, an increased or supergain condition occurs, as discussed in Par. **18-9**. The relative phase velocity $c/v = \lambda/\lambda_z$ to maximize the gain as a function of antenna length is given in Fig. 18-39. A typical phase-velocity variation as a function of specific antenna parameters is given in Fig. 18-40 for the Yagi. Choosing particular antenna parameters so that the optimum phase-velocity conditions are met will result in a good

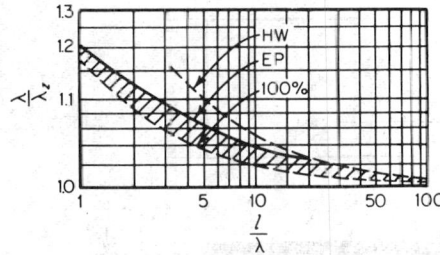

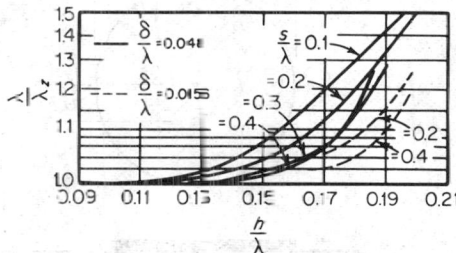

Fig. 18-39. Relative phase velocity $c/v = \lambda/\lambda_z$ for maximum-gain surface-wave antennas as a function of relative antenna length $1/\lambda$. HW = Hansen-Woodward condition; EP = Ehrenspeck-Poehler experimental values; 100% = idealized perfect excitation. (*From Zucker.*[233])

Fig. 18-40. Relative phase velocity on a Yagi antenna (data from Ehrenspeck and Poehler and Frost; see Ref. 232); δ = diameter of wire element, s = spacing between elements, and h = half length of element. (*From Zucker.*[234])

first-cut design. Improved designs require extensive parametric experimental studies where the antenna elements are varied about the initial dimensions. An estimate of how much gain can be expected from surface-wave antennas is given in Fig. 18-41. An excellent summary of surface-wave antenna design and literature has been compiled by Zucker.[233,234]

The surface-wave antenna is a misnomer since surface waves do not exist on finite antennas. Consequently, aside from the use of the general concepts mentioned above, most surface-wave antennas are designed experimentally because of the importance of the feed radiation and end effects. The finite Yagi can be analyzed as an array of dipoles, where one dipole is excited and the rest are shorted, using the wire-antenna method[235] described in Pars. **18-22** and **18-23**. Indeed, all the wire versions of surface-wave antennas can be analyzed and designed in the same manner.

58. Use of Feed Shields. Nearly all surface-wave antennas suffer from side-lobe and beam asymmetry if not fed from special launchers. For example, the helix mounted on a flat ground plane has pattern asymmetry and poor axial ratio (about 3 dB) on axis.[236] Short helices in particular have been found to exhibit this property. A way to improve this helix performance is to use a conical feed shield and to taper

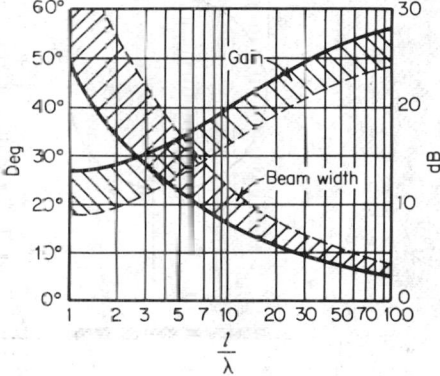

Fig. 18-41. Gain and beam width of surface-wave antenna as a function of relative antenna length $1/\lambda$. For gain (in decibels above an isotropic source) use right-hand coordinate; for beam width left-hand coordinate. Solid lines are optimum values; dashed lines are for low-side-lobe and broadband design. (*From Zucker.*[234])

the beginning and end turns as in the helicone antenna.[237] The feed shield which improves the performance of the cigar antenna is a conical horn[238] or a square cavity or bucket.[239]

MICROSTRIP ANTENNAS

59. General Considerations. In many applications where low-profile antennas are required and bandwidths less than a few percent are acceptable microstrip antennas may have the desired characteristics. Microstrip antennas are constructed on a thin dielectric sheet over a ground plane using printed-circuit-board and photoetching techniques. The most common board is dual-copper-coated Teflon-fiberglass as it allows the microstrip antenna to be curved to conform to the shape of the mounting surface. The antenna itself may be square, rectangular, round, elliptical, etc. The two most common elements (rectangular and round) are illustrated in Fig. 18-42. Circular-polarized radiation can be obtained by exciting the square or round element at two

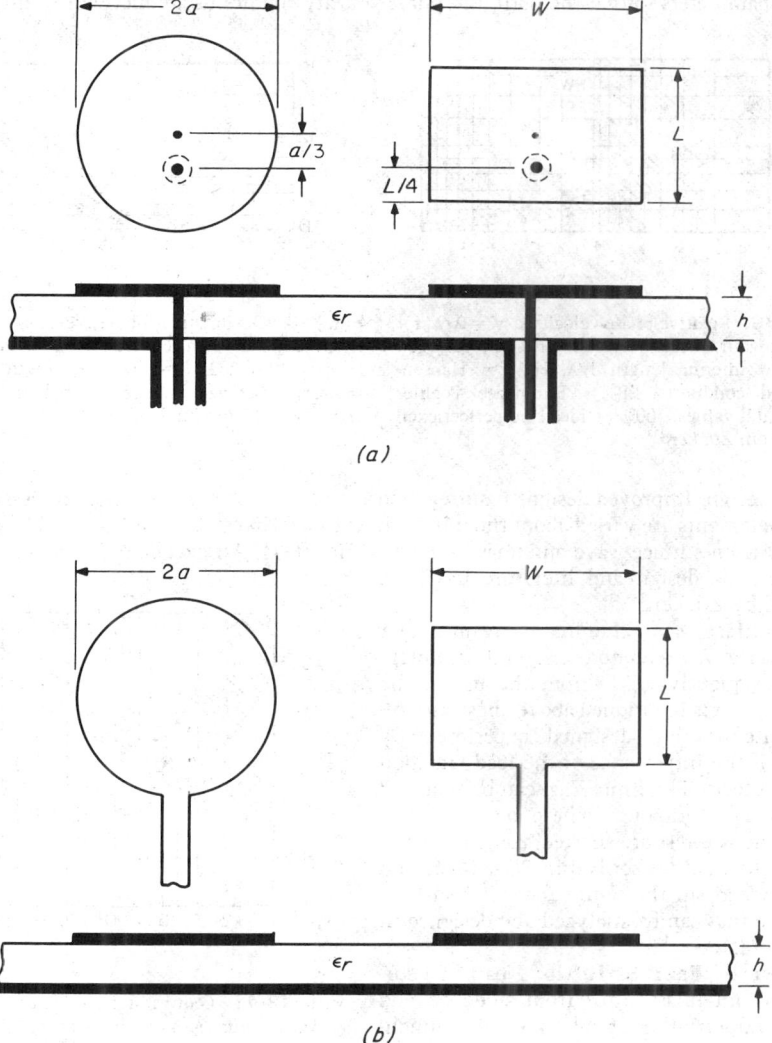

Fig. 18-42. (a) Coaxial-fed microstrip antennas; (b) line-fed microstrip antennas.

feed points 90° apart and in phase quadrature. Circular polarization can also be obtained over a limited frequency range by making the element slightly rectangular ($W/L \approx 1.03$) or slightly elliptical (eccentricity ≈ 0.2) and using a single feed point on the 45° diagonal.

60. Resonant Frequency. The frequency response for the basic elements (rectangular or round) is similar to a resonant tuned circuit or cavity. When viewed as a thin cavity, the rectangular element should be resonant when the length is equal to a half wavelength, and the round element should be resonant when the radius is equal to 0.293 wavelength; however, due to fringing fields at the edges of the element, the actual resonant size is smaller by a few percent, that is, $L_e = 0.5\lambda_e$ and $a_e = 0.293\lambda_e$, where the effective length L_e and effective radius a_e are given by

$$L_e = L + 0.824h[(\epsilon_e + 0.3)/(\epsilon_e - 0.258)][(\alpha L + 0.262h)/(\alpha L + 0.813h)] \quad (18\text{-}58)$$
$$a_e = c[1 + 2h(\pi a\epsilon_r)^{-1} \ln (9.246a/h)]^{1/2} \quad (18\text{-}59)$$

where $\alpha = W/L$, and

$$\epsilon_e = 0.5(\epsilon_r + 1) + 0.5(\epsilon_r - 1)[1 + 12h/\alpha L]^{-1/2} \quad (18\text{-}60)$$

The resonant size (for minimum VSWR) decreases for an increase in the thickness h, as shown in Fig. 18-43 for the square ($W = L$) and in Fig. 18-44 for the round element. The resonant

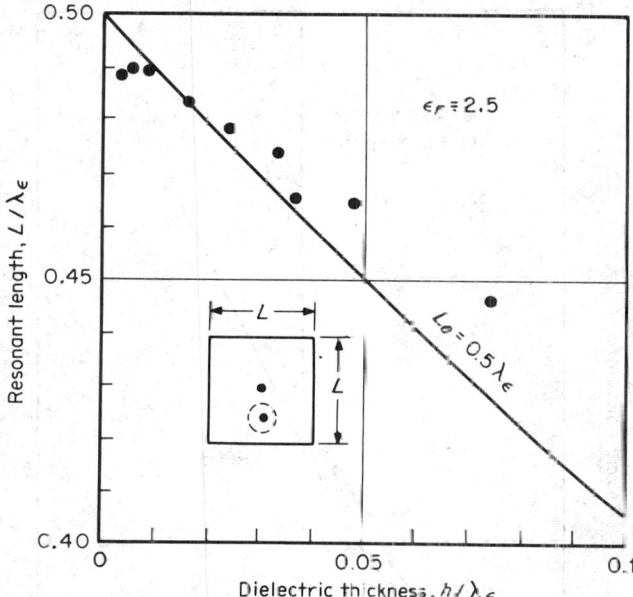

Fig. 18-43. Resonant size of square microstrip antenna.

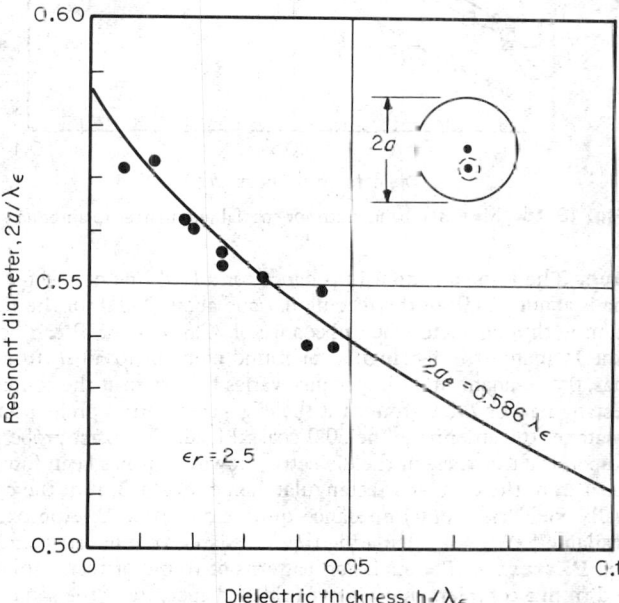

Fig. 18-44. Resonant diameter of round microstrip antenna.

length of the rectangular element also decreases with an increase in the width.[240] A TE- excited strip[241] provides a lower bound on the resonant length of rectangular elements with a large width-to-length ratio.

61. Bandwidth. The bandwidth of microstrip antennas is determined primarily by the thickness of the dielectric, increasing with an increase in thickness, as shown in Fig. 18-45 for square and round elements on Teflon-fiberglass with coaxial feed probes. The data in Fig. 18-45 can be used to select a board thickness to meet the bandwidth requirement, and the resonant size of the element is then determined for this thickness by using the effective length or radius described in Par. **18-60**.

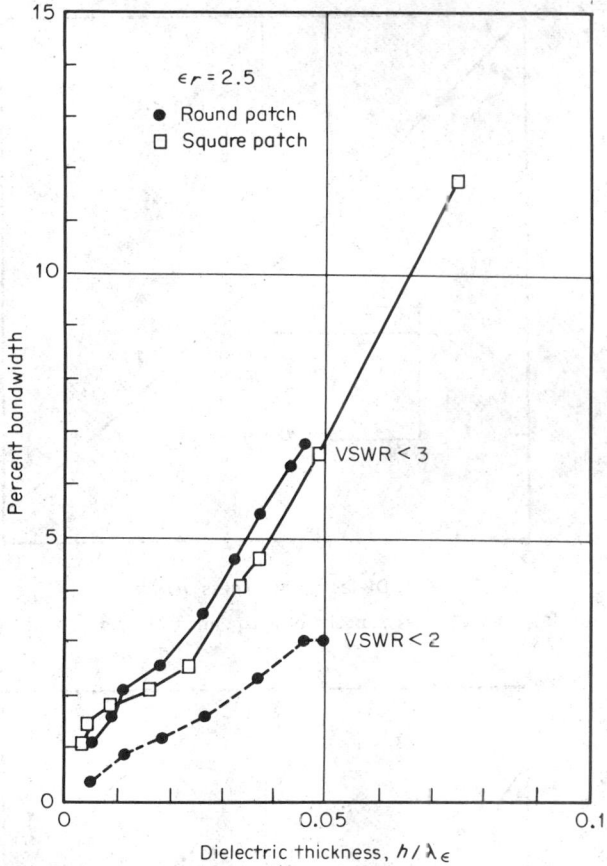

Fig. 18-45. Measured bandwidth for coaxial-fed microstrip antennas.

62. Impedance. The resonant input impedance for a feed line connected at the edge of a microstrip antenna is about 120 Ω for the rectangular and about 300 Ω for the round element. A quarter-wave section with a characteristic impedance of 77.5 or 122.5 Ω can be inserted in the microstrip feed line to match the rectangular or round element to 50 Ω. In coaxial probe-fed microstrip antennas, the resonant input impedance varies from zero at the center of the element to about 120 Ω (rectangular) or 240 Ω (round) at the edge; therefore, a probe position can always be found which matches the antenna to the 50-Ω coaxial feed. The exact probe position for a 50 Ω match depends upon the thickness of the dielectric; however, it has been found that a coaxial probe positioned $L/4$ from the edge of a rectangular element or $a/3$ from the center of a round element will usually yield an input impedance quite close to 50 Ω. Approximate analytical models are also available[240,242-245] for calculating the impedance of microstrip antennas.

63. Radiation Patterns. The radiation patterns of round or rectangular microstrip elements can be found from a complementary-waveguide-fed aperture of the same size as the effective size of the microstrip antenna[246] or from slots located around the perimeter of the element.[247]

64. References on Antennas

1. Stratton, J. A. "Electromagnetic Theory," McGraw-Hill, New York, 1941.

2. Tai, C. T. Kirchoff Theory: Scalar, Vector, or Dyadic?, *IEEE Trans. Antennas Propag.*, January 1972, Vol. AP-20, pp. 114–115.

3. Chu, T.-S. On the Use of Uniform Circular Arrays to Obtain Omnidirectional Patterns, *IEEE Trans. Antennas Propag.*, October 1959, Vol. AP-7, pp. 436–438.

4. Croswell, W. F., et al. A Dielectric-Coated Circumferential Slot Array for Omnidirectional Coverage, *IEEE Trans. Antennas Propag.*, November 1967, Vol. AP-15, pp. 722–727.

5. Croswell, W. F, and C. R. Cockrell An Omnidirectional Microwave Antenna for Use on Spacecraft, *IEEE Trans. Antennas Propag.*, July 1969, Vol. AP-17, pp. 459–466.

6. Ko, H. C. Radio-Telescope Antenna Parameters, *IEEE Trans. Antennas Propag.* December 1964, Vol. AP-12, pp. 891–897.

7. Jelly, J. V., and B. F. C. Cooper An Operational Ruby Maser for Observation at 21 cm with a 60 ft. Radius Telescope, *Rev. Sci. Instrum.*, February 1961, Vol. 32, pp. 166–175.

8. Schuster, D., et al. The Determination of Noise Temperature of Large Paraboloid Antennas, *IEEE Trans. Antennas Propag.*, May 1962, Vol. AP-10, pp. 286–291.

9. Pauling-Toth, I. I. K., et al. The Use of Paraboloid Reflector of Small Focal Ratio as a Low Noise Antenna System, *Proc. IRE*, December 1962, Vol. 50, p. 2483.

10. Jasik, J., and A. D. Bresler A Low Noise Feed System for Large Parabolic Antennas, in E. C. Jordan (ed.), "Electromagnetic Theory and Antennas," Pt. II, pp. 1167–1171, Pergamon, New York, 1968.

11. Degrase, R. W, et al. Ultra Low Noise Antenna and Receiver Combinations for Satellite and Space Communications, *Proc. Natl. Electron. Conf.*, 1959, Vol. 15, p. 370.

12. Crawford, A. B., et al. A Horn Reflector Antenna for Space Communications, *Bell Syst. Tech. J.*, 1961, Vol. 40, pp. 1095–1116.

13. Jones, S. R., and K. S. Kelleher A New Low Noise High Gain Antenna, *IEEE Int. Conv. Rec.*, 1963, Vol. 11, Pt. I, pp. 11–17.

14. Hidy, G. M., et al. Development of a Satellite Microwave Radiometer to Sense the Surface Temperature of the World's Oceans, *NASA Contract. Rep.* NASA CR-1960, February 1972.

15. Otoshi, T. Y. The Effect of Mismatched Components on Microwave Noise-Temperature Calibrations, *IEEE Trans. Microwave Theory Tech.*, September 1968, Vol. MTT-16, No. 9, pp. 675–687.

16. Hollis, J. S., et al. "Microwave Antenna Measurements," Scientific-Atlanta, Atlanta, Ga., 1970.

17. King, R. W. P. Cylindrical Antennas and Arrays, in R. E. Collin and F. J. Zucker, "Antenna Theory," Pt. I, pp. 352–420, McGraw-Hill, New York, 1969.

18. Richmond, J. H Scattering by an Arbitrary Array of Parallel Wires, *Ohio State Univ. Antenna Lab. Rep.* 1522-8, Contract N123(1953)-31663A, April 1964.

19. Richmond, J. H. Scattering by an Arbitrary Array of Wires, *IEEE Trans. Microwave Theory Tech.*, July 1965, Vol. MTT-13, pp. 408–412.

20. Richmond, J. H. A Wire-Grid Model for Scattering by Conducting Bodies, *IEEE Trans. Antennas Propag.*, November 1966, Vol. AP-14, pp. 782–786.

21. Richmond, J. H. Scattering by Imperfectly Conducting Wires, *IEEE Trans. Antennas Propag.*, November 1967, Vol. AP-15, pp. 802–806.

22. Harrington, R. F. Theory of Loaded Scatterers, *Proc. IEE*, April 1964, Vol. 111, pp. 617–628.

23. Harrington, R. F., and J. Mautz Matrix Methods for Solving Field Problems, *Syracuse Univ. Rep.* RADC TR-66-351, Vol. II, August 1966.

24. Harrington, R. F. Matrix Methods for Field Problems, *Proc. IEEE*, February 1967, Vol. 55, pp. 136–149.

25. Harrington, R. F "Field Computation by Moment Methods," Macmillan New York, 1968.

26. Richmond, J. H. Theoretical Study of V Antenna Characteristics for ATS-E Radio Astronomy Experiment, *OSU Electron. Sci. Lab. Rep.* 2619-1, Contract NAS5-11543, February 1969.

27. Richmond, J. H. Computer Analysis of Three-Dimensional Wire Antennas, *Ohio State Univ. Electrosci. Lab. Rep.* 2708-4, Contract DAAD 05-69-C-0031, December 1969.

28. Otto, D. V. A Note on the Induced E.M.F. Method of Antenna Impedance, *IEEE Trans. Antennas Propag.*, January 1969, Vol. AP-17, pp. 101–102.

29. Richmond, J. H. Coupled Linear Antennas with Skew Orientation, *IEEE Trans. Antennas Propag.*, September 1970, Vol. AP-18, pp. 694–696.

30. Richmond, J. H., and N. H. Geary Mutual Impedance between Co-planar-Skew Dipoles, *IEEE Trans. Antennas Propag.*, September 1970, Vol. AP-18, pp. 414–416.

31. Thiele, G. A. "Wire Antennas: Short Course on Computer Techniques for EM and Antennas," University of Illinois, Sept. 28–Oct. 1, 1970.

32. Richards, G. A. Reaction Formulation and Numerical Results for Multiturn Loop Antennas and Arrays, Ph.D. dissertation, Ohio State Univ., 1970.

33. High Frequency Aircraft Antennas, *Ohio State Univ. Electrosci. Lab. Final Rep.* 2235-5, May 1968.

34. Otto, D. V., and J. H. Richmond Rigorous Field Expressions for Piecewise-Sinusoidal Line Sources, *IEEE Trans. Antennas Propag.*, January 1969, Vol. AP-17, p. 98.

35. Harrington, R. F., and J. R. Mautz Straight Wires with Arbitrary Excitation and Loading, *IEEE Trans. Antennas Propag.*, July 1967, Vol. AP-15, pp. 502–515.

36. Harrington, R. F., and J. R. Mautz Electromagnetic Behavior of Circular Loops with Arbitrary Excitation and Loading, *Proc. IEE*, January 1968, Vol. 175, pp. 68–77.

37. Chao, H. H., and B. J. Strait   Computer Programs for Scattering and Radiation by Arbitrary Configurations of Bent Wires, *Syracuse Univ. Rep. 7*, Contract AF 19628-68-C-0180, September 1970.

38. Richmond, J. H., and M. H. Geary   Mutual Impedance between Coplanar Skew Dipoles, *IEEE Trans. Antennas Propag.*, May 1970, Vol. AP-18, pp. 414–416.

39. Kalafus, R. M.   Dipole Design Using the Method of Moments, *IEEE Trans. Antennas Propag.*, November 1971, Vol. AP-19, pp. 771–773.

40. Adams, A. T., and D. E. Warren   Dipole plus Parasitic Element, *IEEE Trans. Antennas Propag.*, November 1971, Vol. AP-19, pp. 536–537.

41. Miller, E. K., et al.   Accuracy-Modeling Guidelines for Integral-Equation Evaluation of Thin-Wire Scattering Structures, *IEEE Trans. Antennas Propag.*, July 1971, Vol. AP-19, pp. 534–536.

42. Bowman, D. F.   Impedance Matching and Broadbanding, in H. Jasik (ed.), "Antenna Engineering Handbook," Chap. 31, pp. 31-22 to 31-31, McGraw-Hill, New York, 1961.

43. Tai, C. T.   Characteristics of Linear Antenna Elements, in H. Jasik (ed.), "Antenna Engineering Handbook," Chap. 3, pp. 3-1 to 3-28, McGraw-Hill, New York, 1961.

44. King, H. E., and J. L. Wong   An Experimental Study of a Balun-Fed Open-Sleeve Dipole in Front of a Metal Reflector, *IEEE Trans. Antennas Propag.*, March 1972, Vol. AP-20, pp. 201–204.

45. Arnold, P. W.   A Circularly Polarized Octave-Bandwidth Unidirectional Antenna Using Conical Dipoles, *IEEE Trans. Antennas Propag.*, September 1970, Vol. AP-18, pp. 696–698.

46. Puttre, R. E.   Study of Small Omnidirectional 250-Mc Antenna for Penetrometer, *Phase I Final Rep., NASA Langley Contract NAS1-4470*, July 1965.

47. Cullen, B. D.   Experimental Evaluation of a UHF, Dual Turnstile Omnidirectional Antenna for a Penetrometer, *Phase I Suppl. Rep. NASA Langley Contract NAS1-4470*, November 1966.

48. Puttre, R. E.   Development and Performance of a Small Crossed-Loop Omnidirectional UHF Prototype Antenna for a Penetrometer, *Final Rep. Wheeler Lab., NASA Contract NAS1-4470*, January 1967.

49. Sengupta, D. L., and V. H. Weston   Investigation of the Parasitic Loop Counterpoise Antenna, *IEEE Trans. Antennas Propag.*, March 1969, Vol. AP-17, pp. 180–191.

50. Sengupta, D. L., and J. E. Ferris   On the Radiation Patterns of Parasitic Loop Counterpoise Antennas, *IEEE Trans. Antennas Propag.*, January 1970, Vol. AP-18, pp. 34–41.

51. Balanis, C. A.   Radiation Characteristics of Current Elements near a Finite-Length Cylinder, *IEEE Trans. Antennas Propag.*, May 1970, Vol. AP-18, pp. 352–359.

52. Balanis, C. A.   Analysis of an Array of Line Sources above a Finite Groundplane, *IEEE Trans. Antennas Propag.*, March 1971, Vol. AP-19, pp. 181–185.

53. Balanis, C. A., and C. R. Cockrell   Analysis and Design of Antennas for Air Traffic Collision Avoidance Systems, *IEEE Trans. Aerosp. Electron. Syst.*, September 1971, Vol. AES-7, pp. 960–967.

54. Balanis, C. A.   Radiation from Conical Surfaces Used for High-Speed Aircraft, *Radio Sci.*, February 1972, Vol. 7, pp. 339–343.

55. Balanis, C. A., and L. Peters   Equatorial Plane Pattern of an Axial-TEM Slot on a Finite Size Groundplate, *IEEE Trans. Antennas Propag.*, May 1969, Vol. AP-17, pp. 351–352.

56. Pauli, W.   On Asymptotic Series for Functions in the Theory of Diffraction of Light, *Phys. Rev.*, December 1938, Vol. 54, pp. 924–931.

57. Hutchens, D. L., and R. G. Kouyoumjian   Asymptotic Series Describing the Diffraction of a Plane Wave by a Wedge, *Ohio State Univ. Electrosci. Lab. Rep.* 2183-3, Contract AF19(628)-5929, 1966.

58. Hutchens, D. L.   Asymptotic Series Describing the Diffraction of a Plane Wave by a Two-Dimensional Wedge of Arbitrary Angle, Ph.D. dissertation, Ohio State Univ., 1967

59. Pathak, P. H., and R. G. Kouyoumjian   The Dyadic Diffraction Coefficient for a Perfectly Conducting Wedge, *Ohio State Univ. Electrosci. Lab. Rep.* 2183-4, Contract AF19(628)-5929, June 1970.

60. Harrington, R. F.   "Time-Harmonic Electromagnetic Fields," pp. 236–237, McGraw-Hill, New York, 1961.

61. Cohen, M. H., et al.   The Aperture Admittance of a Rectangular Waveguide Radiating into a Half-Space, *Ohio State Univ. Antenna Lab. Rep.* 339-22, Contract w33-038 ac 21114, November 1951.

62. Compton, R. T., Jr.   The Aperture Admittance of a Rectangular Waveguide Radiating into a Lossy Half Space, *Ohio State Univ. Antenna Lab. Rep.* 1691-1, NASA Grant NsG-448, September 1963.

63. Compton, R. T., Jr.   The Admittance of Aperture Antennas Radiating into Lossy Media, *Ohio State Univ. Antenna Lab. Rep.* 691-5, NASA Grant NsG-448, March 1964.

64. Croswell, W. F., et al.   The Admittance of a Rectangular Waveguide Radiating into a Dielectric Slab, *IEEE Trans. Antennas Propag.*, September 1967, Vol. AP-15, pp. 627–633.

65. Swift, C. T.   The Input Admittance of a Rectangular Aperture Covered with an Inhomogeneous, Lossy Dielectric Slab, *NASA Langley Tech. Note* D-4197, September 1967.

66. Swift, C. T., and D. M. Hatcher   The Input Admittance of a Rectangular Aperture Antenna Loaded with a Dielectric Plug, *NASA Langley Tech. Note* D-4430, April 1968.

67. Croswell, W. F., et al.   The Input Admittance of a Rectangular Waveguide-Fed Aperture under an Inhomogeneous Plasma: Theory and Experiment, *IEEE Trans. Antennas Propag.*, July 1968, Vol. AP-16, pp. 475–487.

68. Cockrell, C. R.   Higher-Order Mode Effects on the Aperture Admittance of a Rectangular Waveguide Covered with Dielectric and Plasma Slabs, *NASA Langley Tech. Note* D-4774, October 1968.

69. Bailey, M. C., et al.   Electromagnetic Properties of a Circular Aperture in a Dielectric Covered or Uncovered Groundplane, *NASA Langley Tech. Note* D-4752, October 1968.

70. Bailey, M. C., and C. T. Swift Input Admittance of a Circular Waveguide Aperture Covered by a Dielectic Slab, *IEEE Trans. Antennas Propag.*, July 1968, Vol. AP-16, pp. 386–391.

71. Mishustin, B. A. Radiation from the Aperture of a Circular Waveguide with an Infinite Flange, *Sov. Radiophys.*, November–December 1965, Vol. 8, pp. 852–858.

72. Levine, H., and C. H. Papas Theory of the Circular Diffraction Antenna, *J. Appl. Phys.*, January 1951, Vol. 22, pp. 29–43.

73. Hartig, E. O. Circular Apertures and Their Effect on Half-Dipole Impedances Ph.D. thesis, Harvard Univ., 1950.

74. Curtis, W. L. Calculated Values of Admittance for Annular Slot Antenna, D2-20301-1, The Boeing Co., February 1964.

75. Swift, C. T. Input Admittance of a Coaxial Transmission Line Opening onto a Flat Dielectric-Covered Groundplane, *NASA Langley Tech. Note* D-4158, September 1967.

76. Balanis, C. A. Pattern Distortion Due to Edge Diffractions, *IEEE Trans Antennas Propag.*, July 1970, Vol. AP-18, pp. 561–563.

77. Swift, C. T., and J. S. Evans Generalized Treatment of Plane Electromagnetic Waves Passing through an Isotropic Inhomogeneous Plasma Slab at Arbitrary Angles of Incidence, *NASA Langley Tech. Rep.* 172, December 1963.

78. Beck, F. B. Admittance Characteristics of a Circular Waveguide Radiating into a Homogeneous Dielectric Slab and Inhomogeneous Plasma, M.S. thesis, George Washington Univ., June 1971.

79. Collin, R. E. "Field Theory of Guided Waves," pp. 470–476, McGraw-Hill, New York, 1960.

80. Churchill, R. V. "Introduction to Complex Variables and Applications," McGraw-Hill, New York, 1948.

81. Jones, J. Earl The Admittance of a Parallel-Plate Aperture Illuminating a Displaced Dielectric-Dielectric Boundary, *NASA Langley Tech. Note* D-5083, February 1969.

82. Jones, J. Earl The Influence of Air Gap Tolerances on the Admittance of a Dielectric-Coated Slot Antenna, *IEEE Trans. Antennas Propag.*, January 1969, Vol. AP-17 pp. 63–68.

83. Gilreath, M. C. Techniques for Determining the Microwave Properties of Dielectric Materials, M.S. thesis, George Washington Univ., June 1971.

84. Swift, C. T. Admittance of a Waveguide Fed Aperture Loaded with a Dielectric Plug, *IEEE Trans. Antennas Propag.*, May 1969, Vol. AP-17, pp. 356–359.

85. Croswell, W. F., et al. Computations of the Aperture Admittance of an Axial Slot on a Dielectric Coated Cylinder, *IEEE Trans. Antennas Propag.*, January 1972, Vol AP-20, pp. 39–92.

86. Balanis, C. A. Pattern Distortion Due to Edge Diffractions, *IEEE Trans. Antennas Propag.*, July 1970, Vol. AP-18, pp. 561–563.

87. Blass, Judd Slot Antennas, in H. Jasik (ed.), "Antenna Engineering Handbook," Chap. 8, pp. 8-1 to 8-16, McGraw-Hill, New York, 1961.

88. Stevenson, A. F. Theory of Slots in Rectangular Waveguides, *J. Appl. Phys.*, 1948, Vol. 19, pp. 24–38.

89. Watson, W. H. Resonant Slots, *J. IEEE*, Session on Aerials and Waveguides, 1946, Vol. 93, Pt. IIIA, No. 1, pp. 747–777.

90. Oliner, A. A. The Impedance Properties of Narrow Radiating Slots in the Broad Face of Rectangular Waveguide, Pt. I, Theory, *IRE Trans. Antennas Propag.* January 1957, Vol. AP-5, pp. 4–11.

91. Oliner, A. A. The Impedance Properties of Narrow Radiating Slots in the Broad Face of Rectangular Waveguide. Pt. II, Comparison with Measurement, *IRE Trans. Antennas Propag.*, January 1957, Vol. AP-5, pp. 12–20.

92. Bailey, M. C. Design of Dielectric-Covered Resonant Slots in a Rectangular Waveguide, *IEEE Trans. Antennas Propag.*, September 1967, Vol. AP-15, pp. 594–598.

93. Bailey, M. C. The Properties of Dielectric-Covered Narrow Radiating Slots in the Broadface of a Rectangular Waveguide, *IEEE Trans. Antennas Propag.*, September 1970, Vol. AP-18, pp. 596–603.

94. Lawson, R. W., and V. M. Powers Slots in Dielectrically Loaded Waveguide, *Rad o Sci.*, January 1966, n.s., Vol. 1, pp. 31–35

95. Erlich, M. J. Slot Antenna Arrays, in H. Jasik (ed.), "Antenna Engineering Handbook," Chap. 9, pp. 9-1 to 9-18, McGraw-Hill, New York, 1961.

96. Maxum, B. J. Resonant Slots with Independent Control of Amplitude and Phase, *IRE Trans. Antennas Propag.*, July 1960, Vol. AP-8, pp. 384–389.

97. Croswell, W. F., and R. B. Higgins Effects of Dielectric Covers over Shunt Slots in a Waveguide, *NASA Langley Tech. Note* D2158, 1964.

98. Fratila, R., et al. Dielectric Covered Slot Antennas *WADC-OSU Radome Symp.*, Wright Air Dev. Cent. Tech. Rep. 57-314, June 1957.

99. Hanson, R. L., and G. A. Sharp Small Antenna Study, *U.S. Nav. Ordnance Lab.* NAVORD Rep. 4600, July–September 1956.

100. Croswell, W. F., and R. B. Higgins A Study of Dielectric Covered Shunt Slots in a Waveguide, *IEEE Trans. Aerosp.*, April 1964, Vol. AS-2, pp. 278–233.

101. Adams, A. T. Flush Mounted Rectangular Cavity Slot Antennas: Theory and Design, *IEEE Trans. Antennas Propag.*, May 1967, Vol. AP-15, pp. 342–351.

102. Kay, A. F., and A. J. Simmons Mutual Coupling of Shunt Slots, *IRE Trans. Antennas Propag.*, July 1960, Vol. AP-8, pp. 389–400.

103. Simmons, A. J.   Circularly Polarized Slot Radiators, *IRE Trans. Antennas Propag.*, January 1957, Vol. AP-5, pp. 31–36.

104. Bailey, M. C.   Effects of Dielectric Covers over Cross Slots in a Rectangular Waveguide, *NASA Langley Tech. Note* D-4194, October 1964.

105. Getsinger, W. J.   Elliptically Polarized Leaky-Wave Array, *IRE Trans. Antennas Propag.*, March 1962, Vol. AP-10, pp. 165–172.

106. Adams, A. T.   Flush Mounted Rectangular Cavity Slot Antennas: Theory and Design, *IEEE Trans. Antennas Propag.*, May 1967, Vol. AP-15, p. 342–351.

107. Wilkinson, E. J.   A Circularly Polarized Slot Antenna, *Microwave J.*, March 1961, pp. 97–100.

108. Cox, R. M., and W. E. Rupp   Circularly Polarized Phased Array Antenna Element, *IEEE Trans. Antennas Propag.*, November 1970, Vol. AP-18, pp. 804–807.

109. Klein, C. F.   An Equivalent Circuit for the Tee-Fed Slot Antenna, *IEEE Trans. Antennas Propag.*, March 1970, Vol. AP-18, pp. 280–282.

110. Lindberg, C. A.   A Shallow Cavity UHF Crossed-Slot Antenna, *M.I.T. Lincoln Lab. Tech. Rep.* 446, March 1968.

111. Campbell, T. G.   An Extremely Thin, Omnidirectional Microwave Antenna Array for Spacecraft Applications, *NASA Langley Tech. Note* D-5539, November 1969.

112. Frood, D. G., and J. R. Wait   Investigation of Slot Radiators in Rectangular Plates, *Proc. IEE*, January 1956, Vol. 103, pp. 103–110.

113. Swift, C. T., et al.   Radiation Characteristics of a Cavity Backed Cylindrical Gap Antenna, *IEEE Trans. Antennas and Propag.*, July 1969, Vol. AP-17, pp. 467–477.

114. Balanis, C. A.   Radiation from Slots on Cylindrical Bodies Using the Geometrical Theory of Diffraction and Creeping Wave Theory, *NASA Langley Tech. Rep.* R-331, February 1970.

115. Pathak, P., and R. G. Kouyoumjian   Private communication, June 1971.

116. Bailey, M. C., and W. F. Croswell   Pattern Measurements of Slot Radiators in Dielectric-Coated Metal Plates, *IEEE Trans. Antennas Propag.*, November 1967, Vol. AP-15, pp. 824–826.

117. Knop, C. M., and G. I. Cohn   Radiation from an Aperture in a Coated Plane, *J. Res. Natl. Bur. Stand.*, 1964, Vol. 68D, No. 4, pp. 363–378.

118. Sinclair, G.   The Patterns of Slotted-Cylinder Antennas, *Proc. IRE*, December 1948, Vol. 36, pp. 1487–1492.

119. Papas, C. H.   On the Infinitely Long Cylinder Antenna, *J. Appl. Phys.*, May 1949, Vol. 20, pp. 437–440.

120. Papas, C. H.   Radiation from a Transverse Slot in an Infinite Cylinder, *J. Math. Phys.*, January 1950, Vol. 28, pp. 227–236.

121. Silver, S., and W. K. Saunders   The External Field Produced by a Slot in an Infinite Circular Cylinder, *J. Appl. Phys.*, February 1950, Vol. 21, pp. 153–158.

122. Silver, S., and W. K. Saunders   The Radiation from a Transverse Rectangular Slot in a Circular Cylinder, *J. Appl. Phys.*, August 1950, Vol. 21, pp. 745–749.

123. Wait, J. R.   "Electromagnetic Radiation from Cylindrical Structures," Pergamon, New York, 1959.

124. Compton, R. E., Jr., and R. E. Collin   Slot Antennas, in R. E. Collin and F. J. Zucker, "Antenna Theory," Pt. I, pp. 560–620, McGraw-Hill, New York, 1969.

125. Mushiake, Y., and R. E. Webster   Radiation Characteristics with Power Gain for Slots on a Sphere, *IRE Trans. Antennas Propag.*, January 1957, Vol. AP-5, pp. 47–55.

126. Bugnolo, D. E.   Quasi-Isotropic Antenna in Microwave Spectrum, *IRE Trans. Antennas Propag.*, January 1957, Vol. AP-5, pp. 47–55.

127. Bangert, J. T., et al.   The Spacecraft Antennas, *Bell Syst. Tech. J.*, July 1963, Vol. 43, pp. 869–897.

128. Bailin, L., and S. Silver   Exterior Electromagnetic Boundary Value Problems for Spheres and Cones, *IRE Trans. Antennas Propag.*, January 1956, Vol. AP-5, pp. 6–16.

129. Pridmore-Brown, D. C., and G. E. Stewart   Radiation from Slot Antennas on Cones, *IEEE Trans. Antennas Propag.*, January 1972, Vol. AP-20, pp. 36–39.

130. Balanis, C. A., and L. Peters   Analysis of Aperture Radiation from an Axially Slotted Elliptical Conducting Cylinder Using Geometrical Theory of Diffraction, *IEEE Trans. Antennas Propag.*, July 1969, Vol. AP-17, pp. 507–513.

131. Balanis, C. A.   Radiation from Slots on Cylindrical Bodies Using Geometrical Theory at Diffraction and Creeping Wave Theory, Ph.D. thesis, Ohio State Univ., 1969.

132. Yu, C. L., and W. D. Burnside   Elevation Plane Analysis of On-Aircraft Antennas, *Ohio State Univ., Electrosci. Lab. Tech. Rep.* 3188-2, January 1972.

133. Southworth, G. C., and A. P. King   Metal Horns as Directive Receivers of Ultrashort Waves, *Proc. IRE*, 1939, Vol. 27, pp. 95–102.

134. Barrow, W. L., and L. J. Chu   Theory of the Electromagnetic Horn, *Proc. IRE*, January 1939, Vol. 27, pp. 51–64.

135. Chu, L. J., and W. L. Barrow   Electromagnetic Horn Design, *Trans. AIEE*, July 1939, Vol. 58, pp. 333–338.

136. Barrow, W. L., and F. D. Lewis   The Sectoral Electromagnetic Horn, *Proc. IRE*, January 1939, Vol. 27, pp. 41–50.

137. Chu, L. J.   Calculation of the Radiation Properties of Hollow Pipes and Horns, *J. App. Phys.*, 1940, Vol. 11, pp. 603–610.

138. Schelkunoff, S. A. "Electromagnetic Waves," Van Nostrand, New York, 1943.

139. Rhodes, D. R. An Experimental Investigation of the Radiation Properties of Electromagnetic Horn Antennas, *Proc. IRE*, September 1948, Vol. 36, pp. 1101–1105.

140. Schelkunoff, S. A., and H. T. Friis "Antennas Theory and Practice," Wiley, New York, 1952.

141. King, A. P. The Radiation Characteristics of Conical Horn Antennas, *Proc. IRE*, March 1952, Vol. 38, pp. 249–251.

142. Schorr, M. G., and J. J. Beck Electromagnetic Field of the Conical Horn, *J. App. Phys.*, August 1950, Vol. 21, pp. 795–801.

143. Compton, R. T., Jr. and R. E. Collin Open Waveguides and Small Horns, in R. E. Collin and F. J. Zucker, "Antenna Theory," Pt. I, pp. 621–655, McGraw-Hill, New York, 1969.

144. Russo, P. M., R. C. Rudduck, and L. Peters A Method for Computing E-Plane Patterns of Horn Antennas, *IEEE Trans. Antennas Propag.*, March 1965, Vol. AP-13, pp. 219–224.

145. Yu, J. S., R. C. Rudduck, and L. Peters Comprehensive Analysis for E-Plane of Horn Antennas by Edge Diffraction Theory, *IEEE Trans. Antennas Propag.*, March 1966, Vol. AP-14, pp. 138–149.

146. Yu, J. S., and R. C. Rudduck H-Plane Pattern of a Pyramidal Horn *IEEE Trans. Antennas Propag.*, September 1969, Vol. AP-17, pp. 651–652.

147. Thomas, D. T. A Half Blinder for Reducing Certain Side-Lobes in Large Horn, Reflector Antennas, *IEEE Trans. Antennas Propag.*, November 1971, Vol. AP-19, pp. 774–776.

148. Jull, E. V. Reflection from the Aperture of a Long E-Plane Sectoral Horn, *IEEE Trans. Antennas Propag.*, January 1972, Vol. AP-20, pp. 62–68.

149. Potter, P. D. A New Horn Antenna with Suppressed Sidelobes and Equal Beamwidths, *Microwave J.*, June 1963, pp. 71–78.

150. Ludwig, A. C. Radiation Pattern Synthesis for Circular Aperture Horn Antennas, *IEEE Trans. Antennas Propag.*, July 1966, Vol. AP-14, pp. 434–440.

151. Turrin, R. E. Dual Mode Small-Aperture Antenna, *IEEE Trans. Antennas Propag.*, March 1967, Vol. AP-15, pp. 307–308.

152. Caldecott, R. et al. High Performance S-Band Horn Antennas for Radiometer Use, *Ohio State Univ. Electrosci. Lab. Tech. Rep.* 3033-1, NAS 1-10040, May 1972.

153. Love, A. W. The Diagonal Horn Antenna, *Microwave J.*, March 1962, pp. 117–122.

154. Lawrie, R. E., and L. Peters Modifications of Horn Antennas for Low Sidelobes, *IEEE Trans. Antennas Propag.*, September 1966, Vol. AP-14, pp. 605–610.

155. Bohert, W. F., and L. Peters Small-Aperture Small-Flare-Angle Corrugated Horns, *IEEE Trans. Antennas Propag.*, July 1968, Vol. AP-16, pp. 494–495.

156. Peace, G. M., and E. E. Swartz Amplitude Compensated Horn Antenna *Microwave J.*, February 1964, pp. 66–68.

157. LaGrone, H. H., and G. F. Roberts Minor Lobe Suppression in a Rectangular Horn Antenna through the Utilization of a High Impedance Choke Flange, *IEEE Trans. Antennas Propag.*, January 1966, Vol. AP-14, pp. 102–104.

158. Nair, K. G., G. P. Srivastava, and S. Hariharan Sharpening of E-Plane Radiation Patterns of E-Plane Sectoral Horns by Metallic Grills, *IEEE Trans. Antennas Propag.*, January 1969, Vol. AP-17, pp. 91–93.

159. Epis, J. J. Compensated Electromagnetic Horns, *Microwave J.*, May 1961, pp. 84–89.

160. Ajoika, J. S., and H. E. Harry Shaped Beam Antenna for Earth Coverage from a Stabilized Satellite, *IEEE Trans. Antennas Propag.*, May 1970, Vol. AP-18, pp. 322–327.

161. King, H. E., J. L. Wang, and C. J. Zamites Shaped-Beam Antennas for Satellites, *IEEE Trans. Antennas Propag.*, September 1966, Vol. AP-14, pp. 641–643.

162. Tsandoulas, G. N., and W. D. Fitzgerald Aperture Efficiency Enhancement in Dielectrically Loaded Horns, *IEEE Trans. Antennas Propag.*, January 1972, Vol. AP-20, pp. 69–74.

163. Hamid, M. A. K., S. J. Towaij, and G. O. Martens A Dielectric-Loaded Circular Waveguide Antenna, *IEEE Trans. Antennas Propag.*, January 1972, Vol. AP-20, pp. 96–97.

164. Quddus, M. A., and J. P. German Phase Correction by Dielectric Slabs in Sectoral Horn Antennas, *IEEE Trans. Antennas Propag.*, July 1961, Vol. AP-9, pp. 413–415.

165. Satoh, T. Dielectric-Loaded Horn Antenna, *IEEE Trans. Antennas Propag.*, March 1972, Vol. AP-20, pp. 199–201.

166. Chatterjee, J. S., and W. F. Croswell Waveguide Excited Dielectric Spheres as Feeds, *IEEE Trans. Antennas Propag.*, March 1972, Vol. AP-20, pp. 206–208.

167. Walton, K. L., and V. C. Sundberg Broadband Ridged Horn Design, *Microwave J.*, March 1964, pp. 96–101.

168. Sengupta, D. L., and J. E. Ferris Rudimentary Horn Antenna, *IEEE Trans. Antennas Propag.*, January 1971, Vol. AP-19, pp. 124–126.

169. Wong, J. Y. A Dual Polarization Feed Horn for a Parabolic Reflection, *Microwave J.*, September 1962, pp. 188–191.

170. Crawford, A. B., H. C. Hogg, and L. E. Hunt A Horn Reflector Antenna for Space Communication, *Bell Syst. Tech. J.*, 1961, Vol. 40, p. 1095.

171. Teichman, M. Precision Phase Center Measurements of Horn Antennas, *IEEE Trans. Antennas Propag.*, November 1970, Vol. AP-18, pp. 689–690.

172. Wrixom, G. T., and W. J. Welch Gain Measurements of Standard Electromagnetic Horns in the K and K_a Bands, *IEEE Trans. Antennas Propag.*, March 1972, Vol. AP-20, pp. 136–142.

173. Cockrell, C. R.   Reflection Coefficients of Pyramidal and H-Plane Horns Radiating into Dielectric Materials, *NASA Langley Tech. Note* D-5978, September 1970.

174. Jakes, W. C.   Horn Antennas, in H. Jasik (ed.), "Antenna Engineering Handbook," pp. 10-1 to 10-18, McGraw-Hill, New York, 1961.

175. Cottony, H. V., and A. C. Wilson   Gains of Finite-Size Corner-Reflector Antennas, *IEEE Trans. Antennas Propag.*, October 1958, Vol. AP-6, pp. 366–369.

176. Wilson, A. C., and H. V. Cottony   Radiation Patterns of Finite-Size Corner-Reflector Antennas, *IEEE Trans. Antennas Propag.*, March 1960, Vol. AP-8, pp. 144–157.

177. Berry, D. G., R. G. Malech, and W. A. Kennedy   The Reflector Array Antenna, *IEEE Trans. Antennas Propag.*, November 1963, Vol. AP-11, pp. 645–651.

178. Schell, A. C., et al.   An Experimental Evaluation of Multiple Antenna Properties, *IEEE Trans. Antennas Propag.*, September 1966, Vol. AP-14, pp. 543–550.

179. Schell, A. C.   The Multiple Antenna, *IEEE Trans. Antennas Propag.*, September 1966, Vol. AP-14, pp. 550–560.

180. Silver, S.   "Microwave Antenna Theory and Design," M.I.T. Radiation Laboratory Series, Vol. 12, McGraw-Hill, New York, 1949.

181. Gray, C. Larry   Estimating the Effect of Feed Support Member Blocking on Antenna Gain and Sidelobe Level, *Microwave J.*, March 1964, pp. 88–91.

182. Ruze, J.   Lateral-Feed Displacement in a Paraboloid, *IEEE Trans Antennas Propag.*, September 1965, Vol. AP-16, pp. 660–665.

183. Pagones, M. J.   Gain Factor of an Offset-Fed Paraboloidal Reflector, *IEEE Trans. Antennas Propag.*, September 1965, Vol. AP-16, pp. 536–541.

184. Love, A. W.   Spherical Reflecting Antennas with Corrected Line Sources, *IEEE Trans. Antennas Propag.*, September 1962, Vol. AP-13, pp. 529–537.

185. Schell, A. C.   The Diffraction Theory of Large-Aperture Spherical Reflector Antennas, *IEEE Trans. Antennas Propag.*, July 1963, Vol. AP-14, pp. 428–432.

186. Potter, P. D.   The Aperture Efficiency of Large Paraboloidal Antennas as a Function of Their Feed System Radiation Characteristics, *Jet Prop. Lab. Tech. Rep.* 32-149, September 25, 1961.

187. Rusch, W. V. T.   Phase Error and Associated Cross-Polarization Effects in Cassegrainian-Fed Microwave Antennas, *Jet Prop. Lab. Tech. Rep.* 32-610, May 30, 1962.

188. Potter, P. D.   The Application of the Cassegrainian Principle to Ground Antennas for Space Communications, *Jet Prop. Lab. Tech. Rep.* 32-295, June 1962.

189. Potter, P. D.   A Simple Beamshaping Device for Cassegrainian Antennas, *Jet Prop. Lab. Tech. Rep.* 32-214, Jan. 31, 1962.

190. Potter, P. D.   A Computer Program for Machine Design of Cassegrain Feed Systems, *Jet Prop. Lab. Tech. Rep.* 32-1202, Dec. 15, 1967.

191. Rusch, W. V. T.   Edge Diffraction from Truncated Paraboloids and Hyperboloids, *Jet Prop. Lab. Tech. Rep.* 32-113, June 1, 1967.

192. Ludwig, A., and W. V. T. Rusch   Digital Computer Analysis and Design of a Subreflector of Complex Shape, *Jet Prop. Lab. Tech. Rep.* 32-1190, Nov. 15, 1967.

193. Williams, W. F.   High Efficiency Antenna Reflector, *Microwave J.*, July 1965, pp. 79–82.

194. Potter, P. D.   Application of Spherical Wave Theory to Cassegrain-Fed Paraboloids, *IEEE Trans. Antennas Propag.*, November 1967, Vol. AP-15, pp. 727–736.

195. Space Program Summary, *Jet Prop. Lab. Tech. Rep.* 37-50, January 1, 1968–March 31, 1968.

196. The Deep Space Network, *Jet Prop. Lab. Space Programs Summ.* 37-52, Vol. II, July 31, 1968, pp. 78–105.

197. Meeks, M. L., and J. Ruze   Evaluation of the Haystack Antenna and Radome, *IEEE Trans. Antennas Propag.*, November 1971, Vol. AP-19, pp. 723–728.

198. Ingerson, P. G., and W. C. Wong   The Analysis of Deployable Umbrella Parabolic Reflectors, *IEEE Trans. Antennas Propag.*, July 1972, Vol. AP-20, pp. 409–415.

199. Ludwig, A. C.   Conical-Reflector Antennas, *IEEE Trans. Antennas Propag.*, November 1972, Vol. AP-20, pp. 146–152.

200. Sengupta, D. L., and R. E. Hiatt   Reflectors and Lenses, in M. I. Skolnik (ed.), "Radar Handbook," Chap. 10, pp. 10-11 to 10-31, McGraw-Hill, New York, 1970.

201. Rudge, A. W., and M. J. Withers   Design of Flared-Horn Primary Feeds for Parabolic Reflector Antennas, *Proc. IEE*, September 1970, Vol. 117, pp. 1741–1749.

202. Rudge, A. W.   Private communication, Apr. 26, 1972.

203. Rudge, A. W., and M. J. Withers   New Technique for Beam Steering with Fixed Parabolic Reflectors, *Proc. IEE*, July 1971, Vol. 118, pp. 857–863.

204. Ruze, J.   The Effect of Aperture Errors on the Antenna Radiation Pattern, *Nuovo Cimento Suppl.*, 1952, Vol. 9, No. 3, pp. 364–380.

205. Special Issue on Electronic Scanning, *Proc. IEEE*, November 1963, Vol. 56, pp. 1761–2048.

206. Von Aulock, W. H.   Properties of Phased Arrays, *Proceedings of the IRE*, October 1960, Vol. 48, pp. 1715–1727.

207. Hsiao, J. K.   A Broadband Wide-Angle Scan Matching Technique for Large Environmentally Restricted Phased Arrays, *IEEE Trans. Antennas Propag.*, July 1972, Vol. AP-20, pp. 415–421.

208. Wheeler, H. A.   Simple Relations Derived from a Phased-Array Antenna Made of an Infinite Current Sheet, *IEEE Trans. Antennas Propag.*, July 1965, Vol. AP-13, pp. 506-514.

209. Allen J. L.   On Array Element Impedance Variation with Spacing, *IEEE Trans. Antennas Propag.*, May 1964, Vol. AP-12, p. 371.

210. Mallioux, R. J.   Surface Waves and Anomalous Wave Radiation Nulls on Phased Arrays of TEM Waveguides with Fences, *IEEE Trans. Antennas Propag.*, March 1972, Vol. AP-20, pp 160-166.

211. Amitay, N., and V. Galindo   Characteristics of Dielectric Loaded and Covered Circular Waveguide Phased Array, *IEEE Trans. Antennas Propag.*, November 1969, Vol. AP-17, pp. 722-729.

212. Hannan, P. W., and M. A. Balfour   Simulation of a Phased Array Antenna in a Waveguide, *IEEE Trans. Antennas Propag.*, May 1965. Vol. AP-13, pp. 342-353.

213. Balfour, M. A.   Phased Array Simulators in Waveguide for a Triangular Array of Elements, *IEEE Trans. Antennas Propag.*, May 1965, Vol. AP-13, pp. 475-475.

214. Amitay, N., and V. Galindo   Characteristics of Dielectric Loaded and Covered Circular Waveguide Phased Arrays, *IEEE Trans. Antennas Propag.*, November 1966, Vol. AP-17, pp. 722-729.

215. Bailey, M. C.   Analysis of Finite-Size Phased Arrays of Circular Waveguide Elements, *NASA Langley Tech. Rep.* R-408, April 1974.

216. Bailey, M. C.   Mutual Coupling between Circular Waveguide Fed Apertures in a Rectangular Ground-plane, *IEEE Trans. Antennas Propag.*, July 1974, Vol. AP-22, pp. 597-599.

217. Lyon, J. A. M., et al.   Derivation of Aerospace Antenna Coupling-Factor Interference Prediction Techniques, *Univ. Mich. Radiat. Lab. Tech. Rep.* AFAL-TR-66-57, April 1966.

218. Fante, R. L.   Calculation of the Admittance, Isolation, and Radiation Pattern of Slots on an Infinite Cylinder Covered by a Lossy Plasma, *Radio Sci.*, March 1971, Vol. 6, pp. 421-428.

219. Rumsey, V. H.   The Equiangular Spiral, *IRE Nat. Conv. Rec.*, 1957, P. I, pp. 114-118.

220. Dyson, J. D.   The Equiangular Spiral, *IRE Trans. Antennas Propag.*, April 1959, Vol. AP-7, pp. 181-187.

221. Springer, P. S.   End-Loaded and Expanding Helices as Broadband Circularly Polarized Radiators, *Proc. Natl. Electron. Conf.*, 1949, Vol. 5, pp. 161-171.

222. Chatterjee, J. S.   Radiation Characteristics of a Conical Helix of Low Pitch Angles, *J. Appl. Phys.*, March 1955, Vol. 26, pp. 331-335.

223. Duhamel, R. H., and R. E. Isbell   Broadband Logarithmically Periodic Antenna Structures, *IRE Natl. Conv. Rec.*, 1957, Pt. 1, pp. 119-128.

224. Dyson, J. D.   The Unidirectional Equiangular Spiral Antenna, *IRE Trans. Antennas Propag.*, October 1959, Vol. AP-7, pp. 329-334.

225. Mayes, P. E., and R. L. Carrel   Log-Periodic Resonant V-Arrays, *IRE West. Conv.*, 1961.

226. Isbell, D. E.   Log-Periodic Dipole Arrays, *IRE Trans. Antennas Propag.*, May 1960, Vol. AP-8, pp. 260-267.

227. Carrel, R. L.   The Design of Logarithmically Periodic Dipole Antennas, *IRE Natl. Conv. Rec.*, 1961, Pt. 1, pp. 61-75.

228. Bantin, C. C., and K. G. Balmain   Study of Compressed Log-Periodic Dipole Antennas, *IEEE Trans. Antennas Propag.*, March 1970, Vol. AP-18, pp. 195-203.

229. Barbano, N.   Log-Periodic Yagi-Uda Array, *IEEE Trans. Antennas Propag.*, March 1966, Vol. AP-14, pp. 235-238.

230. Mailloux, R. J.   The Long Yagi-Uda Array, *IEEE Trans. Antennas Propag.*, March 1966, Vol. AP-14, p. 128.

231. Collin, R. E.   "Field Theory of Guided Waves," McGraw-Hill, New York, 1960.

232. Brunstein, S. A., and R. F. Thomas   Characteristics of a Cigar Antenna, *Jet Prop. Lab. Q. Rev.*, July 1972, Vol. 1, No. 2, pp. 87-95.

233. Zucker, F. J.   Surface Wave Antennas, in R. E. Collin and F. J. Zucker, "Antenna Theory," Pt. II, Chap. 21, pp. 298-348, McGraw-Hill, New York, 1969.

234. Zucker, F. J.   Surface and Leaky-Wave Antennas, in H. Jasik (ed.), "Antenna Engineering Handbook," Chap. 16, pp. 16-1 to 16-57, McGraw-Hill, New York, 1961.

235. Thiele, G. A.   Analysis of Yagi-Uda Type Antennas, *IEEE Trans. Antennas Propag.*, January 1969, Vol. AP-17, pp. 24-31.

236. Kraus, J. D.   "Antennas," McGraw-Hill, New York, 1950.

237. Angelakos, D. J., and Kajfez Darko   Modifications on the Axial Mode Helical Antenna, *Proc. IEEE*, April 1967, Vol. 55, No. 4, pp. 558-559.

238. Carver, K. R., and B. M. Potts   Some Characteristics of the Helicone Antenna, *1970 IEEE G-AP Symp. Dig.*, pp. 142-150.

239. Croswell, W. F., and M. C. Gilreath   Erectable Yagi Disk Antennas for Space Vehicle Applications, *NASA Langley Tech. Note* D-1401, October 1962.

240. Lo, Y. T., D. Solomon, and W. F. Richards   Theory and Experiment or Microstrip Antennas, *IEEE Trans. Antennas Propag.*, March 1979, Vol. AP-27, pp. 137-145.

241. Bailey, M. C.   Resonant Frequency of Microstrip Antennas Calculated from TE-Excitation of an Infinite Strip Embedded in a Grounded Dielectric Slab, *NASA Langley Tech. Mem.* 30190, November 1979.

242. Derneryd, A. G.   Linearly Polarized Microstrip Antennas, *IEEE Trans. Antennas Propag.*, November 1976, Vol. AP-24, pp. 846-851.

243. Agrawal, P. K., and M. C. Bailey   An Analysis Technique for Microstrip Antennas, *IEEE Trans. Antennas Propag.*, November 1977, Vol. AP-25, pp. 756–759.

244. Carver, K. R.   A Modal Expansion Theory for the Microstrip Antenna, *1979 IEEE AP-S Symp. Dig.*, pp. 101–104.

245. Richards, W. F., and Y. T. Lo   An Improved Theory for Microstrip Antennas and Applications, *1979 IEEE AP-S Symp. Dig.*, pp. 113–116.

246. Bailey, M. C., and F. G. Parks   Design of Microstrip Disk Antenna Arrays, *NASA Langley Tech. Mem.* 78631, February 1978.

247. Hammer, P., D. Van Bauchaute, D. Verschraeven, and S. Van de Capelle   A Model for Calculating the Radiation Field of Microstrip Antennas, *IEEE Trans. Antennas Propag.*, March 1979, Vol. AP-27, pp. 267–270.

Radio-Wave Propagation

BY RICHARD C. KIRBY

FUNDAMENTALS OF WAVE PROPAGATION

65. Introduction: Mechanisms, Media, and Frequency Bands. This part of Sec. **18** deals with radio waves propagated through or along the surface of the earth, through the atmosphere, and by reflection or scattering from the ionosphere or troposphere. The particular propagation mechanism around which a given application is designed depends on the distance to be spanned, the type of information to be transmitted or the service to be provided, and the reliability required. Other propagation mechanisms can affect the performance of the system or lead to interference with and from other systems, depending upon frequency and distance.

Over a line-of-sight path within the nonionized atmosphere, transmission is much as through free space, though atmospheric refraction causes bending, reflection, scattering, and possibly fading. At extremely high frequencies there may be attenuation due to rainfall and absorption by air and water vapor. The conductivity and permittivity (dielectric constant) of the earth are markedly different from those of the atmosphere. A wave mainly diffracted along the surface of the ground encounters increasing loss with increasing frequency. Very low frequency waves are propagated with little attenuation over thousands of kilometers. At high frequencies, losses along the ground become so great that the usefulness of the ground wave is limited to short distances. At medium and high frequencies, ionospheric reflections permit radio communication to great distances. At frequencies much above 30 MHz, ionospheric reflections are not dependable, and most communications depend upon line-of-sight propagation or tropospheric scattering beyond the horizon.

Because of the dependence of propagation characteristics on frequency, much of the discussion of these sections will be in terms of frequency bands, and abbreviations such as VLF for very low frequencies, VHF for very high frequencies, etc., will be used.

The International Telecommunications Union has defined nine frequency bands designated by integer band numbers; for example, 1 MHz is the approximate midband of band 6, etc., as listed in Table 18-3.

TABLE 18-3

ITU band number*	Frequency range (lower limit exclusive, upper limit inclusive)	Corresponding metric subdivision	Abbreviation
4	3–30 kHz	Myriametric waves	VLF
5	30–300 kHz	Kilometric waves	LF
6	300–3,000 kHz	Hectometric waves	MF
7	3–30 MHz	Decametric waves	HF
8	30–300 MHz	Metric waves	VHF
9	300–3,000 MHz	Decimetric waves	UHF
10	3–30 GHz	Centimetric waves	SHF
11	30–300 GHz	Millimetric waves	CHF
12	300–3,000 GHz (3 THz)	Decimillimetric waves	

*Band number N extends from 0.3×10^N to 3×10^N Hz.

Propagation characteristics are discussed according to somewhat different bands, such as 10 to 150 kHz, 150 to 1,500 kHz, etc., corresponding to bands of relatively homogeneous propagation characteristics.

References 1 and 2 provide introductory material; Refs. 3 to 6 are basic texts on electromagnetic propagation; Refs. 7 to 9 offer additional material on tropospheric and ionospheric propagation.* Recent results of theoretical and experimental radio science worldwide are outlined in Ref. 10.

The emphasis here is mainly descriptive, key formulas indicating the behavior of parameters and references to publications providing material for engineering calculations. Maxwell's uniform plane-wave equations are cited only to show the role of the electrical constants and the vector relationships of electric and magnetic field and power flux.

66. Wave Propagation in Homogeneous Media. Electromagnetic radiation is composed of two mutually dependent vector fields, electric and magnetic. The electric field is characterized by the vectors **E**, electric field strength in volts per meter and **D**, dielectric displacement in coulombs per square meter. The magnetic field is characterized by **H**, the magnetic field strength in ampere-turns per meter (or amperes per meter) and **B**, flux density, in webers per square meter. The vector current density **J** is in amperes per square meter.

The relationship between the members of the various pairs of field vectors is characterized by the constitutive parameters, or *electrical constants*, of the medium:

$$\epsilon = \text{dielectric constant, F/m} \qquad \sigma = \text{conductivity, S/m} \qquad \mu = \text{permeability, H/m} \qquad (18\text{-}61)$$

In some cases these constants are functions of the coordinate. Locally, however they are always considered constant. Nearly always the time factor in these sections is $\exp(+i\omega t)$, where ω is the angular frequency $2\pi f$ (f in hertz) and t the time. The *electric field*, then, is the real part of $E \exp i\omega t$.

To explain very basic notation,† a short outline of plane electromagnetic waves in a homogeneous medium is given.

Ohm's law in the complex form is

$$\mathbf{J} = (\sigma + i\epsilon\omega)\mathbf{E} \qquad (18\text{-}62)$$

where $\mathbf{J}$ = current density vector and $\mathbf{E}$ = electric field vector.

The analogous relation for magnetic quantities is

$$\mathbf{B} = \mu\mathbf{H} \qquad (18\text{-}63)$$

In source-free media the above vector quantities are related by

$$\text{curl } \mathbf{E} = -i\mu\omega\mathbf{H} \qquad (18\text{-}64)$$

and

$$\text{curl } \mathbf{H} = (\sigma + i\epsilon\omega)\mathbf{E} \qquad (18\text{-}65)$$

These are Maxwell's equations.

For a homogeneous medium

$$\text{curl curl } \mathbf{E} = \text{grad div } \mathbf{E} - \text{div grad } \mathbf{E} = -i\mu\omega(\sigma - i\epsilon\omega)\mathbf{E} \qquad (18\text{-}66)$$

Since div $\mathbf{E} = 0$,

$$(\nabla^2 - \gamma^2)\mathbf{E} = 0 \qquad (18\text{-}67)$$

where ∇^2 = div grad = Laplacian operator (which operates on the rectangular components of **E**) and $\gamma^2 = i\mu\omega(\sigma + i\epsilon\omega)$. The quantity γ is called the *propagation constant*.

As a simple illustration of the role of the electrical constants of the medium, the field of a wave is assumed to vary only in the z direction in space (time factor $\exp i\omega t$ understood), and the electric field is assumed to have only an x component E_x. For this case Eq. (18-67) reduces to

$$\left(\frac{d^2}{dz^2} - \gamma^2\right) E_x = 0 \qquad (18\text{-}68)$$

and the solutions are $\exp(+\gamma z)$ and $\exp(-\gamma z)$ or, in general,

$$E_x = A \exp \gamma z + B \exp(-\gamma z) \qquad (18\text{-}69)$$

where A and B are constants. The magnetic field then has only a y component given by

---

*For references, see Par. **18-113**.
†From Wait[4] by permission.

$$H_y = -\frac{1}{i\mu\omega}\frac{\partial E_x}{\partial z} = -\eta^{-1}[A\exp(+\gamma z) - B\exp(-\gamma z)] \tag{18-70}$$

where
$$\eta = \left(\frac{i\mu\omega}{\sigma + i\epsilon\omega}\right)^{1/2} \tag{18-71}$$

η is defined as the *characteristic impedance* of the medium for plane-wave propagation. Remembering that the time factor is $\exp i\omega t$, we see that the term $B\exp(-\gamma z)$ is a wave traveling in the positive z direction with diminishing amplitude and the term $A\exp \gamma z$ is a wave traveling in the negative z direction with diminishing amplitude.* The electric and magnetic fields are both transverse to the direction of propagation and orthogonal to each other. Such radiation is termed *plane-polarized*. It is by convention designated as *horizontal* or *vertical* according to the orientation of the plane containing the **E** vector.

The quantity η is equal to the complex ratio of the electric and magnetic field components in the x and y directions, respectively, for plane waves in an unbounded homogeneous medium, i.e.,

$$H_y = (i\omega\epsilon/r)E_x = E_x/\eta \qquad H_x = (-r/i\omega\mu)E_y = -E_y/\eta \tag{18-71}$$

For a perfect dielectric ($\sigma = 0$)

$$\eta = \sqrt{\mu/\epsilon}\ \Omega \qquad \text{and} \qquad r = ik \tag{18-72}$$

where $k = \omega\sqrt{\mu\epsilon} = 2\pi/\lambda$ and λ = wavelength.

The velocity of this wave is

$$v = 1/\sqrt{\mu\epsilon} \qquad \text{m/s} \tag{18-73}$$

v is called the *phase velocity* of the wave. It represents the velocity of propagation of phase or a state and does not necessarily coincide with the velocity with which the energy of a wave or signal is propagated. In fact, v may exceed free-space wave velocity without violating relativity in any way.

The wavelength is defined as the distance the wave propagates in one period

$$\lambda = \frac{2\pi}{\omega\sqrt{\mu\epsilon}} = \frac{v\ (\text{m/s})}{f\ (\text{Hz})} \qquad \text{m} \tag{18-74}$$

For free space

$$\epsilon = \epsilon_0 = 8.854 \times 10^{-12}\ \text{F/m} \qquad \mu = \mu_0 = 4\pi \times 10^{-7}\ \text{H/m} \qquad \sigma = 0$$

Then
$$\gamma = ik = i\omega\sqrt{\mu_0\epsilon_0} = i2\pi/\lambda \tag{18-75}$$

The *velocity of the wave for free space* is

$$v_0 = 1/\sqrt{\mu_0\epsilon_0} \approx 3 \times 10^8\ \text{m/s} \tag{18-76}$$

The *characteristic impedance of free space* is

$$\eta_0 = \sqrt{\mu_0/\epsilon_0} \approx 120\pi\ \Omega \approx 4\pi v_0 \times 10^{-7}\ \Omega \tag{18-77}$$

Energy flow in the electromagnetic field is described by the Poynting vector

$$\mathbf{P} = \mathbf{E} \times \mathbf{H}^* \tag{18-78}$$

where the complex representation of the time-periodic quantities is implied and the asterisk denotes the complex conjugate. The *real part* of the Poynting vector represents the average power flow over a cycle of the time variation per unit area in the direction of transmission

$$\mathbf{P}_{av} = \tfrac{1}{2}\,\text{Re}\,(\mathbf{E} \times \mathbf{H}^*) = E^2/2\eta_0 \qquad \text{W/m}^2 \tag{18-79}$$

$\mathbf{P}_{av}$ is called the *power flux density* or *field intensity*. Note that **E** and **H** are peak values.

The permeability and dielectric constant of any medium relative to free space are called the *relative permeability* and *relative dielectric constant*. These are usually the values given in tables of physical constants; they are dimensionless and designated by μ_r and ϵ_r, respectively.

Additional material on electromagnetic radiation and propagation in waveguides, cavities, and transmission lines is given in Secs. **9** and **25**.

Polarization of the Wave.   Polarization is a term characterizing the orientation of the field vector in its travel. In radio, polarization usually refers to the electric vector. In the simplest case

*The geometry of subsequent paragraphs uses a different convention for x, y, and z directions, shown in figures.

E_z and H_z (field components in the direction of propagation) are zero, and **E** and **H** lie in a plane transverse to the direction of propagation and orthogonal to each other. Such a plane wave is *elliptically polarized* when the electric vector **E** describes an ellipse in the plane perpendicular to the direction of propagation over one cycle of the wave.

When the amplitudes of the rectangular components are equal and their phases differ by some odd integral multiple of $\pi/2$, the polarization ellipse becomes a circle and the wave is *circularly polarized*. It is customary to describe as *right-handed* circularly polarized a clockwise rotation of **E** when viewed in the direction of propagation; counterclockwise rotation is *left-handed* polarization.*

An important case for many radio problems is that in which the polarization is a straight line. The wave is then *linearly polarized*. In *horizontal polarization* the electric vector lies in a plane parallel to the earth's surface.

To obtain maximum transfer of power between two antennas the polarization should match. If the transmitting antenna is horizontally polarized, the receiving antenna must likewise be horizontally polarized. If the transmitting antenna is elliptically polarized with a given degree of ellipticity and a specified direction of rotation, the receiving antenna should have the proper direction of rotation and degree of ellipticity in order to maximize the path antenna gain.

It should be noted that in the process of propagation, except through free space, the polarization may be altered.[10a] Reflections from surfaces can do this. Passage through the ionosphere in the presence of magnetic field is likely to impart elliptical polarization to a plane-polarized incident wave and rotation of the major axis. For MF or HF ionospheric propagation, the downcoming wave may be randomly polarized.

67. Reflection. Most problems in wave propagation involve reflection from a boundary between media of different refractive properties, often between air and the ground or between air and the ionosphere. In general, such a boundary may involve dissipative media (finite conductivity), curvature, finite dimensions, roughness, and stratification.

The *complex index of refraction* for a conducting medium is[11]

$$n^2 = [i\omega\mu(\sigma + i\omega\epsilon)]/[i\omega\mu_0(i\omega\epsilon_0)] \tag{18-80}$$

When $\sigma = 0$,

$$n = \sqrt{\mu_r\epsilon_r} \qquad \text{where } \mu_r = \mu/\mu_0, \ \epsilon_r = \epsilon/\epsilon_0$$

For many applications $\mu = 1$ and $\sigma = 0$, and the index of refraction is simply the square root of ϵ_r.

Figure 18-46 illustrates Snell's law for refraction of plane waves at an infinite plane interface. The angle ϕ between the direction of propagation and the normal to the boundary is called the *angle of incidence*. The angle ψ between the direction of propagation and the boundary, called the *grazing angle* or *elevation angle*, is often more convenient. If the medium of the incident wave is lossy, the angle of incidence is complex and can be defined in various ways.[4] At the boundary, the tangential components of **E** and **H** must be continuous; the phase of the reflected wave is in step with the phase of the incident wave to satisfy this requirement.

Snell's law of refraction for the direction of the transmitted wave toward C is

$$n_1 \sin \phi = n_2 \sin \phi_2 \qquad \text{or} \qquad n_1 \cos \psi_1 = n_2 \cos \psi_2 \tag{18-81}$$

The *penetration depth* δ, or the depth at which the transmitted wave E_t has attenuated to $1/e$ of its incident value (for a conducting medium where $\sigma \gg \omega\epsilon$), is

$$\delta = \frac{1}{\sqrt{\omega\mu\sigma/2}} \qquad \text{m} \tag{18-82}$$

68. Ground Reflection, Reflection Coefficients, Fresnel Zones. A wave incident upon a plane surface can be resolved into two components, one polarized normal and the other parallel to the plane of incidence. The reflection coefficients for the two components differ, and consequently the polarization of the reflected wave depends upon the angle of incidence. Consider an air-earth boundary, taking the media to be nonmagnetic; for the case where the **H** vector is parallel to the ground surface the complex reflection coefficient[11] is

$$R_v = \frac{(\epsilon_r - i60\sigma\lambda) \sin \psi - (\epsilon_r - \cos^2 \psi - i60\sigma\lambda)^{1/2}}{(\epsilon_r - i60\sigma\lambda) \sin \psi + (\epsilon_r - \cos^2 \psi - i60\sigma\lambda)^{1/2}} \tag{18-83}$$

*Right- and left-hand polarization are sometimes misinterpreted with regard to the direction of viewing; the definition given here corresponds to the International Radio Regulations, Geneva, 1979.

where σ = conductivity (S/m), ψ = grazing angle (Fig. 18-46), and ϵ_r = relative dielectric constant of earth to air or free space. $\epsilon_r - i60\sigma\lambda$ is referred to as the *complex dielectric constant*.

If E is parallel to the ground surface and H is in the plane of incidence,

$$\mathbf{R}_h = \frac{\sin\psi - (\epsilon_r - \cos^2\psi - i60\sigma\lambda)^{1/2}}{\sin\psi + (\epsilon_r - \cos^2\psi - i60\sigma\lambda)^{1/2}} \tag{18-84}$$

These are reflection coefficients for vertical and horizontal polarization, respectively. Curves of values for a range of ϵ and σ are given in Refs. 11 and 12.

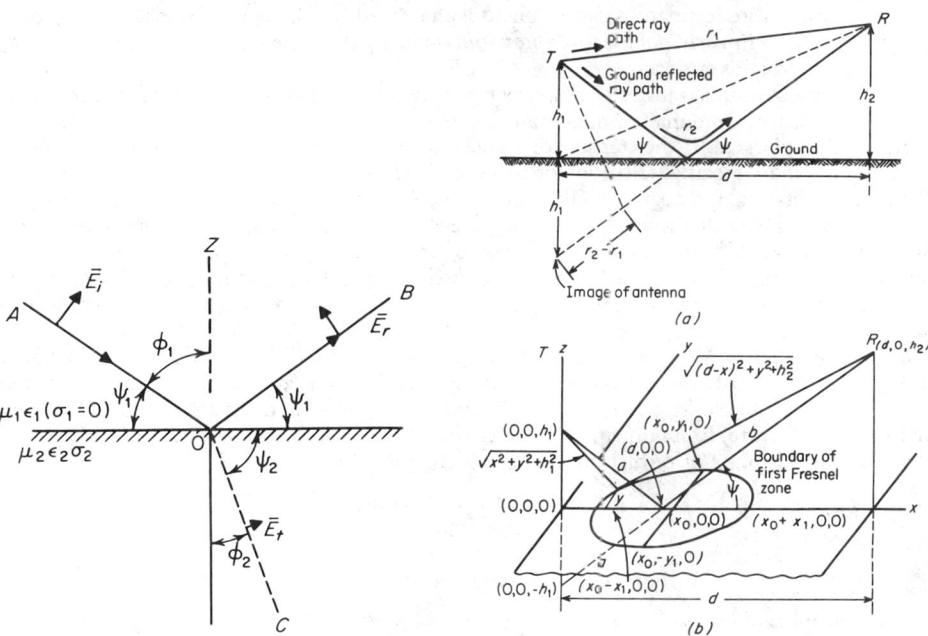

Fig. 18-46. Geometry of reflection and transmission.

Fig. 18-47. Geometry of ground reflection, image antenna, and Fresnel zones for plane earth: (a) ray paths; (b) Fresnel zones.

An important property for vertical polarization is that there exists an angle of incidence for which the reflection coefficient approaches zero (for dielectric media it equals zero). This is the *Brewster angle*, also called the *polarizing angle*, given by

$$\phi_0 = \tan^{-1}\sqrt{\epsilon_1/\epsilon_2} \tag{18-85}$$

It is equal to the angle of incidence for which the reflected and refracted (transmitted) rays are at right angles. If the incidence occurs at the Brewster angle, the reflected wave is polarized entirely in the direction normal to the plane of incidence.

Wave tilt is a property frequently used to determine the electrical constants of the earth.[39] For waves traveling at nearly grazing incidence along the surface of the earth, wave tilt may be interpreted geometrically as the angle between the normal to the wavefront and the tangent to the earth's surface. Wave tilt is defined to be the ratio of the horizontal to the vertical component of the electric field in the air just above the ground:

$$W = E_h/E_v \tag{18-86}$$

Wave tilt is related to the electrical constants of a homogeneous earth by

$$W = \frac{\sqrt{\mu_1/\epsilon_{1c}}}{\sqrt{\mu/\epsilon_0}}\sqrt{1 - \frac{\mu_0\epsilon_0}{\mu_1\epsilon_{1c}}} \tag{18-86a}$$

The subscript 1 refers to earth constants and 0 to free space; ϵ_{1c} = complex dielectric constant = $\epsilon_1 - i\sigma_1/\omega$.

This procedure assumes that $\mu_0 = \mu_1$, generally a valid assumption; if it is not, μ_1 must be determined by some other procedure.

An important consideration in many propagation problems is the interference pattern generated by vector addition of the fields corresponding to the direct ray from an antenna to a point within line of sight plus the ground-reflected ray. In calculating such ground-reflection lobes, the geometry of an image antenna is used,[11,13] as illustrated in Fig. 18-47. Discussion and formulas for ground reflection and Fresnel zones follow Norton and Omberg, Ref. 13, by permission.

A transmitting antenna T is at height h_1 above ear h, and a receiving antenna R or a radar target is at distance d and height h_2. If d is very large, so that $r_1 \approx r_2$, the path difference $r_2 - r_1 \approx 2h_1 \sin \psi$. If e_1 is the free-space direct-path field strength at a unit distance from T in the direction of R and the free-space field in the direction of the ground reflection is also approximately equal to e_1, the resultant field strength at R for horizontal polarization and perfectly reflecting earth is[3]

$$e_R = \underbrace{\frac{e_1}{r_1} \sin \left[\frac{2\pi}{\lambda} \left(\frac{r_1 + r_2}{2} - v_0 t \right) \right]}_{\text{Direct field term}} \underbrace{2 \quad \sin \frac{2\pi h_1 \sin \psi}{\lambda}}_{\substack{\text{Interference term} \\ \text{oscillating between 0 and 2}}} \qquad (18\text{-}87)$$

The angles at which maxima and minima occur, for horizontal polarization, are given by

$$\sin \psi = n\lambda/4h_1 \qquad \text{maxima for } n \text{ odd, minima for } n \text{ even} \qquad (18\text{-}88)$$

For the first maximum to occur at a specified elevation ψ_1,

$$h_1 = \lambda/(4 \sin \psi_1) \qquad (18\text{-}88a)$$

In Fig. 18-47a the ray reflections are shown as though they occurred at a point. Actually, the surface of the earth is illuminated over a wide region corresponding to the radiation patterns of the two antennas and, in accordance with Huygens principle, reradiates elementary wavelets in all directions. In any particular direction, as toward R, these elementary wavelets arrive with strength and phase such that the waves from an elliptical zone in the neighborhood of the ray reflection add nearly in phase. From successive ring areas, similarly bounded by larger ellipses, the waves alternately cancel and add.

These zones of physical reflection are called *Fresnel zones*, since they are closely related to the Fresnel zones of diffraction theory.[*] Most of the energy can be thought of as being reflected from the first Fresnel zone; it is defined, with the aid of Fig. 18-47b, for reflection paths between points such as T and R, as the area from which all the reradiated elementary wavelets arrive, according to geometric optics, within half wavelength of the phase of the direct ray. Thus the length of the geometric ray path at the edge of the nth Fresnel zone is n half wavelengths greater than the geometric ray path.

The cartesian coordinates for the extremities of the major axis of the Fresnel zone[13] are given by $x_0 - x_1$ and $x_0 + x_1$, where

$$x_0 = d_1 \left\{ 1 + \frac{h_2 - h_1}{2h_1[1 + (h_1 + h_2)^2/n\lambda(R + n\lambda/4)]} \right\} \qquad (18\text{-}89)$$

$$x_1 = \frac{(1 + n\lambda/2R)(1 + n\lambda/4R)}{\sin \theta \left[1 + \frac{n\lambda(R + n\lambda/4)}{(h_1 + h_2)^2} \right]} \left\{ \frac{abn\lambda}{a + b + n\lambda/4} \left[+ \frac{n\lambda(R + n\lambda/4)}{4h_1 h_2} \right] \right\}^{1/2} \qquad (18\text{-}90)$$

$$R_n = \sqrt{x^2 + y^2 + h_1^2} + \sqrt{(d - x)^2 + y^2 + h_2^2} \qquad (18\text{-}91)$$

Equation (18-89) determines the centers of Fresnel zones in terms of the location $d_1 = dh_1/(h_1 + h_2)$ of the geometric ground-reflection point.

The minor semiaxis of the ellipse is given by

$$y_1 = \left(1 + \frac{n\lambda}{4R} \right) \left\{ \frac{abn\lambda}{a + b + n\lambda/4} \frac{1 + [n^2(R + n\lambda/4)/4h_1 h_2]}{1 + [n\lambda(R + n\lambda/n)]/(h_1 + h_2)^2} \right\}^{1/2} \qquad (18\text{-}92)$$

For many applications such as radar and air-ground communication, or for radiation aimed at an ionospheric reflection region, $h_2 \gg h_1$, and $h_2 \gg \lambda$. An important use of the above formulas is the determination of the first Fresnel zone under these conditions for transmission at an angle ψ corresponding to the maximum of the first ground-reflection lobe. For this condition,[13] the distances from the antenna T to the near (d_n) and far (d_f) edges of the first Fresnel zone are

$$d_n = x_0 - x_1 \approx 0.688h^2/\lambda \qquad (18\text{-}93)$$

and

$$d_f = x_0 + x_1 \approx 23.3h_1^2/\lambda \qquad (18\text{-}94)$$

For a well-developed ground reflection, the ground should be flat over an area which includes at least the first Fresnel zone. The degree of flatness depends upon the wavelength and angle of incidence; assuming that phase-path changes less than $\lambda/16$ are unimportant, Rayleigh's criterion limits height deviations in terrain from a smooth surface to a magnitude less than $\Delta h = \lambda/(16 \sin \psi)$ over the area of the first Fresnel zone for waves incident at angle ψ. Methods allowing for surface roughness, finite conductivity, and divergence due to the spherical shape of the earth are outlined in Pars. **18-78** to **18-80**.

69. Diffraction and Scattering. The spherical shape of the earth and irregularities of terrain give rise to *diffraction* as an important part of the ground wave and as a mechanism for propagation beyond the optical horizon. Diffraction is included in the methods for calculation given in Pars. **18-76** to **18-82**.

Scattering takes place from the rough surface of the earth and from small-scale irregularities in the index of refraction of the atmosphere or the ionosphere. It is analogous to scattering of light, although the radio problem is complicated by the wide range of relationships of radio wavelength to size of irregularity.

Tropospheric forward scattering is discussed in Par. **18-85** and ionospheric scattering in Par. **18-106**.

70. Reciprocity. Reciprocity in wave propagation means that the source and receiver can be interchanged, with the transmission loss and phase unaffected by direction of propagation. This is an application of the classical reciprocity theorem to radiation, as by J. R. Carson and others. In most radio-wave-propagation problems, with the notable exception of those involving an ionized medium with magnetic field, such reciprocity obtains. As discussed in Par. **18-94**, the refractive index of the ionosphere depends upon magnetic field effects; the direction of propagation of the wave affects attenuation, phase, and bending, and the medium is called *anisotropic*. Thus, especially at very low and medium frequencies propagated via the ionosphere, reciprocity does not obtain. Reciprocity does not in any case imply the same signal-to-noise ratio in both directions. The noise environment may be very different at the transmitting and receiving locations.

71. Transmission Loss; Free-Space Attenuation; Field Strength; Power Flux Density. The *transmission loss* of a radio circuit consisting of a transmitting antenna, a receiving antenna, and the intervening propagation medium, is defined (*CCIR Recomm.* 341)[175] as the dimensionless ratio p'_t/p'_a, where p'_t is the radio-frequency power radiated from transmitting antenna and p'_a is the resultant radio-frequency signal power which would be available from the receiving antenna if there were no circuit losses other than those associated with its radiation resistance. The transmission loss is usually expressed in decibels:*

$$L = 10 \log (p'_t/p'_a) = L_s - L_{tc} - L_c \quad \text{dB} \quad (18\text{-}95)$$

where L_s = system loss and L_{tc}, L_{rc} = losses in transmitting and receiving antenna circuits, respectively, excluding losses associated with antenna radiation resistances; i.e., the definitions of L_{tc} and L_{rc} are 10 log (r'/r), where r' is the resistive component of the antenna circuit and r is the radiation resistance.

For many applications it is convenient to calculate the free-space attenuation between isotropic antennas, i.e., *basic transmission loss*.

The *basic transmission loss* L_b of a radio circuit is the transmission loss expected between ideal, loss-free, isotropic transmitting and receiving antennas at the same locations as the actual transmitting and receiving antennas. At a distance d very much greater than the wavelength λ, the power flux density (field intensity), expressed in watts per square meter, is simply $p'_t/4\pi d^2$ since the power is radiated uniformly in all directions. The effective absorbing area of the isotropic receiving antenna is $\lambda^2/4\pi$, and the available power at the terminals of the loss-free isotropic receiving antenna is given by

$$p'_a = \frac{\lambda^2}{4\pi} \frac{p'_t}{4\pi d^2} \quad (18\text{-}96)$$

Consequently, the *basic transmission loss in free space* can be expressed by

$$L_b = 20 \log \frac{4\pi d}{\lambda} \quad \text{dB} \quad (18\text{-}97)$$

d and λ are expressed in the same units.

*From this point in the text, unless otherwise noted, capital letters are used to denote decibel quantities, and power levels P are referred to a common reference power. Unless otherwise indicated, logarithms are to base 10, and the reference power is 1 W.

A practical form of this equation is

$$L_b = 32.45 + 20 \log f_{MHz} + 20 \log d_{km} \qquad dB \qquad (18\text{-}98)$$

These results of expressions are given in nomogram form in Fig. 18-48a.

The *path antenna gain* G_p is equal to the realized difference in transmission loss between actual antennas used on the circuit and isotropic antennas:

$$G_p = L_b - L \qquad (18\text{-}99)$$

We relate the available power P from the receiving antenna, the total radiated power P_r, the transmission loss L, the basic transmission loss L_b, and the path antenna gain G_p as follows:

$$P = P_r - L \qquad (18\text{-}100)$$
$$L_b = L + G_p \qquad (18\text{-}101)$$

Since the free-space gain of a short lossless electric dipole is $g_t = g_r = 1.5$, the path antenna gain for two optimally oriented short lossless electric dipoles in free space is

$$G_p = G_t + G_r = 3.52 \text{ dB} \qquad (18\text{-}102)$$

Consequently, the transmission loss between two optimally oriented short lossless electric dipoles in free space is

$$L = 10 \log (4\pi d/\lambda)^2 - 3.52 \text{ dB} \qquad (18\text{-}103)$$

The attenuation A relative to free space is defined as

$$A = L_b - L_{bf} \qquad dB \qquad (18\text{-}104)$$

For most radio links it is important to know the attenuation relative to free space A, usually a random variable with time, expressed as a function of frequency, distance, or other path parameters such as elevation angle, taking into account attenuation of the atmosphere or various propagation mechanisms.

In free space, with no absorption, the calculated transmission loss is

$$L = L_{bf} - G_t - G_r \qquad dB \qquad (18\text{-}105)$$

where G_t and G_r are free-space transmitting- and receiving-antenna gains in decibels relative to the gain of an isotropic radiator.

For broadcasting and mobile services, where characteristics and locations of receiving installations vary, it is convenient to calculate *field strength* at some distance from the transmitter by

$$e = \sqrt{30 p_t'}/d \qquad (18\text{-}105a)$$

where $e =$ field strength, rms (V/m), $p_t' =$ effective isotropic radiated power (W), and $d =$ distance (m), or

$$e = 173 \sqrt{p_t'(kW)}/d_{km} \qquad mV/m \qquad (18\text{-}105b)$$

This relation is expressed in the nomogram of Fig. 18-48b.

For microwave and satellite services, power flux density is a convenient term. The *power flux density* (also called *field intensity**) is given by the characteristic relations of a plane wave

$$p = e^2/\eta \qquad W/m^2 \qquad (18\text{-}106)$$

where η is the characteristic impedance of the medium in which the measurement is made ($\eta_0 = 120\pi$ in free space). In free space

$$p = e^2/120\pi \qquad \text{and} \qquad p = 4\pi p_a/\lambda^2 \qquad (18\text{-}106a)$$

where p_a is available power received by an isotropic antenna in the field. This relation is expressed in the nomogram of Fig. 18-48c.

The relations between field strength, power flux density, and available power in the receiving antenna are outlined below.

The absorbing area of a receiving antenna with gain g_r relative to an isotropic antenna can be written

$$a_e = \lambda^2 g_r r_f/4\pi r \qquad (18\text{-}107)$$

where $\lambda =$ wavelength in medium, $r =$ radiation resistance of antenna, and $r_f =$ radiation

*The term power flux density is preferred; field intensity is often confused with field strength.

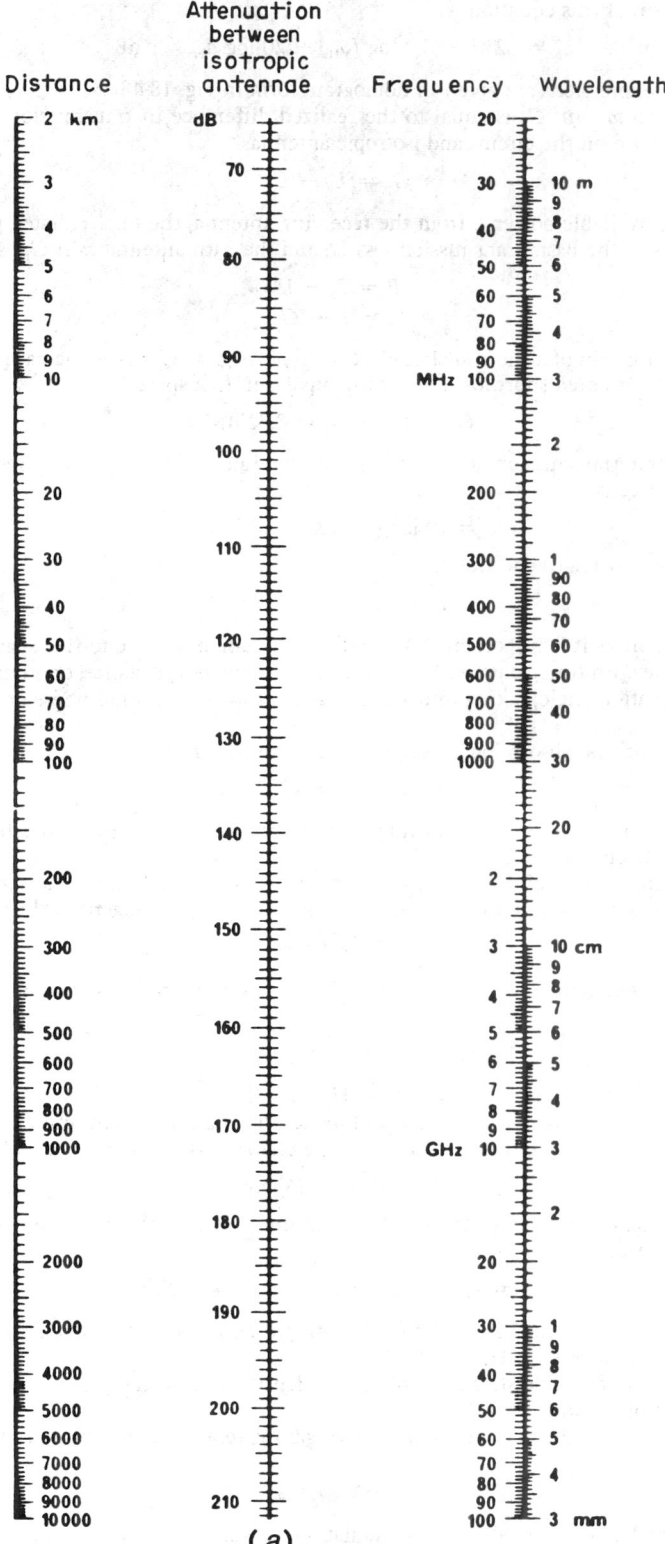

Fig. 18-48. (a) Nomogram for determining basic transmission loss L_{bf} between isotropic antennas in free space. (b) Nomogram for determining free-space field strength. (c) Nomogram for relating characteristics of received plane waves.

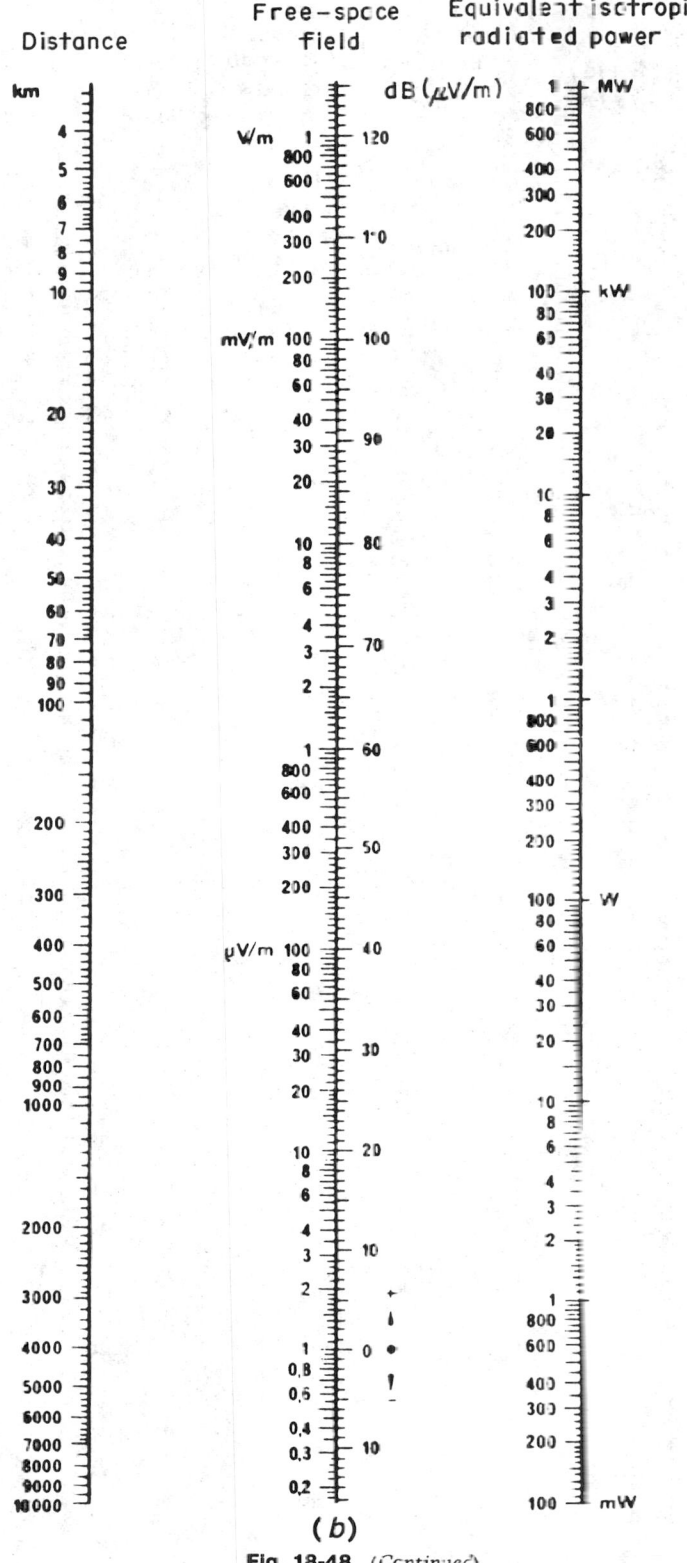

Fig. 18-48. (Continued)

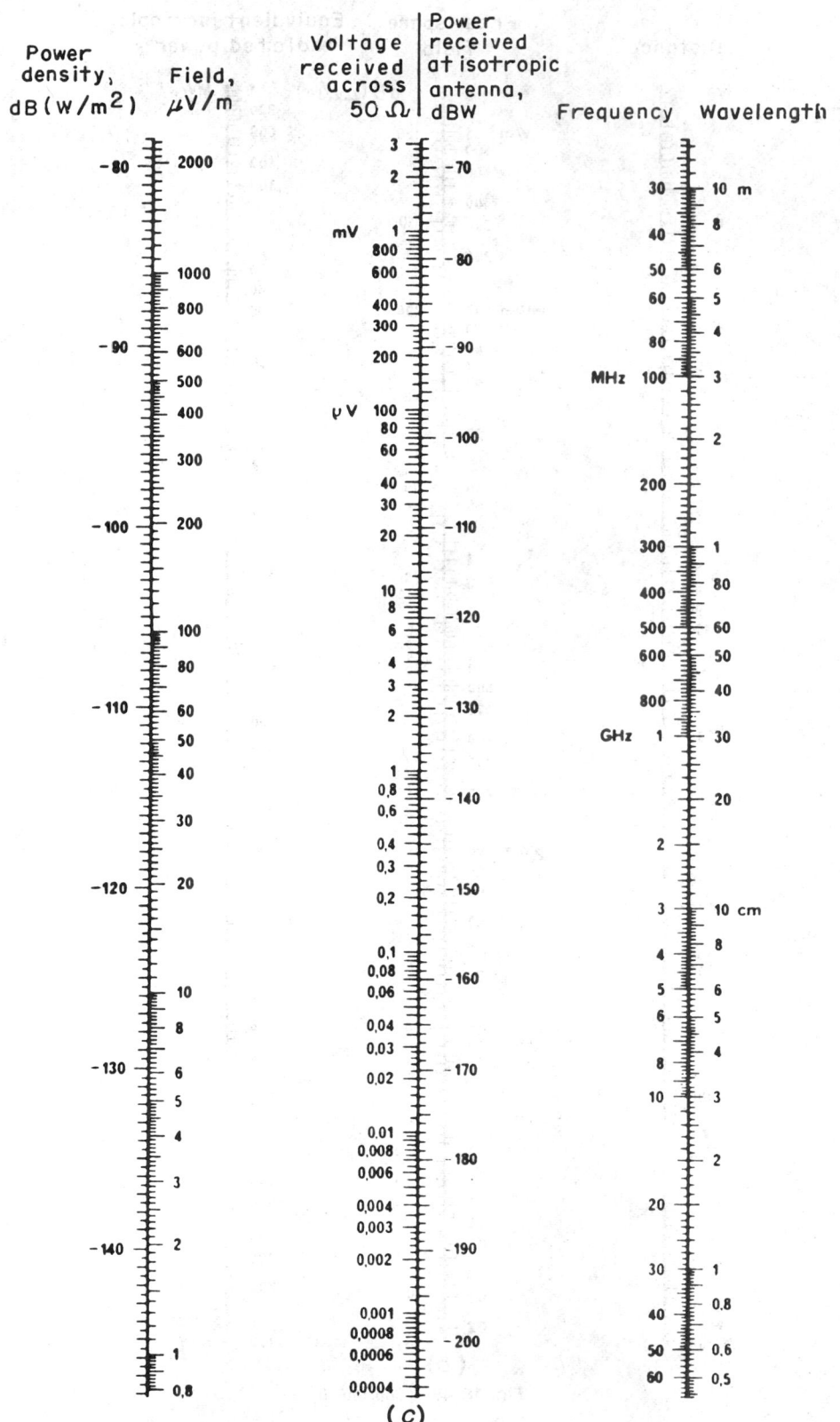

(c)
Fig. 18-48. (Continued)

resistance antenna would have if it were in free space. Combining the above two equations, we find the following formula for the available power p'_a from a lossless receiving antenna:

$$p'_a = e^2\lambda^2 g_r r_f/4\pi\eta r = v^2/4r \qquad (18\text{-}108)$$

The v in this equation denotes the open-circuit voltage induced in the receiving antenna. The field strength is related to the open-circuit voltage by

$$v = e\sqrt{\lambda^2 g_r r_f/\pi\eta} = el \qquad (18\text{-}109)$$

Field-strength meters usually are calibrated in terms of the effective length l of the antenna.

The relation between the available power p_a from the receiving antenna (neglecting losses) and the field strength e can be expressed in decibels as

$$E = 10\log\left[(4\pi\eta_0 P_a \times 10^{12})/\lambda^2 g_r\right] = P_a + 20\log f - G_r + 107.22 \text{ dB}\mu \qquad (18\text{-}110)$$

where f is in megahertz and dBμ means decibels referred to 1 μV/m.

For field strength E in decibels referred to 1 μV/m, referred to 1 kW radiated from a half-wave dipole over perfectly conducting earth, propagation loss is

$$L_p = 139.4 - G_r + 20\log f - E \qquad \text{dB} \qquad (18\text{-}111)$$

If the reference radiation is a short electric dipole, the constant becomes 136.0.

72. Fading; Characterization of Time-Variant Multipath; Channel; Diversity.

Random variations appear in the signal received via various transmission media, especially at frequencies above about 100 kHz when propagation is by the troposphere or ionosphere. Such variation is usually of two types: one is *attenuation*, or *power fading*, which may be quite slow (minute to minute, hour to hour, etc.) and is associated with comparatively large-scale changes in the medium, such as absorption; the other is *variable-multipath* or *phase-interference fading*.

Power fading is usually allowed for in the power margin designed into the system. Phase-interference fading, on the other hand, affects not only the amplitude but also the variable phase-vs.-frequency characteristic of the channel, limiting coherence bandwidth and introducing extraneous fluctuation in received-signal parameters. Alleviation of the effects of variable multipath is possible by diversity techniques, signal design, and signal receiving, processing, and detection operations. Fading media are discussed in Refs. 14 to 19 and CCIR Reps. 238-3, 338-3, 415, 263-4, and 266-4.[175]

The amplitude probability distribution of the fading envelope is usually determined from samples of duration much shorter than the shortest fade duration; observation intervals over which statistical averages are taken are about 1,000 times the reciprocal of the nominal fading rate. The fit of experimental distributions of envelope fading to the Rayleigh distribution is often excellent for ionospheric and tropospheric scatter propagation, and similarly to a Nakagami-Rice[17,18] or Beckmann[19] distribution for situations where a specular component is mixed with scattered components. If long-term (power) fading is mixed with the short-term, the median value of the short-term distribution changes; a log-normal distribution usually represents the long-term variation.

Most theoretical treatments of communication performance in the presence of variable multipath fading resulting from several signal components have been carried out for channels characterized by Rayleigh envelope distribution. The probability density function is given by

$$p(V) = (2V/v^2)\exp\left[-(V/v)^2\right] \qquad (18\text{-}112)$$

where $V = $ fluctuating envelope and $v^2 = $ mean square value of V over distribution. For the Rayleigh fading channel, the probability that the received signal envelope will fall at or below some specified value of V' is given by the cumulative distribution

$$p(V \leq V') = \int_0^{V'} \frac{2V}{v^2}\exp\left[-\left(\frac{V}{v^2}\right)^2\right]dV = 1 - \exp\left[-\left(\frac{V'}{v}\right)^2\right] \qquad (18\text{-}113)$$

The Rayleigh probability distribution function (18-113) is often used in the form

$$p(V \leq V') = 1 - \exp\left[-0.693(V'/V_M)^2\right] \qquad (18\text{-}113a)$$

where V_M is the median value, about 1.6 dB below the rms value. *Fading rate* is important to certain systems. One measure of fading rate is the number of times per second (hertz) the carrier envelope crosses its median with a positive or negative slope. Another measure is the width of the received carrier-envelope spectral density.

For some time, design concern (besides its emphasis on the amplitude variation) has centered on the dispersion and multipath characteristics of the medium, in terms of linear time-variant amplitude and phase- and frequency-distortion parameters, often referred to as *multiplicative noise*, which cannot be overcome by power increase.

A more comprehensive characterization is in terms of the *system function* or *impulse function*. This approach relates the response and excitation of a channel at its input and output terminals.[14,16]

The expressions* for output are formulated in terms of operations on the input time function $x(t)$ or the spectral function (Fourier transform) $X(\omega)$, to produce the output function $y(t)$ or $Y(\omega)$. Each path is characterized by a system function $h(t, \tau)$ that operates on the replica of $x(t)$ traversing it; t is the time at which the observation is made, and τ is the delay or transit time for the path. The spread of delays between the input and output is determined by the system function $h(t, \tau)$, which can be called the *delay-spread system function*. For any particular elemental path x in a distribution of paths that covers some range of delays, the output for a range of delay $\Delta\tau$ centered on τ is given by $h(t, \tau)x(t - \tau) \Delta\tau$, and the total output of the channel is the sum of all such weighted and delayed contributing paths, namely,

$$y(t) = \int_{-\infty}^{\infty} x(t - \tau)h(t, \tau) \, d\tau \tag{18-114}$$

For a Fourier transformable input $x(t)$, the output can be expressed in terms of the input spectral function $X(i\omega)$:

$$x(t) = \int_{-\infty}^{\infty} X(i\omega) \exp (i\omega t) \, d(\omega/2\pi) \tag{18-115}$$

Here we characterize the channel by stating that it modifies the contribution to the structure of $x(t)$ from spectral components in the range $\Delta\omega$ centered at ω by multiplying it by the transfer function $H(i\omega, t)$. The total channel response to $x(t)$ is then

$$y(t) = \int_{-\infty}^{\infty} H(i\omega, t)X(i\omega) \exp (i\omega t) \, d(\omega/2\pi) \tag{18-116}$$

which is the limit of the sum of the channel responses to the components of $x(t)$ from various infinitesimally wide spectral elements. The time-variant, *frequency-dependent transfer function* $H(i\omega, t)$ can be shown to be the *Fourier transform over the delay-spread variable τ* of the *delay-spread function* $h(t, \tau)$.

The randomness of the channel with time is reflected in the treatment of the system functions $H(t, \tau)$ and $H(i\omega, t)$ as sample functions of processes that are random over the space of the time variable t. The autocorrelation function of the channel response process is given by an inverse Fourier transform operation on the product of the spectral density function of the input process and the autocorrelation function of the time-variant, frequency-dependent transfer function $H(i\omega, t)$ of the channel. Further transformations produce a combined time-shift and frequency-shift correlation function and a so-called *scattering function* $S(\tau_0, f_0)$, which has the physical significance of a function that determines the weighting of the signal power as a function of the time delay τ_0 and Doppler shift f_0 incurred in transmission.

On the basis of the above functions, a set of transmission parameters for random time-variant linear filters is defined.[16]

1. The *multipath spread* or *delay spread* is determined by the relative delays of the component paths, or the "duration" of $h(t, \tau)$ over the delay variable τ.

2. The *coherence bandwidth*, the bandwidth over which correlation of amplitude fading (or coherence of phase for some applications) remains to a desired degree is usually defined in terms of specified degradation of error rate, distortion, or other parameter.

3. *Diversity bandwidth* is the frequency separation between two sinusoidal inputs which results in a specified decorrelation of the fluctuating responses, usually taken to be a correlation coefficient of $1/e$.

---

*The outline of channel characterization follows E. J. Baghdady, covered basically in Ref. 16.

4. The *fading rate, fading bandwidth,** *frequency smear,* or *Doppler spread* is a measure of the bandwidth of the received signal when the input to the channel is a stable single-frequency signal.

5. The *diversity time* (or decorrelation time) is a measure of the time separation between input signals to yield correlation of less than $1/e$ between the envelopes of the responses.

These parameters are not all independent; the coherence bandwidth and delay spread are inversely proportional to each other, as are fading bandwidth and decorrelation time.

Most channel characteristics can be measured. The delay-spread response $h(t, \tau)$ can be measured directly by transmitting very short, widely spaced pulses; each received replica will correspond to one path, which can be resolved to examine the relative amplitudes and delays. The amplitude characteristic of the frequency-dependent transfer function $|H(\omega, t)|$ can be measured for short intervals by transmitting a constant-amplitude test signal with repetitive linear sweep covering the desired frequency range. The envelope of the received signal will give a very close approximation to $|H(\omega, t)|$.

Diversity techniques for counteracting short-term fading are used for HF ionospheric communication, forward scatter systems, and increasingly for microwave line-of-sight systems, where high reliability is required. The most common mechanism is to use *spaced antennas*, taking advantage of the fact that fading at one antenna tends to be independent of the signal fluctuation received on another antenna; provision is made to switch between signals or to combine two or more of them. Depending on the propagation mechanism, other kinds of useful diversity include *frequency, angle of arrival, polarization,* and *time* [Refs. 4 and 15 and CCIR Rep. 327-3).[175] *Selection combining* selects the strongest signal; other linear combining methods include *maximal-ratio* and *equal-gain* combining. Diversity operation presumes the availability of *n* independently fading signals, referred to as diversity branches. There is no universal definition of diversity improvement, but it can be indicated for digital error probability.

For Rayleigh fading, maximal ratio combining for coherent diversity reception gives an error probability P_e for digital detection as follows. For one receiver (no diversity)

$$P_{e1} = \tfrac{1}{2} - \tfrac{1}{2}\sqrt{\alpha R/(\alpha R + 1)} \qquad (18\text{-}117)$$

where $\alpha = 1$ for phase-reversal modulation and $\tfrac{1}{2}$ for frequency-shift or on-off keying. R is the normalized signal-to-noise ratio, equal to the ratio of average signal energy to noise energy in each branch. For dual diversity

$$P_{e2} = \tfrac{1}{2} - \tfrac{1}{2}\sqrt{\alpha R(\alpha R + \tfrac{3}{2})/(\alpha R + 1)^3} \qquad (18\text{-}118)$$

The outline of fading and diversity improvement given here has assumed *flat* (nonfrequency-selective) fading and gaussian (white) noise. References 15 and 16 discuss frequency-selective fading. Paragraph **18-73** indicates nongaussian noise effects.

73. Noise: Signal-to-Noise Ratio; Service Probability † Several types of radio noise must be considered in any design, though, in general, one type will be the dominant factor. In broad categories, the noise can be divided into two types: noise internal to the receiving system and noise external to the receiving antenna.

The *noise to the receiving system* is often the controlling noise in systems operating above about 300 MHz. This type of noise is due to antenna losses, transmission-line losses, and the circuit noise of the receiver itself. For receiver noise the instantaneous noise voltage has a gaussian distribution, and the noise envelope is Rayleigh-distributed.[15,20] Its effect on most communication systems can be determined mathematically with high accuracy.[15,21]

The second of the broad categories, *external radio noise,* can be subdivided further. Natural sources of radio noise are (1) atmospheric, (2) galactic, (3) solar noise from antennas pointing at the sun, (4) precipitation (blowing snow or dust), (5) corona, and (6) noise reradiating from any absorbing medium through which the wanted radio signal passes. Very low noise systems used in space communications can be limited by such absorption by clouds, water vapor, and oxygen. Such *sky noise* has the gaussian characteristic of receiver noise. Examples of man-made noise sources are: (1) power lines or generating equipment, (2) automotive ignition systems, (3) fluorescent lights, (4) switching transients, and (5) electrical equipment in general.

Noise power is generally the most significant single parameter in relating the interference value of the noise to system performance. This parameter, however, is seldom sufficient in the

---

* *Fading bandwidth* is also often used in the same sense as *diversity bandwidth,* or *coherence bandwidth,* above.

†This material was contributed largely by R. T. Disney and A. D. Spaulding.

case of impulsive noise, and a more detailed statistical description of the received-noise waveform is generally required.

Figure 18-49 shows the median value of the *available noise-power spectral density* from various sources. F_a is in decibels above Kt_0 ($1 Kt_0 = 3.97 \times 10^{-21}$ W/Hz).* While the solar noise, galactic noise, and sky noise are gaussian, the atmospheric and man-made noises are very impulsive.

CCIR Rep. 322-1 gives detailed definition of F_a, expected atmospheric noise level worldwide, and statistical variation as a function of geographical location, frequency, and time. Envelope probability distributions are given for atmospheric noise. Estimates of the noise-power spectral density from man-made noise expected in business, residential, rural, and quiet rural areas have

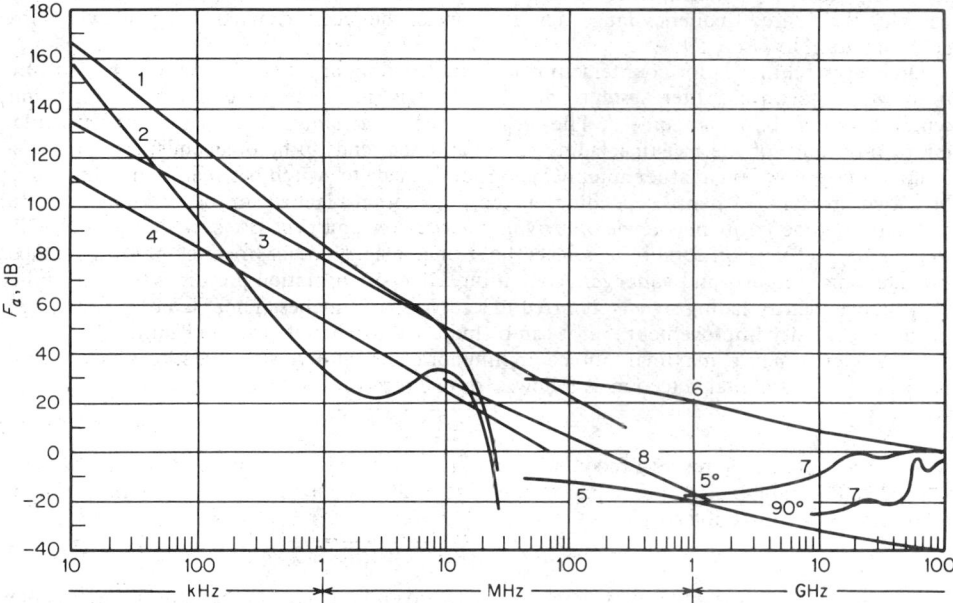

Fig. 18-49. Median radio noise-power spectral density from various sources. Curve 1 = atmospheric noise, summer, 2000 to 2400 h, Washington, D.C., omnidirectional antenna near ground; curve 2 = atmospheric noise, winter 0800 to 1200 h, Washington, D.C., omnidirectional antenna near ground; curve 3 = man-made noise, business area, omnidirectional antenna near ground; curve 4 = man-made noise, quiet rural area, omnidirectional antenna near ground; curve 5 = quiet sun, isotropic (0 dB gain) antenna; curve 6 = disturbed sun, isotropic (0 dB gain) antenna; curve 7 = sky noise, narrow-beam antenna (degrees from vertical); curve 8 = galactic noise, omnidirectional antenna near ground.

been developed from measurements.[22-26] These expected values are the means of a number of location medians, with variation from location to location in each type of area and temporal variation indicated. Generally the noise below 20 MHz is associated with power lines. At 20 MHz and above, automotive electrical systems, especially ignition systems, are the dominant sources in all but rural locations.

Impulsive atmospheric or *man-made noise* disturbs communications in a way quite different from gaussian noise. Figure 18-50 shows the amplitude probability distribution of gaussian noise and a sample of atmospheric noise. The parameter V_d is the ratio in decibels of the rms voltage to the average voltage and is commonly used as an *impulsive index*. The two distributions are plotted relative to their rms level; i.e., both noises shown have the same energy or power. The probability distribution of the noise envelope determines the performance of most basic digital receivers, as indicated by the two error-rate curves for a binary coherent phase-shift-keying system.

Digital receivers are frequently designed for optimum performance in white gaussian noise.

*K is Boltzmann's constant; $t_0 = 300$ K.

Their performance in impulsive man-made or atmospheric noise can be summarized as follows, with comparisons made on the basis of equal noise power:

1. For constant signal, at high signal-to-noise (S/N) ratio, impulsive noise causes more errors than gaussian noise; at lower S/N ratio, gaussian noise causes more errors.

2. For Rayleigh-fading signals, gaussian noise causes more errors at all S/N ratios; flat-fading cases do arise, for which impulsive noise will cause more errors than gaussian noise; for diversity reception, impulsive noise is more harmful.

3. While pairing of errors in differentially coherent phase-shift keying (DCPSK) becomes more unlikely as the S/N ratio increases, pairing of errors increases as the noise becomes more impulsive.

4. For systems with time-bandwidth products on the order of unity, the standard matched filter-receiver is also optimum for impulsive noise.

5. Noise-suppression schemes such as wide-band limiting, smear-desmear, etc., are not particularly effective at high S/N ratios.

6. Receivers especially designed to reject a particular type of impulsive noise perform substantially better than receivers using the "standard" noise-suppression techniques.

The performance of analog voice systems in impulsive noise can be summarized as follows:

1. For a given articulation index, a much lower S/N ratio is required for impulsive noise than for gaussian noise.

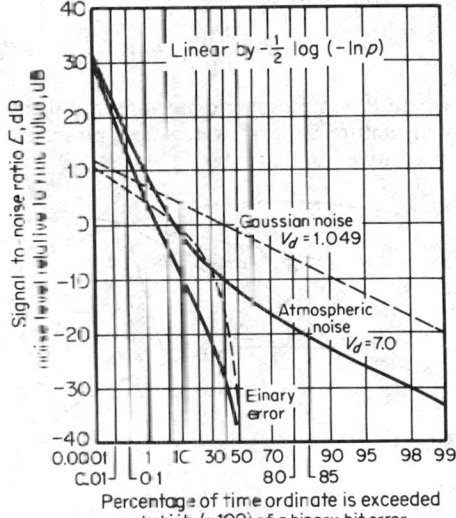

Fig. 18-50. Comparison of noise distribution and error probabilities for gaussian and nongaussian noise (same noise power, coherent phase-shift keying).

2. Various forms of limiting in AM systems (pre-i.f. limiting, i.f. limiting, postdetection limiting) are quite effective in further reducing the required S/N ratio when the noise is impulsive.

Two additional measures (*CCIR Rep.* 322-1) are used to predict long-term performance: the percentage of time a given error rate, or better, will be achieved, termed *time availability*, and the probability that a given system will achieve a specified time availability and error rate, termed *service probability*. The service probability is designed to account for the probable errors or uncertainties in the prediction of the noise and signal distributions.

PROPAGATION OVER THE EARTH THROUGH THE NONIONIZED ATMOSPHERE

74. Introduction. Mechanisms important to propagation over the earth and via the non-ionized atmosphere at frequencies below about 30 MHz include free-space radiation, ground reflection, and diffraction along the surface of the earth, allowing for its finite conductivity, dielectric properties, irregularities, and refractive effects of the atmosphere.

Above about 30 MHz, important mechanisms include free-space radiation, refraction, reflection from elevated atmospheric layers, reflection by the ground and various obstacles, absorption of energy by trees and buildings, diffraction over the surface of the earth and by hills, forward scattering of radio waves from atmospheric irregularities and layers, and scattering and absorption by atmospheric constituents. Terrain and meteorological conditions play important roles in determining the strength and fading properties of transmission through the troposphere.

Above 10 GHz, scattering and absorption effects by water vapor, moisture particles, and oxygen are dominant.

75. Tropospheric Refraction. If a radio wave is propagated in free space (no atmosphere), the path followed by the ray is a straight line. However, in passing through the earth's atmosphere, the vertical gradient of atmospheric refractive index along the trajectory causes the ray path to become curved,[27] as shown in Fig. 18-51. The total angular refraction is τ, bending of the ray. The atmospheric radio refractive index n always has values slightly greater than unity near the earth's surface, for example, 1.0003, and approaches unity (free-space value) with increasing height. Ray paths usually (but not always) have a curvature concave downward, depending upon the gradient dn/dh.

The *radio refractive index* of the atmosphere is

$$n = 1 + N \times 10^{-6}$$

The *refractivity* is

$$N = \frac{77.6}{T} \left(P + \frac{4{,}810e}{T} \right) \quad \text{or} \quad N = \frac{77.6}{T} \left(P + \frac{4{,}810e_s\text{RH}}{T} \right) \tag{18-119}$$

where P = atmospheric pressure (millibars), e_s = saturation water-vapor pressure (millibars at temperature T), e = water-vapor pressure (millibars), T = absolute temperature (K), and RH = relative humidity (%).

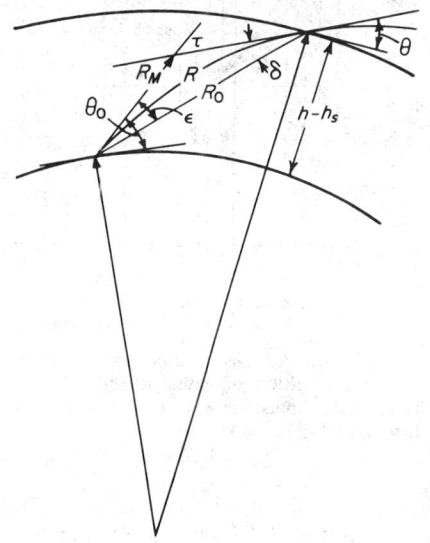

Fig. 18-51. Geometry of the refraction of radio waves. (*Bean et al.*[29])

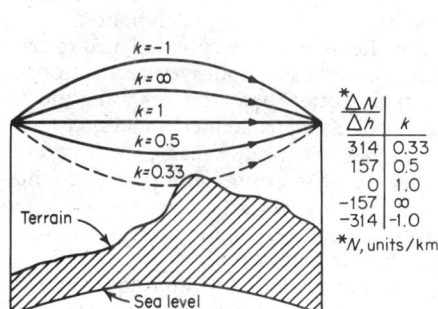

Fig. 18-52. Blending of radio waves for linear gradients. (*Dougherty.*[57])

The elevation-angle error (ϵ in Fig. 18-51) is an important quantity in radar and other positioning or tracking systems.[27,28] It is a measure of the difference between apparent elevation angle to a terminal or target and the true elevation angle:

$$\epsilon = \tan^{-1} \frac{\cos \tau - \sin \tau \tan \theta - n/n_s}{(n/n_s) \tan \theta_0 - \sin \tau - \cos \tau \tan \theta} \tag{18-120}$$

where n_s is the refractive index at the surface

$$\tau_{1,2} = \int_{n_1}^{n_2} \cot \theta \, \frac{dn}{n} \tag{18-121}$$

To evaluate τ, n must be known as a function of height.

Many field-strength, phase, and bending calculations are made assuming a constant gradient of refractive index with height, equivalent to assuming an effective earth radius $a = ka_0$, where a_0 is the real radius and k is determined by

$$k = \frac{1}{1 + (a_0/n)(dn/dh)} \tag{18-122}$$

Widespread practice uses $k = \frac{4}{3}$, corresponding to $dn/dh = -\frac{1}{4}a_0$ or a constant refractivity gradient with height of approximately $-39.3N$ units/km. The gradient is usually expressed in terms of the refractivity, $N = (n - 1) \times 10^6$, so that

$$10^6 \frac{\Delta n}{\Delta h} = \frac{\Delta N}{\Delta h} N \quad \text{units/km} \tag{18-123}$$

and

$$k \approx [1 + (\Delta N/\Delta h)/157]^{-1} \tag{18-124}$$

Ray paths are illustrated in Fig. 18-52 for several values of k and $\Delta N/\Delta h$. The vertical scale is exaggerated relative to the horizontal scale to make the differences in curvature apparent.

For $0 < k < 1.0$, corresponding to positive gradients (subrefractive conditions), the ray curves away from the earth so that the ray joining two terminals passes close to the earth. The ray may even be interrupted by the surface so that the receiving terminal is beyond the radio line of sight. For the common situations $-157 \leq \Delta N/\Delta h \geq 0$, where $\infty \geq k \geq 1.0$, the rays are bent toward the earth's surface. At the critical value $\Delta N/\Delta h = -157 N$ units/km, $|k| = \infty$, and the curvature of the ray path is equal to the curvature of the earth; the rays follow straight paths relative to the earth's surface. For $\Delta N/\Delta h < -157$, the situation is supercritical and the value of k is negative. For negative values the ray paths are bent sufficiently toward the earth's surface for trapping of the radio rays to be possible. $\Delta N/\Delta h < -157$ is commonly referred to as a *trapping* or *ducting* condition.

The use of the constant gradient may lead to errors by overestimating refraction, especially important when one or both terminals are at high altitude, e.g., air to ground, or over long paths, and at frequencies much above VHF; at low frequencies the phase of the ground wave may be in error.

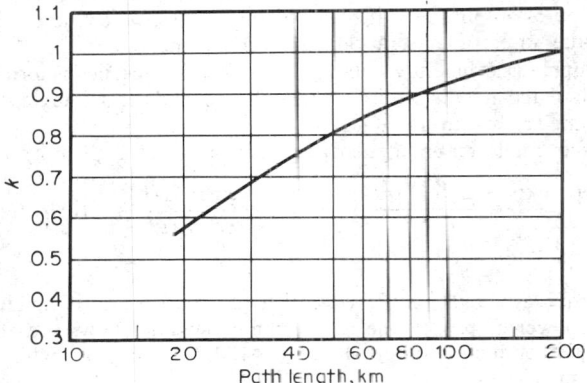

Fig. 18-53. Minimum effective value of k (exceeded 99.9% of the time). (*CCIR Rep. 338-4.*)

A better approximation of the mean refractive index as a function of height is given by the *exponential atmosphere* (Ref. 27, *CCIR Rep. 563-2*):

$$n(h) = 1 + N_s \exp(-bh) \times 10^{-6} \tag{18-125}$$

where N_s = surface value of N, h = height (km), and b is given by

$$\exp(-b) = 1 + \Delta N/N_s \tag{18-126}$$

where ΔN = difference between surface value of N and its value at 1 km height.

ΔN is generally correlated (for long-term means) with surface value N_s, which allows estimating ΔN in the usual case where only surface meteorological data are available:

$$-\Delta N = 7.32 \exp 0.005577 N_s \tag{18-127}$$

(This value is for the United States; other values for other areas are given in *CCIR Rep. 563-2*.) An effective earth's radius a is given by

$$a = a_0(1 - 0.04665 \exp 0.005577 N_s)^{-1} \tag{18-128}$$

Worldwide seasonal charts of N_s and ΔN are published.[28,29] The maps are in terms of N_0 the sea-level value of N, and N_s is obtained from

$$N_0 = N_s \exp(h/7) \tag{18-129}$$

where h = height above sea level (km). From annual mean figures, an average exponential model atmosphere corresponds to an equivalent earth's radius equal to about ⅓ times the actual radius. Generally speaking, this factor k can be derived for any region from either meteorological observations or propagation measurements. The effective value of k also depends upon the path length and geometry. A minimum effective value of k (CCIR Report 718), i.e., a value of k exceeded approximately 99.9% of the time, for quasi-horizontal paths, is shown in Fig. 18-53.

Long-term minimum values of N_0 given in Fig. 18-54 can be used to derive corresponding minimum values of k for particular geographical regions.

Especially at lower frequencies, the effective earth radius also depends upon frequency and polarization. For conditions producing a ⁴⁄₃ earth radius at 100 MHz, typical corresponding values of effective earth's radius at 10, 1, and 0.1 MHz are 1.3, 1.2, and 1.1, respectively.

76. Ground-Wave Propagation over Homogeneous Spherical Terrain. The term ground wave refers to propagation within line of sight as well as by diffraction beyond the horizon, affected by earth conductivity, dielectric constant, and terrain, and by refraction in the lower atmosphere. Useful calculations can be obtained from the theory for a homogeneous, smooth, spherical terrain, although at frequencies much above 10 MHz additional effects of irregular terrain and variable tropospheric refraction must be considered (Pars. **18-78** to **18-86**). At frequencies below about 30 MHz, ionospheric propagation must also be considered (see Pars. **18-92** to **18-106**).

The most informative contemporary treatises on electromagnetic surface waves are those of Wait;[32,32a,32b] the historical development of the theory is traced, and practical computational equations are derived for homogeneous, spherical terrain and for mixed-path propagation. Important additional references are 3 to 5, 33, and 34. A computer program has been documented* which conveniently computes amplitude and phase of the electric field due to electric or magnetic dipoles for any values of earth constants and antenna heights.

A short radial dipole is at height h_T above a spherical, homogeneous terrain of radius a. The electric field is calculated at a distance D and height h_R. The central angle subtended by D is θ, and the arc along the earth's surface is d m.

The vertical electric field for an angular frequency $\omega = 2\pi f$ is given by

$$e_r \approx -\left[300\sqrt{P_r}\, \frac{\exp(-ikd + i\pi/4)}{d} \right] K \int_{\Gamma} \frac{\exp(-iv\theta t)}{W_1'(t)/W_1(t) - q_v}\, H_1(h_R)H_2(h_T)\, dt \qquad \text{V/m}$$

$$(18\text{-}130)$$

The quantity in square brackets is the vertical field over a perfectly conducting plane for an effective radiated power of p_r kW; the rest of the expression shows the effect of propagating along an imperfectly conducting sphere. In Eq. (18-130) Γ is a contour which encloses the poles of the integrand, and

$$K = \theta\sqrt{v/12a \sin\theta} \qquad k = 2\pi/\lambda \qquad q_v = -iv(k/k_2)\sqrt{1 - (k/k_2)^2}$$
$$v = (k_a/2)^{1/3} \qquad k_2 = k\sqrt{\epsilon_r - i60\sigma\lambda}$$

The functions $W_n(t)$ are Airy functions,[4,5] and

$$W_n'(t) = \frac{d}{dt}\, W_n(t)$$

$H_1(h_R)$ and $H_2(h_T)$ are height-gain functions:†

$$H_1(h_R) = W_1(t - y_R)/W_1(t)$$

and

$$H_2(h_T) = \frac{W_2(t - y_T)[W_1'(t) - q_v W_1(t)] - W_1(t - y_T)[W_2'(t) - q_v W_2(t)]}{2i}$$

where

$$y_n = kh_n/v$$

Note that

$$H_1(0) = H_2(0) = 1$$

so that the height-gain functions show the effect of nonzero antenna heights.

At distances beyond the horizon the integral in Eq. (18-130) is more easily calculated by summing the residues at the poles t_s, such that

$$W_1'(t_s) - q_v W_1(t_s) = 0$$

---

*L. A. Berry has contributed the computing formulas here for the ground wave.
†The form of H_1 and H_2 requires that $H_R > H_T$; since propagation is reciprocal, there is no loss of generality and the greater height can be assigned to H_R.

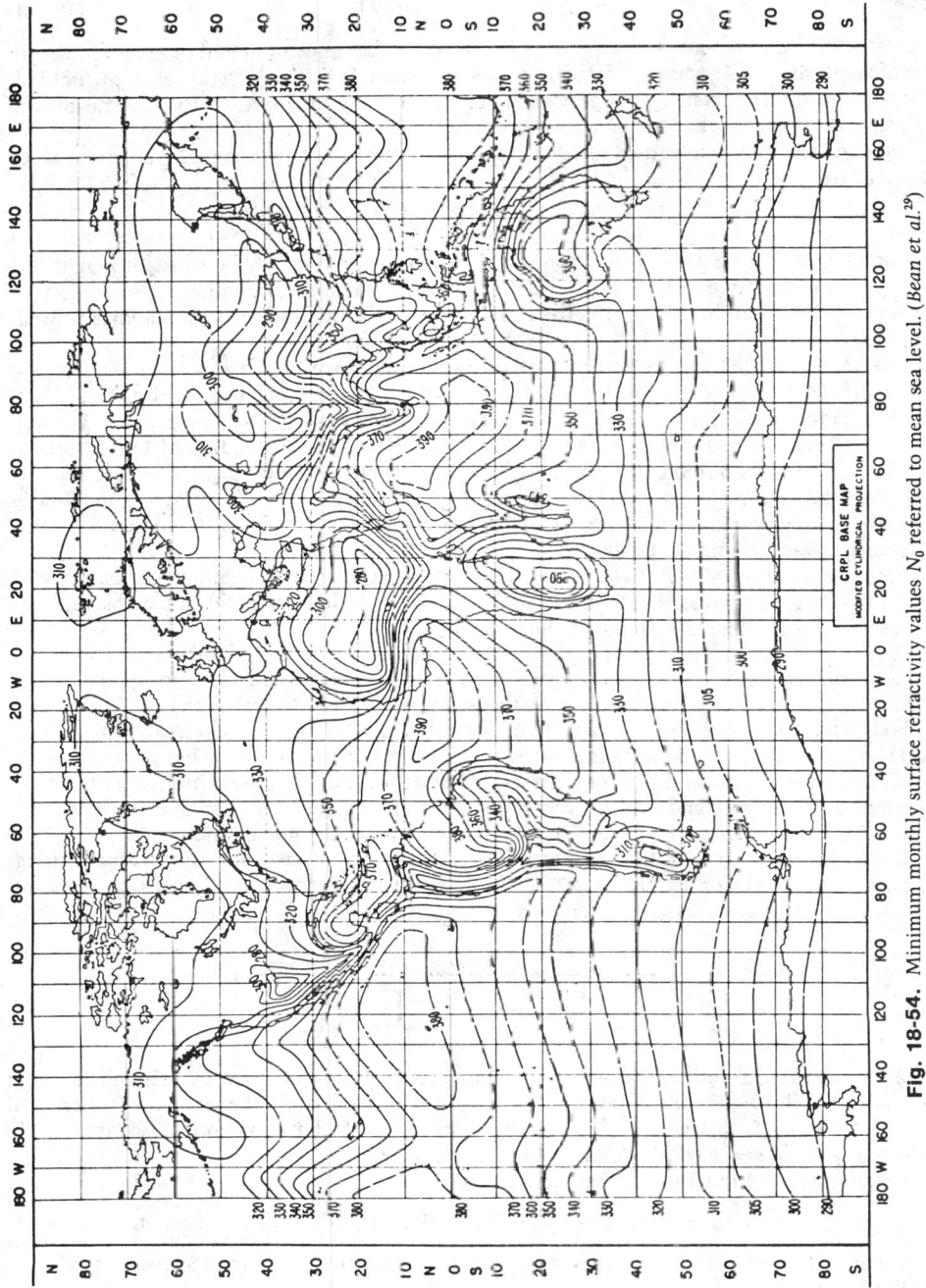

Fig. 18-54. Minimum monthly surface refractivity values N_0 referred to mean sea level. (*Bean et al.*[29])

The classical residue series is then given by

$$e_t \approx -\left[300\sqrt{p_r}\,\frac{\exp(-ikd+i\pi/4)}{d}\right]2\pi iK\sum_{s=0}^{\infty}\frac{\exp(-iv\phi t_s)}{t_s - q_v^2}H(h_R)H(h_T)\qquad \text{V/m}$$

where

$$H(h_n) = W_1(t_s - y_n)/W_1(t_s) \tag{18-131}$$

Because of the slow convergence of Eq. (18-131) when the transmitter and receiver are within line of sight, the saddle-point approximation[3,4] is used within the line-of-sight region, which has been shown to be equivalent to a geometric optical solution. An approximation useful for the diffraction region at higher frequencies is given in Par. **18-78**.

The same formulas can be used to calculate the field strength of a horizontally polarized wave by replacing q_v with

$$q_h = (k_2^2/k^2)q_v$$

Tropospheric refraction is allowed for by use of a linear profile of N, or constant height gradient, using an effective earth radius. The WKB[3,4,32] comparison-equation method, a solution for the linear profile, can be used as the basis for any profile which increases monotonically with height.

CCIR Recomm. 368-4 gives vertically polarized ground-wave field-strength curves from 10 kHz to 300 MHz. Figures 18-55a and b and 18-56a and b give examples for seawater and earth of good conductivity. Curves of phase of the low-frequency ground wave are given in Ref. 35. FCC Rules[36] give field-strength curves for 500 to 1,600 kHz for use in standard AM broadcasting.

Two CCIR Atlases of ground-wave propagation[37,38] give curves for 30 to 10,000 MHz for vertical polarization and elevated antennas, illustrated for 30 and 300 MHz in Fig. 18-56a and b. Field-strength curves for mobile service and broadcasting service, taking into account terrain variations, are shown in Par. **18-82**.

Inhomogeneous or Mixed Paths. Ground-wave propagation over an inhomogeneous path, e.g., a mixture of land areas of different ground constants or mixed water and land paths, requires special (usually graphical) methods of computation.

Wait[32,32a,32b,41] gives amplitude and phase factors for two-section paths for low and medium frequencies; Godzinski[42] gives curves for calculation of attenuation and phase for many-section paths. Widely accepted semiempirical methods are Millington's[43] and Kirke's.[44] Theoretical methods have been developed for high frequencies.[45] Wait[32a,32b] gives recent analytical investigations for mixed paths and applications. A general integral equation is given for the case of smooth boundaries that can be characterized by local surface impedances. A number of practical situations are considered and methods of solution are indicated. References to all important work in mixed path propagation are given. Ott[45] has given an integral equation method.

CCIR recommends the Millington method, as follows. Such mixed paths may be made up of sections S_1, S_2, S_3, etc., of lengths d_1, d_2, d_3, etc., having conductivity and dielectric constant σ_1, ϵ_1; σ_2, ϵ_2; σ_3, ϵ_3, etc., as shown below for three sections:

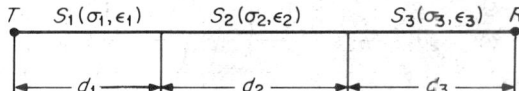

For a given frequency, the curve appropriate to the section S_1 is then chosen and the field $E_1(d_1)$ in decibels (dBμ) at the distance d_1 is then noted. The curve for the section S_2 is then used to find the fields $E_2(d_1)$ and $E_2(d_1 + d_2)$ and, similarly, with the curve for the section S_3, the fields $E_3(d_1 + d_2)$ and $E_3(d_1 + d_2, d_3)$ are found, and so on.

A received field strength E_R is then defined by

$$E_R = E_1(d_1) - E_2(d_1) + E_2(d_1 + d_2) - E_3(d_1 + d_2) + E_3(d_1 + d_2 + d_3)$$

The procedure is then reversed, and calling R the transmitter and T the receiver, a field E_T is obtained, given by

$$E_T = E_3(d_3) - E_2(d_3) + E_2(d_3 + d_2) - E_1(d_3 + d_2) + E_1(d_3 + d_2 + d_1)$$

The required field is given by $\frac{1}{2}(E_R + E_T)$, the extension to more sections being obvious.

The method can in principle be extended to phase changes if the corresponding curves for phase as a function of distance over a homogeneous earth are available. Such information would

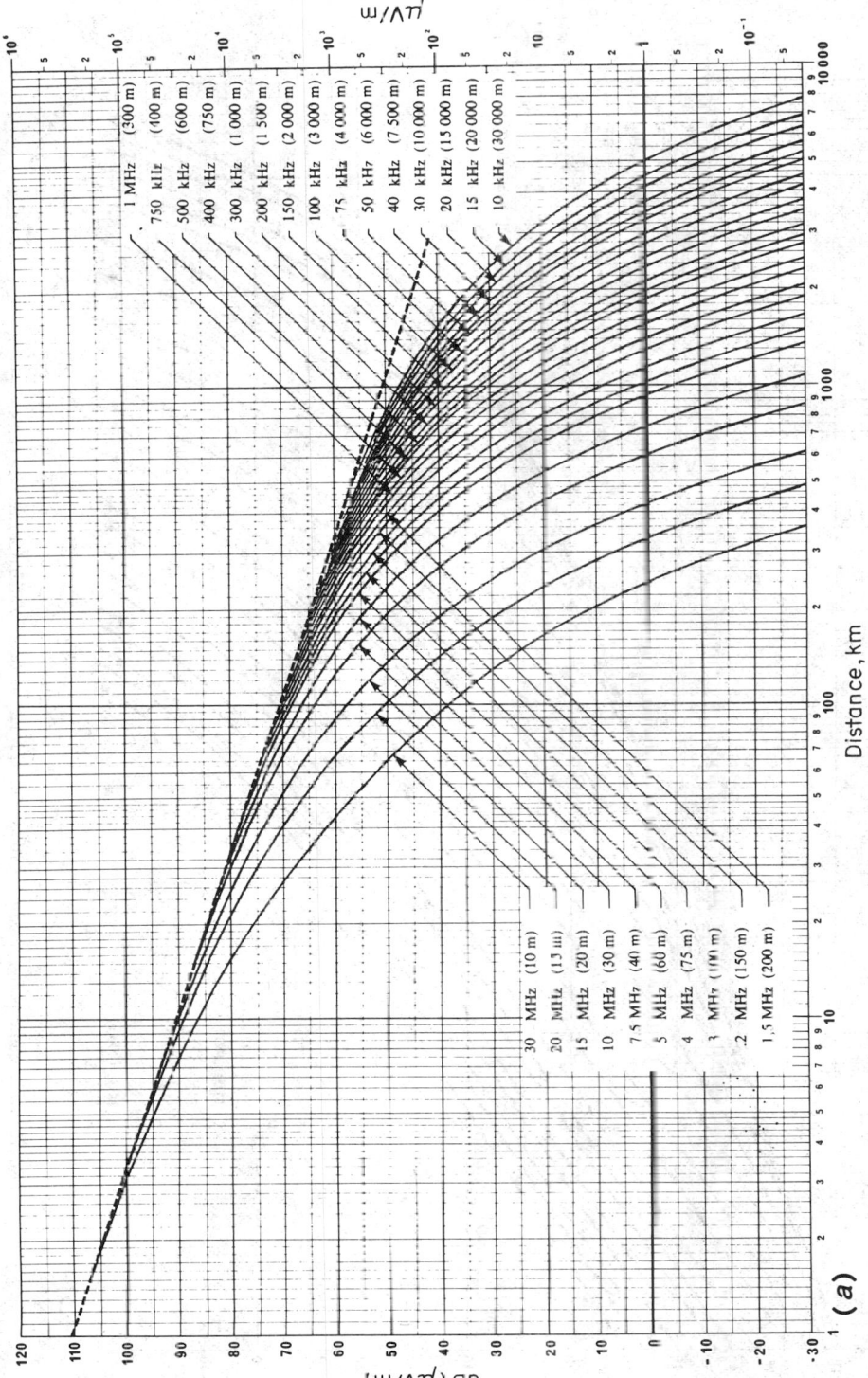

Fig. 18-55a. Ground-wave field strength, reference 1 kW radiated from dipole over perfect earth, that is, 346 mV/m inverse distance field at 1 km. For 10 kHz to 30 MHz, vertical polarization, seawater $\sigma = 5$ S/mm, $\epsilon = 80$.

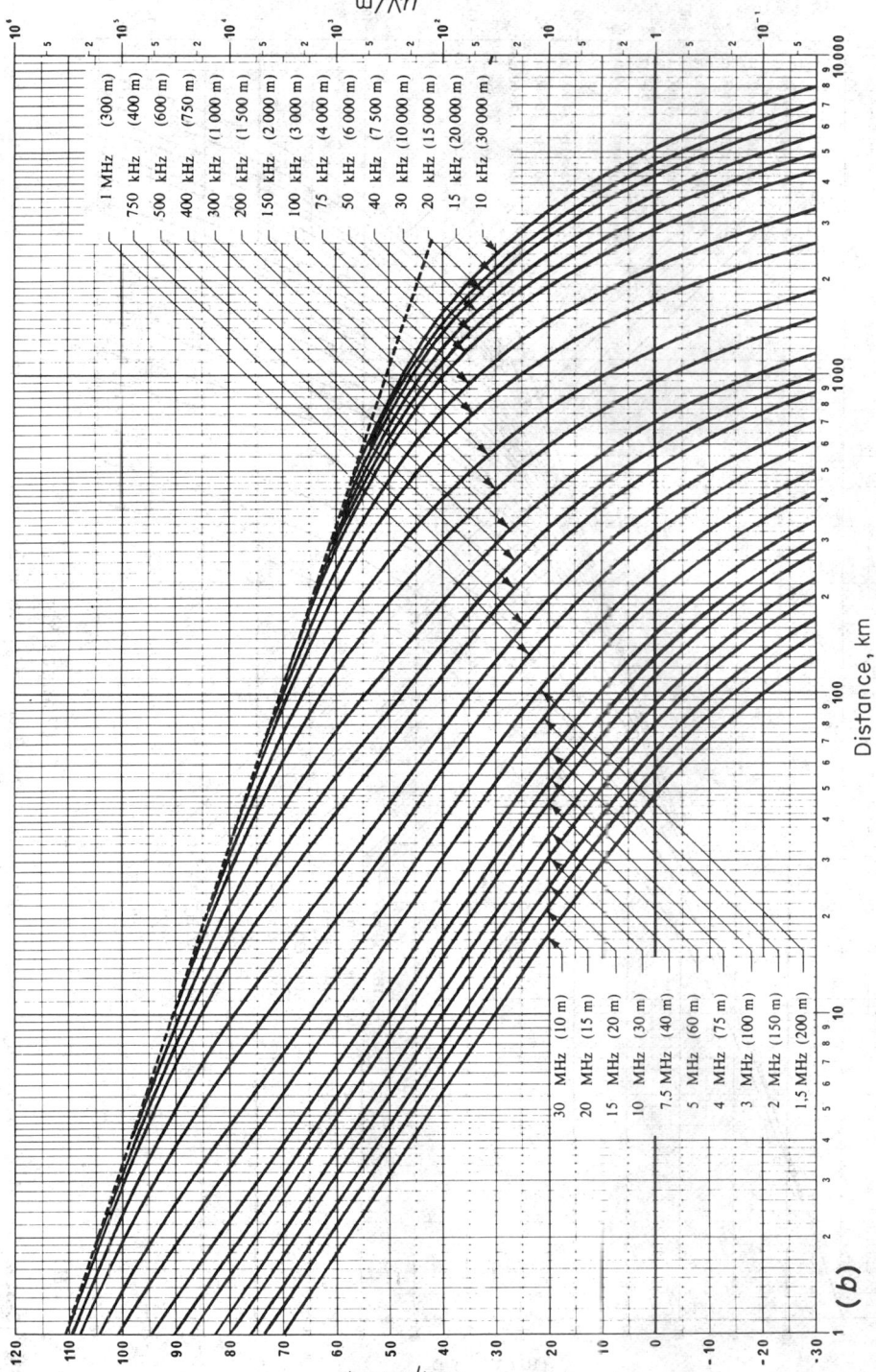

Fig. 18-55b. Same as Fig. 18-55a, for 10 kHz to 30 MHz, vertical polarization, average earth, $\sigma = 1$ ms/m, $\epsilon = 4$.

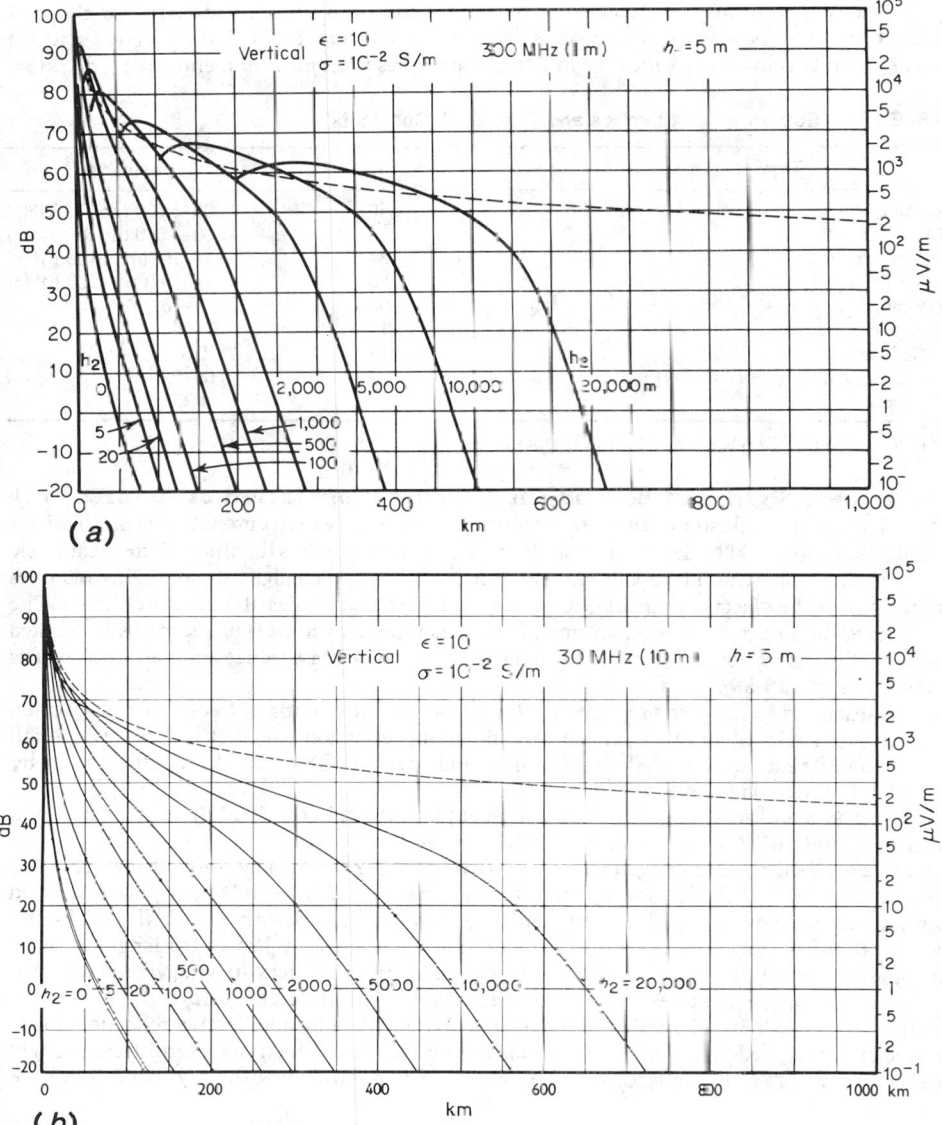

Fig. 18-56. Ground-wave field strength for 1 kW radiated; smooth earth, vertical polarization, good conductivity, $\sigma = 10^{-2}$ S/m, $\epsilon = 10$, h_1, h_2 = antenna heights. (a) 300 MHz; (b) 30 MHz.

be necessary for application to navigational systems. The Millington method is generally easy to use, particularly with a computer.

For planning purposes where the coverage of a certain transmitter is needed, a graphical procedure,[45a] based on the same method, is convenient for a rough and quick estimation of the distance at which the field strength has a certain value.

77. Electrical Characteristics of the Earth. The relative permeability can usually be taken as unity. The relative importance of ϵ, and σ varies with wavelength and can be judged by the ratio $\epsilon_r/60\sigma\lambda$ (λ is in meters). Table 18-4 shows earth constants for various types of surface.

In Table 18-4 the range of conductivity values for a specified type of surface corresponds to differences which exist in different parts of the world. In general, in fertile areas the higher values are applicable while the conductivity of water in lakes and rivers increases with concentration of impurities. The effective values of the ground constants depend on frequency, depth of penetration, lateral spread of the wave, and the geological structure.

FCC Rules[36] give a map of effective ground conductivity in the United States for the AM standard broadcast bands; a similar map is available for Canada. *CCIR Rep.* 717 gives a world atlas of ground conductivities for very low frequencies, based on measurements over long paths.

TABLE 18-4   Some Types of Surface and Their Earth Constants*

Type of surface	ϵ_r	σ, mS/m
Seawater, at 0°C	80	4,000–5,000 (up to 1 GHz)
At 10°C	73	4,000–5,000 (up to 1 GHz)
Fresh water, at 10°C	84	1–10 (up to 100 MHz)
At 20°C	80	1–10 (up to 100 MHz)
Very moist ground	30	5–10
Average ground	15	0.5–5
Arctic land	15	0.5
Very dry ground and large towns (industrial areas)	3	0.05–0.1
Polar ice	3	0.25

*CCIR Recomm. 527 shows the effect of frequency.

78. Line-of Sight and Beyond-the-Horizon Propagation at 30 MHz to 10 GHz. Figure 18-57 illustrates the interrelationships between various propagation mechanisms at frequencies above 30 MHz and the conditions under which each is dominant. The great-circle terrain profile is shown under conditions of normal refraction; the height of mountains and trees and the size of the objects are greatly exaggerated for the amount of earth curvature shown. The legend lists the propagation mechanisms likely to dominate over each of the paths illustrated under conditions of normal refraction (the effect of a ground-based duct and elevated layer is discussed in Par. **18-86**).

Earth-space and space-space propagation, while representing essentially free-space transmission at the higher angles of elevation, encounters important refraction effects at low angles, rainfall attenuation at times at any angle of elevation, and, near 10 GHz and above, attenuation by atmospheric water vapor and oxygen.

The methods outlined below for computation of propagation in the 30-MHz to 10-GHz region are given mainly by Ref. 12 and various CCIR reports.[175]

79. Line-of-Sight Propagation over Smooth Terrain. The simplest ray-optics formulas assume that the field at a receiving antenna is made up of two components, one associated with a direct ray having a path length r_0 and the other associated with a ray reflected from a point on the surface, with equal grazing angles ψ. The reflected ray has a path length $r_1 + r_2$. The field arriving at the receiver via the direct ray differs from the field arriving via the reflected ray by a phase angle which is a function of the path-length difference, $\Delta r = r_1 + r_2 - r_0$, illustrated in Fig. 18-58. The reflected-ray field is also modified by an effective reflection coefficient R_e and associated phase lag $\pi - c$, which depends on the conductivity, permittivity, roughness, and curvature of the reflecting surface. Figure 18-58a shows how the rays bend above the

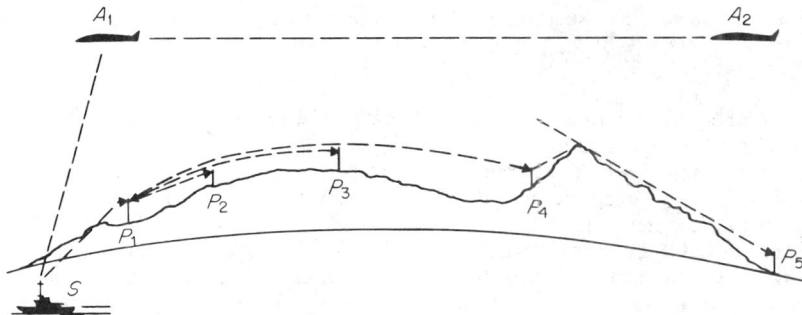

Fig. 18-57. Propagation under conditions of normal refraction; SP_1 = free space plus ground reflection and diffraction over smooth earth; SA_1, A_1A_2 = free space plus ground reflection; P_1P_2 = free space (antenna directivity sufficient to exclude much effect of ground reflection; if not, ground reflection must be considered); P_1P_3 = free space plus ground reflection and diffraction over irregular terrain; P_1P_4 = beyond horizon, mainly diffraction; P_1P_5 = well beyond horizon, forward scatter; P_5A_1, P_5A_4 = knife-edge diffraction.

earth of actual radius a_0, and Fig. 18-58b shows the same rays drawn as straight lines above an effective earth radius a. The effective reflection coefficient R_e is then[46]

$$R_e = DR \exp \{-[(4\pi\sigma_h/\lambda) \sin \psi]^2\} \qquad (18\text{-}132)$$

where the divergence factor D, which allows for the divergence of energy reflected from a curved surface, can be approximated as

$$D = [1 + 2d_1d_2/(ad \tan \psi)]^{-1/2} \qquad (18\text{-}133)$$

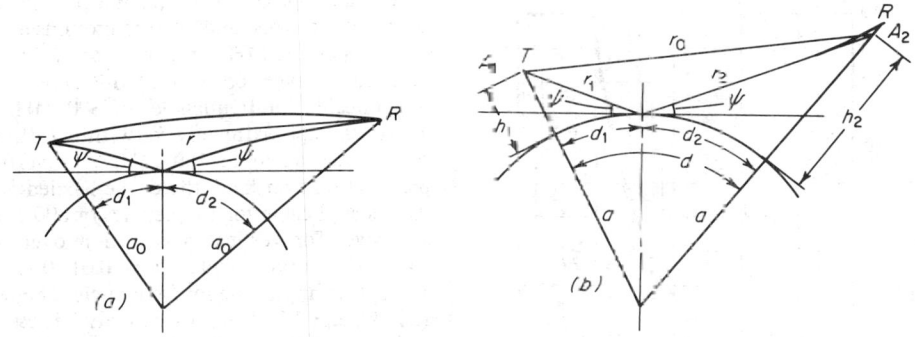

Fig. 18-58. Geometry for line-of-sight paths.

Except for small ψ (less than about 2°) D can be taken as unity. The term R represents the magnitude of the coefficient, $R \exp [-i(\pi - c)]$, for reflection of a plane wave from a smooth plane surface of a given conductivity and dielectric constant (Pars. **18-65** to **18-70** and Refs. 11 and 12). In most cases c can be set equal to zero, and R is very nearly unity. A notable exception occurs for vertical polarization over seawater. The terrain roughness factor σ_h is the rms deviation of terrain elevations.[46]

For line-of-sight transmission involving a single ground reflection, the attenuation relative to free space A can be obtained from

$$A = -10 \log \left[1 + R_e^2 - 2R_e \cos \left(\frac{2\pi \, \Delta r}{\lambda} - c \right) \right] \quad \text{dB} \qquad (18\text{-}134)$$

Over a smooth perfectly conducting surface, $R_e = 1$ and $c = 0$, and

$$A = -6 - 10 \log \sin^2 (\pi \, \Delta r/\lambda) \quad \text{dB} \qquad (18\text{-}135)$$

where $\Delta r = r_1 + r_2 - r_0$ and $\lambda = $ wavelength (m).

For small grazing angles ψ and with antennas h_1 and h_2 km above the earth,

$$\Delta r \approx 2h_1'h_2'/d \qquad (18\text{-}136)$$

where h_1' and h_2' are the heights of the antennas above a plane tangent to the earth at the point of reflection. For equal antenna heights over a spherical earth of effective radius a,

$$\Delta r = d(\sec \psi - 1)$$

Just beyond the radio horizon of a transmitter, the dominant propagation mechanism is usually diffraction. Far beyond the horizon, the dominant mechanism is usually forward scatter.

The curves of Figs. 18-56a and b take into account this geometric-optics approach as well as the effects of diffraction near and beyond the horizon, as discussed in Par. **18-80**.

80. Diffraction over Smooth Terrain at Frequencies above 30 MHz. Paragraphs **18-76** and **18-77** give a method for computing propagation by diffraction over smooth spherical earth which is accurate even to very low frequencies. At frequencies above about 30 MHz, an approximation gives good results. As before, account is taken of atmospheric refraction by use of the equivalent earth's radius equal to $\frac{4}{3}$ the actual radius.

The level A relative to free space beyond the horizon is given [$CCIR$ Rep. 715] by

$$A = G(\chi_0) - F(\chi_1) - F(\chi_2) - 20.5 \quad \text{dB} \qquad (18\text{-}137)$$

where
$$\chi_0 = d_0 B_0 \qquad \chi_1 = d_1 B_0$$

and
$$\chi_2 = d_2 B_0 \qquad B_0 = 670(f/a_e^2)^{1/3}$$

where d_0 = total distance (km), and d_1, d_2 = the distances (km) from each antenna to its radio horizon:*

$$d_1 = \sqrt{2a_e h_1} \qquad d_2 = \sqrt{2a_e h_2}$$

and where h_1, h_2 = antenna heights (km), a_e = effective earth radius (km), and f = frequency (MHz).

The functions $G(\chi_0)$, $F(\chi_1)$, and $F(\chi_2)$ are plotted in Fig. 18-59. The factor k in this figure depends on the frequency and the electrical characteristics of the earth. The curve labeled k

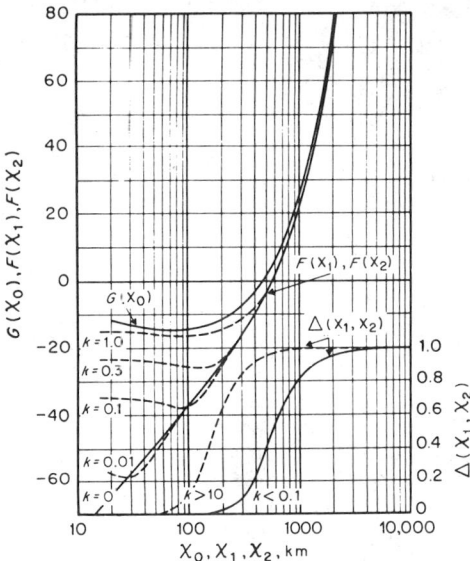

Fig. 18-59. Diffraction (a) by a spherical earth attenuation due to distance and (b) by a spherical earth-height gain.

= 0 is appropriate for horizontal polarization over water or good and average grounds at frequencies of 100 MHz or above. For horizontal polarization over poor ground, this same curve is applicable for frequencies of 600 MHz or above, and approximately so (within 2 dB) for frequencies down to 100 MHz. For vertical polarization over land, the curve labeled k = 0.01 is applicable for frequencies of 600 MHz or above. For vertical polarization over seawater, the curves labeled k = 0.01, 0.1, 0.3, and 1.0 are applicable for frequencies of 3,300, 120, 30, and 7.5 MHz, respectively, or less.

The error in A will be less than 1 dB if

$$\chi_0 - \chi_1 \Delta(\chi_1) - \chi_2 \Delta(\chi_2) > 320 \quad (18\text{-}137a)$$

for horizontal polarization or for vertical polarization on overland paths. For the error in A to be 1.5 dB or less for vertical polarization on oversea paths, the limit indicated as the right-hand side of Eq. (18-137 a) must be 320 for frequencies of 600 MHz or more ($K \leq 0.03$), 335 for 120 MHz ($K = 0.1$), and 115 for 7.5 MHz or less ($K \geq 1.0$). The auxiliary functions $\Delta(\chi_1)$ and $\Delta(\chi_2)$ are given in Fig. 18-59.

Under the same approximation condition (the first term of the residue series is dominant), the calculation can also be made using

$$A = F(d_0) + G(h_1) + G(h_2) \qquad \text{dB} \qquad (18\text{-}137b)$$

The functions F (influence of the distance) and G (height-gain) are given by the nomograms in Fig. 18-59a and 18-59b; F and G have opposite signs.

These nomograms give directly the received level relative to free space for $k = 1$ and $k = \frac{4}{3}$ and for frequencies greater than approximately 30 MHz. k is the effective earth-radius factor. However, the received level for other values of k can be calculated by using the frequency scale for $k = 1$ but replacing the frequency in question by a hypothetical frequency equal to f/k^2 for Fig. 18-59a.

81. Diffraction over Obstacles. Many propagation paths encounter one or several obstacles. Estimates of the losses caused by such obstacles can be obtained by idealizing their form, either assuming a knife-edge or a rounded smooth obstacle with a well-defined radius of curvature. Real obstacles have, of course, more complex forms, so that the idealizations give only approximations, valid mainly at VHF, where the wavelength is small in relation to the size of the obstacle. The method here is outlined from *CCIR Rep.* 715.

In the case of the *knife-edge* obstacle, all the geometrical parameters (Fig. 18-60) are lumped together in a single dimensionless parameter v, which can be found in a number of equivalent forms depending on the geometrical parameters selected

$$v = h \sqrt{(2/\lambda)/(1/d_1 + 1/d_2)} \qquad (18\text{-}138)$$

$$v = \theta \sqrt{\frac{2}{\lambda(1/d_1 + 1/d_2)}} \qquad (18\text{-}138a)$$

*Although metric units are used throughout this section, an approximate and convenient formula for radio horizon distance d in miles for an antenna height of h ft is $d \approx \sqrt{2h}$. This formula allows for $\frac{4}{3}$ earth.

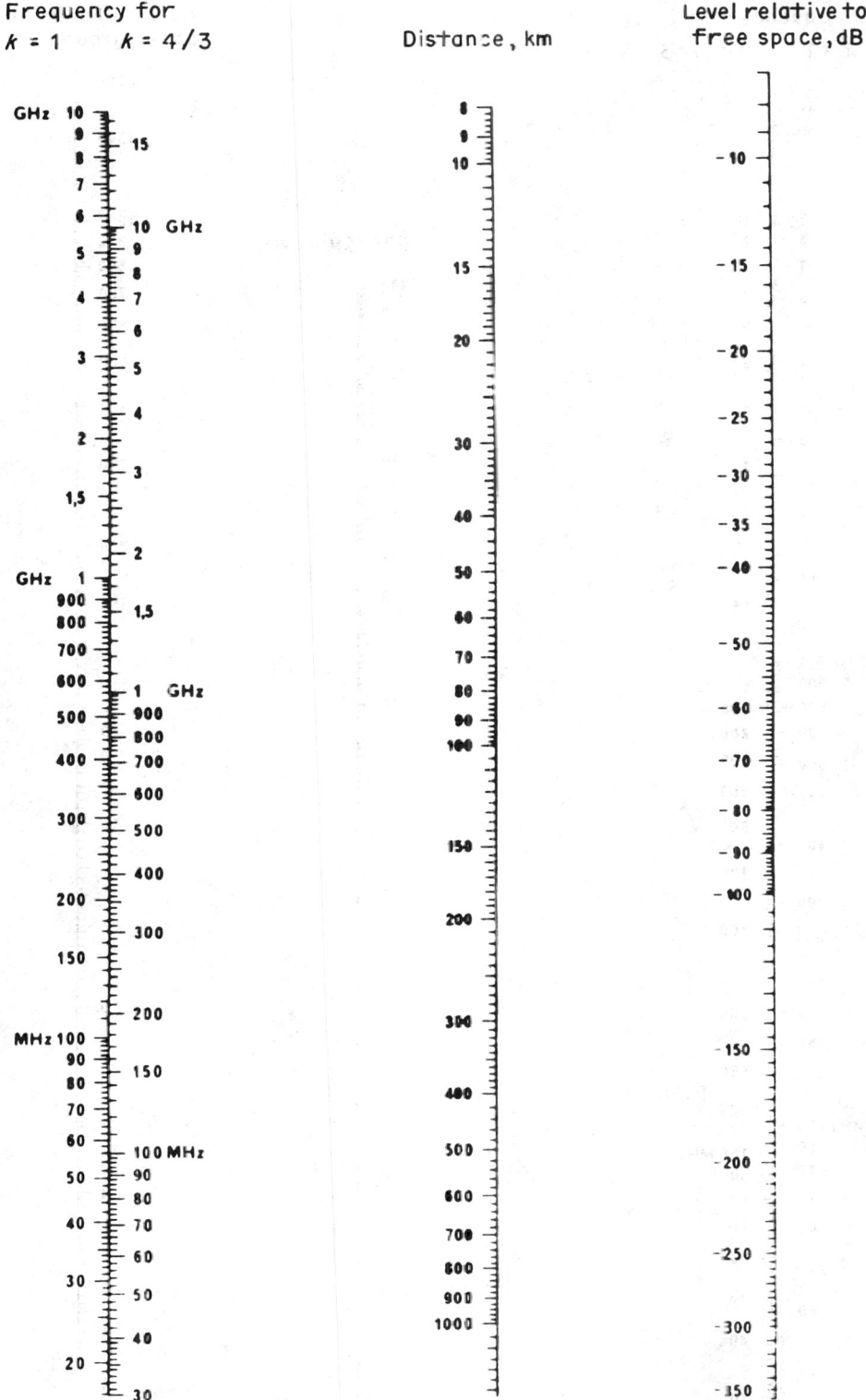

Fig. 18-59a. Influence of distance on free-space signal level.

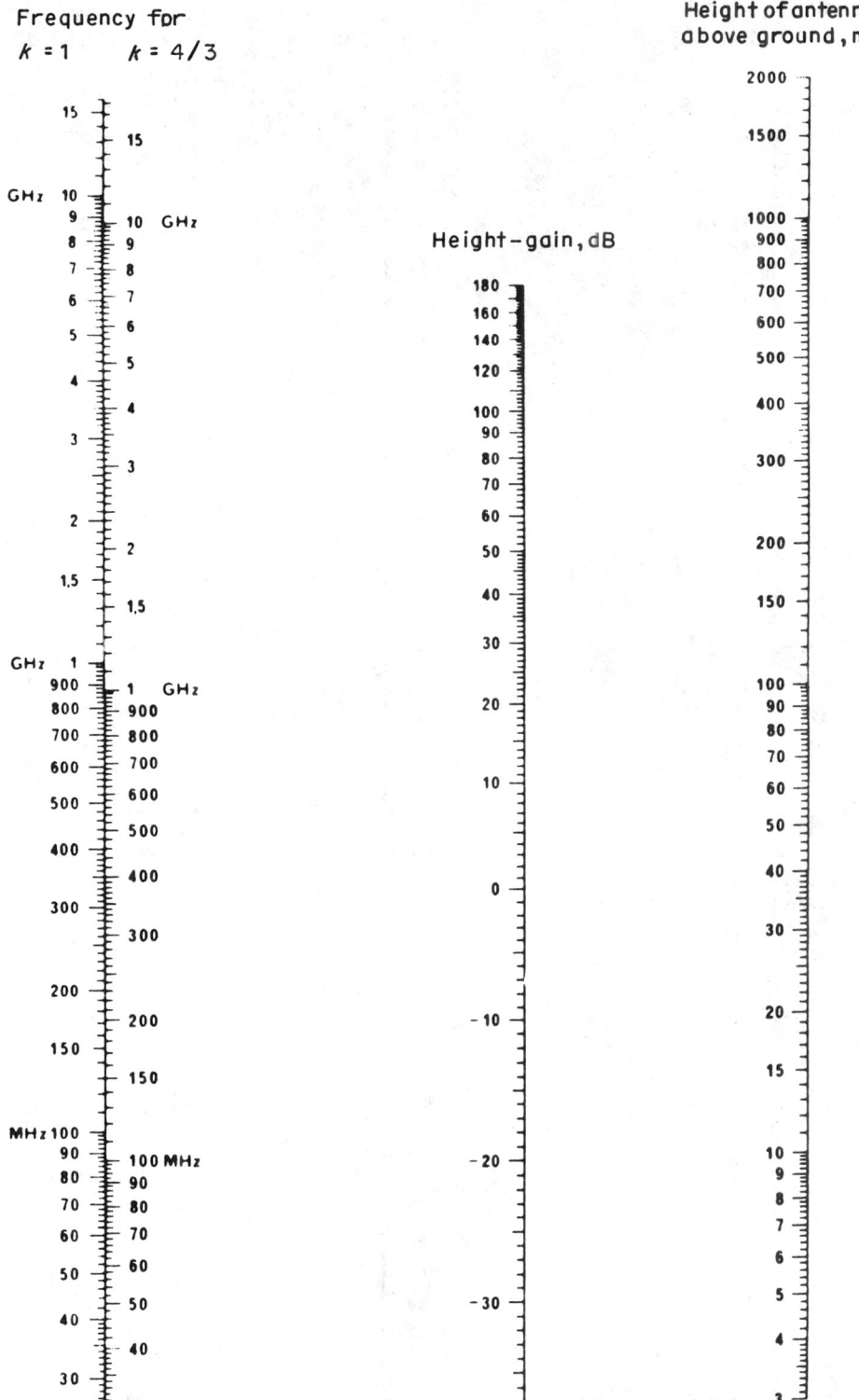

Fig. 18-59b. Height-gain relationship. See Fig. 18-59.

$$v = \sqrt{2h\theta/\lambda} \qquad v \text{ has sign of } h \text{ and } \theta \qquad (18\text{-}138b)$$

$$= \sqrt{(2d/\lambda)\alpha_1\alpha_2} \qquad v \text{ has sign of } \alpha_1 \text{ and } \alpha_2 \qquad (18\text{-}138c)$$

where h = height of top of obstacle above the straight line joining the two ends of the path (if the height is below this line, h is negative); d_1, d_2 = distances of two ends of path from obstacle; $d = d_1 + d_2$ = length of path, θ = angle of diffraction (rad) (its sign is the same as that of h; angle θ is assumed to be less than about 0.2 rad, or roughly 12°); α_1, α_2 = angles between top of the obstacle and one end as seen from other end. α_1 and α_2 are the sign of h in the above equation; h, d, d_1, d_2, and λ are expressed by the same unit.

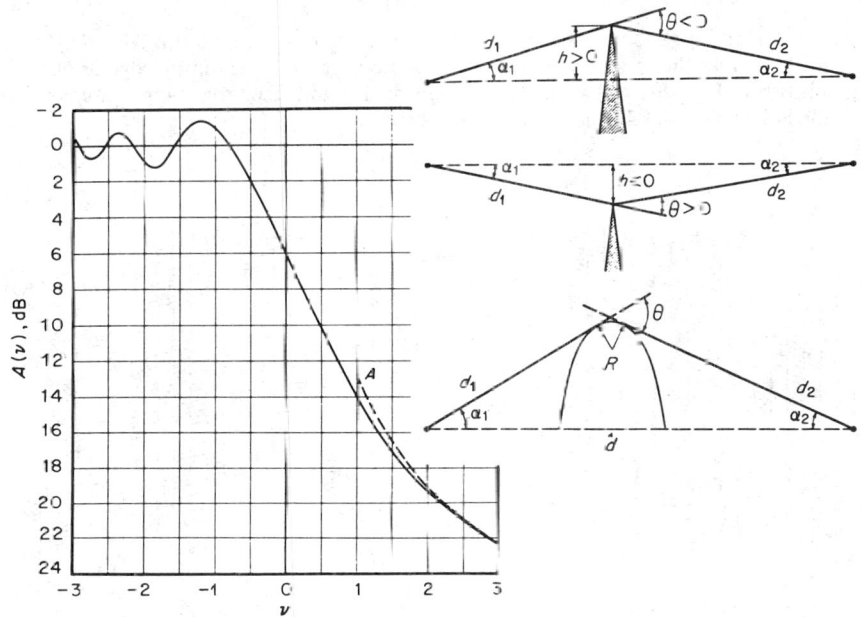

Fig. 18-60. Diffraction by obstacles; attenuation relative to free space. *(CCIR Rep. 715.)*

Figure 18-60 gives the loss in decibels caused by the presence of the obstacle, as a function of v. For v more positive than -1 an approximate value can be obtained from

$$A(v) = 6.4 + 20 \log (\sqrt{v^2 + 1} + v) \qquad dB \qquad (18\text{-}138d)$$

The error is within 0.5 dB.

For many paths the diffraction loss may be 10 to 20 dB greater than the value given by Fig. 18-60 because the obstacle is not a true knife-edge and because of other terrain effects.[47-49a] A propagation path with a single isolated terrain feature, which provides the horizons for both terminals, can often be considered as having a single diffracting rounded knife-edge between the terminals. For $\theta \geq 0$, the diffraction attenuation A (in decibels in excess of free space) can be evaluated from

$$A = F(v) + G(\rho) + E(\chi) \qquad (18\text{-}138e)$$

The *Fresnel-Kirchhoff loss* $F(v)$ is illustrated in Fig. 18-60 as a function of the dimensionless parameter

$$v = 2\left(\sin \frac{\theta}{2} \right) \left[\frac{2(d_a + R\theta/2)(d_b + R\theta/2)}{\lambda d} \right]^{1/2} \qquad (18\text{-}138f)$$

where λ = radio wavelength, d_a, d_b = distances from each terminal to their horizons on terrain feature, and R = effective radius of curvature for terrain feature between horizons as determined by the product of the geometrical radius and the earth-radius factor k. All distances and lengths are in the same units. For $R = 0$, Eq. (18-138f) reduces to Eq. (18-138a).

The loss in decibels for incidence upon the curved surface $G(\rho)$ can be determined[49b] *from*

$$G(\rho) = 7.192\rho - 2.018\rho^2 + 3.63\rho^3 - 0.754\rho^4 \tag{18-138g}$$

where ρ is given by

$$\rho^2 = \frac{(d_a + d_b)/d_a d_b}{(\pi R/\lambda)^{1/3}/R} \tag{18-138h}$$

$E(\chi)$, *the decibel loss for propagation along the surface between the horizons, is given by*

$$E(\chi) = \begin{cases} 12\chi & \text{for } \chi < 4 \\ 17.1\chi - 6.2 - 20 \log \chi & \text{for } \chi \geq 4 \end{cases} \tag{18-138i}$$

where

$$\chi = (\pi R/\lambda)^{1/3}\theta \tag{18-138j}$$

For $R = 0$, ρ and χ go to zero and Eq. (18-138e) reduces to the first term. For $\theta = 0$ and $\chi = 0$ Eq. (18-138e) gives the loss for grazing incidence upon the rounded knife-edge or obstacle.

The solution of Eq. (18-138e) assumes that transmitting and receiving antenna are both remote from their horizon on the diffracting terrain feature.

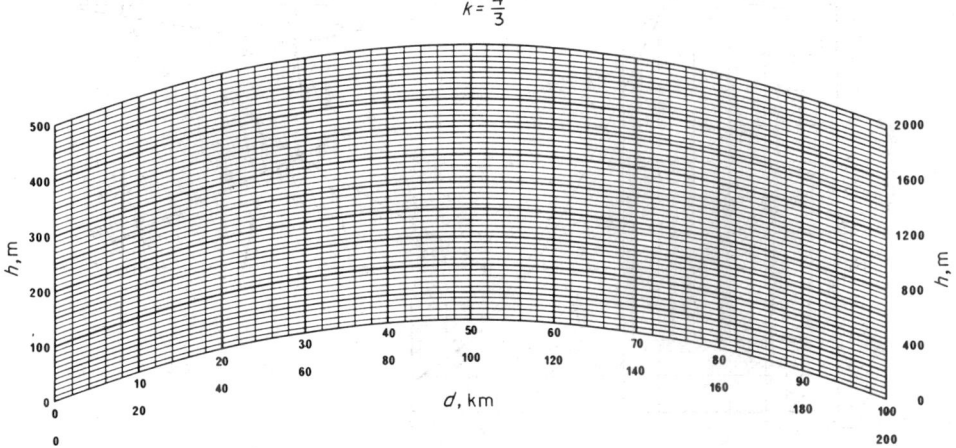

Fig. 18-61. Path profile chart for $k = \frac{4}{3}$.

82. Propagation over Irregular Terrain, 30 to 1,000 MHz. The terrain in most cases is not perfectly smooth. In the VHF and lower UHF regions, propagation for point-to-point, broadcasting, and mobile services usually includes substantial terrain effects, which determine the mean attenuation below free space and affect fading characteristics.

For point-to-point services, it is useful to prepare a path profile scaled to allow for the curvature of the earth modified for refraction by effective earth radius a; this is particularly helpful in determining the distance from a transmitter or receiver to the radio horizon, which may be the bulge of the earth itself or a horizon obstacle, e.g., hills, buildings, or woods. The path profile is determined in the plane of the great circle between the transmitter and receiver (for paths shorter than about 70 km, a rhumb line may be used). Elevations of the terrain are read from topographic maps and tabulated vs. their distances from the transmitting antenna. The recorded elevations should include those of successive high and low points along the path.

An example of a path profile is shown in Fig. 18-61. Special profile paper can be constructed so that a line horizontal from any distance d along the curved surface intersects the height scale at h (same units as d), where $d = \sqrt{2a_eh}$ and a = effective earth radius. The example shown is for normal refraction, using $\frac{4}{3}$ actual earth radius, though planning is often done on the basis of $\frac{2}{3}$ or even $\frac{1}{2}$ actual earth radius, as explained later.

As an alternative to special profile paper, the terrain profile can be plotted on linear graph paper by modifying the terrain elevations to include the effect of the average curvature of the radio-ray path and of the earth's surface. The modified elevation y_i of any point h_i at a distance x_i from the transmitter along a great-circle path is its height above a place which is horizontal at the transmitting-antenna location

$$y_i = h_i - x_i^2/2a \tag{18-139}$$

The vertical scale is exaggerated to represent the detail of terrain irregularities. Plotting terrain elevations vertically instead of radially from the earth's center involves negligible errors and allows use of straight lines for the rays from antennas. For long paths, great-circle distances and bearings are easily calculated from the cosine law for oblique spherical triangles.

If two antennas are mutually visible over the effective earth, geometric optics can be used to estimate the attenuation, provided it is reasonable to fit a straight line or convex curve of radius a to a reflecting portion of the terrain. However, such a procedure is of limited usefulness, being valid mainly at the lower frequencies and with low antenna heights, so that the path-length difference between direct and reflected rays is less than about 60°.

In most cases the terrain is too complex to fit a smooth curve, or only statistical descriptions of irregularities are available. Thus, empirical estimates of propagation are used for broadcasting and mobile services, and these extend well into the beyond-the-horizon region. A computer

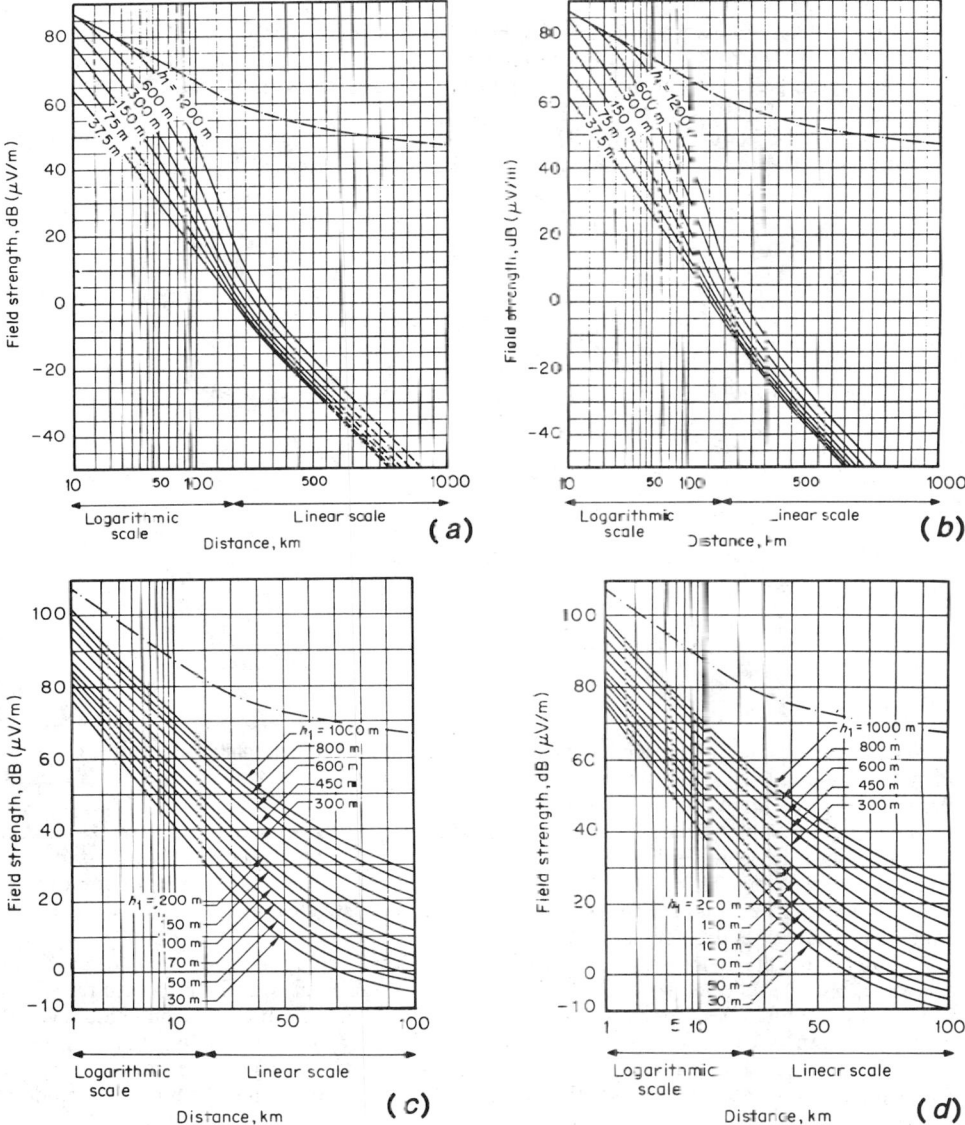

Fig. 18-62. Field strength over irregular terrain, 1 kW effective radiated power; 50% of the time, 50% of locations: (a) broadcasting, 30 to 250 MHz, $h = 10$ m, $\Delta h = 50$ m; (b) broadcasting, 450 to 1,000 MHz, $h = 10$ m, $\Delta h = 50$ m; (c) land mobile, 450 MHz, urban area, $h_2 = 1.5$ m; (d) land mobile, 900 MHz, urban area, $h_2 = 1.5$ m.

method[50] is available and *CCIR Recomm.* 370-3 gives curves of field strength based on a long series of measurements over many paths, principally in the United States and western Europe.

Figure 18-62*a* and *b* gives median field strengths for broadcasting in the 30- to 250- and 450- to 1,000-MHz bands. A parameter Δh used to define the degree of terrain irregularity is the difference in heights exceeded by 10 and 90% of the terrain in the range 10 to 50 km from the transmitter.

The height of the transmitting antenna is that above the local terrain, and the receiving antenna h_2 is taken at 10 m, typical of home-television and broadcast-receiving antennas. For mobile services the height of antenna will be nearer 3 m, and the field strength will be reduced from 4 to 10 dB, depending upon frequency and terrain.

The CCIR Recommendation includes curves for additional frequency bands, corrections for various Δh, and allowances for different heights of receiving antennas. FCC Rules[51] give field strengths for the 88- to 108-MHz FM band in terms of levels exceeded 50% of the time at 50% of locations.

Figure 18-62*c* and *d* gives median field strengths (*CCIR Rep* 567-1) for land mobile services in the 450- and 900-MHz bands, for a receiving antenna height of 1.5 m. Additional data on radio-wave propagation for mobile communications at frequencies from 100 MHz to about 2 GHz are given in Refs. 52 to 54.

An important aspect of VHF and UHF propagation in urban areas and over irregular terrain is the *time-delay spread* due to multipath reflections from buildings and terrain. This is partic-

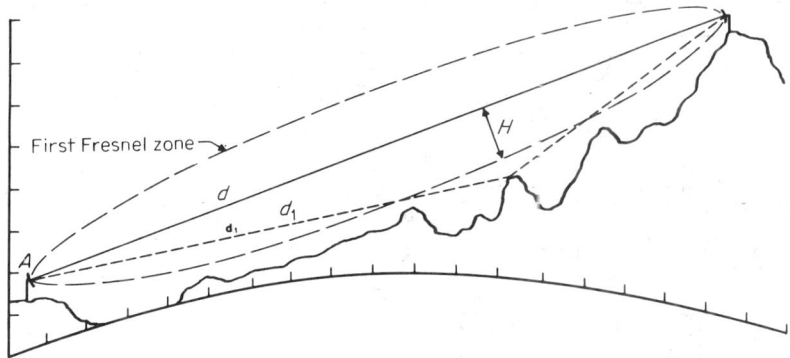

Fig. 18-63. Fresnel-zone clearance of terrain, $k = \%$.

ularly troublesome to FM broadcast and television reception and is a major factor in the design of new mobile radio systems above about 800 MHz. Typical time-delay spread exceeds ½ μs in urban areas and ¼ μs in suburban areas.[54]

83. Line-of-Sight Propagation at Microwave Frequencies, 1 to 10 GHz.
Point-to-point microwave links at frequencies above 1 GHz are designed to achieve essentially free-space attenuation. This is done by the use of narrow-beam antennas and a path-profile chart to engineer the line-of-sight path so that it will have adequate clearance from terrain and surrounding obstacles, as illustrated in Fig. 18-63.

The amount of clearance is usually described in terms of Fresnel zones. All points from which the wave could be reflected with an additional path length of a half wavelength (with respect to the direct path) form an ellipse which defines the first Fresnel zone. If there is no phase reversal on reflection, the second path would partially cancel the direct path. Normally, at grazing angle the phase is reversed so that a delay of an integral number of wavelengths causes cancellation. For any distance d_1 from antenna A, the distance H from the line-of-sight path to the boundary of the first Fresnel zone is approximated by the parabola[56]

$$H = \sqrt{\lambda d_1 (d - d_1)/d} \qquad (18\text{-}140)$$

where λ, d, d_1 are in identical units.

Measurements have shown[56] that the transmission path should pass all obstacles with a clearance of at least 0.6 times the first Fresnel-zone distance and preferably by an amount equal to the first Fresnel-zone distance.

Clearance is required not only to maintain free-space attenuation under normal atmospheric conditions but also to avoid fading problems during abnormal conditions, as discussed in Par. **18-84**. Good engineering practice for such links[55,56] assumes ⅔ effective earth radius as a basis for estimating antenna heights required for terrain clearance rather than the value of ⅔ associated with "standard" propagation. The k = ⅔ provides against subrefractive situations, corresponding to positive refractive-index gradient near the surface and *upward* rather than downward bending of the beam (Fig. 18-65). For adequate terrain clearance *over smooth terrain,* the minimum antenna heights corresponding to the 10-dB curves in Fig. 18-64 are used. This allows for clearance of the terrain by the direct ray over an area of at least 60% of the first Fresnel zone.

Extreme values of the refractive index gradient have a less important effect when averaged over the total path length, so it is possible to use somewhat larger values of k (than ⅔) for paths longer than about 50 km. Figure 18-53 gives estimated minimum effective values of k as a function of path length.

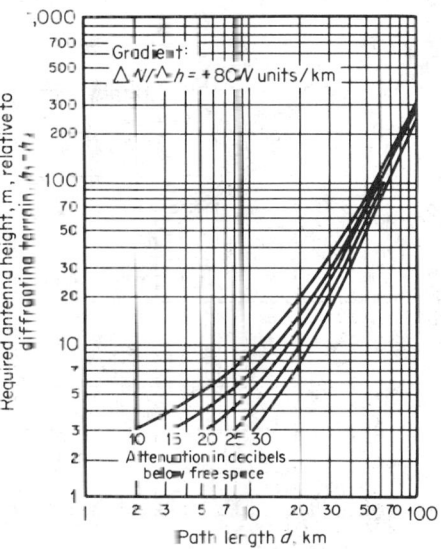

Fig. 18-64. Required antenna height to protect against diffraction attenuation at 2 GHz. (*Dougherty*[57]) The 10-dB curve gives the approximate tower height required for smooth-terrain clearance over six-tenths of the first Fresnel zone for $k = $ ⅔.

84. Fading and Diversity. Most microwave links do not experience serious fading under most meteorological conditions. However, stratification of the atmosphere and other meteorological conditions can cause severe fading. Measurements show that the range of refractivity-index gradient can be very large. At Cape Kennedy, Fla., for example, long-term measurements in the first 100 m near the surface showed that the gradient varied between 230 N/km, the value exceeded 0.5% of the time, and -370 N/km, the value exceeded 99.9% of the time. These values correspond to k values of $+0.4$ and -0.7, respectively (concave earth). A number of situations giving rise to both power fading and multipath fading are illustrated in Fig. 18-65.

Power fading includes results of beam bending, which affects terrain clearance, angle of arrival, trapping or deflection of the beam, and attenuation due to precipitation. Power fading due to loss of terrain clearance, also called *diffraction fading,* may be to depths of 20 to 30 dB. This type can be avoided except for most extreme cases by use of the design criteria of Fig. 18-64 and Ref. 57. Fading is also due to angle-of-arrival variations, up to ±⅔° vertically and 0.1° horizontally. Ducts and layers cause power fades up to 20 dB or more, which may persist for hours

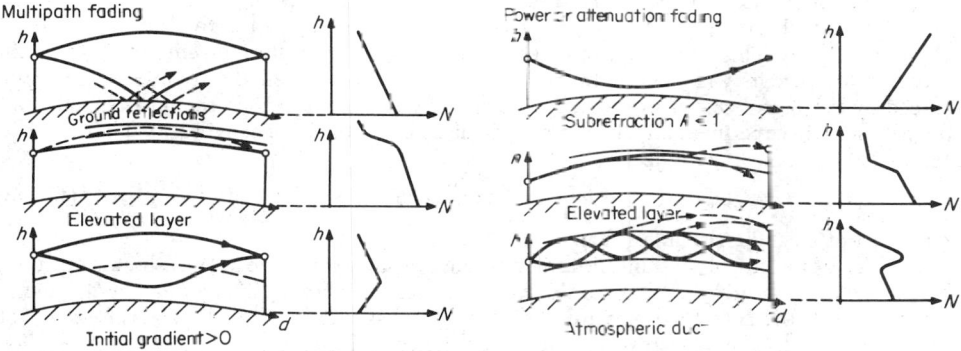

Fig. 18-65. Fading mechanisms for line-of-sight propagation. (*Dougherty*[57])

or days. Precipitation attenuation can be important below 10 GHz but is of principal importance above 10 GHz (see Par. **18-89**).

Multipath fading includes phase-interference effects from ground-reflected and atmospheric paths. As the refractive index varies, interference can occur between the direct wave and the reflection from ground or water surfaces, as well as between the direct wave and partial reflections from atmospheric sheets or elevated layers. Additional direct paths can also be propagated due to surface layers of strong positive refractive gradients or horizontally distributed changes in refractive index encountered with a weather front. The frequency-selective fades extend to 20 or 30 dB below the free space, depending upon the relative amplitudes of the component waves. Specular ground reflection can produce fades persisting for minutes. Proper antenna design and siting can discriminate against terrain reflections.

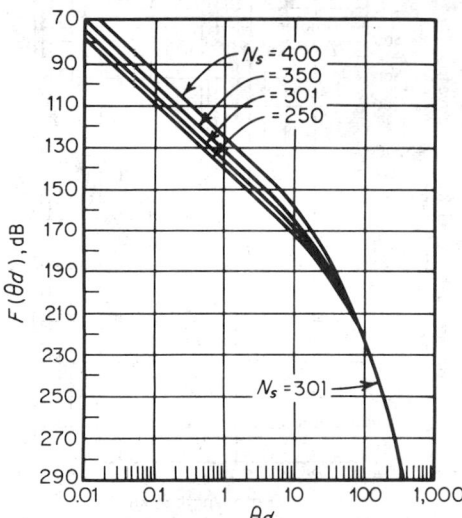

Fig. 18-66. Attenuation function $F(\theta d)$ for forward scatter; d km, θ rad. *(CCIR Rep. 238-4.)*

Since one or two components usually dominate, the amplitude is not generally Rayleigh-distributed but is represented better by the distribution of constant signal plus a random component. Detailed summaries of microwave-fading characteristics and mechanisms, a family of fading distributions, and a bibliography are given in Ref. 57; also important are Refs. 57a to 61a.

To design radio relay systems conforming to CCIR Recommendations, it is necessary to protect against the probability of deep fades for very small percentages of the time, e.g., about 0.0002%. *Space- and frequency-diversity techniques* are used.

A simple, generally effective space-diversity design procedure[59] gives required vertical spacing, center to center:

$$\Delta h = 0.3\sqrt{\lambda d} \qquad (18\text{-}141)$$

where the path length d, wavelength λ, and spacing Δh are all in the same units. Design of diversity separations as a function of frequency, antenna height, path length, and expectation of refractive behavior is given by Ref. 57 for more difficult situations of maritime paths. See also Ref. 61.

85. Forward Scatter. The long-term median transmission loss due to forward scatter is approximately (Ref. 2 and *CCIR Rep. 238-4*)

$$L(50) = 30 \log f - 20 \log d + F(\theta d) - G_p - V(d_e) \qquad \text{dB} \qquad (18\text{-}142)$$

where $F(\theta d)$ is shown in Fig. 18-66. The angular distance θ is the angle between radio horizon rays in the great-circle plane containing the antennas and d_e is the distance between antennas. $V(d_e)$, given by *CCIR Rep.* 238-4 for various climatic regions, may be up to ± 8 dB.

The combined gain of transmitting and receiving antennas may be less than the sum of their plane-wave gains. This apparent drop in gain, termed *gain degradation* or *antenna-to-medium coupling loss*, occurs when the beam widths of the antennas are smaller than the angle subtended by the useful scattering volume. The amount of loss depends on the antenna gain and the path length; experimentally, however, little dependence on distance is observed in the range 150 to 500 km, and an empirical estimate is

$$G_p = G_t - G_r - 0.07 \exp [0.055(G_t + G_r)] \qquad \text{dB} \qquad (18\text{-}143)$$

for values of G_t and G_r each less than 50 dB.

The siting of terminals of transhorizon links requires some care. The antenna beams must not be obstructed by nearby objects, and the basic requirement is that the antennas be directed at the horizon. If the antenna beams are tilted upward by as little as 0.5°, there may be a loss of the order of 10 dB due to decreased scattering efficiency with height.

Theoretical and experimental information on the mechanism and characteristics of tropospheric forward scatter is given in Refs. 62–71.

To estimate *long-term variability* of forward-scatter paths, meteorological information has been used to distinguish between climatic regions (*CCIR Rep.* 238-3), and radio data from more than a thousand paths in various parts of the world provide the basis of prediction of long-term variability about the computed long-term median value in each of these regions.

Short-term *phase interference fading* of a tropospheric scatter signal can usually be approximated by a Rayleigh distribution. The mean fading rate is proportional to carrier frequency and is about 4 Hz at 10 GHz. Fading rate also depends on antenna beam width, being slower for narrow beams because of restriction of the lateral phase-path extremes.

Horizontal spacing of 50 to 100 wavelengths has been found adequate for diversity throughout the frequency and distance range of interest. It is also possible to have angle diversity using multiple feeds and a common reflector.[72]

86. Ducting. When large vertical gradients are sufficient to refract a ray to the same radius of curvature as the earth [vertical decrease of N greater than 157N units/km (Par. **18-74**)], *superrefraction* is said to occur and the wave can be *trapped*. A layer of this type is called a *duct*, and the mode of propagation between the earth and such a layer (or between two layers) is similar to that of a waveguide. Low-loss transmission over great distances is possible via ducts, very distant radar echoes can be observed, and duct propagation constitutes a potential source of interference between satellite-earth stations and other terrestrial radio uses. Climatology of radio ducts is discussed in Ref. 27. The attenuation depends upon refractive-index gradient and thickness of the boundary layers (upper and lower in the case of an elevated atmospheric lower boundary) and how energy is coupled into the duct.[4,73-75] Adequate numerical treatment of propagation via a tropospheric duct is very difficult because of the large number of parameters involved. Reference 74 gives a method for computing an effective cutoff frequency, though the cutoff in practice is not nearly so sharp as for waveguides. Efficient coupling usually occurs only for energy (this may be side-lobe energy) grazing the duct boundary within 1° and when both transmitter and receiver are within the duct. Energy can be scattered from terrain or obstacles into the duct.

One aspect of ducting that is potentially disruptive of microwave links, especially long ones, is the power attenuation associated with an elevated layer, as illustrated in Fig. 18-65. In the example shown, an antenna placed at a normally desirable height may be above an elevated layer, so that the signal is trapped below the height of the antenna. In areas where such ducts are prevalent, as in West Africa (see Fig. 18-67), height diversity can be used.

Experimental studies of ducting over ocean surfaces are given by Refs. 76 and 77. Prediction charts[29] give statistical data on refractivity gradients conducive to ducting. Figure 18-67 illustrates geographical variation of trapping frequency. *CCIR Rep.* 569-2 gives a method for estimating transmission loss under conditions of superrefraction and ducting.

87. Atmospheric Effects above 10 GHz, Millimeter Waves, and Earth-Space Propagation. At frequencies above about 3 GHz the attenuation of radio waves by atmospheric gases and water and refractive scintillation and multipath effects become increasingly important. These factors become dominant considerations in the design of radio-relay or earth-satellite systems at frequencies above 10 GHz.

Growing use of frequencies above 10 GHz is envisaged for earth satellite systems as their information-capacity requirements grow, and the needs involve multiple steerable beams, multiple-access, and high-resolution earth-resource application. Point-to-point video and data-relay systems increasingly require the greater bandwidth capability available above 10 GHz, compared with congested lower-microwave frequencies. Other applications include high-resolution radar, precision guidance, and tracking. New aircraft-landing systems use 15 GHz and higher frequencies to obtain the required resolution. Radio-astronomy studies make use of gaseous resonance lines throughout this region of the spectrum; optical systems are being increasingly used for precise positioning and distance measurement, as well as for high-capacity information transmission. A comprehensive summary of propagation effects on satellite links from 10 to 100 GHz is given in Ref. 79.

Propagation effects of importance are absorption by the clear atmosphere due to molecular resonances; attenuation and scattering by rain or fog; and phase interference and refractive scintillation by atmospheric turbulence, stratification, and terrain effects. Propagation difficulties in this frequency range should not be underestimated. Rain is a most important effect and causes noticeable attenuation at frequencies as low as 3 GHz and interruption of line-of-sight links at frequencies as low as 6 GHz; rain can restrict the reliable use of frequencies much above 30 GHz to terrestrial links of a few kilometers in length or to earth-space links near the zenith.

88. Absorption by the Clear Atmosphere. Transmission through the clear atmosphere is subject to attenuation by molecular oxygen and water vapor (Refs. 78 to 89 and *CCIR*

Rep. 719). An attenuation peak at 22 GHz is due to the single rotational transition of the water molecule; the peak near 60 GHz results from a large number (43) of oxygen absorption lines, effects of pressure broadening in the lower atmosphere playing an important role. Each of these gases has a second absorption region below 300 GHz, oxygen at 118 GHz and water vapor at 184 GHz. Each resonance has an accompanying frequency-dependent phase-velocity, or dispersion, effect. The only serious gaseous absorption is posed by oxygen in the 55- to 76-GHz and the 118-GHz regions and by water vapor above about 125 GHz. Methods for calculating absorption and

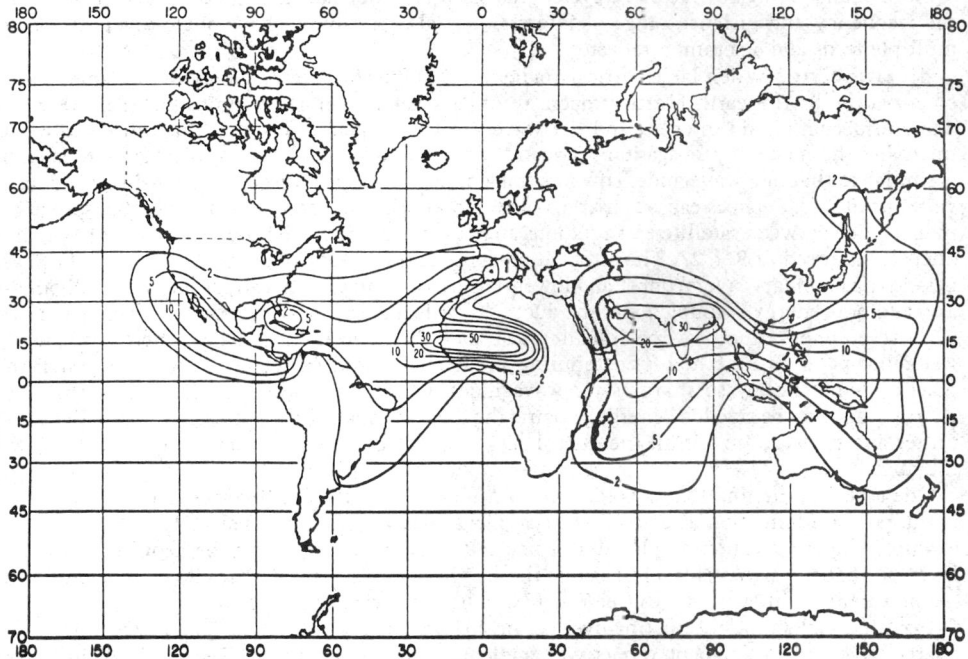

Fig. 18-67. Percentage of time trapping frequency is less than 3 GHz, November. (*Bean et al.*[29])

dispersion as a function of gas density, in terms of line width, strength, and frequency, are given in Refs. 81 and 82, and the parameters are available from laboratory measurements.[83,85] The line width is broadened by molecular collisions depending upon the pressure and gases involved. By modeling the pressure, temperature, and water-vapor profiles it is possible to calculate the attenuation and phase delay through the atmosphere to different heights and at different elevation angles.

It is customary to express the attenuation coefficient (or specific attenuation) in decibels per kilometer for the mean pressure at sea level, 1 atm (1,013.6 millibars), and for a normal humidity value, for example 7.5 g/m³, and for a temperature of 20°C. Figure 18-68 gives a general picture of this absorption from 1 to 350 GHz. Figure 18-69 gives theoretical vertical one-way attenuation from a specified atmospheric height to the top of the atmosphere for a moderate humid atmosphere for frequencies 1 to 150 GHz. Figure 18-70 shows the fine structure of the oxygen absorption lines between 50 and 70 GHz and illustrates the effect of pressure broadening at lower altitudes.[80]

89. Rainfall Attenuation and Scattering.* The attenuation due to rain usually exceeds the combined absorption due to oxygen and water vapor and arises from the absorption of energy in the water droplets and from the scattering of energy out of the beam of the antenna. Attenuation is usually expressed as a function of the rainfall rate R. It depends on both the liquid-water content and the velocity of fall of the drops, which in turn depends on drop size. Rain of a given rainfall rate has various distributions of drop size, and useful estimates of attenuation are obtained empirically.

The total attenuation A_r due to rainfall over a path of length r_0 can be determined by inte-

*This material is taken in large part from *CCIR Rep.* 721.

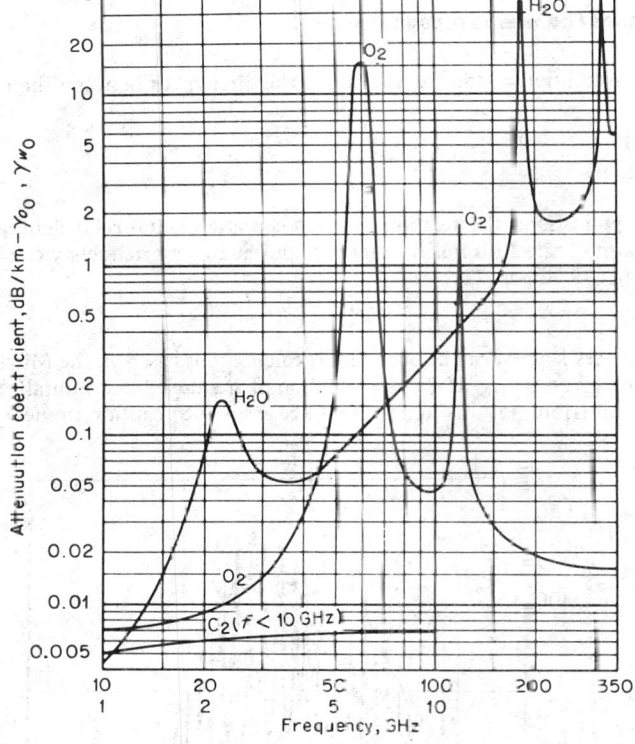

Fig. 18-68. Specific attenuation by atmospheric gases. *(CCIR Rep. 719.)*

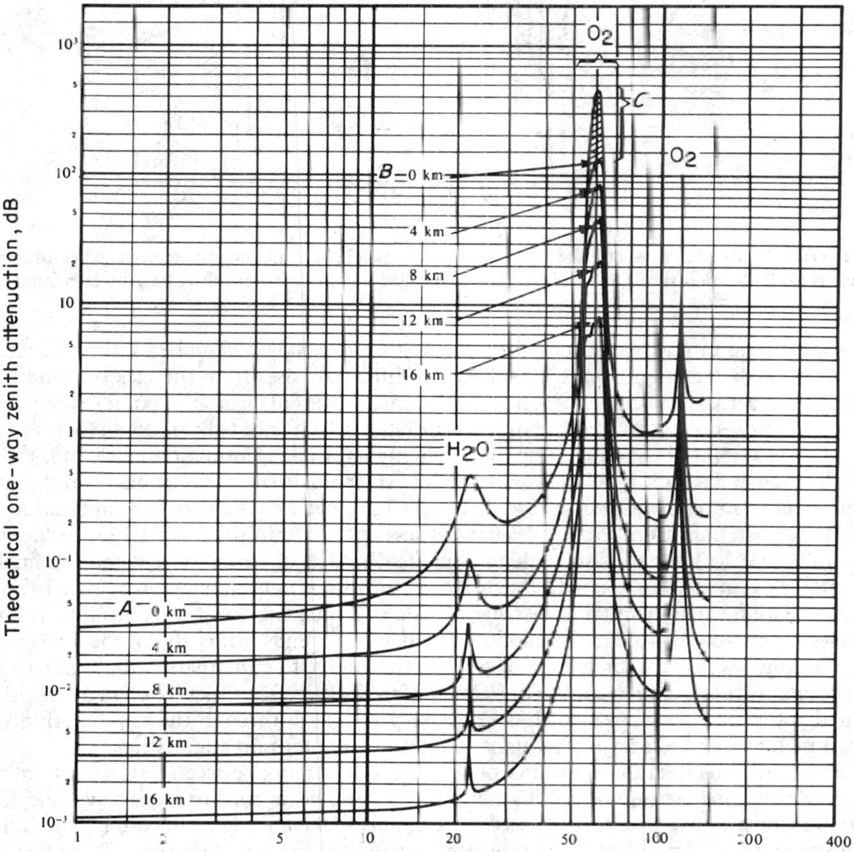

Fig. 18-69. Theoretical one-way attenuation from specified height to top of atmosphere for a moderately humid atmosphere.

grating the rain-absorption coefficient $\gamma_r(r)$ along the direct path between the two mutually visible antennas:

$$A_r = \int_0^{r_0} \lambda_r(r)\, dr \quad \text{dB} \tag{18-144}$$

For practical applications, the relationship between attenuation coefficient γ_r in decibels per kilometer and rainfall rate R in millimeters per hour at a given frequency can be approximated by a power law (CCIR Report 721)

$$\gamma_r = kR^\alpha$$

where the parameters k and α can be obtained by calculations based on the Mie-Ryde theory.[90-91]

Values of γ_r are given in Fig. 18-71 as a function of frequency and rainfall rate R mm/h for a mean drop-size distribution; this figure should be used with caution in view of the tendency

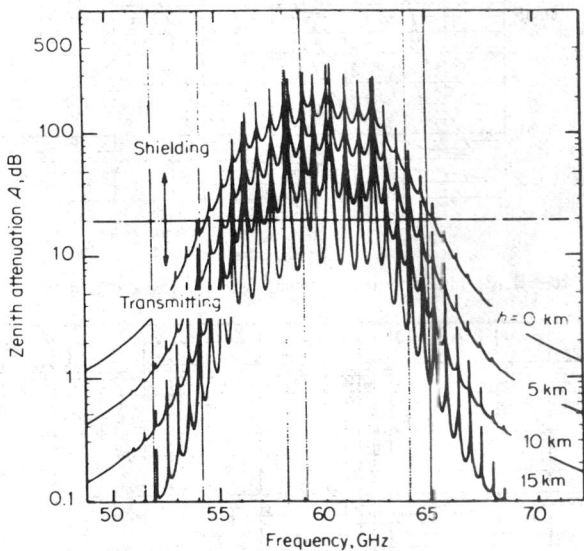

Fig. 18-70. Millimeter-wave attenuation in the 49- to 72-GHz band due to atmospheric oxygen resonances for zenith paths through the U.S. Standard Atmosphere 1962 from different initial heights h to outer space.

of the attenuation measured at some frequencies (between about 20 and 35 GHz) to exceed the maximum predicted attenuations.[90,91] Additional difficulty arises from the lack of uniformity of rainfall over actual transmission paths. Total annual rainfall data at a point, available from weather records, does not usually indicate rain rate statistics needed for attenuation estimates. Nor do surface rainfall data directly provide the high-altitude information needed for estimates of attenuation and scatter for earth-satellite links. An approximate method, based on the concept of a number of rain climatic zones, can be used.[92] This method uses detailed information for 45 United States stations from 1951 to 1960, with less detailed information for 135 worldwide stations from 1931 to 1960, to estimate long-term distributions of 1-min average rainfall rates.

Figure 18-71, in conjunction with the millimeter-per-hour rainfall rate, can be used to estimate rain attenuation for horizontal line-of-sight paths at frequencies below about 60 GHz. At frequencies much above 40 GHz the specific attenuation depends primarily on the concentration of small drops. However, for most drop-size contributions, the attenuation for a given rainfall rate does not increase appreciably with frequency above 100 GHz. Calculation of attenuation for elevated paths requires allowance of an effective path length through the rain.[93] Path diversity on millimeter wave links is an important countermeasure for rain attenuation.[93a,93b]

Scattering by precipitation is a source of potential interference between terrestrial systems and earth-space systems because of the high elevation of the beam for earth-space systems. To calculate transmission via precipitation scatter, one must make allowance for the height and drop-size distribution and an effective path length through the precipitation (Refs. 93 to 95). Figure

18-72 shows calculations of transmission loss at 4, 11, and 30 GHz for scatter from rainfall in intersecting beams on the path "beyond the earth station" as indicated in the geometry. Because the scattering volume is assumed to fill the narrow beams, no account needs to be taken of antenna gain, and transmission loss p_r/p_t is obtained directly.

Coherence bandwidth does not appear to be significantly limited by multipath scattering via rainfall; attenuation usually limits transmission before multipath effects are discernible.[96-98] Important and additional references on rainfall effects on propagation are Refs. 99 and 100.

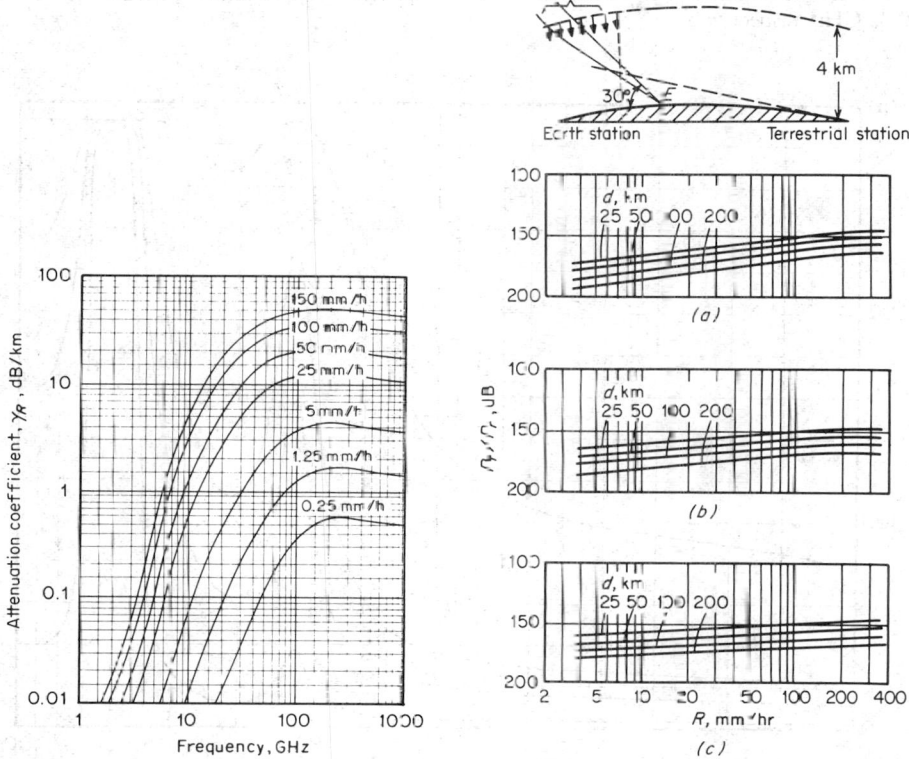

Fig. 18-71. Attenuation coefficient γ_r due to rain.

Fig. 18-72. Scattering by precipitation: transmission loss vs. rainfall for rain on path beyond earth station illuminated by terrestrial microwave signal: (a) $f = 4$ GHz; (b) $f = 11$ GHz; (c) $f = 30$ GHz. (CCIR SJM.)

90. Refraction Errors, Scintillation, Turbulence and Stratification; Fading, Angular, and Bandwidth Limitations.* Ray-tracing methods for refraction errors are outlined in Ref. 27. At low elevation angles (below about 5°) and for horizontal paths, amplitude-, phase-, and angular-scintillation effects can be important. These are caused in the normally turbulent atmosphere by small-scale refractive irregularities associated with random pressure, temperature, and water-vapor variations; similar effects can be produced by phase-interference effects from partial reflection from elevated layers or refractive sheets or in some cases by terrain reflections (especially on long line-of-sight paths). Theoretical methods and experimental results for engineering estimates are given in Refs. 86, 87, 89, and 100a to 103a.

For space-space or earth-space links at high angles of elevation and frequencies below about 20 GHz, small-scale variations in refractive index are generally insignificant. At higher frequencies in some meteorological conditions, their effect may be important. Assuming a source outside the troposphere and a receiving antenna of up to a few meters in diameter, the largest peak-to-

---

*Par. **18-110** discusses scintillations produced by the ionosphere.

peak fades expected in clear air are about ±4 and ±2 dB at 100 and 35 GHz, respectively, for elevation angles exceeding 45°. For angles of elevation of about 10°, however, the fades may occasionally reach ±12 and ±6 dB at 100 and 35 GHz, respectively. The standard deviation of amplitude scintillation at elevation angles below 4° is less than 1 dB even for very large antennas (*CCIR Rep.* 564-1). At frequencies below 10 GHz this scintillation is essentially independent of frequency.

For a distance of 6 m across the wavefront (representative of the diameter of a large millimetric antenna) the rms value of phase differences caused by the troposphere on an earth-space link may sometimes reach 40 and 15° at 100 and 35 GHz, respectively, when the elevation angle exceeds 45°. At about 10° elevation, these values increase to approximately 80 and 30° at 100 and 35 GHz, respectively.

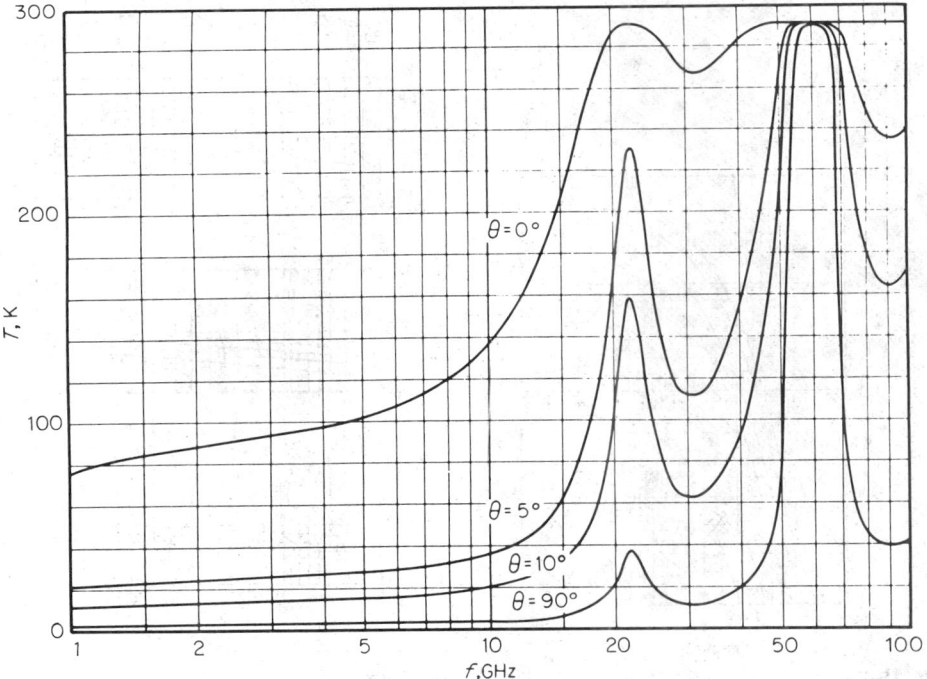

Fig. 18-73. Sky-noise temperature, clear air, for various angles of elevation; surface pressure 1 atm, surface temperature 20°C, water vapor 10 g/m³.

Spectra of phase fluctuations for a horizontal line-of-sight path over the sea at 10 and 35 GHz are available in Ref. 103.

Angle of arrival at 20 to 30° above the horizon shows fluctuation of the order of 0.2 to 0.3 × 10⁻⁴ rad/s (standard deviation for fluctuation of durations greater than 1/10 s). This angle may be considered as the theoretical smallest useful beam for an antenna. At lower elevation the fluctuation will be greater.[27]

Coherence bandwidth is also limited by these scintillations and multipath phenomena, especially at frequencies above 10 GHz for very wide-band systems operating over horizontal (terrestrial line-of-sight) paths. During periods of stable atmospheric stratification, multipath can limit bandwidth coherence to a few tens of megahertz.

A survey of clear-air propagation effects of optical communications is given in Ref. 104.

91. Sky-Noise Temperature Due to Atmospheric Absorption and Precipitation. At frequencies above about 10 GHz, the nonionized region of the atmosphere, as an absorbing medium, is also a source of noise radiation.[105] The effective sky-noise temperature T_s is shown in Fig. 18-73 for various angles of elevation. Rainstorms also provide significant contribution to sky-noise temperature.[106]

PROPAGATION VIA THE IONOSPHERE

92. Introduction. The ionosphere is that part of the atmosphere, at heights above about 50 km, in which free ions and electrons exist in sufficient quantities to affect the propagation of radio waves. At frequencies below about 30 MHz, regular long-distance transmission is possible by way of ionospheric reflections. At frequencies in the 30- to 100-MHz region, regular but weak propagation by ionospheric scattering is obtained, as well as strong intermittent propagation by reflection from sporadic ionospheric layers and meteoric ionization. At frequencies well above 100 MHz auroral echoes are observed, and ionospheric scintillation effects may be important for satellite-earth links.

93. Physical Description of Ionospheric Regions. The various regions of the upper atmosphere are usually described according to the nomenclature of Fig. 18-74, which also illustrates typical daytime densities and temperature.[9,107]

At levels up to about 85 km turbulent mixing keeps the relative chemical composition of the atmosphere essentially the same as at the ground, that is, N_2, O_2, Ar, and CO_2, as well as traces of water vapor, ozone, nitric oxide, and hydrogen. Above about 90 km, O, O_2, and N_2 are the major constituents, and dissociation of O_2 becomes important. The important parameters of radio propagation are the concentration of free electrons (N, electron density) and the rate at which these electrons collide[8,9,106a-108] with neutral particles (ν, collision frequency).

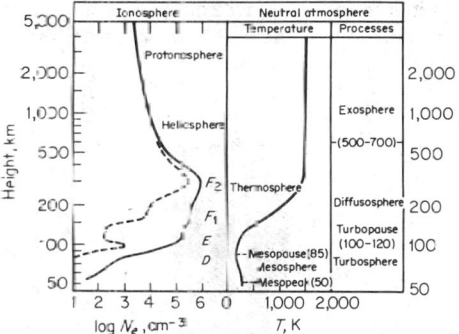

Fig. 18-74. Atmospheric nomenclature and typical daytime ionospheric densities and temperatures. (*After Van Zandt and Knecht,*[107] *by permission of John Wiley & Sons, Inc.*)

Four principal regions, or layers, affect propagation of radio waves. The term layer identifies regions having distinct characteristic processes, heights, and densities.

The *D region* extends from 50 to 90 km height. It is a region of low electron density in collision with neutral gases, mainly causing absorption of radio waves passing through, but sufficiently reflective to provide an upper boundary for VLF and LF propagation and sufficiently irregular and turbulent to scatter waves at VHF. The cause of ionization is generally taken to be solar photoionization of NO by Lyman-α radiation (λ = 121.6 nm). D-region ionization is mainly daytime, though during disturbances and at high latitudes ionization can be caused by particle radiation. The *mesopause* at about 85 km is a region of strong turbulence and wind shear.[9]

The *E region*, at heights 90 to 130 or 140 km, is ionized mostly by solar ultraviolet and x-rays in the daytime with some small nighttime ionization by cosmic rays and meteors. Electron production in the 100- to 140-km height range is ascribed mostly to solar radiation in the 3- to 20-mm wavelength ion,[109] dependent on solar zenith angle and solar cycle. A regular E layer of maximum electron density near 100 km is an important reflecting medium for daytime HF propagation, and at night for MF and IF propagation.

In addition to the regular E layer, irregular cloudlike layers of ionization, called *sporadic E* (or E_s), produce partial reflections and scattering at frequencies up to 150 MHz. There are a number of types and causes of E_s.[110-112] Over the magnetic equator, E_s is a regular daytime phenomenon, attributed to the two-stream plasma instability in the presence of the equatorial electrojet. At midlatitudes E_s is commonly attributed to the concentrating effect of wind shear in the E region on positive ions, thence on electrons; there is a summer peak of occurrence. A high-latitude nighttime E_s is associated with disturbed magnetic conditions and aurora.

The *F region*, above 130 to 140 km, is the most important for high-frequency propagation.

The *F1 layer* exists mainly during daylight at heights of 175 to 220 km; though fairly regular in its characteristics it is not observable everywhere or on all days. F1 depends on daylight solar ionization but is also prevalent during ionospheric disturbance. At night the F1 layer merges into the *F2*.

The *F2 layer*, at heights from 200 to 400 km, is the principal reflecting layer for long-distance high-frequency communication day and night and has the most complex and variable characteristics. Height and electron density vary geographically, diurnally, seasonally, and with solar cycle. Unlike the E and F1 layers, the F2 does not follow solar radiation directly with zenith angle; e.g., the winter midday electron density can be 4 or more times as great as summer midday ionization. Monthly mean maximum electron density is approximately linearly proportional to 12-month running mean sunspot number R_{12}. Ionization depends on the atmospheric model, solar flux, absorption cross section, and ionization efficiency.[108,113]

Knowledge of electron and ion distribution is obtained by a number of probing techniques.[9,114]

Worldwide Sounding. An international program of exchange of ionospheric data, obtained since about 1946 from ground-based radio sounding, has been coordinated by the International Radio Scientific Union (URSI) from the scientific point of view[115] and by International Radio Consultative Committee (CCIR) from the point of view of needs for radio communication (*CCIR Rep* 313-3).

The *ionosonde* is a pulsed (or FM continuous-wave) radar device in which the exploring frequency varies over a wide range from about 1 to 25 MHz or higher. The equipment measures transit time (virtual height) and maximum frequency of reflection, and the results can be interpreted in terms of electron-density profiles from about 100 km to the height of maximum electron density in the F region. A typical vertical ionosonde display (Fig. 18-75) shows E-layer and F1 and F2 reflections.

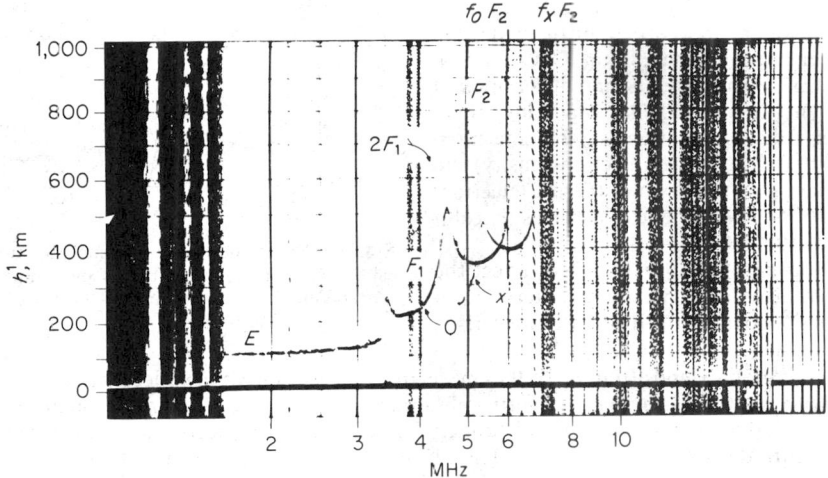

Fig. 18-75. Vertical ionosonde record (Washington, D.C., daytime).

Absorption measurements comparing relative amplitude of pulsed reflections, continuous-wave field-strength recordings, and measurement of galactic-noise attenuation by the ionosphere *(riometer observations)* are used to give information about absorption and electron density in the D region, where the ionosonde gives very little information.

Rockets and Satellites. It is now possible to explore the ionosphere by instrumentation of rockets and satellites. In particular, the region above the F-region maximum electron density, inaccessible to conventional ionosondes, has been studied by satellites. Rockets have provided the most satisfactory measurements of the D region.

"Incoherent Scatter" Radar. High-power VHF radars beamed vertically obtain echoes from electron scattering which can be related to the electron-density profile, providing a means of observing the profile continuously from below the maximum electron density to heights of 1,000 km or more.

94. Refractive Index, Polarization, Reflection, and Critical Frequency. Wave propagation in a magnetoionic medium is discussed in Refs. 8, 9, and 116 to 118. The complex refractive index of the magnetoionic medium (the Appleton formula)[9] is given by

$$n^2 = 1 - \frac{X}{1 - iZ - \dfrac{Y_T^2}{2(1 - X - iZ)} \pm \sqrt{\dfrac{Y_T^4}{4(1 - X - iZ)^2} + Y_L^2}} \qquad (18\text{-}145)$$

In the upper regions of the ionosphere the collision frequency is sufficiently small (for frequencies greater than about 1 MHz) for us to be able to put $Z = 0$; hence the real part of the refractive index[*] is

$$\mu^2 = 1 - \frac{2X(1 - X)}{2(1 - X) - Y_T^2 \pm \sqrt{Y_T^4 + 4(1 - X)^2 Y_L^2}} \qquad (18\text{-}146)$$

In the absence of an imposed magnetic field ($Y_T = Y_L = 0$) and of collisions ($Z = 0$) the refractive index is given by

$$\mu^2 = 1 - X = 1 - (j_N/f)^2 = 1 - kN f^{-2} \qquad (18\text{-}147)$$

where $k = e^2 4\pi^2 \epsilon_0 m = 80.5$ and f is in hertz; f_N is the plasma frequency, where

$$X = \frac{Ne^2}{\epsilon_0 m \omega^2} \qquad Y_L = \frac{eB_L}{m\omega} \qquad Y_T = \frac{eB_T}{m\omega} \qquad \text{and} \qquad Z = \frac{\nu}{\omega}$$

and N = electron density in electrons per cubic meter, e and m = charge and mass of electron, ω = angular frequency, Y = parameter of magnetic field, where B = field (Wb/m²), and T and L = directions of phase propagation, transverse or longitudinal, respectively; Z = parameter associated with collisions with neutral particles, and ν = collision frequency.

An important aspect of propagation in the ionized medium is *polarization*. For the propagation of high-frequency radio waves through the E and F regions of the ionosphere, Z is usually very small and can be neglected. The *wave polarization*[9] is

$$R = \frac{i}{2Y_L} \left[\frac{Y_T^2}{1 - X} \mp \sqrt{\frac{Y_T^4}{(1 - X)^2} + 4Y_L^2} \right] \qquad (18\text{-}148)$$

R is the ratio of the field vectors $H_2/H_3 = E_3/E_2 = P_3/P_2$ in Fig. 18-76. The polarization R gives the amplitude ratio and the phase difference between oscillations of the displacement vector D, the electric field vector E, and the power vector P along the 2- and 3-axes. In general, the tips of these vectors describe ellipses. The H ellipse is similar to the D ellipse (and lies wholly in the 2-3 plane) but is rotated through 90° in the same direction. While the D, B, and H vectors lie in the plane of the wavefront (2-3), the P and E ellipses are tilted forward with respect to the 2-3 plane.

Equation (18-148) tells us that two, and only two, waves can propagate. Analogous to the terminology used for the propagation of light in birefringent crystals, these characteristic waves are called *ordinary* and *extraordinary* waves (o and x, respectively). In the absence of collisions the ellipse of the extraordinary wave can be obtained from the ellipse of the ordinary wave by rotating it through 90° and reversing the sense. For longitudinal propagation the two magnetoionic waves are circularly polarized. For transverse propagation the ordinary wave is polarized with the E and P vectors parallel to the magnetic field, whereas for the extraordinary wave the P and E ellipses lie entirely in the 1-3 plane of Fig. 18-76.

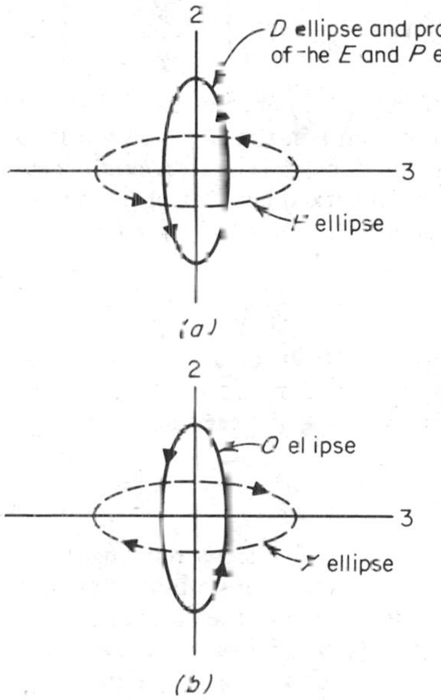

One of the waves (the ordinary) is reflected as in the absence of the magnetic field. The reflection of the other wave (the extraordinary) depends upon the strength (but not the direction) of the imposed magnetic field.

Reflection conditions are as follows. For a radio wave incident at angle ϕ_0 (with respect to normal) on a plane stratified ionosphere (Fig. 18-77) at the bottom of the ionosphere ($N = 0$) μ

Fig. 18-76. Polarization ellipses (no collisions): (a) wave polarization, D and H ellipses; (b) relation between the corresponding ellipses of the ordinary and extraordinary waves. (Davies.[9])

---

[*]Henceforth μ is used for the real part of the ionospheric refractive index and not for magnetic permeability.

is unity. At higher levels of electron concentration, μ falls. If the electron density is sufficiently high, μ will go to zero. Snell's law states that for a wave incident at an angle ϕ_0 to the normal, the angle ϕ with the normal at a level where the refractive index is μ is given by

$$\mu \sin \phi = \mu_0 \sin \phi_0 \qquad (18\text{-}149)$$

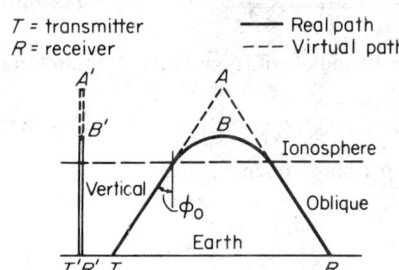

T = transmitter
R = receiver

—— Real path
--- Virtual path

Fig. 18-77. Vertical and oblique equivalents for plane earth and plane ionosphere. (*Davies.*[9])

If the electron density is sufficiently large to reduce μ to zero, a wave (incident normally) will be reflected. Otherwise the wave will penetrate the layer. The highest frequency which can be reflected from the layer at vertical incidence is called the *critical frequency*. The ordinary-wave critical frequency f_0 is associated with the level of maximum electron density; $f_0 F2$ is the critical frequency for the F2 layer, etc, and is related to N by

$$N_{max} = 1.24 \times 10^{10} f_0^2 \qquad \text{electrons/m}^3 \quad (18\text{-}150)$$

where f_0 is in megahertz.

A second critical frequency and reflection level are associated with the extraordinary wave. At frequencies above the gyrofrequency* f_H, the level of reflection is given by $X = 1 - Y$; whereas at frequencies below f_H ($Y > 1$) the condition is $X = 1 + Y$. The critical frequency for reflection of the extraordinary wave is

$$f_x = f_0 + f_H/2 \qquad (18\text{-}151)$$

A number of simplifications[8,9] of the formula for μ are important for special cases of quasi-longitudinal and quasi-transverse propagation, normal incidence, etc.

95. Absorption. Neglecting magnetic field, the *absorption* of the radio wave in passing through a medium of collision frequency ν is

$$\kappa = \frac{\omega}{v_0} \frac{1}{2\mu} \frac{XZ}{1 + Z^2} = \frac{e^2}{2\epsilon_0 m v_0} \frac{1}{\mu} \frac{N_\nu}{\omega^2 + \nu^2} \qquad \text{Np} \qquad (18\text{-}152)$$

Equation (18-152) enables us to distinguish between two types of absorption. *Nondeviative absorption* occurs in the D region of the ionosphere; it predominates when μ is near unity and when the product $N\nu$ is large. *Deviative absorption* occurs near the level of reflection and whenever there is marked bending of the ray, i.e., when μ tends to zero.

For nondeviative absorption, when ω is much greater than ν, we have

$$\kappa \approx \frac{e^2}{2\epsilon_0 m v_0} \frac{N_\nu}{\omega^2} \qquad \text{Np} \qquad (18\text{-}153)$$

Polarization effects on absorption due to the magnetic field are an important factor. The general case is very complex. For a variety of conditions, however, the propagation conditions are quasi-longitudinal, and the nondeviative absorption coefficient is then given, approximately, by

$$\kappa \approx \frac{e^2}{2\epsilon_0 m v_0} \frac{N_\nu}{(\omega \pm \omega_H)^2 + \nu^2} \qquad \text{Np} \qquad (18\text{-}154)$$

The minus and plus signs refer to the ordinary and extraordinary waves, respectively. The ordinary wave suffers less absorption, and the extraordinary suffers more absorption, than the corresponding wave in the absence of a magnetic field.

96. Velocity of Propagation; Phase and Group Velocity. The *phase velocity* of a wave (neglecting magnetic field and collisions) is

$$v = v_0/\mu = v_0(1 - Ne^2/m\epsilon_0\omega^2)^{-1/2} \qquad (18\text{-}155)$$

The relationships between phase velocity, wavelength, and refractive index are

$$v\mu = v_0 \qquad \text{and} \qquad v/\lambda = v_0/\lambda_0$$

When the phase velocity of a wave in a medium is a function of the wave frequency, the medium is said to be *dispersive*.

Group velocity may be regarded as the velocity with which the modulation envelope travels.

97. Propagation at Frequencies below 150 kHz. Transmission in the frequency

*Gyromagnetic resonance frequency $f_H = 2.84 \times 10^{10} B$ MHz ≈ 1.4 MHz.

range below 150 kHz is in the region bounded by the earth and lower ionosphere. ELF waves ($f < 3$ kHz) are of geophysical interest[119] because of their depth of penetration below the earth's surface and are also regarded as a vital means of communication with deeply submerged submarines.[120] VLF is usually employed for very long distance (worldwide) communication and navigation.[121] Limited transmission to points under the sea or beneath the earth is possible at these and lower frequencies. International standard-frequency bands are allocated at 20 and 60 kHz. VLF observations of atmospheric noise give information on worldwide thunderstorm distribution; observations of phase perturbation indicate solar flares and can be used to detect nuclear explosions in the atmosphere.

A number of maritime communication and navigation services operate in the LF region; navigation services include Loran C (90 to 100 kHz), which uses pulses in a hyperbolic system. Because of high levels of atmospheric noise and physical limitations of antennas, high transmitter powers are required for transmission and applications are usually limited to relatively narrow bandwidth. At the lowest frequencies, say below 50 kHz, the dispersive characteristics of the propagation and the high Q of antennas also limit the usable bandwidth. Vertical polarization is employed. Reference 121 gives a good engineering summary.

Waveguide-Mode Theory. At VLF, propagation is usually described in terms of waveguide modes, i.e., the waves propagate between the earth and the lower boundary of the ionosphere. The waveguide-mode theory consists of a full-wave solution that includes the significant effects of diffraction and surface-wave propagation (Refs. 3, 4, 122, and 123 and *CCIR Rep.* 265-4). Such waves are considered to propagate between the earth and the ionosphere as normal waveguide modes, analogous to microwave propagation in a lossy waveguide. At frequencies above about 30 kHz the waveguide is many wavelengths high, and for short distances many propagating modes must be considered; but at VLF, especially at longer distances ($\geq 1,000$ km) only a few modes need be considered. VLF transmitters usually radiate a vertically polarized field.

For long paths, over smooth homogeneous terrain, the vertical field strength on the ground from a transmitter at a distance of d km on the ground (*CCIR Rep.* 265-4 and Ref. 4) is

$$e = \frac{300 \sqrt{p_t}}{\sqrt{a \sin (d/a)}} \frac{\sqrt{\lambda}}{h} \exp \left[-i(kd + \pi/4)\right] \sum_m \Lambda_n \exp \left(-ikS_n d\right) \qquad \text{mV/m} \qquad (18\text{-}156)$$

where $p = $ radiated power (kW) (allowance must be made for antenna efficiency)[4,121], $a = $ radius of the earth, $\lambda = $ free-space wavelength (km), $k = 2\pi/\lambda$, $\Lambda_n = $ excitation factor for the nth mode, $kS_n = $ propagation constant, $d = $ path distance, and $h = $ height of the ionosphere (70 km day, 90 km night).

In general, the terms Λ_n and S_n are complex. The excitation factor Λ_n gives the relative amplitude and phase of each mode of order n excited in the earth-ionosphere waveguide by the source. [Additional height-gain terms $G_r(y)$ and $G_n(y_0)$ for the transmitter antenna are considered when antennas are not located near the ground.] The real part of the propagation constant kS_n contains the phase information for each mode while the imaginary part determines the attenuation rate α. In order to obtain the field strength the contributions of each mode must be summed, with proper attention to the relative phases of the terms. Modification is required near the antipode when $d/a \approx \pi$.

The phase velocity $v_n = v_0/\text{Re}\,(S_m)$, where v_0 is the free-space velocity.

Methods for determining Λ_n and S_r are given in Ref. 4. These factors are related to the wavelength for the ionosphere height, the ground electrical properties, and the spherical reflection coefficient of the ionosphere. In turn the ionosphere reflection coefficients depend on the vertical distribution of electron density and collision frequency, the direction and magnitude of the earth's magnetic field, the frequency, and angle of incidence. The electron-density distribution is a function of latitude, season, solar cycle, time of day, and whether or not ionospheric disturbances are present. Horizontal gradients of electron density, i.e., at sunrise or sunset are also important.[4] Values of Λ, v, and α from Ref. 124 are illustrated in Figs. 18-78 and 18-79. Experimental daytime field strength is illustrated in Fig. 18-80. In summary, the important effects are as follows (*CCIR Rep.* 265-4).

Ground Conductivity. Reducing the ground conductivity increases the attenuation rate of all modes. However, when the conductivity is very low, e.g., polar icecaps, the attenuation rate may approach a maximum and then decrease as the conductivity is lowered further. For moderate conductivities, the magnitude of the excitation factor for the first-order mode usually increases somewhat as the conductivity is reduced and the phase velocity is reduced.

Direction of Propagation with Respect to the Earth's Magnetic Field. The earth's magnetic field causes greater attenuation for propagation to the magnetic west than for propagation to the magnetic east. Propagation in north-south directions gives intermediate attenuation rates in the daytime. At low latitudes, nighttime attenuation may be much greater for north-south paths than for other directions.[125]

Electron-Density Profiles.   The D-region electron-density profiles for VLF calculations can be approximated by[4]

$$\omega_r(z) = \omega_r(h) \exp [\beta(z - h)] \tag{18-157}$$

where $\omega_r = \omega_0^2/\nu$, for $\nu \gg \omega$; ω, ν, and ω_0 = wave angular frequency, collision frequency, and plasma angular frequency, respectively; z = height, and h = reference height at which $\omega_r \approx$

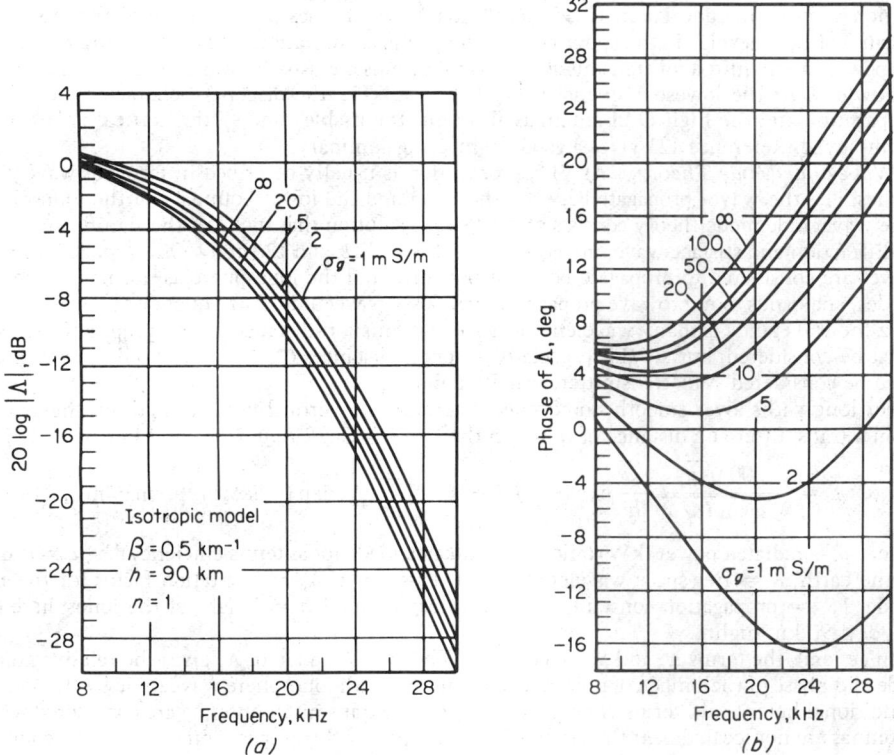

Fig. 18-78. Amplitude (*a*) and phase (*b*) of excitation factor as functions of frequency. (*Wait and Spies.*[124])

2.5×10^5. The term gives the vertical gradient of ω_r. Under daytime conditions, h is about 70 km and $\beta \approx 0.3$ km^{-1}, while at night h is around 90 km and $\beta \approx 0.5$ km^{-1}. These two parameters, h and β, provide a convenient but approximate means of describing the ionosphere for use in VLF calculations. When the magnetic field is considered, the collision-frequency profile $\nu \approx \exp \alpha_0(3 - h)$ must also be considered. Here α_0 is taken to be 0.15 km^{-1}. For a horizontal magnetic field transverse to the propagation path, a magnetic field parameter (see Figs. 18-78 and 18-79) assumes the values -1 and $+1$, corresponding to propagation along the magnetic equator from west to east and east to west, respectively ($\Omega = 0$ corresponds to zero magnetic field).[4]

Among other limitations, the calculations refer to conditions where the path properties are independent of distance. When the electron-density distribution with height along the path is constant but the ground conductivity or magnetic field angle changes, approximate calculations can be made by the methods of Ref. 4 and *CCIR Rep.* 265-4.

At sufficiently low frequencies, i.e., for *ELF propagation* (<3 kHz), the zero-order mode dominates at nearly all distances. In this case only the $n = 0$ term in Eq. (18-156) is needed, and $\Lambda_0 \approx \frac{1}{2}$. In such cases the attenuation rate is of the order of 1 dB per 1,000 km for both day and night models.

Wave-Hop Theory and Geometric-Optical Methods.   Wave-hop theory interprets the resultant field strength and phase at a receiver as the sum of waves (*rays*) that have traveled via the surface wave and one or more earth-ionosphere reflections. Calculations of wave hops using full-wave theory have been shown to be equivalent to waveguide-mode theory.[126,127]

Simpler calculations using geometric-optical formulas for the ionospheric wave,[127–129] condensed here from *CCIR Rep. 265-4*, can be used to approximate the full-wave solution, providing corrections are made for diffraction. The surface-wave field is added to the ionospheric-wave field using the phase difference implied by Fig. 18-81 Application of this method is most prac-

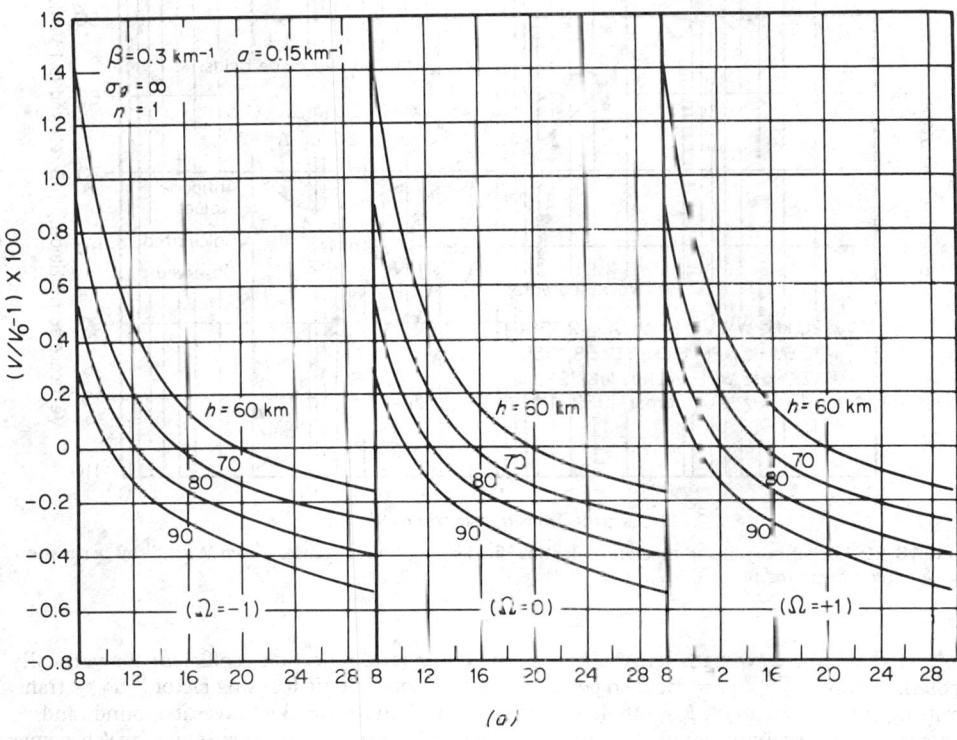

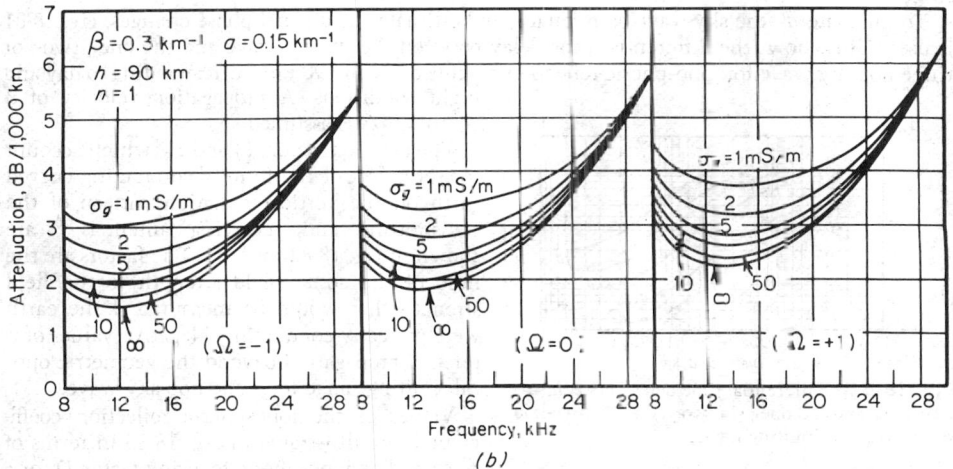

Fig. 18-79. Phase velocity (*a*) and attenuation (*b*) of $n = 1$ mode. (*Wait and Spies.*[124])

tical in the range 50 to 150 kHz. At higher frequencies, the empirical methods of Pars. **18-98** to **18-100** are used. At lower frequencies, the waveguide-mode theory is generally used.

For a vertically polarized transmitting antenna and reception by a small loop antenna located on the surface of the earth, the effective field strength of the sky wave is

$$e_s = 2(300 \sqrt{p_t}/d') \cos \psi \, RDF_T F_R \qquad \text{mV/m} \qquad (18\text{-}158)$$

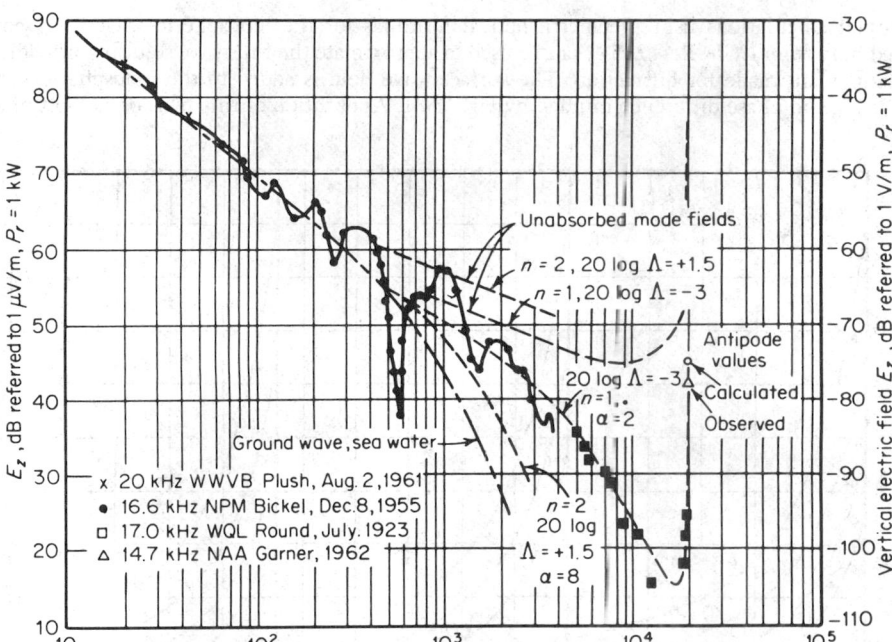

Fig. 18-80. Experimental and theoretical daytime field strength vs. distance. (*From Watt*,[121] *by permission of Pergamon Press, Inc.*)

where d' = sky-wave path length (km), R = ionospheric reflection coefficient for vertically polarized waves,[127-129] p_t = radiated power (kW), D = ionospheric focusing factor,[129] F_T = transmitting-antenna factor,[130] ψ = angle of departure and arrival of sky wave at ground, and F_R = appropriate receiving-antenna factor. For reception by a short vertical antenna cos ψ becomes $\cos^2 \psi$.

To calculate d', the sky-wave path length, and estimate the diurnal phase changes, Fig. 18-81 is used. This shows the differential time delay between the surface wave and the one-, two-, or three-hop sky wave for ionospheric reflection heights of 70 and 90 km, corresponding to day and night conditions. A propagation velocity of 3 $\times$ 10^5 km/s is assumed.

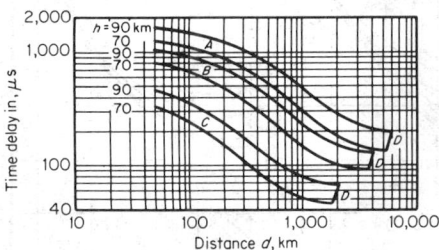

Fig. 18-81. Differential time delay between surface wave and (*A*) one-, (*B*) two-, and (*C*) three-hop sky waves; (*D*) limiting range.

The antenna factors F_T and F_R, which account for the effect of the finitely conducting curved earth on the vertical radiation pattern of the transmitting and receiving antennas,[130] are shown in Fig 18-82 for land. The factors are the ratio of the actual field strength to the field strength that would be measured if the earth were perfectly conducting. Negative values of ψ refer to propagation beyond the geometric optical limiting range for a one-hop sky wave.

Values of the ionospheric reflection coefficient R are illustrated in Fig. 18-83 in terms of $f \cos \phi$. The ionospheric focusing factor D for a spherical earth and ionosphere has a value between 1 and 2.5, depending on path length and frequency. D, R, and F are given in *CCIR Rep. 265-4*.

98. Propagation at 150 to 1,500 kHz. Because of the stable, only moderately attenuated ground wave and efficient nighttime ionospheric propagation, this range of frequencies is used for maritime communications and navigation and for medium-range broadcasting. Aeronautical mobile and fixed communications and radio positioning services are operated in many areas of the world.

Transmission is usually described by summing the *powers of* ground-wave and wave-hop reflections from the lower ionosphere. The description of propagation is similar to that above for frequencies in the 50- to 150-kHz region. However, because the phase path usually involves hundreds of wavelengths and ionospheric reflection characteristics are increasingly variable at higher frequencies, the phase of the downcoming wave is essentially random. This introduces phase-interference fading in the region where ground wave and sky wave are of comparable magnitude. Thus applications are usually designed to take advantage of the ground wave or sky wave separately, avoiding admixture. Polarization effects are important, as discussed in Par. **18-101**.

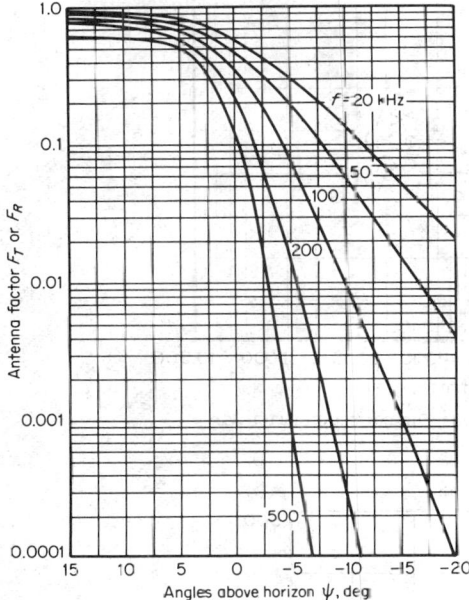

Fig. 18-82. Antenna factors (land): $\epsilon_r = 15$; $\sigma = 2$ ms/m; $a = \frac{2}{3}(6,360)$ km. *(CCIR Rep. 265-2.)*

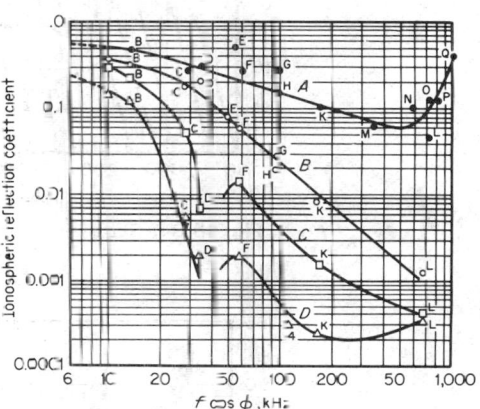

Fig. 18-83. Ionospheric reflection coefficients, so arc cycle maximum. Letters designate vertical incidence measurements: A, night (all seasons); B, day (winter); C day (equinox); D day, summer. *(CCIR Rep. 265-4.)*

During the day, at distances less than 1,000 km, the ground wave is dominant because of intense daylight D-region absorption of the ionospheric wave. At night, ionospheric reflection coefficients can be high, and even multiple reflections are useful. Attenuation varies from one night to the next and with season. Statistical estimates of field strength are obtained empirically. The main bases are two sets of long-term measurements of mf broadcasting stations, one covering the European region by the European Broadcasting Union (EBU)[131] and the International Broadcasting and Television Organization (OIRT), and the other covering United States paths by the Federal Communications Commission (FCC). The European results and analyses are somewhat more comprehensive in frequency and temporal and geographical coverage, and are used by the CCIR. However, there are some differences from the FCC results,[36,13a] which represent measurements over United States paths 2 h after sunset in the 500- to 1,600-kHz band. They have been adopted for official use by North American countries for AM broadcast-frequency-sharing studies and are undoubtedly more accurate for the midcontinental United States.[132a]

99. Estimation of Sky-Wave Field Strength by the CCIR Method (Recomm. 435-3). The field strength is usually represented in terms of an *annual median field strength* for local midnight at the path midpoint, with corrections for antenna, magnetic-dip latitude, local time other than midnight, seasons, and sunspot number.

The annual median sky-wave field strength, assuming reception on a small, vertically polarized antenna, is represented by

$$E = E_0 + \Delta_A \quad \text{dB}\mu \qquad (18\text{-}159)$$

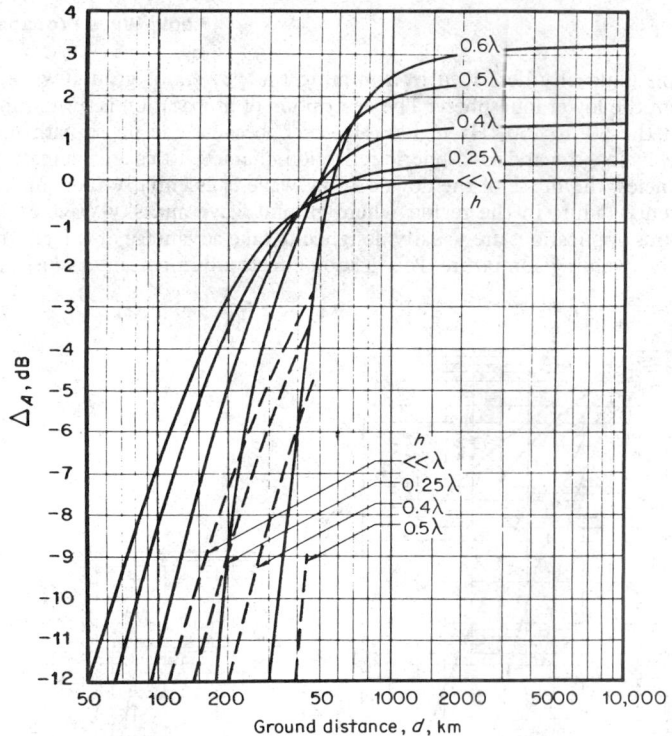

Fig. 18-84. Transmitting antenna gain correction factor Δ_A.

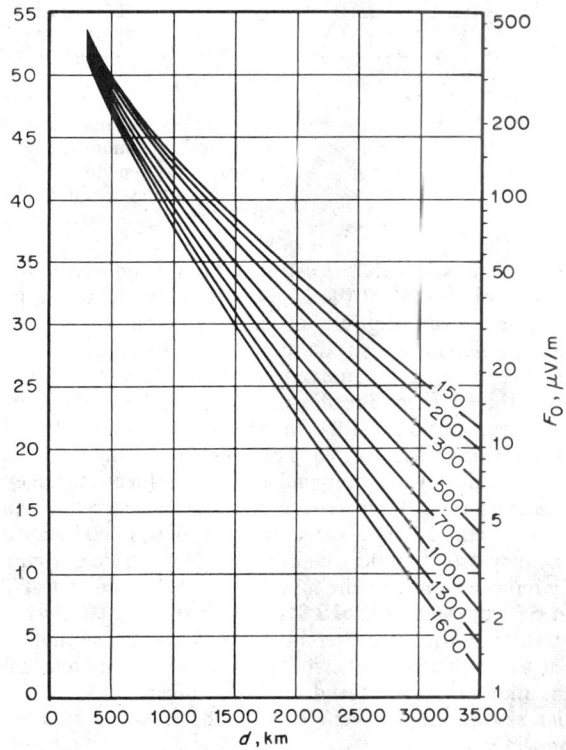

Fig. 18-85. Family of basic curves of annual median field strength E_0 for frequencies 150 to 1,600 kHz for conditions of zero correction for magnetic dip and sunspot number.

where E = annual median value of field strength at a reference hour for power of 1 kW radiated by transmitting antennas; E_0 = annual median value of field strength (dBμ) at reference hour, for reference transmitting antenna giving, over perfectly conducting ground, field strength of 3 $\times 10^5$ μV/m, at a distance of 1 km in all directions above the horizon; Δ_A = 20 log ratio of field strength for a given angle of departure in the vertical plane to field strength of reference antenna above.

Figure 18-84 gives values of Δ_A as a function of the distance between the transmitting and receiving points in the theoretical case of a lossless unloaded vertical antenna of height h/λ placed over perfectly conducting earth and, a reflection of a virtual height of 100 km above a spherical earth, for propagation via the E region of the ionosphere. The discontinuity of the curves at 2,200 km corresponds to the distance beyond which there are at least two hops.

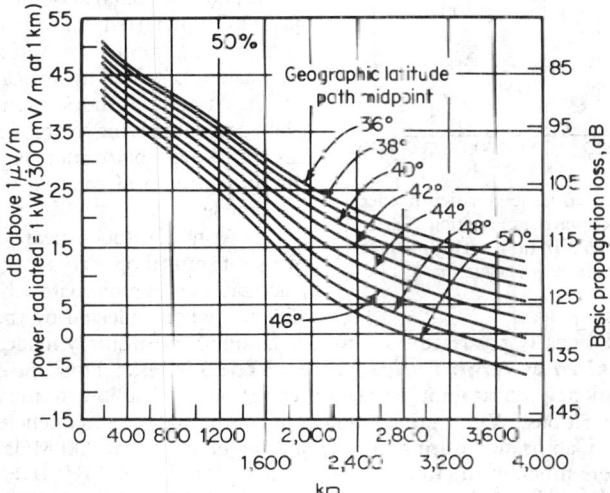

Fig. 18-86. United States sky-wave propagation from FCC data for 50% of nights at 1 MHz (2 h after sunset). (*From Barghausen.*[135])

The value of E_0 is

$$E_0 = 80.2 - 10 \log d - 0.0018 f^{0.26} d \qquad \text{dB}\mu \qquad (18\text{-}160)$$

where d = distance (km) and f = frequency (kHz).

The values of E and E_0 are for the following conditions:

The magnetic dip at the midpoint of the path is 61°.

The sunspot number is $R = 0$

The reference hour is local midnight at the midpoint of the propagation path.

Figure 18-85 shows a family of basic curves of field strength E_0, for $R = 0$, $\Delta_A = 0$, and for a geomagnetic latitude where the correction for magnetic dip is zero. *CCIR Rep* 575-1 gives methods and data for correcting the field for other values of magnetic dip, sunspot number, local time, and percentage of nights of the year.

100. Estimation of Field Strength from FCC Data. The FCC Rules[36] give a series of propagation curves which estimate nighttime sky-wave field strengths exceeded 10 and 50% of the nights (the 50% data are shown in Fig. 18-86). Frequency dependence is not included in the FCC curves, which represent the field strength from 500 to 1,600 kHz, centered at 1,000 kHz. Frequency dependence deduced[132] from FCC data is shown in Fig 18-87 Figure 18-84 can be used to adjust these field strengths for transmitting-antenna factor.

101. Polarization Effects. Paragraphs **18-92** to **18-94** include a description of polarization and absorption effects in the ionosphere. At medium frequencies, because of heavy absorption of the *extraordinary* wave in the E region, propagation depends almost entirely on the *ordinary* wave, which is, in general, elliptically polarized. Polarization coupling losses can occur when vertical transmitting and receiving antennas are used. Though not serious at midlatitude, they may be quite large for east-west paths near the geomagnetic equator, where the ordinary wave tends to be horizontally polarized, and for certain multiple-hop situations involving change of polarization at ground reflection.[132,133] Polarization coupling loss arises because any wave incident on the ionosphere will excite the ordinary mode to the degree that the incident-wave polarization corresponds to the polarization characteristic of that mode. When the two polarizations are orthogonal, e.g., linear polarization at right angles or circular polarizations with opposite sense of rotation, there will be no coupling with the ordinary wave mode. Phillips and Knight[133] identify that serious propagation losses may be encountered in the following situations:

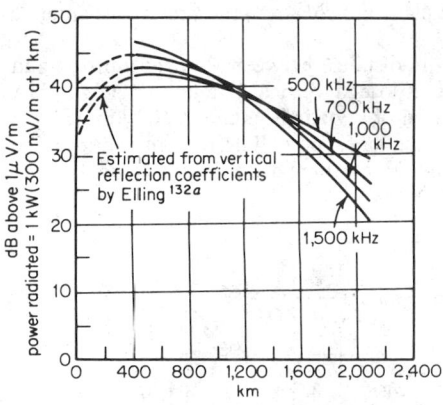

Fig. 18-87. Calculated frequency dependence of United States sky-wave propagation 500 to 1,500 kHz (vertical polarization; nighttime data 2 h after sunset). (*From Barghausen.*[132])

1. At low latitudes, on east-west paths using vertical transmitting and receiving antennas, because the ordinary wave polarization is nearly linear and horizontal. Use of horizontal antennas is to be preferred, for the ionospheric waves, though the ground wave would be appreciably attenuated compared to vertical polarization.

2. At low latitudes, on north-south multihop paths involving sea reflection, because the circular wave polarization after reflection has the wrong sense of rotation to excite the ordinary mode.

3. At midlatitudes (near 45° dip angle), for east-west multihop paths over sea. Here the polarization is approximately linear but tilted at 45° and, when reflected by the sea, emerges at right angles to the linearly polarized wave for excitation of the ordinary mode.

102. Propagation at High Frequencies, 3 to 30 MHz.* From the early 1920s to the present, high-frequency propagation has been used for economical low- to medium-power long-distance communications. While propagation at medium and low frequencies (300 kHz to 3 MHz) suffers heavy absorption during the day and frequencies above 30 MHz are not reflected from the ionosphere much of the time, the frequency range 3 to 30 MHz (HF) usually provides ionospheric reflections day and night. The range of usable frequencies is limited on the upper end by the height and maximum electron density of the controlling layer and on the low-frequency end by absorption in the D region.

HF propagation is characterized by:

1. Variability of propagation conditions, requiring frequent changes in the operating frequency

2. Interruption by ionospheric storms

3. The large number of possible propagation paths and resulting multipath-interference effects

4. Dispersion and frequency distortion

5. Large and rapid phase fluctuations

6. High interference

103. Reflection and Absorption in Oblique Propagation. The reflection process for plane ionosphere is equivalent to mirror-type reflection at a height equal to the virtual height h' of reflection of the equivalent vertical frequency.[9,10] The equivalence theorem must be modified for presence of the earth's magnetic field or for a curved ionosphere.[8,9]

For most purposes, the highest reflection frequency for oblique incident f_{ob} is related to the vertical critical frequency by

$$f_{ob} = k f_v \sec \phi_0 \qquad (18\text{-}161)$$

where $k \approx 1.2$ and is given by transmission curves.[9] A similar equivalence applies for absorption.

The echo structure becomes complex over oblique paths, especially over long distances.

---

*The omission of discussion of propagation at frequencies between about 1.5 and 3 MHz reflects difficulty of adequate methods for treatment and the relatively little use of ionospheric propagation in this frequency range because of intense daytime absorption. For nighttime propagation, one can interpolate between the results obtained by methods for the adjacent higher and lower bands. However, because of the proximity of the operating frequency to the gyromagnetic resonance frequency, attenuation is increased. Theroetical methods for this frequency range are difficult.[9]

Oblique soundings have provided the most revealing insights into the characteristics of long distance HF propagation.[114,134,135]

MUF, originally *maximum usable frequency*, requires some definition. *Classical MUF* is used to designate the highest frequency propagated by ionospheric reflection alone; *operational MUF* denotes the highest frequency permitting operation between points under specified working conditions; it may be higher than the classical MUF as the result of sporadic E, ground scatter, or ionospheric scatter. MOF and LOF refer to highest and lowest *observed frequencies* in the oblique sounding.

Figure 18-88 gives nomenclature for reflections and illustrates the roles of the E and F regions. In the diagram F is used to represent both F1 and F2 layers.

F2 reflections are usually the most important, and during the day the lowest-order (lowest-angle) minimum F ray-path reflection is usually dominant; often even two-hop F reflection is more efficient than one-hop E. The horizon-limited distance for one-hop F2 propagation depends upon layer height but is usually around 4,000 km. Or longer paths, intermediate ground reflection is usually involved, though often the high (Pederson) ray of the first hop F2 reflection is dominant.[134] The effects of the earth's magnetic field, involved in the double-refraction process and the separation and differential absorption of the ordinary and extraordinary rays, depend in a complicated way on path length, the magnetic latitude (dip angle), and the path direction.[8,9,18]

Rays propagated via any of the reflecting layers pass through the absorbing D region, encountering there the principal attenuation relative to free space.

The F1 *layer* is important during daylight hours and during ionospheric disturbances. In summer at high latitudes the principal reflection for paths of 2,000 to 3,500 km is usually via F1.

The E *layer*, because of its low height and low nighttime density, provides useful propagation (mainly daytime) over medium distances to 2,000 km. The ray path of a signal reflected from the F2 layer, however, must penetrate the E layer at one or more points, so that the frequency must be sufficiently high for penetration. The E-layer penetration frequency for one-hop F2 propagation constitutes effectively the lowest useful frequency for many low- to medium-power systems in the distance range to 4,000 km.

Sporadic E (E_s) can also play a dominant role in determining maximum usable frequency. Besides

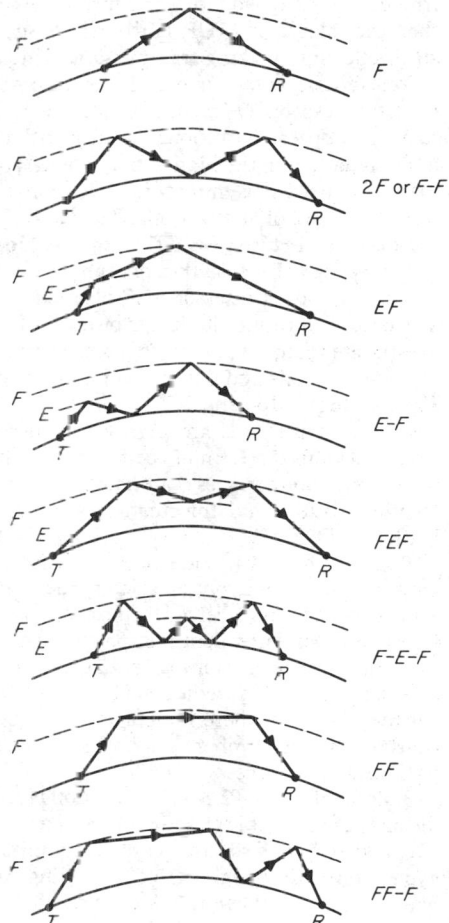

Fig. 18-88. Possible oblique reflection geometries; descriptive nomenclature for oblique ray paths. (*Reprinted by permission from K. Davies,* "Ionospheric Radio Waves," *copyright 1969 Ginn and Co., published by Xerox College Publishing.*)

providing, at times, reflection at distances up to 2,400 km sometimes up to 150 MHz in frequency, E_s can extend the range of F2 propagation, as illustrated in Fig. 18-88. One reflection illustrated provides an extended-range (second-hop) F2 reflection which would otherwise, via the ground, suffer D-region attenuation. Sometimes the operational MUF is higher than the calculated F2 MUF due to the effects of ground scatter, scatter from the ionosphere, sporadic E, or ionospheric tilts.

In the case of *ground scatter* (often called side scatter) the signals are reflected from the ionosphere obliquely to a ground region, scattered from the ground, and propagated again by ionospheric reflection to the receiver. If the transmitter and receiver are collocated, the process is referred to as *backscatter*. This technique has become important as a tool for studying the ionosphere, as well as in over-the-horizon radar systems. If the receiver is located at a distance from the transmitter, the signal arrives from an off-great-circle azimuth and affords transmission under conditions where direct reflections are not possible, because of the more oblique reflections associated with the side-scatter geometry. This phenomenon is allowed for to some extent in estimating extended hours of transmission in international HF broadcasting. Bearing deviations from the great circle due to side scatter are important sources of error in direction-finding and radio-positioning systems.

Ionospheric scatter can also play an important role. In particular, under conditions represented by "spread *F*" on the vertical ionogram, oblique transmissions may be possible at frequencies higher than the calculated MUF. Near the magnetic equator and in auroral regions ionospheric irregularities aligned along the magnetic field can provide returns at much higher frequencies than the calculated MUF. Furthermore, scattering in the *D* region can increase the obliquity of an *F* reflection, making transmission at higher frequencies possible.

Ionospheric tilts can permit two ionospheric reflections without an intermediate ground reflection, especially important for a north-south transmission across the magnetic equator. The signal is propagated to a long distance with only two transits of the absorbing region, thus giving high signal strength; higher frequencies are reflected than with a spherically stratified layer; there is a marked asymmetry in the path, so that the angle of elevation at departure can differ from the angle of arrival (Refs. 9 and 134).

104. Prediction of HF Propagation.   *Atlases of Ionospheric Characteristics; Numerical Mapping.*   Because the ionosphere varies geographically as well as from hour to hour and day to day and with season and solar activity, extensive atlases of ionospheric characteristics are needed to determine the limits of useful frequencies and to perform the calculations necessary to estimate required power, angles of elevation for optimum antenna design, and so on. Comprehensive methods and data for computing high-frequency propagation are given by *CCIR Rep.* 252-2 and Refs. 136 and 137.

Computer methods are highly developed for these predictions. Computer programs, ionospheric data in the form of coefficients for numerical mapping function,[138-139a] and routine computer prediction services[137] are available from the U.S. Department of Commerce, National Telecommunications and Information Administration, Institute for Telecommunication Sciences, Boulder, CO 80302.

The long-term variation of *E*, *F1*, and *F2* parameters is tied closely to the sunspot cycle. Although no completely satisfactory measure of solar activity is available, the 12-month running average sunspot number R_{12} is most widely used. Alternatives such as the ionospheric index I_{F2} (essentially an equivalent sunspot number obtained from ionospheric observations) are useful for short-term predictions (less than 12 months in advance). Solar flux at 10 cm wavelength is especially useful for prediction of *E* and *F1* region characteristics up to 6 months or so in advance. Current values of these indices are published in the *ITU Telecommunication Journal.* Predictions of sunspot number 12 months in advance are also obtainable from the above Department of Commerce address.

Basic worldwide *F2* MUF predictions for 0 and 4,000 km distance and *E* MUF for 2,000 km distance are available (Fig. 18-89). Reference 139 is an atlas of nearly 1,200 charts for $R_{12} = 10$, 110, and 160, representing respectively minimum solar activity, an average maximum of a solar cycle, and a very high period exceeding the peak of the average solar cycle. *CCIR Rep.* 340 contains nearly 600 worldwide charts of *F2* MUF for all months, every 2 h, for $R_{12} = 100$ (Fig. 18-90). Charts and numerical mapping coefficients for the observed probability of occurrence of sporadic *E* are also available[140] (see Fig. 18-91).

Calculation of MUF.   To obtain MUF values for a given path, the great-circle path is drawn on a rectangular coordinate scale corresponding to the prediction map[139] and is used to overlay the radio circuits on the *F2* and *E* MUF maps. For paths shorter than 4,000 km the path midpoint is considered to be the reflection point. Longer paths are divided into an integral number of hops, and the lower MUF values are taken to be controlling. An alternative method for such long paths is more accurate in some cases and less in others: two control points for *F2* transmission are taken, each 2,000 km from the path terminals, and the lower MUF is used.

The computer methods (Ref. 137 and *CCIR Rep.* 252-2) consider additional parameters in some detail, such as the height of the maximum electron density and the height of the bottom of the layer, reconstructing, in effect, an approximate profile of electron density along the path. A parabolic-layer assumption is made to compute the ray geometry along the path. Up to nine ray paths are evaluated in the CCIR program. The ray path must be geometrically possible for a takeoff angle equal to or greater than the minimum value given as input data.

Variability of MUF.   The values of MUF referred to above are monthly medians for a given hour. Day-to-day random variation of *F2* MUF is observed to be distributed according to a chi-squared probability distribution; the constants and computation of this distribution depend on

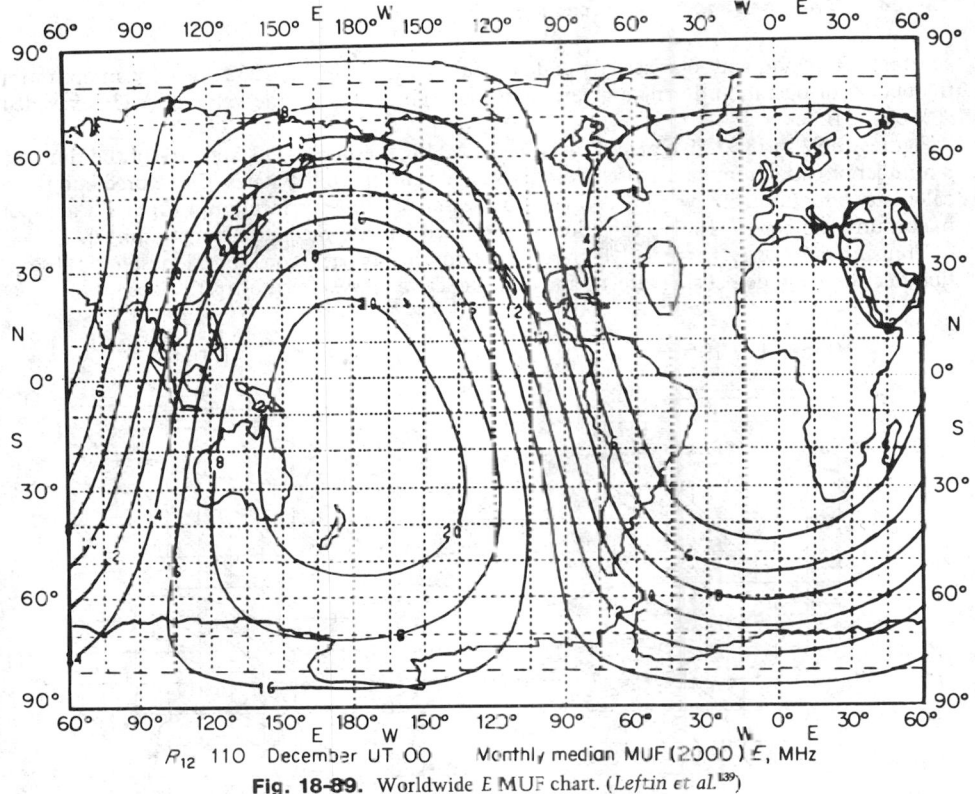

R_{12} 110   December UT 00   Monthly median MUF(2000) E, MHz

Fig. 18-89. Worldwide E MUF chart. (*Leftin et al.*[139])

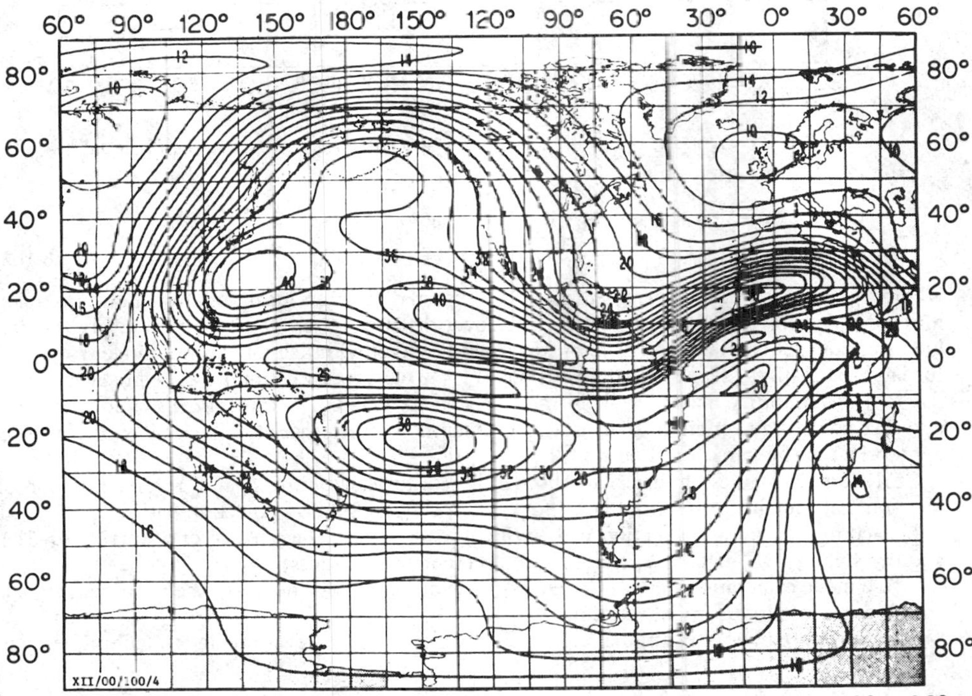

December 00 hours $R_{12}=100$; EJF (4,000) F_2, MHz

XII/00/100/4

Fig. 18-90. Example of worldwide $F2$ MUF chart. To be precise, EJF (estimated junction frequency), or "classical" MUF, is used. (*CCIR Rep.* 340)

geographical region and solar activity and are outlined in *CCIR Rep.* 252. Usually an optimum frequency for operation during a given month at a given hour is the value of MUF exceeded 90% of the time.

Figures 18-92 to 18-97 give predicted values of MUF exceeded 90, 50, and 10% of the time for a number of paths from East Coast, West Coast, and central United States. These predictions are calculated* for summer, winter, and equinox seasons at sunspot minimum (0) and a sunspot maximum number of 110. Interpolation may be used for intermediate sunspot numbers.

Transmission Loss.   Loss relative to free space (for the ray path) consists of ionospheric absorption, focusing and defocusing effects, and earth-reflection losses for multipath hops.

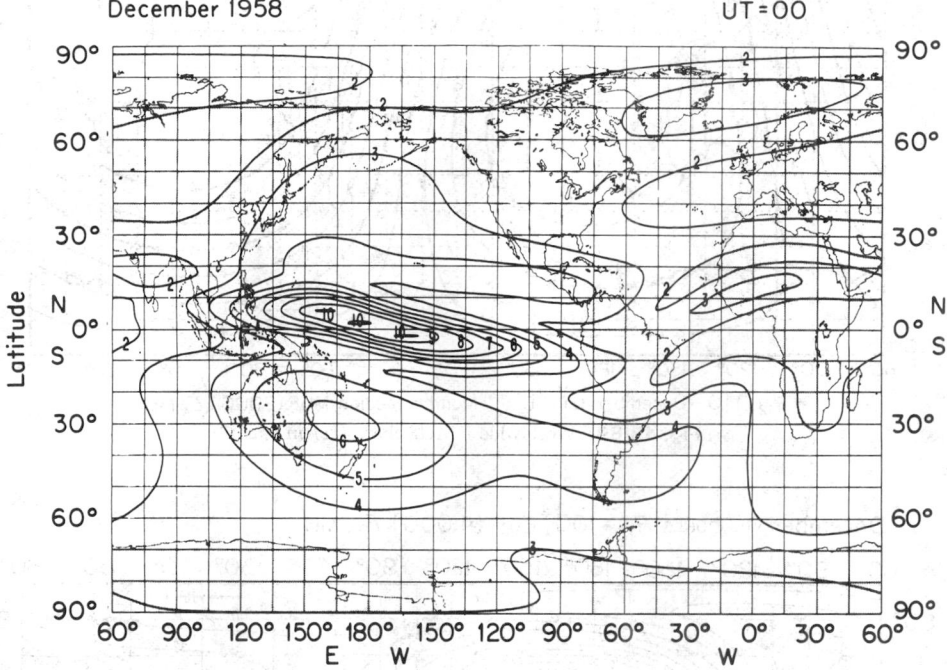

Fig. 18-91. Worldwide map of median value sporadic-E f_oE_s occurrence. (*Leftin et al.*[140])

The total transmission loss is computed according to

$$L = L_{bf} + L_i + L_g + Y_p - (G_t + G_r)   \text{dB} \tag{18-162}$$

where L_{bf} = basic free-space transmission loss expected between ideal, loss-free, isotropic transmitting and receiving antennas in free space over distance of ray path; L_i = monthly median loss from ionospheric absorption; L_g = loss caused by ground reflection (values range from 4 dB for land to 0.2 dB for seawater per reflection); Y_p = excess transmission loss, used to account for day-to-day variability about monthly median; and G_t and G_r = transmitting- and receiving-antenna power gain relative to an isotropic antenna in free space, for the angles of elevation appropriate to the particular hop.

CCIR Rep. 252 gives empirical methods and data for calculating L_i and Y_p as a function of location, path length, season and hour, for each of the ionospheric reflection paths.

L_i is computed for all reflection paths which have a reasonable chance of occurring (>5%), and the smallest value, plus Y_p, is taken to represent the path loss.

These mode calculations also permit an estimate of the relative delays of the multipath signals which can cause intersymbol interference.

The above prediction techniques are for the long term. Short-term prediction techniques gen-

---

*Predictions calculated by U.S. Dept. of Commerce, National Telecommunications and Information Administration, Institute for Telecommunication Sciences, Boulder, Colo.

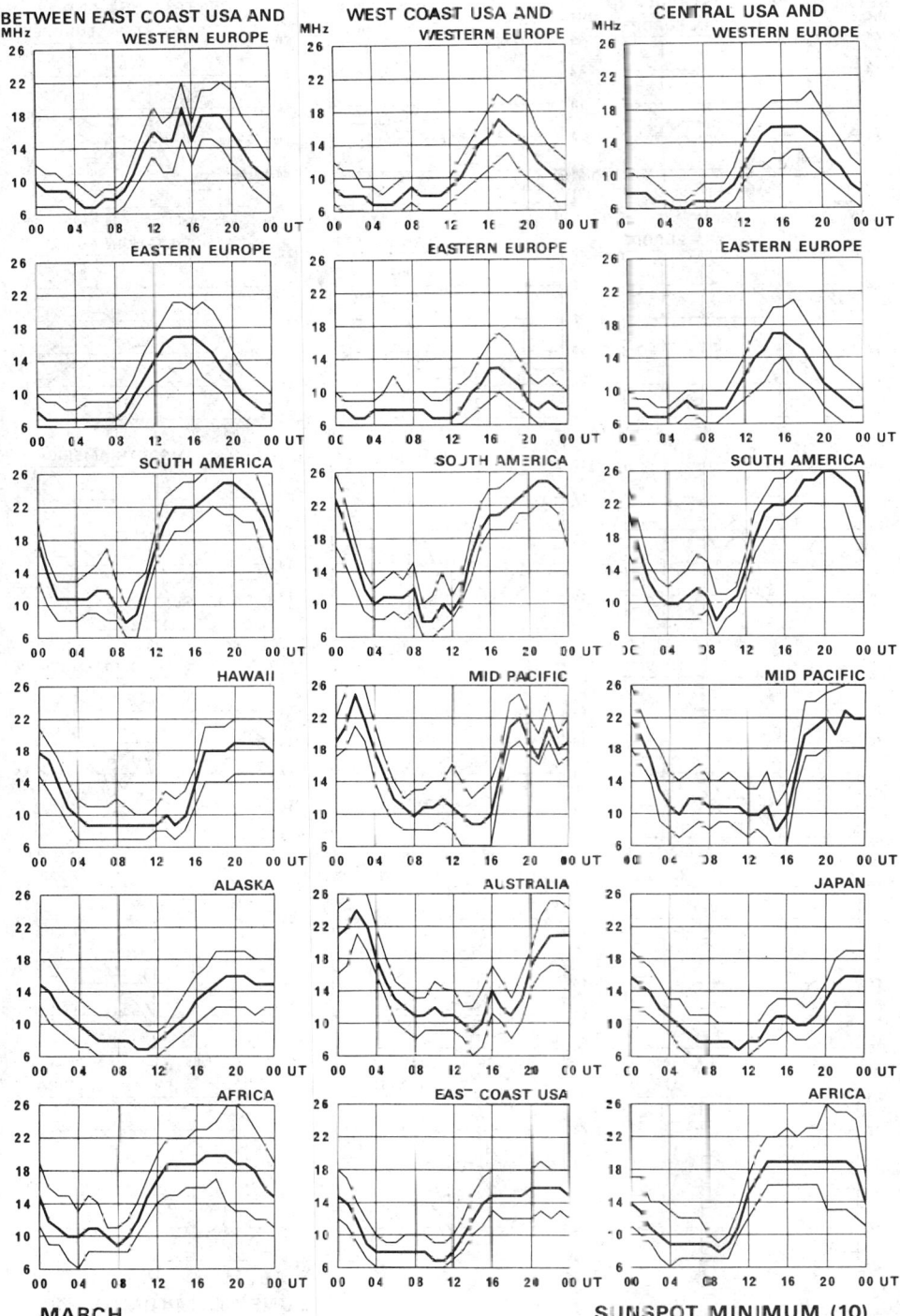

Fig. 18-92. MUF exceeded 90, 50, and 10% of the time; March, sunspot number 10.

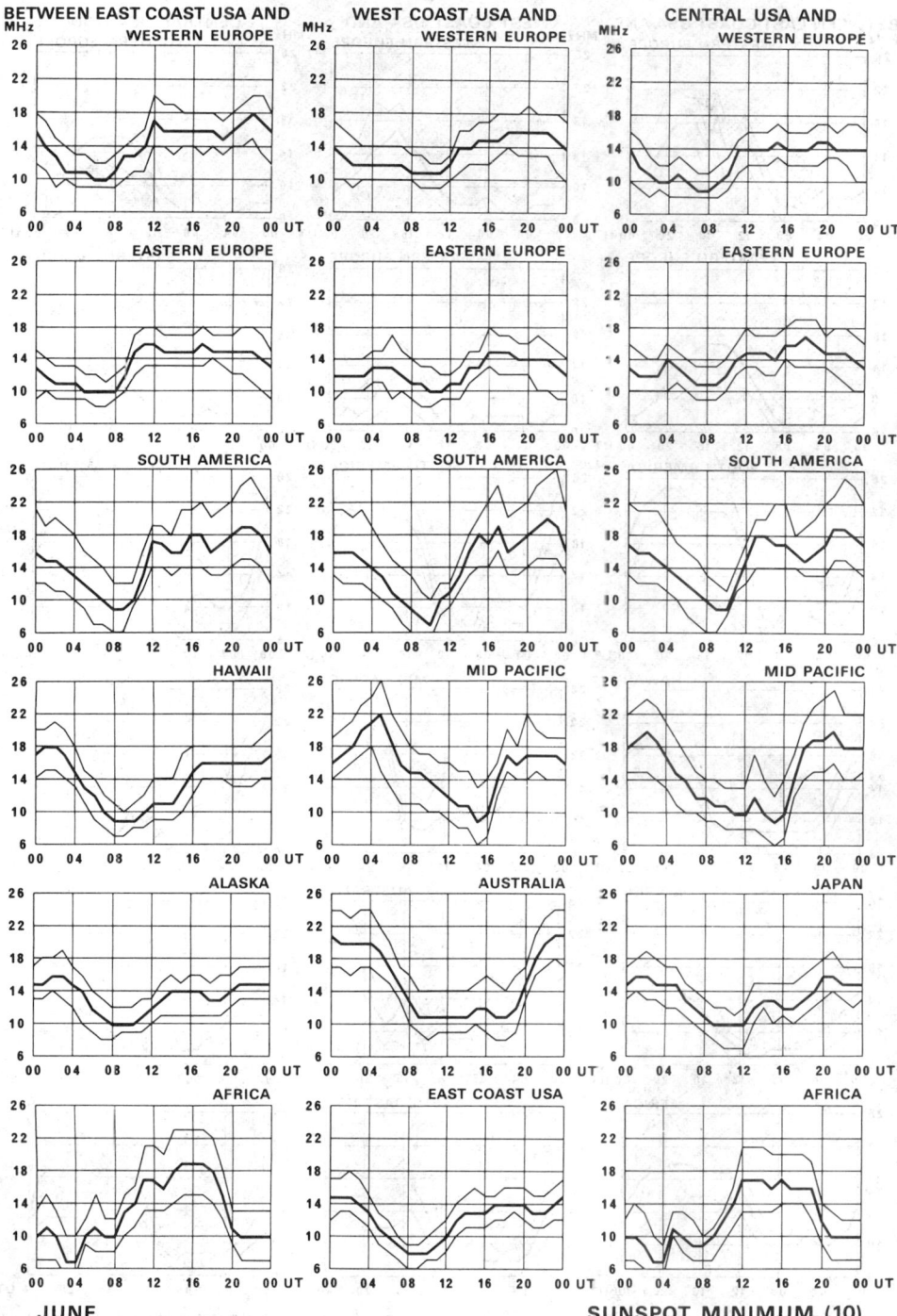

Fig. 18-93. MUF exceeded 90, 50, and 10% of the time; June, sunspot number 10.

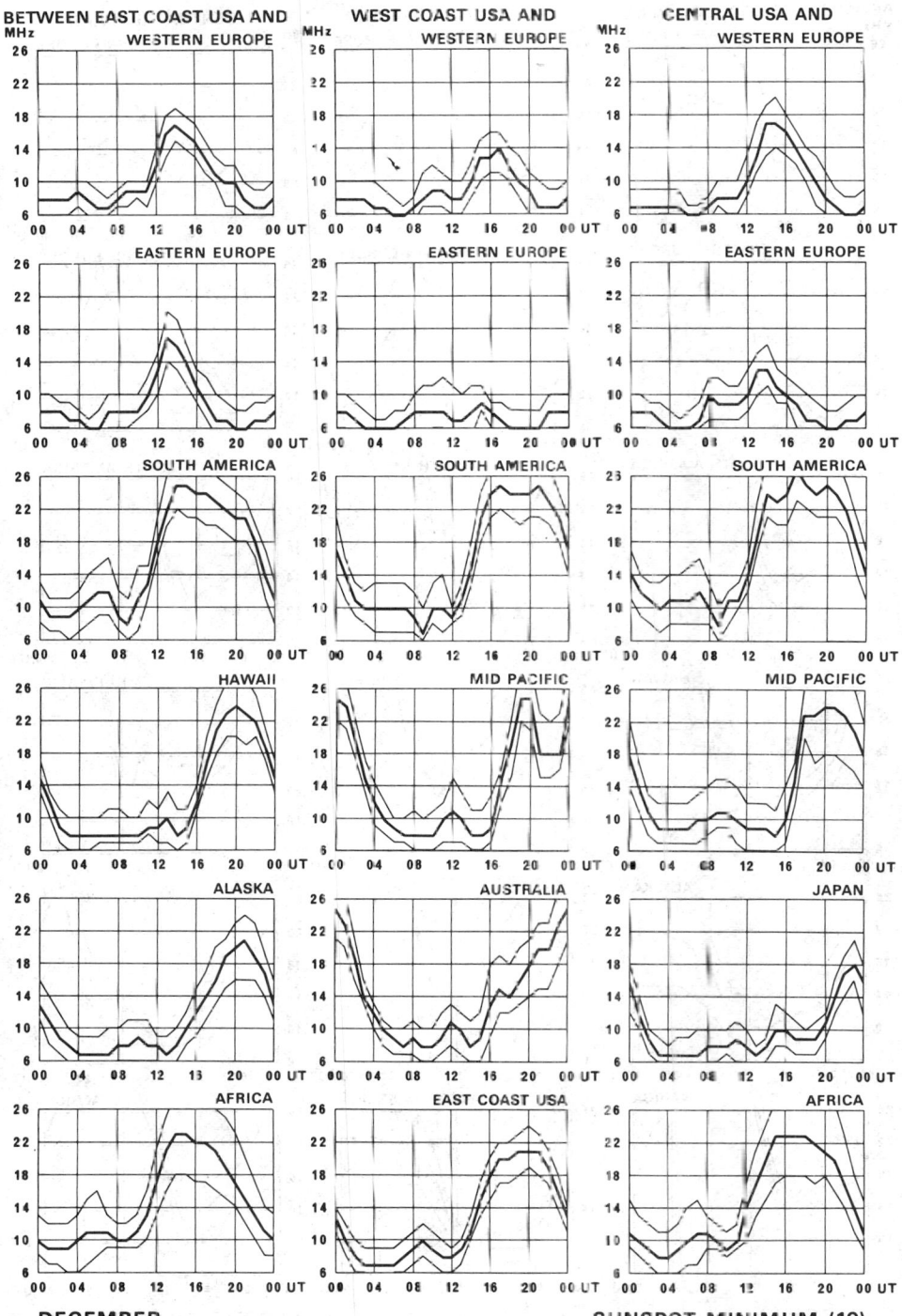

Fig. 18-94. MUF exceeded 90, 50, and 10% of the time; December, sunspot number 10.

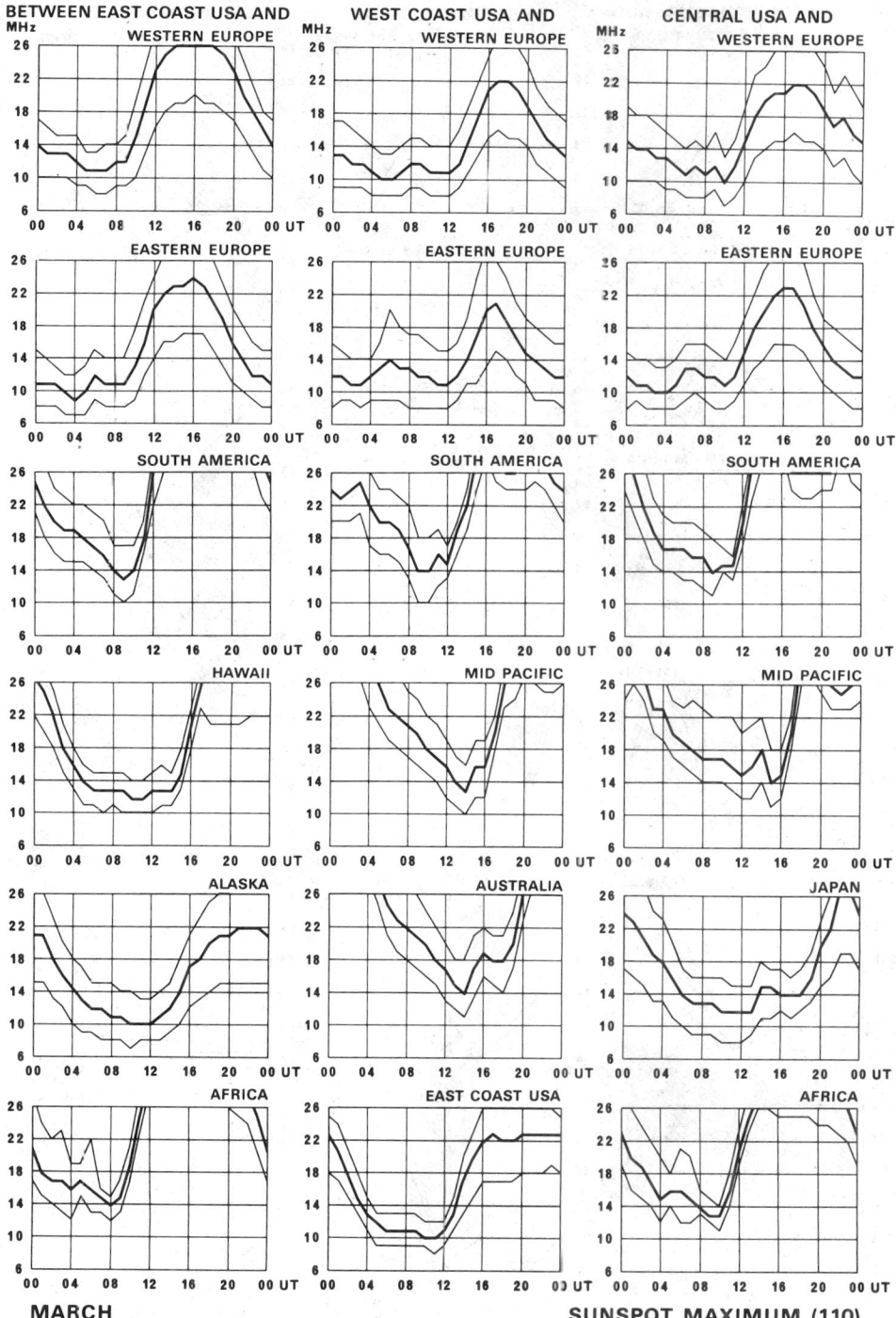

Fig. 18-95. MUF exceeded 90, 50, and 10% of the time; March, sunspot number 110.

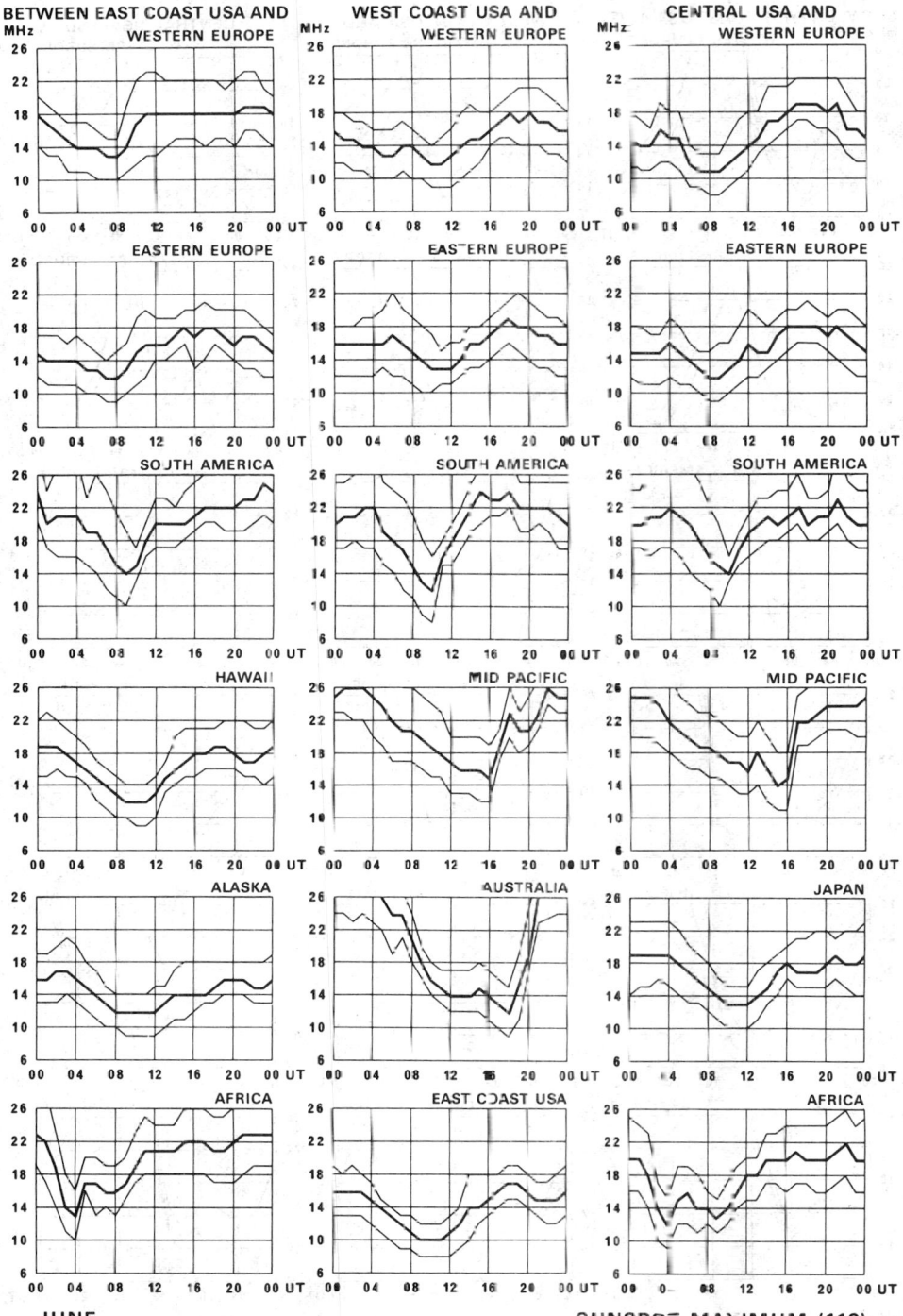

Fig. 18-96. MUF exceeded 90, 50, and 10% of the time; June, sunspot number 110.

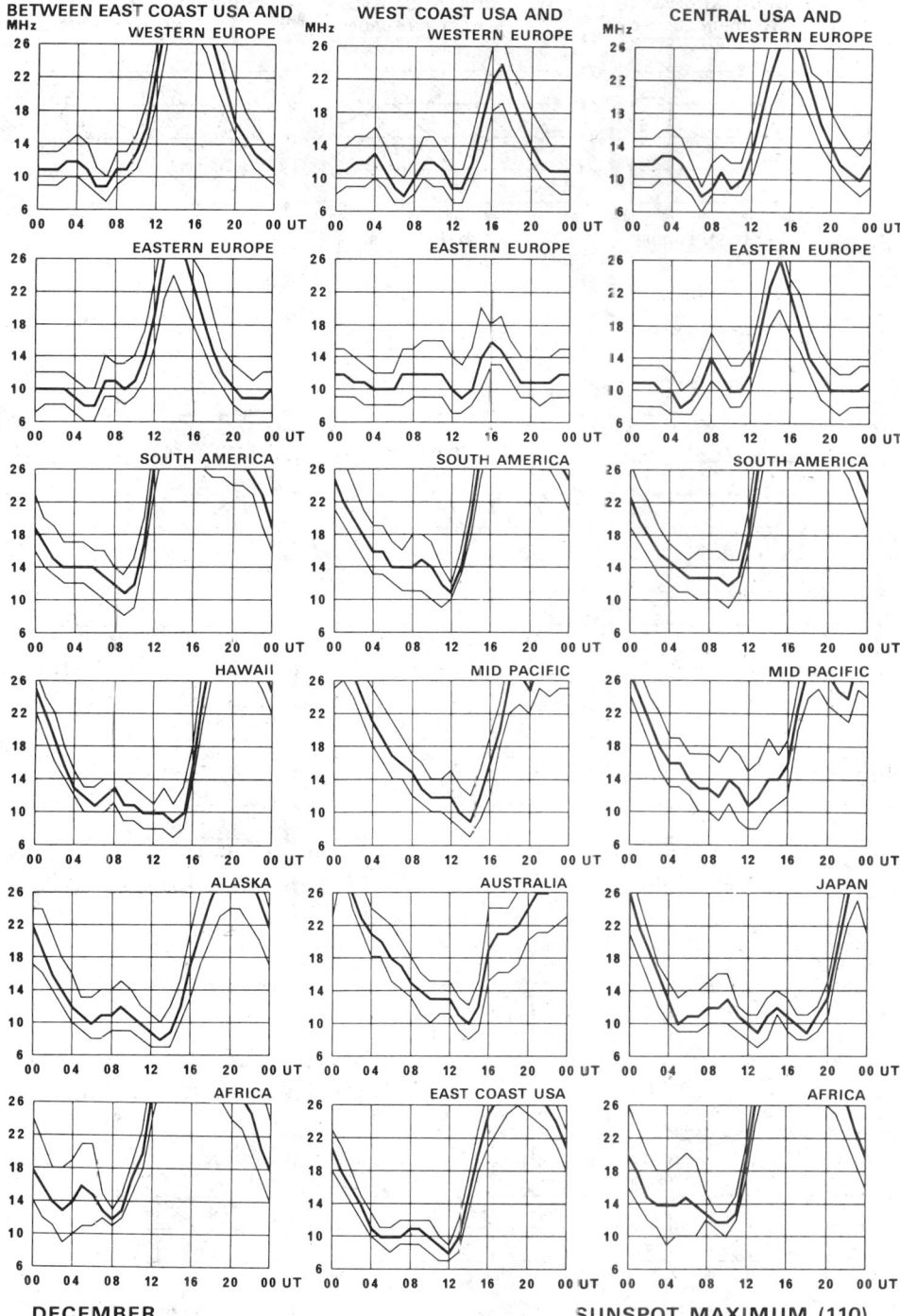

Fig. 18-97. MUF exceeded 90, 50, and 10% of the time; December, sunspot number 110.

erally make use of oblique sounders and provide observations of current conditions rather than predictions in the true sense (Refs. 134, 140, and 140a, and *CCIR Rep.* 249-4).

Siting considerations for HF antennas are outlined in Refs. 141 and 142. For very long paths, the lowest possible angles of radiation are advantageous.

Short-Term Variability, Fading, Multipath. Short-term-fading observations (within periods of about 5 min to exclude the long-term variability) show near Rayleigh distribution most of the time. The fading is somewhat more shallow (Rice-Nakagami distribution) under very strong signal conditions, due to the presence of a strong specular component. Duration of fades may be from 0.1 s to many seconds; the phase variation poses a limitation to phase-keying systems. In equatorial and auroral regions rapid fading is often encountered, with fading rates of 10 to 100 Hz.[143,144]

Under conditions of normal phase-interference fading, a correlation coefficient of less than 0.5 is typical for fading observed at spacings of 15λ, though these parameters depend somewhat on frequency and path length. Diversity spacings of the order of 10λ are useful in overcoming fading of narrow-band signals (bandwidth less than a few hundred hertz). The frequency-selective nature of the multipath fading shows independent fading at frequencies separated by more than a few hundred hertz (Refs. 8, 9, and 143 and 145 and *CCIR Rep.* 266-4). Polarization diversity is also useful for confined spaces, e.g., aboard ship.

Multipath propagation also leads to *intersymbol interference* in digital transmission. Differential propagation delays of several milliseconds are observed between first- and last-arriving significant signal components. Usually less than 5 ms, these relative delays can be mimimized by operating at frequencies sufficiently close to the MUF to reduce multiple reflections. A *multipath reduction factor* has been devised to estimate the significant differential delays for given ratio of operating frequency to prevailing-path MUF.[146] For paths shorter than 1,000 km, multiple reflections can produce multipath delays exceeding 5 ms which are difficult to reduce because the reflection angles are nearly equally steep for all reflections.

105. Prediction of Disturbed Propagation Conditions. Some of the types of ionospheric disturbances which can cause propagation difficulties at HF are:

1. *Sudden ionospheric disturbances* lasting for a few minutes to an hour or more, associated with solar flares and causing intense D-region absorption.

2. *Ionospheric storms*, lasting for a few hours to several days, associated with magnetic disturbances, as a result of solar-particle radiation. The most prominent effects are reduction of F2 MUF and increase in D-region absorption. Aurorae often occur at high latitudes during such storms.

3. *Polar-cap absorption*, lasting for a few hours to several days, associated with high-energy solar protons causing D-region ionization down to heights of the order of 50 km and intense absorption at high latitudes.

The U.S. Department of Commerce, National Oceanic and Atmospheric Administration, operates a Space Environment Forecast Center at Boulder, Colo., which issues current observations and advance forecasts of solar and geomagnetic conditions causing ionospheric disturbances. The U.S. Department of Commerce, National Telecommunications and Information Administration, Institute for Telecommunication Sciences, at Boulder, issues warnings of HF propagation disturbances by broadcast (WWV standard-frequency transmission of the National Bureau of Standards), direct-wire, and time-share computer access. The latter services include estimates of effects on MUF and transmission loss.

106. Scattering from the Ionosphere. At frequencies above about 30 to 300 MHz (VHF) electron densities of the ionosphere are rarely sufficient to reflect waves except for short periods. Transmission is mainly by scattering, although regular F and sporadic E, discussed in the previous paragraphs, do provide reflections often enough to be important sources of long-distance interference to mobile, broadcasting, and other services in this band. It is for this reason the FM broadcasting was reallocated from near 40 MHz to near 100 MHz and channel 1 TV is not used in the United States. Table 18-5 shows the main causes of long-distance interference above 30 MHz.

107. Scattering at VHF from *D*-Region Irregularities. Regular but weak scattering from irregularities in the D region is useful for continuous single-frequency communication in the 30- to 60-MHz frequency range, over paths of the order of 1,000 to 2,000 km (Refs. 70 and 147 to 149). Approximately 1 kW per 100-baud digital channel is required at 35 MHz, depending upon antenna design, modulation, and coding.

The D-region electron-density irregularities are produced by the action of turbulence, wind shears, and overlapping ionization at meteor trails, in the height range 70 to 90 km. Energy is

TABLE 18-5 Causes of Long-Distance Ionospheric Interference to Services Working in the 30- to 100-MHz Region*

Cause of interference	Latitude zone	Period of severe interference	Approximate highest frequency with severe interference, MHz	Approximate frequency above which interference is negligible, MHz	Approximate range of distances affected, km
Regular F-layer reflections	Temperate	Day, equinox winter, solar-cycle maximum	50	60	East-west paths 3,000–6,000 or north-south paths 3,000–10,00
	Low	Afternoon to late evening, solar-cycle maximum	60	70	
Sporadic E reflections	Auroral	Night	70	90	500–4,000
	Temperate	Day and evening, summer	60	85–135	
	Equatorial	Day	60	90	
Sporadic E scatter	Low	Evening up to midnight	60	90	Up to 2,000
Reflections from meteoric ionization	All	Particularly during showers	May be important anywhere in the range		
Reflections from magnetic field aligned columns of auroral ionization	Auroral	Late afternoon and night			Up to 2,000
Scattering in the F region	Low	Evening through midnight, equinox	60	80	1,000–4,000
Special transequatorial effects	Low	Evening through midnight	60	80	4,000–9,000

*From *CCIR Rep. 259-4.*

scattered out of the incident beam, for single isotropic scattering neglecting polarization, according to[149]

$$p_a = p_t r_0^2 l^{-2} b A_r \csc(\gamma/2) S(K)$$
(18-163)

where r_0 = classical electron radius = 2.8×10^{-15} m, l = distance from scattering volume to receiver, b = thickness of scattering volume, A_r = effective area of receiving antenna, γ = angle through which scattering takes place, $S(K)$ = spectrum of turbulent irregularities, p_a = available power at the receiver in same units as p_t. p_t = radiated power, and

$$K = 4\pi/\lambda \sin \gamma/2$$
(18-164)

In the case of turbulent mixing,[149]

$$S(K) = K^{-4}(dN/dh)^2$$
(18-165)

where dN/dh = electron-density gradient.

The relationship of transmission loss to the frequency f and geometry of the propagation path[150] is given by

$$p_t/p_a \propto l^2 f^{n_s}(\sin \gamma/2)^{n} {}^{-1}$$
(18-166)

where n_s ($n_s = n + 2$) is the frequency exponent for scaled antennas (gain constant with frequency). Observations show n_s varies with time but is mostly between 7 and 9½, with a median value of 8.

Figure 18-98 shows distance and frequency dependence of ionospheric scatter compared with tropospheric scattering.[70]

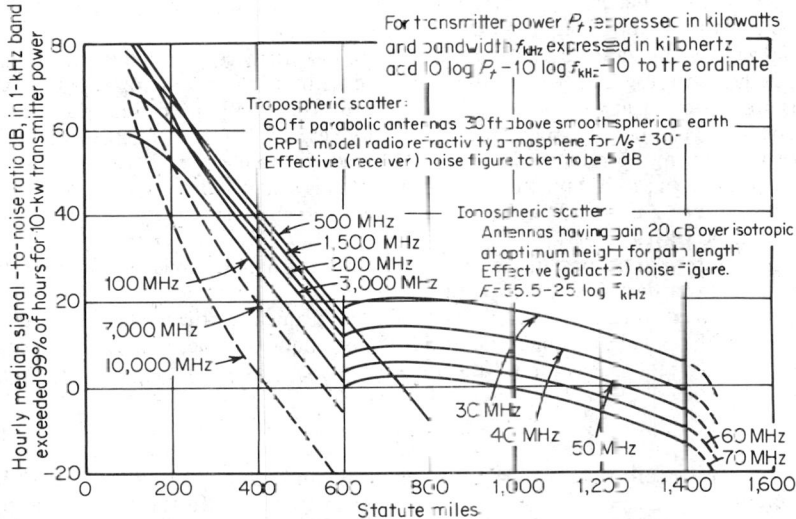

Fig. 18-98. Comparison of distance and frequency dependence of ionospheric and tropospheric scatter. (JTAC.[70])

In the frequency range from 25 to 108 MHz, over distances of from 1,000 to 2,000 km, the transmission loss ranges from about 140 dB to about 210 dB, depending on frequency, antenna, geography, and time.

Short-term fading is Rayleigh-distributed with superimposed bursts from meteor reflections. Fading rates lie between 0.2 and 3 Hz, with a mean value of 1 Hz at 50 MHz and varying approximately proportionately with operating frequency. Space or frequency diversity is applicable, with horizontal spacings about 5 wavelengths (or frequency spacings of about $f \times 10^{-4}$ required, where f is the operating frequency). Simple Yagi and rhombic antennas are satisfactory for some applications, while some large systems have used elaborate corner-reflector arrays. Modulation considerations include the effect of Doppler-shifted meteor reflections and long-delayed F2 multipath.[151]

VHF scatter does not suffer auroral absorption, but auroral scattering may be an important source of multipath; and polar-cap absorption may attenuate the signal at frequencies below 50 MHz. During solar flares, depending upon the intensity, the signal is attenuated at frequencies below about 40 MHz and enhanced at higher frequencies. VHF scatter is presently of little practical importance but may have a potential for reliable low-power data communications, especially in polar regions.

108. Meteor Scatter. Ionized trails are produced by meteors mostly in the height range 80 to 120 km. They diffuse rapidly and usually disappear within a few seconds. These ionized columns reflect VHF waves particularly in the frequency range 20 to 150 MHz. The lower frequency limit is set by the need to be above the regular maximum usable frequency, and the upper limit is set by weakening of the reflections and their shorter duration with increasing frequency (Refs. 152 to 153 and *CCIR Rep.* 251-2).

Meteor-burst propagation is subject to abnormal attenuation during polar-cap absorption events and sudden ionospheric disturbances.

Meteor-burst communications systems have been developed (Refs. 155 and 156 and *CIRR Rep.* 251-2) to operate duplex digital circuits in the 30- to 40-MHz range, over distances of 600 to 1,300 km with transmitter powers of 1 to 3 kW; meteor propagation is currently of little practical importance, but some applications are being suggested for low-power, intermittent data transmission from remote sensors.

109. Other Forms of Scatter. *Equatorial F Scatter.* Occurrence of F scatter is associated with equatorial spread F produced by patches of irregularities located at or below the bottom 50 km or so of the F layer.[9,157] The irregularities are elongated along the magnetic field with longitudinal dimensions of 1,000 m or more and transverse dimensions of the order of 10 m or less. The echoes are aspect-sensitive. While F scatter can permit communication at frequencies as high as 50 MHz (well above the MUF), disastrous for HF communication is the occurrence of flutter fading with rates of 10 Hz or more.[158]

Auroral Scatter. Ionization associated with auroral disturbances gives radio reflections at HF and VHF. Radio scattering is only generally correlated with visible aurora; in general the auroral reflections are aspect-sensitive. Radio amateurs communicate via auroral reflections, and radar observations have been used to study aurora for many years.[159,160] The mechanisms are:[159]

1. Weak scattering from randomly distributed gradients in electron density
2. Strong scattering from randomly distributed clouds of ionization
3. Weak scattering from ordered arrays of gradients in electron density produced by propagating ion-acoustic waves

Incoherent Scatter. The existence of incoherent scatter from electrons (Thomson scattering) in the ionosphere was demonstrated in 1958.[161]

When high-power continuous-wave VHF radar is beamed vertically, weak scattering is observed from all heights in the ionosphere and the technique provides a means of determining the electron-density profile even above the F2 peak, which cannot be seen with conventional ionosondes. The power p_a received[161] from a distance R is

$$p_a = p_t a \sigma v_0 \tau \eta_t^2 \eta_s \eta_A / 8\pi R^2 \tag{18-167}$$

where a = antenna aperture area, σ = cross section per unit volume, v_0 = velocity of light, τ = pulse duration (s), η_t, η_s, and η_A = factors to correct aperture for effects of resistive losses, side lobes, and tapered feed, respectively, and R = range to scattering volume.

p_a is proportional to R^{-2} rather than R^{-4}, as in the case of a single-scatterer. From the divergence of the beam, it can be seen that the scattering volume increases proportional to the square of the range; this removes a factor of R^{-2}.

For thermal equilibrium between ions and electrons, the scattering cross section per free electron is just about one-half the classical Thomson cross section. In the more general case, when the electron and ion temperatures (T_e and T_i, respectively) differ, the relationship between the measured cross section σ_m and the classical value σ is given approximately by

$$\sigma_m = (1/1 + T_e/T_i)\sigma \tag{18-168}$$

Measurements of σ_m give values[161] less than the theoretical value of 5×10^{-29} m² due to D-region absorption during the day and electron-ion temperature relationship at night.

Incoherent scatter is a powerful tool for upper-atmosphere investigations and is currently the subject of much active research.[162-164]

110. Earth-Space Paths via the Ionosphere. Signals from satellites or extraterrestrial sources, upon passing through the ionosphere, undergo refractive bending and scintillation (fading), polarization rotation, absorption, frequency change and dispersion, and some path delay (Refs. 8, 9, and 165 to 168, *CCIR Rep.* 263-4). While the effects are of major concern at frequencies below about 300 MHz, they are of importance ever. at much higher frequencies to some satellite communications and to tracking and radio-astronomy applications.

Formulas for *refraction and ray-path* determination have been given.[169] At very oblique angles the rays may be reflected, but a penetration cone is defined by the semiangles $\phi_0 = \sin^{-1} \mu_m$, where $\mu_m = \sqrt{1 - (f_c/f)^2}$ is the refractive index at the height of maximum electron density (f_c is the critical frequency). The effect of the magnetic field is to produce two images of the source, corresponding to the ordinary and extraordinary ray, each with a different path length and apparent direction of arrival.

Scintillations occur when the waves pass through electron-density irregularities.[166-169e] Phase variations caused by these refractive-index irregularities cause an amplitude-diffraction pattern, focusing, and defocusing. Scintillations can be quite severe and may represent a practical limitation for some types of satellite communications system, especially at VHF. Scintillation characteristics depend very much on the strength and nature of the irregularities and are characteristically different for equatorial, midlatitude, and high-latitude regions. Facing of 20 dB or more is observed at frequencies below 200 MHz at equatorial and high latitudes; fading of as much as 3 to 4 dB at frequencies in the 4- to 8-GHz range has been observed in equatorial regions. Equatorial scintillation peaks near midnight and occurs mainly during equinoxes. High-latitude scintillation is often associated with auroral disturbances.

A useful index for comparing scintillation data is the scintillation index S_4, which is defined as the standard deviation of received power divided by the mean value of received power.[169c] Observations employing 10 frequencies from 138 MHz to 2.9 GHz show a consistent $\lambda^{1.5}$ behavior for S_4 less than about 0.4, where λ is wavelength. Phase scintillation is about proportional to wavelength.

Absorption effects are small (less than about 1 dB) at frequencies above 100 MHz, except in auroral regions. A summary of important effects and their frequency dependence is given in Table 18-6.

TABLE 18-6 Estimated Maximum Ionospheric Effects at 100 MHz for Elevation Angles of About 30°-One-Way Traversal

Effect	Magnitude	Frequency dependence
Faraday rotation	30 rotations	$1/f^2$
Propagation delay	25 μs	$1/f^2$
Refraction	$\leq 1°$	$1/f^2$
Absorption, polar-cap	4 dB	$\sim 1/f^2$
Auroral + polar-cap	5 dB	$\sim 1/f^2$
Midlatitude	<1 dB	$1/f^2$
Dispersion	0.4 ps/Hz	$1/f^3$
Scintillation, midlatitude	≈ 20 dB fade depth	$1/f^{1.5}$
Equatorial	>20 dB fade depth	$1/f^{1.5}$ to $1/f$
High-latitude	>20 dB fade depth	$\sqrt{1/f}$ to $1/f^2$

111. Ionospheric Ducting. Long-distance HF and VHF propagation by ducting along field-aligned ionization[9] has been interpreted from observations of radar and rocket and satellite signals (Refs. 170 to 173 and *CCIR Rep.* 250-4). This type of propagation, depending on the geometry and ionization density, may lead to apparently anomalous signal transmission from satellite to earth at frequencies well above ionospheric-reflection frequencies.

112. Modifying the Ionosphere with Intense Radio Waves. Induced changes in the density, temperature, and distribution of F-region ionization have been demonstrated[174] by beaming powerful ground-based radio-wave emissions upward at near the critical frequency, using 2 MW continuous-wave power, with an antenna gain of about 40. Ambient electron temperature is increased by 35% or more, anomalous reflection structure and attenuation behavior are induced, artificial spread F is generated, and air-glow emission from oxygen is increased. All the induced changes are transitory and self-reversing. Such ionospheric-modification experiments are becoming an important tool for controlled studies of the physics of naturally occurring plasmas.

113. References on Radio-Wave Propagation

1. JTAC [Joint Technical Advisory Council, Institute of Electrical and Electronics Engineers (IEEE) and Electronic Industries Association (EIA)]   "Radio Spectrum Utilization," New York, 1964.

2. Ramo, S., J. R. Whinnery, and T. VanDuzer   "Fields and Waves in Communication Electronics," Wiley, New York, 1965.

3. Bremmer, H.   "Terrestrial Radio Waves: Theory of Propagation," Elsevier, Amsterdam, 1949.

4. Wait, J.R.   "Electromagnetic Waves in Stratified Media," Pergamon, New York, 1962 (2d ed. 1970).

5. Fock, V. A.   "Electromagnetic Diffraction and Propagation Problems," Pergamon, New York, 1965.

6. Born, M., and E. Wolf   "Principles of Optics: Electromagnetic Theory of Propagation: Interference and Diffraction of Light," 3d ed., Pergamon, New York, 1965.

7. duCastel, F.   "Tropospheric Wave Propagation beyond the Horizon," Pergamon, New York, 1966.

8. Davies, K.   "Ionospheric Radio Waves," Blaisdell, Waltham. Mass., 1969.

9. Davies, K.   Ionospheric Radio Propagation, *Natl. Bur. Stand. Monogr.* 80, 1965.

10. "Review of Radio Science 1978-80," *Rep. Int. Union Radio Sci.*, Brussels, 1981.

10a. Beckmann, P.   "The Depolarization of Electromagnetic Waves," Golem, Boulder, Colo., 1968.

11. Booker, H. G.   Developments in the Theory of Radio Propagation, 1900-1950, *Radio Sci.*, Vol. 10., 1975, p. 665.

12. Rice, P. L., A. G., Longley, K. A. Norton, and A. P. Barsis   Transmission Loss Predictions for Tropospheric Communication Circuits, *Natl. Bur. Stand. Tech. Note* 101 (rev.), 1967.

13. Norton, K. A., and A. C. Omberg   Maximum Range of a Radar Set, *Proc. IRE*, 1947, Vol. 35, No. 1, pp. 4-24.

14. Baghdady, E. J.   "Lectures on Communication System Theory," McGraw-Hill, New York, 1961.

15. Schwartz, M., W. R. Bennett, and S. Stein   "Communication Systems and Techniques," McGraw-Hill, New York, 1966.

16. Baghdady, E. J.   Models for Signal Distorting Media, in R. E. Kalman and N. DeClaris (eds.), "Aspects of Network and System Theory," pp. 337-381, Holt, New York, 1971.

17. Nakagami, M.   On the Intensity Distributions and Its Application to Signal Statistics, *Radio Sci. J. Res. Natl. Bur. Stand.*, 1964, Vol. 68D, No. 9, pp. 995-1003.

18. Rice, S. O.   Mathematical Analysis of Random Noise, *Bell Syst. Tech. J.*, 1944, Vol. 23, pp. 282-332; 1945, Vol. 24, pp. 46-156.

19. Beckmann, P.   Rayleigh Distribution and Its Generalizations, *Radio Sci. J. Res. Natl. Bur. Stand.*, 1964, Vol. 68D, No. 9, pp. 927-932.

20. Bennett, W. R.   "Electric Noise," McGraw-Hill, New York, 1960.

21. Middleton, D.   "An Introduction to a Statistical Communication Theory," McGraw-Hill, New York, 1960.

22. Spaulding, A. D., and R. T. Disney   Man-made Noise, Pt. I, *OT Report 44-38*, U.S. Govt. Printing Off., Washington D.C., 1974, and

22a. Spaulding, A. D., R. T. Disney, and A. G. Hubbard   Man-made Noise, Pt. II., *OT Report 75-63*, U.S. Govt. Printing Off., Washington D.C., 1975.

23. Skomal, E. N., An Analysis of Metropolitan Incidental Radio Noise Data, *IEEE Trans. EMC.*, Vol. 15, 1973, pp. 45-57.

24. Report of Joint Technical Advisory Council, Man-Made Noise, Subcommittee 63.1.3, Unintended Radiation, Supple. 9 "Spectrum Engineering," IEEE, March 1968.

25. Middleton, D.   Statistical-Physical Models of Urban Radio Noise Environments, I: Foundations, *IEEE Trans, Electromag. Compat.*, May 1972, Vol. EMC-14, No. 2.

26. Skomal, E. N.,   "Man-made Radio Noise," Van Nostrand Reinhold, New York, 1978.

27. Bean, B. R., and E. J. Dutton   "Radio Meteorology," *Natl Bur. Stand. Monogr.* 92 1966; also Dover, New York, 1968.

28. Bean, B. R., J. D. Horn, and A. M. Ozanich, Jr.   Climatic Charts and Data of the Radio Refractive Index for the United States and the World, *Natl. Bur. Stand. Monogr.* 22, November 1960.

29. Bean, B. R., B. A. Cahoon, C. A. Samson, and G. D. Thayer   A World Atlas of Atmospheric Radio Refractivity, *Environ. Sci. Serv. Admin. Monogr.* 1, 1966.

30. Millington, G.   Propagation at Great Heights in the Atmosphere, *Marconi Rev.*, 1958, Vol. 21, pp. 143–160.

31. Rotheram, D.   Ground Wave Propagation at Medium and Low Frequencies, *Electron. Lett.*, 1970, Vol. 6, pp. 794–795.

32. Wait, J. R.   Electromagnetic Surface Waves in J. Saxton (ed.), "Advances in Radio Research," Academic, New York, 1964.

32a. Wait, J. R.   Theory of Ground Wave Propagation, in J. R. Wait (ed.), "Electromagnetic Probing in Geophysics," Golem, Boulder, Colo., 1971.

32b. Wait, J. R.   Recent Analytical Investigation of Electromagnetic Ground Wave Propagation over Inhomogeneous Earth, *Proc. IEEE*, Aug. 1974, Vol. 62.

33. Norton, K. A.   The Calculation of Ground-Wave Field Intensity over a Finitely Conducting Spherical Earth, *Proc. IRE*, December 1941, Vol. 29, pp. 623–639.

34. Bremmer, H.   The Extension of Sommerfield's Formula for the Propagation of Radio Waves over a Flat Earth to Different Conductivities of the Soil, *Physica'sGrav.*, 1954, Vol. 20, p. 441.

35. Johler, J. R., W. Keller, and L. C. Walters   "Phase of the Low Radio Frequency Ground Wave," *Natl. Bur. Stand. Circ. 573*, 1956.

36. FCC Rules and Regulations, Radio Broadcast Services, Secs. 73.184 and 73.190, March 1980.

37. CCIR "Atlas of Ground-Wave Propagation Curves for Frequencies between 30 and 300 MHz (Vertical and Horizontal Polarization)," International Telecommunications Union, Geneva, 1955.

38. CCIR "Atlas of Ground-Wave Propagation Curves for Frequencies between 30 and 10,000 MHz (Vertical Polarization Only)," International Telecommunications Union, Geneva, 1959.

39. Maley, S. W.   Radio Wave Methods for Measuring the Electrical Parameters of the Earth, Chap. 2 in J. R. Wait (ed.), "Electromagnetic Probing Methods in Geophysics," Golem, Boulder, Colo., 1971.

40. Kirby, R. S., J. C. Harman, F. M. Capps, and R. N. Jones   Effective Radio Ground-Conductivity Measurements in the United States, *Natl. Bur. Stand. Circ. 546*, February 1954.

41. Wait, J. R., and L. C. Walters   Curves for Ground Wave Propagation over Mixed Land and Sea Paths, *IEEE Trans. Antenna Propag.*, 1963, Vol. AP-11, pp. 38–45.

42. Godzinski, Z.   A Comparison of Millington's Method and the Equivalent Numerical Distance Method with the Theory of Ground-Wave Propagation over an Inhomogeneous Earth, *Proc. IEE (Lond.)*, Pt. C, Monogr. 318R, December 1958.

43. Millington, G.   Ground Wave Propagation over an Inhomogeneous Smooth Earth, *J. IEE (Lond).* January 1949, Pt. III, Vol. 96, p. 53.

44. Kirke, H. L.   Calculation of Ground-Wave Field Strength over a Composite Land and Sea Path, *Proc. IRE*, May 1949, Vol. 37, pp. 489–496.

45. Ott, R. H.   An Alternative Integral Equation for Propagation over Irregular Terrain, *Radio Science*, Vol. 5, May 1970, pp. 767–771; see also part 2, April 1971, pp. 429–435.

46. Beckmann, P., and A. Spizzichino   "The Scattering of Electromagnetic Waves from Rough Surfaces," Pergamon, New York, 1963.

47. Dougherty, H. T.   Diffraction by Irregular Apertures, *Radio Sci.*, January 1970, Vol. 5, pp. 55–60.

48. Bachynski, M. P.   Propagation at Oblique Incidence over Cylindrical Obstacles, *J. Res. Natl. Bur. Stand. (Radio Propag.)*, July–August 1960, Vol. 1, 64D, No. 4, pp. 311–345.

49. Wait, J. R., and A. M. Conda   Diffraction of Electromagnetic Waves by Smooth Obstacles for Grazing Angles, *J. Res. Natl. Bur. Stand. (Radio Sci.)*, 1964, Vol. 68D, No. 2, pp. 239–250.

50. Longley, A. G., and P. L. Rice   Prediction of Tropospheric Radio Transmission Loss over Irregular Terrain: A Computer Method, *ESSA Tech. Rep.* ERL 79/ITS 67, 1968.

51. FCC Rules and Regulations, Radio Broadcast Services, Secs. 73.333 and 73.699, March 1980.

52. IEEE Communications Society "Mobile Communications," special issue of *Trans. Commun.*, 1973.

53. Okumura, Y., H. Omuri, T. Kawano, and K. Fukuda   Field Strength and Its Variability in VHF and UHF Land Mobile Radio Service, *Rev. Tokyo Elec. Commun. Lab.*, 1968, Vol. 16, pp. 825–873.

54. Cox, D. C.   Doppler Spectrum Measurements at 910 MHz over a Suburban Mobile Radio Path, *Proc. IEEE*, 1971, Vol. 59, pp. 1017–1018; see also Time and Frequency Domain Characterization of Multipath at 910 MHz in a Suburban Mobile Radio Environment, *Radio Sci.*, December 1972, Vol. 7, pp. 1069–1077.

54a. Cox, D. C.   Multipath Delay Spread and Path Loss Correlation for 910-MHz Urban Mobile Radio Propagation, *IEEE Trans, Veh. Technol.*, 1977, Vol. VT-26.

54b. Jakes, W. C. Jr. "Microwave Mobile Communication," Wiley, New York, 1974.

55. Engineering Considerations for Microwave Communications Systems, Lenkurt Electric Co., Inc., San Carlos, Calif., June 1970.

56. Bell Telephone Laboratories "Transmission Systems for Communications," 4th ed., Western Electric Technical Publishers, Winston-Salem, N.C., 1974.

57. Dougherty, H. T. A Survey of Microwave Fading Mechanisms, Remedies and Applications, *Environ. Sci. Serv. Admin. Tech. Rep.* ERL-69-WPL 4, 1968.

57a. Dougherty, H. T., and R. E. Wilkerson Determination of Antenna Height for Protection against Microwave Diffraction Fading, *Radio Sci.*, 1967, Vol. 2, n.s., pp. 161–165.

58. Ruthroff, C. L. Multipath Fading on LOS Microwave Radio Systems as a Function of Path Length and Frequency, *Bell Syst. Tech. J.*, September 1971, Vol. 50, No. 7, pp. 2375–2398.

58a. Barnett, W. T. Multipath Fading Effects on Digital Radio, *IEEE Trans. Commun.*, December 1979, Vol. COM-27, pp. 1842–1848.

59. Vigants, A. Space-Diversity Performance as a Function of Antenna Separation, *IEEE Trans. Commun. Technol.*, 1968, Vol. COM-16, No. 6, pp. 831–836.

59a. Vigants, A. Space Diversity Engineering, *Bell Syst. Tech. J.*, January 1975, Vol. 54, pp. 103–141.

60. Barsis, A. P., and M. E. Johnson Prolonged Space-Wave Fadeouts in Tropospheric Propagation, *J. Res. Natl. Bur. Stand. (Radio Propag.)*, 1962, Vol. 66D, No. 2, pp. 681–694.

61. Joint Technical Advisory Council IEEE-EIA Microwave Radio Relay System Reliability, Report to the FCC, Mar. 23, 1965.

61a. Bullington, K. Phase and Amplitude Variations in Multipath Fading of Microwave Signals, *Bell Syst. Tech. J.*, July–August 1971, Vol. 50, No. 6, pp. 2039–2053.

62. Booker, H. G., and W. E. Gordon Outline of a Theory of Radio Scattering in the Troposphere, *J. Geophys. Res.*, September 1950, Vol. 55, No. 3, pp. 241–246; see also *Proc. IRE*, April 1950, Vol. 38, No. 4, p. 401.

63. Voge, J. Radioelectricity and the Troposphere, I: Theories of Propagation to Long Distances by Means of Atmospheric Turbulence, *Onde Electr.*, 1955, Vol. 35, pp. 565–581.

64. Megaw, E. C. S. Fundamental Radio Scatter Propagation Theory, *Proc. IEE*, September 1957, Pt. C104, No. 6, pp. 441–455; see also *Monogr.* 236R, May 1957.

65. Hall, M. P. M. Effects of the Troposphere on Radio Communication, *IEE Electromagnetic Waves*, Series 8, London, 1980.

66. Wheelon, A. D. Radio-Wave Scattering by Tropospheric Irregularities, *J. Res. Natl. Bur. Stand. (Radio Propag.)*, September–October 1959, Vol. 63D, No. 2, pp. 205–234; also *J. Atmos. Terr. Phys.*, 1959, Vol. 15, No. 3 and 4, pp. 185–205.

67. Friis, H. T., A. B. Crawford, and D. C. Hogg A Reflection Theory for Propagation beyond the Horizon, *Bell Syst. Tech. J.*, 1957, Vol. 36, pp. 627–644, also published as *Bell Tel. Syst. Monogr.* 2823.

68. duCastel, F., P. Misme, A. Spizzichino, and J. Voge On the Role of the Process of Reflection in Radio Wave Propagation, *J. Res. Natl. Bur. Stand. (Radio Sci.)*, 1962, Vol. 66D, No. 3, pp. 273–284.

69. *Proc. IRE Special Issue on Scatter Propag.*, October 1955, Vol. 43.

70. Joint Technical Advisory Council Radio Transmission by Ionospheric and Tropospheric Scatter, *Proc. IRE*, 1960, Vol. 48, pp. 4–44.

71. Tropospheric Wave Propagation, *IEE, Conf. Publ.* 48, London, 1968 (contains a number of references giving current appraisal of tropospheric propagation).

72. Surenian, D. Experimental Results of Angle Diversity System Tests, *IEEE Trans. Commun. Technol.*, June 1965, Vol. COM-13, No. 2, p. 208.

73. Fock, V. A., L. A. Wainstein, and M. E. Belkina Radiowave Propagation in Surface Tropospheric Ducts, *Radiotechn. Elektron.*, 1958, Vol. 3, No. 12, pp. 1411–1429.

74. Freehafer, J. E. in D. E. Kerr, "Propagation of Short Radio Waves," McGraw-Hill, New York, 1951; reprinted Dover, New York, 1965.

75. Sodha, M. S., A. K. Ghatak, D. P. Tewari, and P. K. Dubez Focusing of Waves in Ducts, *Radio Sci.*, November 1972, Vol. 7, No. 11, p. 1005.

76. Jeske, J., and K. Brocks Comparison of Experiments on Duct Propagation above the Sea with the Mode Theory of Booker and Walkinshaw, *Radio Sci.*, 1966, Vol. 1, n.s., No. 8, pp. 891–895.

77. Pidgeon, V. W. Frequency Dependence of Radar Ducting, *Radio Sci.*, 1970, Vol. 5, No. 3, pp. 541–550.

78. AGARD Telecommunications Aspects on Frequencies 10 to 100 GHz, *Proc. 18th Meet.*

Electromagn. Wave Propag. Panel, 1972 (contains approximately 40 theoretical and experimental papers).

79. NASA   A Propagation Effects Handbook for Satellite System Design, 1980, NASA Communications Division, Washington, D.C.

79a. Thompson, M. C., Jr. L. E. Vogler, H. B. Janes, and L. E. Wood   A Review of Propogation Factors in Telecommunication Applications of the 10 to 100 GHz Radio Spectrum, *U.S. Dept. Commer., Off. Telecommun.* OT/TRER 34, August 1972.

80. Liebe, H. J., and W. M. Welch   Attenuation and Phase Dispersion in the Atmosphere Due to the Microwave Spectrum of Oxygen, *AGARD Proc. 18th Meet. Wave Electromagn. Propag. Panel*, 1972.

81. VanVleck, J. H.   The Absorption of Microwaves by Oxygen, *Phys. Rev.*, April 1947, Vol. 71, No. 7, pp. 413–424.

82. VanVleck, J. H.   The Absorption of Microwaves by Uncondensed Water Vapor, *Phys. Rev.*, April 1947, Vol. 71, No. 7, pp. 425–433.

83. Becker, G. E., and S. H. Autler   Water Vapor Absorption of Electromagnetic Radiation in the Centimeter Wavelength Range, *Phys. Rev.*, 1946, Vol. 70 No. 5 and 6, pp. 300–307.

84. Blake, L. V.   Radar/Radio Tropospheric Absorption and Noise Temperature, *Nav. Res. Lab. Rep.* 7461, October 1972.

85. Liebe, H. J.   Calculated Tropospheric Dispersion and Absorption Due to the 22 GHz Water Vapor Line, *IEEE Trans. Antennas Propag.* 1969, Vol. 17, No. 5, pp. 621–627.

86. Hogg, D. C.   Millimeter-Wave Communication through the Atmosphere, *Science 1968*, Vol. 159, No. 3810, pp. 39–46.

87. Straiton, A. W., and C. W. Tolbert   Factors Affecting Earth-Satellite Millimeter Wavelength Communications, *IEEE Trans.* 1963, Vol. MTT-11, pp. 296–301.

88. Carter, C. J., R. L. Mitchell, and E. E. Reber   Oxygen Absorption Measurements in the Lower Atmosphere, *J. Geophys. Res.*, 1968, Vol. 73, No. 10, pp. 3113–3120.

89. Lane, J. A.   Scintillation and Absorption Fading on Line-of-Sight Links at 35 and 100 GHz, *IEE Conf. Tropospheric Wave Propag. Lond. 1968 Pub* 48.

90. Medhurst, R. G.   Rainfall Attenuation of Centimeter Waves: Comparison of Theory and Experiment, *Trans. IEEE Antennas Propag.*, 1965, Vol. AP-13, No. 4, pp. 550–564.

90a. Olsen, R. L., D. V. Rogers, and D. B. Hodge   The aR^b Relation in the Calculation of Rain Attenuation, *IEEE Trans. Antennas Propag.*, 1978, Vol. AP-26, No. 2, pp. 313–329.

90b. Laws, V. O., and P. A. Parsons   The Relation of Raindrop Size to Intensity, *Trans. Am. Geophys. Union*, 1943, Vol. 24, 165–166.

91. Oguchi, T.   Attenuation of Electromagnetic Waves Due to Rain with Distorted Raindrops, *J. Radio Res. Lab. (Tokyo)*, January 1964, Vol. 11, pp. 19–37.

92. Rice, P. L., and N. R. Holmberg   Cumulative Time Statistics of Surface-Point Rainfall Rates, *IEEE Trans. Commun.*, Oct. 1973, Vol. Com-21, No. 10, pp. 1131–1136.

92a Dougherty, H. T., and E. V. Dutton   Estimating Year-to-Year Variability of Rainfall for Microwave Applications, *IEEE Trans Comm.*, Vol. COM-26, 1978, pp. 1321–1324.

93. Dutton, E. J.   A Meteorological Model for Use in the Study of Rainfall Effects on Atmospheric Telecommunications, *U.S. Dept. Commer., Off. Telecommun.* OT/TRER 24, December 1971.

93a. Ruthroff, C. L   Rain Attenuation and Radio Path Design, *Bell Syst. Tech. J.*, January 1970, Vol. 49, No. 1, pp. 121–135.

93b. Hogg, D. C.   Path Diversity in Propagation of Millimeter Waves through Rain, *IEEE Trans. Antennas Propag.*, 1967, Vol. AP-15, No. 3, pp. 410–415.

93c. Hogg, D. C., and T. S. Chu   The Role of Rain in Satellite Communication, *Proc. IEEE*, 1975, Vol. 63, pp. 1308–1331.

94. Gusler, L. T., and D. C. Hogg   Some Calculations on Coupling between Satellite-Communications and Terrestrial Radio-Relay Systems Due to Scattering by Rain, *Bell Syst. Tech. J.*, 1970, Vol. 49, p. 1491.

95. Crane, R. K.   Propagation Phenomena Affecting Satellite Communications Systems Operating in the Centimeter and Millimeter Wavelength Bands, *Proc IEEE*, 1971, Vol. 59, No. 2, pp. 173–188.

95a. Crane, R. K.   Bistatic Scatter from Rain, *IEEE Trans. Antennas Propag.*, 1974, Vol. AP-22.

95b. Crane, R. K., and D. W. Blood   "Handbook for the Estimation of Microwave Propagation Effects — Link Calculations for Earth-Space Paths," *ERT Tech. Rpt No. 1 NASA Contract NAS5-25341*, NASA, Washington, D.C.

96. Crane, R. K.   Coherent Pulse Transmission through Rain, *IEEE Trans. Antennas Propag.*, 1967, Vol. AP-15, No. 2, pp. 252–256.

97. Gray, D. A.   Transit-Time Variations in Line of Sight Tropospheric Propagation Paths, *Bell Syst. Tech. J.*, July–August 1970, Vol. 49, pp. 1059–1068.

98. Roche, J. F., H. Lake, D. T. Worthington, C. K. G. Tsao, and J. T. deBettencourt   Radio-Propagation 27-40 GHz, *IEEE Trans. Antennas and Propag.*, July 1970, Vol. AP-18, pp. 452–462.

99. Setzer, D.   Computed Transmission through Rain at Microwave and Visible Frequencies, *Bell Syst. Tech. J.*, October 1970, Vol. 49, No. 8, pp. 1873–1892.

100. Ippolito, L. J.   Effects of Precipitation on 15.3- and 31.65-GHz Earth-Space Transmissions with the ATS-V Satellite, *Proc. IEEE*, 1971, Vol. 59, No. 2, pp 189–205.

100a. Tatarski, V. I.   "Wave Propagation in a Turbulent Medium," trans. R. A. Silverman, McGraw-Hill, New York, 1961.

101. Lee, R. W., and A. T. Waterman   Space Correlation of 35 GHz Transmissions over a 28 km Path, *Radio Sci.*, 1968, Vol. 3, pp 135–140.

102. Thompson, M. C., Jr., and H. B. Janes   Measurements of Phase-Front Distortion on an Elevated Line-of-Sight Path, *IEEE Trans. Aerosp. Electron. Syst.* 1970, Vol. AES-6, No. 5, pp. 645–656.

103. Janes, H. B., M. C. Thompson, Jr., D. Smith, and A. W. Kirkpatrick   Comparison of Simultaneous Line-of-Sight Signals at 9.6 and 34.5 HGz, *IEEE Trans. Antennas Propag.*, 1970, Vol. AP-18, No. 4, pp. 447–451.

103a. Mandics, P. A., R. W. Lee, and A. T. Waterman   Spectra of Short-Term Fluctuations of Line of Sight Signals, *Radio Sci.*, 1973, Vol. 8, pp. 185–201.

104. Lawrence, R. S., and J. W. Strohbein   A Survey of Clear-Air Propagation Effects Relevant to Optical Communication, *Proc. IEEE*, 1970, Vol. 58, pp. 1523–1545.

105. Hogg, D. C., and W. W. Mumford   The Effective Noise Temperature of the Sky, *Microwave J.*, March 1960, Vol. 3, pp. 80–84.

106. Decker, M. T., and E. J. Dutton   Radiometric Observations of Liquid Water in Thunderstorm Cells, *J. Amos. Sci.*, 1970, Vol. 27, No. 5, pp. 285–290.

106a. Ratcliffe, J. A.   "Physics of the Upper Atmosphere," Academic, New York, 1960.

107. Van Zandt, T. E., and R. W. Knecht   The Structure and Physics of the Upper Atmosphere, in A. Rosen and D. P. LeGallery (eds.), "Space Physics," Wiley, New York, 1964.

108. Rishbeth, H., and O. K. Garriott   "Introduction to Ionospheric Physics," Academic, New York, 1969.

108a. Hedin, A. E., C. A. Reber, G. P. Newton, N. W. Spencer, H. C. Brinton, H. G. Mayr, and W. E. Potter   A Global Thermospheric Model Based on Mass Spectrometer and Incoherent Scatter Data MSIS-2. Composition, *J. Geophys Res.*, Vol. 82, p. 2148.

109. Sengupta, P. R.   Solar X-Ray Control of the E Layer of the Ionosphere, *J. Atmos. Terr. Phys.*, 1970, Vol. 32, pp. 1273–1282.

110. Smith, E. K., and S. Matsushita   "Ionospheric Sporadic E," Macmillan, New York, 1962.

111. Whitehead, J. D.   Production and Prediction of Sporadic E, *Rev. Geophys. Space Phys.* 1970, Vol. 8, pp. 145–168.

112. Special issue on sporadic E, *Radio Sci.*, March 1972.

113. CIRA "COSPAR International Reference Atmosphere," North-Holland Amsterdam, 1965.

113a. Rawer, K.   Intercomparison of Different Measuring Techniques in the Upper-Atmosphere: the International Reference Ionosphere, *Space Res.*, Vol. XV, 1975, p. 212.

114. Electromagnetic Probing of the Upper Atmosphere; special issue of *J. Atmos. Terr. Phys.*, April 1970, Vol. 32.

115. Piggott, W. R., and K. Rawer   "URSI Handbook of Ionogram Interpretation and Reduction," 2d ed., World Data Center A for Solar Terrestrial Physics, *Report UAG-23*, NOAA, Boulder, Colo., 1972.

116. Ratcliffe, J. A.   "The Magneto-Ionic Theory," Cambridge University Press, Cambridge, 1959.

117. Budden, K. G.   "Radio Waves in the Ionosphere," Cambridge University Press, Cambridge, 1961.

118. Kelso, J. M.   "Radio Ray Propagation in the Ionosphere," McGraw-Hill, New York, 1964.

119. Wait, J. R. (ed.)   "Electromagnetic Probing in Geophysics," Golem Press, Boulder, Colo., 1971.

120. Ocean '72, *IEEE Int. Conf. Eng. Ocean Environ.*, IEEE Pub. 72 CHO 660-1 OCC, 1972.

121. Watt, A. D.   "VLF Radio Engineering," Pergamon, New York, 1967.

122. Budden, K. G.   "The Waveguide-Mode Theory of Wave Propagation," Prentice-Hall, Englewood Cliffs, N.J., 1961.

123. Pappert, R. A.   A Numerical Study of VLF Mode Structure and Polarization below an Anisotropic Ionosphere, *Radio Sci.*, 1968, n.s., Vol. 3

124. Wait, J. R., and K. P. Spies   Characteristics of the Earth Ionosphere Wave Guide for VLF Radio Waves, *Natl. Bur. Stand. Tech. Note* 300, 1964 (and two supplements).

125. Bickel, J. E., J. A. Ferguson, and G. V. Stanley   Experimental Observation of Magnetic Field Effects on VLF Propagation at Night, *Radio Sci.* Vol. 5, No. 1, pp. 19–25.

126. Berry, L. A.   Wave Hop Radio Propagation Theory, *IEE Conf. Publ.* 36, pp. 63–69, 1967.

127. Belrose, J. S., W. L. Hatton, C. A. McKerrow, and R. S. Thain   The Engineering of Communication Systems for Low Radio Frequencies, *Proc. IRE*, May 1959, Vol. 47, pp. 661–680.

128. Piggott, W. R., M. L. V. Pitteway, and E. V. Thrane   The Numerical Calculation of Wave Fields, Reflexion Coefficients and Polarizations for Long Radio Waves in the Lower Ionosphere, II, *Phil. Trans. Soc.*, 1965, Vol. A-257, p. 243.

129. Belrose, J. S.   The Oblique Reflection of Low-Frequency Radio Waves from the Ionosphere, in "Propagation of Radio Waves at Frequencies Below 300 kc," *AGARDOGRAPH* No. 74, pp. 149–165, Pergamon, 1963.

130. Wait, J. R., and A. M. Conda   Pattern of an Antenna on a Curved Lossy Surface, *Trans. IRE (N.Y.)*, 1958, Vol. AP-6, pp. 348–359.

131. Ebert, W.   Ionospheric Propagation on Long and Medium Waves, *EBU Rev.*, Pt. A., Tech. 71 to 73; also *Tech. Monogr.* 3081, 1962.

132. Barghausen, A. F.   Medium Frequency Sky Wave Propagation in Middle and Low Latitudes, *IEEE Trans Broadcast.*, June 1966, Vol. 12, pp. 1–14.

132a. Wang, J. C. H.   Prediction of Medium Frequency Skywave Field Strength in North America, *IEEE Trans. Broadcasting*, Vol. BC-23, 1977, pp. 43–49.

133. Phillips, G. D., and D. Knight   Effects of Polarization on a Medium Frequency Sky Wave Service, Including the Case of Multihop Paths, *Proc. IEE (Lond.)*, January 1965, Vol. 112, pp. 31–39.

134. Hatton, W. L.   Oblique-Sounding and HF Radio Communication, *IRE Trans.*, 1961, Vol. PGCS-9, p. 275.

134a. Neilson, D.   Oblique Sounding of a Transequatorial Path, in P. Newman (ed.), "Spread F and Its Effects upon Radiowave Propagation and Communication," *AGARDOGR.* No. 95, Technivision, 1966, pp. 467–490.

135. Agy, V., and K. Davies   Ionospheric Investigations Using the Sweep-Frequency Pulse Technique at Oblique Incidence, *J. Res. Natl. Bur. Stand.*, September–October 1959, Vol. 63D, pp. 151–174.

136. Bradley, P. A., and C. Bedford   Prediction of HF Circuit Availability, *Electron Lett.*, Vol. 12, 1976, p. 32.

137. Barghausen, A. F., et al.   Predicting Long-Term Operational Parameters of High-Frequency Sky-Wave Telecommunication Systems, *Environ. Sci. Serv. Admin. Tech. Rep.* ERL 110-ITS 78, 1969.

138. Jones, W. B., and R. M. Gallet   Ionospheric Mapping by Numerical Methods, *ITU Telecommun. J.*, December 1960, Vol. 27, No. 12, pp. 280–232.

139. Leftin, M., W. M. Roberts, and R. K. Rosich   "Ionospheric Predictions," 4 Vols., *U.S. Dept. Commer., Off. Telecommun.* OR/TRER 13, 1971.

139a. Jones, W. B., and R. M. Gallet   Methods for Applying Numerical Maps of Ionospheric Characteristics, *J. Res. Natl. Bur. Stand. (Radio Propag.)*, November–December 1962, Vol. 66D, No. 6, pp. 649–662.

139b. Rawer, K.   The Historical Development of Forecasting Methods for Ionospheric Propagation of HF Waves, *Radio Sci.*, Vol. 10, 1975, p. 669.

140. Leftin, M., S. M. Ostrow, and C. Preston   Numerical Maps of f_oE_s for Solar Cycle Minimum and Maximum, *Environ. Sci. Serv. Admin. Tech. Rep.* ERL 73-ITS 63, 1968.

140a. Sandoz, O. A., E. E. Stevens, and E. S. Warren   The Development of Radio Traffic Frequency Prediction Techniques for Use at High Latitudes, *Poc. IRE*, 1959, Vol. 47, p. 681.

141. Utlaut, W. F.   Effect of Antenna Radiation Angles upon HF Radio Signals Propagated over Long Distances, *Natl. Bur. Stand. J. Res.*, Pt. D., March–April 1961.

142. Utlaut, W. F.   Siting Criteria for Hf Communication Centers, *Natl. Bur. Stand. Tech. Note* 139 (PB 16140), April 1962.

143. Koch, J. W., and W. M. Beery   Observations of Radio Wave Phase Characteristics on a High-Frequency Auroral Path, *Natl. Bur. Stand. J. Res.*, 1962, Vol. 66D, pp. 291–296.

144. Koch, J. W., and H. E. Petrie Fading Characteristics Observed on a High-Frequency Auroral Radio Path, *Natl. Bur. Stand. J. Res.* 1962, Vol. 66D, pp. 159–166.

145. Ames, J. The Correlation between Frequency-Selective Fading and Multipath Propagation over an Ionospheric Path, *J. Geophys. Res.*, 1963, Vol. 68, pp. 759–768.

146. Salaman, R. K. A New Ionospheric Multipath Reduction Factor (MRF), *IRE Trans. Commun. Syst.*, June 1962, Vol. CS-10, No. 2, pp. 221–222.

147. Bailey, D. K., R. Bateman, and R. C. Kirby Radio Transmission at VHF by Scattering and Other Processes in the Lower Ionosphere, *Proc. IRE*, October 1955, Vol. 43, pp. 1181–1230.

148. Kirby, R. C. Review of VHF Forward Scatter, *AGARD Proc. 37 Scatter Propag. Radio Waves*, 1968.

149. Wheelon, A. D. Relation of Turbulence Theory to Ionospheric Forward Scatter Propagation Experiments, *J. Res. Natl. Bur. Stand.*, 1960, Vol. 64D, p. 301.

150. Blair, J. C., R. M. Davis, and R. C. Kirby Frequency Dependence of D-Region Scattering at VHF, *J. Res. Natl. Bur. Stand. (Radio Propag.)*, September–October 1961, Vol. 65D, No. 3, pp. 249–263.

151. Koch, J. W. Factors Affecting Modulation Techniques for VHF Scatter Systems, *IRE Trans. Commun. Syst.*, June 1959, Vol. CS-7, pp. 77–92.

152. Manning, L. A., and V. R. Eshelman Meteors in the Ionosphere, *Proc. IRE*, February 1959, Vol. 47, pp. 186–199.

153. McKinley, D. W. R. "Meteor Science and Engineering," McGraw-Hill, New York, 1961.

154. Sugar, G. R. Radio Propagation by Reflection from Meteor Trails, *Proc. IEEE*, February 1964, Vol. 52, p. 116.

155. Forsyth, P. A., E. L. Vogan, D. R. Hansen, and C. O. Hines The Principles of JANET: A Meteor-Burst Communications System, *Proc. IRE*, 1957, Vol. 45, p. 1642.

156. Bartholomé, P. J. Survey of Ionospheric and Meteor Scatter Communications, in K. Folkestad (ed.), "Ionospheric Radio Communications," *NATO Inst. Ionospheric Radio Commun. Artic Proc.* Plenum, New York, pp. 143–154, 1967.

157. Cohen, R., and K. L. Bowles On the Nature of Equatorial Spread F, *J. Geophys. Res*, 1961, Vol. 66, p. 1081.

158. Yeh, K. C., and O. G. Villard A New Type of Fading Observable on High-Frequency Radio Transmission Propagated over Paths Crossing the Magnetic Equator, *Proc. IRE*, 1958, Vol. 46, pp. 1968–1970.

159. Forsyth, P. A. Auroral Scatter, in Scatter Propagation of Radio Waves, Pt. 2, *AGARD Conf. Proc. 37*, August 1968, pp. 34-1 to 34-10.

160. Booker, H. G. Radar Studies of the Aurora, in J. A. Ratcliffe (ed.), "Physics of the Upper Atmosphere," Academic, New York, 1960.

161. Bowles, K. L. Incoherent Scattering by Free Electrons as a Technique for Studying the Ionosphere and Exosphere: Some Observations, and Theoretical Considerations, *J. Res. Natl. Bur. Stand. (Radio Propag.)*, 1961, Vol. 65D, p. 1.

162. Flock, W. L., and B. B. Balsley VHF Radar Returns from the D Region of the Equatorial Ionosphere, *J. Geophys. Res.*, 1967, Vol. 72, pp. 5537–5541.

163. Armistead, G. W., J. V. Evans, and W. A. Reid Measurement of D- and E-Region Electron Densities by the Incoherent Scatter Technique at Millstone Hill, *Radio Sci.*, 1972, Vol. 7, pp. 153–162.

164. Evans, J. V. Theory and Practice of Ionosphere Study by Thomson Scatter Radar, *Proc. IEEE*, 1969, Vol. 57, pp. 496–530.

165. Lawrence, R. S., C. G. Little, and H. J. A. Chivers A Survey of Ionospheric Effects upon Earth Space Radio Propagation, *Proc. IEEE*, 1964, Vol. 52, pp. 4–27.

166. Briggs, H. G. Brief Review of Scintillation Studies, *Radio Sci.*, 1966, Vol. 1, pp. 1163–1167.

167. AGARD Propagation Factors in Space Communications, *Conf. Proc. Symp. Rome, Sept. 21–25, 1965*, Mackay, London, 1967.

167a. Fremouw, E. J., and C. L. Rino An Empirical Model for F-Layer Scintillation at VHF/UHF, *Radio Sci.*, 1973, Vol. 8, pp. 213–222.

167b. Taur, R. R. Ionospheric Scintillation at 4 and 6 GHz, *COMSAT Tech. Rev.*, Vol. 3, 1973 (Spring), No. 1, pp. 145–163.

167c. Briggs, B. H., and I. A. Parkin On the Variation of Radio Star and Satellite Scintillations with Zenith Angle, *J. Atm. Terr. Phys.*, 1963, Vol. 25., pp. 334–365.

167d. Fremouw, E. J., R. C. Livingston, C. L. Rino, M. Cousins, B. C. Fair, and R. L. Leadabrand   Complex Signal Scintillation: Early Results from the DNA-002 Beacon, *Radio Sci.*, 1978, Vol. 13, pp. 167–187.

167e. Crane, R. K.   Ionospheric Scintillation, *Proc. IEEE*, 1977, Vol. 65, pp. 180–199.

168. Evans, J. V.   Propagation in the Ionosphere, Chap. 2, Pt. II, in J. V. Evans and T. Hagfors (eds.), "Radar Astronomy," McGraw-Hill, New York, 1968.

169. Chvojkova, E.   Analytic Formulae for Radio Path in Spherically Stratified Ionosphere, *Radio Sci.*, 1965, Vol. 69D, pp. 453–457.

170. Obayashi, T.   A Possibility of the Long Distance HF Propagation along the Exospheric Field Aligned Ionization, *Rep. Ionos. Spec. Res., Jap.* 1959, Vol. 13, p. 177.

171. Gallet, R. M., and W. F. Utlaut   Evidence of the Laminar Nature of the Exosphere Obtained by Means of Guided High Frequency Wave Propagation *Phys. Rev. Lett.*, 1961, Vol. 6, p. 59.

172. Walker, A.D.   The Theory of Guiding of Radio Waves in the Exosphere, *J. Atmos. Terr. Phys.*, 1966, Vol. 28, p. 1039.

173. Booker, H G.   Guidance of Radio and Hydromagnetic Waves in the Magnetosphere, *J. Geophys. Res.*, 1962, Vol. 67, p. 4135.

174. Utlaut, W. F.   Ionospheric Modification Induced by High-Power HF Transmitters — A Potential for Extended Range VHF-UHF Communications and Plasma Physics Research, *Proc. IEEE*, Vol. 63, 1955, p. 1022.

175. CCIR   See Table 18-7 for a list of Recommendations and Reports (1982) of the International Radio Consultative Committee (CCIR) contained in volumes of the *15th Plenary Assembly, Geneva, 1982*.

TABLE 18-7   Details of Ref. 175

Title	Volume
Recommendations:	
313-3 Exchange of Information for Short-Term Forecasts and Transmission of Ionospheric Disturbance	VI
341 The Concept of Transmission Loss in Studies of Radio System	I
368-3 Ground-Wave Propagation Curves for Frequencies between 10 kHz and 30 MHz	V
370-3 VHF and UHF Propagation Curves for the Frequency Range from 30 MHz to 1000 MHz Broadcasting Services	V
435-3 Prediction of Sky-Wave Field Strength between 150 and 1600 kHz	VI
525 Calculation of Free-Space Attenuation	V
Reports:	
238-4 Propagation Data Required for Trans-Horizon Radio-Relay Systems	V
249-4 Ionospheric Sounding at Oblique Incidence	VI
250-4 Long-Distance Ionospheric Propagation without Intermediate Ground Reflection	VI
251-2 Intermittent Communication by Meteor-Burst Propagation	VI
252-2 CCIR Interim Method for Estimating Sky-Wave Field Strength and Transmission Loss at Frequencies between the Approximate Limits of 2 and 30 MHz	*
263-4 Ionospheric Effects upon Earth-Space Propagation	VI
265-4 Sky-Wave Propagation at Frequencies below 150 kHz with Particular Emphasis on Ionospheric Effects	VI
266-4 Ionospheric Propagation Characteristics Pertinent to Terrestrial Radiocommunication Systems Design (Fading)	VI
322-1 World Distribution and Characteristics of Atmospheric Radio Noise	*
327-3 Diversity Reception	III
338-3 Propagation Data Required for Line-of-Sight Radio-Relay Systems	V
340-3 CCIR Atlas of Ionospheric Characteristics	*
415 Models of Phase-Interference Fading for Use in Connection with Studies of the Efficient Use of the Radio-Frequency Spectrum	*
527 General Use of an Equivalent Antenna Noise Temperature Model for Purposes of Early Coordination	I
563-1 Radiometeorological Data	V
564-1 Propagation Data Required for Space Telecommunication Systems	V
567-1 Methods and Statistics for Estimating Field-Strength Values in the Land Mobile Services Using the Frequency Range 30 MHz to 1 GHz	V

TABLE 18-7   Details of Ref. 175 *(Continued)*

	Title	Volume
575-1	Methods for Predicting Sky-Wave Field Strengths at Frequencies between 150 kHz and 1600 kHz	VI
715	Propagation by Diffraction	V
717	World Atlas of Ground Conductivities	V
718	Effects of Tropospheric Refraction on Radiowave Propagation	V
719	Attenuation by Atmospheric Gases	V
720	Radio Emission Associated with Absorption by Atmospheric Gases and Precipitation	V
721	Attenuation and Scattering by Precipitation and Other Atmospheric Particles	V

*Published separately.

NOTE: At each CCIR Plenary Assembly these reports are updated and the suffixed number advanced. The basic report number (first three digits) remains the primary reference number, to which is added the most recent suffixed number.

Section 19

Sound Reproduction and Recording Systems

DANIEL W. MARTIN *Research Director, Engineering Research Department, Baldwin Piano & Organ Company; Fellow IEEE*

CONTENTS

Numbers refer to paragraphs

STANDARD UNITS FOR SOUND SPECIFICATION[1,2]*

1. Sound Pressure. Airborne sound waves are a physical disturbance pattern in the air, an elastic medium, traveling through the air at a speed which depends somewhat upon air temperature (but not upon static air pressure). The instantaneous magnitude of the wave at a specific point in space and time can be expressed in various ways, e.g., displacement, particle velocity, pressure. However, the most widely used and measured property of sound waves is *sound pressure*, the fluctuation above and below atmospheric pressure which results from the wave.

An *atmosphere* of pressure is typically about 1 million dyn/cm², sometimes called a *bar*. Sound pressure is usually a very small part of atmospheric pressure. The unit of *sound* pressure is 1 dyn/cm², or 1 *microbar* (μbar).

2. Sound-Pressure Level. Sound pressures important to electronics engineering range from the weakest noises which can interfere with sound recording to the strongest sounds a loudspeaker diaphragm should be expected to radiate. This range is approximately 10^6. Consequently, for convenience, sound pressures are commonly plotted on a logarithmic scale called *sound-pressure level* expressed in *decibels* (dB).

The decibel, a unit widely used for other purposes in electronics engineering, originated in audio engineering (in telephony), and is named for Alexander Graham Bell. Because it is logarithmic, it requires a reference value for comparison just as it does in other branches of electronics engineering. The reference pressure for sounds in air, corresponding to 0 dB, has been defined as a sound pressure of 0.0002 μbar. This is the reference sound pressure p_0 used throughout this section of the handbook. Thus the sound-pressure level L_p in decibels corresponding to a sound pressure p is defined by

$$L_p = 20 \log (p/p_0) \qquad \text{dB} \qquad (19\text{-}1)$$

The reference pressure p_0 approximates the weakest audible sound pressure at 1,000 Hz. Consequently most decibel values for sound levels are positive in sign. Figure 19-1 relates sound-pressure level in decibels to sound pressure in microbars.

*Superior numbers correspond to References, Par. **19-93**.

Sound power and sound intensity (power flow per unit area of wavefront) are generally proportional to the square of the sound pressure. Doubling the sound pressure quadruples the intensity in the sound field, requiring 4 times the power from the sound source.

3. Audible Frequency Range. The international abbreviation Hz (hertz) is now used (instead of the former cps) for audible frequencies as well as the rest of the frequency domain. The limits of audible frequency are only approximate because tactile sensations below 20 Hz overlap aural sensations above this lower limit. Moreover only young listeners can hear pure sounds near or above 20 kHz, the nominal upper limit.

Frequencies beyond both limits, however, have significance to audio-electronics engineers. For example, near-infrasonic (below 20 Hz) sounds are needed for classical organ music but can be noise in turntable rumble. Near-ultrasonic (above 20 kHz) intermodulation in audio circuits can produce undesirable difference-frequency components which are audible.

The audible sound-pressure level range can be combined with the audible frequency range to describe an *auditory area,* shown in Fig. 19-2. The lowest curve shows the weakest audible sound-pressure level for listening with both ears to a pure tone while facing the sound source in a free field. The minimum level depends greatly upon the frequency of the sound. It also varies somewhat among listeners. The levels which quickly produce discomfort or pain for listeners are only approximate, as indicated by the shaded and crosshatched areas of Fig 19-2. Extended exposure can produce temporary (or permanent) loss of auditory area at sound-pressure levels as low as 90 dB.

Wavelength effects are of great importance in the design of sound systems and rooms because wavelength varies over a 3-decade range, much wider than is typical elsewhere in electronics engineering. The inverse relation of wavelength to audio frequency is shown in Fig. 19-3 for a sound speed of 344 m/s (1,127 ft/s) at 20°C (68°F). Audible sound waves vary in length from 1 cm to 15 m. The dimensions of the sound sources and receivers used in electroacoustics also vary greatly, e.g., from 1 cm to 3 m.

Sound waves follow the principles of geometrical optics and acoustics when the wavelength is very small relative to object size and pass completely around obstacles much smaller than a wavelength. This wide range of physical effects complicates the typical practical problem of sound production or reproduction.

4. Loudness Level. The simple, direct method for determining experimentally the loudness *level* of a sound is to match its observed loudness with that of a 1,000-Hz sine-wave reference tone of calibrated, variable sound-pressure level. (Usually this is a group judgment, or an average of individual judgments, in order to overcome individual observer differences.)

When the two loudnesses are matched, the loudness *level* of the sound, expressed in *phons,* is defined as numerically equal to the sound-pressure level of the reference tone in decibels. For example, a series of observers, each listening alternately to a machine noise and to a 1,000-Hz reference tone, judge them (on the average) to be equally loud when the reference tone is adjusted to 86 dB at the observer location. This makes the loudness level of the machine noise 86 phons.

At 1,000 Hz the decibel and phon levels are numerically identical, by definition. However, at other frequencies sine-wave tones may have numerically quite different sound- and loud-

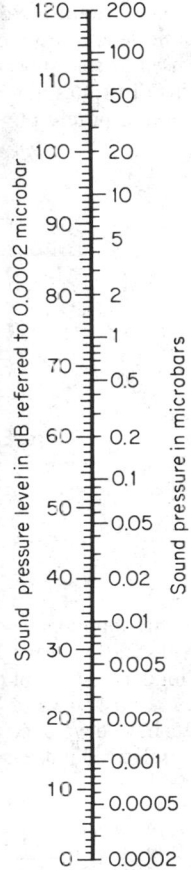

Fig. 19-1. Relation between sound pressure and sound-pressure level.[2]

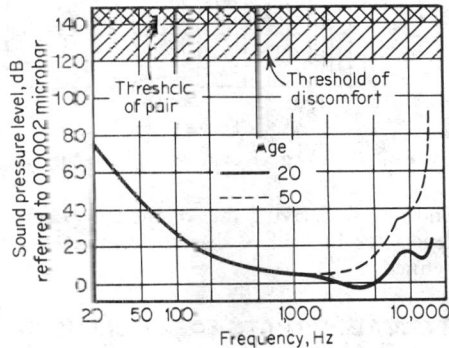

Fig. 19-2. The auditory area.[3]

ness-levels, as seen in Fig. 19-4. The dashed contour curves show the decibel level at each frequency corresponding to the loudness level identifying the curve at 1,000 Hz.[4] For example, a tone at 80 Hz and 70 dB lies on the contour marked 60 phons. Its sound level must be 70 dB for it to be as loud as a 60-dB tone at 1,000 Hz. Such differences at low frequencies, especially at low sound levels, are a characteristic of the sense of hearing. The fluctuations above 1,000 Hz are caused by sound-wave diffraction around the head of the listener and resonances in his ear canal. This illustrates how human physiological and psychological characteristics complicate the application of purely physical concepts.

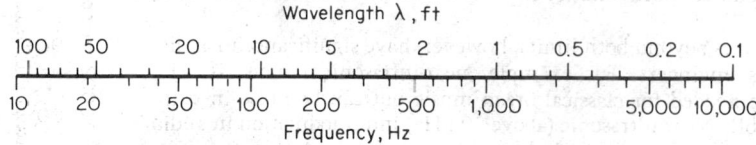

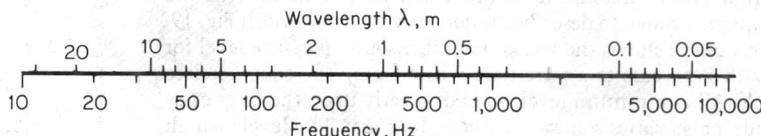

Fig. 19-3. Wavelength vs. frequency in air at 20°C.[2]

Since loudness level is related to 1,000-Hz tones defined physically in magnitude, the loudness-level scale is not really psychologically based. Consequently, although one can say that 70 phons is louder than 60 phons, one cannot say *how much* louder.

5. Loudness. By using the phon scale to overcome the effects of frequency, psychophysicists have developed a true loudness scale based upon numerous experimental procedures involving relative-loudness judgments. *Loudness,* measured in *sones,* has a direct relation to loudness level in phons[5] which is approximated in Fig. 19-5. (Below 30 phons the relation changes slope. Since few practical problems require that range, it is omitted for simplicity.) A loudness of 1 sone has been defined equivalent to a loudness level of 40 phons. It is evident in Fig. 19-5 that a 10-phon change doubles the loudness in sones, which means *twice as loud.* Thus a 20-phon change in loudness level quadruples the loudness.

Another advantage of the sone scale is that the loudness of components of a complex sound are additive on the sone scale as long as they are well separated on the frequency scale. For example (using Fig. 19-5), two tonal components at 100 and 4,000 Hz having loudness levels of 70 and 60 phons, respectively, would have individual loudnesses of 8 and 4 sones, respectively, and a total loudness of 12 sones.

Detailed loudness computation procedures have been developed for highly complex sounds and noises, deriving the loudness in sones directly from a complete knowledge of the decibel levels for individual discrete components or noise bands.[6,7] The procedures continue to be refined.[8]

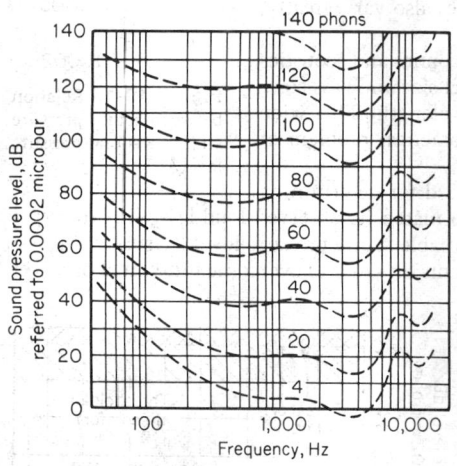

Fig. 19-4. Equal-loudness-level contours.

TYPICAL FORMATS FOR SOUND DATA

Sound and audio electronic data are frequently plotted as functions of frequency, time, direction, distance, or room volume. Frequency characteristics are the most common, in which the ordinate may be sound pressure, sound power, output-input ratio, percent distortion, or their logarithmic-scale (level) equivalents.

6. Sound Spectra. The frequency spectrum of a sound is a description of its resolution into components of different frequency and amplitude. Often the abscissa is a logarithmic frequency scale or a scale of octave (or fractional-octave) bands with each point plotted at the geometric mean of its band-limiting frequencies. Usually the ordinate scale is sound-pressure level. Phase differences are often ignored (except as they affect sound level) because they vary so greatly with measurement location, especially in reflective environments.

Line spectra are bar graphs for sounds dominated by discrete frequency components. Figure 19-6 is an example.

Continuous spectra are curves showing the distribution of sound-pressure level within a frequency range densely packed with components. Figure 19-7 is an example. Unless stated otherwise the ordinate of a continuous-spectrum curve, called *spectrum level*, is assumed to represent sound-pressure level for a band of 1-Hz width. Usually level measurements L_{band} are made in wider bands, then converted to spectrum level L_{ps} by the bandwidth correction

$$L_{ps} = L_{band} - 10 \log (f_2 - f_1) \quad dB \quad (19\text{-}2)$$

in which f_1 and f_2 are the lower- and upper-frequency limits of the band.

When a continuous-spectrum curve is plotted automatically by a level recorder synchronized with a heterodyning filter or with a sequentially switched set of narrow-bandpass filters, any effect of *changing bandwidth* upon curve slope must be considered.

Combination spectra are appropriate for many sounds in which strong line components are superimposed over more diffuse continuous spectral backgrounds. Bowed or blown musical tones and motor-driven fan noises are examples.

Octave spectra, in which the ordinate is the sound-pressure level for bands one octave wide, are very convenient for measurements and for specifications but lack fine spectrum detail. Figure 19-8 is an example.

Third-octave spectra provide more detail and are widely used. Figure 19-9 shows two radically different noise spectra plotted in third-octaves. One-third of an octave and one-tenth of a decade are so nearly identical that substituting the latter for the former is a practical convenience, providing a 10-band pattern which repeats every decade. Placing third-octave band zero at 1 Hz has conveniently made the band numbers equal 10 times the logarithm (base 10) of the band-center frequency; e.g., band 20 is at 100 Hz and band 30 at 1,000 Hz.

Visual proportions of spectra (and other frequency characteristics) depend upon the ratio of ordinate and abscissa scales. There is no universal or fully standard practice, but for ease of visual comparison of data and of specifications, it has become rather common practice in the United States for 30 dB of ordinate scale to equal (or slightly exceed) 1 decade of logarithmic frequency on the abscissa. Available audio and acoustical graph papers and automatic level-recorder charts have reinforced this practice. When the entire 120-dB range of auditory area is to be included in the graph, the ordinate is often compressed 2:1.

7. Response and Distortion Characteristics. Output-input ratios vs. frequency are the most common data format in audio-electronics engineering. The audio-frequency scale (20 Hz to 20 kHz) is usually logarithmic. The ordinate may be sound- or electric-output level in decibels as the frequency changes with a constant electric or sound input; or it may be a ratio of the output to input (expressed in decibels) as long as they are linearly related within the range of measurement.

Visual proportions (Par. **19-6**) are quite important in the publication of such data. Figure 19-10 is an important (but untypical) microphone example. In this case the ordinate scale has been

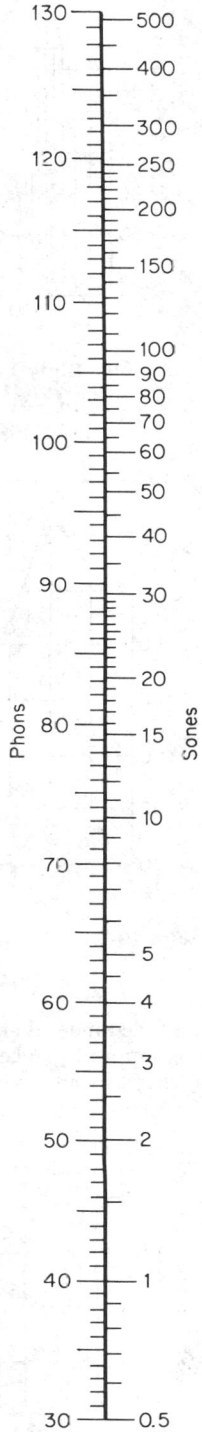

Fig. 19-5. Relation between loudness in sones and loudness level in phons.[6]

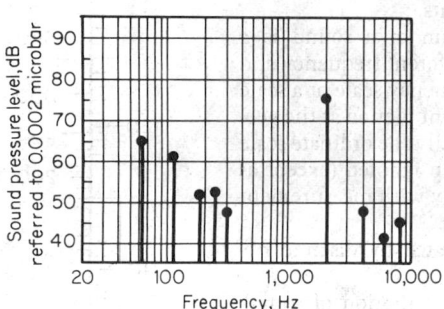

Fig. 19-6. Typical line spectrum.[39]

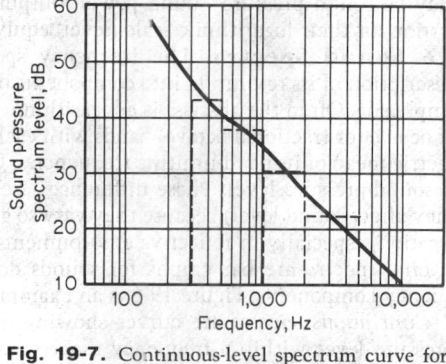

Fig. 19-7. Continuous-level spectrum curve for a motor and blower.[2]

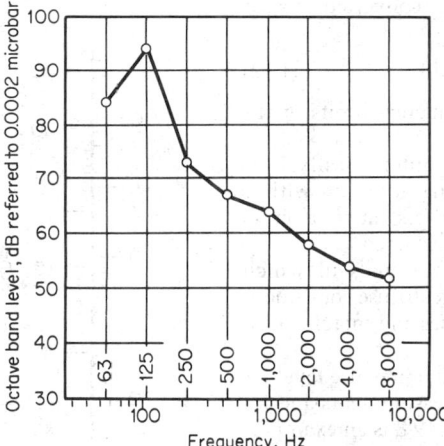

Fig. 19-8. Noise measurement by octave bands.[2]

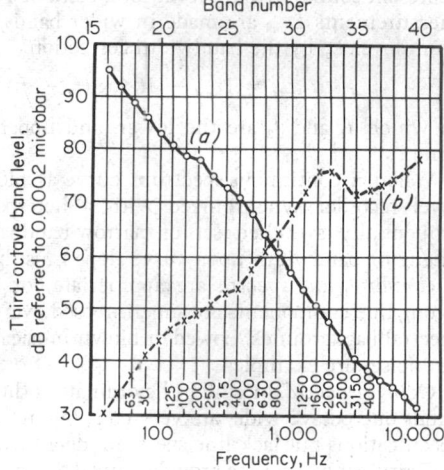

Fig. 19-9. Third-octave band spectra for two noise sources.[2]

expanded more than usual because the response variations are so small. It is necessary for the nonuniformity to be presented accurately here, because this type of microphone has long been a laboratory and industry standard for measurement and comparison.

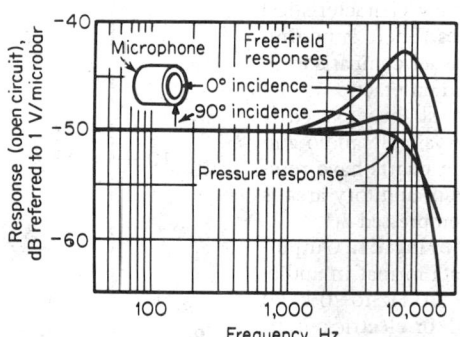

Fig. 19-10. Response vs. frequency of a Western Electric type 640-AA condenser microphone, showing behavior for sounds arriving along the axis and perpendicular to the axis of the microphone, as well as the pressure response.[38]

When the response-frequency characteristic is measured with the input frequency filtered from the output, a distortion-frequency characteristic is the result. It can be further filtered to obtain curves for each harmonic if desired.

8.   Directional Characteristics. Sound sources radiate almost equally in all directions when the wavelength is large compared to source dimensions. At higher frequencies, where the wavelength is smaller than the source, the radiation becomes quite directional. Figure 19-11, an example from musical sound, shows a typical format for plotting sound level vs. angle at different frequencies. On-axis sound level is chosen as a 0-dB reference level of comparison to other angles.

9. Time Characteristics. Any sound property can vary with time. It can build up, decay, or vary in magnitude periodically or randomly. Figure 19-12 is an example showing a high-speed sound-level record of the reverberant sound decay in a room. A reverberant sound field decays rather logarithmically. Consequently the sound level in decibels falls linearly when the time scale is linear. The rate of decay in this example is 33 dB/s.

SPEECH SOUNDS

10. Speech Level and Spectrum.[10,11] Both the sound-pressure level and the spectrum of speech sounds vary continuously and rapidly during connected discourse. Although speech may be arbitrarily segmented into elements called phonemes, each with a characteristic spectrum and level, actually one phoneme blends into another.

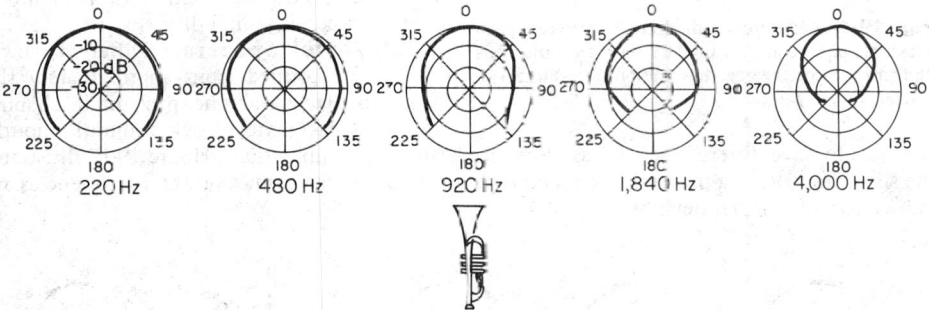

Fig. 19-11. Directional characteristics of a trumpet for five frequencies.[9]

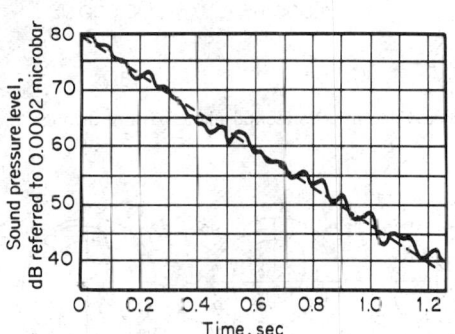

Fig. 19-12. A high-speed level record showing how the sound-pressure level in a room decays with time. Since the sound decays 40 dB in 1.2 s, the reverberation time, i.e., the time to decay 60 dB, is 1.8 s.[36]

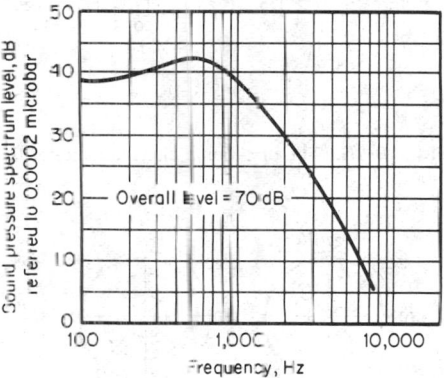

Fig. 19-13. Idealized speech spectrum for male voices at 1 m from the talker's lips. (*After Ref. 10.*)

Different talkers speak somewhat differently, and they sound different. Their speech characteristics vary from one time or mood to another. Yet in spite of all these differences and variations, statistical studies of speech have established a typical "idealized" speech spectrum shown in Fig. 19-13. The spectrum level rises about 5 dB from 100 to 600 Hz, then falls about 6, 9, 12, and 15 dB in succeeding higher octaves.

Overall sound-pressure levels, averaged over time and measured at a distance of 1 m from a talker on or near the speech axis, lie in the range of 65 and 75 dB when the talkers are instructed to speak in a "normal" tone of voice. Along this axis the speech sound level follows the inverse-square law closely to within about 10 cm of the lips, where the level is about 90 dB. At the lips, where communication microphones are often used, the overall speech sound level typically averages over 100 dB.

The peak levels of speech sounds greatly exceed the long-time average level. Figure 19-14 shows the difference between short peak levels and average levels at different frequencies in the speech spectrum. The difference is greater at high frequencies, where the sibilant sounds of relatively short duration have spectrum peaks.

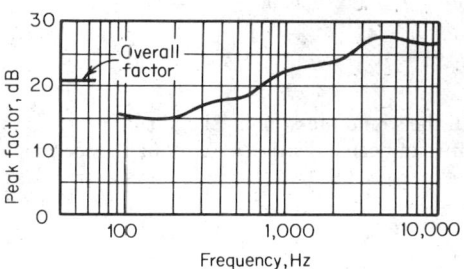

11. Speech Directional Characteristics.[11] Speech sounds are very directional at high frequencies. Figures 19-15 and 19-16 show clearly why speech is poorly received behind a talker, especially in nonreflective environments Above 4,000 Hz the directional loss in level is 20 dB or more, which particularly affects the sibilant sound levels so important to speech intelligibility.

Fig. 19-14. Difference in decibels between peak pressures of speech measured in short (⅛-s) intervals and rms pressure averaged over a long (75-s) interval. *(After Ref. 10.)*

12. Vowel Spectra.[12] Different vowel sounds are formed from approximately the same basic laryngeal tone spectrum by shaping the vocal tract (throat, back of mouth, mouth, and lips) to have different acoustical resonance-frequency combinations. Figure 19-17 illustrates the spectrum filtering process. The spectral peaks are called *formants*, and their frequencies are known as formant frequencies.

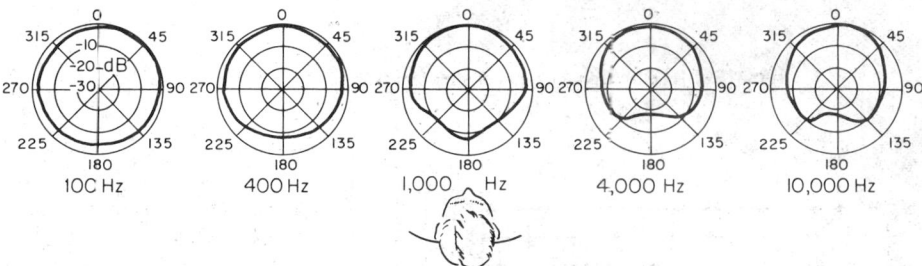

Fig. 19-15. The directional characteristics of the human voice in a horizontal plane passing through the mouth.[22]

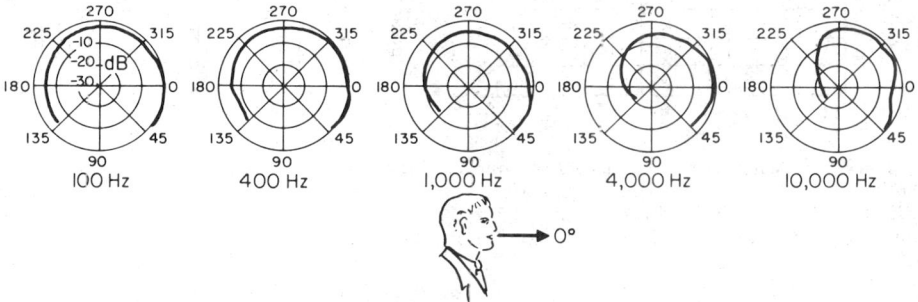

Fig. 19-16. The directional characteristics of the human voice in a bilaterally symmetrical vertical plane passing through the mouth.[22]

Detailed vowel analyses show three or even more formants. In Fig. 19-18 the abscissa is time, the ordinate is frequency, and the darkness of shading is proportional to sound level. The dark bars are formant peaks for different vowel sounds. In diphthongs the vowels and their formants are in transition.

The shapes of the vocal tract, simplified models, and the acoustical results for three vowel sounds are shown in Fig. 19-19. A convenient graphical method[14] for describing the combined formant patterns is shown in Fig. 19-20. Traveling around this vowel loop involves progressive motion of the jaws, tongue, and lips.

13. Speech Intelligibility.[15,16] More intelligibility is contained in the central part of the speech spectrum than near the ends. Figure 19-21 shows the effect upon articulation (the percent of syllables correctly heard) when low- and high-pass filters of various cutoff frequencies are used. From this information a special frequency scale has been developed in which each of 20 frequency bands contributes 5% to a total *articulation index* of 100%. This distorted frequency scale is used in Fig. 19-22. Also shown are the spectrum curves for speech peaks and for speech minima, lying approximately 12 and 18 dB, respectively, above and below the average-speech-spectrum curve. When all the shaded area (30-dB range between the maximum and minimum curves) lies above threshold and below overload, in the absence of noise, the articulation index is 100%.

If a noise-spectrum curve were added to Fig. 19-22, the figure would become an articulation-index computation chart for predicting communication capability. For example, if the ambient-noise spectrum coincided with the average-speech-spectrum curve, i.e., the signal-to-noise ratio is 1, only twelve-thirtieths of the shaded area would lie above the noise. The articulation index would be reduced accordingly to 40%.

Figure 19-23 relates monosyllabic word articulation and sentence intelligibility to articulation index. In the example above, for an articulation index of 0.40 approximately 70% of monosyllabic words and 96% of sentences would be correctly received

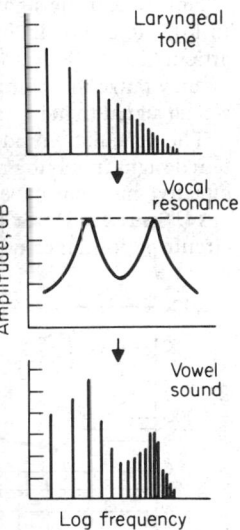

Fig. 19-17. Effects upon the spectrum of the laryngeal tone produced by the resonances of the vocal tract.[2a]

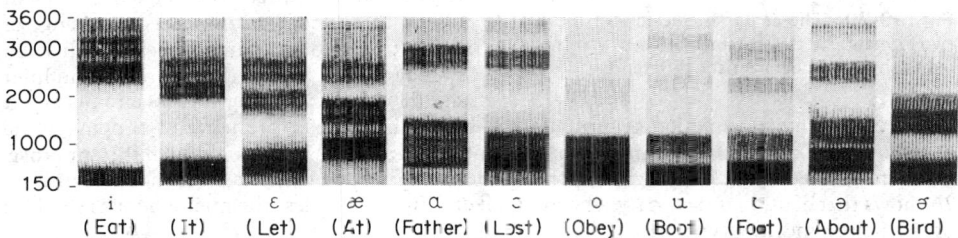

Fig. 19-18. Spectrograms of vowels from a male voice. *(Courtesy of Bell Telephone Laboratories.)*

i (Eat) I (It) ε (Let) æ (At) ɑ (Father) ɔ (Lost) o (Obey) u (Boot) ʊ (Foot) ə (About) ɝ (Bird)

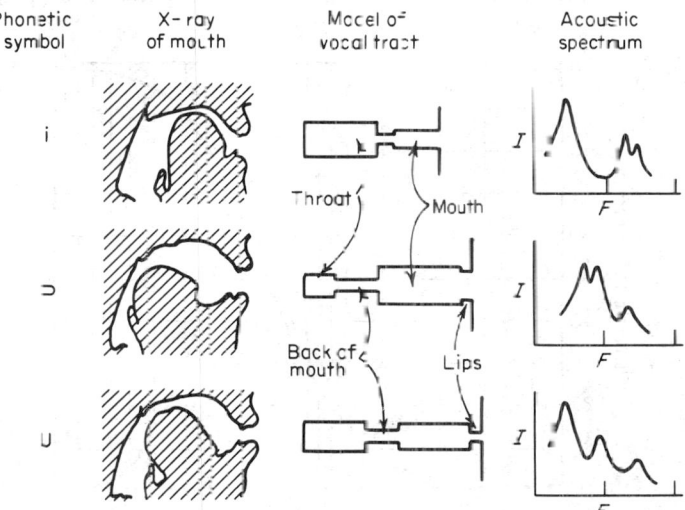

Fig. 19-19. Phonetic symbols, shapes of vocal tract, models, and acoustic spectra for three vowels.[13]

However, if the signal-to-noise ratio were kept at unity and the frequency range were reduced to 1,000 to 3,000 Hz, half the bands would be lost. Articulation index would drop to 0.20, word articulation to 0.30, and sentence intelligibility to 70%. This shows the necessity for wide frequency range in a communication system when the signal-to-noise ratio is marginal. Conversely a good signal-to-noise ratio is required when the frequency range is limited.

The articulation-index method is particularly valuable in complex intercommunication-system designs involving noise disturbance at both the transmitting and receiving stations.[17] Simpler effective methods have also been developed.[18]

14. Speech Peak Clipping. Speech waves are often affected inadvertently by electronic-circuit performance deficiencies or limitations. Figure 19-24 illustrates two types of amplitude distortion, center clipping and peak clipping. Center clipping, often caused by improper balancing or biasing of a push-pull amplifier circuit, can greatly interfere with speech quality and intelligibility. In a normal speech spectrum the consonant sounds are higher in frequency and lower in level than the vowel sounds. Center clipping tends to remove the important consonants.

By contrast peak clipping has little effect upon speech intelligibility[19] as long as ambient noise at the talker and system electronic noise are relatively low in level compared with the speech.

Peak clipping is frequently used intentionally in speech-communication systems to raise the average transmitted speech level above ambient noise at the listener or to increase the range of a radio transmitter of limited power. This can be done simply by overloading an amplifier stage. However, it is safer for the circuits and it produces less intermodulation distortion when back-to-back diodes are used for clipping ahead of the overload point in the amplifier or transmitter. Figure 19-25 shows intelligibility improvement from speech peak clipping when the talker is in quiet and listeners are in noise. Figure 19-26 shows that caution is necessary when the talker is in noise, unless the microphone is shielded or is a noise-canceling type.[17]

Tilting the speech spectrum by differentiation[19] and flattening it by equalization[20] are effective preemphasis treatments before peak clipping. Both methods put the consonant and vowel sounds into a more balanced relationship before the intermodulation effects of clipping affect voiced consonants.

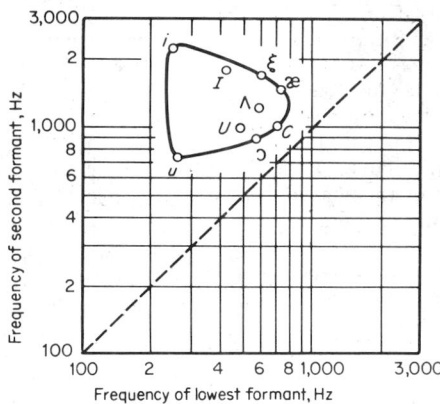

Fig. 19-20. The center frequencies of the first two formants for the sustained English vowels plotted to show the characteristic differences.[14]

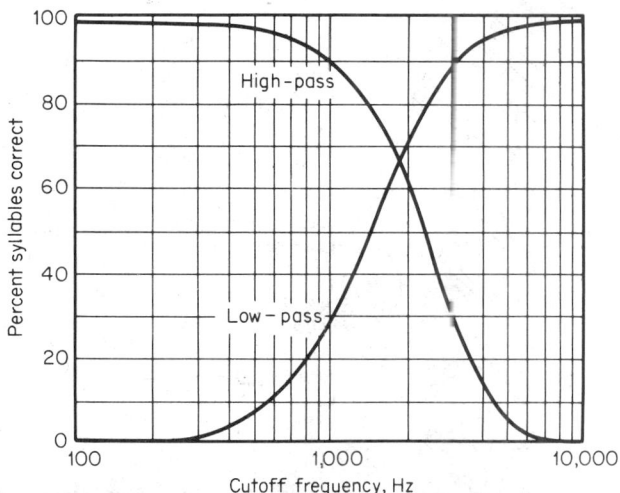

Fig. 19-21. Syllable articulation score vs. low- or high-pass cutoff frequency.[15]

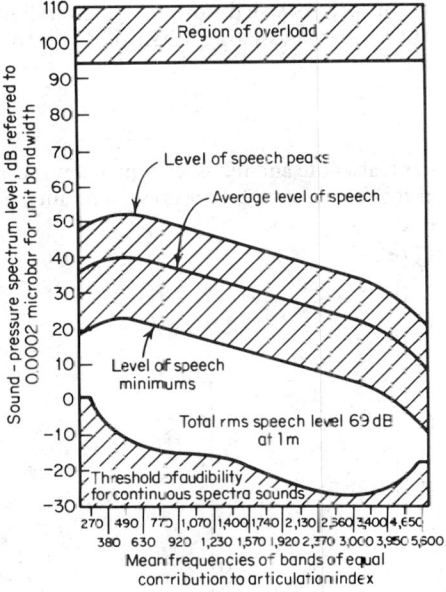

Fig. 19-22. Speech area, bounded by speech peak and minimum spectrum-level curves, plotted on an articulation-index calculation chart.[52]

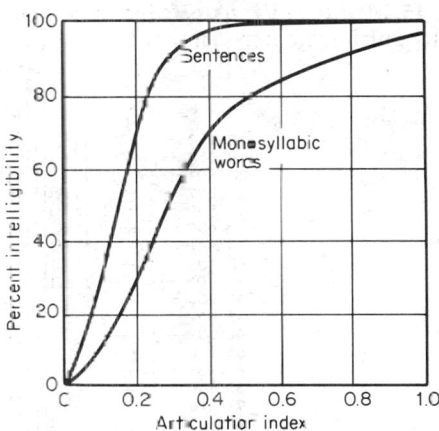

Fig. 19-23. Sentence- and word-intelligibility prediction from calculated articulation index.[17]

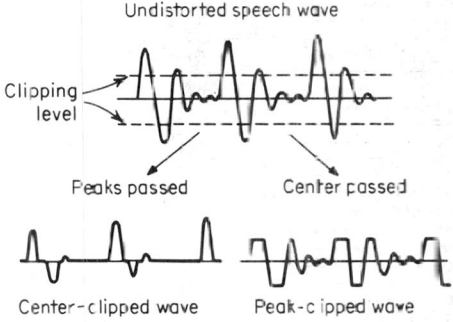

Fig. 19-24. Two types of amplitude distortion of speech waveform.[124]

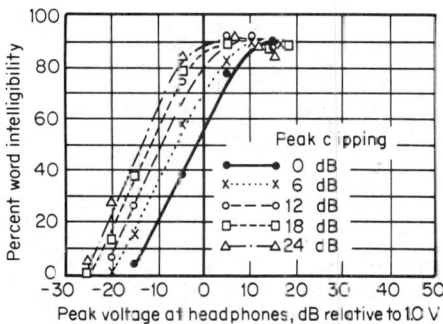

Fig. 19-25. Advantages of peak clipping of noise-free speech waves, heard by listeners in ambient air-craft noise. (After Ref. 17.)

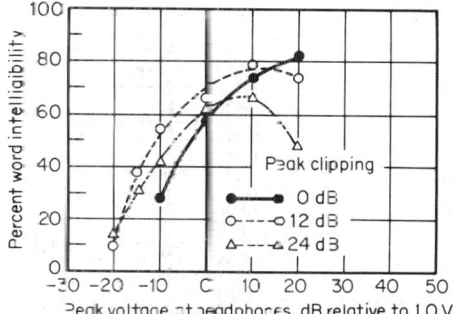

Fig. 19-26. Effects of speech clipping with both the talker and the listener in simulated aircraft noise. Note that excessive peak clipping is detrimental. (After Ref. 17.)

Caution must be used in combining different forms of speech-wave distortion, which individually have innocuous effects upon intelligibility but can be devastating when they are combined.[21]

MUSICAL SOUNDS[22]

15. Musical Frequencies. The accuracy of both absolute and relative frequencies is usually much more important for musical sounds than for speech sounds and noise. The interna-

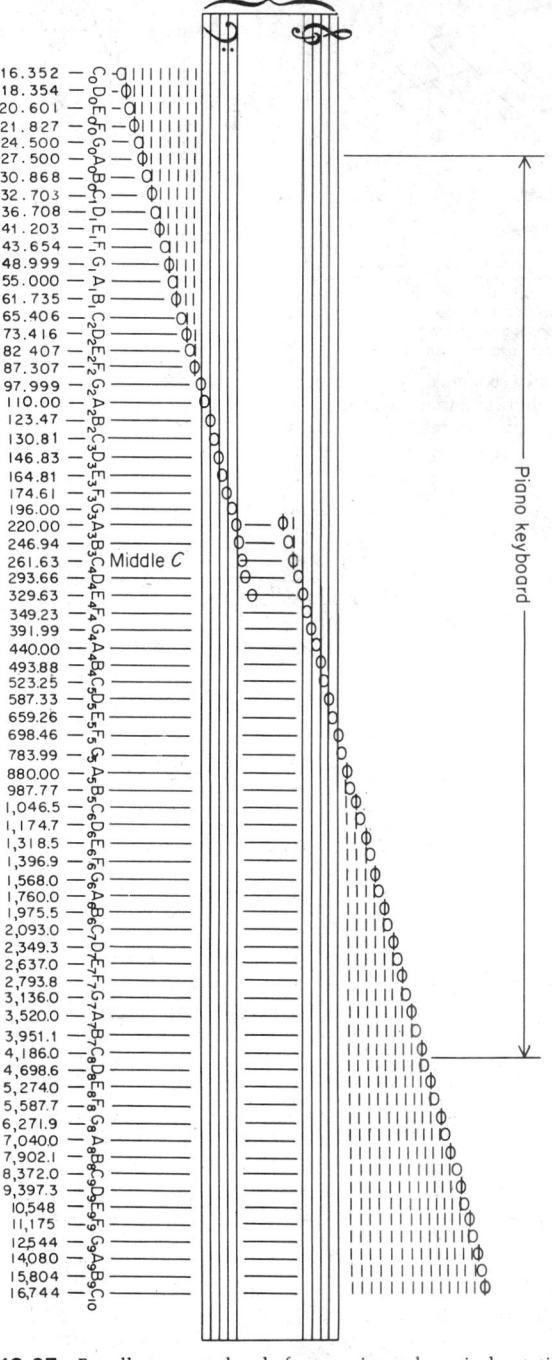

Fig. 19-27. Equally tempered scale frequencies and musical notation.[22]

tional frequency standard for music is defined at 440.00 Hz for A_4, the A above C_4 (middle C) on the musical keyboard. In sound recording and reproduction the disc-rotation and tape-transport speeds must be held correct within 0.2 or 0.3% error (including both recording and playback mechanisms) to be fully satisfactory to musicians.

The mathematical musical scale is based upon an exact octave ratio of 2:1. The subjective octave slightly exceeds this,[23] and piano tuning sounds better when the scale is stretched very slightly.[24]

The equally tempered scale of 12 equal ratios within each octave is an excellent compromise between the different historical scales based upon harmonic ratios.[25] It has become the standard

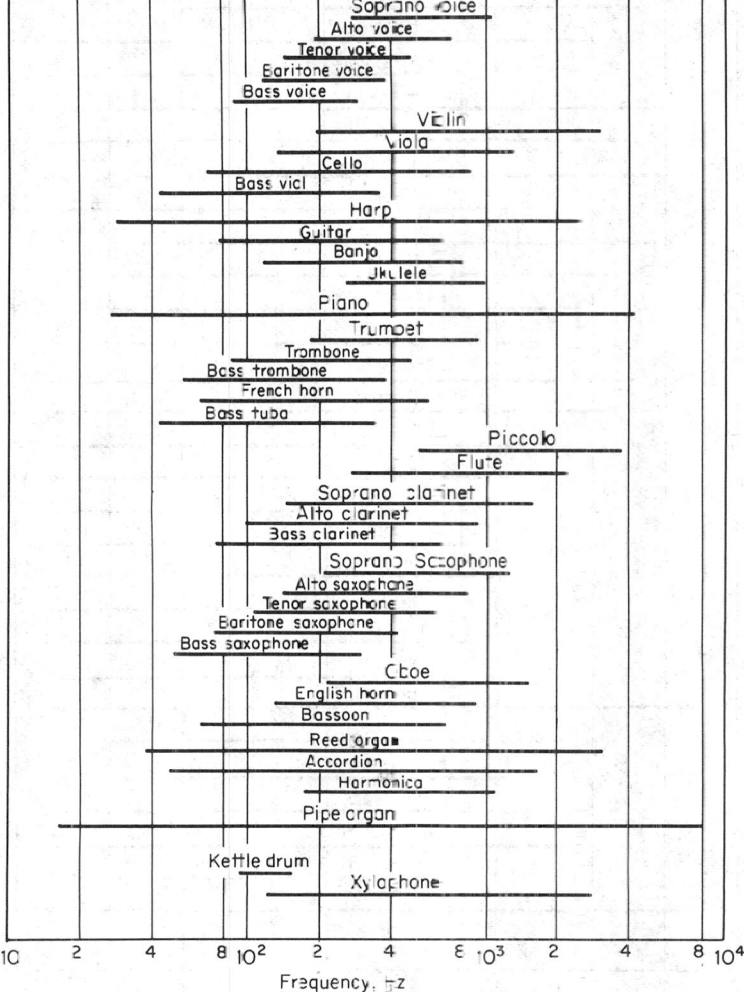

Fig. 19-28. Range of the fundamental frequencies of voices and various musical instruments. *(After Ref. 31.)*

of reference, even for individual musical performances which may deviate from it for artistic or other reasons. Figure 19-27 shows the fundamental frequencies corresponding to ANSI standard[1] musical notation and locations on the musical staff, omitting sharps and flats. The bottom octave is chiefly for the lowest (32-ft) organ pedal tones, and the upper two octaves for the fundamental frequencies of the highest organ stops and mixtures.

Different musical instruments play over different ranges of *fundamental* frequency, shown in Fig. 19-28. However, most musical sounds have many harmonics which are audibly significant to their tone spectra. Consequently high-fidelity recording and reproduction need a much wider frequency range, given in Fig. 19-29. When the musical frequency range must be restricted in recording and reproduction, the result can be estimated from Fig. 19-30.

16. Sound Levels of Musical Instruments. The sound level from a musical instrument varies with the type of instrument, the distance from it, which note in the scale is being played, the dynamic marking in the printed music, the players ability, and (on polyphonic instruments) the number of notes (and stops) played at the same time.

Orchestral Instruments. The following sound levels are typical at a distance of 10 ft in a nonreverberant room. Soft (pianissimo) playing of a weaker orchestral instrument, e.g., violin,

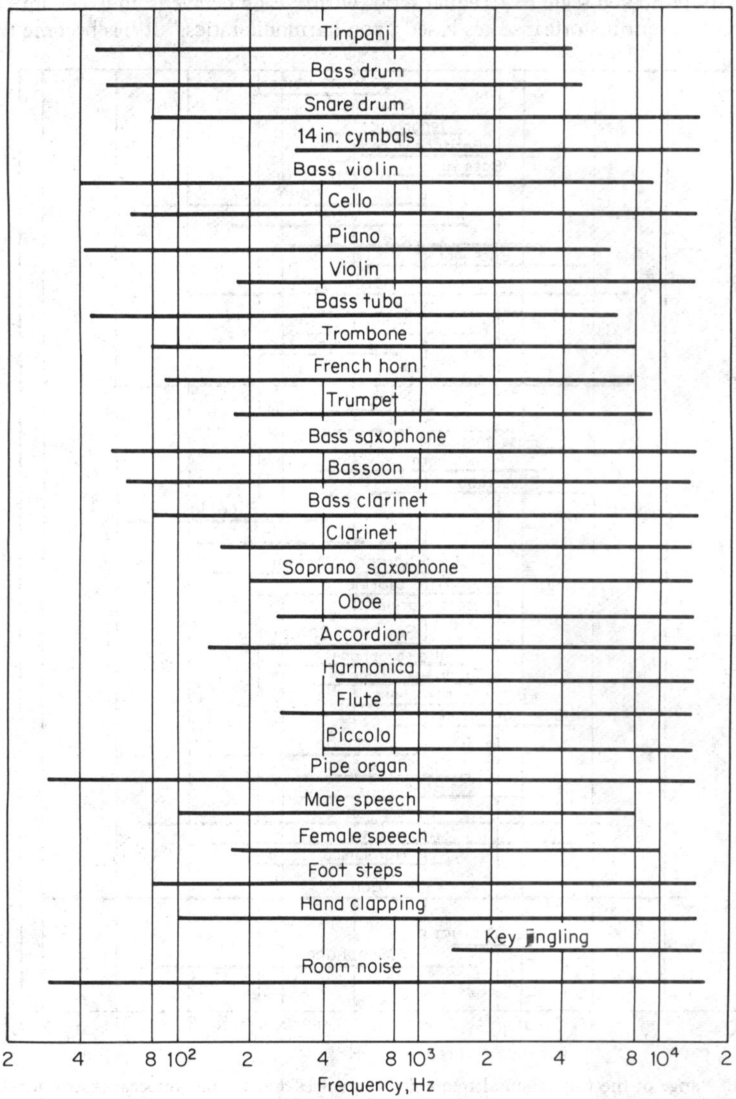
Timpani
Bass drum
Snare drum
14 in. cymbals
Bass violin
Cello
Piano
Violin
Bass tuba
Trombone
French horn
Trumpet
Bass saxophone
Bassoon
Bass clarinet
Clarinet
Soprano saxophone
Oboe
Accordion
Harmonica
Flute
Piccolo
Pipe organ
Male speech
Female speech
Foot steps
Hand clapping
Key jingling
Room noise
2 4 8 10^2 2 4 8 10^3 2 4 8 10^4 2
Frequency, Hz

Fig. 19-29. Frequency ranges required for speech, musical instruments, and noises so that no frequency discrimination will be apparent. *(After Ref. 26.)*

flute, bassoon, produces a typical sound level of 55 to 60 dB. Fortissimo playing on the same instrument raises the level to about 70 to 75 dB. Louder instruments, e.g., trumpet or tuba, range from 75 dB at pianissimo to about 90 dB at fortissimo.[27]

Certain instruments have exceptional differences in sound level of low and high notes. A flute may change from 42 dB on a soft low note to 77 dB on a loud high note, a range of 35 dB. The French horn ranges from 43 dB (soft and low) to 93 dB (loud and high).[27]

Sound levels are about 10 dB higher at 3 ft (inverse-square law) and 20 dB higher at 1 ft. The louder instruments, e.g., brass at closer distances may overload some microphones and preamplifiers.

Percussive Instruments. The sound levels of shock-excited tones are more difficult to specify because they vary so much during decay and can be excited over a very wide range. A bass drum may average over 100 dB during a loud passage with peaks (at 10 ft) approaching 120 dB. By contrast a triangle will average only 70 dB with 30-dB peaks. A single tone of a grand piano played forte will initially exceed 90 dB near the piano rim, 80 dB at the pianist, and 70 dB at the conductor 10 to 15 ft away. Large chords and rapid arpeggios will raise the level about 10 dB.

Instrumental Groups. Orchestras, bands, and polyphonic instruments produce higher sound levels since many notes and instruments (or stops) are played together. Their sound levels are specified at larger distances than 10 ft because the sound sources occupy a large area; 20 ft from the front of a 75-piece orchestra the sound level will average about 85 to 90 dB with peaks of 105 to 110 dB. A full concert band will go higher. At a similar distance from the sound sources of an organ (pipe or electronic) the full-organ (or crescendo-pedal) condition will produce a level of 95 to 100 dB. By contrast the softest stop with expression shutters closed may be 45 dB or less.

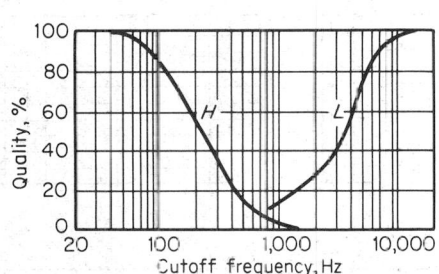

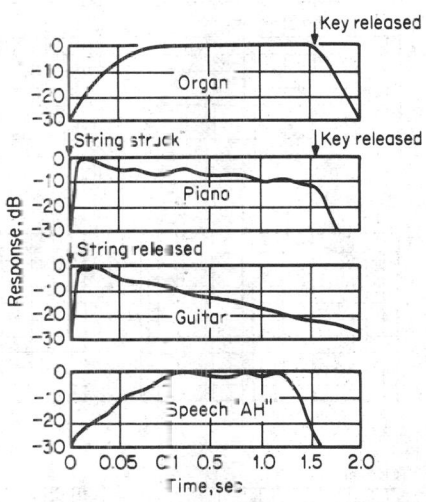

Fig. 19-30. Orchestral musical-quality judgment vs. high-pass (*H*) and low-pass (*L*) filter cutoff frequencies.[26]

Fig. 19-31. The growth and decay as a function of the time of an organ piano and guitar tone and the speech sound "ah."[22]

17. Growth and Decay of Musical Sounds. These characteristics are quite different for different instruments. Examples are given in Fig. 19-31. Piano or guitar tones quickly rise to an initial maximum, then gradually diminish until the strings are damped mechanically. Piano tones have a more rapid decay initially than later in the sustained tone.[28] Orchestral instruments can start suddenly or smoothly, depending upon the musician's technique, and they damp rather quickly when playing ceases. Room reverberation affects both growth and decay rates when the time constants of the room are greater than those of the instrument vibrators. This is an important factor in organ music, which is typically played in reverberant environment.

Many types of musical tone have characteristic transients which influence timbre greatly. In the "chiff" of organ tone the transients are of different fundamental frequency. They appear and decay before steady state is reached. In percussive tones the initial transient is the cause of the tone (often a percussive noise), and the final transient is the result.

These transient effects should be considered in the design of audio electronics such as "squelch," automatic gain control, compressor, and background-noise reduction circuits.

18. Spectra of Musical Instrument Tones.[29] Figure 19-32 displays time-averaged spectra for a 75-piece orchestra, a theater pipe organ, a piano, and a variety of orchestral instruments, including members of the brass, string, woodwind, and percussion families. These vary from one note to another in the scale, from one instant to another within a single tone or chord, and from one instrument or performer to another. For example, a concert organ voiced in a baroque style would have lower spectrum levels at low frequencies and higher at high frequencies than the theater organ shown.

The organ and bass drum have the most prominent low-frequency output. The cymbal and snare drum are strongest at very high frequencies. The orchestra and most of the instruments

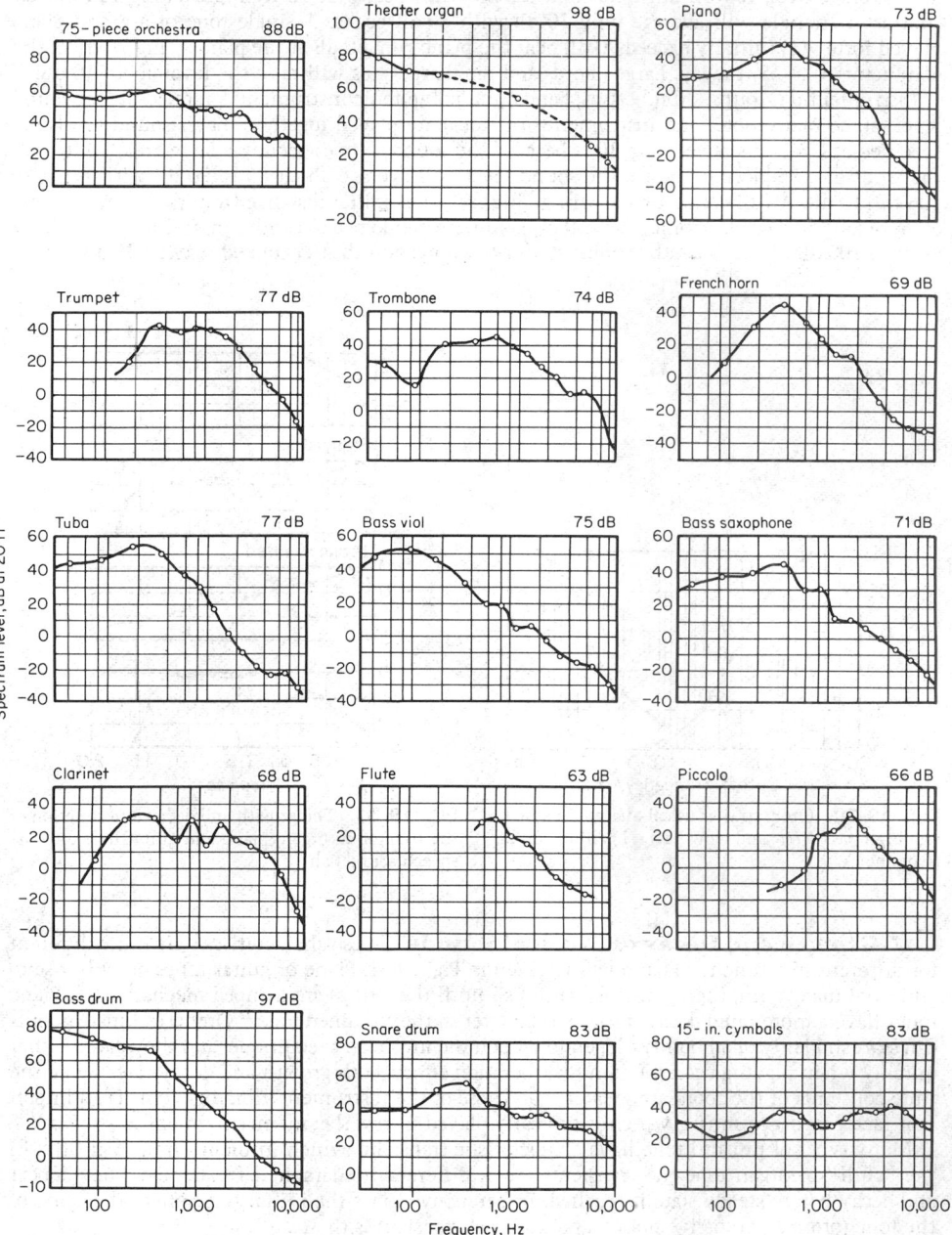

Fig. 19-32. Time-averaged spectra of musical instruments. (*After Ref. 29.*)

have spectra which diminish gradually with increasing frequency, especially above 1,000 Hz. This is what has made it practical to preemphasize the high-frequency components, relative to those at low frequencies, in both disc and tape recording. However, instruments which differ from this spectral tendency, e.g., coloratura sopranos, piccolos, cymbals, create problems of intermodulation distortion and overload.

Spectral peaks occuring only occasionally, for example, 1% of the time, are often more important to sound recording and reproduction than the peaks in the average spectra of Fig. 19-32. The frequency ranges shown in Table 19-1 have been found[29] to have relatively large instantaneous peaks for the instruments listed.

Table 19-1.   Frequency Band Containing Instantaneous Spectral Peaks

Band Limits Hz	Instruments
20–60	Theater organ
60–125	Bass drum, bass viol
125–250	Small bass drum
250–500	Snare drum, tuba, bass saxophone, French horn, clarinet, piano
500–1,000	Trumpet, flute
2,000–3,000	Trombone, piccolo
5,000–8,000	Triangle
8,000–12,000	Cymbal

19. Directional Characteristics of Musical Instruments. Most musical instruments are somewhat directional. Some are highly so, with well-defined symmetry, e.g., around the axis of a horn bell[9] (see Fig. 19-11). Other instruments are less directional because the sound source is smaller than the wavelength, e.g., clarinet, flute. The mechanical vibrating system of bowed string instruments is complex, operating differently in different frequency ranges, and resulting in extremely variable directivity (Fig. 19-33). This is significant for orchestral seating arrangements both in concert halls and recording studios.[30] The large size and complex shape of piano soundboards and bridges and the unsymmetrical shape of grand piano reflecting lids create a complex set of directional characteristics (Fig. 19-34) at high frequencies. Audiences and recording microphones are usually located opposite the inclined reflecting lid.

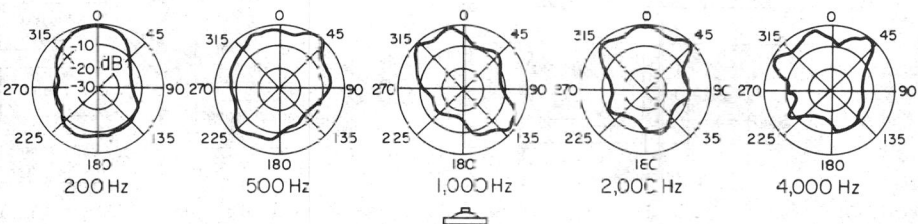

Fig. 19-33. The directional characteristics of a violin for five different frequencies.[22]

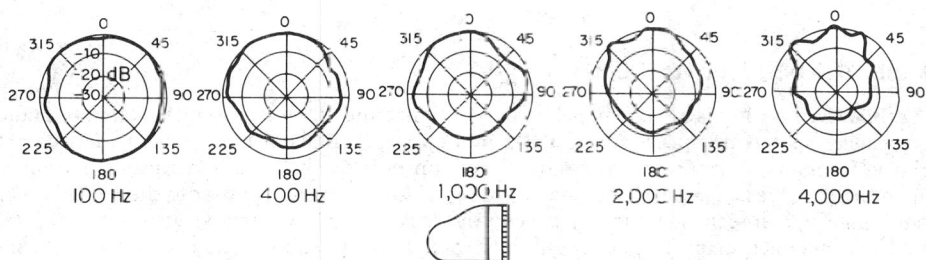

Fig. 19-34. The directional characteristics of a grand piano for five different frequencies.[22]

20. Audible Distortions of Musical Sounds. The quality of musical sounds is more sensitive to distortion than the intelligibility of speech. A chief cause is that typical music contains several simultaneous tones of different fundamental frequency in contrast to typical speech sound of one voice at a time. Musical chords subjected to nonlinear amplification or transduction generate intermodulation components which appear elsewhere in the frequency spectrum.

Difference tones are more easily heard than summation tones because the summation tones are often hidden by harmonics which were already present in the undistorted spectrum and because auditory masking of a high-frequency pure tone by a lower-frequency pure tone is much greater than vice versa.

Figure 19-35 shows perceptible, tolerable, and objectionable amounts of controlled nonlinear distortion for five different system high-frequency cutoffs (shown in the figure) for both speech and music and for two types of distortion. (The pentode distortion type has a more extended harmonic series.) Both speech and music were live (recording and playback were not involved).[31]

When a critical listener controls the sounds heard (an organist playing an electronic organ on a high-quality amplification system) and has unlimited opportunity and time to listen, even lower distortion (0.2%, for example) can be perceived.[32]

Frequency-range division into multiple channels[33] is the successful method for preventing intermodulation distortion between widely separated frequencies (such as 100 and 4,000 Hz).[34] The method has been extended to frequencies which lie within the same octave,[35a] using a larger number of dividing networks and as many as four loudspeaker channels.

By contrast some contemporary forms of musical performance employing electroacoustic or electronic means use a variety of electronic distortion circuits for special and unusual timbre effects.[35b]

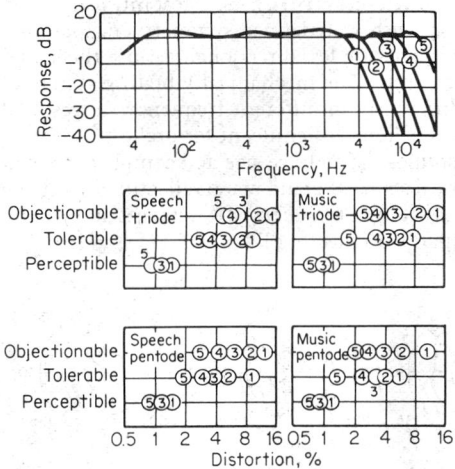

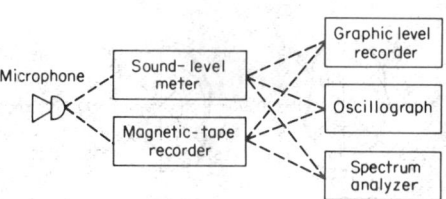

Fig. 19-35. Perceptible, tolerable, and objectionable amounts of controlled nonlinear harmonic distortion of speech and music for various high-frequency cutoffs.[31]

Fig. 19-36. General system for noise measurement.[38]

AMBIENT NOISE AND ITS CONTROL[36,37]

21. Nature of Noise. In audio-electronics engineering, noise is unwanted sound or audio disturbance. Speech or music can be noise if they are crosstalk between audio lines or are received from an adjacent recording studio. The term *noise* is also used sometimes when referring to the very weak spectrum components or to the low-level transients or to the second-order modulations which seem superficially to be in the background of the main signal or the natural sound but which in many cases turn out to be vital although subtle. The treatment here concerns audible-noise measurement, noise criteria for rooms and for hearing conservation, and the reduction and isolation of airborne and structure-borne noise.

22. Noise Measurement.[38,39] Figure 19-36 is a general system for noise measurement. The simplest system is a sound-level meter (with microphone) which indicates sound-pressure level as previously defined [Eq. (19-1)]. Its amplifier output can also be supplied to an oscillograph, a graphic-level recorder, or a sound-spectrum analyzer, or all three. If the noises are to be stored for reference, a magnetic-tape recorder can be used. For later detailed study the tape can be played into the oscillograph, graphic-level recorder, or spectrum analyzer. This assumes that the recorder acts linearly, without automatic gain control, with wide frequency range and low internal-noise level, and is calibrated or has a calibrated reference sound recorded on the tape.

The actual shape of the microphone output wave can be seen on the oscillograph, assuming zero phase shift in the sound-level-meter amplifier. Magnetic-tape recorder-playback phase shift must be considered when tape is used if phase is significant. Special instrumentation tape recorders are available.

An example involving an oscillograph and a graphic-level recorder is shown in Fig. 19-37. The oscillograms on the left have corresponding level graphs on the right. In (a) the single-frequency logarithmic decrement gives a smooth, straight graph of sound-level decay. In (b) the two frequencies beat as shown in both the wavy envelope and the oscillating level decay. In (c), which is more typical of room reverberation, numerous modes contribute a rather random aperiodic modulation to the wave envelope and graph.

23. Sound-Level Meter. A simplified block diagram is shown in Fig. 19-38. The first attenuator prevents the microphone output from overloading the high-gain amplifier when the noise level is very high. The weighting networks are inserted when the response-frequency characteristic is switched from uniform to predetermined standard curves (see Fig. 19-39). The amplified output is available for oscillograph, etc., and is also rectified (full-wave) for the indicating meter (usually rms), which responds logarithmically over a typical range of −4 to 10 dB relative to the switch-selected multiple of 10 dB sound-pressure level. The meter response time is typically 0.5 s. A choice of meter speed is sometimes provided.

The standard overall response-frequency characteristic for different weighting networks is shown in Fig. 19-39. Curve C is essentially unweighted and is used for physical noise measurements. Weighting networks B and A are used for moderate and substantial attenuation of the low-frequency

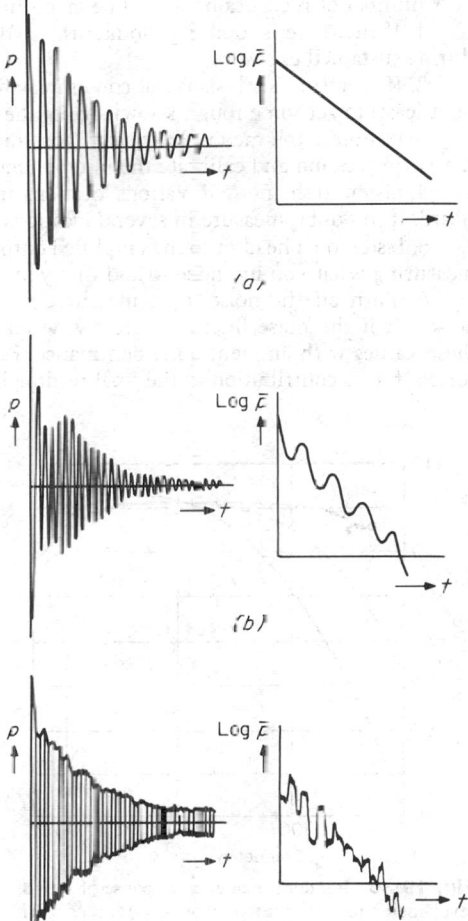

Fig. 19-37. Sound-pressure oscillograms and corresponding graphs of sound-pressure level vs. time: (a) single frequency decay; (b) two adjacent frequencies with the same decay constant; (c) many closely spaced frequencies with the same decay constant.[37]

noise components. Weighting is used when the noise level is low, and the C scale meter reading would correlate poorly with loudness level (see the shape of the 20- and 40-phon equal-loudness contours in Fig. 19-4). The A weighting has also been found useful in connection with hearing-conservation criteria described later.

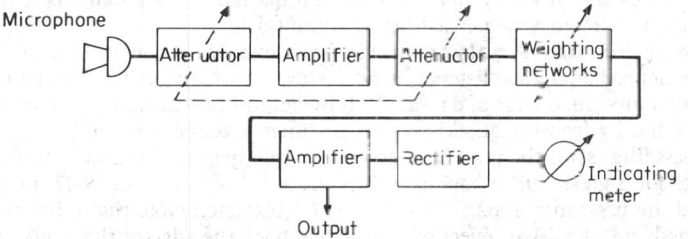

Fig. 19-38. Simplified block diagram of a sound-level meter.[38]

A number of precautions should be taken in making noise-level measurements.

1. If the noise is loud, e.g., louder than your own loud speech sounds, wear ear protection during sustained exposure.

2. If spectrum analysis is not contemplated, make measurements on all three scales (C and A at least) to get some rough knowledge of the noise spectrum.

3. Calibrate the meter electrically (according to the manufacturer's procedure) before each measuring session and calibrate the microphone periodically.

4. Listen to the noise in various locations and measure where the most significant noises are heard. If in doubt, measure in several locations, noting each for later data analysis.

5. Listen on a headset to the amplified output of the sound-level meter to be certain you are measuring what you intended to and that you are not measuring meter noise.

6. Turn off the noise to be measured, if possible, and measure the ambient background noise. Or if the noise fluctuates greatly, watch the lowest values and listen for correlation of these values with ambient-noise dominance. Knowing the level of the ambient noise, you can correct for its contribution to the total reading by using Fig. 19-40.

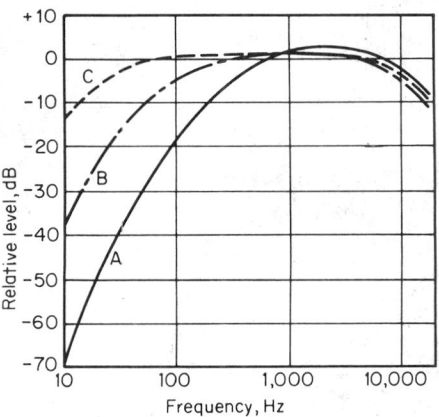

Fig. 19-39. Random incidence response of sound-level meter for different networks. *(Adapted with permission from Figs. 1 to 3 of American National Standard Specification for Sound Level Meters, S 1.4-1971, copyright 1971 by the American National Standards Institute. Copies available for purchase from ANSI, 1430 Broadway, New York, NY 10018.)*

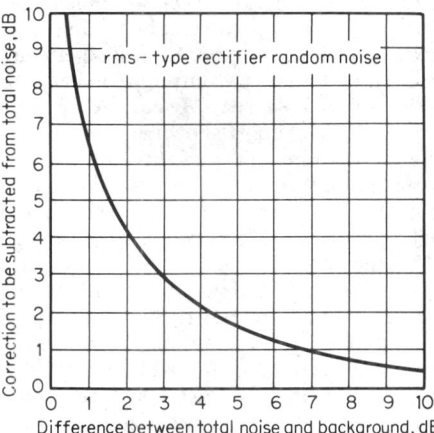

Fig. 19-40. Correction for ambient background level in terms of the difference between the total noise (including the noise being measured) and the ambient noise. The curve indicates the correction where both noises are essentially random in character and the rectifier characteristic is of the rms type.[39]

7. When the noise is dominated by steady components of fixed frequency, especially at low frequencies and in small rooms, the data will probably be position-dependent. If so take enough data for averaging.

8. Note Fig. 19-41. The size and shape of the sound-level meter and the proximity of the observer can influence the reading somewhat. If these differences are important to your data, use a microphone stand and extension cable (with cable correction) or a smaller meter or binoculars to read the meter. In any case hold the meter away from you and stand to one side of the line from the source to the meter.

9. If the noises are caused by sharp or sudden impact, or if they sound percussive or clicky, an impact noise meter[40] may be needed for significant data.

24. Noise-Spectrum Analyzers. The least expensive and most available noise analyzer, an experienced analytical listener, is not a substitute for a good analyzing instrument but can save many hours and dollars of data analysis by helping choose between a narrow- or a broadband analyzer and by focusing quickly on the pertinent frequency range.

The bandpass filter set is the most rapid tool for simple spectrum analysis. Response-frequency characteristics for a good octave bandpass filter set are shown in Fig. 19-42. (Recent standards have centered the passbands at 62.5, 125, 250, 500, 1,000, etc.) Note the uniformity of response within the passband, the 40-dB rejection one octave from the edge of the band, and the greater rejection beyond.

Automatic switching of a level indicator (meter, graphic-level recorder, or oscillograph) to each of the filter outputs in a frequency sequence gives an octave spectrum plot. When the sequence is rapidly cycled, one can monitor time variations of the spectrum. Separate indicators on each filter output allow continuous spectrum monitoring in real time. Half- and third-octave bands are also used for greater spectrum detail.

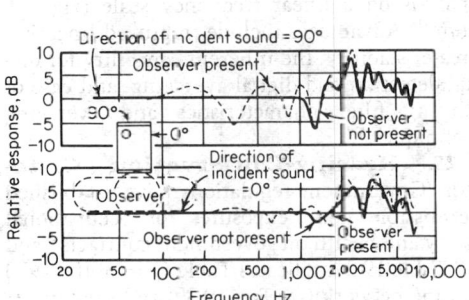

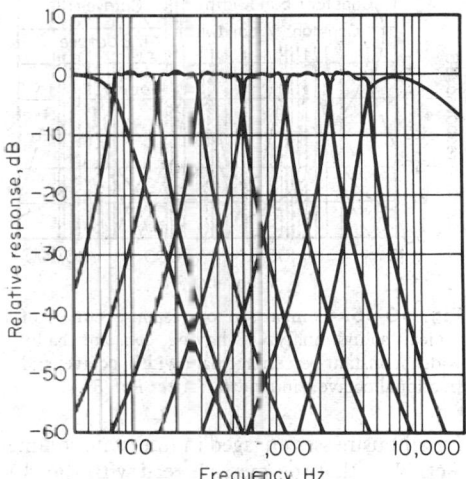

Fig. 19-41. The effect on the frequency response of using the microphone directly on a rectangular sound-level meter, with and without an observer present in a free field.[39]

Fig. 19-42. Response vs. frequency of the eight filters in an octave-band filter set.[38]

Searching for very fine detail on the frequency axis of the noise spectrum is accomplished by a heterodyne analyzer. Figure 19-43 gives a block diagram. The selectivity is determined by an ultrasonic fixed-frequency filter (usually quartz crystal or inductor-capacitor with electronic feedback). The noise signal is mixed (in a balanced modulator) with a variable ultrasonic oscillator signal in order to translate the audio-frequency noise spectrum upward to the filter frequency. Varying the oscillator frequency causes different spectral components to appear at the filter output indicator. Since the filter is fixed, the bandwidth in hertz is constant and spectrum-level data in decibels are automatically provided (after a constant correction for filter bandwidth in hertz). This type of analyzer is very useful in the search for spectral-line components within a continuous-spectrum background.

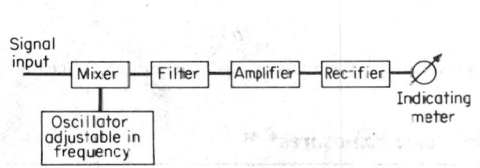

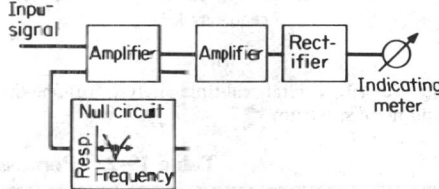

Fig. 19-43. Basic elements of a heterodyne analyzer.[38]

Fig. 19-44. Block diagram of an analyzer using an electric null circuit in a negative-feedback loop.[38]

A good compromise between fractional-octave filters and the heterodyne analyzer is the narrow-band proportional-bandwidth analyzer shown in Fig. 19-44. It uses an electric null circuit of variable frequency in a negative-feedback loop to convert an amplifier stage into a narrow-band-pass filter. Since its bandwidth is proportional to frequency, for example, 2%, the output-level data are of the same type as octave spectra, requiring −3 dB/octave tilt for conversion to spectrum level. This type of analyzer is very useful when the search is for discrete-frequency components at lower frequencies and when such narrow selectivity (in hertz) is neither needed nor desired at high audio frequencies.

Figure 19-45 compares the response-frequency characteristics of all three types of noise-spectrum analyzers in the three principal decades of the audio-frequency range.

A special analyzer[41] for magnetically recorded speech samples has also been found useful for noise. The sample recorded on a loop is played back repeatedly at high speed for heterodyne analysis. The final spectrogram is a plot of frequency vs. time, with spectrum level indicated by shading on a white-black-gray scale (see Fig. 19-18).

Digital real-time analyzers use digital filters for constant percentage bandwidth analysis on a logarithmic frequency scale (Fig. 19-46a); or use fast Fourier transform for constant bandwidth analysis on a linear frequency scale (Fig. 19-46b).[42a] Advantages include improved linearity, greater stability, the inherent capability for digital detection and digital averaging, and ease of varying filter characteristics and averaging techniques.[42b]

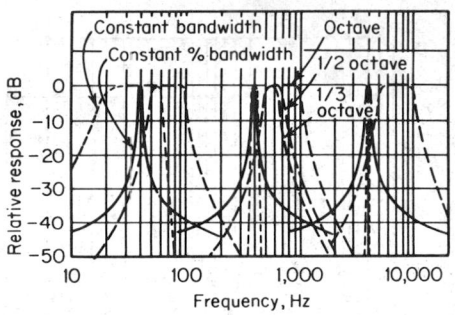

Fig. 19-45. Comparison or response curves of typical sound analyzers having constant bandwidth, constant-percentage bandwidth, octave, and fractional-octave bandwidth. *(After Ref. 39.)*

25. Hearing-Conservation Criteria. Government regulations have established permissible noise exposures for occupational safety and health on government contracts[43] and for all businesses engaged in interstate commerce.[44] Table 19-2 duplicates Table G-16 of the 1970 Act. Note that the meter is read with the *A* weighting network (see Fig. 19-39) and slow meter response.

The effect of the use of the *A* scale is to permit somewhat higher sound-pressure levels at low audio frequencies than at medium and high frequencies.

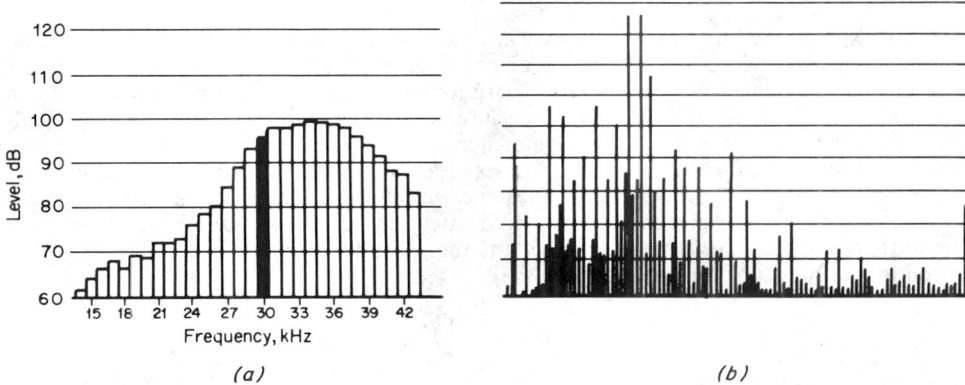

(a)   (b)

Fig. 19-46. Digital real-time analysis: (a) one-third octave-band noise analysis; (b) constant-width narrow-band noise spectrum.[42b]

Table 19-2.   Permissible Noise Exposures* [44]

Duration per Day, h	Sound Level dB(A), Slow Response
8	90
6	92
4	95
3	97
2	100
1½	102
1	105
½	110
¼ or less	115

* When the daily noise exposure is composed of two or more periods of noise exposure at different levels, their combined effect should be considered, rather than the individual effect of each. If the sum of the fractions $C_1/T_1 + C_2/T_2 + \cdots + C_n/T_n$ exceeds unity, the mixed exposure should be considered to exceed the limit value. C_n indicates the total time of exposure at a specified noise level, and T_n indicates the total time of exposure permitted at that level.

There is a separate restriction concerning impact noise peaks, which are not permitted to exceed 140 dB sound-pressure level. Impact-noise peak levels are to be measured only with an impact meter or an oscilloscope.

These regulations are based upon many years of psychoacoustic research, clinical observation, and industrial hygiene records, through the cooperation of scientific, medical, industrial, and government laboratories. The intent is to protect the hearing acuity of the population through protective legislation, industrial cooperation, and government inspection and enforcement.

26. Room-Noise Criteria. In typical background-noise spectra for rooms the octave-band sound-pressure level usually decreases with increasing frequency, for the following reasons:

1. The noise of most machinery in good operating condition is dominated physically by the lower-frequency components.

2. Intervening partitions attenuate the high-frequency noises more than the low-frequency noises.

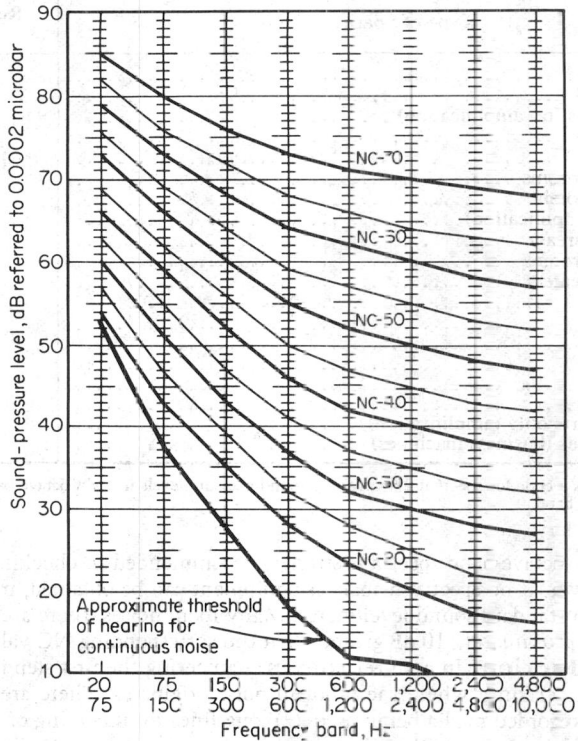

Fig. 19-47. Noise-criterion curves for rooms. Octave-band-level measurements should not exceed the specified criterion curve in any band when the unoccupied room is in normal operation, e.g., ventilation system, business machinery.[37]

3. The sound-absorption coefficient of conventional acoustical materials is greater at medium and high frequencies than at low frequencies. Thus room absorption of sound tends to accentuate the downward spectrum trends of machinery noise.

Fortunately the equal-loudness contours curve upward at low frequencies, especially at low sound levels (see Fig. 19-4). Thus the characteristics of typical building noise and of hearing-response curves tend to complement each other. This convenient (or evolutionary) relationship has led to the widely used set of noise-criterion (NC) curves in Fig. 19-47.

The NC level of the noise in a room is determined by measuring sound-pressure levels in each octave band and then comparing them to the grid of NC curves. The band having the highest interpolated NC value establishes the NC level for the entire spectrum. Consider, for example, a noise having a uniform octave-band level of 40 dB. It would have a rating of NC 43 because at 4.8 to 10 kHz its highest octave band (relative to the NC curves) lies three-fifths of the way between NC 40 and NC 45.

When NC 30 is specified for a room in the planning stage, this means that none of the octave-band levels in the completed room should lie above the NC 30 curve with all normally operating equipment in operation and with the room unoccupied.

Typical recommended noise criteria are given in Table 19-3. These are not invariant, being subject to economic necessity and varying local conditions. In some instances, e.g., offices with poorly attenuating partitions, a higher level of broadband, midfrequency noise is tolerated, or even welcomed for its masking effect upon the intelligibility of distant conversations. However, dependence upon masking to compensate for inadequate construction requires considerable experience, judgment, and care in planning and execution. The recommendations of Table 19-3 are backed by a large amount of acoustical experience and should not be compromised without expert consultation.

Table 19-3.   Typical Recommended Noise Criteria for Rooms

Type of Space	Recommended NC Curve*
Broadcast studios	15–20
Concert halls	20
Legitimate theaters (no amplification)	20–25
Music rooms	25
Schoolrooms	25
Large conference rooms	25
Apartments and hotels	25–30
Assembly halls (amplification)	25–30
Homes (sleeping areas)	25–35
Small conference rooms	30
Motion picture theaters	30
Hospitals	30
Churches	25
Courtrooms	30
Libraries	30
General offices	30–40
Restaurants	45
Coliseums for sports only (amplification)	50
Stenographic offices (business machines)	40–50

* Each NC curve is a code for specifying permissible sound-pressure levels in eight octave bands. The curve should not be exceeded in any band.

Measurement of octave-band spectra is strongly recommended in checking compliance with NC curves. However, if no spectrum-analysis equipment can be obtained, make a rough check on the A scale of a standard sound-level meter. Many room noises (there are exceptions) have a decibel A level *approximately* 10 dB greater than the corresponding NC value.

27. Noise Reduction. In audio-electronics engineering the first step in reducing noise is to determine with certainty where the problem noise originates. There are many possibilities including (1) any recorded media being used, (2) long lines for incoming or outgoing audio signals, (3) the electronic system (either the wiring or the components), (4) input transducers, (5) airborne noise generated within the room, (6) airborne noise generated elsewhere, (7) structure-borne vibration.

In general listening (by high-fidelity headset) to the noise output of an audio system while checking various source possibilities is an important analytical supplement to any visual indicators used such as meters and oscillographs. (This obvious advice is too often ignored by electronics engineers.)

System Noise.   Recorded media are easily checked by stopping them and long lines by disconnecting them temporarily (while replacing them with dummy sources of equivalent nominal impedance).

The *electronic system* probably contains the noise problem if the symptoms persist when incoming and outgoing lines are replaced by dummy-load connectors. However, oscillations from input-output coupling could be caused by wiring proximity at the connectors or lack of shielding near the connectors. Too many ground points, poorly soldered and intermittent ground connections, or lack of a good system ground are common causes of hum and television interference. After volume controls, switches, and disconnects have been tried, some judicious ac shorting of sequential signal points to ground will usually pinpoint which stage of the circuit is the noise source. Then trial component replacement is an obvious remedy.

If the noise problem disappears when the electronic system is isolated and it cannot be found in the incoming and outgoing lines by themselves, the lines may be serving as antennas for the audio electronics at higher than audio frequencies. A good high-frequency oscillograph is a necessity here because listening clues will only be by-products of the main action. Judicious use of decoupling components or networks is suggested if the problem is really at radio frequencies, with full consideration for response and phase conservation at high audio frequencies.

If a noise problem occurs only when the sounds to be recorded, transmitted, or measured are being produced in the room and the problem persists when substitute input transducers are used, try replacing the input transducers with dummy loads and producing the sounds to check for microphonics in the electronic components. These may be caused either by sound or by vibration.

Lifting the equipment or giving it a resilient mechanical support will show whether vibration is the problem. Moving the equipment outside temporarily or enclosing it in a box will check whether the excitation of the electronic components is acoustical.

Input transducers can also be overloaded by excessive sound or recorded levels. Microphones can be affected aerodynamically from air conditioning and close breathing, as well as the wind, and by building vibration transmitted through microphone stands. Microphone lines very high or very low in impedance are also susceptible to electrostatic or magnetic field pickup. Microphone-line impedance of 50 to 600 Ω is a good compromise. Balanced lines with the shield grounded only at the amplifier end are recommended for low noise.

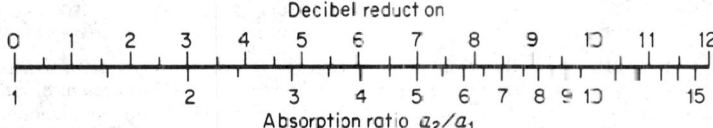

Fig. 19-48. Reduction of generally reflected sound level caused by increasing total sound absorption in the room from a_1 to a_2.[36]

Room Noise. When the noise problem is definitely in the room, not the audio-electronics system, listening is again an obvious analytical aid. If the noise source is within the room and cannot be moved out, e.g., essential operating equipment, it can be dynamically balanced or isolated mechanically to keep the floor or walls from radiating its vibration. Possibly it can be enclosed acoustically, totally or partially. If totally, the possible degree of attenuation is greatly improved. This will be discussed below under noise isolation.

Absorption of noise within a room by the installation of absorptive acoustical treatment is so well known that it is often assumed to be more effective than it is in practice. Absorptive treatment of room boundaries can be quite effective for *reflected* sound but cannot reduce *direct* sound. When the noise source is distant in an acoustically absorptive environment, the noise level decreases approximately 6 dB when the distance is doubled. This occurs in anechoic chambers and is approximated in large-area, low-ceilinged rooms with absorptive acoustical ceilings and carpeted floors. However, when the noise source is nearby, where the direct sound dominates the generally reflected sound, absorptive treatment is not very effective in noise-level reduction. It does reduce reverberation, however, and thereby reduces the spatial impression of noisiness.

The reduction of generally reflected sound in a room is given by

$$\text{Reduction} = 10 \log (a_2/a_1) \qquad \text{dB} \qquad (19\text{-}3)$$

when the total sound absorption in the room is increased from a_1 to a_2. The total absorption is obtained by adding together each surface (or object) area in the room multiplied by its respective absorption coefficient (percentage). For example, doubling the total absorption reduces the built-up noise level 3 dB. Figure 19-48 provides the reduction for other values of the absorption ratio. Bear in mind that this *does not* affect the level of the noise received *directly* without reflection.

Figure 19-49 shows the effect of absorptive treatment upon the loudness of noise buildup (after a noise source is turned on) in a previously reflective room. Reverberation after the noise is turned off is also shown.

In Fig. 19-50 both the direct noise level (dashed line) and the total (direct plus reflected) noise level (solid curves) are shown at different distances from an omnidirectional sound source. The source was arbitrarily selected to produce a direct noise level of 100 dB at a distance of 1 ft. The parameter is total sound absorption in the room in sabins. For example, at 10 ft the total noise is 20 dB higher than the direct noise when the total absorption is only 50 sabins but only 2 dB higher for 10,000 sabins.

28. Noise Isolation. When a noise source in an adjacent space is the problem (even with connecting doors and windows closed) look and listen first for sound-leakage paths. Measure their noise contribution by placing the microphone end of a small sound-level meter as close as possible to the suspected leakage openings. (Use the *A* weighting scale on the meter in this test.)

Small clearance cracks, e.g., under doors, transmit considerable amounts of noise. Their contribution can easily be checked. Close the cracks by any temporary expedient. If this makes an audible or measurable difference, a practical mechanical solution becomes worthwhile.

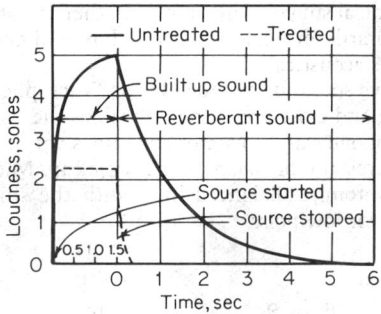

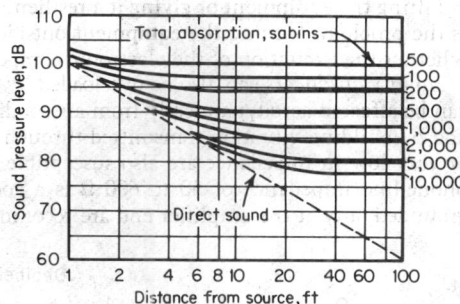

Fig. 19-49. Loudness of sound buildup and reverberant sound in a highly reflective room before and after absorptive acoustical treatment.[36]

Fig. 19-50. Comparison of total noise level with the direct noise level at different distances from the noise source with different amounts of sound absorption.[36]

Partitions.   If noise transmission by the wall itself is suspected, an easy way to check it is to rest one ear on the wall while sealing the other ear with a finger. What appears to be a solid wall, e.g., concrete block, may actually be acoustically transparent if it is highly porous.

Surface porosity can be checked by blowing into the wall with lips sealed against the wall. If air flows freely into the wall surface, it can probably be improved as an acoustical barrier by plastering or by a sealing coat of grout or thick paint. However, if the air escapes between the lips and the wall, or if a back pressure builds up, wall porosity is not the problem.

The *transmission loss* (TL) of a wall is defined by

$$TL = 10 \log (1/\tau) \quad dB \tag{19-4}$$

where τ is the percentage of the incident-sound power which is transmitted. For example, if 1% of the sound reaching a boundary passes through, the TL is 20 dB.

Figure 19-51 shows the two-room test method for measuring the TL of a partition test panel. When the receiving room is anechoic (totally nonreflective acoustically) the measured difference between sound-pressure levels L_1 and L_2 on the source and receiving sides of the panel provides the TL rating. However, when the receiving room is somewhat reverberant, the generally reflected sound will contribute (as in Fig. 19-50) to the sound level at the microphone. Then the equation for determining TL experimentally, including the room-correction term, is

$$TL = L_1 - L_2 + 10 \log (S/a) \quad dB \tag{19-5}$$

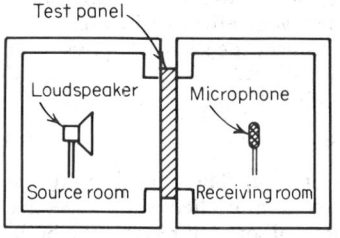

Fig. 19-51. Mounting a partition test panel between two rooms for a sound-transmission-loss test (not to scale).[36]

where S = area of test panel (ft^2) and a = total receiving room absorption (sabins).

The TL of a partition varies with the frequency of the sound, sometimes in a very complex manner. An average of the TL at a number of frequencies was long used as a single-number rating for a partition. Figure 19-52 compares this rating for a wide variety of wall, ceiling, floor, door, and window types. Also shown is a reference line for an empirical mass law, having a slope of 4.4 dB per doubling of the area density. Doors and porous wall materials generally lie below the mass-law line. Double-layer boundaries, with mechanically unbridged air space between, provide higher TL than the mass law.

A mass law also applies to the TL vs. frequency of the noise. Theoretically the TL should increase 6 dB/octave. It usually does so (see Fig. 19-53) up to where the sound at a certain angle of incidence excites a prominent flexural mode of vibration. A broad "coincidence" dip occurs above this frequency, the depth and extent depending upon wall damping and structural design.

Low stiffness and high damping maintain the mass-law trend to higher frequencies. Figure 19-53 applies to simple barriers of well-known materials. The high damping and low stiffness of thin sheet lead excel.

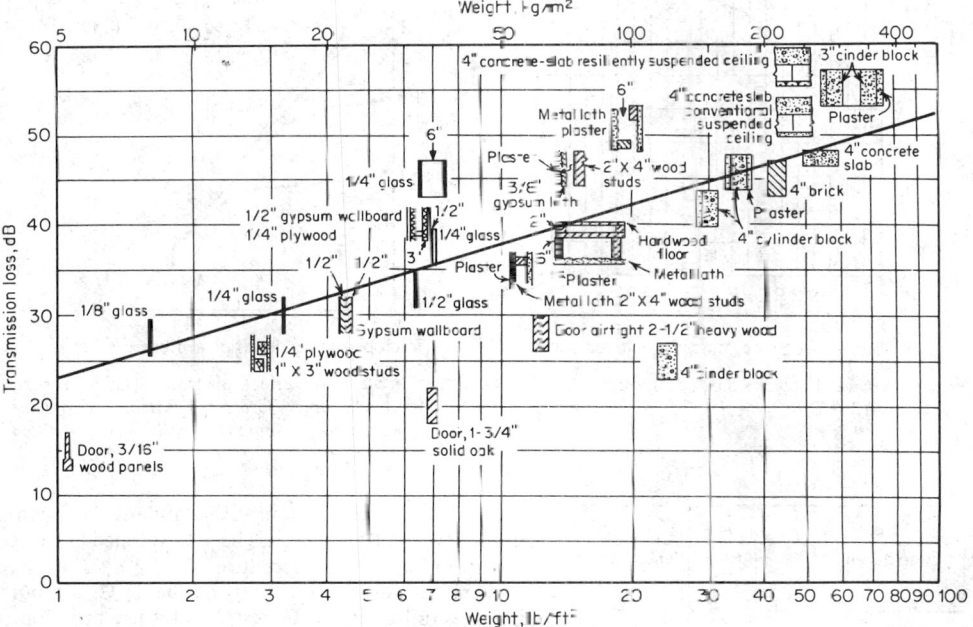

Fig. 19-52. The mass-law relation between average sound-transmission loss and mass per unit area of partition.[36]

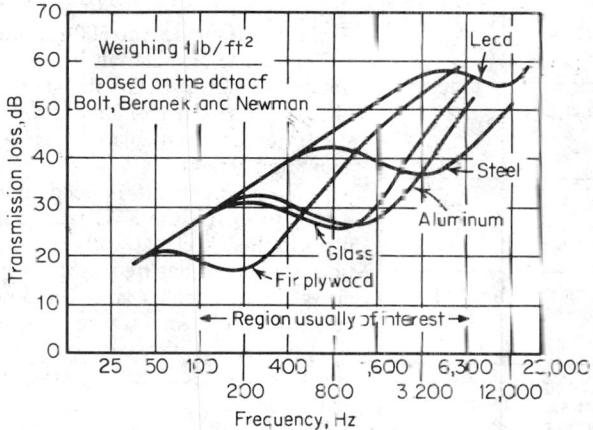

Fig. 19-53. Transmission loss of ideal single-layer barriers having equal density. (*SAE Reprint 833C, Lead Industries Association, Inc., New York, 1964.*)

The use of averaged TL data as a single-value criterion has been largely superseded in recent years by the sound-transmission-class (STC) rating, which is based upon the following considerations:

1. Large TL values at very low frequencies are uncommon and have little value because of the shape of the equal-loudness contours of hearing at low levels and frequencies (see Fig. 19-4).

2. Many boundary materials and constructions provide increasingly large TL values at high

audio frequencies. There is little additional audible advantage beyond the midfrequency range, since typical transmitted noise spectra diminish beyond inaudibility at high frequencies.

3. The most important TL values are at the central frequencies, where walls typically have TL coincidence dips in frequency bands where noise and audibility are both appreciable.

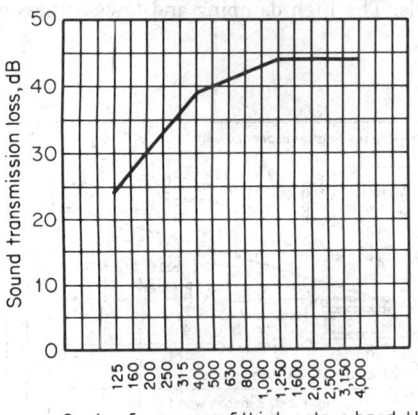

Center frequency of third-octave band, Hz

Fig. 19-54. Typical STC contour (STC-40). *(ASTM E413-70T.)*

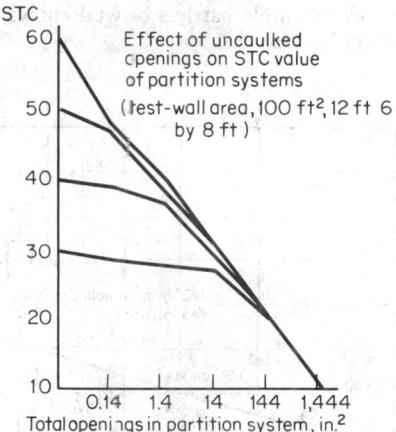

Total openings in partition system, in.²

Fig. 19-55. The effect of leaks. *(L. F. Yerges, "Sound, Noise and Vibration Control," Van Nostrand Reinhold Company, New York, 1969. Reprinted by permission.)*

Based upon these factors, the standard STC contour[45] shown in Fig. 19-54 has been developed. To rate a partition follow this procedure:

1. Place an overlay bearing the STC contour upon the 16-frequency TL test data for the partition.

2. Slide the overlay upward until the sum of the deficiencies (deviations below the contour) equals 32 dB, with no deficiency exceeding 8 dB.

3. Read the STC rating in decibels where the STC contour crosses the 500-Hz ordinate line.

The choice of materials and construction is based upon TL data (or STC values), depending upon the NC criteria selected for a room and the noise or sound spectra expected in adjacent spaces (including those above and below).

Leakage Effect. It is good practice in partition design to match the STC ratings of doors and windows to the associated walls. Tight, careful construction is assumed but frequently is not obtained, especially around doors and windows. Figure 19-55 emphasizes the importance of leakage. The greater the STC value of the partition, the greater the loss of effectiveness produced by the same amount of leakage.

Leaks can best be avoided by care during construction because they may be inaccessible after construction. Figure 19-56 shows the cumulative effect of sequential caulking lines in a multiple-layer partition.

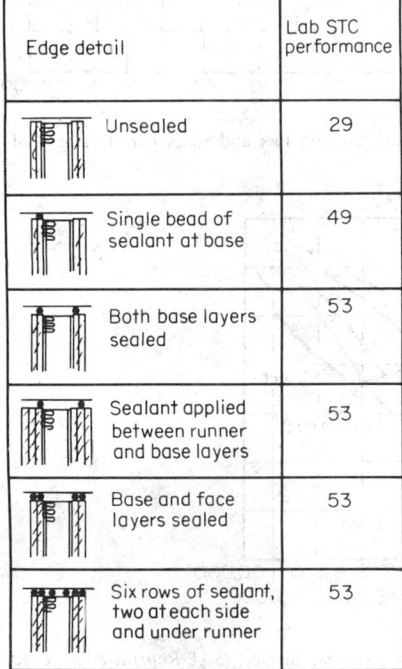

Edge detail	Lab STC performance
Unsealed	29
Single bead of sealant at base	49
Both base layers sealed	53
Sealant applied between runner and base layers	53
Base and face layers sealed	53
Six rows of sealant, two at each side and under runner	53

Fig. 19-56. The effect of caulking. *(United States Gypsum Company.)*

Ventilation Noise.[46,47]   In modern buildings the air-distribution systems often contribute significantly to ambient room noise and sometimes transmit sounds from one space to another through otherwise well-constructed partitions and floors. Figure 19-57 summarizes the acoustical essentials of a ventilation system. Return air ducts are omitted only because their quieting principles are the same as those for supply ducts. Sound travels against airflow, as well as with it.

In Fig. 19-57 the motor and blower are supported on a separate platform mechanically isolated on vibration mounts from the building slab. Canvas sleeves attenuate mechanical vibration transmitted to the plenum and the duct system. The plenum reduces acoustical noise transmission to the duct system, especially at high frequencies and if treated internally. Noise absorption also occurs at bends, especially if the duct is acoustically treated at the bends. Vanes can cause aerodynamic noise to be generated at bends if the air velocity in the duct is too high.

Splitting the duct cross section into parts can increase noise attenuation in the duct if the splitters are also acoustically lined. The duct cross section is often increased where splitting occurs to avoid constriction of the airflow. Special duct attenuators are of this expanded shape.

Many brands of duct liner (having absorptivity in the range of 0.70 to 0.90) are available for combined noise absorption along the duct length and heat insulation through the duct walls. These materials also help damp the vibration of thin duct walls responding mechanically to air turbulence within. Absorptive lining is particularly important in ducts connecting adjacent rooms, but sufficient treated-duct length and absorptive bends are also necessary. Grilles

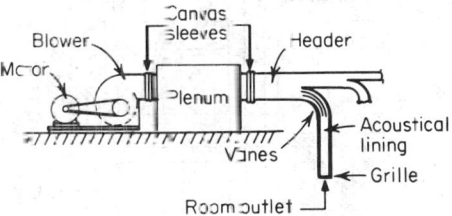

Fig. 19-57. Acoustical essentials of a typical ventilating system. [46]

and diffusers at the openings into the rooms can generate disturbing noise locally if the terminal air velocity is too high. Standards and diffuser ratings to prevent this have been developed.[48,49]

Structure-Borne Noise.[50]   Buildings can have partitions and floors with excellent STC ratings for airborne noise and still have noise problems because of shock and vibration transmitted directly through solid materials and structure. Figure 19-58 illustrates some of the ways impact noise can travel from one room to another. The source of impact in one room may be airborne

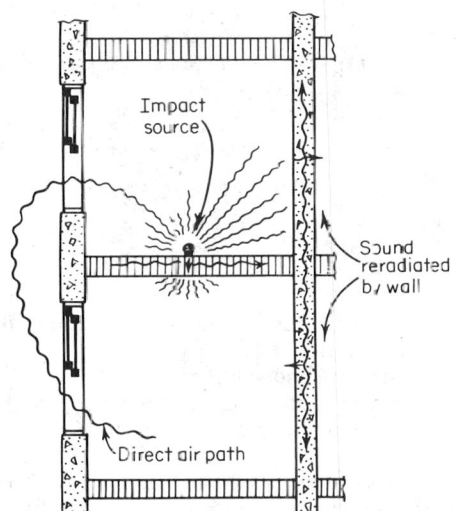

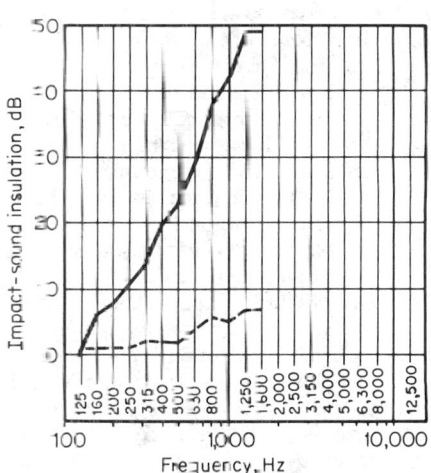

Fig. 19-58. How impact sound travels from one room to another.[50]

Fig. 19-59. Impact-sound insulation provided by two different types of floor finish on a concrete floor: (*a*) ⅛-in linoleum (*dashed line*) and (*b*) ⅜-in Wilton carpet (*solid line*).[50]

to another room through walls or open windows or by multiple reflection. More often impact noise goes through the solid material of a floor and is radiated by ceiling motion in the room below. Even when the lower room has a separate ceiling resiliently supported from the floor above, impact traveling through the structure can shake the walls of the room.

There are structural solutions, but the reduction of impact at the source is most effective. One common solution to the problem is to carpet the floor. Figure 19-59 compares the impact sound insulation of ⅛-in linoleum and ⅜-in carpet on a concrete floor slab. Floating-floor constructions are very effective where a padded surface is impractical.

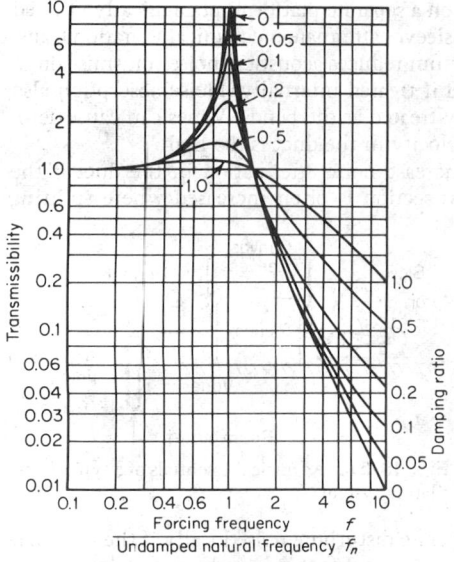

Fig. 19-60. Transmissibility of a vibrating system.

Machinery Quieting.[51]   If a machine is well balanced dynamically and relatively rigid in construction, the principles used in further noise reduction are much the same as for reduction of airborne and structure-borne noise.

To reduce the acoustical noise from purely mechanical or electromechanical machinery, reduce the motion and the area of the exposed surface or enclose it within mechanically isolated and internally acoustically absorptive walls. If the machinery is aerodynamic in operation, reduce the speed of airflow, if possible, and expand the flow cross section through tortuous paths lined with sound-absorbing materials.

Machinery vibration can be isolated from the supporting floor or walls by vibration mounts chosen to resonate (with the machinery mass) at a frequency well below the fundamental frequency of machinery rotation or reciprocation. Figure 19-60 shows the reduction of vibration transmission above the natural resonance frequency of the mass-spring supported (or suspended) system with damping as the parameter.

Several machines are often mounted on a common slab (resiliently supported on vibration isolators) in order to increase the total mass and reduce the number of isolation devices required.

ACOUSTICAL ENVIRONMENT CONTROL

29. Introduction. Acoustical needs within a room vary with its functions. Acoustical conditions in a room depend upon:
1. Internal factors
    a. Room size and shape
    b. Surface types and their locations
    c. Furnishings and their locations
    d. Sound and noise sources within the room
    e. Number and location of auditors
2. External factors
    a. Noise and vibration sources outside the room
    b. Their mechanical and acoustical isolation from the building structure and room walls
    c. The transmission characteristics of the walls and structure
    d. Service-connected equipment, e.g., lighting, heating, and ventilating

The external factors have been described previously. Internal acoustical factors should also be considered because they strongly influence the input program material to audio-electronic systems and the output sound reproduction from them.

30. Acoustical Functions of Rooms. The specialized acoustical purposes of some rooms are clearly indicated by their names, such as lecture rooms, conference rooms, symphony halls, opera houses, recording studios, motion picture theaters, anechoic chambers, reverberation chambers, and audiometric rooms. For such rooms the acoustical conditions which are optimum for the special purpose can usually be specified and obtained by various fixed solutions to the problem.

Other rooms have more varied acoustical purposes, such as convention halls, stadiums, churches, multipurpose auditoriums, and even general-purpose rooms. The optimum internal acoustics for some of the functions of such rooms must either be compromised or provided through sound controls such as variable reflectors, variable absorbers, sound-reinforcement systems, or electroacoustically enhanced reverberation.

31. Internal Acoustical Properties of Rooms. In Fig. 19-61 a pulse wave of sound pressure in a room (24 by 40 ft floor plan) is shown (a) expanding from the source, (b) reflecting from the near (front) wall, (c) reflecting from the side walls, and (d) just after the first reflection from the distant (rear) wall has returned to the source. In less than $\frac{1}{20}$ s a very simple wave in a simple rectangular room has developed a complex reflection pattern.

Room Sound Level.   In most rooms the reflected sound dominates the overall sound level except near the sound source (refer to Fig. 19-50). Figure 19-62 shows how much sound power in milliwatts is required to produce different levels of generally reflected sound, with total sound absorption in the room as the parameter.

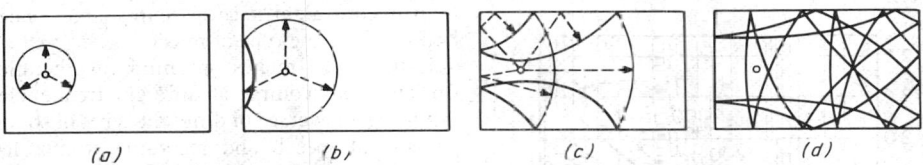

Fig. 19-61.   Progress of a single sound pulse in a closed room: (a) $\frac{2}{200}$ s (b) $\frac{1}{100}$ s, (c) $\frac{2}{40}$ s, (d) $\frac{1}{17}$ s.[36]

The sound powers are quite small. However, in sound-reproduction systems the loudspeaker efficiency ranges from the order of only 0.1% (for some small bookshelf enclosures) to 20% for highly efficient theater-type sound systems. This factor and the 12- to 15-dB peak factors needed for unclipped music reproduction lead to audio-electronic power requirements 80 to 30,000 times the acoustic powers shown.

Room Reverberation.   Reverberation in the room is caused by the room's multiple reflective acoustical properties and is the resulting tendency for sound level in the room to persist after direct sound ceases. The measure of reverberation is the time t_{60} for the sound level to decay 60 dB from its steady-state level after it is turned off. The commonly used Sabine reverberation equation is

$$t_{60} = 0.049V/a \qquad (19\text{-}6)$$

where t_{60} = reverberation time (s), V = room volume (ft^3), and a = total sound absorption in sabins (square-foot units). In metric units

$$t_{60} = 0.161V/a \qquad (19\text{-}7)$$

where V is in cubic meters and a is the total absorption in square meters. The value of the constant in the equations depends to a small degree upon the shape of the room, but this is infrequently considered.

The total sound absorption is calculated from

$$a = \Sigma\alpha_iS_i = \alpha S \qquad (19\text{-}8)$$

where α_i = absorption coefficient (percentage of incident sound absorbed) for each surface area S_i, S = total interior surface area in room, and α = average absorption coefficient of surfaces.

As an example consider (at 500 Hz) a living room 20 ft long, 13 ft wide, and 8 ft high, with a plaster ceiling ($\alpha_1 = 0.02$), a carpeted floor ($\alpha_2 = 0.30$), a wood-paneled side wall ($\alpha_3 = 0.12$), an opposite glass wall ($\alpha_4 = 0.03$), an end wall of medium drapery ($\alpha_5 = 0.40$), and a brick fireplace ($\alpha_6 = 0.02$) for the other end wall. With no additional furnishings or occupants the total sound absorption would be

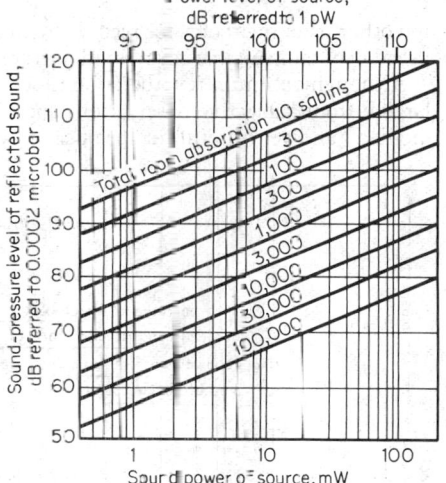

Fig. 19-62.   Sound levels for different sound powers in rooms having different total absorption. (After Ref. 36.)

$$a = (0.02 + 0.30)(260) + (0.12 + 0.03)(160) + (0.40 + 0.02)(104)$$
$$= 151 \text{ sabins} = 0.144(1,048)$$

The average absorption coefficient is 0.14.

The reverberation time at 500 Hz would be approximately

$$t_{60} = 0.049(2,080/151) = 0.68 \text{ s}$$

The optimum reverberation time for a room depends upon room volume, sound frequency, and the type of sound which is most important to the room function, e.g., conversation, recorded music, or instrumental music. Larger rooms need greater reverberation in order to reinforce the loudness of sound at typically greater distances from the sound source. Low-frequency sounds need a longer 60-dB reverberation time than medium- or high-frequency sounds in order to have equivalent audible duration of reverberation, because of the higher threshold of audibility at low frequencies (refer to Fig. 19-2). Speech intelligibility can be degraded somewhat by the same amount of reverberation needed for maximum appreciation of some types of music such as classical organ. Conductors and musicians prefer a crowded rehearsal studio to be less reverberant than an equivalent stage space in a large concert hall for the same type of music.

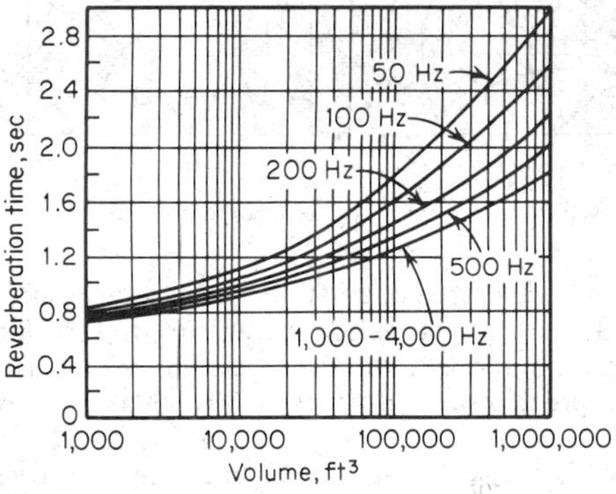

Figure 19-63 relates optimum reverberation time to room volume at different frequencies, assuming an average of different types of sound. For special speech and recording studios less reverberation is desirable. For churches and concert halls more reverberation sounds better.

The increased reverberation in Fig. 19-63 at low frequencies is also a compromise. For speech alone very little increase is desired at low frequencies, and music performance needs nearly twice as much increase as shown.

Figure 19-64 summarizes optimum reverberation (at 500 to 1,000 Hz) for different rooms. The length of the heavy dashed line allows somewhat for different room sizes typical of that function. Note the compromise for combined speech and music requirements. Cinema and other listening rooms for recorded sounds require less reverberation, since the recording will already contain suitable reverberation from the recording environment.

Fig. 19-63. Optimum reverberation for different room sizes at different frequencies (average for speech and music).[22]

Although optimum reverberation times and time-frequency characteristics have been derived largely from subjective observations and empirical evidence, they are well founded and documented[52] and very useful for specification purposes.

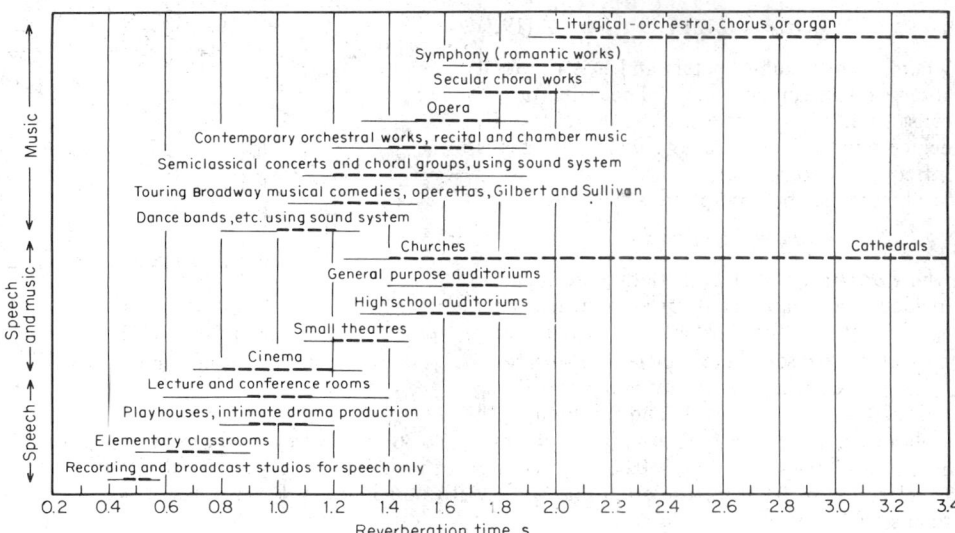

Fig. 19-64. Optimum reverberation (at 500 to 1,000 Hz) for auditoriums and similar facilities for speech and music. (*After Russell Johnson in Ref. 54.*)

Sound Distribution within the Room. Acoustical requirements for sound distribution also depend upon the room function. For example a lecture room needs outstanding one-way speech distribution from a rostrum (usually near one end of the room) to an audience seated toward the other end of the room. Figure 19-61 illustrates this case with reflective surfaces near the lecturer and a need for absorption at the opposite end to minimize reverberation.

By contrast, a conference room has many interchangeable source and receiver locations. A concert hall (Fig. 19-65) is the musical equivalent of combining the conference (a musical ensemble) with chiefly one-way communication from the performing group to the audience. A courtroom is an example having several scattered but well-defined source locations (judge, witness, attorneys) and seated groups of listeners (jury, public).

All these situations require a combination of beneficially shaped sound-reflecting surfaces near the sound source, acoustically absorptive audience areas, and acoustically absorptive or diffusing surfaces beyond the audience areas.

Room echoes are discrete, separately heard sound reflections occurring too late to provide beneficial reenforcement to the direct sound. Beneficial early reflections arrive within about 20 ms of direct-sound arrival. A concentrated echo arriving more than 50 ms late is a serious acoustical defect.

A *flutter echo* is a rapid (usually regular) succession of reflected pulses resulting from a single initial pulse.

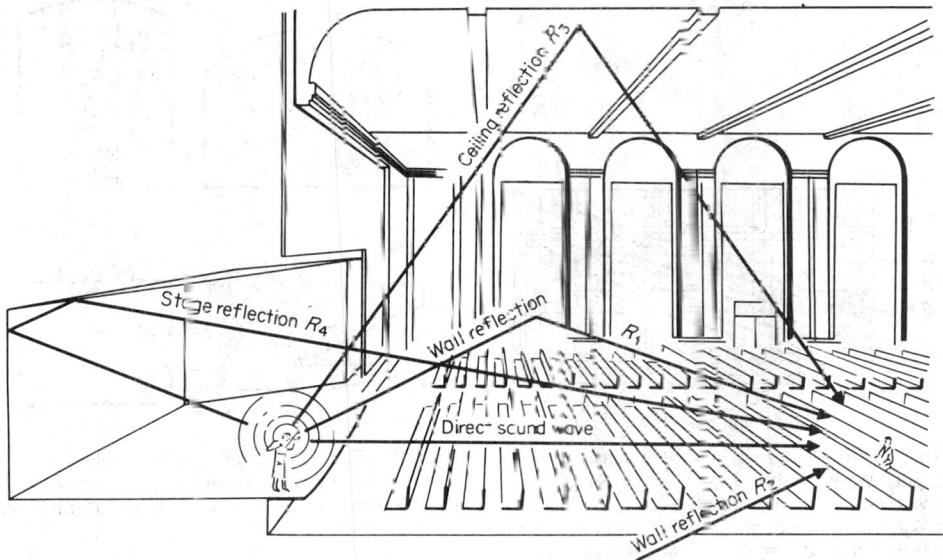

Fig. 19-65. Paths of direct and early reflected sound from an orchestral performer to a listener in a concert hall. *(After Ref. 52.)*

General reverberation, resulting from many superimposed reflections, tends to mask echo effects. Consequently the *reduction* of reverberation, without correcting the cause of an echo, can make the echo *more audible*.

32. Acoustical Shape Factors and Effects.[55] Room boundary shapes can be either beneficial or detrimental to room acoustics. Shape determines the frequency distribution of room resonances and can also produce undesirable sound-focusing effects. Detailed calculation of room resonances and of geometric focusing effects is extremely complicated, but some general understanding of them is very important to room-acoustics design and understanding.

Simple geometric shapes, e.g., spheres, cylinders, and cubes give the least complex room patterns and consequently the most obvious and undesirable acoustical resonances and poorest sound distribution. The scale of dimensions determines where in the audible frequency range the resonance patterns occur. The smaller the room the higher the frequency of the lowest-pitched resonance modes, which are the most noticeable because they are well separated on the frequency scale.

Dimension Ratios. There is no unique set of "perfect" acoustical dimension ratios for a room. However, to help prevent a poor choice and as a guide toward good practice, Fig. 19-66 is an example applicable to rectangular rooms.[56] The three room dimensions are separated by third-octave steps in order to spread the resonance frequencies uniformly on a logarithmic frequency scale. Shifts of one dimension by an octave provide for the needed variety of room types (small, average, low, long). The living room in the reverberation-calculation example above had dimension ratios chosen from the "average" set of Fig. 19-66.

Curved shapes of room boundaries can be classified broadly into convex and concave inward. *Convex* surfaces can be used near sound sources to spread reflected sound more uniformly over a wide listening area. On reflective parallel surfaces convex diffusers are sometimes used to eliminate flutter echo. Convex shapes can be used in place of absorbing material on distant rear walls of auditoriums to prevent an audible rear-wall-to-stage echo when long reverberation is desired, e.g., for music.

Concave shapes are commonly detrimental to good acoustics because they tend to create focused echoes. Figure 19-67 shows examples in which S is a sound source and S' is the point where a listener or a microphone would receive the maximum echo effect. (They are interchangeable.) These are the most difficult and expensive acoustical defects to cure because they usually require a change of surface shape. Absorptive treatment of the concave surface can reduce the echo level as much as 10 dB, but this seldom makes it inaudible enough.

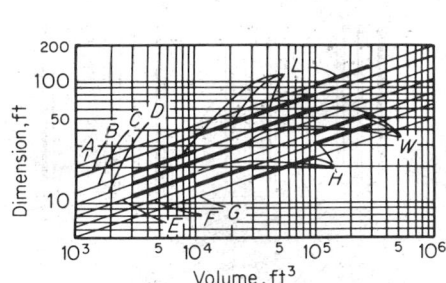

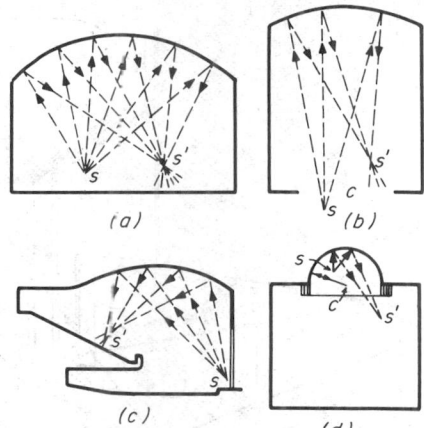

Fig. 19-66. Preferred studio dimensions. H = height, W = width, L = length. Small rooms, $H{:}W{:}L$ = $1{:}1.25{:}1.6$ = $E{:}D{:}C$. Average shape rooms $H{:}W{:}L$ = $1{:}1.6{:}2.5$ = $F{:}D{:}B$. Low-ceiling rooms, $H{:}W{:}L$ = $1{:}2.5{:}3.2$ = $G{:}C{:}B$. Long rooms, $H{:}W{:}L$ = $1{:}1.25{:}3.2$ = $F{:}E{:}A$. *(After Ref. 56.)*

Fig. 19-67. Examples of acoustical defects resulting from concave internal surfaces: (*a*) barreled ceiling; (*b*) concave rear wall; (*c*) domed ceiling; (*d*) cylindrical stage plan.[53]

In some special circumstances concave curvature is not detrimental. When a dome has a radius of curvature small in comparison to the height of its base above the listening space, its focal points will be outside the listening range. A radius of curvature large compared with room dimensions will not focus sounds from remote points, but it can produce whispering-gallery effects at grazing incidence when the source and receiver are both near the curved surface. Close scrutiny of concave curvature is always necessary in acoustical planning.

33. Absorptive Control. The principal design factor for internal acoustics of a room, in addition to shape and size, is the control of sound absorption and reflection. The total absorption required to give optimum reverberation can be calculated from Eq. (19-6) or (19-7). Decisions on the amounts and types of absorptive treatment are based upon Eq. (19-8), i.e., which surfaces are preferably absorptive or reflective, the area of these surfaces, and the selection of acoustical materials which will meet economic, visual, and operational, e.g., fireproof, criteria.

Absorption and Reflection Materials. Table 19-4 shows typical approximate absorption coefficients for a variety of general building materials at three frequencies. Materials having coefficients less than 10% are useful for reflecting sound. Most commercial absorptive materials have coefficients greater than 50%. Moderate absorbers are in the 10 to 50% range.

For more detail see data of individual manufacturers and building materials associations.

A widely used single-number coefficient for commercial acoustical materials is the *noise reduction coefficient* (NRC), which averages the coefficients at 250, 500, 1,000, and 2,000 Hz.

Sound absorption is largely a result of viscous air damping in porous materials, mechanical damping in acoustically driven diaphragms or panels, acoustical damping within acoustical resonators, or combinations of these.

Absorptivity by a porous material is very dependent at low frequencies upon its spacing from

Table 19-4.   Sound-Absorption Coefficients for General Building Materials

Material	125 Hz	500 Hz	2,000 Hz
Brick, unglazed	0.03	0.03	0.05
Painted	0.01	0.02	0.02
Carpet, heavy, on concrete	0.02	0.15	0.60
On heavy padding	0.08	0.50	0.70
Concrete block, porous, unpainted	0.35	0.30	0.40
Pores painted full	0.10	0.10	0.10
Concrete, poured	0.01	0.02	0.02
Covered by floor tile	0.02	0.03	0.03
Floor, wood	0.15	0.0	0.07
Glass, heavy plate	0.18	0.04	0.02
Window	0.35	0.18	0.07
Gypsum board, ½-in. on 2 by 4 in., 16 in. o.c.	0.30	0.05	0.07
Plaster, gypsum or lime, smooth	0.02	0.03	0.04
Plywood paneling, ⅜-in. thick	0.25	0.15	0.10

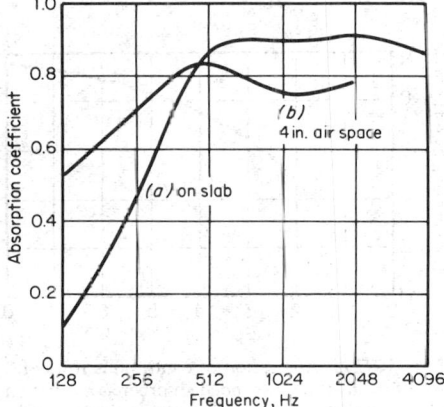

Fig. 19-68. Absorption of a 1-in rock-wool blanket (a) mounted against reflective ceiling and (b) suspended 4 in. below.[53]

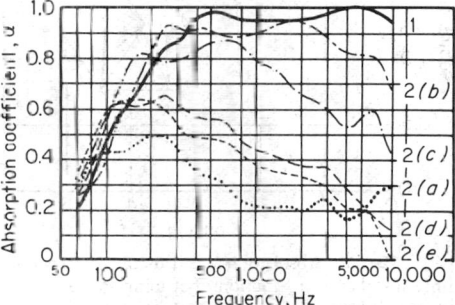

Fig. 19-69. Variation of absorption with spacing of holes in a sheet-metal cover: (1) 3-in rock wool; (2) 3-in rock wool covered with 22 BWG steel sheet (a) unperforated, (b) ⅜-in holes at ⁵⁄₁₆-in centers; (c) ⅜-in centers; (d) 1¼-in centers; (e) 1⅞-in centers. (*From National Physical Laboratories, England.*)

rigid boundaries, where particle velocity is too low for viscous absorption. Figure 19-68 shows the importance of air space behind the material.

Porous absorbing materials may be covered, for mechanical protection or appearance, by thin highly perforated facings without much loss in absorptivity. However, the holes must be closely spaced, as shown in Fig. 19-69.

Absorptive Structures. A widely used mechanical absorber is plywood paneling with air space (Fig. 19-70). This low-frequency absorption can be used to balance or supplement the greater absorption of porous materials at the higher frequencies. Rooms paneled throughout may be lacking in low-frequency reverberation (see Fig. 19-63).

Absorptive structures can also be suspended away from room boundaries for greater absorptive efficiency when it is acceptable visually. Figures 19-71 and 19-72 give the absorption in sabins per structural unit for two different types of porous absorbers. Figure 19-71 refers to a circularly symmetrical unit formed of two cones face to face. Figure 19-72 is for flat 2 by 4 ft absorptive baffles. Note that scattering the units increases absorption per unit and that exposing two sides improves total absorption.

Acoustical resonators are another type of

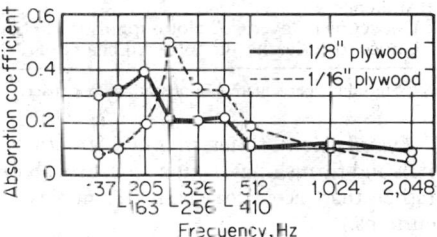

Fig. 19-70. Absorption characteristics of plywood panels, 2 by 9 ft with transverse braces 3 ft apart, with 2¾-in air space. (*From P. Sabine and L. Ramer in Ref. 53.*)

absorptive structure. They can be built into room walls, e.g., hollow tile with circular or slot openings into the room. They can be room furnishings (vases or bottles) but this will seldom be of practical significance. They are best used for selective absorption at their acoustical resonance frequencies.

Seating and Audience Absorption. The audience is one of the largest absorbers of sound in an auditorium. Audience size variation can be the most variable part of the total room absorption. In order to stabilize room acoustics while the audience varies from none to capacity, acousticians often recommend auditorium chairs with upholstered back rests and seat cushions. A person occupying this type of chair covers (subtracts) most of the chair absorption and adds his own absorption. The net increase depends upon his size and attire and the acoustical rating of the chair.

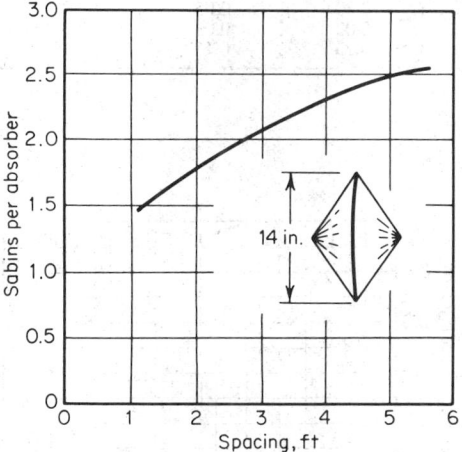

Fig. 19-71. Absorption of illustrated suspended unit absorber as a function of spacing. Absorption is averaged from 250 to 2,000 Hz. *(Johns-Manville Research Center in Ref. 36.)*

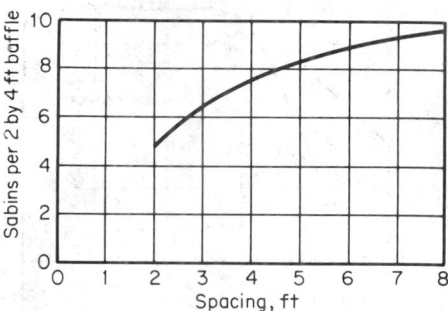

Fig. 19-72. Absorption of continuous rows of 2 by 4 ft baffle-type suspended absorber as a function of spacing. Absorption is averaged from 250 to 2,000 Hz. *(Owens-Corning Fiberglas Corp. in Ref. 36.)*

Published acoustical data on upholstered chairs are useful for chair comparisons, but the values are often larger (in sabins per chair) than the effective absorption in an auditorium installation,[57] for the same reason that spacing affects absorption in Figs. 19-71 and 19-72.

Studies of audience and seating absorption in large halls have led to a recent practice of calculating seating-area absorption, with or without audience, on the basis of sabins per square foot of seating floor space. Evidently the same number of chairs and auditors absorb more sound when spread out than when compactly arranged. Table 19-5 compares seating-area coefficients with audience and without, the latter for two typical chair coverings.[58]

Table 19-5. Sound-Absorption Coefficients* for Audience and Seating Areas

Condition	125 Hz	500 Hz	2,000 Hz
Audience	0.60	0.88	0.93
Unoccupied "average" cloth-upholstered chairs	0.49	0.80	0.82
Unoccupied leather-upholstered chairs	0.44	0.60	0.58

* In sabins per square foot of seating floor space.

Distribution of Materials and Structures. The location and distribution of functional materials and furnishings having acoustical absorptive properties are often determined on a functional rather than acoustical basis. (Examples are seating, carpeting, window draperies, and stage curtains.)

Before deciding where to put supplementary absorption (if any), at least the following factors should be considered:

    1. Acoustical functions of the room
    2. Optimum reverberation time for the functions

3. Probable location of sound sources (talkers, musical instruments, loudspeakers, noisy equipment) and receivers (listeners, microphones)

4. The room shape planned

5. Whether electronic sound reenforcement or reproduction is planned

If the goal is acoustical privacy for individuals or small groups within a large area (library, open-plan school, public dining area), absorptive treatment of the ceiling, the floor, or both is important, and wall treatment is relatively unimportant.

If the goal is airborne communication, either one-way or two-way, an acoustically reflective ceiling is necessary when the sources and receivers are separated very much. Near source locations the walls should usually be reflective.
Near receiver locations wall absorption is desirable. When the communication is definitely one-way, as in a typical lecture room or theater, efficient rear-wall absorption comes first, then rear side walls, and rear ceiling, in that order, if necessary for reverberation control. When opposite surfaces are parallel, the absorptive material can advantageously be distributed or installed in a staggered manner to prevent flutter echo. In a classroom or large conference room some ceiling treatment near the edges provides reverberation control, allowing a central reflective ceiling to aid two-way communication.

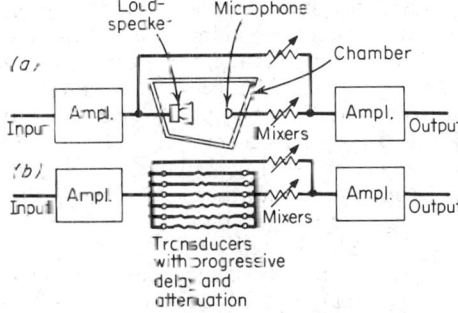

Fig. 19-73. Reverberation-enhancement systems: (a) reverberation chamber; (b delay lines. ("Music, Physics and Engineering," Dover, New York, 1967.)

Because rooms have some acoustical modes of vibration that lie in one direction or plane, it is unwise to concentrate all the sound absorption in one place or along one axis of the room. This allows some isolated modes to reverberate while all the rest are properly damped. A good check is to compare the absorption coefficient averaged over each pair of opposite surfaces (ceiling and floor, front and rear, side walls).

All these suggestions are general and not necessarily the answer to a specific acoustical problem.

Variable Absorbers. In multipurpose rooms variable sound absorbers can provide optimum acoustical conditions for each purpose. However, ease of control is a necessity, and the operation must be understood. Otherwise a good fixed compromise is preferable to random control.

Mechanisms for variable absorption include absorptive areas with reflective hinged covers or rotating shutters, absorptive draperies which can be extended from slot openings, absorptive blankets which roll up into reflective enclosures, and rotating panels with reflective and absorptive sides.

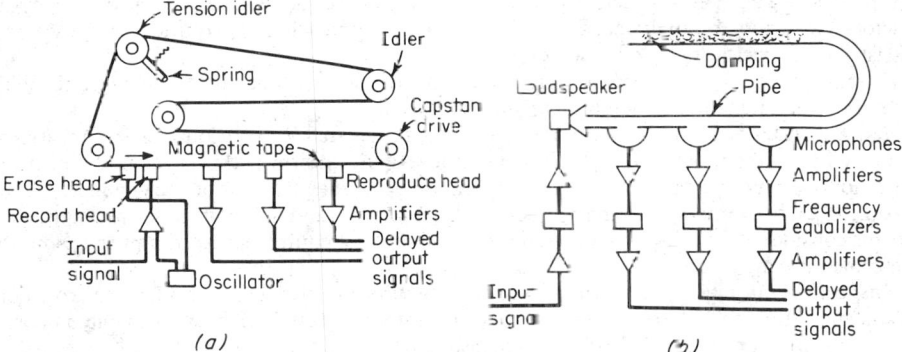

Fig. 19-74. Audio delay systems: (a) magnetic tape; (b) electroacoustic line. ("Music, Physics and Engineering," Dover, New York, 1967.)

Electronic Reverberation Systems. Recordings made in studios lacking sufficient reverberations are often enhanced by the addition of reverberation Figure 19-73 shows two methods. In Fig. 19-73a the audio signal to be recorded or rerecorded is reproduced in a reverberant chamber, picked up, and mixed with direct signal. In Fig. 19-73b a parallel group of delay lines substitutes for the room. Figure 19-74 shows magnetic-tape loop and acoustic-tube delay systems. Metallic plates and springs are also used with appropriate input and output transducers, but all-electronic bucket-brigade delay lines[59a] predominate.

The same electronic systems can be used with an extensive loudspeaker system to enhance the natural reverberation of a room. The loudspeakers must be well distributed over the normally reflective surfaces of the room and oriented for diffuse rather than directional radiation, for the additional reverberation to blend with the natural reverberation.

Acoustical resonators, each containing a microphone and loudspeaker in a stable, controlled acoustical-feedback relation, can be tuned to different frequency ranges and positioned in various parts of an auditorium to extend the natural reverberation time.

34. Room-Acoustics Specifications and Predictions. The complex theories of physical acoustics, the uncertainties in predicting variations in the properties of both natural and manufactured building materials, and the difficulty in controlling installation procedures such as painting, sealing, bracing, furring, and draping all keep the specification and prediction of room acoustics from being an exact science. The frequent need to evaluate the final result by listening rather than by purely quantitative methods gives acoustical planning a somewhat subjective aspect. This is not unique. The solutions to many engineering problems involve human experience and statistical factors.

Examples of *building acoustical properties* often specified include:

1. Noise criteria
2. STC ratings of partitions, doors, and windows
3. Precautions concerning leakage, bridging of double walls, and machinery isolation
4. Ductwork ratings
5. Structural isolation requirements

Examples of *room acoustical properties* often specified include:

1. Reverberation-time goals at various frequencies, room unoccupied and occupied
2. Room size and shape needed for the acoustical functions
3. Room volume needed to obtain the maximum reverberation specified
4. Shape requirements for sound diffusion, for good acoustical dimension ratios (in small rooms) and good sound-level distribution (in large rooms)
5. Shaping precautions to avoid sound focusing and flutter echo
6. Types, amounts, and locations of seating, absorptive and reflective surface materials, and structures expected to fulfill the acoustical goals
7. Means proposed for any final adjustment after construction where the requirements are critical

The adjustment means can be variable absorption or reflection devices, or if variation is not desired, a reserve amount of material, e.g., draperies, which can be omitted or added, depending upon the final result.

A comprehensive and creative study of the dimensional and acoustical properties of concert halls and opera houses[52] includes rating scales intended for correlation of measurable physical properties with expert subjective judgment. Both positive factors, e.g., liveness, and negative factors, e.g., echo, are included. Experience with this approach and evolution of the scales will assist future specification and prediction.[59b]

Electroacoustic system specifications are often included with acoustical specifications. They are discussed later under microphones and loudspeakers.

35. Acoustical Engineering Tests. Tests suggested *before* acoustical specifications are prepared for a room or building include noise levels and spectra at the location of construction or renovation (with noise sources operating), noise data on equipment of the general types to be installed, and reverberation-time frequency characteristics and ambient-noise spectra in similar rooms considered satisfactory by the owner or client. Test equipment for the purposes has been described in Par. **19-22**.

Inspection must be carried out *during construction* for inadvertent bridging (across double walls and vibration isolators), acoustical leakage, improper installation of sound-rated doors and windows, insecure mounting of panels (causing subsequent rattle), omission of the acoustical lining of ducts, and any other items hidden during final inspection. The porosity of acoustical plasters and concrete block should be checked as they are first installed so that materials and installation techniques can be corrected early. Interim acoustical tests of sound-transmission loss and reverberation time (while construction work is shut down) can advantageously be made on completed sample rooms or on incomplete rooms if allowances are made.

Final acoustical tests include compliance with specifications on noise criteria, STC ratings, reverberation characteristics, sound-level distribution, and the effect of any variable absorption.

Echoes are usually detected by listening while traversing the room with a mechanical-impulse generator. Oscillographic methods are helpful in their identification. Maintenance instructions are useful, e.g., on the painting of porous materials the adjustment and repair of sound-rated door mechanisms, and the caulking of any separation or cracks which may occur during settling or use. Operation instructions or schedules are needed for optimum use of variable absorbers.

Articulation Tests.[60a] Both rooms and sound systems can be given a final test by articulation methods. These tests use talkers, listeners, standardized word lists, and simple procedures. However, the results depend somewhat upon talker and listener selection, ability, and training. The articulation score is the average percentage of words correctly heard by all listeners for all talkers. Phonetic spelling is more important than dictionary spelling.

Each word is spoken in a carrier sentence such as "You will write ——— now." An example of a phonetically balanced list of 50 words of one syllable follows: ask, bid, bind, bolt, bored, calf, catch, chant, chew, clod, cod, crack, day, deuce, dumb, each, ease, fad, flip, food, forth, freak, frock, front, guess, hum, jell, kill, left, lick, loot, night, pint, queen, rest, rhyme, rod, roll, rope, rot, shack, slide, spice, this, thread, till, us, wheeze, wig, yeast. Additional lists and rearrangement of words are necessary to prevent memorization.

Speech Interference Level.[60b] An alternative is to use a predictive method for rating the expected speech-interfering aspects of noise, based upon acoustical measurements of the noise.

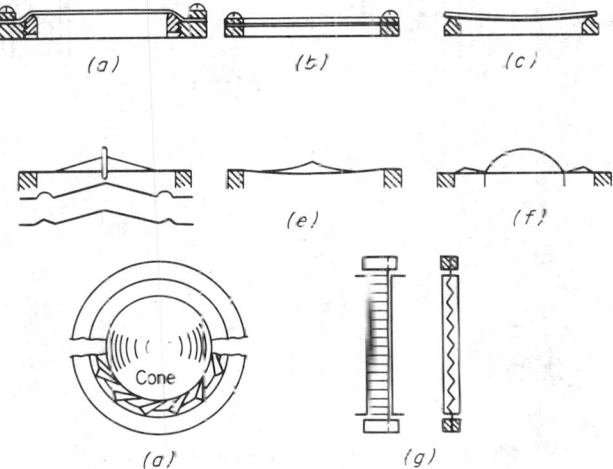

Fig. 19-75. Sound-responsive elements in microphones.[61]

MICROPHONES AND ACCESSORIES[61,62,31]

36. Sound-Responsive Elements. The sound-responsive element in a microphone may have many forms (Fig. 19-75). It may be a stretched membrane (a), a clamped diaphragm, (b), or a magnetic diaphragm held in place by magnetic attraction (c). In these the moving element is either an electric or magnetic conductor, and the motion of the element creates the electric or magnetic equivalent of the sound directly.

Other sound-responsive elements are straight (d) or curved (e) conical diaphragms with various shapes of annular compliance rings, as shown. The motion of these diaphragms is transmitted by a drive rod from the conical tip to a mechanical transducer below.

Other widely used elements are a circular piston (f) bearing a circular voice coil of smaller diameter and a corrugated-ribbon conductor (g) of extremely low mass and stiffness suspended in a magnetic field.

37. Transduction Methods. Microphones have a greater variety of transduction methods currently in use than other types of electroacoustic and electromechanical transducers. The variety is shown in Fig. 19-76.

The loose-contact transducer (Fig. 19-76a) was the first achieved by Bell in magnetic form and later made practical by Edison's use of carbonized hard-coal particles. It is widely used in telephones. Its chief advantage is its self-amplifying function, in which diaphragm amplitude variations directly produce electric resistance and current variations. Disadvantages include noise, distortion, and instability.

Moving-iron transducers have great variety, ranging from the historic pivoted armature (Fig. 19-76b) to the modern ring armature driven by a nonmagnetic diaphragm (Fig. 19-76h). In all these types a coil surrounds some portion of the magnetic circuit. The reluctance of the magnetic circuit is varied by motion of the sound-responsive element, which is either moving iron itself (Fig. 19-76c and d) or is coupled mechanically to the moving iron (Fig. 19-76b, e to h). In some of the magnetic circuits (Fig. 19-76e to h) that portion of the armature surrounded by the coil carries very little steady flux, operating on differential magnetic flux only. Output voltage is proportional to moving-iron velocity.

Electrostatic transducers (Fig. 19-76i) use a polarizing potential and depend upon capacitance variation between the moving diaphragm and a fixed electrode for generation of a corresponding potential difference. The *electret microphone* is a special type of electrostatic microphone which holds polarization indefinitely without continued application of a polarizing potential, an important practical advantage for many applications.[63a,b]

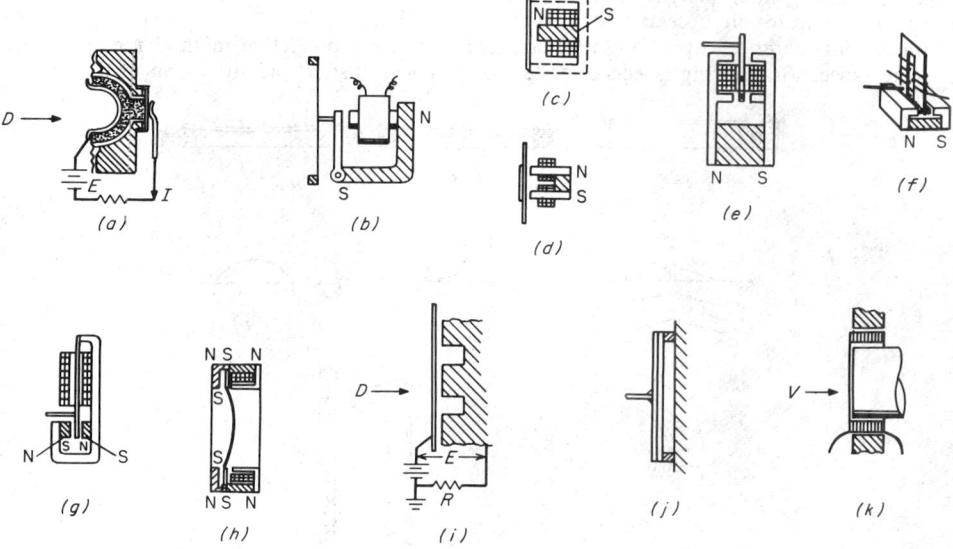

Fig. 19-76. Microphone transduction methods.[61]

Piezoelectric transducers (Fig. 19-76j) create an alternating potential through the flexing of crystalline elements which, when deformed, generate a charge difference proportional to the deformation on opposite surfaces. Because of climatic effects and high electric impedance the rochelle salt commonly used for many years has been superseded by polycrystalline ceramic elements and by piezoelectric polymer.[63c,d]

Moving-coil transducers (Fig. 19-76k) generate potential by oscillation of the coil within a uniform magnetic field. The output potential is proportional to coil velocity.

38. Equivalent Circuits.[64] Electronics engineers understand electroacoustic and electromechanical design better with the help of equivalent or analogous electric circuits. Microphone design provides an ideal base for introduction of equivalent circuits because microphone dimensions are small compared with acoustical wavelengths over most of the audio-frequency range. This allows the assumption of lumped circuit constants.

Figure 19-77 shows equivalent symbols for the three basic elements of electrical, acoustical, and mechanical systems. In acoustical circuits the resistance is air friction or viscosity, which occurs in porous materials or narrow slots. Radiation resistance is another form of acoustical damping. Mechanical resistance is friction. Mass in the mechanical system is analogous to electric inductance. The acoustical equivalent is the mass of air in an opening or constriction divided by the square of its cross-sectional area. The acoustical analog of electric capacitance and mechanical-spring compliance is acoustical capacitance. It is the inverse of the stiffness of an enclosed volume of air under pistonlike action. Acoustical capacitance is proportional to the volume enclosed.

Figure 19-78 is an equivalent electric circuit for a Helmholtz resonator. Sound-pressure and air-volume current are analogous to electric potential and current, respectively. Other analog systems have been proposed. One frequently used has advantages for mechanical systems.[65]

39. Microphone Types and Equivalent Circuits. Different types of microphone respond to different properties of the acoustical *input* wave. Moreover, the electric *output* can be proportional to different internal mechanical variables.

Pressure Type, Displacement Response. Figure 19-79 shows a microphone responsive to the sound-pressure wave acting through a resonant acoustical circuit upon a resonant diaphragm coupled to a piezoelectric element responsive to displacement. (The absence of sound ports in

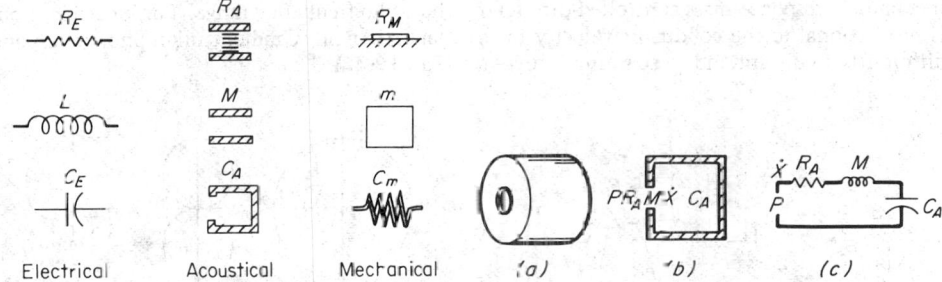

Electrical	Acoustical	Mechanical	(a)	(b)	(c)

Fig. 19-77. Equivalent basic elements in electrical, acoustical, and mechanical systems.[21]

Fig. 19-78. Helmholtz resonator in (a) perspective and (b) in section and (c) equivalent electric circuit.[22]

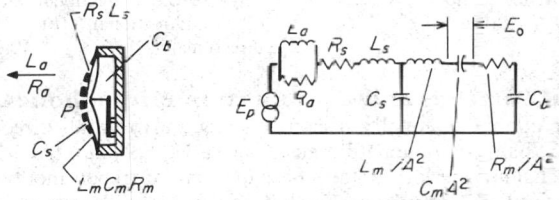

Fig. 19-79. Pressure microphone, displacement response.[61]

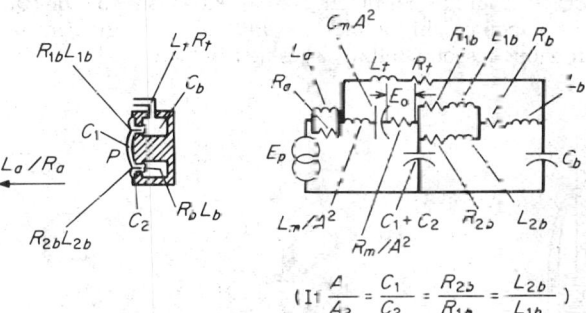

$$\left(\text{If } \frac{L}{A_2} = \frac{C_1}{C_2} = \frac{R_{2b}}{R_{1b}} = \frac{L_{2b}}{L_{1b}} \right)$$

Fig. 19-80. Pressure microphone, velocity response.[61]

the case or in the diaphragm keeps the microphone pressure responsive.) In the equivalent circuit the sound pressure is the generator. L_a and R_c represent the radiation impedance. L_s and R_s are the inertance and acoustical resistance of the holes; C_s is the capacitance of the volume in front of the diaphragm; L_m, C_m, and R_m are the mass, compliance, and resistance of the piezoelectric element and diaphragm lumped together and C_b is the capacitance of the entrapped back volume of air. The electric output is the potential differential across the piezoelectric element. It is shown across the capacitance in the equivalent circuit because microphones of this type are designed to be stiffness-controlled throughout most of their operating range.

Pressure Type, Velocity Response. Figure 19-80 shows a moving-coil pressure microphone which is a velocity-responsive transducer. In this microphone three acoustical circuits lie behind the diaphragm. One is behind the dome and another behind the annular rings. The third acoustical circuit lies beyond the acoustical resistance at the back of the voice-coil gap and includes a leak from the back chamber to the outside. This microphone is resistance-controlled throughout most of the range, but at low frequencies its response is extended by the resonance of the third acoustical circuit. Output potential is proportional to the velocity of voice-coil motion.

Pressure-Gradient Type, Velocity Response. When both sides of the sound-responsive element are open to the sound wave, the response is proportional to the *gradient* of the pressure wave. Figure 19-81 shows a ribbon conductor in a magnetic field with both sides of the ribbon open to the air. In the equivalent circuit there are two generators, one for sound pressure on each side. Radiation resistance and reactance are in series with each generator and the circuit constants of the ribbon. Usually the ribbon resonates at a very low frequency, making its mechanical response mass-controlled throughout the audio-frequency range. The electric output is proportional to the conductor velocity in the magnetic field. Gradient microphones respond differently to distant and close sound sources (see Par. **19-42**).

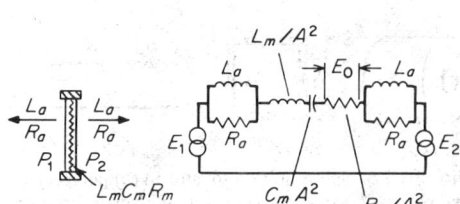

Fig. **19-81.** Gradient microphone, velocity response.[61]

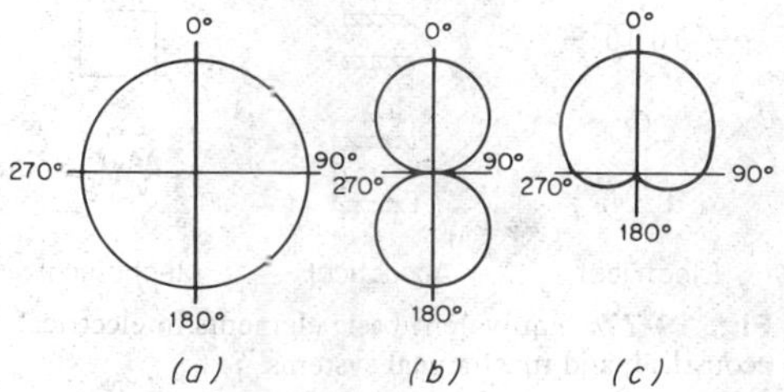

Fig. **19-82.** Directional patterns of microphones: (*a*) nondirectional, (*b*) bidirectional, (*c*) unidirectional.[22]

40. Directional Patterns and Combination Microphones. Because of diffraction, a pressure microphone is equally responsive to sound from all directions as long as the wavelength is larger than microphone dimensions (see Fig. 19-32a). (At high frequencies it is somewhat directional along the forward axis of diaphragm or ribbon motion.)

By contrast a pressure-gradient microphone has a figure-eight directional pattern (Fig. 19-82b), which rotates about the axis of ribbon or diaphragm motion. A sound wave approaching a gradient microphone at 90° from the axis produces balanced pressure on the two sides of the ribbon and consequently no response. This defines the *null plane* of a gradient microphone. Outside this plane the microphone response follows a cosine law.

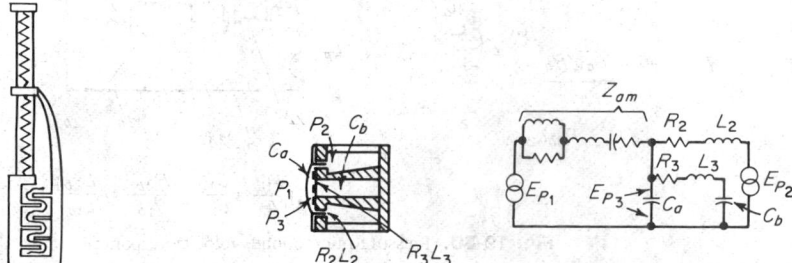

Fig. **19-83.** Combination unidirectional microphone.[61]

Fig. **19-84.** Phase-shift unidirectional microphone.[61]

If the pressure and gradient microphones are combined in close proximity (see Fig. 19-83) and are connected electrically to add in equal (half-and-half) proportions, a heart-shaped cardioid pattern (Fig. 19-82c) is obtained. (The back of the ribbon in the pressure microphone is loaded by an acoustical resistance line.) By combining the two outputs in other proportions other limaçon directional patterns can be obtained.

41. Phase-Shift Directional Microphones. Directional characteristics similar to those of the combination microphones can also be obtained with a single moving element by means of equivalent circuit analysis using acoustical phase-shift networks. Figure 19-84 shows a moving-coil, phase-shift microphone and its simplified equivalent circuit. The phase-shift network is composed of the rear-port resistance R_2 and inertance L_2, the capacitance of the volume under the diaphragm and within the magnet, and the impedance of the interconnecting screen. The microphone has a cardioid directional pattern.

42. Special-purpose microphones include two types which are superdirectional, two which overcome noise, wearable microphones, and one without cables.

Line microphones (Fig. 19-85) use an approximate line of equally spaced pickup points connected through acoustically damped tubes to a common microphone diaphragm. The phase relationships at these points for an incident plane wave combine to give a sharply directional pattern along the axis if the line segment is at least one wavelength.

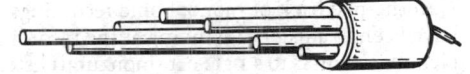

Parabolic microphones face a pressure micro-phone unit toward a parabolic reflector at its focal point, where sounds from distant sources

Fig. 19-85. Line microphone. ("*Music, Physics and Engineering,*" Dover, New York, 1967.)

along the axis of the parabola converge. They are effective for all wavelengths smaller than the diameter of the reflector.

Noise-canceling microphones[66a] are gradient microphones in which the mechanical system is designed to be stiffness-controlled rather than mass-controlled. For distant sound sources the resulting response is greatly attenuated at low frequencies (see Fig. 9-86). However, for a very close sound source the response-frequency characteristic is uniform because the *gradient* of the pressure wave near a point source decreases with increasing frequency. Such a microphone provides considerable advantage for nearby speech over distant noise on the axis of the microphone. In addition, there is an advantage from the effect of the figure-eight directional pattern. Figure 19-87 shows the combined advantage of on-axis close speech over distant randomly incident noise.

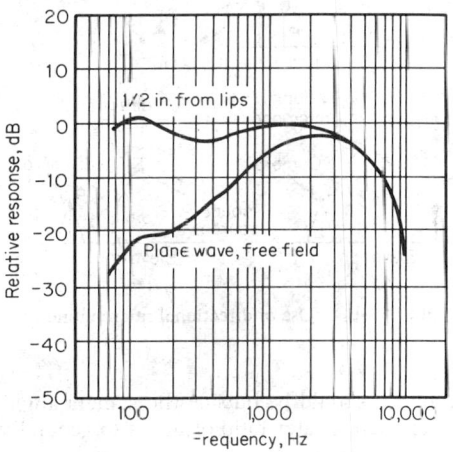
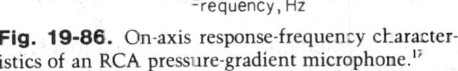

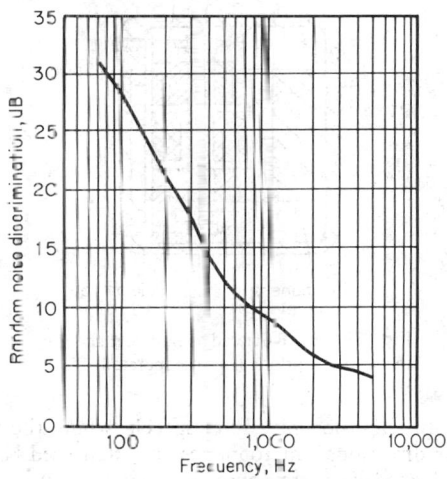

Fig. 19-86. On-axis response-frequency character-istics of an RCA pressure-gradient microphone.[17]

Fig. 19-87. Random-noise discrimination of the microphone in Fig. 9-86.[17]

Contact microphones are used on string and percussion musical instruments, on seismic-vibration detectors, and for pickup of body vibrations including speech. The throat microphone was noted for its convenience and its rejection of airborne noise. Most types of throat microphone are inertia-operated (Fig. 19-88), the case receiving vibration from the throat walls actuated by speech sound pressure in the throat. The disadvantage is a deficiency of speech sibilant sounds received back in the throat from the mouth.[66b]

Lavalier microphones are worn on the chest, suspended by a cord around the speaker's neck. Tie-clasp microphones are worn on the tie. They allow the wearer great flexibility of movement without concern for microphone location. Their fixed position on the talker and their proximity to the speech source improve acoustical feedback margin.

Wireless microphones have obvious operational advantages over those with microphone cords. A wireless microphone contains a small, low-power radio transmitter with a nearby receiver nected to an audio communication system. Any of the microphone types can be so equipped. The potential disadvantage is in rf interference and field effects.

43. Microphone Use in Recordings.[67a] The choice of microphone type and placement greatly affects the sound of a recording. For speech and dialogue recordings pressure microphones are usually placed near the speakers in order to minimize ambient-noise pickup and room reverberation. Remote pressure microphones are also used when a maximum room effect is desired. The farther from the sound source the less direct the pickup and the greater the microphone pickup of generally reflected sound.

In the playback of monophonic recordings room effects are more noticeable than they would have been to a listener standing at the recording microphone position because single-microphone pickup is similar to single-ear (monaural) listening, in which the directional clues of localization are lost. Therefore microphones generally need to be closer in a monophonic recording than in a stereophonic recording.

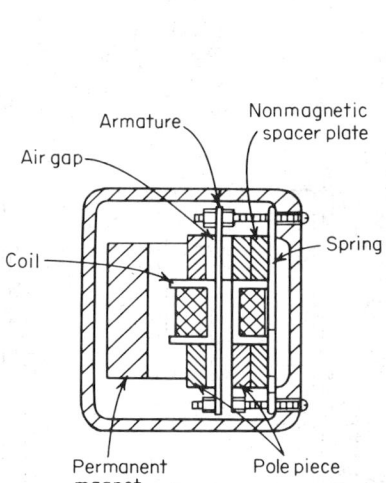

Fig. 19-88. Balanced-armature magnetic throat microphone (inertia-operated).[66b]

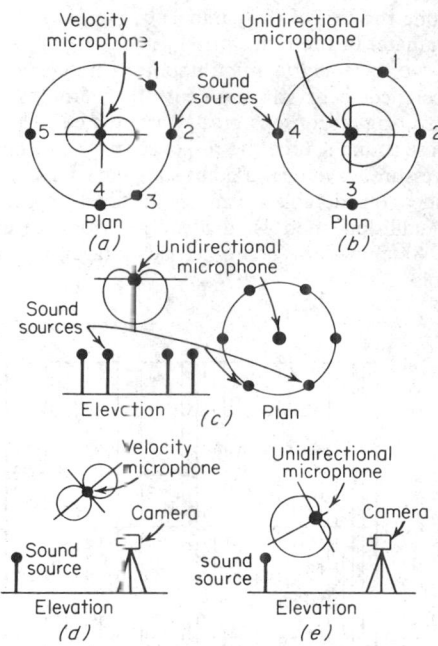

Fig. 19-89. Use of directional microphones.[22]

In television pickup of speech, where the microphone should be outside the camera angle, unidirectional microphones are often used because of their greater ratio of direct to generally reflected sound response.

Both velocity (gradient) microphones and unidirectional microphones can be used to advantage in broadcasting and recording. Figure 19-89a shows how instruments may be placed around a figure-eight directivity pattern to balance weaker instruments 2 and 5 against stronger instruments 1 and 3 with a potential noise source at point 4. In Fig. 19-89b source 2 is favored, with sources 1 and 3 somewhat reduced and source 4 highly discriminated against by the cardioid directional pattern. In Fig. 19-89c an elevated unidirectional microphone aimed downward responds uniformly to sources on a circle around the axis while discriminating against mechanical noises at ceiling level. Figure 19-89d places the camera noise in the null plane of a figure-eight pattern, and Fig. 19-89e shows a similar use for the unidirectional microphone. Camera position is less critical for the cardioid microphone than for the gradient microphone.

Early classical stereo recordings used variations of two basic microphone arrangements. In one scheme two unidirectional microphones were mounted close together with their axes angled toward opposite ends of the sound field to be recorded. This retained approximately the same arrival time and phase at both microphones, depending chiefly upon the directivity patterns to create the sound difference in the two channels.

In the second scheme the two microphones (not necessarily directional) were separated by distances of 5 to 25 ft, depending upon the size of the sound field to be recorded. Microphone axes (if directional) were again directed toward the ends of the sound field or group of sound sources. In this arrangement the time of arrival and phase differences were more important, and the effect of directivity was lessened. Each approach had its advantages and disadvantages.

With the arrival of tape recorders having many channels a trend has developed toward the use of more microphones and closer microphone placement. This offers much greater flexibility in mixing and rerecording, and it largely removes the effect of room reverberation from the recording. This may be either an advantage or a disadvantage depending upon the viewpoint. Reverberation can be added later (see Fig. 19-73).

In sound-reinforcement systems for dramatic productions and orchestras the use of many microphones again offers operating flexibility. However, it also increases the probability of operating error, increased system noise, and acoustical feedback, making expert monitoring and mixing of the microphone outputs necessary.

An attractive alternative for multimicrophone audio systems is the use of independent voice-operated electronic control switches in each microphone channel amplifier, in combination with

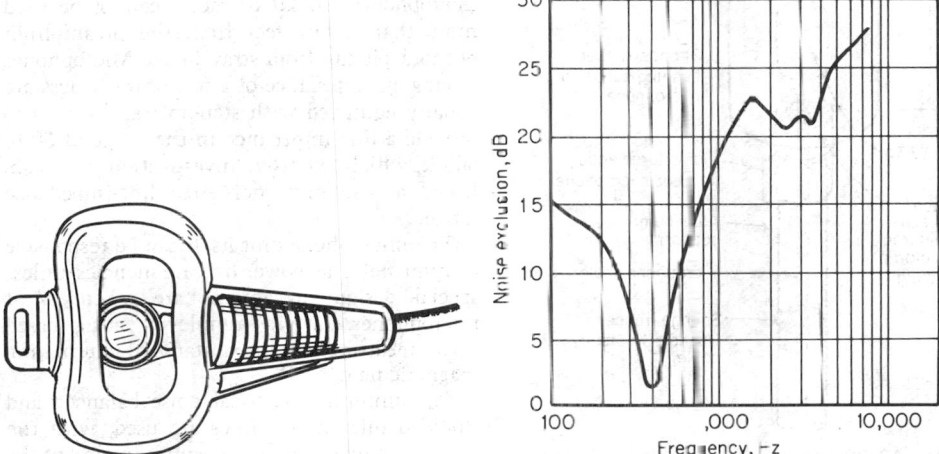

Fig. 19-90. Noise shield on a military pressure-gradient moving-coil RCA microphone (M-34/AIC).[17]

Fig. 19-91. Exclusion of external noise when the shield shown in Fig. 19-90 is worn.[17]

an automatic temporary reduction of overall system gain as more channels switch on, in order to prevent acoustical feedback. Special circuitry has been devised to minimize speech signal dropouts, and to prevent the inadvertent operation of channel control switches by background noises.[67b]

44. Microphone Mounting. On podiums and lecterns microphones are typically mounted on fixed stands with adjustable arms. On stages they are mounted on adjustable floor stands. In mobile communication and in other situations where microphone use is occasional, hand-held microphones are used during communication and are stowed on hangers at other times. For television and film recording, where the microphone must be out of camera sight, the microphones are usually mounted on booms overhead and are moved about during the action to obtain the best speech-to-noise ratio possible at the time. In two-way communication situations which require the talker to move about or to turn his head frequently, the microphone can be mounted on a boom fastened to his headset. This provides a fixed close-talking microphone position relative to the mouth, a considerable advantage in high-ambient-noise levels.

45. Microphone Accessories. Noise shields are needed for microphones in ambient noise levels exceeding 110 dB. Figure 19-90 shows a removable noise shield installed on a military type of hand-held pressure-gradient microphone with a talk switch. The thin rubber flange on the shield is pressed against the face around the mouth during speech. Figure 19-91 shows the noise attenuation by the shield. Noise shields are quite effective at high frequencies, where the random-noise discrimination of noise-canceling microphones diminishes. Noise shields and noise-canceling microphones complement each other. Noise shields are also useful on microphones used for dictation or commentary in situations where the talker's voice would interfere with proceedings.

Windscreens are available for microphone use in airstreams or turbulence. Without them aerodynamically induced noise is produced by turbulence at the microphone grille or openings. Figure 19-92 shows a cylindrical windscreen which lowers wind noise by over 20 dB. Large windscreens are more effective than small ones because they move the turbulence region farther from the microphone.

Figure 19-92 also shows special sponge-rubber mountings for the microphone and cable to reduce extraneous vibration of the microphone. Many microphone stands and booms have optional suspension mounting accessories to reduce shock and vibration transmitted through the stand or boom to the microphone.

46. Special Properties of Microphones. The source impedance of a microphone is important not only to the associated preamplifier but also to the allowable length of microphone cable and the type and amount of noise picked up by the cable. High-impedance microphones (10 kΩ or more) cannot be used more than a few feet from the preamplifier without pickup from stray fields. Microphones having an impedance of a few ohms or less are usually equipped with step-up transformers to provide a line impedance in the range of 30 to 600 Ω, which extensive investigation has established as the most noise-free line-impedance range.

The microphone unit itself can be responsive to hum fields at power-line frequencies unless special design precautions are taken. Most microphones have a hum-level rating based upon measurement in a standard alternating magnetic field.[58]

For minimum electrical noise balanced and shielded microphone lines are used, with the shield grounded only at the amplifier end of the line.

Microphone linearity should be considered when the sound level exceeds 100 dB, a frequent occurrence for loud musical instruments and even for close speech. Close-talking microphones, especially of the gradient type, are particularly susceptible to noise from breath and plosive consonants.

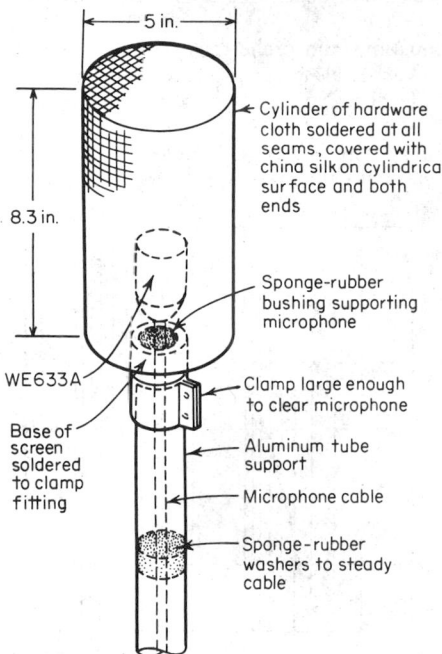

Fig. 19-92. Cylindrical experimental microphone windscreen.[46]

Labels in figure:
- 5 in.
- 8.3 in.
- Cylinder of hardware cloth soldered at all seams, covered with china silk on cylindrical surface and both ends
- Sponge-rubber bushing supporting microphone
- WE633A
- Base of screen soldered to clamp fitting
- Clamp large enough to clear microphone
- Aluminum tube support
- Microphone cable
- Sponge-rubber washers to steady cable

47. Specifications. Microphone specifications typically include many of the following items: type or mode of operation, directivity pattern, frequency range, uniformity of response within the range, output level at one or more impedances for a standard sound-pressure input (for example, 10 dyn/cm²), recommended load impedance, hum output level for a standard magnetic field (for example, 10^{-3} G), dimensions, weight, finish, mounting, power supply (if necessary), and accessories. Table 19-6 compares several microphone types.

48. Microphone Tests. Acoustical tests of microphones are generally run under anechoic conditions, either in an anechoic chamber or on an outdoor tower. This is necessary for the measurement of directional characteristics. Directional characteristics of microphones are plotted on polar-coordinate level recorders while the microphone rotates upon its axis in front of the standard sound source.

Microphone response curves are usually smoother than loudspeaker response curves. For this reason the sound pressure at a microphone under test is sometimes maintained constant by feedback from a nearby standard laboratory microphone to an automatic gain control in the loudspeaker amplifier circuit. This minimizes the effect of loudspeaker variations upon the appearance of the microphone test curve.

LOUDSPEAKERS, EARPHONES, AND ACCESSORIES[62,69,70]

49. Introduction. A loudspeaker is an electroacoustic transducer intended to radiate acoustic power into the air, with the acoustic waveform equivalent to the electrical input wave-

TABLE 19-6 Characteristics of Microphones

Company and model	Transducer	Output, dB re 1 V/μbar	Impedance	Pattern	Freq. range ±1 dB	±3 dB	Length, in	Diameter, in	Weight, lb
Altec 633A	Moving-coil	−90	30 Ω	Circular (no baffle)	200–600	50–5,000	3½	2	%
B & K 4134	Condenser	−58	20 pF	Circular		3–30,000	%	%	On preamp
E-V RE 50	Moving-coil	−83	150 Ω	Circular	120–4,000	80–13,000	7%	2	%
RCA 77 D	Ribbon	−79	250 Ω	Circular, cosine, or cardioid	120–3,000	50–10,000	11%	2½	3
Shure 300	Ribbon	−87	160 Ω	Cosine	50–5,000	40–7,000	9%	1½	1
W.E. 640AA	Condenser	−50	50 pF	Circular	10–10,000	1–12,000	1	1	On preamp

form. An earphone is an electroacoustic transducer intended to be closely coupled acoustically to the ear. Both the loudspeaker and earphone are receivers of audio-electronic signals. The principal distinction between them is the acoustical loading. An earphone delivers sound to air in the ear. A loudspeaker delivers sound indirectly to the ear through the air.

The transduction methods of loudspeakers and earphones are historically similar and are treated together. However, since loudspeakers operate primarily into radiation resistance and earphones into acoustical capacitance, the design, measurement and use of the two types of electroacoustic transducers will be discussed separately.

50. Transduction Methods. Early transducers for sound reproduction were of the mechanoacoustic type. Vibrations received by a stylus in the undulating groove of a record were transmitted to a diaphragm, placed at the throat of a horn for better acoustical impedance matching to the air, all without the aid of electronics. Electroacoustics and electronics introduced many advantages and a variety of transduction methods including moving-coil, moving-iron, electrostatic, magnetostrictive, and piezoelectric (Fig. 19-93).

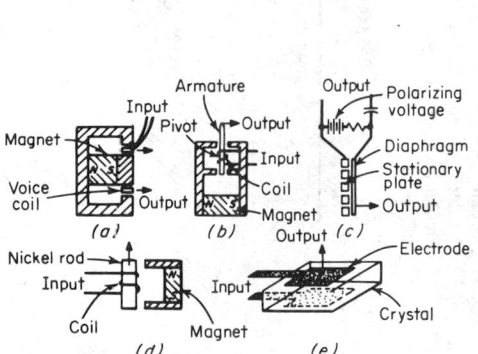

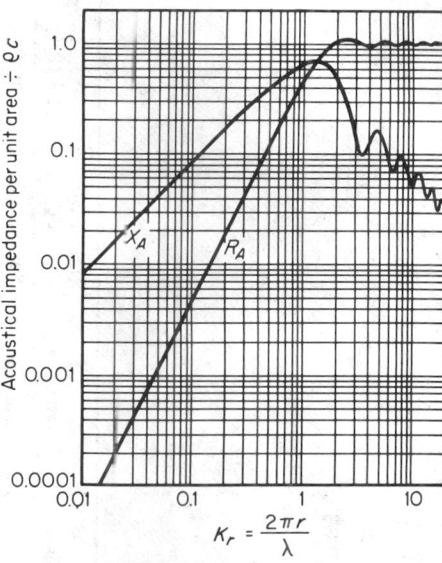

Fig. 19-93. Loudspeaker (and earphone) transduction methods: (a) moving-coil; (b) moving-iron; (c) electrostatic; (d) magnetostrictive; (e) piezoelectric.[69]

Fig. 19-94. Acoustical resistance R_A and reactance X_A per unit area of air load on a piston of radius r in an infinite baffle.[22]

Most loudspeakers are moving-coil type today, although moving-iron transducers were once widely used. Electrostatic loudspeakers are used chiefly in the upper range of audio frequencies, where amplitudes are small. Magnetostrictive and piezoelectric loudspeakers are used for underwater sound. All the transducer types are used in earphones except magnetostrictive.

Moving-Coil. The mechanical force on the moving coil of Fig. 19-93a is developed by the interaction of the current in the coil and the transverse magnetic field disposed radially across the gap between the magnet cap and the iron housing which completes the magnetic circuit. The output force along the axis of the circular coil is applied to a sound radiator.

Moving-iron transducers reverse the mechanical roles of the coil and the iron. The iron armature surrounded by the stationary coil is moved by mechanical forces developed within the magnetic circuit. Moving-iron magnetic circuits have many forms. As an example in the balanced armature system (Fig. 19-93b) the direct magnetic flux passes only transversely through the ends of the armature centered within the two magnetic gaps. Coil current polarizes the armature ends oppositely, creating a force moment about the pivot point. The output force is applied from the tip of the armature to an attached sound radiator. In a balanced-diaphragm loudspeaker the armature is the radiator.[71]

Electrostatic. In the electrostatic transducer (Fig. 19-93c) there is a dc potential difference between the conductive diaphragm and the stationary perforated plate nearby. Audio signals applied through a blocking capacitor superimpose an alternating potential, resulting in a force upon the diaphragm, which radiates sound directly.

Magnetostrictive transducers (Fig. 19-93c) depend upon length fluctuations of a nickel rod caused by variations in the magnetic field. The output motion may be radiated directly from the end of the rod or transmitted into the attached mechanical structure.

Piezoelectric transducers are of many forms using crystals or polycrystalline ceramic materials. In simple form (Fig. 19-93e) an expansion-contraction force develops along the axis joining the electrodes through alternation of the potential difference between them.

51. Sound Radiators. The purpose of a sound radiator is to create small, audible air-pressure variations. Whether they are produced within a closed space by an earphone or in open air by a loudspeaker, the pressure variations require air motion or current.

Pistons, Cones, Ports. Expansion and contraction of a sphere is the classical configuration, but most practical examples involve rectilinear motion of a piston, cone, or diaphragm. In addition to the primary direct radiation from moving surfaces there is also indirect or secondary radiation from enclosure ports or horns to which the direct radiators are acoustically coupled.

Attempts have been made to develop other forms of sound radiation such as oscillating air-streams and other aerodynamic configurations with incidental use, if any, of moving mechanical members (see Par. **19-55**).

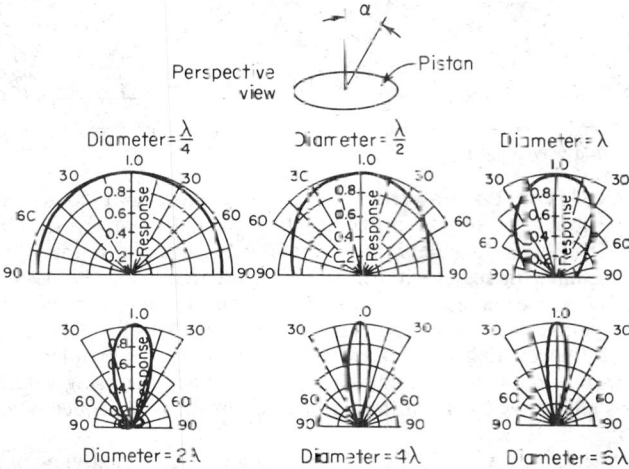

Fig. 19-95. Directional characteristics of rigid circular pistons of different diameters or at different sound wavelengths.[22]

Air Loading. Figure 19-94 shows the resistive and reactive components of the air load on one side of a circular piston of radius r mounted in a very large flat baffle. The abscissa is proportional to frequency, k being $2\pi/\lambda$. (For example, when the diameter equals the wavelength, $kr = \pi$.) The ordinate is mechanical impedance per unit area of piston divided by the characteristic acoustical impedance of air ρc. Note that below $kr = 1.0$ the reactance is directly proportional to frequency and the resistance is proportional to the square of frequency. Beyond $kr = 2$ the resistance is approximately unity, and the reactance falls off inversely with frequency. At low frequencies reactance dominates, and at high frequencies resistance does. These curves are a general basis for understanding the air loading of direct radiator loudspeakers. Not all loudspeaker mountings correspond to a piston in a large flat baffle, but the slopes and critical points of the curves in Fig. 19-94 vary only slightly when the baffle surrounding the piston changes all the way from flat to a long tube extending directly behind the piston.[72]

Directivity. Figure 19-95 shows the directional characteristics of a rigid circular piston for different ratios of piston diameter and wavelength of sound. (In three dimensions these curves are symmetrical around the axis of piston motion.) For a diameter of one-quarter wavelength the amplitude decreases 10% (approximately 1 dB sound level) at 90° off axis. For a four-wavelength diameter the same drop occurs in only 5°. (The beam of an actual loudspeaker cone is less sharp than this at high frequencies, where the cone is not rigid.) Note that all the polar curves are smooth when the single-source piston vibrates as a whole.

Radiator Arrays. When two separate, identical small-sound sources vibrate in phase, the directional pattern becomes narrower than for one source. Figure 19-96 shows that for a separation of one-quarter wavelength the two-source beam is only one-half as wide as for a single piston. At high frequencies the directional pattern becomes very complex. (In three dimensions these curves become surfaces of revolution about the axis joining the two sources.)

Arrays of larger numbers of sound radiators in close proximity are increasingly directional. Circular-area arrays have narrow beams which are symmetrical about an axis through the center of the circle. Line arrays, e.g., column loudspeakers, are narrowly directional in planes containing the line and broadly directional in planes perpendicular to the line.[73]

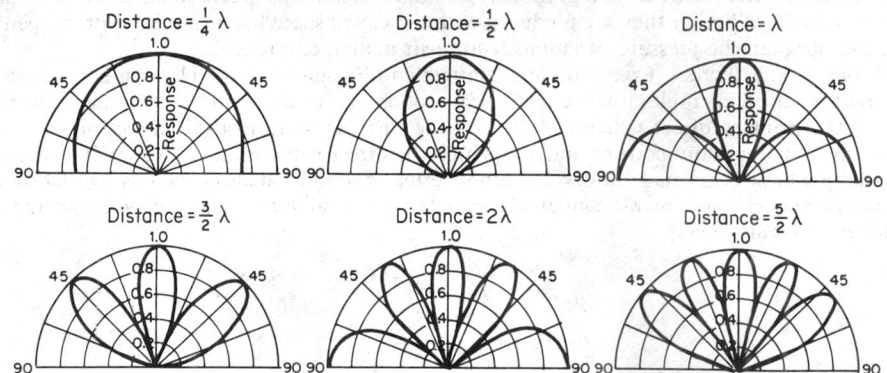

Fig. 19-96. Directional characteristics of two equal small in-phase sound sources separated by different distances or different sound wavelengths.[22]

52. Direct-Radiator Loudspeakers.[74] Most direct-radiator loudspeakers are of the moving-coil type because of simplicity, compactness, and inherently uniform response-frequency trend. The uniformity results from the combination of two simple physical principles: (1) the radiation resistance increases with the square of the frequency (Fig. 19-94), and hence the radiated sound power increases similarly for constant velocity amplitude of the piston or cone; (2) for a constant applied force (voice-coil current) the mass-controlled (above resonance) piston has a velocity amplitude which decreases with the square of the frequency. Consequently a loudspeaker designed to resonate at a low frequency combines decreasing velocity with increasing radiation resistance to yield a uniform response within the frequency range where the assumptions hold.

Equivalent Electric Circuits.[64] Figure 19-97 shows a cross-sectional view of a direct-radiator loudspeaker mounted in a baffle, the electric voice-coil circuit, and the equivalent electric circuit of the mechanoacoustic system. In the voice-coil circuit e is the emf and R_{EG} the resistance of the generator, e.g., power-amplifier output. L and R_{EC} are the inductance and resistance of the voice coil. Z_{EM} is the motional electric impedance from the mechanoacoustic system.

F_M is the driving force resulting from interaction of the voice-coil current field with the gap magnetic field. M_C is the combined mass of the cone and voice coil. C_{MS} is the compliance of the cone-suspension system. R_{MS} is the mechanical resistance. The mass M_A and radiation resistance R_{MA} of the air load complete the circuit.

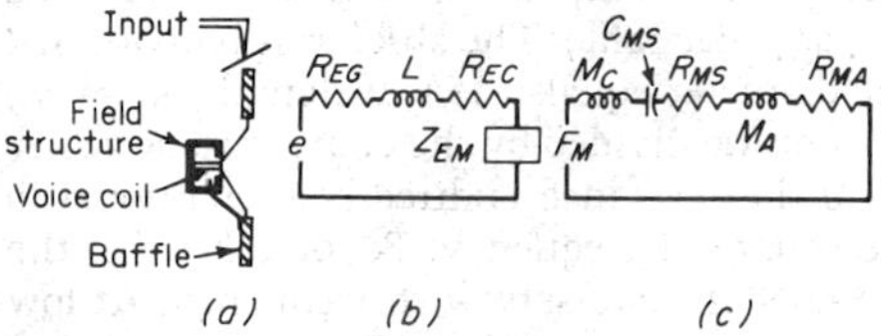

Fig. 19-97. (a) Structure, (b) electric circuit, and (c) equivalent mechanical circuit for a direct-radiator moving-coil loudspeaker in a baffle.[69]

Figure 19-98 summarizes these mechanical impedance factors for a 4-in direct-radiator loudspeaker of conventional design. Above resonance (where the reactance of the suspension system equals the reactance of the cone-coil combination) the impedance-frequency characteristic is dominated by M_C. From the resonance frequency of about 150 Hz to about 1,500 Hz the conditions for uniform response hold.

Efficiency. Since R_{MA} is small compared to the magnitudes of the reactive components, the efficiency of the loudspeaker in this frequency range can be expressed as

$$\text{Efficiency} = \frac{100(Bl)^2 R_{MA}}{R_{EC}(X_{MA} + X_{MC})^2} \text{percent} \tag{19-9}$$

where B = gap flux density (G), l = voice-coil conductor length (cm), and R_{EC} = voice-coil electric resistance (abohms). Since R_{MA} is proportional to the square of the frequency and both X_{MA} and X_{MC} increase with frequency, the efficiency is theoretically uniform.

All this has assumed that the cone moves as a whole. Actually wave motion occurs in the cone. Consequently at high frequencies the mass reactance is somewhat reduced (as shown in the dashed curve of Fig. 19-98), tending to improve efficiency beyond the frequency where radiation resistance becomes uniform.

Magnetic Circuit.   The magnet may be either permanent or field-coil type. However, most magnets now are a high-flux, high-coercive permanent type, either an alloy of aluminum, cobalt, nickel, and iron, or a ferrite of iron, cobalt, barium, and nickel. The magnet may be located in the core of the structure or in the ring, or both. However, magnetization is difficult when magnets are oppositely polarized in the core and ring.

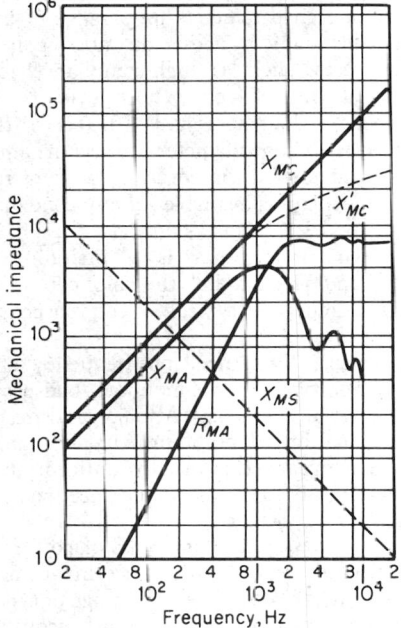

Fig. 19-98. Components of a mechanical impedance of a typical 4-in loudspeaker.[69]

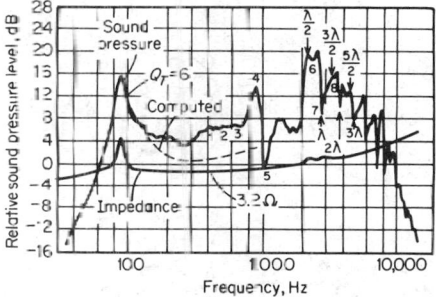

Fig. 19-99. Impedance and response-frequency characteristics for an 8-in-diameter loudspeaker mounted in a very large baffle. Note the response nonuniformity caused by cone breakup.[75a]

Air-gap flux density varies widely in commercial designs from approximately 3,000 to 20,000 G. Since most of the reluctance in the magnetic circuit resides in the air gap, the minimum practical voice-coil clearance in the gap compromises the maximum flux density. Pole pieces of heat-treated soft nickel-iron alloys, dimensionally tapered near the gap, are used for maximum flux density.

Voice Coils.   The voice coil is a cylindrical multilayer coil of aluminum or copper wire or ribbon. Aluminum is used in high-frequency loudspeakers for minimum mass and maximum efficiency. Voice-coil impedance varies from 1 to 100 Ω with 4, 8, and 16 Ω standard. For maximum efficiency the voice-coil and cone masses are equal. However, in large loudspeakers the cone mass usually exceeds the voice-coil mass. Typically the voice-coil mass ranges from tenths of a gram to 5 g or more.

Cones.   Cone diameters range from 1 to 18 in. Cone mass varies from tenths of a gram to 100 g or more. Cones are made of a variety of materials. The most common is paper deposited from pulp upon a wire-screen form in a felting process. For high-humidity environment cones are molded from plastic materials, sometimes with a cloth or fiber-glass base. Some low-frequency loudspeaker cones are molded from low-density plastic foam to achieve greater rigidity with low density.

So far piston action has been assumed in which the cone moves as a whole. Actually at high frequencies the cone no longer vibrates as a single unit. Typically there is a major dip in response resulting from quarter-wave reflection from the circular rim of the cone back to the voice coil. For loudspeaker cones in the range of 8 to 15 in diameter this dip usually occurs in the range of 1 to 2 kHz. Figure 19-99 combines typical impedance and response-frequency characteristics for a baffle-mounted 8-in loudspeaker without edge damping. The quarter-wave dip is at 1,050 Hz.

Other dips are at integral multiples of a wavelength along the cone radius. Peaks occur at odd multiples of a half wavelength.[75]

Figure 19-100 shows corresponding nodal patterns of cone vibration at numbered frequencies of interest in Fig. 19-99. It is this complex pattern of cone breakup that sustains the efficiency of a cone loudspeaker at high frequencies. However, the resulting nonuniformity of response and extreme complexity of directional characteristics make two- and three-way loudspeaker systems with frequency-range dividing networks necessary for high-quality sound reproduction.

Typical Commercial Design Values. Figure 19-101 shows typical values for several cone and voice-coil design parameters for a range of loudspeaker diameters.[62] These do not apply to extreme cases, such as high-compliance loudspeakers or high-efficiency horn drivers. The effective piston diameter (Fig. 19-101a) is less than the loudspeaker cone diameter because the amplitude falls off toward the edges. A range of resonance frequencies is available for any cone diameter, but Fig. 19-101b shows typical values. In Fig. 19-101c typical cone mass is M including the voice coil and M' excluding the voice coil. Figure 19-101d shows typical cone-suspension compliance.

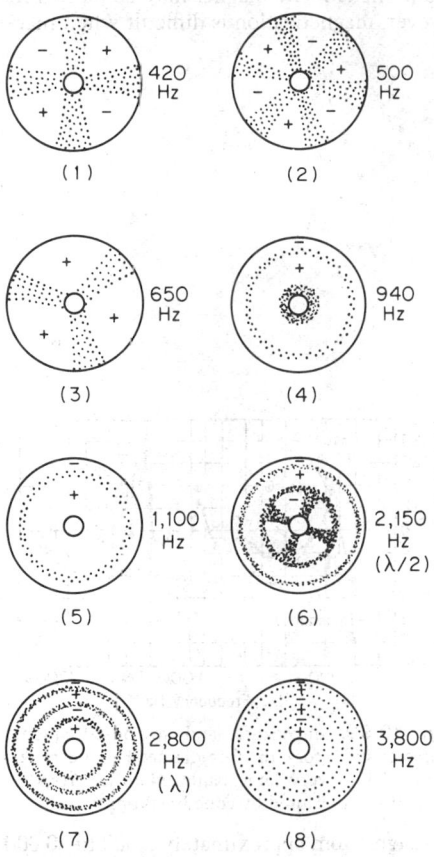

Fig. 19-100. Nodal patterns of the loudspeaker cone of Fig. 19-99 at selected significant frequencies (numbered points along the curve).[75a]

Impedance. The impedance-frequency characteristic of a typical baffle-mounted loudspeaker is also shown in Fig. 19-99. A major peak results from motional impedance at primary mechanical resonance. Impedance is usually uniform above this peak until voice-coil inductance becomes dominant over resistance.

Power Ratings Different types of power rating are needed to express the performance capabilities of loudspeakers. The large range of typical loudspeaker efficiency makes the acoustical power-delivering capacity quite important. The electrical power-receiving capacity (without overload or damage) determines the choice of power amplifier.

Loudspeaker efficiencies are seldom measured but are often compared by measuring the sound-pressure level at 4 ft on the loudspeaker axis for 1-W audio input. High-efficiency direct radiators provide 95 to 100 dB. Horn loudspeakers are typically higher by 10 dB or more, being both more efficient and more directional.

Loudspeakers are also rated by the maximum rms power output of amplifiers which will not damage the loudspeaker or drive it into serious distortion on peaks. Such ratings usually assume that the amplifier will seldom be driven to full power. For example a 30-W amplifier will seldom be required to deliver more than 10 W rms of music program material. Otherwise music peaks would be clipped and sound distorted.

However, in speech systems for high-ambient-noise levels the speech peaks may be clipped intentionally, causing the loudspeaker to receive the full 30 W much of the transmission time. Then the loudspeaker must handle large excursions without mechanical damage to the cone suspension and without destroying the cemented coil or charring the form.

Distortion. Nonlinear distortion in a loudspeaker is inherently low in the mass-controlled range of frequencies. However, distortion is produced by nonlinear cone suspension at low frequencies, voice-coil motion beyond the limits of uniform air-gap flux, Doppler shift modulation of high-frequency sound by large cone velocity at low frequencies, and nonlinear distortion of the air near the cone at high powers (particularly in horn drivers). Methods for controlling these distortions follow.

1. When a back enclosure is added to a loudspeaker, the acoustical capacitance of the

enclosed volume is represented by an additional series capacitor in the mechanical circuit of Fig. 19-97. Insufficient volume stiffens the cone acoustically, raising the resonance frequency and limiting the low-frequency range of the loudspeaker. It is convenient to reduce nonlinear distortion at low frequencies by increasing the cone-suspension compliance and depending upon the back enclosure to provide the system stiffness. Since an enclosed volume is more linear than most mechanical springs, this lowers low-frequency distortion.

2. Distortion from inhomogeneity of the air-gap flux can be reduced by making the voice-coil length either considerably smaller or larger than the gap width. This stabilizes the total number of lines passing through the coil, but it also reduces loudspeaker efficiency.

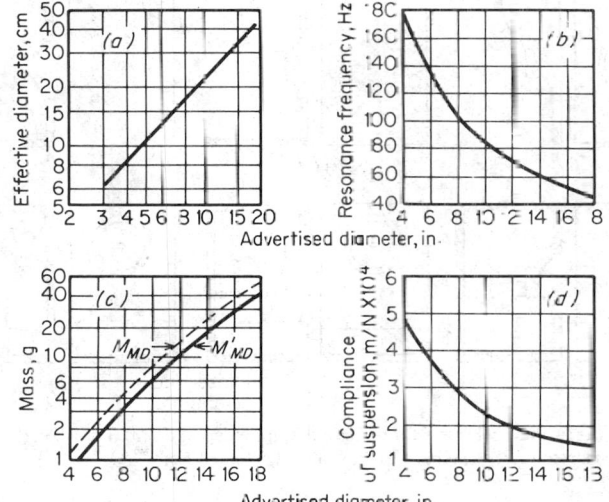

Fig. 19-101. Typical cone and coil design values.[62]

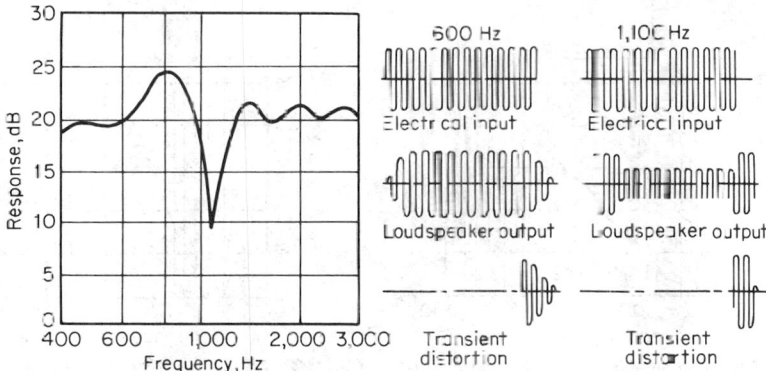

Fig. 19-102. Response-frequency fluctuation in a direct-radiator loudspeaker and transient response to tone bursts at the peak and dip frequencies.[69]

3. Doppler distortion can be eliminated only by separating the high and low frequencies in a multiple loudspeaker system.

4. Air-overload distortion can be avoided by increasing the radiating area.

Transient distortion is a factor near cone resonance frequencies. Figure 19-102 shows the opposite effects at a cone peak and a dip caused by quarter-wave reflection from the cone rim. Repetitive tone-burst testing, combined with digital analysis, storage and averaging, provides cumulative spectra for a three-dimensional display of loudspeaker transients.[25b,c]

53. Loudspeaker Mountings and Enclosures. Figure 19-103 shows a variety of mountings and enclosures. An unbaffled loudspeaker is an acoustic doublet for wavelengths greater than the rim diameter. In this frequency range the acoustical power output for constant cone velocity is proportional to the fourth power of the frequency.

Baffles. In order to improve efficiency at low frequencies it is necessary to separate the front and back waves. Figure 19-103a is the simplest form of baffle. The effect of different baffle sizes is given in Fig. 19-104. Response dips occurring when the acoustic path from front to back is a wavelength are eliminated by irregular baffle shape or off-center mounting.

Enclosures. The widely used open-back cabinet (Fig. 19-103b) is noted for a large response peak produced by open-pipe acoustical resonance. A closed cabinet (Fig. 19-103c) adds acoustical stiffness at low frequencies where the wavelength is larger than the enclosure. At higher frequencies the internal acoustical resonances create response irregularities requiring internal acoustical absorption.

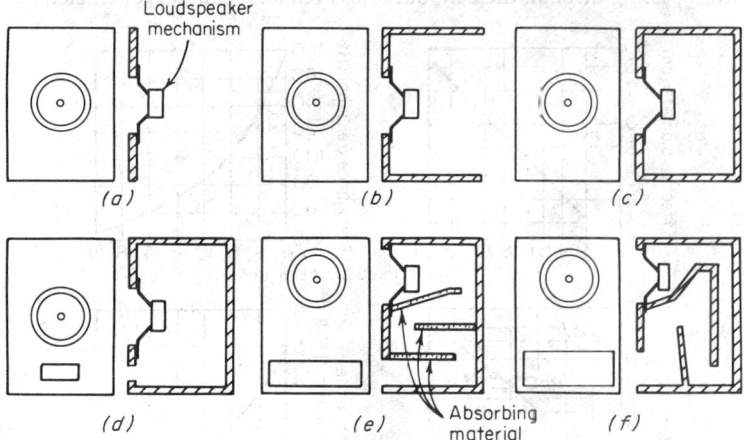

Fig. 19-103. Mountings and enclosures for direct-radiator loudspeaker. (*a*) flat baffle; (*b*) open-back cabinet; (*c*) closed cabinet; (*d*) ported closed cabinet; (*e*) labyrinth; (*f*) folded horn.[69]

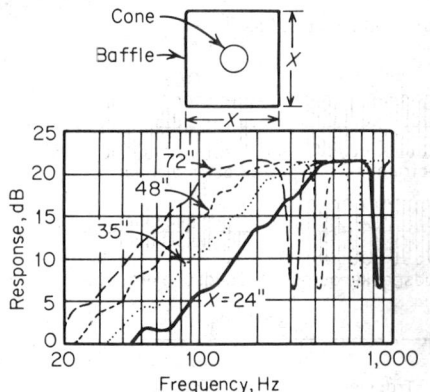

Fig. 19-104. Response frequency for loudspeaker in 2-, 3-, 4-, and 6-ft square baffles.[31]

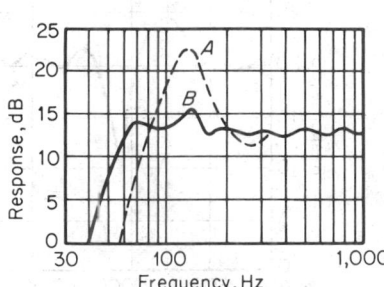

Fig. 19-105. Response frequency for loudspeaker in closed (*A*) and ported (*B*) cabinets.[31]

Ported Enclosures (Fig. 19-103d).   Enclosure volume can be minimized without sacrificing low-frequency range by providing an appropriate port in the enclosure wall. Acoustical inertance of the port should resonate with the enclosure capacitance at a frequency about an octave below cone resonance frequency. *B*, Fig. 19-105, shows that this extends the low-frequency range. This is most effective when the port area equals the cone-piston area. Port inertance can be increased by using a duct. An extreme example of ducting is the acoustical labyrinth (Fig. 19-103e). When duct work is shaped to increase cross section gradually, the labyrinth becomes a low-frequency horn (Fig. 19-103f).

Direct-radiator loudspeaker efficiency is typically 1 to 5%. Small, highly damped types with miniature enclosures may be only 0.1%. Transistor amplifiers easily provide the audio power for domestic loudspeakers. However, in auditorium, outdoor, industrial, and military applications much higher efficiency is required.

54. Horn Loudspeakers.[76a] Higher efficiency is obtained with an acoustic horn, which is a tube of varying cross section having different terminal areas to provide a change of acoustic impedance. Horns match the high impedance of dense diaphragm material to the low air impedance. Horn shape or taper affects the acoustical transformer response. Conical, exponential, and hyperbolic tapers have been widely used. The potential low-frequency cutoff of a horn depends upon its taper rate. Impedance transforming action is controlled by the ratio of mouth to throat diameter.

Horn Drivers. Figure 19-106 shows horn-driving mechanisms and straight and folded horns of large- and small-throat types. A large-throat driver (Fig. 19-106a) resembles a direct-radiator loudspeaker with a voice-coil diameter of 2 to 3 in and a flux density around 15,000 G. A small-throat driver (Fig. 19-106b) resembles a moving-coil microphone structure. Radiation is taken from the back of the diaphragm into the horn throat through passages which deliver in-phase sound from all diaphragm areas. Diaphragm diameters are 1 to 4 in with throat diameters of ¼ to 1 in. Flux density is approximately 20,000 G.

Large-Throat Horns. These are used for low-frequency loudspeaker systems. A folded horn (Fig. 19-106c) is preferred over a straight horn (Fig. 19-106d) for compactness.

Small-Throat Horns. A folded horn (Fig. 19-106e) with sufficient length and gradual taper can operate efficiently over a wide frequency range. This horn is useful for outdoor music reproduction in a range of 100 to 5,000 Hz. Response smoothness is often compromised by segment resonances. Extended high-frequency range requires a straight-axis horn (Fig. 19-106f).

Horn Directivity. Large-mouth horns of simple

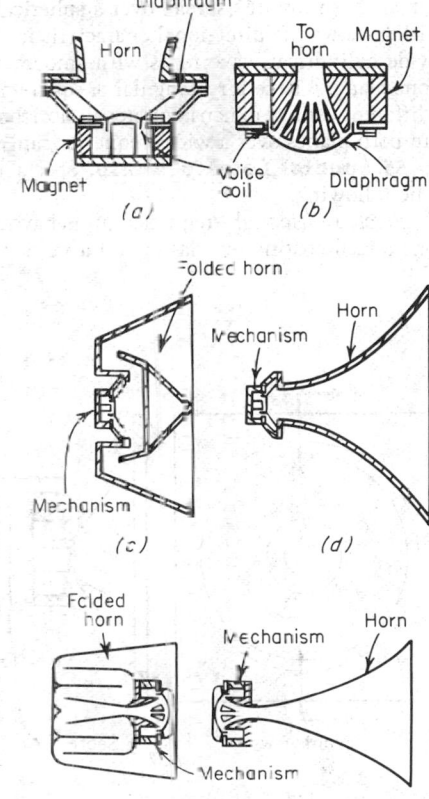

Fig. 19-106. Horns and horn drivers: (a) large throat driver; (b) small-throat driver; (c) folded large-throat horn; (d) straight large-throat horn; (e) folded small-throat horn; (f) straight small-throat horn.[39]

exponential design produce high-directivity radiation which tends to narrow with increasing frequency (as in Fig. 19-95). In applications requiring controlled directivity over a broad angle and a wide frequency range, a horn array (shown in Fig. 19-107a) can be used, with numerous

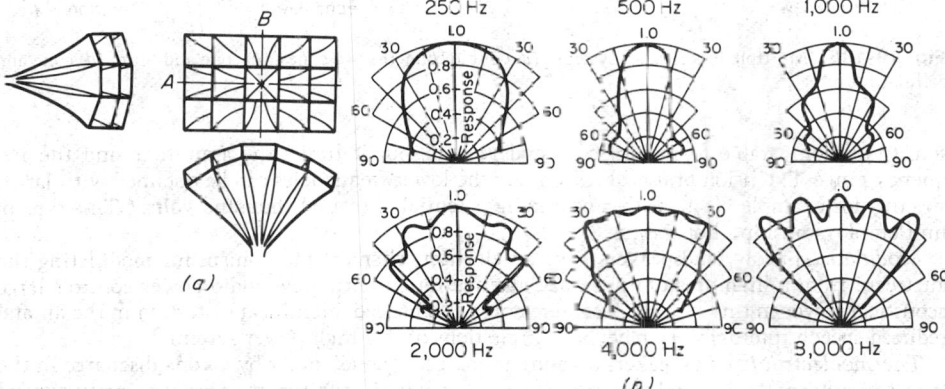

Fig. 19-107. Horn array (cellular) and directional characteristics (a) array; (b) horizontal directional curves.[31]

small horn mouths spread over a spherical surface and throats converging together. Figure 19-107b shows the directional characteristics. Single sectoral horns with radial symmetry can provide cylindrical wavefronts with smoother directional characteristics which are controlled in one plane.[76b] Recent rectangular or square-mouth "quadric" horns designed by computer to have different conical expansion rates in horizontal and vertical planes, provide controlled directivity in both planes over a wide frequency range.[76c]

55. Special Loudspeakers. Special types of loudspeakers for limited applications include the following.

Electrostatic high-frequency units have an effective spacing of about 0.001 in between a thin metalized coating on plastic and a perforated metal backplate.[77] This spacing is necessary for

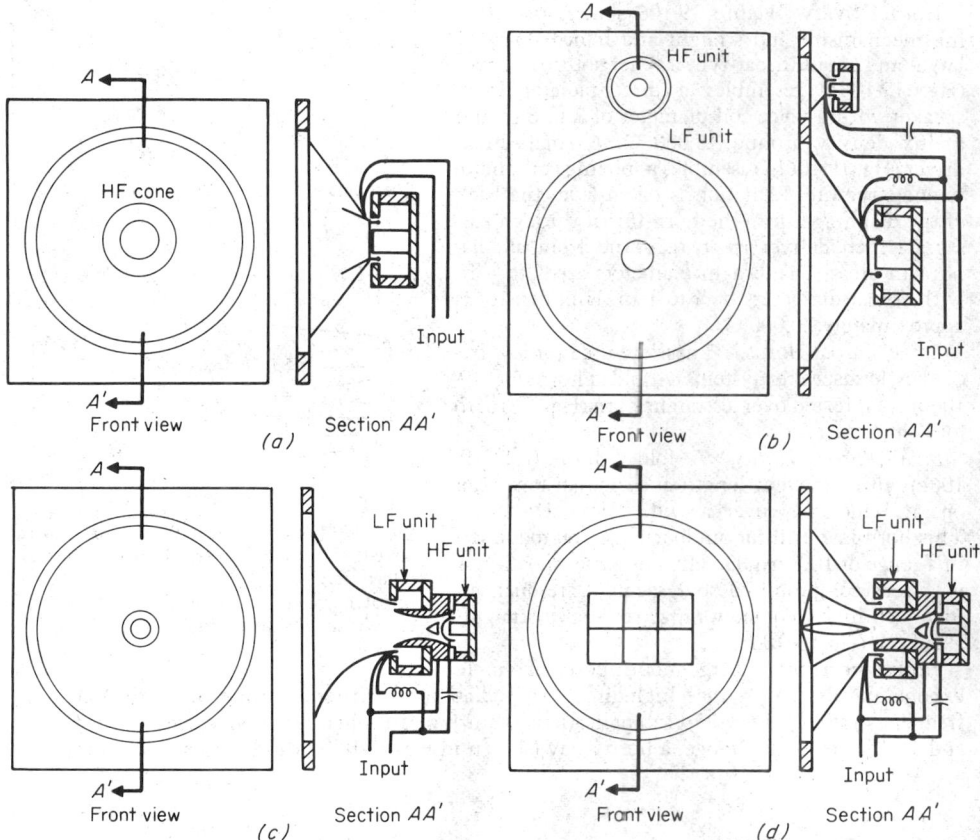

Fig. 19-108. Multiple loudspeaker systems: (a) double-cone; (b) two cones; (c) cone and horn; (d) cone and cellular horn.[69]

sensitivity comparable to moving-coil loudspeakers, but it limits the amplitude and the frequency range. Extension of useful response to the lower frequencies can be obtained with larger spacing, for example, 1/16 in, with a polarizing potential of several thousand volts.[78] This type of unit employs push-pull operation.

Modulated-airflow loudspeakers have an electromechanical mechanism for modulating the airstream from a high-pressure pneumatic source into a horn. Low audio power controls large acoustical power in this system. A compressor is also needed. Nonlinear distortion in the air and reduced speech intelligibility have been limitations of this high-power system.[79]

Thermoelectronic loudspeakers produce an ionized "cloud" of air by corona discharge in the throat of a horn. Audio modulation of the electric field in the throat causes the air to expand and contract accordingly.[80] Large amplitudes of vibration dissipate the concentrated charge, making the device useful chiefly at high frequencies.

56 Loudspeaker Systems. The audio range covers so many octaves that more than one sound radiator is required for the best combination of efficiency, response smoothness, and broad directivity. Dividing the frequency range into parts and assigning each part a suitable sound radiator gives a superior acoustical result.

Frequency-range division is possible either between the power amplifier and the several loudspeaker or driver units (using a low-impedance, high-power dividing network), or ahead of individual power amplifiers for each of the loudspeaker units. The choice depends upon the number of power amplifiers planned. High-power networks in the output are most common, but low-level bandpass filters ahead of the power amplifiers are also used.

Multiple direct-radiator systems are shown in Fig. 19-108a and b. A double cone with a single voice coil is used in Fig. 19-108a. The larger cone radiates the sound at low and medium frequencies. Above a mechanical crossover frequency at the resonance of the small free-edge cone the frequency range is extended by central radiation. This low-cost two way system is used widely although a response dip occurs near the crossover frequency.

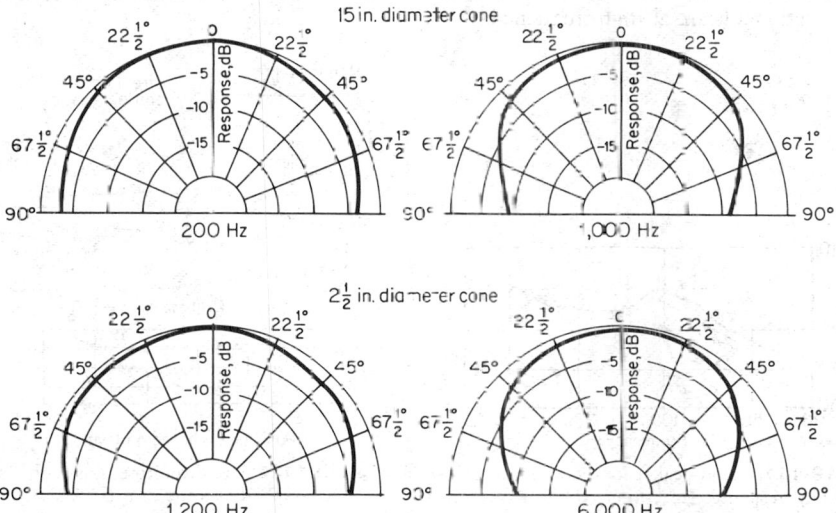

Fig. 19-109. Directional characteristics of two-cone system.[69]

When two separate loudspeakers (Fig. 19-108b) are used with a simple *LC* crossover network, one can get a broad directional characteristic over a wide frequency range. Figure 19-109 shows the beam of a 15-in low-frequency unit getting narrow at crossover frequency just as the broader beam of the 2½-in cone takes over. Doppler distortion in Fig. 19-108a is missing from Fig. 19-108b because of physical separation of the cones.

Direct-Radiator Horn Combination. In Fig. 19-108c a high-frequency horn driver is mounted within the magnet structure of the low-frequency direct-radiator unit. The horn forms part of the magnet structure. The cone of the low-frequency unit is the high-frequency horn. In Fig. 19-108d a small cellular horn radiates the high-frequency sound. These coaxial combinations provide smooth transition in the crossover range with a minimum of distortion from phase difference.

Multiple Horn. When space permits, e.g., behind a movie screen, a two-horn system takes full advantage of horn efficiency. Figure 19-110 shows such a system using a cellular high-frequency horn and a folded low-frequency horn. The frequency-range dividing network shown has a slope of 12 dB/octave outside the passband.

Stereophonic loudspeaker systems use a matched pair of any loudspeaker type. For maximum benefit the stereophonic loudspeaker sys-

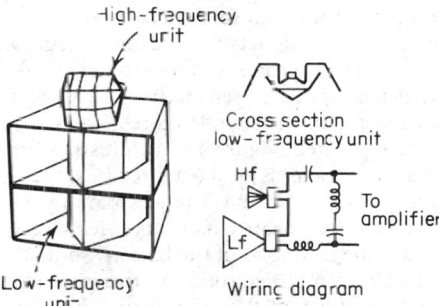

Fig. 19-110. Two-horn theater loudspeaker system.[31]

tems should be separated. When the stereophonic pickup is three-channel, the center channel should seem to come from a phantom loudspeaker between the real loudspeakers.

It is beneficial to aim stereo loudspeaker axes to cross at the rear of the listening area. This allows listeners at one side of the listening area to hear the output from the opposite side better.

57. Loudspeaker Use in Rooms. Although loudspeakers are designed and measured on an idealized basis, they are used in rooms. There are a number of practical considerations.

Location Effects. Loudspeaker response at low frequencies depends upon source location within the room. At specific frequencies response depends upon room dimensions and the associated low-frequency vibration modes. Smoothed loudspeaker response curves of Fig. 19-111 show response trends for the same loudspeaker in four different locations. Maximum low-frequency response is obtained with the loudspeaker in the corner. The next best location is the center of a wall at floor level. A midwall location is next, and suspension in the center of the room provides the least low-frequency sound.

If a high-frequency loudspeaker or a wide-range loudspeaker must be mounted in a location where most listeners will be off the axis, acoustic-lens accessories are available which broaden the directional beam at high frequencies.[81]

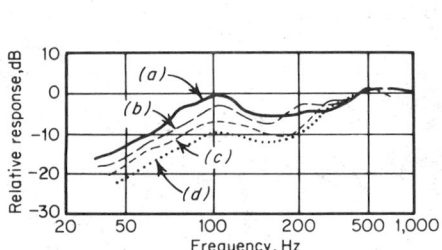

Fig. 19-111. Loudspeaker location effects on low-frequency response: (a) corner; (b) wall center at floor level; (c) center of wall; (d) center of room.[62]

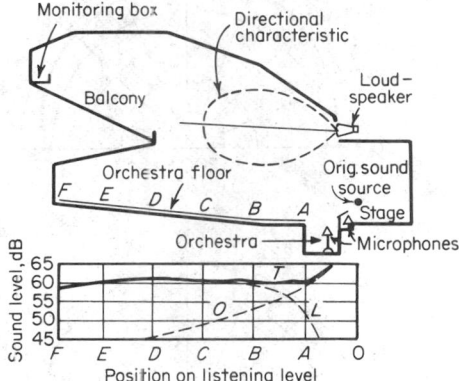

Fig. 19-112. Sound reinforcing system in a theater. The graph shows the direct sound at the points indicated on the orchestra floor: curve O, the original sound; curve L the level due to the loudspeaker; curve T the total sound.[31]

Localization and Orientation. Different sound systems have different purposes. If the intent is to gain attention (announcing systems) or provide realistic sound reproduction (sound motion pictures), the sources should easily be localized by listeners. For such purposes high directivity and point-source radiation are desirable. For background music, the simulation of reverberation, or the simulation of large-area sources it is better to use broad directional patterns, distributed sound sources, and even to direct the loudspeakers toward scattering and reflecting surfaces. Some applications require both types of sound radiation. For example, the reproduction of a studio-recorded orchestra within an auditorium of limited reverberation might use a multichannel group of directional loudspeakers on stage in combination with delayed and artificially reverberated mixed signals from scattered loudspeakers aimed at reflecting surfaces.

Level Distribution. When sound is only reproduced, the distribution of sound level can be predicted from the sensitivity and directional characteristics of the loudspeakers. However, when some of the sound is direct and the remainder is from loudspeakers, a realistic reinforcement is desired. Figure 19-112 illustrates the basic principles for an auditorium with balcony. The loudspeaker is situated directly above the original sound source so that direct and amplified sound arrive at distant listeners simultaneously. The loudspeaker height and its directivity are so chosen that nearby listeners receive chiefly direct sound. The loudspeaker beam is directed toward the front part of the balcony so that the curvature of the loudspeaker distribution pattern minimizes amplified sound in the front rows and increases the amplified sound level gradually toward the rear of the main floor. This gives a combination which is nearly uniform across the entire audience. The geometry of the beam and ceiling reflection of the upper side of the beam assist the sound level at the balcony rear.

Table 19-7. Characteristics of a Variety of Loudspeaker Types

Company	Altec	Altec	Bozak	RCA
Model no.	755C	1505B horn 290D driver	CM-109-23	LC1B
Type	Direct radiator	Cellular horn (3 × 5)	Three-way column	Duo-cone
Sensitivity (at 4 ft for 1 W), dB	95	110	106	95
Frequency range, Hz	40–15,000	300–8,000	65–13,000	25–16,000 (±4 dB)
Impedance, Ω	8	4	8	15
Power rating, W	15	100	20C	20
Distribution angle, deg	90	105 horizontal 60 vertical	90 horizontal 30 vertical	120
Voice-coil diameter, in.	2	2.8	(3 sizes)	(2 cones)
Cone resonance, Hz	52	..	(3 sizes)	22
Crossover frequency, Hz	...	500	800. 2,500	1,600
Diameter, in.	8⅜	18½ high 30½ wide	57 in. high 22¾ w de	17
Depth, in.	2¼	30	15¾	7½
Weight, lb	3¾	43	250	21

58. Loudspeaker Specifications and Measurements. Typical loudspeaker specifications are shown in Table 19-7 for a variety of loudspeaker types.

Loudspeaker impedance is proportional to the voltage across the voice coil when driven by a high-impedance constant-current source. Continuous power ratings are obtained from sustained life tests with typical audio-program material restricted to the frequency range appropriate for the loudspeaker type. Sensitivity, response-frequency characteristics, frequency range, and directivity are most effectively measured under anechoic conditions using calibrated laboratory microphones and high-speed level recorders. However, data so measured should not be expected to be exactly reproducible under room listening conditions.[8c]

Distortion measurements in audio-electronic systems are generally of three types shown in Fig. 19-113. For harmonic distortion a single sinusoidal signal *A* is supplied to the loudspeaker and wave analysis at the harmonic frequencies determines the percent distortion.

Both intermodulation methods supply two sinusoidal signals of different frequency to the loudspeaker. In the older Society of Motion Picture and Television Engineers (SMPTE) method the frequencies are widely separated, and the distortion is expressed in terms of sum and difference frequencies around the higher test frequency. This method is meaningful for wide-range loudspeaker systems.

The CCIF (International Telephone Consultative Committee) method is more applicable to narrow-range systems and loudspeakers receiving input at high frequencies. It supplies two high frequencies to the loudspeaker and checks the low difference frequency.

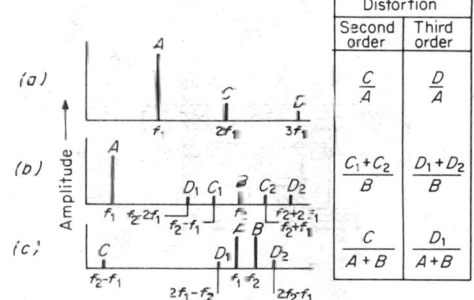

Distortion	Second order	Third order
(a)	$\dfrac{C}{A}$	$\dfrac{D}{A}$
(b)	$\dfrac{C_1+C_2}{B}$	$\dfrac{D_1+D_2}{B}$
(c)	$\dfrac{C}{A+B}$	$\dfrac{D_1}{A+B}$

Fig. 19-113. Methods of measuring nonlinear distortion: (a) harmonic; (b) intermodulation method of SMPTE; (c) intermodulation method of CCIF.[82a]

Transient intermodulation distortion, resulting from nonlinear response to steep wavefronts, is measured by adding square-wave (3.18-kHz) and sine-wave (15-kHz) inputs, with a 4:1 amplitude ratio, and observing the multiple sum- and difference-frequency components added to the output spectrum (Fig. 19-114).

59. Earphones.[84a] The transduction methods are the same as for loudspeakers (Fig. 19-93).

Equivalent Electric Circuits. Figure 19-115 shows a cross section of a moving-coil earphone and the equivalent electric circuit. The voice-coil force drives the voice coil and diaphragm. (Mechanical resonance of earphone diaphragms occurs at a high audio frequency in contrast to loudspeakers.) Diaphragm motion creates sound pressure in several spaces behind the diaphragm and the voice coil and between the diaphragm and the earcap. Inertance and resistance of the connecting holes and clearances combine with the capacitance of the spaces to add acoustical resonances. Z is the acoustical impedance of the ear.

Idealized Ear Loading.   The ear is approximately an acoustical capacitance. However, acoustical leakage adds a parallel resistance-inertance path affecting low-frequency response. At high frequencies the ear canal-length resonance is a factor.

Since the ear is a capacitance, the goal of earphone design is a constant diaphragm amplitude throughout the frequency range. This requires a stiffness-controlled system or a high resonance frequency. The potential across the ear is analogous to sound pressure within the ear cavity. This sound pressure is proportional to diaphragm area and inversely proportional to enclosed volume. Earphone loading conditions are extremely varied for different types of earphone mountings.

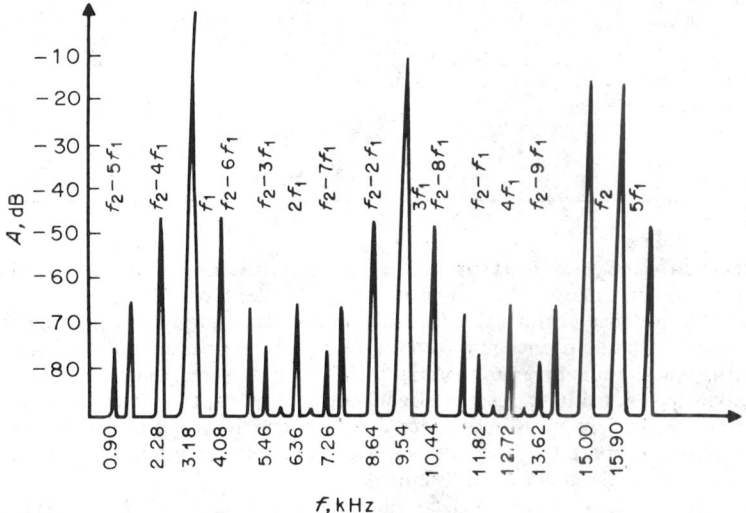

Fig. 19-114 Frequency spectrum of output signal including transient intermodulation distortion components. Input signals include f_1 square wave and f_2 sine wave in 4:1 amplitude ratio.[83]

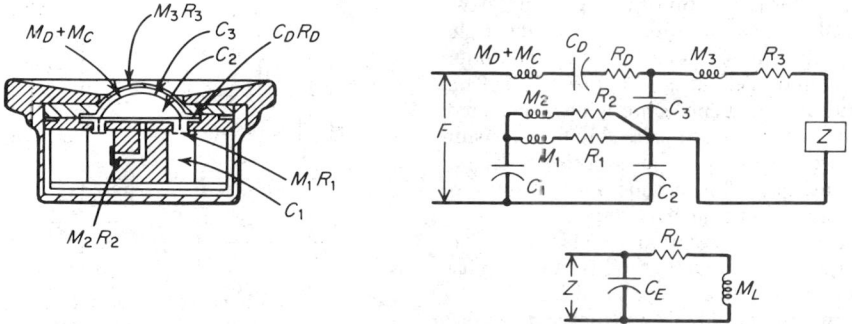

Fig. 19-115. Moving-coil earphone cross section and equivalent electric circuit.[84b]

Earphone Mountings.   The most widely used earphone is the single receiver unit on a telephone handset. It is intended to be held against the ear but is often tilted away, leaving considerable leakage.

Headsets provide better communication than handsets because they supply sound to both ears and shield them. Four types of headset mounting for earphones are shown in Fig. 19-116. A fifth type for the deaf is the insert earphone worn within the outer ear and supported by it. Simulated ears are used for their acoustical measurement.[85]

Ear cushions on a headband (Fig. 19-116a) press against the ear and enclose approximately 6 cm³. Semi-insert phones (Fig. 19-116c) provide a smaller contact area, require less headband force, and enclose approximately 2 cm³. Earmuffs (Fig. 19-116d) have larger contact area, require stronger headbands, and enclose the entire ear (100 to 300 cm³). Military helmets (Fig. 19-116b) may be adapted to any of the ear coupling devices. ANSI Standard S3.7 covers the use of four coupler types.

Insert earphones are attached to individually fitted inserts. There is no applied force. The enclosed volume is 2 cm³ or less.

A remote earphone can drive the ear canal through a small acoustic tube. The length may be an inch or two for hearing aids and several feet for music listening on aircraft.

Efficiency, Impedance, and Driving Circuits. Moving-iron earphones and microphones can be made efficient enough to operate as sound-powered (battery-less) telephones. Efficient magnet structures, minimum mechanical and acoustical damping, and minimum volume of acoustical coupling are required for this purpose. In some earphone applications overall efficiency is less critical, and wearer comfort is important.

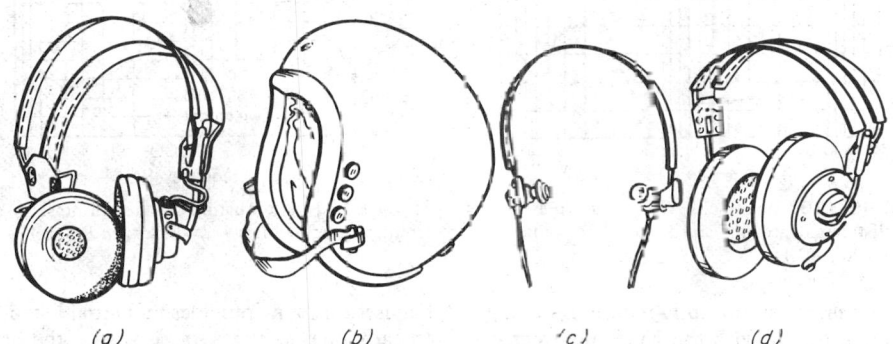

Fig. 19-116 Four types of headset mounting of earphones: (*a*) headband with ear cushions; (*b*) helmet; (*c*) light headband with semi-insert tips; (*d*) headband with circumaural ear muffs. (*After J. Zwislocki, in Ref. 6.*)

Insert earphones need less efficiency than external earphones because the enclosed volume is much smaller; however, they require moderate efficiency to save the amplifier batteries.

Circumaural earphones are frequently driven by amplifiers otherwise used for loudspeakers. Here efficiency is less important than power-delivering capacity.

Typically 1 mW of audio power to an earphone will produce 100 to 110 dB in a standard 6-cm³ coupler. The same earphone will produce less sound level in an earmuff than in an ear cushion and more when coupled to an ear insert.

The shape of the enclosed volume also affects response. The farther the driver is from the eardrum the lower the frequency of standing-wave resonance. Small tube diameters produce high-frequency attenuation.

The response-frequency characteristic of moving-iron or piezoelectric earphones is quite dependent upon source impedance. A moving-iron earphone having uniform response when driven at constant power will have a rising response (with increasing frequency) at constant current and a falling response at constant voltage[85] (Fig. 19-117).

Real-Ear Response. The variety of earphone-coupling methods and the variability of outer-ear geometry (among different listeners) make response data from artificial ears only indicative, not definitive. Out of necessity a real-ear response-measuring technique was developed. A listener adjusts headset input to match headset loudness to an external calibrated sound wave in an anechoic chamber. From matching data at numerous frequencies an equivalent free-field sound-pressure level can be plotted for constant input to the earphone. This curve usually differs from a sound-level curve on a simple earphone coupler. The reason is that probe measurements of sound at the eardrum and outside the ear in a free field differ because of ear amplification and diffraction about the head (Fig. 19-118).

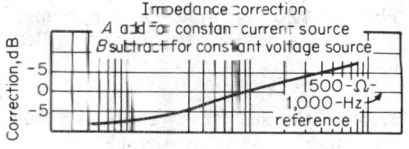

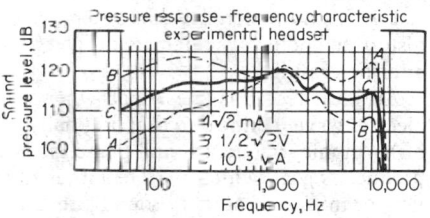

Fig. 19-117. Effect of source impedance upon earphone response curve: (*a*) constant current; (*b*) constant voltage; (*c*) constant power.[86]

Acoustic attenuation by earphones can be measured either by threshold shift or by matching the loudness of tones heard from an external loudspeaker, with and without the headset on. The sound-level difference is plotted as attenuation in decibels (Fig. 19-119). The solid curve is for the headset (Fig. 19-116a), and the dashed curve is Fig. 19-116d. The third curve is for a special earphone socket using wax inside the sealing ring. Even greater attenuation can be obtained below 1 kHz by combining larger mass, larger contact area, a stronger headband, and larger enclosed volume.[88] Attempts to improve attenuation above 1 kHz are defeated by bone conduction to the inner ear unless rigid helmets are worn.

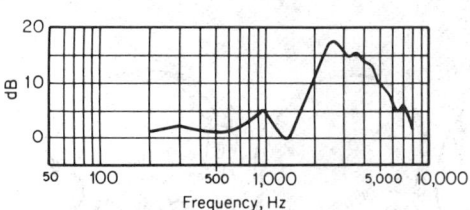

Fig. 19-118. Relative level of sound pressures at the listener's eardrum and in the free sound field. (*After Ref. 87.*)

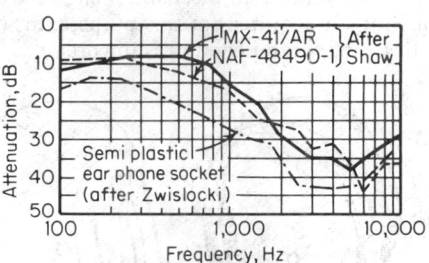

Fig. 19-119. Attenuation of external noise by different headsets. (*After J. Zwislocki in Ref. 6.*)

Monaural, Diotic, and Binaural Listening. A handset earphone provides monaural listening. Diotic listening with the same audio signal in both earphones localizes sound within the head. This is not unpleasant and may actually be an aid to concentration. In natural binaural listening the ears receive sound differently from the same source unless it is directly on the listening axis. Usually there are differences in phase, arrival time, and spectrum (because of diffraction about the head).

Recordings provide true binaural effects only if the two recording microphones are on an artificial head. Stereophonic microphones are usually separated much farther, so that headset listening gives an exaggerated effect. For some listeners this is an enhancement, but for listeners who prefer natural binaural listening a circuit has been developed to convert stereo information into binaural (Fig. 19-120). This circuit produces crosstalk between stereo channels with phase and amplitude differences at each ear to simulate sound diffraction around the head.[89]

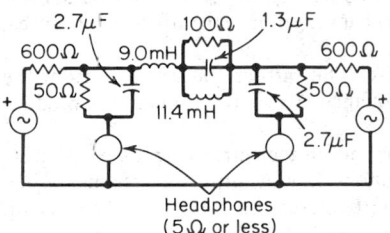

Fig. 19-120. Network for transforming stereophonic playback for binaural headset listening.[89]

DISC RECORDING SYSTEMS

60. Principles and Characteristics of Disc Recording.[90,91] "A sound recording system is a combination of transducing devices and associated equipment suitable for storing sound in a form capable of subsequent reproduction."[1] In mechanical sound recording the vibrations are engraved or embossed in the surface of a material. Embossing deforms the surface by rubbing it and is used chiefly in recording speech dictation for direct reproduction without further processing. Engraving cuts a groove into the surface of the material. It has now been developed to provide the full audio-frequency range and the freedom from distortion and noise which are needed for high-fidelity disc recording and for mass reproduction after processing.

An ideal record is one which when reproduced generates from the output of the pickup an electric wave identical to that originally supplied to the recorder, with a minimum of noise and of harmonic, frequency, and phase distortion. This technical goal is independent of the record producer's goal, which may be either to reproduce or recreate an original sound or to create a new or modified sound or sound effect.

Figure 19-121 shows a stereophonic disc recording system with a choice of source material from either microphones in a studio or a two-channel stereophonic magnetic tape. Other source possibilities include electronic musical instruments and synthesizers. Very often the original recording is made on tape because of the ease of starting, stopping, editing, and playback without damage to the original. Then the finally edited and perhaps enhanced tape original is rerecorded on the disc original. Previous subsections have treated audio program material, studio acoustical environment, microphones, and monitoring loudspeakers (see Pars. **19-72** to **19-86** for magnetic-tape sound recording and reproducing systems). The part of the overall system concerned specifically with disc recording includes the disc material; the turntable and drive, which move the material during recording; the stylus, which cuts the groove; the recording head or cutter, which moves the stylus in a plane perpendicular to the grooves, the drive for the cutter head, and the electronic preemphasis of the electric input wave by the equalizer of Fig. 19-121.

61. Disc Materials. For many years original recordings were made on a wax surface. The actual material was a blend of waxes with metallic soaps, first in the form of a thick disc of wax, and later a thin layer of wax melted and flowed onto a smooth, flat metal base. Before recording began, the solid wax discs were shaved to a highly polished surface while rotating at high speed. For flowed wax the smoothness and flatness of the recording surface depended largely upon the metal base.

Now original recordings are commonly made on a lacquer-coated disc of metal or glass The basic lacquer compound has been combined with new and superior coating techniques to meet the requirements of microgroove stereophonic recording.[92]

62. Drive Mechanisms and Speeds. The National Association of Broadcasters (NAB) standard recording speeds are 33⅓ and 45 r/min (the older 78.26 r/min speed is no longer considered standard). Speeds of 16⅔ and 8⅓ r/min are useful for speech, especially talking books for the blind. The 33⅓ r/min speed can be obtained by 54:1 reduction and 45 r/min by a 40:1 reduction from 1,800 r/min. The NAB standard requires that recording speed be held to 0.1% tolerance. On the time scale this amounts to less than 2 s in ½ h of program, and on the frequency scale less than one-fiftieth of a musical semitone.

Recording Turntables. The NAB accuracy specification necessitates the use of synchronous-driven turntables. Whether the connection between the motor and the turntable is by gears, friction drive, or a belt, mechanical filtering is usually required to prevent transmission of motor vibration to the turntable and to avoid audible flutter. Among the synchronous drives which have been used are single-phase selsyn motors, 3-phase power selsyn motors, and synchronous motors of the variable-reluctance type.

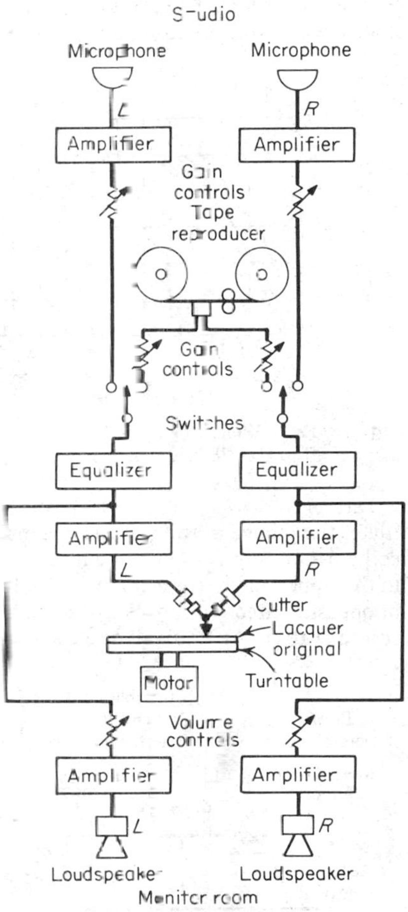

Flutter, Wow, and Rumble. "Flutter is any deviation in frequency of reproduced sound from the original frequency."[1] Wow is a colloquial term commonly applied to flutter frequency of the order of 1 Hz. Rumble is low-frequency noise transmitted

Studio

Fig. 19-121. Stereophonic disc recording system.[22c]

from the motor (or the coupling) through the turntable to the disc and stylus, whether the system has constant speed or not. Since the job of the turntable is to rotate the disc quietly at accurate, constant speed, all these faults must be minimized.

Research has shown[93] that flutter in the wow range is barely audible at 0.04% frequency modulation, the NAB upper limit during recording, but that high-frequency tones with flutter rates of 3 to 10 Hz have a perceptibility threshold an order of magnitude lower. This is reflected in the weighting curve of the IEEE standard 193-1971, Method of Measurement for Weighted Peak Flutter of Sound Recording and Reproducing Equipment, shown in Fig. 19-122 and based on a test frequency of 3,150 Hz. Only the highest-quality recording equipment is free of flutter by these standards.

Cutter Drive Speed. For microgroove recording the lathe screw bearing the cutter head advances it nominally ½₆₀ in per disc revolution. However, between a 4.75-in minimum diameter and a 11.5-in maximum diameter this provides only 22½ min recording time. Variable-pitch recording[94] decreases the cutter-drive speed during low-level passages, extending recording time to 30 min. Special speech recordings (talking books) have up to 650 grooves per inch.

63. Recording-Head Design. In disc recording the cutter must have a high enough mechanical impedance at the top of the stylus to control the recording medium. This requirement leads directly to high stiffness and density for the stylus and indirectly to a resonance centrally located within the frequency range to be recorded. For the stylus velocity to be proportional to the applied force, the resonance peak must be eliminated by mechanical resistance control.

Recording Transduction Methods. Recording heads have operated chiefly on moving-coil, moving-iron, and piezoelectric principles of transduction. In the past the moving-coil principle was usually adopted for vertical (hill-and-dale) recording. Moving-iron and piezoelectric transducers were well suited to lateral groove modulation.

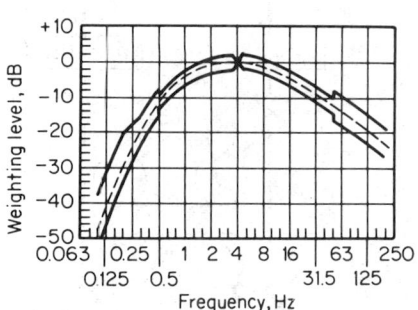

Fig. 19-122. Weighting curve for flutter measurement at 3,150 Hz. *(IEEE Stand. 193-1971.)*

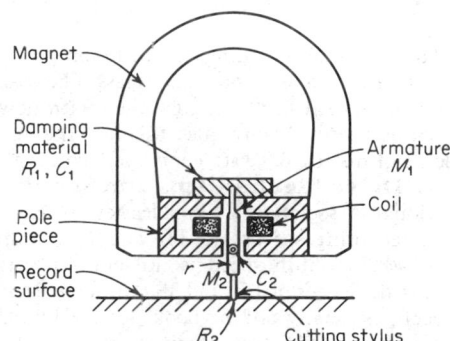

Fig. 19-123. Lateral disc recording head (magnetic).[91]

Lateral Recorder. Figure 19-123 is a sectional view of a magnetic disc recorder which modulates the groove in the plane of the disc. Alternating flux in the armature, corresponding to the coil current, interacts with the balanced permanent-magnetic field, applying an alternating force to the upper end of the armature which is pivoted in the lower gap. Through lever action a force of opposite direction moves the cutting stylus laterally to modulate the groove being cut in the record surface. An equivalent electric circuit of this recording head is shown in Fig. 19-124. For constant-velocity recording the circuit constants must provide a current I_2 which is independent of frequency. This requires the damping resistance to be relatively large.

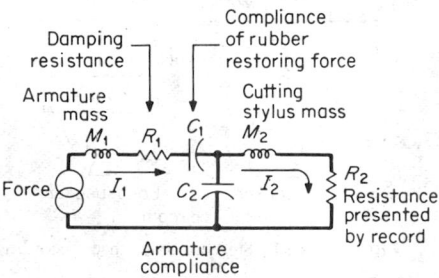

Fig. 19-124. Equivalent electric circuit for Fig. 19-123.[91]

Stereophonic Recorder.[95] The recorder of Fig. 19-125 contains two moving-coil drive units, one for each stereophonic channel. Each unit also contains a feedback coil located in a separate annular branch of the magnetic circuit. Copper shields between the two coils reduce the inductive crosstalk. (The purpose of the feedback coil is given below.) V-shaped beryllium-copper coil-support springs hold the coil assemblies, allowing them to move only along their axes. Motion is transmitted from the coils to a tubular stylus-support member through wire links braced with magnesium sleeves. A single magnet supplies flux to the magnetic circuit through series-parallel gaps, ensuring equal flux densities in the corresponding gaps. The orthogonal drive system controls stylus motion within a plane approximately perpendicular to the axis of the groove. For practical reasons the plane is actually tilted from the vertical at an angle standardized at 15°.

Feedback Recording.[96] The output of each feedback coil is used for negative-feedback control (in its channel amplifier) of resonance in the vibratory system. An advantage of the negative-feedback method is the damping provided to the electromechanical system without resistive losses.

Recording-Stylus Design.[97] Disc recording styli are ground from sapphire. The lack of grain and cleavage planes permits grinding to very acute angles without breakage. Figure 19-126 shows the shape of a cutting stylus, with a magnified view of the stylus point and the principal angles affecting the properties of the recorded groove. The dubbing facet was found necessary for satisfactory cutting of lacquer discs.[98] An inherent loss in high-frequency range, caused by the finite size of the dubbing facet relative to the wavelength, has been overcome by the hot-stylus recording technique,[99] which allows extremely small facets on the cutting edge in combination with a quiet groove and good high-frequency response. The heating coil is wrapped directly around the stylus just above the cutting portion, as shown in Fig. 19-125. The effect of heating upon

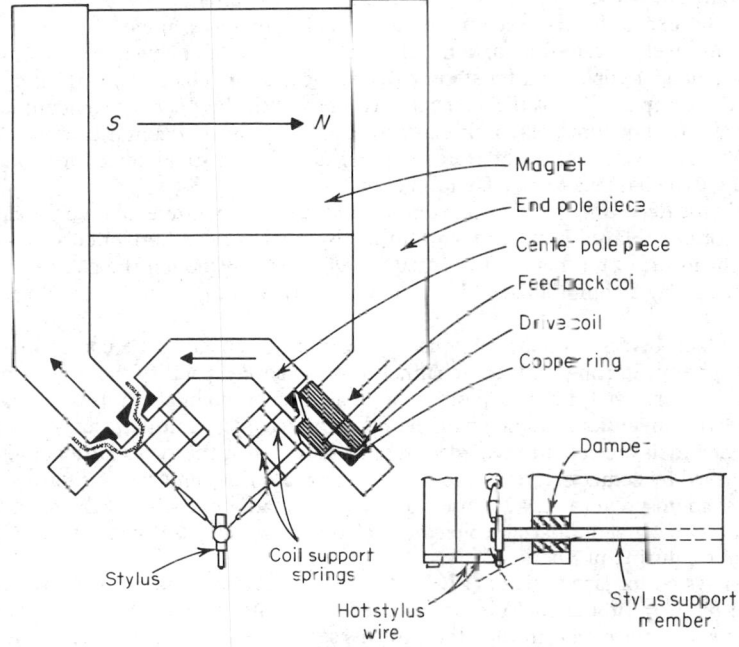

Fig. 19-125. Stereophonic disc recording head.[95]

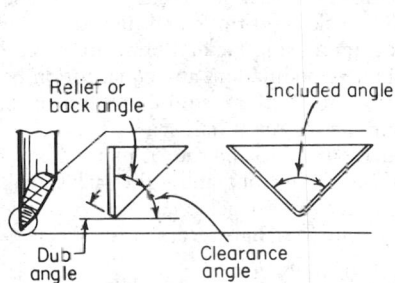

Fig. 19-126. Cutting stylus point.[97]

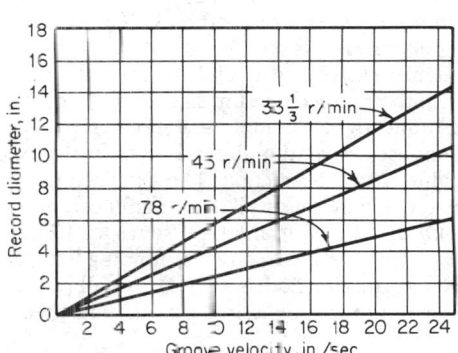

Fig. 19-127. Groove velocity vs. diameter for different record speeds.[100a]

noise can be optimized by current control. Excessive heating produces groove-burning noise. A typical heating coil consists of six turns of No. 40 AWG nichrome wire carrying a current of approximately ½ A.

The clamping of the stylus in its holder is critical. Minute misfitting or looseness can seriously affect recording quality. The grinding of styli is very scientific, but styli use has both technical and interpretive aspects.[100]

64. Groove Modulation. The groove velocity varies with the diameter of the spiral and the rotation rate of the record, as shown in Fig. 19-127. A 6-mil groove width was standard for the old 78 r/min records, with a radius of 1.5 mils at the bottom of the groove. The playback stylus had a radius in the range of 2.5 to 3.0 mils. When 33⅓ and 45 r/min records became standard, the groove width was reduced to 2.6 mils with a bottom radius of 0.2 mil. The playback-stylus radius was reduced to the range of 0.8 to 1.1 mils. Stereophonic recording decreased the stylus radius to 0.7 mil to reduce the tracing distortion associated with the vertical component of stylus motion in the groove.

Stereophonic Modulation. Figure 19-128 compares the theoretical groove shape and dimensions for lateral recording and for stereophonic recording. The solid lines define the unmodulated groove, and the dashed lines show the maximum lateral excursions for monophonic recording and the maximum vertical and lateral excursions for stereophonic recording. The dimensions are very similar except for reduced amplitude of motion in 45-45, amounting to a reduction of 3 dB in output level for each stereophonic channel. Thus the total power remains the same.

The choice of 45-45 operation for stereo (over vertical-lateral channels) was based upon symmetry and the compatibility of the stereophonic mode with the previous standard lateral monophonic mode. Monophonic playback of stereo combines the horizontal components of both channels. The relative phase of the stereophonic channels was standardized for equal in-phase signals in the two channels to give lateral motion.

Quadraphonic Recording.[101-104] In extending the two-channel stereophonic principle a quadraphonic recording, played back through four loudspeakers spaced horizontally around the listeners, sought to surround them with virtual sound sources, although the choice of four channels was necessarily a compromise.[105] Naturally, the goal was to continue recording in a single groove.

Numerous and diverse systems for quadraphony are intended to be compatible with disc record stereophony. Starting with two-channel record-groove capability there were two general approaches. One involved the superposition of an additional modulated ultrasonic carrier upon each of the two conventional audio channels, all being recorded simultaneously. Since four discrete input channels are actually recorded in one groove, with the two additional audio output channels derived by demodulation of the ultrasonic signals, this approach was termed 4-4-4. The other general approach, termed 4-2-4, uses some kind of matrix encoding of the four audio input channels into two conventional stereo-recorded channels, which are decoded back to four audio output channels during playback.

The CD-4 system of Japan Victor (JVC), adopted by RCA, is 4-4-4. It superposes all left-side information (both front and back) in one of the 45-45 components of recording stylus motion and all right-side information in the other. In this sense it is "discrete" or separated. However, for stereo playback compatibility each audio (20-Hz to 15-kHz) channel contains the sum of front and back information; the superposed carrier (30 kHz) is frequency-modulated by the difference between front and back. (Disregarding the carrier, stereo playback recovers the left and right sum signals only.)

Quad playback demodulates the left and right difference signals, for addition or subtraction with their respective sum signals, to yield left-front, left-back, right-front, and right-back signals. Left and right channel separations have the usual compromising factors found in stereo, and front-back separation depends upon perfect modulation, demodulation, and equalization before addition and subtraction. The extended frequency range (to 40 kHz) required improved record material, playback styli, and transducers and recording the master at half speed.

Of the numerous matrix systems for 4-2-4 the SQ method of CBS and the QS method of Sansui have been the most active, combining left-front, left-back, right-front, and right-back signals for later separation. Both SQ and QS use all-pass, phase-shift networks to get quadrature components. This minimizes undesired cancellation effects. The two audio signals recorded for SQ are

$$L_T = L_F - j0.707L_B + 0.707R_B$$
$$R_T = R_F + j0.707R_B - 0.707L_B$$

This allows the two front signals to remain isolated from each other to the degree that they are in stereo systems.

During SQ quadraphonic playback decoding, the L_T and R_T signals are reproduced separately for front channels. Each is also mixed with minus 90° phase-shifted versions of the other to obtain the back channels. All four outputs contain predominantly the appropriate input signal plus two side-effect signals which are symmetrical. The back-channel separation is compromised and the back images are drawn toward each other. Psychoacoustic factors are used to justify this compromise, which assumes that listeners face the front. An advanced version of SQ uses a logic-directed matrix decoder to sense relative levels in signal channels and to adjust relative gains for minimum side effects in the reproduction of highly localized original sources.

The QS system has a symmetrical amplitude and phase matrix defined by

$$L = (L_F + jL_B) \cos \theta + (R_F + jR_B) \sin \theta$$
$$R = (R_z - jR_B) \cos \theta + (L_F - jL_B) \sin \theta$$

Because of its symmetry, QS inherently provides identical front-back and left-right separation. The listener has no preselected listening axis. For any playback channel the crosstalk is down only 3 dB in adjacent channels.

A QMX discrete-matrix disc system by Nippon Columbia is also a 4-4-4 type. It resembles CD-4, but its ultrasonic carrier channels carry matrixed information or low-pass frequency-limited range, reducing system and disc problems above 30 kHz.

"Jury" listening tests conducted by a National Quadraphonic Radio Committee, and later by the Federal Communications Commission, compared different sets of 4-4-4, 4-3-4, and 4-2-4 systems for azimuth localization, for musical preference, and for reduction to stereo. [The 4-3-4 system transmitted a monophonic signal, with an $(L - R)$ signal modulating one subcarrier and an $(L + R)$ signal modulating a similar subcarrier 90° out of phase.] The FCC first and second rankings, respectively, were 4-4-4 and 4-3-4 on localization; 4-4-4 and SQ(4-2-4) on musical preference; and SQ(4-2-4) and either 4-4-4 or 4-3-4 on stereo.

In spite of all of these developments and tests it appears that commercial quadraphonic disc activity has diminished. Nevertheless the technological by-products have improved the art, which may revive through standardization and further research on surround sound.

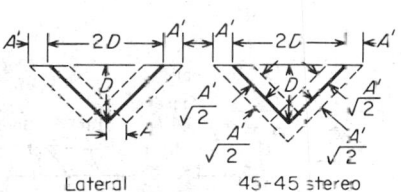

Lateral   45-45 stereo

Fig. 19-128. Comparison of 45-45 stereophonic and lateral monophonic grooves.[95]

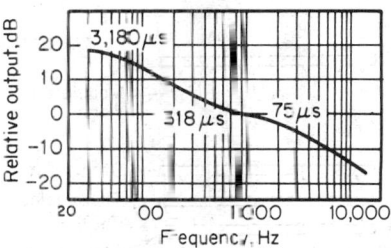

Fig. 19-129. NAB standard disc-reproducing characteristic; relative output level vs frequency for constant velocity input. Tolerance: +2 dB, 50 Hz to 10 kHz; +2 dB, −3 dB below 50 Hz, above 10 kHz. (NAB Disc Recording and Reproducing Standard, 1964.)

65. Preemphasis and Deemphasis. Disc recorders are designed for constant-velocity recording, but typical input spectra (Figs. 19-13 and 19-32) have steep negative slopes. To prevent low-frequency overload and improve high-frequency signal-to-noise ratio the input spectrum is tilted upward before recording, then tilted back down after playback, as shown in Fig. 19-129. The numbers along the curve are network time constants for high-frequency rolloff, low-frequency boost, and very low-frequency rolloff. The standard level for stereophonic discs is 5 cm/s peak at 1 kHz.

66. Electronic Control Circuits.[105] Response-frequency characteristics can be varied automatically with recording level. For example, the dynamic spectrum equalizer boosts low frequencies at low level and the 2- to 8-kHz range at high level to make typical home listening more like a concert hall. Another recording control circuit reduces tracing distortion (see below) by introducing complementary distortion through the use of a tapped delay line in conjunction with sampling gates.[106] Recording overload indicator or control circuits are used with a differentiator to avoid curvature overloading at high frequencies and with an integrator to anticipate groove overload at low frequencies.

67. Practical Recording Techniques.[106, 107] Examples beyond the scope of this book and described in the references include recording stylus mounting and heating, panoramic potentiometers, microphone techniques for smooth response, and monitoring practices.

68. Disc Processing and Specifications. Figure 19-130 shows the steps in the mass production of records. The lacquer original is usually cellulose nitrate with a variety of resins, oils, lacquers, and volatile solvents. The master is produced by spray-silvering the original surface, then plating with nickel and then with copper to a thickness of 0.030 to 0.040 in. The master has ridges where the original had grooves. After a film coating to ensure separability the master is copper-plated to produce the mother, which has grooves like the original. In like manner stampers are made from the mother.

The final record is thermoplastic material pressed between two stampers heated by steam in the press. Disc material properties depend on a correct balance of melt viscosity and melt elasticity.[107b]

The NAB standard specifies a center-hole diameter of 0.286 in + 0.001 − 0.002 in. Hole concentricity should be ±0.005 in. For disc flatness the total indicator reading should not exceed ⅟₁₆ in over the disc surface with a ⅟₃₂ in limit within any 45° sector.

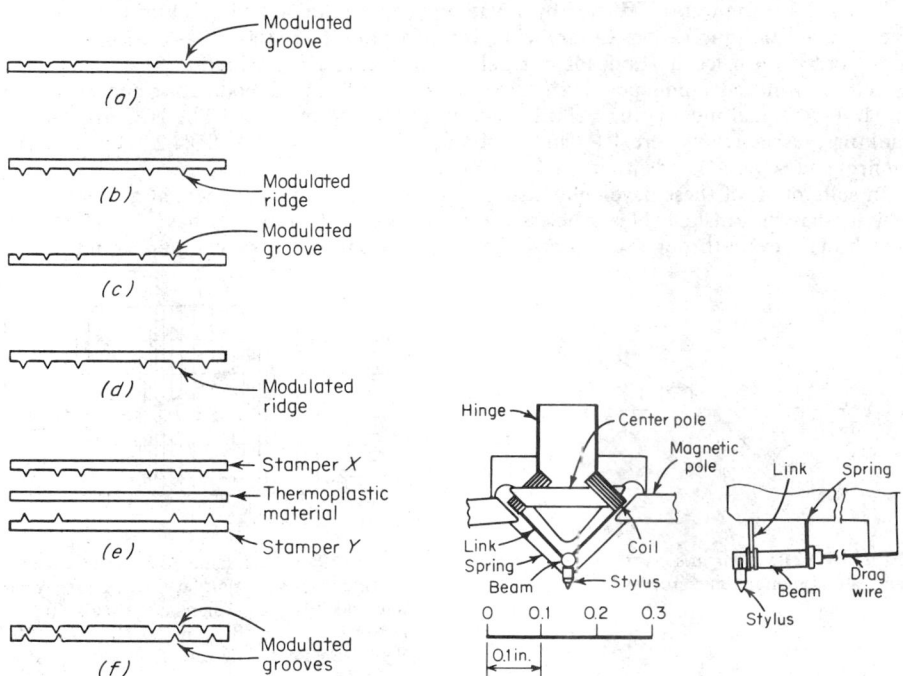

Fig. 19-130. Steps in the mass production process for disc records: (a) original; (b) master; (c) mother; (d) stamper; (e) stamping; (f) record.[22]

Fig. 19-131. Section of moving-coil stereophonic pickup.[95]

DISC REPRODUCTION SYSTEMS

69 Reproducer Heads, Styli, and Arms. A stereophonic system consists of a record on a motor-driven turntable, a stylus on a reproducer pivoted on an arm, a two-channel amplifier with ganged volume contols, each channel including a deemphasis equalizer, a power amplifier, and loudspeaker.

Successful transduction of the groove modulation into an electric output wave depends upon the interactive combination of stylus, pickup, and arm mounting.

Stylus Design.   The playback stylus is a sapphire or diamond cone with a spherical tip of 0.0007-in radius. Although the walls of the record groove wear well under typical tracking forces, there is some stylus penetration of the groove walls because of elastic and plastic deformation.[108] For a disc of Vinylite (a copolymer of vinyl chloride acetate) the pressure from a stylus with a typical tracking force, for example, 1 g, exceeds elastic flow limits and causes some plastic flow. A short duration of stress is important.

Stereophonic Pickups.   Figure 19-131 shows a moving-coil stereophonic pickup. The two coils on Mylar hinges have their axes at 45° from the horizontal. Each coil is linked to a beam bearing the stylus. The beam rear is secured in a flat spring to prevent rotation and to provide uniform stylus compliance for all motions in the plane of groove modulation.

Figure 19-132 shows stereophonic pickups of magnetic and ceramic types. The magnetic pickup has low mechanical impedance. Stylus motion is transmitted through the arm to the magnet. Pole pieces on four sides of the magnet are surrounded by coils, which when connected in proper phase, generate potentials corresponding to the magnet velocity along the $+45°$ and $-45°$ axes.

In the ceramic pickup the stylus displacement bends the ceramic bars along the $+45°$ and $-45°$ axes, generating output potentials. Because the ceramic pickup is displacement-responsive, the response curve approximates the standard reproducing characteristic directly.

Tracing distortion[109] occurs in playback because the spherical stylus does not trace an exact replica of the modulated groove. In Fig. 19-133 the groove modulation is sinusoidal, but the path of the sphere center is not. This type of distortion can be minimized by complementary distortion of the recording stylus path.[106]

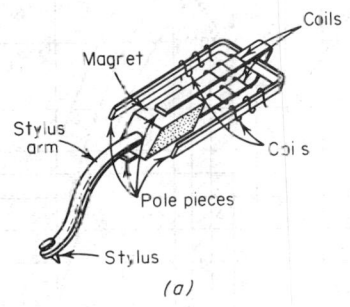

(a)

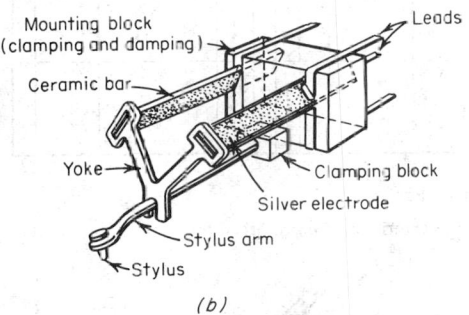

(b)

Fig. 19-132. Stereophonic pickups (perspective): (a) magnetic; (b) ceramic.[22a]

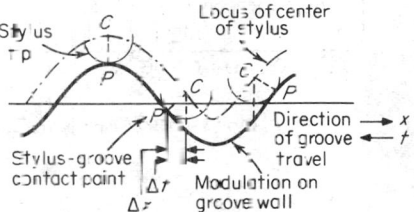

Fig. 19-133. Spherical stylus tracing a sinusoidal modulation in one of the 45-45 stereophonic planes.[136]

Pickup Trackability.[110a] The stylus must not wear the groove unduly but must remain in contact with the record at maximum groove-modulation velocity. The dashed line in Fig. 19-134 is the theoretical maximum limit for recorded velocity (sometimes exceeded). The smooth curves show the velocity to which a specially designed pickup tracks for forces of 1 and ¾ g. Record warpage is, of course, an important factor in trackability.

Arm Design.[111] The arm holds the pickup, applies the stylus tracking force, and provides damping for the resonance of the arm mass with the stylus-support compliance. The dimensions and angle of the arm-pickup-stylus combination determine any tracking error (see Fig. 19-135), which is the angle α between the planes in which the recording and playback styli move. To a first approximation (see Fig. 19-136) this angle can be adjusted for a given arm length l and groove radius R by adjusting (1) the angle β between the pickup axis and the line between the pivot and stylus point and (2) the amount of overhang D by which the stylus passes the record center. The head offset angle creates friction between the stylus and groove, resulting in a tendency to skate toward the record center. Some pivoted arms have an adjustable antiskating compensation weight or spring.

Tracking error can be eliminated by sliding the pickup on a radial arm and using an automatically articulated arm to maintain tangent tracking.

Figure 19-137 shows calculated distortion vs. tracking angle at different radii at 33⅓ r/min.

70. Reproducer Turntables, Drives, and Changers. Record-playing units play either a single disc at a time or a stack of discs in sequence. Either the arm or the changer can be automatic. Turntables have the same basic principles whether recording or reproducing, but design practices differ somewhat.

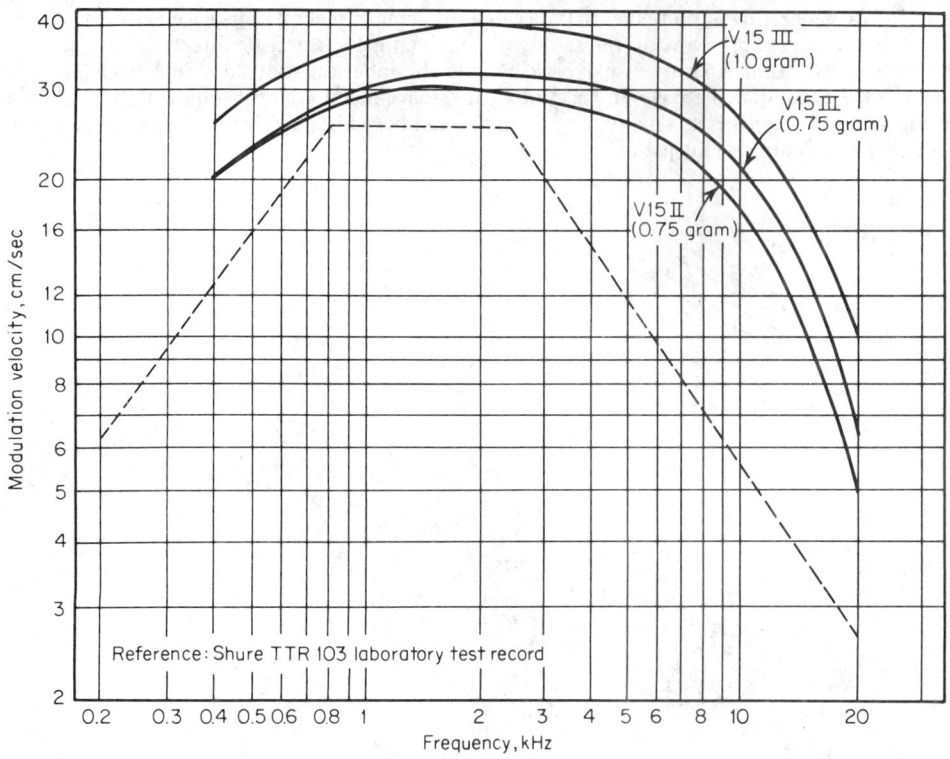

Fig. 19-134. Pickup trackability characteristics. *(Shure Bros., Inc.)*

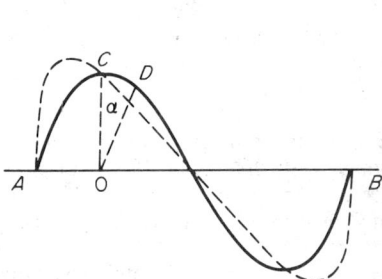

Fig. 19-135. Distortion of laterally recorded sine wave by the error α in tracking angle.[91]

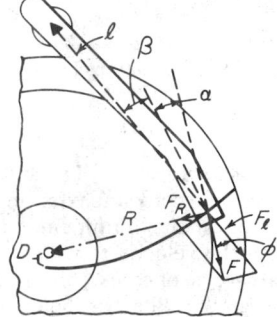

Fig. 19-136. Arm design for minimum tracking error.[111]

Turntables.   The average speed of a reproducing turntable is typically held only to $\pm 0.3\%$, but many turntable drives provide $\pm 6\%$ for fine speed adjustment. This is necessary for musical applications. Wow and flutter are held within 0.05 to 0.1% in playback equipment. Standard center-pin diameter is 0.285 ± 0.0005 in.

The NAB standard for low-frequency noise, or rumble, covers a frequency range of 10 to 500 Hz. Acceptable rumble in a stereo turntable is now -60 dB relative to 1 cm/s peak velocity at 100 Hz. Turntables with good drive systems meet this specification easily. However, the bass boost in the disc reproducing characteristic accentuates the inherent rumble.

Drive Mechanisms.   In most older record changers and single-play turntables an idler wheel presses against the motor shaft and the inside of the turntable rim, transmitting motion from the former to the latter. Idler material and bearings and the design of the disengagement mechanism control the amount of flutter, wow, and rumble transmitted through or contributed by the idler system. Multiple-speed drives may use idlers with stepped diameters or an idler with a conical rather than a cylindrical shape. A conical idler may be moved up and down continuously for fine speed adjustment.

Flexible belt drives are often used between the motor shaft and the turntable. Belt flexibility reduces motor flutter. If the belt is continuous and smooth, it should not contribute noise itself.

Synchronous motors provide good speed accuracy and independence of line-voltage variations. Induction motors have greater torque and inherently less flutter. Hybrid motor designs used now on many record players combine both sets of advantages. DC drive is also being used, both in dc motors coupled directly to the turntable shaft, and in servo electronic circuits, employing quartz phase-lock loop for lower wow and flutter, for example, 0.01 to 0.02%.

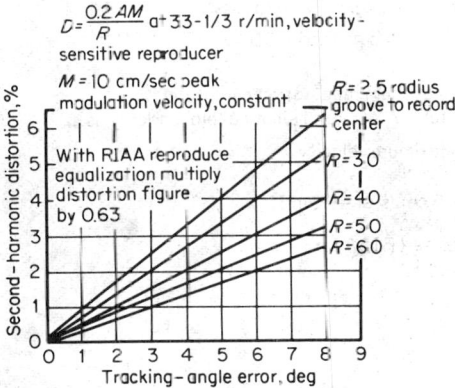

$$D = \frac{0.2\,AM}{R}\ \text{at 33-1/3 r/min, velocity-}$$

sensitive reproducer

$M = 10$ cm/sec peak modulation velocity, constant

$R = 2.5$ radius groove to record center

With RIAA reproduce equalization multiply distortion figure by 0.63

$R = 3.0$
$R = 4.0$
$R = 5.0$
$R = 6.0$

Second-harmonic distortion, %

Tracking-angle error, deg

Reproducer	Record type
10Δ	Vertical
10Δ	Lateral
10Δ	45°
Lateral	Lateral

Percent total harmonic distortion

$F_0 = 400$~z at 2.5 cm/sec per channel
$A = \pm 0.00055$ in

Record diameter, in.

Fig. 19-137. Calculated second-harmonic distortion vs. tracking-angle error at different radii on the disc.[95]

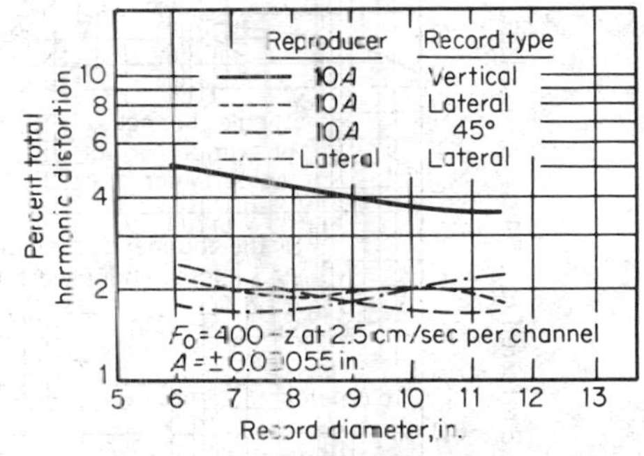

Fig. 19-138. Measured harmonic distortion vs. record diameter for different stereo pickup modes.[95]

Record Changers. Changer mechanisms generally require more power than turntables do and use induction motors. Some systems have two motors, synchronous for the belt drive and induction for the changer mechanism.

Record changers have a special problem in the vertical tracking angle. On a single-play turntable the vertical angle can be corrected by one arm adjustment. On a changer the vertical tracking angle depends upon the number of records in the stack. Generally the angle is compromised and preset for three records in the stack. Some changers with a single-play mode provide one adjustment for the single record and another for a record stack.

A widely used system for changing records has a long center spindle with a retracting member and a small supporting platform at the outer edge of the disc. Other changers use central support only with small retractable members on each side of the spindle, using the disc edge only for diameter sensing in arm control.

Shock and Vibration Isolation. Record players are isolated from external shock and vibration by spring supports which resonate with the changer mass at a very low frequency (see Fig. 19-60). Care must be taken to lock the turntable to its base during transportation; otherwise large excursions at the low-resonance frequency can cause serious damage.

71. Specifications and Measurements. Specifications were previously given for turntable speed, accuracy, wow and flutter, and center-pin dimension, for disc flatness; center-hole diameter and concentricity; groove width, shape, and separation; for stylus tip shape and vertical tracking angle; response-frequency characteristics for the electrical playback system; standard velocity reference levels; minimum rumble level and channel phasing. Another factor is level separation between recorded and unrecorded channels. It should be more than 20 dB (preferably 25 dB) from 100 to 7,500 Hz. Above and below these frequencies the separation should not diminish faster than 6 dB/octave. At 1 kHz the channel balance should match within 0.25 dB.

Record Measurements. Test records can be calibrated by optical-interference patterns[112] in addition to microscopic measurement. Distortion measurements by test recordings[113] are similar to those used for other purposes (see Fig. 19-113). Figure 19-138 is an example of harmonic-distortion measurement for different stereo-pickup modes. Distortion is twice as great for the vertical mode as for the others.

Figure 19-139 is an example of record-noise measurement for a single channel of a Vinylite pressing at 5- and 10-in diameters. The third-octave band spectrum includes the effect of standard playback deemphasis. Also shown (dashed lines) are the single-channel maximum velocity limits set by the physical groove excursion, the maximum slope of a cut, and the minimum radius of curvature of groove modulation. The upper solid curves correspond to the dashed lines after deemphasis. The area between the upper curves and the lower curves defines the available dynamic range for disc recording.

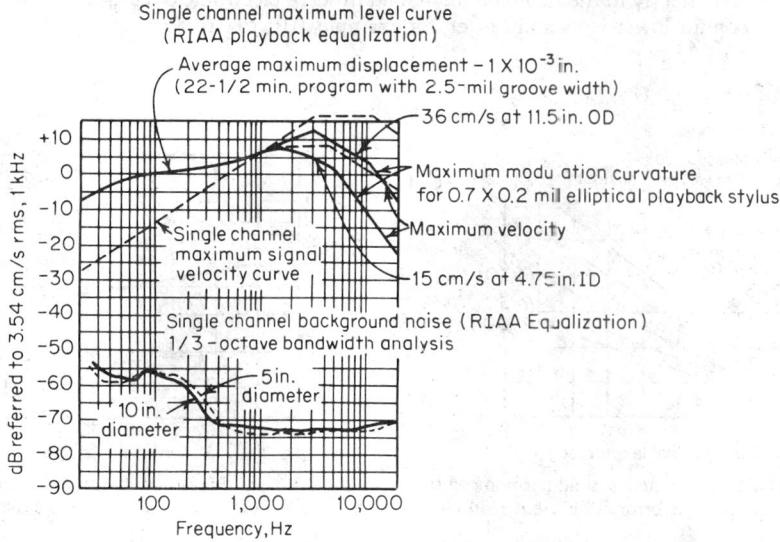

Fig. 19-139. Dynamic range of 33⅓ r/min discs (RIAA equalized) in terms of rms velocity vs. frequency.[114]

MAGNETIC-TAPE SOUND SYSTEMS

72. Principles and Characteristics of Tape Systems.[115–120] In Fig. 19-121 a stereophonic recording system is shown in which the master recording is made on magnetic tape. Facilities are provided either for sound reproduction from the tape or for rerecording upon a disc master. Figure 19-140 shows a typical magnetic-tape sound recorder-reproducer for use in such a system.

In magnetic recording a magnetizable medium (usually tape) from a supply reel is moved at constant speed across a recording head, which induces in the medium a magnetization proportional to the current in the coil of the recording head. Thus the variation of recording current with time is recorded on the tape as a variation of magnetization with distance.

When the tape later passes by the reproducing head, flux from the tape passes through the pickup coil of the reproducing head in proportion to the tape magnetization. Thus during playback (at the same speed) flux variations along the tape induce an electromotive force in the coil of the reproducing head.

In some tape recorders a separate reproducing head is used for continuous monitoring of the recorded result. In other recorders a single head serves as either recorder or reproducer at different times. The erasing head obliterates anything previously recorded and prepares the tape for new magnetization by the recording head.

Constant tape speed, an important requirement, is controlled chiefly by the drive capstan against which the tape is pressed by a roller. Guideposts at each end of the tape segment being driven and recorded isolate the segment from vibrations and variations occurring at the supply and takeup reels, which are generally driven and braked separately to maintain moderate tension in the tape at all times. The reels may be either open or within cartridges or cassettes. When the reels are packaged together, windows are provided (in the cartridge or cassette) where the heads and drive-roller contact the tape segment being recorded or played.

73. Playback Process. The playback process is simpler and easier to understand than the recording process. In Fig. 19-141 the tape record is a multipolar permanent magnet moving past a transducer head. The coil of the head provides an output voltage corresponding to the magnetic field of that part of the magnetic layer in closest proximity to the air gap. Many types of head have been used, but this type with confronting pole heads is the most common.

If the coil has N turns of wire and the recorded flux is ϕ_r, the induced voltage is

$$E = mN \frac{d\phi_r}{dt} \tag{19-10}$$

where m is the fractional part of the recorded flux which threads the coil; m varies with frequency, wavelength λ, gap size g, geometry of the pole pieces, permeability of the head material,

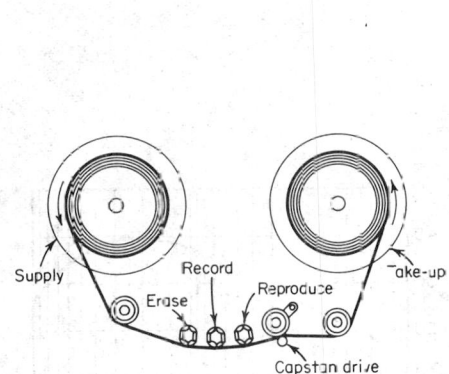

Fig. 19-140. Typical magnetic-tape sound recording and reproducing system.[91]

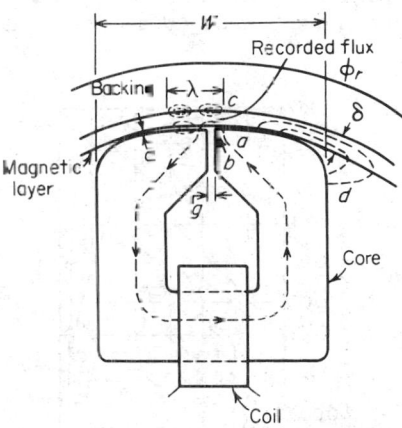

Fig. 19-141. Head-tape configuration.[120]

permeability of the tape, thickness δ of the magnetic layer on the tape, and the clearance α between the head and the tape. Of the various portions of recorded flux shown (Fig. 19-141) d is far from the gap, and c is somewhat removed. Part a, located just outside the gap, threads the pole pieces and coil. When the recorded wavelength is large compared to the width W of the head, the rate of flux change is small and the output is low. As the wavelength approaches the gap size g, the rate and output increase. However cancellation occurs when the wavelength coincides with the gap dimension, as shown in Fig. 19-142. The solid curve is more exact but has a complex function. The approximate dashed curve corresponds to

$$m = \frac{\sin(\lambda/g)}{\lambda/g} \tag{19-11}$$

In practice the output at high frequencies rolls off sooner than predicted by the finite gap width. Other factors include the depth of the magnetic layer and the clearance between the tape and the recording head.

Figure 19-143 shows the effect on the response curve produced by different factors. An ideal tape (curve 1) would have a rising response (6 dB/octave) for constant velocity in which the output is inversely proportional to the wavelength. Finite

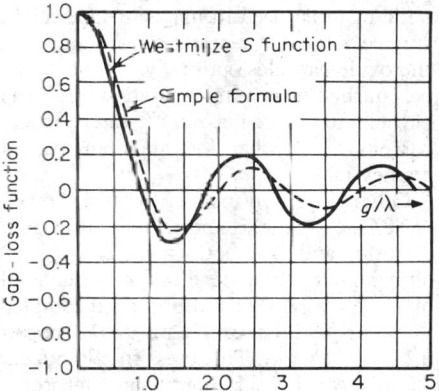

Fig. 19-142. Theoretical output loss at short wavelengths caused by finite gap size.[120]

gap size (curve 4) has its characteristic dip as the wavelength approaches the gap dimension. Curve 2 represents loss caused by the thickness of the layer. Curve 3 shows the effect of small clearance between the tape and head. The combined effects of curves 2, 3, and 4 subtract from the ideal curve 1 to give a typical response-frequency characteristic such as shown in Fig. 19-144 for 7½ in/s tape speed.

74. Tape Materials.[121-124] A review of tape materials and their properties precedes discussion of the recording process. The main function of magnetic tape is to produce at the surface of the reproducing head a magnetic field having the same time variation as that which occurred during the recording process. The tape surface should be smooth and relatively soft to give maximum mechanical contact between the magnetic material and the head. The magnetic layer should be thin to concentrate the magnetic forces near the gap. However, it should be thick enough to prevent amplitude modulation and dropout of the reproduced signal. Magnetic layers on successive wrappings of the tape upon a reel should be separated enough to prevent magnetic interaction known as *print-through*. This layered construction, consisting of a magnetic coating on a nonmagnetic carrier, allows each layer to be optimized separately to fulfill its function.

Polyester films have largely replaced cellulose acetate for magnetic-tape backing because thinner polyester gauges can provide equivalent or greater strength, and polyesters are more stable with regard to humidity, temperature, and time.

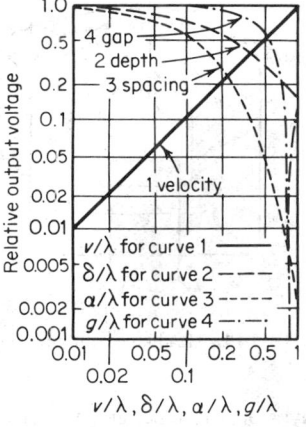

Fig. 19-143. Effect of tape velocity, thickness, spacing, and gap size on output.[120]

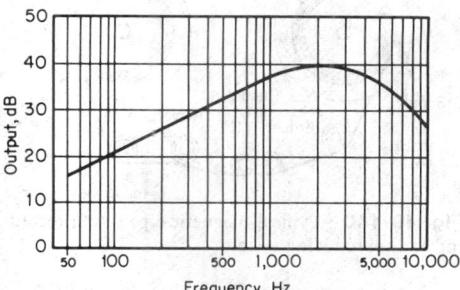

Fig. 19-144. Typical response-frequency characteristic for playback head output from 7½ in/s tape.

Gamma ferric oxide (Fe_2O_3) in the form of needle-shaped crystals has been the principal magnetic material used in magnetic-tape coatings. It is suspended in a liquid binder, coated on one surface of backing material, and (before the liquid dries) passed through a magnetic field to orient the oxide particles optimally. Lubricants, e.g., silicone, can be added, and the tape surface may be polished to minimize friction and head wear. Special tapes of this type have used smaller particles to reduce noise and greater particle density for higher output. Substitution of very small percentages of cobalt for iron atoms has produced higher coercivity and extended the high-frequency range.

Other improved magnetic coatings now widely used include CrO_2, ferrichrome (a thin layer of Fe_2O_3 over CrO_2), and metallic-iron particles[124a] (suspended in a protective lacquer to reduce corrosion and to smooth the recording surface). The resulting higher coercivity and retentivity offer higher output level, high-frequency limit, and S/N ratio but require different recording bias (see Par. **19-76**) and improved complementary head design.

75. Tape Dimensions and Properties. The standard width for nominal ¼-in tape is 0.246 ± 0.002 in. Thickness should not exceed 0.0022 in. Standard base thicknesses are 1.5 and 1.0 mil. A full-track monophonic recording has a width of 0.238 in (Fig. 19-145). For half-track monophonic recordings each track is 0.082 in wide, and the track separation is 0.074 in. These dimensions also hold for two-track stereo, in which the upper track is the left channel. For four-track stereo and four-channel recordings each track is 0.043 in wide, and the spacing between the centerlines of adjacent tracks is nominally 0.067 in. Thus the outside tracks reach the tape edges.

For cassette tape the standard width is 0.150 in (Fig. 19-146) The maximum tape thickness is 0.0008 in. The tape is wound with the magnetic layer facing outward to contact the external heads. The four recorded tracks are 0.025 in. The central blank stripe is wider than the other two separation stripes. When the tape moves from left to right with the magnetic layer facing the observer, the bottom track is numbered 1 and the top 4. Track 1 is the left stereo channel and track 2 the right. (Track 4 left and track 3 right become tracks 1 and 2 when the cassette is turned over.)

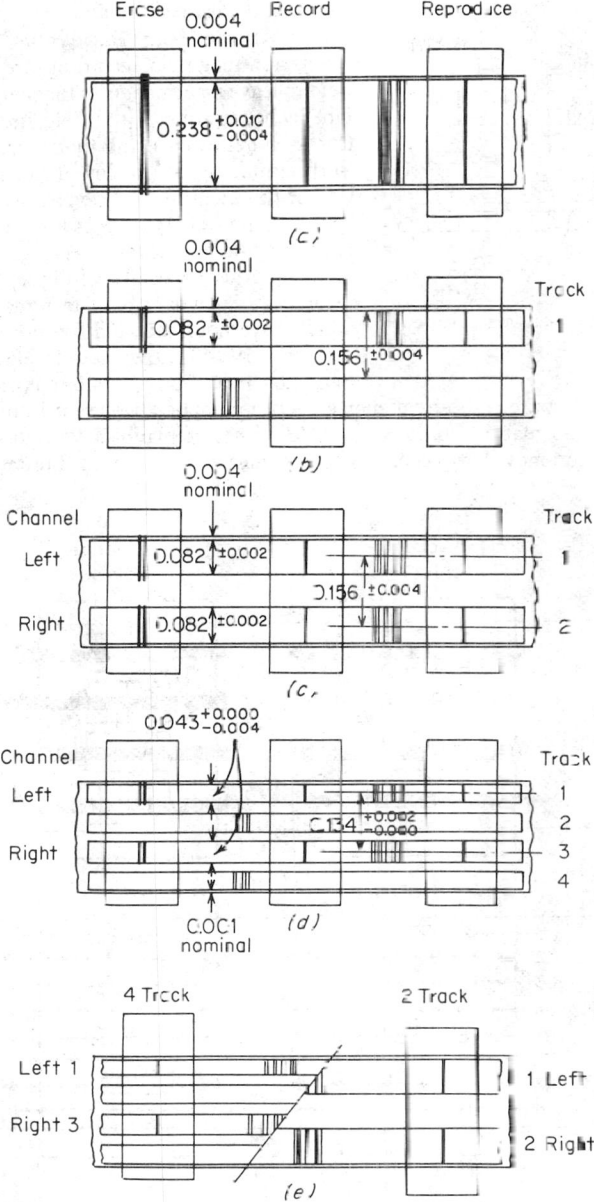

Fig. 19-145. Comparison of track dimensions and locations for different recording modes on ¼-in tape: (a) full track monaural; (b) half-track monaural; (c) two-track stereo; (d) four-track stereo and four-channel; (e) comparison of tracks.

The tape for an eight-track cartridge is standard ¼-in width, and the eight individual tracks are of the same width as in a cassette.

Figure 19-147 summarizes the tracks and playback head locations for open-reel, eight-track cartridge and cassette tapes for both two- and four-channel operation. (In the quadraphonic mode L is left, R is right. F is front, and B is back.)

Educational cassettes generally contain permanent teacher recordings on L1 and L2 and permit recording, erasure and rerecording of student responses on R1 and R2 tracks.

Other mechanical properties[125] affecting tape interaction with mechanical drive systems include static and dynamic coefficients of friction, tape-head abrasion, and adhesion of lubricant where it is used. These factors are especially important to cartridge performance.

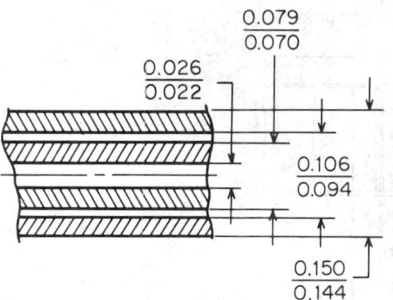

Fig. 19-146. Track dimensions of cassette tape.

76. Recording Process.[126-130] The requirements for magnetic recording are a strong enough field to leave a permanently magnetized record on the tape, concentration of the field into a narrow region (for high definition), and magnetization proportional to the input signal (for low distortion). Proper selection of the medium and reverse use of the head-tape configuration of Fig. 19-141 satisfy the first two requirements. However, meeting the third requirement requires a special technique, because magnetic recording depends upon ferromagnetic materials, which have nonlinear and hysteresis properties.

In Fig. 19-148 curve *abocd* relates the remanent induction in the tape to the maximum magnetizing force in the gap when pure audio-frequency current is supplied to the recording head. The nonlinearity in this transformation would give a highly distorted playback waveform. The ingenious method of high-frequency bias recording overcomes this inherent limitation of magnetic recording.

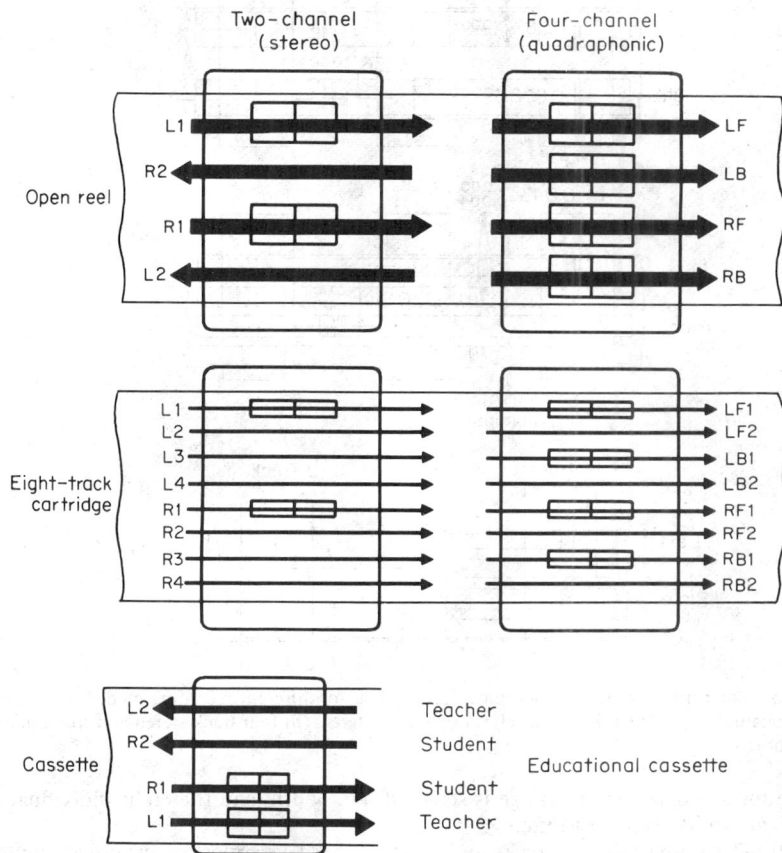

Fig. 19-147. Comparison of track configurations for reel, cartridge, and cassette tapes in two- and four-channel modes.

In this method the unmagnetized tape receives at the recording head a magnetic field which is compounded of the audio signal and an ultrasonic (50- to 150-kHz) component called the *bias*. These two signals are simply added in the recording coil without modulation of one frequency by the other. Electrical addition can be accomplished in various ways. The ultrasonic oscillator output can be connected either in parallel or in series with the audio input to the head, or it can be supplied to a separate winding in the record head. A separate head for the ultrasonic field may also be placed opposite the record head in close proximity to the tape.

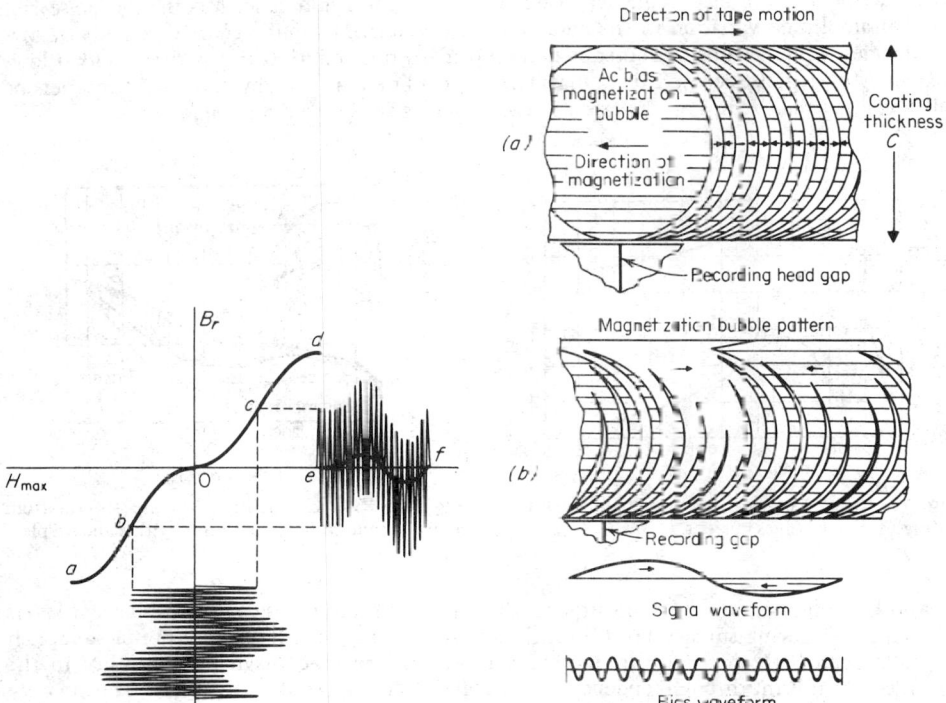

Fig. 19-148. Linear recording by the addition of a high-frequency bias signal.[91]

Fig. 19-149. Model for 100-kHz ac bias recording of 0.3-mil tape coating at 3¾ in/s: (a) zero audio signal; (b) audio signal. (After Ref. 126.)

A simple (incomplete) explanation of bias recording is based on Fig. 19-148. The algebraic sum of the audio and bias signals is applied along the H axis, the bias signal being larger than the audio. This keeps the envelope away from the origin region, in which non linearity occurs. The output variations along the B axis have a similar envelope shape. As the tape leaves the gap, demagnetization takes place at the bias frequency. The remaining net induction is the difference between the positive and negative half cycles of the wave, curve *ef*, which closely resembles the original audio signal. In the absence of sound the tape is almost completely demagnetized for low noise level.

Proper adjustment of the bias amplitude and suitable limitation of the audio amplitude transfer the audio to the tape in the linear portions *b* and *c* of the characteristic curve, minimizing distortion. The shortcomings of this explanation are detailed in the references, which also give alternative explanations in terms of Preisach diagrams and statistical theories and experiments on the fine structure of the recorded magnetic pattern on the tape (Fig 19-149). Understanding of the process is not yet complete.

The optimum amount of ultrasonic bias current depends upon the magnetic properties of the tape used. Figure 19-150 is an example of the variation of tape output level and percent distortion as the bias current is changed. The current is adjusted to give a good compromise between signal-to-noise ratio and distortion.

77. Erasure and Noise.[131,132] Complete reels of tape can be erased by gradual removal or reduction of the magnetic field of a bulk demagnetizer. Also just before recording occurs the erase head of Fig. 19-140 saturates the tape at an ultrasonic frequency (usually the bias frequency to avoid beats). As the tape leaves the gap, the fringes of flux around the edges of the wide erasing gap subject the tape to successive magnetizing cycles of decreasing intensity. Even when bulk-erased, the tape has some background noise because the medium is not continuous.

In the example of Fig. 19-151 the lowest noise spectrum is that generated in the playback equipment when no tape is running. The erased-tape noise is slightly higher in well-designed equipment except at very high frequencies. Introduction of the bias signal increases the noise further, even in the absence of audio. This can result from asymmetry in the bias field or from AM sidebands of the bias frequency occurring within the audio band. Modulation noise is a random amplitude variation in the audio output for constant input, occurring chiefly at low frequencies, which can be attributed to nonuniform distribution of coating particles. Other low-frequency or dc noises come from a magnetized recording head, nearby permanent magnets or solenoids, or small bumps on the tape or head which cause temporary separation.

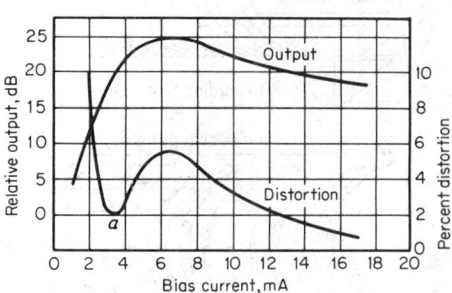

Fig. 19-150. Dependence of output and distortion on high-frequency bias current.[91]

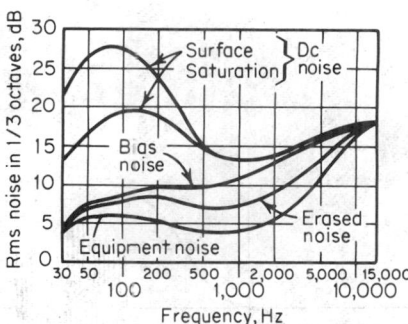

Fig. 19-151. Comparison of noise spectra from audio cassette recorder at 1⅞ in/s with standard playback equalization.[132]

Another common noise is print-through, the printing of recorded signals onto adjacent layers of a tape reel during storage. Print-through is sensitive to time and temperature of storage. It depends upon both proximity to another magnetized layer and magnetic instability in the recorded layer. Print-through tendency can be reduced through control of particle size and crystalline smoothness within the medium.

An important factor in selecting track width is signal-to-noise ratio. Playback signal is proportional to track width, but noise increases with the square root of the width. Thus halving track width reduces the signal-to-noise ratio 3 dB.

78. Tape Standardization and Packaging. The wide variety of tape-recording formats and equipment is based upon a family of standards (from various standards organizations) too numerous to list here.[133] Several standards give detailed support to open-reel, cartridge, and cassette packaging of tapes and associated record-reproduce equipment.[134-136] Tape and track dimensions quoted previously were from these standards. In addition the NAB standards cover hub and reel diameters, winding instructions, recorded level and uniformity, reel dimensions, tape speeds, and the dimensions and shapes of cartridges and their mechanical features. Signal-to-noise ratio (unweighted and weighted), distortion, crosstalk, and stereophonic channel separation are also included. The specified response-frequency characteristics for standard reproducing systems and for standard recording are reviewed in Par. **19-84**.

Special-purpose magnetic-tape cartridges have been developed for continuous-loop operation[137] in which the tape is drawn from the center of the reel. After it passes the heads and pinch roller, it returns to the outside of the reel. The necessity for continuous slippage between adjacent turns of tape requires coating the base material with a lubricant which wears well and will not disintegrate. These are used for eight-track cartridges.

79. Tape Speeds and Related Properties. Standard preferred tape speed is 7½ in/s ± 0.2%. Supplementary tape speeds are 15 and 3¾ in/s with the same tolerance. Recent improvements have expanded the use of 3¾ in/s. Cassette tape speed is 1⅞ in/s.

Flutter. Perceptibility of flutter varies with flutter rate, leading to the weighting curve shown in Fig. 19-122. With a 3-kHz test tone the standard limits for weighted rms flutter are 0.05% at 15 in/s, 0.07% at 7½ in/s, and 0.10% at 3¾ in/s, measured over a flutter frequency range of 0.5 to 200 Hz.

Frequency Limits.   NAB standard tapes contain test tones from 30 Hz (for all tape speeds) to 15 kHz for 7½ and 15 in/s; to 10 kHz for 3¾ in/s; and to 5 kHz for 1⅞ in/s. However, recent tape material and technique improvements extend measurements to 10 kHz and even higher at 1⅞ in/s, making ¹⁵⁄₁₆ in/s a future practical possibility.

80. Tape Transport Mechanisms.[138,139]   The single-capstan system shown in Fig. 19-140 is typical. Figure 19-152 identifies its dynamic elements for theoretical analysis. Semiprofessional machines often use a single motor for the capstan drive. It also operates both the supply and take-up reels through friction idler wheels. Better recorders generally use three motors, a multispeed hysteresis-synchronous motor for the capstan and bidirectional torque motors for the supply and take-up reels.

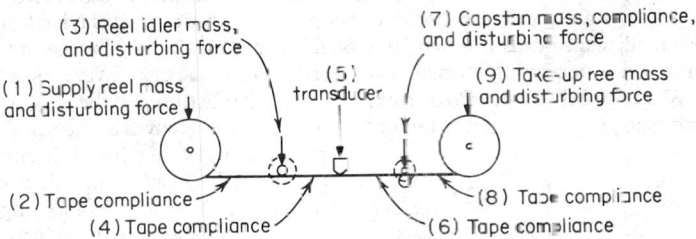

Fig. 19-152. Physical system for a simplified tape transport.[139]

In the record and playback modes the motor on the supply reel (although rotating counterclockwise) exerts a light clockwise torque to provide even tape tension as it unwinds. The take-up motor rotates counterclockwise with just enough torque to supply the correct tension for smooth tape wind. During rewind the supply motor has a high torque in a clockwise direction and the take-up motor a low torque in the opposite direction. For fast forward the situation reverses. Electromechanical brakes on the torque motors stop and hold the reels to prevent spilling the tape.

Another transport system uses a dual-capstan drive, illustrated in Fig. 19-153. Here the capstans tend to turn at slightly different speed, and the tape tension is controlled by the differential action. This system provides better isolation of the active tape segment from the supply and take-up reels. A third type of transport eliminates the pinch-roller completely by using a capstan with high surface friction and a large tape-wrap angle.

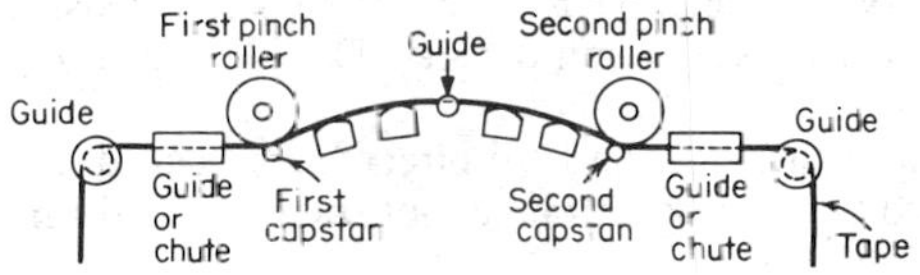

Fig. 19-153.   Dual-capstan tape-transport system.

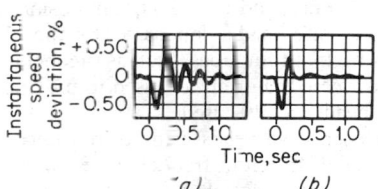

Fig. 19-154. Measured speed deviation vs. time for an impulse at the supply reel, MR-70 tape transport with ½-in-wide tape: (a) undamped flywheel on reel idler, $Q = 4.5$; (b) viscous damper on reel idler, $Q = 1.7$.[139]

In all these systems the goal is constant tape tension, but in some the result is constant torque, so that the tension varies with the radius of the tape winding. In fully professional equipment servo controls have been used to sense tape tension and correct reel torque accordingly.

Mechanical Damping.   Figure 19-152 shows only inertial and compliance elements, omitting resistance elements such as tape friction at heads and guides and tape damping by the rubber puck at the capstan. Without damping, the dynamic system would be unduly excited by mechanical transients caused by tape splices, tape-reel contacts, mechanical imperfections or power fluctuations. Figure 19-154 demonstrates a transient-response improvement produced by addition of a viscous damper to a reel idler.

81. Magnetic-Tape Heads.[140]   Figure 19-141 shows the general shape and design factors. Typically the magnetic circuit consists of two high-permeability laminated cores which are butted together with a thin gap-defining shim between them to form a symmetrical ring-type structure. Balanced coil windings cancel external magnetic fields. After terminals are connected, the assembly is encapsulated. The tape-contacting surface is ground and lapped for curvature and smoothness.

One method of multitrack head assembly precisely stacks individually encapsulated wafer head elements. Selective assembly of tested wafers gives well-matched groups. Shields are selected for thickness to compensate for wafer-thickness variations.

In another assembly method cast case halves are precisely milled to receive shields and laminated core halves which with their coils are epoxied in place. The face of the entire half-head assembly is ground and lapped to achieve a high degree of flatness. A quartz spacer of less than 0.0001 in is evaporated on the pole-tip area before the halves are joined. The laminations are grounded electrically. The entire assembly is shielded for minimum pickup of extraneous fields.

Figure 19-144 shows a typical response-frequency characteristic for a playback head. Typical crosstalk between adjacent heads is down 50 to 60 dB, degraded slightly by the presence of bias. This is compromised in recording formats using very narrow tracks and very close spacing.

82. Head Wear and Tape Contact.[141,142] Abrasive action by the tape oxide surface wears heads and tape guides. If heads extend beyond tape edges, grooves are worn which damage the edges and cause intermittent tape separation from the head. (The effect of separation was shown in Fig. 19-143.) High-frequency losses caused by head wear can be partially regained by relapping the head.

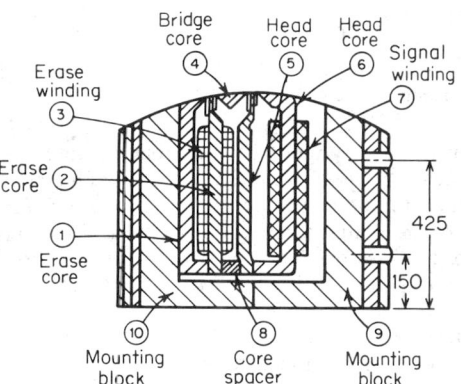

Fig. 19-155. Stereo X-field head. *(After Ref. 143.)*

When tape under tension wraps around a curved surface, a transverse tape curvature is induced. The smaller the wrap curvature and the higher the tension, the greater the tendency for tape cupping, edge separation, and formation of concave grooves.

83. Special Heads.[143–145] Recording resolution can be improved by sharpening the field at the recording gap. When the semicircular field of a conventional head is modified by adding a field normal to the tape thickness, a sharper gradient is produced at one edge of the gap. An X-field head using this principle, shown in Fig. 19-155, uses the erase gap for the X field. Thus only two coils are required per channel. Erase and record gap dimensions are 4 and 0.005 mil, respectively. Elimination of the separate erase head compensates for the additional X-field structure.

For high-output tapes, such as the metal-particle type, special record and playback heads using Sendust core materials are used to provide for the higher saturation density. Double-gap ferrite erase heads are also used.

Flux-sensitive heads using thin semiconductors sensitive to the Hall effect have been developed. Their inherent frequency response is flat into the ultrasonic range, but there are still limitations in the associated head structure.

84. Equalization. The basis for tape equalization is the NAB standard reproducing characteristic shown in Fig. 19-156. This assumes that the recorder circuits will preemphasize low-frequency response and that typical high-frequency losses must be equalized in playback by different amounts, depending upon tape speed. Although curve B is identified with 1⅞ in/s, the standard reproducing characteristic for cassette tapes assumes greater bass preemphasis (1,590-μs time constant) and greater treble loss (120-μs time constant). This corresponds to approximately +10 dB at 30 Hz and −18 dB at 10 kHz. Improved cassette tape material is expected to reduce 120 to 70 μs.

Figure 19-156 assumes constant flux in the reproducing head core, theoretically removing the recorded tape from the measurement. Calibrated standard NAB tapes made under laboratory conditions are available to manufacturers and users. Their test-tone levels are preemphasized so that the output level of a tape reproducer meeting the requirements of Fig. 19-156 should be uniform within the limits of Fig. 19-157.

After the recorder-reproducer designer has met the reproducer limits of Fig. 19-157, the preemphasis circuits of the recorder are designed to give an overall record-reproduce response which is similarly uniform. This equipment specification must then identify the tape type used.

Although the standard reproducing characteristic assumes specific resistance-capacitance networks, preemphasis and postequalization circuits can be designed which are complementary to other system parameters in order to obtain the desired result for a standard tape.

85. Noise-Reduction Circuits.[145–149] At low recording levels the need to suppress background noise from the recording medium has led to various signal-processing systems, some involving both pre- and postprocessing and others acting only on the signal played back.

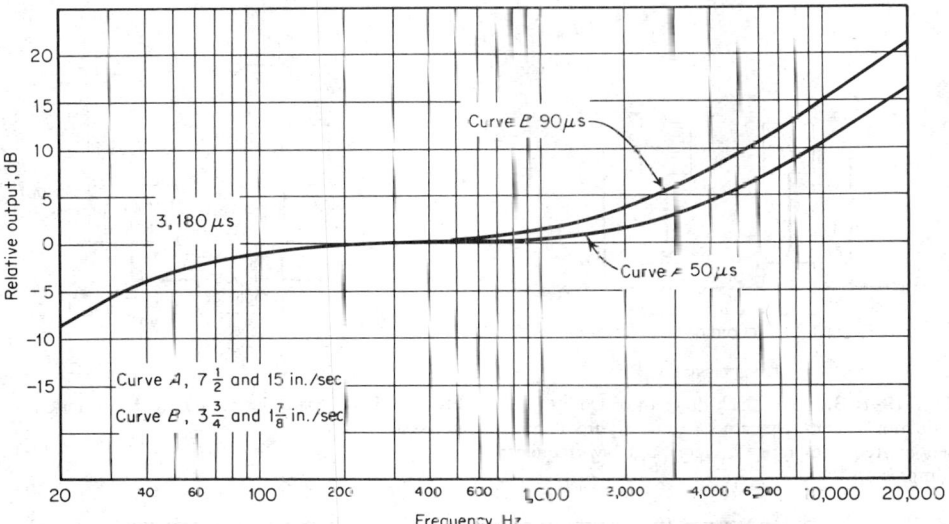

Fig. 19-156. NAB standard reproducing characteristic. Amplifier output for a constant flux in the core of an ideal reproducing head[134]

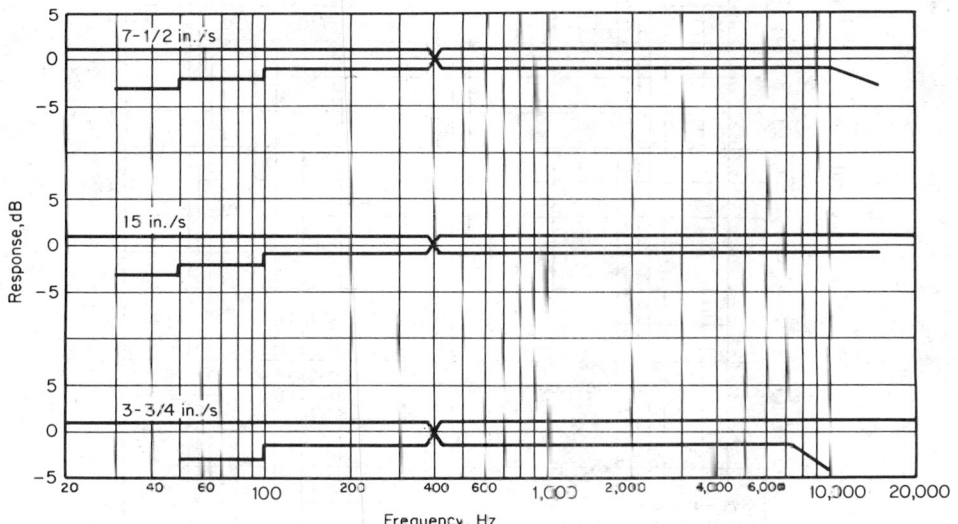

Fig. 19-157. NAB standard reproducing-system response limits.[134]

Figure 19-158 shows a system operating differentially and oppositely on the signal before recording and after playback. Signal multipliers G_1 and G_2 are controlled by the amplitudes and dynamic properties of the signals fed into them. Network G_2 during reproduction sends low-level components back to the subtractor, partially canceling these components in the reproduced signal. To compensate, identical network G_1 adds an identical component ahead of the recording. Originally G_1 and G_2 comprised identical sets of four filters and low-level compressors. At high sound levels all bands were reproduced normally. At low levels the hum, rumble, and hiss ranges were deemphasized. The Dolby type B system uses only two channels with a 1-kHz crossover frequency. Dolby HX (headroom extension) also reduces bias current and noise for bright tone spectra.

Another approach (DBX) compresses the dynamic range about 2:1 at all levels and frequencies before recording, then expands 2:1 before reproduction. A variation (Sanyo Super D) uses less

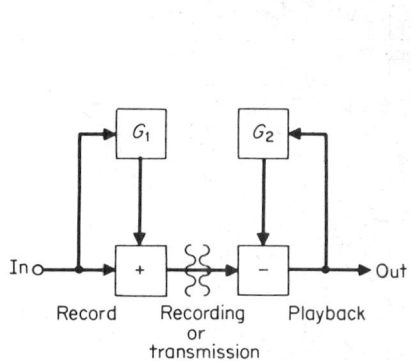

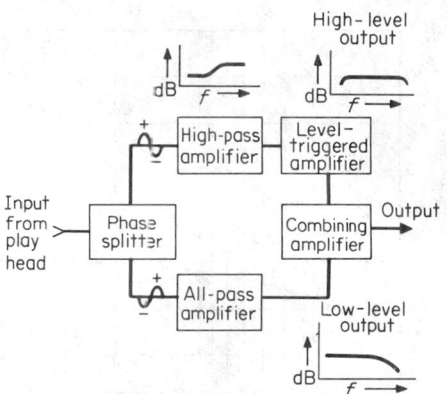

Fig. 19-158. Basic block diagram of Dolby noise-reduction system. Operators G_1 and G_2 are identical sets of frequency-range dividers and low-frequency compressors.[146]

Fig. 19-159. Basic block diagram of hiss-noise-limiter system.

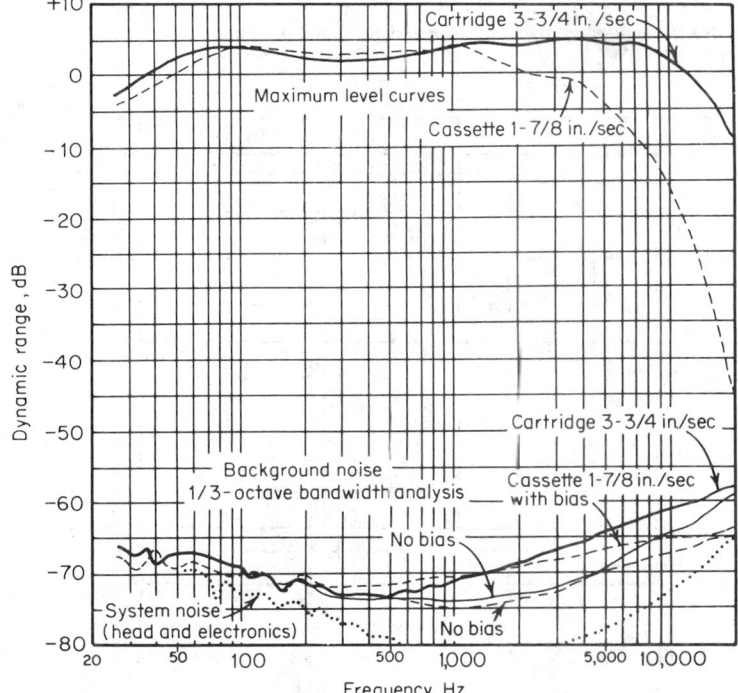

Fig. 19-160. Dynamic range of tape cartridges and cassettes. *(After Ref. 114.)*

compression when both frequency and level are high. Analog recording on improved tape, combined with electronic noise-reduction circuits, approaches the dynamic range possible with digital PCM recording.

The block diagram in Fig. 19-159 shows a noise limiter operating in the playback amplifier after playback equalization. High sound levels are reproduced only through the all-pass amplifier. The phase splitter supplies opposite-phase signals to a 5-kHz high-pass amplifier. When its output falls more than 40 dB below reference level, the level-triggered amplifier output cancels the high-frequency output from the all-pass amplifier. The level-triggered amplifier is normally open during silent passages, effectively canceling interim hiss.

86. Measurements. Knowing how much flux is actually recorded on a tape aids in determining and standardizing reference levels, specifying the properties of magnetic media, and measuring the sensitivity of magnetic heads. Special high-efficiency symmetrical heads have been constructed, calibrated, and verified by magnetometer flux measurements for these purposes.[150]

Overall tape-system measurements are highly complex involving many factors. Their data reflect the state of the art and indicate directions for needed improvements. They also provide guidance for the users of recording equipment. Figure 19-160 is an example of a comprehensive study of the dynamic range of tape cartridges and cassettes.

87. Specifications. The data in the comparison of two tape record-reproducers shown in Table 19-8 were published by independent testing laboratories. Recorder A is open-reel type. Recorder B is an advanced cassette recorder. The price is approximately the same.

TABLE 19-8   Comparison of Tape Recorders

	Recorder A	Recorder B
Type	Open reel	Cassette
Number of channels	2	2
Number of tracks	4	4
Tape width, in	0.246	0.150
Tape speeds, in/s	3¾ 7½	1⅞, ⁵⁄₁₆
Speed accuracy	0.2%	0.4%
Wow and flutter	0.04%	0.06%
Reel size	5 to 10½ in NAB	Cassette
Start time, s	0.1	1.0
Rewind time, s	70 (2,500 ft)	55 (1 800 ft)
Signal-to-noise ratio, dB	66	66
Erasure, dB	70	76
Crosstalk, dB	65	50
Percent third harmonic distortion	<1	<0.8
Record and playback response	20 Hz–20 kHz ± 2 dB	20 Hz–20 kHz ± 3 dB
Head assembly	Erase, record, reproduce	Erase, record, reproduce
Capstan drive motor	Brushless synchronous	
Reel motors	2 direct-coupled	
Drive	Belt to flywheel	
Braking	Electrodynamic	

DIGITAL AUDIO RECORDING AND REPRODUCTION

88. Background. Interaction between different branches of electronic engineering often produces fruitful results. Just as the field of digital computers borrowed storage means from the field of magnetic recording and video recording (both tape and disc) built upon audio recording, a return flow of technology has provided digital recording and reproduction means for potential advancement in the field of sound recording and reproduction. Audio engineers seeking background in digital concepts, terminology and systems, as related to the present topic, are referred to Secs. **4-1** and **4-32** to **4-36,** on information, communication, noise, and interference; Secs. **23-1** to **23-11,** on electronic data processing; Secs. **14-33** to **14-35,** on modulators, demodulators, and converters; Secs. **20-85** through **20-110a,** on video recording systems; and Sec. **22-39,** on communication systems.

It is appropriate to include digital audio recording and reproduction here, before its widespread commercial adoption, because the potential for improvement over the current (1980) analog audio state of the art is an order of magnitude for several operating parameters[151] and because the development progress made so far in digital recording and reproduction for both tape and disc makes its widespread use highly probable. Moreover the digital techniques have already been applied in communication transmission networks.

89. Digital Encoding of the Analog Waveform. Figure 19-161 is a simple example using only 4 bits. The amplitude of the continuous analog audio signal wavetrain A is sampled at each narrow pulse in the clock-driven pulse train B, yielding for each discrete abscissa (time) value a discrete ordinate (voltage) value represented by a dot on or near the analog curve. The vertical scale is subdivided (in this example) into 16 possible voltage levels, each represented by a binary number or "word." The first eight words can be read out either in parallel

$$1000, 1010, 1011, 1011, 1010, 1000, 0110, 0101, \ldots$$

on four channels, or in sequence

$$1000101010111011101010000110010 1 \cdots$$

on a single channel for transmission, optional recording and playback, and decoding into an approximation of the original wavetrain. Unless intervening noise approaches the level of the digit 1, the transmitted or played-back digital information matches the original digital information. The degree to which digitization approximates the analog curve is determined by the number of digits chosen and the sampling rate.

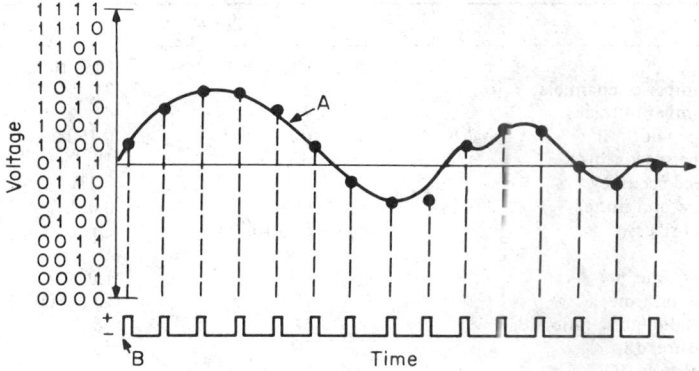

Fig. 19-161. Digital encoding of an analog waveform: (*A*) continuous analog signal wavetrain; (*B*) clock-driven pulse train. At equal time intervals, sample values are encoded into nearest digital word.

90. Record-Playback System.[152,153] Figure 19-162 shows the main electronic blocks of a 5-bit digital recording and playback system without any of the details concerning the actual recording means, recording medium or playback means. The detailed design of much of this circuitry is in the domain of digital-circuit and integrated-circuit engineering, but the selection, specification, and (ultimately) standardization of sampling rate, filter cutoff frequency and rate, number of digits, digitization method, and code error-correction method depend primarily upon the audio recording system engineer, in consultation with video, broadcast, and transmission engineers with whose systems compatibility is desirable.[154,155]

Sampling Rate.[156]   The Nyquist criterion requires the sampling frequency to be at least twice the highest audio frequency to be recorded and reproduced. Since both the input filter, which limits the spectrum to be recorded, and the output filter, which removes the sampling frequency after reconversion, have a finite rate of cutoff, the ratio must exceed the theoretical doubling. The upper frequency limit to be reproduced depends upon system goals, ranging from 4 kHz for telephone systems to 20 kHz for master recordings of music. Corresponding sampling frequencies now in use are 8 kHz for telephone transmission and from 42 to 50 kHz in master recording of music.

Number of Digits.   Although speech-transmission systems need only a 7-digit code (128 identifiable amplitude levels), the digital master recorders for music employ up to 16-digit codes (up to 65,536 identifiable levels), to obtain the extraordinary dynamic range and signal-to-noise ratio which pulse-code modulation (PCM) can theoretically provide with linear digitization. (Other possible signal digitization methods, such as companded and floating-point digitization, have been considered for economic compromise but are not currently in use.)

The dynamic range and signal-to-noise ratio are determined by the number of bits used in quantizing each sample amplitude. Theoretically 14-bit words can provide 84 dB and 16-bit words 96 dB, the low limit being determined by the quantization noise of the least significant

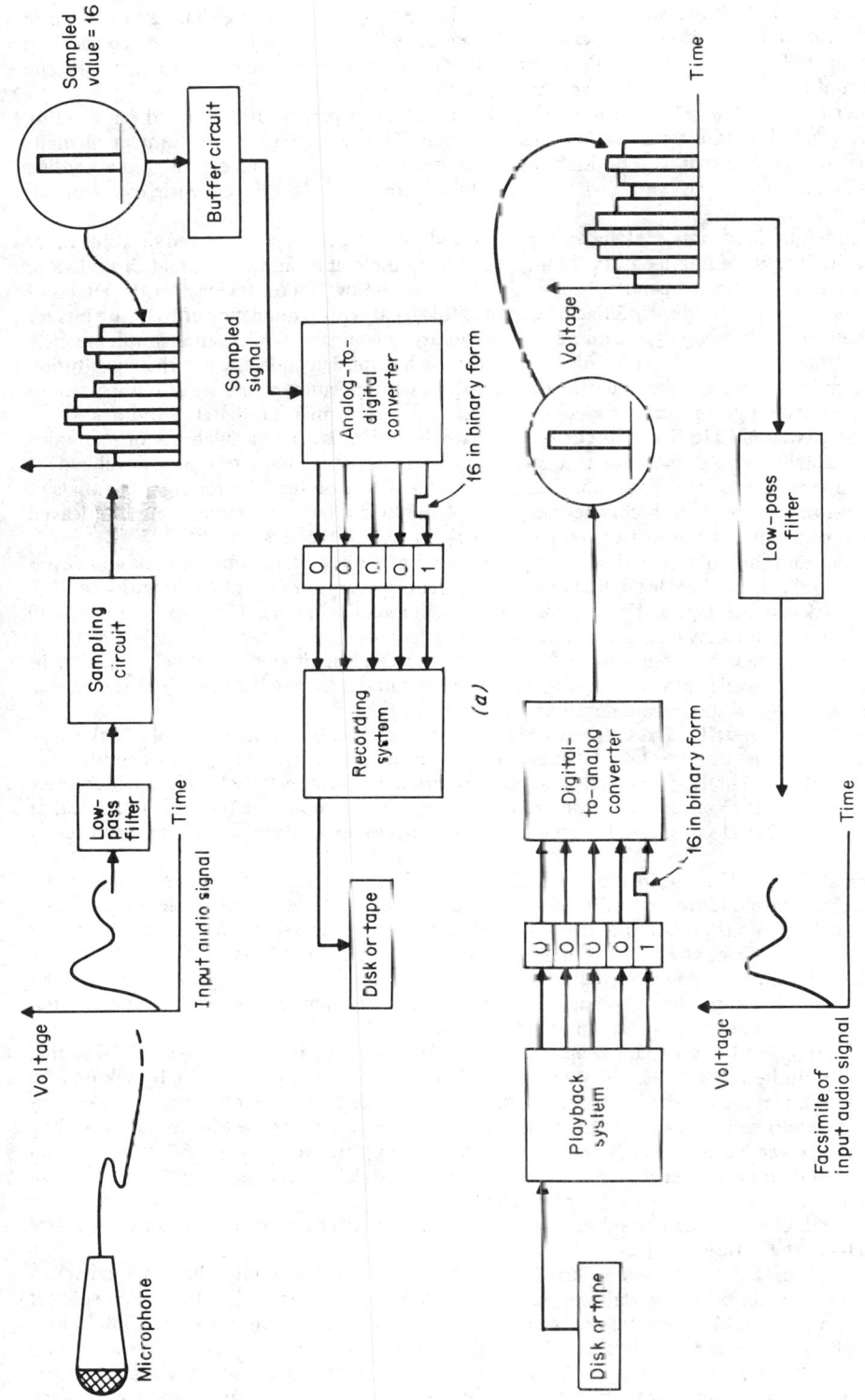

Fig. 19-162. The basic electronic system components for encoding and decoding digital audio signals for (a) recording system and (b) playback system.

digit. These large words require either an equivalent number of recording channels (in the case of tape) for parallel readout or a very high digital recording frequency (for either tape or disc) for sequential readout into a single channel. In the latter case the recording frequency is the product of the sampling rate and the number of digits per sample.

Other Uses for Digits.[157]   Digit space in the recordings is periodically required for synchronization and also for error detection and correction. This compromises the number of digits devoted to signal encoding. The high degree of synchronization possible in digital recording virtually eliminates the wow and flutter problems commonly associated with conventional analog audio recording.

An inherent disadvantage of digital recording and playback is *dropout*, caused by voids in, or damage to, the recording medium. Some dropouts are inevitable, and they could cause loss of synchronization with temporary chaos in the information flow. Protection against the effects of dropouts can be provided by interlacing the encoded signal with redundancy or by using bits for error-detection schemes, e.g., recording sums of words, for comparison with sums simultaneously obtained from playback of the words themselves. Such error detection triggers the substitution of adjacent data, for example, into the dropout gap. Synchronization pulses are also important to the special techniques required for editing of each different format for digital recordings.

91. Digital Audio Tape Recording and Playback. The availability of many different adaptable types of magnetic tape recorder accelerated digital-audio-recording development in the tape medium.[158] Nippon Columbia developed a PCM tape recorder for eight channels of audio information with each channel sampled at 47.25 kHz. Channel samples were interleaved in 13-bit code, plus bits for parity and phase check, in a video-type recorder.[159]

Soundstream, Mitsubishi, 3M, and Ampex have all made important contributions and developments to this field. The 3M digital mastering system, developed in conjunction with the BBC, has 32 tracks on 1-in tape at 45 in/s with one track per audio channel. The sampling rate is 50 kHz. Cyclic redundancy checks are used for error detection.

Digital audio tape recordings have already contributed to the production of better quality in analog discs. They also provide the original program material from which much of the experimentation in digital disc record development is progressing.

92. Digital Audio Disc Recording and Playback. As in video playback discs, there are two general types of digital audio discs under development. One type uses optical laser recording of the digital information in a spiral pattern and optical playback means which track the pattern without contacting the disc directly. The other general type records mechanical or electrical digital tracks along a spiral groove, which provides the guidance for a lightly contacting pickup.

Optical Discs.   The Philips Compact Disc system uses an optical pickup, shown in Fig. 19-163. Light from an aluminum–gallium arsenide diode laser is focused on the surface of the spinning disc. Digital pits pressed into the disc surface, corresponding to patterns recorded by a laser beam on a photosensitive layer on the disc master, modulate the intensity of the beam reflected back into the pickup. The reflected beam is split into two parts for the separate elements of a two-element photodiode for two-channel stereo reproduction from left- and right-channel information carried in alternate words on the record track.

The Sony DAD-1X two-channel optical disc system[160] uses 16-bit linear-encoded PCM with a run-length-limited code termed 3PM (three-position modulation) to achieve a high packing density, for 2½ h playing time on one side of a 30-cm-diameter disc of polyvinyl chloride, coated for reflection, and rotating at 450 r/min. A high-power argon gas laser is used for master recording and a low-power helium-neon laser for playback. Sampling frequency is 44.056 kHz to match the PCM-1600 VTR master recorder. An error-correcting code called *cross interleave correction* is intended to be compatible with various decoder types.

Semiconductor lasers can be substituted for He–Ne lasers to effect great reductions in the size and mass of the optical disc pickup.[161]

Grooved Discs.[162]   AEG-Telefunken and record producer Teldec GmbH have demonstrated a 13.5-cm-diameter polyvinyl chloride audio Mini-Disk, recorded on both sides, which spins at 300 r/min. The PCM system, using 14-bit linear coding with a sampling frequency of 48 kHz for 84 dB dynamic range, records the binary information as mechanical pits in the surface of a metallic plate master for pressing the records. In playback, mechanical readout is by a piezoelectric pickup bearing lightly on the disc for minimum wear. A motor drive system guides the pickup for coarse tracking and tangential scannig, while fine tracking of the stylus is controlled by the groove.

An audio counterpart of the RCA grooved capacitance video disc, in which a metal-backed diamond-tipped stylus picks up the information from electrical capacitance variations pressed into the grooves of a conductive plastic disc, is expected. However technical details for audio have not been announced.

Matsushita and the Victor Company of Japan have proposed a grooveless disc with variable-capacitance signal tracks. A flat metal capacitor pickup rides the disc surface lightly and is guided by tracking signals recorded on the disc instead of by disc groove walls.

It is evident that many commercial factors (especially in the video disc field), in addition to the technological factors will influence both the trend of further development and the standardization[153] needed for full consumer acceptance of digital audio recording.

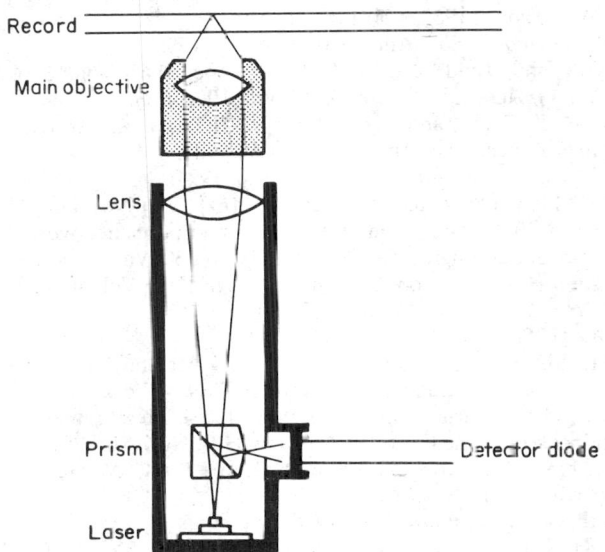

Fig. 19-163. Optical pickup for the digital audio Compact Disc by Philips.

93. References

1. Acoustical Terminology (Including Mechanical Shock and Vibration), SL1-1960, American National Standards Institute, Inc., New York.

2. Young, R. W. Physical Properties of Noise and Their Specification, Chap. 2 in C. M. Harris (ed.), "Handbook of Noise Control," McGraw-Hill, New York, 1957.

3. Robinson, D. W., and R. S. Dadson Br. J. Appl. Phys., 1956, Vol. 7, p. 156.

4. ISO Recommendation 226. International Organization for Standardization, 1961.

5. Expression of the Physical and Subjective Magnitudes of Sound or Noise, ISO Recommendation 131, International Organization for Standardization, 1959 (supplemented by R357-1963).

6. Munson, W. A. The Loudness of Sounds, Chap. 5 in C. M. Harris (ed.), "Handbook of Noise Control," McGraw-Hill, New York, 1957.

7. Method for Calculating Loudness Level, ISO Recommendation 532, International Organization for Standardization, 1966.

8. Stevens, S. S. J Acoust. Soc. Am., 1972, Vol. 51, p. 575.

9. Martin, D. W. J. Acoust. Soc. Am , 1942, Vol. 13, p. 309.

10. Dunn, H. K., and S. D. White J. Acoust. Soc. Am , 1940, Vol. 11, p. 278.

11. Dunn, H. K., and D. W. Farnsworth J. Acoust. Soc. Am., 1939, Vol. 10, p. 184.

12. Potter, Ralph K., G. A. Kopp, and H. C. Green "Visible Speech," Van Nostrand, New York, 1947.

12a. Miller, G. A. "Language and Communication," McGraw-Hill, New York, 1951.

13. Dunn, H. K. J. Acoust. Soc. Am., 1950, Vol. 22, p. 740.

14. Potter, Ralph K., and G. E. Peterson J. Acoust. Soc. Am., 1948, Vol. 20, p. 528.

15. French, N. R., and J. C. Steinberg J. Acoust. Soc. Am., 1947, Vol. 19, p. 90.

16. Beranek, L. L. Proc. IRE, 1947, Vol. 35, p. 880.

17. Hawley, M. E., and K. D. Kryter   Effects of Noise on Speech, Chap. 9 in C. M. Harris (ed.), "Handbook of Noise Control," McGraw-Hill, New York, 1957.

18. Webster, J. C.   *J. Audio Eng. Soc.*, 1970, Vol. 18, p. 114.

19. Licklider, J. C. R.   *J. Acoust. Soc. Am.*, 1946, Vol. 18, p. 429.

20. Martin, D. W.   *J. Acoust. Soc. Am.*, 1950, Vol. 22, p. 614.

21. Martin, D. W., R. L. Murphy, and Albert Meyer   *J. Acoust. Soc. Am.*, 1956, Vol. 28, p. 597.

22. Olson, H. F.   "Musical Engineering," McGraw-Hill, New York, 1952.

22a. Olson, H. F.   "Music, Physics and Engineering," 2d ed., Dover, New York, 1967.

23. Ward, W. D.   *J. Acoust. Soc. Am.*, 1954, Vol. 26, p. 369.

24. Martin, D. W., and W. D. Ward   *J. Acoust. Soc. Am.*, 1961, Vol. 33, p. 582.

25. Martin, D. W.   *Sound*, 1962, Vol. 1, p. 22.

26. Snow, W. B.   *J. Acoust. Soc. Am.*, 1931, Vol. 3, p. 155.

27. Clark, Melville, and David Luce   *J. Audio Eng. Soc.*, 1965, Vol. 13, p. 151.

28. Martin, D. W.   *J. Acoust. Soc. Am.*, 1947, Vol. 19, p. 535.

29. Sivian, L. J., H. K. Dunn, and S. D. White   *IRE Trans. Audio*, 1959, Vol. AU-7, p. 47; revision of paper in *J. Acoust. Soc. Am.*, 1931, Vol. 2, p. 33.

30. Meyer, J.   *J. Acoust. Soc. Am.*, 1972, Vol. 51, p. 1994.

31. Olson, H. F.   "Elements of Acoustical Engineering," Van Nostrand, New York, 1947.

32. Wayne, W. C., A. B. Bereskin, and D. W. Martin, unpublished work.

33. "Motion Picture Sound Engineering," Chaps. 29 and 30, Van Nostrand, New York, 1938.

34. Beers, G. L., and H. Belar.   *Soc. Motion Pict. Eng.*, 1943, Vol. 40, p. 207.

35a. Martin, D. W.   U.S. Patent No. 3,467,758 (1969).

35b. Moog, R. A.   U.S. Patent No. 4,180,707 (1979).

36. Harris, C. M. (ed.)   "Handbook of Noise Control," McGraw-Hill, New York, 1957.

37. Beranek, L. L.   "Noise Reduction," McGraw-Hill, New York, 1960.

38. Peterson, A., and P. V. Bruel   Instruments for Noise Measurements, Chap. 16 in C. M. Harris (ed.), "Handbook of Noise Control," McGraw-Hill, New York, 1957.

39. Scott, H. H.   Noise Measuring Techniques, Chap. 17 in C. M. Harris (ed.), "Handbook of Noise Control," McGraw-Hill, New York, 1957.

40. Peterson, A. P. G.   *Gen. Radio Exp.*, 1956, Vol. 30, p. 1.

41. Koenig, W., H. K. Dunn, and L. Y. Lacy   *J. Acoust. Soc. Am.*, 1946, Vol. 18, p. 19.

42a. Randall, R. B., and R. Upton   *B & K Tech. Rev.*, 1978, Vol. 1, p. 3.

42b. Upton, R.   *B & K Tech. Rev.*, 1977, Vol. 1, p. 18.

43. Paragraph 50-204.10, Walsh-Healey Public Contracts Acts, *Fed. Reg.*, May 20, 1969.

44. Pt. 1910, Tit. 29, Occupational Safety and Health Act of 1970, *Fed. Reg.*, Vol. 36, No. 105, May 29, 1971.

45. Determination of Sound Transmission Class, ASTM Designation E413-70T, 1970.

46. Leonard, R. W.   Heating and Ventilating System Noise, Chap. 27 in C. M. Harris (ed.), "Handbook of Noise Control," McGraw-Hill, New York, 1957.

47. Sound Control chapter, "ASHRAE Guide and Data Book," biennial publication.

48. ASHRAE Standard 36B-63.

49. Air Diffusion Council Test Code 1062R1.

50. Ingerslev, F., and C. M. Harris   Control of Solid-Borne Noise, Chap. 19 in C. M. Harris (ed.), "Handbook of Noise Control," McGraw-Hill, New York, 1957.

51. Harris, C. M., and C. Crede   "Shock and Vibration Hardbook," McGraw-Hill, New York, 1961.

52. Beranek, L. L.   "Music, Acoustics and Architecture," Wiley, New York, 1962.

53. Knudsen, V. O., and C. M. Harris   "Acoustical Designing in Architecture," Wiley, New York, 1950.

54. "Music Buildings, Rooms and Equipment," Music Educators National Conference, Washington, 1966.

55. Morse, P. M.   "Vibration and Sound," Chap. 8, McGraw-Hill, New York, 1948.

56. Volkmann, J. E.   *J. Acoust. Soc. Am.*, 1942, Vol. 13, p. 324.

57. Lane, R. N.   *J. Acoust. Soc. Am.*, 1956, Vol. 28, p. 101.

58. Beranek, L. L.   *J. Acoust. Soc. Am.*, 1960, Vol. 32, p. 661.

59a. Sangster, F. L. J., and K. Teer   *IEEE J. Solid-State Circuits*, 1969, Vol. SC-4, p. 131.

59b. Baxa, D. E., and A. Seireg   *J. Acoust. Soc. Am.*, 1980, Vol. 67, p. 2045.

60a. S3.2-1960, American National Standards Institute, 1960.

60b. S3.14-1977, Acoustical Society of America, 1977.

61. Bauer, B. B. Proc. IRE, 1962, Vol. 50, 50th Anniversary Issue, p. 719.

62. Beranek, L. L. "Acoustics," McGraw-Hill, New York, 1954.

63a. Sessler, G. M., and J. E. West J. Audio Eng. Soc., 1964, Vol. 12, p. 129.

63b. Killion, M. C., and E. V. Carlson J. Audio Eng. Soc., 1974, Vol. 22, p. 237.

63c. Tamura, M., T. Yamaguchi, T. Oyaba, and T. Yoshimi J. Audio Eng. Soc., 1975, Vol. 23, p. 21.

63d. Lerch, Reinhard J. Acoust. Soc. Am., 1979, Vol. 66, p. 952.

64. Olson, H. F. "Dynamical Analogies," Van Nostrand, New York, 1943.

65. Firestone, F. A. J. Acoust. Soc. Am., 1956, Vol. 28, p. 1117.

66a. Olney, B. F. H. Slaymaker, and W. F. Meeker J. Acoust Soc. Am., 1945, Vol. 16, p. 172.

66b. Martin, D. W. J. Acoust. Soc. Am., 1947, Vol. 19, p. 43.

67a. Taylor, C. C. J. Audio Eng. Soc., 1979, Vol. 27, p. 677.

67b. Peters, R. W. U.S. Patent No. 4,149,032, 1979.

68. Anderson, L. J. IRE Trans Audio, 1953, Vol. 1, p. 1.

69. Olson, H. F. Proc. IRE, 1962, Vol. 50, 50th Anniversary Issue, p. 730.

70. Knowles, H. S. Sec. 13 in "Electrical Engineers' Handbook," Wiley, New York, 1950.

71. Hanna, C. R. Proc. IRE, 1925, Vol. 13, p. 437.

72. Levine, H., and J. Schwinger Phys. Rev., 1948, Vol. 73, p. 383.

73. Klepper, D. L., and D. W. Steele J. Audio Eng. Soc., 1963, Vol. 11, p. 198.

74. Rice, C. W., and E. W. Kellogg Trans. AIEE, 1925 Vol. 44, p. 461.

75a. Corrington, M. S. Proc. IRE, 1951, Vol. 39, p. 1021.

75b. Berman, J. M., and L. R. Fincham J. Audio Eng. Soc., 1977, Vol. 25, p. 370.

75c. Suzuki, T., T. Morii, and S. Matsumara J. Audio Eng. Soc., 1978, Vol. 26, p. 511.

76a. Webster, A. G. Proc. Natl. Acad. Sci., 1919, Vol. 5, p. 275.

76b. Keele, D. B. U.S. Patent No. 4,071,112, 1978.

76c. Henricksen, C. A., and M. Ureda J. Audio Eng. Soc., 1978, Vol. 26, p. 629.

77. Janszen, A. A. J. Audio Eng. Soc., 1955, Vol. 3, p. 87.

78. Malme, C. I. J. Audio Eng. Soc., 1959, Vol. 7, p. 47.

79. Fiala, W. T., J. K. Hilliard, J. A. Renkus, and J. J. Van Houten J. Acoust. Soc. Am., 1965, Vol. 38, p. 956.

80. Klein, S. Acustica, 1954, Vol. 4, p. 77.

81. Kock, W. E., and F. K. Harvey J. Acoust. Soc. Am., 1949, Vol 21, p. 471.

82. Schulein, R. B. J. Audio Eng. Soc., 1975, Vol. 23, p. 178.

82a. Beranek, L. L. Proc. IRE, 1962, Vol. 50, p. 767.

83. Leinonen, E., M. Otala, and J. Curl J. Audio Eng. Soc., 1977, Vol. 25, p. 170.

84a. Wente, E. C., and A. L. Thuras J. Acoust. Soc. Am., 1931, Vol. 3, p. 44.

84b. Anderson, L. J. J. Soc. Motion Pict. Eng., 1941, Vol. 37, p. 319.

85. Burkhard, M. D. J. Audio Eng. Soc., 1977, Vol. 25, p. 1008.

86. Martin, D. W., and L. J. Anderson J. Acoust. Soc. Am., 1947, Vol. 19, p. 63.

87. Wiener, F. M., and D. A. Ross J. Acoust. Soc. Am., 1946, Vol. 18, p. 401.

88. Shaw, E. A. G., and G. J. Thiessen J. Acoust. Soc. Am., 1958 Vol. 30, p. 24.

89. Bauer, B. B. J. Audio Eng. Soc., 1961, Vol. 9, p. 148.

90. Bachman, W. S., B. B. Bauer, and P. C. Goldmark Proc. IRE 1962, Vol. 50, 50th Anniversary Issue, p. 738.

91. Frayne, G., and H. Wolfe "Elements of Sound Recording," Wiley, New York, 1949.

92. Jackson, J. E. J. Audio Eng. Soc., 1965, Vol. 13, p 134.

93. Albersheim, W. J., and D. MacKenzie J. Soc. Motion Pict. Eng., 1941, Vol. 37, p. 452.

94. Bachman, W. S. U.S. Patent No. 2,738,385, Mar. 13, 1956.

95. Davis, C. C., and J. G. Frayne Proc. IRE, 1958, Vol. 46, p. 1586.

96. Vieth, L., and C. F. Wiebusch J. Soc. Motion Pict. Eng., 1938, Vol. 30, p. 96.

97. Marcucci, R. J. Audio Eng. Soc., 1965, Vol. 13, p. 130.

98. LeBel, C. J. J. Acoust. Soc. Am., 1942, Vol. 13, p. 265.

99. Bachman, W. S. Audio Eng., 1950, Vol. 34, p. 11.

100. Moura, C. E. J. Audio Eng. Soc., 1960, Vol. 8, p. 228; 1961, Vol. 9, p. 60.

100a. Corrington, M. S., and T. Murakami 1958 IRE Conv. Rec., Pt. 7.

101a. Scheiber, P. J. Audio, Eng. Soc., 1971, Vol. 19, p. 267; also 1971, Vol. 19, 647.

101b. Eargle, J. M. J. Audio Eng. Soc., 1971, Vol 19, p. 552.

101c. Jurgen, R. K. IEEE Spectrum, July, 1972, p. 55.

102a. Inoue, T., N. Takahashi, and I. Owaki J Audio Eng. Soc., 1971, Vol. 19, p. 576.

102b. Owaki, I., T. Muraoka, and T. Inoue J. Audio Eng. Soc., 1972, Vol. 20, p. 361.

102c. Inoue, T., N. Shibata, and K. Goh   *J. Audio Eng. Soc.*, 1973, Vol. 21, p. 166.

102d. Inoue, T., I. Owaki, Y. Ishigaki, and K. Goh   *J. Audio Eng. Soc.*, 1973, Vol. 21, p. 625.

103a. Bauer, B. B., D. W. Gravereaux, and A. J. Gust   *J. Audio Eng. Soc.*, 1971, Vol. 19, p. 638.

103b. Bauer, B. B., G. A. Budelman, and D. W. Gravereaux   *J. Audio Eng. Soc.*, 1973, Vol. 21, p. 19.

103c. Bauer, B. B., R. G. Allen, G. A. Budelman, and D. W. Gravereaux   *J. Audio Eng. Soc.*, 1973, Vol. 21, p. 342.

104a. Itoh, R.   *J. Audio Eng. Soc.*, 1972, Vol. 20, p. 167.

104b. Cooper, D. H., and T. Shiga   *J. Audio Eng. Soc.*, 1972, Vol. 20, p. 346.

104c. Cooper, D. H., T. Shiga, and T. Takagi   *J. Audio Eng. Soc.*, 1973, Vol. 21, p. 614.

104d. Woodward, J. G.   *J. Audio Eng. Soc.*, 1975, Vol. 23, p. 2 and p. 128.

104e. *J. Audio Eng. Soc.*, 1975, Vol. 23, p. 1092.

104f. Woodward, J. G.   *J. Audio Eng. Soc.*, 1977, Vol. 25, p. 843.

104g. Gerzon, M. A.   *J. Audio Eng. Soc.*, 1977, Vol. 25, p. 400.

104h. Bauer, B. B.   *J. Audio Eng. Soc.*, 1979, Vol. 27, p. 866.

105. Camras, M.   *J. Acoust. Soc. Am.*, 1968, Vol. 43, p. 1425.

106a. Olson, H. F.   *J. Audio Eng. Soc.*, 1964, Vol. 12, p. 98.

106b. Fox, E. C., and J. G. Woodward   *J. Audio Eng. Soc.*, 1963, Vol. 11, p. 294.

107a. Eargle, J. M.   *J. Audio Eng. Soc.*, 1969, Vol. 17, p. 276.

107b. Khanna, S. K.   *J. Audio Eng. Soc.*, 1976, Vol. 24, p. 464.

108. Bastiaans, C. R.   *J. Audio Eng. Soc.*, 1967, Vol. 15, p. 389.

109. Pierce, J. A., and F. V. Hunt   *J. Soc. Motion Pict. Eng.*, 1937, Vol. 29, p. 493.

110. Anderson, C. R., and P. W. Jenrick   *J. Audio Eng. Soc.*, 1972, Vol. 20, p. 162.

111. Bauer, B. B.   *Electronics*, 1945, Vol. 18, p. 110; *IEEE Trans.*, 1963, Vol. AU-11, p. 47.

112. Bauer, B. B.   *J. Acoust. Soc. Am.*, 1946, Vol. 18, p. 387; 1955, Vol. 27, p. 586.

113. Roys, H. E.   *J. Audio Eng. Soc.*, 1953, Vol. 1, p. 78.

114. Gravereaux, D. W., A. J. Gust, and B. B. Bauer   *J. Audio Eng. Soc.*, 1970, Vol. 18, p. 530.

115. Poulsen, V.   *Ann. Phys. (Leipz.)*, 1900, Vol. 3, p. 754.

116. Begun, S. J.   "Magnetic Recording," Murray Hill Books, New York, 1949.

117. Wilson, C. F.   *IRE Trans. Audio.*, 1956, Vol. AU-4, p. 53.

118. Westmijze, W. K.   Studies on Magnetic Recording, *Phillips Res. Rep.*, 1953, Vol. 8, pp. 161–183, 245–269.

119. Fujii, M., G. Rehklau, J. G. McKnight, and W. Miltenburg   *J. Audio Eng. Soc.*, 1960, Vol. 8, p. 245.

120. Camras, M.   *Proc. IRE*, Vol. 50, 1962, 50th Anniversary Issue, p. 751.

121. Eilers, D. W.   *J. Audio Eng. Soc.*, 1969, Vol. 17, p. 303; 1970, Vol. 18, p. 540.

122. Nesh, F., and R. F. Brown   *IRE Trans. Audio*, 1962, Vol. AU-10, p. 70; 1964, Vol. AU-12, p. 55.

123. Mee, C. D.   *IRE Trans. Audio*, 1964, Vol. AU-12, p. 72

124. Naumann, K. E., and E. D. Daniel   *J. Audio Eng. Soc.*, 1971, Vol. 19, p. 822.

124a. Van der Giessen, A. A.   *J. Audio Eng. Soc.*, 1978, Vol. 26, p. 838.

125. Finger, R. A., P. Murphy, and E. J. Foster   *J. Audio Eng. Soc.*, 1972, Vol. 20, p. 549.

126. Bauer, B. B., and C. D. Mee   *IRE Trans. Audio*, 1961, Vol. AU-9, p. 139.

127. Woodward, J. G., and E. Della Torre   *J. Appl. Phys.*, 1960, Vol. 31, p. 56.

128. Preisach, F.   *Z. Phys.*, 1935, Vol. 94, p. 277.

129. Eldridge, D. F.   *IRE Trans. Audio*, 1961, Vol. AU-9, p. 155.

130. Radocy, F., and A. Kramer   *J. Audio Eng. Soc.*, 1957, Vol. 5, p. 76.

131. Eldridge, D. F.   *IEEE Trans. Audio*, 1964, Vol. AU-12, p. 100.

132. Daniel, E. D.   *J. Audio Eng. Soc.*, 1972, Vol. 20, p. 92.

133. Audio Standards Listings: Magnetic Tape Sound Recording, *J. Audio Eng. Soc.*, 1970, Vol. 18, p. 319.

134. NAB Standard, Magnetic Tape Recording and Reproducing (Reel-to-Reel), April 1965.

135. NAB Standard, Cartridge Tape Recording and Reproducing, October 1964.

136. Hanson, E. R.   *J. Audio Eng. Soc.*, 1968, Vol. 16, p. 430; 1971, Vol. 19, p. 24.

137. Knox, A.   *J. Audio Eng. Soc.*, 1964, Vol. 12, p. 32.

138. Bixler, O. C.   *IRE Trans. Audio*, 1954, Vol. AU-2, p. 15.

139. McKnight, J. G.   *J. Audio Eng. Soc.*, 1964, Vol. 12, p. 140.

140. Sariti, A. A.   *J. Audio Eng. Soc.*, 1960, Vol. 8, p. 243.

141. Arnold, R. R., L. J. Ananka, and S. V. Marsov   *J. Audio Eng. Soc.*, 1972, Vol. 20, p. 470.

142. Taber, W. D. *J. Audio Eng. Soc.*, 1968, Vol. 16, p. 61.
143. Camras, M. *IEEE Trans. Audio*, 1964, Vol. AU-12, p. 41.
144. Peters, C. J. *IRE Trans. Audio*. 1962, Vol. AU-10, p. 79.
145. Camras, M. *IRE Trans. Audio*. 1962, Vol. AU-10, p. 84.
146. Dolby, R. M. *J. Audio Eng. Soc.*, 1967, Vol. 15, p. 383.
147. Mullin, J. T. *IEEE Trans. Audio*, 1965, Vol. AU-13, p. 31.
148. Olson, H. F. *Electronics*, 1947, Vol. 20, p. 118.
149. Scott, H. H. *Electronics*, 1947, Vol. 20, p. 96.
150. McKnight, J. G. *J. Audio Eng. Soc.*, 1970, Vol. 18, p. 250.
151. Stockham, T. G. Jr. *J. Audio Eng. Soc.*, 1977, Vol. 25, p. 892.
152. Blesser, B. A. *J. Audio Eng. Soc.*, 1978, Vol. 25, p. 739.
153. Bernhard R. *IEEE Spectrum*, 1979, December, p. 28.
154. Busby, E. S. Jr. *J. SMPTE*, 1980, Vol. 89, p. 503.
155. Weisser, A. *J. SMPTE*, 1980, Vol. 89, p. 520.
156. Heaslett, A. *J. Audio Eng. Soc.*, 1978, Vol. 26, p. 66.
157. Stockham, T. G., Jr. U.S. Patent No. 4,202,018, 1980.
158. Myers, J. P., and A. Feinberg *J. Audio Eng. Soc.*, 1972, Vol. 20, p. 622.
159. Iwamura, H., H. Hayashi, A. Miyashita, and T. Anazawa *J. Audio Eng. Soc.*, 1973, Vol. 21, p. 535.
160. Doi, T. T., T. Itoh, and H. Ogawa *J. Audio Eng. Soc.*, 1979, Vol. 27, p. 975.
161. Okada, K., T. Kubo, W. Susaki, and T. Sato *J. Audio Eng. Soc.*, 1980, Vol. 28, p. 429.
162. PCM Audio Disk, *Electronics*, 1979, Nov. p. 68.
163. Digital Audio Technical Committee Report, *J. Audio Eng. Soc.*, 1980. Vol. 28, p. 259.

Television and Facsimile Systems

DONALD G. FINK *Director Emeritus, IEEE; Editor, "Electronics Engineers' Handbook" and "Standard Handbook for Electrical Engineers"; Fellow, SMPTE; Fellow, IEEE*

RENVILLE H. McMANN, JR. *President, Thomson-CSF Laboratories; Fellow, SMPTE; Senior Member, IEEE*

CHARLES W. RHODES *Chief Engineer, Television Product Development, Tektronix, Incorporated; Fellow, SMPTE; Senior Member, IEEE*

KURT F. WALLACE *Television Engineer, Ampex Corporation; Member, SMPTE*

NORMAN W. PARKER *Corporate Staff Scientist, Motorola Incorporated; Fellow, IEEE*

GEORGE M. STAMPS *President, GMS Consulting; formerly Chairman, EIA Facsimile Section; Member, IEEE*

CONTENTS

Numbers refer to paragraphs

Television Fundamentals and Standards*

BY DONALD G. FINK

1. Criteria for Picture Reproduction. In the design of a picture-reproduction system, the basic criteria to be met are that the reproduced images shall be acceptable to the human eye and that the technical details of the system shall not be obtrusively evident to the viewer.

The first quality of a picture judged by the eye is its *sharpness*, or *pictorial clarity*. If the image is out of focus or the details or edges of objects are not clear and sharp, the eye will attempt without success to focus the picture and eyestrain will result. The second quality of importance to the eye is the *contrast* between light and dark areas, and the related *background illumination*. Background lighting affects the contrast observed by the eye and can introduce contrast changes not present in the original scene.

The third quality of importance is *continuity of motion*. Reproduced pictures in motion are created by a succession of still frames, and the illusion of motion is created, in part, by the fact that the human eye briefly retains any image impinging on it. Early work in motion pictures revealed that a rate of 16 pictures per second is sufficient to preserve the sense of continuity of motion. The first movies used 16 pictures per second as a standard, and most amateur movie cameras still do. The standard for motion pictures later became 24 frames per second. In television, the pictures are recreated at 25 pictures per second in most of the countries of the world and at 30 pictures per second in the United States, Japan, and elsewhere (see Table 20-1). This difference originally arose to match the power-system frequencies of 50 and 60 Hz in the respective countries. Table 20-1 lists the television standards in use in the principal countries (for details of the standards, see Tables 20-2 and 20-3).

The fourth picture quality of importance is *flicker*. Even when continuity of motion is preserved, the picture may flicker. To eliminate flicker at useful image brightnesses, the pictures must be presented at a rate considerably greater than that required for continuity of motion. Flicker is a function of picture brightness (see Par. **20-4**). Flicker can cause eyestrain, and it makes the reproduced picture unpleasant to view. Since the visibility of flicker increases sharply with the brightness of the image, it is customary to operate receivers using the 25 picture-per-second standard at lower brightness than can be accommodated at 30 pictures per second. The lower picture rate permits more detail to be resolved in the image, for a given bandwidth occupied by the television signal.

The fifth characteristic is that *color values*, if present, must be accepted as *realistic*. The eye is particularly critical of flesh-tone colors. Color reproduction need not be *accurate* when compared with the original scene, because the colors observed are greatly influenced by the surroundings and illumination and the eye compensates for such variations.

The sixth characteristic is *freedom from extraneous degradations and artifacts*, e.g., those introduced to the reproduced image by noise, interference, signal reflections, and signal-mixing effects.

In establishing the standards of a picture-reproduction system, the problem is to satisfy these requirements of the human eye adequately and economically.

2. Luminance and Brightness. The eye does not respond equally to radiated energy of all visible wavelengths. There is wide variation between observers and the response is also a function of light intensity. From thousands of measurements on human observers, the average eye is considered to respond according to the *luminosity function of the standard observer* (Fig. 20-1).

The *luminance* of a surface is the effect on the average eye of the light emitted by a unit area of the surface. It is the integrated effect of the eye response $y(\lambda)$ (Fig. 20-1) and the visible light

*The editor acknowledges with appreciation the contributions to this subsection in the previous edition by Professor William L. Hughes.

TABLE 20-1   Television Systems by Country

Country	VHF bands Mono	VHF bands Color	UHF bands Mono	UHF bands Color	Country	VHF bands Mono	VHF bands Color	UHF bands Mono	UHF bands Color
Algeria	B	PAL			Jordan	B	PAL		
Argentina	N	PAL	N		Kenya	B	PAL		
Australia	B	PAL			Korea	M	NTSC	M	NTSC
Austria	B	PAL	G	PAL	Kuwait	B	PAL		
Bangladesh	B	PAL			Libya	B	SECAM		
Belgium	B	PAL	H	PAL	Malaysia	B	PAL		
Bermuda	M	NTSC			Mexico	M	NTSC	M	NTSC
Bolivia	N				Morocco	B	SECAM		
Brazil	M	PAL	M	PAL	New Zealand	B	PAL		
Bulgaria	D	SECAM			Norway	B	PAL	G	PAL
Cambodia	M				Pakistan	B	PAL		
Canada	M	NTSC	M	NTSC	Panama	M	NTSC		
Chile	M	PAL			Paraguay	N			
China, P.R.	D	PAL			Peru	M	NTSC		
Columbia	M	SECAM			Philippines	M	NTSC		
Cuba	D	SECAM			Poland	D	SECAM	K	SECAM
Czechoslovakia	D	SECAM	K	SECAM	Portugal	B	PAL	G	PAL
Denmark	B	PAL			Puerto Rico	M	NTSC	M	NTSC
Ecuador	M	NTSC			Rhodesia	B			
Egypt	B	SECAM			Romania	D			
Finland	B	PAL	G	PAL	Saudi Arabia	B	SECAM		
France	E		L	SECAM		M	PAL		
Germany, East	B	SECAM	G	SECAM	Singapore	B	PAL		
Germany, West	B	PAL	G	PAL	South Africa	I	PAL	I	PAL
Ghana	B	PAL			Spain	B	PAL	G	PAL
Great Britain	A		I	PAL	Sri Lanka	B			
Greece	B		G		Sweden	B	PAL	G	PAL
Guatemala	M	NTSC			Switzerland	B	PAL	G	PAL
Holland	B	PAL	G	PAL	Taiwan	M	NTSC		
Honduras	M				Tunisia	B	SECAM		
Hong Kong			I	PAL	Turkey	B	PAL		
Hungary	D	SECAM	K	SECAM	Uruguay	M			
India	B				USA	M	NTSC	M	NTSC
Iran	B	SECAM			USSR	D	SECAM	K	SECAM
Iraq	B	SECAM			Venezuela	M	NTSC		
Ireland	A,I	PAL	I	PAL	Yugoslavia	B	PAL	H	PAL
Israel	B		G		Zaire	K	SECAM		
Italy	B	PAL	G	PAL	Zambia	B			
Japan	M	NTSC	M	NTSC					

Specifications of systems by letter code (see also Tables 20-2 and 20-3)

System code	Lines per frame	Fields per second	Video bandwidth, MHz	Channel bandwidth, MHz	Intercarrier separation, MHz	Modulation polarity	Sound modulation
A	405	50	3	5	3.5	Positive	AM
B	625	50	5	7	5.5	Negative	FM ± 50 kHz
D,K	625	50	6	8	6.5	Negative	FM ± 50 kHz
E	819	50	10	14	11.15	Positive	AM
G,H	625	50	5	8	5.5	Negative	FM ± 50 kHz
I	625	50	5.5	8	6	Negative	FM ± 50 kHz
L	625	50	6	8	6.5	Positive	AM
M	525	60	4.2	6	4.5	Negative	FM ± 25 kHz
N	625	50	4.2	6	4.5	Negative	FM ± 25 kHz

Designation of compatible color systems: NTSC = National Television System Committee standards (see Table 20-2); PAL = phase alternating-line standards (see Table 20-3); SECAM = sequential color with memory standards (see Table 20-3).

source: The data in this table were abstracted from "Broadcasting Engineer's Pocket Book," published by the Independent Broadcasting Authority of Great Britain, and were current in 1978. For current information consult the International Radio Consultative Committee (CCIR), International Telecommunications Union, Geneva, Switzerland. Geneva, Switzerland.

power radiated by the surface $E(\lambda)$, both of which are functions of the wavelength λ. The integration is expressed by

$$\text{Luminance} = 680 \int E(\lambda) \bar{y} \lambda \, d\lambda \qquad \text{lm/unit area}$$

where lm is the abbreviation for lumen and the radiated power $E(\lambda)$ is in watts per unit area. The constant 680 lm/W is the *luminosity* of radiant power at the peak of the luminosity curve, at 546 nm. The *luminous efficacy* is defined as the lumens emitted per watt radiated.

The luminance (commonly "brightness") of a perfectly diffusing surface (in which the luminance flux density falls off as cos Θ) displays no change as the viewing angle Θ changes (Fig. 20-2). The luminance of a surface is expressed in candelas per square meter. The older unit, the footlambert, is equal to 3.42626 cd/m².

A *specular reflector* is one that favors particular directions of reflection. Perfect specular reflectors do not exist, but optically flat mirrors approach this condition. *Diffuse reflection* is used in movie screens so that the audience can view the images adequately over wide areas.

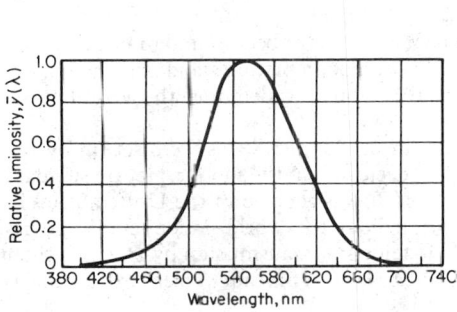

Fig. 20-1. Standard luminosity function.

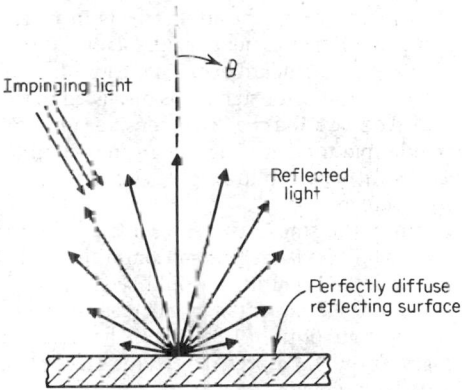

Fig. 20-2. Cosine-law reflector.

3. Contrast (Tonal Range). A characteristic to which the eye is particularly sensitive is *contrast ratio*. This parameter is defined in terms of a diffuse, flat, spectrally neutral reflector, i.e., one that scatters the light falling on it according to the cosine law and reflects all wavelengths of the visible spectrum equally. If two areas in such a reflector have different reflection coefficients, the contrast ratio displayed between the two areas is equal to the ratio of the reflection coefficients. For example, if the reflection coefficients are 80 and 4%, and if both areas are uniformly illuminated, the contrast ratio is

$$0.80/0.04 = 20 \qquad \text{Contrast ratio} = 20\!:\!1$$

When the illumination falling on the two areas is different, the contrast ratio is modified proportionately by the ratio of the illumination. The contrast ratio of that picture is further affected by ambient illumination and by light scattered from one area to another.

In practice in a well-designed, darkened movie theater, contrast ratios of 100:1 or more can be achieved, but in television the situation is quite different. The reflectance of a television picture tube to light falling on it from the room may be 25% or more. Such light is usually present since observation of television in a totally darkened room is not usual and in fact is not recommended. Light from one part of the television image to another is scattered, inside the picture tube and between phosphor grains.

In color television pictures, the definition of contrast ratio, as in monochrome, is the luminance of the brightest area divided by the luminance of the darkest area, independently of whether they are of the same color. High-luminance areas tend to not be highly saturated colors (see Par. **20.16**). Dark areas may have a high color saturation, but they tend to display a dark or blackish appearance.

4. Flicker, Fields, and Frames. The flicker effect has the following characteristics:

1. It is independent of motion in the picture.

2. For a given brightness, it becomes less pronounced as the number of flashes per second is increased.

3. It is a function of brightness. If the large-area highlight brightness is 35 cd/m², the flicker effect disappears at approximately 40 flashes per second. This is a typical condition encountered

in a motion-picture theater. If the large-area brightness is 350 cd/m² (as it may be in a very bright television picture), flicker disappears at 50 or more flashes per second.

4. Small areas have a lower critical flicker frequency than large areas of the same brightness.

For motion pictures, it would be uneconomical to provide enough film for 40 pictures per second merely to overcome flicker, since continuity of motion is preserved at a much lower rate. Consequently 24 still pictures *(frames)* are projected per second but each frame is flashed 2 or 3 times, producing a flash rate of 48 or 72 fields per second. Thus both the continuity of motion and flicker requirements are satisfied with one-half the film consumption.

At the higher brightnesses occurring with television, the 24-frame rate is not entirely adequate. In the United States, a 30-frame–60-field rate has been chosen, originally to minimize hum effects in receivers operated on 60-Hz power. At this high rate, flicker is not evident at brightnesses produced by home television receivers.

In countries where the predominant power frequency is 50 Hz, a 25-frame–50-field rate has been standardized. The permissible (flicker-free) highlight brightness is not as high as at 60 fields, but satisfactory flicker-free performance is obtained.

In recent years, the hum effects in receivers have been minimized to the point where the power frequency is not a major factor. In color television in the United States a field rate of 59.94 Hz has been standardized. The reason for this slight variation is related to the requirements for coding of the color signals, as discussed in Par. **20-21**.

5. Aspect Ratio. In extensive subjective tests, observers have been found to prefer a rectangular picture with slightly greater width than height. In motion-picture standards, the picture *aspect ratio* (width to height) was adopted as 4:3, and this ratio prevailed until the advent of the wide screen.

When the standards for black-and-white television in the United States were set up by the National Television System Committee in 1940, it was decided that the motion-picture standard was valid and that little would be gained by changing it. It was adopted in the United States in 1941 and subsequently in all television systems throughout the world. Wide-screen movies (aspect ratios of from 1.86 to 2.35) have been adapted to television transmission by cropping their edges to the 4:3 aspect ratio. This practice causes some loss of significant subject matter but is usually acceptable.

6. Viewing Distance. In subjective testing to establish moving picture and television standards, considerable attention has been paid to the distance at which viewers choose to view the picture. Most observers prefer to sit at a distance ranging from 4 to 8 times the picture height, and close to the centerline if they sit close to the picture.

This range of preferred viewing distances has a primary effect on the number of scanning lines (and indirectly on the video bandwidth) required in the reproduced pictures. The eye cannot resolve the fine structure of an image viewed from too great a distance. In television images having 525 or 625 lines (the present standards) this limit is reached at a distance of about 10 times the picture height under ideal conditions. At viewing distances less than 4 times the picture height, the line structure may be evident, particularly if the lines are not precisely interlaced (Par. **20-9**).

7. Scanning Patterns and Apertures. It has been established as standard in broadcast television that scanning starts in the upper left corner of the picture and proceeds across to the right and slightly downward. When the right-hand side is reached, the scanning spot retraces

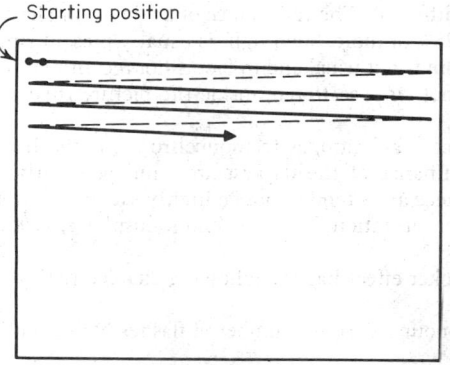

rapidly to a position below its starting position and again proceeds to the right and slightly downward, and so on, ultimately reaching the bottom of the picture (see Fig. 20-3). At that point, the spot returns to the top and repeats the process, except that the lines of the second scanning field fall between the lines of the first field. Thus, successive fields are *interlaced* (Fig. 20-4). This arrangement permits two picture flashes (fields) for each frame and thus greatly reduces the tendency to flicker (see Par. **20-4**) while reducing the required signal bandwidth by a factor of 2.

The electron beams that create the scanning spots are approximately circular, but their intensity is not uniform, their energy falling

Fig. 20-3. Scanning directions and sequence.

off in an error-function distribution, as shown in Fig. 20-5. The *effective width* of the spot is usually taken as the diameter of an equivalent spot of uniform intensity.

8. Number of Scanning Lines. The choice of the number of scanning lines in the image hinges on the resolving capability of the human eye and the viewing distance (Par. **20-6**). From physiological testing it has been determined that if a pair of parallel lines is viewed at such a distance that the angle subtended by them at the eye is less than 2 minutes of arc, the eye sees them as one line. This fact is used to select the number of lines for a television system. If the closest preferred viewing distance is 4 times the picture height, as shown in Fig. 20-6, two parallel lines closer than $d = 0.00232h$ cannot be separately resolved by the observer. The number of lines contained in the picture height at this limit is $1/0.00232 = 431$ lines. Thus, approximately 430 lines is the minimum figure for television scanning. A sharp-eyed observer can resolve the

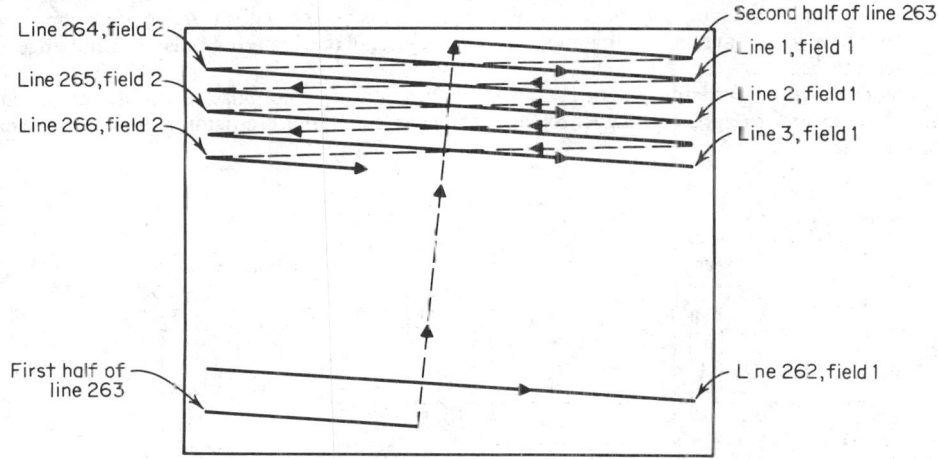

Fig. 20-4. Interlaced scanning pattern (raster).

line structure at a distance of 4 times the picture height, but the average observer cannot. While there is no "correct" number of lines, the choice should be in excess of 400 for reasons of resolution but not too much higher for reasons of spectrum economy.

9. Interlaced Scanning. As stated in Par. **20-4**, large areas of high brightness have a higher critical flicker frequency than small areas of the same brightness. Thus an individual scanning line flickers much less in the critical range than the larger area of the image does as a whole. This fact allows the use of interlaced scanning in television.

Interlaced scanning is achieved by making the horizontal (line-scanning) rate an odd multiple of one-half the vertical (field-scanning) rate. In the United States standards, the horizontal rate is $15,750 = 525(^{60}/_2)$ lines/s. In other words at 30 frames per second, the scanning pattern has 525 lines per frame and 262.5 lines per field (Fig. 20-4).

In color transmissions, the horizontal rate is $15,734.264$ lines/s, and the frame rate is 29.97 per second. An equivalent statement is that interlacing is achieved when the number of lines per

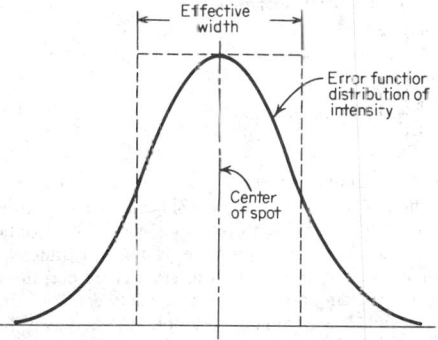

Fig. 20-5. Typical scanning-spot distribution.

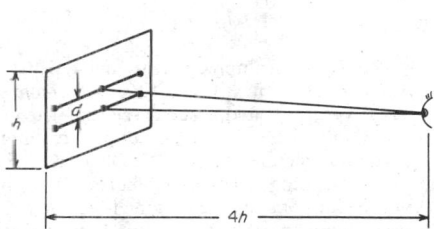

Fig. 20-6. Resolution at 4 times picture height.

frame is an odd number, thus requiring each field to have an even number of lines plus one half line. The half line, left over at the end of a field scan, displaces the next field downward by a full line, and interlacing is achieved. By this scheme, large area flicker is avoided, the number of lines scanned per second is reduced by 2:1, and the resolution is essentially unaffected.

The diagram in Fig. 20-4 is idealized to show the mechanism of interlace. In practice, the vertical-retrace and synchronizing periods occupy the time of several line scans so that less than 500 lines (typically 490) are available for picture information in each frame. At the limit of the tolerances shown in Fig. 20-7, the number of active scanning lines is 484 per frame.

10. Blanking. To prevent the retrace lines from being observed on television receivers and monitors, a blanking pulse is applied to the picture tube, the leading edge of which precedes the leading edge of the horizontal synchronizing pulse. The time relationship of the blanking and synchronizing pulses is illustrated in Fig. 20-8. The region between the leading edge of the blanking signal and the leading edge of the synchronizing pulse is termed the *front porch*, and the region between the synchronizing-pulse trailing edge and the blanking-pulse trailing edge is termed the *back porch*.

Color television standards use the back porch to position the *color burst*, eight or more cycles of color subcarrier (Fig. 20-8b) synchronizing the color-subcarrier oscillator at the end of each scanned line.

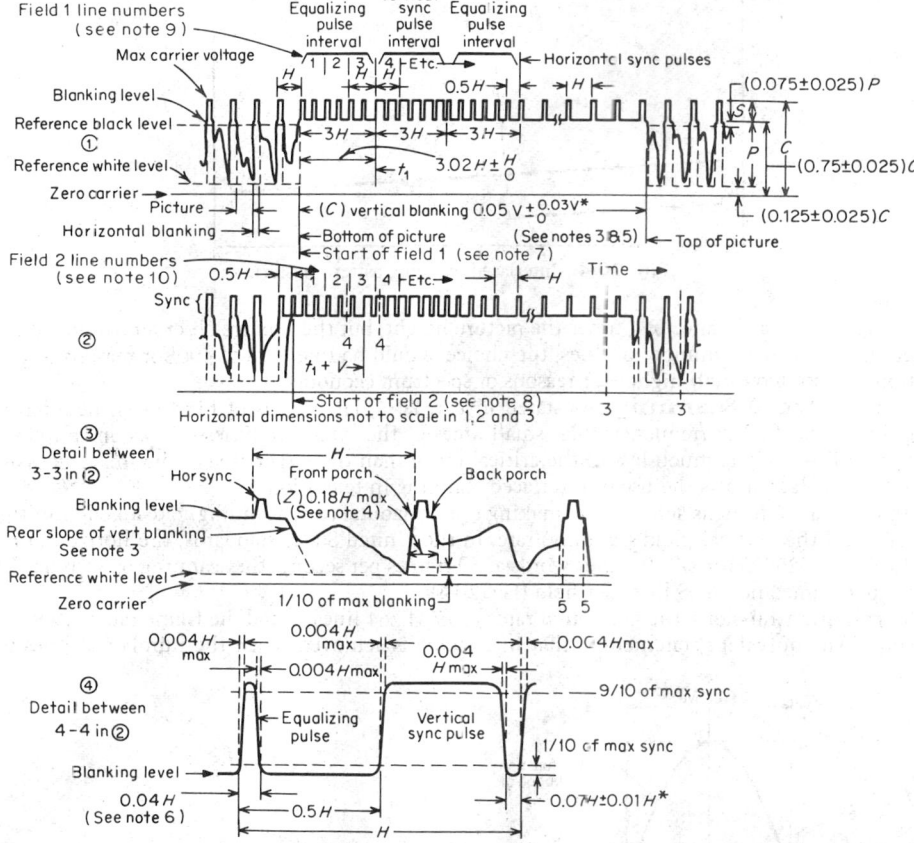

Fig. 20-7. Monochrome synchronizing signal waveform (FCC Standard). *Notes:* (1) H = time from start of one line to start of next line. (2) V = time from start of one field to start of next field. (3) Leading and trailing edges of vertical blanking should be complete in less than $0.1H$. (4) Leading and trailing slopes of horizontal blanking must be steep enough to preserve minimum and maximum values of $(x + y)$ and (z) under all conditions of picture content. *(5) Dimensions marked with an asterisk indicate that tolerances are permitted only for long time variations and not for successive cycles. (6) Equalizing pulse area shall be between 0.45 and 0.5 of the area of a horizontal synchronizing pulse. (7) Start of field 1 is defined by a whole line between the first equalizing pulse and preceding H sync pulses. (8) Start of field 2 is defined by a half line between first equalizing pulse and preceding H sync pulses. (9) Field 1 line numbers start with first equalizing pulse in field 1. (10) Field 2 line numbers start with second equalizing pulse in field 2.

A vertical video blanking pulse is also used, of length = 21 horizonta lines, shown in Fig. 20-7.

11. Black-Level Clamping. The average level of the television signal varies as the picture goes from bright to dark. These dc variations are lost when the signal is passed through any *RC*-coupled amplifier. To provide a reference for reinserting the dc level, it is necessary to employ a clamping circuit, which sets the synchronizing pulses at a constant dc level, independently of picture-brightness variations. The clamping circuit may be keyed by the synchronizing pulses themselves or by specially generated driving pulses. The circuits are variously called clampers, keyed clampers, restorers, etc.

The keyed clamper, operating at the horizontal-line rate, can remove power-line hum interference without disturbing the picture information if the hum is additive to the video signal but not if the signal has become modulated by a nonlinear process.

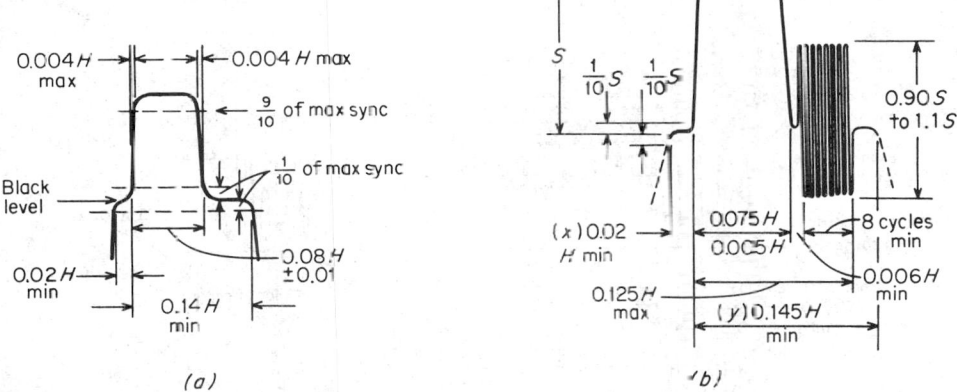

Fig. 20-8. Horizontal blanking and synchronizing pulses: (*c*) black and white; (*b*) National Televison System Committee color.

12. Synchronizing and Blanking Signals. The standard television synchronizing and blanking signals in the United States are as illustrated in Fig. 20-7 for monochrome television and in Fig. 20-9 for color television.

13. Vertical-Interval Signals. (See also Par. **20-114**.) The vertical blanking interval, between successive fields, occupies a time equal to 21 line scans. In the early 1970s the FCC authorized the transmission of test, cue, and control signals during these intervals, beginning with line 17 and continuing through line 20 of each field blanking period. These signals are used to test, control, and monitor the operation of transmitters and networks during broadcasts to the public, the signals not being visible to the audience because they occur during the blanking interval.

In 1975 the FCC designated the nineteenth line interval for so-called vertical-interval-reference (VIR) signals, which are transmitted to control the operation of suitably equipped domestic receivers. The VIR signal contains a chrominance reference (amplitude and phase), a luminance reference, and black-level reference, which are recovered in special receiver circuits and used to control the effective settings of the brightness, contrast and chrominance (hue and saturation) controls to the values intended at the point of program origination.

In 1977 the interval of line 21 of the first field in each frame and the first half of line 21 in the second field were designated by the FCC for the transmission of program-related data signals, e.g., to provide the visual depiction of information presented on the sound channel, as for labeling the picture for the hard of hearing. This latter function is capable of expansion to a wide variety of supplementary services.

14. Colorimetry. Color television standards are based on how the eye perceives colored light. It is a fortunate fact that a wide range of colors can be reproduced, to the satisfaction of the eye, by the addition of only three light sources, e.g., red, green, and blue. The study of colorimetry is based on this property of the eye. Only the elements essential to color television are discussed here.

The three CIE* standard primaries are monochromatic lights of wavelength 700 nm (red),

*Comité International d Éclairage

546.1 nm (green), and 435.8 nm (blue). A *colorimeter* is an optica. system that mixes these lights *additively*, i.e., by superimposing each primary on the the othe- two. Adjacent to the additive area, another area displays monochromatic light at wavelengths selected individually throughout the visible spectrum. The added primaries and monochroma:ic sources are compared, and a match in color and brightness is sought by adjusting the intensity of each of the primaries.

When such matching is performed by thousands of observers, the results are as shown in Fig. 20-10, interpreted as follows: to match a particular monochromatic wavelength, the relative energies required of each primary are shown in the $\bar{r}$, $\bar{g}$, and $\bar{b}$ curves at the wavelength to be matched. If a continuous-spectrum function $I(\lambda)$ is to be matched, the relative primary proportions are computed as follows:

$$R = \int \bar{r}(\lambda) I(\lambda)\, d\lambda \qquad G = \int \bar{g}(\lambda) I(\lambda)\, d\lambda \qquad B = \int \bar{b}(\lambda) I(\lambda)\, d\lambda$$

where λ is the wavelength.

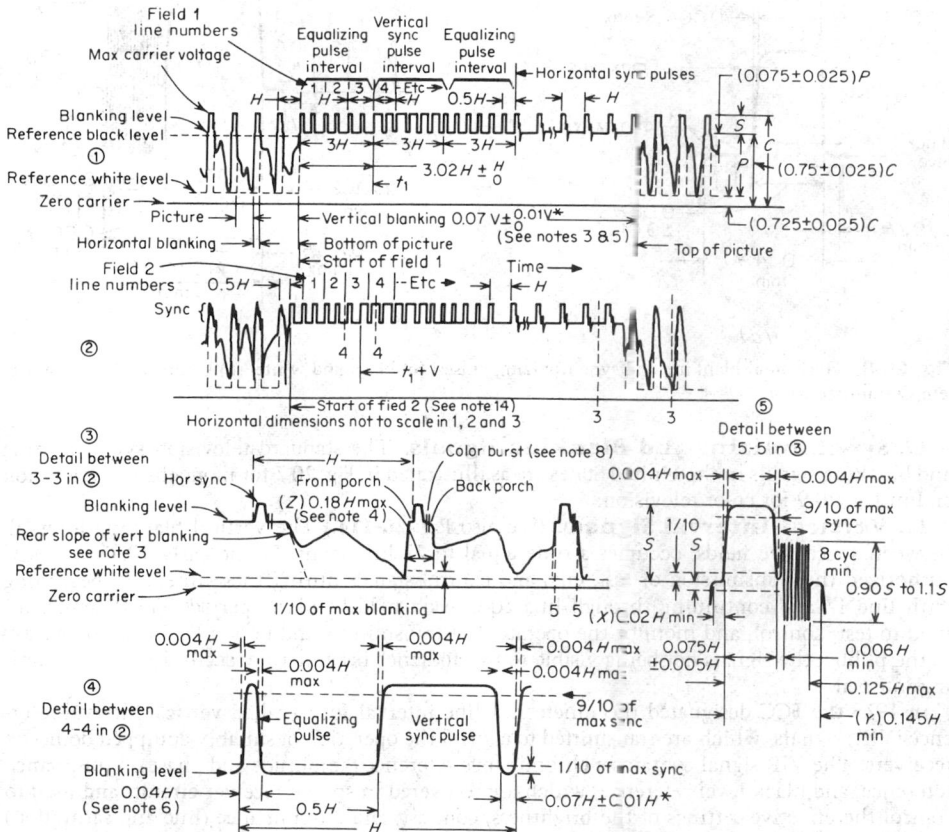

Fig. 20-9. National Television System Committee color synchronizing-signal waveform (FCC Standard). *Notes:* For notes (1) to (6) see Fig. 20-7. (7) Color burst follows each horizontal pulse but is omitted following the equalizing pulses and during the broad vertical pulses. (8) Color bursts to be omitted during monochrome transmission. (9) The burst frequency shall be 3.579545 MHz. The to erance on the frequency shall be ± 10 Hz with a maximum rate of change not to exceed 0.1 Hz/s. (10) The horizontal scanning frequency shall be $\frac{2}{455}$ times the burst frequency. (11) The dimensions specified for the burst determine the times of starting and stopping the burst but not its phase. The color burst consists of amplitude modulation of a continuous sine wave. (12) Dimension P represents the peak excursion of the luminance signal from blanking level but does not include the chrominance signal. Dimension S is the synchron zing amplitude above blanking level. Dimension C is the peak carrier amplitude. (13) Start of field 1 is defir ed by a whole line between first equalizing pulse and preceding H sync pulses. (14) Start of field 2 is defined by a half line between the first equalizing pulse and the preceding H sync pulses. (15) Field 1 line numbers start with the first equalizing pulse in field 1. (16) Field 2 line numbers start with second equalizing pulse in field 2.

The negative values in the $\bar{r}$ curve mean that to obtain a match it is necessary to add the red primary to the monochromatic sample in the colorimeter. For this reason the entire spectrum cannot be matched by the three additive sources, but this limitation is not serious.

15. X, Y, and Z (Nonphysical) Primaries. A linear transformation can be found which transfers the curves of Fig. 20-10 to positive coordinates; that transformation is

$$X = 2.7690R + 1.7518G + 1.1300B$$
$$Y = 1.0000R + 4.5907G + 0.0601B$$
$$Z = 0.0000R + 0.0565G + 5.5943B$$

This transformation has the interesting property that Y is the *luminance* of the monochromatic sources while X and Z have zero luminosity. A plot of the tristimulus values $\bar{x}$, $\bar{y}$, and $\bar{z}$ is given in Fig. 20-11.

Associated with the $\bar{x}$, $\bar{y}$, and $\bar{z}$ tristimulus values are a set of *artificial* (or nonphysical) primaries X, Y, and Z. To match a continuous light source of wavelength vs. energy, represented by a function $I(\lambda)$, the amounts of the primaries X, Y, and Z required are computed from the following integrals:

$$X = \int I(\lambda)\bar{x}(\lambda)\,d\lambda \qquad Y = \int I(\lambda)\bar{y}(\lambda)\,d\lambda \qquad Z = \int I(\lambda)\bar{z}(\lambda)\,d\lambda$$

where λ is the wavelength.

The nonphysical primaries are mathematical transformations to avoid negative numbers. While they cannot exist as light sources, *they can exist as electric signals*. Using the nonphysical primaries, we can define any light source by two quantities, x and y, such that

$$x = X/(X + Y + Z) \qquad \text{and} \qquad y = Y/(X + Y + Z)$$

The third quantity $z = Z/(X + Y + Z)$ is redundant.

16. CIE Chromaticity Diagram. The plot of x and y for all light sources, including monochromatic sources, is the CIE chromaticity diagram (Fig. 20-12). All colors lie within the spectrum line of Fig. 20-12. The CIE chromaticity chart is the world-wide standard method of representing color.

The chromaticity diagram displays the hue and saturation of colors. The *hue* describes the intrinsic nature of the color, i.e., red, green, cyan, purple, etc. *Saturation* is a measure of color intensity, i.e., its pastel vs. vivid quality. Desaturated colors are washed out or whitish. The hue varies on the chromaticity diagram with the angle measured with the white point (illuminant C) as the vertex. Saturation is measured by the radial distance from the white point at the center of the chart.

17. Television Standard Primaries. Since the primaries used in color television reproduction are usually not monochromatic, additional standards must be specified. In setting

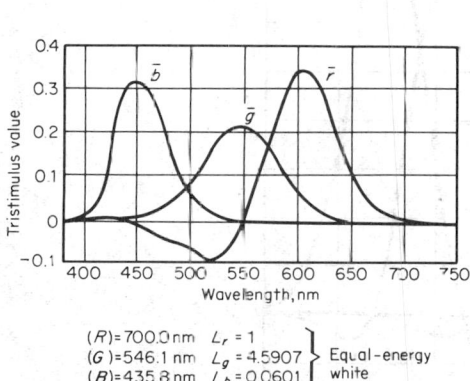

$$(R) = 700.0 \text{ nm} \quad L_r = 1$$
$$(G) = 546.1 \text{ nm} \quad L_g = 4.5907 \Big\} \text{ Equal-energy}$$
$$(B) = 435.8 \text{ nm} \quad L_b = 0.0601 \Big\} \text{ white}$$

Fig. 20-10. Tristimulus values for equal-energy spectrum.

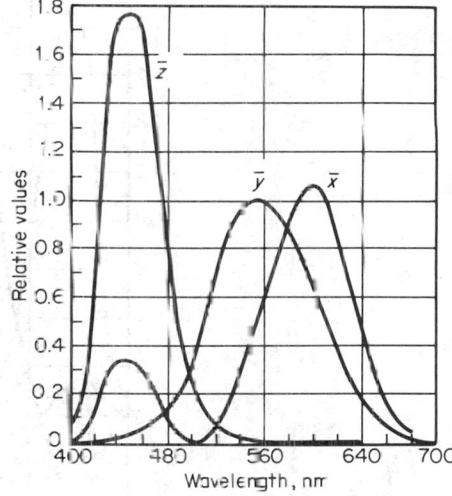

Fig. 20-11. Tristimulus values of CIE nonphysical XYZ primary colors.

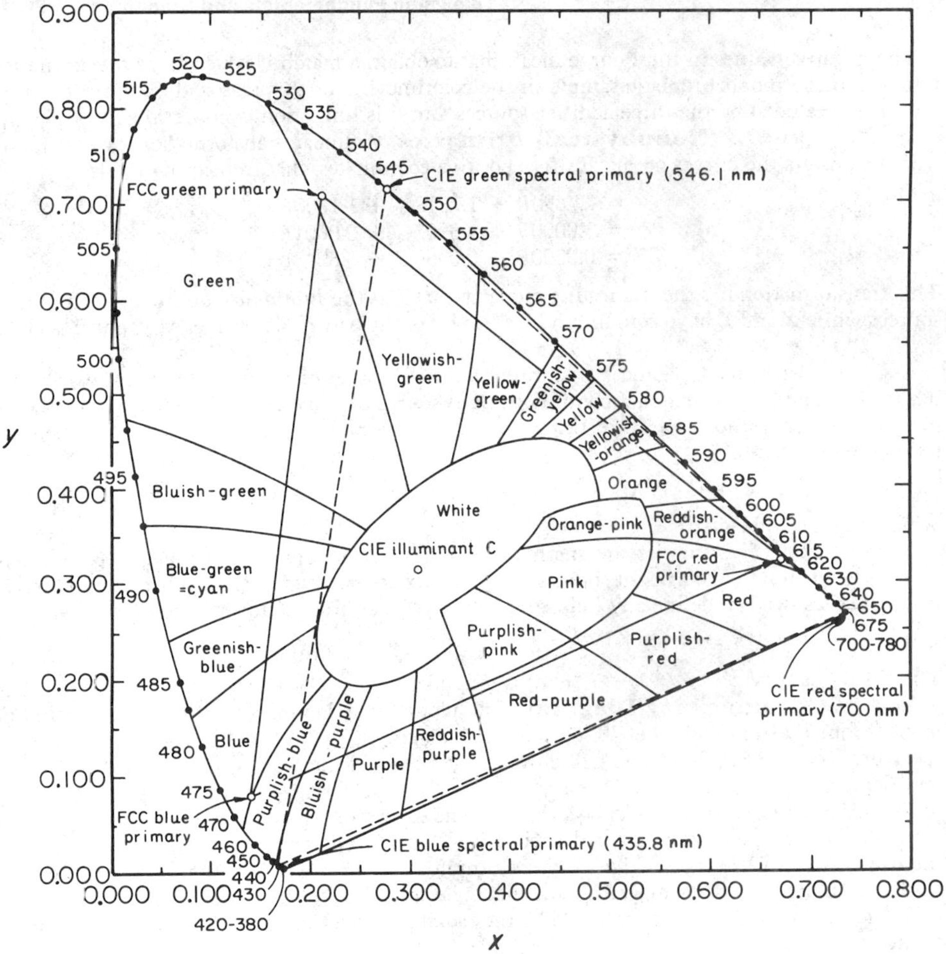

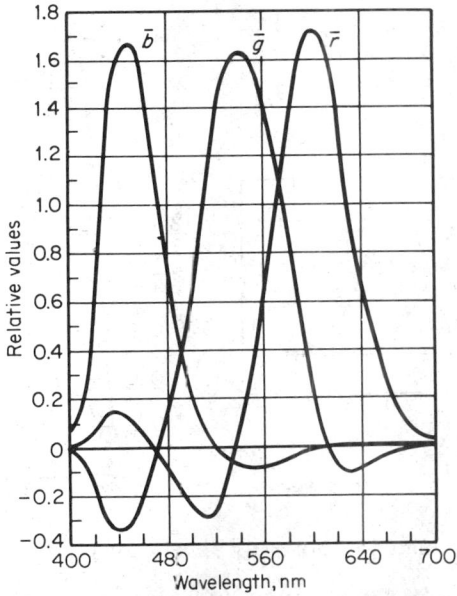

Fig. 20-12. CIE chromaticity diagram.

Fig. 20-13. Tristimulus values of FCC *RGB* primaries.

up the United States color television standards the FCC has specified the x, y coordinates of the standard red, green, and blue primaries as listed in Table 20-2, under Primary colors. These primaries, shown in Fig. 20-12, form a triangle that bounds the color gamut covered by the color system. The transformations to obtain the X, Y, and Z primaries for color television, based on FCC primaries, are

$$X = 0.608R + 0.174G + 0.200B$$
$$Y = 0.299R + 0.537G + 0.114B$$
$$Z = 0.000R + 0.0562G + 1.112B$$

These X, Y, and Z primaries, like those previously defined, are nonphysical and do not represent real colors. They can represent real electric signals but must be electrically transformed (using

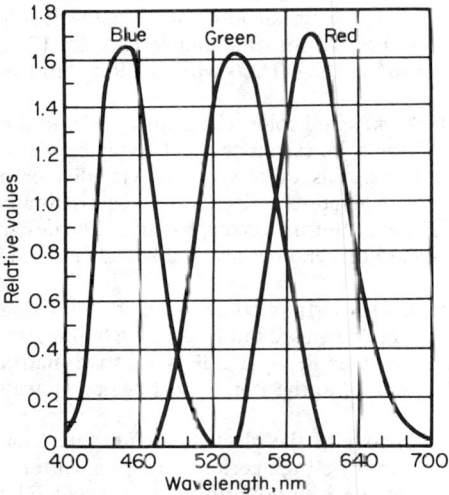

Fig. 20-14. Practical RGB taking sensitivities.

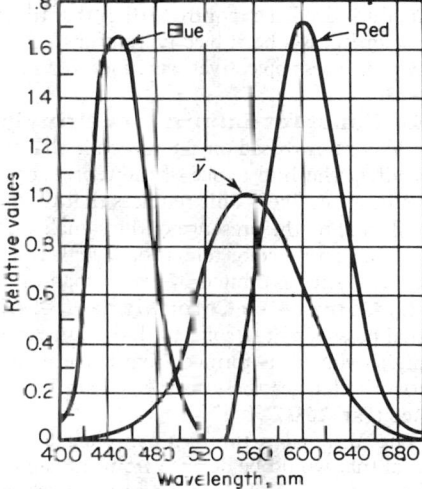

Fig. 20-15. Practical RYB taking sensitivities.

an electrical analog of the transformation equations) to R, G, and B signals before being displayed.

The tristimulus values for the standardized FCC primaries are given in Fig. 20-13. Since negative values occur, an exact match would require a color television camera with six camera tubes, but as a practical matter such cameras are not used. Instead, a compromise set of all-positive *taking primaries* is used. A typical set of such primaries is illustrated in Fig. 20-14. Cameras are also built with a luminance channel and two color channels. Typical taking sensitivities for such cameras are illustrated in Fig. 20-15.

COLOR SYSTEMS

18. Field Sequential Color Systems. The simplest color television system was the first one devised, the field sequential system. This system was developed to a high level of performance by CBS Laboratories and was designated as a standard system in the United States in the early 1950s. It was supplanted as the United States standard by the National Television System Committee (NTSC) system in 1954.

The field sequential system employs a monochrome television camera, with a color-scanning disk mounting near the focal plane. The disk is rotated synchronously with the vertical sweep, such that during one field, light through only a red filter falls on the camera-tube photocathode, during the next field light through only a green filter, and during the next field light through only a blue filter, and so on. The video signal derived from the camera tube thus consists of sequential color fields in the order that the primary light filters appear in front of the camera tube.

The display device was, originally, a black and white cathode-ray tube in front of which another synchronous color filter disk was placed, the camera and receiver color disks being synchronized. Satisfactory color pictures generally resulted.

The most serious disadvantage of the field sequential system for broadcast use is the fact that it is not compatible with the scanning standards of black-and-white sets; i.e., a black-and-white receiver cannot use a sequential color transmission. Another disadvantage is that fast-moving, high-brightness objects, e.g., white gloves and ping-pong balls, have a tendency to display *color breakup;* i.e., they appear as red, green, and blue flashes instead of superimposed colors giving neutral white. After efforts to develop a compatible color television system were successful, these difficulties precluded further development of the sequential system for broadcast purposes.

19. Simultaneous (Compatible) Color System. To be feasible for broadcast use in the United States where the black-and-white receivers in use exceeded 50 million, a compatible color system was required. This requirement was met in the development work of the second NTSC which brought together the talents of the electronic industry in the United States from 1949 to 1954.

The problem of transmitting the three primary-color channels simultaneously, in the same bandwidth, and with essentially similar scanning standards as black-and-white transmissions was difficult indeed. To require further that the images observed on black-and-white receivers during color broadcasts be subjectively undisturbed seemed to pose an insurmountable task. By 1953, however, these objectives were met. For a detailed account of the NTSC's work see Ref. 3 in Par. **20-29**.

20. Constant-Luminance Principle. All the standard color television systems of the world are now based on the principle that if signals in the color-carrier channel do not appreciably affect the luminance of the reproduced picture, the signals are of very low visibility on a television receiver. This requires that the luminance of a given picture area shall be essentially unaffected by the presence of the signals carrying the color information for that area. The design of all composite color television signals is directed toward this end because it takes advantage of characteristics exhibited by the human eye.

21. Composite Color Signals, XYZ System. One form of composite color television signal possessing the constant-luminance characteristic (not now used but illustrative of the principles involved) is formed from three wide-band color signals R, G, and B. A resistive matrix network changes these to the signals X, Y, and Z according to the transformations previously listed (Par. **20-17**).

The Y signal is the luminance signal. It is representative of the black-and-white television signal that would be derived from the same subject matter by a high-performance monochrome camera. The X and Z signals carry the color information, i.e., the *nonluminance content.* The X and Z signals are imposed on a subcarrier in the upper part of the video passband (in the vicinity of 3.6 MHz, using a synchronous, suppressed-carrier quadrature modulation system).

In this arrangement there appear to be several engineering defects.

1. The X and Z color information signals have much less bandwidth than the Y luminance signal.

2. The X and Z signals have high energy content in the upper-frequency part of the luminance channel, which will produce interference with the fine detail of the picture.

3. The upper and lower sidebands of the X and Z signals are not equal. Therefore, when they are detected, quadrature cross talk between them will cause color distortions on sharp edges.

4. To demodulate the X and Z signals, the reference phase of the suppressed carrier frequency must be preserved and reinserted during demodulation.

These problems were in fact solved in the NTSC system by methods described in Table 20-2.

To recover the red, green, and blue signals, an inverse *decoding transformation* is used, as follows:

$$R = 1.191X - 0.532Y - 0.288Z$$
$$G = -0.982X + 2.00Y - 0.0283Z$$
$$B = 0.0585X - 0.119Y + 0.900Z$$

When the decoded R, G, and B signals are displayed on a color receiver tube, even with the limited color information present, acceptable reproduction is obtained. This occurs if the luminance information is kept at full bandwidth because the eye is insensitive to lack of detail in color. Quadrature cross talk does occur, and there are colored halos or fringes at the edges of sharp color transitions.

When the decoded signals are observed on a black-and-white receiver, slanted interference lines occur. The black-and-white picture is acceptable, although the interference effects are objectionable.

22. Reduction of Dot Interference. By selecting the color subcarrier frequency so that it and its sidebands fall at odd multiples of half the line-scanning frequency, successive dots (picture elements) on one line interleave with dots on the next scanning of that line. This reduces the visibility of the slanted interference lines, and the interference, previously noted on the black-and-white receiver, disappears at normal viewing distances. The interference is still actually present, but it appears as a fine dot structure below the threshold of the eye's limiting resolution capability at normal viewing distances. This technique, known as *frequency interleaving*, is used in the design of compatible color television systems (the exception is the French SECAM system, which employs a frequency-modulated color signal, thus precluding the possibility of frequency interleaving).

23. Quadrature Cross Talk. The problem of the colored edges caused by quadrature cross talk has been alleviated by extensive subjective testing with various coded signals. Instead of using the X and Z signals as originally suggested, it was found possible to code the color information in a *luminance signal* Y and two *color-difference signals* R − Y and B − Y. The third color-difference signal G − Y is derived from the Y, R − Y, and B− Y signals at the receiver and need not be transmitted. Early in the NTSC development, in fact, experimentation

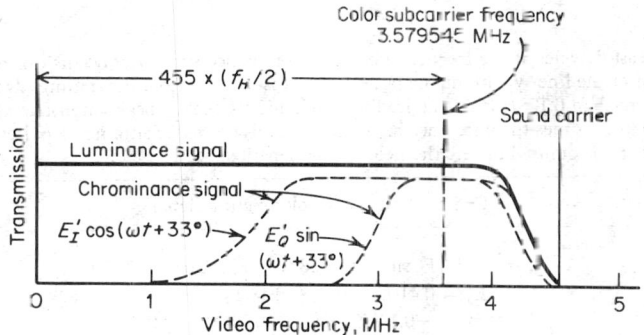

Fig. 20-16. Spectrum pattern of NTSC color signal.

was done with R − Y and B − Y signals rather than with the X and Z signals because the former are easier to derive.

For a neutral color (black, gray, or white), $R = G = B = Y$, and the color-difference signals disappear. Since neutral colors at high brightness are precisely those for which the eye is most critical of objectionable interference, it is good practice for the color signals to go to zero as the saturation of the colors decreases.

It has been found that the cross-talk fringes are reduced to an acceptable level by the following steps:

1. The color subcarrier is precisely synchronized with the scanning rates.
2. Two color signals are used, derived from the color-difference signals as follows:

$$I = 0.74(R − Y) − 0.27(B − Y) \qquad Q = 0.48(R − Y) + 0.41(B − Y)$$

3. The I and Q signals are produced in such a way that the Q signal has a bandwidth of around 0.6 MHz and is transmitted by double sideband suppressed carrier. The I signal has a lower sideband of 1.2 MHz and an upper sideband of 0.6 MHz and is transmitted by suppressed carrier in quadrature with the Q signal. The spectral pattern is shown in Fig. 20-16. The channel arrangement of the U.S. standard (NTSC) system is shown in Fig. 20-17.

This video-signal-coding configuration eliminates half the quadrature cross-talk problem (Q into I) and was selected such that the other half (I into Q) is of low visibility.

24. Subcarrier Phase Recovery. To preserve the subcarrier phase information for synchronous demodulation, a few cycles of the subcarrier frequency (called the *color burst*) are added immediately after each horizontal synchronizing pulse. The burst positioning is shown in Fig. 20-8b.

At the receiver the burst is passed through a gating circuit and thereafter is used to synchronize an oscillator periodically at the end of each line scan. The oscillator signal is used in a synchronous demodulator to decode the color information.

The frequency for the color subcarrier in the United States is 3.579545 MHz ± 10 Hz. This frequency meets two frequency-interleaving requirements. When multiplied by the submultiple $\frac{2}{455}$, the horizontal scanning frequency becomes 15,734.3 Hz and the vertical scanning frequency is 59.94 Hz. These are values so close to the black-and-white standards that monochrome receivers have no difficulty in synchronization. In addition, the beat frequency between 3.579545 MHz and the sound carrier at 4.25 MHz is also frequency-interleaved, making the visibility of the beat frequency very low.

TABLE 20-2 Characteristics of the Color-Television System in Use in the United States

<table>
<tr><td colspan="4" align="center">Scanning and video characteristics</td></tr>
<tr><td>Number of lines per picture (frame)</td><td>525</td><td>System capable of operating independently of power-supply frequency?</td><td>Yes</td></tr>
<tr><td>Field frequency, fields/s</td><td>59.94</td><td></td><td></td></tr>
<tr><td>Interlace</td><td>2:1</td><td></td><td></td></tr>
<tr><td>Picture (frame) frequency, pictures/s</td><td>29.97</td><td>Approximate gamma of picture signal</td><td>0.45</td></tr>
<tr><td>Line frequency, lines/s</td><td>15,734.264</td><td>Nominal video bandwidth, MHz</td><td>4.2</td></tr>
<tr><td>Tolerance, lines/s</td><td>±0.044</td><td>Chrominance subcarrier frequency, MHz</td><td>3.579545</td></tr>
<tr><td>Aspect ratio, width/height</td><td>4:3</td><td>Tolerance, Hz</td><td>±10</td></tr>
<tr><td>Scanning sequence, line</td><td>Left to right</td><td>Hz/s</td><td>< 0.1</td></tr>
<tr><td>Field</td><td>Top to bottom</td><td></td><td></td></tr>
</table>

A burst of at least 8 cycles at the frequency of the chrominance subcarrier occurs during each horizontal blanking period after the line-synchronizing pulse and at least $0.006H$ from the trailing edge of that pulse and lasts until not more than $0.125H$ from the leading level, and its peak-to-peak amplitude about the blanking level is from 0.90 to 1.1 times the difference between the levels of the synchronizing pulses and the blanking level. The color burst is omitted during the field-blanking period. See Fig. 20-9

Composition of the color-picture signal

$$E_M = E_Y' + [E_Q' \sin (\omega t + 33°) + E_I' \cos (\omega t + 33°)]$$

where

$$E_Q' = 0.41(E_B' - E_Y') + 0.48(E_R' - E_Y')$$
$$E_I' = -0.27(E_B' - E_Y') + 0.74(E_R' - E_Y')$$
$$E_Y' = 0.30E_R' + 0.59E_G' + 0.11E_B'$$

For color-difference frequencies below 500 kHz the signal is

$$E_M = E_Y' + \{(1/1.14) [(1/1.78) (E_B' - E_Y') \sin \omega t + (E_R' - E_Y') \cos \omega t] \}$$

where E_M = total video voltage, corresponding to scanning of a particular picture element, applied to modulator of picture transmitter, E_Y' = gamma-corrected voltage of monochrome portion of color picture signal, corresponding to given picture element, E_Q', E_I' = amplitudes of two orthogonal components of chrominance signal corresponding respectively to narrow-band and wide-band axes, E_R', E_G', and E_B' = gamma-corrected voltages corresponding to red, green, and blue signals during scanning of given picture element, and ω = angular frequency = 2π times frequency of chrominance subcarrier.

The portion of each expression between brackets represents the chrominance subcarrier signal which carries the chrominance information.

The phase reference in the E_M equation is the phase of the burst + 180°. The burst corresponds to amplitude modulation of a continuous sine wave.

Bandwidths (equivalent bandwidths assigned before modulation to the color-difference signals E_Q and E_I

Q-channel bandwidth		I-channel bandwidth	
At 400 kHz	< 2 dB down	At 1.3 MHz	<2 dB down
At 500 kHz	<6 dB down	At 3.6 MHz	At least 20 dB down
At 600 kHz	At least 6 dB down		

Primary colors

The gamma-corrected voltages E_R', E_G', and E_B' are suitable for a color picture tube having primary colors with the following chromaticities in the CIE system of specification:

	x	y
Red (R)	0.67	0.33
Green (G)	0.21	0.71
Blue (B)	0.14	0.08

and having a transfer gradient γ (gamma exponent) of 2.2 associated with each primary color. The voltages E_R', E_G', and E_B' may be respectively of the form $E_R^{1/\gamma}$, $E_G^{1/\gamma}$, and $E_B^{1/\gamma}$, although other forms may be used with advances in the state of the art.

TABLE 20-2   Characteristics of the Color-Television System in Use in the United States (Continued)

Subcarrier characteristics

The radiated chrominance subcarrier vanishes on the reference white of the scene. The numerical values of the signal specification assume that this condition will be reproduced as standard illuminant C ($x = 0.310$, $y = 0.316$) of the CIE. E'_Y, E'_Q, E'_I and the components of these signals match each other in time to 0.05 μs. The angle of the subcarrier measured with respect to the burst phase, when reproducing saturated primaries and their complements at 75% of full amplitude, is within $=10°$, and its amplitude is within $\pm 20\%$ of the values specified above. The ratios of the measured amplitudes of the subcarrier to the luminance signal for the same saturated primaries and their complements fall between the limits of 0.8 and 1.2 of the values specified for their ratios.

RF and modulation characteristics

Nominal rf bandwidth	6 MHz
Sound carrier relative to vision carrier	+4.5 MHz
Sound carrier relative to nearest edge of channel	−0.25 MHz
Nominal width of main sideband	<2 MHz
Nominal width of vestigial sideband	0.75 MHz
Type of polarity of vision modulation A5C	Negative
Synchronizing level as a percentage of peak carrier	100
Blanking level as a percentage of peak carrier	72.5–77.5
Difference between black level and blanking level as percentage of peak carrier	2.875–6.75
Reference-white level as percentage of peak carrier	10–15
Type of sound modulation	F3 ±25 kHz, 75μs preemphasis
Ratio of effective radiated powers of vision and sound	10:1–5:1
Polarization of radiated wave	Horizontal or right-hand circular or elliptical

---

Synchronizing signals

Line synchronization	Percent of line period (H)	μs
Line period H	100	63.556
Line-blanking interval	16.5018	10.5–11.4
Interval between time datum H_0 and back edge of line-blanking signal	12.7–16	8.06–10.3
Front porch	>2	>1.27
Synchronizing pulse	6.6–8	4.2–5.1
Buildup time (10–90%), of edges of the line-blanking signal	<0.75	<0.48
Of line-synchronizing pulses	<0.4	<0.25

Field synchronization	Duration
Field period V	16.683 ms
Line period H	63.566 μs
Field-blanking period	1335 ms = 0.08V $\approx$ 21H
Buildup times (10–90%) of edges of field-blanking pulses	<6.36 μs
Duration, of first equalizing pulse sequence	3H
Of synchronizing pulse sequence	3H
Of second sequence of equalizing pulses	3H
Of equalizing pulse	2.29 μs
Of field-synchronizing pulse	28 0 μs
Interval between field-synchronizing pulses	5 μs
Buildup times (10–90%) of edges of synchronizing signals	<0.25 μs

---

25. American Standard (NTSC) Color System. Table 20-2 lists the characteristics of the American (NTSC) Color System.

26. PAL and SECAM Systems of Color Television. In Europe, two color-television systems have evolved, PAL and SECAM. Both operate on the constant-luminance principle. The PAL system is very similar to the NTSC system except that the color-subcarrier phase is reversed every other line and simple color-difference signals are used in place of the I and Q signals.

These modifications make the system less sensitive to subcarrier phase errors and minimize quadrature cross talk. The receiver is somewhat more complicated, however.

The SECAM system transmits one color signal on one line and the other color signal on the subsequent line. The receiver has a delay line, with delay equal to one horizontal-line period, and an electronic switching system. These elements allow the three types of color video information to be displayed simultaneously. The color signals are frequency-modulated on a subcarrier high in the video passband. This arrangement precludes the use of frequency interlace but

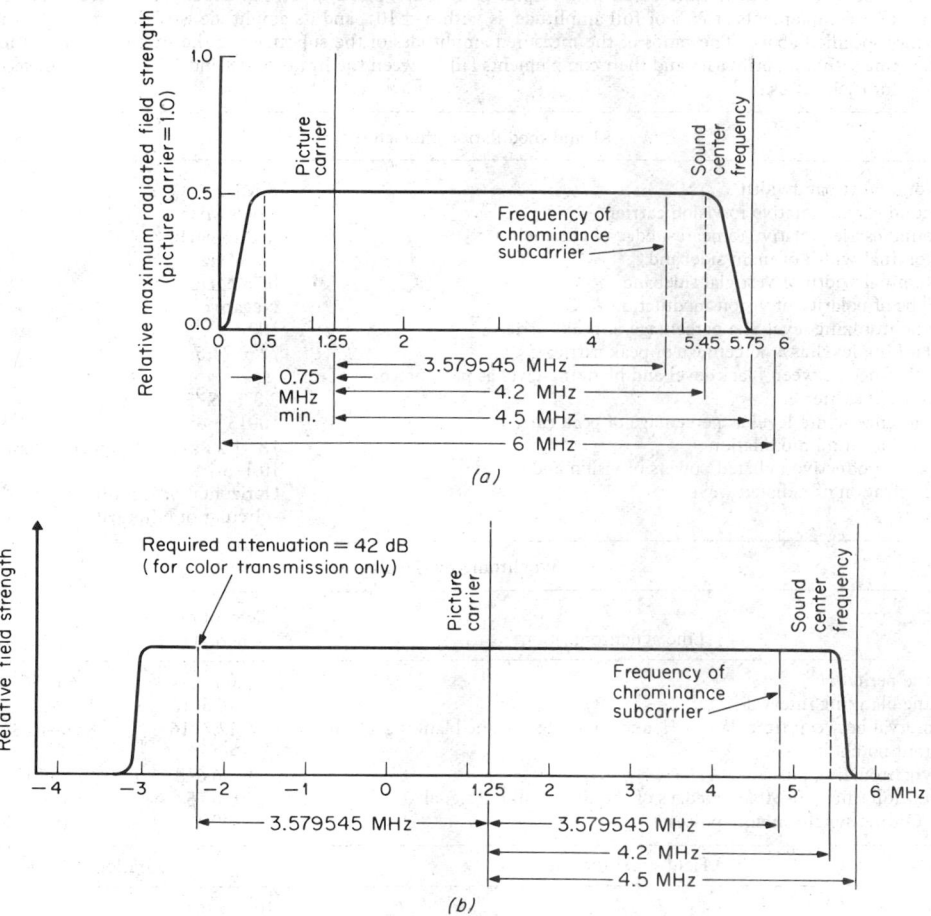

(a)

(b)

Fig. 20-17. Channel arrangement of NTSC transmission (FCC Standard Idealized Transmission Characteristic): (*a*) for all stations except low-power UHF; (*b*) permissible double-sideband channel for stations of peak visual power 1 kW or less, in UHF channels 14 to 83.

makes it possible to ignore subcarrier phase accuracies in recording and in microwave transmission.

These systems operate on 625 lines at 50 fields. Detailed specifications are given in Table 20-3.

27. Digital Video Techniques. As in other forms of electronic communication, digital techniques are being increasingly applied to video engineering. The transfer to digital form occurs in an analog-to-digital converter, in which the analog composite video signal is sampled at a high rate (typically equal to 3 times the chrominance subcarrier frequency), and the amplitude of each successive sample is adjusted (quantized) to the upper or lower limit of the interval between the binary levels. The NTSC composite video color signal is typically quantized into 256 levels [8 bits (b) = 2^8 = 256]. Each quantized sample amplitude is encoded by pulse-code modulation into a binary 8-b number, and the succession of these numbers constitutes a bit stream that represents the video signal in digital form. The analog form is recovered in a digital-to-analog converter.

TABLE 20-3 Specifications of the PAL and SECAM III Systems

Characteristic	Nominal video bandwidth (see also Table 20-1 for associated standards)		
	5 MHz (B,G,H)	5.5 MHz (I)	6 MHz (D,K,L)
1. General specifications			
Luminance component (PAL and SECAM):	Amplitude modulation of the picture carrier		
Chrominance component:			
PAL	Simultaneous pair of components transmitted as amplitude-modulated sidebands of a pair of suppressed subcarriers in quadrature having a common frequency		
SECAM	A pair of components transmitted alternately on successive lines as the frequency modulation of a subcarrier		
2. Color subcarrier f_{sc}			
PAL	4.43361875 MHz ± 5 Hz	4.43361875 MHz ± 5 Hz	4.43361875 MHz ± 5 Hz
SECAM	4.40625 MHz ± 2 kHz (f_{OR}) 4.25000 MHz ± 2 kHz (f_{OB})	4.40625 MHz ± 2 kHz (f_{OR}) 4.25000 MHz ± 2 kHz (f_{OB})	4.40625 MHz ± 2 kHz (f_{OR}) 4.25000 MHz ± 2 kHz (f_{OB})
3. Frequency spectrum of composite color picture and sound signals, MHz			
Vision-to-sound spacing:			
PAL	5.5	5.9996 ± 0.0005	6.5 ± 0.001
SECAM	5.5	5.9996 ± 0.0005	6.5 ± 0.001
Main sideband (luminance):			
PAL	5.0	5.5	6
SECAM	5.0	5.5	6
Vestigial sideband:			
PAL	0.75 (H 1.25)	1.25	1.25 (D, K 0.75)
SECAM	0.75 (H 1.25)	1.25	1.25 (D, K 0.75)
Chrominance sidebands f_{sc}:			
PAL: E_u', F_v' signal	+0.57 −1.3	+1.07 −1.3	...
SECAM: D_R', D_B' (deviation)	+0.350 − 0.506 (OR) +0.506 − 0.350 (OB)	...	+0.350 − 0.506 (OR) +0.506 − 0.350 (OB)

TABLE 20-3 Specifications of the PAL and SECAM III Systems (Continued)

Characteristic	Nominal video bandwidth (see also Table 20-1 for associated standards)		
	5 MHz (B,G,H)	5.5 MHz (I)	6 MHz (D,K,L)
4. Transmitted color-picture-signal waveform			
Color synchronization:			
PAL:			
Subcarrier burst, duration	10 ± 1 cycle		
Start	5.6 ± 0.1 μs after the leading edge of the line-synchronizing pulses		
Amplitude	% of difference between blanking level and peak white level ± 10% (±3% in system I)		
Omission	Omitted during field-blanking periods for 9 lines		
Phase sequence (see Phase reference, below, under 8)	First field (even) starts on line 7, +135° on odd lines		
	Second field (odd) starts on line 319, −135° on even lines		
	Third field (even) starts on line 6, +135° on even lines		
	Fourth field (odd) starts on line 320, −135° on odd lines		
SECAM	Subcarrier signal modulated in frequency and amplitude to correspond to a sawtooth color-difference signal D'_B or D'_R during six lines of each field-blanking period		
Deviation and amplitude	Correspond to maximum D'_B or D'_R		
Duration	Active line period		
Sequence	First field (even), $-D'_B$ on lines 11, 13, 15; $-D'_R$ on lines 10, 12, 14		
	Second field (odd), $-D'_B$ on lines 323, 325, 327; $-D'_R$ on lines 324, 326, 328		
	Third field (even), $-D'_B$ on lines 10, 12, 14; $-D'_R$ on lines 11, 13, 15		
	Fourth field (odd), $-D'_B$ on lines 326, 328; $-D'_R$ on lines 323, 325, 327		
Duration of color subcarrier protection interval	Starts 5.6 ± 0.2 μs after leading edge of line-synchronizing pulse		
5. Luminance component			
Attenuation-frequency characteristics, both	Uniform 0–5.0 MHz	Uniform 0–5.5 MHz (notch filter at f_{sc} permissible)	Uniform 0–6.0 MHz (notch filter at f_{sc} permissible)

6. Scanning

Line-scanning frequency f_{line}: PAL SECAM Gamma, PAL and SECAM	15.625 Hz = $f_{sc}(1,135/4 + 1/625)^{-1}$ 15.625 Hz = $f_{sc}/282 \, (f_{OR})$; $f_{sc}/272 \, (f_{OB})$ Corresponds to display gamma of 2.8

7. Synchronizing and blanking waveforms

Both	Similar to characteristics of monochrome television systems

8. Equation of complete color signal

PAL*	$$E_M = E'_Y + E'_U \sin 2\pi f_{sc} t \pm E'_V \cos 2\pi f_{sc} t$$ where $\quad E'_U = 0.493(E'_B - E'_Y) \quad$ and $\quad E'_V = 0.877(E'_R - E'_Y)$ $\qquad\qquad E'_Y = 0.299 \, E'_R + 0.587 \, E'_G + 0.114 \, E'_B$ E'_U axis. The sign of the E'_V component is the same as that of the subcarrier burst, changing for each line, as under item 4 (color synchronization) above
Phase reference	In large area of color
SECAM*	$$E_M = E'_Y + G \cos (2\pi f_{sc} + E'_C \, \Delta f_{sc})t$$ where E'_C is a color-difference signal D'_R or D'_B and Δf_{sc} is the frequency deviation corresponding to unit amplitude of the preemphasized color-difference signal; D'_R = $-1.9(E'_R - E'_Y)$ and $D'_B = 1.5(E'_B - E'_Y)$. G is a function of $E'_C \, \Delta(2\pi f_{sc})$ and determines the amplitude of the chrominance signal

Equivalent bandwidths of color-difference signals D'_B and D'_R before preemphasis and modulation	1.4 MHz At 1.0 MHz <2 dB 1.4 MHz
	                    down
	                 At 1.5 MHz >5 dB
	                    down
	                 At 2.0 MHz <20 dB
	                    down

9. SECAM color-subcarrier modulating signal

Preemphasis of signal D'_R and D'_B before modulation	Time constant = 1.12 μs (+14 dB at 1 MHz)
Frequency modulation of color carrier Maximum subcarrier deviation	f_{sc} = 230 kHz per unit amplitude of D'_R and D'_B after preemphasis, respectively $f_{sc} \pm 506 \pm 75$ kHz

*E'_M = total video voltage applied to modulator of transmitter; E'_Y = voltage of luminance component of composite signal; E'_R, E'_G and E'_B = gamma-corrected voltages corresponding to red, green, and blue signals.

SOURCE: CCIR "Green Book," 14th Plen. Assem., Kyoto, 1978, Vol. XI, Rep. 6241, International Telecommunications Union, Geneva, 1978.

Since each element of the stream is a bit (0 or 1), it can be stored and/or transmitted with minimal opportunity for introduction of noise or distortion. Thus, many repeaters can be used in tandem or many video-tape duplicates made in succession without visible degradation of the reproduced image. Moreover, the bit stream can be stored for as long as is necessary to perform signal corrections or transformations in the digital form.

One such application is time correction of video tapes made in electronic newsgathering, which need not be rigidly synchronized during the initial "take." The synchronizing signals of the camera tape and the broadcast feed are tied together by reference to the corresponding bits. Storage of video bit streams also permits reduction of noise levels and enhancement of signal resolution. A most important application involves the *frame store*, in which the bit stream for two successive fields is stored, being replaced line by line as successive fields are scanned. This store is addressable; i.e., any particular bit in the stream can be identified and recalled as needed

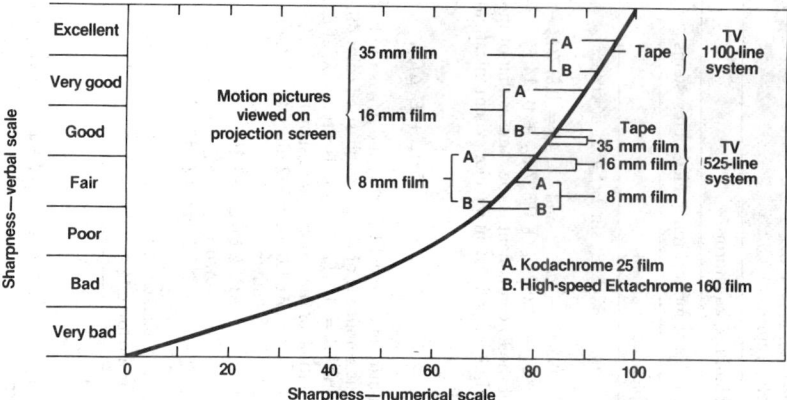

Fig. 20-18. Comparative values of sharpness (definition, judged subjectively) when images are viewed at a distance equal to 5 times the picture height, of motion-picture film projection and directly viewed television images of 525 and 1,100 lines. (*Diagram by Wilmotte, Rose, and Nelson in Fink.*[22])

from the store. Thus it is possible, for example, to store a digitized NTSC signal at 525 lines, 60 fields and convert it to the PAL system at 625 lines, 50 fields or vice versa. Commercial standards-conversion equipment for this purpose was introduced in the late 1970s. Virtually any standards conversion is possible by this means without signal degradation.

The price paid for the advantages of the digital technique is the substantially wider bandwidth occupied by the digitized signal. Thus in the typical 8-b NTSC conversion, with a sampling rate of 3 times the chrominance subcarrier frequency of 3.579 MHz, that is, 10.74 Mb/s, the bit rate is 8(10.74) = 85.9 Mb/s. To avoid aliasing interference the Nyquist theorem requires that the bandwidth be not less than one-half the bit rate, or 47.95 MHz. Additional bits are used for parity checking and housekeeping purposes, so that the typical bandwidth for 8-b NTSC transmission is about 50 MHz.

This wide spectrum requirement precludes the use of digital signals in ground-based broadcasting, although it is potentially feasible in direct satellite broadcasting and in optical-fiber communications. The use of comb filters and differential pulse-code modulation permits narrower bandwidths, i.e., sampling rates lower than those required by the Nyquist theorem. Within the broadcast plant, including video-tape recorders, the wide bandwidth is accommodated by using several parallel channels.

28. High-Definition Television Systems. In the late 1970s research was begun (notably in a program sponsored by the Japan Broadcasting Company, NHK) into the requirements and specifications of a television system that would compete in definition and other qualities with the images projected from 35-mm color motion picture film (see Fig. 20-18). Scanning rates of the order of 1,125 lines per frame at 60 fields per second, with video bandwidths of 20 to 25 MHz, were used in experimental transmission in Tokyo with direct satellite reception in 1979. At that time live cameras capable of resolving 1,125 lines had objectionable lag characteristics, and no available video-tape recorder was capable of recording the 20- to 25-MHz signal at acceptable noise levels and tape-consumption rates. However, rapid progress was being made, and the use of such high-definition systems for projection in theaters and as a substitute for film in the production of 35-mm motion pictures is being actively considered. A definitive report on the state of the art in 1980 has been published.[22]

29. Bibliography

GENERAL REFERENCES ON TELEVISION SYSTEMS AND STANDARDS

1. Fink, D. G. "Television Standards and Practice," McGraw-Hill, New York, 1943.
2. Fink, D. G. "Television Engineering," McGraw-Hill, New York, 1952.
3. Fink, D. G. "Color Television Standards," McGraw-Hill, New York, 1955.
4. Fink, D. G. (ed.) "Television Engineering Handbook," McGraw-Hill, New York, 1957 (microfilm edition, Xerox University Microfilms, Ann Arbor, Mich.).
5. Fink, D. G. Perspectives on Television: The Role Played by the Two NTSC's in Preparing Television Service for the American Public, Proc. IEEE, September 1976, Vol. 64, pp. 1322-133L
6. McIlwain, K., and C. E. Dean "Principles of Color Television," Wiley, New York, 1956.

SCANNING

7. Mertz, P. Television: The Scanning Process, Proc. IRE, October 1941, Vol 29, pp. 529-537.
8. Mertz, P., and F. Gray A Theory of Scanning and Its Relation to the Characteristics of the Transmitted Signal in Telephotography and Television, Bell Syst. Tech. J., July 1934, Vol. 13, pp. 464, 515.
9. Schade, O. H. Electro-Optical Characteristics of Television Systems. Pt. I–IV, RCA Rev., March, June, September, and December 1948.

COLORIMETRY

10. Hardy, A. C. "Handbook of Colorimetry," Technology Press, Cambridge, Mass., 1936.
11. Judd, D. B. The 1931 I.C.I. (C.I.E.) Standard Observer and Coordinate System for Colorimetry, J. Opt. Soc. Am., 1933, Vol. 23, pp. 359-374.
12. Wintringham, W. T. Color Television and Colorimetry, Proc. IRE, October 1951, Vol. 39, pp. 1135-1172.

SEQUENTIAL COLOR SYSTEM

13. Goldmark, P. C., J. N. Dyer, E. R. Piore, and J. M. Hollywood Color Television. Pt. I, Proc. IRE, April 1942, Vol. 30, pp. 162-182.
    Goldmark, P. C., E. R. Piore, J. M. Hollywood, T. H. Chambers, and J. J. Reeves Color Television, Pt. II, Proc. IRE, September 1943, Vol. 31, pp. 465-478.

NTSC COMPATIBLE COLOR SYSTEM

14. Abrahams, I. C. The Frequency Interleaving Principle in the NTSC Standards, Proc. IRE, January 1954, Vol. 42, pp. 81-83.
15. Hirsch, C. J., W. F. Bailey, and B. D. Loughlin Principles of NTSC Compatible Color Television, Electronics, February 1952, Vol. 25, No. 2, pp. 88-95.
16. Hirsch, C. J., and W. F. Bailey Quadrature Cross Talk in NTSC Color Television, Proc. IRE, January 1954, Vol. 42, pp. 84-90.
17. Loughren, A. V. Recommendations of the National Television System Committee for a Color Television Signal, J. Soc. Motion Pict. Telv. Eng., April 1953, Vol. 60, pp. 321-336; May 1953, p. 596.

PAL AND SECAM COLOR SYSTEMS

18. CCIR "Green Book," Recommend. Rep. CCIR, 14th Plen. Assem., Kyoto 1978, Vol. XI, Broadcasting Service (Television), International Telecommunications Union, Geneva, 1978.
19. Pritchard, D. H., and J. J. Gibson Worldwide Color TV Standards – Similarities and Differences, J. Soc. Motion Pic. Telev. Eng., February 1980, Vol. 89, pp. 111-120.

DIGITAL VIDEO TECHNIQUES

20. Davidoff, F., J. Rossi, and C. B. Rubenstein (eds.) "Digital Video," Society of Motion Picture and Television Engineers, Scarsdale, N.Y., 1977.
21. Kennedy, M. C. (ed.) "Digital Video," Vol. 2, Society of Motion Picture and Television Engineers, Scarsdale, N.Y., 1979.
21a. Marcus, R. (ed.) "Digital Video," Vol. 3, ibid, Scarsdale, N.Y., 1980.

HIGH-DEFINITION TELEVISION SYSTEMS

22. Fink, D. G., et al. The Future of High-Definition Television: Report of the SMPTE Study Group on High-Definition Television, J. Soc. Motion Pict. Telev. Eng., February, March 1980, Vol. 89, pp. 89-94, 153-161.
23. Hayashi, K. Research and Development of High-Definition Television in Japan, J. Soc. Motion Pict. Telev. Eng., March 1981, Vol. 90, No. 3, pp. 178-186.
24. Fujio, T., et al. High-Definition Television System-Standard Signal and Transmission, J. Soc. Motion Pic. Telev. Eng., August 1980, Vol. 89, pp. 579-584.

Television Cameras*

BY RENVILLE H. MC MANN JR.

30. Monochrome Cameras. Monochrome TV cameras have been almost entirely superseded for entertainment television by color cameras. Nevertheless, the number of monochrome cameras manufactured far exceeds that of color cameras since the former have been widely accepted for industrial, scientific, domestic, and special-purpose applications. The history of monochrome television cameras covers four distinct phases, each based on the availability of advances in pickup-tube technology. The tubes involved are the image dissector, photoemissive (iconoscope and image orthicon), and photoconductive (vidicon, Plumbicon, and Saticon). Although the photoconductive tube is by far the most common in present practice, the photoemissive tube is still used where maximum sensitivity is required. The image dissector and the iconoscope are no longer manufactured.

31. Monochrome Image-Orthicon Cameras. The image-orthicon tube, invented and developed about 1940, furnished the impetus for a new generation of cameras which monopolized the broadcast field for more than 20 years (Fig. 20-19). The extreme highlight sensitivity and a self-adjusting knee in the exposure-transfer characteristic enable the tube to accommodate to practically all lighting conditions, including outdoor scenes of high contrast.

In operation, lighting and lens openings are sufficient to place the highlight exposure well into the knee portion of the transfer characteristic. This produces a self-stabilized image-electron redistribution around the brightest objects in the scene, outlining them in black and giving an effect of sharpness not actually present. In addition, this effect stretches the low-light information. The result is a video signal with limiting in the highlights and excellent visibility of low-light information. This mode of operation was standard in monochrome broadcasting for many years. It accustomed the public to accept and even expect pictures with unnaturally accented low-frequency resolution. A variation of the image orthicon, the *return-beam image orthicon*, is often coupled with an image intensifier to produce pictures at light levels that approach the quantum limit of sensitivity.

32. Image Orthicon 4½-in Studio Cameras. In the late 1950s the technical quality of monochrome television pictures was greatly improved by the introduction of the 4½-in image orthicon. Detailed comparisons of 3- and 4½-in image-orthicon signal-to-noise ratio, gray-scale characteristic, resolution, and lighting and staging techniques showed that a large improvement in picture quality could be realized by using the larger 4½-in image-orthicon tubes, operated with primary emphasis on correct light exposure. This marked the beginning of the era where correct gray scale and proper tonal rendition were recognized as being extremely important, laying the groundwork for the color cameras to follow.

In the new form of operation, the camera was set up so that normal lens exposure on a gray-scale chart brought the highest highlights just to the knee of the transfer characteristic for close-spaced image orthicons and into the knee for the wide-spaced higher-sensitivity tubes.

*The author and editor wish to acknowledge the contribution to this subsection in the first edition, by the late Henry N. Kozanowski.

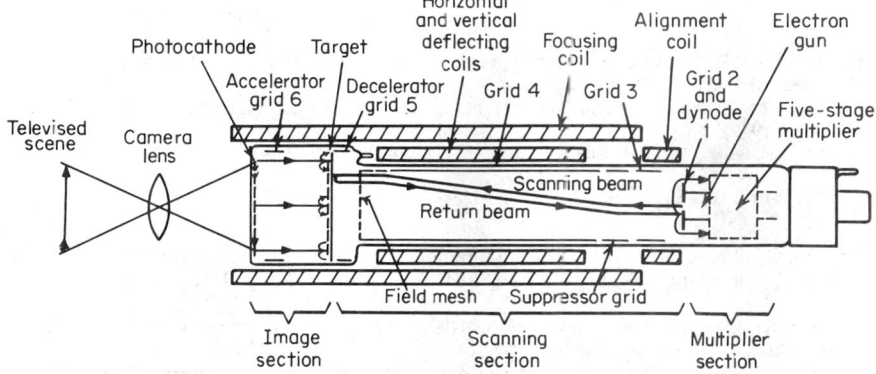

Fig. 20-19. Image-orthicon camera tube with focusing and deflection system.

A further requirement in the new operating techniques was the electronic gamma correction of the signal since the new mode of operation gave substantially a linear gray-scale transfer characteristic. The addition of external gamma correction of 0.7 to 0 5 log-log slope produces pictures that are smooth and pleasing, with excellent face-tone rendition. Because of the greater target area, the signal-to-noise ratio is excellent even after gamma correction.

Cameras designed around the 4½-in image-orthicon tube and intended for operation under controlled exposure conditions were made available by major equipment manufacturers in the early 1960s. These cameras were the first to abandon most vacuum-tube circuits and contained numerous solid-state elements and transistor circuits.

33. Monochrome Vidicon Cameras. Vidicon camera tubes, first introduced in 1951, employ photoconductive pickup. In comparison with the image orthicon, the vidicon stands out in its directness and simplicity of operation. A composite layer of photoconductive material, commonly antimony sulfide, is evaporated onto a transparent conductive substrate sealed to the end of a cylindrical glass tube. An electron gun, with the usual limiting aperture, focus electrodes, and field mesh, completes the essential components of the assembly. An optical image

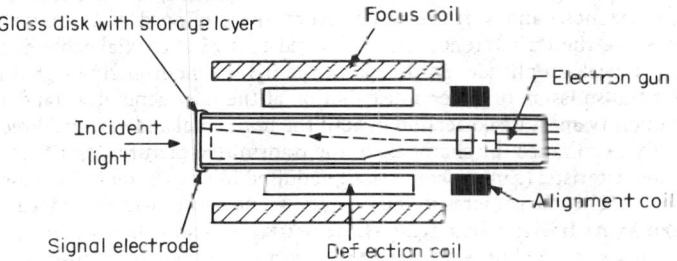

Fig. 20-20. Vidicon camera tube.

focused on the semitransparent photoconductive layer causes the conductivity to vary from point to point in accordance with the light-intensity variations. The charge pattern so developed is scanned by the electron gun (Fig. 20-20). The electrons replaced by this scanning process flow through a load resistor connecting the photoconductive layer to a positive potential source of about 30 V, producing the video signal. For normal operation the peak video-signal current is approximately 300 nA, and the load resistor is approximately 50 kΩ.

Because of the ease of setup, reasonable sensitivity, moderate cost, and simplicity of the circuits, the vidicon has almost completely superseded the image orthicon for industrial and closed-circuit operation.

34. Operating Features of Vidicon Cameras. Vidicon cameras are capable of good operation, with adequate signal-to-noise ratio, at illumination levels of 100 lx or more. Since the gamma (slope of the log-log signal output to light input) is approximately 0.65, the camera is noncritical of the input lighting; i.e., an increase in highlight illumination intensity by a factor 2 will raise the signal by only 50%. This transfer characteristic also improves the visibility of low-light detail, compared with a linear gamma function.

With low values of target voltage, the dark current ("no-light" signal) is negligibly small, producing excellent black-level performance. At larger values of target voltage, the dark current rises rapidly and may become a significant part of the total signal. Since the photoconductive layer is a semiconductor, the dark current doubles in value for every 10°C rise in operating temperature. This can give rise to black-level shading which must be corrected in the video processing circuits.

A very important feature of vidicons having antimony sulfide targets is the increase in effective sensitivity as a function of target potential. The sensitivity varies approximately as the 2.5 power of the target voltage. Thus, automatic gain or sensitivity controls can easily be provided by feedback circuits which increase the target voltage as the incident lighting decreases. A simple device for obtaining such sensitivity control is a bypassed high resistance in series with the target voltage source so that the average target signal determines the target voltage.

35. Lag and Retentivity. Under threshold operating conditions with high target voltages and low scene-light levels, vidicons often show a lag effect with motion in the scene and high retentivity (afterimage burn) on the raster. These effects can be minimized by operating the vidicon at the highest available light level on the photo surface and at the lowest target voltage consistent with generating the required video signal.

While the vidicon camera has been almost universally applied to industrial and other closed-circuit applications, it has had only limited use in broadcast TV work aside from motion-picture pickup (see Par. **20-57**). Even though stationary pictures with excellent signal-to-noise ratio, resolution, and gray scale can be produced at moderate lighting levels, the signal lag, producing smear with motion, is sufficiently troublesome to limit the use of such cameras to programs where the amount of motion is moderate.

The resolution of vidicon cameras can be excellent, the 25-mm-diameter versions frequently exceeding 1,000 lines or more. Recently the 18-mm version has become the most popular for industrial use because it permits lower deflection power and uses physically smaller lenses. The 18-mm tube can readily resolve 600 to 800 lines.

36. Video Processing for Monochrome Camera Signals. The fundamental processes involved in generating picture signals require the use of synchronizing signals, which control the deflection of the scanning raster on the photosensitive pickup tube, and the generation of the video output signal from the scanning of the charge image. The video signal must be compensated for attenuation of high frequencies due to shunt capacitance in the pickup tube.

The millivolt target signal is amplified to a level of a few volts, blanked to clean up the waveforms during horizontal- and vertical-sweep return time, clipped at a reference black level, clamped to preserve the dc reference, and delivered to a 75-Ω coaxial-cable distribution drive amplifier at 0.7 V peak amplitude. At this point the 0.3-V synchronizing signal is added to the video signal for transmission or general distribution at the now generally standard 1 V peak-to-peak level. Frequently aperture correction in both the horizontal and vertical directions is carried out electronically to enhance the sharpness of the transmitted picture. Modification of the gray-scale transfer characteristic (gamma correction), required in color cameras, is now also generally introduced in monochrome cameras to increase the scene contrast ratio they can handle.

37. Camera Synchronizing Requirements. Camera chains require horizontal and vertical drive pulses, a composite blanking waveform, and synchronizing signals for proper operation. These are usually produced in a separate unit to conform to EIA Tentative Standard RS-170A specifications (see Fig. 20-24).

The distribution of the six timing signals within the broadcast plant requires a large investment in coaxial cable, amplifiers, and connectors. Recent installations use simplified single-cable systems in which the timing information is coded in a single complex waveform. At a distribution point this waveform is decoded and reconstituted into the required drive pulses by solid-state circuits.

Industrial camera systems are frequently provided with free-running oscillators or multivibrators to produce the required waveforms, and the vertical-horizontal frequency relationship provides random interlace in the transmitted signal. The free-running characteristic makes the camera pickup tube relatively immune to damage from lack of drive pulses and to deflection-circuit malfunction. The camera is sometimes arranged to accept external "standard" pulses, in which case the output conforms to broadcast practice.

The latest cameras use LSI integrated circuits with the entire sync generator contained on one chip. These one-chip sync generators fully meet EIA RS-170A broadcast standards.

In an elaborate installation involving several cameras, all synchronizing generators are tied together (gen-locked) to achieve system stability in switching from camera to camera. This is very important if the switched signal is to be tape-recorded because the mechanical servos in the tape recorder can be severely upset if the timing of sync abruptly changes. This prevents the tape from being played back without severe disturbances at each switch point.

38. Deflection and Focusing of Camera Tubes. In a magnetically focused tube, such as the vidicon, the motion of an electron from the scanning gun is cycloidal. The axial magnetic-focus field strength and the wall voltage are chosen so that an integral number of cycloidal loops is produced between the gun aperture and the surface being scanned. This produces a well-defined scanning spot for removing the video information.

39. Scanning of Camera Tubes. Linear raster scan requires the generation of a sawtooth of current in the horizontal and vertical deflection windings of the pickup tube yoke. Such waveforms are generated from the horizontal and vertical drive pulses, which periodically discharge suitable capacitor-resistance networks. The vertical output stage is required to furnish a current of 200 mA peak to peak into the vertical deflection winding, which typically has an inductance of about 30 mH and a resistance of 35 Ω. The horizontal deflection winding requires a peak-to-peak sawtooth current of 500 mA. The yoke inductance is typically 100 μH with a dc resistance of a fraction of an ohm.

COLOR CAMERAS

40. Three-Tube Color Cameras. The first generation of color studio cameras used three image-orthicon tubes, which were essentially three identical monochrome camera channels with provisions for superposing the three output-signal rasters mechanically and electrically. The optical system consisted of a taking lens which was part of a four-lens turret assembly. The scene was imaged in the plane of a field lens using a 1.6-in diagonal image format. An alternative arrangement used a 10:1 range zoom lens. The real image in the field lens was viewed by a back-to-back relay-lens assembly of approximately 9 in focal length. At the rear conjugate distance of the optical relay was placed a dichroic-prism beam splitter with color-trim filters.

In this manner, the red, blue, and green components of the scene lens were imaged on the photocathodes of the three image-orthicon tubes. A remotely controlled iris located between the two relay-lens elements was used to adjust the exposure of the image orthicons. This iris was the only control required in studio operation. Only a very few of these cameras are still in use since their operating and setup requirements are complex compared with those of photoconductive cameras.

41. Four-Tube Color Cameras. Four-tube (luminance-channel) cameras were introduced when color receivers served a small fraction of the audience; the viewer of color programs in monochrome became aware of lack of sharpness. Using a high-resolution luminance channel to provide the brightness component in conjunction with three chrominance channels for the R, G, and B components produces images that are sharp and independent of registry errors. The four-tube approach is now largely limited to film cameras where the relatively low cost of the extra vidicon is not as important a factor as in a live camera using expensive Plumbicons or Saticons.

42. Three-Tube Photoconductive Cameras. A color TV camera must produce red, blue, and green video signals which complement the characteristics of the NTSC three-gun–three-phosphor standard additive display tube. For both live and film cameras it is now common to use a camera with three photoconductive pickup tubes with a high-efficiency dichroic light splitter to divide the optical image from the zoom lens into three images of red, blue, and green. The spectral characteristics of the three paths of such a camera are shown in Par. **20-17** (Fig. 20-14).

The light splitting is accomplished either by a prism (Fig. 20-21) or by a relay lens and dichroic system (Fig. 20-22). The prism has the advantage of small size and high optical efficiency but a disadvantage in that the three tubes are not parallel to each other and are thus more susceptible to misregistration produced by external magnetic fields. A more serious problem is that of obtaining a uniform bias light on the face of the tubes. Bias light producing 2 to 10% of the signal current is used in most modern cameras to reduce lag effects. Nonuniformity of the bias light can produce color shading in dark areas of the picture. Careful optical design can minimize this problem, and most new designs now use the prism splitter.

The relay optical system shown in Fig. 20-22 is one or two f stops slower than the prism, but it has several compensating advantages. First, bias-light shading is very slight. Second, since an aerial image is produced in the optical system, a mask can be included around the edge of the picture which will produce a true optical black in the signal. This black level can be used as a clamp reference point to establish true signal black, keeping glare, bias-light, and dark-current changes from upsetting the color balance of the picture. The aerial image is also a convenient point in which to insert points of light via fiber optics, to permit automatic horizontal and vertical centering and size adjustment of the three pickup tubes.

In both the prism and relay light splitters the colorimetry is to some degree a function of the polarization of the light entering the lens. One solution is to include a quarter-wave plate ahead of the splitter so that light of either horizontal or vertical polarization is converted to elliptical polarization. The splitter then sees the same depolarized light no matter what the polarization of the scene illumination. High-quality cameras are currently being produced using both prisms and relay optics.

43. Plumbicon and Saticon Tubes. The Plumbicon (N. V. Philips) and Saticon (Hitachi) camera tubes have almost completely superseded other pickup tubes for live-camera entertainment-quality television. Although superficially of the same construction as a vidicon, the targets operate on quite a different principle. The target of a vidicon is a single-layer photoconductor, usually antimony sulfide, which changes its resistance in accordance with the incident light pattern falling on it.

In contrast, the Plumbicon target (Fig. 20-23) operates as a reversed-bias photodiode. Carefully doped layers of lead oxide serve to produce a reversed bias *pin* semiconductor junction. A potential is produced across the junction by an externally applied target voltage. This bias voltage is usually about 45 V with respect to the cathode.

The scanning-beam side of the target comes to equilibrium at cathode potential. Wherever light strikes the target, carriers are generated which cross the barrier junction. The scanning beam then reestablishes cathode potential and in doing so produces a signal current through the target resistor. Since this process is relatively fast compared with the photoconductive signal-generating process in the vidicon and the Plumbicon, it has much less signal lag.

Lag is measured by applying light to the scanned tube and determining how long the signal remains after the light is removed. A residual 3 to 5% of the initial signal during the third field

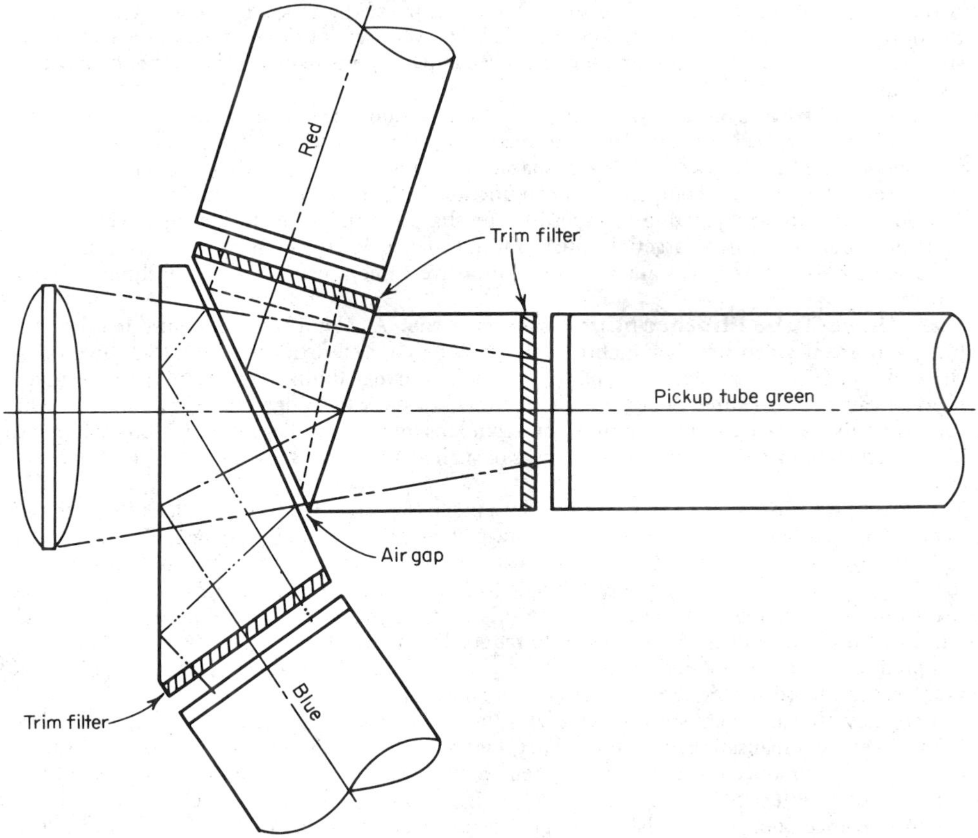

Fig. 20-21. Prism optical system of three-tube Plumbicon camera.

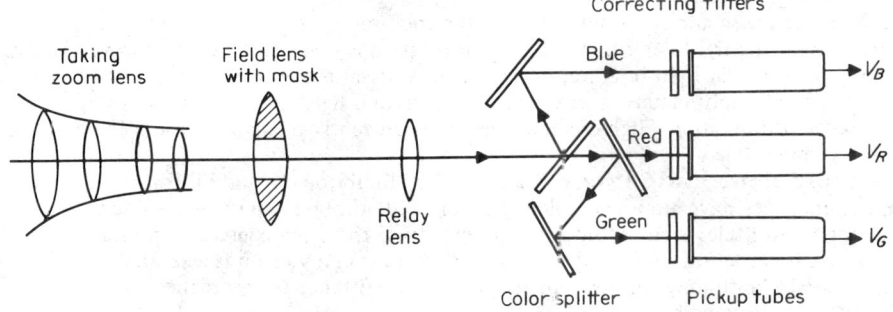

Fig. 20-22. Color separation by relay lens and dichroic filters.

after removal of light is typical for a Plumbicon. The lag is caused by the time constant of the scanning-beam equivalent resistance and the layer capacitance of the target junction.

Making the target layer thinner to improve sensitivity and resolution increases capacitance and hence lag. Lag is particularly bad at low signal currents since under these conditions the scanned side of the target is almost at cathode potential and very little of the scanning beam lands.

Bias light can be applied to the target either from the prism assembly or from a small bulb or light pipe internal to the tube or from an annulus of LEDs placed around the front of the tube. This bias light slightly raises the target potential for dark areas and substantially reduces lag, lowering the effective beam resistance.

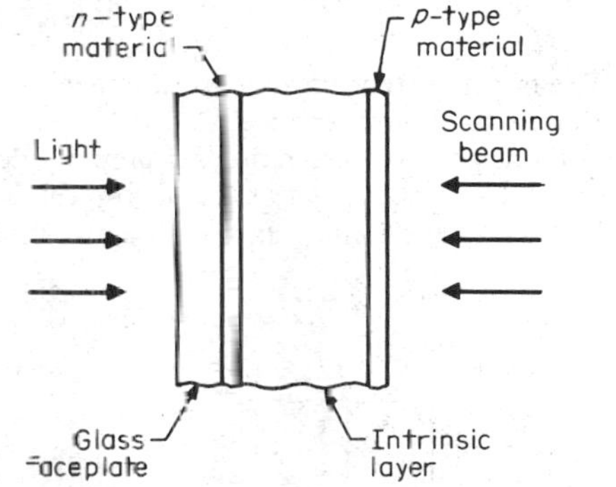

In a new scanning gun, the diode gun (Fig. 20-24) the grid aperture is used as the limiting aperture and is run at positive potential, drawing a current of a few milliamperes. The camera grid-bias circuit must be designed to supply this current with a protection circuit to prevent excess current from overheating the grid aperture. Use of the diode gun reduces velocity differences in the scanning electron beam and thus decreases beam resistance and lag. Diode-gun 18-mm tubes typically have a resolution modulation depth of 60% at 400 TV lines when operated at 750 V accelerating potential and a 70-G axial focusing field.

The Saticon target layer operates as a reversed-bias hetrojunction (Fig. 20-25). In this case the active material is selenium doped by arsenic and tellurium. The lag of the Saticon is considerably higher than that of the Plumbicon without bias light, but when bias light is optically applied, the ultimate lag performance of both tubes is about the same. The resolution of the Saticon layer is independent of the wavelength of light.

44. Preamplification and Sensitivity. Camera sensitivity describes the video-signal output in terms of incident illumination on the scene being viewed. In studio cameras a lens opening of $f/4$ on a 30-mm Plumbicon (raster diagonal of 20 mm) is generally used as a reference base because it offers adequate depth of field. Incident illumination, usually specified in lumens per square foot* (footcandles), is based on the use of a 60% reflectance neutral target as the brightest object in the scene.

The light intensity I_{pc} on the photosensitive raster of the pickup tube required to produce the necessary signal-current level compatible with the desired signal-to-noise ratio is determined by

$$I_s = 4f^2 I_{pc}/TR$$

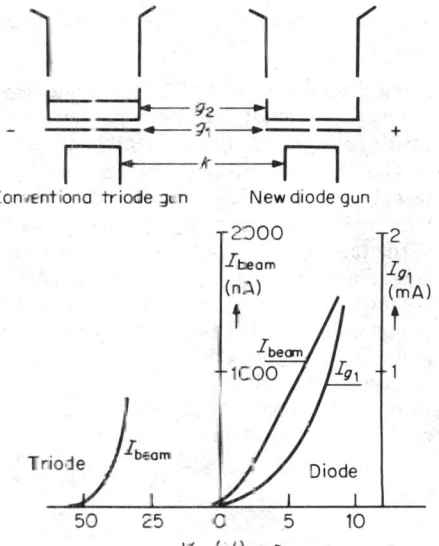

Fig. 20-23. Plumbicon camera-tube detail of photosensitive target layer.

Fig. 20-24. Diode gun of recently developed camera tubes, compared with conventional triode gun.

where I_s = scene illumination (lm/ft²), f = f-number setting of camera-lens aperture, I_{pc} = photocathode illumination of the pickup tube (lm/ft²), T = transmission factor of lens, generally 80% with coated lenses, and R = reflectance of test object in scene (in color pickup this is a matte white card with reflectance of 60%).

For color cameras, an additional factor is required in the denominator to represent the splitter optical transmission coefficients for the red, blue, or green channels.

From this expression it can be determined, for example, that at $f/4$ a three-tube Plumbicon

*The metric unit is the lux = 1 lm/m² = 0.093 footcandle (fc).

camera requires about 150 lm/ft² incident on a 60% neutral reflectance card to develop a 300-n. green video signal with a signal-to-noise ratio of 50 dB. Measurements are conventionally made with a unity gamma transfer characteristic and no aperture compensation or image enhancement. Studio lighting for color TV is furnished by quartz-halogen incandescent-lamp fixtures operating at 3200 K. Because of colorimetry shifts, dimming of lighting fixtures is rarely used in studio practice.

One of the most important criteria for determining the picture quality of a color TV camera is the signal-to-noise ratio. This is measured in decibels according to the formula

$$dB = 20 \log \frac{\text{peak-to-peak video voltage}}{\text{rms noise voltage}}$$

The target current for this measurement is standardized at 300 nA for 30- and 25-mm tubes and 200 nA for 18-mm tubes.

The signal-to-noise ratio of a properly designed camera is set primarily by the preamplifier connected to the target of the pickup tube. A typical circuit for a preamplifier is shown in Fig. 20-26. A cascode input stage with a carefully selected FET transistor feeds a wide-band amplifier with feedback to the gate of the FET. An equalizing network in the feedback path corrects the frequency rolloff due to the target capacitance. The signal-to-noise ratio of such a preamplifier can be shown to be

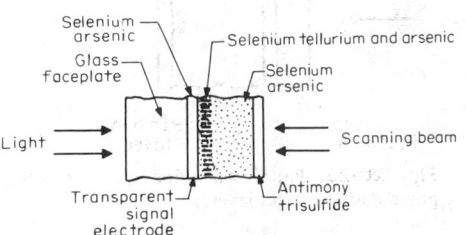

Fig. 20-25. Saticon camera-tube detail of photosensitive target.

$$S/N = \left\{ \frac{I}{4kTf_c\,[(1/R_f) + \frac{4}{3}(C + C_s + C_f)^2/g_m]} \right\}^{1/2}$$

where k = Boltzmann's constant, f_c = bandwidth, T = temperature (K), R_f = feedback resistance, C_s = stray capacitance, C_f = FET input capacitance, g_m = FET transconductance, and C = stray capacitance of pickup tube target. Since it is essential that this capacitance be kept to a minimum, the input FET is frequently put inside the yoke assembly directly adjacent to the target.

The latest Plumbicons and Saticons bring out the target connection via a pin through the faceplate to minimize capacitance to the deflection yoke and focus coil. The output noise spectrum of such a camera is triangular because of the need to compensate for the target capacitance at the rate of 6 dB/octave; i.e., the rms value of noise per unit analyzer bandwidth increases with frequency. This is a desirable characteristic since noise is less visible at higher frequencies.

Since the signal-to-noise performance of the photoconductive tube and preamplifier is high, it is practical to trade improved sensitivity for lower signal-to-noise ratio. Thus by decreasing the scene lighting by a factor of 2 and increasing the video gain by 2, the signal-to-noise ratio is decreased from 50 to 44 dB, which is highly acceptable. Reducing scene lighting by a further factor of 2 reduces the signal-to-noise ratio to 38 dB, which although visibly noisy can be tolerated in many situations.

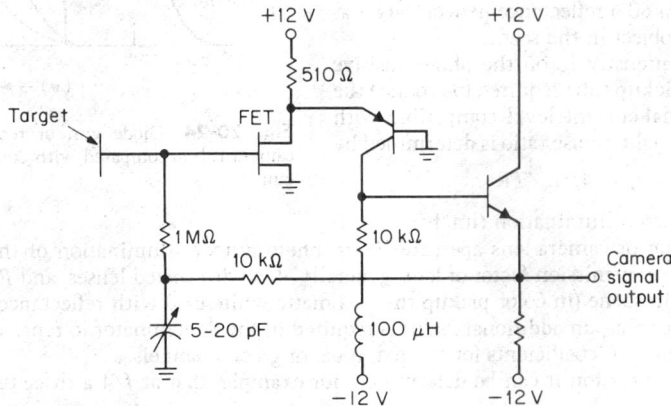

Fig. 20-26. Typical camera preamplifier circuit.

Hence the scene lighting can be reduced from 150 to 37 lm/f² at full video-signal output while keeping the lens aperture at f/4 to obtain adequate depth of field. At this point, while the signal-to-noise is still marginally acceptable, even with bias light the lag or smear of the photoconductive surfaces in picking up motion in the televised scene becomes the dominant factor in picture quality. An optical preamplifier (an image intensifier) is sometimes coupled to the pickup tubes via fiber optics to extend the sensitivity by a factor of 10 to 20 for special purposes such as medical applications and in electronic news gathering. This technique is limited by quantum noise effects at scene illumination below 0.2 fc.

45. Dynamic Control of the Scanning Beam. The scanning beam in a camera must be set high enough to discharge scene highlights on the targets. When the signal currents are typically 300 nA, the beams would be set to 450 to 600 nA to discharge specular reflections in the picture. Unfortunately, the beams cannot be set higher without defocusing and loss of resolution and possible damage to the tube. It would be desirable to set them even lower to improve registration.

Recent cameras use controlled positive feedback to adjust the beams dynamically as they scan, at a level just high enough to discharge the target potentials encountered. Since wide-bandwidth positive feedback is inherently unstable, means must be included to prevent runaway beam currents. One system uses a current-limiting circuit in series with the tube cathode in conjunction with a peak clipper in the feedback amplifier. The loop gain, including the pickup tube transconductance, must be held to slightly less than 1 in this system. This can present a problem as the tube characteristics change with age.

Another camera solves the stability problem by comparing the amplified target current with the cathode current in a difference amplifier and applying an offset voltage to keep the scanning beam current 50 nA greater than the target current at all times. Microcircuits are now available to perform these functions without adding to camera complexity.

A third (historically the first) approach to discharging highlights is to use the horizontal retrace time to limit the potential to which the target can climb. With a special electron gun the cathode is raised a few volts (usually about 4 V) during retrace, and a suitable potential is put on the grid to produce a defocused beam to wipe the target. This method of operation produces large spurious signals which must be accommodated in the camera video signal processing, but it has been successfully accomplished in at least one commercial camera.

46. Gamma Correction. Color picture tubes have a nonlinear transfer characteristic; i.e., the reproduced brightness is a power function of the control-grid video drive. In a log-log graph of the screen brightness plotted against video excitation the slope of the transfer function is about 2.4 over a brightness range of 100 to 1. To produce a pleasing picture in color with consistent colorimetric qualities the transfer characteristic of the pickup camera must exhibit the reciprocal of the picture-tube characteristic, i.e., $1/2.4 = 0.4$ approximately.

Since the Plumbicon is essentially a linear device, i.e., the video signal is directly proportional to the light in the scene, gamma correction in the form of nonlinear signal amplification, to a power of $1/2.4$, must be introduced into the R, G, B channels of the camera before the signal is encoded. The low signal levels are stretched in amplitude and the high signal levels compressed in accordance with the power curve selected. Stretching the black signal by a factor of 2 decreases the signal-to-noise ratio by 6 dB in the blacks, whereas compressing the white signal improves the signal-to-noise ratio. Since significant stretching of blacks must be used, to achieve a gamma of 0.4 the black signal-to-noise may be degraded by 12 dB or more. Various band-limiting and signal-coding techniques to minimize noise effects while stretching the low-light signal have been successfully used in color cameras.

A common technique for producing the wide-band nonlinear video signal is shown in Fig. 20-27. Operation depends on the fact that the voltage across a diode is approximately proportional to the square root of the current through it. By providing two balanced paths a linear signal can also be produced. The setting of potentiometer R produces a mixture of linear and nonlinear signal which can be used to match exactly the slight differences in gamma of the red, blue, and green channels. The high gain of such a gamma circuit for signals near black makes it very susceptible to temperature drift. The use of a feedback clamp circuit operating on the picture black level, after the gamma diode, reduces such drift to a negligible value.

Alternatively, the nonlinear function is sometimes produced by a diode breakpoint arrangement in the path of a feedback amplifier. Both approaches produce satisfactory results.

47. Image Enhancement. The zoom lens, the pickup tube, and, to a lesser extent, the camera circuits all contribute to a loss in resolution at the higher spatial frequencies, both vertically and horizontally. Theoretically, the lens and pickup tube should exhibit a (sin x)/x (phaseless) type of loss. In practice, this proves to be substantially true, but the combination of several rolloffs can produce a response curve which does not follow a simple law.

Aperture correction and *image enhancement* are therefore used in all high-quality color cameras to improve the subjective quality of the pictures. The horizontal aperture corrector is adjusted to make the depth of modulation at a horizontal resolution at 400 lines equal to that

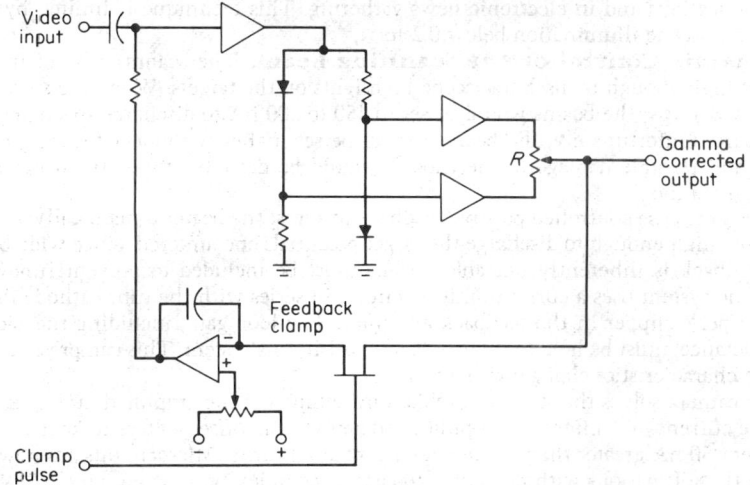

Fig. 20-27. Use of diode for wide-band gamma correction.

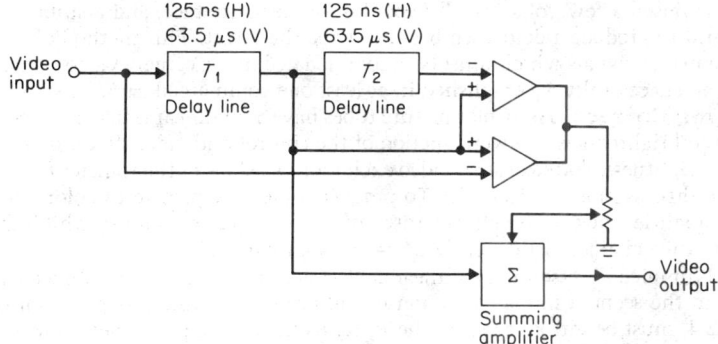

Fig. 20-28. Image enhancement using a transversal filter.

obtained at approximately 50 lines. Transversal delay lines and second-derivative types of corrector are frequently used since they exhibit high-frequency boost without phase shift, thus complementing the $(\sin x)/x$ rolloff, i.e., boosting the high frequencies without introducing ringing.

Once the camera response is flat to 400 lines, an additional correction is applied to increase the depth of modulation in the range of 250 to 300 lines, both vertically and horizontally. This additional correction, known as image enhancement, usually takes the form of a transversal filter (Fig. 20-28). The delay elements T_1 and T_2 are one picture element (or approximately 125 ns) long for horizontal correction and 63.5 μs for vertical correction. Such a corrector produces a correction signal with symmetrical overshoots around transitions in the picture. If overdone, this correction produces an unnatural image, characteristic of the early image orthicons which outlined around midfrequency detail. Image enhancement must therefore be used very sparingly if a natural appearance is to be maintained.

Subjectively the eye is most sensitive to detail in the mid-gray-scale range of the picture. It is therefore beneficial to modulate the aperture and image-enhancement detail correction signals as a function of brightness. In this manner distracting noise in the dark areas of the picture can be eliminated, and the viewer is not aware of the accompanying loss of detail in the shadows.

Coring is sometimes used to slice out the middle of the detail signal so that noise and low level detail signals are removed. This process not only removes noise but also prevents the performers' skin from appearing too rough, while at the same time permitting the highlights in the eyes to pass unattenuated.

48. Masking. The ideal TV camera colorimetric taking characteristics for the NTSC color system (Par. **20-14**) require negative lobes (Fig. 20-13), which cannot be generated optically in the beam splitter or pickup tubes because this would imply negative light, nonexistent in nature. However (Par. **20-17**), negative lobes can be produced electrically in the camera signal processing by a matrix operation called *masking*. One version of this technique balances the matrix for equal red, blue, and green signals, i.e., white, and generates correction signals only when chrominance is present. If the matrix is made polarity-sensitive (Fig 20-29), it is possible to correct individual hues independently to match the NTSC vectors exactly or alternatively to match one camera to another. Masking is also used in film cameras to match the colorimetry of the taking-film dyes to the NTSC display primaries.

In a live camera the matrix operation is usually performed on the linear signals from the pickup tubes before gamma correction. High-quality flying-spot film scanners often are designed with elaborate circuits to achieve mathematical precision in masking. These can correct for color error in the film and can match film segments to a live camera.

49. Portable Cameras. Portable color cameras are extensively used for electronic news gathering and field operations. The quality of pictures obtained from these cameras with diode-gun pickup tubes is now almost the equivalent of the best studio cameras. Weight and power consumption (which can be equated to battery weight) are all important for such use, since the camera may be carried for many hours on the shoulder of the operator. Power consumption of 20 W and 20 lb weight are typical of the latest designs.

Stability under temperature fluctuation is achieved by extensive use of feedback and microcircuits. Microprocessors can adjust setup and black-and-white balance automatically and can adjust centering by correlating edge detail between the red, blue, and green channels, with green as reference. These cameras can be fully operational in 3 s or less if the filaments of the pickup tubes are kept at a low steady voltage while power is removed from the rest of the camera. The use of 18-mm tubes reduces the lens size so that the use of 14:1 zoom lenses weighing less than 4 lb is practicable.

50. Digital Setup—Triaxial-Cable Cameras. A studio camera may have over 100 potentiometers which must be adjusted after changing tubes. Approximately 20 of these may need adjustment on a daily basis to keep the camera in optimum condition. To make many routine adjustments as accessible as possible, studio cameras until recently were designed so that much of the signal processing occurred at the camera control unit (CCU) rather than at the camera head. In this case a CCU-to-camera-head cable might contain five or more individual

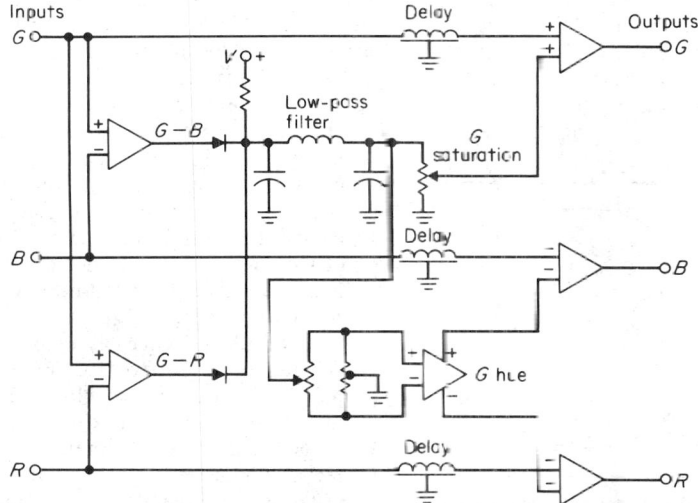

Fig. 20-29. Masking matrix with polarity sensing (matrix shown for G correction only).

cables and 60 to 80 individual wires. Also, maintaining the connectors in such a bulky, heavy camera cable in the field can be a major problem.

Cameras are now available which transmit all the necessary signals and power over a single triaxial cable using digital-multiplexed commands and video subcarriers (Fig. 20-30). When such cameras became available, it was realized that if a microprocessor was used to generate the command signals, suitable software could be used to set up the camera without human intervention. A test-pattern "diascope" projector is included in the camera lens so that the microprocessor can control special setup patterns without the need for stagehands and lighting technicians. The potentiometers are replaced by digital-to-analog converters, which are far more reliable and not subject to misadjustment from vibration or mishaps in the field.

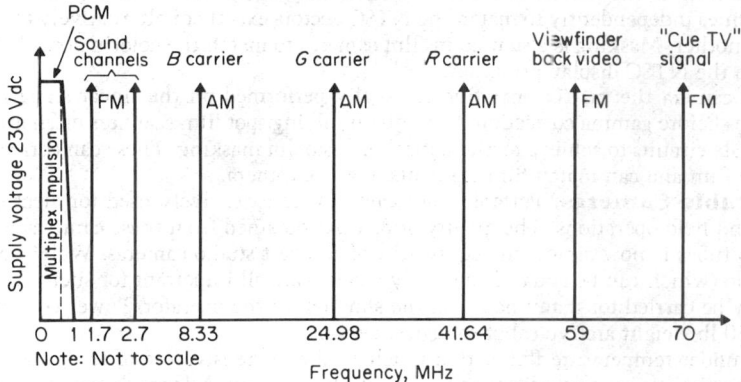

Fig. 20-30. Single triaxial-cable signals for color cameras.

51. Raster Linearity, Registration, and Stability. The starting point for generating color pictures is the optical and electronic superposition of the red-, green-, and blue-tube scanning rasters. In high-quality broadcast service, precision focus and scanning components make possible substantially identical conditions for horizontal and vertical deflection of each camera tube, with precise geometric orientation of the three rasters to the optical picture input.

Solid-state circuits designed for the highest possible linearity and stability are provided to drive the deflection circuits and to maintain magnetic-focus fields constant to within 0.25%. By adjustments made with test charts it is practical to obtain scanning linearity to better than 0.5% within a circle of diameter equal to picture height (zone I, Fig. 20-31) and within 1% in zones II and III. The registration accuracy of the three rasters can be set to closer than 0.1% in zone I and closer than 0.25% in zone II.

Standard test charts for testing linearity and registration test charts (Figs. 20-32 and 20-33) are obtainable as opaque charts or as 2- by 2-in glass-mounted transparencies from Electronic Industries Association, 2001 Eye Street, N.W., Washington DC 20006, and from the Society of Motion Picture and Television Engineers, 862 Scarsdale Ave., Scarsdale NY 10583.

The procedure generally adopted is to use the green channel as a reference, adjust it for correct size and linearity, using the EIA ball chart, and then to match the red and green rasters to it, using both the ball chart and the registration charts. It is important that differential errors in size and linearity be avoided, since they produce color fringing and low resolution in the portions of the raster which do not coincide or register. Small errors in absolute linearity or size go relatively unnoticed.

A technique frequently used in linearity, size, and registry adjustments is that of reversing the polarity of the red and blue video signals with respect to the green video reference. The narrow vertical and horizontal lines which constitute the registration pattern produce pulse waveforms and raster display signals which cancel completely when there is coincidence between red and green

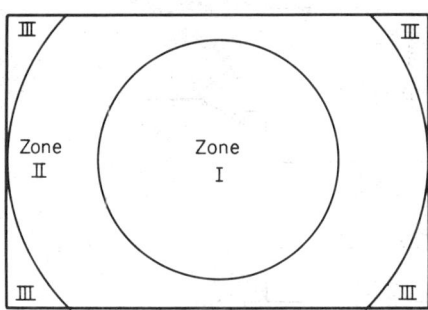

Fig. 20-31. Zone chart for checking scanning linearity and registration of cameras.

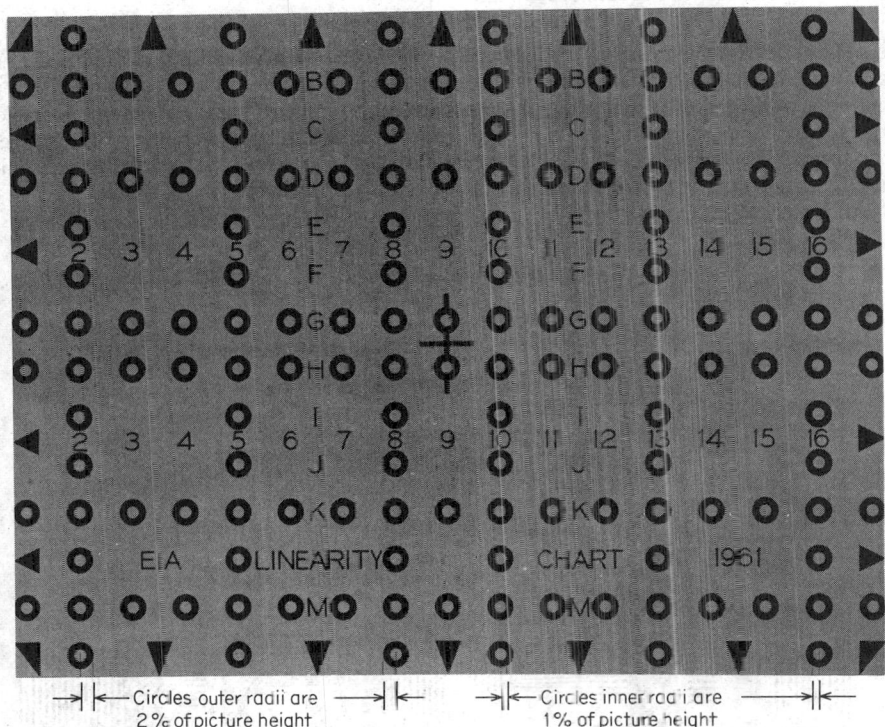

Circles outer radii are 2% of picture height — Circles inner radii are 1% of picture height

Fig. 20-32. EIA linearity test chart. Aspect ratio is 4:3, horizontal blanking 17.5%; electrical grating pattern generator frequencies 315 kHz (horizontal) and 900 Hz (vertical).

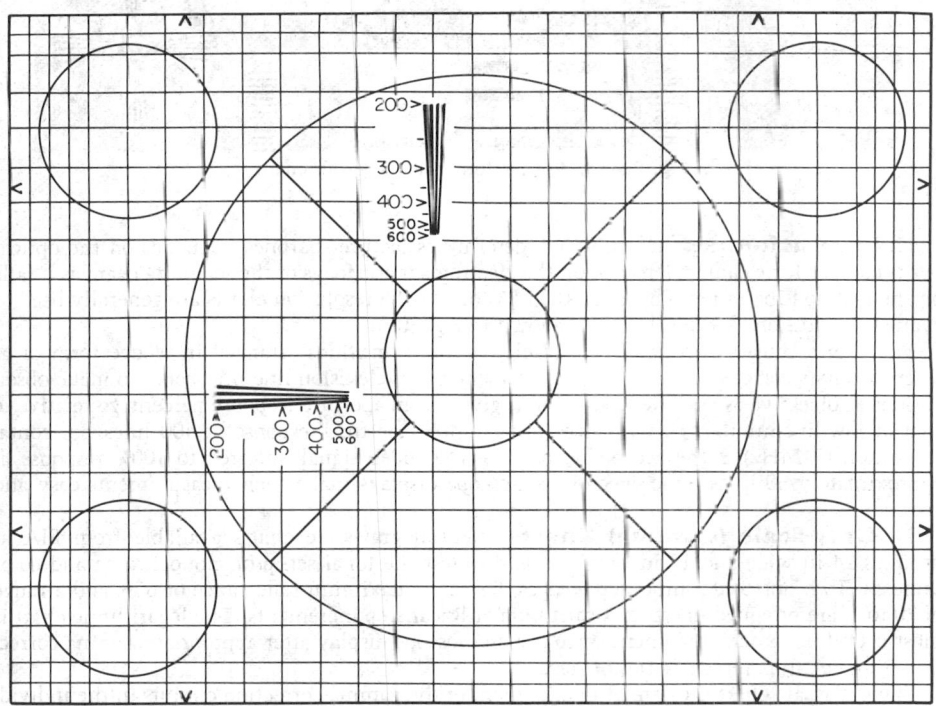

Fig. 20-33. EIA registration test chart.

rasters. The operator's eye is thus drawn to areas where the positive and negative pulses do not cancel and where adjustments in linearity, size, or centering are needed.

The same technique is useful in pickup-tube rotation to obtain parallel scanning of rasters. Such adjustments are necessary on initial setup and require little if any change with change of pickup tubes. Skew control or electrical compensation is provided for residual magnetic and electrical cross talk in the deflection components. A camera initially registered and then turned off will return to its registered condition within 1 to 2 min after being turned on again.

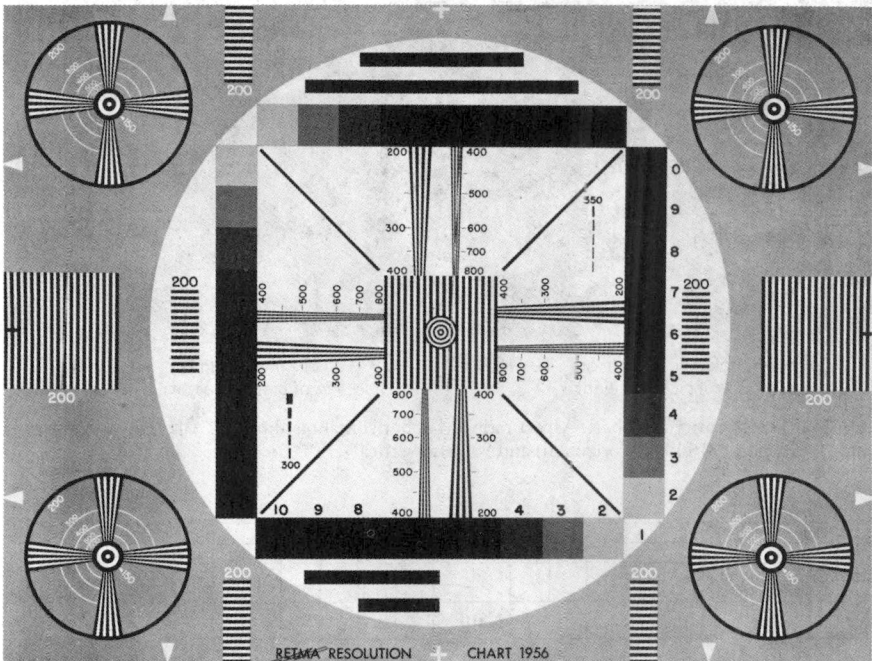

Fig. 20-34. EIA resolution and gray-scale chart.

52. Resolution. Resolution, which determines picture sharpness, depends on the optical performance of the camera lens and on the electromagnetic focus of the scanning beam as it falls on the pickup-tube raster. The EIA (RETMA) or SMPTE resolution charts are generally used to evaluate resolution. A typical chart is shown in Fig. 20-34.

The charts consist of square-wave black-to-white transitions arranged in wedge form or as discrete bursts corresponding to specific frequencies or television line numbers. To make observations as objective as possible, resolution is given as an aperture-response percentage relative to that at low-line-number performance. Thus, a 40% aperture response at 400 lines horizontal resolution (5 MHz) is the acceptable peak-to-peak video signal, referred to 100% response at approximately 50 lines. Line-selector oscilloscope displays make such measurements easy and accurate.

53. Gray-Scale (Gamma) Charts. Neutral gray-scale charts available from EIA as opaques 24 in wide and 18 in high are used to test the tonal setup of monochrome and color cameras. Two horizontal nine-step wedges, having a maximum reflectance of 60% and a range of 20 to 1, are provided in either logarithmic or linear step increments. The logarithmic chart is most useful in camera alignment since the oscilloscope display after appropriate gamma correction is an equal-increment series of steps.

The gray-scale chart is essential in alignment of the gamma-correction circuits in the individual R, B, G color channels. The technique used calls for accurate setting of black level, gain, and gamma correction or differential gain in each channel. The goal is to generate a gray-scale video display in which the color subcarrier vanishes at every gray step level. The waveform monitor becomes a very sensitive device for checking adjustments since any differential variation in the three transfer characteristics will become evident as subcarrier in the display.

The most stringent test of a color camera is its ability to reproduce accurately a neutral or colorless gray scale on a color monitor or receiver. If proper care has been taken in its colorimetric design and adjustment, the production of consistent, accurate color video signals is a necessary consequence.

54. Color Monitors and Receivers. The end use of color television signals is display on color receivers for entertainment, education, or technical application. The *color television monitor* employed by the broadcaster is an evaluation tool to determine how good a picture is being generated.

Color pictures, as specified by NTSC-FCC Rules, are intended to be viewed at a white screen-color temperature of illuminant D, 6500 K. More precisely the intended white of the scene, corresponding to zero subcarrier amplitude, should be reproduced as illuminant D. A close visual approximation to illuminant D is a neutral (white) card illuminated by north-sky daylight. Reference standards for illuminant D are available to the broadcaster in the form of specially controlled phosphor fluorescent lamps or regulated incandescent-lamp sources modified by optical filters. The use of these devices makes it possible to set up a color monitor at any given location to the reference white of illuminant D.

Once this color temperature has been established, all monitors in a broadcasting system must be set to this reference. Monitor setup is time-consuming, tedious, and subject to human error, so the broadcaster often prefers to use a commercial color comparator to measure the ratio of red to green to blue phosphor output. This procedure permits rapid and accurate adjustment of numerous monitors to the illuminant D standard.

The color temperature of the color receiver white has frequently been set at much higher color temperatures, for example 9500 K compared with the broadcast standard of 6500 K. This choice matches monochrome phosphors more closely and offers higher screen brightness. While this practice has no technical basis, experience has shown that satisfactory color pictures can be produced by intuitive adjustment of receiver hue and chrominance controls. Provided that there is no side-by-side comparison with an illuminant D monitor, the viewer is well satisfied with the color picture based on 9500-K white.

55. Flying-Spot Scanners. Flying-spot slide scanners have been used for many years in the reproduction of transparencies. The system uses a high-brightness-kinescope scanning raster optically imaged on a monochrome or color transparency. The light transmitted by the transparency is gathered by phototubes and amplified to form the video signal. Since the raster is generated by a single focused electron spot under the influence of one set of horizontal and vertical deflection fields, raster-size and registration errors do not occur.

The high-brightness, high-resolution raster is generated at second-anode voltages of 25 to 30 kV on 5- or 7- in diameter kinescopes. Special phosphors having short buildup and decay times (less than 0.5 μs) are used.

For color use the phosphors must have high light output in the red, green, and blue portions of the visible spectrum. The light transmitted through the televised transparency is separated into red, green, and blue components by a dichroic mirror system. The respective separated rasters are sufficiently bright to produce simultaneous R, G, B signals with adequate signal-to-noise ratio.

The flying-spot transfer characteristic, relating video-signal output to light input or film transmission, is strictly linear. This requires the use of extensive electronic gamma correction of the output video signal to accommodate the transfer characteristic of the color display picture tube (see Par. **20-46**). A block schematic of a flying-spot slide scanner is shown in Fig. 20-35.

56. Nonphotoconductive Color-Film Cameras. Until recently, the basic incompatibility of the 24-frame-per-second motion picture film standard with the 30-frame television standard has made it impractical to scan motion picture film with a flying-spot scanner. Recently a new technique based on continuous motion of the film has become practical with the advent of digital frame stores. The film is scanned with a noninterlaced raster which tracks the moving film at 24 frames per second. The resulting R, B, G video signals are digitized and stored in a one-frame memory, which is then read out at the 60-field–30-frame-per-second rate.

The quality of these systems is high, but their cost is so great that they are used primarily at production houses for the transfer of film programs to video tape for broadcast.

A variation on this technique is to scan continuously the moving film with a charge-coupled device (CCD) line sensor (see Par. **11-152**) and to put the resulting digitized signal into a frame store. Fixed pattern noise generated by the CCD would show up as vertical lines in the picture but can be removed by a one-line memory which stores a correction signal for each of the 10 to 20 elements in the line array (Fig. 20-36). CCD scanners now commercially available rival the best flying-spot or photoconductive-film cameras in quality.

57. Photoconductive Color-Film Cameras. In the United States, color film is almost always scanned by vidicon photoconductive cameras, although the Plumbicon and Saticon are finding increasing use where high quality is paramount. This differs from the practice in Europe, where flying-spot scanners are widely used, the film being run at the TV standard of 25 frames per second. The vidicon is operated at high light levels and low target voltages, viewing intermittent-motion 16- or 35-mm film transports. Exposure of the raster is carried out during the active scanning sequence and occupies 35% or more of the total scanning time.

Conversion of the standard motion-picture 24-frame-per-second rate to the TV 60-field-per-second display rate is carried out by using a 3:2 intermittent mechanism in the motion-picture projector. One motion-picture frame is held stationary for three television-field raster scans, and the next is held for two field scans. Thus during two film frames (½₂ s at the 24-frame rate) five raster fields are produced in the same time, each at intervals of ⅙₀ s (at 60 fields per second).

The storage characteristics of vidicon film chains are excellent and allow for nonsynchronous operation of the projector relative to the television system. This is especially important in color since the raster is scanned at 59.94 fields per second, i.e., not locked to the 60-Hz power system.

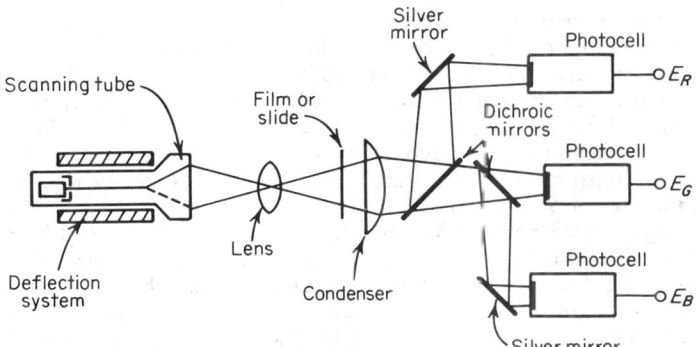

Fig. 20-35. Flying-spot camera for film or slide transparencies.

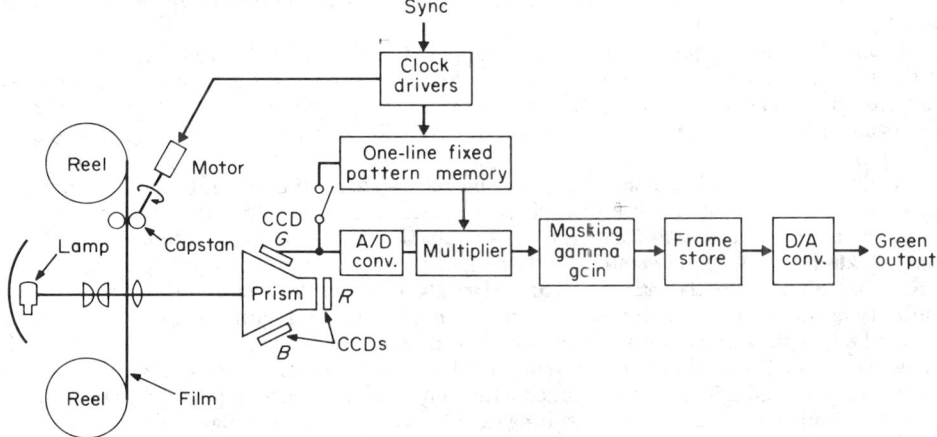

Fig. 20-36. CCD film scanner using frame and line digital memories.

When, as is usual, the motion-picture projector is operated from the local power supply, as long as the light-application time exceeds 30% and the difference between power-line frequency and raster rate does not exceed 0.5 Hz, there are no problems with transition bars in the reproduced picture.

The low dark current, steady black level, inherent low-gamma transfer characteristic, good resolution, and excellent signal-to-noise ratio have made the vidicon very attractive to the broadcaster. Under the operating conditions described, vidicons have low lag and low retentivity and burn characteristics. Because of their long life and moderate initial cost they are the most economical means of reproducing color film. The linear transfer characteristics of the Plumbicon and Saticon can give slightly superior performance, especially from negative film.

Since there are variations in film highlight density which change the peak video output, a film camera is operated to give "constant performance" by increasing the light when dense film

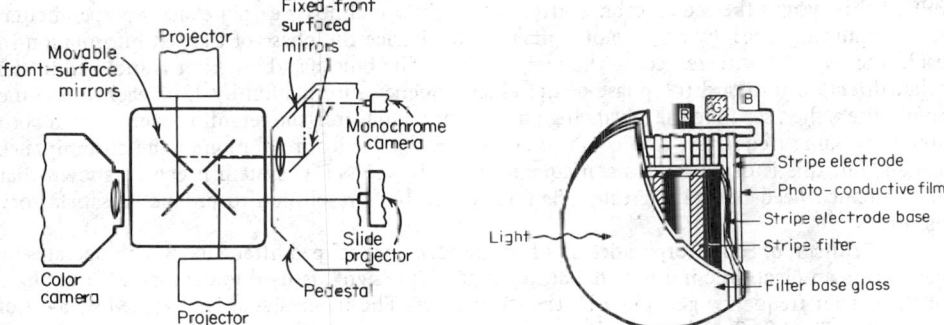

Fig. 20-37. Multiplex use of projectors and scanners (RCA type TP-15).

Fig. 20-38. Stripe-filter single-tube color camera tube with three sets of target electrodes. (*Jpn. J. Electron. Eng.,* April 1976.

is being scanned. A neutral-density optical disk in the projector light path, positioned by a video-level-sensitive servoamplifier feedback loop, uses the light reserve of the motion-picture incandescent lamp to reestablish a reference output video level without any deterioration of signal-to-noise ratio.

For high utilization of film chains, multiplying optical systems are used. Three or four picture sources are arranged on a film island. Movable front-surface mirrors are used to direct any one of the pictures from these sources to the scanner, in the form of a real image at a single field-lens position. The field-lens image is viewed by the vidicon film chain. Thus, 2-by-2 slide projectors and 16- or 35-mm film projectors can be used. A functional diagram of such a multiplexing system is shown in Fig. 20-37.

58. Striped Pickup-Tube Cameras. The problems of registering, equalizing the signals, and providing three matched sets of circuits for a three-tube camera has resulted in a search from the earliest days of color television for a method of obtaining the red, blue, and green signals from a single camera tube.

Cameras now on the market are based on using color-striped filters in front of the pickup tube. Each uses a different method of extracting color information from the video signal generated when scanning the color-striped filter.

An early technique was to put three sets of target electrodes in the tube (Fig 20-38), with one electrode under each color stripe. Then, by using three preamplifiers red, blue, and green signals could be directly obtained from the tube. Capacitive cross talk between the three target leads made this approach unworkable until the advent of very low input impedance feedback preamplifiers. The problems of connecting the three target leads in the tube have been overcome, and tubes with both vidicon and Saticon three-lead targets are now commercially available.

Another approach is to use a pickup tube and a striped filter with a single target lead. Its operation can be understood by considering a striped filter consisting of clear segments interleaved with red absorbing segments placed in front of the target. Such a tube will generate a signal with a frequency component equal to the frequency of the stripes when viewing a scene illuminated by white light. However, when viewing a cyan scene (which has no red component), the stripe-frequency component disappears. The amplitude of the frequency generated by the stripe is therefore a measure of the red content of the scene. A bandpass filter and amplitude detector can extract this signal.

When another striped filter having blue absorption is superimposed upon the first with a slightly different pitch, representing a different frequency, the detected frequency represents the blue content of the scene. The low-pass signal from the tube is then matrixed with the demodulated blue and red signals to obtain green.

For this system to be practical, the spatial frequencies chosen for the striped filters must fall above the luminance-bandwidth limit of 4 MHz. The tube must therefore have sufficient resolution to be able to resolve the stripes over the entire face of the tube without any change in amplitude which would produce color shading. Filter frequencies of 5 and 6 MHz are typical.

In an attempt to reduce the resolution requirements imposed on the pickup tube, varieties of this approach have been devised which make both spatial frequencies identical and which separate the red and blue information by comb filters rather than by frequency differences (Fig. 20-39). In this system the red absorbing stripes are slightly inclined, i.e., not exactly perpendicular to the scanning lines, by an amount sufficient to advance the phase of the red information in each line by 120° with respect to the previous line. The blue-absorbing filter is inclined in the other direction to retard the phase of the blue-frequency information by 120° each line. Line comb filters then separate the advancing phase information from the retarding even though both are at the same frequency. The comb filter reduces the vertical resolution of the chrominance signals, but this is of no serious consequence since in the NTSC system it can be shown that chrominance need not have greater than 40 to 120 lines resolution to produce a satisfactory picture.

The Trinicon of Sony Corporation is a high-performance striped-filter tube which operates by generating an electronic index signal at the target. This signal is used to demodulate the phase of the carrier frequency generated by the stripe filter. The index signal is produced by a set of electrodes (Fig. 20-40), which modulate the target voltage on a line-by-line basis; this in turn modulates the video signal. The index signal can be recovered by a one-line comb filter. The block diagram of such a camera is shown in Fig. 20-41. This tube can closely approach three-tube camera performance.

Since incoming light photons are absorbed in the striped filters, there are fewer photons available to excite the pickup tube, whereas in a three-tube prism camera all the incoming photons are diverted by the dichroic prism surfaces to one of the three tubes. In practice, this effect has limited striped-tube cameras to situations where considerable scene illumination is available.

59. Rotating-Filter Color Cameras. Color systems (not currently intended for broadcast use) have been devised to generate sequentially the video signals corresponding to the red, blue, and green components of a scene being televised. This is done by exposing a single pickup tube to the focused optical information of the scene through a sequence of color-separation filter sectors arranged to rotate synchronously at the vertical scanning rate. Such sequential color cameras have been introduced for industrial and medical use and for the Apollo space program. They

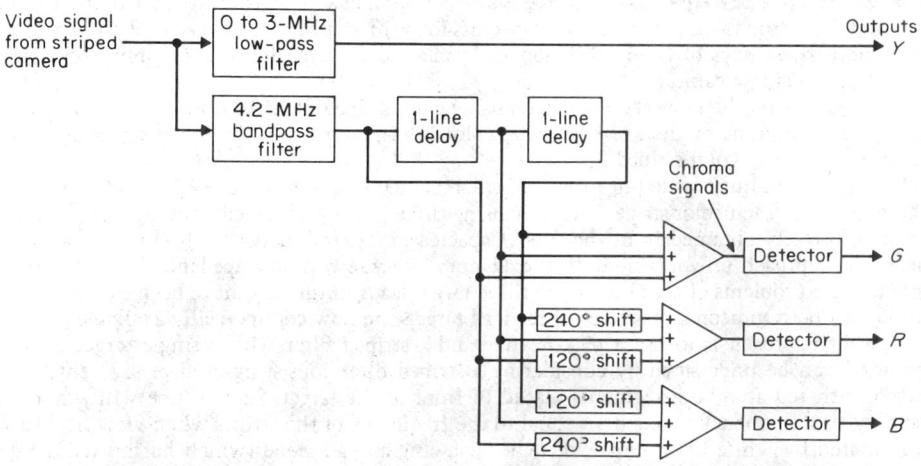

Fig. 20-39. Comb-filter color separation in inclined-stripe single-tube color pickup.

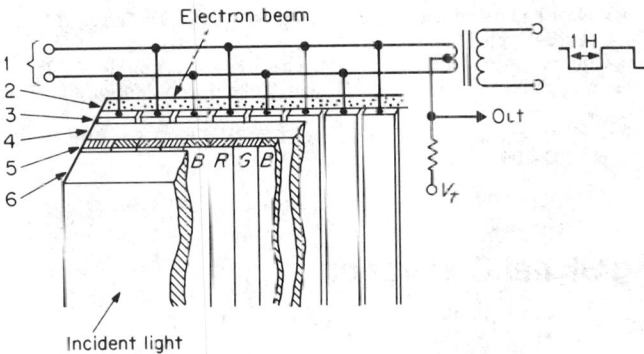

I. ·Busbar   2. Photoconductive layer   3. Electronic indexing electrode
4. Insulating layer   5. Color striped filter   6. Face plate

Fig. 20-40. Production of index signal in Sony Trinicon single-tube color pickup.

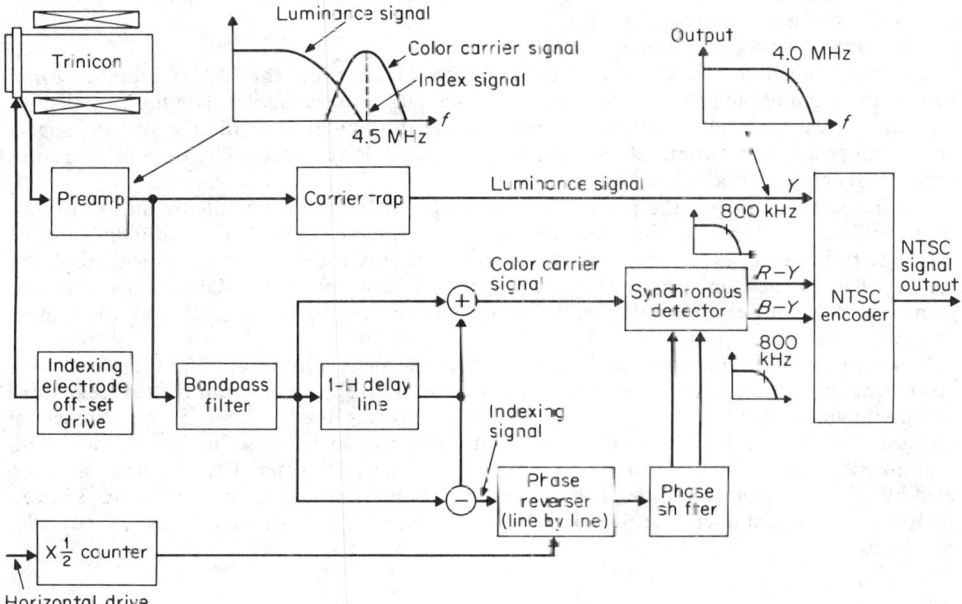

Fig. 20-41. Block diagram of Sony Trinicon camera. *(Sony Corporation.)*

represent the smallest and simplest color cameras that can be produced with available pickup tubes. Conversion of the sequential signals to the NTSC standard by a magnetic-disk scan converter is feasible.

60. Bibliography on Television Camera Tubes

Zworykin, V. K.   The Iconoscope, *Proc IRE*, January 1934, Vol. 22, pp. 16–32.

Farnsworth, P. T.   Television by Electron Image Scanning (Image Dissector), *J. Franklin Inst.*, October 1934, Vol. 218, p. 411.

Iams, H., et al.   The Image Ionoscope, *Proc IRE*, September 1939, Vol 27, pp. 541–547

Rose, A., et al.   The Image Orthicon, *Proc. IRE*, July 1946, Vol. 34, pp. 424–432.

Weimer, P. K., et al.   The Vidicon, *RCA Rev.*, September 1951, Vol. 12, pp. 306–313.

Haan, E. F. de, A. van der Drift, and P. P. M. Schampers   The "Plumbicon," a New Television Camera Tube, *Philips Tech. Rev.*, 1963–1964, Vol. 25, No. 6/7, pp. 131–151.

Haan, E. F. de, F. M. Klassen, and P. P. M. Schampers   An Experimental "Plumbicon' Camera Tube with Increased Sensitivity to Red Light, *Philips Tech. Rev.*, 1965, Vol. 26, No. 2, pp. 49–51

Doorn, A. G. van   The "Plumbicon" Compared with Other Television Camera Tubes, *Philips Tech. Rev.*, 1966, Vol. 27, No. 1, pp. 1–14.

Goto, N., et al.   Saticon, a New Photoconductive Camera Tube with Se-As-Te Target, *IEEE Trans. Electron Dev.*, November 1974, Vol. ED-21, pp. 662–666.

Ehata, S., et al. . Low-Lag Gun for Saticon, *Hitachi Rev.*, March 1978, Vol. 27, pp. 167–171.

Month, A.   A New 30-mm High-Performance Saticon Pickup Tube, *J. Soc. Motion Pic. Telev. Eng.*, July 1980, Vol. 89, pp. 505–507.

Childs, I., and J. R. Sanders   Color Operation of a Line-Array CCD Telecine, *J. Soc. Motion Pic. Telev. Eng.*, February 1980, Vol. 89, pp. 100–106.

Synchronizing Signal Generation

BY CHARLES W. RHODES

61. Performance Requirements. Television sync-pulse generators (SPGs) provide the timing pulses required for the system. In monochrome signals, these pulses are:

1. Composite sync (horizontal sync, vertical serrated sync, and equalizing pulses)
2. Composite blanking (horizontal and vertical combined)
3. Camera horizontal drive pulses
4. Camera vertical drive pulses

The horizontal and vertical blanking pulses are combined within the SPG, and they appear at its output as composite blanking (see Figs. 20-42 and 20-43, upper traces). The horizontal sync, serrated vertical sync pulses, and equalizing pulses are also combined within the SPG and appear at the composite sync output, as in Figs. 20-42 and 20-43, lower traces. The time relationships between sync and blanking are shown.

In early camera designs, the practice was to control the scanning circuits by means of horizontal and vertical drive pulses. These are not combined in the SPG; that is, individual coaxial cables carry these pulses to the camera. The leading edges of the drive pulses coincide with the leading edges of the blanking pulses, as shown in Figs. 20-44 and 20-45. Many modern cameras do not use drive pulses, thus reducing the complexity and weight of the cable from the camera to its camera control unit (CCU).

Monochrome sync generators usually have provision for synchronization to the power line, an internal crystal oscillator, or an external source of 31.5 kHz; or they can be *gen-locked* to a composite video signal. In the NTSC color system, the power-line and internal crystal modes of operation are not feasible. Monochrome SPGs are converted to color use by driving them with a 31.5-kHz signal derived by counting down from the color subcarrier. The subcarrier is generated by a highly stable crystal oscillator held at constant temperature. Such subcarrier sources, including the countdown circuits to produce the locked 31.5-kHz signal, are called *color standards*.

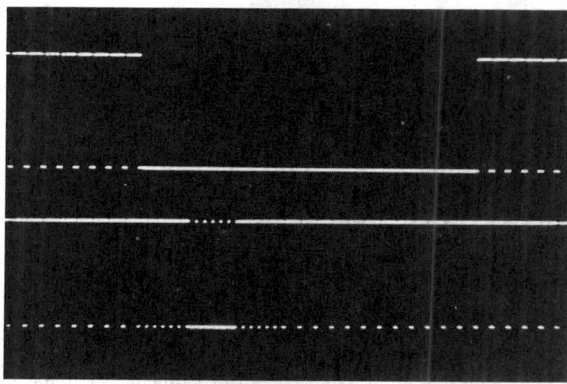

Fig. 20-42. Detail of composite blanking signal *(top)* and composite sync *(bottom)*. True time relationship of vertical blanking and vertical serrated sync pulses is shown.

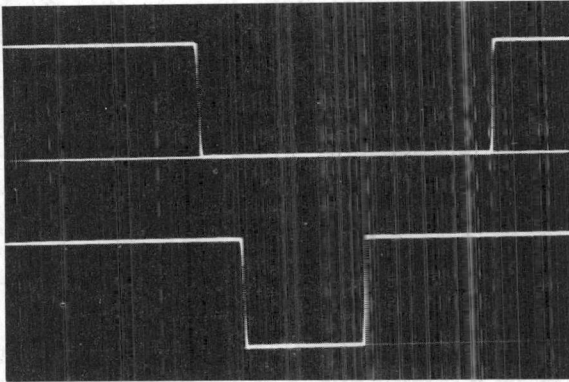

Fig. 20-43. Oscilloscope presentation of composite blanking (top) and composite sync (bottom). True time relationship of horizontal sync and horizontal blanking is shown.

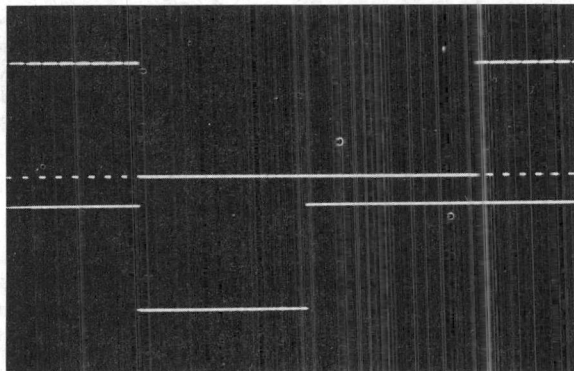

Fig. 20-44. Composite blanking (top) and vertical drive (bottom).

Fig. 20-45. Composite blanking (top) and horizontal drive (bottom).

The FCC Rules require the absolute frequency of the color subcarrier to remain within ± 10 Hz of 3.579545 MHz. The short-time frequency stability (drift rate) is also specified. Better than 3 ppm stability is readily achieved by temperature-stabilized quartz-crystal oscillators. Network operations frequency require a much more stable subcarrier.[1]* Rubidium frequency standards have been developed to provide the color subcarrier and 31.5-kHz signals with a stability of 1 to 5×10^{-11}.

62. NTSC Synchronizing Signal Waveform. Details of the NTSC sync waveform as radiated are shown in Fig. 20-9. This waveform differs from that of the SPG since:

1. The rise and fall times of blanking of video signals are only indirectly controlled by the SPG.

2. The rise and fall times of sync pulses are increased in the transmitter due to its sharply limited (4.2-MHz) bandwidth.

3. The sync-to-burst timing, or breezeway, is altered in the transmitter, principally by the delay precorrection filter, which reduces breezeway by 170 ns.

4. In many cases, the signal timing as initially established by the SPG is altered by video processing amplifiers. This is especially true of video tape recorders and frame synchronizers. Horizontal blanking pulses in particular may increase in duration by 140 ns with each rerecording. In a similar manner, vertical blanking may increase one half line in passing through frame synchronizers, slow-motion recorders, etc.

These factors have been taken into account by Electronic Industries Association's recommendation for a studio output signal (EIA Tentative Standard RS 170-A), as shown in Fig. 20-46. The pulse parameters make allowance for anticipated changes in the signal before it is actually broadcast, so that it is probable that the radiated signal will conform to the FCC rules (Fig. 20-9).

A new timing relationship given in RS 170-A is called S/C-H timing. A specific time interval between the midpoint of the leading edge of the horizontal sync and the first "significant" zero crossing of the color burst is given, and a tight tolerance established. The RS 170-A Standard has in this way defined the four-fields-per-color-frame sequence implicit in the NTSC signal format.

The intent of these specifications is to alleviate the problem of ever-increasing horizontal and/ or vertical blanking durations with video recording and frame synchronizing. These specifications not only help keep the picture from shrinking but also facilitate electronic editing of recorded programming.

It is desirable to control the shape of the sync-pulse *transitions* produced by the SPG, using *sine-squared filters.*[2] These are phase-equalized, low-pass filters which limit the frequency spectrum of these pulses. A sync pulse having such a controlled transition is shown in Fig. 20-47. One particular advantage in band limiting the outputs of an SPG is that ringing and cross talk are reduced, as the signals are distributed within the broadcast plant. This is especially important in current SPG units, in which rise times may be as short as 10 ns. The sine-squared filters have a cutoff frequency approximately the inverse of the step rise time. Rise times are always measured from the 10 to 90% amplitude point on the transition, as shown in Fig. 20-47. Standard RS 170-A (Fig. 20-46) recommends a rise time of 140 $\pm$ 15 ns for sync pulses at studio output. While the usual practice in measuring the pulse durations is to measure between 50% points, the FCC Rules (which apply to transmitter output only) require measurement between the 10% points on the rising and trailing edge of a pulse.

The color burst is not transmitted during the first nine lines in each vertical blanking interval. A nine-line *keyout* of the burst is produced within the SPG by inhibiting generation of the burst-flag pulses. Figure 20-48 compares the burst-flag pulses with the composite video, in the vicinity of the vertical blanking pulse. The absence of the nine color bursts is apparent. The timing of the burst flag relative to both horizontal sync and the burst envelope is shown in Fig. 20-49. The flag precedes the burst, to accommodate the delay in the bandpass filters through which the burst envelope must pass in the encoder.

Black-Burst Technique.   The cost of distributing the individual outputs of an SPG throughout the broadcast plant on multiple coaxial cables has led to the development of the *black-burst signal,* shown in Fig. 20-50. This signal carries all the required timing information from one SPG to another on a single coaxial cable. Many SPGs can thus be slaved to the master SPG via its black-burst output. The slaved SPG units can thus be timed relative to a common point within the plant so that encoded video signals from any source are in precise time synchronism with all other video signals at the production switcher's inputs. This condition, accurate to a few degrees at subcarrier frequency, is essential during switches, fades, wipes, and dissolves. The

---

*Superior numbers correspond to References, Par. **20-67**.

differential pulse cross display can be used to measure the relative timing of two video signals. Waveform monitors also measure relative timing errors, a differential input being required.

A color picture monitor, equipped with both pulse cross display and differential input amplifiers producing signals of opposite polarity, can perform the entire timing operation. The two opposite color bursts will cancel each other when the two bursts are precisely in phase with each other. Oscilloscope techniques can readily measure both the rise time and pulse width. Figure 20-47 shows a rise-time graticule used for this purpose.

63. Pulse Jitter. Pulse jitter must be measured with respect to the color subcarrier. Very stringent specifications of the SPG for jitter are recommended by various broadcasting organizations because SPG jitter reduces the quality of video recordings. Especially with multiple generations of video recording, chrominance noise builds up rapidly due to sync-pulse jitter; 5 ns of jitter is considered excessive in terms of present state of the art.

Jitter can be measured to less than 2 ns by comparing the leading edges of all horizontal sync pulses against the color subcarrier; this can best be done with a *vectorscope*. The time resolution of an NTSC vectorscope is

$$\frac{10^{-6}}{3.58(360°)} = \frac{1 \; \mu s}{1.288 \times 10^3} = 0.77 \; ns/deg$$

The leading edge of horizontal sync pulses generates phase transients which can be observed on the vectorscope, as in Fig. 20-51*. The 2:1 frequency interlace of NTSC causes the double transient display. This figure shows severe jitter.

Another technique for verifying the stability of an SPG is the *pulse cross display* as seen on a picture monitor. Figure 20-52 shows a typical display. The vertical blanking may be expanded for improved resolution in this critical area. Jitter will affect the vertical alignment of the leading edges of horizontal sync pulses and equalizing and serrated pulses. Measurement of jitter is facilitated by using the delayed sweep trigger from a laboratory oscilloscope, the delayed trigger being added to the video signal to modulate the display in intensity. In this way, quantitative measurements can be made with a high degree of accuracy.

64. Burst Timing. Compliance with FCC rules requires a breezeway of 379 ns in the radiated signal. To this must be added the 170 ns introduced in the transmitter. This roughly sets a minimum of 560 ns between the end of horizontal sync and burst start. FCC rules do not specify the effective burst start. Figure 20-53 shows how breezeway varies with burst phase, i.e., the ambiguous start of burst.

EIA Standard RS 170-A has established the effective limits of the burst signal. EIA measures from midpoint of leading edge of sync to the first significant zero crossing of burst. EIA also specifies the sync pulse half-amplitude (50%) duration. EIA does not give any specification of breezeway, but the values chosen by EIA provide operation within the FCC rules.

Standard RS 170-A specifies 5.3 μs (19 cycles of subcarrier) between sync and burst start. The provisional tolerance is ± 0.035 ns ($\pm 45°$), or 1 part in 152. This is well beyond the accuracy of television waveform monitors. Specialized measurement instruments are required, especially where S/C-H timing is to be routinely measured by broadcasters. RS 170-A also defines the effective end of color burst in an unambiguous way.

Table 20-4 summarizes the timing specifications of the NTSC signal as generated (SPG nominal) and as radiated (FCC Rules) †

It is well within the art to generate digitally a composite video signal having correct S/C-H timing inherently. Difference in path lengths of composite sync and subcarrier signals from an SPG can alter this timing relationship. Group-envelope delay or chrominance-luminance delay inequality can also affect the timing relationship in the composite video signal.

65. Organization of SPG Circuits. Interlaced scanning in television systems is achieved by the design of the sync signal. The essential element is that each frame consists of an odd integral number of lines so that each field contains precisely half that number (see Fig. 20-4). Any jitter in the vertical sync with respect to horizontal sync tends to destroy interlace; e.g., a time jitter of 30 μs completely destroys interlace. For this reason, sync generators rely upon the 31.5-kHz pulses to form this leading edge of horizontal sync, vertical serrated, and the equalizing pulses. The functions of the counters (Fig. 20-54) are to select which of the 31.5-kHz pulse edges are to form the sync-pulse components. Within the divide-by-525 counter, it is pos-

---

*Some vectorscopes may blank the beam during sync pulses.

†The PAL system has an eight-field sequence, and SECAM uses a twelve-field sequence.

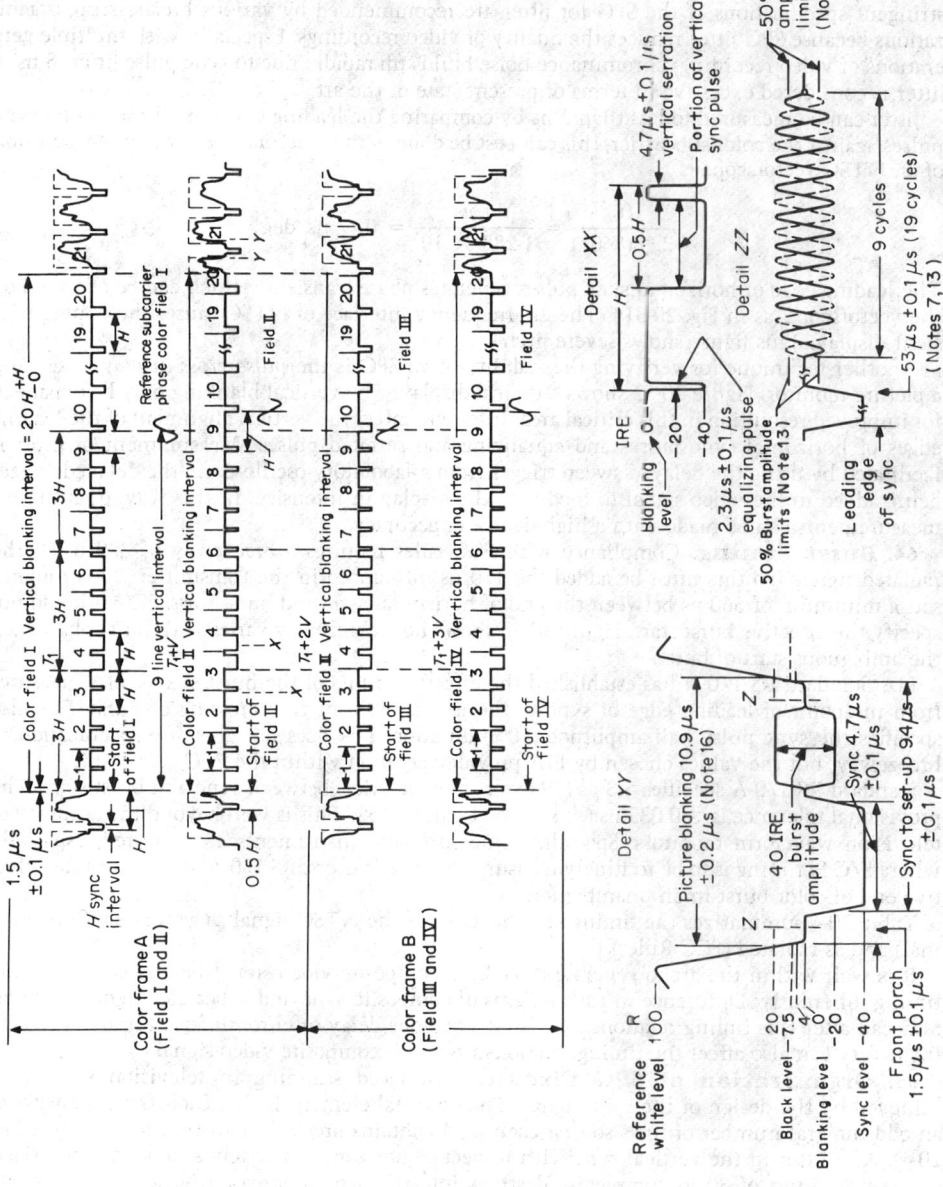

Color frame A
(Field I and II)

Color frame B
(Field III and IV)

Color field I Vertical blanking interval
Color field II Vertical blanking interval
Color field III Vertical blanking interval
Color field IV Vertical blanking interval

Start of field I
Start of field II
Start of field III
Start of field IV

Field II
Field III
Field IV

Reference subcarrier phase color field I

Vertical blanking interval = 20 H_{-0}^{+H}

9 line vertical interval

H sync interval

1.5 μs ±0.1 μs

0.5 H

Detail XX

Detail YY

Detail ZZ

4.7 μs ±0.1 μs vertical serration

Portion of vertical sync pulse

Blanking level 0 —
IRE
−20 —
−40 —

2.3 μs ±0.1 μs equalizing pulse

50% Burst amplitude limits (Note 13)

0.5H

H

50% burst amplitude limits (Note 4)

9 cycles

53 μs ±0.1 μs (19 cycles) (Notes 7,13)

Leading edge of sync

Reference white level 100 —

Black level — 7.5 —
4 —
Blanking level — 0 —
−20 —
Sync level — −40 —

Picture blanking 10.9 μs ±0.2 μs (Note 16)

40 IRE burst amplitude

Sync 4.7 μs ±0.1 μs

Sync to set-up 9.4 μs ±0.1 μs

Front porch 1.5 μs ±0.1 μs

Fig. 20-46. Color television studio picture line amplifier output signal waveforms specified in EIA Tentative Standard RS-170A. *Notes:* (1) Specifications apply to studio facilities. Common carrier, studio to transmitter, and transmitter characteristics are not included. (2) All tolerances and limits shown in this drawing permissible only for long time variations. (3) The burst frequency shall be 3.579545 MHz $\pm$ 10 Hz. (4) The horizontal scanning frequency shall be $\frac{2}{455}$ times the burst frequency [one scan period (H) = 63.556 μs]. (5) The vertical scanning frequency shall be $\frac{2}{525}$ times the horizontal scanning frequency [one scan period (V) = 16,683 μs]. (6) Start of color fields I and III is defined by a whole line between the first equalizing pulse and the preceding H sync pulse. Start of color fields II and IV is defined by a half line between the first equalizing pulse and the preceding H sync pulse. Color field I: that field with positive-going zero crossings of reference subcarrier most nearly coincident with the 50% amplitude point of the leading edges of even-numbered horizontal sync pulses. Reference subcarrier is a continuous signal with the same instantaneous phase as burst. (7) The zero crossings of reference subcarrier shall be nominally coincident with the 50% point of the leading edges of all horizontal sync pulses. When the relationship between sync and subcarrier is critical for program integration, the tolerance on this coincidence is $\pm 40°$ of reference subcarrier. (8) All rise times and fall times unless otherwise specified are to be 0.14 $\pm$ 0.02 μs measured from 10 to 90% amplitude points. All pulse widths are measured at 50% amplitude points unless otherwise specified. (9) Tolerance on sync level, reference black level (setup), and peak-to-peak burst amplitude shall be ± 2 IRE units. (10) The interval beginning with line 17 and extending through line 20 of each field may be used for test, cue, and control signals. (11) Extraneous synchronous signals during blanking intervals, including residual subcarrier, shall not exceed 1 IRE unit. Extraneous nonsynchronous signals during blanking intervals shall not exceed 0.5 IRE unit. All special-purpose signals (VITS, VIR, etc.) when added to the vertical blanking interval are excepted. Overshoot on all pulses during sync and blanking, vertical and horizontal, shall not exceed 2 IRE units. (12) Burst-envelope rise time is $0.3 + 0.2 - 0.1$ μs measured between the 10 and 90% amplitude points. Burst is not present during the nine-line vertical interval. (13) The start of burst is defined by the zero crossing (positive or negative slope) preceding the first half cycle of subcarrier that is 50% or greater of the burst amplitude. Its position is nominally 19 cycles of subcarrier from the 50% amplitude point of leading edge of sync (see detail ZZ). (14) The end of burst is defined by the zero crossing (positive or negative slope) following the last half cycle of subcarrier that is 50% or greater of the burst amplitude. (15) Monochrome signals shall be in accordance with this drawing except that burst is omitted and fields III and IV are identical to fields I and II respectively. (16) Occasionally measurement of picture blanking at 20 IRE units is not possible because of scene content as verified on a picture monitor.

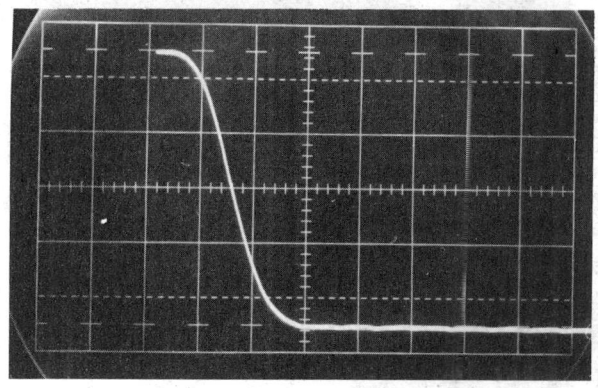

Fig. 20-47. Trailing edge of sine-squared-shaped sync pulse (125 ns between 10 and 90% levels).

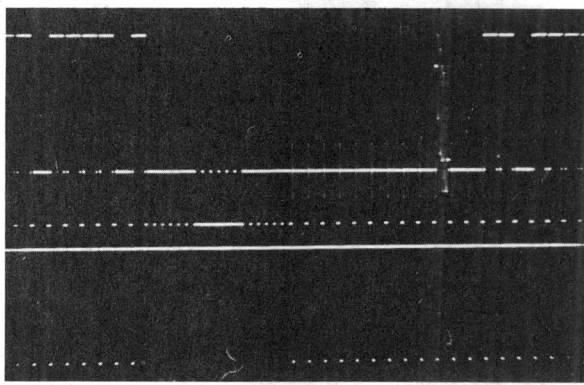

Fig. 20-48. Composite video *(top)*, with burst flag *(bottom)*. During nine lines of the vertical blanking interval, the burst flag is not present due to the nine-line keyout.

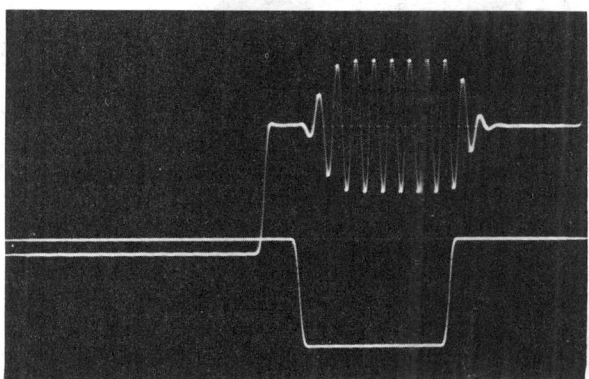

Fig. 20-49. Composite video sync *(top)* and burst flag *(bottom)* during horizontal blanking.

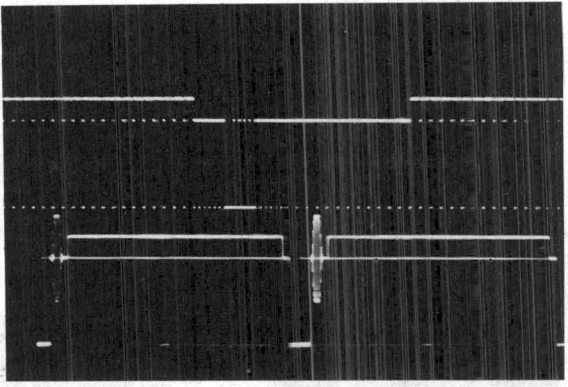

Fig. 20-50. Black burst *(bottom center)*. Vertical blanking detail at top, horizontal blanking detail at bottom.

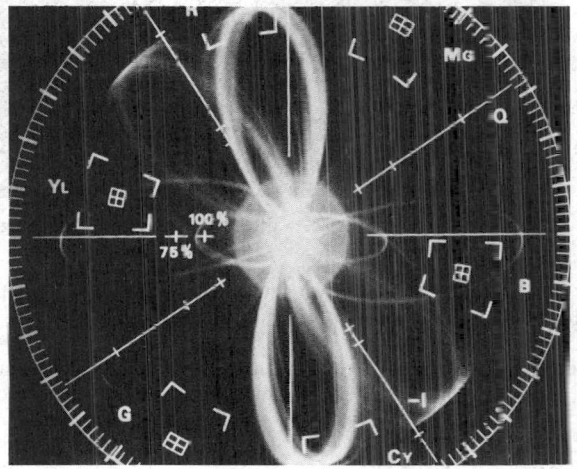

Fig. 20-51. Vectorscope display, showing phase jitter.

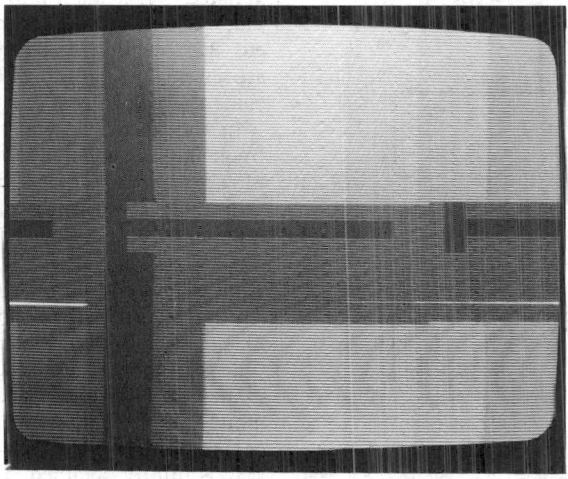

Fig. 20-52. Pulse cross display on a picture monitor

sible, using digital logic, to develop gating pulses at any 31.5-μs interval throughout the television field.

There are three techniques available to time the durations of the sync-pulse components: delay lines, analog-delay pickoffs, and digital counters. The delay-line method is of historical interest only, due to the physical bulk of a suitable delay line. The analog-delay method provides simple and independent adjustments of pulse durations without affecting leading-edge timing.

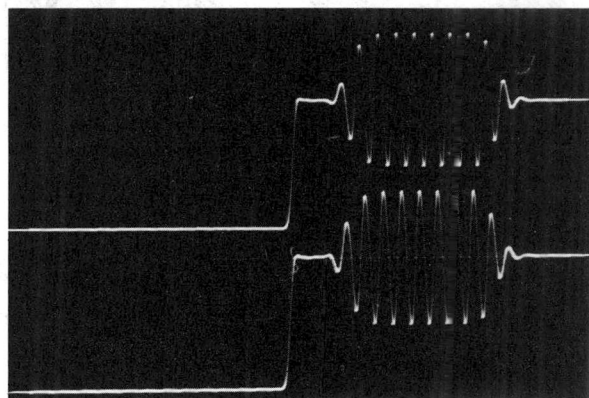

Fig. 20-53. Breezeway display with lower display shifted 180°, showing the change in breezeway with subcarrier phase.

Long-time stability and freedom from jitter or temperature effects can be improved, however, with digital techniques at the expense of pulse durations becoming integer multiples of clock period (70 ns, for example).

66. Digital Techniques. The digital-counting method requires a subcarrier oscillator operating at a harmonic of 3.579545 MHz to provide reasonable time resolution. When the fourth harmonic is chosen, the time resolution of the counter is 70 ns. In practice, this is an appropriate compromise between oscillator stability and time resolution. Higher-frequency quartz oscillators offer less long-term stability. While analog delay is continuously variable over a small range, digital delay provides pulse-width increments of 70, 140 ns, etc. Both techniques have found commercial acceptance, using integrated circuits to overcome the problems of complexity.

In the NTSC system, the color-subcarrier frequency f_{sc} is related to the horizontal-scanning frequency F_h by the ratio

$$f_{sc} = \frac{455(2F_h)}{2 \times 2}$$

Division by 455 and multiplication by 4 satisfies this requirement, i.e., produces $2F_h = 31.5$ kHz. This technique was used in early NTSC sync-generation equipment, but the instability of the times 4 multiplier has led to better techniques. By dividing 4(3.579545 MHz) = 14.3 MHz by 455, the desired $2F_h$ frequency of 31.5 kHz is readily obtained. The color subcarrier is obtained with a divide-by-4 divider driven by the 14.3-MHz crystal oscillator. This technique is used commercially.

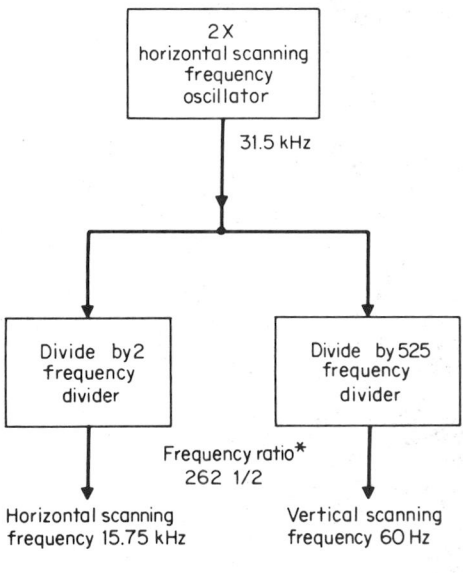

Fig. 20-54. Analog counting technique for generating monochrome sync signals.

TABLE 20-4 Summary of Timing Specifications
(see also Fig. 20-46)

	FCC Rules*	SPG Nominal†
Color subcarrier, 3.579545 MHz	±10 Hz, 0.1 Hz/s	±5 Hz
Pulse widths:		
Horizontal sync	4.49–5.09 μs	4.7 ± 0.1 μs
Equalizing pulses	0.45–0.50 area horizontal sync	2.3 ± 0.1 μs
Serrations in vertical sync	3.81–5.09 μs	4.7 ± 0.1 μs
Rise time	254 ns max	115 ns ± 10%
Amplitude		4 V ± 5% into 75 Ω
Pulse jitter‡		4 ns
Breezeway	379 ns min	759 ± 50 ns
Burst duration	8 cycles min, 11 cycles max	9 cycles
Horizontal blanking	10.48 μs min	10.9 ± 0.2 μs

*Pulse widths measured at 10% point to 90% point.
†Pulse widths measured at half-amplitude duration.
‡Pulse jitter refers to maximum time error between zero crossing of subcarrier cycle and leading edge of any sync pulse in a field.

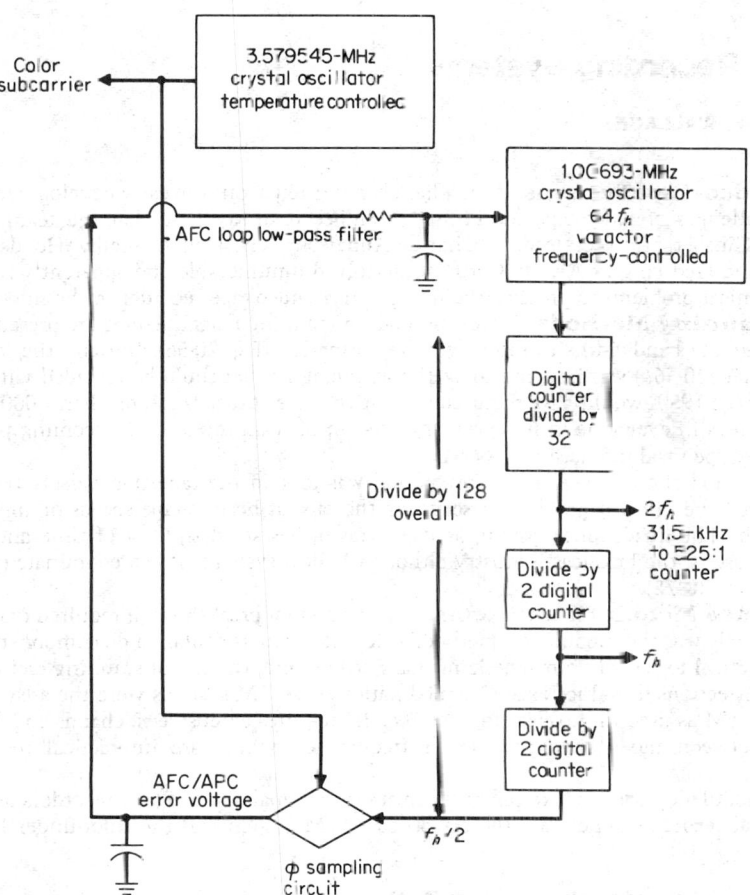

Fig. 20-55. Digital technique for generating NTSC color sync signals.

A second technique avoids both the times 4 multiplier and the crystal oscillator operating at 14.4 MHz in favor of a 3.579545-MHz crystal, which offers improved long-term and temperature stability. This technique, shown in Fig. 20-55, also eliminates the divide-by-455 digital frequency divider, to improve stability. Here, the 1.00693-MHz crystal oscillator (operating at $64F_h$) is digitally divided by 128 to form pulses which sample the subcarrier frequency, a sample being taken once every 455 cycles. The error signal from the sampling gate is the frequency error of the 1.00963-MHz oscillator, with respect to subcarrier frequency. This passes through a low-pass filter to a varactor that controls the frequency of the oscillator with respect to the color subcarrier. The subcarrier frequency is controlled by a proportional-control oven housing the crystal and oscillator circuit. The design of the low-pass filter and the inherent frequency stability of the crystal prevent the 1.00693-MHz crystal from being locked to the wrong multiple of the subcarrier frequency.

Once frequency lock has been established, the system becomes a *phase-locked loop*. It operates to hold the leading edge of the sampling pulses at $F_h/2$ coincident with the positive-going zero crossing of the subcarrier cycles. This defined relationship between the F_h pulses and the subcarrier phase is fundamental to this precise synchronization required in the NTSC system. The timing of leading edges of the horizontal sync pulses, equalizing pulses, and vertical serrated pulses is made absolutely precise by forming the leading edges in the same integrated circuit, in each case triggered by a 31.5-kHz pulse selected by digital logic circuits. Pulse widths are controlled by the reset timing. Pulse widths may be timed using either analog or digital circuits.

67. References on Synchronizing Pulse Generators

1. Davidoff, F.   The CBS Automatic Color Wire-Lock System, *J. Soc. Motion Pict. Telev. Eng.*, August 1969, Vol. 78, pp. 621–625.

2. Kastelein, A.   A New Sine-Squared Pulse and Bar Shaping Network, *IEEE Trans. Broadcast.*, December 1970, Vol. BC-16, No. 4, pp. 84–89.

Video Recording Systems

BY KURT F. WALLACE

68. Video-Tape Systems. Much has been written about the early development and technical challenges of video-tape recording.[1-3]* Suffice it to say here that the team headed by Charles Ginsberg (project leader), Charles Anderson, Ray Dolby, Shelby Henderson, Alex Maxey, and Fred Post, of Ampex Corporation, solved innumerable and apparently insurmountable technical problems to produce the first practical video-tape recorder in the mid-1950s.

69. Scanning Methods. Three methods of scanning magnetic tape are presently used to cover adequate bandwidths to record video information (Fig. 20-56). Initially the longitudinal method (Fig. 20-56a) was impractical, as the wavelengths that could be recorded with the technology of the 1950s would have required moving the tape at speeds greater than 1,000 in/s. After two decades of development, however, this system of scanning is now becoming practical for consumer-type machines (see Par. **20-81**).

The method chosen by Ampex Corporation was to scan the tape transversely (Fig. 20-56b). This solved the difficult problem of scanning the tape at head-to-tape speeds of approximately 145 km/h. The quadruplex system, as it is known, has stood the test of time and is still in common use in the broadcast industry although helical systems now predominate (see Fig. 20-56c and Par. **20-72**).

70. Video Modulation Systems. A fundamental breakthrough required to record television signals was the modulation method. Television signals contain a dc component which it is not practical to record on magnetic media. Furthermore, the use of scanning techniques and tape imperfections introduce excessive modulation noise. FM systems were the answer to these problems. FM as used in broadcasting (see Sec. **21**) requires extensive sidebands and a large separation between modulating and carrier frequencies, which are impractical in wide-band recording.

Two modulation systems are presently in use in all analog video-tape recorders and in discs and optical recorders. These are the direct-record FM system and the color-under heterodyne system.

*Numbers correspond to references in Par. **20-92**.

In the direct-record color high-band recording system a carrier is frequency-modulated with the complete composite video signal so that the sync tip level is deviated to 7.06 MHz, blanking level to 7.9 MHz, and peak white to 10 MHz (in the high-band system, see Table 20-5). Detailed analyses of the sidebands and associated mathematics[4] and the specifications for preemphasis, etc.,[5] are available. Table 20-5 shows super high-band (SHBP), high-band (HB), low-band monochrome (LBM), and low-band color (LBC) frequencies.

The direct-color FM high-band recording system is used in most broadcast video-tape recorders. These systems give excellent-quality reproduction of color video recording with a minimum of cross-modulation products between carrier, video, and subcarrier, known as moiré components. This system suffers the disadvantage of requiring extremely accurate time-base correction to eliminate the errors caused by head-to-tape velocity inaccuracy and rf frequency-response inconsistency in the recording process.

TABLE 20-5 FM Frequencies Used in Video Recording, MHz

	SHBP	HB	LBM	LBC
Peak white	10.7	10.0	6.8	6.5
Blanking	9.9	7.9	5.0	5.79
Sync tip	9.58	7.06	4.28	5.5
Tolerance	±0.02	±0.05	±0.05	±0.05

SOURCE: Extracted from Ref. 5.

ponents. This system suffers the disadvantage of requiring extremely accurate time-base correction to eliminate the errors caused by head-to-tape velocity inaccuracy and rf frequency-response inconsistency in the recording process.

The color-under recording system is shown diagrammatically in Fig. 20-57. The video signal is low-pass-filtered to separate the luminance signal from the chrominance and to roll it off at 2.5 to 3 MHz. In home recording systems the chrominance subcarrier is bandpass-filtered to about ½-MHz bandwidth. One of the common systems (see Table 20-9) records the luminance component on an FM carrier such that peak white is at 4.5 MHz and sync tip at 3.1 MHz. The chrominance component is modulated on a subcarrier at about 688 kHz and added to the rf spectrum. The advantages of the system are simplicity and the fact that time-base errors cause only small changes in subcarrier phase. In practice, some of the low-cost helical recorders, such as U-matic, Betamax, and VHS formats, are used in some special broadcast applications after processing in broadcast-type time-base correction units.

71. Quadruplex Recorders. The basic format of quadruplex recorders is shown in Fig. 20-58. The video head is a four-tipped (quadruplex) scanning assembly rotating at 240 r/s. The drum diameter is 2 in, producing a writing speed (head-to-tape velocity) of 1,560 in/s (38 m/s). One major advantage of the transverse recording system is that errors in tape tension do not in principle affect track position or angle. This means that tracking the transverse recorded information is relatively simple.

A further advantage is that erasing information and subsequent re-recording (electronic editing) is easily achieved by placing a full-width longitudinal erase head, in which the gap is transverse to the tape, upstream of the video head. The erase current must be turned on and off at the appropriate time to produce the edit. Information as short as one TV frame can be erased and re-recorded with new information in a continuous stream of video data.

The audio information is carried on a longitudinal track, shown in Fig. 20-58, with a nominal width of 72 mils (1.7 mm). An optional variation

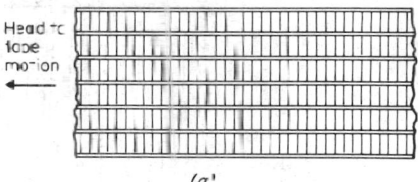

(a)

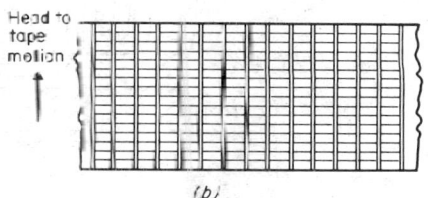

(b)

(c)

Fig. 20-56. Tape-scanning methods.

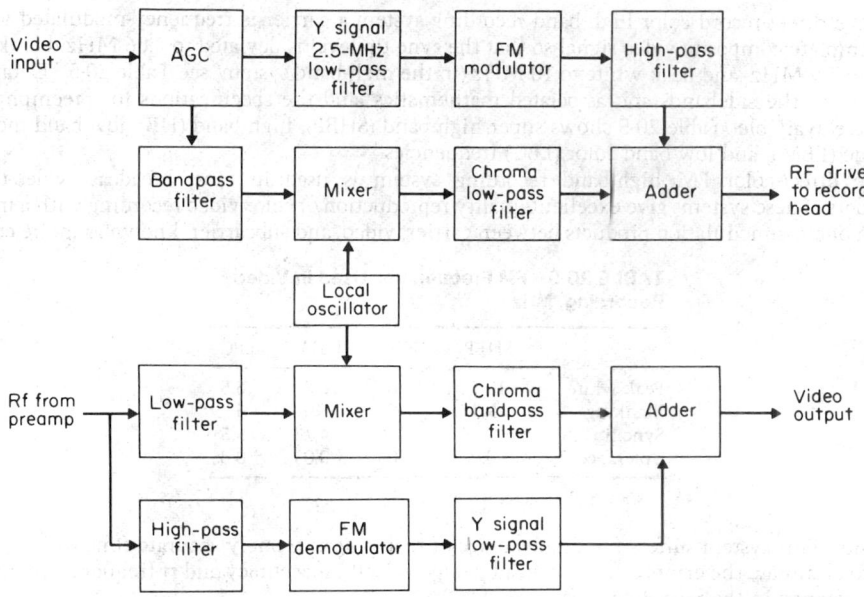

Fig. 20-57. Color-under modulating block diagram.

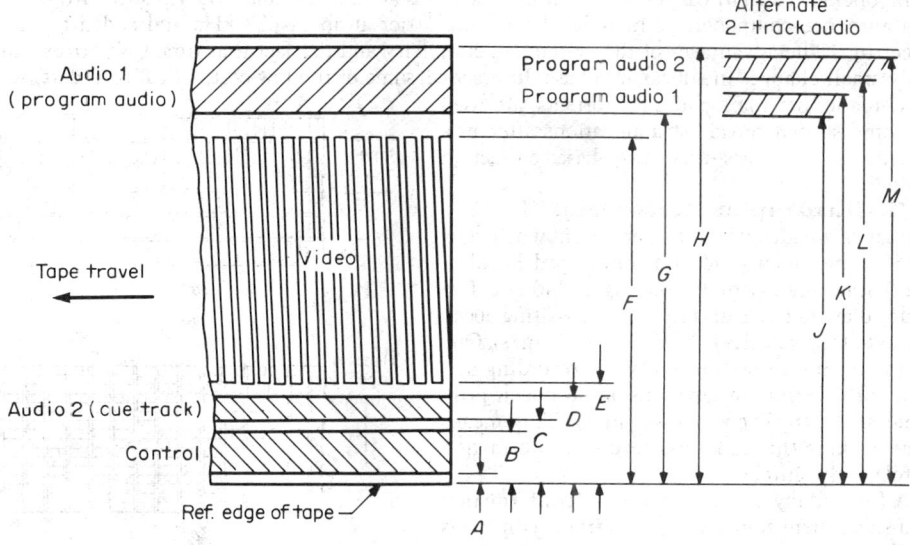

Fig. 20-58. The 2-in quad tape format.[6]

Dimensions	in Min	in Max	mm Min	mm Max	Dimensions	in Min	in Max	mm Min	mm Max
A	0.000	0.004	0.00	0.10	G	1.921	1.930	48.79	49.02
B	0.040	0.049	1.02	1.24	H	1.988	1.996	50.50	50.70
C	0.058	0.062	1.47	1.57	J	1.920	1.928	48.77	48.97
D	0.078	0.085	1.98	2.16	K	1.945	1.951	49.40	49.56
E	0.087	0.094	2.21	2.39	L	1.965	1.971	49.91	50.06
F	1.902	1.914	48.31	48.62	M	1.988	1.996	50.50	50.70

is to split the audio track, for stereo, into two separate tracks measuring 24 mils (0.6 mm) each with a 20-mil (0.5-mm) guard band.[7]

There are two versions of the quadruplex standard with regard to packing density (tape velocity). Most broadcasters still use the original format, which runs at 15 in/s (38 cm/s) longitudinal speed, with a 10-mil video track width separated by a 5-mil guard band. A secondary standard specifies the tape velocity as 7.5 in/s (19 cm/s) with a video track width of 5.5 mils (140 μm) and a track spacing of 2 mils (51 μm).

72. Helical Machines. Helical machines were first considered in the mid-1950s and became practical in the early 1960s. They were welcomed as a solution to the problems of multiple video tips and record-play channels used in quadruplex recorders. The advantage of the helical format is that a single recording and playback channel can be used to record and replay a whole field. This eliminates the troublesome banding problems encountered on quadruplex (and to a lesser extent in the later developed segmented format). A further advantage of recording a single field per head scan is that distinguishable pictures can be obtained when the tape is moved at other than the nominal replaying speed. Thus, with appropriate electronics, pictures can be obtained which are useful for editing purposes at speeds from shuttle velocity down to stop.

73. Automatic Scan Tracking. A recent important development, first used on the Ampex VPR-1 machine, records on 1-in tape with a 360°-wrap format. The technique, known as *automatic scan tracking* (AST), uses a piezoelectric transducer as the mount for the video tip[8] (see Fig. 20-59). The video tip is moved at 90° to the scanning direction by the application of a voltage to the piezoelectric transducer. This motion can be made fast enough to allow the head to jump one track pitch during the vertical blanking interval. This eliminates the interruption caused by skewing across tracks when tape speeds close to zero are used.

Figure 20-59 shows the method used by Ampex. A *dither frequency* is applied to the video tip and to a frequency discriminator, which senses the rf envelope amplitude and detects any minor errors in track curvature or skewing when operating at other than normal speed. Thus the system of moving the video tip laterally produces broadcast-quality pictures from reverse motion to faster than real time without interruptions in the video. The system also compensates for tapes which have been badly stored and have track curvature.

This powerful technique will undoubtedly find widespread use in the future when track widths become even narrower than at present. Philips has announced a scan tracking system on their V2000 recorder which has a track width of 0.9 mil (22.6 μm). See Table 20-8.

74. Broadcast 1-in Helical SMPTE Type C Format. The SMPTE type C format originated at the request of the broadcasters for manufacturers to set up a common 1-in helical standard appropriate to the broadcasters' needs. It was a remarkable achievement on the part of the industry, and especially of the SMPTE Working Groups that agreement between manufacturers and broadcasters was achieved in approximately 1 year.[9] Two proposals had been submitted to the broadcasters: the Ampex format, which had been in production for a number of years and had become the SMPTE 1-in type A format, and an improved version submitted by Sony Corporation. Both formats have almost identical drum diameters, but the methods of tape guid-

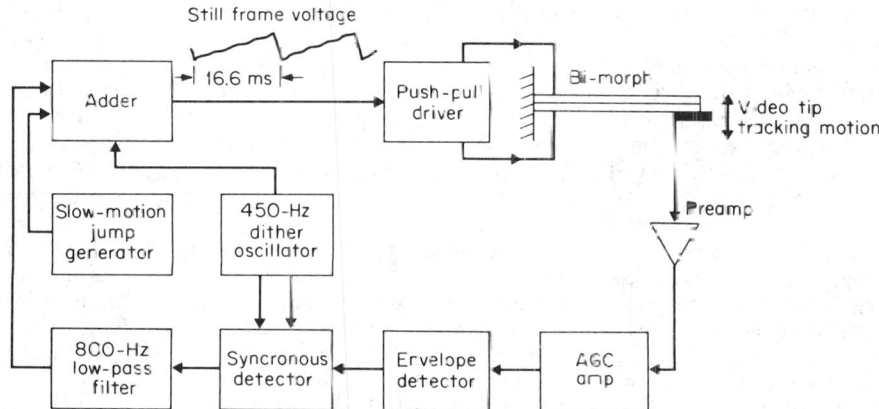

Fig. 20-59. Automatic scan tracking and slow-motion block diagram.[3]

ing and positioning of heads and audio tracks are different. Both proposals share the feature of having one field per head scan and omega-wrap (360°) scan.

The head layout on the drum is shown in Fig. 20-60. It will be noted that the drum has six heads (not one, two, or three as might be expected of a field-per-rotation helical machine). An *erase head*, a *video record head*, and a *video reproduce head* are all that are required to make record and reproduce video images. A second set of heads called *sync heads* are used to record during the vertical sync time. This time has been allocated for digital information and codes such as time-code or station-accounting data.[10]

Figure 20-61 is an extract of the helical video tape type C recording format and specifications from ANSI C98.19.[11] The major advantages of the format lie in its capabilities with regard to slow motion and easy editing. As in all field-per-scan helical machines, pictures can be reproduced which are adequate for identifying scenes at speeds faster than real time and slower, down to stop motion. Broadcast-quality pictures can be obtained from reverse speed to faster than real time using the AST techniques described in Par. **20-73**. Also, because of the head configuration, the tape can be monitored while recording to assure that acceptable pictures are being recorded on satisfactory tape.

A further advantage of single-field helical machines is that only one video tip and one video equalizing channel are required for a whole field. Thus there is no need to match several channels very accurately, as in quadrature and segmented recorders.

75. SMPTE 1-in Type B Format. The SMPTE type B format uses a segmented scan and 180° wrap angle. It was introduced by Bosch-Fernseh. This company developed the format after buying the rights from Echo Arvin (formerly Westel Corporation). The type B format was originally designed for military applications. It is a cross between transverse and helical machines, having many of the advantages and disadvantages of each.

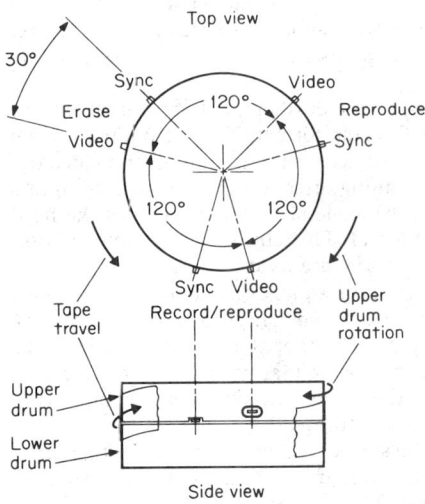

Fig. 20-60. SMPTE 1-in type C head layout.

Two of its advantages are that it is smaller than single-head helical scanners and extremely rugged in the presence of high mechanical acceleration.

Another advantage is that changes in writing speed can be easily introduced by changing the drum rotational speed, rather than changing the size of the drum (as required with helical machines). This flexibility provides the higher writing speed desirable for PAL recording, compared with NTSC machines. To execute the slow-motion effects so popular in helical machines, it is necessary to use a field store and complex tape-acceleration profiles. The video-head scanner and video-head location are shown in Fig. 20-62.

76. Digital Video-Tape Recording. The advances made in the past decade by the semiconductor industry have brought about increasing change in the broadcast industry. The greatest changes have taken place in areas where digital integrated circuits, originally designed for the computer industry, have found their way into the broadcast environment. Field-and-frame stores, used in synchronizers, picture compensators and expanders, noise-reduction systems, and special-effects generators have become practical as a result of the improved performance, small size, and low cost of digital integrated circuits. Digital technology in the broadcast field has been adopted for a reduction in size or cost or for the performance of a function impossible or impractical in the analog domain.

Digital signals recorded on tape must use the tape to its ultimate capacity. The digital signals, although defined in binary code, nevertheless obey all the fundamental laws of tape-recording analog signals, i.e., degradation of amplitude, phase, and signal-to-noise ratio. In digital processing equipment presently used by broadcasters, however, neither bandwidth nor noise presents fundamental limitations. The major advantage of digital video recording is the fact that many generations (copies of copies) can be made with virtually no loss in quality.

To provide a dynamic range comparable to existing analog video-tape machines capable of

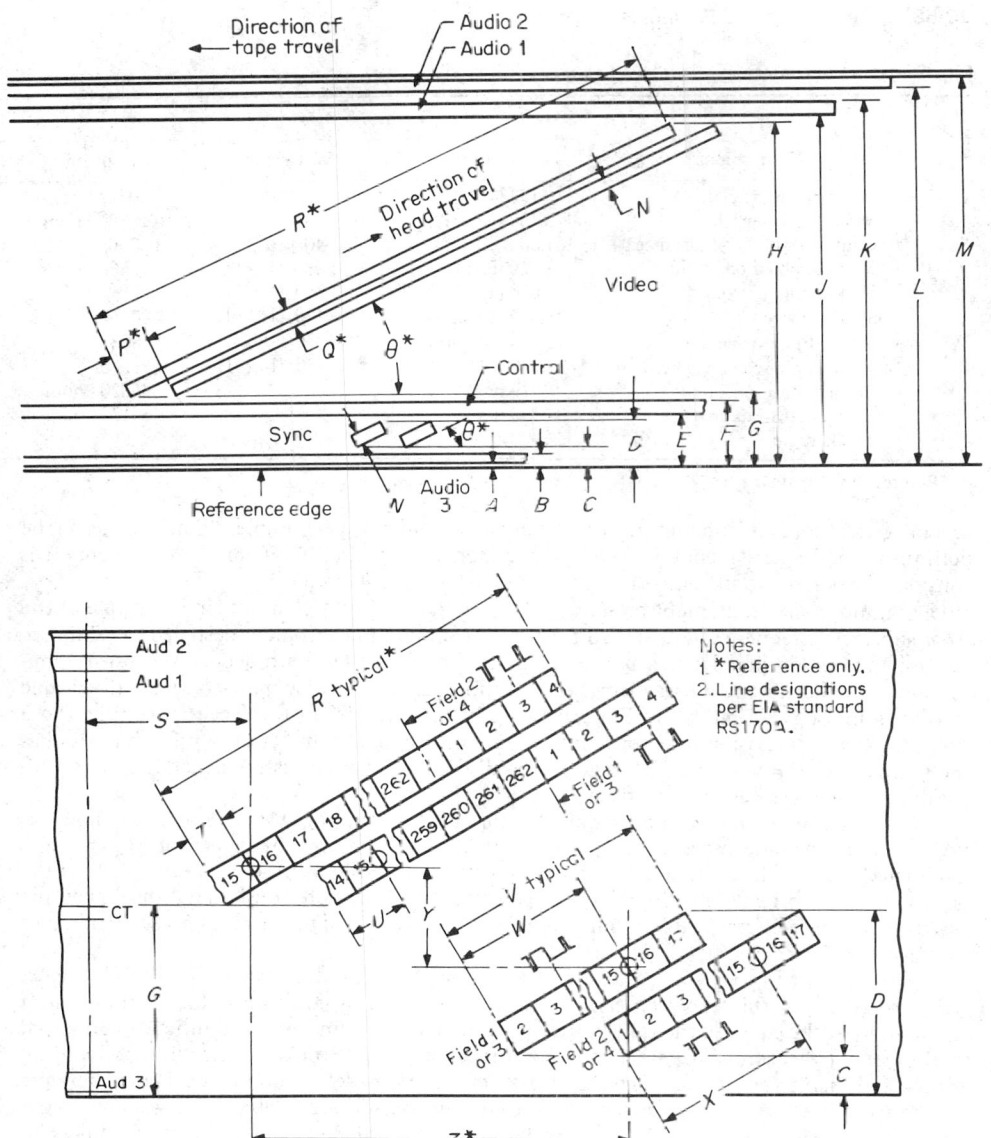

Fig. 20-61. SMPTE 1-in type C format.[11] (See Fig. 20-46.)

	Dimensions	mm		in	
		Minimum	Maximum	Minimum	Maximum
A	Audio 3 lower edge	0.000	0.200	0.00000	0.00787
B	Audio 3 upper edge	0.775	1.025	0.03051	0.04035
C	Sync track lower edge	1.385	1.445	0.05453	0.05689
D	Sync track upper edge	2.680	2.740	0.10551	0.10787
E	Control tract lower edge	2.870	3.130	0.11299	0.12323
F	Control track upper edge	3.430	3.770	0.13504	0.14843
G	Video track lower edge	3.860	3.920	0.15197	0.15433
H	Video track upper edge	22.355	22.475	0.88012	0.88484
J	Audio 1 lower edge	22.700	22.900	0.89370	0.90157
K	Audio 1 upper edge	23.475	23.725	0.92421	0.93406
L	Audio 2 lower edge	24.275	24.525	0.95571	0.96555
M	Audio 2 upper edge	25.100	25.300	0.98819	0.99606
N	Video and sync track width	0.125	0.135	0.00492	0.00531
P	Video offset	4.067 (2.5H) ref		0.16012 nom	

Fig. 20-61 (*Continued*)

Dimensions		mm Minimum	mm Maximum	in Minimum	in Maximu
Q	Video track pitch	0.1823 ref		0.007177 nom	
R	Video track length	410.764 (252.5H) ref		16.17181 nom	
S	Control track head distance	101.60	102.40	4.0000	4.0315
T	Vertical phase odd field	1.220 (0.75H)	2.030 (1.25H)	0.04803	0.0799.
U	Vertical phase even field	2.030 (1.25H)	2.850 (1.75H)	0.07992	0.1122(
V	Sync track length	25.620 (15.75H)	26.420 (16.25H)	1.00866	1.0401(
W	Vertical phase odd sync field	22.360 (13.75H)	23.170 (14.25H)	0.88031	0.9122(
X	Vertical phase even sync field	23.170 (14.25H)	23.980 (14.75H)	0.91220	0.9440!
Y*	Vertical head offset	1.529 nom		0.06020 nom	
Z*	Horizontal head offset	35.350 nom		1.39173 nom	
θ	Track angle	2°34′ ref			

*Reference value only.

several generations of dubbing, it is necessary to record 8 b per sample.[13] The choice of the optimum sampling frequency is a complex matter. For existing NTSC and PAL systems it is considered desirable to lock the sampling rate to the color subcarrier.[14-16]

Digital processing equipment presently on the market operates at an integral multiple of the color subcarrier frequency f_{sc}. A rate 3 times f_{sc} is common in equipment designed to minimize the use and cost of digital stores and recording surface. However, with decreasing cost of semiconductor memories, the use of 4 times the sampling rate has become more common. (See Table 20-6 for sampling and bit rates.) The advantage of this rate lies in the fact that (unlike the 3 times rate) each television line not only has the same integral number of samples but also has the samples vertically aligned. This simplifies all digital signal processing dependent on multiple-line signal delays, as in comb filters.

To solve the problem of vertical-sample lineup in a $3f_{sc}$ sampling system, a method known as PALE is used in some types of equipment. In PALE the sampling frequency of $3f_{sc}$ is phase-modulated with a ½-line-rate square wave. This introduces a 180° shift of sampling phase at the beginning of each television line. The Ampex electronic-still-store (ESS) digital disc recording system, for example, uses PALE to conserve recording surface and to permit digital chrominance separation (see Par. **20-85**).

In the future, fully digitized color television systems will undoubtedly be built. The signal will be digitized at the television camera and remain in digital form up to the final output of the studio and the input to the transmitter. In such digital systems much attention is being paid to *component recording*.[17] Component recording is especially attractive from the standpoint of editing and manipulating the video signal. All frames are by definition identical, and therefore one does not have to circumvent the difficult problems of subcarrier phase changes (over eight fields for the 625-line systems). This is especially true of the SECAM system, which is impossible to edit in the encoded form, as the chrominance information is carried as an FM signal. In the component system several versions have been suggested, all of which entail bandwidths at least as high as the equivalent of the 4 times subcarrier sampling systems.

The problem of recording at these very high bit rates (see Table 20-6) would at first sight appear insurmountable, but to record a given binary level does not require so great a signal-to-noise ratio for a given error rate as an analog signal does[18] (see Table 20-7). Rapid progress has been made in applying digital methods to tape recording, supported by years of experience in the computer industry. Commercially available longitudinal digital machines used in the computer and data recording field have been available for a number of years with bit rates of 100 Mb/s.

The first demonstration of a practical digital video recording system with all the necessary

TABLE 20-6   Sampling and Bit Rates of NTSC, PAL, and SECAM Systems

System	$3f_{sc}$ Sample rate, MHz	$3f_{sc}$ Peak bit rate, Mb/s	$4f_{sc}$ Sample rate, MHz	$4f_{sc}$ Peak bit rate, Mb/s
NTSC	10.7	85.6	14.3	114.4
PAL, SECAM	13.3	106.3	17.7	141.8

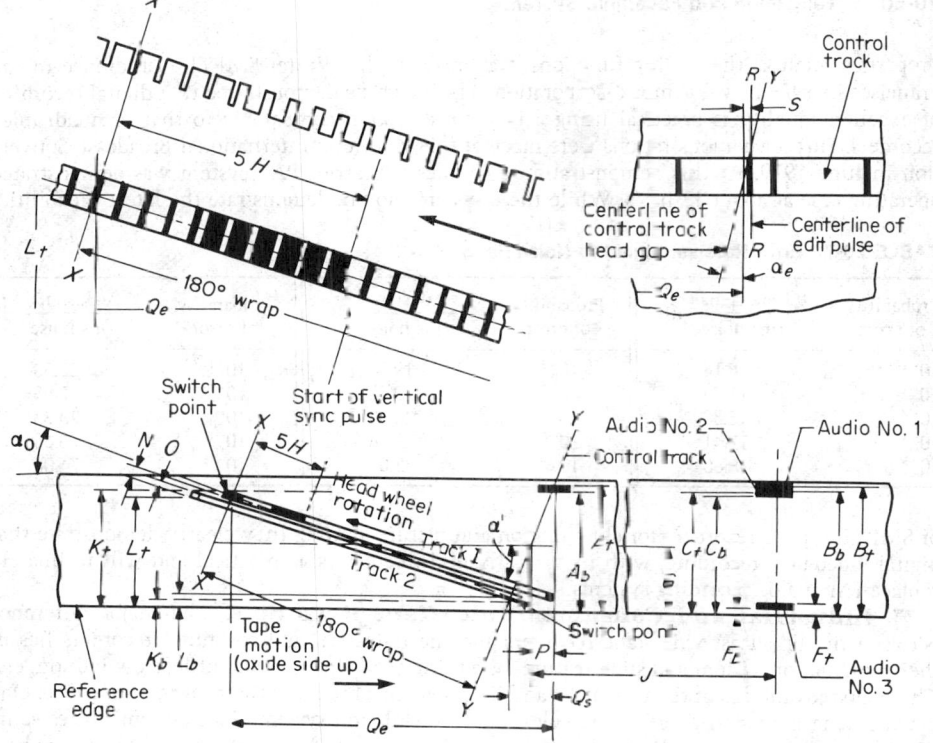

Fig. 20-62. SMPTE 1-in segmented 1-in format.

		mm			in		
	Dimensions*	Min	Max	Ref	Min	Max	Ref
A_b	Control track	23.55	26.65		0.9272	0.9311	
A_t	Control track	23.95	24.06		0.9429	0.9472	
B_b	Audio 1 track	24.35	24.45		0.9587	0.9626	
B_t	Audio 1 track	25.15	25.25		0.9902	0.9945	
C_b	Audio 2 track	22.35	22.45		0.8799	0.8839	
C_t	Audio 2 track	23.15	23.26		0.9114	0.9157	
F_b	Audio 3 track	0.195	0.205		0.00768	0.00807	
F_t	Audio 3 track	0.990	1.01		0.03898	0.0398	
G	Center of video tape			12.70			0.5000
J	Position of audio heads	232.0	233.0		9.134	9.137	
K_b	Full video width	1.18			0.0465		
K_t	Full video width		22.19			0.8736	
L_b	Video width (180°)	1.82			0.0717		
L_t	Video width (180°)		21.55			0.8484	
N	Video track pitch			0.200			0.00787
O	Video track width	0.155	0.165		0.00610	0.00650	
P	Position of control head	2.84	2.88		0.1118	0.1134	
Q_b	Switch point video track 2	82.096	82.121		3.23213	3.23311	
Q_t	Switch point video track 1	5.523	5.533		0.21744	0.21783	
S	Distance between control-track head gap and center edit pulse at 180° switch point			0.040			0.00157
α^0	Scanning angle			14.434°			
α^1	Video track angle (525/60)			4.288°			

*b, t are the dimensions from the reference edge to the bottom and top of the record, respectively.

properties, such as the editing functions, was made at the Winter SMPTE Conference in San Francisco in 1979 by the Ampex Corporation. This machine demonstrated that digital recording of excellent quality is practical using a tape consumption comparable to that of quadruplex recorders. Further demonstrations were made at the Montreux International Broadcast Convention in June 1979. At this demonstration a 4 times subcarrier PAL system was demonstrated operating tape at about 12 in^2/s. While these systems do not demonstrate the latest capabilities

TABLE 20-7   Error Rate vs. Signal-to-Noise Ratio

Probability of error	$\dfrac{\text{Peak signal}}{\text{rms noise}}$, dB	Probability of error	$\dfrac{\text{Peak signal}}{\text{rms noise}}$, dB	Probability of error	$\dfrac{\text{Peak signal}}{\text{rms noise}}$, dB
10^{-1}	8.16	10^{-6}	19.54	10^{-11}	22.53
10^{-2}	13.33	10^{-7}	20.32	10^{-12}	22.95
10^{-3}	15.80	10^{-8}	20.98	10^{-13}	23.33
10^{-4}	17.41	10^{-9}	21.56	10^{-14}	23.68
10^{-5}	18.60	10^{-10}	22.07	10^{-15}	24.00

of SMPTE type C recorders for slow motion and picture shuttle, they clearly demonstrate that digital video-tape recording, with its virtually invisible errors, is practical and will ultimately replace FM analog recording systems.

77. Industrial and Consumer Video-Tape Recorders. The major difference between the broadcast video-tape recorders and the industrial and consumer recorders lies in their modulation system and such features as editing capabilities, fast shuttle, slow motion, etc. The major advantages of the consumer and industrial machines are their low cost and size. One major saving is made by use of the color-under modulation system. This system makes some sacrifice in picture quality, but the performance is adequate and it avoids the elaborate time-base correction needed in direct-color modulation systems.

78. IVC 800 Industrial Helical Recorders. Among the earlier helical recording systems used for industrial applications is the IVC Corporation 800 series. In this helical-scan recorder, described by Reynolds,[19] the tape speed is 6.91 in/s and the head writing speed is 723 in/s. The video signal is recorded on the tape at an angle of 4°45′ and produces a long slanting track across the tape. The track width is 6 mils, and the guard band between tracks is 3.6 mils. The first audio signal (number 1), near the top edge of the tape, is recorded at an angle of 25° and is 39 mils wide. This angle and the angle of the video track make a total of almost 30° angular difference between the two tracks. The center of this audio 1 track is 100 mils in from the edge of the tape. Figure 20-63 shows the format of the IVC helical-scan recording. The control track is near the bottom edge of the tape. Like audio track 1, it is recorded at a 25° angle, 39 mils wide, with the center 100 mils from the edge. Audio track 2 is also at the bottom edge of the tape but is not recorded at an angle. The video is recorded over the audio to conserve tape. This precludes editing the video without rerecording the audio.

79. SMPTE Type E ¾-in (U-matic) Format. One of the most successful series of machines introduced in the early 1970s by Sony Corporation is the U-matic recorder, which has found widespread use in consumer, industrial, and even broadcast applications. Its major advantages for the industrial and broadcast user are its low cost, small size, and low weight compared with all the formats described above.

This format uses a cassette of ¾-in tape running at 3.752 in/s (95.3 mm/s). The head drum diameter is 4.3307 in (110 mm), writing one field per rotation at 404 in/s (10.26 m/s). The mechanical dimensions[20,21] are shown in Fig. 20-64.

80. Consumer Video Recorders. By the latter half of the 70s, a number of different consumer-type video recorders had reached the market. Table 20-8 shows the principal characteristics of available and demonstrated formats. Unlike optical video-disc recording systems (see discussion starting with Par. **20-86**), these machines share the advantage of being able to record as well as playing back color programs. Many versions of the machine are available with such features as microprocessors, which can remember many different times to start and end recordings on several TV channels. This allows preprogramming of recording material as much as 2 weeks in advance. Table 20-8 also lists the cassette dimensions and the recording times in 1979.

81. Longitudinal Video Recorders. The earliest work on video recording was undertaken utilizing the longitudinal principle. One such system was used on the air in Europe by the BBC in 1957 but was abandoned because of the very high tape speed and tape consumption. Also, the first demonstration of quadruplex recording had occurred at the NAB convention in Chicago in 1956.

In 1979, two new longitudinal systems were demonstrated, one at the Consumer Electronics Show in Chicago by Toshiba of Japan and the other by BASF at the Berlin Radio Show. The two approaches are quite different but share the significant advantage of simple mechanical design.

82. Toshiba LVR System. This system[22] uses an endless loop similar to the eight-track audio cartridge. The cartridge is stationary, but the tape is made to rotate inside the cartridge at a speed of 215 in/s (5.5 m/s) (see Table 20-8). The transport layout and tape format are shown in Fig. 20-65. The lowest track, measuring 1 mil (25 μm) wide, is recorded in one pass of the tape in about 25 s. Then the head is stepped one track pitch 1.4 mil (35 μm) to record the next track. The perturbations in tracking which occur during the splice in the continuous-loop tape produce very small errors in video performance.

The TV signals are recorded by the conventional color-under technique, except that the audio signal (which is recorded by the video head) is multiplexed as an FM signal with a carrier frequency of 1.5 MHz. This places the audio carrier between the chrominance subcarrier of 688 kHz and the FM luminance signal, which is at 3.5 MHz on sync tip and 5.4 MHz on peak white.

One of the advantages of this system is that very fast access can be obtained to any track. All that is required is to step the head across the ½-in width of the tape. The transport of the system is very simple, consisting of a capstan, pinch roller and one video tip, mounted on a carriage that allows it to move laterally across the tape. A further advantage of this system is that high-speed duplication of tapes is possible, as the tracks can be recorded with a multiple-head stack, such as a thin-film head assembly. This allows all tracks to be recorded at the same time, reducing the total recording time for one tape by about two orders of magnitude.

83. BASF LVR System. The concept of the BASF LVR system is based on an extremely fast stop-start longitudinal tape transport. The head-to-tape speed is 4 m/s (160 in/s); i.e., it uses the shortest wavelength of any tape recorder to date. The transport reverses at the end of each pass of the tape. During this reversal, stated to be 100 ms, the video level is brought to gray to minimize the interruption. Since the sound is also carried by the video head, an interruption of some 20 to 30 ms results. As in the Toshiba LVR system, rapid mass copying is possible by using multitrack heads, and the transport is less complex than those of conventional helical machines.

84. Broadcast Magnetic Analog Video-Disc System. The earliest commercially available video-disc recording system was introduced by Ampex Corporation for slow-motion

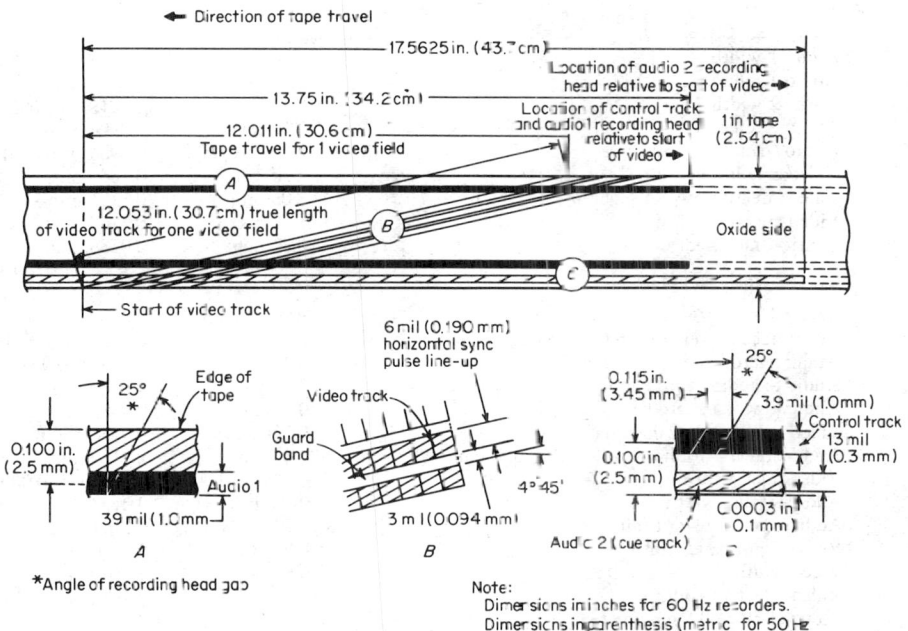

Fig. 20-63. IVC 800 tape format.

playback of sports events. This system employs two 16-in aluminum-substrate discs precision-lapped to a high degree of flatness. The disc is electroplated with a thin layer of nickel-cobalt alloy, which is used as the recording medium. A few microinches of rhodium are plated on the surface to protect it from the atmosphere and to offer a very smooth, hard surface for the heads to ride on.

Two discs are used on each drive, and both sides of each disc are used for recording. The hub is rotated at 3,600 r/min, recording one field per rotation of the spindle for the NTSC system. This gives a storage capacity of approximately 30 s of motion. The heads are so arranged that if head 1 on the top surface of the disc is used to record field 1, during the vertical interval head

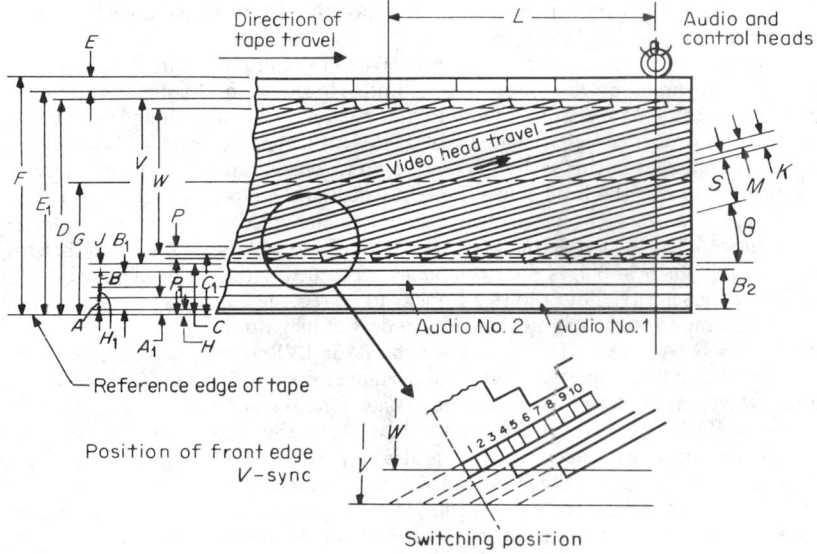

Fig. 20-64. ¾-in SMPTE type F tape format (U-matic).[20]

Dimensions		mm	in
A	Audio 1 width	0.80 ± 0.05	0.0315 ± 0.0020
A_1	Audio 1 reference	1.00 nom	0.0394 nom
B	Audio 2 width	0.80 ± 0.05	0.0315 ± 0.0020
B_1	Audio 2 reference	2.50 nom	0.0984 nom
B_2	Audio track total width	2.30 ± 0.08	0.0906 ± 0.0031
C	Video area lower limit	2.70 min	0.1063 min
C_1	Video effective area lower limit	3.05 min	0.1201 min
D	Video area upper limit	18.20 max	0.7165 max
E	Control track width	0.60 nom	0.0236 nom
E_1	Control track reference	$18.40 \begin{array}{l} +\ 0.28 \\ -\ 0.18 \end{array}$	$0.7244 \begin{array}{l} +\ 0.0110 \\ -\ 0.0071 \end{array}$
F	Tape width	19.00 ± 0.03	0.7480 ± 0.0012
G	Video trace center from reference edge	10.45 ± 0.05	0.4114 ± 0.0020
H	Audio guard band to tape edge	0.2 ± 0.1	0.008 ± 0.004
H_1	Audio-to-audio guard band	0.7 nom	0.028 nom
J	Audio-to-video guard band	0.2 nom	0.008 nom
K	Video track pitch (calculated)	0.137 nom	0.00539 nom
L	Audio and control head position from end of 180° scan	74.0 nom	2.913 nom
M	Video track width	0.085 ± 0.007	0.00335 ± 0.00028
P*	Address track width	0.50 ± 0.05	0.0197 ± 0.0020
P_1	Address track lower limit	2.90 ± 0.15	0.1142 ± 0.0059
S	Video guard band width	0.052 nom	0.00205 nom
Y	Video width	15.5 nom	0.610 nom
W	Video effective width	14.8 nom	0.583 nom
θ	Video track angle, moving tape	4°57′ 33.2″	
	stationary tape	4°54′ 49.1″	

*For reference value only.

TABLE 20-8 Characteristics of Consumer Videotape Recorders*

	U-matic	EAILI	Beta	VHS	Philips V2000	Toshiba LVR	BASF LVR
SMPTE type	E	F	G	H			
Tape width, in	0.75	0.5	0.5	0.5	0.5	0.5	0.33
mm	19	12.65	12.65	12.65	12.65	12.65	8
Head drum diam, in	4.33	4.57	2.95	2.44	2.56		
mm	110	116	75	62	65		
Video track width, mils	3.35	4.33	2.3/1.15/0.77	2.3/1.15	0.89†	1.0	3.94
μm	85	110	58.5/29.2/1.95	58.5/29.2	22.6†	25	100
Head-to-tape speed, in/s	410	437	276	229	200	217	158
m/s	10.4	11.1	7.0	5.8	5.08†	5.5	4
Audio track width, mils	32	39	41	41	26	With video	With video
mm	0.8	1	1.05	1.0	0.65	With video	With video
						Same as head-to-tape	Same as head-to tape
Tape speed, in/s	3.75	7.5	1.5/0.79/.52	1.3/0.66	0.96†	3.9	3.6
cm/s	9.53	19.05	4/2/1.33	3.34/1.67/1.12	2.44	5.4	5.0
FM carrier, MHz, sync	3.8	3.1	3.5	3.4	3.3		
White	5.4	4.4	4.8	4.4	4.8†		
Chrominance subcarrier, kHz	688	767	688	629	625	688	719
Cassette or reel size, in	7.3×4.8×1.2	4.7/25 diam×0.79	6.1×3.8×1	6.4×6.1×1	7.25×4.2×1	5.9×5.2×1.4	4.5×4.2×0.67
mm	186×123×31	120 diam×20	156×96×25	162×104×25	185×107×25	150×132×36	114×106×17
Playing time,‡ min	60	60	60/120/180	20/240	2×240	120	180
Tape thickness,‡ mils	1.1	1.2	0.8	0.8	0.3	0.95	0.33
μm	27	30	20	20	13	24	8.5

* Tape-wrap angle is 180° on all models.
† For the PAL Standards one-half of the tape is recorded in each direction.
‡ Thinner tapes permit longer playing time. The longest time available in the VHS system beginning in 1979 was 6 h of recording.

2 on the second surface is switched to record. Subsequently field 3 is recorded by head 3 and field 4 by head 4.

As each head can be independently moved, the heads are not used to record adjacent tracks on the disc. Instead they are made to jump two track pitches after recording. This means that only alternate tracks are recorded while the heads are stepped towards the center. The interleaved tracks are subsequently recorded when the heads are stepped in the outward direction. This allows a continuous recording of the incoming signal to be made, with a total recorded time of 30 s stored on the disc ready for stop-motion or slow-motion replay as called for by the director.

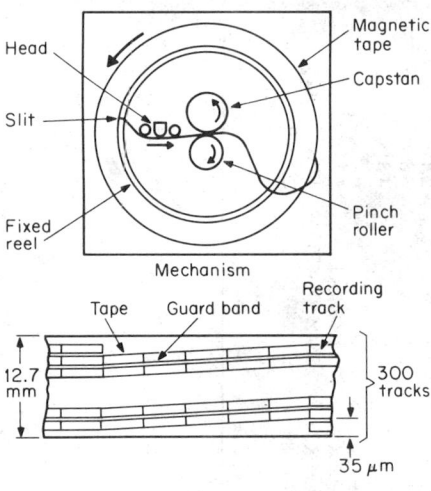

Fig. 20-65. Toshiba longitudinal video-tape-recorder mechanism and format.

The signal system of the HS-100 slow-motion recorder is conventional high-band direct-FM-color recording, as used in most broadcast-quality recorders (see Par. **20-70**).

When a television camera signal source is used with the HS-100 system, displaying a complete frame would be unsatisfactory because of the flicker and serrations in vertical lines caused by motion. To overcome this a single field is displayed twice. The necessary half-line offset of the second replay of the field is obtained by an ultrasonic video delay line.

A fundamental problem of all disc systems which display one field is the change in chrominance subcarrier phase of the PAL and NTSC systems from field to field. In the NTSC system a complete cycle is repeated every four fields, while in the PAL system the complete sequence of subcarrier phase change is repeated every eight fields. All video-disc systems must therefore artificially restore ("reinvent") the subcarrier phase relationships irrespective of whether a field or a frame is being replayed.

85. Broadcast Magnetic Digital Video-Disc System. In recent years the use of computer-type disc recording techniques with heads that ride from 20 to 30 μin above the surface of oxide-coated rigid discs has become practical for video recording purposes. The advantage of digital recording technology is that the recording is virtually transparent; i.e., the input and output of a digitally recording system are identical, with negligible addition of noise or phase errors.

The earliest version of a digital video-disc recording system consisted of a still store which recorded 814 frames with a single-disc drive system, or 1,628 frames with a two-drive system. Subsequently a second generation of machines, called the Ampex ESS 2, has been modified to permit each on-line disc drive to record 27 s of real-time video or 814 still frames, giving a performance very similar to the HS-100 in respect to slow motion and real-time video but with the quality of digital recording.

The computer pack of 10 discs (20 surfaces) employs eight surfaces to record the 8-b digital code used to describe the amplitude of each picture element. The digital sampling rate is 3 times the chrominance subcarrier. This gives an effective signal recording rate of 86 Mb/s.

Sixteen disc surfaces are used to record all the video data. Eight surfaces are used in parallel to record the 8-b video code. One disc surface at the center of the pack is used to record tracking reference signals. The top surface is considered to be too exposed to dirt for satisfactory use. The remaining two disc sides are available for housekeeping or customer data.

86. Nonmagnetic Video-Disc Recording. Two types of practical nonmagnetic disc recorders have been demonstrated. One uses optical record and re-play transducers. The other uses a disc having engraved grooves with mechanical or capacitive readout. Both use the technology of the phonograph industry to produce inexpensive and rapid replication on a plastic recording medium, and both record the video information in FM form in a spiral series of pits or hill-and-dale recordings. Another common characteristic of these systems is their high packing density. Systems presently being marketed record at least 1 h of video on two sides of a 12-in disc.

The advantages of these disc machines, compared with video-tape recorders, lies in the cost of the machine and the size and cost of the recording medium. A further advantage is that very fast access can be made to any part of the recorded program. This is achieved by moving the playback transducer radially across the disc surface, rather than moving many hundreds of feet of tape past the reading head, as in magnetic tape recorders.

Presently none of the consumer disc players offers recording capability. Record and re-play disc machines can be built, but they have been demonstrated only for broadcast and computer digital storage. The problem is that at present they require considerable laser-beam energy for real-time recording[23-25] and (in the case of the mechanical record) are large and expensive.

87. Teldec Mechanical Video-Disc System. One of the earliest video-disc recording systems marketed in Europe was a joint venture of Telefunken and Decca, called Teldec. It attracted attention in the late 1960s and early 1970s. This system is analogous to an audio (phonograph) recording system in that the disc is prepared by cutting a master with a recording-track density of 120 to 140 grooves per millimeter, i.e., about 13 times as many tracks as those of a 33 r/min LP recording.

The readout system, also analogous to a phonograph stylus, is shown in Fig. 20-66. In place of a single stylus, a skid, constructed of diamond, transmits pressure variations to a piezoelectric element which converts the pressure variation into the video signal.

The disc rotation was locked to the European 50-Hz power system rotating at 1,500 r/min. This recording system could not record in real time because mechanical machining of the surface is required. Therefore, some form of video buffering was required to manufacture the master. The modulation system used in the Teldec system was somewhat analogous to color-under magnetic tape recorders, with a chrominance subcarrier of about 1 MHz. Unlike the standard color system, however, the chrominance was line-time-sequential. In the playback unit appropriate time delays were used to reconstruct the color image. This system is no longer in production.

88. RCA Selectavision. A variation of the mechanical recording system is the RCA Selectavision system.[25,27,28] It uses 12-in discs, both sides of which have spiral grooves of V-shaped cross section. Figure 20-67 shows the groove surface of the electrically conductive disc and the diamond tip of the stylus used for tracking the groove. A metallic layer on the rear of the diamond stylus detects capacitance variations as the peaks and valleys of the recorded groove are encountered. To increase the recording time, the disc is rotated at an optimum rate for the recorded wavelength (in this case 450 r/min), which allows four frames to be recorded per disc revolution.

The mastering process originally used electron-beam exposure of an appropriate emulsion.[29] Subsequently lithography techniques were used to produce a metallic master. Later versions have refined the use of mechanical means to form the grooves in a metallic plate which acts as the die for the injection molding process, as in phonograph-record production.

The signal system is frequency modulation, using 4.3 MHz for sync tips and 6.3 MHz for peak white. The chrominance and sound subcarriers are placed below the FM luminance carrier to minimize cross talk. Table 20-9 summarizes the characteristics of the RCA Selectavision system.

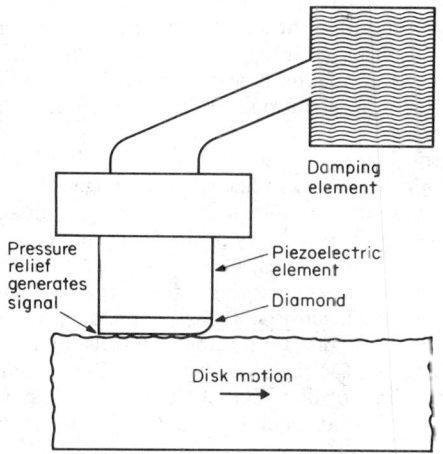

Damping element

Pressure relief generates signal

Piezoelectric element

Diamond

Disk motion

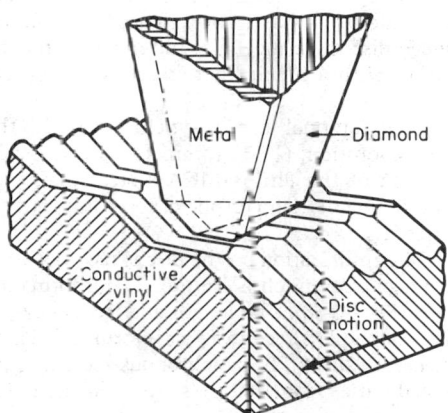

Metal

Diamond

Conductive vinyl

Disc motion

Fig. 20-66. TELDEC disc and pressure transducer.

Fig. 20-67. Selectavision stylus with metal electrode and segment of disc. (RCA.)

89. Matsushita Video Home Disc (VHD). This is a grooveless system that uses a capacitance pickup probe to detect the variations of capacitance of hill-and-dale recording. As the system is grooveless, the tracking is not provided by mechanical means, as in Selectavision (see Par. **20-88**). Two pilot tones, recorded on adjacent tracks, are used to detect tracking errors and to position the flat capacitance stylus to the center of the tracks.[30] The Matsushita disc is run synchronously, allowing the capacitance detector to jump tracks and to produce fast-motion effects. Slow motion and stop motion can be achieved within the limitation of a four-field repetition rate.

TABLE 20-9 Summary of Video Disc Parameters

	Magnavox (Philips) DVA	Thomson CSF	RCA Selectavision	Matsushita VHD
Disc diameter, mm	300	300	300	260
in	12	12	12	10.2
Disc thickness, mm	2.5	0.15	1.8 at inner &	1.2
mil	98	6	70 outer diam.	0.05
Rotation rate, r/min	1,800 (1,500 PAL)	1,500 PAL	450	900
Min/max recording diameter, mm	110/290 (130 PAL)	Not known	145/290	Not known
in	4.3/11.4 (5.1 PAL)		5.7/11.4	
Playing time, min	60 (30 per side)*	60	120 (60 per side)	120 (60 per side)
FM carrier sync/peak white, MHz	7.5–9.2	Not known	4.3–6.3	6.1–7.8
Chrominance bandwidth, MHz	Full NTSC	SECAM Type Mod.	0.5	0.5
Video bandwidth, MHz	4.2	SECAM Type Mod.	3	3.1
Audio carrier, MHz	2.3 & 2.8	Digital in Sync	0.716 & 0.905	3.43 & 3.73
Audio FM deviation, kHz	±100		±50	
Disc turned for second side	Yes	No	Yes	Yes
Sealed recording surface	Yes	No	No	No
Recording/playback	Optical/optical	Optical/optical	Mech./capacitive	Optical/capacitive

*A constant linear velocity (11 m/s) version plays for 120 min (60 min per side).

90. Optical Video Disc. The earliest commercially available optical disc system in the United States was the Philips-MCA video-disc system, called Discovision.[31-33] This system is worth describing in some detail since it differs substantially from other optical disc systems.[34,35] As mentioned in Par. **20-86**, the major advantages of the nonmagnetic disc systems are their high packing density and low cost of replication.

The Philips-MCA video disc, similar in appearance to a phonograph record, consists of a transparent plastic material with a diameter of 12 in (30 cm) and a thickness of 100 mils (2.5 mm) for a double-sided recording. It differs from a phonograph record in the structure of the information. The audio disc has grooves, whose walls move a transducer to reproduce the audio signals. The video disc must meet a requirement of much higher information density, and it therefore has tracks with a much finer structure. The groove spacing is about one-sixtieth that of an audio disc.

The rotational speed is synchronous with the vertical frame rate, to record one complete frame per revolution, (1,800 r/min for NTSC, 1,500 r/min for PAL and SECAM). A longer-playing version of the Philips-MCA system maintains a constant groove-to-pickup velocity, allowing approximately 1 h per side of recording, but eliminating the possibility of slow- and stop-motion effects.

The information on the disc is recorded as a spiral track starting at the inside of the disc. The average track pitch is 1.6 μm. The information in the track consists of small depressions called *pits* (Fig. 20-68). The length and frequency with which the pits are recorded controls the crossings of the FM recording waveform, similar to high-band FM magnetic tape recording systems (see Par. **20-70**). In synchronous recording (constant rate of disc rotation) the track-to-pickup speed varies from 400 in/s (10 m/s) at the inner diameter to about 1,000 in/s (29 m/s) at the outer tracks.

Production Process. The production process for the Discovision video discs is analogous to that used for phonograph recording.[36,37] First a master recording is made on a glass disc covered with a photosensitive layer. The FM coded signal modulates a laser beam, which records the information on the surface of the disc by recording the positive excursions of the waveform in real time. The exposure process is followed by a developing process, which leaves the pattern of pits on the master; from it a series of stampers is produced by a galvanic process. Pressing (injection molding) of a thermoplastic medium produces replicas that conform to the master stampings. A thin aluminum coating of about 1.6 μin (0.04 μm) thickness is deposited on the information side of the disc. A second recording similarly coated with aluminum is mounted back to back to form the 12-in replica, as shown in Fig. 20-68.

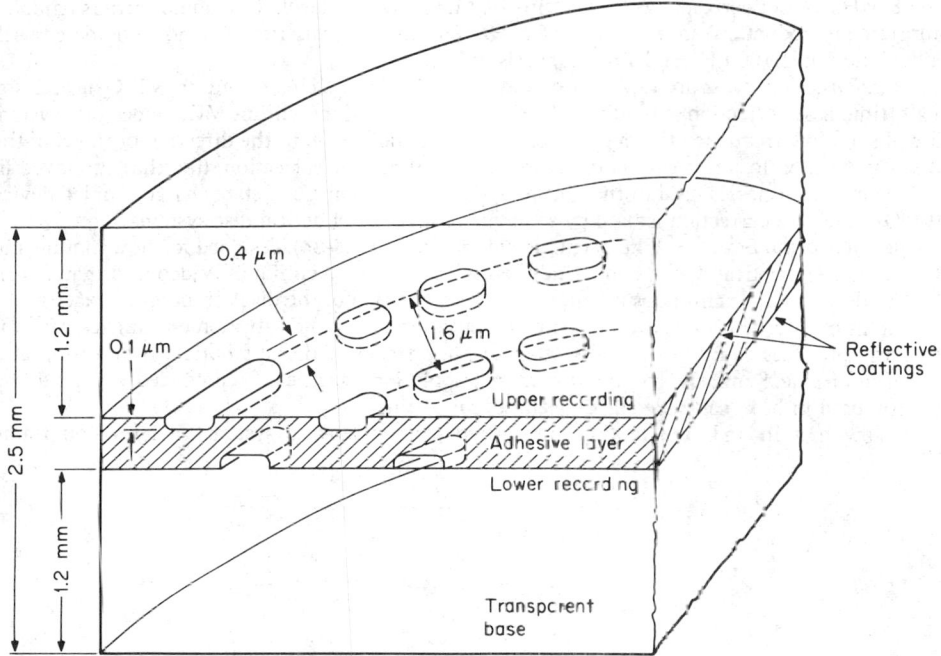

Fig. 20-68. Section of Philips-MCA optical disc.

A feature of the Philips-MCA system is the protective coating which is the base material of the disc itself. As the pits containing the recorded information are extremely small, it is necessary to use a lens of large numeric aperture to form a correspondingly small diffraction-limited readout spot. Thus the focal depth is very short. This disadvantage has been turned into an advantage by using this base material. As the surface of the disc is about 1 mm removed from the recording proper, the surface is well out of focus and dust particles and fingerprints have a negligible effect on the readout.

Focus Servo.   Since the focal depth is so small, great care must be taken to maintain focus. Early attempts at accurately positioning the recording surface used air bearings but all present optical video-disc systems use automatic focus tracking. Figure 20-69 shows a method used in the Philips-Magnavox player. A quadrant detector is used in conjunction with a cylindrical lens. This yields all the information required for correcting any changes in focal distance, as well as providing the rf signal. The optics are so arranged that when the beam is at the correct focal distance, a circular spot is obtained which yields equal output on quadrants A, B, C, and D. When the medium moves too close to the object lens, the spot is elongated, so that more light reaches quadrants A and B than C and D. Similarly, if the medium moves too far from the object lens, the spot is elongated in the direction at right angles, quadrants C and D receiving more light than A and B. Therefore, $(A + B) - (C + D)$ is an error signal which can be fed back to a moving-coil motor (Fig. 20-70), which moves the object lens in the appropriate manner to keep the beam focused. The output $A + B + C + D$ yields the rf frequency-modulated signal.

Radial Tracking.   It is also necessary to track the line of very narrow pits by a closed-circuit servo. To obtain tracking information, two auxiliary light beams are formed which are slightly displaced from the centerline of the track, one on each side of the correct position. Two photodiodes are introduced into the optical paths on either side of the quadrant detectors shown in

Fig. 20-69. The error signal generated by the difference in output of these diodes, after appropriate filtering, is used to deflect a galvanometer mirror and thus to move the beam laterally to correct the error.

Modulation System.   As in magnetic recording systems, the optical video-disc system would produce unwanted variations in dc component as well as unwanted low-frequency components. To avoid these effects, frequency modulation is employed. The frequency response of the Philips-MCA optical video-disc system is such that direct FM recording can be used, allowing a relatively large frequency allocation below the normal FM channel for audio. The rf spectrum allocation of the FM system is similar to the high-band tape system (see Par. **20-70**, Table 20-5). An 8-MHz carrier corresponds to sync tip, 9.2 MHz to white level. Two audio carriers (suitable for stereo reproduction) are at 2.3 and 2.8 MHz. The audio signals are frequency-modulated with a deviation of ± 100 kHz and a preemphasis of 75 μs.

Time-Base Compensation.   As with all direct FM color systems recording NTSC or PAL signals, time-base correction is required. The velocity errors in the Philips-MCA video-disc system are of such low frequency that a galvanometer mirror deflection in the direction of travel of the disc can be used for error correction. No electronic time-base correction, like that employed in video-tape machines, is used in the Philips optical disc system. Solid-state charge-coupled-device (CCD) time-base correction is used by some manufacturers of optical disc systems.

Special Motion Effects.   Like magnetic video discs (Par. **20-84**), designed for slow motion and freeze framing, optical disc systems lend themselves to these techniques. Video information can be laid down in a synchronous manner; i.e., one revolution of the disc can produce exactly two TV fields of information. It is then a simple matter to deflect the galvanometer mirror used for tracking purposes to cause the light beam to jump tracks at the appropriate time to repeat a television frame. Similarly, by jumping tracks fast-motion effects can be produced, similar to the shuttle used in broadcast video-tape machines for editing.

If slow motion and stop motion are desired from video-disc systems, a television frame

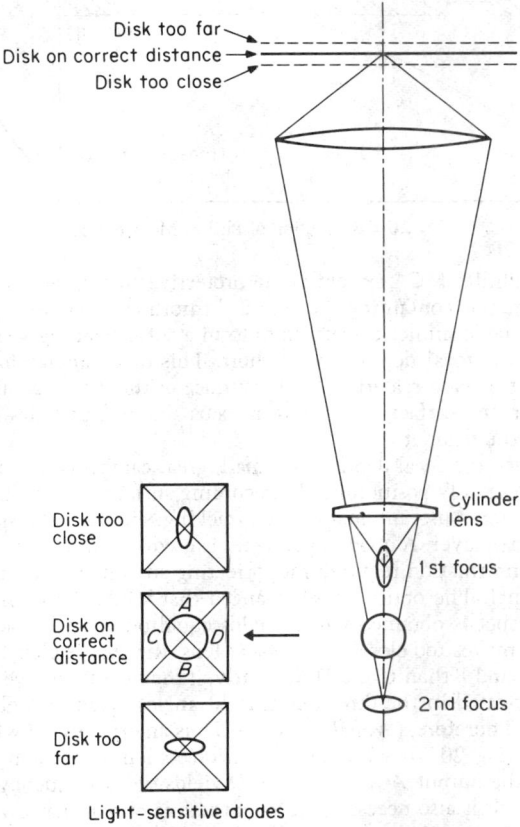

$$(A+B)-(C+D)=\text{focus signal,}\quad A+B+C+D=\text{rf signal}$$

Fig. 20-69. Focus error-detection optics. *(Philips.)*

obtained from a live television camera should not be displayed if motion has taken place, because the television camera, unlike the motion-picture camera, will retan the image of one field and then show a second field with action displaced by the intervening motion. If the original material was shot with a television camera and a complete frame is displayed, objectionable flicker between the two displayed fields will occur if any motion occurred. In the Philips-MCA entertainment system, most of the original material is shot on 35-mm motion-picture film. Care is taken in the player to replay only frames corresponding to the original motion-picture frames in the slow- or stop-motion modes.

91. Thomson-CSF Video Disc. The major differences between the Thomson-CSF video-disc player and the Philips-MCA player lie in the disc itself and in the way the sound is recorded. In the Thomson-CSF disc the medium is transparent, so that both the top and bottom surfaces of the 6-mil (0.15-mm) disc can be read without turning it over.[39] The sound is a digitized audio signal recorded during the sync period. The Thomson-CSF disc also differs from the Philips-MCA disc in having both recording surfaces exposed directly to the atmosphere. The disc is protected in a sleeve. The manufacturing process of the disc is an embossing process which should be significantly less expensive than molding.

The modulation system of the Thomson-CSF disc player is similar to most color-under systems (Par. **20-70**) but uses FM for the chrominance in a manner similar to SECAM. The major features of the system are shown in Table 20-9. This system is presently being marketed with a random-access controller for industrial and educational applications.

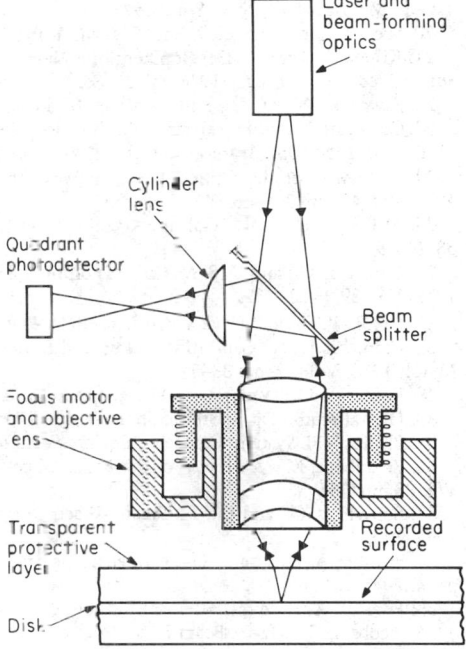

Fig. 20-70. Focus motor and detection optics. (Phuips.)

92. Bibliography and References on Video Recording

1. Robinson, I. "Video Tape Recording Theory & Practice," Hastings House, New York, 1978.

2. White, G. "Video Recording and Replay Systems," Butterworth, London, 1972.

3. Roisen, J. Project Video Tape — 25 Years of World Telev sion, *Int. TV Tech. Rev. Suppl.*, 1961.

4. Felix, M., and H. Walsh FM Systems of Exceptional Bandwidth, *Proc. EE*, September 1965, Vol. 112, No. 9, pp. 1659–1668.

5. Recorded-Carrier Frequencies Pre-emphasis Characteristics, Soc. Motion Pct. Telev. Eng., R.P. 6, *J. Soc. Motion Pict. Telev. Eng.*, December 1978, Vol. 87, p. 840

6. Dimensions of Video & Audio and Tracking of Quadruplex Recorded Tape, *J. Soc. Motion Pict. Telev. Eng.*, November 1979, Vol. 88, pp. 779–780.

7. Proposed Dual Program Audio for 2" Quad, Soc. Motion Pict. Telev. Eng. R.P. 89, *J. Soc. Motion Pict. Telev. Eng.*, June 1978, Vol. 87, p. 396.

8. Hathaway, R. A., and R. Raviso Design Philosophy of VPR1 Helical Scan Machine with Automatic Scan Tracking, *Proc. Int. Telev. Conf.*, Montreux, 1979

9. Alden, A. E. Development of National Standards of One-Inch Helical VTR System, *J. Soc. Motion Pict. Telev. Eng.*, December 1977, Vol. 86, pp. 952–957.

10. Fibush, D. K. SMPTE Type C Helical-Scan Recording Format, *J. Soc. Motion Pict. Telev. Eng.*, November 1978, Vol. 87, pp. 755–760.

11. Soc. Motion Pict. Telev. Eng., R.P.85, R.P.86, *J. Soc. Motion Pict. Telev. Eng.*, March 1978, Vol. 87, pp. 89–91.

12. Soc. Motion Pict. Telev. Eng., Type B, *J. Soc. Motion Pict. Telev. Eng.*, February 78, Vol. 87, pp. 237–245.

13. Cattermole, K. W. "Principles of Pulse Code Modulation," ILIFFE, London, 1969.

14. Oliver, B. M., J. R. Pierce, and C. E Shannon The Philosophy of PCM, *Proc. IRE* November 1948.

15. Devereaux, V. G. Pulse Code Modulation of Video Signals Subjective Study of Coding Parameters, *BBC Res. Rep.* No. 40, 1971.

16. Baldwin, J. L. E. and I. R. Lever Proposed Digital Television Standards for 625 Line PAL Signals, *IBC Conf. No. 166*, London, 1978, pp. 133–136.

17. Sabatier, J., and E. F. Kretz   Sampling the Components of 625 Line Color Television Signals, *EBU Rev.*, October 1978, No. 171, pp. 212–225.

18. Diermann, J., and K. Wallace   Digital Video Tape Recording — An Overview of Design Criteria, *IBC Conf. No. 166, London*, 1978, pp. 127–132.

19. Reynolds, K.   IVC 800 Helical Scan Recorder, in D. G. Fink (ed.), "Electronic Engineers' Handbook," 1st ed., McGraw Hill, New York, 1975.

20. Soc. Motion Pict. Telev. Eng., Type E, R.P. 87, 88, *J. Soc. Motion Pict. Telev. Eng.*, April 1978, Vol. 87.

21. Kihara, N., et al.   Development of a New System of Cassette Type Consumer VTR, *IEEE Trans. Consumer Electron.*, February 1976, pp. 26–36.

22. Sawazaki, N., et al.   Endless Tape Fixed-Head VTR, *Proc. InterMag*, June 1979.

23. Corcoran, J., and H. Ferrier   Melting Holes in Metal Films for Real Time High Density Digital Storage, *SPIE Conf. Proc. Opt. Storage Mat.*, 1972, Vol. 123, pp. 17–31.

24. Bell, A. E., et al.   Optical Recording with the Encapsulated Titanium Trilayer, *RCA Rev.*, September 1979, Vol. 40, pp. 345–362.

25. Miller, R. C., et al.   Galium-Arsenide Laser Facsimile Printer, *Bell Syst. Tech. J.*, November 1979, Vol. 58, No. 9.

26. Keizer, E. O., and D. S. McCoy   Evolution of RCA Selectavision Video Disc System, *RCA Rev.*, March 1978, Vol. 39, pp. 14–32.

27. Fox, L. P.   Conductive Video Disc, *RCA Rev.*, March 1978, Vol. 39, pp. 116–135.

28. Clemens, J. D.   Capacitive Pickup and Buried Subcarrier System Encoding for Video Discs, *RCA Rev.* March 1978, Vol. 39, pp. 33–59.

29. Keizer, E. O.   Video Disc Mastering, *RCA Rev.*, March 1978, Vol. 39, pp. 60–86.

30. Digital Audio Discs, *Studio Sound*, July 1979, pp. 50–58.

31. Philips VPL System (3 papers), *Philips Tech. Rev.*, 1973, Vol. 33, No. 7, pp. 177–193.

32. Broadbent, K.   A Review of the MCA Discovision System, *J. Soc. Motion Pict. Telev. Eng.*, July 1974, Vol. 83, pp. 554–559.

33. Bouwhuis, G., and J. J. M. Braat   Video Disc Player Optics, *Appl. Opt.*, July 1978, Vol. 17, No. 13, pp. 1993–2000.

34. Firester, A. H., et al.   Optical Recording Techniques for RCA Video Discs, *RCA Rev.*, September 1978, pp. 427–471.

35. Firester, A. H., et al.   Optical Readout of RCA Video Discs, *RCA Rev.*, September 1978, pp. 329–426.

36. Jacobs, B. A.   Laser Beam Recording of Video Disc Masters, *Appl. Opt.*, July 1978, Vol. 17, No. 13, pp. 2001–2006.

37. Winslow, J. S.   Mastering and Replication of Reflective Video Discs, *IEEE Trans. Consumer Electron.*, November 1976, pp. 318–326.

38. Dill, J. G., and C. A. Wesdorp   Control of Pit Geometry on Video Discs, *Appl. Opt.*, September 1979, Vol. 18, No. 8, p. 3198.

39. Bricto, C., et al.   Optical Readout of Video Discs, *Proc. IEEE Spring Conf. Consumer Electron.*, April 1976.

Television Image-Reproducing Equipment

BY NORMAN W. PARKER

93. Introduction. In the following paragraphs, the components and circuits used in monitors and receivers to reproduce monochrome and color images are described. The discussion is limited, in general, to those portions of receivers and monitors which follow the second detector, i.e., sync separation, deflection, subcarrier regeneration, dc restoration, color-signal decoding and matrixing, VIR circuits, video amplification, picture tubes, and their auxiliary components. The rf and i.f. aspects of broadcast receivers are treated in Sec. **21**.

94. Synchronizing Signal Separation. The synchronizing signals, transmitted at a signal voltage higher than the active picture content, are separated from the picture content by *amplitude separation* or *clipping*. Since the frequency distribution of the vertical and horizontal signals differs widely, it is possible to separate the vertical, horizontal, and color components by filtering.

The most common method of sync separation generally uses an overloaded amplifier to provide selective transmission of the sync tips, in two steps. The first step provides a signal to the sync separator of sufficient magnitude for the input-signal *operating window* to be less than the magnitude of the sync signals. The second step tracks the sync level and keeps the sync signal in the operating window. The latter step is necessary because the operating level at which the sync signals occur varies at the receiver even with the most sophisticated automatic gain control.

In most sync separators now in use the grid current or base-emitter current provides diode action in a peak-detector circuit which, combined with the overloaded amplifier, provides plate or collector conduction only during the sync pulses. A typical basic circuit of this type is shown in Fig. 20-71.

95. Typical Clipping Circuits (Amplitude Separation). In Fig. 20-71, the base-emitter diode circuit, together with R_1, R_2, and C_1, forms a peak rectifier circuit in which base-emitter current flows only during the sync-pulse interval. During this interval sufficient base current flows to force the transistor into collector-emitter voltage saturation.

The essential design considerations are as follows.

1. The video signal at the base should not exceed the base-emitter breakdown voltage. This limits the video voltage, for most low-power general-purpose transistors, to less than 10 to 15 V, since the clamping action of the input circuit applies the total video as a reverse-bias signal. Although the reverse-bias limitation does not apply to vacuum-tube circuits (which usually have much higher video voltages applied to the grid-cathode diode circuit) vacuum-tube circuits have

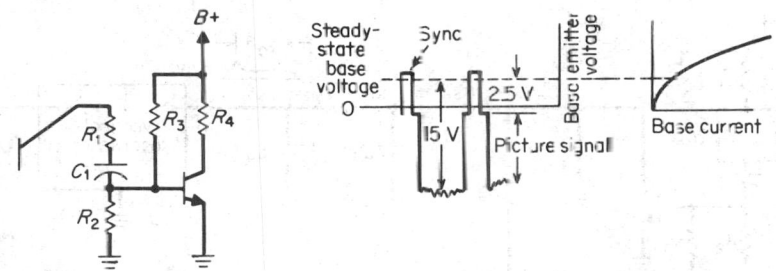

Fig. 20-71. Transistorized sync separator. (Motorola, Inc.)

a larger cutoff-to-zero-bias operating window. This requires much higher video drive to obtain satisfactory separation.

2. The charging time constant of R_1C_1 and the saturation impedance of the base-emitter diode should be much longer than the charging interval of the horizontal sync pulse, and the discharging time constant $R_1R_2C_1$ should be long compared with a horizontal interval. Maintaining appropriately long time constants establishes constant base current and provides symmetrical horizontal pulses, giving a minimum of tilt on the vertical pulses.

3. The resistor R_1 must be chosen large enough to prevent rapid charging of C_1 on short noise pulses coincident with the sync pulses. However, as R_1 is increased, more of the sync pulse appears across R_1 instead of the base-emitter circuit. Making R_1 too large eventually causes the blanking signals to replace the sync pulses.

4. The value of R_3 is chosen to provide some forward bias so that the separator continues to operate when the video goes to black, and the ac signal applied to the base-emitter circuit may be of the order of 1 V. The slight forward bias also helps prevent a reduction in amplitude of the vertical sync signals due to their increased duty cycles reducing the required charging current. The circuit can be operated with the emitter grounded and without temperature-compensating bias since the base current during the conduction period is many times the steady collector-base current and the transistor is switched into saturation during sync pulses.

To prevent noise signals from overcharging C_1 during sync intervals, a second RC network is usually included in the charge path for C_1. This uses a small capacitor which charges up rapidly on large pulses. Since it is shunted by a resistor which discharges the capacitor in approximately one line interval, a single charging pulse is prevented from causing large changes in the voltage on C_1.

96. Vertical Sync Separation. The vertical signal is a series of pulses approximately three lines in duration and occurring at the field-repetition rate. This signal has most of its energy in a frequency band below the spectrum of the horizontal sync pulses. The method in most common use provides simple low-pass filtering, typically two sections of an RC low-pass filter, each designed to cut off at 4 kHz. A separator network of this type is usually referred to as a *vertical integrator* since it places the horizontal pulses on an attenuation characteristic which increases at a rate of 6 dB/octave and can be viewed as integrating the individual pulses to derive the vertical sync pulses (see Fig. 20-72).

Since two sections of 6 dB/octave filtering are used, the circuit provides double integration of the sync signals. A filter of this type passes the first 10 harmonics of the field-frequency signal without attenuation or significant phase shift. This recovered signal is not a replica of the transmitted vertical-pulse envelope, since the three-line vertical-pulse train represents a little over 1% of the frame interval and the energy distribution of the pulse train extends beyond the first terms of the horizontal-envelope signals, having more than one-half its pulse energy above 6 kHz.

97. Vertical Deflection Oscillators. The directly triggered (multivibrator type) relaxation oscillator provides a simple and convenient means of generating vertical sawtooth voltages which give excellent phase accuracy when impulsively triggered by the integrated sync signal. The output of the multivibrator circuit can be used directly to drive the vertical-sweep power amplifier.

The relaxation type of sweep oscillator has the disadvantage that synchronism is accomplished by shortening the relaxation cycle by the action of the sync pulse. The effect is to require that the free-running frequency of the oscillator be lower than the vertical repetition rate. In the absence of sync signals, the video signal precesses with respect to the oscillator and sweep, which

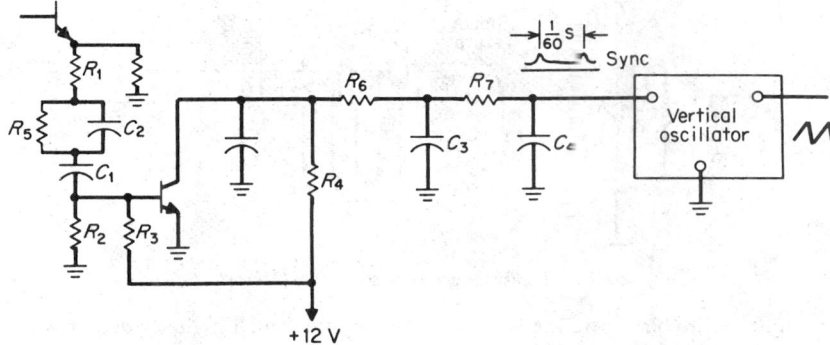

Fig. 20-72. Vertical sync separator. (*Motorola, Inc.*)

produces a continuously rolling picture on the picture tube screen. As shown in Fig. 20-73, the multivibrator becomes increasingly sensitive to pulse synchronism as the pulse time approaches the point where the oscillator spontaneously changes mode. As a result, the circuit possesses some inherent noise-gating effects.

The time required for the system to reach synchronism when a sync signal is applied at random is determined by the rate of precession of the point where the circuit can be triggered. During this time, the picture rolls slowly until the blanking bar reaches the top of the picture, at which time synchronism occurs. The closer the sync rate is to the free-running frequency, the more slowly the picture rolls when sync has been lost. On the other hand, the more rapid the out-of-sync precession is made, the higher the sync-pulse amplitude required to trigger in the proper phase, which generally makes the vertical jitter more violent in the presence of noise pulses accompanying the sync pulse. The sync-pulse level is often adjusted to satisfy the taste of the designer. A typical vertical oscillator and drive circuit are shown in Fig. 20-74.

98. Horizontal Sync Separation. Direct triggering of horizontal scanning, although used in the earliest television receivers, was abandoned because of the high signal-to-noise ratio required to ensure that the horizontal oscillator was sufficiently free of phase modulation. Since the limits of permissible phase modulation are between 3 and 5°, an oscillator with automatic phase control is used universally. The automatic-phase-control loop is characterized by a phase-comparison detector which provides a control voltage proportional to the phase difference between the horizontal oscillator and the horizontal sync signal followed by a low-pass filter that couples the output of the phase comparator to the oscillator. This oscillator is generally of the relaxation type, which can be shifted in frequency by applying a control voltage directly to the oscillator.

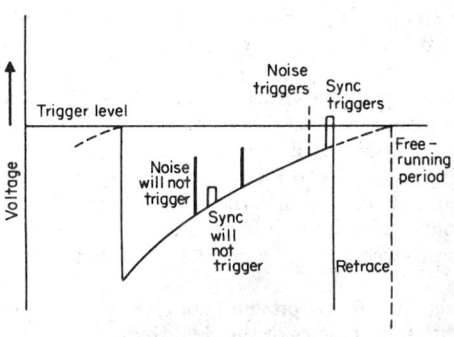

Fig. 20-73. Sync trigger sensitivity.

99. Typical Horizontal Sync Circuits. A common phase detector is the single-ended balanced diode circuit. It has the advantage of using single-ended sync and reference sawtooth voltages while at the same time deriving a control voltage free from offset voltages proportional to either input signal.

A circuit of this type is shown in Fig. 20-75. The flyback pulses are integrated to form a 2.5-V peak-to-peak sawtooth having a voltage ramp during retrace. The sawtooth voltage is applied in series with the synchronizing signals, which are much larger than the sawtooth wave (11 V peak to peak). The voltage developed by diode D_1 appears across capacitors C_{501} and C_{500}. When the automatic-phase-control system is in synchronism, the sync pulses are stationary on the retrace ramp of the sawtooth wave; they produce a voltage on C_{500} and C_{501} which is proportional to the phase between the sync signal and the sawtooth wave derived from the sweep. The positive voltage of the junction of D_1 and D_2 tends to bias D_2 into a cutoff condition, and the positive voltage is transferred to C_{516}, C_{503}, and C_{504} by R_{500}. The negative sync pulses rapidly discharge C_{516}, C_{503}, and C_{504}, keeping the voltage at the automatic-phase-control output at reference ground. The control-voltage variation is limited to the height of the ramp, or ± 1.25 V.

100. Impulse-Noise Protection. The reaction of the sync separator to impulse noise is extremely important in the performance of the deflection oscillators. Impulse noise has two important effects on synchronization: the presence of a noise impulse in the separated sync may cause modulation of the sweep oscillators, and noise impulses higher than sync level may produce temporary errors in the separation levels, causing blocks of sync pulses to be missing. The most critical of these effects is the latter, the loss of sync pulses due to separator charge-up. To

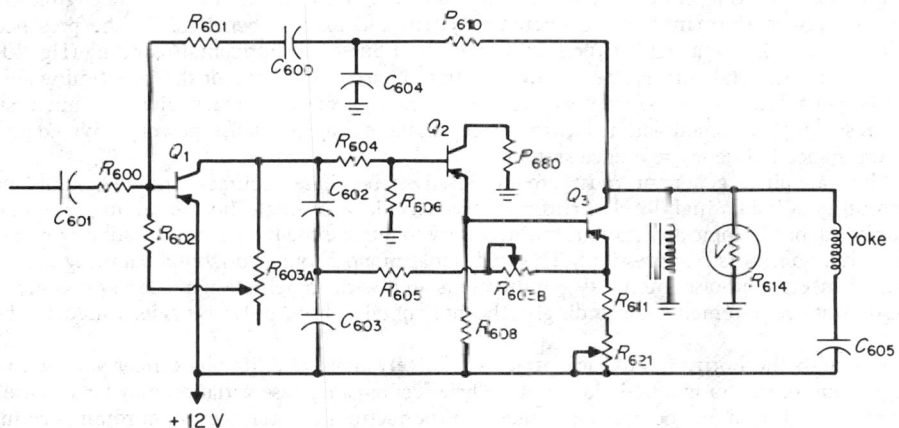

Fig. 20-74. Typical vertical oscillator and driver. (Motorola, Inc.)

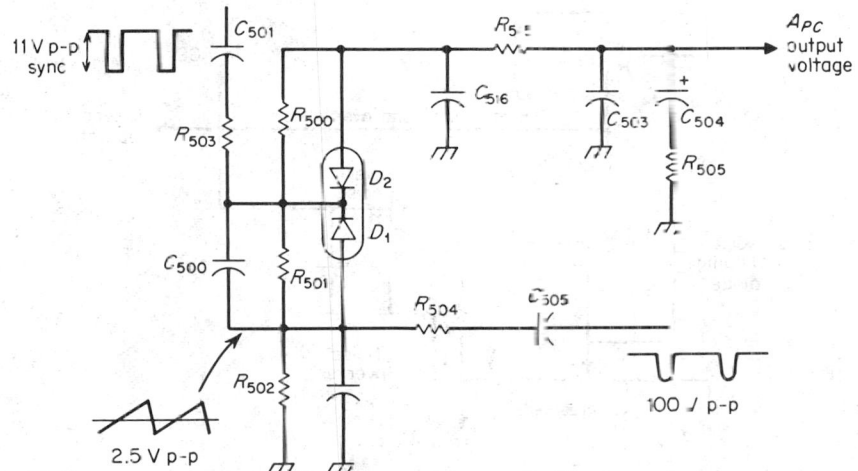

Fig. 20-75. Horizontal automatic-phase-control circuit. (Motorola, Inc.)

prevent this effect it is necessary to ensure that any noise pulses which accompany the sync signal are not appreciably higher in amplitude than the sync signals.

Noise limiting for this purpose can be accomplished by taking the sync signals from a video stage which is designed to limit at a level above sync tips. In addition, a series resistor and a double-time constant prevent sharp pulses of high amplitude but relatively low energy from shifting the bias level on the separator.

In some receivers the sync separator is gated by the sweep-retrace pulses to exclude noise pulses which occur during the active scanning time. This system, however, can have severe difficulties when the receiver is not synchronized. The sync pulses then appear as sampled fractions of the total sync, causing the system to remain locked out of synchronism. As a consequence, gated sync separators are now rarely used.

A more sophisticated system employs the principle of noise inversion. This principle uses a distinct separator which is biased to operate above the sync level and which separates the noise pulses and delivers them to a noise-inverting circuit. The inversion circuit amplifies the noise pulses and subtracts them from the sync signal. The resulting sync signal has sync pulses which are serrated by noise but do not have spikes higher than the inversion level. A circuit of a typical noise inverter is shown in Fig. 20.76.

101. Color-Burst Separation. Separation of the color-synchronizing burst requires time gating. The gate requirements are largely determined by the horizontal sync and burst specifications illustrated in Fig. 20-8b. It is essential that all video information be excluded, and it is desirable that both the leading and trailing edges of the burst be passed so that the complementary phase errors, introduced at these points by quadrature distortion, average to zero.

Widening the gate pulse to minimize the required gate-timing accuracy has negligible effect on the noise performance of the reference system and may be beneficial in the presence of echoes. The $1.2\text{-}\mu s$ spacing between trailing edges of burst and horizontal blanking (Fig. 20-8b) determines the total permissible timing variation. Noise modulation of the gate timing should not be permitted; i.e., noise excursions must not be allowed to encroach upon the burst, since the resulting cross modulation has the effect of increasing the noise power delivered to the chrominance-frequency reference system.

The gate-pulse generator must provide steady-state phase accuracy and reasonable noise immunity. When a high level of chrominance is available, a single-diode disabling gate may be employed, but in some applications it is necessary to remove the burst from the subcarrier before demodulation (burst suppression). The traditional monochrome horizontal-scanning oscillator system meets the noise-immunity requirements and, with some redesign, can approximate the steady-state requirements. Accordingly, the horizontal-flyback pulse is widely used for burst gating.

Although the horizontal-flyback pulse is relatively noise-free, its phase may vary with the adjustment of the horizontal-hold control. The effect of gate-phase variation may be to cause the burst to be clipped, or the gate may slide into the picture; i.e., the picture chrominance infor-

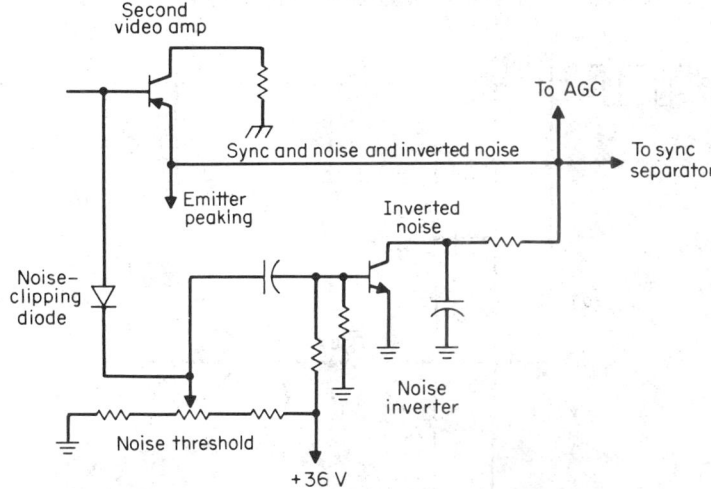

Fig. 20-76. Noise inverter. *(Motorola, Inc.)*

mation tends to serve as the reference phase. For this reason it is desirable to derive a gating signal from a delayed sync pulse which has a fixed time relationship to the burst. A circuit which derives the appropriately delayed sync pulse is shown in Fig. 20-77.

The clipped sync signal is applied to an amplifier which is coupled to the burst-gate stage through a resonant circuit. The resonant frequency of the tuned circuit is adjusted so that the rise time of the translated pulse delays the gate long enough to gate the burst only.

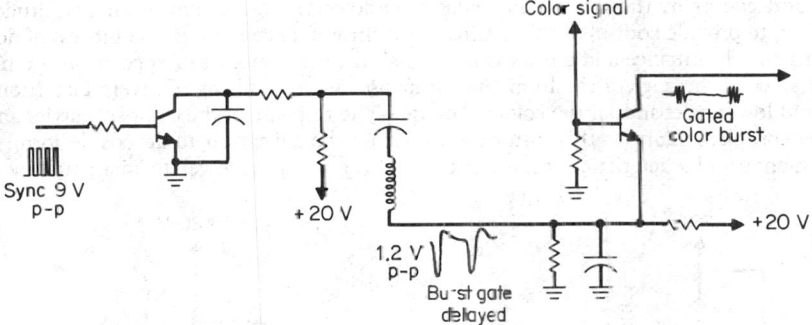

Fig. 20-77. Typical burst gate. (*Motorola, Inc.*)

102. Color Subcarrier Regeneration. The color burst provides a reference phase for correct operation of the synchronous detection process, which in turn restores the color-difference signals to baseband from their quadrature-modulation components in the color subcarrier. The method of recovering an oscillator signal with the correct phase for synchronous detection involves filtering the 3.58-MHz chrominance-subcarrier frequency from the carrier and sidebands that make up the color burst. Two techniques have been used to recover the reference phase oscillator signal. The first uses a narrow-band filter (quartz crystal) which passes the carrier component while substantially attenuating the sidebands of the color burst. The derived carrier component is amplified and limited to give a constant-amplitude reference signal. The second method uses an oscillator in a phase-locked-loop relationship with the burst signal. The phase-locked loop forms an effective filter-limiter combination. The filter width is determined by a low-pass filter in baseband combination with the phase detector and oscillator control element.

While the phase-locked loop was used almost exclusively in early receiver designs, the less complicated *ringing circuit* (crystal filter) is now a popular form of subcarrier regeneration. The ringing circuit may also be used to lock the phase of a free-running oscillator which acts as a regenerative limiter, removing any amplitude modulation which would otherwise result from incomplete attenuation of the sidebands of the burst signal.

In a ringing circuit the burst energy is spread over the entire line period, giving a ringer output voltage of about the average value of the burst level. To maintain a constant amplitude of reference signal it is desirable that the crystal continue to ring over the entire line period, so that a simple limiting amplifier can provide a constant signal. The decay that can be expected from a simple resonant circuit of a given Q, after n cycles of free oscillation, is

$$E_0/E_n = e^{-\pi n/Q}$$

where E_0 = level of voltage at beginning of free oscillation and E_n = level of voltage after n cycles of free oscillation.

At the 3.58-MHz color subcarrier and 63-μs scanning period, the circuit rings freely for about 215 cycles, so that the Q required to maintain 90% of the initial level is about 7,000. To maintain passive circuit Q's of this level requires a crystal filter.

A typical ringing-type subcarrier-regeneration circuit is shown in Fig. 20-78. The burst is applied to a phase-splitter circuit that provides in-phase and reverse-phase burst signals. The reverse-phase burst signal is coupled from the collector and used to cancel the burst signals, through the parallel crystal capacitance. The remaining signal at the junction of the crystal and capacitor C1 is a result of the burst passing through the series resonant circuit made up of the mechanical vibration and piezoelectric effects of the crystal. The low-amplitude continuously ringing signal is amplified in the following transistor stage and coupled to a free-running Colpitts oscillator, which is locked by the amplified ringing signal. The output of the oscillator is a phase-locked reference signal of constant amplitude.

103. Subcarrier Amplitude and Phase Control. Since the color information in the subcarrier contains two separate signals amplitude-modulated on the in-phase and quadrature components of the subcarrier, it is necessary to preserve precisely the phase and amplitude relationships of both the lower sideband and upper sideband of the color subcarrier. Otherwise phase and amplitude distortion of the in-phase and quadrature components will produce cross talk between the components.

The subcarrier amplifier has two principal functions: to restore the subcarrier to a form appropriate for decoding by the synchronous subcarrier detectors and to control the amplitude of the subcarrier, to provide control of the luminance-to-chrominance ratio. In the process of detecting the composite luminance and chrominance signal, the color subcarrier appears on the high-frequency slope at the edge of the luminance response curve, resulting in severe distortion of the upper and lower sidebands of the color subcarrier. The response of the color-subcarrier amplifier must be complementary to this distortion to restore the subcarrier to decodable form. Such a complementary characteristic requires critical tuning of the receiver to maintain the match.

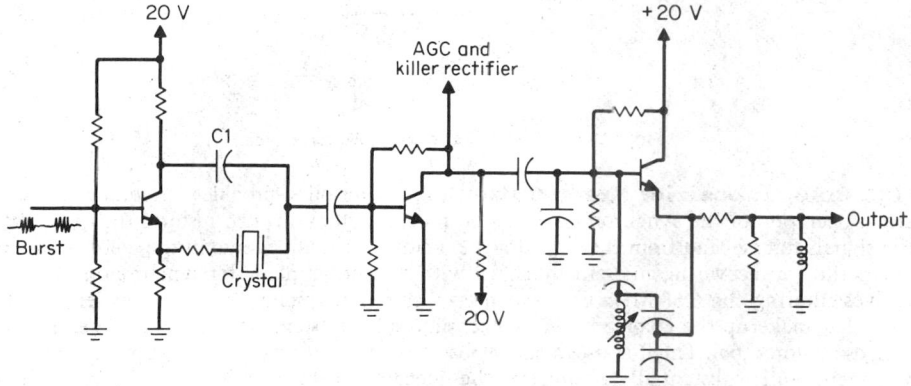

Fig. 20-78. Crystal ringing and subcarrier oscillator circuit. *(Motorola, Inc.)*

However, this technique provides the best compromise since the so-called flat-i.f. response, which does not distort the color subcarrier in the detection process, is also sensitive to tuning adjustments. Variations in the picture carrier level in the latter method cause the received signal to be overmodulated with a resulting rectification of the envelope of the color subcarrier in the video detector and a loss of saturation in the reproduced color picture.

The response of the chrominance bandpass amplifier places the color subcarrier on the rising portion of a sharply peaked response. When combined with the response of the video detector, this produces a flat response around the color subcarrier while operating the video detector with a reduced level of chrominance subcarrier. The bandpass amplifier is also designed with adjustable gain so that the level of the chrominance signal can be adjusted to provide the appropriate color saturation in the color picture. The amplifier gain can also be made to vary in response to the nominally constant level of the burst signal, to provide automatic control of color saturation.

The phase of the demodulation axis controls the hue of the reproduced image. Manual control of the hue can be accomplished by shifting the phase of either the color subcarrier or the subcarrier reference signal. Since it is more difficult to shift the phase of the subcarrier without distorting the sideband components, hue shift is usually accomplished by shifting the phase of the subcarrier reference signal. *RC, RL,* or *RLC* phase shifting may be used. Where the design allows for large shifts in hue, the phase-shift network may introduce amplitude-level shifts, which should be removed by limiting the subcarrier signal.

104. Monochrome Tube Focus and Deflection. In the past, magnetic and electrostatic methods of focusing the electron beam have been used in monochrome picture tubes, but because of its lower cost and inherent simplicity, the electrostatic method is now used exclusively (see Par. **11-97** for details). In deflection of the beam to form the picture raster, both electrostatic and magnetic methods were used, but magnetic means are now used exclusively, a change made necessary to meet the need for wider viewing area with shorter tube length.

If the center of deflection were identical with the center of curvature of the inside glass surface of the picture tube, a rectilinear raster would be obtained. However, the shape of the faceplate is determined by many other factors, and so some compromise is necessary. The *pincushion raster*, produced in wide-angle tubes by the shortened radius of deflection at the edges of the raster, can be corrected by design of the yoke windings. In color, where constraints on the yoke are more critical, separate pincushion-correction circuits are employed.

105. Horizontal Scanning Circuits. The horizontal deflection system provides the linear scanning field in the magnetic-deflection yoke. Since the deflection field is reversed 15,734 times per second, the reactive power which circulates in the yoke may be of the order of 50 W. In a yoke of high Q it is desirable to recover the circulating energy instead of dissipating it during each sweep cycle.

Linearly increasing current (and hence linear deflection) is produced in the yoke by applying constant voltage across the inductive yoke. When the current reaches a value high enough to deflect the electron beam fully, the yoke is, in effect, abruptly disconnected from the constant-voltage source. The stored energy in the yoke then collapses into the distributed capacitance of the yoke winding and into any additional circuit capacitance shunting the yoke. If the resonant frequency of the yoke inductance and these capacitances is of the order of 60 kHz, the energy in the capacitance has sufficient time during the retrace interval to flow back into the yoke. If the voltage source is then, in effect, reconnected, the energy is returned to the yoke during the initial part of the scanning period, while in the last part of the cycle energy is supplied to the yoke. Such a circuit can thus supply large circulating energy for scanning, with little real power dissipated in the process. This type of deflection system is used in all modern receiver designs.

A transistor horizontal deflection circuit of this type is shown in Fig. 20-79. Q_1 saturates to place the battery voltage across the autotransformer input terminals placing a slightly higher, but constant, voltage across the yoke-size coil combination. The current increases linearly in the yoke until the beam reaches the end of the horizontal period, at which time the transistor Q_1 is switched off. The energy stored in the yoke is transferred to the two capacitors and the distributed capacitance by resonance, resulting in a large voltage across Q_1. As the voltage swings back to zero, the current in the yoke reverses and the diode D_1 conducts, returning the energy taken on the last part of the scanning cycle to the 12-V source. The inductance of the deflection coil is chosen to provide a retrace voltage across Q_1 which does not exceed the breakdown voltage of the transistor while at the same time providing sufficient circulating energy for beam deflection.

The transistor Q_1 is chosen to provide as rapid a switch-off characteristic from its saturated condition as possible. Any lag in switch-off causes current in the transistor during the rapidly rising retrace portion of the collector voltage wave, which causes large power losses at the collector of Q_1. The power-supply voltage and yoke impedance are chosen to match the available transistor-switch characteristics.

The output transformer provides a convenient source of obtaining high-voltage power for the picture tube. A power supply of this type has three main advantages: (1) the short retrace-energy exchange results in high pulse voltages which, with a minimum of step-up, can be directly rectified to produce second-anode voltage; (2) since the pulse-repetition rate is high (15,734 Hz line-scanning frequency), high capacitance values of high-voltage filter capacitors, which are both expensive and dangerous when charged, are not required; and (3) the pulse supply is synchronous with the scanning, which avoids interference due to nonsynchronous high-voltage fields.

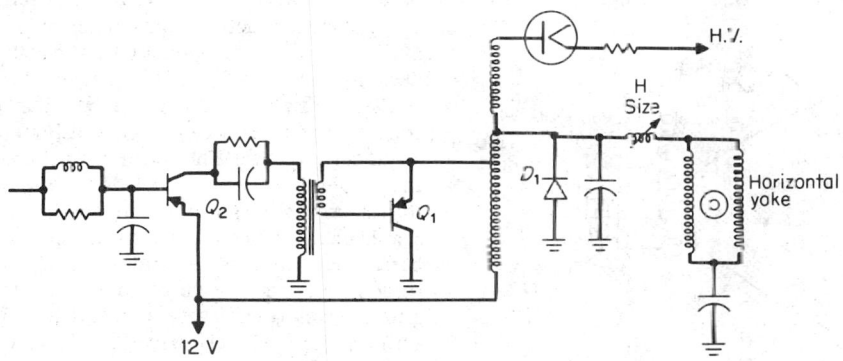

Fig. 20-79. Typical horizontal deflection system. *(Motorola, Inc.)*

106. Vertical Deflection Circuits. Two factors minimize the required scanning power for vertical deflection: (1) the television picture has a 4:3 aspect ratio with the smaller dimension in the vertical direction, which requires a proportionally smaller scanning angle for vertical deflection, and (2) the rate at which reactive energy must be exchanged between the yoke and the source is 1/262.5 of that required for horizontal scanning, with the power reduced accordingly. The result is that vertical scanning reactive power is insignificant. Instead the sweep system is governed by the resistive characteristic of the yoke, which is driven by a waveform derived from a class A power amplifier. A circuit of this type is shown at the right in Fig. 20-74, where the transistors Q_1, Q_2, and Q_3 form an asymmetrical multivibrator with the capacitors in the collector of Q_1 forming a sawtooth signal which drives the base of the power-amplifier output stage Q_3 with the emitter follower Q_2. The yoke is isolated by C_{605} to prevent the dc component caused by the operating point of class A operation from shifting the vertical position of the picture. The rate of voltage rise in the yoke is slow enough to provide essentially resistive loading during the vertical-trace period. However, during retrace an inductive pulse distorts the sawtooth wave at the collector of the power amplifier, and the pulse is resistively damped to limit the peak voltage level that the power transistor must withstand, consistent with short retrace time.

107. Color-Tube Focus and Deflection. Essentially all color tubes in current use have three beams for producing the color picture (see Sec. **11**, Par. **11-114**). Such tubes use electrostatic focus (no external focus elements are required) and magnetic deflection. The operation of the yoke and deflection circuits is similar to that used in black-and-white television, but larger yoke fields are used to accommodate the three beams and to provide a highly uniform field for each. The second-anode voltage supply is higher for color tubes to provide additional beam power to compensate for the beam energy intercepted by the shadow mask or grille.

The requirement that the three beams converge to a single focal point on the screen, equivalent to focusing a single beam of large diameter, severely limits the design parameters of the yoke. To maintain good overall convergence the pincushion correction of the yoke is minimized and additional pincushion-correction circuits, which modulate the sweep amplitude as a function of beam position, provide the necessary raster correction.

The high-voltage supply often includes a shunt regulator to provide a constant load on the horizontal-sweep and high-voltage system. When the load is not held constant, variations in beam current can cause the raster to change size as a function of average picture brightness, and the zero-beam-current high voltage may then exceed the limits of circuit parts and the color tube.

To provide correct colors at all parts of the screen the tube must be capable of providing a uniform field in each of the primary colors when only one beam is present. The purity of the red field is generally most critical. It is adjusted by rotating a weak deflecting field behind the yoke to center the three beams in the yoke-deflecting field. The yoke is also adjusted axially so that the beams arrive at the shadow mask from the proper center of deflection and only the intended phosphor is excited.

To assure that the color picture has a minimum of misregistration (color fringing) the three beams must be *converged* so that they impinge together at all parts of the screen. Static convergence is accomplished by adjusting separate magnets, located so as to affect each beam separately and thus to provide convergence at the center of the picture. In addition, dynamic electromagnetic fields are applied to each beam to ensure convergence at the edges of the picture. The dynamic-field waveforms are derived from the deflection voltages. They control the strength of the convergence action appropriately at all parts of the screen. The components on the color-tube neck are shown in Fig. 20-80.

108. Trinitron Color Tube. The Trinitron is an in-line three-gun tube using electrostatic convergence. The Trinitron differs from the three-gun shadow-mask color tubes in both the screen and the gun. The three in-line closely spaced electron beams are derived from a single-barrel electron lens. A grid structure having three holes, one for each beam, is used, and the grids are connected together. The three separate cathodes are separately driven by the R, G, and B signals. The Trinitron uses low-voltage electrostatic focus (Einzel lens). The focus is

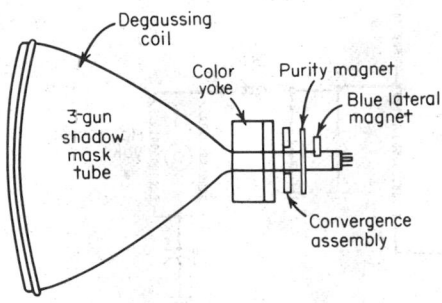

Fig.   20-80. Shadow-mask   color-picture-tube assembly.

adjusted in a manner commonly used in monochrome tubes using Einzel lens focusing; i.e., the focus electrode is returned to any of several low-voltage points. By observing the overall focus, the proper fixed voltage is chosen.

The in-line horizontal-beam arrangment must be maintained as the beam passes through the yoke to the screen if purity and convergence are to be maintained. The fringe fields of the deflection yoke tend to twist the beam since they contain components parallel to the tube axis. The twisting effect is counteracted by a coil on the neck carrying yoke current which produces a countertwist and maintains the horizontal beam alignment as the beam passes to the screen.

The three beams emerging from the spaced grid apertures tend to produce three horizontally spaced spots on the screen. However, the outer beams pass through a pair of deflection plates

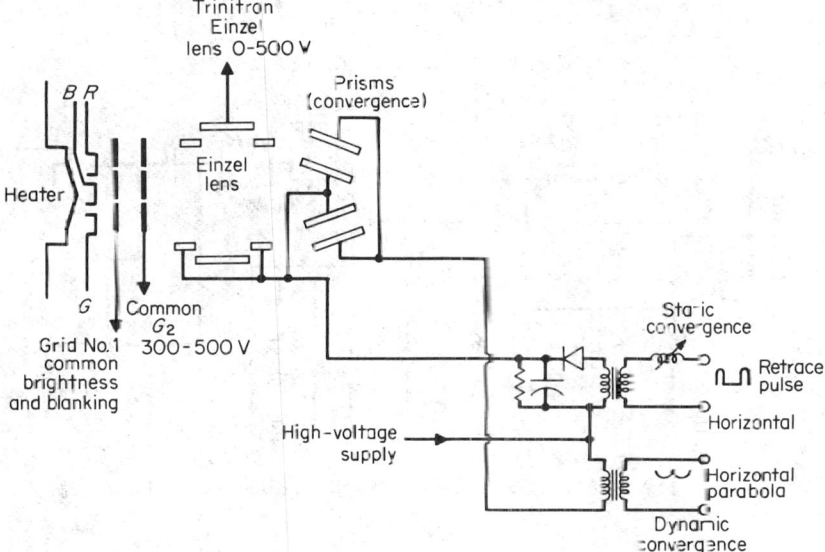

Fig. 20-81. Trinitron gun and convergence circuit.

similar to those used in electrostatically deflected tubes. When the proper voltage is applied, these plates deflect the outer beams to converge with the center beam. To maintain convergence over the tube face, the voltage on the convergence deflection plates is modulated by the scan voltages as the beam is deflected. Figure 20-81 shows gun elements and circuits necessary to operate the Trinitron tube.

Although the Trinitron gun principle can be used with a dot screen or a line screen, current Trinitron tubes use a line screen. An aperture grille, a curved plate of formed slots, provides the shadow apertures for the line screeen.

109. Monochrome Video Amplifiers. A typical television reciever has its signal amplification distributed in three frequency ranges. The tuner, with modest gain, raises the low-level rf signal by 20 to 50 times; the i.f. amplifier, operating at a fixed frequency, raises the signal to several volts; an envelope detector reduces the signal to baseband; finally the video amplifier raises the signal to from 50 to 100 V to drive the picture tube. To minimize the danger of regeneration it is desirable to distribute the gain in these sections as equally as possible, but it is difficult and expensive to increase the gain in the tuner. Since the envelope detector uses a diode which requires signals of at least 1 V to provide reasonable linearity, the largest portion of the gain falls to the i.f. amplifier. Tuners and i.f. amplifiers are treated in Sec. **21.**

The video amplifier may be used to drive the picture tube in either a grounded-grid or grounded-cathode configuration. When the tube is driven in the grounded-cathode circuit, the effects of beam current do not load the video amplifier and the loading is caused only by the capacitive input impedance of the picture tube. With grounded-cathode drive, using a vacuum tube or npn-transistor video-output stage, the operating point of the video stage is near cutoff when the picture tube approaches zero bias (maximum white). Since the drive requirements for the picture tube represent an appreciable portion of the power-supply voltage, the video stage is not linear over the entire range of the output-voltage change. The signal is compressed in the region of cutoff or (for grounded-cathode drive) in the highlights of the picture. At the same time, the picture-tube beam current shunts the video amplifier, tending to minimize white stretching. In this circuit the drive voltage is also effectively applied to the screen grid of the picture tube which reduces the curvature and gamma of the picture tube, resulting in higher average brightness for the same contrast ratio.

Comparisons of the two drive systems show that the cathode-drive circuit provides a brighter picture with less compression of the whites than the grid drive does. Most monochrome receivers use cathode drive.

The video amplifier has adjustable gain to provide customer control of contrast ratio. This is usually done by using a variable degeneration control in the emitter circuit (or cathode of a vacuum-tube circuit). A typical monochrome video amplifier is shown in Fig. 20-82.

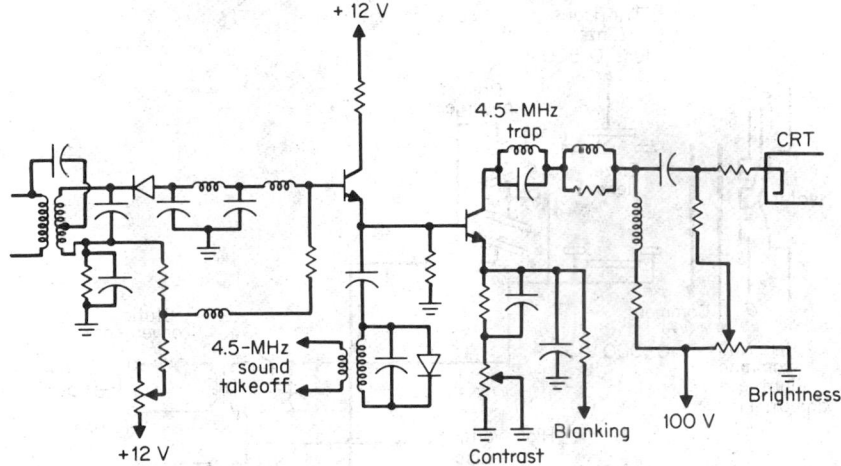

Fig. 20-82. Monochrome video amplifier. *(Motorola, Inc.)*

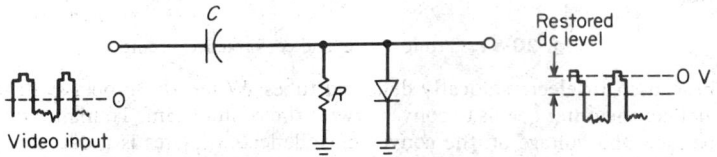

Fig. 20-83. Basic diode dc restoration circuit.

110. DC Restoration. Restoration of the dc component of the video signal can be accomplished in several ways, and it may be omitted entirely for the sake of simplicity of design and low cost. In the latter case, the manual brightness control is used to set the average signal level at the preferred setting.

The simple diode dc restorer shown in Fig. 20-83 is most frequently used. It maintains black level by reinserting the dc component on the sync tips, using the relative values required by signal standards. This type of restorer is very susceptible to error because of noise impulses and is not accurate.

Clamping circuits, although more accurate, faster-acting, and less susceptible to noise, are infrequently used in receivers because of their greater complexity and cost. In the usual circuit, the signal is sampled during the back porch to obtain an output which will bring the level of the sync tips to the correct value. This type of circuit is immune to noise except that present during the sampling interval.

111. Color Decoding and Matrixing Techniques. The color signals corresponding to red, blue, and green in the camera, that is, E_R, E_B, and E_G. are coded before being transmitted. The received signal is in three parts, E_Y, the luminance information, and E_I and E_Q, which must be transformed back into the primary signals. Usually this is done in two steps, first, deriving color-difference signals, $E_Y - E_R$, $E_Y - E_B$, and $E_Y - E_G$, then regaining E_R, E_B, and E_G.

The E_I and E_Q components of the chrominance information are demodulated by synchronous detectors referenced to the phase of the original subcarrier upon which they are modulated in transmission. The luminance and chrominance signals can be recombined in matrix networks to obtain the primary signals, or the composite signal may be directly demodulated.

Four arrangements are available, depending on the combination of the following choices: (1) IQ vs. equiband demodulation; (2) picture-tube vs. pre-picture-tube chrominance and luminance matrixing.

The IQ receiver design utilizes two signals in quadrature. The I signal lags 57° behind the color-burst phase; the Q signal lags by 147°. The required bandwidth for the I signal is 1.25 MHz and for Q, 0.5 MHz. To compensate for the different delays associated with the different bandwidths, a time-delay circuit is necessary in the I channel. Following synchronous demodulation (Par. **20-112**) along the two axes to obtain E_I and E_Q, matrixing is necessary to obtain the color-difference signals. Both polarities of the I and Q signals are necessary for this matrix operation.

Almost all current color receivers employ *equal-bandwidth* designs, using color demodulators which decode the color information along the $R-Y$ and $B-Y$ axes, with the demodulation angles shifted sufficiently to provide acceptable color reproduction.

Vacuum-tube equiband receivers, using matrix recombination to derive the primary signals, employ *picture-tube recombination* of the luminance and chrominance signals, while receivers using solid-state elements generally employ *pre-picture-tube recombination* to provide R, B, and G signals for the picture tube. Picture-tube recombination is generally less complicated, but the color-difference voltages must often exceed twice the luminance voltage, and video-output stages for color must be capable of more than 200 V of video drive. To avoid the use of high-voltage-rating transistors, at the expense of increased complexity, transistor circuits generally use video-output stages that form the R, G, and B drive voltages before application to the picture tube.

112. Synchronous Detection. The synchronous detector is a means of obtaining a vector product of a reference signal (which defines the axis along which the color signal is detected) and the color subcarrier.

$$E_r \cos \omega t \ E_{ct} \cos (\omega t + \phi_c) = E_r E_{ct} \cos \phi_c = (E_R' - E_Y')/2.03$$
$$E_r \sin \omega t \ E_{ct} \cos (\omega t + \phi_c) = E_r E_{ct} \sin \phi_c = (E_R' - E_Y')/1.14$$

The color signal $E_{ct} \cos (\omega t - \phi)$ can be derived from the color-difference signals $E_R - E_Y$ and $E_B - E_Y$, where E_{ct} represents the instantaneous value of color saturation and ϕ_c represents the instantaneous phase of the subcarrier (proportional to the hue of the reproduced image). $E_R - E_Y$ and $E_B - E_Y$ can be derived from the subcarrier by multiplication with unit-amplitude sine and cosine terms.

Large-Signal Demodulation. There are two ways of performing the electrical equivalent of the multiplication process. In the first method a large signal of reference phase is added to the subcarrier color signal. When the reference signal greatly exceeds the value of the color signal, the components which are in phase with the reference signal add directly to its magnitude, while the quadrature components add as the square root of the sum of the squares, which remains constant when one signal dominates. When the combined signal is applied to an envelope detector, the output is proportional to the modulation in the color subcarrier along the reference axis. A balanced-diode synchronous detector using this technique is shown in Fig. 20-84. By using diodes oppositely poled and with the reference signal of the same phase applied to both while opposite-phase color signals are applied to each, the signals from the subcarrier are recovered without an offset voltage from the rectified reference signal.

Product Demodulation. A second method consists of producing a current proportional to one of the signals, i.e., either the color subcarrier or the reference signal, and modulating the developed current with the second signal.

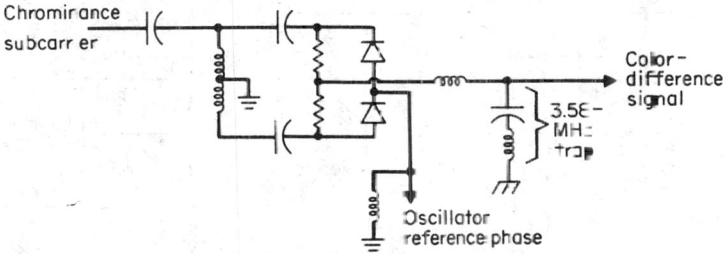

Fig. 20-84. Diode synchronous detector

   In solid-state circuits, since suitable multicontrol-element devices are not readily available, their function is simulated by interconnecting transistors with single control elements. A widely used circuit is shown in Fig. 20-85, where a current which is a function of the chrominance voltage is generated by TR_1. The current from TR_1 provides a constant-current supply for transistors TR_2 and TR_3, which are connected as a differential amplifier. The reference oscillator alternately switches the current from TR_2 to TR_3. The switched output of TR_2 contains the demodulated color signal. To avoid the effects of the switching voltage, the circuits are usually formed in full-wave pairs. An example of a complete solid-state color synchronous detector is shown in Fig. 20-86.

   113. Matrix Circuits. Figure 20-87 shows a block diagram of a decoder using *picture-tube matrixing* to obtain the R, G, and B signals. The synchronous detector may be any of the types previously discussed. The oscillator is similar to the subcarrier-regeneration system described in Par. **20-102**. The phase-shift network is used to obtain the desired axis of demodulation. The matrix provides a resistive mixing of $R - Y$ and $B - Y$ to obtain $G - Y$. Since $G - Y$ is made up of reverse-polarity components of $R - Y$ and $B - Y$, the $G - Y$ video amplifier provides one less phase reversal than the others. The color-difference signals are applied to the grids of the picture tube while the luminance signal is applied to the cathodes. The composite voltage applied between grid and cathode in each case is R, G, and B.

   Figure 20-88 shows a block diagram of a decoder using *pre-picture-tube* matrixing. The functions are essentially the same except that a separate $G - Y$ derivation and separate adders are used to combine the Y signal with the color-difference signals before application to the video-output driver stages.

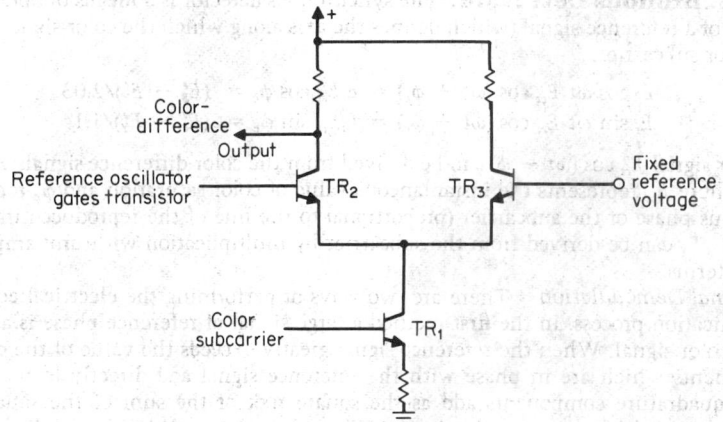

Fig. 20-85. Transistor synchronous detector.

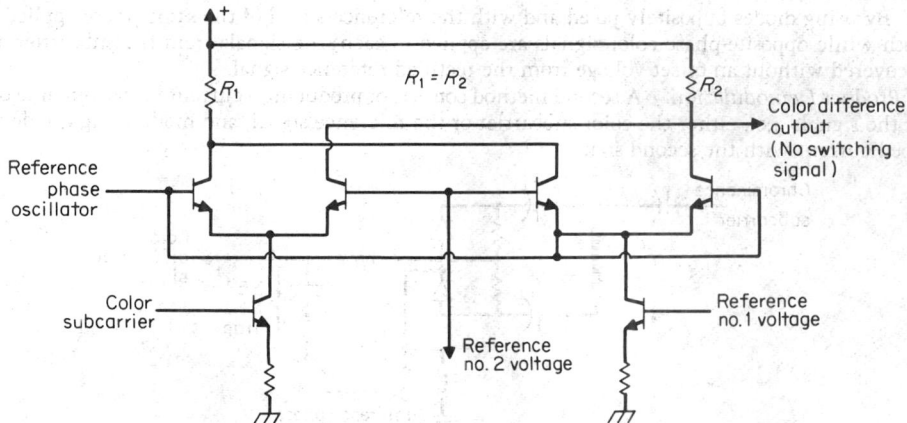

Fig. 20-86. Full-wave transistor synchronous detector.

114. VIR Receiver Circuits. As noted in Par. **20-13**, the nineteenth line retrace time during each vertical blanking period is reserved by the FCC Regulations for the transmission of vertical-interval-reference (VIR) signals that can be used to control the hue, saturation, black level, and contrast of the reproduced image in suitably equipped receivers. Figure 20-89 shows the signal content of the VIR signal; increasing brightness is in the upward direction. The chrominance reference signal is a long sine-wave burst at the chrominance subcarrier frequency. Its amplitude and phase are set at the studio or telecine origination of the broadcast.

When (1) the chrominance reference level and the black reference level are equal at the receiver's $R - Y$ output, the tint (chrominance-phase) setting of the receiver agrees with that of

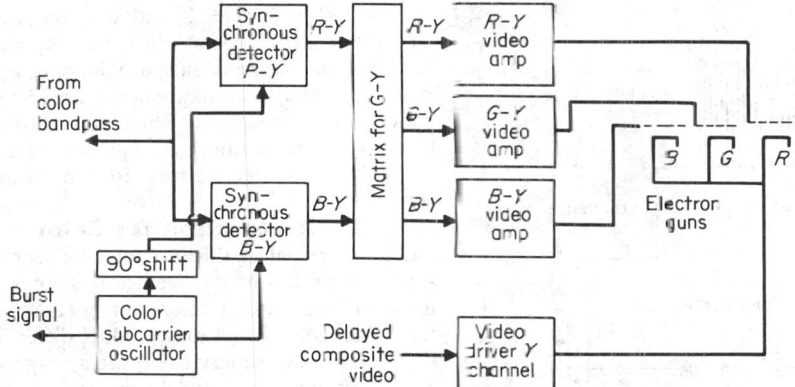

Fig. 20-87. Color-decoding circuit using picture-tube matrixing.

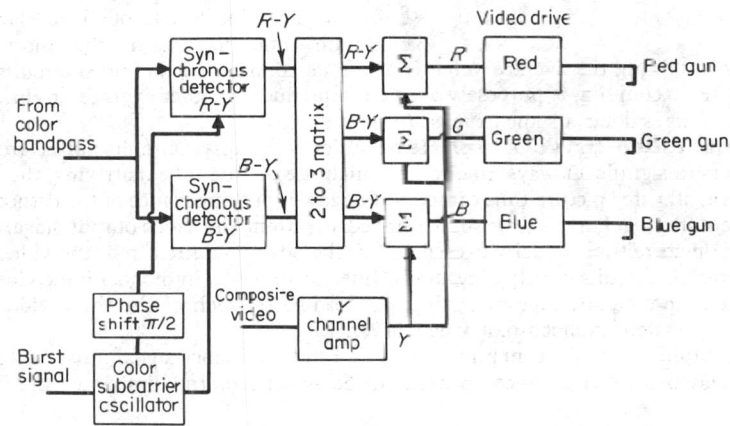

Fig. 20-88. Color decoding using pre-picture-tube matrixing.

the chrominance reference signal, and when (2) the two reference levels are equal at the receiver's blue-drive output, the "color" (chrominance-amplitude setting) agrees with that of the reference signal. If such equality of levels in (1) and (2) is not present, the inequalities serve as error signals (positive or negative) which automatically control the receiver's tint and color controls, respectively. Thus any undesired change in phase or amplitude of the chrominance signal occurring between the point of program origination and the receiver input is automatically corrected. Since the late 1970s most major network stations and many independent stations have routinely broadcast the VIR signal.

Beginning in 1977, receivers containing the VIR circuits for recovery of the correct tint and color (but not luminance level) have been offered to the public. Figure 20-90 shows the block diagram. The VIR receiver model offered in 1977 contained 5 integrated circuits and 30 transistors. Despite this complex and costly addition to the receiver, VIR-equipped receivers have enjoyed substantial popularity.

As shown in the block diagram, the initial functions of the VIR circuit are carried out in the *line recognizer,* which counts the lines in each vertical retrace until the nineteenth line is reached and determines whether the VIR signal is present during that line. If so, another circuit applies voltage to the remaining portions of the VIR chassis and a signal light glows on the front of the receiver. The conventional tint and color control of the receiver are disabled by this action, except that preference controls are available (on the side of the receiver) to make moderate adjustments in hue, tint, and contrast to suit the viewer's personal preference.

The line recognizer also produces keying pulses which are representative of the chrominance reference in the VIR signal and the black reference of that signal. The pulses are fed to the tint controller and color-level controller (Fig. 20-90). Inputs to the controllers are also obtained from the receiver's $R - Y$, $B - Y$, and Y signals, as shown. The two controls operate in effect as error-signal detectors, producing dc output voltages corresponding to the error in chrominance phase and chrominance level, respectively. These dc voltages are fed to the receiver's chrominance-processing circuitry, where they correct errors in tint and color, respectively.

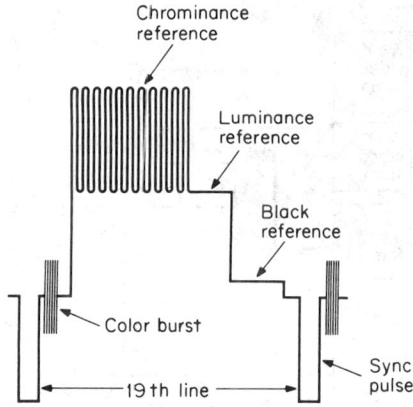

Fig. 20-89. VIR signal for automatic control of receiver hue, saturation, black level, and contrast.

115. DC Restoration for Color. Dc restoration has special significance for color reproduction since the control of the average brightness applies to each of the primary-color components of the picture. Since color scenes may produce differing video signals for each primary color, the average value of brightness may vary widely on each primary. If each dc value is not properly maintained, the primary colors are not properly matched and color errors are introduced. Although dc clamping circuits are sometimes used, the most common method of maintaining the average value is to use dc coupling in the video circuits. In some cases, complete dc coupling is purposely avoided to produce a higher average brightness in the color picture. This is done at some expense to color fidelity.

116. Color Video-Drive Amplifiers. Color video-drive circuits differ from monochrome drivers in significant ways. In circuits which use picture-tube matrixing, the video-output stage drives all three picture-tube cathodes. The low-input impedance of the cathode circuits in parallel requires a relatively low source impedance from the video-output stage. Since the cutoff in color-picture tubes usually exceed 100 V, the power required from the video driver in picture-tube matrix circuits greatly exceeds the video-drive power in monochrome video drivers. In circuits using pre-picture-tube matrixing, the loading on each of the three videodrivers is similar to an equivalent monochrome video driver.

The video amplifier circuits containing the wide-band luminance signal must contain a delay line with a delay of about 1 μs, to compensate the delay between the luminance and the chrom-

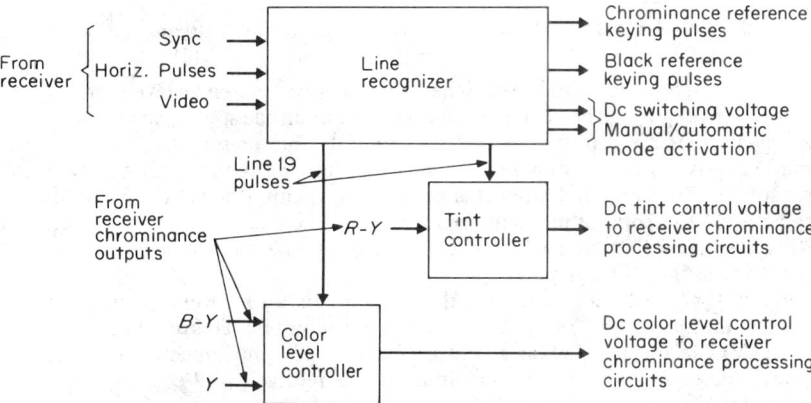

Fig. 20-90. VIR circuit block diagram. *(General Electric Co.)*

inance signals introduced by the bandpass filter. The bandwidth of the luminance channel is a compromise between good resolution and the visibility of the subcarrier. If the bandwidth is extended to provide full resolution, the visible presence of the subcarrier will introduce erroneous luminance information, which desaturates the colors, unless a comb filter is used. The latter was commonly used in receivers from the early 1980s.

117. PAL and SECAM Color Systems. The fundamental premise of the SECAM system is that since the required horizontal resolution in the chrominance components of color pictures is not as high as in the luminance component, the *vertical* resolution devoted to the chrominance components can also be reduced, e.g., by 2 to 1. By imposing the $R - Y$ and $B - Y$ information sequentially on alternate lines, the problem of encoding is substantially simplified; i.e., the carrier is no longer required to be modulated in quadrature by the $B - Y$ and $R - Y$ components of the color signal. The SECAM system uses FM to transmit the color-difference signals. To obtain coherent color signals in a SECAM receiver it is necessary to store the chroma information for a full line period, so that when the $B - Y$ signal is transmitted, it can be combined with the delayed $R - Y$ signal from the previously transmitted line and thus form simultaneously available color signals.

The video amplifier and matrix circuits are similar to those used in the NTSC system. When a shadow-mask tube is used, the high-voltage and convergence circuits are also similar. In the chroma-detector circuits a single discriminator circuit converts the FM subcarrier into sequential $R - Y$ and $B - Y$ signals. A reversing switch activated by color-sync signals selectively alternates the direct and delayed signals to provide continuous color-difference signals. Figure 20-91 shows the block diagram of a SECAM decoder. For details of SECAM and PAL standards, see Table 20-3, Par. **20-26.**

The PAL system uses alternate line averaging and line-period delay but avoids the problems of FM transmission system by using the quadrature-modulation method, similar to the NTSC system. The $R - Y$ and $B - Y$ signals are simultaneous pairs of components transmitted as amplitude-modulated sidebands of a pair of suppressed subcarriers in quadrature, as in the NTSC system. However, the phase of the $R - Y$ signal is reversed on alternate lines. Unlike the SECAM system, in which the subcarrier is always present, the subcarrier of the PAL system disappears on fully desaturated signals. The phase reversal of the $R - Y$ signal on alternate lines causes the $R - Y$ signal to lose interlace with the $B - Y$ signal. Hence the $R - Y$ dot pattern has maximum visibility when $B - Y$ is interlaced at an odd multiple of one-half the line-scanning frequency. As a compromise, the subcarrier is chosen at one-quarter line offset, i.e., at an odd multiple of one-fourth the line-scanning frequency. This compromise does not provide as accurate color interlace as can be obtained with the NTSC signal. The decoder for the PAL system is similar to the NTSC decoder with the addition of a one-line delay and a reversing switch activated by the color-burst signal. A block diagram of a PAL type decoder is shown in Fig. 20-92. The output of the delay line is added to the direct signal to obtain the subcarrier components of $B - Y$ only. The output of the delay line is subtracted from the direct signal to obtain $R - Y$ free of $B - Y$ components. The reinserted color subcarrier is reversed on alternate lines to provide continuous, properly phased $R - Y$ signals. The $R - Y$, $B - Y$, and Y signals are matrixed to provide equivalent $R, G,$ and B signals.

118. Large-Screen Projection Systems. A number of systems have been developed for large-screen projection of television pictures, including the three-CRT, Eidophor, and single-gun Light Valve systems. For domestic use three cathode-ray tubes, with a typical raster diagonal

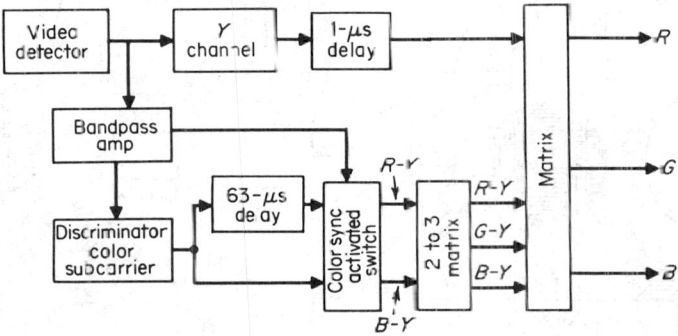

Fig. 20-91. Decoder for SECAM system.

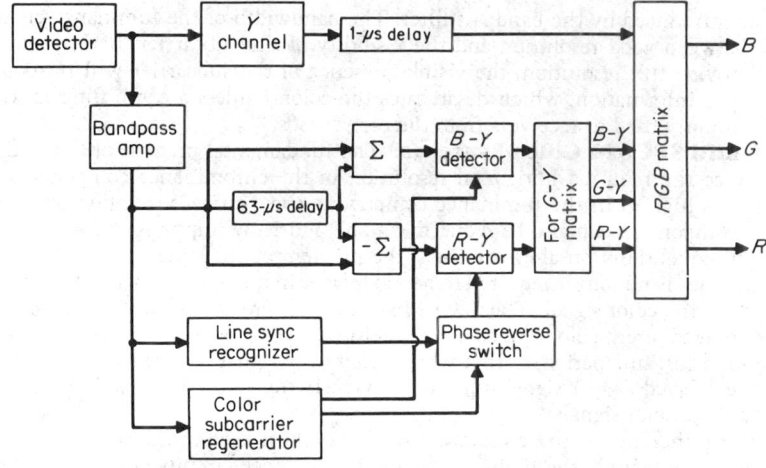

Fig. 20-92. Decoder for PAL system.

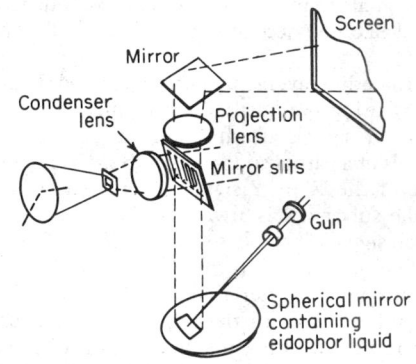

Fig. 20-93. Raster information in Eidophor system. *(After E. Baumann, J. Soc. Motion Pict. Eng., April 1953, Vol. 60, pp. 344–356.)*

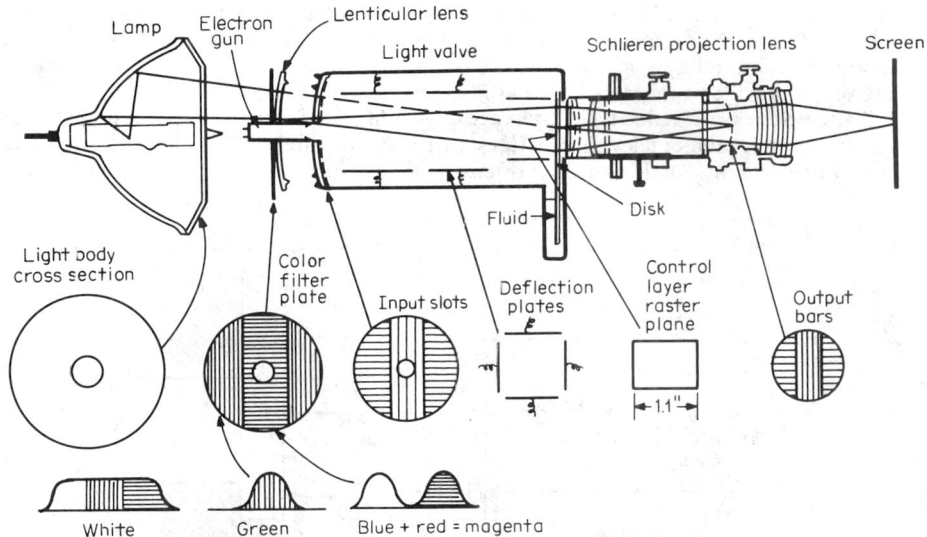

Fig. 20-94. Schematic of single-gun Light Valve projector tube. *(General Electric Co.)*

of 3 in, produce three rasters in red, green, and blue, being driven by the respective R, G, and B drive signals. By careful adjustment of the deflection and orientation of the CRTs the rasters are brought into precise register and geometric congruence and are projected through wide-aperture lenses to a highly reflective screen (screen light gain of 10 times) of 50 to 80 in diagonal. Very high raster brightness is required, and the image brightness is directional, the image being brightest seen directly head on. Projection screen brightnesses of the order of 7 to 86 cd/m² are achieved. The tendency of the three scanning spots to "bloom" at high currents in the highlights limits resolution. Generally speaking, the resolution of this type of system is less than that of a direct-view CRT image of the same subject.

The Eidophor system uses a special form of scanning system, in which an electron beam scans in vacuum the surface of a viscous liquid. This deforms the surface by amounts depending on the video-signal content. The deformation is retained for a time longer than the duration of a field. Figure 20-93 shows that a high-brightness external source of light (typically a 2.5-kW xenon lamp) reaches the liquid through a mirror containing several slits. The slits are so positioned that no light is passed on to the projection screen when the liquid surface is undisturbed, but the deformations of the surface due to the video pattern on each scanning line causes the light to pass through to the projection lens. Three such projectors with filters for each primary color are used with precisely congruent and registered rasters. Using high-wattage xenon lamps as the light sources makes it possible to obtain images of theater size and brightness equal to that of film projection (3000 lm total light output).

The single-gun Light Valve projector is similar in principle but uses only one gun, producing white-light picture elements. Interposed between the light valve and the projection lens is a schlieren optical system (Fig. 20-94). Modulating the vertical width of the scanning lines allows green light to pass through the horizontal slots; magenta (blue and red) passes through vertical slots. Superimposed horizontal deflection of the beam at 16 MHz (for red) and at 12 MHz (for blue) carries the respective primary-color content. Diffraction gratings (Fig. 20-94) break the white light spot into red, green, and blue rasters which automatically fall in precise register and congruence. Both the Eidophor and Light Valve projectors can produce images of very high definition.

Other projection systems, based on light valves using electrooptic crystals, metallic films, and liquid crystals are under development, as is a system using three high-power lasers producing superimposed images in the three primary colors.

Facsimile Systems

BY GEORGE M. STAMPS

SCANNERS AND RECORDERS

119. Introduction. Facsimile combines copying with transmission. An electric signal representing the image of *subject copy* is generated by a *scanner* and transmitted to a *recorder*. The recorder, either nearby or distant, marks a *record medium* in accordance with the received signal, generating a *facsimile copy* (also called *record copy*). The received record copy is a facsimile (image replica) of the original subject copy.

Facsimile is one of the original electrical engineering arts, having been invented by Alexander Bain in 1842. Many characteristics of signaling appeared first in facsimile. In its long development facsimile has served as a mother art, spawning a variety of devices and methods, including the photocell, linear phase filters, adaptive equalizers, compression encoding, the daughter art of television, and the application of transform theory to signals and images.

The facsimile process involves *optical scanning, recording, encoding and decoding, signal processing, modulation and demodulation, record media, and transmission.*

120. Scanning. The facsimile signal corresponds to the diffuse reflectances of light from a sequence of *elemental areas* of the subject copy. The elemental area is defined by an *aperture* through which the light must pass before it induces a signal current in the photocell, photodiode, or other light-sensitive device. The standard elemental area for CCITT group 3 high-resolution facsimile is a rectangle $\frac{1}{208}$ in wide by $\frac{1}{196}$ in high (0.012 by 0.013 cm). Other commonly used elemental areas have dimensions ranging from 20 to 200% of these dimensions. The subject copy may be thought of as being broken up into a rectangular grid of horizontal rows and vertical columns of elemental areas, filling the area to be scanned.

Rectilinear Scan.   In rectilinear scan a single photosensitive device examines an elemental area as it sweeps left to right across a row, reporting the diffuse reflectances encountered as a continuous analog signal.

Array Scan.   In array scan a row of photocells (photodiodes or CCDs) is used, one for each elemental area in the row. Each photocell, left to right in turn, is caused to deliver a current pulse corresponding to the diffuse reflectance in its elemental area. The sequence of output pulses in the array scan is a *discrete-time-series* version of the analog signal.

The scan process is repeated row by row, top to bottom, until the subject copy is completely reported. In rectilinear scan one photocell scans all the rows in series at a constant scan velocity.

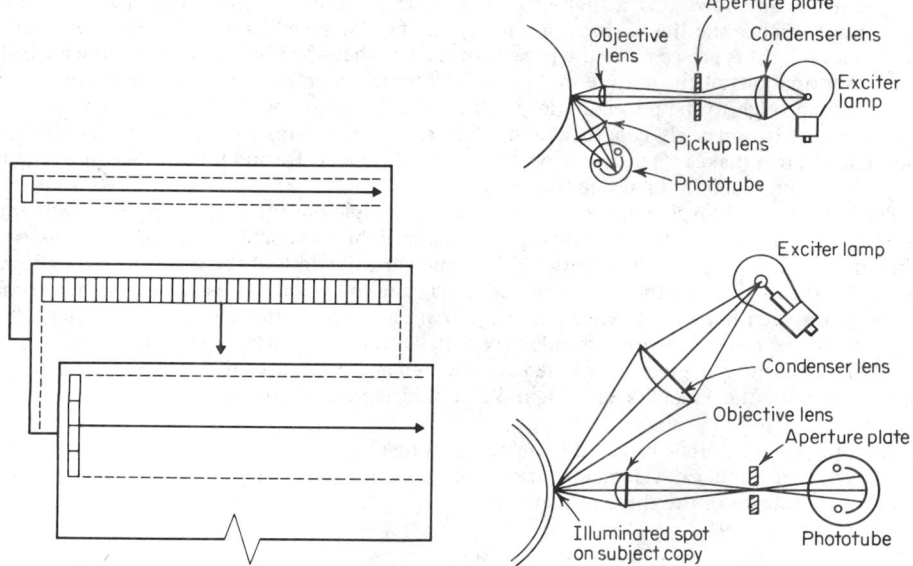

Fig. 20-95. Rectilinear *(top)*, array *(middle)*, and multispot *(bottom)* methods of scan.

Fig. 20-96. Spot projection *(top)* and flood projection *(bottom)*.

In array scan, each of the photocells scans a different vertical column, all the photocells moving in parallel from top to bottom, stepping one row at a time.

The signal corresponding to one elemental area is called a *pixel* (for picture element). When a pixel can only take on the binary values of 1 or 0, the signal represents only black or white on the subject copy and this particular pixel is called a *pel*.

Multispot Scan.   Multispot scan is a variation of rectilinear scan in which *n* consecutive rows are scanned at once by a vertical column of *n* apertures or photocells (spots).

The three methods of scan are illustrated in Fig. 20-95. In multispot it is usual for *n* to be from 2 to 32 spots. To convert the output of *n* multispots to a rectilinear signal, a buffer memory stores *n* lines of scan in parallel and reads them out in series *n* times as fast.

Illumination and Viewing.   Two methods for examining subject copy are illustrated in Fig. 20-96. In *spot illumination (flying spot)* the elemental area is illuminated by a small light spot, created by focusing the image of an illuminated aperture onto the subject copy. The rest of the subject copy is kept dark. A photocell measures light reflected from this elemental area. The pickup lens is located at such an angle that only diffusely reflected light (with no specular reflection or glare light) can reach the photocell.

In *flood illumination (image dissection)* a relatively large area of the subject copy is illuminated and imaged onto an aperture plate. The remaining subject copy need not be kept dark. Only light passing through the aperture (corresponding to light reflected from the elemental area) reaches the photocell. As with spot illumination, the angle between the illumination and pickup optics must assure that only diffusely reflected light reaches the photocell.

121. Mechanical Sweep Methods. The problem in generating mechanical sweeps suitable for facsimile is in converting constant angular velocity (from a motor shaft) into a repetitive (saw tooth) linear motion with constant sweep velocity. This velocity transformation can be made with cams, involutes, Archimedes spirals, helixes, or belts, but the simplest and most common mechanical configuration is that of the *drum* patented as a facsimile device by Frederick Bakewell in 1848).

Drum. Subject copy is wrapped around a drum and held in place by a *clamp bar* (Fig. 20-97). The drum is rotated at constant angular velocity. An *optics carriage*, mounted on a *traverse drive belt*, advances a distance equal to the height of the elemental area for each drum rotation. The signal generated by the photocell is the stream of pixels encountered in the helical scan track shown in Fig. 20-97. For the drum and traverse drive belt directions as shown, the scan is rectilinear. Sometimes the traverse drive is effected with *lead screws*.

It is common practice to use the same mechanical sweep motion for recording, making the mechanism do double duty as both scanner and recorder (called a *transceiver*). In the configuration of Fig. 20-97 a stylus is shown for marking. When used as a recorder, the drum is wrapped with recording paper appropriate for the stylus, i.e., electrosensitive paper for sparking stylus, thermal paper for a thermal nib.

For photographic recording, the photocell is replaced by a modulated light source (typically a *crater lamp*), which causes an illuminated and modulated spot to fall on the drum. Photographic film is wrapped around the drum, and the drum must be shielded from other sources of light.

Other variations of the drum use the optics carriage inside the drum with subject copy on the inner surface of the drum facing inward, and some variations use transparent drums with optical paths penetrating the drum surface.

Polygon Mirrors. Before lasers came into use, the usual polygon mirror scan was configured as in Fig. 20-98. A strip of the subject copy is uniformly illuminated by a tubular lamp. The optical path from subject copy to aperture plate is reflected by one of the plane mirror faces of the polygon. With the elemental area exactly at the focal length of the objective lens, and with the aperture exactly at the focal length of the collimating lens, the optical rays between the lenses form a parallel beam. The focus does not vary with changes in distance between the lenses. Under this condition the aperture remains sharply focused on the elemental areas along a straight line in the plane of the subject copy as the polygon mirror turns (provided the objective lens has a flat field).

For a polygon of n sides, the system will perform n sweeps of the subject copy for each revolution of the polygon mirror. With large n, very fast scanning is possible.

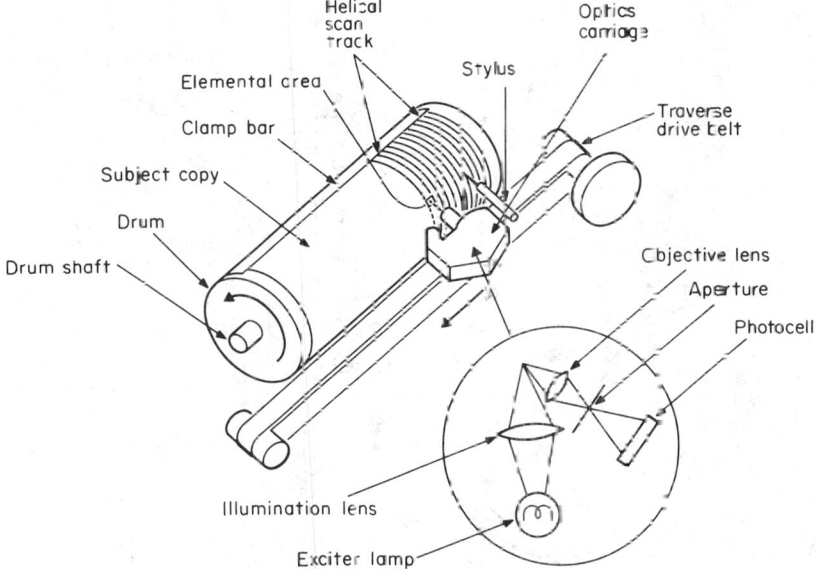

Fig. 20-97. Transceiver drum configuration.

If the photocell of Fig. 20-98 is replaced by a light source, a *flying-spot illumination* of the subject copy can be made, light pickup being performed by a tubular photocell placed near the subject copy. With a modulated light source at the photocell location the polygon mirror system can be used to expose film or a xerographic photoreceptor in a recording system.

The *laser* is the preferred light source for spot-illumination scanning or light-spot recording because the wavefront of its monochromatic coherent light does not spread significantly in the distances involved in polygon-mirror scanning. In scanning applications the laser light is blind to inks of its own color, which appear "white" like the background of typical subject copy. This detriment can be turned into an advantage in printing material on forms used in facsimile transmission which will not be transmitted. The dispatch page for controlling addressing of transmissions in one very high-speed laser-scan facsimile system is printed in the format for mark-sense information in a color the laser cannot see, and the format information is not reproduced.

The nature of the laser satisfies the need for a parallel beam at the mirror surface and at the objective lens (Fig. 20-98). A typical polygon-mirror laser system used to expose a xerographic photoreceptor drum is shown in Fig. 20-99. If the laser beam is fine enough to define the elemental area without further optics, it can shine directly onto the photoreceptor drum (or subject copy in the case of a scanner). If the laser beam is too fine, it is usually broadened by a *beam spreader* consisting of a negative and a positive lens of equal power in series.

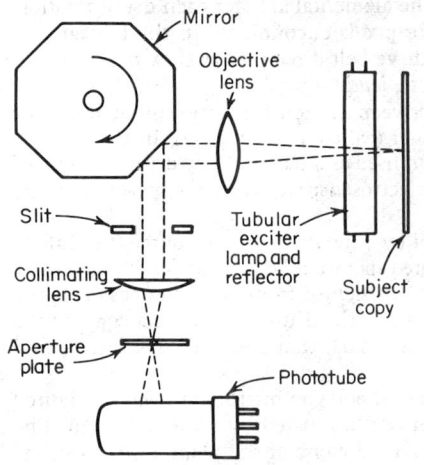

Fig. 20-98. Image-dissecting scanner with rotating mirror.

If d is the distance from the mirror to the midpoint of the sweeping scan, and if θ is the instantaneous angle of the beam to right or left of center, the velocity v of the spot for any given angle θ is

$$v = \omega d \sec^2 \theta$$

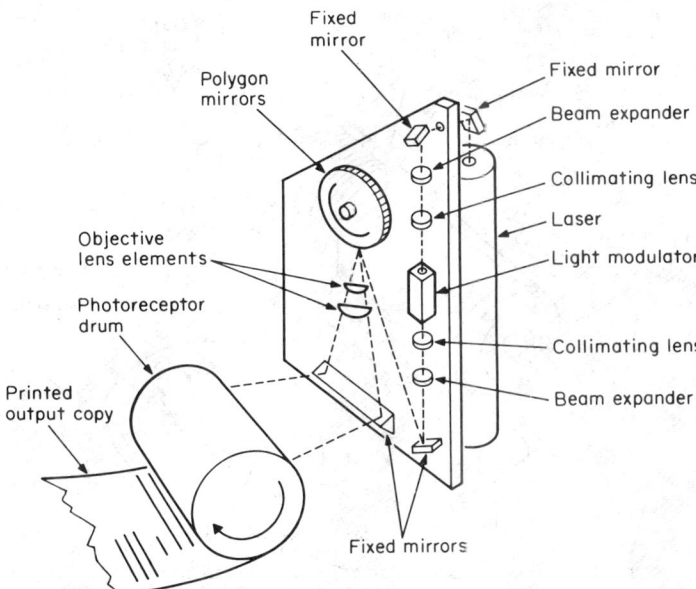

Fig. 20-99. Polygon-mirror laser-scan xerographic facsimile printer.

where ω = angular velocity of prism mirror. The angle over which the beam can sweep is limited to angles for which $\sec^2 \theta$ is a sufficiently good approximation to unity to make the velocity nonlinearity tolerable.

In the xerographic printing system of Fig. 20-99, if the mirror has 18 sides, the maximum value of θ is 10°. Since $\sec^2 10° = 1.031$, the velocity of the spot will vary 3.1% over the length of its sweep. Normally, velocity nonlinearities should not exceed 1.5%.

Optical elements functionally similar to the objective lens of Fig. 20-98 can be used to focus the laser beam to the desired spot size and to reduce the velocity nonlinearity. Special optical elements for this purpose are shown in Fig. 20-99 in the functional location of the objective lens of Fig. 20-98.

Rocking Mirrors.   Two methods of scan are based on a single rocking mirror. The *galvanometer mirror* is a device in which current flow causes rotation of a d'Arsonval movement galvanometer. The polygon mirror of Fig. 20-99 is replaced by a single galvanometer mirror driven with a sawtooth current. Satisfactory operation is possible up to about 70 sweeps per second. The *cam-driven rocking mirror* (Fig. 20-100) uses a sawtooth-motion cam to drive a pivoting spherical mirror to provide sweeping spot illumination of subject copy.

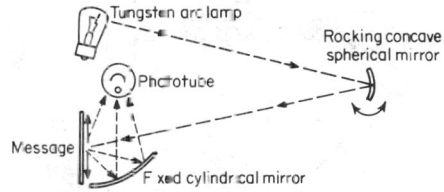

Fig. 20-100. Flat-bed flying-spot scanner using oscillating mirror.

122.   Electronic   Scanners. Above sweep rates of 6 sweeps per second, most facsimile scanning is performed with lasers and polygon mirrors or electronically using techniques based on either the cathode-ray tube or photosensitive arrays.

CRT-Based Methods.   High-resolution vidicons have been applied to facsimile scanning by using them like television cameras to look at subject copy. Conventional CRTs have been used with an objective lens to project flying-spot illumination on subject copy for photocell pickup. Only the following two CRT-based methods, however, stand out in facsimile applications.

The *flat-fiber-optics faceplate CRT* (Fig. 20-101) brings light from the phosphor inside the tube (where the impinging electron beam excites phosphorescence) out to the front surface without blooming, the spreading process which takes place as the light radiates away from the phosphor as it travels through the glass CRT window. Resolution is thus preserved and depends on the size of the electron-beam-focused spot. Brightness is also high, since light otherwise lost to blooming is preserved. In recording, electrofax (zinc oxide–coated) or dry silver photographic paper or a photoconducting xerographic drum is exposed directly by contact feeding against the fiber-optics faceplate.

The *wire-matrix faceplate* recorder (Fig. 20-102) conducts charge from the electron beam inside the CRT to *dielectric paper* in contact outside the CRT. The process is somewhat complex, due to secondary emission (electrons may induce a positive charge on the pins), but the result is that charge is deposited on the dielectric paper in accordance with the CRT signal modulation. Wire-matrix faceplates have been particularly successful in address-label printing applications.

Photosensitive Arrays.   Most current facsimile scanners use photosensitive arrays to scan flood-illuminated copy (Fig. 20-103). An objective lens forms a reduced image of the subject copy exactly on the straight-line array of silicon sensors. For digital facsimile (CCITT group 3) the array has 1,728 sensors in a 1.02-in row to view an 8.5-in-wide subject copy, necessitating an image size reduction of approximately 8.3 times by the objective lens (Fig. 20-104). For the lesser resolution requirement of CCITT group 2 (analog) facsimile, three separate sensor units of 512 photosensors each are commonly used, as in Fig. 20-104 (providing 1,536 pixels per sweep). Although alignment of the mating edges requires high precision, the shorter optical paths save space. Two types of silicon photosensor arrays are in general use.

The *photodiode array* (such as the Reticon device) is a monolithic self-scanning linear array of silicon photodiodes. In each of the 0.00059-in-wide individual cells is a photodiode 0.000275 in wide and 0.00063 in high. Each photodiode is connected to a storage capacitor and an MOS switch. As light shines on the photodiode, it induces a reverse current through the photodiode, allowing capacitor charge to leak off. Every photodiode is sampled by switching in turn to an output. The size of the recharging current pulse to the capacitor measures the integrated light incident on the photodiode for the period since the last sampling.

Since dark current and switching transients are also present, a second complete set of dummy photodiodes, kept dark inside the array chip, is used. The dark diodes and the active phototdiodes are sampled differentially, tending to cancel the dark currents and switching transients and producing an excellent discrete time series of pulses corresponding to the diffuse reflectances of the row of elemental areas on the subject copy being examined.

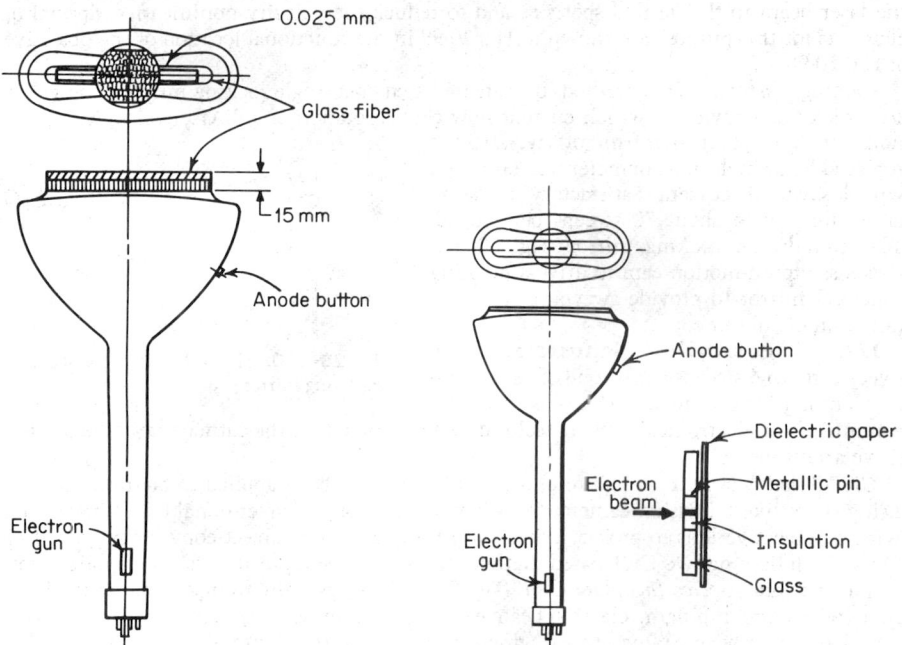

Fig. 20-101. Fiber-optics faceplate CRT.   **Fig. 20-102.** Wire-matrix faceplate recorder.

The *charge-coupled-device (CCD) linear image sensor* (such as made by Fairchild) is functionally similar to the photodiode array. A CCD is a semiconductor device in which isolated charge packets are moved by sequential clocking of an array of gates from one charge-retention cell to the next. Since the amplitudes of the charge packets can take on analog values, a CCD can be used as an analog shift register.

The CCD as a linear image sensor uses silicon sensors (1,728 of them for group 3), as shown in the left center of Fig. 20-105. While light is incident on the silicon, charge accumulates proportional to incident light flux. Transfer gates below the sensors transfer the accumulated charge to a CCD analog shift register for the even-numbered sensors, and transfer gates above the sensors transfer charge to another CCD analog shift register for the odd-numbered sensors. Clocking

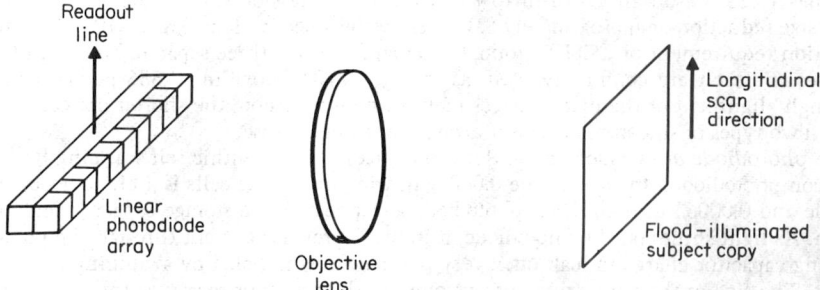

Fig. 20-103. Linear photosensitive array scanner.

signals cause these charges to move down their respective analog shift registers, where they emerge alternately into a preamplifier, the output of which is a discrete time series of analog pulses corresponding to the diffuse reflectances of a row of elemental areas on the subject copy, reported left to right in turn.

123. Recording. The objective in recording is to use the information content of the received signal to control the placement of marks on paper such that a pleasing copy of the original subject copy is reproduced. The psychology of vision plays a primary role in the trade-offs in recorder design. The eye is especially sensitive to *spatial frequencies* and *departures from expected uniformity*. The eye is pleased by smooth curves and regular boundaries. The eye is particularly offended by extraneous marks (specks on white background, holes in solid black, false lines, or streaks).

Figure 20-106 shows the expected alignment of pixels reproducing a narrow vertical line. Horizontal position errors, called *jitter*, and vertical position errors, called *cogging*, are illustrated. Jitter or cogging errors in excess of 0.001 in are usually detectable by the eye, and errors exceeding 0.0025 in are usually disturbing. Mechanical sweep methods are particularly prone to jitter, caused by errors in timing, and to cogging, caused by errors in traverse (feed). Observed jitter and cogging on received copy is the sum of jitter and cogging due to the scanner *and* recorder.

124. Recording Media. Facsimile systems mark record copy by systematically applying electricity, heat, light, ink jet, or

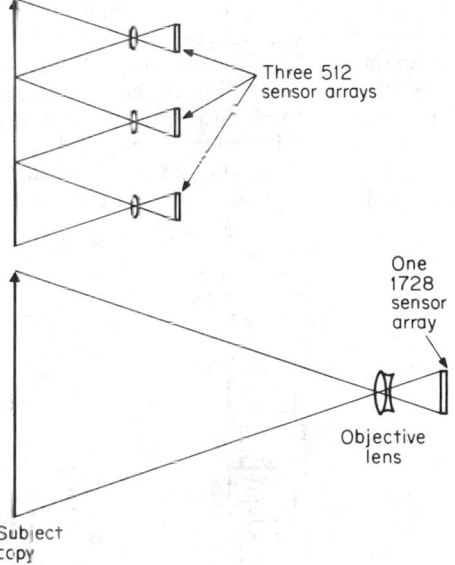

Fig. 20-104. Use of single or separate photosensor arrays for scanning.

pressure to a *recording medium*. Elemental areas of the record medium are marked individually and sequentially by rectilinear, array, or multispot scan (see Fig. 20-95). The tool used to apply the mark to the elemental area is called the *marking transducer* of the facsimile system.

Marking transducers may apply the finished mark directly to the record medium in a *one-step process*, or the marking transducer may create a latent image, which is then rendered visible in a process requiring two, three, or more steps. Except for ink jet, the simplest facsimile

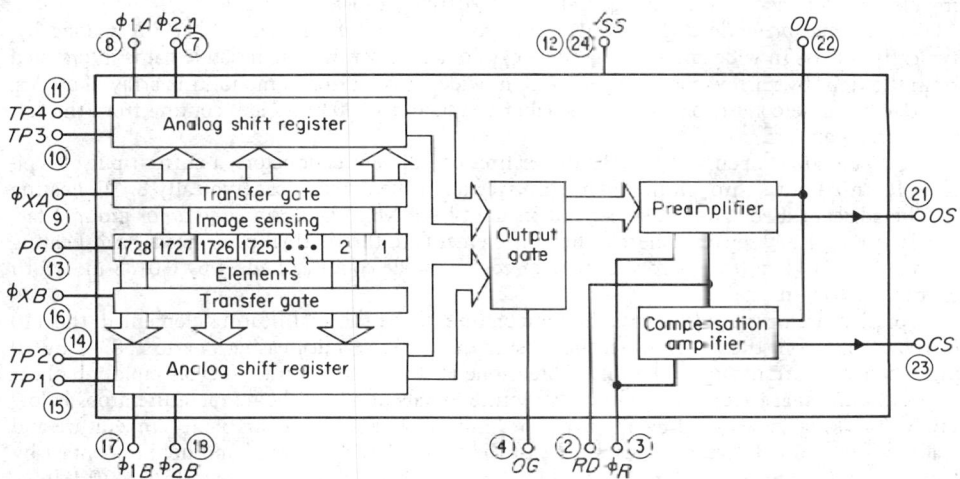

Fig. 20-105. Block diagram of a Fairchild 1,728-element linear image sensor.

machines use a one-step process requiring specially coated recording media. The fastest and most complex facsimile recorders are designed to use plain paper, relying on the paper-cost difference to justify the multistep marking process.

Electrosensitive Paper. Bond paper with a coating of electrically conducting carbon and binder, covered by an off-white high-resistance masking coating of binder and pigments (often zinc oxide) is used as the record medium. Marking is accomplished using a tungsten stylus (0.006 to 0.010 in diameter) lightly loaded (typically 8 g) against the masking surface. The conducting paper is grounded to the metal drum or feed rollers. A regulated dc supply with a 2.2-kΩ series resistor and the stylus side positive is used; current will begin to crater the masking coating at a threshold of 50 V at a spot velocity of 20 in/s.

Black carbon seen through the cratered surface is the mark. Crater width varies with voltage up to a full burnoff of the off-white coating at about 240 V and 50 mA with an optical density of about 1.2. Higher voltage is required at higher spot speeds.

A variation using aluminized conducting paper marks at lower voltage (10- to 50-V range). The typical configuration for facsimile transceivers using electrosensitive papers is the rectilinear

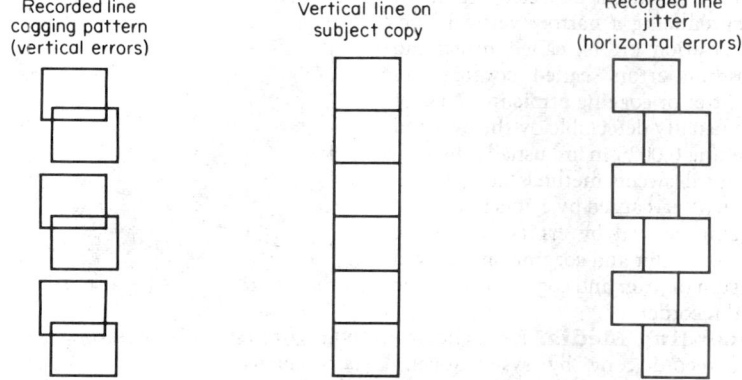

Recorded line cogging pattern (vertical errors)   Vertical line on subject copy   Recorded line jitter (horizontal errors)

Fig. 20-106. Illustration of jitter and cogging.

scan method shown in Fig. 20-97. Electrosensitive papers can support analog gray-scale recording.

Thermal Paper. Bond paper with a thin coating of a colorless leuco dye, a phenol-group color former, and a binder to fuse the coating to the paper is the marking medium. Heating to a temperature in the 110 to 130° C range for 1 ms melts the dye and color former and produces a blue, blue-black, or black indelible mark. Most group 2 and some group 3 facsimile systems employ thermal marking, using array-scan (Fig. 20-95) geometry consisting of one thermal heating element for each elemental area to be resolved along a row

The *thermal print head* (Fig. 20-107) consists of a single film resistive bar 8.3 in long by, typically, 0.0138 in wide and running the length of an array row. The resistive bar is intersected from the right by narrow conductors (0.0028 in wide) coming from a monolithic array of diodes. Interlacing these conductors are even smaller conductors (0.002 in wide) coming from the left from *strobe drivers.*

When a diode is open and a strobe driver fires on adjacent conductors, a current pulse (typically 100 mA) flows through the intervening region of the resistive bar (typically 87 Ω), heating it. The adjacent diode and strobe conductors are typically on 0.0065-in centers for group 2 facsimile, giving 1,279 separate thermal heating elements in the thermal print head. Each heating element is 0.0041 in (from strobe conductor edge to diode conductor edge) by 0.0138 in (width of resistive bar) in size.

Typically, a current pulse raises the temperature of the thermal heating element to the 110 to 130° C range in the first milliseconds, rising thereafter at a decreasing rate to 250° C by 10 ms, when the current stops. The tiny heated zone of the resistive bar then cools rapidly by heat loss to the thermal paper and mounting structure. By about 13 ms the temperature drops below 110°C (marking ceases), and by 17 ms the heating element is back to ambient temperature and ready to fire again. Thermal paper is fed past the thermal printing head and held in contact by a sponge-rubber pressure roller (Fig. 20-108). A typical driver circuit is shown in Fig. 20-109.

The monolithic driver arrays decode a binary facsimile input signal containing a data stream of 1-b pels, causing the correct diode-strobe combinations to fire and heat, melt, and mark elemental areas corresponding to 1s and not mark elemental areas corresponding to 0s. The binary character of thermal paper can provide black-and-white recording only.

Electrostatic Paper. Bond paper whose body has been rendered slightly conducting by the addition of an ionic salt to bring its resistivity to 10^7 Ω-cm or less, and whose surface has been coated to a thickness of about 0.004 in with a dielectric of resistivity 10^{16} Ω-cm or greater, is the marking medium. A *latent image* of electric charge is deposited by means of array scan with a stylus array (typically at 350 V). Opposite charges are attracted through the slightly conducting paper, and the deposited and attracted charges bind in place across the dielectric. The latent charge is rendered visible by *toning* and is fixed (fused) into the paper by heat or pressure (dry toner), or drying (liquid toner). Many group 3 facsimile recorders use dielectric marking.

In the configuration of Fig. 20-110 a standing wave of liquid toner is applied. For dry toning, a toner cascade is employed, followed by the fusing station.

For group 3 facsimile, 1,728 styli (nibs) are used on 0.0048-in centers. When the array row is divided into two interlaced rows in hexagonal close packing, as in Fig. 20-111, signals to the odd styli must be delayed one paper-advance increment after printing the even styli. This creates a continuous overlapping of dots for a more pleasing image and also simplifies the fabrication problem for construction of stylus arrays by increasing the distance between stylus centers. Facsimile recorders with two rows of 0.0048-in center styli interlaced have an effective resolution of 416.7 dots per inch.

The 600 V needed to deposit electrostatic charges on the dielectric surface of dielectric paper is usually applied as -300 V by the *writing stylus* (nib) and $+300$ V by a second *selector electrode* (either at the rear of the paper or mounted as adjacent strip electrodes on the same side), as shown in Fig. 20-110. Charges will not usually deposit below 350 V. The fact that the combined action of both writing and selector electrodes is required for charge deposition is used in *addressing styli.*

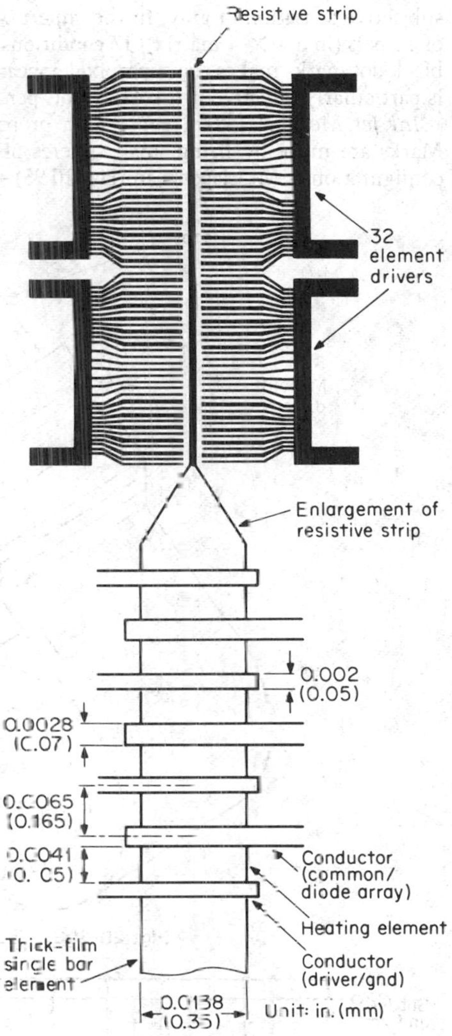

Writing styli are divided into groups of 32 styli each. The nth styli of every group are driven in parallel by the nth output of a single 32-terminal monolithic driver unit. Each selector electrode faces one 32 stylus group. If the nth stylus of the mth group is to print, all nth styli are driven in parallel to -300 V, the mth selector electrode is driven to $+300$ V, and only the nth stylus of the mth group experiences sufficient electric field strength (600 V) to cause charge to deposit.

The number of elemental areas (pels) chosen as standard for group 2 digital facsimile was influenced by its factoring possibilities for decoding, since

$$1{,}728 = 32 \cdot 54 = 2^6 \cdot 3^3 = 4^5 \cdot 3^3 = 12^3$$

In converting array scan to rectilinear scan the 32-stylus driver is clocked to switch from 1 to 32 repeatedly, with the selector electrode driver stepping across its 54 positions once after each thirty-second clock pulse until all 1,728 styli have been addressed serially in turn.

Fig. 20-107. A 128-element section of thermal print head, with enlargement of resistive bar.

When toned, each elemental area is black or white only (no gray scale). Gray scale can be achieved in electrostatic recording by varying the ratio of black area to white area within a single pixel. In the *charge-subtraction method* a single stylus lays down charge, and after a small displacement, the stylus polarity is reversed and some of the charge is reabsorbed by the stylus, leaving only a half-moon mark when toned.

The timing of reabsorption is varied to vary the thickness of the half-moon dot to achieve the subjective sensation of gray. In the *superpixel method* four styli are used to print a superpixel of 16 pels (in a 4 × 4 matrix); 17 conditions of gray ranging from all-white through 16 possible black-dot marks makes the superpixel appear white or various grays up to all black. The method is particularly effective if the 416.7-dots-per-inch stylus density is used.

Ink-Jet Method. Plain paper with properties receptive to the ink chosen is the medium. Marks are made by firing small spheres of ink from a nozzle onto the paper. The preferred configuration is Multispot scan (Fig. 20-95) with a vertical column of nozzles (typically 16). To

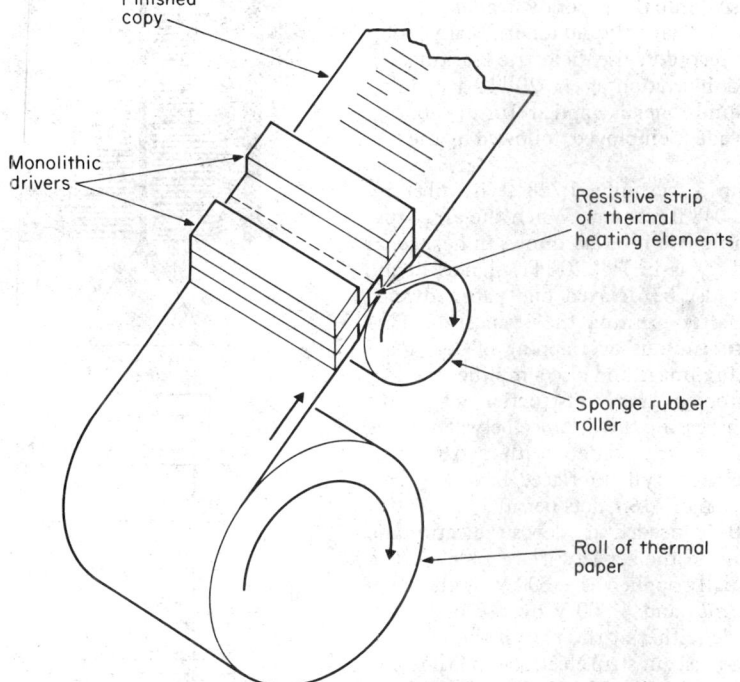

Fig. 20-108. Thermal-printing-head recorder.

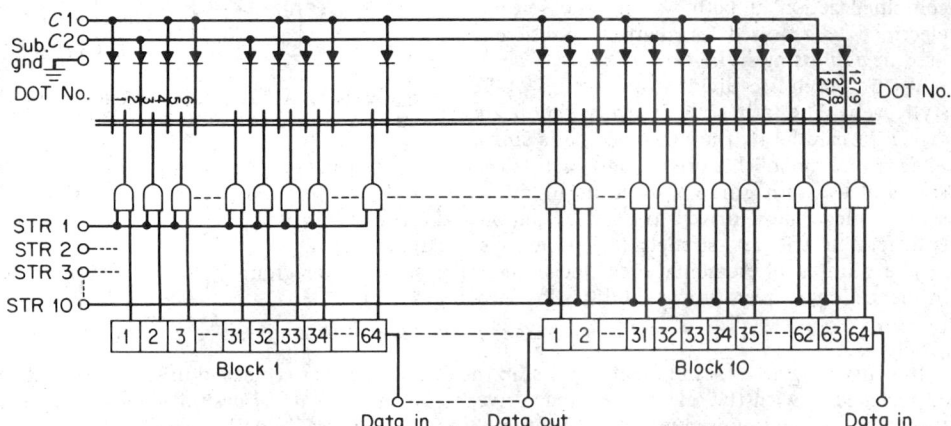

Fig. 20-109. Circuit diagram for thermal print head.

accommodate mechanical constraints the nozzles are sometimes staggered as in Fig. 20-111, except that they are staggered along a column instead of a row. Nozzle firing times are delayed as required to create a single vertical line corresponding to a signal requiring all 16 nozzles to print.

Rectilinear pixel sequences are rendered into Multispot scan by storing 16 rows of signal and then reading each line out in parallel to its nozzle.

Several nozzle methods are used, including methods by C. Hellmuth Hertz and N. G. Erick Stemme. The most common employs an electrostriction contraction to squeeze out a single drop upon applying a voltage across the electrostrictive material in the chamber behind the nozzle. Ink feeds into the nozzle array manifold during marking and is automatically retracted to the ink reservoir during other times to prevent drying and clogging of the nozzles.

Ink jet is used for black-and-white only (pel) marking and has been extended to gray scale through controlling a plurality of ink droplet spheres for one pixel. It is used for color recording by using four jets (three primary colors plus black) fed from four different ink reservoirs and

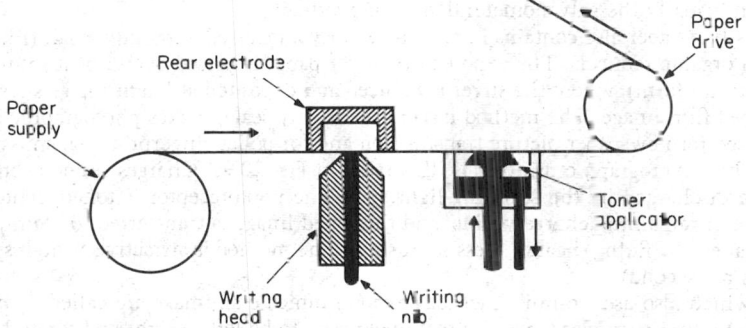

Fig. 20-110. Electrostatic printing with liquid toner (Versatec configuration).

timed to deliver their different colored droplets to reach the same pixel (elemental area) on the record copy.

Pressure Method. Plain paper is the medium. Either an inked ribbon, or a sheet of carbon paper or squeeze-out ink (Plastisol) is used next to the paper. Marks are made when a stylus exerts pressure on the ink carrier, causing ink to transfer.

The first acoustically coupled transceivers used this method with a stylus tip referenced by a guide roller to run lightly over the ink donor sheet. Upon being signaled to mark, a printed coil card pressed the stylus tip as current flowed through the coil card in a strong magnetic field (typically 7 kG = 0.7 Wb/m^2). The stylus squeezed ink onto the paper, the density of the mark being proportional to current in the coil.

A variant uses Multispot scan with an array of styli in a matrix head mechanically swept across an inked ribbon lying against the paper sheet. The usual mode is black-and-white only.

Electrolytic Method. An older method (originally used by Bain in 1842 and improved by Edison) uses paper impregnated with a wet electrolyte containing conducting salts, marking compound, and stabilizers. A mechanically swept electrode (vertical blade) in front of the wet electrolytic paper intersects a stationary horizontal blade electrode behind the paper. A small parallelepiped of paper is squeezed at blade intersection (typically 0.01 in high, due to the horizontal blade thickness; 0.006 in wide, due to the vertical blade thickness, and 0.0025 in thick due to the paper). Usually the rear blade is the cathode and is made of a noble metal (sometimes 83% platinum, 17% iridium). The front blade is the anode, usually stainless steel. Current is passed from anode to cathode through the paper, electroplating Fe^{3+} ions into the paper, where they complex with the marking compound (usually a phenol) to form conjugating chains which are essentially an indelible ink. Typically, at 50 in/s spot speed, a current of 0.25 A will mark the paper to full density, with gray marks at lower currents roughly linear density vs. current.

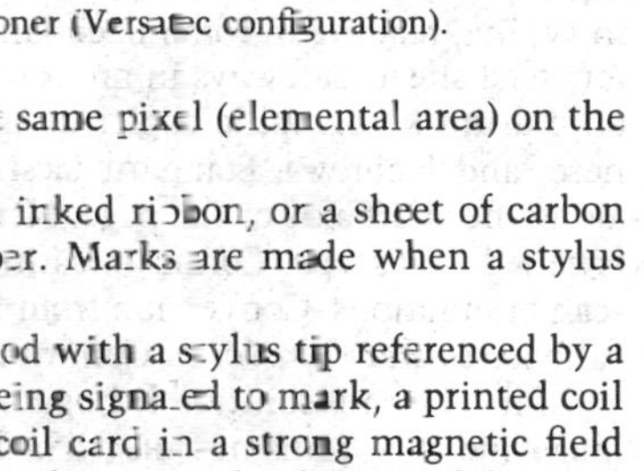

Staggered array

Creates overlapping dots for continuous patterns

Fig. 20-111. Staggered stylus array with delayed printing for odd styli.

The disadvantage is the paper must be kept wet to be usable (approximately 30% moisture by weight). Paper is packaged in sealed bags, and recorders must have humidor chambers for the roll of wet paper. High-wet-strength paper is used, and it must be free of heavy-metal impurities. Shelf life of the paper has been a problem. The anode is consumed and must be regularly replaced.

A favorite configuration uses a helical cathode mounted on a rotating drum to sweep a stationary anode horizontal blade, with paper feeding in between. Some papers require silver anodes.

Current through the paper is modulated, requiring a current driven source. I^2R heat developed in the paper boils the water, which holds the temperature to about 100°C. Moisture must remain or paper will char. Excess water must be driven out by a drier shoe over which the freshly marked paper must pass. The good gray scale permits photographs to be rendered.

Dry Silver Paper. This is a photographic medium containing a silver halide layer capable of forming a latent image upon exposure to light in the blue-green spectrum with an exposure energy required of about 200 ergs/cm² (0.2 J/m²). Blue laser is the usual light source, with rectilinear scan (usually the galvanometer deflection method).

The dry silver paper also contains in close proximity a heat-reducible silver salt (like behenic acid and an organic reducer). The exposed dry silver paper is passed over a heat source (which must be very uniform), where the silver is reduced and deposited at latent image sites, forming the developed film image. The method has excellent gray scale, makes photographic copy, and is widely used for newspaper picture transmission and for police fingerprint transmissions.

Xerography. Xerographic printing is illustrated in Fig. 20-99. Charges resident on a photoreceptor are discharged by the scanning light spot. The photoreceptor is toned, rendering the latent image of remaining charge visible, and the toned image is transferred to plain paper and made permanent by fixing (heat or pressure fusing). The method is attractive at high speeds (0.5 to 2.0 pages per second).

Copiers which also use facsimile methods to communicate an image are called *communicating copiers.* Facsimile standards require rectilinear scan to be left to right and top to bottom (as in typing, reading in romanized languages, and in teletyping and television). Because many copiers feed sheets sideways in order to shorten the paper path, some communicating copiers added onto drum xerographic copier designs scan top to bottom and right to left (as in Chinese, Japanese, and Hebrew). Standard facsimile-scan orientation is called *portrait*, where sweeping is along the short axis of the page. Nonstandard scan with sweeping along the long axis of the page is called *landscape*. Communicating copiers have been designed in both portrait and landscape scan orientations. Conversion from landscape to portrait requires a storage and readout maneuver called a *corner transform*. Communicating copiers typically use 240 or 300 scans per inch.

Film. High-resolution lithographic film is used as the recording medium for newspaper and magazine page transmission from the composing room to the press location. A very high resolution and much larger drum size version of the configuration of Fig. 20-97 is used. The drum must accommodate a full page of a newspaper. Resolution must be adequate to resolve 70-screen halftone dots without moiré pattern. The usual resolutions range from 600 to 1,000 lines/in (where one line is one sweep).

Instead of the viewing optics of Fig. 20-97, a laser or crater lamp source is modulated and imaged onto the film in an otherwise lighttight drum compartment. The exposed film is then developed in the usual way.

A similar drum method is used for transmission, except that the unmodulated spot is used for spot illumination and a photocell is used for pickup.

ENCODING AND TRANSMISSION

125. Classification by Facsimile Groups. The International Telegraph and Telephone Consultative Committee (CCITT), part of the International Telecommunications Union (ITU), an arm of the UN, sets international telecommunications definitions and standards, including definitions of four categories of facsimile equipment and standards for the first three. They can be found in Vol. VII of the Seventh Plenary Assembly CCITT (1980). The categories are called groups, as follows.

Group 1. Group 1 is facsimile apparatus which enables an ISO A4 page (210 by 297 mm = 8.3 by 11.7 in) to be transmitted over a telephone-type circuit in approximately 6 min. Group 1 uses *analog frequency-shift modulation* signaling which will support a gray scale.

Voluntary United States facsimile standards are developed by the Facsimile Systems and

Equipment Engineering Committee TR-29 of the Electronics Industries Association (EIA). Because the United States Direct Distance Dial (DDD) switched-telephone network accommodates greater frequency shifts than the least common denominator of worldwide telephones, U.S. and CCITT Group 1 Standards differ (see Table 20-10). Group 2 and 3 Standards for CCITT and the United States are the same, and the objective is to generate a common set of group 4 standards.

TABLE 20-10   Key Facsimile Group 1 Standards

Parameter	U.S. standard	CCITT standard
Factor of cooperation F*	840 (for 4 min 560)	829 ± 1%
Nominal time/page	6 min (4 min)	6 min
Total scan line L_t	8.75 ± 0.06 in	215 mm (8.46 in)
Usable scan line L_u	8.33 ± 0.03 in	200 mm (7.87 in)
Line-use ratio L_u/L_t	0.952 ± 0.010	0.930
Scan density, traverse (feed)	96 lines/in (4 min, 64 lines/in)	3.85 lines/mm (97.8 lines/in)
Sweep (scan direction)	96 lines/in	97.9 lines/in
Minimum input document size	8.5 × 11.0 in	ISO A4 (210 × 297 mm) (8.3 × 11 7 in)
Scan-line frequency	180 ± 10⁻⁵ m.r.⁻¹	180 ± 10⁻⁵ min⁻¹
FS modulated signal range	1,975 ± 475 Hz	1,700 ± 400 Hz
White frequency	1,500 ± 50 Hz	2,100 Hz
Black frequency	2,425 ± 25Hz	1,300 Hz
Answer tone	(ACK) 1,500 ± 50 Hz for 0.5 to 2.2 s	(Group ID: GI1) 1,650 Hz for 1.5 s repeated after 3 s continuously
End of message	(STOP) 1,100 ± 50 Hz continuous	1,100 ± 3.5% for 3 s
Synchronization (phasing)	15 ± 3 s of scan lines each 96% black and 4 ± 1% white	5 ± 1 s of scan lines each 95% black and 5 ± 1% white
Transmit power	−9 dBm for unmeasured subscriber loop adjusted to 0 dBm by installer	−15 to 0 dBm set by installer
Receive levels	−36 to 0 dBm	−40 to 0 dBm

*F = scan lines/in times total scan line (in); if vertical and horizontal resolutions are the same, F = pixels per total scan line. F also equals index of cooperation times π.

Group 2. Group 2 facsimile apparatus runs at double the speed of group 1 by using *duobinary bandwidth reduction* to halve its Fourier spectrum. Group 2 delivers a page in 3 min over telephone-type circuits.

A group 2 facsimile baseband signal (white = maximum; black = 26 dB less) llustrating black, white, and gray levels and including two isolated black pixels separated by one white pixel is shown in Fig. 20-112a. Signals making a white-black-white reversal trigger a *polarity switch*, which gives the signal a negative sign, making every other white signal of opposite sign. The result is Fig. 20-112b. The dashed lines are the band-limited waveforms The waveforms in Fig. 20-112b have half the frequency of the corresponding waveforms in Fig. 20-112a. The bandwidth-reduced signal cf Fig. 20-112b is used to amplitude modulate a 2,100-Hz carrier frequency f_c. After each polarity switch, and as the baseband signal passes through zero, the phase of the carrier is reversed (shifted 180°), as in Fig. 20-112c. This modulated carrier signal is filtered by the vestigial sideband filter of Fig. 20-113 to create a VSB AM phase-modulated signal for transmission.

Many group 2 facsimile systems record on media which cannot support a gray scale. In such cases the signals of Fig. 20-112a are black and white only, and the modulated VSB AM/PM signal which results is simply a series of carrier bursts of alternating phase, shown in Fig. 20-112d. The highest keying frequency f_k present in the facsimile baseband signal corresponds to an alternating sequence of black and white pixels. For group 2 the nominal f_k = 2,487 Hz. Duobinary reduces the highest modulating frequency by half to 1,243.5 Hz. The position of this information in the telephone channel is as the lower sideband at a nominal 856 5 Hz. For good reception a channel-equalizing filter at the receiver must correct amplitude and phase distortion of the channel so that the energy at 856.5 Hz (nominal) and the energy of the carrier (2,100 Hz) arrive together (minimum differential envelope delay) in their correct relative amplitudes.

Detailed standards for group 2 equipment are in Recommendation T.3 of CCITT, vol. VII (see Table 20-11).

Group 3.   Group 3 is facsimile apparatus which enables a typical A4 (or United States 8.5 by 11.0 in) page to be transmitted by a *digital modem* over a telephone-type circuit in 1 min or less by employing digital-data-compression techniques.

Detailed standards for group 3 equipment are in Recommendation T.4 of CCITT, vol. VII, and in the equivalent U.S. EIA Standard RS-465 entitled Group 3 Facsimile Apparatus for Document Transmission in the General Switched Telephone Network (see Table 20-12).

Common procedures to be followed by all facsimile apparatus (groups 1, 2, and 3) used over the switched telephone network are detailed in Recommendation T.30 of CCITT, Vol. VII, and its equivalent U.S. EIA Standard RS-466 entitled Procedures for Document Facsimile Transmission in the General Switched Telephone Network, which covers procedures for:

1. Call establishment and call release
2. Compatibility checking, status, and control command
3. Checking and supervision of line conditions
4. Control functions and facsimile operator recall
5. Recognized optional functions and nonstandard options

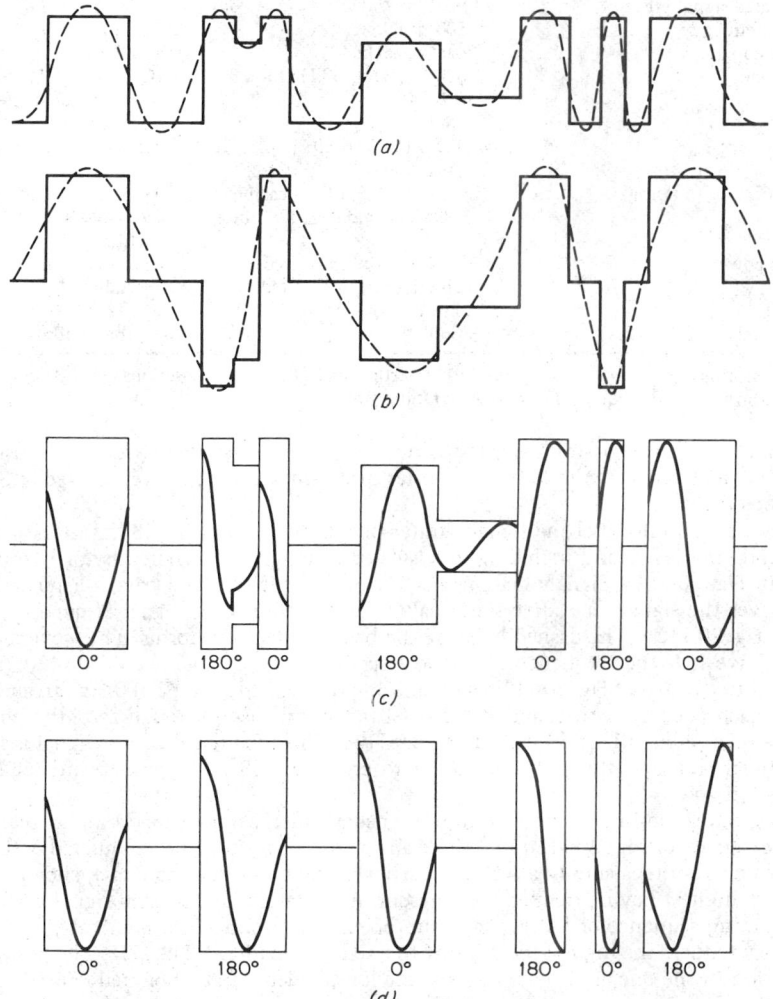

Fig. 20-112. Group 2 waveforms: (*a*) baseband; (*b*) with duobinary inversion; (*c*) modulated carrier, analog response; (*d*) modulated carrier, black-white only.

TABLE 20-11   Key CCITT Group 2 Facsimile Standards

Factor of cooperation	$829 \pm 1\%$
Nominal time/page	3 min (nonstandard 2 min and nonstandard < 2 min with white-space skipping common)
Total scan line L_t	215 mm (8.46 in); up to 222 mm (8.74 in) allowed
Usable scan line L_u	205 mm (8.07 in)
Line use ratio L_u/L_t	0.953
Scan density, traverse (feed)	3.85 lines/mm (97.8 lines/in)
Sweep (scan direction)	Up to 97.9 lines/in
Minimum input document size	ISO A4 (210 × 297 mm = 8.3 × 11.7 in)
Scan-line frequency	360 ± 10^{-5} min^{-1}
Modulation	VS3 AM with phase modulation shifting between 0 and 180°, alternating after each black to white transition; white max, black -26 dB or less
Carrier frequency	$2,100 \pm 10$ Hz
Highest keying frequency	2,487 Hz
Answer tone	Group identification for group 2 = GI2 = 1,850 Hz for 1.5 s repeated after 3 s continuously
Synchronization (phasing)	Line conditioning signal (LCS) 1,00 $\pm$ 50 Hz for 1.5 $\pm$ 0.5 s + 6 $\pm$ 0.5 s of scan lines each 95% black and 5 $\pm$ 1% white; carrier phase reverses each line
Confirmed to receive CFR2	$1,650 \pm 6$ Hz for 3 $\pm$ 15% s indicates receiver has phased and is ready to receive
End of message EOM	$1,100 \pm 38$ Hz for 3 $\pm$ 15% s
Message rec. confirmation MCF2	Repeat of CFR2 starting 0.5 s after end of received EOM
Transmit power	-15 to 0 dBm set by installer
Receive levels	Must operate over -40 to 0 dBm

TABLE 20-12   Key CCITT Group 3 Digital Facsimile Standards

Factor of cooperation	829 standard; 1,658 higher resolution (H and V resolutions not quite same)
Pels per scan line	1,723 (optional 2,048 for 255-mm line)
Nominal time/page	1 min (varies with degree of detail in copy)
Total scan line	215 mm (8.46 in)
Scan density, traverse (feed)	3.85 lines/mm standard (97.8 lines/in) for normal resolution; 7.7 lines/mm higher resolution (195.6 lines/in)
Sweep (scan direction)	Up to 8 lines/mm both modes (204.1 lines/in)
Minimum input document size	ISO A4 (210 by 297 mm = 3.3 × 11.7 in)
Minimum time/scan line	20 ms standard (50 scans/s max)
Options	10 ms with fallback to 20 ms
	5 ms with fallbacks to 10 and 20 ms
	0 ms with fallbacks to 5, 10, and 20 ms plus optional fallback to 40 ms
	40 ms (recognized option)
	10 ms (5 ms higher resolution) with fallback to 20 ms (10 ms higher resolution)
	20 ms standard resolution and 0 ms for higher resolution (this option widely used)
	40 ms (10 ms higher resolution
Coding scheme	1 dimension standard: modified Huffman code*
	2 dimension option: modified READ code
Parameter K	1 dimension code: K = 1
	2 dimension code: normal resolution K = 2
	higher resolution K = 4
Modem	Standard: 2,400 and 4,800 b/s per V.27 ter
	Optional addition: 7,200 and 9,600 b/s per V.29
Premessage controls	Per Recomm. T.30 binary-encoded tonal signaling in HDLC format

*See Tables 20-13 and 20-14.

A facsimile telephone call is made up of five phases, as shown in Fig. 20-114. In phase A (call establishment) the telephone call is placed. In phase B the called station (acting in the identifying mode) responds with signals indicating its capabilities. For groups 1 and 2 the response signal is a series of coded tones. For group 3 the response and further handshake interaction is in binary coded signal with $1 = 1,650 \pm 6$ Hz and $0 = 1,850 \pm 6$ Hz and binary signaling at 300 b/s.

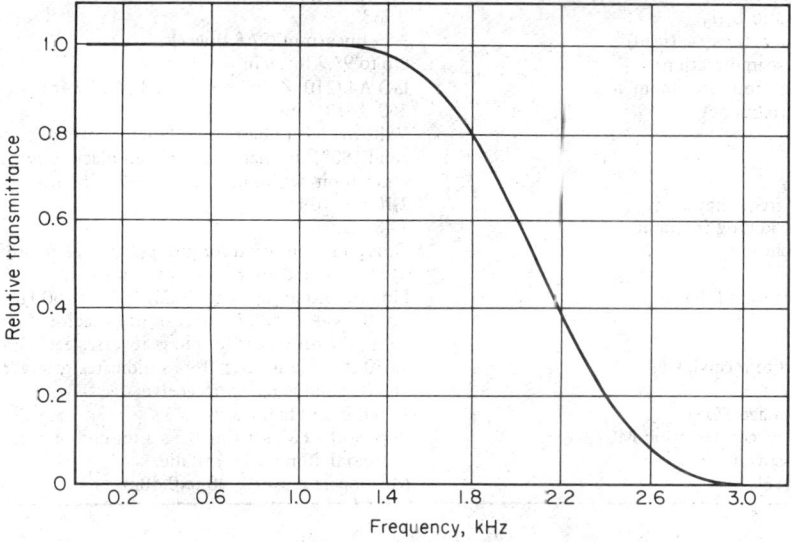

Fig. 20-113. Vestigial sideband filter characteristics for group 2 modulated facsimile signals.

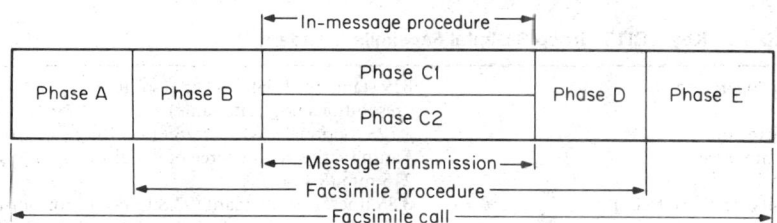

Fig. 20-114. Time-sequence phases of a facsimile telephone call.

The calling station (acting in the command mode) then sends signals stating the mode of operation to be used during the call (group 1, 2, or 3; speed; resolution; special capabilities, etc). If group 1 or 2, a phasing signal is initiated, leading into the message (analog facsimile signal).

For group 3 the procedure is as follows. A *training sequence* is sent, consisting of signals to establish carrier detection, AGC (if required), timing synchronization, to converge (establish transversal equalizer tap settings) the automatic amplitude and phase equalizer with a conditioning sequence, and to synchronize the descrambler (digital modem outputs are scrambled to provide a pseudo-random spectrum).

If the receiver trains and is ready, it acknowledges readiness by sending a confirmation to receive (CFR) signal. Upon receipt of CFR the transmitting station is ready to send the facsimile signal (message). To this point signaling has been at 300 b/s except for the training sequence, which proceeds at the fast data rate of the digital modem. A high-level data-link-control (HDLC) frame structure is used in phase B.

In phase C the facsimile image is encoded and transmitted. Typical facsimile subject copy is 85% white. Group 3 digital facsimile recognizes white or black pels only (no gray scale). The discrete-time-series output from an array scan is 1,728 1s and 0s representing one horizontal line. Since it consists mainly of 0s, the code always assumes that the first pel is a 0. A *white run length* is the number of 0s until a 1 is encountered (run length is zero if the first pel is a 1). Then a *black run length* must follow as the number of 1s until the next 0 is encountered.

Separate tables of white and black run-length words (in binary) describe all possible run lengths from 0 to 1,728. A string of run-length words uniquely describes the line (row) of 1,728 pels.

Run lengths from 0 through 63 are described by terminating codes. Run lengths in equal multiples of 64 from 64 to 1,728 are described by makeup codes, and any run length from 64 through 1,728 can be described by a makeup code plus the appropriate terminating code to add up to the desired run length. The code word 000000000001 signals the end of a line (EOL). It is unique, and even in the presence of errors, any seven consecutive zeros signals EOL.

By choosing the most common run lengths to have the shortest words and encoding them by a procedure due to D. A. Huffman (1952), the subject copy is described with fewer bits than by the full rectilinear scan representation of 1 b for each pel. The ratio of subject-copy pels to encoded bits for one page is called the *compression ratio*. The terminating codes are given in Table 20-13 and the makeup codes in Table 20-14. This run-length scheme used in group 3 facsimile is known as the *modified Huffman code*. Since it compresses only along a line, it is called a *one-dimensional code*.

Compression is optionally extended to the vertical dimension in group 3 by use of the modified READ (relative element address designate) code, which is a two-dimensional code. The technique is to describe the position of an element of the line being transmitted relative to the corresponding element of the previous line. Since successive scan lines tend to be similar, an encoding economy results. In principle the method proceeds as follows:

1. Code the first line using the modified Huffman code.
2. Retain the first line in memory as a *reference line*.
3. If a *transition* (between 0s and 1s) is in the reference line but not in the *new line*, pass it by, using a *passing-mode* code.
4. If the transition in the new line is directly under, or up to 3 pels to the right or left of a transition b_1 in the reference line, describe its position relative to b_1 by a *vertical-mode* code.
5. If none of the above apply, code the run to the next transition in *horizontal mode* using the modified Huffman code.
6. Make the new line the reference line and continue this process until K lines have been encoded.
7. After K lines start afresh by encoding the next line as a new first line using the modified Huffman code again.

To allow time for paper to be fed (advanced) between lines by a stepper motor, a minimum transmission time per total scanning line is standardized at 20 ms (50 scans per second). To avoid inhibiting the use of faster apparatus, options of 0, 5, 10, and 40 ms are also permitted. The fastest available option is identified and selected in phase B.

For lines encoded faster than the minimum transmission time, *fill bits* (0s) are added (essentially a lengthening of EOL) to fill up the remaining minimum transmission time. Together with control signals, fill bits are referred to as *overhead bits*.

Phase D is initiated by sending of the *end-of-transmission* signal consisting of six consecutive EOLs, whose receipt is confirmed. If no more facsimile traffic is to be sent or received, phase E, the termination of the facsimile telephone call, is effected (going on-hook) at both terminals.

Control procedures in T.30 provide for a variety of options, including *escape* procedures for enabling special features incorporated by individual manufacturers, including other private encoding schemes.

Modem signaling speeds of 2,400, 4,800, 7,200, and 9,600 b/s are allowed, with the one-dimensional code and 4,800 b/s, with fall-back to 2,400 b/s, recognized as the standard common-denominator mode for all group 3 facsimile apparatus. The two-dimensional code and 7,200 and 9,600 b/s modem speeds are options, usually furnished because of their higher throughput performance.

Different modem configurations are used for the optional 7,200 and 9,600 b/s speeds from the standard 2,400 and 4,800 b/s speeds. Group 3 facsimile apparatus at 2,400 and 4,800 b/s uses the modulation, scrambler, equalization, and timing signals defined in CCITT Recomm. V.27ter, specifically the preamble sections 2, 3, 8, 10, 11 and the appendix. The V.27ter modem uses a carrier frequency of 1,800 ± 1 Hz. The carrier is amplitude-modulated at 1,600 Bd ± 0.01%. Each baud (Bd) can have eight phase conditions. Each phase condition is 45° apart. Together they correspond to eight combinations of three consecutive bits called *tribits*. The encoded-data-stream output of the scrambler is divided into tribits to modulate the modem. The 1,600 Bd of 3 b/Bd give 4,800 b/s. For the 2,400 -b/s mode the modem is modulated at 1,200 Bd in four phases 90° apart, the bauds being *dibits* (2 bs/Bd).

TABLE 20-13 Modified Huffman Run Length Terminating Codes

White run length	Code word	Black run length	Code word
0	00110101	0	0000110111
1	000111	1	010
2	0111	2	11
3	1000	3	10
4	1011	4	011
5	1100	5	0011
6	1110	6	0010
7	1111	7	00011
8	10011	8	000101
9	10100	9	000100
10	00111	10	0000100
11	01000	11	0000101
12	001000	12	0000111
13	000011	13	00000100
14	110100	14	00000111
15	110101	15	000011000
16	101010	16	0000010111
17	101011	17	0000011000
18	0100111	18	0000001000
19	0001100	19	00001100111
20	0001000	20	00001101000
21	0010111	21	00001101100
22	0000011	22	00000110111
23	0000100	23	00000101000
24	0101000	24	00000010111
25	0101011	25	00000011000
26	0010011	26	000011001010
27	0100100	27	000011001011
28	0011000	28	000011001100
29	00000010	29	000011001101
30	00000011	30	000001101000
31	00011010	31	000001101001
32	00011011	32	000001101010
33	00010010	33	000001101011
34	00010011	34	000011010010
35	00010100	35	000011010011
36	00010101	36	000011010100
37	00010110	37	000011010101
38	00010111	38	000011010110
39	00101000	39	000011010111
40	00101001	40	000001101100
41	00101010	41	000001101101
42	00101011	42	000011011010
43	00101101	43	000011011011
44	00101101	44	000001010100
45	00000100	45	000001010101
46	00000101	46	000001010110
47	00001010	47	000001010111
48	00001011	48	000001100100
49	01010010	49	000001100101
50	01010011	50	000001010010
51	01010100	51	000001010011
52	01010101	52	000000100100
53	00100100	53	000000110111
54	00100101	54	000000111000
55	01011000	55	000000100111
56	01011001	56	000000101000
57	01011010	57	000001011000
58	01011011	58	000001011001
59	01001010	59	000000101011
60	01001011	60	000000101100
61	00110010	61	000001011010
62	00110011	62	000001100110
63	00110100	63	000001100111

The optional 7,200 and 9,600 b/s speeds cannot use the 1,800-Hz carrier. Instead they use a carrier frequency 1,700 $\pm$ 1 Hz to center the power spectrum about the fastest propagation frequency of typical telephone-type circuits. Group 3 facsimile apparatus at 7,200 and 9,600 b/s employs the modulation, scrambler, equalization, and timing signals defined in CCITT Recomm. V.29, specifically sections 1–4, 9, 10, 12, and 13. The 1,700-Hz carrier is modulated at 2,400 Bd $\pm$ 0.01%. The encoded data stream output of the scrambler is divided into bauds of

TABLE 20-14   Modified Huffman Makeup codes

White run length	Code word	Black run length	Code word
64	11011	64	0000001111
128	10010	128	000011001000
192	010111	192	000011001001
256	0110111	256	000001011011
320	00110110	320	000000110011
384	00110111	384	000000110100
448	01100100	448	000000110101
512	01100101	512	0000001101100
576	01101000	576	0000001101101
640	01100111	640	0000001001010
704	011001100	704	0000001001011
768	011001101	768	0000001001100
832	011010010	832	0000001001101
896	011010011	896	0000001110010
960	011010100	960	0000001110011
1,024	011010101	1024	0000001110100
1,088	011010110	1088	0000001110101
1,152	011010111	1152	0000001110110
1,216	011011000	1216	0000001110111
1,280	011011001	1280	0000001010010
1,344	011011010	1344	0000001010011
1,408	011011011	1408	0000001010100
1,472	010011000	1470	0000001010101
1,536	010011001	1536	0000001011010
1,600	010011010	1600	0000001011011
1,664	011000	1664	0000001100100
1,728	010011011	1728	0000001100101
EOL	000000000001	EOL	000000000001

It is recognized that machines exist which accommodate larger paper widths while maintaining the standard horizontal resolution; this option has been provided for by the addition of the makeup code set defined as follows:

Run length (black and white)	Makeup codes	Run length (black and white)	Makeup codes
1,792	00000001000	2,240	000000010110
1,856	00000001100	2,304	000000010111
1,920	00000001101	2,368	000000011100
1,984	000000010010	2,432	000000011101
2,048	000000010011	2,496	000000011110
1,112	000000010100	2,560	000000011111
2,176	000000010101		

four consecutive bits each called *quadbits*. The 2,400-Bd signal can have one of two amplitudes at each of eight phase locations (45° apart). The two amplitude conditions times the eight phase conditions gives the 16 possible combinations of 4 b (quadbits). At 2,400 Bd times 4 b the speed of the V.29 modem is 9,600 b/s. For 7,200 b/s the amplitude is constant and only the phase tribits are used, since 2,400(3) = 7,200 b/s.

If modem internal criteria for successful reception are not met, modem speed is automatically dropped back during phase B. For full modem option apparatus, the V.29 modem can drop back into V.27 ter modem speeds if necessary.

Group 3 facsimile recommended standards and options are summarized in Table 20-12.

Digital facsimile equipment introduced before 1980 is in general not compatible with these standards. Conversion options or conversion services are sometimes available to convert non-standard signals and protocols to other standard or nonstandard formats for forwarding to otherwise incompatible facsimile terminals.

Conversion Services. Automatic store-and-forward conversion services, such as introduced by ITT Faxpak, accept a page of facsimile copy in one signal format and forward it encoded in another, such as acceptance of group 1 U.S. signals (analog FS at 96 lines/in at 6 min/page) and forwarding as group 3 (digital Modified READ Code V.29 modem at 204 lines/in by 98 lines/in in under 1 min) signals. Conversion is effected by decoding the incoming signal to its baseband array of pixels and calculating by interpolation the location and values of pels for the forwarded signal and then encoding them in the forwarded signal system.

Group 4. Group 4 is facsimile apparatus using digital data-compression techniques and interfaced directly to a digital data network in which procedures are incorporated which assure error-free reception.

Under these circumstances no modem is necessary, and the parameter K can equal the number of scan lines in the document (called *infinite K* or *wraparound* because the encoding procedure continues without restart for the equivalent of a single long helical scan line which would "wrap around" the entire drum of Fig. 20-96).

In the early 1980s group 4 standards were being developed by TR-29 and CCITT Study Group XIV. The trend is to employ modified READ code with infinite K within HDLC protocols suitable for common use by Teletex (a CCITT term for a text service successor to Teletype) and digital facsimile traffic. Group 4 facsimile applications imply a store-and-forward character in which the minimum transmission time per total scan line can be zero. With zero minimum time and infinite K the time of transmission for high-resolution pages using modified READ code is cut almost in half compared with group 3 for the same bit rates.

126. Coupling to the Channel. Several methods of coupling facsimile apparatus to transmission channels are used. The group 1 (and some group 2) signals are acoustically coupled to the telephone handset, provided the signal level induced at the tip and ring terminals of the telephone subscriber loop does not exceed -9 dBm and provided no single signal tones exist in the signal-frequency band of $2,600 \pm 150$ Hz. Higher signal levels are allowed for measured loss subscriber loops provided the signal level at the switching central does not exceed -12 dBm.

Signals can be introduced to telephone circuits through a data access arrangement (DAA), which has been certified by the FCC (in the United States). The DAA function can also be built into the facsimile apparatus itself, either by incorporating an isolated approved DAA circuit or by submission of the entire apparatus for FCC approval. The DAA can be connected directly to tip and ring of the telephone circuit. For facsimile equipment conforming to CCITT recommendations the approval procedure for connection to telephone circuits differs in different countries.

Connection to private telephone networks is similar, depending on the network. Connection to special networks depends on the network. For digital interfacing, the group 4 recommendations are being developed.

127. Transform Encoding. For facsimile transmission of gray-scale photographs, the pixel array comprising the subject copy is broken into squares, typically 16×16, 32×32, or 64×64. Each square is transformed into its Fourier, Hadamard, Karhunen-Loeve, slant, or raised cosine transform. These transforms are invertible. The transformed squares can be encoded with a net savings of bits compared with the transmission of a byte for each pixel of the original subject copy. For further details on this approach, see Netravali and Limb; Chen and Smith; and Pratt in the References below.

128. References and Bibliography on Facsimile

Bertine, H. V.   Physical Level Protocols, *IEEE Trans. Commun.*, April 1980, pp. 433–444.

CCITT 7th Plen. Assem., Vol. VII, "Telegraph Technique," International Telecommunications Union, Geneva, 1981.

Chen, W. H., and C. H. Smith   Adaptive Coding of Monochrome and Color Images, *IEEE Trans. Commun.*, November 1977, pp. 1285–1292.

Costigan, D. M.   "Electronic Delivery of Documents and Graphics," Van Nostrand Reinhold, New York, 1978.

Costigan, D. M.   Facsimile Comes Up to Speed, *IEEE Commun. Mag.*, May 1980, pp. 30–34.

Huang, T. S.   Coding of Two Tone Images, *IEEE Trans. Commun.*, November 1977, pp. 1406–1424.

Maeda, N., Y. Kita, M. Koya, K. Ishizuka, and H. Nakazawa   A Facsimile LSI Modem for AM-PM VSB, *ICC 1979 Conf. Rec.*, June 1979, pp. 37.7.1–37.7.5.

Mertz, P., and F. Gray   A Theory of Scanning and Its Relation to the Characteristics of the Transmitted Signal in Telephotography and Television, *Bell Syst. Tech. J.*, July 1932, pp. 464–515.

Mussman, H. G., and D. Preuss   Comparison of Redundancy Reducing Codes for Facsimile Transmission of Documents, *IEEE Trans. Commun.*, November 1977, pp. 1425–1433.

Netravali, A. N., and J. O. Limb   Picture Coding: A Review, *Proc. IEEE*, March 1980, pp. 366–406.

Pratt, W. K.   "Image Transmission Techniques,' Academic, New York, 1979.

"Xerox Telecopier Transceiver Compatibility Specifications," Xerox Business Systems, Dallas, Tex., 1973.

Special Issue on Digital Encoding of Graphics, *Proc. IEEE*, July 1980.

Broadcasting Systems

JOSEPH L. STERN *President, Stern Telecommunications Corporation; formerly Vice President-Engineering, CBS Television Services, Columbia Broadcasting System; Senior Member, IEEE, SCTE and SMPTE*

NORMAN W. PARKER *Corporate Staff Scientist, Motorola, Incorporated; Fellow, IEEE*

CONTENTS

Numbers refer to paragraphs

Broadcast Transmission Practice

BY JOSEPH L. STERN

STANDARD-BROADCAST (AM) PRACTICE

1. Standard-Broadcast Allocations. The band 535 to 1,605 kHz is used for standard amplitude-modulation (AM) sound broadcasting. This band is divided into 107 channels each 10 kHz wide; carrier frequencies are assigned at 10-kHz intervals from 540 to 1,600 kHz.

Table 21-1 lists the standard broadcast channels allocated in the United States by the Federal Communications Commission (FCC) with their service classes.

2. Allocation Standards. Section 73.21 of the FCC Rules establishes in the United States three classes of channels in the standard broadcast band: *clear channels*, for high-powered stations; *regional channels*, for medium-powered stations; and *local channels*, for low-powered stations (see Table 21-1). These stations have three service areas,* primary, secondary, and intermittent (see Pars. **21-8** to **21-10**). Class I stations (Par. **21-3**) render service to all three areas. Class II (Par. **21-4**) stations render service to a primary area, but the secondary and intermittent service areas may be materially limited or destroyed by interference from other stations. Class III and IV (Pars. **21-5** and **21-6**) stations serve primary-service areas, as interference from other stations usually prevents secondary service and may limit intermittent service. See second footnote to Table 21-1.

3. Class I stations are dominant stations operating on clear channels with powers of not less than 10 nor more than 50 kW. These stations are designed to render primary and secondary service over an extended area and at relatively long distances. They are so allocated (Table 21-1) that their primary-service areas are free from objectionable interference from other stations on the same and adjacent channels. Their secondary-service areas are free from objectionable interference from stations on the same channels but not from those on adjacent channels. If it is desired to determine the area in which adjacent-channel ground-wave interference (10 kHz

*Definitions of these service areas are given in Sec. 73.11 of the FCC Rules.

removed) to sky-wave service exists, it may be considered as the area within which the ratio of the desired 50% sky wave of the class I station to the undesired ground wave of a station 10 kHz removed is 1:4.

From an engineering point of view, class I stations can be divided into two groups. In group IA are stations on whose channels (except to the extent provided by that section and by Sec. 73.22, of the FCC Rules) duplicate nighttime operation is not permitted. The power of these stations is required to be 50 kW.

TABLE 21-1   U.S. Standard-Broadcast Carrier Frequencies and Service Classes*

Channel, kHz	Classification†	FCC Class	Channel, kHz	Classification†	FCC Class
540	Clear	II	1,050	Clear	II
550–630	Regional	IIIA, IIIB	1,060–1,140	Clear	I, II
640–680	Clear	I, II	1,150	Regional	IIIA, IIIB
690	Clear	II	1, 60–1,210	Clear	I, II
700–720	Clear	I, II	1,220	Clear	II
730–740	Clear	II	1,230–1,240	Local	IV
750–780	Clear	I, II	1,250–1,330	Regional	IIIA, IIIB
790	Regional	IIIA, IIIB	1,340	Local	IV
800	Clear	II	1,350–1,390	Regional	IIIA, IIIB
810–850	Clear	I, II	1,400	Local	IV
860	Clear	II	1,410–1,440	Regional	IIIA, IIIB
870–890	Clear	I, II	1,450	Local	IV
900	Clear	II	1,460–1,480	Regional	IIIA, IIIB
910–930	Regional	IIIA, IIIB	1,490	Local	IV
940	Clear	I, II	1,500–1,530	Clear	I, II
950–980	Regional	IIIA, IIIB	,540	Clear	II
990	Clear	II	1,550–1,560	Clear	I, II
1,000	Clear	I, II	1,570–1,580	Clear	II
1,010	Clear	II	1,590–1,600	Regional	IIIA, IIIB
1,020–1,040	Clear	I, II			

*For details on special operating procedures and assignment criteria, see FCC Rules, beginning at Sec. 73.25.
†In June 1980, the FCC proposed a change in its Rules that would reduce the protection of clear channels to 750 mi and thereby allow the licensing of more than 200 additional stations in the AM service. At the time of going to press, this proposal was the subject of litigation and had not been put into effect. — ED.

Stations in Group IA are afforded protection as follows:
Daytime. To the 0.1 mV/m ground-wave contour from stations on the same channel and to the 0.5 mV/m ground-wave contour from stations on adjacent channels
Nighttime. To the 0.5 mV/m 50% sky-wave contour from stations on the same channel and to the 0.5 mV/m ground-wave contour from stations on adjacent channels

Stations in group IB are those assigned to channels on which duplicate operation is permitted; i.e., other class I or class II stations operating on unlimited time may be assigned to such channels. During nighttime hours, stations in group IB are protected to the 500 μV/m 50% sky-wave contour and during daytime hours of operation to the 00 μV/m ground-wave contour from stations on the same channel. Protection is given to the 500 μV/m ground-wave contour from stations on adjacent channels for both day and night operation. The operating powers of stations in group IB must not be less than 10 kW or more than 50 kW.

4. Class II stations are secondary stations which operate on clear channels with powers not less than 0.25 kW nor more than 50 kW, except that class IIA stations may not operate during nighttime hours with less than 10 kW. Class II stations are required to use a directional antenna or other means to avoid causing interference within the protected service areas of class I stations or other class II stations. For special rules and standards concerning class IIA stations, see the FCC Rules, Sec. 73.22.

These stations normally render primary service only, the area of which depends on the geographical location, power, and frequency. The service area may be relatively large but is limited by, and subject to, interference from class I stations. It is recommended that class II stations be so located that the interference received from other stations will not limit the service area to greater than the 2.5 mV/m ground-wave contour at night and the 0.5 mV/m ground-wave contour during the day (the values for the mutual protection of this class of stations with other stations of the same class). Class IIA stations are normally protected to their 0.5 mV/m ground-wave contour (daytime) and at night to the limit imposed by the co-channel class IA station.

5. Class III stations operate on regional channels and normally render primary service to the larger cities and to contiguous rural areas. These stations are subdivided into two classes; class IIIA stations, which operate with powers not less than 1 kW nor more than 5 kW, are normally protected to the 2,500 μV/m ground-wave contour at night and to the 500 μV/m ground-wave contour during the day. Class IIIB stations, which operate with powers not less than 0.5 kW nor more than 1 kW at night and 5 kW daytime, are normally protected to the 4,000 μV/m ground-wave contour at night and 500 μV/m ground-wave contour during the day.

6. Class IV stations operate on local channels, normally rendering primary service only to a city or town and the contiguous suburban or rural areas, with power not less than 250 W. The upper limits are 250 W at night and 1 kW daytime. Such stations are normally protected to the 0.5 mV/m contour daytime.

On local channels the separation required for the daytime protection also determines the nighttime separation. Where directional antennas are employed by class IV stations operating with more than 250 W power, the separations must not be less than those necessary to afford protection, assuming nondirectional operation with 250 W.

The actual nighttime limitation must be calculated. An approximate method is based on the assumption of a quarter-wavelength antenna height and 88 mV/m at 1 mi effective field for 250 W power, using the 10% sky-wave field-intensity curve.* Zones defined by circles of various radii specified in the accompanying table are drawn about the desired station, and the interfering 10% sky-wave signal from each station in a given zone is considered to be the value tabulated. The effective interfering 10% sky-wave signal is taken to be the square root of the sum of the squares (rss value) of all signals originating within these zones. Stations beyond 500 mi are not considered.

Zone	Inner radius	Outer radius	10% skywave signal, mV/m
A	. . .	60	0.10
B	60	80	0.12
C	80	100	0.14
D	100	250	0.16
E	250	350	0.14
F	350	450	0.12
G	450	500	0.10

Where the power of the interfering station is other than 250 W, the 10% sky-wave signal should be adjusted by the square root of the ratio of the power to 250 W.

7. Service Reclassification. The class of any station is determined by the channel assignment, the power, and the field-intensity contour to which it renders service free of interference from other stations as determined by the FCC Rules. No station is permitted to change to a class protected to a contour of less intensity than the contour to which the station actually renders interference-free service. Any station of a class normally protected to a contour of less intensity than that to which the station actually renders interference-free service is automatically reclassified by the FCC. Likewise, any station to which the interference is reduced, so that service is rendered to a contour normally protected for a higher class, is automatically reclassified.

8. Signal Levels for Different Service Areas—Primary Service. The signals necessary to render primary service to different types of service areas are as follows:

Area	Field-intensity ground wave,* mV/m
City, business or factory areas	10–50
Residential areas	2–10
Rural, all areas during winter or northern areas during summer	0.1–0.5
Southern areas during summer	0.25–1.0

*Section 73.184 of the FCC Rules gives graphs showing distance to various ground-wave field-intensity contours for different frequency and ground conductivities.

*FCC Rules, Par. 73.190, Fig. 2.

These values are based on the usual noise levels in the respective areas, assuming no objectionable finding or limiting interference from other broadcast stations. The values apply both day and night, but fading or interference from other stations usually limits the primary service at night in rural areas to higher values of field intensity than the values given.

The FCC will authorize a directive antenna for a class IV station for daytime operation only with power in excess of 250 W. In computing the degrees of protection which such an antenna will afford, the radiation produced by this antenna must be assumed to be no less, in any direction, than that which would result from nondirectional operation utilizing a single element of the directional array with 250 W.

Standards are not stated for interference from atmospherics or man-made electric noise, as no uniform method of measuring these effects has been established. In an individual case, objectionable interference from any source, except other broadcast signals can be determined by comparing the actual noise interference reproduced during reception of a desired broadcast signal to the degree of interference that would be caused by another broadcast signal within 20 kHz of the desired signal and having a carrier power ratio of 20:1 with both signals modulated 100% on peaks of usual programs.

9. Secondary Service. Secondary service is delivered in the areas where the sky wave for 50% or more of the time has a field intensity of 500 μV/m or greater. It is not considered that satisfactory secondary service can be rendered to cities unless the sky wave approaches the ground-wave value required for primary service. The secondary service is necessarily subject to interference and fading, whereas the primary-service area of a station is not. Only class I stations are assigned on the basis of rendering secondary service.

Standards have not been established for objectionable fading as such standards necessarily depend on receiver characteristics, which have changed considerably during the years. Only selective fading causing audio distortion and signal fading below the noise level are objectionable in modern receivers, since the automatic volume-control circuits usually maintain the audio output sufficiently constant to be satisfactory during conditions of fading.

10. Intermittent Service. Intermittent service is rendered by the ground wave. It begins at the outer boundary of the primary-service area and extends to the value of signal that has no service value. This limit may extend down to a few microvolts in certain areas and up to several millivolts in areas of high noise level, interference from other stations, or objectionable fading at night. The intermittent-service area may vary widely from day to night and from time to time, as the name implies. Only class I stations are assigned protection from interference from other stations in the intermittent-service area.

11. Objectionable Interference. Objectionable nighttime interference from another broadcast station is defined as the degree of interference produced when at a specified field-intensity contour with respect to the desired station the field intensity of an undesired station (or the rss value of field intensities of two or more stations on the same frequency) exceeds, for 10% or more of the time, the values set forth in the FCC standards.

With respect to the rss values of interfering field intensities (except in the case of class IV stations on local channels) calculation is accomplished by considering the signals in order of decreasing magnitude, adding the squares of the values and extracting the square root of the sum, excluding those signals which are less than 50% of the rss value of the higher signals already included.

The rss value is not considered to be increased when a new interfering signal is added which is less than 50% of the rss value of the interference from existing stations and which at the same time is not greater than the smallest signal included in the rss values of interference from existing stations. The application of this "50% exclusion" method of calculation may result in anomalies. For example, the addition of a new interfering signal or the increase in value of an existing interfering signal may cause the exclusion of a previously included signal and may cause a decrease in the calculated rss value of interference.

In such cases, an alternate method of calculating the rss values of interference should be employed.

As an example of rss interference calculations, assume that the existing interferences are:

Station	Interference, mV/m
1	.8
2	0.40
3	0.39
4	0.38

The rss value from stations 1 to 3 is 1.31 mV/m; therefore interference from station 4 is excluded, for it is less than 50% of 1.31 mV/m.

Station A receives the following interferences:

Station	Interference, mV/m
1	1.0
2	0.60
3	0.59

It is proposed to add a new limitation, 0.68 mV/m. This is more than 50% of 1.31 mV/m, the rss value of stations 1, 2, and 3. The rss value of station 1 and of the proposed station would be 1.21 mV/m, which is more than twice as large as the limitations from station 2 or 3. Under the above provision of a new signal, the three existing interferences are calculated for purposes of comparative studies, resulting in an rss value of 1.47 mV/m. However, if the proposed station is ultimately authorized, only station 1 and the new signal are included in all subsequent calculations for the reason that stations 2 and 3 are less than 50% of 1.21 mV/m, the rss value of the new signal and station 1.

12. Coverage Estimates. For the purpose of estimating the coverage and the interfering effects of stations in the absence of field-intensity measurements, use must be made of Fig. 8 of FCC Rules, Par. 73.190, which describes the estimated effective field for 1 kW power input of simple vertical omnidirectional antennas of various heights with ground systems of at least 120 quarter-wave length radials. Certain approximations, based on the curve or other appropriate theory, may be made when other than such antennas and ground systems are employed, but in any event the effective field to be employed shall not be less than that tabulated:

Class of station	Effective field, mV/m
I	225
II, III	175
IV	150

If a directional antenna is employed by the interfering station, the interference varies in different directions, being greater than the tabulated limiting values in certain directions and less in others. To determine the interference in any direction the measured or calculated radiated field (unabsorbed field intensity at 1 mi from the array) must be used in conjunction with the appropriate propagation curves. The existence or absence of objectionable interference due to sky-wave propagation must be determined by reference to the appropriate propagation curves (Fig. 1a or Fig. 2 of Sec. 73.190 of the FCC Rules).

A summary of FCC regulations applicable to standard broadcast stations appears in Table 21-2.

According to the FCC Rules, Table 21-3 is to be used for determining the minimum ratio of the field intensity of a desired to an undesired signal for interference-free service. For a desired ground-wave signal interfered with by two or more sky-wave signals on the same frequency, the rss value of the latter is used. From Table 21-3 it is apparent that in many cases stations operating on channels 10 and 20 kHz apart may be operated with antenna systems side by side or otherwise in proximity without any indications of interference if the interference is defined only in terms of the permissible ratios listed. As a practical matter, serious interference problems may arise when two or more stations with the same general service area are operated on channels 10, 20, and 30 kHz apart.

13. Ground-Wave Field Intensity. The computed values of ground-wave field intensity as a function of the distance from the transmitting antenna are given in the FCC Rules, Sec. 73.184, Graph 12. The ground-wave field intensity is considered to be that part of the vertical component of the electric field received on the ground which has not been reflected from the ionosphere or the troposphere. These charts were computed for a dielectric constant of 15 for land and 80 for seawater (referred to air as unity) and for the ground conductivities (expressed in millisiemens per meter) given on the curves.

The curves show the variation of the ground-wave field intensity with distance to be expected for transmission from a short vertical antenna at the surface of a uniformly conducting spherical earth with the ground constants shown on the curves; the curves are for an antenna power and

Table 21-2. Summary of FCC Regulations for Standard Broadcast Stations

Class of station	Class of channel used	Permissible power, kW	Signal-intensity contour of area protected from objectionable interference,[a] μV/m		Permissible interfering signal on same channel,[b] μV/m	
			Day[c]	Night	Day[c]	Night[d]
IA	Clear	50	100 same channel 500 adjacent channel	500 same channel (50% sky wave)[e] 500 adjacent channel[c]	5	25[e]
IB	Clear	10-50	100 same channel 500 adjacent channel	500 same channel (50% sky wave) 500 adjacent channel[c]	5	25
IIA	Clear	0.25-50 day 10-50 night	500	500 adjacent channel[c] 500	25	25
IIIB, IIID	Clear	0.25-50	500	2,500ᵍ	25	125
IIIA	Regional	1-5	500	2,500ᵍ	25	125
IIIB	Regional	0.5 1 night 5 day	500	4,000ᵍ	25	200
IV	Local	0.25 night 0.25-1 day	500	ᵍ	25	ᵍ

[a] When a station is already limited by interference from other stations to a contour of higher value than that normally protected for its class, this contour shall be the established standard for such station with respect to interference from all other stations.
[b] For adjacent channel, see paragraph 73.182(w) of FCC Rules.   [c] Ground wave.
[d] Sky-wave field intensity for 10% or more of the time.
[e] Class IA stations on channels reserved for the exclusive use of one station during nighttime hours are protected from cochannel interference on that basis. On the frequency 770 kHz, two class I stations may be assigned.
[f] These values are with respect to interference from all stations except class IB, which stations may cause interference to a field-intensity contour of higher value. However, it is recommended that class II stations be so located that the interference received from class IB stations will not exceed these values. If the class II stations are limited by class IB stations to higher values, then such values shall be the established standard with respect to protection from all other stations.
ᵍ Not prescribed; see paragraph 73.182(a)(4) of FCC Rules.

efficiency such that the inverse-distance field is 100 mV/m at 1 mi. The curves are valid at distances large compared with the dimensions of the antenna for other than short vertical antennas. A typical FCC set of ground-wave field-intensity curves, computed for 1,000 kHz, is shown in Fig. 21-1.

Table 21-3.   Desired-to-Undesired Signal Ratios

Frequency separation of desired to undesired signals, kHz	Desired ground wave to:		Desired 50% sky wave to undesired 10% sky wave
	Undesired ground wave	Undesired 10% sky wave	
0	20:1	20:1	20:1
10	1:1	1:5	*

* The secondary service area of a class I station is not protected from adjacent channel interference. However, if it is desired to make a determination of the area in which adjacent channel ground-wave interference (10 kHz removed) to sky-wave service exists, it may be considered as the area where the ratio of that desired 50% sky wave of the class I station to the undesired ground wave of a station 10 kHz removed is 1:4.

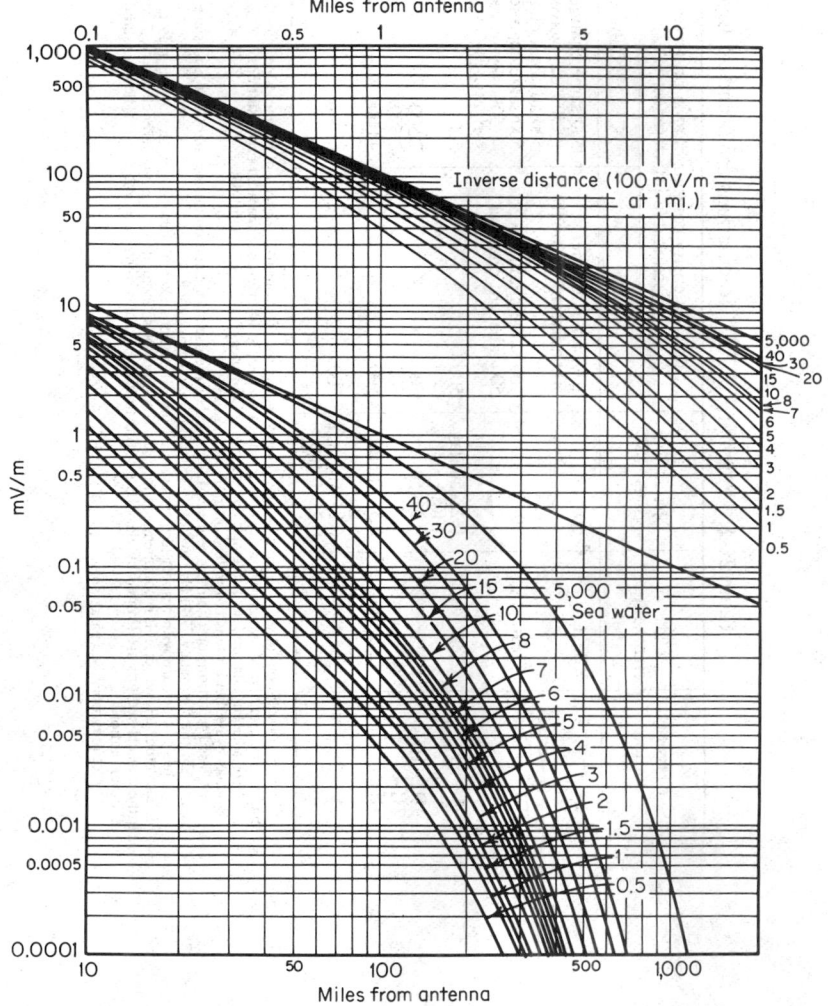

Fig. 21-1. Curves for ground-wave field intensity vs. distance, computed at 1,000 kHz for various values of ground conductivity. The upper curves apply to the upper scale of distance, lower curves to the lower scale. (*From FCC Rules, Sec. 73.174.*)

The *inverse-distance field* (100 mV/m divided by the distance in miles) corresponds to the ground-wave field intensity to be expected from an antenna with the same radiation efficiency when it is located over a perfectly conducting earth. To determine the value of the ground-wave field intensity corresponding to a value of inverse-distance field other than 100 mV/m at 1 mi, the field intensity as given on the charts is multiplied by the desired value of inverse-distance field at 1 mi divided by 100. For example, to determine the ground-wave field intensity for a station with an inverse-distance field of 1,700 mV/m at 1 mi, multiply the values given on the charts by 17.

The value of the inverse-distance field to be used for a particular antenna depends upon the power input to the antenna, the nature of the ground in the neighborhood of the antenna, and the geometry of the antenna.

To allow for the refraction of radio waves in the lower atmosphere due to the variation of the dielectric constant of the air with height above the earth, a radius of the earth equal to four-thirds the actual radius is used in the computations for the effect of the earth's curvature in the manner originally suggested by Burrows; i.e., the distance corresponding to a given value of attenuation due to the curvature of the earth in the absence of air refraction is multiplied by the factor $(4/3)^{2/3} = 1.21$. The amount of this refraction varies from day to day and from season to season, depending on the air-mass conditions in the lower atmosphere. If k denotes the ratio between the equivalent radius of the earth and the true radius, the following table gives the values of k for several typical air masses in the United States.

Air-mass type	k	
	Summer	Winter
Tropical gulf T_c	1.53	1.43
Polar continental P_c.....................	1.31	1.25
Superior S	1.25	1.25
Average	1.32	

STANDARD-BROADCAST TRANSMITTING EQUIPMENT

14. Equipment Functions. The equipment functions of a public broadcast station (which apply to AM, FM, shortwave, and television broadcasting) include the following.

Program Source. The program source consists of inputs from studios, remote locations, disc and tape recordings, and material received from other sources, e.g., network originations, by telephone lines or cables, radio relay, or satellites.

Program Input and Control. The source material, under operator or computer supervision to assure continuity, is fed to the input and control system where it is processed, mixed, switched, and sent to the transmitter plant. The principal functions performed are to raise or lower the volume level, provide equalization for different program sources (or for special effects), to mix multiple sources, to provide switching and continuity and to monitor program activity and quality.

Studio-to-Transmitter Link. For audio sources, the link usually consists of an audio cable connecting the program input and control facilities to the transmitters. If the distance to be covered is large, a high-quality equalized telephone line or radio relay link may be used. In FM broadcasting, off-the-air pickup may be used under favorable reception conditions.

Transmitter Input and Control. These facilities and controls bring the level of the signal from the link up to the point that the transmitter can accept. Included are limiting and clipping amplifiers to maintain a high average signal level, thus keeping the modulation percentage and output power as high as is consistent with good audio quality and the FCC Rules. At the transmitter, facilities are also provided for the measurement of the input level and output monitoring to check the overall quality of the signal, modulation percentage, and carrier-frequency tolerance. The high degree of stability exhibited by modern AM, FM, and TV transmitters has led to widespread use of remote control. In AM stations using nondirectional antennas and FM stations, remote-control facilities provide for an unattended transmitter with all control and monitoring functions performed at the program-source location. Automatic transmission systems providing for fully unattended operation except for manual activation at the beginning of each broadcasting day are permitted for nondirectional AM stations and for FM stations.

15. Transmitter Performance Standards. The Electronic Industries Association (EIA, formerly RETMA) has formulated standards of performance for AM broadcast transmitters, published as Standard TR-101A, Electrical Performance Standards for Standard Broadcast Transmitters. This document, which should be consulted for detailed numerical values and measurement criteria, lists definitions, standards, and methods of measurement of the following items:

Carrier output rating
Carrier-power output capability
Carrier-frequency range
Carrier-frequency stability
Carrier shift
Carrier noise level
Magnitude of rf harmonics
Normal load
Transmitter-output-circuit adjustment facilities
Output voltage and impedance for rf and audio monitor connections
Modulation capability
Audio-input level for 100% modulation
Audio-frequency response
Audio-frequency harmonic distortion
Rated power supply
Power-supply variation
Power input

16. Solid-State Transmitter. The simplified block diagram of a 1-kW solid-state transmitter* is shown in Fig. 21-2. The manufacturer's electrical specifications are given in Table 21-4.

The transmitter has a maximum output of 1,110 W, provided by the combination of five identical modules of four transistors each, operated as class D, full-wave, push-pull amplifiers. The outputs of the five modules are summed in a broadband transformer combining network which provides for balancing the contribution of each of the modules. In the event of a malfunction of

*Sintronics, Type SI-A-IS.

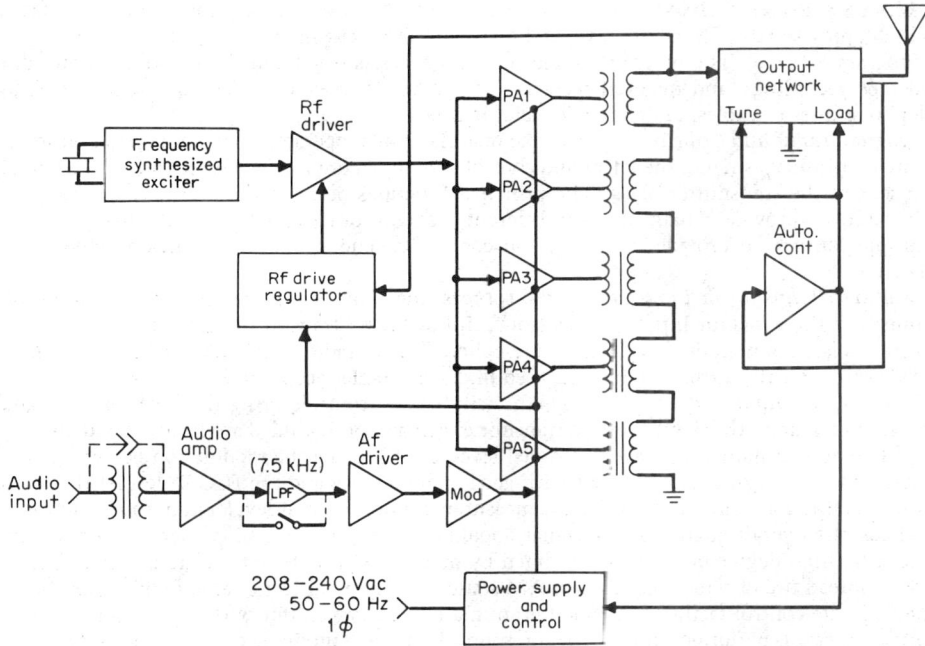

Fig. 21-2. Block diagram of all-solid-state 1-kW transmitter. *(Sintronics, Inc.)*

any one of the modules, operations continue uninterrupted with a 20% reduction in output power. According to the manufacturer, the use of switch-mode class D operation for the power amplifier provides a power-amplifier (PA) efficiency in excess of 95%.

The modulation system is essentially similar to a conventional high-level plate-modulation technique. The PA collector voltage is varied at an audio rate, ranging from zero during 100% negative modulation peaks to twice the carrier-level collector voltage at 100% positive peak modulation. Since the output impedance of the solid-state modulator stage is extremely low, no modulation transformer is required for load-impedance matching. The modulator stage consists of series-parallel transistors operated as a complementary-symmetry cascade power amplifier with an output capability in excess of 1,000 W.

TABLE 21-4   Specifications of a Solid-State AM Transmitter

Performance:	
Power output*	250/500/1,000 W nominal; 1,110 W max
Output impedance	50 Ω, unbalanced
Frequency range	540–1650 kHz
Frequency stability	±5 Hz at 1,000 kHz
Carrier shift	<2% at 100% modulation
Modulation capability	125% positive peaks; 100% negative peaks
Frequency response	±0.5 dB, 50–10,000 Hz (7.5 Hz LPF bypassed)
Harmonic distortion	<1.5% (typically 0.5%) 50–10,000 Hz at 95% modulation
Intermodulation distortion	<2.0% (typically 1.0%; 4:1 ratio)
Audio input impedance	600 Ω, balanced
Audio input level	+10 ± 2 dBm for 100% modulation
AM noise	−60 dB or greater
Mechanical and electrical:	
Power requirements	208–240 V ac, 50/60 Hz single phase
Power consumption (approx.)	Zero modulation 1,800 W; 100% modulation: 3,000 W; average program material: 2,400 W
Maximum altitude	12,500 ft (3,810 m) above mean sea level at 35°C ambient temperature

*Any two power levels available as standard.

A frequency stability of ±5 Hz at 1.0 MHz is maintained through the use of a frequency-synthesized exciter module. Operation of the exciter on any 10-kHz interval within the standard broadcast frequency range is programmable by switch selection on a module card. A 10-kHz reference signal is derived by digital frequency division of the output of a low-temperature-coefficient 4.0-MHz clock-frequency oscillator which operates without an oven. The operating frequency is established by an integrated-circuit voltage-controlled oscillator (VCO). The VCO sample is compared with a 10-kHz clock frequency in a phase comparator, providing a correction voltage which is applied to the VCO and through varactor action locks the VCO to the operating frequency. A square-wave output from the VCO is buffered and amplified to the proper level to feed the rf driver module.

An automatic rf drive-regulator circuit is provided to ensure that the PA drive power is at optimum level throughout the modulation cycle. This regulator samples the modulated rf envelope, and the modulator output signals and provides correction signals through operational amplifiers.

An automatic power-output circuit is also provided which maintains power output by controlling a power-line variable transformer and by adjusting the output-network tuning inductor to accommodate changes in output load reactance and load resistance. Where the automatic control system cannot provide the proper correction, automatic switching to a low-power operating mode is provided. The transmitter is designed for manual operation on all types of remote control systems. Telemetering of voltages and currents is provided. The transmitter is contained in a single cabinet 78 in high, 22 in wide, and 30 in deep.

17. Typical 50-kW AM Transmitters. A typical 50-kW transmitter* uses a high-efficiency linear amplifier with screen-grid modulation. The final (modulated) amplifier consists of

*Continental Electronics Manufacturing Company, Type 317C2

two type 4CX35000C ceramic tetrodes in a high-efficiency linear amplifier configuration. Both tubes are simultaneously screen-grid-modulated from a low-power modulator. In this circuit, the linearity of negative peaks depends mainly on screen-grid linearity and not on control-grid operating conditions. Thus the carrier tube can be operated as a class C amplifier with resulting high plate efficiency.

Since the plate-voltage swing does not increase with positive modulation, the dc plate voltage is much higher than normally used for plate-modulated transmitters. At 16 kV, a plate efficiency of 80% is claimed by the manufacturer. The addition of resonant circuits, giving a trapezoidal shape to the plate-current pulses, yields plate efficiencies of 90% or higher.

Figure 21-3 is a simplified schematic diagram of the output amplifier. The carrier tube V_2 is a conventional grounded-cathode class C amplifier supplying the full 50-kW carrier power without modulation. The screen voltage is maintained at 700 V by a separate low-voltage supply. The

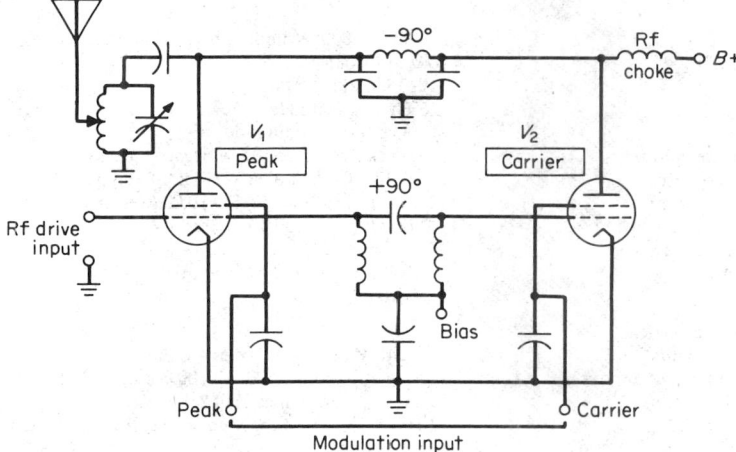

Fig. 21-3. Schematic of a screen-grid-modulated 50-kW final amplifier for standard-broadcast service with 90° phase shift between carrier and peak amplifiers. *(Continental Electronics Manufacturing Company.)*

plate voltage is 16 kV, and rf grid excitation maintains a peak plate swing of 15 kV. This state can be screen-modulated only in a negative direction.

The peak tube V_1 has the same plate voltage and rf grid excitation as the carrier tube V_2. As a positive-going voltage is applied to the peak-tube screen, the tube delivers power into the load until it reaches crest condition, i.e., until the instantaneous screen voltage of V_2 equals the carrier-tube-V_2 screen voltage.

At the crest condition, both tubes are operating with equal-split plate swing, load impedance, screen voltage, and grid drive. Since the carrier-tube load impedance is half what it is at carrier level and the plate swing is the same, the carrier-tube output is doubled from 50 to 100 kW. Since the peak tube is operating under identical conditions, it also provides 100 kW. The combined output is therefore 200 kW on positive peak, as required for 100% amplitude modulation.

Since the voltage contributed by the carrier tube undergoes a 90° phase lag by the time it appears across the load, it is necessary to introduce a 90° phase advance in the carrier-tube grid driving voltage for the power output of both tubes to combine in proper phase. This is accomplished by the leading 90° grid network shown in the figure.

Another 50-kW transmitter* is outlined in the block diagram of Fig. 21-4. This is a high-level plate-modulated air-cooled transmitter using a novel pulse-duration modulation scheme. The transmitter proper uses only five tubes. The pulse-duration modulator operates as follows (see Fig. 21-5):

1. The audio input A is added to a 70-kHz sawtooth wave B to form the threshold input C.

2. The threshold level (power-controlled) determines the point on the sawtooth wave at which the pulse amplifier conducts. After clipping and amplification, squared pulses, which vary in duration with the audio, are formed at D.

*Harris Model MW-50A

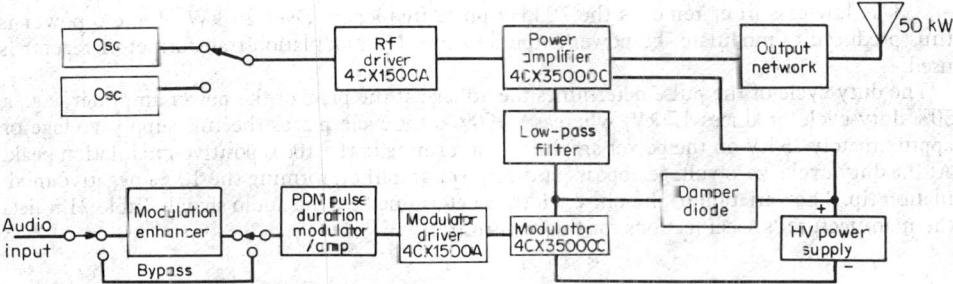

Fig. 21-4. Block diagram of a 50-kW standard-broadcast transmitter using audio-controlled pulse-duration circuits for modulation. *(Harris Corp.)*

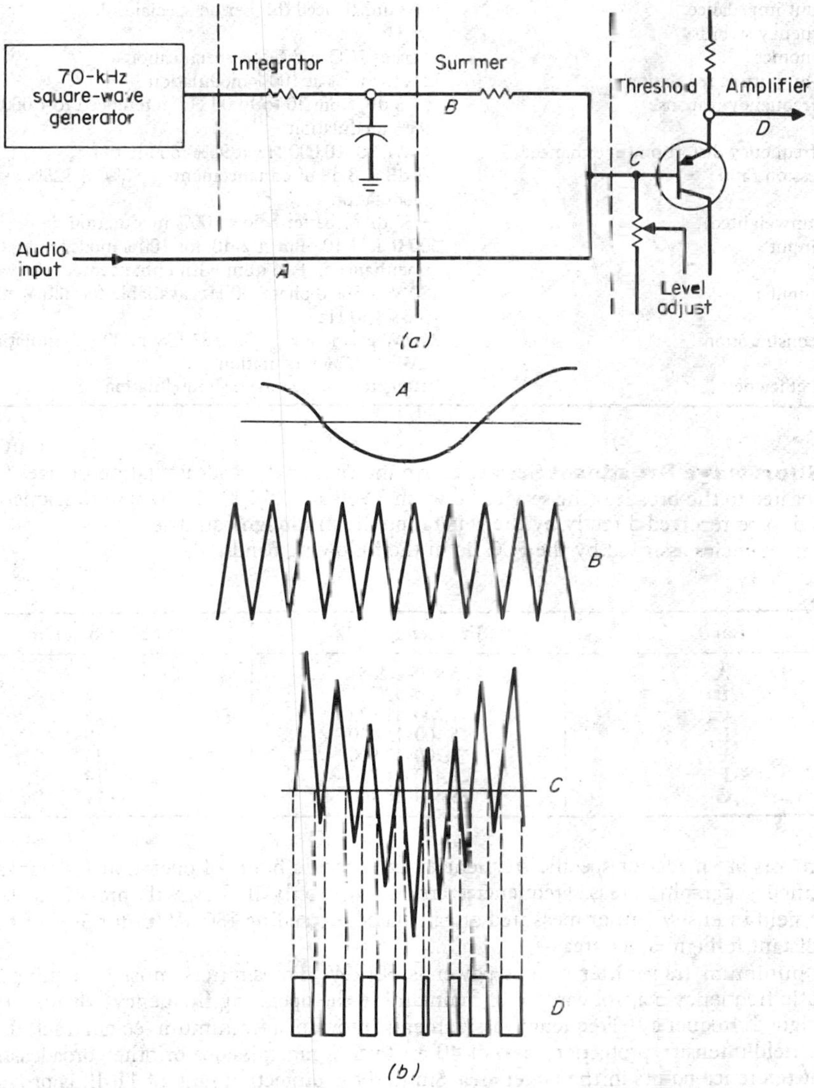

Fig. 21-5. Circuit (*a*) and waveforms (*b*) of the pulse-duration modulation employed in the transmitter of Fig. 21-4. The audio waveform *A*, sawtooth *B*, combined *C*, and pulse duration *D* are described in the text.

3. A low-pass filter removes the 70-kHz pulse frequency. Over 25 kW of audio power is thus produced to modulate the power-amplifier stage. No modulation transformer or reactor is used.

The duty cycle of the pulse determines the voltage at the plate of the power amplifier, e.g., a 50% duty cycle produces 12 kV, whereas a 100% duty cycle places the full supply voltage of approximately 25 kV on the power amplifier, conforming to the 100% positive-modulation peak. At 0% duty cycle, zero voltage appears at the power amplifier, forming the 100% negative-modulation tip. The variation of the pulse width is determined by the audio signal. Table 21-5 lists the manufacturer's specifications for this transmitter.

TABLE 21-5   Specifications of the MW-50A Transmitter

Power output	50 kW (rated), 60 kW (capable); convenient power reduction to 25 or 10 kW
RF frequency range	535–1620 kHz, supplied to frequency as ordered
RF output impedance	50 Ω unbalanced (higher on special order)
RF frequency stability	±5 Hz
RF harmonics	Exceeds FCC and CCIR specifications
Carrier-amplitude regulation	Less than 2% at 100% modulation
Audio-frequency response	±1.5 dB, from 20 to 1,000 Hz, referenced to 1,000 Hz, at 95% modulation
Audio-frequency distortion (unenhanced)	< 3%, 20–10,000 Hz at 95% modulation
Compression ratio	4:1 dB at 3 dB of enhancement; −95%, +125% modulation
Noise (unweighted)	− 57 dB or better below 100% modulation
Audio input	600 Ω at +10 dBm ±2 dB, for 100% modulation, unenhanced; +16 dBm with enhancement activated
Power input	480 V ±5%, 3-phase, 60 Hz; available for 380 V ±5%, 3-phase, 50 Hz
Power consumption	80 kW at 0% modulation, 87 kW at 30% modulation, 110 kW at 100% modulation
Overall efficiency	Better than 60% at average modulation

18. Shortwave Broadcast Service. An international broadcast station utilizes frequencies allocated to the broadcasting service between 5,950 and 26,100 kHz whose transmissions are intended to be received directly by the general public in foreign countries.

The frequencies assigned by the FCC lie in the following bands:

Band	Frequency, kHz	Meter band, m
A	5,950–6,200	49
B	9,500–9,775	32
C	11,700–11,975	25
D	15,100–15,450	19
E	17,700–17,900	16
F	21,450–21,750	14
G	25,600–26,100	11

Assignments are made for specific frequencies and specific hours of operation for transmission to specified geographic areas. Frequencies are assigned only if they will provide a delivered median field intensity, either measured or calculated, exceeding 150 μV/m for 50% of the time at the distant foreign target area.

The minimum transmitter output power is 50 kW. Transmitters must be equipped with automatic frequency control capable of maintaining the operating frequency within 0.003% of the assigned frequency. Frequency assignments provide a minimum co-channel delivered median field-intensity protection ratio of 40 dB to the transmissions of other broadcasting stations, at reference points in the target area. Similarly, a protection ratio of 11 dB is provided for adjacent-channel assignments.

Frequency-Modulation Broadcast Practice

BY JOSEPH L. STERN

19. FM Allocation Standards. The FM broadcast band comprises the radio-frequency spectrum from 88 to 108 MHz, divided into 100 channels of 200 kHz each. The channels available (including those assigned to noncommercial educational broadcasting) are given numerical designations by the FCC, from channel 201 (carrier frequency 88.1 MHz) to channel 300 (107.9 MHz). The channel number N is related to the carrier frequency f in megahertz by

$$N = 5(f - 57.9) \qquad f = N/5 + 47.9 \text{ MHz}$$

Field Strengths in FM Broadcast Service. Figure 21-6 shows estimated field strengths at distances from 1 to 100 mi, with transmitting antenna heights from 100 to 5,000 ft, for 1 kW of effective radiated power (50% of the receiving locations, 50% of the time).

20. Class A. Service. Except as provided in Sec. 73.204 of the FCC Rules, the frequencies in Table 21-6 are designated as class A channels and are assigned to class A stations only.

A class A station is designed to render service to a relatively small community, city, or town and the surrounding rural area. Class A stations are not authorized to operate with effective

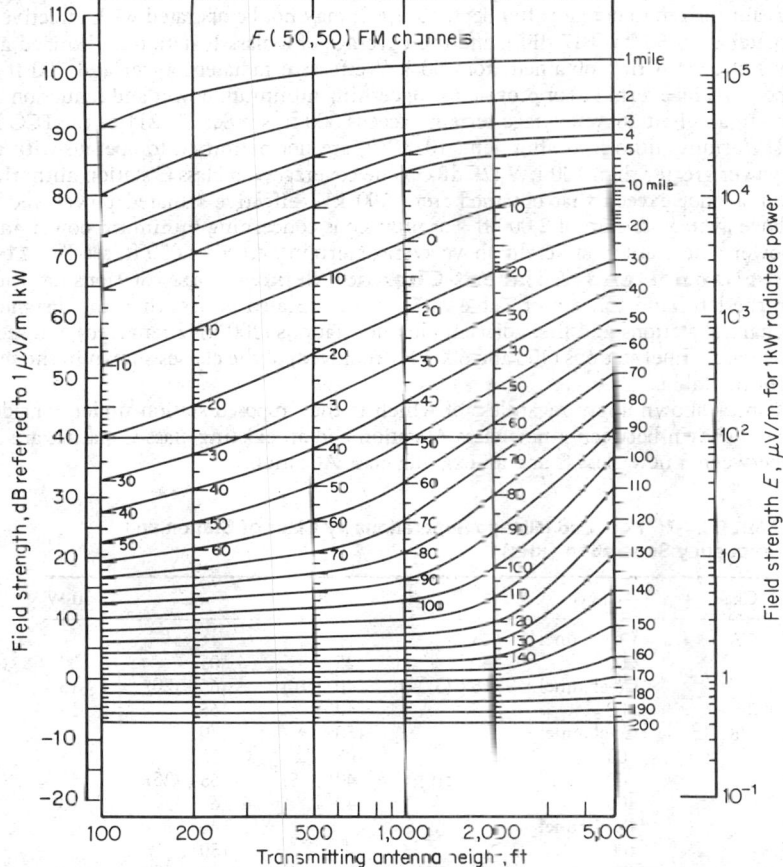

Fig. 21-6. Field-intensity curves vs. antenna height for various distances in FM broadcast service. The $F(50,50)$ designation gives the estimated field strength exceeded at 50% of the potential receiver locations for at least 50% of the time (receiving antenna height 30 ft). (*From FCC Rules, Sec. 73.333.*)

radiated power greater than 3 kW (4.8 dBk), and the coverage of a class A station must not exceed that obtained from 3 kW effective radiated power and antenna height above average terrain of 300 ft. For provisions concerning minimum facilities and reduction in power where antenna height above average terrain exceeds 300 ft, see Sec. 73.211 of the FCC Rules.

TABLE 21-6   Class A Allocation

Frequency, MHz	Channel no.	Frequency, MHz	Channel no.	Frequency, MHz	Channel no.
92.1	221	97.7	249	103.1	276
92.7	224	98.3	252	103.9	280
93.5	228	99.3	257	104.9	285
94.3	232	100.1	261	105.5	288
95.3	237	100.9	265	106.3	292
95.9	240	101.7	269	107.1	296
96.7	244	102.3	272		

21. Class B-C Service. Except for the class A channels listed above, all channels from 222 through 300 (92.3 through 107.9 MHz) are classified as class B-C channels. Subject to the restrictions set forth in Sec. 73.204 of the FCC Rules, they are assigned for use in zones I and 1A by class B stations only and for use in zone II by class C stations only.

Class B and class C stations are designed to render service to a sizable community, city, or town (or to the principal city or cities of an urbanized area) and to the surrounding area. *Class B stations* authorized to operate after Sept. 10, 1962, may not be operated with effective radiated power greater than 50 kW (17 dBk), and the coverage of a class B station authorized after that date may not exceed that obtained from 50 kW effective radiated power and 500 ft antenna height above average terrain. For provisions concerning minimum power and reduction in power where antenna height above average terrain exceeds 500 ft, see Sec. 73.211 of the FCC Rules.

Class C stations authorized after Sept. 10, 1962, are not permitted to operate with effective radiated power greater than 100 kW (20 dBk). The coverage of a class C station authorized after that date may not exceed that obtained from 100 kW effective radiated power and antenna height above average terrain of 2,000 ft. For provisions concerning minimum power and reduction in power where antenna height above average terrain exceeds 2,000 ft, see Par. **21-24.**

22. Co-channel and Adjacent-Channel Separations. Stations of the classes shown in the left-hand column of Table 21-7 must be located no less than the distance shown from co-channel stations and first adjacent-channel stations (200 kHz removed) and second and third adjacent-channel stations (400 and 600 kHz removed) of the classes shown in the remaining columns of the table.

The distances shown apply regardless of which is the proposed station under consideration, e.g., distances shown between a new class A station and an existing class C station are also the distances between a new class C and an existing class A station.

TABLE 21-7   Required Mileage Separations by Class of Station and Frequency Separation (kHz)*

Class	Type	A		B		C		10-W
A	Co-channel	65						
	200	40		25		105		30
	Co-channel	15	(5)	40	(10)	65	(20)	15
	600	15		40		65		15
B	Co-channel			150		170		
	200			105		135		
	400		(10)	40	(15)	65	(25)	40
	600			40		65		40
C	Co-channel					180		
	200					150		
	400		(20)		(25)	65	(30)	
	600					65		

*Stations or assignments separated in frequency by 10.6 or 10.8 MHz (53 or 54 channels) will not be authorized unless they conform to the separation mileages shown in parentheses.

23. Power and Antenna-Height Requirements. *Minimum Power Requirements.* The minimum effective radiated power for FM broadcast stations is:

Class A	100 W (-10 dBk)
Class B	5 kW (7 dBk)
Class C	25 kW (14 dBk)

No minimum antenna height above average terrain is specified. However, if the antenna height exceeds the maximum for each class, the permitted effective radiated power is reduced as noted below.

Maximum Power and Antenna Height. The maximum effective radiated power in any direction and maximum antenna height for equivalence purposes, for the various classes of stations, are given in Table 21-8.

Table 21-8. Maximum Power and Antenna Height

Class	Maximum power		Maximum antenna height, ft above average terrain
	kW	dBk	
A	3	4.8	300
B	50	17.0	500
C	100	20.0	2,000

NOTE: Antenna heights exceeding those specified may be used provided the effective radiated power is reduced. The amount of reduction is specified in Sec. 73.211 and Fig. 3 of Sec. 73.333 of the FCC Rules.

TECHNICAL STANDARDS FOR FM BROADCASTING

24. Definitions.*

General:

Antenna height above average terrain. The average of the antenna heights above the terrain from 2 to 10 mi from the antenna for the eight directions spaced evenly for each 45° of azimuth starting with true north. In general, a different antenna height will be determined in each direction from the antenna. The average of these various heights is considered the antenna height above the average terrain. In some cases fewer than eight directions may be used. Where circular or elliptical polarization is employed, the antenna height above average terrain must be based upon the height of the radiation center of the antenna which transmits the horizontal component of radiation.

Antenna power gain. The square of the ratio of the rms free-space field strength produced at 1 mi in the horizontal plane, in millivolts per meter for 1 kW antenna input power, to 137.6 mV/m. This ratio should be expressed in decibels.

Center frequency. (1) The average frequency of the emitted wave when modulated by a sinusoidal signal, or (2) the frequency of the emitted wave without modulation.

Effective radiated power. The product of the antenna power (transmitter-output power less transmission-line loss) times (1) the antenna power gain or (2) the antenna field gain squared. When circular or elliptical polarization is employed, the term effective radiated power is applied separately to the horizontal and vertical components of radiation. For allocation purposes, the effective radiated power authorized is the horizontally polarized component of radiation only.

Free-space field strength. The field strength that would exist at a point in the absence of reflected waves.

Frequency swing. The instantaneous departure of the frequency of the emitted wave from the center frequency resulting from modulation.

Multiplex transmission. The simultaneous transmission of two or more signals within a single channel. Multiplex transmission as applied to monophonic FM broadcast stations means the transmission of facsimile or other signals in addition to the regular broadcast signals.

Percentage modulation. The ratio of the actual frequency swing to the frequency swing defined as 100% modulation, expressed in percentage. For FM broadcast stations, a frequency swing of ±75 kHz is defined as 100% modulation.

*Adapted from Sec. 73.310 of the FCC Rules.

Definitions Applying to FM Stereophonic Broadcasting:

Crosstalk. An undesired signal occurring in one channel caused by an electric signal in another channel.

FM stereophonic broadcast. The transmission of a stereophonic program by a single FM broadcast station utilizing the main channel and a stereophonic subchannel.

Left (or right) signal. The electrical output of a microphone or combination of microphones placed so as to convey the intensity, time, and location of sounds originating predominantly to the listener's left (or right) of the center of the performing area.

Main channel. The band of frequencies from 50 to 15,000 Hz which frequency-modulate the main carrier.

Pilot subcarrier. A subcarrier serving as a control signal for use in the reception of FM stereophonic broadcasts.

Stereophonic separation. The ratio of the electric signal caused in the right (or left) stereophonic channel to the electric signal caused in the left (or right) stereophonic channel by the transmission of only a right (or left) signal.

Stereophonic subcarrier. A subcarrier having a frequency which is the second harmonic of the pilot subcarrier frequency, employed in FM stereophonic broadcasting.

Stereophonic subchannel. The band of frequencies from 23 to 53 kHz, containing the stereophonic subcarrier and its associated sidebands.

25. FM Broadcast Equipment Standards.* Under FCC regulations, the design of FM broadcast transmitting systems, from input terminals of microphone preamplifier, through the audio facilities at the studio, lines or other circuits between studio and transmitter, audio facilities at the transmitter, and the transmitter must be in accordance with the following principles and specifications.

Modulation Percentage and Bandwidth. The transmitter shall operate satisfactorily in the operating power range with a frequency swing of 75 kHz defined as 100% modulation. The transmitting system shall be capable of transmitting a band of frequencies from 50 to 15,000 Hz.

Pre-emphasis shall be employed in accordance with the impedance-frequency characteristic of a series inductance-resistance network having a time constant of 75 μs. The deviation of the system response from the standard pre-emphasis curve shall lie between two limits (shown in Fig. 2 of Sec. 73.333 of the FCC Rules). The upper of these limits shall be uniform (no deviation) from 50 to 15,000 Hz. The lower limit shall be uniform from 100 to 7,500 Hz, and 3 dB below the upper limit; from 100 to 50 Hz the lower limit shall fall from the 3-dB limit at a uniform rate of 1 dB/octave (4 dB at 50 Hz); from 7,500 to 15,000 Hz the lower limit shall fall from 3-dB limit at a uniform rate of 2 dB/octave (5 dB at 15,000 Hz).

Harmonic Distortion. At any modulation frequency between 50 and 15,000 Hz and at modulation percentages of 25, 50, and 100%, the combined audio-frequency harmonics measured in the output of the system shall not exceed the following rms values:

Modulating frequency, Hz	Distortion, %
50–100	3.5
100–7,500	2.5
7,500–15,000	3.0

Measurements shall be made employing 75-μs de-emphasis in the measuring equipment and 75-μs pre-emphasis in the transmitting equipment (without compression if a compression amplifier is employed). Harmonics shall be included to 30 kHz. It is recommended that none of the three main divisions of the system (transmitter, studio-to-transmitter circuit, or audio facilities) contribute over one-half of these percentages since at some frequencies the total distortion may become the arithmetic sum of the distortions of the divisions.

Noise. The transmitting-system output noise level (frequency modulation) in the band of 50 to 15,000 Hz shall be at least 60 dB below 100% modulation as a reference. The noise-measuring equipment shall be provided with standard 75-μs de-emphasis; the ballistic characteristics of the instrument shall be similar to those of the standard volume-unit meter.

The transmitting-system output noise level (amplitude modulation) in the band of 50 to 15,000 Hz shall be at least 60 dB below 100% modulation as a reference. The noise-measuring

*Section 73.317 of the FCC Rules.

equipment shall be provided with standard 75-μs de-emphasis; the ballistic characteristics of the instrument shall be similar to those of the standard volume-unit meter.

Carrier-Frequency Control. Automatic means shall be provided in the transmitter to maintain the assigned center frequency within the allowable tolerance of 2,000 Hz.

Out-of-Band Emissions. Any emission appearing on a frequency removed from the carrier by between 120 and 240 kHz, inclusive, shall be attenuated at least 25 dB below the level of the unmodulated carrier. Compliance with this specification will be deemed to show the occupied bandwidth to be 240 kHz or less.

Any emission appearing on a frequency removed from the carrier by more than 240 kHz and up to and including 600 kHz shall be attenuated at least 35 dB below the level of the unmodulated carrier.

Any emission appearing on a frequency removed from the carrier by more than 600 kHz shall be attenuated at least 43 + 10 log (power, in watts) decibels below the level of the unmodulated carrier, or 80 dB, whichever is less.

Remote-Control and Automatic Transmission Systems. FM transmitters are permitted to be operated by remote control or fully unattended by means of automatic transmission systems as defined in the FCC rules.

26. Subsidiary FM Communications Authorizations (SCA). An FM broadcast station may be issued a Subsidiary Communications Authorization (SCA) to provide limited types of subsidiary services on a multiplex basis. Permissible uses fall within the following categories:

1. Transmission of programs which are of a broadcast nature but which are of interest primarily to limited segments of the public wishing to subscribe thereto. Examples include background music, storecasting, detailed weather forecasting, special time signals, and other material of a broadcast nature expressly designed and intended for business, professional, educational, religious, trade, labor, agricultural, or other groups.

2. Transmission of signals which are directly related to the operation of FM broadcast stations, e.g., relaying broadcast material to other FM and standard broadcast stations, remote cueing and order circuits, remote-control telemetering functions, and similar uses. SCA operations may be conducted without time restriction so long as the main channel is programmed simultaneously.

Subsidiary communications multiplex operations are governed by the following engineering standards: Frequency modulation of SCA subcarriers must be used. The instantaneous frequency of SCA subcarriers must at all times be within the range 20 to 75 kHz, provided, however, that when the station is engaged in stereophonic broadcasting (Par. **21-27**) the instantaneous frequency of SCA subcarriers must be within the range 53 to 75 kHz

The arithmetic sum of the modulation of the main carrier by SCA subcarriers must not exceed 30%, provided, however, that when the station is engaged in stereophonic broadcasting the arithmetic sum of the modulation of the main carrier by the SCA subcarriers must not exceed 10%. The total modulation of the main carrier, including SCA subcarriers, must meet the requirements of Sec. 73.268 of the FCC Rules. Frequency modulation of the main carrier caused by the SCA subcarrier operation, in the frequency range 50 to 15,000 Hz, must be at least 60 dB below 100% modulation, provided, however, that when the station is engaged in stereophonic broadcasting frequency modulation of the main carrier by the SCA subcarrier operation, in the frequency range 50 to 53,000 Hz, must be at least 60 dB below 100% modulation.

27. Stereo Transmission Standards. (See definitions in Par. **21-24**.) The modulating signal for the main channel consists of the sum of the left and right signals (Fig. 21-7).

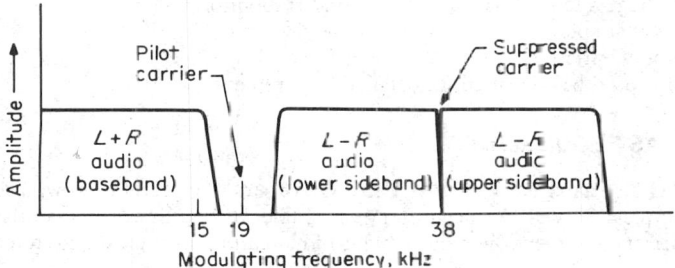

Fig. 21-7. Modulating frequencies for FM stereo transmissions.

The *pilot subcarrier* at 19,000 ± 2 Hz is transmitted and frequency modulates the main carrier between the limits of 8 and 10%.

The *stereophonic subcarrier* is the second harmonic of the pilot subcarrier and must cross the time axis with a positive slope simultaneously with each crossing of the time axis by the pilot subcarrier. Amplitude modulation of the stereophonic subcarrier must be used, and the stereophonic subcarrier must be suppressed to a level less than 1% modulation of the main carrier.

The stereophonic subcarrier must be capable of accepting audio frequencies from 50 to 15,000 Hz, and the modulating signal for the stereophonic subcarrier must be equal to the difference of the left and right signals.

The *pre-emphasis characteristic* of the stereophonic subchannel must be identical with those of the main channel with respect to phase and amplitude at all frequencies.

The sum of the sidebands resulting from amplitude modulation of the stereophonic subcarrier must not cause a peak deviation of the main carrier in excess of 45% of total modulation (excluding SCA subcarriers) when only a left (or right) signal exists; simultaneously in the main channel, the deviation when only a left (or right) signal exists must not exceed 45% of total modulation (excluding SCA subcarriers).

At the instant when only a positive left signal is applied, the main-channel modulation must cause an upward deviation of the main-carrier frequency; and the stereophonic subcarrier and its sidebands signal must cross the time axis simultaneously and in the same direction.

The ratio of peak main-channel deviation to peak stereophonic subchannel deviation when only a steady-state left (or right) signal exists must be within ±3.5% of unity for all levels of this signal and all frequencies from 50 to 15,000 Hz.

The phase difference between the zero points of the main-channel signal and the stereophonic subcarrier sidebands envelope, when only a steady-state left (or right) signal exists, must not exceed ±3° for AM frequencies from 50 to 15,000 Hz.

If the stereophonic separation between left and right stereophonic channels is better than 29.7 dB at AM frequencies between 50 and 15,000 Hz, it will be assumed that the requirements of the preceding two paragraphs have been complied with.

Crosstalk into the main channel caused by a signal in the stereophonic subchannel must be attenuated at least 40 dB below 90% modulation, and crosstalk into the stereophonic subchannel caused by a signal in the main channel must be attenuated a like amount.

28. FM Transmitter Performance Standards. The Electronic Industries Association (EIA, formerly RETMA) has issued standards of performance for FM transmitters, in Standard TR-107, Electrical Performance Standards for FM Broadcast Transmitters (88–108 MHz). This standard lists definitions, standards, and methods of measurements of the following items:

Carrier power-output rating
Carrier power-output capability
Normal load
Rf output coupling-impedance range
Carrier-frequency range
Carrier-frequency stability
Spurious emissions
Modulation capability
Audio-input level for 100% modulation
Audio-frequency response
Audio-frequency harmonic distortion
FM noise level on carrier
AM noise level on carrier
Output voltage and impedance for audio and rf monitoring
Rated power supply
Power supply variation

This standard should be consulted for detailed requirements.

FM BROADCAST EQUIPMENT

29. Typical 2.5- and 3.0-kW FM Transmitters. Figure 21-8 shows a simplified block diagram of a typical 3-kW FM transmitter* and Table 21-9 outlines the manufacturers specifications. It is designed to meet the requirements of monaural, multiplex, and stereo transmission.

*RCA Type BTF-3ESI.

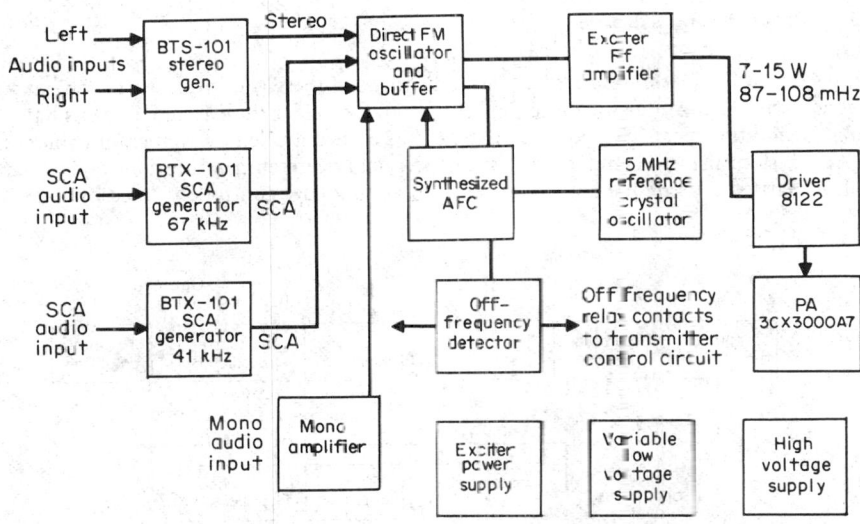

Fig. 21-8. Block diagram of a typical 3-kW FM transmitter. *(RCA Corp.)*

TABLE 21-9   Specifications of Type BTF-3ES1 FM Transmitter

Performance:

Power output	500–3000 W
Output impedance (1% in OD unflanged)	50 Ω
Frequency deviation, 100% modulation	±75 kHz
Modulation capability	±100 kHz
Carrier-frequency stability	±1,000 Hz max
Audio input impedance	600 Ω
Audio input level (100% modulation)	+10 ±2 dBm[a]
Audio-frequency response (30 Hz-15 kHz)	±1 dB max[b]
Pre-emphasis network time constant	0, 25, 50 or 75 μs[c]
Harmonic distortion (50 Hz–15 kHz)	0.3% max[d]
FM noise level (referred to 100% FM modulation)	−68 dB max
AM noise level (referred to 100% AM modulation	−50 dB max
Subcarrier input level (100% modulation)	9 to 30% adjustable
Subcarrier input impedance	Resistive 600 Ω balanced
Subcarrier frequency	20–95 kHz
Main-to-subchannel crosstalk	−50 dB referred to ±6.0 kHz deviation of the subcarrier by a 400-Hz tone; main-channel modulated 70% by a single tone (30–15,000 Hz) and 30% by subcarrier using narrow-band detector
Sub-to-main-channel crosstalk	−60 dB referred to ±75-kHz deviation of the main carrier by a 400-Hz tone; subcarrier modulated (±4.0 kHz) by a single tone (30–5000 Hz), main channel modulated 30% by subcarrier, using narrow-band detector

Electric power requirements:

Line	240/208 V, 50/60 Hz[e]
Combined voltage variation and regulation	±10%
Power consumption (approx)	6 kW
Power factor (approx)	90%

[a]Level measured at input to pre-emphasis network, referred to 400 Hz
[b]Frequency response referred to 75- or 50-μs pre-emphasis curve.
[c]Other time constants available on request.
[d]Distortion includes all harmonics up to 30 kHz and is measured following a standard 75- or 50-μs de-emphasis network.
[e]Ordinarily 3-phase power; unit for single-phase available

Figures 21-9 to 21-11 show simplified block diagrams of the FM exciter, stereo generator, and SCA generators. The exciter shown in Fig. 21-9 is totally solid state. Discrete crystals have been eliminated, and the operating frequency is generated and controlled by a programmable synthesized AFC system using a standard 5-MHz proportional oven-temperature-controlled reference crystal. A Hartley type frequency-modulated oscillator, voltage-controlled, is fed by the divider

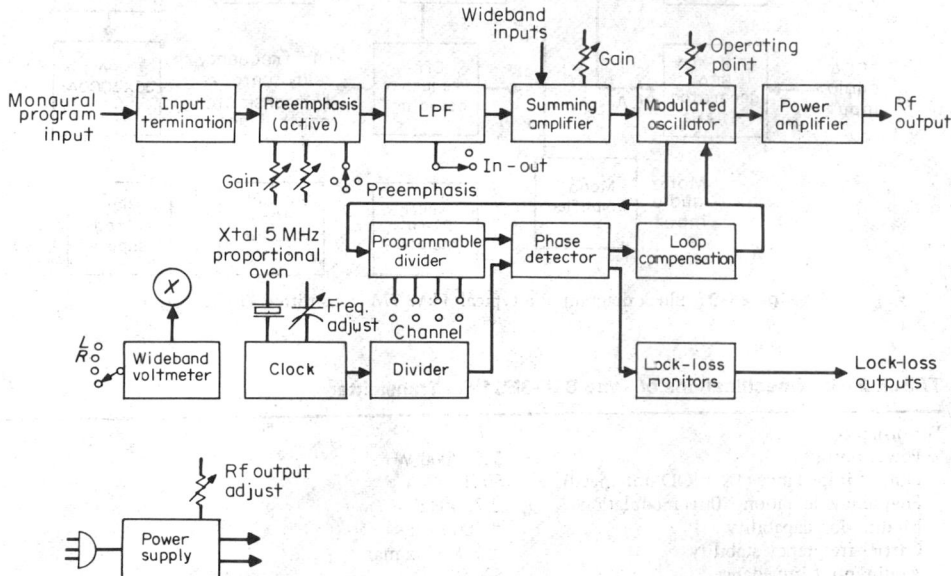

Fig. 21-9. Block diagram of solid-state monophonic FM exciter. *(RCA Corp.)*

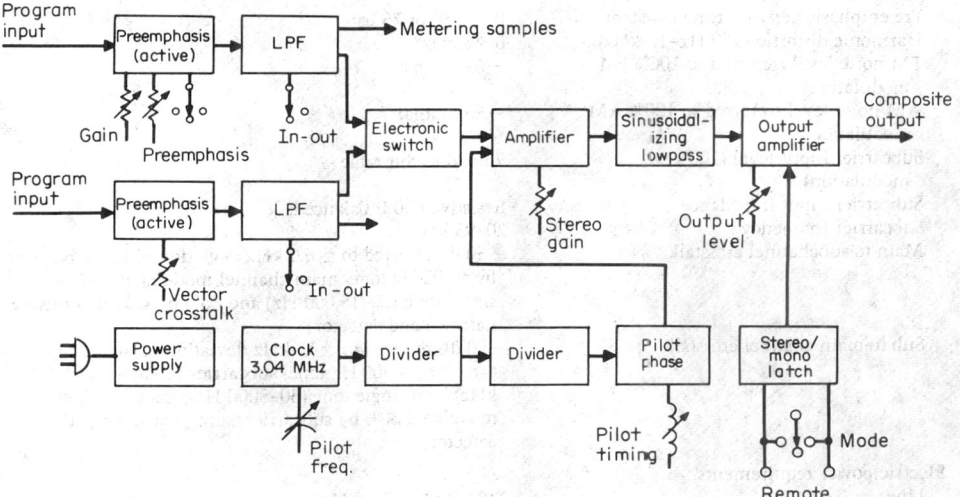

Fig. 21-10. Block diagram of solid-state stereo generator. *(RCA Corp.)*

circuit or can be operated in a free-running position. The buffer and PA amplifiers are wide-band over the range of 88 to 108 MHz with only peak trimming required to achieve 15 W output.

The stereo generator shown in Fig. 21-10 is also completely solid state. Active pre-emphasis is provided, with facilities permitting the use of internal or external audio processing. The pilot frequency of 19 kHz is generated from a 3.04-MHz crystal oscillator through a conventional divider circuit.

The SCA generator shown in Fig. 21-11 provides for automatic muting of the SCA signal in the absence of audio. The unit is provided with flexible control systems to prevent crosstalk into main and stereo channels.

In the transmitter (Fig. 21-8) the exciter output of 15 W is fed to a driver using a type 8122 tetrode and the power amplifier, a grounded grid ceramic type 3CS3000A7. Power output is controlled by means of a variable resistor controlling the screen votage supply of the 8122 tetrode.

The output of the power amplifier feeds through a harmonic filter–directional coupler to the antenna system. The harmonic filter consists of two m-derived half sections in series and three constant-k half sections configured into a lumped-inductance-capacitance network. The m-derived sections at both input and output provide for rapid cutoff in the second harmonic region and provides a 50-Ω impedance at each end of the filter.

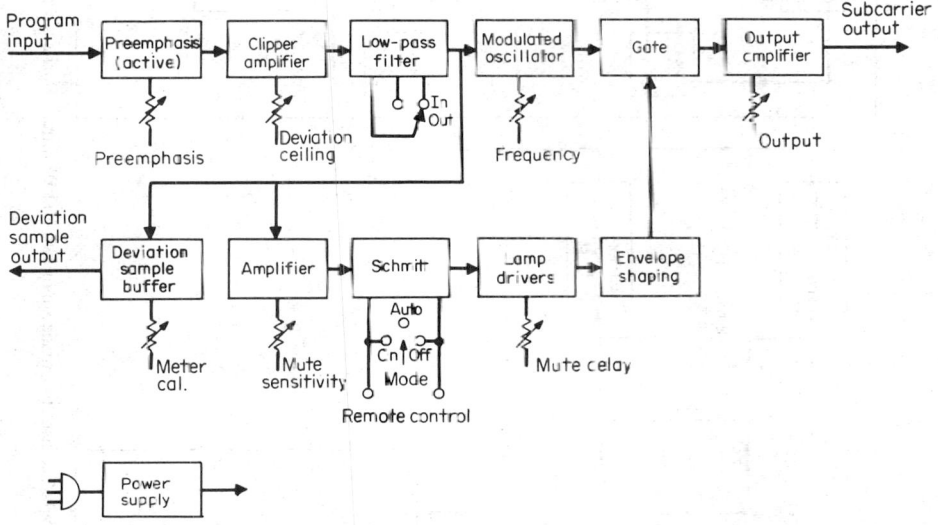

Fig. 21-11. Block diagram of solid-state SCA generator. (RCA Corp.)

Power circuits are protected by magnetic trip breakers as well as overload relays. Automatic sequencing prevents application of plate power before filaments have heated. Stepping relays provide automatic cycling in the event of brief overloads or power interruptions.

30. Typical 40-kW FM Transmitter.* Figure 21-12 shows the simplified block diagram of a 40-kW transmitter made up of two 20-kW transmitters fed by a solid-state exciter, both feeding into a combining network to give the desired 40-kW output. The manufacturer's specifications are shown in Table 21-10. The use of two transmitters offers circuit simplicity, lower parts inventory, and transmission continuity in case one of the units should fail. The output of each 20-kW unit is fed through its own harmonic filter to the combining network. This hybrid network adds the two 20-kW signals to produce a 40-kW output to the transmission line while isolating one transmitter from the other.

Should one transmitter fail, the other continues feeding the combining network and the combiner continues to operate as a power divider, 10 kW being fed to the output transmission line and 10 kW to the dummy load connected to the combiner. This division is necessary to provide isolation between transmitters and to allow the nonoperating unit to be serviced without rf coupling.

The solid-state exciter uses direct carrier-frequency modulation. The modulated oscillator operates at carrier frequency and has an output of 10 W. Separate 10-W isolation amplifiers are used to feed the two 20-kW units. The isolation amplifiers drive a pair of 4CX250B tubes, which produce 400 W of drive for a single 4CX15000 power-amplifier stage. The output tuning of the power-amplifier stage is accomplished with an inductively tuned plate-tank circuit, eliminating the need for vacuum capacitors. The transmitter is equipped for remote operation, including a motor-operated screen voltage-supply control for power-output adjustment.

*Type FM40K, Harris Corporation.

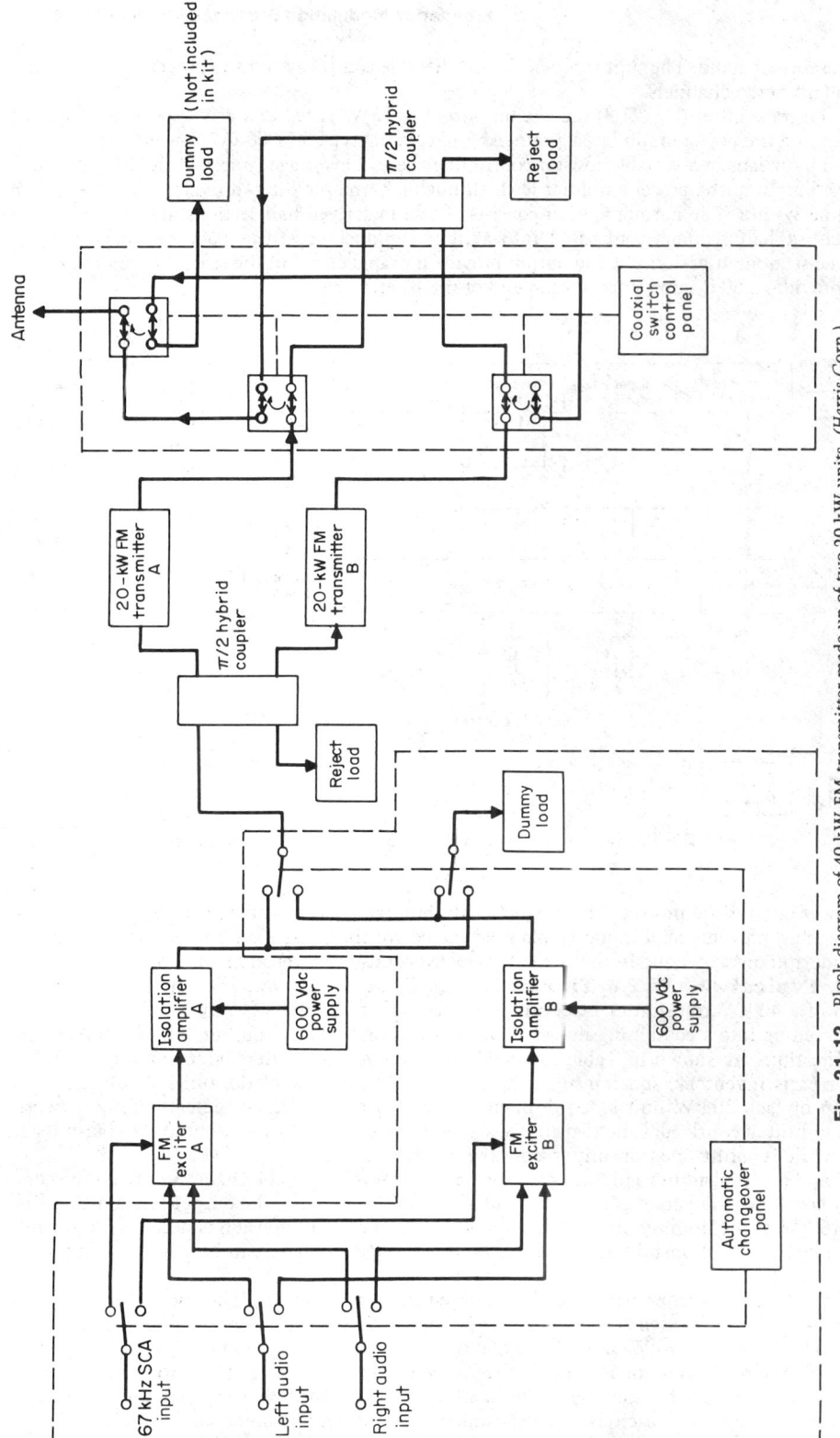

Fig. 21-12. Block diagram of 40-kW FM transmitter made up of two 20-kW units. *(Harris Corp.)*

TABLE 21-10   Specifications for Type FM-2.5K and FM40K Transmitters

Power output:	
FM-2.5K	2.5 kW
FM40K	40 kW
Frequency range	87.5-108 MHz
RF output impedance	50 Ω
Frequency stability	0.00 % or better
Type of modulation	Direct-carrier FM
Modulation capability	±100 kHz
RF harmonics	Suppression meets all FCC requirements
Monaural mode:	
Audio-input impedance	600 Ω balanced
Audio-input level	+10 dBm ±2 dB for 100% modulation at 400Hz
Audio-frequency response	Standard 75-Ω, FCC pre-emphasis curve ± 0.5 dB, 30−15,000 Hz
Harmonic distortion	0.2% or less, 30-15,000 Hz
Intermodulation distortion	0.2%, 60-7,000 Hz, 4:1 ratio
FM noise	68 dB below 100% modulation referred to 400 Hz
AM noise	50 dB below reference carrier AM modulation 100%
Stereophonic mode:	
Pilot oscillator	Crystal-controlled
Pilot stability	19 kHz ±1 Hz
Audio-input impedance	Left and right 500 Ω balanced
Audio-input level	Left and right +10 dBm ±1 dB for 100% modulation at 400 Hz
Audio-frequency response	Left and right standard 75 FCC pre-emphasis curve ± 0.5 dB, 50-15 000 Hz
Harmonic distortion	Left or right 0.4% or less, 50-15,000 Hz
FM noise	Left or right 65 dB min below 100% modulation, referred to 400 Hz
Stereo separation	40 dB min, 50-15,000 Hz
Subcarrier suppression	dB min, 50-15,000 Hz
Crosstalk (main to subchannel or vice versa)	45 dB below 90% modulation
SCA mode	
Modulation	Direct FM
Frequency	41 or 67 kHz programmable
Frequency stability	±500 Hz
Modulation capability	±7.5 kHz
Audio input impedance	600 Ω balanced (ac-coupled) and 2,000 Ω unbalanced (dc-coupled)
Audio input level	±10 dBm ±1 dB for 100% modulation at 400 Hz
Audio-frequency response selectable	41 kHz and 67 kHz 50-μs pre-emphasis ± 1 dB standard; Flat, 50- or 75-μs pre-emphasis
Distortion	Less than 1%, 30−50,000 Hz; ±5 kHz deviation
FM noise (main channel not modulated)	55 dB min (ref 100% = ±5 kHz deviation at 400 Hz)
Crosstalk (SCA to main or stereo subchannel)	−60 dB or better
Main or stereo subchannel to SCA	50 dB below ±5 kHz deviation of SCA, with mono or stereo channels modulated by frequencies 30−15,000 Hz, SCA demodulated with 150-μs de-emphasis
SCA to SAC (41 kHz/67 kHz)	50 dB demodulated with 150-μs de-emphasis
Automatic mute level	Variable, 0 to −30 dBm
Mute delay	Adjustable 0.5 to 20 s
Injection level	1 to 30% of composite, adjustable

Television Broadcasting Practice

BY JOSEPH L. STERN

31. Television Broadcast Allocations. The FCC has authorized 82 6-MHz channels for commercial and educational television broadcasting in the United States. Table 21-11 lists the numerical designations and frequency limits of the channels. Channels 2 to 6 (54 to 88 MHz) are known as the *low-band VHF* channels, 7 to 13 (174 to 216 MHz) as the *high-band VHF* channels, and 14 to 83 (470 to 890 MHz) as the *UHF channels.*

32. Channel Utilization. The FCC has prepared a comprehensive table of the commercial and educational channels assigned to particular communities in the United States, its territories, and possessions. The table appears in Sec. 73.606 of the FCC Rules. Each local channel assignment is identified as to the use of "offset" carrier frequencies. Certain assignments carry the nominal picture and sound carrier frequencies, others are required to operate with carriers 10 kHz above the nominal values (plus offset), and still others with carriers 10 kHz below the nominal values (minus offset). The offset system of carrier assignments substantially reduces the visible effects of co-channel interference.

Table 21-11.   Designations and Frequency Limits of Television Channels in the United States

Channel designation	Frequency band, MHz	Channel designation	Frequency band, MHz	Channel designation	Frequency band, MHz
2	54–60	30	566–572	57	728–734
3	60–66	31	572–578	58	734–740
4	66–72	32	578–584	59	740–746
5	76–82	33	584–590	60	746–752
6	82–88	34	590–596	61	752–758
7	174–180	35	596–602	62	758–764
8	180–186	36	602–608	63	764–770
9	186–192	37	608–614	64	770–776
10	192–198	38	614–620	65	776–782
11	198–204	39	620–626	66	782–788
12	204–210	40	626–632	67	788–794
13	210–216	41	632–638	68	794–800
14	470–476	42	638–644	69	800–806
15	476–482	43	644–650	70	806–812
16	482–488	44	650–656	71	812–818
17	488–494	45	656–662	72	818–824
18	494–500	46	662–668	73	824–830
19	500–506	47	668–674	74	830–836
20	506–512	48	674–680	75	836–842
21	512–518	49	680–686	76	842–848
22	518–524	50	686–692	77	848–854
23	524–530	51	692–698	78	854–860
24	530–536	52	698–704	79	860–866
25	536–542	53	704–710	80	866–872
26	542–548	54	710–716	81	872–878
27	548–554	55	716–722	82	878–884
28	554–560	56	722–728	83	884–890
29	560–566				

33. Co-channel and Adjacent-Channel Separations. Table 21-12 states the minimum co-channel mileage separations set up by the FCC.

The minimum adjacent-channel separations applicable to all zones are 60 mi for channels 2 to 13 and 55 mi for channels 14 to 83. Due to the greater than normal frequency spacing between channels 4 and 5, 6 and 7, and 13 and 14, the minimum adjacent-channel separations specified above are not applicable to these pairs of channels.

The minimum station separations between stations on the UHF channels 14 to 83, inclusive, must meet conditions set forth in Table IV of Sec. 73.698 of the FCC Rules. This table sets up separations based on considerations of intermediate-frequency beat interference, intermodulation, adjacent-channel interference, oscillator interference, sound-image interference, and pic-

TABLE 21-12   Minimum Co-channel Separations in Miles

Zone (see Fig. 21-11)	Channels 2 to 13	Channels 14 to 83
I	170	155
II	190	175
III	220	205

＊ The minimum cochannel mileage separations between a station in one zone and a station in another zone shall be that of the zone requiring the lower separation.

ture-image interference, with separations from 20 to 75 mi, depending on the type of interference. Each UHF channel (14 to 83 inclusive) is paired with another channel or channels, in each category, and the minimum separation for each type of interference must be met or exceeded in each category. Section 73.698 of the FCC Rules should be consulted for details.

34. Definitions Applicable to Television Broadcasting.[*] See also Sec. **20**, Pars. **20-1** to **20-29**.

Antenna electrical beam tilt. The shaping of the radiation pattern in the vertical plane of a transmitting antenna by electrical means so that the maximum radiation occurs at an angle below the horizontal plane.

Antenna height above average terrain. The average of the antenna height above the terrain from 2 to 10 mi from the antenna for the eight directions spaced evenly for each 45° of azimuth starting with true north. (In general, a different antenna height will be determined in each direction from the antenna. The average of these various heights is considered the antenna height above the average terrain.) In some cases fewer than eight directions may be used.

Antenna mechanical beam tilt. The installation of a transmitting antenna so that its axis is intentionally off vertical in order to change the normal angle of maximum radiation in the vertical plane.

Antenna power gain. The square of the ratio of the rms free-space field intensity produced at 1 mi in the horizontal plane, in millivolts per meter for 1 kW antenna input power, to 137.6 mV/m. This ratio should be expressed in decibels. (If specified for a particular direction, antenna power gain is based on the field strength in that direction only.)

Aural center frequency. The average frequency of the emitted wave when modulated by a sinusoidal signal; the frequency of the emitted wave without modulation.

Blanking level. The level of the signal during the blanking interval except the interval during the scanning synchronizing pulse and the chrominance-subcarrier synchronizing burst.

Chrominance. The colorimetric difference between any color and a reference color of equal luminance, the reference color having a specific chromaticity.

Chrominance subcarrier. The carrier which is modulated by the chrominance information.

Effective radiated power. The product of the antenna input power and the antenna power gain. This product should be expressed in kilowatts and the decibels above 1 kW (dBk).

Free-space field intensity. The field intensity that would exist at a point in the absence of waves reflected from the earth or other reflecting objects.

Negative transmission. Transmission in which a decrease in initial light intensity causes an increase in the transmitted power.

Peak power. The power over a radio-frequency cycle corresponding in amplitude to synchronizing peaks.

Percentage modulation. As applied to frequency modulation, the ratio of the actual frequency swing to the frequency swing defined as 100% modulation, expressed as a percentage. For the aural transmitter of television broadcast stations, a frequency swing of ±25 Hz is defined as 100% modulation. See Fig. 21-13.

Polarization. The direction of the electric field as radiated from the transmitting antenna.

Reference black level. The level corresponding to the specified maximum excursion of the luminance signal in the black direction.

Reference white level of the luminance signal. The level corresponding to the specified maximum excursion of the luminance signal in the white direction.

Vestigial-sideband transmission. A system of transmission wherein one of the generated sidebands is partially attenuated at the transmitter and radiated only in part.

Visual-carrier frequency. The frequency of the carrier which is modulated by the picture information.

Visual-transmitter power. The peak power output when transmitting a standard television signal.

35. Television Transmission Standards in the United States. See also Sec. **20**. The following standards have been adopted by the FCC.

Channel Standards. The width of the television broadcast channel is 6 MHz, with the visual-carrier frequency nominally 1.25 MHz above the lower boundary of the channel and the aural center frequency 4.5 MHz higher than the visual-carrier frequency.

The visual-transmission amplitude characteristic is in accordance with Fig. 21-14 (for stations operating on channels 15 to 83 and employing a transmitter with maximum peak visual power

[*]Adapted from Sec. 73.681 of the FCC Rules.

output of 1 kW or less, the visual-transmission amplitude characteristic given in Fig. 5a of FCC Sec. 73.699 may be used).

The chrominance subcarrier frequency is 3.579545 MHz ± 10 Hz with a maximum rate of change not to exceed $\frac{1}{10}$ Hz/s.

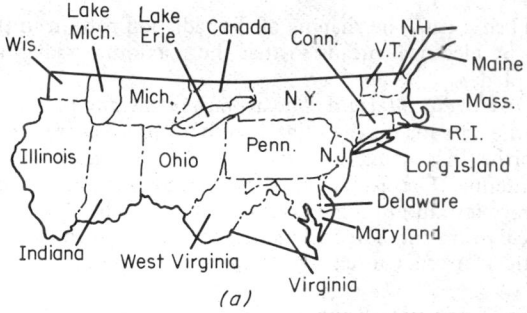

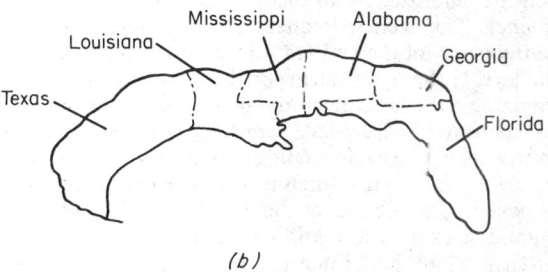

Fig. 21-13. Geographical zones set up by the FCC to take into account different requirements for television co-channel mileage separations. (*a*) zone I, (*b*) zone II. The remainder of the United States is designated as zone III.

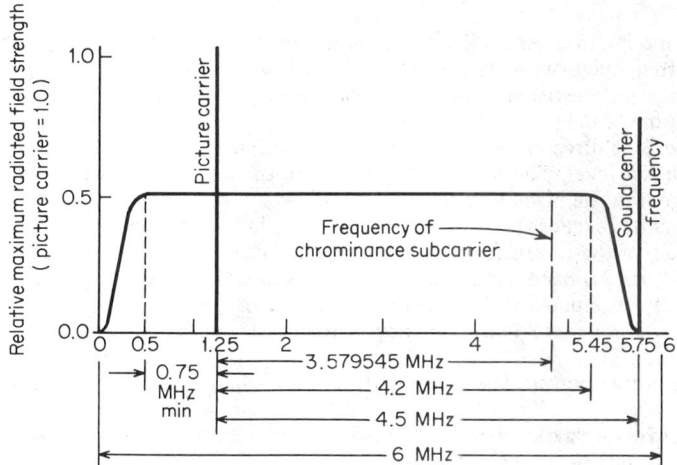

Fig. 21-14. The standard FCC channel for monochrome and color television picture transmissions in the United States (not to scale), shown as an idealized amplitude characteristic of the picture transmission. (*From FCC Rules, Sec. 73.699.*)

For monochrome and color transmissions the number of scanning lines per frame is 525, interlaced two to one in successive fields. The horizontal scanning frequency is $\frac{2}{455}$ times the chrominance subcarrier frequency; this corresponds nominally to 15,750 Hz (the actual value is 15,-734.264 ±0.044 Hz). The vertical scanning frequency is $\frac{2}{525}$ times the horizontal scanning frequency; this corresponds nominally to 60 Hz (the actual value is 59.94 Hz). For monochrome transmissions only, the nominal values of line and field frequencies may be used. Other aspects of the scanning and modulation standards are treated in Sec. **20**, Pars. **20-7** to **20-13**

Frequency Tolerance. The carrier frequency of the visual transmitter must be maintained within ±1,000 Hz of the assigned frequency, while the center frequency of the aural transmitter must be maintained at 4.5 MHz ± 1,000 Hz above the assigned visual-carrier frequency.

The signals radiated are horizontally polarized. The effective radiated power of the aural transmitter must not be less than 10% nor more than 20% of the peak radiated power of the visual transmitter.

The peak-to-peak variation of transmitter output within one frame of video signal due to all causes, including hum, noise, and low-frequency response, measured at both scanning synchronizing peak and blanking level, must not exceed 5% of the average scanning synchronizing peak-signal amplitude (this provision is subject to change but is considered the best practice under the present state of the art).

The waveforms and transfer characteristics set up by the FCC for monochrome and color transmissions are stated in Sec. **20**, Pars. **20-9** to **20-12**.

36. Remote Control. Television stations may be authorized to operate by remote control. When operated in this manner remote monitoring is provided through the use of test signals inserted into the vertical interval of the visual signal. Vertical-interval-test (VIT) signals are defined in the FCC Rules and Regulations. (The FCC is considering rule making to permit the use of automatic transmissions systems for television transmission.)

37. Test and Identification Signals. The interval beginning with the last 12 μs of line 17 and continuing through line 20 of the vertical blanking interval of each field may be used for the transmission of test signals subject to the conditions set forth by the FCC. Test signals include signals used to supply reference modulation levels so that variation in light intensity of the scene viewed by the camera will be faithfully transmitted, signals designed to check the performance of the overall transmission system or its individual components, and cue and control signals related to the operation of the television broadcast station.

The use of test signals must not result in significant degradation of the program transmissions of the television broadcast station or create emission components in excess of those permitted for normal program transmissions.

They may not be transmitted during horizontal blanking.

The intervals within the first and last 10 μs of lines 21 through 28 and 230 through 262 (on a field basis) may contain coded patterns for the purpose of electronic identification of television broadcast programs and spot announcements, provided the coded patterns do not exceed 1 s in duration. The transmission of these patterns must not result in significant degradation of broadcast transmission.

38. Coverage Determinations and Standards. In the authorization of television broadcast stations, two field-intensity contours, specified as grade A and grade B, indicate the approximate extent and area of the coverage over average terrain in the absence of interference from other television stations. Under actual conditions, the true coverage may vary greatly from these estimates when the terrain is different from the average terrain or which the field-strength charts were based. The required field intensities $F(50,50)$ (see Fig. 21-15) in decibels above 1 μV/m (dBμ) for the grade A and grade B contours are given in Table 21-13

In predicting the distance to the field-intensity contours, the $F(50,50)$ field-intensity charts (Figs. 9 and 10 of FCC Sec. 73.699) are used. Figure 21-15 applies to channels 2 to 6 and 14 to 83. The 50% field intensity is defined as that value exceeded for 50% of the time. The $F(50,50)$ charts give the estimated 50% field intensities exceeded at 50% of the locations, in decibels above 1 μV/m. The charts are based on an effective power of 1 kW radiated from a half-wave dipole in free space, which produces an unattenuated field strength at 1 mi of about 103 dB above 1 μV/m (137.6 mV/m).

Depression Angle. In predicting the distance to the grade A and grade B field-intensity contours, the effective radiated power used is that radiated at the vertical angle corresponding to the depression angle between the transmitting-antenna center of radiation and the radio horizon as determined individually for each azimuthal direction concerned. The depression angle is based on radiation above the average terrain and the radio horizon, assuming a smooth spherical earth with a radius of 5,280 mi, determined by

$$A_a = 0.0153 \sqrt{H}$$

where A_h = depression angle (deg) and H = height (ft) of transmitting-antenna radiation center above average terrain of 2- to 20-mi sector of pertinent radial.

The antenna height used with the $F(50,50)$ charts is the height of the radiation center of the antenna above the average terrain along the radial in question. In determining the average elevation of the terrain, the elevations between 2 and 10 mi from the antenna site are used. Profile graphs are drawn for eight radials beginning at the antenna site and extending 10 mi for each

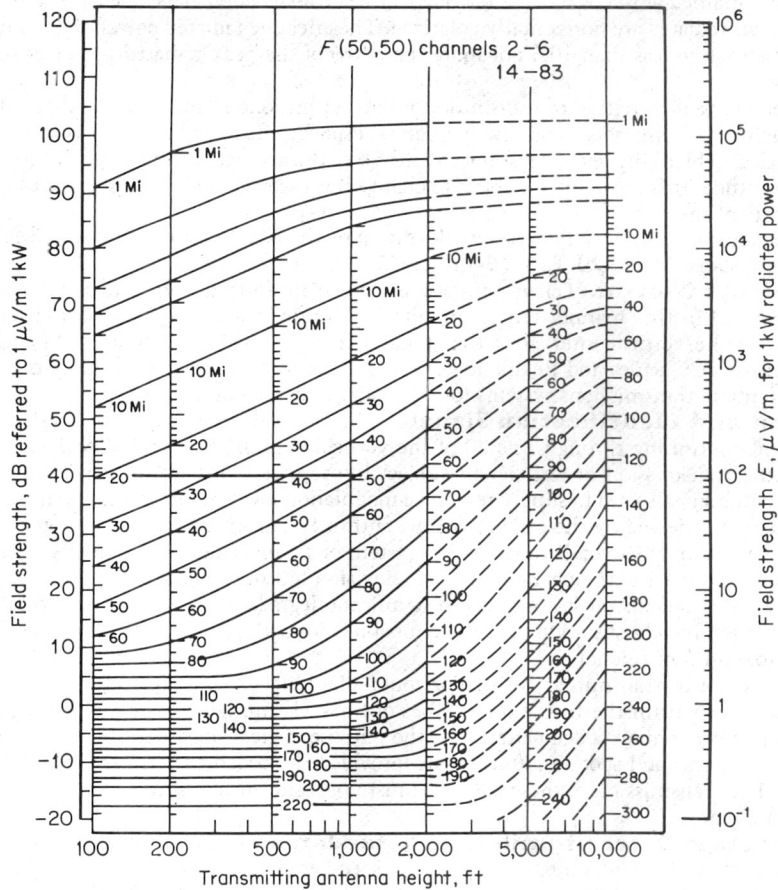

Fig. 21-15. Field-intensity curves vs. antenna height for television transmissions at various distances. For the significance of the designation $F(50,50)$ see Fig. 21-6. Receiver antenna height is 30 ft. *(From FCC Rules, Sec. 73.699.)*

45° of azimuth starting with true north. At least one radial must include the principal community to be served even though it may be more than 10 mi from the antenna site. For further details of coverage predicton, see FCC Rules, Sec. 73.684.

39. Transmitter Location and Antenna System. The transmitter location chosen must provide, on the basis of the effective radiated power and antenna height above average terrain employed, the following minimum field intensity in decibels above 1 μV (dBμ) over the entire principal community to be served: channels 2 to 6, 74 dBμ; channels 7 to 13, 77 dBμ; channels 14 to 83, 80 dBμ.

Table 21-13.   Grades of Television Service

Channels	Grade A, dBμ	Grade B, dBμ
2–6	68	47
7–13	71	56
14–83	74	64

Location of the antenna at a point of high elevation is necessary to reduce to a minimum the shadow effect on propagation due to hills and buildings which may materially reduce the intensity of the station's signals. In general the transmitting antenna of a station should be located at the most central point at the highest elevation available To provide the best degree of service to an area, it is usually preferable to use a high antenna rather than a low antenna with increased transmitter power. The location should be so chosen that line of sight can be obtained from the antenna over the principal community to be served; in no event should there be a major obstruction in this path. The antenna must be constructed so that it is as clear as possible of surrounding buildings or objects that would cause shadow problems. When the shape of the desired service area and population distribution make the choice of a transmitter location difficult, consideration may be given to the use of a directional antenna system, although it is generally preferable to choose a site where a nondirectional antenna can be used.

40. Directional Antennas. An antenna designed or altered to produce a noncircular radiation pattern in the horizontal plane is considered to be a directional antenna. Antennas purposely installed to give mechanical beam tilting of the major vertical radiation lobe are included in this category. Stations operating on channels 2 to 13 are not permitted to employ a directional antenna having a ratio of maximum to minimum radiation in the horizontal plane in excess of 10 dB. Stations operating on channels 14 to 83 and employing transmitters delivering a peak visual power output of 1 kW or less are not limited as to the ratio of directional characteristics.

Television Broadcasting Equipment

BY JOSEPH L. STERN

41. TV Broadcast-Equipment Performance Standards. The Electronic Industries Association (EIA, formerly RETMA) has issued a 21-page set of television-equipment standards, Standard RS-240, Electrical Performance Standards for Television Broadcast Transmitters, Channels 2-6, 7-13, and 14-83. This comprehensive standard should be consulted for the detailed definitions, standards, and methods of measurement it contains. The topics covered fall into four parts: the television transmitter, the visual transmitter, the aural transmitter, and safety standards, as follows:

Television transmitter:
  Television transmitter
  Power rating
  Rated power supply
  Power-supply variation
  Carrier-frequency range
  Transmitter harmonic and subharmonic output
  Cabinet radiation
Visual transmitter:
  Visual transmitter
  Power-output rating
  Peak power-output adjustment
  Variation of output
  Regulation of output
  Carrier-frequency stability
  Amplitude vs. frequency response
  Upper and lower sideband attenuation
  Linearity
  Differential gain
  Differential phase
  Envelope delay
  Transmitter input polarity
  Transmitter input impedance
  Transmission polarity
  Transmitter input level for rated modulation
  Carrier pedestal level

Carrier-reference white level
Output polarity and voltage for composite picture-signal monitor connections
Output voltage for rf monitor connections
Hum and noise
Incidental frequency modulation
Aural Transmitter:
Aural transmitter
Carrier power-output rating
Center-frequency stability
FM noise level on carrier
AM noise level on carrier
Output voltage and impedance for audio and rf monitor connections
Modulation capabilities
Audio-input impedance and input level for 100% modulation
Audio-frequency response
Audio-frequency harmonic distortion
Intermodulation distortion
RF output-coupling impedance range
Safety:
Transmitter enclosure
Grounding
Grounding switches
Grounding sticks
Interlocks
Access to voltages
High-voltage metering devices
Disconnect device
Radiation
Electromagnetic exposure
Mechanical safeguards

42. Signal Processing. The generation of the composite television signal, before its arrival at the transmitter, is treated in Sec. **20**, Pars. **20-61** to **20-66**. The transmitter receives the composite signal via cables, rebroadcast receivers, or microwave relay links. The distortion that can occur, especially to the synchronizing pulses, must be taken into account. To convert the incoming signal into a standard television waveform, a synchronizing regenerator may be introduced at the input of the visual transmitter, to regenerate the synchronizing pulses and stabilize the video level. For color transmission, it is also necessary to comply with two additional requirements: (1) the color burst must be transmitted without distortion, even when using clamping circuits operating during the back-porch interval, and (2) the amplitude of the color subcarrier must be regulated to the nominal value. Either the color-synchronizing burst or a corresponding element in the insertion signal can be used as reference. A typical arrangement for regeneration and stabilization is shown in Fig. 21-16.

43. TV-Transmitter Design Principles. *High- vs. Low-Modulation Levels.*   In high-level modulated transmitters, modulation is carried out in the final rf power stage. Linearity

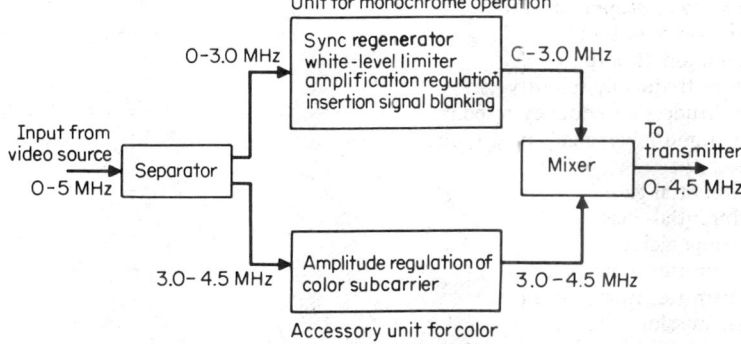

Fig. 21-16. Block diagram of signal-processing circuits for monochrome and color operation.

and high video drive-power requirements are of concern in high-level modulation. In addition, the vestigial-sideband filter must handle the total transmitter-power output. With low-level modulation, one or more linear power amplifiers follow the modulated amplifier stage. These linear amplifiers must be designed for wide-band operation. In this case the vestigial-sideband filter can be located immediately after the modulated amplifier and thus operate at relatively low power.

The visual and aural transmitters are normally connected to the antenna through a combining-unit diplexer. The use of separate antennas for visual and aural transmitters is now rare.

The block diagram of the various stages of a typical VHF transmitter modulated in the power-amplifier stage is shown in Fig. 21-17. The vestigial-sideband filter and diplexer are integrated in a single unit, known as a filterplexer.

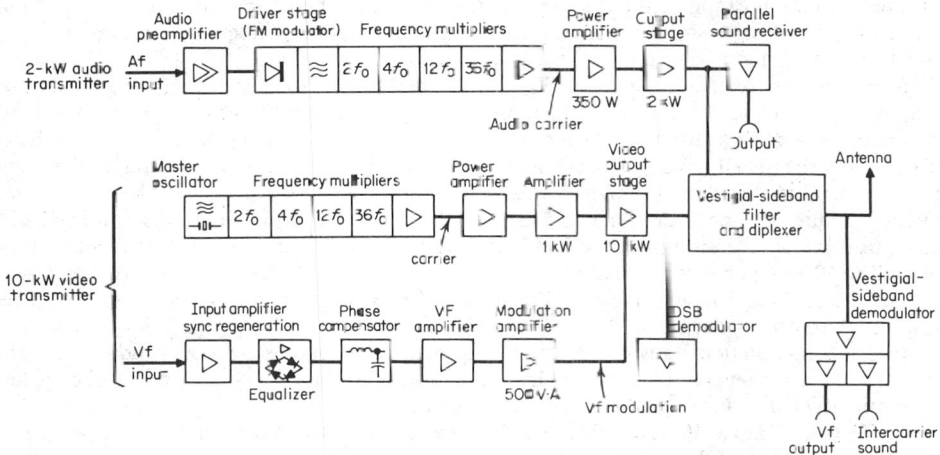

Fig. 21-17. Block diagram of a typical high-level modulated transmitter of 10 kW visual and 2 kW aural output.

Triodes or tetrodes are used principally in VHF transmitter-output stages, tetrodes or klystrons for the UHF band. The tetrode is more efficient than the klystron but three or four tetrode stages are required to equal the gain of a klystron, and the risks of tube failure are greater. The life of a klystron is greater than that of currently available tetrodes. The greater amplification available from the klystron also permits low-power semiconductor drive stages to be used. The cooling system required for a klystron transmitter must handle a heavy cooling load, and the power-supply voltages are considerably higher.

Intermediate-Frequency Modulation. In i.f.-modulated transmitters, however, the video signal is modulated at an intermediate carrier at low power, which is then heterodyned to the output frequency. The advantages of the i.f.-modulated transmitter include the facts that the distortion is reduced, thanks to the low power at which modulation takes place (although the linearity of the subsequent stages must be of a high order), and that group-delay equalization can be applied at the intermediate frequency under optimum conditions.

Direct Modulation. The alternative to i.f. modulation is direct modulation. In the earliest transmitters modulation was applied to the grid of the final rf tube. For large output powers, the video modulation-amplifier requirements were quite onerous. Video voltages in the 400- to 500-V range, with reactive currents approaching 2 A at 5 MHz, were not uncommon.

Improvements in vacuum tubes and the introduction of sweep-frequency techniques for adjustment permitted wide-band rf amplifiers to be designed and high-power transmitters to be made with one or more rf linear amplifiers following the modulated stage.

Improvements in vacuum tubes have continued, and long-life tubes for high-power rf linear amplifier service are available with power gains exceeding 16 dB. This makes possible a single-tube or two-tube linear amplifier to be driven from a solid-state modulated rf drive.

The solid-state drive equipment takes several basic forms, e.g., a low-level modulator followed by transistor linear amplifiers or a higher-level modulator at the power level necessary to drive a vacuum tube. Investigations have also been made into the possible use of PIN diodes and rf transistors for high-power modulators.

UHF transmitters use linear output amplifiers employing multicavity klystrons or traveling-wave tubes. With the high gains provided by these devices, the modulated-drive-level requirements are modest, usually a few watts, and are readily obtained from a solid-state modulator. Because of the low output-power requirements, the continuous-wave (cw) drive level is acceptably small. However, the incidental phase modulation introduced by the modulator is of sufficient magnitude to require prephase modulation of the cw drive signal to cancel incidental phase modulation.

The vestigial-sideband shaping filter may be placed between the modulator and the output linear amplifier. If it is placed at the input to a klystron amplifier, two undesirable effects are introduced: (1) intermodulation products generated in the klystron produce signals in the lower (suppressed) sideband region which, in effect, modify the slope response of the overall amplitude-frequency characteristic; (2) the level dependency of the klystron frequency response introduces differential amplification of the sideband frequencies. As the sidebands are not of equal bandwidth, the resulting distortion makes independent high- and low-frequency (differential gain line-time) correction necessary.

Comparative Circuit Complexity. The circuits required for the visual chain of a transmitter employing i.f. modulation are more complex than for direct modulation. Two high-stability oscillators, one at the visual i.f. carrier frequency and the other at the sum of visual radiated carrier plus the visual i.f. carrier frequency, are required. The heterodyne oscillator chain and the visual mixer have to be provided for translating the visual i.f. signal to the radiated frequency. Solid-state linear amplifiers are used to raise the power from the mixer to a level sufficient to drive a tube amplifier. For VHF signals up to 230 MHz, linear solid-state amplifiers providing 60 W (sync level) output power are available. For UHF transmitters, only 2 or 3 W of output power is required to drive a klystron or traveling-wave amplifier, a level readily obtained from UHF transistor circuits.

For direct-modulation transmitters employing high-power solid-state visual modulators, the carrier-wave drive level is, of necessity, large, and pre- or postphase modulation must be applied to ensure acceptable levels of incidental phase modulation.

44. Typical Low-Power VHF Television Transmitter. Figure 21-18 shows a block diagram of a 1-kW VHF transmitter* using all solid-state circuits, with the exception of

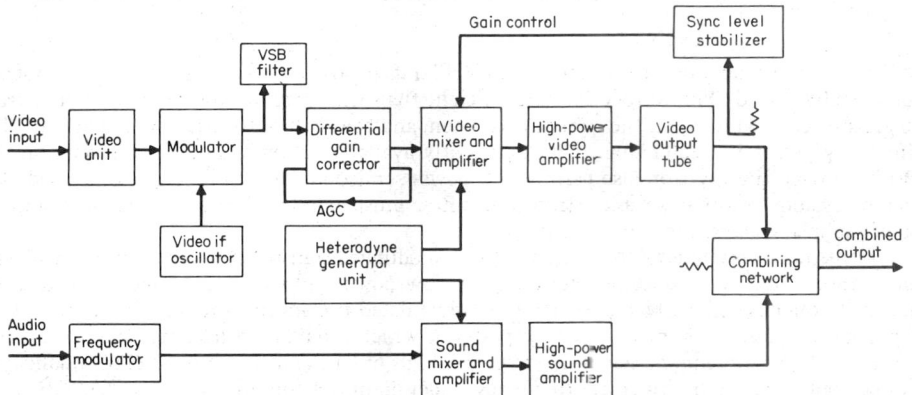

Fig. 21-18. Block diagram of 1-kW VHF television transmitter employing low-level modulation. (*Marconi Broadcasting Division.*)

the final amplifier. The latter is a tetrode in a grounded-grid circuit, with single-tuned cathode input and a double-tuned output The drive power of 70 to 100 W is supplied by a transistorized drive circuit. To achieve this power over the wide band required, six power transistors are used in the driving amplifier, in pairs paralleled by ferrite hybrid transformers. Intermediate-frequency modulation is used at a crystal-controlled vision carrier of 38.9 MHz, which is modulated by the processed picture signal in a circuit employing field-effect transistors.

Frequency modulation for the sound signal is also introduced at an intermediate frequency separated (for U.S. standards) 4.5 MHz from the crystal-controlled vision carrier of 38.9 MHz. The aural power output of 100 W is produced entirely in transistorized circuits and is combined with the visual carrier in a combining network.

*Marconi Broadcasting Division, English Electric.

45. 25-kW VHF Television Transmitter. Figure 21-19 shows the block diagram of a 25-kW VHF transmitter* using low-level i.f. modulation. Less than 1 V of video signal is required to modulate the visual carrier. A highly linear broadband diode-ring modulator is used, with active delay compensation and low-level vestigial-sideband filtering. The transmitter proper is fed by transistorized exciters for the aural and visual inputs. The master oscillator is in the visual exciter.

Ceramic tetrode tubes are used in the final amplifier stages, operating in grounded-grid and grounded-screen configurations. Neutralization is not required. All control functions and metering are arranged for remote-control operation. Table 21-14 lists the specifications of this transmitter.

46. 60-kW Television Transmitter for Low-Band VHF. Figure 21-20 shows the block diagram of a 60-kW transmitter comprising two parallel 30-kW transmitters. Two identical units are used so that one can continue operation if the other fails. The two units provide 13.2 kW of aural carrier output. The circuits use solid-state devices with a driver output of 1600 W

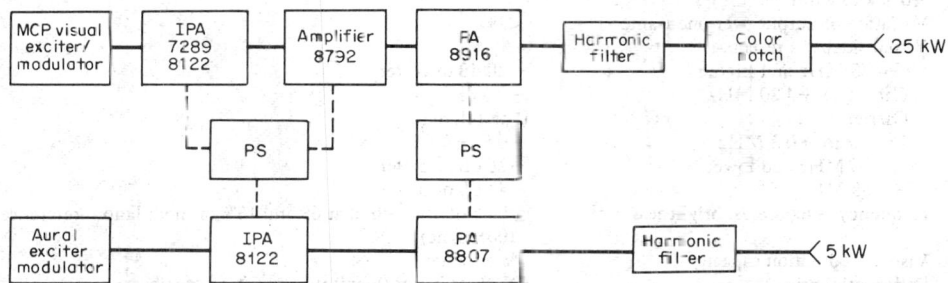

Fig. 21-19. Block diagram of 25-kW VHF television transmitter using low-level modulation. (*Courtesy RCA Corp*).

feeding the visual power amplifier, which is a Type 9007 power tetrode. The aural power amplifier, driven by 75 W, is a Type 8977 tetrode. Two exciters are provided; one drives both transmitters, and the other serves as a standby unit which is automatically switched into operation should the first exciter fail.

The exciter modulator unit also contains video processing functions. Correction for receiver, tuned-amplifier, and notch-diplex group delay is handled in this unit by plug-in equalizer boards. No group-delay correction is required for the phase-linear sideband filter, wide-band rf exciter circuits, or solid-state amplifiers. Differential phase correction is provided in the video domain. Clamping at level is followed by a white clipper with selection provided between luminance, luminance-chrominance, or no clipping action. Motorized controls are provided for separate remotable adjustment of video level and sync gain. The process video signal is applied to a double-balanced modulator which modulates an i.f. signal. An i.f. of 45.75 MHz is used for transmitters operating on FCC (system M) channels while a frequency of 38.9 MHz is used for CCIR systems B, D and K1 (see Table 20-1).

The aural exciter generates a frequency-modulated i.f. signal phase-locked to the reference generator shared with the visual frequency synthesizer An input is provided for a subcarrier input to accommodate optional telemetry of transmitter metering function when operating by remote control.

Picture-signal sideband shaping is accomplished by a surface acoustic wave (SAW) filter, which has a flat-amplitude frequency response over the entire video passband. The SAW filter is inherently free of large band-edged distortions characteristic of lumped-constant sideband filters. The SAW filter is temperature-stabilized by a proportional control heater.

The output of the SAW filter goes to an rf processor which controls the transmitter output power levels and provides for up conversion of the picture and sound signals from intermediate to carrier frequency. The carrier frequency signals are amplified to ½-W level at the rf processor output. Adjustment of transmitter power-output level is provided by controlling the amplitude of the picture and sound i.f. signals by separate pin diode attenuators Linearity correction is performed at the i.f. following power-level control. The up conversion to picture and sound-

*RCA Model BT-25L2/H2.

carrier frequencies is accomplished by mixing each i.f. signal with a phase-locked-loop local-oscillator signal at carrier plus intermediate frequency.

Incidental phase distortion of the visual carrier frequency is corrected in the rf processor by a phase modulator which cancels the distortion by applying modulation in a direction opposite that of the error. Stable picture-power level is maintained by detecting a sample of the final rf

TABLE 21-14   Specifications of the Type BT-25L2/H2 Television Transmitter

Visual:	
Power output	25 kW peak (FCC); 21 kW peak (CCIR B)
Output impedance	50 Ω
Frequency range:	
BT-25L2	48–88 MHz (channels 2–6)
BT-25H2	174–230 MHz (channels 7–13)
Carrier stability	±250 Hz (maximum variation over 30 days)
Regulation of rf output power	<3%
(black-to-white picture)	
Variation of output over one frame	<2%
Visual sideband response	
+4.75 MHz and higher	−20 dB or better
Carrier to +4.20 MHz	+0.5 dB
Carrier	0 dB reference
Carrier to −0.5 MHz	+0.5, −1 dB
−1.25 MHz and lower	−26 dB or better
−3.58 MHz	−42 dB or better
Frequency response vs. brightness	±0.75 dB (measured at 65 and 15% of modulation; reference 100% sync)
Visual modulation capacity	3% or better
Differential gain	0.5 dB or better (maximum variation of subcarrier amplitude from 75 to 10% of modulation; subcarrier modulation percentage 10% peak to peak)
Linearity (low-frequency)	0.5 dB or better
Differential phase	±1° or better (maximum variation of subcarrier phase with respect to burst for modulation percentage from 75 to 10%; subcarrier modulation percentage 10% peak to peak)
Signal-to-noise ratio	−50 dB or better (rms) below sync level
Envelope delay:	
0.05–2.1 MHz	+40 ns*
At 3.58 MHz	±30 ns*
At 4.18 MHz	±70 ns*
Video input	75-Ω system
Harmonic radiation	−80 dB
Aural:	
Power output	5 kW at diplexer output
Audio input	+10 dBm, ±2 dB into 600 Ω
Input impedance	600/150 Ω
Pre-emphasis	75 μs
Frequency response	± 0.5 dB relative to pre-emphasis (30 to 15,000 Hz)
Distortion, after 75 μs de-emphasis	0.5% or less
with ± kHz deviation	
FM noise	−60 dB relative to ±25-kHz deviation
AM noise	55 dB relative to 100% modulation measured after de-emphasis
Output impedance	50-Ω output connector 3⅛-in EIA standard
Frequency stability	±250 Hz (maximum variation over 30 days)

*Reference to standard FCC curve.

output and comparing the peak of the sync signal with a reference voltage. The resulting error signal is applied to the power-control *pin* diode attenuator providing automatic control of power level.

Intermediate power amplification is accomplished by a solid-state broadband amplifier system. The sound signal is amplified by a predriver followed by an output amplifier. The picture signal is amplified in the solid-state predriver then split and fed to a system of intermediate power-amplifier units operating in parallel, each having sync peak power capability of 400 W. Hybrid strip-line splitters and attenuators provide for the power division and isolation. Each 400-W intermediate power-amplifier unit for the low-band transmitter consists of two parallel amplifier circuits in a single module using microstrip splitter-combiner circuitry. The 400-W solid-state amplifiers are cooled by an embedded heat pipe system using a vapor-chamber fluid in a sealed module. The heat pipes are embedded in each amplifier baseplate and transfer the heat from the transistor case to a heat sync at the rear of each module. The heat sync extends into an area where the air-cooling system of the transmitter can remove heat from the heat sync.

47. 55-kW UHF Television Transmitter. Figure 21-21 shows a block diagram of a transmitter* designed for high-power operation in the UHF television band. It uses two identical

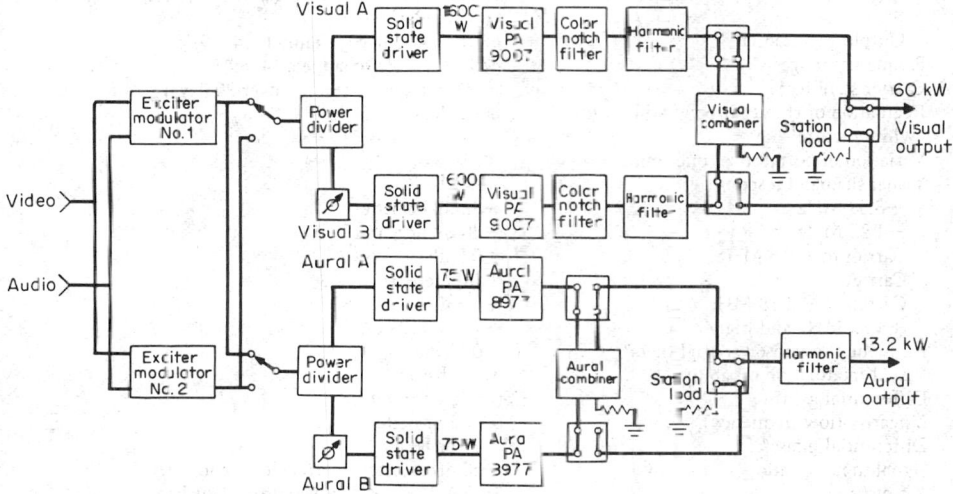

Fig. 21-20. Block diagram of a 60-kW transmitter comprising two parallel 30-kW units. (*Model TTG 30/1, RCA Corp.*)

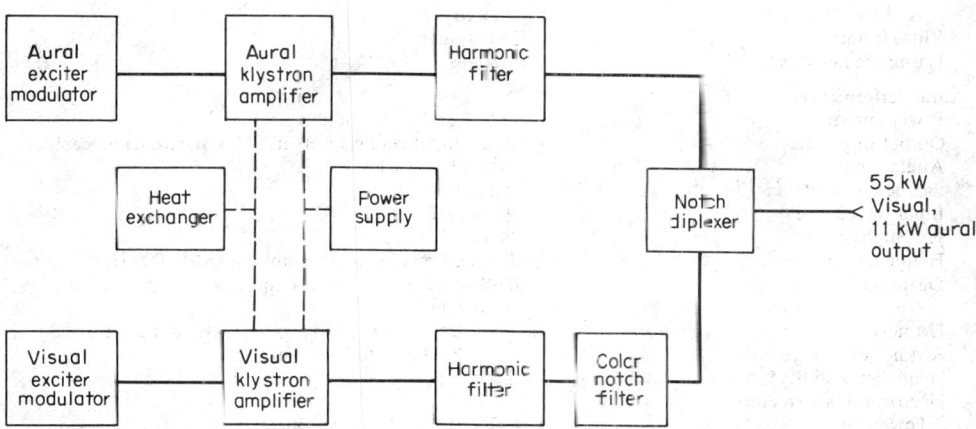

Fig. 21-21. Block diagram of 55-kW UHF television transmitter (*Harris Corp.*)

five-cavity vapor-cooled klystrons as the visual and aural final power amplifiers. These klystrons develop full power output with less than 1 W of drive power. The exciters deliver nominally 2 W of visual drive and 5 W of aural drive. Low-level i.f modulation is used. All circuits except the final amplifiers employ solid-state devices exclusively. The klystrons are controlled by digital logic circuits and operate completely independently, with separate magnet supplies and overload sensors. To remove heat from the klystrons, a unitized heat exchanger is used, containing cooling cores, blower and motor, circulating pump, storage tank, and control devices. The transmitter

*Model BT-55UI, Harris Corporation.

is fully equipped for remote-control operation. Table 21-15 lists the specifications of the BT-55U1 transmitter.

48. 100-W UHF Television Translator. Figure 21-22 shows the block diagram of a 100-W visual 10-W aural output UHF-TV translator. This unit is completely solid state. The receiver–up-converter section contains all circuitry required to provide channel conversion and

TABLE 21-15   Specifications of the Model BT-55UI UHF Television Transmitter

Visual performance:	
Power output	55 kW peak
Output impedance, from cabinet	6⅛-in EIA flanged 50 Ω (channels 14–51); waveguide (channels 52–69)
Output to antenna	6⅛-in EIA flanged (channels 14–69)
Frequency range	470–806 MHz (channels 14–69)
Carrier stability[a]	±500 Hz (max variation over 30 days)
Regulation of rf output power (black to white picture)	3% or less
Variation of output over one frame	<2%
Visual sideband response:	
−3.58 MHz	−42 dB or better
−1.25 MHz and lower	−20 dB or better
Carrier to −0.5 MHz	±0.5 dB
Carrier	0 dB reference
Carrier to +4.18 MHz	+0.5 dB, −2.0 dB
+4.75 MHz and higher	−20 dB or better
Frequency response vs. brightness[b]	±0.75 dB
Visual modulation capability	3% or better
Differential gain[c]	0.5 dB or better
Linearity (low-frequency)	0.5 dB or better
Differential phase[d]	±4° or better
Signal-to-noise ratio	−50 dB or better (rms) below sync level
K factors	2t, 2%, 12.5t, <10% base-line disturbance
Equivalent envelope delay:[e]	
0.05–2.1 MHz	±40 ns[e]
At 3.58 MHz	±30 ns
At 4.18 MHz	±70 ns
Video input[f]	75-Ω system
Harmonic radiation	−80 dB
Aural performance:	
Power output	11 kW at diplexer output
Output impedance	50 Ω; output connector 3⅛-in EIA standard (from cabinet)
Audio input	+10 dBm ± 2 dB
Frequency deviation	±25 kHz
Input impedance	600/150 Ω
Pre-emphasis	75 μs
Frequency response	± 0.5 dB relative to pre-emphasis (30-15,000 Hz)
Distortion	0.5% or less after 75μs de-emphasis with ± 25 kHz deviation
FM noise	−59 dB or better relative to +25 kHz deviation
AM noise[g]	−55 dB relative to 100% modulation
Frequency stability[h]	±500 Hz
Electrical requirements:	
Power input	440/460/480 V, 3-phase 50/60 Hz
Power consumption (typical):	
Channels 14–51	214 kW
Channels 52–69	269 kW
Power factor	>90%

[a]After initial aging of 60 days.

[b]Measured at 65 and 15% of modulation. Reference 100% = peak of sync.

[c]Maximum variation of subcarrier amplitude from 75 to 10% of modulation. Subcarrier modulation percentage: 10% peak to peak.

[d]Maximum variation of subcarrier phase with respect to burst for modulation percentage from 75 to 10%. Subcarrier modulation percentage: 10% peak to peak.

[e]Referenced to standard curve, FCC.

[f]Bridging, loop through input with −30 dB or better return loss up to 5.5 MHz.

[g]After de-emphasis.

[h]Relative to frequency offset by 4.5 MHz (FCC).

AGC functions from VHF or UHF channels. A low-noise preamplifier following the preselection filter sets the system noise figure. The output of the preamplifier is a double-balanced diode mixer which is driven from the input local-oscillator module. The mixer output is at standard TV i.f. frequency of 45.75 MHz.

The i.f. amplifier provides AGC and power-control functions. The AGC is based on a sync gating system, which provides stable peak sync reference for output power independent of signal video content. Power control is obtained from dc control of a pin-ciode attenuator.

A linearizer is provided at i.f. to compensate for the slight amplitude compression that occurs at peak sync levels. The linearizer produces intermocculation and sidebands with the same amplitude but opposite phase from those produced in the amplifier nonlinearity. The linearizer is temperature-compensated.

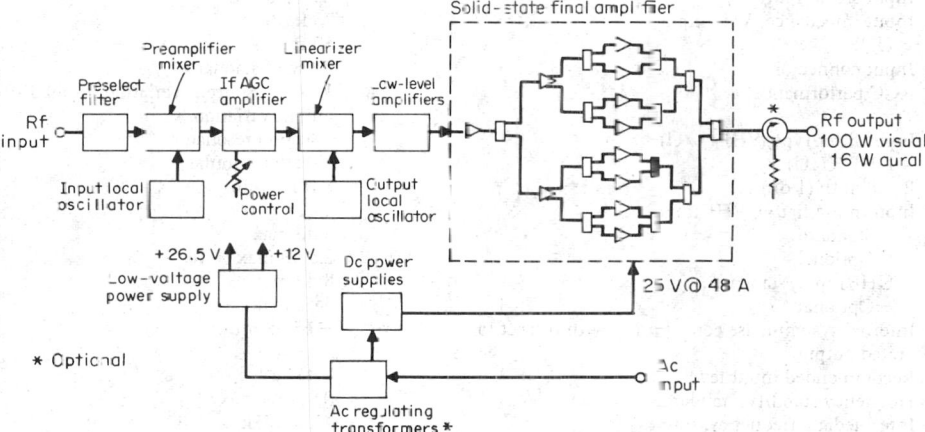

Fig. 21-22. Block diagram of all-solid-state 100-W visual, 10-W aural output translator.

The low-level amplifiers are wide-band solid-state modules which raise the mixer output level to approximately 3 W. The solid-state output amplifier consists of 16 hybrid combining devices on 8 plug-in modules. Each module produces 18 W minimum. The total amplifier, including driver stages, uses 20 transistors. The performance of this unit is shown in Table 21-16.

AM Broadcast Receivers

BY NORMAN W. PARKER

49. Introduction. AM broadcast receivers are designed to receive amplitude-modulated signals between 540 and 1,600 kHz (555 to 185 m wavelength), with channel assignments spaced 10 kHz. To enhance ground-wave propagation the radiated signals are transmitted with the electric field vertically polarized.

AM broadcast transmitters are classified, according to the input power supplied to the power amplifier, from a few hundred watts up to 50 kW. The operating range of the ground-wave signal, in areas where the ground conductivity is high, is up to 200 mi for 50-kW transmitters. During the day the operating range is limited to the ground-wave coverage. At night, refraction of the radiated waves by the ionosphere causes the waves to be channeled between the ionosphere and the earth, resulting in sporadic coverage over many thousands of miles. The night-time interference levels thus produced impose a restriction on the number of operating channels that can be used at night.

The signal-selection system is required to have a constant bandwidth (approximately 10 kHz), continuously adjustable over a 3:1 range of carrier frequencies. The difficulty of designing cascaded tuned rf amplifiers of this type has resulted in the universal use of the superheterodyne principle in broadcast receivers.

A block diagram of a typical design is shown in Fig. 21-23. In this figure the signal is supplied by a vertical monopole (whip) antenna in automobile radio receivers or by a ferrite-rod loop antenna in portable and console receivers. An rf amplifier is used in most automobile designs but not in small portable models.

In some receivers the local oscillator is combined with the mixer, which simplifies the rf portion of the receiver. An intermediate frequency of 455 kHz is used in portable and console

TABLE 21-16   Characteristics of UHF Translator

RF performance:	
Output power, visual	100 W peak
Aural	10 W average
Output frequency	Any UHF channel 14–69
Output connector and impedance	Type N, 50 Ω
Input signal range	35–10,000 μV
Input impedance, VHF input	75 Ω
UHF	50 Ω
Input connector	Type N standard
AGC performance	1 dB max output variation for 50 dB input variation
Input filter type, VHF (V/U)	3 section resonator
UHF (U/U)	5 section tubular
Bandwidth (1 dB)	6 MHz min
Input noise figure, VHF input:	
Standard	5 dB max
Optional	2.5 dB max
UHF input, standard	8 dB max
Optional	5 dB max
Internal system-noise contribution with respect to rated output	−65 dB min
Recommended input level	1,000 μV
Frequency stability, standard	0.0015% max*
Intermediate frequency, sound	41.25 MHz
Visual	45.75 MHz
Harmonic output	60 dB rms min below peak sync
Video performance:	
Differenatial gain	0.5 dB
Differential phase	1° max
Group delay	30 ns max error
2tK factor	2% max
Sync compression	0 dB nominal, 0.5 dB max
Low-frequency linearity	5% max
Service conditions and mechanical:	
Ambient temperature	−30 to 50°C
Ambient relative-humidity range†	0–100%
Altitude	12,000 ft
Power requirements	120 or 240 V ac, single phase, 50/60 Hz
Consumption at black level	1.8 kVA
Dimensions, height	72 in
Width	27 in
Depth	32 in
Weight, 120 V ac	250 lb
240 V ac	325 lb

*Precision stability optional.
†No condensation.

receivers, while 262.5 kHz is common in automobile radio designs. Diode detectors are almost universally used for detection of the i.f. signal. Push-pull class B audio-power amplifiers are used to minimize current drain. A moving coil dynamic speaker is used as the output transducer.

50. Design Categories. AM receiver designs currently fall into three categories:

Portable Battery-Powered Receivers without External Power Supply. These units vary in size from small pocket radios operating on penlite cells to larger hand-carried units using D cells for power. The battery life of the pocket radio depends on the type of cells used and the duty cycle of receiver operation. In a typical pocket radio using carbon-zinc penlite cells with operation of 2 h/day the life expectancy is about 70 h, to an end point at 80% of the initial battery voltage. Failure usually occurs with loss of power output from the audio-power amplifier or from loss of local oscillation. The audio-power output is usually about 75 mW for the pocket receiver. The larger portable units using D cells operate for longer periods of time without shortening the battery life. When the average operation is less than 2 h/day, the battery life is limited by shelf-life degeneration. The power output in the larger portable units is about 250 mW.

Console and Component Type AM Receivers Powered by the Power Line. These units are usually a part of an AM-FM receiver, with high audio-power output capability. The power output ranges from several watts to more than 100 W. Most audio systems use push-pull class B operation. However, since quiescent power drain is not a limitation, some units use single-ended class A power amplifiers to save transistor costs. Most such systems are equipped with two amplifier systems for FM stereo operation.

Automotive Receivers Operated on the 12-V Battery-Generator System of the Automobile. The primary current used in transistorized receivers usually does not exceed 1 A.

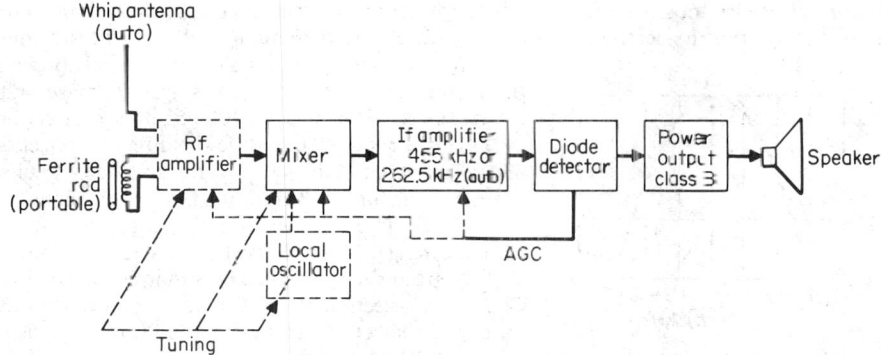

Fig. 21-23. Block diagram typical of AM receivers.

Because of operation in the high ambient noise of the vehicle, the power output required is relatively high (2 to 3 W).

51. Sensitivity and Service Areas. The required sensitivity is governed by the expected operating field strengths. Typical field strengths for primary (ground-wave) and secondary (sky-wave) service are as follows:

	Field strength, mV/m
Primary service:	
Central urban areas (factory areas)	10–50
Residential urban areas	2–10
Rural areas	0.1–1.0
Secondary service	Areas where sky-wave signals > 0.5 mV/m at least 50% of time

Co-channel protection is provided for signals exceeding 0.5 mV/m. The receiver sensitivity and antenna system should be adjusted to provide usable outputs with signals of the order of 0.1 mV if the receiver is to be used over the maximum coverage area of the transmitter.

The required circuit sensitivity is controlled by the efficiency of the antenna system. A car-radio receiver vertical antenna is adjustable to about ̄ m in length. Since the shortest wavelengths are of the order of 200 m, the antenna can be treated as a nonresonant short-capacitive antenna. The open-circuit voltage of such a short monopole antenna is

$$E_a = 0.5 l_{\text{eff}} E_f \quad \text{mV}$$

where l_{eff} = effective length of antenna (m) and E_f = field strength (mV/m).

The radiation resistance of the short monopole is small compared with the circuit resistance of the receiver input circuit, but the antenna is not matched to the input impedance since matching is not critical, adequate antenna voltage being available at the minimum field strength (0.1 mV/m) needed to override noise. The car-radio antenna is coupled to the receiver by shielded cable. This, with the receiver input capacitance, forms a capacitive divider, reducing the antenna voltage applied to the receiver. To ensure adequate operation the receiver should offer 20 dB signal-to-noise ratio when 10 to 15 μV is applied to the input terminals.

Portable and console receivers use much shorter built-in antennas, usually horizontally polarized coils wound on ferrite rods. The magnetic antenna can be shielded from electric field interference. Although the effective length of a ferrite rod is shorter than that of a whip antenna, the higher Q of the ferrite rod and coil provide approximately the same voltage to the receiver. The unloaded Q of a typical ferrite-rod antenna coil is of the order of 200. The voltage at the terminals of the antenna coil is QE_a.

52. Selectivity. Channels are assigned in the broadcast band at intervals of 10 kHz, but adjacent channels are not assigned in the same service area. In superheterodyne receivers, selectivity is required not only against interference from adjacent channels but also to protect against image and direct i.f. signal interference.

The primary adjacent-channel selectivity is provided by the i.f. stages, whereas image and direct i.f. selectivity must be provided by the rf circuits. In receivers using a ferrite-rod antenna and no rf stage, the rf selectivity is provided entirely by the antenna. High Q in the antenna coil thus not only provides adequate signal to override the mixer noise but also protects against image and i.f. interference. With a Q of 200 the image rejection at 1,400 kHz is about 40 dB while the direct i.f. rejection at 600 kHz is about 24 dB. With an rf stage added, the image rejection is about 50 dB and the i.f. rejection about 46 dB.

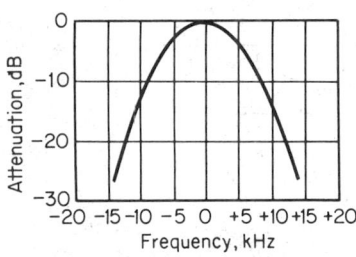

Fig. 21-24. Typical selectivity curve of an AM receiver using a ferrite-rod antenna without rf amplifier.

Since car-radio receivers are subjected to an extreme dynamic range of signal levels, the selectivity must be slightly greater to accommodate strong signal conditions. The image rejection at 1,400 kHz is typically 58 dB in spite of the lower i.f. frequency; the i.f. rejection is typically 50 dB, and the adjacent-channel selectivity is about 20 dB. Figure 21-24 shows the overall response of a typical portable receiver using a ferrite rod without an rf amplifier.

53. High-Signal Interference. When strong signals are present, the distortion in the rf and i.f. amplification stages can generate additional interfering signals. The transfer characteristic of an amplifier system can be expressed in a power series as

$$E_{out} = G_1 E_{in} + G_2 (E_{in})^2 + G_3 (E_{in})^3 + \cdots + G_n (E_{in})^n$$

where E_{out} = output voltage, E_{in} = input voltage (same units) and $G_1, G_2, \ldots, G_n$ = voltage gains of successive amplifier stages.

When the input signal consists of two or more modulated rf signals, E_{in} becomes

$$E_{in} = e_1 \cos \omega_1 t + e_2 \cos \omega_2 t$$

where

$$e_1 = E_1[1 + Ea_1(t)/E_1] \qquad e_2 = E_2[1 + Ea_2(t)/E_2]$$

where E_1 = signal 1 carrier, E_2 = signal 2 carrier, $Ea_1(t)$ = audio modulation first signal, and $Ea_2(t)$ = audio modulation second signal.

I.F. Beat. When two strong signals are applied to the amplifier with carrier frequencies separated by a difference equal to the intermediate frequency, a difference-frequency signal appears which is independent of the local oscillator frequency. Because of the wide frequency spacing, interference of this kind can take place only in the mixer or rf amplifier and only with strong signals (signal strengths of several volts per meter). These signals are derived from the G_2 term:

$$E_{if,beat} = G_2 e_1 e_2 \cos (\omega_1 - \omega_2)t$$

where $e_1 e_2$ and $\omega_1 \omega_2$ are the respective amplitudes and angular frequencies.

Cross Modulation. When a strong interfering signal is present, the modulation on the interfering signal can be transferred to the desired signal by third-order distortion components. This type of interference does not occur at a critical frequency of the interfering signal, provided that it is close enough to the desired signal frequency not to be attenuated by the selectivity of the receiver. This type of distortion can take place in the rf. mixer, or i.f. stages of the receiver. These signals are derived from the G_3 terms:

$$E_{crossmod} = G_3(e_2)^2 e_1/4 \cos \omega_1 t$$

The cross modulation is proportional to the square of the strength e_2 of the interfering signal.

Intermodulation. Another type of interference due to third-order distortion is caused by two interfering carriers. When these signals are so spaced from the desired signal that the first signal is arbitrarily spaced by Δf and the second signal is $2\Delta f$, third-order distortion can create a signal on the desired carrier frequency. These signals are derived from the G_3 terms:

$$E_{intermod} = G_3 e_1^2 e_2/4 \cos (2\omega_1 - \omega_2)t$$

The interference is proportional to the square of the amplitude of the closer carrier times the amplitude of the farther carrier. Intermodulation is sometimes masked by cross modulation; it occurs only when the e_2 signal is stronger than the desired carrier after attenuation by the rf selectivity of the receiver.

Harmonic Distortion. Harmonic-distortion interference usually arises from harmonics generated by the detector following the i.f. amplifier, which radiate back to the tuner input.

54. Choice of Intermediate Frequency. Two intermediate frequencies are used in broadcast receivers: 455 kHz and 262.5 kHz. The 455-kHz i.f. has an image at 1,450 kHz when the receiver is tuned to 540 kHz, thus allowing good image rejection with simple selective circuits in the rf stage. At 540 kHz the selectivity must be sufficient to prevent i.f. feedthrough since the receiver is particularly sensitive at i.f. frequencies because converter circuits typically have higher gain at i.f. than at the converted carrier frequency. The choice of the higher i.f. frequency also makes i.f. beat interference less likely. The second harmonic falls at 910 kHz and the third harmonic at 1,365 kHz.

The 262.5 kHz i.f. has a lower limit of image frequency at 1,065 kHz, which requires somewhat more rf selectivity than is needed when the higher i.f. is used. The second harmonic of the 262.5-kHz frequency falls below the broadcast band (at 525 kHz) and hence does not interfere. On the other hand, there are more higher-order responses in the passband (787.5, 1050, 1,312.5, and 1,575 kHz). Sensitivity to i.f. feedthrough when the receiver is tuned to 540 kHz is greatly reduced by the use of the lower i.f. frequency.

55. Tuners. There are three basic circuit configurations used in AM receiver tuners. The least complicated uses a self-oscillating mixer stage. Figure 21-25 shows a typical circuit of a self-oscillating mixer. The most critical design factor is tracking the oscillator with the antenna tuning to maintain a fixed frequency difference of 455 kHz below the oscillator. The oscillator covers the range from 995 to 2075 kHz as the antenna circuit covers the band from 540 to 1,600 kHz. Since the frequency ratio covered by the oscillator is less than the antenna circuit, the total capacitance change and the rate of capacitance change are less in the oscillator than in the antenna circuit. Since the oscillator and antenna are ganged together on the same control, the plates of the oscillator must be shaped to provide tracking. The shortcoming of this circuit lies in its lack of rf selectivity and the inability to employ automatic gain control (AGC) to control overload. In some circuits a separate mixer is used to allow the application of AGC to the mixer stage.

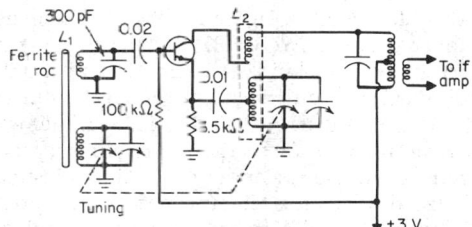

Fig. 21-25. Typical self-oscillating mixer for AM service.

Figure 21-26 shows a circuit with a separate mixer and an rf amplifier stage, using a three-gang variable capacitor for tuning. The capacitors in the base and collector circuit of the rf amplifier are identical. When the trimmer capacitors are adjusted to provide the same effective minimum capacitance in both circuits and the collector coil is adjusted to match the base-circuit resonance, the rf tuning adjustments track. The oscillator relies on capacitor-plate shaping to track the rf stages to produce a constant 455-kHz difference frequency.

The forward bias on the rf transistor is reduced by the negative voltage at the diode detector. Reducing the forward bias reduces transistor beta and reduces the gain to the rf amplifier. Controlling the gain of the rf stage tends to reduce spurious signals which would otherwise occur in the mixer stage when strong signals are received. Since the tuning of the rf circuits relies on the antenna circuit, the Q of this circuit must be as high as possible. Unloaded Q's of more than 200 are desirable. When the loaded Q is adjusted to one-half the unloaded value, the antenna coil is matched for optimum power transfer from the antenna circuit. Since the radiation resistance is a small fraction of the resistance that determines the Q, the antenna source impedance is never matched to the input. Using the highest possible Q with the greatest effective rod length provides the best possible compromise antenna match.

Transistor noise is low enough for satisfactory signal-to-noise ratios to be obtained with 0.1 mV/m when an rf stage is not used.

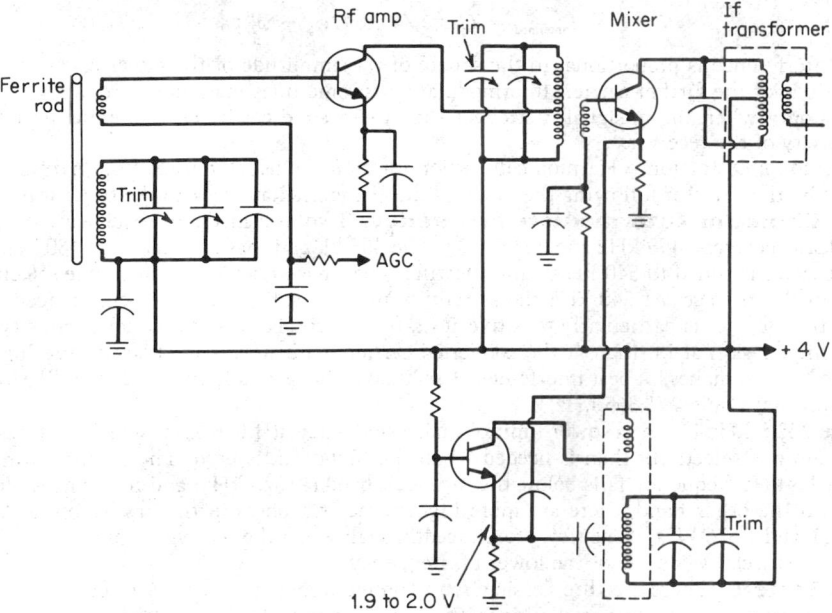

Fig. 21-26. Tuner with rf stage and separate oscillator. *(Motorola, Inc.)*

Automotive radio equipment relies on push-button tuners, which use slide mechanisms that adjust core positions in coils rather than resetting the exact angle on the rotating shaft of a variable capacitor. Figure 21-27 shows a circuit of a typical automotive radio tuner. Since the tuner uses the 262.5 kHZ i.f., the tracking problem is greatly simplified. In the slug-tuned receiver the tracking is accomplished with coil geometry so that a relatively constant frequency difference is maintained throughout the slug travel.

56. I.F. Amplifiers. The i.f. amplifier is primarily responsible for the gain and selectivity of the receiver. Since the i.f. amplifier operates at a fixed frequency, the bandwidth of the receiver is independent of the tuning adjustment over the full tuning range from 540 to 1,600 kHz. Although it is tempting to design essentially all the gain into the i.f. amplifier, this practice may cause instability from regeneration unless expensive shielding is used. Hence total gain is distributed among the rf, i.f., and audio stages.

The i.f. designs usually employ two stages of single-tuned amplification, with about 60 dB gain overall. A typical gain distribution is as follows: 10 to 30 dB in the rf section, 60 dB in the i.f. amplifier, and 30 to 60 dB in the audio section, the total power gain being 100 to 150 dB. Figure 21-28 shows the circuit of a typical two-stage i.f. amplifier with AGC applied to the first stage. The circuit derives its selectivity from three single-tuned transformers with tightly coupled secondary windings to match the output impedance of the collector and the input impedance of the base-emitter circuit of the following amplifier. Ideally, the Q of the circuit is determined by the loading of the amplifiers, but in practical coil designs the attainable Q is limited and some mismatching (2 to 3 dB) of the amplifier occurs.

Significant gain reduction in the first i.f. amplifier occurs when the current drops below ¼ mA and the voltages on base and emitter approach zero. When this occurs, the detector level is slightly over 1 V, the optimum operating detector level. Gain control over 40 dB can be achieved with the single stage. Typical AGC begins to operate with about 1 mV from the mixer and continues to reduce the gain, maintaining a relatively constant output at 200 mV.

Some superheterodyne systems use more than one intermediate frequency. In receivers of this type, the incoming signal is first converted to a relatively high first i.f. chosen to minimize image responses. The output of the first i.f. amplifier is supplied to a second mixer stage, where a fixed-frequency oscillator converts its output to a much lower i.f. frequency, at which high gain and selectivity can easily be obtained.

Mechanically vibrating elements can be designed with Q's much higher than electric circuits, and it is feasible to use mechanical vibrators in i.f. filter circuits to provide excellent selectivity in low-frequency i.f. circuits. There are two basic types of electromechanical filter circuits. (1)

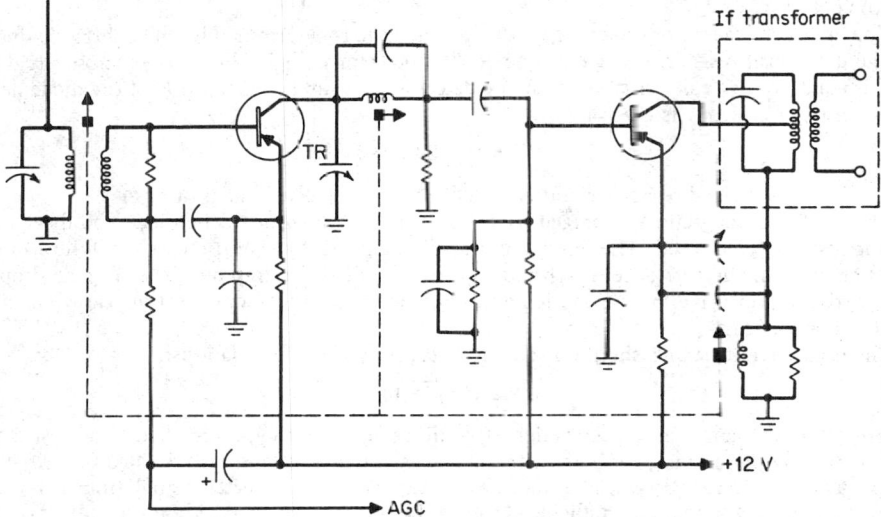

Fig. 21-27. Tuner for automobile AM radio. *(Motorola, Inc.)*

Piezoelectric vibrator plates made of quartz or (for low and medium frequencies) barium titanate form the equivalent of a high-Q coil-and-capacitor combination which can be used singly or in combination to form i.f. filters with steep band-edge selectivity characteristics. The piezoelectric vibrator converts energy from electric to mechanical and back to electric with relatively little loss. (2) Passive mechanical filters are made of mechanically vibrating elements, coupled to form a composite filter of extremely high selectivity. Low-loss filters of this type are often used in communication receivers, where good selectivity is more important than cost. A typical filter of this type uses several lumped masses coupled together by torsion rods. The low-loss rod-and-mass resonant combinations are used as elements in Butterworth or Chebyshev type filters. The signal is converted from electrical form to mechanical form to drive the filter, and the filter output converted back to electrical form by piezoelectic or magnetostrictive converters. The conversions may introduce losses of 20 dB or more.

Barium titanate resonators have been used in some broadcast receiver designs. While they

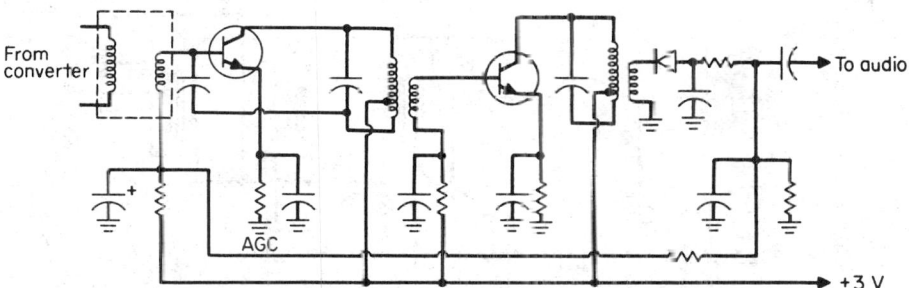

Fig. 21-28. Typical two-stage AM receiver i.f. amplifier. *(Motorola, Inc.)*

have the advantage of small size (at 455 kHz), the disadvantage is the numerous spurious responses caused by multiple modes of vibration. Hence those resonators must be used in combination with coils and capacitors to suppress the spurious responses. The need for supplementary filtering has resulted in limited use of resonators of this type.

57. I.F. Detectors. The series-connected diode detector is almost universally used in detecting the i.f. signal. The diode detector is simple, efficient, and relatively free of distortion when properly used. The ac diode output constitutes the audio signal, while the dc component provides the control bias for AGC on the i.f. (and the rf stage if one is present). To operate the diode at low signal levels it is necessary to provide forward bias to overcome the diode-contact potential. In Fig. 21-25 the AGC circuit is connected to the power supply to forward-bias the first i.f. stage, while a 3,300-Ω resistor connects the detector to the first i.f. stage and thus provides forward bias for the detector.

The effective loading of the diode on the last i.f. stage is determined by the energy dissipated in the diode-load resistor. If the efficiency of the detector is high, the voltage across the diode load is equal to the peak voltage driving the detector. The input resistance R_i of the diode detector at carrier frequency is given by

$$R_i = R_d/2N_0$$

where R_d = diode-load resistance and N_0 = efficiency of detector (maximum 1).

The diode-load capacitor is charged by the i.f. output coil through the diode and discharged by the resistive diode load. The discharge time constant must be short enough to follow downward changes in the carrier level without causing the diode to drop out. When diode dropout occurs, the output wave follows the RC discharge curve and a type of distortion known as *diagonal clipping* occurs.

The capacitive reactance should be chosen to accommodate the relationship

$$Z_m/R_d > m$$

where Z_m = capacitor reactance at highest modulating frequency, R_d = diode-load resistance, and m = modulation factor at highest frequency (modulation percentage divided by 100).

58. Audio Amplifiers. The audio amplifier raises the detected signal from a level of about 100 μW to the required audio-power level. In portable receivers, where the power output is 100 mW, the gain required is about 30 dB with additional reserve gain for low signal levels. In console or component-type receivers, where the audio output may be 100 W, gains in excess of 60 dB are required.

Figure 21-29 shows a typical battery-operated portable-radio audio and power-output circuit. The combined current drain of the first two stages is about 3 mA, with class A biasing. The first two stages provide sufficient power gain to drive the power stages. which supply about 75 mW of audio-power output. Since peak currents of 50 mA would be required to supply a class A audio stage, operation of the output stage would require a large battery or short battery life. To minimize current drain (particularly zero signal-current drain) the audio is operated in push-pull, class B.

In battery-operated portable sets the small speaker enclosures limit the low-frequency response. The tone control consists of a simple RC filter which reduces the response to high-frequency audio signals. In console receivers with wide-band acoustical response, the audio response is varied by more complex tone controls. Figure 21-30 shows a circuit of a tone control used in such receivers. This circuit provides independent attenuation or boost of the bass and

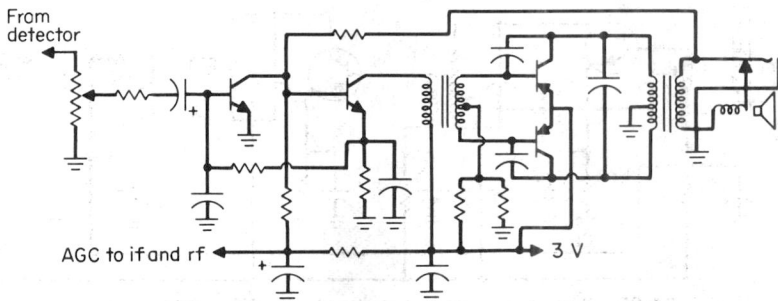

Fig. 21-29. Typical driver in class B audio output stages of a battery-operated portable receiver.

treble responses. In addition to the tone control it is customary to provide a tone-compensated volume control on the larger console or component systems. The tone-compensated volume control provides a bass boost when the volume is reduced, to compensate for the normal drop in the sensitivity of the ear to low levels of the low frequencies. The presence of the 0.009-μF capacitor and series resistor to ground from the tap on the volume control provides progressively increasing bass boost until the control is retarded beyond the tap point, below which the boost remains constant.

59. Loudspeakers. The important element in an audio system affecting the quality of the reproduced sound is the loudspeaker and its enclosure (see also Sec. **19**). Loudspeakers fall into two general categories: direct-radiator piston types, using a moving coil in a permanent magnet field, and horn-type speakers, in which a wide-band acoustical-impedance transformation network (exponential horn) is used to couple a small piston to the free-space radiation impedance.

Although the horn loudspeaker is much more efficient than the direct-radiator type (efficiencies up to about 50%), such high efficiency is available only in horns of very large size if low-

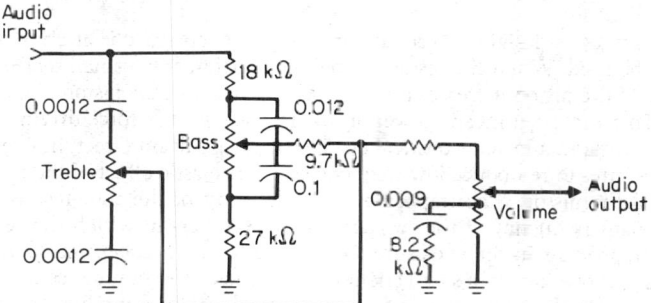

Fig. 21-30. Wide-range tone control and compensated volume control.

frequency signals are to be reproduced. Therefore horn units are used principally in large theater-type installations, where both high power and high efficiency are required.

In radio receivers the direct-radiator speaker is universally used for middle- and low-frequency reproduction. Horn-type tweeter units, designed to reproduce frequencies above 1 kHz only, and hence of moderate size, are used in some combination speaker systems.

In direct-radiator loudspeakers a coil of wire suspended in a magnetic field (voice coil) provides the driving force for a paper-cone piston supported at the edges by a springlike suspension. The force which moves the piston against the spring loading of the suspension is proportional to the voice-coil current. Since the coil is resistive over most of the audio frequencies, the force applied to the piston is constant with frequency for a given voltage applied to the speaker driving element.

A piston enclosed on one side, so that its vibration changes the instantaneous air pressure near the piston, couples acoustical power to the air. For the radiated power to be independent of frequency, the velocity of the piston must be inversely proportional to frequency at low and medium frequencies, where the piston diameter is smaller than the wavelength of the radiated sound. The mechanical system represented by the piston and the elastic suspension is arranged to operate above resonance, in the audio range covered by the speaker. In the region above resonance, the load on the driving force is the mass of the piston, the velocity of which decreases directly with frequency with constant force applied. Below the mechanical resonance the velocity decreases with frequency such that the power output drops 12 dB/octave.

The radiated power is constant for constant drive power from the mechanical resonant frequency up to a frequency at which the distance around the periphery of the piston is equal to the wavelength of the sound in air. Above this point the radiation resistance of the piston is independent of frequency. Since the velocity of the piston is inversely proportional to frequency and the radiated power is proportional to the square of the velocity times the radiation resistance, the output above this point drops 12 dB/octave.

Two effects counter the tendency of the speaker to radiate less power at frequencies where the periphery of the piston exceeds a wavelength. The first effect is the directivity pattern of the radiating piston, above the frequency of constant radiation resistance, which increases the power radiated along the axis perpendicular to the piston. This increase counteracts the loss in total radiated power by concentrating the power radiated directly in front of the speaker.

The second effect which alters the radiated-power pattern at high frequencies is the loss of integrity of the moving piston. The piston is a cone made of light, stiff paper. At high frequencies the cone ceases to move as a single unit and breaks into separate cone resonance patterns. The cone breakup greatly alters the radiation pattern, resulting in substantial changes from the theoretically predicted radiation levels.

Since these effects are undesirable in a wide-range loudspeaker, concentric elastic rings are often built into the cone, which progressively alter the cone motion so that as the frequency is increased, the outer rings of the cone become decoupled and no longer move with the driving force applied at the center of the cone. The result is that the speaker operates over a wide frequency range without reaching a point where the radiation resistance becomes independent of frequency, with the resulting erratic behavior in the sound-power radiation pattern.

In high-quality audio systems it is customary to use several speakers, each covering a separate frequency range, instead of relying on decoupling in the cone. These systems usually employ a bass speaker system using a 12- or 15-in speaker (woofer) to cover the range from 30 to 500 Hz, an 8- to 10-in midrange speaker to cover from 500 to about 2,000 Hz, and a tweeter to cover from 2,000 to 15,000 Hz.

Small speakers may be designed to act as bass sources when space is at a premium and large speakers cannot be used. When the piston is small and enclosed in a small sealed enclosure, the motion required of the piston increases inversely as the square of the piston diameter for a given power output. To radiate significant power at low frequencies the force driving the cone must be increased, resulting in a lower acoustical efficiency. A significant effect, in using small speakers in small enclosures to reproduce low frequencies, is the elastic effect of the trapped air in the enclosure behind the piston. The trapped air provides most of the cone-restoring force, determining the resonant frequency of the speaker system. Speakers in which the restoring force is principally the trapped air in the enclosure are referred to as *air-suspension speaker* systems.

When a speaker is operated in a small enclosure, the enclosure compliance tends to stiffen the suspension. As a result, the lowest usable frequency which can be reproduced is raised. In speakers designed for air suspension, the cone mass is increased to reduce the resonant frequency in the enclosure. The increased mass in the cone requires increased driving force for the same power output and a corresponding loss in efficiency. In battery-operated receivers, little attempt is made to reproduce at full level the frequencies below 150 Hz. In such cases the cone resonance is relied upon to enhance the response at the mechanical resonance of the speaker, which is the lowest frequency where reliable response can be obtained. In this case, the magnetic circuit is purposely weakened to minimize damping of the speaker resonance by the electromagnetic coupling between the mechanical system of the speaker and the source impedance of the power amplifier. This also provides a weight reduction in the speaker since a smaller magnet can be used.

Television Broadcast Receivers*

BY NORMAN W. PARKER

60. General Considerations. Broadcast television receivers are designed to receive signals in two VHF bands and one UHF band according to the United States and Canadian standards (for other standards, see Sec. **20**). The lower VHF band (channels 2 to 6) extends from 54 to 88 MHz in 6-MHz channels, with the exception of a gap between 72 and 76 MHz. The higher VHF band (channels 7 to 13) extends from 174 to 216 MHz in 6-MHz channels. The UHF channels are spaced 254 MHz above the highest VHF channel, comprising 70 6-MHz channels extending from 470 to 890 MHz. The television tuner is thus required to cover a frequency range of more than 16:1. TV tuners use separate units to cover the UHF and VHF bands.

The signal coverage of TV transmitters is generally limited to line-of-sight propagation, with coverage extending from 30 to 100 mi, depending on antenna height and radiated power. The coverage area is divided into two classes of service, depending on the signal level. The service

---

*The following paragraphs on television receivers deal only with the tuner and i.f.-amplifier sections of monochrome and color broadcast receivers. The portions of receivers that handle synchronization, video amplification, color-signal processing, picture tubes, and their auxiliaries (which are common to other forms of television-reproduction equipment) are treated in Sec. **20**, Pars. **20-93** to **20-118**. — ED.

area labeled class A is intended to provide essentially noise-free service and specifies the signal levels shown.

Channels	Peak signal evel, μV/m	Peak open-circuit (half-wavelength) antenna voltage, μV
	Class A Service	
2-6	2,500	3,500
7-13	3,500	11,800
14-83	5,000	700
	Class B Service	
2-6	225	300
7-13	600	300
14-83	1,600	225

For the limiting area of fringe service the signal levels are defined as shown for class B service. The typical level of the sound signal is from 3 to 20 dB below the picture level, due to radiated sound power and antenna gain.

61. Tuners. The TV tuner comprises two sections, one to cover the VHF bands from 54 to 88 MHz and from 174 to 216 MHz, the second to cover the UHF band from 470 to 890 MHz. The VHF tuner supplies gain at the selected transmitter carrier frequency, matches the antenna and impedance, and isolates the receiver local oscillator from the antenna. The UHF tuner consists of input circuits for matching and selectivity together with a crystal mixing circuit and local oscillator to provide an output at the i.f. frequency, which is supplied to the VHF tuner. Because the UHF tuner has a conversion loss instead of a gain, the VHF tuner is designed to provide additional gain at the intermediate frequency. When so operated, the local oscillator in the VHF tuner is turned off and the rf amplifier and mixer are tuned to the i.f. frequency when the receiver is switched to the UHF position.

Most receiver designs use a 300-Ω balanced input to the tuner terminals. The recent trend toward optional operation on closed-circuit television (e.g., CATV) systems, which generally use shielded 72-Ω coaxial cable, has resulted in tuners designed both for 300-Ω balanced and 72-Ω coaxial inputs.

Although balanced 300-Ω input is universally used, the rf amplifier stages are single-ended and so the transmission line must be coupled to the receiver through a balun transformer which converts the balanced input to a matched unbalanced load. The balun transformer usually consists of a two-window ferrite core with two windings of four turns each, coupled to provide a balance-to-unbalance match and an impedance transformation from 300 Ω balanced to 72 Ω unbalanced.

The rf amplifier must have the lowest noise figure possible, consistent with cost, bandwidth, and dynamic range. The typical VHF tuner employs a bipolar transistor in a grounded-base amplifier circuit. To prevent signal overload in the collector circuit and mixer stage, forward-bias AGC is applied to the base of the rf amplifier stage. Channel tuning is accomplished by switching coils in series, on a wafer switch. A typical tuner of this type has four wafers ganged together. The first wafer tunes the input of the rf amplifier, with single tuned circuits. The rf-mixer coupling circuit, with a double-tuned circuit, uses two wafers, and the local oscillator is tuned with the fourth wafer.

The circuit of a complete VHF and UHF tuner of this type is shown in Figs. 21-31 and 21-32. This type of tuner has a typical noise figure of 3 dB and an overall gain of about 30 dB. The intermediate frequency is taken from the collector circuit by a low-impedance coupling network, using a capacitance divider to reduce the collector load impedance by a factor of 170 and feeding the signal through a coaxial coupling cable to the i.f. amplifier. This method of coupling has become widely used since it allows the tuner to be located at a distance from the i.f. amplifier, thus providing flexibility in cabinet and chassis locations; e.g., the channel selector can be located at any convenient place in the cabinet without regard to the location of the i.f. amplifier stages.

62. I.F. Amplifiers. The choice of intermediate frequency is an important factor in determining the ability of the receiver to operate without interference from spurious responses. The local oscillator frequency is positioned above the desired signal frequency such that the frequencies are 45.75 MHz for the picture carrier, 41.25 MHz for the sound carrier, and 42.17 MHz for the color subcarrier. The images on channels 2 to 6 then fall in the gap between channels 6 and 7, while the images for channels 7 to 13 are above channel 13. In the UHF band (channels 14 to 83) each image falls 15 channels higher than the desired channel, so that the FCC UHF channel-assignment plan avoids assigning transmitters with 15-channel separation in the same area.

Numerous other combinations of signals can provide significant interference. The most impor-

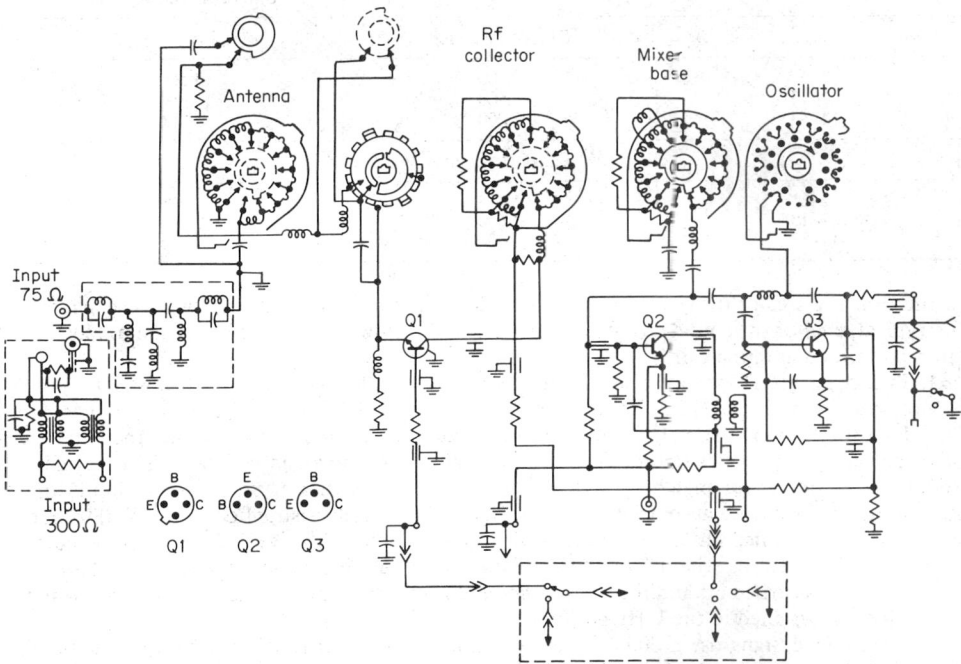

Fig. 21-31. Typical VHF television tuner. *(Motorola, Inc.)*

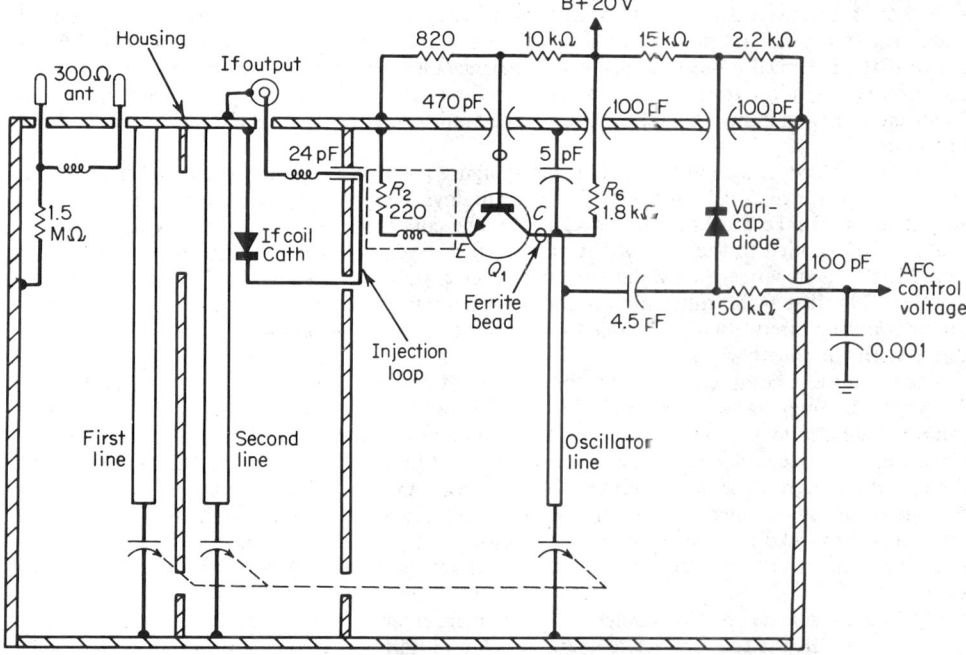

Fig. 21-32. Typical UHF television tuner. *(Motorola, Inc.)*

tant is the i.f. beat, which occurs when the carriers of two strong channels are spaced seven channels apart, the mixture of the two causing a spurious i.f. signal proportional to the product of the two signals. In the UHF band, where seven-channel spacing is possible, the assignment to any one area of channels with seven-channel spacing is avoided. In the VHF band, the beat does not occur, since the channel assignments cannot be spaced seven channels. In the VHF band intermodulation and cross-modulation products are avoided by spacing the channels so that two strong signals do not fall on adjacent channels.

In spite of the care with which the intermediate frequency was chosen, there is at least one significant source of interference. This occurs on channel 6, where harmonics of the carrier generated in the rf amplifier can beat with the local oscillator to produce an interference with no additional signal present. On that channel the second harmonic of the sound channel beats with the local oscillator to produce a 0.75-MHz interfering signal, and the color subcarrier provides additional interference at 1.15 MHz.

The i.f. circuits operate with a vestigial-sideband signal (see Par. 20-25) with the carrier tuned to a point 6 dB below the maximum response point. To avoid possible interference from an adjacent-channel picture carrier, a fixed-tuned trap tuned to 47.25 MHz is used. To prevent non-linearities in the i.f. amplifier from generating spurious signals, selectivity against interference should be introduced at the lowest-level point, preferably between the mixer and the i.f. amplifier. The insertion loss of the filter must be low enough to prevent i.f. noise from contributing to the signal.

A circuit of a typical transistorized i.f. amplifier is shown in Fig. 21-33. The input circuits of the i.f. amplifier provide attenuation of the lower adjacent sound carrier, which appears in the i.f. amplifier above the carrier. An attenuation of the order of 60 dB is required in the adjacent-channel sound trap to avoid visible adjacent-channel sound interference. The interfering signal produced by the sound signal produces a beat with a mean video frequency of 1.5 MHz. Signals falling in this range are visible when the signal is 50 dB below the desired video carrier. More-over, additional attenuation is necessary when the desired signal is weaker than the adjacent-channel interference. The sound signal is attenuated by about 26 dB to ensure that nonlinear effects in the i.f. amplifier do not produce spurious cross-modulation components.

63. Intercarrier Sound Circuits. Before the video detector circuit the intercarrier FM sound signal, with a carrier frequency of 4.5 MHz, is removed from the video signal and is fed to a separate detector for sound only. The separate sound detector is required in color receivers, since the presence of the color subcarrier at 3.58 MHz would produce a 920-kHz beat if nonlin-earities existed in the detector transfer characteristic. Since beats as low as 50 dB below the carrier level are perceptible, the required linearity would be impractical if the sound signal were to be recovered from the video detector.

The separate diode detector is connected to the last i.f. amplifier stage, ahead of the video detector sound trap, for the purpose of developing the 4.5-MHz intercarrier sound beat. The sound-detector load circuit is tuned to 4.5 MHz. Although the intercarrier design makes tuning less critical (since the sound signal is present whenever the composite signal is received), the advent of color has made the tuning adjustment of the picture carrier frequency in the i.f. amplifier more critical, since mistuning causes the color subcarrier to change amplitude rapidly.

In receivers with intercarrier sound, the output of the sound detector is not a function of the tuner oscillator frequency shift, and this method of automatic tuning oscillator control (AFC) cannot be used. To maintain correct tuning in color receivers of intercarrier design, a separate discriminator tuned to the picture carrier provides the control voltage for the tuner AFC. A typical AFC circuit is shown in Fig. 21-34. The sound channel supplies additional gain at 4.5 MHz, bringing the intercarrier sound signal to a level suitable for operation of the FM detector. One of the most popular sound-detector circuits is the ratio detector, since its inherent AM rejec-tion characteristics minimize the need for limiting stages.

In most recent designs, composite amplification and FM detection circuits have been combined on integrated-circuit chips with a minimum number of external connections. In circuits of this type the amplifier are generally of the wide-band type with no interstage selectivity. Instead, the selectivity is placed ahead of the integrated circuit to minimize the number of external connec-tions. The discriminator is also designed to include a minimum number of coils and external connections.

Since in integrated circuits the number of external connections required is more critical than the number of stages used, it is possible to provide additional gain and limiting at 4.5 MHz and to provide an FM detector less complex than the ratio detector, thus reducing the external con-nections. The circuit usually employed is the quadrature detector, which uses a single coil to

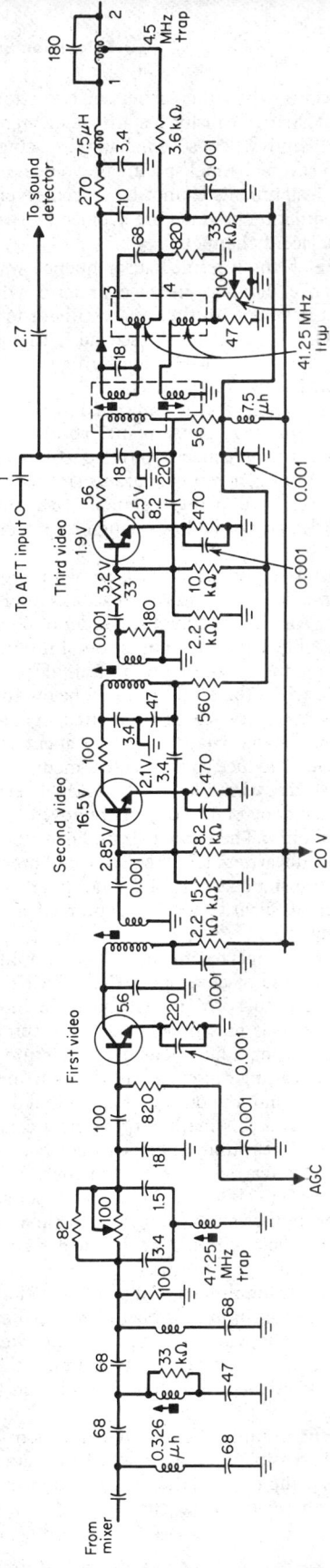

Fig. 21-33. Typical 3-stage television i.f. amplifier. (*Motorola, Inc.*)

provide a quadrature voltage at the carrier frequency. Since the phase of the voltage derived from the tuned circuit varies with frequency, by comparing the tuned-circuit voltage with the voltage obtained directly from the amplifier-limiter, an audio-output voltage proportional to input frequency can be obtained. For additional discussion of circuits used in TV broadcast receivers, see Pars. **20-93** to **20-118**.

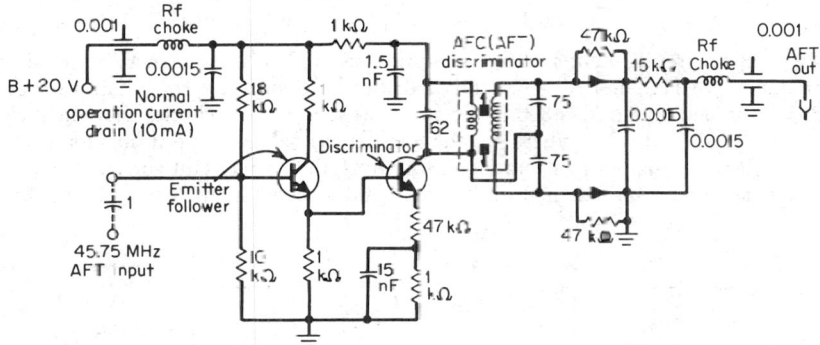

Fig. 21-34. Discriminator for automatic frequency control.

FM Broadcast Receivers

BY NORMAN W. PARKER

64. General Considerations. Broadcast FM receivers are designed to receive signals between 88 and 108 MHz (3.5 to 2.8 m wavelength). The broadcast carrier is frequency-modulated with audio signals up to 15 kHz, and the channel assignments are spaced 200 kHz. The FM band is primarily intended to provide a relatively noise-free radio service with wide-range audio capability for the transmission of high-quality music and speech. The service range in the FM band is generally less than that obtainable in the AM band, especially when sky-wave signals are relied on for extending the AM coverage area.

VHF signals are limited to usable service ranges of less than 70 mi. Since sky-wave signals do not materially affect the transmission of FM signals, there is no equivalent night effect and licenses are not limited to daylight hours only, as with many AM operations. In the past, all FM signals in the United States were horizontally polarized. However, the rules have been changed to allow maximum power to be radiated in both the horizontal and vertical planes.

Unlike AM broadcasting, where the station power is measured by the power supplied to the highest-power rf stage, FM transmitters are rated in *effective radiated power* (ERP), i.e., the power radiated in the direction of maximum power gain. This method of power measurement is used since the transmitting antenna has significant power gain, resulting in an ERP many times the input power supplied to the rf output stage of the transmitter.

Although FM receivers have been principally used in high-fidelity audio installations, more recently small FM receivers have been designed with limited audio-output capabilities. Also, an increasing number of FM broadcast sets have been included in automobile installations.

In 1960 the FCC amended the broadcast rules to allow the transmission of stereophonic signals on FM stations (see Par. **21-27**). This transmission is equivalent to the transmission of two audio signals, each having 15 kHz bandwidth, transmitted on the same carrier as used for monophonic FM signals. Since the FM signal was initially designed to have sufficient additional bandwidth to achieve improved signal-to-noise ratio at the receiver, there is room to multiplex the second component of the stereophonic signal with no increase in the radiated bandwidth. However, the signal-to-noise ratio is reduced when the multiplexing technique is employed. Most console and component-type FM receivers are currently designed for the reception of multiplex stereo.

65. Sensitivity. The field strength for satisfactory FM reception in urban and factory areas is about 1 mV/m. For rural areas, 50 μV/m is adequate signal strength. These signal levels are considerably lower than the equivalent levels for AM reception in the standard broadcast bands. Three effects make satisfactory reception with these lower signal levels possible: (1) the effects of lightning and atmospheric interference (static) are negligible at 100 MHz compared with the interference levels typical of the standard broadcast band; (2) the antenna system at 100 MHz can be matched to the radio-receiver input impedance, providing more efficient coupling between the signal power incident on the antenna and receiver, and (3) the use of the wide-band FM method of modulation reduces the effects of noise and interference on the audio output of the receiver.

The open-circuit voltage of a dipole for any length up to one-half wavelength is given by

$$E_{oc} = E_f(5.59 \sqrt{R_r})/F_s \qquad mV$$

where E_{oc} = open-circuit antenna voltage, E_f = field strength at antenna (mV/m), F_s = received signal frequency (MHz), and R_r = antenna radiation resistance (Ω). For a half-wave dipole R_r = 72 Ω; for a folded dipole R_r is 300 Ω. For antennas substantially shorter than one-half wavelength, $R_r = 8.75 l^2 F_s^2 \times 10^{-3}$, where l is the total length of the dipole in meters. For a folded dipole one-half wavelength long operating at 100 MHz, the open-circuit voltage is $E_{oc} = 0.97 E_f$. The voltage delivered to a matched transmission and receiver input is one-half of this value, $0.48 E_f$.

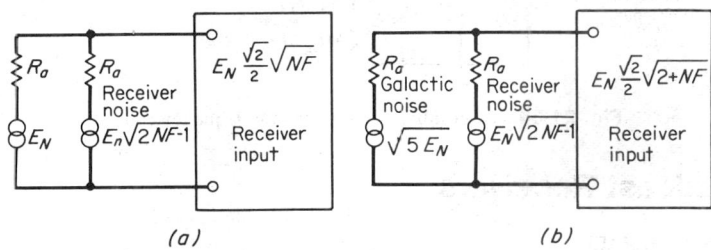

(a) *(b)*

Fig. 21-35. Equivalent sources of receiver noise: (a) with resistive generator; (b) with galactic noise source included.

The noise in the receiver output is caused by the antenna noise plus the equivalent thermal noise at the receiver input. The input impedance generates an excess of noise compared with the noise generated by an equivalent resistor at room temperature. The noise generated (Fig. 21-35) is given by

$$E_{nr} = E_n \sqrt{2NF - 1} \qquad V$$

where E_{nr} = equivalent noise generated in receiver input, E_n = equivalent thermal noise (V) = $(\sqrt{4R_{in}kT\Delta f})$, R_{in} = receiver input resistance, k = Boltzmann's constant = 1.38×10^{-23} J/K, T = absolute temperature = 290 K, Δf = half-power bandwidth of receiver response taken at discriminator, and NF = receiver noise figure. (If the noise figure is given in decibels, $NF_{dB} = 10 \log NF$.)

The generator noise and receiver noise add as the square root of the sum of the squares. Figure 21-35 shows that the equivalent receiver input noise is $0.707 E_n \sqrt{NF}$. For a receiver with 300-Ω input resistance and 200-kHz noise bandwidth, $E_n = 0.984 \mu V$. A typical noise factor for a well-designed receiver is 3 dB, or 2 times power ($\sqrt{2}$ voltage) increase, giving an equivalent noise input of 1.39 μV.

In an AM receiver the signal-to-noise (S/N) ratio at the receiver input is a direct measure of the S/N ratio to be expected in the audio output. In an FM receiver using frequency deviation greater than the audio frequencies transmitted, the S/N ratio in the output may greatly exceed that at the rf input. Figure 21-36 shows typical output S/N ratios obtained with receiver bandwidths adjusted to accommodate transmitted signals with modulation indexes of 1.6 and 5.0, compared with the audio S/N ratio when AM modulation is used and the bandwidth of the receiver is adjusted to accommodate the AM sidebands only.

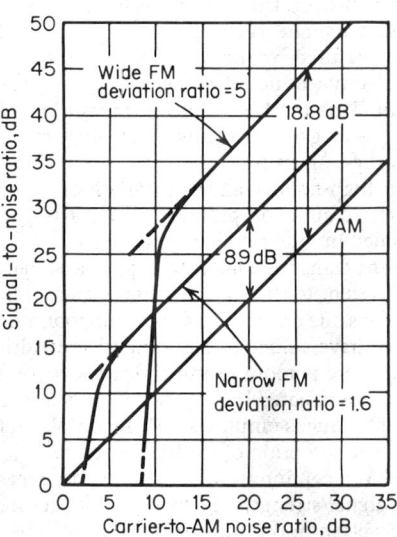

Fig. 21-36. Typical signal-to-noise ratios for FM and AM for deviation ratios of 1.6 and 5.

As shown in this figure, the S/N ratio for a properly designed receiver operating with a modulation index of 5 is 18.8 dB higher than that of an AM receiver with the same rf S/N ratio.

For rf S/N ratios in FM higher than 12 dB, the audio S/N ratio increases 1 dB for each 1-dB increase in rf S/N ratio. For FM S/N ratios lower than 12 dB, the S/N ratio in the audio drops rapidly and falls below the AM S/N ratio at about 9 dB. The point at which the ratio begins to fall rapidly is called the *threshold signal level*. It occurs where the carrier level at the discriminator is equal to the noise level. The threshold level increases directly as the square root of the receiver bandwidth, i.e., approximately the square root of the modulation index. The equation for S/N improvement using FM is

$$\frac{(S/N)_{FM}}{(S/N)_{AM}} = \sqrt{3}\,\Delta$$

where Δ is the deviation ratio. Since broadcast standards in the United States for FM call for a modulation index of 5 for the highest audio frequency, the direct S/N improvement factor is 18.8 dB for rf S/N ratios exceeding 12 dB.

In the FM system a second source of noise improvement is provided by pre-emphasis of the high frequencies at the transmitter and corresponding de-emphasis at the receiver. The pre-emphasis network raises the audio level at a rate of 6 dB/octave above a critical frequency, and a complementary circuit at the receiver decreases the audio output at 6 dB/octave, thus producing a flat overall audio response. Figure 21-37 shows simple RC networks for pre-emphasis and de-emphasis.

The additional S/N improvement using de-emphasis in an FM receiver is

$$\frac{(S/N)_{out}}{(S/N)_{in}} = \frac{f_a^3}{3[f_a f_0^2 - f_0^3 \tan^{-1}(f_a/f_0)]}$$

where $(S/N)_{out}$ = signal-to-noise ratio at de-emphasis output, $(S/N)_{in}$ = signal-to-noise ratio at de-emphasis input, f_a = maximum audio frequency, and f_0 = frequency at which de-emphasis network response is down 3 dB. For f_c = 15 kHz and a 75-μs time constant (f_0 = 2.125 kHz),

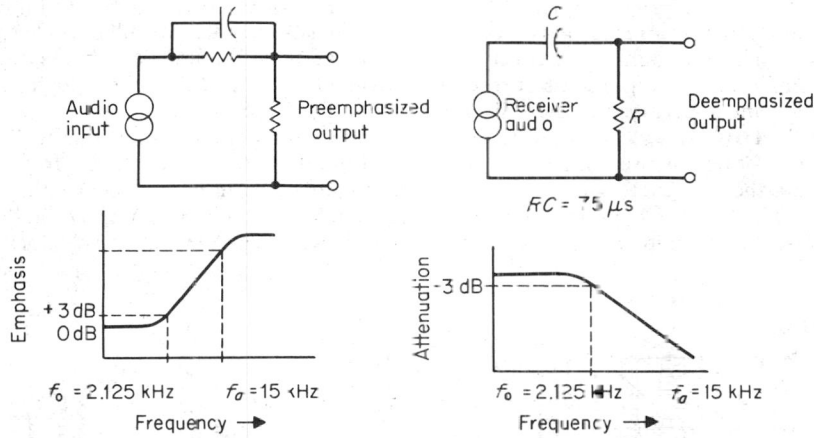

Fig. 21-37. Pre-emphasis and de-emphasis circuits and characteristics.

the S/N improvement is 13.2 dB. The total S/N improvement over AM is 32 dB when the carrier is high enough to override noise plus 12 dB for the 75-kHz deviation used in United States broadcast stations. The minimum coherent S/N ratio is therefore 44 dB.

When a dipole receiving antenna is used, an additional noise component is produced by *galactic noise*, because the dipole pattern does not discriminate against sky signals. The ratio of signal to galactic noise can be improved by using an antenna array with gain in the horizontal direction, i.e., one that discriminates against sky-wave signals. The additional noise is shown in Fig. 21-36. Using the calculated value of E_n and assuming a 3-dB noise factor in the receiver gives an equivalent noise input to the receiver of 1.39 μV. The required field strength to produce a 12-dB ratio at the receiver and a 44-dB S/N ratio at the audio output is 11.5 μV/m with a half-wave dipole.

66. Selectivity. When the FM system uses a high modulation index, the system is not only capable of improving the S/N ratio but will reject an interfering co-channel signal. The

FM signal modulation involves very wide phase excursions, and since the phase excursion which can be imparted to the carrier by an interfering signal is less than 1 rad, the effect of the interference is markedly reduced. The co-channel-interference-suppression effect requires that the interfering signal be smaller than the desired signal, since the larger signal acts as the desired carrier, suppressing the modulation of the smaller signal. This phenomenon is called the *capture* effect since the larger signal takes over the audio output of the receiver.

The capture effect produces well-defined service areas, since signal-level differences of less than 20 dB provide adequate signal separation. Although it is useful in suppressing undesired signals of a level less than the desired signal, the capture effect can also produce an annoying tendency for the receiver to jump between co-channel signals when fading, for example, caused by airplanes makes the desired signal drop below the intefering signal by only a few decibels. This effect also occurs in FM radios used in automobiles when motion of the antenna causes the relative signal levels to change.

67. Tuners. The FM tuner is matched to the antenna input. An rf amplifier is used to override the mixer noise. The mixer provides a 10.7-MHz i.f. signal. Most FM tuners contain a single stage of rf amplification. The mixer may be self-oscillating or employ a separate oscillator. Since the tuning range is from 88 to 108 MHz a range of only 1.23:1, the tuner may be tuned by any convenient means. In auto receivers tuning usually is accomplished inductively, using variable-permeability slugs, suitable for push-button operation. The rf stage must have a low noise figure to reach the minimum threshold-signal level, but its most important requirement is to provide the mixer and i.f. amplifier with signals free of distortion.

When the rf amplifier is overloaded, the signal supplied to the i.f. amplifier may be distorted or suppressed. For single interfering signals there are three significant sources of difficulty: (1) image signals may capture the receiver, suppressing the desired signal; (2) strong signals at one-half the i.f. frequency (5.35 MHz) above the desired signal may capture the receiver; or (3) a strong signal outside the range of the i.f. beat but strong enough to cause limiting in the rf stage may drastically reduce the output of the rf amplifier at the carrier frequency.

When two strong signals are present, three conditions may produce unsatisfactory operation: (1) cross modulation of two adjacent upper- or lower-channel signals may produce an on-channel carrier, (2) two strong signals spaced 10.7 MHz in the rf passband may produce an i.f. beat, or (3) submultiple i.f. beats may be produced by strong signals spaced at i.f. submultiple spacings in the rf band. To minimize the effects of distortion and provide a low noise figure, most FM tuners employ an FET type transistor rf stage. A circuit of a typical FM tuner employing an FET rf amplifier and a separate-oscillator mixer is shown in Fig. 21-38.

68. I.F. Amplifiers. To provide sufficient image separation, a higher intermediate frequency (10.7 MHz) is used in FM than in standard broadcast AM. Since the i.f. frequency is higher and the bandwidth greater (200 kHz), the amplifier design is more demanding in FM receivers. The i.f. amplifier must provide sufficient gain for the noise generated by the rf amplifier to saturate the limiting stages fully if the benefits of wide-band FM are to be obtained at low

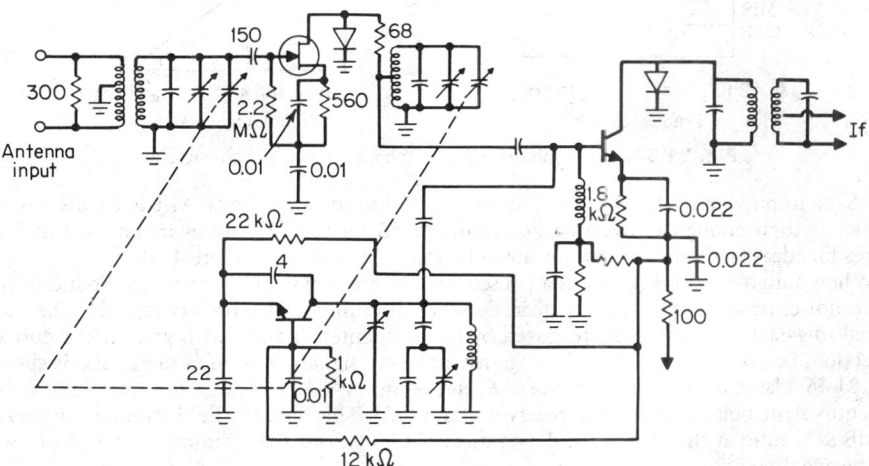

Fig. 21-38. Typical FM tuner using capacitive tuning.

signal levels. The high gain should be supplied with a low noise figure in the first i.f. amplifier, so that the noise introduced by the i.f. is small compared with the noise from the rf amplifier.

One of the most important characteristics of the i.f. amplifier is phase linearity, since envelope-delay distortion in the passband is a principal cause of distortion in FM receivers. Care must also be taken to avoid regeneration since this would cause phase distortion and hence audio distortion in the detected signal.

Although AGC is theoretically unnecessary, it is sometimes used in the rf stage to avoid overload. Such overload, coming before sufficient selectivity is present, could produce cross modulation, causing capture by an out-of-band signal. The requirements of high gain and good phase

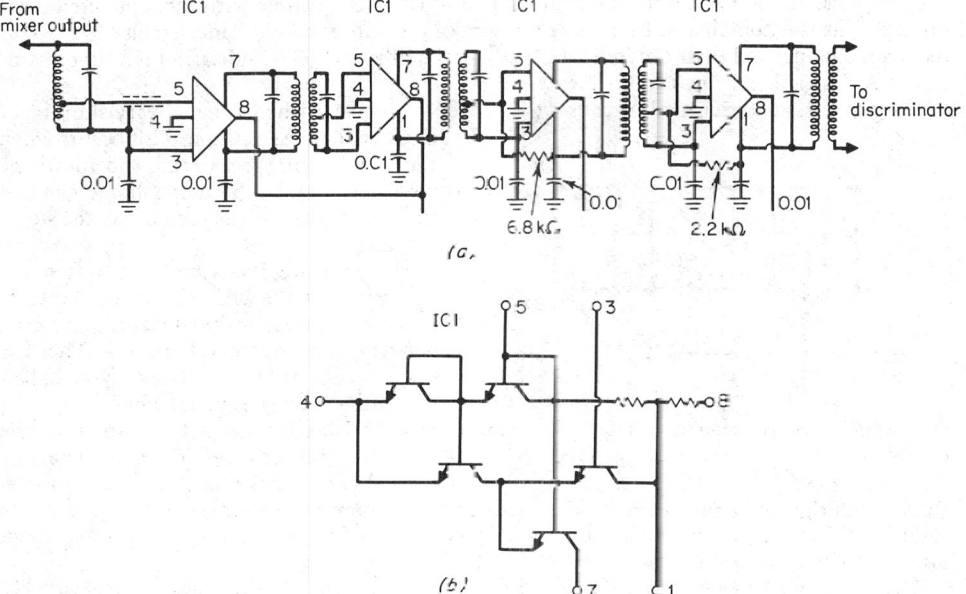

Fig. 21-39. I.F. amplifier for FM receiver using integrated circuits: (a) complete circuit; (b) detail of integrated circuits.

linearity are generally met by using amplifiers with double-tuned circuits adjusted to operate at critical coupling. A circuit of a typical i.f. amplifier is shown in Fig. 21-39. The gain is provided by integrated-circuit differential amplifiers which provide symmetrical limiting.

69. Limiters. The design of the limiter is critical in determining the operating characteristics of an FM receiver. The limiter should provide complete limiting with constant-amplitude signal output on the lowest signal levels which override the noise. In addition, the limiting should be symmetrical, so that the carrier is never lost at low signal levels. This is essential if the receiver is to capture on the strongest signal when there is little difference in signal strengths between the weaker and the stronger signals. Finally, the bandwidth in the output must be wide enough to pass all the significant sideband terms associated with the carrier, to prevent spurious amplitude modulation due to the lack of sufficient bandwidth to provide the constant-amplitude FM signals. The differential amplifier with dc coupling can be made to provide highly symmetrical limiting.

70. FM Detectors. The FM detector should provide an output voltage which changes linearly with frequency. Most balanced detectors provide zero voltage output at the center frequency (when the carrier is unmodulated), and the voltage varies plus and minus as the carrier is modulated above and below the carrier frequency. This provides a means of converting frequency modulation into an audio voltage. The bandwidth of the linear portion of the detector response should be wider than the expected frequency swing to provide protection from demodulation of rapid excursions in frequency generated by interfering signals.

There are five well-known types of FM detectors: (1) the balanced-diode discriminator (Foster-Seeley circuit); (2) the ratio detector using balanced diodes; (3) the slope-detector-pulse-counter circuit using a single diode with an RC network to convert FM to AM; (4) the locked-oscillator circuit, which uses a variation in current as the frequency is varied to convert the output of an oscillator (locked to the carrier frequency) to a voltage varying with the modulation; and (5) the quadrature detector circuit that produces the electrical product of the two voltages, the first derived from the limiter, the second from a tuned circuit which converts frequency variations into phase variations. The output of the product device is a voltage which varies directly with modulation. A typical ratio detector is shown in Fig. 21-40.

71. FM Stereo and SCA Systems. Since FM broadcasting uses a bandwidth of 200 kHz and a modulation index of 5 for the highest audio frequency transmitted, it is possible by using lower modulation indexes to transmit information in the frequency range above the audio. This method of using a supersonic carrier to carry additional information is used in FM broadcasting for a number of purposes, most notably in FM stereo broadcasting.

In broadcasting stereo the main-channel signal must be compatible with monophonic broadcasting. This is accomplished by placing the sum of the left- and right-hand signals $(L + R)$ on the main channel and their difference $(L - R)$ on the subcarrier. The subcarrier is a suppressed-carrier AM signal carrying the $L - R$ signal.

The suppressed-carrier method causes the carrier to disappear when L and R vary simultaneously in identical fashion. This occurs when the $L - R$ signal goes to zero and allows the peak deviation in the monophonic channel to be unaffected by the presence of the stereo subcarrier.

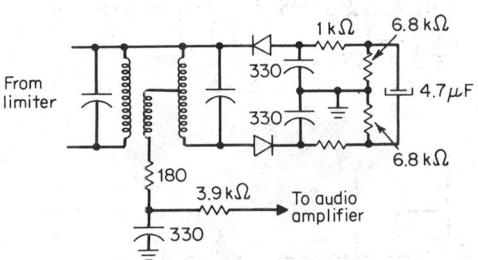

In the receiver it is necessary to restore the subcarrier. In the U.S. Standards, the technique used provides a pilot signal at 19 kHz, one-half the suppressed carrier. The frequency subcarrier is restored by doubling the pilot-signal frequency and using the resulting 38-kHz signal to demodulate the suppressed-carrier $(L - R)$ signal. The suppressed carrier has a peak deviation of less than 2, and the subcarrier is amplitude-modulated. The composite stereo signal thus has a S/N ratio 23 dB below that of monophonic FM broadcasting. The main $(L + R)$ channel is not affected by the stereo subcarrier $(L - R)$ signal.

Fig. 21-40. Ratio-detector circuit for FM demodulation.

The stereo signal can be decoded in two different ways. In the first, the subcarrier is separately demodulated to obtain the $L - R$ signal. The $L + R$ and $L - R$ signals are then combined in a matrix circuit to obtain the individual L and R signals. In the second method, more widely used, the composite signal is gated to obtain the individual L and R signals directly. A circuit of the gated type of stereo decoder is shown in Fig. 21-41. The circuit uses a doubler circuit to obtain the 38-kHz reference signal. The latter signal is added to the composite signal, and the signal is decoded in a pair of full-wave peak rectifiers to obtain the L and R signals directly. The $L - R$ subcarrier sidebands extend from 23 to 53 kHz.

In the SCA system an additional subcarrier is placed well above the stereo sidebands at 67 kHz. This subcarrier is used for auxiliary communication services (Special Communications

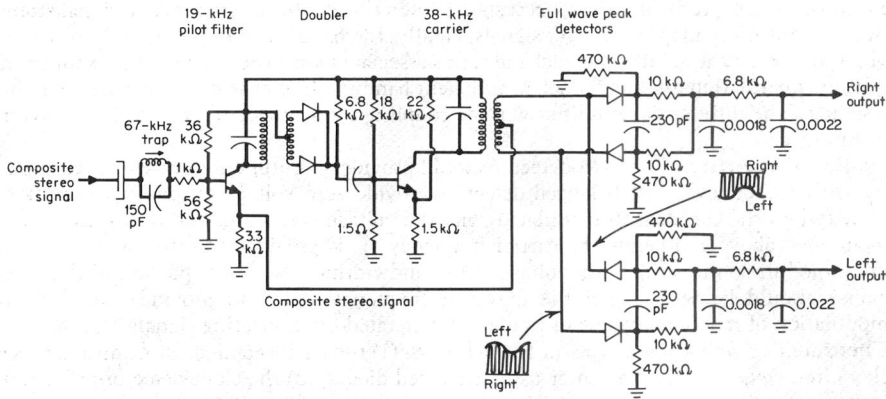

Fig. 21-41. FM stereo decoder using the gating principle.

Authorization, SCA). The level of the subcarrier is kept to about 10% of the peak carrier deviation to minimize its effects on the broadcast service signal. The auxiliary subcarrier is frequency-modulated with 8-kHz peak deviation; the audio is usually limited to 5 kHz. These signals are not decoded by broadcast receivers; special receivers are used at the receiving locations of the special services to decode the subcarrier signal.

Cable Television (CATV) Systems

BY JOSEPH L. STERN

72. General Considerations. Cable television systems, also known as community-antenna television (CATV) systems, use coaxial cable to distribute standard TV signals to homes or establishments subscribing to the service. The program material may originate at a distant VHF or UHF broadcast station, a distant program-origination studio, or a computerized news and information center, or it may be locally generated within the facilities of the CATV system. Signals received off the air are picked up by a specially designed antenna system (community antenna), which provides freedom from noise, interference, and multipath distortion not obtainable directly at the subscriber's homes. In the early 1980s, nearly 4,100 CATV systems were in operation in the United States, serving more than 16 million subscribers in approximately 10,000 communities reaching 25% of all homes equipped for television.

73. Elements of a CATV System. The typical CATV system comprises four main elements: a *head end*, in which the signals are received and processed; a *trunk system*, the main artery carrying the processed signals; a *distribution system*, which is bridged from the trunk systems and carries signals to subscriber areas; and *subscriber drops*, fed from taps on the distribution system to feed into the subscriber's TV receiver. Figure 21-42 shows a diagram of a typical CATV system.

While cable television systems still have the primary goal of bringing high-quality signals to home viewers, another basic function, *interactive communications*, has been developed to allow

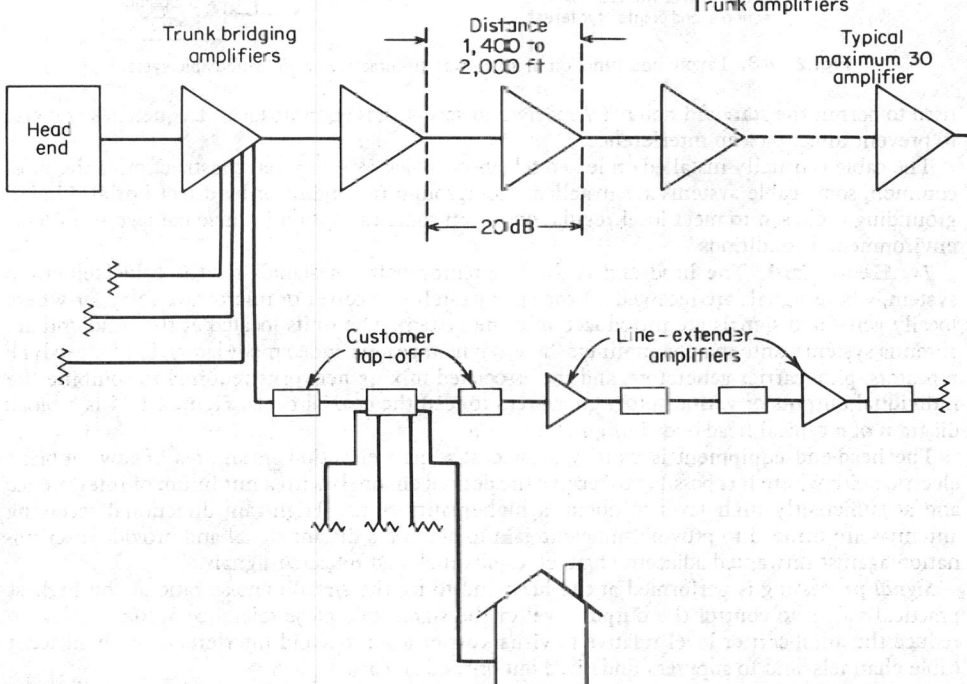

Fig. 21-42. Block diagram of typical CATV system.

the subscriber to interact with the program source to provide or request various types of information capable of being transmitted over the wide-band system. Cable television systems may carry many more channels of television programming than could be received off the air, as well as new communication series. Figure 21-43 illustrates a two-way interactive system and some of its potential uses.

While the primary subscriber device connected to the CATV system is a standard home TV receiver, the distribution of signals occurs within the TV broadcast band as well as on nonbroadcast frequencies. The nonbroadcast frequencies permit carriage of more than 12 channels, and present technology allows up to 54 channels to be distributed. A frequency-conversion device is

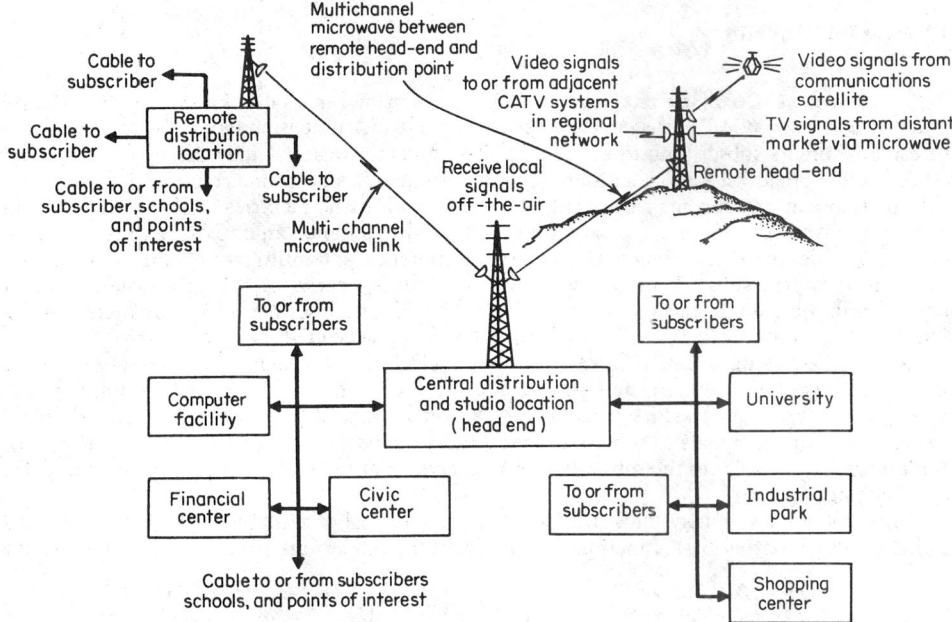

Fig. 21-43. Layout and function of a two-way (interactive) wide-band cable system.

used to permit the standard home TV receiver to accept these nonbroadcast frequencies and also to prevent direct pickup interference.

The cable is usually installed on leased telephone poles. While aerial construction is the most common, some cable systems are installed underground in conduit or by direct burial. Undergrounding is chosen to meet local regulations or, in some cases, to minimize damage from local environmental conditions.

74. Head End. The head end is the originating point of signals for the cable television system, where signals are received off the air, by satellite receiver or microwave relay, or where locally generated signals are introduced into the system. The units located at the head end are antenna systems, antenna preamplifiers, heterodyne repeaters, video modulators, FM heterodyne repeaters, pilot-carrier generators, and the associated mixing networks required to combine the individual outputs of various program sources to feed the coaxial cable. Figure 21-44 is a block diagram of a typical head-end configuration.

The head-end equipment is usually located at a high elevation in an area of low ambient electric noise where it is possible to receive the desired channels with a minimum of interference and at sufficiently high level to obtain a high-quality signal. High-gain directional receiving antennas are utilized to provide sufficient gain to pick up a distant signal and provide discrimination against unwanted adjacent-channel, co-channel, and reflected signals.

Signal processing is performed at the head end to fix the signal-to-noise ratio at the highest practical value, to control the output level of the signal to a close tolerance automatically, to reduce the aural-carrier level relative to visual-carrier level to avoid interference with adjacent cable channels, and to suppress undesired out-of-band signals.

Processing is also used to convert the received signal to a different channel, to convert signals to VHF, to introduce signals received by microwave into the system, and to introduce locally

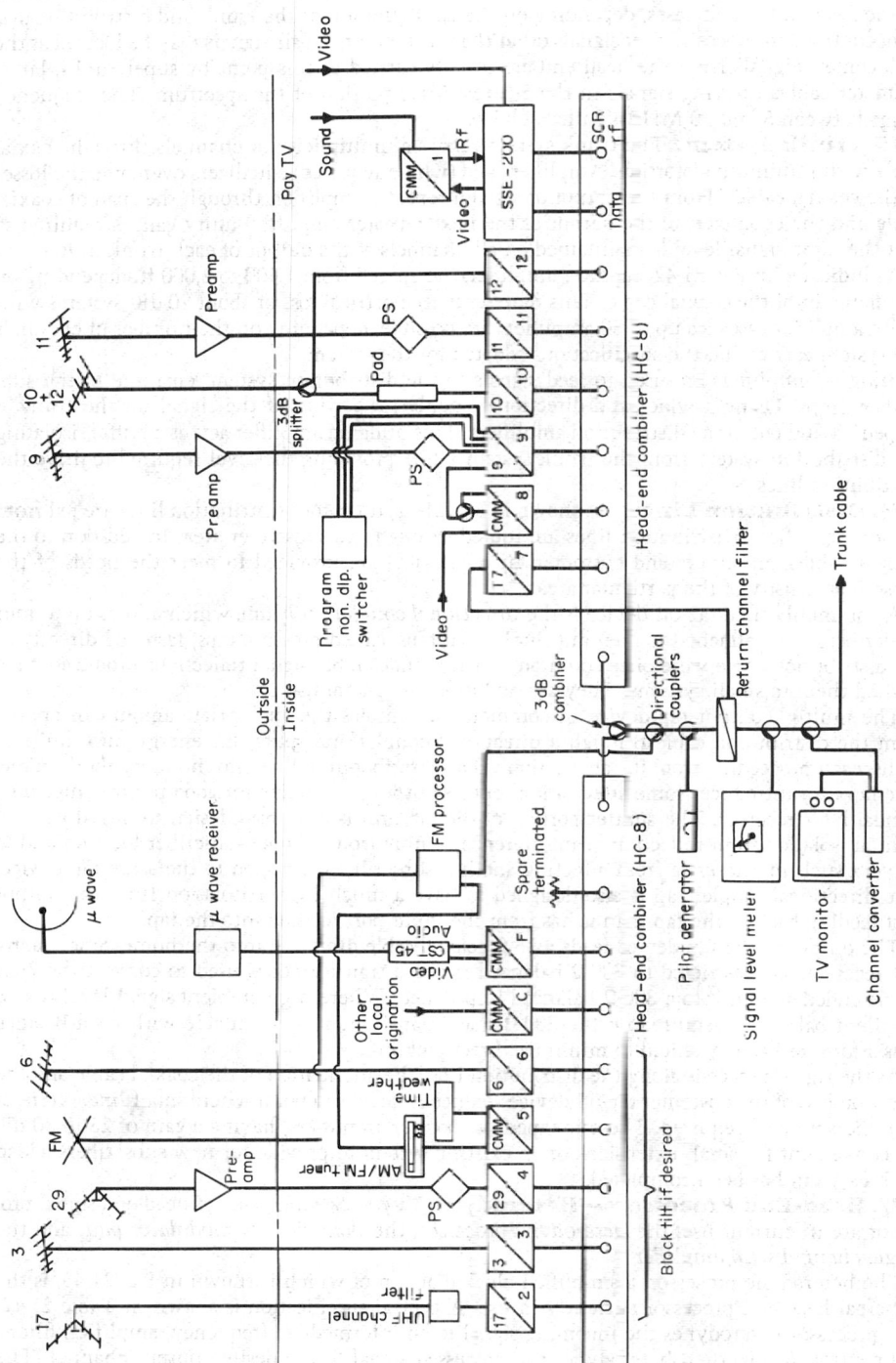

Fig. 21-44. Typical head-end configuration.

originated video services into the cable. FM broadcast signals are converted to respace the FM signals so that the cable system can accommodate as many FM stations as possible and to avoid distortion from direct pickup by the FM receiver.

In some systems the head end feeds a hub center, where locally generated signals are inserted in the system. In such cases, depending on the configuration of the trunk and distribution system, control and processing for signals other than distant off-the-air signals may be located at the hub center. Signals from the head end are usually carried to this point by supertrunks, large-diameter cables carrying signals in the 5- to 95-MHz portion of the spectrum. The frequency range between 5 and 50 MHz is called sub-low.

75. Trunk System. The trunk system carries a multiplicity of channels through coaxial cable with minimum distortion. Amplifiers and (where required) equalizers overcome the losses in the coaxial cables. From the output of a given repeater amplifier, through the span of coaxial cable and the equalizer, to the output of the next repeater amplifier, unity gain is required so that the same signal level is maintained on all channels at the output of each trunk unit.

As indicated in Fig. 21-42, repeater amplifiers are spaced from 1,400 to 2,000 ft, depending on the diameter of the coaxial cable. This represents an electrical loss of about 20 dB. Systems with trunk-amplifier cascades up to 50 amplifiers are possible, depending on the number of channels the system carries and the specifications adopted for the system.

Bridging amplifiers are used to feed signals to the distribution system, en route to the subscriber drops. Using a wide-band directional coupler, a portion of the signal on the trunk is tapped off and fed to the distribution amplifiers. This bridging amplifier acts as a buffer, isolating the distribution system from the trunk system while providing the level required to drive the distribution lines.

76. Distribution Lines. As shown in Fig. 21-42, up to four distribution lines are fed from a bridging station. Distribution lines are routed through the subscriber area. In addition to the coaxial cables, amplifiers and customer tap-off devices are provided to meet the needs of the subscriber density of the particular area.

A commonly used tap-off device is the *directional coupler multitap*, which allows up to four subscribers to be attached to one unit. Individual taps, called pressure taps, fastened directly to the distribution cable were once common, but the undesirable signal reflections produced have limited them to small systems. They are no longer in general use.

The multiple-output tap device in common use samples the appropriate amount of energy from the distribution cable through a directional coupler and splits this energy into multiple paths, each proceeding from its tap location via a subscriber-drop line into the subscriber's home. The tap unit introduces some attenuation, but its output is adequate for good performance on a standard TV receiver. The splitter portion of the tap unit is of hybrid design to introduce substantial isolation from reflection or interference coming from a home-subscriber location and to prevent such interference from affecting another subscriber connected to the same tap device. The directional-coupler tap is also designed to have a much higher isolation from the output port feeding back to the tap than it has from the input port feeding into the tap.

The output of the tap device feeds a 75-Ω coaxial-cable drop line into the home. Since many TV receivers are restricted to 300-Ω balanced input, a transformer is used to convert the 75-Ω single-ended system into a 300-Ω balanced impedance. Where high ambient-signal levels exist, excellent balance is required on the 300-Ω side. Transformers are available with good balance plus a form of Faraday shield to minimize direct pickup.

As the signal proceeds along the distribution line, the attenuation of the coaxial cable and the insertion loss of the customer tap-off devices reduce its level to a point where small *line-extender amplifiers* may be required. These inexpensive booster amplifiers, having a gain of 25 to 30 dB, are convenient for small extensions of an existing system to provide for new subscribers added after a system has been completed.

77. Head-End Processors—Heterodyne Type. Several types of head-end signal processor are in current use: the *heterodyne processor*, the *demodulator-modulator pair*, and the *single-channel strip amplifier*.

The heterodyne processor, a simplified block diagram of which is shown in Fig. 21-45, is the principal head-end processor currently in use. A typical specification is shown in Table 21-17. The processor heterodynes the incoming signal to an intermediate frequency, amplifies, filters, and controls levels, then heterodynes the processed signal to the desired output channel. The following functions are performed:

1. Amplification of the desired signal
2. Rejection of adjacent channels (filtering)

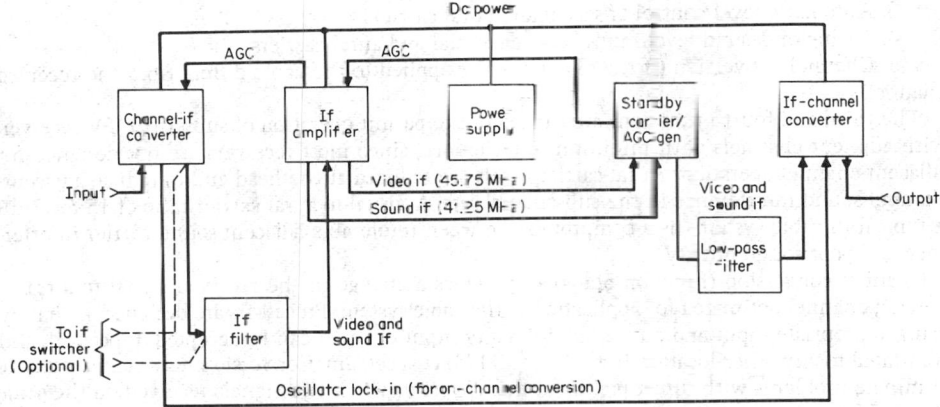

Fig. 21-45. Block diagram of typical heterodyne-type head-end processor. *(Jerrold Manufacturing Co., model CHC-*.)*

TABLE 21-17   Specifications of a Typical Heterodyne Head-End Processor*

Input-frequency range	Single standard VHF TV channel
Output-frequency range	Single standard VHF TV channel
Intermediate frequency	45.75 MHz, video carrier, 41.25 MHz, sound carrier
Frequency response, video	0.75 MHz below video carrier to 4.2 MHz above video carrier
Sound	Sound carrier ± 0.1 MHz
Response flatness	±0.5 dB nominal ±1.0 dB max
Sensitivity	−10 dBm input level to yield at least +60 dBm output level
Adjacent-carrier rejection	50 dB or better
Image rejection	60 dB or better
Input level dynamic range	At least 40 dB (−10 to +30 dBm)
Noise figure, low-band	6 dB maximum at full gain
High-band	7.5 dB maximum at full gain
Input impedance	75 Ω, unbalanced
Output impedance	75 Ω, unbalanced
Output-level range	+50 to +60 dBm
Maximum operational output level, with external filter	+60 dBm, video carrier, +45 dBm, sound carrier; spurious signals down at least 60 dB in any band (standard, sub-, mid-, or mid-high)
Without external filter	+54 dBm, video carrier, +39 dBm, sound carrier; spurious signals down at least 60 dB in standard VHF TV band
AGC regulation	±0.5 dBm maximum output level variation for input level dynamic range of −10 to +30 dBm
AGC type:	
Channel i.f. converter and i.f. amplifier modules	Keyed sync referenced
I.F. channel converter	Sync tip referenced
Sound limiting	10 dB for rated output level with −10 dBm minimum input level
Standby carrier:	
Delay on	~20 s
Delay off	~4 s
Range	Operates in absence of air signal; incorporates facility for manual override
Standby-carrier mode	Unmodulated carrier; carrier modulated with internal 15 kHz, 0 to 37.5%; carrier modulated with video, 0 to 87.5%; carrier modulated with 4.5 MHz

Jerrold model CHC-.

3. Automatic level control of visual and aural carriers

4. Setting of desired level ratios between visual and aural carriers

5. Channel conversion (if the channel to be applied to the cable differs from the received channel)

The third and fourth functions are carried out to permit operation of subscriber TV receivers with adjacent channels with minimum interference, since most receivers are not designed for adjacent-channel operation. Aural-carrier levels are reduced at the head end to reduce adjacent-channel sound interference in the subscribers' sets. A visual-to-aural carrier ratio of 15 to 17 dB is typical in cable systems as a compromise between intolerable adjacent-sound-carrier interference and poor sound quality.

Channel conversion (function 5, above) provides a change in the received signal to a transmission channel optimized for application to the cable system. Processors are designed so that by using appropriate input and output modules any input channel can be accepted, processed, and translated to any other location in the 5- to 400-MHz spectrum. Conversion may be necessary to eliminate problems with direct pickup at the TV set or when two signals received on the same channel from different directions are to be applied to the same cable system. Other applications of this function are to provide for the midband and superband channels of cable systems (see Table 21-18) and to supply the output for subband systems.

The visual-signal i.f. passband of a typical heterodyne converter is shown in Fig. 21-46. Note that this curve is not like that of a TV set, where the visual carrier is set at a point 6 dB down on the response curve. A linear phase characteristic is easier to achieve in the flat portion of the passband. This is one of the reasons the heterodyne processor has better differential phase characteristics than the demodulator processor.

78. Demodulator-Modulator Pair. In a demodulator-modulator pair, the demodulator is basically a high-quality television receiver to the point of video and audio output, whereas the modulator is essentially a low-power television transmitter. The demodulator-modulator pair provides increased selectivity and better AGC control and interconnection flexibility compared with that attainable with a strip amplifier. The demodulator (Fig. 21-47) consists of a tuner, an i.f. amplifier section, a video detector, a 4.5-MHz amplifier, and a discriminator. The tuner uses

TABLE 21-18 CATV Channel Designations and Carrier Frequencies

Channel designation	Visual carrier, MHz	Aural carrier, MHz	Channel designation	Visual carrier, MHz	Aural carrier, MHz
Low band:			Superband:		
2	55.25	59.75	N	241.25	245.75
3	61.25	65.75	O	247.25	251.75
4	67.25	71.75	P	253.25	257.75
5	77.25	81.75	Q	259.25	263.75
6	83.25	87.75	R	265.25	269.75
Midband:			S	271.25	275.75
A-2	109.25	113.75	T	277.25	281.75
A-1	115.25	119.75	U	283.25	287.75
A	121.25	125.75	V	289.25	293.75
B	127.25	131.75	W	295.25	299.75
C	133.25	137.75	Hyperband:		
D	139.25	143.75	AA	301.25	305.75
E	145.25	149.75	BB	307.25	311.75
F	151.25	155.75	CC	313.25	317.75
G	157.25	161.75	DD	319.25	323.75
H	163.25	167.75	EE	325.25	329.75
I	169.25	173.75	FF	331.25	335.75
High band:			GG	337.25	341.75
7	175.25	179.75	HH	343.25	347.75
8	181.25	185.75	II	349.25	353.75
9	187.25	191.75	JJ	355.25	359.75
10	193.25	197.75	KK	361.25	365.75
11	199.25	203.75	LL	367.25	371.75
12	205.25	209.75	MM	373.25	377.75
13	211.25	215.75	NN	379.25	383.75
Superband:			OO	385.25	389.75
J	217.25	223.75	PP	391.25	395.75
K	223.25	227.75	OO	397.25	401.75
L	229.25	233.75	RR	403.25	407.75
M	235.25	239.75			

either a high-quality conventional TV tuner or one with a crystal-controlled local oscillator. It has high gain and a good noise figure and a wide range of AGC control.

The i.f. amplifier is very similar to a high-quality TV receiver. A typical i.f. response curve is shown in Fig. 21-48. Note the point at which the visual carrier is located on the passband. The 4.5-MHz intercarrier sound is taken off before video detection, amplified, and limited to remove video components.

The demodulator must be carefully designed to minimize phase and amplitude distortion. This can be done by linearizing the detector, but quadrature distortion is inherent in a system

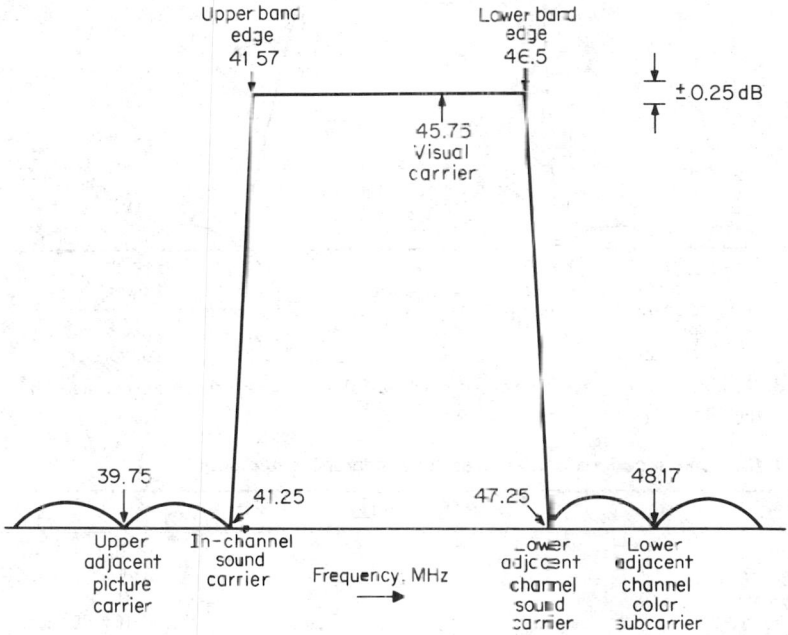

Fig. 21-46. Typical idealized video i.f. response curve of heterodyne processor. Note that the visual carrier is located on the flat-top portion of the curve to permit improved phase response

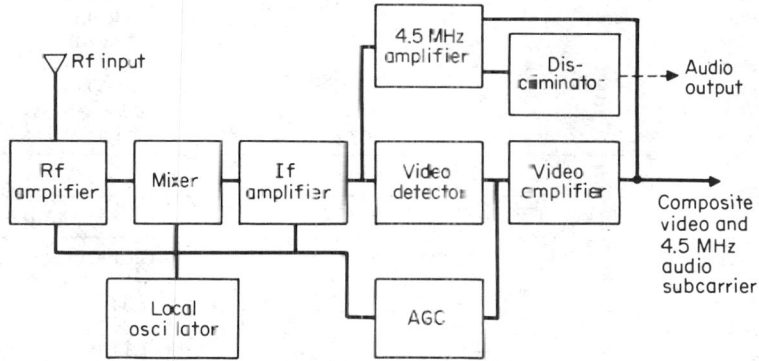

Fig. 21-47. Block diagram of demodulator portion of a demodulator-modulator pair.

using an envelope detector on a vestigial-sideband signal and can be corrected for only by video processing. Typical demodulator specifications are shown in Table 21-19. In the modulator (Fig. 21-49) the composite input is applied to the separation section, where the video and sound subcarriers are separated and the video fed to a video amplifier. From this point the video signal is processed, mixed with a carrier oscillator to obtain the desired output frequency, filtered, and amplified to obtain the necessary power level and remove any undesired products. Following the rf amplifier is the vestigial-sideband filter required to remove most of one sideband and allow

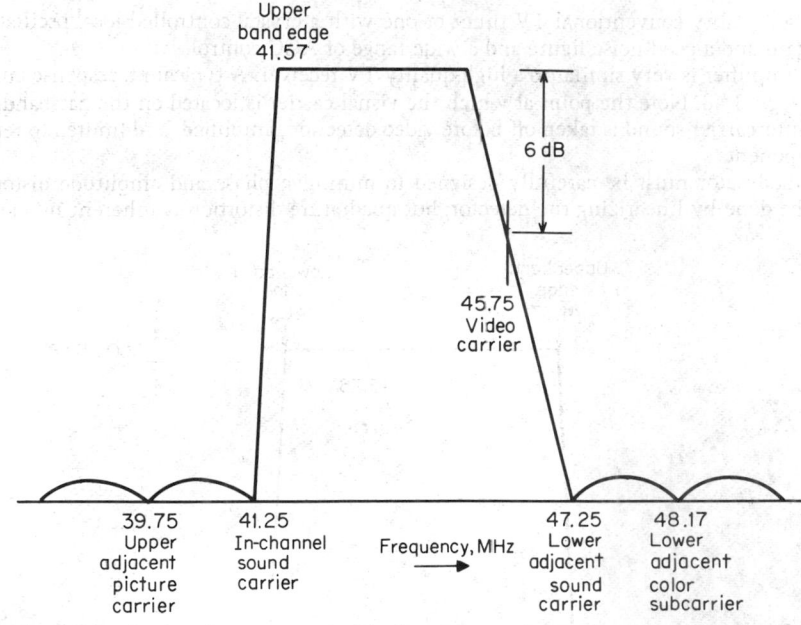

Fig. 21-48. Idealized video i.f. response curve of a demodulator. The visual carrier is located 6 dB below the maximum response.

TABLE 21-19. Specifications of a Typical Demodulator-Modulator Pair

Demodulator	
Input level	−20 to +30 dBm
Input frequency	Any VHF or UHF channel
Input impedance	75 Ω
Noise figure, VHF	VHF 7 dB max
UHF	12 dB max
Image rejection	60 dB min
I.F. response flatness	±0.5 dB 41.6 to 46.5 MHz
Adjacent video-carrier rejection	>40 dB
Adjacent sound-carrier rejection	>40 dB
AGC sensitivity	±0.5 dB
Video output level	1.5 V p-p max
Audio output level	+6 VU
4.5-MHz output level	0.2 V p-p max
Audio-frequency response	±1.5 dB, 50 to 15,000 Hz

Modulator		
	Visual	Aural
Frequency range	Channels 2 to W	Channels 2 to W
Input level	0.5 V p-p	−10 VU for full deviation
Input impedance	75 Ω	600 Ω
Output impedance	75 Ω	75 Ω
Output level	+50 dBm	+40 dBm
Output control range	20 dB	20 dB
Frequency response	30 Hz to 4.0 MHz, ±1 dB	50 Hz to 15 kHz ± 1 dB
		(75 μs pre-emphasis)
Hum and noise	60 dB down at 100%	60 dB down with ±50 kHz
	modulation	deviation
Color response	±2 dB max differential gain	
	±4° max differential phase	
Vestigial-sideband response:		
Picture carrier, −1.25 MHz	−20 dB	
Picture carrier, −3.58 MHz	−42 dB	
Adjacent-channel	−22 dB	

adjacent-channel operation. The characteristics of this filter are designed to meet the same requirements as that of a TV transmitter. Following the filter the carrier-level adjustment occurs, and then the visual carrier is combined with the aural carrier.

In the modulation process, a high percentage of modulation is desirable for high signal-to-noise ratio, but this produces differential gain and phase on the color subcarrier. The vestigial-sideband filter must be optimized to minimize phase distortion near the picture carrier. The compensation required for the demodulator-modulator pair is usually provided in a video equalizer in the modulator. Typical modulator specifications are shown in Table 21-19.

79. Single-Channel Amplifier. Single-channel amplifiers, or *strip amplifiers*, are the simplest head-end processors. They amplify one channel and reject the others. In simplest form they consist of a filter, amplifier, and power supply. More elaborate types provide the above functions plus AGC in some form and usually better input and output filtering. A representative

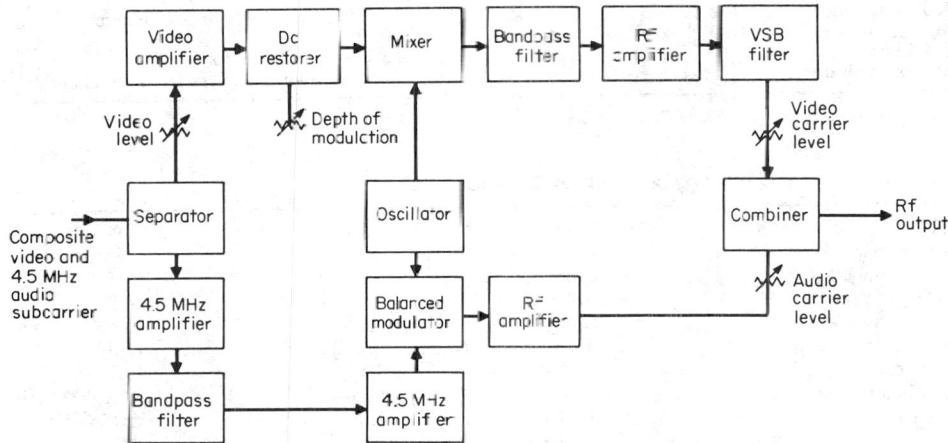

Fig. 21-49. Block diagram of the modulator portion of a demodulator-modulator pair.

type is shown in block-diagram form in Fig. 21-50. Typical specifications are shown in Table 21-20.

Strip amplifiers are used where the desired signal levels are fairly high and the undesired levels low or absent. They do not offer the selectivity of the more complicated heterodyne and demodulator-modulator processors. They also lack such features as independent control of sound- and picture-carrier levels and the ratio between them, separate AGC, and limiting for picture and sound channels, respectively. Their use is restricted to the 12 channels because they cannot be translated to other channels. They are also subject to leading ghost problems if the signal on a given channel can be picked up by sets in the distribution system. The off-the-air signal arrives far enough ahead of the signal through the cable to create a ghost ahead of the cable-transmitted signal.

80. Trunk and Distribution Amplifiers. Since the trunkline is the main artery of the CATV system, the trunk amplifiers must provide minimum degradation to the system. Specifications of a typical trunk amplifier are contained in Table 21-21. Figure 21-51 shows the block

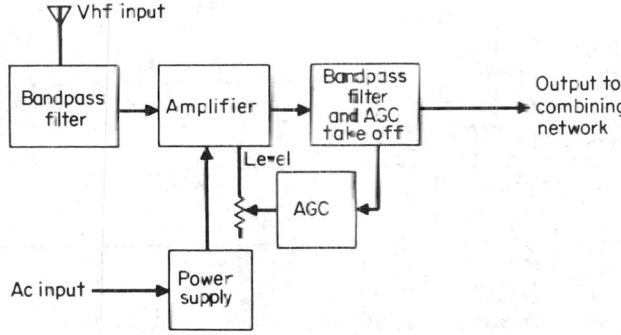

Fig. 21-50. Block diagram of single-channel amplifier (strip amplifier) with AGC.

diagram of a typical trunk-AGC amplifier and the distribution of gain. The first stage is designed for low-noise figure and the remainder for low cross modulation.

81. Cross Modulation. The maximum output level allowable in CATV system amplifiers is almost always determined by cross modulation in the picture signal. Where cross modulation exists, it results in a variation in the peak voltage of an otherwise unmodulated signal substituted for the wanted carrier. The contribution by the third stage to the total cross-modulation distortion is only 1.27 dB; that is, the output stage principally determines the amplifier cross-modulation distortion.

Table 21-20. Typical Single-Channel Amplifier Specifications

Minimum gain	51 dB (channels 2 to 13 and FM)
Maximum output for 0.5-dB gain compression	+66 dBm (2 V)
AGC range	40 dB
AGC capability	±0.5 dB for 40-dB input change
Minimum input for TASO* "excellent" picture	0 dBm
Bandpass	6 MHz, ±0.5 dB for TV
Skirt selectivity	25 dB down ±9 MHz from midchannel

* Television Allocations Study Organization.

TABLE 21-21   Specifications of a Typical Trunk Amplifier

Operating levels,	+10 dBm	Cross modulation	−95 dB
input, channel 13		Second-order beat	−85 dB
Channel 2	+16 dBm	Hum modulation	−60 dB
Output, channel 13	+32 dBm	Return loss, trunk in	18 dB
Channel 2	+27 dBm	Trunk out	18 dB
Control range, recommended	22 dB	Noise figure, channel 2	16 dB
gain, channel 13		Channel 13	10.5 dB
Channel 2	11 dB	Frequency range	50 to 400 MHz
Spacing	22-dB cable to	Ripple, trunk	±0.25 dB
	14-dB cable	Bridger	±0.5 dB
	plus 8 dB flat		
AGC temperature	±3 dB		
compensation, input change			
Output change	± 0.5 dB		

Cross modulation is most likely to occur in the output stage, where the signal levels are high. The cross-modulation distortion products at the output of a typical amplifier (Fig. 21-51) are 93 dB below the desired signal at an output level of +32 dBm. Note from Fig. 21-51 that the gain of the third stage is 8 dB. The output level of the third stage would therefore be

$$P_3 = P_4 - G_4 = +32 - 8 = +24 \text{ dBm}$$

Figure 21-52 illustrates the relationship between cross-modulation distortion and amplifier output level. As shown, a 1-dB decrease in output level produces a corresponding 2-dB decrease in the cross-modulation distortion. Using this relationship, we find the output cross-modulation distortion of the third stage to be −109 dB.

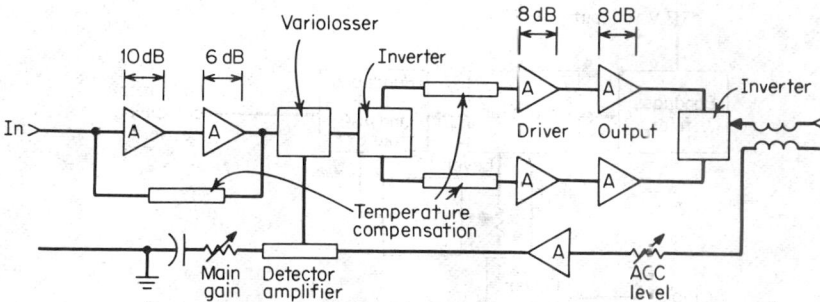

Fig. 21-51. Functional block diagram of a trunk-AGC amplifier and its gain distribution; A = amplifier.

82. Frequency Response and Beat Distortion. The most important design consideration in CATV amplifiers is the frequency response. An amplifier frequency response flat within ±0.25 dB over 40 to 400 MHz is required of an amplifier carrying 50 or more 6-MHz channels to permit a cascade of 20 or more amplifiers. To meet this requirement special attention must be paid not only to the high-frequency parameters of the transistors and associated components but to good high-frequency layout and packaging techniques as well.

The circuit shown in Fig. 21-53 is representative of amplifiers designed to achieve flat response. By properly designing the feedback network comprising C_1, R_1 and L_1, sufficient negative feedback can be used to maintain a nearly constant output over a wide frequency range. The collector transformer T_2 and the splitting transformers T_1 and T_3 play an important part in the amplifier's performance. In a representative amplifier of this type transformer T_2 is bifilar wound on a ferrite core and presents an essentially constant 75-Ω impedance over the entire frequency range. Transformers T_1 and T_3 are similar in construction but have the additional function of providing the required 180° phase shift for the push-pull pair Q_1, Q_2 while maintaining a 75-Ω input-output impedance.

The sweep-response patterns shown in Fig. 21-54 illustrate the wide bandwidth and amplitude linearity necessary for cascade signal processing. The response irregularity indicated in this figure is, by itself, of little concern, but if this small irregularity (perhaps 0.2 dB) occurs at the same frequency in each amplifier, it becomes a response "signature" and accumulates to a magnitude of 6.4 dB at the end of 32 amplifiers. The degree of signature in a high-quality CATV trunk amplifier is typically no more than 0.1 dB.

Second-Order Beat. As the number of CATV television channels has been increased from 12 nonharmonically related to 40 or more harmonically related, the second-order-beat distortion characteristics of an amplifier have become important to CATV equipment manufacturers.

A number of approaches has been used to reduce amplifier second-order distortion, but the most successful is the push-pull configuration. The circuit illustrated in Fig. 21-53 is representative of the 75-Ω push-pull building-block approach. Each push-pull stage is of this basic configuration and differs essentially in component values only. The splitting transformers T_1 and T_3 are the key to the operation of the push-pull amplifier, since for maximum second-order cancellation to occur the proper 180° phase relationship must be maintained over the full amplifier bandwidth. Additionally, it is necessary that the gain be equal in both push-pull halves and that the individual transistors be optimized for maximum second-order linearity.

83. Channel Capacity. CATV system design has advanced to the point where it is feasible to provide 40 or more channels with two-

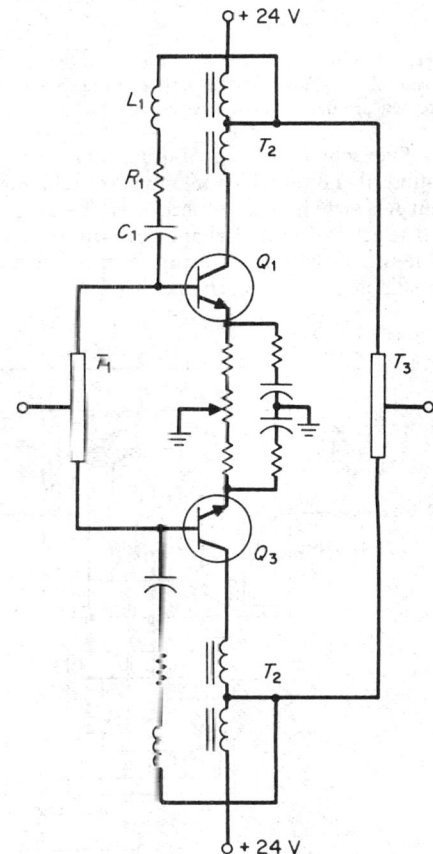

Fig. 21-53. High-performance push-pull wideband amplifier stage for a trunk amplifier.

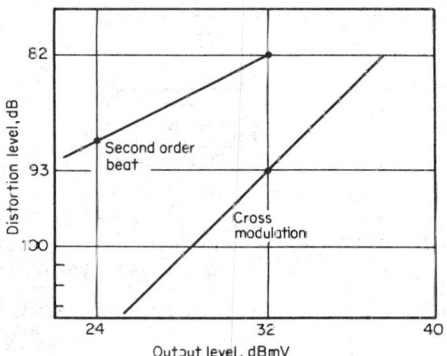

Fig. 21-52. Cross modulation and second-order beat distortion vs. output level of a trunk amplifier.

way capability. The FCC Rules require that CATV systems located in whole or in part within a major television market have at least 120 MHz of bandwidth (the equivalent of 20 television broadcast channels) available for immediate or potential use. In addition, the Rules stipulate that each system maintain a plant having technical capacity for return communication.

A variety of techniques have been proposed for the provision of more than the usual 12 VHF channels, such as use of UHF channels, frequency-division multiplexing of a single cable, dual cables, and switched multiple-cable networks with or without multiplexing. Each of these schemes has technical problems which are much more troublesome than those of the 12-channel systems. The standard 12 TV channels were allocated by the FCC in a manner to minimize second-order distortion products and interference from nonbroadcast signals. As additional channels are utilized, these problems become most significant.

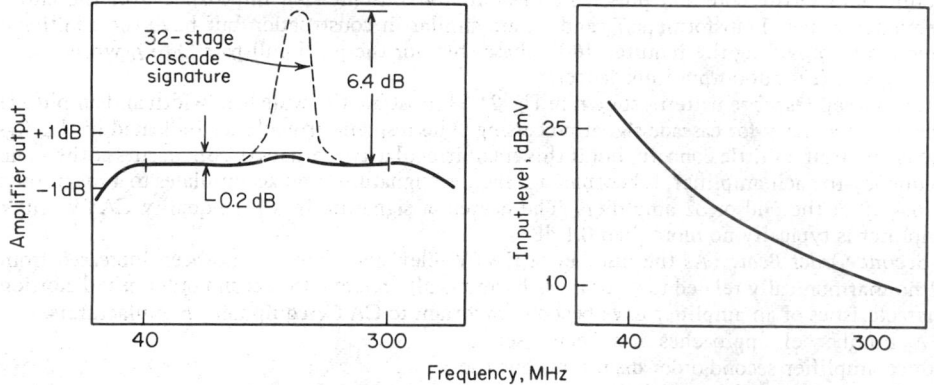

Fig. 21-54. Input and output levels of the amplifier shown in Fig. 21-53 as functions of frequency over the band 40 to 300 MHz. The dashed curve shows the cumulative effect of the 0.2-dB hump when the signal is passed through 32 such stages in cascade.

One scheme for distributing more than 12 channels is to use dual cables. In this case, two subscriber drop cables are brought to the subscriber's TV receiver and fed to the receiver through an A-B switch. The subscriber is able to receive 12 channels on each cable, operating the switch to select the desired channel group. In practice, direct pickup from local TV stations prevents the use of the full 12 channels per cable but generally, in major markets, 14 to 18 channels are available using this scheme.

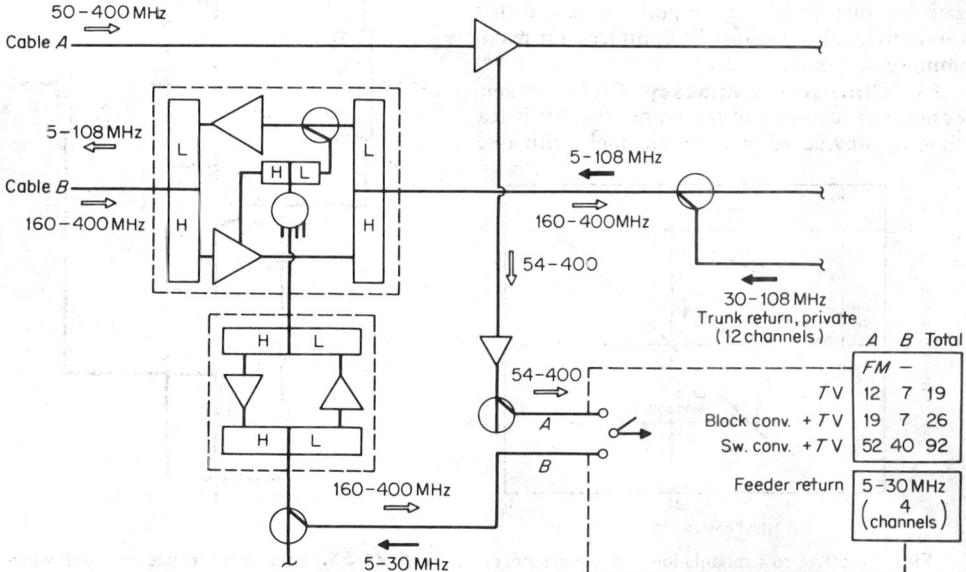

Fig. 21-55. Block diagram of a two-way trunk and feeder return for a dual-cable system.

A second technique to provide more than 12 channels involves the use of a converter. The output of the converter feeds the receiver at some unused VHF channel position. The receiver remains tuned to this unused local channel, and the converter is used to tune in the desired VHF channels. Converters are also used to avoid direct-pickup problems. Converters are commercially available providing for 52 channels. In conjunction with a dual-cable system and an A-B switch, they can provide as many as 104 channels.

UHF distribution on cable has been considered but has not proved satisfactory. The combination of very high cable losses at these high frequencies and the limitations of UHF tuners in commercial TV receivers have made such a distribution scheme impractical.

84. Two-Way Systems. Since most existing CATV systems use a single cable, the majority of conversions to provide additional channel capacity and two-way operation have involved retrofitting the single-cable systems. To do this, the earlier limited-bandwidth single-ended amplifiers have been replaced with broadband push-pull amplifiers and all other components provided with extended frequency ranges. To provide two-way service, filters have been installed to provide frequency separation and feed return signal amplifiers.

A dual-cable trunk and feeder system providing for up to 92 channels (using a switched converter), 12 private return channels, and 4 return channels from subscribers is shown in block form in Fig. 21-55. This system contains no frequency-splitting filters in cable A, minimizing group-delay problems for the main TV channels. It is also useful for private channel distribution in the reverse direction, meeting a number of the requirements outlined in Fig. 21-44.

85. Technical Standards. The Federal Communications Commission has issued definitions and technical standards governing CATV systems. They appear in the FCC Rules, Sec. 76.601-76.617 (subpart K, Technical Standards) and Sec. 76.5 (subpart A, General, Definitions).

Rules and regulations regarding the power levels of specific frequencies used in certain areas by the FAA and the DOD are contained in Sec. 76.610, 76.611 and 76.613. Prior FCC authorization is required for use of some of the frequencies listed therein.

Telecommunications: Point-to-Point and Mobile Systems

A. B. BROWN, JR. *Bell Telephone Laboratories; Senior Member IEEE*

J. C. BAUMHAUER, JR. *Bell Telephone Laboratories*

R. L. CERBONE *Bell Telephone Laboratories*

H. W. EARLE *Bell Telephone Laboratories*

G. C. FRITZ *Teletype Corporation*

J. M. GOTWAY *Bell Telephone Laboratories*

W. E. HOSTETLER *Bell Telephone Laboratories*

R. M. HUNT *Bell Telephone Laboratories*

J. J. JETZT *Bell Telephone Laboratories*

A. E. JOEL, JR. *Bell Telephone Laboratories; Fellow, IEEE*

V. I. JOHANNES *Bell Telephone Laboratories; Member, IEEE*

V. E. MUNSON *Bell Telephone Laboratories*

W. H. NINKE *Bell Telephone Laboratories; Senior Member, IEEE*

PHILIP T. PORTER *Bell Telephone Laboratories; Member, IEEE*

J. J. ROSINSKI *Bell Telephone Laboratories*

J. SALZ *Bell Telephone Laboratories*

C. STOCKBRIDGE *Bell Telephone Laboratories*

R. K. THOMPSON *Bell Telephone Laboratories (Retired)*

G. P. TOROK *Bell Telephone Laboratories*

E. W. UNDERHILL   *Bell Telephone Laboratories*

R. E. WADDELL   *Bell Telephone Laboratories*

D. L. WHITSON   *Bell Telephone Laboratories; Member IEEE*

CONTENTS

Numbers refer to paragraphs

Principal Service Networks

BY A. B. BROWN, JR.

NETWORKS AND OBJECTIVES

1. Service Networks. Communication entails the transfer of information from one point to another; the information may be in any of several forms, including voice, data, video, and facsimile. The facilities used for any particular form of service are known as a *service network*. A service network is made up of *terminals* (Pars. **22-100** to **22-126**), through which the information enters and leaves the network; *transmission facilities* (Pars. **22-40** to **22-70**), which provide the transfer of information from place to place; and *switching* (Pars. **22-71** to **22-99**), which connects the appropriate transmission facilities to cause the information to be delivered to the desired place. Some service networks do not include terminals, and some do not include switching. Terminals are of many types, including telephones, teletypewriters, facsimile terminals, and computer ports.

A service network may be planned and managed by a common carrier a resale organization, or the user of the service. The facilities of which the network is composed may, in turn, be owned by the organization that plans and manages the network or may be provided by another organization. In the latter case, they may be in the form of another service network; e.g., a user may choose to provide his own terminal equipment while obtaining transmission and switching services from a common carrier. The arrangements for maintenance of the network are normally made by the one who organizes and assembles the network; if the facilities are furnished by others, they may perform the maintenance on those facilities.

2. Interfaces. The boundary of the service network, where the user interacts with it (as by means of a telephone) or connects equipment to it is the *interface*. Service networks may provide either an analog or a digital interface to the users; i.e., signals may have a continuum of values or only discrete values. Either digital or analog transmission may be used to furnish either digital or analog service; all four combinations exist. If the interface is digital, the use of digital transmission and switching avoids the need for digital-to-analog (D/A) and analog-to-digital (A/D) conversion of the signals. Some networks that use digital transmission furnish either digital or analog interfaces, according to the service the user wishes.

There is a trend toward communication of various forms of information over the same service network. For example, voice, data, and facsimile are sent over public telephone networks[1]* (Pars. **22-5** to **22-8**). Some of the newer networks (Pars. **22-10** and **22-11** with digital interfaces are intended to be used for digitally encoded video or voice as well as for high-speed data. Some older networks (Par. **22-9**) are used only for data. Reference 2 covers computer communications.

3. Codes and Protocols. Communication of digital data is usually done in the form of characters, each of which is represented as a sequence of bits. Less commonly, the signals may be a bit stream generated and interpreted by the terminal equipment. The number of bits per character and the correspondence between the bit sequence and the character represented is called a *code*. Several codes have been used at various times for telegraphic and computer communications, starting with the Morse dot-dash codes used by telegraphers. The most common code at the present time is International Alphabet no. 5 (known in the United States as the American Standard Code for Information Interchange, ASCII). Other codes have been in use for older types of transmission equipment (teleprinters), and some codes have been introduced by individual manufacturers.

In coded transmission, two methods of maintaining synchronism between the transmitting and receiving points are commonly used. In *start-stop transmission*, the interval between characters is represented by a steady 1 signal; the transmission of a single 0 bit signals the receiving terminal that a character is starting. The information bits follow the start bit and are followed by the stop pulse. It is the same as the signal between characters and has a minimum length that is part of the specification of the terminal: 1.0, 1.42, 1.5, or 2.0 b are commonly used. In the synchronous method, the bits are sent at a uniform rate; if there is no character ready to be sent when it is time to send one, a synchronous idle pattern is used to maintain the timing. The synchronous method is used at higher speeds.

Protocols are standard procedures for the operations of communication; their purpose is to coordinate the equipment and processes at interfaces and at the ends of the communication channel. Protocols are considered to apply to several levels the International Organization for Standardization (ISO) has developed a seven-level Reference Model of Open Systems Interconnection to guide the development of standard protocols.[3] The seven levels and their functions are shown in Table 22-1. The International Telegraph and Telephone Consultative Committee (CCITT) Recommendations in the X series, which apply to public data networks, include X.1, speeds; X.2, user options; X.20, interface for start-stop transmission services; X.21, interface for synchronous operation; X.22, multiplex interface for synchronous terminals; X.24 definitions of interface circuits; and X.121, international numbering plan for public data networks. Additional recommendations apply to packet networks (see Par. **22-10**). Levels of protocol above the transport level apply to the interchange of data between applications and are currently under study in CCITT.

*Superior numbers correspond to References, Par. **22-139**.

Within the United States, the standards of the Electronic Industries Association (EIA), which represent consensus (the result of negotiation) among manufacturers, are in wide use at the physical level. They include RS-232-C and its successor, RS-449, the interface between data terminal equipment and data circuit-terminating equipment employing serial binary data interchange; RS-366-A, interface between data terminal equipment and automatic calling equipment for data communication; and RS-422-A and RS-423-A, electrical characteristics of balanced and unbalanced voltage digital interface circuits.

Some data-link protocols, historically the most common, use characters or combinations of characters to control the interchange of data. Others, including the American National Standards Institute (ANSI) Advanced Data Communications Control Procedures (ADCCP) and its subsets (Par. **22-120**) use sequences of bits in predetermined locations in the message to provide the link control. Reference 4 provides thorough coverage of protocols.

4. Objectives. The objectives that form the basis of network planning are derived from the service the network is to provide. There may, however, be different objectives for a network corresponding to different intended uses by the customers. For example, most of the objectives of the public telephone network (Pars. **22-26** to **22-31**) are derived from the needs of voice com-

TABLE 22-1   ISO Reference Model of Open Systems Interconnection

Layer	Function
Application	Permits communication between applications
Presentation	Presents structured data in proper form for use by application programs
Session	Sets up and takes down relationships between presentation entities; controls data exchange, e.g., dialog control
Transport	Furnishes network-independent transparent transfer of data
Network	Provides network-dependent routing, switching, and transport
Data link	Gives error-free transfer of data over a link
Physical	Covers mechanical and electrical characteristics; meaning of signals

Source: Adapted from Ref. 3. Copyright 1980, IEEE. Used by permission.

munication; however, some objectives, e.g., the limits on delay distortion and impulse noise, concern its use for data communication. In turn, network objectives are the bases for the design objectives of the facilities which will compose the network. In practice networks evolve while they are in service, usually with more stringent objectives to meet the rising expectations of the users and with improved performance made possible by improved technology and designs. Thus, additions to the network may be designed according to objectives different from those of the earlier parts.

Some of the principal service networks are described in the following paragraphs.

PUBLIC TELEPHONE NETWORKS

5. Network Topography. The simplest *local network* is a point-to-point circuit connecting two telephones. Since generally telephone users will wish to call many other users, switching is provided to set up the desired circuits when they are needed. For simple applications, one circuit may be provided that connects all the telephones, but only one conversation can take place at once, and it can be heard by the users of all the telephones. While this kind of circuit has been used on private networks such as railroad dispatchers' telegraph lines, on which many of the messages are intended for more than one addressee, telephone service is provided to the public by means of *switching offices*, which permit connecting any telephone served by the office to any other, allow many simultaneous conversations, and maintain privacy. When the area served is large, it becomes economical to provide more than one switching office, in order to shorten the average length of *loops* (transmission facility from telephone to office). A call from a telephone served by one *local* (or *end*) *office* to a telephone served by another is then routed over a *trunk* connecting the two offices. In this case, the amount of switching equipment provided is increased in order to reduce the cost of transmission facilities. Whether to provide an office and where, in order to reduce the total cost of switching and transmission, depends on the location of the terminals and the cost of transmission and switching.

If the number of offices in the local area is large, it is expensive to provide trunk groups of adequate size between every pair of offices, particularly if some of the trunk groups are small.

As illustrated in Fig. 22-1, the load a trunk group can carry for any given probability of *blocking* (a call cannot be completed because all trunks are busy) increases more rapidly than the number of trunks in the group. The efficiency of a small trunk group providing good service, such as 1% blocking of calls, is low. A more economical arrangement may be to provide a *tandem office*, through which an office can switch calls to any of the other offices. The trunk group from each end office to the tandem office is larger and more efficient than the direct groups would have been, so the total number of trunks is smaller. In addition and more importantly, the busy periods for calls from one end office to others often do not coincide, so the tandem group does not carry the sum of the peak loads but a load that may be considerably less than that sum.

A further saving may result from providing both direct and tandem trunk groups. Figure 22-2 shows such a network. A call is routed over a direct trunk if one is available; otherwise the call is *alternate-routed* via the tandem office. The number of trunks in each direct group is chosen so they will be used efficiently, although the fraction of calls blocked will be larger than

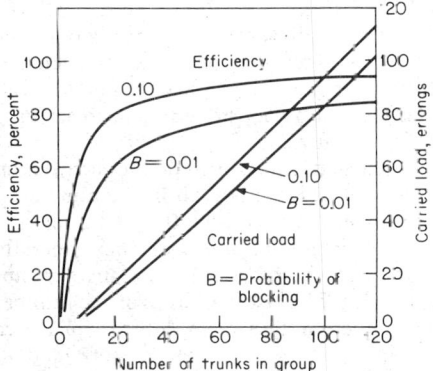

Fig. 22-1. Carried load and efficiency as functions of trunk group size (for first-choice routes). *(Adapted from Ref. 1. Copyright 1977, Bell Telephone Laboratories. Used by permission)*

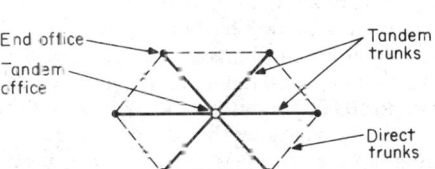

Fig. 22-2. A simple local network. *(Adapted from Ref. 1. Copyright 1977 Bell Telephone Laboratories. Used by permission)*

would be acceptable without alternate routing. Since the group to the tandem office carries the overflow from all the direct groups from that end office, it can be relatively efficient. The size of each direct trunk group is chosen to minimize the total cost of the direct trunk groups and the groups to the tandem office plus the cost of the switching offices. If there is very little traffic between any given pair of offices, direct trunks may not be justified.

Toll networks are arranged to facilitate connections between central offices that are separated geographically, politically, or in community of interest, in the same way that tandem offices facilitate connections between central offices in the same area. The offices in the North American network are arranged in a five-level hierarchy, each office having trunk groups to one or more offices at a higher level. As with local networks, high-use trunk groups are provided between offices at any level where there is enough traffic to justify them. The trade-off between transmission and switching obeys the same principles as in local networks. The arrangement of the network is shown in Fig. 22-3.

Toll and local networks may overlap; in the United States, subscriber loops may be connected to the three lowest levels of office in the toll network, which are thus acting as combined local-toll local offices; a tandem office may also serve as a local office or a toll office.

The plan for selecting a route for a call is

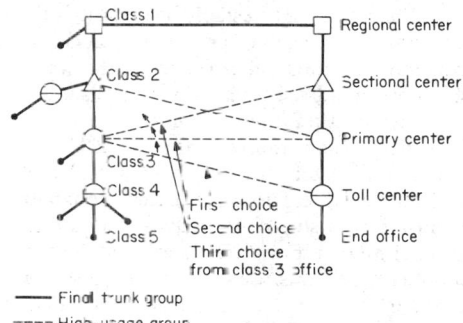

Fig. 22-3. Nominal toll network pattern. *(Adapted from Ref. 1. Copyright 1977, Bell Telephone Laboratories. Used by permission.)*

designed to provide connections economically under normal conditions. Special routing may be introduced as needed. The first-choice route from any office is the trunk group that goes most directly toward the destination, as shown in Fig. 22-3. If that trunk group is busy or out of service, the next choice is toward the next higher office in the destination area; subsequent choices are to higher offices in the originating area. It is a requirement for routing plans that no call may be routed over the same trunk group more than once.

6. Customer Switching. Two types of switching arrangements which provide for the needs of specific users may be considered as adjuncts to the public telephone network.

Private branch exchanges (PBXs) are switching offices used by organizations with a large need for internal communications. They provide simpler numbering for internal calls and provide also for a central attendant to handle calls coming in from the public network. A PBX may be connected to a switching office of the public telephone network; there may also be *tie trunks* to other PBXs of the same user. A variation of PBX service has the switching done by a telephone company switching office. PBXs range in size from about 10 lines to several thousands. Many different features are available with various types of PBX.

Key systems are arrangements used to give individuals access to several lines, using only one telephone (see Pars. **22-111** and **22-112**). A number of individuals may be served by a smaller number of lines to the PBX or central office. The features include the ability to hold one call and talk on another line, to answer a call for a colleague who is not present, and to set up a conference between three persons. Intercom service may be provided.

7. Numbering Plan. A numbering plan for a switched network must define each station uniquely, must be easy for users to understand, and must provide for growth in the number of telephones served. The North American telephone network uses a basic 10-digit numbering plan, in which the digits are grouped $3 + 3 + 4$. The first three are the *area code*; they generally designate nonoverlapping geographical areas. The next three, the *office code*, designate groups of up to 10,000 stations served by a particular switching office. The last four digits of the number identify the station within the central office. A call to a station in the same numbering plan area omits the area code. An initial 1 may be required to obtain access to the toll network, and an extra digit may also be required to reach the public network from a PBX. Outside North America different numbering plans are in use.

With the North American plan, the address dialed by a caller depends only on the destination and not on the route to the called station. Addressing which does not depend on the route to the called station makes possible changes in the network or alternate routing without any need for the users of the network to be informed or to take any action. Addressing that contains the routing of the call is used in some networks.

International numbering is an extension of the national numbering systems; the national numbers are preceded by country codes of one to three digits, which are standardized by the International Telegraph and Telephone Consultative Committee (CCITT) in Recommendation E.161. For customer international dialing, special access codes are required, both to obtain access to the international network and to prepare the local central office to receive called numbers of different lengths. Within North America international codes can be omitted because of the uniform numbering plan.

8. Transmission Plan. A transmission plan for a network is an orderly way of ensuring that the connections established through the network meet the objectives for controlling transmission impairments, such as loss, noise, and distortion (see Pars. **22-26** to **22-31**). The transmission plan must take account of the way the trunks are used in making up a connection. For example, in a local connection there is usually only one trunk, but there may be up to three. In a toll connection, there are normally at least three, and there may be (rarely) as many as nine. In general, each impairment is reduced as far as is consistent with reasonable cost; the impairment is allocated to loops, trunks, and switching offices to minimize cost while meeting the overall objective. Loss is made low enough to give a good received signal level while still meeting echo and near-singing objectives (Pars. **22-26** and **22-27**). It is also desirable to have the received level similar from one connection to another between the same two points, no matter whether a direct trunk was used or the connection was built up over an alternate route. Therefore, the amount of loss in a direct trunk between two offices will be larger than the loss of a trunk from an end office to a tandem.

DIGITAL SERVICE NETWORKS

Paragraphs **22-9** to **22-11** describe networks that furnish digital interfaces; digital facilities are covered in Pars. **22-55** to **22-67**.

9. Low-Speed Data Networks. Low-speed data networks include both switched and private-line types. The former can be derived from the public telephone network or can be special networks; the latter may be either point-to-point or multipoint. Both private-line and switched networks are used to serve teleprinters and computers; the private-line networks are used for alarm, telemetering, and remote-control purposes as well.

The private-line telegraph service in the United States offers speeds up to 150 or 300 bauds (Bd) depending on the carrier. The user interface follows the EIA standard at speeds up to 150 or 300 Bd; at 75 Bd and below, a 20- or 62.5-mA neutral interface is also furnished. "Neutral" means that the current flows in one of the states ("mark") but is zero in the other ("space"). Where metallic facilities are available, the transmission can be at baseband. The 20- and 62.5-mA neutral signals are used for transmission, but their use for this purpose is declining for noise reasons. For short distances, up to about 2 kΩ loop resistance, baseband signals of 6 mA (balanced neutral) or 3 mA (polar) can be used. "Balanced" indicates that the impedance to ground of each of the wires is equal; "polar" means that the current flows in one direction for one state but in the other direction for the other state. Voice-band channels, including those derived from carrier systems (Pars. **22-48** to **22-67**), are used for frequency-shift-keyed (FSK) (see Sec. 14) signals. When a single channel is applied to a voice band, the frequencies are usually chosen to permit simultaneous two-way transmission. To use the transmission facilities more economically, telegraph carrier is often used on the longer circuits. Using frequency-division multiplexing (Par. **22-48**) up to twenty-four 50-Bd, twelve 100-Bd, or six 200-Bd telegraph channels can be applied to a voice channel, according to the CCITT Recommendations. The signals in each telegraph channel are usually FSK. The telegraph circuits are connected at telegraph offices, which provide facilities for measurement of circuit and terminal performance, connection of the parts of multipoint circuits, and regeneration of the signals, where needed, if the signals being used are a commonly used code and speed.

TWX is a switched service for communication between teleprinters. The speeds offered are 60 words per minute, using International Alphabet no. 2,* and 100 words per minute, using International Alphabet no. 5 (ASCII).* The two kinds of terminal can send messages to each other; the connnection uses a central-office-located converter circuit, which translates from one code and speed to the other. Telex is an international service, operating at 50 Bd and using International Alphabet no. 2.

10. Packet Networks. Among the data networks provided up to the mid-1960s, both the circuit-switched and message-switched types were in use. In the first, a channel was assigned full time for the duration of a call. In the second, a message or section of a serial message was transmitted to the next switch if a path (loop or trunk) was available; if not, it was stored until a path became available. The use of trunks between message switches was often very efficient. In many circuit-switched applications, however, data were transmitted only a fraction of the time the circuit was in use. In order to make more efficient use of facilities and to make it possible to charge users only according to the amount of data transmitted, packet networks were developed. In a packet network, a message from one terminal to another is divided into packets of some definite maximum length, often 128 octets (corresponding to bytes in a computer) of data; the packets are sent from the origination point to the destination individually. Each packet contains a header which provides the network with necessary information to handle the packet. The packets transmitted by one terminal to another are interleaved on the facilities between the packets transmitted by other users to their addressees, so that the idle time of one source can be used by another. At the destination switching center, the message is reassembled and formatted before delivery to the called station. In general, a network has an internal protocol to control the movement of data within the network. The internal speed of the network is in general higher than any of the terminal speeds, so that there is a change of speed from source to network and from network to destination.

The same physical interface circuit can be used for communication with more than one other terminal or computer at the same time by the use of *logical channels*. At any given time, each logical channel is used for communication with some particular addressee; each packet includes the identification of its logical channel, and the packets for the various logical channels are interleaved on the physical-interface circuit.

Three methods of handling messages are in common use. *Datagrams* are one-way messages sent from an originator to a destination; the packets are delivered independently, not necessarily in the order they were sent. Delivery and nondelivery notifications may be provided. In *virtual*

---

*See CCITT Recommendations F.1 and V.3, or for example, J. E. McNamara, "Technical Aspects of Data Communication," Digital Equipment Corporation, 1977.

calls, packets may be exchanged between two users of the network; at the destination they are delivered to the addressee in the same order as they were originated. *Permanent virtual circuits* also provide for exchange of packets between two users of the network; each assigns a logical channel, by arrangement with the provider of the network, for exchange of packets with the other. No setup or clearing of the channel is then needed. Some packet networks support terminals that are not capable of formatting the data in packets, by means of a packet assembler and disassembler included in the network.

The earliest major packet network in the United States was ARPANET, set up to connect terminals and host computers at a number of universities and governmental research establishments. An objective was to permit computer users at one location to use data or programs located elsewhere, perhaps in a computer of a different manufacturer. Access to the network itself is through an interface message processor (IMP) at each location, connected to the host computer(s) there and to the IMPs at other locations. The IMPs are not all connected to each other; packets are routed via other IMPs when no direct trunk is provided. At locations where there is no host computer, a terminal interface processor (TIP) is used to provide access for terminals.

Other packet networks have been set up since ARPANET to provide packet communication service to commercial users, in the United States, Canada, France, Spain, Japan, and elsewhere. An early commercial network in the United States provides service at speeds up to 56 kb/s. The service is either common-user or with closed user groups; the latter service prevents communication into or out of a given group of stations, the equivalent of a private network. The network is arranged in a hierarchy of switching nodes, the higher nodes being redundantly connected by 56-kb/s trunks. The transit delay for a packet is less than 200 ms. The network operates in the virtual call mode.

Protocols for packet networks are the subject of several Recommendations of the CCITT, in addition to those listed in Par. **22-3**. Recommendations X.3, X.23, and X.29 cover start-stop mode terminal handling; X.25 covers the interface for packet-mode data terminal equipment, and X.75 covers the interface for packet-network interconnection.

In early packet networks, routing of each packet in a message is independent; each packet carries the address of its destination, as well as a number to permit arranging the packets in the proper order at the destination. Networks designed more recently use a *virtual circuit*, which is set up at the beginning of a call and which contains the routing information for all the packets of that call. The packets after the first contain the designation of the virtual circuit. In some networks, the choice of route is based on measurements, received from all other nodes, of the delay to every other node in the network. Recent results[5] indicate that it is sufficient for a node to measure and transmit to the other nodes the delay to nodes to which it is directly connected.

11. High-Speed Private-Line Networks. Data-transmission service is offered by common carriers using analog transmission channels of group bandwidth and supergroup bandwidth (Par. **22-48**). Digital signals from the information sources are converted to analog signals for transmission by means of *modems* (Par. **22-124**). In some areas, the service at group band may be either private-line or switched; the service at supergroup band is private-line.

With the increased interest in moving large amounts of data by large companies, new forms of high-speed private-line digital networks are being planned and set up, some by the established carriers and others by new carriers organized for the purpose. In addition to moving digital data, the networks are intended to be used for transmission of images, for voice (including secure voice), and for teleconferencing (voice and video). Some of the networks use terrestrial microwave systems to connect the larger cities; others use satellites. Digital transmission systems are used with both satellites and terrestrial systems. The new techniques are supplemented by standard art, especially in reaching user premises from the termini of the long-distance systems.

One of the networks being set up* provides channels for voice, data, or image transmission, arranged in the form of multipoint private lines. The channels are to be provided via satellite, with digital transmission from earth station to satellite and from satellite to earth station. The earth stations are to be located on the users' premises. For analog signals, the A/D conversion is to be at the earth station; bandwidths are to be furnished for carrying data at rates ranging from 2.4 kb/s to 6.3 Mb/s and voice at 32 kb/s. The system will use two spin-stabilized geostationary satellites, with 10 transponders per satellite (see Par. **22-54**). The transponders are to be used in a time-division multiplexed mode, with the ground stations that use the same transponder synchronized via the satellite by a control station. In addition to the time in each frame allocated to

---

*Current descriptions of the networks described in this paragraph can be found in such sources as Datapro Reports on Data Communication and Auerbach Data World.

each ground station, there is an additional pool of time in each frame, which is allocated on a demand basis to meet the needs of the users.

Another digital network being planned to provide nationwide electronic message service, data communication, document transmission, and teleconferencing with audio and freeze-frame video will use satellite channels leased from other carriers, as well as terrestrial microwave, for intercity transmission. Local distribution of signals will use microwave systems in a cellular arrangement, with the distribution area divided into cells of about 6 mi radius. An omnidirectional antenna will be located in the center of each cell for transmission to, and reception from, an antenna located on each customer's roof. Transmission between the local node at the center of the cell and the city node is also to be by microwave. The cellular arrangement will permit reuse of the radio frequencies, in a way similar to that described in Par. **22-20.** The equipment on the user premises will examine the address of the incoming data and will disregard all that is not addressed to that user. Transmission from the user to the local node will be by time-division multiplex, each user being assigned a time slot in which to send data.

A third digital channel service[6] uses terrestrial transmission systems, both radio and cable. Point-to-point and multipoint channels are offered at speeds up to 56 kb/s and at 1.544 Mb/s. Transmission from user premises to the offices is by means of four-wire loops. The interoffice transmission uses digital systems on cable for short distances and microwave radio for longer distances. On the microwave systems, a portion of the frequency band below the voice channels is used for digital signals.

PRIVATE NETWORKS

12. A private network is set up for the exclusive use of a single user (a single company, or an association, such as travel agents or trucking firms). The network may be as simple as one point-to-point circuit or as extensive as a nationwide switched network. The types of application for which a private network may be appropriate are discussed in Pars. **22-34** to **22-39.** Contributory reasons for a private network may be simpler addressing or faster call setup in a simpler network.

A private network may be connected to the public network at one or more points and may achieve the desired advantage of a private network with the flexibility of the public network. One common form of private network used by multilocation companies is a *tandem network*, whose PBXs are connected by tie trunks. These networks permit calls from stations on one PBX to those on another with simpler addressing and switching. The private network may or may not provide for alternate routing of calls during times of equipment trouble or high traffic load. If no alternate paths are provided in the network, they can be provided through the public network by means of arrangements to connect the terminal equipment to lines to the public switching offices.

In local data networks extending not more than a few miles, the short distance may make it practical to use wide-band but expensive (compared with wire pairs) transmission facilities, such as coaxial cable or optical fiber.[7,8] With data rates of a few megabits per second, perhaps 100 stations can be accommodated. A protocol which lets the stations contend for the use of the channel and does not call for a *master station* to control which station will send at any given time reduces the getting-started cost. If the medium is passive (no active circuits in the transmission path), the reliability can be high without need for redundancy. A *gateway* station can be used to transfer messages to or from another local network or a long-distance network. Messages are received by all stations, but only the one to which a message is addressed records it and takes action.

OPERATIONS SYSTEMS

13. Operations Systems. The preceding paragraphs have described the makeup and functions of service networks. Important to the provision of high-quality, economical service are features of the networks which make their support and continued operation possible. Operations may be considered in four parts: engineering the system to provide a network design and the amounts of equipment needed to give the desired quality of service without excess cost, installing the system, maintaining it, and managing it to take account of bad variations or other changes in conditions.

Recent trends, especially in larger systems, have been toward computerized operations systems.[9] Automatic monitoring capability in communication systems is encouraging greater use of centralized operations centers, to which customer-perceived troubles can also be reported. The operations center verifies the trouble, determines its location, and dispatches the maintenance forces. Recent developments have permitted the operations centers to handle new types of communication systems by changes in the software of the operations center; thus, the operations center does not need new equipment for each new type of communication system it monitors. Capabilities of operations systems in use, in addition to those already mentioned, include automatic testing of subscriber loops and changing routing plans to avoid areas where there is congestion caused by equipment trouble or an externally caused load. Paragraphs **22-95** and **22-96** describe operations systems for the support of switching systems.

Land-Mobile Radio Communications Networks

BY PHILIP T. PORTER

14. Regulatory Factors. The Federal Communications Commission has allocated portions of the spectrum in the 2-, 35-, 150-, 450-, and 850-MHz bands to be used for communication with moving or temporarily stationary vehicles on land; other allocations for communication with boats, ships, and airplanes also exist. The allocations have been divided among specific services; each service operates under a subsection (called a Part) of the FCC Rules (Title 47 of the Code of Federal Regulations*). Table 22-2 lists the Part of the FCC Rules applicable to each

TABLE 22-2   Mobile Services Qualifying for FCC Assignments

Part	Service
22	Public mobile (common carriers, public air-to-ground)
81–83	Maritime
87	Aviation (traffic control, advisory, en route)
90	Public safety (police, fire, maintenance, etc.)
	Industrial (utilities, mining, agriculture, construction, business, etc.)
	Transportation (railroads, buses, taxis, towing)
95	Personal (hobbyists, citizens band)

service. The general outline of each of these Parts is: general information, applications, technical standards, operating rules, and specific rules (eligibility and frequencies used).

In most areas of the country, both private ownership and common-carrier subscription are available to potential users of mobile communication. Comparison should be made between these two alternatives with regard to cost, maintenance, coverage area, reliability, blocking probability, etc. The user should be realistic about the coverage to be expected, should recognize that any assigned channel may be shared with other similar users, and should understand how radio calls are made. If owning and operating a private system is indicated, the user should become familiar with the applicable FCC rules and with the type-accepted equipment listed by the various suppliers.

SYSTEM DESIGN FOR RF COVERAGE

15. Propagation. Both the general theory of ideal propagation at all frequencies and a wide variety of practical considerations are discussed in Sec. **18**. Here the focus will be to provide a useful guide to lay out FM systems at 150, 450, and 850 MHz, the most used mobile bands.

Analysis of the fundamental free-space configuration diagrammed in Fig. 22-4, that of a half-wave dipole transmitting antenna and a similar receiving antenna separated by one wavelength, yields a loss of 17.7 dB.[11] Beyond one wavelength, the ratio P_R/P_T, termed *path loss*, is proportional to $1/d^2$ in free space. Thus at a distance of $2^n\lambda$ in free space, the path loss is $17.7 + 6n$ dB.

*Obtainable from the Superintendent of Documents, U.S. Government Printing Office, Washington, 20402.

For example, for λ = 2 m and n = 9, we find the path loss at 1 km to be about 71 dB. However, both large-scale random processes and earth-proximity effects add excess loss to the ideal case, making the estimation of coverage an inexact process but one that can be statistically investigated[10-14] to get at least an order-of-magnitude determination. The following factors enter into this process.

Effective Radiated Power (ERP).   Expressed in dBW, the ERP is the transmitter output power (10 log P_o/1 W) minus the losses in any cavity filters, diplexers, hybrids, or circulators external

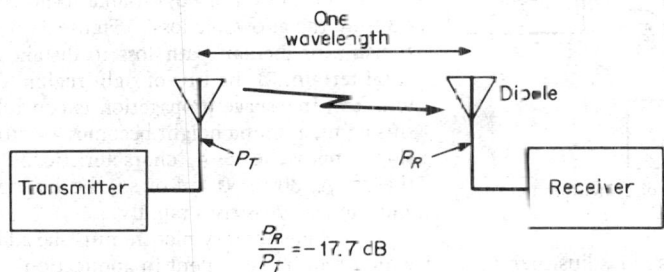

$$\frac{P_R}{P_T} = -17.7\ dB$$

Fig. 22-4. Path loss at one wavelength.

to the transmitter, minus the loss in the transmission line, plus the gain in the transmitting antenna in the direction of interest (relative to a half-wave dipole). Antennas for land mobile use are usually omnidirectional in the horizontal plane (azimuth) and directional in the vertical plane (elevation); however, nearby structures, e.g., the mast, may distort the pattern unexpectedly. Under the FCC rules, each service has its own maximum ERP, but in no case should the ERP exceed what is reasonably necessary for the job at hand.

Sensitivity.   With no rf input signal, the audio output of an FM receiver is random band-limited noise. When an rf signal is introduced, this noise is "quieted." A sensitivity specification quantifies this effect; it is commonly specified either at the 12-dB SINAD point (audio signal + noise + distortion, divided by noise + distortion) for a modulated signal or at the 20-dB quieting point for no modulation, and represents the typical lower limit of reasonable quality. Sensitivity is usually quoted in microvolts at the receiver terminals (see Sec. **21**). To convert this to received power, the familiar E^2/R is used, e.g., for a 50-Ω receiver, 0.5 μV sensitivity is equivalent to −143 dBW. To the specified sensitivity, one adds the transmission line loss and any other loss between the antenna and the receiver and subtracts the gain of the receiving antenna to get the effective receiver sensitivity (ERS) referred to the antenna terminals, i.e., the signal at the antenna required to overcome the receiver noise and produce a usable signal.

Site Noise.   In most locations, external noise, caused by auto ignition, power lines, cochannel interference, spurious emissions from transmitters, etc., rather than receiver sensitivity is the limiting factor at the receiver site. Table 22-3 lists typical levels of rf signal power required for an adequate audio signal for the narrow-band FM channels under discussion. At 35 MHz the situation is 10 to 15 dB worse that at 150 MHz. Of course, careful measurements made at a specific site provide more precise values; see Fig. 22-5 for the recommended test configuration. External noise varies according to the time of day, the seasons, etc. If the noise can be pinpointed to one specific nearby source, a directive antenna with a notch aimed, either in elevation or in azimuth, at the offending source can often provide 2 to 8 dB overall improvement (provided the notch in azimuth can be tolerated).

TABLE 22-3   Required Signal in Typical Noisy Locations

	Required signal for good quality, dBW	
Situation	150 MHz	450 and 850 MHz
In urban or industrial area or in heavy vehicular traffic	−115	−125
In suburban area or in moderate traffic	−125	−135
In rural area or in light traffic (with vehicle ignition well suppressed)	−135	−145

Allowable Path Loss.   This factor is the difference between effective radiated power (ERP) and required signal power (RSP), the latter being the larger of either effective receiver sensitivity or the necessary signal power for the site (measured or estimated from Table 22-3). From this allowable loss and a graph defining the relationship between median path loss and distance, one can predict the range at which 50% of the locations should be usable. (To convert to a criterion of 90% usable locations, subtract 13 dB from the allowable loss.[12]) Figure 22-6 gives the relationship of median path loss to distance over typical rural terrain. In the line-of-sight region (out to about 2 to 3 km) free-space propagation is controlling. Beyond this point, antenna height becomes a factor, and the log slope approaches -4, characteristic of smooth-earth theory. At 30 to 50 km or so, the horizon is reached, and the path loss rises rapidly.

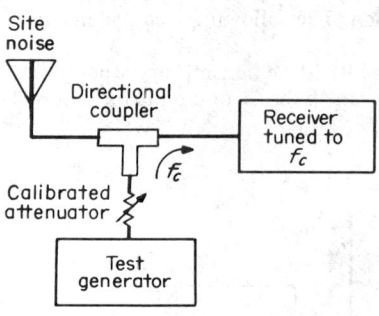

Fig. 22-5. Site-noise measurement technique.

To this elementary picture must be added a series of correction factors or adjustments, all of which require judgment in application:

1. *Antenna height.* For base-station antenna heights (let h = antenna height in meters) other than the 100-m reference of Fig. 22-6 and at distances greater than 2 to 3 km subtract 20 log $(h/100)$ from the values on the y axis. Note that the height to use is the estimated "effective" height, not necessarily the height of the mast; this should be either (a) height above average terrain (if situated on a hilltop) or (b) height above typical surrounding buildings. For mobile antenna heights k other than 2 m, subtract 10 log $(k/2)$; this latter factor is not precise in urban areas, however, since it can be grossly affected by scattering objects within about 100 m of the mobile.

2. *Buildings.* The local environment adds path loss over and above the open rural example of Fig. 22-6. This amount can be estimated using a judgment of local conditions as follows:

	Median added loss, dB		
	150 MHz	450 MHz	850 MHz
Suburban: separated buildings, or low abutting buildings on wide streets	4–8	6–10	8–12
Urban: tall buildings, or narrow streets bordered by abutting three-story buildings or higher	23–25	25–27	27–29

Clearly, there is a continuum of median loss between these two situations. Also, a standard deviation of about 10 dB about the predicted median can be expected within a local area which otherwise appears homogeneous. This correction factor dominates the metropolitan propagation.

3. *Hills.* Shadow losses due to specific hills can be predicted, using Fig. 22-7 plus a profile of the terrain between the two antennas. Once d_1, d_2, and h are known (see sketch which is part of Fig. 22-7), the intersection of d, where $d = \min(d_1, d_2)$, and the curve corresponding to the parameter h in Fig. 22-7 determines the approximate shadow loss to be expected; note that the specific d axis to use depends on the frequency band. In the absence of specific obstructions, 5 to 7 dB should usually be added unless the terrain is unusually flat and open.

An alternative method, suggested by Okumura et al.[14] can be more easily applied where there is a succession of rolling hills. The excess path loss found there is given thus:

Peak-to-valley terrain variation, m	Excess loss, dB		
	At peak	Halfway	At valley
20	0	1	3
50	0	6	10
100	0	10	20

Okumura et al.[14] also proposes correction factors in other, less common situations, for generally sloping terrain and for paths which are partially over water.

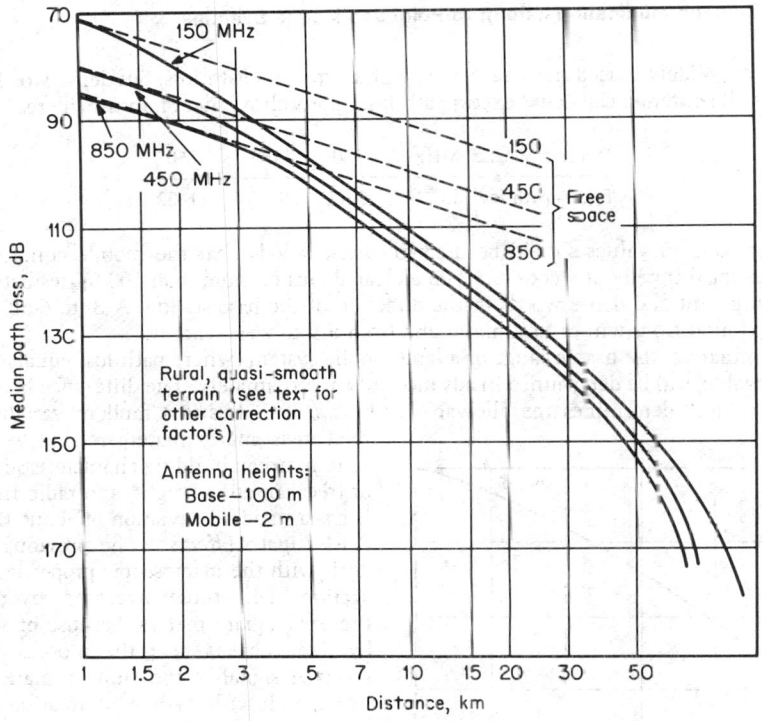

Fig. 22-6. Path loss between dipoles over smooth earth.

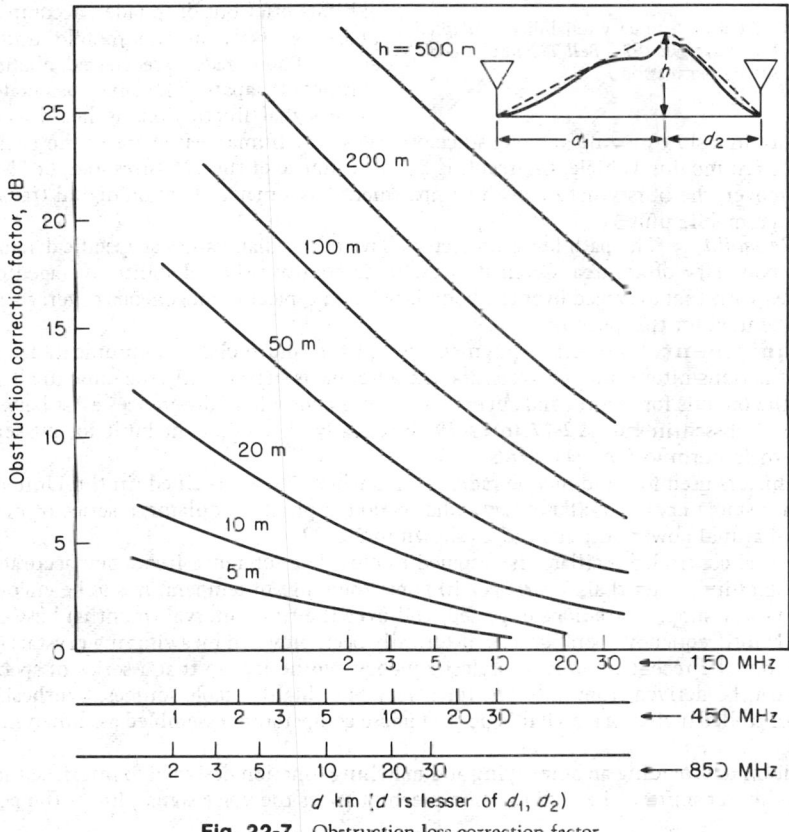

Fig. 22-7. Obstruction-loss correction factor.

4. *Trees.* Widely spaced trees have a negligible effect on path loss, but dense woods close to the low mobile antenna can cause excess path loss approaching that of an urban area:

Frequency band, MHz	150	450	850
Excess path loss, dB	6–12	9–18	11–22

The maximum values should be used for dense woods near the mobile equipment; the lower value for thinned-out woods or when a cleared area of more than 100 m lies between the mobile equipment and dense woods in the direction of the base station. A 3- to 6-dB variation can also be found from winter to summer and from dry to wet weather.

To summarize, the usable range of a land mobile system (where path loss equals the maximum allowable) can be determined in advance only approximately. The difference between the ERP and the RSP determines the allowable path loss. Path loss is a random variable whose median is approximately related to distance, terrain, urban building characteristics, density of trees, antenna height, and radio frequency; it has a standard deviation of about 10 dB.

Multipath Effects. The previous material deals with the macroscopic properties of propagation, i.e., those averaged over a few hundred square meters. Because of scattering by fixed objects near the mobile unit, the received signal is the sum of many components, each with its own attenuation and path length (hence random phase). As a result, the signal follows Rayleigh or Ricean statistics within a local area (see Sec. **18**). The signal exhibits brief but deep fades, accompanied by phase reversals, as the mobile unit moves about. These fades are spaced about a half wavelength apart. The noise associated with these signal fluctuations is heard as discrete events only in a slow-moving vehicle, since otherwise the human ear averages the background noise. In a fast-moving vehicle, the result is a disappearance of the FM threshold, or FM advantage. Moreover, the bursty nature of this impairment has a major effect on digital transmission to or from a mobile unit.

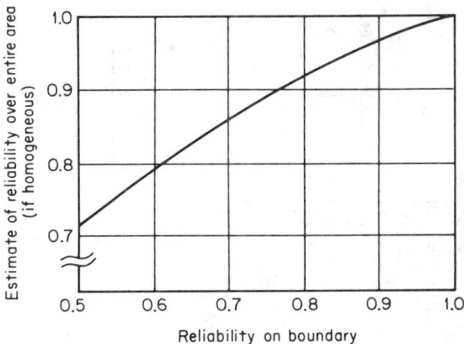

Fig. 22-8. Area vs. boundary reliability. *(Adapted from Ref. 15. Copyright 1974, Bell Telephone Laboratories. Used by permission.)*

Area Reliability. The path losses predicted above are median values at specific distances, i.e., near the boundary of an area. Often it is useful to compare the reliability averaged over the whole area with that averaged over the boundary. For a typical homogeneous coverage area, Fig. 22-8 can be used for this purpose.

16. Equipment. The basic equipment design for land mobile communications, particularly for the transmitters, the receivers, and the antenna, is substantially the same in all services; some of the options for control and supervisory equipment where differences exist between services are discussed in Pars. **22-17** to **22-19**. Practically all equipment built for all services is designed to conform to EIA* standards.

Transmitters used for land-mobile radio service where FM is specified (in the United States) consist of a stable crystal oscillator, an audio section, a phase modulator, a series of multiplier stages, and a final power output stage, as shown in Fig. 22-9.

The crystal-controlled oscillator is designed to provide a reference frequency accurate to the required stability (better than 5 parts per 10^6) over the range of temperatures to be encountered, over the power supply variations expected, and over the time interval (months) between calibrations. Multifrequency operation has historically been provided by switching crystals (one per channel), but the recent trend is to use a frequency synthesizer so that a series of specific frequencies can be derived from only one (or a very few) highly stable sources. Synthesizers use voltage-controlled oscillators, digital logic, and phase comparators, assembled as shown in Fig. 22-10.

In addition to providing an amplifying and matching function designed to interface the microphone to the transmitter, the audio section preemphasizes the voice signal, limits the peak volt-

*Electronic Industries Association, 2001 I St. N.W., Washington, D.C 20006.

age excursions, and deemphasizes the signal, to hold the maximum deviation to within the limits specified by the FCC (typically ±5 kHz, as measured after the multiplying sections). This has the effect of imposing a peak frequency excursion limit on the basic phase-modulated output of the modulator. Since phase modulators are relatively inexpensive and provide the required FM preemphasis automatically, they are universally used in narrow-band FM systems.

Frequency multipliers serve the triple purpose of amplifying the signal from the modulator, multiplying its frequency by an integer (using a highly nonlinear characteristic to generate harmonics and then filtering out the unwanted harmonics), and similarly multiplying the deviation. The final power-amplifier stage, operating in class C, boosts this 1- to 5-W signal up to the required output: 10 to 100 W for mobile units, 30 to 500 W for base stations. The final stage also includes filtering to reduce the amplitude of unwanted harmonics and spurious responses. Circulators (nonreciprocal ferrite devices, see Sec. 9) are also often used to suppress intermodulation products and to protect the final stage against high reflected power because of possible failure of the cable or antenna.

Receivers are also fairly standard in design; see Fig. 22-11. In most designs, the same crystal or synthesizer output which was fed to the transmitter modulator is multiplied and used in the

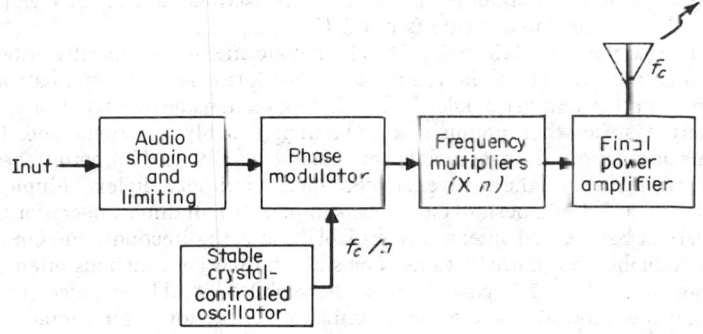

Fig. 22-9. Typical transmitter block diagram.

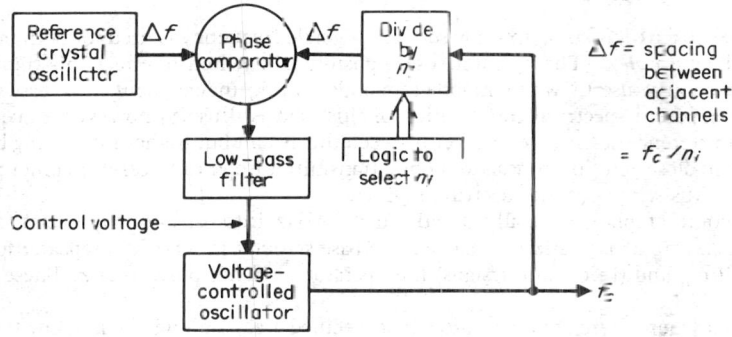

Fig. 22-10. Synthesizer block diagram.

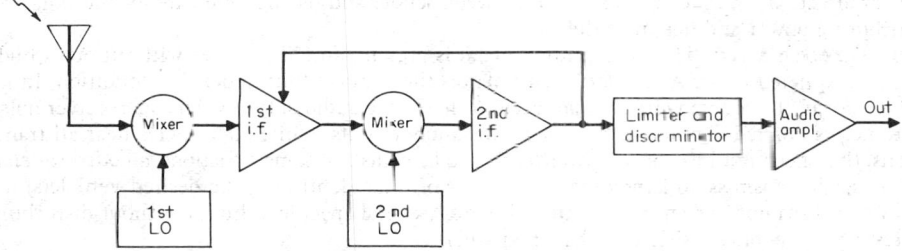

Fig. 22-11. Typical receiver block diagram.

first mixer to demodulate the received signal to a first i.f. at about 5 MHz, typically the separation between the land-transmit and mobile-transmit channel assignments. In other designs (especially at 850 MHz) separate crystals or synthesizer outputs are used for the first mixer.

After amplification at the first i.f., a second mixing process translates the signal to a second i.f. (often about 500 kHz) for further amplification and precise filtering. An FM discriminator and a deemphasis network pass the signal at baseband to the audio output stage for separation of voice and supervisory signals. Typical specifications are 0.5 μV sensitivity, 90-dB adjacent channel selectivity, 60-dB rejection of intermodulation products, and 90-dB rejection of other spurious signals and out-of-band noise.

The use of diversity reception to combat slow fading on point-to-point links dates from the 1920s. More recently, Jakes et al.[15] describe the effect of diversity reception on the multipath fading impairment in modern vehicular systems (see also Sec. **18**). In some situations, diversity will be found to be a cost-effective means either to increase the usable range of a system or to increase the audio signal-to-noise ratio at a specific range. The effectiveness of space diversity depends to a great extent on the separation between the receiving antennas; one-wavelength separation typically provides sufficient decorrelation at the mobile equipment, while seven to ten wavelengths are required at the base site. The expected improvement also depends on the specific diversity-combining technique being used; either maximal-ratio or equal-gain combining is to be preferred over selection diversity (see Sec. **18**).

Antennas are described in detail in Sec. **18**. The mobile antenna is typically either a quarter-wave whip, which uses the roof of the vehicle as a groundplane reflector, or a half-wave coaxial antenna mounted on the bumper, fender, or trunk. From a transmission point of view, the roof mount is preferred, since other mounts tend to be unpredictably directional and thus affected by the vehicle's orientation. A single mobile antenna is generally used for both transmitting and receiving; the transmitter and the receiver are decoupled via either a diplexer (duplex operation) or an rf switch controlled in a push-to-talk mode (simplex or half-duplex operation).

A wide variety of base-station antennas is available, and both directional and omnidirectional antennas are obtainable in a range of gains. Transmit and receive functions often use separate antennas, although combining is possible as in the mobile case. The engineer must also pay attention to lightning protection and to the illumination of the tower for aircraft-warning purposes; this latter is both an FCC and an FAA rule.

SYSTEM DESIGN FOR CONTROL

17. Operational Modes. In some services, a single frequency is used for both talking and listening (called *simplex*). The operators use a push-to-talk switch to effect this transfer. This mode is particularly useful when mobile-to-mobile calling independent of a base station is needed. The apparent spectrum conservation of this plan is illusory, however, since potential cochannel interference at the base receiver between the distant but higher interfering base transmitters and the closer but lower desired mobile transmitter imposes larger cochannel protection distances and thus wipes out the spectrum savings.

The most common plan, especially at 450 and 850 MHz, is to use two frequencies for the two directions. Spectrum conservation is realized, the base station can serve as a repeater for mobile-to-mobile calling, and (in certain designs) the operator need not push to talk. These plans are shown in Fig. 22-12.

In situations where there are many users in a specified area, each with fairly low traffic, such that they can effectively share a common two-frequency channel, a community-repeater arrangement can be used, as diagrammed in Fig. 22-13. In this case, the base stations are merely fixed versions of push-to-talk half-duplex mobile units, transmitting on f_2 or receiving on f_1. The repeater effects the frequency translation between sender and listener and extends the range by contributing power and height to the path.

18. Squelch Circuits. When no rf signal is present, an FM receiver will put out a high level of unquieted noise. A *squelch* circuit blanks the audio output under this condition. In its simplest form, the squelch circuit is adjusted to enable the audio output when the receiver noise drops, i.e., is quieted. Thus in a fleet of several mobile units, while all vehicles hear all transactions, they need not hear noise while waiting to be called. In some situations, a *coded squelch* is used. Each transmission is preceded by a burst of tone identifying the desired vehicle(s). In this way a dispatcher or attendant can call one or several specific vehicles without disturbing the rest of the mobiles assigned to that rf channel.

In full-duplex systems, where a carrier is continuously received squelch circuits are replaced by other forms of selective signaling since each receiver is quieted by the carrier for the duration of the transaction. For example, full-duplex mobile telephone systems use digital signaling at the beginning of every call.

19. Reciprocity and Receiver Selection. In concept, radio transmission is reciprocal; i.e., path loss is independent of direction. However, two factors make mobile-to-land and land-to-mobile transmission design different: (1) antenna height (which provides a different noise environment for the base receiver) and (2) power capability. Comparatively, these two factors often balance; the quieter noise environment of the base receiver antenna tends to compensate for the lower power of the mobile transmitter. Nevertheless in some situations, multiple land receivers are needed to cover the area of interest. Selecting the best receiver output is achieved either manually or by continuous automatic monitoring of noise the automatic means

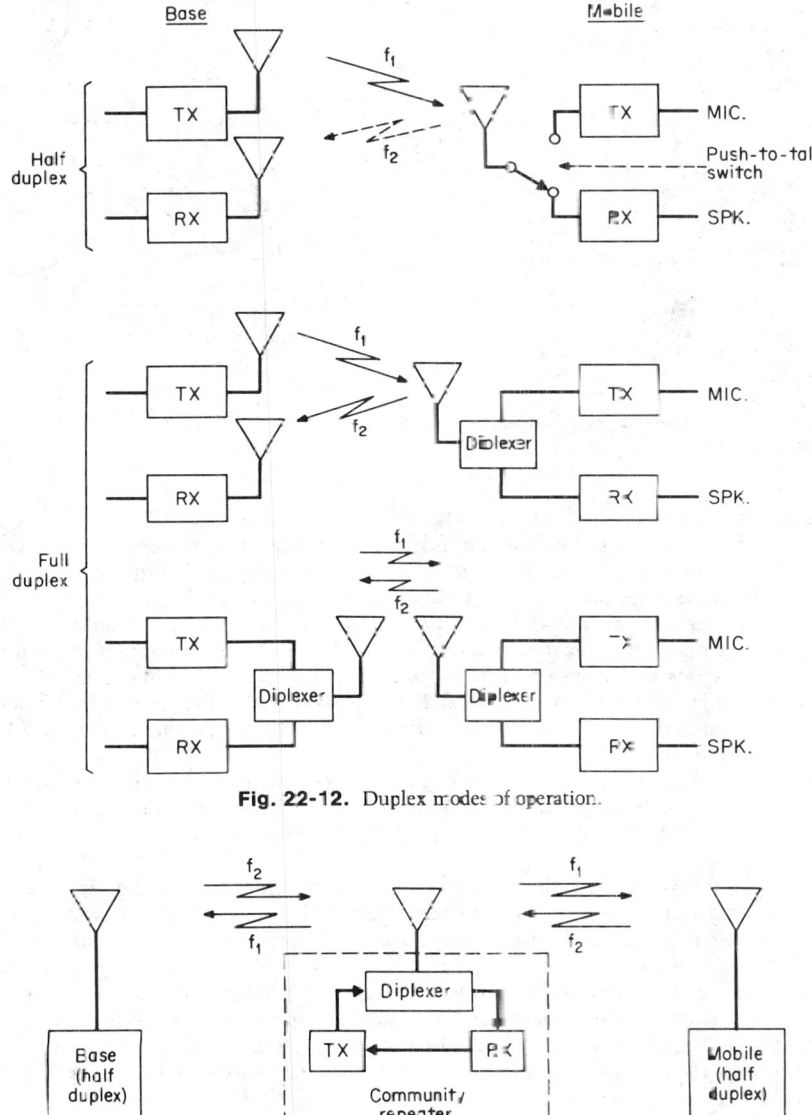

Fig. 22-12. Duplex modes of operation.

Fig. 22-13. A community-repeater system.

are often termed *voting circuits.* The automatic varieties work well, although in some unusual cases the attendant must override.

Use of multiple land transmitters to create a wide coverage area causes other problems. Simultaneous radiation results in beat frequencies which can be annoying and can impair transmission; poor audio delay equalization, phase reversals, and misaligned deviation settings also cause impairments. To use multiple transmitters effectively, high-stability crystal oscillators, delay-equalized connecting facilities, and carefully adjusted deviation settings are required. To overcome these problems, transmitter selection based on the outcome of receiver voting has proved workable in some situations.

20. Mobile Telephone Systems. In privately owned radio systems, there is no automatic connection to the telephone network; an attendant or dispatcher is required. In a modern

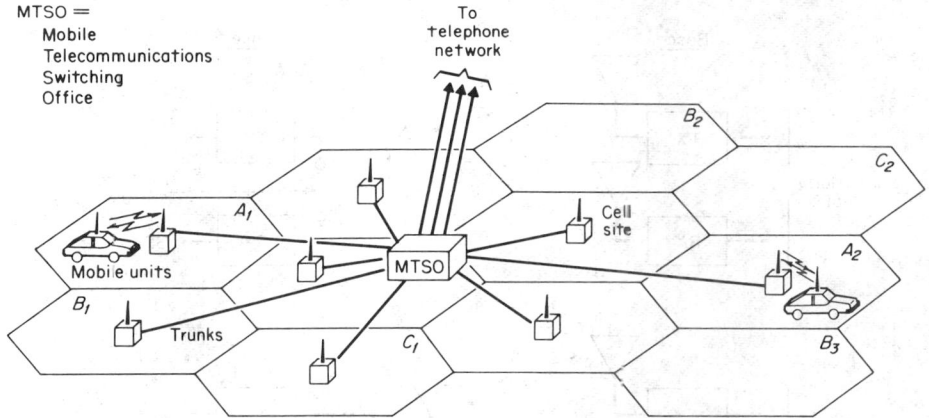

Fig. 22-14. Components of cellular mobile telephone system. *(From Ref. 19. Copyright 1979, American Telephone and Telegraph Co. Used by permission.)*

common-carrier mobile telephone system, the mobile user is provided with a rotary-dial or pushbutton instrument and thus can function much as a land-line telephone user. However, some mobile telephone systems still use an operator and a switchboard to provide a manual interface. In many urban areas, the channels are crowded and waiting lists are long.

Automatic systems at 150 and 450 MHz use a *marked-idle* channel for handling traffic.[16] All mobile units not involved in a call search through the complement of channels available locally and lock onto the one which system logic has currently marked with an idle tone. All incoming and outgoing traffic is routed via this channel; and as soon as the marked channel is seized for use by one subscriber, the tone is automatically moved to another channel (if available) to be ready for another call.

Signaling at the beginning of each call, using tone pulses, identifies the calling or called mobile unit. If *called,* the mobile set then rings; if *calling,* the user is given dial tone as in normal telephone practice. The end of a call is clearly marked with a disconnect tone to prevent incorrect billing.

System logic keeps track of calls for billing and for busy indication; it also identifies itself as required by FCC rules, is self-checking for trouble conditions, and provides the on-hook–off-hook signals to the telephone network. It also administers the protocol for initial seizure of the channel to resolve contentions.

At 850 MHz a new technology[17-19] has been introduced. Known as *cellular,* this radio system plan uses a grid of small service zones (cells) and dictates a specific pattern of channel use within the group of cells so that channels can be reused within a system in a controlled way. This reuse plan, coupled with a larger allocation, permits a great expansion of mobile service and an improved service quality.

In a cellular system, a call is routed via land-line trunks to a cell site in the vicinity of the mobile unit; low-power rf transmission is used for the last few miles only. System logic must locate an active user within the grid so as to hand off control of the call to the proper cell site as the signal strength of the active user changes. This plan is shown in Fig. 22-14; in this example, a unit in cell A_1 can use the same channel as one in cell A_2, with acceptably low probability of interference. Some versions of this plan further improve its spectrum reuse properties by using directional antennas to minimize interference. The cellular plan uses a special group of channels for high-speed binary signaling, thus doing away with the slower marked-idle mode of operation and permitting much greater flexibility of call routing.

21. Paging Systems. Certain of the channels allocated to the specific services mentioned above can be used for one-way signaling to small receivers carried by individuals. Called *paging*, this function can signal an individual selectively to take some prearranged action, e.g., call the office, or can deliver a short message. These variations are known as *tone only* and *tone plus voice*, respectively. The bulk of paging is provided by common carriers, although a number of larger users (hospitals, large plants, etc.) operate private systems.

Planning for rf coverage in paging systems must take into account several factors not important in vehicular systems:

1. In-building coverage is usually required; the need to cover 90% of the area on the ground floors of heavily constructed buildings at the edge of the service area can reduce the allowable path loss by 30 dB.[20,21]

2. The small antenna of the pocket receiver is inefficient (approximately 10 dB worse than a vehicular antenna).

3. Proper antenna orientation cannot be guaranteed. Also the human body becomes a major factor in the field pattern.

For these reasons, paging systems in urban areas typically require several transmitters per system; this in turn leads to a requirement for precise carrier frequencies, audio delay equalization to prevent partial cancellation, and modulation calibration accuracy. Sequencing of transmitters can sometimes obviate this requirement.

Systems using a variety of signaling plans for paging are on the market. Schemes using tones are typically decimal plans; three to five digits are transmitted per page, yielding capacities of 10^3 to 10^5 users per channel and up to five pages per second. Digital schemes typically use binary FSK signaling and error-correcting codes, which permit 10^5 users per system and five pages per second.

The trend in paging is toward enhancement of the basic service to permit more information to be sent or stored. One such feature is an LCD pane on the personal receiver which displays the calling telephone number or some other coded message. An expected enhancement is the use of voice storage at the control terminal to permit message retrieval by the called party after a page. And in some situations (in-building or in-plant systems) the calling and called parties are switched together at the control terminal in a "meet-me" calling arrangement, making paging almost the equivalent of a pocket telephone bell.

22. Trunked Systems. Intermediate in service capability between private and common-carrier systems are *trunked systems*. These plans permit several fleets to use a common group of channels to increase the loading per channel or to decrease the blocking probability. Typically, the channel utilization can be increased to the order of 30% while at the same time the blocking decreases to less than 20% and the queuing delay to less than 10 s. Specific operational details, as of 1981, are just now being released by the suppliers of equipment and service.

23. Personal Portable Systems. The use of hand-held or worn two-way radios is becoming more common as size reductions are made and as their usefulness in specific situations is appreciated. Unfortunately, size reduction tends to come at the expense of battery capacity, antenna efficiency, and other receiver specifications. Design of a system employing personal portables parallels that for a paging system, where building penetration losses, improper orientation, reduced ERP, and increased RSP must be allowed for. Designing with multiple base transmitters and receivers causes increased cost and makes base antenna height a more valuable factor (up to the point where cochannel interference from vehicular units becomes limiting). The most common use is around airports, construction sites, large plants, stadiums, etc. Cellular mobile telephone systems at 850 MHz, with their small-cell design are expected to accommodate portables as that technology matures.

24. Privacy. It must be realized that mobile radio communication is not secure from eavesdropping unless precautions are taken; in certain situations (police, watchmen, etc.), privacy is highly desirable. Thus, more and more use is being made not only of scrambling devices but also of digital encryption. Analog scramblers afford only temporary security against a sophisticated eavesdropper, whereas the digital approach can be virtually unbreakable. Decreased voice quality and/or sensitivity can be expected when a privacy requirement is imposed on an existing system.

Service Network Performance and Design Objectives

BY A. B. BROWN, JR.

NATURE OF PERFORMANCE OBJECTIVES

25. Performance objectives of a communication system must meet the users' needs with a good balance between the quality of service and cost. In practice, needs vary among users, and determination of the appropriate quality of service is subject to judgment. In setting specific performance objectives, the concept of *grade of service* has been found useful. As used here, it is the fraction of users who will rate a service in a given category, such as good or excellent. (Grade of service is also used to describe blocking; see Par. **22-38**.) The grade of service takes into account both the distribution of user opinion for the same quality of service and the distribution of facilities in the plant. In addition to balancing cost and performance, the objectives take into account the trade-offs between various parameters of the service; e.g., increasing the received signal strength may at times also increase the echo.

The numerical values set for performance objectives are based on tests with users,[22] who rate the observed conditions on a scale from unsatisfactory to excellent. For some parameters, the users' response may depend significantly on the values of other parameters, and the test conditions must include variations of them all;[23] for others, there is little interaction, and parameters other than the one under test can be set at a nominal value.

Translation of overall performance objectives to design objectives for trunks and switching equipment involves formulating a model of the connections anticipated in the network.[23] The model describes statistically the numbers and types of trunks and switching offices that may make up a connection. The total value of each parameter is allocated to the parts of the connections according to the feasibility and economy of achieving the resulting objectives for each part.

In order to achieve the performance objectives, the network should be appropriately engineered, taking into account the characteristics of the equipment units and the quantities of each type required. The network must also be properly maintained and properly operated once it is installed and in service.

QUALITY OBJECTIVES FOR ANALOG SERVICE NETWORKS

The following paragraphs cover the principal impairments[24] to be found in voice-band analog service networks and the related quality objectives as they apply to public networks. These objectives may serve as a guide in the planning of private networks, although the balance between users' needs and cost may be different. Similar considerations hold for wide-band (group or supergroup; see Par. **22-48**) channels, but the numerical objectives are different. They are covered in CCITT Recommendations H.14 (group links) and H.15 (supergroup links).

26. Loss. Loss objectives in a voice-band connection are determined principally by considerations of voice communication. For the most part, data circuit-terminating equipments (modems) are designed to operate on connections whose loss has been made satisfactory for voice use.

The strength of a received signal should be such that it is acceptable to listeners (neither too weak nor unpleasantly strong) and is appropriately related to the noise at the receiving end of the channel. The loss of the channel is also used to control echoes (Pars. **22-27** and **22-68**). Subjective tests of listener preferences have shown that most listeners consider -25 to -30 volume units (vu) as good or excellent. A vu meter gives the approximate rms voltage during the talking period. It can be converted to dBm at nominal impedance during the talking period by the relation

$$\text{Average power} = (\text{vu} - 1.4) \text{ dBm}$$

A more recent method of characterizing speech levels during the talking period is the equivalent peak level (EPL).[25] It uses a distribution of levels which is uniform in decibels up to a peak; such a distribution has been found experimentally to be a good approximation to the levels of speech. The EPL is that peak which makes the rms value of the distribution above an arbitrary threshold the same as that of the speech. Both theory and experiment show that the EPL is nearly independent of the threshold over a range of at least 35 dB.

Measurements of 1,000-Hz loss on toll connections, including the end offices but not loops, were made in the United States and Canada in 1969 to 1970.[26] The mean losses for each mileage band ranged from about 6.5 dB for 0 to 180 airline miles to about 7.5 dB for 725 to 2,900 airline miles; standard deviations were about 2 dB. The losses in the two directions often differed; a 5-dB difference was not uncommon.

27. Noise. Noise is any unwanted signal on the communication channel which might interfere with the desired signal. It can be in the form of irregular waveforms that cover the frequency band of the channel, impulses (voltage spikes), crosstalk from another channel, single-frequency tone (usually from another channel), or excessive echo or near-singing of the channel being considered. Measurement of noise, including explanations of reference noise levels and transmisson level points (TLP), is covered in Par. **22-49**.

Message circuit noise is defined as the short-term rms value of the noise voltage, measured with a weighting that approximates the response of telephone sets (C message in North America or, internationally, psophometric; see Par. **22-49** and Fig. 22-32), and with a time constant that approximates that of the human ear. The concept is useful in evaluating the effect of noise on speech.[27] The principal types of message circuit noise are ordinarily thermal noise, power hum, and switching-office noise. Quantizing noise in digital transmission systems may also contribute. The objective for connections for international circuits is given in CCITT Recommendation G.113 as −43 dBm0p, where the reference level is that of the first circuit in the chain of international circuits. The United States objectives are given in Table 22-14 (Par. **22-49**); see also Ref. 28.

Impulse noise consists of the peaks of noise that individually are significantly larger than the rms level of noise on the channel. Its principal effect is on data transmission, where individual noise peaks may mask the data signal for one or more bits. If impulse noise is serious enough, it affects speech. It may result from the operation of switching offices or from other electrical surges coupled into the disturbed channel. Measurement of impulse noise consists of counting the impulses that exceed a given threshold during a given period of time. The derivation of the objectives depends on the properties of the modems expected to be used over the channels under consideration and on the other impairments expected, e.g., amplitude and delay distortion, which may reduce the margin of the data signal against errors caused by impulses.

The objective in the United States network for customer-to-customer connections is not more than 15 counts in 15 min on 80% of the connections at a level 5 dB below the received signal level. This level depends on the end-to-end loss of the connection and can be determined by inserting a tone at the standard level into the channel and measuring its level at the receiving end. This objective is divided into loop and trunk objectives; the loop objective is not more than 15 counts in 15 min at 59 dBrnc, referred to the local switching office. The objective for trunks depends on trunk loss, which is translated into length; the objective is not more than 5 counts in 5 min on 50% of the trunks in a group at thresholds given in Table 22-4.

TABLE 22-4 Impulse Noise Threshold for No More than Five Counts in 5 Min on Carrier Trunk Facilities

Trunk length, mi	0–125	125–1,000	1,000–2,000	Over 2,000
Threshold level, dBrnc0	58	59	61	64

Source: From Ref. 24. Copyright 1970, Bell Telephone Laboratories. Used by permission.

Crosstalk is the effect of coupling the signal of one channel into another. In voice channels, the crosstalk from another voice channel may be intelligible, or it may be so shifted in frequency or inverted that it is not intelligible. In the former case, a loss of privacy can result. In the latter, the spurious signal in the disturbed channel is likely to contain a syllabic rhythm; the listener may think that it could be understood if carefully listened to and may worry about being overheard.

Crosstalk may arise from a number of sources. Inadequate frequency selectivity of frequency-division multiplex (FDM) systems may permit energy from one channel to reach another. Unbalance of the parameters of cables (capacitance and inductance between the wires) can result in a net coupling of the signal on one pair into another. This effect may be made more serious by unbalance in the equipment connected to the ends of the cable. Also, nonlinearities in analog multiplex transmission systems can result in intermodulation distortion, which can result either in unintelligible noise in the disturbed channel or, in some cases, in intelligible crosstalk. Intermodulation is the result of the various frequencies in a multiplexed transmission system being modulated by each other because of nonlinearity in the system. A high-level tone in one channel can act as a local carrier and translate the signal from a second channel into a third channel; in the worst case, the translated signal results in intelligible crosstalk. Any of these effects may be made worse by an excessive signal level in the disturbing channel.

The factors that affect hearing crosstalk involve the volume level of the disturbing signal,[110] the loss in the disturbing circuit to the point of coupling, the coupling loss, the loss in the disturbed circuit from the point of coupling to the receiving point, the hearing acuity of the listener or the sensitivity of the receiving device, and the ambient noise level where the listener is located. Occupancy of the disturbing circuit also affects the probability of hearing crosstalk.

TABLE 22-5 CCITT Crosstalk Loss Recommendations, dB, Referred to Equal-Level Points

	Circuits	Switching offices
Between two circuits	58	70
Between go and return direction of a circuit	43	60

Source: Adapted from CCITT Orange Book, *6th Plenary Assem.*, 1976; Recommendation G.151, Vol. III-1, pp. 74–75, and Recommendation Q.45, Vol. VI-1, p. 58. Reproduced with special permission by the International Telecommunication Union (ITU).

TABLE 22-6 Subjective Reaction to Echo Delay: Round-Trip Loss at Which 50% of Users Rated Channel as Good or Excellent[29]

Round-trip delay, ms	20	40	60	80	100
Round-trip loss, dB	25	32	38	40	42

The CCITT crosstalk objectives are stated in terms of the coupling loss between two circuits, referred to equal level points, as in Table 22-5. The overall performance of the switched telephone network in the United States for trunked connections is about 1% chance of hearing intelligble crosstalk.

Echo and Near-Singing.   Energy is reflected from any impedance mismatch in the transmission channel. If the energy can reach the transmitting point, it will be added to, and interfere with, any other signal being received there. Reflections that occur in the four-wire part of the channel cannot be reflected back toward the talker, but reflections that occur at the four-to-two-wire conversion or in the two-wire part of the channel can reach the talker.[30] If there is another impedance discontinuity, the energy will be reflected again; if it can reach the receiving point, it will interfere with the signal there as well.

For a voice signal being reflected back to the talker, the effect depends on the strength of the reflected signal and on the delay. If the delay is very short, the effect is the same as enhancement of the sidetone; a somewhat longer delay results in a hollow sound of the sidetone, and a long delay produces delayed intelligible speech, which is very disturbing to the talker. Voice signals that undergo a second reflection and reach the listener may also produce an objectionable distortion of the received speech. For data signals, the effect is to spread the energy from one bit period into those coming before and after, producing intersymbol interference.

For either of these uses of a channel, longer delay or increased level causes the echo to be more objectionable. Results of tests of the round-trip loss at which 50% of users rated the channel as good or excellent[29] are given in Table 22-6.

Circuit requirements to reduce echo on terrestrial circuits are set in terms of return loss* between circuits that are connected in the switching offices and of loss of the trunks. Because of the long delay of satellite circuits, special means (Par. **22-68**) are employed to control echo. It is practical to match impedances of one trunk with another, including the transmission facilities and the associated office circuitry, to 20-dB return loss. However, end offices face the loops, which differ in impedance and vary with temperature; the return loss may be as low as 11 dB between 500 Hz and 2,500 Hz. The loss of each trunk is made large enough to make the echo acceptable,[23]

---

*Return loss at an interface is the ratio between the incident and the reflected power, expressed in decibels.

as described in Par. **22-68**. Outside the 500- to 2,500-Hz band, the return loss may not be as much as the 11 dB given above; the objective in the United States network is 6 dB minimum return loss within the band of 250 to 3,200 Hz; the result may be an objectionable hollow sound of the speech or even near-singing. The 2.5-dB loss introduced into each toll-connecting trunk to meet the 500- to 2,500-Hz echo requirements (Par. **22-68**), although not needed for that purpose on short connections, does alleviate the 250- to 3,200-Hz echo problem. Results of measurements of echo performance of toll telephone connections in the United States are given in Ref. 30.

28. Bandwidth. The bandwidth of a connection affects the naturalness of speech and the accuracy of data transmission. CCITT Recommendation G.124B states that, for national networks, circuits should be capable of transmitting 300 to 3,400 Hz between international and local exchanges with the response down not more than 9 dB from the response at 800 Hz. Measurements of the United States and Canadian network are given in the next paragraph.

29. Distortion. *Attenuation distortion* is the variation in response of a channel as a function of frequency within the band. It can occur as a relatively smooth slope, as ripples, or as a combination of the two. The effect of slope on speech is to cause an unnatural sound; if the high frequencies within the voice band are greatly attenuated with respect to the low frequencies, there can be a reduction in intelligibility of the speech. The effect of either slope or ripple on pulse transmission is to distort the shape of the pulses (often including spreading some of the energy into adjoining pulses), thus reducing the margin against errors caused by noise.

Attenuation distortion arises in the filters used in carrier systems (Pars. **22-48** and **22-56**) and in the effects of capacitance between wires in baseband transmission (see Fig. 22-21).

CCITT Recommendation G.132 specifies the limits on attenuation distortion for a chain of four-wire circuits, including both the international and national parts measured with respect to the transmission at 800 Hz, as

2.2 dB	600 to 2,400 Hz
4.3 dB	400 to 3,000 Hz
8.7 dB	300 to 3,400 Hz

Attenuation distortion as measured in the United States and Canadian toll network is shown in Fig. 22-15.

Delay distortion is the difference in the envelope delay at any frequency and the minimum envelope delay in the transmitted band, where envelope delay is the time for a change in the envelope of the wave to pass through the channel. The human ear is relatively insensitive to delay distortion, but the pulses in data and video signals are dispersed in time by it and interfere with the pulses that precede and follow. The result for data is reduced margin against errors caused by noise.

Delay distortion arises principally in the filters used to separate the channels in a frequency-division multiplex transmission system. The steep cutoff required to conserve frequency bandwidth implies delay distortion. In translating overall objectives to objectives for specific equipments, it is necessary to take into account the number of pieces of equipment through which the signals will pass in a connection because the impairments add. CCITT Recommendation G.133 gives the permissible differences of envelope delay for a worldwide chain of 12 circuits, each on a single 12-channel group link, at the lower and upper limits of the frequency band as in Table 22-7.

Typical values of delay distortion on a voiceband network are taken from the results of measurements on the switched telephone network in the United States and Canada. The results on toll connnections[26] are given in Fig. 22-16.

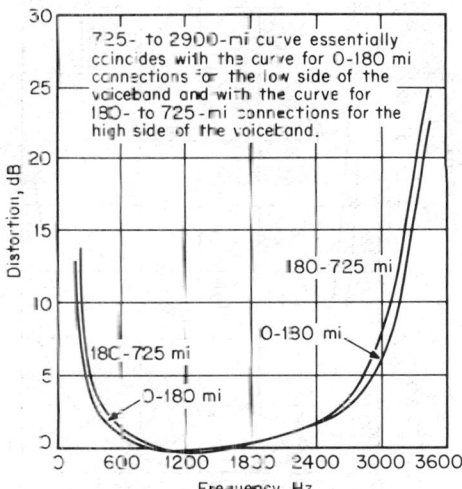

Fig. 22-15. Locus of means for attenuation distortion relative to 1,000 Hz. (Adapted from Ref. 26. Copyright 1971, American Telephone and Telegraph Co. Used by permission.)

Nonlinear distortion is the effect in which the amplitude of the instantaneous output of a channel is not directly proportional to the input. In general, the output can be expressed as a Taylor series, with second-, third-, and higher-order terms as well as the desired linear term. In practice, the terms of order higher than the third can be ignored. The nonlinearities of the channel introduce frequencies in the output that are sums and differences of the input frequen-

TABLE 22-7   Limits on Delay Distortion in 12-Circuit Chain, Allocated to the International Chain and the Two National Four-Wire Extensions

	Delay distortion limits, ms	
	Lower limit of transmitted frequency band	Upper limit of transmitted frequency band
International chain	30	15
Each national four-wire extension	15	7.5
Total	60	30

Source: From CCITT Orange Book, 6th Plenary Assem., 1976; Recommendation G.133, Vol. III-1, p. 64. Reproduced with special permission by the International Telecommunication Union (ITU).

cies. Nonlinear distortion also changes the amplitudes of the desired frequencies (gain compression or expansion); this effect is usually unimportant. If the input frequencies are A and B, second-order nonlinearities will introduce terms whose frequencies are $2A$, $2B$, $A + B$, and $A - B$. Third-order nonlinearities will introduce terms whose frequencies are $3A$, $2A + B$, $2A - B$, $2B + A$, $2B - A$, and $3B$. Many will fall outside the transmission band, but others will be transmitted and interfere with the received signal.

The effect is most serious on transmission that uses tones in various parts of the frequency band, such as voice-frequency telegraph multiplex or full-duplex data transmission in the voice band, where the two directions of transmission use different frequencies. In such cases, the sums or differences of some of the telegraph channels may lie in the frequency bands of others and interfere with them. Nonlinear distortion within a channel is primarily caused by nonlinearities in the transfer characteristic of active devices such as transistors or, to a lesser extent, of coils or transformers, in carrier terminals, or repeaters in a voice-frequency line.

Measurements of nonlinearities in a channel can be made by applying a single tone and measuring the level of harmonics. More recent techniques call for applying two or four tones and measuring the modulation products. Results of measurements in the United States and Canadian telephone network are given in Table 22-8.

30.   Rapid Gain and Phase Changes. Changes in either the level or the phase of the received signal may occur in a small fraction of a second. The changes may either be isolated changes, such as those resulting from switching a radio channel to a spare channel whose carrier supply is at a different phase or whose propagation time is different, or continuous changes resulting from instability in a carrier supply. The principal effect of changes in gain or phase is on data signals; severe incidental phase modulation degrades speech.

For leased circuits intended for special use, such as data transmission, CCITT Recommendation H.12 gives a provisional expectation (subject to further study) that phase jitter will not exceed 15° peak to peak; an objective for

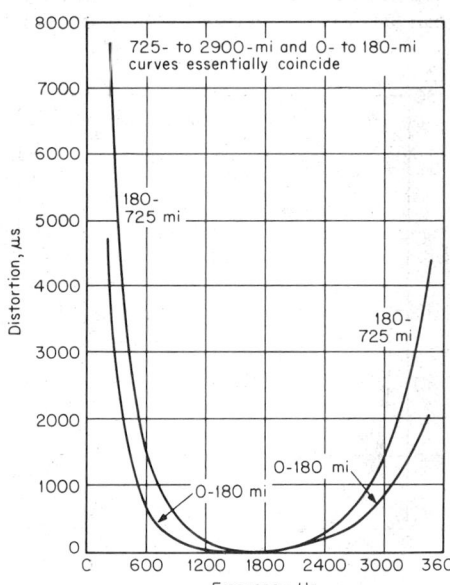

Fig. 22-16. Locus of means for envelope delay distortion relative to 1,700 Hz. (*Adapted from Ref. 26. Copyright 1971, American Telephone and Telegraph Cc. Used by permission.*)

variation of loss with time is given as ∓ 3 dB over a period of a few seconds and ± 4 dB over long periods. Measurements of phase jitter in the United States and Canadian network are given in Table 22-9.

31. Frequency Offset. The frequencies in the received signal may be shifted from those in the transmitted signal. When this effect occurs, the shift is usually the same for all frequencies. The offset usually arises in frequency-division multiplex carrier systems, where the transmitting and receiving carrier supplies may not be at precisely the same frequency. Any difference in the carrier frequencies causes a frequency shift of the same amount in the baseband signal. The effect on voice is negligible for the small shifts that normally occur, but frequency offset prevents baseband data transmission over channels that exhibit it. Since even a small frequency offset would cause severe intersymbol interference, it is necessary to transmit data using a modulation method that is not affected by a few hertz of shift. In practice, the limit on acceptable frequency offset is set by its effect on FSK modulation, especially in a system where the two frequencies are close together, such as a voice-frequency telegraph multiplex. With frequency offset, the received signal is biased toward one of the states; the pulses corresponding to that state are lengthened, and those corresponding to the other are shortened.

CCITT Recommendation G.135 states that, based on the needs of voice-frequency telegraphy, the frequency offset on an international telephone circuit should not exceed 2 Hz. Carrier frequency accuracies to achieve this result are discussed in Par. 22-48. In practice, the frequency offset for a channel whose carrier oscillators are designed to these figures and are properly maintained is rarely more than 2 Hz. Measurements of the frequency offset in the United States and Canadian toll network[26] are given in Table 22-10.

TABLE 22-8   Results for the Ratio of the Total Received Power of a Two-Tone Signal Transmitted at − 10 dBm to Individual Intermodulation Product Powers on Toll Connections for All Mileage Bands

Type	Product A = 1,250 Hz B = 700 Hz Freq., Hz	Power ratio Mean, dB	Standard deviation, dB
$A - B$	550	42.8 ± 2.3	10.6
$2B$	1,400	50.1 ± 2.5	11.5
$A + B$	1,950	43.9 ± 3.0	12.0
$3B$	2,100	54.3 ± 1.2	7.4
$2A$	2,500	48.5 ± 2.2	11.2
$2B + A$	2,650	50.0 ± 2.2	12.3
$2A - B$	1,800	43.0 ± 2.4	10.9
Equivalent $A + B$		42.1 ± 2.2	10.6

Source: Adapted from Ref. 26. Copyright 1971, American Telephone and Telegraph Company. Used by permission.

TABLE 22-9   Results for Peak-to-Peak Phase Jitter on Toll Connections for All Mileage Bands

Frequency band, Hz	Degrees of phase jitter*		
	10%	50%	90%
12–768	0	3.0	7.0
12–24	0	0	2.0
24–48	0	0	1.0
48–96	0	0	2.0
96–192	0.3	0.3	1.3
192–384	0	0.5	1.5
384–768	0	0.6	1.6

*At percentage of measurements shown.
Source: Adapted from Ref. 26. Copyright 1971, American Telephone and Telegraph Company. Used by permission.

TABLE 22-10   Results for Absolute Frequency Offset on Toll Connections

Connection length, airline miles	Offset, Hz*		
	10%	50%	90%
All	0	0	0.2
0–180	0	0	0.1
180–725	0	0	0.3
725–2900	0	0.1	1.1

*At percentage of measurements shown.
Source: Adapted from Ref. 26. Copyright 1971, American Telephone and Telegraph Company. Used by permission.

QUALITY OBJECTIVES FOR DIGITAL SERVICE NETWORKS

The quality objectives that apply to service networks with digital interfaces concern the accuracy of the bits, the proper timing of the bits, and the synchronization of the network, i.e., prevention of the loss or replication of bits. In addition, in packet networks the question of order of delivery of the parts of the message arises. When the interface of the service network is digital, the same objectives apply in general whether the transmission is digital or the digital interface is achieved by applying modems to an analog network.

32. Error Rate. Measurements of the accuracy of transmitted data have been expressed in terms of several parameters. One is the bit-error rate, i.e., the fraction of the received bits that

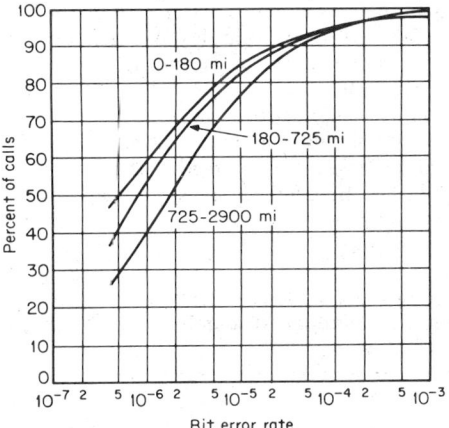

Fig. 22-17. Bit-error rate distributions by mileage strata at 1,200 b/s. *(Adapted from Ref. 31. Copyright 1971, American Telephone and Telegraph Co. Used by permission.)*

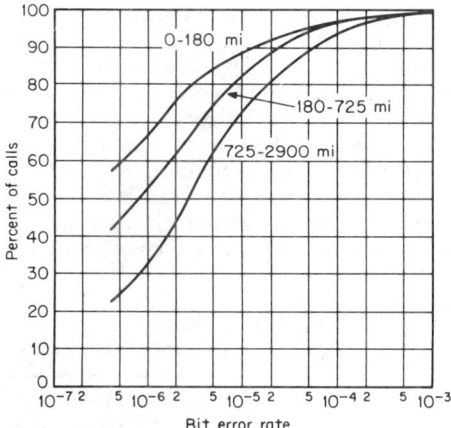

Fig. 22-18. Bit-error rate distributions by mileage strata at 2,000 b/s. *(Adapted from Ref. 31. Copyright 1971, American Telephone and Telegraph Co. Used by permission.)*

are in error. Measurements have shown that the bit errors on the line do not occur as independent random events but tend to be clustered. It is therefore not possible to evaluate an error-control system, either in terms of residual error rate or in terms of the number of retransmissions to be expected, from the bit-error rate alone. A second method of reporting error statistics is the block-error rate; the data are considered to be divided into successive blocks of some given size, and the fraction of those blocks which contain errors is the block-error rate. However, the block lengths in error-control systems are not standard. Recently, accuracy objectives and measurements have been stated in terms of the proportion of errored seconds. Another method of reporting accuracy objectives and results, which is applicable to packet networks, is the packet-error rate. For start-stop transmission, an error on a start bit or stop pulse may result in a loss of character synchronization, and a series of character errors may result from a single bit error. The character-error rate has been used to describe the accuracy of start-stop transmission.

Measurements of errors in data transmission were made on toll connections, excluding loops, in the United States and Canadian public telephone network in 1969 to 1970.[31,32] The effect of omitting loops was studied and found to be minor. The digital interfaces were provided by

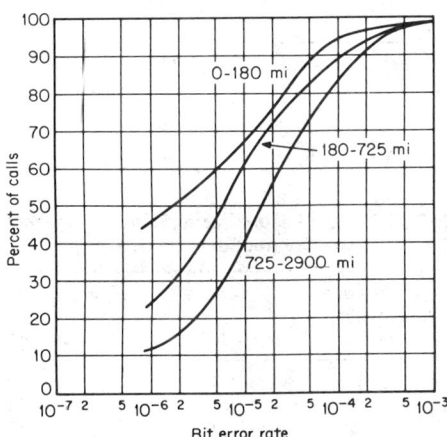

Fig. 22-19. Bit-error rate distributions by mileage strata at 3,600 b/s. *(Adapted from Ref. 31. Copyright 1971, American Telephone and Telegraph Co. Used by permission.)*

commercially available modems, data were transmitted at 150 Bd start-stop and 1,200 b/s synchronously, using FSK modulation; at 2,000 b/s using four-phase modulation, and at 3,600 and 4,800 b/s, using vestigial sideband modulation. Results are shown in Figs. 22-17 to 22-19 for calls in the 0- to 180-, 180- to 725-, and 725- to 2900-mi ranges, at 1,200, 2,000, and 3,600 b/s. At 150 Bd, it was observed that, in addition to character errors, some characters were lost; no bits were received for them. Table 22-11 gives the results of the measurements of 150-Bd transmission. The results varied from one call to another; at all speeds, most of the errors occurred in a small percentage of the calls.

At lower speeds, about 90% of the bursts were less than 25 b long, and about 90% contained fewer than five bit errors. The modems used at 3,600 and 4,800 b/s contained scramblers, which caused each line error to be translated into about three errors at the modem interface. At 3,600 b/s, about 90% of the bursts were less than 75 b long, and about 90% contained fewer than 13 bit errors, as measured at the interface.

The accuracy objectives of digital service networks that use digital transmission can be compared with the results of those using analog transmission.

The error performance objectives of one of the digital service networks being set up (using satellites) are to achieve an error rate on the radio channel of not more than one bit error per 10^6

TABLE 22-11 Character-Error and Lost-Character Statistics at 150 Bd

	All calls	0–180 mi	180–725 mi	725–2,900 mi
Character-error rate	1.46×10^{-4}	1.07×10^{-4}	1.42×10^{-4}	1.90×10^{-4}
Lost-character rate	6.81×10^{-4}	137×10^{-4}	1.98×10^{-4}	5.03×10^{-4}

Source: Adapted from Ref. 32. Copyright 1971, American Telephone and Telegraph Company. Used by permission.

b 95% of the time, not more than one bit error per 10^4 b 99.5% of the time, and not more than one bit error per 10^2 b 99.8% of the time. One bit error per 100 b gives acceptable speech transmission. For data, error control is to be offered that is expected to give an error rate of not more than one bit error per 10^7 b 99.5% of the time. Another of the networks offers 99.5% error-free seconds, end to end. In one of the packet networks, error checking is provided which is expected to result in a bit-error rate within the network of not more than 10^{-12}.

33. Digital-Network Synchronization. In a digital network, the bit timing on trunks is usually controlled by a frequency reference at the transmitting end. Thus, when a message is to be sent over a connection that includes more than one trunk, the bits in it are timed by a different reference on each trunk. To prevent loss or repetition of bits caused by differences in the rates of the frequency references, such differences must either be prevented or counteracted. In some networks, synchronous operation is used, in which a master source of timing provides signals which are distributed to all offices. In other networks such as international networks at the present time, stable and precise timing sources are used in each subnetwork, with provision for accepting slips between the digit streams from time to time.

A third method uses a line bit rate somewhat higher than the rate of information bits plus framing and stuffs extra bits as needed, as described in Par. **22-58**; the stuff bits are removed at the receiving end of the line. Within a synchronous system, there should be no slips, provided that sufficient buffer capacity has been provided to allow for the variation of propagation time from one office to another because of changes in temperature and the like. For international systems operating with separate reference sources in the subnetworks, CCITT Recommendation G.811 includes an objective of not more than one slip on any 64 kb/s channel per 70 days for each exchange. In one commercial digital network, the timing distribution network has the form of a tree, with spare paths that can be substituted if needed. At each major node of the tree is a timing supply that can maintain its frequency if all connection to the master source is lost, with a slip rate that should not exceed one per day.

GENERAL OBJECTIVES

Important service-network design parameters which are determined by other objectives are whether the network should be common user or private line, what the level of security measures should be, what reliability, including redundancy, should be designed into the equipment, and what amount of equipment should be furnished to give the desired level of service.

34. Community of Interest. If the service network is to furnish service to a limited community, a private network or a closed user group on a public network may be appropriate. If the users communicate with a wide variety of other stations, a shared network is likely to be better.

35. One- or Two-Way Communication, Number of Addressees, Message Assurance. If the communication is to be largely one-way from a central station to a number of outlying stations, particularly if the same message is to be sent to many stations, consideration should be given to a multipoint private-line network. Acknowledgment of receipt of messages to multiple addressees is difficult in any network, because of the number of acknowledgment messages.

36. Urgency of Messages. In such networks as those intended for reporting emergencies, a common-user network may make it too likely that all paths to the called party will be busy or that the called party itself will be busy.

37. Security. Business communication systems may be subject to attempts to steal or to alter the information they carry; the level of protection against such attempts properly depends on the sensitivity and value of the information. It may include physical protection of the communication paths, as well as the use of passwords, encryption (Par. **22-62**), and the like, normally implemented in the computer handling the information or in auxiliary terminal equipment. Systems used to report fire and burglar alarms must be secure against deliberate attempts to disable or deceive them. Such attempts may include tampering with the transmission facilities by cutting wires, short-circuiting them, or inserting false signals. The system should be so designed that any such attempt will be detected at the alarm central station. The alarm system should be able to operate in the presence of one such fault; a fault that affects one protected premises should not affect any other; and it is desirable that any fault be easily located from the alarm central station or from a maintenance location.

38. Blocking Probability. Telephone practice does not set an objective for end-to-end blocking. Alternate routing, the sharing of trunk groups by calls from various origins to various destinations, and the differing times of busy periods for various trunk groups make it very difficult to determine the end-to-end performance from measurements on the parts. Objectives for blocking on a trunk group used in automatic operation, without overflow, are given in CCITT Recommendation E.520 as 1% during the mean busy hour (average of the 30 busiest days of the year) and 7% during the busy hour of the 5 busiest days. Recommendations E.521 and E.522 give objectives for trunk groups carrying overflow traffic and for groups from which traffic overflows, respectively. Blocking may also occur in switching systems (Par. **22-74**).

In private networks, the blocking probability can be adjusted to the objectives of the user, with respect to the balance between service and cost. It is frequently allowed to be greater than on a public telephone network.

In analytical studies of queuing it is frequently assumed that calls are submitted to the system independently and at random. If a server is available, the call is served; if not, the call is said to be *blocked* and it either disappears from the system for the time being (blocked calls cleared) or is placed in a queue, to be served when a server becomes available (blocked calls delayed). The first situation occurs in a public telephone network when no trunk is available for a call, and either it is sent by an alternate route or the caller is given a reorder tone. The second situation occurs in a public telephone network when no circuit to receive the dialed digits is available and a caller must wait for dial tone. The probability of finding all servers busy in the ideal blocked-calls-cleared case is approximated by the Erlang B formula. The probability of delay in the blocked-calls-delayed case is given by the Erlang C formula. Curves for each are given in Ref. 33.

The probability distribution of delay for a call, given that it is delayed, depends on the queue discipline, although ideally the average delay and the average number in queue do not. However, long delays may cause callers to abandon calls. Both the probability distribution of call delay and the distribution of the queue lengths are important in system design, the former because of quality of service and the latter because of determining the amount of storage to provide for queues. Several queue disciplines have been studied, including order-of-arrival service, random-order service, and reverse order-of-arrival service. Long waits are minimized by order-of-arrival service. To use this discipline, the order of arrival must be recorded; with computing technology, it can be implemented by electronic memory for the messages and a list of pointers per output line to record the order in which the messages should be transmitted on that line.

39. Availability and Reliability. The availability (the fraction of the time that a user finds a system working) is a parameter that can be adjusted in a private network to achieve a balance with cost, given the user's objectives. In a public network, the availability depends on the carrier's judgment of the subscribers' desires. In one commercial digital service network, the objective is 99.96% availability.

Transmission Systems for Telecommunications

BY V. I. JOHANNES

TRANSMISSION SYSTEM PRINCIPLES

40. Transmission Systems. A telecommunication transmission link can be either a loop, which connects a user with a serving office, or a trunk, which connects two offices. Telephone transmission can be at voice frequency, or a number of voice-frequency channels can be multiplexed together using frequency-division techniques (analog carrier) or time-division techniques (digital carrier). The multiplex signal can then be transmitted over guided wave media, such as wire and optical fibers, or through free space, as in radio systems.

The advantage of carrier techniques over voice-frequency transmission is the greater economy of carrying many channels on a single (albeit more expensive) medium element. While this reduces the cost per channel mile, a cost is incurred in multiplexing the channels together for transmission, so that carrier systems become more attractive as the length and number of channels are increased. Historically, analog carrier was first applied on the longer systems, and it was not until the advent of digital carrier with its lower terminal costs that carrier techniques achieved widespread use on trunks under 20 or 30 mi long. The improvements in digital technology, the extensive use of digital carrier in the exchange plant, and the introduction of digital switching have widened the prove-in range for digital systems so that they are now used for voice-circuit trunks from the shortest to several hundred miles. However, as more voice-frequency circuits can be carried via analog carrier than digital carrier on a microwave radio-relay system or coaxial tube, analog carrier is more economical and still dominates in long-haul applications, both domestic and international. Intelsat, the international satellite carrier, is planning extensive use of digital transmission in the future in order to take advantage of TDMA (time-division multiple-access) techniques and long-haul digital fiber optic systems are planned.

MEDIA FOR TRANSMISSION

41. Paired Cable.[1,24] Paired telephone cables typically consist of from 6 to 1,900 pairs of no. 26 to no. 19 AWG copper wire. The cable is made by stranding together several stranded units (typically 25 pairs) to form the cable core. The core then is protected by applying a sheath, which until 1950 was usually lead. Today most sheaths consist of several layers of aluminum, steel, and polyethylene to provide electric shielding, water resistance, and mechanical protection, as shown in Fig. 22-20. Through the 1930s the insulation material was commonly paper strip, but today the insulation is usually paper pulp formed over each wire or polyethylene or expanded polyethylene extruded over each wire. As further protection against ingress of water into the sheathed cable, air spaces in some newer cables are filled with a greasy substance, and nonfilled cables are often maintained under pressure with dry air after installation. Cables may be run in underground ducts, buried (especially in rural areas), or suspended from poles.

The insertion loss of a typical cable pair in the voice band is shown in Fig. 22-21. The inclusion of series inductors ("loading") with coils of 88 mH at 6,000 ft spacing (H88 loading) can be seen to reduce the loss and improve the in-band frequency response markedly. Pairs can also be used (unloaded) into the megahertz range for carrier transmission, and characteristics of a 6,000-ft length are shown in Fig. 22-22. Above

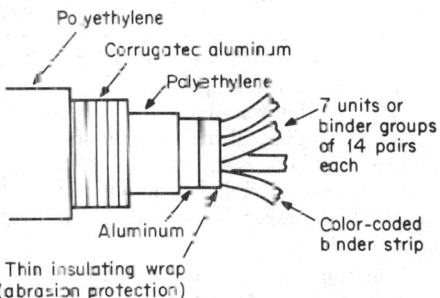

Po yethylene
Corrugated aluminum
Polyethylene
7 units or binder groups of 14 pairs each
Color-coded binder strip
Aluminum
Thin insulating wrap (abrasion protection)

Fig. 22-20. A 104-pair LOCAP cable with ALPETH (aluminum-po yethylene) sheath.

the audio range the loss in decibels is proportional to $\sqrt{f}$ for polyethylene-insulated cables.* (Paper and pulp cables have also an appreciable component of loss proportional to f, while in polyethylene cables the f component of loss is often negligible.) The characteristic impedance of pairs is on the order of 1,000 Ω in the audio band but varies substantially with frequency. At carrier frequencies the characteristic impedance is more constant, typically 100 to 150 Ω.

Crosstalk between pairs in a unit is minimized by giving each pair in the unit a different, carefully selected twist length, but at carrier frequencies crosstalk is usually the factor which limits the performance of transmission systems. Crosstalk from an interferer at the sending end of a cable (near-end crosstalk) (see Fig. 22-22) is more significant than that from an interferer at the far end (far-end crosstalk), as it does not suffer the attenuation along the length of cable

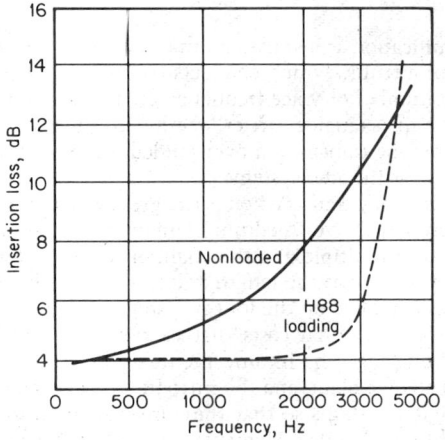

Fig. 22-21. Insertion loss characteristics of 12,000 ft of 26-gauge cable measured between impedances of 900 Ω and 2 mF. *(From Ref. 24. Copyright 1970, Bell Telephone Laboratories. Used by permission.)*

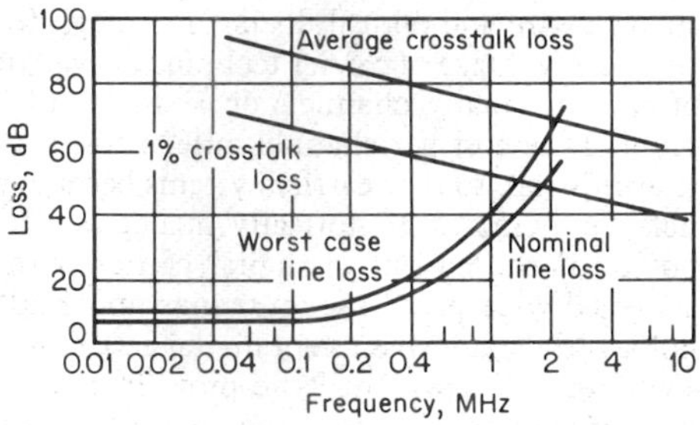

Fig. 22-22. Line loss and near-end crosstalk loss of a 6,000-ft length of 22-gauge exchange-area cable. *(From Ref. 24. Copyright 1970, Bell Telephone Laboratories. Used by permission.)*

suffered by the signal. As a result, some cables are available with an internal shield (*screen*) to separate the directions of transmission, and in some applications a separate cable is used in each direction (two-cable operation).

42. Coaxial Cable.[1,24] A coaxial cable for land telecommunications is shown in Fig. 22-23. Such cables consist of up to 22 coaxial tubes plus some wire pairs (used for maintenance) stranded into a cable and then sheathed (often with lead) and operated under dry air pressure. Each tube consists of an inner conductor (approximately no. 10 AWG copper), 0.375-in polyethylene spacers at about 1-in intervals, which hold an outer copper tube (made from a copper strip with a longitudinal serrated seam), and a protective steel-strip outer wrap. (Similar ¼-in tubes have also been used.)

The loss of an air dielectric coaxial tube in decibels is proportional to $\sqrt{f}$ over most of its useful frequency range. For 0.375-in tubes, the loss at 1 MHz is about 3.84 dB/mi. Periodic irregularities in the cable caused by stranding can cause ripples in the frequency response which can be significant above 150 MHz. The loss also increases with temperature (±0.38 dB/mi at 20 MHz and ±0.67 dB/mi at 60 MHz with $\pm20°$F change from 55°F, which is about the maximum temperature range for a cable buried at 4 ft).

Coaxial cable for submarine transoceanic telephone use consists of a single tube with solid dielectric and a steel strength member inside the inner conductor of copper (Fig. 22-24). Where the cable is laid in shallow water (up to 500 or 1,000 fathoms), protection from anchors and trawls may be provided by an outer layer of steel-armor wires or by burying. Earlier cables were similar but smaller and accordingly had higher loss.

---

*The propagation constant of a transmission line is $\gamma = \alpha + j\beta = \sqrt{(R + j\omega L)(G + j\omega C)}$, where R and L are the series resistance and inductance and G and C are the shunt conductance and capacitance between conductors per unit length. For air- and polyethylene-insulated cables (pair or coaxial) $G \approx 0$, and L and C are constant.

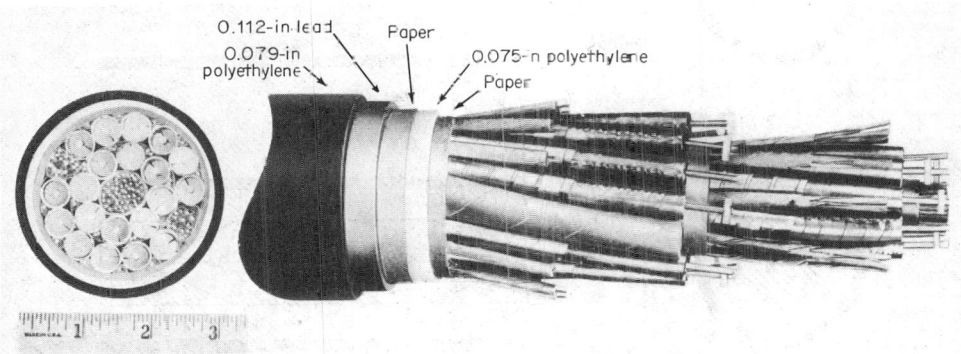

Fig. 22-23. Coaxial cable with twisted pairs. *(From G. H. Duvall. and L. M. Rackson, Bell Syst.Tech. J., 1969, Vol. 48, p. 1968. Copyright 1969, American Telephone and Telegraph Co. Used by permission.)*

The loss of the cable of Fig. 22-24 is $K\sqrt{f}$, where K varies (due to the dielectric) from 1.38 dB/nmi $\cdot$ MHz$^{1/2}$ at low frequencies to 1.46 at 30 MHz. The loss increases about 0.3 to 0.5% per thousand fathoms of submersion, due to compression of the cable.

43. Optical-Fiber Cables.[34] Optical fibers for telecommunications are typically about 5 mils thick and consist of a glass core, a glass cladding of lower index of refraction, and a protective coating. Figure 22-25 shows three types of fibers.

Loss in fibers is a result of both scattering in the glass one component of which decreases with increasing wavelength, and absorption due to impurities. Quite low losses can be achieved in practical cables, as shown in Fig. 22-26.

The signal bandwidth a fiber can support is primarily limited by the dispersion due to the difference of group velocity between the modes supported by the fiber (modal dispersion) and the difference in velocity of the various wavelengths of light being transmitted (chromatic dispersion). In the single-mode fiber (see Fig. 22-25a) the core is sufficiently narrow to permit only one mode to propagate, but the small core makes splicing and coupling difficult compared with the fiber of Fig. 22-25b, which can support many modes. Modal dispersion in multimode fibers can be reduced by using a graded index fiber, in which the index of refraction in the core varies with distance from the center (see Fig. 22-25c).

Thus the rays which travel farthest have a higher average speed, giving the various modes about the same group velocity. Chromatic dispersion can be reduced by using a light source with high spectral purity, such as a laser. The interplay of these factors, based on typical 0.8-μm light sources and pulse transmission with 40 dB loss, is shown in Fig. 22-27.

Fiber cables must protect the fibers both from externally induced mechanical damage and from excessive tensile strain (especially during installation). Unlike that for copper, the tensile strain on fibers should usually be significantly less than 1%. A compliant coating on each fiber protects the fiber from short-period bending which could cause increased loss (microbending). Several fibers can be

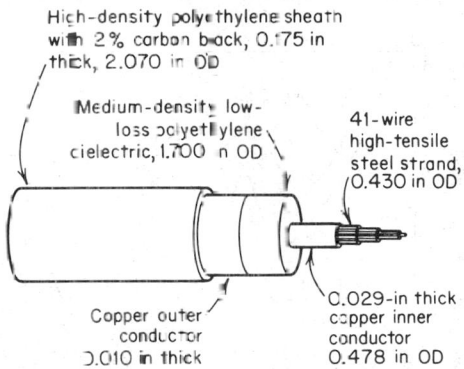

Hich-density polyethylene sheath with 2% carbon black, 0.175 in thick, 2.070 in OD

Medium-density low-loss polyethylene dielectric, 1.700 in OD

41-wire high-tensile steel strand, 0.430 in OD

Copper outer conductor 0.010 in thick

0.029-in thick copper inner conductor 0.478 in OD

Fig. 22-24. Armorless SG ocean cable. *(From S. G. Morse et al., Bell Syst. Tech. J., 1978, Vol. 57, p. 2437. Copyright 1978, American Telephone and Telegraph Co Used by permission.)*

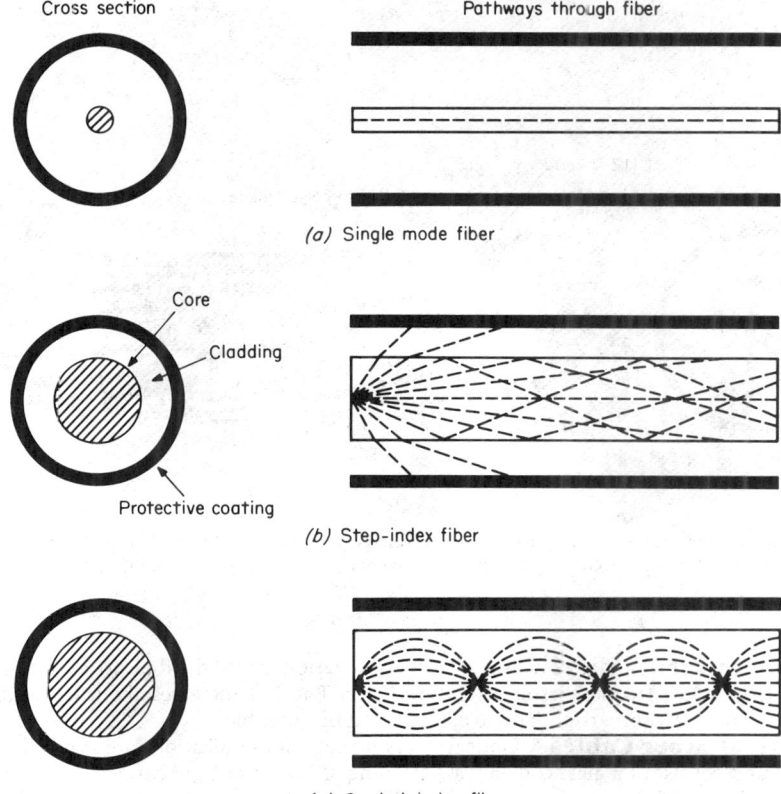

Fig. 22-25. Fiber profiles and modes. (From Ref. 34. Copyright © 1979, Bell Telephone Laboratories, Incorporated. Reprinted with permission of Bell Laboratories RECORD.)

embedded in a single ribbon of compliant material. A cable also requires sheathing and strength members. Cables with stranded fibers which resemble sheathed twisted-pair cables, with central or outer strength members, and cables which have wire-reinforced plastic sheaths with the optical-fiber ribbons "floating" inside have been used.

Splicing of optical fibers in the field has been done by fusing the fibers together and by various mechanical arrangements which hold the ends of fibers together closely and in good alignment. Both fused and mechanical splices can have losses in the 0.1- to 0.2-dB range. Mechanical arrangements have been made for connecting individual fibers,[35] a technique useful for patching (shown in Fig. 22-28) and for handling groups of fibers for cable splicing.

While fibers have been used to transmit analog signals such as TV (using a form of FM, for example), interest for telecommunications applications has been primarily in digital transmission.

44. Radio-Propagation Effects. Radio propagation is discussed in detail in Sec. **18**. In the microwave range, where most present telecommunications transmission systems function, propagation is primarily line-of-sight, and at typical relay tower spacings of 15 to 30 mi the free-space loss will be about 140 dB in the 4- and 6-GHz bands. With antenna gains of about 40 dB, the normal loss of a section from antenna input to output will be some 60 dB. Loss well above

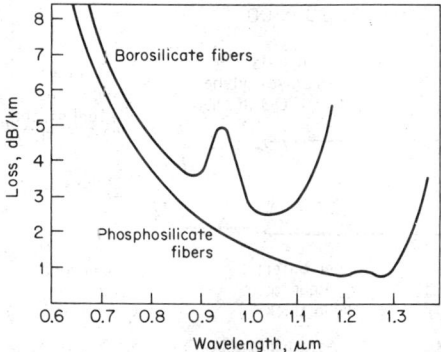

Fig. 22-26. Loss of optical fibers. (From Ref. 34. Copyright © 1979, Bell Telephone Laboratories, Incorporated. Reprinted with permission of Bell Laboratories RECORD.)

the normal can occur, as discussed in Sec. **18**. At microwave frequencies greater than 10 GHz, the additional attenuation resulting from heavy rainfall is often a limiting factor. Path lengths 10 to 15 mi in the 11-GHz band and 1 to 3 mi in the 18-GHz band are typical. The deployment of additional microwave systems is often limited by the need to avoid interference with existing systems which use the same frequency band.

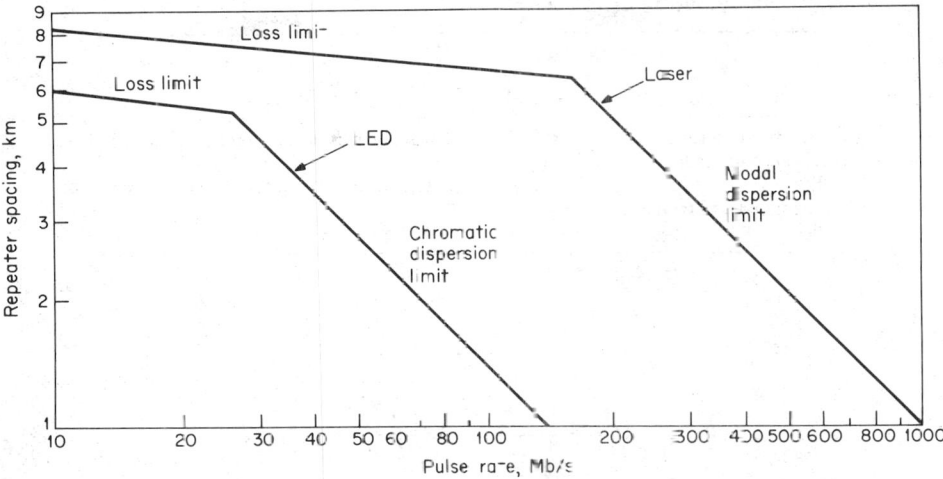

Fig. 22-27. Typical repeater spacing for 0.8-μm graded-index fiber (40 dB loss). (*From Ref. 34. Copyright © 1979, Bell Telephone Laboratories Incorporated. Reprinted with permission of Bell Laboratories RECORD.*)

VOICE-BAND TRANSMISSION[1,24]

45. Application. Voice-band (~200 to ~3,500 Hz) transmission over a pair of wires in a cable is widely used to connect telephone instruments to the local telephone exchange (loops) and for trunks used to interconnect nearby exchanges.

For lengths of a few miles or more, the cable is often loaded (Par. **22-41**) to improve voice-band performance. The performance of trunks is generally controlled more tightly than that of loops.

46. Loops. In a loop, wire gauge and loading are normally selected to keep the loop resistance below 1,300 Ω, which permits proper operation of the supervisory and transmission circuits. Loops up to about 3 mi can be 26-gauge pairs. Loops up to 5 mi can be 24 gauge, and longer loops are often of mixed gauges with the thinner wire closer to the office. An alternative approach referred to as *long-route design* (*unigauge*) permits extensive or exclusive use of 26-gauge wire and up to 2,800 Ω loop resistance. In the range from 1,300 to 2,800 Ω the additional loss is compensated for by audio gain and equalization, as well as increased signaling power, at the central office. These functions are provided by a range extender, or range extender with gain, which can in some switching arrangements be inserted by the switch, so that it is not necessary to have one for every loop served.[36]

47. Trunks. Voice-band trunk design is usually determined by voice-transmission performance rather than supervisory signaling. Typical values are given in Table 22-12.

Repeaters for two-wire voice-band circuits may be varieties of series and/or shunt negative-impedance converters, shown in composite form in Fig. 22-29. This approach requires use of line-impedance build-out networks (LBOs) to establish a fixed controlled-impedance interface to the negative-impedance elements. Alternatively, amplifiers may be used, with the direction of transmission split by hybrid coils, as shown in Fig. 22-30a. Both

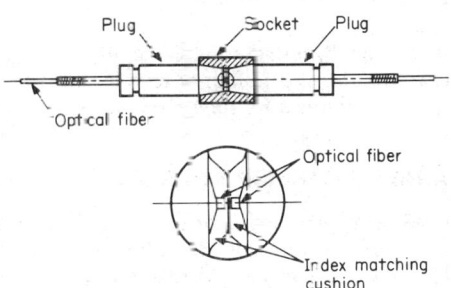

Fig. 22-28. Cross section of the single-fiber connector detail showing elastomeric index matching cushions. (*From Ref. 35. Copyright 1978, American Telephone and Telegraph Co. Used by permission.*)

methods require careful impedance matching including selection and adjustment of hybrid balance networks and/or LBOs at each repeater to avoid instability (singing). A maximum of about 12 dB gain can be achieved in practice, typically limited by cable crosstalk.

Four-wire repeatered operation is shown in Fig. 22-30b. Since the only feedback path is from hybrid to hybrid, overall balance problems are generally less severe, permitting lower loss and

TABLE 22-12   Voice-Band Trunks

	Loss, dB		
Type*	Avg	Max	Typical realization
Local office to local office	5	7	2-wire, usually loaded, often repeated
Local office to tandem office	3	4	2-wire, usually loaded, often repeated
Tandem office to tandem office	0.5	1.5	4-wire, repeated

*These terms are explained in Par. **22-5**.

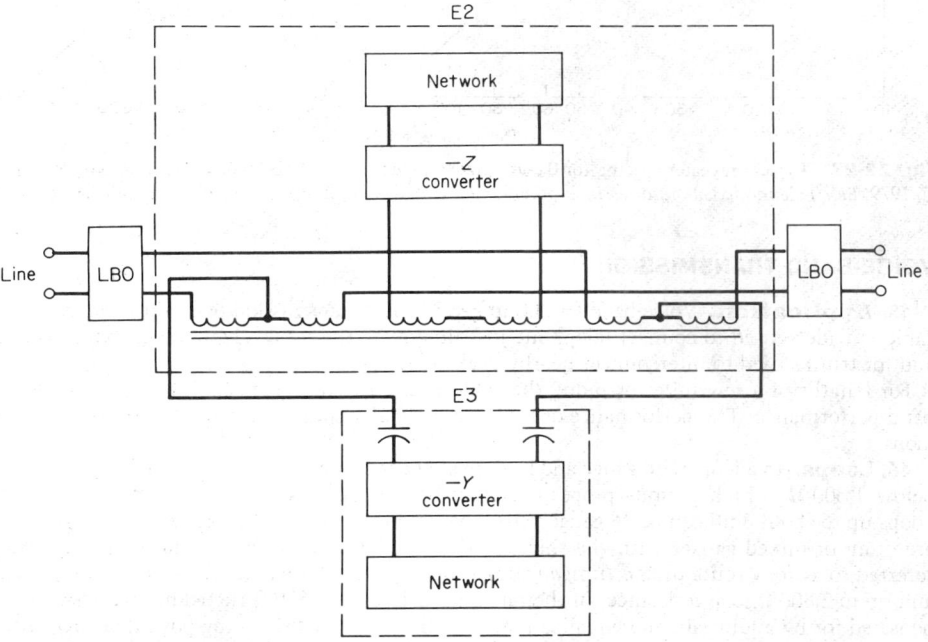

Fig. 22-29. Block schematic of negative-impedance repeater. *(From Ref. 24. Copyright 1970, Bell Telephone Laboratories. Used by permission.)*

better-equalized facilities. Generally, the overall trunk loss is controlled by echo considerations (Par. **22-68**). While four-wire voice-band trunks of any length are possible, carrier systems are more economical for longer trunks.

ANALOG CARRIER SYSTEMS[24]

48. Frequency-Division Multiplex.[1,24] Multiples of voice channels are combined, by amplitude modulation plus filtering, into signals for transmission over carrier systems. The hierarchy for analog transmission in North America is shown in Table 22-13. Several variants to this hierarchy exist. In countries outside of North America, a 300-channel master group is used. In transoceanic cable systems, a 200- to 3,050-Hz voice channel with 3-kHz carrier spacing is sometimes used. Within the United States, the N-carrier arrangements shown in the table are used in short-haul applications where multiplex costs are dominant. Furthermore, the N terminals include companding (*compression-expansion*). In this process, incoming signals below a certain level (typically 5 dBm0) are increased in amplitude before transmission, i.e., the range is compressed, and restored to their original amplitude, i.e. the range is expanded, at the receiving terminal, thus reducing the noise amplitude as well. In the absence of speech, line noise is reduced by the maximum expandor loss. (Similar techniques are used in digital channel banks.)

The master group is the common unit in long-haul transmission. In addition to the voice channels, single-frequency tones (pilots) are provided (for monitoring and carrier-system adjustment) in the group, supergroup, and master-group spectra. Combinations of master groups, spe-

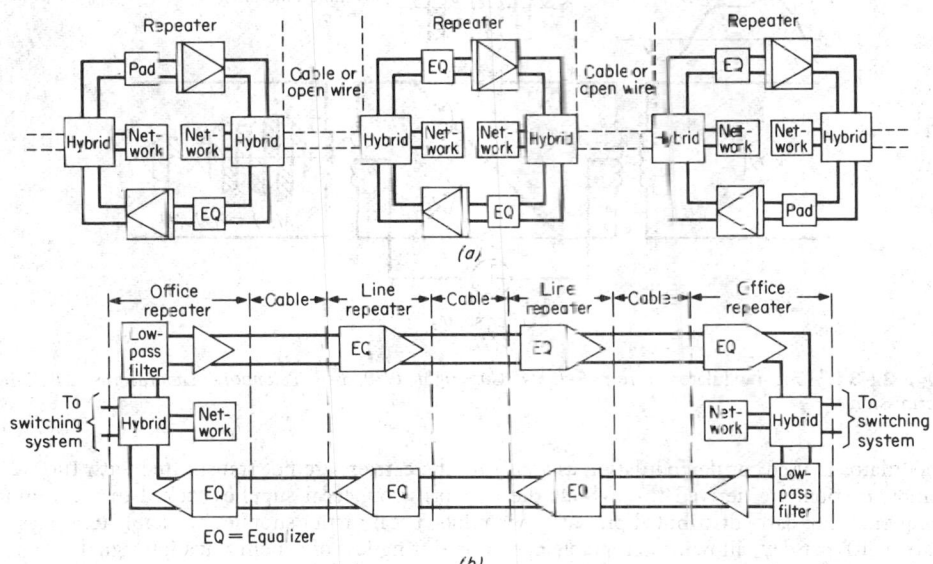

Fig. 22-30. Two- and four-wire methods of operation. (*From Ref. 24. Copyright 1970. Bell Telephone Laboratories. Used by permission.*)

TABLE 22-13 Analog Hierarchy (North America)

Signal unit	Content	Allocated frequency band, kHz	Formed in
Voice channel	Voice-band signal	0–4	Telephone or voice-band data set
Group	12 voice channels multiplexed or 50 kb/s data	60–108	Channel bank or 303 type data set
Super group	5 groups (60 voice channels)	312–552	Group bank
L600 master group	10 supergroups (600 voice channels)	60–2,788	Supergroup bank
U600 master group	10 supergroups (600 voice channels)	564–3,084	Supergroup bank
N1, N2 group*†	12 voice channels	36–140 164–268	N terminals; low-frequency group and high-frequency group in opposite direction on paired cable
ON, N3, N4 terminal group*‡	24 voice channels	36–132 172–268	

*Used directly for short-haul transmission on paired cable. Cannot be multiplexed directly into a supergroup.
†Double-sideband-transmitted carrier.
‡Single-sideband-transmitted carrier.

cialized to the carrier system involved, are also used. They are derived either from specialized multiplexes akin to the banks of Table 22-13 or by combining a number of master-group translators, each of which modulates a single master group to and from the desired frequency range. Modulation in the long-haul systems is single-sideband suppressed-carrier performed in balanced modulators such as Fig. 22-31, followed by a filter to select the desired sideband(s). (The balanced

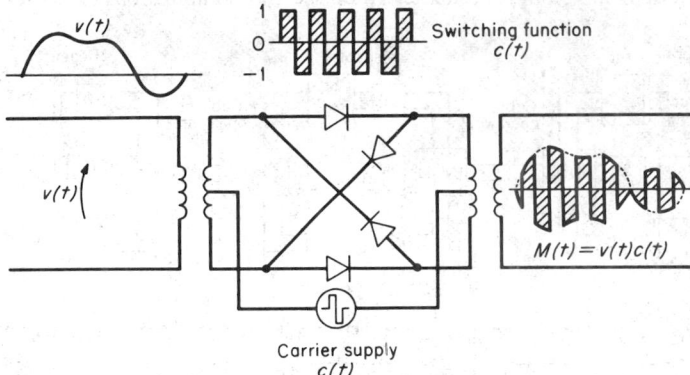

Fig. 22-31. Ring modulator. *(From Ref. 24. Copyright 1970, Bell Telephone Laboratories. Used by permission.)*

modulator is used for demodulation as well.) As the carriers are not transmitted with the sidebands, carriers are derived throughout the network from local supplies locked to a common frequency standard, distributed primarily via pilots. Local carrier supplies are stable to at least 1 part in 10^8 per day, allowing acceptable operation during loss of synchronization signal.

49. Analog Carrier on Cable, Noise Measures.[1,24] In an analog carrier system, the frequency-division-multiplexed signal is amplified at regular intervals along the cable by *repeaters*. The repeaters include amplification closely approximating the loss in the preceding section and may contain equalization to correct the variation of loss with frequency in that (and possibly previous) sections. Design of such a system must account for and balance the effects of:

Output power available from the repeater
Section loss (which must be overcome by repeater gain)
Cable frequency response (which must be equalized)
Thermal noise (in repeater input stage primarily)
Crosstalk from other systems in the cable
Impulse noise from switching transients and other interferers
Intermodulation distortion in the repeater (output stage primarily)

The effects of noise, intermodulation, and crosstalk are usually expressed in terms of the total noise power in a 4-kHz bandwidth, with a frequency weighting which approximates the subjective effect of the noise as heard through the telephone instrument. The C weighting function used in the United States and the psophometric function used by the CCITT are shown in Fig. 22-32. To allow meaningful direct addition of the weighted noise powers from various points in a system which do not have the same transmission level (consider noise injected at the input and output of a repeater, for example), the measure used is the noise power multiplied by the system gain to a unity-gain reference point (zero transmission level point, 0 TLP). Transmission levels throughout the analog system are also given as the ratio (expressed in decibels) of the signal power to the power of the same signal at 0 TLP. The 0 arises from the 0-dB level of the reference point.

In Europe, the customary noise measure is picowatts, psophometrically weighted, referred to 0 TLP, and abbreviated pW0p or pWp0. In North America, where C message weighing is used, the result is expressed in decibels referred to a refer-

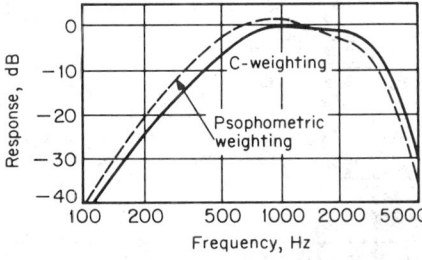

Fig. 22-32. Comparison of noise-weighting characteristics. *(From Ref. 40. Copyright 1977, American Telephone and Telegraph Co. Used by permission.)*

ence noise (rn) of 1 pW and referred to 0 TLP, abbreviated dBrncC (dBrnc0 $\approx$ 10 log pWp0).

When noise and/or intermodulation products from different sources in an analog system are combined (as from two repeaters), the resultant noise power is the sum of the two constituent noise powers.* (So if the two repeaters contribute equal noise, the sum noise, as measured in dBrnc0, is 3 dB more than the noise of either repeater.)

A comparison of objectives for short- and long-haul analog cable systems is given in Table 22-14. Typical loss variation over the 300- to 3,000-Hz band is about ±1 dB for carrier-derived trunks and ±3 dB for voice-band trunks.

TABLE 22-14   Message-Circuit Noise Objectives for Short- and Long-Haul Cable Systems

System type	Usual medium	Typical objective* message-circuit noise, dBrnc0	Typical factors determining limit on	
			Repeater gain	Repeater output power
Short haul (to 250 mi)	Pairs	28 mean up to 60 mi	Crosstalk	Repeater overload
Long haul (above 250 mi)	Coax	40 mean at 4,000 mi	Thermal noise	Intermodulation

*Objective changes by 3 dB per double distance, 60 to 4,000 mi.

50. Analog Carrier on Paired Cable. Analog carrier is used on paired cable for systems up to 200 mi, although the usual system is much shorter. (The higher capacity of a coaxial tube makes it the economic choice for longer cable systems.) In such short-haul systems the cost of the terminals is significant (if not dominant), and it is often worth sacrificing line capacity to reduce terminal cost. Several generations of short-haul terminals have been designed, the later terminals being smaller, costing less, and dissipating less power. The N type systems are designed for use with about 5-mi repeater spacing on multipair cable (open-wire carrier systems are obsolescent) and use a separate pair for each direction of transmission. Frequency-separation filters allowing transmission of both directions on the same pair are used to permit additional systems to be added to cables filled to capacity with four-wire systems. To reduce the effects of crosstalk, the two directions are also carried in different frequency bands (see Table 22-13), and the high and low groups are interchanged at each repeater (frequency frogging). This interchange also provides first-order equalization for the cable loss characteristic.

51. Analog Carrier on Coaxial Cable. Although some short-haul applications exist, analog coaxial systems are generally designed for long-haul use. Therefore, thousands of repeater sections may be in tandem, and provision must be made in system design to limit the misalignment (deviation from nominal or ideal signal levels) both with distance along the system and with frequency across the transmitted band. As an example, note that if the cable loss of a 4,000-mi system varied only 0.1 dB/mi with temperature at the highest transmitted frequency, an unequalized system would have a top frequency misalignment of 400 dB and the signal would be completely lost, either in noise or in overload, depending on whether the deviation is loss or gain. To limit misalignment, a combination of fixed and adjustable equalizers is typically placed at periodic intervals along the system (including the terminals). Adjustable equalizers may be manually adjustable to compensate for misalignment fixed in time but unique to a particular link or automatically regulated to compensate for time-varying misalignment. Regulation may be open-loop, based on sensing the temperature variations which are the major causes of misalignment, or closed-loop, based on responding to variations in signal amplitude of one or more pilot tones. The major parameters of some current United States and European systems are given in Table 22-15.

Modern systems rely on ultralinear transistor feedback amplifiers and extremely precise adjustment to achieve the desired performance.

As an example of a modern terrestrial system, Fig. 22-33 shows the major elements of the L5 system.[37] The main station includes an automatic protection switch, which monitors transmit and receive pilots as well as received signal power. If either is out of limits, communications between the protection switches at opposite ends of the switching span (via an FSK signal at 68.76 and 68.78 MHz) results in transfer of traffic to a protection line, provided on a 1:10 ratio.

*In some cases, intermodulation products add coherently, and the resultant is the sum of the voltages rather than the powers.

The station also applies 910 mA at up to $\pm 1,150$ V to the center conductor of each coaxial for powering the line repeaters. The basic repeater, located at 1 mi nominal spacing, contains a fixed-gain amplifier (32 dB at 66 MHz). Every 7 mi (maximum) a regulating repeater is used. This includes amplification as in the basic repeater, plus pre- and postregulation. The preregulator adjusts gain up to ± 2.4 dB from nominal in accord with temperature variations sensed in the earth near the manhole. This compensates for about half of the cable-loss variations with temperature in a 7-mi section. The postregulator senses the level of a temperature pilot at 42.88 MHz and compensates for the remaining variation of cable loss, as well as for any other loss deviation present at this frequency. Gain adjustments have $\sqrt{f}$ shape to equalize the cable loss, and are implemented using Bode equalizers with a thermistor as the variable element. Additional equalization is provided as indicated in Table 22-16.

TABLE 22-15   Parameters of Coaxial Systems

System	Typical multiplexing arrangement	Voice channels	Frequency band, kHz	Noise objective message circuit	Nominal repeater spacing
L4	6 master groups	3,600	564–17,548	40 dBrnc0, 4,000 mi	2 mi
L5	3 jumbo groups of 3 master groups each	10,800	3,124–60,556 (gain within ± 4 dB over message band)	40 dBrnc0, 4,000 mi	1 mi
L5E	3 multi-master groups, 2 consisting of 7 and 1 of 8 master groups	13,200	3,224–64,844 (gain within ± 4 dB over message band)	40 dBrnc0, 4,000 mi	1 mi
12 MHz CCITT G.332	3 super-master groups of 3 300-channel master groups each	2,700	300–12,435	3 pWp0/ km	4.5 km
60 MHz CCITT G.333	12 super-master groups as above	10,800	4,287–61,160	3 pWp0/ km	1.5 km

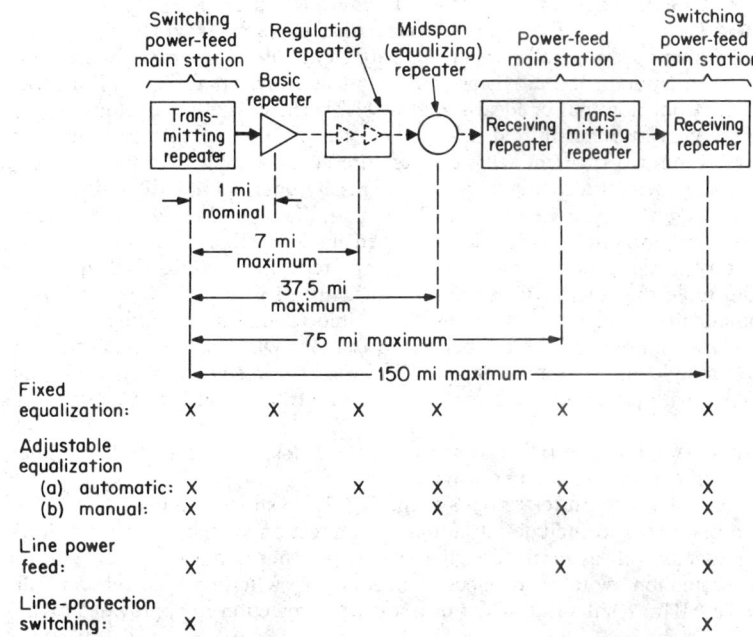

Fig. 22-33. Main features of an L5 switching span. *(From Ref. 37. Copyright 1974, American Telephone and Telegraph Co. Used by permission.)*

TABLE 22-16 Equalization Plan for L5/L5E

Equalization type	Location	Equalization range, dB	Comment
Fixed $\sqrt{f}$	All repeaters	26 at 42.88 MHz	Approximates nominal cable to ±0.1 dB
Fixed (deviation)	Midspan and receiving repeaters	About ±5	Matches difference between nominal cable and fixed $\sqrt{f}$ equalizer design
Line build-out ($\sqrt{f}$) selected at installation	All repeaters as required	Equivalent to cable in steps of 0.1 mi from 0.1–0.5 mi	Builds out loss to be equivalent to 1 mi of cable where manholes are more closely spaced
Manually adjusted, E1	Transmitting and receiving main stations, equalizer manhole	±4.5	10 adjustable 4.5 dB bumps plus 1 temperature-pilot bump
E2	Transmitting and receiving main stations	±3.5	18 adjustable 3.5-dB bumps; with E1, switching section gain adjusted flat within ±0.4 dB 1.6 to 66 MHz
Adjustable $\sqrt{f}$	Regulating and main station repeater	±3.5	Corrects for cable temperature variations
Automatic (fix). L5E uses 2.976 and 66.048-MHz pilots; L5 uses 20.992- and 42.880-MHz pilots also	Receiving main station	About ±5	Holds switching-section gain deviations within ±0.4 dB with variations in temperature, aging, etc.

52. Submarine Cable Systems. A tabulation of the characteristics of the major trans-oceanic systems is given in Table 22-17. Terminals for these systems use 3-kHz channel spacing and special multiplexing arrangements designed to conserve bandwidth.

Long submarine cable systems usually carry both directions on a single coaxial tube, using different frequency bands for the two directions; they are then equivalent four-wire. This can

TABLE 22-17   Transatlantic Cables

System designation	Service date	VF circuits*	Cable diameter (dielectric), in	Nominal top frequency, MHz	Top-frequency repeater gain, dB	Active element	Line current, mA	Repeater, V	Repeater spacing, nmi
TAT-1 (SB)	1956†	50	0.62	0.164	60	Vacuum tubes	225	124	37
TAT-3 (SD)	1963	140	1	1.1	50	Vacuum tubes	389	60	20
TAT-5 (SF)	1970	820	1.5	6	40	Germanium mesa transistor	136	13.1	10
TAT-6 (SG)[108]	1976	4,200	1.7	29.5	41	Silicon planar transistor	657	12	5

*Without use of TASI (time-assignment speech interpolation).
†Retired 1979.

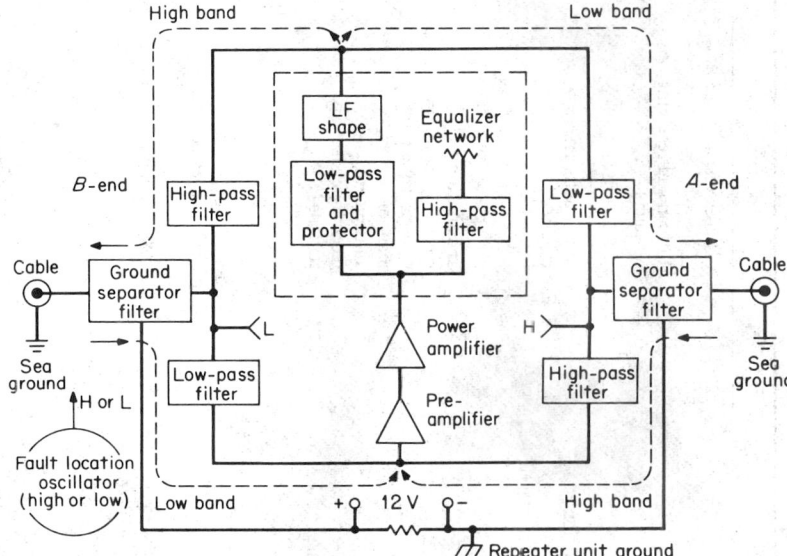

Fig. 22-34. SG repeater block diagram. *(From Ref. 1. Copyright 1977, Bell Telephone Laboratories. Used by permission.)*

be accomplished with either one or two amplifiers and appropriate high- and low-pass filters. A one-amplifier version is shown in Fig. 22-34. A complete repeater is shown in Fig. 22-35. The inaccessibility of the repeaters for repair dictates high reliability (21 failures in 10^9 h for the repeater shown) and generally longer repeater spacing than is common with terrestrial systems. Equalization for the nominal variation of cable loss with frequency is provided in each repeater, and a manually set passive equalizer is usually located along the system at intervals (every 98 nmi for TAT-6).* This equalizer is adjusted (or in some cases designed and built) on shipboard just before being laid. In addition, some systems have equalizers in the cable which can be

*TAT-6 refers to trans-Atlantic telephone cable no. 6.

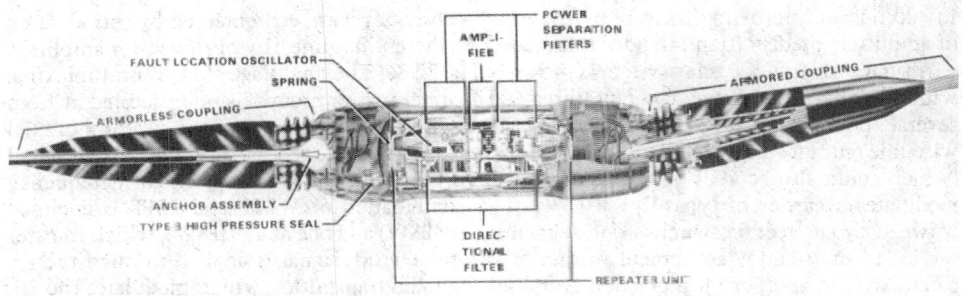

Fig. 22-35. SG repeater. (*From C. D Anderson et al., Bell Syst. Tech. J., 1978, Vol. 57, p. 2359. Copyright 1978, American Telephone and Telegraph Co. Used by permission.*)

adjusted from shore to compensate for aging of cable, etc. TAT-6 includes four such shore-controlled equalizers, which use mechanical switches and relays to switch equalizing networks, with a total system adjustment capability of ±75 dB at the top frequency. Additional manually adjustable equalizers are provided at the terminals on shore.

53. Microwave Radio.[1,24] Analog microwave radio relay systems carry the bulk of long-haul telecommunications in the United States and many other countries. The major common-carrier bands and their application are shown in Table 22-18. The history of microwave relay has been one of increasing channel capacity and lowering cost, mostly by use of improved solid-state devices and circuits, as illustrated in Table 22-19. Until the introduction of single-sideband amplitude modulation in this application, the typical modulation technique had been low-index FM with total bandwidth of 2 or more times 4 kHz for each voice channel. With the single-sideband system, however, the bandwidth limit of the radio channel has been approached. The single-sideband system requires a very high degree of linearity in its amplifying circuits, which,

TABLE 22-18 Selected United States Common-Carrier Microwave Frequency Allocations

Band, GHz	Allotted frequencies, MHz	Bandwidth, MHz	Application
2	2,110–2,130	20	Limited, due to small bandwidth
	2,160–2,180	20	
4	3,700–4,200	500	Major long-haul microwave relay band
6	5,925–6,425	500	Long and short haul
11	10,700–11,700	1,000	Short haul
18	17,700–19,700	2,000	Short haul, limited use
30	27,500–29,500	2,000	None to date (1981)

TABLE 22-19 Analog Microwave Radio Relay Systems: Technological Progress

Year	4 GHz, circuits per 20-MHz channel	6 GHz, circuits per 30-MHz channel
1950	480	
1953	600	
1960	*	1,800
1966	900	
1968	1,200	
1970		1,800
1972	1,500	
1979	1,800	2,400
1980		6,000

*In 1960 the number of usable 20-MHz channels was doubled by introducing the horn reflector antenna, which allowed the use of vertical and horizontal polarization.

in addition to improving linearity by the usual techniques, has been obtained by introduction of amplitude predistortion intended to cancel the inherent nonlinearity of the power amplifier.

A typical microwave relay system is shown in Fig. 22-36. The final stage of FDM multiplexing, which might, for example, combine three 600-channel master groups, is often located at from several thousand feet to a few miles from the antenna site, necessitating the use of a coaxial wire-line entrance line (WLEL) which may include intermediate repeaters. The baseband signal (which could also be a TV signal) is applied to an FM terminal (FMT) on which it frequency modulates a carrier of typically 70 MHz. The combination of WLEL and FMT is enclosed between two protection switches, one at baseband (BSBSW) and one at i.f. (IFSW), which transfer service to the standby equipment in case of failure. The i.f. signal is applied through the i.f. patch bay and another i.f. protection switch to the radio transmitter, which modulates the 20-MHz-wide channel to its assigned slot in the 4-GHz band. The i.f. patch bay serves for access to the regular and standby channels for emergency use, e.g., to restore service normally carried over another route. The second i.f. switch, in combination with an i.f. switch at the receiving main station, automatically protects against equipment failures and frequency-selective fades on the radio route. Communication between the switches takes place over the auxiliary channel (AUXCHAN), which also collects alarms for the entire system. This auxiliary channel is often realized over a narrow-band radio system. The repeater station shown here demodulates only to the i.f. level.

54. Communications Satellite Systems.[38] Communications satellites (except for the Russian Molniya Constellation) are located in geostationary orbit 22,300 mi above the equator. They receive signals from an earth station and include transponders, which amplify and translate the signal in frequency and retransmit it to the receiving earth station, thus making effective use of the line-of-sight microwave bands without requiring erection of relay towers. The transponders are powered from solar cells, with batteries for periods of eclipse. *Spin-stabilized* satellites are roughly cylindrical and spin at about 60 r/min, except for a "despun" portion, including the antennas, which is pointed at the earth. *Three-axis-stabilized* satellites have internal high-speed rotating wheels for stability and solar cells on appendages which unfold after they are in orbit. Adjustments in the position and orientation of a satellite in orbit are accomplished under control of the telemetry tracking and control (TTC) station on the earth, and the exhaustion of fuel for this purpose is the normal cause of end of life of the satellite. The characteristics of the Intelsat satellites, used for international communication, are shown in Table 22-20.

Most civilian communication satellites have used the common-carrier bands of 5,925 to 6,425 MHz in the uplinks, and the 3,700 to 4,200-MHz band in the downlinks. Future plans involve the 12- and 14-GHz bands for down- and uplinks respectively. Since the 12- and 14-GHz bands are not so widely used in terrestrial microwave relay, interference problems should be less although rain attenuation is much higher. All frequencies are subject to sun-transit outage when the satellite is directly between the sun and the receiving earth station, so that the receiving antenna is pointing directly at the noisy sun. This occurs for periods of up to ½ h/d for several days around the equinoxes. The effect can be avoided by switching to another distant satellite during the sun-transit period.

Some satellites use each frequency band twice, once in each polarization. Earth antenna directivity permits reuse of the same frequencies by different satellites as long as satellites are not too close in orbit (4° is the present norm for domestic United States satellites in the 4- to 6-GHz band). Further frequency reuse is possible in a single satellite by using more directive satellite antennas which direct separate beams to different earth areas. Most present satellite antennas have beam widths covering upward of 1,000 mi on earth. The ease of launching larger antennas promised by the space shuttle and the use of the 12- to 14-GHz bands will encourage more directive antennas. A simplified transponder block diagram is given in Fig. 22-37.

Transponder Utilization and Multiple Access.   A transponder can be used for a single signal (single carrier operation) which may be either frequency- or time-division-multiplexed. Such signals have included TV, two 600-channel analog master groups multiplexed together and used to frequency modulate a carrier, and a digital signal with rate up to 64 Mb/s used to modulate a carrier using quaternary phase-shift keying (QPSK). Single-carrier operation can be either point-to-point (as for normal telecommunication) or broadcast (as for distributing TV programs). Transponders can also be used in either of two multiple-access modes in which the same transponder carries (simultaneously) signals from several locations.

In FDMA (frequency-division multiple access) the frequency band of each transponder is subdivided and portions assigned to different earth stations. Each station can then transmit continuously in its assigned frequency band without interfering with the other signals. All earth sta-

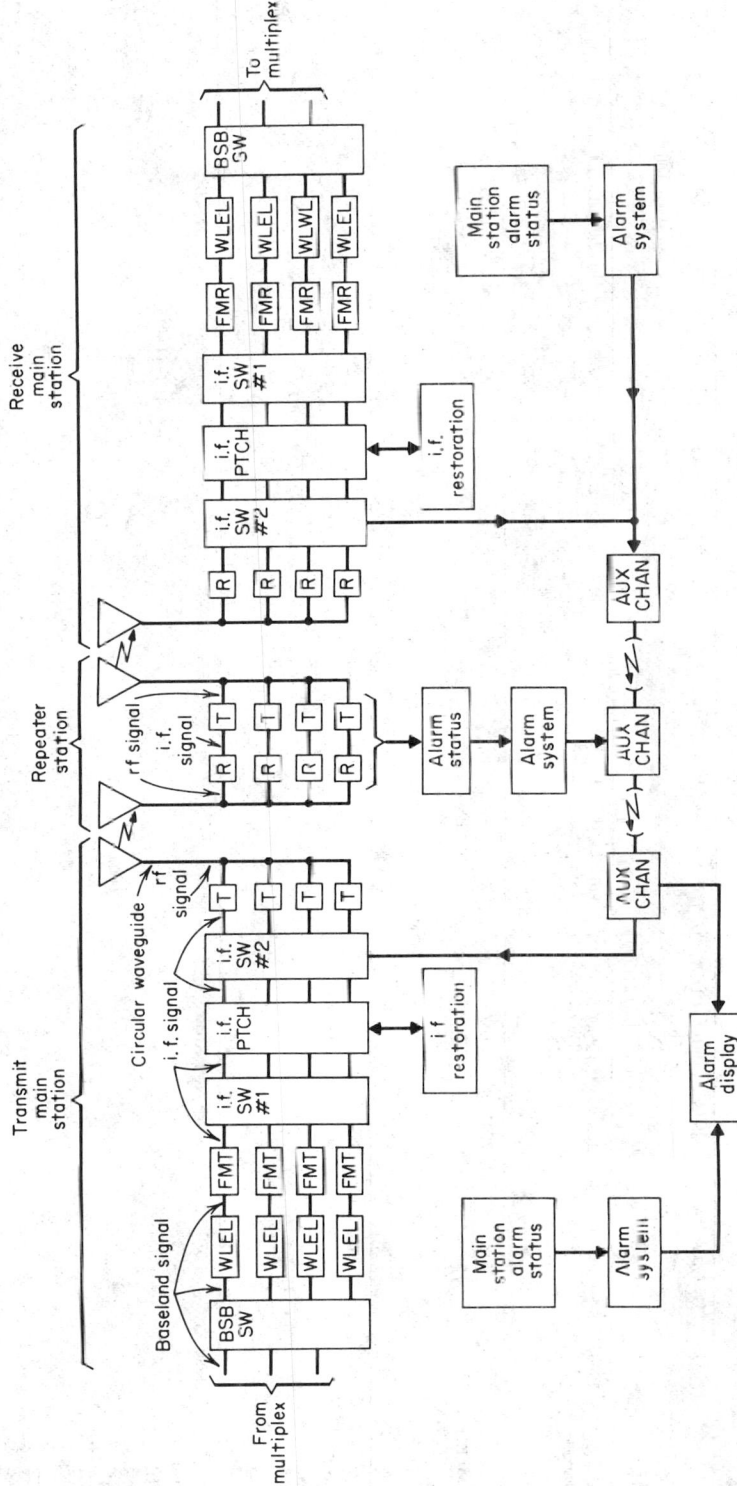

Fig. 22-36. Typical block diagram (for one direction of transmission) of radio system composed of three regular channels and one protection channel. (*Adapted from Ref. 1. Copyright 1977, Bell Telephone Laboratories. Used by permission.*)

TABLE 22-20 Technological Development of Intelsat I to V

Model	First launch	Height overall, cm	Mass in orbit, kg	Launch vehicle	Primary power, W	Transponders	Bandwidth per transponder, MHz	Coverage	EIRP[a] per beam, dBW	No. of telephone circuits	Design life, yr	Annual cost per circuit
I Intelsat 1	1965	59.6	38	Thor-Delta	40	2	25	Northern hemisphere	11.5	240[b]	1.5	$30
II Intelsat II	1967	67.3	86	Improved Thor-Delta	75	1	150	[c]	15.5	240	3	10
III Intelsat III	1968	104	152	Long-tank Thor-Delta	120	2	225	[c]	23	1,200	5	2
IV Intelsat IV	1971	528	700	Atlas/Centaur	400	12	36	[c,d]	22.5[c] 33.7[d]	4,000	7	1.2
IVA Intelsat IV-A	1975	590	790	Atlas/Centaur	500	20	32–36	[c,e]	22[c] 29[d]	6,000	7	1.1
V Intelsat V	1980	1,570	976	Atlas/Centaur	1,200	[f]	36–241	[c–e]	23.5–26.5[c,g] 26–29[d,g] 41.1–44.4[d,g]	12,000[h]	7	0.8

[a] Effective isotropic radiated power. [b] No multiple access. [c] Global. [d] Spot beam. [e] Hemibeam. [f] Using dual polarization. [g] Switched arrangement with reception in 6- and 14-GHz bands and transmission in 4- and 11-GHz bands. [h] Plus two TV.

Source: Adapted from B. I. Edelson et al, *IEEE Commun. Soc. Mag.*, January 1977, Vol. 15, No. 1, and B. I. Edelson, Global Satellite Communications, *Scientific American*, February 1977, Copyright © 1977 by Scientific American, Inc.; all rights reserved, reprinted by permission 1977 IEEE and copyright.

tions receive all signals but demodulate only signals directed to that station. In the limit of subdivision, one voice channel can be placed on a single carrier (single channel per carrier or SCPC). As the high-power amplifiers (HPA) in the earth station and the satellite are highly nonlinear, power levels must be reduced considerably ("backed off") below the saturation level

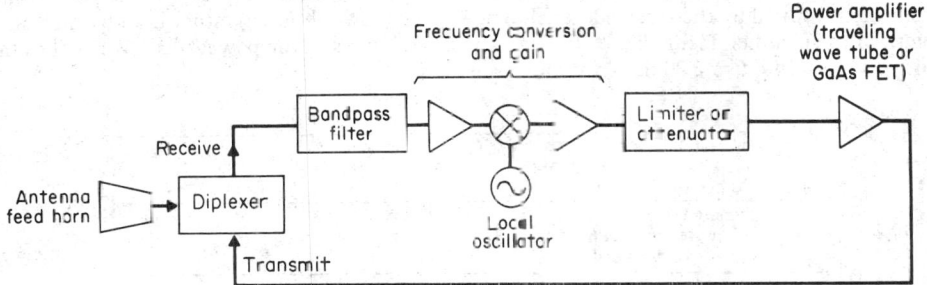

Fig. 22-37. Satellite transponder.

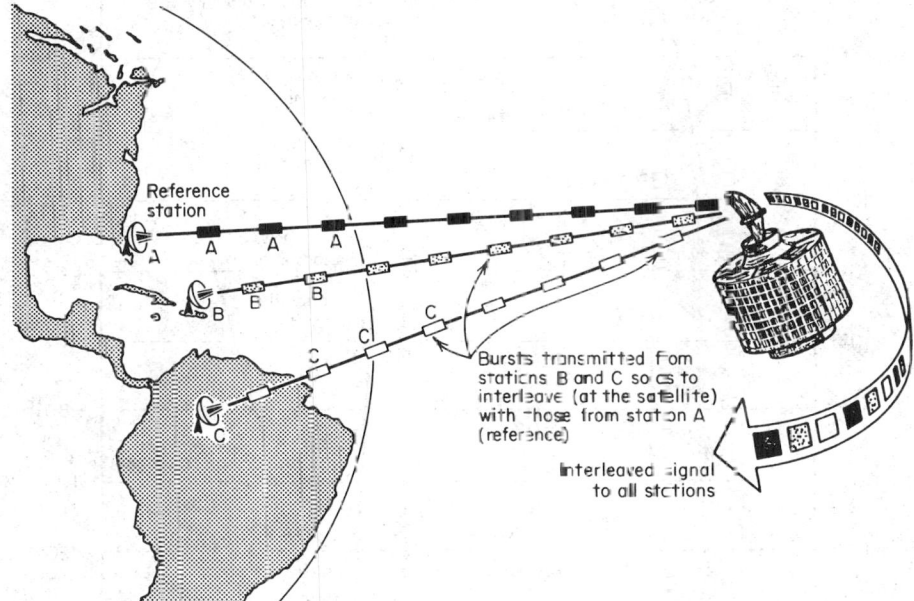

Fig. 22-38. Satellite time-division multiple access. *(From Digital Communications Corporation, "TDMA Concepts." Used by permission.)*

to reduce intermodulation distortion between the several carriers. It is also possible to use *demand assignment* in which a given frequency slot can be reassigned among several earth stations as traffic demands change.

In TDMA (time-division multiple access) each earth station uses the entire bandwidth of a transponder for a portion of the time, as illustrated in Fig. 22-38. This arrangement implies digital transmission (such as QPSK) with buffer memories at the earth stations to form the bursts. A synchronization arrangement which controls the time of transmission of each station is also required. As at any given time only a single carrier is involved, less backoff is required than with FDMA, allowing an improved signal-to-noise ratio. Demand assignment can be realized by reassigning burst times among the stations in the network.

Transmission Considerations. The free-space loss between a satellite and the earth is about 200 dB. To overcome this large loss, earth stations customarily use large parabolic antennas (usually 10 to 30 m in diameter, although 5- and 7-m antennas* have been used in some low-band-

*Smaller receive-only antennas are used in some applications.

width applications), high output power (up to several kilowatts), and low-noise receiving amplifiers (cryogenically cooled in some cases). Transponder output power is typically only about 5 W, and, therefore, downlink thermal noise often accounts for most of the system noise, with intermodulation in the transponder power amplifier often the next largest contributor. Because of the expense of increasing transmitted power from the satellite, the capacity of a satellite channel is often limited by the received signal-to-noise ratio (power-limited) rather than by the bandwidth of the channel. Figure 22-39 illustrates gains and losses in the power of a TV signal transmitted via satellite using 30-m earth-station antennas.

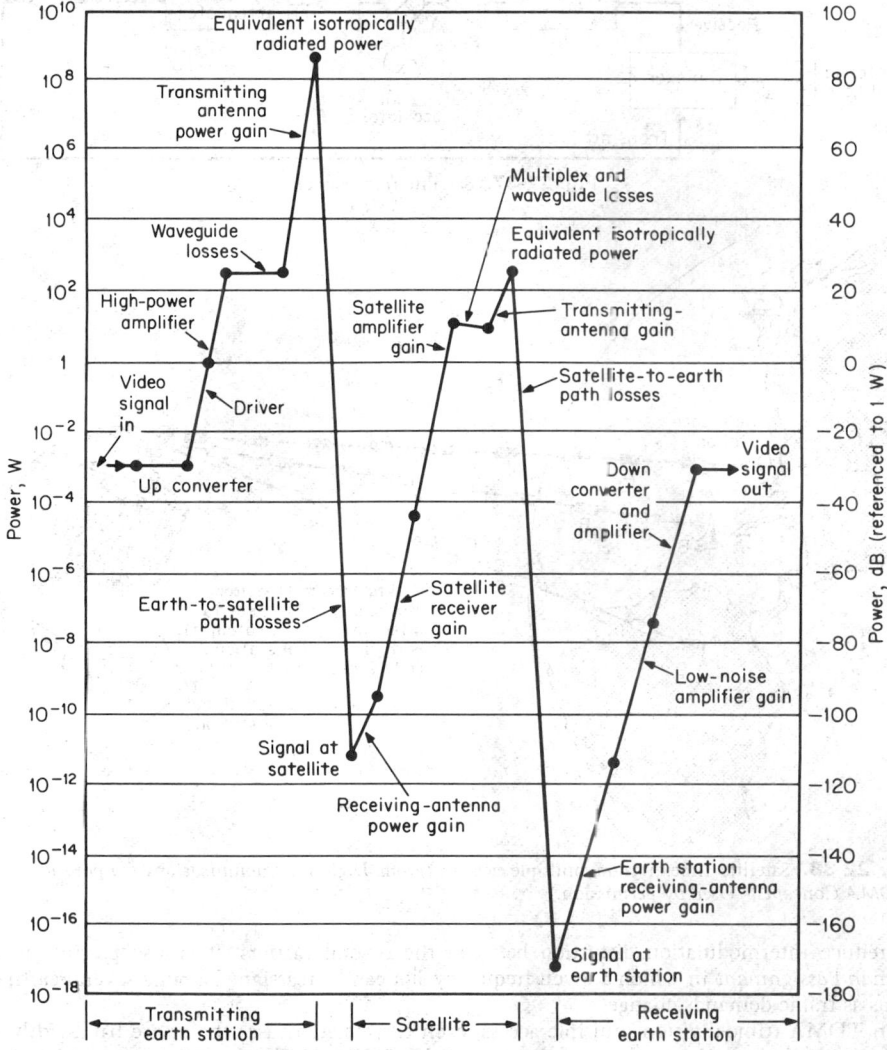

Fig. 22-39. Power levels in transmission of a TV signal via satellite *(From B. I. Edelson, Global Satellite Communications, Scientific American, February 1977. Copyright © 1977 by Scientific American, Inc. All rights reserved.)*

DIGITAL CARRIER SYSTEMS[1,24,40,41]

55. Network Factors.[42] The first digital carrier was the 24-voice channel 1.544 Mb/s T1 system, introduced in the United States in 1962 for short-haul (up to 50 mi) application. The major advantage of T1 over short-haul analog systems was the lower-cost terminals made possible by the ease of handling signaling information (e.g., on-hook, off-hook) on a digital system, the sharing of a single codec (coder-decoder) over 24 channels, and the economy of time-division multiplexing. T1 has been widely deployed throughout the United States, Canada, and Japan and, although supplemented by higher-speed systems, is still being installed in large quantities. In the late 1960s, Europe standardized on a 30-channel 2.048 Mb/s system, with a coding technique also differing somewhat from that ultimately adopted by the United States, Canada, and Japan.

The major components and rates of the digital transmission networks in the United States are shown in Fig. 22-40. The basic 1.544 Mb/s digital signals may come from a digital channel bank which encodes voice signals, a digital switch (such as the no. 4 ESS, or data-connecting arrangements as shown at the bottom of the figure. Provision for an FDM 300-channel master group to be coded directly and inserted at the DSX3 has been made. The multiplexes combine several lower-speed signals into a higher-rate signal (and correspondingly demultiplex the higher-speed

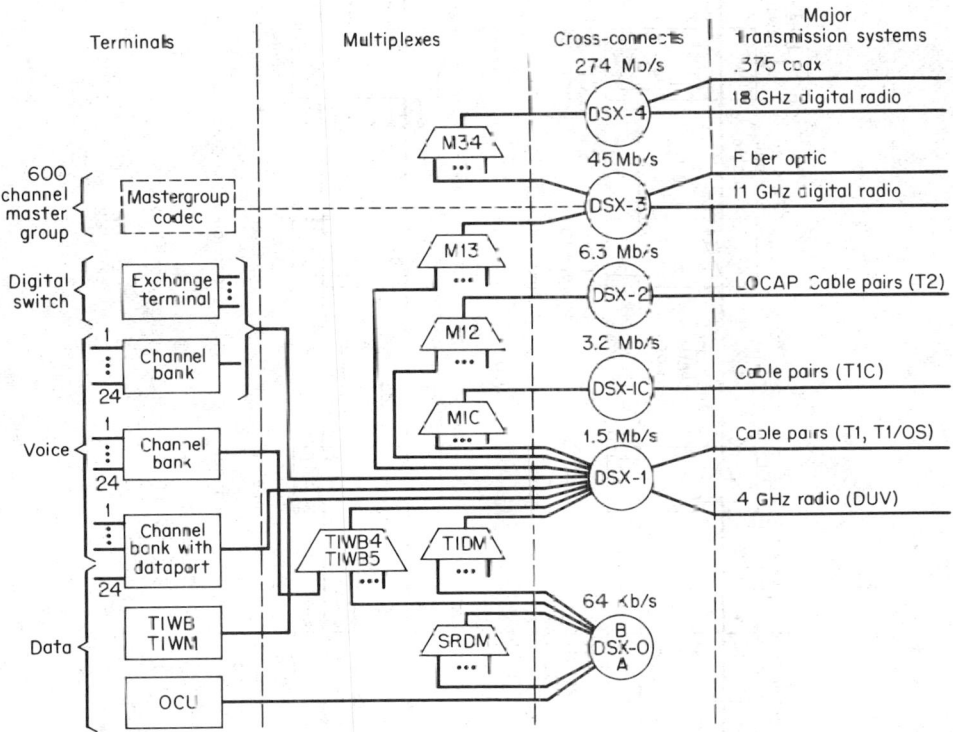

Fig. 22-40. Bell System digital transmission network.

received signal). The crossconnects are standard-level connection points at which all signals of a given rate in an office appear, facilitating rearrangements and testing. The major transmission systems associated with each rate are also indicated.

The hierarchy in Canada is similar, except that the 3.152 Mb/s rate is not used, and multiplexes* connecting the DSX-2 and DSX-3 are used in place of an M13 to reach the higher rates. (Since the M13 is, in effect, an M12† plus M23, it can connect with such arrangements.) In Japan the 64 kB/s and 1.544 and 6.312 Mb/s rates are used, along with 32.064 Mb/s (suitable for encoding the 300-channel master group used in Japan) and 97.723 and 397.200 Mb/s rates. The European hierarchy uses 64 kB/s and 2.048, 8.448, 34.368 (in some countries), and 139.264 Mb/s. Systems of about 560 Mb/s have been explored in North America and Europe, as have 800 Mb/s systems in Japan; they may constitute the next step.

A block diagram of a simple 24-channel T1 system showing the actual equipment items involved is given in Fig. 22-41a. The voice-frequency connection to the switch requires one pair

*A multiplex consists of a multiplexer and demultiplexer with any associated maintenance equipment.
†Read "M one-two."

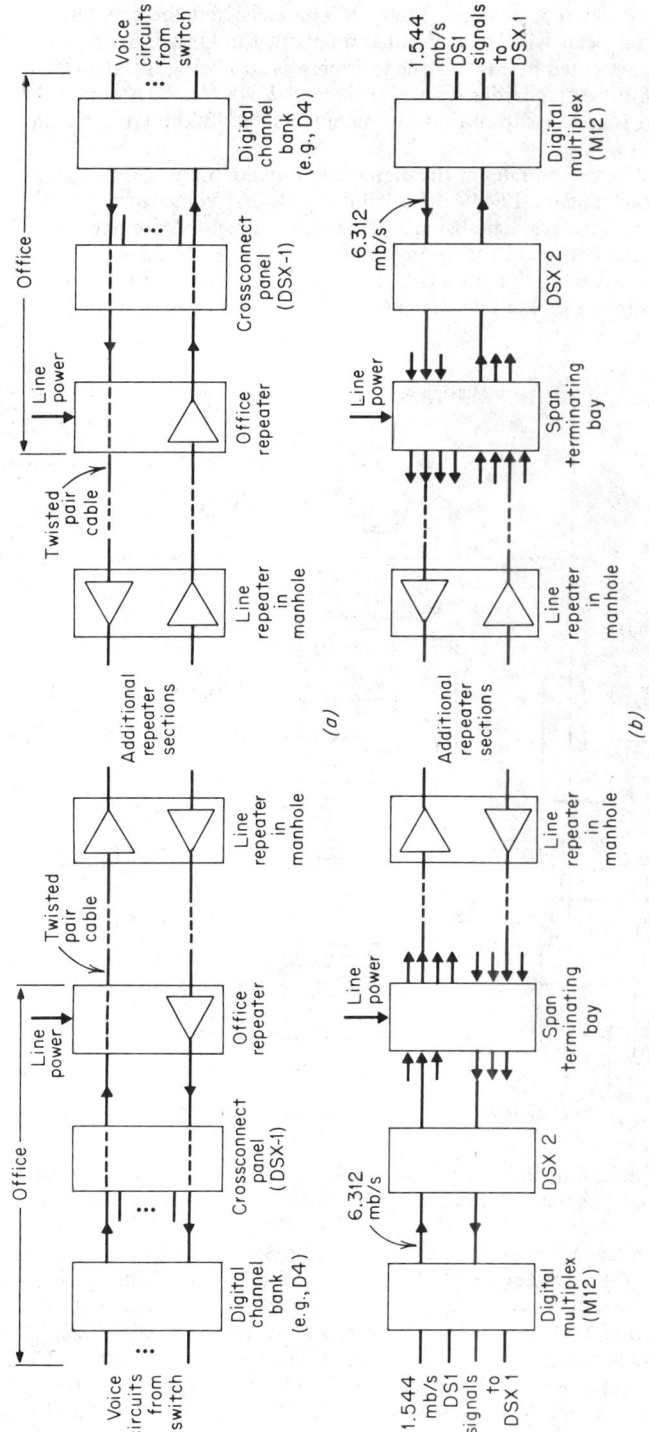

Fig. 22-41. (a) Basic T1 system; (b) T2 system.

in each direction, and an additional pair may be provided to carry signaling information. In the event of a line failure, a spare line is patched in at the DSX-1 crossconnect. A representative block diagram of a higher-speed system with automatic protection switching and requisite multiplexing is shown in Fig. 22-41b. (Protection switches are also sometimes used with T1.)

56. Voice Encoding.[24] To prepare a voice signal for digital transmission, it is band-limited to about 3,500 Hz and sampled at 8 kHz; each sample is encoded into 8-b PCM, producing a 64-kb/s signal. The quantization of the signal to one of $2^8 = 256$ levels which is inherent in the coding process produces noise, termed *quantizing noise* or *quantizing distortion*, in the decoded signal. (This is the major impairment suffered by a voice signal transmitted digitally.)

To improve the signal-to-distortion ratio at small signal levels, a logarithmic encoding law is employed in order to use a greater portion of the available levels for weak signals. In this way, a certain *percentage* change in the instantaneous signal corresponds closely to a change of one encoding level, and the signal-to-distortion ratio is approximately constant over a wide range of talker volume. This process can be thought of as compressing the signal in amplitude according to one of the curves of Fig. 22-42 and then performing a linear A/D conversion. In current practice, however, a piecewise linear approximation to the logarithmic curve is incorporated in the coder proper and, of course, in the decoder as well.

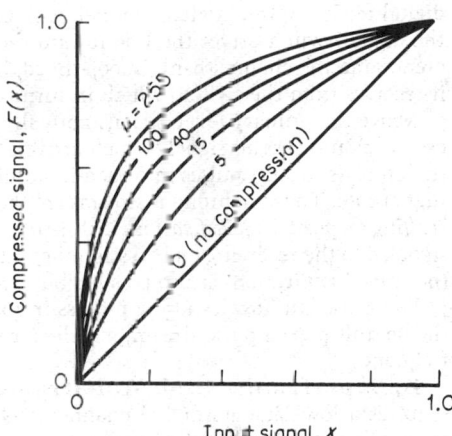

Fig. 22-42. Logarithmic compression characteristics. *(From Ref 24. Copyright 1970, Bell Telephone Laboratories. Used by permission.)*

The two approximations in current use are shown in Table 22-21. Both meet the CCITT requirement of 33-dB signal-to-quantizing–distortion ratio for sine waves from 0 to −30 dBm0, with overload point at about 3 dBm0, but the μ-law coder is a few decibels better at low signal levels. Idle channel noise for the μ-law coder is specified as less than 23 dBrnc0. (In this, as in other sine-wave specifications of coders, exact submultiples of the 8-kHz sampling frequency are excluded.) In the μ-law code, the all-zero code word is not used, slightly increasing the quantizing noise.[43]

Many coding schemes for voice which use less than 64 kb/s have been proposed, and a few, particularly delta modulation, have found limited application.

57. Time-Division Multiplexing; Framing.[124] If two or more synchronous digital signals are to be multiplexed, they can be adjusted in phase and pulse width and combined directly, with some method for identification of the signals for later demultiplexing. A common method of identification is to group a number of input bits into a frame with an additional framing bit at the beginning of the frame. For example, in the 24-channel PCM channel bank, twenty-four 8-b words, each representing one sample, constitute 192 b of a 193-b frame, and the 193d bit is the framing bit. If no other features were desired, the framing bit could be a simple pattern such as 101010, which would allow frame to be recovered at the receiving channel bank by finding a position in the received pulse stream which alternates between 0 and 1 at intervals

TABLE 22-21   CCITT Recommended Coding Laws for Voice

Name	Basic law $0 \leq |x| \leq 1$	Parameter value	Segments in linear approximation	Primary multiplex rate, Mb/s
μ law	$y = \dfrac{\ln(1 + \mu|x|)}{\ln(1 + \mu)}$	$\mu = 255^*$	15	1.544
A law	$\dfrac{1 + \ln A|x|}{1 + \ln A} \quad \dfrac{1}{A} \leq x \leq 1$ $\dfrac{A|x|}{1 + \ln A} \quad 0 \leq x \leq \dfrac{1}{A}$	$A = 87.6$	13	2.048

$^*\mu = 100$ was used in original T1 system.

of 193 b. A more complex framing pattern is used in modern banks, however, to identify groups of twelve 8-b code words in each channel, as shown in Fig. 22-43, to allow robbing the eighth bit of every sixth code word in each channel to carry signaling, e.g., on or off hook, information.

58. Synchronization, Pulse Stuffing. Multiplexing to the primary level (1.544 or 2.048 Mb/s) is synchronous, and entire 8-b words are multiplexed. In channel banks the locally encoded voice signals are synchronous and appear 8 b at a time, so that this is a natural mode of operation. Where 64-kb/s signals from more than one source are to be intermingled, e.g., in a digital switch, they are synchronized to a central reference which is distributed via designated digital facilities. In a digital channel bank connected to a digital switch, e.g., via a digital line, the digital switch drives the line toward the channel bank at a rate synchronous with the reference and the channel bank is loop-timed; i.e., it derives its transmit frequency from the signal it receives from the switch, which in turn is timed from the network reference.

Above the primary level, signals normally arrive at the multiplex a bit at a time, and it is not easy to assure that all signals which arrive at a multiplex will be synchronous. Present practice, therefore, is to stuff pulses into each input digital stream to synchronize them all to a common, higher rate. The synchronous streams are then multiplexed bit by bit. This process, called *pulse stuffing* or *positive justification*, is illustrated in Fig. 22-44. The location of the stuffed pulses is signaled to the receiving end on still other bits added to the frame for the purpose. Pulse deletion (negative justification) is also possible but has found little use.

The pulse stuffing-destuffing process introduces some jitter (undesired phase modulation) in the demultiplexed pulse stream, which is usually not troublesome as it is of low amplitude and frequency.

59. Superframes and Multiframes. The synchronization scheme described above requires a low-data-rate digital channel between multiplexer and demultiplexer to signal the presence or absence of stuffed time slots. Low-data-rate channels may also be required to provide communication between terminals for maintenance functions such as switching to a spare or for parity bits for detection of transmission errors. These low-data-rate channels are provided by adding bits which are located by establishing a superframe (also called multiframe) structure encompassing many of the information-bit frames discussed above.

An example of a superframe is given in Fig. 22-45, where the superframe consists of 24 frames. Each frame consists of 48 information bits (shown as [48] in the figure) plus one service bit (shown as M, F, or C). The service bit provides framing, superframe location, and pulse-stuffing information. The 48 information bits are taken one at a time from each of the four input signals in turn, for a total of 12 b from each input. This basic frame of 49 b is identified by the F_0 and F_1 framing bits (which are 0 and 1 respectively). The 0111 pattern of the M_0 and M_1 "marker" bits (again 0 and 1 respectively) permits the superframe, which here consists of 24 basic frames, to be identified.

The format of this example is used in a multiplex which combines four 1.544-Mb/s (T1) signals (I, II, III, IV) into a 6.312-Mb/s stream. It is therefore necessary to have four distinct stuff–no-stuff indications, which are provided by the bits C_I, C_{II}, C_{III}, and C_{IV}. Each of these bits is sent three times, as shown in the figure, to allow correcting a single transmission error. When a stuff is to take place in T1 signal number IV for example, the three C_{IV} bits of a superframe are sent as 1s and the stuff is performed toward the end of the superframe, in the slot indicated in the figure. Thus only one stuff per channel per superframe is possible.

60. Digital Carrier on Cable: System Factors. Digital (like analog) carrier systems consist of terminals connected by cables with regularly spaced repeaters. In digital systems, however, the repeater does not merely amplify the receiving signal but detects, regenerates, and retimes it, thus delivering a clean, noise-free output. Design of such a system must consider the same factors as listed for analog cable, except that intermodulation distortion is not among them. Further, noise and distortion do not accumulate along a digital line as they do in an analog system.

Thanks to the regenerative process, the only impairments introduced by a digital repeatered line are errors and jitter (unwanted phase modulation). Error performance can be characterized in terms of average error rate (number of errors divided by time of observation), and this is a satisfactory measure if the errors are more or less randomly distributed.

In many cable and radio systems, enough design margin against thermal noise and crosstalk (the normal causes of randomly distributed errors) is provided for the residual errors to occur in infrequent bursts, as a result of impulse noise or other disturbances. Performance of such systems is better described in terms of the percentage of seconds which are error-free, or of 1-min intervals which individually have better than 10^{-6} average error rate, as illustrated in Table 22-22.

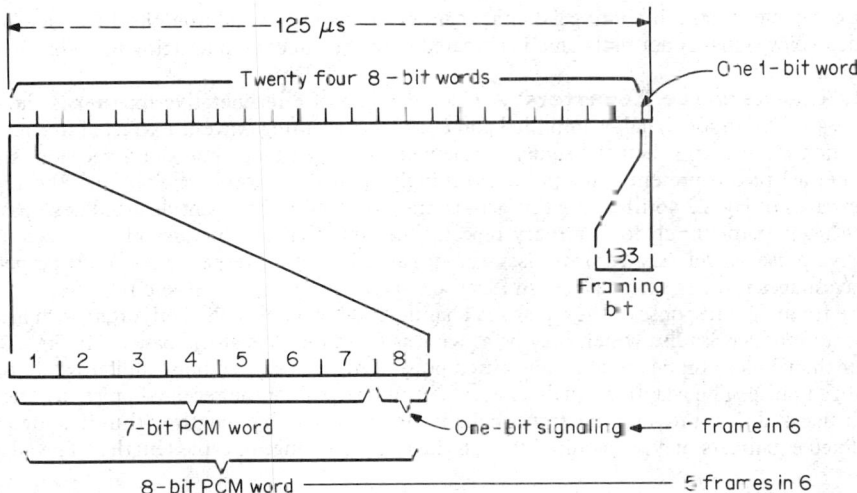

Fig. 22-43. A 24-channel channel-bank framing format. (From Ref. 43. Copyright 1972, American Telephone and Telegraph Co. Used by permission.)

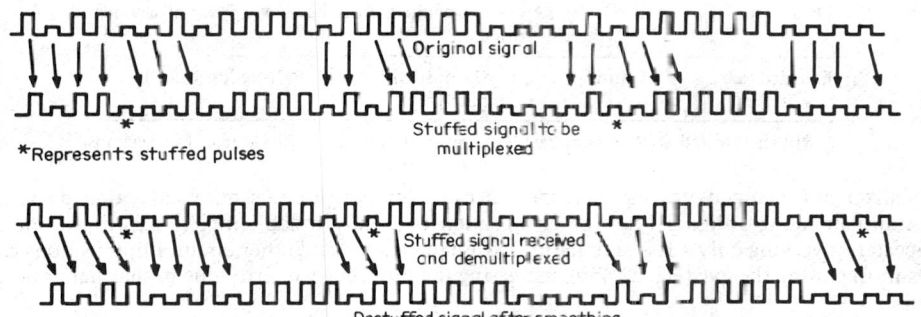

Fig. 22-44. Pulse-stuffing synchronization. (From Ref. 24. Copyright 1970, Bell Telephone Laboratories. Used by permission)

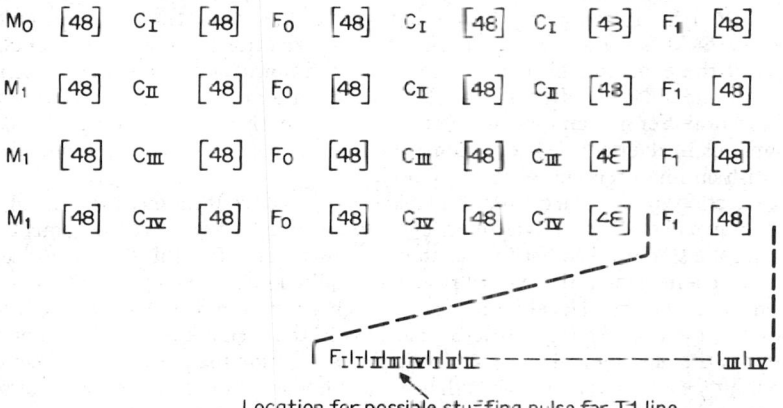

Fig. 22-45. M12 multiplexer format. (From Ref. 24. Copyright 1970, Bell Telephone Laboratories. Used by permission.)

For a complete system, including a digital channel bank and a digital line, the effect of errors on a coded voice signal is normally small compared with the effect of quantizing noise in the channel bank.

61. Regenerative Repeaters. A block diagram of a regenerative repeater is shown in Fig. 22-46. The input signal is amplified and equalized, a timing wave is extracted in the clock-extraction circuit, and then the signal is regenerated; i.e., in each time slot a decision is made whether a 1 or 0 is present and a pulse accordingly applied (or not) to the output. The stylized waveshapes of Fig. 22-46 illustrate this action, and Fig. 22-47a shows an idealized "eye pattern" at the regenerator input, for a ternary repeater, i.e., one that can produce either a positive or negative pulse as well as a zero, as discussed in Par. **22-62**. An eye pattern is the superposition of waveshapes resulting from all possible pulse sequences $(0+0, -+-,$ etc.).

The frequency response of the equalized channel is deliberately rolled off, often with a cosine shape, to produce a pulse which may be as wide as two time slots at the base (as in Fig. 22-47a) rather than a close replica of the transmitted pulse. This reduces the noise and crosstalk on the equalized pulse. The adaptive equalizer acts to bring the peak of the received pulse to a standard amplitude and thus provides the proper equalization for any cable length within its range. (Several fixed equalizers may be required to span the complete range of cable lengths.) The adaptive

TABLE 22-22   CCITT Provisional End-to-End Error-Performance Specification for Long International Connection at 64 kb/s

Intended application	Interval	Requirement
Data	1 s	A 64-kb/s component to have over 95%* of seconds error-free
Encoded voice	1 min	Over 90% of such intervals to have less than 1 error per 10^6 b

*In the United States commercial systems guarantee up to 99.5% error-free seconds.

equalizer also compensates for variations in cable temperature. As misequalization does not accumulate along a digital line, the equalization is usually much simpler than in an analog repeater, involving only a few singularities. Misequalization and other circuit imperfections can result in closing the eye (Fig. 22-47b), increasing the probability of error due to thermal or other noise.

Jitter in a repeatered line results from pattern variations in the signal transmitted. As a result of imperfect equalization and other idiosyncrasies of the repeater, these pattern variations appear as phase variations applied to the timing extraction filter (see below), and any components of the phase variation at a frequency within the bandpass of this filter appear as jitter on the timing signal and hence on the repeater output. The rms jitter of a line is approximately proportional to the square root of the number of repeaters. (For T1, for example, with a repeater tank Q of 80, the rms jitter for 10 repeaters is about 3°.) Jitter in the amounts usually encountered has little effect on 64-kb/s encoded voice signals. Jitter (unlike errors) can be reduced or eliminated completely at the endpoints by writing the receiving information into a buffer memory and reading it out under control of a stable clock. Thus the major concern in practice is to assure an adequate size of buffer memory where dejitterization intentionally or inadvertently takes place. (An example of inadvertent dejitterization would be the connection of a repeatered line to a terminal with an input repeater of higher Q.)

Clock extraction for a ternary repeater involves full-wave rectification of the signal, filtering the result in an LC tank or equivalent, in order to obtain a sine wave at the symbol rate, and then shaping the sine wave to obtain a sampling pulse or edge. The full-wave rectification produces a strong component at the symbol rate. The required C of the tank circuit, or equivalent, depends on the line code selected. In any case, a higher Q in the filter reduces the jitter of the timing wave, which results from pattern variations in the digital stream. This jitter produces output jitter of the repeater and dynamically displaces the sampling pulse from the center of the eye, reducing repeater margin. On the other hand, a higher Q increases the static offset of the sampling pulse due to temperature variations of the frequency of the tank. Arrangements used have included an LC tank with Q of 80 (T1), a monolithic crystal filter, and a phase-locked loop with crystal-controlled VCO (voltage-controlled oscillator) in some higher-speed systems (see Par. **22-62**).

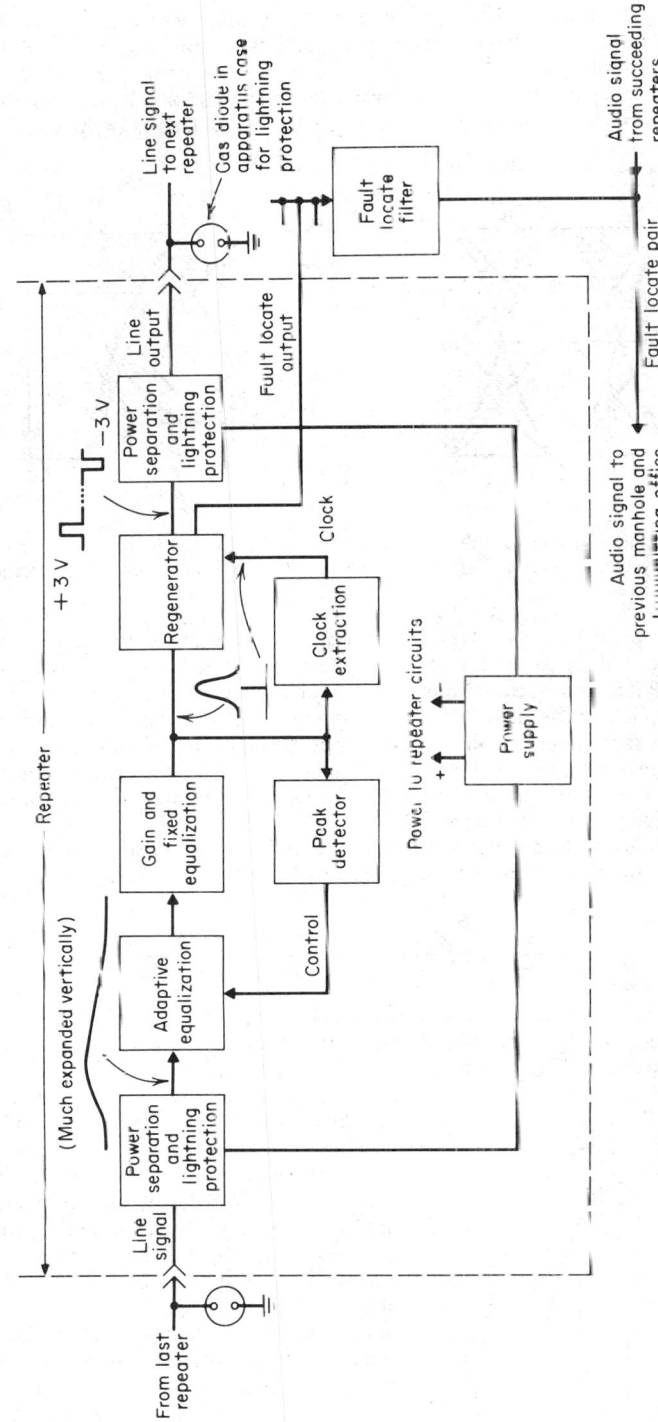

Fig. 22-46. Representative regenerative repeater for cable system with associated external components.

The regenerator proper is a clocked flip-flop or similar arrangement with carefully controlled threshold. The output pulse is often clocked to about 60% of the time slot, but NRZ (non-return-to-zero) is also used in some cases.

Most regenerative repeaters have some provision for fault location, i.e., determining from an office which of the many repeaters in a failed line is at fault. This is often accomplished, as

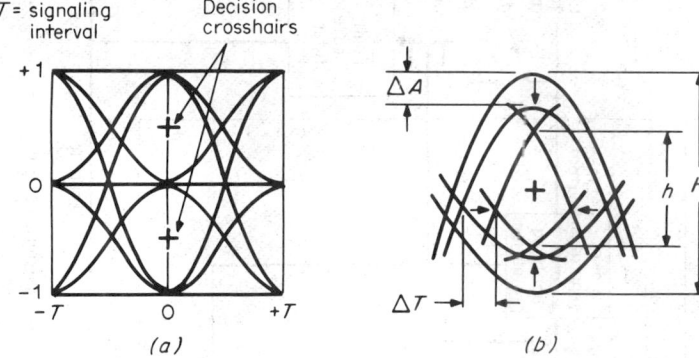

Fig. 22-47. Eye diagrams for a ternary repeater: (a) ideal eye; (b) closing the eye to account for practical degradations (upper eye only). *(From Ref. 24. Copyright 1970, Bell Telephone Laboratories. Used by permission.)*

shown in Fig. 22-46, by feeding a small part of the output back to the originating station through a filter which passes only a particular audio frequency assigned to that manhole. To locate a faulty repeater, a digital pattern containing the particular audio frequency associated with a certain manhole is applied to the line. If the audio frequency then appears on the fault-locating pair, the repeater in that manhole and all repeaters upstream of that line are known to be operating. The procedure is then repeated with other audio frequencies until the defective repeater is located. All repeaters at a given manhole share the same audio filter, and all manholes share the same voice-frequency fault-locate pair.

The probability that an ideal digital repeater with gaussian noise added to its input will make an error is the probability that the noise will exceed half the eye height, which for m-level transmission is

$$P_E = \frac{m-1}{m} \, \text{erfc} \left| \frac{1}{(m-1)\sqrt{2}} \frac{\text{peak signal}}{\text{rms value of noise}} \right|$$

This is plotted in Fig. 22-48. It can be seen that in the range of usual interest, 10^{-6} or better, variation of a few decibels of signal-to-noise ratio produces a variation of many orders of magnitude in error probability. This is generally true in practice as well, even though crosstalk or some other interference, rather than thermal noise, may be the major cause of errors.

The features described above are typical of regenerative repeaters for use on cable systems. Since such cable repeaters normally are powered from a direct current carried on the same conductors as the signals, they also require a power-separation filter, as shown in Fig. 22-46. Regenerative repeaters for radio use require modulators and demodulators as well, and these are usually QPSK (quaternary phase-shift-keyed) or 8PSK (eight-phase-shift-keyed) which may be applied to the i.f. or directly to the rf signal. Optical-fiber repeaters require a light source and optical detectors, as well as the fibers described in Par. **22-43**. Both solid-state lasers and LEDs (light-emitting diodes) of gallium–aluminum arsenide with output wavelength of 0.82 to 0.85 μm are available. (Generally LEDs are less expensive but provide less light

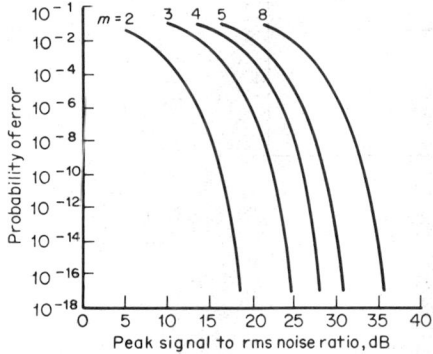

Fig. 22-48. Probability of error vs. peak signal to rms gaussian noise for random m-level polar transmission. *(From Ref. 24. Copyright 1970, Bell Telephone Laboratories. Used by permission.)*

power to the fiber.) As fiber loss is less at longer wavelengths, sources with longer wavelength are being sought. Detectors are typically silicon avalanche photodiodes.

62. Scramblers and Line Codes for Digital Systems. Normally it is not possible to transmit an arbitrary digital stream directly as a binary signal because of the possibility that some patterns might cause the line system to misfunction. For example, too long a string of zeros

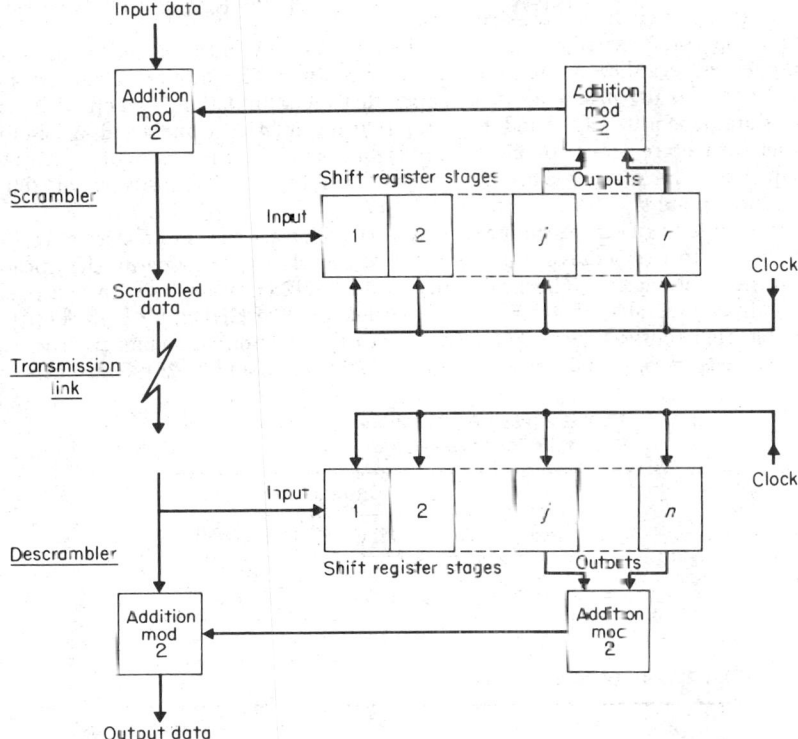

Fig. 22-49. Self-synchronized scrambler and descrambler block diagram. *(From Ref. 24. Copyright 1970, Bell Telephone Laboratories. Used by permission.)*

could cause loss of timing signal in the repeaters, repetitive patterns could produce strong discrete frequencies in the output of radio systems which might violate emission limits, and a variation of the average density of 1s would cause a wander of the baseline of the signal after passing through a transformer or power-separation filter. To counter these effects, the signal is usually scrambled or coded (sometimes both).

A digital signal is scrambled by adding to it, in mod 2,* a long predetermined pseudo-random pattern of 1s and 0s. It can be unscrambled at the receiving end by adding (mod 2) the same pattern, in the same phase. The pseudo-random pattern is usually obtained from a shift register clocked at the signal rate with some taps added and fed back to the input. Appropriately selected taps produce a maximal-length pattern $2^N - 1$ long, where N is the number of stages in the register. The phase of the sequence can be synchronized to the information signal by some external means, such as a framing pattern. If this is not convenient, a self-synchronized scrambler like that of Fig. 22-49 can be used. It can be seen that after a settling period the output data of the descrambler are identical to the input data, and the appropriate selection of length and taps produces, in effect, a maximal-length scrambling sequence. The disadvantage of the self-synchronized scrambler is that in descrambling it multiplies any line errors by 3.

Scrambling may reduce the probability (but cannot *guarantee* the absence) of any unfortunate pattern such as a long string of zeros. Therefore, if scrambling is to serve this purpose, high-Q timing recovery circuits such as phase-locked loops are usually required. Another use of scram-

*In mod 2 (modulo 2) addition $0 + 0 = 0, 1 + 0 = 1, 1 + 1 = 0$.

bling is encryption of a digital signal to maintain its confidentiality. This can be accomplished by scrambling the signal with a very long sequence known only to the parties involved. This practice was formerly largely confined to government communications, but heightened commercial interest in privacy has led to a standard method of encrypting, the DES (data encryption standard) promulgated by the National Bureau of Standards, and reasonably priced silicon integrated circuits to implement it. Scrambling for encryption is, of course, best applied at the source and destination of the signal to be encrypted, while scrambling to control signal statistics is applied within a particular transmission line.

Coding, as opposed to scrambling, adds enough redundancy to the signal to guarantee some desired properties, regardless of the pattern to be transmitted. The most prevalent code is bipolar (also called AMI, for alternate mark inversion) with all zero limitation, used on T1. In this code alternate 1s are transmitted as $+$ and $-$ pulses, assuring dc balance and avoiding baseline wander. Further, an average density of one pulse in eight slots with a maximum of 15 zeros between ones is required.* This maintains timing signal sufficient for the inexpensive, low-Q (80) timing circuits in the repeaters.

Another scheme to guarantee timing density with a bipolar code is to replace strings of zeros with a pattern with two successive pulses of the same polarity, allowing its identification and removal at the receiving end. The arrangement, called BNZS (for bipolar with N zero substitution) is also in considerable use. In Europe, it is called HDB or CHDB, for high-density bipolar or compatible high-density bipolar. BNZS codes all carry one information bit per ternary symbol, and codes with $N = 3$, 4, and 6 are in use (see Tables 22-23 and 22-25). Greater information

TABLE 22-23   B3ZS Choice of Sequences to Substitute for Three Zeros

Last pulse transmitted	Last substitute sequence used	
	$00+$ or $+0+$	$00-$ or $-0-$
$+$	$-0-$	$00+$
$-$	$00-$	$+0+$

TABLE 22-24   Simple 4B3T Code

Binary word	Ternary words*		Binary word	Ternary words*	
	$+$	$-$		$+$	$-$
0000	$0 - +$	$0 - +$	1000	$0 + -$	$0 + -$
0001	$- + 0$	$- + 0$	1001	$+ - 0$	$+ - 0$
0010	$- 0 +$	$- 0 +$	1010	$+ 0 -$	$+ 0 -$
0011	$+ - +$	$- + -$	1011	$+ 0 0$	$- 0 0$
0100	$0 - +$	$0 - -$	1100	$+ 0 +$	$- 0 -$
0101	$0 + 0$	$0 - 0$	1101	$+ + 0$	$- - 0$
0110	$0 0 +$	$0 0 -$	1110	$+ + -$	$- - +$
0111	$- + +$	$+ - -$	1111	$+ + +$	$- - -$

*Use $+$ column if sum of pulses transmitted so far is $-$, otherwise use $-$ column.

capacity can be obtained using block codes such as 4B3T (each block consists of 4 b on three ternary symbols). Generally, as the information capacity is increased, the reduction in redundancy results in poorer timing and baseline-wander performance. A simple 4B3T example is given in Table 22-24. There are many variants, but all have spectra which are zero at both zero frequency and the symbol rate, as do the bipolar and BNZS codes.

Another type of code is epitomized by the *duobinary* code, which is a 1-b-per-symbol ternary code in which adjacent pulses of opposite polarity cannot occur. Thus the redundancy is used to reduce the transmitted energy at high frequencies rather than to control baseline wander and assure timing. Scrambling, high-Q tanks, and circuit design to minimize baseline wander are normally required. The duobinary spectrum is maximum at zero frequency and zero at half the symbol rate. The reduction of high-frequency energy makes it attractive for sharply band-lim-

*This is readily obtained in voice-band coding by simply not using the all-zero word.

ited media such as radio channels or, on wire pairs, to minimize crosstalk. Variants with larger numbers of transmitted levels have been dubbed *partial-response codes* and can also limit the transmitted spectrum in a similar fashion.

Various pseudo-ternary codes are illustrated by example in Table 22-25. The redundancy

TABLE 22-25 Coding Examples

Code name	Coding rule	Example
Binary		1 0 1 1 1 0 0 1 1 0 0 0 0 0 0 0 1 1 1 1
Bipolar, AMI	Invert alternate 1s	+0−+−0 0+−0 0 0 0 0 0 0+−+−
B3ZS, CHDB2	Invert alternate 1s, but for 3 0s substitute according to Table 22-23	+0−+−0 0+−0 0−+0 +0−+−+
4B3T	See Table 22-24	+0 0 +−0 0+−0−+−−−
Duobinary	From binary input sequence a first form sequence* C $C_n = a'_n \oplus C_{n-1}$ then output = $C_n + C_{n-1} - 1$	0 1 1 1 1 0 1 1 1 0 1 0 1 0 1 0 0 0 0 0 −0+++0 0++0 0 0 0 0 0−−−−

* $\oplus$ = addition modulo 2, and a'_n is the logical inverse of a_n; that is, substitute 0 for 1 and vice versa. The first step avoids a long string of decoding errors resulting from a single transmission error.

added in any of these codes is sufficient to allow reasonably accurate estimates of errors in transmission by detection of violations of the constraints of the particular code. In uncoded systems parity bits may be added to accomplish this purpose.

Forward-acting error-correcting codes, in which redundancy is added to allow error detection at the receiving end, are used only in special applications, as, in general, the additional bandwidth required for forward error correction would improve performance more if used to increase system margins. The special applications in which forward error correction is used include satellite digital transmission, in which the transmitted rate is often limited by the available power on the satellite rather than available bandwidth.

63. Digital Transmission Systems.[1,24,40,41] Table 22-26 lists the major digital transmission systems in use in the United States and Canada with their usual application. Radio systems using regenerative repeaters as well as cable systems are included; digital transmission on radio systems using analog repeaters is discussed below. For some years, there was worldwide interest in digital systems employing circular waveguide of about 2 in diameter, operating in the 40- to 110-GHz range, and it appeared that over 200,000 two-way voice circuits could be carried on such a system.[44] (Circular waveguide, rather than rectangular waveguide, was contemplated due to the lower losses in the millimeter-wave region.) This interest waned, however, as optical fibers approached practicality. Similarly, optical-fiber systems may displace 560-Mb/s systems for 0.375-in coaxial cable, which have been designed but not yet deployed.

While T1 was originally designed for short trunks (up to 50 mi) it has been extended to longer applications by addition of a protection switch and somewhat more stringent engineering rules. It has also been used in carrier systems for the loop plant as well. In this application the terminals are not standard toll-channel banks but may use delta modulation or other low-bit-rate coding schemes, as well as concentration by switching.

Optical-fiber transmission up to 1,000 Mb/s has been explored. Complete systems at 44.736 Mb/s have been announced by several manufacturers, and extensive deployment of systems at this rate appears likely (see Table 22-26). There is considerable interest in developing longer-wavelength light sources to allow greater lengths without repeaters and higher-rate systems.

64. Broadband Coding and Connection between Digital and Analog Carrier Systems. To handle an analog voice circuit on a digital switch or carrier system the individual voice signal is encoded into digital form, as discussed above. Wider-bandwidth signals, such as TV or an analog master group, could also be encoded and enter the digital hierarchy at higher levels, the original wide-band analog signal being recovered at the receiving end. (The 44.736-Mb/s level was chosen to allow connection of an encoded 600-channel master group, and equipment to accomplish this has been described.)[15] Systems encoding TV have been proposed with bit rates ranging from 1.5 to 108 Mb/s. The higher-rate systems use 8-b-per-sample PCM encoding with sampling at about 12 MHz and were directed at very high quality for network distribution. The lower-rate systems make use of frame-redundancy techniques, in which only

TABLE 22-26 Major North American Digital Systems

Line rate, Mb/s	64 kb/s voice-channel capacity	Bell System or widely used designation	Line formats	Usual medium	Major design limit	Typical repeater spacing, mi	Typical section loss,† dB	Protection switch	Comment
1.544	24	T1, T1OS (T1-outstate)	Bipolar, with one-in-eight 1s density, 15 0s maximum	Wire pairs in single cable (but two directions in different units)	Near-end crosstalk	1 (on 22 gauge)	32	Special application only	Intended for exchange trunks. Most widely used system
3.152	48	T1C, T1D, T148	Bipolar, 4B3T; duobinary; modified duobinary	As for T1	Near-end crosstalk	1 (on 22 gauge)	48	Special application only	Displacing T1 in many new installations, usually installed with repeater housing (apparatus case) with separately shielded transmit and receive cables to reduce crosstalk[109]
6.312	96	T2	B6ZS	LOCAP paired cable, two-cable operation	Far-end crosstalk	3	55	Yes	Intended for intercity use
44.736	672	3ARDS	QPSK scrambled, with parity*	11-GHz radio; 40-MHz channel can be used in both polarizations simultaneously for capacity of 2×44.736 Mb/s	Rain fades	10–20	...	Yes	
2×44.736	1344	DRG, DR11	8PSK*, 16 QAM (quadrature amplitude modulation)*	6- and 11-GHz radio	Multipath fades at 6 GHz, rain fades at 11 GHz	10–25 at 6 GHz, 10–20 at 11 GHz	...	Yes	
44.736	672	FT3	Binary scrambled with parity*	Optical fiber (light wavelength = 0.82 μm)	Optical shot noise	4	40	Yes	Pilot systems installed, major deployment starting
274.176	4032	T4M	Polar binary with synchronous scrambler parity (United States) B3ZS (Canada)	0.375-in coaxial cable	Thermal noise and circuit limitations	1	55	Yes	
274.176	4032	LD4 DR18	QPSK with synchronous scrambler, parity	18-GHz radio (220-MHz channel)	Rain fades	2–5	50	Yes	

*The standard 44.736-Mb/s signal (DS3) which appears at a telephone company crossconnect is B3ZS, with prescribed framing, parity and communication bits. This is converted to the line code in the line-terminating equipment, and, in the case of 2 × 44.736 Mb/s digital radio equipment, multiplexed there with another such signal as well.
†At half the line rate for cable systems.

the differences between successive picture frames are transmitted. There is of course some loss of quality in this process, and the lowest-rate systems are directed toward visual conferencing or other applications where this is acceptable.

Conversely, digital (data) signals can be carried on an analog facility using a modulator-demodulator (modem), as described below. A third distinct possibility of connection arises when it is desired to connect an equal number of voice channels between the analog and digital hierarchies. This can be done by connecting analog and digital channel banks back to back, as shown in Fig. 22-50. A single equipment which incorporates the function (dotted lines in the figure) is called a *transmultiplexer*, and transmultiplexers built along these lines are considerably simpler and less expensive than using three channel banks. Experimental transmultiplexers in which A/D conversion is applied directly to the wide-band analog signal (converted to and from the appropriate digital signals by digital processing) have also been constructed.

65. Digital Interfaces for Switches. In recent years it has proved desirable to perform the telephone switching function on digital signals (see Par. **22-77**). The interface between such a *digital* switch and *analog* trunks resembles a digital channel bank in that a number of circuits are converted between the analog and digital forms. An actual channel bank is not generally suitable, however, as the number of channels, bit rate, and format for the switch do not usually correspond to the line format. Furthermore, the requirements for maintenance and for handling signaling information are quite different in the two cases. This has led to the development of specialized A/D converters for this purpose. For example, the voice interface frame, for connection to the no. 4 ESS, handles 120 voice channels, and the digital signal (on the switch side) has a 128-channel format, with the additional channels used for maintenance.

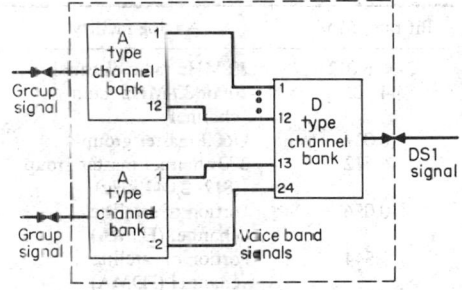

Fig. 22-50. Transmultiplexer.

To connect a *digital* switch to a *digital* transmission system requires an interface unit (exchange terminal), which performs the required digital format conversion and multiplexing to convert 24-channel DS1 (or 30-channel 2.048 Mb/s) which appears on the line to the digital format of the switch. Unlike line or facility multiplexers, the multiplexing is synchronous as the switch deals only with 8-b words. This requires a memory of about two frames in the interface unit, which aligns the frame of the received line signal and dejitterizes it.

An interface between a *digital* switch and an *analog* carrier system can be obtained economically by using an exchange terminal connected to the analog carrier system via a transmultiplexer.

The channel bank is the major interface used today between *analog* switches and *digital* carrier systems. Another possibility is to include as part of the switch a terminal which interfaces directly with the digital carrier system and provides analog voice circuits to be switched. This *digital carrier trunk* is attractive for an analog switch with stored program control, as it avoids the translation of signaling information from digital form to contact closures, which otherwise must be performed in the switch and again in the channel bank. As a consequence, it also reduces the cross-office wiring.

66. Hybrid and Modem-Type Digital Systems, Terrestrial and Satellite. The systems listed in Table 22-26 were designed to be economical for voice traffic in the intended application. The systems carry only digital signals and use only regenerative repeaters. (Hybrid systems with all digital traffic but with intermediate linear repeaters between regenerative receivers have been studied but never deployed.)

Mixed systems using modems (modulators-demodulators) typically arise when it is desired to carry digital traffic on an existing analog facility. Some unit of analog bandwidth is allocated to the digital signal, which is appropriately modulated for transmission. In such systems, both digital and analog signals are carried simultaneously through linear repeaters, with regenerative repeaters appearing only at infrequent intervals or perhaps only at the ends. The most common examples of this technique are the widely used voice-band data modems. Group band modems (50 kb/s) and terrestrial systems above this rate are also available. Table 22-27 lists some of them.

Since present satellites do not include regenerative repeaters, they fall in the category described above when used to transmit digital signals. Table 22-27 includes examples of modems for this application.

67. Maintenance Techniques in Digital Systems. Faults in the basic T1 system (Fig. 22-41*a*) are detected by the inability of a channel bank to find framing pulses in its 1.544-Mb/s input signal. When such a condition persists long enough (perhaps 2 s), the channel bank causes the trunks it serves to be declared "busy," lights a red alarm light, and sounds an office alarm. It also transmits a special code which causes the other channel bank to take the trunks out of service, light a yellow alarm light, and sound an alarm. Maintenance personnel then clear the trouble, perhaps by patching in a spare T1 line, and proceed with fault location and repair. Banks can be checked by looping, i.e., connecting digital output and input. Repeatered line-fault location techniques are discussed in Par. **22-61.**

In higher-speed systems, such as in Fig. 22-41*b*, automatic line and multiplex protection switching is often provided. A typical line-protection switch monitors and removes violations of the redundancy rules of the line signal on the working line at the receiving (tail) end. When

TABLE 22-27   Modem-Based Systems for Carrying Digital Signals on Wide-Band Analog Facilities

Bit rate, Mb/s	Analog facility	Line format	Comment
3 × 6.312	20-MHz radio channel	QPSK	In some use since 1970
44.736	20- or 30-MHz radio channel	QPSK	Available, but little field experience
6.312	U600 master group	16 QAM	Under development in U.S.
6.312	300-channel master group (812–2,044 kHz)	12-level vestigial sideband	Under development in Japan
0.056	Portion of satellite channel (FDMA)	QPSK	In service with several companies
1.544	Portion of satellite channel (FDMA)	QPSK	In limited use
60–64	36-MHz satellite channel (TDMA)	QPSK	In limited use

violations in excess of the threshold are detected, a spare line is bridged on at the transmitting (head) end and if a violation-free signal is received on this line, the tail-end switch to spare is completed. If the spare line also has violations, there is probably an upstream failure and no switch is performed. A multiplex-protection switch is typically based on a time-shared monitor which evaluates each of several multiplexers and demultiplexers in turn by pulse-by-pulse comparison of the actual output with correct output based on current input. Multiplex monitors also usually check the incoming high- and low-speed signals using the line-code redundancy.

ECHO CONTROL, SIGNALING, AND SPEECH INTERPOLATION

68. Echo Control.[26] The causes and effects of echo are discussed in Par. **22-27**. If the round-trip delay is less than 45 ms, acceptable echo performance can be achieved by designing all trunks to have a loss which increases the return loss of the echo. The net loss of a connection required presently is

$$\text{Net loss} = (0.1)(\text{round-trip delay, ms}) + (0.4)(\text{number of trunks}) + 5.0 \text{ dB}$$

The 5.0 dB is provided in the toll-connecting trunks and associated equipment, and the other terms (known as via net loss or VNL) are included in the design of all trunks. (With the advent of digital systems, which do not lend themselves to the VNL administration, fixed-loss plans have been proposed.) Above 45 ms delay, VNL results in excessive loss, and echo suppressors or echo cancelers are used to reduce the required loss.

An *echo suppressor* is a pair of voice-operated switches which insert a loss of 35 dB or more in the reverse direction while one subscriber is talking. If both subscribers talk simultaneously, only the louder talker is heard. An echo suppressor can be located at or near only one end of a connector and control both directions (full echo suppressor) or be split and provided at both ends.

While echo suppressors control echo effectively, the time necessary to recognize speech and actuate the switches (2 to 50 ms in various types) can clip the beginning of words, producing another impairment. Since this effect would be very detrimental for many voice-band data signals, echo suppressors are equipped with a tone disabler, which disconnects the suppressor upon receipt of a short burst of a sine wave of frequency between 2,010 and 2,240 Hz, which is provided by the modem. Certain toll switches are also equipped to disable echo suppressors on attached trunks, to avoid more than one suppressor in a connection.

Where satellite transmission is used, the much greater round-trip delay (0.6 s) aggravates the echo problem. This has led to the development of the echo canceler, which internally adjusts taps on a delay line to duplicate the effect of the echo and then subtracts the resultant signal, canceling the echo. This internal adjustment is based on measuring the echo and is done rapidly and continuously. Echo cancelers are much more complicated than echo suppressors, but an LSI

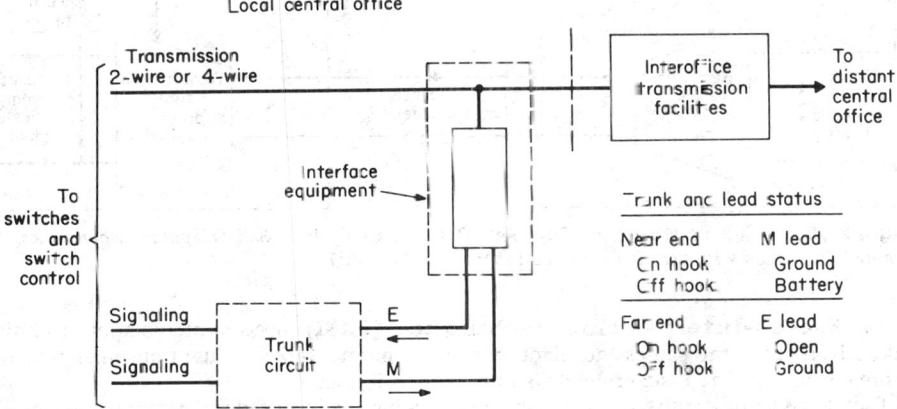

Fig. 22-51. E and M leads. *(From Ref. 40. Copyright 1977, American Telephone and Telegraph Co. Used by permission.)*

(large-scale-integration) version which provides the function for digitized signals on a single chip incorporating 35,000 devices has been developed.[46]

69. Signaling on Transmission Systems. As explained in detail in Pars. **22-72** and **22-84** to **22-91**, signaling refers to transmission of information controlling a call, such as on-hook–off-hook (supervision) and address (number being called). Where this information is channel-associated and is to be carried by the transmission system, it normally appears in the switch as relay-contact closures.

One widely used interface technique, E and M, is shown in Fig. 22-51. Other systems have similar block diagrams. The interface extension equipment required depends on the interoffice transmission facilities used. Where voice-band trunks with dc continuity are involved, various arrangements based on the application of either continuous and/or pulsed battery or ground signals to the wires may be used. Where dc continuity is not present, as in analog carrier systems, the usual signaling systems are the single-frequency system and multifrequency pulsing system. The single-frequency (SF) system places a 2,600-Hz tone on the trunk to signal the on-hook condition. This is removed when the trunk goes off-hook. The multifrequency (MF) pulsing system is used only to transmit address information. It operates by having the switch send simultaneously two (of six possible) frequencies in the voice band to represent each digit of the address. While this does not require any special transmission equipment, it must be supplemented by SF or a similar system to carry off-hook–on-hook information. A related address-signaling mode used on loops is dual-tone multifrequency, which uses combinations of two out of eight audio tones. It is used in conjunction with battery and ground-supervisory signaling.

In digital systems the interface equipment is part of the channel bank, and the E and M lead status is signaled to the far end by making the robbed bits (Fig. 22-43) 1 or 0. As 2 b are available per channel, four states can be signaled in each direction. MF can, of course, be carried over digital systems as well.

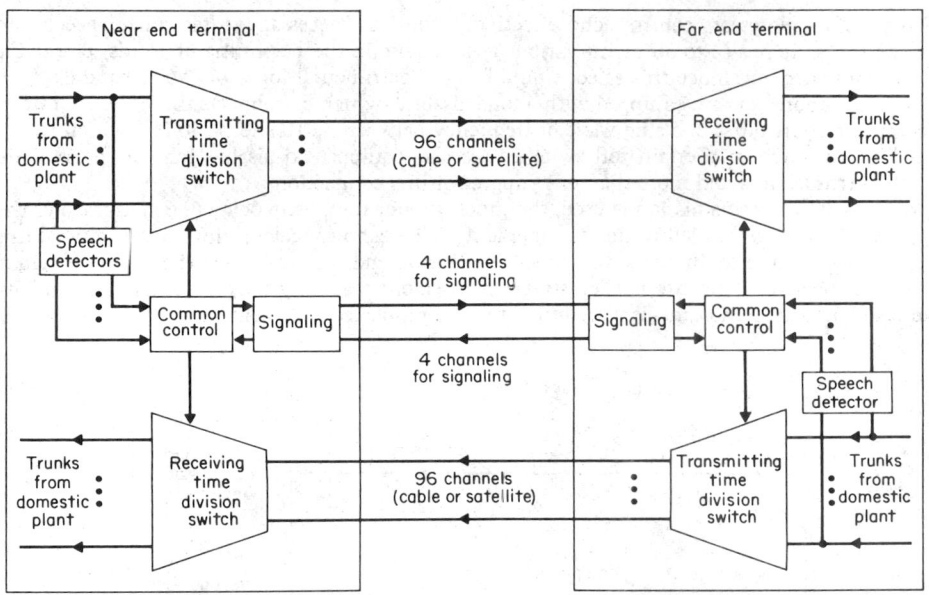

Fig. 22-52. TASI-B block diagram. *(From Ref. 47. Copyright © 1970, Bell Telephone Laboratories, Incorporated. Reprinted with permission of Bell Laboratories RECORD.)*

70. Speech-Interpolation Techniques (TASI). Speech-interpolation techniques take advantage of the pauses and silent intervals in normal speech to use transmission facilities more efficiently. The basic principle is illustrated in Fig. 22-52.

Each transmitting trunk has a speech detector which indicates to the common control when a voice connection is needed. The common control consists of a central processor made of typical computer building blocks. These blocks register, store, and compare information. When a connection is to be established or taken down, the common control signals the other end. Signaling includes other commands and responses in addition to connection information. The specific numbers in the figure are for TASI-B[47] (time-assignment speech interpolation), but the original TASI system and subsequent systems (also called DSI, for digital speech interpolation, in some cases) are similar in principle. The number of trunks is limited by the loss of speech due to the delay in recognizing the presence of speech and establishing a connection (*processing clip*) and the loss of speech when more channels are needed than are available (*competitive clip*).These parameters for TASI-B are given in Table 22-28.

TABLE 22-28   TASI-B Parameters

Trunks connected to TASI	274
Average trunks off hook	235
Speech activity	37% of off-hook time
Processing clip	1.5%
Competitive clip	0.5%

The processing and competitive clips cause subjective degradation in speech and can cause severe problems with some types of data signals, requiring tone disablers as for echo suppressors. To avoid the appearance of a dead circuit, any listener not connected to a talker is fed noise approximating that which exists on the circuit.

The expense of the speech-interpolation equipment means that this principle has found its major application on long intercontinental trunks, originally on submarine cable, and more recently on satellite links as well. Where the trunks are digital, or can be digitized, the switches and speech detectors can take advantage of digital integrated circuits for more economical realization, and some use of TASI within the United States has been reported.

Switching Systems

BY A. E. JOEL, JR.

A telecommunication service that includes directing a message from any input to a selected output requires a switching system. The terminals are connected to the switching system by *loops*, which together with the terminals are known as *lines*. The switching systems at nodes of a network are connected to each other by channels called *trunks*. This section deals primarily with systems that provide *circuit switching*, i.e., provision of a channel that remains for the duration of a call. Other forms of switching are noted in Par. **22-83**. Switching systems find application throughout a communication network. They range from small and simple key telephone systems or PBXs to the largest local and toll switching systems.

SWITCHING FUNCTIONS

71. Introduction. A switching system performs certain basic functions plus others that depend upon the type of service being rendered. Generally switching systems are designed to act on each message or call, although there are some switches that perform less often, e.g., to switch spare or alternate facilities. Each function is described briefly here and in greater detail in specific paragraphs devoted to each function.

The basic function of a telecommunication switching system is connection by the *switching network*,* the transfer of communication from a source to a selected destination. Vital to this basic function are the additional functions of *signaling* and *control* (call processing) (Fig. 22-53). Other functions are required to *operate*, *administer*, and *maintain* the system.

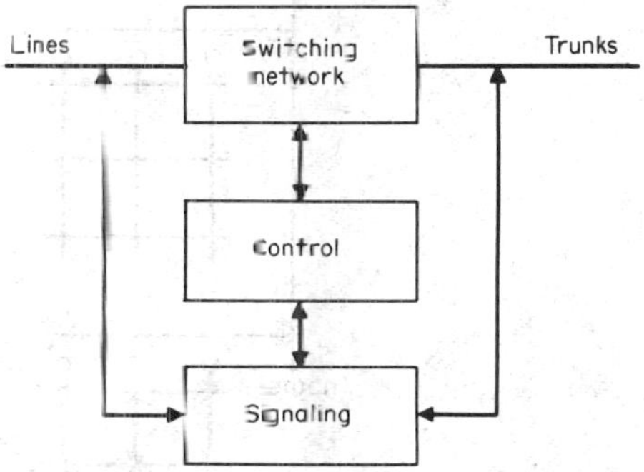

72. Signaling. Almost all switching today is remote-controlled switching. Transfer of control information from the user to the switching office and between offices requires electrical

Fig. 22-53. Basic switching functions in circuit switching

technology and a format. This is known as signaling, and it can be thought of as a special form of data communication.

Originally, signaling was developed to accommodate the type of switching technology used for local switching. Most of these systems used dc electric signals. Later, with the advent of longer distances, signaling using single- and multiple-frequency tones in the voice band was developed. Most recently, long-distance signaling using digital signals has been introduced over dedicated networks, distinct from the talking channels.

As dialing between more distant countries became feasible, specific international signaling standards were set. The standards were necessarily different from national signaling standards since it was necessary to provide for differences in call handling such as requests for language assistance or restrictions in routing.

73. Control. Control of switching systems and their application is called *system control*, the overall technique by which a system receives and interprets signals to take the required actions and to direct the switching network to carry them out.

In the past, the control of even the most complex switching systems was accomplished by logic circuits using relays and other electromechanical switches. Today, virtually all new systems employ stored-program control (SPC). By changing and adding to a program one can modify the behavior of a switching system faster and more efficiently than w th wired logic control. Several manufacturers provide SPC to augment or replace the wired logic of electromechanical switching systems.

*The term *switching network* will be used in these paragraphs to identify the implementation of the connection function within a switching system. The term *communications network* will refer to the collection of switching systems and transmission systems which constitute a communication system.

74. Switching Networks. The switching network provides the function of connecting channels within a circuit-switching system. Store-and-forward or packet-switching systems do not need complex switching networks but do require connecting networks such as a bus structure.

In the past, switching systems have generally derived their names or titles from the type of switching technology used in the switching network, e.g., step-by-step, panel, and crossbar. These devices constitute the principal connective elements of switching networks. Portions of the device that change in electrical impedance are known as *crosspoints*. Typically electromechanical crosspoints are metallic and go from almost infinite to zero impedance; electronic

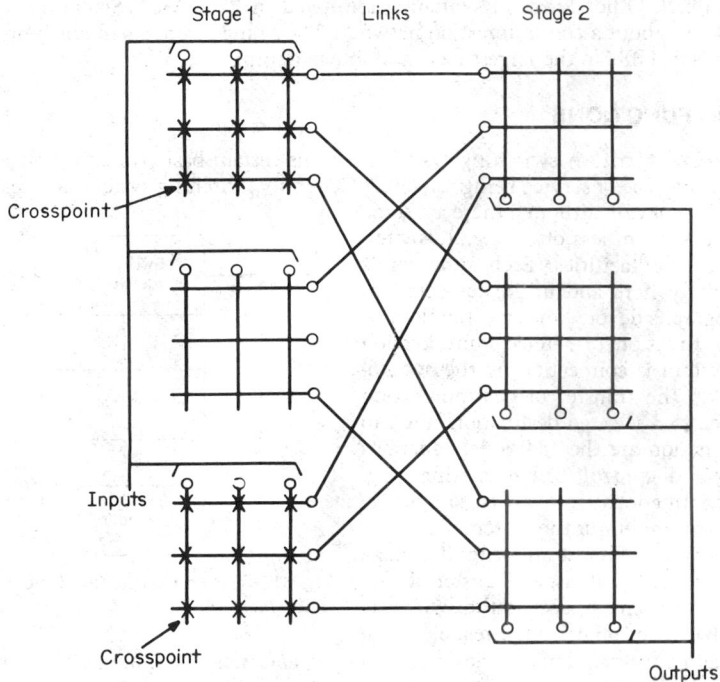

Fig. 22-54. Two-stage switching network.

crosspoints change impedance by several orders of magnitude. The off-to-on impedance ratio must be great enough to keep intelligible signals from passing into other paths in the network (crosstalk). A plurality of crosspoints accessible to or from a common path or *link* is known as a *switch* or, if a rectangular array, a *switch matrix*. Each crosspoint may contain more than one gate or contact. The number depends on the information switched and the technology.

Generally a number of stages of switches are used to provide a network in order to conserve the total number of crosspoints required. A simple example of a two-stage network is shown in Fig. 22-54. Instead of a single 9×9 switch with 81 crosspoints, the two-stage network uses six 3×3 switches requiring 54 crosspoints. An output of each first-stage switch is connected to an input of a second-stage switch via a link. There is a connectable path for each and every input to each and every output. Since each input has access to every output, the network is characterized as having *full access*. However, two paths may not simultaneously exist between two inputs on the same first-stage switch and two outputs of a single-output-stage switch. (There is only one link between any first- and second-stage switch.) A second call cannot be placed, and this network is said to be a *blocking network*. By making the switches bigger and adding links to provide parallel paths, the chance of incurring a blocking condition is reduced or eliminated. The design of most practical switching networks includes a modest degree of blocking in order to provide an economical design.

Large central-office switching networks may have more than 100,000 lines and trunks to be interconnected and provide tens of thousands of simultaneous connections. Such networks typically require six to eight stages of switches and are built to carry loads which result in less than 2% of call attempts in the busy hour being blocked.

75. Network Control. While the switching system as a whole requires a control (Par. 22-73), the control required for a switching network may be separated in part or in its entirety from the system control function. The most general network control accepts the address of the input and output(s) for which an interconnection is required and performs all the logic and decision functions associated with the process of establishing (and later releasing) connections. The control for some networks may be common to many switches or, as in the step-by-step system, individual to each switch.

Some form of memory is involved with all networks. It may be intimately associated with the crosspoint device employed, e.g., to hold it operated, or it may be separated in a bulk memory. The memory keeps a record of the device in use and of the associated switch path. (In some electronic switching systems it may also designate a path reserved for future use.)

76. Operation, Administration, and Maintenance (OAM). The functions introduced in the preceding paragraphs are required to form the basic foundation of a switching system, and they are essential. Generally when switching systems are to be used by the public, a high-quality, continuous service day in and day out over every 24-h period is required.

A system providing such reliable service requires additional functions and features. Examples are continuity of service in the presence of device or component failure and capability for growth while the system is in service.

Separate maintenance and administrative functions are introduced into systems to monitor, test, and record and to provide human control of the service-affecting conditions of the system. These functions together with a human input/output (I/O) interface constitute the basic maintenance functions needed to detect, locate, and repair system and component faults.

In addition to specific maintenance functions, *redundancy* in the switching system is usually necessary to provide the desired quality of service. Complete duplication of an active system with a standby system will protect against one or more failures in one system but presents severe recovery problems in the event of a simultaneous failure of both systems. Judicious subdivision of the system into parts that can be reconfigured (e.g., either of a pair of central processors may work with either of a pair of program memories) can greatly increase the ability of the system to continue operation in the presence of multiple faults.

Where there are many switching entities in a telecommunications network and as systems have become more reliable and training more expensive, the centralization of maintenance has become a more efficient technique. It ensures better and more continuous use of training and can also provide access to more extensive automated data bases that benefit from more numerous experiences.

For public operation, a basic subset of administration and operation features has become accepted as required features. These include the collecting of traffic data, service-evaluation data, and data for call charging.

SWITCHING NETWORKS

Three different aspects will be considered in the design of switching networks: (1) the types of switching network, (2) the technology of the devices, and (3) the topology of their interconnection.

77. Types of Switching Networks. The three types of switching networks are known by the manner in which the information passes through the network.

In *space-division* networks the electric signals representing the message pass through a succession of operated crosspoints that are assigned to the call for all or most of its duration. Even with relatively slow crosspoints, the control interval is small compared with the message time. Therefore space-division networks can be constructed of metallic electromagnetically or electromechanically operated crosspoints or semiconductor crosspoints. The signals are expected to pass through the network without distortion. The network elements need not be linear if the only signals transmitted are tolerant of nonlinearity. The crosspoints and the wiring between them, *links*, influence the allowable bandwidth of the signals that can be carried. Pulses that vary in amplitude or digitized sample pulses can be transmitted through space-division networks. Space-division networks are generally used where the message signals require wide bandwidth or considerable power.

Time-division networks operate wih streams of electric pulses representing interleaved signals from a number of inputs. Voice signals may be sampled, and the information retained as short pulses. Other voice signals are similarly sampled, and the pulses are interleaved or *multiplexed* (Fig. 22-55). For voice communication the sampling is generally at a rate of 8 kb/s, which allows the eventual reconstruction of speech (to 3,500 Hz) with acceptable fidelity.

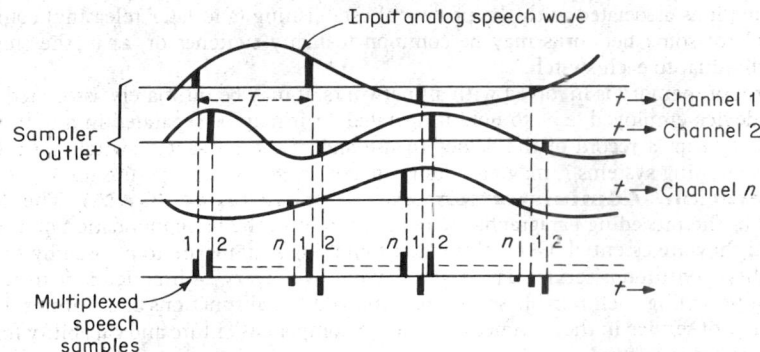

Fig. 22-55. Sampling and multiplexing. *(Adapted from Ref. 1. Copyright 1977, Bell Telephone Laboratories. Used by permission.)*

Figure 22-55 shows *n time slots* or voice-band *channels*, repeating every T μs ($T = 125$ in most time-division networks). One such array of *n* time slots represents one *frame*. The number of time slots contained in a frame depends on the speed with which the electronic gates can respond.

Figure 22-56 shows a simple time-division switch comprising a bus and electronically operated gates. By controlling the time of operation of selected gates the switch makes the requisite connections. For example, during a first time slot, switches $x(1)$ and $y(m - 1)$ are closed and information is conveyed from input $I(1)$ to output $O(m - 1)$.

In the next time slot switch $x(m - 1)$ and $y(2)$ can be operated to interconnect $I(m - 1)$ and $O(2)$, and so on. The pattern of control pulses is repeated in successive frames. This is accomplished by storing the connection pattern in memory and reading the contents of that memory in sequential circular fashion to obtain the corresponding pattern of control signals.

In the example shown, the amplitude of the pulse contains the information. This information can also be contained by varying the width of the pulse or by encoding the value of the amplitude in a sequence of binary digits. (See Par. **22-56** for details of voice encoding.) Time-division networks which switch information in this last format are referred to as *digital time-division networks* (DTDN).

When such networks are designed to work with the format of digital carrier systems either in the loop (digital pair-gain systems, or line concentrators) or in the trunk side (T carrier in North America), the multiplexed bit streams can be directly applied to the inputs and no demultiplexing of the carrier bit stream is required before switching. This represents a cost advantage of such a switching arrangement. Standard frames of 24 voice channels (mainly in North America and Japan) and 30 voice channels (elsewhere) are most commonly used. As digital carrier systems become more prevalent, digital time-division switching systems become more attractive.

Operating administrations are looking forward to an integrated digital network where, at some future time, the customer's telephone will provide the digital encoding. Analog voice signals will be found only within the customer's station equipment. An integrated digital network would accommodate voice or data at 56 or 64 kb/s. Such an evolution is depicted in Fig. 22-57.

In DTDNs there are two kinds of time-division switching elements, referred to as *space* switches and *time* switches (or time-slot interchanges, TSI). The space switch (also known as a time-multiplexed switch, TMS), shown in Fig. 22-58, operates like the normal space-switch matrix described in Par. **22-74**, but with each new time slot the

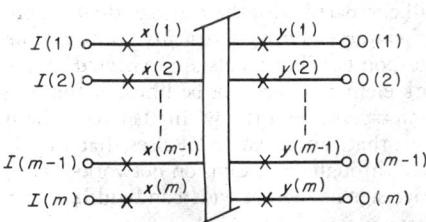

Fig. 22-56. Time-division bus system.

electronic gates are reconfigured to provide a new set of input and output connections. The two-dimensional space switch now has an added third dimension of time.

The time-slot interchange uses a buffer memory, into which one frame of input information is stored. Under direction of the contents of the control memory, transfer logic reorders the sequence of information stored in the buffer, as shown in Fig. 22-59. To avoid race conditions the input and output access of the buffer memory must be interleaved in time. The TSI necessarily creates delays in handling the information stream. To meet overall transmission-delay requirements, the number of TSI stages that can be used in a DTLN is limited.

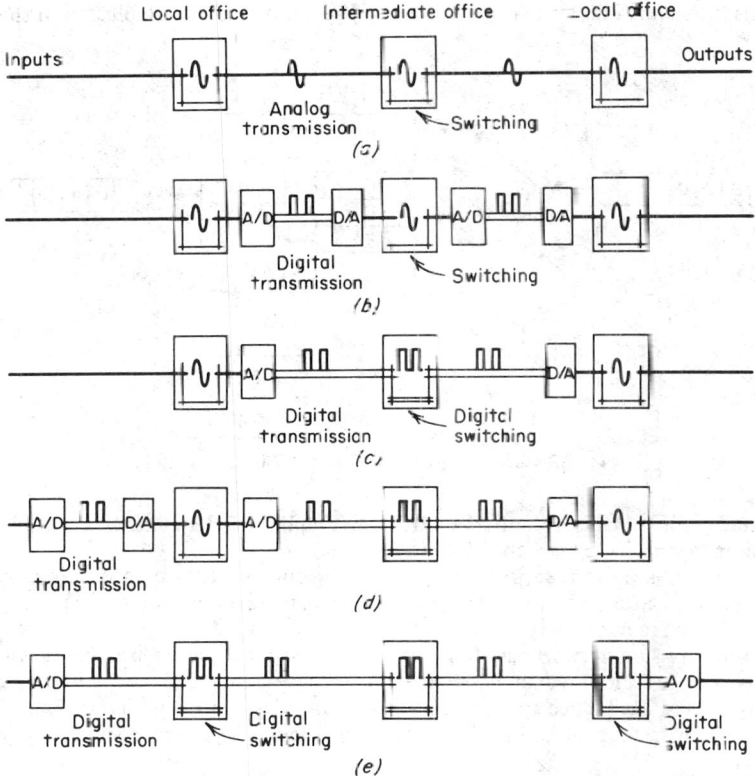

Fig. 22-57. Evolution of an integrated digital network: (*a*) integrated analog switching and transmission, (*b*) growth of digital trunk transmission, (*c*) digital intermediate office and trunk transmission, (*d*) digital pair gain, (*e*) digital local office with A/D conversion at the central office (*right*) or at or near the station (*left*). (*From A. E. Joel, Jr., Digital Switching: How It Has Developed, IEEE Trans. Commun., July 1979, Vol. COM-27, No. 7, pp. 948-959. Copyright 1979 IEEE. Used by permission.*)

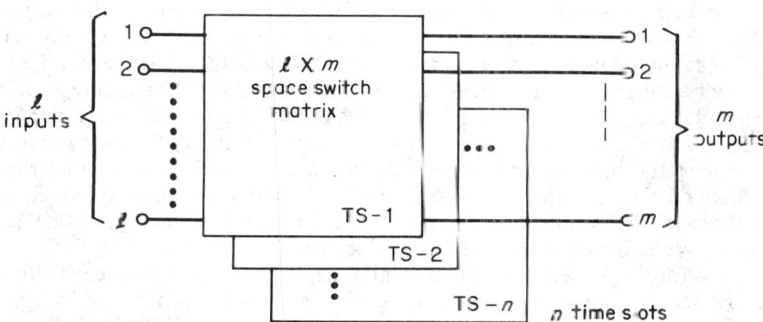

Fig. 22-58. Time-multiplex switch (TMS); space switch.

Channels arriving at the switch in time-multiplexed form can be further multiplexed (and demultiplexed) into frames of greater (or lesser) capacity, i.e., at a higher rate and with more time slots. This function is generally used before using TSI so that channels from different input multiplexes can be interchanged.

Time-division switches are designated by the sequence of time and space stages through which the samples pass, e.g., TSST.

Analog samples can be switched in both directions through bilateral gates. An efficient and accurate transfer of the pulse is effected by a technique known as *resonant transfer*. For most analog and all digital time-division networks, however, the two directions of signals to be switched are separated. Therefore two reciprocal connections or what is known as four-wire connections (the equivalent of two wires in each direction) must be established in the network.

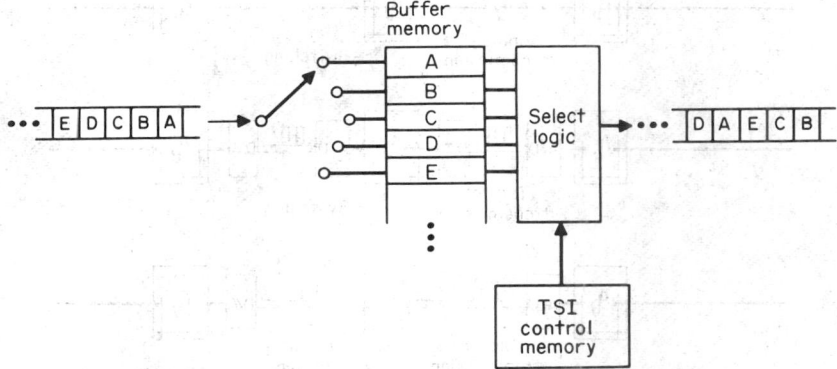

Fig. 22-59. Time-slot interchange (TSI): time switch.

When connections transmit in only one direction, amplification and other forms of signal processing can more readily be switched into the network.

If multiplexing is performed in such a way that samples from an incoming circuit can be assigned arbitrarily to any of a number of time slots on an outgoing circuit, time-slot interchange and multiplexing are effectively achieved in a single operation.

Digital time-division networks are becoming economically attractive because they can be constructed with large-scale integrated components. However, the need for four-wire connections requires hybrids in local office applications when connected to two-wire loops. The hybrid must be balanced with respect to the impedance of the particular loop served since loop impedances vary widely.

The number of time slots provided for in a time-division network depends upon the speed employed. Typically in a voice-band network there may be from 32 to 1,024 time slots. The coded information in digitized samples may be sent serially (typically 8 b per sample for voice signals), in parallel, or combinations. Extra bits are sometimes added as they pass through the switch for checking parity (Pars. **22-134** to **22-136**), for other signals, or to allow for timing adjustments (Par. **22-58**).

Figure 22-60 shows the block diagram of the switching network for a no. 4 ESS, a large digital time-division switching system presently being deployed mainly in North America. Like most such DTDNs, this switching network is symmetrical. Analog signals are multiplexed into frames of 120 channels, and incoming digital T carrier streams (five T1 lines with 24 channels each) are further multiplexed to frames of 120. The information is buffered in registers to permit the time-slot interchange. The TSIs on the right side of the figure reverse the order of selecting and buffering; selected input sequences driven by a control memory (not shown) and sequentially gated out of the buffer attain the desired interchange in time. Note that the network shown is unilateral (left to right); the actual network includes a second unilateral network to carry the right-to-left portion of the conversation. This network can accommodate over 47,000 simultaneous conversations with essentially no blocking.

Frequency Division.   Since frequency-multiplex carrier has been used successfully for transmission, its use for switching has been proposed. Connections are established by assigning the same carrier frequency to the two terminals to be connected. Generally to achieve this requires

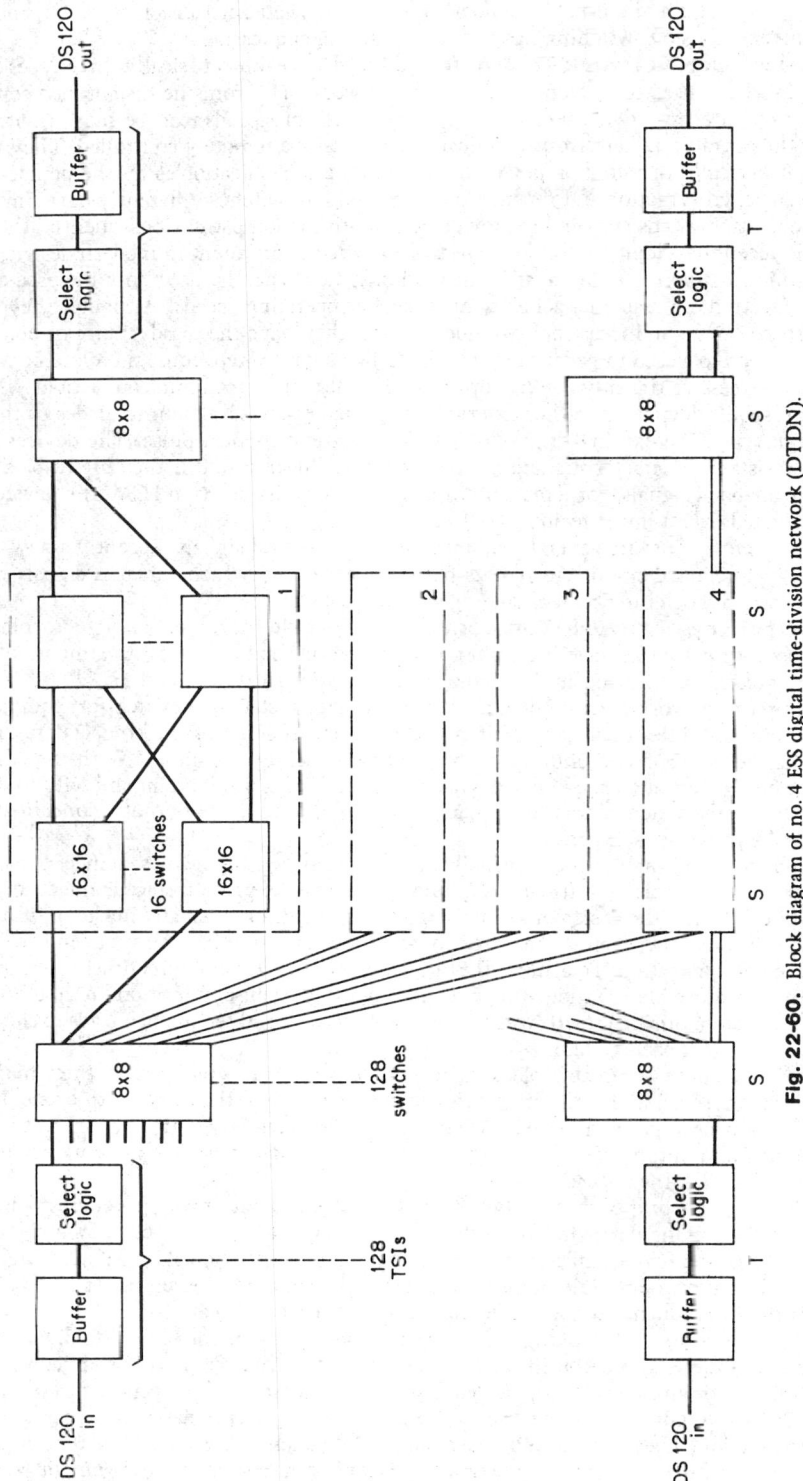

Fig. 22-60. Block diagram of no. 4 ESS digital time-division network (DTDN).

a tunable modulator and a tunable demodulator to be associated with each terminal, and therefore frequency-division switching has had little practical application.

78. Switching-Network Technology. Broadly speaking, basically three types of technology have been used to implement switching networks. (1) From the distant past comes the *manually operated switch*, where wires, generally with plug ends, can be moved within the reach of the operator. (2) *Electromechanical* switches can be remotely controlled. They may be *electromagnetically operated* or *power-driven*. Another classification is by the contact movement distance, *gross* motion and *fine* motion. Gross-motion switches inherently have limitations in their operating speeds and tend to provide noisy transmission paths. Consequently, they have seen little recent development. (3) The *electronic* switch is prevalent in modern design.

Electronic Crosspoints. Gross- and fine-motion switches can be used only in space-division systems. Electronic crosspoints achieve much higher operating speeds. Although they can be used in space-, time- and frequency-division systems, they have the disadvantage of not having as high an open-to-closed impedance ratio as metallic contacts. Steps must therefore be taken to ensure that excessive transmission loss or crosstalk is not introduced into connections.

The crosspoint devices are either externally triggered or are self-latching diodes of the four-layer *pnpn* type. The external trigger may be an electric or optical pulse. The devices have a negative resistance characteristic and are operated in a linear region if they are to pass analog voice or wide-band signals. For fixed-amplitude pulse transmission, as in PCM, the devices need not be operated over a linear region.

Electronic crosspoints are generally designed to pass low-level signals. Recently a new class of high-energy integrated-circuit crosspoints has been developed which can pass signals used in telephone circuit switching such as ringing and coin control.

79. Switching-Network Topology. Of all the switching functions, the topology and traffic aspects of networks have been most amenable to normal analytical treatment, although many less precise engineering and technology considerations are also involved.[48,49]

The simplest network is one provided by a single-stage rectangular switch (or, equivalently, a TSI) so that any idle input can reach any idle output. If some of the contacts are omitted, *grading* has been introduced and not every input can reach every output. With the advent of electronic crosspoints and time division, grading has become less important and will not be pursued further here. When inputs to a rectangular switch exceed the outputs, *concentration* is achieved; the converse is *expansion*.

A telephone switching network is usually arranged in stages. Input lines connect to a concentration stage, several stages of distribution follow, and a last expansion stage connects to trunks or other lines. Within the design of a switching system, provision is usually made for installation of switches in only the quantity required by the traffic and number of inputs and outputs of each particular application. To achieve this, the size of each stage and sometimes the number of stages is made adjustable. Consideration of control, wiring expense, transition methods (for rearranging the system during growth without stopping service), and technology leads to the configurations selected for each system.

In order to achieve acceptable blocking in networks that are smaller than their maximum designed size more parallel paths are provided from one stage to the next. In this case, because the distribution need is also reduced, the connections between stages are rewired so that those switch inputs and outputs which are required for distribution in a large network are used for additional parallel paths instead.

Nonblocking networks of the Clos type[50] can be constructed with many fewer crosspoints than the single-stage rectangular switches to which they are equivalent. Clos networks use expansion in the first stage to provide sufficient parallel paths in the distribution stages to guarantee a path from any idle input to any idle output under all traffic conditions. Even fewer crosspoints are needed if the internal connection can be *rearranged* for each new call.[51]

Switching Networks. In talking about network topologies it is convenient to divide the network into groups of stages according to the direction of the connection. (Calls are considered as flowing from an originating circuit, associated with the request for a connection, to a terminating circuit.) Local interoffice telephone trunks, for example, are usually designed to carry traffic in only one direction. The trunk circuit appearances at a tandem office are then either originating (incoming) or terminating (outgoing). Figure 22-61a illustrates such an arrangement where the whole network is *unidirectional*.

Telephone lines are usually *bidirectional*: they can originate or terminate calls. For control or other design purposes, however, they can be served by unidirectional stages, as shown in Fig. 22-61b. Concentration and expansion are normally used with line switching to increase the internal

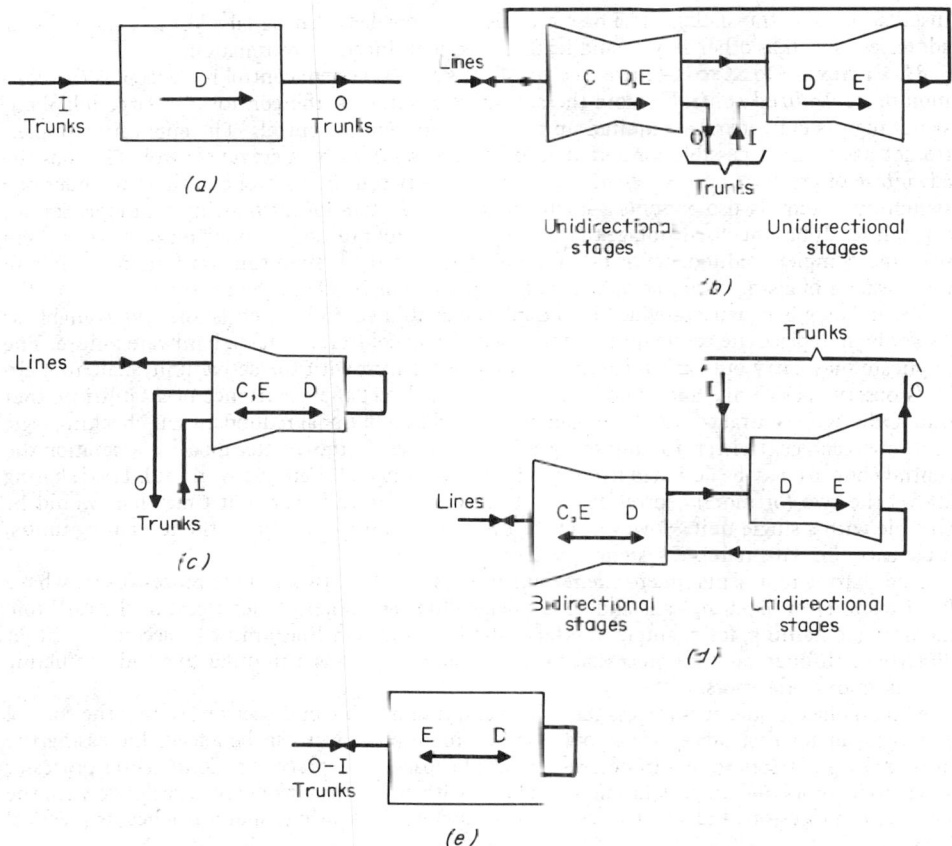

Fig. 22-61. Switching-center networks: (a) unidirectional network (tandem), (b) unidirectional network (local), (c) bidirectional network (local), (d) combined network (local-toll), (e) bidirectional network; line stages: C = concentration, E = expansion, D = distribution; trunk stages O = outgoing I = incoming. Arrows indicate direction of progress of setup of call.

network occupancy above that of lines. In smaller systems a bidirectional network can serve all terminal needs: lines, trunks, service circuits, etc. (Fig. 22-61c). When interconnection between trunks, as in a combined local-tandem office, is required, line stages can be kept bidirectional while trunk stages are unidirectional (Fig. 22-61d). When the majority of trunks are bidirectional, as may occur in a toll office, a bidirectional switching network is used (Fig. 22-61e). Many other configurations are possible.

SYSTEM CONTROLS

80. Stored Program Control. As discussed in Par. 22-73, most modern systems use some form of general-purpose stored-program control (SPC). Full SPC implies a flexibility of features, within the capability of existing hardware, by changes in the program.

SPC system controls generally include two memory sets: one for the program and other semipermanent memory requirements and one for information that changes on a real-time basis, such as progress of telephone calls, or the busy-idle status of lines, trunks, or paths in the switching network. These latter writable memories are *call stores* or *scratch-pad memories*. The two memories may be in the same storage medium, in which case there is a need for a nonvolatile backup store such as disc or tape. Sometimes the less frequently used programs are also retrieved from this type of bulk storage when needed.[52]

Nonprogram semipermanent memory is required for such data as *parameters* and *translations*. A switching system is generally designed to cover a range of applications; memory, network and other equipment modules are provided in quantities needed for each particular switching office. Parameters define for the program the actual number of these modules in a particular installation. The translations data base provides relations between signal addresses and physical addresses as well as other service and feature class identification information.

81. Central Control. *Single Active.* The simplest system control in concept is the *common* or *centralized control.* Before the advent of electronics, the control of a large telephone switching system generally required up to 10 or more central controls. The application of electronics has made it possible for a system to be fully serviced by a single control. This has the advantage of greatly simplifying the access circuits between the control and the remainder of a switching system. It also presents a single point in the system for introducing additional service capabilities. It has the disadvantage that a complete control must be provided regardless of system size, and complete redundancy is often required so that the system can continue to operate in the presence of a single trouble or while changes are being made in the control.[53]

Redundancy is usually provided by a duplicate central control which is idle and available as a standby to replace the active unit (the unit actually in control) if it has a hardware failure. The duplicate may carry out each program step in synchronism with the active unit; matching circuits observe both units and almost instantaneously detect the occurrence of a fault in either unit. Otherwise central-control faults can be detected by additional redundant self-checking logic built into each central control unit or by software checks. In these latter modes of operation the central controls may be designed to operate independently and share the workload. *Load sharing* allows the two (or more) central controls to handle more calls per unit time than would be possible with a single unit. However, in the event of a failure of one unit, the remaining unit(s) must carry on with reduced system capacity.

Load-sharing represents *independent multiprocessing* where two or more processors may have full capability of handling calls and do not depend on each other. At least part of the call store memory (containing, for example, the busy-idle indications of lines) must be accessible, either directly or through another processor, to more than one processor in order to avoid conflicting actions among processors.

A small office would require less than the maximum number of processors, so that the control cost is lower for that office; as the office grows, more processors can be added. Increasing the number of processors results in decreasing added capacity per processor. Conflicts on processor access to memory and other equipment modules, with accompanying delays, accelerate with the number of processors. Independent multiprocessing or load sharing rapidly reaches its practical limit.

Functional Multiprocessing. Another way to allocate central-control workload is to assign different functions to different processors. Each carries out its task; together they are responsible for total capability of the switching system. This functional or *dependent* multiprocessing arrangement can also evolve from a single central control. A small office can start with the entire program in one processor. When one or more functional processing units are added, the software is modified and apportioned on a functional basis. As in load sharing, the mutually dependent processors must communicate with each other directly or through common memory stores.

In handling calls, each processor may process a portion and hand the next step to a succeeding processor, as in a factory assembly line. This *sequential multiprocessing* has been used in wired-logic electromechanical switching systems. Virtually all SPC-dependent multiprocessing arrangements are *hierarchical.* A master processor assigns the more routine tasks to subsidiary processors and maintains control of system.

The one or more subsidiary processors may be *centralized* or *distributed.* If the subsidiary processors are centralized, they have full access to network and other peripheral equipment. Distributed controls are dedicated to segments of the switching network and associated signaling circuits. As network and associated signaling equipment modules are added, the control capability is correspondingly enlarged. Most newer switching systems use distributed controls.

TYPES OF SWITCHING SYSTEMS

In the preceding paragraphs the various switching functions were described. A variety of switching systems can be assembled using these functions. The choice of system type depends on the environment and the quantity of the services the system is required to provide. Combining the various types of systems within one embodiment is also possible.

82. Circuit Switching. Circuit switching is generally used where visual, data, or voice messages must be delivered with imperceptible delay (<0.050 s) and are relatively long. For these applications connections are established through the switch that will pass the required bandwidth. The Nyquist criterion (see Sec. 4) is used in time-division networks by choosing the sample rate to be at least twice the desired maximum bandwidth.

Circuit switching usually (but not always) implies message dialogue, i.e., reciprocal communication in both directions.

Circuit switching has the advantage of being able to switch a very broad spectrum of signal rates. For example, analog or digital video signals can be switched using either metallic or nonmetallic crosspoints.

83. Systems Other than Circuit Switching. Services that provide one-way transmission and more or less deferred delivery of hard-copy messages have been available in both public and private versions for many years; the switching involved is sometimes called *message switching*. Messages are stored in the switching system until a transmission channel is available, thus giving efficient use of channels, or until it is convenient to deliver the message to the recipient's station, thus giving efficient use of stations. Further, messages can be provided from the source to the communications system in bulk, or over a single high-volume *port*, which permits efficiencies in the nature and operation of such a source. Semipermanent record storage and message numbering can be provided by the switching system. Message switching lends itself particularly well to *multiple-address messages* It can also be used to provide special features, e.g., help in the preparation of messages.

The implementation of automatic message switching started with teletypewriterlike equipment and has evolved to systems that are computer-based For example, the Bell System put into operation in 1969 (for internal traffic) a service employing the standard large electronic local telephone switching system augmented with stores, special input-output equipment, and special programs. The airlines industry has long been a user of computer-based electronic message switching for applications ranging from operations traffic to agent traffic. With time, the concepts of message switching have been extended from hard copy to transient text display, voice, and graphics.

To get some of the benefits of traditional message switching and add the benefits of two-way conversational mode plus, if desired, minimal delay, a scheme called *packet switching* was introduced (Par. **22-10**). The switching system typically employs line- and trunk-terminating units, processing units, stores, and buses. The architecture of the switching system is adapted to derive from the header of a packet the information corresponding to signaling in a circuit-switching system.

SIGNALING

84. Basic Purposes. Signaling has three basic purposes: *supervising* the call or message, *addressing* the call or message, and conveying *supplementary* information relative to the call. Supervision consists of indicating a call origination answer or end, and station alerting. The application of these signals may be between the stations and the switch, in which case it is known as *station signaling*, or between switching offices, when it is known as *interoffice signaling*. There are two basic approaches to signaling, *per channel* and *common channel*. Per channel signaling may be in or out of the bandwidth used to carry communications. In the case of digital time-division multiplex transmission, signals may occur in each channel time slot (*in slot*) or in a separate signaling channel (*out of slot*.) Common-channel signaling, on the other hand, is carried over one or more channels dedicated to signaling which usually serve many trunks. The electrical characteristics of signals are divided into three categories. dc, ac, and *digital signals.*

85. Standards and Compatibility. Only by following standards for the transmission and use of signaling can the various elements of a telecommunications network function together. Signaling standards for public networks are generally established by the CCITT. The deliberations leading to the setting of these standards involve telecommunication administrations and recognized private operating agencies with the assistance of scientific and industrial organizations. Over the years, as new technology and requirements have appeared, new and revised signaling standards have followed.

National systems are more dependent upon the type of signaling required by the local switching systems, and conditions vary widely throughout the world.

86. Station Signaling. There are few varieties of station signaling in public networks. One is the universal dc loop supervision and dial pulsing. Other signals sent over the loop are those for ringing, coin control, party identification, toll denial (signals indicating limitation of access to the DDD network of a particular station), metering, etc. Worldwide they vary in voltage, frequency, and how ground is used as a conductor. An important attribute is the distance or electrical range over which each signal functions satisfactorily.

One standard ac station-address signaling using two-out-of-eight frequencies (twice one-out-of-four) is known as *dual-tone multifrequency* (DTMF).[54] For purposes of interconnection to the United States network an Electronics Industries Association standard[55] has been issued on station signaling as applied to PBXs.

87. Distributed Switching. The range of dc signaling has been particularly important in station signaling since it has determined the location and the number of wire centers required to serve a given area. Various signaling arrangements were developed to extend the dc loop and ringing range of central switching systems. The lower cost of electronics for small transmission and switching systems has made it economical to extend the range further and to decentralize switching.

88. Interoffice Signaling. In the past many forms of dc interoffice signaling have been developed and used. They were designed to accommodate specific electromechanical switching systems. Reference 56 describes those used in the United States in general terms.

Signaling systems designed for use between offices, particularly over long distances and between countries, must be designed to be transmitted over carrier systems either analog or digital. For analog carrier trunks international standards have been set involving codes of two-out-of-six frequencies (multifrequency code, MFC) for address signaling and one or two frequencies for supervisory signaling. In digital carrier systems a means of carrying encoded supervisory signals in the bit stream is provided to be used with the MFC address signaling. These systems are summarized in Table 22-29.

TABLE 22-29   CCITT Signaling Systems

	No. 3 (obsolete)	No. 4	No. 5	No. 6	No. 7	R1 Analog	R1 Digital	R2 Analog	R2 Digital
In-band signaling	X	X	X			X		X	
Out-band signaling								X	
Common-channel signaling				X	X				
Analog two-frequency signaling		X	X						
Analog multifrequency signaling			X			X	X	X	X
Digital				X	X		X		X
Suitable for operation over satellites?	No	No	Yes	Yes	Yes	Yes	Yes	No	No
Suitable for operation with TASI circuits?	No	No	Yes	Yes	Yes	No	No	No	No
Recommended for operation between SPC Exchanges?	No	No	No	Yes	Yes	No	No	No	No

The per channel signaling systems have a limited number of signals that can be transmitted and usually lengthen the time of use of the transmission paths. To overcome these limitations and provide other advantages, common-channel signaling (CCS) systems have been designed. Common-channel signaling systems consist of a full-time data link between two signaling points and the necessary terminal equipment. By appropriately encoding the data stream, an almost unlimited number of signals of both address and supervisory types can be transmitted at higher speed between SPC switches. If the CCS link carries the signals for the group of trunks between two offices only, it is said to be operating in the *associated* mode. To signal between two offices with too few trunks to justify an associated link, it is possible to signal over two or more signaling links in tandem via intermediate signaling points. These act as packet switches, routing messages by means of the label information contained in each message. In this case, since the links are carrying the signals for more than one trunk group, the links are not associated with any one and are said to be operating in the *nonassociated* mode (see Fig. 22-62).

For common-channel signaling, two international standards have been adopted, one optimized for analog transmission (CCITT signaling system 6) and one optimized for use with digital data links (CCITT signaling system 7) (see Table 22-29). In the United States, *common-channel* interoffice signaling (CCIS), based on system 6, was first introduced in 1976. As long-distance digital transmission becomes available, CCIS will be converted to system 7.[57]

89. Signaling Networks. In the United States, as CCIS becomes the basic signaling method, its signaling links become a separate *signaling network*. The switching systems appear as users on this network.

Each office eventually will rely upon this network for all its interoffice or internodal signaling needs. *Signal transfer points* (STP) serve offices in a geographical *signaling region*. Where the number of trunks justifies an associated signaling link between two offices at any level of the hierarchy it can be added just as high-use trunks are used where appropriate in the trunk network. To ensure service reliability, each office has access to two or more STPs (see Fig. 22-63). The message switches that serve as STPs employ redundancy of their own and achieve the same quality of service continuity as the switching offices.

Separate data links can also be used for signaling from customer premises, e.g., for passing the calling PBX station identification to the central office in what is known as automatic identified outward dialing (AIOD). Eventually digital loop signals will be standardized internationally.

90. Protocols. Communication signaling

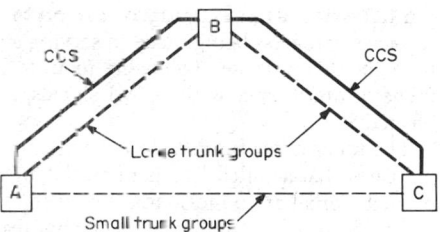

(AB – associated)   (AC – Nonassociated via B)
(BC – associated)

Fig. 22-62. Associated and nonassociated common-channel signaling modes. *(Adapted from C. A. Dahlbohm, ISS Conf. Rec., 1972, p. 421. Copyright 1972 IEEE. Used by permission.)*

as described above has been developed and standardized for the control of telephone switching systems. With the growth of data communications, user terminals (including computers), known in CCITT parlance as data terminal equipment (DTE), must be connected to the telephone network or to some kind of data network and must deal with the network to bring about the functions of addressing and control. These functions include those found in telephone station signaling plus specialized functions needed for data transmission and switching. Examples of data functions are bit synchronization and error control which may be needed on a private channel or a circuit-switched channel or with packet or message switching. A particularly complex example found in packet switching is the case of a terminal, generally a computer, which pro-

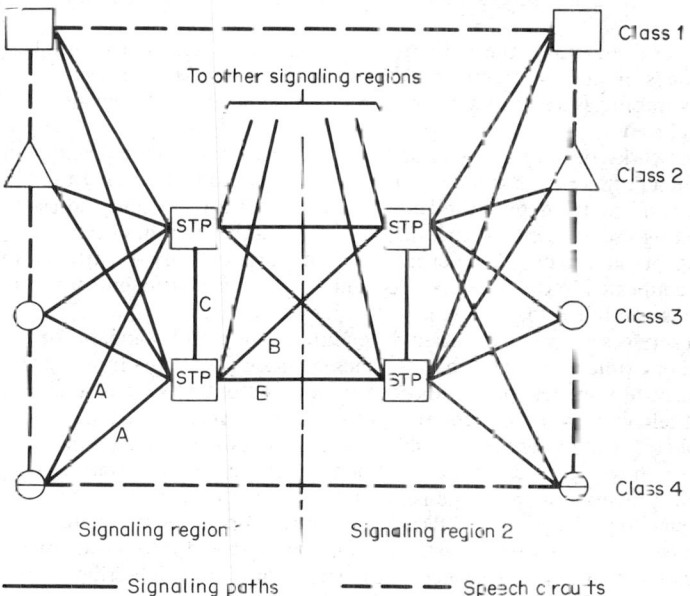

Fig. 22-63. Signaling-network concept. From R. C. Nance and B. Kaskey, ISS Conf. Rec., 1976, p. 413-2-1. Copyright 1976 Inst. of Elec. and Comm. Engrs. of Japan. Used by permission.

duces a stream of packets of information, each packet including routing information directing it to a specific terminal on the network.

To standardize interfaces between DTEs and networks, protocols in the form of a hierarchy of rules and formats have been developed. Noteworthy are CCITT Recommendation X.21, dealing with synchronous transmission, and CCITT Recommendation X.25, dealing with terminals using the packet mode. Reference 4 is strongly recommended for insight into such interfaces and standards (see also Par. **22-3**). The development of switched data communication networks is facilitated by the existence of terminal interface standards.

91. Tones and Announcements. Switching systems need to inform the users of the service of progress being made in serving a particular call. Tones and verbal announcements are used for this purpose. To enable tones to be automatically detected by call originators and to prevent interference with DTMF signals, a precise (frequency) tone plan is being adopted in the United States.[58]

The tones range from *dial tone*, which prompts the caller to start dialing, to *busy tone*, which indicates that a called line is unavailable. Similarly, where more than symbolic information is needed, verbal announcements are given to the caller.

Generally tones and announcements reach the caller through a regular switching network connection. For tones, one or more network terminals are used. If the same source supplies all terminations, low impedance is employed to prevent crosstalk between calls simultaneously reaching the same tone source.

Announcements may be composed from a recorded vocabulary; e.g., intercept systems announce number changes by concatenating a limited vocabulary of phrases and digits.[59]

SERVICES AND OAM FEATURES

92. A *service* is what the user perceives as being delivered by a switching system. OAM *features* of central-office telephone switching systems are those uses of system functions which are needed to operate, administer, and maintain the switching, customer terminal, and transmission equipment to provide services.

For example, to provide telephone service (see below) a directory number or address is assigned to a line. Because directory numbers are not permanently associated with switching-network terminations, an administrative feature is required to store this association as a translation in system-control memory.

93. Telephone Service. The switching functions required to set up a telephone call are listed in Par. **22-71**. Other services also use these functions, plus such additional functions as may be needed.

Telephone service includes the ability to place calls in a telephone network. Users reach other users' telephones by dialing a sequence of digits and can similarly be reached. To respond and alert users, switching systems in a network provide the appropriate signals: dial tone, ringing, busy tone, and so on.

In many networks, operators provide additional services: assisting in placing toll calls (person-to-person), calculating charges and collecting coins in certain calls originating from public coin telephones, providing telephone numbers for users given a name and address (directory assistance), informing users whenever they have dialed a number that has been changed or is not in service (intercept), and so on. These operator services may be fully or partially automated with SPC systems connected to the network. Depending on the operating entity and the service provided, users may be billed for the use of such operator services.

Additional services are provided (again depending on the country and/or the operating entity) for additional one-time or periodic charges. These include DTMF dialing, key telephone or PBX services, or private user networks embedded within the public switched communication network, special telephone sets for decorative purposes, and special calling services.

An example of a calling service available in the North American network where SPC switching systems are in service is *call waiting*. When a user who has call-waiting service is engaged in a telephone conversation, a distinctive tone will be received if a new call arrives. The user can be connected to the new call by "flashing," or briefly operating the switch hook. The system responds by placing the existing connection in a hold state and reconnects the subscribing user to the new call. By subsequent flashes of the switch hook the subscribing user can alternate connections between the two calls as needed.

A user wishing to purchase or subscribe to services such as the above makes appropriate arrangements with the business office of the telephone administration or operating company. Some of the other services may presently be activated, used, and deactivated by dialing directly into an SPC switching system. As the connectivity of SPC systems pervades a communication network, more and more services become available to meet user needs and the trend is toward further automation of the direct control by the user in obtaining such services.

94. New Services. With the introduction of centralized SPC and associated bulk memory, many new services have been devised for switching offices. Usually they can be implemented with little or no additional hardware except for the memory to store the additional program. Switching systems generally include a dynamic set of service capabilities. For many years public coin telephone service in the United States in large cities required the deposit of a coin before dial tone was given to the caller. More recently this has been changed to *dial tone first* so that certain destinations, such as the operator and emergency service bureaus, could be called without a coin deposit. This is an example of the changing requirements for customer services.

Typically telephone systems must provide many capabilities. They serve party and mobile telephones, private branch exchanges, and many different rate categories (flat rate or measured service, coin, etc). Some central office systems include capabilities of serving intramural business needs directly to avoid the need for key and PBX systems.

95. OAM Features. *Network Management.* Automatic and manual routing control is provided in large modern networks to route calls around portions of the network that are temporarily congested or where disasters or other types of problems have reduced or eliminated the ability to reach or receive traffic from portions of the network. This is known as *network management.* Network management also includes turning back calls to reduce congestion to a particular destination and the temporary augmentation of facilities to serve overloads. Further, it includes threshold measurements and alerting when offered calls do not appear to be reaching their intended specific destinations.

Traffic Measurement. To observe how the switching system is operating, an administration needs indications of what loads are being carried by the system components. Typical are counts of calls, measurements of call delays, counts of call dispositions, duration of all circuits busy, and circuit use. The last measurement is best reported in terms of call hours per hour, or *erlangs.* It is often made by periodic counting of busy circuits and reporting the average or by summing the actual hours of circuit use for all circuits in a group. Output of data may be directly printed out by a switching system or may be transmitted to a centralized support system.

Charging. Broad allocation of costs to various users of a network can be determined by traffic measurements. In public service systems, more detailed information is needed. Two basic methods are used, *bulk billing* and *detailed billing.* In bulk billing, charge units are allocated to call setup, duration of call, and distance called. In Europe, the pulse metering system is used, where the local office generates pulses on a per line basis. In North America detailed billing is used for toll calls. Call details are recorded, either centrally or locally, and charges are later computed in centralized data processing centers where bills are prepared. This process, called *automatic message accounting* (AMA), has the advantage of more flexibility and full reporting of charges to the customer at the expense of more data processing than is required of bulk billing.

For coin telephones, with pulse metering systems, the charge pulses can be used to control coin collection directly. With detailed billing systems, operators or centralized charge calculation and control capabilities are required to quote charges and handle the collection of coins.

Calls in public data networks are usually billed on the basis of duration of call or number of packets plus a network access charge. The rates may change with the time of day and priority required.

96. Maintenance. The place of the switching system in a communication network puts unusually severe requirements on its maintenance. Since loss of service for any but very short periods of time is unacceptable, detection, recovery, diagnosis, and repair of trouble must be carried out while the system continues to process calls. Central processor design therefore incorporates a considerable amount of attention to internal trouble detection and the ability to reassign faulty system elements automatically so that processing can continue without loss of calls. Tests diagnosing the nature of the trouble may be also automatic or subject to request by maintenance personnel.

Alarms and diagnosis results must be generated; most indications appear as lighted lamps or typewritten characters. A standard *user-machine language* is being adopted by the CCITT.[60]

Contents of translation and other semipermanently stored data bases change daily in a large public system. Provision is made to change this information locally or remotely. Care is needed to ensure that changes are free from error and that the data bases will not be lost. In some systems changes are made in two steps. The information is first stored in a temporary *recent change* location and later relocated to a regular data base address.

Another class of feature deals with the cutover of a new system, recovery of a failed system, and change of hardware or software in a working system.

In addition to the tests internal to the switching system, separate test sets, consoles, display boards, and test access switchboards are used to varying degrees and constitute maintenance features. The switching system is also used for connecting test circuits to remote switching and transmission systems.

In the design of systems, a critical factor is the objective in-service time. For public switching systems for two-way voice service, an objective of 2 h in 40 years downtime has been used.

Centralized Operation Support. In the past, operational features have been largely contained within the design of the switching system. There is a trend toward providing these features, as well as maintenance and administrative facilities, at a centralized point where they can service a number of switching systems. The centralized systems are known as *operations support systems* (OSS). Operations includes maintenance and administration as well as operation. The design of switching system hardware and/or software includes features needed to provide and interact with OSSs.

In some centralized support systems, programs and other information for infrequently used real-time call processing can be accessed at a centralized call processor or OSS designed with the required reliability objective.

APPLICATIONS

The principles of switching system or node designs have been given in Pars. **22-71** to **22-96**. This section gives the reader an understanding of typical switching systems.

To determine whether a given switching system meets a specific application need one must first know the basic dimensions and capacity requirements.

97. Dimensions. Limiting the size of a system are the number of physical terminations for lines and trunks, the call-carrying capacity of the control, the traffic capability of the network(s), and the address range of the memories.

The *terminations* are those lines, trunks, service and similar circuits which are principal service inputs and outputs of the system. The service circuits do not extend out of the system but are used to provide a function such as ringing, or call signal receiving or transmitting.

Small data and PBX switching systems may serve tens or hundreds of terminations while large central-office voice-switching systems can have a termination capacity of over 150,000. Most switching systems are designed so that they can grow over a range of terminations, typically an order of magnitude or more. Ideally a switching system should be able to grow over a greater range, and new technology continues to expand this range.

System capacity depends on many factors, including not only the properties of the switching system but also customer traffic characteristics and customer expectation of grade of service. Grade-of-service criteria are usually defined for the average busy season busy hour (ABS) as well as for the 10 highest busy hours and for the highest busy hour. Examples of ABS criteria are 1.5% of calls delayed over 3 s in receiving dial tone and 2% of incoming calls blocked in the switching network. Capacity must be determined for each central office installation, as it depends on the amount of equipment provided.

Call Capacity. The limit to the maximum office size is frequently designed to be the capacity of the central processor(s). Processor capacity is usually stated in terms of busy hour originating plus incoming calls, although any particular office capacity calculation must consider the proportions of all types of calls. In determining engineered capacity, consideration must be given to the maximum capacity of the central processor. Usually this corresponds to almost 100% use of the time available for call processing. An allowance must then be made for safety (typically 5%) on a high-day engineering basis, and beyond that for the ratio of high day to average business day. (A common ABS utilization is 70% of available processor time.) Then all delay criteria must be checked; if any are not met, the capacity must be lowered further.

Systems have been built with distributed microprocessors serving as many as 9,000 busy-hour call attempts (BHCA), and with a central SPC using multiprocessors and high-speed integrated-circuit technology of up to 750,000 BHCA.

Data-message or packet-switching capacity is measured in terms of maximum rate of *throughput*. The capability of the control includes overhead as well as message handling. The control capacity is generally considered to be the major factor in throughput. Processing capacity can be limited by congestion delays within a system delivering or receiving calls for processing.

Switching-Network Capacity.   The switching-network capacity is expressed in the number of erlangs which can be carried at an objective blocking. Networks for smaller offices are partially equipped; each arrangement will have its own capacity. If the average call duration is long, the erlang capacity may limit call capacity below that of the processor

$$\text{Network calls/h} = \frac{\text{erlang capacity}}{\text{call holding time (h)}}$$

Because processors are not partially equipped and networks can be, designers choose to design networks with a smaller chance of being the limiting dimension of system capacity.

Memory Capacity.   The total memory requirements depend upon many components. An important factor is whether there is only one storage subsystem or separate storage subsystems are used for different storage needs. Typically, separate subsystems might be used for program and call data storage. In a message-switching system separate subsystems might be used for call processing and message storage.

One limit on memory size is the amount of memory that can be directly addressed. By having larger program words (more bits per word), more memory can be accessed, but this means that larger, more expensive program memories are required.

As in most SPC systems (Par. **22-73**), software techniques can be used to extend the address range at the expense of real time. One address can refer to a table containing another range of addresses.

Physical memory modules are used to form a storage subsystem. Only as many memory modules are used as are needed to provide for the memory requirements for a particular installation.

98. Typical Systems.   Table 22-30 shows some of the important dimensions of a few typical circuit-switching systems in common use in North America. These systems were introduced in the late 1970s and all have SPC control. The trend toward solid-state networks and digital time-division switching is evident.

The dimensional limits represent the individual maxima in each case (call attempts per hour, erlangs, and terminations). Depending on the specific environment, any one of these may limit further growth of the system, and the remaining limits would be unattainable. These limits represent approximate values, which in themselves depend on assumptions of the system environment, e.g., the ratio of intraoffice to interoffice calls in a local central office. The limits change with time as additional features and services are added to a system or when improvements in hardware or software are introduced.

Two basic switching entities are used at points distant from the host central wire center. One is the *remote line concentrator* (RLC) and the other the *remote switching unit* (RSU). Both the RLC and RSU are used to concentrate traffic at a point closer to the lines they serve. The RLC provides only for remoting this network function. Generally, in the central office an equivalent expansion function is provided, and consequently each RLC line is given a central office appearance. Some RLC systems eliminate the need for this expansion by connecting RLC trunks to links within the switching network. Generally the line range is not extended with the use of RLCs.

Should all the trunks between the RLC and the central office be busy or the facilities carrying the trunks be severed, calls to and from the RLC lines cannot be served. Since the lines served are in a confined area, they are subject to higher traffic variation. Concentrators are generally small, serving no more than a few hundred lines.

With the advent of microprocessors not only the network but also the control can be remoted. This means that more intelligence can be designed into the remote switching. Also, the remote switch may be able to complete intra-RSU calls without using trunks to the host office. The RSUs have been developed with SPC and limited call-processing capability so that they can provide basic service (Par. **22-93**) if the link to the host is lost. This is known as *stand-alone* capability.

99. Trends in Switching Systems.   Switching systems have made the transition from those accomplished with relays and other forms of electromechanical switches to all-electronic networks and controls. Depending on the administration, this change was begun during the 1960s and 1970s. The major ingredient to stimulate this transition has been stored-program control. SPC has provided the flexibility and power to add to and change switching system capabilities easily. As administrations make the transition to national and international networks of SPC systems, capabilities can be extended further. The advent of common-channel signaling in

TABLE 22-30 Typical Circuit Switching Systems (North America)

System name	Application	Year introduced	Control		Network		Terminations	
			Call capacity, peak attempts per hour	Type	Type	Traffic capacity, erlangs	Lines or Stations	Trunks and service circuits
No. 1A ESS	Local, tandem, toll	1976	240,000	2 processors, single active	Space division magnetic latching reed	10,000	128,000	32,000
DMS-10	Local	1977	13,000	2 processors, single active	DTDN T-S-T	850	8,300 lines plus trunks	
No. 3 EAX	Toll	1978	360,000	4 processor pairs, distributed	DTDN S-T-T-S	16,700	----	61,400
No. 4 ESS	Toll	1976	500,000	Hierarchical and distributed	DTDN T-S-S-S-T	47,000	----	107,000
Horizon	Key telephone system	1978	*	1 microprocessor	Space division pnpn crosspoints	15	79	32
GTD-120	PBX	1976	*	2 microprocessors, single active	DTDN T	30	120	28
Dimension 400	PBX	1975	2,000	1 microprocessor	Analog time division	45	436/312	64/128

*Not limiting factor.

the late 1970s has provided a separate fast communication network between the processors of SPC switching systems, further increasing the capability of telecommunication networks. New service concepts that are possible within an SPC office may be extended to the entire communication network.

Telecommunication networks are also making the transition, begun in the late 1970s, from the switching of analog signals to digital signals. Space-division networks are yielding to digital time-division networks. The switching, transmission, and station equipment is evolving toward an integrated digital network, greatly increasing the ways in which the telecommunication users' needs can be met.

Terminal Equipment

TELEPHONES

BY W. E. HOSTETLER, R. M. HUNT, E. W. UNDERHILL, J C. BAUMHAUER, JR., D. L. WHITSON, AND J. M. GOTWAY

100. Introduction. Telephone equipment ranges from the familiar telephone set to the versatile and specialized equipment of the information age. The merging of telecommunications and computer technologies makes use of the entire spectrum of voice, video, data, text, and color graphics.

101. The Telephone Set. A typical common-battery telephone set is shown schematically in Fig. 22-64. The functional elements consist of a carbon transmitter, electromagnetic receiver, switch hook, dial, loop-equalizer circuit, hybrid transformer antisidetone circuit, and an electromagnetic ringer to alert the user to an incoming call. The loop equalizer and antisidetone circuits are collectively known as the *network*.

Sound waves impinge on the diaphragm of the carbon transmitter, resulting in variations in its resistance. An alternating current, an analog of speech is generated as the varying resistance modulates the direct current provided by the central-office (CO) battery.

When the varying component of a modulated current generated by a far-end talker reaches the near-end listener, the receiver winding on a permanent magnet is activated. The resulting varying magnetic field causes movement of the receiver diaphragm and generates sound waves corresponding to those impinging on the far-end transmitter.

The telephone set must be able to function with various switching systems (panel, step-by-step, crossbar, PBX, electronic) over varying loop lengths (typically up to 15,000 ft of 26-gauge cable, or 1,300 Ω) and to provide the desired levels of transmitted and received signals and sidetone under these conditions.

Sidetone is that portion of the transmitted signal which is heard n the receiver. In modern station sets the sidetone network generally provides a sidetone signal at about the same level as a nominal received signal. If the sidetone level were too high in the receiver, it would cause the talker to speak too softly for good transmission. Sidetone is subjectively desirable because it provides a live quality to the telephone conversation.

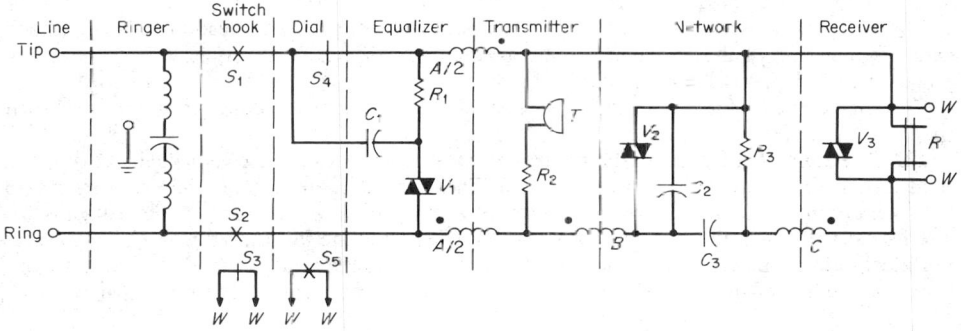

Fig. 22-64. Typical common-battery telephone set.

When the handset is taken off the hook, the switch hook contacts S_1 and S_2 (Fig. 22-64) close and direct current is provided to the transmitter, line, and balance varistors over the metallic loop from the CO battery. The switch hook contacts S_3 prevent switching transients from being heard in the receiver. Dial pulsing (DP) is accomplished by operation of the rotary dial (contacts S_4), which provides controlled interruptions (nominally 10 pulses per second) of the loop current. Further details are given in Par. **22-104**.

Resistor R_1 and varistor V_1 constitute the loop-equalizer network. The higher levels of current on short loops lower the resistance of the current-sensitive varistor V_1, thereby reducing the transmit and receive levels on short loops. The lower levels of current on long loops cause V_1 to remain at a relatively high resistance, so that transmit and receive levels are not significantly changed and good performance is obtained. The effect of the loop equalizer on transmission level is shown in Fig. 22-65a.

The combination of a three-winding hybrid transformer and impedance balancing circuitry

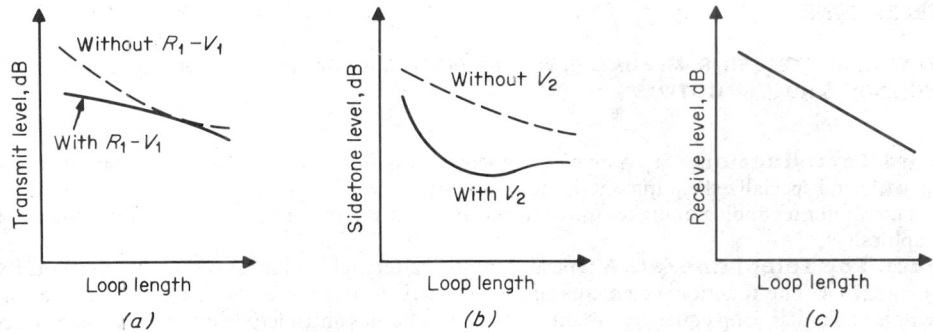

Fig. 22-65. Effect of (a) loop equalizer, (b) antisidetone network, and (c) overall receiver power-loop length relationship.

provides the means of coupling the transmitter and receiver to the loop efficiently. This is called an antisidetone network.

Incoming ac signals from the telephone loop cause induced voltages in the three windings of the transformer. The induced voltages are such that most of the incoming signal power is delivered to the receiver with little power to the balance network. Alternating currents originating in the transmitter induce voltages in the three windings, so that most of the signal power is divided between the balance network and loop impedances with little to the receiver. The choice of impedances and turns ratios provides a compromise in sidetone balance and impedance matching to the telephone loop.

Impedance matching of the set to the loop is important because it influences return loss* and therefore transmission quality. In Fig. 22-64 capacitor C_2, varistor V_2, capacitor C_3, and resistor R_3 make up the antisidetone balance-network impedance. Varistor V_2 compensates for changes in the impedance of the loop and varistor V_1. The effect of varistor V_2 on the sidetone levels is shown in Fig. 22-65b. Varistor V_3 is placed across the receiver to limit received signals or noise to an acceptable level to protect the ear. The receiver does not draw any dc power, and receiver ac power varies little with loop lengths, as indicated in Fig. 22-65c.

The ringer is shown for bridged ringing, connected across the loop ahead of the switchhook contacts. The user is alerted when a nominal ringing voltage of 86 V rms at 20 Hz is superimposed on the telephone loop at the central office.

The characteristics for a typical telephone set, with a typical talker and connection, are shown in Table 22-31.

102. Electromechanical Ringers. In most telephones an electromechanical ringer is used to alert the customer to an incoming call. The example shown in Fig. 22-66 consists of two bells which produce a distinctive alerting sound when struck by the clapper.[61] The ringer, with a series capacitor, is usually connected across the telephone line for bridged ringing. The capacitor blocks direct current and resonates with the ringer inductance at about 20 Hz. Other ringer-connecting schemes are used, however, so that customers can be selectively rung on party lines.[62]

*See footnote, Par. **22-27**.

Selective ringing schemes involve the connection of the ringer between the tip or ring conductor to ground, ringers tuned to different frequencies, or positive or negative dc voltage combined with the ac ringing voltage.

Miniaturization of the bell ringer has been a recent international trend.[63] Continued design improvement has resulted in a compact single-bell ringer with performance comparable to that of the larger two-bell ringer.[64]

103. Tone Ringers. Electronic tone ringers are being used for some customer alerting applications. Typically, the tone ringer consists of a detector circuit to distinguish between valid ringing signals and transients on the telephone line and a tone-generator circuit which drives an efficient electroacoustic transducer. The tone-ringer circuit is capacitor-coupled to the telephone line and provides the proper input impedance. Tone ringers generally produce fewer spectrum components than a bell ringer, but various field trials and human-factor studies have shown equivalent effectiveness and acceptability.

A new scheme for telephone alerting, using laser-generated optical power delivered over a fiber lightguide, has been proposed for future telephones.[55] The replacement of conventional local telephone lines with optical fibers is possible if this research leads to a practical method of telephone alerting.

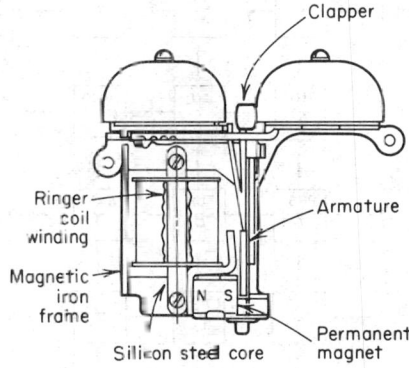

Fig. 22-66. Typical ringer used in telephone set.

104. Rotary Dials. The rotary dial interrupts the telephone line current with a series of breaks. The number of breaks in a string represents the number being dialed, one break being a 1 and ten being a 0. As shown in Table 22-31, these breaks occur at a nominal rate of 10 pulses per second. The ratio of the time the line current is broken to the total pulse time (percent break) is 61 ± 3%. As shown in Fig. 22-64, the dial pulsing contacts S_4 are protected by the capacitor-resistor circuit (C_1 and R_1). Dial contacts S_5 mute the receiver to minimize clicks to the user's ear during dialing. The mechanical rotary dial is generally of the single-lobe cam-and-pawl type design, which is driven by a spring motor wound up by the user as each digit is dialed. The return motion is controlled by a governor mechanism to maintain the proper dial pulsing rate.

Electronic dials perform the rotary dial function but use a pushbutton key pad similar to that used in the dual-tone multifrequency dials described in Par. **22-105** (see Fig. 22-74). These electronic dials interrupt the line current with transistors or relays. Since the user can enter the number into the dial faster than the number can be pulsed out, a first-in, first-out memory is generally used to store the number to be dialed. The timing is generated using electronic circuitry. Power for the circuitry is derived either from the telephone line or from commercial power.

105. Dual-Tone Multifrequency (DTMF) Dials. Multifrequency dialing consists of two simultaneously transmitted audio frequencies. On the standard 4 × 3 dial format, each column and each row is associated with a different frequency, as shown in Fig. 22-67. This method of signaling permits faster dialing for the user and more efficient use of the switching systems. Some special service dials use a 4 × 4 matrix, which provides four extra frequency

TABLE 22-31   Characteristics for Typical Telephone Set with a Typical Talker and Connection

Characteristic	Short loop	Long loop
Line current, mA	90	20
Impedance, Ω	600	900
Transmit level, dBV	−15	−25
Receiver level, dB re 20 μPa	83	78
Sidetone level, dB re 20 μPa	79–87	74–82
Rotary dial	8–11 pulses/s	
	(10 pulses/s nominal)	
Percent break	61 ± 3%	

pairs for additional signaling capability. Since the frequencies are in the audio band, they can be transmitted through the telephone network from one user to another after a call has been set up and used for data communications.

Several methods are commonly used to generate these frequencies. The first method is shown in Fig. 22-68. This inductor-capacitor oscillator uses two independently tuned transformers A and B to obtain two-frequency operation. Different values of inductance are switched into the circuit to obtain different frequencies.

Another method of generating multifrequency signals is a resistor-capacitor oscillator (Fig. 22-69). Here a twin-tee-notch filter is placed in the feedback loop of a high-gain amplifier to obtain the desired frequency. Two of these amplifier-filter combinations are used in each dial, one for each group of frequencies.

A new method uses digital synthesis techniques to generate the frequency signals. As shown in Fig. 22-70, the key-pad information is combined with the master clock to generate the frequency to be produced. This information is fed to a D/A converter, whose output is a stair-step waveform. This is then filtered and fed to a driver circuit, which provides the desired sine-wave frequency signals to the telephone line.

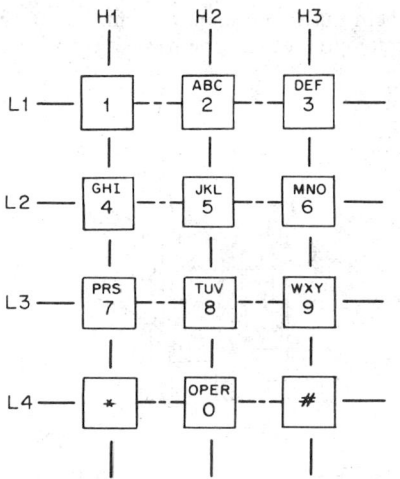

Fig. 22-67. Basic arrangement of pushbuttons used for dual-frequency dialing.

106. Transmitters. The use of granular carbon transmitters[66,67] dates back 100 years to the birth of telephony. Sound striking the diaphragm imparts a pressure fluctuation on the carbon aggregate (Fig. 22-71a). Since granule contact force and dc resistance R_0 are inversely related, a modulation of the telephone loop current I_0 results.[68,69]

While carbon transmitters offer 20 to 25 dB of inherent signal-power gain and low cost, new electronic telephone networks require a unit that consumes much less power. The electret[70-72] is a capacitive transmitter with low sensitivity to mechanical vibrations and a small power requirement. An effective bias voltage V_0 depends on a polymer diaphragm's fixed electret charge[73] (Fig. 22-71b). Noncarbon units require electronic amplification but generally yield longer service life and smaller size.

107. Receivers. The receiver[74,75] converts the analog electric signal back into acoustic vibrations. An electromagnetic (moving-iron) receiver uses voice coil currents to modulate the dc flux, which produces a variable force on the armature (Fig. 22-72a).[76] Certain hearing aids can inductively couple to the leakage flux from some receivers.[77]

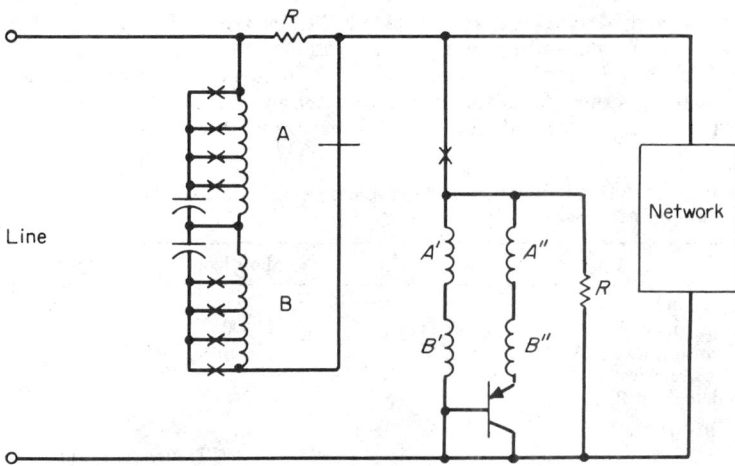

Fig. 22-68. LC dial circuit for dual-frequency operation.

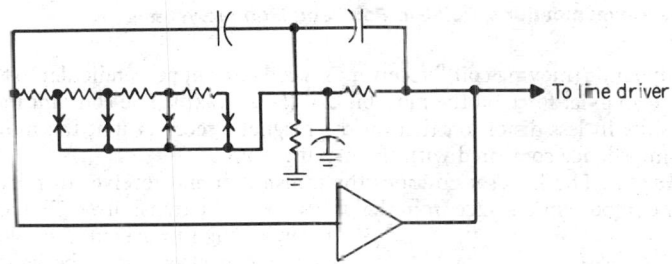

Fig. 22-69. Typical RC dial circuit.

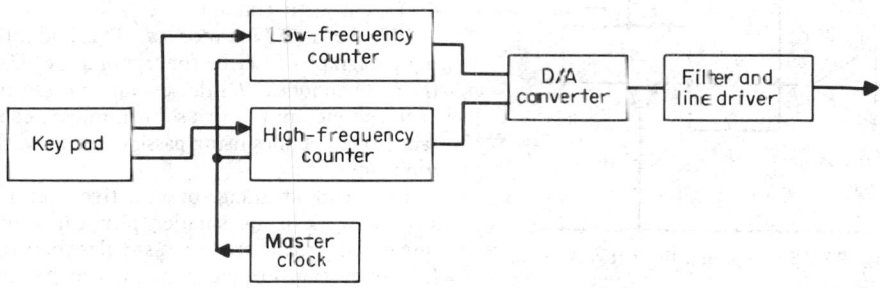

Fig. 22-70. Digital-synthesis circuit.

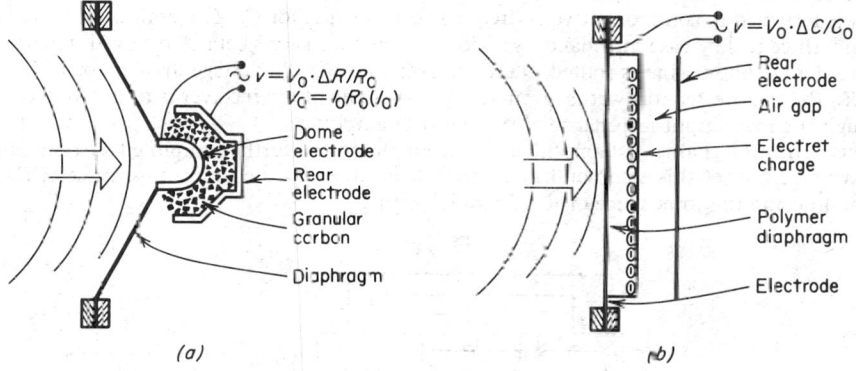

Fig. 22-71. Transmitters: (a) carbon and (b) electret.

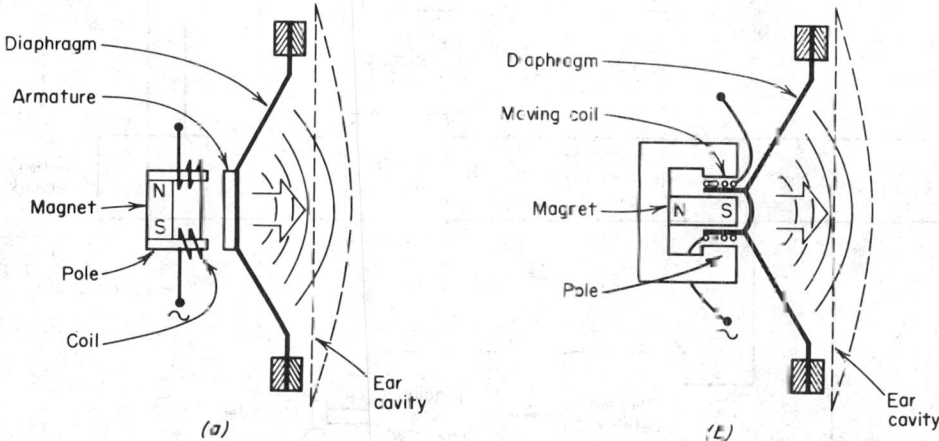

Fig. 22-72. Receivers: (a) central armature magnetic (moving iron) and (b) dynamic (moving coil). [(a) From Ref. 77. Copyright 1951, American Telephone and Telegraph Co. Used by Permission.]

The electrodynamic (moving-coil)[78] receiver uses coil current perpendicular to the dc magnetic field to generate an axial force on the movable coil (Fig. 22-72b). The constant reluctance of the coil air gap results in less distortion than in the magnetic receiver, but the moving coil offers limited input impedance compared with the moving iron.

108. Handsets. The handset positions the transmitter and receiver in relation to the ear, determining the important distance from the user's lips to the transmitter. The handset must be heavy enough to operate the telephone switch-hook when placed on it, be light enough to be comfortably held for long periods, provide an acoustic seal to the ear, and often interface with data-set acoustic couplers.

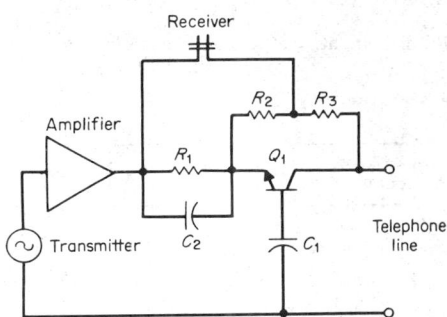

Fig. 22-73. Typical active-network circuit.

109. Active Networks. This and following paragraphs describe the use of active circuitry in telephones. While several of these types of telephone are not in use in numbers comparable to telephones using passive circuitry, their use is increasing.

The main advantage of an active over a passive network is its smaller physical volume; integrated circuits are much smaller than voice-frequency transformers that must carry direct current. An active network also provides power gain, thus allowing the use of passive transmitters such as electrets.

A typical active network presently used in some telephones is shown in Fig. 22-73. The base of Q_1 is returned to common (at voice frequencies) by capacitor C_1. The emitter of Q_1 is virtual ground, since its low base-impedance is divided by the transistor's beta. A received signal appearing on the telephone line is routed to the receiver through the voltage divider consisting of R_2 and R_3; R_2 is connected to a virtual ground. The other end of the receiver is returned to common through the low output impedance of the transmit amplifier.

The transmit signal is first amplified by the amplifier and further amplified by transistor Q_1. The voltage gain of this common-base stage is determined by the input impedance of the telephone line and the impedance of R_1 in parallel with C_2.

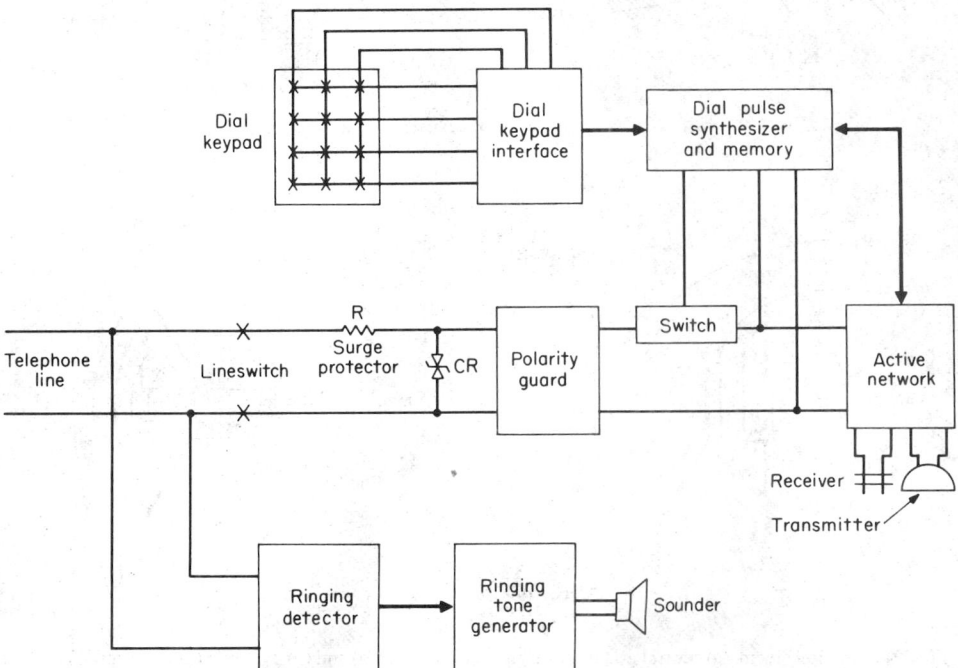

Fig. 22-74. Electronic dial-pulse telephone.

The sidetone balance is achieved by adjusting the voltage divider (R_2 and R_3) to compensate for the gain of the common-base stage, leaving about the same potential at both ends of the receiver. Capacitor C_2 is added to minimize any phase shift through this stage caused by the capacitance of the telephone line.

110. Electronic Telephones. In electronic telephones several conventional components, such as the bell ringer, the transformer-coupled speech network, and the rotary or DTMF dial, are replaced by active electronic devices that are incorporated into large-scale integrated (LSI) circuits.[79] Several advantages result: the overall reliability of the telephone is increased; efficient manufacturing is possible using highly automated assembly; telephone-set weight and volume are reduced, permitting a wide variety of physical designs,[80,81] and finally, overall transmission performance of the telephone set is improved.

A diagram of a typical dial-pulse electronic telephone set is shown in Fig. 22-74. It contains a tone ringer, an active network, an electronic dial, and a linear transmitter. A bridge rectifier is provided to guarantee proper tip and ring polarity for the electronic elements. To protect the set from high voltage surges such as lightning, the combination of a current-limiting resistor R and a surge-protection diode CR are connected across the telephone line.

The tone alerter (Par. **22-103**) in an electronic telephone replaces the conventional electromechanical clapper-and-gong type of bell ringer. The dial in an electronic telephone uses a push-button key pad to provide either DTMF signaling or dial-pulse signaling (Pars. **22-104** and **22-105**). Electronic logic performs a variety of common switching functions, such as switching out the transmitter and lowering the gain to the receiver during dialing, functions that are performed by mechanical switches in conventional telephone sets.

The speech circuit in an electronic telephone is an integrated active network (Par. **22-109**) rather than a transformer-coupled passive network. Electronic telephones typically use either electret microphones or electrodynamic microphones, which provide high-fidelity speech transmission and have good aging characteristics. In the active network the gain of the transmit and receive amplifiers is automatically adjusted, depending upon the loop current, providing compensation for transmission losses which vary with loop length.

TELEPHONES WITH ADDED FUNCTIONS

BY R. K. THOMPSON, R. E. WADDELL, J. M. GOTWAY, R. L. CERBONE, AND V. E. MUNSON

111. Key Telephone Sets (Electromechanical Key Systems). Additional features are required of a telephone set when more than one central-office (CO) or PBX line is used. Identification of the line in use, switching from one line to another, and the ability to hold one or more lines are desirable features. Further capabilities include multiline conferencing, visual and audible signals to indicate specific line status, or alerting, and provision for auxiliary equipment.

Line selection is typically provided by adding button-actuated switches (keys) to the telephone, thus creating a key telephone set. Telephones are available that access from two to a few tens of lines (see Fig. 22-75). Separately mounted groups of keys can provide access to over 100 lines.

Key sets can be used alone to access CO or PBX lines; these do not require auxiliary circuitry or commercial power. More commonly, additional logic and relay line circuits are used to add hold, visual signals, and a variety of audible signals to the overall system. When such circuitry is used, the multibutton station sets feature a single common hold button, regardless of the num-

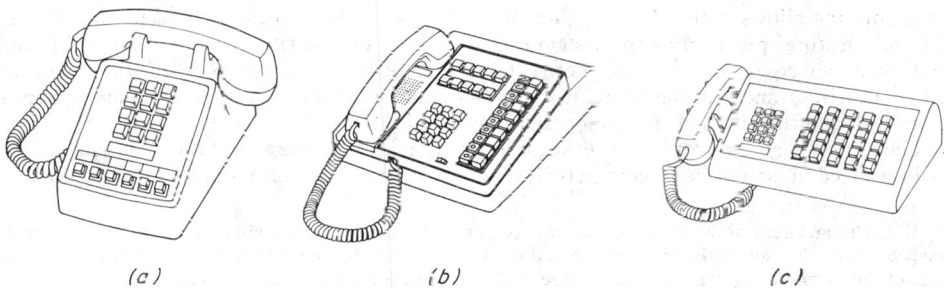

(a)   *(b)*   *(c)*

Fig. 22-75. Typical key sets: (a), 6-button (b), 10-button with DSS button array, and (c) 30-button.

ber of line appearances. Also, with such circuitry, a holding bridge applied to a line by one station will be released by any station which selects that line and goes off-hook on it.

Many multibutton station sets are arranged to offer visual (lamp and LED) and audible signals to aid in user interpretation of line status in the system. A 60-interruptions-per-minute flashing signal indicates the presence of an incoming call on a line. This can be augmented by interrupted audible signals if needed. A steady indicator signifies a line in use or a held line. Another rhythmical wink or flutter signal is often used to identify the held-line condition.

The ringer in most multiline sets can be associated with one of the lines appearing at the station, to operate as a bridged ringer for that line. However, the ringer is usually connected as a common ringer, activated from local circuitry, to indicate incoming calls on any line or lines appearing at the station. Voice signaling on intercom calls via a DSS (direct-station-selection) button array is a feature of some newer systems.

Many installations incorporate one or more local intercommunicating paths between stations to relieve traffic on CO or PBX lines. These paths range in complexity from communal circuits (with separate pushbutton and buzzer signaling) to more elaborate DSS button fields or dial-selective facilities. They may have one or more talking links, options for visual and automatic ringing features, and the ability to conference other stations.

Key sets and key mountings are generally arranged for two-wire operation, although increasing use is being made of optional four-wire line-terminating arrangements in the same set for improved transmission on either switched or private lines.

Each line pickup makes use of dedicated conductors in the cabling system. These, coupled with conductors assigned to other features, require conductor counts in the cables and cords serving the telephones that range from 30 to 200 or more. Multibutton telephone-set mounting cords are provided with one or more multicontact plugs, which allow for rapid installation or change.

112. Electronic Key Telephone Sets. Electronic key sets are telephones containing active electronic circuitry that is used at an interface with electronic-key-system control equipment. The normal telephone functions of ringing, dialing, and transmitting and receiving of speech signals may be performed in the conventional manner or with active circuitry. Using active circuitry, the electronic key equipment and telephones exchange control and status information via a data channel; by switching at the key equipment, only one voice pair to the telephone is needed.

As indicated in Par. **22-111**, telephones used in relay-circuitry-control-type key systems require large multipair cords. Electronic key sets by contrast have significantly fewer pairs (generally less than four).

Neither the number of conductors in the set cord nor the functional definition for conductor pairs in set cordage is standardized. One conductor pair in a two-pair electronic key telephone cord might be used as a conventional tip-ring speech pair. The second pair could serve as a full duplex data channel to the key-system equipment serving the telephone. Power for the electronic key telephone control circuits could be supplied from the key equipment over either pair or phantomed* over both. Variations in this arrangement requiring a three-pair cord occur where the data channel is realized with two conductor pairs instead of one. Phantoming of power can then be accomplished via the data pairs, leaving the tip-ring pair undisturbed. A fourth pair can be added to the set cord to provide for auxiliary services or additional power.

Data rates on the data channel range from several thousand to over 200,000 b/s. Information received by the set might activate audible or visual signaling devices to denote set and system status. Information transmitted from the set can inform the control equipment of switch-hook position and buttons pushed at the station. The buttons allow the station user to control the pickup and hold functions on lines serving the key system. In many electronic key systems, programming allows button line assignments at the station to be readily changed. In addition, different features provided by the system can easily be made available via the different buttons and signal devices of the set. Thus there is no need to rewire the inside of an electronic key set when the lines and features to which it has access are changed. System software is reprogrammed, and the labels of the appropriate buttons on the set are updated.

The circuitry an electronic key set incorporates to receive, process, and transmit data is typically realized in large-scale-integrated (LSI) semiconductor devices. Either microprocessors or cus-

---

*A phantom circuit is derived from two wire pairs that carry conventional circuits. A transformer, center-tapped on the line side, is inserted at each end of each pair; the phantom circuit is connected to the center taps of the transformers, and each wire pair serves as one conductor of the phantom circuit.

tom-designed LSI devices are used to enhance features, minimize cost, and reduce mechanical operations in the set.[82]

A few types of electronic key sets use speech encoding to facilitate digital switching in the associated system control equipment. Some systems have combined the encoded speech with the system status and control data. This reduces the number of conductor pairs needed in set cordage and consolidates electronic circuitry.

113. Multifunction Electronic Telephones. The use of active electronic circuits in many telephone components makes it possible for the telephone to provide additional features and services at low incremental costs. For example, the addition of a small amount of silicon-chip memory makes automatic dialing of a stored number possible with a single button push. Inclusion of an electronic display panel allows the telephone user to verify stored numbers or to display numbers as they are dialed. If a clock (timer) circuit is provided in the telephone, the user can view on the display the elapsed time for a long distance call. A residential hold feature can make it possible to transfer a call from one extension phone to another. Other electronic circuits can be added to provide various types of radio telephony.

The architecture of a multifunction telephone typically includes some custom LSI circuits, such as tone-generator chips, clock (timer) chips, and display-driver chips. Depending upon the number and the kinds of features provided, the telephone may also incorporate a microprocessor which controls the operation of the various custom LSI circuits.[83]

A diagram of a typical microprocessor-controlled multifunction electronic telephone is given in Fig. 22-76. The microprocessor receives information from the ringing detector, the dial key pad, other control buttons, the line switch, and the active network. The microprocessor controls such items as the ringer tone generator, the dial DTMF synthesizer, the display, and the speech network. Interfacing with both random-access memory (RAM) and read-only memory (ROM) units, the microprocessor can either store received information, such as telephone numbers entered from the key pad, or retrieve previously stored information for processing.

114. Repertory Dialers. A repertory dialer is a device that permits the user to program several telephone numbers and to dial automatically from a directory of stored numbers.[84] They are used for both address signaling and for some end-to-end signaling applications. Early products typically used cards, magnetic tapes, or drums as the storage media but virtually all current products use solid-state memories with capacities of anywhere from a few to hundreds of telephone numbers.

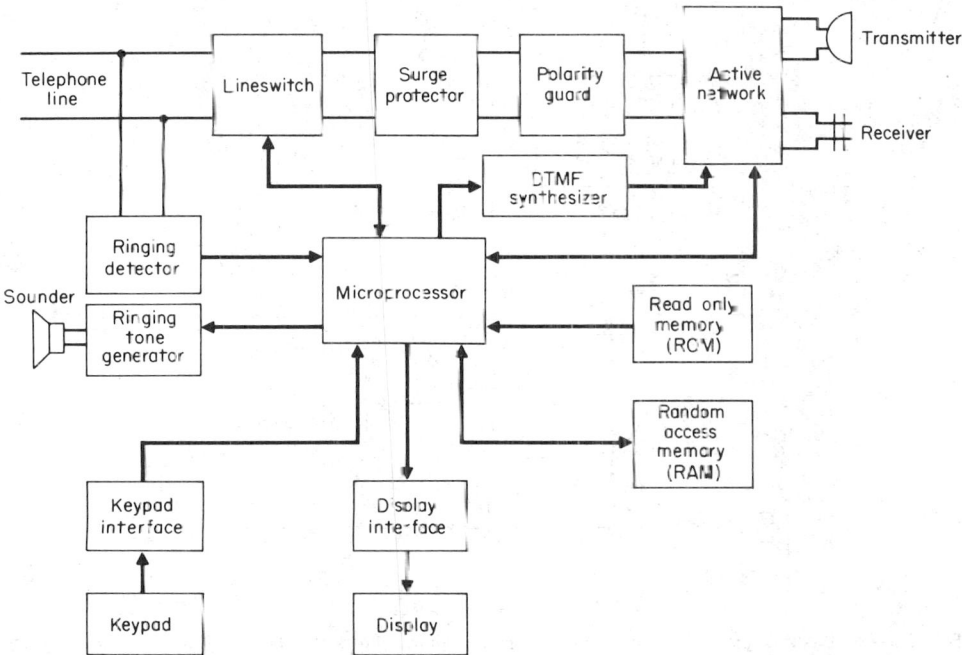

Fig. 22-76. Multifunction electronic telephone.

Both dual-tone multifrequency (DTMF) designs and dial-pulse designs are available. Some manufacturers have provided a 20-dial-pulse-per-second option for use with central offices that accept that higher rate. Common repertory features are last number dialed; dial-tone detectors; one-touch dialing (turns on a speakerphone, waits for dial tone, and dials with only a single button push); and numeric displays of the dialed digits (with a query mode where the number may be displayed without dialing).

Repertory dialers have evolved from large, rather expensive business units (usually requiring bulky power supplies) to small inexpensive designs for business and residences using low-power microcomputers and RAMs. Some dialer designs are powered from the telephone line. Figure 22-77 is a block diagram of a typical microcomputer-based design.

115. Speakerphone. A speakerphone is basically the transmission portion of a telephone set in which the handset has been replaced by a microphone and loudspeaker, typically located within a few feet of the user. This arrangement provides the user freedom to move about with hands free during a telephone conversation, as well as some reduction in fatigue on long calls. It also facilitates small-group participation on a telephone call and can be of benefit in special cases involving hearing loss and physical handicaps.

The lengthened transmitting and receiving acoustical paths in the speakerphone arrangement, compared with those of a conventional telephone, introduce loss, which can be on the order of 20 dB or more in each path. This requires that gain be added to both the transmit and receive channels over that provided in a conventional telephone. The amount of gain which can be added in each channel to compensate for the loss in the acoustical paths is limited by a "singing" problem and room echos which are returned to the distant talker. A signal from the microphone reaches the loudspeaker via the sidetone path and returns to the microphone through acoustic coupling in the room. Singing can occur when too much gain is added in this loop. Even before

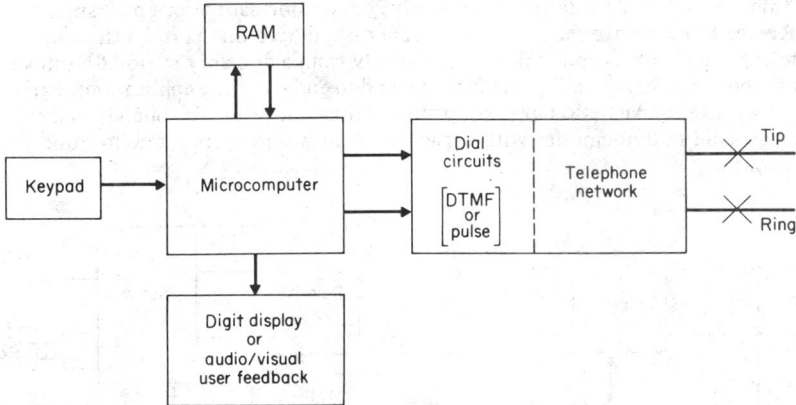

Fig. 22-77. Repertory dialer block diagram.

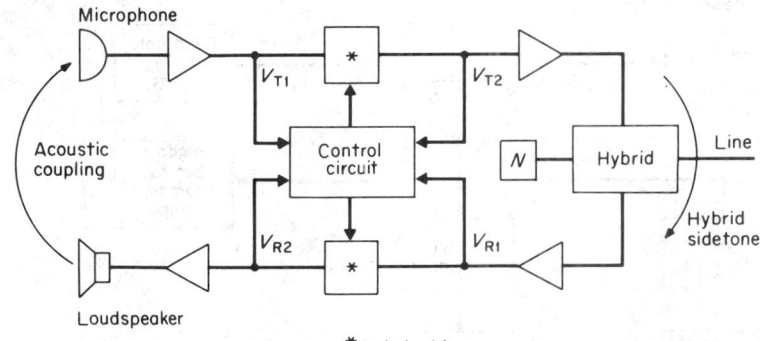

*Switched loss

Fig. 22-78. Block diagram of a voice-switched speakerphone. *(From Ref. 86. Copyright 1960, American Telephone and Telegraph Co. Used by permission.)*

reaching this condition, however, the return of room echos to the distant talker can become highly objectionable. Echos occur when coupling from loudspeaker to microphone causes the incoming speech to be returned to the distant party with delay.

A solution to these problems can be found in *voice switching*,[85,86] in which only one direction of transmission is fully active at a time. With voice switching, a switched-loss or switched-gain element is provided in both the transmit and receive channels, which operate in a complementary fashion. In this manner, full gain is realized in the chosen direction of transmission, while margin is provided against singing and distant talker echo. Voice switching, however, results in one-way-at-a-time communication. Additionally, there can be a problem of clipping of at least some portion of the speech, since control of the voice-switching operation is derived from the speech energy itself.

A functional diagram of the essential elements for a voice-switched speakerphone is shown in Fig. 22-78, in which a measure of the speech energy is provided to the control circuit from four distinct locations in the transmission paths. Signals V_{T1} and V_{R1}, which are a measure of the speech energy in the transmit and receive paths, respectively, are compared to determine the direction of transmission to be enabled. Signals V_{T2} and V_{R2} are used by the control circuit to compensate for the speech energy transmitted via the hybrid sidetone circuit and room acoustics, respectively, which might otherwise result in false activation of the voice switch. Many speakerphone designs do not use all four control signals directly, but equivalent functions are generally provided by other means.

While voice switching is effective in eliminating the problems of singing and distant-talker echo, it does not relieve the higher transmitted levels of room ambient noise and reverberant speech caused by increased gain in the transmit channel.

TELECONFERENCE TERMINALS

BY G. P. TOROK, C. STOCKBRIDGE, AND J. J. JETZT

116. Handwriting Transmission Systems. Graphic transmission terminals can send handwriting over conventional 3-kHz-bandwidth voice-grade telephone lines. Such systems can be used for remote teaching[87,88] and teleconferencing. By using telephone conference bridges a large number of locations can be interconnected for multipoint interactive communication.

A handwriting-transmission system consists of an input terminal which detects the position of the writing instrument (pen, chalk, light pen), a suitable processor for sending and receiving the information, and a display terminal which reproduces the transmitted information (see Fig. 22-79).

Unlike facsimile systems, where a graphic image is dissected and sent in a scanned fashion, handwriting transmission can be accomplished in a time-sequential manner in real time. The inertia of a human hand and the limitations of the nervous system cause the information to be generated quite slowly with handwriting. Figure 22-80 shows a typical relationship between the peak-to-peak amplitude of one spatial coordinate of handwriting on a blackboard and frequency. It can be seen that the highest frequency is less than 16 Hz. This frequency can easily be transmitted over ordinary telephone lines, even when high reproduction quality is desired.

Commercial systems use AM, FM, and various digital transmission methods. In the simplest systems the X and Y coordinates are transmitted as a pair of tones. One frequency band, that is, 600 to 1,200 Hz, corresponds to X and another band, 1,500 to 2,100 Hz, corresponds to Y. Such a system has reasonable performance on quiet lines but is susceptible to jitter and distortion on other lines. Better performance can be obtained with FM transmission, where even a stationary XY coordinate is modulated on an FM carrier, thus spreading the energy of the signal over a frequency band. The best performance can be obtained through digital transmission. In a digital system the coordinates can be sampled at a rate more than twice the highest frequency, that is, > 2 times 16 Hz, and then reconstructed using techniques similar to the ones used in processing sampled voice signals (Par. **22-56**). Alternatively, the information can be transmitted point by point consistent with the resolution of the display.

The input terminal can be a graphics tablet using capacitive or magnetostrictive technology for determining the position of the pen on a tablet. Light pens have been used in conjunction with a CRT, but this method is slow and has limited resolution. In one system the input terminal is a 4- by 5-ft blackboard on which the pressure of a conventional piece of chalk causes two

Fig. 22-79. Electronic blackboard.

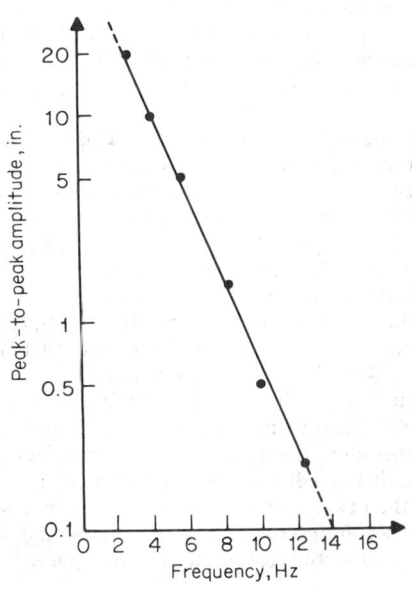

Fig. 22-80. Handwriting frequency response. *(From Ref. 87. Copyright 1977 IEEE. Used by permission.)*

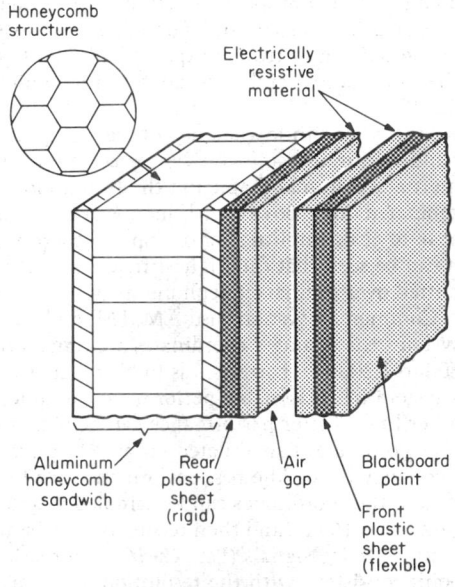

Fig. 22-81. Electronic blackboard construction. *(From L. E. O'Boyle et al., Bell Laboratories Record, October 1979, Vol. 57, p. 257. Copyright © 1979, Bell Telephone Laboratories, Incorporated. Reprinted with permission of Bell Laboratories RECORD.)*

resistively coated surfaces to make physical contact. The system consists of two large potentiometers, and the position is sensed by voltage division (see Figs. 22-81 and 22-82).

Traditional display terminal systems use a pen in a mechanical arm which is positioned over a paper or transparent film by a pair of servo motors. Resolution s very good, but inertia and reliability have presented problems. Some experimental systems use laser beams which write on ultraviolet-sensitive film or on infrared-sensitive smectic liquid crystals. Storage CRTs have also been used but have limited brightness and size. The most widely available systems use a frame memory for storage and a conventional video monitor for display. The advantage of this combination is that the video signal is standard and can be distributed to numerous monitors. As processing and storage become less expensive, new features such as color, retransmission of stored graphics, and combined handwriting, computer graphics, and facsimile will become cost-effective.

117. Audio Teleconference Terminals. To some, teleconferencing means full-motion, studio-quality interactive television and to others, third party add-on using telephone handsets. Both are correct; they are simply the opposite ends of the spectrum of teleconferencing

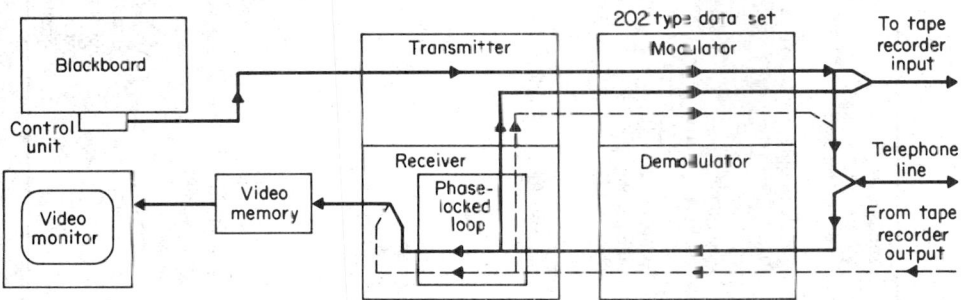

Fig. 22-82. Block diagram of electronic blackboard system. (*From L. E. O'Doyle et al., Bell Laboratories Record, October 1979, Vol. 57, p. 258. Copyright © 1979, Bell Telephone Laboratories, Incorporated. Reprinted with permission of Bell Laboratories RECORD.*)

possibilities.[89-92] Conferencing between two or more groups using hands-free telephone equipment together with supplementary electronic graphics represents an intermediate view.

The simplest audio terminals are the speakerphones described in Par. **22-115**. These teleconference terminals were designed for general-purpose hands-free telephony, tele-education and small ad hoc business conferences. When the microphones are used within the critical acoustic distance of the speaker's mouth, they perform well. The transmitter unit should be placed centrally in a group of people, not more than 3 ft from the talkers. The loudspeaker should be more than 18 in from the transmitter; its position is otherwise not critical. For best performance and minimum intrusion of the switching action between the listening and talking modes on the conversation, the volume control should be set as low as will give an adequate listening level.

Other terminals are specifically designed for teleconferencing. One such set connects up to eight table or lavaliere microphones and multiple loudspeakers, usually mounted in the ceiling; it is intended for permanent installation in conference rooms, general-purpose meeting rooms, executive board rooms, classrooms, and auditoriums. Network interface objectives for audio teleconferencing equipment are as follows: (1) they should transmit normal speech (71 dBSPL at 18 in) at a level of -18 vu on tip and ring, and conversely, (2) they should transduce received signals at -32 vu to a level of 65 dBSPL at the listener's ears (dBSPL refers to decibels with respect to reference sound-pressure level, which, at 1,000 Hz, is 20 μPa or 0.0002 dyn/cm^2 rms).[78]

In any teleconference using microphones within the critical acoustic distance room acoustics plays a major role in determining the quality of the speech.*

Table 22-32 lists representative audio equipment for various conference applications as a function primarily of the number of people who wish to talk. It cannot be overemphasized that room acoustics has a significant effect on the perception of the quality with which any of these systems perform. Thus a *quiet* room will assure that the ratio of transmitted speech to ambient room

*Guidelines for "Teleconference Room Acoustics" are available from Engineering Director, Customer Equipment System, American Telephone and Telegraph Company, 295 North Maple Avenue, Basking Ridge, NJ 07920, as Publ. 42901.

TABLE 22-32 Audio Teleconference-Equipment Selection Table

Equipment	Microphone coverage	Loudspeaker coverage	Telephone set	Teleconference application
Bell System:				
4 A speakerphone system	Up to 6 people, omnidirectional microphone	20 people in quiet room	Yes, separate	Small business conferences, 2–6 people in offices or meeting rooms
50A1 conference set	6 people on each of 2 external omnidirectional microphones and one internal cardioid	20–30 people	Yes, built-in	Telelectures or similar conferences of 20–30 people
WA7400038 group audio teleconference set	Expandable, 20 people typical	Expandable, 20 or more people	Yes, and extension to another room	Permanent conference rooms and auditoriums 20–500 people

Locally engineered systems used on an experimental basis for the indicated applications:

Equipment	Microphone coverage	Loudspeaker coverage	Telephone set	Teleconference application
50A1 plus auxiliary audio equipment	Expandable, 20 people typical	Expandable, 20 or more people	Yes, built-in	Permanent installation, 20 or more people
WA7400038 plus auxiliary audio equipment	Expandable, 20 or more people	Expandable, 50 or more people	Yes, see above	Special conference rooms and auditoriums
Non-Bell:				
Darome, Inc., convenor ser. 610 and 611	4–8 close-talking cardioid microphones, 8–20 people	20–300 people with auxiliary loudspeakers	No, requires telephone set for dialing and ringing	Telelectures with classroom participating; model 611 for private dedicated 4-wire systems only
Precision Components PC-50B	Cardioid, 3 people on each of 2 external microphones	20–30 people	Built-in	Conferences of 12 or more people
Bell Northern Conference 2000 adjunct to Bell Canada speakerphone	Omnidirectional below loudspeaker up to 12 people at round table	20 people in quiet room, speaker faces up	Yes, separate	Business conferences in small conference room
Konferens telefon Swedish PTT	10 individual headsets with close-talking microphone	1 earphone per participant	Yes, separate excellent pickup and voice quality	Up to 10 people in noisy environments

noise is at least 20 dB, which is minimal for conference telephony, while a *nonreverberant* room will provide a ratio of direct speech to indirect reverberant speech of at least 6 dB, which again is minimal.

It is to be expected that a conferencing system which integrates audio, interactive writing, graphics display with pointing, and slow-scan TV will be most effective. Other visual adjuncts to audio teleconferencing are reviewed in Ref. 93. Table 22-33 lists some of them.

TABLE 22-33   Visual Capabilities of Telewriting, Televideo, and Facsimile Systems

Capability	Example	Device
	Hand-drawn graphics	
Instant transmission; shown on monitors or permanent paper and acetate copies; stored on audio tape	Writing, drawings, outlines, equations, diagrams, graphs	Telewriters (electromechanical pens, electronic blackboard, light-pen video writers, graphics tablets)
	Hand-drawn and computer-drawn graphics	
Instant or delayed transmission; shown on monitors; stored on tape, discs, or in memory; reproduced on printers and plotters	Free-hand drawings, traced diagrams, computer-drawn symbols (diagrams, graphics, schematics, charts), alphanumerics, graphics functions (zoom, reduce, label)	Computer graphics, e.g., TOPES-WECo; storage oscilloscopes; video projectors associated with alphanumeric keyboards
	Graphics and pictures	
Delayed transmission; shown on monitors; stored on audio tape; reproduced on paper	Hand-drawn graphics, prepared graphics, writing, monochrome pictures, color pictures	Slow-scan televideo (Robot, Colorado Video, N.E.C., others)
	Paper reproductions	
Delayed transmission; permanent paper copies	Typewritten pages, documents, prepared graphics, pictures	Facsimile (Xerox, Graphics Sciences, many others)

Source: Adapted from C. Olgren, Ref. 93. Used by permission.

118. Video Teleconference Arrangements. A video teleconference room is a specially equipped room that provides two-way audio and video communication with participants in another similar room (see Fig. 22-83). The audio facility provides hands-free talking and listening. The facility provides face-to-face communication. By selective switching, the system can transmit pictures of the participants in the room. In addition the video facility provides a graphics capability sufficient to resolve about one-half of an 8½ by 11-in page of typewritten material. Acoustically, the room should have low ambient noise and a short reverberation time.

NTSC television standards (see Sec. 20) have been found satisfactory in a video teleconference room, with conventional television cameras producing the outgoing video signal. These cameras must have the proper sensitivity for the typically low lighting level (approximately 100 fc). Voice-activated controls may switch the scene of the outgoing video by the selection of one of several cameras aimed at different sets of participants.

The color rendition by these cameras must be consistent as the scene changes. User-activated controls may select and manipulate the various auxiliary facilities used for the transmission of graphics, which include cameras aimed at front- and backlighted surfaces, a slide-chain mechanism (arrangement for projecting a slide into a camera), and an adjustable camera for miscellaneous purposes. If digital picture processing is used for the transmission of the output video, or if special effects such as split-screen presentations are employed, the cameras must be run from a common synchronization source.

Video teleconference rooms may use conventional television monitors to display the outgoing and incoming video. In addition, the auxiliary cameras may have associated preview monitors to allow setting up graphics material without disturbing the conference.

Conventionally installed ceiling loudspeakers provide proper coverage of the incoming audio. Conventional lavaliere microphones provide good-quality audio pickup from individual participants as well as inputs to the voice-activated camera-switching circuitry. The use of lavalieres and voice-voting circuitry at the microphone preamplifiers minimizes the acoustic feedback in the room. Nevertheless, the audio system requires echo suppression to prevent excessive echo to the talker. If there is significant transmission delay between the video teleconference rooms, echo suppression becomes more critical for satisfactory service. Voice-activated switching, similar to that provided by speakerphone terminals, is often used for echo suppression.

Video teleconference rooms transmit and receive signals on video, audio, and control channels.

Fig. 22-83. Video teleconference room.

The transmission may be digital or analog via wide-band facilities. If it is digital, picture processing may be used to reduce the required bit rate.

Video teleconference service may include multilocation video teleconferencing, where three or more rooms may have a simultaneous conference. The audio will be mixed so that each room will hear all the others. If there is only one incoming video signal at a room, however, a bridging arrangement must actively select the proper picture to be seen by each location.

DATA TERMINALS

BY H. W. EARLE, G. C. FRITZ, J. J. ROSINSKI, W. H. NINKE, AND J. SALZ

119. Transaction Telephones and Terminals. A *transaction telephone* is a point-of-sale terminal specifically designed for use on the public telephone network to improve the efficiency of the credit card authorization or similar procedure. The terminal may serve as the user's main station telephone in addition to providing facilities for automatically dialing a credit service center or data base, for transmitting an inquiry message using DTMF or FSK (frequency-shift keying) data, and for receiving a voice, tone, or FSK data response. The inquiry message might contain terminal information, merchant data, customer credit card data, and the amount of sale. *Transaction terminals* are available for use on polled-access channels in data networks. These terminals transmit inquiries and receive responses using FSK data, and hence, are generally not equipped with facilities for dialing or voice.

Features may include the following: (1) a prompting system, (2) means for reading information from plastic cards, (3) a control panel, (4) means for providing data-base response outputs, and (5) an auxiliary key pad.

User prompting is generally accomplished by sequenced lamps, backlighted panels, multi-segment (or dot) character displays, or CRT displays.

Information is read from plastic cards by a magnetic stripe reader for ABA track 2, an optical scanner for characters or bar codes, an embossed character reader, or a punched-hole reader.

Control panels contain pushbutton keys for terminal control and function selection, a keyboard for manual data entry, and a display for verifying manually entered data.

Data-base response outputs are provided by a handset or loudspeaker for audio responses, by lamps activated by special tones or FSK data, by an alphanumeric display panel to display authorization codes or dollar amounts or further instructions, and by a printer for printing the authorization of checks or sales and deposit receipts.

Some applications may require a separate key pad located several feet from the terminal to permit private entry of a personal identification number

For flexibility, more recent terminal designs incorporate microprocessor control systems. Specific application routines may be stored locally in ROM or downloaded into a RAM from a remotely located controller. Buffer memory is provided for storing input data before transmission and for storing response messages before display or printout.

To improve efficiency, dialing codes, merchant data, and special system instructions can be encoded on plastic cards for quick loading into the terminal via the card reader. Such data may be stored temporarily or permanently in the terminal to eliminate the necessity for reentering the data when transactions are continually made to the same data center.

Transaction telephones may operate with key telephone systems, and automatic dialing features may include the capability to operate behind a PBX with two-part dialing when a second dial tone wait is required. Split mode dialing by transmitting the first part of a two-part number in dial pulses followed by a second part in DTMF, or vice versa, may also be provided.

A comparison of commercially available terminals, features, and operating protocols is given in Refs. 94 and 95.

120. Teletypewriter Data Terminals. Data terminals are used in a large variety of data-communication systems, which generally fall into two categories: terminal-to-terminal and terminal-to-computer. Both are used to communicate messages. Computer-based data-communication systems can also be used for such applications as inquiry-response, data entry, data collection, record update, remote batch, and message switching. Each application requires specific terminal features.

The service networks used in data communications are ordinarily switched networks or private lines, and a different type of terminal is required for each. For example, in *multipoint private-line selective-calling* systems, the terminal must be able to recognize specific commands from the computer conditioning it to send or receive, and a switched-network terminal must be able to access and be accessed by the network.

Data-communication facilities, including terminals, may be half or full duplex. With *half-duplex* terminals, data can be sent and received but not simultaneously. With *full-duplex* terminals, transmission and reception can occur at the same time, providing greater throughput.

Data terminals are made up of combinations of the following modules: keyboard, CRT display, printer, paper-tape reader or punch, floppy-disc drive, electronic buffer, magnetic-tape drive, card punch or reader, and controller. These modules can be organized into the categories of operator *input-output*, terminal *control* and *storage*. A typical data terminal is shown in Fig. 22-84.

Keyboards are available in a number of formats, including typewriter, keypunch, and display with edit. In some cases, two formats are provided on the same keyboard, side by side or integrated, e.g., typewriter with a numeric pad. Future designs call for multifunction keyboards whose designations or functions can be easily changed by the user, because in many cases differently trained operators will be using the same terminal to enter different types of data for different applications. There is also a trend toward special-application keyboards, such as the cash register keyboard used by a fast food chain which has the name of each item on the menu on a proximity-switch plastic overlay.

CRT displays allow the operator to enter, view, and edit information; editing information with a CRT display is typically much faster than if a mechanical printer were used. The most popular CRT displays exhibit 24 lines of up to 80 characters each. Other size variations include 12 lines of up to 80 characters and a full page display (approximately 66 lines of up to 80 characters). Other features found on CRT displays include blinking, half-intensity upper- and lowercase character sets, foreign-character sets, variable character sizes, graphic (line-drawing) character sets, and multicolor displays (see also Par. **22-122**).

Printers for data terminals are classified in two ways, according to the methods they use to (1) receive and print information (character at a time or line at a time) and (2) to form the printed characters (impact or nonimpact). *Character-at-a-time* printers are efficient for low-speed message printouts or brief computer communications such as those found in inquiry-response and record-update applications. *Line-at-a-time* printers are efficient for high-volume, high-speed printouts such as those found in remote-batch applications. However, if a storage module such as a buffer is placed ahead of a character-at-a-time printer, it can be used in higher-speed applications, since the buffer can receive at one speed and send at another. This permits a low-speed character-at-a-time printer to be used where a more expensive high-speed line-at-a-time printer would otherwise be required.

Impact printers strike characters onto paper via a print ribbon. *Nonimpact* printers typically burn (thermal) or otherwise chemically react with special (electrostatic) paper to produce what appears to be a conventional printout. Still others print with a stream of ink droplets which is directed toward the paper and electrostatically deflected to trace characters. The speed of the nonimpact type can approach that of a line-at-a-time printer, and their operation is nearly silent. However, they produce only one copy of the printout and, in some cases, provide less legibility.

Character-at-a-time impact printers are available in whole-character and dot-matrix versions. *Whole-character* printers print via typewheel, typebox, daisy wheel, ball, and many other print-head methods, with varying results in print quality and speed. *Dot-matrix* printers are inherently somewhat faster and can be easily altered through the electronic circuits to produce a variety of printing densities, character sizes, type styles, and languages, whereas whole-character printers must often be mechanically altered at least by changing the print head. Legibility was once a limiting factor of dot-matrix printers, but it has rapidly improved.

The most popular media for *storage modules* include paper (card, tape), magnetic (floppy diskettes, tape), and electronic (integrated-circuit RAM, bubble). Data stored on magnetic or electronic media can be erased or replaced with other data. Data stored on paper media such as a punched card or paper tape can only be re-recorded on virgin media in order to be edited. Access to stored information can be sequential, e.g., tape, or random (diskette). Typically, information can be retrieved from random-access storage orders of magnitude faster than from sequential access storage.

Controllers interconnect the channel interface and the various terminal components and make them interact to perform the terminals' specific functions. They also perform other functions, such as recognizing and acting on protocol commands, e.g., polling and selecting sequences in selective calling systems; code translation; and error detection and correction. The trend in controller design is toward the use of solid-state electronics, especially microprocessors, to increase their functions, including programmability, which greatly increases terminal versatility.

Standardized codes, protocols, and interfaces provide a uniform framework for the transmission and reception of data by data terminals. Codes and some character-oriented protocols are discussed in Par. **22-3**. Interfaces are discussed in Pars. **22-2** and **22-9**.

IBM's binary synchronous communication (bisynch) protocol has been implemented widely. It accommodates half-duplex transmission and is character-oriented. Bit-oriented protocols that provide both half- and full-duplex transmission are coming into wider use. The American National Standards Institute's (ANSI) Advanced Data Communications Control Procedures (ADCCP), and the International Organization for Standardization's High Level Data Link Control (HDLC) are the current standards for bit-oriented protocols. Although there are differences, they are generally compatible, and the trend is for other bit-oriented protocols to establish compatibility with them.

121. Teletext and Videotex Systems. Teletext* and Videotex are information communication systems which provide users access to visual (text or graphics) information.

Teletext[96,97] is based on a broadcast capability (airwave or cable). Frames (each frame being one screen of information) are transmitted as coded data rather than in video form to suitably equipped television receivers. A large number of these frames are continuously transmitted from an information-storage facility in cyclic order, during the vertical blanking interval of an in-use television channel or in an entire unused video channel. A user typically selects the frame or frames desired for viewing by entering frame-identification numbers on a key pad. The receiver extracts the data signal associated with the desired frame from the overall bit stream. The data

---

*Not to be confused with the CCITT Teletex service.

signals are then decoded, and information is stored in the Teletext receiver to permit display on a television set. Some Teletext systems provide an additional communications link to allow the user to transmit messages back to the information storage facility e.g., for a voting-from-home application.

Videotex[98,99] is based on a two-way switched telecommunications system network (telephone or data). Information can be displayed on modified television sets or on special-purpose data-display terminals. Information is transmitted as modulated data bidirectionally between the terminal and the information source, which consists of a computer system and associated information data bases. A user selects the desired information via the terminal's input device, typically a key pad or a keyboard. User-to-user communications are possible with Videotex thanks to the use of telecommunications networks for transmission.

Although Teletext service has a greater transmission bandwidth and hence greater data-transmission speed than Videotex service, the total number of frames available from the information

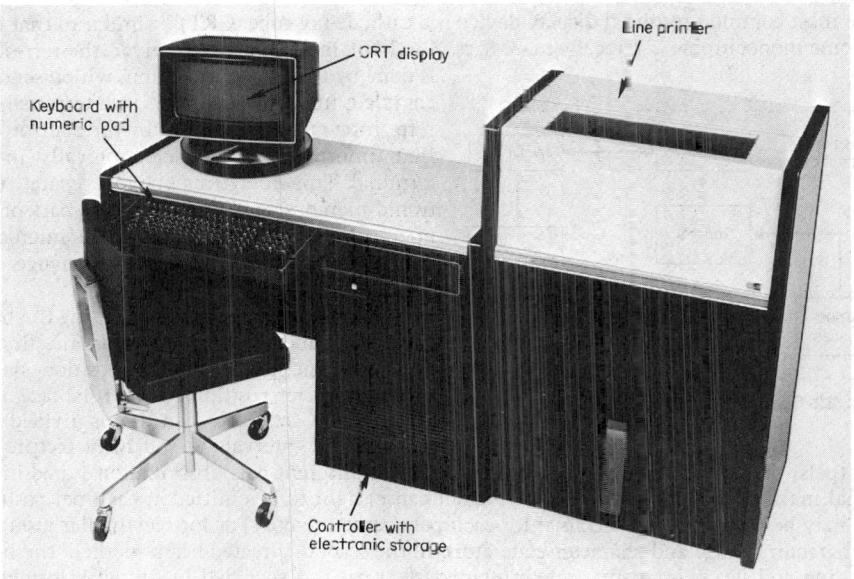

Fig. 22-84. Teletype 4540 data terminal. *(Teletype Corporation.)*

source is much smaller, due to the limited time a user is willing to wait for the desired frame to be "captured" from the continuously transmitted stream of frames. On the other hand, Teletext allows simultaneous information access by a virtually unlimited number of users, while Videotex service has a limited simultaneous user capability due to loading on the information source unless additional computers are provided.

The various countries and companies implementing Videotex systems are considering alternative methods for communicating graphics information. One method mosaic graphics, involves the juxtaposition of symbols, coded as discrete characters for transmission, on the screen. The set of display symbols can be permanently stored in the terminal or downloaded to it from the information storage facility. A second method, geometric graphics, involves transmitting position and size parameters to the terminal, which then executes stored programs for drawing points, lines, arcs, and polygons on the display. A third method, photographic graphics, is analogous to facsimile in that individual picture elements (pels) or groups of pels are coded and transmitted individually.

122. Display Terminals (Text and Graphics). Terminals for the entry and electronic display of textual and or graphic information generally are used for communicating between people and computers and, increasingly, between people. Typical application areas include computer program preparation, data entry, inquiry-response, inventory control, word processing, financial transactions, computer-aided design, and electronic mail. A terminal is divided functionally into control, display, user input, and communication-line-interface portions (see Fig. 22-85). The first three portions are detailed in the following sections. Interface to a communication line is typically through either a directly coupled or acoustically coupled modem (Par. **22-124**).

Terminals have differing amounts of control or decision-making logic built into them. A popular classification scheme is based on terminal control capabilities.[100] A nonintelligent ("dumb") terminal is a basic I/O and control device. Although it may have some information buffering, it must rely on the computer to which it is connected for most processing of any entered information and editing of output displays. A "smart" terminal has both information entry and editing capabilities, and although it too is generally connected to a computer, it can perform information processing locally. The terminal usually contains a microcomputer (see Sec. **8**) which is programmed by the terminal manufacturer to meet the general and special needs of a user. An "intelligent" terminal has a microcomputer that can be programmed by a user to meet user-specific needs. Terminals can be connected to a supporting computer either directly or through a cluster controller that supervises and services the data-communication needs of a number of terminals.

The most common terminal display device is a cathode-ray tube (CRT)[101] similar to that used in a home monochrome TV receiver (see Secs. **7** and **20**). In a home TV receiver, the refreshing is done by the broadcast station, which sends 30 complete images per second. In a terminal, the refreshing must be provided by the control logic from information stored electronically in the terminal. This storage may be in a separate electronic memory or memory that is part of the microcomputer. In either case the microcomputer can address, enter, or change the information.

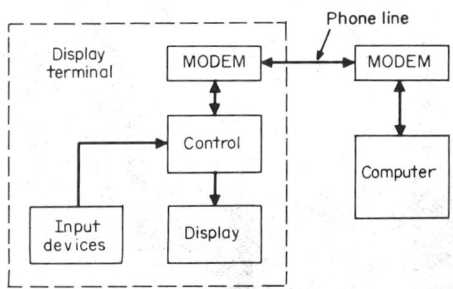

Fig. 22-85. Display-terminal organization.

Two methods exist for coordinating the beam positioning and beam intensity. In the first, or raster-scan method, the beam position sweeps through a raster position on the tube face as on a regular TV set. Each scan line is divided into a number of intervals which form picture elements (pels). As the beam position passes through the different pels, information stored in the terminal makes the beam current cause (or not cause) light to be emitted at each pel position. There may be storage in the terminal for each pel (frame memory) or for rectangular groups of pels (character storage and character generator). In the second, directed-beam method, the beam positioning and beam intensity are simultaneously varied as specified by stored information. Images are "drawn" out much as a pencil writes on paper.

Some terminals use special storage CRTs in which an image is stored as a charge pattern deposited by the electron beam on an internal mesh screen or special faceplate layer (see Sec. 7). Refreshing is not needed, and there is no flicker. To change an image, however, the entire charge pattern must be erased by an electric signal generated by the control and a new image totally reformed. For very complicated images this can be a slow process. Standard refreshed CRT displays reform a new image every 1/30 s. Since to make a small change, only a small portion of the electronic storage need be changed (a rapid process), refreshed displays are usually better for presenting rapidly changing images and for providing feedback to a user through mechanisms such as blinking or intensity inversion.

Display terminals that can present images in several colors have recently been introduced. A color CRT, like that of a home TV set, is used (see Secs. 7 and 20). The extra dimension of color allows information association to be presented easily, allows special attention to be directed at certain information, and also allows some esthetically pleasing effects.

Displays constructed using flat plates of glass are now appearing. These displays are more compact than CRTs and permit design of smaller terminals. Gas discharge[102] is the most common flat-panel technology. In it, light is generated from the cathode glow of the breakdown or discharge of a gas, usually neon. The gas is sealed in a flat-glass-panel envelope. Conductor patterns on the glass define the locations where discharges can be formed.

Two types of displays are made, dc and ac. In dc gas displays, the conductors are in contact with the gas, and a pulsed single-polarity voltage is used to cause breakdown. In existing commercial displays, the breakdowns occur in a time-division fashion. External semiconductor memory is used to store what is to be displayed. The memory repetitively controls the drive circuits to refresh the image on the display at a rate well above the flicker rate of the eye, as in a refreshed CRT. In the ac gas display, or plasma panel, conductors are isolated from the gas by a thin dielectric film. An alternating polarity-sustaining voltage causes a discharge site, once broken down, to discharge anew with each alternation of the sustaining voltage. Special selection circuitry is used to cause an initial breakdown. A display thus has inherent memory like a storage CRT and need not be refreshed from external memory.

Most display terminals have a typewriterlike keyboard and a few extra buttons as input devices for the entry of textual and control information. Text is entered at a position on the display indicated by a special symbol called a cursor. The control moves this cursor along much as a typewriter head moves along as text is entered via the keyboard. Different manufacturers use different cursors (underlines, blinking squares, video inverted characters).

To allow editing of existing images or flexible entry of additions, auxiliary control devices are provided to change the cursor position. The simplest is a set of five buttons, four of which move the cursor up, down, right, or left one character position from the current position. The fifth indicates that the cursor is to be "homed" to the upper left-hand corner of the image, where textual entry usually starts. Another popular position-controlling device is a joystick, a lever mounted on gimbals. Movement of the top of the lever is measured by potentiometers attached to the gimbals. The control senses the potentiometer outputs and moves the cursor correspondingly.

Terminals intended for display of complex graphic information e.g., for computer-aided design, generally have either tablet-stylus devices or light pens[103] for input of new information or indication of existing information. One popular tablet-stylus has a surface area under which magnetostrictive waves are alternately propagated horizontally and vertically. A stylus with a coil pickup in the tip senses the passing of a wave under the stylus position. Electronic circuitry measures the time between launching of a wave and its sensing by the stylus and computes the position of the stylus from that time and the known velocity of the wave.

A light pen senses when light is within its field of view. In a raster-scan display, the time from the start of a displayed image until the light-pen signal is received indicates where the pen is in the raster-scan pattern and, with suitable scaling factors, gives the position of the pen over the image. In a directed-beam display, pen-position locating is more complicated. The centering of a special tracking pattern under the pen is sensed and through a feedback arrangement controlled by the terminal computer the pattern is moved to keep it centered. The position of the pattern is then the same as the pen position.

Although most current users of display terminals are divorced from accompanying voice communication, the future promises an incorporation of extensive display functions into the ordinary voice-telephone terminal. Much special information retrieval now being planned for Videotex services (Par. **22-121**) will be possible through these "telephones." Graphic-image communication, now requiring special audiographic teleconferencing terminals (Par. **22-117**), will also be done through future telephones. Future telephones probably will consist of voice input and output transducers, a general-purpose flat-panel display, some control buttons, and a microprocessor. Terminal functions will be defined by programs in the microprocessor, some permanent and some downloaded from storage at a telephone central office. There will thus be display telephones whose functions are defined by software.

123. Data Transmission on Analog Circuits. The devices discussed in Pars. **22-119** to **22-122** and Sec. **23** generate baseband signals[104,105] (see Sec. **4**). These signals are not compatible with the voice circuits of the public telephone network, partly because of frequency range but more importantly because these circuits are likely to produce a small frequency offset in the transmitted signal. The offset causes drastic changes in the received waveform. This effect, while quite tolerable in voice communications, destroys the integrity of individual pulses of a baseband signal. Compatible transmission is obtained by modulating a carrier frequency within the channel passband by the baseband signal in a modem (modulator-demodulator).

For a band-limited system, Nyquist showed that the maximum rate for sending noninterfering pulses is two pulses (usually called symbols) per second per hertz of bandwidth. The bit rate depends upon how these pulses are encoded. For example, a two-level system transmits 1 b with each pulse. A four-level system transmits 2 b per pulse; an eight-level system transmits 3 b; etc. Unfortunately, we cannot go to an arbitrarily large number of levels because, assuming that the total power is limited, the levels will become so closely spaced that the random disturbances in the transmission medium will make one level indistinguishable from the next. Shannon's fundamental result states that there is a maximum rate, called the *channel capacity*, up to which one can send information reliably over a given channel. This capacity is determined by the random disturbances on the channel. If these random disturbances can be characterized as white gaussian noise, the channel capacity C is given by

$$C = W \log_2 (1 + S/N)$$

where W is the bandwidth of the channel and S and N are the average signal and noise powers respectively.

With today's more elaborate commercial modulation techniques it is possible to transmit data at approximately one-half the capacity of the channel. With the most complex modulation and coding techniques, it is possible to get very close to the capacity of the channel, but this is achieved at the expense of considerable processing complexity and delay.

Data transmission employs all three modulation methods (amplitude, frequency, and phase) plus combinations of them. A description of these methods is given in Sec. **14**.

An on-off amplitude-modulated (AM) signal is shown in Fig. 22-86a. This modulation scheme is little used for voice-band modems but is used in wide-band modems with raised cosine pulse shaping. Envelope-detection schemes are not common in data transmission, since the envelope is often severely distorted by the transmission channel. A variation of the method in which one sideband and a portion of the other are transmitted, called vestigial-sideband (VSB) AM, is sometimes used to give increased transmission speed over double-sideband AM while permitting carrier recovery for detection purposes.

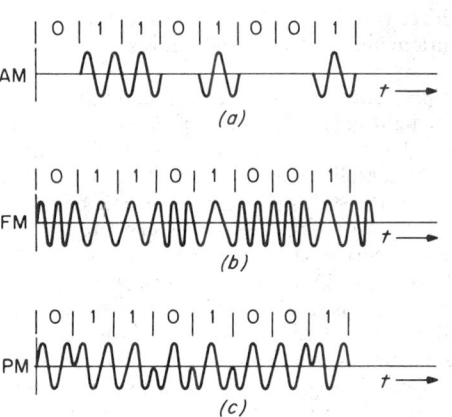

(a)

(b)

(c)

Fig. 22-86. Binary (a) amplitude-, (b) frequency-, and (c) phase-modulated carrier waves.

An example of a binary frequency-modulated (FM) carrier wave, sometimes called frequency-shift keying (FSK), is shown in Fig. 22-86b. While FM or FSK requires somewhat greater bandwidth than AM for the same symbol rate, it gives much better performance in the presence of impulse noise and gain change (Pars. **22-27** and **22-30**). It is used extensively in low- and medium-speed (voice-band) telegraph and data systems.

A binary phase-modulated (PM) carrier wave is shown in Fig. 22-86c, where a phase change of 180° is depicted. However, this modulation method is usually employed with a four-phase and sometimes an eight-phase change, which must be accommodated within a range of $\pm 180°$. In a four-phase-change system, the binary bits are formed in pairs called *dibits*. The dibits determine the phase change from one signal element to the next. The four phases are spaced 90° apart, but several different translations from dibit to phase change are in use. CCITT Recommendations include two: one translates the dibits 00, 01, 11, 10 into 0, 90, 180, and 270°, respectively; the other, into 45, 135, 225, and 315°, respectively. In effect, this method employs a four-state signal, and such a system is inherently capable of greater transmission speeds for the same bandwidth, as is obvious from the Nyquist-Shannon criteria stated above. With improvement of voice channels, the four-phase (quaternary-phase-modulated) scheme is being employed increasingly for medium-speed (voice-band) data transmission systems to give higher transmission speeds than FM for the same bandwidth. The system is useful only in synchronous transmission.

Table 22-34 gives comparative data for a number of systems (both binary and multilevel) on a fixed-bandwidth basis. Signal-to-noise ratio is given on the basis of both fixed maximum steady-state power and average power. This table indicates theoretical performance of perfectly implemented systems with optimum signal shaping and filtering. Actual systems offer poorer performance by 1 dB or more.

124. Data-Circuit-Terminating Equipment (DCE). The DCE (incorporating a modulator-demodulator, or modem) accepts the telegraph or data baseband signal, modulates it, transmits it over the communication system, and performs the reverse when receiving the data- or telegraph-modulated signal. It also incorporates some type of interface for control of the telegraph or data user equipment.

Choice of Modulation System   In present-day modems FSK is the preferred modulation technique for bit rates below 1,800 b/s. At 2,400 b/s, the commonly used modulation technique is PSK using four phases, and at 4,800 b/s it is PSK using eight phases. The latter requires the use of an adaptive equalizer, an automatically adjustable filter that minimizes the distortion of the transmitted pulses resulting from the imperfect amplitude and phase characteristics of the chan-

TABLE 22-34   Comparison of Data Systems under Limitations of Average and Maximum Steady-State Power*[104]

System	No. of states	Speed per hertz of bandwidth, b/s	Signal-to-noise ratio for 10⁻⁴-b error rate	
			Avg. signal power	Max. steady-state signal power
Unipolar baseband	2	2	14.4	17.4
	4	4	22.8	26.9
Bipolar baseband	2	2	14.4	17.4
Polar baseband	2	2	11.4	11.4
	4	4	13.3	20.8
	8	5	24.3	28.0
	16	3	30.2	34.4
Full-carrier AM:				
Envelope detection	2		11.9	14.9
Coherent detection	2		11.4	14.4
	4	2	15.8	23.8
	8	3	26.5	31.0
Suppressed-carrier AM, coherent detection	2	1	8.4	8.4
	4	2	15.3	17.8
	8	3	21.3	25.0
	16	4	27.2	31.4
PM, coherent detection	2	1	8.4	8.4
	4	2	11.4	11.4
	8	3	16.5	16.5
	16	4	22.1	22.1
	32	5	28.1	28.1
	64	6	34.1	34.1
Differential detection	2	1	9.3	9.3
	4	2	13.7	13.7
	8	3	19.5	19.5
FM	2	1	11.7	11.7
	4	2	21.1	21.1
	8	3	28.3	28.3
VSB, 50% modulation	2	2	16.2	17.9

*For VSB, suppressed-carrier, coherent detection: see figures for polar baseband for quadrature AM, suppressed-carrier, coherent detection: see figures for polar baseband.

nel. At 9,600 b/s, the preferred modulation technique is quadrature amplitude modulation, a combination of amplitude and phase modulation.

125. Data-Circuit-Terminating-Equipment (DCE) Trends. Most modern businesses of significant size depend heavily upon their data-communications networks. Large businesses usually have a staff of specialists whose job it is to manage the network. To assist in this network-management function, DCE manufacturers have provided various testing capabilities in their products. Sophisticated DCEs now are capable of automatically monitoring their own "health" and reporting it to the centralized location where the network management staff resides. These new capabilities are frequently implemented through the use of microprocessors. Some DCEs can also establish whether a trouble is in the modem itself or the interconnecting channel. If it is the channel that is in trouble, equipment is available that automatically sets up dialed connections to be used as backup for the original channel. Another capability is to send in a trouble report automatically when a malfunction is detected.

FCC REGULATIONS

BY A. B. BROWN, JR.

126. Certain characteristics of communications terminals are governed by Parts 15 and 68 of the FCC Regulations (obtainable from Superintendent of Documents, U.S. Government Printing Office, Washington, 20402). Part 15 applies to computing devices (defined as devices that generate timing signals or pulses at rates of more than 10,000 pulses per second and use digital techniques) and covers radiation of rf energy into space and its conduction along the power line, with resulting harmful interference with radio communications. Different radiation and conduction limits are given for computing devices that are intended for use in a commercial, industrial, or business environment and devices that are marketed for use in a residential environment. Requirements are also given for test methods, labeling the equipment, and information to be furnished to the user.

Part 68 applies to equipment that is to be directly connected to the telephone network and covers properties that could cause harm to the service provided by the network or to people working on it. Part 68 applies to all terminal equipment that is directly connected to the telephone network for use with services other than certain private-line services, party-line service, and coin service.

The Part 68 requirements cover the insulation of the terminal equipment, specified in terms of leakage currents and voltage levels appearing on leads or surfaces, to prevent unsafe voltages or currents from appearing anywhere that might cause injury or harm. The requirements also cover the signal-power levels the terminal equipment may apply to the telephone line and the longitudinal balance it presents to the line, to prevent overload of carrier systems and crosstalk in cables. The impedance the equipment presents to the line in both the on-hook and off-hook states is also specified, to provide for proper supervision and maintenance of the loop. The range of ringing signals to which ringing detectors must respond is also specified, to prevent inefficient use of network facilities. Timing requirements are given for the transmission of energy when the equipment has initially gone off-hook, to provide for proper billing of the call.

The electrical requirements are to be satisfied in all operational states of the equipment, both before and after the application of specified environmental stresses, involving vibration, temperature and humidity, shock, and metallic and longitudinal surges.

Illustrative Applications

BY A. B. BROWN, JR.

BUSINESS COMMUNICATIONS SYSTEMS

The opportunity to increase the productivity of office workers and managers has led to greatly increased use of information-handling equipment. People have used this equipment for interactive communication with their associates (teleconferencing), for noninteractive communication with their associates (electronic mail), and for interactive communication with stored information. The use of computer terminals and the channels described in previous paragraphs to retrieve and process information according to the needs of the worker is done in ways that can be inferred from this section and Sec. **23**. Electronic mail and corporate message switching are described in the next paragraphs and teleconferencing in Pars. **22-116** to **22-118**.

127. Electronic Mail. Much business communication consists of transferring information in one direction, with interaction, if any, occurring later, e.g., after the recipient has had time to assimilate the information and prepare comments or other response. The communication may be within the business organization or to another company. Electronic mail has not yet acquired a precise definition through consensus; it may mean transferring information that is coded in characters or the image of a document; it may mean transferring hand-drawn information. Many vendors are offering packages to implement these functions. The U.S. Post Office is working on transferring information electronically to the destination city and delivering it to the addressee through the usual mail channels.

Requirements for electronic mail differ from those of interactive information transfer; time delays of minutes or hours are acceptable, and delivery to multiple addressees and translation of a single reference into a list of addressees may be desired. The recipient should be informed of the arrival of the mail, should be able to see or hear it, should be able to get a permanent copy, and should be able to discard it. In some applications, the ability to store received messages and to recall them by some characteristic, e.g., keyword or date, may be needed.

An electronic mail system is composed of terminals, transmission, and switching facilities as described in earlier paragraphs. A very simple system may consist of a number of communicating word processors with a controller; it may, on the other hand, involve communication via large computers and a large network.

Uses of an intracompany electronic mail system include the transmission of informal notes or formal memoranda or leaving messages for people away from their work station. Intercompany mail may include letters, purchase orders, or invoices. Comments on a draft report by several members of a committee can be distributed to each member and serve as the basis for further comments without routing the draft to each member in sequence.

128. Corporate Message Switching. Closely related to electronic mail is the process of sending messages to many people at different locations of a multilocation company. For efficient use of transmission facilities, it may be desirable, when the routes to several addressees coincide for part of their length, to send only one copy as far as the point where the routes diverge and to initiate separate copies at that point.

ALARM-REPORTING SYSTEMS

129. Functions and Requirements. Systems that are intended to report an intrusion or a fire to the police or fire department or to an alarm company must be reliable, operate within less than a minute, and, in the case of intrusion, withstand attempts by the intruders to disable the alarm. Systems that are intended to report a mechanical failure, such as that of a food freezer in a supermarket, may take longer to make the report and are not subject to attempts to disable them. Systems requiring high reliability are designed to operate through any single fault (open, short, or ground) on the transmission facilities, either through the circuit design of the equipment or through the operating practices, as described below. The trouble situation is detected by means which are provided by the user of the premises or by the alarm company: heat or smoke alarm, intrusion detection, freezer temperature, and the like. The trouble sensor is used to activate a transmitter which sends the alarm to the appropriate central point and, if needed, identifies the premises and the nature of the trouble by a coded signal.

130. Dial-Up Systems. The simplest systems, which are appropriate for the less demanding types of alarm, are those which report alarms automatically by use of the switched telephone network. The reporting terminal has stored within it the telephone number of the agency to which the alarm is to be reported; when activated by the alarm sensor, it goes off-hook, waits for dial tone, passes the number to the central office, and transmits the alarm signal. The alarm signal is often in the form of a recorded voice message, which identifies the location and nature of the emergency. The terminal may detect the dial tone, or it may simply wait a specified time; it may not require an answer signal but may transmit the alarm message several times after completion of dialing. Such a system is relatively inexpensive and requires no special transmission facilities, but it is subject to congestion in the telephone network and to deliberate attempts to disable it.

131. Voice-Band Private-Line Systems. A second type of system uses private-line voice-band channels, either point-to-point or multipoint, for transmission of the alarms. A variety of terminal equipment is used on such systems; in general, the signals do not depend on dc continuity of the channel. Such a system is free from network congestion problems, attempts to disable one of the transmitters can be detected by polling the stations from the central point and looking for a response from each station. However, noise or jamming introduced at any point of the network can affect the signals from all stations.

132. McCulloh Loop Systems. A McCulloh loop is a metallic circuit, starting at the alarm central station, going to each of the protected premises in turn, and returning to the alarm central station. The wire pairs are usually provided by the telephone company. At each protected premises, one or more transmitters are attached to the circuit in series. At the alarm central station, a dc voltage is applied to the circuit and between the circuit and ground. An alarm transmission consists of sending a coded series of pulses, each of which involves a momentary opening of the series circuit and momentary application of a ground. Relays at the central station detect the opens and grounds, and the series of pulses is recorded, to give the location of the emergency. An open in the series circuit is detected at the central station, and any alarm before it is repaired is detected by the grounding pulses applied by the transmitter. Similarly, a ground on the circuit is detected and an alarm is transmitted by the momentary opens applied by the transmitter. If both sides of a wire pair leading to a given premises are cut or grounded, that location can send no alarm signals, but the other locations are still protected. Maintenance procedures in the telephone offices through which the loop passes locate the trouble for repair.

The McCulloh loop system can be used to protect a single premises; there is then no problem of overlapping alarm signals from various locations, and any disturbance of the circuit can be related to the single protected premises. The system does require dc continuity of the transmission facility, even between telephone company offices. In some areas, such a facility is scarce and expensive.

133. Switched Private Systems. A recently developed alarm system uses switched private voice-band facilities, which do not need dc continuity. The switch is controlled from the alarm central station and provides a voice-band path to one protected premises at a time. The alarm central station polls each transmitter, to verify that the circuit and the transmitter are in good order or to receive an alarm. The polling arrangement, number of stations connected, and length of transmission are chosen so that the entire polling cycle is completed within the time permitted for an alarm to be reported. The arrangement prevents any attempt to interfere from one location with an alarm to be reported from another location.

ERROR-FREE TRANSMISSION

134. Principles. While completely error-free transmission can not be achieved, errors that may be introduced into data during transmission can be greatly reduced (several powers of 10) by the use of redundancy in the transmitted bits. Before transmission, additional bits are introduced into the message; they are calculated from the bits that carry the information. The principle is basically similar to the redundancy that exists in written English; if a character is typed in error, the reader can usually make out the intended meaning from the context. In data transmission, redundancy is often used to detect one or more errors in a section of the message, and retransmission of that section is requested in order to correct the error. In other cases, when there is no means of requesting a retransmission, sufficient redundancy is used to identify the bit or bits that are in error; they can then be corrected by logic at the receiving terminal.

135. Types of Codes. The simplest form of error-detecting code involves one extra bit per character; in International Alphabet no. 5, which contains seven information bits per character, this is an eighth bit. It is chosen so that the total number of 1 bits in the character is even (or, in certain applications, odd). This is known as even (or odd) *parity*. If any 1 is changed to a 0 or vice versa, the total number of 1 bits will be odd instead of even, i.e., the parity will not check; the error can be detected. If two bits in the character are in error, the parity still checks and the error is not detected. The code can be made more powerful by the use of extra characters, called a *longitudinal redundancy check*, at intervals in the message. These characters are derived from the other characters, either by setting each bit according to the parity of the corresponding bits of the other characters or, if several longitudinal check characters are used, by one of the more powerful error-checking codes. In either case, two bit errors in one character will be detected; several errors distributed through the section of message may be detected.

One of the better-known codes was developed by Hamming;[106] it is shown in its simplest form in Fig. 22-87. The message is divided into blocks of n bits each, where $n = 2^c - 1$, of which c are check bits and $k (= n - c)$ are information bits. Each check bit is a parity bit over $(n + 1)/2$ bits. Of the 2^n possible sequences of n bits, only 2^k are used, corresponding to the possible sequences of the k information bits. Each sequence used differs from every other sequence used in at least three bit positions; two bit errors can always be detected. Alternatively, a single bit error can be corrected. By permuting their bits appropriately, the Hamming codes can be

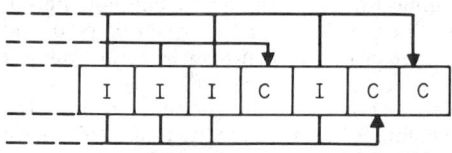

Fig. 22-87. Hamming code: I = information bit C = check bit. The block can be longer, with additional check bits.

shown to be equivalent to members of the class of *cyclic codes*, in which every code word can be converted into another by permuting the bits cyclically. These codes are particularly easy to encode and decode using shift registers.

The codes described above are block codes; i.e., the message is divided into blocks, and the coding of each block is independent. *Convolutional codes*, in which the check bits are generated by a parity pattern that slides along the message, have also been studied but have been less used than block codes. Reference 107 gives a thorough coverage of codes.

136. Implementation. Error control may be applied over each link of a network[3] (see Par. **22-3**) or end to end through the channel or network. The choice of a code for a particular application depends on the rate and distribution of errors on the line, on the desired accuracy of the data delivered to the recipient, on the desired throughput of the system, and on whether a reverse channel and storage of data at the source make error correction by retransmission possible. In general, an error detection and retransmission system will require less redundancy to achieve a given final error rate than a system using an error-correcting code. The rate of errors in the delivered data can be estimated from measurements of transmission facilities such as those discussed in Par. **22-32** by identifying the most likely error patterns that will be undetected or improperly corrected by the code. The block error rate is the probability that such an error pattern will occur; the bit error rate is that probability times the number of resulting errors in the block divided by the number of bits in the block. Each of these must be averaged over the various bit-error patterns that will be undetected or improperly corrected.

In an error-detection and retransmission system, the transmitting terminal may wait for each block to be acknowledged before transmitting the next; this gives the simplest procedures and requires the least storage but reduces the throughput, especially if the round-trip propagation time is long. A second method, applicable where there is a simultaneous reverse channel, is for the transmitting terminal to continue sending while waiting for an acknowledgement; if it is negative, or if none is received within a stated time, the message is retransmitted starting with the block in trouble. A third method provides for retransmitting only the block that was not correctly received; the receiving terminal must be able to insert a block that was first received in error and later received correctly into the message in its proper place. The throughput is highest because the material retransmitted is least.

VOICE-MESSAGE STORAGE

137. Answering and Recording of Messages. A user may wish to record telephone messages that arrive when they cannot be answered and to listen to them later. Customer-premises equipment is available which will detect the alerting signal, go off-hook, play a message recorded by the user that invites the caller to leave a message, record the message, and go on-hook.

A second method of recording messages, a telephone answering service, uses equipment that bridges to the customer's line at the telephone central office. The equipment is used by attendants who answer the calls and record messages manually, which are read to the user of the service on request. The service is used by physicians, for example; the service enables the attendant to take action in an emergency, if appropriate. The caller can be referred to a hospital or to another physician who is covering for the user during a vacation.

Call answering can also be provided using digital storage of voice in the central office. In one implementation, the storage medium is disc and the control is by closely associated peripheral processors in a stored-program-control switching office. In such a centralized implementation, a call arriving when the line is busy can be routed to the answering equipment; more than one caller can record a message for the same person at the same time. If the message was recorded while the telephone was busy, the user is notified when the telephone stops being busy; otherwise, the notification is by a modified dial tone the next time a call is originated.

138. Storage of Voice Messages for Later Delivery. A user may wish to leave a message for someone who cannot be reached at the time and who has no provision for recording incoming messages. For example, one may wish to inform someone of a change of travel plans or to leave a message for someone in a different time zone.

Such a service can be implemented by means of the central-office voice-storage equipment described above. The user dials an access code, dials the number of the telephone to which the message is to be delivered, and records the message. The delivery may be at a time specified (via the dial) by the user or after a standard interval from the recording of the message. In either case, repeated attempts to deliver the message will be made if the first one does not succeed.

139. References

1. Bell Telephone Laboratories, Inc., Murray Hill, N.J.   "Engineering and Operations in the Bell System," 1977.

2. Special issue on Computer Communications, *IEEE Trans. Commun.*, January 1977, Vol. COM-25, No. 1.

3. Zimmermann, H.   OSI Reference Model: The ISO Model of Architecture for Open Systems Interconnection, *IEEE Trans. Commun.*, April 1980, Vol. COM-28, pp. 425–432.

4. Special Issue on Computer Network Architectures and Protocols, *IEEE Trans. Commun.*, April 1980, Vol. COM-28.

5. McQuillan, J. M., I. Richer, and E. C. Rosen   An Overview of the New Routing Algorithm for the ARPANET, *Sixth Data Commun. Symp.*, November 1979, pp. 63–68.

6. Snow, N. E., and N. Knapp, Jr.   Digital Data System: System Overview, *Bell Syst. Tech. J.*, 1975, Vol. 54, pp. 811–832.

7. Metcalfe, R. M., and D. R. Boggs   Ethernet: Distributed Packet Switching for Local Computer Networks, *Commun. ACM*, July 1976, Vol. 19, pp. 395–404.

8. Tokoro, M., and K. Tamaru   Acknowledging Ethernet, *COMPCON 77 Fall*, September 1977, pp. 320–325.

9. Buchner, M. M., Jr.   Planning for the Evolving Family of Operations Systems, *Bell Lab. Rec.*, 1979, Vol. 57, pp. 118–124.

10. Bullington, K.   Radio Propagation at Frequencies above 30 Megacycles, *Proc. IRE*, October 1947, Vol. 35, p. 1122.

11. Bullington, K.   Radio Propagation Fundamentals, *Bell Syst. Tech. J.*, May 1957, Vol. 36, p. 593.

12. Young, W. R., Jr.   Comparison of Mobile Radio Transmission at 150, 450, 900, and 3700 MC, *Bell Syst. Tech. J.*, 1952, Vol. 31, p. 1068.

13. Egli, J.   Radio Propagation above 40 MC over Irregular Terrain, *Proc. IRE*, 1957, Vol. 45, pp. 1383–1391.

14. Okumura, Y., E. Ohmori, T. Kawano, and K. Fukuda   Field Strength and Its Variability in VHF and UHF Land Mobile Service, *Rev. Elec. Commun. Lab.*, 1968, Vol. 16, p. 825.

15. Jakes, W. C. (ed.)   "Microwave Mobile Communications," Wiley, New York, 1974.

16. Nylund, H. W., and R. M. Swanson   Improved Mobile Dial Telephone Service, *IEEE Trans. Vehic. Commun.*, 1963, Vol. VC-12, pp. 32–36.

17. Special Joint Issue on Radio Communications, *IEEE Trans. Commun.*, November 1973, Vol. COM-21, No. 11.

18. Special Issue on Emerging 900 MHz Technologies, *IEEE Trans. Vehic. Technol.*, November 1978, Vol. VT-27, No. 4.

19. Special Issue on Advanced Mobile Phone Service, *Bell Syst. Tech. J.*, January 1979, Vol. 58, No. 1.

20. Rice, L. P.   Radio Transmission into Buildings on 35 and 150 Mc, *Bell Syst. Tech. J.*, 1959, Vol. 38, p. 197.

21. Durante, J. M.   Building Penetration Loss at 900 MHz, *IEEE Conf. Pap.* 73CH0817-7VT-B-6, October 1973.

22. Cavanaugh, J. R., R. W. Hatch, and J. L. Sullivan   Models for the Subjective Effects of Loss, Noise, and Talker Echo on Telephone Connections, *Bell Sys. Tech. J.*, 1976, Vol. 55, pp. 1319–1371.

23. Spang, T. C.   Loss-Noise-Echo Study of the Direct Distance Dialing Network, *Bell Syst. Tech. J.*, 1976, Vol. 55, pp. 1–36.

24. Bell Telephone Laboratories, Inc., Murray Hill, N.J.   "Transmission Systems for Communications," 4th ed., 1970.

25. Brady, P. T.   Equivalent Peak Level: A Threshold-Independent Speech Level Measure, *J. Acoust. Soc. Am.*, 1968, Vol. 44, pp. 695–699.

26. Duffy, F. P., and T. W. Thatcher, Jr.   Analog Transmission Performance on the Switched Telecommunications Network, *Bell Syst. Tech. J.*, 1971, Vol. 50, pp. 1311–1347.

27. Aikens, A. J. and D. A. Lewinski   Evaluation of Message Circuit Noise, *Bell Syst. Tech. J.*, 1960, Vol. 39, pp. 879–909.

28. Andrews, F. T., Jr., and R. W. Hatch   National Telephone Network Transmission Planning in the American Telephone and Telegraph Company, *IEEE Trans. Commun. Technol.*, June 1971, Vol. COM-19, pp. 302–314.

29. Sullivan, J. L.   Is Transmission Satisfactory? Telephone Customers Help Us Decide, *Bell Lab. Rec.*, 1974. Vol. 52, pp. 90–98.

30. Duffy, F. P., G. K. McNees, I. Nasell, and T. W. Thatcher, Jr.   Echo Performance of Toll Telephone Connections in the United States, *Bell Syst. Tech. J.*, 1975, Vol. 54, pp. 209–243.

31. Balkovic, M. D., H. W. Klancer, S. W. Klare, and W. G. McGruther   High-Speed Voiceband Data Transmission Performance on the Switched Telecommunications Network, *Bell Syst. Tech. J.*, 1971, Vol. 50, pp. 1349–1334.

32. Fleming, H. C., and R. M. Hutchinson, Jr.   Low-Speed Data Transmission Performance on the Switched Telecommunications Network, *Bell Syst. Tech. J.*, Vol. 50, 1971, pp. 1385–1405.

33. Cooper, R. B.   "Introduction to Queueing Theory," Macmillan, New York, 1972.

34. Jacobs, I.   Lightwave Communications Begins Regular Service, *Bell Lab. Rec.*, 1979, Vol. 57, pp. 298–304.

35. Runge, P. K., and S. S. Cheng   Demountable Single-Fiber Optic Fiber Connectors and Their Measurement on Location, *Bell Syst. Tech. J.*, 1978, Vol. 57, pp. 1771–1790.

36. Cieselka, A. J., and N. G. Long   New Technology for Loops A Plan for the 80's, *IEEE Trans. Commun.*, July 1980, Vol. COM-28, pp. 923–930.

37. Kelcourse, F. C., and F. J. Herr   L5 System: Overall Description and System Design, *Bell Syst. Tech. J.*, 1974, Vol. 53, pp. 1901–1934.

38. Spilker, J. J. Jr.   "Digital Communications by Satellite," Prentice-Hall, Englewood Cliffs, N.J., 1977.

39. Rusch, R. J., J. T. Johnson, and W. Baer   INTELSAT V Spacecraft Design Summary, *AIAA Conf. Satell. Commun.*, San Diego. 1978.

40. American Telephone and Telegraph Co., New York,   "Telecommunications Transmission Engineering," Vol. 1, "Principles," 2d ed., 1977.

41. American Telephone and Telegraph Co., New York,   "Telecommunications Transmission Engineering," Vol. 3, "Networks and Services," 2d ed., 1977.

42. Johannes, V. I.   The Evolving Digital Network, *Bell Lab. Rec.*, 1976, Vol. 54, pp. 268–273.

43. Henning, H. H., and J. W. Pan   D2 Channel Bank System Aspects, *Bell Syst. Tech. J.*, 1972, Vol. 51, pp. 1641–1658.

44. Alsberg, D. A., J. C. Bankert, and P. T. Hutchison   The WT4/WT4A Millimeter Wave Transmission System, *Bell Syst. Tech. J.*, 1977, Vol. 56, pp. 1829–1848

45. Andrews, H. W., V. I. Johannes, and W. E. Woodzell   A 44.736 Mbits/s Codec for U600 Mastergroups, *IEEE Trans. Commun.*, February 1977, Vol. COM-25, pp. 264–271.

46. Duttweiler, D. L., and Y. S. Chen   A Single-Chip VLSI Echo Canceller, *Bell Syst. Tech. J.*, 1980, Vol. 59, pp. 149–160.

47. Leopold, G. R.   TASI-B: A System for Restoration and Expansion of Overseas Circuits, *Bell Lab. Rec.*, 1970, Vol. 48, pp. 299–306.

48. Benes, V.   "Mathematical Theory of Connecting Networks and Telephone Traffic," Academic, New York, 1965.

49. Feiner, A., and J. G. Kappel   A Method of Deriving Efficient Switching-Network Configurations, *Proc. Natl. Electron. Conf.*, Chicago, 1970, pp. 818–823.

50. Clos, C.   A Study of Nonblocking Switching Networks, *Bell Syst. Tech. J.*, 1953, Vol. 32, pp. 406–424.

51. Paull, M. C.   Reswitching of Connection Networks, *Bell Syst. Tech. J.*, Vol. 41, 1962, p. 833.

52. Tokita, Y., et al.   ESS Software Architecture for Multiprocessing System, *3d Int. Conf. Software Eng. Telecommun. Switching Syst.*, Helsinki, 1978, pp. 132–130.

53. Briley, B. E., and W. N. Toy   Telecommunications Processors, *Proc. IEEE*, 1977, Vol. 65, p. 1305.

54. Battista, R. N., C. G. Morrison, and D. H. Nash   Signaling System and Receiver for Touch-Tone Calling, *IEEE Trans. Commun. Electron.*, March 1963, pp. 9–17.

55. EIA Standard RS-464 PBX Switching Equipment for Voiceband Applications, December 1979.

56. Breen, C., and C. A. Dahlbom   Signaling Systems for Control of Telephone Switching, *Bell Syst. Tech. J.*, 1960, Vol. 39, pp. 1381–1441.

57. Schlanger, G. G.   Planning for the Application of CCITT No. 7 in the CCIS Network, *Natl. Telecommun. Conf.*, Houston, Tex., December, 1980.

58. Notes on The Network, Sec. 5, American Telephone and Telegraph Co., New York, 1980.

59. Special Issue on Automatic Intercept System, *Bell Syst. Tech. J.*, January 1974, vol. 53, No. 1.

60. Bourgonjon, R. H.   A High Level Programming Language for SPC Software Systems, *Int. Switching Symp. Proc., Kyoto, Japan, 1976,* pp. 222–231.

61. Inglis, A. H., and W. L. Tuffnell   An Improved Telephone Set, *Bell Syst. Tech. J.,* 1951, Vol. 30, pp. 239–270.

62. Ott, Henry W.   Ringing Problems on Long Subscriber Loops, *Telephony,* June 24, 1974, Vol. 186, No. 25, pp. 33–40.

63. Yamozaki, S., and I. Fujimoto   Miniaturization of Magneto Ringers, *Keng Kyo Sitso Yo Kahoko,* 1971, Vol. 20 No. 2, pp. 303–340.

64. Hunt, R. M., and J. W. Nippert   Computer-Aided Magnetic Circuit Design for a Bell Ringer, *Bell Syst. Tech. J.,* 1978, Vol. 57, pp. 179–203.

65. Deloach, B. C., R. C. Miller, and S. Kaufman   Sound Alerter Powered over an Optical Fiber, *Bell Syst. Tech. J.,* 1978, vol. 57, pp. 3309–3316.

66. Frederick, H. A.   The Development of the Microphone, *Bell Tel. Q.,* 1931, Vol. 10, pp. 164–188.

67. Gayford, M. L.   "Electroacoustics," Sec. 2.2, American Elsevier, New York, 1971.

68. Kett, R. W.   Carbon Microphones for Communications, *Proc. IREE Australia,* April 1964, pp. 250–256.

69. Means, D. R.   T1 Carbon Transmitter Model for Use in Computer-Aided Analysis of Telephone Set Transmission Characteristics, *Bell Syst. Tech. J.,* 1975, Vol. 54, pp. 1301–1318.

70. Sessler, G. M., and J. E. West   Electret Transducers, A Review, *J. Acoust. Soc. Am.,* 1973, Vol. 53, pp. 1589–1599.

71. Reedyk, C. W.   Noise-Canceling Electret Microphones for Lightweight Head Telephone Sets, *J. Acoust. Soc. Am.,* 1973, Vol. 53, pp. 1609–1615.

72. Baumhauer, J. C., Jr., and A. M. Brzezinski   The EL2 Electret Transmitter: Analytical Modeling, Optimization and Design, *Bell Syst. Tech. J.,* 1979, Vol. 58, pp. 1557–1578.

73. Wintle, H. J.   Introduction to Electrets, *J. Acoust. Soc. Am.,* 1973, Vol. 53, pp. 1578–1588.

74. Ebel, H., and E. Martin   State of the Art of Telephone Transmitters and Receivers: A Century after Bell's Invention, *Telefon Report,* 1976, Vol. 12, Iss. 1, pp. 30–39.

75. Gayford, M. L.   "Electroacoustics," Sec. 2.6, American Elsevier, New York, 1971.

76. Murakami, M., and M. Tobita   Optimum Design of Electromagnetic Receivers, *Electron. Commun. Jpn.,* 1969, Vol. 52-A, No. 10, pp. 10–18.

77. Mott, E. E., and R. C. Miner   The Ring Armature Telephone Receiver, *Bell Syst. Tech. J.,* 1951, Vol. 30, pp. 110–140.

78. Beranek, L. L.   "Acoustics," Chap. 7, Mc-Graw Hill, New York, 1954.

79. Luff, P. P.   The Electronic Telephone, *Sci. Am.,* March 1978, Vol. 238, No. 3, pp. 58–64.

80. Boeryd, A., et al.   Electronic Push-Button Telephone Set: Ericofon 700, *Ericsson Rev.,* 1976, Vol. 53, No. 3, pp. 118–133.

81. Ruffer, W. J.   Designing an Electronic Phone with Special Market Appeal, *Telephony,* Feb. 19, 1979, Vol. 196, No. 8, pp. 63–72.

82. Audette, J., R. Hawkins, and B. Voss   The User Interface: SL-1 Terminals and Peripheral Equipment, *Telesis (Tech. J. Bell-Northern Res. Ltd., Ottawa, Canada),* Fall 1975, Vol. 4, No. 3, pp. 84–90.

83. Krepick, W.   Smart Phones Aren't Coming: They're Here, *Telephony,* Feb. 26, 1979, Vol. 196, No. 9, pp. 45–52.

84. Prince, T. B.   Touch-A-Matic Telephone, *Proc. IEEE Int. Conf. Commun., Minneapolis,* 1974, Pap. 21/3.

85. Clemency, W. F., and W. D. Goodale, Jr.   Functional Design of a Voice-Switched Speakerphone, *Bell Syst. Tech. J.,* 1961, Vol. 40, pp. 649–668.

86. Busala, A.   Fundamental Considerations in the Design of a Voice-Switched Speakerphone, *Bell Syst. Tech. J.,* 1960, Vol. 39, pp. 265–294.

87. Torok, G. P.   Electronic Blackboard: Have Chalk Will Travel, *Int. Conf. Commun., Chicago,* June 1977, Vol. 2, p. 19.1–19.22. © 1977 IEEE.

88. Dahl, D. A., and J. W. Seyler   Educational Applications of the Electronic Blackboard, *Int. Conf. Commun., Chicago,* June 1977, Vol. 2, p. 19.2–19.26.

89. Stockbridge, C.   Multilocation Audiographic Conferencing, *Telecommun. Policy,* June 1980, Vol. 4, No. 2, pp. 96–107.

90. Johansen, R., J. Vallee, and K. Spangler   "Electronic Meetings: Technical Alternatives and Social Choices," Addison-Wesley, Reading, Mass., 1979.

91. Bolsky, M. I., and C. Stockbridge   "The Audio Teleconference: What Is It; Why Do It; How to Do It," April 1980, Bell Telephone Laboratories, Holmdel, N.J.

92. Parker, L. A., and M. K. Monson "Teletechniques — An Instructional Model for Interactive Teleconferencing," Educational Technology Publications, Englewood Cliffs, N.J., 1980.

93. Olgren, C. Visual Systems for Teleconferencing: Telewriting, Televideo, and Facsimile, in "Technical Design for Audio Teleconferencing," University of Wisconsin Extension, Center for Interactive Instructional Programs, Madison, 1978, pp. 81–108.

94. Datapro Research Corp. A Buyer's Guide to Credit Card Authorization Systems, Rep. B51-010-101, 1978.

95. Bell System Tech. Reference Switched Network Transaction Telephone System Interfacing with Transmission Control Units, Pub 41805, April 1977.

96. Cawkell, A. E. Developments in Interactive On-Line Television Systems and Teletext Information Services in the Home, Inst. Sci. Inf. On-Line Rev. (Gr. Br.), March 1977, Vol. 1, No. 1, pp. 31–38.

97. Schmedel, S. R. Video Frontier: TV Systems Enabling Viewers to Call Up Printed Data Catch Eye of Media Firms, Wall Street J., July 24, 1979, p. 40.

98. Fedida, S. Viewdata: The Post Office's Textual Information and Communications System: Background and Introduction, Post Office Res. Cent, Wireless World (Gr. Br.), February 1977, Vol. 83, No. 1494, pp. 32–36.

99. Viewdata/Videotex Report, part of The Home Inf. Rep. (Link, 215 Park Ave. South, New York 10003) December 1979, Vol. 1, No. 2.

100. Newman, W. M., and R. F. Sproull "Principles of Interactive Computer Graphics," 2d ed., pp. 211–289, McGraw-Hill, New York, 1979.

101. Oess, F. G. CRT Considerations for Raster Dot Alpha Numeric Presentations, Proc. Soc. Inf. Display, 1977, Vol. 20, No. 2, pp. 81–88.

102. Jackson, R. N., and K. E. Johnson Gas Discharge Displays: A Critical Review, Chap. 35, in Adv. Electron. Electron Phys. 1974.

103. Ritchie, G. J., and J. A. Turner Input Devices for Interactive Graphics, Int. J. Man-Machine Stud., No. 7, 1975, pp. 639–660.

104. Bennett, W. R., and J. R. Davey "Data Transmission," McGraw-Hill, New York, 1965.

105. Lucky, R. W., J. Salz, and E. J. Weldon, Jr. "Principles of Data Communication," McGraw-Hill, New York, 1968.

106. Hamming, R. W. Error Detecting and Error Correcting Codes, Bell Syst. Tech. J., 1950, Vol. 29, pp. 147–160.

107. Peterson, W. W., and E. J. Weldon, Jr. "Error-Correcting Codes," 2d ed., MIT Press, Cambridge, Mass., 1972.

108. Brewer, S. T., R. L. Easton, H. Soulier, and S. A. Taylor SG Undersea Cable System: Requirements and Performance, Bell Syst. Tech. J., 1978, Vol. 57, pp. 2319–2354.

109. Graczyk, J. F., E. T. Mackey and W. J. Maybach T1C Carrier: The T1 Doubler, Bell Lab. Rec., 1975, Vol. 53, pp. 256–263.

110. Park, K. I. Intelligible Crosstalk Performance of Voice-Frequency Customer Loops, Bell Syst. Tech. J., 1978, Vol. 57, pp. 3001–3029.

Section 23

Electronic Data Processing

RICHARD E. MATICK   *Research Staff Member, IBM Thomas J. Watson Research Center; Member, IEEE*

BERTON D. MOLDOW   *Senior Staff Member, IBM Systems Research Institute*

CLAUDE E. WALSTON   *Assistant for Software Technology, IBM Federal Systems Division; Senior Member, IEEE*

CONTENTS

Numbers refer to paragraphs

The authors acknowledge the assistance of Joseph T. Ma, Herbert B. Michaelson, George C. Stierhoff, and Donald T. Tang in the preparation of this manuscript. The manuscript was typed by Katherine I. Chandri, Janis T. Riznychok, Billie A. Sykes, and Ann R. Tartaglia.

PRINCIPLES OF DATA PROCESSING

1. Definition of a Computer. A computer is a device that determines the solution of some problem by calculation, i.e., by the application of a set of *logical* or *mathematical operations* to *data* or *information* to achieve a result called a *solution.* An example of a computer is the slide rule. Numerical values are laid out on the slide rule so as to be proportional to the logarithm of the number shown. When the operator adds the lengths on the two scales, the sum of the logarithms of the numbers is obtained. Since the sum of the logarithms of two numbers is the logarithm of the product, the sum of the lengths indicates the product on the rule. The accuracy is limited by the length of the rule and by the ability of the operator to set and read the scales.

Another type of computer is the abacus. Each of the beads on the abacus is assigned a numerical value. A multiplace decimal number is entered by pulling down (or pushing up) appropriate beads in successive columns. Other numbers are added to the first entry by displacing appropriate beads successively in each column and by following rules for the *carry operation* whenever the sum of the digits of the first and second numbers exceeds the number of bead values in a column. The accuracy of the abacus is determined by the number of columns used to represent a quantity. For example, a 20-column abacus can represent a decimal digit to an accuracy of 20 places. If appropriate procedures are followed, the abacus can be used to perform a wide variety of arithmetic operations.

The slide rule and abacus are examples of the two basic classes of computational machines, *analog* and *digital*, respectively. In an analog device, a substitution of a physical quantity for a number occurs, the substituted quantity (length, current, etc.) being used to represent the number. In the digital computer, computations are carried out upon objects that represent a *code* for numerical values. The coded quantity can represent any quantifiable entity.

2. Desk Calculator. A desk calculator is another type of digital computer. Conceptually, it is only slightly more complex than an abacus. A keyboard, with buttons for each of the 10 decimal digits, is used for entry of numerical information into mechanical or electrical *registers* arranged for *display*. Addition is accomplished by successively entering numbers into the keyboard and depressing an add key, which initiates a sequence of electrical or mechanical actions corresponding directly with the motions of an abacus.

Desk calculators capable of multiplication and division have a *shifting function*, such that an upper carriage is shifted to the left or right relative to the lower carriage. Multiplication is performed by adding the multiplicand a number of times equal to a digit of the multiplier. Then, after the carriage is shifted left, the multiplicand is added the number of times specified by the next multiplier digit, and so on.

For multiplication, the machine follows two prescribed procedures, shifting and addition, to realize the product. Division takes place by a similar routine of subtractions and shifts.

3. Stored-Program Methods. A punched paper tape or similar device can be used to control the sequence of buttons pressed on a desk calculator, thus allowing more complicated computations, e.g., extracting a square root or other such functions, to take place automatically. Such a tape is said to contain a *program* for the solution of the given problem. Modern desk calculators incorporate magnetic or electronic storage that operates in this way.

The next step in automating a calculator is to provide a second paper tape for the automatic *entry of information* in the keyboard and a third tape to collect and record the *computed result*. In digital-computer language, the desk calculator has been provided with *input* and *output* equipment.

A calculator equipped with storage tapes can realize any algebraic equation whose solution depends on a straightforward sequence of the fundamental operations of arithmetic. However, certain functions cannot be so formed. For example, if the absolute value of a number is desired, different machine actions are implied depending upon whether the number is positive or negative. Such an operation means that the machine must operate as the result of a *decision*, depending upon a specific result at an intermediate stage of computation. A single paper tape, operated in strictly sequential fashion, cannot determine such an operation, since it specifies only one specific sequence. But if *two* control tapes are available so that depending upon the outcome of a specific operation specified on the first tape, the second can be activated to contribute an alternative program, the absolute value of a number can be realized and its sign identified. In the terminology of digital computers, the two program tapes provide a *branch* operation.

The control and input-output functions associated with such tapes can be initially read into a unit, called a *store* (or *memory*), prior to the initiation of procedures for a solution. Furthermore, if the store can be readily altered during the course of operation, intermediate results of complex operations can be stored temporarily for use in later sequences of a program.

4. Internal Modification of Stored Programs. If a suitable alterable store is used in conjunction with a calculator, and if provision is made for programming variable sequences of operations and for automatic input and output of data, a powerful device is available for the solution of a wide class of computational problems. In addition, a fundamental factor in the power of the modern digital computer is its ability to modify its stored program *within the computer itself*.

To accomplish this, a numerical code stored in the computer memory is used to specify the particular operation of the calculator at any one time. For example, the code number 1 may be used to specify addition, the code number 2 to specify subtraction, and these codes may be followed by a numerical code to determine what data are to be operated upon. This latter numerical code, called an *address*, specifies the location of data in the physical store. The operation at different parts of the program can be changed from addition to subtraction by adding the number 1 to the code that specifies addition.

Similarly, if the data on which the operation is performed are to be varied during the computation, the numerical field that specifies the location of the information associated with the instruction need only be changed from one value to another by adding a suitable constant. Such operations imply that the machine is capable of treating its own program as *data subject to manipulation*.

Suppose, for example, that a problem calls for the execution of a series of arithmetic steps on each of a list of a thousand numbers. These numbers can be stored sequentially in storage. The problem is to perform the sequential operations 1,000 times, without having to provide 1,000 separate program steps, each differing only in the location of data upon which it operates. An option is to provide a means, at the termination of each sequence of operation, for the data portion of each instruction in the program to be modified to call upon data at successive locations. The ability of the computer to operate numerically upon its program permits such a procedure.

5. Memory, Processing, and Control Units. A desk calculator performs arithmetic operations, but it can be arranged to receive *alphabetic* as well as *numeric* information by assigning internal codes to represent the alphabetic characters (letters). The letters, in turn, can be manipulated by applying the rules of logic or arithmetic to the codes that represent them. For example, a calculator could compare letters and determine which precedes the other and thus could place in alphabetical order a set of letters from A to Z. It is characteristic of modern computers that the arithmetic section can manipulate more general codes than purely arithmetic data. This section is therefore commonly called the *arithmetic logic unit* (ALU). Another name applied to the ALU is the *processing unit*.

Implicit in the above discussion of the operation of the desk calculator is a mechanism by which the program is sequenced. In computer language the section that translates the stored program codes into an appropriate sequence of operations in the ALU, memory, or input-output equipment is called the *control unit*.

In summary, the basic subsystems in a computer are the input and output sections, the store, the ALU, and the control section. Each of these units is described in detail in this section of the handbook. Generally, a computer operates in the following way. An external device such as a magnetic tape, card reader, or disc file delivers a program and data to specific locations in the computer store. Control is then transferred to the stored program, which manipulates the data (and the program itself) to generate the output. These output data are delivered to a device, such as a tape, printer, or display, where the information is used in accordance with the purpose of the computation.

6. Historical Background. In the early 1800s, almost 150 years before their widespread use, stored-program digital computers were conceived and partially implemented by a remarkable Englishman, Charles Babbage.[3]* Babbage's conception included a unit (corresponding to the ALU) that he called the *mill*, an internal storage device, control units, and input-output devices. Babbage understood the requirements for transfer of control depending upon the results of intermediate computations and the need for program storage. Furthermore, he realized the potential and power of the computer's treating its own program as data to modify the procedure as the calculation proceeded. In his work Babbage was aided and supported, especially in the area of programming, by Byron's daughter Augusta Ada, the Countess of Lovelace. Babbage's ideas were

*Numbers correspond to references in Par. **23-146**.

never fully reduced to practice because of the primitive technology that existed at that time.[2] There were no electronics, and the mechanical devices used were neither precise enough nor reliable enough to realize a practical machine.

In the 1890s Hollerith developed card-tabulating equipment to cope with the growing demands of the United States census. His work laid the basis for the subsequent development of electromechanical methods that were used in commerce and industry during the early part of the twentieth century.

Starting in the 1930s, many contributions were made that set the stage for modern computer systems. The high-speed electromechanical relay was developed for application to card-tabulating equipment and to telephone and telegraphic switch gear. The development of radio and other electronic equipment contributed to the widespread availability of electronic components and methods. In particular, cosmic-ray and radar research led to pulsed and digital circuits.

During the early 1940s at the Moore School of Electrical Engineering at the University of Pennsylvania, at Harvard, and at IBM, a number of projects were initiated. At Harvard the Mark I machine was developed in conjunction with IBM.[1] This was an electromechanical sequence calculator programmed by paper tape and mechanical switches. The Electronic Numerical Integrator and Calculator (ENIAC)[5] at the Moore School and the IBM card-programmed calculator were built with electronic components to achieve high-speed operation. At the Moore School of Engineering, a team fortunate enough to have the services of John von Neumann, of the Princeton Institute of Advanced Study, proposed a fully programmable Electronic Discrete Variable Automatic Computer (EDVAC). This machine incorporated most of the elements of modern systems. The description of the machine written at the time discussed in detail many problems of digital computers

Fig. 23-1. A large-scale-integrated (LSI) memory circuit chip that can hold up to 64,000 b, roughly equivalent to 1,000 eight-letter words. The chip dimension is ¼ in on each side.

that are still pertinent.[5a,6] Besides von Neumann others from Princeton contributed to the Moore School program under the sponsorship of the U.S. Army Ordnance Department.

Thus, technologically, there has been a line of development of mechanical calculator components, beginning with Babbage and leading to a variety of mechanical desk and larger mechanical calculators. Another line of development has used relays as computing circuit elements. Today's computers have benefitted from these lines of development, but especially they are based on electronic components, the vacuum tube and the transistor. The transistor, first described by Shockley, Bardeen, and Brattain in 1947, began a line of development that is today characterized by the miniaturization and low-power operation of large-scale integration (LSI) (see Sec. 8).

LSI permits the interconnection of large numbers of computing elements by means of microscopic layered structures on a semiconductor substrate (usually silicon) or chip sometimes as small as ¼ in square. Since the entire arithmetic and logic circuit of a computer can be built on a single chip ("microprocessor"), computers incorporating large-scale integration are often called minicomputers or microcomputers. An LSI chip containing more than 155,000 components—some 85,000 field-effect transistors and 70,000 capacitors—is shown in Fig. 23-1.

7. Price-Performance Ratio and Utilization. Digital computers are required to operate at extremely high speed, in part because of economic requirements. This need is expressed in the price-performance ratio. Figure 23-2 illustrates the improvement in this ratio of a series of computers produced by one manufacturer from 1952 to 1979 in terms of decreasing storage rental cost and increasing processor speed.

Other measures of computer efficiency are reliability and accessibility, which together result in utilization. Reliability is the ability of a functional unit to perform its intended function under stated conditions for a stated period of time. Availability is the degree to which a computer is ready for use. A typical guideline for a current *central processing unit* (CPU) *utilization* in an 8-h prime period is 90%, that is, 7.2 h of CPU time available. These improvements have been made possible by the enormous decrease in cost and increase in reliability of components in the progression from relay logic to vacuum-tube logic to large-scale integrated circuits.

8. Binary and Decimal Numbers. Most transistors display random variations of their operating parameters over relatively wide limits. Similarly, passive circuit elements experience a considerable degree of variation, and noise and power-supply variations, etc., limit the accuracy

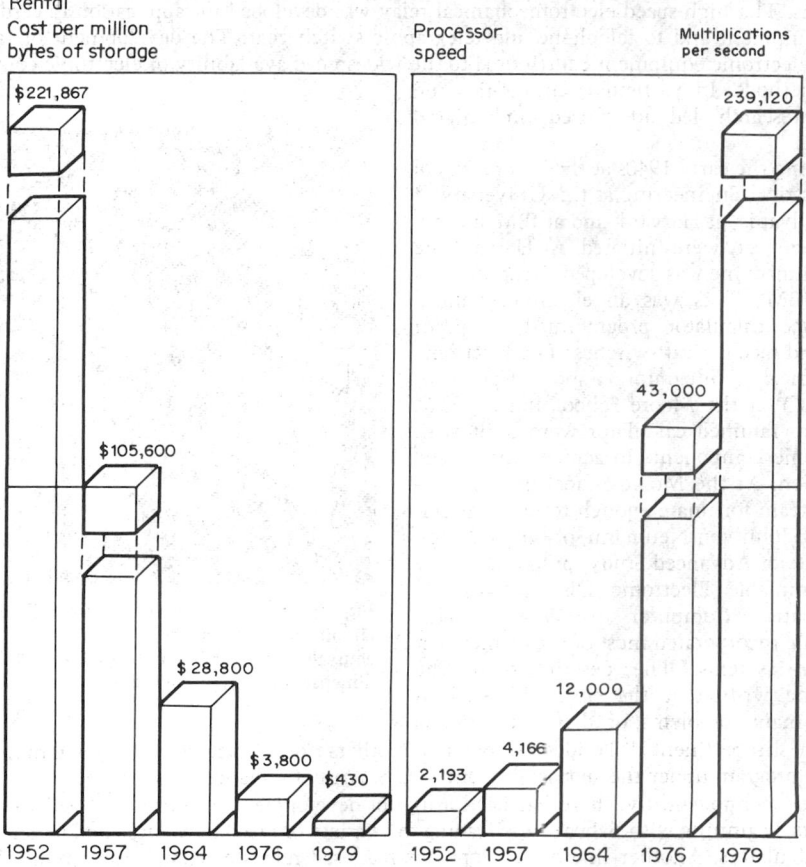

Fig. 23-2. Improvement in the price-performance ratio between 1952 and 1979.

with which quantities can be represented. As a result, the preferred method is to use each circuit in the manner of an on-off switch, and representation of quantities in a computer is thus almost always on a *binary* basis. Figure 23-3 shows the binary numbers equivalent to the decimal numbers between 1 and 10. Figure 23-4 shows the addition of binary 6 to binary 3 to obtain binary 9.

The process of addition can be dissected into digital, logical, or boolean operations upon the binary digits, or bits (b). For example, a first step in the procedure for addition is to form the so-called EXCLUSIVE-OR addition between bits in each column. This function of two binary numbers is expressed in Fig. 23-5a in tabular form. This table is called a *truth table*. In Fig. 23-5b is the table used to generate the *carries* of a binary bit from one column to another. This latter function of two binary numbers is variously called the AND function, *intersection*, or *product*. The entries at each intersection in each table are the result of the combination of two binary numbers in the respective row and column. Figure 23-5 also shows the decimal addition tables. They illustrate the relative simplicity of the binary number system.

The names truth table and logical function arise from the fact that such manipulations were first developed in the *sentential calculus*, a subsection of the calculus of logic, dealing with the truth or falsity of combinations of true or false sentences.

Binary Encoding. Information in a digital processing machine need not be restricted to numerical information since a different specific numeric code can be assigned to each letter of the alphabet. For example, A in the EBCDIC code is given by the binary sequence 11000010. When alphanumeric information is specified, such a code sequence represents the symbol A, but in the numeric context the same entry is the binary number equal to decimal 194.

9. Internal and External Information. The fact that a data processing system operates internally in binary form raises the problem of communication between the system and

Decimal	Binary
0	0
1	1
2	10
3	11
4	100
5	101
6	110
7	111
8	1000
9	1001
10	1010

Decimal	Binary
6	110
+3	+ 11
9	1001

Fig. 23-3. Decimal and binary numbers between 0 and 10.

Fig. 23-4. Addition of 6 and 3 in decimal and binary.

the operators who provide input/output and programming. It is difficult for most users to write or recognize numbers or codes presented in binary but the machine can use no other type of material.

A number of approaches to the problem of internal-external communication with machines have been followed. For example, a 4-bit (4-b) binary code can be used inside a machine to represent the 10 decimal numbers. By assigning suitable weights to each bit, it is a relatively straightforward task for an operator to translate such a code into decimal equivalents. Such computers have been called binary-coded decimal (BCD) machines. The use of BCD in a machine implies underutilization of the circuits and storage elements in the system.

A second option is for the machine to work internally in binary fashion but at each juncture of input or output to translate the codes, either by programming or by hardware, so as to accept and offer decimal numeric information. Such systems are complicated, and failures in the coding and decoding system can prevent interaction with the program.

The third approach is a compromise between the human requirement for a decimal system and the machine requirement for binary. An example is the use of the base-16 (hexadecimal) system, which is relatively amenable to human recognition and manipulation.

10. Error-Correction Codes. Though the circuits in modern data processing systems have reached degrees of reliability undreamed of in the relatively recent past, errors can still arise. Hence it is desirable to detect and, if possible, correct such errors. As discussed in Par. **23-23**, it is possible by appropriate selection of binary codes to detect errors. For example, if a 6-b code is used, a seventh bit can be added to maintain the number of 1 bits in the group of 7 as an odd number. When any group of 7 with an even number of 1s is found by appropriate circuits in the machine, an error is detected. Such a procedure is known as *parity checking*. Although these error-control coding schemes were originally developed for noisy transmission channels, they are also applicable to storage devices in data processing systems.

	0123456789
0	0123456789
1	1234567890
2	2345678901
3	3456789012
4	4567890123
5	5678901234
6	6789012345
7	7890123456
8	8901234567
9	9012345678

Decimal addition table
excluding the carry

	0	1
0	01	
1	10	

Binary addition table
excluding the carry
(a)

	0123456789
0	0000000000
1	0000000001
2	0000000011
3	0000000111
4	0000001111
5	0000011111
6	0000111111
7	0001111111
8	0011111111
9	0111111111

Decimal addition
carry generation

	0	1
0	00	
1	01	

Binary addition
carry generation
(b)

Fig. 23-5. Addition tables for decimal and binary numbers. The binary addition table (a) is called the EXCLUSIVE-OR or module-2 truth table; the carry table (b) performs the AND or *intersection* operation.

11. Boolean Functions. Figure 23-6 illustrates truth tables (Par. **23-8**) for functions of one, two, and three binary variables. Each x entry in each table can be either 0 or 1. Hence for one variable x four functions $f(x)$ can be formed; for two variables x_1 and x_2, 16 functions $f(x_1,x_2)$ exist; for three variables, 256 functions, etc. In general, if $f(x_1, \ldots, x_n)$ is a function of n binary variables, 2^{2^n} such functions exist.

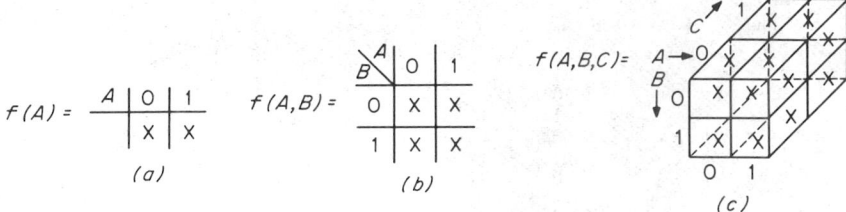

Fig. 23-6. Binary functions of (*a*) one, (*b*) two, and (*c*) three binary variables.

For functions of one variable, the most important is the *inverse function*, defined in Fig. 23-7*a*. This is called NOT A or $\overline{A}$, where A is the binary variable. Also illustrated in Fig. 23-7 are the two most important functions of two binary variables, the AND (*product* or *intersection*) and the OR (*sum* or *union*). If A and B are the two variables, the AND function is usually represented as AB and the OR as $A + B$.

Figure 23-8 shows how the products AB, $\overline{A}B$, $A\overline{B}$, and $\overline{A}\overline{B}$ are summed to yield any function of two binary variables. Each of these products has only one 1 in the four positions in their respective truth tables, so that appropriate sums can generate any function of two binary variables. This concept can be expanded to functions of more than two variables, i.e., any function of n binary variables can be expanded into a sum of products of the variables and their negatives. This is the general theorem of boolean algebra. Such a sum is called the *standard sum* or the *disjunctive normal form*.

The fact that any binary function can be so realized implies that mechanical or electrical simulations of the AND, OR, and NOT functions of binary variables can be used to represent any binary function whatever.

12. Logical-Function Realization. Relays or toggle switches are logical devices. Figure 23-9 shows how the contacts on a relay or switch are interconnected to form a logical function. The inverse boolean[4a] function of one variable (NOT) is realized by using a *normally closed* contact of a switch or relay in a given circuit. The AND function of two variables is realized by a series connection of contacts and the OR function by a parallel connection.

In the example of the implementation of the EXCLUSIVE-OR function given in Fig. 23-9*e* no

$f(A) = \overline{A} =$

A	0	1
	1	0

(*a*)

$f(A,B) = A + B = A \cup B = A \vee B$

B \ A	0	1
0	0	1
1	1	1

(*b*)

$f(A,B) = AB = A \cap B = A \wedge B$

B \ A	0	1
0	0	0
1	0	1

(*c*)

Fig. 23-7. Significant functions of one and two binary variables: (*a*) the negation (NOT) function of one binary variable; (*b*) the OR function of two binary variables; (*c*) the AND function of two binary variables.

$AB =$

B \ A	0	1
0	0	0
1	0	1

$A\overline{B} =$

B \ A	0	1
0	0	0
1	1	0

$\overline{A}B =$

B \ A	0	1
0	0	1
1	0	1

$\overline{A}\overline{B} =$

B \ A	0	1
0	1	0
1	0	0

$A\overline{B} + \overline{A}B = A \oplus B =$

B \ A	0	1
0	0	1
1	1	0

Fig. 23-8. The four products of two binary variables (top). The realization of the EXCLUSIVE-OR function is shown below.

current flows in the left branch of the circuit unless relay A is off and relay B is on. Similarly, no current flows in the right branch unless relay A is on and relay B off. Since either branch can carry current, the condition under which current is transmitted is $A\bar{B} + \bar{A}B$, where the symbols refer to the state of current in the relay coils. For mechanical switches, the same considerations apply except that the position of the switch is determined by a mechanical force. With relays or switches, the logical-function operations are expressed by interconnecting wires, the variables being associated with the physical contacts.

13. Electronic Realization of Logical Functions. Logical functions can be realized using electronic circuits. Figure 23-10 illustrates the realization of the OR function and Fig. 23-11 the realization of the AND circuit using diodes. Each input lead is associated with a boolean variable, and the upper level of voltage represents the logical 1 for that variable; in the OR circuit, any input gives rise to an output. Thus for a three-variable input, the output is $A + B + C$. With the AND function no output is realized unless all inputs are positive; the output function generated is ABC.

The inverse function (NOT) of a boolean variable cannot be readily realized with diodes. The circuit shown in Fig. 23-12 uses the inverting property of a grounded-emitter transistor amplifier to perform the inverse function. Also shown in Fig. 23-12 is an example of how the OR function and the NOT function are combined to form the NOT-OR (NOR) function. In this case, since the transistor circuit provides both voltage and current gain, the signal-amplitude loss associated with transmission through the diode can be compensated, so that successive levels of logic circuits can be interconnected to form complex switching nets.

Figure 23-13 illustrates the realization of the EXCLUSIVE-OR function. Note that the variables are represented by the wiring of interconnected circuit blocks while the function is realized by the circuit blocks themselves.

14. Levels of Operation in Data Processing. A detailed sequence of operations is generally required in a data processing system to realize even simple operations. For example, in carrying out addition, a machine typically performs the following sequence of operations:

1. Fetch a number from a specific location in storage.
   a. Decode the address of the program instruction to activate suitable memory lines. Such decoding is accomplished by activating appropriate AND and OR gates to apply voltage to the lines in storage specified by the instruction address.

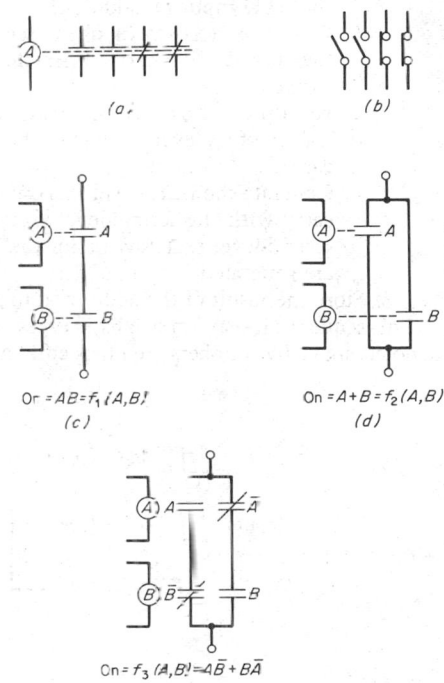

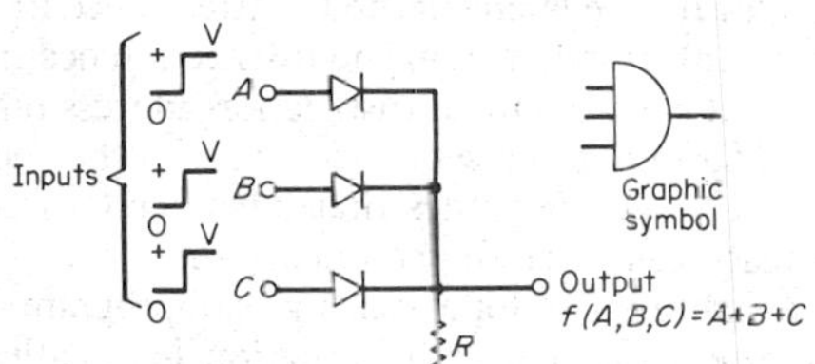

Or $= AB = f_1(A,B)$
(c)

On $= A+B = f_2(A,B)$
(d)

On $= f_3(A,B) = A\bar{B} + B\bar{A}$
(e)

Fig. 23-9. Relay or switch contacts for realization of logical functions. (a) Relay and (b) switch with normally open and closed contacts. A normally off contact corresponds to an inverse function of one variable. A series connection (c) realizes the AND function and a parallel connection (d) the OR. (e) The EXCLUSIVE-OR circuit. The contacts correspond to the binary variables, while the interconnections express the functions realized.

Fig. 23-10. Diode realization of an OR circuit. A positive input on any line produces an output.

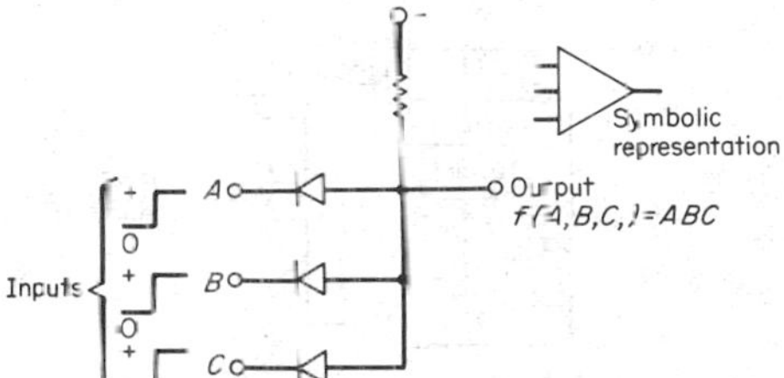

Inputs { A, B, C — Graphic symbol — Output $f(A,B,C) = A+B+C$

Fig. 23-11. Diode realization of an AND circuit. All inputs must be positive to produce an output.

Inputs { A, B, C — Symbolic representation — Output $f(A,B,C) = ABC$

    b. Sequence storage to withdraw the information and place it in a storage output register.

    c. Transmit information from the storage output register into the appropriate ALU.

  2. Withdraw a number from storage and add it to the number in the ALU. These operations break down into:

    a. Decode the instruction address, activate storage lines, and transmit the information to the ALU input for addition.

    b. Form the EXCLUSIVE-OR of the second number with the number in the ALU to form the sum less the carry. Form the AND of the two numbers to develop the first-level carry.

    c. Form the second-level EXCLUSIVE-OR sum.

    d. AND the first-level carry with the first-level EXCLUSIVE-OR sum to form the second-level carry.

    e. Generate the third-level EXCLUSIVE-OR by forming the EXCLUSIVE-OR of the second-level carry with the second-level EXCLUSIVE-OR sum. AND the second-level carries with the second-level EXCLUSIVE-OR for the third-level carry and so forth until no more carries are generated.

  3. Store the result of the addition into a specified location in storage.

This sequence illustrates two basic types of operation in a data processing machine. Operations denoted above by numbers are of specific interest to the programmer, since they are concerned

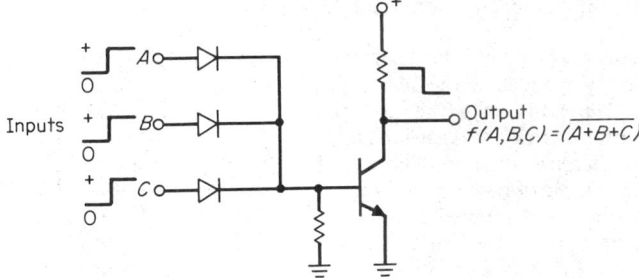

Fig. 23-12. Use of a transistor circuit for inverting a function. The circuit shown forms the NOT-OR (NOR) of the inputs.

with the data stored and the operations performed thereupon. The second level, denoted above by letters, are operations at the logical-circuit level within the machine. These operations depend upon the particular configurations of circuits and other hardware in the machine at hand.

If only the higher-level (numbered) instructions are used, some flexibility in machine operation is lost. For example, only an add operation is possible at the higher level. At the lower-level (lettered) operations the AND or EXCLUSIVE-OR of the data words can be formed and placed in storage.

The organization of current digital computers follows the lines of these two divisions (numbered and lettered, above). The *macroinstruction set* associated with each machine can be manipulated by the programmer. These instructions are usually implemented in a numerical code. For example, the instruction "load ALU" might be 01 in binary. "Add ALU" might be given by 10 and "store ALU" by 11. Similarly, each instruction has an associated storage address to provide source data. The microinstruction set comprises a series of suboperations that is combined in various sequences to realize a given macroinstruction.

Two methods of realizing the sequence of suboperations specified by the operations portion of the instruction have been used in machine design. In one such method a direct decoding of the information from the instruction occurs when it is placed in an instruction address register. Specific clock sequences turn on the successively required lines that have been wired in place to realize the action sought.

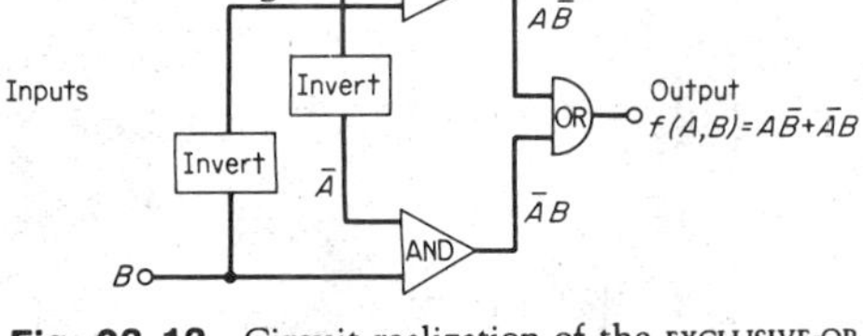

An alternative for actuating a subprogram is to store a number of information bits, called *microinstructions*, that are successively directed to the appropriate control circuits to

Fig. 23-13. Circuit realization of the EXCLUSIVE-OR function.

activate selectively and sequentially individual wires to gate sequential actions for the realization of the requisite instruction.

The first method of computer design is called *hard-wired*, and the second is the *micropro-grammed*. The microprogram essentially specifies a sequence of operations at the individual circuit level to specify the operations performed by macroinstruction. Microprogramming is preferred in modern computer design.

15. Types of Computer Systems. There is a wide variety of computer-system arrangements, depending upon the type of application. One type of installation is that associated with *batch processing*. A computer in a central job location receives programs from many different sources and runs the programs sequentially at high speed. An overall supervisory program, called an *operating system*, controls the sequence of programs, rejecting any that are improperly coded and completing those which are correct.

Another type of system, the *time-shared system*, provides access to the computer from a number of remote input-output stations. The computer scans each remote station at high speed and accepts or delivers information to that location as required by the operator or by its internal program. Thus a small terminal can gain access to a large high-speed system.

Still another type of installation, the *minicomputer*, involves an individual small computer that, though limited in power, is dedicated to the service of a single operator. Such applications vary from those associated with a small business, with limited computational requirements, to an individual engaged in scientific operations.

These three types of systems are representative of the major classifications. Other computers are used for dedicated control of complex industrial processes. These are individual, once-programmed units that perform a *real-time* operation in *systems control*, with sensing elements that provide the inputs. Highly complex interrelated systems have been developed in which individual computers communicate with and control each other in an overall major *systems network*. Among the first of such systems was the SAGE network, developed in the 1950s for defense against missile or aircraft attack.

Computers that are interconnected to share workload or problems are said to form a *multiprocessing system*. A computer system arranged so that more than one program can be executed simultaneously is said to be *multiprogrammed*.

16. Internal Organization of Digital Computers. The internal organization of a data processing system is called the *system architecture*. Such matters as the minimum addressable field in memory, the interrelations between data and instruction word size, the instruction format and length or lengths, parallel or serial (by bit or set of bits) ALU organization, decimal or binary internal organization, etc., are typical questions for the system architect. The answers depend heavily upon the application for which the computer is intended.

Two broad classes of computer systems are *general-purpose* and *special-purpose* types. Most systems are in the general-purpose class. They are used for business and scientific purposes. General-purpose computers of varying computer power and memory size can be grouped, sharing a common architecture. These are said to constitute a *computer family*.

A computer scientifically designed for, and dedicated to the control of, say, a complex refinery process is an example of a special-purpose system.

A number of design methods have been adopted to increase the speed and functional range for a small increase in cost. For example, in the instruction sequence, the next cell in storage is likely to be the location of the next instruction. Since an instruction can usually be executed in a time that is short compared with storage access, the store is divided into subsections. Instructions are called from each subsection independently at high speed and put into a queue for execution. This type of operation is called *look-ahead*. If the instructions are not sequential, the queue is destroyed and a new queue put in its place.

Since instructions and data tend to be clustered together in storage, it is advantageous to provide a small, high-speed store (local store) to work with a larger, slower-speed, lower-cost unit. If the programs in the local store need information from the larger store, a least-used piece of the local store reverts to the larger store and a batch of data surrounding the information sought is automatically brought into the high-speed unit. This arrangement is called a *hierarchical memory* and the high-speed store is often called a *cache* (Par. **23-62**).

NUMBER SYSTEMS, ARITHMETIC, AND CODES

17. Representation of Numbers. A distinction must be drawn between the concept of a *number*, an abstraction concerning the quantity or order of elements in a given set, and the *numerals* that are used to *name* these quantities. Considerable latitude exists in the selection of names, so that a choice can be made according to a definite purpose. Though the word "numeral" should properly be used whenever the names of numbers are being discussed, in common practice "number" is acceptable unless a distinction is necessary.

A set of codes and names for numbers that meets the requirements of extendability and convenience of operation can be obtained using the following power series:

$$N = A_n X^n + A_{n-1} X^{n-1} + \cdots + A_1 X + A_0 + A_{-1} X^{-1} + \cdots + A_{-m} X^{-m} \quad (23\text{-}1)$$

Here the number is represented by the sum of the powers of an integer X, each having a coefficient A_i. A_i may be an integer equal to or greater than zero and less than X. In the decimal system, X equals 10 and the coefficients A_i range from 0 to 9.

Note that Eq. (23-1) can be used to represent X^{m+n+1} numbers ranging between 0 and $X^{n+1} - X^m$ with an accuracy limited by X_m. Thus m and n must be of reasonable size to be useful in most applications. A useful property of the power series is the fact that its multiplication by X^k can be viewed as a shift of the coefficients of any given term by the number of positions specified by the value of k. These results are independent of the choice of X in the series representation.

There is little reason to write the value of the number in the form shown in Eq. (23-1) since complete information on the value can be readily deduced from the coefficient A_i. Thus, a number can be represented merely by a sequence of the values of the coefficients. To determine the value of the implied exponents on X, it is customary to mark the position of the X_0 term by a period immediately to the right of its coefficient. The power series for a number represented in the decimal system ($X = 10$) and its normal decimal notation are

$$3 \times 10^3 + 0 \times 10^2 + 2 \times 10^1 + 4 \times 10^0 + 6 \times 10^{-1} + 2 \times 10^{-2} = 3{,}024.62 \quad (23\text{-}2)$$

The value of X is called the *radix* or *base* of the number system. Where ambiguity might arise, a subscript to indicate the radix is attached to the low-order digit, as in $1000_2 = 8_{10} = 10_8$ (1000 binary equals 8 decimal equals 10 octal). The power series for a number in base 2 and its representation in binary notation is

$$1 \times 2^4 + 1 \times 2^3 + 0 \times 2^2 + 1 \times 2^1 + 1 \times 2^0$$
$$+ 0 \times 2^{-1} + 1 \times 2^{-2} + 1 \times 2^{-3} = 11011.011 \quad (23\text{-}3)$$

18. Number-System Conversions. Since computer systems, in general, use number systems other than base 10, conversion from one system to another must be carried out frequently. Equation (23-4) shows the integer N represented by a power series in base 10 and base 2

$$\sum_{i=0}^{n} A_i 10^i = \sum_{j=0}^{m} B_j 2^j \quad (23\text{-}4)$$

The problem is to find the correlation between the coefficients A_i and B_j. In the binary series, if N is divisible by 2, then B_0 must be 0. Similarly, if N is divisible by 4, B_1 must be 0, and so forth. Thus if the decimal coefficients A_i are given, successive divisions of the decimal number by 2 will yield the binary number, the binary digits depending on the value of the remainder of each successive division. This process is shown in Fig. 23-14.

The conversion of a binary integer to a decimal integer is

$$100011011 = 1 \times 2^8 + 0 \times 2^7 + 0 \times 2^6 + 0 \times 2^5$$
$$+ 1 \times 2^4 + 1 \times 2^3 + 0 \times 2^2 + 1 \times 2 + 1 \times 2^0$$
$$= 283$$

In the case of conversion of an integer in binary to an integer in decimal, the powers of 2 are written in decimal notation and a decimal sum is formed from the contribution of each term of the binary representation. For conversion from a binary fraction to a decimal fraction, a similar procedure is used since the value of terms as multiplied by the A_i can be added together in decimal form to form the decimal equivalent.

The conversion of a decimal fraction to a binary fraction is defined by

$$0.5764_{10} = A_{-1} 2^{-1} + A_{-2} 2^{-2} + \cdots + A_n 2^{-n} \quad (23\text{-}5)$$

To determine the values of the A_i, first multiply both sides of Eq. (23-5) by 2 to give

$$1.1528_{10} = A_{-1} + A_{-2} 2^{-1} + \cdots + A_{-n} 2^{n-1} \quad (23\text{-}6)$$

Since the position of the decimal point (more accurately called the *radix point*) is invariant, and since in a binary series each successive term is at most half of the maximum value of the preceding term, the leading 1 in the decimal number in Eq. (23-6) indicates that A_{-1} must have been 1. A second multiplication by 2 can similarly determine the coefficient A_{-2}. This process of conversion of a base-10 fraction to a base-2 fraction is illustrated in Fig. 23-15.

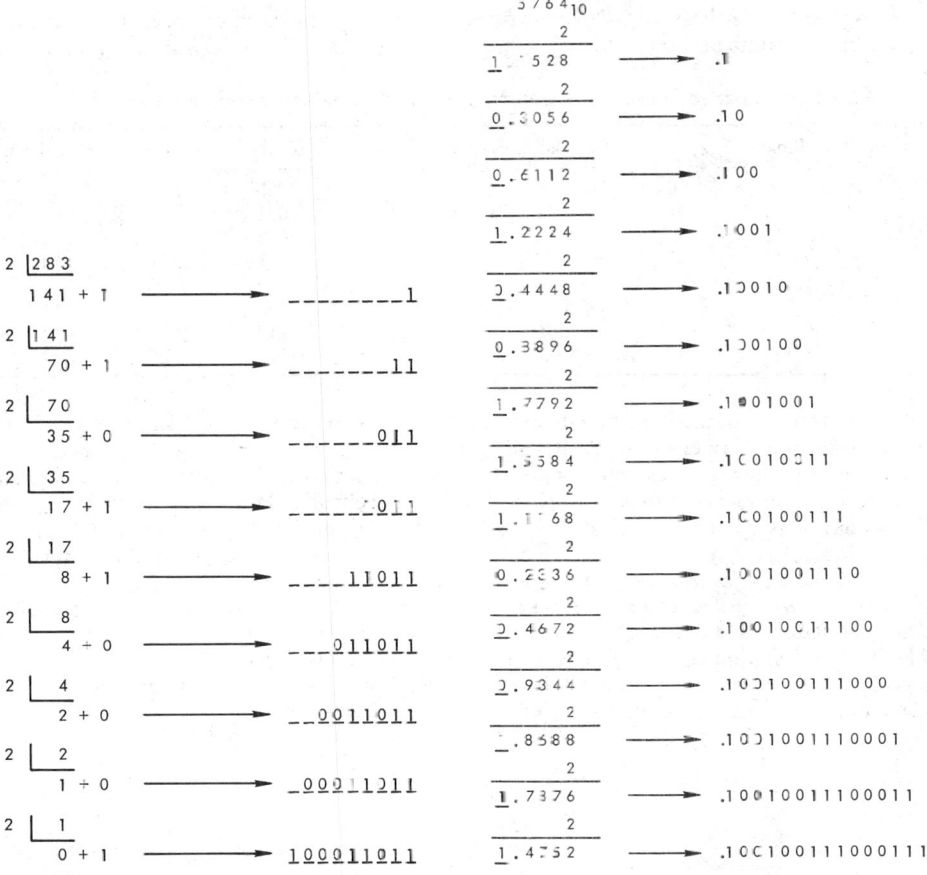

Fig. 23-14. Conversion from a decimal to binary by repeated division of the decimal integer. At each division the remainder becomes the next higher-order binary digit.

Fig. 23-15. Conversion of a decimal fraction into a binary fraction. At each stage the number to the right of the decimal is multiplied by 2. The resulting number to the left of the decimal point is entered as the next available lower order position of the binary fraction to the right of the binary radix point.

Conversion from binary integers to octal (base 8) and the reverse can be handled simply since the octal base is a power of 2. Binary to octal conversion consists of grouping the terms of a binary number in threes and replacing the value of each group with its octal representation. The process works on either side of a decimal point. The octal-to-binary conversion is handled by converting each octal digit, in order, to binary and retaining the ordering of the resulting groups of three bits.

Since there are not enough symbols in decimal notation to represent the 16 symbols required for the hexadecimal system, it is customary in the data processing field to use the first six letters of the alphabet to complete the set.

Conversions from decimal to octal or hexadecimal can proceed indirectly by first converting decimal to binary and then binary to octal or hexadecimal. Similarly, a reverse path from octal or hexadecimal to binary to decimal can be used. Direct conversions, however, between hexadecimal and octal and decimal exist and are widely used. In going from hexadecimal or octal to decimal, each term in the implied power series is expressed directly in decimal, and the result is summed. In converting from a decimal integer to either hexadecimal or octal, the decimal is divided by either 16 or 8, respectively, and the remainder becomes the next higher-order digit in the converted number. Examples of four common number representations are shown in Table 23-1.

19. Binary-Arithmetic Operations. Figure 23-16 shows an example of the addition of two binary numbers, 1001 and 1011 (9 and 11 in decimal). The rules for manipulation are

TABLE 23-1   Comparison of Decimal, Binary, Octal, and Hexadecimal Numbers

Decimal	Binary	Octal	Hexadecimal	Decimal	Binary	Octal	Hexadecimal
0	0	0	0	8	1000	10	8
1	1	1	1	9	1001	11	9
2	10	2	2	10	1010	12	A
3	11	3	3	11	1011	13	B
4	100	4	4	12	1100	14	C
5	101	5	5	13	1101	15	D
6	110	6	6	14	1110	16	E
7	111	7	7	15	1111	17	F

similar to those in decimal arithmetic except that only the two symbols, 1 and 0, are used and the addition and carry tables are greatly simplified.

Figure 23-17 shows an example of binary multiplication with a multiplication table. This process is also simple compared with that used in the decimal system. The rule for multiplication in binary is as follows: if a particular digit in the multiplier is 1, place the multiplicand in the product register; if 0, do nothing; in either case shift the product register to the right by one position; repeat the operations for the next digit of the multiplier.

Figure 23-18 shows an example of binary subtraction and the subtraction and borrow tables. The subtraction table is the same as the addition table, a feature unique to the binary system. The borrow operation is handled in a fashion analogous to that in decimal. If a 1 is found in the preceding column of the subtrahend, it is borrowed, leaving a 0. If a 0 is found, an attempt is made to borrow from the next higher-order position, and so forth.

An example of binary division is

$$
\begin{array}{r} 110 \\ 101\ \overline{)11110} \\ 101 \\ \hline 101 \\ 101 \\ \hline 0 \end{array}
\qquad
\begin{array}{r} 6 \\ 5\ \overline{)30} \end{array}
$$

The procedure is as follows:

1. Compare the divisor with the leftmost bits of the dividend.

2. If the divisor is greater, enter a 0 in the quotient and shift the dividend and quotient to the left.

3. Try subtraction again.

4. When the subtraction yields a positive result, i.e., the divisor is less than the bits in the dividend, enter a 1 in the quotient and shift the dividend and the quotient left one position.

5. Return to step 1 and repeat.

Binary division, like binary multiplication, is considerably simpler than the decimal operation.

20. Subtraction by Complement Addition. If subtraction were performed by the usual method of borrowing from the next higher order, a separate subtraction circuit would be required. Subtraction can be performed, however, by the method of *adding complements* (or adding 1's complements, as the method is also called). By this method, the *subtrahend*, i.e., the number that is to be subtracted, is *inverted*, changing the 0s to 1s and the 1s to 0s. Then the inverted subtrahend is added to the *minuend*, i.e., the number that is to be subtracted from, and an additional 1 is added to find the difference. As an example, consider the subtraction 1101 − 1001. The subtrahend (1001) is first inverted to form the complement (0110). The difference is formed by adding the minuend and the complement of the subtrahend (plus 0001) as follows: $1101 - 1001 = 1101 + 0110$ (complement) $+ 000 = (1)0011 + 0001 = 0100$. Note that in subtraction by complement addition, a leading 1 (in parentheses) in the result (difference) must be suppressed, and that 1 must be added to obtain the result. The result of this operation can be verified by observing that the decimal equivalent of this operation is $13 - 9 = 3 + 1 = 4$.

21. Floating-Point Numbers. In a computer having a fixed number of bits that define a word, the bits represent the maximum size of a numerical value. For example, if 40 bit positions are provided for a word, the maximum decimal number that can be represented is of the order of 1.009×10^{12}. Though this number is large, it does not suffice for many applications, especially in science, where a greater range of magnitudes may be routinely encountered. To extend the range of values that can be handled, numbers are represented in floating-point notation. In floating point the most significant digits of the number are written with an assumed radix point immediately to the left of the highest-order digit. This number is called the *fraction*. The intended position of the radix point is identified by a second number, called the *characteristic*, which is appended to the fraction. The characteristic denotes the number of positions that the assumed radix point must be shifted to achieve the intended number. For example, the number 146.754 in floating point might be 146754.03 where 146754 would be equivalent to 0.146754 and the .03 would denote a shift of the decimal point three places to the right. In binary notation the number 11011.011 (27.375 in decimal) might be represented in floating point as 11011011.101 with the fraction again to the left of the decimal and the characteristic to the right.

With floating-point addition and subtraction, a shift register is required to align the radix points of the numbers. To perform multiplication or division, the fraction fields are appropriately multiplied or divided and the exponents summed or subtracted, respectively. As with fixed-point addition or subtraction, provision is usually made to detect an overflow condition in the characteristic fields. In some systems provision is made to note when an addition or subtraction occurs with such widely differing characteristics that justification destroys one of the two numbers (by shifting it out the end of a shift register).

Binary		Decimal
1011	=	11
+ 1001	=	9
10100	=	20

Fig. 23-16. Binary addition and corresponding decimal addition.

Binary	Decimal
1011	11
×1001	× 9
1011	99
101100	
1100011	

Fig. 23-17. Binary multiplication. The binary multiplication table is the AND function of two binary variables. The process of multiplication consists of merely replicating and adding the multiplicand, as shown, if a 1 is found in the multiplier. If 0 is found, a single 0 is entered and the next position to the left in the multiplier is taken up.

22. Numeric and Alphanumeric Codes. The numeric codes used to represent numerical values previously discussed include the hexadecimal, octal, binary, and decimal codes. In many applications the need arises for the coding of nonnumeric as well as numeric information, and such coding must use the binary scheme. A code embracing numbers, alphabetic characters, and special symbols is known as an *alphanumeric code*.

A widely used code with its roots in the past is the telegraph code (the Baudot code). Figure 23-19 illustrates this code, which is still used in some major communication networks. Other alphanumeric codes have been devised for special purposes. One of the most significant of these, because of its present use and its contribution to the design of other codes, is the Hollerith code, developed in the 1890s. Hollerith's equipment contributed to the development of electromechanical accounting machines that provided the foundation for electronic computers.

In the Hollerith code bits (holes) located in rows 0, 11 or 12 of a punched-card column are used in conjunction with one or more holes in rows 1 to 9 of the same column to represent an alphabetic or special character. The Hollerith code is generally used for character sets consisting

Binary	Decimal	$A - B$, less borrow	$A - B$ borrow
100110	38		
1001	9		
11101	29		

Fig. 23-18. Binary subtraction and corresponding decimal subtraction. The subtraction table is the same as the addition table. The borrow operation is handled analogously to decimal subtraction.

	Code signals						Lowercase	Uppercase			
Start	1	2	3	4	5	Stop		CCITT standard international telegraph alphabet 2	United States teletype commercial keyboard	A T & T fractions keyboard	Weather keyboard
	●	●				●	A	-	-	-	↑
	●			●	●	●	B	?	?	$\frac{5}{8}$	⊕
		●	●	●		●	C	:	:	$\frac{1}{8}$	○
	●			●		●	D	Who are you?	S	S	↗
	●					●	E	3	3	3	3
	●		●	●		●	F	Note 1	!	$\frac{1}{4}$	→
		●		●	●	●	G	Note 1	&	&	↘
			●		●	●	H	Note 1	#		↓
		●	●			●	I	8	8	8	8
	●	●		●		●	J	Bell	Bell	'	↙
	●	●	●	●		●	K	(	(	$\frac{1}{2}$	←
		●			●	●	L	)	)	$\frac{3}{4}$	↖
			●	●	●	●	M	.	.	.	.
			●	●		●	N	,	,	$\frac{7}{8}$	⊕⊕
				●	●	●	O	9	9	9	9
		●	●		●	●	P	0	0	0	∅
	●	●	●		●	●	Q	1	1	1	1
		●		●		●	R	4	4	4	4
	●		●			●	S	'	,	Bell	Bell
					●	●	T	5	5	5	5
	●	●	●			●	U	7	7	7	7
		●	●	●	●	●	V	=	;	$\frac{3}{8}$	⊕⊕
	●	●			●	●	W	2	2	2	2
	●		●	●	●	●	X	/	/	/	/
	●		●		●	●	Y	6	6	6	6
	●				●	●	Z	+	"	"	+
						●	Blank				-
	●	●	●	●	●	●	Letters shift				↓
	●	●		●	●	●	Figures shift				↑
			●			●	Space				▮
				●		●	Carriage return				<
		●				●	Line feed				=

● Denotes positive current

Fig. 23-19. The Baudot telegraphers' code, a 5-b code. The code can be extended by using a shift character.

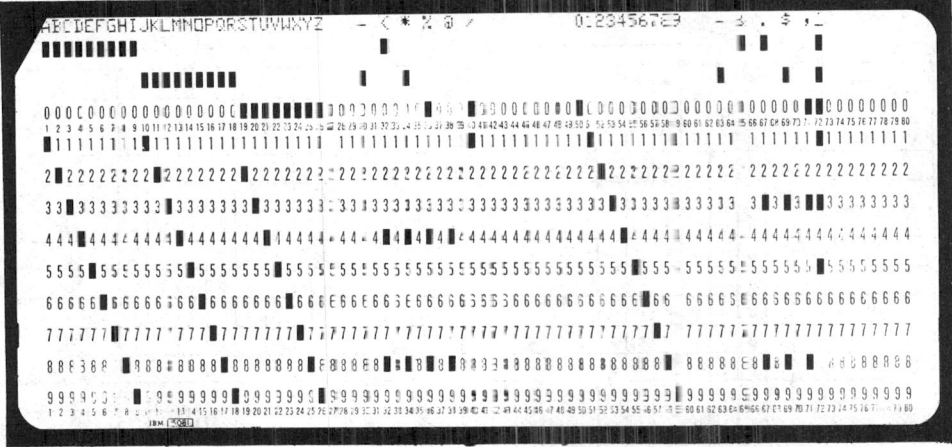

Fig. 23-20. A Hollerith code punched card, first developed in the 1890s and still widely used. The 0, 11, and 12 punches are called *zone punches*; 1 through 9 are the numeric field.

of 48 alphabetic, numeric, and special characters. Figure 23-20 shows a tabulating card encoded in Hollerith.

Figure 23-21 shows the binary-coded decimal (BCD) code, an outgrowth of the Hollerith code. In this 6-b code, the four lower-order bits are a binary representation of the numeric portion of a Hollerith coded character. The two higher-order bits correspond to the presence or absence of holes in the 0, 11, or 12 rows of the Hollerith code. For example, the letter Q is coded in BCD as 101000, whereas in Hollerith it is represented by holes in rows 8 and 11. With the development of more powerful computers an extension of the BCD code was required, so that more symbols could be represented. Such a code is shown in Fig. 23-22. Another code of importance in the United States is the American Standard Code for Information Interchange (ASCII) (see Fig. 23-23).

This code, developed by a committee of the American National Standards Institute (ANSI), has the advantage over most other codes of being contiguous, in the sense that the binary combination used to represent alphanumeric information is sequential. Hence alphabetic sorting can be easily accomplished by arithmetic manipulation of the code values.

Codes used for data transmission generally have both data characters and *control characters*. The latter perform control functions on the machine receiving information. For example in the Baudot code (Fig. 23-19) the characters for space, carriage return, and line feed do not generate a character but operate mechanisms associated with the receiving printer. In more sophisticated codes, such as ASCII, these control functions are greatly extended and hence are applicable to machines of different design.

By sending special characters called *escape characters* the mode of operation of the receiving machine can be changed to generate a different character set. Such characters or groups of such characters can extend the scope of any coding sys-

	00	0	10	11
0000	blank	b	—	&+
0001	1	/	J	A
0010	2	S	K	B
0011	3	T	L	C
0100	4	U	M	D
0101	5	V	N	E
0110	6	W	O	F
0111	7	X	P	G
1000	8	Y	Q	H
1001	9	Z	R	I
1010	0	‡	1	?
1011	⧣=	,	$	.
1100	@'	% (	*	□)
1101	:	ɣ	]	[
1110	>	\	;	<
1111	√	+++	△	⧧

identification of
control codes and special meanings

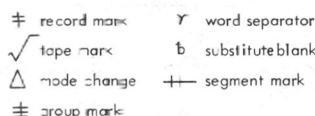

‡ record mark   ɣ word separator
√ tape mark   b substitute blank
△ mode change   +++ segment mark
⧧ group mark

Fig. 23-21. The IBM BCD interchange code, a 6-b code based upon the Hollerith code. The last 4 b represent a numeric portion on a Hollerith card in BCD, while the first 2 b represent the zone punches

Data characters

Normal shift

Character	S	B	A	8	4	2	1	Parity
1	0	0	0	0	0	0	1	0
2	0	0	0	0	0	1	0	0
3	0	0	0	0	0	1	1	1
4	0	0	0	0	1	0	0	0
5	0	0	0	0	1	0	1	1
6	0	0	0	0	1	1	0	1
7	0	0	0	0	1	1	1	0
8	0	0	0	1	0	0	0	0
9	0	0	0	1	0	0	1	1
0	0	0	0	1	0	1	0	1
a	–	1	1	0	0	0	1	0
b	0	1	1	0	0	1	0	0
c	0	1	1	0	0	1	1	1
d	0	1	1	0	1	0	0	0
e	0	1	1	0	1	0	1	1
f	0	1	1	0	1	1	0	1
g	0	1	1	0	1	1	1	0
h	0	1	1	1	0	0	0	0
i	0	1	1	1	0	0	1	1
j	0	1	0	0	0	0	1	1
k	0	1	0	0	0	1	0	1
l	0	1	0	0	0	1	1	0
m	0	1	0	0	1	0	0	1
n	0	1	0	0	1	0	1	0
o	0	1	0	0	1	1	0	0
p	0	1	0	0	1	1	1	1
q	0	1	0	1	0	0	0	1
r	0	1	0	1	0	0	1	0
s	0	0	1	0	0	1	0	1
t	0	0	1	0	0	1	1	0
u	0	0	1	0	1	0	0	1
v	0	0	1	0	1	0	1	0
w	0	0	1	0	1	1	0	0
x	0	0	1	0	1	1	1	1
y	0	0	1	1	0	0	0	1
z	0	0	1	1	0	0	1	0
.	0	1	1	1	0	1	1	0
$	0	1	0	1	0	1	1	1
,	0	0	1	1	0	1	1	1
/	0	0	1	0	0	0	1	1
'	0	0	0	1	0	1	1	0
&	0	1	1	0	0	0	0	1
-	0	1	0	0	0	0	0	0
@	0	0	1	0	0	0	0	0

Upper shift

Character	S	B	A	8	4	2	1	Parity
=	1	0	0	0	0	0	1	
c	1	0	0	0	0	1	0	1
;	1	0	0	0	0	1	1	1
:	1	0	0	0	1	0	0	0
°C	1	0	0	0	1	0	1	1
'	1	0	0	0	1	1	0	0
-	1	0	0	0	1	1	1	0
+	1	0	0	1	0	0	0	1
(	1	0	0	1	0	0	1	0
)	1	0	0	1	0	1	0	0
A	1	1	1	0	0	0	1	1
B	1	1	1	0	0	1	0	1
C	1	1	1	0	0	1	1	0
D	1	1	1	0	1	0	0	1
E	1	1	1	0	1	0	1	0
F	1	1	1	0	1	1	0	0
G	1	1	1	0	1	1	1	1
H	1	1	1	1	0	0	0	1
I	1	1	1	1	0	0	1	0
J	1	1	0	0	0	0	1	0
K	1	1	0	0	0	1	0	0
L	1	1	0	0	0	1	1	1
M	1	1	0	0	1	0	0	0
N	1	1	0	0	1	0	1	1
O	1	1	0	0	1	1	0	1
P	1	1	0	0	1	1	1	0
Q	1	1	0	1	0	0	0	0
R	1	1	0	1	0	0	1	1
S	1	0	1	0	0	1	0	0
T	1	0	1	0	0	1	1	1
U	1	0	1	0	1	0	0	0
V	1	0	1	0	1	0	1	1
W	1	0	1	0	1	1	0	1
X	1	0	1	0	1	1	1	0
Y	1	0	1	1	0	0	0	0
Z	1	0	1	1	0	0	1	1
.	1	1	1	1	0	1	1	1
!	1	1	0	1	0	1	1	0
,	1	0	1	1	0	1	1	0
?	1	0	1	0	0	0	1	0
±	1	0	0	1	0	1	1	1
+	1	1	1	0	0	0	0	0
-	1	1	0	0	0	0	0	1
*	1	0	1	0	0	0	0	1

(a)

Fig. 23-22. The extended IBM BCD 1-b code used for data transmission: (a) data characters and (b) control characters.

tem. For example, with the Baudot code shown in Fig. 23-19, transmission of the up-shift code ↑ causes the machine to print in uppercase until a down-shift ↓ code is received. The addition of these two escape characters almost doubles the character set of the device.

Other Numeric Codes. Not all numeric information is represented by binary numbers. Other codes are also used for numeric information in special applications. Figure 23-24 shows a

Control characters	(Either shift) (Either setting of ? bit)					
Backspace	1	0	1	1		0
End of transfer	0	0	1	1		1
Delete	1	1	1	1	1	1
Down-shift	1	1	1	1	1	0
Carriage return	1	0	1	1	0	1
Prefix	0	1	1	1	1	1
Idle	1	0	1	1	1	1
Reader stop	0	0	1	1	0	1
Space	0	0	0	0	0	0
End of block	0	1	1	1	1	0
Up-shift	0	0	1	1	1	0
Line feed	0	1	1	1	0	1
Tab	1	1	1	1	0	1
Restore	1	0	1	1	0	0
Bypass	0	1	1		0	0
End of heading	0	0	1	0	1	1
Punch on	0	0	1	1	0	0
Punch off	1	1	1	1	0	0

(b)

Fig. 23-22. (continued).

widely used code called the *reflected* or *Gray* code. It has the property that only 1 b is changed between any two successive values, irrespective of number size. This code is used in digital-to-analog systems since there is no need for propagation of carry integers in sequential counting as in a binary code.

23. Error Detection and Correction Codes. The integrity of data in a computer system is of paramount importance because the serial nature of computation tends to propagate errors. Internal data transmission between computer-system units takes place repeatedly and at high speed. Data may also be sent over wires to remote terminals, printers, and other such equipment. Because imperfections of transmission channels inevitably produce some erroneous data, means must be provided to detect and correct errors whenever they occur.

A basic procedure for error detection and correction is to design a code in which each word contains more bits than are needed to represent all the symbols used in a data set. If a bit sequence is found that is not among those assigned to the data symbols, an error is known to have occurred.

One such commonly used error-detection code is called the *parity check.* Suppose that 8 b are used to represent data and that an additional bit is reserved as a check bit. A simple electronic circuit can determine whether an even or odd number of 1 bits is included in the eight bit positions. If an even number exists, a 1 bit can be inserted in the check position. If an odd number of 1s exists, the check position contains a 0. As a result all code words must contain an odd number of 1 bits. If a 9-b sequence is found to contain an even number of 1s, an error can be presumed.

There are limitations in the use of the simple parity check as a mechanism for error detection, since in many transmission channels and storage systems there is a tendency for a failure to produce simultaneous errors in two adjacent positions. Such an error would not be detected since the parity of the code word would remain unchanged.

To increase the power of the parity check, a list of code words in a two-dimensional array can be used, as shown in Fig. 23-25. The code words in the horizontal dimension have a parity bit added, and the list in the vertical dimension also has an added parity bit, in each column. If one bit is in error, errors appear in both the row and the column. If simultaneous errors occur in two adjacent positions of a code word, no parity error will show up in that row, but the column checks will detect two errors. This code can detect any 3-b errors.

	000	001	010	011	100	101	110	111
0000	NULL	①DC₀	♭	0	@	P		
0001	SOM	DC₁	!	1	A	Q		
0010	EOA	DC₂	"	2	B	R		
0011	EOM	DC₃	#	3	C	S		
0100	EOT	DC₄ (STOP)	$	4	D	T		
0101	WRU	ERR	%	5	E	U		
0110	RU	SYNC	&	6	F	V		
0111	BELL	LEM	'	7	G	W	Unassigned	
1000	FE₀	S₀	(	8	H	X		
1001	HT SK	S₁	)	9	I	Y		
1010	LF	S₂	*	:	J	Z		
1011	V TAB	S₃	+	;	K	[		
1100	FF	S₄	,	<	L	\		ACK
1101	CR	S₅	-	=	M	]		②
1110	SO	S₀	.	>	N	↑		ESC
1111	SI	S₇	/	?	O	←		DEL

Identification of control symbols and some graphics

NULL	Null/idle	V$_{TAB}$	Vertical tabulation	S₀–S₇	Separator (information)
SOM	Start of message	FF	Form feed	♭	Word separator (space, normally nonprinting
EOA	End of address	CR	Carriage return		
EOM	End of message	SO	Shift out	<	Less than
EOT	End of transmission	SI	Shift in	>	Greater than
WRU	"Who are you?"	DC₀	Device control 1 Reserved for data link escape	↑	Up arrow (exponentiation)
RU	"Are you...?"			←	Left arrow (implies/ replaced by)
BELL	Audible signal	DC₁–DC₃	Device control	\	Reverse slant
FE₀	Format effector	DC₄ (STOP)	Device control (stop)	ACK	Acknowledge
HT	Horizontal tabulation	ERR	Error	②	Unassigned control
SK	Skip (punched card)	SYNC	Synchronous idle	ESC	Escape
LF	Line feed	LEM	Logical end of media	DEL	Delete/idle

Fig. 23-23. The ASCII code has a contiguous alphabet, so that numeric ordering permits alphabetic sorting.

Decimal	Gray code
0	0000
1	0001
2	0011
3	0010
4	0110
5	0111
6	0101
7	0100
8	1100
9	1101
10	1111
11	1110
12	1010
13	1011
14	1001
15	1000

Fig. 23-24. The Gray code, used in analog-to-digital encoding systems. There is only a 1-b change between any two successive integers.

It is possible to design codes that can detect directly whether errors have occurred in two bit positions in a single code word. Figure 23-26 shows such a code, an example of a *Hamming* code. The code positions in columns 1, 2, and 4 are used to check the parity of the respective bit combinations. Two code words of a Hamming code must differ at three or more bit positions, and therefore any 2-b error patterns can be detected. The pattern formed by the particular parity bits that show errors indicates which bit is in error in the case of a single bit failure.

In general, if two code words must differ at D or more bit positions, the code can detect up to $D - 1$ bit errors. For $D = 2t + 1$, the code can detect $2t$ bit errors or correct t bit errors.

Word parity	Binary code
1	000000
0	000001
0	000010
1	000011
0	000100
1	000101
1	000110
0	000111
0	001000
1	001001
1	001010
0	001011
1	001100
0	001101
0	001110
1	001111
0	010000
1	010001
1	010010
0	010011
	010100
0	010101
0	010110
1	010111
1	011000
0	011001
0	011010
1	011011
0	011100
1	011101
1	011110
0	011111
0	100000
1	100001
1	100010
0	100011
1	100100
0	100101
0	100110
1	00111
1	01000

C 1 0 1 1 1   List parity

Fig. 23-25. Two-dimensional parity checking, in which a single error can be corrected and a triple error detected.

COMPUTER ORGANIZATION AND ARCHITECTURE

24. Introduction. The terms computer architecture and computer organization have been given a variety of meanings by different authors and have even been used interchangeably. Brooks[56a] defines *computer architecture* as the complete and detailed specification of the user interface. For the computing system it is the union of the manuals the user must consult to do the job. The architecture describes what the system is to do. How the architecture is implemented determines how it does it. *Computer organization* describes the structure of the implementation. The *realization* of a computer is the actual construction of the machine with a specific component technology.

Over the past 25 years great progress has been made as component technology has moved from vacuum tubes to solid-state devices to large-scale integration (LSI) (see Par. **23-6**). This has been achieved as a result of increased understanding of semiconductor materials along with improvements in the fabrication processes. The result has been significant enhancement in the performance of the logic and memory components used in computer construction along with significant reductions in cost and size. Figure 23-27, for example, indicates the reduction in volume of main memory that has occurred in the last 25 years. The volume required to store 1 million characters has been reduced by a factor of 6,400 in that period. Similarly, during that same period, the system cost to execute a specific mix of instructions has also decreased by a factor of 130. At the same time, the time required to execute that instruction mix has decreased by a factor of 125 as a result of increased processing performance. This combined reduction amounts to a cost-performance improvement of 16,250 over this time span. There is no reason to expect that this progress will not continue into the future as new technology improvements continue to occur.

These advances in component technology have also had a major impact on computer organization and its realization. Functions and features that were too expensive to be included in earlier designs are now feasible, and the trade-offs between software and hardware need to be reevaluated as hardware costs continue to decrease. New approaches to computer organization must also be considered as technology continues to improve. The advances in component technology also have had a major impact on such aspects of computer realization as packaging, coding, and power. For more information on the component technology itself see Sec. **8**.

25. Basic Computer Organization. The basic organization of a digital computer is shown in the block diagram of Fig. 23-28. This structure was proposed in 1946 by von Neumann.[5a] It is a tribute to his genius that this design, which was intended for use in solving differential equations, has also been applicable in solving other types of problems in such diverse areas as business data processing and real-time control. Von Neumann recognized the value of maintaining both data and computer instructions in storage and in being able to modify instructions as well as data. He recognized the importance of branch or jump instructions to alter the sequence of control of computer execution. His contributions were so significant that the vast

Decimal digit	Position						
	1	2	3	4	5	6	7
0	0	0	0	0	0	0	0
1	1	1	0	1	0	0	1
2	0	1	0	1	0	1	0
3	0	0	0	0	0	1	1
4	1	0	0	1	1	0	0
5	0	1	0	0	1	0	1
6	1	1	0	0	1	1	0
7	0	0	0	1	1	1	1
8	1	1	1	0	0	0	0
9	0	0	1	1	0	0	0

Parity checks   8, 4, 2, 1 code

Fig. 23-26. A Hamming code. The parity bit in column 1 checks parity in columns 1, 3, 5, and 7; the bit in column 2 checks 2, 3, 6, and 7; and the bit in column 4 checks 4, 5, 6, and 7. The overlapping structure of the code permits the correction of a single error or the detection of single or double errors in any code word.

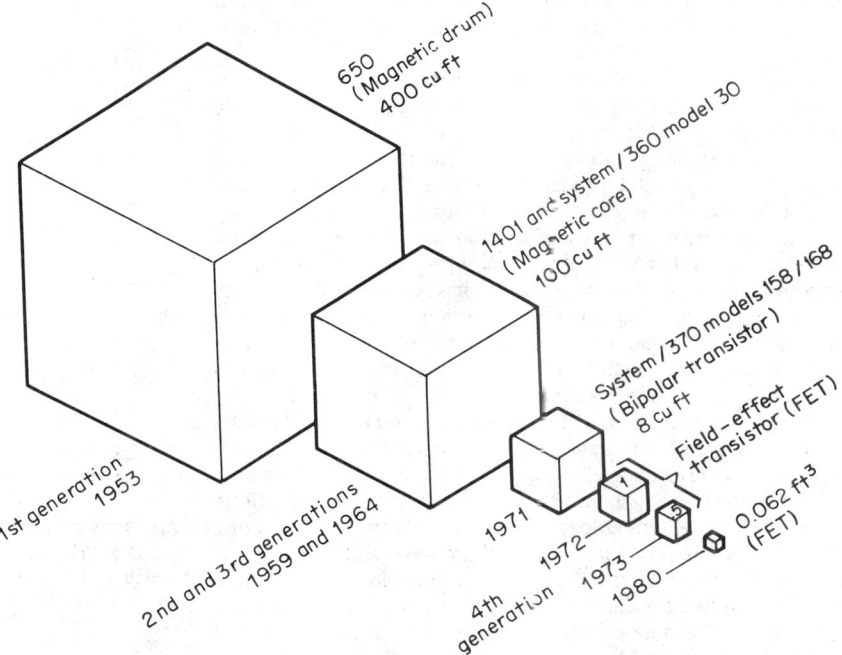

Fig. 23-27. The reduction by a factor of 6,400 in memory size from the first- to the fourth-generation families of IBM computers.

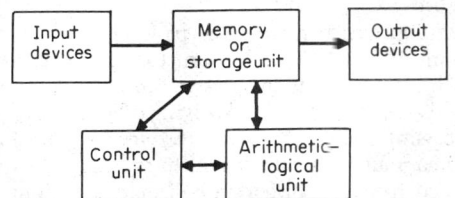

Fig. 23-28. Block diagram of a digital computer illustrating the main elements of the von Neumann architecture.

majority of computers in use today are based on his design and are called von Neumann computers.

The four basic elements of the digital computer are its *main storage, control unit, arithmetic-logic unit* (ALU), and *input/output* (I/O). These elements are interconnected as shown in Fig. 23-28. The ALU, or *processor*, when combined with the control unit, is referred to as the *central processing unit* (CPU).

Main storage provides the computer with directly addressable fast-access storage of data. The storage unit stores programs as well as input, output, and intermediate data. Both data and programs must be loaded into main storage from input devices before they can be processed. (See the discussion on computer storage beginning with Par. **23-62**.)

The control unit is the controlling center of the computer. It supervises the flow of information between the various units. It contains the sequencing and processing controls for instruction

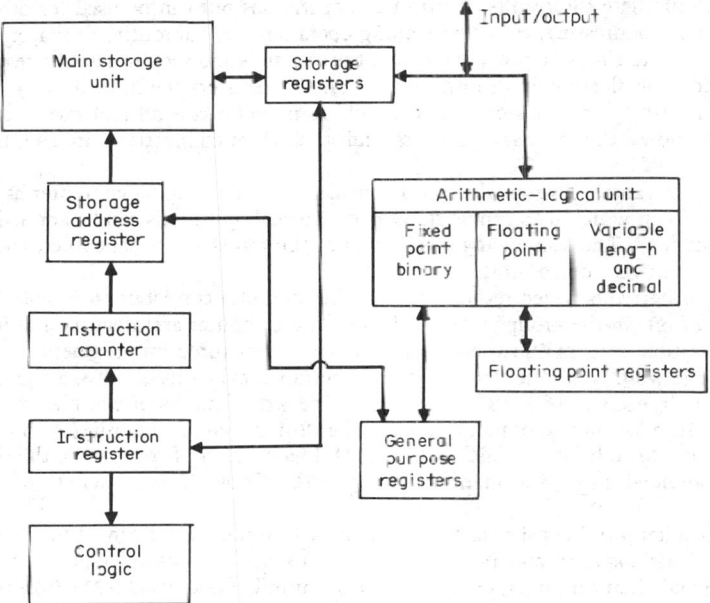

Fig. 23-29. Basic structure of a digital computer showing a typical register organization and interconnections.

decoding and execution and for handling interrupts. It controls the timing of the computer and provides other system-related functions.

The ALU carries out the processing tasks specified by the instruction set. It performs various arithmetic operations as well as logical operations and other data processing tasks.

Input/output devices, which permit the computer to interact with users and the external world, include such equipment as card readers and punches, magnetic-tape units, disc storage units, display devices, keyboard terminals, printers, teleprocessing devices, and sensor-based devices. (See the discussion on input/output beginning with Par. **23-80**.)

26. Detailed Computer Organization. The block diagram of Fig. 23-29 provides an overview of the basic structure of the digital computer. Computer systems are complex, and block diagrams cannot describe the computer in sufficient detail for most purposes. One is therefore forced to go to lower levels of description. There are at least five levels[66a] that can describe the implemention of a computer system:

1. Processor-memory-switch (block-diagram) level
2. Programming level (including the operating system)
3. Register-transfer level
4. Switching-circuits level
5. Circuit or realization level

Each of these levels is an abstraction of the levels beneath it. A number of computer-hardware description languages have been developed to represent the components used in each level along with their modes of combination and their behavior. These languages are discussed under hardware design, starting with Par. **23-60**. The programming level is not considered here but is described later under software, beginning with Par. **23-119**.

A *register* is a device capable of receiving information, holding it, and transferring it as directed by control circuits. The actual realization of registers can take a number of forms depending upon the technology used.

Registers are found in every element of the computer system They are an integral part of main storage, being used as storage registers to contain the information being transferred from memory (read) or into memory (write) as well as storage address registers (SAR) to hold the address of the location in storage involved in the information transfer. In the control unit, the instruction (or program) counter contains the storage address of the instruction to be executed while the instruction register holds the instruction being decoded and executed.

In the ALU internal registers are used to hold the operands and partial results while arithmetic and logical operations are being performed. Other ALU registers, called *general-purpose registers*, are used to accumulate the results of arithmetic operations but can be used for other purposes such as indexing, addressing, counting looping operations, or subroutine linkage. In addition, floating-point registers may be provided to hold the operands and accumulate the results of arithmetic operations on floating-point numbers. Of all the registers mentioned, only the general-purpose and floating-point registers are accessible to program control and to the programmer. Figure 23-29 shows the primary registers and their interconnections in the basic digital computer.

At the register transfer level of abstraction one can describe a digital computer as a collection of registers between which data can be transferred. Logical operations can be applied to the data during the transfers. The sequencing and timing of the transfers are scheduled and controlled by logic circuits in the control unit.

The data transferred between registers within the computer consist of groups of binary digits. The number of bits in the group is determined by the computer architecture and in particular by the organization of its main storage. Main storage is structured into segments, called *words*, and each storage word is uniquely identified by a number, called its *address*, assigned to it. The number of bits in each word is its *word length*. The word lengths of computers vary widely, ranging from 16 b for minicomputers, 32 b for the IBM System/370 family, 36 b for the UNIVAC 1100 series, to 60 b for the CDC Cyber series. In a number of computers, the storage word is further subdivided into 8-b segments called *bytes* (B). Thus, the word length of the System/370 is 4 B.

Since a computer word consists of binary digits, a 16-b (2-B) word can contain $65,536 = 2^{16}$ different combinations of 0s and 1s. These bit patterns can be interpreted as: (1) a pure binary word, (2) a signed binary number, (3) a floating-point number (see Par. **23-21**), (4) a binary-coded decimal number, (5) data characters, or (6) an instruction word.

In a signed binary number the high-order (leftmost) bit indicates the sign of the number. If the bit is 0, the number is positive; a 1 indicates it is negative. Thus

$$0bbbbbbb \text{ represents a positive 7-bit number,}$$
$$1bbbbbbb \text{ represents a negative 7-bit number}$$

A negative number is carried in 2's-complement (inverted) form. For example,

$$11111110_2 = 2_{10}$$

A binary-coded decimal code uses 4 b to represent a decimal digit. It uses the binary digit combinations for 0 to 9; combinations greater than 9 are not allowed. Thus

$$0101_2 = 5_{10} \qquad 1010_2 = \text{illegal}$$

The sign of the decimal number can be indicated in several ways. One technique uses the low-order bits to indicate the sign; for example,

$$bbbbbbbbbbbb1100 \text{ represents a positive 3-digit number}$$
$$bbbbbbbbbbbb1011 \text{ represents a negative 3-digit number}$$

Thus, $00010101001 11011_2 = -153_{10}$.

For external communication, as well as text processing and other nonnumeric functions, the digital computer must be able to handle character sets. The byte has been accepted as the data unit for representing character codes. The two most common codes, described earlier, are the American Standard Code for Information Interchange (ASCII) and the Extended Binary Coded Decimal Interchange Code (EBCDIC). The 16-b word 1100011111001010 coded in EBCDIC represents the two-letter word "go." (See Par. **23-22** and Figs. 23-21 to 23-23.)

The *instruction word* is composed of two major parts, an *operation part* and an *operand part*. The length of the instruction word is determined by the computer architecture. Some computers have a single format for their instruction words and thus a single length whereas other computers have several formats and several different lengths. The operation part consists of the operation code that describes the particular operation to be performed by the computer as a result of executing that instruction. The operand part usually contains the addresses of the two operands involved in the operation. For example, the RR (register-to-register) instruction format in the System/370 is

Op Code	Reg 1	Reg 2
bbbbbbbb	bbbb	bbbb

0   7   8 11   12 15

The instruction 1AB4 (in hexadecimal), for example, instructs the computer to add the contents of register 11 to the contents of register 4 and to put the resulting sum in register 11, replacing its original contents. (See Par. **23-18** and Table 23-1.)

Two facts should be noted. First, the discussion thus far may have implied that digital computers can deal only with fixed-length words. That is true for some computers, but other families of computers can also deal with variable-length words. For these, the operand part of the instruction contains the address of the first digit or character of each variable-length word plus a measure of its length, i.e., the number of characters it contains. The second fact is that it is impossible to distinguish between the various data representations when they are stored. For example, there is nothing to indicate whether a word of memory contains a binary number or a binary-coded decimal number. Programmers must make the distinction in the programs they develop and not attempt meaningless operations such as adding a binary number to a decimal number. The only way the computer distinguishes an instruction word from other data words is by the time when it is read from storage into the control unit. This is discussed in Par. **23-27**.

27. Instruction Execution. The digital computer operates in a cyclic fashion. Each cycle is called a *machine cycle* and consists of two main subcycles, the *instruction* (I) *cycle* (sometimes called the *fetch cycle*) and the *execution* (E) *cycle*. During the machine cycle, the following basic steps occur in sequence (see Fig. 23-29):

1. The cycle begins with the I cycle:
   a. The contents of the instruction counter are transferred to the storage address register (SAR). (The instruction counter holds the address of the next instruction to be executed.)
   b. The specified word is transferred from storage to the instruction register. (The control unit assumes that this storage word is an instruction.)
   c. The contents of the instruction register are decoded by logical circuits in the control unit. This identifies the type of operation to be performed and the locations of the operands to be used in the operation.
2. At this point, the E cycle begins
   a. The specified computer operation is performed using the designated operands, and the result is transferred to the location indicated by the instruction.
   b. The instruction counter is advanced to the address of the next instruction in the sequence. (If a branch, or change in execution control sequence, is to occur, the contents of the instruction counter are replaced by an address as directed by the instruction currently being executed.)
3. At this point, the I cycle is repeated.

To indicate in more specific terms what happens in the CPU during instruction execution it is necessary to go to the switching level of description. The following paragraphs describe the operations of the ALU and the control section in more detail.

28. Arithmetic Logic Unit (ALU). The ALU performs arithmetic and logical operations between two operands, such as OR, AND, EXCLUSIVE-OR, ADD, MULTIPLY, SUBTRACT, or DIVIDE. The unit may also perform operations such as INVERT on only one operand, and it tests for minus or zero and forms a complement.

Adders and multipliers are at the heart of the ALU. In Fig. 23-30a one bit position of an ALU is shown as part of a wider data path. One latch (part of a register A) feeds an AND circuit that is conditioned by a CONTROL A. The output feeds INPUT A of the adder circuit. One latch of register B is also ANDed with CONTROL B and feeds the other input into the adder. A true-complement circuit is shown on the B line. This latter circuit has to do with subtraction and can be assumed to be a direct connection when adding. Each adder stage is a combinatorial circuit that accepts a carry from the stage representing the next lower digit in the binary number (assumed to be on the right). The collection of outputs from all adder stages is the sum. This sum is ANDed into register D by CONTROL D.

All bit positions of each of the registers are gated by a single control line. If the gate is closed (control equal to 0), all outputs are 0s. If the gate is open (control equal to 1), the bit pattern

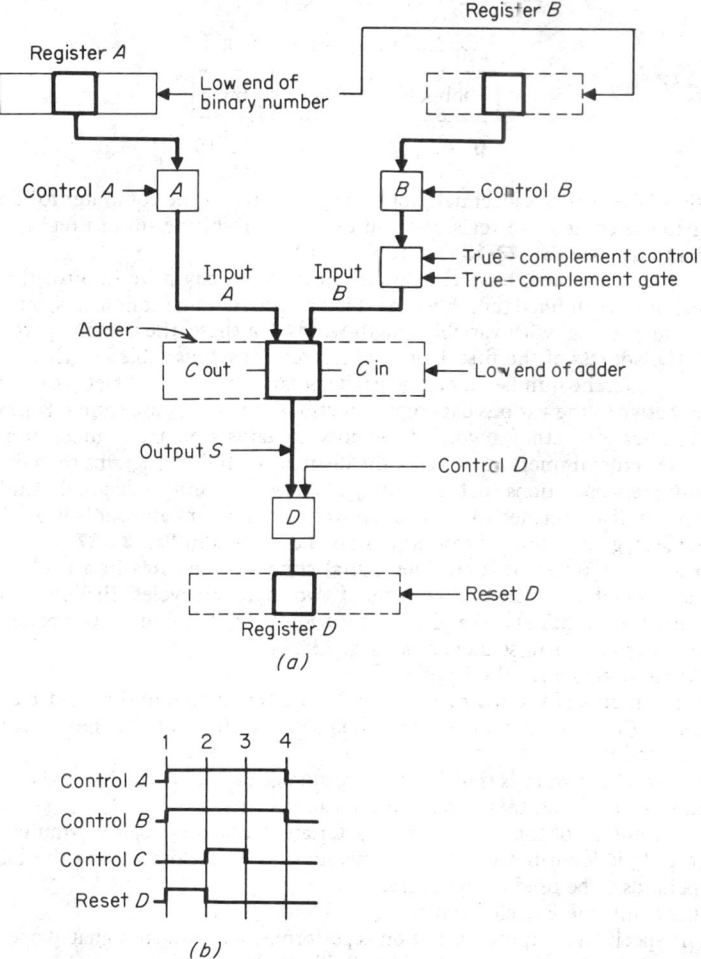

Fig. 23-30. Basic addition logic. The heart of the ALU is the adder circuit shown in functional form in (a). The control section applies the appropriate time sequence of pulses (b) on the control to perform addition. Heavy lines indicate a repeat of circuitry in each bit position to form a machine adder.

appearing in the register is transmitted through the series of AND circuits to the input of the adders. Thus, a gate is a two-way AND circuit for each bit position. The diagram of Fig. 23-30 illustrates all positions of an n-position adder since all positions are identical. In such a case heavy lines, as shown, indicate that this one line represents a line in each bit position.

29. Binary Addition. At the outset of an addition, it is assumed that registers A and B (Fig. 23-30a) contain the addends. An addition is performed by pulsing the control lines with signals originating in the control section of the CPU. Time is assumed to be metered into fixed intervals by an oscillator (clock) in the control section. These time slots are numbered for easier identification (Fig. 23-30b). At time 1 the inputs are gated to the adder, and the adders begin to compute the sum. At the same time, register D is *reset* (all latches set to 0). At time 2 the outputs of the adders have reached steady state, and control line D is raised, permitting those bit positions for which the sum was 1 to set the corresponding latches in register D. Between times 2 and 3, the result is latched up in register D, and at time 3, control D is lowered. Only after the result is locked into D and cannot change, may control A and B be lowered. If they were lowered earlier, the change might propagate through the adder to produce an incorrect answer.

The length of the pulses depends on the circuits used. The times from 2 to 3 and 3 to 4 are usually equal to a few logic delays (time to propagate through an AND, OR, or INVERTER). The time from 1 to 2 depends on the length of the adder and is proportional to the number of positions in a parallel adder due to potential carry-propagation times. This delay can be reduced by *carry look-ahead* (sometimes called *carry bypass*, or *carry anticipation*).

30. Binary Subtraction. Subtraction can be accomplished using the operation of addition, by forming the complement of a number. Negative numbers are represented throughout the system in complement form. To subtract two numbers such as B from A, a set of logic elements may be put into the line shown in Fig. 23-30c as input B. Using 2's complement, the sign of a number is changed by complementing each bit and adding 1 to the result. The inversion of the bit is performed by the logic element interposed on the input B line in Fig. 23-30a known as a *true-complement* (T/C) *gate*. This unit gates the unmodified bit if the control is 0 and inverts each bit if the control is 1. The boolean equation for the output of the T/C gate is

$$\text{Output} = \overline{T/C} \cdot B + T/C \cdot \overline{B}$$

The T/C gate is a series of EXCLUSIVE-ORs with one leg, common to all bit positions, connected to the T/C control line. The other leg of each EXCLUSIVE-OR is connected to one bit of the circuit containing the number to be complemented.

The T/C gate produces the 1's complement; a 1 must be added in the low-bit position to produce the true complement. The low stage of an adder may be designed to have an input for a carry-in, designed to accommodate the 1 bit automatically produced from the high-order position of the T/C gate. Such a logical interconnection accomplishes the required 1 input for a true-complement system when a positive number B is subtracted from a positive number and is called an *end-around carry*. Consistency of operation is obtained by entering the appropriate high-order T/C gate into the low-order carry position.

31. Decimal Addition. In some systems the internal organization of a computer is such that decimal representations are used in arithmetic operations. In binary-coded decimal (BCD), a conventional binary adder can be used to add decimal digits with a small amount of additional hardware. Adding two 4-b binary numbers produces a 4- or 5-b binary result. When two BCD numbers are added, the result is correct if it lies in the range 0 to 9. If the result is greater than 9, that is the resulting bit pattern is 1010 to 10010, the answer must be adjusted by adding 6 to that group of 4 b, this number being the difference between the desired base (10) and the actual base ($2^4 = 16$). The binary carry from a block of 4 b must also be adjusted to generate the appropriate decimal carry.

The circuits to accomplish decimal addition are shown in Fig. 23-31. A test circuit generates an output if the binary sum is 1010 or greater. This output causes a 6 to be added into the sum and also is ORed with the original binary carry to produce a decimal carry. The added circuits needed to perform decimal additions with a binary adder represent one-half to two-thirds of the circuits of the original binary adder.

32. Decimal Subtraction. In most computers that provide for decimal operation, decimal numbers are stored with a sign and magnitude. To perform subtraction the true complement is formed by subtracting each decimal digit in the number from 9 and adding back 1. Once the complement is formed, addition produces the desired difference.

In a machine that provides for decimal arithmetic a *decimal true-complement* switch may be incorporated in each group of four BCD bits to form the complement. As with binary, provision must be made for the addition of an appropriate low-order digit and for the occurrence of overflows.

33. Shifting. All computers have shift instructions in their instruction repertoire. Shifting is required, for example, for multiply and divide operations.

The minimum is a shift of one position left or right, but most computers have circuits permitting a shift of one or more bits at a time, that is, 1, 4, or 8.

In shifting, a problem arises with the bit(s) shifted out at the end of the register and the new (open) bit position(s). Those shifted out at one end can be inserted in the other end (referred to as *end-around shift* or *circulate*), or they can be discarded or placed in a special register. The newly created vacancies can be filled with all 0s, all 1s, or from another special register.

In a typical computer, shifting is not performed in a shift register but with a set of gates that follow the adders. The outputs of the shift gates are connected to the output register, with an offset by providing a separate gate for each distance and direction of shift. One bit position of the output register can therefore be fed from several adder outputs, but only one gate is opened at a time, as in Fig. 23-32. The pattern shown is repeated for every bit position.

34. Multiplication. Figure 23-33 shows a possible data-flow system for multiplication, i.e., the data flow of the adder shown in Fig. 23-30 with the addition of register C to hold the multiplier and the shift registers as shown in Fig. 23-32. An extra register E holds any extra bits

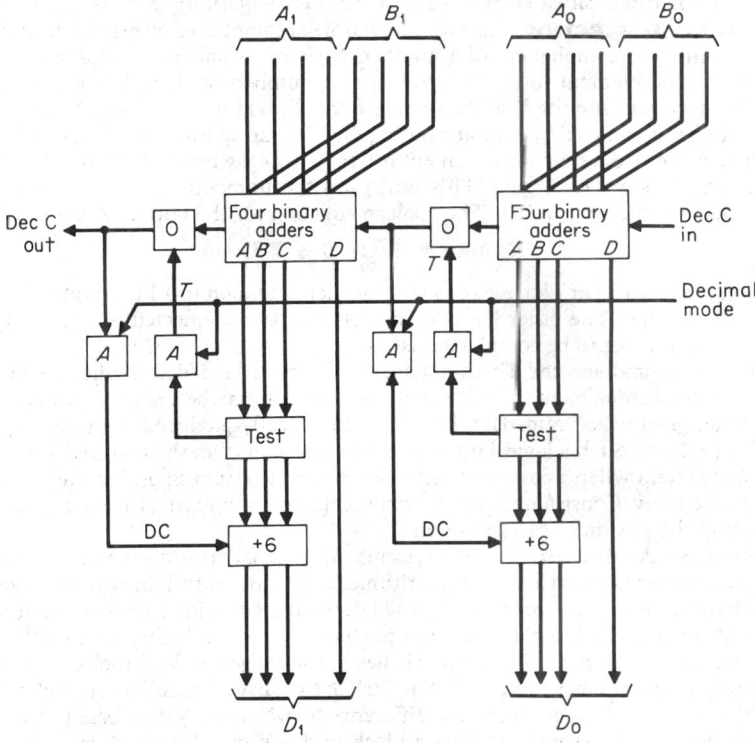

Fig. 23-31. Binary adder with decimal-add feature. If the input of the binary add is 1010 or greater, the addition of a binary 6 produces the correct result by modulo-10 addition.

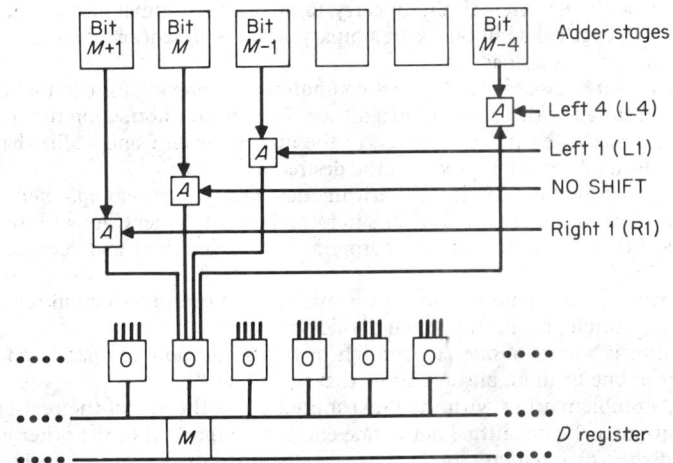

Fig. 23-32. Shift gates. In many systems shifting is accomplished in conjunction with the output of the adder; i.e., position shifts can be accomplished with only one circuit delay.

that might be generated in the process of multiplication. Register E in the particular system shown also receives the contents of C after transmission through C's shift register.

The process of binary multiplication involves decoding successive bits in the multiplier, starting with its lowest-order position. If the bit is a 1, the multiplier is added into an accumulating sum; if 0, no addition takes place. In either case the sum is moved one position to the right and the next higher-order bit in the multiplier considered.

In Fig. 23-33 the multiplier is stored in register C, the multiplicand in A, and all others are

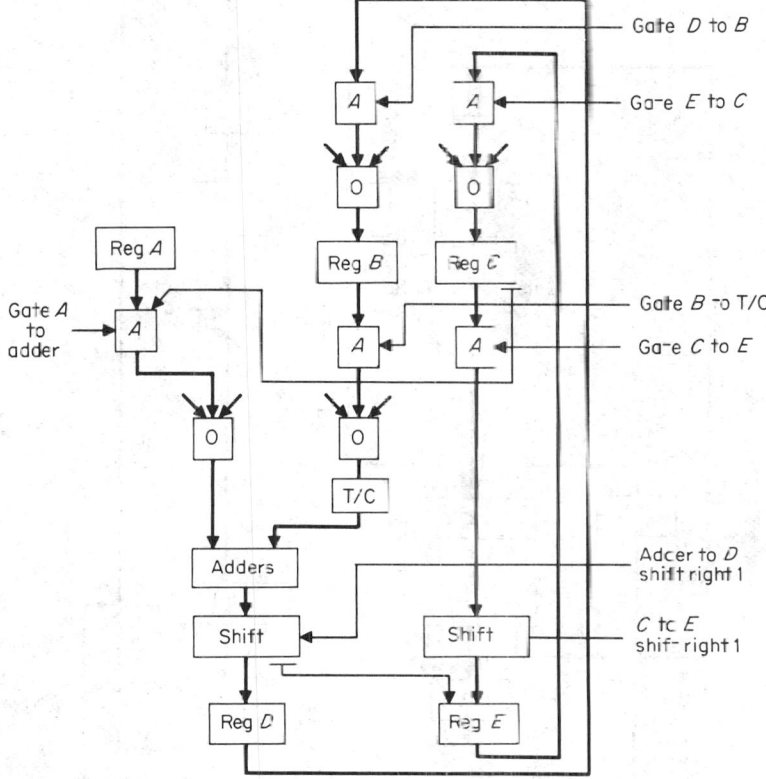

Fig. 23-33. Data flow for multiplication.

reset to zero. If the low-order position of C is a 1, the contents of B and C are added and shifted one position to the right into register D. After the addition, register C is shifted one position to the right and stored in register E. If the low-order bit in C had been a zero, only register B would have been gated into the adders; i.e., the addition of the contents of A would have been suppressed, but subsequent operations would have remained the same.

Each add and shift operation subsequent to an add may generate low-order bits that may be shifted into register E, since as the contents of C are shifted to the right, unused positions become successively available. After the add and shift cycles, registers D and E are transferred into B and C, respectively, and the process is repeated until all positions of the multiplier are used. The content of D and E is the product.

35. Division. To provide for the division of two numbers the functions of the registers in Fig. 23-33 must be rearranged as shown in Fig. 23-34. A gating loop is provided from the shift register to register A. Initially the divisor is placed in register B and the dividend in A. The T/C gate is used to subtract B from A, and if the result is zero or positive, a 1 is placed in the low-order position of register E. If the result is negative, a 0 is placed in E and the input from B ungated to reset to the contents of A. The shift register then shifts the output of the adder one position to the right and gates it back to A. E is gated through C and shifted on to the right. The whole process is repeated until the dividend is exhausted. E contains the quotient and A the remainder.

36. Floating-Point Operations. In some applications, it is convenient to represent numerical values in floating-point form. Such numbers consist of a *fraction* that represents the number's most significant digits, in a portion of the field of a word in the machine, and a *characteristic* in the remaining portion. The characteristic denotes the decimal-point position relative to the assumed decimal point in the characteristic field. In floating-point addition and subtrac-

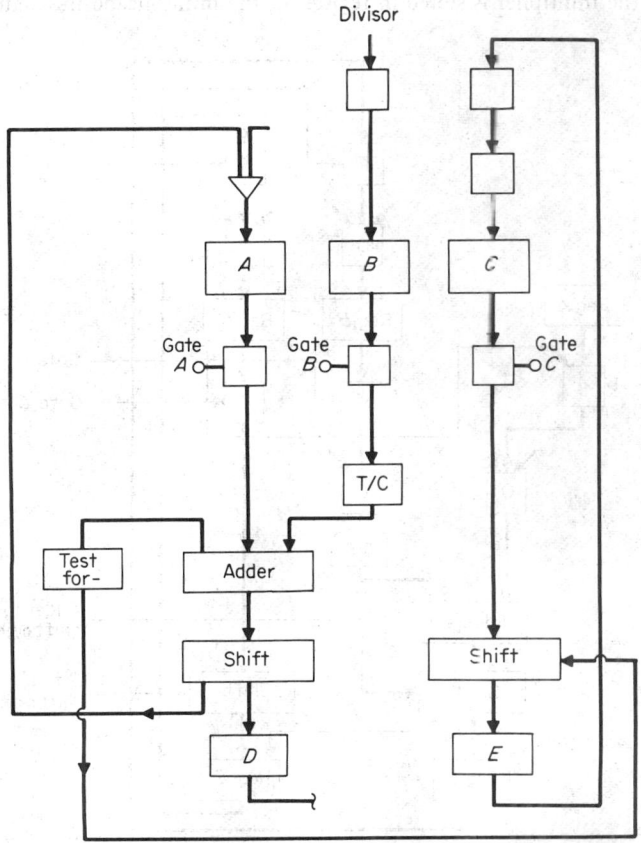

Fig. 23-34. Data flow for division. B holds the divisor and A the dividend. A trial subtraction is made of the high-order bits of B from A, and if the results are 0 or positive, a 1 is entered into the low-order bit of E and the result of the subtraction shifted left and reentered into A (with gate A closed). If the result had been negative the B gate would have closed to negate the subtraction, and a 0 would have been entered into E. The output of the adder would then have been shifted left one position and reentered in A.

tion, justification of the fractions according to the contents of the characteristic field must take place before the operation is performed; i.e., the decimal points must be lined up.

In an ALU such as in Fig. 23-33, the operation proceeds in the following way. Two numbers A and B are placed in registers A and B, respectively. The control section then gates only the characteristic fields into the adder, in a subtract mode of operation. The absolute difference is stored in an appropriate position of the control section. Controlled by the sign of the subtraction operation, the fraction of the least number is placed through the adder into the shift register. The control section then shifts the fraction the required number of positions, i.e., according to the stored difference of characteristic fields, and places the result back in the appropriate register. Addition or subtraction can then proceed according to the generating machine instruction.

This procedure is costly of machine time. Instead of using the ALU adder for the characteristic-field difference, provision can be made for subtraction of the characteristics in the control section. In such a case, only the characteristics A and B need be entered into registers A and B and the lesser number can be placed in B so that shifting can be accomplished by the circuits normally used in multiplication.

37. Control Section. The control section is the part of a CPU that initiates, executes, and monitors the system's operations. It contains clocking circuits and timing rings for opening and closing gates in the ALU. It fetches instructions from main storage controls execution of these instructions, and maintains the address for the next instruction to be executed. The ALU also initiates I/O operations.

The underlying design concepts of a control section are not so well developed as those of the ALU. The control contains about the same amount of logic circuit for a medium-size parallel machine (say 30 b) but contains considerably more hardware than the ALU for machines with narrower data paths. There is no typical design approach for a control section, though there are some common practices. To exemplify one of these practices a control section is described that is compatible with the ALUs discussed in the previous paragraphs.

38. Basic Timing Circuits. Basic to any control unit is a continuously running oscillator. The speed of this oscillator depends on the type of computer (parallel or serial), the speed of the logic circuits, the type of gating used (the type of registers), and the number of logic levels between registers. The number of logic delays for an average oscillator cycle time is typically between 6 and 15. The oscillator pulses are usually grouped to form the basic operating cycle of the computer, referred to as *machine cycles*. In this example four pulses are combined into such a group.

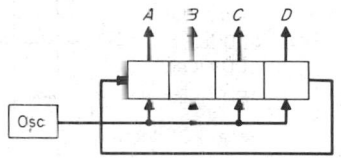

In Fig. 23-35 an oscillator is shown that drives a four-stage ring. At any one time, only one stage is on. Suppose an addition is to be performed between registers A and B and the result is to be placed back in B. The addition circuitry described in Fig. 23-33 uses three registers, two for the addends (registers A and B), and one for temporary storage (register D). The operation to be performed is to add the content of register A to the content of register B and store the result in register B. Register D is required for temporary storage since if B were connected back on itself, an unstable situation would exist; i.e., its output (modified by A) would feed its input.

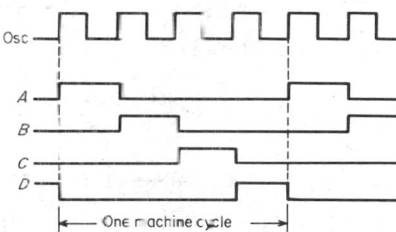

The operation of the four-stage clock ring in controlling the addition and transfer is shown in Fig. 23-36. Action is initiated by the coincidence of an ADD signal and clock pulse *A*, which starts the *add ring*. This latter ring has two stages and advances to the next stage upon occurrence of the next *A* clock pulse. The timing chart in Fig. 23-36 describes one sequence of actions required and the gates needed for the addition. The circuit diagram shows a realization of these gates. Each register is reset before it is reused, and all pulses are derived from the four basic clock pulses shown in Fig. 23-35. The add ring initiates the add by opening the gates between registers A and B and the adder. The add latch is then reset, D is reset, the ring stage transferred, and so forth. An ADD-FINISHED signal is furnished, and may be used elsewhere in the system.

In the timing diagram, it is assumed that the time required for transmission through the adder is about one machine cycle and that one clock cycle is sufficient for the signal to propagate through the necessary gating and set the information in the target register. These times must include the delay in the circuits and any signal-propagation delay.

39. Control of Instruction Execution. The following approach for the design of a control section is straightforward but extravagant of hardware. For each instruction in the computer a timing diagram is developed similar to the one shown for the add operation in Par. **23-38**. These timings are implemented in rings that vary in length, according to the complexity of the instruction. The concept is simple and has been widely used, but it is costly To reduce cost, rings are used repeatedly within one instruction and/or by several instructions. Subtraction, for example, might use the ADD ring, except for an extra latch that might be set at ring time 1, clock time *A* (denoted 1.*A*) and reset at 2.*C*. This new latch feeds the T/C gates. Another latch that might be added to denote decimal arithmetic would also be set at 1.*A* time and reset at 2.*C*

Fig. 23-35. Oscillator and ring circuit. Many control functions can be performed by using an oscillator in conjunction with a timing ring that sequentially sends signals on separate lines, in synchronism with the oscillator.

time. The addition of two latches and a few additional ANDs and ORs permits the elimination of three two-stage rings and associated logic. Further reductions in the number of required circuits can be achieved by considering the iterative nature of some instructions, such as multiplication, division, or multiple shifting.

40. Controls Using Counters (Multiplication). To exemplify this approach multiplication is considered. Multiplication can be implemented as many add-shift cycles, one per

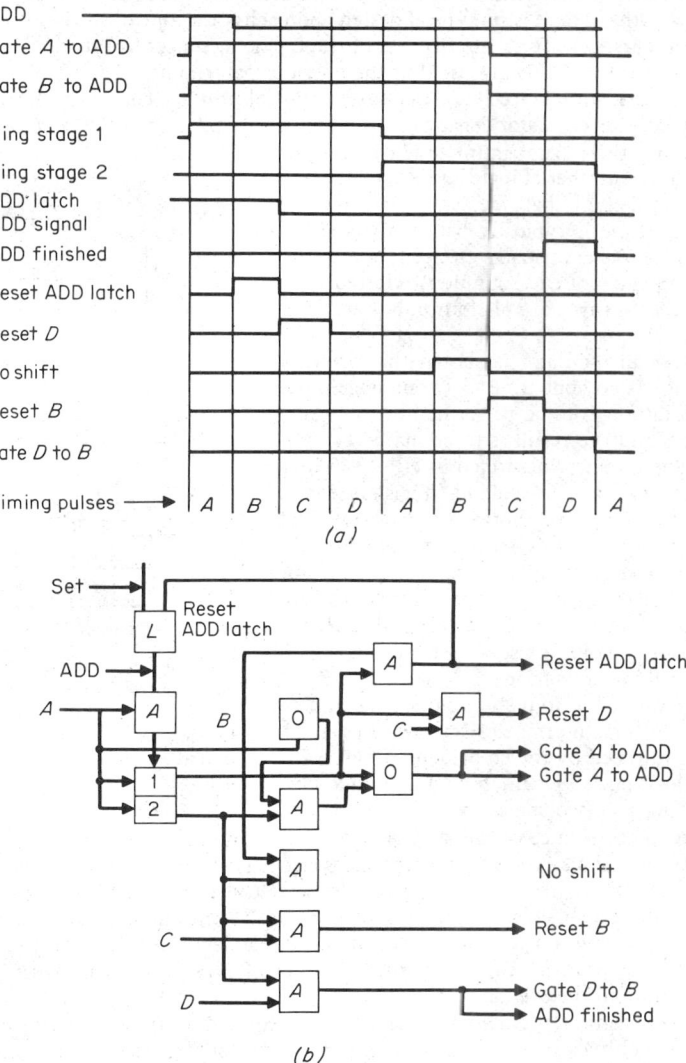

Fig. 23-36. Control circuit (b) and timing chart (a) for addition. The timing ring shown in Fig. 23-35 is used to control the adder circuit shown in Fig. 23-33 with the help of additional switching circuits.

digit in the multiplier (say N). The control for such an instruction can be implemented using one $2N + 1$ position ring, the first position initializing and the next $2N$ positions for N add-shift cycles (two positions are needed per add cycle). Such an approach requires not only an unnecessarily long ring but is relatively inflexible for other than N-b multipliers.

The alternate approach, requiring considerably less hardware, uses the basic operation in multiplication, an add and a shift. Therefore a multipy can be implemented by using the controls for the add-shift instruction, plus a binary counter, plus some assorted logic gates to control the gates unique to the multiply, as in Fig. 23-37. In the figure some of the less important controls needed for multiply have not been included to simplify the presentation. For example, the NO SHIFT signal for add must be conditioned with a signal that multiply is not in progress, and during multiplication an ADD-FINISHED signal should be ignored (nor start an I cycle), the reset B also resets C, the reset D also resets E, gate D to B also gates E to C, gating of A to the adder is conditioned on the last bit of C, etc.

In Fig. 23-37 the action is started by raising the MPY line that sets a *multiply* latch. This in turn sets the binary counter to the desired cycle count. Also set is the ADD-LATCH, and the

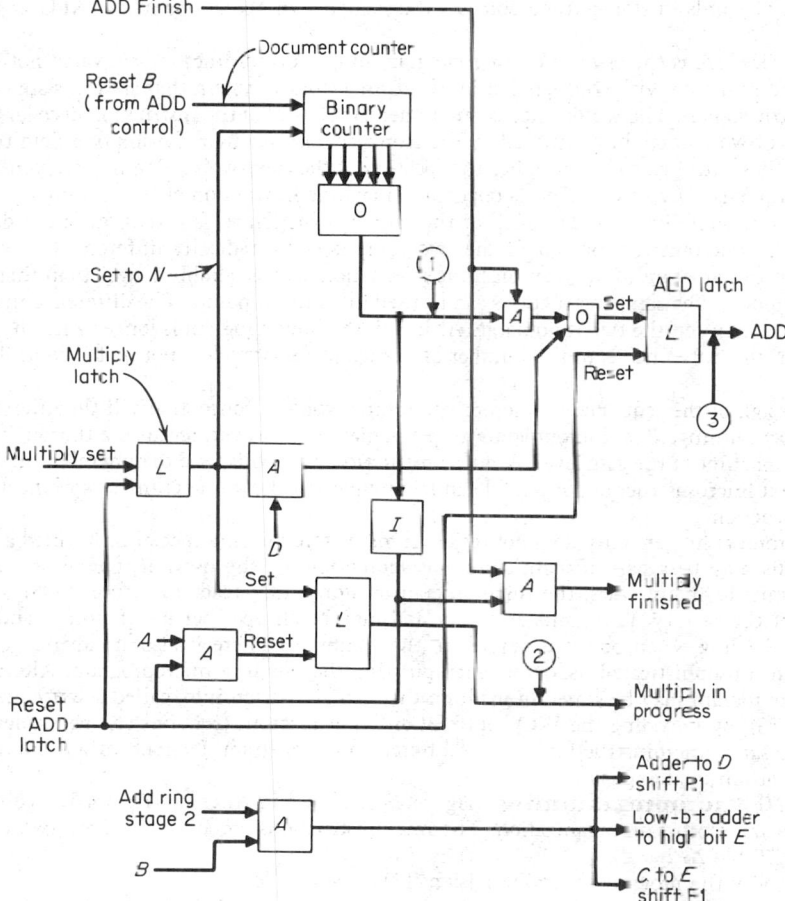

Fig. 23-37. Control circuit for multiplication. Repetitive operations such as add and shift are combined in cyclical fashion. The timing diagram of Fig. 23-36a is assumed.

addition cycles start. When the counter goes to 0, the ADD-LATCH is no longer set and the MULTIPLY-FINISHED signal is raised.

41. Microprogramming. The control of the E cycle, in the preceding descriptions, is performed by a set of sequential circuits designed into the machine. Once an execution is initiated, the clocking circuits complete the operation, through wired paths in a specific manner. An alternative method of design for a control unit is *microprogramming*. The concept is not sharply defined but has as its objective implementation of the control section at low cost and with high flexibility.

In many cases it is desirable to design compatible computers, i.e., units with the same instruction set, with widely varying performance and cost. To provide a slower version at a lower cost, the width of the data path is reduced to lower circuit counts in the ALU On the other hand, to operate with the same instruction set, the reduced data path width usually implies more iterative operations in the control section, at added cost.

Considerable investment is required for programming development, and normally such systems run only on computers with identical instruction sets. If appropriate flexibility is provided, one computer can mimic the instruction set of another. Microprogramming provides for such operations. The process by which one system mimics another is called emulation.

In the control-section operation for an add given above in the example of a wired system (Figs. 23-30 and 23-31), the ring circuit delivers a series of electric signals in time sequence on specific logical control lines. An alternative approach follows. Upon decoding the instruction code, a sequence of words is called forth from a special storage unit and put into a device that converts the bit positions of each word directly into signals that selectively activate the control wire. The sequence of words in storage then controls the sequence of operations in the ALU at the gate level.

This procedure is the essence of microprogramming. Control lines are activated not by logic gates in conjunction with counters but by words in a storage system that are translated directly into electric signals. The words selected are under the control of the instruction decoder, but the sequence of words may be controlled by the words themselves by provision of a field that does not directly control gates but specifies the location of the next word. The name given to these control words is *microinstruction* as opposed to machine instruction or instruction.

When two computers are designed for the same instruction set but with different data-path widths, the microinstruction sets of the two computers are radically different. For the small computer, the program for a given machine instruction is considerably longer than that for the large computer. The same instructions can be used in both computers. The difference in control-system cost between the two is not large. Although the microprogram is longer with the smaller computer, the difference is in the number of storage places provided, not in the control-section hardware.

The design of the sequence of microprogramming words is conceptually little different from other programming. The microprogram implementation, however, requires a thorough knowledge of a machine at the gate level. A *microinstruction counter* is used to remember the location of the next microinstruction, or a field can be provided to allow branching or specification of a next instruction.

A microprogram generally does not reside in main store but in a special unit called a *control store*. This may be a part of main store, not addressable by the user, or it may be a separate storage unit. In many cases, the microprogram is stored in a read-only store (ROS); i.e., it is written at the factory. ROS units are faster and may be cheaper per bit of storage and do not require reloading when power is applied to the computer. There is also an advantage in preventing an unsophisticated user from manipulating the bits in a microprogram. Alternatively the microprogram may be stored in medium that can be written into, called a *writable control store* (WCS). By reloading the WCS, entirely different macroinstructions can be implemented, using the same microinstruction set in a different microprogram. By such means emulation is achieved at minimal expense.

42. CPU Microprogramming. Figure 23-38 shows an ALU and Fig. 23-39 a control section with microprogram organization. The microinstructions embodied in these two units are shown in Table 23-2.

To simplify the program several provisions have been made:

1. Each microinstruction contains the control-store address of the next microinstruction. If omitted, the address in one microinstruction is the current address of the one written just below it. It may *not* be next in numeric sequence.

2. An asterisk at the beginning of a line in the program indicates a comment. This means that the entire line contains information about the program and does not translate into a microinstruction.

3. To simplify the drawing, a gate in a path is indicated by placing an X in the line and omitting from the drawing any control lines controlling these gates. It is also assumed that where two lines join, OR circuits are implied.

4. Rather than listing a numeric value for each field, a shorthand description for the desired action is invented. All actions not so described are assumed to be zero. For example, A to ADD implies that register A is gated to the adder, or T/C means raise T/C gate. These changes do not in any way modify the concept of microprogramming but make the result more readable.

43. Instruction FETCH. In the preceding paragraphs the execution of instructions (E cycles) is discussed. In these cases, operations are initiated by setting an appropriate latch for the function to be performed. The signals that set the latch are in turn generated by circuits that interpret the information of the operation-code part of an instruction cycle (I cycle).

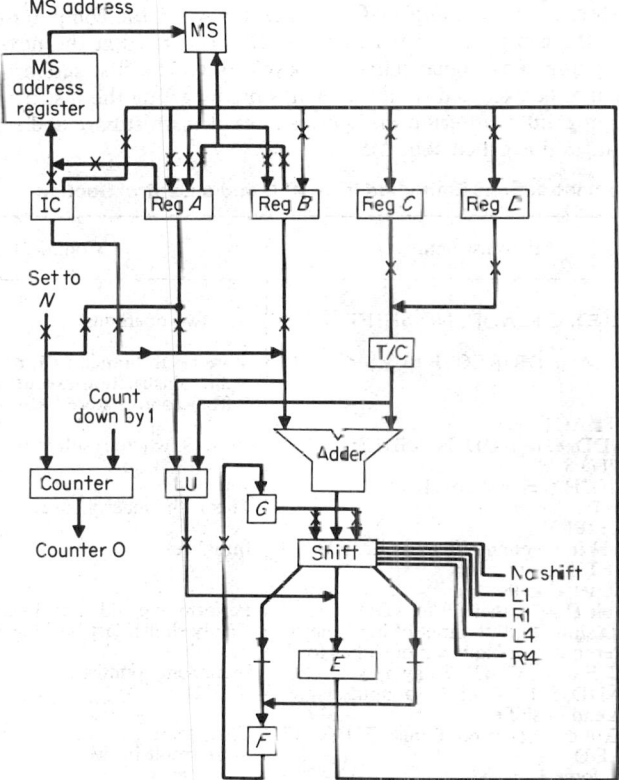

Fig. 23-38. Microprogrammed ALU.

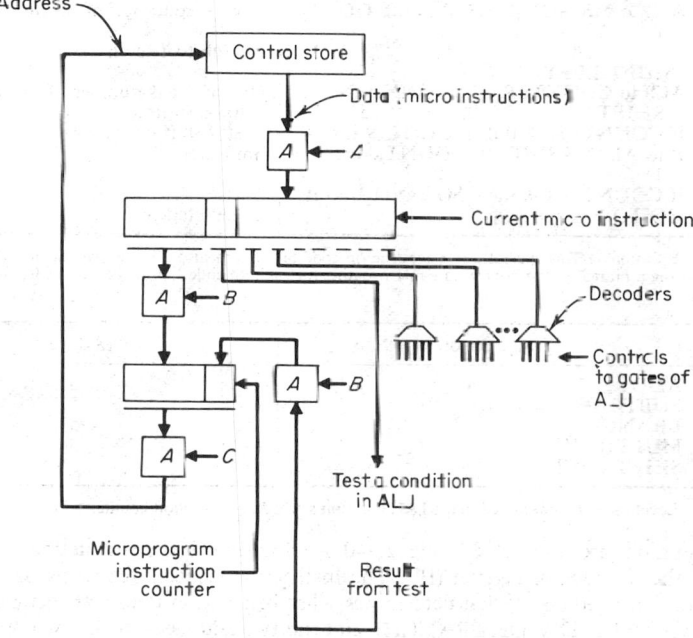

Fig. 23-39. Microprogrammed control unit.

Whenever an instruction has completed execution (or when the computer operator presses the start button on the console), a *start I-cycle* signal is generated at the next *A* time of the master-clock timing ring. This signal starts the I-cycle ring. The first action of this ring is to fetch the instruction to be executed from the main store by gating the instruction counter (IC) (sometimes called *program counter*) to the address lines of main storage and initiating a main-store cycle by pulsing a line called *start MS*.

TABLE 23-2   Microinstructions Embodied in an ALU and a Control Section

Current address	Microinstruction	Comment
	*ADD	
51	B to ADD, C to ADD, NO-SHIFT	Add two operands
8	E to B	
9	A(2) to MS-ADR-REG, B to MS, GO TO 1	Store result branch to 1, next microinstruction executed to be taken from control store location 1
	*SUBTRACT	
52	B to ADD, C to ADD, NO-SHIFT T/C, GO TO 8	Go to 9, where result is stored
	*BRANCH (unconditional)	
53	A(3) to IC, GO TO 1	This is the macroinstruction branch
	*MULTIPLY	
54	Set an N into counter, C to ADD, NO-SHIFT	Initialize
17	E to D, set C to 0s	
18	If last bit D = 1, then (B to ADD), C to ADD shift-R1, 0 to input of high end of shifter, output of low end of shifter to F	Perform one add shift if last bit D = 1, only shift if last bit D = 0.
19	E to C, F to G, COUNT down by 1	Increment counter
20	D to ADD, SHIFT-R1, G to input of high end of shifter	Shift D
21	E to D, if counter is not 0 then GO TO 17	Close loop
22	C to ADD	Store result in two MS locations
23	E to B, force 1 into C	
24	A(2) to ADD, C to ADD NO-SHIFT; A(2) to MS-ADR-REG B to MS	Store first half of result, increment result address
25	E to A(2)	
26	D to ADD, NO-SHIFT	
27	E to B	
28	A(2) to MS-ADR-REG, B to MS GO TO 1	Store second half of result
		Branch to 1-fetch
	*SHIFT LEFT	
55	A(2) to COUNTER, C to ADD, NO-SHIFT	Operand 1 is number of bits operand 2 is to be shifted
10	If COUNTER = 0 then GO TO 9 E to B	Test if shift count was 0
11	B to ADD, SHIFT L1, COUNT down by 1	Shift loop
12	If COUNT not 0, then GO TO 11 E to B	
13	GO TO 9	Completed shift

NOTE 1:   (Location 7) This special test places the op-code bit pattern into the low part of the address for the next instruction, causing a branch to the appropriate microroutine for each op code.   Branches are as follows:

Op code	Instruction	Address
1	ADD	51
2	SUBTRACT	52
3	BRANCH	53
4	MULTIPLY	54
5	SHIFT LEFT	55

NOTE 2:   (Location 1) It is assumed that START button sets microinstruction counter to 1.

These operations are illustrated in Fig. 23-40. At ring time 2, the instruction arrives back and is placed in the instruction register (IR). The instruction typically contains three main fields: the *operation code* (op code), that determines what instruction is to be executed (ADD, SUB-TRACT, MULTIPLY, DIVIDE, BRANCH), and the two addresses of the two operands partici-pating in the operation. For certain classes of instruction, the operands must be delivered to appropriate locations before the E cycle begins.

During ring time 3 and 4, the first operand is fetched from main store and stored in A. During ring time 5, this operand must be transferred from A to an alternate location, depending upon the nature of the instruction being executed. During ring time 5 and 6, the second operand is fetched and stored in A. Ring time 6 is also used to gate the op code to the *instruction decoder*, which is a combinatorial logic circuit accepting the *operation code*, or *instruction code*, from the P b in the op code. The decoder has 2^P input wires, one for each unique input combination.

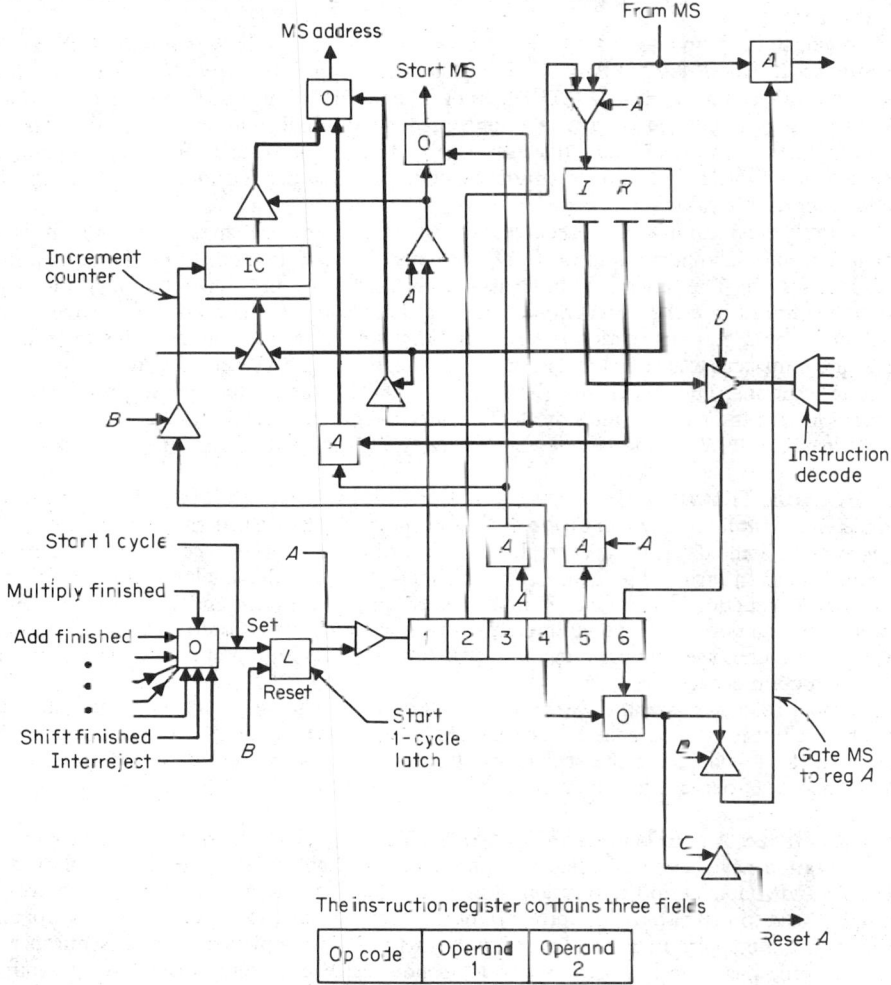

The instruction register contains three fields

Op code	Operand 1	Operand 2

Fig. 23-40. Implementation of equipment for an I cycle in a two-address machine. The instruction is first brought from main store and deposited in the instruction register. The operation proceeds by successively gating the information associated with the two address fields into register A. The first is moved out of A while the second is being sought from main store.

Thus for each bit combination entering the decoder, one output line is activated. These signals represent the start of an E cycle with each wire initiating the execution of one instruction by setting some latch, e.g., an add or multiply latch, or by initiating an appropriate microprogram.

Some op-code bit combinations may not be valid instruction codes. The outputs of the decoder are ORed together and fed into a circuit that ultimately interrupts the normal processing of the computer and indicates the invalid condition. At the beginning of the I cycle, the content of the instruction counter (IC) points to the instruction to be executed next. An *increment-counter* signal is generated during the I cycle in order to increment the counter, so that the address stored in IC points to the next sequential instruction in the program.

44. Instruction- and Execution-Cycle Control. In normal program operation, I and E cycles alternate. The I cycle brings forth an instruction from storage, sets up the ALU for execution, resets the instruction counter for the next I cycle, and initiates an E cycle. A number of conditions serve to interrupt this orderly flow, as follows:

1. When no more work is to be done, the end of a program is reached and the computer goes to a WAIT state. Such a state is reached, for example, by a specific instruction that terminates operations at the end of the I cycle in question, by a signal from the instruction counter when a predetermined limit is reached.

2. A STOP button may prevent the next setting of the *Start I cycle* latch. A START button resets this latch.

3. When starting up after a shutdown, e.g., in the morning, activity is usually initiated by depressing an INITIAL PROGRAM LOAD button on the operator's console (the name varies from system to system, e.g., IPL, LOAD, START). The button usually performs three functions: a reset of all rings, latches, and registers to some predefined initial condition; a read-in operation from an I/O device into main store (usually the first few locations) of a short program; and an initiation of an I cycle. Program execution generally starts at a fixed location so that the IC initially is set to this value.

4. In multiprogramming, i.e., concurrent operations upon more than one program, only one program at a time is in operation in the CPU, but transfers occur from one to another as required by I/O accesses, etc. The program to be transferred is handled by an *interrupt*. Under interrupt, an address is forced into the instruction counter so that upon completion of the E cycle of the current program, a new instruction is referenced that starts the interruption. This instruction initiates program steps that store the data of the old program, e.g., in special registers or in special main-store locations. The contents of the IC, part of the IR, the contents of any registers, and the reason for the interruption are stored. The collection of these fields together is called a *program status word* (PSW). It can be referenced to reinitiate action at a later time on the program interrupted.

45. Branch Instructions. Two kinds of instructions permit change in program sequence: *conditional* and *unconditional*. The purpose of such instructions is to permit the system to make some decision, so as to alter the flow of program, and to continue execution at some point not in the original sequence. In systems that use two address fields as described above, the program instruction to be branched to is usually contained in operand 2. The execution of the branch instruction thus involves merely moving operand 2 from the IR to the IC. In nonconditional branches the original program instruction provides for a branch whenever the particular instruction occurs.

Conditional branches take the extra step of determining if some condition to be tested has been satisfied. Either the op code or the operand 1 field normally defines the test and/or the condition to be tested for. If the specified test has been satisfied, the branch is executed as described above. Otherwise no action is taken, and the next normally sequenced instruction is executed.

46. Advanced Architectural Features. The basic structure of the digital computer and its operation have been described in the previous paragraphs. This structure has proved to be flexible and adaptable to the solution of many different applications. There is, however, a continuing need to increase the performance of the computer and to make it easier to program so that even more applications can be handled. This has been achieved through a number of different approaches. One has been to develop sophisticated operating systems and to couple them closely to the hardware design of the computer. The net effect is that a programmer viewing the computer does not distinguish between the hardware and the operating system but perceives them as an integrated whole.

A second solution has been the development of architectural features that permit overlapped processing operations within the computer. This is in contrast to earlier computers that were strictly sequential and resulted in reduced performance and low throughput because valuable computer resources could remain idle for relatively long periods of time. The newer architectural features include such concepts as data channels, storage-organization enhancements, and pipelining, described briefly in the following paragraphs. Another totally different approach to improved computer performance has been the development of computers with non-von Neumann architecture, described in Par. **23-52**.

47. Data Channels. An increase in computer performance can be achieved by overlapping input/output (I/O) operations and processor operations. *Channels* have been introduced to permit concurrent reading, writing, and computing. The channel is in effect a special processor which acts as a data and control buffer between the computer and its peripheral devices. Figure 23-41 shows the organization of the computer when channels are introduced. Each channel can accommodate one or more I/O device control units (see Pars. **23-82** to **23-84**). The channel is designed with a standard interface which is designed to permit a standard set of control status signals and data signals and sequences to be used to control I/O devices. This permits a number of different I/O devices to be attached to each channel by using I/O device control units that also meet the standard interface requirements. Each device control unit is usually designed to function with only one I/O device type, but one control unit may control several devices.

The channel functions independently of the CPU. It has its own program that controls its operations. The CPU controls channel activity by executing I/O instructions. These cause it to send control information to the channel and to initiate its operation. The channel then functions independently by being given a channel program to execute. This program contains the commands to be executed by the channel as well as the addresses of storage locations to be used in the transfer of data between main storage and the I/O devices. The channel in turn issues orders to the device control unit, which in turn controls the selected I/O device. When the I/O operation is completed, the channel interrupts the CPU by sending it a signal indicating that the channel is again free to perform further I/O operations. Several channels can be attached to the CPU and can operate concurrently.

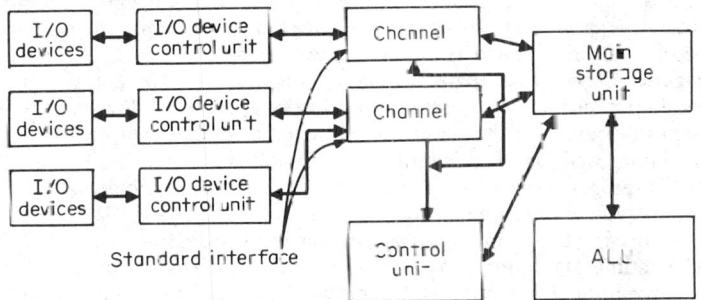

Fig. 23-41. Computer organization with channels, separate logical processors that permit simultaneous input, output, and processing.

48. Cache Storage. *Cache storage* was introduced to achieve a significant increase in the performance of the CPU at only a modest increase in cost. It is, in effect, a very high speed storage unit that is added to the computer but is designed to operate in a unique way with the main storage unit. It is transparent to the program at the instruction level and can thus be added to a computer design without changing the instruction set or requiring modification to existing programs. Cache storage was first introduced commercially on the IBM System/360 model 85 in 1968. For more details on its operation, See Pars. **23-62** and **23-76**.

49. Virtual Storage. Properly using and managing the memory resources available in the computer has been a continuing problem. The programmer never seems to have enough high-speed main storage and has been forced to use fairly elaborate procedures such as overlays to make programs fit into main storage and run efficiently. *Virtual-storage* systems were introduced to permit the programmer to think of memory as one uniform *single-level storage* unit but to provide a *dynamic address-translation unit* that automatically moves program blocks on pages between auxiliary storage and high-speed storage on demand. Virtual-storage operation is described in Pars. **23-75** and **23-76**.

50. Pipelining. A further improvement in computer performance was achieved through the use of *pipelining*. This technique consists of decomposing repetitive processes within the computer into subprocesses that can be executed concurrently and in an overlapped fashion.

Instruction execution in the control unit of the CPU lends itself to pipelining. As discussed earlier, instruction execution consists of several steps that can be executed relatively independently of each other. There is instruction fetch, decoding, fetching the operands, and then execution of the instruction. Separate units can be designed to perform each one of these steps. As each unit finishes its activity on an instruction, it passes it on to the next succeeding unit and begins to work on the next instruction in succession. Even though each instruction takes as long to execute overall as it does in a conventional design, the net effect of pipelining is to increase the overall performance of the computer. For example, under optimal conditions, once the pipeline is full, when one instruction finishes, the next instruction is only one unit behind it in the pipeline. In this four-unit example, the net effect would be to increase the speed of instruction execution by a factor of 4.

This approach can be carried to the point where 20 or 30 instructions are at various stages of execution at one time. This type of processing is called *pipeline processing*. No difficulties arise during uninterrupted processing. When an interrupt does occur, however, it is difficult to determine which instruction has caused the interrupt since the interrupt may arise in a subsystem sometime after the IC has initiated action. In the meantime, the IC may have started a number of subtasks elsewhere by stepping through subsequent cycles. Operands within subunits may not be saved in an arbitrary intermediate state, since information is in the process of being generated for return to the main program. Because of the requirement that no further I cycles be started, interrupt is signaled when the pipeline is empty. At the time the interrupt is signaled, the IC does not point at the instruction causing the interrupt but somewhat past it. This type of interrupt is called *imprecise*.

51. Firmware Engineering. The concept of microprogramming was described in Par. **23-42** as a substitute for hard-wired internal control logic in the CPU. Microprogramming, however, can also be used to improve computer system performance by migrating certain software primitives, e.g., read and write buffer operations, or selected segments of operating system and applications programs into microcode so that they can be executed directly in hardware. In a number of cases this has resulted in a significant improvement in overall system performance. The whole process of specifying, designing, implementing, and testing the microcode which is to be developed has become known as *firmware engineering*.

52. Advanced Organizations. In addition to the von Neumann computer and its enhancements, a number of other computer organizations have been developed to provide alternative approaches to satisfying the needs for improved computational performance, increased throughput, and improved system reliability and availability. In addition, certain unique architectures have been proposed to solve specific problems or classes of problems.

Flynn[62a] has categorized computer organizations according to the number of procedure (instruction) streams and the number of data streams processed. He describes four categories of computer organization: (1) a single-instruction-stream–single-data-stream (SISD) organization, which is the conventional computer; (2) a multiple-instruction-stream–multiple-data-stream (MIMD) organization, which includes multiprocessor or multicomputer systems; (3) a single-instruction-stream–multiple-data-stream (SIMD) organization; this uses a single control unit that executes a single instruction at a time, but the operation is applied across a number of processing units each of which acts in a synchronous, concurrent fashion on its own data set (parallel and associative processors fall into this category); and (4) a multiple-instruction-stream–single-data-stream (MISD) organization. (Pipeline processors fall into this category.)

53. Multiprocessor (Multicomputer) (MIMD) Organizations. One way to achieve an improvement in performance and to improve reliability at the same time is to use multiprocessors. The American National Standard Vocabulary for Information Processing defines a multiprocessor as "a computer employing two or more processing units under integrated control." Enslow[61a] amplifies this definition by pointing out that a multiprocessor contains two or more processors of approximately comparable capabilities. Furthermore, all processors share access to common storage, to I/O channels, control units, and devices. Finally, the entire system is controlled by one operating system that provides interaction between processors and their programs.

A number of different multiprocessor system organizations have been developed. Enslow[61a] classifies them into three major types, according to the method used in interconnecting their system elements: (1) time-shared, or common-bus, systems that use a common communication path to connect all functional units, (2) crossbar-switch systems that use a crossbar switching matrix to interconnect various system elements, and (3) multiport storage systems in which the switching and control logic is concentrated at the interfaces to the memory units.

54. Distributed Processing. One of the newest concepts in computer organizations is that of *distributed processing*. Unfortunately, the term distributed processing has not yet been precisely defined, and it has been loosely applied to any computer system which has any degree of decentralization. The consensus seems to be that a distributed processing system consists of a number of processing elements (not necessarily identical) which are interconnected but which operate with a distributed, i.e., decentralized, control of all resources. With the advent of the less expensive micro- and miniprocessors, distributed processing is receiving much attention since it offers the potential for organizing these processors so that they can handle problems that would otherwise require more expensive supercomputers. Through resource sharing and decentralized control, distributed processing also provides for reliability and extensibility since processors can be removed or added to the system without disrupting system operations.

At present there is no standard or generally agreed upon taxonomy for classifying distributed process systems. Anderson and Jensen[53a] have developed a taxonomy based upon the interconnection structure and describe configurations in terms of the processing elements, paths, and switching elements. They describe 10 system architecture types that arise from their taxonomy: (1) loop, (2) complete interconnection, (3) central memory, (4) global bus (5) star, (6) loop with central switch, (7) bus with central switch, (8) regular network, (9) irregular network, and (10) bus window. Distributed systems have been designed that fall into each of these categories. Hybrid forms use combinations of two or more of these architectural types.

There are a number of problems that need further research and development work before distributed process system can be more generally applied. One key area is that of implementing truly decentralized operating systems. Almost all those in current operation require that the entire global state of the system be available in a central location. Another key area is that of implementing distributed data bases; their successful implementation will require the solution to such data management problems as query processing when the data requested are not in one location, update processing, and data-base integrity and recovery in the event of failures.

55. Parallel Processors (SIMD). A number of large problems require high throughput rates on structured data, e.g., weather forecasting, nuclear-reactor calculations, pattern recognition, and ballistic-missile defense. Problems like these require high computation rates and may not be solved cost-effectively using a general-purpose (SISD) computer. Parallel processors, which are SIMD organization types, were designed to address problems of this nature. A parallel processor consists of a series of process elements (cells) each having data memories and operand registers. The cells are interconnected. A central control unit accesses a program, interprets each program step, and broadcasts the same instructions to all the processing elements simultaneously.

56. Stack Computers. In the CPUs discussed thus far, the instructions store all results, so that the next time an operand is used it has to be fetched. For example, a program to add A, B, and C and put result into E appears as

MOVE A to E   1 fetch, 1 store
ADD E to B store in E   2 fetch, 1 store
ADD E to C store in E   2 fetch, 1 store

In languages such as PL/I or FORTRAN (Par. **23-128**) this program might be written as the single statement

$$E = A + B + C$$

This equation describes a sequence of actions, as in the case of the program, but the specific sequence is not described. Since addition is commutative, a correct result is achieved by $E = ((A + B) + C)$, $E = (A + (B + C))$, or $E = ((A + C) + B)$, each step occurring in any order. The computer, however, uses a specific program in achieving a result so that the method of writing the equation must generate a specific sequence of actions.

A method of writing an equation that specifies the order of operation is called *Polish notation*. For the above example of addition, one possible Polish string would be

$$AB + C + E =$$

In this string, the system would find A and B and, as determined by the plus sign *following* the two operands, add them. The result is then combined with C under addition called for by the second plus sign. The E = symbols indicate that the result is to be stored in E. The plus sign appears *after* the A and B, and the specific string shown is called *postfixed*. An equivalent convention could place the operator first and would be called *prefixed*.

Any complex expression can be translated into a Polish string. For example in PL/I language, the statement

$$M = (A + B)*(C + D*E) - F;$$

means evaluate the right-hand side of the equation and store the result in the main-store location corresponding to variable M (asterisks indicate multiplication). The Polish string translation for this statement is

$$AB + DE*C +*F - M =$$

In translation from the types of expression permitted by higher-level languages (Pars. **23-126** to **23-132**) a machine can be programmed to analyze successively an arithmetic expression of the types shown above. In so doing, first, the outermost expressed or implied parentheses are aggregated and successively broken down until no more quantities remain. The first such quantities analyzed are generally the last computed, so that in the development of a Polish string from an algebraic expression, a first-in, last-out situation prevails.

57. Stacks. Evaluation of a Polish string in a machine is best performed using a *stack* (pushdown list). A stack has the property that it holds numbers in order. A PUSH command places a value on the stack; i.e., it stores a number and an operation at the top of the stack and, in the process, lowers all previous items by one position. Numbers are retrieved from the stack by issuing a POP command. The number returned by the stack on a POP command is the most recently PUSHED one. The following example illustrates the behavior of a stack. The value in parentheses is the value *placed* on the stack for PUSH and returned by the stack for POP (assume the stack is initially empty):

PUSH (A)	stack contains	A
PUSH (B)	stack contains	B A
POP (B)	stack contains	A
PUSH (C)	stack contains	C A
POP (C)	stack contains	A
POP (A)	stack contains	nothing

Such a stack lends itself very well to the evaluation of Polish strings. The rules for evaluation are:

1. Scan the string from left to right.
2. If a variable (or constant) is encountered, fetch it from main store and place its value on the stack.
3. If an operator is encountered, POP the operands and PUSH the result.
4. Stop at the end of the string. If executed correctly, the stack is in the same state at the end of execution as it was at the start.

The advantage of using a stack is that intermediate results never need storing and therefore no intermediate variables are needed. In sequences of instructions where there are no branches, the operations can be stored in a stack. A program becomes a series of such stacks put together between branches.

This approach is called *stack processing*. In stack processing, a program consists of many Polish strings that are executed one after another. In some cases the entire program may be considered to be one long string.

Stacks are implemented by using a series of parallel shift registers, one per bit of the character code. The input is placed into the leftmost set of register positions. PUSH moves an entry to the right, and POP moves it to the left. The length of the shift registers is finite and fixed. The stack, however, usually must appear to the user as though it were infinitely deep. The stack is thus implemented so that the most active locations are in the shift register, and if the shift register overflows, the number shifted out at the right on a PUSH is placed in main storage. There the order is maintained by hardware, microprogramming, or a normal system program.

58. New Trends in Computer Organization. Several new trends in computer architecture and organization may have a significant impact on future computer systems. The first of these are the *data-flow computers*, which are data-driven rather than control-driven. In the data-flow computer, an instruction is ready for execution when its operands have arrived; there is no concept of control flow, and there is no program counter. A data-flow program can feature concurrent processing since many instructions can be ready for execution at the same time. In another area *capability systems* are receiving increased attention because their inherent protection facilities make them ideal for implementing secure operating systems. A capability is a protected token (key) authorizing the use of the object named in the token. One approach to implementing capabilities is through a *tagged* architecture, where tag bits are added to each word in storage and to each register. This tag specifies whether the contents represent a capability or not.

59. Special-Purpose Processors. Certain classes of problems require unique processing capabilities not found in general-purpose computers. Special-purpose processors have been designed to solve these problems. In some cases they have been designed as stand-alone processors, but often they are designed to be attached to a general-purpose computer that acts as the host. One such class of special-purpose processors is the *associative processor*. It uses an associative store. Unlike the storage units described earlier, which require explicit addresses, an associative store retrieves data from memory locations based upon their content. The associative store does its searching in parallel over its entire storage in approximately the same time as required to access a single word in a conventional storage unit.

In digital signal processing, many repetitive mathematical operations must be performed, e.g., fast Fourier transforms, and require a large number of multiplications and summations. The *array processor* has been designed for these types of operations. It has a high-speed arithmetic processor and its own control unit and can operate on a number of operands in parallel. It attaches to a host CPU, from which it receives its initiation commands and data and to which it returns the finished results of its computation.

The *hybrid processor* uses a host digital CPU to which is attached an analog computer. These systems operate in a digital-analog mode and provide the advantages of both methods of computation.

HARDWARE DESIGN

60. Computer Hardware Description Languages (CHDL). There are a number of basic levels at which the implementation of a computer system must be described. The problem has been to describe the design of the system at each level so that it can be communicated between computer engineers, software engineers, and those who have to implement the design in hardware. Graphical representations (drawings and schematics) have long been used for this purpose, but in recent years hardware description languages have a so been developed. These languages have additional advantages in that they permit the analysis of proposed system designs and can serve as the input to a simulation program so that further testing of the design can be done. At the moment there are a number of CHDLs in existence, but not much standardization has occurred. Among the leading languages* that have emerged are (1) Computer Design Language (CDL), which encompasses both a design methodology and a language for expressing the design; (2) Digital System Design Language (DDL), which is a block-oriented language whose blocks are intended to correspond to the subsystems, assemblies, parts, etc, of the hardware systems; (3) a Processor-Memory-Switch language (PMS) and an Instruction Set Processor Language (ISP), which were developed to describe the physical structure of a computer and the programming level of a computer precisely; and (4) a Hardware Programming Language (AHPL), which is a hardware description language based on the notational conventions of the programming language APL.

61. Design and Packaging of Computer Systems. The basic techniques, construction, and processing of integrated circuits and microelectronic components are treated in Sec. **8**. Computers, perhaps more than other devices, depend upon microcircuitry. Even a small computer may have as many as 10,000 such circuits, whereas a large system may have as many as a million.

Nanosecond circuit speeds are common in high-performance computer systems. Since light travels approximately ⅓ m in 1 ns, system configurations must be kept small to take advantage of the speed potential of available circuits. Thus the emphasis is on the use of microcircuit fabrication and packaging techniques. A computer system contains from 10^5 to 10^7 interconnections between circuits, depending on its size and power. The layout of the system must minimize the length and complexity of these interconnections and must be realized, without error, from detailed manufacturing instructions. To permit these requirements to be met a *basic* circuit package must be available. The upper limit to the size of such a basic unit, e.g., an integrated circuit, is set by the number of crystal defects per unit area of silicon. If these defects are distributed at random, the selection of too large a chip size results in some noperative circuits on a majority of chips. There is thus an economic balance between the number of circuits that can be fabricated in an integrated circuit and the yield of the manufacturing process.

Another limit on the size of the basic package is set by the number of interconnections between the integrated circuits. An empirical relation ((*Rent's rule*) exists between the number *m* of circuits in a package and the number *n* of external connections required. The relation is $n = km^a$, where *a* is a constant in the range of 1.4 to 1.6 and *k* is a constant depending on the fan-in and fan-out characteristics of the particular circuit family used.

*For descriptions of these languages see the December 1974, issue of *Computer*, Vol. 7.

COMPUTER STORAGE

62. Basic Concepts. The main memory attached to a processor represents the most crucial resource of the computing system. In most instances, the storage system determines the bulk of any general-purpose computer architecture. Once the word size and instruction set have been specified, the rest of computer architecture and design deals mainly with optimization of the attached memory and storage hierarchy. Early computers were designed by first choosing the main memory hardware. This not only specified much of the remaining architecture, e.g., serial or parallel machine, but also dictated the *processor cycle time*, which was chosen equal to the *memory cycle time*. As computer technology evolved, the logic circuits and hence processor cycle time* improved dramatically. This improved speed demanded more main-memory capacity to keep the processor busy, but the need for increasingly larger capacity at higher speed placed a difficult and usually impossible demand on memory technology. In the early 1960s, a gap appeared between the main-memory and processor cycle times, and the gap grew with time. Fundamentally, it is desirable that main-memory cycle time be approximately equal to the processor cycle time, so this gap could not continually widen without serious consequences. In the late 1960s, this gap became intolerable and was bridged by the "cache" concept, introduced by IBM in the System/360 Model 85. See Ref. 85.

The cache concept proved to be so useful and important that by the late 1970s it became quite common in small, medium, and large machine architectures. The *cache* is a relatively small high-speed random-access memory that is paged out of main memory and holds the most recently and frequently used instructions and data. The same fundamental concepts that provide the basis for cache design apply equally well to "virtual memory" systems; only the methods of implementation are different.

In terms of implementation method, there are five types of storage systems used in computers.

1. *Random-access memory* is one for which any location (word, bit, byte, record) of relatively small size has a unique, physically wired-in addressing mechanism and is retrieved in one memory-cycle time interval. The time to retrieve from any given location is made to be the same for all locations.

2. *Direct-access storage* is a system for which any location (word, record, and so on) is not physically wired in and addressing is accomplished by a combination of direct access to reach a general vicinity plus sequential searching, counting, or waiting to find the final location. *Access time* depends on the physical location of the record at any given time; thus access time can vary considerably, from record to record and to a given record when accessed at a different time. Since addressing is not wired in, the storage medium must contain a certain amount of information to assist in the location of the desired data. This is referred to as *stored addressing information*.

3. *Sequential-access storage* designates a system for which the stored words or records do not have a unique address and are stored and retrieved entirely sequentially. Stored addressing information in the form of simple interrecord gaps is used to separate records and assist in retrieval. Access time varies with the record being accessed, as with direct access, but sequential accessing may require a search of every record in the storage medium before the correct one is located.

4. *Associative (content-addressable) memory* is a random-access type of memory that in addition to having a conventional wired-in addressing mechanism also has wired-in logic that makes possible a comparison of desired bit locations for a specified match for all words simultaneously during one memory-cycle time. Thus, the specific address of a desired word need not be known since a portion of its contents can be used to access the word. All words that match the specified bit locations are flagged and can then be addressed on subsequent memory cycles.

5. *Read-only memory (ROM)* is a memory that has permanently stored information programmed during the manufacturing process and can only be read and never destroyed. There are several variations of ROM. *Postable* or *programmable* ROM (PROM) is one for which the stored information need not be written in during the manufacturing process but can be written at any time, even while the system is in use; i.e., it can be posted at any time. However, once written, the medium cannot be erased and rewritten. Another variation is a *fast-read, slow-write* memory for which writing is an order of magnitude slower than reading. In one such case, the writing is done much as in random-access memory but very slowly to permit use of low-cost devices. Another version of slow-write memory is one with a changeable or replaceable storage medium, e.g., magnets on a card, wires, or metal plates (capacitors) punched with holes. These

---

*Processor cycle time is roughly 10 logic-gate delays with appropriate circuit and package loads.

are read-only memories that are programmable at any time but require minutes to hours to change.

The reason for the large variety of storage types is cost, which is related to the access time. A short access time can be obtained only at a high cost. Conversely, inexpensive memories have slower access times. Approximate rules of thumb for cost and access time comparisons are as follows (where T = access time):

Cost or price:

$$\text{Cache cost} = 10 \times \text{main cost} = 10^4 \times \text{disc cost} \begin{cases} = 10^7 \times \text{tape cost} & \text{off line} \\ \approx 10^5 \times \text{tape cost} & \text{on line} \end{cases} \quad (23\text{-}7)$$

where all costs are in cents per byte

Access time:

$$T_c = 10^{-1}T_m = 10^{-6}T_d = 10^{-9}T_t \quad (23\text{-}8)$$

where T_c = cache time, T_m = main time, T_d = disc time, and T_t = tape time.

The large gaps between main memory and discs as well as disc and tapes produce large gaps in access time; i.e., access time is sacrificed to achieve economy.

63. Storage-System Parameters. In any storage system the most important parameters are the capacity of a given module, the access time to any piece of stored information, the data rate at which the stored information can be read out (once found), the cycle time (how frequently the system can be accessed for new information), and the cost to implement all these functions.

Capacity is simply the maximum number of bits (b), bytes (B), or words that can be assembled in one basic self-contained operating module. For example, 64K B of integrated-circuit memory and 29M B on a disc pack are typical modules.*

Access time can vary depending on the type of storage. For random-access memory the *access time* is the time from the instant a request appears in an address register until the desired information appears in an output register, where it can subsequently be further processed. For non-random-access storage, the access time is the time from the instant an instruction is decoded asking for information until the desired information is found but not read. Thus, access time is a different quantity for random- and nonrandom-access storage. In fact, it is the access time that distinguishes the two, as is evident by the definitions above. Access time is made constant on random-access memory whereas on nonrandom storage access time may vary substantially, depending on the location of information being sought and the current position of the storage system relative to that information.

Data rate is the rate (usually bits per second, bytes per second, or words per second) at which data can be read out of a storage device. *Data transfer time* for reading or writing equals the product of the data rate and the quantity of the information being transferred. Data rate is usually associated with nonrandom-access storage where large pieces of information are stored and read serially. Since an entire word is then read out of random-access memory in parallel, data rate has no significance for such memories.

Cycle time is the rate at which a memory can be accessed, i.e., the number of accesses per unit time, and is applicable primarily to random-access storage. It does not necessarily equal the access time for various reasons. If a random-access memory works in the destructive readout mode, the information must be regenerated before another access can be made. This causes a wide disparity between access and cycle time. Even if nondestructive readout is used, there are often transients that must be allowed to die out. Drivers or sense amplifiers must recover from large transients that drive them into saturation, and ringing, which is caused by multiple pulse reflections on the array lines, must be allowed to die out. Thus, cycle time is often substantially larger than access time. Cycle time has little meaning with nonrandom serial storage. Cycle time is essentially the access time plus data-transfer time, both of which can vary widely on a given storage system as a function of time and of the data being accessed.

64. Fundamental System Requirements for Storage and Retrieval. In order to be able to store and subsequently find and retrieve information, a storage system must have the following four basic requirements:

Medium for storing energy

---

*Although K and M are read "kilo-" and "mega-," they are not the SI prefixes k- and M- (meaning 10^3 and 10^6) but the binary approximation of them. That is, K stands for 1,024, and M = K². To emphasize the distinction the K or M is attached to the number rather than the unit (b or B).

Energy source for writing the information, i.e., write transducers on word and bit lines

Energy sources and sensors to read, i.e., read and sense transducers

Information addressing capability, i.e., address-selection mechanism for reading and writing

The fourth requirement implicitly includes some coincidence mechanisms within the system to bring the necessary energy to the proper position on the medium for writing and a coincidence mechanism for associating the sensed information with the proper location during reading. In random-access memory, it is provided by the coincidence of electric pulses within the storage cell whereas in nonrandom-access storage, it is commonly provided by the coincidence of an electric signal with mechanical position. In many cases, the write energy source serves as the read energy source as well, thus leaving only sense transducers for the third requirement. Nevertheless, a read energy source is still a basic requirement.

The differences between storage systems lie only in how these four requirements are implemented and, more specifically, in the number of transducers required to achieve these necessary functions. Here a *transducer* denotes any device (such as magnetic head, laser, transistor circuits)

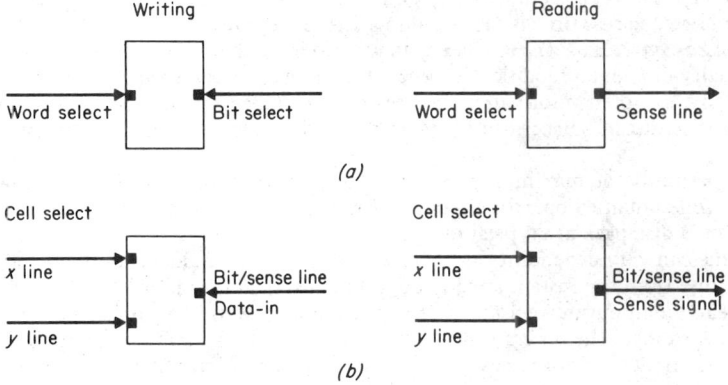

Fig. 23-42. Reading and writing operations for random-access memory cells having two and three functional terminals: (a) cell with two functional terminals, and (b) cell with three functional terminals.

that generates the necessary energies for reading and writing, senses stored energy and generates a sense signal, or provides the decoding for address selection.

Since random-access memory is so crucial to processor architecture, we discuss this type first and continue in the order defined above. Virtual memory is considered in Par. **23-76** and a brief discussion of charge-coupled devices (CCDs) is given in Par. **23-78** and magnetic bubbles in Par. **23-79**.

The organization of random-access memory systems is intimately dependent on the number of functional terminals inherent in the memory cells. The cell and overall array organization are so interwoven that a discussion of one is difficult without the other. We discuss first the fundamental building block, the memory cell, and subsequently build the array structure from this.

65. Random-Access Memory Cells. To be useful in random-access memory, the cell must have at least two independent functional terminals consisting of a word line and a common bit/sense line, as shown in Fig. 23-42a. For writing into the cell, the coincidence of a pulse on the word-select line with the desired data on the bit-select line places the cell into one of two stable binary states. For reading, only a pulse on the word-select line is applied, and a sense signal indicating the binary state of the cell is obtained on the sense line. It should be apparent that for reading no coincidence selection is possible, which limits the use of the cell to either a 2- or 2½-dimensional (2D or 2½D) organization described later. A more versatile but more complex cell is one having three functional terminals, as in Fig. 23-42b. A coincidence of pulses on both the x and y lines is necessary to select the cell for both reading and writing. The appearance of data on the bit/sense line puts the cell in the proper binary state. If no data are applied, a sense signal indicating the binary state of the cell is obtained. The use of coincidence for reading as well as writing allows this cell to be used in more complex array organizations such as the 3D and 2½D (described later) but only at the expense of a more complex and costly cell.

Magnetic ferrite cores, which played a key role in the early history of computers, were first introduced as three-terminal cells. In fact, the first ferrite core cells used four wires through each core (separate bit and sense lines) to achieve a three-functional terminal cell as in Fig. 23-42b. Refinements gradually allowed the bit and sense line to be physically combined to give a three-wire cell and eventually even a two-wire two-terminal cell.

The introduction of *integrated-circuit cells* in the early 1970s for random-access memory led to the disuse of ferrite cores. There are two general types of integrated-circuit cells, static and dynamic. *Static cells* remain in the state they are set until they are reset or the power is removed and are read nondestructively; i.e., the information is not lost when read. A *dynamic cell* gradually loses its stored information and must be refreshed periodically. In addition, the cell state is destroyed when read, and the data must be rewritten after each read cycle. However, dynamic cells can be made much smaller than static cells which gives a greater density of bits per semiconductor chip. The resulting lower cost more than compensates for the additional complexity.

66. Static Cells. All static cells use the basic principles of the Eccles-Jordan[82] flip-flop circuit to store binary information in a cross-coupled transistor circuit like that shown in Fig. 23-43. Either *junction (bipolar)* or *field-effect transistors (FET)* can be used in the same configu-

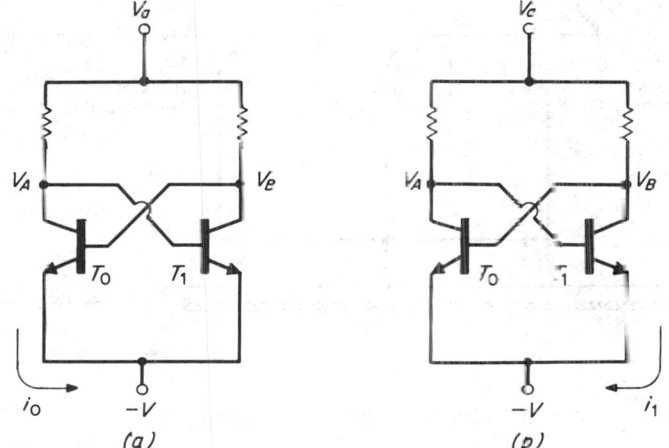

Fig. 23-43. Transistor flip-flop circuits: (a) stored 0, T_0 conducts, $V_A = 0$, $V_c = 1$ V; (b) stored 1, T_1 conducts, $V_A = 1$ V, $V_B = 0$.

ration; only the voltages, currents, and load resistors will be different. To store a 0, transistor T_0 is turned on and T_1 turned off. A stored 1 is just the opposite condition, T_0 off and T_1 on. To achieve these states, it is necessary to control the node voltages at A and B. For instance, if node A voltage V_A is made sufficiently low, the base of T_1 will be low enough to turn T_1 off, causing node B to rise in voltage. This turns the base and hence the collector of T_0 on and holds node A at a low voltage so that a 0 is stored in the flip-flop. If the voltage at node B is made sufficiently low, the opposite state occurs, giving a stored 1. For reading, it is only necessary to sample the voltages at nodes A or B to see which is high or low. This gives a nondestructive read cell since the state is not changed, only sampled. Hence, nodes A and B are the access ports to the cell for both reading and writing.

The basic difference between all *static cells* is in how nodes A and B are accessed. Note that although the flip-flop of Fig. 23-43 has two access nodes A and B, it is functionally only a one-terminal cell since nodes A and B are not independent but operate in a cooperative mode. To make a cell suitable for a random-access array, at least another functional terminal must be added. This can be achieved by the addition of another FET to each node, A and B, as shown in Fig. 23-44. Although the two transistors provide a total of four physical connections to the cell, that is, T_2 provides one gate g_2 and one drain d_2 for node A and T_3 provides g_3 and d_3 for node B, the circuit operates in a symmetrical, balanced mode. Since these four terminals are not independent, only two functional terminals are present. The operation of the cell can be understood by tracing through the pulsing sequence of Fig. 23-44. The peripheral circuits used to obtain these pulse sequences are not shown. An equivalent two-terminal static cell can be achieved by using junction transistors with some changes, but the operating principles remain basically the same.

A two-terminal static cell that was very popular in early integrated-circuit memories used multiemitter transistors. The cell required high power, however, a condition that greatly limits chip density.

Higher density and lower power can be obtained with the Schottky diode flip-flop cell of Fig. 23-45, a very popular cell for high-speed, low-power junction transistor memories. Resistors R_c are part of the internal collector contact resistance of the devices and are much smaller than R_L.

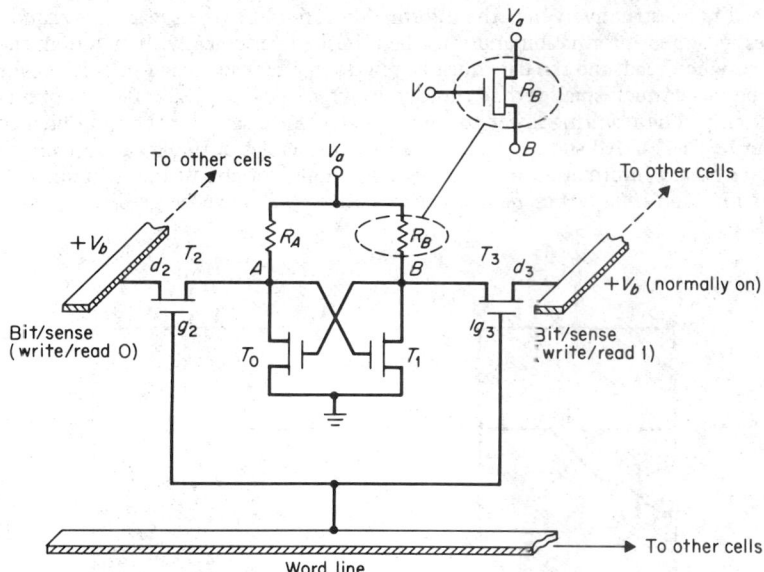

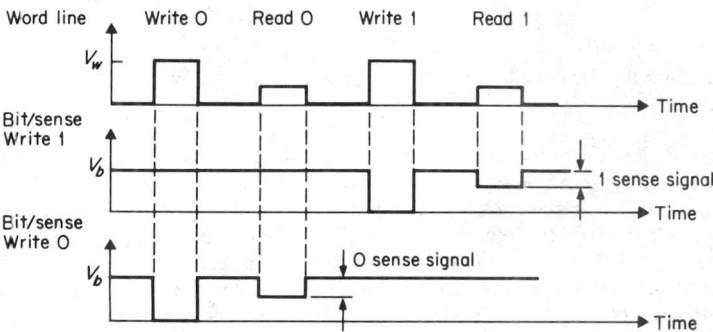

Fig. 23-44. MOSFET two-terminal storage cell.

The pulses on the word and bit/sense lines are used to forward- or back-bias the diodes D_0 and D_1, thus controlling or sensing the voltages and nodes A and B, as can be seen by tracing through the pulse sequence shown.

67. Dynamic Cells. Since the static cells described above operate in a balanced, differential mode, it would seem reasonable that a nondifferential mode would be able to achieve at least a twofold reduction in component count per cell. This is indeed the case, and by sacrificing other properties of static cells, such as their inherent nondestructive read capability, a very significant reduction in cell size is possible. The most widely used such cell is the one-FET dynamic cell shown in a nonintegrated form in Fig. 23-46. The essential principle is simply the storing of charge on capacitor C, for a 1 and no charge for a 0. A single capacitor by itself is sufficient to accomplish this, but an array of such devices requires a means of selecting the desired capacitors for reading and writing and of isolating nonselected capacitors from the array lines. If isolation is not provided, the charge stored on half-selected capacitors may inadvertently be removed, making the scheme inoperable. The isolation and selection means is provided by the simple one-

FET device in series with the capacitor, as shown. For operation in an array, the terminal c can be either at ground or $+V_c$, depending on the technology and details of the cell. Assume that c is at $+V_c$, as indicated. To write either state, the word line is pulsed high to turn the FET on. If the bit/sense line is at ground, V_c will charge C_s to a stored 1 state. However, if the bit/sense line is at $+V_c$, any charge on C_s will be removed or no charge is stored if none was there originally, giving a stored 0. For reading, the word line is pulsed high to turn the FET on with the

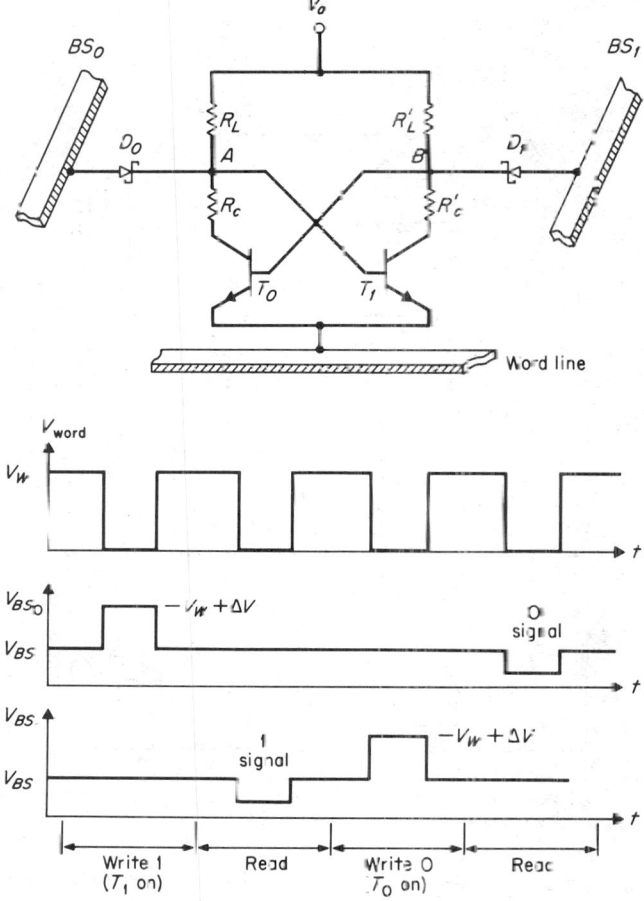

Fig. 23-45. Schottky diode storage cell.

sense line at a normally high voltage. If there was charge on the capacitor, it will discharge through the bit/sense line to give a signal as shown. If there was no stored charge, no signal would be obtained.

Note that the reading is destructive since a stored 1 is discharged to a 0 and requires regeneration after each read operation in an array. Note also that the FET must carry current in both directions; i.e., the current charging C_s during writing is in the opposite direction of the current during reading. This cell has one further disadvantage: in an integrated structure, the charge on C_s will unavoidably leak off in a time typically measured in milliseconds. Hence the cell requires periodic refreshing, which necessitates additional peripheral circuits and complications. This feature gives rise to the term *dynamic cell*. Despite these disadvantages, this technique allows a very substantial improvement in cell density and cost, at very adequate cycle times and has become very popular. Densities of 64K b per chip are used commercially, and 128K b or higher are possible.

68. Random-Access Memory Organization. There are four types of memory organization: one type of 3D organization; two types of 2½D; and one type of 2D organization. The two types of 2½D are distinguished by the fact that one uses coincident selection at the lowest, e.g., cell, level for both reading and writing in a manner resembling a 3D organization. The second type of 2½D organization uses coincident selection at the lowest (cell) level only for writing. An important and fundamental concept that should be kept in mind throughout this discussion is that these methods of organization apply at all levels of organization — cells, chips, cards, boards, memory modules, memory systems, etc. In fact a different organization can be used simultaneously at different levels of the same memory system. For instance, it is quite

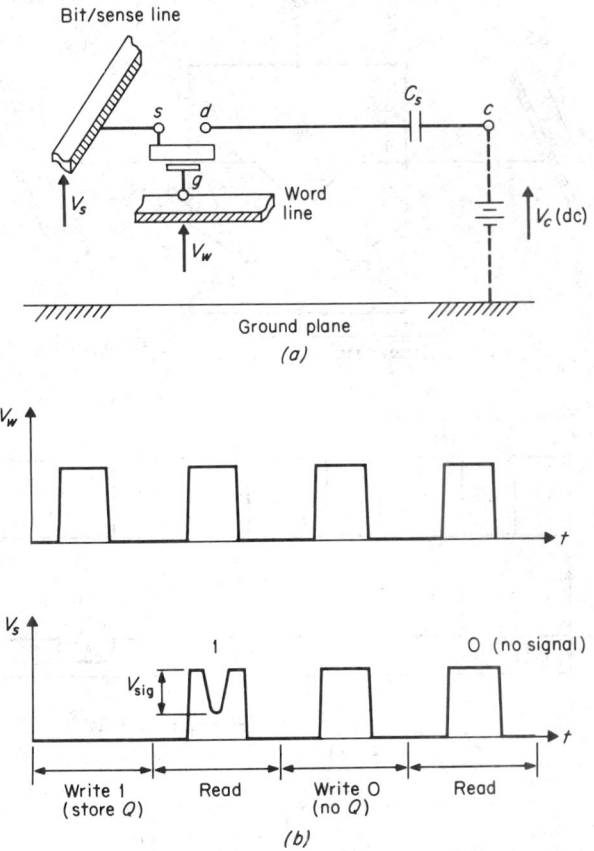

Fig. 23-46. One-FET-device dynamic storage cell: (a) general equivalent circuit, (b) pulsing sequence.

common for integrated-circuit memories to use a 2½D organization for selection of cells on a chip and a 2D organization of the chips that constitute the memory.

In any memory system, if E units are to be selected, an address length of N bits is required which satisfies

$$2^N = E$$

In most cases of interest E is equal to W, the number of logical words in the system.

69. 3D Organization. The unit being selected, i.e., cell, chip, module, must have at least three functional terminals; it can be connected in several different physical arrangements depending on the electrical characteristics of the unit being selected and the peripheral circuits. For an array of W logical words by b bits per word, a simple 3D scheme using integrated-circuit chips is shown in Fig. 23-47. Here it is assumed that the three-terminal cells have common bit/sense line, and on-chip word line decoders are employed in a square array of $W/2$ x-select word lines by $W/2$ y-select word lines. With the arrangement shown, only 1 b on each chip can be selected by a coincidence of an x and y signal, so a 1 b/chip organization is realized. This obviously requires b such chips to provide the full logical word. This is the pure form of 3D selection and was used extensively in early magnetic-core storage technology but with some modifications.*

A 3D on-chip selection scheme was also used in early integrated-circuit memories but is seldom used today because the added cell complexity means lower density and higher cost.

70. 2½D Organization. If we continue to use a three-terminal cell, we can rearrange the 3D array of Fig. 23-47 and obtain one form of a 2½D organization. For instance, we could put pairs of adjacent y-word lines in series so there would be only $W/4$ rather than $W/2$ independent lines in the y direction for selection. But now this would select 2 b for each x-selected word line. Likewise, there would have to be two sets of independent bit/sense lines, as illustrated in Fig. 23-48. Note that the same size memory requires the same number of chips as previously but

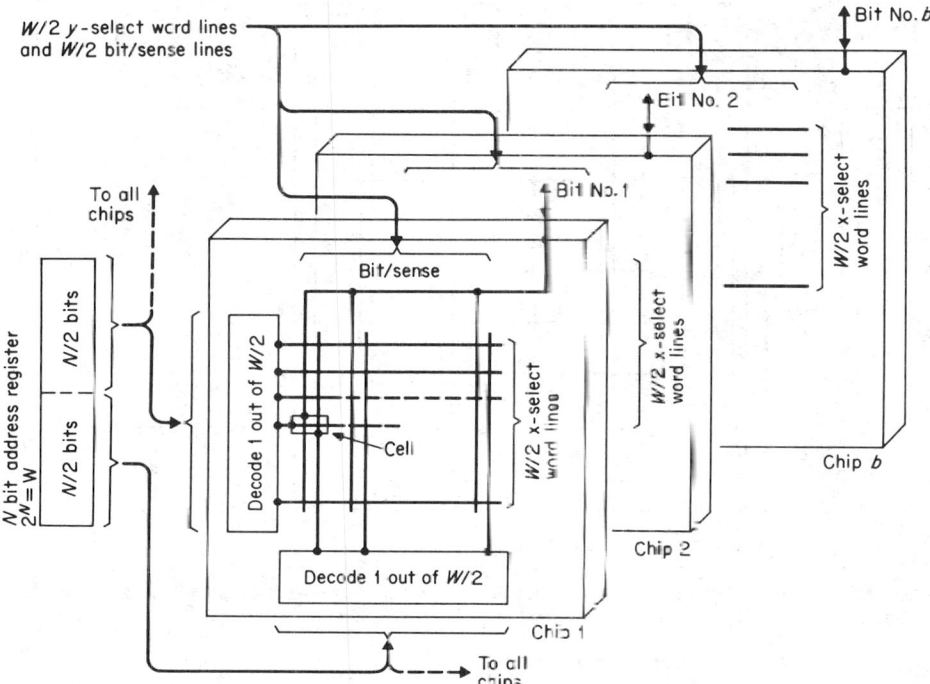

Fig. 23-47. Simple 3D organized integrated-circuit array using 1 b/chip and on-chip decoders.

each chip gives twice as many bits as before, or a total of $2b$ bits. But only b bits are needed for each logical word. Note also that since there are only one-half as many y-select word lines as previously, one less address bit, or $N/2 - 1$ is used for the y decoding on each chip. The excess address bit is now used to select one-half of all the b chips, i.e., select either the first set of $b/2$ or the second set of $b/2$ chips. This is done by the ENABLE or chip-select signal.

This selection of the proper group or segment of bit/sense lines must be done during both reading and writing. This bit/sense line selection could also be done in other ways. For instance, instead of selecting a group of chips, we could allow each chip to provide 1 b/chip by selecting one of the two bit/sense lines, as shown in Fig. 23-49. Hence we see the number of address bits remains the same but they are used differently. In Fig. 23-47 all chips are accessed on all memory cycles so no chip selection is necessary and only 3D selection is used. In Fig. 23-48 the bits on the chip are 2½D-selected, but the chips themselves are also selected by a noncoincident, direct ENABLE (which is fundamentally 2D selection). Hence the bits are 2½D-selected whereas the chips are 2D-selected, giving a mixed organization. Mixed organizations are common in integrated-circuit memories, often mixing 2½D on-chip selection with 2D for chip selection. Another form of 2½D selection using two-terminal rather than three-terminal cells is possible.

---

*There were no on-module decoders, and bit lines were arranged differently for noise suppression. Also, corresponding x lines between modules were often connected in series, as were the y lines.

71. 2½D Organization Using Two-Terminal Cells. In order to achieve higher density and lower cost, two-terminal cells are desirable. They can be used in a modified form of the above 2½D organization or 2D organization discussed later. Since a two-terminal cell cannot be read in a coincident selection scheme, we cannot use both x- and y-select-word lines; i.e., one of these must be eliminated. Suppose we eliminate the y word lines in Fig. 23-48 or 23-49. For a 1

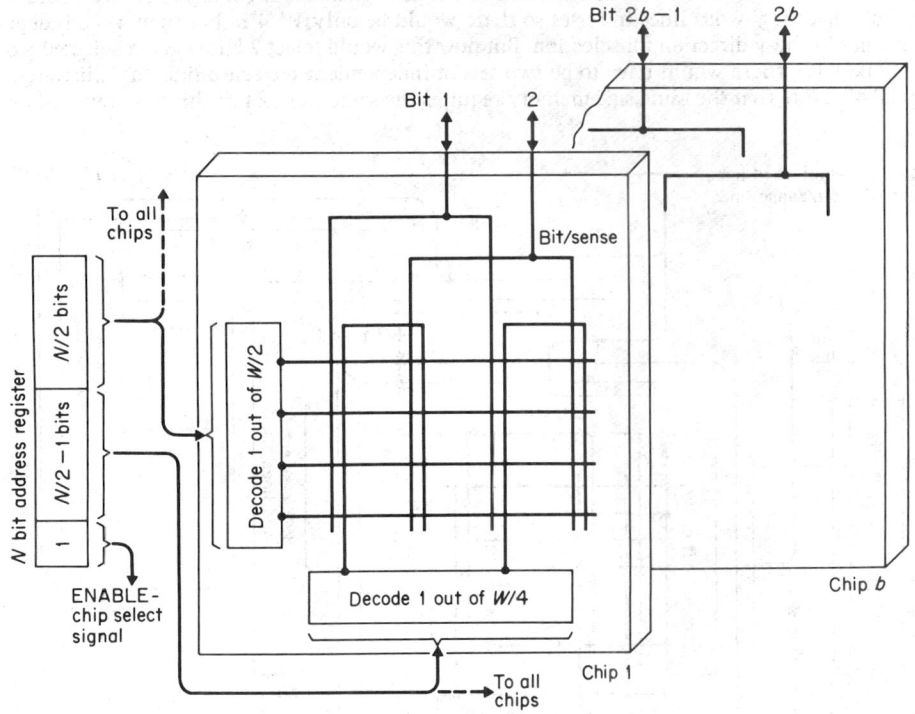

Fig. 23-48. Variation of the simple 3D organization using 2 b/chip with 2½D on-chip selection and 2D selection.

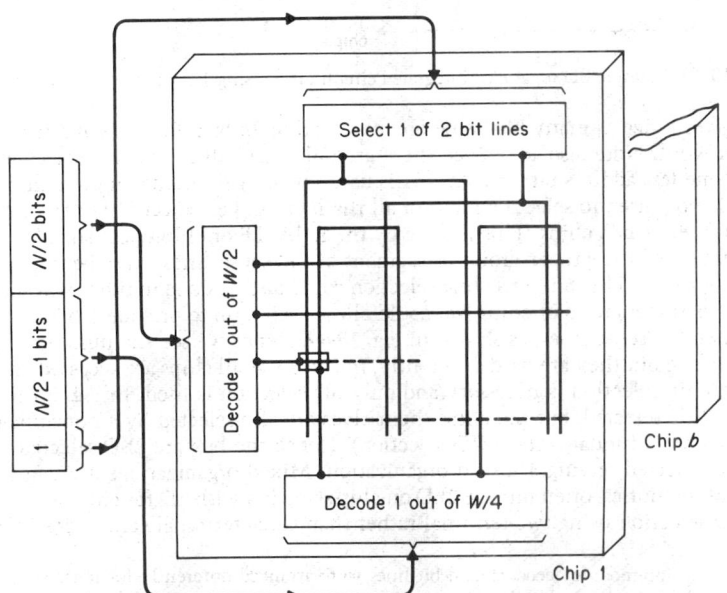

Fig. 23-49. Example of 2½D organization with 1 b/chip.

b/chip organization, the scheme of Fig. 23-50 is a typical example. For both reading and writing, the n_1 and n_2 address bits must be decoded on each of six chips to select one of the $W = 2^{n_2}$ word lines and one of $2^{n_1} = b$ bit/sense lines. During writing, all bits along the selected word line are energized, and the selected bit line on each chip has the proper data applied to store the correct binary state. All other bit lines are held in a quiescent state so that all other cells along the energized word line receive a so-called *half-select disturb pulse*. Cells must be capable of withstanding such disturbances. During reading of the memory of Fig. 23-50 one word line is energized and reads all bits along its length. This produces sense signals on all bit/sense lines, but the bit/sense line decoder selects only one of these, at the corresponding position of each chip, giving a total of b bits, as required.

Note that the total number of words possible in such a scheme is equal to the number of bits per chip. If we have available a technology to produce chips of 128K or even 256K b, a memory of 128K or 256K words is possible. The number of bits per word, typically 16 to 72, is, of course,

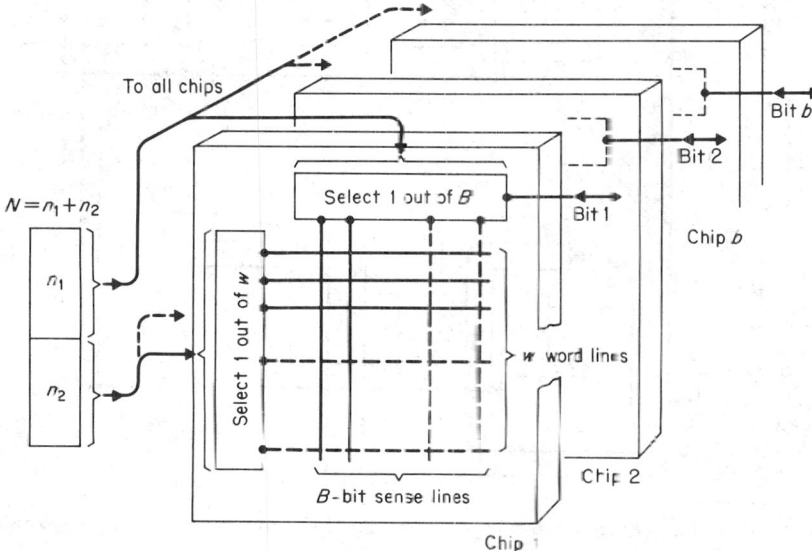

Fig. 23-50. Example of 2½D organization using two-terminal cells in a 1 b/chip organization.

determined by the number of chips stacked in parallel. It is often desirable to produce a memory with a much larger number of words and corresponding larger address. The organization of Fig. 23-50 can still be used on the chip but with provisions to allow for more chips and additional decoding of the additional address bits. One method of doing this, shown in Fig. 23-51, consists essentially of groups of b chips, each identical to Fig. 23-50 but with an additional ENABLE or chip-select capability. The address bits n_1 and n_2 are sent to all chips in all groups and attempt to read or write all chips of the array. The additional address bits are decoded separately and are used to select one of the 2^{n_3} groups of chips. The bit/sense lines from corresponding chips of each group must be tied in parallel either directly through a dot or, as shown, or by an OR gate. Obviously, any size memory can be made with the same chip organization. The organization in Fig. 23-51 represents a 2½D selection on chip and 2D or direct selection of the groups of chips. Many variations of this scheme are possible, such as a 3D selection of the groups of chips. Since this requires an additional coincident selection circuit on each chip to replace the off-chip decoder plus more address lines to each chip, it is undesirable and seldom used.

The selection can be reorganized on the chip rather than off the chip. Instead of bringing out or in only one data bit per chip, consider placing all b bits of a word on a chip. Since the number of bits per physical word line on a chip is often larger than the number of logical bits per word, it is necessary to use several segments of b bit/sense lines per segment.

The selection of 1 out of 2^{n_2} words on each chip is the same as previously discussed. Selection of the bit/sense lines now requires only the selection of one out of s segments with each segment containing b bits. For the same size memory as before a smaller value of the n_1 address bits is required. Since the total number N must remain the same, obviously the additional bits became a part of n_3 and are used for additional chip-select decoding. Thus only one chip is enabled on any one read or write access.

This type of organization was used in some older core memories but is seldom used today. The 1 b/chip type of organization of Fig. 23-51 is commonly used in integrated-circuit memories for reliability reasons, which can be understood as follows. Error-correcting codes that correct all single-bit errors and detect all double-bit errors occurring in a word are easily implemented. The memory organization must be one that uses the error-correcting code to protect against the most serious failure condition. While failure modes can and do vary considerably among technologies

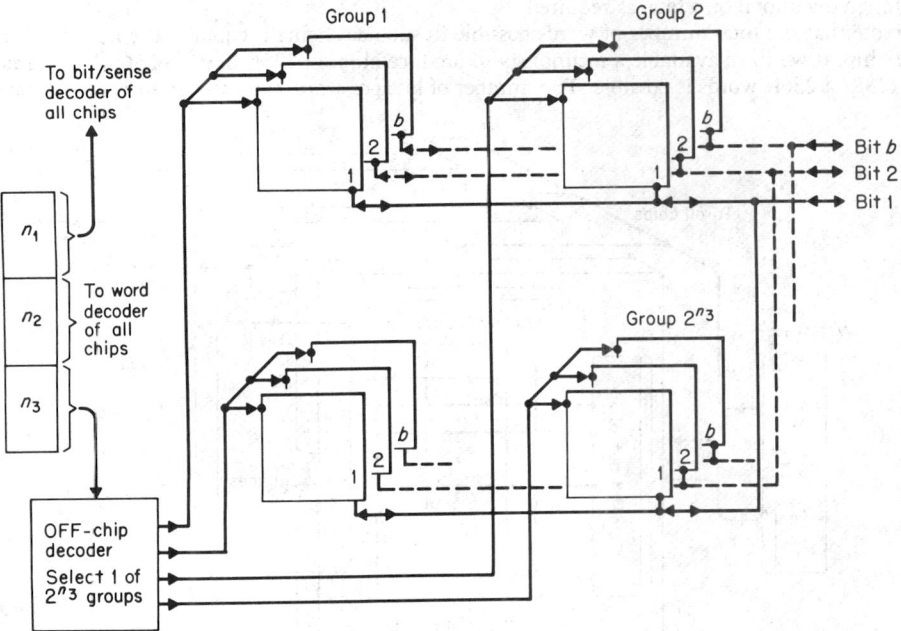

Fig. 23-51. Mixed memory organization using 2½D on-chip and 2D (direct) chip-ensemble accessing 1 b/chip.

as well as manufacturers, a total chip failure, i.e., chip killer, often represents 40 to 60% of the failures. If 2 b or more of any accessed word belong to one chip, single-bit error correction cannot correct a chip failure. If, however, only 1 b/chip is ever accessed, a chip failure deletes only 1 b from a number of different words. Thus single-bit error correction is effective against this common type of failure.

72. 2D Organization. This scheme makes use of two-terminal storage cells with selection performed only on the word lines. Such being the case, there must be b bits along each word line, as shown in Fig. 23-52. Since b is usually small (16 to 72 b) and W is usually large (256K to 4M b), this requires a long, narrow array, which is undesirable. Such a configuration could be obtained by connecting many chips together in the long direction. Nevertheless such a scheme still has many disadvantages, such as requiring at least 2^N decoders and word drivers. In addition, more than 1 b/chip is used, which prevents effective use of error-correction circuits. For small arrays which can be made nearly square, this scheme is simple and fast. It is hardly ever used on chip, but the basic idea is used to perform the chip-select ENABLE function, as in Fig. 23-51. The off-chip decoder and the individual ENABLE signal lines of Fig. 23-51 are equivalent respectively to the on-chip decoder and individual word lines of Fig. 23-52. In both cases, one and only one entity, i.e., either a group of chips or a word line, is directly selected by one level of decoding, using all the address bits available for that level of organization. It should be apparent that there can be additional levels of a memory with additional levels of decoding, which is often the case.

Integrated-circuit memories were introduced in about 1973 with 16 to 64 b/chip. By 1980, chip density had risen to 64K b/chip with prospects of 128K and 256K in the early 1980s.

73. Digital Magnetic Recording. Magnetic recording is attractive for data processing since it is inexpensive, easily transportable, unaffected by normal environments, and can be reused many times with no processing or developing steps. Since magnetic recording is nonrandomly accessed, no wired-in hardware, i.e., no wired cell or array configuration, is required.

The essential parts of a simplified but complete magnetic recording system are shown in Fig. 23-53. They consist of a *controller* (sometimes a large computer) to perform all the logic functions as well as write-current generation and signal detection; seria-to-parallel conversion registers; a read-write head with an air gap to provide a magnetic field for writing and sensing the flux during reading; and finally the medium. The wired-in cells, array, and transducers of random-access memory have been replaced by one read-write transducer, which is shared by all

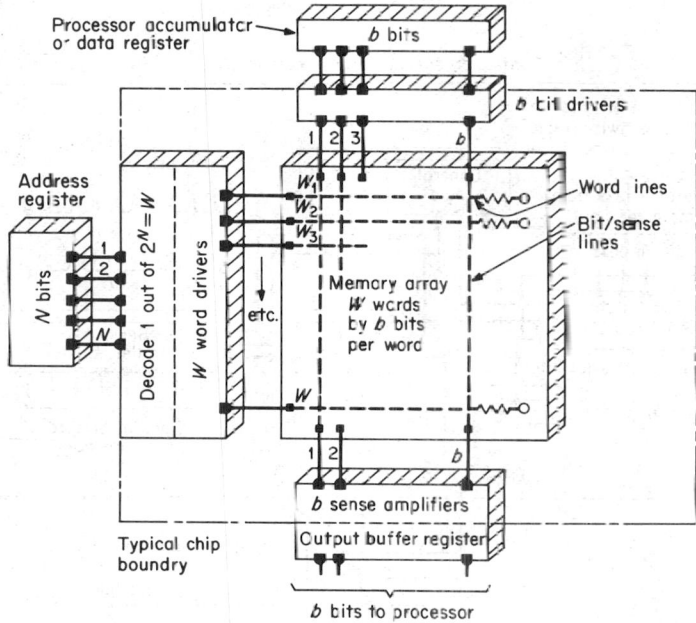

Fig. 23-52. 2D memory organization showing peripheral circuits typically included on chip.

stored bits, and a shared controller. Coincident selection is still required for reading and writing, and this is obtained by the coincidence of electric signals in the read-write head with physical positioning of the medium under the head.

The recording medium is very similar in principle to ferrite material used in cores. The common material for digital magnetic recording is ferric oxide, Fe_2O_3. It has remained essentially unchanged for many years except for reductions in the particle size, smoother surfaces, and thinner, more uniform coating, all necessary for high density. This material remained the sole medium for discs and tapes until the late 1970s, when NiCo (nickel cobalt) was introduced. These new materials have a higher coercive force and can be deposited thinner and smoother, allowing higher recording densities. Operation of these media requires a reasonably rectangular magnetic hysteresis loop with two stable residual states $+B_r$ and $-B_r$ for storing binary information. The media must be capable of being switched an infinite number of times by a magnetic field produced by the write head, which exceeds the coercive force. Stored information is sensed by moving the magnetized bits at constant velocity under the read head to provide time-changing flux and hence an induced sense signal.

The essence of magnetic recording consists of being able to write very small binary bits, to place these bits as close together as possible, to obtain an unambiguous read-back voltage from these bits, and to convert this continuously varying voltage into discrete binary signals. The write head is not a major factor in determining density since the writing is done by the trailing edge of the write field

The minimum size of one stored bit is determined by the minimum transition length required within the medium to change from $+B_r$ to $-B_r$ without self-demagnetizing. The smaller the transition length, the larger the self-demagnetizing field. The minimum spacing at which adjacent bits can now be placed with respect to a given bit is governed mainly by the distortion of the sense signal when adjacent bits are too close, referred to as *bit crowding*. This results from the overlapping of the fringe field from adjacent bits when they are too close and this total, overlapped magnetic field is picked up in the read head as a different induced signal from that produced by a single transition. Conversion of the analog read-back signal to digital form requires accurate clocking; this means that clocking information must be built into the coded information, particularly at higher densities.

Neglecting clocking and analog-to-digital conversion problems for the moment, the signals obtained during a read cycle are just a continuous series of 1s and 0s. A precise means of identifying the exact beginning and end of the desired string of data is necessary, and furthermore,

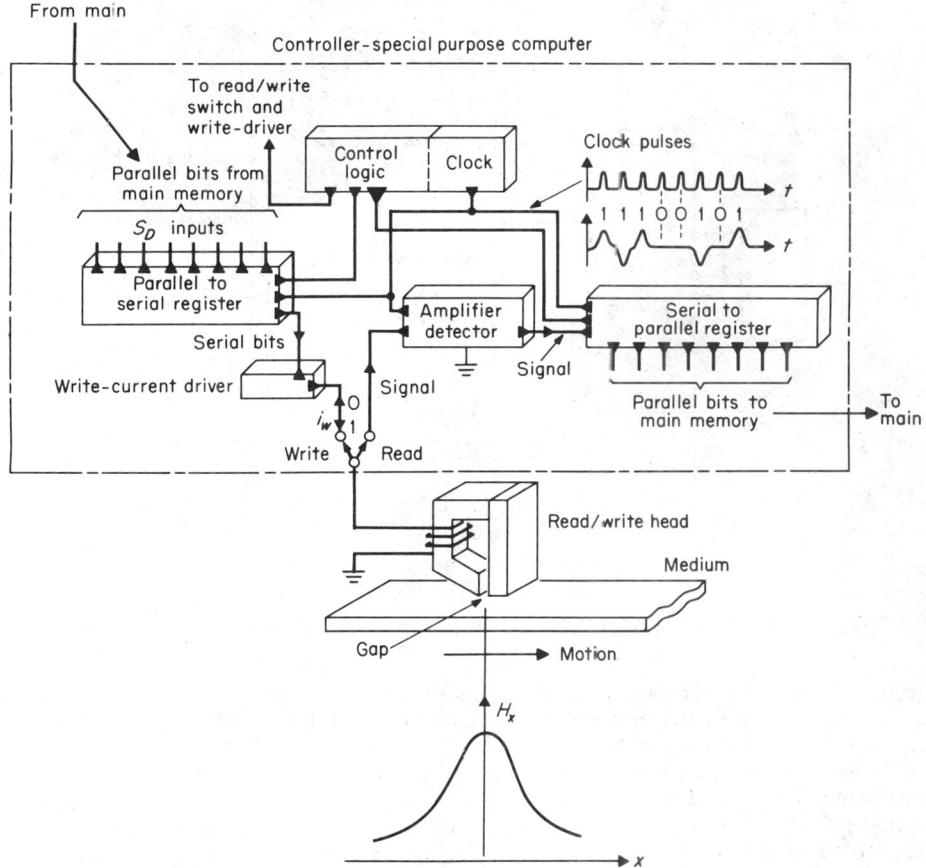

Fig. 23-53. Schematic of simplified complete magnetic recording system.

some means for identifying specific parts within the data string is often desirable. Since the only way to recognize particular pieces of stored information is through the sequence of pulse patterns, special sequences of patterns such as gaps, address markers, and numerous other coded patterns are inserted into the data. These can be recognized by the logic hardware built into the controller, which is a special-purpose computer attached to the storage unit. These special recorded patterns, along with other types of aids, are referred to as the *stored addressing information* and constitute at least a part of the addressing mechanism.

Coding schemes are chosen primarily to increase the linear bit density. The particular coding scheme used determines the frequency content of the write currents and read-back signals. Different codes place different requirements on the mode of operation and frequency response of various parts of the system such as clocking techniques, timing accuracy, head time constant, medium response, and others. Each of these can influence the recording density in different ways, but in the overall design the trade-offs are made in the direction of higher density. Thus special coding schemes are not fundamentally necessary but only meet practical needs. For instance, it is possible to store bits by magnetizing the medium over a given region where say $+M_r$ (magnetization) is a stored 1 and $-M_r$ is a stored 0, as in Fig. 23-54a. The transition region in between 1s and 0s is assumed to have $M = 0$ except for the small regions of north and south poles on the edges as shown. As the medium is moved past the read head, a signal proportional to dM/dt or dM/dx is induced, so that the north poles induce, say, a positive signal and south poles a negative signal as shown (polarity arbitrary).

This code is known as *return to zero* (RZ) since the magnetization returns to zero after each bit. Each bit has one north- and one south-pole region so that two pulses per bit result. Not only are two pulses per bit redundant, but considerable space is wasted on the medium for regions

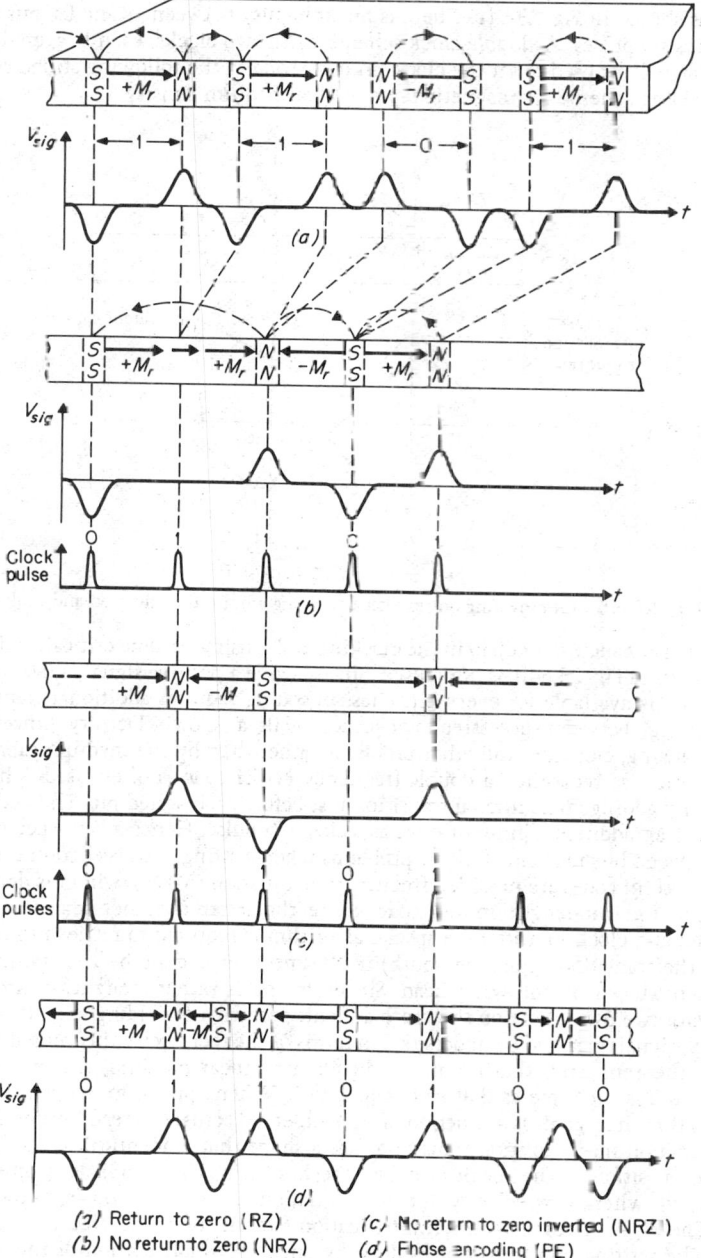

(a) Return to zero (RZ)   *(c)* No return to zero inverted (NRZ')
(b) No return to zero (NRZ) *(d)* Phase encoding (PE)

Fig. 23-54. Cross-sectional view of magnetic recording medium showing stored bits, signal patterns, and clocking for various codes

separating stored bits. It is possible to push these bits closer together, as in Fig. 23-54b, so that the magnetization does not return to 0 when two successive bits are identical, as shown for the first two 1s; hence the name *nonreturn to zero* (NRZ). The result is then only one transition region and one signal pulse per bit. By adjusting the clocking pulses to coincide with the signal peaks, we have a coding scheme in which 0s are always negative signals and 1s are always positive (Fig. 23-54b). The difficulty is that only a change from 1 to 0 or 0 to 1 produces a pulse; a string of only 1s or 0s produces no signals. This requires considerable logic and accurate clocking in the controller to avoid accumulated errors as well as to separate 1s from 0s.

One popular coding scheme is a slightly revised version of the above, namely *nonreturn to zero inverted* (NRZI), in which all 1s are recorded as a transition (signal pulse) and all 0s as no transition, as shown in Fig. 23-54c. There is no ambiguity between 1s and 0s, but again a string of 0s produces no pulses. A *double-clock* scheme, with two clocks, each triggered by the peaks of alternate signals, is used to set the clock timing period to the following strobe point. NRZI is a common coding scheme for magnetic tapes used at medium density.

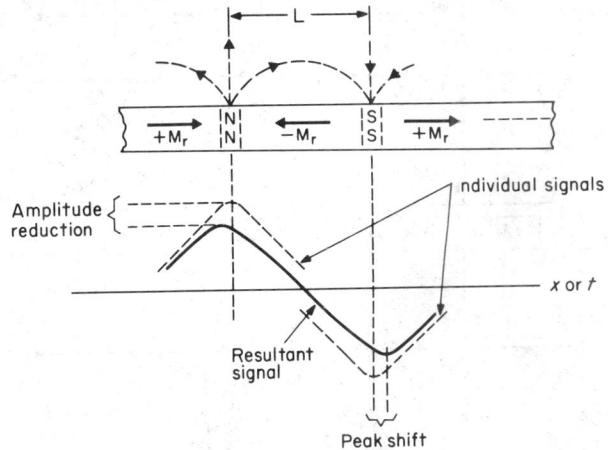

Fig. 23-55. Bit crowding on read-back showing amplitude reduction and peak shift.

For high density such as 1,600 b/in the clocking and sensing become critical, and *phase encoding* (PE), shown in Fig. 23-54d, is often used. Since 1s give a positive signal and 0s give a negative signal, a signal is available for every bit. Phase encoding requires additional transitions within the medium, e.g., between successive 1s or successive 0s, as shown. Density, however, is usually limited by sensing, clocking, and other problems rather than by the medium capability.

For magnetic-disc recording, a double-frequency NRZI code is often used. This is obtained from NRZI by adding an additional transition just before each stored bit. The additional transition generates an additional pulse to serve as a clocking pulse. Hence a well-specified window is provided between bits to avoid clocking problems when a string of 0s is encountered. Two other popular and useful codes are modified frequency-modulation (MFM), which is derived from the DF code, and the run-length-limited code. Since the latter does not maintain a distinction between data and clock transitions, a special algorithm is required to retrieve the data.[85]

Writing the transitions (north or south) in the medium is done by the trailing edge of the fringe field produced by the write head. Since writing is rather straightforward and is not a limiting factor, we dwell here on the more difficult problems of reading and clocking.

Read-back signals can best be understood in terms of the reciprocity theorem of mutual inductance. This theorem states that for any two coils in a linear medium, the mutual inductance from coil 1 to 2 is the same as that from coil 2 to 1. When applied to magnetic recording, the net result is that the signal, as a function of time observed across the read-head winding induced by a step-function magnetization transition, has a shape that is identical to the curve H_x vs. x for the same position of the medium below the head gap (Fig. 23-53). It is only necessary to replace x by vt, where v = velocity, for the translation of the x scale on H_x to the time scale on V_{sig} vs. t. The H_x-vs.-x curve with a multiplication factor is often referred to as the *sensitivity function*. The writing of a transition is done by only one small portion of the H_x-vs.-x curve, whereas the sense signal is determined by the entire shape of H_x vs. x; that is, the signal is spread out.

This fact gives rise to *bit crowding*, which makes the read-back process more detrimental in limiting density than the writing process. To understand bit crowding, suppose there are two step-function transitions of north and south poles separated by some distance L, as in Fig. 23-55. When these transitions are far apart, their individual sense signals shown by the dashed lines appear at the read winding. However, as L becomes small, the signals begin to overlap and, in fact, subtract from each other,* giving both a reduction in peak amplitude and a time shift in the peak position as shown. This represents, to a large extent, the actual situation in practice. The transitions can be written closer together than they can be read.

Clocking or strobing of the serial data as they come from the head to convert them into digital characters is another fundamental problem. If perfect clock circuits with no drift and hence no accumulated error could be made, the clocking problem would disappear. But all circuits have tolerances, and as the bit density increases, the time between bits becomes comparable to the

TABLE 23-3   Typical Parameters of Common Tape Systems

Tape speed, in/s	Data rate, B/s	Linear density, b/in	Rewind time, s	Recording code
18.75	15K	800	Minutes	NRZI
37.5	30K	800	Minutes	NRZI
75	120K	1,600	45–100	PE
	60K	800	45–100	PE
	41.5K	556	45–100	NRZI
	15K	200	45–100	NRZI
100	160K	1,600	72	PE
	80K	800	72	NRZI
112.5	180K	1,600	55–97	PE
	90K	800	55–97	NRZI
125	200K	1,600	55	PE
	100K	800	55	NRZI
200	320K	1,600	45–60	PE
	160K	800	45–60	NRZI
	111.2K	556	45–60	NRZI
250	800K	3,200	45	PE
200	1,250K	6,250	45	

drift in clock cycle times. Since the drift can be different during reading and writing, serious detection errors can result. For high density, it is necessary to have some clocking information contained within the stored patterns as in the PE and double-frequency NRZI codes discussed previously.

74. Magnetic Tape. The most common tape consists of ½-in-wide Mylar, about 1 mil thick, coated with about 0.5-mil-thick magnetic oxide, about 2,400 ft long and wound on a 10½-in-diameter reel.

Since there are either seven or nine tracks written across the width of the tape, either seven or nine read-write heads are required to store one complete character or byte at a time. Tape tracks and hence bit-cell widths are approximately 0.04 in wide, but the bit spacing along a track is much smaller. The latter is approximately the reciprocal of the linear density, or 0.00125 for an 800 b/in system. The actual transition lengths are generally about half this bit-cell spacing. In many systems there are separate read and write gaps in tandem to check the reliability of the recording by reading immediately after writing. The tape is mechanically moved back and forth in contact with the heads, all under the direction of a controller. Both the tape and heads wear from abrasive contact and must be replaced periodically. The important operational parameters are tape speed, linear density, and rewind time. Some typical values are given in Table 23-3.

The stored addressing information in tapes is relatively simple, consisting of specially coded bits and tape marks in addition to *interrecord gaps* (IRG). The latter are blank spaces on the tape to provide space to accelerate and decelerate between records since reading and writing can only be done at constant velocity. The common gap sizes are 0.6 and 0.75 in for specifications shown in Table 23-4A. Typically, a tape recorded at 800 b/in with eight tracks plus parity storing records of 1K B each and a gap of 0.6 in between each record can hold over 10^8 b of data. Even

---

*Linear superposition is possible since the air gap makes the read head linear.

though this represents a large capacity, the gap spaces consume nearly 50% of the tape surface, a rather extravagant amount. In order to increase efficiency, records are often combined into groups known as blocks. Since the system can stop and start only at an interrecord gap, the entire block is read into main memory for further processing during one read operation. The highest-density tapes can hold nearly eight times the above amount.

TABLE 23-4A   Tape Interrecord Gaps (IRG)

Tracks	Density, b/in	IRG,* in
7	200, 556, 800	0.75
9	800, 1,600,	0.6
	6,250	0.3

*Standard accepted by most of the industry.

TABLE 23-4B   Disc Parameters

Linear density	1,100–6,000⁺ b/in
Track density	100–400 REM tracks/in
Rotation speed	1,200–3,600 r/min
Arm movement, between adjacent tracks	50 ms
Across all tracks	100 ms
Data rates	0.5 to 10×10^3 b/s

75. Direct-Access Storage Systems—Discs. There are two major types of direct-access storage systems: movable-head (arm) systems and fixed, one-head-per-track, systems. The latter includes both discs and drums, as shown in Fig. 23-56. We concentrate on the first. It is more common since it is less expensive as a result of a greater sharing of the read-write heads.

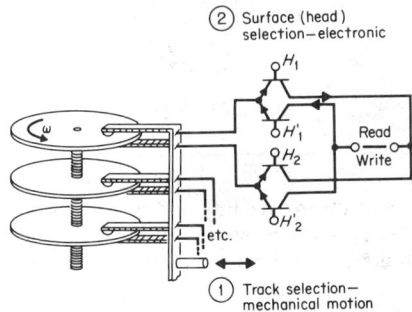

(a) Movable–arm system

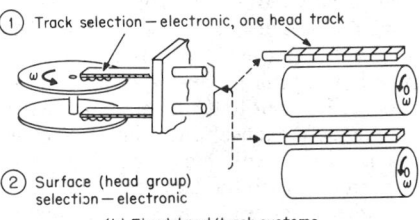

(b) Fixed-head/track systems

Fig. 23-56. Essential features of direct-access storage systems.

The recording head usually consists of one gap, which is used for both reading and writing. The head is "flown" on an air cushion above the disc surface at separations in the neighborhood of 5 to 100 μin, depending on the system. A well-controlled separation is vital to reliable recording.

For discs, the medium consists of very thin coatings of about 10 μin or less of the same magnetic material used on tapes but applied to polished aluminum discs. Several discs are usually mounted on one shaft, all rotated in unison. Each surface is serviced by one head, as in Fig. 23-56a. The arms and heads are moved mechanically along a radial line, and each fixed position sweeps out a track on each surface, the entire group of heads sweeping out a cylinder. A typical bit cell is a rectangle 0.005 in wide by 0.0005 in long for a 2,000 b/in linear density. The transition length is about half this size. See Table 23-4B for typical disc parameters.

The fundamental difference between various disc systems centers on the stored addressing information and addressing mechanisms built into the system. Some manufacturers provide a rather complex track format permitting the files to be organized in many different ways, using keys, identifying numbers, stored data, or other techniques for finding a particular word. Thus, the user can "program" the tracks and records to retrieve a particular word "on the fly." This provides a very versatile system but only with additional cost, since considerable function must be built into the controller. Other systems use a very simple track format consisting mainly of gaps and sector marks that do not permit the user to include programmable information about the data. This scheme is more suitable for well-organized data, such as scientific data; it still can be used in other applications with more user involvement.

Floppy discs are finding wide applications as inexpensive high-density, medium-speed peripherals. Floppy discs often consist of the same flexible medium (hence the name floppy) as magnetic tape but cut in the form of a disc 5 to 7½ in in diameter. Such discs straighten out when spun. The read-write mechanism is usually identical to that above except that the head is in contact with the medium, as in tape. This causes wear that is more significant than in ordinary discs, requiring frequent replacement of the disc and occasional replacement of the head, particularly when heavily used.

Although some floppy-disc systems use only one track recorded as a spiral as on phonographic recording, most use the more sophisticated system with movable heads and optical track-following servo. Many newer systems record on both sides. The major deviations from flying-head discs are much smaller track density and data rate. The linear bit densities are quite comparable to those of tape. Typical parameters are 77 to 150 tracks per disc surface, 1,600 to 6,800 b/in, rotation from 90 to 3,600 r/min.

76. Virtual Memory Systems. Virtual memory is a term usually applied to the concept of paging a main memory out of a disc or drum.[85] This concept makes the normal-sized main memory appear to the user as large as the virtual-address space* while still appearing to run at essentially the speed of the actual memory. Virtual memories are particularly useful in multiprogrammed systems. On the average, virtual memory provides for better management of the memory resource with little wasted space due to fragmentation, which can otherwise be quite severe.

To understand *fragmentation*, suppose that a number of programs to be processed require small, medium, and large amounts of memory on a multiprogrammed machine. Suppose further that a small and a large program are both resident in main memory and that the small one has been completed. The operating system attempts to swap into memory a medium-sized program to be processed, but it cannot fit in the space freed by the small program. If none of the other waiting programs is small enough to fit, this memory space is wasted until the large program has been completed. It can be seen that with many different sized programs, fitting several of them into available memory space becomes difficult and leads to much unusable memory at any one time, i.e., fragmentation. Virtual memory avoids this by breaking all programs into *pages* of equal size, usually 2K or 4K B, and dynamically paging them into main memory on demand. The identical concepts used in a virtual memory are applicable to a cache paged out of main memory; only the details of implementation are different, as shown later. We discuss first the basic concepts as applied to a main memory paged out of a disc and indicate, whenever possible, the differences applicable to a cache.

All virtual memories start with a *virtual address* that is larger than the address of the available main memory. Such being the case, the desired information may not be resident in the memory. Hence it is necessary to find out *if*, in fact, it is present. If the information is resident, it is necessary to determine *where* it is residing because the physical address cannot bear a one-to-one correspondence to the virtual address. In fact, in the most general case, there is no relationship between the two, so that an address translation scheme is necessary to find *where* the information does reside.

If the requested page is not resident in memory, a *page fault* results. This requires a separate process to find the page on the disc, remove some page from memory and bring the new page into this open spot, called a *page frame* Which page to remove from main storage is determined by a page-replacement algorithm which replaces some page "not recently used." The address-translation and page-replacement functions are shown conceptually in Fig. 23-57.

Thus there are at least three fundamental requirements: (1) a mapping function to specify how pages from the disc are to be mapped into physical locations in memory, (2) an address translation function to determine *if* and *where* a virtual page is located in main memory, and (3) a replacement algorithm to determine which page in memory is to be removed when the *if* translation is a "no," i.e., when a page fault occurs. These are the three fundamental requirements needed to implement a virtual memory system, either a virtual main memory as here described, or a cache.

To make the system efficient and practical, there are other desirable features, the most important being data clustering. In the process of writing programs, users tend to keep related pieces of data and instructions close together. Only infrequently is logically related information physically separated by large distances. Exceptions occur in subroutines or data arrays, but they are closely structured within themselves. This is no different from what an engineer does in solving, say, an analytical problem: books and papers related to the problem are kept close at hand for

*Represented by a register holding the virtual address.

easy access. Only rarely is it necessary to travel to the library archives to fetch new information. When this is necessary, the book or paper is then brought as a unit back to the office for local processing. If there is not enough room in the office for the new book, an old one, most likely one not recently used, will be sent back to the library to make room. In a virtual memory, a completely analogous situation arises. Pages required for the current processing are placed in the memory. As the processing proceeds, most of the data are clustered in one of several pages so no interruption occurs. Eventually data or an instruction is required that is not in memory. These data or this instruction and other closely related data are then brought into memory upon

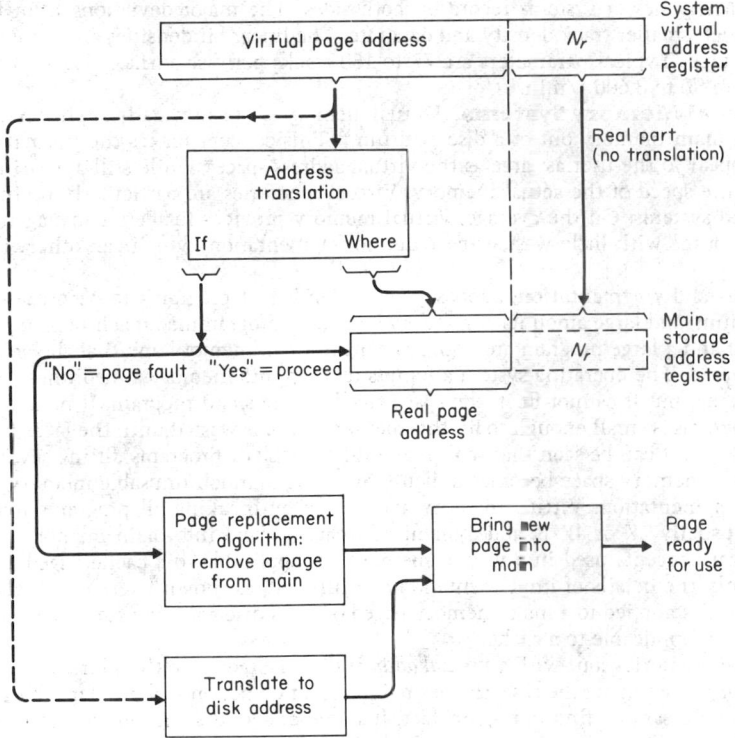

Fig. 23-57. Block diagram of address-translation and page-replacement process.

demand (*demand paging*). If there is no available page frame *then* any page not recently used is removed to make room. Processing continues until another page fault occurs. The clustering of data is usually expressed in terms of the hit ratio H_R, defined as

$$H_R = \frac{\text{references found in memory}}{\text{total memory references}}$$

This ratio varies as a function of time and problem type. An average value over a long time is used to specify the amount of data clustering. For a typical cache paged out of main memory, hit ratios of 90 to 98% are common, whereas a main memory paged out of disc in a multiprogrammed system can have over 99% hit ratios. The hit ratio varies with the relative size of the members of the storage hierarchy and the problem type.

77. Mapping Function and Address Translation. The mapping function is a logical construct, whereas the address-translation function is the physical implementation of the mapping function. Mapping functions cover a range from *direct mapping* to fully *associative mapping*, with a continuum of *set-associative* mapping functions in between. A very simple way to understand maps is to consider the example of building one's own personal telephone directory "paged" out of larger telephone books. Assume that the personal directory is of fixed size, say 4(26) = 104 names, addresses, and associated telephone numbers. A direct mapping from the large books to the personal directory is an alphabetical listing of the 104 names. Given any name, we can go directly to the directory entry, if present. Such an address translation could be hard-wired if desired. There are two difficulties with direct mapping: (1) it is very difficult to change, and (2) suppose we allow one entry for Jones. If later we wish to include another Jones, there is no room. If both Joneses are needed, there is a conflict unless we restructure the entire directory. Because of such conflicts, direct maps are seldom used.

At the other end of the spectrum is a *fully associative directory* in which 104 names in any combinations are placed in any positions of the directory. This directory is very easily changed because a name not frequently used can simply be removed and a new name entered in its place without regard to the logical (alphabetical) structure of the two names. For instance, if the directory is full and we wish to make a new entry Zeyer, we first find a name not used much recently. If Abas is in position 50 of the directory, remove Abas and replace the entry with Zeyer. The major difficulty, obviously, is in searching the directory. If we wish to know the number for Smith, we must associatively search the entire directory (worst case) to find the desired information. This is very time-consuming and impractical in most cases.

Imagine the usual telephone directory that is associatively organized. There are several ways to resolve the fundamental conflict between ease of search and ease of change. The fully associative directory can be augmented with a separate, directly organized and accessed table that contains a list of all names. However, the only other piece of data is a number indicating the entry this name now occupies in the associative directory. If a directory entry is changed, the new entry number must be placed in this table. If we wish to access a given name, a direct access to the table gives the entry number. A subsequent direct access to this entry number gives the desired address and telephone number.

The penalty is the two accesses plus the storage and maintenance of the translation table. Nevertheless, this is exactly the scheme used in all virtual main memories paged out of disc, drum, or tape. This table is typically broken into a hierarchy of two tables called *segment table* and *page table* to facilitate the sharing of segments among users and to allow the tables to also be paged as units of 2K or 4K B. These tables are built, manipulated, and maintained by supervisory programs and generally are invisible to the user. Although these tables can consume large amounts of main memory and of system overhead, the saving greatly exceeds the loss.

An example of such a virtual memory with a fully associative mapping using a two-level table-translation scheme is illustrated in Fig. 23-58. Each user has a separate segment and page table (the two are, in principle, one table) stored in main memory along with the user's data and programs. When an access is required to a virtual address, say N_vN, as in Fig. 23-58, a sequence of several accesses is required. The user ID register bits μ give a direct address to the origin of that user's segment table in main memory. The higher-order segment-index bits (SI) of the virtual address (typically 4 to 8 b) specify the index (depth) into this table for the required entry. This segment table entry contains a flag specifying *if* this entry is valid or not, and a *where* specifying the origin of that user's page table in memory, as shown. The lower-order page index bits (PI) of the virtual address specify the index into the page table as shown. The page-table entry so accessed contains an *if* bit to indicate whether the entry is valid (*if* page is present in main memory) and a *where* address that gives the real main-memory address of the desired page. The lower-order N_r bits of the address, typically 11 or 12 b for 2K- or 4K-B pages, are real, representing the word or byte of the page and hence do not require translation.

The principles described above for the two-level table translation are those used in the IBM System/360 and System/370 virtual-memory systems and the Amdahl 470 system, although considerably more detail is required. The Honeywell MULTICS system uses a similar scheme but with a three-level table translation. It should be apparent that all these schemes require at least two accesses to main memory just to do the address translation followed by a final access to the required data. Obviously such a system could not run at the speed of a nonvirtual memory system if a table translation were required for every access. To circumvent this slow translation process, a small, fast, set-associative directory is provided in high-speed hardware. This directory* is typically similar to that of Fig. 23-60, discussed later for a cache, with two notable exceptions: (1) it is only a partial directory, containing address-translation information for only some of the more recently used pages; (2) a two-way set associativity is sufficient to provide large hit ratios. When a page is first accessed, the slow translation tables must be used to obtain the real address. This information is then entered into the partial directory (DLAT), and any subsequent references to that page are quickly translated via the DLAT. Such a scheme is quite efficient and typically can give hit ratios of 90 to 95%.

In cache memories, speed is so important that a fully associative mapping with a table-trans-

---

*Generally called a translation look-aside buffer (TLB) or dynamic look-aside table (DLAT).

lation scheme is impractical. Instead a set-associative mapping and address translation is implemented directly in hardware. Although it is not generally realized, set-associative mapping and translation is commonly used in everyday life. It is a combination of a partially direct mapping and selection, with a partially associative mapping and translation. A simple example is the most common type of personal telephone directory as shown in Fig. 23-59, where a small index knob is moved to the first letter of the name being sought. Suppose there is one known position on

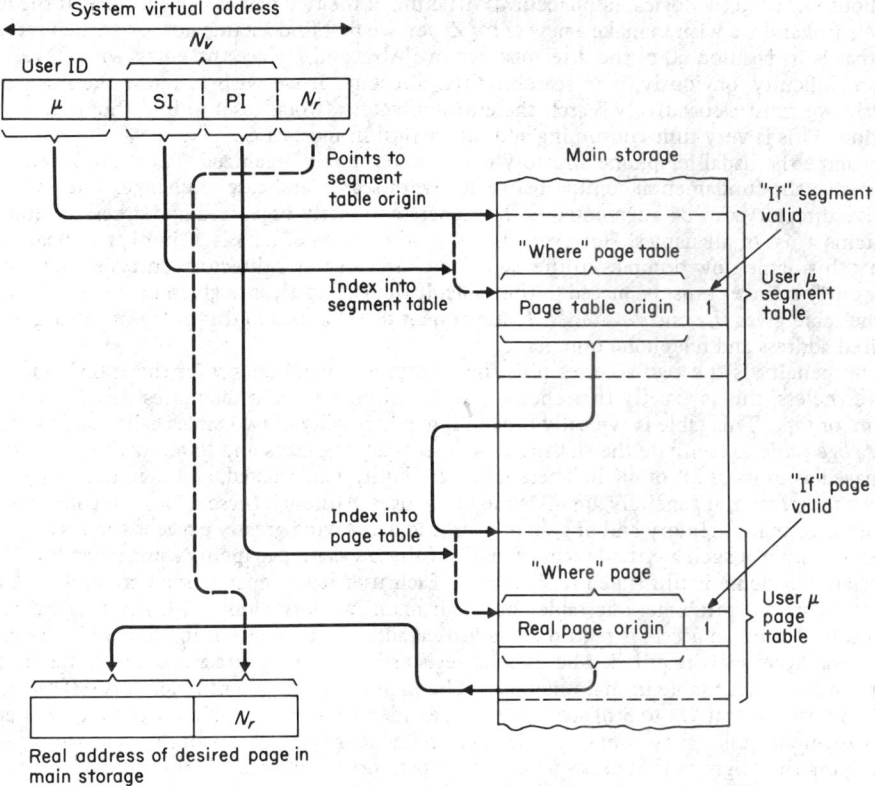

Fig. 23-58. Virtual-storage-address translation using a two-level table (segment and page) for each user.

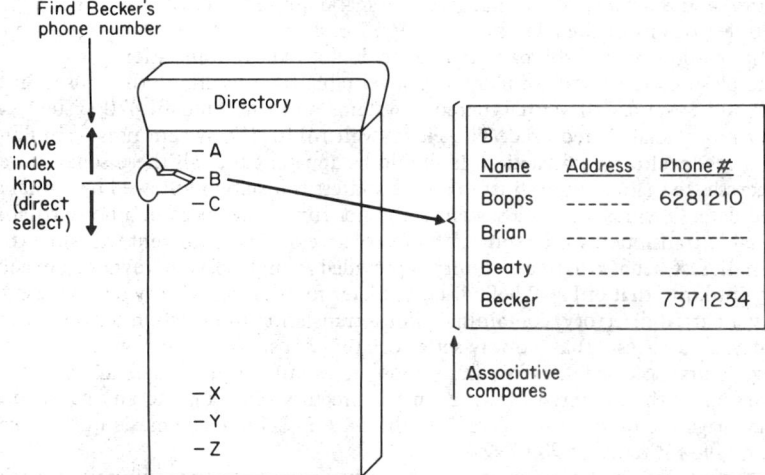

Fig. 23-59. Fundamentals of a general set-associative directory showing direct and associative parts of the addressing process.

the directory for each letter of the alphabet and we organize it with a set associativity of four. This means that for each letter there are exactly four entries or names possible, and these four can be in any order, as shown by the insert of Fig. 23-59. To find the telephone number for any given name such as Becker, we first do a direct selection to the letter B followed by an associative search over the four entries. Thus it is apparent that a set-associative access is a combination of the two limiting cases, namely part direct and part associative. Many different combinations are possible with various advantages and disadvantages.

This set-associative directory (Fig. 23-59) with four names or entries per set is exactly the same fundamental type used in many cache-memory systems. The directory is implemented in a random-access memory array as shown in Fig. 23-60 where there are 128(4) entries total, requiring 9 total virtual page address bits. Each set of 4 is part of one physical word (horizontal) of the

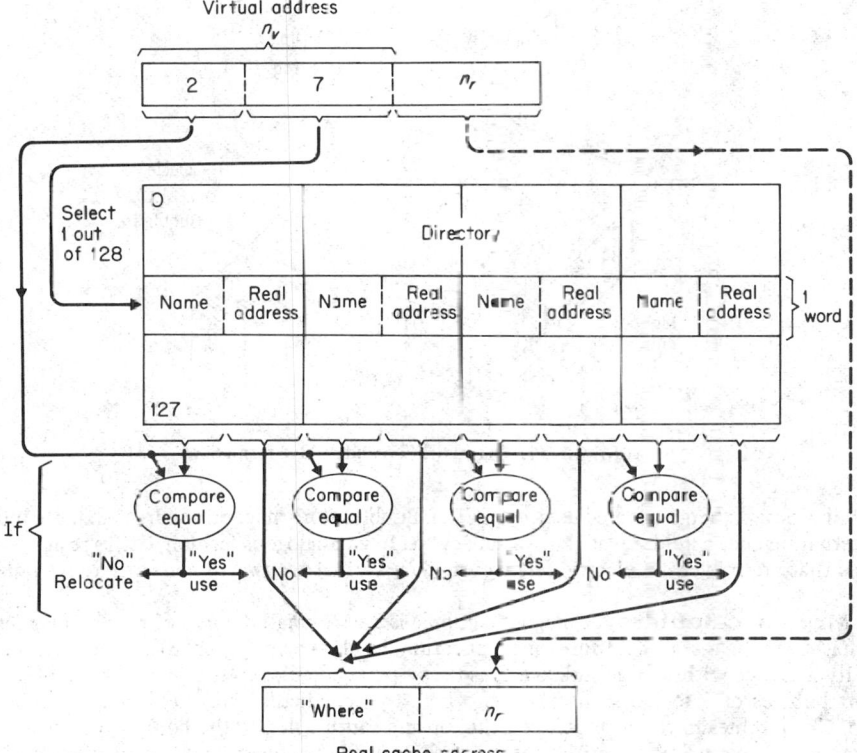

Fig. 23-60. Cache directory using 4-way set-associative organization

random-access array, so there are 128 such words, requiring 7 address bits. The total virtual address $n_v = 9$ must be used in the address translation to determine if and where the cache page resides. As before, the lower-order bits n_r, which represent the byte within the page, need not be translated. Seven virtual bits are used to select directly one of the 128 sets as shown. This is analogous to moving the index knob in Fig. 23-59 and reads out all four names of the set. The NAME part is 2 b long, representing one of the four of the set. All four are compared simultaneously with the 2 b of the virtual address. If one of these gives a "yes" on compare equal, then the correct "real" address of the page in the cache, which resides in the directory with the correct NAME, is gated to the "real" cache-address register.

The latter is used on a subsequent cycle to obtain the correct information from the cache array (not shown). This feature of storing only the "real" location of the data within the directory is the one essential difference between Figs. 23-60 and 23-59 and is done for practical reasons. We could actually have the desired information stored within the directory, but the overall system would be less efficient in a total hierarchy. Cache pages are typically only 64 to 128 B long, whereas main-memory pages are 2K to 4K B. Cache pages are often referred to as *blocks* or *lines*. The latter term is very confusing to hardware designers since a "line" has other meanings, e.g., a cache-line fault is quite ambiguous.

In a three-level virtual-memory hierarchy containing a main memory paged out of disc and a cache paged out of main memory the same principles described above apply, but some changes in detail are introduced for practical reasons. In such a case, the DLAT is used to do part of the address translation for the cache directory for two reasons: (1) it allows the directory to be smaller, faster, and cheaper, and (2) if a miss occurs to the cache, an access to that entry of the DLAT has to be done anyway. Thus by parallel address translation the entire system is faster and more efficient.

78. Electronic Shift-Register Storage. In magnetic discs and tapes, information bits are fixed in the recording medium and are accessed by mechanically moving the medium in a type of mechanical shift register. An electronic shift register is one in which the medium is held stationary and the bits are shifted electronically in the medium. Two common schemes are mag-

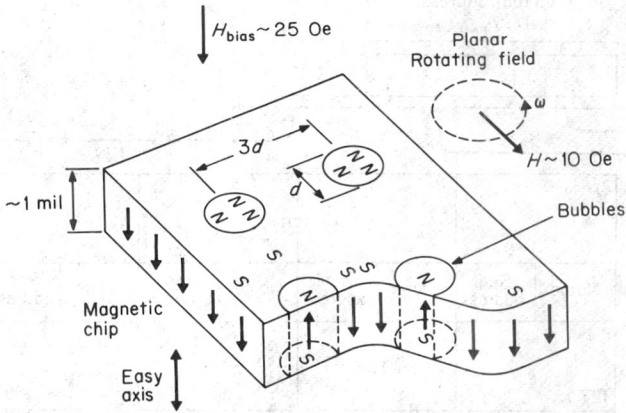

Fig. 23-61. Magnetization and bubble formation.

netic bubbles and charge-coupled devices (CCDs). Bubbles store magnetic charges that are moved by external magnetic fields created in various ways. In an analogous fashion, CCDs store electric charges that are moved by electric fields created in various ways. We concentrate on bubbles here.

79. Magnetic Bubbles. A magnetic bubble is a very small domain of reverse magnetization that exists in a sea of uniform magnetization, as shown in Fig. 23-61. A magnetic chip, typically a garnet with a large uniaxial anisotropy perpendicular to the plane of the chip, can support bubbles of 1 to 5 μm in diameter when a perpendicular bias field H_{bias} is applied as shown. The bubbles have north poles on the top and south poles on the bottom surface and are very mobile. Of course, all polarities reverse if the bias field is reversed. A small magnetic field of south poles near the top surface of the bubble attracts it, causing it to move. This principle can be used in practical devices provided there is a means for generating, annihilating, and sensing bubbles. One method for performing all these functions makes use of soft magnetic permalloy overlays on the top surface, along with printed-circuit wiring, as shown in Fig. 23-62. An externally applied planar, rotating magnetic field H_p over the entire chip induces time-changing north and south poles in the permalloy overlays to give the shifting, generation, and annihilation as follows. The dimensions of the T and I bars are chosen such that the long dimension is about three bubble diameters and the short dimension is less than one bubble diameter. The south poles induced along any long dimension of a T or I bar attract any nearby bubble, while induced north poles repel a bubble. The north and south poles induced across a short dimension of any T or I bar are too close together to have any net effect on a bubble because they cancel each other. Thus any nearby bubble is attracted to, and remains at, the induced south poles of the leftmost I bar of Fig. 23-62a. Shifting of bubbles from left to right occurs as the planar field rotates to successive 90° positions, as shown in Fig. 23-62b. South poles are successively induced in I and T bars, the south poles moving from left to right. The north poles on the top surface of the bubble follow the south poles, thus creating an electronic shift register in a medium that is stationary.

Generation of a new bubble is obtained by stretching or elongating one permanent, nonshifting bubble and then snipping off one section to form a new bubble. The permanent bubble resides on the circular generator overlay, as shown in Fig. 23-62a. This bubble rotates around the generator as the planar field rotates. However, for the 90° position just before Fig. 23-62a the planar field H_p is horizontal, pointing to the left. This induces south poles on that protrusion of the generator so that the bubble sits here, very close to the first I bar. As H_p rotates, the bubble rotates toward the top of the generator, but it is also attracted to the south poles of the I bar. It is thus elongated, as shown dashed in Fig. 23-62a. As the field continues to rotate, the stretching exceeds the elastic limit of the bubble and a new bubble breaks off. However, the new bubble

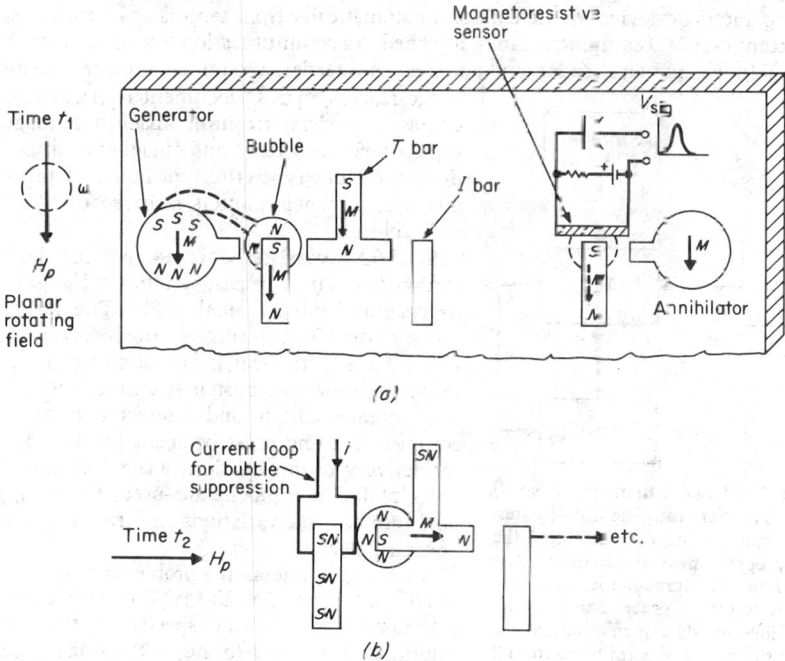

Fig. 23-62. Essential features of a magnetic-bubble shift register using T and I bars for propagation.

does not make the permanent bubble any smaller, so that an infinite number of new bubbles can be generated. The new bubble can represent, say, a binary 1. In that case, a binary 0 is the absence of a bubble, which requires suppression of a new bubble. Suppression is obtained by applying a current in the loop of wire around the first I bar to cancel the south poles, as illustrated in Fig. 23-62b. In this case, the permanent bubble on the generator is not attracted to the I bar, and hence no bubble is generated for this cycle.

The presence of a bubble is sensed as it passes under the magnetoresistive sensor on the right side of Fig. 23-62a. The magnetic field of the bubble causes a change in the resistance of the sensing element of several percent, which generates a small voltage pulse as shown, indicating a 1. If no bubble is present, no pulse is obtained, indicating a 0. Annihilation of bubbles is the exact inverse of generation. In fact, if the direction of rotation of H_p is reversed, bubbles are generated on the right side and propagate toward the left. A continuous-loop shift register can be made by the proper arrangement of TI bars, as well as switching, decoding, and logic functions.[81] All these can be placed on the same chip with the same technology. Bubbles offer densities well over 10^6 b/in^2 but only at moderate speeds of generally less than 1-MHz shift rates. As such they are attractive for fixed-head file types of applications, typically giving an access time in the range of 1 ms, depending on the size and organization of the system. See Chap. 5 of Ref. 86 for methods of organization.

INPUT/OUTPUT

80. Input/Output (I/O) Equipment. I/O equipment includes printers, console typewriters, card readers and punches, paper tapes, keyboards, character-recognition units, process-control sensors, and cathode-ray tubes, or other display devices. Such equipment presents a wide range of characteristics that must be taken into account at their interface with the central processing unit (CPU) and the arithmetic logic unit (ALU). A magnetic-tape unit operates serially by byte, writing or reading information of varying record length. A printer ordinarily uses information one line at a time. Different types of I/O gear operate over widely different speed ranges, from a few bytes per second to millions of bytes per second. In addition to these variables, magnetic drums, printers, card equipment, etc., present differences in access and start-up times, and the operational speed and capabilities of CPUs vary over several orders of magnitude.

For telecommunications an ever-growing array of I/O equipment can be attached to communications lines, including satellites for transmitting data to a computer. Data are entered by human operators or devices or are collected automatically from sensors or instruments.

Most computer I/O equipment can be attached to a communication line or satellite. The *input devices* include paper-tape readers and punches, card readers and punches, magnetic card readers, badge readers, optical document readers, magnetic-ink character readers, facsimile machines, magnetic-tape units, tape cassettes, and magnetic discs. *Output devices* include typewriters, printers, teleprinters, CRT displays, gas panels, plotters, strip recorders, and facsimile machines.

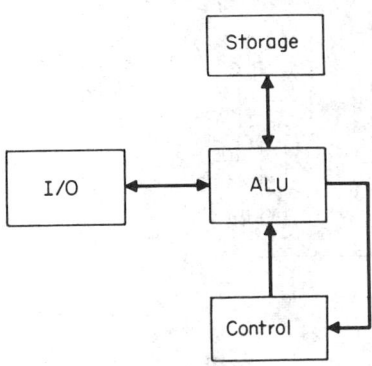

81. I/O Configurations. Figure 23-63 shows a configuration used in past systems and presently found in systems having a small CPU. The program gains access to the I/O gear through the logic circuitry of the ALU. Thus to transfer information to the I/O equipment the program must first extract the information from storage, edit it, and interact with the receiving equipment. Time must be spent by the CPU waiting for delivery of information to the I/O gear. If several types of I/O equipment are used, the program must take into account variations in formatting and control of each type.

A method of alleviating problems at an I/O interface is *I/O buffering* (Fig. 23-64). The buffer accumulates information at machine speeds, so that subsequent information transfers to the I/O gear can take place in an orderly fashion without holding up the CPU. A buffer system is most appropriate when the average

Fig. 23-63. Organization of a small machine. Transfers into and out of external equipment take place through the ALU. A program must first transfer information from the store to the ALU and from there to the I/O gear. The organization provides for data manipulation and editing in the ALU and permits the control unit to handle special I/O requirements.

I/O equipment information rate is less than the program rate of delivery. Such a system lacks flexibility in the types of I/O equipment accommodated and must use some of the capability of the ALU for control and formatting.

82. I/O Memory-Channel Methods. The arrangements shown in Figs. 23-63 and 23-64 involve information transfer from the I/O gear into the ALU and thence to main store. With a modularized main store, I/O data can be directly entered in, and extracted from, main storage. Direct access to storage implies control interrelationships at the interface between the I/O and the CPU to ensure coordination of accesses between the two units. The basic configuration used for direct access of I/O into storage is shown in Fig. 23-65. The connecting unit between the

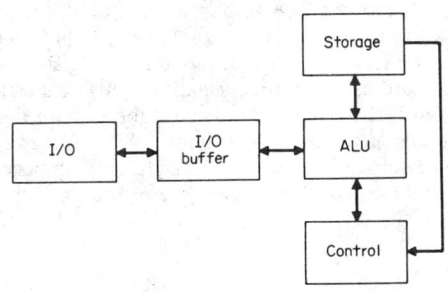

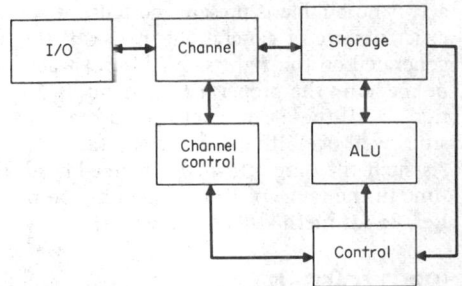

Fig. 23-64. Machine organization with an I/O buffer. Data can be accumulated in the buffer at machine speeds and transferred at a different ratio to the I/O gear, and vice versa.

Fig. 23-65. Machine organization with a channel, a separate logical device that can be used to help solve problems associated with maintaining CPU speed when working with I/O gear.

main store and the I/O gear is called a *channel*. A channel is not merely a data path but a logical device that incorporates control circuits to fulfill the relatively complex functions of timing, editing, data preparation, I/O control, etc.

83. Channel Functions. When a program calls for writing into a specific piece of equipment, the program instruction first initiates a channel address. The channel in turn brings forth a sequence of commands by means of its own control circuits so that it can specify the location of the information to be transferred from main storage, specify the commands for the initiation of attachment of the device sending or receiving information, and edit, modify, or otherwise

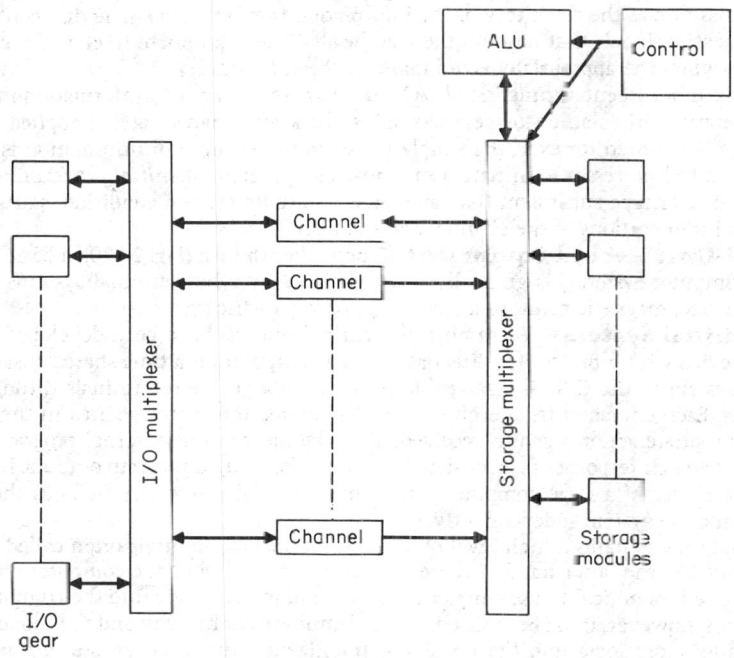

Fig. 23-66. Multiple multiplexed channels. In large systems a set of channels may be used in conjunction with multiplexers for storage and I/O gear. By such means the CPU can initiate multiple I/O operations and pursue other internal programs while waiting for an I/O response.

correlate the data into a format suitable for transmission. In medium-sized systems a channel may have limited computation facilities, but in a large system the channel may have a complexity approaching that of a medium-sized data processing machine.

Figure 23-66 shows an arrangement in which several channels are incorporated in a system. The principal I/O signal paths are multiplexed into a sequence of channels that in turn can be multiplexed into the CPU. Once the CPU initiates action upon a particular channel, it is free to pursue other programs that may initiate other channel actions. In particular, a CPU may operate upon a program until the program specifies an I/O access. Command is then transferred to a specific channel, and a new program is initiated in the CPU until the information transfer for the previous program has occurred. If more than one channel is available, the second program can initiate action on a new channel and a third program can be undertaken until the results are available in a previous program.

Multiprogramming is the ability of a system to operate on more than one program at a time. Multiprogramming permits maintenance of the internal speed of operation of the CPU and the ALU while accommodating the slower speed of the I/O equipment. The flexibility of channel architecture also permits many different types of I/O gear to be incorporated without unduly complicating the programming system of the CPU

Channels may operate in either the *burst mode* or the *unit-message mode*. In the burst mode, the total message is sent, the information being emitted continuously until the message is completed. In the unit-message mode, information may be sent separately to different devices on the same channel, one at a time.

Some channels (*selector channels*) can attach to only one I/O device at a time. Other channels can attach simultaneously to a number of subchannels on a time-shared multiplexed basis. Such channels are called *multiplexer channels*. Selector channels operate in the burst mode to be efficient.

84. Channel Attachment. Many pieces of I/O equipment generate electric signals to control specific operations. For example, in one type of printer (Par. **23-91**) an image of a printing unit, called a *chain*, must be maintained and periodically matched in time with the contents of the information to be printed. Such functions, specific to the device addressed, usually are handled by a *controller* that interfaces between the channel and the I/O gear itself.

In the operation of a direct-access disc storage device, the channel may call for a specific track on a particular disc as the repository of the information from the CPU. The disc controller generates the electric signals that position the read head at the appropriate track and receives feedback signals when the appropriate record mark has been found.

A specialized but essential function of I/O gear is to provide enough information for the CPU to begin operation. In volatile storage, the store is blank when power is first applied. Since program storage is required for even the simplest operations, an input of information is needed to permit the central processor to initiate operations. The problem of *initial program load* (IPL) is solved by a disc or tape subsystem that, subsequent to the power-on condition, starts an initial program load into portions of the channel and the main store.

85. Card-Unit Record Equipment. The Hollerith card (Fig. 23-20), a basic unit input record to computer systems, is gradually being replaced by other terminal systems that use a key-entry device, magnetic cards, or a key-to-tape or key-to-disc system.

86. Terminal Systems. A number of *terminal systems* have been developed that use a key-entry device with a printer for direct access to a computer on a time-shared basis. The programming system in the CPU is arranged to scan a number of such terminals through a set of subchannels. Each character from each terminal is entered into a storage area in the computer until an appropriate action signal is received. Then by means of an internal program the computer generates such responses as error signals, requests for data, or program outputs. By terminal systems, the power of a large computer can be made available to many users as though each were operating the system independently.

A terminal that contains a high level of logic operational capability is often called an *intelligent terminal*. On the other hand, if there are many terminals and one computer on a system, it may be more economical to use simpler terminals and put the logic into the computer. Other considerations, however, must be assessed, e.g., communication-line cost and the economic trade-off of building more logic into the terminals. Intelligent terminals may create a better user-machine interface and lessen the number of errors. They also minimize inconvenience when a computer or communication line fails.

Logic capability in the terminal used to be expensive and unreliable. The cost of logic, however, is decreasing fast, and its reliability, with the advent of LSI circuitry, has become very high. Buffering and logic can reside on a single mass-produced silicon chip in the terminal. Like other logic devices, micro- and minicomputers are decreasing in cost, leading to increasingly greater use of intelligent terminals and intelligent I/O systems.

87. Process-Control Entry Devices. In process control, e.g., chemical plants and petroleum refineries, inputs to the computer generally come from sensors (see Sec. **10**) that measure such physical quantities as temperature, pressure, rate of flow, or density. These units operate with a suitable analog-to-digital converter that forms direct inputs to the computer. The CPU may in turn communicate with such devices as valves, heaters, refrigerators, pumps, etc., on a time-shared basis, to feed the requisite control information back to the process-control system. See also Sec. **24**.

88. Magnetic-Ink Character-Recognition Equipment. A number of systems have been developed to read documents by machine. *Magnetic-ink character recognition* is one. Magnetic characters are deposited upon paper or other carrier materials in patterns designed to be recognized by machines and by operators. A change in reluctance, associated with the presence of magnetic ink, is sensed by a magnetic read head. The form of the characters is selected so that each yields a characteristic signature.

89. Optical Character Recognition. Information to be subjected to data processing comes from a variety of sources, and often it is not feasible to use magnetic-ink characters. To avoid retranscription of such information by key-entry methods, devices to read documents optically have been developed.

Character recognition by optical means occurs in a number of sequential steps. First, the characters to be recognized must be located and initial starting points on the appropriate characters found. Second, the characters must be scanned to generate a sequence of bits that represents the character. The resulting bit pattern must then be compared with prestored reference patterns to identify the pattern in question.

The scanning function may be performed in a number of ways. In one method, an image of the character is projected on the face of an image orthicon tube. An electron beam scanning across the tube derives a bit pattern in accordance with the image on the tube. In another method, the character image is translated with respect to a linear photodiode array, and the output of the array is sampled at periodic intervals to determine the bit pattern. In a third method, a scanning mirror is used in conjunction with a light-emitting diode (LED) array. A laser may also be used in conjunction with a photodiode.

Finding the initial character in the group to be recognized and sequencing from that character involve control of the character locations on the document. In another technique a *dark-space search* locates an upper left dark area printed on the document.

Another character-recognition technique involves computation of a *correlation function* between the input and the stored patterns. Optical spatial filtering may be used, although digital techniques are more common.

Another method of data entry is through the encoding of characters into a digital format, permitting scan by a wand. Typical of this encoding scheme is the bar code, known as the *universal-product code* (UPC), which has been standardized in the supermarket industry by industry agreement to the code selection by the Uniform Grocery Product Code Council, Inc. Besides being readable by a wand reader, the UPC can also be scanned and read by a laser, as in supermarket checkout terminals.

90. Batch-Processing Entry. One computer can be used to enter information into another. For example, it is often advantageous for a small computer system to communicate with a larger one. The smaller system may receive a high-level language problem, edit and format it, and then transmit it to a larger system for translation. The larger system, in turn, may generate machine language that is executed at the remote location when it is transmitted back.

Remote systems may receive data, operate upon them, and generate local output, with portions transmitted to a second unit for filing or incorporation into summary journals. In other cases, two CPUs may operate generally upon data in an independent fashion but may be so interconnected that in the event of failure one system can assume the function of the other. In other cases, systems may be interconnected to share the work load. Computer systems that operate to share CPU functions are called *multiprocessing* systems.

91. Printers. Much of the output of computers takes the form of printed documents, and a number of *printers* have been developed to produce them. The two basic types of printer are *impact printers*, which use mechanical motion of type slugs driven against a carbon ribbon or paper, and *nonimpact* printers, which use various physical materials to produce the characters.

Impact printers are divided into *line printers* and *serial printers*. Line printers print a full line at one time and operate at from 50 to 2,000 lines/min. Serial printers include typewriters and other such devices that print one character at a time. Some serial printers use engraved type for the formation of solid character images. Others use a matrix of wires to print dots that form a character. The latter are called *matrix printers*.

Impact printers generally use a high-speed electromechanical hammer to drive a print slug into contact with a ribbon and paper. In the *front printer* the hammer drives the print slug against the ribbon and paper. In the *back printer* the hammer drives the paper and ribbon against the type slug.

In a *drum printer* the drum is engraved with a set of characters for each column to be printed. As the drum rotates, the full set of characters is successively presented at each print position. In a back printer the drive hammers are mounted behind the paper, and the paper is driven against the drum at appropriate times to print the characters desired. A full rotation of the drum is required in order to print a line. The paper stays stationary during printing.

Figure 23-67 shows the basic configuration of a back-printing chain printer. The chain moves characters horizontally across the paper. Hammers at the print positions behind the paper are fired at the appropriate times to print the required character as the belt moves by. Chain printers have the advantage over drum devices in that any mistiming of the hammer results in a horizontal displacement of the character printed. Such a displacement is less discordant to the eye than the vertical displacement produced by a drum printer. The *train printer* is similar to the chain device except that the type slugs are pushed through a guide instead of being pulled on a belt.

Two basic types of print hammers are used in high-speed printers. One operates by means of an electromagnetic circuit with a movable iron armature. The armature itself forms the hammer-drive mechanism. The other uses the current in a coil in a magnetic field as a motive mechanism and operates in a manner similar to the voice coil of a loudspeaker.

Carriage movement in printer systems is provided by hydraulic actuators that may step the carriage from one line to the next in times as short as 4 ms, resulting in speed of paper movement ranging up to 100 in/s. Line printers are equipped to skip at high speed past lines not to be printed so as to increase the *throughput*, i.e., the total length of paper printed per unit time.

Electronic circuits, such as power drivers for the carriage and hammers, are usually contained in a separate control unit. In chain or drum devices, an electronic image of the physical position of the chain or drum is stored and compared with the characters desired at each location. When a match is found, the hammer is fired and the appropriate character is printed. The logical devices required for such operations are often contained in a separate control unit but may be located in the CPU box.

92. Serial-Impact Printers. A number of mechanisms have been developed for serial-impact printing. One consists of a ball covered with characters that is rotated and tilted before it is moved forward to imprint a character. This device is widely used in typewriters. Ordinary

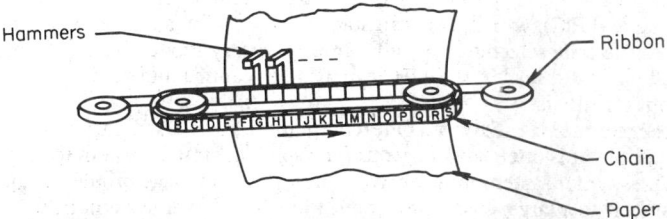

Fig. 23-67. A chain printer. The characters, on an endless belt, are moved transversely with respect to the paper. A hammer at each print position is fired as the character to be printed at that position appears.

electric typewriters, using type bars and horizontal carriage motion, are also used as serial printers.

The most widely used mechanism for serial-impact printers uses a matrix of wires to print a configuration of dots that form a character. The configuration for "T", for example, would be

$$\begin{matrix} \circ\;\circ\;\circ\;\circ\;\circ \\ \circ \\ \circ \\ \circ \\ \circ \end{matrix}$$

93. Nonimpact Printers. Nonimpact printers use various mechanisms to imprint characters on a page. There are two major classes of such devices, those using *special paper* and those using *ordinary paper*. In one type of special-paper device, the paper is coated with an opaque waxy substance on a contrasting background. When the paper is heated, an image appears. Characters are formed by contact with selective heated wires or resistors that form a matrix character. Another type of special paper is coated with a thin aluminum or other metal film that can be evaporated with a laser light. The laser beam is switched on and off under electronic control while being scanned across the paper by means of a mirror. The metal film can also be eroded away by means of a spark discharge from a suitably actuated array of wires.

In other systems, electrophotographic methods are used, with papers coated or filled with a photoconductive substance such as zinc oxide. The sheet is charged by corona wires and is selectively discharged by the optical image of the character. Small electrostatically charged particles, called *toners*, can be made to adhere to the paper at the positions of discharge and fixed by heat or pressure or both. Another method of nonimpact printing on special paper called *electrography* uses a dielectric-coated paper that is directly charged by ion generation. The charged paper can be toned and fixed as in electrophotography. Nonimpact printing using chemically treated optically sensitive papers is also available. The characters are generated by exposure to shaped light beams. In diazo systems, for example, the paper is optically illuminated by the character images and developed in a bath of gaseous ammonia.

Nonimpact printing can also be accomplished using plain paper, an advantage in a large computer installation, where the cost of paper is a major factor in the overall cost of printing. In *transfer electrophotography*, used with plain paper, a photoconductive surface (such as selenium or certain organic materials) is charged and selectively discharged as in electrophotography. A toner image is developed directly from the photoconductor, but this image is subsequently electrostatically transferred to plain paper, where it is then fused.

Recently announced printers combine laser and electrophotographic techniques to produce high-speed computer printout. The printing process uses a light-sensitive photoconductive material wrapped around a rotating drum. The photoconductor is electrically charged and then

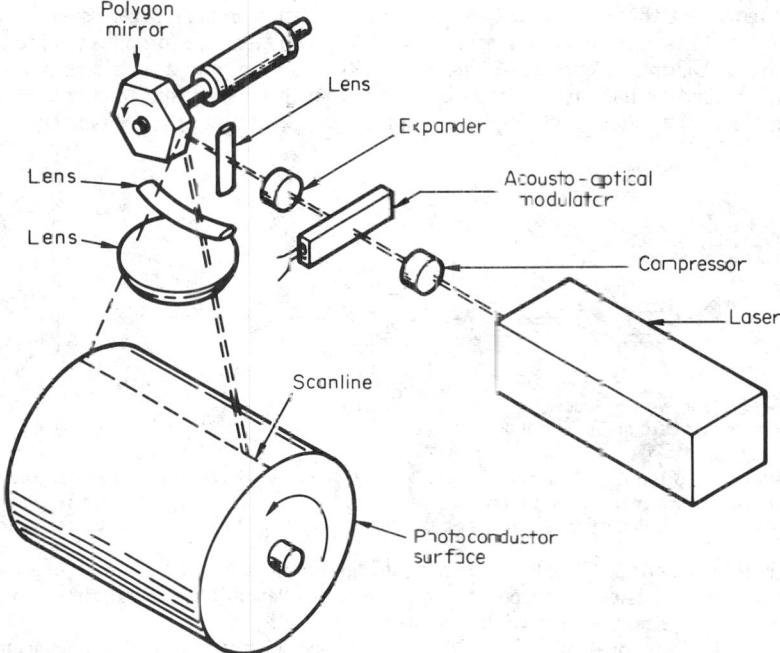

Fig. 23-68. A laser printing subsystem, showing the polygonal mirror and scanning method.

exposed to light images of alphanumeric information. These images selectively discharge the photoconductor where there is light and leave it charged where there is no light. A powdered black toner material is then distributed over the photoconductor where it adheres to the unexposed areas and does not adhere to the exposed areas, thus forming a dry powder image of the information. The image is then transferred to paper where the dry black toner is fixed to the paper by fusing it with heat.

Two exposure techniques are used. In one method, an optional full-page mask containing fixed information is placed close to the photoconductor surface and exposed once per page by a flash lamp. The second technique is to expose the photoconductor by using a focused laser beam. One way to scan parallel to the drum axis is to use a rotating polygonal mirror. An acoustooptical modulator can be used to turn the beam on and off during the scan. Figure 23-68 illustrates a laser printing subsystem. A helium-neon laser provides the energy for exposing the photoconductor mounted on the rotating drum.

94. Image Formation. Electrophotography, transfer electrophotography, and thermal methods require that optical images of characters be generated. In some types of image generation, the images of the characters to be printed are stored. Other devices use a linear sweep arrangement with an on-off switch. The character image is generated from a digital storage device that supplies bits to piece together images. In the latter system any type of material can be printed, not just the character sets stored; the unit is said to have a noncoded-information (NCI) capability.

One method of optical character generation uses a cathode-ray tube. A character mask in the tube selectively intercepts the electron stream. The resulting image on the tube face can be projected upon a suitable nonimpact printing receptacle. In another method, the electron beam undergoes linear scanning and is switched on and off by an electronic character generator. Each line of the sweep produces a section of a line of characters. Repeated sweeps are used to generate a full line of information.

Characters can be also generated using a laser and a rotating mirror, either by switching the laser off and on or by projecting through a standard or holographic character mask. Sets of light-emitting diodes can also be used for character generation.

The output terminals for digital image processing are high-resolution monochrome or color monitors for real-time displays. For off-line processing, a high-resolution printer is required. The inputs to digital image-processing requirements are generally devices such as the vidicon, flying-spot scanner, or color facsimile scanner.

95. Ink Jets. Another method of direct character formation on plain paper uses ink droplets. When a stream of ink emerges from a nozzle vibrated at a suitable rate, droplets tend to form in a uniform, serial manner. Figure 23-69 shows droplets emerging from a nozzle and being electrostatically charged by induction as they break off from the ink stream. In subsequent flight through an electrostatic field, the droplets are displaced according to the charge they received

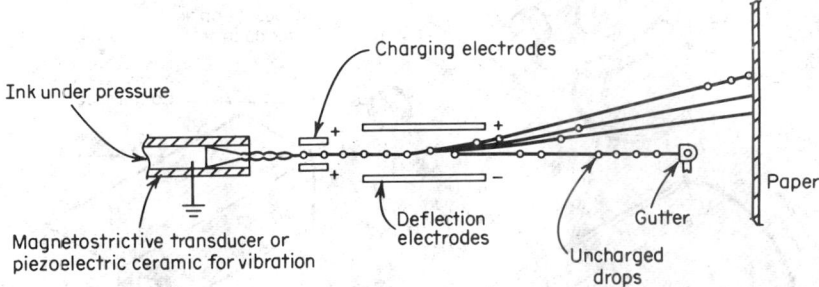

Fig. 23-69. Ink-jet printing. Ink under pressure is emitted from a vibrating nozzle, producing droplets which are charged by a signal applied to the charging electrodes. After charging, each drop is deflected by a fixed field, the amount of deflection depending on the charge previously induced by the charging electrodes.

from the charging electrode. Droplets generated at high rates (up to 500,000 droplets per second) are guided in one dimension and deposited upon untreated paper. The second dimension is furnished by moving the nozzle relative to the paper.

A set of nozzles, as shown in Figure 23-70, can be used with vertical paper displacement to deposit characters. Each nozzle prints sections of characters in sequential fashion under electronic control in the horizontal direction while paper movement completes character formation in the vertical direction. Ink-jet systems have been developed using very fine matrices for character generation, producing high document quality.

There are three main variations of ink-jet printers, *continuous-droplet, impulse,* and *electrostatic ink-jet* printers. Continuous-droplet ink-jet printing has been developed for lower-quality high-speed printing and medium- to high-quality serial character printing. Impulse ink-jet printing tends to give lower-quality data output. A significant feature is its overall simplicity and low cost, which have made it attractive for telecommunications output and facsimile. Electrostatic ink-jet printing is characterized by low print speed and low output quality. A typical application is facsimile printers.

96. Visual-Display Devices. Visual-display devices associated with a computer system range from console lights that indicate the internal state of the system to cathode-ray-tube displays that can be used for interactive problem solving. In the cathode-ray tube, a raster scan, in conjunction with suitable bit storage, generates the output image. Fixed character sets can also be generated by masks in the cathode-ray tube. The latter devices require less stored information, since a coded input signal can be used to display the appropriate character. Such displays are often used with a keyboard for information entry, so that the operator and computer can operate in an interactive mode.

Gas panel displays (Fig. 23-71) use a dot matrix to form characters by means of vertical and horizontal fine wire conductors that serve as the anode and cathode of miniature gas discharge "tubes" at each crosspoint of the dot matrix. A pattern is formed by energizing appropriate horizontal and vertical electrodes to create a matrix of light dots.

Gas panel displays are versatile devices because they are capable of displaying a wide range of patterns. They are expensive, however, and require relatively complex character-generation logic to control the horizontal and vertical components of the image. High voltage and the need to select a number of dots to form one character add to the high cost.

One of the most promising displays under development, with a potential for low cost, low power, and high performance, is the *liquid-crystal display* (LCD). Liquid crystals differ from most other displays in that they depend on external light sources for visibility. This type of

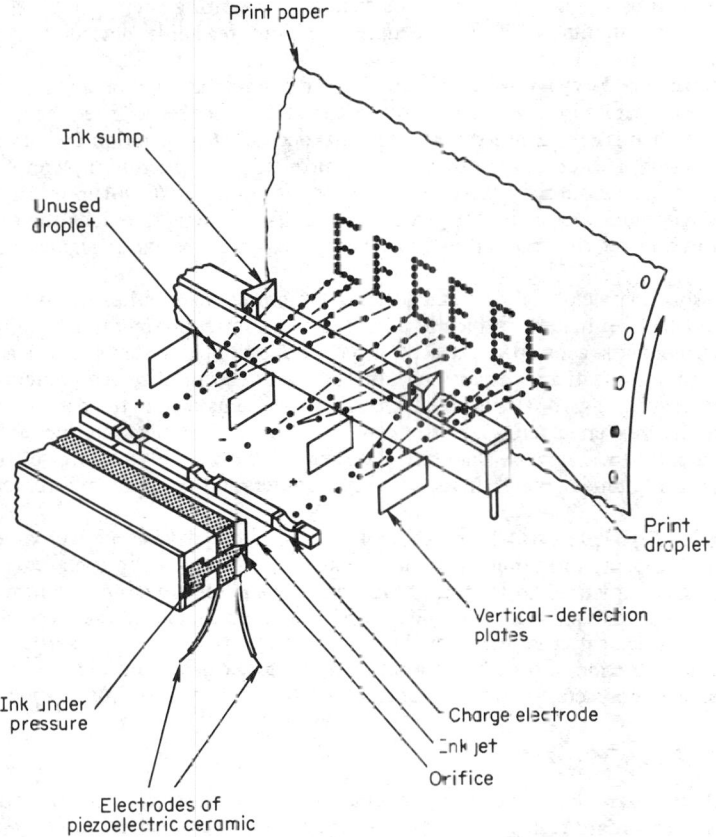

Fig. 23-70. A printing system using ink jets. Individual nozzles are used to print characters in sets of two. Horizontal electrostatic deflection forms line sections of each character, while the paper movement supplies the vertical dimension.

display is composed of two parallel glass plates with conductive coatings on their inner surfaces and a liquid-crystal compound of the nematic variety sandwiched between them. In the dynamic scattering display, the clear organic material becomes opaque and reflective when subjected to an electric field. As in other displays, the characters are usually built up from segments.

97. Program Interrupt. Remote access to a computer system requires that an external inquiry be able to interrupt a program in process. In most systems operating in a multiprogramming mode, accesses from multiple inputs are queued for attention so that as a program in progress is completed or as an interrupt is made mandatory, the computer can store its current program and give attention to the external inquiry. When the inquiry is satisfied, the system resumes operation on its original program until the next interrupt occurs.

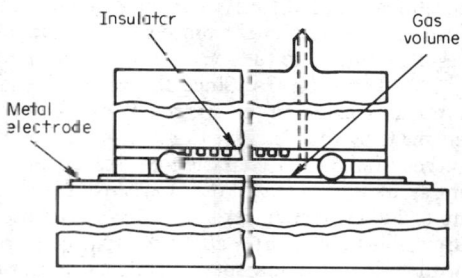

Fig. 23-71. Schematic representation of an ac display panel.

98. Channel Extension. In systems with multiple inputs and extensive communication links, channel control systems may be extended. For example, an intermediate storage mechanism for each input channel can be introduced through a drum or fixed-head file. A CPU subsystem is then used to control the information flow from external equipment to the file and from the file to the CPU.

99. Paper-Tape I/O. Paper-tape drives have been used as I/O gear. Such devices usually operate by punching or sensing the presence of holes, for writing and reading, respectively. A sprocket maintains alignment. Paper-tape units are slow, read-only devices that are usually inexpensive.

100. Computer-Output Microfilm. Record keeping in such organizations as finance and insurance companies involves massive quantities of data that must be retained, even though individual reference is rare. Since the cost of paper-document preparation and storage in such applications is high, it is customary to store the information on microfilm. Microfilm printers developed for computer output, called *computer output microfilm* (COM) printers, use nonimpact printing techniques for character generation and lens systems to reduce the optical image for storage directly on the microfilm. Laser beams may also be used, scanning directly on microfilm.

The increasing complexity of data and information processing requires the successful use of computers and I/O equipment to handle all the processing requirements in environments of many configurations of centralized, decentralized, or distributed systems. With the advances made in data communications, networking, and the use of decentralized computers, distributed processing systems have gradually evolved. Distributed systems, where the data storage and processing functions are shared across a mix of computers and communication lines and nontrivial operations are performed at more than one place, have grown rapidly. A large distributed processing system can include several thousand minicomputers and terminals and *remote job-entry* (RJE) stations.

101. Audio Equipment. In some applications, audio devices deliver a recognizable speech signal in response to an inquiry to the computer. The simplest approach is to use recorded words or phrases, sequenced to form messages. More complex devices can synthesize speech sounds into words and phrases. Equipment has also been developed that can discriminate between human voice patterns. Words are broken down into frequency bands, and the intensity of the bands as a function of time forms a recognizable pattern. Recognition of such patterns is accomplished as in character recognition, and the word information is thus extracted.

TELECOMMUNICATIONS

102. Background. Early computer systems were tightly integrated with locally attached I/O facilities, such as card readers and punches, tape readers and punches, tape drives, and discs. Data had to be moved to and from the site of the programming system physically. For certain systems, the turnaround time in such processes was considered excessive and it was not long before steps were taken to interconnect *remotely located* I/O devices to a computer site via telephone and telegraph facilities.

This *interconnection* is accomplished in one of two ways. The *off-line* approach involves the interconnection of two devices via a communications line. Communication of the data is completely separate from the processing computer. It is typically in batch form, and data enter the processing system in the same way other locally acquired information enters. The *on-line* approach is to use the computer as the receiving device. Here, data are transmitted via communications media directly into the computer and are then either stored on an auxiliary storage medium for subsequent processing (remote job entry, RJE) or processed immediately and the results returned to the source in *real time*, i.e., *conversationally* or *interactively*.

103. Terminals. Since the early advent of telecommunications, many terminal device types have come on the market. The earliest and most popular devices were the keyboard printers made by the Teletype Corporation and others. These electromechanical machines operate at speeds from 5 to 11 characters per second. Such devices were also used in the early data processing systems and are still used today. More recently electronic terminals with buffers that operate up to 120 characters per second have been used. New low-cost electronic technology has made it possible to offer *error-checking, code conversion*, and *editing* facilities. The major application of these low-cost, low-speed terminals is interactive communications via the *switched communications facilities* offered by the common carriers.

The second most popular type of terminal device is the cathode-ray-tube (CRT) *visual display* and keyboard; it operates at speeds from 1,200 to 9,600 b/s across common-carrier telephone facilities or directly attached to the local channel of a processor. Displays generally are alphanumeric or graphic, the preponderance of terminals being alphanumeric. Many such displays can be attached to a single control unit, which can greatly reduce the overall costs of the electronic devices required to drive the multiple devices. The control unit can also reduce communications cost by allowing the multiple locally connected devices to share a communications line to a remote *host computer* by a technique called *concentration* (Par. **23-114**). Some controllers support the attachment of hard-copy printers and have sufficient logic to permit printout of information on a display screen without communicating with a host computer. Many of the new display controllers are, in reality, small *mini-* or *microcomputers* that contain logic to support such functions as paging, error detection and correction, buffering, code conversion, and data compaction or compression. Displays are generally used in conversational or interactive modes of operation.

A third category is the *batch-entry terminal*. These devices generally permit the attachment of card readers and punches, line printers, and other terminal devices. They are designed to permit the batching of data at a remote site and then to transmit bulk stored data to a host, usually via switched facilities of the common carrier, or to receive bulk data from a host for later output to a selected device on the batched terminal. For storage, tape units, discettes, and tape cassettes are attached to such terminals. Many of these terminals are programmable and can thus perform many sophisticated I/O functions.

The many specialized *transaction-type terminals* that are being designed for specific industry applications form a fourth category of terminals. Special point-of-sale terminals for the retail industry, bank teller and cash dispensers, and airline ticketing terminals are examples of transaction-type terminals.

A category of terminals that incorporates many of the features of previous categories, the *intelligent work station*, is really a remote mini- or microcomputer system that permits the distribution of application program functions to remote sites. Such *intelligent terminals* are capable of supporting a diversity of functions from interactive processing to remote batched job entry. Some of these terminals contain such a high level of processing capability that they are, in fact, true host processors.

104. Communications Media. Several media by which data are transported are wire, microwave, coaxial cable, optical-fiber cable, and radio and satellite links. Within a single enterprise location, communication media may be provided by the using organization. However, when a public thoroughfare has to be crossed, the user must acquire approval from a regulatory body, usually a state utility commission, or acquire the service of a recognized common carrier. In countries other than the United States and Canada the carrier that provides telecommunication services generally also has the regulatory power. A general term used to describe this entity is *postal, telephone,* and *telegraph company* (PTT).

The rapidly growing demand for data-transmission capacity has fostered the development of many new forms of communications services. The *Federal Communications Commission* (FCC) is the agency of the United States government responsible for regulating interstate and international communications originating or terminating in the United States. In recognizing the growing needs of the data communications industry, the FCC has opened the industry to competition. Thus, within the past 10 years, new classes of carrier have entered the United States market.

For example, data users can now acquire communications services from organizations classified as specialized common carriers, value-added carriers, and satellite carriers. (Besides data transmission, a *value-added carrier* may provide such services as storage and error detection and correction.) Each new type of carrier has introduced new forms of media with characteristics more suited to the transmission of data than voice. The competition has been met by the conventional carriers with competitive offerings, so that data users have a number of options to satisfy their communications needs. Outside the United States, the PTTs may offer similar services, but they represent the only source of telecommunications service within their countries.

Telecommunications services are characterized by their *technology, connectivity, capacity, error characteristics, propagation delays,* and *availability*.

105. Technology. Information can be transmitted over media in either an *analog* or a *digital mode*. Analog transmission implies a continuous spectrum of frequencies, whereas the digital mode involves a stream of discrete pulses very much like the movement of data within computer circuits. Obviously, when data are transmitted between digital logic devices such as terminals or computers, digital tranmission facilities are preferable because the data are more compatible with those media. However, the preponderance of information transfer today is via facilities designed to carry analog voice signals.

Thanks to the rapidly declining cost of digital circuitry, there is a growing trend toward digital communications facilities, but because the existing capital investment in analog facilities is well over 100 billion dollars in the United States, it is expected to be many years before digital facilities completely replace analog facilities. Thus, in most current teleprocessing environments, the digitally encoded data of the computer must be transformed into analog signals at the source to match the characteristics of the transporting medium and then retransformed to the digital form at the receiving end.

This is accomplished by using a paired set of devices called the *modulator* at the sending end and *demodulator* at the receiving end. The modulator and demodulator equipment are collectively termed *modems*. The sending modem modulates the amplitude, frequency, or phase of an analog carrier signal in accordance with the value of the digital signal (0 or 1), and the receiving modem converts the modulated carrier back to the original bit stream, as shown in Fig. 23-72.

In communications parlance, the equipment attached to the communications system is called *data terminating equipment* (DTE), and the device to which the DTE attaches is called *data circuit terminating equipment* (DCE). Historically, the carriers have provided the modem for switched network use. (Outside the United States many carriers continue to be the sole source

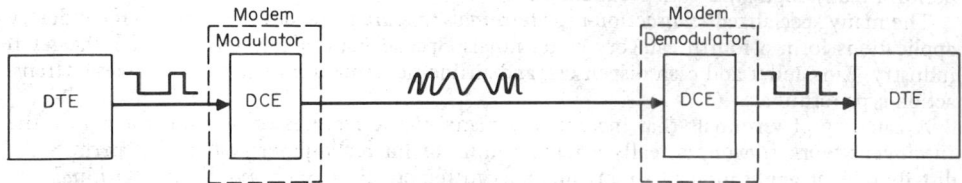

Fig. 23-72. Flow of digital data via analog communications facilities.

of this equipment for both switched and leased circuits.) Thus, the term *modem* has become synonymous with DCE. Still another synonym frequently used in the United States for a modem provided by the carrier is *data set*.

Digital transmission facilities do not require costly modems for data transmission, but do require a *signal converter* between the communications system and the user device. The signal converter matches the signal levels and the clock rate required by the digital transmission facilities. Thus, a signal converter may also be called a DCE.

106. Connectivity. Connectivity can be described in terms of its permanence, direction, and shape.

Permanence. Connections may be temporary (switched or dialed) or permanent. The charge for common-carrier switched facilities normally varies with the length of time a connection is held. Thus when data volumes between known DTEs demand a long connect time, it may be more economical for users to lease permanent circuits. The economic crossover point is a function of the carrier's pricing (tariff) structure. However, the decision to use leased or switched facilities involves more than tariff considerations alone. The decision must also take into consideration the *shape* or configuration (topology) of the components being interconnected and the directionality, capacity, and integrity requirements of the communicating components.

Directionality. Communications channels may be unidirectional *(simplex)*, bidirectional nonconcurrent (two-way alternate or half duplex), or *bidirectional concurrent* (two-way simultaneous or full duplex). Typically, the analog-switched facilities provided by common carriers are two-way alternate (TWA). The familiar telephone circuits are of a TWA type; two persons cannot talk at the same time.

Two-way alternate channels impose additional delays in communication since it becomes necessary to reverse the direction of transmission or to "turn the line." In data communications, this involves reversing modems so that a sending modem shuts its carrier and assumes a receiving role and vice versa. With voice-grade switched media, TWA involves reversing echo suppressors inherent in the circuit that prevent concurrent signal flow in both directions. Almost all leased analog and digital facilities are inherently two-way simultaneous circuits.

Shape. DTEs connected via a common-medium facility are generally said to be physically adjacent to each other. It is shown later that this physical connectivity does not imply anything regarding the ability to communicate between such devices. Such a capability also depends on the existence of higher levels of logic contained in the communicating devices.

DTEs may connect to each other either in point-to-point or multipoint configurations. *Point-to-point* connections imply the existence of a single circuit between two DTEs. The circuit may be permanent or switched and may be real or virtual. A *real circuit* implies the dedication of resources to effect the connection between the two endpoints by means of either a *frequency spectrum* or *time-division* slots on a channel.

A *virtual circuit* implies that there is awareness of connectivity at both terminating ends but no fixed end-to-end allocation of resources. Resources are made available only upon the occurrence of data at the interference between the DTE and DCE. In addition, when resources are allocated for a virtual circuit, they may exist only over a segment of the path between the origin and destination points of the connection. This type of communications technology is sometimes referred to as *statistical time-division multiplexing* or *store and forward* (Pars. **23-113** to **23-116**).

Multipoint refers to the connection of more than two DTEs to a common shared communications channel. The shape of a multipoint facility is that of a star, bus, or ring, as shown in Fig. 23-73. Star and bus configurations are generally bidirectional; a ring is generally unidirectional.

Sharing such channels can be achieved either by providing a control mechanism in the communications facility or by a *protocol agreement* (set of rules) between the DTEs attached. The protocol approach usually implies that the channel between DTEs is static and merely exists to transport information bits. In common-carrier dial facilities, connections between DTEs are made by switching controls contained inside the carrier's communications facilities.

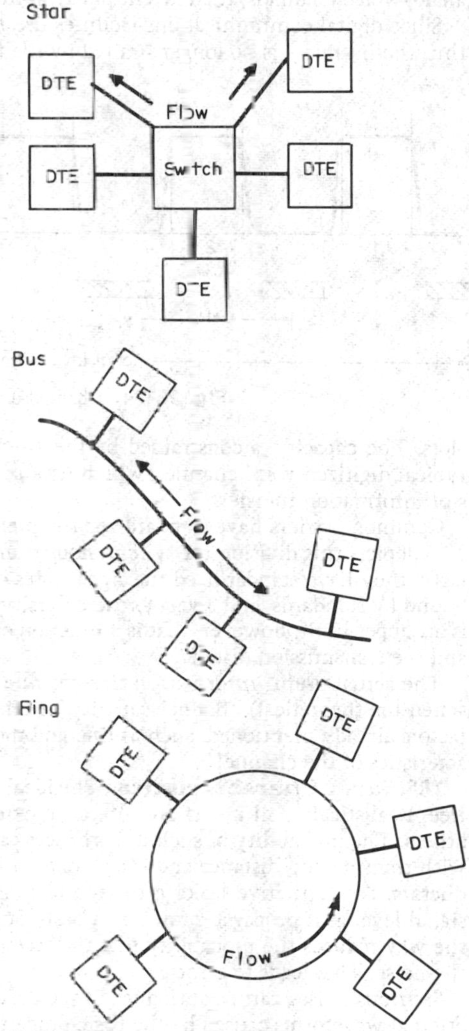

107. Capacity. The inherent information-carrying capacity of most media is quite large. For example, some systems today use a copper wire carrying 1.544×10^5 b/s. Dedicating a single-wire conductor to a voice signal is very wasteful of the medium because effective voice communications requires less than 4 kHz of frequency spectrum. Thus, common carriers have developed a technique that permits the division of the potential frequency spectrum of the medium into multiple 4-kHz slots, so that many voice conversations can share one physical connection. This technique of channelization is called *frequency-division multiplexing*. Each 4-kHz slot is called a *channel* (see Fig. 23-74).

Fig. 23-73. Shapes of shared communications channels.

Historically, the capacity of a data-communications channel has been specified in bauds (Bd), named after Baudot, an early worker in the communications field). This term is usually applied to analog communications and is defined as the maximum number of analog signal transitions per second (signal rate) that can occur on a channel of a given medium. This is determined by the bandwidth, which defines the range of frequencies a channel can successfully pass, and by the type of modulation used. Thus a channel designed to pass a frequency bandwidth up to 2,400 Hz is designated as a 4,800-Bd line.

In digital technology, a more useful measure of communications capacity is *bits per second*.

The number of information bits per second that can be transmitted on existing analog facilities depends on the signaling techniques employed by the paired modems. Thus, it is quite possible to transmit many more bits per second than the baud rate of a line implies.

Work by Shannon shows that the theoretical maximum limit of information transfer on a voice-grade channel with 2,500-Hz bandwidth is approximately 24,900 b/s. This assumes that the only disturbance on the channel is that introduced by Brownian (molecular) noise. With analog voice channels, the current practical limit is 9,600 b/s.

Since digital communications facilities use *time-division multiplexing,* the bandwidth — and thus the bit rate — is no longer limited by the filters required to divide the spectrum into 4-kHz

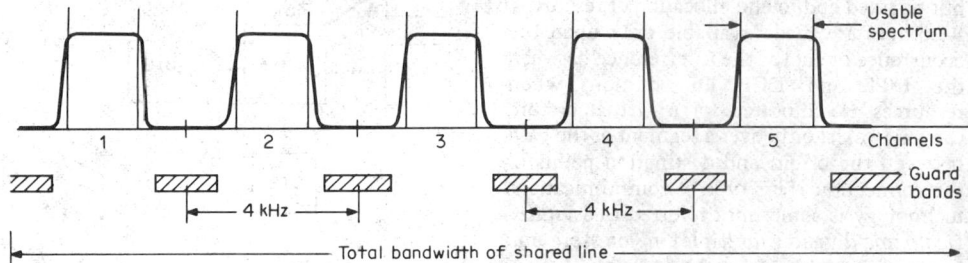

Fig. 23-74. Channel utilization for voice signals.

slots. The capacity is constrained by the *frame size* and *clock rate* imposed by the carrier. A typical digitized voice channel, which uses *pulse-code modulation,* can handle up to 56,000 b/s of information transfer.

Common carriers have standardized the preferred signal rates of their analog channels. Table 23-5 depicts this division for *synchronous transmission* on leased telephone-type services. Similarly, they have standardized the signal rates of their digital channels. Private facilities are not bound by standards, and a very broad diversity of bandwidths and bit rates can be found. There is an upper limit, however, that is a function of the electrical properties of the selected medium and the transmission distance.

The actual useful *information transfer rate* across a communication channel is less than the stated (or theoretical) channel capacity in bits per second. The lower practical rate is due to factors already mentioned, such as line and modem turnaround time, as well as the error characteristics of the channel.

108. Error Characteristics. The ideal medium for data transmission is one that is error-free. Realistically, all media are subject to external and internal disturbances, some more than others. The probability of such disturbances causing errors in the transmission of data increases with transmission distance and is a characteristic of each medium. For example, with wire conductors, the capacitive leakage to ground increases with distance. This attenuates the source signal level as it propagates toward its destination. Any introduction of noise to the signal along the way reduces the probability that the receiver can distinguish between the information and the noise. This leads to errors.

Spurious noises can be introduced by external electromagnetic impulses or distortion of the original waveforms caused by the basic properties of the medium itself. To reduce the effects of

TABLE 23-5   Standardization of Data Signaling Rates for Synchronous Data Transmission on Leased Telephone-Type Circuits

Preferred range of data signaling rates, b/s	
600*	4,800*
1,200*	7,200
2,400*	9,600*
3,600	

*Modems for use on leased telephone-type circuits at these data signaling rates. See *CCITT Recomm.* V.23, V.27, V.27 bis, and V.29, respectively.

attenuation, *repeaters* are placed in the medium to reamplify the signal levels. In analog facilities, any noise introduced between repeaters is amplified along with the signal.

Digital communications repeaters, on the other hand, can both amplify the signal and remove the effects of many low-level noise impulses by reshaping the pulses. Thus, the error characteristics of digital transmission facilities are usually orders of magnitude better than those of analog transmission facilities.

Furthermore, leased analog transmission is generally better than switched transmission since the physical channel is dedicated and identifiable. This makes it possible for the carrier to take steps to overcome the effects of medium parameters that introduce undesired distortion. Thus

TABLE 23-6 Typical Average Channel Bit Error Rates (BER)

Channel	BER
Subvoice grade	10^3–10^4
Voice grade, switched	10^3–10^5
Leased	10^4–10^5
Digital, terrestrial	10^6–10^7
Satellite	10^7–10^9
With forward error control	10^9–0^{12}
Computer channel	10^2

leased facilities can be acquired with different *grades of service*. This grading implies that the carriers have conditioned each medium to reduce it to a specific level of distortion. Certain carriers also attempt to reduce the apparent error rate of their media by adding redundancy to the signals to permit error detection and correction at the receiving end of the carrier facility. Such additions are transparent to the attached DTEs and are termed *forward error correction* (FEC).

Typical error characteristics to be found on various types of media are shown in Table 23.6. Errors are usually expressed in expected average *bit error rates* (BER).

109. Propagation Delay. Propagation delay pertains to the time required to transfer a signal from a sending DTE to a receiving DTE and is a function of the physical distance the signal must travel. On terrestrial circuits this delay is usually measured in milliseconds and has no significant influence on performance.

Satellite circuits, however, introduce delays of the order of 260 ms on a simplex circuit, since the space segment of the link is approximately 45,000 mi one way, i.e., the distance from sender to satellite plus distance from satellite to receiver. Since the normal procedure for handling errors in a medium is to request retransmission from the origin, it is apparent that such long propagation delays could greatly influence the effective data transmission rate of a satellite channel. To reduce the effect of delay on performance, satellite channels normally use FEC.

110. Availability. Circuits between DTEs can fail for many reasons. Generally, the reliability of common carrier facilities is extremely high (99%+). However, *availability* is a more important attribute than *reliability*. As users continue to move toward real-time and on-line applications and the operation of their business becomes more dependent on communications, the availability of these services becomes critical. *Leased services* may offer lower availability than switched service since there is no direct backup. (Alternate dial backup can be purchased to overcome this shortcoming.) In many instances, a failure of a *switched-circuit* connection can by overcome by redialing, or automatic switchover within the carrier's facilities can be used. New switched circuits can be established very quickly, and the failing circuit can be isolated. Users of a switched service know that the resource is only temporarily allocated to them and that they may never use the same circuit twice.

Leased multipoint is generally lower in availability than leased point-to-point since it becomes much harder to isolate the source of failure when multiple DTEs are connected to a common channel.

111. Line Control. The establishment of a connection the transmission of data across the connection, and the termination of a connection are three major phases of operation that require coordination between a DCE and DTE. The carriers and business equipment manufacturers, through international standards organizations, have defined several standard recommendations for the definition of the interface between a DCE and a DTE. The most widely implemented recommendations for communication between DTEs attached to public switched networks are

CCITT Recomm. V.24 and V.25 and the related United States standards, such as EIARS 232C and EIARS 366, respectively.

There are numerous other interfaces, but the key requirement is that the DTE contain the logic and electrical and physical characteristics to permit it to talk to the DCE. This logic is termed *line control.* Some examples of line-control protocol information exchanged between the DTE and DCE are shown in Table 23-7.

TABLE 23-7   Typical DTE-DCE Interchange Circuits

	Data		Control	
	From DCE	To DCE	From DCE	To DCE
Transmitted data		×		
Received data	×			
Request to send				×
Ready for sending			×	
Data set ready			×	
Connect data set to line				×
Data terminal ready				×
Data channel received line signal detector			×	

112. Link Control. The term *link* has a logical rather than a physical connotation. Whereas line control is concerned with protocol between a DTE and DCE, *link control* is concerned with protocols between DTEs communicating across the medium. The main functions of link control are to initiate and terminate contact between communicating DTEs, establish and maintain synchronization of data frames being exchanged, enhance the apparent integrity of the link by providing error detection and correction procedures, and in certain cases to control access to avoid possible collision of data frames when the line is being shared. If the detected data-frame error rate is determined to be excessive, link control may also initiate diagnostic procedures in an attempt to isolate the source of trouble.

Asynchronous, i.e., start-stop, mode was the earliest form of link control and was used in terminals provided by the Teletype Corporation and others. Since these devices were not buffered, the link controls were designed with no error-detection codes and were asynchronous by character. Each character transmitted was enveloped by a start bit and a stop bit (not necessarily of the same duration). The early character-code representation was the 5-b Baudot code (Fig. 23-19). Subsequent enhancements have led to the 7-b ASCII code (Fig. 23-23). Asynchronous transmission proves to be quite wasteful of an expensive resource. In the Baudot code, each 5-b character is framed by 2.42 character time periods for an overhead of 33%. In addition, the asynchronous rate of data entry introduces very high rates of idle time on the communications line.

113. Link-Control Methods. Possibly excessive idle time and synchronization-bit overhead were recognized very early by the computer industry and led to the development of *synchronous link protocols.* The most popular of these protocols, termed *binary synchronous communication* (BSC), depends on the existence of buffered communicating devices that disassociate the rate of transmission of information from the rate of entry to the device. Thus, data are blocked and forwarded in bursts. The overhead introduced by the delineating characters is now lessened and depends on the size of the data frame. It is true that with a single key-entry terminating device line, idle time is still evident with BSC, but by attaching multiple devices to a common line in a multipoint configuration the line can effectively be shared, reducing the idle time. For this purpose further logic must be added to the link control to control access to the link. Nevertheless, even considering the cost of such logic, significant savings result.

In the configuration shown in Fig. 23-75 one station on the line is designated as a primary or master station, and all other stations are designated as secondary. The primary station polls each secondary station, asking whether it has anything to send. If there is nothing to send, a negative acknowledgment is returned. If there are data, the

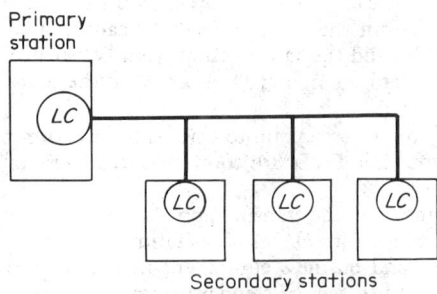

Primary station

Secondary stations

Fig. 23-75. Multipoint configuration with primary and secondary stations and link control (LC).

station proceeds to transmit. The primary selects the secondary station when it has information to send. Note that in this configuration the secondary can communicate only with the primary and not with another secondary.

Early versions of BSC were highly code-sensitive, and certain characters were reserved for control. Such bit configurations could not appear in data streams because control signals might be confused with data. This limitation was overcome in later versions of BSC with introduction of the *data-link escape* character (DLE). The presence of a DLE signals the receiving station that all data to follow are not to be interpreted as control until another DLE appears. Thus data char-

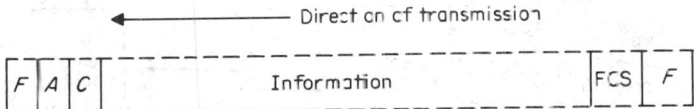

Fig. 23-76. Frame structure for high-level data link control (HDLC) in which F is a flag, A is address, C is control, and FCS is frame-check sequence

acters with the same bit configuration as control characters can be transmitted without ambiguity.

DLE still requires the sending station to be sensitive to the content of data fields in the bit stream to make sure that data do not contain a DLE character representation. When such a situation occurs, the sending station adds a DLE character to the data stream and the receiving station upon detecting two concurrent DLEs discards the added character and treats the remaining DLE as data. This mode, defined as a *transparent mode of operation*, permits the user to send any bit configuration as data.

The syntax and semantics of the BSC protocols in use today are relatively standard, but there is a wide variation in (1) the circumstances under which control information can appear in a frame and (2) the nature of an acceptable sequence of protocol exchanges. Thus, many so-called BSC DTEs are not really compatible and cannot communicate with each other effectively.

Error control is provided by adding a *block-count-sum field* that employs a *polynomial*. The receiving station regenerates the block sum using the same algorithm as the sending station. If it is not the same, an error in transmission is assumed and a negative response (negative acknowledgment, NAK) is sent back, signaling the sender to retransmit the frame. Otherwise a positive response (acknowledgment, ACK) signals the sender to proceed with the next frame. This form of error correction is termed automatic request for repeat (ARQ).

To overcome the incompatibilities of existing BSC protocols a newer and more efficient *synchronous* bit-oriented protocol, called *high-level data link control* (HDLC), has been developed by the *International Standards Organization* (ISO). This protocol is based on a fixed frame structure with a rigid definition of bit positions in control headers and trailers (Fig. 23-76). Start-stop and BSC protocols use characters to signify control information, whereas HDLC uses bits. The control overhead is thus significantly reduced. The protocol is code-transparent and highly flexible and thus easily adaptable for the transmission of noncoded information over media with widely varying attributes, e.g., terrestrial and satellite communications. One mode of HDLC supports nonpolled communications and permits a large number of multiple frames to be transmitted before requiring a response at the sending stations. This permits the efficient use of satellites as wide-band, long-propagation-delay links.

As one would expect, the high cost of a communication medium over a long distance leads to the development of many techniques to optimize its use. Multiplexers and multipoint interconnection for this purpose have been mentioned. But with reduced cost of computing and memory, it has become feasible to consider using these mechanisms as *store and forward* switching facilities to effect a higher degree of sharing of communications media. This has led to the development of *concentrators, message switches,* and commercial *packet switching.*

114. Concentrators. Traffic from many low-speed devices can be accumulated in the memory of a processor and can share a common outgoing line by being queued for service. Since, statistically there is low probability that all attached devices will be active concurrently, and since it is even less probable that all devices will be sending at the same exact instant, the capacity of the outgoing line can be significantly less than the sum of the capacities of the incoming low-speed lines, as shown in Fig. 23-77. Should there be concurrent inputs while one message is being serviced on the outgoing line, the other messages can be stored in the processor buffers awaiting their turn for link service.

Several types of concentrators exist. The terminal concentrator was described in Par. **23-103**. A second type of concentrator, termed a *front-end processor,* moves communications line and link control functions, as well as other higher-level functions, from a host into a separate processor. This frees host CPU cycles for applications functions. A third type of concentrator is a user-provided switching node (DTE) interconnected by carrier leased-line facilities that replace the switching functions of the carrier. Concentrators combining sets of these functions can also be found. The switching technologies used are either *message switching* or *packet switching.*

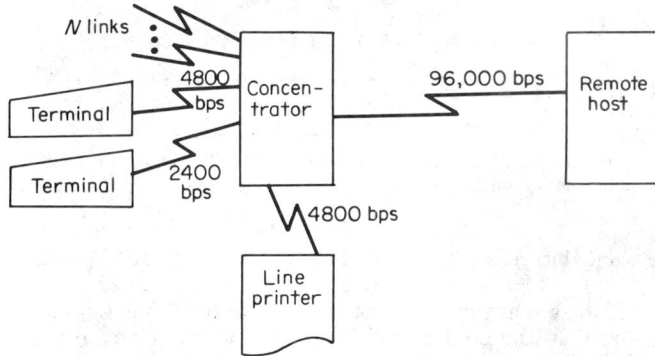

Fig. 23-77. Effect of a concentrator showing that the sum of the line speeds between the terminals and the concentrator is greater than the link speed from the concentrator to the host.

115. Message Switching. The ability to store a message and later forward it to its addressed destination is an early computer application. It replaced torn paper tape, by which messages sent from an origin and containing destination addresses in their headers are sent to a collection center and rerouted to the addressed destinations. The large auxiliary storage devices and the logic capabilities of a computer are ideally suited for the automation of this function.

Message switching supports communications between two DTEs that need not be concurrently connected. The sender forwards a message to the switch (typically a host processor) that stores it on a recoverable medium. When the message is correctly stored, the sender is informed and may then disconnect. At some later time when the receiver is available, the switch establishes a connection, retrieves the message from its file, and delivers it, as shown in Fig. 23-78.

116. Packet Switching. *Packet switching* draws from both message switching and concentrator technology. It uses computer nodes (concentrators) that support the boundary interconnection of local terminals via low-cost lines or computers via high-speed lines. These concentrator nodes are in turn connected to each other in a limited mesh configuration, as shown in Fig. 23-79. There is generally provision for at least two points of connection to each node to provide increased availability of paths between DTEs. The interconnected packet nodes form a *backbone network* between DTEs.

The availability of more than one path between any origin-boundary and destination-boundary node permits the selection of alternate paths by *routing functions* in the concentrators. Selection of a specific route may be made at the time a circuit is established, i.e., *fixed-path routing.* It can be dynamically selected for each data unit to be routed, based upon the switching node's current awareness of the resistance to flow experienced on any path between itself and the designated destination, i.e., *adaptive routing.* Many routing strategies exist that can be categorized in the context of one of these two approaches.

Data units are forwarded from node to node by the routing mechanism. They are stored in node memory rather than in the auxiliary file storage employed by message switching. Because the paths taken for messages between a paired set of end-user elements (programs and terminals) are not dedicated solely to this pair and may not be dedicated for the duration of their logical connection, the circuit

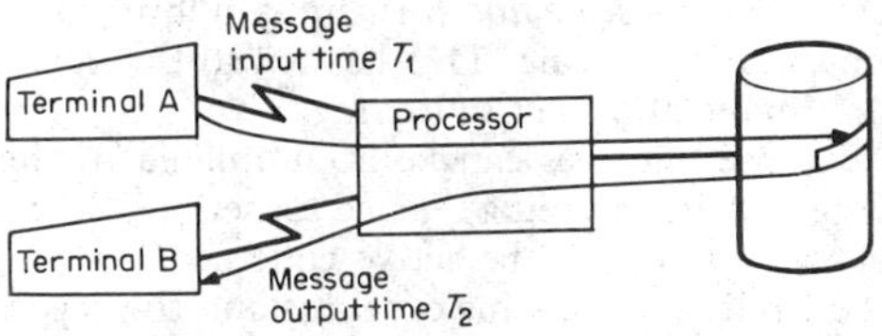

Fig. 23-78. Message switching where $T_1 \geq T_2$. Terminal B need not be connected to the processor when terminal A sends its message.

between the communicating elements is said to be *virtual*. The term *packet* relates to the maximum upper limit in message size permitted to transverse the *store-and-forward* nodes. Certain packet-switching nodes permit DTEs to enter messages longer than the packet size specified by segmenting the large message into multiple packets and reassembling them at the destination boundary before passing them on. The prime reason for packeting is to reduce the transit delay in the store-and-forward nodes. This is achieved through *pipelining* and reduced *average queue wait time*.

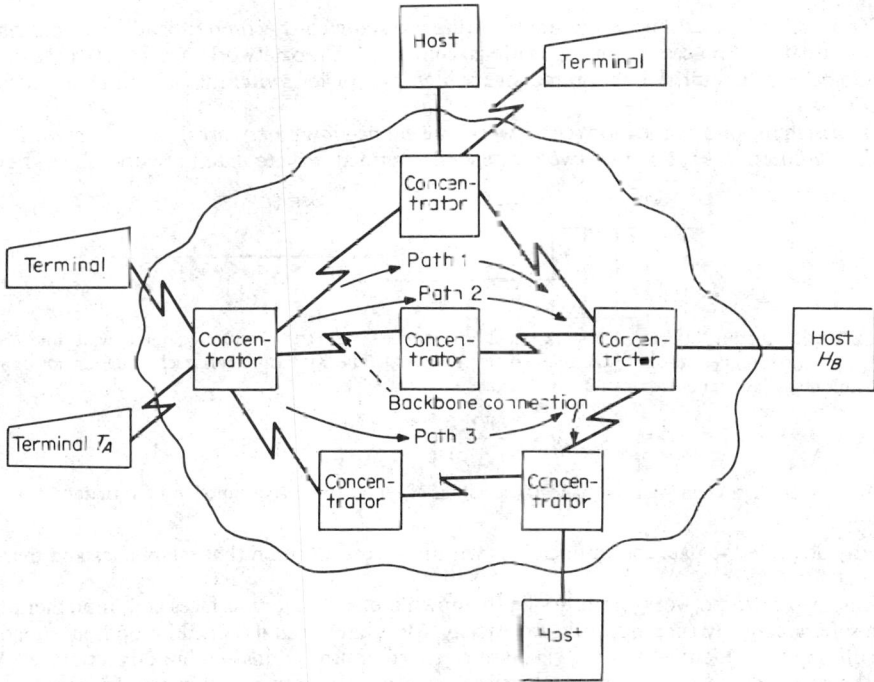

Fig. 23-79. Limited-mesh interconnection of packet-switching nodes in which there are three paths from T_A to H_B.

Pipelining in packet networks is similar to that in parallel processors. Portions of a message can traverse several cascaded links concurrently, thereby providing parallelism of link service times (Fig. 23-80).

To discuss *average queue wait time* we begin with the fact that establishing a maximum packet size imposes a limit on the variability of message size. Since packet nodes receive data traffic from many unrelated sources the arrival pattern of data to links other than those at a boundary is random. The Khintchine-Pollaczek formulation for the average wait time in a single-server queue shows a relationship between the time a message waits for service and the variability of message length (Fig. 23-81). From this is should be apparent that elimination of very long messages significantly reduces the average packet delay across the network. By invoking an appropriate queueing discipline, interactive traffic can be given priority and thus lower response time can be provided for interactive traffic while the network is being shared with batch or bulk transmissions.

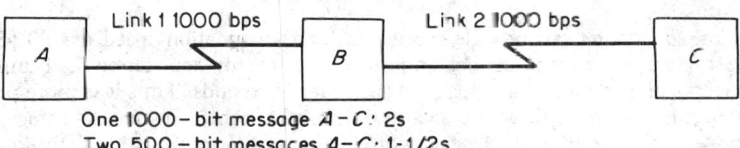

One 1000-bit message $A-C$: 2s
Two 500-bit messages $A-C$: 1-1/2s

Fig. 23-80. Effect of pipelining: because of overlapping times for the two 500-b case, the first 500-b message is transferred on link 2 while the second 500-b message traverses link 1 in the same ½ s.

117. Communications Networks. A *communications network* is defined as a grouping of *nodes*, such as concentrators, interconnected by media and cooperatively providing communications services. Cooperation is achieved through the exchange of control messages. The rules of cooperation are defined as a protocol. The error detection and correction procedures effected between two adjacent nodes by *link control* are examples of the exchange of control information in accordance with a protocol. The rules for routing data units through store-and-forward nodes are another example of protocol. The boundary of what constitutes a network is defined by the commonality of the protocols and the commonality of control of the network elements.

Communications data networks are structured to share costly communications resources and the terminals and processors (or applications) connected to the network. These networks use the technologies of circuit switching, message switching, packet switching, or combinations of all three.

The three major classifications of networks are *home-grown networks*, *vendor networks*, and *public data networks*. The first two are generally termed *private data networks* since they are

Queue   Server

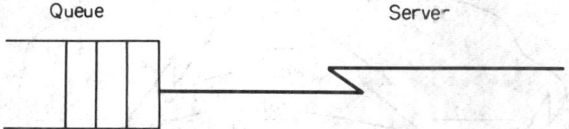

Fig. 23-81. Queuing effects: $E(t_s)$ = expected link service time; $E(t_q)$ = expected queue wait time = $E(t_w)$ + $E(t_s)$; $E(t_w)$ = expected total link delay = $E(t_s)$ + $E(t_q)$. The Kintchine-Pollaczek equation for expected total link delay, assuming exponential arrival rate, is

$$E(t_w) = \frac{\rho\, E(t_s)}{2(1-\rho)} \left[1 + \left(\frac{\sigma\, t_s}{E(t_s)} \right)^2 \right]$$

where ρ = utilization and $\sigma\, t_s$ = standard deviation of service time as a function of variation of message length.

usually dedicated to data communications within the organization that establishes and manages the network.

In home-grown networks users design their own protocols and interfaces and implement their own software. Many such networks exist today. Most have been designed for optimal support of specific application environments, e.g., interactive communications and bulk data transfer. Very few home-grown networks can support the diverse set of communication requirements needed to handle a broad spectrum of applications.

Vendor networks are those marketed by manufacturers of computing equipment to meet the growing need for networks. Such networking offerings typically provide for a system of architectural protocols, interfaces, software, and products. Capabilities include the sharing of programs, processors, data, and communications lines. These networks also provide the basic support for a new concept of computing referred to as *distributed data processing* (DDP). The media-interconnection responsibility in such networks rests with the purchaser of the network components.

There are two types of *public data networks*, digital circuit switching and packet switching. *Digital networks* can be used to provide either leased or switched digital communications service. The significance of digital switching is the rate at which a circuit can be established or disconnected. In normal voice-grade networks *(public switched networks)* circuit setup times can require tens of seconds to establish. If one assumes the attachment of a buffered terminal on a 9,600 b/s circuit with an average message length of 100 characters (8 b each), it is obvious that the transmission time is significantly shorter than the time required to establish and terminate the circuit. Therefore, it is very inefficient to use a switched voice-grade circuit for short intermittent (bursty) transmissions, regardless of whether the connection is retained or reestablished for each transmission.

Digital switched circuits can provide connection and termination speed of 100 to 200 ms. Whereas the typical voice-grade service minimum charge for connection is 1 min, digital switched networks offer a minimum charge of the order of seconds. Thus it is more practical to reestablish a circuit each time there are data to transmit. Another obvious advantage of digital public data networks is the inherent integrity offered by digital transmission techniques.

Packet switched public data networks also offer the equivalent of the leased and switched service alternatives found in digital switched networks. Unlike circuit switched facilities, where users pay for the circuit as long as it is maintained, regardless of the amount of data moved across the facility, packet-switching tariffs are data-volume-sensitive. The user pays for the quantity of data packets transmitted. Suppliers of this type of network service are sometimes termed *value-added carriers* (VAC) because they enhance the availability and integrity of the end-to-end communications services by using the store-and-forward concentration techniques. Public data networks in general assume responsibility for network management. Like vendor networks, they also support the concept of a closed user group, which provides each customer with the appearance of having a private network. International standards recommendations for connection to both forms of public data networks are X.21 for digital circuit switching and X.25 for packet switching.

118. Higher Levels of Logic. Most of this discussion of teleprocessing has centered on aspects of transport mechanisms for moving data from one location to another. This includes the media, line and link control, and store-and-forward switching for origins and destinations that are not physically adjacent.

But *communications*, defined as the transfer of thought or information, implies that attention

TABLE 23-8   Possible Communication Access Method Services

Data line control	Data editing
Data link control	Data sequence control
Flow control DTE to DTE (pacing)	Data compaction
DTE to DCE	Data compression
Session (DTE to DTE connection):	Data recovery
Establishment	Packetization
Recovery	Transport network management
Termination	Diagnostics
Dialogue exchange control DTE to DTE	Statistics
Code translation	Debugging aids
Data	Trace
Device control	Address translation
Display formatting	

must be paid to achieving a common understanding between two attached users. The world of devices and computer systems is highly heterogeneous. Machines talk different codes, e.g., ASCII, EBCDIC, and WORDS. Even the same machines appear heterogeneous when they contain different operating systems or subsystems. Many functions beyond the transport mechanism can be useful in improving communications between heterogeneous connecting elements.

Some of these functions can handle the details of supporting the logical interface between DTE and DCE; others can provide for recovery of information in case of failures or handle code differences by providing translation aids; others can reduce the volume of transferred bits further by compacting or compressing the data representation. The list of functions that can be relegated to such higher levels of logic is extensive. A potential subset is summarized in Table 23-8.

In early computer systems the responsibility for providing these higher levels of logic fell upon the application program. It did not take long to realize that there was a high degree of commonality of communications services required across a large number of applications programs. Logic, or *communication access methods*, has evolved to address this need, much in the same way *operating systems* have evolved to provide a set of shared computer-resource management services.

Many of these access methods offer options to the application programs so that each application can selectively tailor the services required to specific needs. These access methods are very similar in construction to the access methods provided in operating systems to support communications between auxiliary storage media or locally attached devices. In many of the newer operating systems, access methods have been subsumed by the operating systems.

SOFTWARE

119. Nature of the Problem. Even though hardware costs have been declining dramatically over the past 10 years, the overall cost of developing and implementing new data processing systems and applications has not decreased. Primarily because of the costs of developing software, a predominantly labor-intensive effort, overall costs have been increasing. Furthermore, the problems being solved by software are becoming more and more complex. This creates a real challenge to achieve intellectual and management control over the software development process to eliminate schedule slippages and budget overruns and to deliver quality products to the user.

Because of these problems, a great deal of effort is being devoted to the software development process and its management. In these paragraphs the software development process is described along with technologies developed to overcome the problems mentioned above. First we must define some terms in widespread use. In the past, the focus has been on the computer *program*, which can be defined as an organized sequence of computer instructions and associated data designed to provide a specific function or service or to solve a specific problem. As the problems to be solved have become larger and more complex, the term *software* has come into vogue. Initially, software was used synonymously for programs, but it has now come to refer to not

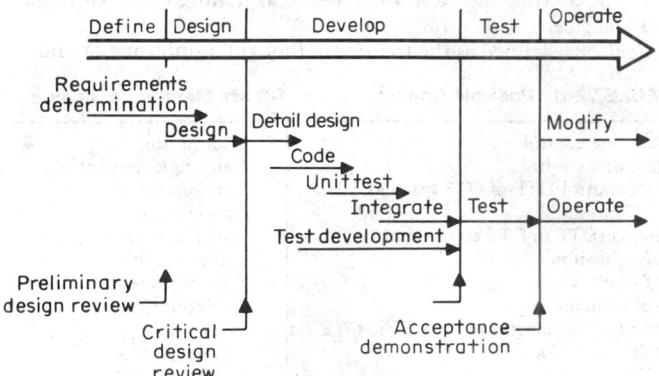

Fig. 23-82. Traditional model of the software life-cycle process showing its five major phases.

only the programs themselves but also the associated data, documentation, and operational procedures necessary to develop, operate, and maintain those programs.

The successful development of software requires discipline and rigor coupled with appropriate management control arising from adequate visibility into the development process itself. This has led to the rise of *software engineering*, defined as the application of scientific knowledge to the design and construction of computer programs and the associated documentation.

120. The Software Life-Cycle Process. In the earlier history of software the primary focus was on its development, but it has become evident that many programs are not one-shot consumables but are tools intended to be used repetitively over an extended time. As a result, it is obvious that the entire software life cycle must be considered. The software life cycle is that period of time over which the software is defined, developed, and used. Figure 23-82 shows the traditional model of the software life-cycle process and its five major phases. It begins with the *definition phase*, which is the key to everything that follows. In some cases, the life cycle may be considered to begin at an earlier point with a *conceptual phase*, which is concerned with a problem definition study that results in a systems concept and with the preparation of a proposal. During the definition phase, the system requirements to be satisfied by the system are developed and the system specifications, both hardware and software, are developed. These specifications describe *what* the software product must accomplish. At the same time, test requirements should also be developed as a requisite for systems acceptance testing.

The *design phase* is concerned with the design of a software structure that can meet the requirements. The design describes *how* the software product is to function. During the *development phase*, the software product is itself produced, implemented in a programming language, tested to a limited degree, and integrated. During the *test phase*, the product is extensively tested to show that it does in fact satisfy the user's requirements. The *operational phase* includes the shipment and installation of the data processing system in the user's facility. The system is then employed by the user, who usually embarks on a maintenance effort, modifying the system to improve its performance and to satisfy new requirements. This effort continues for the remainder of life of the system.

New approaches to software design, especially *top-down design* and *structured programming*, are having a major influence on how software systems are developed and managed. In the traditional approach, the process is viewed and managed as being composed of distinct sequential phases, but when software engineering is used, the distinctions between the design, develop, and test phases disappear. These phases are no longer sequential but in effect become a continuum in which they are recognized as specifying activites. This is shown in Fig. 23-83. The work products resulting from these activities are shown in Fig. 23-84.

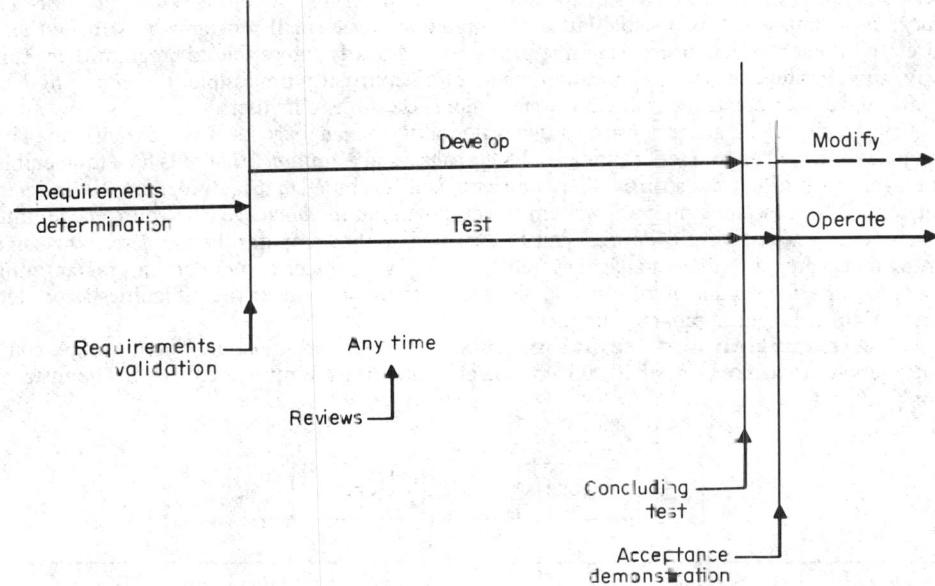

Fig. 23-83. The impact of software engineering on the traditional software life-cycle process.

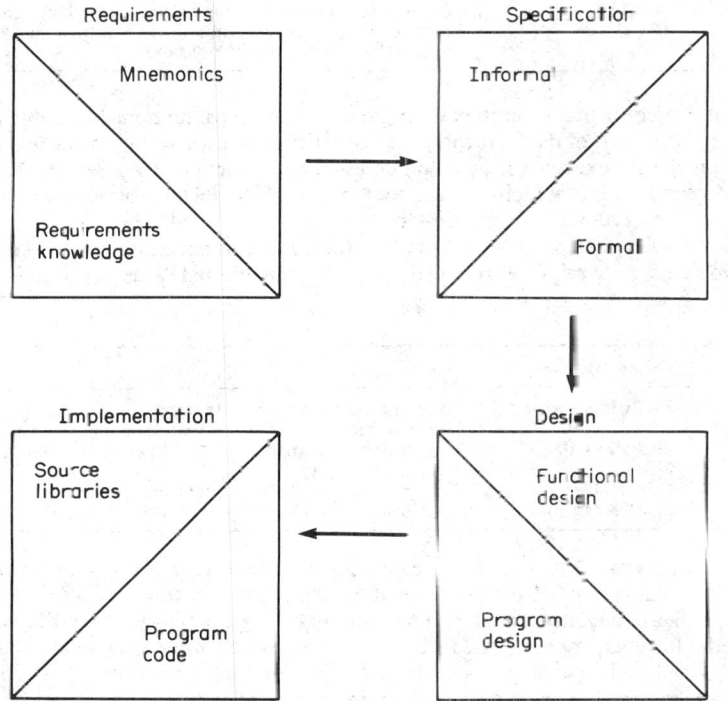

Fig. 23-84. Some of the major software products developed during the software life cycle.

121. Machine-Language Programming. When a stored-program digital computer operates, its storage contains two types of information: the data being processed and program instructions controlling its operations. Both types of information are stored in binary form. The control unit accesses storage to acquire instructions; the ALU makes reference to storage to gain access to data and modify it. The set of instructions describing the various operations the computer is designed to execute is referred to as a *machine language*, and the act of constructing programs using the appropriate sequences of these computer instructions is called *machine-language programming*. It is possible but not desirable to create small programs by writing them directly in machine languages. For large programs it is nearly impossible to program them this way, and maintenance and modification would also be virtually impossible. *Programming languages* have been created to make the system more accessible to its users.

A programming language consists of two major parts: the language itself and a translator. The language is described by a set of *symbols* (the *alphabet*) and a *grammar* that tells how to assemble the symbols into correct strings. The *translator* is a machine-language program whose main function is to translate a program written in the *programming language* (the *source code*) into machine language (*object code*) that can be executed in the computer. Before describing some of the major programming languages currently in use, we consider two important programming concepts, *alternation* and *iteration*, and also see by example some of the difficulties associated with machine-language programming.

122. Alternation and Iteration. These techniques are illustrated here using a computer whose storage consists of 10,000 words each containing 4 B numbered 1 to 4. The instruction format is

1	2	3	4
op code	0	Address	

Op code	Name	Description
01	LOAD	Loads value from addressed word into data register
02	COMP	Compares value of addressed word with data-register value
03	ADD	Adds value of addressed word to data register
08	STORE	Copies value of data register into addressed storage word
20	BRLO	Branches if data-register value from last previously executed COMP was less than comparand

The computer used in this simplified example contains a separate nonaddressable data register than contains one word of data. Further, each instruction is accessed at an address 1 more than that of the previously executed instruction unless that instruction was a BRLO instruction with a low COMP condition, in which case the address part of the BRLO instruction is the address at which the next instruction is to be accessed.

Consider the following program instructions (beginning at address 0100) to select the lower value of two items (in words 0950 and 0951) and place the selected value in a specific place (word 0800):

Address	Instruction	Effect
0100	01000950	Place first item value in data register
0101	02000951	Compare second-item value with data-register value
0102	20000104	Branch to next instruction at address 0104 if data-register value was lower
0103	01000951	Place second item value in data register
0104	08000800	Store lower value in result (word 0800)

123. Flowcharts. One way to depict the logical structure of a program graphically is by the use of *flowcharts*. Flowcharts are limited in what they can convey about a computer program, and with the advent of modern programming design languages they are becoming less widely used. However, they are used here to portray these simple programs graphically. The program of the preceding example is depicted by the flowchart shown in Fig. 23-85.

The flowchart contains boxes representing processes (rectangular boxes) and decisions (alternations — diamond-shaped boxes). The arrows connecting the boxes represent the paths and sequences of instruction execution. An alternation represents an instruction (or a sequence of instructions) with more than one possible successor depending on the result of some processing test (this is commonly a conditional branch). In the example, instruction 0103 is or is not executed depending on the values of the two items.

If the example is extended to require finding the least of four item values, the flowchart is that shown in Fig. 23-86. If the example is further extended to find the largest value of 1,000 items (in locations 0336 through 0790 inclusive in hexadecimal), the flowchart and the corresponding program become very large if analogous extensions of the flowcharts are used.

The alternative is to use the technique known as the *program loop*. A program loop for this latter example is:

Address	Instruction	Effect
0100	01000950	Move first item as initial value of result (ITEMHI)
0101	08000800	
0102	01000900	Initialize loop to begin with item 2
0103	08000104	
0104	(00000000)	Loop: Nth item to data register
0105	02000800	Compare with prior ITEMHI value
0106	20000108	Branch to 108 if Nth item value low
0107	08000800	Store Nth item value as ITEMHI
0108	01000104	Increment value, of N by 1
0109	03000901	
010A	08000104	
010B	02000902	Compare against N = 1,001
010C	2000104	Branch for looping if N < 1,001
010D	end	
0900	01000951	Load item 2; initial instruction
0901	00000001	Address increment of 1
0902	010003E9	Limit test; load 1,001st item

The corresponding flowchart appears in Fig. 23-87. The loop proper (instructions 0104 to 010C) is executed 999 times. The instruction at 0104 accesses the Nth item and is altered each time the program flows through the loop so that on successive executions successive words of the item table are obtained. After each loop execution, a test is made to determine whether processing is complete or a branch should be made back to the beginning of the loop to repeat the loop program.

The loop proper is preceded by several instructions that *initialize* the loop, presetting ITEMHI and the instruction 0104 value for the first time through. A loop customarily has a *process* part, an *induction* part to make changes for the next loop iteration, and an *exit test* or termination to determine whether an additional iteration is required.

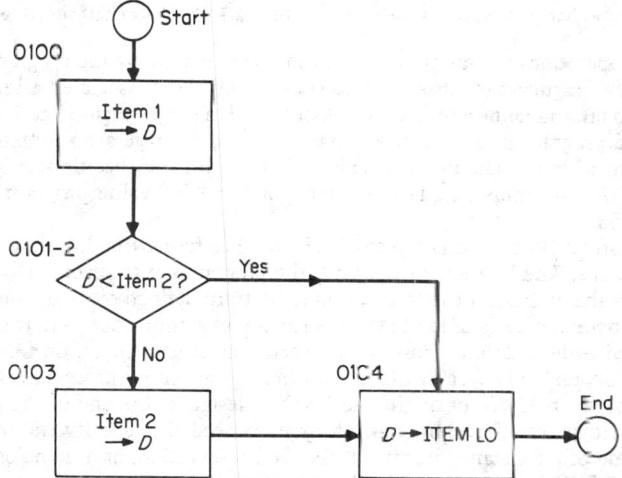

Fig. 23-85. Flowchart of a simple program. The boxes represent processes; the diamonds represent decisions.

124. Symbolic Assembly Languages. The previous example illustrates the difficulty of preparing and understanding even simple machine-language programs. One help would be the ability to use a symbolic (or mnemonic) representation of the operations and addresses used in the program. The actual translation of these symbols to specific computer operations and addresses is a more or less routine clerical procedure. Since computers are well suited to perform-

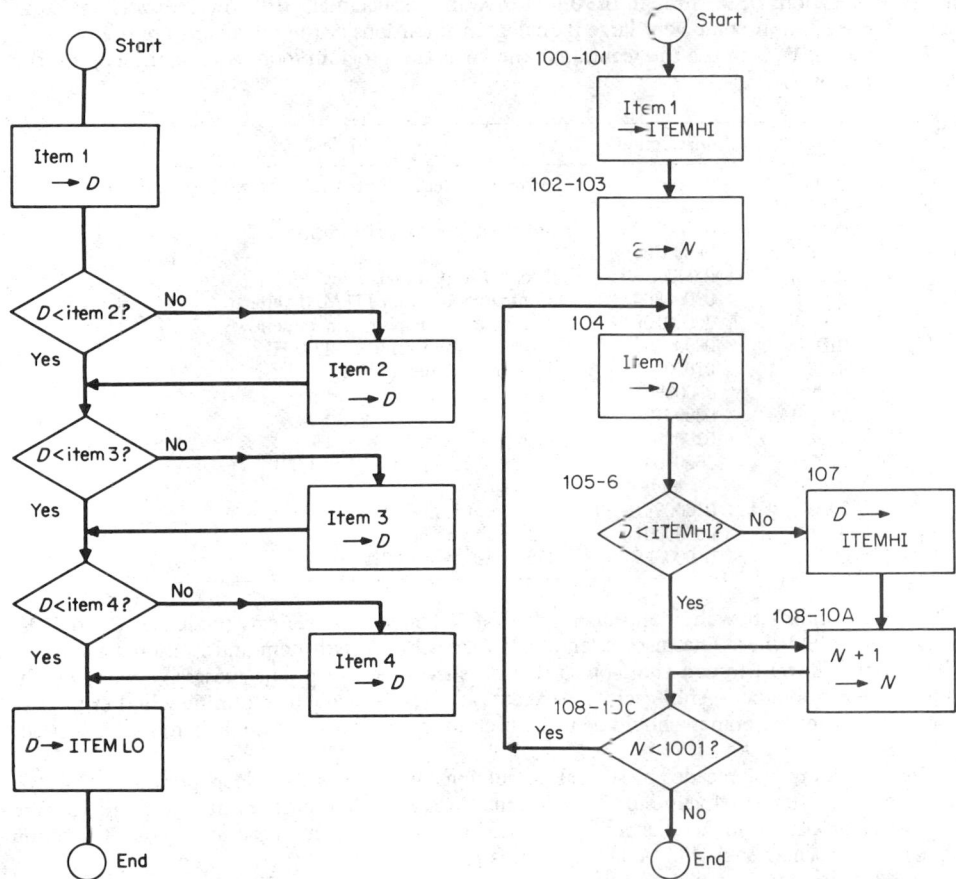

Fig. 23-86. Flowchart of a repetitive task.   **Fig. 23-87.** Flowchart showing a program loop.

ing such routine operations, it was quite natural that the first automatic programming aids, the *symbolic assembly languages* (or *assembly languages*) and their associated assembly programs, were developed to take advantage of that fact. Assembly languages permit the critical addressing interrelations in a program to be described regardless of the storage arrangement, and they can produce therefrom a set of machine instructions suitable for the specific storage layout of the computer in use. An assembly-language program for the 1,000-value program of Fig. 23-87 is shown in Fig. 23-88.

The program format illustrated is typical. Each line has four parts: location, operation, operand(s), and comments. The location part permits the programmer to specify a symbolic name to be associated with the address of the instruction (or datum) defined on that line. The operation part contains a mnemonic designation of the instruction operation code. Alternatively, that line may be designated to be a datum constant, a reservation of data space, or a designation of an assembly *pseudo operation* (a specification to control the assembly process itself). Pseudo operations in the example are ORG for origin and END to designate the end of the program.

The operand field(s) give the additional information needed to specify the machine instruction, e.g., the name of a constant, the size of the data reservation, or a name associated with a pseudo operation. The comment part serves for documentation only; it does not affect the assembly-program operation.

After a program is written in assembly language, it is entered as input data to its *associated assembly program* (or assembler), either by direct keyboard entry or indirectly via punched cards, paper tape, or magnetic tape. The assembly program reads the symbolic assembly-language input and produces (1) a machine instruction program with constants, usually in a form convenient for subsequent program loading, and (2) an assembly listing that shows in typed or printed

LOC	OP	Operand	Comment
	ORG	100	Start at ADDR 100
START	LOAD	ITEM 1	Move first item value to ITEMHI
	STORE	ITEM HI	
	LOAD	CONST 1	Set LOOP to start with second item
	STORE	LOOP ST	
LOOP ST	CONST		* Load NTH item
	COMP	ITEM HI	Compare with previous HI
	BRLO	LOOP INC	Skip if lower
	STORE	ITEM HI	Store on equal or high
LOOP INC	LOAD	LOOP ST	Increment N or 1
	ADD	ONE	
	STORE	LOOP ST	Modify storage
	COMP	CONST 2	Exit test
	BRLO	LOOP ST	Repeat if $N < 1001$
			(End of example)
	ORG	700	
CONST 1	LOAD	ITEM 1 + 1	Initial LOOP ST instruction
ONE	CONST	− 1	Incrementation constant
CONST 2	LOAD	ITEM 1 + 1000	LOOP end test constant
	ORG	800	
ITEM HI	RESRV	1	Word where high value left
	ORG	950	
ITEM 1	RESRV	1	First item value
	RESRV	999	Second through thousandth items

Fig. 23-88. An assembly program. The program statements are in a one-for-one correspondence with machine instructions. Hence the procedure is fully supplied by the programmer according to the particular macroinstruction set of the system. The assembly language alleviates housekeeping routines, such as specific assignments, and makes user-oriented symbols possible instead of numeric or binary code.

form each line of the symbolic assembly-language input, together with any associated machine instructions or constants produced therefrom.

The assembly pseudo operation ORG specifies that the instructions and/or constant entries for succeeding lines are to be prepared for loading at successive addresses, beginning at the specified load origin (value of operand field of ORG entry). Thus the 13 symbolic instructions following the initial ORG line in Fig. 23-88 are prepared for loading at addresses 0100 through 010C inclusive, with the following symbolic associations established:

Location symbol	(Local) address
START	100
LOOP ST	104
LOOP INC	108

Four instructions of this group of 13 contain the symbol LOOP ST in the operand field, and the corresponding machine instructions will contain 0104 in their address parts.

The operation of a typical assembly program therefore consists of (1) collecting all location symbols and determining their values (addresses), called *building the symbol table*, and (2) building the machine instructions and/or constants by substituting op codes for the OP mnemonics and location symbol values for their positions in the operand field. The symbol table must be formed first since, as the first instruction in the example shows, a machine instruction may refer to a location symbol that appears in the location field near the program end. Thus most assembly programs process the program twice; the *first pass* builds the symbol table and the *second pass* builds the machine-language program. Note in the example the use of the operation RESRV to reserve space (skipping in the load-address sequence) for variable data.

Assembly language is specific to a particular computer instruction repertoire. Hence, the basic unit of assembly language describes a single machine instruction (so called one-for-one assembly process).

Most assembly languages have a *macroinstruction* facility. This permits the programmer to define *macros* that can generate desired sequences of assembly-language statements to perform specific functions. These macro definitions can be placed in macro libraries, where they are available to all programmers in the facility.

125. Routines. The term *routine* (also *subroutine, procedure,* and *subprogram*) is used to refer to a group of instructions that perform some particular function used repeatedly in essentially the same *context*. The quantities that vary between contexts may be regarded as parameters (or arguments) of routine. The method of adaptation of the routine determines whether it is an *open* or *closed* routine.

An open subroutine is adapted to its parameter values during code preparation (assembly or compilation) in advance of execution, and a separate copy of the subroutine code is made for each different execution context. A closed subroutine is written to adapt itself during execution to its parameter values; hence, a single copy suffices for several execution contexts in the same program. The open subroutine executes faster since tailoring to its parameter values occurs before execution begins. The closed subroutine not only saves storage space, since one copy serves multiple uses, but is more flexible, in that parameter values derived from the execution itself can be used.

A closed subroutine must be written to determine its parameter values in a standard way (including the return point after completion). The conventions for finding the values and/or addresses of values are called the subroutine linkage *conventions.* Quite commonly, a single address is placed in a particular register, and this address in turn points to a consecutively addressed list of addresses and/or values to be used. Subroutines commonly use (or *call*) other closed subroutines, so that there are usually a number of levels of subroutine control available at any point during execution. That is, one routine is currently executing, and others are waiting at various points in partially executed condition.

126. High-Level Programming Languages. On general-purpose digital computers, *high-level programming languages* have largely superseded assembly languages as the predominant method of describing application programs. Such programming languages are said to be *high-level* and *machine-independent.* High-level means that each program function is such that several or many machine instructions must be executed to perform that function. Machine-independent means that the functions are intended to be applied to a wide range of machine-instruction repertoires and to produce for each a specific machine representation of data.

The high-level language translator is known as a *compiler,* i.e., a program that converts an input program written in a particular high-level language (*source program*) to the machine language of a particular machine type (*object program*).

The advantages of high-level languages, as opposed to assembly languages, are that it is possible to use the same program for execution on several different machine types with few if any changes; fewer errors are to be expected, and checkout is simpler because the higher-level functions require less detailed coding; programming is easier, faster, and more economical; programming skills in a particular language are transferable to another machine; and better programs result since compilers use specialized techniques that it is not economical for an assembly-language programmer to employ on an extensive basis.

127. High-Level Procedural Languages. Most of the high-level programming languages are said to be *procedural.* The programmer writing in a high-level procedural language thinks in terms of the precise sequence of operations, and the program description is in terms of sequentially executed *procedural statements.* Most high-level procedural languages have statements for *documentation, procedural execution, data declaration,* and various compiler and execution *control specifications.*

The program in Fig. 23-89, written in the FORTRAN high-level language (Par. **23-128**), describes the program function given in Fig. 23-88 in assembly language. The first six lines are for documentation only, as indicated by C in the first column. The DIMENSION statement defines ITEM to consist of 1,000 values. The assignment statement ITEMHI = ITEM (1) is read as "set the value of ITEMHI to the value of the first ITEM." The next statement is a loop-control statement meaning: "do the following statements through the statement labeled 1 for the variable N assuming every value from 2 through 1,000." The statement labeled 1 causes a test to be made to "see if the Nth ITEM is greater than .GT. the value of ITEMHI, and if so, set the ITEMHI value equal to the value of the Nth item."

128. FORTRAN. The high-level programming languages most commonly used in engineering and scientific computation are FORTRAN, ALGOL, BASIC, and APL. FORTRAN, the first to appear, was developed during 1954 to 1957 by a group headed by Backus.[97a] of IBM. Based on algebraic notation, it allows two types of numbers: integers (positive and negative) and floating point. Variables are given character names of up to six positions. All variables beginning with the letters I, J, K, L, M, or N are integers; otherwise they are floating point. Integer constants are written in normal fashion, 1, 0, −4, etc. Floating-point constants must contain a decimal point, 3.1, −0.1, 2.0, 0.0, etc. For example, 6.02×10^{24} is written 6.02E24. This standard notation was adopted to accommodate the limited capability of computer input-output equipment.

READ and WRITE statements permit values of variables to be read into or written from the ALU, from or to input, output, or intermediate storage devices. The latter may operate merely by transcribing values or may be accompanied by conversion or editing specified in a separate

IBM — FORTRAN Coding Form — GX28-7327-6 U/M 050 Printed in U.S.A.

FORTRAN STATEMENT

```
C     THIS FORTRAN PROGRAM SELECTS
C       THE LARGEST VALUE OF 1000
C       ITEMS AND PUTS IT IN ITEMHI
C
C     THE FOLLOWING DECLARES 'ITEM'
C     IS A VECTOR WITH 1000 VALUES
      DIMENSION ITEM(1000)
C     NEXT SET ITEMHI TO VALUE OF
C       1ST ITEM
        ITEMHI = ITEM(1)
C     NEXT BEGIN LOOP THAT EXTENDS
C         THRU STATEMENT 1, N STARTS
C         AT 2 AND GOES TO 1000
      DO 1   N=2, 1000
    1   IF (ITEM(N).GT.ITEMHI) ITEMHI=ITEM(N)
          END
```

Fig. 23-89. An example of a FORTRAN program, corresponding to the flowchart of Fig. 23-87 and assembly program of Fig. 23-88.

FORMAT statement. Some idea of the range of operations provided in FORTRAN is shown by the following value-assignment statement:

ROOT = (−(B/2.0) + SQRT ((B/2.0) ** 2 − A*C))/A

This is the formula for the root of a quadratic equation with coefficients A, B, and C. The asterisk indicates multiplication, / stands for division, and ** exponentiation.

The notation: *name (expression)* and *name (Expression, expression)* etc., is used in FORTRAN with two distinct meanings depending on whether or not the specific name appears in a DIMENSION statement. If so, the *expression*(s) are subscript values; otherwise the *name* is considered to be a function name, and the expressions are the values of the arguments of the function. SQRT((B/2.0)**2 − A*C) in the preceding assignment statement requires the expression (B/2.0)**2 − A*C to be evaluated, and then the function (square root here) of that value is determined. Square root and various other common trigonometric and logarithmic functions and their respective inverses are standardized in FORTRAN, typically as closed subroutines.

The same notation may be employed for a function defined by a FORTRAN programmer in the FORTRAN language. This operation is performed by writing a separate FORTRAN program headed by the statement

FUNCTION name (arg 1, arg 2, etc.)

where arg represents the name that stands for the actual argument value at each evaluation of that function. Similarly, any action or set of actions described by a closed FORTRAN subroutine is called for by "CALL subroutines (args)" together with a defining FORTRAN subroutine headed by "SUBROUTINE subroutine name (args)."

A FORTRAN program, with all necessary subprograms (SUBROUTINE and FUNCTION programs), defines an execution process. However, it has not been customary to compile the entire executable program in one step. It is usual instead to compile each program or subprogram as a separate process into a *relocatable* form and to use a relocating loader (also called *linking loader, link editor,* etc.) to combine the relocatable forms to the absolute machine-language form. The relocatable form permits all symbolic-address references to be adjusted relative to a small number of address values determined by the loader in combining the appropriate cross-references between the relocatable programs, one from each compilation.

Two advantages follow from this procedure: (1) during checkout, each correction involves changes to only one (or a small number of) programs that must then be recompiled, and (2) if the assembly program is designed to produce output in the same relocatable format, particular functions and/or subroutines need not be necessarily written in FORTRAN. This procedure allows the use of processes not expressible in FORTRAN or ones that are more efficient in assembly language.

ALGOL was developed by a committee appointed jointly by the Association for Computing Machinery in the United States and its counterparts in France, Germany, Britain, and the Netherlands during 1957 to 1960. ALGOL includes most of the properties of FORTRAN but has a number of features new to programming languages. Probably the most distinctive feature of ALGOL is its *block* structure. An ALGOL program, as written, consists of blocks of contiguous statements. Two contiguous blocks either have no common statements or one block is wholly contained in the other. Declarations apply to a particular block (and to any blocks nested within it). Names of variables, statements, and procedures are declared (defined) and have that declared meaning only for statements within that block.

The Revised Report on the Algorithmic Language (ALGOL 60[109a]) presents a new method of specifying the syntax of ALGOL, introducing a formal notation to eliminate ambiguities, inconsistencies, or misunderstandings. Although not part of ALGOL proper, the use of such a description of ALGOL syntax serves both to demonstrate the underlying unity of the ALGOL language and to show the usefulness of such descriptive methods in dealing with the syntax of programming languages. ALGOL is more widely used in Europe than the United States, probably because FORTRAN had become firmly established in the United States at an early date.

ALGOL 68 is also block-structured language. It provides a rich collection of data types (called modes), and it also provides facilities for programmers to construct their own data types from primitive and old modes. It supports arrays and structures. ALGOL 68 is an expression language. This means that every construction in the language yields values and in principle can appear on the right-hand side of an assignment. It supports procedures with fully typed parameters and pointers. However, in contrast to most programming languages, ALGOL 68 considers procedures to be objects, complete with values and modes.

In addition to expressing sequential processing, ALGOL 68 can also be used to describe independent, noncommunicating processes that run in parallel through the use of collateral clauses. If synchronization between processes is necessary, semaphores are provided as synchronization primitives. ALGOL 60 was widely criticized for its lack of I/O capabilities. ALGOL 68 has overcome this by providing a powerful and flexible set of I/O procedures.

BASIC is a high-level programming language based on algebraic notation that was developed for solving problems at a terminal; it is particularly suitable for short programs and instructional purposes. The user normally remains at the terminal after entering his program in BASIC, while it compiles, executes, and types the output, a process that typically requires only a few seconds.

BASIC is similar to FORTRAN. Many of the important differences relate to the mode of use i.e., immediate compilation and execution, with the terminal serving as the sole input-output device. Hence subroutines are treated differently: BASIC compiles a single program to produce a self-contained, complete object program. Every statement in BASIC is numbered, and the statements are arranged in ascending order, before compilation. This feature facilitates program correction during checkout. For example, if more than one statement is entered with the same number, only the last is retained. The other statement is deleted by entering a statement number only (a null statement replaces the statement being deleted). Statements are inserted by using a statement number between the two statements at which insertion is to be made. Since statement numbers need not be consecutive (in ascending order only) it is good programming practice in BASIC to leave gaps in the initial statement numbering to allow for later insertion.

129. APL. APL (*a programming language*) is high-level language that is widely used because it is easy to learn and has an excellent interactive programming system supporting it. Its primitive objects are arrays (lists, tables, etc). It has a simple syntax, and its semantic rules are few. The usefulness of the primitive functions is further enhanced by operators that modify their behavior in a systematic manner. The sequence control is simple because one statement type embraces all types of branches and the termination of the execution of any function always returns control to the point of use. External communication is established by variables shared between APL and other systems.

130. PASCAL. The most recent of the major high-level programming languages is PASCAL, developed by Niklaus Wirth. It has had widespread acceptance and use since its introduction in the early 1970s. The language was developed for two specific purposes: (1) to make available a language to teach programming as a systematic discipline and (2) to develop a language that supports reliable and efficient implementations. PASCAL provides a rich set of both control statements and data structuring facilities. Six control statements are provided: BEGIN-END, IF-THEN-ELSE, WHILE-DO, REPEAT-UNTIL, FOR-DO, and CASE-END.

In addition to the standard scalar data types, PASCAL provides the ability to extend the language via user-defined scalar data types. In the area of higher-level structured data types, PASCAL extends the array facility of ALGOL 60 to include the record, set, file, and pointer data types. In addition to these, PASCAL contains a number of other features that make it useful for programming and teaching purposes. In spite of this, PASCAL is a systematic language and modest in size, attributes that account for its popularity.

131. Other Scientific Languages. Research is continuing in the development of new languages that incorporate the concepts and ideas arising out of modern software technology development. One of the many new languages to be introduced during the coming decade is Ada, named after Lord Byron's daughter. This language perhaps has the greatest potential for success among the new languages. It is being developed by the United States Department of Defense to be the single successor to a number of high-level languages currently in use by the Armed Forces of the United States. Ada is a PASCAL-based language but incorporates a number of features designed to make it suitable for real-time and concurrent processing applications.

132. High-Level Commercial Languages. High-level programming languages used for business data processing applications emphasize description and handling of files for business record keeping. Two widely used programming languages for business applications are COBOL (common business-oriented language) and RPG (report program generation). Compilers for these languages, with generalized sorting programs, form the fundamental automatic programming aids of many computer installations primarily used for business data processing. COBOL and RPG have comparable file, record, and field-within-record descriptive capabilities, but the specification of processing and sequence control derive from basically different concepts.

COBOL is a procedure-oriented language like the scientific programming languages FORTRAN and BASIC. A COBOL program has a *procedure division* part in which executable statements control not only computation and summarization but also input, output, loop, and branching (GO TO) statements. Thus the COBOL programmer specifies the precise execution sequence, and the COBOL language in the *procedure division* reflects the notion of what might be called a procedure statement instruction counter.

RPG reflects the concepts and uses the terminology of punched-card accounting-machine methods. Four different forms are used to describe different aspects of the application: the files involved, the fields of each input record by file, the fields of working storage and how their contents are established, and the output record (and its editing) in terms of which input and/or working storage fields are to be copied.

After an RPG programmer has completed the forms describing the application, their content is entered on punched cards to an RPG compiler. This compiles an object program to perform the programmed task.

COBOL was developed by the Conference on Data Systems Languages, beginning in 1959. Besides a procedure division with GO TO, IF, and loop control statement (analogous to FORTRAN or ALGOL), COBOL has an *environment division* and a *data division* which specifies fields in input records, output records, and working storage.

133. Operating Systems. There are many reasons for developing and using an *operating system* for a digital computer. One of the main reasons is to optimize the scheduling and use of computer resources, so as to increase the number of jobs that can be run in a given period. Creation of a multiprogramming environment means that the resources and facilities of the computing system can be shared by a number of different programs, each written as if it were the only program in the system.

Another major objective for an operating system is to provide the full capability of the computing system to the user while minimizing the complexity and depth of knowledge of the computer system required. This is accomplished by establishing standard techniques for handling system functions like *program calling* and *data management* and providing a convenient and effective interface to the user. In effect the user is able to deal with the operating system as an entity rather than having to deal with each of the computer's features. As indicated in Fig. 23-90, each user will be dealing with an extended machine that may be thought of conceptually as a unit consisting of both the hardware and the programs and procedures that make up the operating system.

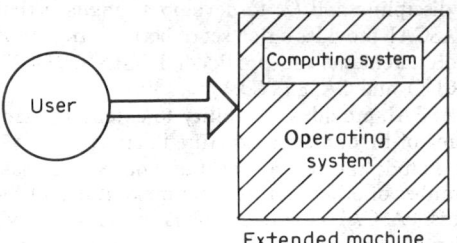

Fig. 23-90. The user's view of the operating system as an extension of the computing system yet an integral part of it.

134. General Organization of an Operating System. There are many ways to structure operating systems, but for the purpose of this discussion the organization shown in Fig. 23-91 is typical. The operating system is composed of two major sets of programs, control (or supervision) programs and processing programs. *Control programs* supervise the execution of the support programs (including the user application programs), control the location, storage, and retrieval of data, handle interrupts, and schedule jobs and resources needed in processing. *Processing programs* consist of language translators, service programs, and user-written application programs, all of which are used by the programmer in support of program development.

The work to be processed by the computer can be viewed as a stack of jobs to be run under the management of the control program. A *job* is a unit of computational work that is independent of all other jobs concurrently in the system. A single job may consist of one or a number of *steps*. Certain essential information about each job and its steps must be supplied to the operating system by the programmer. The form and structure in which this information is supplied is determined by the operating system and is called the *job control language* (JCL).

135. Types of Operating Systems. There are five basic types of operating systems: serial-batch system, multiprogramming system, time-sharing system, real-time system, and multiprocessing system. The *serial-batch system* was the first operating system to be developed. It uses sequential job scheduling and runs jobs one at a time. The multiprogramming system permits two or more batch jobs to be running in the system concurrently. Time sharing permits a number of users to access the system simultaneously. Even though each user is sharing the computer system, the time-sharing system gives the appearance of having the exclusive use of the computer's resources. The users may be geographically dispersed, accessing the computer via remote terminals and a telecommunications network. The real-time system must be designed to ensure that the processing functions are executed within rigid time constraints. For example, it may be performing control functions where sensor signals are read and processed and commands are transmitted to actuating devices, or it may provide the facility to query, retrieve, and update a centralized bank of data.

The multiprocessing system must schedule and control the execution of jobs that are distributed across two or more coupled processors. These processors may share a common storage, in which case they are said to be *tightly* (or *directly*) *coupled*, or they may have their own private storage and communicate via other means such as sending messages over networks, in which case they are said to be *loosely coupled*.

136. Job-Management Function. *Job management* is responsible for setting up the environment and arranging the external conditions necessary for running the job. One of the things that must be done is to schedule the jobs to be run. Job management maintains a queue of jobs to be run and then schedules them for execution. In addition to sequential scheduling, many operating systems permit priority scheduling so that jobs can be initiated, based on their job class and on user-assigned priorities. Job management is also responsible for handling communications with the console operator. Job management is responsible for allocating the necessary I/O devices. It initiates the job and then turns control of the system to the *task management* (or *system management*) *function*.

137. Task-Management Function. This function, sometimes called the *supervisor*, controls the operation of the system as it executes units of work known as *tasks* or *processes*. (The performance of a task is requested by a job step.) The distinction between a task and a program should be noted. A *program* is a static entity, a sequence of instructions, while a *task* is a dynamic entity, the work to be done in execution of the program. Task management initiates and controls the execution of tasks. If necessary it controls their synchronization. It allocates system resources to tasks and monitors their use. In particular it is concerned with the dynamic allocation and control of main storage space.

Task management handles all interrupts occurring in the computer, which can arise from five different sources: (1) supervisor call interrupts occur when a task needs a service from the task-management function, such as initiating an I/O operation; (2) program interrupts occur when unusual conditions are encountered in the execution of a task; (3) I/O interrupts indicate that an I/O operation is complete or some unusual condition has occurred; (4) machine-check interrupts are inititated by the detection of hardware errors; and (5) external interrupts are initiated by the timer, by the operator's console, or other external devices.

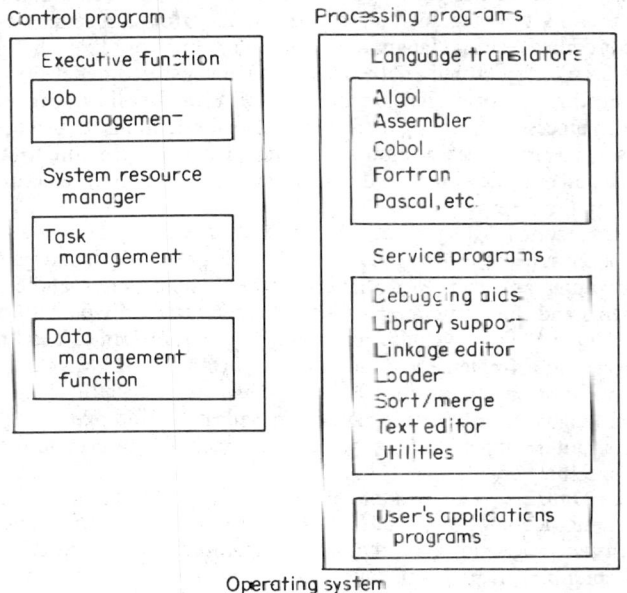

Fig. 23-91. A typical operating system and its constituent parts

138. Data Management. This function provides the necessary I/O control system services needed by the operating system and the user application programs. It frees the programmer from the tedious and error-prone details of I/O programming and permits standardization of these services. It constructs and maintains various file organization structures, including the construction and use of index tables. It allocates space on disc (auxiliary) storage. It maintains a directory showing the locations of data sets (files) within the system. It also provides protection for data sets against unauthorized entry.

139. Operating System Security. One of the major concerns in the design of operating systems is to make certain that they are reliable and that they provide for the protection and the integrity of the data and programs stored within the system. Work is under way to develop secure operating systems. These systems use the concept of a security kernel — a minimal set of operating system programs that are formally specified and designed so that they can be proved to implement the desired security policy correctly. This assures the correctness of all access-controlling operations throughout the system.

140. Software-Development Support. The last 15 years have seen great strides in software engineering technology. Out of the research and development efforts in universities, industry, and government have emerged a number of significant ideas and concepts that can have significant and long-lasting influence on the way that software is developed and managed. These concepts are just now starting to find their way into the software-development process but should become more widely used in the future. They are briefly reviewed below.

141. Requirements and Specifications. This has been one of the problem areas through the years. Analysis and design errors are, by far, the most costly and crucial types of errors, and a number of attempts are being made to develop methods for recording and analyzing software requirements and developing specifications. Currently, most requirements and specifications documents are recorded in English narrative form, which introduces problems of inconsistency, ambiguity, and incompleteness. Research activities have the goals of developing a requirements methodology with a more structured representational form. The approaches being taken seem to fall into one of two categories: a graphical-notation or graphic-language approach and a formal-language approach. The graphic approaches include HIPO, SADT,* and structured analysis. The formal-language approach includes PSL/PSA and SPECIAL.

HIPO (hierarchy plus input-process-output) was originally developed as a software design and documentation technique. A HIPO diagram primarily shows functions and data flow. Each diagram has, from left to right, an input, a process, and an output section. Each diagram describes a single function and shows its decomposition into steps that are listed in the process section.

SADT, structured analysis and design technique, is a graphics-based requirements-definition methodology. It uses a set of diagrams to describe a system from a specific viewpoint and for a specific purpose. It embeds natural language in a graphic framework like a blueprint and is both modular and hierarchic. *Structured analysis* is a methodology used to develop a special kind of functional specification, a structured specification. It uses data-flow diagrams showing the sources of data, the processes operating on data, and data destinations. It also uses a data dictionary and presents a narrative specification statement written in structured English, a form of English structured and embedded within the basic control structures of structured programming to achieve clarity and to improve rigor.

The *problem statement language/problem statement analyzer (PSL/PSA)* system provides a very useful computer support for the systems analyst. It provides a language (PSL) to describe the requirements and an analyzer (PSA) that can perform completeness and consistency checks on the requirements and then generate a variety of reports that provide both management and analyst support. PSL/PSA was developed under the direction of Daniel Teichroew at the University of Michigan and is implemented on a variety of computers.

SPECIAL is a specification and assertion language developed at SRI, International. It is a mathematically formal language for describing system specifications and expressing assertions about them. A set of computer support tools is available to check for correctness the syntax and the assertions of the specification.

142. Software Design. The work of Dijkstra, Hoare, and Mills has had a major influence on software-design methodology by introducing a number of concepts that have led to the development of *structured programming*. Structured programming is not the absence of GO TOs (unconditional branching) in programs, but it is the introduction of a methodology based on mathematical rigor. It introduces the concept of top-down design and implementation by describing the program at a high level of abstraction and then expanding (refining) this abstraction into more detailed representations through a series of steps until sufficient detail is present for implementation of the design in a programming language to be possible. This process is called *stepwise refinement*. The design is represented by a small, finite set of primitives, such as shown in Fig. 23-92. These three primitives are adequate for design purposes but for convenience several others have been introduced; namely the *indexed alternation* or *case*, the *do-until*, and the *indexed sequence* or *for-do* structure.

The initial concentration of methodology development was on control structures, but data are also key elements in software design. It was soon recognized that the organization and representation of software systems can be made cleaner if certain data and the operations permitted on those data are organized into data abstractions. The internal details of the organization of the data are hidden from the user.

The result of applying this methodology is the organization of a sequential software process into a hierarchical structure through the use of stepwise refinement. The software system structure is then defined at three levels: the *system*, the *module*, and the *procedure*. The system (or *job*) describes the highest level of program execution. The system is decomposed into modules. The module is composed of one or more procedures and data that persist between successive invocations of the module. The procedure is the lowest level of system decomposition, the executable unit of the stored program. Dijkstra[98a] showed how to extend this concept into an even more powerful design technique by organizing an entire system as a cooperating society of sequential processes, using process synchronization to ensure cooperation.

Another important aspect of the design process is its documentation. There is a critical need to record the design as it is being developed, from its highest level all the way to its lowest level of detail, before its implementation in a programming language. What is needed is a language

---

*SADT is a registered trademark of SOFTECH, Inc.

that can be used not only to communicate software designs between specialists in software development but also between specialists and nonspecialists in rigorous logical terms. A class of languages has been developed and are in use for this purpose, called process, or programming, design languages (PDL).

143. Software-Development Facilities. A number of software tools have been developed to support the various activities that are an essential part of the software-development process. They are intended to support the individual programmer as well as the project management team who are concerned with monitoring and controlling the entire process.

One key tool is the programming support library (PSL), or program production library (PPL), as it is also called. This is more than just a file system. It is designed to support all the facets and phases of the software project. Programming is fundamentally a specifying process. The specifi-

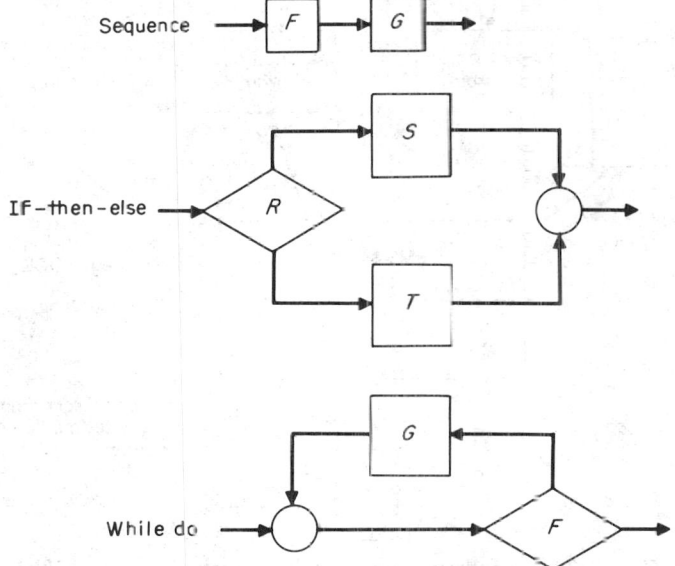

Fig. 23-92. Basic set of structured control primitives.

cations are the products of this process and range from the product of the requirements-analysis activity, which is a requirements specification, through the design specification in PDL, to the implementation specification in source and object code. The software development plan can also be viewed as a specification in this context.

These specifications consist basically of data elements and the relationships that exist between them and can therefore be stored in the programming support library for control and manipulation. It provides management visibility into the process and can support other critical efforts such as cross checking between requirements and design specifications and analysis of change requests. The PSL is hierarchically structured. It supports individual programmers on their program and code segments. It controls the migration of program units and modules from one level to the next within the development library, and it supports the project manager in controlling the integration and delivery of the master system. It provides management reports and statistics to ensure management visibility into the process.

Important elements of the software development process are reviews, walk-throughs, and inspections, which can be applied to the products of the software-development process to ensure that they are complete, accurate, and consistent. They are applied to such areas as the design (design inspections), the software source code (code inspections), documentation, test designs, and test-results analysis. The goals of the software review and inspection process are to ensure that standards are met, to check on the quality of the product, and to detect and correct errors at the earliest possible point in the software life cycle. Another important value of the review process is that it permits progress against development-plan milestones to be measured more objectively and rework for error correction to be monitored more closely.

144. Testing. An important activity in the software-development cycle that is often ignored until too late in the process is testing. It is also important to note what testing can and cannot do for the software product. Quality cannot be tested into the software; it must be designed into it. Test planning begins with the requirements analysis at the beginning of the project. Requirements should be testable; i.e., they should be stated in a form that permits the final product to be tested to assure that it satisfies the requirements. Test planning and test designs should be developed in parallel with the design of the software. Test requirements, test designs, and the analyzed results of the tests themselves should all be kept in the PSL.

145. Data-Base Management Systems. Data processing facilities have to handle a large variety of files. In the past each file was treated as an independent entity even though elements of its data structure might be identical to those in another file or two files might contain data that were related in some fashion. There were a multiplicity of applications programs writ-

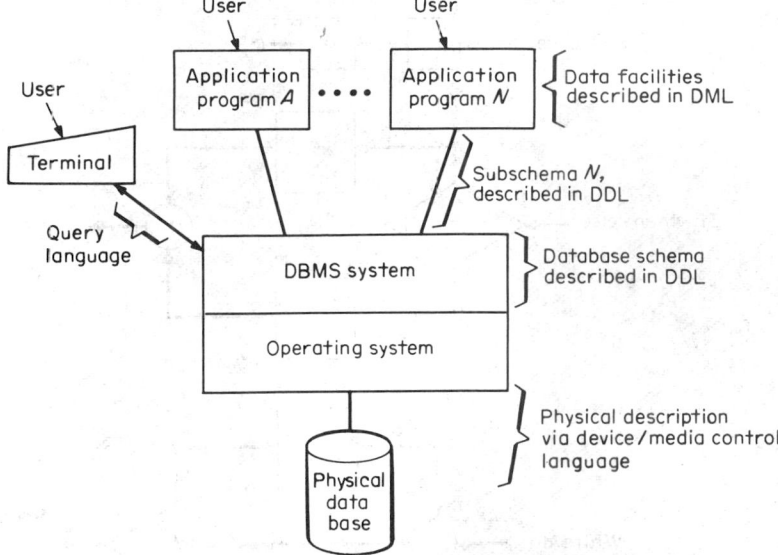

Fig. 23-93. Overview of a typical data-base management system (DBMS).

ten to process data from the files. If a data record had to be changed, the impact would affect a number of existing application programs in ways that were not obvious. The effect of hardware changes, such as adding new disc drives, could be devastating to the existing programs. In response to problems like these, the need for integrating data into data bases and for building generalized data-handling and data-management facilities became apparent. The result was the creation of software systems that became known as data-base management systems (DBMS).

The DBMS design has to provide a number of features. Concurrency of operations and avoidance of deadlocks are essential, since many users may have to access the same data at the same time. Data integrity and data security are of great concern because the data base is a very valuable commodity to its owners and users. Data independence, the ability to make logical changes in the data base or physical changes in the hardware on which it resides without significantly affecting the users and their applications programs, is also important.

An overview of a typical DBMS is shown in Fig. 23-93. As can be seen, there are three levels of data description in the system. The highest level is the conceptual view of information and its structure as seen by the user. This must be mapped into a logical structure, which is the second level of description, according to the structure of the data model chosen and the constraints of the DBMS. This logical structure is called the *schema* and is described in a *data definition language* (DDL) by a *data-base administrator* (DBA). Each application program usually has access to only a subset of the data base, called a *subschema* and defined in a DDL. The third level of description is the actual layout of the data on the physical devices on which it is to reside in the computing system. This description is made using a device-media control language that can take a number of forms, depending on the computing system actually selected. Application programs request data from the DBMS by using a special *data-manipulation* language (DML). Some data-base management systems permit users to retrieve and manipulate the data through the use of a special nonprocedural language called a *query language*.

The DBMS is identified by the data-base model it uses to represent the logical structure. There are three main types: *hierarchic* model, *network* model, and *relational* model. The hierarchic model is a tree structure in which a *parent* record type can have one or more *child* record types but a child record type may not have more than one parent record type; i.e., it does not permit a many-to-many relationship between two record types. In the network model, any record type can be related either as a child or parent record type to any number of record types; i.e., a child record can have more than one parent record. The relational model is made from a set of flat tables called *relations*. The relation is a set of *tuples*; i.e., if the table has n columns, the relation is said to be of degree n. Mathematical notation drawn from relational algebra and calculus is used to describe relations and operations on them.

146. References and Bibliography

PRINCIPLES OF DATA PROCESSING

1. Aiken, A. H., and G. M. Hopper   The Automatic Sequence Controlled Calculator, *Elec. Eng.*, 1948, Vol. 65, p. 384.

2. Babbage, H. P.   "Babbage's Calculating Engines," Spon, London, 1889

3. Babbage, Charles   "Passages from the Life of a Philosopher," Longmans, London, 1864.

4. Bartee, T.C.   "Digital Computer Fundamentals," 4th ed., McGraw-Hill, New York, 1977.

4a. Boole, G. "The Mathematical Analysis of Logic," 1847; reprinted Blackwell, Oxford, 1951.

5. Brainerd, J. G., and T. K. Sharpless   The ENIAC, *Elec. Eng.*, February 1948, pp. 163–172.

5a. Burks, A. W., H. H. Goldstine, and J. von Neumann   Preliminary Discussion of the Logical Design of an Electronic Computing Instrument, P. I II, *Datamation*, September 1962, Vol. 8, No. 9, pp. 24–31, October, No. 10, pp. 36–41.

6. Burks, A. W., H. H. Goldstine, and J. von Neumann   "Preliminary Discussion of the Logical Design of an Electronic Computing Instrument," Institute for Advanced Study, Princeton, N.J., 1946.

7. Caiannello, E. R. (ed.)   "Automatic Theory," Academic, New York, 1966.

8. Caldwell, S. H.   "Switching Circuits and Logical Design," Wiley, New York, 1958.

9. Cardenas, A. F. (ed.)   "Computer Science," Wiley-Interscience, New York, 1972.

10. Dertouzos, M. L.   "Threshold Logic: A Synthesis Approach," M.I.T Press, Cambridge, Mass., 1965.

11. Dolotta, T. A., M. I. Berstein, R. S. Dickson, Jr., N. A. France, B. A. Rosenblatt, D. M. Smith, and T. B. Steel, Jr.   "Data Processing in 1980–1985," Wiley, New York, 1976.

12. Feldman, J. A.   Programming Languages, *Sci. Am.*, December 1979, Vol. 241, No. 6, pp. 94–116.

13. Fogel, L. J., A. J. Owens, and M. J. Walsh   "Artificial Intelligence through Simulated Evolution," Wiley, New York, 1966.

14. Forsythe, A. I., T. A. Keenan, E. I. Organik, and W. Stenberg   "Computer Science: A First Course," 2d ed., Wiley, New York, 1975.

15. Gill, A.   "Introduction to the Theory of Finite-State Machines," McGraw-Hill, New York, 1962.

16. Ginsburg, S.   "An Introduction to Cybernetics," Addison-Wesley, Reading, Mass., 1962.

17. Gluskov, V. M.   "Introduction to Cybernetics," Academic, New York, 1966.

18. Goldstine, H. H.   "The Computer from Pascal to von Neumann," Princeton University Press, Princeton, N.J., 1972.

19. Gray, H. J.   "Digital Computer Engineering," Prentice-Hall, Englewood Cliffs, N.J., 1963.

20. Gschwind, H. W., and E. J. McCluskey   "Design of Digital Computers," 2d ed. Springer-Verlag, New York, 1976.

21. Harrison, M. A.   "Introduction to Switching and Automata Theory," McGraw-Hill, New York, 1965.

22. Hartmanis, J., and R. E. Sterns   "Algebraic Structure of Sequential Machines," Prentice-Hall, Englewood Cliffs, N.J., 1963.

23. Hellerman, H.   "Digital Computer System Principles," 2d ed. McGraw-Hill, New York, 1973.

24. Hennie, F. C., III,   "Iterative Arrays of Circuit Logic," M.I.T. Press, Cambridge, Mass., and Wiley, New York, 1961.

25. Hu, S. T.   "Threshold Logic," University of California Press, Berkeley, Calif., 1965.

26. Humphrey, W. S., Jr.   "Switching Circuits with Computer Applications," McGraw-Hill, New York, 1958.

27. Hurley, R. B.   "Transistor Logic Circuits," Wiley, New York, 1961.

28. Keonjian, E. "Microelectronics: Theory, Design, and Fabrication," McGraw-Hill, New York, 1963.

29. Kreiger, M. "Basic Switching Circuit Theory," Macmillan, New York, 1967.

30. Lewis, P. M., and C. L. Coates "Threshold Logic," Wiley, New York, 1967.

31. Lynn, D. K., S. Meyer, and D. J. Hamilton "Analysis and Design of Integrated Circuits," McGraw-Hill, New York, 1967.

32. Maley, G. A., and J. Earle "The Logic Design of Transistor Digital Computers," Prentice-Hall, Englewood Cliffs, N.J., 1963.

33. McCluskey, E. J., Jr. "Introduction to the Theory of Switching Circuits," McGraw-Hill, New York, 1965.

34. Marcus, M. P. "Switching Circuits for Engineers," Prentice-Hall, Englewood Cliffs, N.J., 1962.

35. "Microelectronics, a *Scientific American* Book," Freeman, San Francisco, 1977.

36. Miller, R. E. "Switching Theory," Vol, 1, "Combinatorial Circuits," Wiley, New York, 1965.

37. Minsky, M. "Computation: Finite and Infinite Machines," Prentice-Hall, Englewood Cliffs, N.J., 1967.

38. Moore, E. F. (ed.) "Sequential Machines: Selected Papers," Addison-Wesley, Reading, Mass., 1964.

39. Motorola, Inc. "Integrated Circuits: Design Principles and Fabrication," McGraw-Hill, New York, 1965.

40. Motorola, Inc. "Analysis and Design of Integrated Circuits," McGraw-Hill, New York, 1967.

41. Nichols, J. E. "The Structure and Design of Programming Languages," Addison-Wesley, Reading, Mass., 1975.

42. Phister, M., Jr. "Logical Design of Digital Computers," Wiley, New York, 1958.

43. Prather, M., Jr. "Introduction to Switching Theory: A Mathematical Approach," Allyn & Bacon, Boston, 1967.

44. Randell, B. (ed.) "The Origins of Digital Computers, Selected Papers," Springer-Verlag, New York, 1973.

45. Richards, R. K. "Arithmetic Operations in Digital Computers," Van Nostrand, Princeton, N.J., 1965.

46. Sammet, J. E. "Programming Languages: History and Fundamentals," Prentice-Hall, Englewood Cliffs, N.J., 1969.

47. Shannon, C. E., and J. McCarthy (eds.) "Automatic Studies," Princeton University Press, Princeton, N.J., 1956.

48. Sippl, C. J. "Calculator Users Guide and Dictionary," Matrix, Champaign, Ill., 1976.

49. Tang, D. T., and R. T. Chien Coding for Error Control, *IBM Syst. J.*, 1969, Vol. 8, No. 1, pp. 48–86.

50. Tippett, J. T. D. A., L. C. Berkowitz, L. C. Clapp, C. J. Koester, and A. Vanderburgh, Jr. "Optical and Electro-Optical Information Processing," M.I.T. Press, Cambridge, Mass., 1965.

51. Warfield, J. N. "Principles of Logic Design," Ginn, Boston, 1965.

52. Xlander, M. "Fundamentals of Reliable Circuit Design," Vols. I and II, ILIFFE Books, London, 1966.

53. Young, J. Z. "A Model of the Brain," Clarendon, Oxford, 1964.

COMPUTER ORGANIZATION AND ARCHITECTURE

53a. Anderson, B. A., and E. D. Jensen Computer Interconnection Structures: Taxonomy, Characteristics and Examples, *Comput. Surv.*, December 1975, Vol. 7, No. 4, pp. 197–213.

54. Awad, E. M. "Business Data Processing," 4th ed., Prentice-Hall, Englewood Cliffs, N.J., 1975.

55. Arbib, M. "Brains, Machines, and Mathematics," McGraw-Hill, New York, 1964.

56. Bloch, E., and D. Galage Component Progress: Its Effects on High-Speed Computer Architecture and Machine Organization, *Computer*, April 1978, Vol. 11, pp. 64–76.

56a. Brooks, F. P., Jr. "The Mythical Man-Month," Addison-Wesley, Reading, Mass., 1975.

57. Bulman, D. M. Stack Computers: An Introduction, *Computer*, May 1977, Vol. 10, pp. 18–28.

58. Cypser, R. J. "Communications Architecture for Distributed Systems," Addison-Wesley, Reading, Mass., 1978.

59. Davidson, S., and B. D. Shriver An Overview of Firmware Engineering, *Computer*, May 1978, Vol. 11, pp. 31–31.

60. Denning, P. J.   Third Generation Computer Systems, *ACM Comput. Surv.*, December 1971, Vol. 3.

61. Dennis, J.   The Varieties of Data Flow Computers, *Proc. 1st Int. Conf. Distrib. Syst. IEEE Comp. Soc.*, Oct. 1-5, 1979, p. 430–439.

61a. Enslow, P. H., Jr.   Multiprocessor Organization — A Survey, *Comput. Surv.*, March 1977, Vol. 9, No. 1, pp. 103–129.

62. Feigenbaum, E. A., and J. Feldman (eds.)   "Computers and Thought," McGraw-Hill, New York, 1966.

62a. Flynn, M. J.   Some Computer Organizations and Their Effectiveness, *IEEE Trans. Comput.*, September 1972, Vol. C-21, No. 9, pp. 948–960.

63. Gear, C. W.   "Computer Organization and Programming," 2d ed., McGraw-Hill, New York, 1978.

64. Hill, F. J., and G. R. Peterson   "Digital Systems: Hardware Organization and Design," 2d ed., Wiley, New York, 1978.

65. Husson, S. S.   "Microprogramming: Principles and Practices," Prentice-Hall, Englewood Cliffs, N.J., 1970.

66. Kartashev, S. I., and S. P. Kartashev   Dynamic Architectures: Problems and Solutions, *Computer*, July 1978, Vol. 11, pp. 26–40.

66a. IEEE Computer Society Task Force on Computer Architecture   A Course of Study in Computer Hardware Architecture, *Computer*, December 1975, Vol. 3, No. 12, pp. 44–57.

67. Kimbleton, S. R., and G. M. Schneider   Computer Communications Networks: Approaches, Objectives, and Performance Considerations, *ACM Comput. Surv.*, September 1975, Vol. 7, pp. 129–173.

68. Lipovski, G. J., and K. L. Doty   Development Directions in Computer Architecture, *Computer*, August 1978, Vol. 11, pp. 54–67.

69. Mano, M. M.   "Digital Logic and Computer Design," Prentice-Hall, Englewood Cliffs, N.J., 1979.

70. Nilsson, N. J.   "Learning Machines," McGraw-Hill, New York, 1965.

71. Reddi, S. S., and E. A. Feustel   A Conceptual Framework for Computer Architecture, *ACM Computing Surv.*, June 1976, Vol. 8, pp. 227–300.

72. Slotnick, D. L., and J. K. Slotnick   "Computers: Their Structure, Use, and Influence," Prentice-Hall, Englewood Cliffs, N.J., 1979.

73. Sondak, N., and E. Mallach   "Microprogramming," Artech House, Dedham, Mass., 1977.

74. Special Issue: Computer Systems Architecture   *ACM Comput. Surv.*, December 1975, Vol. 7.

75. Special Issue: Parallel Processors and Processing, *ACM Comput. Surv.*, March 1977, Vol. 9.

76. Su, S. Y. H. (guest ed.)   Hardware Description Language Applications, *Computer*, June 1977, Vol. 10, pp. 10–49.

77. Thurber, K. J.   "Large Scale Computer Architecture: Parallel and Associative Processors," Hayden, Rochelle Park, N.J., 1976.

78. Wall, H. M. (guest ed.)   Design Automation, *Computer*, April 1975, Vol. 8, pp. 18–51.

79. Uhr, L. (ed.)   "Pattern Recognition," Wiley, New York, 1966.

COMPUTER STORAGE

80. Becker, J., and R. M. Hayes   "Introduction to Information Storage and Retrieval: Tools, Elements, Tolerances," Wiley, New York, 1963.

81. Chang, H., J. Fox, D. Lu, and L. Rosier   A Self-Contained Magnetic Bubble Domain Memory Chip, *Solid State Circ. Conf., Philadelphia*, February 1971.

82. Eccles, W. H., and F. W. Jordan   A Trigger Relay Utilizing Three-Electrode Thermionic Vacuum Tubes, *Radio Rev.*, 1919, Vol. 1, p. 143 (copies available at a service charge from the Engineering Societies' Library, 345 E. 47 Street, New York, N.Y. 10017).

83. Hamming, W. R.   Error Detecting and Error Correcting Codes, *Bell Syst. Tech. J.*, 1947, Vol 29.

84. Hsiao, M. Y.   A Class of Optimal Minimum Odd-Weight Column SEC-DED Codes, *IBM J. Res. Dev.*, 1970, Vol. 14, No. 4, pp. 395–410.

85. Matick, R. E.   "Computer Storage Systems and Technology," Wiley, New York, 1977.

86. Stone, H.   "Introduction to Computer Architecture," 2d ed., Science Research Associates, Chicago, 1980.

87. Salton, G.   "Automatic Information Organization and Retrieval," McGraw-Hill, New York, 1968.

88. Weiderhold, G.   "Data Base Design," McGraw-Hill, New York, 1977.

89. Wegner, P.   "Programming Languages, Information Structures and Machine Organization," McGraw-Hill, New York, 1968.

Telecommunications

90. Davies, D. W., D. L. A. Barber, W. L. Price, and C. M. Solomonides   "Computer Networks and Their Protocols," Wiley, New York, 1979.
91. Doll, D. R.   "Data Communications—Facilities, Networks, and System Design," Wiley, New York, 1978.
92. Housley, T.   "Data Communications and Teleprocessing Systems," Prentice-Hall, Englewood Cliffs, N.J., 1979.
93. Martin, J.   "Future Developments in Telecommunications," 2d ed., Prentice-Hall, Englewood Cliffs, N.J., 1977.
94. Martin, J.   "System Analysis for Data Transmission," Prentice-Hall, Englewood Cliffs, N.J., 1972.
95. Martin, J.   "Teleprocessing Network Organization," Prentice-Hall, Englewood Cliffs, N.J., 1970.
96. Special Issue on Packet Communications Networks, *Proc. IEEE*, November 1978, Vol. 66, No. 11, 1301–1588.

Software

97. American National Standards Institute   Standards X3.9-1966, and X3.10-1966, American National Standards Institute, New York, 1966.
97a. Backus, J. W., et al.   The FORTRAN Automatic Coding System, *Proc. West. Joint Comput. Conf., 1957,* Vol. II, p. 188.
98. Cardenas, A. F.   "Data Base Management Systems," Allyn and Bacon, Boston, 1979.
98a. Dijkstra, E. W.   *Commun. Ass. Comput. Mach.,* 1968, Vol. 11, No. 5, p. 341.
99. Dijkstra, E. W.   "A Discipline of Programming," Prentice-Hall, Englewood Cliffs, N.J., 1976.
100. Hansen, P. B.   "The Architecture of Concurrent Programs," Prentice-Hall, Englewood Cliffs, N.J., 1977.
101. Husson, S. S.   "Microprogramming Principles and Practice," Prentice-Hall, Englewood Cliffs, N.J., 1970.
102. Jensen, K., and N. Wirth   "PASCAL User Manual and Report," Springer-Verlag, New York, 1978.
103. Kernighan, B. W., and P. J. Plauger   "The Elements of Programming Style," McGraw-Hill, New York, 1974.
104. Lanciaux, D. (ed.)   "Operating Systems: Theory and Practice," North-Holland, Amsterdam, 1978.
105. Ledgard, H. D., J. F. Hueras, and P. A. Nogin   "PASCAL with Style, Programming Proverbs," Hayden, Rochelle Park, N.J., 1979.
106. Madnick, S., and J. Donavan   "Operating Systems," McGraw-Hill, New York, 1974.
107. Martin, J.   "Computer Data Base Organization," Prentice-Hall, Englewood Cliffs, N.J., 1975.
108. Martin, J.   "Design of Real-Time Computer Systems," Prentice-Hall, Englewood Cliffs, N.J., 1967.
109. Mills, H. D.   Software Engineering, *Science,* Mar. 18, 1977, Vol. 195, pp. 1199–1204.
109a. Naur, P. (ed.)   Revised Report on the Algorithmic Language ALGOL 60, *Commun. Ass. Comput. Mach.,* 1963, Vol. 6, p. 1.
110. Organick, A. I., A. I. Forsythe, and R. P. Plummer   "Programming Language Structure," Academic, New York, 1978.
111. Rosen, S.   "Programming Systems and Languages," McGraw-Hill, New York, 1967.
112. Sammet, J. E.   "Programming Languages: History and Fundamentals," Prentice-Hall, Englewood Cliffs, N.J., 1969.
113. Special Issue: Data Base Management Systems, *ACM Comput. Surv.,* March 1976, Vol. 8.
114. Tanenbaum, A. S.   A Tutorial on ALGOL 68, *ACM Comput. Surv.,* June 1976, Vol. 8, pp. 155–190.
115. Wegner, P. (ed.)   "Introduction to System Programming," Academic, New York, 1965.

116. Wegner, P. (ed.) "Research Directions in Software Technology," M.I.T. Press, Cambridge, Mass., 1979.

117. Wirth, N. "Algorithms + Data Structures = Programs," Prentice-Hall, Englewood Cliffs, N.J., 1976.

118. Yeh, R. T. (ed.) "Current Trends in Programming Methodology," Prentice-Hall, Englewood Cliffs, N.J., Vol. 1, Software Specification and Design, 1977; Vol. 2, Program Validation, 1977; Vol. 3, Software Modeling, 1978; and Vol. 4, Data Structuring 1978.

Section **24**

Electronics in Processing Industries

W. E. VANNAH *Technology Staff, The Foxboro Company; Senior Member, IEEE*

P. D. HANSEN *Senior Technical Consultant, The Foxboro Company; Member, ASME*

D. A. RICHARDSON *Principal Development Engineer, The Foxboro Company*

C. M. PATEL *Research Coordinator, Electronic Systems, The Foxboro Company; Member, IEEE*

M. PRASAD *Product Assurance Consultant, The Foxboro Company*

W. CALDER *Manager, Corporate Quality Assurance, The Foxboro, Company; Senior Member, Instrument Society of America*

CONTENTS

Numbers refer to paragraphs

Overview

BY W. E. VANNAH

1. Applications Requirements. This section presents requirements for electronic measurement and control equipment applied to process industries. Factors are stressed which impose design constraints differing from those imposed by other areas of application.

The process industries include chemical, petroleum, power, food, textile, paper, and metallurgical, among others. These industries continuously or semicontinuously process gases, liquids, or solids. Electronic equipment measures, indicates, and controls their flow, pressure, temperature, level, and composition.

Measurement and control applications for process variables range from indication and/or regulation of a single process variable to the optimization of the kinetics and throughput of hundreds of variables in an entire plant.

A pronounced, continuing trend of process management is to provide maximum profit per unit time of process operation. This emphasis demands larger aggregates of equipment assembled in systems configurations in order to:

1. Monitor more data simultaneously
2. Make more efficient control of interactive variables possible
3. Present more information "by exception" (see Par. **24-24**)
4. Ensure a high level of availability of the process and continuity of process operation
5. Allow for lower-cost expansion, both "vertically" (compatibility with higher-level computing equipment) and "horizontally" (system-size expansion)
6. Facilitate entry of more exogenous data

An Application Is Unique. Each process control system is designed as a one-of-a-kind application. Uniqueness occurs because:

1. Factors determining process dynamics vary widely between industries and between plants within industries, e.g., vessel capacities, transport lags.
2. The relative importance of system functions depends on the needs of the individual user, who may determine that either control enforcement through regulatory control or continuous operator information is more significant and who may trade reliability against system cost.
3. Environmental requirements vary dramatically.

Although different control strategies emphasize different performance characteristics of the system, the functions normally required of the control system include:

1. Process data acquisition
2. Alarm for abnormal conditions
3. Display and recording of process measurement, set point, and output values
4. Single-variable control using standard feedback algorithms
5. Multivariable control including cascade, ratio, feedforward, and interacting configurations
6. Supervisory control and reporting of management information

Flexibility Results. Process-control equipment has evolved continually over the last 30 years in response to the variety of functional requirements described. Rapid advances in electronic component technology and design techniques have facilitated innovative development of signal processing subsystems and electronic devices. Incorporated in distributed information processing and measurement equipment, they have paced control system precision and flexibility. The results include:

1. Opportunity for coordinated control strategies through control-center consolidation.

2. Remote and distributed communication subsystems. Figure 24-1 traces their standardized field transmission signals between process variables and regulator controllers (see discussion

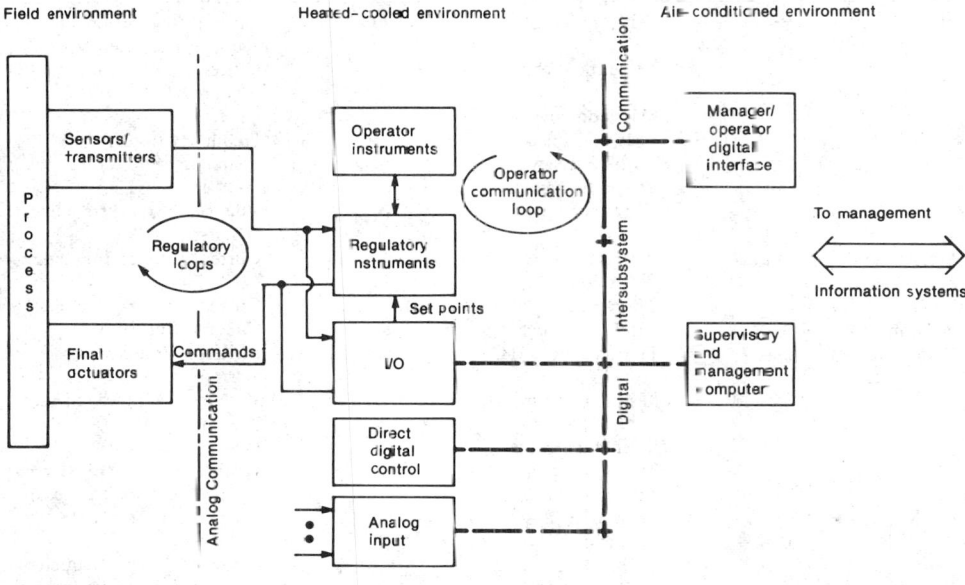

Field environment   Heated-cooled environment   Air-conditioned environment

Fig. 24-1. Communication subsystems to provide interface between process, operator, and plant management.

beginning with Par. **24-9**); and digital communication, common between data acquisition, regulatory control, and management computing.

3. Computers designed for process control.

4. Reliability of system performance and designs of equipment which are safe for operating personnel as well as for operation in combustible environments.

5. Broad flexibility for configuring application systems and adaptability to changes in processing and regulatory-agency requirements.

2. System Architecture. Structuring of functions and equipment varies widely. Paragraph **24-19** identifies a comprehensive architecture. The three major interfaces displayed in Fig. 24-1 and identified in Pars. **24-3** to **24-5** are common to it and other accepted architectures. Distributed microprocessors are used in the digital subsystem interfaces the operator interface, supervisory control computers, and direct regulatory control computers.

Paragraph **24-21** demonstrates a facility, offered by a common intersubsystem communication network, to distribute regulatory control functions geographically or by level.

3. Process Interface. Paragraph **24-6** discusses the major issues involved with measurement and transmission of process variables. The signal levels involved have become a worldwide standard. Table 24-1 indicates some common types of process measurement sensors, transmission signals, and command signals.

4. Operator Interface. The window through which the operator views the process is critical and variable. The discussion starting with Par. **24-24** delineates some of the options presently available.

5. Digital Subsystem Interface. The increasing use of digital computers in process control justifies a less costly and more efficient interface than the earlier generations offered. Computers are used as supervisory devices (providing optimizing calculations and outputting setpoint information to regulatory controllers), direct-digital-control computers (comparing process measurements with set points, calculating the required control action, and outputting commands to final actuators), and as communications processors. Paragraph **24-23** presents current trends in providing a more compatible interface to distributed digital computers.

Service conditions in plant locations dictate other specifics of designing and applying electronic equipment to process measurement and control. These locations are described in standards pub-

TABLE 24-1 Typical Sensors and Transmission Signals

Measurement	Sensor type	Transmission signal*
Flow	Orifice place or other head meter	Current (proportional to flow squared), frequency
	Target flowmeters	Current (proportional to flow squared)
	Magnetic flowmeters	Current
	Positive-displacement meters	Frequency (linear)
	Turbine flowmeters	Frequency (linear)
	Vortex flowmeters	Frequency (linear)
	Ultrasonic flowmeters	Voltage, current, frequency (linear)
Pressure and differential pressure	Force balance, strain, or motion detection	Current (linear), frequency
	Differential-pressure transmitter	Current (linear), frequency
Position and level	Displacers, force detector	Current (linear)
Temperature	Thermocouple (TC)	Millivolts (nonlinear), current
	Resistance temperature detector (RTD)	Variable resistance, current (linear)
Speed	Magnetic pickups	Frequency (linear)
	Photodiode pickups	Frequency (linear)
	Tachometers	AC and dc voltage (linear)
Force	Force balance, strain, or motion detection	Current (linear)
pH	Electrochemical	Current, voltage, frequency
Ion-selective	Electrochemical	Current, voltage, frequency
Oxidation-reduction potential (ORP)	Electrochemical	Current, voltage, frequency
Conductivity, resistivity	AC impedance detector	Current, voltage, frequency
Composition	Chromatograph	Current, voltage, frequency
	Coulometer	Current, voltage frequency
	Spectrophotometer	Current, voltage frequency
Power, current, and voltage	Hall effect transducers	Current (linear)
	Thermal converters	Millivolts (linear)
	Transformers	AC volts (linear)

*Currents range from 10 to 50, 4 to 20, or 1 to 5 mA. Millivolt values range from $< \pm 55$ mV (thermocouples) up to $\pm 1,000$ mV (ORP). Voltages range from 0 to 1 to 0 to 10 V. Frequency signals range up to 4 kHz.

lished by the International Electrotechnical Commission and include ranges of temperature, humidity, explosive contaminants, and electromagnetic effects. Paragraph **24-30** presents these important qualifying conditions.

Process Signal Transmission

BY P. D. HANSEN

6. Transmission of Measurements and Commands. Centralized control of chemical plants, petroleum refineries, and other process industries may require transmission of hundreds of measurement and command signals over many hundreds of meters. The cost of the installed transmission system is significant, often exceeding that of the measuring instruments and the final actuator command converters. Furthermore, the signal transmission system can significantly affect plant safety, control reliability, and product quality. Paragraphs **24-7** to **24-**

12 outline many factors that should be considered in designing and specifying instruments and cable for electrical process signal transmission.

7. Types of Measurements and Commands. In an electronic sensor a process variable modulates a parametric impedance, admittance, transformer coupling, voltage, current, or frequency. A transmitter, usually located near the process function (in the field), provides suitable excitation to the sensor and converts the parametric modulation into a form which can be transmitted up to hundreds of meters over wire pairs to a control center. Table 24-1 lists some common process variables measured, their sensors or transmitters, and associated signals.

At the control center an analog regulatory controller and/or a digital input-output (I/O) converter processes the transmitted measurement signal (see Pars. 24-14 and 24-19). The regulatory controller or the digital I/O converter transmits commands back to a field converter, which in turn modulates power to position a final actuator (control valve or motor drive). Table 24-2 lists I/O converters, field converters, and their transmission signals.

8. Interference. Most process variables change very slowly, their measurements containing little useful information at frequencies beyond 1 Hz. In an analog system, pickup from power-line and higher-frequency sources is usually filtered adequately by signal converters, controllers, recorders, and indicators, as well as by the process. As a result, special precaution is required only for low-level signals or when nonlinear operations are involved.

When signals are periodically sampled, Fourier components of the original signal with frequencies greater than half the sampling frequency cannot be distinguished from lower-frequency components. In particular, any signal component that is an integral harmonic of the sampling frequency appears after sampling as if it were constant. Consequently, it is much more important for high-frequency signal components in digital signal processing to be smaller than those employed in analog signal processing. This is accomplished in several ways: through the use of shielding (e.g., twisted leads and other measures to minimize pickup), by filtering at the receiver, and by use of transmission signal forms less corruptible by pickup.

Radio-frequency interference (electromagnetic, EMI) is particularly difficult to eliminate with conventional discrete-component analog filters. To prevent problems with EMI, each electronic enclosure should provide adequate grounded shielding. In addition, any nonshielded signal or power-lead penetration of an enclosure should be made with a feedthrough capacitor designed to shunt the external rf signal components to the enclosure-grounded shield while allowing the lower-frequency signal or power component to penetrate the enclosure shield.

9. Measurement Voltage Signal. Low-frequency and steady-state voltage signals are conveyed with a pair of leads (Fig 24-2). Each lead has a potential with respect to local earth ground. The average of these potentials is called the *common-mode voltage*. The difference in potentials is the *signal*, or *normal-mode voltage*. Twisting the leads causes nearly identical electromagnetic (inductively coupled series) voltage pickup on each lead. Also, electrostatic (capacitively coupled) current pickup from an external high-voltage source tends to be the same in both leads. If the impedances are matched from each lead to earth ground, the lead voltages resulting from these currents will be equal. Consequently, balanced-impedance twisted leads minimize inductive and possible capacitive normal-mode pickup but do not affect common-mode pickup. Therefore, a receiver with high common-mode rejection (10^6 or better) is essential for low-level-voltage measurement. Capacitively coupled pickup can be further reduced by shielding. Resistively coupled pickup is the result of a common conduction path. When two circuits have a common node, there should be minimum length of shared conductor. Also, a conductor

TABLE 24-2 Some Converters and Signals

Converter	Input	Output
Positioner	4–20 or 10–50 mA	Valve position
Actuator	4–20 or 10–50 mA	Pressure 20–100 kPa (3–15 lb/in²)
A/D	As in Table 24-1	12-b value
D/A	10- or 12-b value	4–20 or 10–50 mA
Check	Contact closure denoting status or interrupt	1-b value
Digital output	1-b value	Contact closure or 80 mA at 48 V dc, 2 mA at 240 V ac
Pulse input	<12,800 pulses/s	12 b
Pulse output	1 b	Pulse train

loop should be avoided because magnetically induced circulating current, together with the loop's series resistance, may cause voltage variations around the loop.

Many voltage sources (e.g., thermocouples and ion-selective electrodes) are grounded in the field. Because of possible differences between local ground potentials, their transmission lines

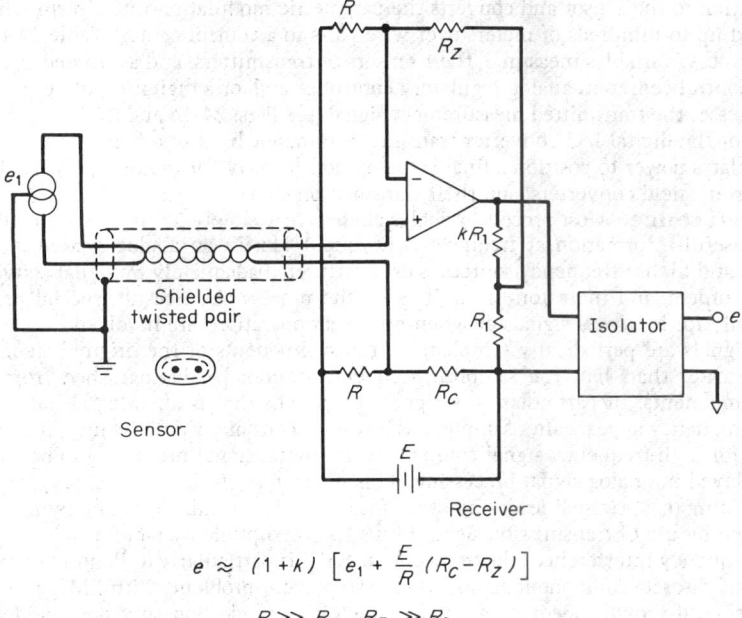

$$e \approx (1+k)\left[e_1 + \frac{E}{R}(R_C - R_Z)\right]$$

$$R \gg R_C, R_Z \gg R_1$$

Fig. 24-2. Low-level voltage-signal transmission.

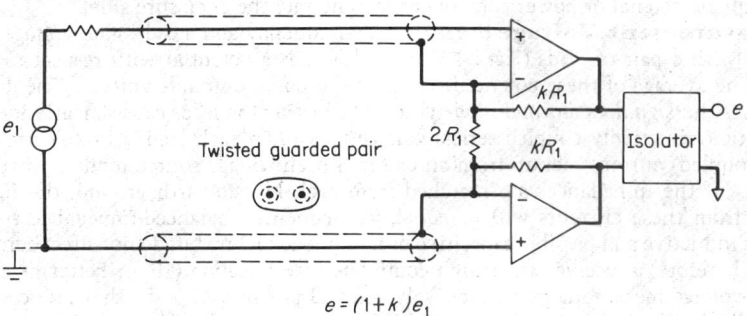

$$e = (1+k)e_1$$

Fig. 24-3. High-source-impedance voltage transmission.

should be isolated by high impedance from the control-center ground to prevent current flow through the signal leads.

The normal-mode current flow must be kept small to prevent errors caused by cable series resistance. The receiver should have a high normal-mode input impedance and low normal-mode input current. Consequently, power to operate a voltage-signal field transmitter should not be supplied through either signal wire. Often, a very small normal-mode current (10 nA) is intentionally introduced at the amplifier input to cause the output to saturate if the measurement circuit opens. In a thermocouple application, this feature is called *burnout indication*. An open thermocouple then causes the amplifier output to go full high or low, depending on the direction of the normal-mode bleed current.

Very high source-impedance voltage measurement (Fig. 24-3) requires independent guarding of each input lead, as well as a receiver with much higher input impedance and high common-mode rejection. Such precautions are commonly employed at the inputs of pH and ORP (oxidation-reduction potential) sensors and also with magnetic flowmeters.

10. Measurement Impedance Signal. As with voltage measurement, the simplest method of conveying an impedance modulation to the control center is over direct-connected lines. A transmitter may be interposed since the nonideal characteristics of the connecting leads, principally series resistance and shunt capacitance, may affect the measurement's calibration and its sensitivity to service conditions.

Temperature measurement with a resistance thermal detector (RTD) offers an example of direct connection. This resistance measurement uses a three-wire circuit which requires current flow through two wires (see Figs. 24-4 and 24-5).

If the wire resistances match, voltage drops cancel when the sensor bridge is balanced. This compensates for the effect on zero of the cable resistance change with ambient temperature

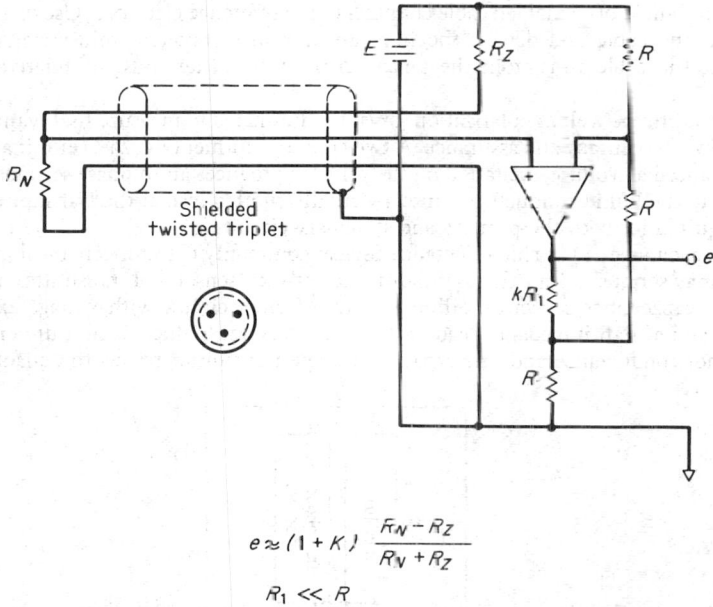

$$e \approx (1 + K) \frac{R_N - R_Z}{R_N + R_Z}$$

$$R_1 \ll R$$

Fig. 24-4. Series-excitation transmission from a nickel RTD.

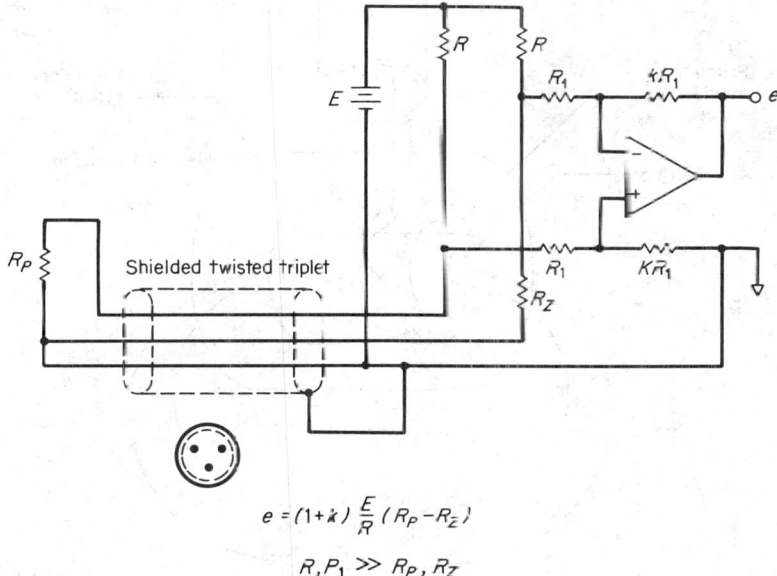

$$e = (1 + k) \frac{E}{R} (R_P - R_Z)$$

$$R, P_1 \gg R_P, R_Z$$

Fig. 24-5. Parallel excitation transmission from a platinum RTD.

(copper resistivity changes 20%/50°C, 40%/100°F). With additional receiver circuits (or by using more wires) it is possible to achieve first-order cancellation of the effect of cable resistance on span, as well. The wires should be twisted to eliminate normal-mode inductive pickup. A shield or grounded conduit will minimize capacitive pickup. Also, normal-mode capacitive pickup can be reduced by placing the reference resistor R_z near the sensor, instead of near the amplifier. This nearly balances the impedances to earth of the two critical lines but may subject the reference resistor to greater swings in ambient temperature. A field-mounted transmitter should be employed when transmission distances are greater than a few hundred meters.

Other impedance-measurement examples are the measurement of water resistivity and the conductivity of ionized chemical process streams. An electrode sensor with ac excitation to minimize polarization is used when fluid resistance is large compared with lead and electrode contact resistance. The shield of a shielded cable connects to the reference electrode. Use of a transmitter which drives the cable and detects the in-phase current component minimizes cable error. Nevertheless, the cable runs from the sensor to the transmitter must be relatively short to achieve high accuracy.

Electrode fouling, as well as polarization, must be eliminated to measure high values of liquid conductivity with accuracy and assurance. A two-core transformer (Fig. 24-6) eliminates the electrodes. A balanced ac voltage excitation on the primary produces an in-phase secondary current, proportional to the liquid conductance, into a virtually shorted load at the transmitter. Shielded wires are required for both the primary and secondary circuits.

Variable-capacitance or variable-reluctance devices commonly detect mechanical motion. Line capacitance may seriously limit the useful distance between sensor and transmitter, particularly for a variable-capacitance sensor. A differential transformer circuit with voltage excitation on the primary and a high impedance load on the secondary can reduce quadrature errors caused by transformer conductance and line capacitance. The transmitter rejects the quadrature com-

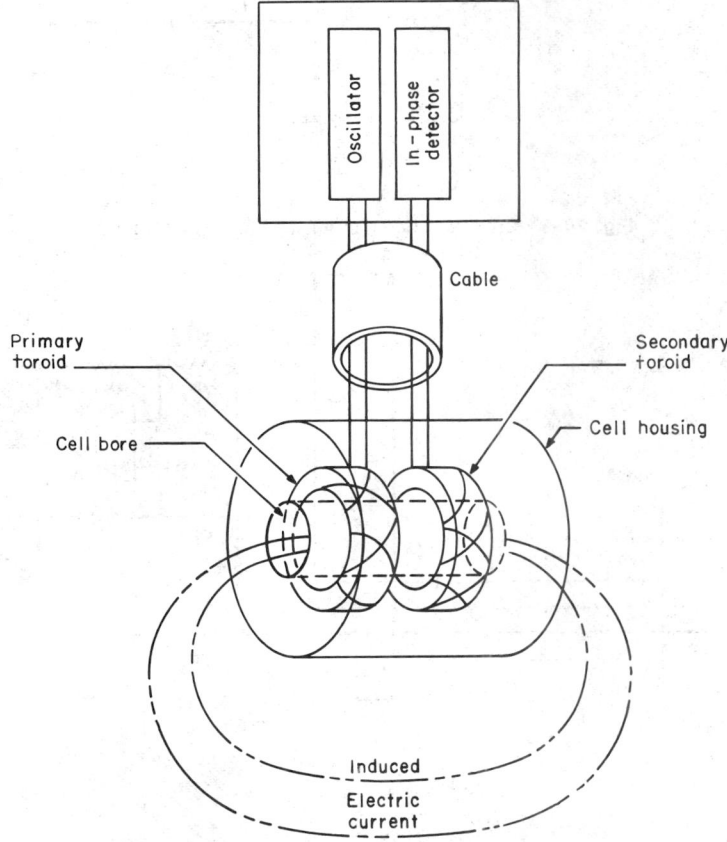

Fig. 24-6. Electrodeless conductivity sensor.

ponents by sensing only the secondary voltage component in phase with the primary voltage. Both the primary and secondary wire pairs should be independently twisted to reduce inadvertent coupling.

11. Current Signal. In order to power field equipment and avoid errors resulting from two-wire transmission-line series resistance, direct current is used to convey a measurement signal from a field transmitter to a central data-conditioning and control site or to convey a command signal from that site to a field converter. The most commonly used current range is 4 to 20 mA. Since the current and the voltage drop across the field instrument are always positive, a moderate amount of continuous power (at least 50 mW) is available for powering field trans-

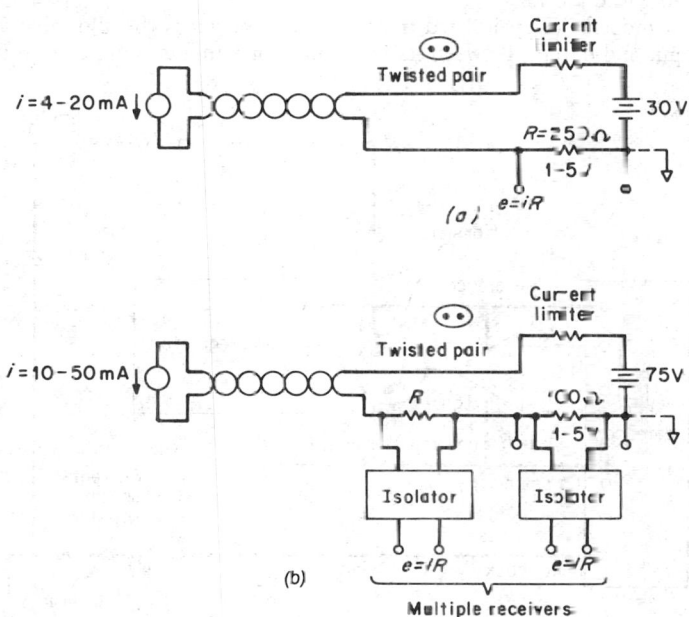

Fig. 24-7. High-level current transmission communicates (a) measurement and (b) command signals.

mitters and converters and for detection of opens or shorts. If a measurement-signal transmission line were to open-circuit, the receiver would overrange low. A short circuit would cause the receiver to overrange high. The voltage across a command-signal transmission line would be abnormally high for an open circuit and low for a short circuit.

A current signal can be conveyed through a single conductor (Fig. 24-7). The return path can be shared with other circuits without significant error, but independent return paths are used to avoid excessive and unpredictable line drop that would occur in a shared common path. Also, because a pair of wires forms a loop of small area tending to intercept a small number of magnetic flux lines, a twisted pair may further help eliminate inductive coupling. The inductive voltage pickup divides between the transmitter and receiver according to the ratio of their impedances. (Consider half the cable capacitance in shunt with each.) Thus, if the transmitter output impedance at power-line frequency is 25 kΩ and the receiver impedance is 250 Ω, 1% of the inductive pickup appears at the receiver.

Capacitively coupled current pickup is generally small compared with the signal-current span. Unusually large pickup may result from running signal cables together with power lines. Nevertheless, this practice is followed without adverse effect in some analog systems. For example, for the peak-to-peak capacitively coupled current from a 115-V 60-Hz power line to be less than ¼% of span (16 mA) after passing through a 10-Hz first-order low-pass filter, the unbalanced mutual capacitance must be less than 1.7 nF. If the unbalanced coupling capacitance is 30 pF/m, power and signal cables can be run together for 60 m.

Cable shunt conductance (approximately 0.30 pS/m for PVC-insulated cable) limits the length for accurate dc current transmission in unguarded cable. For example, a 4- to 20-mA circuit with a 30-V dc supply has a maximum leakage current of 1 μA/km, or 0.006% of span per kilometer. A 10- to 50-mA circuit with a 75-V dc supply has a maximum leakage current of 2.3 μA/km;

also 0.006% of span per kilometer. By using a driven shield or guard, this error can be reduced by one to four orders of magnitude, depending on the current range.

If the current signal is to be transmitted to several locations, the receivers may be connected in series between the transmitter and the power supply (Fig. 24-7b). To prevent the presence or absence of one receiver from affecting the signal at another, each receiver output should be isolated with high impedance from its input, other inputs, the transmission-circuit power supply, and ground. Shunt paths that could otherwise cause signal current to bypass some of the receivers are thus eliminated. If the transmission-circuit power supply is also isolated from ground and other transmission circuits, one fault, a short circuit between two circuits or ground, may be tolerated without error.

An important property of an isolated transmitter or receiver is the allowable voltage range between its input and output, between its input and other inputs, and between its input and

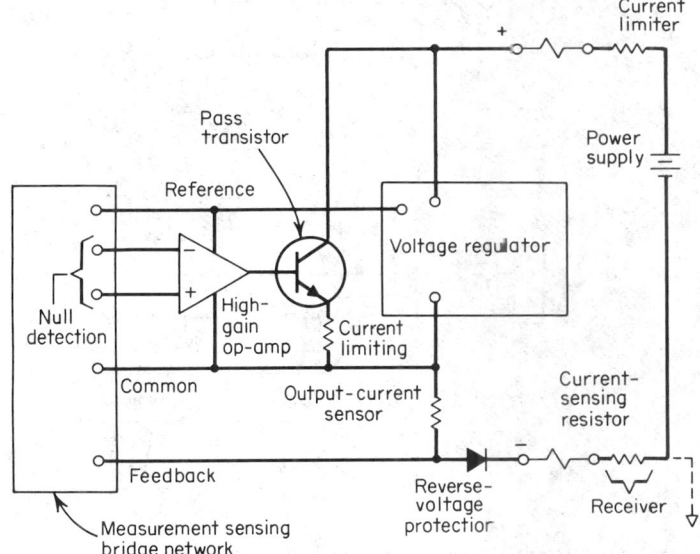

Fig. 24-8. Two-wire current output with voltage feedback.

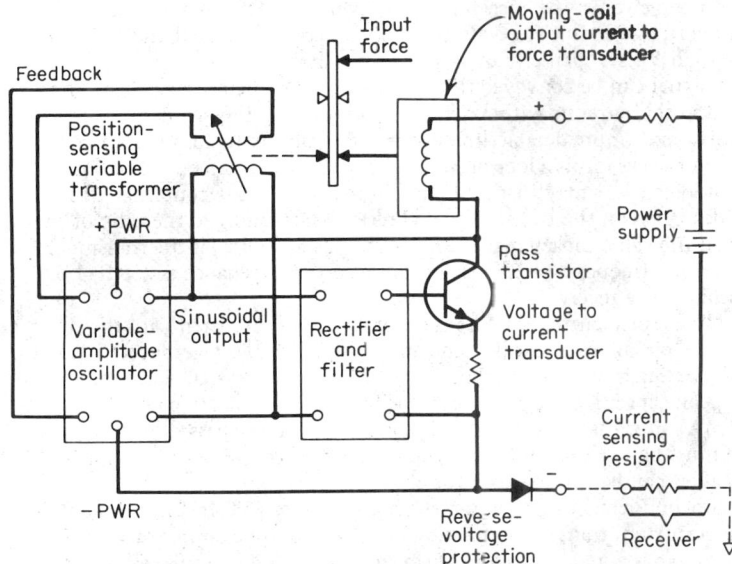

Fig. 24-9. Two-wire current-feedback force-balance transmitter.

ground. Some applications require rejection of 500 V or more. To achieve this, the signal must
be transmitted across a nonconducting (galvanically isolated) barrier separating the input and
output. This is commonly done with an isolation transformer or magnetic amplifier, requiring
that the primary-side dc input signal modulate an ac carrier and that the secondary ac signal be
demodulated or rectified to achieve the dc output. Digital or frequency signal transmission across
an isolation barrier can be performed with an optical coupler or pulse transformer.

Figures 24-8 and 24-9 show two common two-wire transmitter concepts. In Fig. 24-9 a series
diode blocks current flow in the reverse direction to prevent damage should power supply leads

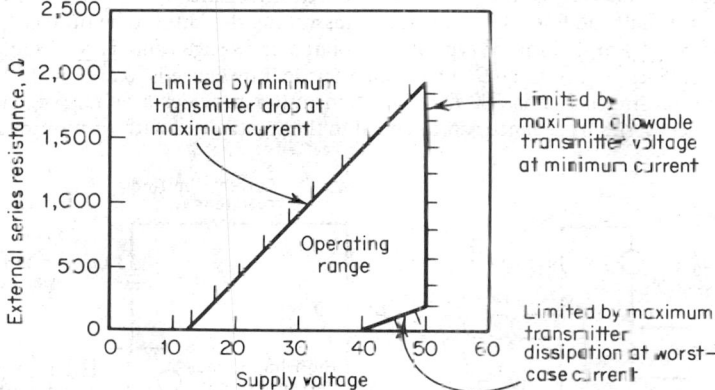

Fig. 24-10. Output operating limits of a 4- to 20-mA transmitter.

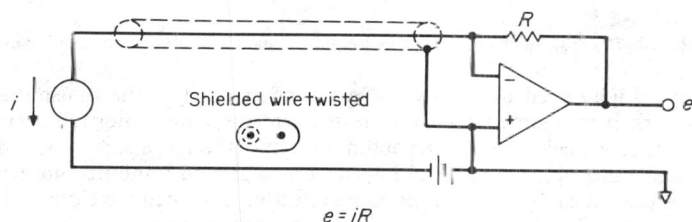

$$e = iR$$

Fig. 24-11. Current signal from a photomultiplier.

be reversed. In Fig. 24-8 the output current is sensed resistively and fed back as a voltage to
rebalance a bridge network. Accuracy depends upon the stability of the reference voltage, the
sensing resistor, and the bridge resistances.

In Fig. 24-9, the output is passed through a voice coil, a moving coil in a permanent magnetic
field. This action generates a force that rebalances a mechanical lever system. The balance is
sensed with a differential transformer that controls the amplitude of a sine-wave oscillator,
which in turn controls the output current. Accuracy depends on the current-force relationship
of the voice coil and the mechanical stability of the lever system. Both these circuits can have
a very high output impedance and can operate over a wide range of power-supply voltage and
external series resistance. Typical operating limits are shown in Fig. 24-10. Limited dc power and
high internal frequency (requiring small internal energy storage incapable of igniting an explosive mixture under fault conditions) permit application of these transmitters to intrinsically safe
circuits.

Three- or four-wire transmitters are used when more than 50 mW quiescent power is required.
Examples include the magnetic flowmeter, which requires high-current field excitation, and the
pH or selective-ion sensor, which may use a dedicated ultrasonic or motor-driven cleaner to
prevent electrode fouling. Both use four wires, two to transmit the signal to the receiver and two
to connect directly to the ac mains. This arrangement permits galvanic isolation of the signal
from earth and the ac mains.

Low-level current measurement (Fig. 24-11) requires shielding to reduce capacitive pickup and
leakage current due to cable conductance. The receiver input resistance is made virtually zero
to reduce inductive pickup as well as cable leakage.

12. Frequency Signal. Frequency is the only property of an ac wave that is not altered by transmission. Amplitude and duty cycle are affected by attenuation and dispersion. An audio-frequency signal can readily be transmitted through a galvanic isolator. Also, it can be converted into a digital value by counting zero crossings long enough to achieve the desired resolution. A faster conversion can be made by measuring period.

At the same time that a variable-frequency voltage signal is transmitted in one direction, a low-frequency current signal can be transmitted in the opposite direction while dc power is supplied to the frequency transmitter over a single pair of wires (Fig. 24-12).

For minimum distortion of a signal with high-frequency content, the terminal impedance at one end of the line should equal the line's characteristic impedance $\sqrt{L/C}$ (neglecting dielectric dissipation) minus half the line's series resistance (assuming the latter to be lumped half at each end). For example, 1 km (0.6 mi) of typical two-conductor 19-gage cable is represented by $L = 0.6$ mH, $C = 0.06$ μF, and $R = 60$ Ω. The characteristic impedance is $\sqrt{L/C} = 100$ Ω, and the optimum terminal impedance is 500 Ω. If the transmitter provides a voltage signal with low source impedance, series impedance can be added to the transmitter with minimum attenuation

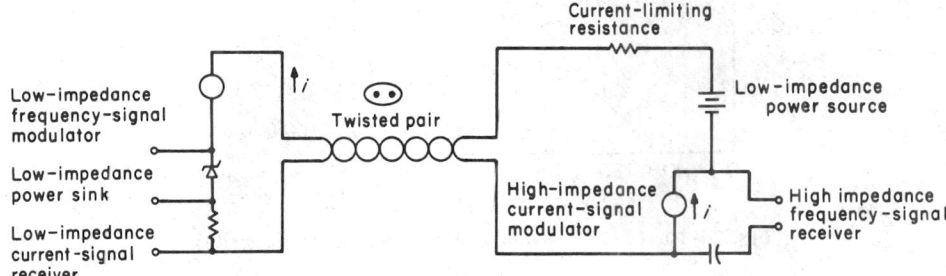

Fig. 24-12. Two-wire simultaneous frequency and current signal transmission.

of the signal at a high-impedance receiver. On the other hand, if the transmitter provides a current signal with high source impedance, least attenuation and distortion occurs when the receiver's impedance is made equal to the optimum terminal impedance. A twisted pair is used to reduce inductive pickup, and matched impedances to ground minimize normal-mode capacitive pickup. A coaxial cable may be even more effective at reducing pickup and also provide a higher characteristic impedance because of lower shunt capacitance.

13. Annotated Standards. *Transmission.*   Standards of the International Electrotechnical Commission (IEC) and U.S. national standards describe accepted practice for current and voltage transmission in industrial process plants. The standards specify signal current level, the number of wires required for the transmission of signal and field power, the interdependence of load resistance and power-supply voltage, and the isolation characteristics of the transmitter and receiver. Adherence to these standards will provide interchangeability between subsystems and equipment. The standards do not apply to direct connection between a sensor in the field and a transmitter at a control center because requirements are usually dictated by the peculiar characteristics of the sensor and its extension wires. The transmission standards are:

IEC Publ. 381 (1971) Analogue DC Current Signals for Process Control Systems.

IEC Publ. 381A (1975) First Supplement (a proposed supplement may reduce power supply voltage to 20 V)

IEC Publ. 381-2 (1978) Part 2: Direct Voltage Signals

IEC Publ. 382 (1971) Analogue Pneumatic Signal for Process Control Systems

Electrochemical Measurement.   Standards on electrochemical measurement are:

SAMA AI-1.1-1974 Recommended Practice for Glass pH and Reference Electrodes

Draft IEC standards specify signal ranges, test procedures, ranges of service conditions, and formats for reporting performance:

Expression of Performance of Electrochemical Analyzers, Part I: General, Part II: pH Value, and Part III: Electrolytic Conductivity.

Standards and recommended practices referenced are subject to periodic change. Latest editions of IEC Publications are available from American National Standards Institute, Inc., 1430 Broadway, New York, NY 10018, except as indicated here.

Scientific Apparatus Makers Association (SAMA), 1101 6th Street, N.W., Washington, DC 20036

National Fire Protection Association (NFPA), 470 Atlantic Ave., Boston MA 02210

Underwriters Laboratories Inc. (UL), 333 Pfingsten Rd., Northbrook, IL 60062

Canadian Standards Association (CSA), 178 Rexdale Blvd., Rexdale, Ontario, Canada M9W 1R3

VDE, refer to DKE, Stresemannallee 21, D-6000, Frankfurt 70, Federal Republic of Germany

Comité Européan de Normalisation Electrotechnique (CENELEC), Rue Brederode, 2, Bte 5, 1000, Bruxelles, Belgique

American Bureau of Shipping, McGraw-Hill, Inc., 1221 Avenue of the Americas, New York, NY 10020.

Data and Control

BY D. A. RICHARDSON

14. Functions. A control center contains the functions which process measurement signals and generate command signals transmitted to final actuators. Data-conditioning functions include scaling, linearizing, shaping, algebraic computing, and dynamic compensation. Control functions operate on the process measurement signals with a variety of steady-state and dynamic control modes.

Regulatory control is the first level of control in a process control hierarchy. Here, control functions hold measured variables at predetermined values (set points) by manipulating associated controlling elements. Also, state transfer, adaptive, and information display elements establish and monitor performance of a process unit.

15. Data Conditioning. In many control applications, a direct measurement of a process variable is not adequate. It may be necessary to linearize the signal, take its root, or characterize it on some empirical engineering scale. Where direct measurement of a variable, say energy transfer, is not practical, the system designer may specify computation of the value from several variables. Dynamic compensation, to match process dynamics, is common.

Equipment producers supply a number of computational and characterizing elements to modify and combine readily available measurements. Each element is usually supplied as a single-function device with standardized input and output signal levels. The elements provide standard continuous computing functions the system designer can use to configure complex measuring techniques. Typically, they can be scaled to provide proper spans for different process requirements. The functions are provided in pneumatic, analog, or digital electronic elements. Typical functions provided by most equipment manufacturers include:

Square-root extraction.

Summing, having two to eight inputs, for addition and subtraction. Scaling is provided for input spans and biasing.

Multiplying-dividing, often in the form $D = AB/C$ with all variables scaled to suit the spans and relationships of process variables.

Selecting, to select the highest, median, or lowest of several variables.

Signal characterization, primarily adjustable lead and lag transfer functions or curve fitting for nonlinear functions.

Time-function generation, such as ramp generating or signal programming, for batch operations.

The following applications are provided.

Linearization. Where flow rate is determined by measuring the differential pressure across an orifice plate, the measurement signal is proportional to the square of the flow rate. In some cases, as when the desired measurement is the sum of several flows, the square root of each differential pressure signal must be taken before the flow signals are added. Each requires a square-root-extracting function that solves the equation $E_{out} = 10\sqrt{E_{in}}$. The function requires no scaling because the measurement is always zero-based and transmitter scaling provides proper span.

Computing. Where the variable to be controlled is not easily measurable, a computing function may calculate the value. A measurement of mass flow, for instance, can be calculated from

$$M = K\sqrt{hp/T}$$

where M = mass flow, h = absolute pressure, p = differential pressure across orifice plate, T = temperature, and K = constant.

The calculation requires a multiplying-dividing function, with each input scalable to suit the flow equation. The selection of transmitter spans can greatly influence the accuracy of the calculated result.

Another application of computing functions occurs when the controlled variable is dependent on the measurement of an independent variable. A typical example is the control of composition by the introduction of additives. Here, the amount of additive (controlled flow) is to be maintained in a specific relationship to the main ("wild") flow rate. A ratio element sets the proper proportioning in the form $E_{out} = RE_{in}$, where R is the ratio of controlled to wild flow.

16. Regulatory Control. In an industrial process in which the value of a variable is to be enforced at a set point, a controller compares the measurement signal with the set point and calculates a command to minimize the difference. Although controllers range from the simple ON-OFF type to those employing complex mathematical functions, all have certain functions in

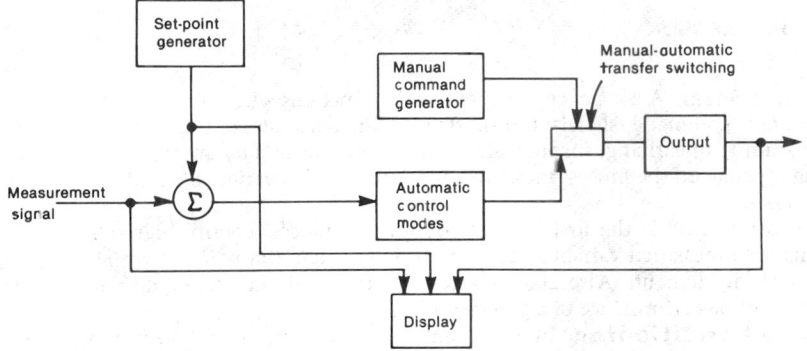

Fig. 24-13. Typical functions of a process controller.

common. The control functions can be implemented in many media, such as pneumatic, mechanical, hydraulic, electronic analog, and digital. The principles are the same for all. Typical functions or sections of a process controller are shown in Fig. 24-13. Means for generating the set point include a manual knob but may also include switching between several sources, such as a computer or external signal source.

Stepless switching between manual and automatic states permits process startup and troubleshooting without process upsets. Switching functions include an automatic technique for balancing the controller output with manual output.

Display functions present measurement, set-point, and output command signals. The measurement and set-point displays are usually so arranged that the plant operator can easily see their difference, which is the primary indication of control effectiveness.

Control Modes. A control mode counteracts the deviations between measurement and set point. This section discusses a number of the commonly used modes.

Two-Position (ON-OFF). This is the simplest control mode; in forward action, when the measurement is below the set point the output is ON, and when the measurement is above the set point, the output is OFF. It is used primarily for large single-capacity processes such as tank level or room heating. ON-OFF controllers often incorporate an adjustable dead band, called a differential gap, to reduce the cycle rate inherent in this type of control.

Proportional Mode (P). Proportional mode is the oldest and simplest form of continuous control. Output E_0 is proportional to deviation in the form $E_0 = (100/PB)e + B$, where e = deviation between measurement and set point, B = bias, and PB = proportional band, defined as

$$PB = \frac{1}{\text{proportional gain}} \times 100\%$$

The low gain (wide PB), often necessary to maintain process stability, allows large offsets induced by process load changes. For this reason, proportional mode is most commonly used for applications, such as surge tanks, where tight control is neither necessary nor desirable, or on essentially single-capacity processes which have a maximum phase shift of 90° and thus will be stable even at very narrow proportional band.

Proportional plus Integral (P + I). The most commonly used control function is a combination of proportional plus integral modes in the form

$$E_0 = \frac{100}{PB}\left(e + \frac{1}{R}\int e\, dt\right)$$

where e = deviation, PB = proportional band, R = integral time, and t = time.

The combination provides a wide dynamic proportional band to achieve process stability and a high static gain to minimize load-induced offset at a rate tuned to process dynamics. $P + I$ is used on nearly all flow control loops and many level and pressure loops. Proportional bands range from 500 to 2 and integral times from 1 s to 60 min. Figure 24-14 shows the magnitude ratio of a $P + I$ function's frequency response.

Proportional plus Integral plus Derivative (P + I + D). Some processes having a number of capacities in series can be more successfully controlled by the addition of a derivative or phase-lead mode to the $P + I$ function. Temperature control often requires the $P + I + D$ function to compensate for the sequential lags (time constants) of heaters, vessels, and sensor thermal wells. The ideal form for this control function would be

$$E_0 = \frac{100}{PB}\left(e + \frac{1}{R}\int e\, dt + D\frac{de}{dt}\right)$$

where e = deviation, PB = proportional band, R = integral time, D = derivative time, and t = time.

Figure 24-15 shows the magnitude ratio of a practical $P + I - D$ function's frequency response. "Rate gain" represents the increment of magnitude ratio, in the frequency range for the derivative mode, over that for the proportional mode. By design, the increment is restrained, as shown, to inhibit response to electrical noise.

17. Applied Regulatory-Control Functions. *State-transfer control* changes control modes sequentially or in response to changes in process throughput, measured variables, or system state.

A batch cooking process offers an example of a time-sequenced state-transfer control. For this process, a fill valve is opened for a period of time and the temperature set point held at a minimum. At the termination of the fill period, the fill valve is closed and the temperature set point is raised to the desired cooking temperature. After the cooking period the set point is reduced to zero and a drain valve is opened. Instrumentation for this process includes a timing mechanism and contact outputs to effect the required valve state transfers at the appropriate times, as well as a means of changing the temperature set point during the cooking period.

A safety override demonstrates a simple state-transfer control driven by measurement and control signals. When an upstream blockage occurs in a flow line, a low-pressure signal automatically overrides the flow-control signal, which would otherwise cause the valve to try to maintain flow. Such a system prevents pump damage from cavitation caused by low line pressure.

Cascade Control. Many processes have an intermediate variable that responds both to the manipulated variable and to process disturbances. In Fig. 24-16 temperature is controlled by manipulating fuel-flow control.

Pressure changes in the fuel line (intermediate variable) may occur much faster than the time constant of a temperature change. Appreciable fuel-rate changes could occur before the temperature controller could take corrective action. The

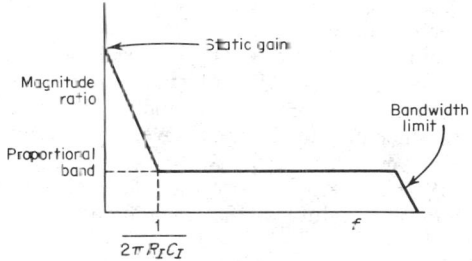

Fig. 24-14. Frequency response of a $P + I$ control function.

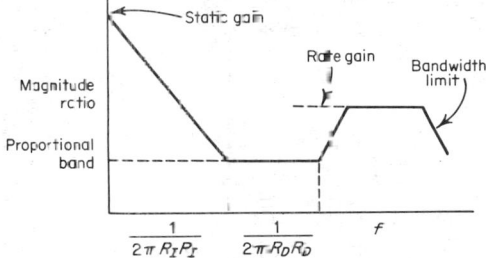

Fig. 24-15. Frequency response of a $P + I + D$ control function.

addition of a secondary loop that directly controls fuel flow minimizes fuel-pressure effects and provides significant improvement in loop stability.

Adaptive Control.   Nonlinearities of process gain and transient response occur in many processes. Fixed controller proportional band and integral time settings will be correct for control-loop dynamics over only a limited range of operation. Adaptive control automatically adjusts a control mode in accordance with known relationships between a measurable variable and the process gain or time constants.

Relatively few control loops have adaptive control, either because the nonlinearities encountered in the normal operating range are not severe or because proper mode settings cannot be obtained from available measured variables.

Feedforward Control.   In a feedback control system, a deviation between measurement and set point must exist before control action will occur. If the major sources of load variation can be measured, it is possible to act directly on the manipulated variable to compensate for the load variations before their effect is felt by the controlled variable and a deviation ensues between the set point and the controlled variable. This action is known as feedforward control.

The feedforward function is an approximate model of the process. Although it is theoretically possible to use feedforward control only, it is far more realistic to use a combination of feedforward and feedback control. The feedforward elements thus reduce the deviation seen by the feedback system, which then has only to correct for the imperfections in the feedforward process model. This technique is particularly applicable to processes having significant dead time.

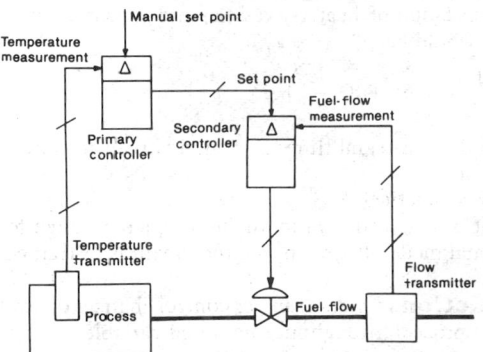

Fig. 24-16. Temperature control in cascade with flow control.

18. Analog Implementation. The very long time constant required to match the dynamics of physically large process vessels, heat exchangers, and reactors dictate the use of extremely high-gain high-impedance amplifiers with very low offset deviations. The typical design life of 10 years in industrial atmospheres further requires that extreme care be taken in selecting and derating components, packaging techniques, and circuit reliability. For these reasons, chopper-stabilized amplifiers are often used, and high-input-impedance modulators using varactors or FETs are common.

P + I.   Figure 24-17 illustrates a generalized schematic of a $P + I$ controller. It is available with the proportional bands and integral times given in Par. **24-16**. Note the automatic-manual switch and manual output drive, commonly used for automatic process startup and some emergency manual procedures.

Startup.   Any control function which includes an integral mode is subject to saturation (integral windup) when a deviation between measurement and set point persists. Although this

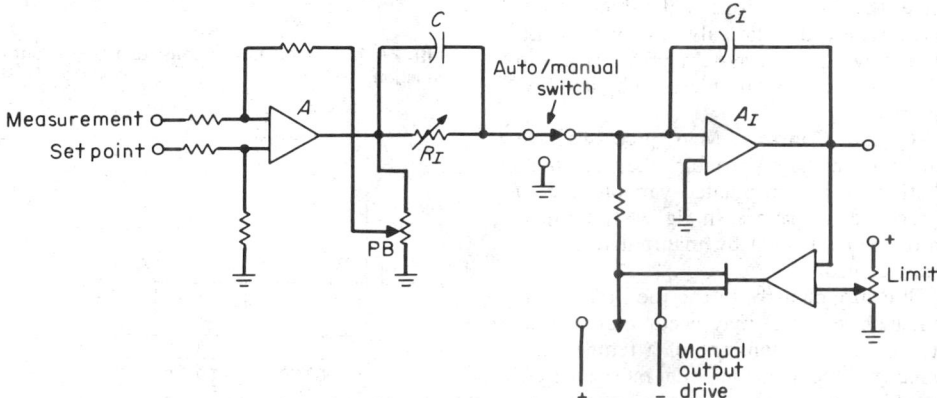

Fig. 24-17. Implementation schematic of a $P + I$ control function.

situation normally does not occur in a continuous process under automatic control, it must be prevented on automatic shutdown and startup of batch processes. If saturation is not prevented, no automatic control occurs on startup until the measurement signal reaches the set-point value. Measurement overshoots the set point by an appreciable amount, as illustrated in Fig. 24-18.

The schematic in Fig. 24-17 includes a section which prevents integral windup and measurement overshoot. During shutdown, the section holds or manipulates controller output to a limit value which is well within either minimum or maximum saturated value. Thus, when the sequence of process operations calls for automatic startup, all control modes renew action within proportioning limits. Overshoot is eliminated if the integral time exceeds the time constant of the measured process variable.

$P + I + D$. A derivative mode is usually implemented in the feedback of a controller. A better method (Fig. 24-19) applies derivative mode to the measurement signal only and isolates it from the integral mode. This implementation, although not available in all analog controllers, is a significant feature, particularly for startup under step-by-step commands from a supervisory computer.

19. Digital implementation takes several forms which implement functions as blocks:

1. A microprocessor-based signal processor for a *single regulatory loop*. It duplicates all the analog data-conditioning functions including scaling, linearization, algebraic computing, and real-time dynamic compensation. The signal processor includes the P, I, and D modes described in Par. **24-16**. It adds a variety of conditional logic and adaptive functions.

2. A computer-based direct digital control subsystem incorporating *multiple regulatory loops*. System capability includes flexibility to adapt to wide variations in process configuration. Either the control

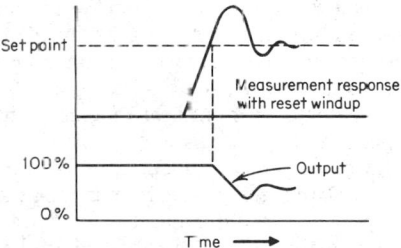

Fig. **24-18.** Characteristic response of $P + I$ control without integral antiwindup function.

subsystem or separate microprocessor subsystems implement I/O functions. Logic and calculation capability allows implementation of loop checking and of statistical functions such as fast Fourier transforms and single-point variance-distribution algorithms. Assurance considerations, including self-checking and redundant backup, determine the scope of plants governed by the subsystem.

3. User-programmable configuring of data-conditioning and regulatory control functions. Generally available programming languages based on FORTRAN and BASIC implement the selection of functions and the construction of system configurations.

All three digital implementations perform their functions through software. The first two generally allow the algorithm software to reside in read-only memory (ROM), while the last implementation calls for software resident in random-access (read-write) memory (RAM). In all cases, the user, by changing data in writable memory, may change individual parameters in a control algorithm and also interconnect (where allowed) different algorithms to carry out a complex control strategy.

20. Standards Relating to Data and Control

IEC Publ. 546 (1976) Methods of Evaluating the Performance of Controllers with Analog Signals for Use in Industrial Process Control

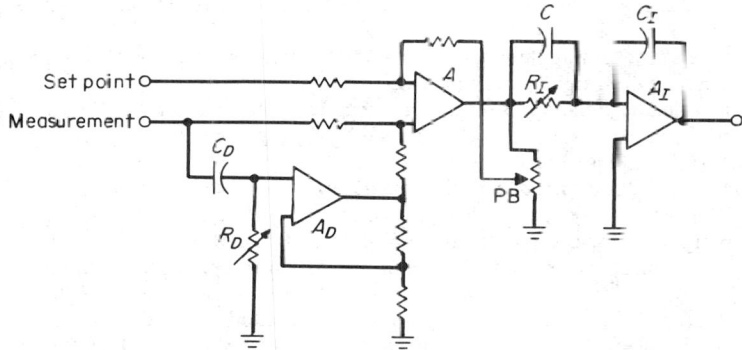

Fig. 24-19. A $P + I + D$ controller takes the first derivative of only the measurement signal.

ISO 3511/1-1977(E) Process Measurement (and) Control Functions and Instrumentation — Symbolic Representation — Part 1: Basic Requirements

ISO standards and IEC publications are subject to periodic change. Latest editions are available from American National Standards Institute, Inc., 1430 Broadway, New York, NY 10018.

Computer-System Architectures

BY C. M. PATEL

21. Major Functions. Process-control-system architectures consist of a number of functionally partitioned subsystems, interconnected by digital data busses (serial or parallel). Software executes process control functions in real time. The commonly used programming languages are Assembly Language or higher-level languages based on BASIC and FORTRAN. Up through the supervisory control level, all functions interact in real time with the operator or supervisor of a process unit. Beyond that level, strategic planning and optimizing of plant operations use techniques such as nonlinear programming which require mathematical modeling of process unit, plant, and operations.

Common major functions in a computer-system architecture are:

Data logging. Data from process measurement signals are collected, analyzed, tabulated, and reported, either on request or periodically. The computer converts raw measurement signals into engineering units and checks alarm limits. People use the resulting information to control the process directly.

Supervisory control. Performance calculations and optimization calculations provide tactical set points for regulatory control.

Direct digital control (DDC). A computer gathers data from process measurements and status signals and uses them to solve equations which are equivalent to analog data conditioning and control functions. Computer output signals command the positions or velocities of final actuators. Either a human operator or a supervisory control computer enters set points for regulatory control. The advantage of DDC over analog control lies in its adaptability for implementing sophisticated control concepts. DDC offers a wider selection of control algorithms. Through adaptive tuning of control modes it can improve dynamic response of control loops to nonlinear process gain changes caused by large process upsets.

DDC consolidates regulatory control functions for distributed process units. Where the consequences of failure are of sufficient concern, analog backup for each critical control loop provides assurance of continued process operation.

The computer system implements a variety of other process-related functions:

Process scheduling. The computer can efficiently handle scheduling tasks involving selection of raw material and processing equipment, the sequencing of process states, and blending from base-stock components. On-line linear programming implements these tasks.

Process unit optimization. Optimization techniques using nonlinear programming can be used for strategic planning and optimizing of plant and process operations. Such techniques involve use of mathematical models of the plant, process unit, operation, etc., with live and historical sensor-based data and/or manually entered data for investigation, analysis, planning, and control.

Batch processing. A computer system can implement automatic startup, normal production state, shutdown, and cleanup of batch processes.

Information management and reporting. A computer can perform numerous tasks involving the collection, analysis, storage, preparation, and printing or display of operating information for quality control and managerial reports.

22. Equipment Required for Centralized Process Control. Centralized computer control consists of major subsystems shown in Fig. 24-20: central processor unit, bulk storage devices, peripheral devices, power isolation unit, CRT and other operator interfaces, process I/O interfaces, communications interface, and analog control interface. These are described below.

The *central processor unit* (CPU) performs all the computation and data-processing functions for the control of a plant. Depending on the complexity of the process, the CPU can be a 16-b minicomputer or a 24- or 32-b high-performance processor. The main memory, which may be semiconductor or core, is used for program and data storage. Memory size can vary from 24,000 to several hundred thousand words. The processor communicates with all other subsystems through a parallel bus structure.

Bulk storage devices store programs and data for loading into main memory as needed, thereby reducing main memory size and facilitating execution of many different types of programs on an as-needed basis. The bulk storage devices can be moving-head discs, head-per-track discs, floppy discs, bulk core, etc., depending on performance and reliability requirements. Storage ranges from ½ million words to several million words.

Peripheral devices are similar to those used by data-processing computers (see Sec. **23**). These devices may consist of line printers, video terminals, keyboard-printers, video copiers, etc.

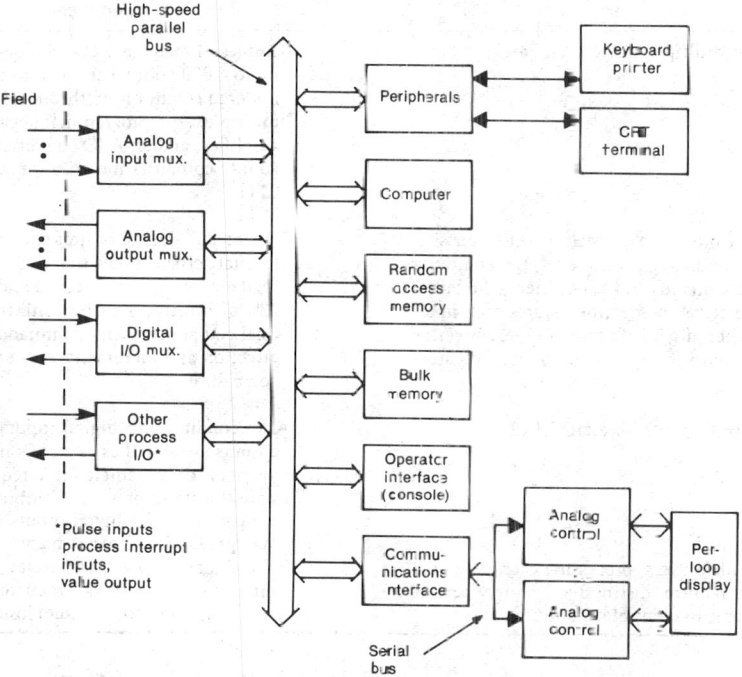

Fig. 24-20. Centralized process control system.

Power-Isolation Unit. The power used in most industrial plants is subject to transients and electromagnetic interference from a variety of sources. A power-isolation unit provides transient-free and galvanically isolated power to the computer system.

CRT consoles and other operator interfaces provide a means by which a process operator can interact with the process. The discussion beginning at Par. **24-24** describes in detail this critical operator facility.

Process I/O interfaces connect between a variety of analog and/or digital sensors, actuators, and the process computer (see Figs. 24-21 and 24-22). Diverse signals and elements need to interface to the CPU. They include thermocouples, resistance temperature detectors (RTD), 4- to 20-mA and 10- to 50-mA signals from transmitters and to final actuators, other voltage signals between −200 mV and +10 V dc (input and output), contact inputs, process-interrupt inputs, pulse-counter inputs, solid-state digital outputs (steady-state and momentary), and relay outputs (steady-state and momentary).

Process sensor-transmitter and actuator signals penetrate a control center through "field termination" racks. The termination racks provide a means for plant electricians to perform all plant wiring before final connection to control center equipment.

Table 24-3 describes multiplexers which interface between the major subsystems and signals at the termination racks.

Communications Interfaces. The main communication interface, shown in Fig. 24-20 for a centralized process control system, is a parallel bus which connects all the major subsystems to the CPU. This bus consists of data lines (16, 24, or 32, depending on the computer used), address lines to address memory locations and I/O devices, and control lines which control information transfer between the processor and other subsystems. The information transfer rate on the bus is quite high (a rate of 1 to 2 million words per second is typical).

Other communications interfaces consist of:

1. Serial ASCII character-oriented communication (via EIA RS-232-C or 20-mA loop) to keyboard-printers, video terminals, etc. Appropriate modems provide paths to other computers.

2. Bi-sync (binary synchronous) or SDLC (synchronous data-link control) modules which communicate to other computers, e.g., an IBM model 370. Such communication is generally

TABLE 24-3   Multiplexers for Process I/O Interfaces

Type	Comment
Analog input multiplexer (see Fig. 24-21)	Samples as many as 2,000 analog field signals up to 4,000 points per second; signals undergo conditioning through a multiplexer, a programmable gain amplifier, and an A/D converter and then go into computer memory for action by the CPU
Analog outlet multiplexer; performs the opposite function of an analog multiplexer; the computer calculates commands and stores them; the output multiplexer scans these values, stores each in a latch, and then directs them to a D/A converter; after conditioning they emerge to final actuators	Two major types of output commands issue to final actuators: position and "velocity"; final actuator position command might be 23% of full travel, 90% of full travel, etc.; final actuator velocity command signifies direction and increment, e.g., up 12%, down 20%
Digital I/O multiplexer (see Fig. 24-22)	After conditioning, digital inputs and outputs are stored as 1s and 0s in computer memory; only a single bit is required to store the status of a digital input or output; as many as 1,500 digital inputs and outputs may prevail in a process plant
Pulse I/O multiplexers; processing each type of pulse signal requires distinctive circuitry before representing it in computer memory	Pulse signals may be of constant width and variable height, voltage or current, variable duration, trains, or contact closure trains

required to link a process control computer to a large computer used for processing management information, e.g., inventory control, corporate accounting, and financial analysis.

3. A vital communications interface is the serial bus between a centralized system and an analog regulatory control subsystem (Fig. 24-21). Bus security is essential, since the analog subsystem assures integrity of process operation by sustaining critical regulatory loops should the CPU fail. Bus security pervades the plant through the location of analog regulatory control subsystems hundreds of meters from the CPU. Figure 24-23 shows typical formats of data words transmitted. Note that for every data word transmitted a 5-b check, e.g., Bose-Chandhuri, code and status bits for the control loop are also sent for secure communication. Control-loop state may include remote, local, or computer set point and automatic, manual, or computer control.

Analog Control Interface.   Until recently, interfacing methods treated each controller as a separate entity, connecting it to the computer through dedicated input and output multiplexer points. This required a multiplicity of wires between the CPU and each analog control loop (typically 8 to 10 wires).

Merging of the input and output multiplexing functions into the regulatory control can provide a digital communication medium common to groups of analog control loops. Figure 24-24 shows a group serviced by a serial communication bus. The bus accommodates multiple groups, and the merging permits drastic reduction in the number of wires required. It also enables a controller to accept directly from the CPU either positional or incremental updates of set points or outputs or commands to change status from automatic to manual, or the reverse.

Analog Control Subsystem.   Locations for the subsystem divide into two areas shown in Fig. 24-25, display and nest. The *display area* provides per loop displays which can include recorders, indicators, and trend-recording selector panels. An operator uses the display area to manipulate each control loop. The discussion starting at Paragraph **24-24** elaborates on display.

A *nest area* contains signal-conditioning, control, and computing cards. A typical nest handles 16 controllers and their associated communication medium. The signal-conditioning cards convert field signals into a common internal signal, for example, 0 to 10 V dc, for use by the control and computing cards. Failure of any one card can fault only one regulatory loop. Other loops maintain a level of subsystem integrity sufficient to minimize the consequences of fault and thus assure continued process operation.

23. Equipment for Distributed Process Control. In centralized systems (Par. **24-22**) large numbers of field signals from sensors and final control elements connect to I/O multiplexers at the control center. As the transmission distances can be as long as 300 m, the cost of laying cables can be quite high.

Decreasing costs of processors and memory in recent years have allowed cost-effective use of multiple processors. System designers can now dedicate a microprocessor to the control of each

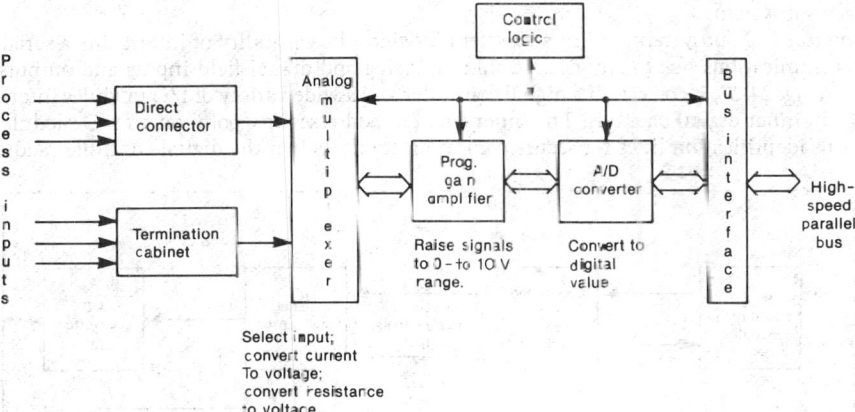

Fig. 24-21. Analog input multiplexer.

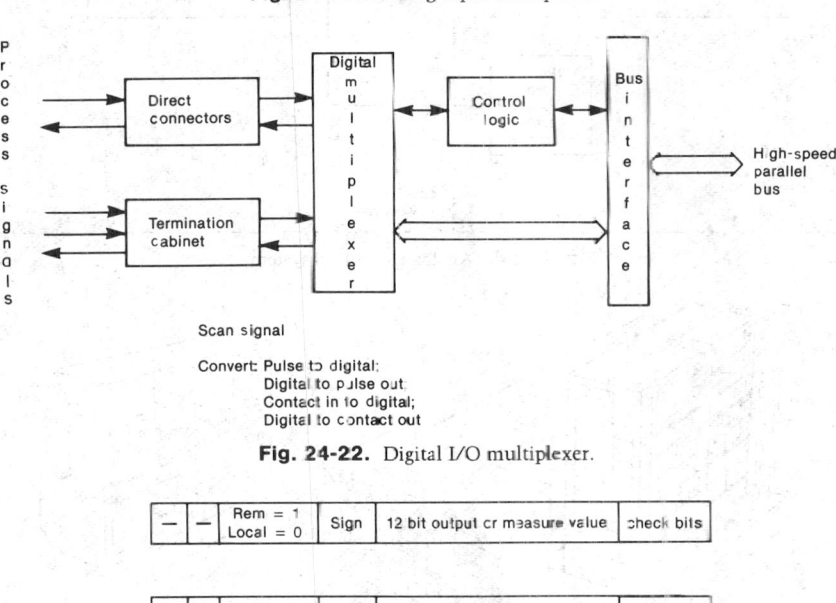

Fig. 24-22. Digital I/O multiplexer.

Fig. 24-23. Format of every update word transmitted over a serial communications bus includes status checks.

separable plant unit. Decentralized computer control, with processors distributed to match the geographical distribution of plant units, is now feasible and cost-effective. The cost of field-signal cabling further reduces as the distance between plant equipment and processors shrinks. Each processor contains its own I/O modules for interfacing with field signals. A serial communication bus links the distributed processors.

The functions of distributed control (Fig. 24-26) essentially duplicate those of centralized control. Differences center on implementation and integrity. Proximity to a local plant unit and connection of unit systems to the serial communication bus allow greater local autonomy. A unit system continues to perform its assigned tasks even if the common communication link is broken.

A distributed process control system may consist of one or more of the following subsystems linked by a serial communications bus: universal I/O subsystem, analog I/O subsystem, multiloop DDC subsystem, analog control subsystem, supervisory computer, and/or main process interface subsystem.

Universal I/O Subsystem.   This subsystem provides the capability of interfacing a serial digital communications bus to any desired mix of analog and digital field inputs and outputs. As shown in Fig. 24-27, it consists of a digital controller and a wide variety of I/O modules interfaced to it by an inner digital data bus. The inner data bus addresses any point on an I/O module and maintains identification lines for secure communications. When the digital controller addresses

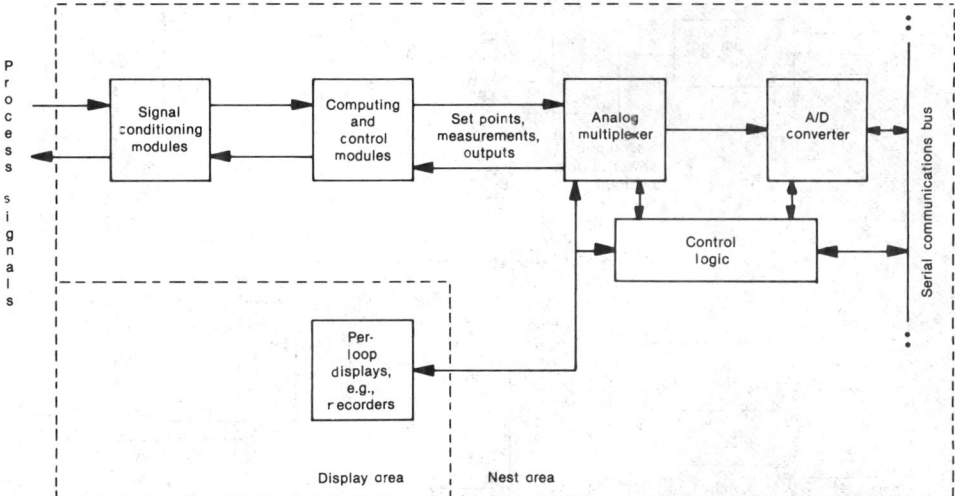

Fig. 24-24. Analog control subsystem.

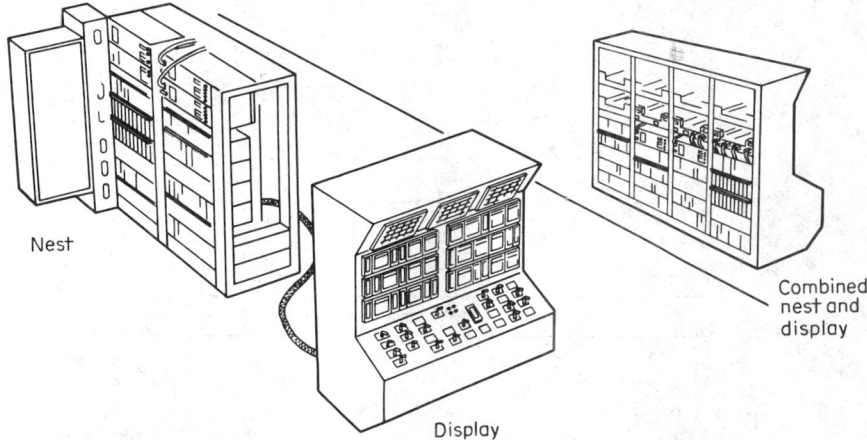

Fig. 24-25. Display and nest areas of an analog control subsystem.

an I/O module, a code identifies the type of I/O module at that address. An interrupt line handles process interrupt inputs, such as change of status of a contact.

A typical universal I/O subsystem may offer 30 slots for I/O modules. Typical I/O modules handle: two analog inputs or outputs, eight digital inputs or outputs, four digital outputs, four analog outputs, four pulse inputs, and eight process interrupt inputs or unique input-outputs.

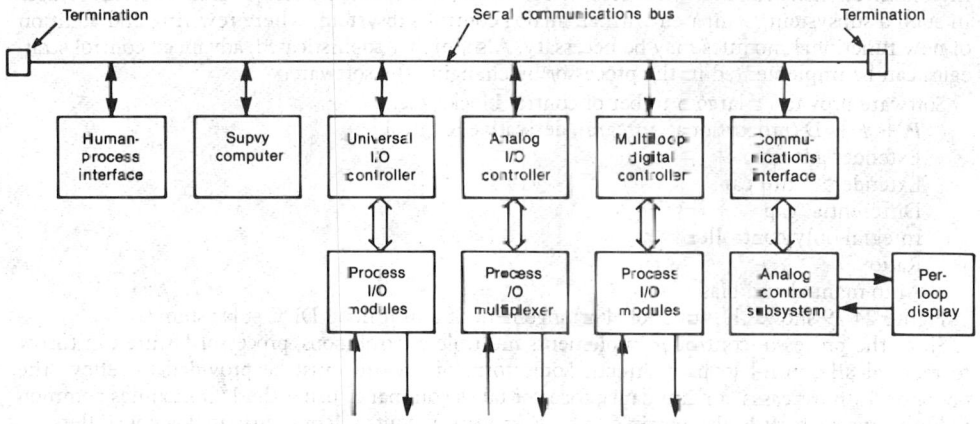

Fig. 24-26. Serial communications bus for distributed process control subsystems.

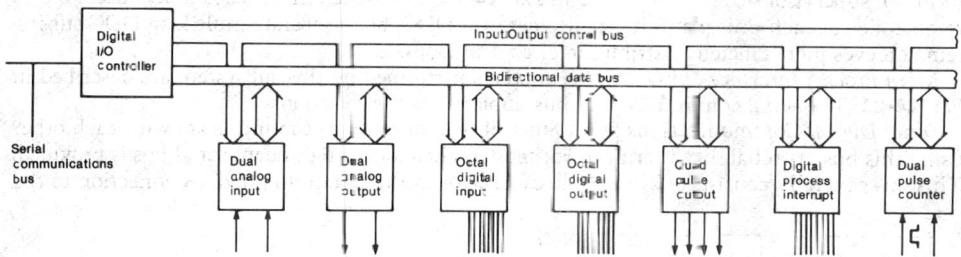

Fig. 24-27. Universal I/O subsystem.

Each I/O module interfaces directly to process signals and contains appropriate signal conditioning circuitry.

The digital I/O controller consists of a processor program, and data memory. It scans all I/O modules and stores their data and identifications in memory. Output modules read data from memory at regular intervals. This memory is also available via the serial communications bus for communication with other subsystems of the distributed process control system. For example, the operator-process subsystem may read the digital I/O controller's memory for input values and output values and display them on the CRT screen

Analog I/O Subsystem. For applications which require a large number of inputs, the universal I/O subsystem described above may not be cost-effective. For example, where a process control application presents a large number of thermocouple inputs, a subsystem specifically designed to handle a large number of analog inputs and outputs is specified Such a system is similar to the universal I/O subsystem except that each I/C module may be required to accept as many as eight inputs. Also, instead of 30 slots for I/O modules, the subsystem can be designed to accept as many as 120 card slots, requiring multiple racks. A single subsystem may process as many as 1,000 thermocouple inputs Figure 24-28 illustrates a generalized analog input subsystem. Usually, far fewer analog outputs than inputs are required.

The digital I/O controller in such a subsystem performs input functions such as linearization and reference-junction compensation. Like the universal I/O subsystem, the digital I/O controller scans all I/O modules (at least once per second) and stores data for access by other subsystems via the serial communications bus.

Multiloop DDC Subsystem. As the cost of processor and memory comes down, it is feasible to dedicate a processor to perform DDC on a small number of loops located in a unit plant. Such a subsystem requires essentially similar types of I/O modules as in the universal I/O subsystem, but the digital controller, instead of performing I/O functions, is programmed to perform DDC. The control functions (often called blocks) performed by the digital controller are basically the same as those performed by analog circuitry in an analog control system. Software makes the interconnections between control blocks. Therefore, it is relatively easy to change control stages in such a subsystem, compared with an analog control subsystem, where rewiring and addition of new functional modules may be necessary. Also, highly sophisticated, advanced control strategies can be implemented in the processor by changing the software.

Software provides a large number of control blocks, e.g.,

$P + I + D$ (proportional, integral, derivative) control
Extender-adaptor
Extender, nonlinear
Differential gap
Integral-only controller
Ratio
Auto-manual and bias

Figure 24-29 shows the functional arrangement of a multiloop DDC subsystem.

Since the processor-controller implements multiple control loops, processor failure can throw to manual all control loops it directs. Some form of backup must be provided to relieve the operator. In most cases, a redundant processor takes command, using the I/O modules common to both processors. It is also possible to back up critical control loops with analog controllers.

Analog Control Subsystem.   The subsystem is similar to the subsystem described in Par. **24-22**. The system designer now has a choice between analog control, DDC, or any mix of the two.

Computer Subsystem.   The computer connected to the serial communications bus can perform the supervisory control described in Par. **24-22**. The basic difference is that process I/O is remote, i.e., at each unit plant. It can also perform DDC, but a separate multiloop DDC subsystem achieves more efficient distribution of control loops.

Main Process Interface Subsystem.   Functions performed by this subsystem are described in Par. **24-22**. The serial-communications bus implements the functions.

Serial-Digital-Communications Bus.   Since all the subsystems communicate with each other using this bus, its reliability is critical. For most applications, a redundant serial bus is provided. This bus generally consists of a coaxiable cable with multiple taps to provide connection to the

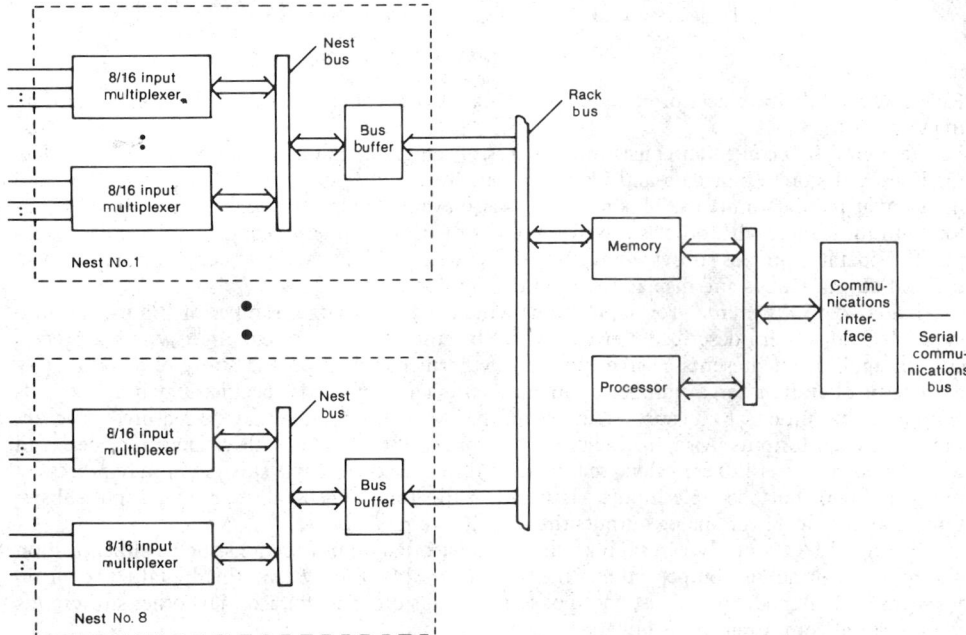

Fig. 24-28. Generalized analog input subsystem.

subsystems. Some busses may use a fiber-optics cable. Serial data rate on this bus can be as high as 1 Mb/s. Some means must be provided to arbitrate between access requests from the subsystem since all subsystems share this bus.

The bus transmits information as a sequence of message frames. Figure 24-30 shows the format of a typical message frame. Each message frame consists of a number of fields. The flag fields

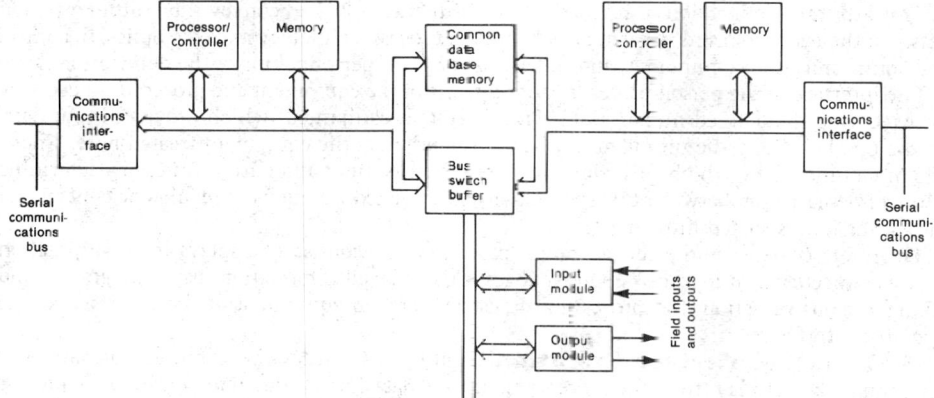

Fig. 24-29. Multiloop DDC subsystem with redundant digital controller.

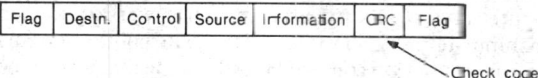

Check code

Fig. 24-30. Message frame of serial-digital-communications bus or distributed control.

define the start and end of the message frame. The source and destination fields define the addresses of the subsystems involved in the communication. The control field directs the message. The information field defines the message, and the CRC field is used for error checking.

The types of messages sent across the communications bus depend upon the functions of a distributed system. The basic types are network control, file transfer, initiate tasks, print messages, initiate and respond to control functions, and perform process I/O operations.

Future Prospects. Functions listed in Par. **24-21** will endure. Architectures will vary between centralized control, distributed control, and combinations of the two. The combinations will depend upon the limits of local autonomy the processing industries will allow. Industry expects standardization at the serial-communications bus. National committees and the IEC are developing standard functions for the bus. Implementation of the bus functions with broadly available microprocessor technology will pace the changes in autonomy.

Person-Process System Interface

BY M. PRASAD

24. Tasks frame people's perception of the process system. From this viewpoint, the process system consists of:

Processing equipment (tanks, reactors, pipes, and prime movers)

In-line measurement and actuation equipment (flowmeters, temperature sensors and transmitters, valves, pumps, and motors)

Control equipment (individual regulatory controllers, direct digital control, and supervisory control computers)

Communications equipment (interface subsystems, serial and parallel links, and modems)

Related communications stations (remote computers and enquiry points)

A person thinks of, senses, and manipulates the process equipment described in performing his or her tasks. The tasks vary with functional responsibilities of the person's job. Typical jobs and their functions are:

Operator. Operational supervision of the process and its regulatory control functions

Equipment engineer. Identification and diagnosis of malfunctions of control and communications equipment

Manager. Adherence to legal and financial restrictions

This arbitrary delineation of responsibilities is intended only to convey some differentiation between the tasks involved in each job. Changes in process technology and economics, in control and communications equipment, and in training of plant personnel make the delineation fluid.

The interface of the person-process system consists of the equipment and procedures necessary for executing the defined human tasks. Process system equipment, which provides input data, on the one hand, and the means for acting, on the other, is the equipment treated here. Tradeoffs in techniques by which this equipment implements input and output functions determine a line of demarcation between two such closely interrelated functions as regulatory control and the means for associated human action.

The interface between the person and the process system consists of displays and manipulation devices. Since each human task requires digesting displayed information, using judgment, and taking responsive action, the processing industries refer to equipment at the interface as the *operator interface.*

25. Operator-Interface Evolution. Figure 24-31 traces the evolution of functions common to data display (from the process system) and data entry (human response) to the process system. Major developments are as follows:

1. Progress in *transmitter technology* allows transmission of sensor signals to operator stations, which may be either centralized or distributed.

2. Increased *computational capability,* with smaller power requirements and physical size, makes it feasible to manipulate raw data more closely to its point of origin before transmission.

3. Earlier *means of manipulation* consisted mostly of mechanical elements. Progress in *keyboard technology* increasingly points to keyboards as data-entry devices, the computer providing extensive checking of entered data. The economics of *final control elements* (control valves and pumps) have not offered the incentive to adopt the digital technology incorporated in other elements of the operator interface. Consequently, converters transform keyboard data into a positional or incremental form accepted by pneumatic actuators and positioners (see Table 24-2).

4. The *cathode-ray tube* (CRT) has increasingly supplanted other display devices. Alphanumeric and graphic capability together make the CRT a convenient device for information display in a wide variety of applications. Advances in storage techniques coupled with availability of computational power provide almost unlimited display flexibility.

Modern control strategy emphasizes the minimization of the effects of upsets while optimizing normal steady-state operation. This, in combination with the rapid changes in electronic technology, prescribes the following characteristics in an operator interface:

1. Greater responsibility for one person.
2. High information density.
3. Supervision by exception.
4. Flexible, multifunction, multipurpose display and data-entry equipment.
5. Transformation of raw data into understandable information.
6. Diagnostic information for pinpointing malfunctions of process control system and communications equipment.
7. Methods of transferring various units of related equipment to redundant or backup mode.
8. Ability to configure control and communications systems as well as operator displays.
9. Expansion and interpretation of raw data into easier to understand forms of representing relationships. Relationship displays can be combinations of alphanumeric, graphic, geometric shapes, time-based CRT or chart-recorder plots, or color-coded schemes.

The developments listed have one feature in common: they imply the use of computational techniques in transforming data and presenting information to the operator.

26. Functional Requirements. Points of human access (physical and informational) are usually sheltered. The process, however, consists of equipment located in environments ranging from desert dry heat to subarctic conditions. Paragraph **24-30** defines the variety of locations and their associated service conditions. Control centers designed to incorporate computers require controlled levels of temperature, humidity, and chemical contaminants.

The physical size of process units dictates the geographic spread of a process. Transportation facilities for receipt of raw materials (solids, liquids, or gases), transfer of intermediate products, and storage of final products determine plant layout. Sensors and final actuators mounted to the process units follow the extremes of layout and spread. Control equipment (rack-mounted electronics and computers) is concentrated in enclosures because it demands a controlled environment and because people must have access to it.

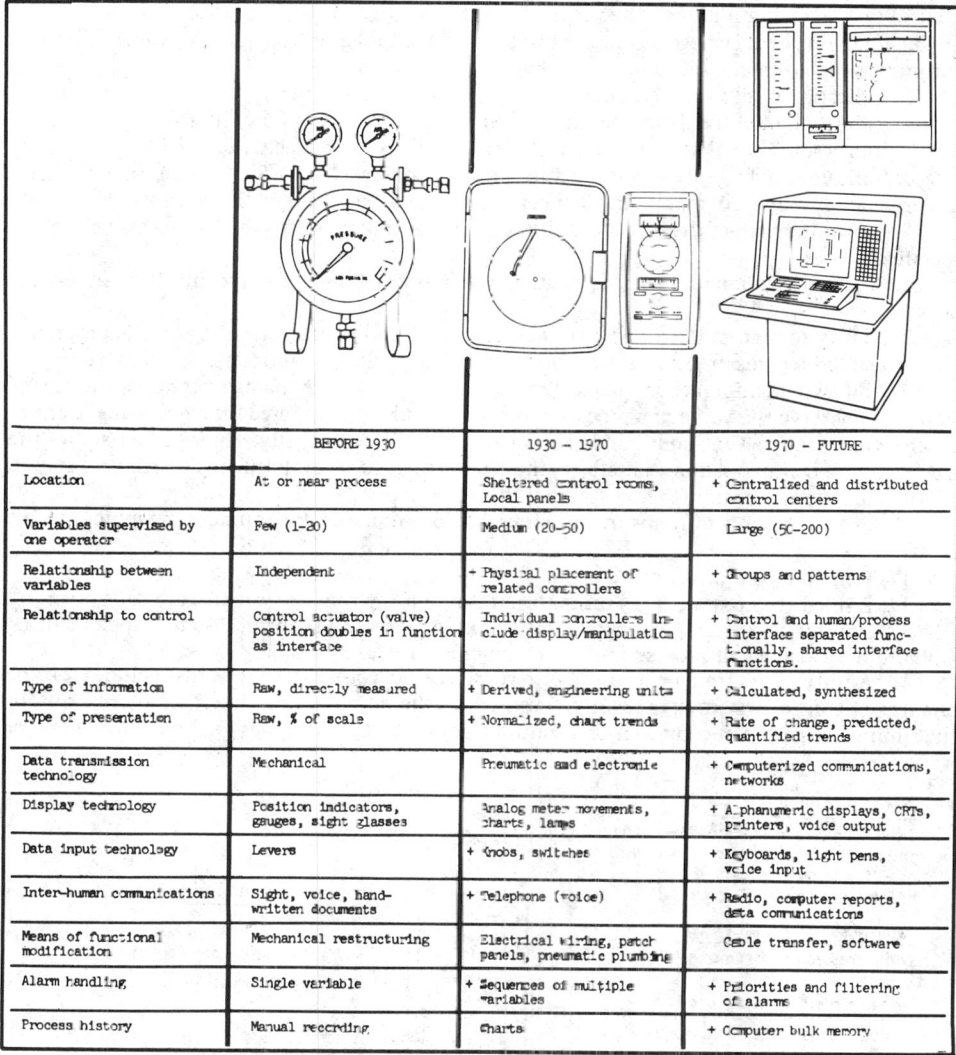

	BEFORE 1930	1930 – 1970	1970 – FUTURE
Location	At or near process	Sheltered control rooms, Local panels	+ Centralized and distributed control centers
Variables supervised by one operator	Few (1–20)	Medium (20–50)	Large (50–200)
Relationship between variables	Independent	- Physical placement of related controllers	+ Groups and patterns
Relationship to control	Control actuator (valve) position doubles in function as interface	Individual controllers include display/manipulation	+ Control and human/process interface separated functionally, shared interface functions.
Type of information	Raw, directly measured	+ Derived, engineering units	+ Calculated, synthesized
Type of presentation	Raw, % of scale	+ Normalized, chart trends	+ Rate of change, predicted, quantified trends
Data transmission technology	Mechanical	Pneumatic and electronic	+ Computerized communications, networks
Display technology	Position indicators, gauges, sight glasses	Analog meter movements, charts, lamps	+ Alphanumeric displays, CRTs, printers, voice output
Data input technology	Levers	+ Knobs, switches	+ Keyboards, light pens, voice input
Inter-human communications	Sight, voice, hand-written documents	+ Telephone (voice)	+ Radio, computer reports, data communications
Means of functional modification	Mechanical restructuring	Electrical wiring, patch panels, pneumatic plumbing	Cable transfer, software
Alarm handling	Single variable	+ Sequences of multiple variables	+ Priorities and filtering of alarms
Process history	Manual recording	Charts	+ Computer bulk memory

Fig. 24-31. Evolution of interface functions.

Regardless of the degree of automation, careful thought has to be given to the degree and modes of *interaction* between the process system and the human being. Interaction may occur at several levels:

The plant manager must know (and be able to affect) production levels of different process entities constituting the plant.

The process manager may need access to data from each of the units making up the process.

The unit supervisor needs access to individual pieces of equipment that are part of a process unit.

The degree and frequency of detail necessary vary from level to level. For example, access at the plant manager's level may consist of written or printed daily reports and change orders. At the unit level the supervisor needs to know instantaneous values of process variables and be able to change the values manually either by varying the set point of a regulatory controller or by manually changing the signal level to a final control element.

At any of these levels, an *audit trail*, i.e., a permanent record, may be necessary for future analysis, fault detection, and/or location. Regulatory-agency requirements may also necessitate such records. The proliferation of computers at all levels necessitates an even greater number of audit trails to allow people to retain control over and above the autonomy of the computer. The flexibility offered by computers also requires means for adapting functions when changes in control system configuration are specified.

The following broad categories of data display and manipulation make up the human interface for process management and control:

1. Indication of physical variables (flow, pressure, temperature, etc.).
2. Indication of derived variables (chemical composition, mass flow, etc.).
3. Indication of final actuator positions and/or signals to final actuators.
4. Indication of target values (set points, production targets, etc.). Such values may be direct or derived, i.e., manually manipulated or computationally generated by the control system itself.
5. Alarm indication, the alarm condition resulting from comparison of a process variable to specified values.
6. Binary indication of process or control system states (pump ON-OFF, controller AUTOMATIC-MANUAL, etc.).
7. Ability to manipulate or change the items displayed in categories 1 to 6, as appropriate.
8. Printed records of items in categories 1 to 6 (numeric, alphanumeric, graphic, etc.).
9. Ability to manipulate or change items in categories 3 to 6. A change might require several steps; e.g., observe the value of a process variable, observe the computed set-point value, transfer control status to local set point, and then change set point manually. Figure 24-32 shows the faceplate of a control station that allows these operations to be performed. This control station is dedicated to one control loop.
10. Printed records of items in categories 1 to 6 (numeric, alphanumeric, graphic, or time plot).
11. Printed record of alarm transitions.
12. Printed record of changes made to and through the control system, as in category 8.
13. Graphic or semigraphic representation of the process and control system, supplementing alphanumeric display of process and control system states and variable values.
14. Ability to manipulate, change, or reconfigure the control system. This includes routine changes of control-system parameter values such as the duration of a timer, an alarm limit, or the tuning constant of a controller or a control algorithm. It also implies total reconfiguration of

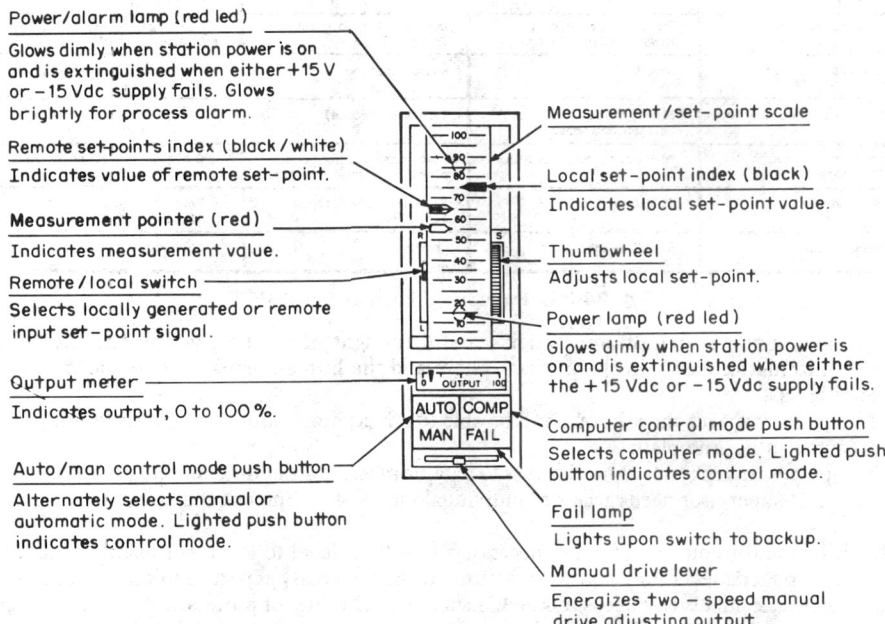

Power/alarm lamp (red led)
Glows dimly when station power is on and is extinguished when either +15 V or −15 Vdc supply fails. Glows brightly for process alarm.

Measurement/set-point scale

Remote set-points index (black/white)
Indicates value of remote set-point.

Local set-point index (black)
Indicates local set-point value.

Measurement pointer (red)
Indicates measurement value.

Thumbwheel
Adjusts local set-point.

Remote/local switch
Selects locally generated or remote input set-point signal.

Power lamp (red led)
Glows dimly when station power is on and is extinguished when either the +15 Vdc or −15 Vdc supply fails.

Output meter
Indicates output, 0 to 100 %.

Computer control mode push button
Selects computer mode. Lighted push button indicates control mode.

Auto/man control mode push button
Alternately selects manual or automatic mode. Lighted push button indicates control mode.

Fail lamp
Lights upon switch to backup.

Manual drive lever
Energizes two − speed manual drive adjusting output.

Fig. 24-32. Control-station faceplate details.

a complex control system without having to shut the process down. The last requirement usually implies regulatory control done by computer. Figure 24-33 shows the configuration display for a control algorithm in a distributed control scheme. The interface, including the CRT, is located remotely from the control device.

15. Displays and printed records of data which might have only remote or no direct relationship to the actual process control function. Results of process modeling or process simulation

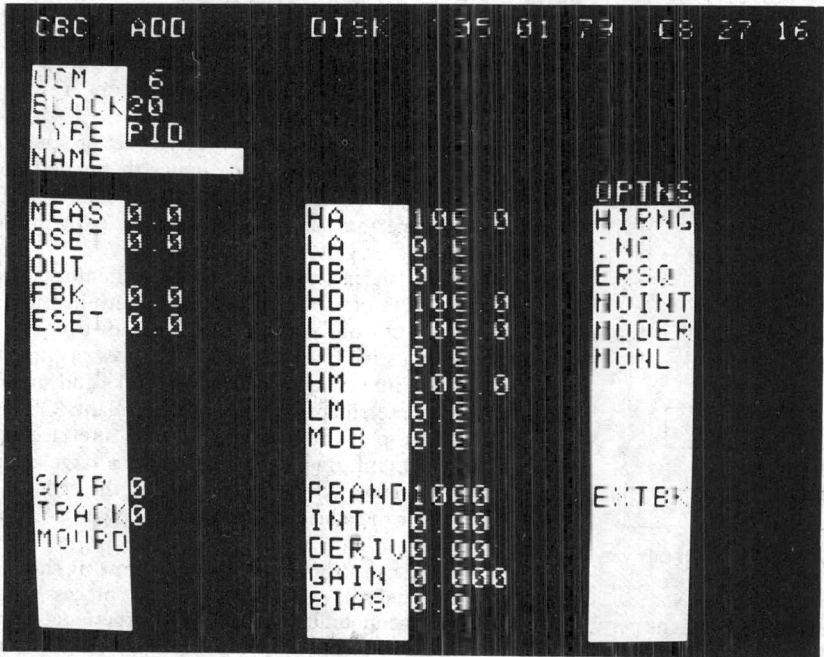

Fig. 24-33. Configuration display for distributed control.

fall in this category. Means of accessing computer software and user programs are obviously implied.

16. Ability to change interface functions. A plug-in membrane switch keyboard which alters function based on the insert, for instance, vastly increases the utility of the CRT-keyboard combination.

17. Ability to simulate process and/or control system for training purposes. The simulated process is itself transparent to the operator. All display and manipulation requirements would be as in a live situation.

27. Human Characteristics and Limitations. The sole beneficiary of the operator interface is the human element of the system. Human response to a specific interface design includes a large measure of subjective, possibly emotional, factors. To some extent, the operator's previous experience (knowledge of the process and familiarity with data-display–data-entry concepts) governs these factors. Attention to human engineering details can help the operator identify the interface as part of the solution, rather than as part of the problem.

The interface must distinguish between abnormal conditions in the process system and malfunctions in the control or communications system. To the extent possible, the interface must indicate its own local malfunctions. For example, the interface can sometimes provide notification of equipment faults such as erratic memory (if it has built-in error correction) or communication retries (if the necessary counters are designed in).

In all cases, the fault notification must be clear and unambiguous. This is true for all process-related notification because of its impact on assurance of process operation. For all related control, communications, and interface equipment, a vague notification of a fault would reduce the operator's confidence in the interface. Such confidence is of the utmost importance in helping the operator to distinguish and identify a catastrophic process emergency from a minor equipment fault.

Process-alarm philosophy is a major area of importance and emphasis, both in existing interface systems and in continuing design work toward future systems. An alarm structure that limits itself to notifying the operator of an off-normal condition at a finite element level is attractive both in terms of ease of implementation and in consistency between different elements. (Simple alarming of an off-normal variable value is a case in point.) However, this structure tends to overwhelm the operator with a multiplicity of individual notifications in case of a major process upset. Such emergencies usually involve a number of related events, and with this structure the operator is forced to sort the significant events from the secondary ones, based solely on human judgment. Such judgment is often a distillation of previous experiences in other emergencies involving pattern recognition by the operator. Faced with a new pattern, the operator is left with two possible courses of action: (1) approximate the new pattern to a known earlier pattern and react accordingly or (2) guess the correct action required under unfamiliar circumstances.

Given the major economic and/or safety-related penalties for an incorrect action, the operator is pushed to the limits in coming to a conclusion. Decisions made under these conditions are often wrong and also differ between operators, as in the case of two persons, one on each shift, reacting differently to a similar emergency.

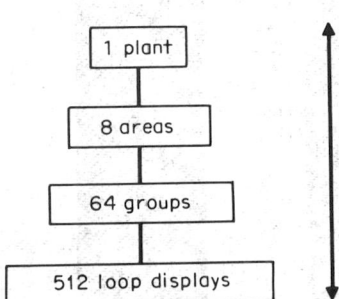

Fig. 24-34. Hierarchical view of process-plant relationships.

A second alarm structure, one that attempts to recommend a course of action based upon a built-in pattern recognition, is obviously more desirable. However, this calls for an exhaustive knowledge of the process, coupled with up-to-the-minute knowledge of the quirks and nuances of every single element of the physical plant. Often such detail is simply not available to the interface system. Where such information is available, a large degree of expensive customizing is necessary and the economic justification for the added computational power is not always clear.

A third alarm structure is based on allowing the operator to group the display data from different process segments. Such customized designs permit the grouping to be modified as operational experience increases. While retaining some aspects of pattern recognition, this structure circumvents the economic and reliability penalties imposed by the added computational requirements necessary for more comprehensive pattern recognition.

A hierarchical view of a plant, as shown in Fig. 24-34, allows the operator to home in to an alarm at a detailed level, starting at a macroscopic level. In this particular example, the plant consists of eight areas, each consisting of eight groups. Each group is made up of eight loops. Configuration software allows the lowest element (loop) to be used as often as desired, in combination with other elements. Patterns can thus be defined, based on operational considerations. Each level has a unique display, allowing greatest detail at the lowest level as arranged in Fig. 24-35.

Data entered by the operator must result in quick response from the interface. When rejected by the interface, invalid data input (which may represent either function or value) must result in nonambiguous notification. Some designs (see Par. **24-29**) offer ways of making it impossible for the operator to make a functional mistake.

Ability to correct an unintentional (though valid) value change is another important requirement. Simplicity and consistency in the data-entry procedure ensures safe operator reaction to emergency conditions. Key-locked access at several levels, corresponding to the levels of human authority, is usually included in the data-entry scheme.

Where an operator has to deal with multiple interface systems simultaneously, e.g., different design philosophies are placed in close physical proximity, it is important for the interface design coordinator to achieve some degree of commonality in the data-entry and display schemes.

Operators tend to judge displays subjectively. Consequently, they frequently participate in the design of such displays. Broader availability of computing capability and memory capacity have opened the way for displays customized to the experiences and preferences of operators.

28. Operator-Interface Technology. Data-entry and data-display devices, supplemented by auxiliary equipment, make up the interface.

At present, all communications to the human being from the process system are normally designed to depend almost exclusively on the sense of sight. The only exception is an audible alarm which notifies the operator of a situation in need of closer examination. The operator might use one or more of the other senses to help diagnose a particular situation. The difficulty in quantifying data for such sensing, coupled with extremely wide variations in the corresponding human perception, does not permit designing for this. Even the first two senses, sight and hearing, are used to complement each other to overcome individual limitations such as color blindness or hearing deficiency. While it has long been recognized that in some situations people use all senses to aid in pattern recognition, there are at present no known ways of designing for this fact.

Data Entry. Conventional devices include knobs, switches, and levers. Physical position of the entry device or a related display such as a driven pointer or a lamp indicates an entered value. Pushbuttons enter binary data, e.g., pump ON-OFF. Keyboards of somewhat greater complexity call for the assembly of entering numerical data, one numeral at a time. Associated display is alphanumeric.

Standardized typewriterlike keyboards have achieved nearly universal use in conjunction with CRT displays. Other variations in keyboards depend on the availability of a computer. Examples include multifunction keyboards, where the computer reads the identity of a plug-in or slide-in keyboard to set the function of each key by context and code. Keyboards using membrane switches with audible feedback and backlighting lend themselves to such flexible use. Combining multifunction keyboards with microprocessors provides practically unlimited diversity in function.

An interesting variation of a functional keyboard is the variable-function keyboard shown in Fig. 24-36. Here the computer assigns a function to a specific key based on the context of the associated display. The current function of each key is identified on the CRT associated with the keyboard. Though initially the concept seems complex, experience has shown that operators actually appreciate its simplicity. This method greatly reduces the clutter of keys necessary to service the multiplicity of functions. Functions which need to be quickly accessed, i.e., through one keystroke, are allocated dedicated keys.

Data Display. Conventional devices include lamps, electromechanical counters, and meter movements with pointers. While these devices do not easily lend themselves to being shared by multiple points of data generation, their legibility makes them attractive in simple applications and dirty conditions.

Highly readable gas-discharge displays provide bar-graph indication. Used in conjunction with microprocessors, they are attractive because of their ease of calibration and maintenance. Flashing segments of the neon glow may display alarms. Figure 24-37 shows two control-station faceplates, one using a gas-discharge display and the other using a conventional meter movement.

Light-emitting-diode (LED) and liquid-crystal-display (LCD) techniques are used for alphanu-

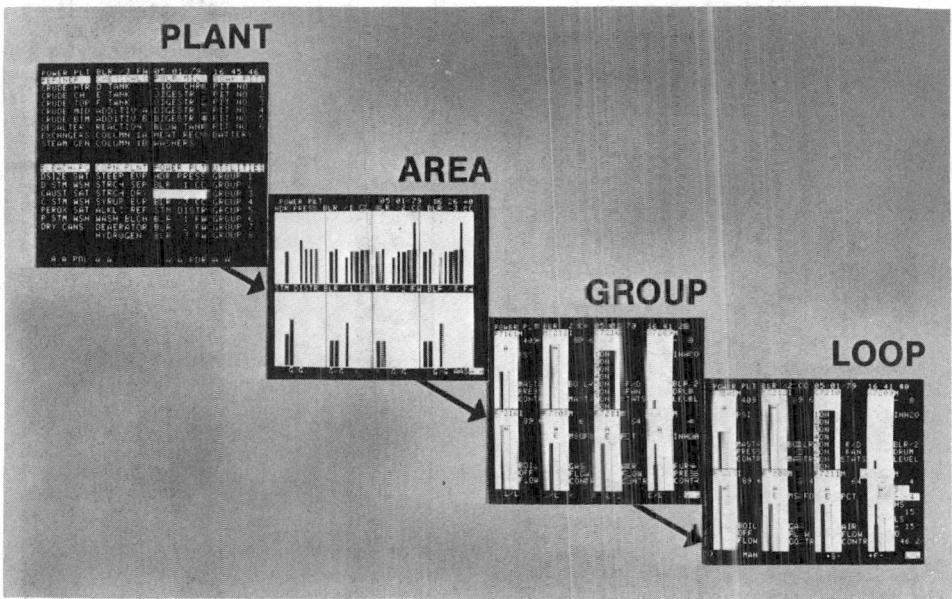

Fig. 24-35. Levels of detail in hierarchical display.

meric characters, as are plasma displays, but most of these technologies offer only monochromatic light.

The CRT, in both monochromatic and color versions, has gained wide acceptance. In addition to providing extremely flexible formats, combinations of alphanumeric displays with graphics yield attractive functional combinations, as pictured in Fig. 24-38. Character and bit-map graphics provide the flexibility needed for displaying graphic representations of the process: bar graphs to show variable values and curve plotting to show process variable history. A wide variety of CRTs, differing in screen size, resolution, refresh rates, and graphics capability, have become part of the process-control work station.

Chart recorders furnish hard-copy records of one or more process variables as a function of time. They use a variety of round and strip charts, often with custom-printed scales. Inking pens are common; thermal marking has less acceptance. Some recorders have multiple pens using

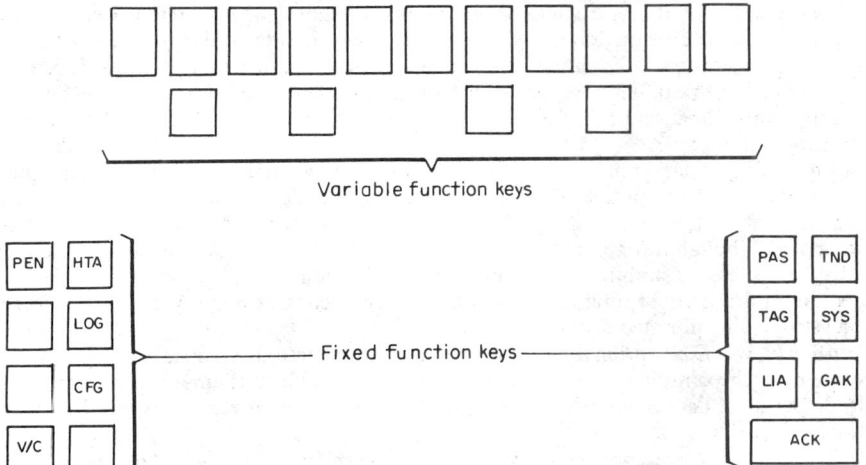

Fig. 24-36. Variable-function keyboard.

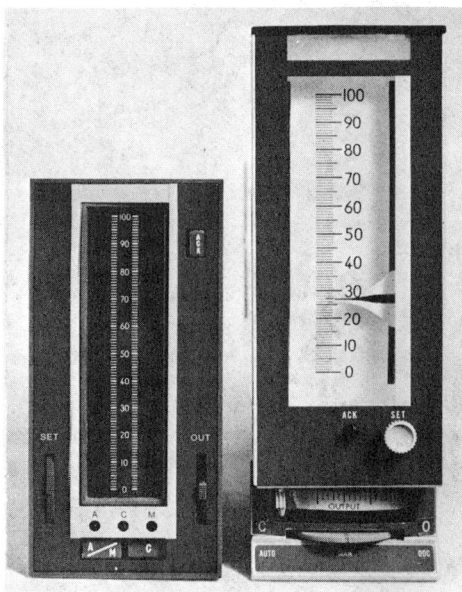

Fig. 24-37. Control stations.

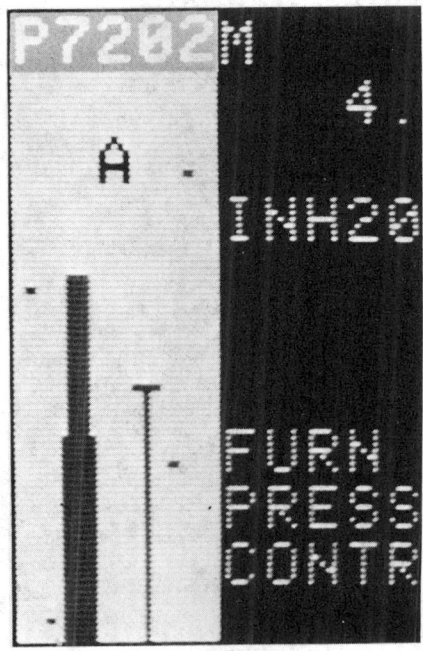

Fig. 24-38. CRT faceplate.

different colors of ink, while other recorders use an input multiplexer to record multiple variables. Used in conjunction with data recorded on a computer's bulk memory, such as a disc, chart records are highly useful for process analysis. Regulatory-agency rules sometimes dictate retention of chart records for specified periods of time.

Serial typers and line printers are common adjuncts to process-control computers. Used both for alarm and periodic logging, they serve as versatile output devices for microprocessor-based systems.

Video copiers, driven from CRT screens, are especially useful for obtaining hard copies of graphic displays. A multiplexer allows the screen displays from several remotely located CRTs to be directed to one video copier.

Alarm annunciator subsystems, self-contained devices, are also used to indicate process alarm conditions. By grouping variables, the operator is provided a simple means of alarm notification and acknowledgement. Such subsystems supplement other alarm-notification schemes, such as flashing or changed color displays on CRTs, and provide additional security as needed.

Graphic panels consist of instruments, such as controllers, recorders, indicators, push-button switches, and status lamps, directly mounted on a pictorial representation of the process, with positional significance. Such panels tend to take up space but may be dictated by the preferences of operating personnel.

Projectors, backlighted displays, and placards are sometimes used to meet individual application requirements and preferences. Microfilm and slide projectors in combination with CRTs are the focal points of modern control rooms.

Voice input-output, presently in the experimental stage, is beginning to find limited application.

29. Typical Operator Interfaces. This discussion covers the functional detail of two interface systems. The first, termed *distributed operator interface*, is a stand-alone system designed to service a large number of stations dedicated either to regulatory process control or to data gathering, all communicating over a serial communications link. The second system, termed *centralized operator interface* is designed to work in conjunction with a process-control computer, using the same serial link as the first system. The fundamental difference between the two systems is one of design philosophy. The first system tends to maximize the number of display functions available, while the second one acts as a flexible, intelligent terminal, whose functions are determined either by user programs or by application software in the process-control computer.

Specific data-entry and -display techniques, the amount and mix of data in each display, means of alarm notification, the volume of hard-copy records (printed reports and chart records), and a host of other features are determined by multiple trade-offs between system requirements such as:

1. Present application requirements
2. Application expansion plans
3. Age and economics of process and related control system
4. Division of security and administrative responsibility in a given process plant
5. Compatibility and ease of communication with control system
6. Interface supplier trade-offs in system design

Availability of computational power, either remote (shared) or local (dedicated), further increases the options available for any one particular installation. Overall economics, coupled with security and maintainability restrictions, determine the system features in most cases. Fundamental technical restrictions can usually be traced to the lack of availability of primary data rather than to interface technology.

Operator interfaces described here are typical of equipment available to meet broad functional requirements. The uniqueness of a given system depends on the mix of interface technology and human engineering.

Distributed Operator Interface. This system is designed to provide most of the functions contained in a conventional control panel, illustrated in Fig. 24-39. Conventional panels are sometimes extremely large, and association between two instruments located physically far apart tends to be difficult. Figure 24-40 shows a modern interface incorporating current technologies.

The major subsystems provided in this interface, for use with a distributed computer control system, are as follows:

1. *High-speed serial communications bus.* Process measurements and related data are acquired over the link from equipment dedicated to measurement data generation. The bus also transmits control data (set points, regulatory control status, valve-output signals, etc.) to and from

regulatory control subsystems. All devices on the bus, including supervisory computers, transmit status information via the bus. The bus also monitors its own status.

2. *Recorder subsystem.* Multipoint strip-chart recorders record data received directly from process I/O and control subsystems. Thumbwheel switches specify the multiplexed data for a specific recorder; one position on each switch allows the recorder to receive data from the interface.

3. *Annunciator subsystem.* Switches and alarms important to process and overall system security are grouped in a central panel for effective alarm management. Annunciator lamps and

Fig. 24-39. Conventional control panel.

Fig. 24-40. Interface for distributed control.

a variety of pushbutton switches are arranged in custom groupings. Inputs are wired directly to field contacts, and manual switch contacts are brought directly to the field. Such functional independence from the rest of the interface enhances the security of critical alarms.

4. *Display subsystem.* Consisting of one or more color or monochromatic CRTs (each with a separate variable function keyboard), coded alarm status lamps, and a general-purpose keyboard, this subsystem is the nucleus of the interface system. A video copier allows hard copies of the CRT displays to be made.

5. *Processor subsystem.* While seemingly transparent to the operator, this subsystem consists of a medium-sized minicomputer which provides all the data for the display subsystem and receives operator-entered data from the associated keyboards. Bulk memory (such as floppy diskettes) provides long-term storage capacity for the specialized software, which provides all the interface functions such as displays, alarming, communications, etc. For greater reliability, this

software resides strictly in random-access memory (RAM) when the interface is on line. This subsystem (together wih the full display data base) drives an alarm-logging printer for hard copy.

Other than the few functions which are kept independent for security reasons, software implements almost all the functions of the interface. Major functions are as follows:

1. *Display of process status.* Displays are presented in a hierarchy of levels — plant, area, group, and loop Starting at a plant level (consisting of several hundreds of control loops), each step down in the hierarchy presents progressively greater detail on a smaller portion of the plant. Displays consist of alphanumeric identifiers, bar-graph, and/or alphanumeric representation of values associated with control loops, e.g., set point, measurement, output, high- and low-value alarms, deviation alarms, high and low valve-position alarms, alarm dead bands, and loop-tuning parameters. Flashing and intensified (or color-coded) areas of the CRT screen present alarm information. The operator homes in on a particular parameter by repeated use of the variable-function keyboard. A pair of these keys then allows modification of parame ers such as a set point or a loop-tuning parameter. Functionally invalid operations during any part of this procedure are made virtually impossible because only the valid keys are identified and enabled.

2. *Display of process trends.* The video screen may display either real time process data or historical-trend data collected and stored on a diskette For comparative analysis, multiple variables can be examined simultaneously.

3. *Trend recording.* As in the display of process trends, process variables can be recorded on strip-chart recorders.

4. *Dedicated alarm display.* Color-coded alarm status lamps indicate the status of plant sections. Audible alarms at each of several consoles in the interface system also help to draw attention to the alarm condition.

5. *Hard-copy record of alarm.* A printer using two-color ribbon automatically enters a description of an alarm condition on the log sheet.

6. *Logging of process data.* The operator can obtain printed alarm summaries on demand.

7. *System security.* A family of displays at multiple levels presents information related to the security status of the interface system components such as memory, printer, CRTs, and communications. A second set of displays relates to the status of the I/O devices and control devices which provide the basic process data. A third set of displays is dedicated to the communication subsystem, providing both alarm data and statistics related to bus-traffic analysis.

8. *Display configuring.* An on-line program allows configuring of various types of display loops, including data necessary for showing display data in engineering units

9. *Control configuring.* This on-line program allows the system engineer to configure the arrangement of remote control devices for implementing specific control strategies. A record of the configuration is kept on the bulk memory device and can be simultaneously transmitted to the control device. This feature permits rapid on-line configuration and commissioning of control strategies among a group of geographically separated control devices.

10. *Interface system configuring.* One off-line program is used to configure the interface system itself, such as number of CRTs, communication ports, memory, etc. A second program allows specification of key-lock access to some process data. A key-locked parameter can be examined but not modified unless the access protection is defeated through the key lock. A third program allows custom tailoring of the color scheme of the displays

11. *Utilities.* A series of off-line programs provides utility functions used in maintaining the interface system software. Expansion of this distributed interface (within its design constraints) consists of a combination of two steps: (a) reconfiguring the system using the appropriate programs and (b) addition of modular functions such as CRTs and annunciator lamps and pushbuttons as appropriate.

Expansion of the interface scope (beyond the design limits of one system) consists of duplicating the interface onto the communications bus. Distribution of the functions among multiple operator interfaces rests with the judgment of the system designer.

Centralized Operator Interface.   Figure 24-41 shows a CRT-keyboard combination used with a small process-control computer. Both standard alphanumeric keys and special-function keys are used.

As noted earlier, dependence upon application software in a powerful process-control computer and performance as a smart terminal distinguish the centralized operator interface from a distributed one. It can be physically located as part of the distributed operator interface since it uses the same high-speed communications bus as other process I/O subsystems. Figure 24-42 shows one such system.

The interface consists of a console with a color or monochromatic CRT and a series of versatile keyboards using flexible-membrane switches. Different edge-coded keyboard overlays allow adaptation of the operator keyboard for specific use, corresponding to software resident in the process-control computer. An annunciator keyboard provides alarm capability. A key indicates an alarm condition; when pressed, it calls up a unique display. A removable standard typewriter keyboard is used for data entry. Another removable function keyboard allows functional exten-

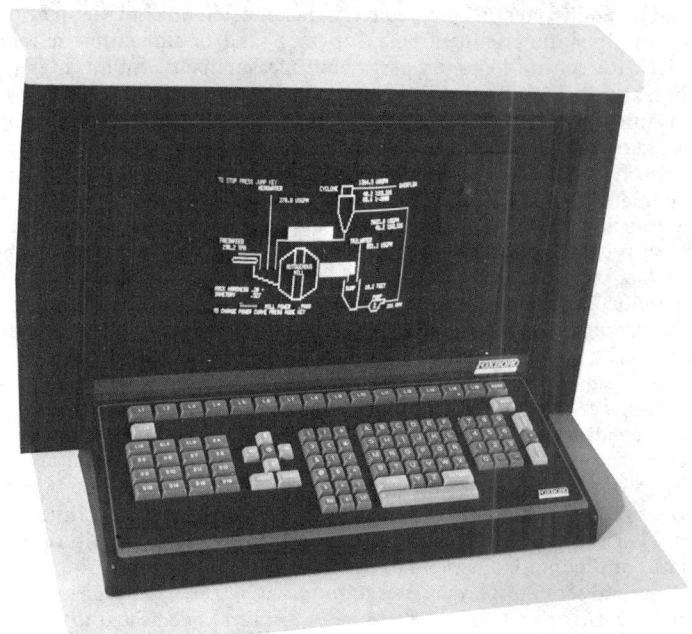

Fig. 24-41. CRT console for small process-control computer.

Fig. 24-42. Operator interface for centralized control.

sion of the operator keyboard. Keys in this last keyboard allow for easy customizing to meet special needs through user programming. Display capability includes alphanumeric characters, character graphics, and bit-map graphics. Character display may flash or be protected. A video copier attached to the interface can provide hard copies of displays

The process-control computer has several peripheral devices local to it, such as the keyboard-printers, video terminals, and card input-output. They fulfill the data processing requirements, as differentiated from process-control needs.

Several application packages for supervisory and regulatory control, each with its own data-entry and data-display requirements, reside in computer memory. Each data base must be designed to avoid element redundancy. The method for entry and later modification of such a data base is an integral part of the design, down to the keyboard level. Examples of such packages are as follows:

1. *Process analysis and control.* On-line entry and modification of data-base information, such as I/O specification, scan frequencies, engineering units, and control-algorithm specification. An alarm condition flashes a dedicated-action push-button key; when pressed, it calls up a unique display. Other keys call up displays of varying complexity. Displays combine alphanumeric data with graphic data. Table 24-4 lists the entry and display parameters and is indicative of the complexity of the data base.

2. *Process management and reporting.* The operator requests trend displays by specifying the desired number of variables, graph scaling, and the time-span–data-interval combination. The resulting graphic presentation is used for monitoring inputs or for other process-analysis tasks.

3. *Sequential process control and monitoring.* Plant startup and shutdown cycles and sequential processes are controlled by recipes or lists of sequential time- and/or event-driven actions. Display of large numbers of dissimilar simultaneous operations does not lend itself to easy codification. Recipe building and modification and process-status displays use combinations of graphics and alphanumeric data.

Effects of Service Conditions

BY W. CALDER

30. Integrity of equipment performance is subject to service conditions which are peculiar to equipment locations in process plants. Integrity depends on the equipment designer's ability to select components and circuits that will tolerate or compensate for effects of these conditions. A successful equipment design stabilizes the net effects within limits that assure operation to performance specifications.

Types of service conditions that affect equipment performance in the processing industries include the following:

1. *Natural environmental conditions.* Temperature and humidity variations, seismic disturbances, and surges from lightning strikes

2. *Generated disturbances.* Mechanical vibration and electromagnetic interference from other equipment, released contaminants, and the presence of flammable materials in the atmosphere

Requirements which do not necessarily affect performance but do call for elegant design are

3. *Safety requirements*

4. *Reliability considerations*

31. Standards Programs Establish Service Profiles. Known effects of damp heat and energy stressing, established for component qualification, set the base of a service-condition framework. Trade associations such as the Scientific Apparatus Makers Association (SAMA) and technical societies such as the Institute of Electrical and Electronics Engineers (IEEE) and the Instrument Society of America (ISA) have gathered application experience on other natural conditions and generated disturbances. Their collective experience is filling in the framework. At the same time the International Electrotechnical Commission (IEC) Technical Committees 65 (Industrial Process Measurement and Control), 66 (Electronic Measuring Equipment), 31 (Electrical Apparatus for Explosive Atmospheres), and 56 (Reliability and Maintainability) have estab-

TABLE 24-4   Standard Display Content and Operator-Modifiable Parameters (Data Base)

Parameter or information type	Block†	Loop	Unit	Meas.	Trend	BAM†	Supv. library
Time and date	D	D	D	D	D	D	D
Three-character ID			D	D			
Display title			D	D			
Block identification	D	D	D	D	D	D	
Block type	D	D				D	
Block description	D					D	
Set point	M	M	M			M	
Input (up to 2, excluding measurement)	D					D	
Output (up to 3)	M	M	M				
Input source ID		D	D				
Measurement input	M	M	M	D		D	
Engineering units	D	D	D	D		D	
Application description				D			
On-off control status	M	M	M	D		M	
DDC/SPC	M	M	M			M	
Remote or local status	M	M	M			M	
Open-close set-point status (cascade)	M	M	M			M	
Auto-manual mode status	M	M	M			M	
Switch status	M	M	M				
High-low absolute alarm limits	M					M	
High-low deviation alarm limits	M					M	
Rate-of-change alarm limit	M					M	
Absolute or deviation dead band	D					M	
Block alarm status	D	D	D	D			
Scan frequency	D						
Block diagram of loop		D					
Loop identification			D			D	
Loop alarm status		D					
On-off scan status	M	M	M			M	
Recorder number (up to 2)					D		
Pen number (up to 3 per recorder)					D		
Variable type number					D		
High-low engineering units range					D		
Variable type-number legend					D		
Trend-pen-assignment instruction					D		
Block off-trend instruction					D		
Suppress alarm						M	
Conversion information						M	
Algorithm type						D	
Filter index and constant						M	
Tuning parameters						M	
Range increment						M	
Scaling constants						M	
Output limits						M	
Integration indicator						M	
Error squared constant and range						M	
Library number							D
Program name							D
Program description							D
Program on-off status							D
Execution period							D
Next scheduled run time							D
Foreground or background type							D
Paging instructions							D

*D = display only; M = display and modify.
†Type of information displayed depends on block type. BAM = block adder and modifier.

lished a consensus on similar service condition data. The result is a series of profiles published or planned for publication by the IEC.

32. Atmospheric Service Conditions. Table 24-5 compiles from several sources the ranges of natural and generated atmospheric influences which may affect the functional performance of process measurement and control equipment (Par. **24-39**). The compilation recognizes that process-plant design limits the conditions in some equipment locations by heating, cooling, and air conditioning. A pragmatic consensus now exists for ranges of service conditions in four classes of location. The location classes for operation and storage range from A through D, defined below, in increasingly broader ranges of temperature, humidity, and barometric pressure. The classification also establishes limits of concentration of several airborne contaminants.

Class A: Air-Conditioned Areas. Temperature and humidity are controlled within narrow limits. A force-filtered air-circulation system reduces the quantity of particulate and chemical contaminants. In the event of air-conditioning equipment failure, conditions may approach the limits described for class B areas.

Plant designers provide class A areas for process control computers and operator-interface subsystems. Major control centers for critical operations, e.g., nuclear power station safety shutdown, also normally occupy class A areas.

Faulty operation of air conditioners or introduction of spares brought in from storage areas kept below the dew point of the air-conditioned area may cause condensation. Condensation occurs only for brief intervals.

Class B: Heated and/or Cooled Enclosed Areas. The atmospheric conditions of an enclosure are maintained within specified limits, but automatic maintenance of temperature and/or

TABLE 24-5   Range of Atmospheric Temperature, Humidity, and Barometric Pressure and Airborne Contaminants Classified by Location[1,2]

	Environmental class			
	A	B	C_1	D_2
	Operating conditions, comparable to IEC Publ. 654-1 (1979)			
Temperature, °C	15 to 30	5 to 50	−25 to +55	−40 to +85
Relative humidity, %	20 to 80	5 to 95	5 to 100	5 to 100
Max water content, kg/kg dry air	0.022	0.028	0.028	0.050
Atmospheric pressure, kPa		70 to 108		
	Storage conditions			
Temperature, °C	−40 to +70	−40 to +85	−40 to +85	−55 to +100
Relative humidity, %		0 to 100*		
Atmospheric pressure, kPa		20 to 108		
	Selected contaminants			
	Field appraisal†	Field appraisal†	Fed. Reg.‡	Field appraisal†
Hydrogen sulfide (H_2S), ppm	≤0.1	≤1.0	≤100	<25
Sulfur dioxide (SO_2), ppm	≤0.1	≤1.0	≤5.0	<10
Chlorine (Cl_2), ppm	≤0.01	≤0.1	≤1.0	<5.0
Ammonia (NH_3), ppm	≤0.05	≤1.0	≤500	<100
Oxides of nitrogen (NO, NO_2, N_2O_2), ppm	≤0.01	≤1.0	≤5.0	<10
Oxidants (O_3 and others), ppm	≤0.05	≤0.05	≤0.1	<0.5
Carbon monoxide (CO), ppm	≤10	≤25	≤50	<100
Hydrocarbons, ppm	§	§	≤50	<25
Dust, $\mu g/m^3$		≤100	≤200	≤500

*Conditions producing sustained condensate are not allowed.
†The Foxboro Company.
‡1971, Vol. 36, No. 105, Pt. 2.
§To be determined.

humidity may not exist. Condensation may occur for brief periods, particularly when the relative humidity is high.

Class B areas often house equipment requiring frequent operator surveillance and use. Manufacturers recommend storage of measurement and control equipment in these areas.

Class C: Sheltered Areas. These areas, which are typical for process operating equipment (and often for storage sites and transportation containers), furnish protection from direct sunlight, wind, rainfall, and precipitation of plant materials. In most instances, neither heating nor cooling is provided. Ventilation (if any) is natural. Minimum temperatures approach low ambient; maximum temperatures may exceed high ambient due to radiant heating of the shelter. Limited wind-driven precipitation, dripping water, and spray may penetrate class C areas. Occasional condensation occurs.

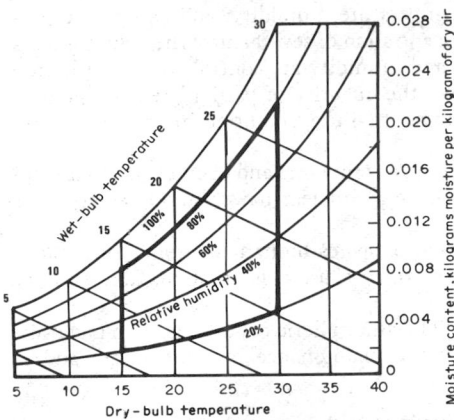

Fig. 24-43. Psychrometric chart for air-conditioned location of electronic equipment.

"Shacks" for transmitters, valve converters, and local indicators often occupy this class of environment. Equipment stored in the shacks is subject to the same unpurged environment.

Class D: Outdoor Areas. Equipment is totally unprotected from atmospheric conditions, including sunshine, wind, precipitation and airborne plant contaminants. Sensors, transmitters, final control elements, actuators, and local indicators are often located in class D areas, adjacent to plant vessels and piping.

Rapid changes of air temperature may occur. Equipment temperature changes may occur even more rapidly. For instance, apparatus exposed to direct sunlight may suddenly be cooled by rainfall. Sustained condensation should be considered normal.

Table 24-5 and these descriptions of location classes suggest that the plant engineer, like the system designer, has significant responsibilities. The designer attempts to furnish equipment which tolerates broad ranges of atmospheric service conditions but is restrained by the effects on specific elements of a measurement and control system. The plant engineer, however, may provide an environment which falls within limits established for an element or an instrument. Elements rated for class A and B service conditions would call for an equipment room or control center having environmental controls to limit the excursions of temperature and humidity and to filter expected contaminants from incoming air.

Psychrometric charts define temperature and humidity limits conveniently for both the system designer and the plant engineer. Figure 24-43 is a psychrometric chart with class A environmental service conditions outlined. A system designer would expect the atmospheric conditions for equipment installed in class A areas to fall within the outline.

33. Nonatmospheric Service Conditions. In conjunction with the development of standard ranges of atmospheric conditions, exploration of limits for other conditions has progressed as well. With the exception of seismic effects, they are disturbances generated by plant equipment rather than by nature. They include electrical transients, electromagnetic interference, vibration, and shock. Because they occur without regard to location class, their effect is general. Although these effects are pervasive, experience of equipment manufacturers and system designers reveals patterns and levels. As with atmospheric conditions, trade associations, technical societies, and multinational organizations have compiled profiles and procedures for evaluating the susceptibility of electronic equipment. Since the coupling of the generated disturbances is diffuse, guidelines have evolved for estimating propagation through circuits and structures.

34. Electromagnetic Interference. Technological advance in design of control-system and other plant equipment has resulted in the emergence of electromagnetic interference (EMI) as a compelling service condition. Control-equipment influences are mutual: (1) at reduced operating levels, signals are more susceptible to interference, and (2) switching power supplies and CRT monitors introduce sources of interference which are radiated and conducted to control equipment.

Other plant equipment includes maintenance communication systems, such as hand-held radio transceivers, that generate local interference.

Standards prescribing EMI protection levels remain obscure, but experience in field strength levels is providing concrete guidance. The most significant factors are emitted power, frequency ranges, and distance from the emitter to leads and signal wires. The sophistication of EMI susceptibility measurement techniques further complicates design assurance. Nevertheless, SAMA has produced a standard[5] which defines test procedures and three levels of EMI. Table 24-6 summarizes the levels and factors.

Other phenomena include *electrostatic discharge*, at 15 to 20 kV with energy levels in the 16-

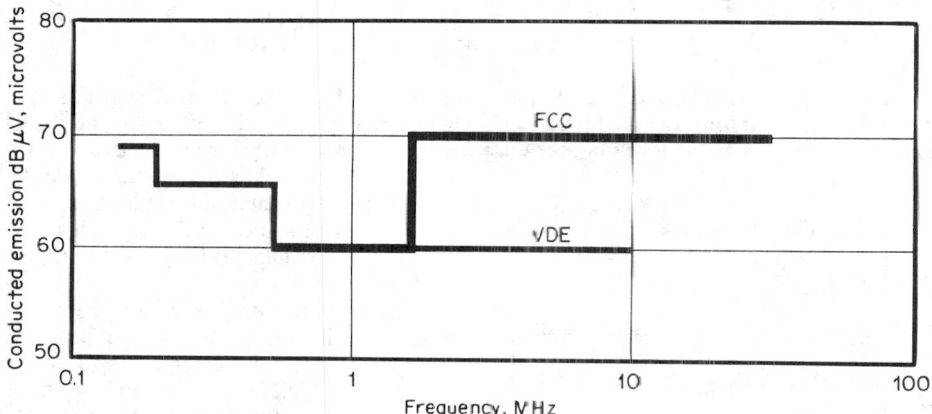

Fig. 24-44. Conducted EMI ranges identified by VDE[6] and FCC[3].

or 17-mJ range, between operator and equipment, and *showering arc* due to circuit inductance and relay or contact bounce. A "spark-gap" test determines these effects.

IEC Technical Committee 65/Working Group 4 is considering appropriate levels of the EMI phenomena described. Tests that it will evaluate may include lightning impulse, common- and normal-mode interference at power-line frequency, power interruption,[3,4] and mains-voltage depression.[3,4]

EMI technology is inexact and subject to diverse technical opinion. Wide acceptance of requirements and test procedures will take exposure and time. Federal Communications Commission (FCC) rules and Verbande Deutscher Electrotechniker (VDE) standards offer adequate challenge for the time. Figures 24-44 and 24-45 interpret the requirements established by VDE and FCC. The interpretations are subject to change, in anticipation of eventual applications of 1979 FCC rules.

35. Vibration and shock are mechanical influences. They occur in three application categories: industrial plants, nuclear power plants, and marine.

Study of standard mechanical influences for industrial plants is incomplete, but standards exist for nuclear and marine applications.

1. In *nuclear plants*, IEEE 344-1975, Recommended Practices for Seismic Qualifications of Class 1E Equipment for Nuclear Power Generating Stations, establishes some criteria and test procedures for seismic effects. Two bases have evolved for specifying levels: (a) identification of equipment-mounting response spectra for a given nuclear facility site and (b) regeneration of earthquake motion which is generic to anticipated sites

Figure 24-46 shows generic earthquake motions in the form of the seismic required-response spectra (RRS), identified by multiple resonant-frequency detectors rigidly attached to the same surfaces to which equipment is expected to be mounted. Below the crossover frequency of vibra-

TABLE 24-6   Radiated EMI Levels Identified by SAMA PMC 33.1-1973

EMI level	Field intensity	Frequency range, MHz		
		20–50	50–300	300–1,000
I	3 V/m	1a	1b	1c
II	10 V/m	2a	2b	2c
III	Not specified	3a	3b	3c

tion-test equipment, amplitude of motion is programmed to a segment of random frequency to generate motions which envelop the RRS. Nuclear licensees must meet two sets of criteria: (a)

Nuclear licensees must meet two sets of criteria: (a) equipment intended for use in class 1E safety-related systems must meet a manufacturer's performance specifications before, during, and after seismic vibration tests, and (b) equipment not destined for a safety-related control loop, while subjected to the same test levels, need only maintain structural integrity. There are no requirements for functional performance during or after tests.

2. For *marine applications*, insurance underwriters such as Lloyd's Registry of Shipping (UK), Bureau Veritas (France), and American Bureau of Shipping have identified standard mechanical influences. These organizations have led in the development of several IEC standards.

While shipboard vibrations are not as intense as those in land-based nuclear plants, they combine sustained rolling and pitching of a ship with rotation of machinery, shafts, and propellers. Lloyd's specifies rising levels of acceleration up to 0.7 g over a frequency range of 1 to 13.2 Hz and a constant 0.7 g acceleration out to 100 Hz. Even though the levels are low, sustained tests at each resonance for 2 h on each axis, to simulate continuous cruising conditions, can be destructive.

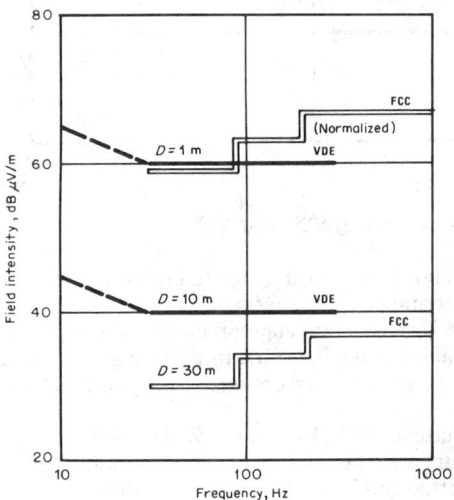

Fig. 24-45. Radiated EMI.

3. In *processing plants* (land-based and offshore), field audits are incomplete. Reported experience converges on acceleration ranging from 0.02g to 3g, depending on the level of service. Vibration is independent of the location classes defined for atmospheric service conditions. Nevertheless three levels of vibration are observed for similar locations: level I: control centers; level II: moderate field mounting; and level III: process pipe mounted. Table 24-7 gives values of acceleration and frequency for each of the three levels.

Actual equipment levels depend on the design of mounting structures and on the nature of vibration sources, such as piping and resonating machinery. Plant engineers and control system designers should consider preferred installation sites and mounting techniques.

As the values in Table 24-7 suggest, the levels reflect test techniques in which amplitude is constant up to a crossover frequency of 50 Hz, when acceleration becomes constant. Figure 24-47 shows the patterns of the three levels. Equipment test procedures call for a low-intensity resonant search, followed by sustained vibration at each resonant frequency to determine effects on performance.

Industry has not established shock levels, although field-mounted equipment frequently experiences 30 to 50g shock. Test procedures simulate these levels by dropping equipment from specified heights and by using impact hammers. Shock specifications remain a subjective issue, due to uncertain levels, inconsistent results at levels suggested, and the amount of damping associated with impact hammers.

Transportation influences occur when equipment is not operating. Levels for packaging

TABLE 24-7   Vibration Levels Observed in Processing Industries

Vibration level	Amplitude,* 10–50 Hz		Acceleration 50–2,000 Hz	
	mm	in	m/s²	g
I	0.050	0.004	4.9	0.5
II	0.10	0.008	9.8	1.0
III	0.30	0.024	29.4	3.0

*Amplitudes in millimeters are for single amplitude (peak values), the common practice in Europe and elsewhere, and in inches are for double amplitude (peak to peak), the common practice in North America.

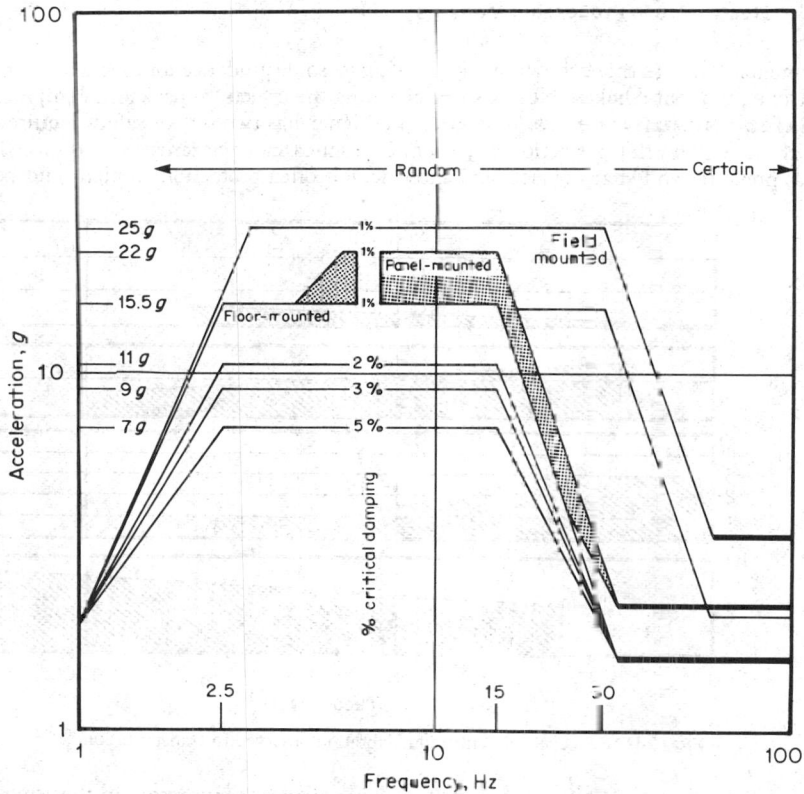

Fig. 24-46. Typical generic required response spectra (RRS) for floor-mounted, panel-mounted, and field-mounted instruments.

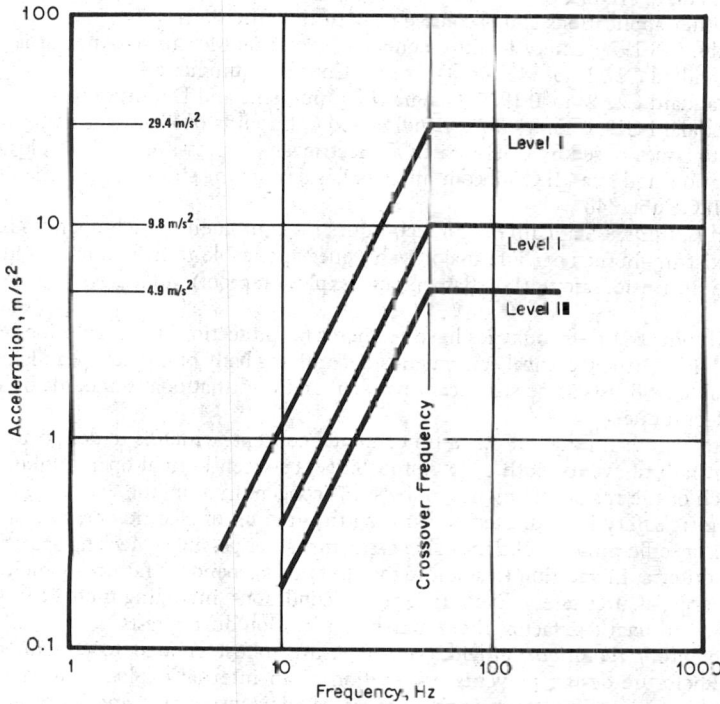

Fig. 24-47. Acceleration vs. frequency for three levels of vibration observed in the processing industries.

requirements,[10] such as those shown in Fig. 24-48, give some guidance for mechanical influences on control equipment. Shake tables exist for checking the effects on packaged equipment.

36. Safety Requirements. The system designer has two sets of safety requirements to satisfy: those which offer protection to personnel from casualty hazards such as electrical phenomena, pressure, and sharp objects and those which offer protection to plant and personnel

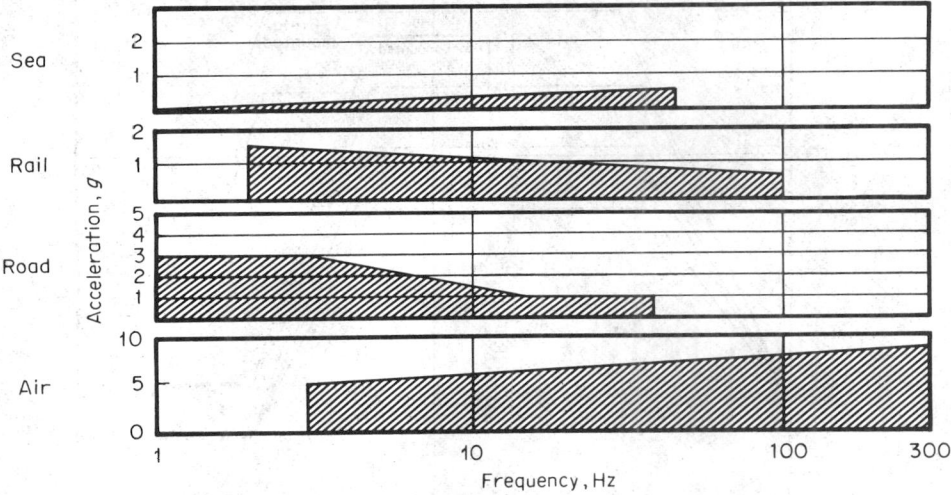

Fig. 24-48. Levels of continuous vibration observed in transportation.[10]

from explosion hazards which exist when flammable materials are present in the atmosphere in concentrations capable of ignition.

Casualty Hazards. Because the processing industries have recognized such hazards for years, standardization of design for safety is in an advanced state. The most comprehensive standard is the 1980 version of ANSI C39.5.[11] This standard addresses recognized hazards of industrial electronics for other applications and is consistent with the provisions of:

IEC Publ. 348-1978, Safety Requirements for Electronic Measuring Apparatus
CSA Standard C22.2 no. 142 (draft), Process Control Equipment
CSA Standard C22.2 no. 0-1975, General Requirements and Definitions
CSA Standard C22.2 no. 04-1972, Bonding and Grounding of Electrical Equipment

It responds to issues raised by OSHA, such as electric shock, pressure, CRT implosion or explosion, rf radiation, and heat. IEC Subcommittee 66E is developing similar requirements in a major revision of IEC Publ. 348.

Explosion Hazards. Standard design techniques for equipment which operates in flammable materials exist in abundance. The major techniques applicable to instruments and instrument systems are intrinsic safety (IS), flameproof (explosion-proof) housings, and pressurization (purging).

Nearly all industrialized countries have standards for protection from explosion hazards, as do IEC and CENELEC (the electrical equipment standardizing body of the European Economic Community). Table 24-8 lists some significant national and multinational standards by number and protection technique.

Each protection technique follows a fundamental concept, which is to design for at least two independent failure events, both of low probability, between normal operation and a potential disaster. Each of the major techniques follows this concept in a unique way.

1. *Intrinsic safety* is predicated on limiting the amount of electric energy available so that ignition of a specific mixture of flammable gas in air is not possible. Meeting specified construction or test criteria, in addition to assuming up to two independent failures which increase the available energy, assures safety. Even under these conditions, including increase of the available energy by an artificial test factor, there may be no ignition during tests.

2. *Flameproof technique* provides an enclosure robust enough to contain any internal explosion. Enclosure design prevents propagation of an internal explosion to the outside and ignition of the surrounding atmosphere. Again, specific construction and/or test criteria prove

the enclosure, its fittings, and its joints. The enclosure itself is one of the elements in the concept of having two levels of protection.

The second level of protection is subtle. First, the enclosure, by design, is extremely tight. A flammable atmosphere would have to persist outside for an extended time for a flammable mixture to accumulate inside the enclosure. In addition, if there are ignition-capable sparks or hot surfaces inside the enclosure, they will burn off any flammables which do enter before they

TABLE 24-8 Electrical-Equipment Standards for Explosion Hazards

Source	IS	Flameproof	Pressurization
United States	NFPA 493	UL 698	NFPA 496
Canada	CSA C22 2 no. 157	CSA C22.2 no. 30	NFPA 496
United Kingdom	SFA 3012		
West Germany	VDE 0170/0171	VDE 0170/0171	
IEC	Publ. 79-11	Publ. 79-1	Publ. 79-2
CENELEC	EN 50020-/EN 50014	EN 50018/EN 50014	EN 50016/EN 50014

*NFPA = National Fire Protection Association, UL = Underwriters Laboratories, CSA = Canadian Standards Association, SFA = British Approvals Service for Electrical Equipment n Flammable Atmospheres Standards, VDE = Verbande Deutscher Elektrotechniker, EN = European Norm.

reach an explosion-producing level. Second, if no ignition sources exist normally inside the enclosure, a failure of the enclosed electrical equipment is required to cause ignition. Conformance to these conditions, in conjunction with an intact enclosure, has provided an acceptable level of protection experience.

3. *Pressurization technique* prevents entry of flammable material into an enclosure by maintaining internal pressure higher than the outside (barometric) pressure. There are construction requirements and test criteria. The two levels of protection vary with equipment design. Overpressuring provides the first level. If the enclosure contains ignition-capable equipment, the second level of protection interlocks the power source with the pressurization to remove power on loss of overpressure. Should the enclosure contain equipment which in normal operation is not ignition-capable with power intact, no power interlock is required. In this case, only an alarm to indicate loss of pressurization is required.

No matter which technique is used, design and installation to existing standards assures safe operation. But satisfying all the standards listed poses a cost-benefit design challenge. While the principles of the multiple standards are consistent for each technique, significant deviations in detail exist. Satisfying such deviations with options on a single fundamental design may lead to overdesign elegance. A technically sound effort is under way to reduce national deviations. IEC and CENELEC projects lead the way.

37. Reliability has come to the forefront as an objective engineering discipline only during the last decade (see Sec. **28**). During this same time, process-control equipment has become more complex and has encompassed more measurement, control, and management functions. The consequences of failure can be significant to process operating economics and to protection of life and property. Equipment and system designers have become much more aware of reliability considerations.

Standards for estimating and reporting the reliability of general electronics exist. MIL Handbook 217C[26] is the most widely recognized source of general data and estimation techniques. IEC Technical Committee 56, Reliability and Maintainability, has developed reliability testing procedures.[31,32] But reliability standards specifically for electronics in the processing industries have not yet reached a public consensus. Nevertheless, SAMA is developing promising PMC drafts[27,28,29,30] and has issued a publication on terminology. A glance at the SAMA and IEC documents serves to demonstrate acceptable techniques of determining module and subsystem reliability, as well as system availability.

Generation of *mean-time-between-failure* (MTBF) data begins with a unit module which performs a specific function such as analog-to-digital conversion. Five methods are available: parts count, stress analysis using preset stress, stress analysis using actual stresses, demonstration based upon life tests performed by the manufacturer, and demonstration based upon plant experience.

Accuracy (but also engineering cost) increases from the parts-count method to the plant-demonstration method. SAMA drafts span the parts-count and stress-analysis methods for unit modules. IEC TC56 publications address both demonstration techniques.

Service condition	Electronic-equipment design alternatives	Plant design and operation alternatives
Temperature, humidity	Compensate for temperature in circuitry; encapsulate and seal for humidity	Air-condition, by design
Corrosive contaminants	Use connectors and seal-to-metal conductors with noble-metal surfaces	Use active filters, by administrative rule
Radiated and conducted EMI	Line filters with capacitances to ground shielding	Restrain EMI emitters to >1 m away, by administrative rule
Surge withstand	Provide components to discharge surges to ground	
Vibration	Secure connectors; fasten racks and slides; use static components; reduce motions and increase torques	Stiff mountings to concrete rather than to plant piping
Seismic effects	Reduce mass-to-stiffness ratio of cantilevered electromechanical components; furnish stiff mounting hardware for field-mounted equipment; stiffen mechanical structure and lower the centers of gravity of racks and panels	Provide solid bases for mounting racks and panels
Safety of operating personnel[25]	Limit energy levels; prevent casual human contact; eliminate sharp edges; cover and warn; insulate to withstand dielectric tests; design process-wetted parts to withstand test pressures	Instruct operating personnel
Explosion hazards:[25] Intrinsic safety (IS)	Limit IS circuit voltage with isolating transformers, zener diodes, and silicon controlled rectifiers; limit current with resistors and semiconductors; separate IS circuits from other circuits with spacing and mechanical barriers; printed-circuit layout and routing of interconnecting wiring is critical; keep maximum surface temperatures below ignition level by derating components and by encapsulation	Adhere to manufacturers' instructions for installation and maintenance, to preserve integrity of IS circuits
Explosion-proof	See standards referred to in Par. **24-36**, which cover (1) providing an enclosure robust enough to withstand pressures up to 4 times explosion pressure, (2) designing enclosure joints to meet requirements for gap width and length of flame path, (3) ensuring that all openings in an enclosure can be sealed off to prevent propagation of explosion, (4) designing to keep external enclosure temperatures from exceeding ignition temperatures of the gas, vapors, and/or dust to which the enclosure may be exposed	Install per ANSI/NFPA 70, National Electrical Code; provide routine inspection and maintenance to assure integrity of explosion-proof enclosure
Pressurization	Provide (1) an enclosure tight enough to maintain an internal pressure at least 0.1 in H_2O greater than the external barometric pressure with an economic flow of protective gas, usually filtered air; (2) a switch to interlock electric-supply shutoff upon loss of internal pressure; (3) an alarm switch which sounds on loss of pressurization if internal parts are not normally ignition-capable or internal parts are normally ignition-capable and external atmosphere is not normally hazardous; (4) an exhaust port to carry internal flammable material away during initial and subsequent purges	Install per ANSI/NFPA 70, National Electrical Code; provide routine inspection and maintenance to assure integrity of explosion-proof encosure

TABLE 24-9   Equipment-Design and Plant-Design Alternatives for Meeting Service Conditions (*Continued*)

Service condition	Electronic-equipment design alternatives	Plant design and operation alternatives
Reliability	Keep the design simple; the more components, the more chances for failure; use the highest quality components and materials; derate components to reduce stress — electrical mechanical, and thermal select materials for the nature of airborne contaminants; use "burned-in" electrical components to reduce or eliminate infant failures test; select spare parts carefully	Know the plant consequences of equipment performance degradation, as well as loss of equipment function; follow a schedule of checking, maintenance, and service
	For system configurations Specify redundancy where practicable; consider available service and spares; conduct comprehensive testing	

The same techniques yield system MTBF values, using already developed values for the modules which make up a system.

Availability numerics are calculated for all levels of industrial electronic equipment but most frequently for subsystems and power supplies. The numerics predict the percentage of time during which equipment will be available for its intended function. Techniques for determining availability include reliability block diagrams, a more sophisticated version which considers mean time to repair, and state transition rate analysis, which includes repair. Again, accuracy and engineering cost increase in the order of listing the techniques.

A more recent concern highlights the data security of communication busses. Called *communications reliability*, it deals with signal integrity and the probability that a given signal from a sending device will get to the proper receiving device, when a multitude of devices are linked to the same data bus. Public development of reliability-evaluation techniques for communication busses is under active consideration.

38. Response to Service Conditions by Design of Equipment and Installation. An equipment designer must respond to the potential effects of service conditions with features that go well beyond required function. Each of the service conditions described in Pars. **24-30** to **24-37** prompts specific design features which increase equipment performance. Should plant designers neglect opportunities to limit service conditions by engineering installation sites appropriately, extravagant equipment performance features will result.

While the equipment designer may experience only intermittent opportunity to limit service conditions, the plant designer has this opportunity in every proposed plant. During the life cycle of the equipment, both the equipment designer and the plant designer gain insight to the variable costs of (1) features of measurement and control equipment which respond to severe service conditions and (2) air conditioning, equipment mountings free of seismic effects, secure power sources, and even administrative rules for operation of other equipment which may generate part of the service conditions.

Surely, experience with both sets of variable costs will be convincing during a decade of equipment life. Management challenge both equipment and plant designers to convince each other during equipment design. Table 24-9 sketches the major design alternatives facing them. All the alternatives affect the installed cost of equipment. A significant few reduce measurement and control equipment-design extravagance.

39. Standards*

ATMOSPHERIC SERVICE CONDITIONS

1. IEC Publ. 654-1 (1979), Operating Conditions for Industrial-Process Measurement and Control Equipment, Part 1: Temperature, Humidity and Barometric Pressure.

2. *Fed. Reg.*, 1971, Vol. 36, No. 105, Pt 2.

*Standards, rules, norms, and recommendations are subject to periodic change. Latest editions are available from American National Standards Institute, Inc., 1430 Broadway, New York, NY 10018, except as indicated here.

ELECTROMAGNETIC INTERFERENCE

3. IEC Publ. 654-2 (1979), Operating Conditions for Industrial-Process Measurement and Control Equipment, Pt. 2: Power.

4. IEEE Standard 446 (1980), Recommended Practice for Emergency and Standby Power Systems for Industrial and Commercial Applications.

5. SAMA Standard PMC 33.1 (1978), Electromagnetic Susceptibility of Process Control Instrumentation.

6. VDE 0871/6.78, Specification for Equipment that Generates or Processes RF.

7. VDE 0875/6.77, Regulation for Equipment (Including in Industrial Areas) Exposed to Unintentional RF.

8. FCC 79-555, First Report and Order—Technical Standards for Computing Equipment.

VIBRATION, SHOCK, AND SEISMIC EFFECTS

9. American Bureau of Shipping, "Rules for Building and Classifying Steel Vessels," 1980. These rules are updated annually by the American Bureau of Shipping, and the latest edition should be consulted to learn the requirements in effect at the time of application.

10. Dummer, G. W. A., and N. B. Griffin, "Environmental Testing Techniques for Electronics and Materials," Pergamon, New York, 1962.

SAFETY REQUIREMENTS

Casualty Protection

11. ANSI C39.5 (1980 draft), Safety Requirements for Electrical and Electronic Measuring and Controlling Instrumentation.

Explosion Protection

12. NFPA 493 (1978), Intrinsically Safe Apparatus for Use in Division 1, Hazardous Locations.

13. NFPA 496 (1974), Purged and Pressurized Enclosures for Electrical Equipment in Hazardous Locations.

14. UL 698 (1973), Industrial Control Equipment for Use in Hazardous Locations. Class I, Groups A, B, C, and D and Class II, Groups E, F, and G.

15. CSA C22.2 no. 157 (1979), Intrinsically Safe and Non-incendive Equipment for Use in Hazardous Locations.

16. CSA C22.2 no. 30 (1970), Explosion-Proof Enclosures for Use in Class 1 Hazardous Locations.

17. VDE 0170/0171 (1979), Specification for the Construction and Testing of Electrical Apparatus for Use in Explosive Gas Atmospheres for the Mining Industry (VDE 0170); for Industries Other than Mining (VDE 0171).

18. IEC Publ. 79-11 (1976), Electrical Apparatus for Explosive Gas Atmospheres, Pt. 11: Construction and Test of Intrinsically-Safe and Associated Apparatus.

19. IEC Publ. 79-1 (1971), Electrical Apparatus for Explosive Gas Atmospheres, Pt. 1: Construction and Test of Flameproof Enclosures of Electrical Apparatus.

20. IEC Publ. 79-2 (1975), Electrical Apparatus for Explosive Gas Atmospheres, Pt. 2: Pressurized Enclosures.

21. CENELEC EN 50014 (1977), Electrical Apparatus for Potentially Explosive Atmospheres, Pt. 1: General Requirements.

22. CENELEC EN 50016 (1977), Electrical Apparatus for Potentially Explosive Atmospheres, Pt. 3: Pressurized Apparatus 'p'.

23. CENELEC EN 50018 (1977) Electrical Apparatus for Potentially Explosive Atmospheres, Pt. 5: Flameproof Enclosure 'd.'

24. CENELEC EN 50020 (1977), Electrical Apparatus for Potentially Explosive Atmospheres, Pt. 7: Intrinsic Safety 'i.'

Casualty and Explosion Protection

25. Magison, E. C. "Electrical Instruments in Hazardous Locations," 3d ed., Instrument Society of America, 1978.

RELIABILITY CONSIDERATIONS

26. MIL Handbook 217C (1979).

27. SAMA Standard PMC 32.0 (1980 draft), Process Instrumentation Reliability Techniques: Preface.

28. SAMA Standard PMC 32.1 (1976), Process Instrumentation Reliability Techniques: Terminology.

29. SAMA Standard PMC 32.2 (1980 draft), Process Instrumentation Reliability Techniques: Full Module Level MTBF Predictions.

30. SAMA Standard PMC 32.3 (1980 draft), Process Instrumentation Reliability Techniques: System Level MTBF Predictions.

31. IEC Publ. 605-1 (1978), Equipment Reliability Testing, Pt. 1: General requirements.

32. IEC Publ. 605-7 (1978), Equipment Reliability Testing, Pt. 7: Compliance Plans for Failure Rate and Mean Time between Failures Assuming Constant Failure Rate.

EVALUATION

33. SAMA PMC 31.1 (1980), Generic Test Methods for the Testing and Evaluation of Process Measurement and Control Instrumentation.

34. IEC Publ. 546 (1976), Methods of Evaluating the Performance of Controllers with Analog Signals for Use in Industrial Process Control.

Section 25

Radar, Navigation, and Underwater Sound Systems

DAVID K. BARTON Consulting Scientist, Missile Systems Division, Raytheon Company; Fellow, IEEE

HAROLD R. WARD Consulting Scientist, Equipment Division, Raytheon Company; Fellow, IEEE

SVEN H. DODINGTON Avionics Consultant, International Telephone and Telegraph Company; Fellow, IEEE

JAMES F. BARTRAM Consulting Engineer, Submarine Signal Division, Raytheon Company; Senior Member, IEEE

STANLEY L. EHRLICH Consulting Engineer, Submarine Signal Division, Raytheon Company; Senior Member, IEEE

DONALD A. FREDENBURG Principal Engineer, Submarine Signal Division, Raytheon Company

JACK H. HEIMANN Principal Engineer, Submarine Signal Division, Raytheon Company; Senior Member, IEEE

JOSEPH A. KUZNESKI Senior Engineer, Submarine Signal Division, Raytheon Company; Senior Member, IEEE

PAUL SKITZKI Technical Consultant, Submarine Signal Division, Raytheon Company; Member, IEEE

CONTENTS

Numbers refer to paragraphs

Radar Principles*

By DAVID K. BARTON

1. Basic Functions. The basic functions of radar are inherent in the word, which stands for *radio detection and ranging*. Measurement of target angles has been included as a basic function of most radars, and doppler velocity is often measured directly as a fourth basic quantity. Resolution of the desired target from background noise and clutter is a prerequisite to detection and measurement, and resolution of surface features is essential to mapping or imaging radar. The radar resolution cell is a four-dimensional volume, bounded by antenna beamwidths, width of the processed pulse, and bandwidth of the receiving filter. Within each such resolution cell, a decision may be made as to presence or absence of a target, and if a target is present, its position may be interpolated to some fraction of the cell dimensions.

The block diagram of a typical pulsed radar is shown in Fig. 25-1. The equipment has been divided arbitrarily into seven subsystems, corresponding to the organization of the paragraphs on radar technology (Pars. **25-59** to **25-83**). Radar operation is initiated by a synchronizer, which controls the time sequence of transmissions, receiver gates and gain settings, signal processing, and display. When called for by the synchronizer, the modulator applies a pulse of high voltage to the rf amplifier, simultaneously with an rf drive signal from the exciter. The resulting high-

---

*Section 25 is a contribution of the IEEE Aerospace and Electronic Systems Society. The editor is indebted to David B. Dobson, editor of *IEEE Transactions on Aerospace and Electronic Systems*, for his assistance in planning its contents and arrangements with the contributors. — D.G.F.

power rf pulse is passed through a transmission line or waveguide to the duplexer, which connects it to the antenna for radiation into space. The antenna shown is of the reflector type, steered mechanically by a servo-driven pedestal. A stationary array may also be used, with electrical steering of the radiated beam.

After reflection from a target, the echo signal reenters the antenna, which has been connected to the receiver preamplifier or mixer by a duplexer. A local-oscillator signal furnished by the

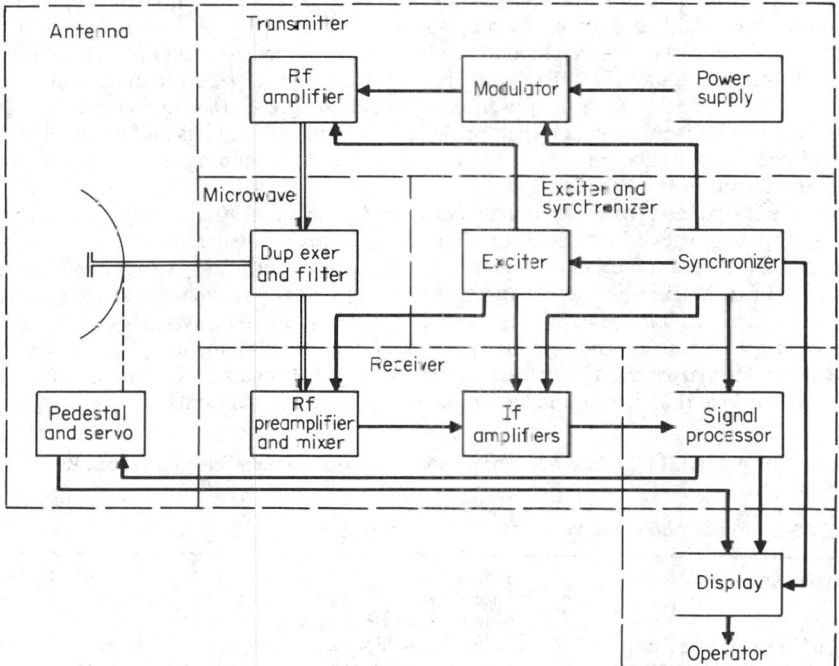

Fig. 25-1. Block diagram of typical pulsed radar.

exciter translates the echo frequency to i.f., which can be amplified and filtered in the receiver prior to more refined signal processing. The processed i.f. signal is passed through an envelope detector and displayed, with or without video processing. Data to control the antenna steering and to provide outputs to an associated computer are extracted from the time delay and modulation on the video signal.

2. Radar Parameters. A radar system can be described in terms of several basic parameters listed in Table 25-1, which determine its performance as a detection and measurement device. Other parameters and specifications are needed to describe subsystem design characteristics (antenna type, signal stability, input power source, etc.), but those shown in the table establish fundamental constraints on performance. From these parameters, the maximum operating range R_{max} on typical targets can be found, resolution properties can be described, and the signal-to-noise and signal-to-clutter ratios which constrain detection and measurement performance can be calculated for different environments.

Table 25-1.   Radar Parameters

Subsystem	Parameters
Transmitter	Frequency, wavelength, peak and average power, pulse width, pulse energy, pulse repetition frequency, duty ratio, signal bandwidth
Antenna	Size, effective aperture area, beamwidths, scan rate and coverage
Receiver and signal processor	Effective input temperature, bandwidth, input noise power, output pulse width, MTI improvement factor, video integration gain
System performance	System losses, received power and energy on standard target, time on target, signal-to-noise ratios in power and energy, measurement slope factors, data rate, instrumental errors

3. Target Characteristics. Given a set of radar parameters, the detection and location performormance will depend also on target characterisics: (a) target cross section (average or median) and fluctuation with time and frequency; (b) target size, shape, distribution of scatterers, aspect angle, and rate of turn; (c) target position in range, azimuth, and elevation and its velocity and acceleration in all three coordinates.

4. Noise and Background. The input noise level, determined by the receiving-system noise factor under standard conditions, is not an accurate measure of radar performance. When connected to the antenna, the receiver sees a low-noise background of empty space, modified by surrounding terrain or sea surfaces and atmosphere, at about 290 K, the sun at several thousand kelvins, galactic noise (at low frequencies), and various man-made sources of interference. Noise received through antenna side lobes must be added to internal receiver noise and main-lobe noise. Finally, the background of unwanted echoes from the earth's surface and atmosphere (including rain, birds, insects, and man-made objects such as buildings and automobiles) must be considered. These echoes may be either targets or clutter, depending on the radar application; they are described in Par. **25-12**.

5. Radar Applications. A complete catalog of radar applications would extend for many pages, with new entries added each year. The major fields of application, however, remain as shown in Table 25-2. Numerous miscellaneous applications, not readily categorized, can also be mentioned: intrusion alarms, monitoring of bird migration, ground-vehicle control, rendezvous of space vehicles, etc. The basic principles of radar apply to all these systems with suitable definition of target parameters and resolution or measurement requirements.

6. Radar Frequencies.[9]* In essence, there are no fundamental bounds on radar frequency. Any device that detects and locates targets by radiating electromagnetic energy and uses

*Courtesy of Dr. Merrill I. Skolnik. Superior numbers correspond to numbered references, Par. **25-161**.

TABLE 25-2   Radar Applications

Air surveillance	Long-range early warning, ground-controlled intercept, acquisition for weapon system, height finding and three-dimensional radar, airport and air-route surveillance
Space and missile surveillance	Ballistic missile warning, missile acquisition, satellite surveillance
Surface-search and battlefield surveillance	Sea search and navigation, ground mapping, mortar and artillery location, airport taxiway control
Weather radar	Observation and prediction, weather avoidance (aircraft), cloud-visibility indicators
Tracking and guidance	Antiaircraft fire control, surface fire control, missile guidance, range instrumentation, satellite instrumentation, precision approach and landing
Astronomy and geodesy	Planetary observation, earth survey, ionospheric sounding

TABLE 25-3   Radar Frequency Bands (IEEE Standard 521-1976)

Name	Frequency range	Radiolocation bands based on ITU assignments in region II
VHF	30–300 MHz	137–144 MHz
		216–225 MHz
UHF	300–1,000 MHz	420–450 MHz
		890–940* MHz
P band†	230–1,000 MHz	
L band	1,000–2,000 MHz	1,215–1,400 MHz
S band	2,000–4,000 MHz	2,300–2,550 MHz
		2,700–3,700 MHz
C band	4,000–8,000 MHz	5,255–5,925 MHz
X band	8,000–12,500 MHz	8,500–10,700 MHz
K_u band	12.5–18 GHz	13.4–14.4 GHz
		15.7–17.7 GHz
K band	18–26.5 GHz	23–24.25 GHz
K_a band	26.5–40 GHZ	33.4–36 GHz
Millimeter	>40 GHz	

*Sometimes included in L band.
†Seldom used nomenclature.

the echo scattered from the target can be classed as a radar, no matter what its frequency. Radars have been operated at wavelengths of 100 m (short waves) or longer to wavelengths of 10^{-7} m (ultraviolet) or shorter. The basic principles are the same at any frequency, but the implementation is widely different. In practice, most radars operate within the microwave-frequency range, but there are many notable exceptions.

A set of letter designations exists for the frequency bands commonly used for radar (see Table 25-3). The original code letters, (P, L, S, X, and K) were introduced during World War II. After the need for secrecy no longer existed, these designations remained. Others were added later (C, K_u, and K_a) as new bands were opened, and some were seldom used (P and X). There have been attempts to subdivide the entire microwave-frequency spectrum by letter codes and to extend the letter nomenclature to the millimeter-wave region, but this has not gained wide acceptance.

7. Radar Propagation. *Atmospheric Attenuation.* The frequency bands used for radar were selected to minimize the effects of the atmosphere while achieving adequate bandwidth, antenna gain, and angular resolution. Attenuation is introduced by the air and water vapor, by rain and snow, by clouds and fog, and (at some frequencies) by electrons in the ionosphere.

Attenuation in the clear atmosphere is seldom a serious problem at frequencies below 16 GHz (Fig. 25-2). The initial slope of the curves shows the sea-level attenuation coefficient k in decibels

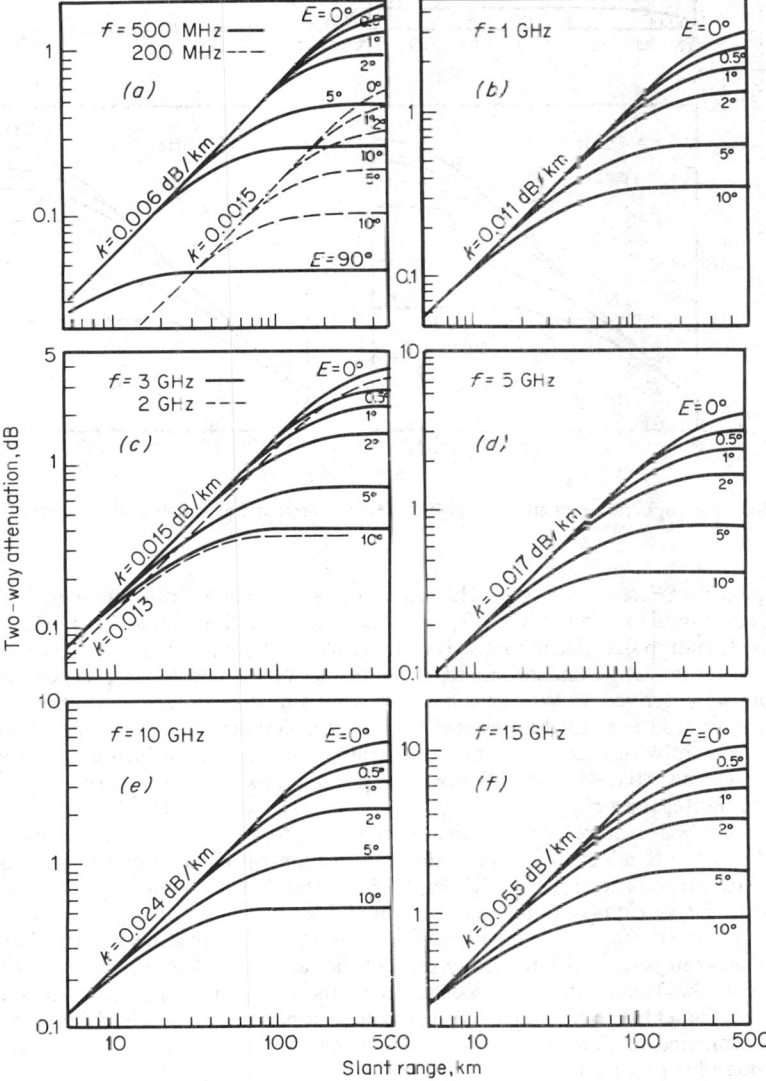

Fig. 25-2. Atmospheric attenuation (0.2 to 15 GHz) vs. range and elevation angle E. (*Data from Ref. 10.*)

per kilometer, falling off as the path reaches higher altitude. Above 16 GHz, atmospheric attenuation is a major factor in system design (Fig. 25-3). The absorption lines of water vapor (at 22 GHz) and oxygen (near 60 GHz) are broad enough to restrict radar operations above 16 GHz in the lower troposphere to relatively short range, even under clear-sky conditions. Attenuation vs. frequency for two-way paths through the entire atmosphere is shown in Fig. 25-4.

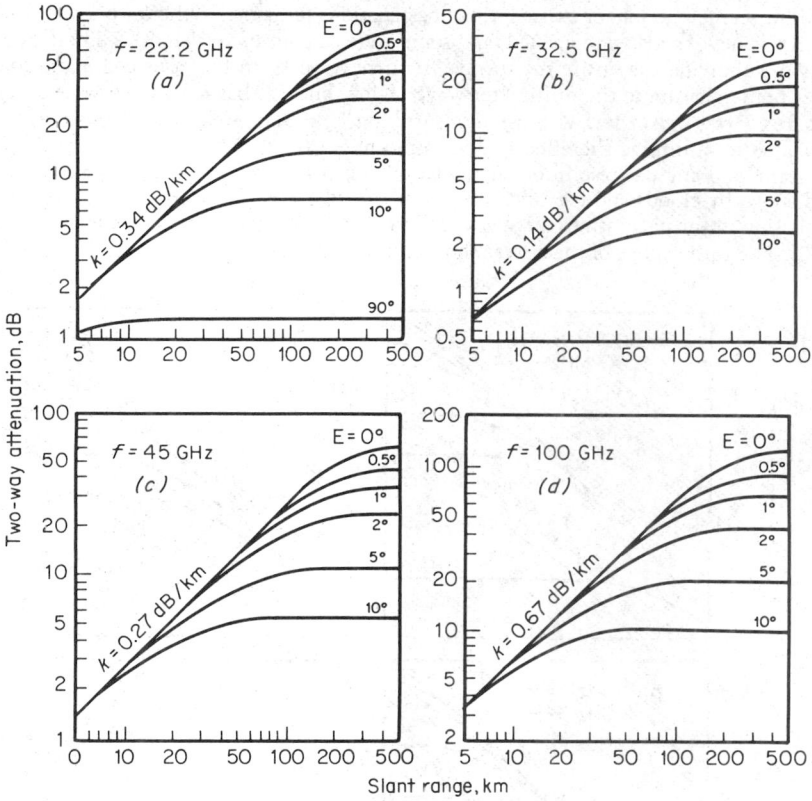

Fig. 25-3. Atmospheric attenuation (20 to 100 GHz) vs. range and elevation angle E. (*Data from Ref. 10.*)

Precipitation Effects.   Above 2 GHz, rain causes significant attenuation, with k roughly proportional to rainfall rate r and to the 2.5 power frequency. The classical data[1] on rain attenuation[11,12] were based on drop-size distributions given by Ryde and Ryde,[13] which gave generally accurate results, except for a 40% underestimate of the loss between 8 and 16 GHz, at low rainfall rates. Later data were derived by Wexler and Atlas[14] from a modified Marshall-Palmer distribution.[15] At high rates (100 mm/h) the loss coefficient k/r is doubled between 8 and 16 GHz, giving better agreement with measurements and matching the estimates of Medhurst[16] above 16 GHz. The Wexler and Atlas data provide the most satisfactory estimates for general use, and these were used in preparing Fig. 25-5.

Very small water droplets, suspended as clouds or fog, can also cause serious attenuation, especially since the affected portion of the transmission path can be tens or hundreds of kilometers. Attenuation is greatest at 0°C (Fig. 25-6). Transmissions below 2 GHz are affected more seriously by heavy clouds and fog than by rain of downpour intensity.

Water films that form on antenna components and radomes are also sources of loss. However, such surfaces can be specially treated to prevent the formation of continuous films.[17,18]

Apparent Sky Temperature.   Associated with the atmospheric loss is a temperature term, which must be added to the radar receiver input temperature (Par. **25-73**). Figure 25-7 shows this loss temperature T'_a as a function of frequency. Also included in T'_a is the galactic background noise for $f \leq 1$ GHz.

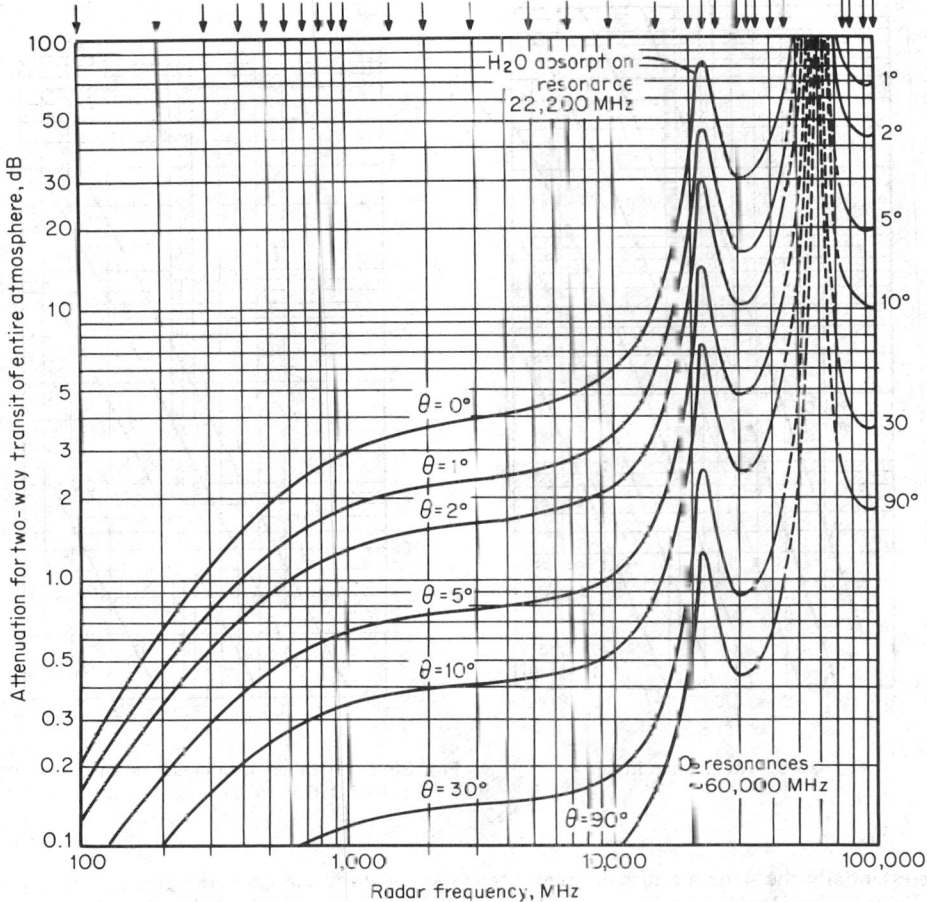

Fig. 25-4. Absorption loss for two-way transit of the entire troposphere, at various elevation angles.[10]

Ionospheric Attenuation. In the lowest radar bands, the daytime ionosphere may introduce noticeable attenuation.[19] However, above 100 MHz, this attenuation seldom exceeds 1 dB.

8. Surface Reflections. The radar antenna illuminates the target with direct rays and indirectly with energy reflected from the surface (Fig. 25-8a). The direct and reflected rays combine to produce a pattern of lobes and nulls (Fig. 25-8b). The effect of specular reflection from a plane surface is described most conveniently by assuming an image antenna below the reflecting plane, tilted downward by the same angle as the real antenna tilts upward. If the free-space voltage pattern of the antenna, as a function of elevation angle E, is $f(E)$, the pattern when operated over the reflecting plane is

$$f'(E) = f(E) + \rho(E)f(-E) \exp\left[-j\left(\frac{4\pi h_r}{\lambda}\sin E + \phi\right)\right] = Ff(E) \qquad (25\text{-}1)$$

where ρ = magnitude of surface reflection coefficient, h_r = antenna height, λ = radar wavelength, ϕ = phase angle of reflection coefficient, and F = pattern-propagation factor.[10] Equation (25-1) is valid for targets at long range $(R \gg h_r^2/\lambda)$, as is normally the case in land-based or shipborne radar.

At zero elevation, where $f(0) = f(-0)$, then $\rho \approx 1.0$ and $\phi \approx 180°$ for either polarization, leading to a total null in the pattern. Partial nulls with gains $f(E) - \rho f(-E)$ will occur when $\sin E$ is an even multiple of $\lambda/4h_r$, over elevations where $\phi \approx 180°$. Figure 25-9 shows the magnitude of ρ for vertical and horizontal polarizations and for different surfaces. Reflection from seawater is a function of frequency below 2 GHz, but freshwater and ground reflections

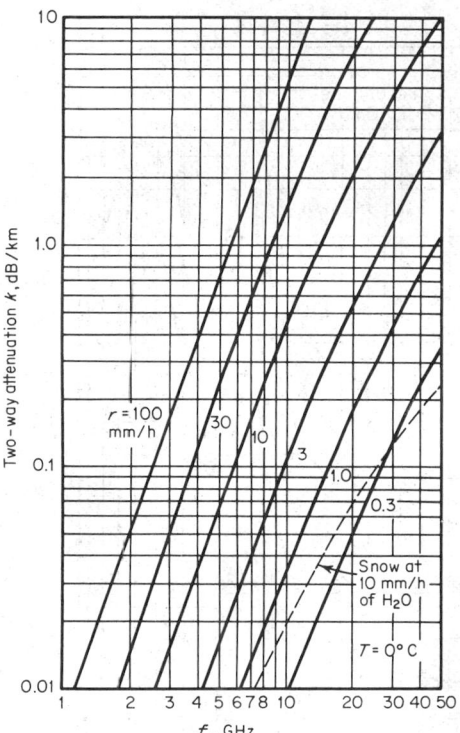

Fig. 25-5. Attenuation in rain.

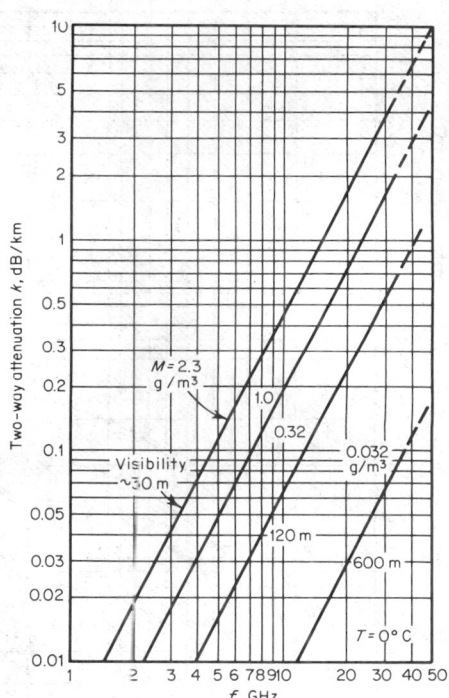

Fig. 25-6. Attenuation in clouds or fog (values of k are halved at $T = 18°C$).

are essentially the same for all radar frequencies. As shown in Fig. 25-8, the grazing angle γ is equal to target elevation angle E_t for a long-range target over a level surface. For all practical purposes, $\phi \approx 180°$ for horizontal polarization at all grazing angles and for vertical polarization below the angle of minimum ρ (the Brewster angle). More precise data for all frequencies down to 30 MHz are given in Ref. 20.

Rough-Surface Reflection. In the preceding discussion, an idealized smooth reflecting surface was assumed. Actual land and water surfaces are irregular, multiplying the smooth-surface specular reflection coefficient (Fig. 25-9) by a factor

$$\rho_s = \exp\{-2\,[(2\pi\sigma_h \sin \gamma)/\lambda]^2\} \tag{25-2}$$

as shown in Fig. 25-10. Here, σ_h is the rms deviation in height of the surface from its average plane. The energy lost from the specular reflection is scattered at other angles, including backscatter to the source.

9. Tropospheric Refraction. The refractive index of the troposphere, for all radar frequencies, can be expressed as

$$N \equiv (n - 1) \times 10^6 = \frac{77.6}{T}\left(P + \frac{4{,}810p}{T}\right) \tag{25-3}$$

where T = temperature (K), P = total pressure (mbar), p = partial pressure of water-vapor component, n = refractive index, and N = scaled-up deviation in n, known as the refractivity. Dry air at sea level can have a value of N as low as 270, but normal values lie between 300 and 320.

The Central Radio Propagation Laboratory (CRPL) of the National Bureau of Standards established a family of exponential approximations to the normal atmosphere in which the average United States conditions are represented by[22]

$$N(h) = 313.0e^{-0.14386h} \tag{25-4}$$

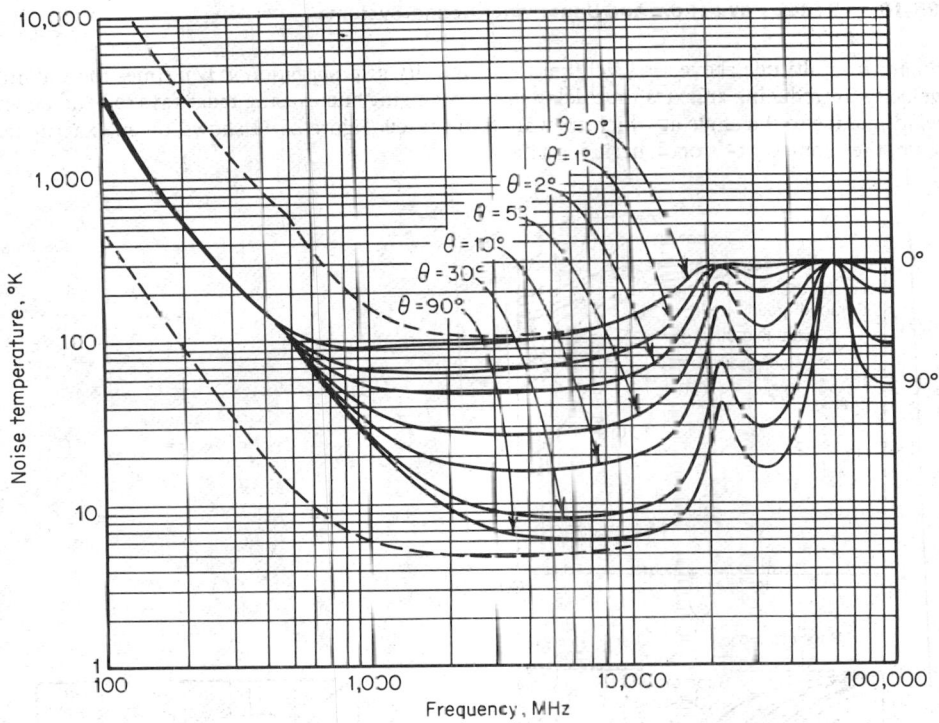

Fig. 25-7. Noise temperature of an idealized antenna (lossless, no earth-directed side lobes) located at the earth's surface, as a function of frequency, for a number of beam elevation angles. Solid curves are for geometric-mean galactic temperature, sun noise 10 times quiet level, sun in unity-gain side lobe, cool-temperature-zone troposphere, 2.7-K cosmic blackbody radiation, zero ground noise. Upper dashed curve is for maximum galactic noise (center of galaxy, narrow-beam antenna), sun noise 100 times quiet level, zero elevation angle, other factors the same as the solid curves. Lower dashed curve is for minimum galactic noise, zero sun noise, 90° elevation angle.[10]

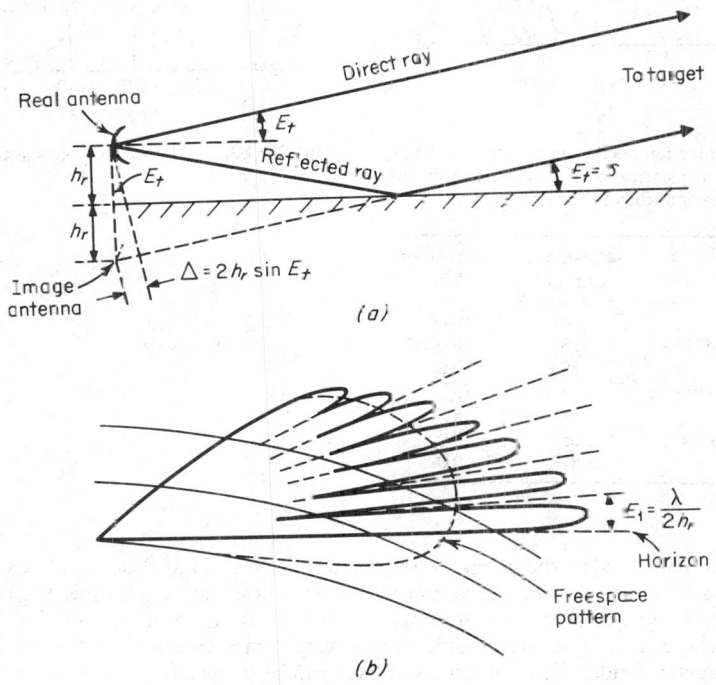

(a)

(b)

Fig. 25-8. Effect of surface reflections.[3]

where h = altitude above sea level (km). The velocity of propagation is $1/n$ times the vacuum velocity, introducing an extra time delay in radar ranging and causing radar rays to bend downward relative to the angle at which they are transmitted. Figure 25-11 shows, on an exaggerated scale, the geometry of tropospheric refraction.

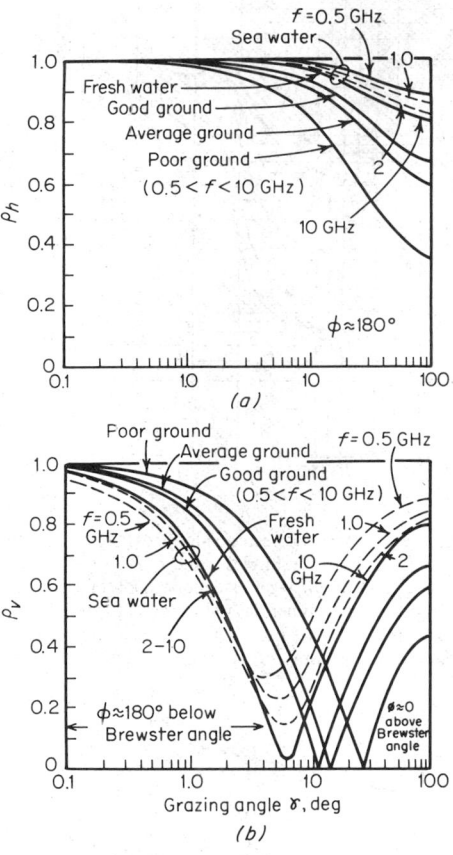

Fig. 25-9. Reflection coefficient vs. grazing angle for (a) horizontal polarization and (b) vertical polarization. Surface conditions defined as follows:

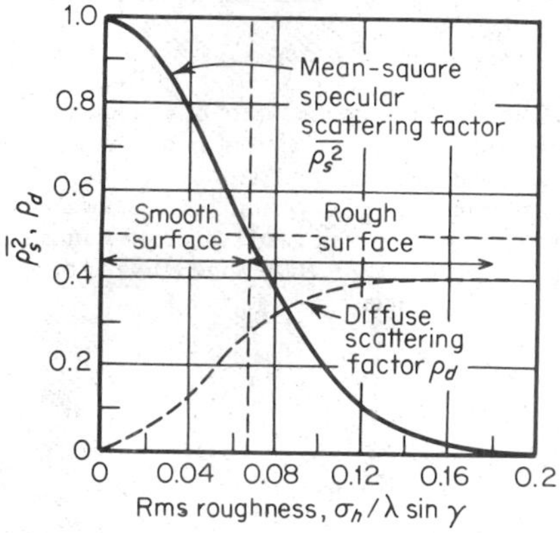

Fig. 25-10. Scattering factors vs. roughness.[8]

	Dielectric constant	Conductivity S/m
Poor ground	4	0.001
Average ground	15	0.005
Good ground	25	0.02
Seawater, 10 GHz	65	15
0.5 GHz	81	5
Fresh water, 10 GHz	65	10
0.5 GHz	81	0.01

For air-surveillance radar, the effects of tropospheric refraction are adequately expressed by plotting ray paths as straight lines above a curved earth whose radius is ⅘ times the true radius.

A special problem can arise when the ray is transmitted at elevations below 0.5° into an atmosphere whose refractivity has a rapid drop, several times greater than the standard 45 N units per kilometer. Under these conditions, the ray can be trapped in a surface duct (Fig. 25-12) or in an elevated duct bounded by layers of rapidly decreasing N. The result is a great increase in radar detection range for targets within the duct (surface targets and clutter, in most cases) at the expense of coverage just above the ducting layer. Although there is some leakage of energy from the top of the duct, increasing at lower frequencies, a duct will usually trap all radar frequencies, leaving a coverage gap just above the horizon.

10. Tropospheric Fluctuations. Deviations from average atmospheric conditions at a given site will fluctuate over a continuous spectrum from several hertz down to one cycle per

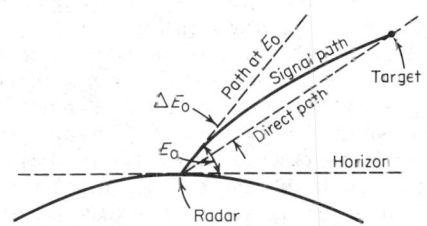

Fig. 25-11. Geometry of tropospheric refraction.

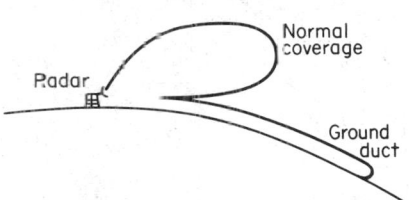

Fig. 25-12. Low-angle ducting effect.[2]

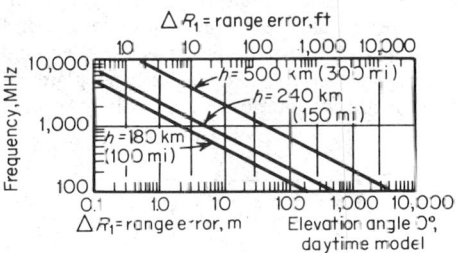

Fig. 24-13. Ionospheric range error vs. frequency.[24]

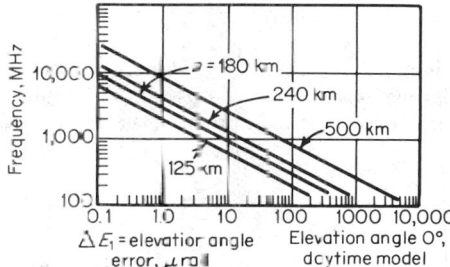

Fig. 25-14. Ionospheric angle error vs. frequency.[24]

year, with increasing spectral density at lower frequencies.[23] The average spectral density $W(f)$ can be expressed[8] in parts per million (ppm) of the path length as

$$W(f) = \begin{cases} 32/f \text{ ppm}^2/\text{Hz} & 10^{-8} \leq f \leq 10^{-5} \text{ Hz} \\ 10^{-6}/f^{2.5} \text{ ppm}^2/\text{Hz} & f \geq 10^5 \text{ Hz} \end{cases} \qquad (25\text{-}5)$$

Integration of this spectrum over frequencies which can be observed during a measurement interval t_0 (i.e., down to about $1/t_0$) gives the variance of the path length in ppm^2 occurring during that interval.

11. Ionospheric Refraction. The refractivity of the ionosphere at radar frequencies is given by

$$N_i = (n - 1) \times 10^6 = -(40N_e/f^2) \times 10^6 = (-1/2)(f_c/f)^2 \times 10^6 \qquad (25\text{-}6)$$

where N_e = electron density (m^{-3}) and f_c = critical frequency (Hz) ($f_c \approx 9\sqrt{N_e}$). Since f_c seldom exceeds 14 MHz, the refractivity at 100 Mhz is less than $10^6 N$ units, and above 1 GHz it does not exceed 100 N units. Figures 25-13 and 25-14 give the range and elevation-angle errors for normal ionospheric conditions for targets at different altitudes. Ionospheric errors are not significant for radars operating in the gigahertz region but can dominate the error analysis in the 200- to 400-MHz band.

12. Targets and Clutter. *Radar Cross Section.* The primary parameter describing a radar target is its radar cross section, or backscattering coefficient, defined as 4π times the ratio of reflected power per unit solid angle in the direction of the source to the power per unit area in the incident wave. If a target were to scatter power uniformly over all angles, its radar cross section would be equal to the area from which power was captured from the incident wave. A large sphere (whose radius $A \gg \lambda$) captures power from an area πa^2, scattering it uniformly in solid angle, and hence has a radar cross section $\sigma = \pi a^2$ equal to its projected area.

The variation of sphere cross section with wavelength (Fig. 25-15) illustrates the division into three regions of the spectrum, for any smooth target:

1. The optical region ($a \gg \lambda$), where cross section is essentially constant with wavelength.
2. The resonant region ($a \approx \lambda/2\pi$), where cross section oscillates about its optical value, due to interference of the direct reflection with a creeping wave, propagated around the circumference of the object.
3. The Rayleigh region ($a \ll \lambda/2\pi$), where the cross section drops rapidly below its optical value, varying as (a/λ^4).

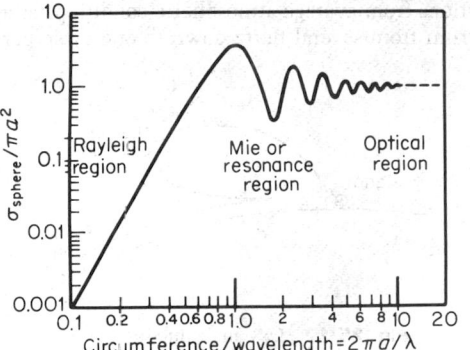

Fig. 25-15. Radar cross section of a sphere; a = radius, λ = wavelength.[2]

Although the sphere cross section varies with wavelength, it is constant over all aspect angles, and hence can serve as a reference for radar testing and evaluation. The cross sections of all other objects vary with aspect angle, requiring more complex descriptions: amplitude probability distributions, fluctuation frequency spectra (or time correlation functions), and radar frequency-correlation functions. For relatively simple objects, it is possible to write analytical expressions for cross section as a function of wavelength and geometry (see Table 25-4). Reflectivity patterns can be plotted for such objects as cylinders and flat plates, and lobe widths can be found as for antenna patterns. For cylinders and plates of length L (measured normal to the radar beam and to the axis of rotation), there will be a main

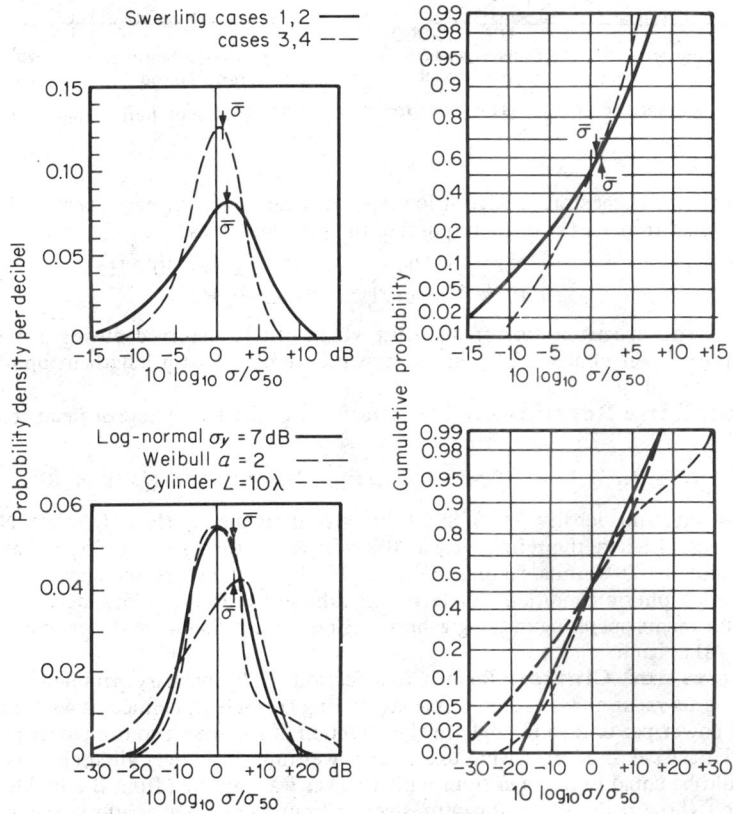

Fig. 25-16. Cross-sectional amplitude distributions.

(specular) reflection lobe whose half-power width is

$$\Delta = 0.44\lambda/L \qquad \text{rad}$$

whose side lobes are of width $\Delta_s = \lambda/2L$ and whose null-to-null main-lobe width is $2\Delta_0 = \lambda/L$.

13. Amplitude Distributions. When the target aspect angle is unknown or the target is too irregular to permit cross section to be expressed as a function of geometry, the amplitude distribution can be used to describe cross section statistically. Figure 25-16 plots the probability density and cumulative probability functions for several types of target, on decibel scales. The first two plots are the Swerling fluctuation models:[25]

$$dP = \begin{cases} (1/\bar{\sigma})e^{-\sigma/\bar{\sigma}}\,d\sigma & \sigma \ge 0, \text{ cases 1 and 2} \qquad (25\text{-}7) \\ (4\sigma/\bar{\sigma}^2)e^{-2\sigma/\bar{\sigma}}\,d\sigma & \sigma \ge 0, \text{ cases 3 and 4} \qquad (25\text{-}8) \end{cases}$$

where $\bar{\sigma}$ = arithmetic mean of the distribution. The median value σ_{50} is used as the center of each plot. Cases 1 and 3 describe slowly fluctuating targets such that all pulses integrated within a single scan are correlated but successive scans give uncorrelated values. Cases 2 and 4 describe fast (pulse-to-pulse) fluctuation.

Table 25-4. Radar Reflectivity Characteristics of Simple Bodies

Object	σ_{max}	σ_{min}	Number of lobes	Major lobe width
Sphere	πa^2	πa^2	1	2π
Ellipsoid ($k = a/b$)	πa^2	$\dfrac{\pi b^2}{k^2}$	2	$\sim \dfrac{b}{a}$
Cylinder	$\dfrac{2\pi a L^2}{\lambda}$	Null	$\dfrac{3L}{\lambda}$	$\dfrac{\lambda}{L}$
Flat plate	$\dfrac{4\pi A^2}{\lambda^2}$	Null	$\dfrac{8L}{\lambda}$	$\dfrac{\lambda}{L}$
Dipole	$0.88\lambda^2$	Null	2	$\dfrac{\pi}{2}$
Infinite cone (half-angle α)	$\dfrac{\lambda^2 \tan^4\alpha}{16\pi}$	Null		
Convex surface	$\pi a_1 a_2$			
Square corner reflector	$\dfrac{12\pi a^4}{\lambda^2}$	$\cdots$	4	$\dfrac{\pi}{4}$
Triangular corner reflector	$\dfrac{4\pi a^4}{3\lambda^2}$	$\cdots$	4	$\dfrac{\pi}{4}$

SOURCE: D. K. Barton, "Radar System Analysis," copyright 1964. Reprinted by permission of Prentice-Hall, Inc., Englewood Cliffs, N.J.

The lower plots in Fig. 25-16 show two analytical models used to describe clutter or targets with a broad spread of amplitudes, such as the distribution of a cylinder. Physically, the Swerling case 1 or 2 model corresponds to any target composed of more than three scatterers of comparable size (including aircraft at most aspect angles and at frequencies above a few hundred megahertz, rain clutter, and surface clutter viewed with grazing angles larger than about 5°). The case 3 or 4 model corresponds to a target in which one constant-amplitude scatterer dominates the echo but is modulated by two or more smaller scatterers. Weibull and lognormal distributions describe

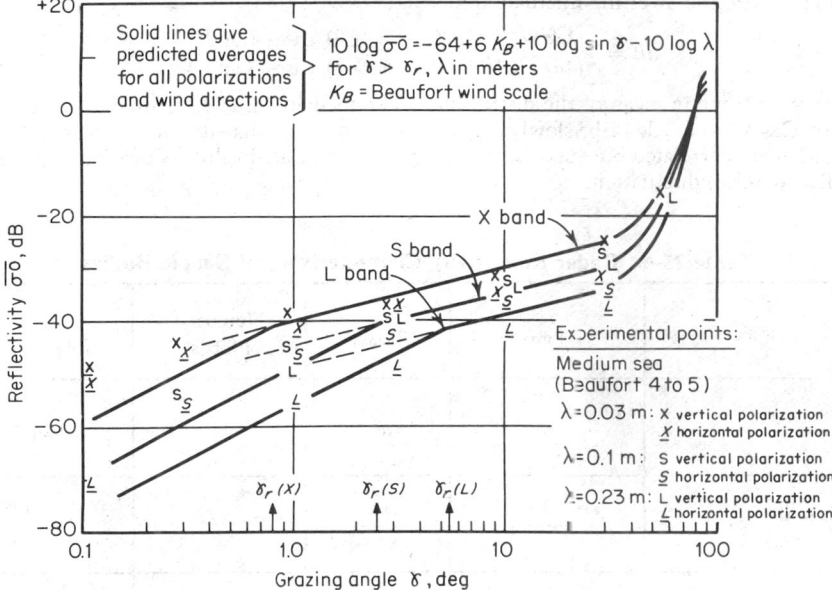

Fig. 25-17. Sea clutter reflectivity vs. grazing angle.

the statistics of surface clutter viewed at low grazing angles by medium- or high-resolution radar.[6,26,27,28]

14. Spectra and Correlation Intervals. Radar performance is sensitive to the time scale of cross-sectional fluctuation, which is related to the rate of rotation in aspect angle and to internal vibration or rotation of target components. Considering a rigid target whose aspect angle changes at a rate ω rad/s, the width of the fluctuation power spectrum will be proportional to the product ωL_x, where L_x is the spread of scatterers normal to the radar line of sight and the axis of rotation. If the scatterers are distributed uniformly over L_x, the spectrum will be uniform over a band of width $f_{max} = 2\omega L_x/\lambda$. Nonuniform scatterer distributions will produce corresponding-shaped spectra,[8] but any object can be approximated closely by an equivalent spread L_x and bandwidth B, analogous to the equivalent noise bandwidth of a filter. The corresponding correlation interval in time is $t_c = 1/f_{max} = \lambda/2\omega L_x$. The bandwidth can be increased (and t_c decreased) by internal vibrational and rotational components such as aircraft propellers or turbines. Typical values of f_{max} for aircraft-body echoes are on the order of a few hertz at X band.

Spread of target scatterers over a length L_r along the radar line of sight leads to frequency sensitivity of cross section, with a correlation interval $f_c = c/2L_r$. Radars using frequency diversity can take advantage of this to obtain independent samples of target echo at different frequencies, all transmitted and received within one target correlation time interval t_c. Thus the Swerling case 1 (or 3) target model (Fig. 25-16), applicable when $t_c \geq t_0$, may be changed to case 2 (or 4) by rapid change in radar frequency, reducing the echo correlation time to a new value $t_c' \ll t_0$. Typical aircraft targets, for which $L_r \approx 10$ m, give correlation frequency intervals on the order of 15 MHz, so that many independent samples are available in the tuning band of a microwave radar.

15. Target Glint and Scintillation. Targets composed of multiple scattering elements whose varying phase relationships cause fluctuations in signal amplitude are subject to errors in radar position measurement.[29] The apparent source of the composite echo signal wanders back and forth across the target, and at times the signal appears to originate from points well beyond the physical spread of the target itself. In principle, the variance in position measurement is infinite, for a measuring system with unlimited dynamic range and bandwidth. However, for practical systems this "glint" error is closely approximated by a gaussian distribution with standard deviation $\sigma_r = 0.35\ L_r$ (in range), or $\sigma_\theta = 0.35\ L_x/R$ rad (in angle).[8] On typical aircraft targets, the distribution of scatterers is equivalent to a uniformly scattering object with L from one-third to two-thirds of the physical span, leading to rms errors from 0.1 to 0.25 times the tip-to-tip aircraft dimensions. The apparent doppler spread of target echoes (frequency glint) is a scaled replica of the cross-range glint, with $\sigma_f = 0.35 L_x(2\omega/\lambda)$ for a target aspect angle

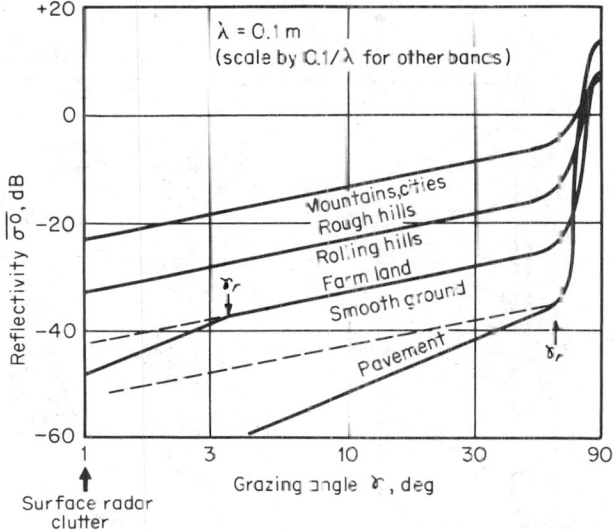

Fig. 25-18. Land clutter reflectivity vs. grazing angle.

rate ω rad/s (or 0.1 to 0.25 times the tip-to-tip difference in doppler frequency). The fact that the peak-to-peak doppler error exceeds the band of frequencies actually present in the signal is the result of nonlinearities inherent in the measurement process.

16. Clutter Echoes. Unwanted radar echoes (clutter) may originate from land or sea surfaces (characterized by a dimensionless surface reflectivity σ^0), from weather, from chaff occupying a volume of space (with a volume reflectivity η_v in m^2/m^3, or from discrete objects described by cross sections in square meters. The cross section of surface clutter is found by multiplying σ^0 by the surface area A_c within the radar resolution cell, while for volume clutter, η_v is multiplied by the resolution volume (see Par. **25-47**). These parameters have statistical distributions and spectra describing variation in space and time. Sea and land clutter as functions of grazing angle are shown in Figs. 25-17 and 25-18.

Rain and snow reflectivity, based on the classical (and generally accurate) Gunn and East data,[12] is plotted in Fig. 25-19. Although rainfall statistics averaged over periods of an hour show negligible probability of encountering rain rates greater than 50 mm/h, instantaneous (1-min) values at the surface and sustained densities aloft may reach 100 to 300 mm/h.

17. Detection. *Signal and Noise Statistics.* Most radar signals are sinusoids with narrowband modulation (in amplitude or phase) superimposed by the transmitter, antenna, and target. Random noise is also constrained approximately to the signal bandwidth when it reaches the detection circuits after passage through the receiver. The instantaneous i.f. output noise voltage V is described by a gaussian probability distribution

$$dP_v = (1/\sqrt{2\pi N})e^{-V^2/2N}\, dV \tag{25-9}$$

where dP_v = probability that voltage lies between V and $V + dV$, and N = mean-square noise voltage (average noise power). The video noise voltage E_n out of a linear envelope detector has a Rayleigh distribution

$$dP_e = (E_n/N)e^{-E_n^2/2N}\, dE_n \qquad E_n \geq 0 \tag{25-10}$$

The output voltage of a square-law detector follows the exponential distribution of i.f. noise power,

$$dP_\psi = (1/N)e^{-\psi/N}\,d\psi \qquad \psi \geq 0 \qquad (25\text{-}11)$$

These noise distributions are shown in Fig. 25-20. The probability that the noise envelope will exceed a given threshold voltage level E_t is the shaded area to the right of E_t in Fig. 25-21

$$P_n = \int_{E_t}^{\infty} (E_n/N)e^{-E_n^2/2N}\,dE_n = e^{-E_t^2/2N} \qquad (25\text{-}12)$$

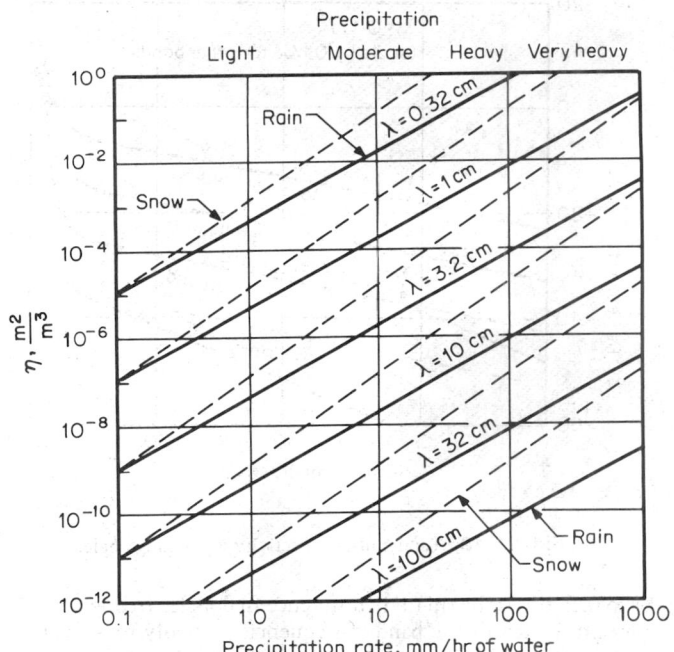

Precipitation

Fig. 25-19. Radar reflectivity of rain and snow.[3]

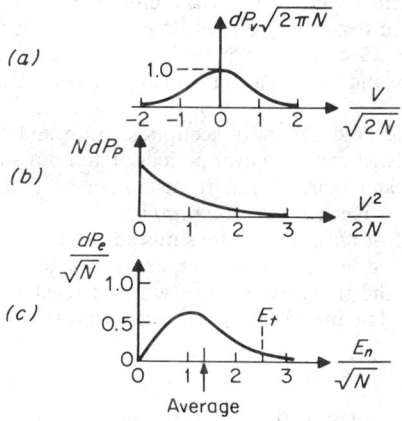

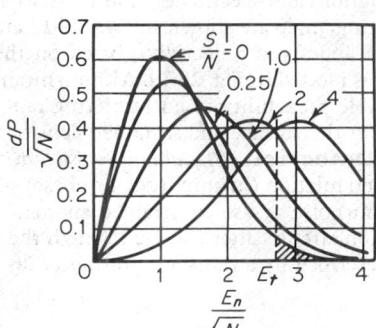

Fig. 25-20. Probability distributions of noise: (a) gaussian distribution of noise voltage; (b) exponential distribution of noise power; (c) Rayleigh distribution of detected envelope.[3]

Fig. 25-21. Probability distributions of envelope of signal plus noise.[3]

If a sinusoidal signal of peak amplitude E_s is present with the noise, the distribution of envelope voltage E_r is Rician,

$$dP_s = (E_r/N)e^{-(E_s^2+E_r^2)/2N} I_0 (E_n E_r/N)\, dE_r \qquad (25\text{-}13)$$

where I_0 = Bessel function with imaginary argument. The probability of detection P_d for a sample taken at the peak of the signal is the area under this curve and above the threshold E_t, as shown in Fig. 25-21 for different signal-to-noise power ratios $S/N = E_s^2/2N$.

Single-Sample Detection Probability   When a single sample is available for detection, the threshold is set to give the desired false-alarm probability according to Eq. (25-12), and the detection probability follows from the integral of Eq. (25-13). The results are plotted in Fig. 25-22,

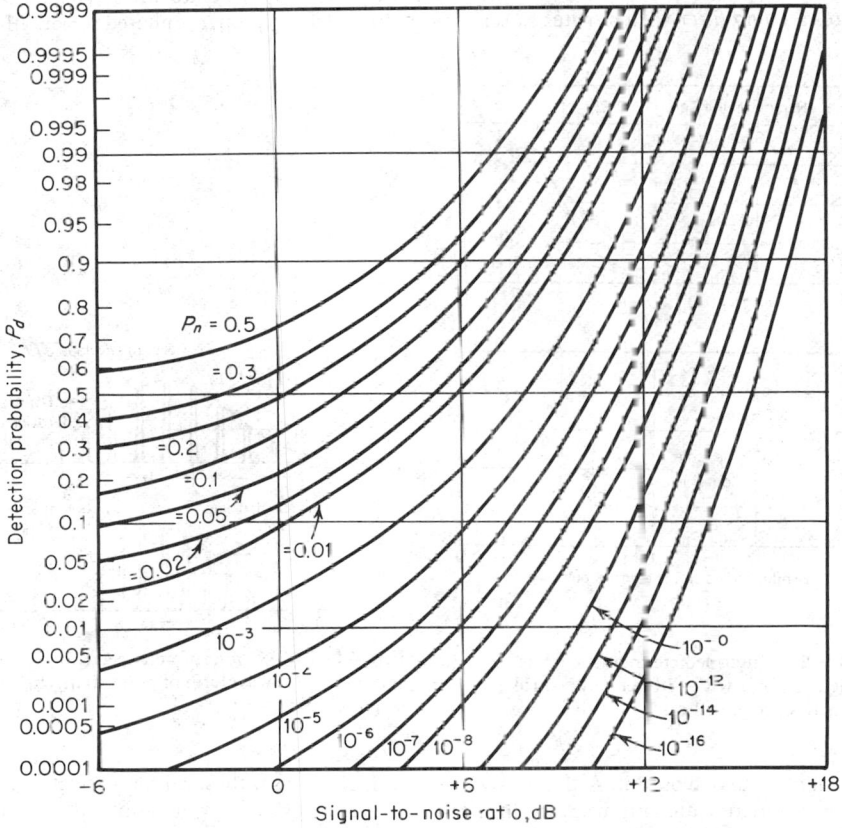

Fig. 25-22. Detection probability vs. signal-to-noise ratio.

where each curve gives P_d vs. S/N for the threshold setting corresponding to the P_n value shown. The value of S/N required to achieve a given P_d with fixed P_n is denoted by $D_0(1)$, the single-pulse detectability factor for the steady-signal case. Data from this figure are used as the basis for establishing detectability factors in multiple-pulse systems with different target types.

18. Filters and Signal Spectra. Detection performance for a single sample has been shown to depend on the S/N ratio at the i.f. output. Noise power at this point is a function of the receiver gain, the bandwidth, and the noise spectral density in the early stages of the receiver, before bandwidth is established. It is customary in detection analysis to assume an ideal (noise-free) receiver of unity gain, and to replace all actual noise sources with an equivalent broadband source of spectral density N_0, added to the input signal. The output noise power is then $N = N_0 B_n$, where B_n is the equivalent noise bandwidth of the receiver. The signal output power will also depend upon the receiver bandpass characteristics, and on the input energy.

The maximum output S/N will be

$$(S/N)_{\text{max}} = E/N_0$$

where E = total signal energy and $(S/N)_{max}$ is measured at the maximum of the signal envelope out of a matched filter. Practical filters can approach, but never exceed, the performance of a matched filter.

The matched filter for a single, uncoded pulse of duration τ can be approximated by a conventional bandpass filter whose bandwidth is $B \approx 1/\tau$. A matching loss is defined for this case as

$$L_m = \frac{E/N_0}{S/N}$$

(see Fig. 25-23). Pulses with internal coding or modulation require more complex filters (see Par. **25-74**).

A train of pulses produces a spectrum consisting of many separate lines (Fig. 25-24) and requires a *comb filter* with a matched series of response bands, properly phased to add all signal

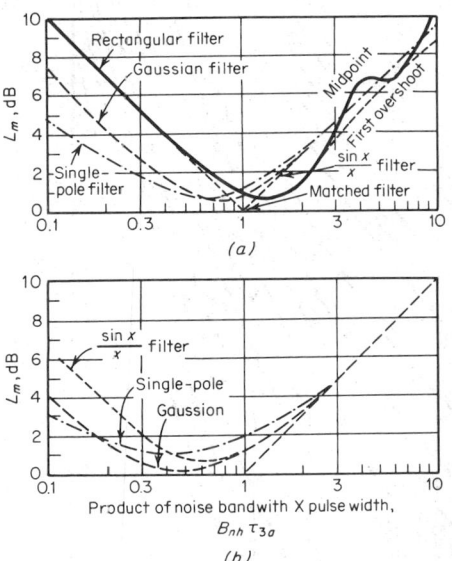

Fig. 25-23. Intermediate-frequency filter loss: (a) rectangular pulse with different filters; (b) gaussian pulse with different filters.[8]

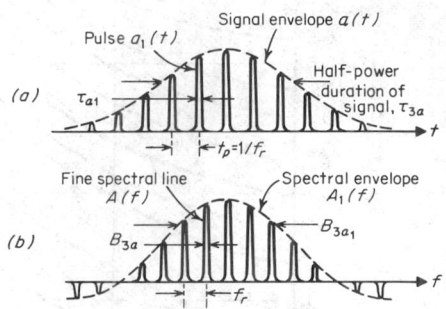

Fig. 25-24. Waveform and spectrum of coherent pulse train: (a) waveform of pulse train; (b) spectrum of pulse train.[8]

components into one output. Although such a filter is readily synthesized for a selected point in the time-frequency domain, using a range gate and narrow-band filter, it is difficult to match to signals of unknown time delay and doppler shift. Hence the integration of pulse trains is usually carried out at video (after envelope detection). The i.f. filter is matched approximately to a single pulse, giving $S/N = (E_1/N_0)/L_m$ at the envelope detector, where $E_1/N_0 = (E/N_0)/n$ is the average energy in one pulse of the n-pulse train. Successive video pulses are then added on the radar display or in some other range-ordered storage device before threshold detection.

19. Video Integration. Pulse-train integration at video reduces the required energy of each pulse but introduces a loss in performance relative to predetection (matched-filter) integration. This integration loss is defined as the increase in total signal energy required for a given P_d and P_n, relative to that needed with predetection integration. The detectability factor for n-pulse video integration as defined by Blake[10] and the IEEE[30] is

$$D_0(n) = \frac{L_i D_0(1)}{n} = \frac{E_1/N_0}{L_m} \tag{25-14}$$

Curves of integration loss L_i vs. n are shown in Fig. 25-25, with $D_0(1)$ as a parameter. For example, if $P_d = 0.90$ and $P_n = 10^{-6}$ are needed, $D_0(1)$ is 13.1 dB from Fig. 25-22, and L_i is 5.7 dB for $n = 100$ pulses. From these two figures, the steady-target data can be generated with an accuracy of about 0.1 dB.

20. False-Alarm Time. The D_0 curves of Fig. 25-22 correspond to a given probability P_n of a false alarm from each independent sample applied to the threshold. If the detection circuits are operative for $t_g \leq t_p$ seconds in each repetition interval t_p, there will be $\eta = t_g B_n$ independent samples applied to the threshold during the integration interval nt_p, producing an average of ηP_n false alarms. The ratio of total time to average number of alarms is the false-alarm time. In some

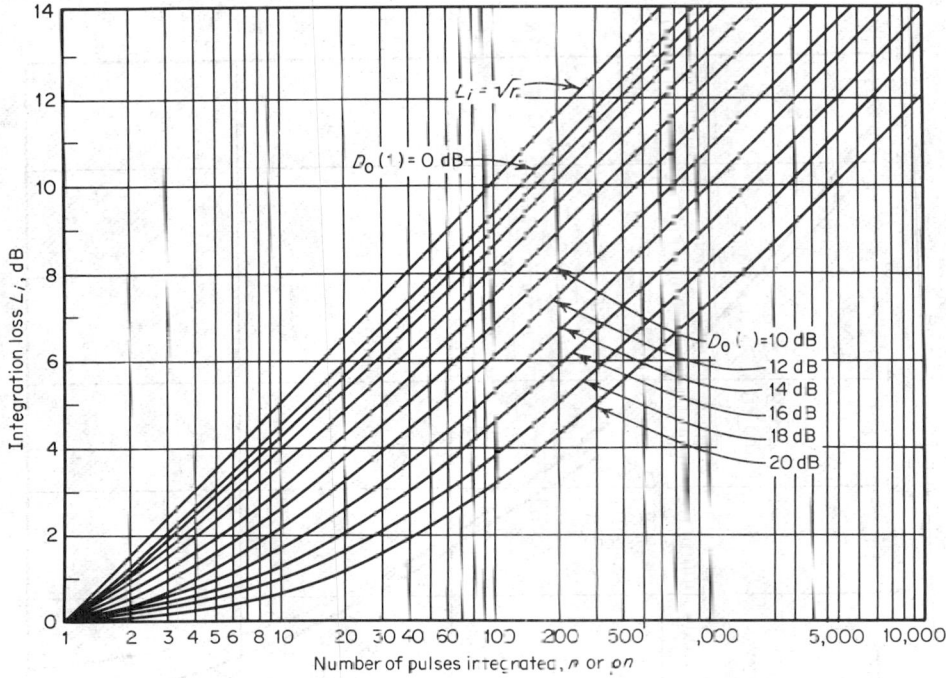

Fig. 25-25. Integration loss vs. number of pulses [33]

discussions, following Marcum's usage,[31] the false-alarm time t_{fa} is defined as the interval in which the probability of at least one false alarm is 0.50, and a false-alarm number n' is defined as the number of independent samples applied to the threshold during t_{fa}. Since $t_{fa} = 0.69 \overline{t_{fa}}$, Marcum's false-alarm number n' is

$$n' = 0.69 B_n t_g f_r \overline{t_{fa}}/n = 0.69/P_n \tag{25-15}$$

21. Collapsing Loss. Practical radars seldom preserve their full rf signal resolution through the integration and thresholding processes, where insufficient video bandwidth or broadened range gates may prove economical. The n video samples of signal plus noise are then integrated, along with m extra samples of noise alone, giving a collapsing ratio ρ

$$\rho = 1 + m/n \tag{25-16}$$

The effect, when using a square-law detector, is the same as if the signal energy were redistributed over $\rho n = m + n$ pulses, leading to the larger L value (Fig. 25-25) associated with integration of ρn pulses. The additional loss is referred to as the collapsing loss L_c

$$L_c(\rho, n) = L_i(\rho n / L_i n) \tag{25-17}$$

Formulas for ρ in different cases are given in Par. **25-42**, with rules for determining whether the number of independent threshold samples remains constant or varies inversely with ρ. In the latter case, P_n may be allowed to increase, and the energy added to overcome collapsing effects need not be as great.

22. CRT Integration. Signal detection by human operators, using cathode-ray-tube displays, cannot be characterized by a definite false-alarm probability. Instead, a *visibility factor* V_0 is used to give approximate requirements for E_1/N_0 under optimum bandwidth, display, and viewing conditions. Figure 25-26 shows V_0 for 0.50 detection probability using PPI and A-scope displays, assuming rectangular pulses and approximately matched rectangular filters ($B\tau \approx 1.2$). The visibility factor V_0 cannot be compared directly with D_0, but it is approximately equal to $L_m L_c D_0$ for $P_n = 10^{-4}$, assuming a minimum product $L_m L_c = 1.7$, or 2.3 dB, and an averaging time of 3 s for the combination of A-scope and operator. An operator can be expected to achieve the performance shown in Fig. 25-26 only under ideal conditions involving short attention spans and at least some a priori knowledge of where the signal should be expected on the display.

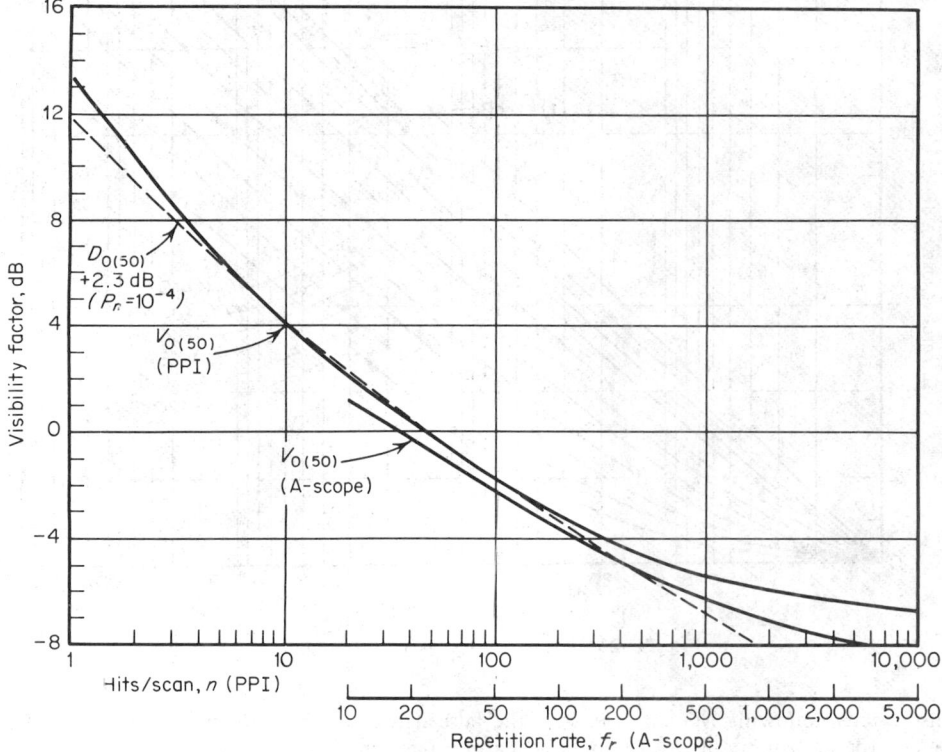

Fig. 25-26. CRT display visibility factors. *(After Ref. 10.)*

Signal energy required for initial detection of an unexpected target during an extended watch may be higher by several decibels (operator loss).

23. Detection of Fluctuating Signals. The single-pulse detectability factor $D_1(1)$ for Swerling case 1, or case 2 (Fig. 25-16), is plotted in Fig. 25-27, showing a considerable increase in the required S/N ratio for high detection probabilities.

The difference in decibels between $D_0(1)$ and $D_1(1)$ for a given P_d and P_n is known as *fluctuation loss* L_f for the case 1 target, and L_f remains essentially constant for all values of n on this target. However, when n pulses from a case 2 target are integrated, the loss in decibels is $1/n$ times as great as for case 1. In general, when n pulses containing n_e independent Rayleigh target samples are integrated, the detectability factor $D_e(n,n_e)$ can be found by adding to $D_0(n)$ the loss L_f/n_e decibels. A system with n_e-fold diversity will then have a *diversity gain* of $L_f (1 - 1/n_e)$. The case 3 target has an equivalent $n_e = 2$ for all values of n, while case 4 has $n_e = 2n$. Values of n_e for intermediate cases are computed[33] using ratios of the observation time t_0 to target correlation time t_c and of tuning bandwidth Δf to correlation frequency f_c.

24. Detection with Lognormal Statistics. In response to a growing awareness of actual signal and clutter distributions which are broader than the gaussian and Rayleigh models, detection curves have been derived for lognormal signals in gaussian noise[34] and for steady and Rayleigh signals in lognormal clutter.[35]

The defining equation for lognormal distribution of signal or noise power ψ is

$$dP_\psi = \frac{1}{\sqrt{2\pi}\,\sigma_\psi} \exp\{- [\ln (\psi/\psi_m)]^2/2\sigma^2\}\,d\psi \qquad (25\text{-}18)$$

where ψ_m = median power and σ = standard deviation of $\ln \psi$. The width of the distribution is also described by the standard deviation in decibels, $\sigma_y = 4.34\sigma$. The ratio of average to median power in the lognormal distribution increases rapidly with σ or σ_y.

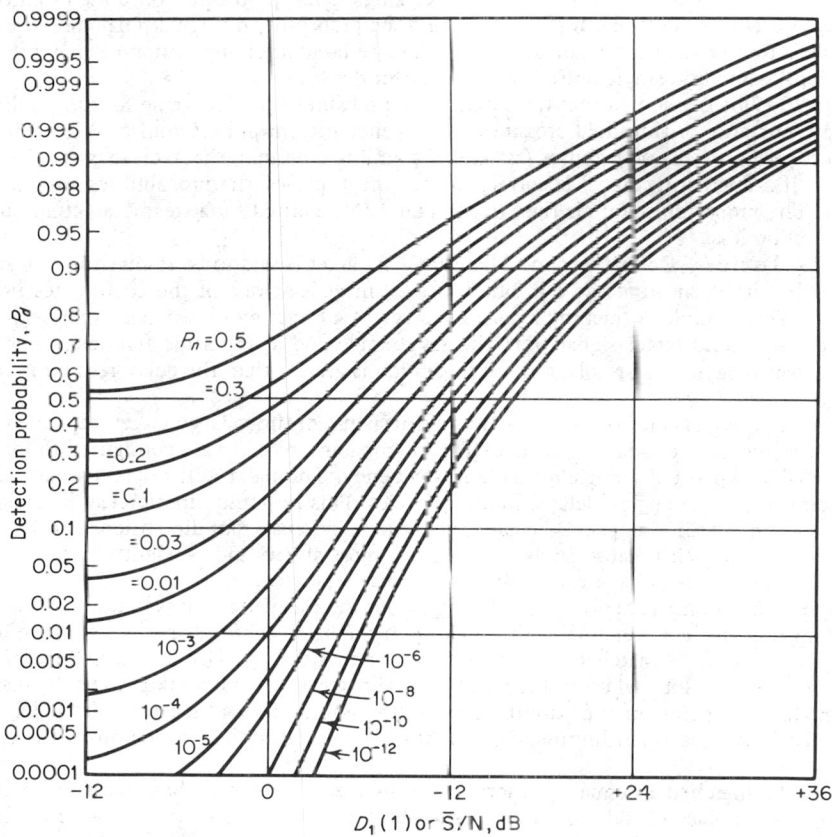

Fig. 25-27. Detectability factor for Rayleigh target (Swerling case 1).

Because a large portion of the average power is contributed by the low-probability tail of the distribution, detection of lognormal signals is characterized by a large fluctuation loss. This loss can be reduced substantially by integration of independent signal samples, but the diversity gain is not as large as given by Ref. 33 for the Swerling models.

The large ratios of average and peak power to median power also affect detection of steady and fluctuating signals in lognormal clutter.[28,35] In this case, the threshold must be set well above the average clutter level to achieve low false-alarm rates. Since the average clutter power can be many decibels above the median, the required ratio of signal to median clutter can be several tens of decibels. The solution is to use a *median detector*, of which the binary integrator is an example, with as many independent clutter samples as possible.

25. Binary Integration. An integration procedure which has found wide use in modern radar systems is the binary, or double-threshold, integrator. In this system, the receiver output is applied to a threshold, generating a sequence of binary ones and zeros within each repetition interval. These binary signals are summed for each range cell over n repetition intervals, and an alarm is generated when at least k out of n ones are accumulated in a cell. If the probability of a threshold crossing is p for a single interval, the probability of exactly j threshold crossings in

n intervals is given by the binomial distribution

$$P(j) = \frac{n!}{j!(n-j)!} p^j (1-p)^{n-j} \tag{25-19}$$

The probability that j will equal or exceed the second threshold k is the sum $P(k) + P(k+1) + \cdots + P(n)$. The median detector is instrumented by setting $k/n \approx \frac{1}{2}$.

The performance of the optimum video integrator is approximated (with about 1 dB extra loss) for $0.25 \le k/n \le 0.75$, depending upon the type of signal fluctuation. For a given value of k/n, Eq. (25-19) is used to calculate the required single-pulse probability on noise or clutter to give the required low false-alarm probability, and the probability on signal plus noise or clutter to give detection probability. Figure 25-22 can then be used (after application of an appropriate fluctuation loss) to find single-pulse S/N required for detection.

In Eq. (25-19) it is assumed that the single-pulse probability p is the same for each pulse but that the occurrence of threshold crossings is independent from pulse to pulse. When the S/N ratio varies over a train of n pulses (as with a scanning antenna), the average p can be used. However, if a fluctuating signal is correlated over the n pulses, the probabilities are not independent. The proper procedure is then to compute S/N required for a steady target and to add the fluctuation loss.

26. Resolution. *Definitions and Measures.* A target is said to be resolved if its signal is separated by the radar from those of other targets, in at least one of the coordinates used to describe it. For example, a tracking radar may describe a target by two angles, time delay, and frequency. A second target signal from the same angle and at the same frequency but with different time delay can be resolved if the separation is greater than the delay resolution of the radar.

Resolution, then, is determined by the relative response of the radar to targets separated from the target to which the radar is matched. The antenna and receiver are configured to match a target signal at a particular angle, delay, and frequency. The radar will respond with reduced gain to targets at other angles, delays, and frequencies. This *response function* can be expressed as a surface in a four-dimensional coordinate system. Because four-dimensional surfaces are impossible to plot, and because angle response is almost always independent of the delay-frequency response, these pairs of coordinates are usually separated.

In angle, the response function $\chi(\theta, \phi)$ is simply the antenna pattern. It is found by measuring the system response as a function of the angle from the beam center. It has a main lobe in the direction to which it is matched and side lobes extending over all visible space. Angular resolution, i.e., the main-lobe width in the θ and ϕ coordinates, is generally taken to be the distance between the 3-dB points of the pattern. The width, amplitude, and location of the lobes are determined by the aperture illumination (weighting) functions in the two coordinates across the aperture.

Because the matched antenna is uniformly illuminated, its response has relatively high side lobes, which are objectionable in most radar applications. To avoid these, the antenna illumination may be mismatched slightly, with resulting loss in gain and broadening of the main lobe.

Time delay and frequency can also be viewed as if they were two angular coordinates, i.e., as a two-dimensional surface which describes the filter response to a given signal as a function of the time delay t_d and the frequency shift f_d of the signal relative to some reference point. Points on the surface are found by recording the receiver output voltage while varying these two target coordinates. The response function $\chi(t_d, f_d)$ is given, for any filter and signal, by

$$\chi(t_d, f_d) = \int_{-\infty}^{\infty} H(f) A(f - f_d) e^{j2\pi f t_d} \, df \tag{25-20}$$

or

$$\chi(t_d, f_d) = \int_{-\infty}^{\infty} h(t_d - t) a(t) e^{j2\pi f_d t} \, dt \tag{25-21}$$

where the functions $A(f)$ and $a(t)$, $H(f)$ and $h(t)$, are Fourier transform pairs describing the signal and filter, respectively.

The transform relationships, Eqs. (25-20) and (25-21), governing the time-frequency response function are similar to those which relate far-field antenna patterns to aperture illumination. Hence data derived for waveforms can be applied to antennas, and vice versa, by interchanging analogous quantities between the two cases. There is a significant difference between waveform and antenna response functions and constraints, however, because the two waveform functions (in time delay and frequency) are dependent upon each other through the Fourier transform. The two antenna patterns (in θ and ϕ coordinates) are essentially independent of each other, depending on aperture illuminations in the two spatial coordinates x and y. Further differences arise from the two-way pattern and gain functions applicable to the antenna case.

The ability of the radar to form distinguishable response peaks on closely spaced targets can be illustrated by plotting the composite waveform of two equal targets with varying rf phase difference (Fig. 25-28). With 0.5-pulse-width sep-aration, separate peaks appear only when the phase difference is near 180°. Separations between 1.0 and 1.25 give distinguishable peaks over most phase angles, and at 1.5 pulse widths the two tar-gets are always distinguishable (separately detect-able and measurable with small interaction). For the smoothly shaped pulse with no side lobes, shown in the figure, the targets become resolvable at separations between 1.25 and 1.5 times the half-power pulse width. In angle, with a scanning two-way antenna pattern, the corresponding resolu-tion is between 0.9 and 1.1 times the half-power (one-way) beamwidth.

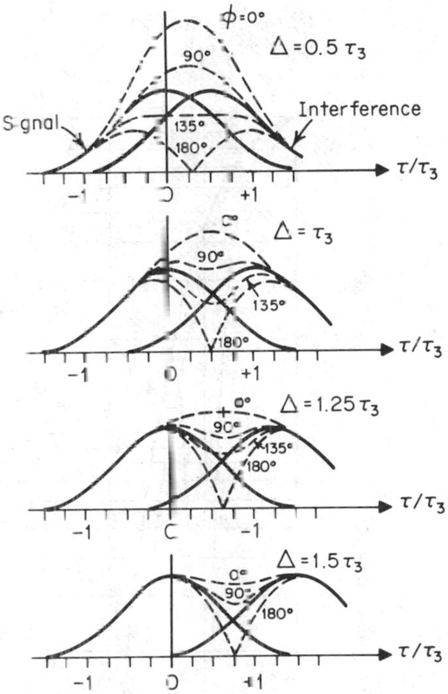

When a large target is present, greater separa-tion is needed to resolve a smaller target. How-ever, because of the steep skirts attainable in most radar response functions, the separation need sel-dom exceed twice the pulse width, or 1.5 times the beamwidth, unless the side-lobe response from the larger target obscures the smaller one. Thus the half-power width of the response func-tion can provide a measure of resolution, subject to modification by a factor near unity to account for targets of different amplitudes. Other mea-sures of resolution, such as Woodward's time and frequency resolution constants[36] (extended, by analogy, to angle) will also be shown to be closely related to the half-power widths.

Fig. 25-28. Resolution between adjacent tar-gets with separation Δ and rf phase difference ϕ.

27. Antenna and Waveform Func-tions. The resolution properties of response functions generated by continuous apertures and waveforms can be summarized by a few simple parameters. In a coordinate z, which may represent time delay, frequency, or angle, the response function $\chi(z)$ can be described by its width at significant levels below the peak response χ_m:

z_3, the half-power (3-dB) width

z_6, z_{10}, z_{20}, at the 6-, 10-, or 20-dB points

Figure 25-29 shows plots of three typical waveform functions, from which these widths can be measured.

Other measures of resolution are

Effective width for noise or random clutter: $\quad z_n \equiv \dfrac{1}{\chi_m^2} \displaystyle\int_{-\infty}^{\infty} \chi^2(z)\, dz$ $\qquad$ (25-22)

rms response width: $\quad z_{\text{rms}} \equiv 2\pi \left[\dfrac{\displaystyle\int_{-\infty}^{\infty} z^2\chi^2(z)\, dz}{\displaystyle\int_{-\infty}^{\infty} \chi^2(z)\, dz} \right]^{1/2}$ $\qquad$ (25-23)

Woodward resolution constant: $\quad z_w \equiv \left[\displaystyle\int_{-\infty}^{\infty} |R(z)|^2\, dz \right] \Big/ \chi_m^2$ $\qquad$ (25-24)

Equation (25-24) is a measure of $R(z)$, the total amount of response (or ambiguity) of the auto-correlation of χ in the z coordinate, where

$$R(z) = \int_{-\infty}^{\infty} \chi(s)\chi^*(s + z)\, ds$$

Table 25-5 lists the resolution widths for several functions in terms of waveform coordinates t and f. These coordinates can be interchanged, or antenna coordinates θ and x/λ can be substituted.

If the first three functions of Table 25-5 are excluded, a remarkable similarity between the several tapered functions can be seen. These have almost identical main lobes (at least out to the

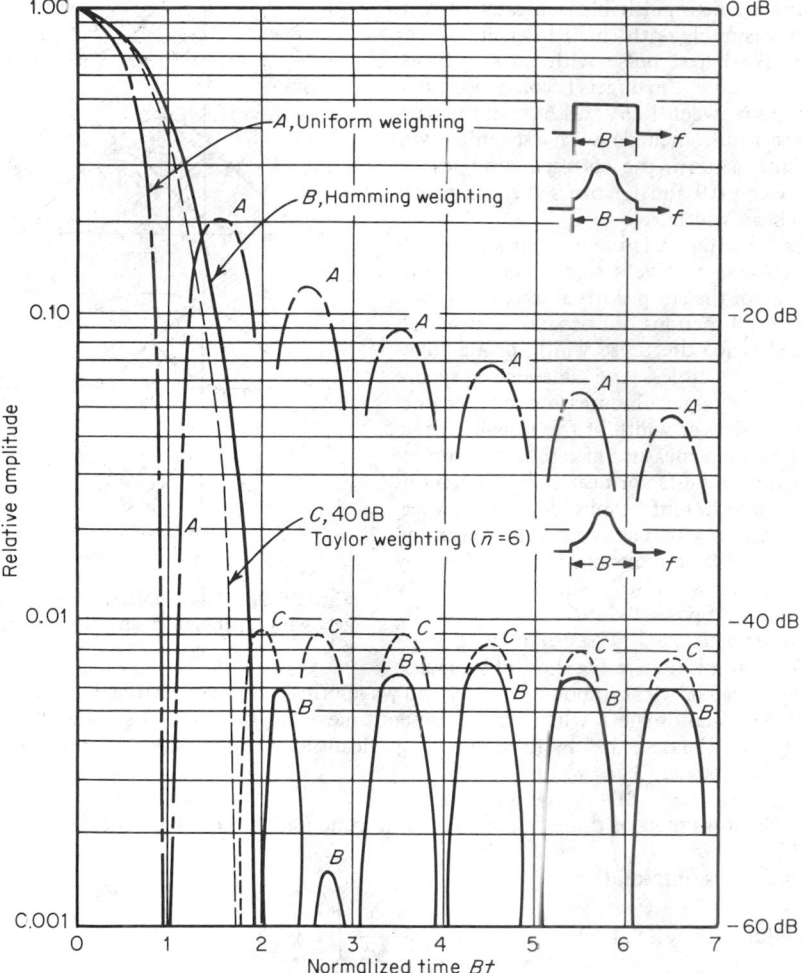

Fig. 25-29. Comparison of waveform functions.[38]

−10-dB points), with $\tau_n \approx 1.05\tau_3$ and $\tau_w \approx 1.3\tau_3$. The significant differences are side-lobe levels and presence of extended lobes in frequency or time domains (or equivalent antenna coordinates).

28. Ambiguity Functions of a Single Pulse. The time-frequency response function for a system whose filter is matched to the waveform is the ambiguity function for that waveform. This function describes the response to objects at different time delays and frequencies (doppler shifts). The ambiguity function of a simple pulse of width τ (without phase modulation or coding) consists of a single main lobe whose base width is 2τ along the time axis and whose

Table 25-5. Resolution Properties of Different Waveforms

Waveform	Spectrum	Side-lobe level, dB	$\tau_3 B_3$	$\tau_3 B$	τ_6/τ_3	τ_{10}/τ_3	τ_{20}/τ_3	τ_n/τ_3	$\alpha/\tau_3 = \tau_{rms}/t_3$	τ_w/τ_3
Rectangular (width τ)	$\dfrac{\sin \pi f \tau}{\pi f \tau}$	∞	0.89	∞	1.00	1.00	1.00	1.00	1.81	0.67
$\dfrac{\sin \pi t B}{\pi t B}$	Rectangular (width B)	14	0.89	0.89	1.36	1.66	*	1.13	∞	1.13
$\dfrac{\sin^2(\pi t B/2)}{(\pi t B/2)^2}$	Triangular (width B)	26	0.37	1.28	1.39	1.72	2.3	1.05	2.70	1.41
$\dfrac{\cos(\pi t/\tau)}{1 - 4 f^2 \tau^2}$ $(t_3 = 0.5\tau)$	$\dfrac{\cos \pi f \tau}{1 - 4 f^2 \tau^2}$	∞	0.59	∞	1.33	1.60	1.87	1.00	2.27	1.07
$\dfrac{\cos^2(\pi t/\tau)}{(\tau_3 - 0.37\tau)}$	$\dfrac{\sin \pi f \tau}{(1 - f^2 \tau^2)\pi f \tau}$	∞	0.53	∞	1.37	1.70	2.18	1.02	2.42	1.30
$\dfrac{\cos \pi t B}{1 - 4 t^2 B^2}$	$\cos(\pi f/B)$	23	0.59	1.19	1.40	1.70	2.18	1.05	2.66	1.27
$\dfrac{\sin \pi t H}{(1 - t^2 H^2)\pi t B}$	$\cos^2(\pi f/B)$	32	0.53	1.44	1.39	1.73	2.28	1.08	2.52	1.36
Taylor	Taylor weighted (max. width B)	30	0.52	1.12	1.40	1.74	2.26	1.05	∞	1.34
Taylor	Taylor weighted (max. width B)	40	0.50	1.24	1.40	1.76	2.35	1.05	∞	1.39
Hamming	$0.54 + 0.46 \cos\left(\dfrac{2\pi f}{B}\right)$	43	0.51	1.12	1.43	1.77	2.35	1.06	∞	1.39
Gaussian $\exp(-t^2/2\sigma_t^2)$	Gaussian $\exp(-f^2/2\sigma_f^2)$	∞	0.44	∞	1.41	1.82	2.58	1.06	2.67	1.50

* Side-lobe level less than 20 dB down.

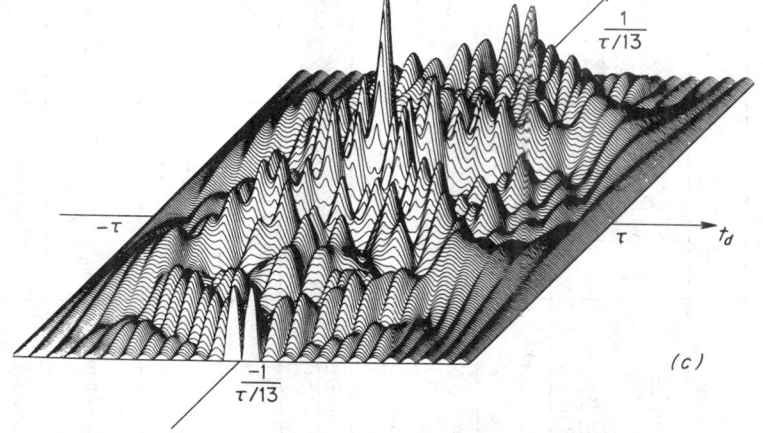

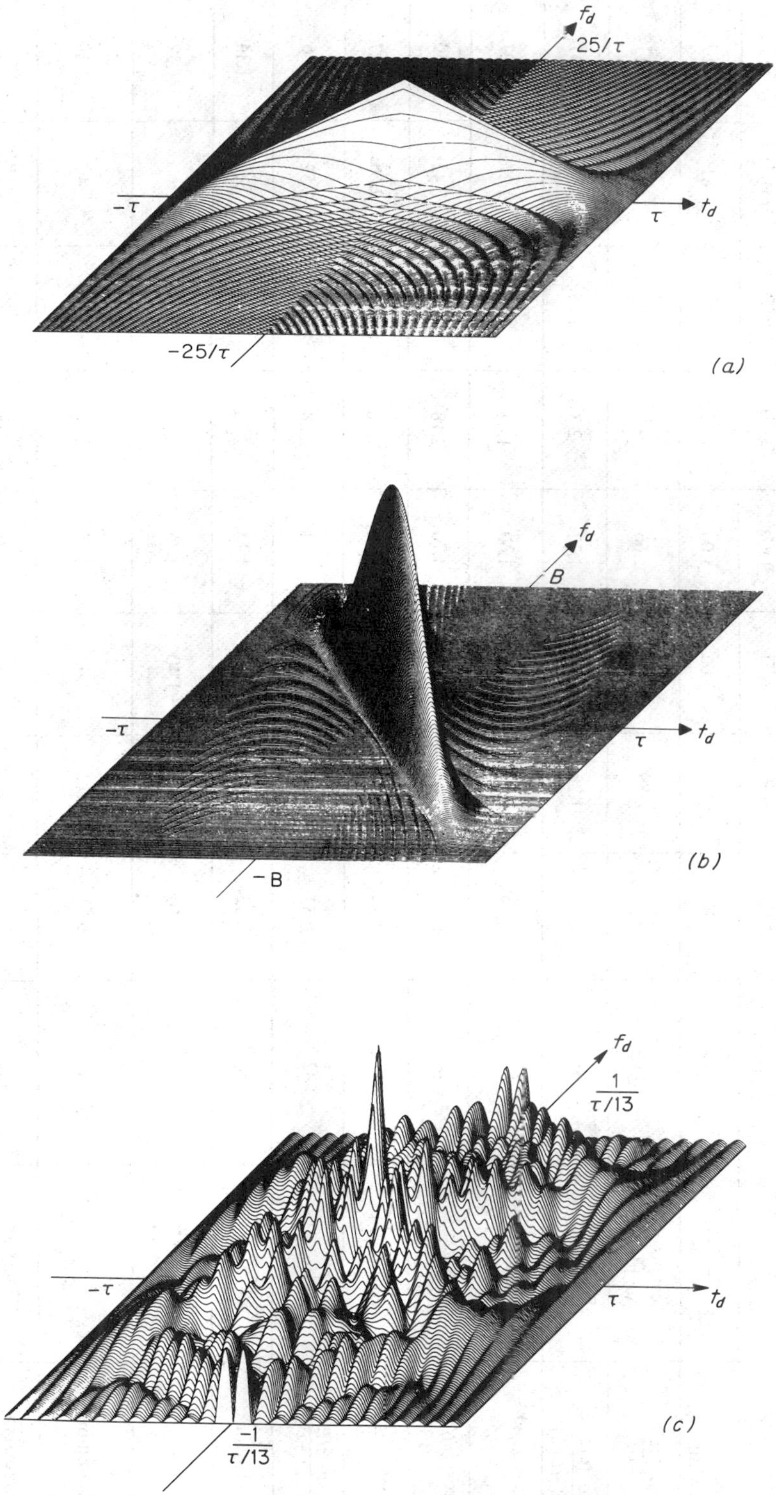

Fig. 25-30. Ambiguity functions of single pulses: (a) response to constant carrier pulse with a rectangular envelope; (b) chirp response for bilateral Hamming weighting; (c) response for 13-element Barker code.[37]

frequency response has a lobe structure dependent on the discontinuities in the time function or its derivatives (Fig. 25-30a).

Introduction of phase modulation during the pulse broadens the frequency spread of the function and narrows the response along the time axis. This is the principle of pulse compression, of which the most common form is linear FM (or chirp), shown in Fig 25-30b. With the linear FM function, very low side lobes can be obtained in regions both sides of the main, diagonal response ridge. Along this ridge, however, the response falls off slowly from its central value, and targets separated by almost one transmitted pulse width will be detected if they are offset in frequency by the correct amount. Pseudo-random phase coding can generate a single narrow spike in the center of the ambiguity surface (Fig. 25-30c), at the expense of large-amplitude side lobes elsewhere on the ambiguity surface.

The plots of Fig. 25-30 illustrate an important property of the ambiguity function: the total volume under the surface describing the square of the function, $|\chi(-_d, f_d)|^2$ is equal to the signal energy (or to unity, when normalized) for all waveforms. Compression of the response along the time axis must be accompanied by an increase in response elsewhere in the time-frequency plane, either along a well-defined ambiguity ridge (for linear FM) or in numerous random lobes (for random-phase codes). For mismatched filters, the response function (cross-ambiguity function) has similar properties, although the central signal peak may be reduced in amplitude.

29. Ambiguity Functions of Pulse Trains. The principle of constant volume under the squared ambiguity function is also applicable to pulse trains. A train of noncoherent pulses of width τ, with interpulse period $t_p = 1/f_r \gg \tau$, merely generates a repeating set of surfaces similar to Fig. 25-30a at intervals t_p in time. The added volume equals the energy of the additional pulses, and if the energy is normalized to unity (by division among the several pulses), the amplitude of each response peak is reduced proportionately. The location of peaks and associated side lobes is shown in Fig. 25-31a. When the signal is coherent over an observation time $t_0 = nt_p$, the time response of the matched filter stretches the ambiguity function to $\pm t_0$ along the time axis.

Within the central lobes, this response is concentrated in spectral lines separated by f_r and approximately $1/t_0 = f_r/n$ wide in frequency (Fig. 25-31b). Near the ends of the ambiguity function, where the matched-filter impulse response overlaps only $n' < n$ pulses of the received train, the lines broaden to width f_r/n', and at the end of the ambiguity function, where $n' = 1$, no line structure remains. If the repetition rate is increased, holding constant the pulse width and number of pulses in the train, t_0 is decreased and the ambiguity volume is redistributed into a smaller number of broader lines. A decrease in pulse width, such that t_0 is restored to its original value and n is increased, leads to a broader overall ambiguity function, with the original number of lines in frequency but with narrower and more numerous response bands along the time axis (Fig. 25-31c).

30. Resolution with CW Transmissions. A cw radar, observing a target over a time interval t_0, can be regarded as having transmitted a single pulse of width $\tau = t_0$. Generally, this interval is sufficiently long so that all targets and clutter sources within the beam are included within the main response lobe in time, and frequency resolution is relied upon to distinguish targets. Introduction of a linear FM sweep produces a single ambiguity ridge of the type shown in Fig. 25-30c, with a very long time span and a correspondingly narrow frequency width. Time resolution, along the axis, is roughly the reciprocal of the total frequency sweep. Repeated FM sweeps at intervals $t_p \ll t_0$ produce a diagonal-line structure.

31. Resolution of Targets in Clutter. The choice of radar waveform is often dictated by the need to resolve small targets (aircraft, buoys, or projectiles) from surrounding clutter. The clutter power at the filter output is found by integrating the response function over the clutter region, with appropriate factors for variable clutter density, antenna response, and the inverse-fourth-power range dependence included in the integrand. Signal-to-clutter ratio (S/C) for a target on the peak of the radar response is then given by

$$S/C = \frac{\sigma G_t(0) G_r(0) |\chi(0, 0)|^2}{\int_v \eta_v(\theta, \phi, f_d, t_d) G_t(\theta, \phi) G_r(\theta, \phi) |\chi(f_c, t_c)|^2 (R/R_c)^4 \, dv} \tag{25-25}$$

where σ = target cross section, η_v = clutter reflectivity, R/R_c = target-to-clutter range ratio, and v = four-dimensional volume containing clutter. The usual equations for S/C ratio (Par. **25-41**) are simplifications of Eq. (25-25) for various special cases, e.g., surface clutter and homogeneous clutter filling the beam. Clearly, the S/C ratio is improved by choosing a waveform and filter such that $\chi(f_d, t_d)$ is minimized in clutter regions while maintaining a high value $\chi(0, 0)$ for all potential target positions. In a search radar, a two-dimensional bank of filters would be constructed to cover the intervals in delay and doppler occupied by targets, and the clutter power for each of these filters would then be evaluated using Eq. (25-25).

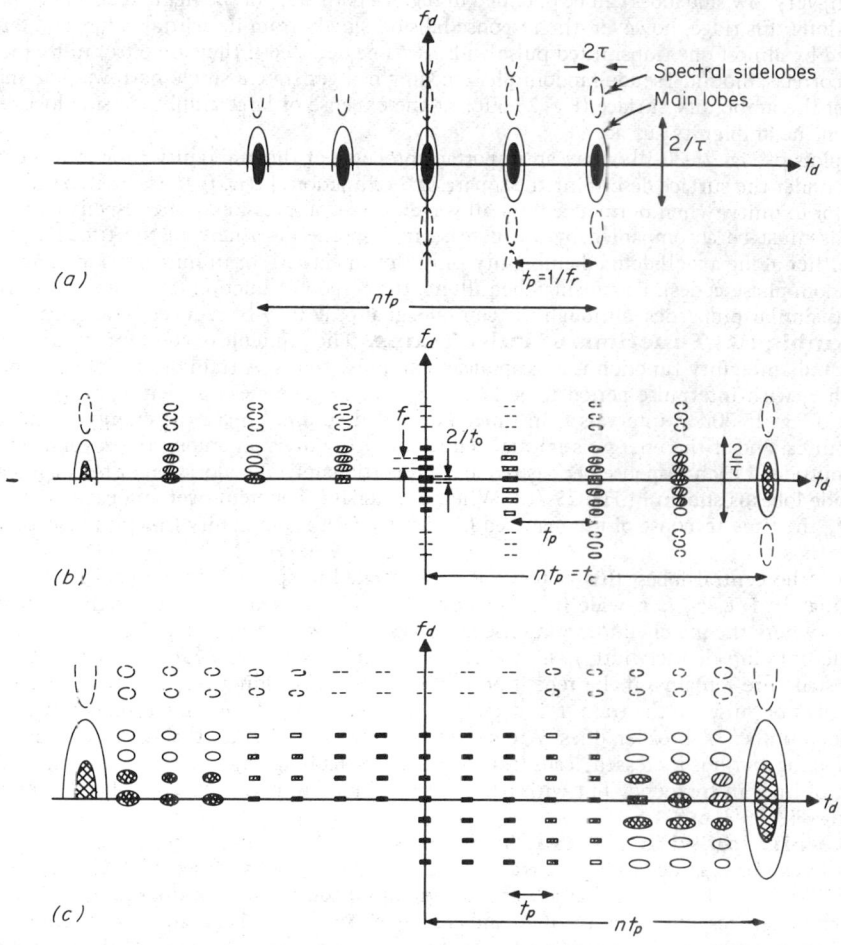

Fig. 25-31. Location of response lobes for uniform pulse trains: (a) noncoherent pulse train; (b) coherent pulse train; (c) coherent, reduced τ, higher f_r.

32. Basic Process of Measurement. A radar measures target position in a given coordinate by identifying one or more resolution cells containing detectable target signals and then interpolating to refine the estimate of location. A very simple estimator consists of a bank of contiguous resolution cells (range gates, doppler filters, or beam positions) in which the center of the cell containing the strongest signal above a detection threshold is identified as the target location. Although noise may lead to identification of the wrong cell in cases where a target is in a region of overlapping response, the error seldom exceeds $\pm$ one-half cell width, and an rms error of 0.2 to 0.3 times the 3-dB cell width can be expected on targets strong enough to be detected reliably.

More accurate estimates are made by interpolating between overlapping cells, or by placing the peak of the response function at the target location. Ideally, this is done by forming a function which is the derivative of the response function in the measured coordinate, and adjusting its position until the target gives a null output. If this difference channel null occurs within the region of peak response in the main response, or Σ, channel, the target must lie exactly at the peak of the Σ channel, where the derivative is zero. Since the derivative is an S-shaped (odd) function with a linear region each side of the null, small deviations in target positions can be estimated directly from the difference-channel output. In an idealized monopulse angle estimator, two offset beam patterns (Fig. 25-32a) are formed by feed networks. Outputs of these beams are combined in a hybrid to produce an on-axis Σ pattern and a Δ pattern which is the derivative of the Σ (Fig. 25-32b). When the Δ voltage is normalized to Σ, a calibrated curve for Δ/Σ vs. target displacement θ is produced (Fig. 25-32c) which gives estimates of θ independent of signal strength. Noise or interference will introduce errors in the estimate which are inversely proportional to the slope of this curve and to the voltage ratio $\sqrt{S/N}$ or $\sqrt{S/I}$.

33. Ideal Estimators. The ideal estimator for noise achieves an output signal-to-noise ratio equal to the ratio of twice the signal energy reaching the aperture to input noise power density, $\mathcal{R}_0 = 2E/N_0$, using a matched filter and uniform aperture illumination for the Σ channel. At the same time, it maximizes the Δ slope separately in each coordinate, using a Δ channel identical with Σ in three coordinates but following the derivative of Σ in the measured coordinate. The maximum Δ slope is proportional to the second derivative at the peak of the Σ channel, which is determined by the rms width of the transform of the response function.

34. Practical Monopulse Estimators. The high side-lobe level of the uniformly illuminated antenna is seldom satisfactory in radar, and horn-fed apertures must employ tapered illumination to avoid excessive spillover loss. The use of taper reduces the maximum gain from F_0^2 to $F_m^2 = \eta_a F_0^2$, reduces the energy extracted from the incident wave from $\mathcal{R}_0$ to $\mathcal{R}_m = \eta_a \mathcal{R}_0$, reduces the Δ slope from K_0 to $K = K_r K_0$, reduces the Σ and Δ side-lobe levels, and increases the Σ beamwidth.

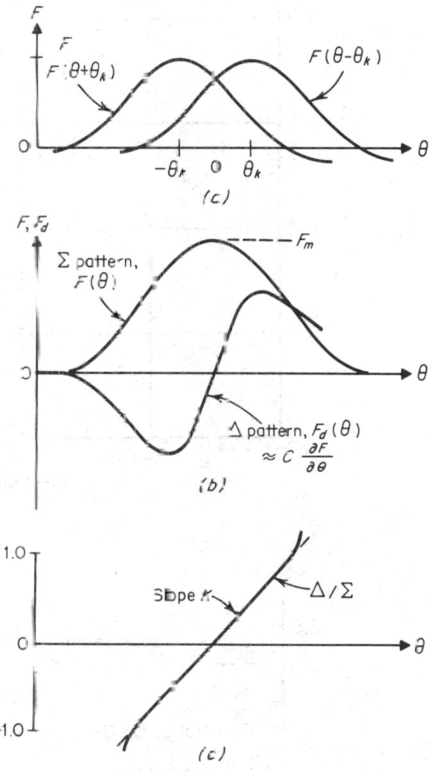

Tables of these parameters, for many different illumination functions, are available.[8,39] Cases of common interest are summarized in Fig. 25-33 and Table 25-6. The parameters plotted in Fig. 25-33 are the relative difference slope $K_r = K/K_0$ and the monopulse slope normalized to Σ-channel gain and beamwidth.

Table 25-6 shows the performance characteristics of practical horn-fed apertures, in terms of H-plane (x-coordinate) and E-plane (y-coordinate) slopes and both Σ- and Δ-channel side-lobe ratios. Use of rectangular apertures is assumed, but the absolute levels of performance are essentially unchanged if the corners are removed to produce circular apertures (η_a and K_r values will be increased because these are referred to the reduced potentialities of the smaller circular aperture). Additional noise errors are produced if the Σ beam is not kept exactly on the target.

35. Scanning Systems. Angle estimates can also be made by scanning a single beam around (conical scan) or across (linear scan) the target obtaining sequentially the offset samples which are gathered simultaneously in monopulse radar. Two-dimensional measurement by sequential scanning is much less efficient than monopulse, because the target is not fully illuminated by the transmitting and receiving Σ patterns, the difference slope is lower, and the energy must be shared between two orthogonal measurements. Figure 25-34 shows values of conical-scan errors for one- and two-way conical-scan systems, using gaussian and (sin x)/x beam patterns, as functions of offset angle.

The one-dimensional linear-scan case, encountered in search and height-finding radars, gives the following accuracy, expressed as a function of actual signal energy ratio, number of hits per beamwidth, and on-axis signal-to-noise ratio $(S/N)_m$:

$$\sigma_\theta = \theta_3/(k_p \sqrt{\mathcal{R}}) \approx \theta_3/2\sqrt{n(S/N)_m} \tag{25-26}$$

The constant k_p varies from 1.18 for one-way operation with a gaussian pattern to 1.76 for two-way operation with a (sin x)/x pattern, but the associated beam-shape loss varies so that the

Fig. 25-32. Basic angular measurement process. (a) Offset beam patterns; (b) sum and difference patterns, (c) calibration curve for Δ/Σ.

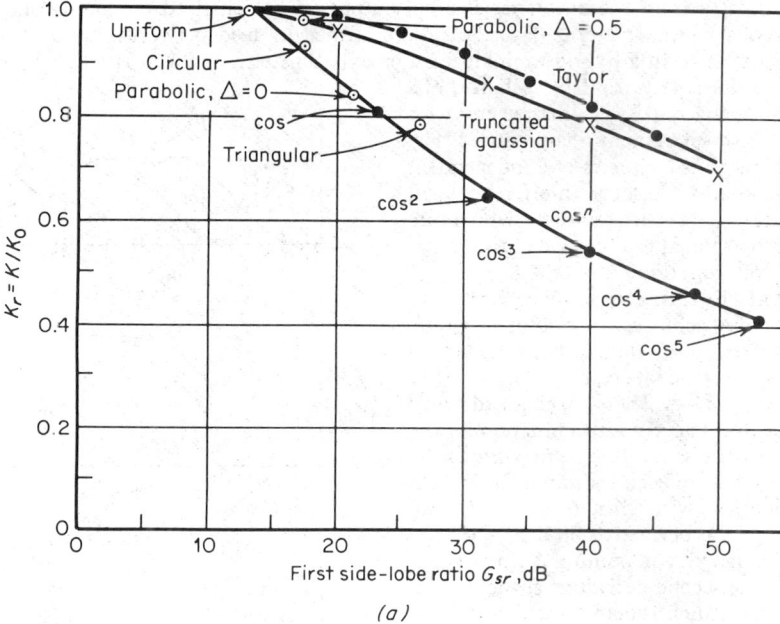

(a)

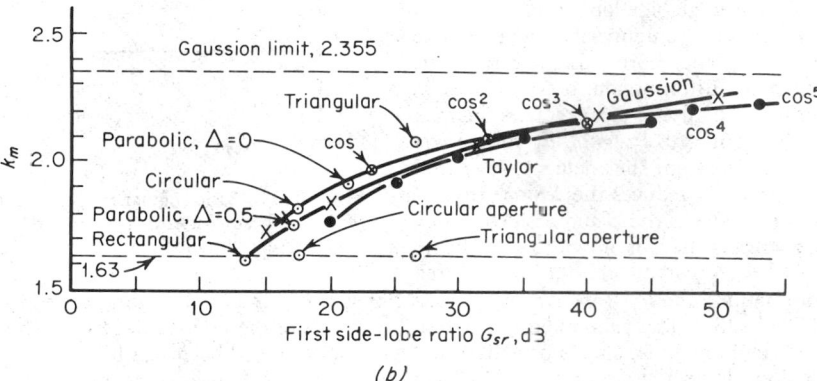

(b)

Fig. 25-33. Difference slopes vs. Σ-channel side-lobe ratio: (a) relative-difference slope; (b) normalized monopulse slope.[8]

approximation shown is accurate to within 15% in σ_θ for all cases. It is assumed that an optimum estimation process is used on the received pulse train and that n is large (≥ 10).

36. Range Estimators. Radar-range R measurements are made by estimating the round-trip time delay t_d between transmission and reception of a signal, and converting to range, using the velocity of light in vacuum c

$$R = t_d c/2$$

$$c = \begin{cases} 2.997925 \times 10^8 \text{ m/s} \\ 9.9835692 \times 10^8 \text{ ft/s} \\ 1.618750 \times 10^5 \text{ nmi/s} \end{cases}$$

Thus the delay per unit range is

$$t_d/R = 2/c = \begin{cases} 6.671281 \text{ ns/m} \\ 2.033410 \text{ ns/ft} \\ 12.35521 \text{ }\mu\text{s/nmi} \end{cases}$$

Table 25-6. Monopulse Feed Horn Performance

Type of horn	η_a	H-plane $K_r\sqrt{\eta_y}$	H-plane k_m	E-plane $K_r\sqrt{\eta_x}$	E-plane k_m	G_{Σ}, dB	G_{Δ}, dB	Feed shape
Simple four-horn	0.58	0.52	1.2	0.48	1.2	19	10	
Two-horn dual-mode	0.75	0.68	1.6	0.55	1.2	19	10	
Two-horn triple-mode	0.75	0.81	1.6	0.55	1.2	19	10	
Twelve-horn	0.56	0.71	1.7	0.67	1.6	19	19	
Four-horn triple-mode	0.75	0.81	1.6	0.75	1.6	19	19	

SOURCE: D. K. Barton and H. R. Ward, "Handbook of Radar Measurement," copyright 1969. Reprinted by permission of Prentice-Hall, Inc., Englewood Cliffs, N.J.

According to the ideal process described earlier, the arrival time of a signal should be estimated by passing the signal through a matched filter, differentiating it, and measuring the point at which the derivative passes through zero. The accuracy will then be determined by the energy ratio and the rms bandwidth. Matched filters for many waveforms and spectra can be closely approximated by practical receivers, with resulting bandwidths, 3-dB pulse widths, and spectral widths, as shown in Table 25-7.

In pulse-compression radar, the transmission is often a rectangular pulse of width τ and bandwidth $B \gg 1/\tau$, with an approximately rectangular spectrum. Weighting is used in the receiver to reduce side lobes, giving a mismatched filter with reduced S/N and Δ slope. The range error with mismatched filter is

$$\sigma_t = 1/K\sqrt{\mathcal{R}} = \tau'/k_t \sqrt{2S/N} = (\sigma_t)_{min}/K_r \qquad (25\text{-}27)$$

where τ' is the 3-dB width of the output waveform, and the slope factors are the waveform analogs of the corresponding angular slopes shown in Fig. 25-33. Thus, for the special case of uniform-spectrum pulse compression, k_t values between 1.6 and 2.3 can be used to describe the difference slope normalized to the 3-dB width of the compressed pulse.

Another case of special interest is that of the band-limited (but uncoded) rectangular pulse (Ref. 2, p. 468) for which the rms bandwidth shown in Table 25-7 is $\beta = \sqrt{2B/\tau}$. Then the range error is

$$\sigma_t = \sqrt{\tau/2B\mathcal{R}} = \tau/\sqrt{2B\tau\mathcal{R}} \qquad (25\text{-}28)$$

Other studies[40] have shown a theoretical limit for unrestricted bandwidth, given by

$$\sigma_t = \sqrt{2\tau/\mathcal{R}} \qquad (25\text{-}29)$$

Table 25-7. Waveform and Spectrum Measurement Parameters

Input-signal description		Half-power widths		Rms widths	
Spectrum	Waveform	B_3	τ_3	β	α
$\dfrac{\sin \pi f\tau}{\pi f\tau}$ $(\|f\| < B/2)$	Band-limited rectangular (width τ)	$\dfrac{0.89}{\tau}$	τ	$\sqrt{\dfrac{2B}{\tau}}$	$1.81\,\tau$
Time-limited rectangular (width B)	$\dfrac{\sin \pi tB}{\pi tB}$ $(\|t\| < \tau/2)$	B	$\dfrac{0.89}{B}$	$1.81\,B$	$\sqrt{\dfrac{2\tau}{B}}$
$\dfrac{\sin^2(\pi f\tau/2)}{(\pi f\tau/2)^2}$	Triangular (width τ)	$\dfrac{1.27}{\tau}$	$0.29\,\tau$	$\dfrac{3.45}{\tau}$	$1.28\,\tau$
Triangular (width B)	$\dfrac{\sin^2(\pi tB/2)}{(\pi tB/2)^2}$	$0.29\,B$	$\dfrac{1.27}{B}$	$1.28\,B$	$\dfrac{3.45}{B}$
$\dfrac{\cos \pi f\tau}{1 - 4f^2\tau^2}$	$\cos \dfrac{\pi t}{\tau}$	$\dfrac{1.18}{\tau}$	$0.50\,\tau$	$\dfrac{3.14}{\tau}$	$1.14\,\tau$
$\cos \dfrac{\pi f}{B}$	$\dfrac{\cos \pi tB}{1 - 4t^2B^2}$	$0.50\,B$	$\dfrac{1.18}{B}$	$1.14\,B$	$\dfrac{3.14}{B}$
$\dfrac{\sin \pi f\tau}{(1 - f^2\tau^2)\pi f\tau}$	$\cos^2\dfrac{\pi t}{\tau}$	$\dfrac{1.43}{\tau}$	$0.37\,\tau$	$\dfrac{3.65}{\tau}$	$0.89\,\tau$
$\cos^2\dfrac{\pi f}{B}$	$\dfrac{\sin \pi tB}{(1 - t^2B^2)\pi tB}$	$0.37\,B$	$\dfrac{1.43}{B}$	$0.89\,B$	$\dfrac{3.65}{B}$
Gaussian $\exp(f^2/2\sigma_f^2)$ $= \exp(-2\pi^2 f^2\sigma_t^2)$	Gaussian $\exp(-t^2/2\sigma_t^2)$ $= \exp(-2\pi^2 t^2\sigma_f^2)$	$1.66\,\sigma_f$	$.66\,\sigma_t$	$\begin{array}{l}4.45\,\sigma_f \\ = 2.67\,B_3\end{array}$	$\begin{array}{l}4.45\,\sigma_t \\ = 2.67\,\tau_3\end{array}$

To realize the optimum accuracy of a band-limited rectangular pulse with $B\tau \gg 1$, the receiving system must adapt its bandwidth to the actual energy ratio.[8]

Practical range estimators, especially those operating on pulse trains with low single-pulse energy and S/N, take the form of a split-gate tracker (see Par. **25-56**). The multiplication of a waveform $a(t)$ by the split-gate function, followed by averaging, is a form of differentiation, although the composite filter-differentiator function may be quite different from the ideal matched-filter estimator.

Figure 25-35 shows the performance of rectangular split gates, of varying width τ_g, in terms of the product $K\tau_{3e}$, where τ_{3a} refers to the 3-dB width of the input waveform to the receiver and

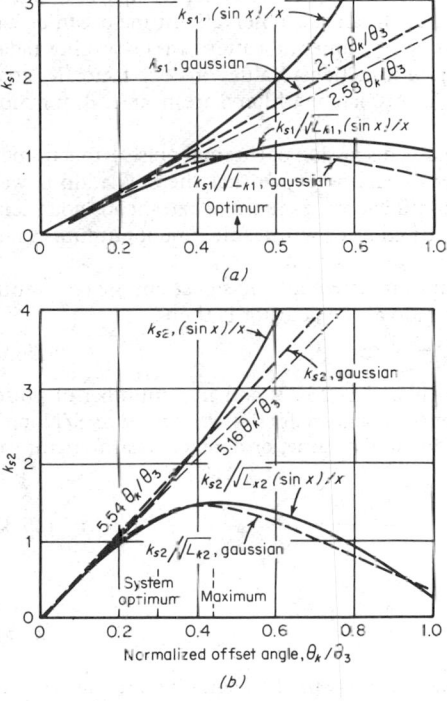

(a)

(b)
Normalized offset angle, θ_k/θ_3

Fig. 25-34. Conical-scan error slopes: (a) one-way case; (b) two-way case.[3]

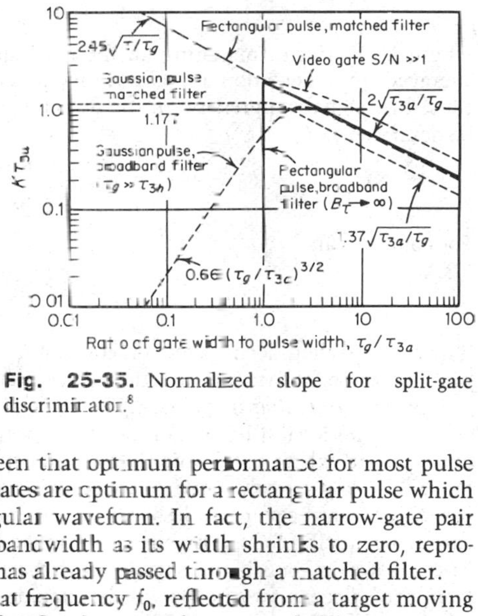

Fig. 25-35. Normalized slope for split-gate discriminator.[8]

τ_g is the total width of the split-gate pair. It is seen that optimum performance for most pulse shapes requires $\tau_g \approx 1.4\tau_{3a}$ but that very narrow gates are optimum for a rectangular pulse which has been optimally filtered to produce a triangular waveform. In fact, the narrow-gate pair approaches an ideal differentiator of unlimited bandwidth as its width shrinks to zero, reproducing the ideal estimator for any signal which has already passed through a matched filter.

37. Doppler Estimators. A transmission at frequency f_0, reflected from a target moving with radial velocity v_r, will be received at $f_0 + f_d$. The change in frequency f_d is known as the doppler shift:

$$f_d = f_0\left(\frac{c - v_r}{c + v_r} - 1\right) = \frac{-2f_0 v_r}{c}\left(1 - \frac{v_r}{c} + \frac{v_r^2}{c^2} - \cdots\right)$$

$$\approx -\frac{2f_0 v_r}{c} = -\frac{2v_r}{\lambda} \tag{25-30}$$

A measurement of f_d can be translated to radial velocity

$$v_r = -\frac{f_d c}{2f_0}\left(1 - \frac{f_d}{2f_0} + \frac{f_d^2}{4f_0^2} - \cdots\right) \approx -\frac{f_d c}{2f_0} = -\frac{f_d \lambda}{2} \tag{25-31}$$

In most cases, the signal bandwidth is small enough relative to f_0 for the doppler shift to be regarded as a simple displacement of the spectrum relative to that transmitted.

The spectrum of a typical coherent pulse train is shown in Fig. 25-24. The spectral envelope is determined by the waveform of an individual pulse, and can provide a coarse frequency estimate if applied to an i.f. discriminator. This discriminator would ideally be matched to the derivative of the spectral envelope $A_1(f)$, in which case the error in frequency estimate would be

$$\sigma_f = 1/\alpha\sqrt{\mathcal{R}}$$

with α representing the rms width of an individual pulse. Frequency error on rectangular pulses with mismatched filters can be estimated using the analogy to antennas, as described in connection with Eq. (25-27) for time measurements, but with time and frequency interchanged.

Of greater practical significance is the measurement of doppler shift on the fine-line spectrum, where the line width is a function of a pulse-train duration t_0 and its envelope shape, or of phase-stability factors in the radar equipment. For stable radar, an rms observation time α can be calculated using the pulse-train envelope (as determined by antenna pattern in a scanning radar, for instance). Table 25-7 shows values of α for typical functions. Values of slope factor K_f, analogous to K in angular or range measurement with mismatched filters, are near $1/B_3$ for most signals and discriminators.

38. Signal-Processing Losses. The preceding discussion of thermal-noise errors in measurement has been based on the optimum use of signal energy, so that the S/N ratio is well above unity before envelope detection or similar nonlinear processing, and extraneous noise samples are excluded from the averaging process. Increased error will result if nonoptimum conditions apply, as in the following cases.

Detector Loss. In most radars, it is impractical to integrate all the signal energy coherently in a matched filter before the detectors. The S/N ratio at the detector is then

$$S/N = \mathcal{R}_1/2L_m = \mathcal{R}/2nL_m \tag{25-32}$$

where L_m = loss in matching i.f. filter to single pulse (Fig. 25-23) and n = number of pulses averaged after detection in performing measurement. The reduction in effective S/N ratio caused by signal suppression in the detector for monopulse, time, or doppler measurement can be described as L_x:

$$L_x = \frac{(S/N) + 1}{S/N} = \frac{S + N}{S} \tag{25-33}$$

For conical scan,

$$L_x = \frac{2(S/N) + 1}{S/N} = \frac{2S + N}{2S} \tag{25-34}$$

Thus, even when a postdetection integrator or data filter is used to combine all the received samples, the effective energy ratio will be reduced by L_x.

Matching or Collapsing Loss. The loss L_m in Eq. (25-32) implies that the i.f. filter passes more noise than an ideal filter would relative to peak signal. Usually, the i.f. bandwidth is wider than optimum, in which case more than one independent sample of noise error is available on each pulse. Use of a narrow video filter following the detector, or of a matched range gate and low-pass filter, can recover this loss in data, except to the extent L_m has contributed to detector loss through reducing S/N. The effective number of measurement samples becomes $nB\tau$ for these cases. However, if B is too narrow, or if the range gate is wider than the pulse, noise samples from the adjacent range cells will be included in the output data, and error will increase proportionately. The average S/N ratio during each output sample should be used in the error expressions for these cases: $S/N = \mathcal{R}_1/L_m$.

Time Sharing of Signals. It was noted that the energy available in conical-scan measurements of each angular coordinate was only half the received energy. A similar situation exists in some monopulse configurations, where a single Δ channel is shared between the two angular-error signals and $\mathcal{R}$ should be reduced accordingly.

39. Other Sources of Error. Thermal noise is only one component of error in radar measurement, although it has received much attention because it lends itself to mathematical analysis. Other errors are sometimes random and noiselike but may also appear as fixed bias, slow drifts from a calibrated setting, sinusoids, or other functions of time or target motion.

Errors from random clutter and noiselike interference can often be analyzed as though the interfering signals were thermal noise. Care must be taken to ensure that the interference is approximately normally distributed (Rayleigh amplitude distribution) and homogeneous over the resolution cells surrounding the target. Also, since clutter may be correlated over time intervals longer than one repetition period, the number of independent samples may be less than the number of pulses averaged n. When a target is observed in land clutter or certain types of interference, signal-to-interference ratio S/I may follow a broad, lognormal distribution and large peak errors are possible.[41] Special editing procedures may be necessary to maintain tracking even when the median S/I is large.

Multipath errors, caused by reflection of target signals from ground or sea surfaces, become a serious problem for low-angle targets. Primarily affecting elevation angle, these reflections may also cause significant errors in other coordinates of precision tracking systems.[8]

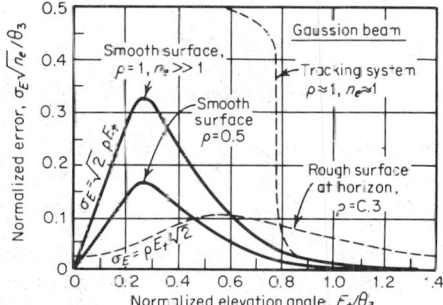

Fig. 25-36. Elevation multipath error vs. target elevation.[8]

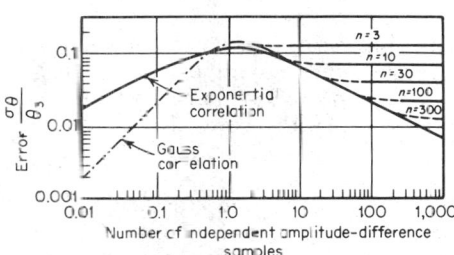

Fig. 25-37. Scintillation error in a scanning radar. Horizontal lines on the right indicate limits derived by Swerling for the case 2 target with different values of n.[8]

For targets within a beamwidth of the horizon, and for any interfering target near enough in range and angle to be unresolvable, the angular error is

$$\sigma_\theta = \theta_i / \sqrt{2(S/I)n_e} \qquad (25\text{-}35)$$

where θ_i = angular separation of interference (image or reflection region) from target and S/I = signal-to-interference power ratio. Figure 25-36 shows the low-angle elevation error over smooth and rough reflecting surfaces, for a gaussian beam and derivative Δ pattern. Below $0.8\theta_3$ over a smooth surface, the tracking systems become unreliable because they may at times lock on the image and will generally oscillate randomly between target and image. However, over rough or absorbing surfaces, tracking remains possible, with large elevation errors, to the horizon.

Errors from target glint (Par. **25-15**) are described in detail in the literature.[8,29] While glint is a factor in all measurements of extended targets, scintillation error depends upon the sequential sampling procedure and its relationship to target correlation time. The scintillation error on a Rayleigh target is

$$\sigma_\theta \approx \frac{\theta_3}{k_s} \sqrt{\frac{\beta_n}{2\pi^2 t_c f_s^2}} = \frac{0.225\theta_3}{f_s k_s} \sqrt{\frac{\beta_n}{t_c}} \qquad (25\text{-}36)$$

For typical values (f_s = 30 Hz, k_s = 1.5, β_n = 4 Hz, t_c = 0.1 s) the error is about $0.03\theta_3$. In search radar, the error depends on the ratio of time on target to correlation time, as shown in Fig. 25-37. The same curves apply to frequency-scanned antennas, where the ratio of frequency shift Δf, required for one-beam-width scan, to correlation frequency f_c is substituted for t_0/t_c.

40. Combination of Errors. An estimate of overall radar accuracy in any coordinate can be made by calculating separately the several error components discussed above, along with instrumental errors in the equipment and lag errors caused by target motion, and combining them in root-sum-square fashion:

$$\sigma^2 = \sigma_1^2 + \sigma_2^2 + \cdots$$

The process is repeated for each measured coordinate, and the major axes of an ellipsoid of error are described in terms of a radial error $\sigma_r = \sigma_t c/2$ and two orthogonal components $\sigma_n = R\sigma_\theta$ for elevation and traverse errors. Where smoothing or differentiation of radar outputs is performed, it may be necessary to form a composite error spectrum for each coordinate, so that the effect of the data filters on bias and noise at different frequencies can be evaluated.

41. Radar-Range Equations. *Signal-to-Noise Ratio.* The *signal power S* received by the radar antenna is calculated from transmitted power P_t and a series of factors that describe, basically, the geometry of the radar beam and target:

$$S = P_t G_t A_r \sigma/(4\pi)^2 R^4 \tag{25-37}$$

where G_t = transmitting antenna gain, A_r = effective receiving aperture area, σ = target cross section, and R = target range. Since the receiving aperture and gain are related by

$$A_r = G_r \lambda^2/4\pi \tag{25-38}$$

the signal power can also be given as

$$S = P_t G_t G_r \lambda^2 \sigma/(4\pi)^3 R^4 \tag{25-39}$$

In pulsed radar, P_t and S are the so-called peak power levels (actually the mean power over an rf cycle at the peak of the pulse envelope).

Radar noise is a combination of receiver-generated and environmental noise, extending over a broadband with a power density N_0:

$$N_0 = kT_s = kT_0 F_{n0} \tag{25-40}$$

where k = Boltzmann's constant = 1.38×10^{-23} W/(Hz·K), T_s = system noise temperature referred to the antenna terminals, T_0 = standard temperature (290 K) used in measuring noise factor, and F_{n0} = operating noise factor ($F_{n0} \equiv T_s/T_0$). The system noise temperature[10] is calculated from receiver noise factor F_n, line losses L_r, and antenna temperature T_a:

$$T_s = T_a + T_r + L_r T_e \tag{25-41}$$
$$T_a = (0.876 T_a' - 254)/L_a + 290 \tag{25-42}$$
$$T_r = T_{tr}(L_r - 1) \tag{25-43}$$
$$T_e = T_0(F_n - 1) \tag{25-44}$$

where T_r = temperature contribution of loss L_r, T_a' = sky temperature from Fig. 25-7, L_a = antenna ohmic loss, T_{tr} = physical temperature of the receiving line (290 K), and T_e = temperature contribution of the receiver. For $T_a \approx 290$ K, the operating noise factor $F_{n0} \approx F_n$, and $T_s \approx 290 F_n$. For antennas directed into space, F_{n0} can be much less than F_n, and T_s much less than 290 K.

The noise power at the i.f. output of the receiver will depend upon receiver bandwidth and gain, but this noise power is equivalent to an input power at the antenna terminals of

$$N = N_0 B = kT_s B_n$$

where B_n = noise bandwidth of i.f. filter. For a wide-band filter ($B_n\tau \gg 1$) the signal peak is not affected by the filter and $S/N = S/kT_s B_n$. In general, however, the S/N ratio at the receiver i.f. output is calculated from the ratio of received pulse energy $S\tau$ to noise density.

Ideal energy ratio for single pulse:

$$E_1/N_0 = P_t \tau G_t G_r \lambda^2 \sigma/(4\pi)^3 R^4 kT_s \tag{25-45}$$

Intermediate-frequency power ratio for single pulse:

$$S/N = E_1/N_0 L_m = P_t \tau G_t G_r \lambda^2 \sigma/(4\pi)^3 R^4 kT_s L_m \tag{25-46}$$

where L_m = i.f. filter matching loss shown in Fig. 25-23 and τ = pulse width. For a cw or coherent pulse radar which integrates over an observation interval t_0 in a predetection filter, the i.f. output S/N ratio is

Intermediate-frequency power ratio over interval t_0:

$$S/N = E/N_0 L_m = P_{av} t_0 G_t G_r \lambda^2 \sigma/(4\pi)^3 R^4 kT_s L_m \tag{25-47}$$

where P_{av} = average transmitter power ($P_t\tau f_r$ for pulsed radar) and L_m = matching loss of the filter to the entire waveform over t_0 s.

42. Loss Factors. Equations (25-41) to (25-47) consider free-space transmission conditions and ideal radar operation. In practice, a number of other factors must be included.

(a) *Signal Attenuation before Receiver.* Transmission line loss L_t; antenna losses (included in G_t, G_r, T_s); receiving line and circuit losses at rf (included in T_s); atmospheric attenuation L_a (from Figs. 25-2 to 25-6); atmospheric noise (included in T_s for clear air; $T_a' \to 290$ for larger values of L_a).

(b) *Surface Reflection-Diffraction Effects.* Pattern-propagation factor F, calculated from Eqs. (25-1) to (25-3) with data from Figs. 25-9 and 25-10, appears as F^4 in the numerator ($F > 1$ implies extra gain); details for the diffraction case appear in Ref. 10.

(c) *Antenna Pattern and Scanning.* For tracking and searchlighting case, G_t and G_r are defined for beam axis; for one-coordinate scan at ω rad/s, a reference energy is calculated using $t_0 = \theta_3/\omega$, where θ_3 is one-way half-power beamwidth in radians, and gains G_t and G_r at the point in the scan nearest the target; the effective energy is this reference level reduced by a beam-shape loss $L_{p1} \approx 1.45$ (or 1.6 d3); for two-coordinate scan, the reference energy is based on maximum gains and $t_0 = t_s\theta_a\theta_e/\psi_s$, where θ_a and θ_e are azimuth and elevation beamwidths, and ψ_s is the solid angle searched in time t_s; effective energy is this reference reduced by $L_{p2} \approx L_p^2 \approx 2.1$ (or 3.2 dB).

(d) *Signal-Processing Losses.* For noncoherent integration of $n = t_0 f$, pulses, effective energy is reduced by $L_i(n)$ (Fig. 25-25) relative to the matched-filter value in Eq. (25-47); for loss of resolution in signal processing, a collapsing loss L_c defined by Eq. (25-17) is included; collapsing ratio ρ is found from Table 25-8; losses from nonoptimum threshold settings or operator factors are also included as a factor L_x. These several factors are incorporated into the radar equation for S/N ratio as a factor F^4/L, where L is the product of loss factors for a given case, e.g.,

Intermediate-frequency power ratio for single pulse, in tracking radar:

$$S/N = P_t\tau G_t G_r \lambda^2 \sigma F^4/(4\pi)^3 R^4 k T_s L_m L_t L_a \tag{25-48}$$

Effective energy ratio for noncoherent search radar:

$$\frac{E}{N_0} = P_{av}t_0 G_t G_r \lambda^2 \sigma F^4/(4\pi)^3 R^4 k T_s L_m L_t L_a L_p L_i L_c L_x \tag{25-49}$$

43. Calculation of Detection Range. When the requirement for signal-to-noise ratio in power or energy is known, Eqs. (25-45) to (25-49) can be solved for maximum radar range R_m. In Par. **25-19** the S/N ratio required for each of n equal signal pulses in a train was defined by Eq. (25-14) as the detectability factor $D_0(n)$. For fluctuating targets the required average S/N,

TABLE 25-8 Equations for Collapsing Ratio $\rho = (m + n)/n$

Cases for which P_n/ρ remains constant	
1. Restricted CRT sweep speed s, where d = spot diameter and τ = pulse width	$\rho = \dfrac{d - s\tau}{s\tau}$
2. Restricted video bandwidth B_v, where $B = 1/\tau$ = i.f. signal bandwidth	$\rho = \dfrac{2B_v - 1/\tau}{2B_v} = \dfrac{2B_v + B}{2B_v}$
3. Collapsing of coordinates onto the display, where $2\Delta_t/c$ = time-delay interval displayed per display cell, $\omega_e t_v$, $\omega_a t_v$ = elevation and azimuth scans during integration time t_v, and θ_e, and θ_a = beamwidths	$\rho = \dfrac{2\Delta_t}{c\tau}$ or $\rho = \dfrac{\omega_e t_v}{\theta_e}$ or $\rho = \dfrac{\omega_a t_v}{\theta_a}$
Cases for which P_n remains constant	
4. Excessive i.f. bandwidth $B_n > 1\tau$ followed by matched video	$\rho = \dfrac{B + 1/\tau}{B}$ (use L_c in place of L_m)
5. Receiver outputs mixed at video, where M = number of receivers	$\rho = M$
6. I.F. filter followed by gate of width τ_g and by video integration	$\rho = \dfrac{1}{B\tau} + \dfrac{\tau_g}{\tau}$

computed using $\bar{\sigma}$ in Eq. (25-46) is denoted by D_1, D_2, D_3, D_4, or D_ϵ, depending on the target case and number of diversity samples n_e. For a given case 1, 2, 3, 4, or ϵ, relationships for $D_e(n)$, $L_i(n)$, and fluctuation loss L_{fe} are shown in Figs. 25-25 to 25-27. The maximum range equation can be written

$$R_m^4 = P_t \tau G_t G_r \lambda^2 \bar{\sigma} F^4/(4\pi)^3 kT_s D_e(n,n_e) L_m L_t L_\alpha L_p L_c L_x \tag{25-50}$$
$$R_m^4 = P_{av} t_0 G_t G_r \lambda^2 \bar{\sigma} F^4/(4\pi)^3 kT_s D_0(1) L_s \tag{25-51}$$

where $L_s = L_m L_t L_\alpha L_p L_c L_x L_i L_{fe}$ and fluctuation loss L_{fe} = function of target case, n_e, and P_d.
   The definition of fluctuation loss is

$$L_{fe} = D_e(n,n_e)/D_0(n) \tag{25-52}$$

Equation (25-50) is essentially the form given by Hall in his classic 1956 paper,[42] while Eq. (25-51) is the more universal form used in a later work.[43]
   If detection is performed visually on a CRT display, the product $D_0 L_m$ for a steady target in Eq. (25-50) can be replaced by $V_0 C_B$, where $V_0(n)$ is the visibility factor from Fig. 25-26 and C_B is a bandwidth correction factor[10]

$$C_B = (B\tau/4.8)(1 + 1.2/B\tau)^2 \tag{25-53}$$

Since $C_B = 1$ for the optimum $B\tau = 1.2$, an allowance for minimal matching and collapsing loss $L_m L_c \approx 1.7$ is included within V_0. Curves for V_0 are available only for $P_d = 0.50$ at an effective $P_n \approx 10^{-4}$ on steady targets. Adjustments to other cases can be made by assuming that $V_e \approx 1.7 D_e$ for optimum viewing conditions.
   A worksheet for maximum-range calculation, devised by L. V. Blake of the Naval Research Laboratory (Blake Chart) has been widely accepted as a means of standardizing such calculations. The procedure is based on Eq. (25-50), with λ replaced by c/f and $D_0 L_m$ replaced by $V_0 C_B$. It can be applied to any target, with or without diversity, if $D_e(n,n_e)$ is entered in place of V_0, $\bar{\sigma}$ in place of σ, and L_m in place of C_B. Radar and target parameters are converted into decibel form relative to common engineering units, and the conversion constants are combined into a single constant. The Blake Chart appears in *NRL Rep.* 6930, published by the Naval Research Laboratory, and in Ref. 10.
   For a tracking radar, the Blake chart or Eq. (25-50) can be used to calculate maximum range for a given performance level by finding the required single-pulse S/N ratio and entering it in place of D_e or V_0. Alternatively, if the requirement for doubled energy ratio $\mathcal{R}$ is known, Eq. (25-51) can be used with $D_0(1) L_{fe} = \mathcal{R}/2$.
   44. Search-Radar Equation. The potential performance of a search radar can be determined from its average power, receiving aperture, and system temperature, without regard to its frequency or waveform. The steps in deriving optimum search performance from Eq. (25-51) are as follows.
   1. *Assume uniform search, without overlap,* of an assigned solid angle ψ_s in a time t_s using a rectangular beam whose solid angle is

$$\psi_b = \theta_a \theta_e = 4\pi/G_t L_n \ll \psi_s = A_m(\sin E_m - \sin E_0) \tag{25-54}$$

where θ_a, θ_e = 3-dB beamwidths, A_m = azimuth sector searched, and E_m, E_0 = upper- and lower-elevation search limits.
   2. *Express the observation time t_0 for a target as*

$$t_0 = t_s \psi_b/\psi_s = 4\pi t_s/G_t \psi_s L_n \tag{25-55}$$

and assume that all signal energy reaching A_r during t_0 is integrated for one detection decision. Note that the definition of two-coordinate beam-shape loss, which is included in L_s, is consistent with Eq. (25-55).
   3. *Substitute Eqs. (25-38), (25-54), and (25-55) into Eq. (25-49) and assume $F = 1$, to obtain the search-radar equation*

$$R_m^4 = P_{av} A_r t_s \bar{\sigma}/4\pi \psi_s kT_s D_0(1) L_s L_n \tag{25-56}$$

Neither frequency nor waveform appears directly in Eq. (25-56) although frequency and aperture must permit Eq. (25-54) to be satisfied to concentrate energy within ψ_s, and the loss terms will vary with frequency, waveform, and scan procedure. The new loss term L_n appearing in Eq. (25-54) is an antenna beam loss which accounts for energy outside the main lobe and not available for integration. Spillover and side-lobe energy, for example, reduce G_t without adding to the beam angle ψ_b or time t_0. The conventional gain expression $G = 25,000/\theta_a\theta_e$, with θ in degrees, corresponds to a 40% loss of useful energy, or $L_a = 2.0$ dB. Reduced directivity caused by illumination taper does not appear in L_a but enters the search equation through A_r. Practical minimum values for L_sL_n are about 10 dB on steady targets (for $P_d \approx 0.90$) and 11 to 13 dB on fluctuating targets with optimum diversity and scan procedure.

45. Cumulative Probability of Detection. Search and acquisition radars normally scan their assigned volumes more than once in order to achieve high detection probability. If a single-scan probability P_1 is obtained on each of k scans, the cumulative detection probability P_c is

$$P_c = 1 - (1 - P_1)^k \tag{25-57}$$

This procedure gives earlier detection of some penetrating targets, and minimizes effects of fluctuation, pattern lobing, and patches of large clutter. However, the distribution of energy into k scans without scan-to-scan integration is less efficient than matched-filter integration (or even postdetection integration). The loss in effective energy relative to the matched filter can be expressed[3] as a scan distribution loss L_d

$$L_d \equiv \frac{kD_0(1), \text{ for } P_d = P_1}{D_0(1). \text{ for } P_d = P_c} \tag{25-58}$$

For example, to obtain $P_c = 0.90$ in $k = 4$ scans, $P_1 = 0.44$ and $D_0(1) = 12.7$ dB (at $P_n = 2.5 \times 10^{-9}$). The same result is obtained in one scan with $D_0(1) = 14.2$ dB (at $P_n = 10^{-8}$), giving $L_d = 4.5$ dB. All or a portion of L_d is recovered if the target fluctuates, because L_f is lower at $P_1 = 0.44$ than at $P_d = 0.90$; also, using postdetection integration, $L_i(n/k)$ is less than $L_i(n)$. The optimum k usually approximates 4.

46. Beacon-Range Equations. One-way transmission from a radar to a beacon gives the interrogation power level S_b at the beacon receiver

$$S_b = P_tG_tG_b\lambda^2/(4\pi)^2R^2L_tL_{al}L_b \tag{25-59}$$

where G_b = beacon antenna gain in radar direction, L_{al} = one-way atmospheric loss, and L_b = loss between beacon antenna and receiver. On the return link, a beacon peak power P_b determines the signal power at the radar antenna terminal.

The beacon response power is

$$S = P_bG_bG_r\lambda^2/(4\pi)^2R^2L_rL_{al} \tag{25-60}$$

from which the single-pulse signal-to-noise ratio is

$$S/N = P_b\tau G_bG_r\lambda^2F^2/(4\pi)^2R^2kT_sL_mL_bL_{al} \tag{25-61}$$

47. Radar Range in Clutter. The echo power returned by clutter is found from Eq. (25-37) or (25-39) when the clutter range R_c and cross section σ_c are used in place of target σ. The cross section of homogeneous clutter is found by multiplying reflectivity by the area or volume of the resolution cell.

$$\sigma_c = \begin{cases} A_c\sigma^0 = \dfrac{R_c\theta_a}{L_p}\dfrac{\tau_n c}{2}\sigma^0 & \text{surface clutter} \tag{25-62} \\[2mm] V_c\eta_v = \dfrac{R_c^2\theta_a\theta_e}{L_p^2}\dfrac{\tau_n c}{2}\eta_v & \text{volume clutter} \tag{25-63} \end{cases}$$

where θ_a, θ_e = 3-dB beamwidths in azimuth and elevation, τ_n = effective (noise) width of processed pulse (Table 25-5), and $L_p \approx 1.45$ = beam-shape loss. In writing expressions for signal-to-clutter ratios, the approximation $\tau_n \approx 1/B$ will be used, and an improvement factor I will be included to describe the increase in signal-to-clutter (S/C) output, due to MTI or doppler processing

$$I \equiv \frac{(S/C)_{out}}{(S/C)_{in}} = \frac{\sigma_c}{\sigma}\left(\frac{R}{R_c}\right)^\epsilon (S/C)_{out} \tag{25-64}$$

In systems without range ambiguity, the clutter competing with a target is at the same range; so $R/R_c = 1$. The processed clutter or clutter residue will be assumed noiselike, so that required S/C ratios are given approximately by V_0, D_0, or D_e. If the output clutter power competing with the signal is correlated from pulse to pulse, an increased integration loss term will be needed to describe the reduced number of independent samples: $L_{ic} \geq L_i$. Thus the subclutter visibility (SCV) is defined and related to D_e and D_0 by

$$\text{SCV} = (C/S)_{in} \qquad \text{when } (S/C)_{out} = D_e(n)$$
$$= 1/D_e(n) = 1n/D_0(1)L_f L_{ic} \qquad (25\text{-}65)$$

Substituting Eq. (25-62) into (25-65), with $\tau_n = 1/B$, $n = t_0 f$, and unambiguous range $R_u = c/2f_r = t_0 c/2n$, the maximum range becomes

$$R_m = \frac{B l \sigma t_0 (R_c/R_m)^3}{R_u \theta_a \sigma^0 D_0(1) L_f L_{ic} L_c L_x} \qquad (25\text{-}66)$$

(losses L_p, L_f, and L_x have been applied to the target). The equivalent form for volume clutter is

$$R_m^2 = \frac{B l \sigma t_0 (R_c/R)^2}{4\pi R_u \psi_b \eta_v D_0(1) L_f L_{ic} L_c L_x} \qquad (25\text{-}67)$$

48. Search-Radar Techniques. *Control of Coverage Patterns.* The most basic problem in search radar is to establish reliable detection coverage over the assigned volume (e.g., for

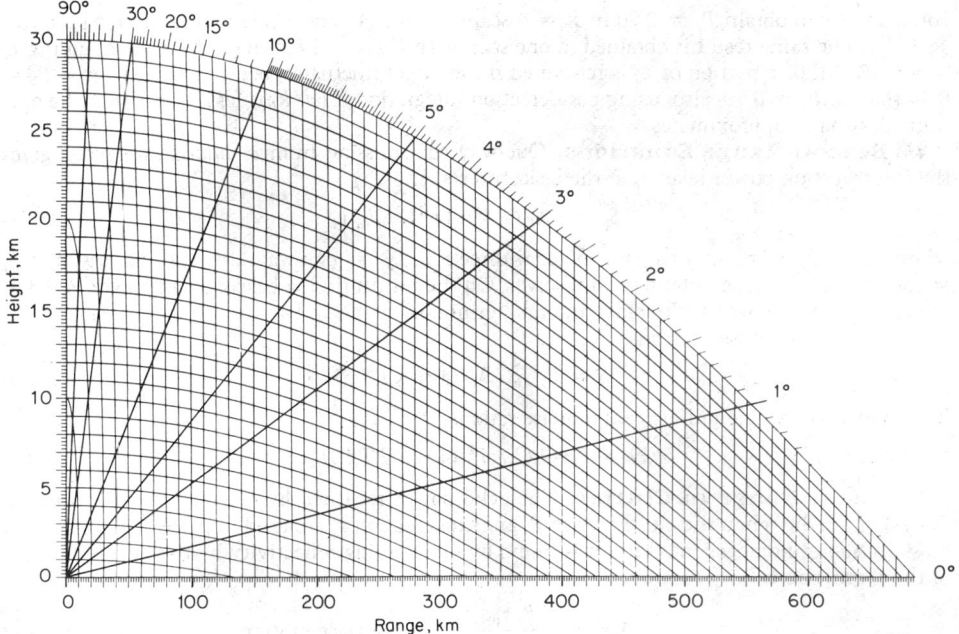

Fig. 25-38. Range-height-angle chart for aircraft.[45]

air surveillance, a volume extending from the horizon to a given altitude within a maximum range). Above 100 MHz, propagation paths are reliable only above the horizon, restricting coverage on targets in the troposphere.

The horizon range limit is

$$R_h \approx \begin{cases} 4.15(\sqrt{h_r} + \sqrt{h_t}) & \text{km } (h \text{ in meters}) \\ 1.23(\sqrt{h_r} + \sqrt{h_t}) & \text{nmi } (h \text{ in feet}) \end{cases} \qquad (25\text{-}68)$$

where h_r = radar antenna height and h_t = target height. These are the conventional values based on the ⅘-earth's-curvature approximation. More accurate coverage estimates for targets at all altitudes can be made using range-height-angle charts such as Figs. 25-38 and 25-39, based on ray tracing through the exponential reference atmosphere.[2,22,45]

Search-radar vertical coverage is conventionally plotted on linear scales as in Fig. 25-38, with contours showing R_m vs. E_t for different values of P_d. A value of R_π for arbitrary E_{t0} is computed from Eq. (25-50) or (25-51), and this is scaled at other elevations:

$$\frac{R_m(E_t)}{R_m(E_{t0})} = \frac{F(E_t)}{F(E_{t0})} \left(\frac{G_t G_r / L_a \text{ at } E_t}{G_t G_r / L_a \text{ at } E_{t0}} \right)^{1/4} \tag{25-69}$$

For example, as shown in Fig. 25-40a, an air-surveillance radar provides coverage to 30,000 ft (10 km) at 80 mi (150 km), with a cosecant-squared antenna pattern to maintain approximately constant-altitude coverage at shorter range.

$$G_t = G_r = G(3°) \csc^2 E_t$$

$$\frac{R_m(E_t)}{R_m(3°)} = \frac{F(E_t)}{F(3°)} \left[\frac{L_a(3°) \csc^4 E_t}{L_a(E_t) \csc^4 3°} \right]^{1/4} \approx \frac{\csc E_t}{\csc 3°} \tag{25-70}$$

since F and L_a are approximately constant for $E_t > 3°$. Below $3°$ the antenna gain peaks and falls off to its half-power point approximately at the horizon. A small increase in atmospheric

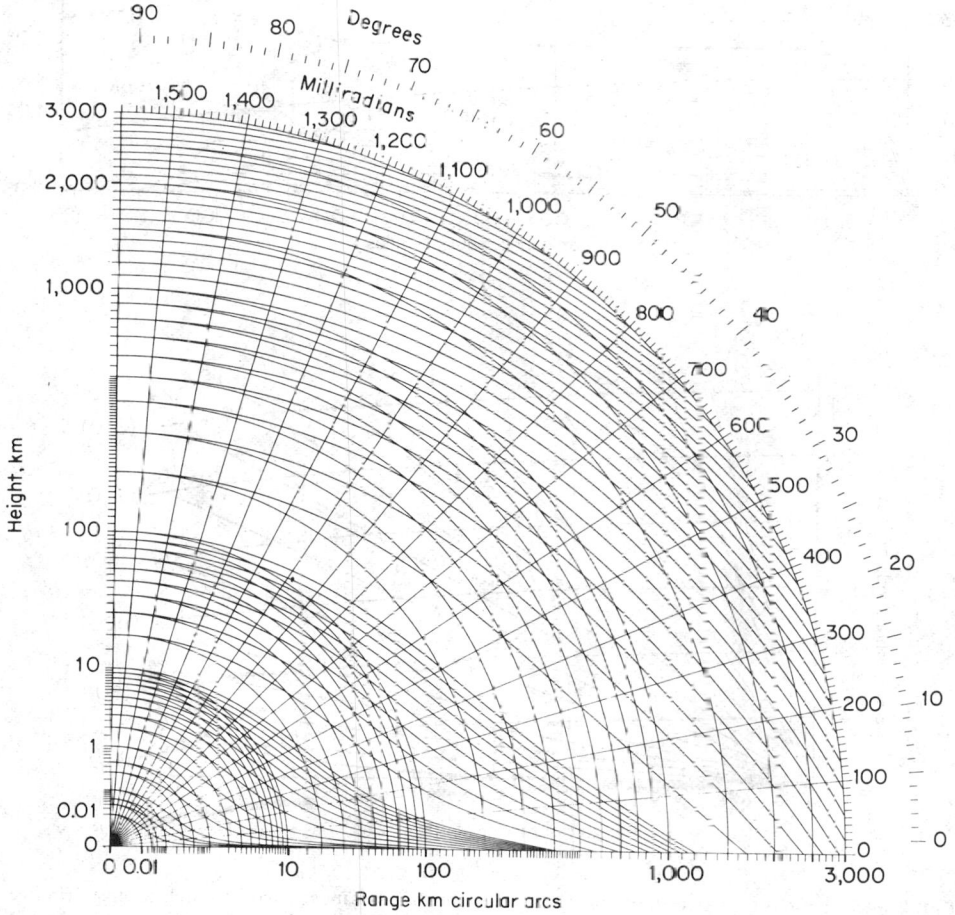

Fig. 25-39. Range-height-angle chart for missiles and satellites.[45]

attenuation is included in the average ($F = 1$) pattern at low angle, and the effect of lobing is shown as a series of lobes whose spacing depends on antenna height, as in Eq. (25-1). The average range near the horizon can be increased by directing the beam axis nearer the horizon, but the depth of nulls will be greater and they will extend further into the high-angle coverage.

Apart from lobing problems, heavy illumination of the ground can introduce intolerable clut-

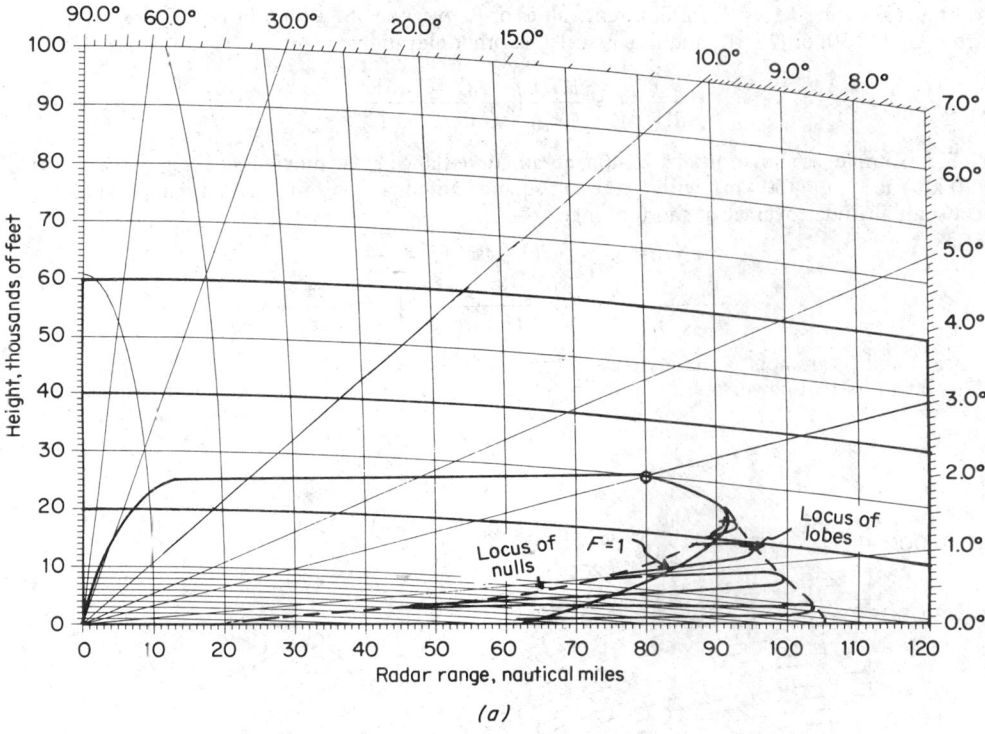

(a)

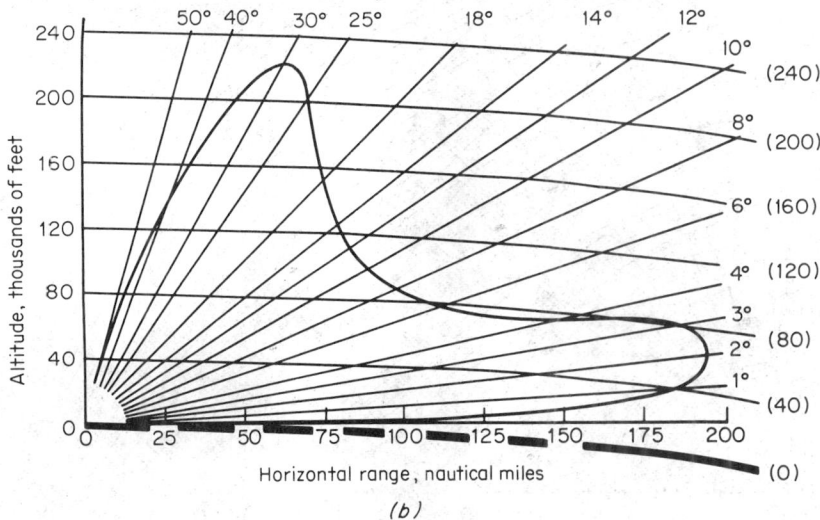

(b)

Fig. 25-40. Search-coverage patterns for aircraft detection: (a) cosecant-squared air-search coverage; (b) free-space coverage diagram for the ARSR-2 antenna; (c) two-way coverage diagram for adjustable receiving beam.[47]

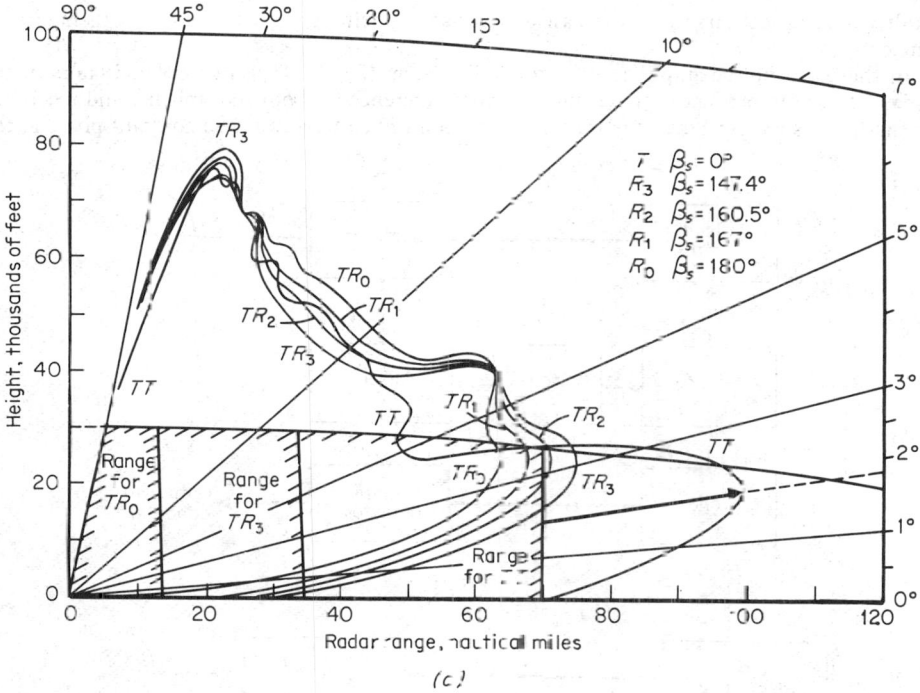

(c)

Fig. 25-40 (continued).

ter at short range, both from fixed objects and from moving clutter such as birds and insects. To maintain visibility of aircraft at high elevation angles it may be necessary to direct substantial energy into these regions while using a pattern which cuts off rapidly at and below the horizon (Fig. 25-40b). Horn-fed reflector systems have been designed that vary the receiving pattern as a function of range delay in order to obtain the benefits of good high-angle visibility and low response to short-range clutter without sacrificing long-range coverage near the horizon (Fig. 25-40c).

Search coverage can be controlled even more completely when a narrow pencil beam is scanned in a raster or when several such beams are stacked one above the other to cover an elevation sector with multiple receiver channels. With the agile beam, both transmitted energy and receiver gates can be varied to select the desired coverage in each beam position, but the number of beam positions may not permit long enough dwells for MTI or doppler processing.

49. Search-Radar Detection. Early radar depended entirely on CRT displays with human operators for target detection, and this procedure remains one of the most efficient and adaptable. The curves for visibility factor (Fig. 25-26) show signal integration performance near the optimum limit set by information theory. Such performance cannot be expected under field conditions, where the operator may be fatigued or distracted by the surroundings. However, it remains true that trained operators, using their experience to recognize targets and reject interference on the basis of complex visual patterns and scan-to-scan memory, may outperform the most sophisticated automatic detector in a difficult environment. To overcome the factors of fatigue and inattention in early-warning applications, as well as minimizing effects of random interference and collapsing on a display, the CRT-operator integration may be replaced by a video sweep integrator (Par. **25-80**). The output of the integrator provides a bright blip on an otherwise uncluttered display, and may also operate an audible alarm to attract attention in early-warning use. Performance is restricted by the resolution and stability of the integrator memory device and by ability to set and hold the threshold precisely relative to random noise (see CFAR discussion, Par. **25-53**).

50. Moving-Target Indication. A moving-target indicator (MTI) is a device which limits the display of radar information primarily to moving targets. The sensitivity to moving targets is provided primarily through their doppler shifts although *area* MTI systems have been built which cancel targets on the basis of overlap of their signal envelopes in both range and angle.

In the usual pulsed-amplifier coherent MTI system (Fig. 25-41a), two cw oscillators in the radar are used to produce a phase and frequency reference for both transmitting and receiving, so that echoes received over the train of pulses from fixed targets have a constant phase at the

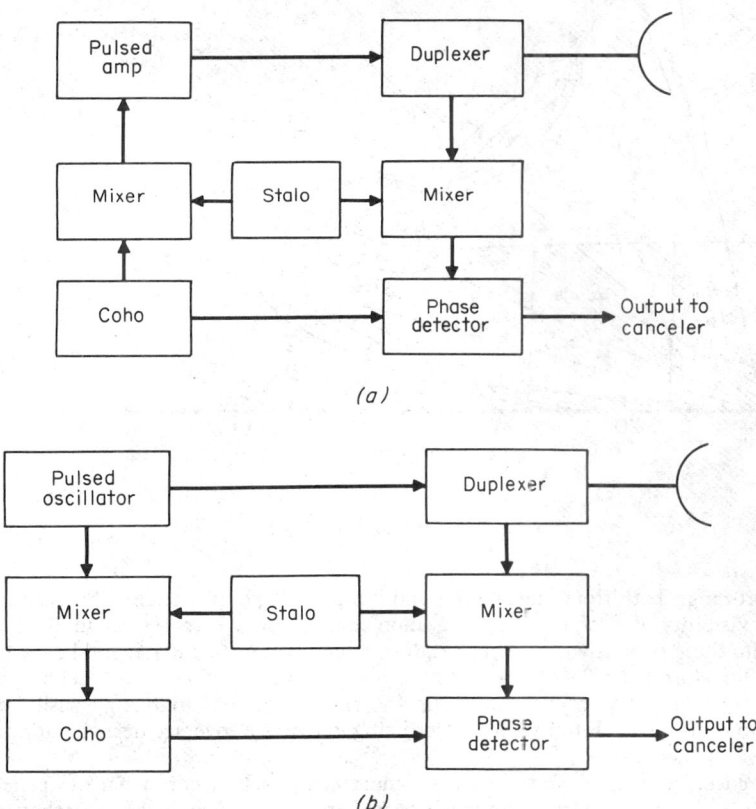

Fig. 25-41. Block diagrams of MTI radars: (a) simplified block diagram for power amplifier; (b) pulsed oscillator.

detector. These echoes will be canceled, leaving in the output only those signals whose phase varies from pulse to pulse. Coherent MTI can also be implemented with a pulsed-oscillator transmitter (Fig. 25-41b), in which the coherent oscillator is locked in phase to remember the random phase with which each successive pulse is transmitted. Although the transmitted signal is noncoherent, the i.f. signal is coherent and has the same line structure as in the pulsed amplifier system. Both systems attenuate targets in a band centered at zero radial velocity, the depth and width of the rejection notch depending on design of the canceler and stability of the received signals.

Two variations on the coherent MTI are available for rejection of clutter with nonzero radial velocity. In the clutter-locked MTI, the average doppler shift of a given volume of clutter is measured and used to control an offset frequency oscillator in the receiver, shifting the clutter into the rejection notch. Short- or long-term averages can be used to obtain rapid adaptation to varying clutter velocity (as in weather clutter) or better rejection of selected parts of a complex clutter background. The alternative is *noncoherent MTI*, in which the clutter surrounding a target provides the phase reference with which the target signal is mixed to produce a doppler signal. Although simpler to implement, noncoherent MTI does not cancel as completely and may lose target signals when the clutter is too small (as well as when it is too large).

Cancelers for MTI radar are designed to pass as much of the target spectrum as possible while rejecting clutter. Since search-radar MTI must cover many range cells without loss of resolution, canceling filters are implemented with delay lines, with multiple-range gates feeding bandpass filters, or with range-sampled digital filters which perform both these functions. The response of several typical cancelers is shown in Fig. 25-42. A wide variety of response shapes is available through use of feedback and multiple, staggered repetition rates.[45] In particular, through proper use of stagger (Fig. 25-42*d*), it is possible to maintain detection of most targets with nonzero radial velocities, even those which would fall in one of the blind speeds v_{bj} (ambiguous rejection notches) of an MTI with a single repetition rate:

$$v_{bj} = j\lambda f_r/2 \qquad j = \pm 0, 1, 2, \ldots \tag{25-71}$$

51. Performance of MTI. The basic measure of MTI performance is the MTI improvement factor I, defined in Eq. (25-64). This is equal to the clutter attenuation when the canceler gain is set to unity on noise or targets uniformly distributed in velocity. The basic relationships between I and radar or clutter parameters can be expressed in terms of the ratio of rms clutter spread to repetition rate or blind speed:

$$z = 2\pi\sigma_f/f_r = 2\pi\sigma_v/v_{bl} = 4\pi\sigma_v/\lambda f_r \tag{25-72}$$

where σ_f = standard deviation of clutter power spectrum (Hz), f_r = repetition rate, σ_v = standard deviation in (m/s), and v_{bl} = first blind speed from Eq. (25-71). For a scanning gaussian beam

$$z = 2\sqrt{\ln 2}\,(\omega/f_r\theta_3) = 1.665/n \tag{25-73}$$

For a (sin x)/x beam, the constant 1.665 should be changed to 1.760 giving slightly poorer MTI performance.

In general, the rms spread should be calculated using the sum of components σ_s due to scanning, σ_i due to internal motion of the clutter, σ_m due to relative motion between the radar and the clutter, and σ_x due to low-frequency instabilities in the radar.

52. CW Radar. A cw transmission has no velocity ambiguity, and so cw radar equipment can easily be designed to provide 80 to 100 dB rejection of fixed or moving clutter. Coherent integration of target signals in selected doppler bands is also provided by narrow-band filters of relatively simple construction. Three problems, however, restrict the performance and usefulness of cw radar for search:

(a) *Isolation of Receiver from Transmitter.* Direct feedthrough of transmitter power to the receiver must be minimized, requiring separate antennas in high-power systems and careful design in all systems to avoid receiver saturation.

(b) *Magnitude of Short-Range Clutter Echo.* The echo power received from clutter in a cw radar is the integrated product of the reflectivity, (range)$^{-4}$, and antenna gain factors of the radar equation (25-39) over the common volume of the transmitting and receiving beams. In both volume and surface clutter, the echo power is controlled by the clutter at the shortest range in the common volume, and the effective clutter cross section is a function of beamwidth and range R_c to the point where the beams substantially overlap. The required clutter improvement for a cw radar may therefore be very high because of the $(R/R_c)^4$ term.

(c) *Transmitter Noise.* Both the direct feed-through from transmitter to receiver and the echoes from short-range clutter will contain random-noise components from the transmitter. Special circuits may be designed to cancel the direct feed-through and low-frequency components of reflected noise, but the higher-frequency components will appear with phase shift from the range delay and cannot be canceled completely. Subclutter visibility in cw systems is generally controlled by these noise components.

53. Control of False Alarms. The signal-detection statistics previously described are based on normally distributed random noise (Rayleigh envelope distribution) with fixed (and known) rms level. The false-alarm probability for that case, given in Eq. (25-12), is a function of threshold setting. In the real environment, however, the background against which targets must be detected has a varying rms level and may also depart from the Rayleigh assumption. This is especially true of clutter, and even the residue after MTI filtering may be non-Rayleigh and partially correlated from pulse to pulse. Many techniques are used to minimize false alarms under these conditions, generally (if inaccurately) described as *constant false-alarm rate* (CFAR) techniques.

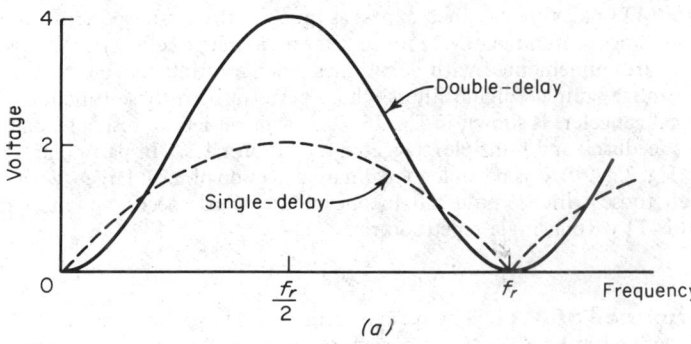

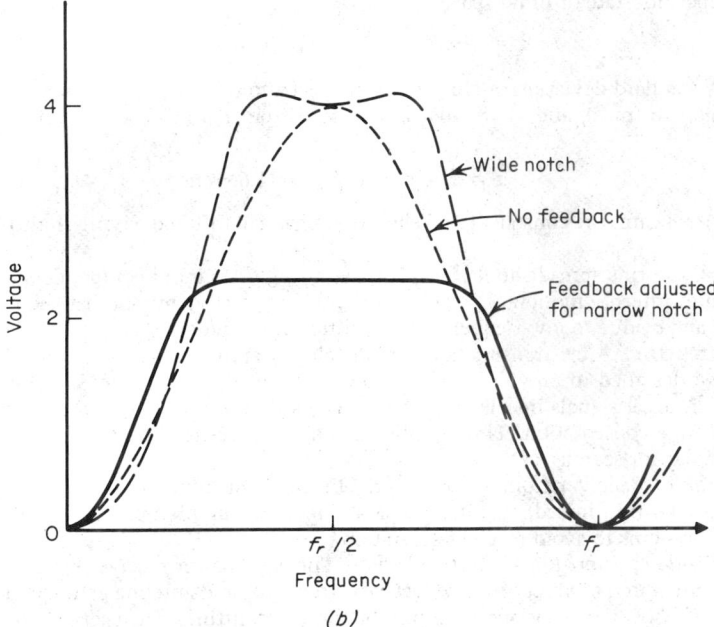

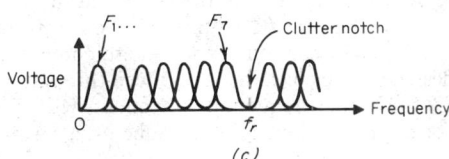

Fig. 25-42. Frequency response of MTI filters: (*a*) single and double delay without feedback; (*b*) double delay with feedback; (*c*) range-gated filter bank; (*d*) double and triple delay with staggered pulse repetition rate and feedback.[46]

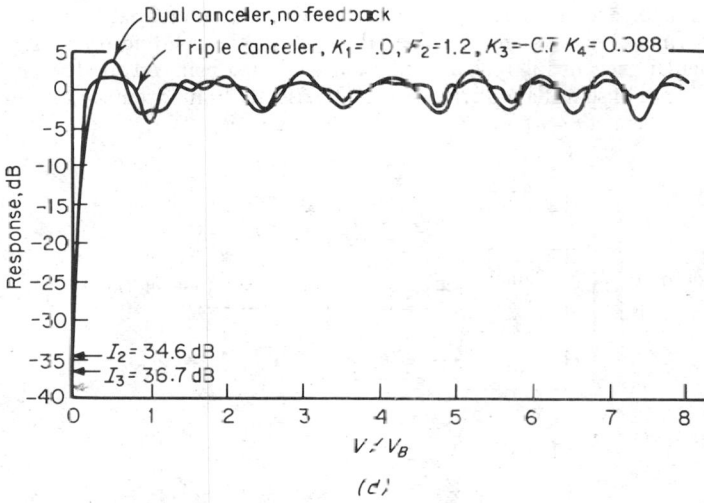

Fig. 25-42. (continued)

Early radars used a simple programmed variation in receiver gain, as a function of range, to reduce alarms from small short-range objects. This procedure, known as *sensitivity time control* (STC) remains an effective means of rejecting echoes from birds and insects and of minimizing saturation on larger clutter sources. If gain is programmed to maintain a constant signal level for targets near the center of the beam, the noise level is sharply reduced at short range, and false alarms in that region should result only from clutter. Targets not in the main beam (e.g., aircraft flying at constant altitude into the cosecant-squared portion of search coverage) will also be lost. Some compromise in gain program and antenna pattern will normally be used to give increased target-detection probability at short range without excessive clutter alarms.

Another well-established technique is the log-FTC receiver, in which a logarithmic i.f. or video response is followed by a video differentiator of fast time constant. This is effective in compressing the range of output fluctuation from random clutter whose rms level is varying slowly (as in rain-cloud echoes or clutter from hilly terrain). Target echoes in the clutter region are also suppressed, but targets which exceed the local noise or clutter fluctuation by the visibility factor V_0 remain detectable. This technique is especially useful in non-MTI systems where clutter would otherwise saturate the display or video processor elements. It is compatible with frequency diversity and video integration techniques and can also be used with MTI outputs. A moderate loss (1 or 2 dB) in detectability is usually incurred.

Examples of CFAR techniques are:

(a) *Guard-Band Receivers.* Filter channels adjacent to the signal spectrum are used to estimate the broadband noise level (Fig. 25-43a).

(b) *Dicke Fix or Wide-Band Limiter.* As with example (a) the level over a band broader than the signal spectrum is used to control i.f. gain, using in this case a limiter to hold total output constant (Fig. 25-43b). The effective number of samples is the ratio of total bandwidth to signal bandwidth. In coherent systems, this procedure can be applied within the ambiguous doppler interval, f_r, which is wider than the spectral-line width of an individual target.

(c) *Range-Averaged AGC.* Automatic gain control based on a local average of range cells near the detection cell can provide a measure of noise plus clutter within a single repetition interval (Fig. 25-43c). It may be combined with time averages over several repetition intervals to increase the accuracy of the estimate or increase the rate at which the estimate will follow local changes in clutter as a function of range.

(d) *Side-Lobe Blanking.* Averages in the angular coordinates may be used to recognize side-lobe response to strong clutter or active sources. These averages are generally taken with an auxiliary, broad-beam antenna and receiving channel (Fig. 25-43d).

54. Search-Radar Measurements. The theoretical limits on angular measurement, applicable to scanning-search radar, are summarized in Fig. 25-32. The optimum beam-splitting process can be approached closely by differentiating the range-gated output of a video integrator to find the peak of the beam-pattern envelope. This process is the angular analog of the optimum time-delay estimator, which consists of a matched filter and differentiator. As with time-delay measurement, approximations are available which locate the centroid of the signal envelope using split angular gates or interpolation between a selected point on the leading and trailing edges. The presence of scintillation error (Fig. 25-32) usually limits accuracy to a level near 0.05 beamwidth, regardless of S/N ratio or processor efficiency.

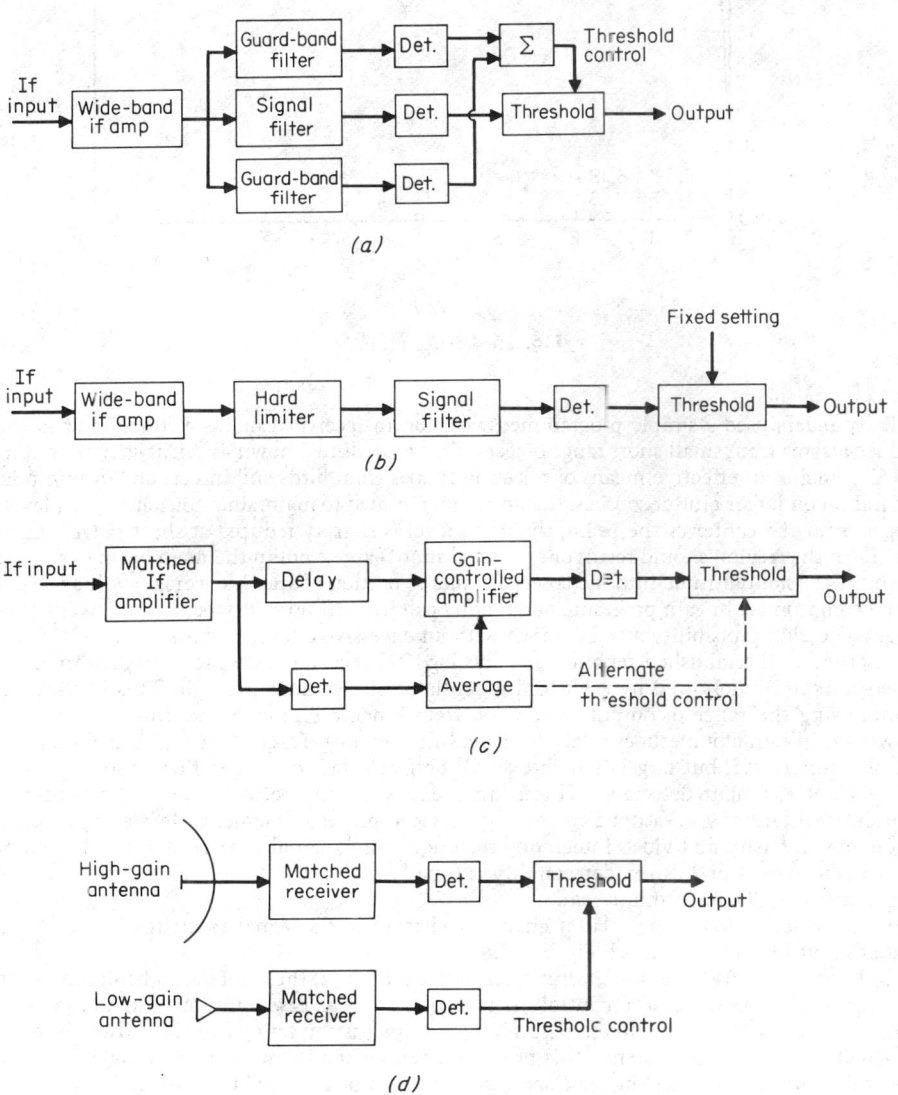

Fig. 25-43. CFAR techniques: (a) guard-band system; (b) Dicke fix system; (c) range-averaged AGC; (d) side-lobe blanker.

An important function in some surveillance systems is target-height measurement on multiple targets detected by the search radar. Since range is known to high accuracy, the elevation angle error becomes the dominant problem. One of the following procedures may be used to measure elevation.

A separate height finder, designated to target azimuth by the two-dimensional search radar, scans a narrow beam in elevation (Fig. 25-44a) with accuracy expressed by Eq. (25-26) and Fig. 25-37 relative to the elevation beamwidth. One or two such height finders are sequenced among many targets to achieve a low data rate, e.g., one reading per 30 s.

Separate Receivers. The search-radar elevation coverage is divided among several stacked receiving beams feeding separate receivers (Fig. 25-44*b*). With sufficient overlap to give smooth coverage, the receiver outputs may be compared (at r.f. or video) to produce a monopulse estimate. The number of receivers is minimized at some cost in accuracy if integrated video outputs are compared. Scintillation does not affect accuracy.

The V-beam approach can be used,[8] wherein two broad beams scan in azimuth, one of them tilted to produce an output whose delay (in the azimuth scan cycle) is proportional to elevation angle (Fig. 25-44*c*). The elevation error is proportional to the difference between the two azimuth

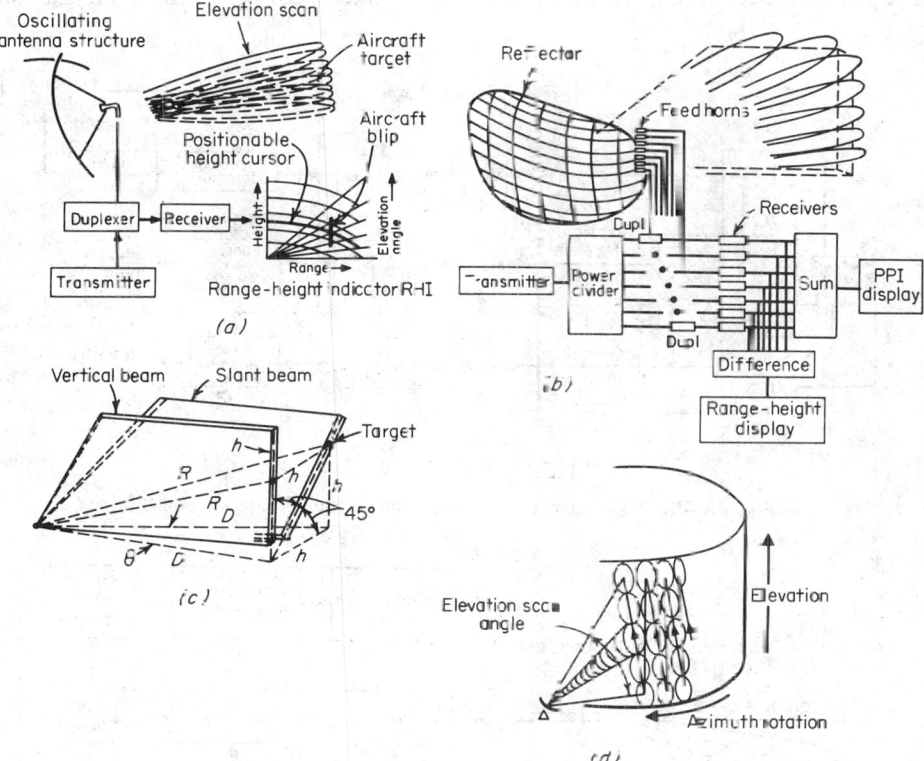

Fig. 25-44. Height finding in surveillance radar system: (*a*) nodding-height-finder representation; (*b*) stacked-beam-radar representation; (*c*) V-beam geometry; (*d*) azimuth-elevation raster produced by a three-dimensional scanning radar.[49]

readings multiplied by the cotangent of the tilt angle. The azimuth readings are subject to independent scintillation errors, leading to rather large errors in the elevation estimate.

A scanning pencil beam (Fig. 25-44*d*) can be used to measure both angular coordinates, with monopulse instrumentation if scintillation error is to be avoided. If frequency scan is used in elevation, the scintillation error is found from Fig. 25-37, using the correlation frequency f_c of the target and the frequency shift Δf per beamwidth to give $n_e = \Delta f/f_c$.

For low-elevation targets ($E_t < 1.5\theta_e$), the multipath error from Fig. 25-36 will be a significant contributor to height-finder error. In a V-beam system, the effective elevation beamwidth for evaluation of multipath depends on the vertical aperture used to generate the tilted beam.

TRACKING-RADAR TECHNIQUES

55. Angle Tracking. In tracking radar, a narrow (and generally circular) beam is directed at a selected target, either continuously (with a mechanically steered reflector or lens) or with a time-shared array beam. The electromechanical or electrical servo loop is controlled to minimize the angle errors, as measured by an error-sensing antenna and receiver system. The error-sensing sensitivity of monopulse antennas is described in Par. **25-34** in terms of the normalized slope of the Δ/Σ pattern ratio, while conical-scan sensitivity depends on the fractional modulation of the pattern per beamwidth of target displacement.

In a typical *monopulse radar* (Fig. 25-45), Σ and Δ patterns are formed with a four-horn feed and hybrid network before conversion of signals to intermediate frequency. Normalization (formation of the Δ/Σ ratio) is performed using a common automatic gain control (AGC) voltage, derived from the range-gated Σ-channel output and applied to Σ and Δ receivers in parallel. Phase-sensitive error detectors, using the Σ output as a reference, produce bipolar video outputs proportional to the Δ/Σ ratios, and hence to the off-axis error angles in each coordinate. After range gating to select the target at the range of interest, these video signals can be stretched and

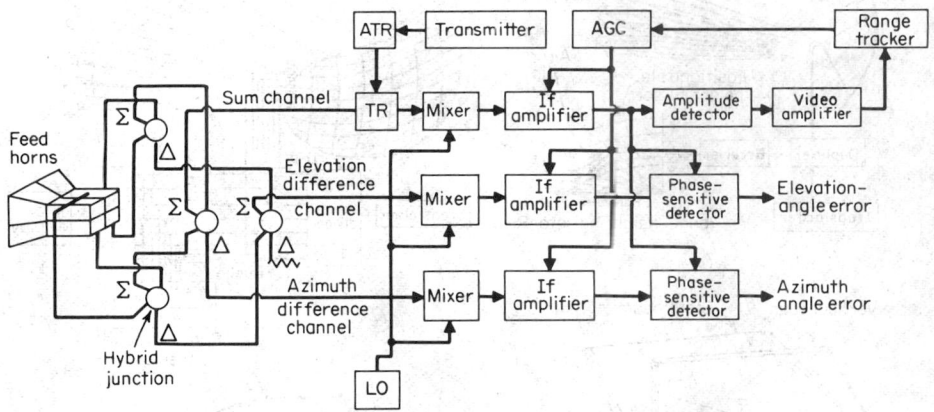

Fig. 25-45. Block diagram of a conventional monopulse tracking radar.[9]

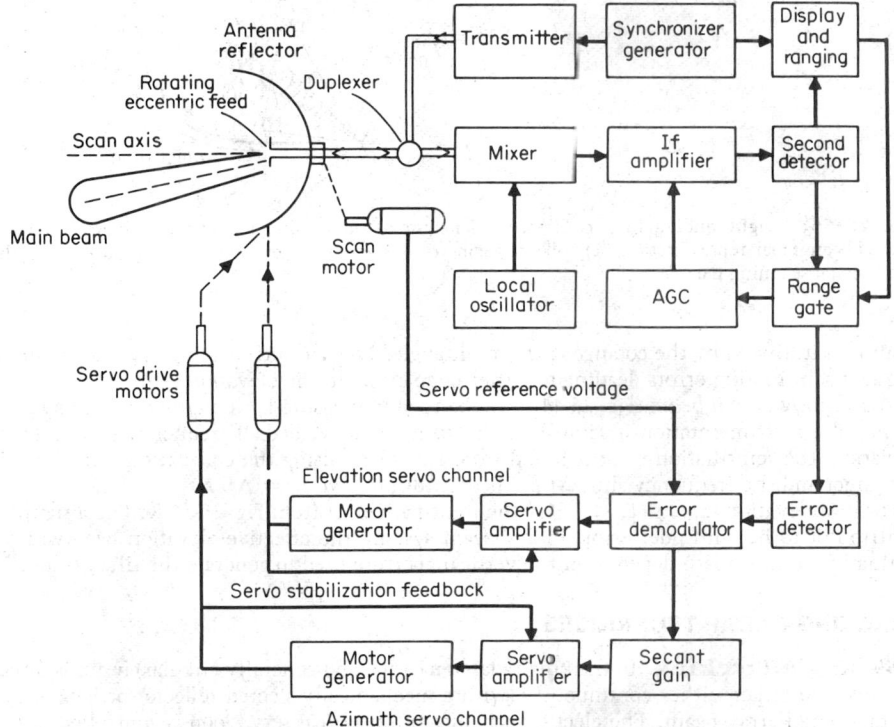

Fig. 25-46. Block diagram of a conical-scan radar.[29]

smoothed to produce dc error signals for the servo loop. Apart from the choice of monopulse feeds and receiver designs, the major variations between monopulse radar techniques appear in the procedure for normalization, error detection, and calibration.

The *common-AGC* normalization technique works well for a single-target tracker, where pulse-to-pulse signal variations are relatively small and longer-term fluctuations can be followed by a fast AGC loop.[50] Multiple-target trackers require single-pulse normalization, which can be provided by IAGC, logarithmic receivers, or limiter systems. Details on these different error sources and effects will be found in the literature.[8,9]

The *conical-scan tracker* (Fig. 25-46) samples the beam positions around the tracking axis sequentially, obtaining a train of amplitude-modulated pulses (Fig. 25-47) from which elevation and azimuth (traverse) errors are extracted. Only a single receiver channel is used, with AGC to normalize the error signals. In place of the mechanically rotating or nutating feed, some systems use electronic scan to produce higher scan rate or to scan on receive only. This latter approach is sometimes referred to as pseudo monopulse because it employs a fixed four-horn feed with Σ and Δ outputs, but its sensitivity to noise and scintillation is the same as given for conical scan with one-way patterns (Fig. 25-34a). The primary performance differences between conical-scan and monopulse trackers are the increased tracking noise from target scintillation (amplitude noise) in conical scan and the greater thermal noise error at long range.

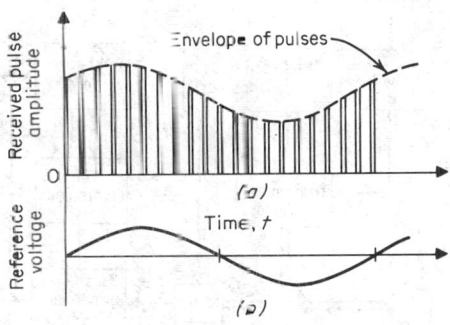

Fig. 25-47. Conical-scan signal modulation: (a) angle error in envelope of received pulses; (b) reference signal derived from drive of scan feed.[29]

56. Range Tracking. A simple video-signal range tracker is shown in Fig. 25-48. Video pulses are applied to a time discriminator which measures the position of the centroid, peak, or leading edge of the signal relative to the delay stored in the tracker. Any difference between these two positions is used as the servo error signal to reposition the tracker output, after suitable smoothing and filtering to control servo bandwidth and lag. The range delay output is used to gate the angle tracker and AGC circuit, so that ideally all three coordinates are controlled by the single selected target. The techniques used in the time discriminator and the delay generator determine accuracy and sensitivity of the range tracker. The split-gate discriminator process is similar to the use of two offset beams in monopulse in that the difference in voltage between the offset gates is found on each pulse, and normalized to the total signal in the main range gate (Σ channel).

An alternative process forms a signal which approximates the derivative of the video pulse, and locates its zero crossing by placing an error detector gate to obtain zero average output. The

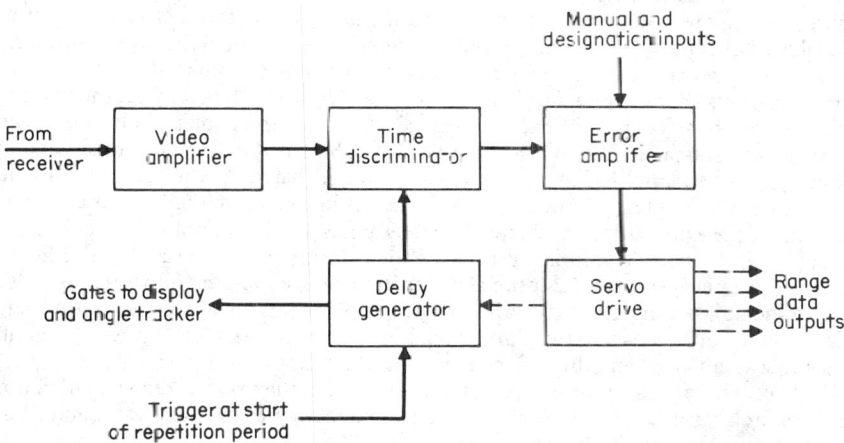

Fig. 25-48. Basic elements of automatic range tracker.[51]

leading-edge discriminator consists of two differentiators in cascade, the first producing a short pulse width $\tau' \approx 1/B$ on the leading edge of an extended echo, and the second locating the peak of this pulse (hence the point of greatest slope on the original input signal). Any of these techniques will work well on strong signals, and the first two procedures can be used on signals near or below noise level if the error-detecting gate outputs are smoothed over many pulses.

57. Doppler Tracking. In cw or high-pulse-repetition-rate pulsed radar, doppler tracking is used instead of range tracking as a primary means of resolving targets. It is also used in low- and medium-pulse-repetition-rate pulsed radar to supplement range resolution in clutter environments and to measure radial velocity with high accuracy. Used only for measurement, the

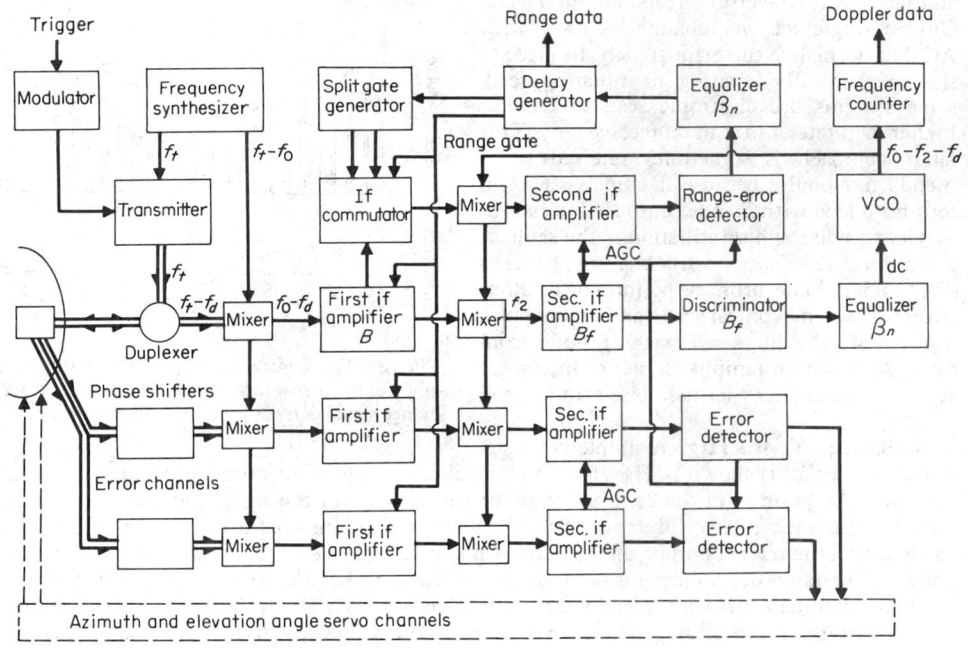

Fig. 25-49. Four-dimensional tracker, monopulse type.[24]

doppler tracker can operate in parallel with (and independently of) the range- and angle-tracking loops, which do not require coherent data. If the doppler resolution is needed to obtain adequate S/C or S/N ratio, however, all four loops must be interdependent, with coherent signals being filtered and processed in a common Σ channel and four Δ channels.

Figure 25-49 shows such a system, in which the monopulse Σ channel feeds a discriminator for doppler error sensing as well as providing reference inputs for AGC and error detectors in range, azimuth, and elevation. The discriminator, operating on the selected fine line of the signal spectrum, controls the voltage-controlled oscillator (VCO), which heterodynes the first i.f. signal into the center of the doppler filter at the second i.f. To obtain doppler resolution in the range Δ channel, the split-gate function is applied as an i.f. phase reversal (commutator), and the narrow-band second i.f. amplifier averages the two reversed halves of the signal to produce zero output with a centered gate and opposite-polarity error outputs on either side of center.

58. Tracker-Acquisition Procedure. Targets for high-resolution trackers must be designated with reasonable accuracy if they are to be acquired when near the threshold of sensitivity. In the absence of four-dimensional designation, one or more coordinates must be scanned to acquire the target, and sensitivity will be degraded. Since it is possible to observe all range and doppler cells simultaneously, a two-dimensional bank of acquisition cells is usually established (as in some search radars) and the region of angular uncertainty is scanned sequentially until the target is detected. Targets strong enough to be acquired can then be tracked with high accuracy, since the search losses associated with overlap and straddling of multiple gates and filters, and with beam scanning, are eliminated in track.

Radar Technology

By HAROLD R. WARD

Radar development, since its beginning during World War II, has been paced by component technology. As better rf tubes and solid-state devices have been developed, radar technology has advanced in all its applications. This subsection presents a brief overview of radar technology. It emphasizes the components used in radar systems that have been developed specifically for the radar application. Since there are far too many devices for us to mention, we have selected only the most fundamental to illustrate our discussion.

In this subsection, the sequence in which each subsystem is discussed parallels the block diagram of a radar system. Pictures of various radar components give an appreciation for their physical size, while block diagrams and tabular data describe their characteristics. The material for this subsection was taken by permission largely from Ref. 51, to which we refer the reader for more detail and references.

59. Radar Transmitters. The requirements of radar transmitters have led to the development of a technology quite different from that of communication systems. Pulse radar transmitters must generate very high power, pulsed with a relatively low duty ratio.

The recent development of high-power rf transistors capable of producing a few hundred watts of peak output power at L band has made solid-state radar transmitters feasible. A corporate combiner is needed to sum the outputs of many devices to obtain the power levels required of medium- and long-range search radars. Such solid-state transmitters offer the following advantages over tube transmitters: low-voltage operation (typically 36 V), no modulator required, and reliable operation through the redundant architecture. While the solid-state transmitter still costs more than its tube equivalent, the solid-state technology is developing more rapidly.

Tube-type power oscillators and power-amplifier stages consist of three basic components: a power supply, a modulator, and a tube. The power supply converts the line voltage into dc voltage of from a few hundred to a few thousand volts. The modulator supplies power to the tube during the time the rf pulse is being generated. Although the modulation function can be applied in many different ways, it must be designed to avoid wasting power in the time between pulses. The third component, the rf tube, converts the dc voltage and current to rf power. The devices and techniques used in the three transmitter components are discussed in the following paragraphs.

60. RF Tubes. The tubes used in radar transmitters are classified as *crossed-field*, *linear-beam*, or *gridded* (see Sec. **9**). The crossed-field and linear-beam tubes are of primary interest because they are capable of higher peak powers at microwave frequencies. Gridded tubes such as triodes and tetrodes are sometimes used at UHF and below. Since these applications are relatively few, gridded tubes will not be described here (see Pars. **9-12 to 9-23**).

61. Modulators. If a pulsed radar transmitter is to obtain high efficiency, the current in the output tube must be turned off between pulses. The modulator performs this function by acting as a switch, usually in series with the anode current path. Some rf tubes have control electrodes that can also be used to provide the modulation function. There are three kinds of modulators in common use today: the line-type modulator, magnetic modulator, and active-switch modulator. Their characteristics are compared in Table 25-9.

The *line-type modulator* is the most common and is often used to pulse a magnetron transmitter. A typical circuit including the high-voltage power supply and magnetron is shown in Fig. 25-50. Between pulses, the pulse-forming network (PFN) is charged. A trigger fires the thyratron V_1, shorting the input to the PFN, which causes a voltage pulse to appear at the transformer T_1. The PFN is designed to produce a rectangular pulse at the magnetron cathode, with the proper voltage and current to cause the magnetron to oscillate. The line-type modulator is relatively simple but has an inflexible pulse width.

Active-switch modulators are capable of varying their pulse width within the limitation of the energy stored in the high-voltage power supply. A variety of active-switch cathode pulse modulators is shown in Fig. 25-51. Active-switch modulators using a vacuum tube free of gas but capable of passing high current and holding off high voltage are called *hard-tube modulators*.

The *magnetic modulator*, a third type of cathode pulse modulator (Fig. 25-52) has the advantage that no thyratron or switching device is required. Its operation is based on the saturation characteristics of inductors L_2, L_3, and L_4. A long, low-amplitude pulse is applied to L_1 to charge

Table 25-9. Comparison of Modulators

Modulator	Fig.	Flexibility		Pulse-length capability		Pulse flatness	Crowbar required		Modulator voltage level
		Duty cycle	Mixed pulse lengths	Long	Short		Load arc	Switch arc	
Line-type: Thyratron/SCR	25-50	Limited by charging circuit	No	Large PFN	Good	Ripples		No	Medium/Low
Magnetic modulator	25-52	Limited by reset and charging time	No	Large C's and PFN	Good	Ripples		No	Low
Hybrid SCR– magnetic modulator	...	Limited by reset and charging time	No	Large C's and PFN	Good	Ripples		No	Low
Active switch: Series switch	25-51a	No limit	Yes	Excellent; large capacitor bank	Good	Good	Maybe	Yes	High
Capacitor-coupled	25-51b	Limited	Yes	Large coupling capacitor	Good	Good	Maybe	Yes	High
Transformer-coupled	25-51c	Limited	Yes	Difficult: XF gets big; large capacitor bank	Good	Fair	Maybe	Yes	Medium-high
Modulator anode	...	No limit	Yes	Excellent; large capacitor bank	OK, but efficiency low*	Excellent	Yes	Yes	High
Grid	...	No limit	Yes	Excellent; large capacitor bank	Excellent	Excellent	Yes	...	Low

* Unless ON and OFF tubes carry very high peak current or unless modulator anode has high mu.   After Weil. Ref. 52. Par. 25-161.

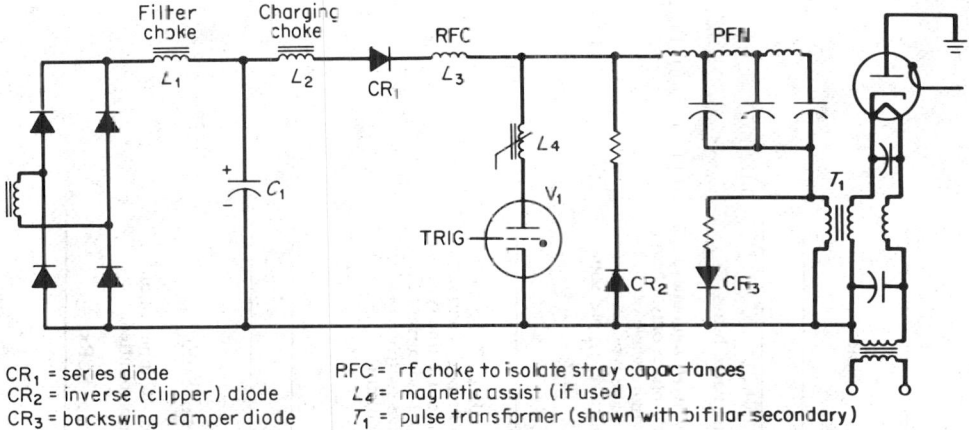

CR$_1$ = series diode
CR$_2$ = inverse (clipper) diode
CR$_3$ = backswing camper diode

RFC = rf choke to isolate stray capacitances
L_4 = magnetic assist (if used)
T_1 = pulse transformer (shown with bifilar secondary)

Fig. 25-50. Line-type modulator.[52]

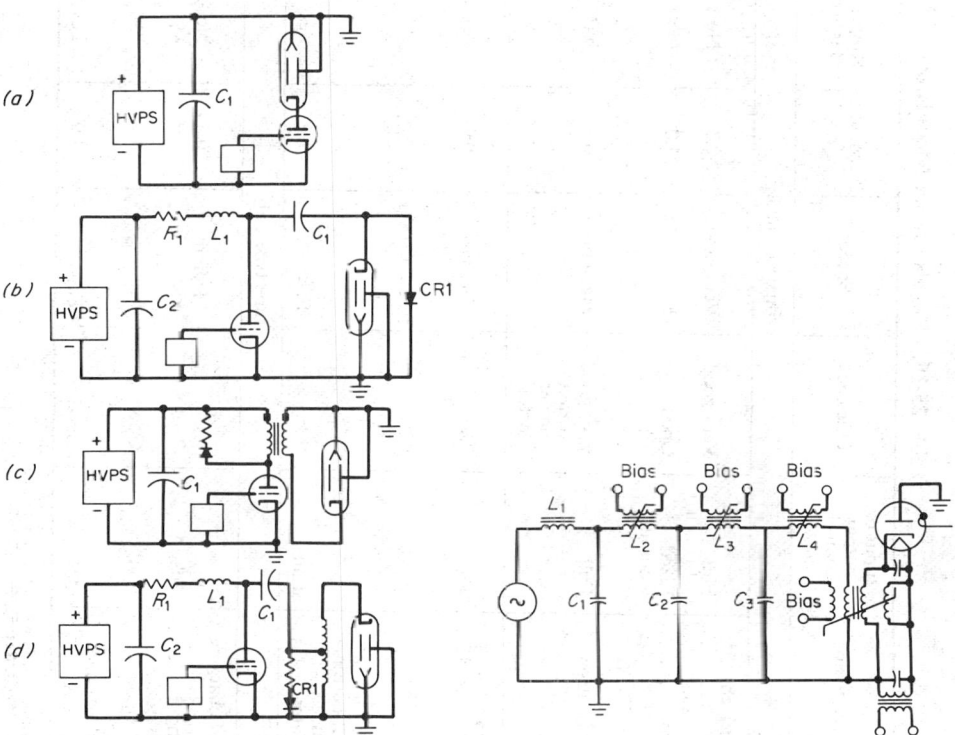

Fig. 25-51. Active-switch cathode pulsers. (*a*) direct-coupled; (*b*) capacitor coupled; (*c*) transformer coupled; (*d*) capacitor- and transformer-coupled.[52]

Fig. 25-52. Magnetic modulator.[52]

Table 25-10. Regulators for Modulators

Type of regulator	Input or output control range	Efficiency	Speed	Accuracy, %	Ripple reduction	Notes
Primary control						
Motor-driven Variac*	Full	Very good	Very slow	0.3	No	Moving parts, heavy.
Ferroresonant regulator	Fixed output	Good	Fast	1.0	No	Heavy, fixed frequency, single phase only. May be square-wave or sinewave type.
SCR or ignitron ..	Full	Very good (lowers power factor)	Medium	0.1	No	Small, raises ripple, some RFI.
Dc series regulator	Full	Poor	Very fast	0.01	Yes	Uses tubes. Tube must handle peak charging current (in line type) or peak load current (in active-switch modulator).
Constant-current hard-tube modulator	Full	Poor	Fast	1.0	Yes	
Line type only						
Damped charging choke	30%	Poor	Medium	0.5	No	Tubes.
De-Q-ing	30%	Poor	Fast	1.0	Yes	Tubes or SCRs.
Series triode and return diode	Full	Good	Fast	1.0	Yes	Tubes. Clipper circuit can be omitted.
Series diode and return SCR	30%	Good	Fast	1.0	Yes	Tubes or SCRs.

* Or equivalent Powerstat, Inductrol, etc.
SOURCE: Weil, Ref. 52, Par. 25-161.

C_1. When C_1 is nearly charged, L_2 saturates, and the energy in C_1 is transferred resonantly to C_2. The process is continued to the next stage, where the transfer time is about one-tenth that of the stage before. The energy in the pulse is nearly maintained so that at the end of the chain a short-duration high-amplitude pulse is generated.

62. Power Supplies and Regulators. The power supply converts prime power from the ac line to dc power, usually at a high voltage. The dc power must be regulated to remove the effects of line-voltage and load variation. A comparison of various regulators is given in Table 25-10.

Protective circuitry is usually included with the high-voltage power supply to prevent the rf tube from being damaged in the event of a fault. Improper triggers and tube arcs are detected and used to trigger a crowbar circuit that discharges the energy stored in the high-voltage power supply. The crowbar is a triggered spark gap capable of dissipating the full energy of the power supply. Thyratrons, ignitrons, ball gaps, and triggered vacuum gaps are used.

Stability. Radar systems with moving-target-indicator (MTI) place unusually tight stability requirements on their transmitters. Small changes in the amplitude, phase, or frequency from one pulse to the next can degrade MTI performance. In the transmitter, the MTI requirements appear as constraints on voltage, current, and timing variations from pulse to pulse. The relation between voltage variations and variation in amplitude and phase shift differs with the tube type used. Table 25-11 lists stability factors for the various tube types used with a high-voltage power supply (HVPS).

63. Radar Antennas. The great variety of radar applications has produced an equally great variety of radar antennas. These vary in size from less than a foot to hundreds of feet in diameter. Since it is not feasible even to mention each of the types here, we shall discuss the three basic antenna categories, search antennas, track antennas, and multifunction array antennas, after first reviewing some basic antenna principles.

A radar antenna directs the radiated power and receiver sensitivity to the azimuth and elevation coordinates of the target. The ability of an antenna to direct the radiated power is

Table 25-11. Stability Factors

	Frequency- or phase-modulation sensitivity	Impedance ratio	Current or voltage change for 1% change in HVPS voltage	
		Dynamic static	Line-type modulator	Low-impedance* hard-tube modulator, or dc operation
Magnetron	$\frac{\Delta f}{f} = \left(\begin{matrix}0.001\\ \text{to}\\ 0.003\end{matrix}\right)\frac{\Delta I}{I}$	0.05–0.1	$\Delta I = 2\%$	$\Delta I = 10\text{–}20\%$
Stabilotron or stabilized magnetron	$\frac{\Delta f}{f} = \left(\begin{matrix}0.0002\\ \text{to}\\ 0.0005\end{matrix}\right)\frac{\Delta I}{I}$	0.05–0.1	$\Delta I = 2\%$	$\Delta I = 10\text{–}20\%$
Backward-wave CFA	$\Delta\phi = 0.4$ to $1°$ for 1% $\Delta I/I$	0.05–0.1	$\Delta I = 2\%$	$\Delta I = 10\text{–}20\%$
Forward-wave CFA	$\Delta\phi = 1.0$ to $3.0°$ for 1% $\Delta I/I$	0.1–0.2	$\Delta I = 2\%$	$\Delta I = 5\text{–}10\%$
Klystron	$\frac{\Delta\phi}{\phi} = \frac{1}{2}\frac{\Delta E}{E} \quad \phi \approx 5\lambda$ $\Delta\phi \approx 10°$ for 1%$\Delta E/E$	0.67	$\Delta E = 0.8\%$	$\Delta E = 1\%$
TWT	$\frac{\Delta\phi}{\phi} \approx \frac{1}{3}\frac{\Delta E}{E} \quad \phi \approx 15\lambda$ $\Delta\phi \approx 20\%$ for 1%$\Delta E/E$	0.67	$\Delta E = 0.8\%$	$\Delta E = 1\%$
Triode or tetrode	$\Delta\phi = 0$ to $0.5°$ for 1%$\Delta I/I$	1.0	$\Delta I = 1\%$	$\Delta I = 1\%$

* A high-impedance modulator is not listed because its output would ideally be independent of HVPS voltage.

SOURCE: Weil, Ref. 52, Par. 25-161.

described by its antenna pattern. A typical antenna pattern is shown in Fig. 25-53. It is a plot of radiated field intensity measured in the far field (a distance greater than twice the diameter squared divided by the wavelength from the antenna) and is plotted as a function of azimuth and elevation angle. Single cuts through the two-dimensional pattern, as shown in Fig. 25-54, are more often used to describe the pattern. The principle of reciprocity assures that the antenna pattern describes its gain as a receiver as well as transmitter. The gain is defined relative to an isotropic radiator.

The gain used as a defining parameter is the gain at the peak of the beam or main lobe (see Fig. 25-54). This is the one-way power gain of the antenna

$$G_p = 4\pi A\eta_a/\lambda^2$$

where A = area of antenna aperture (reflector area for a horn-fed reflector antenna), λ = radar wavelength in units of A, and η_a = aperture efficiency, which accounts for all losses inherent in the process of illuminating the aperture. Tapered-aperture-illumination functions designed to produce low side lobes also result in lower aperture efficiency η_a and larger beamwidth θ_3, as shown in Fig. 25-55. A second gain definition sometimes used is directive gain. This is defined

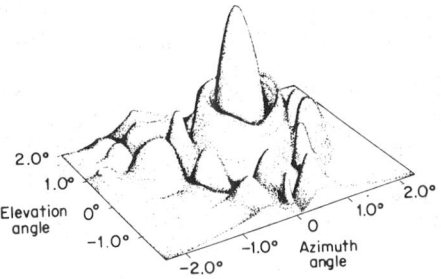

Fig. 25-53. Three-dimensional pencil-beam pattern of the AN/FPQ-6 radar antenna. *(Courtesy of D. D. Howard, Naval Research Laboratory.)*

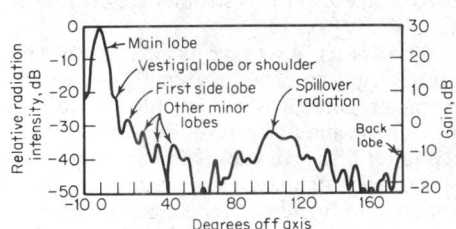

Fig. 25-54. Radiation pattern for a particular paraboloid reflector antenna illustrating the main-lobe and side-lobe radiation.[53]

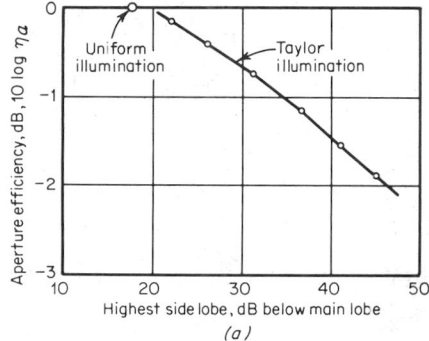

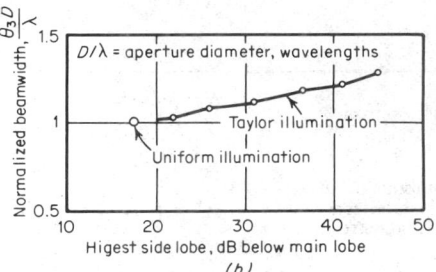

Fig. 25-55. Aperture efficiency and beamwidth as a function of highest side-lobe level for a circular aperture: (a) aperture efficiency; (b) normalized beamwidth.[55]

as a maximum radiation intensity, in watts per square meter, divided by the average radiation intensity, where the average is taken over all azimuth and elevation angles.

Directive gain can be inferred from the product of the main-lobe widths in azimuth and elevation, over a wide range of tapers (including uniform). For example, an array antenna, with no spillover of illumination power, gives

$$G_d = 36,000/\theta_{3a}\theta_{3e} \qquad (25\text{-}73a)$$

where θ_{3a} = 3-dB width of the main lobe in the azimuth coordinate (degrees) and θ_{3e} = 3-dB main-lobe width in the elevation coordinate (degrees). For horn-fed antennas the constant in Eq. (25-73a) is about 25,000.

64. Search Antennas. Two examples of search-radar antennas illustrating the variety of shapes and sizes are shown in Fig. 25-56 (an airborne weather radar antenna used to locate storms in the path of the aircraft) and Fig. 25-57 (a conventional air-search radar antenna).

Conventional surface and airborne search radars generally use mechanically scanned horn-fed reflectors for their antennas. The horn radiates a spherical wavefront that illuminates the reflector. The shape of the reflector is designed to cause the reflected wave to be in phase at any point on a plane in front of the reflector. This focuses the radiated energy at infinity. Mechanically

Fig. 25-56. RDR-1200 airborne weather radar system. (*Bendix Corporation.*)

Fig. 25-57. AN/TPN-19 airport surveillance radar. (*Raytheon Co.*)

scanning search radars generally have fan-shaped beams that are narrow in azimuth and wide in the elevation coordinate. In a typical surface-based air-search radar the upper edge of the beam is shaped to follow a cosecant-squared function. This provides coverage up to a fixed altitude. Figure 25-58 illustrates the effect of cosecant-squared beam shaping on the coverage diagram as well as on the antenna pattern. In the horn-fed reflector the shaping can be achieved by either the reflector or the feed, and the gain constant in Eq. (25-73a) is reduced to about 20,000.

65. Tracking-Radar Antennas. The primary function of a tracking radar is to make accurate range and angle measurements of a selected target's position. Generally, only a single target position is measured at a time, as the antenna is directed to follow the target by tracking servos. These servos smooth the errors measured from beam center to make pointing corrections. The measured errors, along with the measured position of the antenna, provide the target-angle information.

Tracking antennas like that of the AN/FPS-16 use circular apertures to form a pencil beam about $1°$ wide in each coordinate. The higher radar frequencies (S, C, and X band) are preferred because they allow a smaller aperture for the same beamwidth. The physically smaller antenna can be more accurately pointed. In this section we discuss aperture configurations, feeds, and pedestals.

One of the simplest methods of producing an equiphase wavefront in front of a circular aperture uses a parabolic reflector. A feed located at the focus directs its energy to illuminate the reflector. The reflected energy is then directed into space focused at infinity. The antenna is inherently broadband because the electrical path length from the feed to the reflector to the plane wavefront is the same for all points on the wavefront.

Locating the feed in front of the aperture is sometimes inconvenient mechanically. It also produces spillover lobes where the feed pattern misses the reflector. The Cassegrain antenna shown in Fig. 25-59 avoids these difficulties by placing a hyperbolic subreflector between the

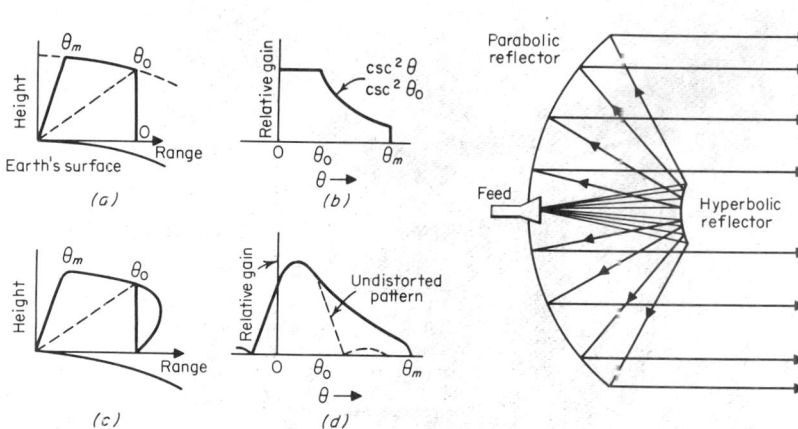

Fig. 25-58. Elevation coverage of a cosecant-squared antenna: (a) desired elevation coverage; (b) corresponding antenna pattern desired; (c) realizable elevation coverage with pattern shown in (d); (d) actual cosecant-squared antenna pattern.[58]

Fig. 25-59. Schematic diagram of a Cassegrain reflector antenna.[59]

parabolic reflector and its focus. The feed now illuminates the subreflector, which in turn illuminates the parabola and produces a plane wavefront in front of the aperture.

Lenses can also convert the spherical wavefront emanating from the feed to a plane wavefront over a larger aperture. As the electromagnetic energy passes through the lens, it is focused at infinity (see Fig. 25-60). Depending on the index of refraction n of the lens, a concave or convex lens may be required. Lenses are typically heavier than reflectors, but they avoid the blockage caused by the feed or subreflector.

A single feed providing a single beam is unable to supply the angle-error information necessary for tracking. To obtain azimuth and elevation-error information, feeds have been developed than scan the beam in a small circle about the target (conical scan) or that form multiple beams about the target (monopulse). Conical scanning may be caused by rotating a reflector behind a dipole feed or rotating the feed itself. It has the advantage compared with monopulse that less hardware is required in the receiver, but at the expense of somewhat less accuracy.

Modern trackers more often use a monopulse feed with multiple receivers. Early monopulse feeds used four separate horns to produce four contiguous beams that were combined to form a reference beam and azimuth and elevation-difference beams. More recently, multimode feeds have been developed to perform this function more efficiently with fewer components (see Fig. 25-45).

Shaft angle encoders quantize radar pointing angles through mechanical connections to azimuth and elevation axes. The output indicates the angular position of the mechanical bore-site axis relative to a fixed angular coordinate system. Because these encoders make an absolute measurement, their outputs contain 10 to 20 bits of information. A variety of techniques is used, the complexity increasing with the accuracy required. Atmospheric errors ultimately limit the number of useful bits to about 20, or 0.006 mrad. In less precise tracking applications, synchros attached to the azimuth and elevation axes indicate angular position within a fraction of a degree.

66. Multifunction Arrays. *Array antennas* form a plane wavefront in front of the antenna aperture. These points are individual radiating elements which, when driven together, constitute the array. The elements are usually spaced about 0.6 wavelength apart. Most applications use planar arrays, although arrays conformal to cylinders and other surfaces have been built.

Phased arrays are steered by tilting the phase front independently in two orthogonal directions called the *array coordinates*. Scanning in either array coordinate causes the beam to move along a cone whose center is at the center of the array. The paths the beam follows when steered in the array coordinates are illustrated in Fig. 25-61, where the z axis is normal to the array. As the

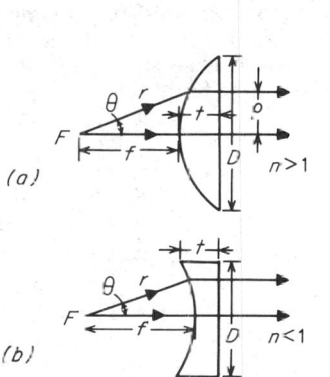

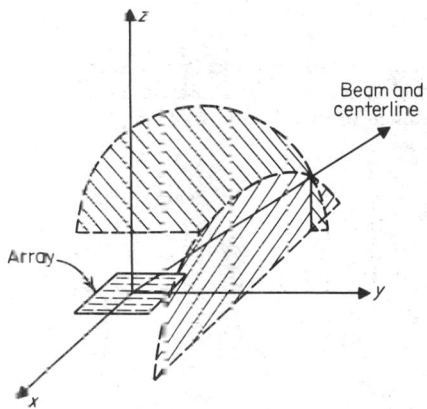

Fig. 25-60. Geometry of simple converging lenses: (a) $n > 1$; (b) $n < 1$.[59]

Fig. 25-61. Beam-steering contours for a planar array.

beam is steered away from the array normal, the projected aperture in the beam's direction varies, causing the beamwidth to vary proportionately.

Arrays can be classified as either active or passive. *Active arrays* contain duplexers and amplifiers behind every element or group of elements; *passive arrays* are driven from a single feed point. Only the active arrays are capable of higher power than conventional antennas.

Both passive and active arrays must divide the signal from a single transmission line among all the elements of the array. This can be done by an optical feed, a corporate feed, or a multiple-beam-forming network. The *optical feed* is illustrated in Fig. 25-62. A single feed, usually a monopulse horn, illuminates the array with a spherical phase front. Power collected by the rear elements of the array is transmitted through the phase shifters that produce a planar front and steer the array. The energy may then be radiated from the other side of the array, as in the lens, or be reflected and reradiated through the collecting elements, where the array acts as a steerable reflector.

Corporate feeds can take many forms, as illustrated by the series-feed networks shown in Fig. 25-63 and the parallel-feed networks shown in Fig. 25-64. All use transmission-line components to divide the signal among the elements. Phase shifters can be located at the elements or within the dividing network.

Multiple-beam networks are capable of forming simultaneous beams with the array. The Butler matrix shown in Fig. 25-65 is one such technique. It connects the N elements of a linear array to N feed points corresponding to N beam outputs. It can be applied to two-dimensional arrays by dividing the array into rows and columns.

The phase shifter is one of the most critical components of the array. It produces controllable phase shift over the operating band of the array. Digital and analog phase shifters have been developed using both ferrites and *pin* diodes. Phase shifter designs always strive for a low-cost, low-loss, and high-power-handling capability.

The Reggia-Spencer phase shifter consists of a ferrite inside a waveguide, as illustrated in Fig. 25-66. It delays the rf signal passing through the waveguide. The amount of phase shift can be controlled by the current in the solenoid, through its effect on the permeability of the ferrite. This is a reciprocal phase shifter which has the same phase shift for signals passing in either direction. Nonreciprocal phase shifters (where phase-shift polarity reverses with the direction of propagation) are also available. Either reciprocal or nonreciprocal phase shifters can be locked or latched in many states by using the permanent magnetism of the ferrite.

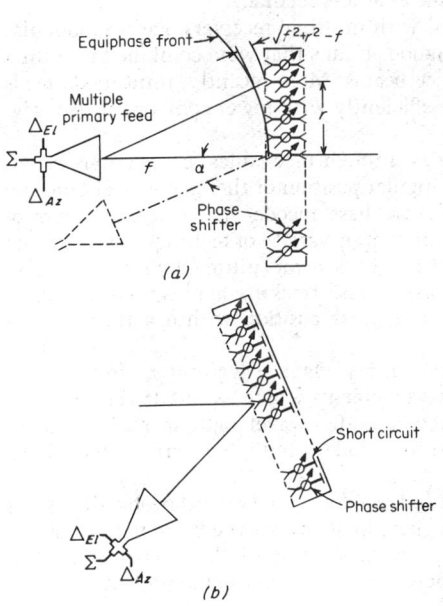

Fig. 25-62. Optical-feed systems: (a) lens; (b) reflector.[61]

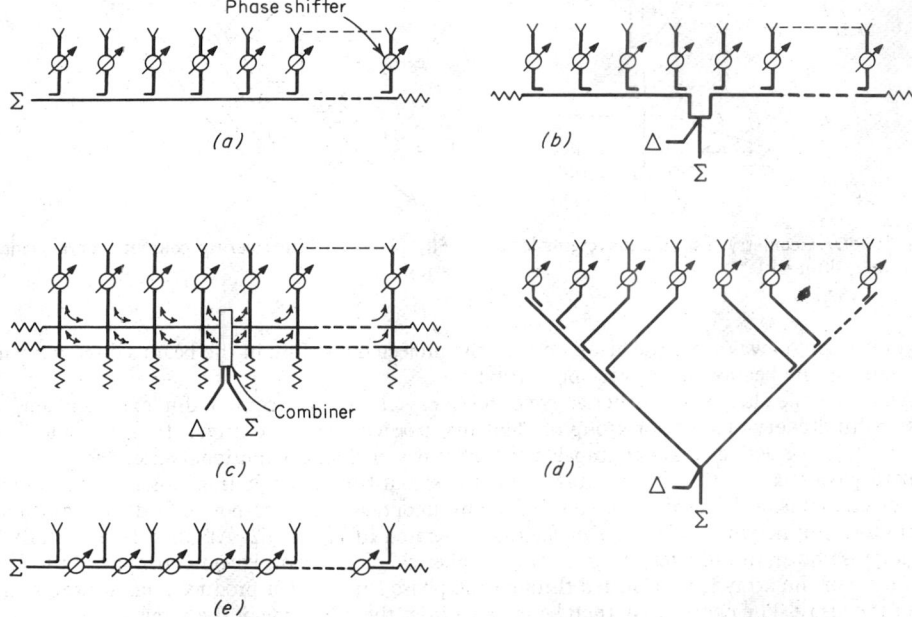

Fig. 25-63. Series-feed networks: (a) end feed; (b) center feed; (c) separate optimization; (d) equal path length; (e) series phase shifters.[61]

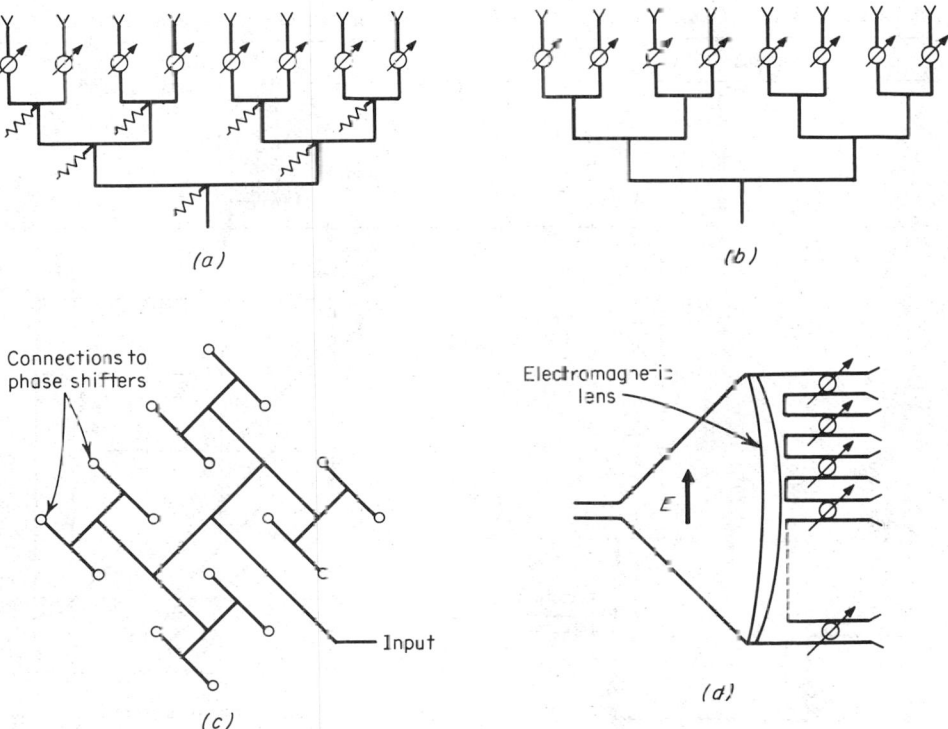

Fig. 25-64. Parallel-feed networks: (a) matched corporate feed; (b) reactive corporate feed; (c) reactive strip-line; (d) multiple reactive divider.[61]

Phase shifters have also been developed using pin diodes in transmission-line networks. One configuration, shown in Fig. 25-67, uses diodes as switches to change the signal path length of the network. A second type uses pin diodes as switches to connect reactive loads across a transmission line. When equal loads are connected with a quarter-wave separation, a pure phase shift results.

When digital phase shifters are used, a phase error occurs at every element due to phase quantization. The error in turn causes reduced gain, higher side lobes, and greater pointing errors. Gain reduction is tabulated in Table 25-12 for typical quantizations. Figure 25-68 shows the rms

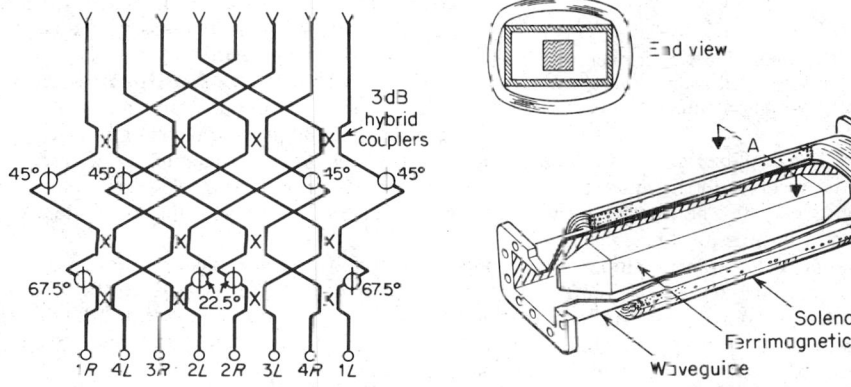

Fig. 25-65. Butler beam-forming network.[61]   **Fig. 25-66.** Typical Reggia-Spencer phase shifter.[62]

Table 25-12. Gain Loss in a Phased Array with m-Bit Digital Phase Shifters

Number of Bits, m	Gain Loss, dB
3	0.228
4	0.057
5	0.0142
6	0.00356
7	0.0089
8	0.0022

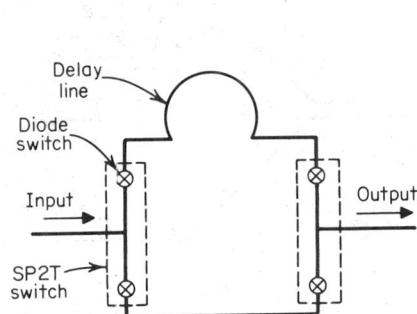

Fig. 25-67. Switched-line phase bit.[62]

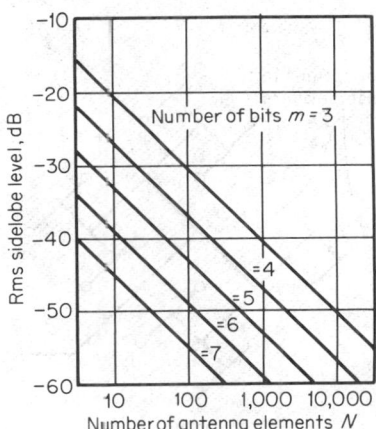

Fig. 25-68. RMS side lobes due to phase quantization.[61]

side-lobe levels caused by phase quantization in an array of N elements. The rms pointing error relative to the 3-dB beamwidth is given by

$$\sigma_\theta/\theta_3 \approx 1.12/2^m \sqrt{N}$$

where m = number of bits of phase quantization and N = number of elements in array.[55]

Frequency scan is a simple array-scanning technique that does not require phase shifters, drivers, or beam-steering computers. Element signals are coupled from points along a transmission line as shown in Fig. 25-69. The electrical path length between elements is much longer than the physical separation, so that a small frequency change will cause a phase change between elements large enough to steer the beam. The technique can be applied only to one array coordinate, so that in two-dimensional arrays, phase shifters are usually required to scan the other coordinate.

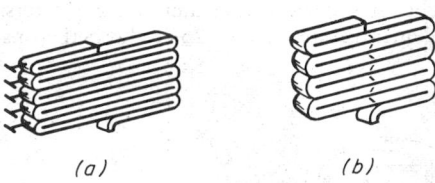

Fig. 25-69. Simple types of frequency-scanned antennas: (*a*) broad-wall coupling to dipole radiators; (*b*) narrow-wall coupling with slot radiators.[64]

67. Microwave Components. The radar transmitter, antenna, and receiver are all connected through rf transmission lines to a duplexer. The duplexer acts as a switch connecting the transmitter to the antenna while radiating and the receiver to the antenna while listening for echoes. Filters, receiver protectors, and rotary joints may also be located in the various paths. See Sec. **9** for a description of microwave devices and transmission lines.

A variety of other transmission-line components are used in a typical radar. Waveguide bends, flexible waveguide, and rotary joints are generally necessary to route the path to the feed of a rotating antenna. Waveguide windows provide a boundary for pressurization while allowing the microwave energy to pass through. Directional couplers sample forward and reverse power for monitoring, test, and alignment of the radar system.

68. Duplexers. The duplexer acts as a switch connecting the antenna and transmitter during transmission and the antenna and receiver during reception. Various circuits are used that depend on gas tubes, ferrite circulators, or *pin* diodes as the basic switching element. The duplexers using gas tubes are most common. A typical gas-filled TR tube is shown in Fig. 25-70. Low-power rf signals pass through the tube with very little attenuation. Higher power causes the gas to ionize and present a short circuit to the rf energy.

Figure 25-71 shows a balanced duplexer using hybrid junctions and TR tubes. When the transmitter is on, the TR tubes fire and reflect the rf power to the antenna port of the input hybrid.

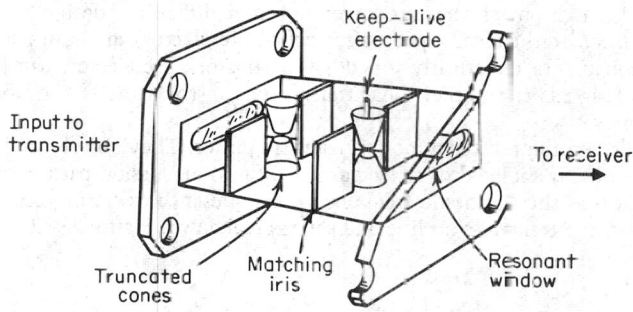

Fig. 25-70. Typical TR tube.[53]

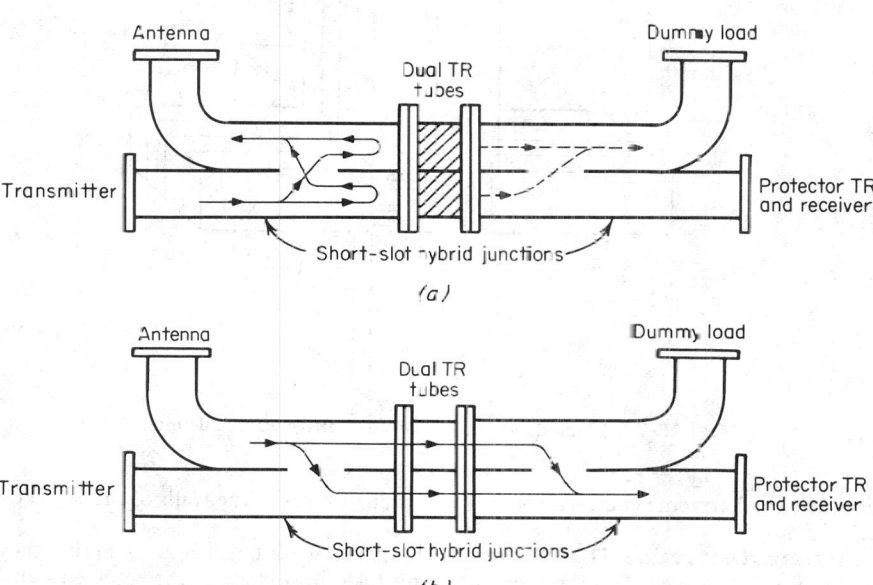

Fig. 25-71. Balanced duplexer using dual TR tubes and two short-slot hybrid junctions: (*a*) transmit condition; (*b*) receive condition.[53]

On reception, signals received by the antenna are passed through the TR tubes and to the receiver port of the output hybrid.

69. Circulators and Diode Duplexers. Newer radars often use a ferrite circulator as the duplexer. A TR tube is required in the receiver line to protect the receiver from the transmitter power reflected by the antenna due to an imperfect match. A four-port circulator is generally used with a load between the transmitter and receiver ports so that the power reflected by the TR tube is properly terminated.

In place of the TR tubes *pin* diode switches have been used in duplexers. These are more easily applied in coaxial circuitry and at lower microwave frequencies. Multiple diodes are used when a single diode cannot withstand the required voltage or current.

70. Receiver Protectors. TR tubes with a lower power rating are usually required in the receive line to prevent transmitter leakage from damaging mixer diodes or rf amplifiers in the receiver. A keep-alive ensures rapid ionization, minimizing spike leakage. The keep-alive may be either a probe in the TR tube maintained at a high dc potential or a piece of radioactive material. Diode limiters are also used after TR tubes to further reduce the leakage.

71. Filters. Microwave filters are sometimes used in the transmit path to suppress spurious radiation or in the receive signal path to suppress spurious interference. Because the transmit filters must handle high power, they are larger and more difficult to design.

Narrow-band filters in the receive path, often called *preselectors*, are built using mechanically tuned cavity resonators or electrically tuned YIG resonators. Preselectors can provide up to 80 dB suppression of signals from other radar transmitters in the same rf band but at a different operating frequency.

Harmonic filters are the most common transmitting filter. They absorb the harmonic energy to prevent it from being radiated or reflected. Since the transmission path may provide a high standing-wave ratio at the harmonic frequencies, the presence of harmonics can increase the voltage gradient in the transmission line and cause breakdown. Figure 25-72 shows a harmonic

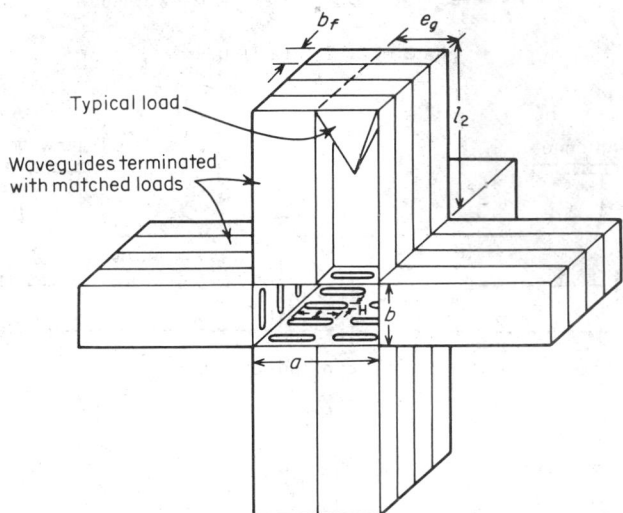

Fig. 25-72. Typical construction of a dissipative waveguide filter.[65]

filter where the harmonic energy is coupled out through holes in the walls of the waveguide to matched loads.

72. Radar Receivers. The radar receiver amplifies weak target returns so that they can be detected and displayed. The input amplifier must add little noise to the received signal, for this noise competes with the smallest target return that can be detected. A mixer in the receiver converts the received signal to an intermediate frequency where filtering and signal decoding can be accomplished. Finally, the signals are detected for processing and display.

73. Low-Noise Amplifiers. Because long-range radars require large transmitters and antennas, these radars can also afford the expense of a low-noise receiver. Considerable effort has been expended to develop more sensitive receivers. Some of the devices in use will be described here after a brief review of noise-figure and noise-temperature definitions.

Noise figure and noise temperature measure the quality of a sensitive receiver. Noise figure is the older of the two conventions and is defined as

$$F_n = \frac{\text{S/N at output}}{\text{S/N at input}}$$

where S = signal power, N = noise power, and the receiver input termination is at room temperature. Before low-noise amplifiers were available, a radar's noise figure was determined by the first mixer, which would be typically 5 to 10 dB. For these values of F_n it was approximately correct to add the loss of the waveguide to the noise figure when calculating signal-to-noise ratio. As better receivers were developed with lower noise figures, these approximations were no longer accurate, and the noise-temperature convention was developed.

Noise temperature is proportional to noise-power spectral density through the relation

$$T = N/kB$$

where k = Boltzmann's constant and B = bandwidth in which the noise power is measured. The noise temperature of an rf amplifier is defined as the noise temperature added at the input of the amplifier required to account for the increase in noise due to the amplifier. It is related to noise figure through the equation

$$T = T_0(F_n - 1)$$

where T_0 = standard room temperature = 290 K.

The receiver is only one of the noise sources in the radar system. Figure 25-73 shows the receiver in its relation to the other important noise sources. Losses, whether in the rf transmis-

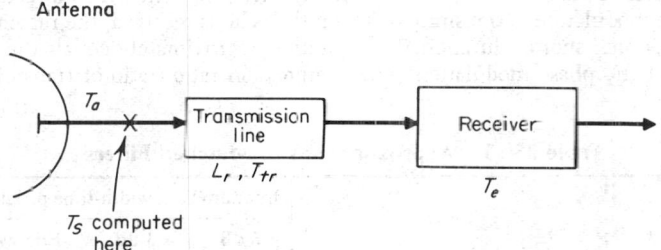

Antenna

T_a

Transmission line

L_r, T_{tr}

Receiver

T_e

T_s computed here

Fig. 25-73. Contributions to system noise temperature.

sion line, antenna, or the atmosphere, reduce the signal level and also generate thermal noise. The load presented to the rf transmission line by the antenna is its radiation resistance. The temperature of this resistance, T_a, depends on where the antenna beam is pointed. When the beam is pointed into space, this temperature may be as low as 50 K. However, when the beam is pointed toward the sun or a radio star, the temperature can be much higher. All these sources can be combined to find the system noise temperature T_s, according to the equation

$$T_s = T_a + (L_r - 1)T_{tr} + T_e L_r$$

where T_a = temperature of the antenna (see Par. **25-7**), L_r = transmission-line loss defined as ratio of power in to power out, T_{tr} = temperature of transmission line, and T_e = receiver noise temperature.

Figure 25-74 shows the noise temperature as a function of frequency for radar-receiver front ends. All are noisier at higher frequencies. Transistor amplifiers and uncooled parametric amplifiers are finding increased use in radar receivers. *Transistor amplifiers* have been improved steadily, with emphasis on increased operating frequency. Although the transistor amplifier is a much simpler circuit than the parametric amplifier, it does not achieve the parametric amplifier's low noise temperature.

The *parametric amplifier* is unusual in that it

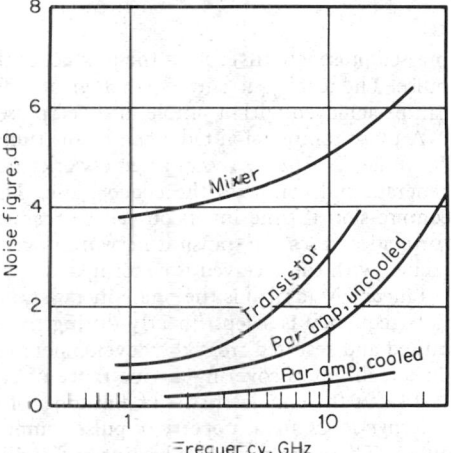

Fig. 25-74. Noise characteristics of radar front ends.

requires a high-frequency pump. Figure 25-75 shows a parametric amplifier circuit. A varactor diode acts as a capacitance, varying at the pump frequency, typically K band for a microwave amplifier. In this condition, the varactor presents a negative resistance to the circulator port 2. This in turn produces an amplified reflected signal from that port which leaves the circulator from port 3. Gains of 10 to 20 dB per stage and bandwidths up to 10% are typical.

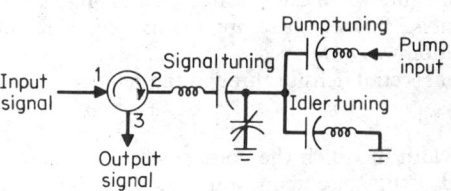

Fig. 25-75. Single-port parametric amplifier.[66]

A balanced mixer is often used to convert from rf to i.f. Balanced operation affords about 20 dB immunity to amplitude noise on the local-oscillator signal. Intermediate frequencies of 30 and 60 MHz are typical, as are 1.5 to 2 dB intermediate-frequency noise figures for the i.f. preamplifier. Double conversion is sometimes used with a first i.f. at a few hundred megahertz. This gives better image and spurious suppression.

The *matched filter* (Par. **25-18**) is usually instrumented at the second i.f. frequency. This filter is an approximation to the matched filter and therefore does not achieve the highest possible signal-to-noise ratio. This deficiency is expressed as mismatch loss. Table 25-13 lists the mismatch loss for various signal-filter combinations when the optimum bandwidth is used.

74. Pulse Compression. Pulse compression is a technique in which a rectangular pulse containing phase modulation is transmitted. When the echo is received, the matched-filter output is a pulse of much shorter duration. This duration approximately equals the reciprocal of the bandwidth of the phase modulation. The compression ratio (ratio of transmitted to com-

Table 25-13. Approximations to Matched Filters

Pulse shape	Filter	Optimum bandwidth-time product			Mismatch loss, dB
		6 dB	3 dB	Energy	
Gaussian	Gaussian bandpass	0.92	0.44	0.50	0
Gaussian	Rectangular bandpass	1.04	0.72	0.77	0.49
Rectangular	Gaussian bandpass	1.04	0.72	0.77	0.49
Rectangular	5 synchronously tuned stages	0.97	0.67	0.76	0.50
Rectangular	2 synchronously tuned stages	0.95	0.61	0.75	0.56
Rectangular	Single-pole filter	0.70	0.40	0.63	0.88
Rectangular	Rectangular bandpass	1.37	1.37	1.37	0.85
Rectangular chirp	Gaussian	1.04 × 6 dB width of equivalent $(\sin x)/x$ pulse, $(0.86 \times$ width of spectrum)			0.6

Source: Taylor and Mattern, Ref. 66, Par. **25-161**.

pressed pulse lengths) equals the product of the time duration and bandwidth of the transmitted pulse. The technique is used when greater pulse energy and range resolution are required than can be achieved with a simple uncoded pulse.

A pulse-compression radar can be instrumented in three basic ways, as illustrated in Fig. 25-76. In Fig. 25-76a two conjugate networks are used, one on transmit and the other on receive to generate and compress the coded pulse. The same network can be used for generation and compression if time inversion can be realized, as shown in Fig. 25-76b. The third technique, correlation, uses the transmit network to generate a replica of the transmitted pulse that is correlated with the received waveform, as shown in Fig. 25-76c.

Linear FM (chirp) is the phase modulation that has received the widest application. The carrier frequency is swept linearly during the transmitted pulse. The wide application has both caused and resulted from the development of a variety of dispersive analog delay lines. Delay-lines techniques covering a wide range of bandwidths and time durations are available. Table 25-14 lists the characteristics of a number of these dispersive delay lines.

Range lobes are a property of pulse-compression systems not found in radar using simple cw pulses. These are responses leading and trailing the principal response and resembling antenna side lobes; hence the name range lobes. These lobes can be reduced by carefully designing the phase modulation or by slightly mismatching the compression network. The mismatch can be described as a weighting function applied to the spectrum.

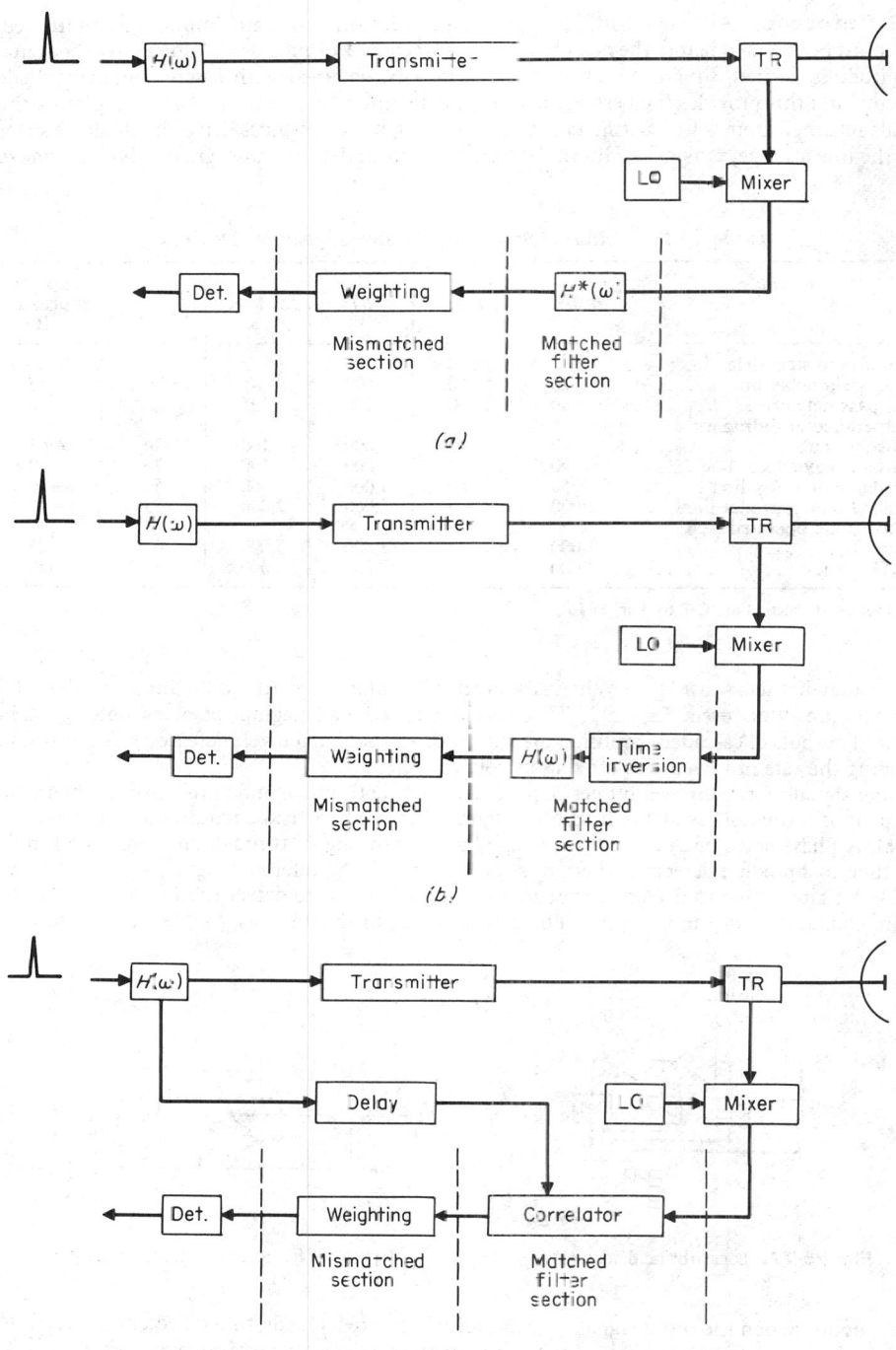

Fig. 25-76. Pulse-compression radar using (a) conjugate filters, (b) time inversion, and (c) correlation.[67]

75. Detectors. Although bandpass signals on an i.f. carrier are easily amplified and filtered, they must be detected before they can be displayed, recorded, or processed. When only the signal amplitude is desired, square-law characteristics may be obtained with a semiconductor diode detector, and this provides the best sensitivity for detecting pulses in noise when integrating the signals returned from a fluctuating target. Larger i.f. signal amplitudes drive the diode detector into the linear range, providing a linear detector. The linear detector has a greater dynamic range

Table 25-14.   Characteristics of Passive Linear-fm Devices

	B, MHz	T, μs	BT	f_0, MHz	Typical loss, dB	Typical spurious, dB
Aluminum strip delay line ...	1	500	200	5	15	-60
Steel strip delay line	20	350	500	45	70	-55
All-pass network	40	1,000	300	25	25	-40
Perpendicular diffraction delay line	40	75	1,000	100	30	-45
Surface-wave delay line	40	50	1,000	100	20	-50
Wedge-type delay line	250	65	1,000	500	50	-50
Folded-tape meander line	1,000	1.5	1,000	2,000	25	-40
Waveguide operated near cutoff	1,000	3	1,000	5,000	60	-25
YIG crystal	1,000	10	2,000	2,000	70	-20

SOURCE: Farnett et al., Ref. 67, Par. 25-161.

with somewhat less sensitivity. When still greater dynamic range (up to 80 dB) is required, log detectors are often used. Figure 25-77 shows the functional diagram of a log detector. The detected outputs of cascaded amplifiers are summed. As the signal level increases, stages saturate, reducing the rate of increase of the output voltage.

Some signal-processing techniques require detecting both phase and amplitude to obtain the complete information available in the i.f. signal. The phase detector requires an i.f. reference signal. A phase detector can be constructed by passing the signal through an amplitude limiter and then to a product detector, where it is combined with the reference signal, as shown in Fig. 25-78. An alternative to detecting the amplitude and phase is to detect the in-phase and quadrature components of the i.f. signal. The product detector shown in Fig. 25-78 can also provide

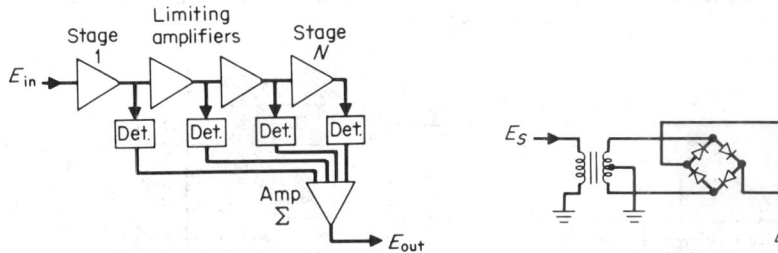

Fig. 25-77. Logarithmic detector.[66]   **Fig. 25-78.** Balanced-diode detector.[66]

this function when the input signal is not amplitude-limited. Quadrature detector circuits differ only in that the reference signal is shifted by 90° in one detector relative to the other.

76. Analog-to-Digital Converters. Digital signal processors require that the detected i.f. signals be encoded by an analog-to-digital converter. A typical converter may sample the detected signal at a 1-MHz rate and encode the sampled value into a 12-bit binary word. Encoders operating at higher rates have been built, but with fewer bits in their output. Encoders typically have errors that about equal the least significant bit.

77. Exciters. Two necessary parts of any radar system are an *exciter* to generate rf and local-oscillator frequencies and a *synchronizer* to generate the necessary triggers and timing pulses.

The components used in exciters are oscillators, frequency multipliers, and mixers. These can be arranged in various ways to provide the cw signals needed in the radar. The signals required depend on whether the transmitter is an oscillator or a power amplifier.

Transmitters using power oscillators such as magnetrons determine the rf frequency by the magnetron tuning. In a noncoherent radar, the only other frequency required is that of the local oscillator. It differs from the magnetron frequency by the i.f. frequency, and this difference is usually maintained with an automatic frequency control (AFC) loop. Figure 25-79 shows the

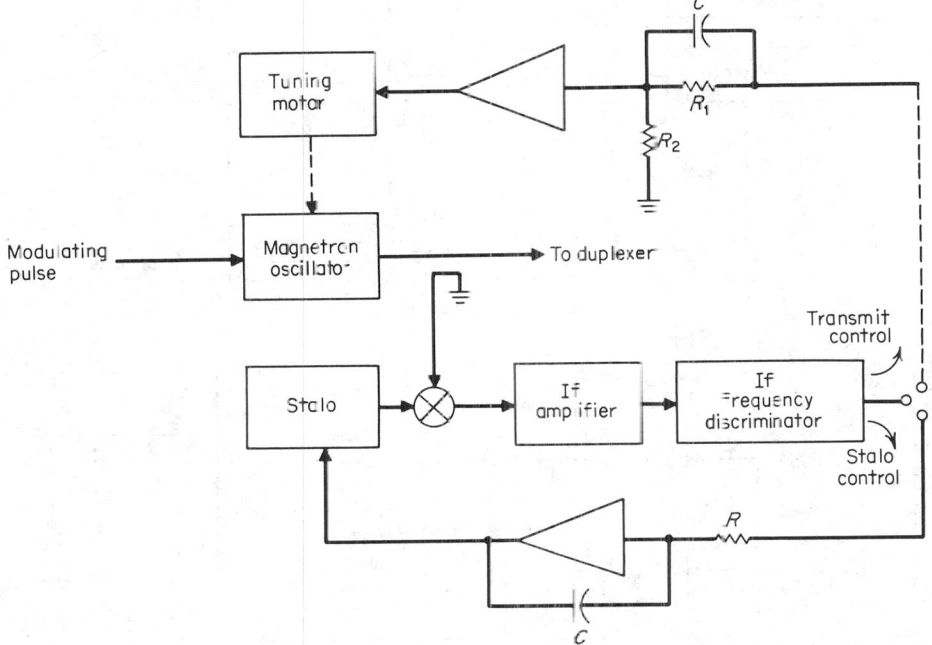

Fig. 25-79. Alternative methods for AFC control.[66]

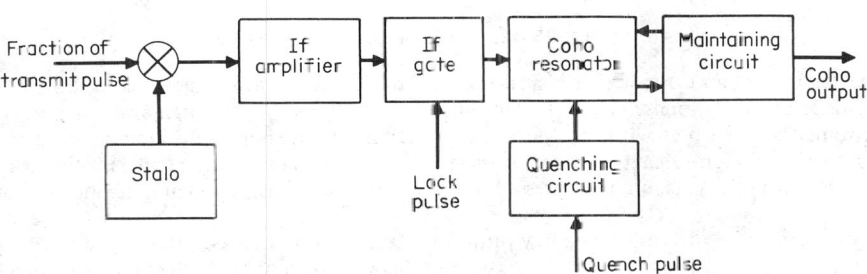

Fig. 25-80. Keyed coho.[66]

circuit of a simple magnetron radar, illustrating the two alternative methods of tuning the magnetron to follow the *stable local oscillator (stalo)* or tuning the stalo to follow the magnetron.

If the radar must use coherent detection (as in MTI or pulse doppler applications), a second oscillator, called a *coherent oscillator (coho)*, is required. This operates at the i.f. frequency and provides the reference for the product detector. Because an oscillator transmitter starts with random phase on every pulse, it is necessary to quench the coho and lock its phase with that of the transmitter on each pulse. This is accomplished by the circuit shown in Fig. 25-80.

When an amplifier transmitter is used, coho locking is not required. The transmit frequency can be obtained by mixing the stalo and coho frequencies, as shown in Fig. 25-81. The stalo and coho are not always oscillators operating at their output frequency. Figure 25-82 shows an exciter using crystal oscillators and multipliers to produce the rf and local-oscillator frequencies. Crystals may be changed to select the rf frequency without changing the i.f. frequencies.

The stability required of the frequencies produced by the exciter depends on the radar application. In a simple noncoherent radar a stalo frequency error shifts the signal spectrum in the

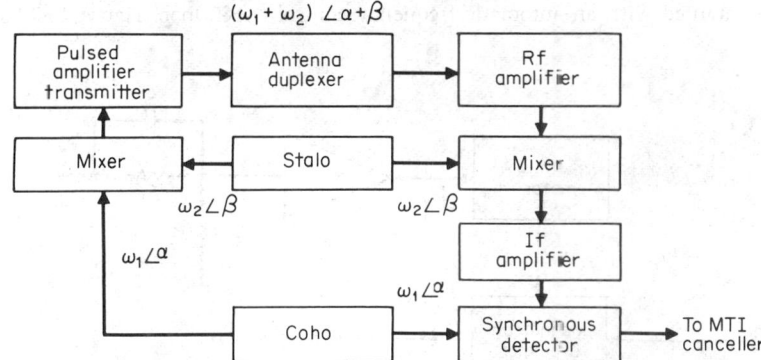

Fig. 25-81. Coherent radar.[66]

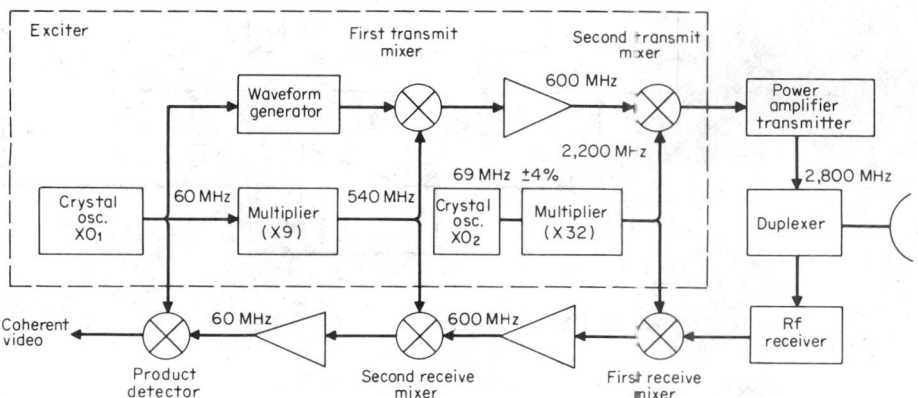

Fig. 25-82. Coherent-radar exciter.

i.f. passband, and an error which is a fraction of the i.f. bandwidth can be allowed. In MTI or pulse doppler radars, phase changes from pulse to pulse must be less than a few degrees. This requirement can be met with crystal oscillators driving frequency multipliers or fundamental oscillators with high-Q cavities when sufficiently isolated from electrical and mechanical perturbation. Instability is often expressed in terms of the phase spectrum about the center frequency.

Crystal oscillators driving frequency multipliers are finding increased use as stalos. A typical multiplier might multiply a 90-MHz crystal oscillator frequency by 32 to obtain an S-band signal. This source has the long-term stability of the crystal oscillators, but with degraded short-term stability. This is because the multiplier increases the phase modulation on the oscillator signal in proportion to the multiplication factor; i.e., each doubler stage raises the oscillator sidebands 6 dB. Frequency may be varied by tuning the crystal oscillator (about 0.25%) or by changing crystals.

78. Synchronizers. The synchronizer delivers timing pulses to the various radar subsystems. In a simple marine radar this may consist of a single multivibrator that triggers the transmitter, while in a larger radar 20 to 30 timing pulses may be needed. These may turn on and off the beam current in various transmitter stages; start and stop the rf pulse time attenuators; start display sweeps; etc.

Timing pulses or triggers are often generated by delaying a pretrigger with delays that may be either analog or digital. New radars are tending toward digital techniques, with the synchronizer incorporated into a digital signal processor. A diagram of the delay structure in a digital synchronizer is shown in Fig. 25-83. A 10-MHz clock moves the initial timing pulse through shift registers. The number of stages in each register is determined by the delay required. Additional analog delays provide a fine delay adjustment to any point in the 100-ns interval between clock pulses.

The synchronizer will also contain a range encoder, in radars where accurate range tracking is required or where range data will be processed or transmitted in digital form. Range is usually quantized by counting cycles of a clock, starting with a transmitted pulse and stopping with the

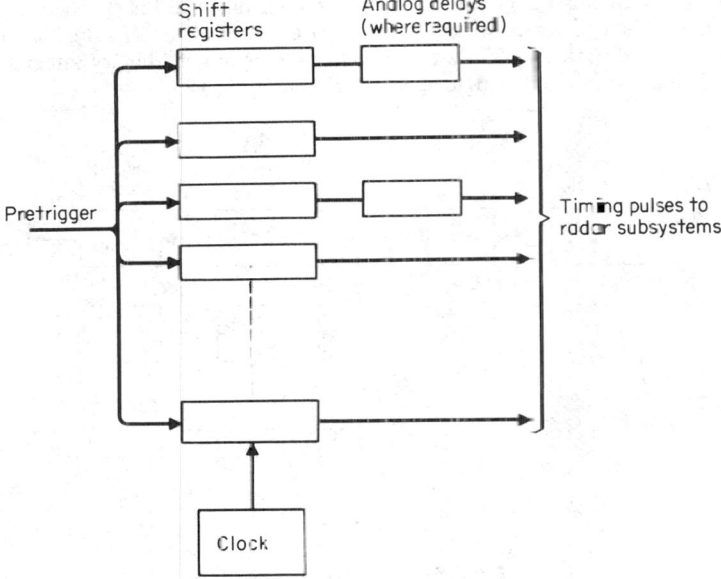

Fig. 25-83. Digital synchronizer.

received echo. Where high precision is required, the fine range bits are obtained by interpolating between cycles of the clock, using a tapped delay line with coincidence detectors on the taps.

79. Signal Processing. The term *signal processing* describes those circuits in the signal path between the receiver and the display. The extent to which processing is done in this portion of the signal path depends on the radar application. In search radars, postdetection integration, clutter rejection, and sometimes pulse compression are instrumented in the signal processor. The trend in modern radar has been to use digital techniques to perform these functions, although many analog devices are still in use. The following paragraphs outline the current technology trends in postdetection integration, clutter rejection, and digital pulse compression.

80. Postdetection Integration. Scanning-search radars transmit a number of pulses toward a target as the beam scans past. For best detectability, these returns must be combined before the detection decision is made. In many search radars the returns are displayed on a plan-position indicator (PPI), where the operator, by recognizing target patterns, performs the postdetection integration. When automatic detectors are used, the returns must be combined electrically. Many circuits have been used, but the two most common are the video sweep integrator and binary integrator.

The simplest video integrator uses a single delay line long enough to store all the returns from a single pulse. When the returns from the next pulse are received, they are added to the attenuated delay-line output. Figure 25-84 shows two forms of this circuit. The second (Fig. 25-84b) is preferred because the gain fac-

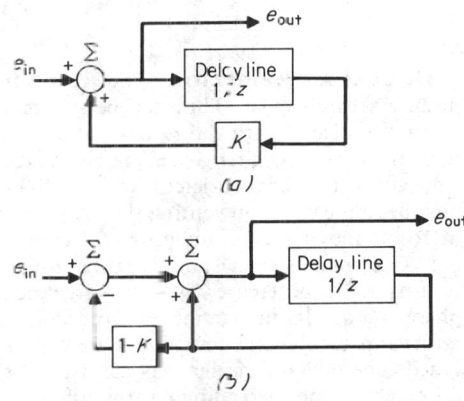

Fig. 25-84. Two forms of sweep integrator.[68]

tor K is less critical to adjust. The circuit weights past returns with an exponentially decreasing amplitude where the time constant is determined by K. For optimum enhancement

$$K = 1 - 1.56/N$$

where N = number of hits per one-way half-power beamwidth. By limiting the video amplitude into the integrator, the integrator eliminates single-pulse interference. The delay may be analog or digital, but the trend is to digital because the gain of the digital loop does not drift.

The binary integrator or double-threshold detector is another type of integration used in automatic detectors. With this integrator, the return in each range cell is encoded to 1 bit and the last N samples are stored for each range cell. If M or more of the N stored samples are 1s, a detection is indicated. Figure 25-85 shows a functional diagram of a binary integrator. This integrator is also highly immune to single-pulse interference.

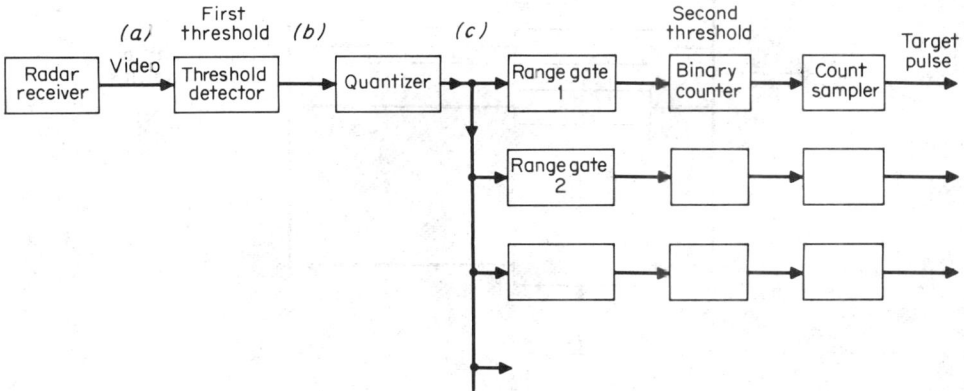

Fig. 25-85. Binary integrator.[53]

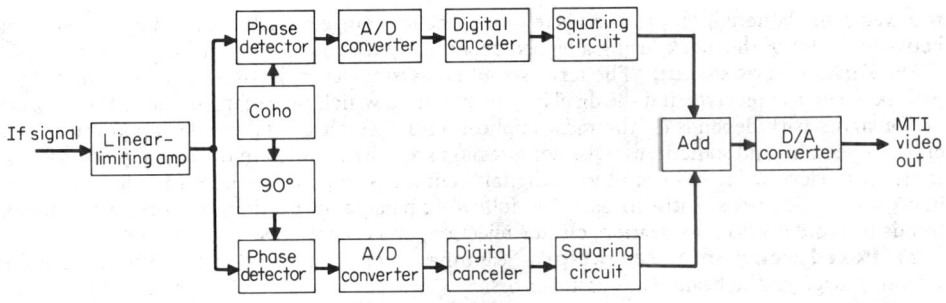

Fig. 25-86. Digital vector canceler.[68]

81. Clutter Rejection. The returns from land, sea, and weather are regarded as clutter in an air-search radar. They can be suppressed in the signal processor when the spectrum is narrow compared with the radar's pulse repetition rate (prf). Filters that combine two or more returns from a single-range cell are able to discriminate between the desired targets and clutter. This allows the radar to detect targets with cross section smaller than that of the clutter. It also provides a means of preventing the clutter from causing false alarms. The two classes of clutter filters are moving target indicator (MTI) and pulse doppler.

MTI combines a few pulse returns, usually two or three, in a way that causes the clutter returns to cancel. Figure 25-86 shows a functional diagram of a digital vector canceler. The in-phase and quadrature components of the i.f. signal vector are detected and encoded. Stationary returns in each signal component are canceled before the components are rectified and combined. The digital canceler may consist of a shift register memory and a subtractor to take the difference of the succeeding returns. Often only one component of the vector canceler is instrumented, thereby saving about half the hardware, but at the expense of signal detectability in noise.

A pulse doppler processor is another class of clutter filter where the returns in each range resolution cell are gated and put into a bank of doppler filters. The number of filters in the bank approximately equals the number of pulse returns combined. Each filter is tuned to a different frequency, and the passbands contiguously positioned between zero frequency and prf. Figure 25-87 shows a functional diagram of a pulse doppler processor. The pulse doppler technique is most often used in either airborne or land-based target-tracking radars, where a high ambiguous

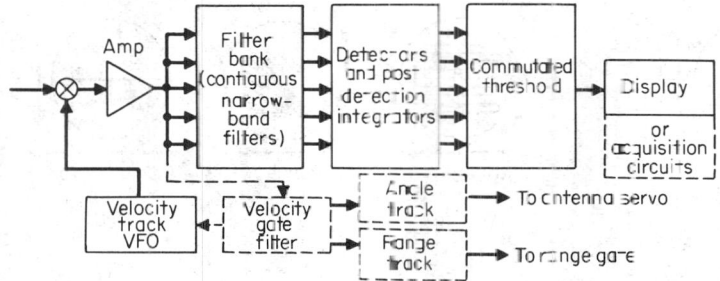

Fig. 25-87. Typical pulse doppler processor.[69]

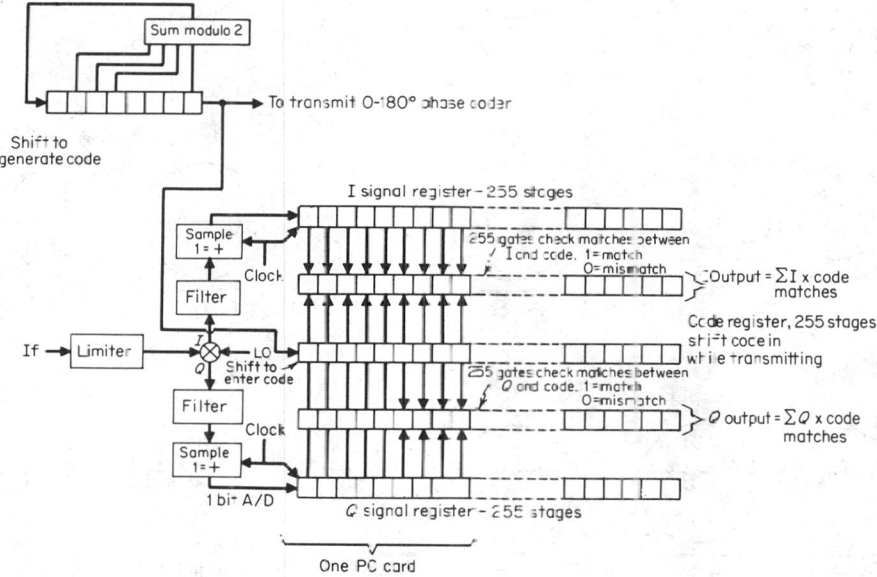

Fig. 25-88. Digital pulse compressor using 0 to 180° binary-phase-coded modulation.[70]

prf can be used, thus providing an unambiguous range of doppler frequencies. The filter bank may be instrumented digitally by a special-purpose computer wired according to the fast Fourier transform algorithm.

82. Digital Pulse Compression. Digital pulse compression performs the same function as the analog pulse compression described in Par. **25-74**, except that it is instrumented in the signal processor. Digital pulse compression is easiest to instrument with binary-phase-coded waveforms. The transmitted pulse is divided into equal-length subpulses, where the phase of each subpulse is set at 0 or 180° according to the phase code, and the number of subpulses equals the compression ratio. Figure 25-88 shows a functional diagram of a digital pulse compression processor. The quadrature components of the i.f vector are encoded to 1 bit, and shift registers, having one stage per subpulse, store a length of data equal to the transmitted pulse. Each of the subpulse bits is summed after reversing the bits corresponding to the positions of 180° phase

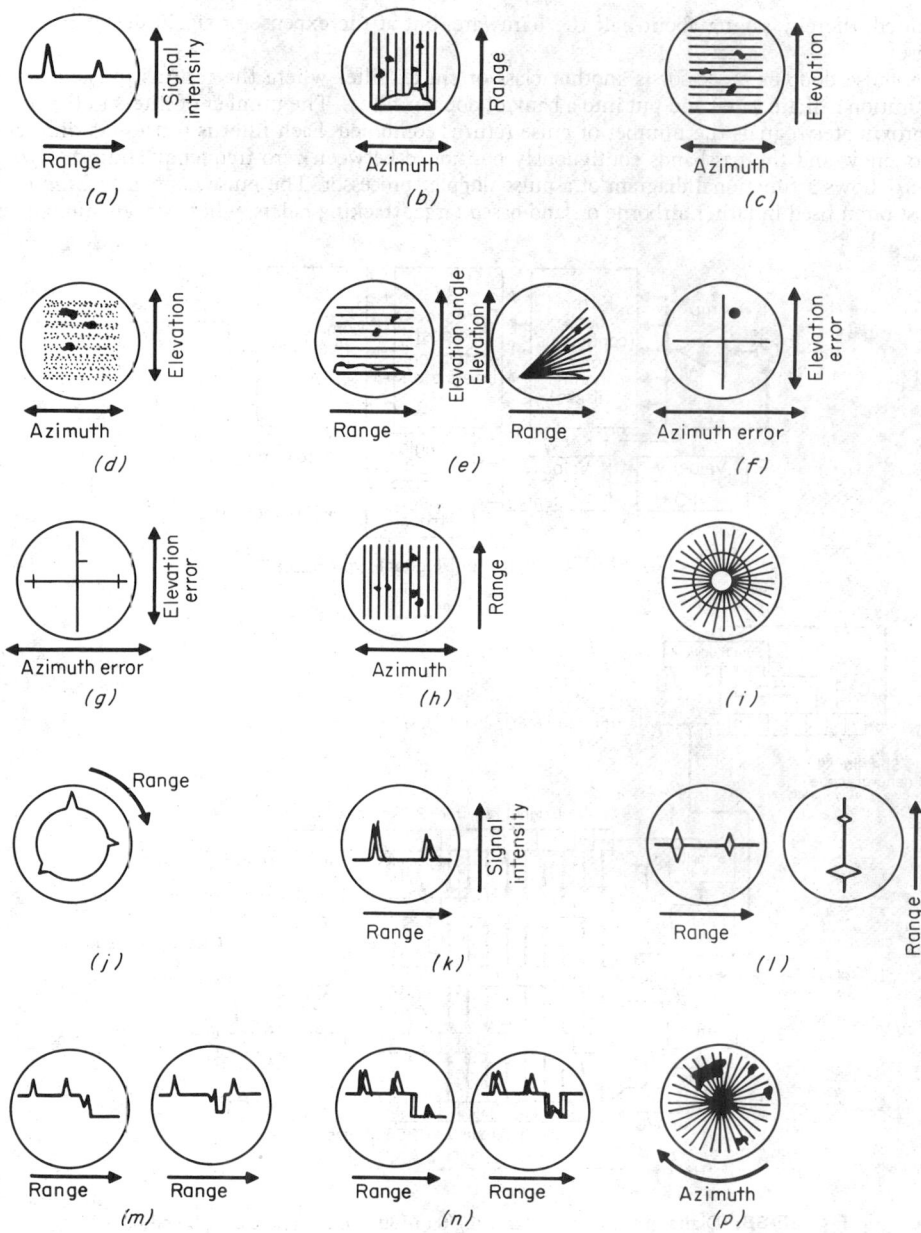

Fig. 25-89. Types of radar data presentation: (*a*) A presentation; (*b*) B presentation; (*c*) C presentation; (*d*) coarse-range information provided by position of signal in broad azimuth trace; (*e*) range-height indication; (*f*) single signal only; in the absence of a signal, the spot may be made to expand into a circle; (*g*) single signal only; signal appears as a winged spot, the position of which gives azimuth error and elevation error; length of wings is inversely proportional to range; (*h*) signal appears as two dots; left dot gives range, and azimuth of target; relative position of right dot gives rough indication of elevation; (*i*) antenna is conical; signal appears as circle, the radius of which is proportional to the range; the brightest part of the circle indicates the direction from the axis of the cone to the target; (*j*) same as type (*a*) except that the time base is circular and signals appear as radial deflections; (*k*) type (*a*) with antenna switching; spread voltage displaces signals from two antennas; when pulses are of equal size, antenna assembly is directed toward target; (*l*) same as type (*k*), but signals from two antennas are placed back to back; (*m*) type (*a*) with range step or range notch; when pulse is aligned with step or notch, range can be read from dial or counter; (*n*) a combination of types (*k*) and (*m*); (*p*) plan position; range is measured radially from the center.[71]

shift on transmission. A maximum output is obtained when the expanded pulse just fills the shift register. Shifting the pulse in the register by one subpulse destroys the match between the signal and the tap coding, causing the output to drop to a low value. This form of digital pulse compression has an advantage in that the code is easier to change than in analog pulse compression, where the coding is determined by a dispersive network.

83. Displays. Radar indicators are the coupling link between radar information and a human operator. Radar information is generally a time-dependent function of range or distance. Thus the display format generally uses a displacement of signal proportional to time or range. The signal may be displayed as an orthogonal displacement (such as an oscilloscope) or as an intensity modulation of brightening of the display. Most radar signal presentations have the intensity-modulated type of display where additional information such as azimuth, elevation angle, or height can be presented.

A common format is a polar-coordinate, or plan-position, indicator (PPI) which results in a maplike display. Radar azimuth is presented on the PPI as an angle, and range as radial distance. In a cartesian coordinate display, one coordinate may represent range and the other may represent azimuth elevation, or height. Variations may use one coordinate for azimuth and the other coordinate for elevation and gate a selected time period or range to the display. The increasing use of processed radar data can provide further variations. In each case, the display technique is optimized to improve the information transfer to the operator. Fifteen display formats with their scanning patterns are shown in Fig. 25-89.

The types J, K, L, M, and N formats are modifications of the type A display and use deflection modulation for signal presentation. Type R is an expanded-scale version of type A for precision range measurement. The types C, D, E, H, and I formats are modifications of the basic type B and type P (PPI) displays and are intensity-modulated. The type E format shows two variations: the left is elevation angle vs. range, and the right is elevation height vs. range. The latter format is generally used for range-height indicators (RHI). The PPI display can be presented either as a centered display or in various combinations of off-centered and sector scans for specific requirements.

Dual displays can be incorporated on a single cathode-ray tube. The precision approach radar (PAR) in ground-control approach (GCA) systems commonly has a type B scan located above a type E scan. The range axis of both formats is horizontal, the vertical scale being azimuth and elevation angle, respectively.

Marker signals may be inserted on the displays as operator aids. These can include fixed and variable range marks, strobes, and cursors as constant-angle or elevation traces. Alphanumeric data, tags, or symbols may be incorporated for operator information or designation of data.

84. Cathode-Ray Tubes. The cathode-ray tube (CRT) is the most common display device used for radar indicators. The cathode-ray tube is used most because of its flexibility of performance, resolution, dynamic range, and simplicity of hardware relative to other display techniques. Also, the cathode-ray tube has varied forms, and parameters can be optimized for specific display requirements (see Secs. 7 and 11).

Cathode-ray tubes using charge-storage surfaces are used for specialized displays (see Sec. 7). The direct-view storage tube is a high-brightness display tube. Other charge-storage tubes use electrical write-in and readout. Such tubes may be used for signal integration, for scan conversion so as to provide simpler multiple displays, for combining multiple sensors on a single display, for increasing viewing brightness on an output display, or for a combination of these functions.

Electronic Navigation Systems

By SVEN H. DODINGTON

85. Introduction. Some form of electronic navigation is used by all commercial airlines, by most military and general-aviation aircraft,[76] and by most ships.[77] In addition, electronic position-fixing systems are used in surveying, particularly in connection with offshore oil prospecting.[78]

While the known speed of propagation of radio waves allows good accuracies to be obtained in free space, multipath effects along the surface of the earth are the main enemies of practical airborne and shipborne systems, and there are consequently many different systems in use, none simultaneously satisfying the requirements for high accuracy, large service area, and low cost.

86. Categories and Terminology of Navigation Systems. *Service Area.* In both aviation and marine services, it is customary to divide systems into three classes:

1. Long-range, above 200 mi, mainly transoceanic
2. Medium-range, 20 to 200 mi, mainly above populated land masses and along coasts
3. Short-range, below 20 mi, in connection with approach, docking or landing

Cooperative or Self-Contained. Cooperative systems depend on transmission, one- or two-way, between one or more ground stations and the vehicle. These are capable of providing the vehicle with a fix, independent of its previous position. Self-contained systems are entirely contained in the vehicle and may be radiating or nonradiating. In general, they measure the distance traveled and have errors that increase with time or distance.

Data Rate. For map making, it may be acceptable to take a multiplicity of readings at one spot hours apart on different days, to achieve the highest possible redundancy and accuracy. At the other extreme, a fast-moving aircraft, during the approach and landing phase, may need positional updating at a rate of 10 times a second.

Cost. Systems designed for small, privately owned boats or aircraft have vehicular equipment costs similar to those of household electrical appliances, while those designed for high-accuracy military-weapons delivery often have vehicular equipment costs in the hundreds of thousands of dollars. In cooperative systems, it is generally the object to maximize the ground equipment cost and minimize the vehicular equipment cost, for a given total cost.

Accuracy. For transoceanic navigation, errors of a few miles may be acceptable, while for docking and landing, the required accuracies are measured in feet.

Safety. It has long been standard design philosophy that when a navigation aid fails, it shall give *no* information rather than false information. Designs are therefore favored which, as far as possible, automatically fail safely. Where this cannot be achieved, elaborate monitoring and alarm circuits become mandatory. These are sometimes more costly than the device they monitor.[80]

Vehicle-Derived or Ground-Derived. While the majority of navigation aids produce a direct readout on the vehicle, an important exception is a class of aids in which the position of the vehicle is read out on the ground and the resulting information transmitted to the vehicle by ordinary communication (voice, telegraphy, or data link). While this results in low vehicular cost (since the communication link is already available), such systems have not found favor, in part due to complexity, time delay, division of responsibility, and allocation of cost. However, in earth-controlled space navigation it has, so far, been standard practice to track the vehicle from the ground and then to send various corrective instructions to the vehicle.

87. Standardizing Agencies. Since aircraft and ships may travel to any part of the world and obviously do not wish to carry different equipment for each country, a large degree of international standardization is in effect for those navigation aids which depend on cooperation between a shore and a vehicular station. These standards, typically, take a decade to become established, and then remain in effect for several decades. Major agencies are:

International Civil Aviation Organization (ICAO, MONTREAL, Canada), an agency of the United Nations. Defines the signal characteristics, but not the hardware, for standard civil aviation systems, worldwide.

Federal Aviation Agency (FAA), Washington, D.C. Operates ground-based navigation aids and traffic control systems in the United States, for both civil and military aircraft.[80]

U.S. Coast Guard (USCG), Washington, D.C., operates shore-based navigation aids in the United States for shipping.

Radio Technical Commission for Aeronautics (RTCA), Washington, D.C., sets minimum performance standards for airborne equipment.

European Civil Aviation Electronics organization (Eurocae), Paris, sets standards for ground and airborne equipment, working closely with RTCA.

In each country, the military services operate additional navigation aids. Some of these are compatible with the international civil aids. For surveying, the systems are largely of a proprietary nature, each firm supplying the complete system.

88. Glossary of Navigation Terms

Accelerometer. A device that senses the force per unit mass along a given axis, due to acceleration of vehicle.

Angle of cut. The angle at which two lines of position intersect, preferably a right angle.

Approach path. The portion of the flight path between start of descent and touchdown.

Azimuth. The angle in the horizontal plane with respect to a fixed reference, usually true North, measured clockwise (refer to definition of bearing below).

Back course. In ILS, the course located on the opposite end of the runway behind the localizer.

Base line. The line joining two points between which electrical time is compared. Large base lines produce high instrument accuracy but may also introduce instrument ambiguities.

Bearing. An angle in the horizontal plane with respect to a reference, usually expressed in degrees measured clockwise from the reference. *Relative* bearing is to some arbitrary reference; *absolute* bearing is to North, usually magnetic North in navigation systems.

Bend. A departure of a course line from a straight line. It is usually oscillatory and generally caused by interference between direct and multipath signals (refer to definition of scalloping below).

Bore sighting. The process of aligning a directional antenna system, often by optical means.

CEP. Circular error probability. In a two-dimensional error distribution, the radius of a circle encompassing half the errors.

Chain. A network of stations operating as a group.

Clearance sector. In ILS, the sector from the course to the back course, in which sector it is desirable to maintain the left-right needle off scale.

Coherent pulses. Pulses used in navigation systems in which the phase of the radio-frequency cycles within the pulse is retained for measurement purposes (as in Loran C, as contrasted with Loran A, which uses only the envelope).

Cone of ambiguity. In VOR and Tacan, the conical volume of airspace above the beacon in which bearing information is below specification.

Cone of silence. The conical volume above an antenna where field strength is relatively low.

Course. The intended direction of travel. Also the direction defined by a navigation aid.

Course-line computer. A vehicle-carried device which converts navigation signals into courses not generated directly by the signals themselves (for example, hyperbolic to straight-line, rho-theta to straight-line).

Course softening. Intentional decrease in course sensitivity as the navigation aid is approached.

CPE. Circular probable error. Same as CEP

Crab angle. Correction angle to compensate for wind drift. The angular difference between course and heading.

Dead reckoning. Determination of position at one time with respect to known position at a previous time, by the application of course and distance information derived without reference to external aids.

Decision gate. In ILS, the point at which the pilot must decide to land or to execute a missed-approach procedure.

Drift angle. Angle between heading and track, due to effect of wind or water currents.

Error, attitude. Varies with attitude of vehicle Often related to polarization error.

Error, flight technical. Error due to failure of aircraft to follow prescribed path.

Error, instrument. Error caused by the equipment itself.

Error, polarization. Varies with polarization of antenna at one or both ends of signal path. Often related to attitude error.

Error, propagation. Error caused by variations in the propagation medium.

Error, readout. Error caused by failure of navigator to read his instrument properly.

Error, site. Error caused by reflections from obstructions close to the site of the navigation aid. Of major concern in directional systems.

Fix. A position determined without reference to a former position.

Flag alarm. An indicator on a navigation instrument to show when a reading is unreliable.

Flare-out. That part of the approach path which rapidly decreases the glide angle by nosing up the aircraft at touchdown.

GDOP. Geometrical dilution of precision. Loss of accuracy when angle of cut is not a right angle.

Geoid. The shape of the earth as defined by the hypothetical extension of mean sea level through all land masses.

Heading. The horizontal direction in which a vehicle is pointed with respect to a reference, often magnetic North, usually expressed in degrees, clockwise from the reference.

Homing. The process of approaching a desired point by directing the vehicle toward that point.

Instrument approach. An approach using navigation instruments rather than direct visual reference to the terrain.

LOP. Line of position. A line plotted on the earth's surface representing the locus of constant indication of navigational information (in VOR, straight radial lines; in DME, circles; in Decca, Loran, and Omega, hyperbolas).

Most probable position. A computed position based on several lines of position, all adjusted to a common time and weighted in accordance with their estimated probable errors.

Night effect. An error occurring mainly at night, when ionospheric reflection is at a maximum. A major limitation to the useful range of continuous-wave systems in the lf/mf bands.

Octantal error. A bearing error, usually due to departure of an antenna pattern from an ideal shape. It varies sinusoidally throughout the 360° and has four positive and four negative maxima.

Pitch. The angular displacement between the longitudinal axis of the vehicle and the horizontal.

Quadrantal error. An error in measured bearing, frequently due to antenna or goniometer characteristics, which varies sinusoidally throughout the 360° and has two positive and two negative maxima.

Reciprocal bearings. The opposite direction to a bearing (bearing ± 180°).

Rho-rho. A generic term for navigation systems that derive position by measurement of distance to two stations (DME/DME).

Rho-theta. A generic term for navigation systems that derive position by measurement of distance and bearing from a single station (VOR/DME, Tacan).

Roll. The angular displacement between the transverse axis of the vehicle and the horizontal.

RVR. Runway visual range. The forward distance visible along the runway during a landing approach.

Scalloping. Oscillatory-course bends occurring at a rate higher than can be followed by the vehicle.

Slant distance. Distance between two points not at the same elevation. Also called *slant range.*

Theta-theta. Generic term for navigation systems that derive position by measurement of bearing from two stations (VOR/VOR).

Track. Actual path traveled.

Yaw. Angular displacement between the normal axis of the vehicle and the course line.

89. Navigation Techniques. *Direction Finding.* Direction finding (DF) is the oldest and most widely used form of navigation aid. The direction of a transmitter may be determined by comparing the arrival time of its transmission at two or more known points. In the simplest practical system, these two points are the vertical arms of a loop antenna connected to a receiver. As the loop is rotated, the received signals cancel each other when the plane of the loop is at right angles to the direction of the station. Two such nulls, 180° apart, exist if the loop is less than half a wavelength wide. At greater widths, additional nulls appear. These ambiguities can be resolved by use of additional antennas.[81]

The weakness of direction finding systems is their susceptibility to site errors (see Fig. 25-90). As shown, an obstruction is located close to the DF site in such a way as to reflect a signal having 90° error. When direct and reflected signals are equal and in phase, the resulting error is 45°. If, on the other hand, the receiving and transmitting stations were reversed, the error would be

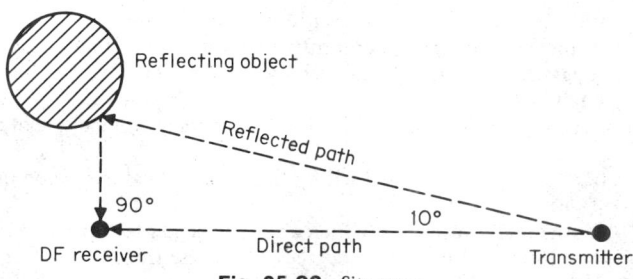

Fig. 25-90. Site error.

5°. This error is therefore associated with the *site* of the DF antenna, and is particularly serious on ships and aircraft, where large reflecting objects (funnels, masts, vertical stabilizers, wings) are present.

The chief weapon against site error is the use of large DF antenna aperture. In most cases, a multiplicity of antennas, suitably combined, can be made to favor the direct path and discriminate against indirect paths. However, most vehicles cannot afford the space for such large antenna arrays. A reverse form of DF is therefore employed whereby the directional antenna system is at the transmitter and a simple omnidirectional antenna is at the receiver. The site error then occurs at the transmitter, but there is usually sufficient space there to allow reduction by the use of large antenna aperture. This principle is used in instrument landing systems and in omnidirectional ranges, such as VOR and Tacan.

Two-Way Distance Ranging.   For automatic distance measuring, it is customary to use a *transponder* on the desired target (Fig. 25-91). This device receives the interrogator pulse and replies

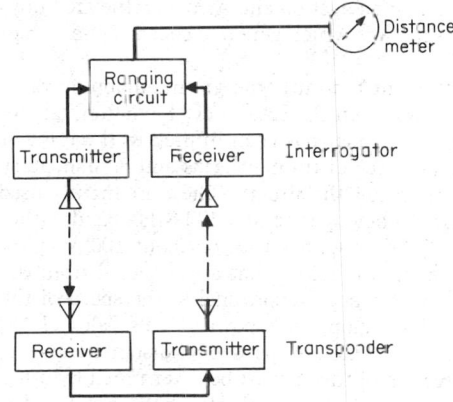

Fig. 25-91. Two-way distance ranging.

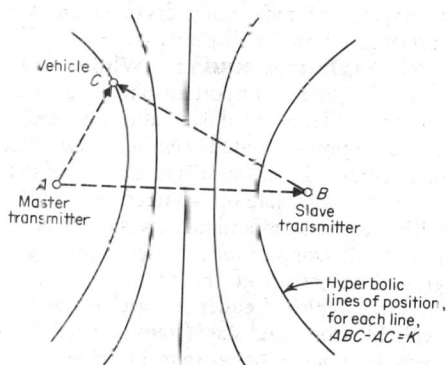

Fig. 25-92. Differential distance ranging (hyperbolic).

Fig. 25-93. One-way distance ranging.

to it with a much stronger pulse, usually on a different frequency. The distance in meters between interrogator and transponder is then equal to 150 times the elapsed round-trip time in microseconds (less any built-in fixed equipment delays). Various codes can be employed to limit responses to a single target or class of target. While pulse-type distance measuring is the most common, there are also cw systems in which received phase is compared with transmitted phase, ambiguities being solved by the use of different subcarriers and subsubcarriers. These latter can produce high instrumental accuracies, but usually with only one target at a time.

Differential Distance Ranging.   Two-way ranging requires a transmitter at both ends of the link. To avoid carrying a transmitter on the vehicle, two transmitters are placed on the ground. One is a master, and the other a slave repeating the master, as shown in Fig. 25-92. The receiver measures the *difference* in the arrival of the two signals. For each time difference, there is a hyperbolic line of position, and such systems are therefore called *hyperbolic systems.* They may be either pulsed, using a common-carrier frequency, or continuous-wave (cw), using different (but related) carrier frequencies. At least two pairs of stations are needed to produce a fix.

One-Way Distance Ranging.   If both stations are provided with stable synchronized clocks, the distance between them can be established by a one-way transmission whose elapsed time is measured with reference to the two clocks (see Fig. 25-93). The distance in meters is equal to 300 times the elapsed time in microseconds (less any built-in equipment delays). An error of approximately 1 mi per elapsed hour occurs for a frequency difference between the clocks of 1 part in 10^9. This method is extensively used in VLF navigation (as a backup to Omega). It shows promise for the Joint Tactical Information Distribution System and the Global Positioning System (Pars. **25-122** and **25-120**, respectively). In these systems clock stabilities are of the order of 1 part in 10^{12}.

90. Radar Navigation Systems. *Ground-Based.* Ground-based radar navigation stations are positioned along airways and at the entrance to harbors. In airways they are primarily for the purpose of air traffic control[82] and not for navigation, although aircraft in terminal areas are usually "vectored" by command from the air-traffic controller, the latter, in effect, temporarily navigating the aircraft. In the harbors, the radar PPI display is sometimes televised to ships, allowing them to see themselves in relation to other traffic, even though they do not have radar of their own. This has also frequently been proposed for the airways but has not found favor due in part to the extensive interpretation problems involved.

Ship-Based.[77] Radar is a very valuable tool when it allows mapping of coastlines and detection of other ships; in those circumstances it becomes a major navigation aid. Also of value are buoys equipped with radar reflectors or *racons*, small transponders which generate coded replies when interrogated by the ship's radar.

91. Airborne Radar.[83] While it was once thought that airborne ground-mapping radar would become an important navigation aid, this has not been the case in civil aviation, largely due to problems of interpretation. Instead, civil airborne radar has been limited to three major areas: (1) nose-mounted *weather radar* detects the presence of thunderheads and is mandatory on airliners. (2) *Altimeters* read distance to the terrain below the aircraft. These are mainly used to improve the flare-out characteristic during landing. They operate in 4,200-MHz band, either with frequency-modulated cw signals at less than 1 W or with pulses of about 100 W peak power. (3) *Doppler radar*[84] uses relatively low-power transmissions that are reflected from the ground and return to the aircraft with a frequency difference proportional to the speed of the aircraft. Typical frequencies are in the 10- to 20-GHz region, with power levels below 1 W, available from solid-state (transistorized) equipment. Integration of speed gives distance traveled, to an accuracy of better than 1%. However, the direction of travel must be determined by other means. The system is much used by transoceanic airlines and by military helicopters and is substantially less expensive than inertial systems.

92. Other Systems. *Inertial Navigation.*[87,112] This is the most widely used "self-contained" technology. Accelerometers constantly sense the vehicle's movements and convert them, by double integration, into distance traveled. To reduce errors caused by vehicle attitude, the accelerometers are mounted on a gyroscopically controlled stable platform. Accuracy is limited by gyroscopic drift and is typically 1 mi per elapsed hour. Since this is the most expensive vehicular navigation technology, it is used only by military aircraft and transoceanic airlines.

The *laser gyro*,[85] by eliminating moving parts, should reduce the cost of maintenance. An important by-product of inertial navigation is the improved attitude reference provided to other airborne instruments.

Celestial Observation.[88] Position fixing with reference to the stars has been standard marine practice for centuries and has been aided in recent years by electronic star trackers and by vehicular computers. Ships, however, still have to cope with a weather problem when using this technique, and while high-flying aircraft are less troubled by this problem, the considerable human judgment needed has limited the use of this technology.

Earth Satellites.[89] The limitation of celestial observation by the weather has led to consideration of artificial satellites using radio, and one such system has seen several years of operational use. More recently, the Global Positioning System (GPS), also called Navstar, has been under development.[86] The chief stumbling blocks have not been technical but political, e.g., what agency should provide and maintain the satellite system.

93. Radio Frequencies Used in Electronic Navigation. Virtually all radio frequencies have been used in navigation systems. At the low-frequency end, systems are limited by the massive antenna systems needed and by low data rates. Above 10 GHz there are limits on the amount of rainfall that can be penetrated.[15] With few exceptions, frequencies and technologies have been chosen to avoid dependence on ionospheric reflection; while such reflection has been of much value in communication, it is deliberately used only in navigation systems where its effects are well understood and predictable. Table 25-15 lists the principal frequency bands used for navigation.

94. Internationally Standardized Systems. The systems described in Pars. **25-95** to **25-107** are used by hundreds of thousands of vehicles, throughout the world. Standardization refers principally to the radiated signal characteristics. Each country, manufacturer, or user decides individually on the detailed equipment design which best fits its needs, market, and resources. These systems have been in use for at least two decades and can be expected to remain in use for at least two more decades.

95. Direction Finding. *Shipboard DF and Coastal Beacons.*[91]   While the more than 10,000 broadcasting stations in the world can be used as "targets" for a shipboard direction finder, special coastal beacons operate in the 285- to 325-kHz band, specifically for the benefit of ships. This frequency band provides ground-wave coverage over seawater to about 1,000 mi, and the assigned channels are such that interference between beacons is minimized out to that distance. The beacons also transmit notices of interest to mariners. Power varies from 100 W to 10 kW.

The simplest shipboard antenna consists of a loop about 2 ft in diameter which can be rotated around its vertical axis, by hand (Fig. 25-94a). It is connected to a receiver whose output shows a sharp null when the plane of the loop is at right angles to the direction of the transmitter. If the resulting 180° ambiguity causes doubt (it rarely does), a "sense" antenna is switched in; in conjunction with the loop, this generates a cardioid pattern (Fig. 25-9cb) having a single null but less sharp.

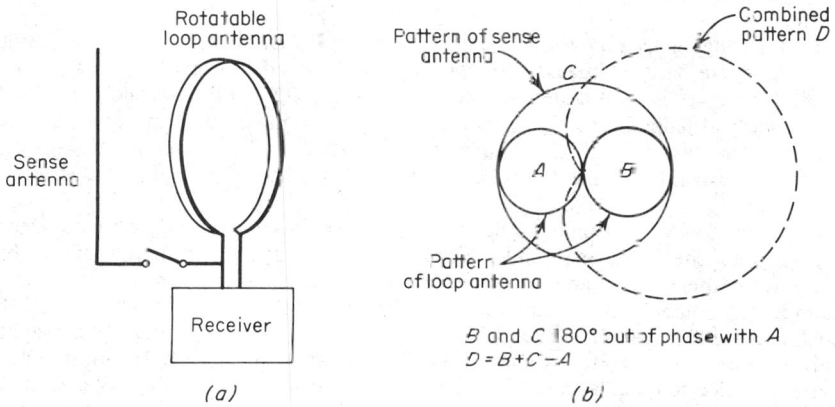

Fig. 25-94. Sense antenna applied to direction finding: (a) antennas; (b) patterns.

TABLE 25-15   Radio Frequencies Used in Electronic Navigation

System	Frequency band	No. of stations	No. of vehicles
Omega	10–13 kHz	8	10,000
VLF Comm.	16–24 kHz	10	5,000
Decca	70–130 kHz	150	30,000
Loran-C/D	100 kHz	50	10,000
Lf range	200–400 kHz	*	*
ADF/NDB†	200–1,600 kHz	4,000	106,000
Coastal DF†	285–325 kHz	1,000	100,000
Consol	250–350 kHz	15	5,000
Loran A	2 MHz	*	*
Marker beacon†	75 MHz	2,500	150,000
ILS localizer†	108–112 MHz	1,200	150,000
VOR†	108–118 MHz	2,000	250,000
ILS glide slope†	329–335 MHz	1,200	150,000
Transit	150,400 MHz	6	5,000
DME†, Tacan	960–1,215 MHz	2,000	70,000
ATCRBS†	1,030, 1,090 MHz	800	250,000
GPS	1,227, 1,575 MHz	‡	‡
Altimeter	4,200 MHz	. . .	5,000
Talking beacons	9 GHz	3	1,000
MLS	5 GHz	‡	†
Weather radar	5, 9 GHz	. . .	10,000
Doppler radar	10–20 GHz	. . .	5,000

*Obsolescent.
†Internationally standardized systems.
‡In development.

The chief disadvantages of this system are site errors caused by the ship's superstructure, ionospheric reflections (sky waves) which may not always travel along the vertical plane joining the transmitter and the receiver, atmospheric noise, and polarization errors. The latter occur when the polarization of the arriving signal is not precisely vertical. The relatively slow speed of ships, and the relatively long integration time that can be employed, allow a well-designed DF system to show an accuracy of about ±2° under typical conditions.[92]

While the single rotating loop is the simplest system, most large ships employ two fixed loops, at right angles to each other, with a rotatable goniometer below decks feeding their outputs to the receiver. More recent designs use digitally controlled electronic goniometers and digital displays. While these ease the task of the operator, they do little to alleviate the propagation effects previously listed.

Airborne ADF and Ground NDB.[93]   Early aircraft used the same system as described for ships. Since, on most aircraft, the navigator is also the pilot, a greater degree of automation became necessary, culminating around 1935 in the *automatic direction finder*, abbreviated ADF. Typically two crossed ferrite-core loops are used, projecting about 1 in from the skin of the aircraft, with an electronic goniometer generating a direct-reading display of bearing to the selected transmitter.

The tuning range is usually 200 to 1,600 kHz, taking in not only the broadcast stations and maritime coastwise beacons but also specially designed inland beacons operating from 200 to 1,600 kHz. These are known as *nondirectional beacons* (NDB). They radiate from 10 W to 2 kW, into a vertical radiator usually at least 100 ft high. An important feature of these beacons is the sharp reduction in signal strength obtained directly over the beacon due to vertical polarization of both transmitting and receiving antennas. This effect is frequently used to obtain a position fix.

The low ground-wave attenuation occurring over seawater is not present over land. Depending on soil conditions, the effective ground-wave range may be only a few hundred miles, while site error, sky-wave interference, and atmospheric and polarization errors are as serious as in the shipboard case.[94] Consequently, accuracy is usually not better than ±5°, and the useful range not much above 200 mi at 200 kHz and 50 mi at 1,600 kHz. This range may be reduced to zero during heavy atmospherics. Nevertheless, the ADF navigation aid is one of the most widely used of all airborne electronic aids, and NDBs are to be found in practically every country of the world.

96. Instrument Landing System (ILS).[95]   Early work on instrumental landing in Germany and England before World War II was consolidated into a standard U.S. system in 1942 and became an ICAO standard in 1947. The ground equipment is made up of three separate elements: the *localizer*, giving left-right guidance; the *glide slope*, giving up-down guidance; and the *marker beacons*, which define progress along the approach course. Using line-of-sight frequencies, ILS is free of atmospherics and sky-wave effects, but, it is still much subject to site effects.

The *localizer* operates on 40 channels spaced 50 kHz apart, 108 to 112 mHz, radiating two antenna patterns which give an equisignal course on the centerline of the runway, the transmitter being located at the far end of the runway. The left-hand pattern is amplitude-modulated by 90 Hz, the right-hand pattern by 150 Hz. The airborne receiver detects these tones, rectifies them, and presents a left-right display on a zero-center dc meter in the cockpit. The accuracy is better than ±0.1°.

Minimum ICAO performance calls for the airborne meter to remain hard left or hard right to a minimum of ±35° from the centerline; i.e., there must be no ambiguous or "false" courses within this region (Fig. 25-95a). More sophisticated systems exist in which a usable "back course" (with reverse sense) is obtained, and a separate transmitter, offset by about 10 kHz, provides "clearance" (Fig. 25-95b), so that no ambiguities exist throughout ±180°. Total rf power is of the order of 25 W. The localizer may be voice-modulated, but this feature is seldom used.

The *glide-slope transmitter*, of about 7 W power, is located at the approach end of the runway and up to about 500 ft to the side. It operates in the 329 to 335-MHz band, each channel being paired with a localizer channel. In the airborne receiver, both channels are selected by the same control. Two antenna patterns are radiated, giving an equisignal course about 3° above the horizontal. The lower pattern is modulated with 150 Hz, and the upper pattern by 90 Hz (Fig. 25-95c). The airborne receiver filters these tones, rectifies them, and presents the output on a horizontal zero-centered meter mounted in the same instrument case as the localizer display, the two together being called a *cross-pointer display*. When one needle is horizontal and the other vertical, the aircraft is exactly on course. The accuracy is better than ±0.1°.

In this frequency band it has not been found possible to generate the required antenna patterns without use of either an excessively tall array (100 ft or more, which would be dangerous to aircraft) or of deliberate ground reflection. The glide slope therefore suffers from course bends due to the terrain in front of the array, and is generally not depended on below 50 ft of altitude.

Marker beacons operate at a fixed frequency of 75 MHz, and radiate about 2 W upward toward the sky with a fan-beam antenna pattern whose major axis is across the direction of flight (Fig. 25-95d). At each ILS installation there is an "outer" marker about 5 mi from touchdown, and a "middle" marker about 3,500 ft from touchdown. A few runways also have "inner" markers just before touchdown. Each type is modulated by audio tones which are easily recognized as the aircraft passes through their antenna pattern. Alternatively, differently colored lamps are set

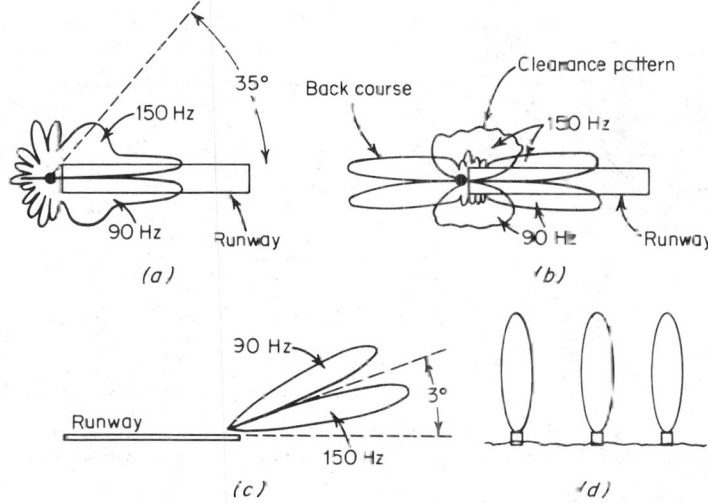

Fig. 25-95. Instrument landing system: (a) minimum ICAO localizer pattern; (b) localizer pattern with back course and clearance; (c) glide-slope pattern; (d) marker beacons.

to light in the cockpit as each marker is passed. Eventually, marker beacons may be replaced by DME, particularly for overwater approaches.

ICAO has established categories of ILS performance, dependent on the quality of the installation and the qualifications of the air crew. These place the following minimum limits on how close an aircraft may approach the touchdown point:

Category I: 200-ft ceiling and ½-mi visibility (about 1,200 runways in the world)
Category II: 100-ft ceiling and ¼-mi visibility (fewer than 200 runways in the world)
Category III: zero ceiling and 700-ft visibility (as of 1980, only 20 runways in the world)

ILS installations are usually dual (except antenna), with elaborate monitoring and immediate changeover to the standby unit when the course is found to deviate excessively. If the standby unit also fails, the facility is shut down.

97. VOR System. The very high frequency omnidirectional range, often called Omni-Range or just Omni, was standardized in the United States in 1946 and became an ICAO standard in 1949 for en route flying. Like ILS, it uses line-of-sight frequencies and is thus free of atmospherics and sky-wave distortions. Also, like ILS, it places the directional burden on the ground, rather than in the aircraft, where more extensive means can be employed to alleviate site errors. Line-of-sight limits its service area to about 200 mi for high-flying aircraft, and some stations are intended for only 25-mi service to low-flying aircraft. There are more than 1,000 stations in the United States and about an equal number in the rest of the western world. There are two variations: conventional and doppler (which provides increased site-error reduction).

Conventional VOR[96] operates on 40 channels, 100 kHz apart, between 108 and 112 MHz (interleaved between ILS localizer channels), and on 120 channels, spaced 50 kHz, between 112 and 118 MHz. The airborne receiver is frequently common with the airborne localizer receiver and may use the same airborne antenna. Power output from the ground transmitter varies from 25 to 200 W, depending on antenna design and on the desired service area.

The ground-antenna pattern forms a cardioid (Fig.25-96a) in the horizontal plane which is rotated 30 times per second. The cw transmission is amplitude-modulated by a 9,960-Hz tone which is frequency-modulated ±480 Hz at a rate of 30 Hz. This latter 30-Hz "reference" tone, when extracted in the airborne receiver, is compared with the 30-Hz amplitude modulation provided by the rotating antenna. The phase angle between these two 30-Hz tones is the bearing of the aircraft, with respect to North (Fig. 25-96b).

Instrument accuracy of the complete system is of the order of $\pm1°$. However, site errors may often degrade this to $\pm3°$ or worse. Sectors may be declared unflyable between certain bearings and below certain altitudes when these errors exceed certain limits.

Doppler VOR[97] reduces site error about tenfold by using a large-diameter antenna array at the ground station. This array constitutes a 44-ft-diameter circle of antennas. Each antenna is sequen-

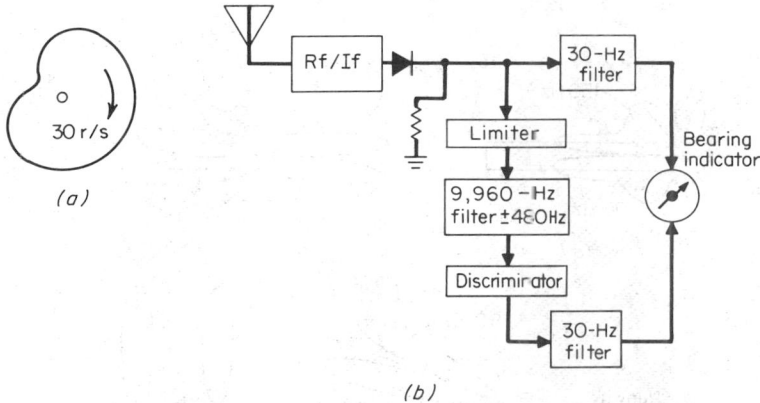

(a)

(b)

Fig. 25-96. VOR system: (a) conventional VOR antenna pattern; (b) receiver for conventional and doppler VOR.

tially connected to the transmitter in a manner to simulate the rotation of a single antenna around a 44-ft-diameter circle at 30 r/s. The receiver sees an apparent doppler shift in the received rf of ±480 Hz at a 30 r/s rate, and at a phase angle proportional to the receiver's bearing with respect to North. This signal is therefore identical with the conventional VOR reference tone. It remains merely to transmit at 30-Hz AM tone, separated 9,960 Hz, as the doppler reference, to radiate an identical signal to the conventional one, receivable in an identical receiver but benefiting from a tenfold increase in ground-antenna aperture.

Since the doppler VOR costs substantially more than the conventional VOR, it is used only at sites than cannot be properly served by the conventional VOR. In the United States this currently amounts to about 3% of the sites, with plans to reach about 10%. In some other countries, the doppler system is being installed at all new sites, its extra cost being somewhat compensated by lower site cost. (The conventional VOR needs a flat, cleared area of at least 1,500-ft radius and even then can be seriously disturbed by buildings a mile or more away.) VOR can be voice-modulated, though little use of this is made except for identification purposes. The VOR presentation in the cockpit is usually on a radio-magnetic indicator (RMI), on which the angle between needle and rotating card is the VOR bearing, while the angle between rotating card and the frame is the compass output. The needle thus represents the angle between heading and station if the magnetic deviation is the same at the station as at the aircraft.

98. Distance-Measuring Equipment.[98] DME is an interrogator-transponder two-way distance-ranging system (Fig. 25-91). It became an ICAO standard in 1959. Some 2,000 ground stations and 70,000 pieces of airborne equipment are in use in the western world.

The airborne interrogator transmits 1-kW pulses of 3.5-μs duration, 30 times a second, on one of 126 channels 1 MHz apart, 1,025 to 1,150 MHz. The ground transponder replies with similar pulses on another channel 63 MHz above or below the interrogating channel. (This allows both equipments to use the transmitter frequency as the receiver local-oscillator frequency if the intermediate frequency is 63 MHz.) All channels are crystal-controlled, and a single antenna is used at both ends of the link. To reduce interference from other pulse systems (e.g., ATCRBS) and to allow addition of future functions by pulse coding, paired pulses are used in both directions, their spacing being 12, 30, or 36 μs. The fixed delay in the ground transponder is 50 μs.

In the airborne set, the received signal is compared with the transmitted signal, their time difference derived, and a direct digital reading of miles is displayed. The typical accuracy is ±0.2 mi. Ground transponders are arranged to handle interrogation from up to 100 aircraft simultaneously, each aircraft recognizing the replies to its own interrogation by virtue of the pulse repetition frequency being identical with the interrogation. Early analog models required some 20 s for this identity to be initially established (after which a continuous display was provided), but newer digital models perform the search function in less than 1 s.

In ICAO practice, the DME is nearly always associated with a VOR, the two systems forming the basis for a rho-theta area navigation system; 80 DME frequencies are paired with 160 VOR channels, the same selector in the cockpit operating both sets; 20 DME frequencies are paired with 40 ILS frequencies.

DME is particularly immune to site and propagation errors, and better accuracy is readily obtainable if required. For this reason, DME is seriously being considered for addition to the new Microwave Landing System (MLS) (Par. **25-121**).

99. TACAN[99,100] is a NATO military system which adds a bearing function to DME, on the same frequencies, allowing greater portability of the ground station than with ICAO's VOR/DME, particularly on aircraft carriers.

A standard ICAO DME beacon is arranged to operate at "constant duty cycle"; i.e., when 100 sets of interrogations are lacking, the gain of its receiver is increased until an equivalent set "squitter" pulses (from receiver noise) is generated. The DME antenna is then replaced by rotating directional antenna generating two superimposed patterns (Fig. 25-97). One of these is a cardioid, as in VOR, but rotating at 15 Hz. The other is a nine-lobe pattern, also rotating at 15 Hz. The squitter pulses and replies are amplitude-modulated as the antenna rotates. Reference pulses are transmitted at 15 and 135 Hz.

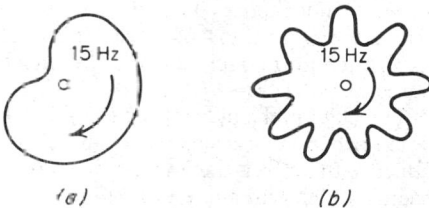

(a)   *(b)*

Fig. 25-97. Tacan antenna pattern.

In the aircraft, a coarse phase comparison can then be made at 15 Hz, supplemented by a fine comparison at 135 Hz, the overall instrumental accuracy being on the order of ±0.2°. Since the DME frequency is about 10 times the VOR frequency, the antenna system is about 10 times smaller than that of the VOR and thus more portable.

A major problem in Tacan has been the design of sufficiently good vertical directivity in the ground antenna to preclude the generation of vertical nulls by ground reflections.

100. Vortac. In NATO countries having a common air traffic control system for the civil and the military (e.g., the United States, Germany), the ICAO rho-theta system is implemented by the use of Tacan rather than DME. Tacan transponders are colocated with VOR stations, and civil aircraft get their DME service from the Tacan station, as shown in Fig. 25-98. In the United States, about 700 VORs have colocated Tacan transponders. About 40,000 airborn Tacan sets have been built.

101. IFF (Identification of Friend from Foe) Systems. To distinguish friend from foe, early radars employed an interrogator-transponder system operating at a different set of frequencies, the "friend" being transponder-equipped and the "foe" not. The interrogator was pulsed at about the same time as the radar, and the airborne transponder produced coded replies

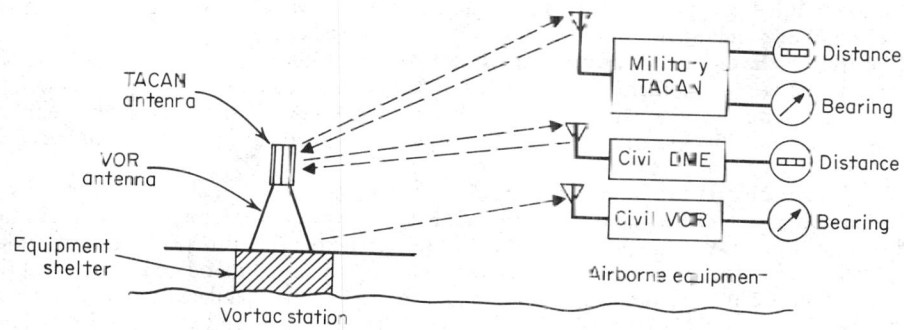

Fig. 25-98. Vortac system.

shortly after the direct radar reply from the aircraft skin, as shown in Fig. 25-99. In theory, even if the foe used the same transponder equipment, he would not know the code of the day. This identification of friend from foe system became known as IFF, and after World War II, became stabilized among NATO countries as a system in which all interrogation takes place at 1,030 MHz and all replies at 1,090 MHz. Typical pulse powers are 500 W, with 1-μs length for interrogation and 0.5-μs length for reply. The ground antennas are typically 20 ft long and rotate at 15 r/min.

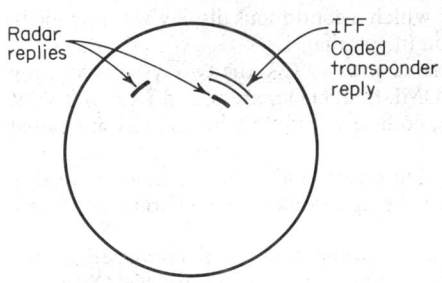

Fig. 25-99. PPI display with IFF.

102. ATCRB System. In 1958 IFF became an ICAO standard, known as *SSR* (secondary surveillance radar) or *ATCRBS* (air-traffic-control radar beacon system). It is used by air-traffic-control authorities to track aircraft and is not primarily intended for navigation. In the United States, there are over 500 ATCRBS interrogators and 200,000 pieces of airborne equipment.

Paired pulses are used for interrogation, and a third pulse between the two is radiated omnidirectionally to reduce triggering by side lobes (Fig. 25-100). The airborne transponder replies only when the directional pulses are stronger than the omnidirectional pulse. The reply pulses comprise a train of up to 14 pulses, lasting 21 μs, which can be combined into 4,094 codes. These can be used to identify the aircraft or to communicate its altitude to the ground controller (height-finding radar has been found impractical for this purpose).

The chief problem besetting the ATCRBS system is the interference (garbling) which occurs when two or more aircraft are at about the same azimuth and distance from the interrogator. To alleviate this effect, the FAA has plans in the United States to institute a system of interrogation coding which will allow each aircraft to be addressed by a discrete code, and thus only "speak when spoken to." This system will be compatible with the present ATCRBS, to allow an orderly

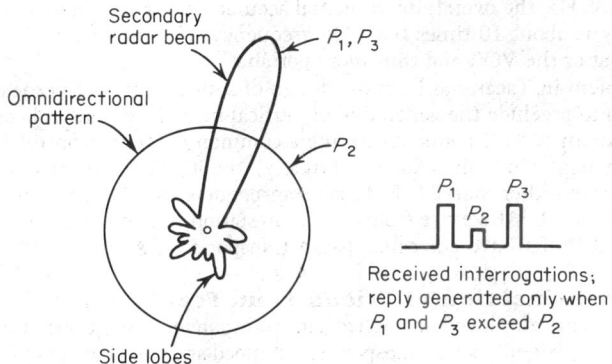

Fig. 25-100. Side-lobe suppression.

transition.[101] It is known as DABS for *discrete address beacon system*. A similar system is being developed in England, known as ADSEL for *address selective beacon system*.

103. Mechanical vs. Electronic Modulation. Most of the systems just described depend in one way or another on the amplitude modulation of a carrier or on the rotation of a directional antenna pattern. While these objectives might have been reached by electronic means, the unreliability of early vacuum-tube designs caused many designers to turn to *mechanical* means. Two of these, still in wide use, are described below.

104. ILS Mechanical Modulation. For both localizer and glide slope, it is necessary to modulate the carrier with 90- and 150-Hz waves. Here the carrier is generated in a single source and then split into two equal parts by an Alford bridge. The main feature of this bridge, which has a 180° phase shift in one arm, is that energy entering any one corner does not appear at the opposite corner. Thus the two carrier portions can be separately modulated without effect on each other. This modulation is performed by two paddlewheels, one with three paddles and one with five, both rotated at 30 r/s by a constant-speed motor. The two signals are then combined in another Alford bridge to give the signals needed by the antenna system, carrier plus sidebands and sidebands only.

The advantages for this system are that variations in the rf generator and in the motor speed affect the 90- and the 150-Hz components equally, and therefore the course can be shifted only by variations in the modulators, which, being mechanical, can remain fixed for years at a time.

105. VOR Mechanical and Electronic Modulation Systems. The *Mechanical VOR System* calls for a rotating cardioid pattern. One way to generate this has been by a mechanically rotating dipole, which generates a figure-eight pattern, superimposed on an omnidirectional pattern. On the same shaft as the rotating dipole is also mounted the 30-Hz reference generator. A variation uses a fixed array of the four loop antennas but feeds them from a mechanically rotating goniometer whose shaft also drives the reference generator.

In both cases, phase lock between the two 30-Hz signals is maintained because they both are generated by the same motor shaft.

The *VOR electronic-modulation system* uses solid-state technology to produce the modulation electronically and to achieve the reliability formerly associated with mechanical modulation. In one design, the modulation occurs at low level to preserve linearity of modulation; the sidebands are then separated from the carrier and from each other, amplified in class C power stages, and recombined and fed to a pair of fixed crossed dipoles, to generate a rotating figure eight. As in ILS, the result is lower first cost, less power consumption, and easier maintenance.

106. Tacan Modulation. This calls for a cardioid superimposed on a nine-lobe pattern, both rotating at 15 Hz. Nearly all operational stations in the world achieve this by radiating the pulsed carrier omnidirectionally and then mechanically rotating a series of parasitic dipoles around the central radiator. The cardioid is generated by a single parasite at about 3 in radius, and the nine-lobe pattern is generated by nine parasites at about 15 in radius. On the same shaft are generators for the reference signals.

An electronically rotated version uses a large number of fixed parasites (over 100 for the nine-lobe array) and switches them on and off by pin diodes controlled from a multiphase square-wave generator. By suitable programming of this generator, the granularity of the on-off modulation is held to a level below that imposed by the discrete DME pulses on which the Tacan bearing signal is modulated.

107. ATCRBS Modulation. The ground-based interrogator for this system uses a 20-ft directional antenna which rotates mechanically at about 15 r/min. This speed is a compromise; if it were higher, to give a higher data rate, the antenna would have to be smaller, have less aperture, and provide less azimuth resolution.

108. Nonstandard Cooperative Systems. The following systems have existed, at least in developmental form, for a decade or more, and some have seen extensive use. They have failed to win international standardization for a variety of reasons.

109. Ground-Based Direction Finding.[92] The site-error limitations of vehicular direction finding (DF) led to the concept of placing the direction-finding station at a good site on the ground, taking bearings on the vehicle's transmitter, and then communicating the result to the vehicle. As recently as 1961, this concept was being seriously proposed for a national air-navigation system in one country, aircraft position to be obtained by triangulation from a multiplicity of ground stations connected by telephone lines. The system obviously has great economy as far as the vehicle is concerned. The chief disadvantages appear to be low data rate and relatively expensive ground operations, including much manpower. The principle is still in use, but mainly as an aid in the location and rescue of vehicles whose other navigational functions have been lost. Major technologies used are as follows:

LF-MF. The "crossed-loop" principle, but using vertical elements only, spaced several hundreds yards apart, in a so-called *Adcock array.*[102]

HF. The *Wullenweber system*, introduced in Germany during World War II. Here, a 1,000-ft-diameter circle of antennas feeds a rotating commutator which connects each antenna to an appropriate delay line so that all signals coming from a given direction reach the receiver with the same phase, while those from other directions arrive with random phase. A British version, known as CADF (for commutated antenna DF), operates in a similar manner. Instrumental accuracy is better than 1° but is adversely affected by the errors caused by ionospheric reflection. Systems of this type are in use by military and intelligence agencies for nonnavigational purposes.

VHF-UHF. Here the doppler principle, used in the doppler VOR, is most effective. A circle of antennas, about 10 ft in diameter, is commutated in such a manner that the receiver sees, in effect, a single antenna rotating around that circle. As the receiver moves away from the signal, the frequency appears to be lowered, and as it moves toward the signal, the frequency appears to be raised. An FM discriminator at the output of the receiver delivers a sine wave whose phase depends on the direction of the transmitter. The display is usually on a cathode-ray tube with circular scan, radial lines being produced in the direction of received signals. The FAA has about 200 such installations.

The Wullenweber, CADF, and doppler systems are basically schemes for using a wide antenna aperture, without ambiguity, and thereby reducing site errors.

110. LF Four-Course Range.[93] This system was introduced in the United States in 1927, reached its zenith during World War II, and was overtaken by the VOR before it could achieve international standardization. Since about 1950, it has gradually been phased out of existence, one of the very few radio navigation aids to have suffered this fate. Four 125-ft towers are arranged in a 400-ft square, and each diagonal pair fed with about 200 W in the 200- to 400-kHz band. The rf is modulated with a 1,020-Hz tone, which is keyed on a diagonally located pair of antennas with the Morse code letter A and, on the other pair, with the letter N, the phases of these Morse codes being such that, when they are received at equal signal strength, a constant 1,020-Hz tone results. Four equisignal courses are thus generated, each course being characterized by the reception of A on one side, N on the other side, and a pure tone on course. The system is frequently referred to as the *A-N range.*

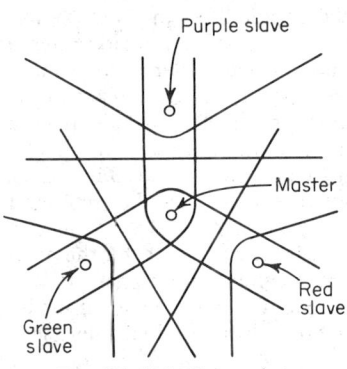

Fig. 25-101. Decca chain.

Aside from the disadvantage of only four courses, the LF range suffers from sky-wave reflections and atmospherics, limiting its service range to no more than is obtained with line-of-sight systems.

111. Decca.[103] This system has been developed in the United Kingdom starting in 1939. It is a continuous-wave hyperbolic system operating in the 70- to 130-kHz band. A typical chain comprises four stations, one master and three slaves, separated about 70 mi, arranged as in Fig. 25-101. Each station is fed with a signal which is an accurately phase-controlled multiple of a base frequency f, in the 14-kHz region, the master at $6f$, the "red" slave at $8f$, the "green" slave at $9f$, and the "purple" slave at $5f$. At the receiver, these four signals are received, multiplied, and phase-compared, as shown in Fig. 25-102. Each of the three phase meters, called *decometers*, thus provides a position along a hyperbolic line, and by plotting these positions on a map a unique fix is obtained. Considerable ambiguity exists with this system, since equiphase hyperbolic lines are obtained as close as 1 mi apart. Additional complexity is added to resolve the ambiguity.

There are about 25 Decca chains in Europe and about 20 elsewhere in the world. The system is used by about 30,000 ships, half of which are fishing vessels.[104] Accuracy varies from under 100 yd on a summer day, close to the chain, to several miles on a winter night 200 mi from the chain. The chief advantage of Decca is that it is non-line-of-sight and that the simplest receiver is relatively inexpensive. Disadvantages are those common to other cw LF aids: sky-wave propagation and atmospherics. From time to time, attempts have been made to justify Decca for airborne applications. These have not been successful, a major difficulty being the high cost of converting the hyperbolic readouts to information directly usable by an aircraft pilot.

112. Loran A.[93] The original Loran system (Loran A) was developed in the United States during World War II and has since been maintained by the U.S. Coast Guard as an aid to ships and aircraft. The name is an abbreviation of *long range navigation.* It is a hyperbolic system operating around 2 MHz and using pulses, to allow discrimination against sky waves. Receivers use a cathode-ray-tube display and the earlier arrival of the ground wave can readily be observed.

A chain normally comprises a master and two slaves, spaced 200 mi on each side of the master along a coastline, for the benefit of ships and aircraft crossing the ocean. Discrimination between chains is by use of four radio frequencies, 50 kHz apart, and 24 sets of pulse-repetition frequencies in the 20 to 35 pulses/s region. The pulse rise time is 21 μs.

The peak power is about 100 kW into an antenna about 100 ft high. At the receiver, the pulse positions are matched on a cathode-ray oscilloscope to an accuracy of about 1 μs, and navigational accuracy (which depends on the angle of cut of the hyperbolic lines), is typically 1,000 ft between the stations and about ½ mi at the extreme ground-wave range.

At one time there were about 25 chains, but by 1981 they were to have been shut down and replaced by Omega and Loran C.

113. Loran C.[106] This is the officially designated United States marine system in the coastal-confluence zone. It is a hyperbolic system, with chains maintained by the Coast Guard. It uses pulses at a carrier frequency of 100 kHz and, unlike Loran A, matches the individual rf cycles within each pulse, thereby gaining added resolution and accuracy. The present signal characteristics were established around 1960. Since all stations operate on the same frequency, discrimination between chains is by the pulse-repetition frequency. A typical chain comprises a master and two slaves, about 600 mi from the master, along a coastline. Each antenna is 1,300 ft high and is fed 5-MW pulses, which build up to peak amplitude in about 50 μs and then decay to zero in about 100 μs. The slow rise and decay times are necessary to keep the radiated spectrum

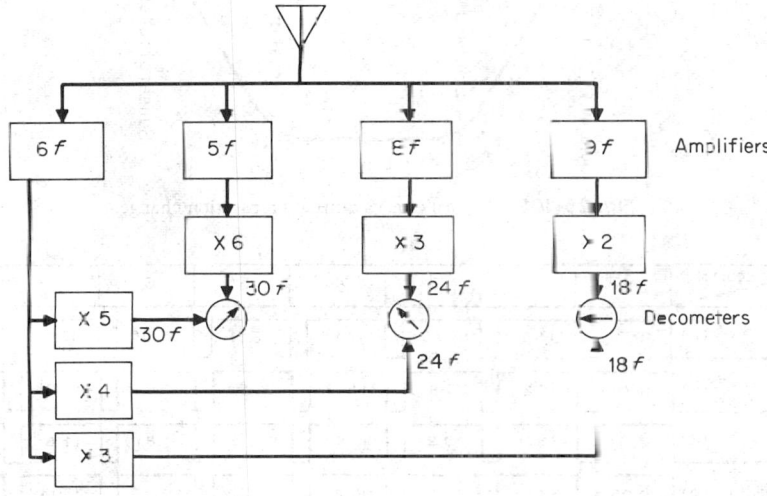

Fig. 25-102. Decca receiver.

within the assigned band limits of 90 to 110 kHz. At the receiver, the first three rf cycles are used to measure the time of arrival. At this point, the pulse is at about half amplitude. The rest of the pulse is ignored, since it may be contaminated by sky-wave interference.

To obtain greater average power at the receiver without resorting to higher peak power, the master transmits groups of nine pulses, 1,000 μs apart, and the slaves transmit eight pulses, also 1,000 μs apart. These groups are repeated at rates ranging from 10 to 25 per second. Within each pulse, the rf phase can be varied for communication purposes.

At the receiver, phase-locked loops track the master and slave signals and present their time differences on a digital display, from where they can be transferred to a map on which hyperbolic lines of position have been printed. Alternatively, at the cost of doubling the size and price of the equipment, a digital computer can provide direct readouts in latitude and longitude or in left-right steering information and distance to go. To reduce interference from the numerous cw stations in the band, narrow notch filters are employed at the front end of the receiver. These automatically track and eliminate the strongest interference; another filter then tackles the second strongest; and so on.

The advantages of Loran C are that it extends beyond the line of sight and has long range (up to 1,000 mi) and high accuracy (better than 0.1 mi). The disadvantages are high cost, slow acquisition of the signal (up to 10 min), inconvenient readout, and lack of worldwide coverage.

Coverage extends to all United States coastal areas, plus certain areas of the North Pacific, North Atlantic, and Mediterranean, provided by about 50 transmitters.

114. Loran D is a tactical version of Loran C, with shorter base lines, lower power, and smaller masts. To compensate for the necessarily reduced performance, 16 pulses are employed per group, and measurements are made at the peak of the pulse, rather than at the 50% point. This is justified by the reduced sky wave interference at short ranges. Receivers are the same as for Loran C.

115. Omega.[107] This is a system aimed at providing worldwide coverage from only eight stations. It is a hyperbolic system, using the VLF band at 10- to 13 kHz. At this low frequency, sky-wave propagation is relatively stable. There is, however, a marked difference in propagation between day and night, as shown in Fig. 25-103, but this is predictable and can be compensated if the observer at the receiver has a rough idea of his location. Overall accuracy is consequently of the order of 1 mi, even at ranges of 5,000 mi. There are no masters or slaves, each station transmitting according to its own standard. The signal format is shown in Fig. 25-104. Each

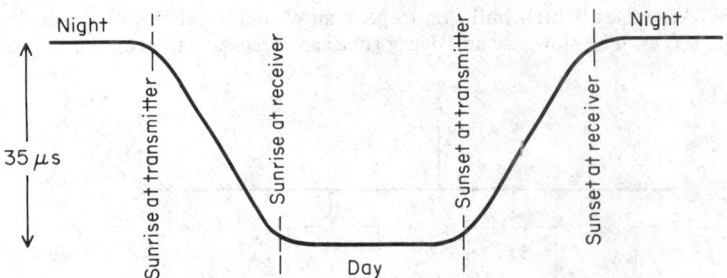

Fig. 25-103. Typical Omega diurnal propagation change.

Segment / Station	1	2	3	4	5	6	7	8
Norway (A)	10.2	13.6	11 1/3	12.1†	12.1†	11.05	12.1†	12.1†
Liberia (B)	12.0†	10.2	13.6	11 1/3	12.0†	12.0†	11.05	12.0†
Hawaii (C)	11.8†	11.8†	10.2	13.6	11 1/3	11.8†	11.8†	11.05
North Dakota (D)	11.05	13.1†	13.1†	10.2	13.6	11 1/3	13.1†	13.1†
La Reunion (E)	12.3†	11.05	12.3†	12.3†	10.2	13.6	11 1/3	12.3†
Argentina (F)	12.9†	12.9†	11.05	12.9†	12.9†	10.2	13.6	11 1/3
Australia (G)	11 1/3	13.0†	13.0†	11.05	13.0†	13.0†	10.2	13.6
Japan (H)	13.6	11 1/3	12.8†	12.8†	11.05	12.8†	12.8†	10.2

Transmission interval — 0.9 — 0.2 — 1.0 — 0.2 — 1.1 — 0.2 — 1.2 — 0.2 — 1.1 — 0.2 — 0.9 — 0.2 — 1.2 — 0.2 — 1.0 — 0.2
—10 s—

Fig. 25-104. Omega-system signal-transmission format (frequencies in kHz). Frequencies marked † are the unique frequencies for the respective stations.

station transmits on one frequency at a time, for a minimum of about 1 s, the cycle being repeated every 10 s. These slow rates are necessitated by the high Q's of the necessarily inefficient transmitting antennas.

The simplest type of receiver receives only the 10.2-kHz signals, comparing those from one station against those of another by use of a medium-stability internal oscillator. The phase differences are transferred to a map with hyperbolic coordinates.

At the transmitted VLF frequencies, lane ambiguity occurs every 8 mi or so (half a wavelength). However, by using the beats between the VLF frequencies, these ambiguities can be extended to 24 mi for a two-frequency receiver and 72 mi using a three-frequency receiver.

A large proportion of Omega receivers also are able to use the VLF communication stations.[108] There are about 10 of these in the world, operating between 16 and 24 kHz with powers between 50 and 1,000 kW and with frequency stabilities of 1 part in 10.[12] This allows one-way DME to be accomplished to a number of stations, with an overall accuracy of about 1 mi/h or about the same as inertial. Since the cost is far less, in 1980 a number of the world's transoceanic airlines were converting at least some of their equipment from the inertial system to Omega/VLF.

116. Transit.[109] This is a U.S. Navy satellite-based system and is the only satellite system in operational use. Its first test was in 1959, and in 1967 it was released for civil use.

In its simplest form, shown in Fig. 25-105, one satellite in polar orbit at 600 mi altitude circles the earth every 1¾ h, radiating two cw frequencies near 150 and 400 MHz. As it passes an observer on the surface of the earth, these frequencies undergo a doppler shift, the magnitude of which depends on the distance to the satellite, as shown in Fig. 25-106. From the rate of change (slope) of this doppler shift, the distance to the satellite can be computed, and since the satellite position is predicted in published tables, the observer's position can be determined. For a single satellite at least four such fixes are obtained per day for all positions on the earth. The system was therefore the first to provide worldwide coverage, albeit at a rather low data rate.

While a single radiated frequency would allow an accuracy of about 1 mi, two frequencies allow errors due to ionospheric refraction to be reduced. Other errors are caused by air friction and variations in the earth's gravitational field. These errors are observed by four ground-based tracking stations,

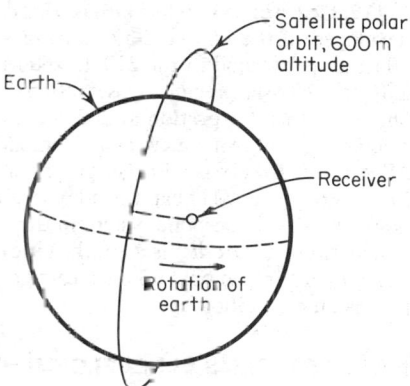

Fig. 25-105. Transit orbit.

and correctional notes are sent to the satellite, which rebroadcasts them. Total system error on board a ship traveling at known speed is below 0.25 mi. Even with several satellites in orbit (typically five), the data rate is too low to be of interest in most airborne applications. It continues to be a popular marine aid, the chief complaint in the 1980s being that the satellites have been allowed to become bunched, with correspondingly long waits between passes in some parts of the world.

117. Consol.[110] This system, originally known as Sonne, was developed in Germany for guidance of submarines during World War II. It operates in the 300-kHz region and consequently is subject to atmospherics and to sky-wave propagation effects. Three vertical antennas are stationed in a line about 2,700 m apart, the central one being fed cw at about 3 kW and the outer ones cw at 750 W each. The phases of the signals from the outer ones are varied in time steps of 1 s for 30 s with alternate transmission of the Morse letters E and T, producing a multilobe rotating pattern.

The observer, using a simple receiver with audio output, can estimate his direction by counting the number of E's and T's he has heard since the start of the transmission, which repeats every minute. The range over seawater is of the order of 1,000 mi, and the accuracy better than 1° under favorable conditions. Although the system has frequently been declared obsolete, the simplicity of the receiving equipment is attractive, and as recently as 1971, the Norwegian government installed four new sets of stations for the benefit of fishing vessels. Its low data rate and indirect readout have precluded its use in airborne applications.

118. Talking Beacons. Since the advent of highly directional microwave beams, it has frequently been proposed that a simple directional system could be devised in which such a directional beam rotated slowly while constantly announcing, by recorded voice, the direction in which it was pointing. The U.S. Coast Guard started several such developments in the early

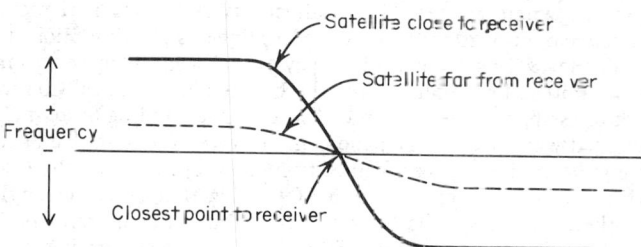

Fig. 25-106. Doppler shift of Transit signal.

1960s, but they were unsuccessful, due mainly to mistakes in hardware design. However, since 1966 Japan has had a successful three-station system in operation in the Straits of Tsushima.[111] Stations are at least 100 m above sea level, transmit 7-kW pulses at 9,300 MHz, and are pulse-duration-modulated by voice. The pulse repetition frequency is 10 kHz, and rotational rate is ⅛ r/min, with three beams per station 120° apart. The voice modulation is recorded on magnetic drums which are geared to the antenna rotation. Antenna beamwidth is 2°, and bearing announcements are made every 2°. The accuracy is about 1°, and the range about 70 mi.

119. Cooperative Systems Used for Surveying.[78] *Shoran* is a two-way pulse system developed in the United States during World War II. It transmits 0.25-μs interrogation pulses, 10 times per second in the 210- to 260-MHz region, which are replied to by transponders in the 290- to 320-MHz region. Observation is on an oscilloscope. *Trident* is a French variation of Shoran with interrogation at 230 MHz and reply at 270 MHz. *Raydist* is a United States proprietary system using a cw phase comparison in two-way and hyperbolic modes in the 1.6- to 5-MHz band. *Hi/Fix* is a British proprietary system (owned by the Decca Co.) using bursts of cw in the 1.6- to 2.6-MHz band. Sea-Fix is a higher-power version with shorter transmissions. *Toran* is a French cw hyperbolic system in the 1.4- to 2-MHz band. Major French ports have permanent installations. *Hydrodist* is a South African proprietary two-way system using cw phase comparison, superimposed on a 36-GHz carrier, with parabolic dishes to reduce multipath effects. Display is on an oscilloscope.

MAJOR SYSTEMS IN DEVELOPMENT

120. Satellite Systems. Many believe than only a system using earth satellites can provide worldwide navigational coverage with the accuracy associated with line-of-sight frequencies. Aside from Transit, whose data rate is considered too low, many other schemes have been proposed, many relying on the use of synchronous satellites, successfully used for communication.[112] The problems are primarily political, stemming chiefly from the question of the country or agency responsible for the high costs of satellite manufacture, launch, and maintenance.

Because of the relatively inaccurate navigation aids and the absence of direct surveillance in midocean, transatlantic aircraft have long been forced to fly on parallel paths that are 120 mi apart at the same altitude, compared with similar paths 5 mi apart in areas having VOR/DME service. This forces some aircraft either to take circuitous routings or to fly at uneconomic altitudes. However, the advent of larger aircraft has tended to keep the number of aircraft fixed, despite traffic growth. Thus the pressure to solve this problem has not been great. The airlines would like better communication with transoceanic aircraft but are not particularly concerned with navigation or surveillance. The FAA is primarily interested in surveillance. Only the military, so far, have shown much interest in satellite navigation.

One such system, the Global Positioning System (GPS), also known as Navstar,[86] has received much funding and publicity (Fig. 25-107). When completed in the late 1980s, it will use 24 satellites, each circling the earth twice a day, at about 10,000 mi altitude. Each satellite will transmit cw, phase-modulated with a pseudo-random code, which will allow measurement of distance in much the same way as at VLF but without atmospheric disturbances since the frequencies will be 1,227 and 1,545 MHz. Depending on angles of cut, integration time, and other factors, accuracies of tens of feet are claimed.

The supporters of GPS claim that once the system is installed for military purposes, civil uses will also be found for it.

121. Microwave Landing System.[113] While the present ICAO ILS has served well for over 30 years, requirements for the future are believed to necessitate more channels, more flexible approach paths, and greater freedom from site effects. These can readily be obtained at microwave frequencies where greater antenna directivity and a wider frequency spectrum are available. Since a range of only 20 mi or so is needed, line-of-sight limitations pose no problem. Towards the end of the 1960s a number of new systems began to appear, operating between 1 and 16 GHz. In an effort to halt proliferation, the U.S. Radio Technical Commission for Aeronautics undertook to establish a single standard Microwave Landing System (MLS). The work began in 1967, elicited worldwide participation, and received the tentative blessing of ICAO in 1978. It calls for angular guidance to be obtained from fan-shaped beams which scan the airspace, using 200 radio frequencies between 5,031 and 5,091 MHz. The perceived angle in the aircraft is proportional to the time it takes the beam to pass through the aircraft first in one direction and then in the other. The scanning rate is 50 μs/degree. The system is thus known as time-referenced scanning beam (TRSB); it was expected to get the full approval of ICAO in 1980.

122. Joint Tactical Information Distribution System (JTIDS).[114] While chiefly a military communication system, it is mentioned here because it uses the same frequency band as DME and Tacan and because it has received support of the same financial magnitude as GPS and MLS. It employs pulses which are simultaneously hopped in frequency, time, and phase to produce marked protection against wide-band noise.

123. Collision-Avoidance System.[101] Since about 1955 the scheduled airlines have been actively looking for a system that would protect against midair collisions. Most of the schemes proposed, while recognizing the threat of another approaching aircraft, have suffered

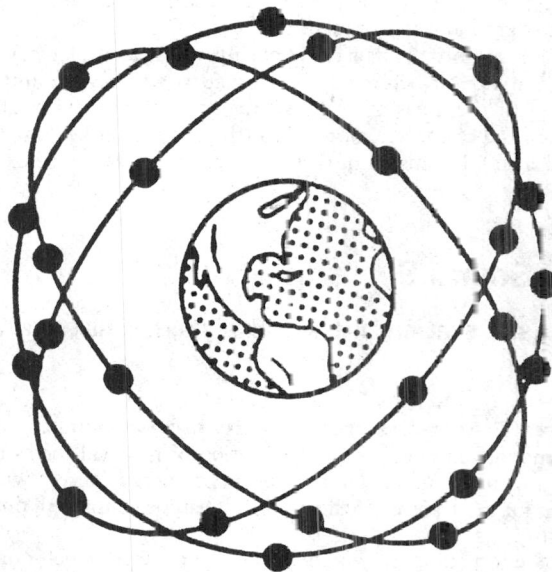

Fig. 25-107. GPS satellite coverage (eight satellites in each of three orbit planes).

from excessive false alarms; this is not surprising when one considers that two aircraft on parallel tracks may pass each other very closely but with complete safety. A major difficulty has been that no practical airborne antenna has been available with sufficient directional discrimination to distinguish between a collision course (no change of angle) and a passing course (small change of angle). A further difficulty has been that all practical systems have required equipment on all aircraft, not just those willing and able to pay for the protection. Thus low cost has been an essential requirement.

The most promising solution appears to be one that makes use of the altitude-coded ATCRBS transponder already carried by large numbers of aircraft. This is known as B-CAS (beacon-derived collision avoidance system). This will at least allow an aircraft equipped with B-CAS to see all other transponder equipped aircraft in the neighborhood. However, in the terminal area, where most accidents occur, this system may cause overload.

Not the least of the problems is what happens to other traffic in a dense environment when two aircraft take evasive action, unknown to the air traffic controller on the ground. While alleviating one problem, it may generate others. A growing school of thought consequently believes that the only long-term solution is an accurate and complete traffic-control system in which all aircraft are under surveillance and control by a single authority on the ground.

124. Possible Future Trends. At the end of World War II, in 1945, few ships and aircraft were equipped with electronic navigation, aside from direction finders, and about 100 different systems were in development or being proposed. In 1980, as shown in Table 25-15, hundreds of thousands of vehicles are equipped, and only two radically new systems are in development, GPS and MLS. All the other new systems such as DABS, Adsel, B-CAS, and JTIDS are outgrowths of existing systems and provide compatibility with those they replace. Thus, during the transition phase from old to new, users of these latter systems do not have to carry both the old and the new sets; the new sets accommodate the old functions. It is believed that this trend toward compatibility will continue.

Nearly all the systems described have been *vehicle*-derived; i.e., the computation is done on board the vehicle. In the past, there were many good reasons for this arrangement, but the advent of reliable computers may be inducing a change to *ground*-derived systems.

An example of a new system which combines both these concepts is Germany's DAS, or DME-derived *a*zimuth *s*ystem, which is being proposed as a possible adjunct to MLS. In this system, the interrogations from an existing standard airborne DME are received on the ground and their direction of arrival determined, pulse by pulse, by a special high-speed computerized direction finder. The resulting azimuth angle is then telemetered back to the originating aircraft via the normal DME reply link. Thus, the aircraft obtains its bearing at practically no extra cost over that of getting distance alone, and the bearing accuracy is primarily a function of how large an antenna array the ground station can afford.

We have considered mainly the navigation of ships and aircraft. However, at a radionavigation users' conference[115] in 1979 it became apparent that there is a large untapped *land*-navigation market, particularly in the areas of *site registration* (the business of accurately defining the location of a vehicle) and of *vehicle monitoring* (the business of keeping fleet owners advised of where their vehicles are). Expansion in these and related fields can be expected in the 1980s.

Underwater Sound Systems

By J. F. BARTRAM, S. L. EHRLICH, D. A. FREDENBURG, J. H. HEIMANN, J. A. KUZNESKI, AND PAUL SKITZKI

125. Principles, Functions, and Applications. Sound energy travels in water as a result of particle motion initiated by the application of physical forces to the particles from a vibrating diaphragm, collapsing air bubbles, or other energy sources with sufficient mechano-acoustical coupling for the transfer of the energy. It can be controlled, directed, and transmitted for many useful purposes.[121–125]

Water is an excellent medium in which to transmit compressional sound waves. Liquids have higher specific acoustic impedances by several orders of magnitude than gases. The high acoustic impedance of water (1.5 MN·s/m³ for seawater) makes it possible to design transducers whose internal mechanical impedance approaches the radiation load impedance, with conversion efficiency on the order of 50% over a band of an octave or over 80% over a narrow band.[121]

The transmission and reception of underwater sound can be controlled and directed to perform the functions of communications, navigation, detection, tracking, classification, etc., which in aerospace are accomplished with electromagnetic energy.[122] The wavelengths of underwater sound systems and radar systems are of the same order of magnitude, since the frequencies employed differ by the ratio of the speed of sound to the speed of electromagnetic waves. The term *sonar*, derived from sound navigation and ranging, is used synonymously with *underwater sound* and *underwater acoustics*.

The applications of underwater sound for defense purposes, both pro- and antisubmarine, have advanced with development of the nuclear submarine and other platforms. In military applications, underwater sound is used for depth sounding; navigation; ship and submarine detection, ranging, and tracking (passively and actively); underwater communications; mine detection; and or guidance and control of torpedoes and other weapons. Most systems are monostatic, but bistatic systems are also employed.

Civilian applications of underwater sound are numerous and are continuing to increase as attention is focused on the hydrosphere, the ocean bottom, and the subbottom. These applications include depth sounding; bottom topographic mapping, object location; underwater beacons (pingers); wave-height measurement; doppler navigation; fish finding; subbottom profiling; underwater imaging for inspection purposes; buried-pipeline location; underwater telemetry and control; diver communications; ship handling and docking aid; antistranding alert for ships; current flow measurement; and vessel velocity measurement.

PROPAGATION

126. Propagation of sound in water can be represented by the sound pressure, the sound particle velocity, and/or the sound intensity as a function of position, time, and frequency.

Because the sound pressure can usually be measured more nearly directly, it is the preferred parameter for most experimental data.[135] The *sound-pressure amplitude* p in water is expressed in pascals.* The logarithmic unit of a *sound-pressure level* L_p is expressed in decibels with respect to the reference sound pressure amplitude, that is, 1 μPa $= p_0$, where $L_p = 20 \log (p/p_0)$. The phase of the sound pressure is expressed in degrees or in radians with respect to a specified reference.

The difference between the sound-pressure level at the reference position and the sound-pressure level at a point in the sound field is called the *propagation loss* N_W for that point. For a

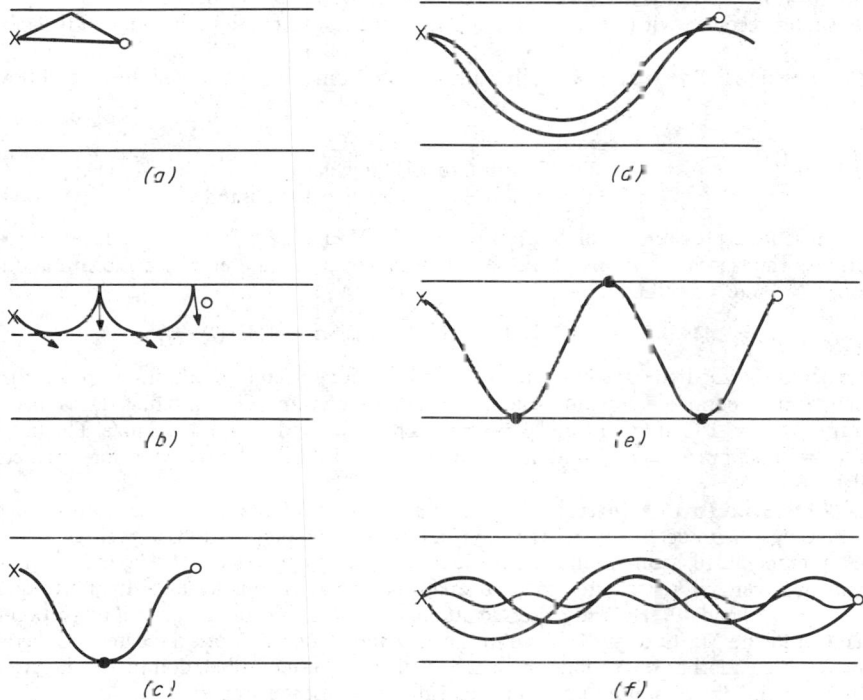

(a)

(c)

(b)

(e)

(c)

(f)

Fig. 25-108. Propagation paths.

small sound source, the reference position may be at a standard distance of 1 m in the direction of the maximum response. For a larger source, far-field data may be extrapolated back to the reference distance r_0.

The propagation loss may be considered[129-132] to consist of two basic components, one due to the spreading of sound energy with increasing radial distance r from the sound source N_{spr} and the other to attenuation of sound as it propagates through the medium N_{att}

$$N_W = N_{spr} + N_{att}$$

The *spreading loss* is given by $10n \log (r/r_0)$, where n is dependent on the spreading law and is equal to 2 for the theoretical case of *spherical spreading*. The *attenuation loss* is given by $10^{-3}\alpha r$, where α, the *attenuation coefficient*, is as discussed in Par. **25-128**.

Common propagation paths are illustrated[129] in the ray diagrams of Fig. 25-108. The paths of the direct ray and a ray with a single surface reflection in water with constant sound speed, are shown in Fig. 25-108a. In b a surface-layer sound channel confines the ray near the surface, with leakage rays due to diffraction and reflected waves from a rough surface. A ray which experiences a single bottom reflection is shown in c, while two bottom reflections with an intermediate surface reflection are shown in e. A pair of rays that diverge and return to a crossover in the convergence zone is shown in d. The three diverging rays, trapped in a deep sound channel and

*1 pascal $= 1$ N/m^2 $= 10$ dyn/cm^2 $= 10$ μbars.

crossing each other several times before converging at the receiver, are shown in f. The reliable acoustic path, RAP (not shown), exists between a source at moderate depth and a surface receiver.

The theoretical treatment of sound propagation in water depends on assumptions used to simplify the mathematical formulation. Typical assumptions used by various authors include combinations of one or more of the following:[129,132,139]

One sound source with constant frequency and spherically isotropic radiation

Propagation medium with linear transmission characteristics and sound speed dependent only on depth

An ideal horizontal plane sea surface with zero acoustic impedance

An ideal horizontal plane sea bottom with infinite acoustic impedance

A sound receiver with zero rise time, flat frequency response, and spherically isotropic reception

127. Speed of Sound. Essentially, all seawater can be represented by the following conditions:

$$\text{Temperature } T = -3 \text{ to } +30°C$$
$$\text{Depth } d = 0 \text{ to } 10,000 \text{ m}$$
$$\text{Salinity } S = 33 \text{ to } 37 \text{ parts per thousand}$$

with atmospheric pressure (absolute pressure) 0.1013 MPa at zero depth at sea level. For these conditions, the speed of sound c in seawater is given by an empirical formula due to Wilson,[126-128] here simplified to

$$c = 1449.3 + 4.572T - 0.0445T^2 + 0.0165d + 1.398(S - 35) \quad \text{m/s}$$

At zero depth the accuracy of c is about 3 m/s, or 0.2%; for extreme conditions with comparable accuracy a more complete equation, including higher-order and cross-product terms, becomes necessary. At $T = 0°C$, $d = 0$ m, and $S = 35$ parts per thousand, $c = 1,449.3$ m/s. The nominal value $c = 1,500$ m/s, corresponding to about $T = 13°C$, is convenient for engineering calculations.

128. Attenuation of Sound. The attenuation of sound in seawater has been studied by many investigators to determine its variation in different frequency bands as well as its dependence on temperature, salinity, and depth. At frequencies greater than 1 MHz, the attenuation mechanism is generally attributed to shear and dilatational viscous losses.[132] In the frequency band between 10 and 40 kHz, the increased attenuation is almost solely due to a relaxation-type mechanism in the $MgSO_4$ salts dissolved in the seawater.[138] Recent work indicates that between 0.1 and 1.0 kHz another relaxation-type mechanism, not yet identified, dominates the attenuation.[133] Below 50 Hz other attenuation mechanisms are of greater importance.

An expression for the attenuation coefficient of seawater from 0.1 kHz to 100 MHz includes three components resulting from the three main attenuation mechanisms,[129,132,133,136] multiplied by a depth-dependent term:[137]

$$\alpha = \left[\frac{0.11f^2}{1 + f^2} + \frac{0.70f_T f^2(S/35)}{f_T^2 + f^2} + \frac{0.03f^2}{f_T} \right] (1 - 65 \times 10^{-6}d) \quad \text{dB/km}$$

where f = frequency (kHz), f_T = relaxation frequency[136] (kHz) = $21.9 \times 10^{6-1,520/(T+273)}$, and T = temperature (°C), S = salinity (parts per thousand), d = depth below air-water boundary surface (m).

At $T = 4°C$ ($f_T \approx 71$ kHz), $d = 0$ m, and $S = 35$ parts per thousand, the equation simplifies to

$$\alpha = \frac{0.11f^2}{1 + f^2} + \frac{50f^2}{5,000 + f^2} + 0.0004f^2 \quad \text{dB/km}$$

Further simplification is possible for the above listed conditions, at frequencies below about 20 kHz, to approximately

$$\alpha = \frac{0.11f^2}{1 + f^2} + 0.010f^2 \quad \text{dB/km}$$

and above 200 kHz to approximately

$$\alpha = 50 + 0.0004f^2 \quad \text{dB/km}$$

Measured values of attenuation in seawater include absorption losses, scattering losses due to random internal inhomogeneities, and interaction losses with the bottom boundary, the subbottom, and upper surface boundary.

129. Reflection and Refraction. Reflection and refraction of sound are normally in accordance with Snell's law, which states that $(\cos \theta_i)/c_i$ is a constant, where θ_i is the angle between the direction of propagation and the horizontal plane and c_i is the sound velocity at the point i of the ray. At a boundary where the sound velocity is discontinuous, the angle θ_i is the angle of incidence with respect to the plane tangent to the boundary, which is not necessarily horizontal.

A special case of interest occurs in a region with a constant *sound speed gradient* ∇c. This results in a circular sound ray path with radius equal to $c_i/(\nabla c \cos \theta_i)$. The center of the circle corresponds to a position where c_i becomes zero. In a surface layer, the resultant upward refraction leads to shadow zones which theoretically contain no propagated sound energy.

A second special case occurs at a boundary where the incident sound speed is lower than the refracted sound speed c_r, such as a plane interface between water and air. The well-known phenomenon of total internal reflection results when θ_i is less than a critical angle θ_{crit}, at which $c_i/(\cos \theta_{crit})$ is equal to c_r.

Another important case occurs when the sound in water is propagated by a direct path and also by a single reflection from a boundary of slower sound speed, such as air. Because of a 180° phase shift in the sound pressure at such a reflecting boundary, the resultant sound pressure exhibits maxima and minima as a function of the position of the receiver. This is called the *Lloyd mirror*, an image interference effect, since the condition can be represented by an additional sound source of equal amplitude and opposite phase as a virtual image that provides the constructive and destructive interference with the real sound source.

Other important cases include a multiple-layered medium where the sound field at each discontinuity must be accounted for; irregular and nonstationary boundaries which tend to randomize the reflection and refraction properties; moving boundaries which tend to modify the frequency of the sound energy because of the doppler effect; and intentional discontinuities introduced in transducer and array designs.

130. Reverberation. Reverberation of sound in water produces energy usually unwanted at the receiver. It is caused by scattering, i.e., reflection and refraction of sound from discontinuities other than those of primary interest. When the sound source and sound receiver are at the same location, reverberation is produced primarily by backscattering. When the scatterers are boundaries of the medium, the effect is called *surface reverberation*, which may be subdivided into *sea-surface* and *sea-bottom* reverberation. When the scatterers are contained within the medium, the effect is called *volume reverberation*, which may be due to fine particles, fish, or other inhomogeneities, including the structure of the sea.

NOISE

131. Background Noise. Underwater sound systems operate in a medium which has a very low acoustic-noise level under quiet conditions. Stimulation from natural and human causes, however, can generate and propagate acoustic noise at various levels and frequencies. The acoustic-noise background consists of an *ambient-acoustic-noise level* and a *self-noise level* caused by the sonar-platform presence and its movement. The factors contributing to the acoustic-noise background are shown in Fig. 25-109. Passive sonar systems detect target-generated noise, while active sonar systems detect target echoes in the presence of the noise background.

132. Ambient Noise. The main contributions to ambient noise are shown in Fig. 25-110. Tides, waves, and seismic disturbances predominate at the very low frequencies, in the region of 1 Hz. At frequencies used for sonar systems, the main sources are sea surface agitation due to meteorological effects, noise from marine life, and human noises from shipping and other activities.[140,141] Of these, surface agitation due to wind and wave action is most significant. The acoustic spectrum levels for various sea states are shown in Fig. 25-111

At the high frequencies (above 100 kHz), thermal molecular motion of the water is the principal noise contributor but does not limit sonar performance at lower frequencies. It has a spectrum level that rises with frequency at a rate of 6 dB/octave. The mean-square sound pressure in a 1-Hz band is

$$p_T^2 = 4\pi k T_\rho c/\lambda^2$$

where k = Bolzmann's constant, T = temperature of water (K), ρc = specific acoustic impedance, and λ = wavelength. Acoustic thermal noise is consistent with its more familiar electrical
counterpart. It is far below sea state levels at frequencies below 10 kHz. At higher frequencies
than 100 kHz, it sets a threshold on the minimum detectable pressure levels in the medium.[142]
The ambient-noise level, as a sonar parameter, is the intensity of the noise background as measured with a nondirectional hydrophone, referred to the intensity of a plane wave having an
rms pressure of 1 μPa.

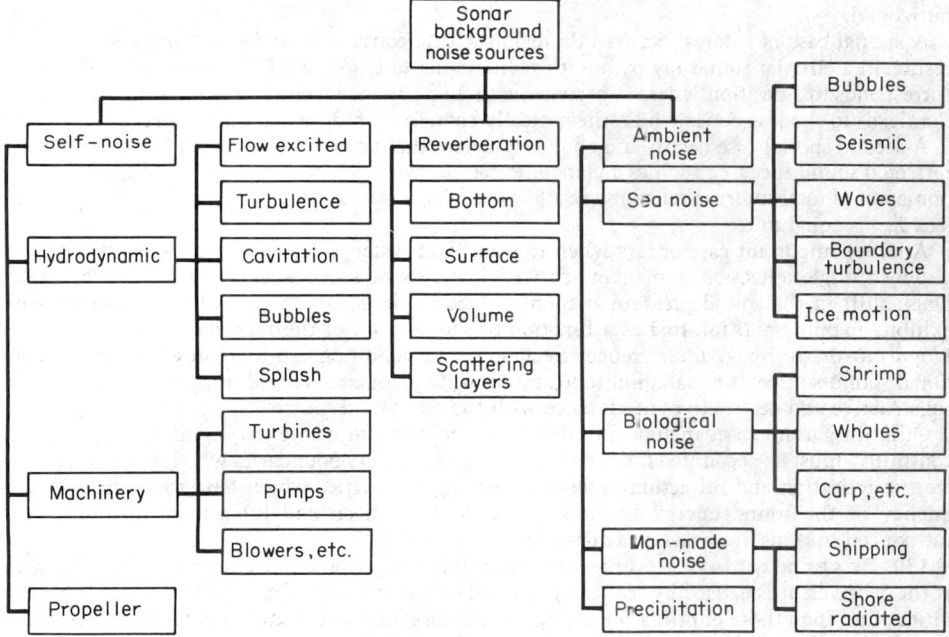

Fig. 25-109. Sonar background-noise sources.

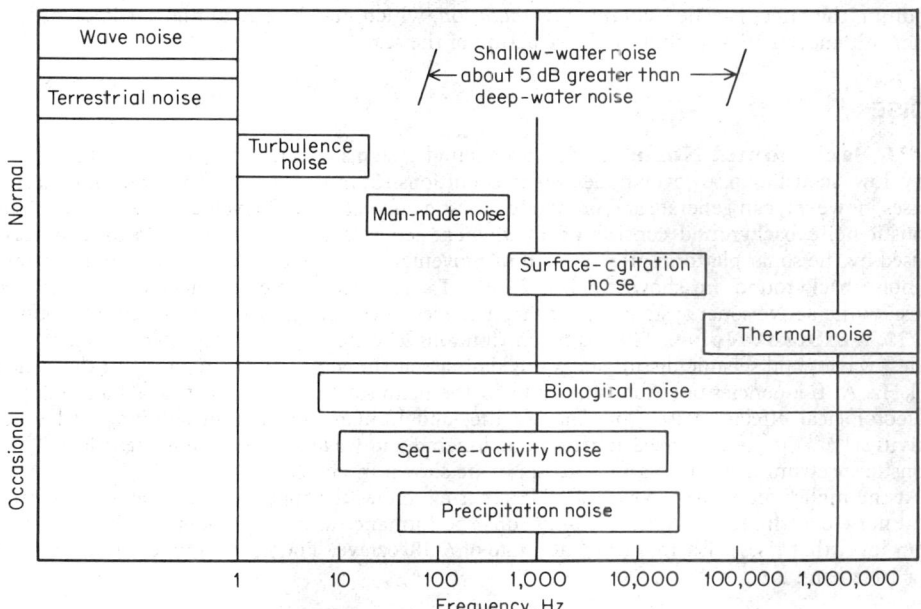

Fig. 25-110. Principal ambient-noise sources.

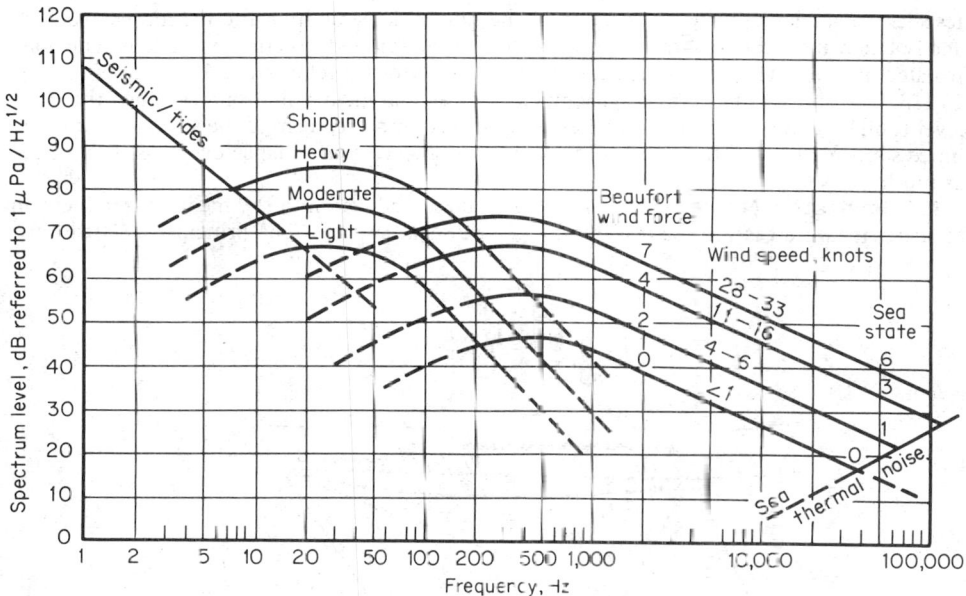

Fig. 25-111. Average deep water noise.

133. Platform Noise. Platform noise is a degrading factor in the performance of underwater sound systems, particularly in the case of mobile platforms such as ships, submarine, aircraft, torpedoes, and other sonar-carrying vehicles. Fixed-position platforms, including sonobuoys, moored structures, bottom-mounted structures, mines, etc., are also plagued with platform-noise problems, primarily induced by hydrodynamic flow or motion. Good sonar performance with mobile platforms at any speed above approximately 10 kn is achievable only by most careful attention to platform-noise reduction and by design of the sonar system for minimum susceptibility to local noise. Figure 25-112 shows the dominant sources and relative levels at various ship speeds.

Platform noise may enter the sonar system via radiation in the medium or by conduction through the platform structure. Generally, conducted noise can be reduced below the level arriving at the array via the medium. The techniques involve sensor design for minimum

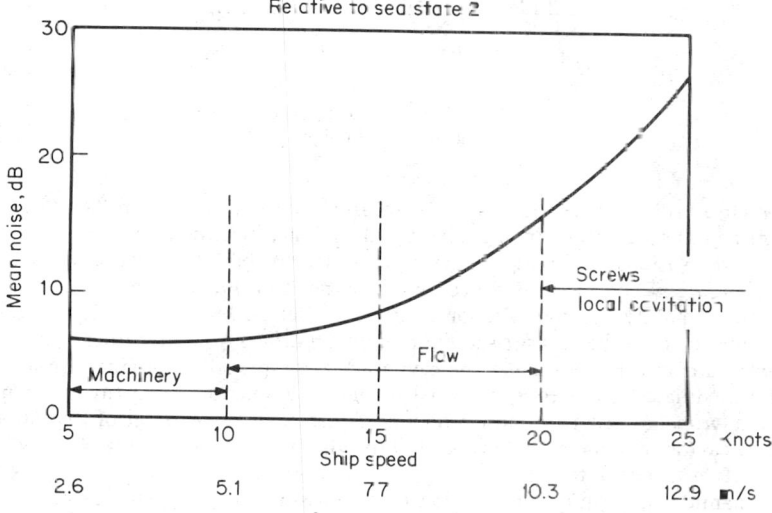

Fig. 25-112. Self-noise behavior.

response to acceleration forces, isolation of the sensor elements from the mounting structure, and isolation and location of the array away from hull-borne and structural vibrations. The noise radiated into the medium may reach the array directly or via reflected paths, as shown in Fig. 25-113. The ship's propellers and machinery are the dominant noise sources. Since the noise level is highest on stern bearings, baffles are generally employed behind the sonar array to minimize stern noise, even though this results in loss of sonar performance over a portion of the azimuth.

134. Radiated Noise. Platform-generated noise, radiated into the medium, produces an acoustic signature that can be detected by passive sonar systems. The principal radiated noise

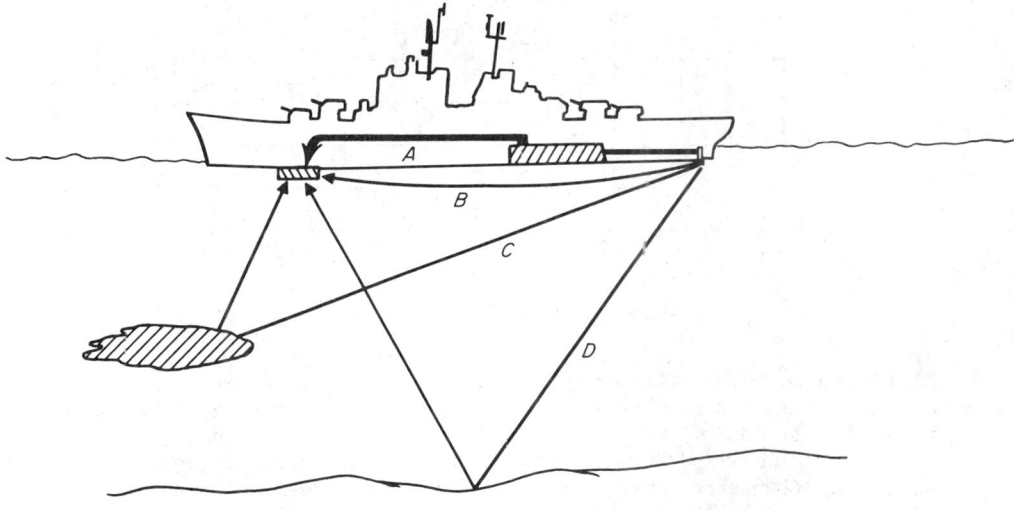

Fig. 25-113. Paths of self-noise.

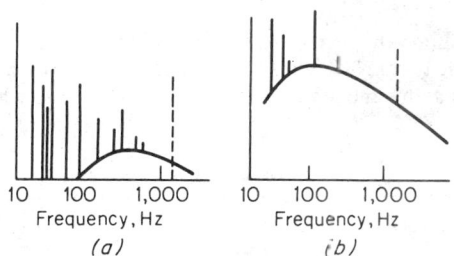

10   100   1,000   10   100   1,000
Frequency, Hz   Frequency, Hz
(a)   (b)

Fig. 25-114. Diagrammatic spectra of submarine noise: (a) low speed; (b) high speed.

sources on ships, submarines, and torpedoes are listed in Table 25-20 (Par. **28-149**). Machinery noise is defined as noise caused by propulsion and auxiliary machinery on the vessel. The noise produced by the various machines, generators, pumps, actuators, etc., travels by diverse paths to the hull structure, where it is introduced into the medium. Propeller noise originates outside the hull and is mainly due to cavitation at the propeller blades. Cavitation-produced bubbles generate acoustic noise, the acoustic spectrum of which differs from machinery noise and varies with speed and depth. In addition to the cavitation noise, propellers produce amplitude-modulated noise modulated at a frequency equal to the shaft rotation speed times the number of blades. Such *propeller beats* are most pronounced just beyond the onset of cavitation and are swamped by cavitation noise at higher speeds. Propellers may also produce a "singing" noise due to vibrational resonance of the blades.[141]

Hydrodynamic noise results from the flow of fluid past the moving platform. It increases with hydrodynamic structural irregularities of the platform and the fluid flow rate. Breaking bow and

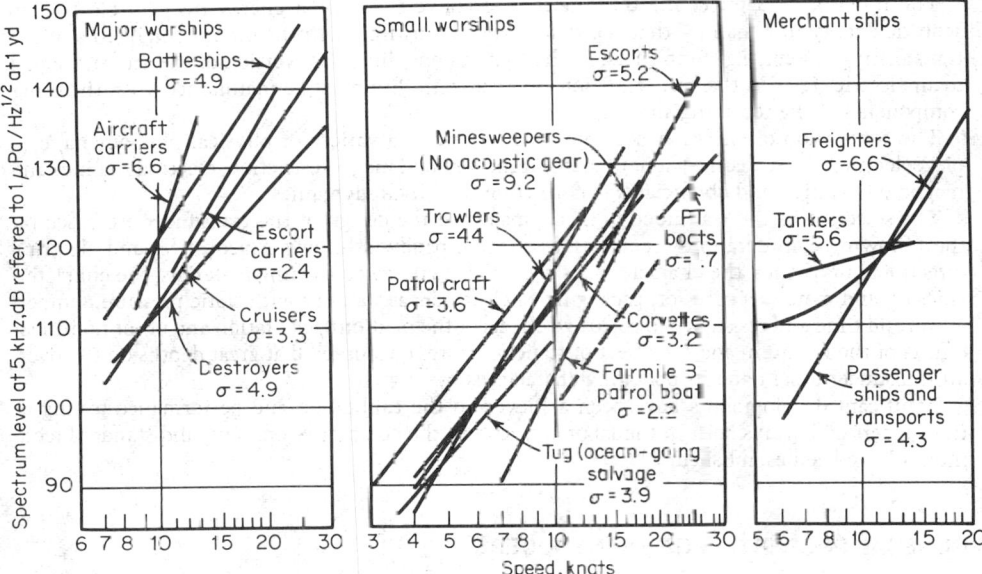

Fig. 25-115. Average radiated spectrum levels of ships.

stern waves can excite the hull or structural members. Hydrodynamic noise is a minor contributor to the platform-radiated noise, and is usually swamped by machinery and propeller noise. However, it is an important element in consideration of self-noise for underwater sound systems associated with the platform.

135. Radiated Noise Levels. Radiated noise consists of broadband noise and tonal noise (line components). Measurements of radiated noise are made at some distance, say 200 yd, from the vessel and reduced to source spectrum level values by correction for the test distances and the measurement bandwidths. Tonals are determined by fine-grain spectral analysis. Figure 25-114 illustrates broadband and tonal noise from a submarine at two speeds. In Fig. 25-114a the broadband noise from cavitation at the propeller begins to appear, and the tonals from machinery noise are predominant. In Fig. 25-114b the broadband noise has increased as a result of higher speed, and

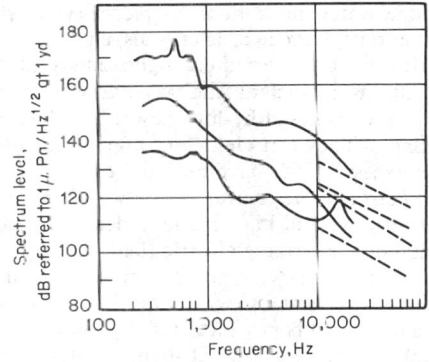

Fig. 25-116. Noise spectra of torpedoes.

many tonals are masked by the broadband noise while other tonals are changed in amplitude or frequency.[141] Figure 25-115 shows average radiated-noise spectra for several classes of ships, and Fig. 25-116 illustrates noise from running torpedoes.

Transducers and Arrays

136. Transducers for underwater sound applications perform the functions of generating a sound wave in the medium or detecting the existence of a sound wave and its properties (e.g., amplitude and phase) in the medium. In the generating case, the transducer is commonly referred to as a *source*, or *projector*, and in the detecting case, as a *hydrophone*. Often the transducer is required to perform both functions. Single transducers or arrays of transducers may be designed to control the directional properties (i.e., directivity or beam pattern) of the generated acoustic energy, and to discriminate against the noise in the receiving case.

The function of a projecting transducer is to convert the input energy (usually electric) to acoustic energy in a manner that is efficient and compatible with the other components of the transmitting system, e.g., amplifiers. In the hydrophone, linear conversion of the acoustic signal to an electric signal is the basic function, and compatibility must be maintained with the other components of the receiving subsystem.

The conversion of energy is accomplished by any of a variety of physical phenomena, e.g., piezoelectricity and electrostriction; piezomagnetism and magnetostriction; electrodynamics and magnetodynamics; and chemcial transformations and hydrodynamics.

The selection of the transduction mechanism and the design of the transducer are based on the following considerations: operating frequency; bandwidth; power (acoustical and electric); directional properties; the characteristics of available energy-converting materials; the characteristics of, and materials used for, packaging (such factors as stability with static pressure, temperature, and time and resistance to corrosive effects of the medium); cavitation and other nonlinear effects of the medium; and the effect of static pressures encountered at great depths on the overall transducer design and its operating characteristics.

Significant developments have been achieved in the calibration and performance testing of transducers and arrays both in the laboratory and in the ocean environment, and standard techniques have been established.[146,147]

GENERAL PROPERTIES OF TRANSDUCERS

137. Types of Projectors. There are two main transducer types used as underwater sound sources: those operating with a continuous-wave or modulated (amplitude, frequency) input and those operating as impulse sources. The former are used for most military and many commercial applications; the latter are used mainly for oceanographic and geophysical applications.

138. CW and Modulated Sources. Transducers designed for cw or modulated-input underwater applications use piezoelectric, electrostrictive, or magnetostrictive energy-conversion materials. Piezoelectric crystals, e.g., quartz, have a linear relationship between strain and electric field. However, their application is limited by low dielectric constant, low electromechanical coupling coefficient (the ratio of the converted energy to the total input energy in a transducer), narrow bandwidth, low power-handling capability, and limited availability of geometrical shapes. They can yield very high conversion efficiencies (quartz transducers having efficiencies in excess of 90% have been built).[148]

Ferroelectric crystals, in either single-crystal or polycrystalline ceramic form, are often electrostrictive and have higher-order nonlinear properties in their natural state. With the application of a polarizing electric field, the electromechanical processes in these materials can be linearized over a wide range of operating conditions. These materials have the advantage of high dielectric constant (which results in low impedance), high electromechanical coupling coefficients, broad bandwidth, and high power-handling capability when properly used, and are available in a wide variety of shapes (plates, cylinders, rings, spherical zone sections). Efficiencies in excess of 70% can be achieved. Operating frequencies from less than 1 Hz to more than 10 MHz can be achieved.

The most widely used materials in the ferroelectric class are modified lead zirconate titanate and modified barium titanate. Idealized transmitting current and voltage responses of a transducer that utilizes these materials are shown in Fig. 25-117a and b, where f_0 is the frequency of mechanical resonance of the transducer. Figure 25-118 shows a typical impedance characteristic.

Magnetostrictive transducers depend upon the interchange of energy between magnetic and mechanical forms. In an unpolarized state, such transducers are nonlinear, frequency-doubling devices. However, in a polarized state (achieved by the use of permanent magnets, direct current, or operation at remanence) they are linear (i.e., piezomagnetic). Commonly used materials include various nickel alloys and ferrites.

Properly designed magnetostrictive transducers can achieve radiated acoustic powers up to several kilowatts with efficiencies in excess of 50%. Operating frequency is usually limited to frequencies below 100 kHz. Figure 25-119 shows a typical idealized transmitting current characteristic.

Electrodynamic sources have been used in underwater acoustics for low-frequency applications. One noteworthy application is the low-frequency *standard projector.*[149]

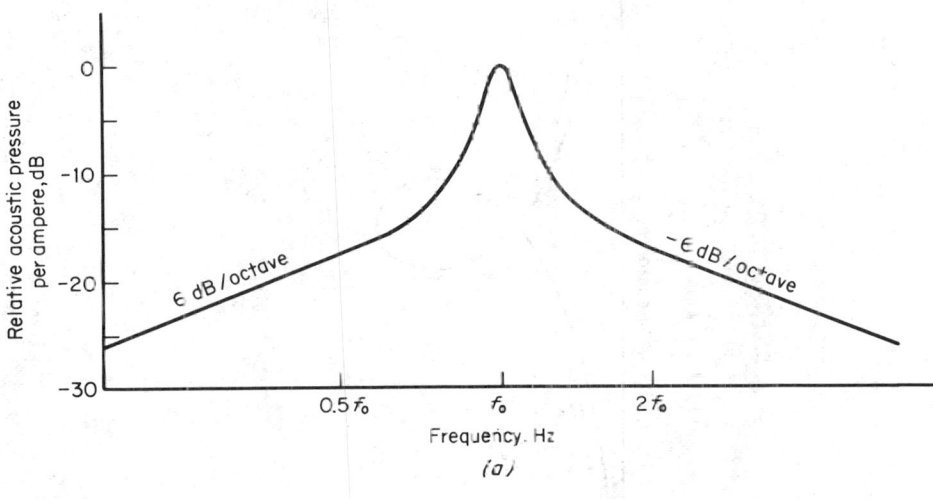

(a)

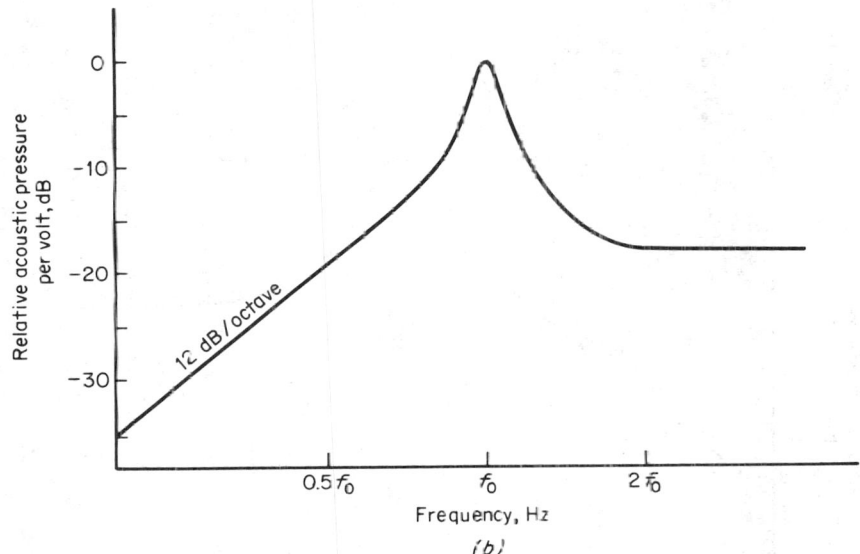

(b)

Fig. 25-117. Idealized transmitting current and voltage responses for a piezoelectric transducer: (a) current response; (b) voltage response.

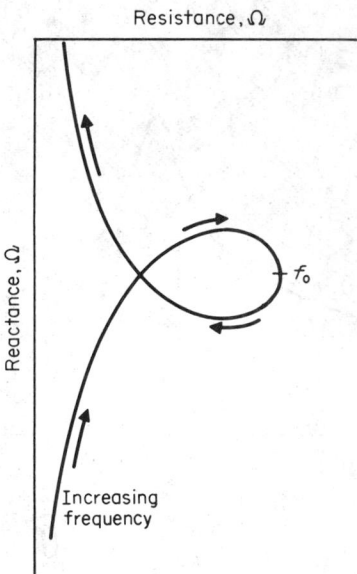

Fig. 25-118. Idealized impedance locus for a piezoelectric transducer.

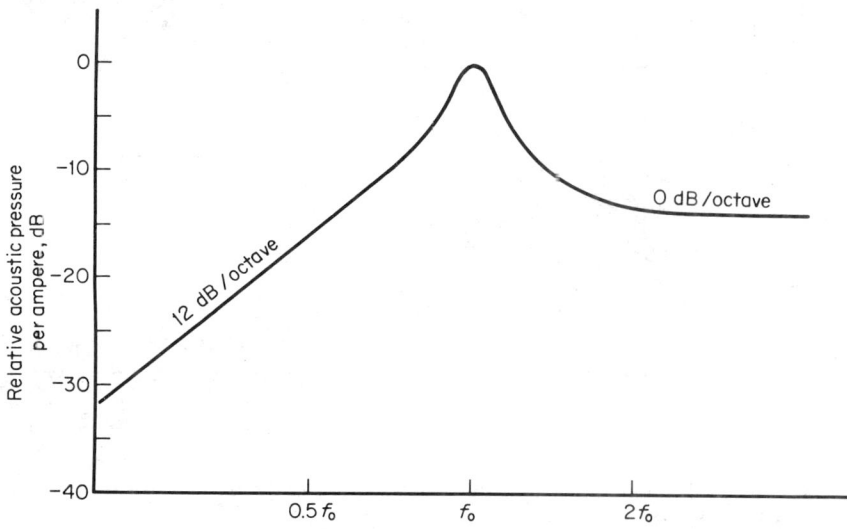

Fig. 25-119. Idealized transmitting current response for a magnetostrictive transducer.

139. Impulse Sources. Impulse sound sources produce a short-time duration, high-amplitude transition of more or less regular waveform. For example, explosives (TNT or other high-burning-rate chemical) with provision for hydrostatic, electrical, or fused detonation produce in the ocean medium a pressure wave that is initially steep-fronted, displays approximately exponential decay, and is followed by a sequence of bubble pulses. The shock wave is usually so intense that the resulting finite-amplitude effects are appreciable. Figure 25-120 shows a typical pressure-time characteristic.

Empirical formulas yield $p_0 = 4.22 \times 10^8 (w^{1/3}/r)^{1.13}$ to $36.4 \times 10^{-6} w^{1/3} (r/w^{1/3})^{0.22}$ and $T = 3.35 w^{1/3}/(0.0305d + 10.1)^{5/6}$, where d is depth (m), r is range (m), and w is equivalent yield of TNT (kg).

Charges ranging in weight from 1 oz to 50 lb are in common use. A 4-lb charge of TNT will produce a pressure level at 1 km of 4 MPa (that is, 252 dB re 1 μPa). Other impulse type

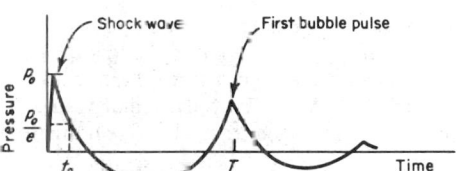

Fig. 25-120. Pressure-time characteristic for an explosive source. [50]

sources include implosive devices, spark-gap generators, and pneumatic- and mechanical-impact mechanisms.

140. Hydrophones. The receiver in a sonar system employs a hydrophone or hydrophone array coupled to an amplifier. Hydrophone elements generally use piezoelectric energy-conversion materials, although magnetostrictive and electrodynamic mechanisms are sometimes used. Typical hydrophone sensitivities are on the order of -180 to -200 dB re 1 V/μPa. Proper impedance termination and suitable amplification are necessary to obtain useful electrical levels. Figure 25-121 shows a typical idealized open-circuit receiving response for a piezoelectric hydrophone.

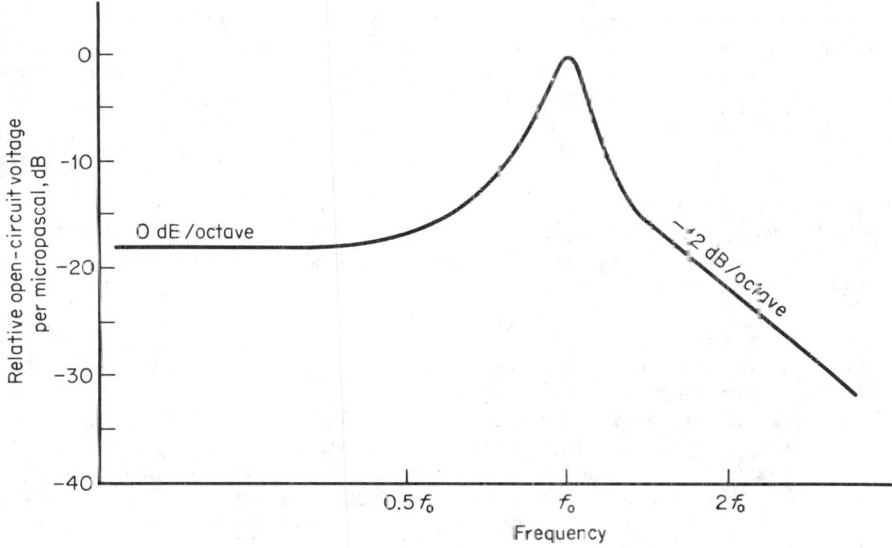

Fig. 25-121. Idealized open-circuit receiving response for a piezoelectric transducer.

PRODUCTION OF SOUND FIELDS

141. Acoustic Principles. The production of a sound field in the medium involves an electrical input that is converted by the transducing mechanism into the motion of a surface in contact with the medium. The motion initiated by the moving surface of the transducer is communicated to the adjacent water particles, and a sound wave is propagated from its surface.

The *far-field sound pressure* produced by the source can be described in terms of the radiated acoustic power P_a by

$$p^2(r,\phi,\theta) = \rho c P_a D_i R(\phi,\theta)/4\pi r^2 \qquad (25\text{-}74)$$

where $p^2(r,\phi,\theta)$ = mean-square acoustic pressure (Pa); r,ϕ,θ = spherical coordinates, r in meters; ρc = product of density and speed of sound of medium, i.e., specific acoustic impedance of medium ($N \cdot s/m^3$); $R(\phi,\theta)$ = normalized pattern function; and D_i = directivity factor of source.

The directivity factor is defined as

$$\frac{1}{D_i} = \frac{1}{4\pi r_0^2} \int \int \frac{p^2(r,\phi,\theta)}{p_0^2} dS \qquad (25\text{-}75)$$

where p_0 = pressure at distance r_0 in direction of maximum response and dS = element of surface area on sphere having radius r_0.

An example of a normalized beam pattern is shown in Fig. 25-122. This figure shows a plot of $10 \log p^2(\theta,\phi_k)$ vs. θ for a particular value of $\phi = \phi_k$. Dividing Eq. (25-74) by the square of the input current to the transducer I^2 and rearranging gives

$$\frac{p^2(r,\phi,\theta)}{I^2} = \frac{\rho c D_i R(\phi,\theta) R_e}{4\pi r^2} \frac{P_a}{P_e} \qquad (25\text{-}76)$$

where P_e = electrical input power to transducer (W) and R_e = electrical input resistance of transducer (Ω).

The current transmitting response ($20 \log S_0$) is given by

$$20 \log S_0 = 10 \log R_e + 10 \log D_i + 10 \log \eta_{ea} + 170.8 \qquad \text{dB re 1 } \mu\text{Pa/A at 1 m} \qquad (25\text{-}77)$$

The term $10 \log D_i$ is referred to as the *directivity index* N_{DI}, or gain, of the transducer; $\eta_{ea} = P_a/P_e$, or electrical efficiency.

The free-field open-circuit receiving response M_0 in volts per micropascal is related to the current transmitting response for a reciprocal transducer by

$$20 \log M_0 = 20 \log S_0 - 20 \log f - 294 \qquad \text{dB re 1 V}/\mu\text{Pa} \qquad (25\text{-}78)$$

where f = frequency (Hz) and nominal conditions in the water are assumed.

In determining the reaction of the medium on the moving surface of the transducer, it is assumed that the vibrating surface of the souce has a velocity u, and that the surface exerts a force F_r on the water, and the force exerted by the water on the moving surface of the source is $-F_r$. The radiation impedance Z_r is expressed as[151]

$$Z_r = -F_r/u = R_r + jX_r \qquad (25\text{-}79)$$

where R_r = radiation resistance and X_r = radiation reactance.

In a linear system consisting of a continuous source, the value of Z_r is frequency-dependent, but is a constant at constant frequency. If the radiation impedance is known, calculating the acoustic power P_a of the source is greatly simplified, since

$$P_a = \tfrac{1}{2} u_{peak}^2 R_r = u_{rms}^2 R_r \qquad (25\text{-}80)$$

Table 25-16 lists the radiation impedances for various radiating surfaces, and Fig. 25-123 shows plots of radiation impedance for typical surfaces.

142. Nonlinear Acoustic Principles. The equations governing the propagation of sound in water are in fact nonlinear. The conventional linear relationships are really approximations, strictly valid only in the limit of infinitesimal sound amplitudes. Although the observable effects of this fundamental nonlinearity when relatively intense sound amplitudes are employed have normally been viewed as a source of performance degradation, e.g., distortion and loss, beneficial practical applications have been found.

In particular, if two relatively intense sound waves of nearly equal frequency are caused to propagate through water in the same direction, the nonlinearity causes them to interact with each other to form waves having the sum and difference frequencies. The original primary waves are eventually absorbed in propagating through the water, and the sum-frequency wave, being at a higher frequency, is absorbed even sooner. The difference-frequency wave, being at a lower frequency, persists and has an independent existence.

The far-field sound pressure (all pressures expressed in pascals) of the difference-frequency wave p_d can be described in terms of the radiated sound pressures, p_1, p_2 of the two collimated primary waves by the following basic equation,[151a] which is subject to various refinements discussed below:

$$p_d = \frac{(1 + \tfrac{1}{2}B/A) \, \omega_d^2 p_1 p_2 S}{8\pi r \rho c^4} \frac{1}{j\alpha + k_d \sin^2 (\theta/2)} \qquad (25\text{-}80a)$$

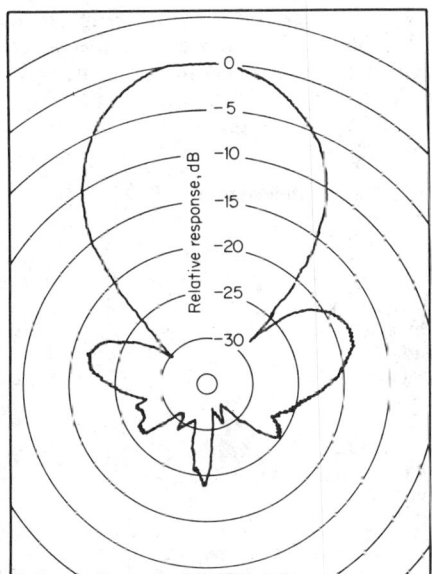

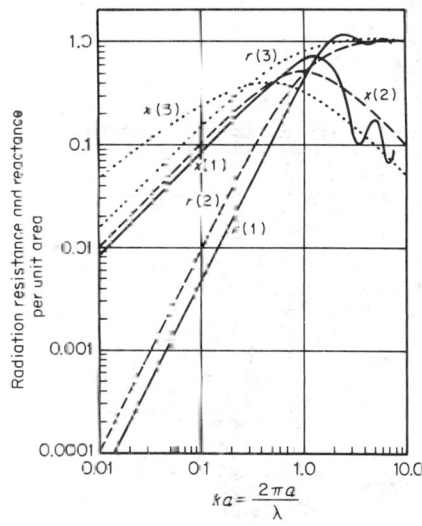

$$ka = \frac{2\pi a}{\lambda}$$

Fig. 25-122. Typical beam pattern of an underwater sound transducer.

Fig. 25-123. Radiation resistance and reactance per unit area divided by ρc as a function of ka (a = radius) for (1) a circular piston in a rigid baffle; (2) a pulsating sphere; (3) a pulsating cylinder of infinite length.

Table 25-16.   Radiation Impedance for Simple Geometries

Type of Radiator		Radiation Impedance
Rigid circular piston in infinite baffle		$Z = \pi a^2 \rho c \left[1 - \frac{J_1(2ka)}{ka} + j \frac{S_1(2ka)}{ka} \right]$ where J = Bessel function S = Struve function
Vibrating Strip of infinite length in an infinite baffle		Per unit length: $Z = 2\rho c a \left[2\Lambda(2ka) - H_1^{(2)}(2ka) + \frac{j}{\pi ka} \right]$ where $\Lambda(x) = \frac{1}{2} \int_0^x H_0^{(2)}(x)\,dx$ H = Hankel function
Sphere		
Pulsating		$Z = \frac{4}{3}\pi a^2 \rho c \frac{(ka)^2 + jka}{1 + (ka)^2}$
Oscillating		$Z = \frac{4}{3}\pi a^2 \rho c \frac{(ka)^4 + jka(1 + k^2 a^2)}{4 + (ka)^4}$
Pulsating cylinder of infinite length		Per unit length: $Z = 2\pi a \rho c j \dfrac{J_0(ka) - jN_0(ka)}{J_1(ka) - jN_1(ka)}$ where J = Bessel function N = Neumann function

where B/A = parameter of nonlinearity (Table 25-17 lists values of this parameter for various liquids), ω_d = frequency of difference-frequency wave (rad/s), S = cross-sectional area of collimated primary beams (m²), r = distance to measuring point (m), ρ = density of medium (kg/m³), c = speed of sound (m/s), α = attenuation coefficient (np/m), k_d = wave number of difference-frequency wave (m⁻¹), θ = angle to measuring point with respect to axis of beam.

TABLE 25-17   Values of B/A Except Where Indicated at Atmospheric Pressure[151b]

Substance	T,°C	B/A	Substance	T,°C	B/A
Distilled water	0	4.2	Benzyl alcohol	30	10.2
	20	5.0	Diethylamine	30	10.3
At 1 atm	30	5.2	Ethylene glycol	30	9.7
At 200 kg/cm²		6.2	Ethyl formate	30	9.8
At 4,000 kg/cm²		6.2	Heptane	30	10.0
At 8,000 kg/cm²		5.9	Hexane	30	9.9
	40	5.4	Methyl acetate	30	9.7
	60	5.7	Cyclohexane	30	10.1
	80	6.1	Nitrobenzene	30	9.9
	100	6.1	Mercury	30	7.8
Seawater (3.5%)	20	5.25	Sodium	110	2.7
Methanol	20	9.6	Potassium	100	2.9
Ethanol	0	10.4	Tin	240	4.4
	20	10.5	Indium	160	4.6
	40	10.6	Bismuth	318	7.1
n-Propanol	20	10.7	Monatomic gas	20	0.67
n-Butanol	20	10.7	Diatomic gas	20	0.40
Acetone	20	9.2	Methyl iodide	30	8.2
Benzene	20	9.0	Sulfur	121	9.5
Chlorobenzene	30	9.3	Glycerol (4% H_2O)	30	9.0
Liquid nitrogen	bp	6.6	1,2-Dichlorohexa-fluorocyclopentene (DHCP)	30	11.8

Figure 25-124 is an example of the difference-frequency wave beam pattern. It will be noticed that this beam has no side lobes and is quite narrow. θ_d, the value of θ for which the intensity is down 3 dB, is that which causes the real and imaginary parts of the denominator of Eq. (25-80a) to be equal. Because this value of θ is small, it is approximately

$$\theta_d = 2\sqrt{\alpha/k_s} \qquad (25\text{-}80b)$$

The beamwidth, which is twice this angle, is narrower than that achievable when a wave of the same (difference) frequency is generated directly using a transducer of the same size. Also, this beamwidth does not vary significantly over a wide frequency range.

Figure 25-125 is a nomogram for finding the far-field source level and beamwidth of the difference-frequency wave when the medium is seawater. The dashed lines indicate how the nomogram is to be used. As an example, assume primary waves with frequencies of 95 and 105 kHz. The mean primary frequency is 100 kHz, and the difference frequency is 10 kHz, resulting in a half beamwidth of 1.1°. A mean primary source level minus directivity index of 170 dB re 1 μPa results in a difference-frequency source level of 143 dB at 1 m re 1 μPa at 1 m.

It will be observed that there is an adverse conversion loss between primary and difference-frequency source levels, particularly if the primary directivity index is taken into account. This is the price paid for the benefits gained in terms of the narrow beamwidth.

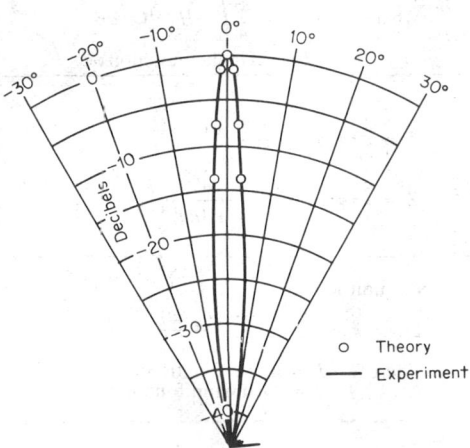

Fig. 25-124. Difference-frequency beam pattern.

Several practical factors serve to modify the above relationships.

1. Equation (25-80a) assumes collimated plane primary waves and essentially zero primary beamwidth. More realistic nonzero primary beamwidths require a modification. Reference 151e provides useful curves for taking the shape and dimensions of the transducer into account.

2. As the intensity of the primary waves increases, a point is reached where a shock wave is formed, at which point a different set of relationships govern the behavior. This phenomenon is still a subject of research; a good current reference is 151f.

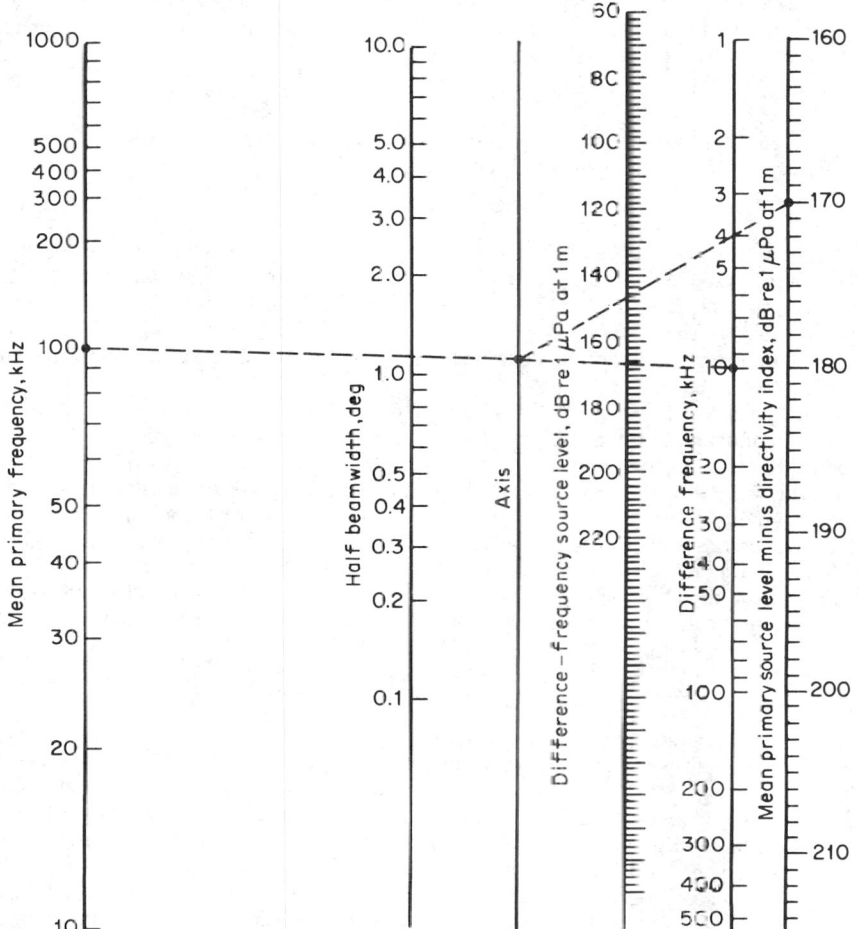

Fig. 25-125. Nomogram for determining difference-frequency source level and half beamwidth of a parametric endfire array in seawater.[151d]

In addition to the transmitting-array application described above, it is also possible to create a highly directive receiving array by placing a high-frequency high-intensity acoustic source, called a *pump*, at a relatively great distance from the receiver, in line with a relatively low-frequency plane wave arriving from a distant source. Again waves at the sum and difference frequencies are generated by the nonlinear interaction in the medium. One or both of these sidebands are separated from the pump frequency by filtering. This application and the governing relations are presented in Ref. 151g.

TRANSDUCER MATERIALS

143. Piezoelectric Materials. Since 1950, man-made piezoelectric ceramics as transducing materials have reached maturity by achieving reproducibility for a given composition and by diversification.[156]

The electromechanical nature of these materials in a polarized state is described by linear equations. With stress and electric field as the independent variables,

$$S = s^E T + dE \qquad D = dT + \epsilon^T E \tag{25-81a}$$

With stress and charge density (electric displacement) as the independent variables,

$$S = s^D T + gD \qquad E = -gT + \beta^T D \tag{25-81b}$$

where S = strain, T = stress, E = electric field, D = electric displacement, s^E = elastic compliance at constant electric field, S^D = elastic compliance at constant electric displacement, ϵ^T = dielectric constant at constant stress, and β^T = dielectric impermeability at constant stress; d and g are piezoelectric constants defined as

$$d = (\partial S/\partial E)_T = (\partial D/\partial T)_E \qquad g = (-\partial E/\partial T)_D = (\partial S/\partial D)_T \tag{25-82}$$

The electromechanical coupling coefficient is an important measure of the effectiveness of the energy conversion mechanism and is defined as

$$k = U_m/U_e U_d = d/\epsilon^T s^E \tag{25-83}$$

where U_m = mutual elastic and dielectric energy density, U_e = elastic self-energy density, and U_d = dielectric self-energy density. This is a quasi-static parameter that can be related to the fundamental material constants for one-dimensional transducers. The definition is not necessarily applicable to all geometries.

A more complete treatment of the piezoelectric equations and the measurement of the various constants can be found in Refs. 154 to 160.

Although piezoelectric ceramics can be operated in various modes, two types are of major importance to underwater sound transducers. In the parallel, or *33-mode* type, the stress, strain, and electric field are in the same direction. In the transverse, or *31-mode* type, the stress and strain are the same direction but orthogonal to the electric field. Each of these mode types is characterized by its associated constants, for example d_{31} or d_{33}, g_{31} or g_{33}, k_{31} or k_{33} respectively. The dielectric constant of interest is in the direction of the electric field for both cases and is denoted ϵ_{33}. For the above constants, the direction of the electric vectors are denoted by the 3 direction, and the direction of the mechanical variable, if different, is the other subscript. An additional electromechanical coupling coefficient k_p is useful for some applications. It is related to k by

$$k_p = \sqrt{2/(1 - \sigma^E)}\, k_{31} \tag{25-84}$$

where σ^E = Poisson's ratio at constant electric field.

The most important piezoelectric ceramic materials for underwater sound transducers are the modified lead zirconate titanate (PZT) compositions and to lesser extent modified barium titanate compositions. Table 25-18 lists some more important properties of these materials. A more comprehensive table of properties is available in Ref. 156, Chap. 3. These materials are characterized as being "very hard" lead zirconate titanate, "hard" lead zirconate titanate, and "soft" lead zirconate titanate. Progress has been made in specifying and classifying the various ceramic compositions.[161] The properties of the piezoelectric ceramics vary as functions of time, static stress, stress cycling, and electric field strength.[156]

144. Magnetostrictive Materials. Magnetostrictive materials offer certain advantages for some underwater sound transducer applications. One example is that of a large low-frequency source that is submerged to a great depth and must operate unattended for long periods of time. Two forms of magnetostrictive materials are used, metal alloys and ceramic compositions.

The physical quantities regarding the magnetic and mechanical state of a polarized material are the magnetic field strength H, the magnetic flux density B, the mechanical stress T, and the mechanical strain S. For a sinusoidal variation, they are related by

$$S = s^H T + dH \qquad B = dT + \mu^T H \tag{25-85a}$$

or

$$T = c^B S - hB \qquad H = -hS + v^S B \tag{25-85b}$$

where s^H = elastic compliance at constant magnetic field strength, c^B = elastic stiffness at constant magnetic flux density, μ^T = permeability at constant stress, and v^S = reluctivity at constant strain; d and h are the piezomagnetic constants, which are defined as

$$d = (\partial S/\partial H)_T = (\partial B/\partial T)_H \qquad h = -(\partial T/\partial B)_S = -(\partial H/\partial S)_B \tag{25-86}$$

More detailed information regarding the magnetostrictive equations and constants can be found in Refs. 153, 156, 162, and 163.

The electromechanical coupling factor k of magnetostrictive materials has the same physical meaning as for piezoelectric materials, with the same limitations.

Eddy currents can play an important role in the efficiency of magnetostrictive materials. For this reason, magnetostrictive assemblies are often constructed from thin laminations cemented together, usually in an annealed state. Eddy current losses can be taken into account by multiplying the permeability by a complex eddy current factor χ. This factor depends on the geometry. With a modification of the analysis in Ref. 163, a skin-effect parameter m^2 can be defined as

$$m^2 = j\omega\mu(0)/\rho_e \qquad (25\text{-}87)$$

where $\mu(0)$ = permeability with no eddy current effect, ω = angular frequency, and ρ_e = resistivity.

The apparent permeability is given by

$$\mu = \mu(0) \frac{\tanh (mt/2)}{mt/2} \qquad (25\text{-}88)$$

where t = thickness of sheet of material.

Figure 25-126 shows a plot of the magnitude and phase angle of the complex correction factor χ vs. $mt/2$.

Table 25-19 contains values of important properties of a number of magnetostrictive materials.[156] Reference 156 contains a table that includes more materials but is limited in the number of characteristics shown.

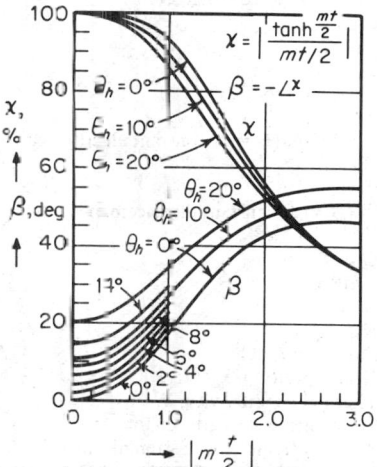

Fig. 25-126. Eddy-current loss factor, magnitude, and phase angle of magnetic materials; $\vartheta_e < \mu(0)$.[163]

TRANSDUCER ARRAYS

145. Beam Formation. Transducers and hydrophones can be arranged individually or in arrays to possess omnidirectional or directional characteristics, depending upon effective aperture dimensions, geometrical shape, and vibrational modes used. At high frequencies, since the wavelengths are short, highly directional individual units can be designed. At lower frequencies, multiple transducers or hydrophones are used in arrays of planar, cylindrical, spherical, or volumetric configuration.

Directionality is highly desirable in underwater sound detection systems because it makes both directional transmission and the determination of the direction of arrival of a signal possible. As in directional radar or communications antennas, this reduces the noise relative to the signal from other directions. Arrays can be steered mechanically by physical rotation or electrically by phasing or time delay networks. The direction of maximum sensitivity of a plane array of elements can be rotated into a direction lying at angle θ_0 to a reference direction by delaying differentially the output of each element. In this way, an irregular array can be effectively converted into a line array.

In their simplest form, arrays are arranged with elements along a line or distributed along a plane. The acoustic axis of such line or plane arrays, when unsteered, lies at right angles to the line or plane. The beam pattern of a line array may be visualized as a doughnut-shaped figure having supernumerary attached doughnuts formed by the side lobes of the pattern. The three-dimensional pattern of a plane array is a searchlight type of figure with rotational symmetry about the perpendicular to the plane plus side lobes.

146. Lines of Equally Spaced Elements.[150] The beam pattern of a line of equally spaced, equally phased (i.e., unsteered) elements is derived as follows. Let a plane sinusoidal sound wave of unit pressure be incident at an angle θ to a line of n such elements, each spaced from the next a distance d. The output of the mth element relative to that of the zeroth element is delayed by the time necessary for sound to travel the distance $l_m = md \sin\theta$. The corresponding phase delay for sound of wavelength λ, at frequency $\omega = 2\pi f$, is

$$u_m = mu$$

where the phase delay between adjacent elements, in radians, is

$$u = (2\pi d/\lambda) \sin\theta$$

TABLE 25-18 Characteristics of Commonly Used Piezoelectric Ceramics, Low-Signal Properties at 25°C

Quantity	PZT-4*	PZT-5*	PZT-8*	95% wt $BaTiO_3$, 5% wt $CaTiO_3$
k_p	0.58	0.60	0.51	0.36
k_{31}	0.334	0.344	0.30	0.212
k_{33}	0.70	0.705	0.64	0.50
$\epsilon_{33}^T/\epsilon_0$ (T = at constant stress)	1,300	1,700	1,290	1,700
$\epsilon_{33}^S/\epsilon_0$ (S = at constant strain)	635	830	580	1,260
tan δ	0.004	0.02	0.004	0.006
d_{33}, pC/N	289	374	225	149
d_{31}	-123	-171	-97	-58
g_{33}, mV·m/N	26.1	24.8	25.4	14.1
g_{31}	-11.1	-11.4	-10.9	-5.5
s_{11}^E, pm²/N (E = at constant electric field)	12.3	16.4	13.5	8.6
s_{33}^E	15.5	18.8	11.5	9.1
s_{11}^D (D = at constant displacement)	10.9	14.4	10.4	8.3
s_{33}^D	7.90	9.46	8.0	7.0
Q_M	500	75	1,000	400
ρ, 10^3 kg/m³	7.5	7.75	7.6	5.55
N_1, Hz·m†	1,650	1,400	1,700	2,290
N_3‡	2,000	1,770	2,070	2,740
Curie point, °C	328	365	300	115
Heat capacity, J/kg·°C	420	420	420	500
Thermal conductivity, W/m·°C	2.1	2.1	2.1	3.5
Static tensile strength, lb/in²	13,000	13,000	12,000	12,000
Rated dynamic tensile strength, lb/in²	6,000	4,000	7,000	7,500

*Trademark, Vernitron Piezoelectric Division.
†N_1 = frequency constant of a thin bar with electric field perpendicular to length, $f_r l$.
‡N_3 = frequency constant of a thin plate with electric field parallel to thickness, $f_r t$.

TABLE 25-19 Properties of Magnetostrictive Materials[156]

	Nickel	Alfenol	Ferroxcube 7A1	Ferroxcube 7A2
k_{33} (opt)	0.15 to 0.31	0.25 to 0.31	0.25 to 0.30	0.21 to 0.25
d_{33} (opt), nWb/N	~-3.1	~7.1	-2.8 to -4.4	-1.6 to -2.9
μ_{33}^S/μ_0 (opt)	22	58	15 to 25	8 to 15
$1/S_{33}^H$, GPa	~200	~140	151	161
$Q_M H$	50 to 250	. . .	2,500 to 5,000	2,500 to 5,000
tan δ	. . .	. . .	0.001 to 0.002	0.001 to 0.002
H_{opt}, A/m*	700 to 1000	700 to 1000	1500 to 2400	1100 to 1900
B_{bias}, Ta†	. . .	. . .	0.22 to 0.24	0.22 to 0.24
B_R, Ta	~0.4	~0.6	0.11 to 0.16	0.15 to 0.17
μ_{33}^S/μ_0, remanence	~20	. . .	30 to 45	30 to 50
k_{33}, remanence	~0.14	. . .	0.15 to 0.20	0.15 to 0.19
d_{33}, remanence, nWb/N	~-1.5	. . .	-2.3 to -3.8	-2.2 to -3.7
Resistivity, Ω·m	7×10^{-8}	9×10^{-7}	>10	>10
Curie point, °C	358	~500	530	530
H_c, kA/m	0.03	~0.01	0.25 to 0.5	0.2 to 0.4
ρ, 10^3 kg/m³	8.8	6.5	5.35	5.35
ν^B, km/s	~5.0	~4.8	~5.65	~5.75
ν^H, km/s	~4.85	~4.55	~5.45	~5.6

*Field required for highest k_{33}.
†B_{opt} = 0.7 B_{sat}.

The output voltage of the mth element of voltage response R_m is

$$V_m = R_m \cos(\omega t + mu)$$

and the array voltage is the sum of such terms:

$$V = R_0 \cos \omega t + R_1 \cos(\omega t + u) + \cdots + R_n \cos(\omega t + mu) - \cdots + R \cos(\omega t + nu)$$

In the complex notation, the array voltage will be

$$V = (R_0 + R_1 e^{iu} + R_2 e^{2iu} + \cdots + R_n e^{(n-1)u})e^{i\omega t}$$

If the array elements all have unit response ($R = 1$),

$$V = (1 + e^{iu} + e^{2iu} + \cdots + e^{(n-1)iu})e^{i\omega t}$$

Multiplying by e^{iu} and subtracting gives

$$V = (e^{inu} - 1)/(e^{iu} - 1)e^{i\omega t}$$

Neglecting the time dependence, this becomes

$$V = [\sin(nu/2)]/[\sin(u/2)]$$

Finally, expressing u in terms of θ, the beam pattern, the square of the function-normalized to unity at $\theta = 0$ is

$$b(\theta) = \left(\frac{V}{n}\right)^2 = \left[\frac{\sin(n\pi d \sin\theta/\lambda)}{n \sin(\pi d \sin\theta/\lambda)}\right]^2 \qquad (25\text{-}89)$$

147. Continuous-Line and Plane Circular Arrays.[150] When the array elements are so close together that they may be regarded as adjacent, the array becomes a continuous-line transducer and the beam pattern is found by integration rather than by summation. For this case, let the line transducer be of length L and have a response per unit length of R/L. The contribution to the total voltage output produced by a small element of line length dx located a distance x from the center is (neglecting the time dependence)

$$dv = (R/L)e^{i2\pi x/\lambda \sin\theta}dx$$

The beam pattern, the square of V normalized so that $b(\theta) = 1$, is

$$b(\theta) = \left(\frac{V}{R}\right)^2 = \left[\frac{\sin[(\pi L/\lambda)\sin\theta]}{(\pi L/\lambda)\sin\theta}\right]^2 \qquad (25\text{-}90)$$

In a similar manner, the beam pattern of a circular-plane array of diameter D of closely spaced elements can be shown to be

$$b(\theta) = \left[\frac{2J_1[(\pi D/\lambda)\sin\theta]}{(\pi D/\lambda)\sin\theta}\right]^2 \qquad (25\text{-}91)$$

where $J_1[\pi D/\lambda) \sin\theta]$ is the first-order Bessel function of argument $(\pi D/\lambda)\sin\theta$.

Generalized beam patterns for continuous-line and circular-plane arrays are drawn in Fig. 25-127 in terms of the quantities $(L/\lambda)\sin\theta$ and $(D/\lambda)\sin\theta$.

Figure 25-128 is a nomogram for finding the angular width between the axis and the -3-dB and -10-dB points of the beam pattern of continuous-line and circular-plane arrays. The dashed lines indicate how the nomogram is to be used. Thus a circular-plane array of 500 mm diameter at a wavelength of 100 mm (corresponding to a frequency of 15 kHz at a sound speed of 1,500 m/s) has a beam pattern 6° wide between the axis of the pattern and the -3-dB points.

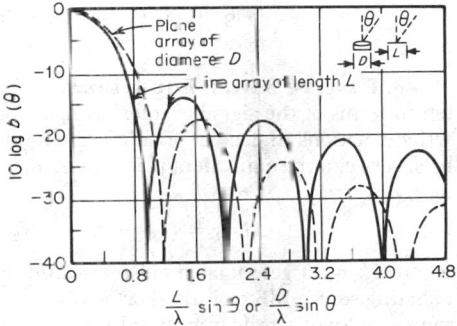

Fig. 25-127. Beam patterns of a line array of length L and a circular plane of diameter D.

PASSIVE SONAR SYSTEMS

Passive sonar systems, also referred to as listening sonar systems, are designed to respond to acoustic energy radiated by sources in the band of the sonar receiver. These systems are designed to accentuate the response to wanted signals while suppressing unwanted background noise. Passive systems are designed to maximize the signal-to-noise ratio to the degree that the characteristics of the signal and noise are known or can be predicted.

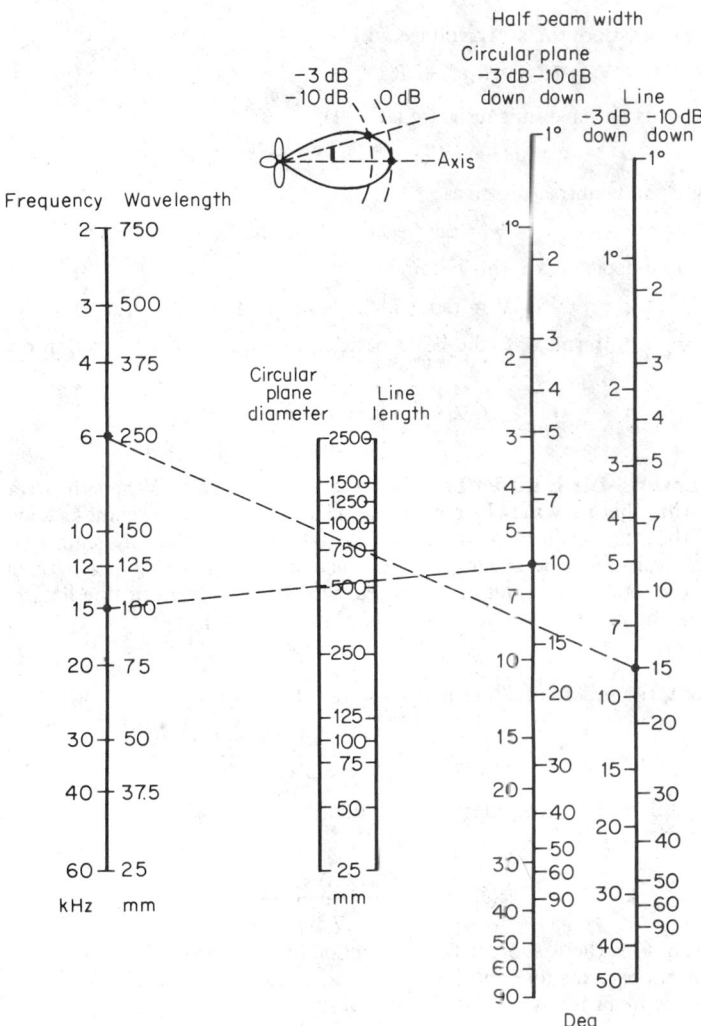

Fig. 25-128. Nomogram for finding width of beam pattern.

148. Passive Sonar Equations. The fundamental relation of passive sonar can be written in terms of the signal-to-interfering-background ratio. Since in sonar the usual practice is to write equations in decibel notation, the ratio is defined as the *signal differential* $\Delta L_{S/N}$, in decibels, between the equivalent plane-wave levels of signal and interfering noise at the passive sonar receiving array:

$$\Delta L_{S/N} = (L_{Sf} - N_w + N'_{BW}) - [(L_{Nf} - N_{DI} + N_{BW}) + N_{\delta}] \qquad (25\text{-}92)$$

where L_{Sf} = target-radiated-noise spectrum level vs. frequency at 1 m from effective center of radiating source (dB re 1 μPa/Hz$^{1/2}$), N_w = one-way acoustic propagation loss between radiating source and passive sonar array (dB), N'_{BW} = 10 log (signal bandwidth) = signal-bandwidth-level correction, L_{Nf} = equivalent-plane-wave interfering-noise spectrum level vs. frequency at passive sonar array resulting from summation of noise from all sources (d3 re 1 uPa/Hz$^{1/2}$), N_{DI} = 10 log D, where D = effective directivity of the passive sonar array and beam former against isotropic noise (dB) (if the noise background has directional components, the effectiveness of the array in decreasing background noise is modified), N_{BW} = 10 log (noise bandwidth) = (noise-bandwidth-level correction, and N_δ = receiving deviation loss (dB).

The value of the signal differential required at the array output varies with the application, e.g., detection, classification, or localization.

The signal differential can be considered to be the sum of two terms: the detection threshold N_{DT}, defined as that value of the signal differential required to just detect the signal, and the signal excess N_{SE}, which is the amount in decibels by which $\Delta L_{s/n}$ exceeds N_{DT}. A detection threshold N_{DT} adjustment in the system is usually set to a value consistent with a sufficiently low false-alarm rate on the display in use. Substituting this sum into Eq. (25-92) gives

$$N'_{SE} = (L_{Sf} - N_W + N'_{BW}) - [(L_{Nf} - N_{Di} + N_{BW}) - N_\delta + N_{DT}] \qquad (25\text{-}93)$$

Equation (25-93) is arranged to show that signal excess in decibels is the differential between a set of signal terms and a set of effective-noise terms, i.e.,

$$N_{SE} = L_i - L_{MD} \qquad (25\text{-}94)$$

where L_i = incident signal level and L_{MD} = minimum detectable signal level, given by

$$L_{MD} = L_N - N_{DI} + N_{EW} + N_\delta + N_{DT} \qquad (25\text{-}95)$$

Another useful measure is the figure of merit N_{FM}, defined as the maximum allowable one-way propagation loss under the condition of zero signal excess. From Eq. (25-93),

$$N_{FM} = (L_{Sf} + N'_{BW}) - [(L_{Nf} - N_{Di} + N_{BW}) + N_\delta - N_{DT}] \qquad (25\text{-}96)$$

and from Eq. (25-95), the figure of merit can also be written

$$N_{FM} = L_{Sf} + N'_{EW} - L_{MD} \qquad (25\text{-}97)$$

An example of the use of the passive sonar detection equation is detailed in Fig. 25-129.

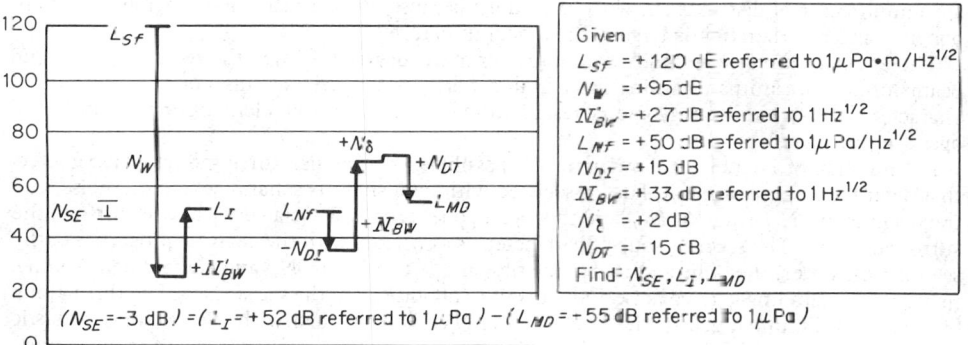

Given

L_{Sf} = +120 dB referred to 1 μPa•m/Hz$^{1/2}$
N_W = +95 dB
N'_{BW} = +27 dB referred to 1 Hz$^{1/2}$
L_{Nf} = +50 dB referred to 1 μPa/Hz$^{1/2}$
N_{DI} = +15 dB
N_{BW} = +33 dB referred to 1 Hz$^{1/2}$
N_δ = +2 dB
N_{DT} = −15 dB

Find: N_{SE}, L_I, L_{MD}

$(N_{SE} = -3$ dB$) = (L_I = +52$ dB referred to 1 μPa$) - (L_{MD} = -55$ dB referred to 1 μPa$)$

Fig. 25-129. Passive sonar system analysis.

149. Sonar Parameters. The radiated-noise level of a target is usually composed of broadband and narrow-band noise from the propeller, machinery, and possibly echo-ranging pings from the target, as well as from other ships. Since this radiation can vary considerably in frequency and transmitted intensity as a function of time, the signal excess, and consequently the maximum detection range, can also vary over a wide range.

Acoustic propagation loss, also called *transmission loss,* varies according to an applicable spreading law modified by absorption, refraction, and reflection. In deep water, spherical spreading applies, and the propagation loss N_W in decibels is given by

$$N_W = 20 \log R + aR + 60 \qquad (25\text{-}98)$$

where R = range (km); $\alpha(R)$ = absorption (dB/km), which varies with frequency f; and 60 represents the conversion from meters to kilometers. For shallow sources in deep oceans when a surface layer is present, spherical spreading applies for the first kilometer and cylindrical spreading thereafter, and the propagation loss then becomes

$$N_W = 10 \log R + aR + 60 \tag{25-99}$$

The interfering background noise places a limit on detectability, but depending on the character of the noise, on the array interelement correlation, and on the signal processing used, the threshold of detection can be lowered considerably below the rms noise band level. Background noise is composed of self-noise, ambient noise, and, when present, reverberation. The noise components in these categories are listed in Table 25-20.

TABLE 25-20   Sources of Background Noise

A. Self-noise	B. Ambient noise	3. Man-made
1. Hydrodynamic	1. Sea	a. Shipping
a. Flow-excited	a. Bubbles	b. Shore-radiated
b. Turbulence	b. Seismic	4. Precipitation
c. Cavitation	c. Waves	C. Reverberation
d. Bubbles	d. Boundary turbulence	1. Bottom
e. Splash	e. Ice motion	2. Surface
2. Machinery	2. Biological	3. Volume
a. Turbines	a. Shrimp	4. Scattering layers
b. Pumps	b. Whales	
c. Blowers etc.	c. Carp etc.	
3. Propeller		

Self-noise is generated by all those sources associated with the listening vessel and its interaction with the surrounding water. *Ambient noise* is caused by both natural and man-made sources, while *reverberation* results from reflections of sonar pings by the ocean surface, bottom, volume, and scattering layers.

Self-noise has many directional characteristics (near-field effects); ambient noise generally has an omnidirectional distribution, with exceptions because of man-made and precipitation components; and reverberation is largely directional in nature.

The *effective directivity index* is an indicator of the degree to which the receiving array and beam former discriminates against background noise, assuming that this noise is isotropic in character and that the array geometry results in low correlation from element to element in the operating frequency band.

Examination of Eq. (25-96) shows that it is possible to distinguish three groups of parameters that determine figure of merit. First, associated with own ship, its sonar, background noise, and the operator are N_{DI}, N_{BW}, N_δ, N_{DT}, and those components of L_N associated with own ship and the surrounding sea. The second group of parameters is associated with the acoustic properties of the sea (or freshwater) medium and its boundaries, and it consists of N_W and the ambient components of L_N, including surface noise and reflections of noise from the sea surface. The third group of parameters, summarized in the terms L_S and N'_{BW}, is concerned with the radiating acoustic source or target, i.e., the description of the characteristics of the target in terms of radiated-noise spectrum level as a function of frequency, including both broadband and narrow-band spectra, and the bandwidth of the radiated noise.

150. System Configuration and Parameters. A generalized passive sonar detection model, shown in Fig. 25-130, depicts all the parameters associated with a passive sonar system and its operator. To account for system hardware and operator gains and losses, terms must be added to Eqs. (25-92), (25-93), (25-95), and (25-96). While these equations are applicable at the output of the idealized beam former, the modified equations apply at the operator decision point indicated in Fig. 25-130. Specifically, at the decision point the signal excess is

$$N'_{SE} = (L_{Sf} + N_S - N_W + N'_{BW} + N_P)$$
$$- [(L_{Nf} - N_{DI} + N_{BW}) + N_M + N_\delta + N_T + N_{SD}] \tag{25-100}$$

The added terms are N_S, to account for more than one ray path from the acoustic source to the sonar N_P, for signal processing gain; N_M, for hardware design margin (loss); N_p, for signal processing loss, as in a clipping processor; N_T for the effect of the threshold level, and N_{SD}, to allow for the signal differential required for signal recognition or the particular display used.

The sonar detection system shown in Fig. 25-131 is placed between the hydrophone array and the decision point. It consists of a receiver, a visual and/or aural display, and an operator. The detection threshold N_{DT} is defined[164] as the ratio in decibels of the signal power S in the receiver band to the noise power N_0 in a 1-Hz band, measured at the receiver terminals; i.e.

$$N_{DT} \equiv 10 \log (S/N_0) \tag{25-101}$$

The signal and signal-plus-noise values are taken in the receiver band, and N_{DT} is computed for use in the sonar equation. Noise backgrounds are expressed as power spectrum levels, i.e., as powers in 1-Hz bands.

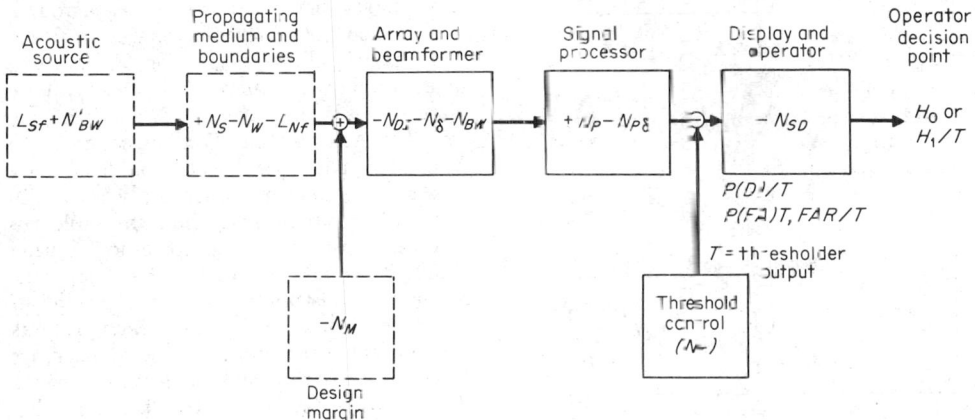

At operator decision point, N_{SE} is given in Eq. (25-100)

Fig. 25-130. Passive sonar system detection model.

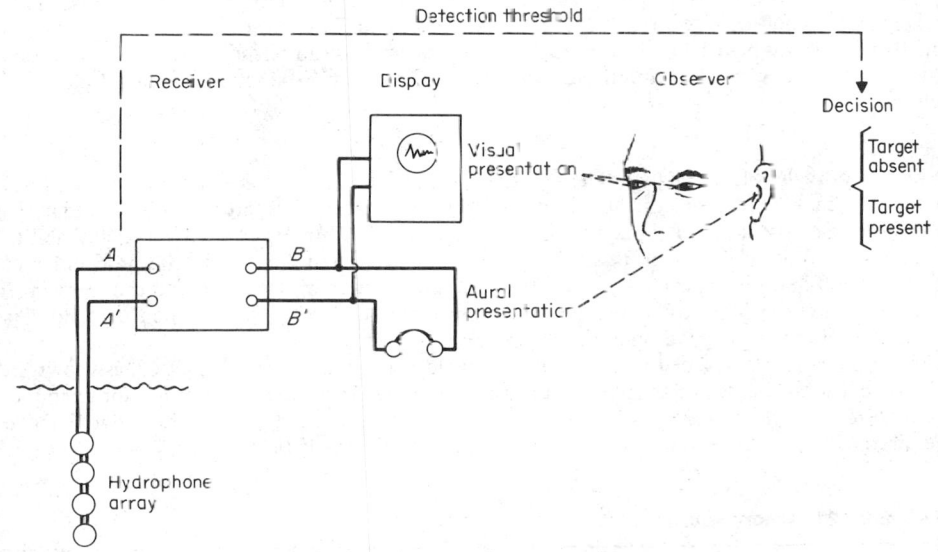

Fig. 25-131. Elements of a sonar receiving system.

Detection decisions are binary in nature; i.e., signal is present or signal is absent. But since a signal can actually be present or actually absent at the receiver input, there are four possible situations, summarized in Table 25-21. The correct decisions are shown on the diagonal of the matrix, namely, the detection and null decisions, with probabilities $P(D)$ and $1 - P(FA)$, respectively, where $P(D)$ is the probability of detection and $P(FA)$ is the probability of false alarm. False-alarm and miss decisions can occur with respective probabilities $P(FA)$ and $1 - P(D)$.

To implement the detection threshold criterion, it is necessary to apply a threshold voltage to a comparison circuit, at the receiver output. The threshold voltage is either fixed at a level corresponding to the desired N_{DT} or it can be controlled and calibrated in terms of N_{DT}. Whenever the level of the waveform at the comparison point exceeds the threshold voltage, the decision of signal present is made, unless a rule is adopted or circuit implemented, which, for example, counts the number of times the threshold is exceeded during a fixed time interval and bases the detection decision on this count.

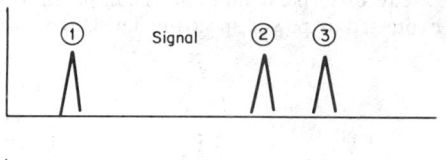

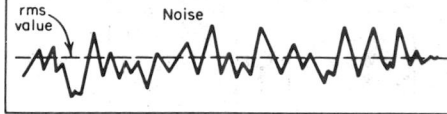

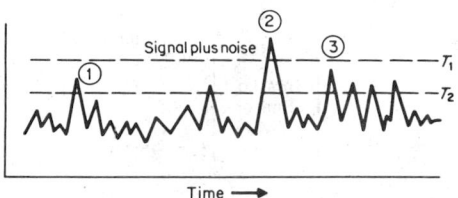

Fig. 25-132. Signal and noise at two threshold settings.

The effects of setting the threshold voltage at various levels are shown in Fig. 25-132, at three target signals in a noise background, with two possible threshold voltages. High settings like T_1 allow only strong targets to be detected and extremely few, if any, false alarms; consequently, both $P(D)$ and $P(FA)$ are low. Low settings like T_2 allow many more possible signals to be detected and also many false alarms to occur; consequently both $P(D)$ and $P(FA)$ are high. Both these threshold settings are unsatisfactory. A threshold setting should be found that produces a display with a uniformly distributed (gray) background in the presence of noise only and which prevents the buildup of the number of noise markings as a function of time and the decay rate of the display storage system in use; i.e., the false-alarm rate must be bounded.

Detection decisions depend upon the independent probabilities $P(D)$ and $P(FA)$, and consequently on the distributions of noise and signal-plus-noise at the receiver output. Figure 25-133 shows the probability density of noise alone and signal-plus-noise plotted as a function of amplitude[164] a. The curves are assumed to have a gaussian distribution with variance σ^2, the first with mean noise amplitude $M(N)$ and the second with mean signal-plus-noise amplitude $M(S + N)$. Then the parameter detection index d is defined as

$$d \equiv (M_{S+N} - M_N)^2/\sigma^2 \tag{25-102}$$

which is equivalent to the signal-to-noise ratio of the envelope of the receiver output effectively at the terminals where the threshold voltage T is established. The area (integral) under the probability density curve of signal-plus-noise to the right of T in Fig. 25-133 is the probability that an amplitude in excess of T is due to signal-plus-noise and is equal to the probability of detection $P(D)$. Similarly, the area under the probability curve of noise alone to the right of T is equal to the probability of false alarm $P(FA)$. Since $P(D)$ and $P(FA)$ vary as the threshold T is changed, their values depend upon the parameter d.

The probabilities associated with various threshold voltage levels can be plotted, as shown in Fig. 25-134, with the detection index d as a parameter. Thus, for high-T settings, for example, the corresponding low $P(D)$, $P(FA)$ points are plotted in the lower left portion of the figure above the diagonal. These are known as receiver operating characteristic (ROC) curves.

TABLE 25-21   Binary Decision Matrix

	And the decision is:	
When at the input:	Signal present	Signal absent
Signal present	Correct decision: detection with probability $P(D)$	Incorrect decision: miss with probability $1 - P(D)$
Signal absent	Incorrect decision: false alarm with probability $P(FA)$	Correct decision: null with probability $1 - P(FA)$

If the curves of Fig. 25-133 are imagined to refer to the receiver input, the likelihood ratio for an input sample of amplitude a is

$$L = B/A \tag{25-103}$$

Thus, the ROC curves are related to the signal-to-noise ratio at the receiver input required for detection.

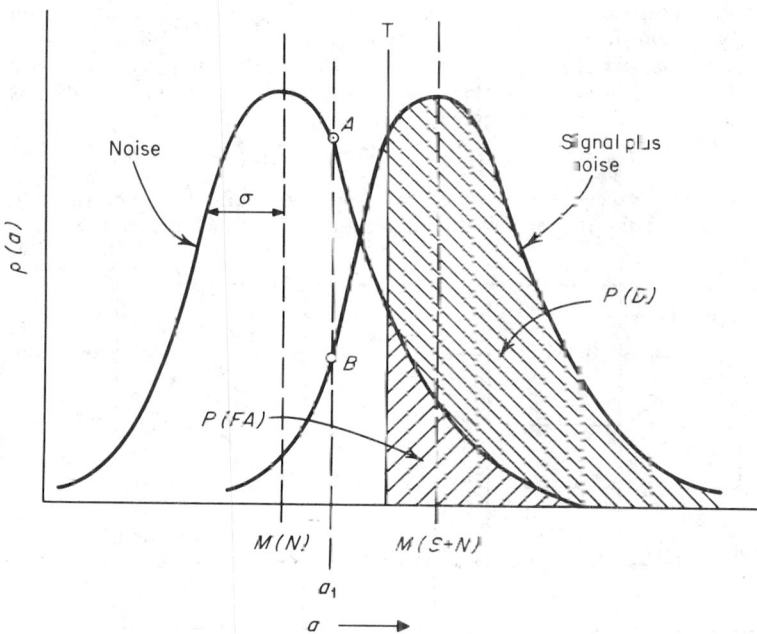

Fig. 25-133. Probability density distributions of noise.

Following the detector, a postdetection averaging filter is used to smooth out fluctuations due to detector output noise. The optimum integration time T of this filter is the time equal to the signal duration t. In passive sonar systems t can be as long as it is operationally useful to integrate the incoming signal in the noise and/or reverberation background. Here, factors such as stabilization of own ship's receiving beams (for rotating or preformed beams), rotating receiving beam RPM (which determines the time on target during each rotation), receiving beamwidth, possible range of the target, and possible speed of the target must be considered.

ACTIVE SONAR SYSTEMS

151. Introduction. Active sonar systems, like radar, make use of *reflected energy reception*, referred to as *echo ranging*. Active sonar systems are used primarily to determine target range and bearing. In addition, active systems may be used for determining target depth, aspect angle, course, and speed. Target motion may be computed from successive echo returns or measurement of target-generated doppler on each ping.

A wide variety of signal waveforms, pulsed or continuous, may be used in echo-ranging operations. Pulsed continuous waves of 1 ms

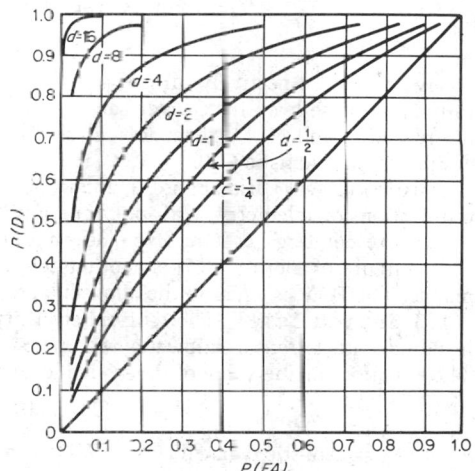

Fig. 25-134. Receiver operating characteristic (ROC) curves.

to several seconds duration are commonly used, short pulses providing maximum range resolution and long pulses providing maximum energy return from the target, hence greater detection range. Pulsed FM waves with up or down sweeps and pseudo-random noise waves are used to obtain better target discrimination in a reverberant background. Combinations of the foregoing waveforms may be used to reduce mutual interference.

Continuous transmission systems (cw or FM) are employed in short-range work, such as navigation sonars. Continuous transmission is not used in long-range detection systems because of the difficulty of acoustic isolation between the receiving and transmitting arrays.

152. Active Sonar Equation. The active sonar equation, corresponding to the model shown in Fig. 25-135, is used for performance prediction. The following equation is commonly used for applications with two-way propagation loss:

$$N_{SE} = L_S - (N_{\delta T} + N_{\delta R}) + N_{TS} - 2N_W - L_B - N_{DT} \qquad (25\text{-}104)$$

where N_{SE} = signal excess (dB), L_S = source level (dB re 1 μPa at 1 m) on transmit beam axis, $N_{\delta T}$, $N_{\delta R}$ = transmit and receive deviation losses due to target ray being off axis of transmit and

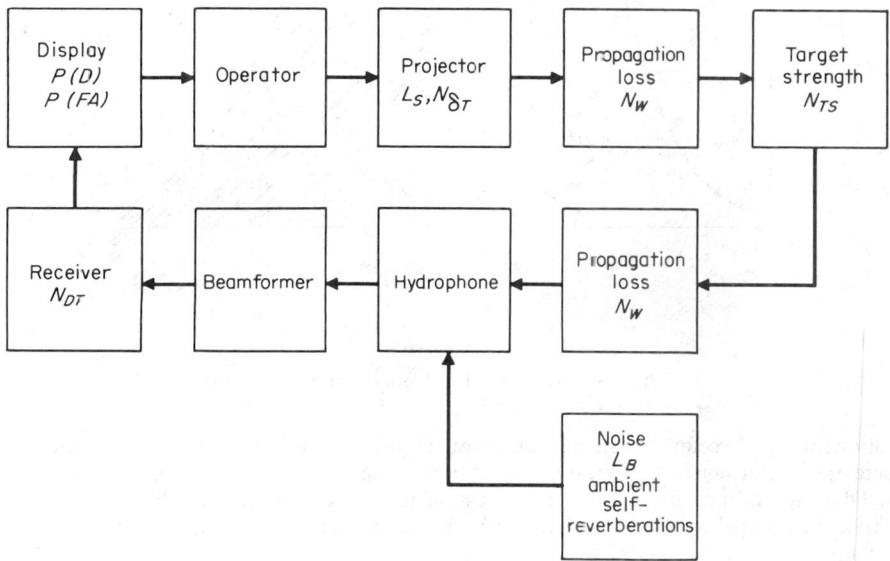

Fig. 25-135. Echo-ranging block diagram.

receive beams, respectively (dB), N_{TS} = target strength (dB), N_W = one-way propagation loss (dB), L_B = noise level in receiver band, including ambient noise, self-noise, and reverberations (dB re 1 μPa), and N_{DT} = detection threshold in receiver band and at receiver input to produce desired display statistic (dB).

Equation (25-104) also applies to applications with one-way propagation losses, such as communications or telemetry, with only one propagation loss taken into account and background noise corresponding to the receive platform only.

The figure of merit (FOM) is another commonly used criterion for evaluating sonar performance. The FOM is equal to the allowable propagation loss when the signal excess is zero.

153. Source Level. The radiated acoustic powers of shipboard sonars[164] range from a few hundred watts to tens of kilowatts, with transmitting directivity indexes between 10 and 30 dB. The equation for the value of the source level L_S is

$$L_S = 170.8 + 10 \log P_A + N_{DI} - N_D \qquad (25\text{-}105)$$

where P_A = acoustic radiated power (W), N_{DI} = directivity index (dB), and N_D = dome loss (dB) where applicable.

A plot of L_S vs. P_A is shown in Fig. 25-136 with N_{DI} as a parameter. The efficiency of converting electric to acoustic power varies with the transducer design, the match between transmitter source and transducer load impedance, and array-mutual-interaction effects.

There are two major acoustic limitations that limit achievable source level; cavitation and array mutual interaction. Cavitation limits occur[16] when power applied to a transducer is such that bubbles begin to form on the face and just in front of the transducer. These bubbles are a manifestation of cavitation and are caused by the rupture of water through negative pressure of the generated sound field. Cavitation causes a deterioration of transducer performance and life by erosion of the transducer surface; loss of radiated acoustic power in absorption and scattering by cavitation bubbles; deterioration in beam pattern; and reduction of the acoustic impedance into which the transducer must radiate. Cavitation threshold occurs at the point of departure from linearity of the input-output power curve and is a function of frequency, pulse length, and operating depth.

The effect of increased depth of operation is to increase the cavitation threshold

$$P_c(h) = P_c(0)(1 + 0.1h)^2 \tag{25-106}$$

where $P_c(h)$ = cavitation threshold (W/m²) at operating depth h m. The cavitation threshold increases with decreasing pulse lengths below 5 ms, as shown in Fig. 25-137.

Unequal loading effects occur in arrays of transducer elements because of mutual radiation impedances[173,174] between independent sound sources. These effects cause hot spots in the array

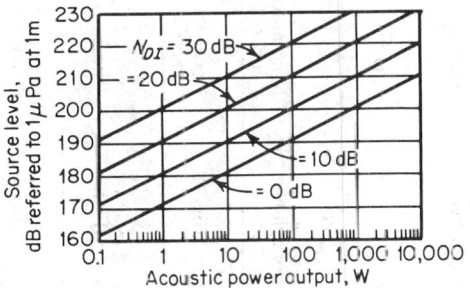

Fig. 25-136. Plot of Eq. (25-105).

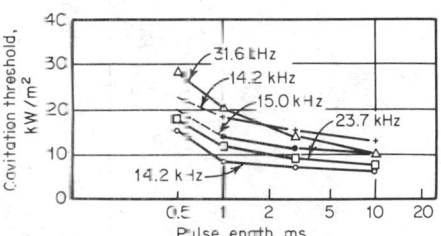

Fig. 25-137. Cavitation threshold vs. pulse duration.

(i.e., one element receives many times the average power per element), so that it may be driven to destruction. Some elements can also absorb the acoustic output of others even when all elements are driven in like manner. Other effects are a gradual deterioration of beam pattern (or directivity index) and reduction of electrical power into the array, with corresponding loss of L_s. Mutual-interaction effects can be controlled by proper spacing between array elements, tuning the transducer element to reduce mutual-radiation impedance, increasing the self-radiation impedance and controlling individual transducer velocity and phase.

Sonar domes may have a significant transmission loss. Expressions for the transmission loss and specular reflections have been obtained theoretically and verified experimentally. Both increase with frequency and thickness, as well as density of the dome wall.

154. Target Strength. Target strength relates the echo intensity returning from a target to the incident intensity. The target strength, a concept equally useful for submarine, mine detection, and fish location, is defined as 10 times the logarithm to the base 10 of the ratio of the intensity of the sound returned by the target, at a distance of 1 m from its acoustic center to the incident intensity from a distant source. The theoretical target strength of a number of geometric shapes and forms is shown in Table 25-22. Nominal values of measured target strength for typical underwater targets are tabulated in Table 25-23. Target strength is a function of aspect angle, frequency, and pulse length.

155. Deviation and Propagation Losses. *Deviation loss* results when the target is not on the maximum response axis of either the transmitting or receiving beam or both. Deviation loss varies from 0 dB on the maximum response axes to several decibels off the axis, depending upon the beam pattern responses at the specific angles involved.

The propagation loss is a highly variable function, and only average values can be predicted. Average propagation-loss data have been published as a function of ocean model, propagation path, wind state, and season.[164,168] In the absence of empirical data, several theoretical models may be used: spherical or cylindrical spreading loss as a function of range, sound absorption loss as a function of range and frequency, and sound wave or ray propagation.

TABLE 25-22 Target Strength of Simple Forms

Form	Target Strength $t = 10 \log t$	Symbols	Direction of incidence	Conditions
Any convex surface	$\dfrac{a_1 a_2}{4}$	$a_1 a_2$ = principal radii of curvature r = range $k = 2\pi/$wavelength	Normal to surface	$ka_1, ka_2 \gg 1$ $r > a$
Sphere Large	$\dfrac{a^2}{4}$	a = radius of sphere	Any	$ka \gg 1$ $r > a$
Small	$61.7\dfrac{V^2}{\lambda^4}$	V = volume of sphere λ = wavelength	Any	$ka \ll 1$ $kr \gg 1$
Cylinder Infinitely long Thick	$\dfrac{ar}{2}$	a = radius of cylinder	Normal to axis of cylinder	$ka \gg 1$ $r > a$
Thin	$\dfrac{9\pi^4 a^4}{\lambda^2}r$	a = radius of cylinder	Normal to axis of cylinder	$ka \ll 1$
Finite	$\dfrac{aL^2}{2\lambda}$	L = length of cylinder a = radius of cylinder	Normal to axis of cylinder	$ka \gg 1$ $r > L^2/\lambda$
	$\dfrac{aL^2}{2\lambda}\left(\dfrac{\sin\beta}{\beta}\right)^2 \cos^2\theta$	a = radius of cylinder $\beta = kL\sin\theta$	At angle θ with normal	
Plate	r^2		Normal to plane	

Finite Any shape	$\left(\dfrac{A}{\lambda}\right)^2$	A = area of plate L = greatest linear dimension of plate l = smallest linear dimension of plate	Normal to plate	$r > \dfrac{L^2}{\lambda}$ $kl \gg 1$
Rectangular	$\left(\dfrac{ab}{\lambda}\right)^2 \left(\dfrac{\sin\beta}{\beta}\right)^2 \cos^2\theta$	a, b = side of rectangle $\beta = ka\sin\theta$	At angle θ to normal in plane containing side a	$r > \dfrac{a^2}{\lambda}$ $kb \gg 1$ $a > b$
Circular	$\left(\dfrac{\pi a^2}{\lambda}\right)^2 \left(\dfrac{2J_1(\beta)}{\beta}\right)^2 \cos^2\theta$	a = radius of plate $\beta = 2ka\sin\theta$	At angle θ to normal	$r > \dfrac{a^2}{\lambda}$ $ka \gg 1$
Ellipsoid	$\left(\dfrac{bc}{2a}\right)^2$	u, b, c = semimajor axes of ellipsoid	Parallel to axis of a	$ka, kb, kc \gg 1$ $r \gg a, b, c$
Conical tip	$\left(\dfrac{\lambda}{8\pi}\right)^2 \tan^4\psi \left(1 - \dfrac{\sin^2\theta}{\cos^2\psi}\right)^{-3}$	ψ = half angle of cone	At angle θ with axis of cone	$\theta < \psi$
Average overall aspects Circular disk	$\dfrac{a^2}{8}$	a = radius of disk	Average overall directions	$ka \gg 1$ $r > \dfrac{(2a)^2}{\lambda}$
Any smooth convex object	$\dfrac{S}{16\pi}$	S = total surface area of object	Average overall directions	All dimensions and radii of curvature large compared with λ
Triangular corner reflector	$\dfrac{L^4}{3\lambda^2}(1 - 0.00076\theta^2)$	l = length of edge of reflector	At angle θ to axis of symmetry	Dimensions large compared with λ

SOURCE: Urick, Ref. 164, Par. **26-161.**

The propagation from the target to the transducer is the same as that from the transducer to the target but may include several propagation paths. Propagation paths of interest for active sonar include in-layer direct path, across-layer direct path, bottom bounce, and convergence zone. Separate elevation angles may be required for the different propagation paths, depending on the ocean model.

156. Background Noise. Background noise in an active sonar system[164,168] is the random addition of ambient and self-noise as well as reverberations. Ambient and self-noise are assumed to be stationary during the ping cycle, while reverberations vary with detection range. Ambient and self-noise are usually assumed to be isotropic in azimuth and elevation for performance predictions (i.e., in the active sonar equation), so that the ambient-noise band level L_B at the receiver input is

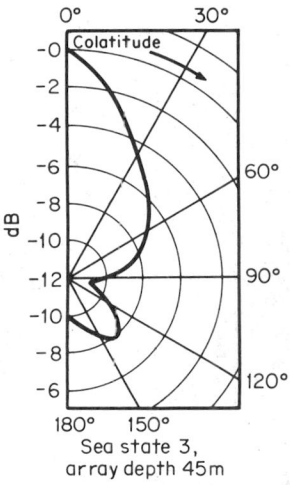

Fig. 25-138. Vertical directivity of ambient sea noise.

$$L_B = L_{Nf} + N_{BW} - N_{DI} \qquad \text{dB} \qquad (25\text{-}107)$$

where L_{Nf} = equivalent isotropic spectrum level of ambient or self-noise (dB re 1 μPa) in 1-Hz band, N_{BW} = receiver input band level (dB re 1 Hz), and N_{DI} = directivity index (dB).

Neither the ambient nor the self-noise is in fact isotropic, so that performance predictions based on equivalent isotropic noise levels predict average conditions. Ambient noise varies with elevation angle, as shown in Fig. 25-138, and varies in azimuth as well as with the maximum noise arriving from the direction of the wave swells. In addition to being anisotropic, ambient noise varies with the ocean configuration, wind, current, diurnal heating and cooling cycles, and widely varying velocity structures.

Self-noise is also highly directional, and, as expected, the highest self-noise on ships and submarines originates at the propeller. Self-noise can be separated into platform and equipment noise. Sonar systems are usually designed so that equipment noise is small compared with ambient or platform noise.

The receiving transducer may be the same as the transmitting transducer or can be completely independent. When the same transducer is used, a transmit/receive switch connects the transducer to either the transmitter or beam former as required. The receive transducer (or set of transducers) accepts the echo plus noise and furnishes these to a beam former. Beam formers reject noise and have either an adaptive or fixed configuration. Adaptive beam formers are designed to reject variable directional noise sources such as self-noise and reverberations by monitoring the noise field and adjusting internal parameters to minimize the background noise. Fixed beam formers are designed to reject isotropic noise and take advantage of the directional property of the echo.

157. Reverberation.[164] Reverberation often limits performance on modern high-power sonars. The boundaries of the ocean at the surface and bottom and inhomogeneities in the ocean, such as schools of fish, layers of air bubbles near the surface, the deep scattering layer, as well as pinnacles and seamounts on the sea bed, all reradiate a portion of acoustic energy incident on them. This reradiation of sound is called *scattering*, and the sum-total acoustic energy at the

TABLE 25-23   Nominal Values of Target Strength

Target	Aspect	Target strength, dB
Submarines	Beam	+25
	Bow-stern	+10
	Intermediate	+15
Surface ships	Beam	+25 (highly uncertain)
	Off-beam	+15 (highly uncertain)
Mines	Beam	+10
	Off-beam	+10 to −25
Torpedoes	Bow	−20
Fish of length L, ft	Dorsal view	$-31 + 30 \log L$

SOURCE: Urick, Ref. 164, Par. 25-161.

receiving transducer from all the scatterers is called *reverberation*. There are many different types of reverberations, and it is important to visualize the kinds of reverberations occurring during the ping cycle and what azimuth and elevation angle these will be coming from.

Certain characteristics of reverberations can be used to good advantage by the sonar designer. Knowledge of the directions of reverberations may be used to reduce their effect on sonar operations. For instance, with a direct propagation path, a narrow vertical transmit-and-receive beam pattern will first decrease the incident acoustic energy on the bottom, and second, reject reverberations coming from the bottom outside the main lobe of the receive beam pattern.

A knowledge of the reverberation spectrum can also be used to minimize the effect. The sonar system then becomes self-noise or ambient noise limited with a significant drop in the background-noise level and corresponding improvement in system performance. Long cw pulses have reverberation spectrums which are small compared with the receiver input bandwidth and lend themselves readily to this scheme. Own-ship doppler nullification (ODN) circuits compensate for platform velocity-induced shifts in center frequency of the reverberation spectrums. The notch filter needed for rejection of the reverberation has to be wide enough for the transmitted pulse spectrum convolved with the scatterer motion. Homogeneous scatterer motion causes center-frequency shifts, and random scatterer motion causes frequency spreading. For echo-ranging applications, it is usually possible to reject the reverberations with a notch filter with a bandwidth corresponding to a ± 0.25 m/s random scatterer motion.

The doppler shift of the center frequency of the reverberations from that of the transmitted pulse spectrum can be used to measure ship speed with respect to the immediate surrounding water.

158. Detection Threshold. The detection threshold N_{DT} is the signal-to-noise ratio S/N required in the receiver input band at the receiver input to produce the desired display statistics.[164] The term N_{DT} is usually applied to surveillance displays and refers to the desired single-ping probability of detection and false alarm. For other sonar modes of operation, such as fine-grained measurement of target range, doppler range rate, bearing, and elevation angle, the target has already been detected, and the required S/N at the receiver input is related to the desired standard deviation of the estimator.

The desired display statistics for a surveillance display are usually a 50% probability of detection and an acceptably low number of false alarms on the display. A typical surveillance display is a B scan with linear presentation of range along the vertical axis and linear presentation of bearing along the horizontal axis. The number of independent range cells is usually taken as the echo-ranging time required for the display range gate, in seconds, times the receiver output bandwidth in hertz. The number of independent bearing cells is usually the number of beams required to give continuous bearing coverage for the desired azimuthal sector. The false-alarm probability of any one of the independent range-bearing cells is just the number of allowed false alarms on the display divided by the total number of such range-bearing cells.

Once the desired probability of detection P(D) and probability of false alarm P(FA) have been established, the required S/N at the display input can be derived through the use of receiver operating curves (ROC) as described in Par. **25-150**.

The N_{DT} at the receiver input can be established through the use of S/N transfer functions for a given signal processor.[164,175–177] It should be noted that the desired P(FA) of independent range-bearing cells is usually also a function of the signal processor because of different receiver output bandwidths for different signal processors.

A sonar receiver for surveillance has to work over a wide range of conditions, which, because of the nature of the targets, are generally unknown. A *robust receiver* (i.e., one that works well over all the expected variations of input parameters) or a combination of several receivers to cover such variations is usually desired. The "ideal" N_{DT} for any particular receiver degrades with target doppler, multipath arrivals, and nonstationary noise background.

The selection of the receiver for a particular application cannot be based on the lowest N_{DT} alone, but has to take many other receiver characteristics into account, such as cost, reliability, maintainability, logistics, power consumption, cooling requirement, operator training, weight, space, plus additional items peculiar to the application.

The optimum receiver for a signal known exactly, with white, stationary, gaussian noise, is a matched filter. The signal processing gain for this type of receiver is 10 times the logarithm to the base 10 of the time-bandwidth product of the signal, and the resulting N_{DT} is the lower limit. Cross-correlators and comb-filter receivers are matched-filter-type receivers.

Incoherent receivers are optimum for unknown signals, with white, stationary, gaussian noise.

The signal processing gain for this type of receiver is 5 times the logarithm to the base 10 of the time-bandwidth product. Energy and envelope detectors, as well as autocorrelators and spatial correlators, are examples of this type of receiver.

Since the echo for an active sonar is neither known exactly nor completely unknown, semi-coherent receivers are sometimes used which have matched-filter features for those echo characteristics which are relatively well known and incoherent receiver characteristics for those echo characteristics which are likely to change significantly. The postdetection pulse-compression receiver is an example of this kind of receiver. The signal processing gain of this type of receiver is between that of the matched-filter type and the incoherent receiver.

159. Surveillance Receivers. A typical active surveillance display is shown in Fig. 25-139. This is a rectangular B-scan format with linear presentation of range along the vertical axis

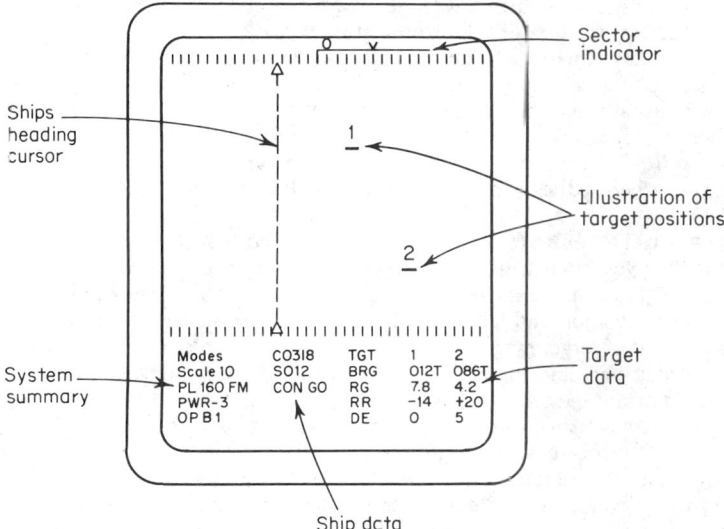

Fig. 25-139. B-scan surveillance display.

and linear presentation of bearing along the horizontal axis. The center bearing of the B-scan display can usually be selected to be true North or ship's bearing by the operator. It is driven by a thresholded output of a multichannel surveillance receiver. The signal excess over the threshold controls the display intensity in each bearing-range cell. The ability to threshold relies on noise-power normalization, so that changes in background-noise levels caused by variations in ambient noise and/or self-noise, as well as reverberations, do not cause an excessive number of false alarms.

A number of different surveillance receivers are available to drive the display: linear or clipped correlators, autocorrelator, spatial correlator, comb-filter receivers (with or without reverberation suppression), energy or envelope detectors, and pulse compression receivers. Performance prediction for the various types of receivers is covered in the literature on signal processing techniques. It is important to remember that these receivers should be compared on the basis of equal display statistics and nonstationary reverberations, self-noise, and ambient background noise, as well as "ideal" conditions.

Noise-power variations may be reduced by clipping, nonlinear amplification such as logarithmic amplifiers, or variable linear amplification such as time variable gain (TVG), automatic gain control (AGC), or step gain control (SGC) ahead of the receiver. Clipping is inexpensive but has the disadvantage of possible capture of the clipper by strong interfering noises. Logarithmic or other nonlinear amplifications can hold the noise power constant but cause frequency distortion and spreading. TVG is generally used early in the ping cycle to prevent system saturation right after transmission and then to increase the gain vs. time in a predetermined manner as a function of the transmitted source level and pulse length. AGC is closed-loop gain control where the noise power is estimated and the gain inversely controlled to maintain constant noise power. SGC is like AGC except that the gain is controlled in steps, so that inverse proportional gain control is not needed. Both AGC and SGC are sometimes referred to as reverberation gain control (RGC).

Another type of active surveillance display is the plan position indicator (PPI) display. The PPI format has the ship at the display center, with bearing linearly presented as display angle and the ship's heading presented vertically upward. Range is linearly presented as the radius from the display center.

Track and localization displays are usually presented in A-scan format, with a range gate linearly presented along the horizontal axis and either azimuth or elevation angle gate or doppler

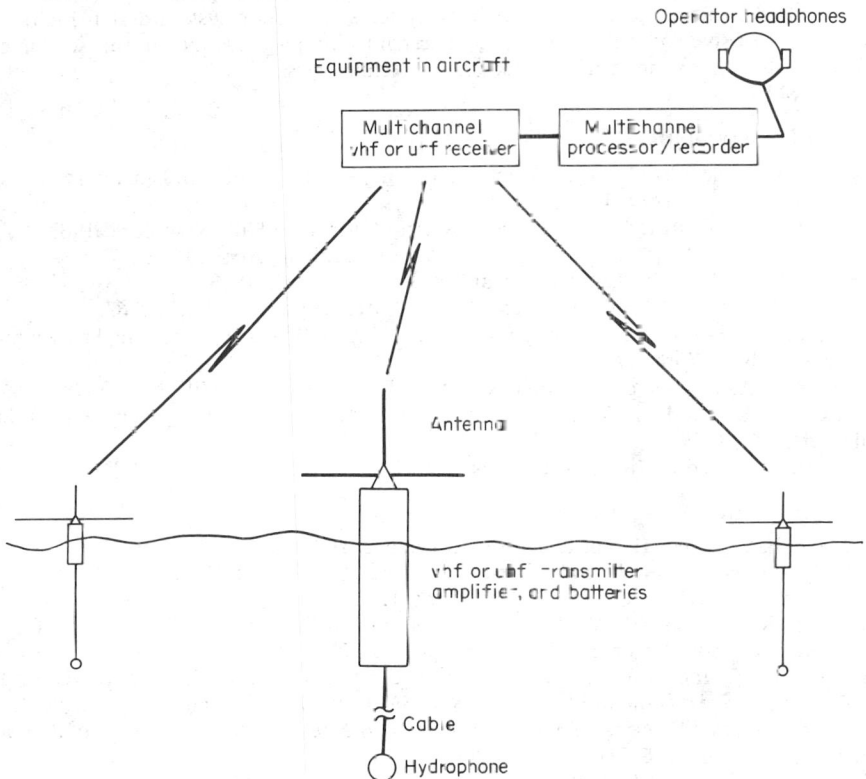

Fig. 25-140. Passive sonobuoy system.

range rate linearly presented along the vertical axis. Localization and track displays and receivers are generally rated according to the resolution in range, bearing, elevation, and doppler range rate available to the operator as a function of S/N.

160. Sonobuoy Systems. Sonobuoy systems are miniature passive and/or active sonar systems designed for deployment from aircraft, making use of VHF or UHF radio channels for transmission of information from the buoy to the aircraft. In its simplest configuration, a passive sonobuoy consists of a cable-connected omnidirectional hydrophone which is released from the buoy after water entry and which sinks to a predetermined operating depth; an amplifier to raise the level of the hydrophone output; a VHF or UHF transmitter; antenna system; and batteries for power. These items are often contained in a cannister less than 4 ft long and 5 in in diameter, fitted with a rotochute or other device for control of descent after release from the aircraft. The buoy is designed to float so as to maintain its antenna system above the water surface. On the aircraft the sonobuoy signal is extracted from the radio channel and applied to a processor-recorder and operator's headphones. Multichannel receiving and processing equipment is used to monitor the outputs of several sonobuoys simultaneously (Fig. 25-140).

More complex sonobuoy systems result when directional arrays are used in lieu of the simple omnidirectional hydrophone, necessitating the transmission of array orientation and target-bearing information via the radio link. Greater complexity, posing a severe challenge to the sonobuoy designer, results from combination of passive and active systems and control of the sonobuoy functions from the aircraft via a command radio link. Because of the advances made in micro-miniaturization, integrated circuits, multiplexing, miniature accustic arrays, digital signal processing, and high-density packaging, it is now possible to design sophisticated multifunction sonobuoy systems.

The design of such systems includes heavy emphasis on sonobuoy performance, cost, size, weight, life, and reliability factors — in view of the end use of the buoy as an expendable item. System partitioning, in which the functions to be performed in the buoy and in the aircraft equipment are defined, is an important element of the performance per cost consideration, particularly for the more sophisticated sonobuoy systems. Sonobuoy system design includes the application of passive and active sonar system design techniques described in this section of the Handbook, plus aerodynamic and hydrodynamic technologies.

161. Bibliography

REFERENCE BOOKS ON RADAR

1. Ridenour, L. N. (ed.)   "Radar System Engineering," M.I.T. Radiation Laboratory Series, vol. 1, McGraw-Hill, New York, 1947.
2. Skolnik, M. I.   "Introduction to Radar Systems," McGraw-Hill, New York, 1962.
3. Barton, D. K.   "Radar System Analysis," Artech, Dedham Mass., 1976.
4. Berkowitz, R. S. (ed.)   "Modern Radar," Wiley, New York, 1965.
5. Cook, C. E., and M. Bernfeld   "Radar Signals," Academic, New York, 1967.
6. Nathanson, F. E.   "Radar Design Principles: Signal Processing and the Environment," McGraw-Hill, New York, 1969.
7. Rihaczek, A.   "Principles of High Resolution Radar," McGraw-Hill, New York, 1969.
8. Barton, D. K., and H. R. Ward   "Handbook of Radar Measurement," Prentice-Hall, Englewood Cliffs, N.J., 1969.
9. Skolnik, M. I. (ed.)   "Radar Handbook," McGraw-Hill, New York, 1970.

OTHER REFERENCES ON RADAR PRINCIPLES

10. Blake, L. V.   Prediction of Radar Range, Chap. 2 in Ref. 9.
11. Goldstein, H.   Attenuation by Condensed Water, Sec. 8.6 in D. E. Kerr (ed.), "Propagation of Short Radio Waves," McGraw-Hill, New York, 1951.
12. Gunn, K. L. S., and T. W. R. East   The Microwave Properties of Precipitation Particles, *Q. J. R. Meteorol. Soc.*, October 1954, Vol. 80, pp. 522–545.
13. Ryde, J. W., and D. Ryde   Attenuation of Centimetre and Millimetre Waves by Rain, Hail, Fogs and Clouds, *General Electric Co. Rep.* 8670, Wembly, England, 1945.
14. Wexler, R., and D. Atlas   Radar Reflectivity and Attenuation of Rain, *J. Appl. Meteorol.*, April 1963, Vol. 2, pp. 276–280.
15. Marshall, J. S., and W. M. Palmer   The Distribution of Raindrops with Size, *J. Meteorol.*, August 1948, Vol. 5, pp. 165–166.
16. Medhurst, R. G.   Rainfall Attenuation of Centimeter Waves: Comparison of Theory and Measurement, *IEEE Trans.*, July 1965, Vol. AP-13, No. 4, pp. 550–564.
17. Blevis, B. C.   Losses Due to Rain on Radomes and Antenna Reflecting Surfaces, *IEEE Trans.*, January 1965, Vol. AP-13, No. 1, pp. 175–176.
18. Ruze, J.   More on Wet Radomes, *IEEE Trans.*, September 1965, Vol. AP-13, No. 5, pp. 823–824.
19. Millman, G. H.   Atmospheric Effects on VHF and UHF Propagation, *Proc. IRE*, August 1958, Vol. 46, No. 8, pp. 1492–1501.
20. Rice, P. L., A. G. Longley, K. Norton, and A. P. Barsis   Transmission Loss Predictions for Tropospheric Communication Circuits, *Natl. Bur. Stand. Tech. Note* 101 (Vol. 2), 1967, rev. GPO.
21. Beckman, P., and A. Spizzichino   "The Scattering of Electromagnetic Waves from Rough Surfaces," Macmillan, New York, 1963.
22. Bean, B. R., and G. D. Thayer   CRPL Exponential Reference Atmosphere, *Natl. Bur. Stand. Monogr.* 4, Oct. 29, 1959.
23. Thompson, M. C., H. B. Jones, and R. W. Kirkpatrick   An Analysis of Time Variations in Tropospheric Refractive Index and Apparent Radio Path Length, *J. Geophys. Res*, January 1960, Vol. 65, No. 1, pp. 193–201.
24. Pfister, W., and T. J. Keneshea   "Ionospheric Effects on Positioning of Vehicles at High Altitudes," *Air Force Surv. Geophys.* 83, Cambridge Research Center, Cambridge, Mass., March 1956 (DDC document AD 98 777).

25. Swerling, P.   Probability of Detection for Fluctuating Targets, *IRE Trans.*, April 1960, Vol. IT-6, No. 2, pp. 269–308.

26. Boothe, R. R.   "The Weibull Distribution Applied to the Ground Clutter Backscatter Coefficient," U.S. Army Missile Command, Redstone Arsenal, Ala., Rep. RE-TR-69-15, June 12, 1969 (DDC document AD 691 109).

27. Barton, D. K.   Radar Equations for Jamming and Clutter, November 1967, *IEEE Trans.*, Vol. AES-3, No. 5, pp. 340–355 (EASCON Suppl.).

28. Trunk, G. V., and S. F. George   Detection of Targets in Non-Gaussian Sea Clutter, *IEEE Trans.*, September 1970, Vol. AES-6, No. 5, pp. 620–628.

29. Dunn, J., and D. D. Howard   "Target Noise," Chap. 21 in Ref. 9.

30. IEEE Standard Radar Definitions, No. 686, November 1977.

31. Marcum, J. I.   A Statistical Theory of Target Detection by Pulsed Radar, *IRE Trans.*, April 1960, Vol. IT-6, No. 2, pp. 59–267.

32. DiFranco, J. V., and W. L. Rubin   "Radar Detection," Prentice-Hall, Englewood Cliffs, N.J., 1968.

33. Barton, D. K.   Simple Procedures for Radar Detection Calculations, *IEEE Trans.*, September 1969, Vol. AES-5, No. 5, pp. 837–846.

34. Heidbreder, G. R., and R. L. Mitchell   Detection Probabilities for Log Normally Distributed Signals, *IEEE Trans.*, January 1967, Vol. AES-3, No. 1, pp. 5–13.

35. Trunk, G. V.   Detection of Targets in Non-Rayleigh Sea Clutter, *IEEE EASCON 1971 Rec.*, pp. 239–245.

36. Woodward, P. M.   "Probability and Information Theory with Applications to Radar," Pergamon, New York, 1953.

37. Rihaczek, A. W.   "Principles of High Resolution Radar," McGraw-Hill, New York, 1969.

38. Farnett, E. C., T. B. Howard, and G. H. Stevens   Pulse-Compression Radar, Chap. 20 in Ref. 9.

39. Hannan, P. W.   Optimum Feeds for All Three Modes of a Monopulse Antenna, *IRE Trans.*, September 1961, Vol. AP-9, No. 5, pp. 444–461.

40. Manasse, R.   Range and Velocity Accuracy from Radar Measurements, *M.I.T. Lincoln Lab Rep.* 312–326, Feb. 3, 1955 (DDC document AD 236 236).

41. Barton, D. K   Radar Measurement Accuracy in Log-Normal Clutter, *IEEE EASCON 1971 Rec.*, pp. 246–251.

42. Hall, W. M.   Prediction of Pulse Radar Performance, *Proc. IRE*, February 1956, Vol. 44, No. 2, pp. 224–231.

43. Hall, W. M.   General Radar Equation in "Space/Aeronautics R and D Handbook," 1962–1963.

44. Barton, D. K. and W. W. Shrader   Interclutter Visibility in MTI Systems, *IEEE EASCON 1969 Rec.*, pp. 294–297.

45. Blake, L. V.   Radio Ray (Radar) Range-Height-Angle Charts, *Microwave J.*, October 1968, Vol. 4, No. 10, pp. 49–53.

46. Shrader, W. W   MTI Radar, Chap. 17 in Ref. 9.

47. Winter, C. F.   Dual Vertical Beam Properties of Doubly Curved Reflectors, *IEEE Trans.*, March 1971, Vol. AP-19, No. 2, pp. 174–180.

48. Ward, H. R., and W. W. Shrader   MTI Performance Degradation Caused by Limiting, *IEEE 1968 EASCON Rec.*, pp. 168–174.

49. Brown, B. P.   Radar Height Finding, Chap. 22 in Ref. 9.

50. Dunn, J. H., and D. D. Howard   The Effects of Automatic Gain Control of Performance on the Tracking Accuracy of Monopulse Radar Systems, *Proc. IRE*, March 1959, Vol. 47, No. 3, pp. 430–435.

REFERENCES ON RADAR TECHNOLOGY

51. Skolnik, M. I. (ed.)   "Radar Handbook," McGraw-Hill, New York, 1970.

52. Weil, T. A.   Transmitters, Chap. 7 in Ref. 51.

53. Skolnik, M. I.   "Introduction to Radar Systems," McGraw-Hill, New York, 1962.

54. Sherman, J. W.   Aperture-Antenna Analysis, Chap. 9 in Ref. 51.

55. Barton, D. K., and H. R. Ward   "Handbook of Radar Measurement," Prentice-Hall, Englewood Cliffs, N.J., 1969.

56. Ashley, A., and J. S. Perry   Beacons, Chap. 38 in Ref. 51.

57. Croney, J.   Civil Marine Radar, Chap. 31 in Ref. 51.

58. Freedman, J.   Radar, Chap. 14 in "System Engineering Handbook," McGraw-Hill, New York, 1965.

59. Sengupta, D. L., and R. E. Hiatt Reflectors and Lenses, Chap. 10 in Ref. 51.

60. Dunn, J. H., D. D. Howard, and K. B. Pendleton Tracking Radar, Chap. 21 in Ref. 51.

61. Cheston, T. C., and J. Frank Array Antennas, Chap. 11 in Ref. 51.

62. Stark, L., R. W. Burns, and W. P. Clark Phase Shifters for Arrays, Chap. 12 in Ref. 51.

63. Kefalas, G. P., and J. C. Wiltse Transmission Lines, Components, and Devices, Chap. 8 in Ref. 51.

64. Hammer, I. W. Frequency-Scanned Arrays, Chap. 13 in Ref. 51.

65. Matthaei, G. L., L. Young, and E. M. T. Jones "Microwave Filters, Impedance Matching Networks, and Coupling Structures," McGraw-Hill, New York, 1964.

66. Taylor, J. W., and J. Mattern Receivers, Chap. 5 in Ref. 51.

67. Farnett, E. C., T. B. Howard, and G. H. Stevens Pulse-Compression Radar, Chap. 20 in Ref. 51.

68. Shrader, W. W. MTI Radar, Chap. 17 in Ref. 51.

69. Mooney, D. H., and W. A. Skillman, Pulse-Doppler Radar, Chap. 19 in Ref. 51.

70. Nathanson, F. "Radar Design Principles: Signal Processing and the Environment," McGraw-Hill, New York, 1969.

71. Berg, A. A. Radar Indicators and Displays, Chap. 6 in Ref. 51.

REFERENCES ON ELECTRONIC NAVIGATION

72. Kayton, M., and W. Fried "Avionics Navigation Systems," Wiley, New York, 1969. (Includes chapters on radio, doppler, inertial, radar, celestial, and satellite navigation systems, together with other chapters on computers, ILS, air traffic control, and displays.)

73. Sandretto, P. C. "Electronic Avigation Engineering," International Telephone and Telegraph Corp., New York, 1958.

JOURNALS ON ELECTRONIC NAVIGATION

74. *Navigation*, published quarterly by the Institute of Navigation, Washington.

75. *IEEE Transactions on Aerospace and Electronic Systems*, published six times a year by the IEEE, New York.

REFERENCES ON ELECTRONIC NAVIGATION

76. Anderson, E. W. Air Navigation Techniques, *Navigation*, Spring 1971, Vol. 18, No. 1.

77. Dunlap, G. D. Major Developments in Marine Navigation During the Last 25 Years, *Navigation*, Spring 1971, Vol. 18, No. 1.

78. French Institute of Navigation "Étude comparative des systèmes hyperboliques et circulaires pour la navigation et la localisation," Paris, April 1970.

79. Braverman, N. Aviation System Design for Safety and Efficiency, *Navigation*, Fall 1971, Vol. 18, No. 3.

80. Jackson, W. "The Federal Airways System," *IEEE*, New York, 1970.

81. Keen, R. "Wireless Direction Finding," Iliffe, London, 1938.

82. Astholtz, P. Air Traffic Control, in Ref. 72.

83. Wiley, C. Radar Navigation, in Ref. 72.

84. Fried, W. Doppler Navigation in Ref. 72.

85. Laser Navigators for Boeing 757/767. *Flight International*, Dec. 16, 1978.

86. Special Issue on G.P.S., *Navigation*, Summer 1978, Vol. 25, No. 2.

87. Kayton, M. Inertial Navigation, in Ref. 72.

88. Quasius, G. Celestial Navigation, in Ref. 72.

89. Duncan, R. C. Satellite Navigation, in Ref. 72.

90. Hawkins, H., et al. Radar Performance Degradation in Fog and Rain, *IRE Trans. ANE*, March 1959, Vol. ANE-6, No. 1.

91. Jansky, C. M. The Current State of the Science of Marine Navigation, *Navigation*, Spring 1965, Vol. 12, No. 1.

92. Hopkins, H. G., et al. Current D/F Practice, *Proc. IEE*, March 1959, Vol. 105, Pt. B, No. 9.

93. Sandretto, P. "Electronic Avigation Engineering," International Telephone & Telegraph Corp., New York, 1958.

94. Busignies, H. Evaluation of Night Errors in Aircraft Direction Finding, *Electr. Commun.*, June 1946, Vol. 23, No. 2.

95. Kayton, M. Landing Guidance, in Ref. 72.

96. Popp, H. Solid-State VOR, *Elect. Commun.*, December 1969, Vol. 44, No. 4.

97. Anderson, S., et al.   The CAA Doppler Omnirange, *Proc. IRE*, May 1959, Vol. 47, No. 5.

98. Poritsky, S. (ed.)   Special issue on VOR/DME, *IEEE Trans. ANE*, March 1965, Vol. ANE-12, No. 1.

99. Colin, R., et al.   Principles of Tacan, *Electr. Commun.*, March 1956 Vol. 33, No. 1.

100. Dodington, S.   Recent Developments of the Tacan Navigation System, *Electr. Commun.*, December 1969, Vol. 44, No. 4.

101. Special Issue on Collision Avoidance,   *ICAO Bulletin*, March 1979, Vol. 34, No. 3.

102. Adcock, F.   British Patent No. 130490, 1919.

103. Powell, C.   The Decca Navigator System for Ship and Aircraft Use, *Proc. IEE*, March 1958, Vol. 105, Pt. B, No. 9.

104. Rörholt, B. A.   Electronic Aids to Navigation for Fishing Vessels and Other Open Sea Users, *Navigation*, Fall 1969, Vol. 16, No. 3.

105. Kuebler, W.   Marine Electronic Navigation Systems: A Review, *Navigation*, Fall 1968, Vol. 15, No. 3.

106. Van Etten, J. P.   Loran C System and Product Development, *Electr. Commun.*, June 1970, Vol. 45, No. 2.

107. Swanson, E. R.   Omega, *Navigation*, Summer 1971, Vol. 18, No. 2.

108. Ontrac II, Versatile Navigator, *Flight International*, Apr. 10, 1975.

109. Stansell, T.   Transit, the Navy Navigational Satellite System, *Navigation*, Spring 1971, Vol. 18, No. 1.

110. Brown, A. H.   Consol Navigation System, *J. IEE*, 1947, Pt. 3A, Vol. 94, No. 15.

111. Tadaro, T., et al.   Talking Beacon System in Japan, *Navigation*, Winter 1969, Vol. 16, No. 4.

112. Hirst, M. Ten Years of Airline INS, *Flight International*, July 29, 1979.

113. RTCA, SC-139, Minimum Performance Standards for Airborne MLS, 1980.

114. JTIDS/TIES Consolidated Tactical Communications, *E-W Rev.* September–October 1977.

115. U.S. Department of Transportation   *Rep. Radionavigation Users Conf.*, Oct. 9–11, 1979.

116. Pogust, F.   The Status of Microwave Scanning Beams Landing System Developments, *IEEE EASCON 1969 Rec.*

117. Earp, C. W., et al.   Doppler Scanning Guidance System, *Electr. Commun.*, December 1971, Vol. 46, No. 4.

118. Edwards, J. A.   Microwave Landing System, *IEEE EASCON 1971 Rec.*

119. Special issue on CAS, *Trans. IEEE Group AES*, March 1968, Vol. AES-4, No. 2.

120. Borrok, M. J., et al.   Results of ATA CAS Flight Test Program, *Navigation*, Fall 1970, Vol. 17, No. 3.

REFERENCES ON UNDERWATER SOUND-DETECTION SYSTEMS: PRINCIPLES AND FUNCTIONS

121. Batchelder, L. B.   Sonics in the Sea, *IEEE Proc.*, October 1965, Vol. 53 No. 10.

122. Becken, B. A.   Sonar, *Adv. Hydrosci.*, 1964.

123. Horton, J. W.   "Fundamentals of Sonar," U.S. Naval Institute, Annapolis, Md., 1959.

124. Gray, D. E.   "American Institute of Physics Handbook," 2d ed., McGraw-Hill, New York, 1963.

125. Urick, R. J.   "Principles of Underwater Sound," McGraw-Hill, New York, 1975.

PROPAGATION

126. Wilson, W. D.   *J. Acoust. Soc. Am.*, 1960, Vol. 32, p. 1357.

127. Ibid., p. 641, 1960.

128. Ibid., p. 866, 1962.

129. Urick, R. J.   "Principles of Underwater Sound," McGraw-Hill, New York, 1975.

130. Tolstoy, I., and C. S. Clay   "Ocean Acoustics," McGraw-Hill, New York, 1966.

131. Albers, V. M.   "Underwater Acoustics Handbook," Penn State University Press, University Park, Pa., 1960

132. Horton, J. W.   "Fundamentals of Sonar," U.S. Naval Institute, Annapolis, Md., 1957.

133. Thorp, W. H.   *J. Acoust. Soc. Am.*, 1967, Vol. 42, p 270.

134. Batchelder, L.   *Proc. IEEE*, 1965, Vol. 53, p. 10

135. Bobber, R. J.   "Underwater Electroacoustic Measurements," Naval Research Laboratory, Washington, D.C., 1970.

136. Schulkin, M., and H. W. Marsh   *J. Acoust. Soc. Am.*, 1962, Vol 34, p. 864.

137. Fisher, F. H. *J. Acoust. Soc. Am.*, 1958, Vol. 30, p. 442.

138. Liebermann, L. N. *Phys. Rev.*, 1949, Vol. 76, p. 1520.

139. Stephens, R. W. B. (ed.), "Underwater Acoustics," Wiley-Interscience, London, 1970.

NOISE

140. Becken, B. A. Sonar, *Adv. Hydrosci.*, 1964.

141. Urick, R. J. "Principles of Underwater Sound," McGraw-Hill, New York, 1975.

142. Batchelder, L. Sonics in the Sea, *IEEE Proc.*, October 1965, Vol. 53, No. 10.

143. Bartberger, C. L. Lecture Notes on Underwater Acoustics, U.S. Naval Air Development Center, Johnsville, Md., May 17, 1965.

144. Bobber, R. J. "Underwater Electroacoustic Measurements," Naval Research Laboratory, Washington, 1970.

145. "Introduction to Sonar Technology," NavShips Publ. 0967-129-3010, Navy Dept., Bureau of Ships, Washington, 1965.

TRANSDUCERS AND ARRAYS

146. American National Standards Institute ANSI S1.20-1972, Procedures for Calibration of Underwater Electroacoustic Transducers, New York.

147. Bobber, R. J. "Underwater Electroacoustic Measurements," Naval Research Laboratory, Washington, 1970.

148. Hueter, T. F., and R. H. Bolt "Sonics," Wiley, New York, 1955.

149. Sims, C. C. High-Fidelity Underwater Sound Transducers, *Proc. IRE*, 1959, Vol. 47, p. 866.

150. Urick, R. J. "Principles of Underwater Sound," McGraw-Hill, New York, 1975.

151. Kinsler, L. E., and A. R. Frey "Fundamentals of Acoustics," Wiley, New York, 1962.

151a. Westervelt, P. J. *J. Acoust. Soc. Am.*, 1963, Vol. 35, p. 535.

151b. Beyer, R. T. "Nonlinear Acoustics," Naval Sea Systems Command, Department of the Navy, 1974.

151c. Muir, T. G. An Analysis of the Parametric Acoustic Array for Spherical Wave Fields, *Univ. Tex. Tech. Appl. Res. Lab. Rep.* 71-1, 1971.

151d. Lockwood, J. C. Nomographs for Parametric Array Calculations, *Univ. Tex. Tech. Appl. Res. Lab. Mem.* 73-3, 1973.

151e. Berktay, H. O. and D. J. Leahy *J. Acoust. Soc. Am.*, 1974, Vol. 55, p. 539.

151f. Moffett, M. B. and R. H. Mellen *J. Acoust. Soc. Am.*, 1977, Vol. 61, p. 325.

151g. Barnard, G. R., J. G. Willette, J. J. Truchard and J. A. Shooter *J. Acoust. Soc. Am.*, 1972, Vol. 52, p. 1437.

152. Olson, H. F. "Acoustical Engineering," Van Nostrand, Princeton, N.J., 1957.

153. Kikuchi, Y. "Ultrasonic Transducers," Corona, Tokyo, 1969.

154. Cady, W. G. "Piezoelectricity," Vols. 1 and 2, Dover, New York, 1964.

155. Mason, W. P. "Piezoelectric Crystals and Their Application to Ultrasonics," Van Nostrand, Princeton, N.J., 1950.

156. Berlincourt, D. A., D. R. Curran, and H. Jaffee "Piezoelectric and Piezomagnetic Materials and Their Function in Transducers," in W. P. Mason (ed.), "Physical Acoustics," Vol. I, Pt. A, Academic, New York, 1964.

157. Piezoelectric Crystals, IEEE Standard 176, 1949.

158. Definitions and Methods of Measurements of Piezoelectric Vibrators, IEEE Standard 177, 1966.

159. Piezoelectric Crystals: Determination of the Elastic, Piezoelectric, and Dielectric Constants: The Electromechanical Coupling Factor, IEEE Standard 178, 1958.

160. Measurement of Piezoelectric Ceramics, IEEE Standard 179, 1961.

161. Piezoelectric Ceramic for Sonar Transducers, MIL-STD-1376 (SHIPS).

161. Magnetostrictive Materials: Piezomagnetic Nomenclature, IEEE Standard 319, 1971.

163. Kikuchi, Y. Magnetostrictive Metals and Piezomagnetic Ceramics as Transducer Materials, in O. E. Mattiat (ed.), "Ultrasonic Transducer Materials," Plenum, New York, 1971.

PASSIVE SONAR SYSTEMS

164. Urick, R. J. "Principles of Underwater Sound for Engineers," McGraw-Hill, New York, 1975.

165. Peterson, W. W., and T. G. Birdsall   The Theory of Signal Detectability, *Univ. Mich. Eng. Res. Inst. Rep.* 13, 1953.

166. Peterson, W. W., T. G. Birdsall, and W. C. Fox   The Theory of Signal Detectability, *Trans. IRE*, vol. PGIT-4, p. 171, 1954.

167. Lawson, J. L., and Uhlenbeck   Threshold Signals, M.I.T. Radiation Laboratory Series, vol. 24, McGraw-Hill, New York, 1950.

ACTIVE SONAR SYSTEMS

168. *Adv. Hydrosci.*, vol. 1, 1964.

169. Special Issue on Detection Theory and Its Applications, *Proc. IEEE*, May 1970.

170. Schwartz, Misha   "Information Transmission, Modulation and Noise," 2d ed., McGraw-Hill, 1970.

171. Broch, J. T.   Effects of Spectrum Non-linearities upon Peak Distributions of Random Signals, *Brüel Kjaer Tech. Rev.*, no. 3, 1963.

172. Croney, J.   Clutter on Radar Displays: Reduction by Use of Logarithmic Receivers, *Wireless Eng.*, 1956, Vol. 33.

173. Sherman, C. H., and D. F. Kass   Radiation Impedances, Radiation Patterns, and Efficiency for Large Array on a Sphere, *U.S. Navy Underwater Sound Lab Res. Rep.* 429, Fort Trumbell, New London, Conn., July 17, 1959.

174. Carson, D. L.   Diagnosis and Cure of Erratic Velocity Distributions in Sonar Projector Arrays, *J. Acoust. Soc. Am.*, September 1962, Vol. 34, No. 9.

175. Skolnick, M. J.   "Introduction to Radar System," McGraw-Hill, New York, 1962.

176. Davenport, W. B., and W. L. Root   "Introduction to Random Signals and Noise," McGraw-Hill, New York, 1958.

177. Gerlach, A. A.   Theory and Application of Statistical Wave-Period Processing, *Cook Electric Co., Rep.*, Chicago.

Section 26

Electronics in Medicine and Biology

PETER W. CHEUNG Associate Professor of Biomedical Engineering, Case Western Reserve University

DUDLEY CHILDRESS Director, Prosthesis Research Laboratory, Northwestern University; Member IEEE

PETER G. KATONA Associate Professor of Biomedical and Electrical Engineering, Case Western Reserve University; Member IEEE

RAYMOND S. KIRALY Chief Engineer, Research Division, Department of Artificial Organs, Cleveland Clinic Foundation

WEN H. KO Director, Engineering Design Center, Professor of Electrical Engineering, Professor of Biomedical Engineering, Case Western Reserve University; Fellow, IEEE

PAUL S. MALCHESKY Department of Artificial Organs, Research Division, Cleveland Clinic Foundation

JAMES D. MEINDL Professor of Electrical Engineering Stanford Electronics Laboratories, Stanford University; Fellow IEEE

FLORO MIRALDI Professor of Engineering, Case Western Reserve University and Chief, Nuclear Medicine, Cleveland Metropolitan General Hospital

J. THOMAS MORTIMER Associate Professor of Biomedical Engineering, Case Western Reserve University

MICHAEL R. NEUMAN Associate Professor of Biomedical Engineering in Reproductive Biology, Case Western Reserve University; Member IEEE

ROBERT PLONSEY Professor of Biomedical Engineering, Case Western Reserve University; Fellow IEEE

CONTENTS

Numbers refer to paragraphs

Biomedical Electronics

BY WEN H. KO

   1. Biomedical electronics applies electronic theory, technology, instrumentation, and computing systems to biological research and medical problems; it is a major area within the new discipline of biomedical engineering. For the last half decade, along with the advances of solid-state electronic devices, integrated circuits, and microcomputers, the field has grown with great speed both in sophistication and diversification. The major contributions made and sub-areas established are (1) physiological monitoring and research, (2) diagnostic instruments and imaging techniques, (3) automated clinical laboratories, (4) patient monitoring for the critically ill, (5) prosthetic and orthotic devices, including artificial organs, (6) computer data processing and management, and (7) implant instrumentation systems.

   This section of the Handbook presents a summary of selected topics in these areas with references, but a survey of biomedical transducers is included in place of item 3, the clinical laboratory, since in clinical laboratories the electronic involvement is centered on transducers and the well-established control systems discussed in other sections.

Electronic theory, technology, and design procedures of devices, circuits, and systems can be matched and applied to biomedical problems with little or no modification, but two special topics are common to all biomedical electronics: (1) electrical safety and (2) packaging and material. A brief summary of them follows.

2. Electrical and Electromagnetic Radiation Safety. The safety standard of leakage current on biomedical instruments from dc to 1 MHz accepted in the United States is the American National Standard Safe Current Limits for Electromedical Apparatus, ANSI/AAMI SCL 12/78, prepared by the Association for the Advancement of Medical Instrumentation

TABLE 26-1 DC to 1-kHz Risk-Current Limits in rms Microamperes

Electrical apparatus	Patient connection	Core-connected grounding conductor or exposed metal*	Permanently connected per National Electrical Code, grounding conductor or exposed metal, ground open test
With isolated patient connection	10†	100	5,000
With nonisolated patient connection	50	100	5,000
Likely to contact patient	Not applicable	100	5,000
No patient contact	Not applicable	500	5,000

*Measured to ground from exposed metal or from a 200-cm² foil over insulating enclosures. For insulating enclosures, the current is also measured in the grounding conductor. The limit for double-insulated apparatus is one-half of the values given. However, if the functional or supplementary insulation is effectively bypassed during a test, the limit is as given. The microampere limit in the grounding conductor applies to any apparatus or group thereof with a single cord connection to the electrical supply.

†The allowed sink risk current is 20 μA for isolated electromedical apparatus with patient cables when measured at the patient end of the cable.

SOURCE: ANSI/AAMI SCL 12/78.

and approved by the American National Standards Institute, Inc., in January 1978. This publication sets risk-current limits (RCL) and measuring techniques for risk currents of electromedical apparatus as a function of the frequency, the characteristics of the apparatus, and the nature of the intentional contact with the patient.

Table 26-1 summarizes the risk-current limits for four types of instrumentation between dc to 1 kHz. For frequencies from 1 kHz to 1 MHz, the RCL is as shown in Fig. 26-1. Both these RCLs are based on immediate fibrillatory and sensation thresholds and do not consider long-term physiological effects.

In some countries smaller current limits have been set with consideration of physiological effects, but no uniform standard has been agreed upon at this time.

The standard test load for testing medical instruments is shown in Fig. 26-2, which can be used to represent the patient in the risk-current measurement. Both source and sink currents should be measured, as shown in Figs. 26-3 and 26-4.

The leakage current from 60-Hz power lines can be reduced by:

1. Using the isolation transformer with Faraday shield between the primary and secondary winds and grounding the shield
2. Using primary or rechargeable batteries as the power source
3. Using a dc-dc inverter-converter
4. Using energy converters with nonelectromagnetic coupling:
   a. Mechanical: motor-generator
   b. Optical: light bulb solar cells
   c. Ultrasonic or vibrational coupling

For high frequencies, from 10 MHz to 100 GHz, the radiation standard is based on the heating effect on the most sensitive part of the body. The United States standard recommended[9]* is (1) power density 10 mW/cm² averaged over a 0.1-h period; (2) energy density 1 mWh/cm² for both

*Superior numbers correspond to references in Par. **26-55.**

partial and whole-body irradiation. The safety levels of some other countries may be as low as 0.01 mW/cm^2 for $f > 300$ MHz.[10-12] On the other side of safety is the medical use of electro-magnetic waves. Many applications have been developed, including microwave hyperthermia for cancer therapy.[13]

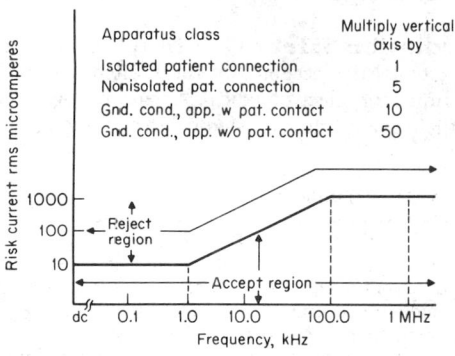

Fig. 26-1. Source risk-current limits. *(ANSI/AAMI SCL 12/78.)*

Fig. 26-2. AAMI standard test load. *(ANSI/AAMI SCL 12/78.)*

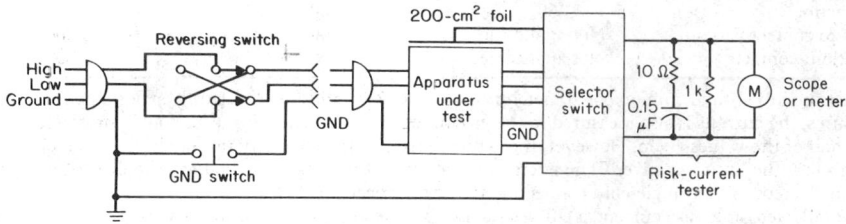

Fig. 26-3. Source risk-current test circuit (includes the battery charger of apparatus with rechargeable batteries). The oscilloscope must be ungrounded when the ground switch is open. *(ANSI/AAMI SCL 12/78.)*

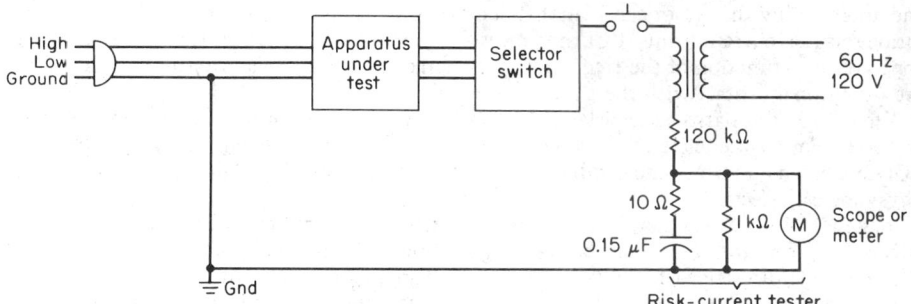

Fig. 26-4. Sink risk-current test circuit. *(ANSI/AAMI SCL 12/78.)*

3. Packaging and Materials. Packaging electronic devices, instruments, and complex systems for biomedical uses must satisfy the following considerations in addition to electrical safety and ordinary industrial requirements:

1. The mechanical structure must be strong for use in a medical environment by nonengineering personnel.

2. Biocompatibility of the material must be ensured in order to protect the body in contact with the device from possible toxic effect.

3. The electronics must be protected from the body environment, which is much more corrosive than saline solution or seawater.

4. Size and weight limitations for implant devices must be met.

The structure of medical devices and instruments should enable them to withstand use in a medical environment where emergency situations arise routinely. Instruments for operating rooms and intensive-care units are often required to be explosion-proof. Since the devices must be failure-safe and reliable, foolproof design needs to be considered.

Biocompatibility is essential, and the material in contact with tissue must therefore be non-toxic. Many materials have been tested for biocompatibility.[14] Metals such as 316 stainless steel, cast Co–Cr alloy, wrought Co–Cr alloy, Co–Cr–Mo alloy (vitallium), titanium, Ti-6 AL-4V, tantalum, and MP-35N are currently used as surgical implants.[15] For polymers, the medical grade silastic 382 RTV is the best-known inert material.

Other commonly used stable polymers are dacron, high-density polyethylene, polyether urethanes, polypropylene, silicon rubber and Teflon. Pyrolytic carbon, ceramics, and glass are generally inert to body tissues, but variation in the manufacturing process and impurity content in these materials may greatly alter their biocompatibility. Care should be taken when selecting specific blends of material with known impurity content for biocompatibility.

The molding technique, fabrication procedures, and surface condition of medical implants can often influence the resulting body reaction to these materials, and such effects should also be considered.

To protect them from corrosive and highly conductive body fluid, electronic circuits must be enclosed in hermetic enclosures or other protective coatings. Metal flatpacks, hermetically sealed by resistive welding or laser welding, are preferred for long-term applications. Pure tin soldering of the lid to the flatpack, gold plating over the soldered area, and a Lysol epoxy and 382 RTV coating on the outside may be used for shorter-term applications. Macor, machinable ceramic, and glass encapsulation can be used for long-term applications, in which electromagnetic waves need to be transmitted through the encapsulation. Material and packaging present major problems in prosthetic and orthotic devices, artificial organs, and implant instrumentation. Special attention must be directed to this problem to ensure proper operation.

4. Future Directions. Researchers in the field of biomedical engineering believe that the future development in medical technology[18] will be in the following areas:

1. Care of patients with heart disease, cancer, stroke, kidney failure, and chronic respiratory disease

2. Care of infants and children, related to congenital abnormalities, prematurity, birth injury, respiratory distress, nutritional disorders, and disabilities resulting from incomplete recovery from these problems, such as motor paralysis, mental retardation, and other residual effects

3. Prevention and care of injuries from accidents

4. Preventive and therapeutic measures

5. New and automatic drug-delivery systems

6. Home-based health care

Electronics will definitely play an important role in all these areas and will contribute to the program on improved health care with new technology.

Electrocardiography and Biopotentials

BY ROBERT PLONSEY

5. Electrophysiology of the Heart. In electrocardiography (ECG), the electric signals of interest are generated by the heart itself. For this reason the ECG has diagnostic value, since it reflects the biological (hence, clinical) behavior of the heart. While the following pertains specifically to the heart, it reflects basic bioelectric principles applicable to other organ systems, e.g., the electroencephalogram (EEG).

The human heart (Fig. 26-5) is composed primarily of a specialized muscle (cardiac muscle). Individual muscle cells are around 15 μm in diameter and 100 μm long and are stacked together rather like bricks. Collectively, their long dimensions lie parallel to each other, and in the aggregate they constitute bands of muscle fibers which make a broad spiral in forming the heart walls.

Individual heart cells are surrounded by a plasma membrane which separates the intracellular from the extracellular space. For a typical heart cell, metabolic energy is used to create an internal environment rich in potassium but low in sodium compared with an extracellular composition of high sodium and low potassium. Because of this existing imbalance, a resting potential exists across the cell membrane, the inside being some 90 mV negative relative to the outside. When the cell is stimulated (by passing an electric current which momentarily increases the transmembrane potential), the membrane properties go through a cyclic change, the first phase of which is a greatly elevated permeability to sodium. A large (early) inward sodium current results from the diffusion and electrical gradients.

This inflow constitutes precisely a generation of electric current, and during the transient the cell behaves essentially like a current dipole source. This transient sodium current is responsible for (and is part of) a local circuit current which links with adjoining cells and serves as a stimulus to initiate their activity. In this way activity (once started) spreads contiguously to adjacent cells. When the membrane recovers (returns to resting properties), the *action potential* of the cell is ended, and it is once again at rest and capable of being restimulated.

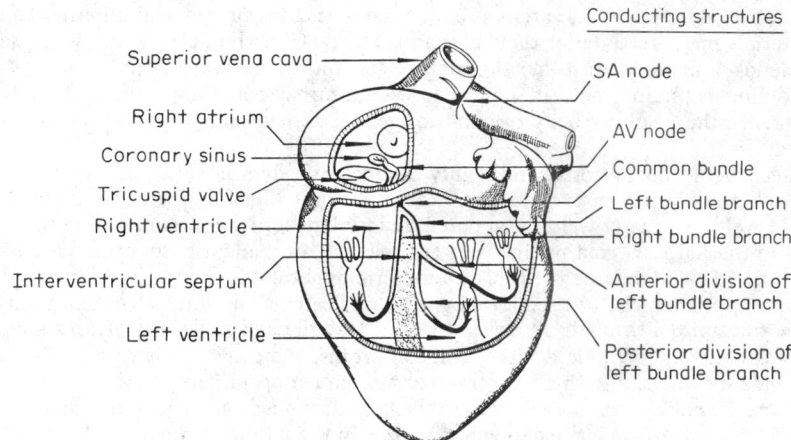

Fig. 26-5. Cardiac muscle: structure and function.

Figure 26-6 shows a typical ventricular action potential. The *upstroke* corresponds to activation *depolarization*, and during this period the cell constitutes a source of emf representable as a dipole. During the plateau (phase 2 in Fig. 26-6) the cell is quiescent (as an electrical source). Because recovery (phase 3) proceeds at a relatively slow rate, the cell generates a relatively weak emf during the prolonged period.

In addition to muscle cells of the type described, specialized *pacemaker cells* (P cells) exist in the SA nodal region of the heart and in the AV node (see Figs. 26-5 and 26-6). While the resting potential can be maintained indefinitely in ordinary muscle cells, for the pacemaker cells the transmembrane potential spontaneously increases until the threshold for excitation is reached, giving rise to an action potential, as previously described. Thus pacemaker cells are self-excitatory. Their oscillations communicate to neighboring cells by local circuit currents, and in this way cyclic excitation is established in all portions of the heart.

An additional cell type which forms "conductive" tissue (where conduction velocity is from 2 to 10 times that of ordinary muscle) is known as the *Purkinje cell*. These cells are found in specific bundles (tracts) which exist between the SA and AV node (internodal tracts), through the AV node (common bundle, bundle of His), and from the ventricular part of the AV node to the inner walls of the ventricles. The latter constitutes the His-Purkinje system (Purkinje network).

As noted the SA node causes a cyclical phenomenon, the frequency being the heart rate. In disease states when the SA node is nonfunctional, the AV nodal pacemakers may take over. The heart rate then will be lower, since ordinarily to drive the AV node the SA

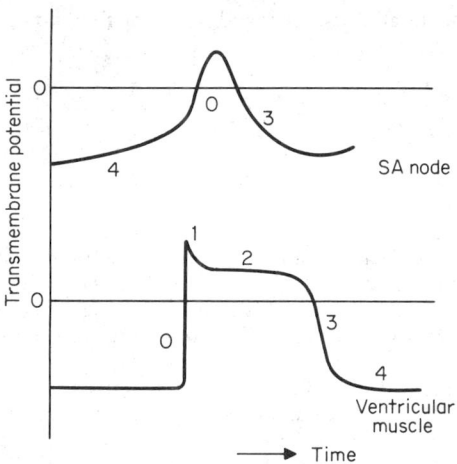

Fig. 26-6. Transmembrane action potentials. For ventricular muscle, phase 0 is the upstroke due to activation of the cellular membrane. Phase 2 is a prolonged period of little potential change (plateau). Phase 3 is that of rapid recovery, and phase 4 is the stable resting condition. In contrast, the pacemaker's phase 4 shows a progressive rise in potential until, at the break, an action potential ensues. There is no phase 1 or 2 in this case.

node must have a higher natural frequency. Ordinary heart muscle, under certain conditions, can depolarize spontaneously and become a pacemaker (ectopic focus), leading to a premature ventricular contraction (PVC).

The activation of all cells in the heart, initiated by the SA node, forms the basis for the ECG. As noted, cells undergoing activation (depolarization) act as dipole current sources. The resulting local currents not only contribute to the spread of activity but also result in the presence of currents everywhere in the torso (which behaves like a passive resistive medium). During recovery, each cell again behaves like an emf source, though the magnitude is much less and the duration much longer than during activation.

6. The Electrocardiogram. If the electric potential measured between the right and left arm is amplified, the ECG signal will have the appearance shown in Fig 26-7. Each of the three wave complexes, P, QRS, and T, corresponds to a particular electrophysiological event. Initiation of the electric cycle takes place by the pacemaker cells of the SA node. Their spontaneous "firing" is communicated by conduction throughout the atrial tissue. Thus, the first major event is depolarization of the right and left atria, and this collective electrical activity gives rise to the P wave.

Activity is prevented from spreading directly from the atria to the ventricles by the presence of a ring of nonconducting fibrous tissues separating these two regions except in the vicinity of the AV node. At the AV node, a pathway for conduction exists; however, a very low velocity is characteristic of this region. Thus, initiation of activity in the ventricles is delayed by the time required for activity to propagate first from the SA to AV node and then through the common bundle of the AV node. The interval gives rise to the designated P-R interval, which is a measure of the composite delay.

Once the ventricular conduction (Purkinje) system is activated, rapid spread occurs and general ventricular activity is initiated. This gives rise to the electrocardiographic QRS complex. The QRS waveform reflects the composite depolarization of ventricles, and its duration measures the total time required. As noted in Fig. 26-7, the Q wave is the first downward deflection before any upward deflection, the R wave is the first upward deflection, and the S wave is the first downward deflection after an upward deflection. One or more parts may be missing.

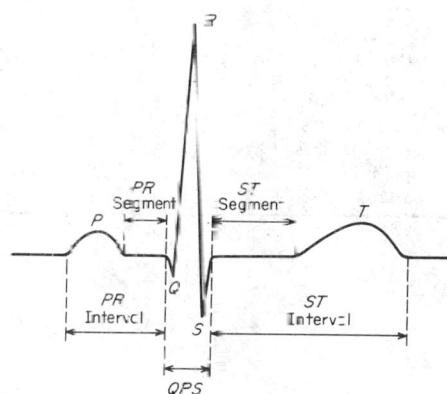

Fig. 26-7. Standard scalar electrocardiogram.

Recovery is also a process that generates a potential field, generally of lower magnitude. Ventricular recovery produces the T wave. The S-T segment is ordinarily at the base line as determined by successive TP segments; however, in the case of cell injury this may be elevated or depressed. Recovery in the atrial tissue also results in an electrocardiographic wave (TP); it is of low magnitude, however, and ordinarily masked by the QRS, with which it overlaps.

Normal values of electrocardiographic intervals, segments, and waveform durations are given in Table 26-2.

In recording the ECG, a standard-time and potential-amplitude base are used. Corresponding to a paper speed of 25 mm/s, major 5-mm divisions represent intervals of 0.2 s. Amplification is adjusted ordinarily to 0.1 mV/mm. Peak ECG signals are generally under 5 mV.

7. Mechanical Activity of the Heart. The function of the heart is to pump blood. Electrical activity is important only because it initiates muscular contraction. Abnormal conduction pathways or heart rates that are too high or low can cause reduced cardiac output and may produce clinical symptoms. Conversely, diseases affecting the strength of muscular contraction may be reflected in abnormalities in the ECG. Examples of the former are conduction defects and atrial flutter or tachycardia, while the latter may arise from loss of blood flow to portions of the heart, resulting in areas of tissue necrosis and subsequent replacement with fibrous (noncontractile) tissue (myocardial infarction).

Under normal conditions, the correlation between electrical and mechanical events is as depicted in Fig. 26-8. Mechanical and circulatory values connected with the heart's behavior as a pump are given in Table 26-3.

TABLE 26-2   Normal Electrocardiographic Values for Adults*

P-R interval	0.12–0.20 s	P duration	Lead II ≤ 0.11 s
QRS interval	0.07–0.10 s	R amplitude	V_1–V_6 ≤ 2.7 mV
P amplitude	Lead II ≤ 0.25 mV	T amplitude	Lead II ≤ 0.8 mV

Heart rate	Q-T interval, s	S-T segment, s
60	0.33–0.43	0.14–0.16
70	0.31–0.41	0.13–0.15
80	0.29–0.38	0.12–0.14
90	0.28–0.36	0.11–0.13
100	0.27–0.35	0.10–0.11

*See Fig. 26-9 for definition of leads.

TABLE 26-3   Normal Heart Parameters

Dimensions, cm:		Blood flow, L/min	5.5
Left ventricular wall thickness	1.2	Heart weight, g	250–390
Mean heart height	9.7	Volume, mL:	
Mean heart width	10.7	Stroke (resting)	75
Duration, s:		Left ventricular end diastolic	135
Atrial systole	0.11	Left atrial (max.)	60
Atrial diastole	0.72	Heart	700–750
Isometric contraction	0.06	Heart rate (resting), min^{-1}	69
Isometric relaxation	0.05	Pressure, mmHg:	
Ventricular systole	0.27	Peak aortic	120
Ventricular diastole	0.56	Peak pulmonary	25
Rapid inflow	0.16	End diastolic, aorta	80
Diastasis	0.23	Pulmonary aorta	10

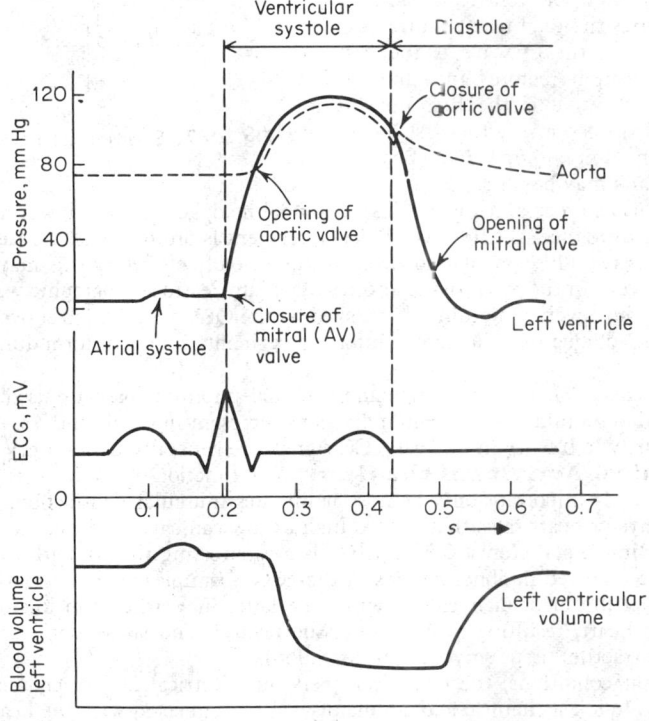

Fig. 26-8. Diagram of hemodynamic and ECG events in the cardiac cycle.

8. Electrocardiographic Lead Systems. Figure 26-9 shows the placement of electrodes on the body for standard clinical ECG leads. Voltages are recorded between selected pairs of electrodes, for example,

$$V_I = \text{(potential at left arm)} - \text{(potential at right arm)}$$
$$V_{II} = \text{(potential at left leg)} - \text{(potential at right arm)}$$
$$V_{III} = \text{(potential at left leg)} - \text{(potential at left arm)}$$

These leads (electrode connections) are referred to as the extremity or limb, leads. Note that because of Kirchhoff's voltage law, $V_I + V_{III} = V_{II}$, only two are independent.

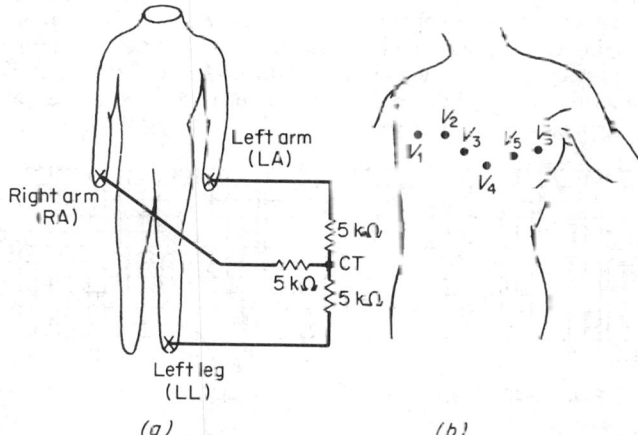

Fig. 26-9. Location of leads for 12-lead electrocardiogram (further details in Table 26-5); CT = Wilson central terminal: (a) limb leads, (b) precordial leads.

A synthetic lead used as reference is formed by connecting the limbs through 5-kΩ resistors to a common point called the *Wilson central terminal* (CT). Precordial leads (V_1 to V_6) are voltages between designated chest electrodes (see Fig. 26-9) and the CT. Augmented leads result from measuring each limb lead against the central terminal, except that the respective limb contribution to the CT is disconnected. (This results in augmentation of the signal by a factor of 1.5.)

The resultant 12 leads are conventionally used in clinical electrocardiography. A summary of the connections is given in Table 26-4. Figure 26-10 shows a typical normal record from each of the 12 leads.

TABLE 26-4   Standard Electrocardiographic Leads

	Standard or limb leads	
$V_I = \Phi(LA) - \Phi(RA)$	$V_{II} = \Phi(LL) - \Phi(RA)$	$V_{III} = \Phi(LL) - \Phi(LA)$
where $\Phi(LA)$ = potential of left arm	$\Phi(RA)$ = potential of right arm	$\Phi(LL)$ = potential of left leg

Precordial leads

Wilson central terminal (CT) is the junction of three 5-kΩ resistances, each connected to a limb lead; precordial leads use CT as reference; precordial leads are located as follows:
- V_1 Fourth right intercostal space at sternal edge
- V_2 Fourth left intercostal space at sternal edge
- V_4 Fifth left intercostal space at the mid-clavicular line
- V_3 Midway between V_2 and V_4
- V_5 Same level as V_4 at anterior axillary line
- V_6 Same level as V_4 at midaxillary line

Augmented limb leads

- aV_L Left arm with respect to junction of 5-kΩ resistors, one to RA, the other to LL
- aV_R Right arm with respect to junction of 5-kΩ resistors, one to LA, the other to LL
- aV_F Left leg with respect to junction of 5-kΩ resistors, one to RA, the other to LA

To the extent that the heart behaves like a dipole, all 12 leads can be synthesized from the dot product of the heart dipole (vector) and a vector reflecting the geometry of the lead (lead vector). In fact, the locus of the heart vector (vector loop) is the object of specialized corrected orthogonal lead systems of *vectorcardiography*. The dipole nature of the heart, as an electric generator, appears to be a good approximation. Current research has shown that significant non-dipolar behavior is exhibited during portions of the QRS. Practical measurement and clinical utilization of higher-order (multipole) terms are still under investigation.

9. The ECG Signal. Typical signals in electrocardiography are shown in Fig. 26-10 for the 12-lead ECG. Under normal conditions the waveform is periodic at the heart rate (normally around 70 beats per minute). A single cycle is on the order of 0.85 s, and the QRS duration is around 0.09 s, while the R-wave spike might cover an interval of 0.03 s. Signal theory suggests a high-frequency content up to perhaps 60 Hz, due mainly to the QRS complex. The low-frequency content might be estimated from the periodicity itself, which is approximately 0.8 Hz.

For diagnostic purposes the time intervals and segments, as well as amplitudes, are used. In addition, the waveform of the P, QRS, and T waves can be important. These waveforms may be abnormal if an ectopic pacemaker initiates a cardiac cycle, i.e., a PVC, because the pathway of

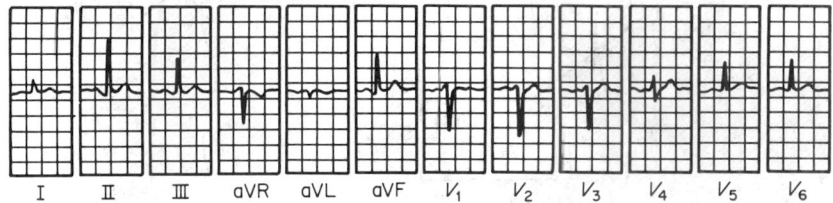

| I | II | III | aVR | aVL | aVF | V_1 | V_2 | V_3 | V_4 | V_5 | V_6 |

Fig. 26-10. Twelve-lead electrocardiogram from normal heart.

activation is abnormal; hence the temporal development of cardiac sources and their potential fields are quite different.

Several successful approaches for computer analysis and diagnosis of electrocardiograms have been developed; algorithms are based on waveform recognition and both amplitude and time intervals of the component waves.

10. ECG Instrumentation. The essential part of a typical ECG machine is a high-input-impedance differential amplifier. The amplifier input may be ac-coupled with a dc voltage balance control to balance off the difference in dc base-line voltages of each electrode.

One of the most persistent problems in ECG measurements is the reduction or elimination of ac interference. If the body is considered as a conducting sphere of 0.5 m radius, its capacitance to infinity is given by

$$C = 4\pi\epsilon_0(0.5) = 55 \text{ pF}$$

The capacitance to surrounding power-line sources varies with proximity but is approximated at 100 pF. At the power-line frequency of 60 Hz, this constitutes an impedance of 26 MΩ. If the right leg is connected to the common ground of the amplifier through an electrode with contact resistance of, say, 2 kΩ, a power-line potential of 120 V will be reduced by the ratio of 2,000/(26 × 10^6) to a value of around 10 mV as a noise input. This is a substantial signal, in excess of the ECG itself. However, by using differential amplifiers, this signal will be mostly eliminated, because normally each lead develops essentially the same ac signal.

A second major problem in ECG recording is that of *base-line drift*, which may arise from physiological tension or poorly connected electrodes. The patient should be lying down when recording ECGs, and freedom from noise or other distractions should be ensured. Such conditions also reduce interference from muscle noise. In spite of this care, an *RC* network is necessary to block residual direct current. A time constant of 3 s corresponds to a low-frequency half-power point of 1/20 Hz, which appears low enough to retain normal waveshape. Another solution to this problem is to use a *base-line clamper*.

A third important problem is electrical safety during continuous monitoring in the hospital.

Inadequate high-frequency response can result in waveform distortion. In particular, the amplitudes and durations of a pertinent wave can be in error. A high-frequency response between 60 and 500 Hz is generally accepted as the desirable range. Studies on characteristic ECGs have shown that amplitude errors in Q, R, and S of less than 0.1 mV require cutoff frequencies above 100 Hz. (Reduction of amplitude errors below 0.05 mV would require an even higher cutoff frequency of perhaps 200 Hz.) Timing errors are less affected and appear satisfactory at cutoffs of 50 to 80 Hz.

The current Recommendations for Standardization of Leads and of Specifications for Instruments in Electrocardiography and Vectorcardiography by the American Heart Association are available. Table 26-5 summarizes some salient features.

TABLE 26-5   Direct-Writing Electrocardiographs

Input impedance between any two electrodes	5 MΩ for frequencies up to 60 Hz
Maximum leakage current to patient	10 μA
Frequency response:	
0.14–25 Hz	Flat to ±0.5 dB
0.05–0.14 Hz	<3 dB loss in response at 0.05 Hz (from 0.14 Hz); requires time constant >3.2 s
25–100 Hz	For 5 mm peak-to-peak output at 25 Hz, <3 dB decrease in strength up to 100 Hz
Noise level	At 10 mm/mV <0.1 mm rms (subject simulated with 25 kΩ between leads)
Electrode impedance	Skin-to-electrode impedance (with appropriate preparation and use of paste or jelly) <50 kΩ at 10 Hz for current <10 μA
System performance	<5% deviation from linearity for peak-to-peak amplitudes under 5 mm deviation not to exceed 0.25 mm (input signal frequency components 0.05–100 Hz)

SOURCE: Committee on Electrocardiography of the American Heart Association, Selected Summary of Recommendations for Standardization of Leads and Specifications for Instruments in Electrocardiography and Vectorcardiography, *Circulation*, 1975, Vol. 52, pp. 11-31. At this writing AAMI/ANSI Standards for Electrocardiographic devices are being developed.

11. Other Electrophysiological Systems. In addition to cardiac muscle, striated (skeletal) muscle, smooth muscle, and nerve are also capable of undergoing electrical activity. As in the heart, an action potential can be elicited from each such excitable cell. During the time-varying phase the cell acts as a dipole source of electricity, and surface or implanted electrodes record the collective contribution to the electric potential field from all such sources (gross activity). Since these measurements reflect the electrophysiological character of the responsible cells, diagnostic information can be deduced concerning them.

Electroencephalography (EEG) is concerned with the measurement of electric signals on the scalp which arise from the underlying neural activity in the brain (including synaptic sources). Such signals are therefore a reflection of the state of the central nervous system and prove useful in the diagnosis of epilepsy, brain tumors, and other disease states. Table 26-6 lists a number of physiological systems in which muscle or nerve generates a waveform that has been identified and characterized. The problem of noise is more critical in electroencephalography since the desired signal is of the order of 50 μV (Fig. 26-11). Bandwidth requirements for electromyography range from 20 to 500 Hz, while EEG requires direct current to 150 Hz. Not all signals are available at the body surface, and some type of deep electrode is necessary. Electrodes specially adapted to each system under study have been developed which provide good contact, correct geometrical location, and low impedance (see Table 26-6).

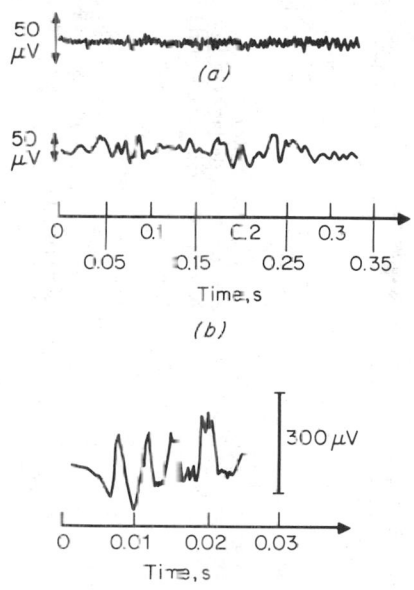

Fig. 26-11. Biopotential waveforms: (a) EEG under excited conditions; (b) EEG under sleep conditions; (c) EMG wave.

TABLE 26-6 Electrode Characteristics

Type of bioelectric phenomena	Type and placement of electrodes	Biological source	Physical characteristics of signal	Magnetic induction amplitude, pT
Electroencephalography	Small metal disk, ~7 mm (preferably Ag–AgCl) placed on scalp; needle electrodes may be inserted into subcutaneous tissue [array of 16 electrodes over top of skull roughly at intersection of four longitudinal and four transverse columns (bipolar electrodes)]	Gross electrical activity of brain nerve cells	Amplitude under 100 μV, average 10–50 μV; frequency range dc to 150 Hz (see text)	1
Electrocardiography	Metal plate–silver electrodes 3.5 by 5 cm placed on extremities, concave suction-cup type for chest (4.75 cm diam)	Gross electrical activity of heart	Amplitude under 10 mV, average 1–5 mV; frequency range 0.14–100 Hz (see text)	50
Electromyography	Surface electrodes superficial to muscle; Ag–AgCl type; needle electrodes, concentric steel outer sheath 0.65 mm, platinum-wire core 0.04 mm diam beveled (bipolar or unipolar; use wire core only); inserted into muscle bundle.	Extracellular recording of one or more motor units (groups of muscle fibers) due to their action potentials arising from muscle contraction	Amplitude under 3 mV, usually 0.1–1.0 mV; frequency range 40–3,000 Hz (see text)	2
Electroretinography	One electrode consists of chlorided silver wire contacting the physiological saline solution held by contact lens placed on cornea; indifferent electrode is plate-type (Ag–AgCl) on forehead (use skin preparation and electrode paste)	Summation of receptor potentials generated by light-sensitive cells (rods and cones) in retina due to light stimulation	Amplitude under 1 mV; frequency range dc to 20 Hz	0.2
Electrooculography	Electrode pairs equidistant from pupil above and below and/or laterally and medially (at outer and inner canthi); electrodes small (10-mm diam), plate silver (preferably chlorided silver) for surface contact (use skin preparation and electrode paste)	Eyebulb in orbit establishes dc current dipole of galvanic type	Amplitude 0.05–3.5 mV (0.2–0.4 mV/ 20° rotation); frequency range dc to 125 Hz	10

Biomagnetism. The electric currents generated by bioelectric sources are themselves sources of a magnetic field. The study of such magnetic fields of biological origin is known as *biomagnetism.* Since the magnetic field from even the strongest of such sources (the heart) is many times smaller than the normal ambient magnetic field, practical measurement systems became possible only with recent technological developments. Based on units of magnetic induction in teslas (1 T = 10^4 G), typical biological field magnitudes are in the picotesla (1 pT = 10^{-12} T) and femtotesla (10^{-15} T) range. Typical values for several basic biological systems are given in Table 26-6.

The detection of such weak magnetic fields ordinarily requires extensive magnetic shielding from external sources of noise and/or the use of differential detection and signal averaging. The most successful detector is the SQUID (superconducting quantum interference device), which has a noise level of 10 fT/$Hz^{1/2}$. This compares advantageously with the fluxgate magnetometer, whose rms noise level is 20 pT/$Hz^{1/2}$. Because of the need for liquid helium, the SQUID presents greater measurement difficulties than electrical measurements. On the other hand, this device opens a number of advantages including dc measurements that are free from contact artifact and a different weighting of the sources, in contrast to electrical detection, which may provide new aspects of the source.

Therapeutic and Diagnostic Radiology

BY FLORO MIRALDI

12. Production of X-Rays. Radiology consists of several major divisions from which, for our purposes, we shall select diagnostic radiology, therapeutic radiology, and special radiologic imaging systems. The main function of the diagnostic radiologist is to produce and interpret shadow images of internal organs of the body using x-radiation. Therapeutic radiology concerns itself primarily with the treatment of disease by the destruction of tissue with high-energy radiation. Special imaging radiology includes nuclear techniques, computed axial tomography (CAT), and ultrasound. Computed tomography and ultrasound are purely diagnostic modalities. Nuclear medicine has both diagnostic and therapeutic aspects but differs from the others in that the radiation is usually internally distributed in the body and arises from the decay of radioactive materials which have been ingested or injected.

In *diagnostic radiology* and for superficial therapy purposes, the energy spectrum of radiation varies from about 10 to 100 keV. In *therapeutic radiology* and in *nuclear medicine*, the energies of interest range from about 100 to 10,000 keV. Corresponding wavelengths range from 0.1 to 0.0001 nm and constitute the x-ray region of the electromagnetic spectrum.

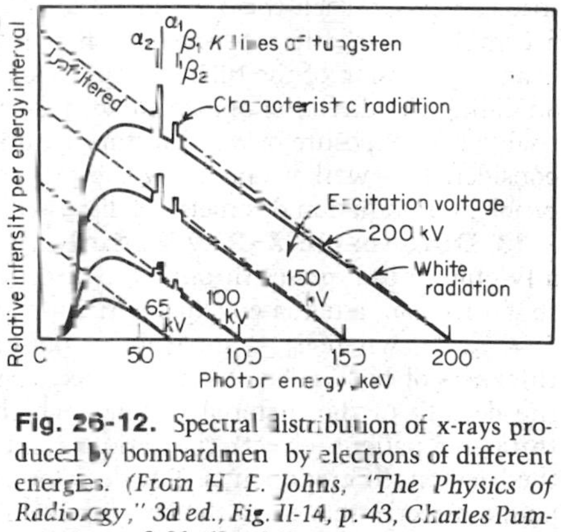

Production of x-rays is accomplished by bombarding material with high-speed electrons, using

Fig. 26-12. Spectral distribution of x-rays produced by bombardment by electrons of different energies. *(From H. E. Johns, "The Physics of Radiology," 3d ed., Fig. II-14, p. 43, Charles Pummis. Springfield, Ill., 1969.)*

the principle that accelerated charged particles radiate.[33,34] Figure 26-12 shows spectral distributions of x-rays produced by electron bombardment at different energies. The continuous distribution which results is called *bremsstrahlung* or *white radiation.* The superimposed peaks on the white radiation are called *characteristic radiation* because they are characteristic of the atoms in the target. Also shown in Fig. 26-12 are characteristic radiations of tungsten, the most common target material. Even in the most favorable cases the useful radiation output is only a few percent of the total electron energy; i.e., most of the energy is dissipated in the target as collisional energy or heat.

Characteristic radiation occurs when electrons from higher-atomic-energy orbits go into lower-energy orbits. This occurs when the lower-energy electrons in an atom are ejected by a collisional process.[35] Hence *k* radiation is produced when the *k* electron is ejected from the atom; *l* radiation when the *l* electron is ejected from the atom; etc. The principal emission lines for tungsten are shown in Table 26-7.

The production of x-rays is straightforward, but the practical requirements of constructing instruments to deliver x-rays of specific quality and quantity are quite severe. A conventional x-ray tube consists of an anode-and-cathode assembly placed in an evacuated glass envelope. The anode is usually a massive piece of copper in which a small tungsten target is placed. The cathode assembly usually consists of a filament of tungsten wire in a shallow focusing cup. The hot tungsten filament provides the source of electrons, which are accelerated toward the anode by applying a high voltage between the anode and cathode. Although such tubes are usually excited by an alternating voltage, the electron currents through the tube can flow only when the anode is positive with respect to the cathode. Such tubes are *self-rectifying*, and the current through the tube consists of a half wave.

Under severe electron bombardment, the tungsten target of the cathode may reach a temperature at which emission from its surface begins. On the half cycle in which the filament is positive, electrons can then travel from anode to filament. If conduction occurs during the inverse cycle, the filament will be destroyed by the electron bombardment. To avoid this, reverse electron flow is prevented by rectifiers.

Because of the low efficiency of conversion of the electron energy into radiation energy, the heat load of the anode becomes severe in high-current machines. Since most diagnostic machines

TABLE 26-7.   Principal Emission Lines for Tungsten

Transition	Symbol	Energy, keV
$N_{II}N_{III}$–K	$K\beta_2$	69.089
M_{III}–K	$K\beta_1$	67.236
L_{III}–K	$K\alpha_1$	59.310
L_{II}–K	$K\alpha_2$	57.972

Source: H. E. Johns, "The Physics of Radiology," 3d ed., Table II-2, p. 51, Charles Pummis, Springfield, Ill., 1969.

operate with tube currents of the order of hundreds of milliamperes, at 10^4 to 10^6 V, the power load can easily reach the order of 10^4 to 10^5 W. Thus, cooling of the target is required, which is accomplished in different ways, such as rotating the anode to expose different surfaces to the beam, immersion of the tube in liquid coolant, requiring intermittent use of the machine, etc.[36] Shielding and direction of a beam with the use of diaphragms and ports are required to obtain a usable film exposure or a satisfactory treatment.[37] Problems of radiation tissue damage must be considered, as well as delivering the proper amounts of radiation where they are needed. The problem of radiation dosimetry is discussed by Hine and Brownell.[38]

13. Diagnostic X-Ray Radiology. In this application a beam of x-rays, which is usually quite broad, passes through the object and impinges on a detector. Use is made of the fact that x-rays are attenuated exponentially in passing through a medium according to the formula $I = I_0 e^{-\mu x}$, where I_0 = intensity of impinging beam, μ = attenuation coefficient, and x = thickness of body. The attenuation coefficient is a strong function of both atomic number and the density of the material through which the beam is passing. Thus, when a beam passes through a patient (object), the various components of the body yield a different attenuation for the beam at different points. The detector then receives varying amounts of radiation at various points directly related to the varying attenuation of the body. In this way a two-dimensional representation of the attenuation to the body is obtained.

Photographic film is by far the most common detector for a number of reasons. It produces an image or picture which is easily viewed and studied, it provides a permanent record, and it is relatively sensitive and has high resolution capabilities. Since in modern x-ray facilities, resolution capabilities are of the order of 0.1 mm, the quality of the images obtained can be very high.

The main disadvantage of this system is the production of a two-dimensional representation of a three-dimensional body. Thus images contain a large amount of ordered noise. The other disadvantage is that because the attenuation coefficients of the various soft tissues of the body are very nearly the same, it is very difficult to obtain an image which differentiates one soft-tissue organ from another.

Both these disadvantages have been overcome by *tomography*. In conventional tomography the x-ray source (x-ray tube) and detector (film) are moved in opposite directions along a preassigned path, which may be linear, circular, ellipsoidal, or any other, as determined by the manufacturer. Because of the particular geometry and speed of the relative source and detector, only one plane in the body will be in focus at all times; i.e., structures in a thin slice centered about a particular plane will project onto the film in the same relative positions throughout the entire examination. All other planes above and below will project in different positions of the film at different times through the motion and will therefore be blurred.

The end result is an image in which the structures in the focused plane are seen with sharp boundaries, whereas structures in all other planes are blurred on the film. This yields a smearing of the ordered noise into a more uniform noise background for the image, except for an overlying sharper image of the structures in the focused plane.

This form of tomography (looking at a thin slice of the body) is a widely used technique. It solves to a degree the disadvantage noted above. It does not solve (or address itself to) the problem of small variations in attenuation coefficient of soft-tissue structures in the body. This latter problem is handled more effectively by computerized tomography (Par. 26-14).

The second most widely used detector is fluorescent screens. They can be viewed directly, and the radiologist can observe the object being visualized by watching the image on a fluorescent screen. The efficiency of the screens is good, but the light output is quite low and requires dark-room adaptation of the eyes to see structures adequately. In modern machinery, the fluorescent screens are attached to image amplifying systems, which then allow the signal strength to be boosted so that the images can be displayed on conventional television screens. In this way the dark adaptation is no longer necessary and the visual signal strength can be improved. Image amplifier systems also allow brightness control and contrast control. Direct visualization of the fluorescent screens is rarely used in modern installations today. Image amplification systems are discussed elsewhere in this Handbook (see Sec. 11).

14. Computerized Tomography. Computerized tomography is a modification of the radiographic technique described above. In this procedure a very narrow x-ray beam is passed through the body in a plane perpendicular to the axis. The total attenuation of the beam is then measured by a detector. This measurement is repeated thousands of times with different rays across the body in the same plane. The result is thousands of attenuation measurements across a narrow slice of the body in a plane perpendicular to the axis.

Use is made of the fact that the attenuation of a narrow beam along a path can be expressed as a product of the attenuation through each segment of that path. Since the attenuation is exponential, the total attenuation argument μx can be expressed as the sum of the individual arguments for each segment:

$$\mu x = \mu_1 x_1 + \mu_2 x_2 + \mu_3 x_3 + \cdots + \mu_n x_n$$

With this concept one envisions the actual slice in question as being composed of a number of small volume elements, called voxols. Each measurement of the attenuation of the narrow beam across the slice is thus a measurement of the attenuation of those voxols in the path of that measurement. Accordingly, it is possible to set up a set of linear equations for the solution to the attenuation in each voxol element.

To obtain reasonable resolution it is necessary to have small voxol elements, which means a large grid. Present grid sizes are of the order of a 300×300 matrix, and the number of measurements presently obtained is of the order of several hundred thousand. The efficient solution to such a large set of equations requires a relatively sophisticated software and relatively large computer systems for rapid calculations.

The final result is a calculation of the attenuation coefficient for each voxol element. The attenuation coefficients are then ranked in magnitude according to one of several schemes which uses a normalization of zero for water and some other numbers, such as 1000, for a dense material such as calcium or aluminum. A gray scale is then attached to this range of numbers, and the computed attenuation coefficients are displayed on a screen in terms of the gray-scale ranking. In this way a gray-scale image of the attenuation coefficients throughout this slice of the body is obtained. Because of the very fine grid and high degree of accuracy attained in this determination of the attenuation coefficients, the small differences in tissue characteristics throughout the body are quite well defined. The images so produced thus tend to have a very close resemblance to the anatomy of the body across such an axial slice.

This technique has received overwhelming success because the images bear such a close resemblance to the visualized anatomy. The ability to detect lesions or variations from a normal anatomy has also been phenomenal and makes this modality very useful.

This particular technique eliminates the disadvantages described above for standard x-ray radiography, but has its own disadvantages, primarily that of viewing numerous axial slices rather than visualizing an entire region easily. The technique has also introduced a higher radiation dose to the patient, but this problem is rapidly diminishing as more efficient techniques are developed.

There are many variations on a theme in this technique. The first consists of motion of the x-ray source and detector in a linear fashion across the body, followed by rotation of the entire unit and a repeat linear sweep. This is reiterated many times through many angles of rotation. Other techniques rotate the source and detectors about the body, and still others rotate the source with stationary detectors placed in a circumference about the body. All the techniques have some advantages and some disadvantages.

Detectors used range from ionization chambers with high-density gas, such as xenon, to scintillation crystals, to solid-state devices. Again, there are disadvantages and advantages to all the various systems. At present commercially available units use various combinations of technique and detectors to achieve relatively similar end results.

15. Radioisotope Imaging. In the area of nuclear medicine the most important procedure is that of organ imaging. The patient ingests a gamma-emitting radioisotope in appropriate pharmaceutical form which allows the concentration of the isotope in a particular organ system (or conversely, the exclusion of an isotope from a particular organ). Pathological lesions, variations in blood perfusion, or anatomical variations will produce an abnormal accumulation or exclusion of the isotope in particular organs. Accordingly, mapping the distribution of the isotope via detection of the emitted gamma rays yields an image which provides clues of the presence of pathological lesions.

Mapping the time distribution usually requires following the activity over one or several selected areas. The devices for this are the simplest types of radiation detector systems and generally use scintillation crystals because they are the most efficient for gamma detection. The basic configuration of such a probe consists of a collimator to limit the field of view, followed by a scintillator, which in turn is followed by a photomultiplier with associated electronics for recording. Output curves are handled in many ways, such as oscilloscope display, use of scalers, count-rate meters, or strip-chart recorders.

Spatial mapping, or *imaging,* of an organ is somewhat more complex. At present two basic approaches are used, rectilinear scanning and scintillation-camera scanning.

16. Rectilinear Scanning. The rectilinear scanner is primarily a scintillation probe which is moved over the region of interest in a linear motion. The system is a point-by-point detection method which places a dot on film or paper for each incident recorded at a particular point. Most rectilinear scanners use a focusing collimator to limit the field of view without undue sacrifice of sensitivity.

The most significant disadvantage of the rectilinear system is that, for high resolution, the sensitivity is small and scanning time is long. The low sensitivity is partly the result of observing only a small region at a time. Depending on the organs scanned, times required may range from 15 min to several hours. A second major disadvantage of these systems is that they tend to be tomographic. Obviously, in some cases the tomographic feature can be used to advantage.

The sensitivity of these instruments, measured as the number of data points displayed per gamma ray emitted in the body during the scan, is typically of the order of 10^{-4} to 10^{-6}. Increasing the amount of radioisotopes to increase the speed with which a scan may be performed has its limitations because of radiation dose delivered. Consequently, for better sensitivity and better scans, other approaches are needed.

17. Scintillation Cameras. The scintillation camera is an attempt to bypass the problem of low sensitivity. In this concept, a large field is viewed simultaneously. Obviously, if 10 detectors were placed in parallel, the speed would increase by an order of magnitude. A multiplicity of detectors of the rectilinear scanner type ganged to run in parallel paths has been proposed, and such an instrument is available commercially.

A second approach which employs a matrix of scintillation crystals covering the entire field of interest, each with its own collimator system, has been suggested and developed. When systems like this do not involve motion over the body, the system is called a camera, e.g., the autofluoroscope of Bender and Blau. It suffers the disadvantage that, to cover a reasonably sized field, the number of crystals and photomultipliers required is large, resulting in a rather complex system.

The most promising camera system to date, and also the most widely used, is the scintillation-camera system of Anger[38a,38b]. This instrument consists of a large scintillation crystal behind a collimator and backed by an array of photomultipliers. At present the number of photomultipliers ranges from approximately 19 to approximately 100, depending upon the manufacturer.

Using a passive network to triangulate on the position of a scintillation as viewed by several photomultipliers, this system is capable of resolving scintillations well enough for most clinical applications. It is many times faster and more sensitive than the rectilinear scanners because it exposes much more crystal to the field of view than the other systems. Its sensitivity appears to be approximately 4 to 7 times faster than the rectilinear scanner for comparable resolution. It suffers the disadvantage that its resolution becomes poor for gamma rays below about 100 keV, and at high energies the sensitivity drops rapidly. This system is of great interest at present because many pharmaceuticals can be tagged with Tc-99m, which emits a gamma ray of 140 kV.

18. Positron Scanning Devices. The elements involved in most biological materials, hydrogen, carbon, nitrogen, and oxygen, would appear to be the best tags if one wished to trace various chemicals or natural biological materials in the body. Unfortunately for nuclear-imaging methods, isotopes of these elements with gamma emissions or reasonable half-lives do not exist. Carbon, nitrogen, and oxygen, however, do have isotopes that are positron emitters with short but usable half-lives. In particular, carbon 11, nitrogen 13, and oxygen 15 fall into this category. Compounds containing these elements can be made with the normal or stable element replaced by such positron emitters. One then has a natural compound with a detectable tag and a usable half-life. The radiation detected in the positron emitters is the annihilation gammas which occur when the positron unites with an electron to produce two 0.511-meV gamma rays traveling at 180° to each other. If one uses an array of gamma detectors, such as scintillation counters, around a body with appropriate coincidence-counting techniques, one can determine the line along which the two gammas travel and thus the line containing the source of the annihilation.

A number of such devices have been constructed, but the most successful to date has been the axial-tomographic device described by Ter-Pogossian[9a]. It contains a ring of detectors which surrounds the body and determines, by the coincidence method, the paths of multiple disintegrations. An image is constructed by techniques similar in concept to the method of computer tomography. Because of the similarity to the computed-axial-tomography system, such a device is called an E-CAT device, which stands for emission computed axial tomography. The beauty of this system is its ability to use natural compounds that are not altered by attaching a tag to them as is the case with most compounds used in nuclear radiology. Thus the configuration of the molecule is not changed. In essence this technique allows one to make in vivo physiological tests not heretofore possible.

The major disadvantage is the relative short half-life of these isotopes (20.4 min for carbon 11, 9.9 min for nitrogen 13, and 118 s for oxygen 15). Such short half-lives require that the isotopes essentially be produced at the site of use and that chemical manipulations to substitute the isotope into the appropriate compound be rapid. The entire procedure from production to administration to completion of the examination must be accomplished in a matter of minutes to a few hours at most.

This requirement forces the production facility to be associated with the hospital and scanning device and is not an inexpensive venture. In particular, one requires a high-energy accelerator for the production of the isotopes, usually a cyclotron. At present small commercially available cyclotrons are being produced for this purpose, demonstrating that even such restrictions do not prevent the use of an interesting, viable technique.

19. Other Devices for Radiation Mapping. Interest has also developed in the multiple-wire proportional counter, in which the anode of a conventional gas-proportional counter is replaced by a grid. By blanking the wires of the grid in timed sequence the position of an interaction in the photomultiplier tube can be determined. In this manner, a position-dependent proportional-counter system is obtained. The resolution is extrememly high and can be of the order of millimeters. Sensitivity is also high for very-low-energy photons but becomes completely inadequate at energies above about 30 kV.

Another system which has been developed uses an image-amplifier tube situated behind a large scintillation crystal or an array of scintillation crystals. The potential resolution of these systems is very high, but they also suffer sensitivity problems inherent in the image-amplifier portion of the system.

A recent innovation is a hybrid scanner which combines both rectilinear and camera concepts to produce a line detector. Using a bar of scintillation crystal, it has been shown that the light output from the ends of the bar falls off exponentially with distance from the end. A photomultiplier situated at the end of the bar, followed by logarithmic amplifiers, then has an output which is directly proportional to position measured from the end of the bar. The bar detector is moved across the field of view, and the position of a scintillation is determined by noting the position of the bar and position along the bar where the scintillation occurred.

Although all these devices have been demonstrated and their technical potential proved, they have not been widely accepted, probably because their advantages are not overwhelming compared with those of instruments presently available and highly developed both technically and commercially.

20. Ultrasound. Ultrasound systems have recently been adapted to medical problems for the visualization of boundaries and different tissue densities, particularly in the abdomen. The general principles are those of sonar (see Sec. **25**). In essence, medical ultrasound is an adaptation to a particular use of the general principles developed in the sonar field.

Artificial Organs

BY PAUL S. MALCHESKY AND RAYMOND J. KIRALY

21. Kidney Substitutes. The kidneys' function is to maintain the chemical and water balance of the body by removing waste materials from the blood. The kidneys do this by sophisticated mechanisms of filtration and active and passive transport. In addition, the kidneys support such physiological processes as red blood cell production, bone metabolism, and blood-pressure control. Renal failure occurs when the kidneys are damaged to the extent that they can no longer function to detoxify the body. The failure may be acute or chronic. In chronic renal failure and in the absence of a successful transplant, the patient must be maintained on dialysis. About 50,000 people in the United States and 150,000 worldwide are on dialysis. In dialysis solutes pass from one solution to another through a semipermeable membrane, as a result of a concentration gradient.

Hemodialysis. Hemodialysis is literally the dialysis of blood. The patient's blood is anticoagulated with heparin for its extracorporeal treatment. Blood access is usually from a fistula permanently made in the forearm. The blood is drawn from the body at a flow of 50 to 350 mL/ min, depending on patient conditions, and pumped with a roller pump through the extracorporeal circuit that includes the dialyzer. In the dialyzer, which contains the semipermeable membrane, the transport process between the blood and the dialysate takes place. The dialysate is a solution of electrolytes of about the same concentration as normal plasma.

The membranes being used are almost exclusively cellulosic and have an average pore size of 2.5 nm. Such membranes permit the low-molecular-weight solutes such as urea, creatinine, uric acid, electrolytes, and water to pass freely but prevent the passage of high-molecular-weight proteins and blood cellular elements.

Dialyzers use either hollow fibers or film membranes in sheet or flat tubular form. Many variations of these two basic designs are commercially available. Generally, blood and dialysate flows are countercurrent.

Water removal in hemodialysis is accomplished primarily by ultrafiltration. Under a hydraulic pressure gradient between the blood and the dialysate compartments, convective flux of water and its contained solutes takes place. The functions of the dialysate delivery system are to prepare and deliver dialysate of the required chemical makeup for use in the hemodialyzer. Monitoring and control equipment is included as part of the system to ensure that the dialysate composition is correct and ready for use by the hemodialyzer.

Methods of preparing and delivering dialysate are either batch or continuous. In a continuous dialysate-supply system, concentrate and processed tap water are continuously mixed during the course of the dialysis and delivered to the dialyzer. This type of system eliminates the space required for mixing an entire batch of dialysate. However, to be effective, the system must be closely controlled, since any malfunction can result in an improperly mixed dialysate. Dialysate can be recirculated, although fresh dialysate is most effective because of its high concentration gradient with the blood. In a single-pass system, the flow of dialysate is usually kept low (about 500 mL/min) to limit the amount of dialysate required.

Dialysate delivery systems also include monitoring, controlling, and safety equipment. These devices range from simple components to automated systems capable of operating without an attendant. Monitoring equipment includes flowmeters, temperature and pressure gages, dialysate-conductivity probes, and display meters. Control equipment includes thermostat-controlled heaters or mixing valves for regulating dialysate temperatures and composition, valves for regulating flow rates, and adjustable high and low limits on various safety-monitoring devices. Water conditioning and treatment equipment is usually available separately from the dialysate delivery system. Safety equipment includes devices designed to indicate or correct any factor in dialysis which exceeds the established limits for safe operation. This includes audio and/or visual alarms and failsafe shutdown sequences that would be accomplished during the course of dialysis.

Hemodialysis is generally performed for periods of from 3 to 6 h, 3 times a week and is used by about 97% of the dialysis population. Hemodialysis may be performed in the hospital, a dialysis center, or at home. Equipment requirements vary, depending upon where and by whom the dialysis is performed. Figure 26-13 schematically shows the circuitry with on-line dialysate preparation, and Table 26-8 outlines the most common factors monitored during hemodialysis with reference to the site of monitoring shown in Fig. 26-13.

TABLE 26-8   Factors Most Commonly Monitored during Hemodialysis

Factor	Equipment position (Fig. 26-13)	Operation	Remarks
Extracorporeal blood pressure	1, 2, and/ or 3	Measures pressure in drip chambers by mechanical or electronic manometer; abnormal pressure indicates any one of several malfunctions (increased line resistance, clotting, blood leak) and has high and low alarm limits	Installation in location 3 is considered mandatory and provides most meaningful information with respect to change in flow (clots) or blood-line leak
Blood-leak detector	4	Photoelectric pickups in effluent dialysate line detect optical transmission changes due to presence of blood	Detection threshold is adjustable with alarm circuit to shut off blood pump or bypass the dialysate flow
Dialysate pressure	5 and/or 4	Measures pressure of dialysate inlet and/or outlet by mechanical or electronic manometer; usually has high and low alarm limit; abnormal operation can result in membrane rupture or improper ultrafiltration	Transmembrane pressure may be displayed with possible control with blood-side pressure
Dialysate temperature	5	Thermostatic measurement generally used to control electric heaters; dial thermometer or thermocouple gage readout; out-of-range operation can result in patient discomfort or fatal blood damage	In central dialysate delivery systems, dialysate is centrally heated with trimming at individual stations
Dialysate flow rate	5	Measures and displays flow in a rotameter; unless extremely low, cannot result in undue harm to patient	In through-flow systems, dialysate is normally used at rate of 500 mL/min
Dialysate concentration	5	Measures and displays electric conductivity of dialysate; improper concentration can result in blood cell and central nervous system damage	Continuous concentration measurement is a necessity in delivery system using continuous proportioning of dialysate

Peritoneal dialysis is dialysis carried out in the peritoneal cavity of the patient. The peritoneum is a serous membrane lining the abdominal cavity and covering the abdominal organs, essentially forming a closed sac. Through a cannula placed through the skin or a catheter permanently implanted, dialysate solution (about 2 L in an adult) is infused, allowed to dwell for a designated period of time, and drained. This process is repeated according to the needs of the patient. The peritoneum is a semipermeable membrane permitting the solutes in the blood, which perfuses the peritoneum, to pass into the dialysate. Peritoneal dialysis may be performed intermittently or continuously.

In intermittent peritoneal dialysis a single cycle for dialysate solution infusion, dwell, and drainage is generally accomplished in less than 30 min and the procedure continued for 8 to 12 h per treatment. Dialysate is available in commercially prepared bags or bottles or can be made on site, as in hemodialysis, from dialysate concentrate and water. Because of the infusion of the dialysate solution into the body, it must be sterile and, therefore, extra precautions must be taken with its preparation compared with the dialysate used in hemodialysis.

Final water treatment is generally done by reverse osmosis. Monitoring and control equipment for the preparation and delivery of the dialysate is generally automated and capable of operating without an attendant; it includes automatic timers, pumps, electrically operated valves, thermostat-controlled heaters, conductivity meter, and alarms.

In continuous peritoneal dialysis the infused dialysate is allowed to reside for a few hours. Generally, five to six exchanges are made per day. The difference between this technique and that of intermittent peritoneal dialysis and hemodialysis is that body chemistries are more stable and not fluctuating between the extremes of the pre- and postdialysis periods.

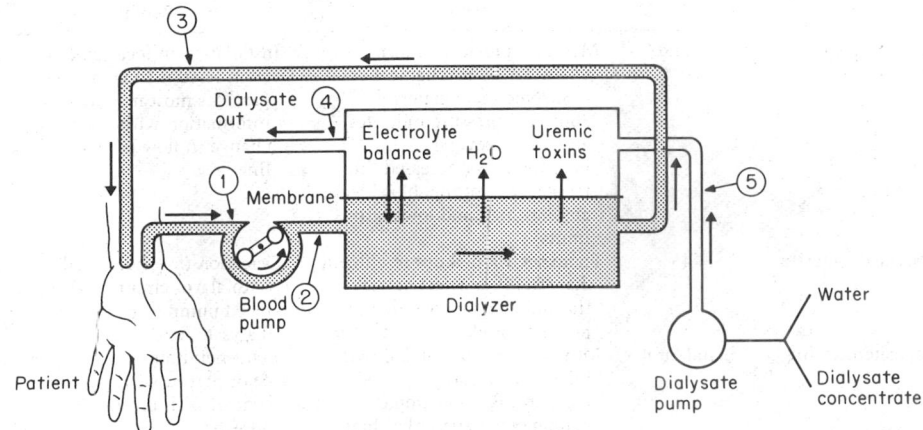

Fig. 26-13. Schematic of hemodialysis circuitry with on-line dialysate preparation. Numbered locations refer to instrumentation listed in Table 26-8.

In peritoneal dialysis since direct blood contact with dialysis equipment does not take place, anticoagulants are not required. As hydraulic pressure cannot be created across the peritoneal membrane, water removal can be accomplished only by osmotic pressure. High concentrations of dextrose in the dialysate are used to achieve the desired water withdrawal.

The low volume of dialysate, the low degree of agitation of the dialysate in the peritoneal cavity, and blood-flow dynamics in the peritoneum mean that the removal of small solutes is less efficient than in hemodialysis. As with hemodialysis, peritoneal dialysis can be performed in the home. Figure 26-14 shows peritoneal dialysis schematically.

Hemofiltration is a process under investigation as an alternate to dialysis for the treatment of renal failure. Solute and water removal is accomplished strictly by convective flux. Blood is pumped into a filtering device that allows the passage of water and solute molecules. High-flux membranes (water flux 5 to 10 times greater than hemodialysis membranes) are used. Due to the high flux, large volumes of sterile infusion solutions must be given back. Either pre- or postfilter addition schemes may be used. Due to the high physiological sensitivity to changes in circulating volume, highly sensitive monitoring equipment and infusion pumps must be used to maintain the fluid flow balances accurately, as required for a net removal of fluid from the patient.

Sorbent Systems.   Sorbents are substances that remove solutes from fluids by physical and/or chemical attraction and, as such, can be employed to detoxify blood. Because of the wide range of chemical abnormalities associated with chronic renal failure, no one sorbent alone provides adequate treatment. One system, produced by Organon Teknika of Oklahoma City, uses multiple sorbents including activated charcoal, cation and anion exchange resins, and the enzyme urease for the breakdown of urea in the regeneration of dialysate solution. This system eliminates the need for a large volume of dialysate. The complete deionization of the recirculated solution makes the infusion of electrolytes necessary. Monitoring and control equipment similar to that used in hemodialysis is also required, but no water pretreatment is necessary.

22. Lung Substitutes and Support. The lungs' primary functions are to transport oxygen to blood and remove carbon dioxide. Failure of the lungs to carry out these functions requires artificial respiratory support, in the form of mechanical assistance or substitution.

A ventilator provides mechanical assistance in breathing. A ventilator can be of two types, volume- or pressure-controlled. The cycling of the ventilator may be controlled by electronic or electromechanical timing units. Alarm signals may be incorporated to indicate failure, the disconnection of the circuit, or the onset of respiratory obstruction. In most cases ventilators are set to operate continuously and independently of the patient's inspiratory efforts.

In substitution, a blood-gas exchanger or so-called artificial lung or oxygenator is used. Artificial lungs are required for open-heart surgery. In the United States there are approximately 120,000 cases per year. Artificial lungs are of two types, direct blood-contacting and membrane.

Direct Blood-Contacting Oxygenators. There are two kinds. In *bubble-type oxygenators,* oxygen gas is bubbled through the blood. The large surface area of the bubbles and their intimate

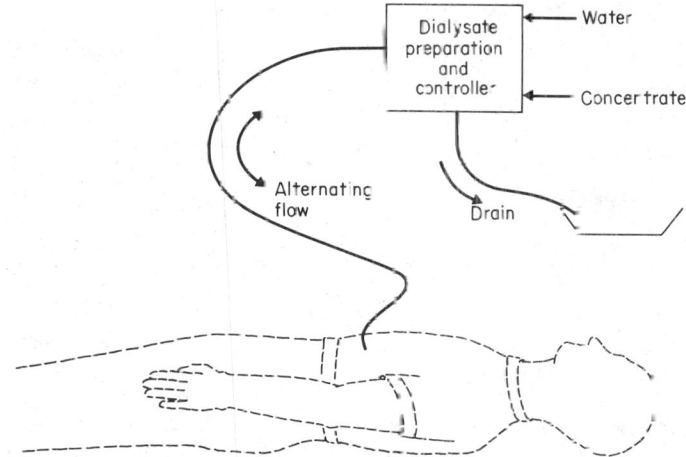

Fig. 26-14. Schematic of intermittent peritoneal dialysis with on-site dialysate preparation.

contact with blood promote high gas-transport rates. Since the blood cells can be damaged by the mixing action, this type is restricted to short-term use for routine cardiac surgery. Bubble oxygenators are used in about 80% of all open-heart surgeries.

In *film-type oxygenators,* the gas contacts the blood, which is spread or distributed in films. These devices generally incorporate a mechanical mixing device so that the blood film is continually renewed. Since such complex devices are not made in disposable form, they are no longer employed clinically.

Membrane Oxygenators. In this type the blood and gas are separated by a membrane. The membrane may be of two types, diffusion (as silicone rubber) or microporous (as Teflon or polypropylene) and may be in the form of film or hollow fibers. In the membrane oxygenator the gas transport is similar to that in the natural lungs, where a thin tissue layer separates the gas from the blood. Damage to blood in membrane oxygenators is generally considered to be less than in direct-contacting devices, allowing them to be used for extended periods of time (up to weeks), although such extended uses are rare at present. Membrane oxygenators are used in about 20% of all open-heart surgeries. Membrane oxygenators are also used in organ preservation to provide oxygen and remove carbon dioxide from the perfusion fluid circulating through the organ.

In open-heart surgery an oxygenator is used in conjunction with a roller pump ("heart-lung machine"). Such equipment is primarily manually controlled. Figure 26-15 shows the heart-lung machine components.

23. Red-Cell Substitutes (Artificial Blood). Fluorocarbon liquids have a high solubility for oxygen (3 times greater than red blood cells) and carbon dioxide and therefore have been studied as artificial blood. An emulsion of the fluorocarbon liquid is necessary, since the density of the liquid is nearly twice that of water and it is immiscible in water. The emulsion preparation serves only as a gas-transport medium and lacks cells, such as platelets and leucocytes, which are important in clotting and defense from foreign invasion. Such a preparation is universal and unlike blood does not need to be typed. At present this fluorocarbon emulsion is being investigated clinically in emergency situations where the blood of the proper type is not available.

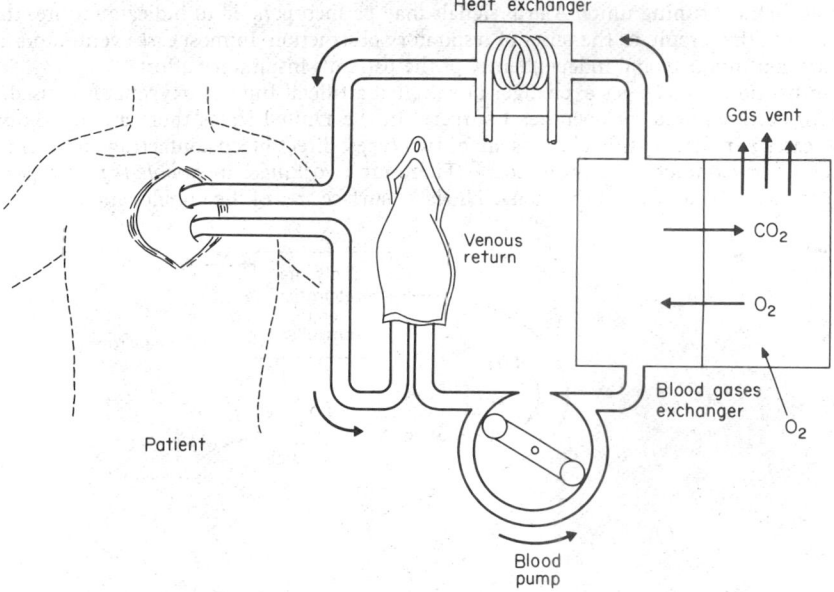

Fig. 26-15. Circuitry for heart-lung bypass.

24. Liver Support Systems. The liver is a complex organ; its many functions can be classified as detoxification, metabolism, and synthesis. By chemically modifying certain toxins the liver makes it possible for them to be excreted by the kidneys. The liver is, in a sense, the major chemical factory of the body, controlling the metabolism and synthesis. Present clinical activities have concentrated primarily upon supporting the function of detoxification. As many of the toxins are protein-bound, conventional dialysis is not generally successful; thus techniques of sorption and high-flux filtration are employed. Direct perfusion over sorbents such as activated charcoal or nonionic resins are particularly useful in cases of drug overdose and have met limited success in cases of hepatic failure. The circuit is quite simple, as shown in Fig. 26-16.

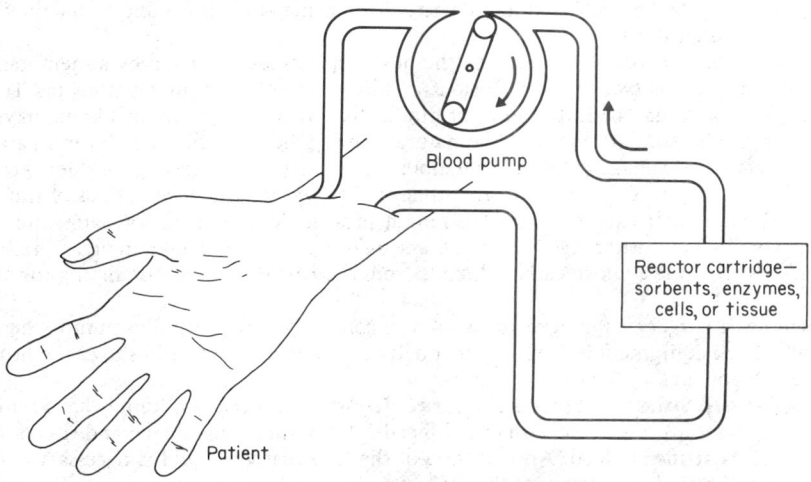

Fig. 26-16. Circuitry for reactor perfusion.

As in renal failure, hepatic failure is associated with a wide range of chemical abnormalities, and the use of multiple sorbent systems is preferred. To prevent the detrimental interaction between the blood cellular elements and the sorbents, plasma is separated from whole blood on line, using blood-cell separators or membrane devices; the plasma is perfused over the sorbents for detoxification; the plasma is then reunited with the bulk blood flow and returned to the body, as shown in Fig. 26-17. Treatment times for hepatic failure patients are much longer (up

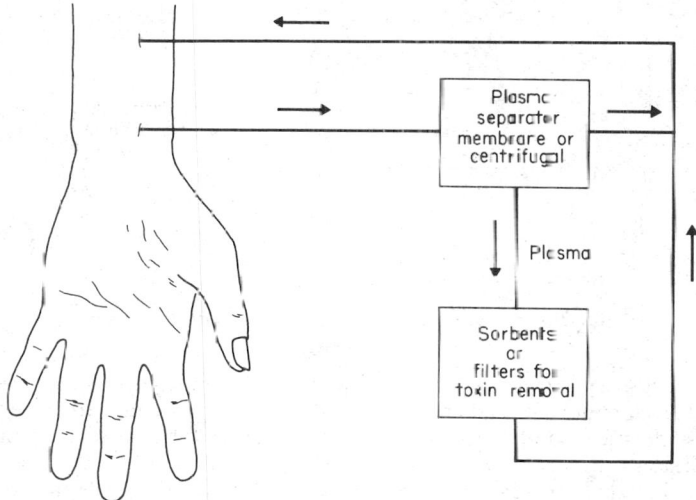

Fig. 26-17. Schematic of plasmaphoresis with on-line plasma detoxification.

to 12 h) than for hemodialysis. At present, the extracorporeal treatment consists of manual control of the necessary equipment, such as blood, plasma, and infusion pumps. For the treatment to become universal, control, safety, and monitoring equipment must be integrated.

25. Pancreas Support Systems. The pancreas secretes digestive enzymes and produces the hormones insulin and glucogen, which regulate carbohydrate metabolism. The major area of pancreatic support involves glucose control by insulin infusion Since biologic control by transplantation or islet-tissue infusion has related immunologic problems, the emphasis for clinical applications has been on the nonbiologic techniques of insulin infusion. Mechanical nonbiologic systems are classified as either closed- or open-loop. In the open-loop system, insulin infusion is based on a preprogrammed rate independent of the blood-glucose concentration. The system consists of a slow infusion pump, power source, controller, and the insulin storage reservoir. While the pump is generally set to deliver at a given rate, it can be actuated to deliver a bolus injection to augment the daily basal dose. Present open-loop systems can be made portable.

The closed-loop system includes a glucose sensor, controller, and glucose and/or insulin pumps, as shown in Fig. 26-18. The controller uses physiological information on the glucose-insulin homeostasis. The present long-term unreliability of the glucose sensor means that closed-loop systems are not yet portable.

26. Other Blood-Detoxification Systems. In most diseases, biochemical abnormalities exist which can be alleviated by blood detoxification. Many neurologic, renal, rheumatic, and hematologic disorders have associated plasma macromolecules which, if removed, cause remission of the symptoms of the disease. For the removal of these macromolecules, plasmaphoresis is performed. Plasmaphoresis can be carried out by the centrifugal separation of the plasma from blood collected by venipuncture with the subsequent reinfusion of the blood cells. More commonly, on-line membrane separators or centrifugal separators are used. Such on-line systems include control, monitoring, and safety equipment.

A growing interest may be noted in the applications of these techniques and in making the applications more universal. Combining plasmaphoresis with on-line detoxification systems such as membrane filters and sorbents (as shown in Fig. 26-17) make it possible to remove those toxic substances from the plasma without replacing fluids.

27. Artificial Heart Systems. Artificial hearts are blood pumps used to assist or replace the function of a diseased or impaired natural heart. The complete system consists of the blood pump itself, which must be connected into the cardiovascular system, as well as the energy

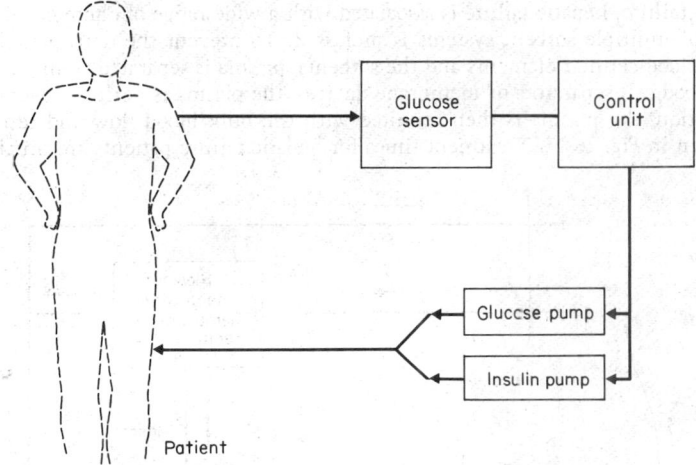

Fig. 26-18. Schematic of closed-loop insulin delivery system.

system with its associated controls. The blood pump must be designed to achieve the required physiological levels of pressure and flow with very high reliability and with proper design of the flow passages and selection of the materials to avoid damage to the blood cells and clotting on the pump surfaces.

The natural human heart has two pumping chambers, or ventricles. The right ventricle is fed from the venous system at near ambient pressure. Contraction of the heart pumps blood from the right ventricle through the pulmonary artery into the lungs. The blood, receiving oxygen and giving up CO_2 in the lungs, returns to the left ventricle through the pulmonary veins. Pressure in the pulmonary arteries is normally 20 mmHg; pulmonary venous pressure is near atmospheric. The left ventricle's contraction pumps blood into the aorta for circulation in the body. Aortic pressure ranges from 60 to 150 mmHg. The natural heart beats in the range of 60 to 140 beats per minute while pumping 5 to 14 L/min.

Four types of pumps are used in artificial heart systems, as shown in Fig. 26-19: (1) The roller pump has received widespread use for heart-lung bypass during cardiac surgery. (2) The centrifugal pump is a compact and simple device, which produces steady flow and pressure. (3) The collapsing-bladder-type pump uses a flexible bladder contained within a rigid container. Check valves are required at the inlet and outlet of the pump, and alternating fluid pressure from an external source provides the pumping action. (4) The pusher-plate pump is also a pulsatile pump using a piston or pusher plate attached to a flexible diaphragm to achieve pumping motion. This type of pump is most easily integrated with mechanical or electrical systems for driving the pump. Current developments aimed at long-term, totally implantable systems are based mainly on the pusher-plate-pump concept.

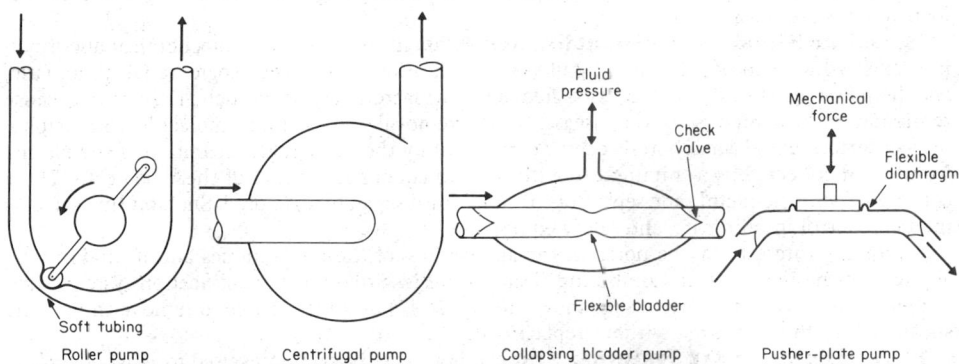

Fig. 26-19. Four common pumps used as blood pumps in artificial heart systems replacing or assisting the natural heart's pumping capacity.

Applications of artificial heart systems can be classified according to duration and purpose. The blood pump is classified as an assist device if it is used in conjunction with a diseased or impaired natural heart. The assist devices can be used for short-term as well as permanent applications. Assist devices are being used clinically for patients requiring natural heart assistance temporarily following cardiac surgery. In these applications the pump is connected in parallel with the natural heart's left ventricle and used for a few hours to a few days while the natural heart recovers.

If the natural heart does not recover or the disease state is such that it cannot be surgically corrected (assuming that the right ventricle function is satisfactory), a left ventricular assist pump might be implanted in the body as a permanent device.

If the natural heart has irreparable damage and an assist device is insufficient, a total replacement pump might be considered. For this application the diseased natural heart is removed and replaced with a two-ventricle pump which can be employed as a temporary or permanent device.

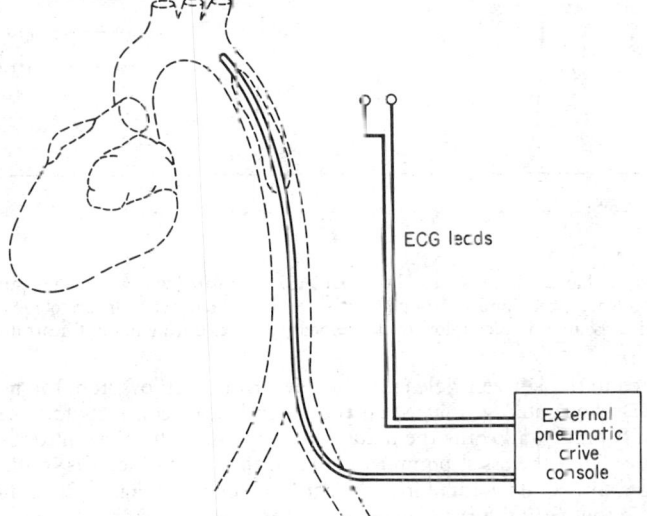

ECG leads

External pneumatic drive console

Fig. 26-20. Schematic of intraaortic balloon pump. Balloon is inserted into the aorta via femoral artery. An ECG is used to synchronize balloon with cardiac cycle.

A temporary device would be used if a transplant of a natural heart from a cadaver donor would be feasible. The artificial heart supports the patient while the cadaver heart is being obtained and proper tissue matching testing is being conducted. Artificial hearts can also be envisioned as an alternative to transplantation where a totally implantable artificial heart system could be put into a patient with expectations of reliable operation for many years meeting the patient's physiological requirements.

Assist Devices.   An assist device with current widespread clinical use is the intraaortic balloon pump (IABP). This device is technically not a pump but an elongated balloon that is inserted into the patient's aorta generally through the femoral artery, as shown in Fig. 26-20. This balloon is inflated and deflated from an external pneumatic control console. When the natural heart contracts (systole) to eject the blood into the aorta, the balloon is rapidly deflated, momentarily reducing the pressure in the aorta, allowing the natural heart to eject against the lower pressure. After the natural heart has completed its ejection and is in diastole, or filling phase, the balloon is inflated, elevating the aortic pressure during this phase of the cardiac cycle. This action of the balloon pump has the effect of reducing the workload on the natural heart as well as increasing the aortic pressure while the natural heart is in diastole, thereby increasing the blood flow and oxygen to the heart muscle through the coronary arteries, as shown in Fig. 26-21.

The balloon pump is used as a temporary device, generally with patients needing assistance following cardiac surgery. It is a simple, effective device which is easily inserted, simply operated, and easily removed with minimal complications. The balloon pump can apply moderate assistance to the natural heart and is effective only if the balloon is accurately synchronized with the natural cardiac cycle. The ECG is used for synchronization and triggering of the balloon pump, while the arterial pressure display is used to set the proper adjustment for phasing the balloon pump with respect to the cardiac cycle.

Pumps used as left ventricular assist devices (LVAD) can take over virtually the entire load of the natural heart. The LVAD is connected to the natural heart with the inlet to the pump either

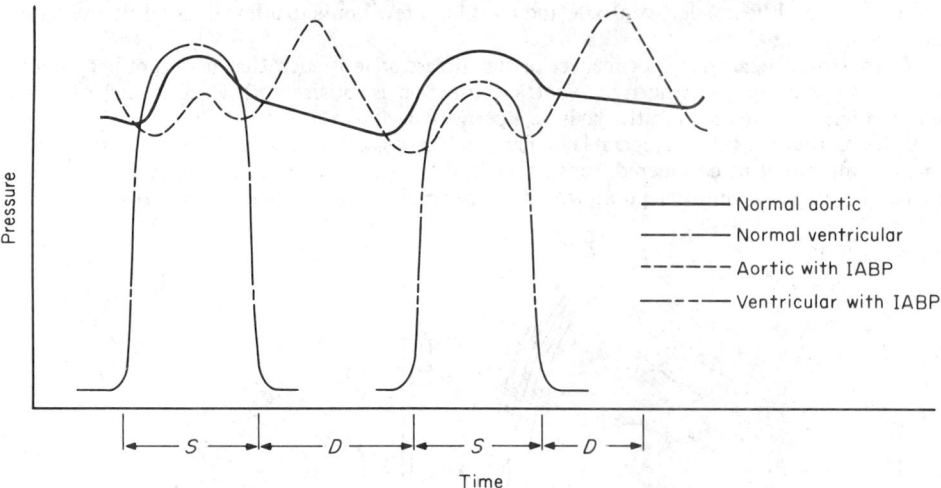

Fig. 26-21. Pressure characteristics with and without IABP. Compared with normal pressures without IABP, the use of IABP decreases aortic and ventricular pressures during natural heart systole (*S*) while increasing aortic pressure and consequently blood flow to the coronary arteries during natural heart diastole (*D*).

directly connected to the left ventricle or to the left atrium; outflow from the pump goes to the aorta. The LVAD is operated so that when the natural heart contracts for ejection, the assist pump is essentially empty, allowing the natural heart to pump the blood into the assist pump at very low pressure. Once the assist pump is filled and the natural heart is in diastole, the assist pump is energized to eject its volume into the aorta, as shown in Fig. 26-22. Synchronization of the assist pump to the natural heart is generally desired but not essential.

The assist pump can be controlled using feedback information from the pusher-plate position, as might be obtained with a Hall-effect sensor and a magnet attached to the pusher plate. The

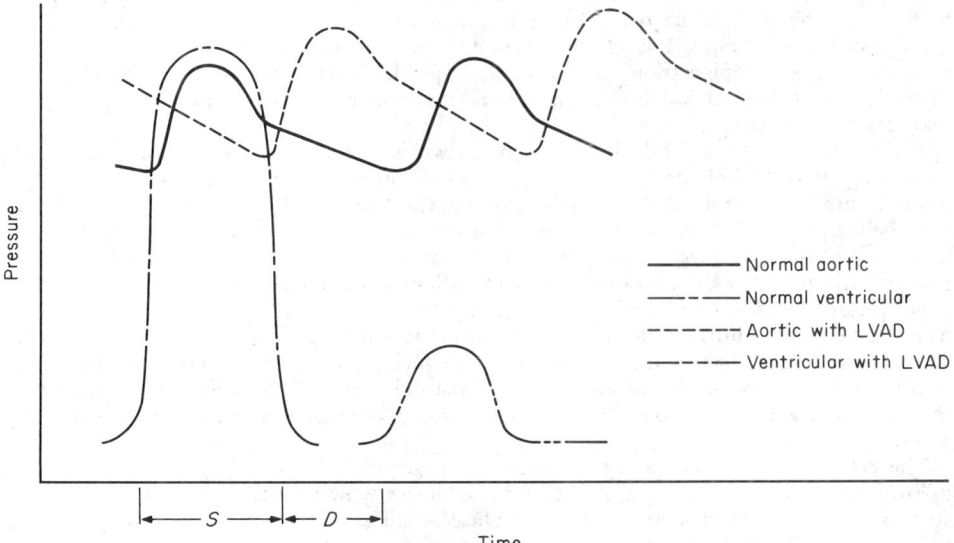

Fig. 26-22. Pressure characteristics showing the effect of the LVAD. It significantly reduces left ventricular pressure during natural heart systole (*S*) and shifts the aortic pressure out of phase with the natural heart.

control system would sense that the natural heart has ejected into the assist pump, as detected by the motion of the pusher plate. After an appropriate delay, the pump ejects the blood into the aorta and then remains idle waiting for the next beat of the natural heart.

LVADs have been used clinically for short-term assistance for patients not recovering from cardiac surgery. Not only pulsatile pumps but also the roller pump and centrifugal pump have been used for these applications. Biventricular assist devices are also used temporarily in patients whose right ventricle performance is sufficiently impaired to require a second pump to assist or take over the function of the right heart. For long-term or permanent application of the LVAD, the pump would be implanted in the patient's body in conjunction with an integrated energy system, as discussed later. Figure 26-23 shows one of several system concepts being developed for permanent assistance.

Total Artificial Hearts. Total artificial hearts can be used either as temporary devices in a patient awaiting a transplant or when a completely implanted system is available to take over the total function of the natural heart. Although a short-term artificial heart implantation has already been used clinically in the case of a transplant, the use of permanent total artificial heart systems will depend upon the development of very highly reliable and efficient systems. Many researchers worldwide have demonstrated the ability to maintain animals alive for several months on an artificial heart.

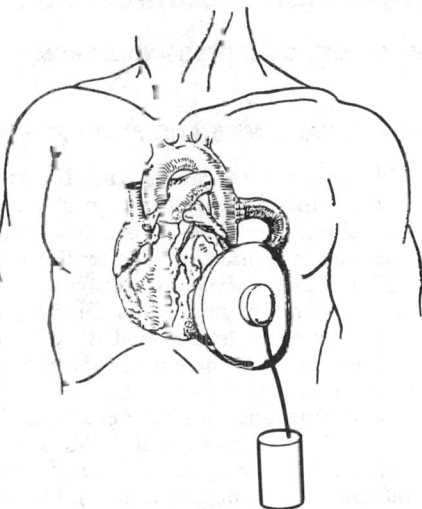

Fig. 26-23. One conceptualization of left ventricular assist pump permanently implanted inside a patient's chest. The pump is driven by a high-pressure hydraulic system powered by an electric motor or a heat engine implanted in the abdomen.

The total artificial heart comprises two pumps, a right and left ventricle implanted in the chest. Both pumps might be actuated by a common drive system or be completely independent. Various control schemes have been tried, most of which involve attempts to maintain either a set arterial pressure or atrial pressure.

Some interaction between the two pumps can take place. For example, the left pump operates to maintain a set aortic pressure, while the right pump maintains the left atrial pressure. Some of the simpler systems operate both right and left pumps independently, each attempting to maintain set atrial pressures. Most of the systems being developed and evaluated in experimental animals involve a collapsing-bladder type of artificial heart. These pumps are connected to external pneumatic systems allowing a simple pump using easily controlled pneumatics to actuate the implanted device.

Energy Systems. The energy required to pump blood under normal physiological conditions in an adult person is approximately 3 to 5 W. At present, the only practical concept for a totally implantable system, without requiring maintenance or recharging, would be to use a radioisotope heat source to drive a heat engine converting the thermal energy into mechanical energy to drive the pump. Systems currently under development show an overall efficiency of nearly 20%, but long-term, reliable operation must be demonstrated before they can be applied clinically. The engines generally use a Stirling thermodynamic cycle system.

A number of electric motor systems are also being developed to power total artificial hearts or LVADs for permanent applications. Some of these electrical systems use rotary motors coupled by cams or direct solenoid actuation of the pusher plate. Systems are also being developed which use electric motors to pump hydraulic fluid to actuate the pusher plate. The electric motors are of various shapes, depending upon the blood-pump configuration to which they are to be mated. Overall efficiencies near 50% are achievable with the electrical systems.

All electrical systems require transmission of the electric energy from a wearable battery pack to the motor coupled to the pump implanted inside the patient. Two modes of energy transmission are being examined and developed. The simplest is a percutaneous connection system using a skin-to-material interface being developed to be stable, infection-free, and permanent. The second method being evaluated is the transmission of electric energy across the skin, using electromagnetic transformer systems. The latter method eliminates skin penetration, which increases the danger of infection.

All the components of these systems must be brought to a high level of development and demonstrated reliability before they can be accepted as permanent devices in clinical patients. Generally, high reliability can be achieved by redundancy in components or by sensing and warning of incipient failures. These requirements combined with high efficiency, anatomically compatible size and shape, and availability at reasonable costs makes the artificial heart system a continuing engineering challenge.

Myoelectric Control and Functional Stimulation

BY DUDLEY CHILDRESS AND J. THOMAS MORTIMER

ARTIFICIAL LIMBS AND ASSIST DEVICES

28. Myoelectricity (muscle electricity) is the electric activity generated when a muscle contracts. Generally "myoelectricity" is used to refer to the electricity generated by skeletal (voluntary) muscles, although smooth muscle and cardiac muscle also produce an electric signal during contraction. The cardiac signal, however, is usually called the electrocardiogram (ECG, Par. **26-6**). Myoelectric activity, in its recorded form, is called an electromyogram (EMG), and this term is often used to refer to myoelectric potentials in general.

The electromyogram is used as a diagnostic tool for muscle pathologies and to assess muscle involvement when there is peripheral or central nervous system impairment due to trauma or disease. Processed myoelectricity has also been used by amputees as a control-signal input to artificial limbs and, to a lesser extent, as a control signal for other types of rehabilitation equipment. This is commonly called EMG control.

The myoelectric signal resulting from contraction of a muscle can be used to control a powered limb prosthesis (limb replacement). This control can be natural if the powered limb replaces the original function controlled by the contracting muscle. In a similar way an orthotic device, e.g., a hand brace or splint, can be controlled by a partially paralyzed person through the myoelectric signal from an intact muscle.

29. Myoelectric potentials result from the depolarization of individual muscle fibers that are normally polarized at about 80 mV (inside negative). The depolarization is in response to chemical release of acetylcholine at the myoneural junction when an action potential arrives from the motor neuron.[51] The depolarization wave propagates away from the initial point toward the ends of the fiber at about 5 m/s. The resulting electric field has been modeled.[52]

Each motor neuron (axon plus cell body) sends collaterals to several muscle fibers; this combination is called a single motor unit. The number of fibers belonging to a single motor unit varies from 5 to 10 in extraocular muscle, 200 to 300 in extrinsic muscles of the hand, and 1,000 to 2,000 in large muscles of the leg.[53] It is estimated that there are approximately 500 to 1,000 motor units in an average muscle and from 10^4 to 5×10^5 muscle fibers. The fibers of a motor unit are not bundled together but commingled with other motor units over a region of the muscle. An action potential from the motor neuron will activate all the fibers of a single motor unit more or less synchronously, and the resulting potential is the motor-unit potential (MUP). This almost simultaneous activity of many fibers causes potentials that may be sufficiently large to be detected at considerable distance, even on the surface of the skin.

Myoelectric potentials can be picked up with intramuscular electrodes or with surface electrodes. Monopolar or bipolar electrodes may be used, although monopolar methods may require the use of a Faraday cage to avoid noise interference.[54]

Intramuscular microelectrodes (25 to 30 μm in diameter) can be used to pick up activity of individual fibers.[53] These electrodes, particularly if bipolar (<200 μm apart) are highly selective. They have high impedance and require amplifiers with high input impedance and low input capacitance. Single-fiber action potentials have durations of about 1.0 ms, peak amplitudes of 5 to 10 mV, significant spectral components from 100 Hz to 20 kHz, with peak energy at around 2 kHz. Microelectrodes and single-fiber recording are generally used in studies of neuromuscular function.

Intramuscular macroelectrodes (200 to 1000 μm) are used to detect MUPs and gross muscle myoelectric activity. They generally have low impedance (1 to 2 kΩ). Various designs have been described,[54] but the bipolar type is most often used in prosthetic and orthotic applications.[55]

MUPs have lower-frequency compcnents than individual fibers because of electrode position and geometry, because high frequencies are attenuated by the tissues, and because the potential is due to the activity of several fibers that do not depolarize at precisely the same time. MUPs have durations of approximately 5 ms and significant spectral components ranging between 10 and 1,000 Hz, with peak energies around 100 to 200 Hz.

Active motor units of the forearm generally fire at 5 to 20 pulses per second. As muscular contraction increases, more motor units are recruited and the firing rate may also increase. The asynchronous activity of many motor units results in a myoelectric potential that takes on the appearance of an interference pattern. This gross myoelectric potential has a frequency spectrum determined principally by the frequency spectrum of the MUPs.[35,56] The peak amplitude is generally below 10 mV, and in unfatigued muscle the amplitude has a probability density function that is gaussian.

It is difficult to use intramuscular electrodes for prosthetic or crhotic applications because wires coming through the skin are a nuisance, may become infected, and may break. Electro-magnetically coupled implants are usually expensive and require surgery. Consequently, most

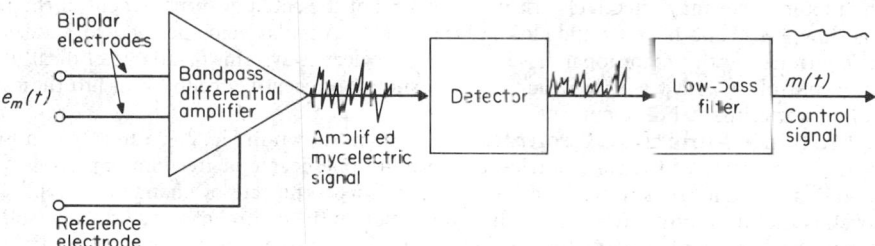

Fig. 26-24. Typical processing of myoelectric signals; amplification, detection, and smoothing.

practical applications of the myoelectric signal as a control input have been limited to surface electrodes.

Surface electrodes are usually bipolar (2 to 3 cm apart) and connected to differential amplification. Silver–silver chloride electrodes used with a paste electrolyte give the most stable electrode. However, dry electrodes are usually required for prosthetic or crhotic work because paste is a nuisance to use on a daily basis and may cause skin irritation. Stainless steel has been shown empirically to have small motion artifacts.[57] If low-frequency components (below 100 Hz) are attenuated, fluctuations of the half-cell potential are rendered unimportant. Electrode impedance of stainless-steel electrodes at the higher frequencies is as good as for other metals.

The surface electromyogram is remarkedly similar to the intramuscular myoelectric signal except that the amplitude is lower, the frequency spectrum shifted slightly lower, and single motor units more difficult to detect. The amplitude has a gaussian distribution and zero mean (nonfatigued muscle, capacitively coupled) for a given level of contraction.[58] A moderate level of contraction typically produces an rms value of 100 μV. As contraction level increases, the variance and hence the rms value increases.

The rms value of the surface EMG is, in general, a nonlinear function of muscle force, although linear relationships are sometimes obtained for isometric (constant-length) contractions of small muscles of the hand and for other muscles where the range of force is limited.[59] Muscle length, velocity (shortening or lengthening), and rate-of-force change all influence the rms value of the surface EMG, as does the EMG from nearby synergistic muscles. Therefore, great caution must be used in attempting to relate EMG to muscle force.

30. Myoelectricity as a Control Signal. For control purposes the amplified myoelectric signal need not be of instrumentation quality. Bandpass gains cf 10^4 to 10^5 with 100- and 400-Hz cutoff frequencies are usually sufficient for low-noise operation. The low-frequency cutoff attenuates 60-Hz signals as well as low-frequency electrode noise. Relatively little energy is present in the surface EMG above 400 Hz, and the upper-frequency cutoff reduces noise. After amplification the signal is usually detected and smoothed, as shown in Fig. 26-24.

Filter time constants greater than 0.2 s make the system too sluggish for most prosthetic and orthotic applications ($T_c \leq 0.1$ s is desirable). Various other processing schemes are possible in the generation of a control signal from the raw myoelectric signal.[60,61]

The impedance of dry electrodes is moderately high and changes with time. Perspiration lowers electrode impedance but may do so differently at each electrode. Also, perspiration may cause unbalanced surface impedances between differential electrodes and the reference electrode. This will cause common-mode signals to be converted into differential-mode signals and will nullify the common-mode rejection characteristics of the differential amplifier. For this reason, the reference electrode should be located some distance from the bipolar electrodes.

Differential input impedance of the amplifier should be moderately high (1 MΩ or higher) to minimize signal reduction caused by high source impedance. More importantly, the amplifier's common-mode impedances must be as large as possible to minimize the effect of unequal electrode impedances. As with the surface-impedance effect, low common-mode impedance can result in conversion of common-mode signals into differential-mode signals.

31. Clinical Applications. The most common application of myoelectric control is in control of electric hand prostheses for below-elbow amputees. Myoelectric signals are taken from two muscle sites to control opening and closing the electric hand. Figure 26-25 illustrates the control concept. A similar arrangement can be used to control an elbow. Figure 26-26 is a photograph of the system as clinically applied for below-elbow amputation. Powered prostheses of this type are used on a modest scale in North America.[62]

The motor driver may effectively use on-off control of the actuator if movement of the limb part is slow, e.g., hand opening and closing $\approx$ 1.0 rad/s. Angular velocities of 2 to 3 rad/s are better controlled with proportional drive. Present myoelectric systems (available clinically) are velocity-controlled with visual feedback. Sensory and proprioceptive feedback in prostheses are still under experimental development.

32. Multiple Functional Control. The system shown in Fig. 26-25 uses two muscles to control one degree of freedom. Single-site control of one degree of freedom is possible with three-state approaches sensitive to the level of $m(t)$[63] or to the rate of change of $m(t)$.[64] This approach is useful in prosthetics and also has application in the control of hand orthoses (splints) by persons with neuromuscular deficits (assuming they have some residual muscle control).

When the hand is moved, a constellation of muscles within the hand and forearm work synergistically to produce the desired motion. Also, when the arm is moved, there is coordinated activity of muscles, particularly those of the upper arm and shoulder. Multiple electrodes and pattern-classification techniques have been used to identify intended motion from the myoelectric signals of muscles remaining in the forearm[65] or from muscles remaining in the arm and shoulder.[66]

A unified theory for myoelectric control of multiaxis arms has also been recently put forward.[67] In addition, it has been proposed that the identification can be achieved from a single EMG site. Because the myoelectric activity at a single electrode site is some combination of activity from a constellation of nearby muscles, it appears possible to identify intended motion by detailed analysis of the single EMG.[68]

Despite recent advances, it is still to be determined how extensively myoelectric control will be used in prosthetics and orthotics, beyond control of hand prehension and elbow flexion. Other control methods for complex systems must still be considered. Microcomputer technology will soon have a strong influence on this field.

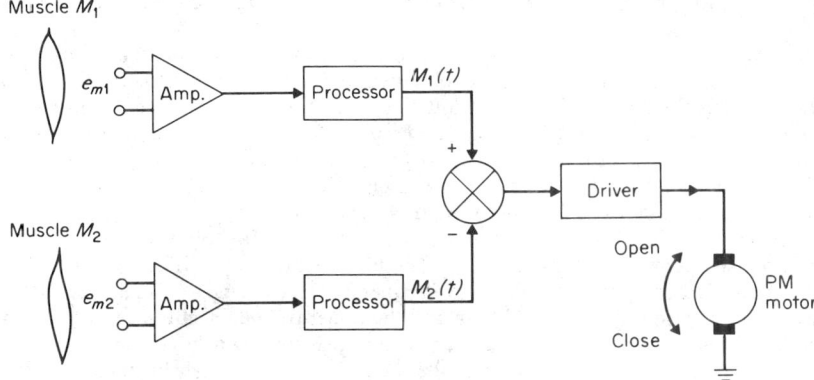

Fig. 26-25. Typical two-site control of a dc motor with myoelectric signals.

ELECTRICAL STIMULATION OF EXCITABLE TISSUE

Electrical excitation of neural tissue is a technique which offers an opportunity to restore missing or impaired body function. For instance, by the application of electrical stimulation to the appropriate regions a visual sensation can be elicited in the blind, an auditory sensation can be evoked in the deaf, and paralyzed limbs can be made to move.

In this text, general guidelines to the practical aspects of stimulation will be given rather than the specifics of applications. To this end, tissue damage will be considered first, followed by a discussion of some of the characteristics of the electrical-excitation phenomenon. A more thorough treatment of these subjects can be found in Ref. 72.

33. Tissue Damage. Metal electrodes provide the most common means of localizing an electrical stimulus. In the leads and the electrode, charge flow is supported by electron migration, while in the body spaces it is supported by ion migration. The conversion from electron to ion migration occurs at the surface of the electrode. On the basis of what we have observed to date, it appears that tissue damage, incurred during stimulation, is related to the surface or electrochemical reactions occurring during this conversion process.

Electrochemical Aspects of Stimulation. When the surface of a metal electrode and an ionic solution come into contact a potential difference develops, reflecting the thermodynamic forces which operate to bring them into equilibrium. The application of an external potential during stimulation alters the surface equilibrium potential of the electrode. The new potential is accommodated by various reaction processes which depend on the metal, the electrode potential V_E, and the ionic species in the surrounding electrolyte.

For an implanted electrode like those in current use, the electrode potential and hence the reaction processes are directly proportional to the charge moved through the system and inversely proportional to the surface area of the electrode (the charge density Q/A). This relationship is illustrated in Fig. 26-27.

Three regions are distinguished here; the irreversible region to the right of point I, corresponding to the extremes of anodic stimulation; the middle region where the reaction processes are fully reversible; and the irreversible region to the left of point II, corresponding to the extremes of cathodic stimulation. On the basis of current observations it is strongly recommended that stimulation be restricted to the reversible region where no inevitable reaction products are generated.

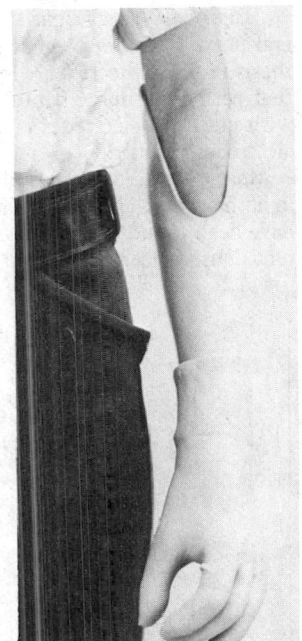

Fig. 26-26. A below-elbow amputee wearing a myoelectrically controlled hand prosthesis similar to the system of Fig. 26-25. Stainless-steel electrodes over remaining finger extensor muscles pick up the signal that opens the hand. Finger flexor muscles activate the hand closing. Closing force (max 100 N) is proportional to the level of the myoelectric signal; 250-mAh (12-V) nickel-cadmium batteries power the hand. Quiescent current drain is less than 1 mA. Motor stall current is about 800 mA.

To restrict the electrode potential to the reversible region, a balanced-charge biphasic stimulus waveform must be used, and the charge density per phase must be within the limits of the reversible region for that particular electrode material. When monophasic stimulation is applied to an electrode, the electrode potential will be driven eventually into the irreversible region, where potentially toxic by-products will be generated or metal corrosion will occur.[73]

Brain Stimulation. If great care is taken during surgery, surface electrodes can be placed on the brain with minimal trauma to the underlying cells. Platinum, the most commonly used material, has been found to be "safe" with charge densities (per pulse) less than or equal to 0.3 μC per square millimeter of real area.[74] (Real area is distinguished from geometric area by taking into account irregularities of the surface that would yield a larger value than geometric measurements would.[75]) An alternative material, Ta_2O_5, looks very promising.[76–78] Early studies of this material used the porous slug of a tantalum capacitor as the electrode. The oxide coating eliminates irreversible reactions, and the porous structure provides a large real surface area for a very small geometric area (approximately 100:1). *A note of caution:* like the tantalum capacitor, the Ta_2O_5 electrodes must be operated with an anodic bias.

Peripheral-Nerve Stimulation. Systematic studies of the effect of stimulus current on peripheral nerve have not been reported. Work in progress through the NIH-NINCDS Neural Prosthesis Program, however, should be complete in the early to mid-1980s. Available data indicate that peripheral nerve stimulation can be applied safely.[79-81] The most likely cause of trauma with cuff-type electrodes is mechanical. The damage appears to be greatest when close-fitting electrodes are used.

Muscle Stimulation. Coiled-wire intramuscular electrodes[82] have been used for over a year in both animals and human beings. Electrodes of this configuration, formed of 316 stainless steel, have been found to be safe with pulsed monophasic cathodic stimuli, provided the average current density does not exceed 10 μA per geometric square millimeter.[83] With balanced charge biphasic stimulation (recommended) the "safe" limits are 0.4 μC per geometric square millimeter. If this limit is exceeded, the electrode may also corrode.

Balanced Biphasic Stimulator. Perfect charge balance with two rectangular (positive and negative) pulses is almost impossible to achieve. Any long-term imbalance will force the electrode to operate in the irreversible region. A relatively simple method of achieving balanced charge with a regulated current stimulus is illustrated in Fig. 26-28.

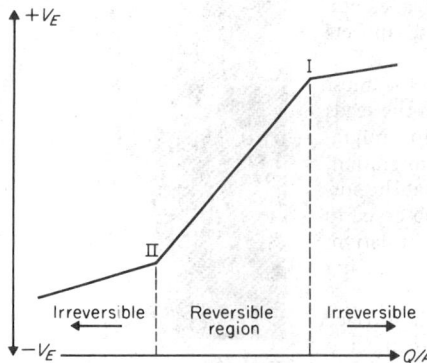

Fig. 26-27. Idealized representation of the relationship between electrode potential V_E and charge density (charge per unit of real electrode area). Charge injection in the reversible region involves processes which are completely reversible and do not result in a net change in chemical species. Charge injection in the regions to the right of point I or left of point II involve electrochemical reactions which cannot be reversed by driving current in the opposite direction.[72]

34. Properties of Electrically Excited Nerve.
A nerve is a hollow tube formed from a membrane. In the resting or unexcited state, it is polarized, the potential on the outside being several tens of millivolts greater than on the inside. In the excited state the potential difference decreases, and at the peak it actually reverses. The practical duration of the excited state is approximately 1 ms. The local current flowing during the excited state is sufficient to produce an excited state in adjacent resting membrane, which results in a propagated action potential along the length of the nerve.

The action potential is one mechanism through which one part of the body exerts control on another body part. The excited state can be induced locally by a brief externally applied electric field. The orientation of the field must be in a direction to depolarize the nerve membranes, i.e., lower the potential of the outside with respect to the inside. In general (but not always) the proper field orientation can be effected with the least current (for a given pulse width) when the stimulus pulse is cathodic (negative). Exceptions are detailed in Ref. 72.

Amplitude-Duration Relationship. Experimentally it has been found that the stimulus amplitude required to effect a propagated action potential I_{th} increases as the pulse width decreases (Fig. 26-29). The curve can be described with reasonable accuracy by

$$I_{th} = I_R/(1 - e^{-kt})$$

where I_R = rheobase current, t = pulse width, and k = constant.

In the life science literature this relationship is referred to as the strength-duration curve. The relation differs for different nerve fibers (diameter) and tissue (muscle). The I_R term is difficult to measure in practice because, for very long pulse widths, the threshold stimulus actually begins to increase rather than decrease. Further, I_R depends on the spacing between the electrode and the nerve (or tissue of interest). For these reasons convention has led to the use of the term chronaxie t_C, which is the pulse duration required for a threshold pulse twice the magnitude of I_R (Fig. 26-29).

Charge-Duration Relationship. In Par. **26-33** it was noted that the charge injection should be minimized to avoid damage (if a small electrode size is important). The threshold charge injection Q_{th} can be found by multiplying both sides of the strength-duration equation by the pulse width t. Theoretically, Q_{th} can be shown to decrease as the pulse duration decreases (confirmed experimentally)[84] and reaches a minimum I_R/k, at $t = 0$. Since I_{th} increases as t decreases, there are practical limits to the stimulus current and duration. For most practical systems pulse widths between 50 and 200 μs are recommended. At a pulse width equal to the chronaxie, the excess charge over the theoretical minimum is 38%.[7]

Threshold Dependence on Nerve-Fiber Diameter and Tissue. Small nerve fibers have a higher threshold than large nerve fibers under the same geometrical conditions.[85] Furthermore, the separation between thresholds I_{th} increases as the pulse width decreases. These properties make it possible to activate one size group of nerve fibers without exciting the smaller group, even though they are located in the same physical space. This is important since nerves subserving different functions sometimes have nerve fibers of grossly different size. As an example, nerve fibers controlling muscle contraction are much larger than those fibers conveying sensa-

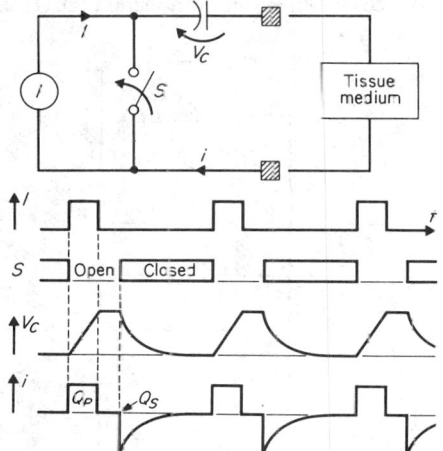

Fig. 26-28. Schematic diagram for practical balanced-charge biphasic stimulator with timing diagrams. I = stimulator current; S = switch (closed during indicated period), V_C = voltage developed across capacitor C, and i = electrode current.[72]

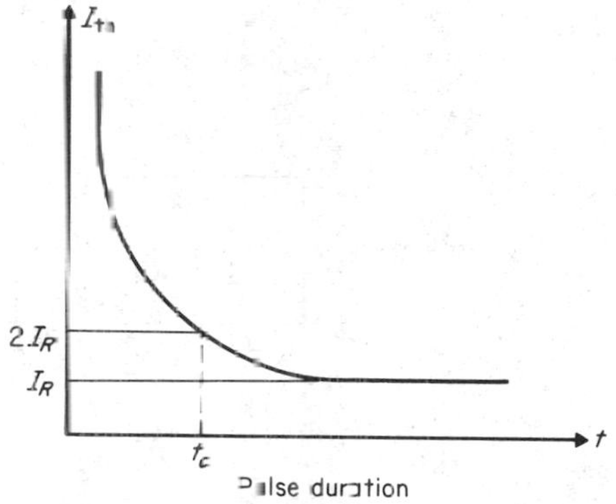

Fig. 26-29. Pulse-width-amplitude relationship for excitable tissue. See text for parameter definition.

tions of pain. Therefore, using narrow pulse widths may make it possible to excite muscle without undue discomfort from coincident pain-fiber stimulation.

The strength-duration curve for direct muscle stimulation increases much more rapidly than that for muscle nerves as the stimulus pulse width decreases. The two curves can be readily distinguished by comparing chronaxie times. The chronaxie for muscle is greater than 10 ms and for muscle nerve less than 10 ms (probably less than 1 ms). This property allows one to activate muscle through its nerve supply at a much lower stimulus amplitude or to excite muscle nerves without directly exciting the muscle fibers with electrodes placed within the muscle.

Electronic Patient Monitoring

BY MICHAEL R. NEUMAN

35. Patient monitoring is concerned with the continuous observation of seriously ill patients, including appearance, physical examination, records of physiologic variables, and intervention and administration of therapy when necessary. Electronic devices can monitor, display, record, and make elementary decisions concerning a patient's hospital course. A generalized scheme for an electronic monitoring system is shown in Fig. 26-30. The transducers convert physiologic variables into electric signals, which in turn are manipulated by the signal processor to allow qualitative or quantitative display and recording. Computing algorithms for pattern recognition can be applied to data so that abnormalities are recognized and the clinical staff is alerted.

Electronic patient-monitoring systems are applied in many areas of the hospital, including medical, surgical, and pediatric intensive-care units, operating and recovery rooms, labor and delivery rooms, and the newborn intensive-care nursery. With the advent of radio telemetry and miniature recording devices, monitoring devices need not be limited to any one area of the hospital but can be applied to ambulatory patients as well.

36. Coronary-Care Monitoring. Many deaths occur in heart disease as a result of cardiac arrhythmias (disturbances in the normal sequence of the heartbeat). For example, a myocardial infarct resulting from blockage of blood flow to a portion of the heart muscle can cause other heart muscle fibers to contract randomly. This is known as fibrillation; if it is not rapidly detected and corrected, it results in the patient's death. Electronic monitoring devices can be applied to detect fibrillation and some of its precursors.

A typical cardiac monitoring system is shown in Fig. 26-31. The electrocardiogram (ECG), as

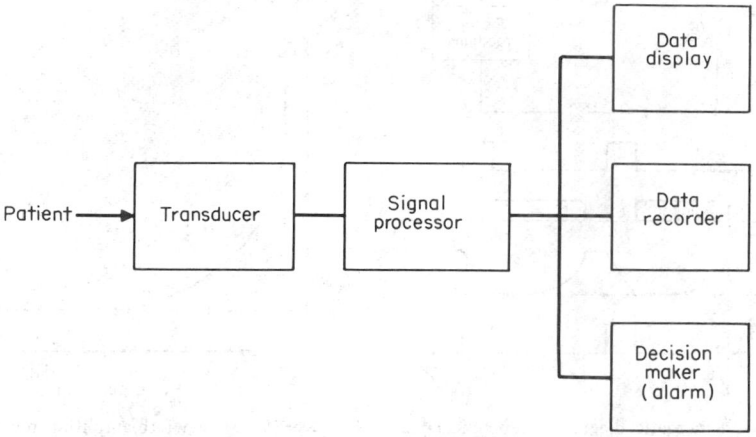

Fig. 26-30. Generalized electronic monitoring system.

sensed by electrodes on the anterior chest wall, is the main variable monitored (see Pars. **26-6** to **26-10**). The signal is amplified through a high-input-impedance amplifier (5 to 10 MΩ). The frequency response of the amplifier should range from 0.1 to 100 Hz with 3 dB for an undistorted signal; however, amplifiers with narrower bandwidth often are used to reduce artifact.

The amplified signal is displayed on a cardioscope with a long-persistence phosphor or with a continuously updated memory. A recording oscillograph may also be used to provide a permanent record. The amplifier output is also fed to a short-term memory loop which stores the ECG of the most recent 15 s. If the operator or the alarm circuitry detects an arrhythmia, the chart recorder can be switched to the memory loop to record the ECG for 15 s leading up to the observed arrhythmia. The memory loop is usually a shift-register-type solid-state memory.

Heart rate is calculated by cardiotachometer circuits, either of the averaging or the beat-to-beat type. The former determines average heart rate over a period of several seconds; the latter determines heart rate as the reciprocal of the time interval between individual beats. Output from the cardiotachometer is fed to analog or digital display devices and a heart-rate alarm circuit to set off an alarm when the heart rate exceeds a maximum or falls below a minimum, as determined by the operator. In many coronary-care monitors, the alarm also connects the recording oscillograph to the memory loop to record automatically the event which set off the alarm.

Most coronary-care units have displays at the patient's bedside as well as at a central monitoring station. Many monitoring units can also continuously monitor other physiologic variables such as blood pressure, peripheral pulse pressure, blood gases, and respiration.

Several problems are common to all cardiac monitoring instruments, the most serious being interference from the power mains due to improper connection of electrodes or poor grounding of the patient. Poor-quality or improperly prepared electrodes can produce electrical noise and/ or dc offset potentials. Motion of the electrodes with respect to the patient's skin can produce sharp transients of 100 mV or greater or give drifting base lines. Electrolyte paste may evaporate or leak out from the electrode, causing loss of signal. Finally, the electrodes, the electrolyte paste, or the adhesive which holds the electrode against the skin may cause skin irritation. These problems can result in false or missed alarms.

A 60- or 50-Hz interference from the power mains can also be a result of improper electrode attachment. Dissimilar electrode-to-body impedances of a differential pair of electrodes can result in a voltage of power-line frequency at the monitor input caused by displacement currents between the power line and the electrode lead wires being returned to ground through the patient.

Similarly, displacement current from a power line to the patient which is returned to ground through the monitor will give a differential input voltage if the sum of the monitor input impedance and electrode impedance for each electrode is different. Finally, high-common-mode input

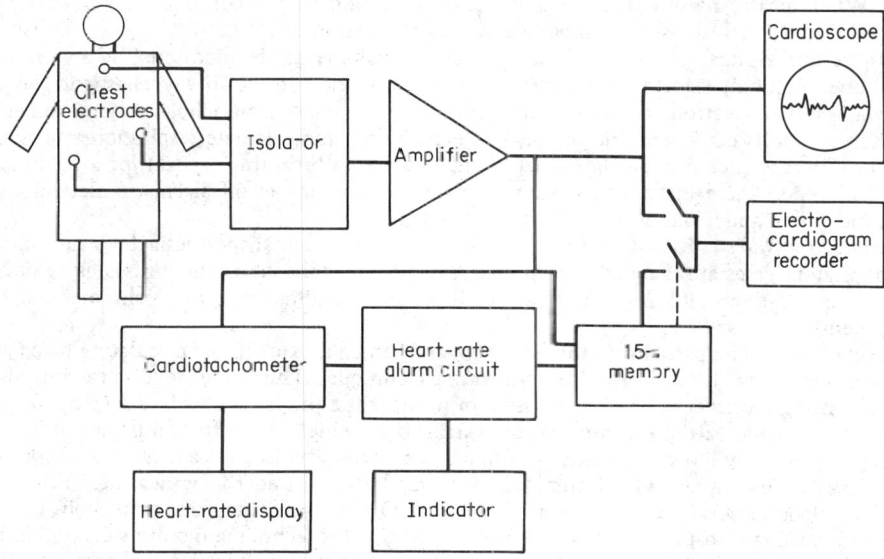

Fig. 26-31. A typical cardiac monitoring system.

voltages at the power-line frequency resulting from poor grounding of the patient can produce interference if the monitor's common-mode rejection ratio is not extremely high.

Radio telemetry and computers are used in cardiac monitoring. A radio transmitter, worn by the patient and having a range of several hundred feet, can broadcast ECG and other physiologic variables to a receiver at a central station, allowing greater patient mobility. In this system, the patients are completely isolated from the monitoring apparatus connected to the power mains, thereby eliminating the electric-shock hazard from the monitor and reducing the power-mains artifact. On-line computers have been introduced to recognize various cardiac arrythmias, the regularity of beat-to-beat intervals, the presence of abnormal beats, the frequency of these occurrences, and other characteristics of ECG waveforms.

37. Intensive-Care, Operating-Room, and Recovery-Room Monitoring. Besides the coronary-care unit, patients with other acute illnesses, in shock, or suffering from severe burns, as well as those undergoing and recovering from surgery, often require intensive monitoring. In these patients several vital signs must be monitored continuously. Some of the methods of monitoring them are listed below.

Heart Rate and ECG. See Par. **26-36**.

Arterial and Venous Blood Pressure. The direct method involves the cannulation of an artery or vein with a miniature pressure transducer or a fluid-filled catheter for hydrostatically coupling the vessel to an external pressure transducer. The indirect method uses either an automatic sphygmomanometer cuff, which determines the systolic and diastolic pressures as in manual operation, or uses the cuff to obtain sphygmograms by inflating the cuff to a pressure intermediate between systolic and diastolic levels and continuously recording the pressure within the cuff.

Peripheral Pulse Pressure. Changes in peripheral pulse pressure can be determined by plethysmographic methods. A finger plethysmograph, consisting of a small cuff, fits over the finger and determines changes in finger volume due to the blood-pressure pulse. The electric impedance across the finger will also change with blood volume, but this effect is more prone to artifact than the pressure method. Optical plethysmographs also have been used; a light source is placed on one side of the earlobe with a photodetector on the opposite side. The amount of light passing through the lobe decreases with increasing blood volume. Similar variations in reflected light from the finger or other skin surface can also be used to determine relative peripheral pulse. All these plethysmographic methods give only a relative reading that can be used to observe changes but not absolute pressure.

Peripheral and Core Temperatures.   Temperatures in the rectum and the esophagus and skin temperatures at various locations can be monitored by thermistors and associated electronic circuits. When measuring skin temperatures, the use of sensors with small thermal mass and good thermal contact with the skin is important for reducing errors.

Respiration; Apnea.   Respiration rate and relative volume can be monitored by a variety of techniques including transthoracic electric impedance, gas flow in the airway, chest and abdominal-wall motion, electromyogram from the muscles of respiration, and whole-body movements. Electronic circuits can determine the peak of each breath and calculate respiration rate. Some monitors are specifically designed to detect the absence of breathing (apnea) for a minimum period of time. These instruments, known as apnea monitors, set off alarms to alert clinical personnel when apnea has persisted beyond the set time.

Respiratory Gases.   For patients undergoing surgery with general anesthesia, both inspiratory and expiratory gases are sampled. Continuous analysis is conducted on these gases using either oxygen and carbon dioxide analyzers, small mass spectrometers, or gas-chromatographic equipment.

Blood Gases.   The partial pressure of oxygen and carbon dioxide in the circulating blood can be monitored using either invasive or noninvasive techniques. The former require the introduction of sensing electrodes into the bloodstream itself, while the latter involves sensing oxygen and carbon dioxide partial pressures on the skin surface which result from diffusion from the underlying capillary blood. The latter technique is known as transcutaneous blood-gas monitoring. It uses a sensor probe which dilates the skin capillary bed under it by heating, so that the capillary blood closely approximates arterial blood. Oxygen saturation of hemoglobin can be determined using transmission or reflectance oximetry. This technique monitors color changes in the blood by means of fiber-optic catheters, direct contact with the blood, or light transmission for reflectance through tissue such as the earlobe.

Weight.   Load-cell-instrumented chairs or beds can be used to monitor a patient's weight. This is especially important for following the uptake or loss of water in patients undergoing dialysis.

Electroencephalogram.   Scalp electrodes with amplifying and recording or displaying instruments are used. Computers have been used to analyze the complex EEG waveforms and to identify specific features.

38. Perinatal Intensive-Care Monitoring. Patient-monitoring techniques have been applied to the intensive management of mother, fetus, and newborn in the period surrounding birth to detect and manage traumatic labors in high-risk pregnancies and premature or severely ill newborns. Obstetrical monitoring involves observing fetal heart rate and uterine contractions during labor. In the indirect method, sensors are attached to the maternal abdomen. Fetal heart activity can be detected through phonocardiographic, ultrasonic, or electrocardiographic methods. In the first method a sensitive microphone is placed on the maternal abdomen to detect fetal heart sounds. In the ultrasonic method, Doppler frequency-shifted ultrasonic beam reflections from the fetal cardiovascular system in the vicinity of the heart are detected and processed to determine the fetal heart cycle. A weak fetal electrocardiogram can be sensed on the maternal abdominal wall. This signal must be separated from the stronger maternal electrocardiogram to determine fetal heart rate.

Uterine contractions are detected indirectly by means of a tokodynamometer, usually consisting of a linear differential transformer or strain gages coupled to a plunger which is pressed against the abdominal wall and, hence indirectly, the uterus. Direct methods of obstetrical monitoring require access to the uterine cavity and fetus either transabdominally or through the vagina when labor has progressed far enough to allow rupture of fetal membranes.

The direct method of obtaining the fetal ECG is by means of an electrode attached to the fetus with a reference electrode in contact with the maternal vaginal mucosa. Fetal heart rate is then determined by an instantaneous cardiotachometer. Direct uterine pressure recordings can be made by coupling the amniotic fluid in the uterine cavity to an external pressure transducer through a fluid-filled catheter or by placing a miniature pressure sensor in the uterine cavity itself.

In the typical obstetrical monitoring system, beat to-beat fetal heart rate is displayed on one channel and uterine contractions are shown on the other of a two-channel chart recorder. The clinician examines the chart for magnitude, duration, interval or frequency, and base line of uterine contractions and compares this information with the base line, accelerations, and decelerations in fetal heart rate during and immediately following uterine contractions.

Monitoring of premature, postoperative, and seriously ill newborns involves methods similar to those used in adult intensive care. Variables commonly monitored include (1) ECG and heart rate, (2) arterial and venous blood pressures, (3) surface, core, and environmental temperatures, (4) respiration rate and apnea, (5) environmental oxygen, and (6) blood gases. Special attention must be directed to the small size of the patient in neonatal monitoring Sensors must be much smaller than for the adult, and care must be taken to ensure that they are not interfering with neonatal physiological processes themselves.

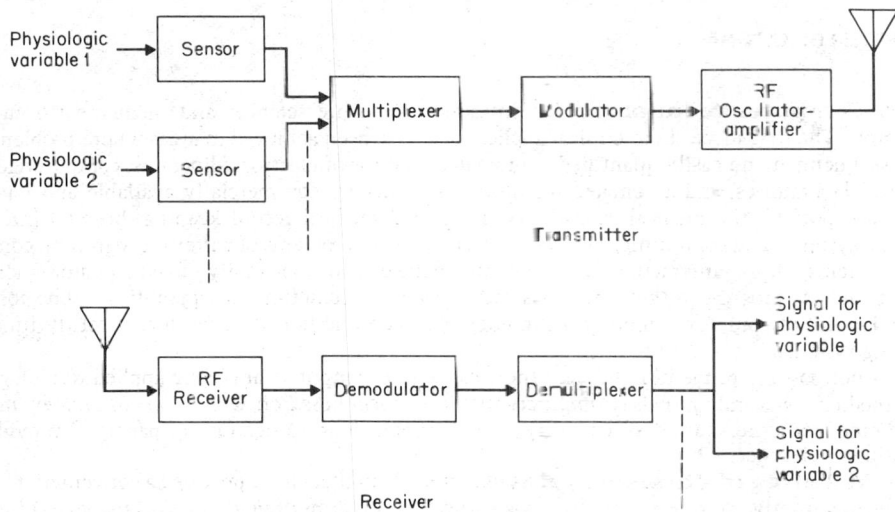

Fig. 26-32. Block diagram of a radio-telemetry system

39. Ambulatory-Patient Monitoring. Electronic monitoring of the patient not confined to bed is carried out by one of three basic techniques radio telemetry, miniature wearable recorders, or miniature wearable monitors. Single- or multiple-channel radio-telemetry systems, in which a miniature transmitter with appropriate signal sensors is worn by the patient, have the general arrangement shown in Fig. 26-32. Signals corresponding to physiologic variables which are sensed by appropriate transducers are combined by a multiplexer, and this signal is used to amplitude-, frequency-, or pulse-modulate a conventional radio-frequency transmitter circuit. The receiver system demodulates the radio-frequency signal, and a demultiplexer separates the signals corresponding to the original physiologic variables. Single-channel operation can be achieved by eliminating the multiplexer and demultiplexer. Small transmitters worn by patients can have ranges in excess of 200 m.

Miniature battery-powered magnetic-tape recorders or solid-state memory circuits, which can be worn by patients and interfaced to appropriate signal gathering sensors, are used in diagnostic patient monitoring where randomly occurring events such as cardiac arrhythmias need to be detected. The tape recorders, sometimes known as Holter monitors, can store 24 h of three channels of information on a small tape cassette. Solid-state memory units cannot store as much data but through intelligent selection can store only those data which are pertinent to the diagnosis; therefore they can be used over extended periods of time. As the available memory chip becomes higher in capacity and lower in cost, portable solid-state memory can be used to replace the tape recorder in the near future.

Large-scale-integrated (LSI) microelectronic devices have made it possible to miniaturize patient monitoring systems, like that shown in Fig. 26-31, to such an extent that they are small enough to be worn by a patient. These devices, usually cardiac monitors, can determine when abnormalities of the electrocardiogram are occurring and alert the patient to seek medical help. Some of these devices incorporate memories as well, so that the event that precipitated the alarm can be played back to the physician treating the patient. This area is relatively new to patient monitoring and will no doubt be the subject of expanded development in the coming years.

Although electronic patient monitoring devices do not have the same power of observation, data handling, and interpretation as their human counterparts, they are capable of accurate, continuous, routine monitoring of some physiologic variables. These devices also can alert the clinical staff when significant changes occur, thereby freeing the staff to carry out other patient management functions.

The use of computers to analyze and recognize patterns and trends in the data has reduced data management demands on the clinician (see Pars. **26-40** to **26-46**).

Computer Applications

BY PETER G. KATONA

40. Computer technology was introduced into life sciences and medicine through research. The first successes in clinical applications have been achieved in areas where problems are well defined and easily quantified. Automated analysis of electrocardiograms, computerized clinical laboratories, and automated monitoring systems are commercially available and have been accepted by the medical community. Diagnosis, medical record keeping, hospital information systems, and community health-care facilities pose problems of pattern recognition, complex systems, human interaction, and economics, none of which are easily solved by automation. Consequently, progress in these areas has been slower and sometimes disappointing.[86] The cost-effectiveness of complex computer technology has been questioned[87] even for smoothly functioning systems.

It is not possible to mention here all the areas where computers are being applied to biology and medicine. Several journals (*Computers and Biomedical Research, Computers in Biology and Medicine*, and others) are devoted solely to such applications. Collections of papers[88,89] provide additional references.

41. Modeling of Physiological Systems. Computers have provided a convenient tool for the quantitative description of physiological systems in terms of mathematical models. Before this tool is used, however, the purpose of the model building must be specified and clearly understood. The models are almost never unique, and an appropriate model can be selected only if there are objective criteria with which the models can be evaluated.

"Black-box" models describe the input-output relationship of a physiological system but disregard the underlying physiological mechanisms. They are not generally useful for understanding physiology, but they find applications in the design of systems that involve interactions between people and machines.

The goal of most simulation studies is a better understanding of the physiological system. It can be achieved by (1) interpreting the model structure and/or parameters in terms of physiologically significant quantities, (2) using the model to suggest further experiments on the physiological system, or (3) constructing the model to serve as a component within a quantitative description of a more encompassing physiological system.

Modeling starts with the assumption of a structure for the model, including inputs and outputs. If possible, this structure, usually consisting of an interconnection of compartments or network elements, should correspond to known or hypothesized physiological mechanisms operating within the system. Linear components are characterized by transfer functions (or equivalently, linear differential equations or impulse responses). The most common nonlinearities are multiplicative interaction of variables and input-output relationships showing both threshold and saturation. Thresholds are especially significant in physiological systems since chemical concentrations and frequencies of nerve firings cannot assume negative values.

Models usually contain unknown parameters which must be adjusted in such a way that outputs of the model best approximate outputs of the real system for the same inputs. If the outputs do not match within a certain tolerance, the model structure may have to be changed and new parameters computed. This iteration is continued until structure and model parameters are obtained which yield appropriate outputs. The model obtained must be verified by applying inputs that were not used in determining the model.

Although linear systems, in principle, can be characterized by using almost any input waveform, pseudo-random binary inputs are increasingly used as test signals. These computer-generated signals switch between two levels at apparently random time intervals, and the mean switching rate can be adjusted so that the power spectrum of the signal is flat over the frequency response of the system tested. Cross-correlating the input with the output yields the impulse response, even if the output measurement is corrupted by noise that is not correlated with the input.

Optimization models are alternatives to the more traditional structural models for understanding physiological control mechanisms. Optimization models are based on the assumption that the system performs its function in a way which minimizes an associated cost. For example, the respiratory system has been analyzed assuming that the airflow pattern is such that the work required to produce a given ventilation is minimized.[1] The lack of uniqueness of solutions is a major difficulty of this approach.

The complexity of physiological systems poses a variety of difficulties. Foremost is the variability of data caused by inability of the experimenter to control the large and usually unknown number of variables that continuously influence the system under investigation. The model builder may construct a model to fit the averaged data or use a model in which parameters are allowed to vary from one experiment to the next. An alternative approach is to use a model which accounts for only gross characteristics of the response. For steady-state analysis, such a model assigns ranges of possible outputs to ranges of possible inputs.

Most modeling of physiological systems is now performed using digital rather than analog computers. Digital computers provide an orderly and objective way of determining model parameters using standard optimization techniques, while analog computers usually require manual adjustment of parameters and a subjective judgment of "best fit." Hybrid computers are sometimes useful in real-time applications. In these computers setting the parameters and evaluating the model are done digitally, but the model output is generated by analog means.

Several books deal with modeling physiological systems in general.[92-94] Others contain collections of articles describing models of neural, cardiovascular, respiratory, and renal-endocrine systems.[95] The development of a complex model of blood-pressure regulation is especially instructive.[96]

42. Signal Analysis. The need to analyze neural activity motivated the development of the first laboratory minicomputer in the early 1960s. Since then minicomputers and microcomputers have become standard tools for the analysis of physiological signals, in both laboratory and clinical settings.

Neurophysiology. The activity of neurons is described using a variety of histograms which characterize the spontaneously occurring activity of a single neuron, the response of the neuron to a stimulus, and the relationship between the activity of several simultaneously recorded neurons. A single recording containing the activity of several neurons can be processed to characterize the aggregate activity or to separate the contributions of the individual neurons. The timing of impulses can be determined either by software after analog-to-digital conversion or by using Schmitt triggers to generate interrupts when neural impulses occur.

Continuous signals originating from the nervous system that can be recorded noninvasively include sensory-evoked potentials, eye movements, and electroencephalograms (EEG). All these signals are used clinically to diagnose neurologic abnormalities, but computerized analysis of the most commonly used signal, the EEG, has not yet been widely accepted. The analysis is complicated by the presence of artifacts, lack of objective criteria for classification, and the need to consider up to 16 simultaneous channels. The background activity is usually described by the frequency spectrum using fast Fourier transforms, while the occasional transient events are most often detected using pattern-recognition techniques in the time domain.[97]

In addition to these traditional analyses of time and waveform, computers have also been used to describe and track three-dimensional neuronal structures.[98] This application is closely related to *imaging,* discussed below.

ECG Analysis.[99-101] ECG analysis provided one of the first clinically accepted applications of computerized signal processing. ECG analysis may be performed as diagnosis to determine whether any cardiac abnormalities are present in a short stretch of recording or as part of continuous monitoring of a patient requiring intensive care. Both types of analysis require sampling of the ECG signal recorded from the patient as one, three, six, or twelve waveforms. Since standard electrocardiographic amplifiers have a bandwidth of 100 Hz, a sampling rate of 250 to 500 per second is recommended.

Diagnosis requires the simultaneous examination of the shape of waveforms on several leads. It starts with identification of QRS complexes (see Par. **26-5**, accomplished by detection of the largest rates of change in the waveform, and followed by a search for the location of the P, Q, R, S, and T waves. Magnitudes, slopes, and time intervals are determined to characterize morphologic and timing features of the ECG. After obtaining required parameters, a sequence of branching decisions is generally followed to arrive at a suggested diagnosis. Multivariate analysis has also been used. Present commercial services generally use a special cart which transmits the ECG to a central computer by telephone; however, microprocessor-based stand-alone systems are expected to become more common. An example of computer-generated ECG analysis is shown in Fig. 26-33.

```
TELEMED CORPORATION
COMPUTER PROCESSED ELECTROCARDIOGRAM
MSDL APPROVED VERSION D 41-42-25-11

PATIENT 007986205
DATE     111770
TIME     0905
CODE     1216
         I    II   III  AVR  AVL  AVF  V1   V2   V3   V4   V5   V6
PR      .29  .25  .13  .00  .00  .14  .13  .00  .00  .00  .00  .00
QRS     .11  .10  .11  .08  .12  .11  .08  .09  .10  .10  .14  .10
QT      .33  .32  .35  .27  .34  .40  .47  .39  .38  .44  .48  .33
RATE     66   66   65  102  102  104   81   77   81   64   64   64

CODE    4A   4A   4A   2A   4A   2A    2    3    2    2    2    2

   AXIS IN   P  QRS  T   Q    R   S  STO              ST-T QRS-T
   DEGREES     -66 222     17 -63 243                  21   72

2631 QRS VARIABLE, P ABSENT    . ATRIAL FIBRILLATION
8311 QRS AXIS RANGE -30 TO -90 . LEFT AXIS DEVIATION
6413 ST DEPRESSION, -.10MV OR  . EXCLUDE DIGITALIS EFFECT
     MORE NEGATIVE             .
6212 NEGATIVE T WAVES          . CONSISTENT WITH ISCHEMIA
                               . ABNORMAL ECG

                               . ------------ M.D.
```

Fig. 26-33. Computer-generated analysis of a 12-lead ECG. *(TELEMED Corporation.)*

While diagnostic ECG analysis can be performed off line, analysis of ECG in monitoring applications must be performed in real time. In coronary-care units, for example, a single-lead ECG is continuously scanned to detect any occurrence of premature ventricular contractions. Such an event, identified by an abnormality in the shape and time of occurrence of the QRS complex, may precede fatal irregularities in cardiac rhythm and usually requires therapeutic intervention. Algorithms to detect abnormal beats are based either on correlation or feature-extraction techniques. Commercially available minicomputer-based systems typically monitor up to 32 patients.

A considerable challenge is the automated analysis of one- or two-lead ECG recorded on miniature tape recorders for 24 h. These recordings are played back at 60 to 120 times real time, and extensive analog or digital preprocessing is usually required before the ECG is analyzed. A hierarchical approach in which digital preprocessing, QRS detection, and beat classification are performed in microprocessors operating in parallel appears especially promising.

Patient Monitoring. In addition to automating routine monitoring functions (see Par. **26-35**), computers also provide a display of trends in selected variables. A desirable feature for such a display is an indication of the reliability of collected data as a function of time. Variability in readings or the percentage of lost data are such possible indicators. Computers can also derive clinically significant information by considering the relationship between recorded variables. Respiratory mechanics, for example, are often partially characterized by compliance and resistance values. Both can be computed, according to a simple model, from proper measurements of respiratory pressure and flow. It is generally hoped that indirect measurements and mathematical models will eventually help predict the status of the monitored patient.

Imaging. Computers are used in image analysis in at least two fundamentally different ways. (1) They scan, enhance, and analyze two-dimensional images that are presented directly. Examples include chromosome analysis, automated cell recognition, and two-dimensional scanning of the lung after the inhalation of a radioactive isotope; some of this activity is often referred to as pattern recognition.[97] (2) They reconstruct cross-sectional images of the body from its projections. These projections can be obtained using x-ray transmission, ultrasound transmission or reflection, or the emission of photons or positrons from a radioisotope substance introduced into the body. The challenge is to improve the image, reduce the radiation dose needed, and reduce the time necessary for the reconstruction. A major breakthrough in reconstruction time would allow several cross-sectional images of the heart to be taken during cardiac contraction. Radiologic imaging is discussed in Pars. **26-14** to **26-16** and in recent reviews.[102]

43. Closed-Loop Control. In the physiological laboratory computers are increasingly used not only for signal acquisition and analysis but also in running the experiments. Computers are used to trigger electrical stimulators, to present acoustical or visual stimuli, to set the initial length of stimulated muscles, or to keep physiological parameters within specified limits.

In a clinical setting, computer-aided dosage planning of digitalis and other toxic drugs relies on the periodic estimate and determination of blood serum concentration of the drug. Data describing the patient are entered into the computer, which then computes the recommended dosage required on the basis of a pharmacokinetic model. Since the response differs with the patient and may change with time even in one particular patient, a comparison of actual and previously predicted drug concentration can be used to individualize the pharmacokinetic model before computing the next suggested dosage. Such systems have been shown to outperform physicians.[103,104]

Direct computer control of therapy is a relatively new development, although at least one system has been in clinical service since the early 1970s.[105] The system monitors patients after open-heart surgery and infuses blood on the basis of a simple control algorithm which considers left atrial pressure as the primary variable. The mean arterial pressure is lowered by a vasodilator drug using a proportional-derivative controller. To protect the patient, the maximum rate of infusion is limited regardless of the blood pressures. Automated computer-controlled infusion of insulin for diabetic patients is now also possible using continuous-flow glucose analysis at the bedside.[106]

44. Medical Diagnosis.[107,108] The essence of the diagnostic process is to consider data (symptoms) obtained from the patient and to decide, according to these data and accumulated knowledge and experience, which of many possible diseases, if any, the patient has. Since in most cases, the logical steps by which the physician arrives at a diagnosis are obscure, computer programs should not necessarily attempt to mimic human thought processes but should explore algorithms suitable for automated computation.

The simplest algorithm is a decision tree which is followed according to the symptoms of the patient until a diagnosis is reached. This approach can be very successful for relatively narrow medical problems but is not suitable for medical decision making in general. Statistical approaches based on Bayes' theorem have been extensively explored, but necessary assumptions about independence of symptoms and mutual exclusiveness of diseases limit its usefulness.

Rules that link diseases and symptoms and heuristic algorithms that determine how these rules must be applied form the basis of the newest and most promising approaches. An advantage of these approaches is that the rules can be easily updated without having to alter the algorithms that govern the method of invoking the rules. Automated diagnosis programs are still primarily in the research stages, and some have been more widely accepted as teaching aids than as tools for clinical decision making.

45. Medical Information Systems. *Medical Record.* The core of medical information systems is the patient record. The record may be that of an essentially healthy person, or it may belong to the acutely ill patient. Traditionally, the patient record has consisted of a chronologic series of hard-to-read notes, impressions, and test results without logical connection. Yet, patient care often depends on physicians who exchange information primarily through such a record. The major problem in developing computer-based medical records is that while computers are effective in dealing with well-structured information, traditional medical records do not have a well-defined structure.

The problem-oriented medical record is organized around a list of the patient's problems. All entries in the patient's record, including the data gathered, treatment plans, and physician's impressions, must be related to specific problems. In its prototype implementation the allowed entries are highly structured, and the patient's record consists of answers to some 0.5 to 1 million frames.[107,109] The elaborate system, consisting of a minicomputer network, is designed to allow scaling down to a more modest size.

Since analyzing narrative information is a formidable task, medical information should be captured in a structured format at a time of entry regardless of the organization of the record. This is facilitated by a question-and-answer dialog between user and computer, using carefully designed interactive terminals. Information is entered by responding to questions displayed by the computer. Response is made by pushing a function button, typing an answer on a keyboard, touching a portion of the screen, or pointing to a spot on the display by a light pen (see Fig. 26-34). The computer acknowledges the answer and provides immediate feedback if the entered information is conflicting or is in the wrong format. Frame sequencing generally depends on the answers provided by the user.

Fig. 26-34. Use of an interactive display terminal. *(Courtesy of Dr. G. Octo Barnett.)*

Hospital Information Systems.[110,111] It has been estimated that over 25% of a hospital budget is spent on information processing, indicating a large potential for automation. Initially, computers were used off line for accounting purposes only. Current emphasis is on on-line systems, with terminals located throughout the hospital. In these systems the computer not only acts as an information processor but also as the center of a communications network.

Assume, for example, that the physician orders blood tests to be performed on a particular patient. A clerk enters this information into the computer at the nursing station. When the sample is drawn, the computer generates a label that is attached to the test tube before it is sent to the clinical laboratory, where tests are performed by automated analyzers. Results are stored in the patient's file. If the label indicates an emergency, results are immediately reported to the nursing station; otherwise they are reported at the end of the day as part of a summary. Simultaneously, appropriate charges for the tests are entered into the patient's account.

Hospital information systems maintain a census on patients and beds in the hospital and contain a record for each patient. Ideally, the entire medical record of the patient should be integrated with the hospital information system, but generally only a portion of the record is kept by the computer. This computerized record may show test results (x-ray findings, ECG interpretation, urinalysis), medications ordered and taken, and data such as periodic blood pressure and temperature readings. Properly designed systems should reduce administrative workload and improve patient care by providing up-to-date information upon which diagnostic and therapeutic procedures are based. However, reductions in administrative costs and improvements in patient care are very difficult to demonstrate.

Progress in developing such systems has been slower than expected. A major problem stems from a lack of understanding of the complexities involved in developing a viable system serving the entire hospital. The typical hospital system is characterized by complexity and inertia; thus, practicality often limits development of an all-inclusive information system in existing hospitals.

In new institutions, automated information processing can be designed into the physical plant and administrative organization. The most promising solution for existing hospitals is a stepwise, modular development of computer systems. Automated information systems for specific and limited goals are developed first. Once thoroughly tested, individual modules are interconnected for a gradually expanding system.

Clinical laboratories were among the first hospital units to be automated. The discovery of new test procedures and the development of continuous-flow analyzers have led to the development of commercially available systems that collect large volumes of data from highly automated analytic instruments. To facilitate the distribution of test results, the laboratory computer should be integrated with the rest of the hospital information system. This integration is hindered by the lack of a convenient and economical method of linking the readings of the automated instruments with the patient whose sample produced the result. Dedicated computer systems in the pharmacy are likewise commercially available. These systems maintain a record of medications dispensed and alert the physician if interaction of the prescribed drugs that may have been overlooked would endanger the patient.

Another major problem in hospital information systems is capturing manually entered input data in a reliable yet convenient manner. Sensitivity to the capabilities, motivations, and time limitations of the user is of extreme importance. Consider, for example, a case when the computer requires the user to specify the required dosage for a particular drug. Instead of expecting a typed answer, it is desirable to display the most commonly used dosages for that drug and allow the user to choose by pointing at or touching the desired portion of the display.

The success of computerized hospital information systems is critically dependent on the choice of the software operating system. It is essential that systems analysts and programmers concentrate on the logical relationships in the information system and not on the details of applications programming. Widely available high-level problem-oriented languages with convenient file-manipulating capability are a prerequisite for the efficient development of systems which can be maintained, expanded, and used in an environment other than that in which the system was originally created.

Ambulatory Care Systems.[12]   A significant proportion of medical care is delivered by physicians in a group practice outside of a hospital. In such a setting the patient may be seen by any of a number of nurses and physicians who derive their primary information about the patient from the medical record. Computers are acquiring a major role in such ambulatory care systems, and at least one system has been developed to such an extent that the entire patient record is computer-based. These records do not approach the complexity of records found in hospitals, but their computerization allows potentially significant improvement in the care provided. For example, patients with abnormal test results automatically receive the appropriate followup.

46. Minicomputers vs. Microcomputers. The trend in biomedical applications of computers is toward decentralization and distributed computing. In place of a large central processor, individual minicomputers or a network of minicomputers are being increasingly used in hospital information systems. For example, separate but interconnected systems may be used for the pharmaceutical clinical laboratory and census operations of the hospital. In applications in which, traditionally, a single minicomputer was used, such as in patient monitoring, networks of microcomputers are expected to supplant them increasingly. For example, separate microcomputers at the bedside may feed a central mini- or microcomputer at the nursing station. Alternatively, different microcomputers may be dedicated to the acquisition and analysis of specific physiological signals.

Decentralization of computing takes advantage of falling hardware prices while trying to reduce the increasing cost of software development. The effort necessary to create and debug a complex program grows faster than linearly with the length of the program; it is thus advantageous to break a complex task down into several smaller ones. Modularity in both software and hardware facilitates maintenance, reduces the need for big backup systems, and makes disastrous breakdowns unlikely.

In many applications the distinction between minicomputers and microcomputers has been blurred. With the availability of 16-bit microcomputers, sophisticated operating systems, high-level languages, and increasingly standardized interfaces, the convenience and power of microprocessor systems approach those of minicomputer systems of just a few years ago.

Implantable Instrumentation

BY JAMES D. MEINDL

47. Functions. All electronic instruments are intended for use "by" people. The distinctive feature of biomedical instruments is that they are intended for use "on" people (and other living systems). The various types of biomedical instruments used on people and animals can be classified according to their orientation relative to the subject and the generic function which they perform, as shown in Table 26-9. In this representation each matrix element is a specific example of a particular type of instrument. Implantable instruments installed during surgery are represented by the "subcutaneous" row of the matrix and also are taken to include devices such as ingestible telemetry pills entirely enclosed by inner surfaces of the body.

Research.[114] Animal models of human disease are indispensable tools in medical research for a host of ethical, legal, scientific, and economic reasons. Implantable telemetry systems offer an invaluable appendage to animal models since implants provide the means for acquisition of data not available from the surface of the body, when the animal is not anesthetized, restrained, or interfered with in any way, over prolonged periods of many months of normal life using automated techniques for 24-h data collection. Totally implantable telemetry and telestimulation systems offer an absolute minimum of interference with and by the subject and no risk of infection from percutaneous wires in physiological, pharmacological, and pathological studies of animals. In addition, they provide an essential step in the evolution of new implantable instruments for use in human beings.

Diagnosis.[115] The ideal diagnostic instrument provides definitive data on the condition of a patient, causes no harm or discomfort, and is convenient, reliable, and economical for a medical practitioner to use. For obvious reasons, these criteria are not readily satisfied by implantable instruments. Consequently, they are not ordinarily used for diagnosis. Ingestible telemetry pills, for example, represent an exception.[116]

Monitoring. In addition to fulfilling all the requirements of an ideal diagnostic instrument, an ideal monitoring instrument imposes the additional stringent requirement of virtually total freedom from the need for a human operator.[115] Because of their invasive character, implants are most frequently used for monitoring in postsurgical patients. For instance, an implant can obviate periodic skull surgery for measurement of intracranial fluid pressure in a hydrocephalic child following surgery for installing a shunt.[117,118]

Therapy. Therapeutic uses of implants include chemical, electrical, and mechanical stimuli to perform remedial functions. Typical examples are small electric currents to accelerate bone healing,[119] strain gages to measure axial forces in Harrington rods for correction of scoliosis,[120] and dorsal column electrical stimulation for controlled interruption of pain transmission to the brain.[121]

Prostheses. The most common implantable electrical prosthesis is the artificial cardiac pacemaker, a device which is vital to many patients.[122] It is estimated that over 100,000 are installed annually in the United States alone. Other prostheses now under investigation include an artificial pancreas,[123] an auditory prosthesis for the profoundly deaf,[115] and control systems for skeletal muscle using electrical stimulation.[124]

The various functions performed by implantable instruments imply a unique set of generic performance requirements. The foremost of these are small size and weight, low energy consumption, low supply voltage (often a single-cell battery), long operating life, high reliability, unusual sensors and transducers, and biological compatibility. Custom integrated circuits in hybrid assemblies offer the most promising approach to fulfilling these requirements, although they may cost more than units assembled with standard components.

48. Research Telemetry and Telestimulation. A block diagram of a totally implantable multichannel telemetry system is shown in Fig. 26-35.[114] The portion of the system inside the dashed rectangle is implanted in the body of a laboratory research animal. The remainder is externally located, typically within a range of 3 to 10 m. The signal outputs are displayed on a cathode-ray tube and recorded on magnetic tape or a paper-strip chart.

Internally, the six input-stage preamplifiers, the 10- to 15-kHz current-controlled oscillator, and its output stage are implemented in a custom integrated circuit designed to accept input signals from a variety of transducers for measuring pressure, electric potentials, temperature, and similar physiological parameters. The rf telemetry transmitter is a custom chip which transmits at frequencies up to 125 MHz, using either frequency or pulse modulation. The CMOS counter is a commercial 5-by-8 decoder. The command receiver is an rf-controlled elapsed-time power switch operating in the 27-MHz citizens' band. Its standby power drain is 7 μW.

Upon activation by a 0.5-s rf burst from the external command transmitter, the power switch serves to connect the single lithium iodide cell power source to the telemetry electronics for a preselected self-timed data collection interval of several minutes, following which the switch automatically disconnects the power source. A low-duty cycle greatly prolongs the useful operating life of the implanted unit. Location of the command receiver in the sealed assembly with the battery prevents leakage current in the cable to the telemetry electronics during quiescent periods.

TABLE 26-9   Biomedical Instrument Matrix: Orientation vs. Function[113]

Type	Research	Diagnostic	Monitoring	Therapeutic	Prosthetic
Subcutaneous	Totally implantable telemetry		Cerebral-pressure telemetry	Nerve block for pain relief	Cardiac pacemaker
Supercutaneous	Animal backpack telemetry	Ingestible pH telemetry capsule	Ambulatory care ECG telemetry	Nerve block for pain relief	Hearing aid
Percutaneous	Implantable transducer with external leads	Catheter-tip blood-gas sensor	Catheter-tip pressure sensor	Therapeutic muscle stimulation with percutaneous electrode	Functional muscle stimulation with percutaneous electrode
Transcutaneous	X-ray	Computerized x-ray tomography	Ultrasonic imaging	Defibrillator	Electronic reading aid for the blind
Extracutaneous	Electron microscope	Mass spectrometer	Gas chromatograph		

The supply-voltage range of the multichannel systems is 2.3 to 2.8 V. The nominal current drain of the signal-processing electronics is 400 μA. Input-signal variations of 66 dB, ranging from 2.0 μV to 400 mV in amplitude and from 0 to 250 Hz in bandwidth, can be accommodated on six channels.

A photograph of an implantation-ready unit encapsulated in medical grade silastic rubber and equipped with one titanium diaphragm piezoresistive pressure sensor and three cardiac electrodes with coiled cables is shown in Fig. 26-36. The cylindrical object is the battery pack and command receiver. The telemetry transmitter dipole antenna leads extend to the right and left of the electronics package, which is hermetically sealed to prevent leakage of body fluids. Table 26-10 is a summary of physiological signals which can be telemetered by this system, with certain exceptions, when equipped with suitable transducers.[114,125, 26]

The major physiological parameters not yet telemetered with multichannel systems, such as illustrated in Fig. 26-35, are blood flow and dimensions using ultrasonic transducers. (Totally implantable electromagnetic blood flowmeters are precluded in many applications by the very large current and power requirements of the transducer field coils as well as base-line instability.[127]) The principal obstacle is the large bandwidths of the "video" signals present in these instruments before demodulation, which is handled externally because of the complexity and bulk of the necessary circuitry.[114]

To achieve sufficient range resolution in the measurement of either blood-velocity profiles with a pulsed Doppler ultrasonic flowmeter or dimensions via transit time of ultrasonic pulses, typical pulse lengths must be less than 1.0 μs, which corresponds to a distance in tissue of 1.5 mm. These short pulses give rise to 500- to 1,000-kHz video signals with 60-dB amplitude variations. Such signals are more practical to transmit across the skin via a short-range inductively coupled data link than an rf telemetry link. In the future, through more extensive use of integrated circuits, internal demodulation to reduce signal bandwidth to the 0.1- to 20-Hz frequency

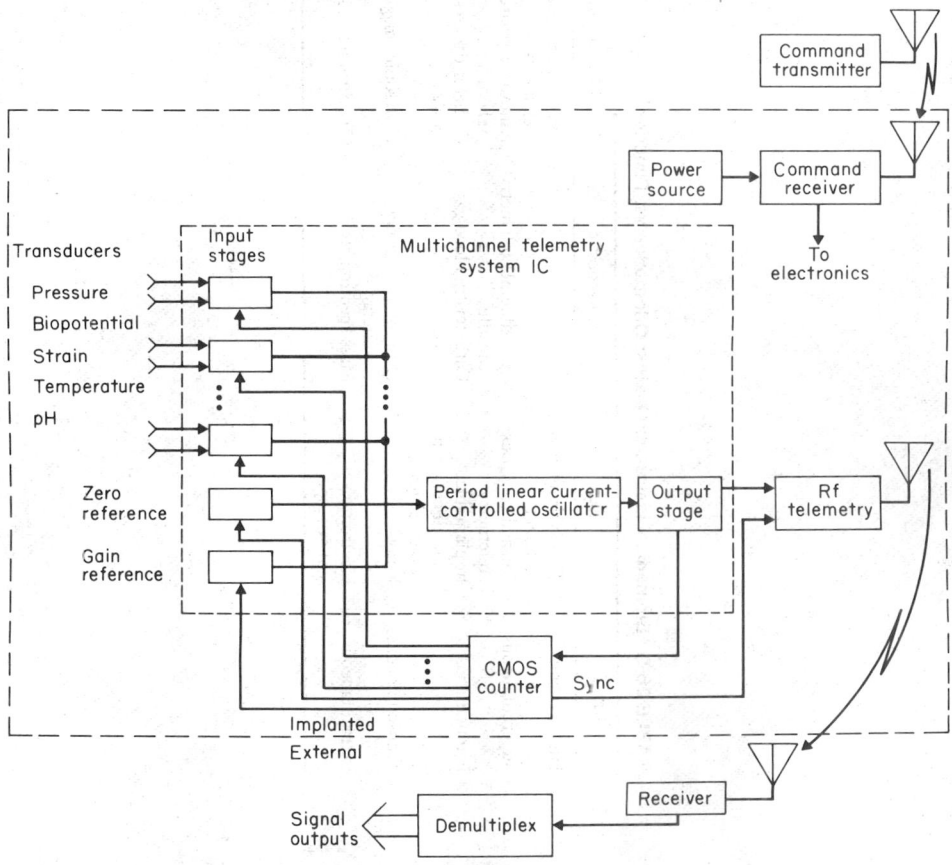

Fig. 26-35. Multichannel-telemetry-system block diagram.[114]

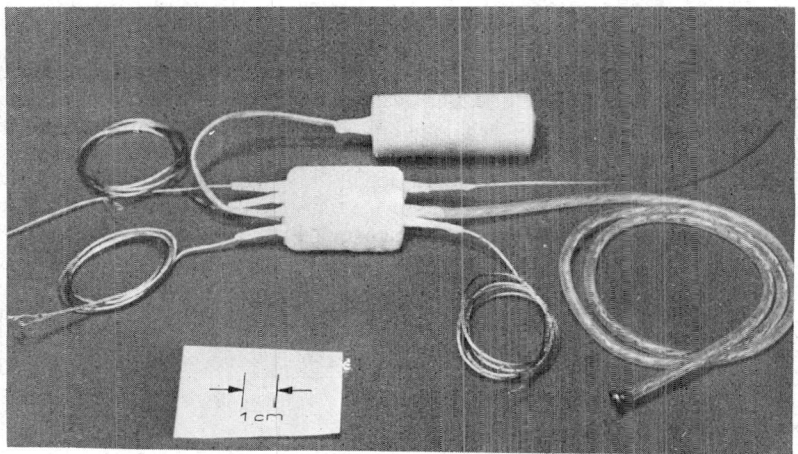

Fig. 26-36. Multichannel telemetry system.

TABLE 26-10 Signal Characteristics of Telemetered Parameters

Physiological parameters	Transducer	Amplitude range	Frequency range
Acceleration	Silicon cantilever strain gage	0.001–50g	0–100 Hz
Bladder pressure	Strain gages	0–100 cmH$_2$0	0–10 Hz
Blood flow	Ultrasonic flowmeter:		
	Continuous-wave Doppler	1–300 cm/s	1–20 Hz flow
	Pulsed Doppler		20–40 kHz Doppler
	Electromagnetic flowmeter		500–1,000 kHz video
Blood pressure	Piezoresistive and capacitive pressure diaphragms; strain gages coupled to diaphragms or in cuffs	0–400 mmHg	0.5–100 Hz
Dimensions	Ultrasonic transit time	1–30 mm	0–10 Hz; 500–1,000 kHz video
Electrocardiogram (ECG)	Electrodes	0.005–4 mV pulse	0.1–100 Hz
Electroencephalogram (ECG)	Electrodes	10–75 μV	0.5–200 Hz
Electrogastrograph	Surface electrodes	10–350 μV	0.05–0.2 Hz
Electromyogram (EMG)	Electrodes	0.1–4 mV pulse	2–10^5 Hz (10–500 Hz clinical)
Eye potential, EOG or ERG	Electrodes	500 μV	0–250 Hz
Gastrointestinal pressure	Variable inductance	20–100 cmH$_2$0	0–10 Hz
Intestinal forces	Strain gages	1–40g	0–1 Hz
Ion concentrations, e.g., pH	Ion-selective electrodes on a metal-insulator-semiconductor FET, pH glass electrodes	3–13	0–0.1 Hz
Nerve potentials	Electrodes	3 mV peak	Up to 1,000 pulses/s, rise time 0.3 s
Respiration rate	Electrode impedance; piezoelectric devices; pneumograph	· · ·	0.15–6 Hz
Temperature	Integrated-circuit band-gap reference, diodes, thermistors	90–100°F	0–0.1 Hz
Tidal volume	Impedance pneumograph	50–100 mL/breath	0.15–6 Hz

range of physiological parameters should permit narrow-band rf telemetry as used for the multichannel system of Fig. 26-35.

A block diagram of a continuous-wave bidirectional Doppler ultrasonic blood flowmeter is shown in Fig. 26-37. A 6-MHz exciter oscillator drives a transmitting piezoelectric transducer contained in a cuff surrounding the blood vessel. A second receiving transducer located in the cuff opposite the transmitter receives ultrasonic signals Doppler-shifted in frequency by an amount proportional to the velocity of blood flowing in the sample volume, defined by the intersection of the near-field acoustic-beam patterns of the transducers. The Doppler frequency shift is given by $f_d = f_0(v/c) \cos \theta$, where f_0 = exciter frequency, v = blood velocity, c = sonic

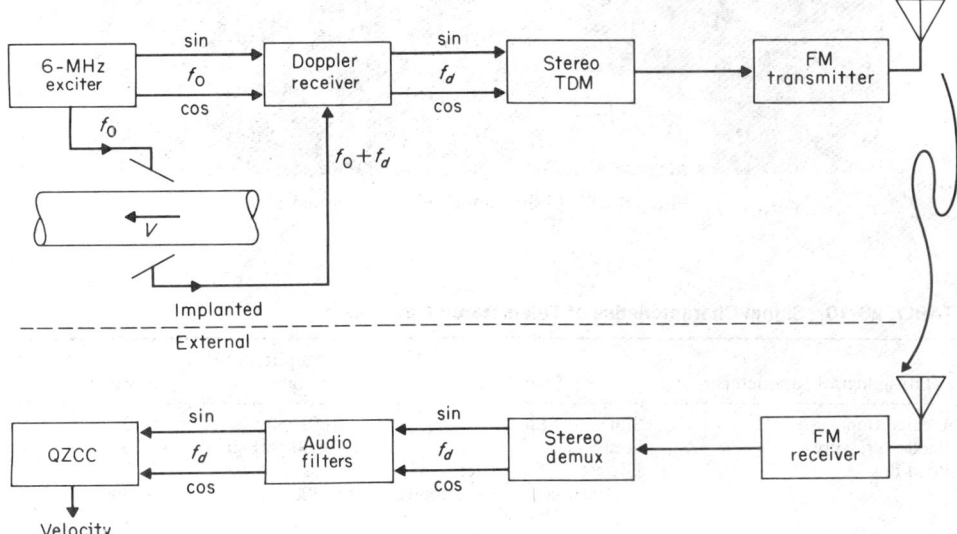

Fig. 26-37. Continuous-wave Doppler ultrasonic blood flowmeter.[114]

velocity in blood, and $\cos \theta$ = the angle between transducer beam and blood velocity. Directional sensing is retained by the characteristics of the sine and cosine functions $\sin(-2\pi f_d t) = -\sin 2\pi f_d t$ and $\cos(-2\pi f_d t) = \cos 2\pi f_d t$. The bandwidth allowed for these two baseband Doppler signals is approximately 100 to 10,000 Hz each. A dedicated rf telemetry link is convenient for use in this case. Long-term base-line stability is excellent thanks to the very small changes in f_0, $\cos \theta$, and c with time, and power drain is less than 30 mW.

Restrictions on size and weight mean that implantable telemetry transmitters generally are not crystal-controlled and are limited to inefficient antenna configurations,[128,129] which transmit low-level signals to reduce power drain and to comply with FCC regulations. Consequently, special-purpose external telemetry receivers incorporating a high degree of automatic frequency and gain control, along with high sensitivity, are helpful.

Energy sources for implantable instruments include primary batteries,[114] secondary batteries recharged by inductive coupling,[127] nuclear batteries,[130] inductive coupling only,[131-133] and biological sources.[134] Primary batteries are effective for uncaged animals without backpacks. They can be replaced by minor surgery if the implantable unit is properly designed.[114] Inductive coupling is most useful for small caged animals. It can cause interference with low-level signals. Nuclear batteries and biological sources are infrequently used in research animals.

49. Diagnostic and Monitoring Telemetry. Ingestible telemetry pills are used for measurement of temperature, pressure, pH, and orientation in the gastrointestinal tract of human patients and animals.[116,135-137] A block diagram of an ingestible pill for temperature and motion measurement is shown in Fig. 26-38. This pill uses a semicustom integrated circuit, a gravity-sensitive switch, a thermistor, and a single 1.35-V mercury power cell whose current drain is 50 μA. The size (14 mm diameter and 4.5 mm height) and buoyancy are constrained to be similar to those of an aspirin tablet. A multiantenna, phase-locked receiver system gathers motion information by tracking the rotation of the pill. The phase-locked technique reduces by 90% interfering noise occurring between rf bursts.

A block diagram of an implantable system for long-term monitoring of intraventricular pressure following neurosurgery is illustrated in Fig. 26-39. Inductive power at 3.5 MHz is beamed into a 3.5- by 2.0- by 0.5-cm package, and pressure and temperature modulated rf is returned at 120 MHz. The receiving system uses phase-locked-loop detectors and demodulators and provides both analog and digital readout of intracranial pressure and transducer temperature (to provide compensation for temperature changes).

50. Therapeutic and Prosthetic Implants. Perhaps the best known and clinically useful chronic implant at this time is the cardiac pacemaker.[122,138] This electronic device has been used in patients for treatment of chronic heart block since 1960. The earliest artificial pacemakers were simple regenerative pulse oscillators which applied fixed rate asynchronous electrical stimulation to the heart. A simplified schematic diagram of such a pulse-generator circuit is

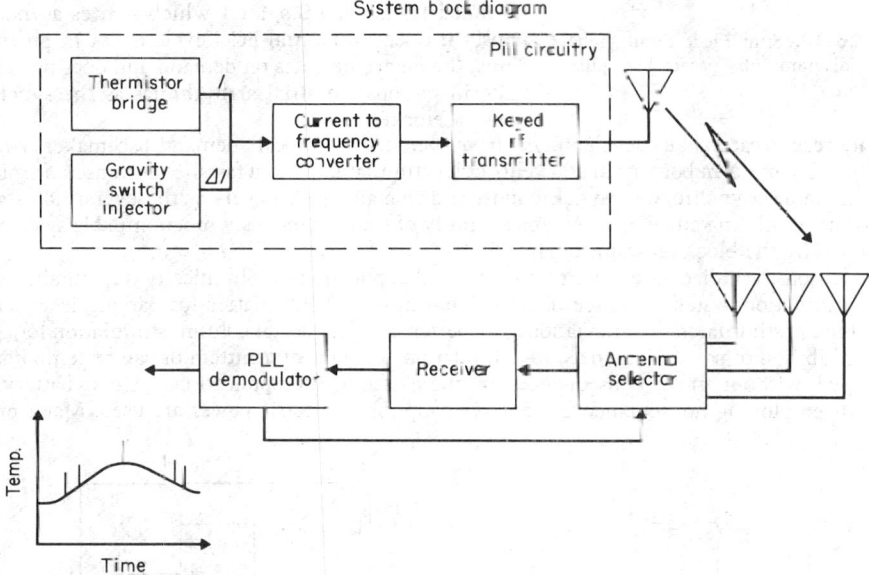

Fig. 26-38. Ingestible telemetry-pill block diagram.[13]

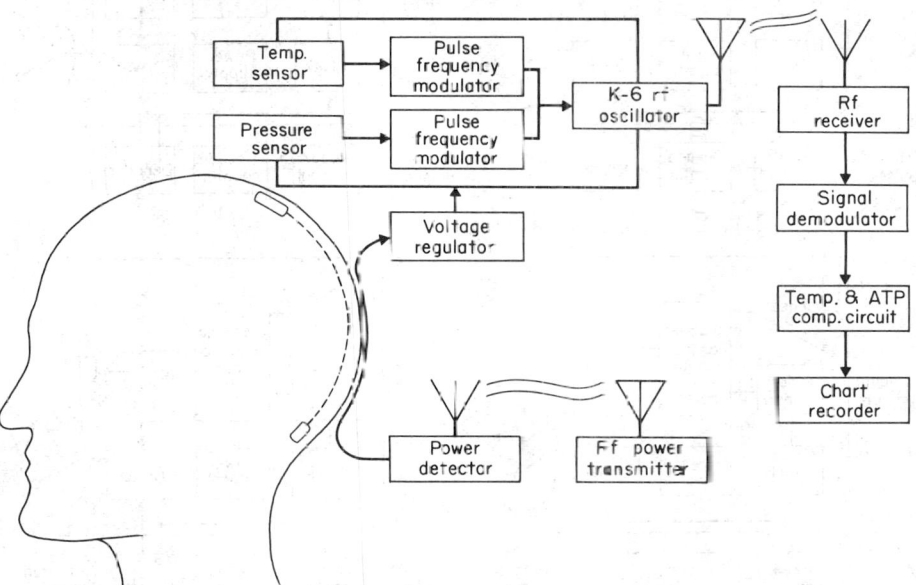

Fig. 26-39. Block diagram of the modules of implant, power, and receiver systems for intraventricular pressure telemetry.[118]

shown in Fig. 26-40. An early extension of this basic *asynchronous system* was the provision for rate adjustment by use of a magnet held against the skin. Following this, the *synchronous pacemaker* was introduced. Essentially, it is an ECG preamplifier plus a pacemaker pulse generator that uses an extra sensing electrode. The ECG preamplifier senses a central nervous system signal at the right atrium and responds by triggering the pulse generator, causing it to stimulate the ventricle synchronously.

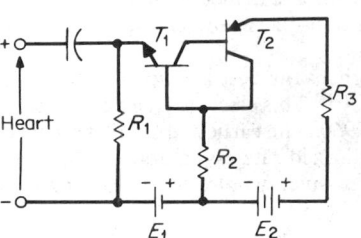

Fig. 26-40. Simplified circuit schematic diagram for pacemaker pulse generator.[139]

The *demand pacemaker* represents a third approach to pacing. It senses the presence of a natural ECG signal. Should this signal exist, the demand pacemaker becomes inhibited for about 1.0 s, after which it fires asynchronously if a second natural ECG cycle is not in progress. Thus, the device operates on demand and does not stimulate in response to atrial arrhythmias as the synchronous pacemaker may.

More recent pacemaker developments have included a "bifocal" demand pacemaker, which is designed to perform both atrial and ventricular stimulation,[140] and families of transcutaneously programmable asynchronous, synchronous, and demand pacemakers with several adjustable rates and stimulating currents.[141] A typical family of transcutaneously programmable systems is illustrated by the block diagrams of Fig. 26-41.

In addition to cardiac pacemakers, other clinical applications of chronically implantable electrical stimulation systems include bladder[142] and muscle[124] stimulators for paraplegic patients, carotid-nerve stimulators for alleviation of hypertension, and dorsal column stimulation for electrical inhibition of pain.[143] The need for stimulation is either intermittent or can be temporarily suspended without grave consequences in these stimulator applications. Thus, batteryless implants employing transcutaneous inductive coupling of electric power are used. Major prob-

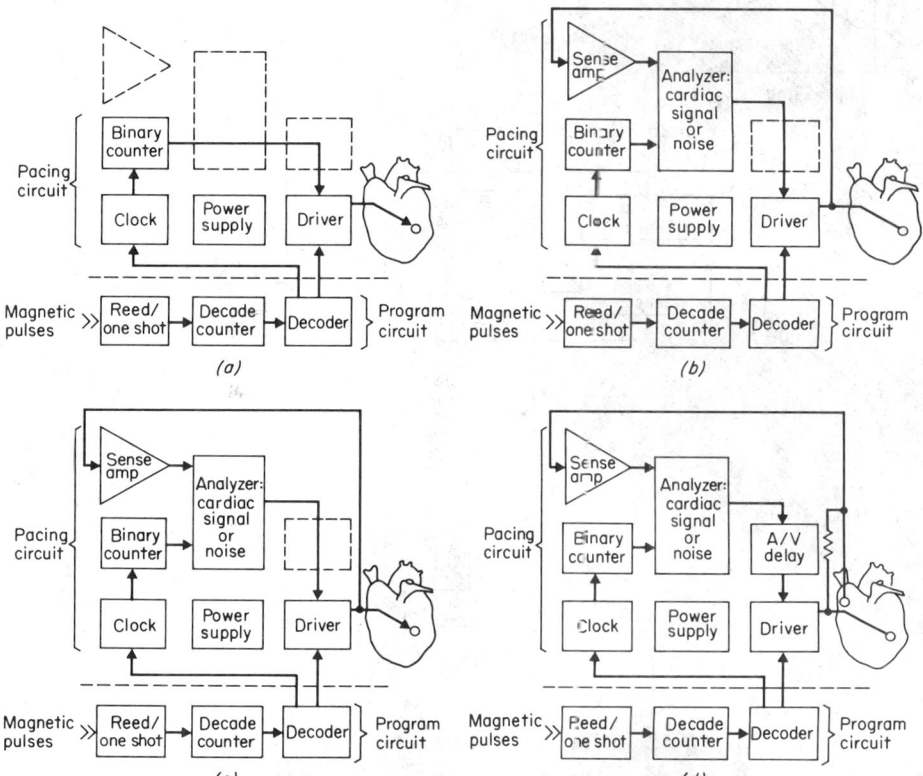

Fig. 26-41. Transcutaneously programmable pacemaker block diagrams: (a) asynchronous system, (b) synchronous (R-wave) system, (c) demand (R-wave-inhibited) system, (d) synchronous (P-wave) system.[138]

lems in such stimulators include the electrodes and tissue damage. Requirements for electrodes include minimal corrosion in the body, good fatigue resistance, and minimum body reaction to limit the increase of threshold.

Single- or multistrand wires of gold, platinum, stainless steel, and Elgiloy are used. Coils or helical-wound multistrand wires provide better flexibility and less fatigue. Elgiloy is reported to have the best fatigue resistance but corrodes when used as an anode. Platinum electrodes can be successfully used either as anode or cathode.

Tissue damage from electrical stimulation using "nontoxic electrodes" can be caused by gas generation due to large current density; heat generated at the electrode site; toxic material generated by electrode-chemical reaction at the electrode site; and mechanical stress caused by the electrode assembly.

Opportunities for advanced electrical stimulation systems to serve as neural prostheses can be enhanced substantially by using integrated-circuit technologies to produce microelectrode arrays.[144,145] Such arrays are being applied to a cochlear prosthesis for the profoundly deaf.[146,147]

TABLE 26-11   Stimulating Parameters

	Pulse width, ms	Pulse rate, s^{-1}	Pulse voltage, V	Pulse current, mA
Heart	0.1–10	0.5–3.33*	0–9	4–45
Bladder	0.1–10	15–35	0–10	50
Nerve	0.05–1	5–100	0–10	0.1–10
Indirect muscle (via nerve):				
Intramuscular	0.1–10	1–50	0–10	0.5–10
Surface	0.1–10	1–50	10–60	5–50

*30–200 pulses per minute.

This implantable system uses the unusual approach of an rf link for transcutaneous power transmission and an ultrasonic link for information transmission in order to reduce interference between the two links.[146]

Feasibility studies of a visual prosthesis incorporating multielectrode stimulation of the visual cortex have been reported.[148,149] Substantial advances on many fronts are required for such a prosthesis to become practical. A summary of electrical stimulation parameters is given in Table 26-11.

In addition to neural prosthetic devices, a variety of other types using electronics is being pursued. Interesting cases include artificial hip-joint telemetry devices,[150] the artificial heart,[151] and the artificial pancreas.[123,152]

Transducers for Biomedical Applications

BY WEN H. KO AND PETER W. CHEUNG

51. Biomedical Transducers. The transducer is the key to clinical measurement and diagnostic instruments. With the advance of integrated circuits and microprocessors, many measurement and instrument functions can be admirably performed by available integrated circuits or microprocessors in the field of biomedical electronics, just as they can for industrial, consumer, and aerospace applications. Present transducers cannot match the versatility and performance of integrated circuits and microprocessors, however, and a need for development and research exists.

Furthermore, for biomedical applications, several special requirements must be considered, e.g., (1) variations of size and weight limitations with the subjects (patients) dealt with; (2) smaller temperature range (20 to 45°C); (3) better stability and reliability requirements for long-term experiments and patient care; (4) magnitude and range of quantities measured, which are usually small compared with those of other applications, requiring higher sensitivity; and (5) the environment in which the transducer will be used, i.e., the biocompatibility and protection of the transducer from body fluids.

For biomedical electronics, both input and output transducers are needed. At the input end, transducers are used to convert biomedical quantities into electric signals for the detection of body condition and behavior. At the output end, transducers are generally used to convert electric signals into some form of stimuli of the body tissues to elicit a desired response. The physical principles, design considerations, and structure of biomedical transducers are similar to those used for other applications discussed in Sec. **10**. A summary of biomedical transducers as commonly used is presented here.

With the advance of solid-state technology, many transducers have been designed with reduced size and weight specifically for medical applications. They have the potential of being used for implants for long-term study. A survey of these solid-state physical and chemical transducers follows.

52. Physical Transducers. Quantities measured by these devices and the common principles involved can be outlined as follows.

Temperature. Body and environment temperatures are measured with various transducers, including thermal-resistive devices (thermistor), thermoelectric devices (thermocouple), *pn* junction diodes, temperature-sensitive resonant circuits, infrared radiation, and chemical devices (liquid crystals and others).

Position, Displacement, and Motion. Position and displacement are measured with potentiometric devices, strain gages, linear-variation differential transformers, capacitive displacement devices, and ultrasonic devices. Velocity is measured with ultrasonic or optical Doppler effects,

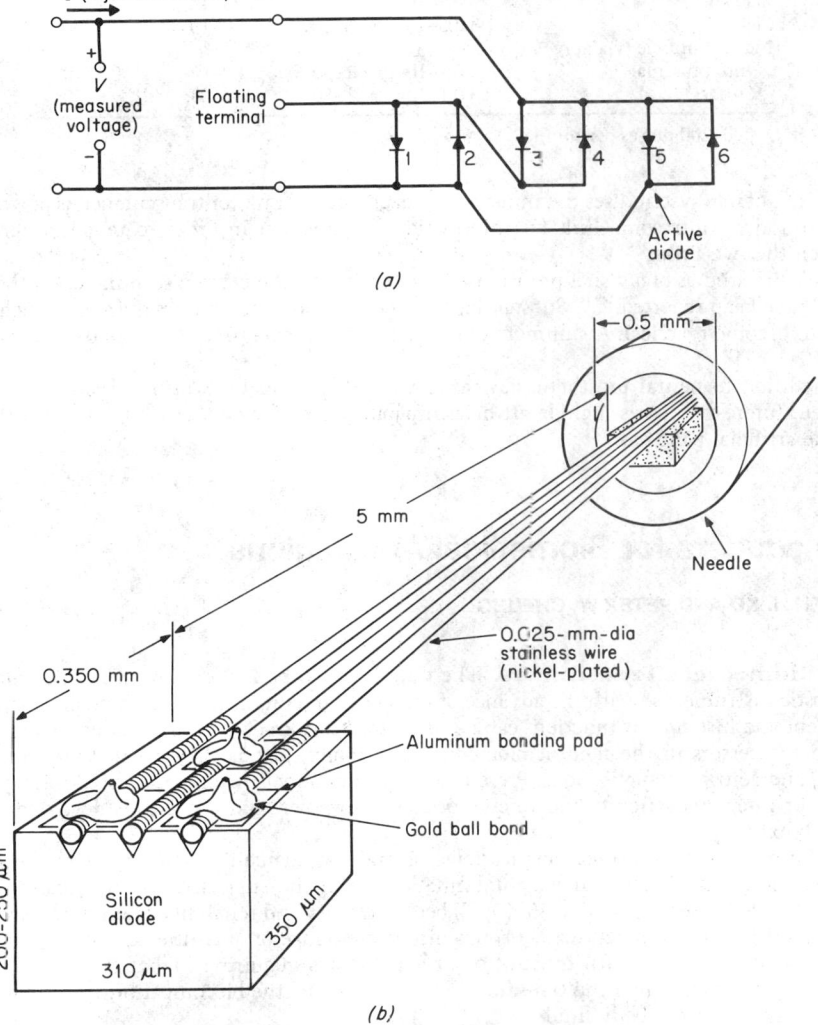

Fig. 26-42. Six-diode array of temperature sensors: (*a*) circuit design; (*b*) structure.

electromagnetic devices, and integration and differentiation of displacement or acceleration measurements. Acceleration is measured by mass and strain-gage systems and piezoelectric devices.

Force. Forces are measured by strain gages and devices balancing the unknown force against an electromagnetic force, elastic element, or a fluid-pressure element

Pressure. Various indirect methods are used to measure body-fluid pressures, e.g., the cuff method of measuring blood pressure and force-balance method used in tonometry. Direct methods include measuring mechanical elastic properties of tubes and diaphragms and using piezoresistive, capacitive, and optical methods to convert mechanical deformation of metal, glass, and silicon diaphragms into electric signals representing the pressure.

Fluid Flow. Fluid flow can be measured in the body with electromagnetic methods, devices using the principles of ultrasonic Doppler, pressure gradient, thermotransport, plethysmography, and the Fick and rapid-injection indicator-dilution methods.

53. Solid-State Physical Transducers. With the development of solid-state electronic techniques, miniature transducers can be made with accurate dimensions while retaining the same or better performance as those of regular size. Various physical and chemical transducers have been developed using the micromachining technique. A selected summary of these transducers designed specifically for biomedical applications follows.

Temperature. Figure 26-42 shows an interesting connecting scheme using *pn* junction arrays to measure the temperature profile in the body. A change of 2 to 3 mV/°C across the junction can be obtained. The change of energy band gap with temperature in semiconductor materials can be measured by the optical absorption spectrum using fiber optics. This provides a means of measuring body temperature with nonconductive devices during microwave-hyperthermia treatment of cancer.

Position and Displacement. A two-dimensional position detector in the form of a 6- to 8-mm rectangle has been fabricated on silicon. This device can determine the two-dimensional coordinates of a light spot on the surface of a device with good linearity.[155] As shown in Fig. 26-43,

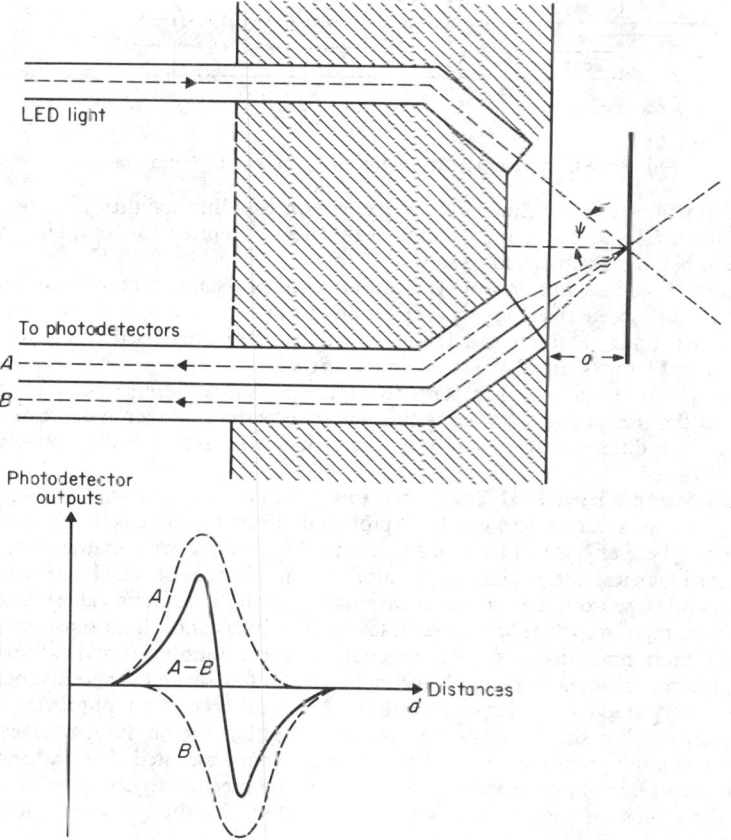

Fig. 26-43. Principle of optical position detector.

the displacement and position of a refracting surface can be measured accurately by optical reflection. An analog signal can be obtained at the output with an accuracy of a few micrometers.

Velocity and Acceleration. Ultrasound transducer arrays have been used for velocity-profile measurements. Figure 26-44 shows a mass and spring accelerometer, 2 by 3 by 0.6 mm in size, weighing less than 0.02 g, which has been reported with 10^{-1} to 10^{-3} m/s^2 sensitivity and 1,000 Hz frequency response.[159]

Force and Pressure. Piezoresistivity has been used to fabricate miniature force and pressure transducers. Both gage and absolute pressure devices were designed about several millimeters

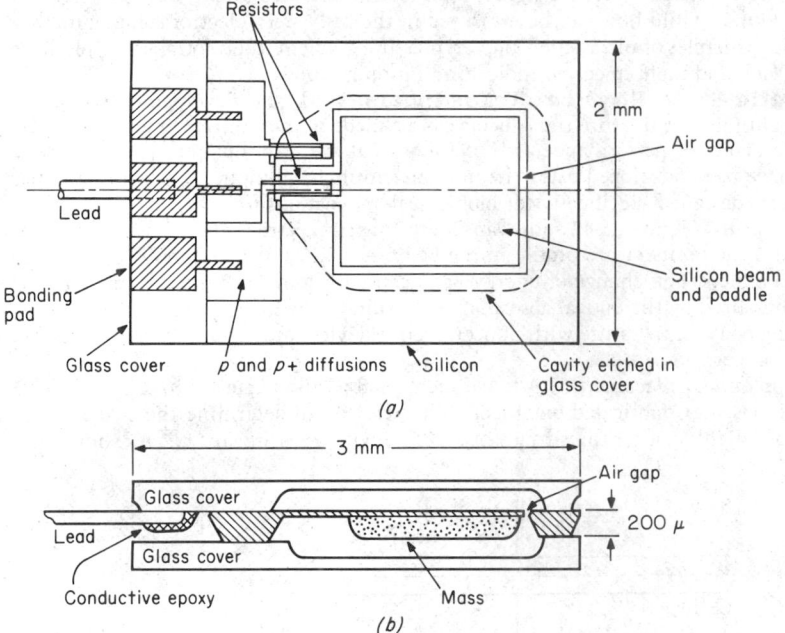

Fig. 26-44. Accelerometer: (*a*) top view; (*b*) centerline cross section.[162]

square and fractions of a millimeter thick. Long-term base-line stability of these devices is a critical problem. The problem of packaging of this device to protect it from body-fluid seepage and resultant bias drift remains to be solved.

Miniature silicon capacitor pressure transducers with first-stage processing circuits integrated on the chip have been made to measure fluid pressures. The optical-reflection technique described above under Position and Displacement has also been used to convert diaphragm deflection caused by pressure change into an electric signal.

Osmometer. With the two-chamber method, the pressure difference between the reference chamber and the measuring chamber with a selective permeable membrane can be measured with a solid-state differential pressure transducer that will give an indication of the osmotic pressure in the body.[160]

54. Solid-State Chemical Transducers. The chemisorption of gases and ions on the surface of semiconductor devices can have profound effects on the electrical characteristics of these devices. These effects can be used for chemically sensitive semiconductor devices (CSSD) especially for biological and medical applications where needs exist which are not met by conventional chemical sensors. The potential advantages of solid-state chemical transducers include miniature size, rapid response, low current, low output impedance, high sensitivity, low noise, and reduced interference due to short connections between transducers and external circuits.

The application of integrated-circuit technology can offer several other advantages, e.g., (1) multiple or array sensors for single or multiple chemical species for improved accuracy and reliability; (2) on-chip signal processing and compensation for nonideal characteristics, e.g., effects of temperature, nonlinearity, and interference from unwanted chemical species; and (3) potential low cost by volume production via efficient and precise processing techniques. A brief survey of the solid-state chemical transducers and systems for the measurement of ions, gases and biomolecules follows.

Humidity Sensor. Humidity is an important parameter in pulmonary and environmental medicine. One solid-state humidity sensor uses a silicon chip on top of a Peltier cooling element.[164] In this design the silicon chip contains a periodic structure whose capacitance changes abruptly when water vapor starts to condense on top of the chip as the temperature is lowered.

Another design uses a thin porous oxide layer sandwiched between a base metal and an extremely thin overlay gold electrode permeable to water vapor. The device acts as a leaky capacitor whose impedance is related to the humidity in the environment. The charge-flow transistor[165] (described under Gas Sensor below) also offers an interesting alternative for the humidity sensor.

Ion Sensor. The rapid measurement and sometimes continuous monitoring of electrolytes H^+, K^+, Na^+, Ca^{2+}, etc., in the blood and body fluids can provide important information for

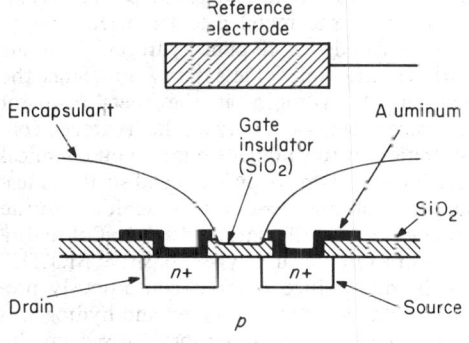

Fig. 26-45. Structure of silicon dioxide gate ISFET.

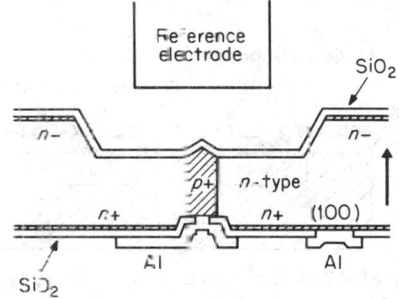

Fig. 26-46. Structure of gate-controlled diode.[170]

effective diagnosis and therapeutic management of critically ill patients. The ion-sensitive field-effect transistor (ISFET) (Fig. 26-45) is essentially an insulated gate field-effect transistor (IGFET) without its metal gate. The operation of the ISFET is similar to that of IGFET if one considers the reference electrode and the electrolyte as the modified gate.

The interfacial potential at the electrolyte-insulator interface produced by the net surface charge due to the ionization and complexation with the ions in solution will affect the channel conductance of the ISFET in the same way as the external gate voltage applied to the reference electrode. The drain current of the ISFET is therefore a function of the electrolytes in solution, for constant drain-source voltage.

Various materials have been investigated for gate insulators, such as SiO_2, Si_3N_4, and Al_2O_3. For pH sensors, Si_3N_4 and Al_2O_3 are reported to give better performance. ISFETs for other ions, such as K^+, Na^+, and Ca^{2+}, have been reported in which the gate insulator is coated with a layer of valinomysin in PVC, aluminosilicate, and dedecyl phosphonate, respectively.[166-168]

The concept of multiarray sensors has been demonstrated to be useful through a 10-channel pH sensor on a chip. Improved accuracy and reliability were obtained by incorporating majority-logic signal-processing schemes with an external microprocessor.

Chemical sensors can also be fabricated by hybrid technology. A miniature pH electrode was made by bonding a FET, connected as a source follower, adjacent to a pH-sensitive glass membrane on a common ceramic substrate, using thick-film hybrid technology.[169] A similar approach yields a fluoride-sensitive ISFET using lanthinum fluoride (LaF_3) as the ion-sensitive material.

A different type of CSSD referred to as a gate-controlled diode (GCD) is shown in Fig. 26-46. The GCD is biased in the inversion region by the reference electrode. At high frequencies the capacitance–gate-voltage response becomes frequency-dependent. If the GCD is maintained at constant capacitance, the frequency response of the GCD is a function of the interfacial electrochemical potential developed on the gate insulator. A pH-sensitive GCD has been reported using SiO_2 as the gate insulator.[170] This device can be extended to other ions by altering the chemical sensitivity of the gate insulator.

Gas Sensor. The measurement of partial pressure of gases such as oxygen and carbon dioxide in blood and tissues is significant in physiology and patient monitoring. The measurements of other gases such as hydrogen, nitrogen, argon, nitrous oxide, etc., are required in assessing tissue perfusion and pulmonary gas exchange using various washout techniques. The electrical characteristics of catalytic metal-insulator-semiconductor structures and bulk semiconductors are altered as their surfaces are exposed to various gases. These phenomena were explored for semiconductor gas sensors for biomedical, industrial, and environmental applications.

In some bulk semiconductors such as ZnO and SnO_2, the electron concentration near the surface can be varied by several orders of magnitude by absorption and reaction of gases at elevated temperature. The change in electric conductance can be used to measure the detected gas concentration. Although these semiconductor gas sensors generally offer a very high sensitivity, practical applications are limited by the high temperature required and problems of selectivity, reproducibility, and calibration. An example is the zinc oxide oxygen sensor.[171]

Another type of solid-state gas sensor is based on the principle that the chemisorption of gases at catalytic metal surfaces can alter the work function of the surfaces and thus modulate the space-charge layer adjacent to the surface by field effect. The metal-insulator-semiconductor structure (MIS) can be used to capitalize on this phenomenon for solid-state gas sensors. A hydrogen-sensitive MOSFET using a thin layer of palladium as the metal gate has been demonstrated.[171] The results showed that at 150°C, the device can detect 40 ppm hydrogen gas in air with response times of less than 2 min. Since the amount of hydrogen at the metal-insulator interface depends not only on the hydrogen concentration in the ambient but also on chemical reactions taking place on the metal surface, measurement of other gases can be achieved in the presence of a small controlled amount of hydrogen. For example, an oxygen-sensitive MOSFET can be made since water is continuously produced in an ambient of oxygen and hydrogen.

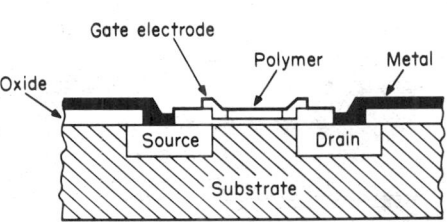

Fig. 26-47. Cross-sectional view of the charge-flow transistor.[165]

Another type of gas sensor is based on the potential generated across an electrochemical cell built with solid electrolytes. Solid-state oxygen sensors have been made using a number of solid electrolytes that exhibit a high oxygen-ion conduction in the solid electrolyte. These sensors offer high sensitivity and rapid response but have the disadvantages of instability, high operating temperature, long-term drift, and degradation in sensitivity due to aging effects.

A new device, the charge-flow transistor (CFT), has been developed for gas and humidity sensors.[165] A thin film of polymeric material is incorporated into the gate structure of the CFT, as shown in Fig. 26-47, such that the time delay between the application of gate-to-source voltage and the appearance of a complete channel depends on the resistivity of the thin film which senses the gas and moisture in the environment.

Semiconductor technology has also been applied to improve the conventional polarographic oxygen electrode. A monolithic silicon oxygen electrode with a multicathode array has been reported which promises to minimize the flow sensitivity of the oxygen cathode.[172]

Biomolecular Sensors and Integrated Chemical Systems. The application of enzyme-substrate reactions to detect the concentration of enzyme and other protein molecules has received much attention in biomolecular sensor development, especially with the advance of enzyme-immobilization techniques. It has been demonstrated that polyacrylamide gel immobilization of an enzyme onto a platinum-screen matrix using thick-film techniques can yield working electrodes for potentiometric measurement of substrate concentration. For example, L-lactate and glucose can be measured using immobilized lactate dehydrogenase and glucose oxidase. This technique can be incorporated in existing CSSD structures such as the ISFET to form a new class of enzyme sensors.

The solid-state ion and gas sensors described above can be used together with immobilized enzymes to sense the by-products, such as H^+, H_2S, O_2, H_2, NH_3, etc., for the development of a wide variety of enzyme sensors.

An immunochemically sensitive FET is based on the principle that when dissolved antigen combines with antibody immobilized on the gate insulator of the ISFET, the net charge of the antibody changes and can be sensed by the ISFET.

The potential of semiconductor technology in chemical sensors was illustrated by Terry and coworkers in a miniaturized gas-chromatography system on a silicon wafer.[173] Photolithography and chemical etch were used to fabricate a gas-injection valve and a 1.5-m-long gas-chromotography column, together with a thermal-conductivity sensor, on a 5-cm-diameter silicon wafer, reducing the size of the instrument by three orders of magnitude.

The future of adopting solid-state electronics technology for chemical sensors is believed to be vast and is limited only by the creative approach and careful implementation of these concepts.

55. Bibliography and References

GENERAL REFERENCES

1. Cromwell, L., et al.   "Medical Instrumentation for Health Care," Prentice-Hall, Englewood Cliffs, N.J., 1976.

2. Webster, J. G. (ed.)   "Medical Instrumentation: Application and Design," Houghton Mifflin, Boston, 1978.

3. Welkowitz, W. and S. Deutsch   "Biomedical Instruments: Theory and Design," Academic, New York, 1976.

4. *IEEE Transactions on Biomedical Engineering; Biomedical Engineering Journal,* Culver City, Calif.; *Journal of American Association of Medical Instrumentation- BioMedical Engineering,* United Trade Press, London; *Biomedical Engineering (Transactions of Meditsinskaya Tekhnike); Proceedings of Annual Conference of Engineering in Medicine and Biology* (since 1959).

5. *Biophysics, Bioengineering and Medical Instrumentation,* International Medical Abstracting Service, sec. 27, Excerpta Medica, Amsterdam; *Medical Electronics and Communications Abstract,* MultiScience Publishing Company, Essex, England; *Index Medicus,* U.S. Dept. of H.E.W., Public Health Service, National Library of Medicine (monthly).

6. Dummer, G. W. A., and J. M. Robertson (eds.)   "Medical Electronic Equipment," Pergamon, New York, 1970.

7. Weyer, E. M. (ed.)   Materials in Biomedical Engineering, *Ann. N.Y. Acad. Sci.,* 1968, Vol. 146, Prt. 1.

8. Levine, S. N. (ed.)   *J. Biomed. Mater. Res.,* March 1967.

BIOMEDICAL ELECTRONICS

9. ANSI   Safety Level of Electromagnetic Radiation with Respect to Personnel, ANSI C95.1-1974, IEEE, New York. 1974.

10. Neukomm, P. A.   Health Hazards from Telemetry RF Exposure, in Fryer, Miller, and Sandler (eds.), "Biotelemetry III," pp. 41–44, Academic, New York, 1976.

11. Schwan, H. P.   Microwave Radiation: Biophysical Considerations and Standards Criteria, *IEEE Trans. Biomed. Eng.,* 1972, Vol. BME 19, p. 304.

12. Gandhi, O. P.   State of the Knowledge for Electromagnetic Absorbed Dose in Man and Animals, *Proc. IEEE,* January 1980, Vol. 68, No. 1, pp. 24–32.

13. Special Issue on Biological Effects and Medical Applications of Electromagnetic Energy, *Proc. IEEE,* January 1980, Vol. 68, No. 1.

14. Rostoker, W., and J. O. Galante   Materials for Human Implantation, *Trans. ASME,* February 1979, Vol. 101, pp. 2–14.

15. Fraker, A. C., and A. W. Ruff   Metallic Surgical Implants: State of the Art, *J. Metals,* May 1977, pp. 22–28.

16. Boretos, J. W.   "Concise Guide to Biomedical Polymers," Thomas, Springfield, Ill., 1973.

17. Lefaux, R.   "Toxicology of Plastics," CRC Press, Boca Raton, Fla., 1968.

18. Rushmer, R. F.   Future Horizons for Technology in Health Care Delivery, *Adv. Biomed. Eng.,* 1976, Vol. 6, pp. 99–153.

ELECTROCARDIOGRAPHY, BIOPOTENTIALS, AND PATIENT SAFETY

19. Hoffman, I. (ed.)   "Vectorcardiography," Vol. 2, North-Holland, Amsterdam, 1971.

20. Pozzi, L.   "Basic Principles in Vector Electrocardiography," Thomas, Springfield, Ill., 1961.

21. Plonsey, R.   "Bioelectric Phenomena," McGraw-Hill, New York, 1969.

22. Geselowitz, D. B., and O. H. Schmitt   Electrocardiography, in H. P. Schwan (ed.), "Biological Engineering," McGraw-Hill, New York, 1969.

23. McFee, R., and G. M. Baule   Research in Electrocardiography and Magnetocardiography, *Proc. IEEE,* 1972, Vol. 60, pp. 290–321.

24. Kossmann, C. E. (chairman)   Recommendations for Standardization of Leads and of Specifications for Instruments in Electrocardiography and Vectorcardiography, *Circulation,* 1967, Vol. 35, pp. 583–602.

25. Hurst, J. W., and R. B. Logue (eds.)   "The Heart," McGraw-Hill, New York, 1970.

26. Walter, C. W. (ed)   "Electric Hazards in Hospitals," National Academy of Sciences, Washington, 1970.

27. Geddes, L. A., and L. E. Baker "Principles of Applied Biomedical Instrumentation," Wiley, New York, 1968.

28. Brazier, M. "The Electrical Activity of the Nervous System," Macmillan, New York, 1958.

29. Donchin, E., and D. B. Lindsley (eds.) "Average Evoked Potentials," NASA, Washington, 1969.

30. Pinelli, P. (ed.) "Progress in Electromyography," Elsevier, Amsterdam, 1962.

31. Wulfsohn, N. L., and A. Sances, Jr. "The Nervous System and Electric Currents," Plenum, New York, 1970.

RADIOLOGY

32. Glasser, O., et al. "Physical Foundations of Radiology," 2d ed., Hoeber, New York, 1952.

33. Evans, R. D. "The Atomic Nucleus," McGraw-Hill, New York, 1956.

34. Johns, H. E., and J. S. Laughlin Interaction of Radiation with Matter, Chap. 2 in Hine and Brownell (eds.), "Radiation Dose Symmetry," Academic, New York, 1956.

35. Heitler, W. "Quantum Theory of Radiation," 3d ed., Oxford University Press, Fair Lawn, N.J., 1954.

36. Johns, H. E. "The Physics of Radiology," 3d ed., Charles Pummis, Springfield, Ill., 1969.

37. Glasser, O., E. H. Quimby, L. S. Taylor, and J. L. Weatherwax "Physical Foundations of Radiology," Hoeber, New York, 1954.

38. Hine, G., and G. Brownell "Radiation Dosimetry," Academic, New York, 1956.

38a. Bender, M. A. The Autofluoroscope, in Bland (ed.), "Nuclear Medicine," 2d ed., McGraw-Hill, New York, 1971.

38b. Anger, H. O. Gamma Ray and Positron Scintillation Camera, *Nucleonics*, 1963, Vol. 21, p. 56.

39. Wagner, H. B. "Principles of Nuclear Medicine," Saunders, Philadelphia, 1968.

39a. Ter-Pogossian, M. A., N. A. Mullani, J. Hood et al. A Multiposition Positron Emission Computed Tomograph (PETT IV) Yielding Transverse and Longitudinal Images, *Radiology*, 1978, Vol. 128, pp. 477–484.

40. Blahd, W. H. "Nuclear Medicine," 2d ed., McGraw-Hill, New York, 1971.

41. Maynard, D. C. (ed.) "Clinical Nuclear Medicine," Lea & Febiger, Philadelphia, 1969.

42. Gross, W., S. Feitelberg, and E. H. Quimby "Radioactive Nuclides in Medicine and Biology: Basic Physics and Instrumentation," 3d ed., Lea & Febiger, Philadelphia, 1970.

43. Early, P. J., M. A. Razzak, and D. B. Sodee "Textbook of Nuclear Medicine Technology," Mosby, St. Louis, 1969.

ARTIFICIAL ORGANS

44. *Artificial Organs*, official journal of the International Society for Artificial Organs.

45. *Transactions of the American Society of Artificial Organs.*

46. Nose, Y. "Manual on Artificial Organs," Vol. 1 'The Artificial Kidney," Mosby, St. Louis, 1969.

47. Nose, Y. "Manual on Artificial Organs," Vol. 2, "The Oxygenator," Mosby, St. Louis, 1973.

48. *Devices Technol. Branch Contract. Meet. Proc. 1978*, National Institutes of Health, Publ. 79-1670, July 1979.

49. Galletti, P. M., and G. A. Brecher "Heart-Lung Bypass," Grune & Stratton, New York, 1962.

50. *11th Annu. Contract. Conf. Artif. Kidney Program NIAMDD Proc.*, January 1978, DHEW Publication (NIH) 79-1442.

MYOELECTRIC CONTROL AND FUNCTIONAL STIMULATION

Myoelectric Control

51. Katz, B. "Nerve, Muscle, and Synapse," McGraw-Hill, New York, 1966.

52. Ruch, T. C., and H. D. Patton "Physiology and Biophysics," Saunders, Philadelphia, 1971.

53. Stalberg, E., and J. Trontelj "Single Fibre Electromyography," Mirvalle, 1979.

54. Basmajian, J. "Muscles Alive," Williams & Wilkins, Baltimore, 1967.

55. Scott, R. N., P. A. Parker, and V. A. Dunfield Myoelectric Control, *IEEE Med. Electron. Monogrs.* 7-12, 1974.

56. Agarwal, G., and G. Gottlieb An Analysis of the Electromyogram by Fourier, Simulation and Experimental Techniques, *IEEE Trans. Biomed. Eng.*, 1975, Vol. 22, No. 3, pp. 225–229.

57. Bergey, G., R. Squires, and W. Sipple   Electrocardiogram Recording with Pasteless Electrodes, *IEEE Trans. Biomed. Eng.*, 1971, Vol. 18, pp. 206–211.

58. Roesler, H.   Statistical Analysis and Evaluation of Myoelectric Signals for Proportional Control, in Herberts et al. (eds.), "The Control of Upper-Extremity Prostheses and Orthoses," pp. 44–57, Thomas, Springfield, Ill., 1974.

59. Bouisset, S.   EMG and Muscle Force in Normal Motor Activities, in Desmedt (ed.), "New Developments in Electromyography and Clinical Neurophysiology," Vol. 1, pp. 547–583, Karger, Basel, 1973.

60. Kreifeldt, J., and S. Yao   A Signal-to-Noise Investigation of Nonlinear Electromyographic Processors, *IEEE Trans. Biomed Eng.*, 1974, Vol. 21, No. 4, pp. 298–308.

61. Childress, D., D. Holmes, and J. Billock   Ideas on Myoelectric Prosthetic Systems for Upper-Extremity Amputees, in Herberts et al. (eds.), "The Control of Upper-Extremity Prostheses and Orthoses," pp. 86–106, Thomas, Springfield, Ill., 1974.

62. Childress, D.   Powered Limb Prostheses: Their Clinical Significance, *IEEE Trans. Biomed. Eng.*, 1973, Vol. 20, No. 3, pp. 200–207.

63. Dorcas, D., and R. Scott   A Three-State Myoelectric Control, *Med. Biol. Eng.*, 1966, Vol. 4, pp. 367–370.

64. Childress, D.   A Myoelectric Three-State Controller Using Rate Sensitivity, *Proc. 8th Intl. Conf. Med. Biol. Eng.*, 1969, pp. 4–5.

65. Herberts, P., C. Alstrom, R. Kadefors, and P. Lawrence   Hand Prosthesis Control via Myoelectric Patterns, *Acta Orthop. Scand.*, 1973, Vol. 44, pp. 389–409.

66. Wirta, R., D. Taylor, and R. Finley   Pattern Recognition Arm Prosthesis: A Historical Perspective — A Final Report, *Bull. Prosthet. Res.*, (B.P.R. 10–30) 1978, pp. 8–35.

67. Jacobsen, S., and R. Jerard   Laboratory Evaluation of a Unified Theory for Simultaneous Multiple Axis Artificial Arm Control, *ASME Publ.* 79-WA/Bio-8, December 1979.

68. Graupe, D., J. Magnussen, and A. Beex   A Microprocessor System for Multifunctional Control of Upper-Limb Prostheses via Myoelectric Signal Identification, *IEEE Trans. Autom. Control*, 1978, Vol. AC23, 4, pp. 538–544.

Functional Stimulation

69. Brindley, G. S., and W. S. Lewin   The Sensations Produced by Electrical Stimulation of the Visual Cortex, *J. Physiol.*, 1968, Vol. 106, pp. 479–493.

70. Dobelle, W. H., S. S. Stensaas, M. G. Mladejovsky, and J. B. Smith   A Prosthesis for the Deaf Based on Cortical Stimulation, *Ann. Otol., Rhinol. Laryngol.*, 1973, Vol. 32, pp. 445–464.

71. Peckham, P. H., and J. T. Mortimer   Restoration of Hand Function in the Quadriplegic through Electrical Stimulation, pp. 83–95, in Hambrecht and Reswick (eds.), "Functional Electrical Stimulation, Applications in Neural Prostheses," Dekker, New York, 1977.

72. Mortimer, J. T.   Motor Prostheses, Chap 39 in Brooks (ed.), "APS Handbook and Motor Control," American Physiological Society, in press.

73. McHardy, J., D. Geller, and S. B. Brummer   An Approach to Corrosion Control During Electrical Stimulation, *Ann. Biomed. Eng.*, 1977, Vol. 5, pp. 144–149.

74. Pudenz, R. H., L. A. Bullara, D. Dru, and A. Talalla   Electrical Stimulation of the Brain, II: Effects on the Blood Brain Barrier, *Surg. Neurol.*, 1975, Vol. 4, pp. 265–270.

75. Brummer, S., and M. J. Turner   Electrical Stimulation with Pt Electrodes, I: A Method for Determination of "Real" Electrode Areas, *IEEE Trans. Biomed. Eng.*, 1977, Vol. BME 25, No. 5, pp. 436–439.

76. Guyton, D. L., and F. T. Hambrecht   Capacitor Electrode Stimulates Nerve or Muscle Without Oxidation-Reduction Reactions, *Science*, July 6, 1973, Vol. 131, pp. 74–76.

77. Guyton, D. L., and F. T. Hambrecht   Theory and Design of Capacitor Electrodes for Chronic Stimulation, *Med. Biol. Eng.*, September 1974, pp. 613–620.

78. Bernstein, J. J., L. L. Hench, P. F. Johnson, W. W. Dawson, and G. Hunter   Electrical Stimulation of the Cortex with Tantalum Pentoxide Capacitive Electrodes, pp. 465–477, in Hambrecht and Reswick (eds.), "Functional Electrical Stimulation, Applications in Neural Prostheses," Dekker, New York, 1977.

79. Glenn, W. L., W. G. Holcomb, J. F. Hogan, T. Kaneyuki, and J. Kim   Long-Term Stimulation of the Phrenic Nerve for Diaphragm Pacing, pp. 97–112 in Hambrecht and Reswick (eds.), "Functional Electrical Stimulation, Applications in Neural Prostheses," Dekker, New York, 1977.

80. Waters, R.   Electrical Stimulation of the Peroneal and Femoral Nerves in Man, pp. 55–64 in Hambrecht and Reswick (eds.), "Functional Electrical Stimulation, Applications in Neural Prostheses," Dekker, New York, 1977.

81. Hershberg, P. I., D. Shon, G. P. Argawal, and A. Kantrowitz Histologic Changes in Continuous, Long-Term Electrical Stimulation of a Peripheral Nerve, *IEEE Trans. Biomed. Eng.*, 1967, Vol. BME 14, pp. 109–114.

82. Caldwell, C. W., and J. B. Reswick A Percutaneous Wire Electrode for Chronic Research Use, *IEEE Trans. Biomed. Eng.*, September 1975, Vol. BME 22, No. 5, pp. 429–432.

83. Mortimer, J. T., D. Kaufman, and U. Roessmann Tissue Reaction to Intramuscular Electrical Stimulation, *Proc. FASAB*, April 1974.

84. Crago, P. E., P. H. Peckham, J. T. Mortimer, and J. P. Van Der Meulen The Choice of Pulse Duration for Chronic Electrical Stimulation via Surface, Nerve, and Intramuscular Electrodes, *Ann. Biomed. Eng.*, 1974, Vol. 2, pp. 252–264.

85. McNeal, D. R. Analysis of a Model for Excitation of Myelinated Nerve, *IEEE Trans. Biomed. Eng.*, 1976, Vol. BME 23, pp. 329–337.

COMPUTER APPLICATIONS

86. Barnett, G. O. Computers in Patient Care, *New Engl. J. Med.*, 1968, Vol. 279, pp. 1321–1327.

87. Schroeder, S. A. Commentary: Why the Reservations about Medical Technology?, *Proc. IEEE*, 1979, Vol. 67, pp. 1337–1339.

88. Stacy, R. W., and B. D. Waxman "Computers in Biomedical Research," Vols. I–IV, Academic, New York, 1965–1974.

89. Special Issue on Pattern Recognition and Image Processing, *Proc. IEEE*, 1979, Vol. 67, pp. 707–856.

90. Marmarelis, P. Z., and V. Z. Marmarelis "Analysis of Physiological Systems: The White Noise Approach," Plenum, New York, 1978.

91. Yamashiro, S. M., and F. S. Grodins Optimal Regulation of Respiratory Airflow, *J. Appl. Physiol.*, 1971, Vol. 30, pp. 597–602.

92. Grodins, F. S. "Control Theory and Biological Systems," Columbia University Press, New York, 1963.

93. Blesser, W. B. "A Systems Approach to Biomedicine," McGraw Hill, New York, 1969.

94. Jones, R. W. "Principles of Biological Regulation," Academic, New York, 1973.

95. Brown, J. H. U., and D. S. Gann (eds.) "Engineering Principles in Physiology," Vols. I and II, Academic, New York, 1973.

96. Guyton, A. C., T. G. Coleman, A. W. Cowley, J. F. Liard, R. A. Norman, and R. D. Manning Systems Analysis of Arterial Pressure Regulation and Hypertension, *Ann. Biomed. Eng.*, 1972, Vol. 1, pp. 254–281.

97. Barlow, J. S. Computerized Clinical Electroencephalography in Perspective, *IEEE Trans. Biomed. Eng.*, 1979, Vol. BME-26, pp. 377–391.

98. Lindsay, R. D. (ed.) "Computer Analysis of Neuronal Structures," Plenum, New York, 1977.

99. *IEEE Proc. Comput. Cardiol. Conf. 1974*.

100. Van Bemmel, J. H., and J. L. Willens (eds.) "Trends in Computer-Processed Electrocardiograms," North-Holland, Amsterdam, 1977.

101. Thomas, L. H., K. W. Clark, C. N. Mead, K. L. Ripley, B. F. Spenner, and G. C. Oliver Automated Cardiac Disrhythmia Analysis, *Proc. IEEE*, 1979, Vol. 67, pp. 1322–1337.

102. Kak, A. C. Computerized Tomography with X-ray Emission, and Ultrasound Sources, *Proc. IEEE*, 1979, Vol. 67, pp. 61–71.

103. Jelliffe, R. W., and F. J. Goicoechea Computerized Dosage Regimens for Highly Toxic Drugs, *Am. J. Hosp. Pharm.*, 1974, Vol. 31, pp. 61–71.

104. Scheiner, L. B., M. Halkin, C. Peck, B. Rosenberg, and K. L. Melmon Improved Computer-Assisted Digoxin Therapy, *Ann. Intern. Med.*, 1975, Vol. 82, pp. 619–627.

105. Sheppard, L. C. The Computer in the Care of Critically Ill Patients, *Proc. IEEE*, 1979, Vol. 67, pp. 1300–1306.

106. Albisser, A. M. Devices for the Control of Diabetes Mellitus, *Proc. IEEE*, 1979, Vol. 67, pp. 1308–1320.

107. Schoolman, H. M., and L. M. Bernstein Computer Use in Diagnosis, Prognosis and Therapy, *Science*, 1978, Vol. 200, pp. 976–931.

108. Shortliffe, E. H., B. G. Buchanan, and E. A. Feigenbaum Knowledge Engineering for Medical Decision Making: A Review of Computer-Based Clinical Decision Aids, *Proc. IEEE*, 1979, Vol. 67, pp. 1207–1224.

109. Schultz, J. R., and L. Davis   The Technology of PROMIS, *Proc. IEEE*, 1979, Vol. 67, pp. 1237–1244.

110. Collen, M. F. (ed.)   "Hospital Computer Systems," Wiley, New York, 1974.

111. Barnett, G. O., and R. A. Greenes   Interface Aspects of a Hospital Information System, *Ann. N.Y. Acad. Sci.*, 1969, Vol. 161, pp. 756–768

112. Barnett, G. O., N. S. Justice, M. E. Somand, B. Adams, P. D. Leaman, M. S. Parent, F. R. Van Deusen, and J. K. Greenlie   COSTAR: A Computer-Based Medical Information System for Ambulatory Care, *Proc. IEEE*, 1979, Vol. 67, pp. 1226–1237.

IMPLANTS

113. Meindl, J. D.   Integrated Electron Devices in Medicine, *Tech. Dig. 1977 Int. Electron Devices Meet., Washington*, pp. 1A–1D.

114. Kimmich, H. P., and J. W. Knutti (eds.)   Implantable Telemetry Systems Based on Integrated Circuits, *Biotelem. Patient Monitoring*, Special Issue, 1979, Vol. 6, No. 3, pp. 95–106.

115. White, R. L., and J. D. Meindl   The Impact of Integrated Electronics in Medicine, *Science*, Mar. 18, 1977, Vol. 153, No. 4283, pp. 1119–1124.

116. Davis, P. R., D. A. Stubbs, and J. E. Ridd   Radio Pills: Their Use in Monitoring Back Stress, *J. Med. Eng. Technol.*, July 1977, pp. 209–212.

117. Walker, A. E., L. J. Viernstein, and J. G. Chubbuck   Intracranial Pressure Monitoring in Neurosurgery, pp. 69–77 in D. G. Fleming, W. H. Ko, and M. R. Neuman (eds.), "Indwelling and Implantable Pressure Transducers," CRC Press, Cleveland, Ohio 977.

118. Lorig, R. J., E. M. Cheng, and W. H. Ko   Systems for the Long-Term Monitoring of Intraventricular Pressure in Neurosurgery, pp. 79–84 in D. G. Fleming, W. H. Ko, and M. R. Neuman (eds.), "Indwelling and Implantable Pressure Transducers," CRC Press, Cleveland, Ohio, 1977.

119. Watson, J.   The Electrical Stimulation of Bone Healing, *Proc. IEEE*, September 1979, Vol. 67, pp. 1339–1352.

120. Nachemson, A., and G. Elfstrom   Results with Intravital Wireless Telemetry of Forces in the Harrington Distraction Rod, *Intravit. Telem.* June 1973, Vol. 9, No. 6, pp. 779–786.

121. Myelostat Dorsal Column Stimulator, Medtronics, Inc., Minneapolis, Minn., 1972.

122. Roy, O. Z.   The Current Status of Cardiac Pacing, *CRC Crit. Rev. Bioeng.*, June 1975, pp. 259–327.

123. Spencer, W. J.   For Diabetics: An Electronic Pancreas, *IEEE Spectrum*, June 1978, pp. 38–42.

124. McNeal, D. R. and J. B. Reswick   Control of Skeletal Muscle by Electrical Stimulation, *Adv. Biomed. Eng.*, 1976, Vol. 6, pp. 209–256.

125. *IEEE Trans. Electron Devices*, 1979, Vol. ED-26, No. 12, Special Issue on Solid-State Sensors, Actuators, and Interface Electronics, pp. 1861–1983.

126. McCutcheon, E. P. (ed.)   "Chronically Implanted Cardiovascular Instrumentation," Academic, New York, 1973.

127. Fryer, T. B., H. Sandler, and W. Freund   A Multichannel Implant Telemetry System for Cardiovascular Flow Pressure and ECG Measurement, *Biotelem.*, 1974, Vol. 2, pp. 40–42.

128. Ko, W. H., R. Plonsey, and S. Kang   The Radiation from an Electrically Small Circular Wire Loop Implanted in a Dissipative Homogeneous Spherical Medium, *Ann. Biomed. Eng.*, 1972, Vol. 1, pp. 135–145.

129. Ko, W. H., and S. Kang   The Radiation from an Electrically Small Off-Centered Loop in a Dissipative Homogeneous Spherical Medium, *Ann. Biomed. Eng.*, 1974, Vol. 2, pp. 321–325.

130. Ko, W. H.   Implant Evaluation of a Nuclear Battery—Betacel, *Proc. 3d Int. Conf. Med. Phys., Goteborg*, 1972, p. 30.8.

131. Fryer, T. B., G. F. Lund, and B. A. Williams   An Inductively Powered Telemetry System for Temperature, EKG, and Activity Monitoring, *Biotelem. Patient Monitoring*, 1978, Vol. 5, pp. 53–76.

132. Ko, W. H., and S. P. Liang   RF Powered Cage System for Implant Telemetry, *Proc. 30th Annu. Conf. Eng. Med. Biol., Los Angeles*, 1977, p. 260.

133. Bettice, J. A., A. Leung, R. J. Lorig, Y. Machtey, and W. H. Ko   Telemetric Intracranial Pressure Monitoring in Normal Goats, *Proc. 31st Annu. Conf. Eng. Med. Biol., Atlanta*, 1978, p. 324.

134. Sandler, H.   Biotelemetry: Its First 50 Years, *Biotelem.*, 1976, Vol. 3.

135. Pope, J. M., T. B. Fryer, and H. Sandler   An Ingestible Temperature Transmitter, *Proc. 24th Annu. Conf. Eng. Med. Biol., Las Vegas*, 1971, p. 321.

136. Ko, W. H., J. Hynecek, C. W. Poon, and E. Greenstein   Ingestible Telemetry System for Low Frequency Signals, *Proc. 26th Annu. Conf. Eng. Med. Biol., Minneapolis*, 1973, p. 5.

137. Lefferts, R. B., and J. D. Meindl   A Miniature Ingestible Telemetry System for Gastrointestinal Temperature Measurement, *Biotelem.*, 1978, Vol. 4, pp. 65–68.

138. Meindl, J. D.   Integrated Electronics in Medicine, *Adv. Biomed. Eng.*, 1976, Vol. 6, pp. 45–98.

139. Raillard, H.   Development of an Implantable Cardiac Pacemaker, *Dig. Tech. Pap.*, 1962 *IEEE Int. Solid State Circuits Conf., Philadelphia*, 1962, pp. 88–89.

140. Berkovits, B. V.   Bifocal Demand Pacing, *Proc. 9th Int. Conf. Eng. Med. Biol., Melbourne, Australia*, 1971, p. 75.

141. Terry, R.   Development of a Programmable Cardiac Pacing System, *Proc. 26th Annu. Conf. Eng. Med. Biol., Minneapolis*, 1973, p. 210.

142. Godec, C., A. S. Cass, and G. F. Ayala   Electrical Stimulation for Incontinence, *Urology*, April 1976, Vol. 7, pp. 388–397.

143. Cook, A. W., J. K. Taylor, and F. Nidzgorski   Functional Stimulation of the Spinal Cord in Multiple Sclerosis, *J. Med. Eng. Technol.*, January 1979, Vol. 3, pp. 18–23.

144. Wise, K., J. Angell, and A. Starr   An Integrated Circuit Approach to Extracellular Microelectrodes, *IEEE Trans. Biomed. Eng.*, July 1970, Vol. BME-17, pp. 238–247.

145. Wise, K., and J. Angell   A Microprobe with Integrated Amplifiers for Neurophysiology, *Dig. Tech. Pap., 1971 IEEE Int. Solid-State Circuits Conf., Philadelphia*, 1971, pp. 100–101.

146. Gheewala, T. R., R. Melen, and R. L. White   A Two Channel Approach to the Stimulation of Implantable Multielectrode Arrays, *Proc. 27th Annu. Conf. Eng. Med. Biol., Philadelphia*, 1974, p. 428.

147. White, R. L., R. G. Mathews, and G. A. May   An Implantable Cochlear Prosthesis for the Profoundly Deaf: Principles and Performance Requirements, *IEEE 1979 Frontiers Eng. Health Care, Denver*, 1979, p. 211.

148. Dobelle, W. H., et al.   Artificial Vision for the Blind: Electrical Stimulation of Visual Cortex Offers Hope for a Functional Prosthesis, *Science*, February 1974, Vol. 183, pp. 440–444.

149. Lin, W. C., et al.   Feasibility Study of the Engineering Problems in a Multi-Electrode Visual Cortex Stimulation System, *Med. Biol. Eng.*, 1972, Vol. 10, pp. 365–375.

150. Carlson, C. E., R. W. Mann, and W. H. Harris   A Radio Telemetry Device for Monitoring Cartilage Surface Pressures in the Human Hip, *IEEE Trans. Biomed. Eng.*, July 1974, Vol. BME-21, pp. 257–264.

151. *Devices Technol. Branch Contract. Meet. Proc. 1978*, Sec. 4, Blood Pumps: Total Artificial Heart, NIH Publ. 79-1670, pp. 53–61, July 1979.

152. Albisser, A. M.   Devices for the Control of Diabetes Mellitus, *Proc. IEEE*, September 1979, Vol. 67, pp. 1308–1320.

BIOMEDICAL TRANSDUCERS

153. Scheibner, E. J.   Solid State Physical Phenomena and Effects, I–IV, *IRE Trans. Components Parts*, December 1961, Vol. CP-8, No. 4, pp. 133–151; March 1962, Vol. CP-9, No. 1, pp. 19–32; June 1962, Vol. CP-9, No. 3, pp. 61–74; September 1962, Vol. CP-9, No. 3, pp. 119–141.

154. Meindl, J. D., and K. D. Wise (eds.)   Special Issue on Solid State Sensors, Actuators, and Interface Electronics, *IEEE Trans. Electron Devices*, December 1979, Vol. ED-26, No. 12.

155. Cobbold, R. S. C.   "Transducers for Biomedical Measurements," Wiley, New York, 1974.

156. Geddes, L. A., and L. E. Baker   "Principles of Applied Biomedical Instrumentation," Wiley, New York, 1968.

157. Fleming, D., W. H. Ko, and M. R. Neuman   "Implantable and Indwelling Pressure Transducers," CRC Press, Cleveland, Ohio, 1977.

158. Cheung, P., D. Fleming, W. H. Ko, and M. R. Neuman   "The Design and Biomedical Applications of Solid State Chemical Sensors," CRC Press, Cleveland, Ohio, 1978.

159. Middelhoek, S., J. B. Angell, and D. J. W. Noorlag   Microprocessors Get Integrated Sensors, *IEEE Spectrum*, February 1980, Vol. 17, No. 2, pp. 42–46.

160. Czulewicz, A. J., I. Greber, and M. R. Neuman   Continuous Measurement of Tissue Oncotic Pressure, *13th Annu. Meet. Ass. Adv. Med. Instrum., Washington*, 1978.

161. Sensor and Transducers Special Report, *EDN*, Mar. 20, 1980, pp. 122–146.

162. Roylance, L. M., and J. B. Angell   A Batch-Fabricated Silicon Accelerometer. *IEEE Trans. Electron Devices*, December, 1979, Vol. ED-26, No. 12, p. 1911.

163. Neuman, M. R., D. G. Fleming, W. H. Ko, and P. W. Cheung   'Solid State Physical Sensors for Biomedical Application," CRC Press, Boca Raton, Fla 1979.

164. Regtien, P. P. L., and H. K. Makkink   A Capacitive Dew-Point Sensor, *Delft Prog. Rep.*, 1978, Vol. 3, p. 107.

165. Senturia, S. D., C. M. Sechen, and J. A. Wishnensky   The Charge-Flow Transistor: A New MOS Device, *Appl. Phys. Lett.*, 1977, Vol. 30, No. 2, pp. 106–08.

166. Cheung, P. W., W. H. Ko, D. J. Fung, and S. H. Wong   "Theory, Design, and Biomedical Applications of Solid State Chemical Sensors," pp. 91–118, CRC Press, West Palm Beach, Fla., 1978.

167. Abe, H., M. Zsashi, and T. Matsuo   ISFET's Using Inorganic Gate Thin Films, *IEEE Trans. Electron Devices*, 1979, Vol. ED-26, No. 12, pp. 1939–1944.

168. Moss, S. D., C. C. Johnson, and J. Janata   Hydrogen, Calcium, and Potassium Ion-Sensitive FET Transducers: A Preliminary Report, *IEEE Trans. Bio. Med Electron.*, 1978, Vol. BME-25, No. 1, pp. 49–54.

169. Afromowitz, M. A., and S. S. Yee   Fabrication of pH-Sensitive Implantable Electrode by Thick Film Hybrid Technology, *J. Bioeng.*, 1977, Vol. 1, pp. 55–60.

170. Wen, C. C., T. C. Chen, and J. N. Zemel   Gate-Controlled Diodes for Ionic Concentration Measurement, *IEEE Trans. Electron Devices*, 1979, Vol. ED-26, No. 12, pp. 1945–1951.

171. Lundström, I., S. Shivaraman, C. Svensson, and L. Lundkrist   A Hydrogen-Sensitive MOS Field-Effect Transistor, *Appl. Phys. Lett.*, 1975, Vol. 26, No. 2, pp. 55–57.

172. Siu, W. M., and R. S. C. Cobbold   Characteristics of a Multicathode Polarographic Oxygen Electrode, *Med. Biol. Eng.*, 1976, Vol. 14, pp. 109–121.

173. Terry, S. C., J. H. Jerman, and J. B. Angell   A Gas Chromatographic Air Analyzer Fabricated on a Silicon Wafer, *IEEE Trans. Electron Devices*, 1979, Vol. ED-26, No. 12, pp. 1880–1886.

Computer-Aided Design of Electronic Circuits

M. R. LIGHTNER *Assistant Professor of Electrical Engineering, University of Illinois; Member, IEEE*

S. W. DIRECTOR *U. A. and Helen Whitaker Professor of Electronics and Electrical Engineering, Carnegie-Mellon University; Fellow, IEEE*

CONTENTS

Numbers refer to paragraphs

1. Introduction. Circuit designers of previous generations found it necessary to build breadboards, fit them with worst-case or limit devices, i.e., active devices whose characteristics were at the high or low specification limits, and then see whether the circuit performed satisfactorily. The profession has come a long way since then, and over the past several decades the complexity of the systems engineers are called upon to design has increased at a rapid pace. This increase in complexity is due both to advances in technology and to the need for meeting a number of simultaneous design specifications. As an example of increase in complexity due to rapid advances in technology, consider the evolution of very large-scale integrated circuits (VLSI), where the designer now has to contend with circuits that contain hundreds of thousands of transistors. Power-system design represents an example of increased complexity due to the number of design objectives such as minimum cost but high reliability. The difficulty in meeting multiple objectives is that they often compete with each other, so that meeting one of the objectives requires a degradation in performance for some other objective. The increase in the complexity of the design process has forced the designer to turn to the computer.

Because of rapidly evolving computer technology, it is now possible to use the computer not just as a means of simulating a particular system design but also to model the effects of statistical variations in the manufacturing process, to optimize parameters in a given design to meet some set of design objectives better, and also to explore various design alternatives to determine the relative trade-off between the various competing objectives.

As a specific example of the use of the computer as a design tool, consider the design of VLSI circuits. During the course of the design process, an engineer might use one computer program to model the two-dimensional effects of semiconductor devices, another computer program to model the fabrication process for manufacturing integrated circuits to determine the effects of process variation, a logic simulator to verify the logical operation of the design, a layout program to help with the placement of the many thousands of transistors of the VLSI circuit on a chip, and a circuit simulator to determine the actual electrical functioning of a circuit.

Unfortunately, a detailed discussion of the general area of computer-aided design (CAD) or even of the techniques used in the typical VLSI circuit-design process would require much more space than is available. To keep within the space limitations, we focus on the techniques used for the simulation of the electrical behavior of a given circuit. The reason for this choice is that since circuit simulators were one of the earliest CAD tools to be developed, circuit simulation is one of the most stable branches of computer-aided design. Furthermore, the techniques we present in this section are likely to be employed in circuit simulators for the forseeable future, whereas many of the techniques being used in the more developed CAD tools are likely to change.

The development of a simulator for electronic circuits requires a blending of techniques from many disciplines: semiconductor physics, numerical analysis, circuit theory, and computer science. We assume that an adequate model of the active devices in the circuit is available and begin our discussion with generating a set of equations to describe the electrical behavior of the circuit of interest.

These equations may describe the frequency-domain, dc, or transient behavior of the circuit. Although circuit theory provides us with many possible ways of generating the equations associated with a circuit, for computer simulation only two methods have emerged. They are discussed in Pars. **27-2** to **27-4**.

Once the equations describing a circuit have been generated, they must be solved. For the circuits of interest an analytical solution is generally not possible and a numerical solution must be generated. Finding a numerical solution requires the solution of a set of simultaneous complex equations in the frequency domain, a set of simultaneous nonlinear (or linear) equations for a dc solution, and a set of simultaneous nonlinear differential equations for a transient solution. The numerical techniques for solving these various sets of equations form the core of every circuit simulator. Although many techniques have been developed for performing each of these tasks, certain methods have proved best suited for the simulation of circuit equations. These methods, their special characteristics, and special implementation considerations are discussed in Pars. **27-5** on.

Paragraph **27-19** discusses some sources for simulation programs and briefly covers some applications and new techniques being developed for the simulation of large-scale electronic circuits. The computer-science aspect of circuit simulators is the one area we do not discuss. Since circuit simulators tend to be very large programs, the proper use of data structures and (in industrial settings) data bases is important for an efficient implementation. Furthermore, as the circuits being simulated become larger and more complex, a high-quality user-machine interface, including graphics, becomes imperative for every general-purpose program. The detailed discussion of these issues requires a background in computer science and is also very machine-dependent. The interested reader is referred to the computer-science literature.

EQUATION FORMULATION

2. Considerations in Equation Formulation for CAD. Many techniques exist for writing a consistent set of equations describing an electronic network (see Sec. **3**, Circuit Principles). However, when determining which form to use in a CAD program a number of points must be considered:

1. The equations must be generated automatically and efficiently from a description of the network.

2. The equation formulation should be able to handle all types of elements.

3. The equations should allow easy calculation of any network variable in which the user is interested.

4. The numerical characteristics of the formulation should be considered, including sparsity, number of equations, total storage, etc.

Based upon the preceding considerations two equation-formulation methods emerge as predominant in CAD packages today. The most common is the *modified nodal approach.*[1]* The second, used in practice and also an important theoretical tool, is the *tableau approach.*[2]

3. Modified Nodal Formulation. The modified nodal approach (MNA) is a hybrid equation-formulation method that allows both voltage and current variables to be unknowns. The MNA can be considered as an extension of nodal analysis. In fact, if the network contains only linear conductances and independent current sources, the modified nodal equations reduce to the nodal equations $YV_n = J$, where Y is the nodal admittance matrix, V_n the node voltages (excluding the datum), and J the current source vector. By allowing current variables to be unknowns the MNA is able to accept all four types of controlled sources, independent voltage and current sources, any type of two-terminal nonlinearity, and linear R, L, and C elements.

The modified nodal equations are generated by considering Kirchhoff's current law at each node; i.e., the sum of the currents leaving the node is zero. However if a current is to be considered as an unknown, the *branch relationship* (BR) of the element is added to the set of node equations. The result, for linear networks, is a set of equations of the form

$$\begin{bmatrix} Y_r & B \\ C & D \end{bmatrix} \begin{bmatrix} V_n \\ I_b \end{bmatrix} = \begin{bmatrix} J \\ F \end{bmatrix} \tag{27-1}$$

where $Y_r = (n-1) \times (n-1)$ nodal admittance matrix, $B = (n-1) \times b$ matrix taking into account the certain unknown branch currents, e.g., voltage source and inductor currents, leaving a node; $C = b \times (n-1)$ matrix expressing certain branch relationships, e.g., voltage-source relationships; $D = b \times b$ matrix accounting for certain controlled source branch relationships; and J, $F = (n-1)$- and b-dimensional vectors, respectively, which are the corresponding right-hand-side (RHS) entries. The modified nodal equations are easily formulated by considering a *stamp* for the various circuit elements.

We shall indicate the kth columns of Y and C by V_k, the kth column of B and D by I_k, the kth row of Y and B by V_k, and the kth row of C and D by i_k. Hence, if the kth branch is a conductance G_k, if it is connected between nodes i and j, and if the current through the conductance is not a desired output variable, the stamp would be

Rows ↓	V_i	V_j	LHS	← Columns
V_i	G_k	$-G_k$	0	
V_j	$-G_k$	G_k	0	

The stamps indicate how each element affects the entries in the matrices defined by Eq. (27-1). Specifically, the stamp shown above indicates that a conductance increased the terms Y_{ii} and Y_{jj} by G_k and Y_{ij} and Y_{ji} by $-G_k$. If the current through the conductance is desired as an output, the stamp becomes

Rows ↓	V_i	V_j	I_G	RHS	← Columns
V_i	0	0	1	0	
V_j	0	0	-1	0	
I_G	G_k	$-G_k$	-1	0	

*Superior numbers correspond to references in Par. **27-20**.

where the last row is a statement of the branch relationship of the conductance.

The general rules for deriving element stamps can be stated as follows:

1. The branch current of a voltage source and the branch current of an inductor are always included as variables.
2. For current sources, resistors, conductances, and capacitors the current is added as a variable if:
   a. The current through the specific element is desired as an output.

TABLE 27-1   Element Stamps for the MNA for the General Node Shown in Fig. 27-1

Element type		Current not output				Branch current is output			
		V_i	V_1	RHS		V_i	V_1	I_G	RHS
G	V_i	G	$-G$		V_i			1	
	V_1	$-G$	G		V_1			-1	
					I_G	G	$-G$	-1	
						V_i	V_2	I_V	RHS
V					i			1	
					2			-1	
					1	1	-1		V
		V_i	V_3	RHS		V_i	V_3	I_C	RHS
C	V_i	$C\dfrac{d}{dt}$	$-C\dfrac{d}{dt}$		V_i			1	
	V_3	$-C\dfrac{d}{dt}$	$C\dfrac{d}{dt}$		V_3			-1	
					I_C	$C\dfrac{d}{dt}$	$-C\dfrac{d}{dt}$	-1	
						V_i	V_4	I_L	RHS
L					i			1	
					4			-1	
					1	1	-1	$-L\dfrac{d}{dt}$	
		V_i	V_5	RHS		V_i	V_5	I_I	RHS
I	V_i			$-I$	V_i			1	
	V_5			I	V_5			-1	
					I_I			1	I
		V_i	V_6	RHS		V_i	V_6	I_F	RHS
$f(v)$	V_i		$-f(V_6 - V_i)$		V_i			-1	
	V_6		$f(V_6 - V_i)$		V_6			1	
					I_F		$f(V_6 - V_i)$	-1	
						V_i	V_7	I_{CCV}	I_G	RHS
CCV					V_i			-1		
					V_7			1		
					I_G	-1	1			$-\beta$

b. Or there are other elements in the circuit, either controlled sources or nonlinearities, that depend on the current.

We now present two examples, the first a general node and the associated stamps and the second a simple network and the associated modified nodal equations (see Figs. 27-1 to 27-3 and Table 27-1).

In the actual solution of a set of modified nodal equations the derivative operator would be discretized and the nonlinearities linearized, yielding a set of linear equations. These concepts will be elaborated upon in later paragraphs.

Modified nodal equations are very easy to generate on a computer directly from the network description and the desired outputs. The equations also have the property of being sparse (relatively few nonzero entries in each row) and numerically well conditioned (see Par. **27-8**). The modified nodal method is at present the preferred technique for formulating circuit equations in circuit-simulation programs.

4. Tableau Formulation. When formulating equations for solution by hand, an obvious objective is to write as few equations as possible. However, the opposite objective, to write as many equations as possible, is the goal of the tableau approach. The idea behind the tableau

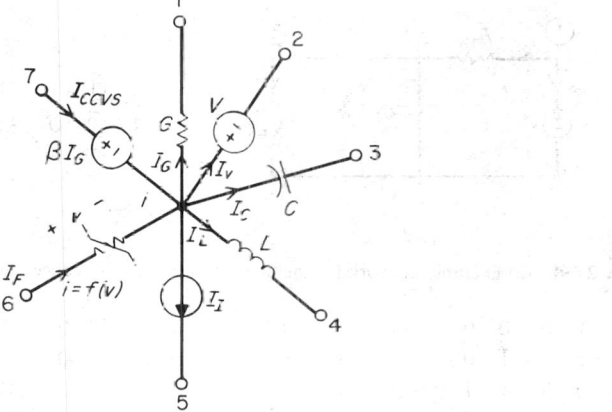

Fig. 27-1. A general node used in describing the modified nodal approach to equation formulation.

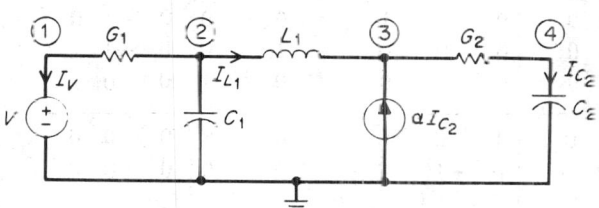

Fig. 27-2. Network used as an example of the modified-nodal-equation method.

G_1	$-G_1$	0	0	1	0	0	V_1		0
$-G_1$	$G_1 + C_1 \dfrac{d}{dt}$	0	0	0	1	0	V_2		0
0	0	G_2	$-G_2$	0	-1	$-\alpha$	V_3		0
0	0	$-G_2$	G_2	0	0	1	V_4	$=$	0
1	0	0	0	0	0	0	I_v		V
0	1	-1	0	0	$-L_1 \dfrac{d}{dt}$	0	I_ℓ		0
0	0	0	$C_2 \dfrac{d}{dt}$	0	0	-1	I_{C_2}		0

Fig. 27-3. Modified nodal equations for the circuit of Fig. 27-2.

approach is to write a large number of very sparse (few nonzero entries per row) equations and then use advanced sparse-matrix techniques when solving the equations on the computer. The nature of sparse-matrix techniques is such that a large sparse set of equations can be solved more efficiently than a smaller, denser set of equations.

The tableau formulation is based on collecting the Kirchhoff current law, Kirchhoff voltage law, and branch-relationship equations into one *large sparse* system of equations. Let A be the reduced incidence matrix for the network (see Par. **27-3**) and R_i, G_v the branch relation operators. Then the tableau is

$$\begin{bmatrix} A & 0 & 0 \\ 0 & I & -A^T \\ R_i & G_v & 0 \end{bmatrix} \begin{bmatrix} i \\ v_b \\ v_n \end{bmatrix} = \begin{bmatrix} 0 \\ 0 \\ S \end{bmatrix} \qquad (27\text{-}2)$$

where $i = b$ branch currents, $v_b = b$ branch voltages, $V_n = n - 1$ node voltages excluding datum, S = vector depending on the branch relationships, and I = identity matrix. This is a set of $2b + n - 1$ equations in $2b + n - 1$ unknowns and is very sparse. Furthermore, unlike

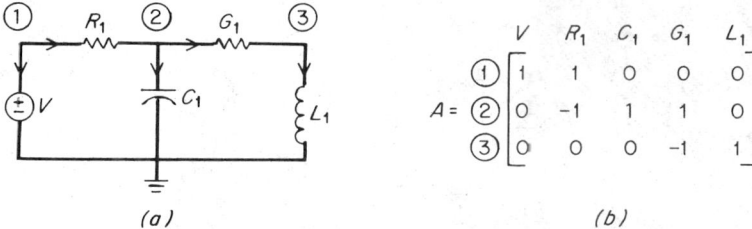

Fig. 27-4. (a) Example network for the tableau approach; (b) reduced nodal incidence matrix for (a).

$$\left[\begin{array}{ccccc|ccc|ccccc} 1 & 1 & 0 & 0 & 0 & 0 & 0 & 0 & 0 & 0 & 0 & 0 \\ 0 & -1 & 1 & 1 & 0 & 0 & 0 & 0 & 0 & 0 & 0 & 0 \\ 0 & 0 & 0 & -1 & 1 & 0 & 0 & 0 & 0 & 0 & 0 & 0 \\ 0 & 0 & 0 & 0 & 0 & 1 & 0 & 0 & 0 & -1 & 0 & 0 \\ 0 & 0 & 0 & 0 & 0 & 0 & 1 & 0 & 0 & -1 & 1 & 0 \\ 0 & 0 & 0 & 0 & 0 & 0 & 0 & 1 & 0 & 0 & -1 & 0 \\ 0 & 0 & 0 & 0 & 0 & 0 & 0 & 0 & 1 & 0 & 0 & -1 & 1 \\ 0 & 0 & 0 & 0 & 0 & 0 & 0 & 0 & 0 & 1 & 0 & 0 & 1 \end{array}\right]$$

	1	1	0	0	0	0	0	0	0	0	0	0	0	i_V		0
	0	-1	1	1	0	0	0	0	0	0	0	0	0	i_{R_1}		0
	0	0	0	-1	1	0	0	0	0	0	0	0	0	i_{C_1}		0
	0	0	0	0	0	1	0	0	0	-1	0	0	0	i_{G_1}		0
	0	0	0	0	0	0	1	0	0	-1	1	0	0	i_{L_1}		0
	0	0	0	0	0	0	0	1	0	0	-1	0	0	v_V		0
	0	0	0	0	0	0	0	0	1	0	0	-1	1	v_{R_1}		0
	0	0	0	0	0	0	0	0	0	1	0	0	1	v_{C_1}	$=$	0
	0	0	0	0	0	1	0	0	0	0	0	0	0	v_{G_1}		V
	0	R_1	0	0	0	0	-1	0	0	0	0	0	0	v_{L_1}		0
	0	0	1	0	0	0	0	$-C_1\dfrac{d}{dt}$	0	0	0	0	0	$v_{\textcircled{1}}$		0
	0	0	0	1	0	0	0	0	$-G_1$	0	0	0	0	$v_{\textcircled{2}}$		0
	0	0	0	0	$L_1\dfrac{d}{dt}$	0	0	0	0	-1	0	0	0	$v_{\textcircled{3}}$		0

Fig. 27-5. Tableau equations for the network shown in Fig. 27-4(a).

many other formulation methods, all network variables are available after solution of the equations.

In order to illustrate the tableau formulation consider the simple network shown in Fig. 27-4. The tableau equations for this network are shown in Fig. 27-5.

Obviously, very powerful computational techniques are necessary if the equations in Fig. 27-5 are to be solved efficiently. It can be shown that solving the tableau in various ways, i.e., reordering the equations before solution, can lead to any of the standard equation formulation methods such as the nodal, modified-nodal, and state-variable methods. Thus, all other formulation techniques can be viewed as specifying, a priori, a certain way of solving the tableau equations. However, these alternate solution techniques do not take into account the sparsity of the tableau equations. When modern sparse-matrix techniques are applied to the tableau equations, a solution order specifically tailored to the problem at hand is chosen. Therefore, solution of the tableau equations can be much more efficient than solution of other formulations.

There are, however, several drawbacks to the tableau approach. First, a modern, highly efficient sparse-matrix program must be available. Second, for some networks with indeterminate solutions or no solutions the tableau equations will be singular. When the solution process terminates because of the singularity, the cause of the problem in the network is difficult to isolate. Regardless of these drawbacks, the tableau formulation has proved to be computationally viable and an important theoretical tool in computer-aided circuit analysis.

NUMERICAL SOLUTION OF LINEAR EQUATIONS

5. Introduction. At the heart of every circuit-simulation program is an algorithm to solve n simultaneous linear equations in n unknowns

$$Ax = b \tag{27-3}$$

For linear networks, every equation-formulation method results in a set of linear equations. For nonlinear dc networks the standard solution technique requires the sequential solution of sets of linear equations. Thus, the need for efficient methods to solve Eq. (27-3) is clear.

6. Gaussian Elimination. The first technique we consider is gaussian elimination. This method essentially eliminates variables successively from each equation until one equation in one variable remains. The single variable is solved for and then the remaining variables found by reversing the elimination process. This forward elimination and back substitution is illustrated in the following example. Consider the set of equations

$$
\begin{aligned}
x_1 + 3x_2 + 2x_3 &= 1 \\
2x_1 + x_2 + 4x_3 &= 0 \\
3x_2 + x_3 &= 1
\end{aligned}
\tag{27-4}
$$

or in matrix notation

$$
\begin{bmatrix} 1 & 3 & 2 \\ 2 & 1 & 4 \\ 0 & 3 & 1 \end{bmatrix}
\begin{bmatrix} x_1 \\ x_2 \\ x_3 \end{bmatrix} =
\begin{bmatrix} 1 \\ 0 \\ 1 \end{bmatrix}
\tag{27-5}
$$

We proceed by eliminating x_1 from the second and third equation to yield

$$
\begin{bmatrix} 1 & 3 & 2 \\ 0 & -5 & 0 \\ 0 & 3 & 1 \end{bmatrix}
\begin{bmatrix} x_1 \\ x_2 \\ x_3 \end{bmatrix} =
\begin{bmatrix} 1 \\ -2 \\ 1 \end{bmatrix}
\tag{27-6}
$$

Next, we use the second equation in Eq. (27-6), to eliminate x_2 from the third equation, obtaining

$$
\begin{bmatrix} 1 & 3 & 2 \\ 0 & 1 & 0 \\ 0 & 0 & 1 \end{bmatrix}
\begin{bmatrix} x_1 \\ x_2 \\ x_3 \end{bmatrix} =
\begin{bmatrix} 1 \\ \frac{2}{5} \\ -\frac{1}{5} \end{bmatrix}
\tag{27-7}
$$

Now we find that $x_3 = -\frac{1}{5}$, $x_2 = \frac{2}{5}$, and $x_1 = \frac{1}{5}$.

If we assume that at the ith stage of the elimination process the diagonal element a_{ii}, associated with the variable to be eliminated from the remaining equations, is nonzero, we can write the gaussian elimination procedure as follows.

Forward Elimination $i = 1$

(1) $a_{ij} \leftarrow a_{ij}/a_{ii}$ $j = i, i + 1, \ldots, n$
     $b_i \leftarrow b_i/a_{ii}$
     If $i = n$, go to *Back Substitution*
     for $k = i + 1, \ldots, n$
(2) $a_{kj} \leftarrow a_{kj} - a_{ki} a_{ij}$ $j = i, i + 1, \ldots, n$
     $b_k \leftarrow b_k - a_{ki} b_i$
     $i \leftarrow i + 1$
     Go to (1)

[At the ith stage of the elimination a $(n - i + 1) \times (n - i + 1)$ submatrix is modified.] At the end of the forward elimination the original system has been reduced to an upper triangular matrix with 1s on the diagonal

$$
\begin{bmatrix}
1 & * & * & \cdots & * \\
0 & 1 & * & & * \\
\cdot & & & & \\
\cdot & & & & * \\
\cdot & & & & \\
0 & & & & 1
\end{bmatrix}
\begin{bmatrix}
x_1 \\
x_2 \\
\cdot \\
\cdot \\
\cdot \\
x_n
\end{bmatrix}
=
\begin{bmatrix}
* \\
* \\
\cdot \\
\cdot \\
\cdot \\
*
\end{bmatrix}
$$

where the * represent possible nonzero entries.

Back Substitution $x_n = b_n$

$$\text{For } i = n - 1, n - 2, \ldots, 1$$

$$x_i = b_i - \sum_{j=i+1}^{n} a_{ij}x_j$$

where the a_{ij} are the entries of the upper triangular matrix.

If at the ith stage the a_{ii} term, the *pivot element*, is zero, the algorithm fails. However, if the original set of equations has a unique solution, a reordering of equations (row pivoting) or variables (column pivoting) will allow the process to continue. For example, consider the set of equations

$$
\begin{bmatrix}
0 & 2 \\
2 & 1
\end{bmatrix}
\begin{bmatrix}
x_1 \\
x_2
\end{bmatrix}
=
\begin{bmatrix}
2 \\
1
\end{bmatrix}
\tag{27-8}
$$

Gaussian elimination as stated will not solve this set of equations, but if we row pivot to obtain the equivalent

$$
\begin{bmatrix}
2 & 1 \\
0 & 2
\end{bmatrix}
\begin{bmatrix}
x_1 \\
x_2
\end{bmatrix}
=
\begin{bmatrix}
1 \\
2
\end{bmatrix}
\tag{27-9}
$$

or column pivot to obtain

$$
\begin{bmatrix}
2 & 0 \\
1 & 2
\end{bmatrix}
\begin{bmatrix}
x_2 \\
x_1
\end{bmatrix}
=
\begin{bmatrix}
2 \\
1
\end{bmatrix}
\tag{27-10}
$$

the elimination algorithm will obtain the correct solution. Notice that at the ith step, pivot candidates can be considered only from the remaining $(n - i + 1) \times (n - i + 1)$ submatrix. Also, since the matrix entries change at every step of the elimination, it is generally impossible to specify a pivot or elimination order a priori. More will be said about pivoting later.

7. LU Decomposition. We often want to solve the system of equations $Ax = b$ with different right-hand sides. For example, if different sources were applied to the same network, A would remain constant and only b would change. Alternately, certain portions of A may change in such a way that we do not have to solve the set of equations again completely. LU decomposition is a solution method that is equivalent to gaussian elimination if the set of equations is solved once but which gains computational advantage if repeated solutions are required.

Consider a decomposition of the matrix A into the product of a *lower and an upper triangular matrix*

$$Ax = LUx = b \tag{27-11}$$

where L is lower triangular and U is upper triangular with 1s on the diagonal. Now we let

$$Ux = y \tag{27-12}$$

where

$$Ly = b \tag{27-13}$$

Thus, Eq. (27-12) can be solved by a forward elimination for y, and Eq. (27-13) can be solved by a back substitution to find x. Notice that if only b changes, we simply need a forward elimination and back substitution in order to find the new solution; that is, A does not have to be factored again.

The question now arises: How much work do we actually save with a procedure like LU decomposition? The traditional measure of work in gaussian elimination is the number of multiplications and divisions. If the matrix is $n \times n$ and full, generating the upper triangular matrix in gaussian elimination takes approximately $n^3/3 + n^2/2$ operations and the back substitution approximately $n^3/2$ operations. To generate a LU decomposition takes approximately $n^3/3$ operations, and the forward and backward substitutions require a total of n^2 operations. Thus for a single solution, LU decomposition and gaussian elimination require the same amount of work; but if the equations are to be solved with a new right-hand side, LU decomposition requires only n^2 operations compared with $n^3/3 + n^2$ for another gaussian elimination.

There are several algorithms for generating an LU decomposition of an $n \times n$ matrix A. We discuss an algorithm due to Crout. We again assume that no pivoting is necessary. The Crout algorithm calculates a column of L and then a row of U successively until the factorization is complete. The algorithm proceeds as follows.

Given a $n \times n$ matrix A:

(1) Set $l_{k1} = a_{k1}$, $k = 1, 2, \ldots, n$; set $u_{ii} = 1$, $i = 1, \ldots, n$

(2) Set $u_{1k} = \dfrac{a_{1k}}{l_{11}}$, $k = 2, \ldots, n$

(3) $j \leftarrow 2$

(4) $l_{kj} = a_{kj} - \displaystyle\sum_{m=1}^{j-1} l_{km} u_{mj}$, $k = j, j+1, \ldots, n$

(5) If $j = n$, stop; otherwise go to (6)

(6) $u_{jk} = \dfrac{\displaystyle\sum_{m=1}^{j-1} l_{jm} u_{mk}}{l_{jj}}$, $k = j+1, j+2 \ldots, n$

(7) $j \leftarrow j+1$ go to (4)

Note that since the diagonal elements of U are 1, the entire LU decomposition can overwrite the A matrix, requiring no extra storage. Consider the example

$$\begin{bmatrix} 1 & 3 & 2 \\ 2 & 1 & 4 \\ 0 & 3 & 1 \end{bmatrix} \begin{bmatrix} z_1 \\ z_2 \\ z_3 \end{bmatrix} = \begin{bmatrix} 1 \\ 0 \\ 1 \end{bmatrix} \tag{27-14}$$

The LU decomposition algorithm generates the following sequence of L and U matrices:

$$L_1 = \begin{bmatrix} 1 & 0 & 0 \\ 2 & 0 & 0 \\ 0 & 0 & 0 \end{bmatrix} \qquad U_1 = \begin{bmatrix} 1 & 3 & 2 \\ 0 & 1 & 0 \\ 0 & 0 & 1 \end{bmatrix} \tag{27-15}$$

$$L_2 = \begin{bmatrix} 1 & 0 & 0 \\ 2 & -5 & 0 \\ 0 & 3 & 0 \end{bmatrix} \qquad U_2 = \begin{bmatrix} 1 & 3 & 2 \\ 0 & 1 & 0 \\ 0 & 0 & 1 \end{bmatrix} \tag{27-16}$$

$$L_3 = \begin{bmatrix} 1 & 0 & 0 \\ 2 & -5 & 0 \\ 0 & 3 & 1 \end{bmatrix} \qquad U_3 = \begin{bmatrix} 1 & 3 & 2 \\ 0 & 1 & 0 \\ 0 & 0 & 1 \end{bmatrix} \tag{27-17}$$

Notice that U_3 is the same as the upper triangular matrix found using gaussian elimination. This is always true. Furthermore, if we solve

$$L_3 y = \begin{bmatrix} - \\ 0 \\ - \end{bmatrix} \tag{27-18}$$

we find

$$y = \begin{bmatrix} - \\ \frac{2}{5} \\ \frac{1}{5} \end{bmatrix} \tag{27-19}$$

which is the right-hand side found during gaussian elimination. Again, this is always true.

A zero pivot element during LU factorization, as in gaussian eliminations, is handled by row or column pivoting.

8. Pivoting for Accuracy. When solving a set of linear equations using gaussian elimination or LU decomposition, it may prove necessary to alter the order of elimination, i.e., to pivot. If a zero element is detected as the coefficient of the next variable to be eliminated, we must pivot in order to continue the elimination. We now address the issue of how to pick a suitable pivot. In this paragraph we discuss pivoting in order to maintain accuracy, while in the next pivoting to maintain sparsity is discussed.

Consider the set of equations

$$
\begin{bmatrix}
-0.000125 & 1 & 1 \\
1 & 0.635 & 0 \\
1 & 0 & 0
\end{bmatrix}
\begin{bmatrix}
x_1 \\
x_2 \\
x_3
\end{bmatrix}
=
\begin{bmatrix}
4\ 999875 \\
2.270 \\
1
\end{bmatrix}
\tag{27-20}
$$

Recognizing the triangular nature of these equations, we find

$$
x_1 = 1 \qquad x_2 = 2 \qquad x_3 = 3
\tag{27-21}
$$

If we use gaussian elimination with five significant digits and the pivot order (1, 3), (2, 2), (3, 1), we find the same result, but if we use gaussian elimination along the diagonal, using five significant figures, we find

$$
x_1 = 2 \qquad x_2 = 1.6668 \qquad x_3 = 3.3333
\tag{27-22}
$$

which is clearly wrong. Thus, even if we do not have a zero pivot element, an erroneous solution can be found as a result of round-off error in the specific pivot order used. Furthermore, it is not simply a matter of small pivot elements inducing numerical inaccuracies.[3,4]

The general, although certainly not foolproof, rule used in pivoting for accuracy is the following. At the ith stage of the elimination find the largest element, $a_{ji}, j = i, i + 1, \ldots , n$. Use this element as the pivot element. This type of pivoting is known as *partial pivoting* or *partial row pivoting*. An alternate but more expensive pivoting scheme involves searching at the ith stage for the largest element in the remaining unreduced submatrix and pivoting on this element. This is known as *full pivoting* (in all schemes ties are broken arbitrarily). There is also the possibility of *partial column pivoting*. [Partial column pivoting works for Eq. (27-20) whereas partial row pivoting does not.]

Neither partial nor full pivoting is guaranteed to lead to an accurate solution. The question thus arises: Can we ever be sure that pivoting is unnecessary to maintain numerical accuracy? It has been shown[3] that if a coefficient matrix A is symmetric positive definite or if A is diagonally dominant, i.e.,

$$
|a_{ii}| > \sum_{j \neq i} |a_{ij}| \qquad \text{for } i = 1, \ldots , n
\tag{27-23}
$$

it is not necessary to pivot in order to maintain numerical accuracy. In fact, in either one of these situations pivoting can degrade the accuracy of the solution. The importance of this condition in the solution of network equations is that network equations tend to be diagonally dominant, e.g., node equations, or almost diagonally dominant, e.g., modified nodal equations. Thus, if a nodal or modified nodal approach is used to formulate network equations, pivoting for accuracy is generally not necessary. (For special rules for modified nodal equations see Ref. 6.) In some cases, especially in frequency-domain problems, pivoting for accuracy may still be important. However, in most modern circuit-simulation packages pivoting is done for reasons of sparsity and not for numerical accuracy.

9. Sparse-Matrix Techniques. In our previous discussions of equation formulation and linear-equation solution we have mentioned the concept of sparsity. A matrix is said to be sparse if all but a small number, say 30%, of its entries are zero. Network equations are generally very sparse. The main idea behind sparse-matrix techniques is to recognize the zero elements in a matrix and not perform any arithmetic operations using them. Sparse techniques also include the ideas of not storing any zero-valued elements and in choosing a pivot order to maintain the sparseness of the matrix during factorization.

In order to understand the importance of sparsity consider the solution of the following set of equations:

$$
\begin{bmatrix}
* & & & * \\
& * & & * \\
& & * & * \\
& & * & * \\
* & * & * & * & *
\end{bmatrix}
\begin{bmatrix}
x_1 \\
\cdot \\
\cdot \\
\cdot \\
x_5
\end{bmatrix}
=
\begin{bmatrix}
* \\
* \\
* \\
* \\
*
\end{bmatrix}
\tag{27-24}
$$

where the * represent nonzero elements. If we store only nonzero elements, the total storage is 18 instead of 30 for a full matrix. Futhermore, if we solve this system using gaussian elimination and carrying out only nontrivial operations, i.e., operations which involve only nonzeros, only 21 operations are required. However, if this matrix were handled as a full matrix, 67 operations would be required. Thus, considerable gains can be made in the solution of linear equations by taking the sparsity of the equations into consideration.

Finally, consider a solution of Eq. (27-24) with the rows and columns pivoted

$$
\begin{bmatrix}
* & * & * & * & * \\
* & * & & & \\
* & & * & & \\
* & & & * & \\
* & & & & *
\end{bmatrix}
\begin{bmatrix}
x_5 \\ x_4 \\ x_3 \\ x_2 \\ x_1
\end{bmatrix}
=
\begin{bmatrix}
* \\ * \\ * \\ * \\ *
\end{bmatrix}
\tag{27-25}
$$

The first elimination step results in

$$
\begin{bmatrix}
* & * & * & * & * \\
0 & * & \circledast & \circledast & \circledast \\
0 & \circledast & * & \circledast & \circledast \\
0 & \circledast & \circledast & * & \circledast \\
0 & * & \circledast & \circledast & *
\end{bmatrix}
\begin{bmatrix}
x_5 \\ x_4 \\ x_3 \\ x_2 \\ x_1
\end{bmatrix}
=
\begin{bmatrix}
* \\ * \\ * \\ * \\ *
\end{bmatrix}
\tag{27-26}
$$

A matrix that was sparse is now full because the elimination process caused zero elements to become nonzero. These new nonzero elements are denoted by $\circledast$ and called *fill* or *fill-ins*. In this paragraph we discuss the storage and pivot strategies associated with sparse matrices. (Note that there are many details, techniques, and programming tricks that cannot be discussed in this section.)[2,3,5,6]

The first consideration in dealing with a sparse matrix is how to store and access only the nonzero elements of the matrix. The technique most often used is that of a *pointer system*. Three linear arrays are used: (1) the *integer row pointer array* IR has $n + 1$ entries; (2) the *integer column pointer array* IC has a list of the column indices of the nonzero elements in each row; (3) the *element array* VAL contains the values of the nonzero elements and has the same number of entries as IC. Consider the example shown in Fig. 27-6.

The ith entry in IR points to the beginning of ith row column data in IC and value data in VAL. The number of nonzero elements in the ith row is given by $IR[i + 1] - IR[i]$. Thus the third row of A has $IR[4] - IR[3] = 2$ nonzero elements, in positions (3, $IC[IR[3]]$) = (3, 3) and (3, $IC[IR[3] + 1]$) = (3, 4) having values $VAL[IR[3]] = 8$ and $VAL[IR[3] + 1] = 12$, respectively.

$$
A = \begin{bmatrix}
1 & 0 & 7 & 4 & 0 \\
0 & 1 & 0 & 0 & 3 \\
0 & 0 & 8 & 12 & 0 \\
10 & 0 & 0 & 4 & 1 \\
0 & 0 & 0 & 3 & 4
\end{bmatrix}
$$

$$
IR = \begin{bmatrix} 1 \\ 4 \\ 6 \\ 8 \\ 11 \\ 12 \end{bmatrix}
\quad IC = \begin{bmatrix} 1 \\ 3 \\ 4 \\ 2 \\ 5 \\ 3 \\ 4 \\ 1 \\ 4 \\ 5 \\ 4 \\ 5 \end{bmatrix}
\quad VAL = \begin{bmatrix} 1 \\ 7 \\ 4 \\ 1 \\ 3 \\ 8 \\ 12 \\ 10 \\ 4 \\ 1 \\ 3 \\ 4 \end{bmatrix}
$$

Fig. 27-6. Example of pointer-system storage for a sparse matrix

The pointer system just described requires 2NZ $+ N + 1$ storage locations for an $n \times n$ matrix with NZ nonzero entries. A full matrix requires n^2 storage locations. Thus if a matrix is sparse, the compact pointer system can offer great savings in storage. However, performing LU factorization and taking fill-in into consideration on a matrix stored in pointer system requires careful and clever programing.[5]

As shown in Eqs. (27-24) and (27-25), the order in which the equations are solved plays a key role in maintaining sparseness during the factorization. An *optimal ordering* for a set of equations is a pivot order chosen to minimize the number of fills during the factorization. Generally, the ordering schemes do not consider numerical accuracy at all. This omission can sometimes lead to roundoff error, but for most network problems it is not a problem if a few precautions are taken.[2]

Since a pivot order is generally selected before the actual factorization is performed, a simulation of the zero-nonzero pattern of a factorization is required. A truly global optimal order would require a complete simulation of the factorization for all possible pivots at all stages of the factorization. This global ordering can be more expensive than computing a matrix inverse (n^3 operations). Thus, the usual approach is to consider a locally optimal or locally suboptimal ordering strategy. A locally optimal ordering would require a simulation to find the best pivot at each stage of the elimination. This can also be very expensive. A locally suboptimal order tries to approximate, in a computationally efficient manner, a locally optimal order.

Typical rules for choosing a locally suboptimal order are the following:

1. Any rows or columns with only one entry should be ordered first since they will cause no fill-in.

2. At any stage, let r_i be the number of nonzeros in the ith row and c_j the number of nonzeros in the ith column of the remaining unfactored submatrix. Choose as the next pivot the ij element with min $(r_i - 1)(c_j - 1)$. Ties are broken arbitrarily.

The first rule is an obvious one. The second rule, known as the *Markowitz criterion*,[7] is one of the best and easiest schemes to implement that results in a suboptimal ordering. Rules 1 and 2 (with some modifications[5]) are typical of those used in modern circuit-simulation programs. An example of this ordering is now given. If a matrix has the zero-nonzero pattern

$$
\begin{bmatrix}
* & 0 & * & * & 0 & 0 \\
* & 0 & 0 & 0 & 0 & * \\
0 & 0 & * & 0 & * & 0 \\
* & 0 & 0 & 0 & 0 & 0 \\
* & * & * & 0 & 0 & 0 \\
0 & * & 0 & * & 0 & *
\end{bmatrix}
\tag{27-27}
$$

rules 1 and 2 would yield the pivots circled below, where the F represent fill-in during a factorization:

$$
\begin{bmatrix}
* & 0 & * & \circledast^4 & 0 & 0 \\
* & 0 & 0 & 0 & 0 & \circledast^3 \\
0 & 0 & * & 0 & \circledast^2 & 0 \\
\circledast^1 & 0 & 0 & 0 & 0 & 0 \\
* & \circledast^5 & * & 0 & 0 & 0 \\
0 & * & \circledast{F} & * & 0 & *
\end{bmatrix}
\tag{27-28}
$$

Only one fill-in has occurred (fill-ins are eligible as pivot candidates).

The use of modern sparse-matrix techniques is one of the major reasons that large-scale integrated circuits (LSI) can be simulated in a cost-effective manner by circuit-simulation programs.

NUMERICAL SOLUTION OF NONLINEAR EQUATIONS

10. Introduction. In the analysis of modern electronic circuits it is necessary to consider the solution of simultaneous nonlinear equations which result from models used for diodes, bipolar transistors, field-effect transistors, etc. As mentioned in the paragraphs on equation formulation, there is in general no difficulty in writing a set of nonlinear equations describing a network—only in solving them.

11. Newton-Raphson Iteration. The standard method in circuit-analysis programs for solving nonlinear equations is the Newton-Raphson method. Consider the nonlinear equation $f(x) = 0$. If x^0 is an initial guess at the solution, we can expand $f(x)$ about x^0 using two terms of the Taylor series

$$
f(x^0 + \Delta x) = f(x^0) + \frac{\partial f}{\partial x}\bigg|_{x=x^0} \Delta x
\tag{27-29}
$$

If $x^1 = x^0 + \Delta x$ is the solution to $f(x) = 0$, we can find

$$
f(x^1) = 0 = f(x^0) + \frac{\partial f}{\partial x}\bigg|_{x^0} \Delta x
\tag{27-30}
$$

or

$$
\Delta x = -\left(\frac{\partial f}{\partial x}\bigg|_{x^0}\right)^{-1} f(x^0)
\tag{27-31}
$$

Alternatively,

$$x^1 = x^0 + \Delta x = x^0 - \left(\frac{\partial f}{\partial x}\bigg|_{x^0}\right)^{-1} f(x^0) \tag{27-32}$$

The Newton-Raphson iteration consists in repeated applications of Eq. (27-32)

$$x^{i+} = x^i - \left(\frac{\partial f}{\partial x}\bigg|_{x^i}\right)^{-1} f(x^i) \tag{27-33}$$

This *iterative* procedure continues until some stopping or convergence criteria, for example, $|\Delta x| \leq \epsilon$ for ϵ small, is met. Thus for nonlinear equations we only find an approximate solution. It is useful to consider a geometric interpretation of the Newton-Raphson iteration. The term $\partial f/\partial x\big|_{x^i}$ is the slope of the nonlinearity at x^i. The iteration of Eq. (27-33) is presented graphically in Fig. 27-7.

The equation of each straight line in Fig. 27-7 is given by

$$\bar{f}(x) = \frac{\partial f}{\partial x}\bigg|_{x^i} x - \left[f(x^i) - \frac{\partial f}{\partial x}\bigg|_{x^i} x^i\right] \tag{27-34}$$

Solving for $\bar{f}(x^{i+1}) = 0$ yields

$$x^{i-1} = x^0 - \left(\frac{\partial f}{\partial x}\bigg|_{x}\right)^{-1} f(x^i) \tag{27-35}$$

which is the Newton iteration.

The extension of the Newton-Raphson method to several equations in several unknowns is straightforward. Consider

$$\mathbf{f(x)} = \mathbf{0} \tag{27-36}$$

where $\mathbf{f}$ is a set of n nonlinear equations in the n unknowns $\mathbf{x}$. The Newton-Raphson iteration for Eq. (27-36) is

$$\mathbf{x}^{i+1} = \mathbf{x}^i - \left(\frac{\partial \mathbf{f}}{\partial \mathbf{x}}\bigg|_{\mathbf{x}^i}\right)^{-1} \mathbf{f(x^i)} \tag{27-37}$$

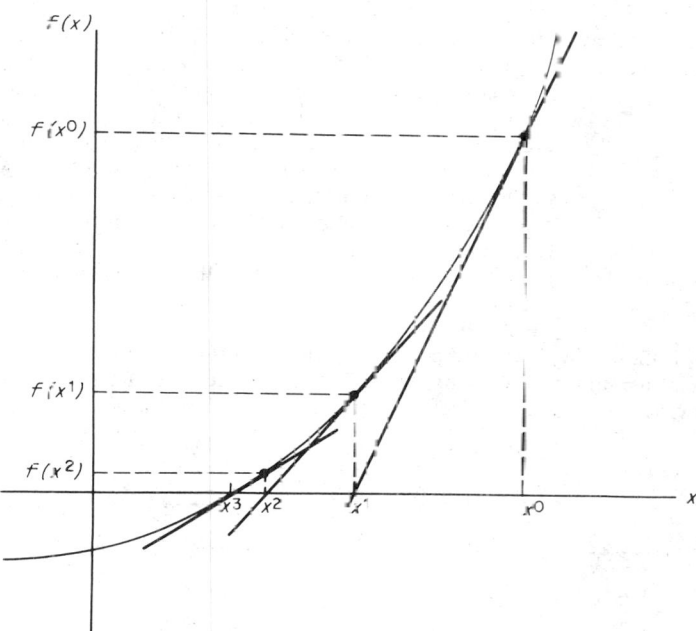

Fig. 27-7. Geometric interpretation of the Newton-Raphson iteration.

where $\partial\mathbf{f}/\partial\mathbf{x}|_{\mathbf{x}^i}$ is the $n \times n$ Jacobian matrix. Notice that Eq. (27-37) can be rewritten as

$$\left.\frac{\partial\mathbf{f}}{\partial\mathbf{x}}\right|_{\mathbf{x}^i} \Delta\mathbf{x}^i = -\mathbf{f}(\mathbf{x}^i) \qquad (27\text{-}38)$$

which is a set of *linear* equations. Thus, the solution of nonlinear equations using the Newton-Raphson method requires the repeated solution of sets of linear equations.

12. Companion Models. We now examine the meaning of the Newton-Raphson iteration for circuit simulation. First we consider all voltages and currents in the network as unknowns and the set of network equations as a set of nonlinear equations. Without loss of generality we group all nonlinearities as $\mathbf{i} = \mathbf{f}(\mathbf{v})$. Thus, the tableau equations become

$$F(i_b, v_b, v_n) = \begin{bmatrix} A & 0 & 0 \\ 0 & I & -A^T \\ I & -f(v_b) & 0 \\ R_i & G_v & 0 \end{bmatrix} \begin{bmatrix} i_b \\ v_b \\ v_n \end{bmatrix} = \begin{bmatrix} \ \\ \ \cdot \\ \ \end{bmatrix} \qquad (27\text{-}39)$$

where R_i and G_v represent the linear branch relationships as in Eq. (27-2). The Jacobian of this set of equations, $\partial F/\partial(i_b, v_b, v_n)$, is

$$\begin{bmatrix} A & 0 & 0 \\ 0 & I & -A^T \\ I & \dfrac{\partial\mathbf{f}}{\partial\mathbf{v}_b} & 0 \\ R_i & G_v & 0 \end{bmatrix} \qquad (27\text{-}40)$$

Thus the Jacobian of a set of nonlinear network Eqs. (27-40) has the same form as Eqs. (27-39) except that the nonlinearities have been linearized.

There is, however, a much more informative way of looking at the application of Newton-Raphson to nonlinear network equations, the companion-model approach.[8] Consider the diode branch shown in Fig. 27-8, where I_s = saturation current, q = charge of electron, k = Boltzmann's constant, and T = temperature (K). Let us expand the diode branch relationship in a two-term Taylor series about v_D^i, i_D^i:

$$i_D^{i+1} - i_D^i = I_s \frac{q}{kT} e^{\frac{qv_D^i}{kT}} (v_D^{i+1} - v_D^i) \qquad (27\text{-}41)$$

This, of course, is just a linearization of the diode characteristic about v_D^i, i_D^i. A network-theoretic interpretation of Eq. (27-41) is shown in Fig. 27-9. That is, at the ith iteration, the diode can be replaced by a conductance in parallel with a current source. This linearization can be carried out for each nonlinearity and the equations of the linearized (i.e., companion) network written. These equations will be exactly the same as if we had applied Newton-Raphson to the original nonlinear equations. However, with the companion-model approach a stamp (Par. **27-3**) for each nonlinear element can be generated and the techniques for linear networks can easily be used for nonlinear networks. We illustrate this procedure by writing the linearized network at the mth iteration of the circuit shown in Fig. 27-10.

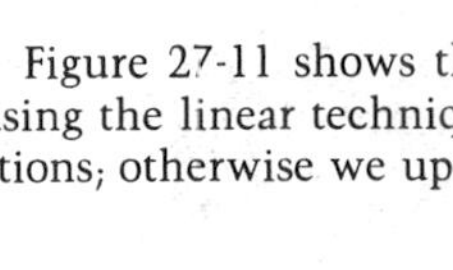

$$i_D = I_s(e^{qv_D/kT} - 1)$$

Fig. 27-8. Diode; symbol and functional description.

Figure 27-11 shows the linearized companion circuit. This network can easily be analyzed using the linear techniques already presented. If convergence criteria are met, we stop the iterations; otherwise we update the model and perform another linear analysis.

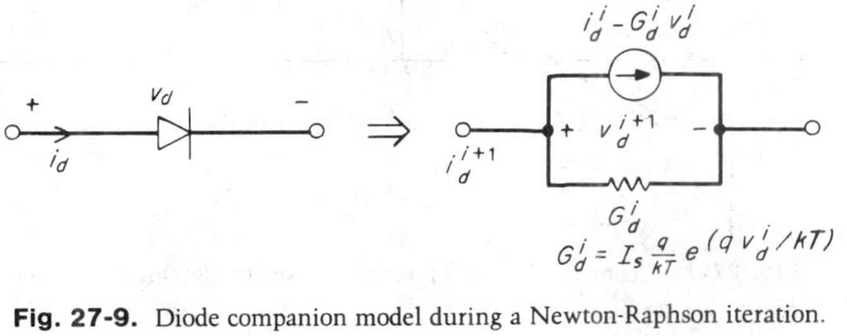

$$G_d^i = I_s \frac{q}{kT} e^{(qv_d^i/kT)}$$

Fig. 27-9. Diode companion model during a Newton-Raphson iteration.

A flow diagram of a nonlinear circuit simulation program is shown in Fig. 27-12. (For a discussion of some aspects of the figure see the discussion of convergence, Par. **27-13**.)

13. Convergence. After each iteration of the Newton-Raphson algorithm a check is made to see whether we have reached a solution. If $\mathbf{x}^k$, $\mathbf{x}^{k+1}$ represent the kth and $(k+1)$th estimates of the solution, a reasonable test for convergence is to check whether

$$|x_i^{k+1} - x_i^k| \leq \epsilon_{ai} + \epsilon_{ri} \min (x_i^{k+1}, x_i^k) \qquad \text{for all } i \qquad (27\text{-}42)$$

The term ϵ_{ai} represents the absolute allowable error for the ith component. For example, $\epsilon_{ai} = 10^{-6}$ would represent an error of 1 μV if x_i was a branch voltage. The ϵ_{ri} is a relative-error term

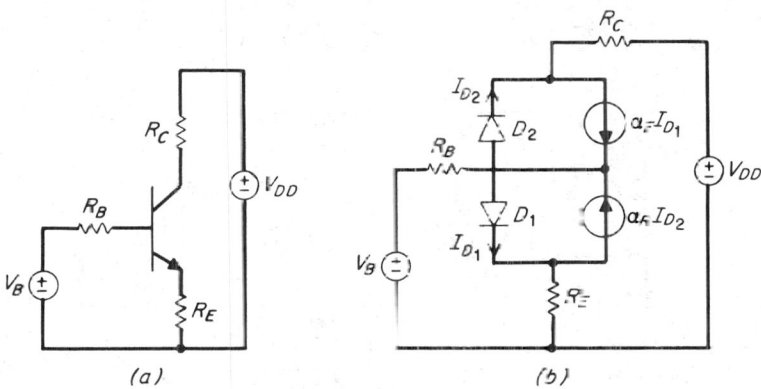

(a)   (b)

Fig. 27-10. Companion-model example: (a) example circuit, (b) circuit including a simplified transistor model.

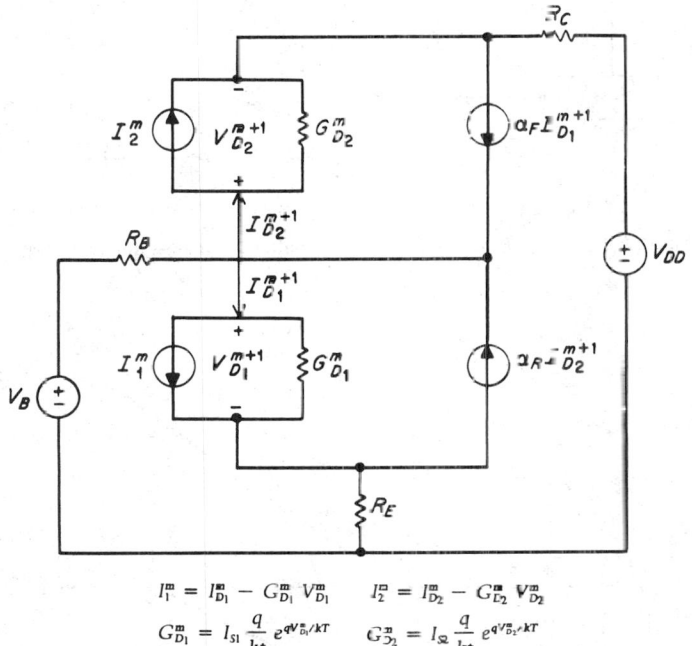

$$I_1^m = I_{D_1}^m - G_{D_1}^m V_{D_1}^m \qquad I_2^m = I_{D_2}^m - G_{D_2}^m V_{D_2}^m$$

$$G_{D_1}^m = I_{S1} \frac{q}{kt} e^{qV_{D_1}/kT} \qquad G_{D_2}^m = I_{S2} \frac{q}{kt} e^{qV_{D_2}/kT}$$

Fig. 27-11. Companion network associated with a Newton-Raphson solution of the circuit in Fig. 27-10

which takes the magnitude of the iterates into account. A typical choice for the relative error is 10^{-3} or 10^{-4}.

However, the error criterion given in Eq. (27-42) can sometimes give false results when the nonlinearity $f(x)$ changes rapidly for a small change in x, as may be the case for a diode nonlinearity. The usual fix for this problem is to apply a convergence criterion of the form shown in

Eq. (27-42) to *all* voltages and currents in the nonlinear network being solved. These two convergence checks are typical in modern circuit-simulation programs.

Independent of the convergence checks just discussed there exists the problem of the initial guess. The one failing of the Newton-Raphson method is that it will not always converge to the correct solution of a set of nonlinear equations. If the nonlinear function is continuously differentiable, for a suitable initial guess the iteration will converge to *a* solution, but the Newton iteration might cycle or converge to a wrong solution, depending on the initial starting point.

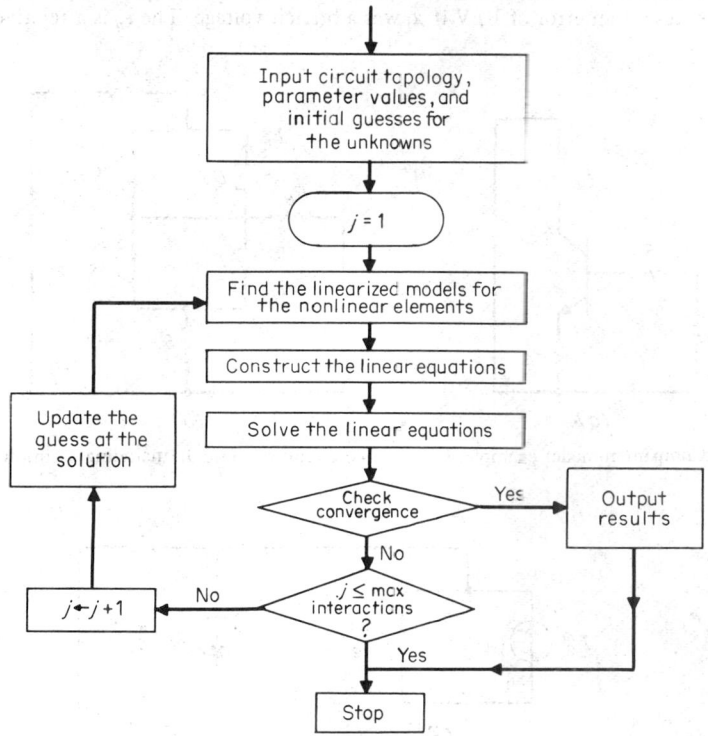

Fig. 27-12. Simplified flow diagram of a nonlinear circuit-simulation program.

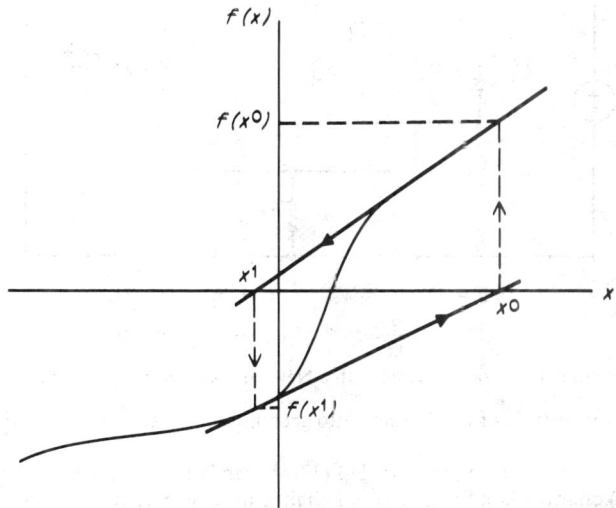

Fig. 27-13. Example of cycling in a Newton-Raphson iteration.

Some possible effects of poor initial guesses on the Newton-Raphson iteration are illustrated in Figs. 27-13 and 27-14.

The first alteration of Newton-Raphson, as indicated in Fig. 27-12, is to control the maximum number of iterations. This is a commonsense technique in any iterative procedure. For most electronic circuits a maximum of 50 iterations is sufficient.

The main computational difficulty in solving nonlinear networks using Newton-Raphson arises from the exponential nature of the nonlinearity associated with the pn junction. For example, at high forward-bias or reverse-bias conditions the conductance associated with the linearization of the diode can be very large or very small, respectively. Thus, a typical modifi-

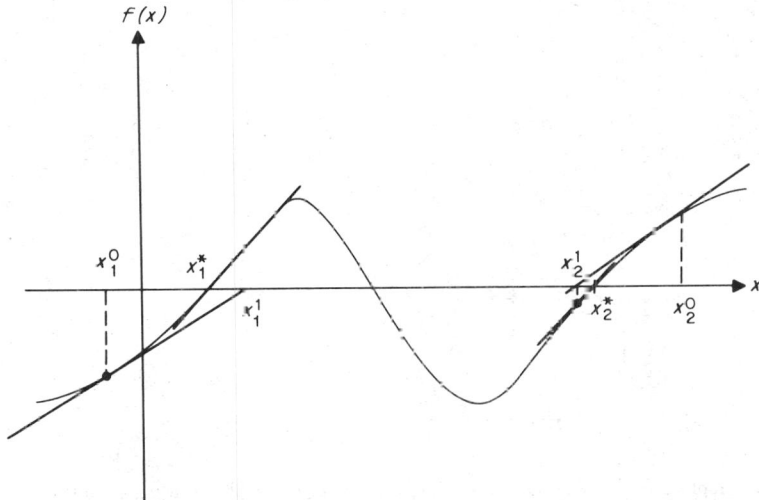

Fig. 27-14. Multiple solutions found using the Newton-Raphson iteration depending on the initial guess.

cation, when using Newton-Raphson on networks containing pn junctions, is to limit the maximum and minimum values of the conductance associated with the diode. Typical limits are 10^{12} and 10^{-12} S.

An alternative to limiting the conductance of a linearized diode is to limit the upper and lower allowable voltages that can appear across the diode. The more usual case, however, is to limit the minimum conductance (10^{-12} S) and the maximum forward-bias voltage (2 V). This scheme is quite effective in controlling numerical difficulties that arise when working with pn junction nonlinearities.

An extension of the limiting schemes discussed above is the idea of limiting the actual step size Δx^k taken by the algorithm. This step-size limitation can be useful when the initial guess is far from the solution. A typical scheme is

If $|x_i^{j+1} - x_i^j| > \Delta x_{max}$

then $\hat{x}_i^{j+1} = x_i^j + \Delta x_{max} \qquad x_i^{j+1} > x_i^j$

          $x_i^j - \Delta x_{max} \qquad x_i^{j+1} < x_i^j$

If $|x_i^{j+1} - x_i^j| \le \Delta x_{max}$

then $\hat{x}_i^{j+1} = x_i^{j+1}$

where Δx_{max} = maximum allowable step, x_i^{j+1} = new estimate of solution given by Newton-Raphson, and $\hat{x}_i^{j+1}$ = actual estimate of solution used in next iteration. A step-size-limiting algorithm was used in the simulation program CANCER[9] and has been shown to be reasonably reliable and fast for bipolar electronic circuits.

The previous modifications to the Newton-Raphson algorithm are quite useful, but one modification has proved exceptionally powerful and should be included in all nonlinear circuit-simulation programs. This modification is known as a *current-voltage iteration*. It uses the fact that most nonlinearities involved in electronic circuits are invertible: i.e,

If $i = j(v)$

then $v = g(i) = f^{-1}(i)$

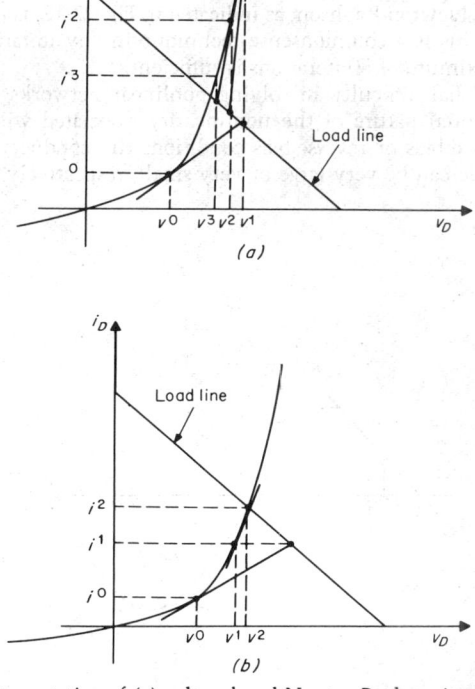

Fig. 27-15. Graphical interpretation of (a) voltage-based Newton-Raphson iteration and (b) current-based Newton-Raphson iteration.

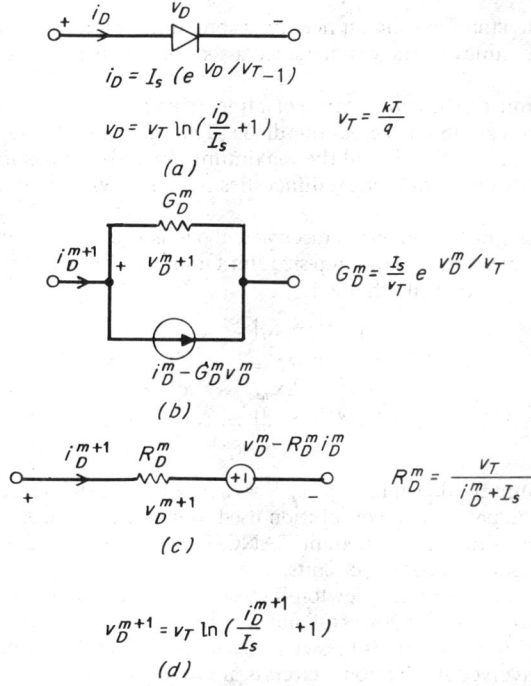

$$i_D = I_s \, (e^{\, v_D/v_T} - 1)$$

$$v_D = v_T \ln\left(\frac{i_D}{I_s} + 1\right) \qquad v_T = \frac{kT}{q}$$

(a)

$$G_D^m = \frac{I_s}{v_T} \, e^{\, v_D^m/v_T}$$

(b)

$$R_D^m = \frac{v_T}{i_D^m + I_s}$$

(c)

$$v_D^{m+1} = v_T \ln\left(\frac{i_D^{m+1}}{I_s} + 1\right)$$

(d)

Fig. 27-16. (a) Voltage and current description of a diode; (b) voltage-iteration companion model; (c) current-iteration companion model; (d) usual implicit method of implementing a current iteration while using the companion model of (b).

The choice of form, i.e., voltage-dependent or current-dependent, in which to express the nonlinearity has a significant effect on the convergence of the Newton-Raphson iteration. For example, Fig. 27-15 illustrates a voltage iteration and a current iteration on a diode nonlinearity and a linear load line. If v^0 is the initial guess and the diode is written $i = f(v)$, the iterations yield very small steps and slow convergence (Fig. 27-15a); but if the diode is written $v = g(i)$ and the iteration is performed on the current, the convergence is very fast (Fig. 27-15b). The opposite situation holds when the diode is reverse-biased. Thus, the idea of the voltage-current iterations is to switch between the different representations of the diode nonlinearity in order to speed convergence. The details of the current-voltage iteration are given in Fig. 27-16. This mixed representation can be done implicitly and does not require a new equation formulation.

The voltage-current iteration version of the Newton-Raphson algorithm is exceptionally effective, but there is as yet no theoretical basis for choosing when to switch from one representation to another. A reference voltage of 0.3 V is used in some analysis programs.[10]

The modifications to the Newton-Raphson iteration presented in the previous paragraphs when incorporated in a circuit-simulation program have proved effective in speeding convergence and overcoming various numerical difficulties. The modifications of principal importance seem to be the current-voltage iteration combined with a lower limit on conductance and upper limit on diode forward voltage. This set of modifications should be used in any circuit simulator that solves nonlinear dc or transient circuits containing pn junction devices.

TRANSIENT SIMULATION OF ELECTRONIC CIRCUITS

14. Preliminaries and Definitions. The simulation of electronic circuits in the time domain requires the use of numerical techniques for solving differential equations since the branch relationships of the capacitor and inductor are differential equations

$$i = C\frac{dv}{dt} \qquad v = L\frac{di}{dt}$$

Thus the equations for a network containing capacitors and inductors would have the form

$$\mathbf{f}(\mathbf{x}, \dot{\mathbf{x}}, t) = \mathbf{0} \qquad 0 \le t \le T \qquad \mathbf{x}(0) = \mathbf{x}_0 \tag{27-43}$$

where $\mathbf{x}$ represents the voltages, currents, charges, and fluxes in the network and $\dot{\mathbf{x}} = d\mathbf{x}/dt$. The form of Eq. (27-43) is typical of those appearing in circuit-simulation programs. However, for purposes of discussion and to be compatible with numerical analysis texts the form

$$\dot{\mathbf{x}} = \mathbf{f}(\mathbf{x}, t) \qquad 0 \le t \le T \qquad \mathbf{x}(0) = \mathbf{x}_0 \tag{27-44}$$

is also used. [Of course, the $\mathbf{f}$'s in Eqs. (27-43) and (27-44) are different.]

When solving differential equations on the computer it is typically not possible to find a closed-form analytic solution. Instead an approximation $\hat{\mathbf{x}}(t_k)$ to the true solution $\mathbf{x}(t_k)$ is found for a discrete set of time points t_k, $k = 1, \ldots, n$. $0 \le t_k \le T$ for all k. In discussing the algorithms used to solve for this approximation we shall be concerned with the accuracy of the solution and the computational expense of generating it

In our discussion of the numerical solution of differential equations we shall always assume that a solution exists and, for any given initial condition, is unique. A discussion of the existence and uniqueness of the solution of differential equations appears elsewhere.[11]

Although the literature on numerical solution of differential equations is vast, for discussing the transient simulation of electronic circuits we can limit the presentation to those techniques and concepts used in practice. In order to have a common vocabulary and for purposes of reference we now present a list of definitions and concepts associated with the numerical solution of differential equations.

Initial-Value Problem. A differential equation of the form (27-44), where a solution $\mathbf{x}(t)$ is to be found which matches a given initial condition, $\mathbf{x}(0) = \mathbf{x}_0$, is called an initial-value problem.

Step Size. The approximate solution $\hat{x}(t_k)$ of Eq. (27-44) is found at discrete time points. The difference $t_{k+1} - t_k$ is called the *time step* or *step size* at t_k. In many cases the step size h is constant, and $t_{k+1} = t_k + h$.

Order. Consider differential equations whose solutions $x(t)$ are polynomials. If p is the highest-degree polynomial that can be exactly solved by a differential-equation solution method, the method is said to be of order p or be a pth-order method. (This definition assumes infinite-precision arithmetic, or equivalently neglects roundoff error.)

   Truncation Error.   The numerical solution of a differential equation is usually not exact. Truncation error is one way of discussing the error associated with a specific technique. If $\hat{x}(t_k)$ is the approximate solution to Eq. (27-44) at t_k and $x(t_k)$ the exact solution, then $x(t_k) - \hat{x}(t_k)$ is the global truncation error at t_k. If, however, the solution technique starts from the exact solution for $t_j < t_k$ and generates $\hat{x}(t_k)$ based upon these exact data, then $x(t_k) - \hat{x}(t_k)$ is called the local truncation error. Generally a local error is the only type that can easily be estimated and thus is used to control various parameters during the solution of a differential equation.

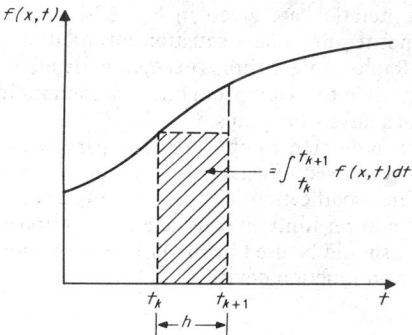

Fig. 27-17. Graphical representation of integration using forward Euler.

   Roundoff Error.   The accuracy of any computer solution is limited by the number of significant digits (number of bits) in the computer representation of a number. The error made by approximating a number with a large number of significant digits by fewer significant digits is called roundoff error. As with truncation error, both a local and global roundoff error can be discussed. Also local roundoff error is typically the only type of roundoff error estimated. However, in general, roundoff error is not a major concern in circuit-simulation programs because all computations are typically performed in double precision.

   Numerical Stability.   A solution technique with the property that the local roundoff error decreases with an increasing number of time steps is said to be numerically stable.

   Stability.   If a solution technique has the property that both local truncation and local roundoff error remain bounded as the number of time steps increases, it is said to be stable.

   Convergence.   If a solution technique has the property that local truncation and local roundoff error go to zero uniformly for $0 \le t \le T$ as the number of time steps becomes arbitrarily large, it is said to be convergent.

   Stiff Systems.   In electronic and other systems it is common to have widely different time constants. Thus the solution of these networks has high-frequency and low-frequency components. Systems with widely separated time constants are said to be stiff and require special consideration for efficient numerical solution.

   Implicit Methods.   In solving differential equations of the form (27-44) it is common to use a recursion of the form

$$\hat{x}(t_{k+1}) = g(\hat{x}(t_{k+1}), \hat{x}(t_k), t_{k+1})$$

where $g(\cdot, \cdot)$ is generally nonlinear and related to $f(\cdot, \cdot)$. Thus, finding $\hat{x}(t_{k+1})$ requires solution of a set of equations. Solution techniques with this form, i.e., with $\hat{x}(t_{k+1})$ on both sides of the equal sign, are called implicit.

   Explicit Methods.   If the recursion used in solving a differential equation has the form

$$\hat{x}(t_{k+1}) = g(\hat{x}(t_k), t_k)$$

the solution technique is called explicit.

   Multistep Methods.   A solution technique of the form

$$\hat{x}(t_{k+1}) = g(\hat{x}(t_{k+1}), \hat{x}(t_k) \ldots , \hat{x}(t_{k-m}), t_{k+1}) \qquad m \ge 1$$

is called an m-step implicit multistep method. If $m = 1$, the method is a one-step method. Explicit multistep methods have the form

$$\hat{x}(t_{k+1}) = g(\hat{x}(t_k), \hat{x}(t_{k-1}), \ldots , \hat{x}(t_{k-m}), t_k) \qquad m \ge 0$$

$m = 0$ being a one-step method.

   Predictor-Corrector Methods.   In using implicit methods to solve a differential equation a nonlinear equation

$$\hat{x}(t_{k+1}) = g(\hat{x}(t_{k+1}), \hat{x}(t_k), t_{k+1}) \qquad (27\text{-}45)$$

generally needs to be solved. An alternative to solution of this equation is to use an explicit method

$$\hat{x}^{(1)}(t_{k+1}) = \bar{g}(\hat{x}(t_k), t_k) \tag{27-46}$$

to predict $\hat{x}(t_{k+1})$ and then use Eq. (27-45) to correct the value

$$\hat{x}^{(2)}(t_{k+1}) = g(\hat{x}^{(1)}(t_{k+1}), \hat{x}(t_k), t_{k+1}) \tag{27-47}$$

The iteration of Eq. (27-47) can be continued until a desired accuracy is obtained. Alternatively, Eq. (27-46) can be used as an initial guess for a Newton-Raphson solution of Eq. (27-45). (This last method is common in circuit simulation programs.)

Companion Model. The typical set of network equations has the form

$$\mathbf{f}(\mathbf{x}, \dot{\mathbf{x}}, t) = 0$$

and not

$$\dot{\mathbf{x}} = \mathbf{f}(\mathbf{x}, t)$$

However, as with nonlinear network elements, it is possible to apply every technique for solving differential equations to the individual branch relationships. Thus there is a companion model for each energy-storage element. This companion model gives rise to a companion network which must be solved to find the response at t_{k+1}. To find the response at the next time point the companion models are updated and the associated companion network resolved.

15. Basic Methods. Three elementary, although useful, techniques for solving differential equations are now presented. Although, these techniques are used in circuit-simulation programs, the main reason for discussing them, independent of the general techniques of the next section, is to obtain a clear presentation of the essential concepts associated with the numerical solution of differential equations. For each method we present the analytical and graphical description of the algorithm, discuss the truncation error and stability, and finally present its companion model.

Forward Euler. Forward Euler is a first-order single-step explicit method. If we are solving

$$\dot{x} = f(x, t) \qquad 0 \le t \le T \tag{27-48}$$
$$x(0) = x_0$$

the forward Euler method is

$$\hat{x}(t_{k+1}) = \hat{x}(t_k) + hf(\hat{x}(t_k), t_k) \tag{27-49}$$

where $h = t_{k+1} - t_k$ is the step size.

If we express the solution of Eq. (27-48) from t_k to t_{k+1} in integral form we have

$$x(t_{k+1}) = x(t_k) + \int_{t_k}^{t_{k+1}} f(x, t) \, dt \tag{27-50}$$

Plotting $f(x, t)$ vs. t, we have the graphical representation of Fig. 27-17. If we make the approximation

$$\int_{t_k}^{t_{k+1}} f(x, t) \, dt \approx (t_{k+1} - t_k)f(x(t_k), t_k) = hf(x(t_k), t_k)$$

we have the forward Euler method. Alternatively, with $\dot{x}(t_k) = f(x(t_k), t_k)$, Eq. (27-49) can be written

$$\frac{x(t_{k+1}) - x(t_k)}{h} \approx \dot{x}(t_k) \tag{27-51}$$

A graphical interpretation of Eq. (27-51) is given in Fig. 27-18.

From Fig. 27-18 we see that the forward Euler method is a first-order method. Thus if $x(t)$ is linear, i.e., a polynomial of degree 1 or less, forward Euler gives the exact solution (neglecting roundoff error) to Eq. (27-51).

The local truncation error associated with forward Euler is rather easy to estimate[12] and is given by

$$x(t_{k+1}) - \hat{x}(t_{k+1}) = (-h^2/2) \ddot{x}(\tau) \qquad t_k < \tau < t_{k+1} \tag{27-52}$$

That is, the local truncation error is proportional to the product of the step size squared h^2 and the second derivative of $x(t)$ at some point $\tau \in (t_k, t_{k+1})$. Thus if $x(t)$ is linear, the truncation error is zero. For more complicated functions the local truncation error can be controlled by changing the step size h. This should be clear from Figs. 27-17 and 27-18, where the smaller the

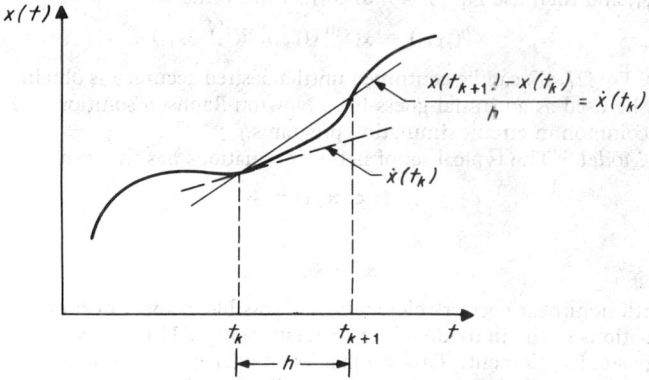

Fig. 27-18. Graphical representation of differentiation using forward Euler.

step size, the better the approximation. (We again remind the reader that local truncation error assumes that the correct solution is available at t_k.)

The main drawback of forward Euler is its stability properties. It is unstable for too large a step size. This can be easily seen by examining the test equation

$$x = -\lambda x \qquad \begin{matrix} 0 \le t \le T \\ x(0) = x_0 \end{matrix} \qquad (27\text{-}53)$$

where λ is a real positive constant. Applying forward Euler to Eq. (27-53) yields

$$\begin{aligned} x_1 &= x_0 - h\lambda x_0 \\ x_2 &= x_1 - h\lambda x_1 = (1 - \lambda h)^2 x_0 \\ &\cdots \\ x_{n+1} &= x_n - h\lambda x_n = (1 - h\lambda)^n x_0 \end{aligned} \qquad (27\text{-}54)$$

assuming a constant step size h. Notice that if $1 - h\lambda > 1$, $x_n \to \infty$ as $n \to \infty$. Thus to ensure numerical stability we require

$$1 - h\lambda < 1 \qquad (27\text{-}55)$$

or

$$h < 1/\lambda \qquad (27\text{-}56)$$

This type of analysis can be extended to systems of equations such as (27-44) with the result that the step size in forward Euler is constrained by the smallest time constant of the network.

Although forward Euler is not a numerically stable algorithm, it is finding use at present in mixed-level[13] and timing simulators,[14] where the explicit nature of the algorithm is important. In general-purpose simulators, however, it is used only as a predictor in a predictor-corrector scheme.

The companion model associated with the forward Euler method is straightforward to generate. As an example consider a capacitor branch relation $i = Cv$. We have

$$v(t_{k+1}) = v(t_k) + \frac{1}{C} \int_{t_k}^{t_{k+1}} i(t)\, dt \qquad (27\text{-}57)$$

or, using forward Euler,

$$\hat{v}(t_{k+1}) = \hat{v}(t_k) + (h/C)\hat{i}(t_k) \qquad (27\text{-}58)$$

Thus we have the companion model for forward Euler illustrated in Fig. 27-19; that is, the companion model of a capacitor using the forward Euler method is simply a voltage source. This is also illustrated in the simple network example of Fig. 27-20. The companion network can be solved for $\hat{i}_c(t_{k+1})$ using any standard analysis method. Then the companion network for t_{k+2} can be found and solved, and so on until we have generated an approximation to the entire transient response of the network.

Backward Euler.   Backward Euler is a first-order single-step implicit method whose recursion is given by

$$\hat{x}(t_{k+1}) = \hat{x}(t_k) + hf(\hat{x}(t_{k+1}), t_{k+1}) \tag{27-59}$$

The method is implicit because $\hat{x}(t_{k+1})$ appears on both sides of the equals sign.

If we write

$$x(t_{k-1}) = x(t_k) + \int_{t_k}^{t_{k+1}} f(x(t), t)\, dt$$

backward Euler is characterized by a piecewise constant approximation to the integral

$$hf(x(t_{k+1}), t_{k+1}) \approx \int_{t_k}^{t_{k+}} f(x(t), t)\, dt \tag{27-60}$$

This is illustrated in Fig. 27-21. Alternatively we write

$$\hat{x}(t_{k+1}) = \hat{x}(t_k) + hf(\hat{x}(t_{k+1}), t_{k+1}) = \hat{x}(t_k) - h\hat{x}(t_{k+1}) \tag{27-61}$$

so that

$$\hat{x}(t_{k+1}) \approx \frac{\hat{x}(t_{k+1}) - \hat{x}(t_k)}{h} \tag{27-62}$$

This interpretation is illustrated in Fig. 27-22.

The local truncation error for backward Euler can be shown to be

$$x(t_{k+1}) - \hat{x}(t_{k+1}) = (h^2/2)\hat{x}(\tau) \qquad t_k < \tau < t_{k+1} \tag{27-63}$$

which is quite similar to the truncation error for forward Euler. Both methods are first order and both have local truncation errors that depend on the square of the time step.

Despite this similarity, the important difference between forward and backward Euler methods is in their stability properties. Consider the test equation

$$\dot{x} = -\lambda x \qquad x(0) = x_0$$

Applying backward Euler results in the recursion

$$x_n = \frac{1}{(1 + a\lambda)^n} x_0 \tag{27-64}$$

Noting that for all $h > 0$ and λ a positive real constant, Eq. (27-64) implies $x_n \to 0$ as $n \to \infty$, we see that backward Euler is a numerically stable algorithm. This stability property is exceptionally important and makes backward Euler the simplest numerical-integration method to be considered for use in a circuit simulator.

The companion model for a capacitor using a backward Euler integration scheme is found as follows:

$$i = C\dot{v}$$

Using Eq. (27-62) gives

$$\dot{v}(t_{k+1}) \approx \frac{v(t_{k+1}) - v(t_k)}{h} \tag{27-65}$$

Thus

$$i(t_{k+1}) \approx \frac{C}{h} v(t_{k+1}) - \frac{C}{h} v(t_k) \tag{27-66}$$

which has the circuit interpretation given in Fig. 27-23. Of course, the Thevenin equivalent of this companion model can also be used.

A simple network and its associated companion network at t_{k+1} are shown in Fig. 27-24. The companion network can be solved for the response at t_{k+1} using any technique. Next the companion model for the capacitor is updated and a new companion network solved. This continues until the desired transient response is obtained.

We note in passing that although stable, the backward Euler method may require a very small time step to be accurate. As shown in Eq. (27-63) the time step depends on the second derivative of the signal and the desired bound on the local truncation error.

Trapezoidal Method.   The methods we have discussed so far have been first-order methods,

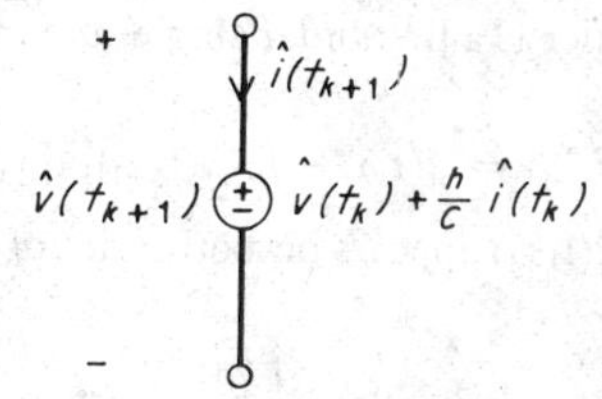

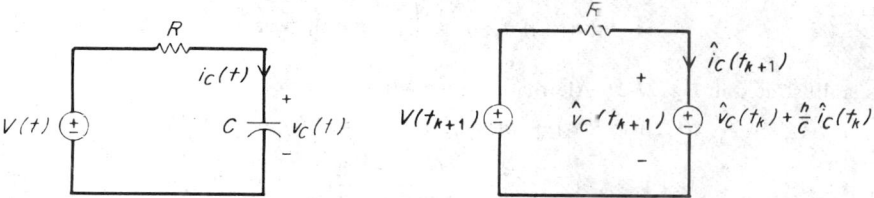

Fig. 27-19. Capacitor companion model for forward Euler.

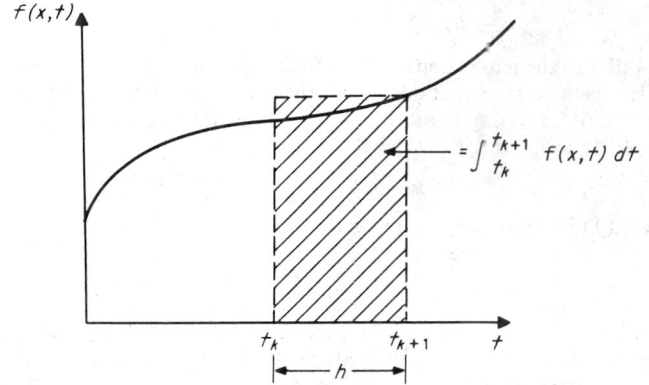

Fig. 27-20. A simplified circuit and the associated companion model when using forward Euler.

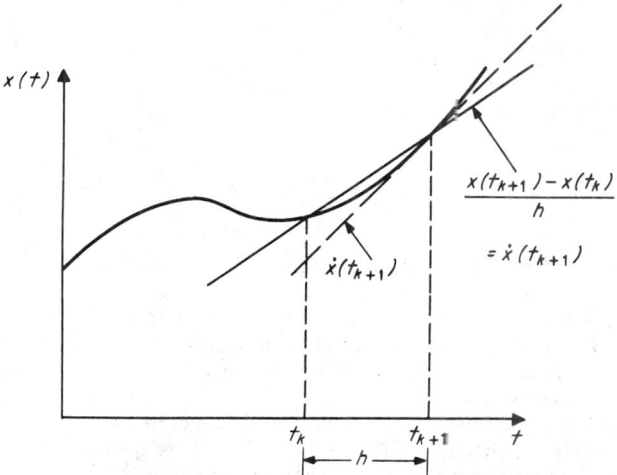

Fig. 27-21. Graphical interpretation of backward Euler integration.

$$= \int_{t_k}^{t_{k+1}} f(x,t)\, dt$$

Fig. 27-22. Graphical interpretation of differentiation using backward Euler.

$$\frac{x(t_{k+1}) - x(t_k)}{h} = \dot{x}(t_{k+1})$$

i.e., accurate for polynomials of order 1 or less. The trapezoidal method is a second-order single-step implicit method whose recursion is given by

$$\hat{x}_n(t_{k+1}) = \hat{x}(t_k) + (h/2)[f(\hat{x}(t_{k+1}), t_{k+1}) + f(x(t_k, t_k)]$$ (27-67)

In integral form this method yields

$$\int_{t_k}^{t_{k+1}} f(x(t), t) \, dt \approx \frac{1}{2} h[f(x(t_{k+1}), t_{k+1}) + f(x(t_k), t_k)$$ (27-68)

This is illustrated in Fig. 27-25, which shows that if $x(t)$ is linear, the trapezoidal method is exact. Thus if $x(t)$ is a polynomial of order 2 or less, it will be found exactly by the trapezoidal method.

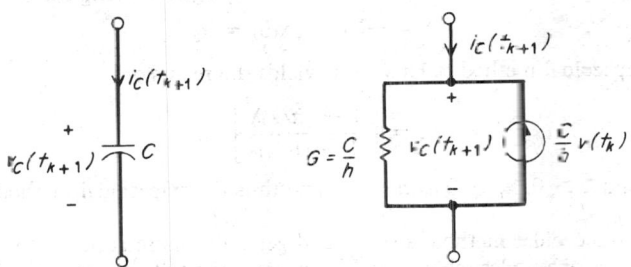

Fig. 27-23. Capacitor companion model for backward Euler.

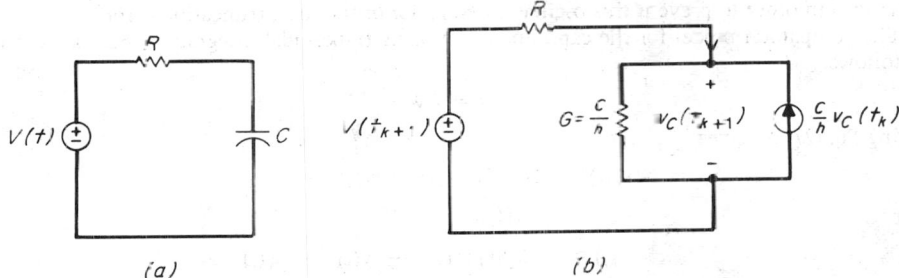

(a) (b)

Fig. 27-24. (a) Example circuit; (b) associated companion network using backward Euler.

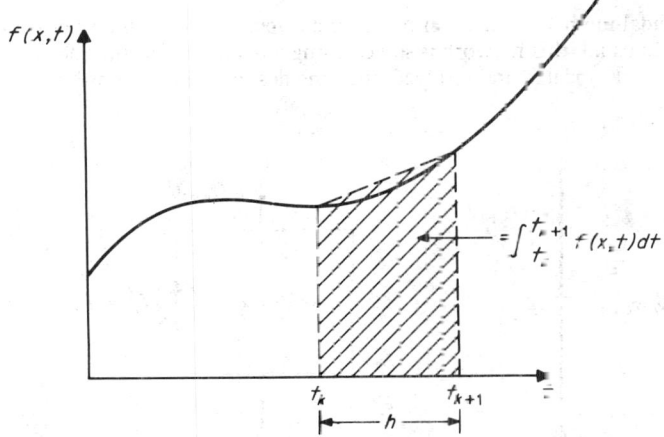

Fig. 27-25. Graphical interpretation of integration using the trapezoidal method.

The trapezoidal method can also be interpreted as

$$\hat{x}(t_{k+1}) = \hat{x}(t_k) + (h/2)[\hat{\dot{x}}(t_{k+1}) + \hat{\dot{x}}(t_k)]$$

or

$$\dot{x}(t_{k+1}) = (2/h)[\hat{x}(t_{k+1}) - \hat{x}(t_k)] - \hat{\dot{x}}(t_k)$$

(27-69)

Since the trapezoidal method is a second-order method, we would expect the local truncation error to be smaller than was for forward or backward Euler. The local truncation error for trapezoidal integration is given by

$$x(t_{k+1}) - \hat{x}(t_{k+1}) = (-h^3/12)\,\dddot{x}(\tau) \qquad t_k < \tau < t_{k+1}$$

(27-70)

For a fixed function and fixed step size, trapezoidal integration should be more accurate than forward or backward Euler. Thus the trapezoidal rule (or some variant on a second-order implicit single-step method) is the most common integration technique used in circuit simulators today.

We examine the stability of the trapezoidal method by again making use of the test equation

$$\dot{x} = -\lambda x \qquad x(0) = x_0$$

(27-71)

Applying the trapezoidal method to Eq. (27-71) yields the recursion

$$x_n = \left[\frac{1 - (h/2)\lambda}{1 + (h/2)\lambda}\right]^n x_0$$

(27-72)

For any $h > 0$ and $\lambda > 0$, $x_n \to 0$ as $n \to \infty$, and thus the trapezoidal method is numerically stable.

Although the trapezoidal method is stable and generally more accurate than the backward Euler method, it has one undesirable numerical characteristic: The sign of the local truncation error can change from one time point to the next for large step sizes. Thus the integration scheme may predict a small oscillatory behavior in the circuit response that is strictly a function of the solution technique and not the circuit.[15] The step size must be $h < 2/\lambda$ (for the test equation) in order to prevent this oscillatory behavior of the local truncation error.

The companion model for the capacitor when using trapezoidal integration method is found as follows:

$$i = C\dot{v}$$

Using Eq. (27-69) gives

$$\dot{v}(t_{k+1}) \approx (2/h)[v(t_{k+1}) - v(t_k)] - \dot{v}(t_k)$$

but

$$\dot{v}(t_k) = i(t_k)/C$$

so

$$i(t_{k+1}) \approx (2C/h)[v(t_{k+1}) - v(t_k)] - i(t_k)$$

(27-73)

The capacitor companion model for trapezoidal integration is presented in Fig. 27-26. The one drawback of this model is that the capacitor current must be calculated at every time point. This requires extra coding in a nodal-analysis scheme or an extra equation in a modified-nodal-analysis method.

The trapezoidal-method companion model and associated companion network are illustrated in Fig. 27-27. As usual, this network is solved using any analysis technique, and the companion model and network updated and resolved until the desired time response has been generated.

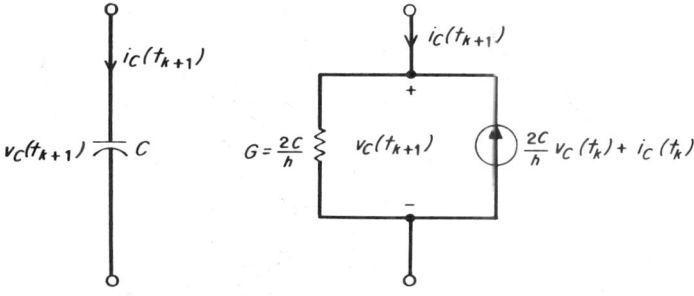

Fig. 27-26. Capacitor companion model using trapezoidal integration.

The companion models for an inductor and capacitor using forward Euler, backward Euler, and trapezoidal integration methods are compiled in Table 27-2.

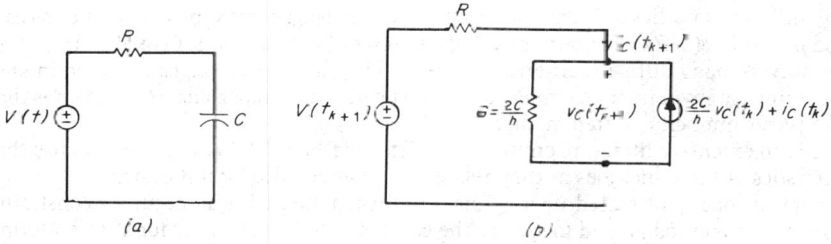

Fig. 27-27. (*a*) Example circuit; (*b*) associated companion network using the trapezoidal method.

TABLE 27-2   Capacitor and Inductor Comparison Models for Forward and Backward Euler and the Trapezoidal Integration Methods

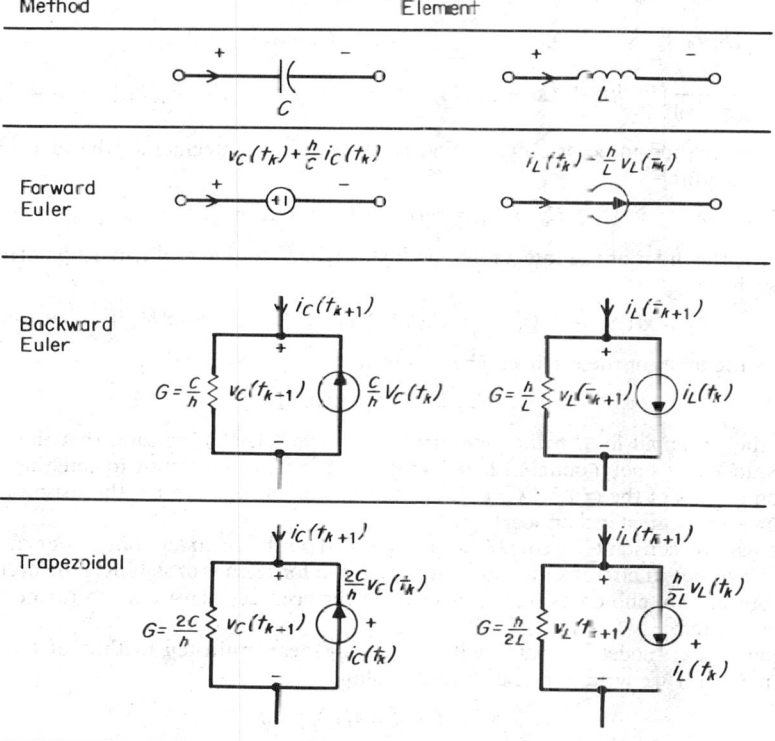

16. General Linear Multistep Methods.

The integration techniques discussed in Par. **27-15** are members of the class of general linear multistep formulas. Given

$$\dot{\mathbf{x}} = \mathbf{f}(\mathbf{x}, t) \qquad \mathbf{x}(0) = \mathbf{x}_0 \qquad 0 \le t \le T$$

the general linear multistep formula is

$$\sum_{j=0}^{k} \alpha_{n+1-j}\,\hat{\mathbf{x}}(t_{n+1-j}) = h \sum_{j=0}^{k} \beta_{n+1-j}\,\mathbf{f}\,\hat{\mathbf{x}}(t_{n+1-j}, t_{n+1-j}) \tag{27-74}$$

where h is the time step, assumed constant for the k steps used in the formula. We can relate Eq. (27-24) to the previously discussed methods as follows:

$$k = 1 \qquad \alpha_{n+1} = 1 \qquad \alpha_n = -1 \qquad \beta_{n+1} = 0 \qquad \beta_n = 1 \Rightarrow \text{forward Euler}$$
$$k = 1 \qquad \alpha_{n+1} = 1 \qquad \alpha_n = -1 \qquad \beta_{n+1} = 1 \qquad \beta_n = 0 \Rightarrow \text{backward Euler}$$
$$k = 1 \qquad \alpha_{n+1} = 1 \qquad \alpha_n = -1 \qquad \beta_{n-1} = \beta_n = \tfrac{1}{2} \Rightarrow \text{trapezoidal}$$

Thus, all the methods presented so far have been single-step methods. Further, examining Eq. (27-74) shows that for a linear multistep method to be explicit we require $\beta_{n+1} = 0$; otherwise the method is implicit.

Many different families of integration techniques can be generated by different choices of the α_i's, and β_i's in Eq. (27-74). For normalization purposes α_{n+1} is always set equal to 1, that is, $\alpha_{n+1} = 1$. However, many different criteria, such as stability, accuracy, etc., can be used in choosing the remaining parameters. For example, it can be shown[12] that trapezoidal integration is the most accurate linear implicit one-step method.

There is an extensive literature on linear multistep methods.[12,16] We can present only the basic characteristics of these methods as they relate to computer-aided circuit design.

The typical constraint placed upon a linear multistep method is an accuracy constraint; i.e., the order of the method is used to specify the constants in Eq. (27-74). If for a k-step formula as in Eq. (27-74) we define

$$
\begin{aligned}
C_0 &= \alpha_{n+1} + \alpha_n + \alpha_{n-1} + \cdots + \alpha_{n+1-k} \\
C_1 &= (k\alpha_n + (k-1)\alpha_{n-1} + \cdots + 2\alpha_{n-k} + \alpha_{n+1-k}) \\
&\quad - (\beta_{n+1} + \beta_n + \beta_{n-1} + \cdots + \beta_{n+1-k}) \\
C_q &= \frac{1}{q!}[k^q\alpha_n + (k-1)^q\alpha_{n-1} + \cdots + 2^q\alpha_{n-k} + \alpha_{n+1-k}] \\
&\quad - \frac{1}{(q-1)!}[k^{q-1}\beta_n + (k-1)^{q-1}\beta_{n-1} + \cdots + 2^{q-1}\beta_{n-k} + \beta_{n+1-k}] \qquad q = 2, 3, \ldots
\end{aligned}
$$

(27-75)

then for the method to be of order p, that is, to be able to integrate a pth-order polynomial exactly, we require[12]

$$
C_0 = C_1 = \cdots C_p = 0 \qquad C_{p+1} \neq 0 \tag{27-76}
$$

As before, the order of the integration method is related to the local truncation error. In particular, we have

$$
x(t_{n+1}) - \hat{x}(t_{n+1}) = C_{p+1} h^{p+1} f^{(p+1)}(x(t_n), t_n) + O(h^{p+2}) \tag{27-77}
$$

where p is the order of the method. The first term,

$$
C_{p+1} h^{p+1} f^{(p+1)}(x(t_n), t_n) \tag{27-78}
$$

is called the *principal local truncation error*. The term $O(h^{p+2})$ indicates that the remaining term is bounded by a polynomial in h of degree $p + 2$.[12,16] It is interesting to note that the *global* truncation error is of the order $O(h^p)$. The accumulation of local error is the reason that global truncation error is greater than local error.

The remaining constants in Eq. (27-74) not specified by the accuracy constraint can be chosen in many ways, but in circuit-simulation programs (and for reasons of stability) the order and the number of steps are chosen so that after meeting the accuracy constraint no further constants need be evaluated.

The companion model associated with a general linear multistep method of Eq. (27-74) is easily generated if we write the branch relationship as

$$
\dot{x}(t_{n+1}) = f(x(t_{n+1}), t_{n+1}) \tag{27-79}
$$

Thus Eq. (27-79) can be rewritten as

$$
\dot{x}(t_{n+1}) = \frac{1}{\beta_{n+1}h}\sum_{j=0}^{k} \alpha_{n+1-j}\hat{x}(t_{n+1-j}) - \frac{1}{\beta_{n+1}}\sum_{j=1}^{k} \beta_{n+1-j} f(\hat{x})(t_{n+1-j}), t_{n+1-j}) \tag{27-80}
$$

Using the capacitor branch relation as an example, we have

$$
\dot{v}(t_{n+1}) = (1/C)i(t_{n+1})
$$

or from Eq. (27-80)

$$
\frac{\alpha_{n+1}}{h\beta_{n+1}} \hat{v}(t_{n+1}) + \frac{1}{\beta_{n+1}}\left[\frac{1}{h}\sum_{j=1}^{k} \alpha_{n+1-j}\hat{v}(t_{n+1-j}) - \sum_{j=1}^{k} \beta_{n+1-j}\frac{1}{C}\hat{i}(t_{n+1-j})\right] = \frac{1}{C}\hat{i}(t_{n+1}) \tag{27-81}
$$

Notice that the term in square brackets is completely known at t_{n+1}. Equation (27-81) yields

$$
\hat{v}(t_{n+1}) = \frac{h\beta_{n+1}}{C\alpha_{n+1}} \hat{i}(t_{n+1}) - \frac{1}{\alpha_{n+1}}\left[\sum_{j=1}^{k} \alpha_{n+1-j}\hat{v}(t_{n-1-j}) - \frac{h}{C}\sum_{j=1}^{k} \beta_{n+1-j}\hat{i}(t_{n+1-j})\right] \tag{27-82}
$$

Defining

$$G_{n+1} = \frac{h}{C} \frac{\beta_{n+1}}{\alpha_{n+1}} \tag{27-83}$$

$$I_{n+1} = \frac{1}{\alpha_{n+1}} \left[\sum_{j=1}^{k} \alpha_{n+1-j} \hat{v}(t_{n+1-j}) - \frac{h}{C} \sum_{j=1}^{k} \beta_{n+1-j} \hat{i}(t_{n+1-j}) \right] \tag{27-84}$$

we have the companion model for the capacitor given in Fig. 27-28, which is structurally the same as the companion model for the trapezoidal rule represented by Eq. (27-73). The companion model can be used to generate a companion network which is solved for the response at t_{n+1}. Then the model and network are updated and another solution point generated.

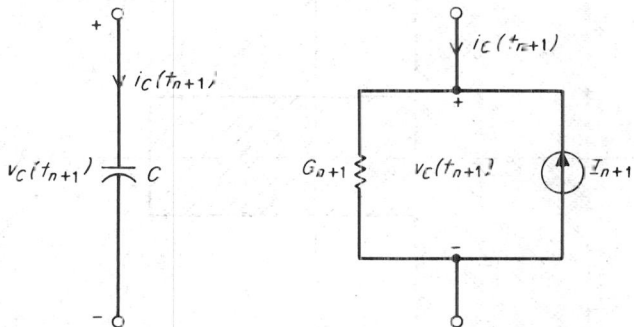

Fig. 27-28. Capacitor companion model for the general linear multistep method.

In order to discuss the convergence and stability of general linear multistep methods we must introduce a few definitions.

A linear multistep method is *consistent* if it has order $p \geq 1$. The *first* and *second* characteristic *polynomials* of linear multistep method of Eq. (27-74) are

$$\rho(z) = \sum_{j=0}^{k} \alpha_{n+1-j} z^{k-j} \tag{27-85}$$

$$\sigma(z) = \sum_{j=0}^{k} \beta_{n+1-j} z^{k-j} \tag{27-86}$$

Note that consistency requires

$$\rho(1) = 0 \qquad \rho'(1) = \sigma(1) \tag{27-87}$$

We can now define the concept of *zero stability*. A linear multistep method is said to be *zero stable* (or *D-stable*) if all the roots of the first characteristic polynomial of Eq. (27-85) have modulus ≤ 1 and every root with modulus 1 is simple.

We now state the principal convergence criteria for linear multistep methods:

A linear multistep method is convergent if and only if it is consistent and zero-stable.

Any integration technique used in a circuit simulator should be convergent. Thus we are interested in methods with maximum accuracy (order) which are also convergent. The following result presents the limits on the maximum number of steps for a zero-stable method:

A zero-stable linear multistep method of step number k has a maximum order $k + 1$, for k odd, and $k + 2$ for k even.

For dealing with many systems the concept of zero stability is not sufficient. Several alternate forms of stability arise frequently in the literature. Instead of using two characteristics polynomials, as in Eqs. (27-85) and (27-86), we now consider one characteristic or stability polynomial defined as

$$P(Z, h\lambda) = \rho(Z) - h\lambda\sigma(Z) \tag{27-88}$$

where $\rho(\cdot)$ and $\sigma(\cdot)$ are given in Eqs. (27-85) and (27-86), h is the step size, and λ is the constant associated with the test equation (27-71) or the largest eigenvalue of a system of differential equations (Note that λ is now complex.) We now use the stability polynomial to define some other forms of stability:

A linear multistep method is said to be *absolutely stable* for $h\lambda$ if all roots r_i of (27-88) for that $h\lambda$ satisfy $|r_i| < 1$ i.e., are within the unit circle in the complex plane. The interval (a, b) of the real line is called an *interval of absolute stability* if the method is absolutely stable for all $h\lambda \in (a, b)$.

The concept of absolute stability is specialized in the following definitions:

A linear multistep method is said to be *A-stable* if its interval of absolute stability contains the entire left half plane Re $h\lambda < 0$. (In this case the test equation is $\dot{x} = \lambda x$ for λ complex, Re $\lambda < 0$.)

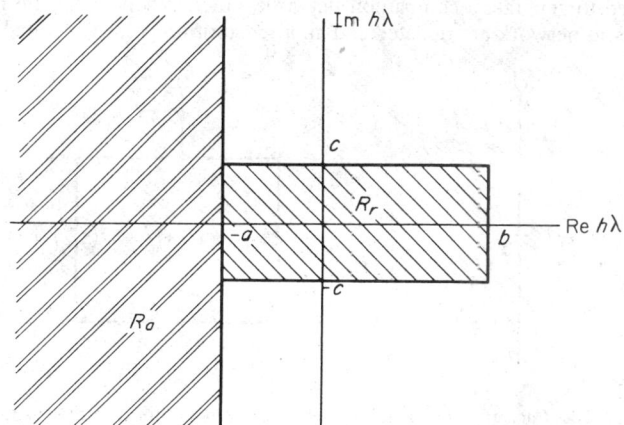

Fig. 27-29. Desirable stability regions R_a, R_r for a stiffly stable integration algorithm.

A weakened version of A stability is useful for stiff systems:

A linear multistep method is called *stiffly stable* if its region of absolute stability contains R_a and R_r and it is accurate for all $h \in R_r$ (for the test equation $\dot{x} = \lambda x$, λ complex, Re $\lambda < 0$), where

$$R_a = \{h\lambda \,|\, \text{Re } h\lambda < -a\}$$
$$R_r = \{h\lambda \,|\, -a \le \text{Re } h\lambda \le b, -c \le \text{Im } h\lambda \le c\}$$

for positive constants a, b, and c (see Fig. 27-29).

The concepts of stability associated with the linear multistep methods are complex, and most, for practical purposes, are not important for circuit simulation. However, many electronic circuits have widely separated time constants, i.e., are stiff, and thus a stiffly stable algorithm is important for circuit simulation. A major drawback of linear multistep formulas[17] is that the trapezoidal method is the most accurate stiffly stable multistep algorithm. (Backward Euler is also stiffly stable.)

Gear's Algorithm. In order to develop stiffly stable algorithms Gear proposed the recursion

$$\dot{x}(t_{n+1}) = a_n(k)\dot{x}(t_n) + a_{n-1}(k)\dot{x}(t_{n-1}) + \cdots + a_{n-k+1}(k)\dot{x}(t_{n-k+1})$$
$$+ h[b_{n+1}(k)f(\dot{x}(t_{n+1}), t_{n+1})] \quad (27\text{-}89)$$

as the kth-order implicit method with the coefficients $a_n(k)'s$, $b_{n+1}(k)$ dependent upon the order k. The coefficients are chosen so that the solution is exact for all polynomials of order k or less. It can be shown that Gear's formulas for orders 1 to 6 are stiffly stable. Details of Gear's algorithm can be found elsewhere.[16] We shall not dwell on these methods because an alternate formulation more suited to circuit simulation will be presented in the next section.

The key feature of Gear's algorithm or the linear multistep methods is that the order and step size of the method can be tailored to the problem being solved. In fact, the most efficient use of these algorithms calls for the possibility of varying the time step and order during the course of a simulation. The main idea is to control local truncation error (in hopes of controlling global error) while at the same time maximizing the step size. Thus the changing order is used to maintain accuracy while the time step is increased in order to decrease the total computational cost associated with the transient analysis. Practical details of implementing a numerical-integration scheme in a circuit simulator are discussed in the next section.

17. Variable Order Variable-Step-Size Methods for Circuit Simulation.

The method proposed by Gear for deriving a family of stiffly stable algorithms does not lend itself to easy implementation in a circuit-simulation program. However, circuit simulation, especially of analog and digital integrated circuits, requires the use of stiffly stable algorithms and the dynamic changing of the order and step size of those algorithms. This situation led to the development of a family of stiffly stable methods well suited to circuit simulation.[18]

The general form of the equations describing a nonlinear dynamic electronic circuit is

$$\hat{f}(x, \dot{x}, t) = 0 \qquad 0 \le t \le T \tag{27-90}$$

For simplicity of notation we consider a single equation in a single variable. The extension to a system of equations is obvious. The technique we now discuss is based upon an implicit backward-difference approach to approximating $\dot{x}(t_{n+1})$

$$\hat{x}(t_{n+1}) = -\frac{1}{h} \sum_{i=0}^{k} \alpha_i \hat{x}(t_{n+1-i}) \qquad 1 \le k \le 6 \tag{27-91}$$

Thus there is a family of six different formulas given in Eq. (27-91). The term $\dot{x}$ in Eq. (27-90) is replaced by Eq. (27-91) and the resulting set of nonlinear equations solved by Newton-Raphson. The initial guess for the Newton-Raphson iteration is based upon a predictor of the form

$$\hat{x}^P(t_{n+1}) = \sum_{i=1}^{k+1} \gamma_i \hat{x}(t_{r+1-i}) \tag{27-92}$$

This predictor is also useful in estimating and controlling the local truncation error. For purposes of controlling roundoff error Eqs. (27-91) and (27-92) are sometimes written using backward differences $\Delta x_n = \hat{x}(t_{n+1}) - \hat{x}(t_n)$

$$\hat{x}(t_{n+1}) = -\frac{1}{h} \sum_{i=0}^{k} \hat{\alpha}_i \, \Delta x_{n-i} \tag{27-93}$$

$$\hat{x}^P(t_{n+1}) = \hat{x}_n + \sum_{i=1}^{1} \gamma_i \, \Delta x_{n-i} \tag{27-94}$$

It can be shown[18] that both representations require the same computational effort.

In addition to being stiffly stable the primary merit of Eq. (27-91) is that the local truncation error can easily be calculated. Local truncation error for Eq. (27-91) is given for a k-step method by

$$E_k = \frac{h}{t_{n+1} - t_{n-k}} [\hat{x}(t_{n+1}) - \hat{x}^P(t_{n+1})] + O(h^{k+2}) \tag{27-95}$$

where $h = t_{n+1} - t_n$ and $\hat{x}^P(t_{n+1})$ is given by Eq. (27-92). Thus neglecting the term in h^{k+2}, we can easily calculate the local truncation error at any specific time point. Furthermore, because the local truncation error is easy to find, it can be used in a scheme to control the order and step size of the integration during the course of the calculation.

The coefficients α_i, γ_i, ($\hat{\alpha}_i$, $\hat{\gamma}_i$) are found using the typical accuracy constraint: if a k-step method is used, it should provide the exact solution when $x(t)$ is a polynomial of order k. Details for calculating these coefficients as well as an investigation of the computational expense of the calculations can be found in Ref. 18.

The main logic necessary for a successful implementation of Eq. (27-91) for circuit simulation is the method of changing step size and order. It is clear that initially (at $t = t_0$) only a first-order method can be used. However, after six time steps have been taken, the order can be adjusted from 1 to 6 in order to optimize the accuracy and computational-expense trade-off involved in solving stiff systems. Again the main idea is to take as large a time step as possible, reducing overall computation, while adjusting the order to maintain the local truncation error within certain bounds.

A typical scheme for using a variable-order variable-step-size method for the kth time point is as follows. With order p and step size h

(1) Set $t_{k+1} = t_k + h$.
(2) Use Eq. (27-92) to calculate $\hat{x}^P(t_{k+1})$.
(3) Use $\hat{x}^P(t_{k+1})$ as the initial guess and solve Eq. (27-90) using a Newton-Raphson iteration with $\dot{x}(t_{k+1})$ replaced by Eq. (27-91).

(4) Compute the local truncation error E_p using Eq. (27-95).

(5) If E_p is too large, reject the solution (typically the step size is reduced, say by a factor of 2 and the algorithm restarted from $t = t_k$). Alternatively, if k steps have been taken at the present order, the tests outlined below can be performed to calculate the maximum step size at a reduced order and this order and step size are used if desired.[18]

(6) If E_p is less than a user-specified error, choose a new step size and order in order to maximize the step size while maintaining the desired accuracy.

The main heuristic part of this scheme is step 6, where the order and step size are adjusted. The user initially specifies a maximum allowable global truncation error E_T. Thus for $0 \leq t \leq T$ the allowable error per unit time is E_T/T, and the allowable error for a time step h is hE_T/T. The heuristic attempts to maximize h subject to

$$E_p \leq hE_T/T \qquad (27\text{-}96)$$

The order p is allowed to change only if $p + 1$ steps have been taken using that order. Also, the order is allowed to increase or decrease by only 1 within the bounds of $1 \leq p \leq 6$. The changing of step size (and order) is based upon calculating a time-step ratio

$$\eta_p = \frac{t_{k+2} - t_{k+1}}{t_{k+1} - t_k} = \frac{\Delta t_{new}}{h} \qquad (27\text{-}97)$$

If only the step size is being changed, the time-step ratio is given by

$$\eta_p = \left(\frac{hE_T}{E_p T} \right)^{1/p} \qquad (27\text{-}98)$$

and the new step size is given by

$$\Delta t_{new} = h\eta_p \qquad (27\text{-}99)$$

However, if a change in order and step size is being considered, the time-step ratio for each order is calculated

$$\eta_i = \left(\frac{hE_T}{E_i T} \right)^{1/i} \qquad i = p - 1, p, p + 1 \qquad (27\text{-}100)$$

We note that the local truncation error E_i for $i + p - 1, p, p + 1$ requires the calculation of the predicted value $\hat{x}^p(t_{n+1})$ for each order. Letting $\hat{x}(t_{n+1})$ be the value found in step 3 of the algorithm, we calculate the local truncation errors $E_i, i = p - 1, p + 1$, from Eq. (27-95). The new order i is taken as the index of

$$\eta^* = \max_{i \in (p-1, p, p+1)} \eta_i \qquad (27\text{-}101)$$

The new step size is given by

$$\Delta t_{new} = h\eta^* \qquad (27\text{-}102)$$

with

$$t_{n+2} = t_{n+1} + \Delta t_{new} \qquad (27\text{-}103)$$

The details for efficient implementation of these tests can be found in Ref. 18.

The variable-order variable-step-size scheme outlined above has proved very effective in the simulation of a large class of electronic circuits and is considered the state of the art. We note, however, that if several variables are being approximated by the backward-difference formula (27-91), a time-step ratio η_i^j must be calculated for each of the j variables. The actual value used for η_i in Eq. (27-100) is then

$$\eta_i = \min_j \eta_i^j \qquad i = p - 1, p, p + 1 \qquad (27\text{-}104)$$

where the minimum is over all variables being integrated. This is also the case if only the step size is being varied, as in Eq. (27-98). Thus, we always choose the maximum step size so that the maximum error [minimum time-step ratio, see Eq. (27-98)] is within the user-specified bounds. [Note that it is also possible to specify a separate global truncation error for each variable, leading to obvious modifications in the test equations (27-96).]

18. Practical Considerations for Nonlinear Transient Simulation. Although the variable-order variable-step-size method discussed in Par. **27-17** is excellent for circuit simulation, some practical heuristics have proved to be effective in the simulation of nonlinear circuits in the time domain.

Step Size. It is expedient to select a minimum step size Δt_{min}. This should be related to the highest-frequency component the user expects in the response of the circuit, e.g.,

$$\Delta t_{min} \leq \frac{2\pi}{2f_{max}}$$

If in the reduction of step size a step $\Delta t_{new} < \Delta t_{min}$ is calculated from Eq (27-102), Δt_{new} is set equal to Δt_{min}, $\Delta t_{new} = \Delta t_{min}$. It has been noted[18] that even if $\Delta t_{new} = \Delta t_{min}$ and the local truncation error is not satisfied, it is important to allow the order to change as suggested in Eq. (27-101). This freedom allows the local truncation error to come within bounds much sooner than if the order were set to $p = 1$ and $\Delta t = \Delta t_{min}$.

Order. Although the Gear and backward-difference formulas are stiffly stable for orders $1 \leq p \leq 6$, experimentation suggests that the maximum useful order is $p = 5$. Furthermore, many simulation programs have been written where the order was restricted to $p \leq 3$ or even $p \leq 2$. The consensus seems to be that for a general-purpose simulator the ability to vary the order $1 \leq p \leq 3$ is probably sufficient.

Newton Iterations (Corrector Iterations). The set of nonlinear equations that result after the derivatives $\dot{x}_{n+1}$ have been replaced using Eq. (27-91) are typically solved using a Newton-Raphson algorithm. This can also be viewed as a corrector iteration in a predictor-corrector scheme.[2] However, these corrector iterations are expensive and, if the predictor of Eq. (27-92) was not accurate, may require many iterations for convergence. Thus it is typical to put a maximum limit on the number of corrector iterations (typically ≤ 5). If convergence has not been attained in the specified number of iterations, the step size is reduced (usually by a factor of 2) and a new predictor-corrector iteration begun.

Time Cusps. The main cause of difficulty in a numerical-integration scheme is when one of the variables is changing rapidly or there is a discontinuity in the derivative of a variable. In electronic circuits these conditions are generally caused by a rapid change in an input signal or the switching of a nonlinear device. If the occurrence of such conditions can be predicted by the user, i.e., at $t_{C1}, t_{C2}, \ldots, t_{Cm}$, then if $t_k < t_{Ci}$ and $t_{k+1} = t_k + \Delta t_k \geq t_{Ci}$, we set $t_{k+1} = t_{Ci}$, set the order to 1 and the step size to Δt_{min}, and proceed with the iteration. The use of time-cusp strategies has proved useful in the simulation of switching circuits[18]

In order to consolidate the information on simulation of electronic circuits in the time domain we present in Fig. 27-30 a flowchart of the variable-order variable-step-size integration method. The flowchart assumes that we are solving

$$\mathbf{f}(\mathbf{x}, \dot{\mathbf{x}}, t) = \mathbf{0} \qquad 0 \leq t \leq T$$

and that more than six time steps have been taken. The present order is assumed to be p, $1 \leq p \leq 6$, the present step size h, the time t_{ki}; we assume that there are n variables $\dot{x}_i$, $i = 1, 2, \ldots, n$, being approximated by the backward-difference formula. The user has specified a global truncation error E_T, which is the same for all variables, the maximum number of Newton iterations per time step is five, the minimum step size is Δt_{min}, and we assume that no time cusps have been specified. Further, we let NSLC be the number of time steps taken since the last change in the order of the integration formula. Also at the time step t_k we assume that the truncation errors for each variable for the order p, E_p^i, $i = 1, 2, \ldots, n$, have been calculated. In short, we have all the information necessary to take the next time step. Finally, we let $N = \{1, 2, \ldots, n\}$.

SOURCES OF SIMULATION PROGRAMS AND NEW DEVELOPMENTS

19. Program Sources. Many circuit-simulation programs have been written during the past 15 years. These programs have been developed by universities, industry, and commercial software companies. In general however, there are very few widely accepted, easily available general-purpose simulation programs. Probably the most widely used, and modified, simulation program is SPICE, Simulation Program with Integrated Circuit Emphasis, developed, maintained, and distributed by the University of California at Berkeley.[10] There are other programs available free from universities or at moderate to exorbitant cost from various companies. An excellent discussion of these various programs can be found in Ref. 19.

Since new programs and modifications to existing programs are being developed constantly, readers interested in obtaining a simulation program should get help from an authority and/or consult the literature in order to determine the best program currently available. The state of the art in simulation programs is highly developed. Besides the techniques discussed in this chapter, many techniques from computer science are used in the input, output (user-machine interface), and data-base aspects of a simulation program. Thus for the general user we recommend the use of existing packages, with special modifications if necessary, rather than developing simulation programs from scratch. For many fine papers on simulation see Refs. 20 and 21.

In this section we have not discussed the many uses of circuit simulation. The obvious use is the evaluation of the electrical functioning of a circuit, but it is also possible to calculate the sensitivity of any particular circuit response with respect to any circuit parameters.[22] Thus a simulator can be used to find sensitive parameters. The sensitivity information can also be used to help design circuits. The design can be accomplished using manual techniques or the techniques and algorithms of optimization theory.[23]

In the manufacture of integrated circuits the various circuit parameters cannot be controlled exactly. Thus the manufactured circuit performance will not be that specified in the design. If the actual response is too far from the specifications, the circuit is a failure. The yield of a design is the number of circuits that meet the specifications divided by the total number manufactured. One prime economic concern in the integrated-circuit industry is to keep the yield as high as

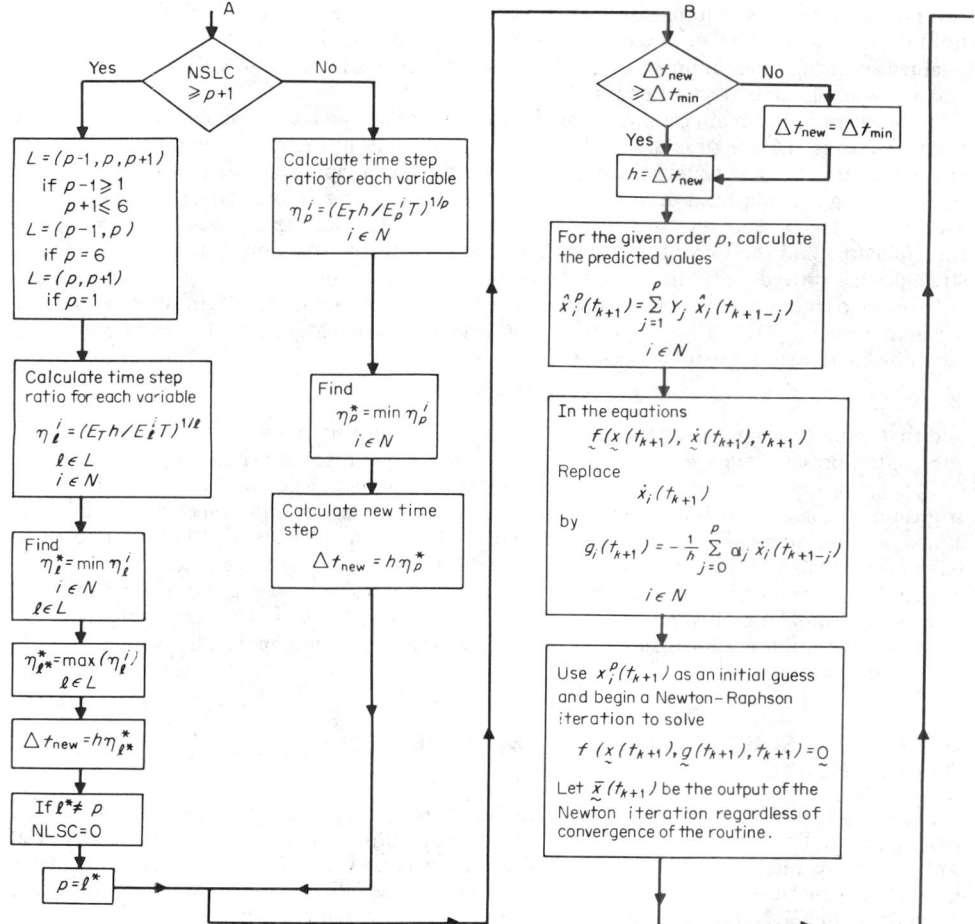

Fig. 27-30. Flow chart for a variable-order variable-step-size integration method for solving nonlinear dynamic electronic circuits.

possible. This has led to the use of simulators to estimate, and even increase, the yield of a design.[24]

The evolution of the integrated-circuit industry has led to the production of large-scale and very large scale integrated circuits (LSI and VLSI), chips which may contain as many as 50,000 to 100,000 transistors. Although this huge number of devices is beyond the economic capacity of the best circuit simulator and fastest computer, these VLSI chips still need to be simulated. This need is leading to the development of new concepts in simulation. We briefly discuss two of these ideas, timing simulation and mixed-mode simulation.

When simulating many digital circuits, especially MOSFET circuits, it is not necessary to obtain an exact solution to the circuit equations. In most cases, a timing simulation can be performed. Timing simulation is midway between the expensive but informative circuit simulation and a cheap but sometimes inadequate logic simulation. Timing simulation is accomplished by three techniques: (1) a simplified model of the MOS transistor and MOS logic gate is used; (2) an explicit integration scheme is used which obviates the need for solving nonlinear equations and essentially decouples each gate so that it can be treated separately and (3) instead of using an analytic description of the device model, a tabular description accompanied by an efficient table-look-up scheme is used to model the active devices. Timing simulation has proved to be a cost-effective design tool and is achieving rapid acceptance in the integrated-circuit industry.[14]

In VLSI circuits it is common to be concerned with the circuit response of one part of the chip, the timing behavior of another, and only the logical functioning of a large portion of the

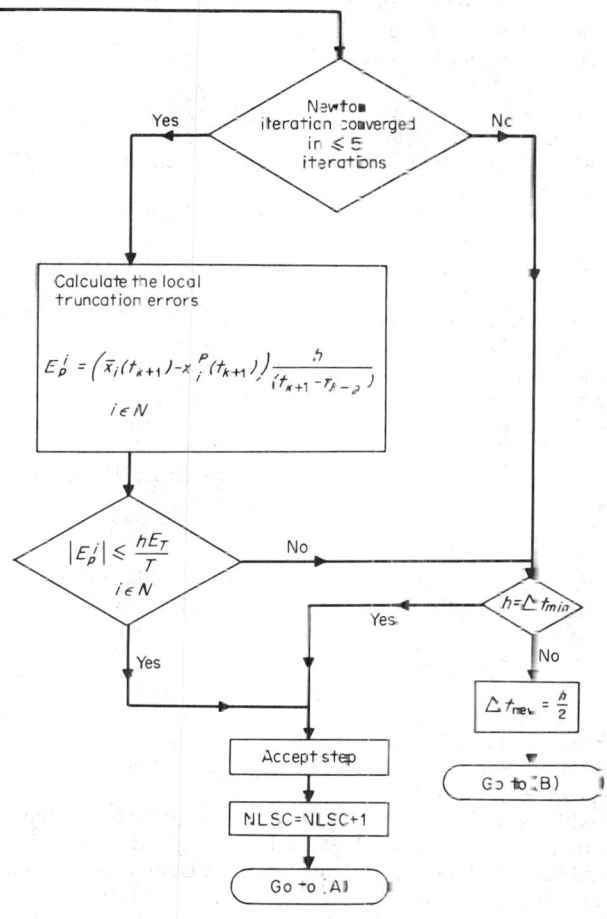

chip. This has led to the development of mixed-mode simulators capable of simulating an entire chip by using different descriptions of various portions of the chip.[13,25] The concept of mixed simulators is relatively new; although the concepts discussed in this section are applicable to these simulators, they will require intensive development and testing before they reach the level of acceptance of circuit simulators.

Thus, although the techniques discussed here are common to circuit simulators, many new techniques in simulation are being developed to deal with the advent of VLSI technology. The techniques discussed here will remain valid and useful as circuit-design aids for many years, but we refer the reader to the current literature for development of new concepts and techniques in VLSI simulation.

20. References

1. Ho, C. W., A. E. Reuhli, and P. A. Brennan   The Modified Nodal Approach to Network Analysis, *IEEE Trans. Circuits Syst.*, June 1975, Vol. CAS-22, pp. 504–509.

2. Hachtel, G. D., R. K. Brayton, and F. G. Gustavson   The Sparse Tableau Approach to Network Analysis and Design, *IEEE Trans. Circuit Theory*, 1971, Vol. CT-18, pp. 101–113.

3. Jennings, A.   "Matrix Computations for Engineers and Scientists," Wiley, New York, 1977.

4. Stewart, G. W.   "Introduction to Matrix Computation," Academic, New York, 1973.

5. Hajj, I. N., P. Yang, and T. Trick   Avoiding Zero Pivots in the Modified Nodal Approach, *Proc. 1980 IEEE Int. Symp. Circuits Syst.*, pp. 797–800.

6. Duff, I. S.   A Survey of Sparse Matrix Research, *Proc. IEEE* April 1977, Vol. 65, pp. 500–535.

7. Markowitz, H. M.   The Elimination Form of the Inverse and Its Applications to Linear Programming. *Manage. Sci.*, 1957, Vol. 3, pp. 255–269.

8. Calahan, D. A.   "Computer-Aided Network Design," rev. ed., McGraw-Hill, New York, 1972.

9. Nagel, L. W., and R. A. Rohrer   Computer Analysis of Nonlinear Circuits, Excluding Radiation (CANCER), *IEEE J. Solid-State Circuits*, August 1971, Vol. SC-6, pp. 166–182.

10. Nagel, L. W.   SPICE2: A Computer Program to Simulate Semiconductor Circuits, Ph.D. dissertation, University of California, May 1975.

11. Coddington, E. A., and N. Levinson   "Theory of Ordinary Differential Equations," McGraw-Hill, New York, 1955.

12. Lambert, J. D.   "Computational Methods in Ordinary Differential Equations," Wiley, New York, 1973.

13. Agrawal, V. D., A. K. Bose, P. Kozak, H. N. Nham, and E. Facas-Skewes   A Mixed-Mode Simulator, *Proc. 17th Design Autom. Conf., June 1980*, pp. 618–625.

14. Chawla, B. R., H. K. Gummel, and P. Kozak   MOTIS: An MOS Timing Simulator, *IEEE Trans. Circuits Syst.*, December 1975, Vol. CAS-22, pp. 901–910.

15. Chua, L. O., and P. M. Lin   "Computer-Aided Analysis of Electronic Circuits: Algorithms and Computational Techniques," Prentice-Hall, Englewood Cliffs, N.J., 1975.

16. Gear, C. W.   "Numerical Initial Value Problems in Ordinary Differential Equations," Prentice-Hall, Englewood Cliffs, N.J., 1971.

17. Dahlquist, G.   A Special Stability Problem for Linear Multistep Methods, *BIT*, 1963, Vol. 3, pp. 27–43.

18. Brayton, R. K., F. G. Gustavson, and G. D. Hachtel   A New Efficient Algorithm for Solving Differential-Algebraic Systems Using Implicit Backward Differentiation Formulas, *Proc. IEEE*, 1972, Vol. 60, pp. 98–108.

19. Kaplan, G.   Computer-Aided Design, *IEEE Spectrum*, October 1975, Vol. 12, no. 10, pp. 40–47.

20. Director, S. W., and B. J. Karafin (eds.)   *IEEE Trans. Circuit Theory*, Special Issue on Computer-Aided Design, November 1973, Vol. CT-20.

21. Director, S. W., and R. K. Brayton (eds.)   *IEEE Trans. Circuits Syst.*, Special Issue on Computational Methods, September 1979, Vol. CAS-26.

22. Director, S. W., and R. A. Rohrer   The Generalized Adjoint Network and Network Sensitivities, *IEEE Trans. Circuit Theory*, August 1969, Vol. CT-16, pp. 318–323.

23. Director, S. W., and R. A. Rohrer   Automated Network Design: The Frequency Domain Case, *IEEE Trans. Circuit Theory*, August 1979, Vol. CT-16, pp. 330–337.

24. Polak, E., and A. Sangiovanni-Vincentelli   Theoretical and Computational Aspects of the Optimal Design Centering, Tolerancing, and Tuning Problem, in Ref 21.

25. Newton, A. R.   Techniques for the Simulation of Large-Scale Integrated Circuits, in Ref. 21.

Section 28

Reliability of Electronic Components and Systems

RONALD T. ANDERSON Reliability Manager, IIT Research Institute; Member, IEEE

STANISLAW KUS Research Engineer, IIT Research Institute

HENRY C. RICKERS Research Engineer. Reliability Analysis Center RADC (RBRAC); Member, IEEE

JAMES W. WILBUR Research Engineer, Reliability Analysis Center RADC (RBRAC); Member, IEEE

GERALD D. NETZBAND Components Engineering Manager, Martin Marietta Aerospace, Orlando Division

DWIGHT E. DAVIS Components Engineer, Martin Marietta Aerospace, Orlando Division

H. BENNETT DREXLER Components Engineer, Martin Marietta Aerospace, Orlando Division

KURT E. GONZENBACH Components Engineer, Martin Marietta Aerospace, Orlando Division

NEIL V. OWEN Components Engineer, Martin Marietta Aerospace, Orlando Division

EDWIN W. KIMBALL Staff Reliability Engineer, Martin Marietta Aerospace, Orlando Division

CLARENCE M. BAILEY, JR. Supervisor, Data Reporting and Failure Analysis, Bell Telephone Laboratories; Member, IEEE

S. SUGIHARA, Manager, Systems Effectiveness Analysis Section, The Aerospace Corp.

CONTENTS

Numbers refer to paragraphs

Reliability Design and Engineering*

BY RONALD T. ANDERSON, STANISLAW KUS, HENRY C. RICKERS, AND JAMES W. WILBUR

RELIABILITY CONCEPTS AND DEFINITIONS

1. The intrinsic reliability of a system, electronic or otherwise, is based upon its fundamental design, but its reliability is often less than its intrinsic level due to poor or faulty procedures at three subsequent stages: manufacture, operation, or maintenance.

*Section 28 was coordinated by the Associate Editor, who wishes to express his appreciation for support and advice to Joseph B. Brauer, D. S. Peck, C. H. Zierdt, Jr., and James J. Egan. — D. C.

2. Definitions. The definition of reliability involves four elements: *performance requirements, mission time, use conditions, and probability.* Although reliability has been variously described as "quality in the time dimension" and "system performance in the time dimension," a more specific definition is *the probability that an item will perform satisfactorily for a specified period of time under a stated set of use conditions.*

Failure rate, the measure of the number of malfunctions per unit of time, generally varies as a function of time. It is usually high but decreasing during its *early life,* or *infant-mortality* phase. It is relatively constant during its second phase, the *useful-life period.* In the third, *wear-out* or *end-of-life,* period the failure rate begins to climb due to the deterioration that results from physical or chemical reactions: oxidation, corrosion, wear, fatigue, shrinkage, metallic-ion migration, insulation breakdown, or, in the case of vacuum tubes or batteries, an inherent chemical reaction that goes to completion.

The failure rate of most interest is that which relates to the useful life period. During this time, reliability is described by the single-parameter exponential distribution

$$R(t) = e^{-\lambda t} \qquad (28\text{-}1)$$

where $R(t)$ = probability that item will operate without failure for time t (usually expressed in hours) under stated operating conditions, e = base of natural logarithms = 2.7182, and λ = item failure rate (usually expressed in failures per hour) = const for any given set of stress, temperature, and quality level conditions. It is determined for parts and components from large-scale data-collection and/or test programs.

When values of λ and t are inserted in Eq. (28-1), the *probability of success,* i.e., reliability, is obtained for that period of time.

The reciprocal of the failure rate $1/\lambda$ is defined as the *mean time between failures* (MTBF). The MTBF is a figure of merit by which one hardware item can be compared with another. It is a measure of the failure rate λ during the useful life period.

3. Reliability Degradation. *Manufacturing Effects.* To assess the magnitude of the reliability degradation due to manufacturing, the impact of manufacturing processes (process-induced defects, efficiency of conventional manufacturing and quality-control inspection, and effectiveness of reliability screening techniques) must be evaluated. In addition to the latent defects attributable to purchased parts and materials, assembly errors can account for substantial degradation. Assembly errors can be caused by operator learning, motivational, or fatigue factors.

Manufacturing and quality-control inspections and tests are provided to minimize degradation from these sources and to eliminate obvious defects. A certain number of defective items escaping detection will be accepted and placed in field operation. More importantly, the identified defects may be overshadowed by an unknown number of *latent defects,* which can result in failures under conditions of stress, usually during field operation. Factory screening tests are designed to apply a stress of given magnitude over a specified time to identify these kinds of defects, but screening tests are not 100% effective.

Operational Effects. Degradation in reliability also occurs as a result of system operation. Wear-out, with *aging* as the dominant failure mechanism, can shorten the useful life. Situations also occur in which a system may be called upon to operate beyond its design capabilities because of an unusual mission requirement or to meet a temporary but unforeseen requirement. These situations could have ill effects on its constituent parts.

Operational abuses, e.g., rough handling, extended duty cycles, or neglected maintenance, can contribute materially to reliability degradation, which eventually results in failure. The degradation can be a result of the interaction of personnel, machines, and environment. The translation of the factors which influence operational reliability degradation into corrective procedures requires a complete analysis of functions performed by personnel and machines plus fatigue and/or stress conditions which degrade operator performance.

Maintenance Effects. Degradation in inherent reliability can also occur as a result of maintenance activities. Studies[1]* have shown that excessive handling from frequent preventive maintenance or poorly executed corrective maintenance, e.g., installation errors, degrades system reliability. Several trends in system design have reduced the need to perform adjustments or make continual measurements to verify peak performance. Extensive replacement of analog by digital circuits, inclusion of more built-in test equipment, and use of fault-tolerant circuitry are representative of these trends.

These factors, along with greater awareness of the cost of maintenance have improved ease of

---

*Superior numbers correspond to the references in Par. 28-63.

maintenance, bringing also increased system reliability. In spite of these trends, the maintenance technician remains a primary cause of reliability degradation. The effects of poorly trained, poorly supported, or poorly motivated maintenance technicians on reliability require careful assessment and quantification.

4. Reliability Growth. Reliability growth represents the action taken to move a hardware item toward its reliability potential, during development or subsequent manufacturing or operation. During early development, the achieved reliability of a newly fabricated item or an off-the-board prototype is much lower than its predicted reliability because of initial design and engineering deficiencies as well as manufacturing flaws. The reliability growth process, when formalized and applied as an engineering discipline, allows management to exercise control of, allocate resources to, and maintain visibility of, activities designed to achieve a mature system before full production or field use.

Reliability growth is an iterative test-fail-correct process with three essential elements: detection and analysis of hardware failures, feedback and redesign of problem areas, and implementation of corrective action and retest.

5. Glossary

Availability. The availability of an item, under the combined aspects of its reliability and maintenance, to perform its required function at a stated instant in time.

Burn-in. The operation of items before their ultimate application to stabilize their characteristics and identify early failures.

Defect. A characteristic which does not conform to applicable specification requirements and which adversely affects (or potentially could affect) the quality of a device.

Degradation. A gradual deterioration in performance as a function of time.

Derating. The intentional reduction of stress-strength ratio in the application of an item, usually for the purpose of reducing the occurrence of stress-related failures.

Downtime. The period of time during which an item is not in a condition to perform its intended function.

Effectiveness. The ability of the system or device to perform its function.

Engineering reliability. The science that takes into account those factors in the basic design which will assure a required level of reliability.

Failure. The inability (more precisely termination of the ability) of an item to perform its required function.

Failure analysis. The logical, systematic examination of an item or its diagram(s) to identify and analyze the probability, causes, and consequences of potential and real failures.

Failure, catastrophic. A failure that is both sudden and complete.

Failure mechanism. The physical, chemical, or other process resulting in a failure.

Failure mode. The effect by which a failure is observed, e.g., an open or short circuit.

Failure, random. A failure whose cause and/or mechanism makes its time of occurrence unpredictable but which is predictable in a probabilistic or statistical sense.

Failure rate. The number of failures of an item per unit measure of life (cycles, time, etc.); during the useful life period, the failure rate λ is considered constant.

Failure, wear-out. A failure that occurs as a result of deterioration processes or mechanical wear and whose probability of occurrence increases with time.

Hazard rate $Z(t)$. At a given time, the rate of change of the number of items that have failed divided by the number of items surviving.

Maintainability. A characteristic of design and installation which is expressed as the probability that an item will be retained in, or restored to, a specified condition within a given time when the maintenance is performed in accordance with prescribed procedures and resources.

Mean maintenance time. The total preventive and corrective maintenance time divided by the number of preventive and corrective maintenance actions during a specified period of time.

Mean time between failures (MTBF). For a given interval, the total functioning life of a population of an item divided by the total number of failures in the population during the interval.

Mean time between maintenance (MTBM). The mean of the distribution of the time intervals between maintenance actions (preventive, corrective, or both).

Mean time to repair (MTTR). The total corrective-maintenance time divided by the total number of corrective-maintenance actions during a given time.

Redundancy. In an item, the existence of more than one means of performing its function.

Redundancy, active. Redundancy in which all redundant items are operating simultaneously rather than being switched on when needed.

Redundancy, standby. Redundancy in which alternative means of performing the function are inoperative until needed and are switched on upon failure of the primary means of performing the function.

Reliability. The characteristic of an item expressed by the probability that it will perform a required function under stated conditions for a stated period of time.

Reliability, inherent. The potential reliability of an item present in its design.

Reliability, intrinsic. The probability that a device will perform its specified function, determined on the basis of a statistical analysis of the failure rates and other characteristics of the parts and components which constitute the device.

Screening. The process of performing 100% inspection on product lots and removing the defective units from the lots.

Screening test. A test or combination of tests intended to remove unsatisfactory items or those likely to exhibit early failures.

Step stress test. A test consisting of several stress levels applied sequentially for periods of equal duration to a sample. During each period, a stated stress level is applied, and the stress level is increased from one step to the next.

Stress, component. The stresses on component parts during testing or use which affect the failure rate and hence the reliability of the parts. Voltage, power, temperature, and thermal environmental stress are included.

Test-to-failure. The practice of inducing increased electrical and mechanical stresses in order to determine the maximum capability of a device so that conservative use in subsequent applications will increase its life through the derating based upon these tests.

Time, down (downtime). See Downtime.

Time, mission. The part of uptime during which the item is performing its designated mission.

Time, up (uptime). The element of active time during which an item is alert, reacting, or performing a mission.

Uptime ratio. The quotient determined by dividing uptime by uptime plus downtime.

Wear-out. The process of attrition which results in an increase of hazard rate with increasing age (cycles, time, miles, events, etc., as applicable for the item).

RELIABILITY THEORY AND PRACTICE

6. Exponential Failure Model. The life-characteristic curve (Fig. 28-1) can be defined by three failure components which predominate during the three periods of an item's life. The shape of this curve suggests the usual term *bathtub curve*. The components are illustrated in terms of an *equipment hazard rate* $Z(t)$. The hazard rate is the conditional probability of failure. The failure components (see Fig. 28-1) include:

1. *Early failures* due to design and quality-related manufacturing, which have a decreasing hazard rate.

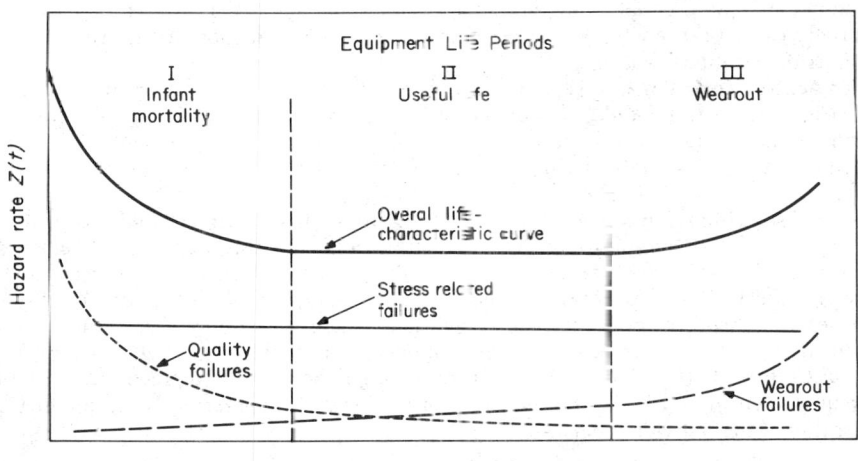

Fig. 28-1. Life-characteristic curve, showing the three components of failure.

2. *Stress-related failures* due to application stresses, which have a constant hazard rate.

3. *Wear-out failures* due to aging and/or deterioration, which have an increasing hazard rate.

From Fig. 28-1 three conclusions can be drawn: (1) that the *infant-mortality* period is characterized by a high but rapidly decreasing hazard rate that comprises a high quality-failure component, a constant stress-related failure component, and a low wear-out-failure component. (2) The *useful-life* period is characterized by a constant hazard rate comprising a low (and decreasing) quality-failure component, a constant stress-related-failure component, and a low (but increasing) wear-out-failure component. The combination of these three components results in a nearly constant hazard rate because the decreasing quality failures and increasing wear-out failures tend to offset each other and because the stress-related failures exhibit a relatively larger amplitude. (3) The *wear-out period* is characterized by an increasing hazard rate comprising a negligible quality-failure component, a constant stress-related-failure component, and an initially low but rapidly increasing wear-out-failure component.

The general approach to reliability for electronic systems is to minimize early failures by emphasizing factory test and inspection and to prevent wear-out failures by replacing short-lived parts. Consequently, the useful life period characterized by stress-related failures is the most important period and the one to which design attention is primarily addressed.

Figure 28-1 illustrates that during the useful life period the hazard rate is constant. A constant hazard (or failure) rate is described by the exponential-failure distribution. Thus, the exponen-

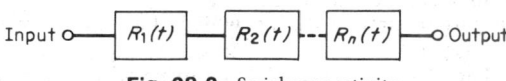

Fig. 28-2. Serial connectivity.

tial-failure model reflects the fact that the item must represent a mature design whose failure rate, in general, is primarily due to stress-related failures. The magnitude of this failure rate is directly related to the stress-strength ratio of the item.

The validity of the exponential reliability function, Eq. (28-1), relates to the fact that the hazard rate (or the conditional probability of failure in an interval given at the beginning of the interval) is independent of the accumulated life.

The use of this type of "failure law" for complex systems is judged appropriate because of the many forces that can act upon the item and produce failure. The stress-strength relationship and varying environmental conditions result in effectively random failures.

The approach to randomness is aided by the mixture of part ages which results when failed elements in the system are replaced or repaired. Over time the system hazard rate oscillates, but this cyclic movement diminishes in time and approaches a stable state with a constant hazard rate.

Another argument for assuming the exponential distribution is that if the time hazard rate is essentially constant, the exponential represents a good approximation of the true distribution over a particular interval of time.

7. System Modeling. To evaluate the reliability of systems and equipment, a method is needed to reflect the *reliability connectivity* of the many part types having different stress-determined failure rates that would normally make up a complex equipment. This is accomplished by establishing a relationship between equipment reliability and individual part or item failure rates.

Before discussing these relationships, it is useful to discuss system reliability objectives. For many systems, reliability must be evaluated from the following three separate but related standpoints: reliability as it affects personnel safety, reliability as it affects mission success, and reliability as it affects unscheduled maintenance or logistic factors. In all these aspects of the subject, the rules for reliability connectivity are applicable. These rules imply that failures are stress-related and that the exponential failure distribution is applicable.

Serial Connectivity. The serial equipment configuration can be represented by the block diagram, shown in Fig. 28-2. The reliability of the *series configuration* is the product of the reliabilities of the individual blocks

$$R_s(t) = R_1(t)R_2(t) \cdots R_i(t) \cdots R_n(t) \tag{28-2}$$

where $R_s(t)$ = series reliability and $R_i(t)$ = reliability of ith block for time t.

The concept of constant failure rate allows the computation of system reliability as a function of the reliability of parts and components:

$$R(t) = \prod_{i=1}^{n} e^{-\lambda_i t} = e^{-\lambda_1 t} e^{-\lambda_2 t} \cdots e^{-\lambda_n t} \tag{28-3}$$

This can be simplified to

$$R(t) = e^{-(\lambda_1 t + \lambda_2 t + \cdots + \lambda_n t)} = e^{-(\lambda_1 + \lambda_2 + \cdots + \lambda_n) t} \tag{28-4}$$

The general form of this expression can be written

$$R(t) = \exp\left[-t \sum_{i=1}^{n} \lambda_i\right] \tag{28-5}$$

Another important relationship is obtained by considering the jth subsystem failure rate λ_j to be equal to the sum of the individual failure rates of the n independent elements of the subsystems such that

$$\lambda_j = \sum_{i=1}^{n} \lambda_i \tag{28-6}$$

Revising the MTBF formulas to refer to the system rather than an individual element gives the *mean time between failures* of the system as

$$\text{MTBF} = \frac{1}{\lambda_j} = \frac{1}{\displaystyle\sum_{i=1}^{n} \lambda_i} \tag{28-7}$$

Successive estimates of the jth subsystem failure rate can be made by combining lower-level failure rates using

$$\lambda_j = \sum_{i=1}^{n} \lambda_{ij} \qquad j = 1, \ldots, m \tag{28-8}$$

where λ_{ij} = failure rate of the ith component in jth-level subsystem and λ_j = failure rate of jth-level subsystem.

Parallel Connectivity. The more complex configuration consists of equipment items or parts operating both in series and parallel combinations, together with the various permutations. A parallel configuration accounts for the fact that alternate part or item configurations can be designed to ensure equipment success by redundancy. A two-element parallel reliability configuration is represented by the block diagram in Fig. 23-3. To evaluate the reliability of parallel configurations, consider, for the moment, that a reliability value (for any configuration) is synonymous with probability, i.e., probability of successful operation, and can take on values ranging between 0 and 1. If we represent the reliability by the symbol R and its complement $1 - R$, that is, unreliability, by the symbol Q, then from the fundamental notion of probability,

$$R + Q = 1 \qquad \text{and} \qquad R = 1 - Q \tag{28-9}$$

From Eqs. (28-9) it can be seen that a probability can be associated with successful operation (reliability) as well as with failure (unreliability). For a single block (on the block diagram) the above relationship is valid. However, in the two-element parallel reliability configuration shown in Fig. 28-3, two paths for successful operation exist, and the above relationship becomes

$$(R_1 + Q_1)(R_2 + Q_2) = 1 \tag{28-10}$$

Assuming that $R_1 = R_2$ and $Q_1 = Q_2$, that is, the blocks are identical, this can be rewritten as

$$(R + Q)^2 = 1 \tag{28-11}$$

Upon expansion, this becomes

$$R^2 + 2RQ + Q^2 = 1 \tag{28-12}$$

We recall that reliability represents the probability of successful operation. This condition is represented by the first two terms of Eq. (28-12) Thus, the *reliability of the parallel configuration* can be represented by

$$R_p = R^2 + 2RQ \tag{28-13}$$

Note that either both branches are operating successfully (the R^2 term) or one has failed while the other operates successfully (the $2RQ$ term).

Substituting the value of $R = 1 - Q$ into the above expression, we obtain

$$R_p = (1 - Q)^2 + 2(1 - Q)Q = 1 - 2Q + Q^2 + 2Q - 2Q^2 = 1 - Q^2 \quad (28\text{-}14)$$

To obtain an expression in terms of reliability only, the substitution $Q = 1 - R$ can be made, which yields

$$R_p = 1 - (1 - R)(1 - R) \quad (28\text{-}15)$$

The more general case where $R_1 \neq R_2$ can be expressed

$$R_p = 1 - (1 - R_1)(1 - R_2) \quad (28\text{-}16)$$

By similar reasoning it can be shown that for n blocks connected in a parallel reliability configuration, the reliability of the configuration can be expressed by

$$R_p(t) = 1 - (1 - R_1)(1 - R_2) \cdots (1 - R_n) \quad (28\text{-}17)$$

The series and parallel reliability configurations (and combinations of them), as described above in Eqs. (28-5) and (28-17), are basic models involved in estimating the reliability of complex equipment.

Redundancy. The serial and parallel reliability models presented in the preceding paragraphs establish the mathematical framework for the reliability connectivity of various elements. Their application can be illustrated to show both the benefits and penalties of redundancy when considering safety, mission, and unscheduled maintenance reliability. Simplified equipment composed of three functional elements (Fig. 28-4) can be used to illustrate the technique.

Elements 1 and 2 are identical and represent one form of functional redundancy operating in series with element 3.

Reliability block diagrams can be defined corresponding to *nonredundant serial, safety, mission,* and *unscheduled maintenance* reliability. The block diagrams show only those functional elements which must operate properly to meet that particular reliability requirement. Figure 28-5 depicts the various block diagrams, reliability formulas, and typical values corresponding to these requirements. It indicates that the use of redundancy provides a significant increase in safety and mission reliability above that of a serial or nonredundant configuration; however, it imposes a penalty by adding an additional serial element in the scheduled maintenance chain.

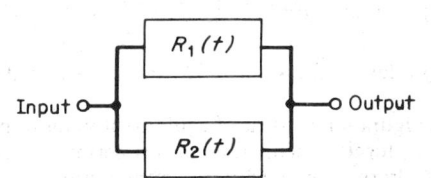

Fig. 28-3. Parallel connectivity.

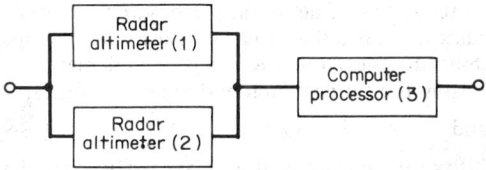

Fig. 28-4. Serial and parallel connectivity.

8. Part-Failure Modeling. The basic concept which underlies reliability prediction and the calculation of reliability numerics is that system failure is a reflection of *part failure.* Therefore, a method for estimating part failure is needed. The most direct approach to estimating part-failure rates involves the use of large-scale data-collection efforts to obtain the relationships, i.e., models, between engineering and reliability variables. The approach uses controlled test data to derive relationships between design and generic reliability factors and to develop factors for adjusting the reliability to estimate field reliability when considering application conditions.

These data have been reduced through *physics-of-failure techniques* and are included in MIL-HDBK-217[2] in a form suitable for estimating stress-related failure rates. MIL-HDBK-217 provides guidance during design and allows individual part-failure rates to be combined within a suitable system reliability model to arrive at an estimate of system reliability.

Although part-failure models (Fig. 28-6) vary with different part types, their general form is

$$\lambda_{\text{part}} = \lambda_b \pi_E \pi_A \pi_Q \cdots \pi_n \quad (28\text{-}18)$$

where λ_{part} = total part-failure rate, λ_b = base or generic failure rate, and π's = adjustment factors. The value of λ_b is obtained from reduced part-test data for each generic part category, where the data are generally presented in the form of failure rate vs. normalized stress and temperature factors. The part's primary-load stress factor and its factor of safety are reflected in this basic failure-rate value. As shown in Fig. 28-6, the value of λ_b is generally determined by the anticipated stress level, e.g., power and voltage, at the expected operating temperature. These values of applied stress (relative to the part's rated stress) represent the variables over which design control can be exercised and which influence the item's ultimate reliability.

π_E is the environmental adjustment factor which accounts for the influences of environments

Reliability requirement	Reliability block diagram	Calculated values
1. Serial (nonredundant) reliability	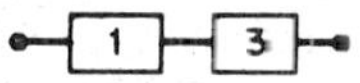	$R = P_1 R_3 = 0.842$ MTBF = 575 hr
2. Safety (or mission) reliability	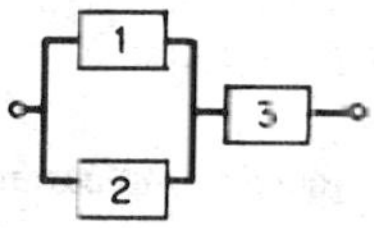	$R = \left[2R_1 - R_1^2\right] R_3$ $= 0.97$
3. Unscheduled maintenance reliability	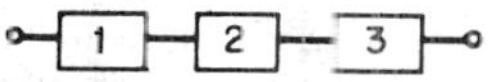	$R = R_1 R_2 R_3 = 0.715$ MTEF = 298 hr

$$R_{\text{Serial}} = R_1 \cdot R_2 \cdots R_n$$

where

$$R_n = e^{-\lambda_n t} \qquad\qquad R_1 = 0.85$$
$$\text{MTBF} = \frac{1}{\lambda_n} \qquad\qquad R_2 = 0.85$$
$$\qquad\qquad\qquad\qquad R_3 = 0.99$$
$$R_{\text{Parallel}} = 1 - (1 - R)(1 - R) \qquad t = 100 \text{ h}$$
$$= 2R - R^2$$

Fig. 28-5. Calculations for system reliability

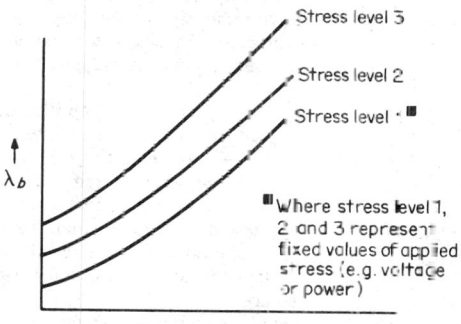

Stress level 3

Stress level 2

Stress level 1

Where stress level 1, 2 and 3 represent fixed values of applied stress (e.g. voltage or power)

λ_b

Temperature ⟶

Failure-rate adjustment factor	π_E	π_A	π_B	. . .	π_n
Value	x	y	. . .	. . .	. . .

$$\lambda_{\text{part}} = \lambda_b \pi_E \pi_A \pi_Q \cdots \pi_n$$

Fig. 28-6. Conceptual part-failure model.

other than temperature; it is related to the operating conditions (vibration, humidity, etc.) under which the item must perform. These environmental classes have been defined in MIL-HDBK-217. Table 28-1 defines each class in terms of its nominal environmental conditions. Depending upon the specific part type and style, the value of π_E may vary from 0.2 to 120. The missile-launch environment is usually the most severe and generally dictates the highest value of π_E. Values of π_E for monolithic microelectronic devices have been added to Table 28-1 to characterize this range for a particular part type.

π_A is the application adjustment factor. It depends on the application of the part and takes into account secondary stress and application factors considered to be reliability-significant.

π_Q is the quality adjustment factor, used to account for the degree of manufacturing control with which the part was fabricated and tested before being shipped to the user. Many parts are covered by specifications which have several quality levels. Several parts have multilevel quality specifications.[2] Values of π_Q relate to both the generic part and its quality level.

π_N is the symbol for a number of additional adjustment factors which account for cyclic effects, construction class, and other influences on failure rate.

The data used as the basis of MIL-HDBK-217 consisted of both controlled test data and field data. The controlled test data directly related stress-strength variables on a wide variety of parts and were suitable for establishing the base failure rates λ_b.

TABLE 28-1   Environmental Symbols and Adjustment Factors[2]

Environment	π_E symbol	Nominal environmental conditions	π_E value*
Ground, benign	G_B	Nearly zero environmental stress with optimum engineering operation and maintenance	1.0
Fixed	G_F	Conditions less than ideal to include installation in adequate racks with adequate cooling air, maintenance by military personnel and possible installation in unheated buildings	2.5
Mobile	G_M	Conditions more severe than those for G_F, mostly for vibration and shock; cooling-air supply may also be more limited and maintenance less uniform	4.0
Space, flight	S_F	Earth orbital; approaches G_B conditions without access for maintenance; vehicle neither under powered flight nor in atmospheric reentry	1.0
Naval, sheltered	N_S	Surface ship conditions similar to G_F but subject to occasional high shock and vibration	4.0
Unsheltered	N_U	Nominal surface shipborne conditions but with repetitive high levels of shock and vibration	5.0
Airborne, inhabited, transport	A_{IT}	Typical conditions in transport or bomber compartments occupied by aircrew without environmental extremes of pressure, temperature, shock, and vibration and installed on long-mission aircraft such as transports and bombers	3.5
Fighter	A_{IF}	Same as A_{IT} but installed on high-performance aircraft such as fighters and interceptors	7.0
Uninhabited, transport	A_{UT}	Bomb bay, equipment bay, tail, or wing installations where extreme pressure, vibration, and temperature cycling may be aggravated by contamination from oil, hydraulic fluid, and engine exhaust; installed on long-mission aircraft such as transports and bombers	4.0
Fighter	A_{UF}	Same as A_{UT} but installed on high-performance aircraft such as fighters and interceptors	8.0
Missile, launch	M_L	Severe conditions of noise, vibration, and other environments related to missile launch and space-vehicle boost into orbit, vehicle reentry, and landing by parachute; conditions may also apply to installation near main rocket engines during launch operations	10.0

*Values for monolithic microelectronic devices.

MIL-HDBK-217 completely describes failure-rate models, failure-rate data, and adjustment factors to be used in estimating the failure rate for the individual generic part types. Table 28-2 presents a tabulation of several models, their base failure rates λ_b, associated π factors, and failure-rate values for several representative part types. The specific procedures for deriving the failure rates differ according to part class and type.

TABLE 28-2 Representative Part-Failure-Rate Calculations

Value	Monolithic bipolar microelectronic device $\lambda_p = \pi_L \pi_Q(C_1 \pi_T + C_2 \pi_E)$	Fixed resistor $\lambda_r = \lambda_b \pi_E \pi_R \pi_Q$	Fixed capacitor $\lambda_p = \lambda_b \pi_E \pi_{cv} \pi_Q$
λ_b	....	0.0015	0.003
π_E	6.0	8.0	24.0
π_Q	5.0	5.0	1.0
π_L	1.0		
π_{T2}	1.9		
C_1	0.006		
C_2	0.002		
π_R	....	1.6	
π_{cv}	...	...	2.0
$\lambda_p \times 10^{-6}$	0.115	0.096	0.144

RELIABILITY EVALUATION

9. Summary. Reliability prediction, failure modes and effects analysis (FMEA), and reliability growth techniques represent prediction and design evaluation methods that provide a quantitative measure of how reliably a design will perform. These techniques help determine where the design can be improved. Since specified reliability goals are often contractual requirements which must be met along with functional performance requirements, these quantitative evaluations must be applied during the design stage to guarantee that the equipment will function as specified for a given duration under the operational and environmental conditions of intended use.

10. Prediction Techniques. *Reliability prediction* is the process of quantitatively assessing the reliability of a system or equipment during its development, before large-scale fabrication and field operation. During design and development, predictions serve as quantitative guides by which design alternatives can be judged for reliability. Reliability predictions also provide criteria for reliability growth and demonstration testing, logistics cost studies, and various other development efforts.

Thus, reliability prediction is a key to system development and allows reliability to become an integral part of the design process. To be effective, the prediction technique must relate engineering variables (the language of the designer) to reliability variables (the language of the reliability engineer).

A prediction of reliability is obtained by determining the reliability of the item at the lowest system level and proceeding through intermediate levels until an estimate of system reliability is obtained. The prediction method depends on the availability of accurate evaluation models that reflect the reliability connectivity of lower-level items and substantial *failure data* that have been analyzed and reduced to a form suitable for application to low-level items.

Various formal prediction procedures are based on theoretical and statistical concepts that differ in the level of data on which the prediction is based. The specific steps for implementing these procedures are described in detail in reliability handbooks. Among the procedures available are parts-count methods and stress-analysis techniques. Failure data for both methods are available in MIL-HDBK-217.[2]

Parts-Count Method. The parts-count method provides an estimate of reliability based on a count by part type (resistor, capacitor, integrated circuit, transistor, etc.). This method is applicable during proposal and early design studies where the degree of design detail is limited. It involves counting the number of parts of each type, multiplying this number by a generic failure rate for each part type, and summing up the products to obtain the failure rate of each functional circuit, subassembly, assembly, and/or block depicted in the system block diagram.

The advantage of this method is that it allows rapid estimates of reliability to determine quickly the feasibility (from the reliability standpoint) of a given design approach. The technique uses information derived from available engineering information and does not require detailed part-by-part stress and design data.

Stress-Analysis Method. The stress-analysis technique involves the same basic steps as the parts-count technique but requires a detailed part models plus calculation of circuit stress values for each part before determining its failure rate. Each part is evaluated in its electric-circuit and mechanical-assembly application based on an electrical and thermal stress analysis. Once part-failure rates have been established, a combined failure rate for each functional block in the reliability diagram can be determined.

To facilitate calculation of part-failure rates, worksheets based on part-failure-rate models are normally prepared to help in the evaluation. These worksheets are prepared for each functional circuit in the system. When completed, these sheets provide a tabulation of circuit part data, including part description, electrical stress factors, thermal stress factors, basic failure rates, the various multiplying or additive environmental and quality adjustment factors, and the final combined part-failure rates. The variation in part stress factors (both electrical and environmental) resulting from changes in circuits and packaging is the means by which reliability is controlled during design. Considerations for, and effects of, reduced stress levels (derating) which result in lower failure rates are treated in Pars. **28-15** to **28-19**.

11. Failure Analysis. *Failure mode and effects analysis* (FMEA) is an iterative documented process performed to identify basic faults at the part level and determine their effects at higher levels of assembly. The analysis can be performed with actual failure modes from field data or hypothesized failure modes derived from design analyses, reliability-prediction activities, and experience of how parts fail. In their most complete form, failure modes are identified at the part level, which is usually the lowest level of direct concern to the equipment designer. In addition to providing insight into failure cause-and-effect relationships, the failure mode and effects analysis provides the disciplined method for proceeding part by part through the system to assess failure consequences.

Failure modes are analytically induced into each component, and failure effects are evaluated and noted, including severity and frequency (or probability) of occurrence. As the first mode is listed, the corresponding effect on performance at the next higher level of assembly is determined. The resulting failure effect becomes, in essence, the failure mode that affects the next higher level.

Iteration of this process results in establishing the ultimate effect at the system level. Once the analysis has been performed for all failure modes, each effect or symptom at the system level usually may be caused by several different failure modes at the lowest level. This relationship to the end effect provides the basis for grouping the lower-level failure modes.

Using this approach, probabilities of the occurrence of the system effect can be calculated, based on the probability of occurrence of the lower-level failure modes, i.e., modal failure rate times time. Based on these probabilities and a severity factor assigned to the various system effects, a *criticality number* can be calculated. Criticality numerics provide a method of ranking the system-level effects derived previously and the basis for corrective-action priorities, engineering-change proposals, or field retrofit actions.

Fault-Tree Analysis. Fault-tree analysis (FTA) is a tool that lends itself well to analyzing failure modes found during design, factory test, or field data returns. The fault-tree-analysis procedure can be characterized as an iterative documented process of a systematic nature performed to identify basic faults, determine their causes and effects, and establish their probabilities of occurrence.

The approach involves several steps, among which is the structuring of a highly detailed logic diagram which depicts basic faults and events that can lead to system failure and/or safety hazards. Then follows the collection of basic fault data and failure probabilities for use in computation. The next step is the use of computational techniques to analyze the basic faults, determine failure-mode probabilities, and establish criticalities. The final step involves formulating corrective suggestions which, when implemented, will eliminate or minimize faults considered critical. The steps involved, the diagrammatic elements and symbols, and methods of calculation are shown in Fig. 28-7.

This procedure can be applied at any time during a system's life cycle, but it is considered most effective when applied (1) during preliminary design, on the basis of design information and a laboratory or engineering test model, and (2) after final design, before full-scale production, on the basis of manufacturing drawings and an initial production model.

The first of these (in preliminary design) is performed to identify failure modes and formulate general corrective suggestions (primarily in the design area). The second is performed to show

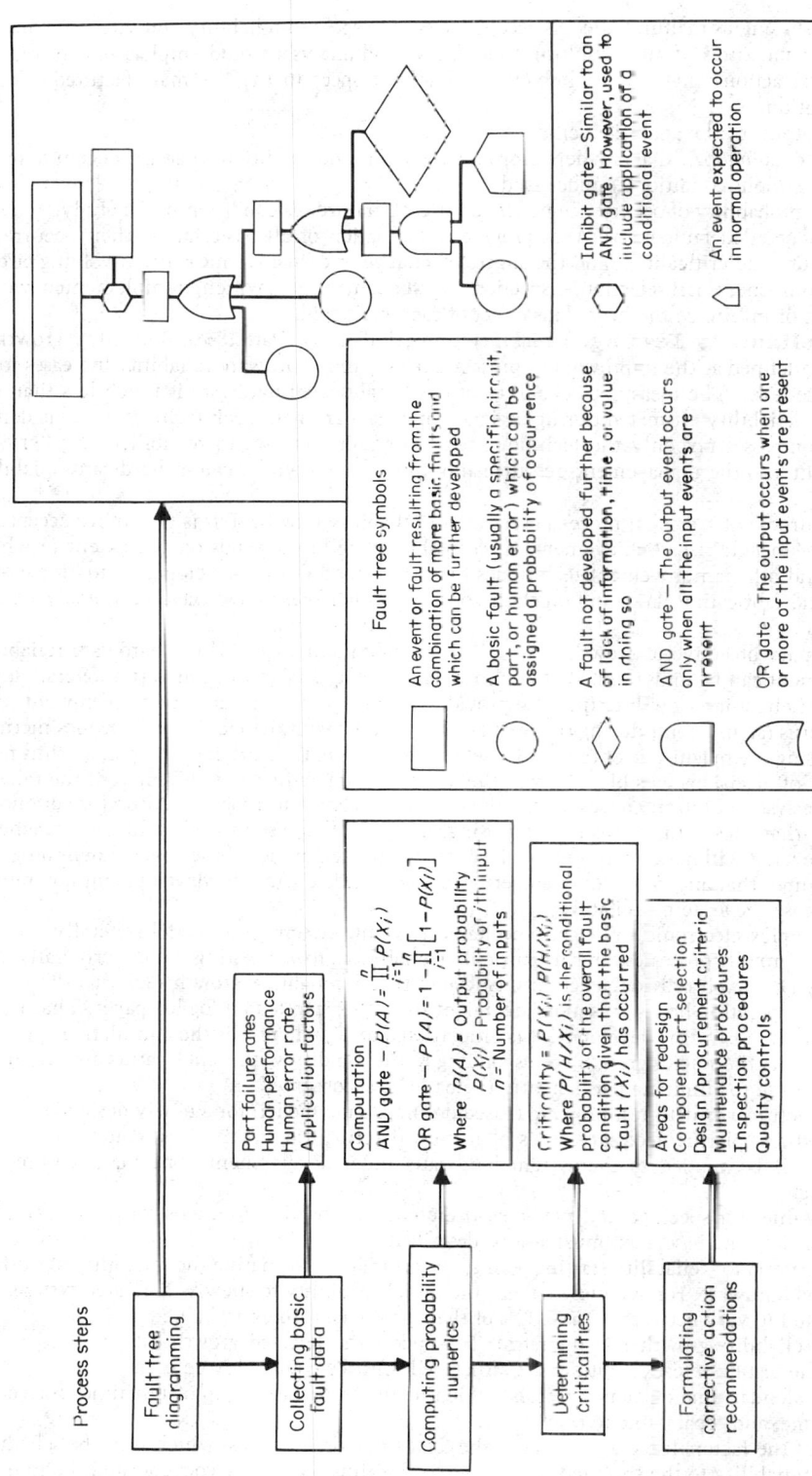

Fig. 28-7. Fault-tree analysis.

that the system, as manufactured, is acceptable with respect to reliability and safety. Corrective actions or measures, if any, resulting from the second analysis would emphasize controls and procedural actions that can be implemented with respect to the "as manufactured" design configuration.

The outputs of the analysis include:

1. A detailed logic diagram depicting all basic faults and conditions that must occur to result in the hazardous condition(s) under study.

2. A probability-of-occurrence numeric for each hazardous condition under study.

3. A detailed fault matrix that provides a tabulation of all basic faults, their occurrence probabilities and criticalities, and the suggested change or corrective measures involving circuit design, component-part selection, inspection, quality control, etc., which, if implemented, would eliminate or minimize the hazardous effect of each basic fault.

12. Reliability Testing. *Reliability Growth* (See also Par. **28-4**). Reliability growth is generally defined as the improvement process during which hardware reliability increases to an acceptable level. The measured reliability of newly fabricated hardware is much less than the potential reliability estimated during design, using standard handbook techniques. This definition encompasses not only the technique used to graph increases in reliability, i.e., "growth plots," but also the management–resource-allocation process which causes hardware reliability to increase.[3]

The purpose of a growth process, especially a reliability-growth test, is to achieve acceptable reliability in field use. Achievement of acceptable reliability depends on the extent to which testing and other improvement techniques have been used during development to "force out" design and fabrication flaws and on the rigor with which these flaws have been analyzed and corrected.

A primary objective of growth testing is to provide methods by which hardware reliability development can be dimensioned, disciplined, and managed as an integral part of overall development. Reliability-growth testing also provides a technique for extrapolating the current reliability status (at any point during the test) to some future result. In addition, it provides methods for assessing the magnitude of the test-fix-retest effort before the start of development, thus making trade-off decisions possible. Many of the models for reliability growth represent the reliability of the system as it progresses during the overall development program. Also, it is commonly assumed that these curves are nondecreasing; i.e., once the system's reliability has reached a certain level, it will not drop below that level during the remainder of the development program. This assumes that any design or engineering changes made during the development program do not decrease the system's reliability.

For complex electronic and electromechanical avionic systems, the model generally used for reliability-growth processes, and in particular reliability-growth testing, is one originally published by Duane.[4] It provides a deterministic approach to reliability growth such that the system MTBF vs. operating hours falls along a straight line when plotted on log-log paper. That is, the change in MTBF during development is proportional to T^d, where T is the cumulative operating time and d is the rate of growth corresponding to the rapidity with which faults are found and changes made to eliminate permanently the basic faults observed.

To structure a growth test program (based on the Duane model) for a newly designed system, a detailed test plan is necessary. This plan must describe the test-fix-retest concept and show how it will be applied to the system hardware under development. The plan requires the following:

1. Values for specified and predicted (inherent) reliabilities. Methods for predicting reliability (model, data base, etc.) must also be described.

2. Criteria for reliability starting points, i.e., criteria for estimating the reliability of initially fabricated hardware. For avionics systems, the initial reliability for newly fabricated systems has been found to vary between 10 and 30% of their predicted (inherent) values.

3. Reliability-growth rate (or rates). To support the selected growth rate, the rigor with which the test-fix-retest conditions are structured must be completely defined.

4. Calendar-time efficiency factors, which define the relationship of test time, corrective-action time, and repair time to calendar time.

Each of the factors listed above affects the total time (or resources) which must be scheduled to grow reliability to the specified value. Figure 28-8 illustrates these concepts and the four elements needed to structure and plan a growth test program.

1. *Inherent reliability* represents the value of design reliability estimated during prediction studies; it may be greater than that specified in procurement documents. Ordinarily, the contract

specifies a value of reliability that is somewhat less than the inherent value The relationship of the inherent (or specified) reliability to the starting point greatly influences the total test time.

2. *Starting point* represents an initial value of reliability for the newly manufactured hardware, usually falling within the range of 10 to 30% of the inherent or predicted reliability. Estimates of the starting point can be derived from previous experience or based on percentages of the estimated inherent reliability. Starting points must take into account the amount of reliability control exercised during the design program and the relationship of the system under development to the state of the art. Higher starting points minimize test time

3. *Rate of growth* is depicted by the slope of the growth curve, which is, in turn, governed by the amount of control, rigor, and efficiency by which failures are discovered, analyzed, and

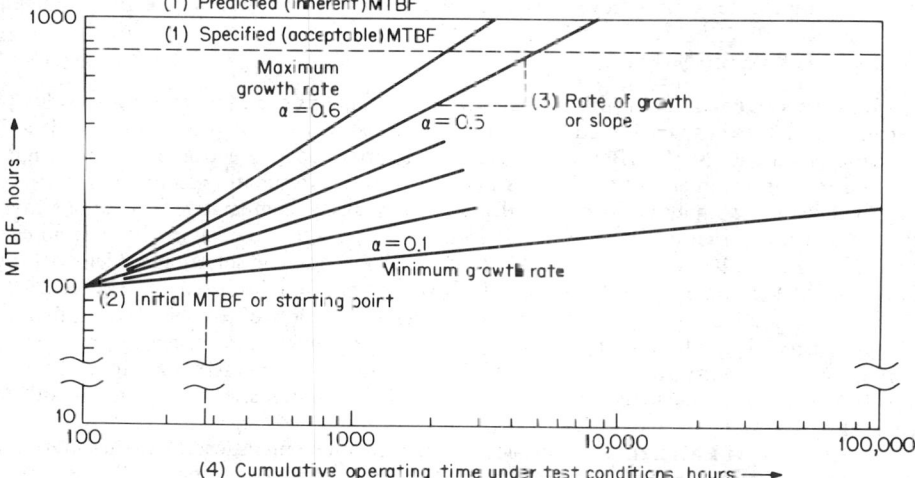

Fig. 28-8. Reliability-growth plot.

corrected through design and quality action. Rigorous test programs which foster the discovery of failures, coupled with management-supported analysis and timely corrective action, will result in a faster growth rate and consequently less total test time.

4. The ratio of *calendar time to test time* represents the efficiency factors associated with the growth test program. Efficiency factors include repair time and the ratio of operating and nonoperating time as they relate to calendar time. Lengthy delays for failure analysis, subsequent design changes, implementation of corrective action, or short operating periods will extend the growth test period.

Figure 28-8 shows that the value of the growth-rate parameter can vary between 0.1 and 0.6. A growth rate of 0.1 can be expected in programs where no specific consideration is given to reliability. In those cases, growth is largely due to solution of problems affecting production and to corrective action taken as a result of user experience. A growth rate of 0.6 can be realized from an aggressive reliability program with strong management support. Such a program must include a formal stress-oriented test program designed to aggravate and force defects and vigorous corrective action.

Figure 28-8 also shows the requisite hours of operating and/or test time and continuous effort required for reliability growth. It shows the dramatic effect that the rate of growth α has on the cumulative operating time required to achieve a predetermined reliability level. For example, for a product whose MTBF potential is 1,000 h it shows that 100,000 h of cumulative operating time is required to achieve an MTBF of 200 h when the growth rate is $\alpha = 0.1$. A rate of 0.1 is expected when no specific attention is paid to reliability growth. However, if the growth rate can be accelerated to 0.6, only 300 h of cumulative operating time is required to achieve an MTBF of 200 h.

Reliability Demonstration. Reliability-demonstration tests are designed to prove a specific reliability requirement with a stated statistical confidence, not specifically to detect problems or for reliability growth. The test takes place after the design is frozen and its configuration is not permitted to change. However, in practice, some reliability growth may occur because of the subsequent correction of failures observed during the test.

Reliability demonstration is specified in most military-system procurement contracts and often involves formal testing conducted per MIL-STD-781. This standard defines test plans, environmental exposure levels, cycle times, and documentation required to demonstrate formally that the specified MTBF requirements of the equipment have been achieved. Demonstration tests are normally conducted after growth tests in the development cycle using initial production hardware.

Reliability-demonstration testing carries with it a certain statistical confidence level; the more demonstration testing, the greater the confidence. The more reliability-growth testing performed, the higher the actual reliability. Depending on the program funding and other constraints, system testing may follow one of two options. The first option maximizes growth testing and minimizes demonstration testing, resulting in a high MTBF at a low confidence. The second option minimizes reliability growth testing with a resultant lower MTBF at higher confidence.

RELIABILITY DESIGN DATA

13. Data Sources. Reliability design data are available from a number of sources. Both the parts-count and stress-analysis methods of predicting reliability rely on part-failure-rate data. One source of such data is MIL-HDBK-217, but not all parts used in electronic system design are covered in that document. Other sources may be sought, or estimating techniques using comparative evaluations may be used. Provided similarity exists, comparative evaluations involve the extrapolation of failure data from well-documented parts to those having little or no failure data. The "Reliability Design Handbook"[5] contains application and selection guidelines for various types of electronic components, along with generic failure rates based on MIL-HDBK-217.

Publications containing up-to-date experience data for a variety of parts, including digital and linear integrated circuits, hybrid circuits, and discrete semiconductor devices, are available through the Reliability Analysis Center, RADC/RBRAC, at Griffiss Air Force Base, Rome, NY 13440. The publications include malfunction through distributions, screening fallout, and experienced failure rates.

14. Physics of Failure. The physical or chemical phenomena leading to the deterioration or failure of electron devices or components in storage or under operating conditions is termed *physics of failure* or *reliability physics*. A major source of information on reliability-physics phenomena of electron devices is the *Annual Proceedings of the International Reliability Physics Symposium* (IRPS). This symposium, which began in 1962, was originally called Physics of Failure in Electronics. The first four symposia (1962 to 1965) were jointly sponsored by Rome Air Development Center (RADC) and IIT Research Institute (IITRI). The fifth (1966) was jointly sponsored by RADC and Battelle Memorial Institute. In 1967 the name was changed to Reliability Physics Symposium; joint sponsorship by IEEE's Electron Devices Society and IEEE's Reliability Group began and has continued to the present.

A search and retrieval index[6] to the *IRPS Proceedings*, 1968 to 1978, available from the Reliability Analysis Center, consists of four different types of indexes. For the researcher or failure analyst searching for leads to interpret failure phenomena, there is an index of considerable detail (3,080 terms). For work reported by a given person or from a specific company, an author index and a corporate index have been provided. An alphabetical list of the detailed index terms facilitates scanning for applicable terms. A chronological listing of all the papers is provided, and in this presentation the detailed index terms appear with each listing to give an overview of the general intent and depth of each paper.

Failure Modes.   A knowledge of the physics of device failure is helpful in predicting and avoiding device failure. Prevalent failure modes are identified in a number of publications, besides the *IRPS Proceedings*. Other sources include the RAC "Reliability Design Handbook,"[5] MIL-HDBK-217C,[2] and MIL-STD-1547(USAF).[7]

Suspect Devices.   In selecting parts for a particular application and in designing screens to identify potential early-life failures, it is helpful to be aware of failure-suspect device designs. A standard intended for the procurement of "space quality" piece parts for space missions, MIL-STD-1547(USAF), includes an identification of reliability-suspect items. Clearly the identification of such parts *does not suggest their inapplicability for all types of electronic systems.*

Examples of reliability suspect devices listed in the aforementioned standard include *capacitors:* all non-ER (established-reliability) types except hermetic styles of MIL-C-14409, wet-slug tantalum capacitors except style CLR-79, glass-encased axial-lead ceramic capacitors, and miniature solid tantalum (style CSR-09); *resistors:* carbon film, all non-ER types; *diodes:* plastic encapsulated, whisker; *transistors:* plastic encapsulated; *microcircuits:* plastic-encapsulated; *thyristors:* plastic-encapsulated.

Derating Factors and Application Guidelines

BY G. D. NETZBAND, D. E. DAVIS, H. B. DREXLER, K. E. GONZENBACH, N. V. OWEN, AND
E. W. KIMBALL

15. Introduction. The following derating guidelines were developed and adopted for use in designing equipment manufactured by Martin Marietta Aerospace, Orlando Division, and it must be recognized that they may be too strict for use by designers of equipment for other markets, such as nonspace or nonmilitary applications. It is also recognized that to achieve higher projected reliability, customers for certain specialized equipment may impose derating factors even more severe than those given here. Nevertheless, the principles underlying the idea of derating are useful in all applications.

Derating is the reduction of electrical, thermal, mechanical, and other environmental stresses on a part to decrease the degradation rate and prolong its expected life. Through derating, the margin of safety between the operating stress level and the permissible stress level for the part is increased, providing added protection from system overstresses unforeseen during design.

The criteria listed in this section indicate maximum application stress values for design. Since safety margins of a given part at failure threshold and under time-dependent stresses are based on statistical probabilities, parts should be derated to the maximum extent possible consistent with good design practice.

When derating, the part environmental capabilities defined by specification should be weighed against the actual environmental and operating conditions of the application. Derating factors should be applied so as not to exceed the maximum recommended stresses.

For derating purposes the *allowable application stress* is defined as the *maximum allowable percentage of the specified part rating at the application environmental and operating condition.* Note that ambient conditions specified by the customer usually do not include temperature rise within a system that results from power dissipation. Thus, a thermal analysis must be performed early in the development phase to be used in the derating process.

RESISTOR DERATING AND APPLICATION GUIDELINES

16. Resistor Types. Variable and fixed resistors are of three types, composition, film, or wire-wound (see Sec. 7). The composition type is made of a mixture of resistive materials and a binder molded to lead wires. The film type is composed of a resistive film deposited on, or inside, an insulating cylinder or filament. The wire-wound type consists of a resistance wire wound on an appropriate structural form.

General Applications. For ordinary military uses *established-reliability* (ER) part types are contractually required as preferred parts.

1. MIL-R-39005, RBR (fixed, wire-wound accurate) Higher stability than any composition or film resistors, where high-frequency performance is not critical. Operation is satisfactory from dc to 50 kHz. Relatively high cost and large size.

2. MIL-R-39007, RWR (fixed wire-wound power type). Select for large power dissipation and where high-frequency performance is relatively unimportant. Generally satisfactory for use at frequencies up to 20 kHz, but the reactive characteristics are uncontrolled except for available "noninductive"-type windings at reduced resistance ranges. Wattage and working voltage must not be exceeded. Power derating begins at 25°C ambient.

3. MIL-R-39008, RCR (fixed, composition-insulated). Select for general-purpose resistor applications where initial tolerance need be no closer than about $\pm 8\%$ and long-term stability no better than $\pm 20\%$ at room temperature under fully rated operating conditions. RF characteristics in resistance values higher than about 500 Ω are unpredictable. These parts are generally capacitive.

4. MIL-R-39009, RER (fixed, wire-wound power type, chassis-mounted) Relatively large power dissipation in given unit size. RF performance is limited. Chassis area for heat dissipation is essential to reach rated wattage. Not as good as RWRs in low-duty-cycle pulsed operation where peaks exceed steady-state rating.

5. MIL-R-39015, RTR (variable, wire-wound, lead-screw-actuated) Use for matching, balancing, and adjusting circuit variables in computers, telemetering equipment, and other critical applications. Requires special consideration in severe environments. Should be used with fixed resistor, if possible, in a circuit designed to reduce sensitivity to movable contact shift.

6. MIL-R-39017, RLR (fixed, metal film). These film resistors (mostly thick film) have semi-precision characteristics and small size. These sizes and wattage ratings are comparable to those of MIL-R-39008, and stability is between that of MIL-R-39008 and MIL-R-55182. Design-parameter tolerances are looser than those of MIL-R-55182, but good stability makes them desirable in most electronic circuits. RF characteristics in values above 500 Ω are much superior to composition types. Initial tolerances are 2 and 1%.

7. MIL-R-39035, RJR (variable, non-wire-wound, lead-screw-actuated). Use for matching, balancing, and adjusting circuit variables in computers, telemetering equipment, and other critical applications. Use of potentiometers in severe environments requires special consideration. Should be used with fixed resistors, if possible, in a circuit designed to reduce sensitivity to movable contact shift.

8. MIL-R-55182, RNR (fixed, film, high stability). Use in circuits requiring higher stability than provided by composition resistors or thick-film, insulated resistors and where ac frequency requirements are critical. These thin-film resistors provide the best high-frequency characteristics available unless special shapes are used. Metal films are characterized by low temperature coefficient and are usable for ambient temperatures of 125°C or higher with small degradation.

17. Mounting Guide. Since improper heat dissipation is the predominant contributing cause of wear-out failure for any resistor type, the lowest possible resistor surface temperature should be maintained. The intensity of radiated heat varies inversely with the square of the distance from the resistor. Maintaining maximum distance between heat-generating components serves to reduce cross-radiation heating effects and promotes better convection by increasing airflow. For optimum cooling without a heat sink, small resistors should have large-diameter leads of minimum length terminating in tie points of sufficient mass to act as heat sinks. All resistors have a maximum surface temperature which must not be exceeded. Resistors should be mounted so that there are no abnormal hot spots on the resistor surface. Most solid surfaces, including insulators, are better heat conductors than air.

18. Rating Factors. The permissible power rating of a resistor is another factor that is initially set by the use to which the circuit is put, but it is markedly affected by the other conditions of use. It is based on the hot-spot temperature the resistor will withstand while still meeting other requirements of resistance variation, accuracy, and life.

Self-generated heat in a resistor is equal to I^2R. It is a usual practice to calculate this value and to use the next larger power rating available in conjunction with the derating guides.

Ambient Conditions vs. Rating.   The power rating of a resistor is based on a certain temperature rise from a specified ambient temperature. If the ambient temperature is greater than this value, the amount of heat the resistor can dissipate is even less and must be recalculated.

Accuracy vs. Rating.   Because all resistors have a temperature coefficient of resistance, a resistor expected to remain near its measured value under conditions of operation must remain relatively cool. For this reason, all resistors designated as "accurate" are very much larger, physically, for a certain power rating than ordinary "nonaccurate" resistors. In general, any resistor, accurate or not, must be derated if it is to remain very near its original measured value when it is being operated.

Life vs. Rating.   If especially long life is required of a resistor, particularly when "life" means remaining within a certain limit of resistance drift, it is usually necessary to derate the resistor, even if ambient conditions are moderate and if accuracy by itself is not important. A good rule to follow when choosing a resistor size for equipment that must operate for many thousands of hours is to derate it to one-half of its nominal power rating. Thus, if the self-generated heat in the resistor is ⅛ W, do not use a ½-W resistor but a 1-W size. This will automatically keep the resistor cooler, will reduce the long-term drift, and will reduce the effect of the temperature coefficient.

In equipment that need not live so long and must be small, this rule may be impractical, and the engineer should adjust his dependence on rules to the circumstances at hand. A "cool" resistor will generally last longer than a "hot" one and can absorb transient overloads that might permanently damage a "hot" resistor.

Pulsed Conditions and Intermittent Loads.   When a resistor is used in circuits where power is drawn intermittently or in pulses, the actual power dissipated with safety during the pulses can sometimes be much more than the maximum rating of the resistor. For short pulses the actual heating is determined by the duty factor and the peak power dissipated. Before approving such a resistor application, however, the design engineer should be sure of the following:

1. The maximum voltage applied to the resistor during the pulses is never greater than its permissible maximum voltage.

2. The circuit cannot fail in such a way that continuous excessive power can be drawn through the resistor.

3. The average power being dissipated is well within the rating of the resistor.

4. Continuous steep wavefronts applied to the resistor do not cause malfunctions because of electromechanical effects of high voltage gradients.

19. Resistor Derating. The resistor derating factors shown in Table 28-3 require the application of the principles shown in the following sections. The applicable percentages or ratios should be applied to the characteristics or ratings, taking into consideration the actual temperature and frequency of operation.

Power Derating.   The objective of power derating is to establish the worst-case hot-spot temperature for the resistor. The power dissipated by a resistor causes the temperature to rise above ambient by an amount directly proportional to the amount of power dissipated. The maximum allowable power can vary due to applied voltage and temperature.

Computations of derated power apply to the maximum power permissible under conditions of voltage and ambient temperature. The derating percentage is applied after the permissible

TABLE 28-3   Resistor Derating Factors

Type	Military specifications (MIL-R-)	Style	Maximum permissible percentage of military specification stress rating		
			Power	Voltage*	Current
Fixed, wire-wound,					
accurate 1.0%	39005	RBR	50	80	70
0.1%			25	80	70
Power	39007	RWR	50	80	70
Chassis-mounted	39009	RER	50	80	
Composition, insulated	39008	RCR	50	80	
Film, metal	39017	RLR	50	80	
High-stability	55182	RNR	50	80	
Variable, lead-screw-					
actuated, wire-wound	39015	RTR	50	80	70
Non-wire-wound	39035	RJR	50	80	70

*Voltage applied should be no more than the smaller of V_{d1} or V_{d2} (Par. **28-19**).

power is determined from the specification rating when all conditions and recommendations are observed. For instance, chassis-mounted resistors are designed to conduct most of the heat through the chassis. Thus, power ratings require knowing the thermal resistivity of the mounting surface and its temperature. MIL-STD-199[8] defines chassis areas upon which power ratings are based.

Voltage Derating.   The voltage should be derated to a percentage of the maximum allowable voltage as determined for the specification rating. This voltage may be limited by derated power as well as by the maximum voltage of the resistor. The derated voltage should be the smaller of

$$V_{d1} = C_v V_r \quad \text{and} \quad V_{d2} = P_d R \tag{28-19}$$

where V_d = derated voltage, P_d = derated power, C_v = derating constant = (percent derating)/100, V_r = rated voltage, and R = resistance value.

CAPACITOR DERATING FACTORS AND APPLICATION GUIDELINES

20. Capacitor Types. Electrostatic capacitors, widely used in electronic equipment, include mica, glass, plastic film, paper-plastic, ceramic, air, and vacuum. Electrolytic types are aluminum and tantalum foil and wet or dry tantalum slug.

21. Environmental Factors. A capacitor may fail when subjected to environmental or operational conditions for which the capacitor was not designed or manufactured. Designers must understand the safety factors built into a given capacitor, the safety factors they add of their own accord, and the numerous effects of circuit and environmental conditions on the parameters. It is not enough to know only the capacitance and the voltage ratings. It is important to know to what extent the characteristics change with age and environment

Temperature Variations. Temperature variations have an effect on the capacitance of all types of capacitors. Capacitance change with temperature is directly traceable to the fact that the dielectric constant of the materials changes with temperature. The capacitance of polarized dielectrics is a complex function of temperature, voltage, and frequency; nonpolarized dielectrics exhibit less change than polarized materials. Many dielectrics exhibit a very large decrease in capacitance with a relatively small decrease in temperature. The increased power factor at this temperature may raise the dielectric temperature sufficiently to recover lost capacitance. When

TABLE 28-4   Capacitor Derating Factors

Capacitor type	Mil. spec. MIL-C-	Style	Maximum permissible percentage of military specification stress rating[a]			
			Voltage[b]	Current[c]	ac ripple	Surge
All tantalum, slug	39006/22	CLR79	80	80[d]		
Fixed, aluminum (ER)	39018	CUR	80 min 95 max	75		
Ceramic, temperature-compensating	20	CC	50	70		
Chip	55681	CDR	60	70		
General-purpose (ER)	39014	CKR	60	70	70	70
Electrolytic	39003	CSR[e]	50	70	70	
Tantalum, non-solid (ER)	39006	CLR	50	70		
Feedthrough	11693	CZR	70	70		
Film plastic (ER) polycarbonate	83421	CRH	60	70	70	70
Glass (ER)	23269	CYR	75	70	70	70
Metallized paper-film (ER)	39022	CHR	50	70	70	70
Mica (ER)	39001	CMR	80	70	70	70
Paper-plastic (ER) and plastic film (ER)	19978	CQR	70	70	70	70
Tantalum, solid (ER)	55365	CWR	50	70		

[a]Manufacturer's derating factors must be applied before applying these factors.
[b]Voltage equals instantaneous total of dc, ac, surge, and transient voltage.
[c]Rated current is defined as $I_R = \sqrt{P_{MAX}/R_{MAX}}$ and by limiting the current to 0.70 times rated current, power is limited to 0.50 maximum.
[d]Package for maximum thermal dissipation.
[e]Limit to 85°C ambient temperature.

a capacitor is initially energized at low temperatures, the capacitance will be a small percentage of its nominal value, and if the internal heating is effective, the thermal time constant of the capacitor must be considered. A change in the distance between the conductors and the effective areas of the conductor due to thermal expansion will also cause a change in capacitance.

Insulation resistance decreases as the temperature increases and varies linearly with the inverse of the capacitance. The time of electrification is most critical in the determination of insulation resistance. The effect of the insulation-resistance value is quite critical in many circuit designs and can cause malfunction if its magnitude and variation with temperature are not considered. The *dielectric strength* decreases as the temperature increases.

The life of the capacitor decreases with an increase in temperature. As a rule of thumb, life decreases by a factor of 2 for each 10°C rise in temperature.

The *operating temperature* and changes in temperature also affect the mechanical structure in which the dielectric is housed. The terminal seals, using elastomeric materials or gaskets, may leak due to internal pressure buildup. Expansion and contraction of materials with different thermal-expansion coefficients may also cause seal leaks and cracks in internal joints. Electrolysis effects in glass-sealed terminals increase as the temperature increases.

If the capacitor is operated in the vicinity of another component operating at high temperature, the flashpoint of the impregnant should be considered.

Moisture. Moisture in the dielectric decreases the dielectric strength, life, and insulation resistance and increases the power factor of the capacitor. Capacitors operated in high humidities should be hermetically sealed.

Aging. The extent and speed of aging of a capacitor depend on the dielectric materials used in its construction. Aging does not affect glass, mica, or stable ceramic capacitors. The most common capacitors with significant aging factors are the medium-K and hi-K ceramic type (CKR series) and aluminum electrolytic types. Detailed aging and storage life data are given in MIL-STD-198.[9]

External Pressure. External pressure is not usually a factor to be considered unless it is sufficient to change the physical characteristics of the container housing the capacitor plates and the dielectric. Certain high-density CKR types demonstrate piezoelectric effects.

Shock and Vibration. The capacitors and mounting brackets, when applicable, must be designed to withstand the shock and vibration requirements of the particular application. Internal capacitor construction must be considered in selecting a capacitor for a highly dynamic environment.

22. Capacitor Derating. The capacitor derating factors (Table 28-4) should be applied after all derating (stated or implied by the MIL-SPEC or manufacturer) has been applied in the circuit design. The table shows the maximum allowable percentage of voltage and current.

Precautions. The following checklist will help achieve high reliability.

1. Do not exceed the current rating on any capacitor, taking into account the duty cycle. Provide series resistance or other means in charge-discharge circuits to control surge currents. In particular, solid-tantalum types should have an effective series impedance of at least 3 Ω/V for highest reliability. Reliability derating factors are available for lower impedances.

2. Include dc, superimposed peak ac, peak pulse, and peak transients when calculating the voltage impressed on capacitors.

3. The MIL-SPEC or manufacturer's recommendations for frequency, ripple voltage, temperature, etc., should also be followed for further derating.

SEMICONDUCTOR DERATING FACTORS AND APPLICATION GUIDELINES

23. General Considerations. Semiconductor device derating should be applied after all deratings stated or implied by the part MIL-SPEC have been used in the circuit design.

For designs using silicon active components, transistors and diodes, a junction temperature of 110°C must not be exceeded for ground and airborne applications and 140°C for missile-flight applications (see Table 28-5).

TABLE 28-5   Transistor Derating Factors

Parameter	Derating factor*
Voltage (V_{CEO}, V_{CBO}, V_{EBO})	0.75
Current	0.75
Junction temperature:	
Ground and airborne use	110°C
Missile-flight use	140°C
Allowing for:	
Increase in leakage (I_{CBO} or I_{CEO})	+100%
Increase in h_{FE}	−50%
Decrease in h_{FE}	−50%
Increase in $V_{CE(SAT)}$	−10%

*Derating factor (applicable to all transistor types)

$$\frac{\text{Maximum allowable stress}}{\text{Rated stress}}$$

A maximum power rating on any semiconductor device is by itself a meaningless parameter. The parameters of value are maximum operating junction temperature, thermal resistance, and/ or thermal derating (reciprocal of thermal resistance). For all semiconductor devices, the mechanism for removal of heat from a junction is usually that of conduction through the leads, not convection. For all silicon transistors and diodes, the maximum operating junction temperature should be 110 or 140°C, respectively.

The method for calculating device junction temperature is

$$T_J = T_A + \theta_{J-A} P_D \tag{28-20}$$

where T_J = junction temperature, T_A = maximum ambient temperature at component, θ_{J-A} = thermal resistance from junction to air, and P_D = power dissipated in device. Where heat sinks are used, the expression is expanded to

$$T_J = T_A + (\theta_{J-C} + \theta_{C-S} + \theta_{S-A})P_D \qquad (28\text{-}21)$$

where

$$\theta_{J-C} + \theta_{C-S} + \theta_{S-A} = \theta_{J-A}$$

and where θ_{C-S} = thermal resistance between case to heat sink (usually includes mica washer and heat-sink compound), θ_{S-A} = thermal resistance of heat sink, and θ_{J-C} = thermal resistance from junction to case. Examples for calculation of junction temperature for various conditions follow.

24. Thermal Resistance and Power Calculations. Using the types of inputs described in Par. **28-23**, the following examples demonstrate the ease with which thermal calculations can be performed.

Example 1. Given: 1N753 reference diode; find: θ_{J-A}. Specifications: $\overline{P}_D$ = 400 mW (max power), $\overline{T}_J$ = 175°C (max junction), T_A = 25°C (max ambient).

The manufacturer does not give θ_{J-A}, but it can be calculated using the above data. The maximum power dissipation is calculated from specified maximums at room temperature:

$$T_J = T_A + \theta_{J-A}P_D$$
$$175°C = 25°C + (\theta_{J-A})(0.4)$$
$$\theta_{J-A} = 375°C/W$$

TABLE 28-6 Contact Thermal Resistance of Insulators

Insulator	Thickness, in	θ_{C-S}, °C/W
No insulation	. . . . .	0.4
Anodized aluminum	0.016	0.4
	0.125	0.5
Mica	0.002	0.5
	0.004	0.65
Mylar	0.003	1.0
Glass cloth (Teflon-coated)	0.003	1.25

Example 2. Determine the thermal resistance of a heat sink required for a 2N3716 power transistor which is to dissipate 14 W at an ambient temperature of 70°C.

$$\theta_{J-C} = 1.17°C/W \qquad \text{from 2N3716 specifications}$$
$$\theta_{C-S} \approx 0.5°C/W$$

(θ_{C-S} = 0.5 ° C/W for mica washer; Table 28-6)

$$110°C = 70°C + \theta_{J-A}P_D$$
$$\theta_{J-A} = (110 - 70)\frac{1}{P_D} = \frac{40}{14} \approx 2.9°C/W$$
$$\theta_{J-A} = \theta_{J-C} + \theta_{C-S} + \theta_{S-A}$$
$$2.9 = 1.17 + 0.5 + \theta_{S-A}$$
$$\theta_{S-A} \leq 1.23°C/W$$

(Heat-sink thermal resistance required; note Table 28-7 for thermal resistance of some common commercially available heat sinks.)

Example 3. Determine the maximum power that the 2N3716 can dissipate without a heat sink in an ambient of 70°C:

$$\theta_{J-A} = 35°C/W$$

(Not given on Motorola data sheets, but for almost all TO-3 devices θ_{J-A} = 35°C/W; see Table 28-8.)

$$T_J = T_A + \theta_{J-A}P_D$$
$$P_D = \frac{T_J - T_A}{J - A} = \frac{110 - 70}{35} = 1.14 \text{ W}$$

Example 4. Determine the maximum power that can be dissipated by a 2N2222 transistor (missile-flight use) with an ambient temperature of 70°C. Derating given: 3.33 mW/°C for 2N2222, TO-18. Therefore

$$\theta_{J-A} = \frac{1°C}{3.33 \text{ mW}} = 300°C/W$$

and

$$P_D = \frac{T_J - T_A}{\theta_{-A}} = \frac{140 - 70}{300} = 233 \text{ mW}$$

25. Semiconductor Derating. *Power Derating.* The objective of power derating is to hold the worst-case junction temperature to a value below the normal permissible rating. The typical diode specification for thermal derating expresses the change in junction temperature with power for the worst case. The actual temperature rise per unit of power will be considerably less, but this is not a value which can readily be determined for each unit.

Junction-Temperature Derating Junction-temperature derating requires the determination of ambient temperature or case temperature. The worst-case ambient temperature or case temperature for the part is established for the area and for the environmental conditions which will be encountered in service. The ambient temperature for a device which does not include some

TABLE 28-7   Thermal Resistance of Heat Sinks

| Shape | Surface area, in | Volume displacement | | | | w, g | Finish | Thermal resistance, °C/W |
		L, in	W, in	H, in	Vol, in			
			Extrusion					
Flat-finned	65	3.0	3.6	1.0	10.3	114	Anod black	2.4
							Bright alum	3.0
							Gray	2.8
	60	3.0	4.0	0.69	8.3	123	Anod black	2.8
	95	3.0	4.0	1.28	15.3	180	Anod black	2.1
	64	3.0	3.8	1.3	15.0	155	Black paint	2.2
	83	3.0	4.0	1.25	15.0	140	Anod black	2.2
	44	1.5	4.0	1.25	7.5	75	Anod black	3.0
	137	3.0	4.0	2.63	31.5	253	Anod black	1.45
	250	5.5	4.0	2.63	58.0	46	Anod black	1.10
	130	6	3.6	1.0	21.5	253	Anod black	1.75
	78	3.0	3.8	1.1	12.5	190	Anod gray	2.9
	62	3.0	3.8	1.3	15.0	170	Anod gray	2.2
	78	3.0	4.5	1.0	13.5	140	Gold alodine	3.0
			Machined casting					
Cylindrical fins, horizontal	30	1.75		0.84	2.0	40	Anod black	8.5
	50	1.75		1.5	3.6	67	Anod black	7.1
	37	1.75		1.5	3.6	48	Anod black	6.65
			Casting					
Cylindrical fins, vertical	7.5	1.5		0.9	4.4	38	Anod black	8.1
	12	1.5		1.4	6.9	51	Anod black	7.0
	25	1.5		2.9	14.2	112	Anod black	5.6
	35	1.5		3.4	16.7	132	Anod black	5.1
	32	2.5		1.5	7.4	94	Anod black	4.5
	20	2.5		0.5	2.45	48	Anod black	6.6
Flat-finned	23	1.85	1.86	1.2	4.15	87	Anod black	5.06
			Sheet metal					
Vertical fins, square	12	1.7	1.7	1.0	2.9	19	Anod black	7.4
Cylindricals	15	2.31*		0.81	3.35	18	Black	7.1
Horizontal fins, cylindrical	6	1.81*		0.56	1.44	20	Anod black	9.15
	55	2.5		1.1	5.4	115	Gold irridate	7.9

*Diameter.

means for thermal connection to a mounting surface should include the temperature rise due to the device, adjacent devices, and any heating effect which can be encountered in service.

Voltage Derating. The voltage rating of a semiconductor device can vary with temperature, frequency, or bias condition. The rated voltage implied by the tabulated rating is the voltage compensated for all factors determined from the manufacturer's data sheet. Derating consists of the application of a percentage figure to the voltage determined from all factors of the rating. Three distinct deratings cover the conditions which can be experienced in any design situation.

1. *Instantaneous peak-voltage derating* is the most important and least understood derating. It is required to protect against the high-voltage transient spike which can occur on power lines as a result of magnetic energy stored in inductors, transformers, or relay coils. Transient spikes

TABLE 28-8   Thermal Resistances of Packages, °C/W*

Package type	Still air		θ_{J-A}, soldered to PCB board
	θ_{J-A}	θ_{J-C}	
TO-3	30–50	1.2–2.5	
TO-5	100–300	30–90	
TO-18	400–500	150–250	
TO-66	30–50	4.0–7.5	
TO-99, TO-100	197	. . .	185
Flat pack	187	. . .	165
Plastic DIP	150	. . .	145
Ceramic DIP	115	. . .	110

*Final estimates should be based on military specification or vendor values, whichever θ_J is higher.

also can result from momentary unstable conditions which cause high amplitude during switching turn-on or turnoff.

Transient spike or oscillating conditions in test sets or life-test racks or resulting from the discharge of static electricity will cause minute breakdown of surface or the bulk silicon material.

Lightning transients, which enter a circuit along power lines or couple from conducting structural members, are a frequent cause of failure or of damage which increases the probability of failure during service.

2. The *continuous peak voltage*, on the other hand, is the voltage at the peak of any signal or continuous condition which is a normal part of the design conditions.

3. The *design maximum voltage* is the highest average voltage. This is essentially the dc voltage as read by a dc meter. The ac signals can be superimposed on the dc voltage to produce a higher peak voltage, providing the continuous peak voltage is not exceeded.

26. Transistor Guidelines. The major failure modes are the degradation of h_{FE} and increased leakage with prolonged use at elevated temperatures. Depending on the application, this parameter degradation can result in a catastrophic failure or decrease in system performance outside the bound of the worst-case design. It is necessary to maintain the junction operating temperature below 110°C for ground and airborne use (140°C for missile-flight use). The principal design criteria for silicon transistors are shown in Table 28-5. This derating is applicable to all transistor types.

Transistor Application Information. These general guidelines apply:

1. h_{FE} has a positive temperature coefficient. This criterion may not be valid for some power transistors operating at high current levels where h_{FE} decreases with temperature.

2. If a maximum leakage is specified at 110°C for ground and airborne use (140°C for missile use), double the value for end of life.

3. The ratings of Table 28-5 apply for operating junction temperature, not just ambient temperature.

4. Typical thermal resistances for common case sizes are described in Table 28-8.

Thermal Resistance and Heat Sinks. Table 28-6 lists contact thermal resistance for various insulators, and Table 28-7 gives the thermal resistance for various heat sinks. Figure 28-9 is used to calculate θ_{S-A} for solid copper or aluminum plates. This is helpful in determining the thermal capabilities of metal chassis. Note that a vertically mounted plate has better thermal properties than a horizontally mounted plate. A list of approximate thermal resistances for various package sizes is given in Table 28-8.

27. Diode Guidelines. The junction-temperature limits specified in Par. **28-23** apply to all diodes. For non-heat-sinked components, a quick calculation can be made to determine the power a given device may dissipate. The calculations described in Par. **28-24** can be used to determine this parameter. The derating for silicon diodes is given in Table 28-9.

For zener diodes, the best worst-case end-of-life tolerance that can be guaranteed is $\pm 1\%$. This places a limitation on the final accuracy of any analog system end of life at some value greater than 1%.

Zener-Diode Voltage-Variation Calculation. The change in zener voltage over a specified operating range is primarily a function of the zener temperature coefficient (TC) and dynamic

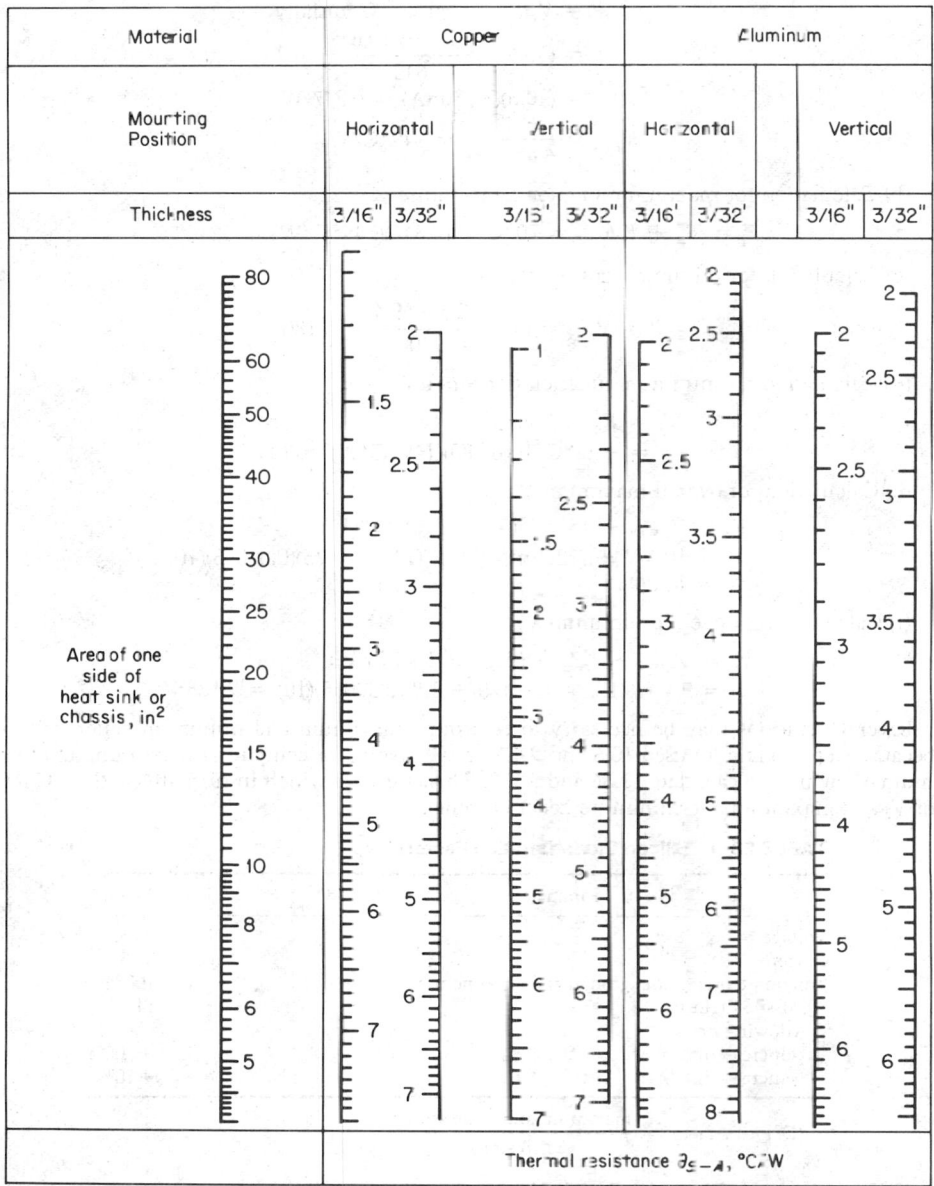

Fig. 28-9. Thermal resistance as a function of heat-sink dimensions.

Instructions for use: Select the heat sink area at left and draw a horizontal line across the chart from this value. Read the values of θ_{S-A} depending on the thickness of the material, type of material, and mounting position.

resistance. The temperature to be used for the TC is the junction temperature of the device in question.

Given: 1N3020B; 10-V zener ±5% (add 1% end of life)

$$P = 1 \text{ W} \qquad \text{derate } 6.67 \text{ mW/°C}$$
$$Z_z = 7 \text{ } \Omega \qquad \text{at } 25 \text{ mA} = I_{zT}$$
$$\text{TC} = 0.055\%/°C \qquad V_{cc} = 30 \text{ V} \pm 5\%$$

The ambient range is -25 to $70°C$, V_{cc} (applied through a resistor R) $= 30 \text{ V} \pm 5\%$, and $R = 800 \pm 20\%$

(a) Calculation for the worst-case maximum power of zener and θ_{J-A}

$$P = V_z I_z \qquad \text{neglect TC initially}$$
$$= (10.6 \text{ V}) \frac{31.5 - 10.6}{640}$$
$$= (10.6)(32.7 \text{ mA}) = 0.347 \text{ W}$$
$$\theta_{J-A} = \frac{1°C}{6.67 \text{ mW}} = 149°C/W$$

(b) Calculation for maximum junction temperature

$$T_J = T_A + P_D\theta_{J-A} = 70°C + (0.347)(149°C/W) = 121.7°C$$

(c) Calculation for minimum power in zener

$$P = V_z I_z = (9.4) \frac{28.5 - 9.4}{960} = 0.180 \text{ W}$$

(d) Calculation for minimum junction temperature

$$T_J = T_A + P_D\theta_{J-A}$$
$$= -25°C + (0.180)(149°C/W) = 1.8°C$$

(e) Calculation for overall maximum V_z

$$V_z = V_{z,MAX} + I_z Z_z + (T_J - T_{amb})(TC)(V_z)$$
$$= 10.6 \text{ V} + (32.7 \text{ mA})(7) + (121.7 - 25)(0.00055)(10)$$
$$= 11.36 \text{ V}$$

(f) Calculation for overall minimum V_z

$$V_z = V_{z,MIN} - I_z Z_z - (T_{amb} - T_J)(TC)(V_z)$$
$$= 9.4 - (19.1)(7) - (25 - 1.8)(0.00055)(10) = 9.138 \text{ V}$$

Several iterations may be necessary to determine maximum and minimum zener voltages, because steps (a) and (c) used 10.6 and 9.4 V, respectively, for computing maximum and minimum V_z, while we calculate 11.36 and 9.138. This affects P, which in turn affects the TC term of V_z and maximum and minimum zener currents.

TABLE 28-9 Silicon Diode Derating Factors*

Parameter	Factor*
Voltage	0.75
Current	0.75
Junction temperature, ground and airborne use	100°C
Missile-flight use	140°C
Allowing for:	
Increase in leakage I_R	+100%
Increase in V_F	+10%

$$*\text{Derating factor} = \frac{\text{maximum allowable stress}}{\text{rated stress}}$$

28. Integrated-Circuit Guidelines. Derating is a process that improves in-use reliability of a component by reducing the life stress on the component or making numerical allowances for minor degradation in the performance of the component. This technique is applied to integrated circuits in two separate and distinct ways.

The first is to specify a derating factor in the application of the component. Derating factors are applied to the voltage, current, and power stresses to which the integrated circuit is subjected during operation. Derating factors must be applied knowledgeably and singly; i.e., they must be applied only to a degree which improves reliability, and they must be applied only once throughout the entire cycle which stretches from the design of the integrated circuit to its application in a system.

From the outset, integrated circuits are designed to a set of conservative design-rating criteria (see Sec. 8). The currents that flow through the conductors and through the wire bonds on a chip, the voltages applied to the semiconductor junctions, and the overall power stress on the entire chip are conservatively defined during the design of an integrated circuit. Therefore it usually is not appropriate to derate the integrated circuit further in its application. Derating power consumption of a digital integrated circuit is, in fact, usually not possible in the design cycle, since the circuit must operate at a specified level of power-supply voltage for maximum performance. However, some linear circuits designed to operate over an extended range of power-supply voltages and power dissipations may accept some degree of derating when it is appropriately applied.

Thus the main area of derating in integrated circuits is not in derating the *stresses* applied to the circuit but in derating the expected and required *performance*. The designer must fully recognize potential performance degradation of integrated circuits over their life. This parametric degradation can require using a digital circuit at less than its full fanout. It can mean designing for an extra noise margin, sacrificing some of it to the degradation of the integrated circuit. It can also mean applying the integrated circuit at performance levels below those guaranteed by the circuit's characterization-specification sheet.

Establishment of derating factors for integrated-circuit parameters must be made after careful analysis of each particular parameter in the circuit. Parameters depending directly on transistor beta, resistor value, or junction leakage are most prone to shift during life. Parameters depending directly on the saturation voltages of junctions and on the ratios of resistors are most likely to remain stable. Device fanout should be derated by a factor of 20%, and logic noise-margin levels should be derated by a factor of 10%.

The severity of the application further establishes the degree of proper derating. It is not customary to derate ac parameters, such as delay times or rates, as these parameters do not vary greatly over the life of an integrated circuit. Allowances should be made, however, for unit-to-unit variation within a given integrated-circuit chip. The delay times of separate gates within one integrated-circuit package can vary greatly. These parameters are usually not measured on 100% of the units.

Although one may be able to derate an integrated circuit for reliability in specified special cases, one cannot take advantage of the derating designed into the integrated circuit and use it beyond its rating or specified capability.

TRANSFORMER, COIL, AND CHOKE DERATING

29. General Considerations. The ratings and deratings of transformers, chokes, and coils are covered in the following paragraphs. Transformers are frequently designed for a particular application and can become a major source of heat. Two major considerations result: derating of transformers must include consideration of their heating effects on other parts; and transformer derating requires control of ambient- plus winding-temperature rises.

30. Voltage Derating. Winding voltages are fixed voltages and cannot be derated to any significant degree as a means of improving reliability. The voltages present between any winding and case or between any winding and shield, as specified, should be derated in accordance with the voltage derating factors of Table 28-10.

31. Power Derating. The power dissipated in a transformer should be derated to control the winding temperature to the maximum derated temperature under full load conditions which are normal to the worst-case service conditions.

Temperature rise is determined for service conditions by measurement of winding resistance using the procedure of MIL-T-27B.[10]

The insulation grade of a transformer is rated for a maximum operating temperature. Deratings shown in Table 28-10 are allowances of temperature to be subtracted from the rated temperature to determine derated temperature. All considerations of frequency, hot-spot temperature, and other factors included in the manufacturer's data must be allowed for before applying this reliability derating temperature.

32. Current Derating. The maximum current in each winding should be derated in accordance with the percentage deratings shown in Table 28-10. The derated current should be considered as the largest current which can flow in the winding under any combination of operating conditions.

In-rush transient currents should be limited to the maximum allowable in-rush or surge rating of the transformer, as shown in Table 28-10. The current in all windings combined should not cause a power dissipation or temperature in excess of the derated temperature requirements.

TABLE 28-10   Coil, Transformer, and Choke Derating Factors

| | Maximum permissible percent of manufacturer's stress rating | | | |
| | Insulation breakdown voltage | | | |
Type	Maximum	Transient	Operating current, A	Allowable winding temp rise, °C
	Coil			
Inductor, saturable reactor	60	90	80	30
General	60	90	80	30
RF, fixed	60	90	80	35
	Transformer			
Audio	50	90	80	35
Pulse, low-power	60	90	80	30
RF	60	90	80	30
Saturable-core	60	90	80	30

SWITCHES AND RELAYS

33. Switch Considerations. Switches are to be applied in circuits with operating current loads and applied voltages well within the specified limits of the type designated. A major problem is contamination, which includes particles and contaminant films on contacts. The storage life on switches exceeds 10 years if they are hermetically sealed. The cycle life for switches can be in excess of 100,000 cycles.

34. Derating. The contact power (volt-amperes) for general-purpose switches (1 to 15 A) should be derated as shown in Table 28-11 from the maximum rated contact power within the maximum current and voltage ratings.

Temperature. Switches should not be operated above rated temperature. Heat degrades insulation, weakens bonds, increases rate of corrosion and chemical action, and accelerates fatigue and creep in detent springs and moving parts. The derating factor in Table 28-11 is defined as the maximum allowable stress divided by the rated stress.

For capacitor loads, capacitive peak in-rush current should not exceed the derated limit. If the relay-switch specification defines inductive, motor, filament (lamp), or capacitive load ratings, they should be derated 75% instead of the derating specified in Table 28-11.

35. Relay Considerations. Relays are to be used in circuits with operating current load and applied coil and contact voltage well within specified ratings. The application for each device should be reviewed independently.

Relay cycle life varies from 50,000 to more than 1 million cycles depending on the relay type, electrical loads of the contacts, duty cycle, application, and the extent to which the relay is

TABLE 28-11   Relay and Switch Derating Factors

Type of load	Switching current, A
Resistive	0.75
Inductive	0.4
Motor	0.2
Filament	0.1
Capacitive	0.75

derated. The storage life of hermetically sealed relays, with proper materials and processes employed to eliminate internal outgassing, is over 10 years.

The chief problem in electromechanical relays is contamination. Even if cleaning processes eliminate all particulates, the problem of internal generation of particles, due to wear, is still present.

36. Relay Derating. The contact power (volt-amperes) should be derated as shown in Table 28-11 from the maximum rated stress level for loads of 1 to 15 A.

Temperature. Relays should not be operated above rated temperature because of resulting increased degradation and fatigue.

CIRCUIT BREAKERS, FUSES, AND LAMPS

37. Circuit Breakers. Circuit breakers should be sized for each application to protect the circuit adequately from overvoltage or overcurrent. For optimum reliability *thermal circuit breakers* should not be subjected to operation in environments where temperatures vary from specified rating(s) because the current-carrying capability of the sensing element is sensitive to temperature variations. For optimum reliability *magnetic circuit breakers* should not be subjected to dynamic environments exceeding the device limitations.

Derating should depend upon the type of circuit breaker or circuit being protected. Normally, circuit breakers of the magnetic type, which are relatively insensitive to temperature, should not be derated except for the interrupting capacity, which should be derated up to 75% of maximum rated interrupting capacity. Derating of standard circuit breakers used in high-reactance-type circuits may be required to avoid undesired tripping from high in-rush current.

Thermal-sensitive circuit breakers used outside of specified ratings should be derated to compensate for effects of operating ambient temperatures.

33. Fuses. Fuses should have ratings which correspond to those of the parts and circuits they protect. These fuse ratings should be compatible with starting and operating currents as well as ambient temperatures. Current-carrying capacity may vary with temperature. An example is given in MIL-F-23419/9.[11]

Fusing should be arranged so that fuses in branch circuits will open before fuses in the main circuit.

39. Indicator Lamps. Indicator lamps should be protected from voltage or current surges above ratings.

Derating. The applied voltage should be derated to 90% of the maximum rated lamp voltage. When strength, reliability, shock, or vibration is of primary importance, lamps rated for 6.3 V or less should be used. Where applicable, long-life lamps should be used.

Semiconductor Reliability

BY C. M. BAILEY, JR.

40. General Considerations. Three terms are typically used to describe the life of equipment and some devices: infant mortality, constant-failure-rate life (long-term reliability), and wear-out. In semiconductor devices, the *infant mortality* and *constant-failure-rate* regions are typically the areas of concern. In normal environments the *wear-out mechanisms* are usually far enough away in time to have little or no effect on the reliability of semiconductor devices in normal operating life.

Although long-term failure rates for semiconductor devices are relatively low, they still can be troublesome and costly. Considering the significant cost and reliability impact of unreliable devices, one must have a technique for predicting and controlling long-term reliability. This involves the recognition of a definable statistical failure distribution.

For semiconductor devices, life distributions have been found to be lognormal, and this is the distribution used for modeling long-term failure rates. Since failure rates depend on stress, *accelerated-stress testing* provides the basic technique for predicting and controlling long-term reliability. The Arrhenius relationship [Eq. (28-22) Par. **28-44**] is used for determining the effect of temperature as an accelerating stress. There are other accelerating environments besides temperature, depending on the failure mechanism being exercised. Accelerated-stress testing provides the basis for lot-acceptance testing to control long-term reliability, as well as for qualification to demonstrate that the required reliability objectives can be met.

The infant-mortality aspects of semiconductor-device reliability have now become a matter of concern. Recent experience shows that infant-mortality defects, though small, can have a significant impact on cost and reliability. Defects are in the range of 0.3 to 2% and include both *dead-on-arrival* and *device-operating failures*. Failure mechanisms are variable in both nature and degree, resulting in very high initial failure rates. Infant-mortality failure rates can be modeled by either lognormal or Weibull statistics [Par. **28-49,** Eq. (28-28)]. Screening to reduce infant-mortality defects is by accelerated-stress tests. A model for estimating the effects of one type of screen, dynamic device burn-in, is discussed in Par. **28-51**.

41. Measuring and Predicting Failure Rate. The quantitative and qualitative measure of reliability is failure rate. Techniques are available for predicting semiconductor-device failure rates in advance and for controlling those rates in the future. We could, of course, simply put semiconductor devices into service and wait to see some pattern of how they fail (how rapidly, at what intervals, etc.). Unfortunately, this data-collection process generally takes too long, and too much equipment is committed in the meantime. Considering that, in early life, semiconductor-device failure rates change with time, we must also think in terms of failure-rate change with time (or *instantaneous failure rate*) rather than average failure rate.

Devices can fail in one of two ways — by degrading gradually or by suffering sudden, catastrophic failure. Most semiconductor devices in general use today fail catastrophically. (The exceptions tend to be devices on the forefront of new technology. For example, early light-emitting diodes exhibited failure largely through diminishing of light output with time; the low level of catastrophic failures due to the breakage of weak wire-bonds would not be recognized until the technology decreased the rate of degradation.)

We are thus limited here to a study of the catastrophic failures, which are typical of the mature semiconductor devices, typically a silicon integrated circuit or transistor. Treatment of degradation, where it occurs in state-of-the-art devices, is not discussed.

Basically, the material covers the techniques for prediction and control of semiconductor-device failure rates during the various phases of device life, with particular emphasis on the use of accelerated-stress testing.

Bathtub Curve. In most cases there are two distinct intervals in the life of a semiconductor device, infant mortality and "constant"-failure rate. The infant-mortality region is characterized by an initially high but rapidly decreasing failure rate. The duration of the infant-mortality period cannot be exactly defined but generally it has its greatest impact in the first year of life. Failure rates in the infant-mortality period may be substantially higher than in those expected in later life. For example, a device with a long-term failure rate of failures per interval of time (FITs)* could have an initial rate (at the first hour) of 10^4 or 10^5 FITs.

The long-term region is characterized by a constant (or nearly constant) failure rate. It is to this region that one refers when the failure rate is not further qualified, e.g., a 10-FIT device.

A third interval, wear-out, is typically used, along with the infant-mortality and long-term intervals, to describe the life of equipment and some devices (the so-called bathtub curve, Sec. **28-6** and Fig. 28-1). For most semiconductor devices in normal environments the wear-out period is far enough away to have little or no impact on the reliability of the devices through normal equipment operating life. An exception, among others, is light-emitting diodes, for which there is essentially no constant-failure-rate region. Despite improvements, these devices still exhibit a wear-out mechanism over their life-span which results in gradual degradation of light output. For most semiconductor devices, however, device life is very long compared with equipment life, and one is concerned with the infant-mortality and long-term but not wear-out regions of the curve.

LONG-TERM RELIABILITY

42. Useful-Life Period. The constant-failure-rate region is sometimes referred to as the *useful-life period*. Although failure rates of semiconductor devices during the useful-life period are relatively low, they still can be troublesome and costly. One then needs a technique for predicting and assuring long-term failure rates. Inherent in the prediction and assurance of long-

---

*FITs are defined as failures in 10^9 device hours of operation. Thus, for example, a 100-FIT device will have a rate of 10^{-7} failures per hour, or 0.01% per thousand hours.

term failure rates is knowledge of the contributing failure modes* and failure mechanisms* and how they can be modeled and controlled. The basic technique involves using accelerated-stress testing coupled with design and process control. The following paragraphs discuss *life distributions, accelerated-stress effects,* the associated *reliability models* and the other factors relative to long-term reliability.

43. Life Distribution. The rational handling of life data and particularly the use of those data to predict what is likely to happen in the future or at some other operating condition requires recognition of a definable statistical failure distribution. The mathematical "definition" should offer formulas describing critical functions of the distribution and should allow the use of properly prepared plotting papers to simplify the use of the distribution.

For semiconductor devices, the lognormal distribution fits more data collections than any other and is assumed to be the proper distribution for semiconductor life in ordinary environ-

TABLE 28-12   Lognormal Mathematical Functions

Distribution functions

Probability distribution function $f(t) = \dfrac{1}{\sigma t \sqrt{2\pi}} \exp\dfrac{1}{2}\left(\dfrac{\ln t - \mu}{\sigma}\right)^2$

Cumulative density function $F(t) = \dfrac{1}{\sigma \sqrt{2\pi}} \displaystyle\int_c^t \dfrac{1}{x} \exp -\dfrac{1}{2}\left(\dfrac{\ln x - \mu}{\sigma}\right)^2 dx$

Instantaneous failure rate, hazard rate $\lambda(t) = \dfrac{f(t)}{1 - F(t)}$

Significant distribution properties

Median (50% failure) $t = t_{50\%} = e^{\mu}$ Mean (average) $t = e^{\mu + \sigma^2/2}$
Mode [highest $f(t)$]: $t = e^{\mu - \sigma^2}$ Location parameter e^{μ}
Shape parameter σ
   S, estimate of σ, can be calculated as $\ln (t_{50\%}/t_{16\%})$

ments. When the choice is not obvious from the data, the lognormal should still be used. The Weibull distribution, however, can be used to model infant mortality [see Par. **28-49,** Eq. (28-29)].

The lognormal distribution logically results from the multiplicative interaction of the many semiconductor failure mechanisms which involve contaminant quantities and diffusion constants, current densities and temperature gradients, temperature variations and voltage gradients, etc.

Table 28-12 lists the pertinent mathematical functions for the lognormal distribution. Figure 28-10 shows the distribution for several values of σ. The curves shown are for constant median life e^{μ}. The term σ refers to the standard deviation of the natural logarithm of time to failure. As shown in Figure 28-11, plotting paper has been prepared for the lognormal distribution so that data, using real time-to-failure measurements which fit that distribution, will plot as a straight line.

It is important that when different failure mechanisms occur within a sample, they be treated independently, as they may be accelerated differently and extrapolations from combined data can be very misleading. (This emphasizes the need for failure analysis to the extent of identifying the failure mechanism.) Hence, in practice, only one failure mechanism should be shown in one plot.

Even within a common failure mechanism, some failures may occur substantially earlier than the bulk of the population on test. These early-life test failures, often called *freaks* or *sports,* result from some extreme combination of effects in individual devices which result in failure much earlier than would be consistent with the major population.

Evaluation of various device data suggests that freak populations also appear reasonably to follow lognormal distributions. Such a combination is shown in Fig. 28-12. This example would suggest that 10% of the population is included in a freak population. The resulting S-shaped (bimodal) curve is regularly seen in life tests of silicon semiconductor devices. The inflection

---

*Failure mode is defined as the effect by which a failure is observed, e.g., an open or short circuit. Failure mechanism is the physical, chemical, or other process resulting in a failure.

point in the curve defines the percentage contribution of the freak distribution and the time on test needed to remove those freaks.

Once the median life and standard deviation have been determined for the main population, it is possible to use a set of normalized lognormal failure-rate curves to estimate the expected failure rate over time (see Fig. 28-13). Note that for the lognormal distribution, failure rates begin low, rise up to peak, and then decrease again at a later time. The curves of Fig. 28-13 represent the failure rates over time assuming no replacement of failed devices. Since, in normal equipment life, the percentage of device failures is quite small (a few tenths of 1%), these curves remain useful and accurate in most practical situations.

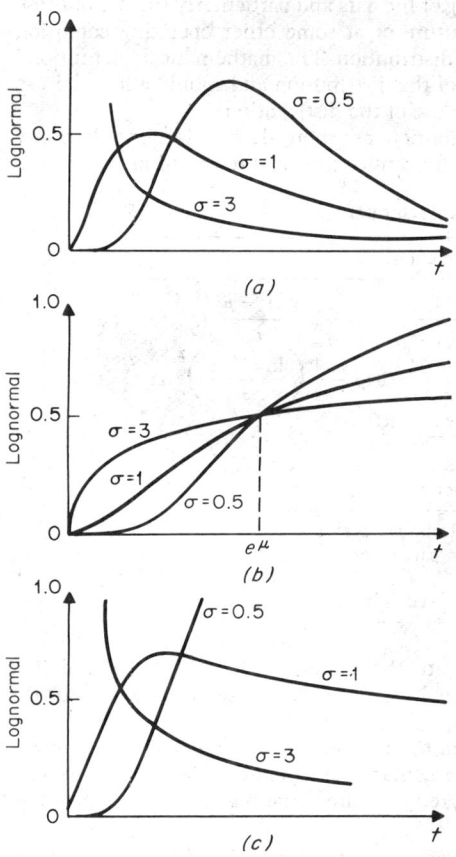

Fig. 28-10. Lognormal distribution.

If there are freaks in the population, they can be separated from the main distribution and handled by themselves. In that case the failure rates for the freaks and the main population can be plotted separately (from Fig. 28-13) and added to provide the total failure-rate estimate.

44. Time-Temperature Relationship. A major basis for accelerated testing lies in the fact that many of the chemical and physical processes leading to failure are accelerated by temperature in a way that can readily be modeled and reproduced. The basic nature of the relationship between the reaction rate of these processes and temperature can be defined by the Arrhenius equation

$$R = R_0 e^{-E_A/kT} \qquad (28\text{-}22)$$

where $R_0 = $ const, $E_A = $ activation energy (eV), $k = $ Boltzmann's constant $= 8.6 \times 10^{-5}$ eV/K, $T = $ absolute temperature (K).

The reaction rate will be related to failure time through some undefined function. Regardless of the specific function, the failure time should follow the temperature dependence of the Arrhenius equation. Given this effect of temperature, we can visualize the change of a device from its normal state to a failed state when some parameter exceeds its functional limit, as in Fig. 28-14, where R_1 is rate of reaction at temperature T_1 and R_2 is the rate at temperature T_2 ($T_1 > T_2$).

If the assumption is made that the reaction is linear in time $R_1 t_1 = R_2 t_2$. Then with $t_f = $ time to failure

$$R_{t_f} = \text{const} \qquad t_f \propto \frac{1}{R} \propto e^{E_A/kT} \qquad \text{and} \qquad \ln t_f = C + \frac{E_A}{kT} \qquad (28\text{-}23)$$

To calculate activation energy from equivalent times at two temperatures

$$\ln t_{f_1} = C + E_A/kT_1 \qquad (28\text{-}24)$$
$$\ln t_{f_2} = C + E_A/kT_2$$

Subtracting gives

$$\ln t_{f_1} - \ln t_{f_2} = \frac{E_A}{k}\left(\frac{1}{T_1} - \frac{1}{T_2}\right) \qquad (28\text{-}25)$$

or

$$E_A = k \ln \frac{t_{f_2}}{t_{f_1}}\left(\frac{1}{1/T_2 - 1/T_1}\right) \qquad (28\text{-}26)$$

The acceleration factor which relates failure times at two different temperatures, when the activation energy is known, is

$$A = \exp\left[\frac{E_a}{k}\left(\frac{1}{T_2} - \frac{1}{T_1}\right)\right]$$

(28-27)

The possibility that the failure mechanism depends on t^n rather than linearly on t cannot be determined from device-failure data. In that case, $\ln t_f = c + E_A/nkt$, and the apparent activation energy is E_A/n, or $1/n$ times the true activation energy of the process leading to failure. This presents no problem because the apparent activation energy is the quantity of practical interest.

If one plots $1/T$ vs. $\ln t_f$, one arrives at a straight-line plot where the slope is related to k/E_A.

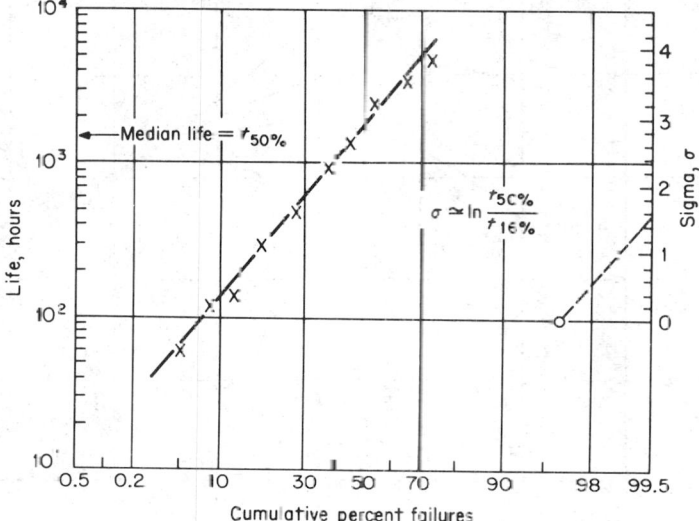

Fig. 28-11. Lognormal plot.

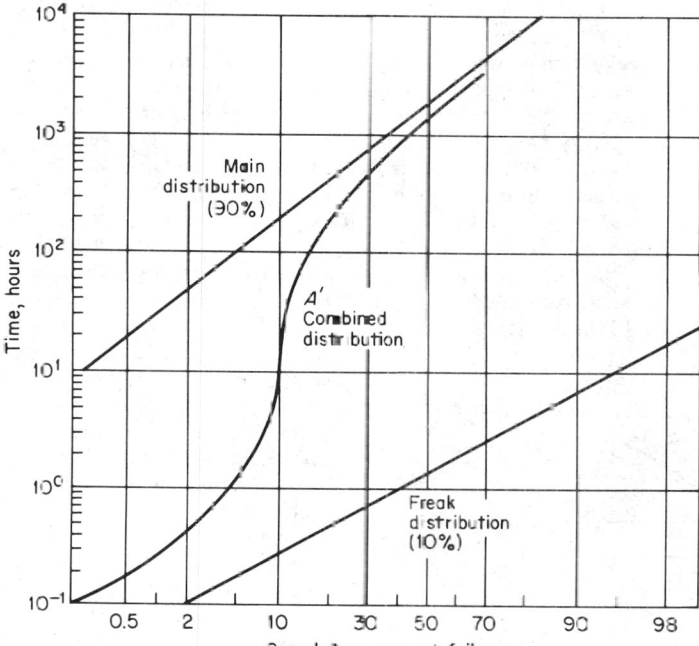

Fig. 28-12. Freak-distribution separation

Such an Arrhenius plot is shown in Fig. 28-15. Note that this plot can be used to determine the apparent activation energy of the failure process or to determine the acceleration factor A.

45. Accelerated Stress Testing. To determine the relationship of stress and time for a given product, a group of tests can be arranged, on a homogeneous run of product, so that several samples are life-tested at different temperatures and the results can be compared. A life distribution will result for each sample, and the results are compared by the median life estimates. These data are plotted on an Arrhenius plot. Freaks and infant-mortality failures are removed statistically so that the median lives of the main failure mechanism are the numbers used in the Arrhenius plot for that mechanism. The plots of Fig. 28-13, with median life suitably

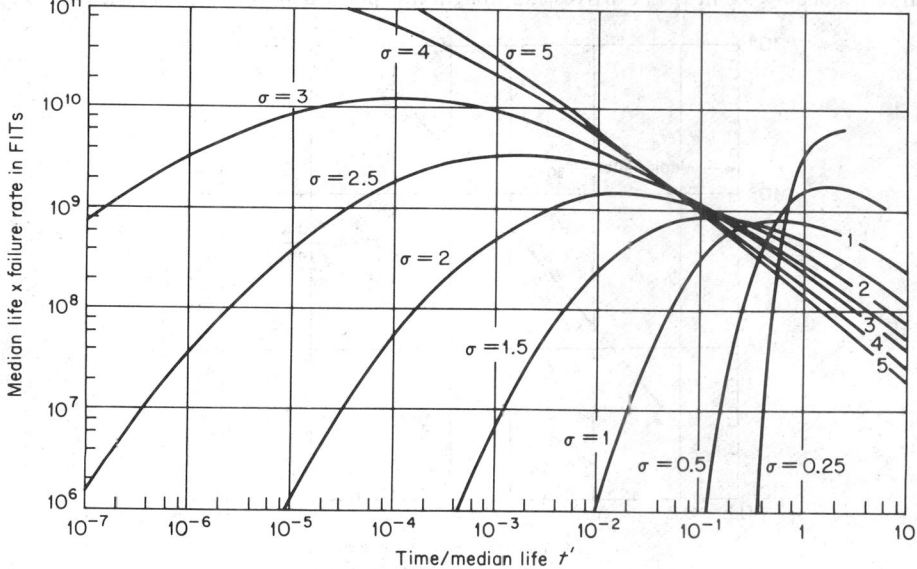

Fig. 28-13. Normalized lognormal failure.

translated for the acceleration between normal-use stress and the accelerated stress, are then used to determine the failure rate to expect.

Example. To illustrate these techniques, consider a simple example based on actual data. Assume that we are qualifying a CMOS integrated circuit and wish to predict its long-term failure rate in normal use (say 50°C, operating temperature). Failure analysis identifies the main failure mechanism as parasitic transistors due to surface contamination. The lognormal plots show some evidence of freaks. After the freaks have been statistically separated, the lognormal plots show median lives of 150 h at 200°C, 600 h at 175°C, and 2,600 h at 150°C. We estimate a σ of approximately 2 for each sample (Fig. 28-15). An Arrhenius plot of the median lives, as in Fig. 28-15, indicates an activation energy E_A of 1 eV and an extrapolated median life at 50°C of approximately 1×10^7 h. Referring to the normalized failure-rate curves of Fig. 28-13, we find 5×10^8 as the product of median life and failure rate, for $\sigma = 2$ and time divided by median life = 1×10^3 (1,000/1×10^7). Since median life = 1×10^7, the failure rate is 50 FITs expected at 10,000 h.

This example involves extrapolation from accelerated-stress results to normal-stress expectations. The development of adequate life-test requirements to protect the reliability requirements of normal-stress application requires a reverse process, summarized as follows:

1. The reliability required of a device by

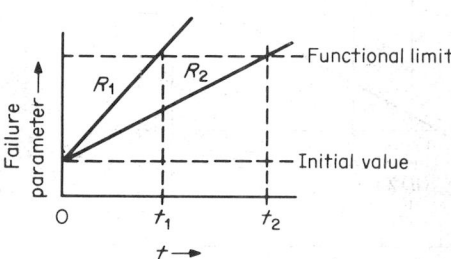

Fig. 28-14. Failure reaction related to time and temperature.

the system (for a given system lifetime) can be translated into a minimum median life t_m given an estimate of σ.

2. This t_m requirement can be related to a choice of t_m at accelerated-stress levels, depending on life-test time choices, etc., and given a knowledge of the activation energy for the probable failure mechanism.

3. The accelerated stress t_m, together with the σ assumed in step 1, provides a distribution from which one can choose a combination of time and percent defects allowed (PDA) that will be satisfactory to both user and supplier.

Failure Mechanisms in Accelerated Testing. There are many accelerating environments besides temperature, depending on the failure mechanism being exercised. Some of the common stresses of interest in addition to temperature are *temperature cycling, power cycling, humidity, voltage,* and *current density.* All are subject to the same basic treatment described here, requiring identification of the equation that associates the effect of stress on time to failure and also requiring that the distribution of failure times be known or identifiable by test.

A number of specific failure mechanisms inherent in semiconductor devices vary in both nature and degree with design and process features. To attempt to list them all is impractical. From the standpoint of accelerated testing, some mechanisms can be grouped by the type of accelerated conditions which affect them, as follows:

1. *Accelerated temperature without* applied electrical bias can test for diffusion effects, chemical actions, decomposition of materials, etc., sometimes in combination with other environments but sometimes alone. One example is chemical reaction in contact areas, where contact metals can react with the semiconductor material. Another is the growth of intermetallic materials at the bonds of dissimilar metals such as Au and Al.

2. *Cycling of temperature* results in a fatigue stressing of materials. This can reveal weak internal wire bonds, mis-

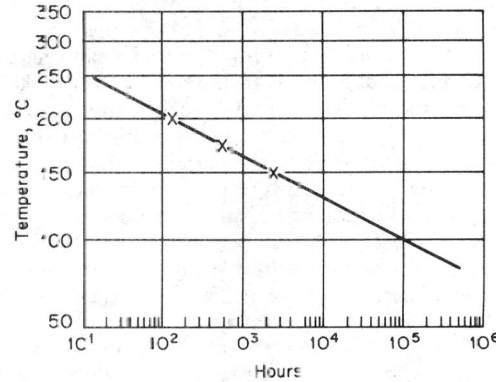

Fig. 28-15. Arrhenius plot. The case for $E_A = 1$ eV.

matches of seal materials in hermetic packages, and mismatch of the silicon-device chip and the package in the die-attach region, for example.

3. If temperature is likely to release mobile ions from surrounding materials or to make contaminating ions on the device active surface mobile, the additional presence of voltage will accelerate the effect of ion movement on surface-dependent characteristics. Hence, *temperature and voltage* stress can screen for mechanisms such as surface inversion, surface-charge movement, dielectric breakdown, etc. Figure 28-16 illustrates how surface-related problems can occur in some oxide-protected integrated circuits. Here the presence of positive charges in (or on) the oxide inverts the p-type base material to n-type. Similarly the presence of negative charges

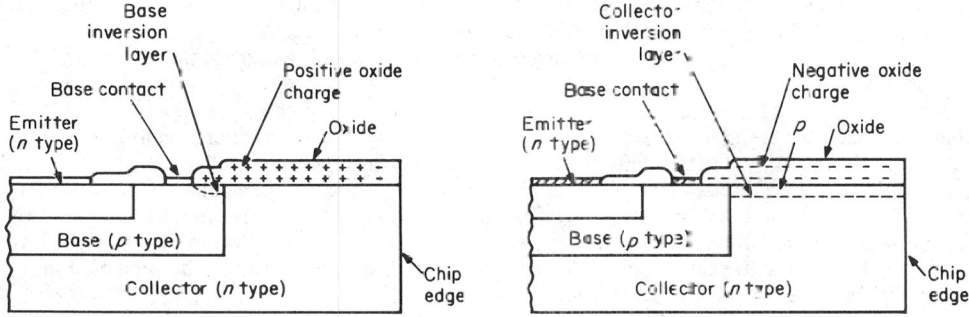

Base and collector inversion layers

Fig. 28-16. Surface inversion results from ionic contamination.

inverts the *n*-type collector material to *p*-type. These inversions result in marked increases in leakage current (shunt) paths, decreases in breakdown voltages, and hence failure of the device.

4. *Temperature, humidity, and voltage* stress is the combination that causes particular trouble with materials subject to corrosion, aided by some contamination to provide an electrolyte. Electrolyte corrosion of aluminum or gold exposed to humidity and normal circuit bias is a good example. Corrosion of aluminum typically produces open metallization lines; gold dendritic growths result in shorts from one line to the next. External leads can corrode off in extreme conditions.

5. *Lowering of the temperature* is an unusual acceleration condition, but at least two failure mechanisms can be observed at a higher failure rate at reduced temperature. For example, moisture sealed in a package will condense at a temperature dependent on the amount of moisture present. When condensation occurs, surface conductivity and leakage increase. Also, the possibility of corrosion is increased since, when condensation occurs, a better electrolyte is provided. The second mechanism involves the phenomenon of hot-electron-induced degradation of *n*-channel MOS devices. The degradation has a higher failure rate at a lower temperature than at a higher temperature.

6. *High temperature and current* can accelerate electromigration of metallization lines. In this mechanism, metal ions are swept by the impact of current flow and cause hillocks of metal at one end of a line and voids at the other end, until an open circuit finally develops.

7. *Naturally occurring α particles* can cause soft errors in small-cell MOS devices such as dynamic random-access memories. Existing ceramic packaging materials contain trace, though significant, levels of materials such as thorium and uranium, which decay, emitting α particles. When an α particle penetrates the chip surface, it can create enough extra electrons to cause a random, singe-bit error. The errors are not permanent; i.e., no physical defect is associated with the failed bit. This failure mechanism depends to a large degree on the physical memory-cell size (or to the amount of charge being stored). Accelerated testing is done with a high-flux α-particle source.

8. Where *high voltage gradients* exist through an insulator, the voltage level has been seen to be an accelerating factor for dielectric failures in MOS circuits and in on-chip capacitors of linear circuits.

Table 28-13 summarizes many of the failure mechanisms in silicon devices, the process, relevant factors, acceleration factors, and the activation energy.

46. Qualification and Lot Acceptance Testing. In the present context qualification refers to the steps taken to demonstrate that the required long-term performance and reliability objectives of a new semiconductor device or technology can be met, with particular emphasis on the use of accelerated-stress testing for that purpose.

Typically, qualification testing is conducted on samples of devices which incorporate the design features, materials, and processes to be used in production. The basic purposes of this testing are to confirm or identify the main failure mechanism and to provide data to predict or confirm the required long-term reliability and establish the lot-acceptance criteria. Enough samples must be tested to provide statistical confidence in the long-term failure-rate predictions. The accelerated-stress tests used will vary in nature and degree with the device technology, the failure mechanisms, and the reliability objectives. For example, a new device technology will need more extensive qualification than extension of an existing one.

Example. We can illustrate a satisfactory qualification procedure by defining those followed for Bell Laboratories integrated-injection-logic (I^2L) devices fabricated using bipolar technology (beam-lead sealed-junction). Accelerated stress tests performed were:

1. *Operation* of the chip, with no moisture protection, *in salt water* for more than 1,000 h without failure.

2. *Temperature-bias testing* at 300°C on 100 units. The test data indicated a median life of 600,000 h and a σ of approximately 4. Using a conservative activation energy estimate of 1.02 eV, a failure rate of less than 1 FIT for surface inversion was predicted at 60°C operation.

3. *Temperature-humidity bias* (85°C/85% RH) testing on 75 units. Median life was 100,000 h with a σ of about 2.3. Using an electrolysis model and a 10°C rise in junction temperature above ambient, the failure rate for the typical condition (25°C/50% RH) was predicted to be less than 10 FITs over a 40-year life for this mechanism. For a severe condition, the predicted maximum failure rate was 50 FITs at the end of a 40-year life.

4. *Current gain.* An additional 25 units were subjected to *temperature-bias step testing.* (Step testing is an accelerated-stress technique where the samples are exposed to successively higher steps of stress at constant times at each stress, in order to determine the range of the samples' reasonable response to that stress.) Average current gain vs. temperature was measured.

TABLE 28-13 Failure Mechanisms in Silicon Devices

Device association	Process	Relevant factors	Accelerating factors	Acceleration	
Silicon oxide and silicon–silicon oxide interface	Surface-charge accumulation	Mobile ions, voltage, temperature	Temperature	$E_A = \begin{cases} 1.0\text{–}1.05 \text{ eV} \\ 1.2\text{–}1.35 \end{cases}$	bipolar MOS
	Dielectric breakdown	Electric field, temperature	Electric field		
	Charge injection	Electric field, temperature, Q_{ss}	Electric field, temperature	$E_A = 1.3 \text{ eV (slow trapping)}$	
Metallization	Electromigration	Temperature, current density, area, gradients of temperature and current density, grain size	Temperature, current density	$E_A = 0.5\text{–}1.2 \text{ eV}, j$ to j^4	
	Corrosion (chemical, galvanic, electrolytic)	Contamination, humidity, voltage, temperature	Humidity, voltage, temperature	Strong humidity effect $E_A \approx$ 0.3–0.6 eV for electrolysis	
	Contact degradation	Temperature, metals, impurities	Varied	Voltage may have thresholds	
Bonds and other mechanical interfaces	Intermetallic growth	Temperature, impurities, bond strength	Temperature	For Al–Au $E_A = 1.0\text{–}1.05 \text{ eV}$	
	Fatigue	Temperature cycling, bond strength	Temperature extremes in cycling		
Hermeticity	Seal leaks	Pressure, differential atmosphere	Pressure		

The data indicated that an initial gain specification 10% above the end-of-life requirement is conservative in providing adequate gain during life.

An important aspect of qualification is in-circuit testing of prototype units. This is frequently needed to prove the design and to assure that the models used for qualification are correct from an applications standpoint. A new MOS memory, for example, may undergo considerable testing in the actual equipment circuit before formal qualification testing is done. Lot-acceptance testing refers to the tests and criteria specified on semiconductor-device products to provide assurance that the reliability and performance objectives for the device, once qualified, will continue to be

TABLE 28-14   Typical Specification Requirements for Integrated-Circuit Packages

Test	Reliability 100 FITs		Reliability 200–1,000 FITs	
	Ceramic	Plastic	Ceramic	Plastic
100% screens				
Temperature storage at 300°C, h	0.25	. . . .	0.25	
Burn-in, h, static, at 150°C	168	168		
Dynamic, at 150°C	168	168		
At 125°C	168	168	168	168
Temperature cycle, c, at −40 to 150°C	10	. . . .	10	
At 0 to 100°C	. . . .	50	. . . .	50
Hot test	. . . .	100°C*		
Centrifuge leak tests, cm³/s	10^{-8}	. . . .	10^{-8}	
Sampled tests				
Temperature-bias, life, h, static at 200°C	100, 10% PDA†	100		
Dynamic, at 150°C	1,000, 10% PDA	1,000, 10% PDA		
At 125°C	. . . .	. . . .	1,000, 10% PDA	1,000, 10% PDA
Humidity life, at 85°C, 85% RH, h	. . . .	300, 5% PDA	. . . .	300, 10% PDA
Hot test	. . . .	. . . .	. . . .	100°C‡ 70°C§

*Continuity or parametric.   ‡Continuity
†Percent defective allowable.   §Functional.

realized in production. Here the requirements are established on the basis of the known failure mechanisms and their probable incidence. Table 28-14 shows some typical lot-acceptance requirements for integrated circuits specified by Bell Laboratories for outside-procured integrated circuits.

EARLY-LIFE RELIABILITY

47. Infant mortality includes two major elements, *dead-on-arrivals* (DOAs) and *device operating failures* (DOFs). DOAs are devices that fail when initially tested after shipment or incorporation in the next assembly level. For example, incoming inspection failures, first-circuit-pack failures, and equipment turn-on failures after installation are all DOAs. Device operating failures refer to devices that fail after some period of operating time. It is important to note that while semiconductor device DOA levels may range from about 0.3 to 1.0% of the product, DOF levels observed in normal factory testing and equipment burn-in are typically smaller, ranging from 0.01 to 0.5%. The length of the infant-mortality period cannot be exactly defined, but generally it has its greatest impact in the first year of device use.

In the past the infant-mortality aspect of semiconductor device reliability has been somewhat neglected due to a lack of recognition of its impact and understanding of its nature and degree. It has now become a matter of concern, particularly for equipment which goes directly into service with the final user. Recent experience has shown that infant-mortality defect levels, though small, can have a significant impact on cost and reliability.

Total semiconductor infant-mortality levels, including DOAs and DOFs, are in the range of

0.3 to 2%. These levels, if not corrected, can result in an unacceptable equipment performance. For example, although semiconductor-device removals in burn-in of 500-part equipment average less than 0.1%, they result in failures of almost half the equipment during the burn-in. As an example of the possible impact of infant mortality, consider a circuit board containing approximately 100 integrated circuits. If the DOA percentage for these devices averages about 0.5%, then 40% of these boards will fail, on the average, at first board test resulting in an appreciable testing and repair effort.

Cost due to infant-mortality failures is incurred in the equipment factory and in the field. Costs include those due to failures at incoming test, circuit-pack test, system test and burn-in, field installation and prove-in, and service failures. Lowest costs will be associated with drop outs at incoming-device tests. Equipment repair costs will increase at each stage, the lowest repair cost being after failure at first circuit pack and the highest repair cost after failure of in-service units. In the final analysis, however, total costs due to DOAs are likely to be equal to those due to DOFs.

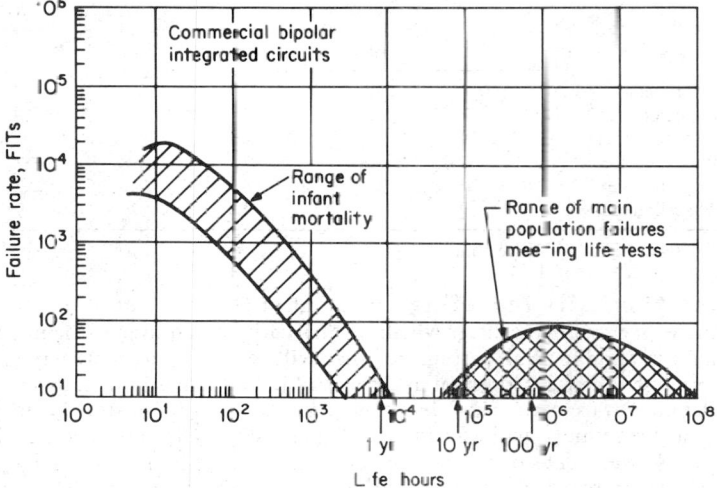

Fig. 28-17. Operating failure rate for infant-mortality and long-term mechanisms.

Finally, to put the infant-mortality problem in perspective, consider Fig. 28-17. This shows the typical infant-mortality failure rate in digital bipolar integrated circuits compared with that resulting from the failure mechanism typical of the main population. The range of long-life main population is suggested to be within realistic expectation, based on accelerated-stress tests on a variety of integrated-circuit types. It represents product which could pass a life test of 100 h at 125°C. What is demonstrated is that even if the long-life reliability of an integrated circuit were quite satisfactory, infant mortality could provide failure rates which would be undesirable for much present-day equipment, for times which are significant with respect to equipment life.

48. Infant-Mortality Defects. Infant-mortality defects stem from a variety of device design, manufacturing, handling, and application-related causes. The device failure mechanisms causing infant mortality may differ in type and degree from product to product and from lot to lot in the same product. The failures are generally caused by gross manufacturing defects. Such defects as oxide pinholes, photoresist or etching defects resulting in near opens or shorts, conductive debris, scratches, weak bonds, or partially cracked chips or ceramics can lead to infant-mortality failures. In some device designs, infant-mortality failures can also result from surface-inversion problems, probably associated with gross passivation defects.

Some of the defects are workmanship related, but many are beyond operator control. Some are inherent in design rules and limitations, some are due to practical limitations, and some are due to limitations of process and material control. Small manufacturing variations or perturbations can produce device lots with infant-mortality characteristics significantly different from the norm. Conversely, some systematic mechanisms due to assignable causes can be corrected by changes in design or in fabrication techniques or in handling the devices during manufacture.

In some cases infant-mortality failures may be caused by assembly operations, e.g., mechanical damage, or are application-dependent (overstress, misapplication, or inadequate specification of device performance requirements). Failures of integrated circuits due to electrostatic discharge during assembly operations have become more significant as technological advances have produced devices with smaller geometries and thinner dielectrics. CMOS and linear integrated circuits are the most sensitive.

Table 28-15 lists the percentage of certain device failures due to various failure mechanisms. It emphasizes the variability of infant mortality from product to product and, by implication, from time to time. For example, there is no good reason why T²L circuits should have a greater percentage of bond problems than CMOS circuits and no reason why CMOS circuits should be free of surface problems. These results were found in a particular sample of field failures, and a sample taken at a different time could show a significantly different mix of the defects which typically make up infant mortality. The removals from overstress shown in this table demonstrate the greater vulnerability of MOS devices to electrostatic-discharge effects.

TABLE 28-15   Infant-Mortality Percentage Failure for Various Mechanisms

	Commercial			Western Electric	
	T²L	CMOS	Memory	T²L*	Memory†
Overstress	4	60	17	35	9
Oxide defects	2	1	51	. . .	53
Surface defects	18	. . .	24		
Bonds, beams	37	5	7	39	27
Metallization	30	34	. . .	4	2
Miscellaneous	9	. . .	1	22	9

*Beam-lead.   †Wire-bonded.

49. Infant-Mortality Modeling. In the infant-mortality part of the total population we are dealing with a small percentage which will fail early with respect to the main population. Since they fail early and in small percentages, they will be completely replaced in a time period of practical interest. Their contribution to the overall failure rate is diluted by their proportion in the population. For example, if we have 1% of infant mortality devices having a 1,000-FIT failure rate at a given time, it will contribute 10 FITs (1,000 × 1%) to the overall failure rate for the population. As those defective devices are displaced in an operating system by good devices, the percentage of defective devices continually decreases, as does their contribution to the overall failure rates.

This relative effect can be accounted for by modification of the normalized failure-rate curves of Fig. 28-13 by introducing the original proportion p of infant-mortality devices and using t_{mf} as the median life of the infant-mortality failures. The overall failure rate of the population remains p, due to the infant-mortality proportion assuming that the failure rate of the rest of the population is negligible, or is added to this if it is known. Figure 28-18 shows these curves for several values of σ. These curves are used like Fig. 28-13 except that the median life t_{mf} is that of the infant-mortality proportion, P_f is the infant-mortality proportion, and the net result is the total failure rate of the population ascribed only to infant mortality.

Another technique for modeling infant mortality involves the Weibull distribution. The instantaneous failure rate as a function of time for one form of the Weibull model can be represented as

$$\lambda = \lambda_0 t^{-\alpha} \tag{28-28}$$

where λ_0 and α are parameters describing the scale and shape of the failure-rate plot. This plot has the favorable feature that when $\ln \lambda$ is plotted vs. $\ln t$, the plot becomes a straight line, as shown in Fig. 28-19. λ_0 is the intercept of the plot at $t = 1$ h and α is the slope of the projection.

In many cases, either the lognormal model or the Weibull model can be used equally well to treat the data. The Weibull is often used because it is more convenient. Figure 28-20 is a plot, from field data, of instantaneous failure rate vs. time, showing a nearly linear slope but with a deviating curvature at very short times. Using the Weibull model to describe this infant-mortality failure rate will result in a somewhat pessimistic estimate, because it ignores the curvature.

While the foregoing provides the basis for modeling, it must be noted that large sample sizes are generally required to provide enough data to characterize infant mortality. It is often impractical during the device qualification to obtain enough samples to characterize the expected infant-mortality experience. In many cases one must await the accumulation of valid field data.

However, reasonably early estimates can be made based on similarity to other technology, knowledge of the failure mechanisms, the limited accelerated-stress test data, and early production.

50. Infant-Mortality Screening. *Screening* describes the application of some test to 100% of the product in order to remove (or reduce) defective or potentially defective units. In the broad sense, electrical parametric or functional testing is a screen, but in eliminating infant-mortality defects one must concentrate on the use of accelerated-stress tests. In some cases such screens can be aimed at long-term failures as well as infant mortality. Freaks of the main population can also appear in the infant-mortality region. Screening conditions typical of accelerated testing of semiconductor devices involve temperature, temperature and bias, etc. The selection of the proper accelerated-stress condition for the screen depends, to a large measure, on the

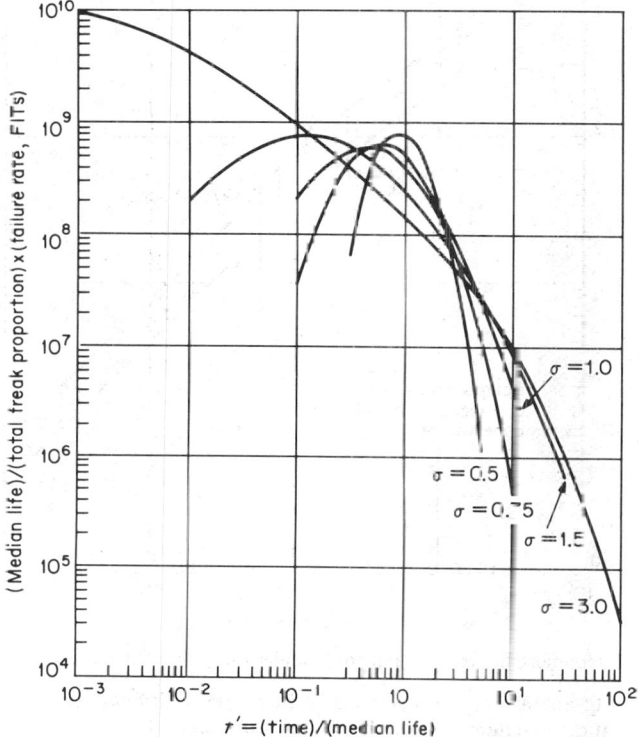

Fig. 28-18. Normalized lognormal failure rates for freaks.

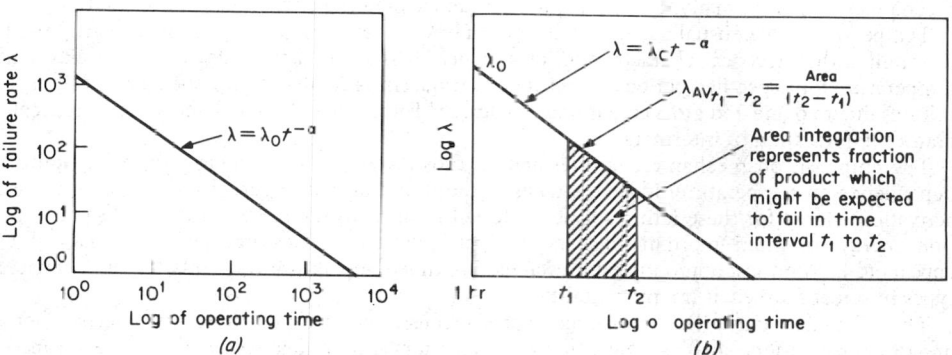

Fig. 28-19. Weibull model. Failure-rate characteristic for integrated circuit type XYZ.

nature and degree of the failure mechanisms contributing to the infant-mortality defects and how they are accelerated by the particular screen.

For example, temperature cycling is a commonly used screen for semiconductor devices. In many integrated circuits, mechanical defects are the major contributor to infant mortality. Temperature cycling followed by a 100°C continuity and gross functional test has proved quite effective in screening out many of the mechanical defects. Temperature cycling will screen for poor wire bonds, die-attach defects, and other mechanical problems. It will accelerate defects due to the formation of intermetallics in Au–Al wire-bond systems. For such systems, it is usually followed by a centrifuge test to detect bonds weakened by intermetallics. When followed by leak tests, temperature cycling is an effective screen for poor seals of hermetic devices. In plastic-encapsulated devices, defects due to fatigue stressing of wire bonds and beam leads by plastic materials surrounding the lead and the bond area are accelerated.

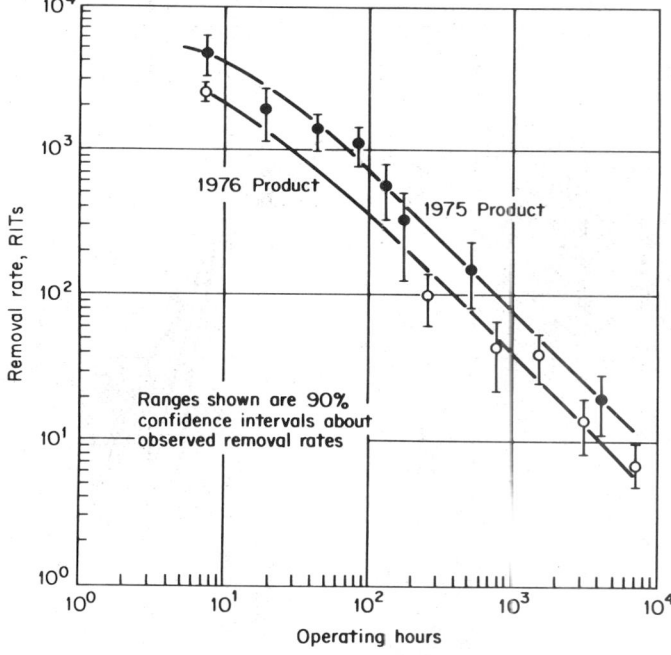

Fig. 28-20. Infant-mortality removal rate for beam-lead sealed-junction T²L integrated circuits.

The modeling of temperature-cycling effects on infant-mortality defects (and, for that matter, long-term defects) and the determination of the optimum conditions is a complicated problem. The failure mechanisms in temperature cycling are likely to reflect the peculiarities of the specific device or structure, and there may be several competing mechanisms in one device. This makes for a complicated analysis and generally requires separate consideration of each failure mechanism. Usually, many data are required to understand the true effects of acceleration.

Temperature alone, usually called *stabilization bake* or *temperature storage*, is primarily used to stabilize the electrical characteristics of a device. It involves storing devices at an elevated temperature for a specified period of time. The thermal stress applied during the bake accelerates failures due to oxide and gross contamination defects. For the Au–Al wire-bond systems it accelerates the formation of intermetallics.

The emphasis on mechanical failures and on those largely due to voltage suggest a limited dependence on temperature. Indeed, studies of the infant-mortality period indicate a very low activation energy for these failures. Data collected from many sources indicate a 0.4-eV activation energy for infant-mortality defects in bipolar and MOS integrated circuits. That value appears to be the most appropriate for establishing time-temperature trade-offs for these device types in screens aimed at infant mortality.

The discussion above illustrates some of the screens used for infant mortality defects. There are, of course, others, such as *power aging* of transistors and *voltage stressing* of on-chip capacitors in linear integrated circuits. As in temperature and temperature cycling, they are oriented toward predominant infant-mortality failure mechanisms in the specified device types.

Typical screens are shown in Fig. 28-12. One screen, burn-in, deserves special consideration, and is discussed separately in the following paragraph.

51. Device Burn-In. Device burn-in has been proved to be an effective means for screening out defects contributing to long-term reliability as well as infant mortality. This screen typically combines electrical stresses with temperature and time. The accelerated stresses thus applied are intended to activate the temperature-dependent failure mechanisms in a relatively short time.

Burn-in can be characterized as static or dynamic. In *static burn-in* a dc bias is applied to the device at an elevated temperature. The bias is applied in order to reverse-bias as many junctions as possible within the device. In *dynamic burn-in*, the devices are operated in order to exercise all the circuit, simulating actual system operation.

Static burn-in is particularly effective for defects resulting from corrosion or contamination. For surface problems, the elevated temperature makes the contaminants more mobile, and the electric field created by the bias causes ions to drift preferentially into critical areas. It is also effective for accelerating intermetallic formation in Au–Al wire-bond systems. The electric fields stress defective oxides to the point of failure.

Dynamic burn-in does not appear to be as effective as static burn-in screening for surface problems, because of the lack of a constant electric field to provide preferential drift of ions. On the other hand, dynamic operation produces higher current densities, power dissipation, and chip temperature than static burn-in. Dynamic burn-in provides more complete access and exercise of internal device elements. For complex LSI devices especially, dynamic burn-in probably screens for more of the infant-mortality failure mechanisms than static burn-in does.

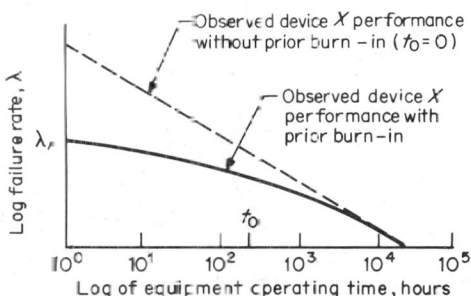

Fig. 28-21. Effect of device burn-in before use in equipment.

The choices of the type of burn-in, static or dynamic, and the specific stress conditions are a function of the device technologies and the reliability requirements. (As in any screen or life test, there are also cost trade-offs, not considered here.)

Dynamic-Burn-In Model. A model based on the assumption of the Weibull distribution for infant mortality has been proposed to model the effects of dynamic device burn-in. Since operation of devices in equipment does lead to reduced device replacement rates, it seems reasonable to assume that such improvements, due to device aging, should also result from dynamic operation in burn-in boards simulating use in the system. The model thus assumes that devices follow monotonically decreasing failure-rate curves, irrespective of where operation takes place. Operation during device burn-in will then produce a certain amount of aging and will result in a reduced failure rate when the devices are assembled into a system. Subsequent operation in equipment will begin at the reduced failure rate and continue down the failure-rate curve with additional operating time.

As shown in Fig. 28-21, when observed from an equipment operating-time perspective, burned-in devices should no longer follow a straight-line log-log failure-rate-vs.-time plot. Devices will then follow a form $\lambda = \lambda_0 (t + t_0^{-\alpha})$, where t_0 represents effective operating time incurred in device burn-in. The observed device performance should begin at a lower failure rate and gradually approach the curve for non-burned-in devices. The improvement attained through device dynamic burn-in thus equates to a gain in early-life reliability.

Since there is an accelerating effect of temperature, dynamic burn-in at a high temperature is equivalent to operation for a longer time at a lower temperature. The effective time of burn-in (equivalent time at normal operating temperature) is related to the actual burn-in time by the temperature-dependent acceleration factor A, so that

$$A = \exp\left[\frac{E_A}{k}\left(\frac{1}{T_n} - \frac{1}{T_{BI}}\right)\right] \qquad (28\text{-}29)$$

where T_n = normal operating temperature (K), T_{BI} = burn-in temperature (K), and E_A = 0.4 eV for infant mortality. We can illustrate this model by the following example.

Suppose a device is burned in at 150°C for 100 h (its normal operating temperature is 50°C). We wish to calculate the effect of the burn-in. Using E_A = 0.4 eV, k = 8.6 × 10^{-5}, T_{BI} = 423 K, and T_n = 323 K, we find

$$A = \exp\left(\frac{0.4}{8.6 \times 10^{-5}} \frac{1}{323} - \frac{1}{423}\right) \approx 30$$

Therefore, the effective burn-in time is 100 (30) = 3,000 h.

If we now want to find the effect of the burn-in on the device reliability during the first month of service, it is fairly simple. The failure rate can be represented as

$$\lambda = \lambda_0 t^{-\alpha} \qquad \text{where } \lambda_0 \text{ is in FITs} \tag{28-30}$$

The percentage failing in a given time interval from t_1 to t_2 is given by

$$\text{Percentage failing} = 10^{-7} \int_{t_1}^{t_2} \lambda_0 t^{-\alpha} \, dt$$

$$= 10^{-7} \frac{\lambda_0}{1-\alpha} \left(t_2^{1-\alpha} - t_1^{1-\alpha} \right)$$

If the device has α = 0.8 and λ_0 = 2 × 10^4, for the first month.
Without burn-in:

$$t_1 = 0 \qquad t_2 = 720 \qquad \text{Failing (\%)} = 3.72 \times 10^{-2} \approx 0.04\%$$

With burn-in:

$$t_1 = 3{,}000 \qquad t_2 = 3{,}720 \qquad \text{Failing (\%)} = 2.17 \times 10^{-3} \approx 0.002\%$$

In this example, the burn-in reduces the number of devices that would fail in the first month by a factor greater than 10. Average device failure rate in the first month of operation can also be calculated

$$\lambda_{av}, t_1 - t_2 = \frac{\int_{t_1}^{t_2} \lambda_0 t^{-\alpha}}{t_2 - t_1} = \frac{[\lambda_0/(1-\alpha)] \, t_2^{1-\alpha} t_1^{1-\alpha}}{t_2 - t_1} \tag{28-31}$$

Without burn-in: $\qquad t_1 = 0 \qquad t_2 = 720 \qquad \lambda_{av}$ = 518 FITs
With burn-in: $\qquad t_1 = 3{,}000 \qquad t_2 = 3{,}720 \qquad \lambda$ = 30 FITs

52. Infant-Mortality Failure Rates. Failure rates during the infant-mortality period are sometimes expressed in terms of the Weibull model [Eq. (28-28), Par. **28-49**]. This assures that, on the average, the devices will exhibit that failure rate. Since the device failure mechanisms causing infant mortality are usually variable, there will be a range of failure rates for a type of product as well as for a given product. Listed in Table 28-16 for several well-controlled device types are typical ranges of cumulative percent device defects along with typical slopes (α from the Weibull model), representing about 1 year of operation. Rates for the same devices in other equipments may differ from these values.

TABLE 28-16   Infant-Mortality Rates and Weibull Slopes of Typical Semiconductor Devices

Device type	Typical range of cumulative % DOF (1 year)	Typical slope α
Transistors	0.02–0.04	0.6
Diodes	0.01–0.02	0.6
Bipolar TTL ICs*	0.05–0.07	0.8–0.9
Linear ICs	0.10–0.18	0.75
Digital CMOS ICs*	0.05–0.07	0.75
MOS memory ICs	0.07–0.40	0.7–0.8

*SSI/MSI.

Reliable System Design

BY RONALD T. ANDERSON, STANISLAW KUS, HENRY C. RICKERS, AND JAMES W. WILBUR

53. General. System reliability can be enhanced in several ways. A primary technique is through the application of *redundancy*. Other methods include *design simplification, degradation analysis, worst-case design, overstress analysis,* and *transient analysis.*

54. Redundancy. Depending on the specific applications a number of approaches are available to improve reliability through redundant design. They can be classified on the basis of how the redundant elements are introduced into the circuit to provide a parallel signal path. There are two major classes of redundancy.

In *active redundancy* external components are not required to perform the function of detection, decision, or switching when an element or path in the structure fails. In *standby redundancy* external elements are required to detect, make a decision, and switch to another element or path as a replacement for a failed element or path.

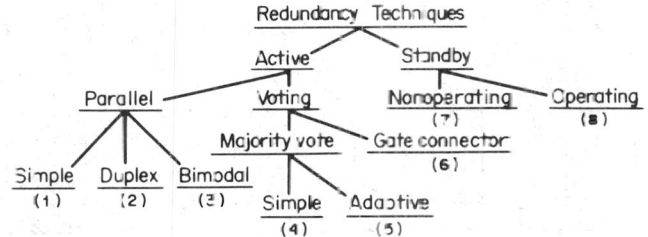

Fig. 28-22. Redundancy techniques.

Techniques related to each of these two classes are depicted in the simplified tree structure of Fig. 28-22. Table 28-17 further defines each of the eight techniques.

Redundancy does not lend itself to categorization exclusively by element complexity. Although certain of the configurations in Table 28-17 are more applicable at the part or circuit level than at the equipment level, this occurs not because of inherent limitations of the particular configuration but because of such factors as cost, weight, and complexity.

Another form of redundancy can exist within a normal nonredundant design configurations. Parallel paths within a network often are capable of carrying an added load when elements fail. This can result in a degraded but tolerable output. The allowable degree of degradation depends on the number of alternate paths available. Where a mission can still be accomplished using an equipment whose output is degraded, the definition of failure can be relaxed to accommodate degradation. Limiting values of degradation must be included in the new definition of failure. This slow approach to failure, *graceful degradation,* is exemplified by an array of elements configured to an antenna or an array of detectors configured to a receiver. In either case, individual elements may fail, reducing resolution, but if a minimum number operate, resolution remains good enough to identify a target.

The decision to use redundant design techniques must be based on a careful analysis of the trade-offs involved. Redundancy may prove the only available method when other techniques of improving reliability have been exhausted or when methods of part improvement are shown to be more costly than duplications. Its use may offer an advantage when preventive maintenance is planned. The existence of a redundant element can allow for repair with no system downtime. Occasionally, situations exist in which equipment cannot be maintained, e.g., spacecraft. In such cases, redundant elements may prolong operating time significantly.

The application of redundancy is not without penalties. It increases weight, space, complexity, cost and time to design, and maintenance cost. The increase in complexity results in an increase of unscheduled maintenance actions; safety and mission reliability is gained at the expense of logistics MTBF.

In general, the reliability gain for additional redundant elements decreases rapidly for additions beyond a few parallel elements. As illustrated in Fig. 28-23 for simple parallel redundancy, there is a diminishing gain in reliability and MTBF as the number of redundant elements is increased. As seen for the simple parallel case, the greatest gain achieved through addition of the first redundant element, is equivalent to a 50% increase in the system MTBF. In addition to

TABLE 28-17 Redundancy Techniques

Simple parallel redundancy

In its simplest form, redundancy consists of a simple parallel combination of elements; if any element fails open, identical paths exist through parallel redundant elements

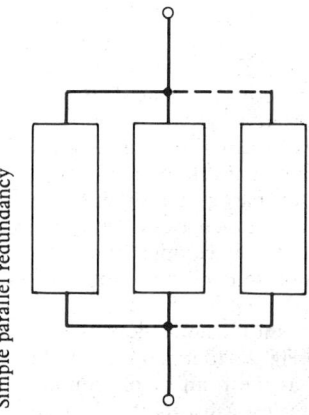

Duplex redundancy

This technique is applied to redundant logic sections, such as A_1 and A_2, operating in parallel; it is primarily used in computer applications where A_1 and A_2 can be used In duplex or active redundant modes or as a separate element; an error detector at the output of each logic section detects noncoincident outputs and starts a diagnostic routine to determine and disable the faulty element

Majority-voting redundancy

Decision can be built into the basic parallel redundant model by inputting signals from parallel elements into a voter to compare each signal with remaining signals; valid decisions are made only if the number of useful elements exceeds the failed elements

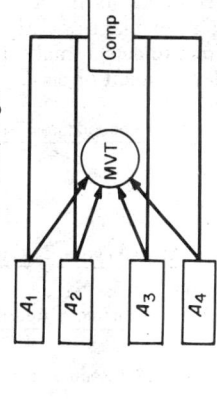

Adaptive-majority logic

This technique exemplifies the majority-logic configuration with a comparator and switching network to switch out or inhibit failed redundant elements

Gate-connector redundancy

Similar to majority voting; redundant elements are generally binary circuits; outputs of the binary elements are fed to switchlike gates, which perform the voting function; the gates contain no components whose failure would cause the redundant circuit to fail; any failures in the gate connector act as though the binary element were at fault

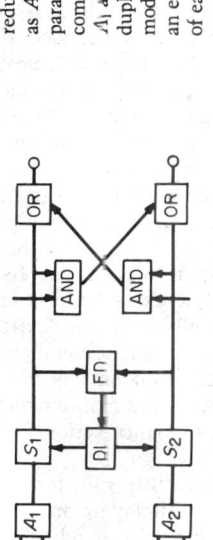

Bimodal parallel-series redundancy

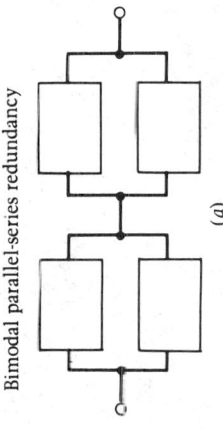

(a)

Bimodal series-parallel redundancy

(b)

A series connection of parallel redundant elements provides protection against shorts and opens; direct short across the network due to a single element's shorting is prevented by a redundant element in series; an open across the network is prevented by the parallel element; network (a) is useful when the primary element-failure mode is open; network (b) is useful when the primary element-failure mode is short

Standby redundancy

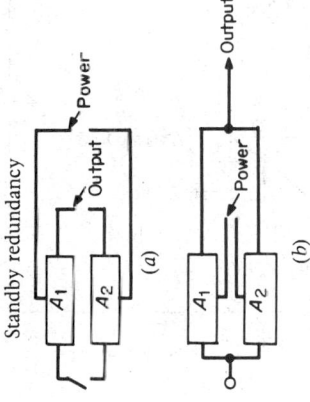

(a)

(b)

A particular redundant element of a parallel configuration can be switched into an active circuit by connecting outputs of each element to switch poles; two switching configurations are possible: (a) the element may be isolated by the switch until switching is completed and power applied to the element in the switching operation; (b) all redundant elements are continuously connected to the circuit and a single redundant element activated by switching power to it

Operating redundancy

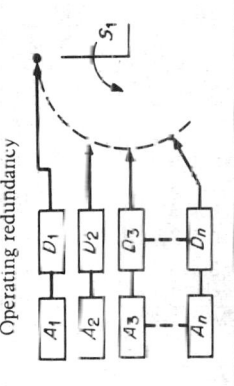

In this application, all redundant units operate simultaneously; a sensor on each unit detects failures; when a unit fails, a switch at the output transfers to the next unit and remains there until failure

28-47

maintenance-cost increases due to repair of the additional elements, reliability of certain redundant configurations may actually be worse. This is due to the serial reliability of switching or other peripheral devices needed to implement the particular redundancy configuration (see Table 28-17).

The effectiveness of certain redundancy techniques, especially standby redundancy, can be enhanced by repair. Standby redundancy allows repair of the failed unit (while operation of the unfailed unit continues uninterrupted) by virtue of the switching function built into the standby redundant configuration. The switchover function can also provide an indication that failure has occurred and that operation is continuing on the alternate channel. With a positive

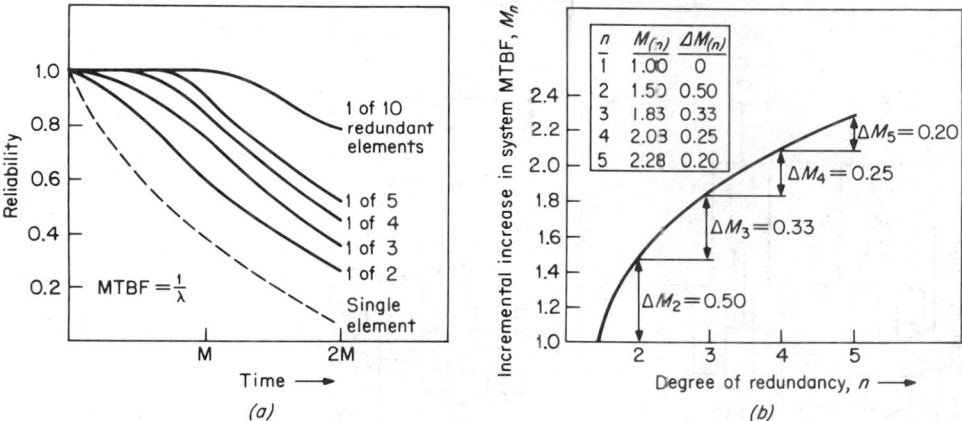

Fig. 28-23. Decreasing gain in reliability as number of active elements increases: (a) simple active redundancy for one of n elements required and (b) incremental increase in system MTBF for n active elements.

failure indication, delays in repair are minimized. A further advantage of switching is related to built-in test (BIT) objectives. Built-in test can be readily incorporated into a sensing and switchover network.

An illustration of the enhancement of redundancy with repair is shown in Fig. 28-24. The achievement of increased reliability through redundancy depends on effective isolation of redundant elements. Isolation is necessary to prevent failures from affecting other parts of the redundant network. The susceptibility of a particular redundant design to failure propagation can be assessed by application of failure-mode-effects analysis (see Par. **28-11**).

Interdependence is most successfully achieved through standby redundancy, as represented by configurations classified as *decision with switching*, where the redundant element is disconnected until a failure is sensed. However, design based on such techniques must provide protection against switching transients and must consider the effects of switching interruptions on system performance.

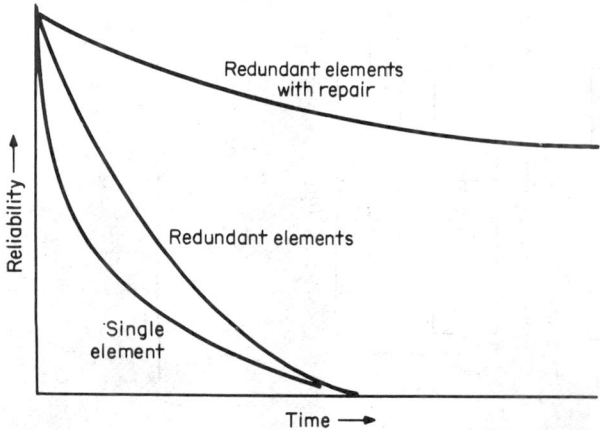

Fig. 28-24. Reliability gain for repair of simple parallel redundant elements upon failure.

Furthermore, care must be exercised to assure that reliability gains from redundancy are not offset by increased failure rates due to switching devices, error detectors, and other peripheral devices needed to implement the redundancy configurations.

55. Design Simplification. Many complex electronic systems have subsystems or assemblies that operate serially. Many of their parts and circuits are in series so that only one

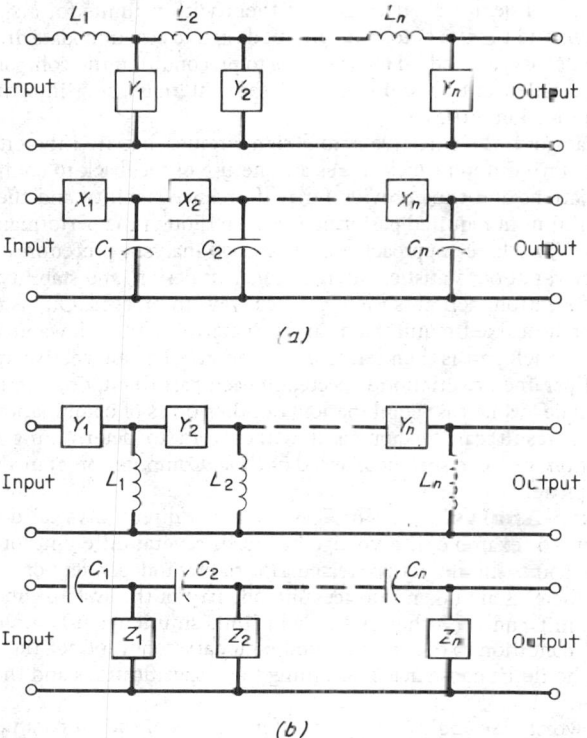

Fig. 28-25. Alternative filter designs: (a) low-pass (b) high-pass.

need fail to stop the system. This characteristic, along with the increasing trend of complexity in new designs, tends to add more and more links to the chain, thus greatly increasing the statistical probability of failure.

Therefore, one of the steps in achieving reliability is to simplify the system and its circuits as much as possible without sacrificing performance. However, because of the general tendency to increase the loads on the components that remain, there is a limiting point to circuit simplification. This limit is the value of electrical stress that must not be exceeded for a given type of electrical component. Limit values can be established for various types of components as determined by their failure rates. In addition, it is also clear that the simplified circuit must meet performance criteria under application conditions, e.g., worst-case conditions.

Design simplification and substitution involves several techniques: the use of proved circuits with known reliability, the substitution of highly reliable digital circuitry where feasible, the use of high-reliability integrated circuits to replace discrete lumped-constant circuitry, the use of highly reliable components wherever individual discrete components are called for, and the use of designs that minimize the effects of catastrophic failure modes.

The most obvious way to eliminate the failure modes and mechanisms of a part is to eliminate the part itself. For instance, a digital design may incorporate extraneous logic elements. Minimization techniques, e.g., boolean reduction, are well established and can be powerful tools for incorporating reliability in a design through simplification. Simplification can also include the identification and removal of items that have no functional significance.

In addition, efforts should also be directed toward the reduction of the critical effects of component failures. The aim here is to reduce the result of catastrophic failures until a degradation in performance occurs. As an example, consider Fig. 28-25 which illustrates the design of filter circuits. A low-pass design, shown in Fig. 23-25a, can involve either series inductances or shunt capacitances. The latter are to be avoided if shorting is the predominant failure mode peculiar to the applicable capacitor types, e.g., solid tantalum, since a catastrophic failure of the filter could result. Similarly, in the high-pass filter of Fig. 28-25b, using a shunt inductor is superior to using a series ceramic capacitor, for which an open is the expected failure mode. Here, the solid-tantalum capacitor, if applicable to the electrical design, could be the better risk, since its failure would result only in noise and incorrect frequency, not a complete loss of signal.

56. Degradation Analysis. There are basically two approaches to reduce part variation due to aging: control of device changes to hold them within limits for a specified time under stipulated conditions and the use of tolerant circuit design to accommodate drifts and degradation time. In the first category, a standard technique is to precondition the component by burn-in. In addition, there is detailed testing and control of the materials going into the part, along with strict control of fabrication processes.

In the second category, the objective is to design circuits that are inherently tolerant to part parameter change. Two different techniques are the use of feedback to compensate electrically for parameter variations and thus provide for performance stability and the design of circuits that provide the minimum required performance, even though the performance may vary somewhat due to aging. The latter approach makes use of analyses procedures such as worst-case analysis, parameter variation, statistical design, transient design, and stability analysis.

In the design of electronic circuits there are two ways to proceed. One is to view the overall circuit specification as a fixed requirement and to determine the allowable limits of each part parameter variation. Each part is then selected accordingly. The alternative approach is to examine the amount of parameter variation expected in each part (including the input) and to determine the output under worst-case combination, or other types of combination, e.g., rms or average deviation. The result can be appraised with regard to determining the probability of surviving degradation for some specified period of time. Computer programs are helpful in both types of design approach.

57. Worst-Case Analysis. In worst-case analysis, direct physical dependence must be taken into account. For example, if a voltage bus feeds several different points, the voltages at each of the several points should not be treated as variables independent of each other. Lifewise, if temperature coefficients are taken into account, one part of the system should not be assumed to be at the hot limit and the other at the cold limit simultaneously unless it is physically reasonable for this condition to occur. A general boundary condition for the analysis is that the circuit or system should be constructed according to its specifications and that the analysis proceeds from there.

In the *absolute worst-case analysis*, the limits for each independent parameter are set without regard to other parameters or to its importance in the system. The position of the limits is usually set by engineering judgment. In some cases, the designer may perform several analyses with different limits for each case to assess the result before fixing the limits.

Modified worst-case analyses are less pessimistic than absolute worst-case analysis. Typically, the method for setting the limits is to give limits to critical items as in absolute worst-case analysis and to give the rest of the items limits of the purchase tolerance.

In any worst-case analysis, the values of the parameters are adjusted (within the limits) so that circuit performance is as high as possible and then readjusted so it is as low as possible. The values of the parameters are not necessarily set at the limits; the criterion for their values is to drive the circuit performance to either extreme. The probability of this occurring in practice depends on the limits selected by the engineer at the outset, on the probability functions of the parameters, and on the complexity of the system being considered.

Computer Analyses. Computer routines are available for performing these analyses on electronic circuits. Generally speaking, the curve of circuit performance vs. each independent parameter is assumed to be monotonic, and a numerical differentiation is performed at the nominal values to see in which direction the parameter should be moved to make circuit performance higher or lower. It is also assumed that this direction is independent of the values of any of the other parameters as long as they are within their limits. If these assumptions are not true, a much more detailed analysis of the equations is necessary before worst-case analysis can be performed. This involves generation of a response surface for the circuit performance which accounts for all circuit parameters.

58. Overstress and Transient Analysis. Semiconductor circuit malfunctions generally arise from two sources: *transent circuit disturbances* and *component burnout*. Transient upsets are generally overriding because they can occur at much lower energy levels.

Transient Malfunctions. Transients in circuits may prove troublesome in many ways. Flip-flops and Schmitt triggers may be triggered inadvertently, counters may change count, memory may be altered due to driving current or direct magnetic field effect, one-shot multivibrators may pulse, the transient may be amplified and interpreted as a control signal, switches may change state, and semiconductors may latch up in undesired conducting states that require rest. The effect may be caused by transients at the input, output, or supply terminals or combinations of these. Transient upset effects can be generally characterized as follows:

1. Circuit threshold regions for upset are very narrow; i.e., there is a very small voltage-amplitude difference between the largest signals which have no probability of causing upset and the smallest signals which will certainly cause upset.

2. The dc threshold for response to a very slow input swing is calculable from the basic circuit schematic. This can establish an accurate bound for transients that exceed the dc threshold for times longer than the circuit propagation delay (a manufacturer's specification).

3. Transient upsets are remarkably independent of the exact waveshape, depending largely on the peak value of the transient and the length of time over which the transient exceeds the dc threshold. This waveform independence allows relatively easy experimental determination of circuit behavior with simple waveforms such as a square pulse.

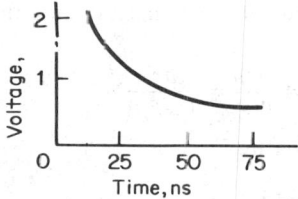

Fig. 28-26. Square-pulse trigger voltage for typical low-level integrated circuit.

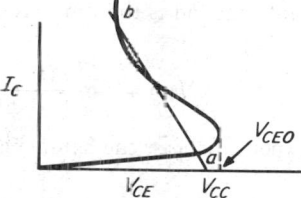

Fig. 28-27. Latch-up response.

4. The input leads or signal reference leads are generally the ones most susceptible to transient upset.

Standard circuit handbook data can often be used to gauge transient upset susceptibility. For example, square-pulse triggering voltage is sometimes given as a function of pulse duration. A typical plot for a low-level integrated circuit is shown in Fig. 23-26.

Latch-Up.   There are various ways in which semiconductors may latch up in undesired conducting states that require rest (power removal). One common situation is shown in Fig. 28-27, which shows an open-base transistor circuit with collector current as a function of collector-emitter voltage and the load line for a particular collector resistance. The collector current is normally low (operating point *a*), but a transient can move the operating level to point *b*, where the circuit becomes latched up at a high current level. The signal required to cause this event can be determined by noting that the collector-emitter voltage must be driven above the V_{CEO} (collector-emitter breakdown) voltage.

Another mode of latch-up can occur when a transistor is grown in a semiconductor substrate, e.g., an *npn* transistor in a doped *p*-substrate. Under usual voltage or gamma-radiation stress, the device can act like an SCR device, latching into conduction.

Transient Overstress.   Although various system components are susceptible to stress damage, the most sensitive tend to be semiconductor components.

Transient Suppression.   There are many techniques available for transient suppression. Some of these are illustrated in Ref. 5 and apply in the following areas: transistors, SCRs, CMOS, ITL, and diode protection. These techniques are representative of generally applicable methods and are not intended as an exhaustive list.

System Modeling

BY S. SUGIHARA

59. General Discussion.   The probability of survival or the reliability function denoted by R(t) is the probability that no failures occur during the time interval (0, t). The most commonly used reliability function is the exponential given by

$$R(t) = e^{-\lambda t} \qquad \begin{matrix} \lambda > 0 \\ t \geq 0 \end{matrix} \qquad (28\text{-}32)$$

The failure density is the negative derivative of the reliability function (when it exists). For the exponential case the failure density is

$$f(t) = -R(t) = \lambda e^{-\lambda t} \qquad \begin{matrix} \lambda > 0 \\ t \geq 0 \end{matrix} \qquad (28\text{-}33)$$

The cumulative distribution $F(t)$ is the probability of failure up to time t and is simply given by $1 - R(T)$. For the exponential case where the density exists we have

$$F(t) = 1 - R(t) = \int_0^t f(z)\,dz = 1 - e^{-\lambda t} \qquad \begin{matrix} \lambda > 0 \\ t \geq 0 \end{matrix} \qquad (28\text{-}34)$$

The concept of the hazard function $h(t)$ or the *instantaneous failure rate* is also useful. The failure rate at time t is defined by

$$z(t) = \frac{R(t) - R(t+h)}{hR(t)} \qquad h > 0 \qquad (28\text{-}35)$$

The hazard function is the limit (assuming it exists) of Eq. (28-35) as h approaches the limit zero, and thus

$$h(t) = \lim_{h \to 0} \frac{R(t) - R(t+h)}{hR(t)} = \frac{-R(t)}{R(t)} = \frac{f(t)}{R(t)} = \frac{f(t)}{1 - F(t)} \qquad (28\text{-}36)$$

For the exponential case the failure rate and hazard function are

$$z(t) = \frac{1 - e^{-\lambda h}}{h} \qquad h(t) = \lambda \qquad (28\text{-}37)$$

Thus for the exponential case we have the unique property of having a constant hazard function.

In describing the reliability of a complex system the exponential function is rarely appropriate. The reliability function is generally nonexponential because (1) the system is composed of redundant elements (even if each element were itself exponential the system is no longer exponential), and (2) the system experiences burn-in and wear-out properties so that the hazard function is nonconstant. During burn-in the hazard function generally decreases, and during wear-out the hazard function generally increases.

For systems formulated as (1) above, one is given the element reliabilities so that the system reliability can be obtained as a function of them. The particular function obtained depends on the type of redundancy used.

For systems formulated as (2) above, one is given the hazard function so that reliability is obtained as the solution of the linear differential equation

$$R(t) = -h(t)R(t) \qquad \begin{matrix} R(0) = 1 \\ t \geq 0 \end{matrix} \qquad (28\text{-}38)$$

The solution of Eq. (28-38) is

$$R(t) = \exp\left(-\int_0^t h(z)\,dz\right) \qquad t \geq 0 \qquad (28\text{-}39)$$

In general, $h(t)$ need only be piecewise continuous. If failure time is truncated at $t = L$, then $R(t)$ is continuous in the half-open interval $[0,L)$ and the hazard function is not defined at $t = L$.

Reliability for a system is defined here as the product of individual module or branch reliabilities, where each module or branch reliability is made up of reliability functions described in categories (1) and/or (2).

60. Module-Reliability Models. This section describes various reliability models currently available. The models are developed at the module level, and the system reliability is the product of these module and branch reliabilities. Branch reliabilities are described in Par. **28-61**. The set of module reliabilities is not exhaustive; others may be developed.[5]

Model 1: Active Exponential Redundant Reliability. There are n elements in parallel; each element is active with an identical exponential reliability. The module reliability is then

$$R(t) = 1 - (1 - e^{-\lambda t})^n \qquad (28\text{-}40)$$

Equation (28-40) is simply 1 minus the probability that all n elements fail. The symbol λ is the failure rate of the exponential reliability model and is used in the models that follow whenever the exponential reliability model is assumed.

Model 2: Standby Exponential Redundant Reliability. There are n elements in parallel, each element in standby until called upon to operate. Each element has an identical exponential reliability and does not experience any degradation while on standby. The module reliability is then

$$R(t) = \sum_{x=0}^{n-1} \frac{\epsilon^{-\lambda t}(\lambda t)^x}{x!} \tag{28-41}$$

Equation (28-41) is derived as a special case in Ref. 12.

Model 3: Binomial Exponential (Active) Redundancy Reliability. There are n elements of which c are required to function properly for the module. Each element is assumed active with an identical exponential reliability. The module reliability is given by the binomial sum

$$R(t) = \sum_{x=c}^{n} \binom{n}{x} (\epsilon^{-\lambda t})^x (1 - e^{-\lambda t})^{n-x} \tag{28-42}$$

In Eq. (28-42) the variable x is interpreted as the number of successes. If we define N ($N = c$, $c + 1, \dots$) as the number of trials until c successes occur, we have the negative binomial sum for module reliability

$$R(t) = \sum_{N=c}^{t} \binom{N-1}{c-1} (e^{-\lambda t})^c (1 - e^{-\lambda t})^{N-c} \tag{28-43}$$

If we further define $y = N - c$ ($y = 0, 1, 2, \dots$) as the number of failures until c successes occur, we have an alternate form of the negative binomial sum for module reliability

$$R(t) = \sum_{y=0}^{n-c} \binom{c+y-1}{y} (e^{-\lambda t})^c (1 - e^{-\lambda t})^y \tag{28-44}$$

Model 4: Standby Exponential Redundant Reliability with Exponential Standby Failure. There are n elements in parallel standby as in model 2, but it is further assumed that the elements in standby have an exponential standby reliability with failure rate λ. Thus each element has in addition to an identical exponential operational reliability, an identical exponential standby reliability. The module reliability is then

$$R(t) = e^{-\lambda t} \sum_{x=1}^{n} \frac{(1 - e^{-\mu t})^{x-1} \Gamma(\beta + x - 1)}{\Gamma(x)\Gamma(\beta)} \tag{28-45}$$

where $\beta = \lambda/\mu$

The special case for $\mu = 0$ (no standby failure) given as Eq. (28-41) is obtained from Eq. (28-45) by taking its limit as $1/\beta \to 0$.

Model 5: Open-Close Failure-Mode Exponential Redundant Reliability. There are n elements in series in each of m parallel lines, as shown in Fig. 28-28. The reliability of the module illustrated is

$$R(t) = (1 - Q_b^n)^m - (1 - R_a^n)^m \tag{28-46}$$

where
$$Q_b = (1 - R_b) = q_b(1 - e^{-\rho t}) \qquad Q_a = (1 - R_a) = q_a(1 - e^{-\rho t})$$

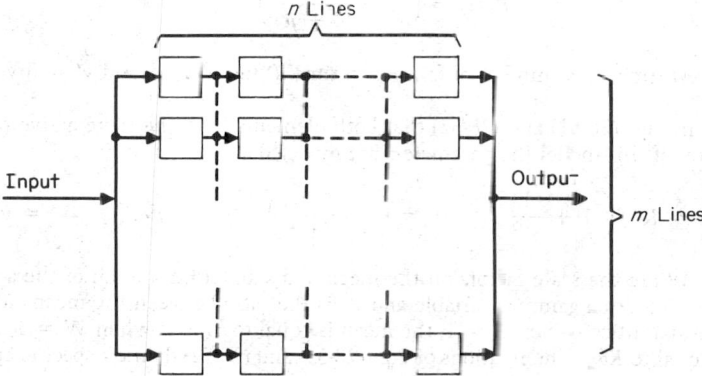

Fig. 28-28. General open-close failure system

and where q_b = conditional probability of failure to close (valve) or failure to open (switch) given system failure = μ/ρ, q_a = conditional probability of failure to open (valve) or failure to close (switch) given system failure = λ/ρ, and $\rho = \lambda + \mu$. There is a total of nm elements in the module. A configuration dual to the module illustrated in Fig. 28-28 is the "bridged" module, which is the same as that in Fig. 28-28 except that the dashed vertical lines are now made solid so that the module units are bridged. For this dual module the reliability is

$$R(t) = (1 - Q_a^m)^n - (1 - R_b^m)^n \qquad R_b \geq Q_a \qquad (28\text{-}47)$$

Model 6: Gaussian Redundant Reliability. There are n identical gaussian elements in standby redundancy, so that if t_i is the failure time of the ith element, the module failure time is

$$t = \sum_{i=1}^{n} t_i \qquad (28\text{-}48)$$

The individual elements are independent gaussian with mean μ_0 and variance σ_0^2. The module then is also gaussian with mean and variance

$$\mu = n\mu_0 \qquad \sigma^2 = n\sigma_0^2$$

The module reliability is therefore given by

$$R(t) = \frac{c}{\sqrt{2\pi}\sigma} \int_t^\infty e^{-(1/2)\sigma^2(x-\mu)^2} \, dx \qquad t \geq 0 \qquad (28\text{-}49)$$

where

$$c = 1/R(0)$$

Model 7: Bayes Reliability for Inverted Gamma Function. The module reliability for a single standby element is given by

$$R(t) = R_1(t) + \int_0^t f_1(t_1) R_2(t - t_1) R_{2s}(t_1) \, dt_1 \qquad (28\text{-}50)$$

where $R_i(t)$ = reliability of ith element, $i = 1, 2$; t_1 = failure time of first element ($i = 1$); $f_1(t)$ = failure density of first element $[= -\dot{R}_1(t)]$, and $R_{2s}(t)$ = reliability of second element in standby mode.

Equation (28-50) is a general formulation of the two-element standby reliability in the sense that the element reliabilities are general. In particular, for the case where the reliabilities are exponential, Eq. (28-50) reduces to Eq. (28-45) for $n = 2$. For the case considered here it is assumed that the reliabilities are bayesian estimates of reliability where the mean-time-to-failure parameter of the exponential has an inverted gamma density. The bayesian model is described in Ref. 13. The reliabilities are given by

$$R_j(t) = \left(1 + \frac{t}{T + \mu}\right)^{-(r+v)} \qquad j = 1, 2 \qquad (28\text{-}51)$$

and

$$R_{2s}(t) = \left(1 + \frac{t}{T + \mu'}\right)^{-(r+v')} \qquad (28\text{-}52)$$

where T = test time, r = number of failures in time T and μ, v, μ', and v' = inverted gamma parameters.

It is noted in Eqs. (28-51) and (28-52) that both elements have the same active reliability. In the application of this model the parameters are modeled as follows:

$$\mu = K\theta_A\left(1 + \frac{1}{W^2}\right) \qquad v = \left(2 + \frac{1}{W^2}\right) \qquad \mu' = \mu/K_1 \qquad v' = v \qquad (18\text{-}53)$$

where K and W are the scale factors on the mean and standard deviation of the a priori mean time to failure (inverted-gamma) variable and K_1 is the ratio between the means for the active and standby reliabilities. When $K = 1$, the mean is equal to θ_A, and when $W = 1$, the standard deviation is equal to $K\theta_A$. The relations of Eqs. (28-51) and (28-52) define a specific application of Eq. (28-50).

Model 8: Reliability Model for Hazard Function Composed of Three Piecewise Continuous Functions. The general reliability model considered here is Eq. (28-36) for the hazard function

$$h(t) = \begin{cases} h_1(t) & 0 \le t < t_1 \\ h_2(t) & t_1 \le t < t_2 \\ h_3(t) & t_2 \le t < \infty \end{cases} \qquad (18\text{-}54)$$

The particular case considered here is where h_1 and h_3 are Weibull hazard functions and h_2 is the exponential hazard function. Further, it is assumed that the hazard function is continuous at the joins t_1 and t_2 so that we have

$$\begin{aligned} h_1(t) &= \alpha_1\lambda_1 t^{\alpha_1-1} & 0 \le t < t_1 \\ h_2(t) &= \lambda & t_1 \le t < t_2 \\ h_3(t) &= \alpha_2\lambda_2 t^{\alpha_2-1} & t_2 \le t < \infty \end{aligned} \qquad (18\text{-}55)$$

where $\qquad \alpha_1, \alpha_2, t_1, t_2,$ and λ are input parameters

$$\lambda_i = \frac{\lambda}{\alpha_i t_i^{\alpha_i-1}} \qquad i = 1, 2$$

Model 9: Tabular Reliability Model. An arbitrary reliability function can be evaluated as an input table of reliability vs. time. For each output time, reliability is obtained by interpolation of the input table. This capability is useful as an approximation to a complex reliability model and is also useful in evaluating a reliability function which is available as a plot or a table for which no mathematical model is available.

Model 10: Active Exponential Redundant Reliability with Different Redundant Failure Rate. There are n elements in parallel. The original c elements are exponential each with failure rate λ. The $n - c$ remaining exponential elements in active parallel with the original elements each have failure rate μ. The module reliability is

$$R_i(t) = 1 - (1 - e^{-\lambda t})^c (1 - e^{-\mu t})^{n-c} \qquad (28\text{-}55)$$

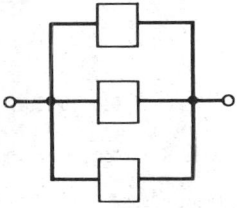

Fig. 28-29 Branch system a consisting of a model 3 module.

61. Branch-Reliability Models. A branch-reliability model is a model which evaluates reliability (at each instant in time) for a set of modules arranged in branches. Each branch individually consists of a set of modules. In particular, a branch may be a module itself. Some typical branches are shown in Figs. 28-29 to 28-31 and are identified as branches a, b, and c respectively.

Branch system a is simply a module. Branch system c is a branch consisting of two modules in series and can be evaluated as two separate modules. Branch system b is made up of two subbranches in parallel.

A more complex branch system consisting of subbranch systems a, b, and c is shown in Fig. 28-32. Branches a and b are in series in active redundancy to branch system c.

Although the reliability functions for particular branch models are not developed here, clearly it is possible and a generalized computer program based on both module and branch models can be written.

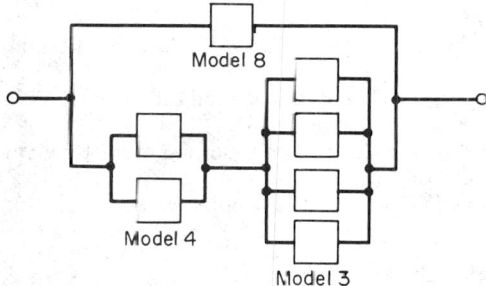

Fig. 28-30. Branch system b consisting of models 3, 4, and 8.

Model 8

Model 4

Model 3

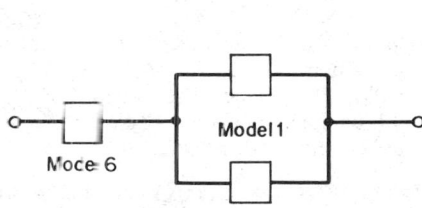

Model 6

Model 1

Fig. 28-31. Branch system c consisting of models 1 and 6.

62. The Systems Model. The system reliability model is obtained as the product of the individual branch reliabilities. In addition, the branch reliabilities are multiplied by a system exponential reliability factor and a constant factor; thus,

$$R_s(t) = P_0 e^{-\lambda_s t} \prod_{i=1}^{K'} R_i(t)$$ (28-57)

where λ_s = exponential hazard function for system elements not included in branches, P_0 = system reliability factor, $R_i(t)$ = ith branch reliability, $i = 1, 2, \ldots, K'$, and K' = number of branches in series.

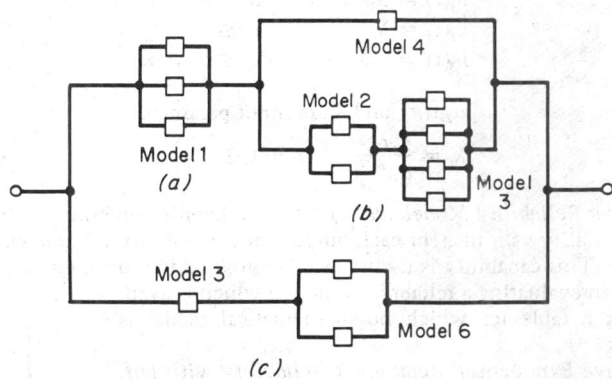

Fig. 28-32. System consisting of branch systems shown in Figs. **28-29** to **28-31**.

63. References and Bibliography

REFERENCES

1. General Electric Co.   "Research Study of Radar Reliability and Its Impact on Life Cycle Costs for the APQ-113, -114, -120 and -144 Radar Systems," Aerospace Electronic Systems Dept., Utica, N.Y., August 1972.
2. U.S. Dept. of Defense   "Military Standardization Handbook, Reliability Prediction of Electronic Equipment," MIL-HDBK-217C, April 1979.
3. Selby, J., and S. Miller   Reliability Planning and Management, *Symp. Reliability Maintainability Technol. Mechanic. Syst.*, April 1972.
4. Duane, J. T.   Learning Curve Approach to Reliability Monitoring, *IEEE Trans. Aerospace*, 1964, Vol. II.
5. Reliability Analysis Center, RADC/RBRAC   "Reliability Design Handbook," RDH 376, March 1976.
6. Reliability Analysis Center, RADC/RBRAC   "Search and Retrieval Index to IRPS Proceedings, 1968–1978."
7. U.S. Dept. of the Air Force   Military Standard Technical Requirements for Parts, Materials, and Processes for Space and Launch Vehicles, MIL-STD-1547 (USAF), December 1980.
8. U.S. Dept. of Defense   "Resistors: Selection and Use of," MIL-STD-199B, March 1977.
9. U.S. Dept. of Defense   "Capacitors: Selection and Use of," MIL-STD-198D, Naval Publications and Forms, Philadelphia, November 1976.
10. U.S. Dept. of Defense   "Transformer Audio Frequency," MIL-T27B, Naval Publications and Forms, Philadelphia, September 1963.
11. U.S. Dept. of Defense   Fuse Instrument Type Style FMO9 (Non-indicating), MIL-STD-F23419/9, U.S. Naval Publications and Forms, Philadelphia, August 1977.
12. Thompson, R. J.   "A Generalized Expression for the Reliability of a System Utilizing Standby Redundancy," Aerospace Corp., Rep. TOR-1001 (2303)-1, August 1966.
13. Bhattacharya, S. K.   Bayesian Approach to Life Testing and Reliability Estimation, *J. Am. Statist. Asso.*, March 1967.

PRINCIPAL GENERAL DOCUMENTS AND REFERENCES (See also Refs. 1, 4, and 6)

14. IEEE *Proc. Annu. Reliability Maintainability Symp.*, 1979.
15. IEEE *Proc. Annu. Reliability Maintainability Symp.*, 1980.
16. IEEE *Proc. Int. Reliability Phys. Symp.*, 1979.

17. IEEE *Proc. Int. Reliability Phys. Symp.*, 1980.

18. Von Alven, W. (ed.) "Reliability Engineering," Prentice-Hall, Englewood Cliffs, N.J., 1964.

19. Kapur, K. C., and L. R. Lamberson "Reliability in Engineering Design," Wiley, New York, 1977.

20. Cunningham, C. E., and W. Cox "Applied Maintainability Engineering," Wiley, New York, 1972.

21. Blanchard B. S., Jr., and E. E. Lowery "Maintainability: Principles and Practices," McGraw-Hill, New York, 1969.

22. Arsenault, J. E., and J. A. Roberts (eds.) "Reliability and Maintainability of Electronic Systems," Computer Science Press, Potomac, Md., 1980.

23. Bazovsky, I. "Reliability Theory and Practice," Prentice-Hall, Englewood Cliffs, N.J., 1961.

24. Special Issue on Reliability of Semiconductor Devices, *Proc. IEEE*. February 1974.

FAILURE DATA AND ANALYSIS

25. Nicholls, D. B. Digital Failure Rate Data, Reliability Analysis Center, RADC/RBRAC, Griffiss Air Force Base, N.Y., July 1979.

26. Klein, M. R. Memory/LSI Data, Reliability Analysis Center, RADC/RBRAC, Griffiss Air Force Base, N.Y., July 1979.

27. Reliability Analysis Center, RADC/RBRAC Hybrid Circuit Data, 1978, MDR-9, July 1978.

28. Reliability Analysis Center, RADC/RBRAC Digital Evaluation and General Failure Analysis Data, MDR-10, October 1978.

29. Reliability Analysis Center, RADC/RBRAC Linear/Interface Data MDR-11, May 1979.

30. Reliability Analysis Center, RADC/RBRAC Transistor/Diode Data, DSR-3, 1979.

31. Reliability Analysis Center, RADC/RBRAC Nonelectronic Parts Reliability Data, NPRD-1, August 1978.

32. Reliability Analysis Center, RADC/RBRAC Microcircuit Wire Bond Reliability, TM72-1.

33. Goth, G. F. Quantification of Printed Circuit Board (PCB) Connector Reliability, Reliability Analysis Center, RADC/RBRAC, RADC-TR-77-433, January 1973.

34. Cottrell, D. F., and T. E. Kirejczyk Crimp Connector Reliability, Reliability Analysis Center, RADC/RBRAC, RADC-TR-78-15, January 1978.

35. Libore, C. Rectangular Flat-Pack Lids under External Pressure: Improved Formulas for Screening and Design (Revised), Reliability Analysis Center, RADC/RBRAC, RADC-TR-79-138, June 1979.

36. Kobini, K., R. W. Perkins, and C. Libore Thermal Stress Analysis of Glass Seals in Microelectronic Packages Under Thermal Shock Conditions, Reliability Analysis Center, RADC/RBRAC, RADC-TR-79-201, July 1979.

37. Thomas, R. W. Moisture, Myths, and Microcircuits, *IEEE Trans. Parts, Hybrids, Packag.*, September 1976, Vol. PHP-12, No. 3.

38. Reliability Analysis Center Hybrid Microcircuit Failure Rate Predictions, RADC-TR-78-97, April 1979, ADA 055 656.

39. Reliability Analysis Center, RADC/RBRAC Impact Stresses in Wire and Wire Bonds: Upper Limit Analysis, RADC-TR-78-113, May 1978, ADA 055 270.

40. Reliability Analysis Center, RADC/RBRAC Impact Stresses in Flat-Pack Lids and Bases, RADC-TR-78-98, April 1978, ADA 054 947.

41. Reliability Analysis Center, RADC/RBRAC Flat-Packs under Thermal Shock: A Simplified Analysis of Flexural Stress in the Lid-to-Wall Seal, RADC-75-308, December 1975, ADA 021 253.

42. U.S. Dept. of Defense Military Standard, Electrostatic Discharge Control Program for Protection of Electrical and Electronic Parts, Assemblies and Equipment (Excluding Electrically Initiated Explosive Devices), DOD-STD-1636, May 1980.

43. U.S. Dept. of Defense Electrostatic Discharge Control Handbook for Protection of Electrical and Electronic Parts, Assemblies and Equipment (Excluding Electrically Initiated Explosive Devices), DOD-HDBK-263, May 1980.

MAINTAINABILITY (See also Refs. 20 and 21)

44. Electronic Industries Association "A Mathematical Derivation of Maintainability Requirements from Operational Readiness Goals," May 1960.

45. Electronic Industries Association "The Techniques of Maintainability Applied to System Effectiveness Quantification," September 1969.

SCREENING AND TESTING

46. Jones, E. R. "A Guide to Component Burn-In-Technology," Wakefield Engineering Inc., 1972.
47. Zemel, J. N. (ed.) "Nondestructive Evaluation of Semiconductor Materials and Devices," Plenum, New York, 1978.
48. Reliability Analysis Center, RADC/RBRAC Microcircuit Screening Effectiveness, TRS-1.
49. Reliability Analysis Center, RADC/RBRAC Critique for the Centrifuge as a Stressing Device, RADC-TR-78-102, May 1978, ADA 054 947.

MODELING

50. Sugihara, S. Burn-In Screening Based on Mixture Models, Aerospace Corp. Rep. ATR-73(7254-01)-1, March 1973.
51. Sugihara, S. General System Reliability Evaluation, Aerospace Corp. Rep. TOR-0158 (3303)-2, May 1968.
52. Sugihara, S. Module Reliability Monitoring, Aerospace Corp. Rep. TOR-00073(3409-01)-52, November 1972.
53. Reliability Analysis Center, RADC/RBRAC LSI/Microprocessor Reliability Prediction Model Development, RADC-TR-79-97, March 1979, ADA 068 911.

STATISTICS AND MATHEMATICS

54. ISO Information Center "Statistical Methods," American National Standards Institute, New York, 1979.
55. Hollander, M., and D. A. Wolfe "Nonparametric Statistical Methods," Wiley, New York, 1973.
56. Conover, W. J. "Practical Nonparametric Statistics," Wiley, New York, 1971.
57. Kalbfleisch, J. D., and R. L. Prentice "The Statistical Analysis of Failure Time Data," Wiley, New York, 1980.
58. Laha, R. G., and V. K. Rohatgi "Probability Theory," Wiley, New York, 1979.
59. Randles, R. H., and D. A. Wolfe "Introduction to the Theory of Nonparametric Statistics," Wiley, New York, 1979.

DESIGN

60. Gansler, J. S., and G. W. Sutherland A Design to Cost Overview, *Defense Manage. J.,* September 1974.

Index

1